The VLSI Handbook

Second Edition

The Electrical Engineering Handbook Series

Series Editor
Richard C. Dorf
University of California, Davis

Titles Included in the Series

The Handbook of Ad Hoc Wireless Networks, Mohammad Ilyas
The Avionics Handbook, Second Edition, Cary R. Spitzer
The Biomedical Engineering Handbook, Third Edition, Joseph D. Bronzino
The Circuits and Filters Handbook, Second Edition, Wai-Kai Chen
The Communications Handbook, Second Edition, Jerry Gibson
The Computer Engineering Handbook, Vojin G. Oklobdzija
The Control Handbook, William S. Levine
The CRC Handbook of Engineering Tables, Richard C. Dorf
The Digital Avionics Handbook, Second Edition Cary R. Spitzer
The Digital Signal Processing Handbook, Vijay K. Madisetti and Douglas Williams
The Electrical Engineering Handbook, Second Edition, Richard C. Dorf
The Electric Power Engineering Handbook, Leo L. Grigsby
The Electronics Handbook, Second Edition, Jerry C. Whitaker
The Engineering Handbook, Third Edition, Richard C. Dorf
The Handbook of Formulas and Tables for Signal Processing, Alexander D. Poularikas
The Handbook of Nanoscience, Engineering, and Technology, William A. Goddard, III,
    Donald W. Brenner, Sergey E. Lyshevski, and Gerald J. Iafrate
The Handbook of Optical Communication Networks, Mohammad Ilyas and Hussein T. Mouftah
The Industrial Electronics Handbook, J. David Irwin
The Measurement, Instrumentation, and Sensors Handbook, John G. Webster
The Mechanical Systems Design Handbook, Osita D.I. Nwokah and Yidirim Hurmuzlu
The Mechatronics Handbook, Robert H. Bishop
The Mobile Communications Handbook, Second Edition, Jerry D. Gibson
The Ocean Engineering Handbook, Ferial El-Hawary
The RF and Microwave Handbook, Mike Golio
The Technology Management Handbook, Richard C. Dorf
The Transforms and Applications Handbook, Second Edition, Alexander D. Poularikas
The VLSI Handbook, Second Edition, Wai-Kai Chen

The VLSI Handbook

Second Edition

Edited by

Wai-Kai Chen

University of Illinois
Chicago, USA

CRC Press
Taylor & Francis Group
Boca Raton   London   New York

CRC Press is an imprint of the
Taylor & Francis Group, an informa business

CRC Press
Taylor & Francis Group
6000 Broken Sound Parkway NW, Suite 300
Boca Raton, FL 33487-2742

International Standard Book Number-10: 0-8493-4199-X (Hardcover)
International Standard Book Number-13: 978-0-8493-4199-1 (Hardcover)

Library of Congress Cataloging-in-Publication Data

The VLSI handbook / edited by Wai-Kai Chen.—2nd ed.
    p. cm.—(Electrical engineering handbook series; 38)
  Includes bibliographical references and index.
  ISBN 0-8493-4199-X
  1. Integrated circuits—Very large scale integration. I. Chen, Wai-Kai, 1936- II. Title. III. Series.

TK7874.75.V573 2006
621.39′5—dc22
                                                      2006050477

Visit the Taylor & Francis Web site at
http://www.taylorandfrancis.com

and the CRC Press Web site at
http://www.crcpress.com

Preface

We are most gratified to find that the first edition of *The VLSI Handbook* (2000) was well received and is widely used. Thus, we feel that our original goal of providing in-depth professional-level coverage of VLSI technology was, indeed, worthwhile. Seven years is a short time in terms of development of science and technology; however as this handbook shows, momentous changes have occurred during this period, necessitating not only the updating of many chapters of the handbook, but more startling, the addition and expansion of many topics. Significant examples are low-power electronics and design, testing of digital systems, VLSI signal processing, and design languages and tools to name a few of the more prominent additions.

Purpose

The VLSI Handbook provides in a single volume a comprehensive reference work covering the broad spectrum of VLSI technology. It is written and developed for practicing electrical engineers in industry, government, and academia. The goal is to provide the most up-to-date information in integrated circuits (IC) technology, devices and their models, circuit simulations, low-power electronics and design, amplifiers, analog and logic circuits, memory, registers and system timing, microprocessor and ASIC, test and testability, design automation, VLSI signal processing, and design languages and tools. The handbook is not an all-encompassing digest of everything taught within an electrical engineering curriculum on VLSI technology. Rather, it is the engineer's first choice in looking for a solution. Therefore, full references to other sources of contributions are provided. The ideal reader is a BS level engineer with a need for a one-source reference to keep abreast of new techniques and procedures as well as review standard practices.

Background

The handbook stresses fundamental theory behind professional applications. To do so, it is reinforced with frequent examples. Extensive development of theory and details of proofs have been omitted. The reader is assumed to have a certain degree of sophistication and experience. However, brief reviews of theories, principles, and mathematics of some subject areas are given. These reviews have been done concisely with perception. The handbook is not a textbook replacement, but rather a reinforcement and reminder of material learned as a student. Therefore, important advancement and traditional as well as innovative practices are included.

Since most of the professional electrical engineers graduated before powerful personal computers were widely available, many computational and design methods may be new to them. Therefore, computers and software use are thoroughly covered. Not only does the handbook use traditional references to cite sources for the contributions, but it also contains all *relevant* sources of information and tools that would assist the engineer in performing his/her job. This may include sources of software, databases, standards, seminars, conferences, etc.

Organization

Over the years, the fundamentals of VLSI technology have evolved to include a wide range of topics and a broad range of practice. To encompass such a wide range of knowledge, the handbook focuses on the key concepts, models, and equations that enable the electrical engineer to analyze, design, and predict the behavior of very large-scale integrated circuits. While design formulas and tables are listed, emphasis is placed on the key concepts and theories underlying the applications.

The information is organized into 13 major sections, which encompass the field of VLSI technology. Each section is divided into chapters, each of which is written by a leading expert in the field to enlighten and refresh knowledge of the mature engineer, and to educate the novice. Each section contains introductory material, leading to the appropriate applications. To help the reader, each article includes two important and useful categories: defining terms and references. *Defining terms* are key definitions and the first occurrence of each term defined is indicated in italic type in the text. The *references* provide a list of useful books and articles for further reading and for additional information on the topic.

Locating Your Topic

Numerous avenues of access to information contained in the handbook are provided. A complete table of contents is presented at the beginning of the book. In addition, an individual table of contents precedes each of the 13 sections. Finally, each chapter begins with its own table of contents. The reader is urged to review these tables of contents to become familiar with the structure, organization, and content of the book. For example, see Section VIII: Microprocessor and ASIC, then Chapter 64: Microprocessor Design Verification, and then Section 64.2: Design Verification Environment. This tree-like structure enables the reader to move up the tree to locate information on the topic of interest.

A combined subject and author index has been compiled to provide means of accessing information. It can also be used to locate definitions; the page on which the definition appears for each key defining term is given in this index.

The VLSI Handbook is structured to provide answers to most inquiries and to direct inquirer to further sources and references. We trust that it will meet your needs.

Acknowledgments

The compilation of this book would not have been possible without the dedication and efforts of the section editors, the publishers, and most of all the contributing authors. I wish to thank all of them and also my wife, Shiao-Ling, for her patience and understanding.

Wai-Kai Chen
Editor-in-Chief

Editor-in-Chief

Wai-Kai Chen is professor and head emeritus of the Department of Electrical Engineering and Computer Science at the University of Illinois at Chicago. He received his BS and MS in electrical engineering from Ohio University, where he was later recognized as a distinguished professor. He earned his PhD in electrical engineering from the University of Illinois at Urbana-Champaign.

Wai-Kai Chen

Professor Chen has extensive experience in education and industry and is very active professionally in the fields of circuits and systems. He has served as visiting professor at Purdue University, University of Hawaii at Manoa, and Chuo University in Tokyo, Japan. He was editor of the *IEEE Transactions on Circuits and Systems, Series I and II*, president of the IEEE Circuits and Systems Society, and is the founding editor and editor-in-chief of the *Journal of Circuits, Systems and Computers*. He received the Lester R. Ford Award from the Mathematical Association of America, the Alexander von Humboldt Award from Germany, the JSPS Fellowship Award from Japan Society for the Promotion of Science, the National Taipei University of Technology Distinguished Alumnus Award, the Ohio University Alumni Medal of Merit for Distinguished Achievement in Engineering Education, the Senior University Scholar Award and the 2000 Faculty Research Award from University of Illinois at Chicago, and the Distinguished Alumnus Award from the University of Illinois at Urbana-Champaign. He is also the recipient of the Golden Jubilee Medal, the Education Award, the Meritorious Service Award from IEEE Circuits and Systems Society, and the Third Millennium Medal from the IEEE. He has also received more than a dozen honorary professorship awards from major institutions in Taiwan and China.

A fellow of the Institute of Electrical and Electronics Engineers and the American Association for the Advancement of Science, Professor Chen is widely known in the profession for his published works which include *Applied Graph Theory* (North-Holland), *Theory and Design of Broadband Matching Networks* (Pergamon Press), *Active Network and Feedback Amplifier Theory* (McGraw-Hill), *Linear Networks and Systems* (Brooks/Cole), *Passive and Active Filters: Theory and Implements* (John Wiley), *Theory of Nets: Flows in Networks* (Wiley-Interscience), *The Circuits and Filters Handbook* (CRC Press), and *The Electrical Engineering Handbook* (Elsevier Academic Press).

Contributors

Ramachandra Achar
Department of Electronics
   Carleton University
Ottawa, Ontario, Canada

Arshad Ahmed
DSP R&D
Texas Instruments, Inc.
Dallas, Texas

Jonathan A. Andrews
Department of Electrical and
   Computer Engineering
Virginia Commonwealth
   University
Richmond, Virginia

James H. Aylor
School of Engineering and
   Applied Science
University of Virginia
Charlottesville, Virginia

R. Jacob Baker
Department of Electrical and
   Computer Engineering
University of Idaho at Boise
Boise, Idaho

Andrea Baschirotto
Department of Innovation
   Engineering
University of Lecce
Lecce, Italy

Charles R. Baugh
C. R. Baugh and Associates
Seattle, Washington

Magdy Bayoumi
The Center for Advanced
   Computer Studies
University of Louisiana
Lafayette, Louisiana

David Blaauw
Department of Electrical
   Engineering and Computer
   Science
University of Michigan
Ann Arbor, Michigan

Victor Boyadzhyan
Jet Propulsion Laboratory
Pasadena, California

Alison Burdett
Toumaz Technology Ltd.
Abingdon, UK

Wai-Kai Chen
University of Illinois
Chicago, Illinois

Kuo-Hsing Cheng
Tamkang University
Tamkang, Taiwan

Bi-Shiou Chiou
Department of Electronics
   Engineering
National Chiao Tung University
Hsinchu, Taiwan

John Choma, Jr.
Department of Electrical
   Engineering/Electrophysics
University of Southern California
Los Angeles, California

Amy Hsiu-Fen Chou
National Tsing-Hua
   University
Hsinchu, Taiwan

Moon Jung Chung
Department of Computer
   Science
Michigan State University
East Lansing, Michigan

David J. Comer
Department of Electrical and
   Computer Engineering
Brigham Young University
Provo, Utah

Donald T. Comer
Department of Electrical and
   Computer Engineering
Brigham Young University
Provo, Utah

Daniel A. Connors
Department of Computer
   Science
University of Colorado
Boulder, Colorado

Donald R. Cottrell
Silicon Integration
   Initiative, Inc.
Austin, Texas

John D. Cressler
School of Electrical and
   Computer Engineering
Georgia Institute of Technology
Atlanta, Georgia

Sorin Cristoloveanu
Institute of Microelectronics,
   Electromagnetism and
   Photonics
Grenoble, France

Wouter De Cock
Katholieke Universiteit Leuven
Leuven-Heverlee, Belgium

Abhijit Dharchoudhury
Motorola, Inc.
Austin, Texas

Robert P. Dick
Department of Electrical
   Engineering and Computer
   Science
Northwestern University
Evanston, Illinois

Vassil S. Dimitrov
Department of Electrical and
   Computer Engineering
University of Calgary
Calgary, Alberta,
   Canada

Donald B. Estreich
Microwave Technology
   Division
Agilent Technologies
Santa Rosa, California

John W. Fattaruso
Texas Instruments, Inc.
Dallas, Texas

Ayman A. Fayed
Texas Instruments, Inc.
Dallas, Texas

Eby G. Friedman
Department of Electrical
   and Computer Engineering
University of Rochester
Rochester, New York

Shantanu Ganguly
Intel Corporation
Austin, Texas

Aman Gayasen
Department of Computer
   Science and Engineering
Pennsylvania State University
University Park, Pennsylvania

Jan V. Grahn
School of Information
   and Communication
   Technology
KTH, Royal Institute of
   Technology
Kista, Sweden

Flavius Gruian
Department of Computer
   Science
Lund University
Sweden

Maria del Mar Hershenson
Stanford University
Stanford, California

Charles Ching-Hsiang Hsu
National Tsing-Hua
   University
Hsinchu, Taiwan

Jen-Sheng Hwang
National Science Council
Taipei, Taiwan

Wen-mei W. Hwu
University of Illinois at
   Urbana-Champaign
Urbana, Illinois

Kazumi Inoh
Center for Semiconductor
   Research and
   Development
Semiconductor Company
Toshiba Corporation
Yokohama, Japan

Ali Iranli
Electrical Engineering
   Department
University of Southern
   California
Los Angeles, California

K. Irick
Department of Computer
   Science and
   Engineering
Pennsylvania State
   University
University Park,
   Pennsylvania

M. J. Irwin
Department of Computer
   Science and Engineering
Pennsylvania State University
University Park, Pennsylvania

Hidemi Ishiuchi
Center for Semiconductor
   Research and Development
Semiconductor Company
Toshiba Corporation
Yokohama, Japan

Mohammed Ismail
Department of Electrical
   and Computer
   Engineering
Ohio State University
Columbus, Ohio

Hiroshi Iwai
Frontier Collaborative
   Research Center
Tokyo Institute of Technology
Yokohama, Japan

Vikram Iyengar
IBM Microelectronics
Essex Junction, Vermont

W. Kenneth Jenkins
Department of Computer
   Science and Engineering
Pennsylvania State University
University Park, Pennsylvania

Jeff Jessing
Department of Electrical
   and Computer
   Engineering
Boise State University
Boise, Idaho

Niraj K. Jha
Department of Electrical
   Engineering
Princeton University
Princeton, New Jersey

Graham A. Jullien
Department of Electrical
   and Computer
   Engineering
University of Calgary
Calgary, Alberta, Canada

Dimitri Kagaris
Department of Electrical
    and Computer Engineering
Southern Illinois University
Carbondale, Illinois

Steve M. Kang
University of California at
    Santa Cruz
Santa Cruz, California

Nick Kanopoulos
Atmel Corporation
Morrisville, North Carolina

Naghmeh Karimi
Electrical and Computer
    Engineering
University of Tehran
Tehran, Iran

Tanay Karnik
Strategic CAD Labs
Intel Corporation
Hillsboro, Oregon

Yasuhiro Katsumata
Engineering Planning Division
Semiconductor Company
Toshiba Corporation
Kawasaki, Japan

Ali Keshavarzi
Circuit Research Labs
Intel Corporation
Hillsboro, Oregon

Heechul Kim
Department of Computer
    Science and Engineering
Hankuk University of Foreign
    Studies
Yongin, Kyung Ki-Do, Korea

Jihong Kim
School of Computer
    Science and Engineering
Seoul National University
Seoul, Korea

Hideki Kimijima
System LSI Division II
Semiconductor Company
Toshiba Corporation
Kitakyushu, Japan

Robert H. Klenke
Department of Electrical
    and Computer
    Engineering
Virginia Commonwealth
    University
Richmond, Virginia

Ivan S. Kourtev
Department of Electrical and
    Computer Engineering
University of Pittsburgh
Pittsburgh, Pennsylvania

Seok-Jun Lee
DSP R&D
Texas Instruments, Inc.
Dallas, Texas

Thomas H. Lee
Stanford University
Stanford, California

Harry W. Li
Formerly with the University
    of Idaho at Boise

Yijun Li
The Center for Advanced
    Computer Studies
University of Louisiana
Lafayette, Louisiana

Chi-Sheng Lin
Department of Electrical
    Engineering
National Cheng Kung University
Tainan, Taiwan

Frank Ruei-Ling Lin
National Tsing-Hua University
Hsinchu, Taiwan

Bin-Da Liu
Department of Electrical
    Engineering
National Cheng Kung
    University
Tainan, Taiwan

John Lockwood
Department of Computer
    Science and Engineering
Washington University
St. Louis, Missouri

Stephen I. Long
Department of Electrical and
    Computer Engineering
University of California
Santa Barbara, California

Ashraf Lotfi
Enpirion, Inc.
Bridgewater, New Jersey

B. Gunnar Malm
School of Information and
    Communication Technology
KTH, Royal Institute of
    Technology
Kista, Sweden

Mohammad Mansour
Department of Electrical and
    Computer Engineering
American University of Beirut
Beirut, Lebanon

Diana Marculescu
Department of Electrical and
    Computer Engineering
Carnegie Mellon University
Pittsburgh, Pennsylvania

Radu Marculescu
Department of Electrical and
    Computer Engineering
Carnegie Mellon University
Pittsburgh, Pennsylvania

Martin Margala
Electrical and Computer
    Engineering Department
University of Massachusetts
Lowell, Massachusetts

Shin-ichi Minato
NTT Network Innovation
    Laboratories
Kanagawa, Japan

Shahrzad Mirkhani
Electrical and Computer
    Engineering Department
University of Tehran
Tehran, Iran

Sunderarajan S. Mohan
Stanford University
Stanford, California

Hisayo S. Momose
Center for Semiconductor
    Research and Development
Semiconductor Company
Toshiba Corporation
Yokohama, Japan

Eiji Morifuji
System LSI Division I
Semiconductor Company
    Toshiba Corporation
Yokohama, Japan

Toyota Morimoto
Memory Division
Semiconductor Company
Toshiba Corporation
Yokohama, Japan

Saburo Muroga
University of Illinois at
    Urbana-Champaign
Urbana, Illinois

Roberto Muscedere
Research Centre for Integrated
    Microsystems (RCIM)
University of Windsor
Ontario, Canada

Akio Nakagawa
Discrete Semiconductor Division
Semiconductor Company
Toshiba Corporation
Kawasaki, Japan

Yuichi Nakamura
NEC Corporation
Kawasaki, Japan

Michel S. Nakhla
Department of Electronics
Carleton University
Ottawa, Ontario, Canada

Zainalabedin Navabi
Nanoelectronics Center
    of Excellence
School of Electrical and
    Computer Engineering
University of Tehran
Tehran, Iran

Philip G. Neudeck
NASA Glenn Research Center
Cleveland, Ohio

C. Nicopoulos
Department of Computer
    Science and Engineering
Pennsylvania State University
University Park, Pennsylvania

Hideaki Nii
System LSI Division I
Semiconductor Company
Toshiba Corporation
Yokohama, Japan

Umit Y. Ogras
Department of Electrical
    and Computer
    Engineering
Carnegie Mellon University
Pittsburgh, Pennsylvania

Tatsuya Ohguro
Center for Semiconductor
    Research and Development
Semiconductor Company
Toshiba Corporation
Yokohama, Japan

Mikael Östling
School of Information and
    Communication
    Technology
KTH, Royal Institute of
    Technology
Kista, Sweden

Seok-Bae Park
Department of Electrical
    and Computer
    Engineering
Ohio State University
Columbus, Ohio

Alice C. Parker
Department of Electrical
    Engineering-Systems
University of Southern
    California
Los Angeles, California

Massoud Pedram
Electrical Engineering
    Department
University of Southern
    California
Los Angeles, California

Patrick Reynaert
Elektrotechniek, ESAT-MICAS
Katholieke Universiteit Leuven
Leuven-Heverlee, Belgium

Mahsan Rofouei
Nanoelectronics Center
    of Excellence
School of Electrical and
    Computer Engineering
University of Tehran
Tehran, Iran

J. Gregory Rollins
Antrim Design Systems
Scotts Valley, California

Saeed Safari
Nanoelectronics Center
    of Excellence
School of Electrical and
    Computer Engineering
University of Tehran
Tehran, Iran

Kirad Samavati
Stanford University
Stanford, California

Naresh R. Shanbhag
Electrical and Computer
    Engineering Department
University of Illinois at
    Urbana-Champaign
Urbana, Illinois

Li Shang
Department of Electrical
    and Computer Engineering
Queen's University
Kingston, Ontario, Canada

Rick Shih-Jye Shen
National Tsing-Hua University
Hsinchu, Taiwan

Bing J. Sheu
Taiwan Semiconductor
    Manufacturing Company
Taiwan

Dongkun Shin
Samsung Electronics Co., LTD.
Suwon, Korea

Muh-Tian Shiue
Department of Electrical
    Engineering
National Central University
Chung-Li, Taiwan

Hamid Shojaei
Electrical and Computer
    Engineering Department
University of Tehran
Tehran, Iran

Bang-Sup Song
Department of Electrical and
    Computer Engineering
University of California
San Diego, California

Michiel Steyaert
Katholieke Universiteit
    Leuven
Leuven-Heverlee, Belgium

Earl E. Swartzlander, Jr.
Department of Electrical
    and Computer
    Engineering
University of Texas
Austin, Texas

Haruyuki Tago
Toshiba Semiconductor
    Company
Saiwai, Kawasaki, Japan

Naofumi Takagi
Department of Information
    Engineering
Nagoya University
Nagoya, Japan

Emil Talpes
Advanced Micro Devices, Inc.
Sunnyvale, California

Baris Taskin
Department of Electrical
    and Computer
    Engineering
Drexel University
Philadelphia, Pennsylvania

Donald C. Thelen
American Microsystems, Inc.
Bozeman, Montana

T. Theocharides
Department of Electrical and
    Computer Engineering
University of Cyprus, Cyprus

Yosef Tirat-Gefen
Department of Electrical
    Engineering-Systems
University of Southern
    California
Los Angeles, California

Chris Toumazou
Institute of Biomedical
    Engineering
University of London
London, UK

Spyros Tragoudas
Department of Electrical
    and Computer
    Engineering
Southern Illinois University
Carbondale, Illinois

Yuh-Kuang Tseng
Industrial Research and
    Technology Institute
Hsinchu, Taiwan

N. Vijaykrishnan
Department of Computer
    Science and Engineering
Pennsylvania State University
University Park, Pennsylvania

Suhrid A. Wadekar
Department of Electrical
    Engineering-Systems
University of Southern
    California
Los Angeles, California

Chorng-Kuang Wang
Department of Electrical
    Engineering
National Taiwan University
Taipei, Taiwan

R. F. Wassenaar
Department of Electrical
    Engineering
University of Twente
The Netherlands

Louis A. Williams III
Texas Instruments, Inc.
Dallas, Texas

Wayne Wolf
Department of Electrical
    Engineering
Princeton University
Princeton, New Jersey

Chung-Yu Wu
Department of Electronics
    Engineering
National Chiao Tung University
Hsinchu, Taiwan

Evans Ching-Song Yang
National Tsing-Hua University
Hsinchu, Taiwan

Kazuo Yano
System LSI Research
    Department
Central Research Laboratory
Hitachi Ltd.
Kokubunji, Tokyo, Japan

Ko Yoshikawa
CAD Department
Computers Division
NEC Corporation
Fuchu, Tokyo, Japan

Kuniyoshi Yoshikawa
Quality Promotion Center
Semiconductor Company
Toshiba Corporation
Yokohama, Japan

Takashi Yoshitomi
System LSI Division I
Semiconductor Company
Toshiba Corporation
Tokyo, Japan

Min-Shueh Yuan
Department of Electrical
    Engineering
National Taiwan University
Taipei, Taiwan

C. Patrick Yue
Stanford University
Stanford, California

Table of Contents

SECTION I   VLSI Technology

SECTION II Devices and Their Models

SECTION III Low Power Electronics and Design

SECTION IV Amplifiers

SECTION V   Logic Circuits

SECTION VI   Memory, Registers and System Timing

SECTION VII   Analog Circuits

SECTION VIII   Microprocessor and ASIC

SECTION IX   Testing of Digital Systems

SECTION X   Compound Semiconductor Integrated Circuit Technology

SECTION XI   Design Automation

Section I

VLSI Technology

John Choma, Jr.
University of Southern California

1

Bipolar Technology

B. Gunnar Malm
Royal Institute of Technology

Jan V. Grahn
Royal Institute of Technology

Mikael Östling
Royal Institute of Technology

CONTENTS

1.1 Introduction

The development of a bipolar technology for integrated circuits (ICs) went hand in hand with the steady improvement in semiconductor materials and discrete components during the 1950s and 1960s. Consequently, silicon bipolar technology formed the basis for the IC market during the 1970s. As circuit dimensions shrink, the MOSFET (or MOS) has gradually taken over as the major technological platform for silicon ICs. The main reasons are the ease of miniaturization and high yield for MOS compared with bipolar technology. For VLSI circuits the low standby power of complementary MOS (CMOS) gates is a significant advantage compared with integrated bipolar circuits.

The evolution of MOS technology has followed the famous Moore's law that predicts a steady decrease in gate length. Bipolar technology has also benefited from the progress in lithography and is currently fabricated using deep UV tools with feature sizes close to 100 nm. The scaling has led to a significant performance improvement and is further illustrated in Figure 1.1, where the reported gate delay time for emitter coupled logic (ECL) and current mode logic (CML) circuits is plotted for a 10-year period. In addition to the reduced dimensions, the introduction of SiGe epitaxy for the base region has further pushed the performance limits. SiGe bipolars are now considered a mature technology and is mainly offered as a high-speed complement to the low-power MOS in the so-called BiCMOS technology. By adding a small amount of carbon to the SiGe epitaxial base, better profile control and compatibility with MOS process flows have been obtained [1].

A mature Si bipolar technology with implanted base at the 0.25 μm MOS technology node offers 12 ps gate delay and can be used to realize 10 Gb/s ICs [2]. The continuous performance increase owing to reduced dimension is illustrated in Table 1.1, where several generations of a commercial BiCMOS technology are compared [3]. As the dimensions are reduced, the traditional local oxidation of silicon (LOCOS) isolation technology is replaced by shallow and deep trenches to increase the packing density, and also to optimize the process flow by getting a more planar structure. As seen in the table, the epitaxial SiGe base markedly improves the device performance at the same technology node.

Apart from high-speed performance, the bipolar transistor is recognized by its excellent analog properties which feature high linearity, superior low- and high-frequency noise behavior as well as very high transconductance [4]. Such properties are highly desirable for many RF applications, both for narrow-band and broad-band circuits [5]. The high current drive capability per unit silicon area makes the bipolar transistor suitable for input/output stages in many IC designs (e.g., in fast SRAMs). The disadvantage of bipolar technology is the low transistor density, combined with large power dissipation.

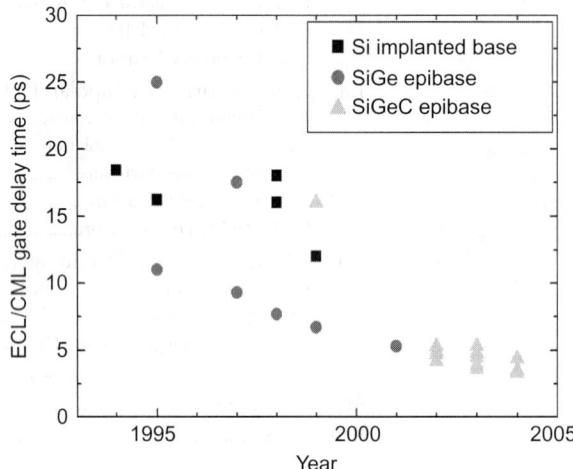

FIGURE 1.1   Reported gate delay time for bipolar ECL and CML circuits versus year.

TABLE 1.1   Technology Evolution and Performance Trends of Commercial BiCMOS Technologies

	Si Implanted Base				SiGe Epibase
Emitter lithography (μm)	1.0	0.7	0.5	0.25	0.25
f_T (GHz)	13	20	30	40	70
f_{MAX} (GHz)s			60	90	>100
Isolation		LOCOS		STI/DTI	
CMOS gate length (μm)	0.78	0.70	0.42	0.14	0.14
Metal layers	3	3	4	5/6	5/6

Source: Modified from Deixler P. et al., *IEEE Bipolar/BiCMOS Circuits Technol. Meeting Tech. Dig.*, 201, 2002.

High-performance bipolar circuits are therefore normally fabricated at a modest integration level (MSI/LSI). By using BiCMOS design, the benefits of both MOS and bipolar technology are utilized [6]. One example is mixed analog/digital systems, where a high-performance bipolar process is integrated with high-density CMOS [7]. This technology forms a vital part in several system-on-a-chip designs for telecommunication and wireless circuits.

In this chapter, a brief overview of bipolar technology is given with an emphasis on the integrated silicon bipolar transistor. The information presented here is based on the assumption that the reader is familiar with bipolar device fundamentals and basic VLSI process technology. Bipolar transistors are treated in detail in well-known textbooks by Ashburn [8] and Roulston [9]. Section 1.2 will outline the general concepts in bipolar process design and optimization. Three generations of integrated devices representing state-of-the-art bipolar technologies for the 1970s, 1980s, and 1990s will be presented in Sections 1.3, 1.4, and 1.5, respectively. Finally, some future trends in bipolar technology are outlined.

1.2 Bipolar Process Design

The design of a bipolar process starts with the specification of the application target and its circuit technology (digital or analog). This leads to a number of requirements formulated in device parameters and associated figures of merit. These are mutually dependent, and a parameter trade-off must therefore be made, making the final bipolar process design a compromise between various conflicting device requirements.

1.2.1 Figures of Merit

In the digital bipolar process, the cutoff frequency (f_T) is a well-known figure of merit for speed. The f_T is defined for a common-emitter configuration with its output short circuit when extrapolating the small signal current gain to unity. From a circuit perspective, a more adequate figure of merit is the gate delay time (t_d) measured for a ring-oscillator circuit containing an odd number of inverters [10]. The t_d can be expressed as a linear combination of the incoming time constants weighted by a factor determined by the circuit topology (e.g., ECL) [10, 11]. Alternative expressions for t_d calculations have been proposed [12]. Besides speed, power dissipation can also be a critical issue in densely packed bipolar digital circuits, resulting in the power-delay product as a figure of merit [13].

In the analog bipolar process, the DC properties of the transistor are of utmost importance. This involves minimum values on common-emitter current gain (β), Gummel plot linearity (β_{max}/β) breakdown voltage (BV_{CEO}), and early voltage (V_A). The product $\beta \times V_A$ is often introduced as a figure of merit for the device DC characteristics [14]. Rather than f_T, the maximum oscillation frequency $F_{MAX} = \sqrt{F_T/(8\pi R_B C_{BC})}$, is preferred as a figure of merit in high-speed analog design, where R_B and C_{BC} denote the total base resistance and the base-collector capacitance, respectively [15]. Alternative figures of merit for speed have been proposed in the literature [16, 17]. Analog bipolar circuits are often crucially dependent on a certain noise immunity, leading to the introduction of the corner frequency and noise figure as figures of merit for low-frequency and high-frequency noise properties, respectively [18].

1.2.2 Process Optimization

The optimization of the bipolar process is divided between the intrinsic and extrinsic device design. This corresponds to the vertical impurity profile and the horizontal layout of the transistor, respectively [10]; see example in Figure 1.2, where the device cross section is also included. It is clear that the vertical profile and horizontal layout are primarily dictated by the given process and lithography constraints, respectively.

Figure 1.3 shows a simple flowchart of the bipolar design procedure. Starting from the specified DC parameters at a given bias point, the doping profiles can be derived. The horizontal layout must be adjusted for minimization of the parasitics. A (speed) figure of merit can then be calculated. An implicit relation is thus obtained between the figure of merit and the processing parameters [11, 19]. In practice,

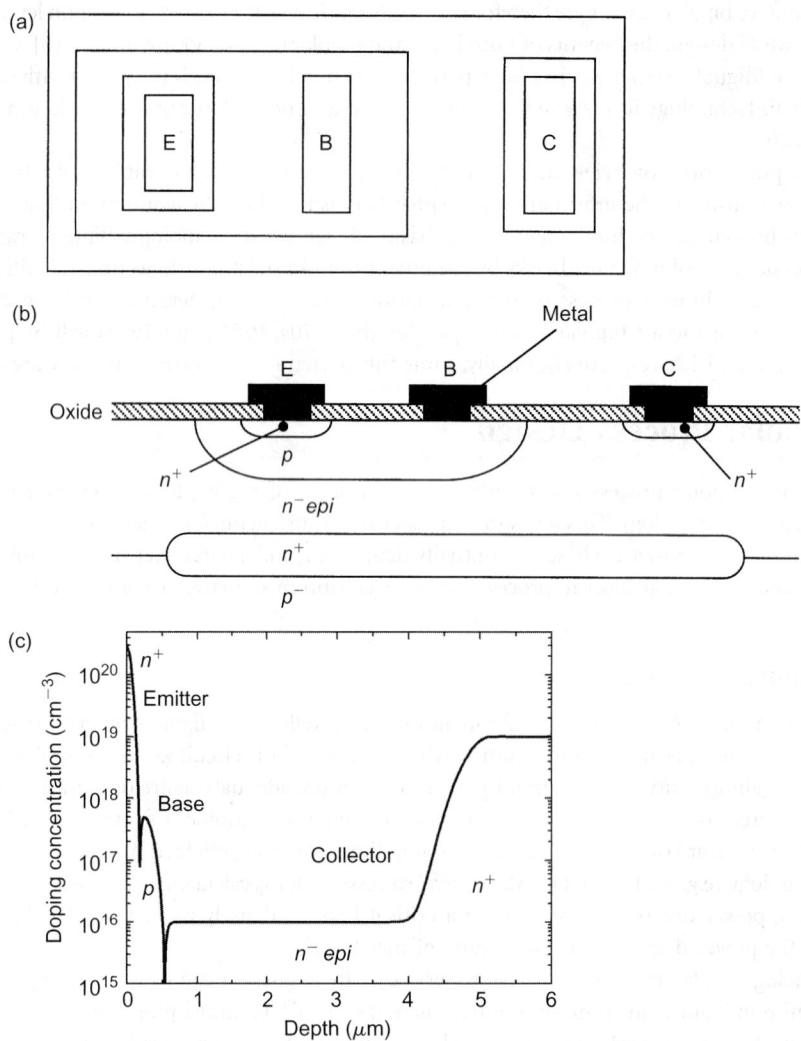

FIGURE 1.2   (a) Layout, (b) cross section, and (c) example of impurity profile through emitter window for an integrated bipolar transistor (E, emitter; B, base; C, collector).

several iterations must be performed in the optimization loop to find an acceptable compromise between the device parameters. This procedure is substantially alleviated by one- or two-dimensional process simulations of the device fabrication as well as finite-element physical device simulations of the bipolar transistor [20, 21]. For optimization of a large number of device parameters, the strategy is based on screening out the unimportant factors, combined with a statistical approach (e.g., response surface methodology) [22, 23].

1.2.3   Vertical Structure

The engineering of the vertical structure involves the design of the collector, base, and emitter impurity profiles. In this respect, f_T is an adequate parameter to optimize. For a modern bipolar transistor with suppressed parasitics, the maximum f_T is usually determined by the forward transit time of minority carriers through the intrinsic component. The most important f_T trade-off is against BV_{CEO}, as stated by the Johnson limit for silicon transistors [24], the product $f_T \times BV_{CEO}$ cannot exceed 200 GHz V. A more detailed calculation taking into account realistic doping profile predicts values of >500 GHz V) [25]. In

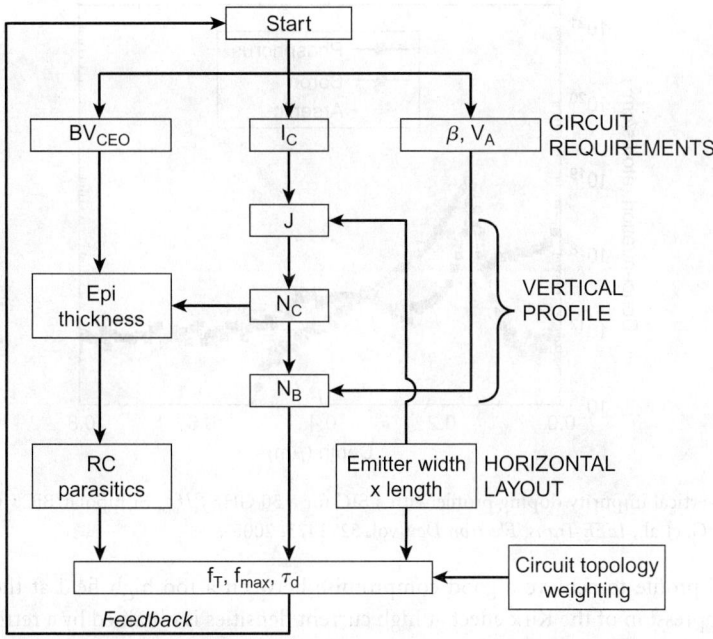

FIGURE 1.3 Generic bipolar device optimization flowchart.

fact, recent experimental results for high-speed SiGeC bipolar transistors have shown a value of 510 GHz V for a 300 GHz f_T technology with 1.7 V BV_{CEO} [26].

1.2.4 Collector Region

The vertical n-type collector of the bipolar device in Figure 1.2 consists of two regions below the p-type base diffusion: a low or moderately doped n-type epitaxial (epi) layer, followed by a highly doped n+ subcollector. The thickness and doping level of the subcollector are noncritical parameters; a high arsenic or antimony doping density between 10^{19} and 10^{20} cm^{-3} is representative, resulting in a sheet resistance of 20–40 Ω/sq. In contrast, the design of the epilayer constitutes a fundamental topic in bipolar process optimization.

To first-order, the collector doping in the epilayer is determined by the operation point (more specifically, the collector current density) of the component (see Figure 1.3). A normal condition is to have the operation point corresponding to maximum f_T, which typically means a collector current density of the order of 2–4 × 10^4 A/cm². As will be recognized later, bipolar scaling results in increased collector current densities. Above a certain current level, there will be a rapid roll-off in current gain and cutoff frequency. This is due to high-current effects, primarily the base pushout or Kirk effect, leading to a steep increase in the forward transit time [27]. Since the critical current value is proportional to the collector doping [28], a minimum impurity concentration for the epilayer is required, thus avoiding f_T degradation (typically around 10^{17} cm^{-3} for a high-speed device). Usually, the epilayer is doped only in the intrinsic structure by a selectively implanted collector (SIC) procedure [29]. An example of such a doping profile from an advanced 0.25 μm BiCMOS technology is seen in Figure 1.4. Such a collector design permits an improved control over the base-collector junction, that is, shorter base width as well as suppressed Kirk effect. The high collector doping concentration, however, may be a concern for both C_{BC} and BV_{CEO}. The latter value will therefore often set a higher limit on the collector doping value.

The SIC technology provides a simple way of creating a high-BV_{CEO} device by masking the implantation. The reduced collector doping in the SIC-free device will also reduce the pinch-base resistance and C_{BC}, which is in favor of high f_{max} [3].

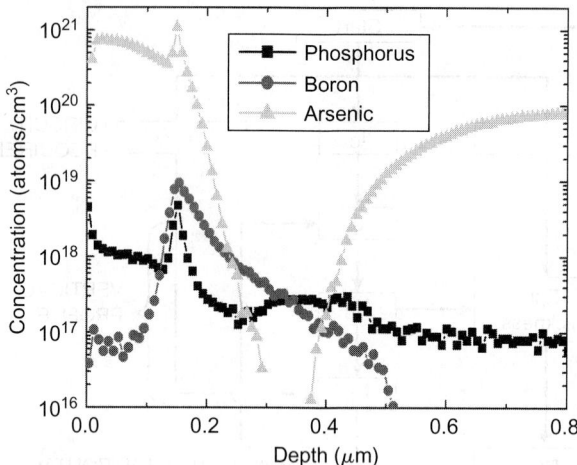

FIGURE 1.4   Vertical impurity doping profile with a SIC for a 50 GHz f_T/f_{max} Si-bipolar/BiCMOS technology (from Malm, B. G. et al., *IEEE Trans. Electron Dev.* vol. 52, 1423, 2005).

The preferred profile to achieve a good compromise between a too high field at the base-collector junction and suppression of the Kirk effect at high current densities is obtained by a retrograde collector profile [30]. For this profile the SIC implantation energy is chosen to obtain a low impurity concentration near the base-collector junction and then increasing toward the subcollector.

The thickness of the epilayer exhibits large variations among different device designs, extending several micrometers in depth for analog bipolar components, whereas a high-speed digital design typically has an epilayer thickness around 1 μm or below, thus reducing the total collector resistance. As a result, the transistor breakdown voltage is sometimes determined by reach-through breakdown (i.e., full depletion of penetration of the epicollector). The thickness of the collector layer can therefore be used as a parameter in determining BV_{CEO}, which in turn is traded off against f_T.

In cases where f_{max} is of interest, the collector design must be carefully taken into account. Compared with f_T, the optimum f_{max} is found for thicker and less doped collector epilayers [32, 33]. The vertical collector design will therefore, to a large extent, determine the trade-off between f_T and f_{max}.

1.2.5   Base Region

The width and peak concentration of the base profile are two of the most fundamental parameters in vertical profile design. In a conventional Si bipolar process the base width is limited by the implantation energy and to some extent the collector doping, since an implanted profile will have a Gaussian tail toward the collector. The base width W_B is normally in the range 0.1–1 μm, whereas a typical base peak concentration lies between 10^{17} and 10^{18} cm^{-3}. In contrast to this, base widths of <100 Å [34] with peak doping close to 10^{20} cm^{-3} can be achieved by SiGe epitaxy including a small amount of carbon for added profile control. The integral of the base doping over the base width is known as the Gummel number. The current gain of the transistor is determined by the ratio of the Gummel number in the emitter and base. In an SiGe transistor, the current gain is also strongly (exponentially) dependent on the Ge concentration in the base and therefore a higher base doping can be used without sacrificing gain. Usually, a current gain of at least 100 is required for analog bipolar transistors, whereas in digital applications, a β value around 20 is often acceptable. A normal base sheet resistance (or pinch resistance) for conventional bipolar processes is of the order of 100 Ω/sq, whereas the number for high-speed devices (implanted or epitaxial base) typically is in the interval 1 to 10 kΩ/sq [3]. This is due to the small W_B (<0.1 μm) necessary for a short base transit time. However, a too narrow base will have a negative impact on f_{max} because of its R_B dependence. As a result, f_T and f_{max} exhibit a maximum when plotted against W_B [35] or base doping [36].

The base impurity concentration must be kept high enough to avoid punch-through at low collector voltages; that is, the base-collector depletion layer penetrates across the neutral base. In other words, the base doping level is also dictated by the collector impurity concentration. Punch-through is the ultimate consequence of base width modulation or the early effect manifested by a finite output resistance in the I_C–V_{CE} transistor characteristic [37]. The associated V_A or the product $\beta \times V_A$ serves as an indicator of the linear properties for the bipolar transistor. The V_A is typically at a relatively high level (>30 V) for analog applications, whereas digital designs often accept relatively low V_A (<15 V).

A limiting factor for high base doping numbers above 5×10^{18} cm^{-3} is the onset of forward-biased tunneling currents in the emitter-base junction leading to nonideal base current characteristics [38]. Since the tunneling current is dependent on mid-bandgap states induced by process steps such as implantation, significantly lower tunneling has been reported for epitaxial base devices [39]. Therefore, base dopings well above 10^{19} cm^{-3} can be used in SiGe HBTs, with no significant nonideality observed in the base current.

The shape of the base profile has some influence on the device performance. The final base profile is the result of an implantation and diffusion process and, normally, only the peak base concentration is given along with the base width. Nonetheless, there will be an impurity grading along the base profile (see Figure 1.2 and Figure 1.4), creating a built-in electrical field and thereby adding a drift component for the minority carrier transport [40]. For very narrow base transistors, the uniform doping profile is preferable when maximizing f_T [41, 42]. This is also valid under high injection conditions in the base [43]. Uniformly doped base profiles are common in advanced bipolar processes using epitaxial techniques for growing the intrinsic base.

1.2.6 Emitter Region

The conventional metal-contacted emitter is characterized by an abrupt arsenic or phosphorus profile fabricated by direct diffusion or implantation into the base area (see Figure 1.2) [44]. In keeping emitter efficiency close to unity (and thus high current gain), the emitter junction cannot be made too shallow (~1 μm). The emitter doping level lies typically between 10^{20} and 10^{21} cm^{-3} close to the solid solubility limit at the silicon surface, hence providing a low emitter resistance as well as a large emitter Gummel number required for keeping current gain high. Bandgap narrowing, however, will be present in the emitter, causing a reduction in the efficient emitter doping [45].

When scaling bipolar devices, the emitter junction must be made shallower to ensure a low emitter-base capacitance. When the emitter depth becomes less than the minority carrier recombination length, the current gain will inevitably degrade. This precludes the use of conventional emitters in a high-performance bipolar technology. Instead, polycrystalline (poly) silicon emitter technology is utilized. By diffusing impurity species from the polysilicon contact into the monocrystalline (mono) silicon, a very shallow junction (<0.2 μm) is formed; yet gain can be kept at a high level and even traded off against a higher base doping [46]. A gain enhancement factor between 3 and 30 for the polysilicon compared with the monosilicon emitter has been reported (see also Section 1.4) [47, 48].

1.2.7 Horizontal Layout

The horizontal layout is carried out to minimize the device parasitics. Figure 1.5 shows the essential parasitic resistances and capacitances for a schematic bipolar structure containing two base contacts. The various RC constants in Figure 1.5 introduce time delays. For conventional bipolar transistors, such parasitics often limit device speed. In contrast, the self-alignment technology applied in advanced bipolar transistor fabrication allows for efficient suppression of the parasitics.

In horizontal layout, f_{max} serves as a first-order indicator in the extrinsic optimization procedure because of its dependence on C_{BC} and (total) R_B. These two parasitics are strongly connected to the geometrical layout of the device. The more advanced t_d calculation takes all major parasitics into account under given load conditions, thus providing good insight into the various time delay contributions of a bipolar logic gate [49].

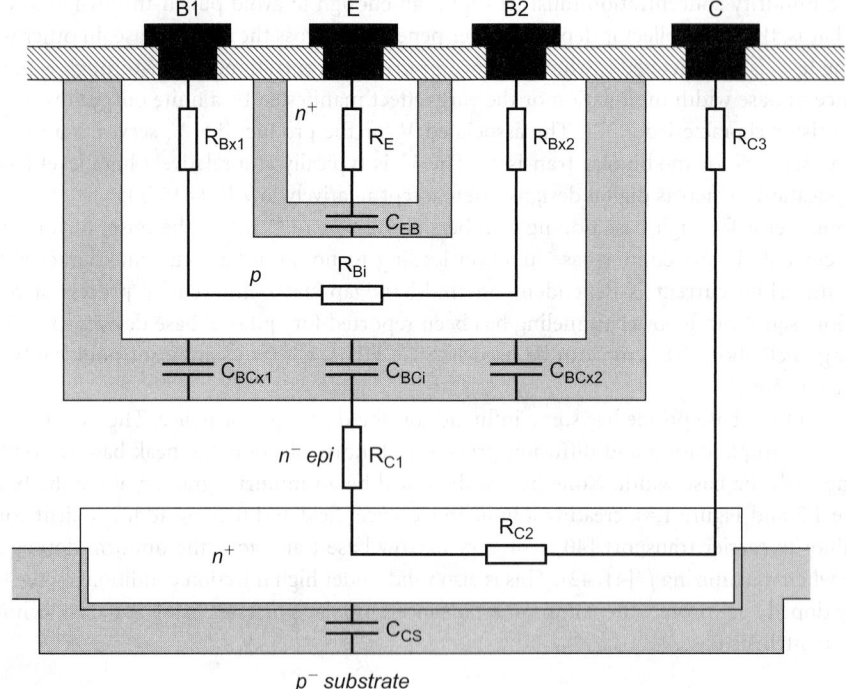

FIGURE 1.5 Schematic view of the parasitic elements in a bipolar transistor equipped with two base contacts. R_E, emitter resistance; R_{Bi}, intrinsic base resistance; R_{Bx}, extrinsic base resistance; R_C, collector resistance; C_{EB}, emitter-base capacitance; C_{BCi}, intrinsic base-collector capacitance; C_{BCx}, extrinisic base-collector capacitance; C_{CS}, collector-substrate capacitance. Gray areas denote depletion regions. Contact resistances are not shown.

From Figure 1.5, it is seen that the collector resistance is divided into three parts. Apart from the epilayer and buried layer previously discussed, the collector contact also adds a series resistance. Provided the epilayer is not too thick, the transistor is equipped with a deep phosphorus plug from the collector contact down to the buried layer, thus reducing the total RC.

As illustrated in Figure 1.5, the base resistance is divided into intrinsic (R_{Bi}) and extrinsic (R_{Bx}) components. The former is the pinch-base resistance situated directly under the emitter diffusion, whereas the latter constitutes the base regions contacting the intrinsic base. The intrinsic part decreases with the current owing to the lateral voltage drop in the base region [50]. At high current densities, this causes current crowding effects at the emitter diffusion edges. This results in a reduced onset for high-current effects in the transistor. The extrinsic base resistance is bias independent and must be kept as small as possible (e.g., by utilizing self-alignment architectures). By designing a device layout with two or more base contacts surrounding the emitter, the final R_B is further reduced at the expense of chip area. Apart from enhancing f_{max}, the R_B reduction is also beneficial for device noise performance.

The layout of the emitter is crucial since the effective emitter area defines the intrinsic device cross section [51]. The minimum emitter area, within the lithography constraints, is determined by the operational collector current and the critical current density, where high-current effects start to occur [52]. Eventually, a trade-off must be made between the base resistance and device capacitances as a function of emitter geometry; this choice is largely dictated by the device application. Long, narrow emitter stripes, meaning a reduction in the base resistance, are frequently used. The emitter resistance is usually non-critical for conventional devices; however, for polysilicon emitters, the emitter resistance may become a concern in very small-geometry layouts [53].

Of the various junction capacitances in Figure 1.5, the collector-base capacitance is the most significant. This parasitic is also divided into intrinsic (C_{BCi}) and extrinsic (C_{BCx}) contributions. Similar to R_{Bx}, the C_{BCx}

is kept low by using self-aligned schemes. For example, the fabrication of an SIC causes an increase only in C_{BCi}, whereas C_{BCx} stays virtually unaffected. The collector-substrate capacitance C_{CS} is one of the minor contributors to f_T; the C_{CS} originates from the depletion regions created in the epilayer and under the buried layer. C_{CS} will become significant at very high frequencies owing to substrate coupling effects [54].

1.3   Conventional Bipolar Technology

Conventional bipolar technology is based on the device designs developed during the 1960s and 1970s. Despite its age, the basic concept still constitutes a workhorse in many commercial analog processes where ultimate speed and high packing density are not of primary importance. In addition, a conventional bipolar component is often implemented in low-cost BiCMOS processes.

1.3.1   Junction-Isolated Transistors

The early planar transistor technology took advantage of a reverse-biased pn junction in providing the necessary isolation between components. One of the earliest junction-isolated transistors, the so-called triple-diffused process, is simply based on three ion implantations and subsequent diffusion [55]. This device has been integrated into a standard CMOS process using one extra masking step [56]. The triple-diffused bipolar process, however, suffers from a large collector resistance owing to the absence of a subcollector, and the npn performance will be low.

By far, the most common junction-isolated transistor is represented by the device cross section of Figure 1.6, the so-called buried-collector process [55]. This device is based on the concept previously shown in Figure 1.2, but with the addition of an n⁺-collector plug and isolation. This is provided by the diffused p⁺ regions surrounding the transistor. The diffusion of the base and emitter impurities into the epilayer allows relatively good control of the base width (more details of the fabrication is given in the next section on oxide-isolated transistors).

The main disadvantage of the junction-isolated transistor is the relatively large chip area occupied by the isolation region, thus precluding the use of such a device in any VLSI application. Furthermore, high-speed operation is ruled out because of the large parasitic capacitances associated with the junction isolation and the relatively deep diffusions involved. Indeed, many of the conventional junction-isolated processes were designed for doping from the gas phase at high temperatures.

1.3.2   Oxide-Isolated Transistors

Oxide isolation permits a considerable reduction in the lateral and vertical dimensions of the buried-layer collector process. The reason is that the base and collector contacts can be extended to the edge of the isolation region. More chip area can be saved by having the emitter walled against the oxide edge. The principal difference between scaling of junction- and oxide-isolated transistors is visualized in Figure 1.7. The device layouts are Schottky clamped, i.e., the base contact extends over the collector region. This hinders the transistor from entering the saturation mode under device operation.

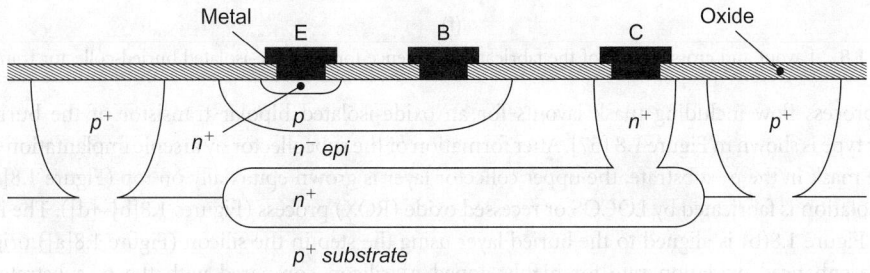

FIGURE 1.6   Cross section of the buried-collector transistor with junction isolation and collector plug.

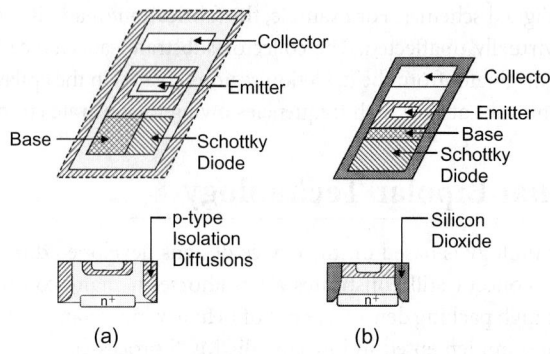

FIGURE 1.7 Device layout and cross section demonstrating scaling of (a) junction-isolated and (b) oxide-isolated bipolar transistors.

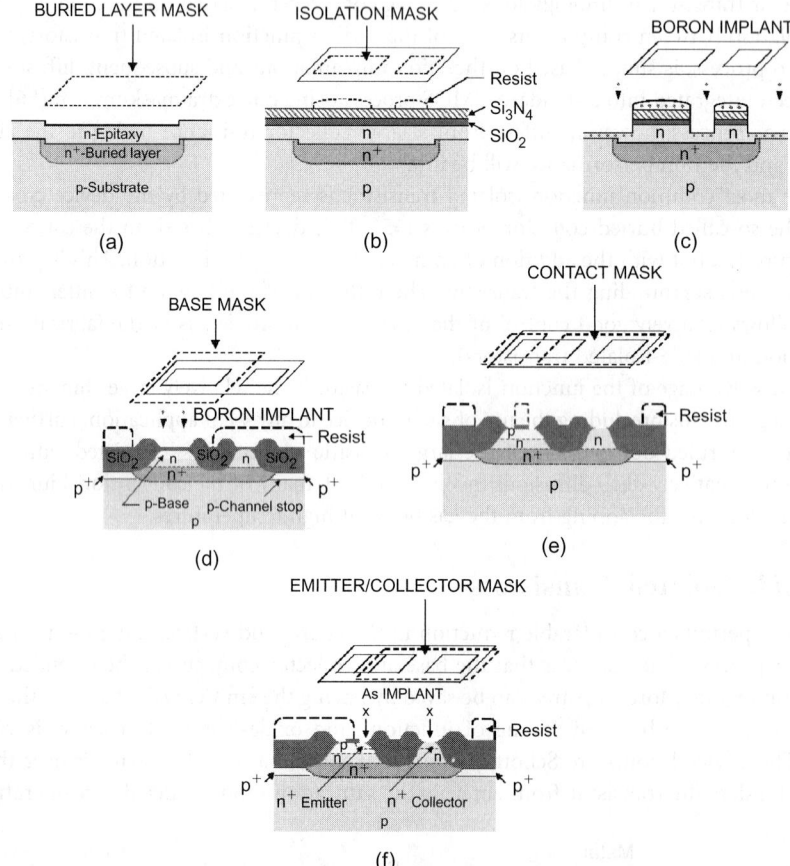

FIGURE 1.8 Layout and cross section of the fabrication sequence for an oxide-isolated buried-collector transistor.

The process flow including mask layouts for an oxide-isolated bipolar transistor of the buried-layer collector type is shown in Figure 1.8 [57]. After formation of the subcollector by arsenic implantation through an oxide mask in the p⁻ substrate, the upper collector layer is grown epitaxially on top (Figure 1.8[a]). The device isolation is fabricated by LOCOS or recessed oxide (ROX) process (Figures 1.8[b]–[d]). The isolation mask in Figure 1.8(b) is aligned to the buried layer using the step in the silicon (Figure 1.8[a]) originating from the enhanced oxidation rate for highly doped n⁺ silicon compared with the p⁻ substrate during activation of the buried layer. The ROX is thermally grown (Figure 1.8[d]) after the boron field implantation

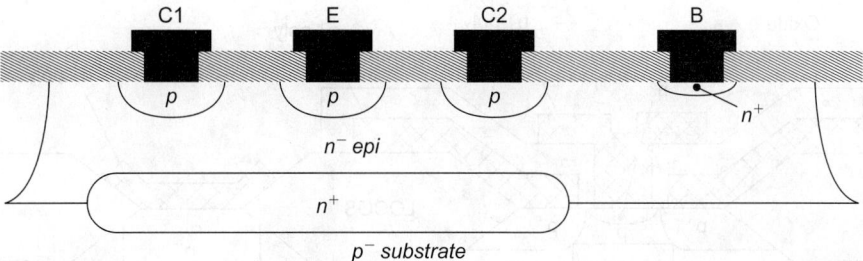

FIGURE 1.9 Schematic cross section of the lateral pnp transistor.

(or channel stop) (Figure 1.8[c]). This p$^+$ implant is necessary for suppressing a conducting channel otherwise present under the ROX. The base is then formed by ion implantation of boron or BF$_2$ through a screen oxide (Figure 1.8[d]); in the simple device of Figure 1.8, a single base implantation is used; in a more advanced bipolar process, the fabrication of the intrinsic and extrinsic base must be divided into one low dose and one high dose implantation, respectively, adding one more mask to the total flow. After base formation, an emitter/base contact mask is patterned in a thermally grown oxide (Figure 1.8[e]). The emitter is then implanted using a heavy dose arsenic implant (Figure 1.8[f]). An n$^+$ contact is simultaneously formed in the collector window. After annealing, the device is ready for metallization and passivation.

Apart from the strong reduction in isolation capacitances, the replacement of a junction-isolated process with an oxide-isolated process also adds other high-speed features such as thinner epitaxial layer and shallower emitter/base diffusions. A typical base width is a few thousand angstroms and the resulting f_T typically lies in the range of 1–10 GHz. The doping of the epitaxial layer is determined by the required breakdown voltage. Further speed enhancement of the oxide-isolated transistor is difficult due to the parasitic capacitances and resistances originating from contact areas and design-rule tolerances related to alignment accuracy.

1.3.3 Lateral pnp Transistors

The conventional npn flow permits the bipolar designer to simultaneously create a lateral pnp transistor, to be used, for example, as a bandgap reference. This is made by placing two base diffusions in close proximity to each other in the epilayer, one of them (pnp collector) surrounding the other (pnp emitter) (see Figure 1.9). In general, the lateral pnp device exhibits poor performance since the base width is determined by lithography constraints rather than vertical base control as in the npn device. In addition, there will be electron injection from the subcollector into the p-type emitter, thus reducing emitter efficiency.

1.4 High-Performance Bipolar Technology

The development of a high-performance bipolar technology for ICs signified a large step forward, both with respect to speed and packing density of bipolar transistors. A representative device cross section of a so-called double-poly transistor is depicted in Figure 1.10. The important characteristics for this bipolar technology are the polysilicon emitter contact, the advanced device isolation, and the self-aligned structure. These three features are discussed here with an emphasis on self-alignment where the two basic process flows are outlined—the single- and double-poly transistor.

1.4.1 Polysilicon Emitter Contact

The polysilicon emitter contact is fabricated by a shallow diffusion of n-type species (usually arsenic) from an implanted n$^+$-polysilicon layer into the silicon substrate [58] (see emitter region in Figure 1.10). The thin oxide sandwiched between the poly- and monosilicon is partially or fully broken up during contact formation. The mechanism behind the improved current gain is strongly related to the details

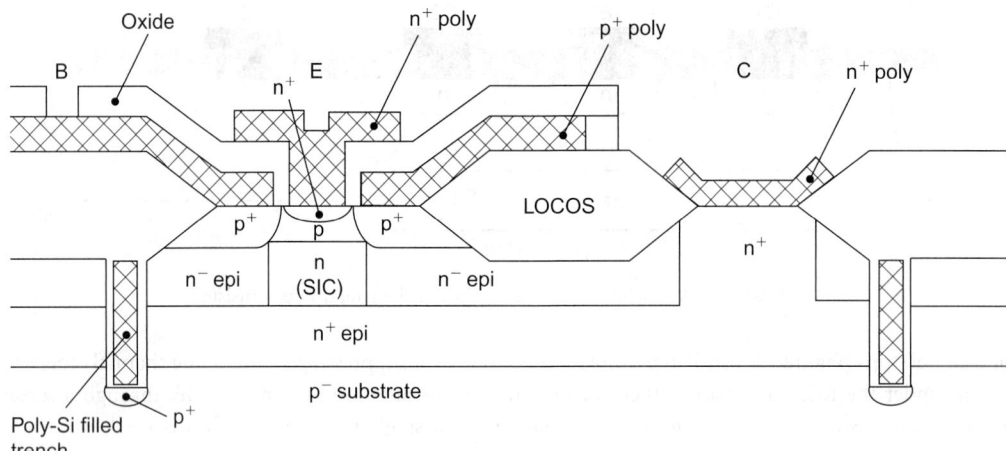

FIGURE 1.10   A double-poly self-aligned bipolar transistor with deep-trench isolation, polysilicon emitter, and SIC. Metallization is not shown.

of the interface between the polysilicon layer and the monosilicon substrate [48]. Hence, the cleaning procedure of the emitter window surface before polysilicon deposition must be carefully engineered for process robustness. Otherwise, the average current gain from wafer to wafer will exhibit unacceptable variations. The emitter window preparation and subsequent drive-in anneal conditions can also be used in tailoring the process with respect to gain and emitter resistance.

From a fabrication point of view, there are further advantages when introducing polysilicon emitter technology. By implanting into the polysilicon rather than into single-crystalline material, the total defect generation as well as related anomalous diffusion effects are strongly suppressed in the internal transistor after the drive-in anneal. Moreover, the risk of aluminum spiking during the metallization process, causing short-circuiting of the pn junction, is strongly reduced compared with the conventional contact formation. As a result, some of the yield problems associated with monosilicon emitter fabrication are, to a large extent, avoided when utilizing polysilicon emitter technology.

1.4.2   Advanced Device Isolation

With advanced device isolation, one usually refers to the deep trenches combined with LOCOS or shallow trenches [59] as seen in Figure 1.10. Before trench etching the collector region has been formed by a buried-layer implantation followed by epitaxy or a double-epitaxial layer (n$^+$–n) grown on a much lower doped p$^-$ substrate. The deep trench must reach below the buried layer, meaning a high-aspect ratio reactive-ion etch. Hence, the trenches will define the lateral extension of the buried-layer collector for the transistor.

The main reason for introducing advanced isolation in bipolar technology is the need for a compact chip layout. Quite naturally, the bipolar isolation technology has benefited from the trench capacitor development in the MOS memory area. The deep-trench isolation allows bipolar transistors to be designed at the packing density of VLSI.

The fabrication of a deep-trench isolation includes deep-silicon etching, channel-stop p$^+$ implantation, an oxide/nitride stack serving as isolation, intrinsic polysilicon fill-up, planarization, and cap oxidation [60]. The deep-trench isolation is combined with a LOCOS or shallow-trench isolation, which is added before or after deep-trench formation. The most advanced isolation schemes take advantage of shallow-trench isolation rather than ordinary LOCOS after the deep-trench process; in this way, a very planar surface with no oxide lateral encroachment ("birds beak") is achieved after the planarization step. The concern regarding stress-induced crystal defects originating from trench etching requires careful attention so as not to seriously affect yield.

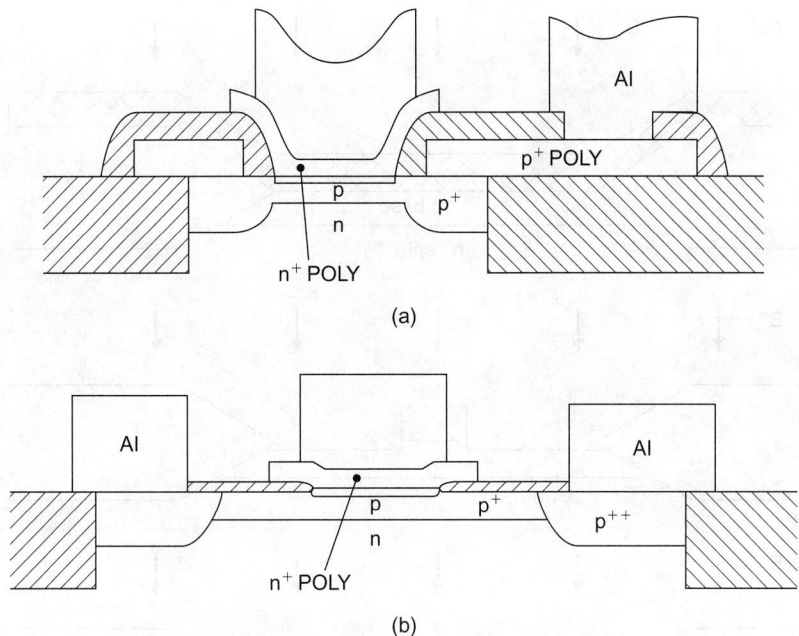

FIGURE 1.11 (a) Double-poly structure and (b) single-poly structure. Buried layer and collector contact are not shown (after Ref [61], copyright © 1989, IEEE).

1.4.3 Self-Aligned Structures

Advanced bipolar transistors are based on self-aligned structures made possible by polysilicon emitter technology. As a result, the emitter-base alignment is not dependent on the overlay accuracy of the lithography tool. The device contacts can be separated without affecting the active device area.

The self-aligned structures are divided into single-polysilicon (single-poly) and double-polysilicon (double-poly) architectures, as visualized in Figure 1.11 [61]. The double-poly structure refers to the emitter polysilicon and base polysilicon electrode, whereas the single-poly only refers to the emitter polysilicon. From Figure 1.11, it is seen that the double-poly approach benefits from a smaller active area than the single-poly one, manifested in a reduced base-collector capacitance. Moreover, the double-poly transistor in general exhibits a lower base resistance. The double-poly transistor, however, is more complex to fabricate than the single-poly device. On the other hand, by applying inside spacer technology for the double-poly emitter structure, the lithography requirements are not as strict as in the single-poly case, where more conventional MOS design rules are used for definition of the emitter electrode.

1.4.4 Single-Poly Structure

The fabrication of a single-poly transistor has been presented in several versions, more or less similar to the traditional MOS flow. An example of a standard single-poly process is shown in Figure 1.12 [62]. After arsenic emitter implantation (Figure 1.12[a]) and polysilicon patterning, a so-called base link is implanted using boron ions (Figure 1.12[b]). Oxide is then deposited and anisotropically etched to form outside spacers, followed by the heavy extrinsic base implantation (Figure 1.12[c]). Shallow junctions (including emitter diffusion) are formed by rapid thermal annealing (RTA). A salicide or polycide metallization completes the structure (Figure 1.12[d]).

Another variation of the single-poly architecture is the so-called quasi-self-aligned process (see Figure 1.13) [63]. A base oxide is formed by thermal oxidation in the active area and an emitter window

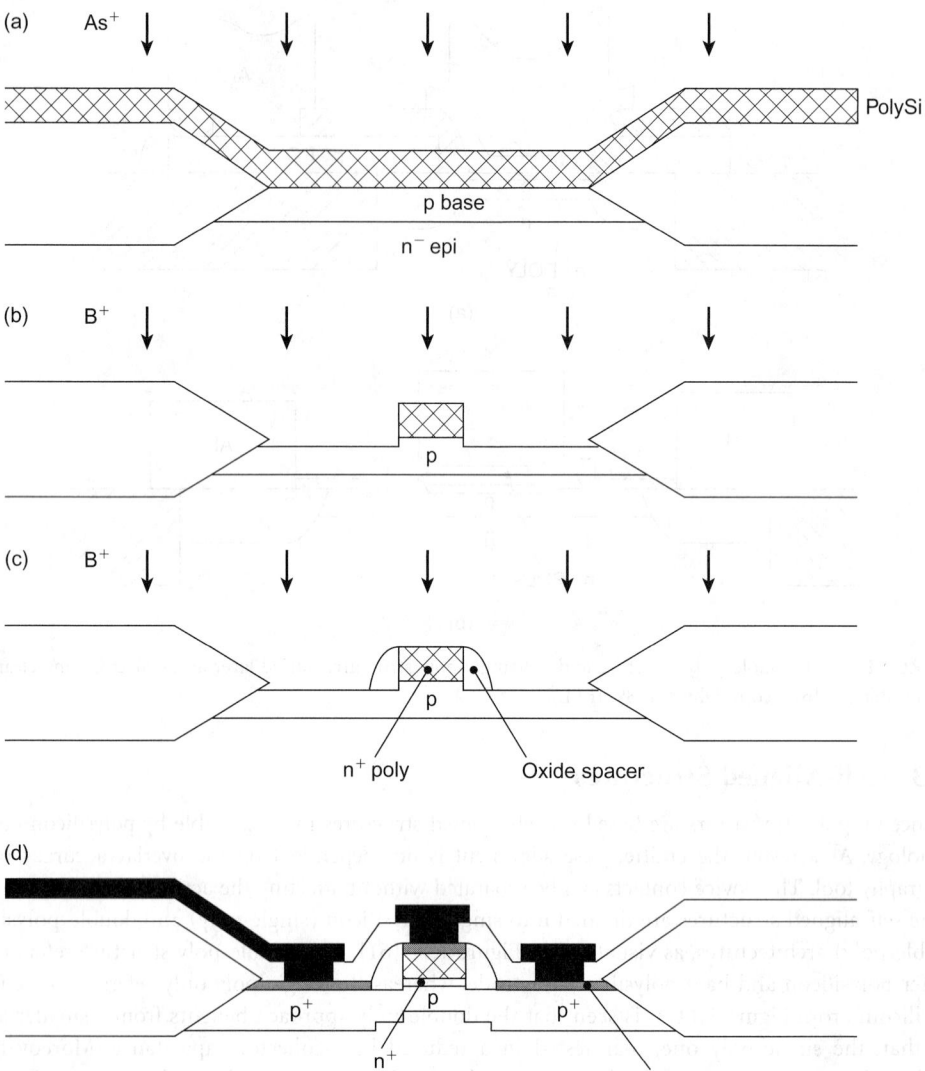

FIGURE 1.12 The single-poly, self-aligned process: (a) polyemitter implantation, (b) emitter etch and base-link implantation, (c) oxide spacer formation and extrinsic base implantation, and (d) final device after junction formation and metallization.

is opened (Figure 1.13[a]). Following intrinsic base implantation, the emitter polysilicon is deposited, implanted, and annealed. The polysilicon emitter pedestal is then etched out (Figure 1.13[b]). The extrinsic base process, junction formation, and metallization are essentially the same as in the single-poly process shown in Figure 1.13. Note that in Figure 1.13, the emitter-base formation is self-aligned to the emitter window in the oxide, not to the emitter itself, hence explaining the term "quasi-self-aligned." As a result, a higher total base resistance is obtained compared with the standard single-poly process.

The boron implantation illustrated in Figure 1.12(b) is an example of the so-called base-link engineering aimed at securing the electrical contact between the heavily doped p+-extrinsic base and the much lower doped intrinsic base. A poor base link will result in high total base resistance, whereas a base link with a too strong diffusion may create a lateral emitter-base tunnel junction leading to nonideal base current characteristics [64]. Furthermore, a poorly designed base link jeopardizes matching between individual transistors since the final current gain may vary substantially with the emitter width.

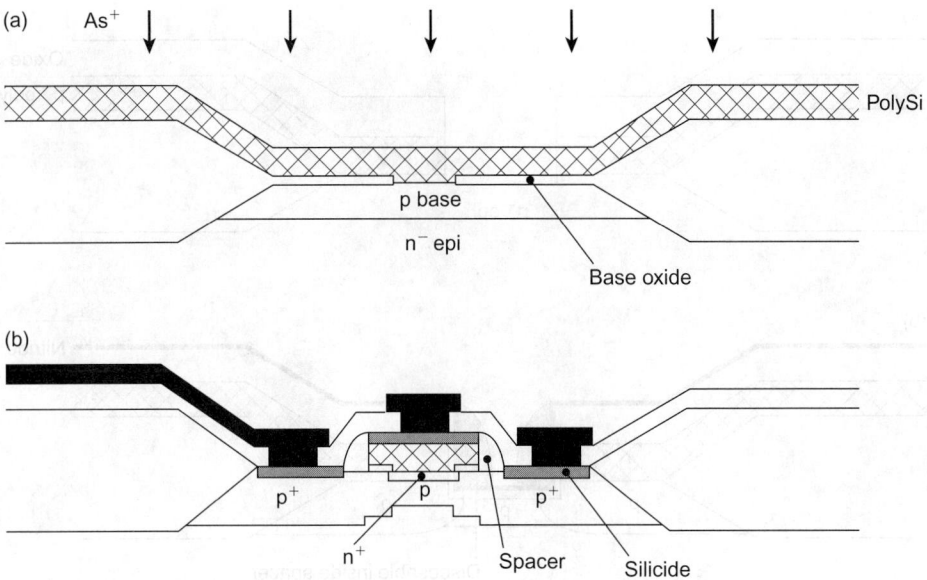

FIGURE 1.13   The single-poly, quasi-self-aligned process: (a) polyemitter implantation and (b) final device.

1.4.5 Double-Poly Structure

The double-poly structure originates from the classical IBM structure presented in 1981 [65]. Most high-performance commercial processes today are based on double-poly technology. The number of variations are less than for the single-poly one, mainly with different aspects on base-link engineering, spacer technology, and SIC formation. One example of a double-poly fabrication is presented in Figure 1.14. After deposition of the base polysilicon and oxide stack, the emitter window is opened (Figure 1.14[a]) and thermally oxidized. During this step, p^+ impurities from the base polysilicon diffuse into the monosilicon, thus forming the extrinsic base. In addition, the oxidation repairs the crystal damage caused by the dry etch when opening the emitter window. A thin silicon nitride layer is then deposited, the intrinsic base is implanted using boron, followed by the fabrication of amorphous silicon spacers inside the emitter window (Figure 1.14[b]). The nitride is exposed to a short dry etch, the spacers are removed, and the thin oxide is opened up by an HF dip. Deposition and implantation of the polysilicon emitter film is carried out (Figure 1.14[c]). The structure is patterned and completed by RTA emitter drive-in and metallization (Figure 1.14[d]). The emitter will thus be fully self-aligned with respect to the base. Note that the inside spacer technology implies that the actual emitter width will be significantly less than the drawn emitter width.

The definition of the polyemitter in the single- and double-poly process leads to some overetching into the epilayer; see Figure 1.12(b) and Figure 1.14(a), respectively. This can be alleviated by inserting, for example, a thin thermal oxide as an etch-stop layer. After emitter patterning this oxide can be removed by a short wet etch, selective to the underlying silicon. This situation is of no concern for the quasi-self-aligned process where the etch of the polysilicon emitter stops on the base oxide.

Also, vertical pnp bipolar transistors based on the double-poly concept have been demonstrated [66]. Either boron or BF_2 is used for the polyemitter implantation. A pnp device with an f_T of 35 GHz has been presented in a classical double-poly structure [67].

1.5 Advanced Bipolar Technology

This section treats state-of-the-art production-ready bipolar technologies, with focus on optimized BiCMOS process for mixed-signal and RF applications. Examples are given of BiCMOS integration at the 0.25 µm CMOS technology node. Alongside the traditional down-scaling in design rules, efforts have

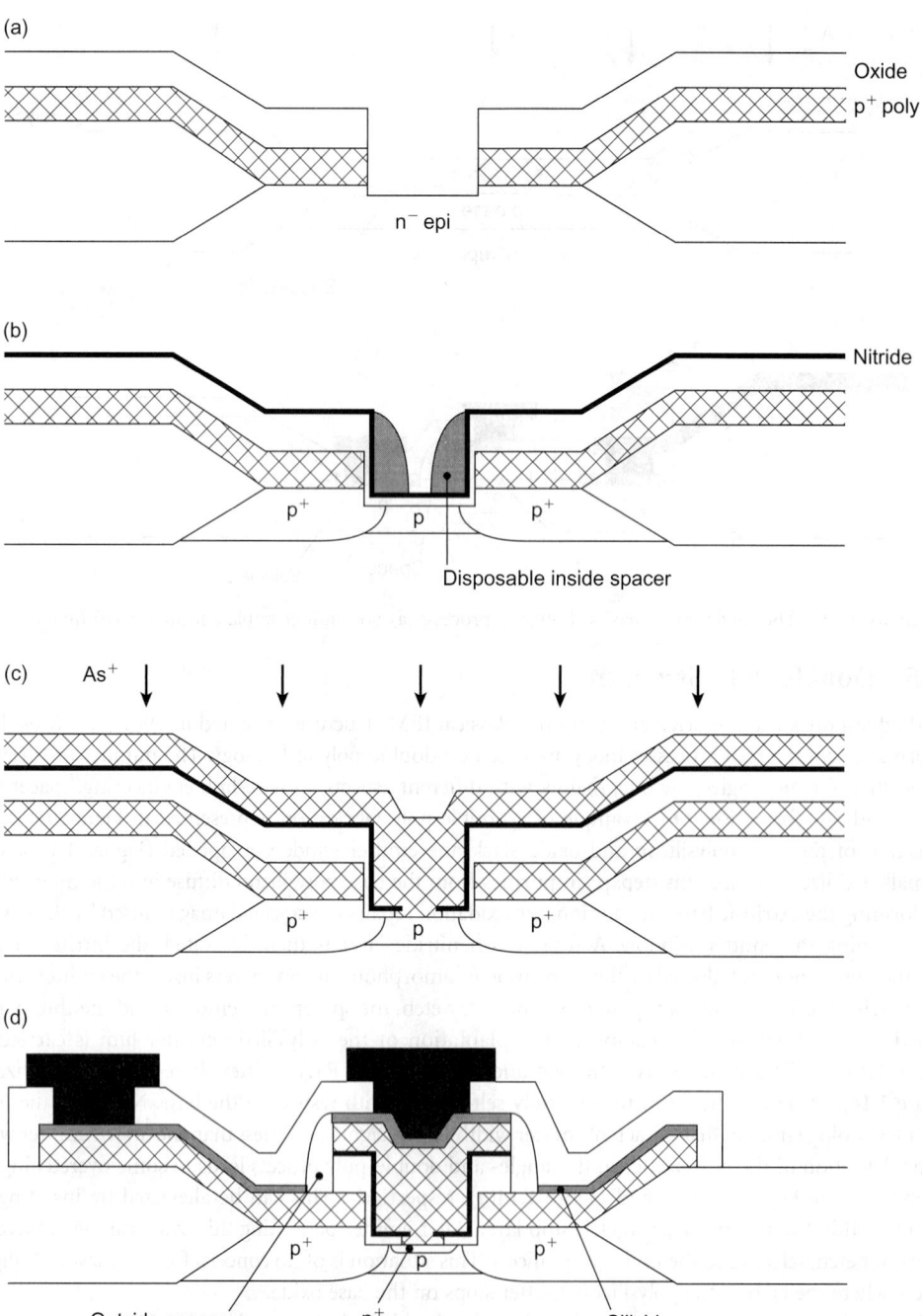

FIGURE 1.14 The double-poly, self-aligned process: (a) emitter window etch, (b) intrinsic base implantation through thin oxide/nitride stack followed by inside spacer formation, (c) polyemitter implantation, and (d) final device after emitter drive-in and metallization.

focused on new innovations in emitter and base-electrode fabrication. A key issue has been the integration of epitaxial SiGe intrinsic base into the standard npn process flow. For the emitter polysilicon electrode, in situ doping using both arsenic and phosphorus has been implemented for improved profile control and reduced influence of small emitter dimensions.

Bipolar integration on silicon-on-insulator (SOI) substrates is discussed with emphasis on thermal effects and self-heating. An example of process optimization resulting in reduced low frequency is given. This section concludes with an outlook on the future trends in bipolar technology after 2005.

1.5.1 Implanted Base

Today's most advanced commercial processes are specified with an f_T ~50 GHz [2, 7]. In these optimized processes, an in situ doped polysilicon emitter is used combined with rapid thermal processing for the final activation and drive-in of the emitter profile. In this way, an intrinsic base width of 500 nm with 12.3 kΩ pinch base resistance can be achieved. The major development has been carried out recently using double-poly technology, although similar performance has also been reported for quasi-self-aligned single-poly architectures [63, 68].

Since ion implantation is associated with a number of drawbacks such as channeling, shadowing effects, and crystal defects, it may be difficult to reach an f_T >50–60 GHz based on such a technology. Also, the emitter implantation can be removed by utilizing in situ doped emitter technology (e.g., arsine [AsH$_3$] gas during polysilicon deposition) [69]. Two detrimental effects are then avoided; namely, emitter perimeter depletion and the emitter plug effect [70]. The former effect causes a reduced doping concentration close to the emitter perimeter, whereas the latter implies the plugging of doping atoms in narrow emitter windows causing shallower junctions compared with larger openings on the same chip.

Arsenic came to replace phosphorus as the emitter impurity during the 1970s, mainly because of the emitter push-effect plaguing phosphorus monosilicon emitters. The phosphorus emitter has, however, experienced a renaissance in advanced bipolar transistors by introducing the so-called in situ phosphorus-doped polysilicon (IDP) emitter [71]. One motivation for using the IDP technology is the reduction in final emitter resistance compared with the traditional arsenic polyemitter, in particular for aggressively downscaled devices with very narrow emitter windows. In addition, the emitter drive-in for an IDP emitter is carried out at a lower thermal budget than the corresponding arsenic emitter owing to the difference in diffusivity between the impurity atoms.

Base electrode engineering in advanced devices has become an important field in reducing the total base resistance, thus improving the f_{max} of the transistor. One straightforward method in lowering the base sheet resistance is by shunting the base polysilicon with an extended silicide across the total base electrode. It is possible to realize f_{max} > 90 GHz in a production-ready double-poly process using TiSi$_2$ [7].

1.5.2 Epitaxial Base

By introducing epitaxial film growth techniques for the intrinsic base formation, the base width is readily controlled up to the order of some hundred angstroms. In combination with the high current gain, which is obtained by the SiGe base-emitter heterojunction, this allows a very high base doping to be used. Hence, epitaxial base transistors feature comparable pinch-base resistance to implanted base transistors even though the base width is significantly reduced. Note that epitaxial base growth is not used for silicon-only transistors, owing to the added process complexity. In SiGe bipolar transistors, both selective and nonselective epitaxial growth (SEG and NSEG, respectively) are currently used. In a self-aligned double-poly approach, the SEG base epitaxy replaces the base implanation after emitter opening. This is schematically illustrated in Figure 1.15(a), shown directly after base epitaxy and prior to emitter poly deposition. This type of structure relies on a good thickness control and uniformity of the epitaxial growth to form the base link between the p$^+$-poly overhangs and the SEG intrinsic base.

A self-aligned base-emitter structure can also be obtained for an NSEG base, which is formed in a blanket deposition, after completed device isolation. The purpose of the self-aligment is to minimize parasitics, such as the extrinsic base resistance, compared with the quasi-self-aligned structure, which has been widely used for this type (NSEG) of epitaxial base. In the 350 GHz transistor from IBM [72], the so-called raised extrinsic polysilicon base is formed self-aligned to a sacrificial emitter mandrel (see Figure 1.15[b]). The emitter-base dielectric isolation is also formed self-aligned to the mandrel.

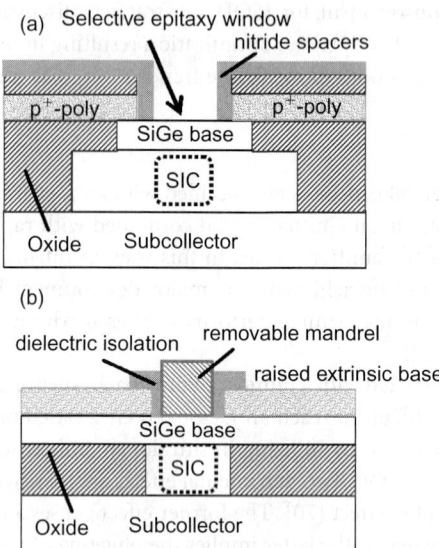

FIGURE 1.15   (a) Schematic of SEG base growth in the emitter window of a double-poly structure, after Ref. [74] (b) Structure with a self-aligned extrinsic base to emitter mandrel and NSEG base growth, after Ref. [75].

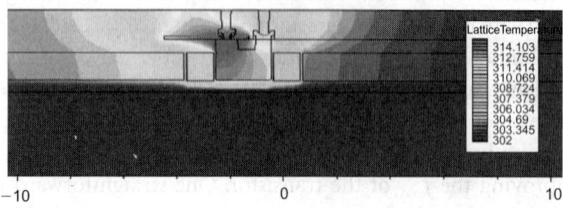

FIGURE 1.16   Simulated temperature distribution in a fully oxide insulated device using SOI substrate and deep trenches.

After removal of the mandrel the polysilicon emitter can then be formed. Another way to form a raised extrinsic base with low sheet resistance is by using selective epitaxy after the poly-emitter definition [73].

1.5.3   Bipolar Integration on SOI

Silicon-on-insulator substrates, with a thin silicon layer on top of a buried oxide, is an important process option, especially fow low-power and mixed-signal applications. The integration of MOS transistors on SOI is relatively straightforward, while bipolar integration presents some challenges. High-speed bipolar devices are operated at relatively high current density and the poor thermal conductivity of the buried oxide, compared with silicon, might lead to an increased junction temperature. For an advanced isolation scheme, using both shallow and deep trenches on an SOI substrate, the active device will be completely surrounded by oxide. In Figure 1.16, the simulated temperature distribution of a high-speed 0.25 µm bipolar transistor is shown. Fully coupled electrothermal simulations were used and the thermal boundary conditions at both the substrate and the metal interconnects were accounted for.

Future-generation MOS technologies, starting approximately from the 130 nm node, will use thin, fully depleted substrates. In this case, the traditional buried-collector layer cannot be used. The solution is to use a lateral collector structure as demonstrated by several groups [76, 77]. A high collector doping must be used to reduce saturation owing to high resistance in the undepleted part of the lateral collector. This leads to a trade-off between a low breakdown voltage and high f_T and f_{max} values.

1.5.4 Future Trends

The advances in silicon bipolar technology during the last 10-year period have been remarkable. While mixed-signal circuits for RF applications in the frequency range 1–10 GHz is still one of the main application areas, the outstanding performance of SiGeC epitaxial base bipolar transistors can be utilized for very demanding applications. State-of-the art technologies with cutoff frequencies above 300 GHz, gate delays below 4 ps are available and can be used to realize circuits operating at 100 GHz. Very high bit rates of 40–80 Gbit/s can be achieved and used for optical applications. In this area, SiGe technology will challenge high-performance III/V, (InP) bipolar technology. Integration of SiGe bipolars have been demonstrated in BiCMOS technologies at the 90 nm node and it is expected that the bipolar development will continue to follow the CMOS road map closely, where the BiCMOS solutions will be offered one or two technology nodes behind pure CMOS. The close connection to CMOS technology is seen, for example, in the choice of the metallization system. For contact metallization, nickel silicide is required in the sub-100 nm CMOS dimensions and has also been demonstrated in the SiGeC processes. An interesting application of nickel silicide technology is the demonstration of a metal emitter, developed to reduce the emitter resistance compared with the conventional polysilicon emitter technology, which has been used for more than 25 years. Bipolar technology also benefits directly from the recent transition to copper metallization, since this allows the devices to be scaled to higher current densities without concern for electromigration.

Among the most important goals for future bipolar technologies is to achieve high performance at low power dissipation. Operating at low current is also key to reduce the noise in very sensitive amplifiers. An important growing application area is automotive electronics, which has higher voltage requirements than the conventional ICs. The devices should also operate at elevated temperatures without performance degradation. A good uniformity and matching of device parameters is a traditional benefit of bipolar technology and will still make bipolar technology very competitive to deep submicron CMOS.

Acknowledgments

The authors wish to thank Susanna Norell for providing new illustrations for the second edition of this chapter. The support from the High-Frequency Electronics and Photonics Program by VINNOVA and the High-Frequency Silicon Program by the Swedish Foundation for Strategic Research is greatly acknowledged.

References

1. Osten, H. J., Knoll, D., Heinemann, B., Rücker, H., and Tillack, B., "Carbon Doped SiGe Heterojunction Bipolar Transistors for High-Frequency Applications," *IEEE Bipolar/BiCMOS Circuits Technol. Meeting, Tech. Dig.*, 109, 1999.
2. Böck, J., Knapp, H., Aufinger, K., Wurzer, M., Boguth, S., Schreiter, R., Meister, T. F., Rest, M., Ohnemus, M., and Treitinger, L., "12 ps Implanted Base Silicon Bipolar Technology," *Int. Electron Devices Meeting Tech. Dig.*, 553, 1999.
3. Deixler, P., Colclaser, R., Bower, D., Bell, N., de Boer, W., Szmyd, D., Bardy, S., Wilbanks, W., P, B., van Houdt, M., Paaschens, J. C. J., Veenstra, H., van der Heijden, E., Donkers, J. J. T. M., and Slotboom, J. W., "QUBiC4G: A f_T/f_{max} = 70/100 GHz 0.25 µm Low-Power SiGe-BiCMOS Production Technology with High-Quality Passives for 12.5 Gb/s Optical Networking and Emerging Wireless Applications up to 20 GHz Networking," *IEEE Bipolar/BiCMOS Circuits Technol. Meeting Tech. Dig.*, 201, 2002.
4. Barber, H. D., "Bipolar device technology challenge and opportunity," *Can. J. Phys.*, vol. 63, 683, 1985.

5. Baltus, P., "Influence of Process- and Device Parameters on the Performance of Portable RF Communication Circuits," *Proc. 24th Europ. Solid State Device Res. Conf.,* 1994.

6. Burghartz, J. N., "BiCMOS process integration and device optimization: Basic concepts and new trends," *Electrical Eng.,* vol. 79, 313, 1996.

7. Szmyd, D., Brock, R., Bell, N., Harker, S., Patrizi, G., Fraser, J., and Dondero, R., "QuBiC4: A Silicon Rf-BiCMOS Technology for Wireless Communication ICs," *IEEE Bipolar/BiCMOS Circuits Technol. Meeting, Tech. Dig.,* 60, 2001.

8. Ashburn, P., *SiGe Heterojunction Bipolar Transistors,* Wiley, Chichester, 2003.

9. Roulston, D. J., *Bipolar Semiconductor Devices,* McGraw-Hill, New York, 1990.

10. Tang, D. D., and Solomon, P. M., "Bipolar transistor design for optimized power-delay logic circuits," *IEEE J. Solid-State Circuits,* vol. SC-14, 679, 1979.

11. Chor, E.-F., Brunnschweiler, A., and Ashburn, P., "A propagation-delay expression and its application to the optimization of polysilicon emitter ECL processes," *IEEE J. Solid-State Circuits,* vol. 23, 251, 1988.

12. Stork, J. M. C., "Bipolar Transistor Scaling for Minimum Switching Delay and Energy Dissipation," *Int. Electron Devices Meeting Tech. Dig.,* 550, 1988.

13. Wilson, G. R., "Advances in bipolar VLSI," *Proc. IEEE,* vol. 78, 1707, 1990.

14. Prinz, E. J. a. S., J. C., "Current gain—Early voltage products in heterojunction bipolar transistors with nonuniform base bandgaps," *IEEE Electron. Dev. Lett.,* vol. 12, 661, 1991.

15. Kurishima, K., "An analytical expression of f_{max} for HBT's," *IEEE Trans. Electron Dev.,* vol. 43, 2074, 1996.

16. Taylor, G. W. and. Simmons, J. G., "Figure of merit for integrated bipolar transistors," *Solid State Electronics,* vol. 29, 941, 1986.

17. Hurkx, G. A. M., "The relevance of f_T and f_{max} for the speed of a bipolar CE amplifier stage," *IEEE Trans. Electron Dev.,* vol. 44, 775, 1997.

18. Larson, L. E., "Silicon Bipolar Transistor Design and Modeling for Microwave Integrated Circuit Applications," *Bipolar Circuits Technol. Meeting Tech. Dig.,* 142, 1996.

19. Fang, W., "Accurate analytical delay expressions for ECL and CML circuits and their applications to optimizing high-speed bipolar circuits," *IEEE J. Solid-State Circuits,* vol. 25, 572, 1990.

20. Johnson, J. B., Stricker, A., Joseph, A. J., and Slinkman, J. A., "A Technology Simulation Methodology for AC-Performance Optimization of SiGe HBTs," *Int. Electron Devices Meeting, Tech. Dig.,* 489, 2001.

21. De Vreede, L. C. N., De Graaff, H. C., Willemen, J. A., van Noort, W., Jos, R., Larson, L. E., Slotboom, J. W., and Tauritz, J. L., "Bipolar transistor epilayer design using the MAIDS mixed-level simulator," *IEEE J. Solid-State Circuits,* vol. 34, 1331, 1999.

22. Alvarez, A. R., Abdi, B. L., Young, D. L., Weed, H. D., Teplik, J., and Herald, E. R., "Application of statistical design and response surface methods to computer-aided VLSI device design," *IEEE Trans. Comp.-Aided Design,* vol. 7, 272, 1988.

23. Haralson, E., Malm, G., and Östling, M., "Device design for a raised extrinsic base SiGe bipolar technology," *Solid-State Electronics,* vol. 48, 1927, 2004.

24. Johnson, E. O., "Physical limitations on frequency and power parameters of transistors," *RCA Rev.,* vol. 26, 163, 1965.

25. Ng, K. K., Frei, M. R., and King, C. A., "Reevaluation of the ftBVCEO limit in Si bipolar transistors," *IEEE Trans. Electron Dev.,* vol. 45, 1854, 1998.

26. Khater, M., Rieh, J.-S., Adam, T., Chinthakindi, A., Johnson, J., , Krishnasamy, R., Meghelli, M., Pagette, F., Sanderson, D., Schnabel, C., Schonenberg, K. T., Smith, P., Stein, K., Stricker, A., Jeng, S.-J., Ahlgren, D., and Freeman, G., "SiGe HBT Technology with f_{max}/f_T = 350/300 GHz and Gate Delay below 3.3 ps," *Int. Electron Devices Meeting, Tech. Dig.,* 247, 2004.

27. Kirk, C. T., "A theory of transistor cut-off frequency falloff at high current densities," *IEEE Trans. Electron Dev.,* vol. ED9, 164, 1962.

28. Roulston, D. J., *Bipolar Semiconductor Devices,* McGraw-Hill, New York, 1990, p. 257.

29. Konaka, S., Amemiya, Y., Sakuma, K., and Sakai, T., "A 20 ps/G Si Bipolar IC using Advanced SST with Collector Ion Implantation," presented in *Ext. Abstracts 19th Conf. Solid-State Dev. Mater.*, Tokyo, 1987.

30. Crabbé, E. F., Meyerson, B. S., Stork, J. M. C., and Harame, D. L., "Vertical profile optimization of very high frequency epitaxial Si- and SiGe-base bipolar transistors," *Int. Electron Devices Meeting Tech. Dig.*, 83, 1993.

31. Malm, B. G., Haralson, E., Johansson, T., and Östling, M., "Self-heating effects in a BiCMOS on SOI technology for RFIC applications," *IEEE Trans. Electron Dev.*, vol. 52, 1423, 2005.

32. Kumar, M. J., Sadovnikov, A. D. and Roulston, D. J., "Collector design tradeoffs for low voltage applications of advanced bipolar transistors," *IEEE Trans. Electron Dev.*, vol. 40, 1478, 1993.

33. Kumar, M. J. and Datta, K., "Optimum collector width of VLSI bipolar transistors for maximum f_{max} at high current densities," *IEEE Trans. Electron Dev.*, vol. 44, 903, 1997.

34. Tominari, T., Wada, S., Tokunaga, K., Koyu, K., Kubo, M., Udo, T., Seto, M., Ohhata, K., Hosoe, H., Kiyota, Y., Washio, K., and Hashimoto, T., "Study on Extremely Thin Base SiGe:C HBTs Featuring sub 5 ps ECL Gate Delay," *IEEE Bipolar/BiCMOS Circuits and Technology Meeting, Tech. Dig.*, 107, 2003.

35. Roulston, D. J. and Hébert, F., "Optimization of maximum oscillation frequency of a bipolar transistor," *Solid-State Electronics*, vol. 30, 281, 1987.

36. Martinet, B., Baudry, H., Kemarrec, O., Campidelli, Y., Laurens, M., Marty, M., Schwartzmann, T., Monroy, A., Bensahel, D., and Chantre, A., "100 GHz SiGe:C HBTs using Non Selective Base Epitaxy," *Proc. 31 ESSDERC*, 97, 2001.

37. Early, J. M., "Effects of space-charge layer widening in junction transistors," *Proc. IRE*, vol. 42, 1761, 1954.

38. del Alamo, J. and Swanson, R. M., "Forward-biased tunneling: A limitation to bipolar device scaling," *IEEE Electron. Dev. Lett.*, vol. 7, 629, 1986.

39. Matutinovic-Krstelj, Z., Prinz, E. J., Schwartz, P. V., and Sturm, J. C., "Reduction of p^+–n^+ junction tunneling current for base current improvement in Si/SiGe/Si heterojunction bipolar transistors," *Electron Device Lett., IEEE*, vol. 12, 163 1991.

40. Roulston, D. J., *Bipolar Semiconductor Devices*, McGraw-Hill, New York, 1990, 220 pp.

41. Van Wijnen, P. J. and Gardner, R. D., "A new approach to optimizing the base profile for high-speed bipolar transistors," *IEEE Electron. Dev. Lett.*, vol. 4, 149, 1990.

42. Suzuki, K., "Optimum base doping profile for minimum base transit time," *IEEE Trans. Electron Dev.*, vol. 38, 2128, 1991.

43. Yuan, J. S., "Effect of base profile on the base transit time of the bipolar transistor for all levels of injection," *IEEE Trans. Electron Dev.*, vol. 41, 212, 1994.

44. Kerr, J. A. and Berz, F., "The effect of emitter doping gradient on f_T in microwave bipolar transistors," *IEEE Trans. Electron Dev.*, vol. ED-22, 15, 1975.

45. Slotboom, J. W. a. d. G., H. C., "Measurement of bandgap narrowing in silicon bipolar transistors," *Solid-State Electronics*, vol. 19, 857, 1976.

46. Cuthbertson, A. and Ashburn, P., "An investigation of the tradeoff between enhanced gain and base doping in polysilicon emitter bipolar transistors," *IEEE Trans. Electron Dev.*, vol. ED-32, 2399, 1985.

47. Ning, T. H., and Isaac, R. D., "Effect of emitter contact on current gain of silicon bipolar devices," *IEEE Trans. Electron Dev.*, vol. ED-27, 2051, 1980.

48. Post, R. C., Ashburn, P., and Wolstenholme, G. R., "Polysilicon emitters for bipolar transistors: A review and re-evaluation of theory and experiment," *IEEE Trans. Electron Dev.*, vol. 39, 1717, 1992.

49. Ashburn, P., *Design and Realization of Bipolar Transistors, chap. 7.* Wiley, Chichester, 1988.

50. Lary, J. E. and Anderson R. L., "Effective base resistance of bipolar transistors," *IEEE Trans. Electron Dev.*, vol. ED-32, 2503, 1985.

51. Rein, H.-M., "Design considerations for very-high-speed Si-bipolar ICs operating up to 50 Gb/s," *IEEE J. Solid-State Circuits*, vol. 8, 1076, 1996.

52. Schröter, M. and Walkey D. J., "Physical modeling of lateral scaling in bipolar transistors," *IEEE J. of Solid-State Circuits,* vol. 31, 1484, 1996.

53. Warnock, J. D., "Silicon bipolar device structures for digital applications: Technology trends and future directions," *IEEE Trans. Electron Dev.,* vol. 42, 377, 1995.

54. Pfost, M., Rein, H.-M., and Holzwarth, T., "Modeling substrate effects in the design of high-speed Si-bipolar IC's," *IEEE J. Solid-State Circuits,* vol. 31, 1493, 1996.

55. Lohstrom, J., "Devices and circuits for bipolar (V)LSI," *Proc. IEEE,* vol. 69, 812, 1981.

56. Wolf, S., *Silicon Processing for the VLSI Area,* vol. 2, Lattice Press, Sunset Beach, 1990, p. 532.

57. Parrillo, L. C., *VLSI Process Integration in VLSI Technology,* Sze, S. M., Ed., McGraw-Hill, Singapore, 1983, p. 449.

58. Ashburn, P., "Polysilicon Emitter Technology," *Bipolar Circuits Technol. Meeting Tech. Dig.,* vol. 1989, 90, 1989.

59. Forsberg, M., Johansson, T., Liu, W., and Vellaikal, M., "A shallow and deep trench isolation process module for RF BiCMOS," *J. Electrochem. Soc.,* vol. 151, G839, 2004.

60. Li, G. P., Ning, T. H., Chuang, C. T., Ketchen, M. B., Tang, D.D., and Mauer, J., "An advanced high-performance trench-isolated self-aligned bipolar technology," *IEEE Trans. Electron Dev.,* vol. ED-34, 2246, 1987.

61. Tang, D. D.-L., Chen, T.-C., Chuang, C. T., Cressler, J. D., Warnock, J., Li, G.-P., Polcari, M. R., Ketchen, M. B., and Ning, T. H., "The design and electrical characteristics of high-performance single-poly ion-implanted bipolar transistors," *IEEE Trans. Electron Dev.,* vol. 36, 1703, 1989.

62. De Jong, J. L., Lane, R. H., de Groot, J. G., and Conner, G. W., "Electron Recombination at the Silicided Base Contact of an Advanced Self-Aligned Polysilicon Emitter," *Bipolar Circuits Technol. Meeting, Tech. Dig.,* 202, 1988.

63. Niel, S., Rozeau, O., Ailloud, L., Hernandez, C., Llinares, P., Guillermet, M., Kirtsch, J., Monroy, A., de Pontcharra, J., Auvert, G., Blanchard, B., Mouis, M., Vincent, G., and Chantre, A., "A 54 GHz f_{max} Implanted Base 0.35 μm Single-Polysilicon Bipolar Transistor," *Int. Electron Devices Meeting, Tech. Dig.,* 807, 1997.

64. Tang, D. D., Chen, T.-C., Chuang, C.-T., Li, G. P., Stork, J. M. C., Ketchen, M. B., Hackbarth, E., and Ning, T. H., "Design considerations of high-performance narrow-emitter bipolar transistors," *IEEE Electron. Dev. Lett.,* vol. EDL-8, 174, 1987.

65. Ning, T. H., Isaac, R. D., Solomon, P. M., Tang, D. D.-L., Yu, H.-N., Feth, G. C., and Wiedmann, S. K., "Self-aligned bipolar transistors for high-performance and low-power-delay VLSI," *IEEE Trans. Electron Dev.,* vol. ED-28, 1010, 1981.

66. Maritan, C. M. and Tarr, N. G., "Polysilicon emitter p-n-p transistors," *IEEE Trans. Electron Dev.,* vol. 36, 1139, 1989.

67. Warnock, J., Lu, P.-F., Cressler, J. D., Jenkins, K. A., and Sun, J. Y. C., "35 GHz/35 psec ECL pnp Technology," *Int. Electron Devices Meeting, Tech. Dig.,* 301, 1990.

68. Chantre, A., Gravier, T., Niel, S., Kirtsch, J., Granier, A., Grouillet, A., Guillermet, M., Maury, D., Pantel, R., Regolini, J. L., and Vincent, G., "The design and fabrication of 0.35 μm single-polysilicon self-aligned bipolar transistors," *Jpn. J. Appl. Phys.,* vol. 37, 1781, 1998.

69. Burghartz, J. N., Megdnis, A. C., Cressler, J. D., Sun, J. Y.-C., Stanis, C. L., Comfort, J. H., Jenkins, K. A., and Cardone, F., "Novel in-situ doped polysilicon emitter process with buried diffusion source (BDS)," *IEEE Electron. Dev. Lett.,* vol. 12, 679, 1991.

70. Burghartz, J. N., Sun, J. Y.-C., Stanis, C. L., Mader, S. R., and Warnock, J. D., "Identification of perimeter depletion and emitter plug effects in deep-submicrometer, shallow-junction polysilicon emitter bipolar transistors," *IEEE Trans. Electron Dev.,* vol. 39, 1477, 1992.

71. Shiba, T., Uchino, T., Ohnishi, K., and Tamaki, Y., "In situ phosphorus-doped polysilicon emitter technology for very high-speed small emitter bipolar transistors," *IEEE Trans. Electron Dev.,* vol. 43, 889, 1996.

72. Rieh, J.-S., Jagannathan, B., Chen, H., Schonenberg, K. T., Angell, D., Chinthakindi, A., Florkey, J., Golan, F., Greenberg, D., Jeng, S.-J., Khater, M., Pagette, F., Schnabel, C., Smith, P., Stricker, A.,

Vaed, K., Volant, R., Ahlgren, D., Freeman, G., Stein, K., and Subbanna, S., "SiGe HBT's with cut-off frequency of 350 GHz," *Int. Electron Devices Meeting, Tech. Dig.*, 771, 2002.

73. Rücker, H., Heinemann, B., Barth, R., Bolze, D., Drews, J., Haak, U., Höppner, W., Knoll, D., Köpke, K., Marschmeyer, S., Richter, H. H., Schley, P., Schmidt, D., Scholz, R., Tillack, B., Winkler, W., Wulf, H.-E., and Yamamoto, Y., "SiGe:C BiCMOS Technology with 3.6 ps Gate Delay," *IEDM Tech. Dig.*, 121, 2003.

74. Meister, T. F., Schäfer, H., Aufinger, K., Stengl, R., Boguth, S., Schreiter, R., Rest, M., Knapp, H., Wurzer, M., Mitchell, A., Böttner, T., and Böck, J., "SiGe Bipolar Technology with 3.9 ps Gate Delay," *Proceedings of IEEE BCTM*, 103 2003.

75. Jagannathan, B., Khater, M., Pagette, F., Rieh, J.-S., Angell, D., Chen, H., Florkey, J., Golan, F., Greenberg, D. R., Groves, R., Jeng, S. J., Johnson, J., Mengistu, E., Schonenberg, K. T., Schnabel, C. M., Smith, P., Stricker, A., Ahlgren, D., Freeman, G., Stein, K., and Subbanna, S., "Self-aligned SiGe NPN transistors with 285 GHz f_{MAX} and 207 GHz f_T in a manufacturable technology," *IEEE Electron Device Lett*, vol. 23, 258, 2002.

76. Cai, J., Ajmera, A., Ouyang, C., Oldiges, P., Steigerwalt, M., Stein, K., Jenkins, K., Shahidi, G., and Ning, T., "Fully-Depleted-Collector Polysilicon-Emitter SiGe-Base Vertical Bipolar Transistor on SOI," *Tech. Dig. VLSI Symp.*, 172, 2002.

77. Avenier, G., Chevalier, P., Vandelle, B., Lenoble, D., Saguin, F., Frégonese, S., Zimmer, T., and Chantre, A., "Investigation of fully and partially depleted self-aligned SiGeC HBTs on thin film SOI," *Proc. 35th ESSDERC*, 133, 2005.

2
CMOS/BiCMOS Technology

Yasuhiro Katsumata

Tatsuya Ohguro

Kazumi Inoh

Eiji Morifuji

Takashi Yoshitomi

Hideki Kimijima

Hideaki Nii

Toyota Morimoto

Hisayo S. Momose

Kuniyoshi Yoshikawa and
Hidemi Ishiuchi
*Semiconductor Company,
Toshiba Corporation*

Hiroshi Iwai
Tokyo Institute of Technology

CONTENTS

2.1　Introduction

Silicon large-scale integrated circuits (LSIs) have progressed remarkably in the past 30 years. In particular, complementary metal oxide semiconductor (CMOS) technology has played a great role in the progress of LSIs. By the downsizing of MOS field effect transistors (FETs), the number of transistors in a chip increases, and the functionality of LSIs is improved. At the same time, the switching speed of MOSFETs and circuits increases and the operation speed of LSIs is improved.

In contrast, system-on-chip technology has come into widespread use [1], and, as a result, the LSI system requires several functions such as logic, memory, and analog functions [2]. Moreover, the LSI system sometimes needs an ultra-high-speed logic or an ultra-high-frequency analog function [3]. In some cases, bipolar-CMOS (BiCMOS) technology is very useful.

In Section 2.2, we focus on CMOS technology, which is the major LSI process technology, including embedded memory technology. In Section 2.3, we describe BiCMOS technology. Finally, we introduce the recent process issues.

2.2   CMOS Technology

2.2.1   Device Structure and Basic Fabrication Process Steps

Complementary MOS (CMOS) was first proposed by Wanlass and Sah in 1963 [4]. Although the CMOS process is more complex than the NMOS process, it provides both n-channel (NMOS) and p-channel (PMOS) transistors on the same chip and CMOS circuits can achieve lower power consumption. Consequently, the CMOS process has been widely used as an LSI fabrication process.

Figure 2.1 shows the structure of a CMOS device. Each "FET" consists of gate electrode, source, drain and channel, and gate bias controls carrier flow from source to drain through channel layer.

Figure 2.2 shows the basic fabrication process flow. The first process step is the formation of p tub and n tub (twin tub or twin well) in silicon substrate. Because CMOS has two types of FETs, NMOS is formed in p tub and PMOS in n tub.

The isolation process is the formation of field oxide to separate each MOSFET active area in the same tub. After that, impurity is doped into the channel region to adjust the threshold voltage, V_{th}, for each type of FET. Gate insulator layer, usually silicon dioxide (SiO_2), is grown by thermal oxidation, because the interstate density between SiO_2 and silicon substrate is small. Polysilicon is deposited as gate electrode material and gate electrode is patterned by reactive ion etching (RIE).

Gate length, L_g, is the critical dimension, because L_g determines the performance of MOSFETs and it should be small to improve the device performance. Impurity is doped in source and drain regions of MOSFETs by ion implantation. In this process step, gate electrodes act as a self-aligned mask to cover channel layers. After that, thermal annealing is carried out to activate the impurity of diffused layers.

In the case of high-speed LSI, the self-aligned silicide (salicide) process is applied for the gate electrode and source and drain diffused layers to reduce parasitic resistance. Finally, the metallization process is carried out to form interconnect layers.

2.2.2   Key Process Steps in Device Fabrication

2.2.2.1   Starting Material

Almost all silicon crystals for LSI applications are prepared by the Czochralski crystal growth method [5], because it is advantageous for formation of large wafers. (100) orientation wafers are usually used for MOS devices, because their interstate trap density is smaller than those of (111) and (110) orientations [6]. The light doping in the substrate is convenient for the diffusion of tub and reduces the parasitic capacitance

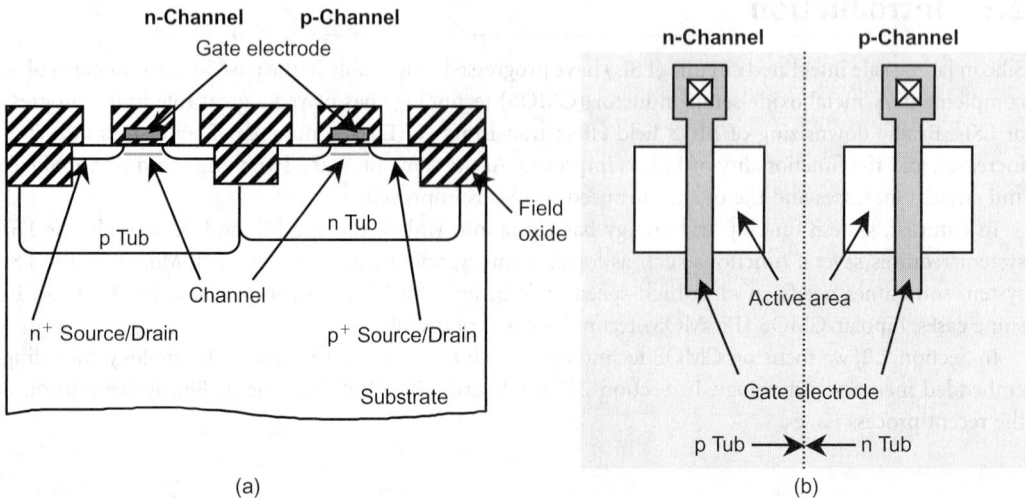

FIGURE 2.1   Structure of CMOS device. (a) Cross-sectional view of CMOS. (b) Plain view of CMOS.

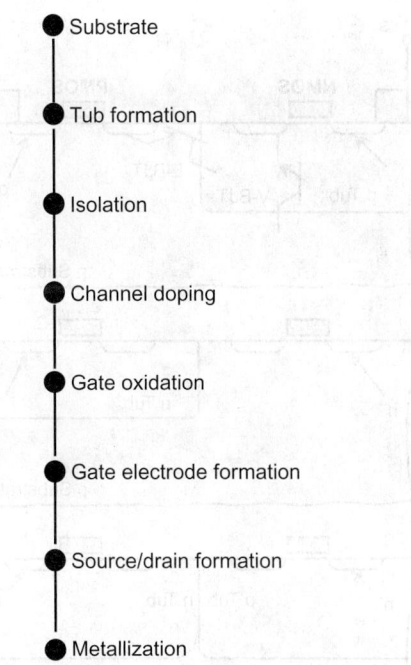

FIGURE 2.2 Basic process flow of CMOS.

between the silicon substrate and tub region. As a starting material, lightly doped ($\sim 10^{15}$ atoms/cm^3) p-type substrate is generally used.

2.2.2.2 Tub Formation

Figure 2.3 shows the tub structures, which are classified into six types: p tub, n tub, twin tub [7], triple tub, twin tub with buried p$^+$ and n$^+$ layers, and twin tub on p-epi/p$^+$ substrate. In the case of the p tub process, NMOS is formed in p diffusion (p tub) in the n substrate, as shown in Figure 2.3(a). The p tub is formed by implantation and diffusion into n substrate at a concentration that is high enough to over-compensate the n substrate.

The other approach is to use an n tub [8]. As shown in Figure 2.3(b), NMOS is formed in the p substrate.

Figure 2.3(c) shows the twin-tub structure [7] that uses two separate tubs implanted into the silicon substrate. In this case, doping profiles in each tub region can be controlled independently, and thus neither type of device suffers from excess doping effect.

In some cases, such as mixed-signal LSIs, deep n tub layer is sometimes formed optionally, as shown in Figure 2.3(d), to prevent the cross-talk noise between digital and analog circuits. In this structure, both n and p tubs are electrically isolated from the substrate and other tubs on the substrate.

To realize high packing density, tub design rule should be shrunk; however, an undesirable mechanism, the well-known latch-up, might occur.

Latch-up, i.e., the flow of high current between V_{DD} and V_{SS}, is caused by parasitic lateral pnp bipolar junction transistor (L-BJT) and vertical npn bipolar junction transistor (V-BJT) actions [9] as shown in Figure 2.3(a), and it sometimes destroys the functions of LSIs. The collectors of each of these bipolar junction transistors feed each others' bases and together make up a pnpn thyristor structure. To prevent latch-up, it is important to reduce the current gain, h_{FE}, of these parasitic bipolar junction transistors, and the doping concentration of tub region should be higher. As a result, device performance might be suppressed because of large junction capacitances.

To avoid this problem, several techniques have been proposed, such as p$^+$ or n$^+$ buried layer under p tub [10] as shown in Figure 2.3(e), the use of high-dose, high-energy boron p tub implants [11,12],

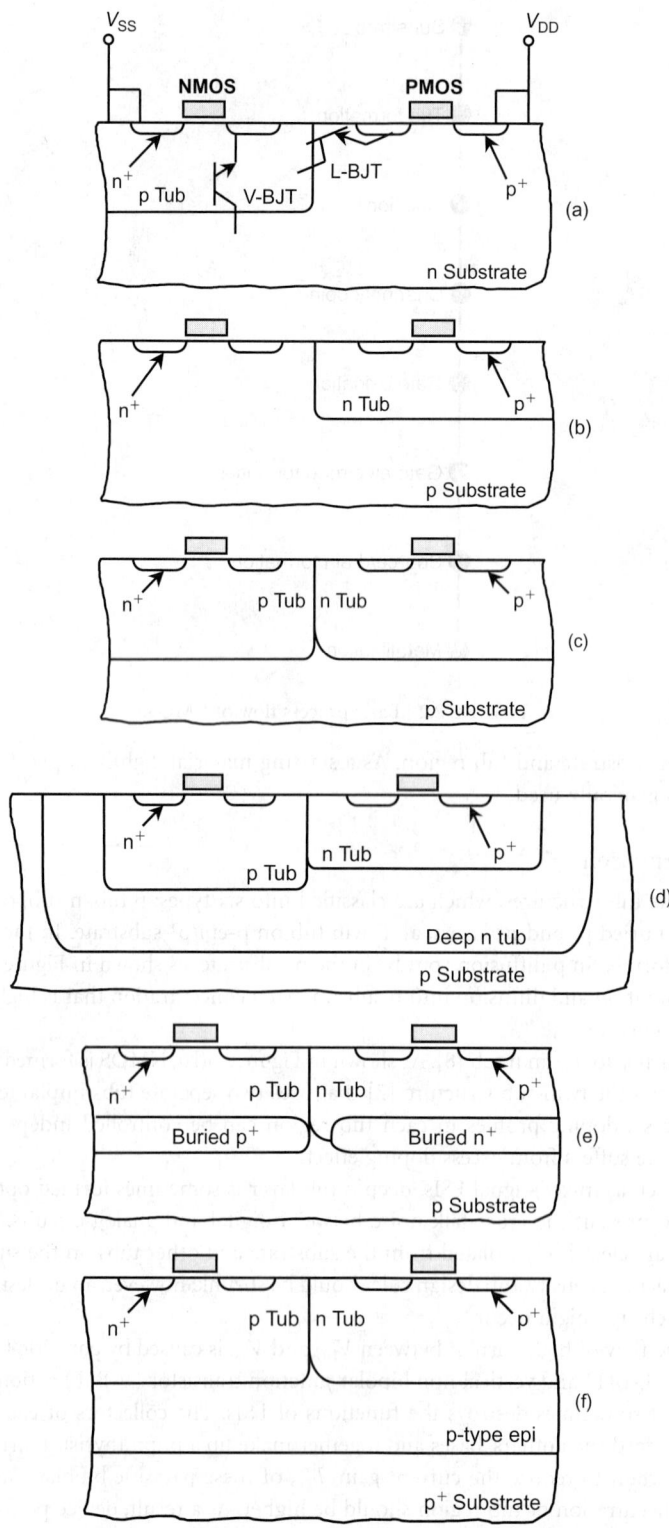

FIGURE 2.3 Tub structures of CMOS. (a) p tub. (b) n tub. (c) Twin tub. (d) Triple tub. (e) Twin tub with buried p$^+$ and n$^+$ layers. (f) Twin tub on p-epi/p$^+$ substrate.

and the shunt resistance for emitter–base junctions of parasitic bipolar junction transistors [9,13,14]. It is also effective to provide many well contacts to stabilize the well potential and hence to suppress the latch-up. Recently, substrate with p epitaxial silicon on p^+ substrate, as shown in Figure 2.3(f), is also used to stabilize the potential for high-speed logic LSIs [15].

2.2.2.3 Isolation

Local oxidation of silicon (LOCOS) [16] is a widely used isolation process, because this technique can allow channel-stop layers to be formed self-aligned to the active transistor area. It also has the advantage of recessing about half of the field oxide below the silicon surface, which makes the surface more planar.

Figure 2.4 shows the LOCOS isolation process. First, silicon nitride and pad oxide are etched for the definition of active transistor area. After channel implantation as shown in Figure 2.4(a), the field oxide is selectively grown, typically to a thickness of several hundreds of nanometers.

A disadvantage of LOCOS is that the involvement of nitrogen in the masking of silicon nitride layer sometimes causes the formation of a very thin nitride layer in the active region, and this often impedes the subsequent growth of gate oxide, thereby causing low gate breakdown voltage of the oxides. To prevent this problem, a sacrificial oxide is grown and then removed before the gate oxidation process after stripping the masking silicon nitride [17,18].

In addition, the lateral spread of field oxide (bird's beak) [17] poses a problem regarding reduction of the distance between active transistor areas to realize high packing density. This lateral spread is suppressed by increasing the thickness of silicon nitride and decreasing the thickness of pad oxide. However, there is a trade-off with the generation of dislocation of silicon.

Recently, shallow trench isolation (STI) [19] has become a major isolation process for advanced CMOS devices. Figure 2.5 shows the process flow of STI. After digging the trench into the substrate by RIE as shown in Figure 2.5(a), the trench is filled with an insulator such as silicon dioxide as shown in Figure 2.5(b). Finally, by planarization with chemical mechanical polishing (CMP) [20], filling material on active transistor area is removed, as shown in Figure 2.5(c).

STI is a useful technique for downsizing not only the distance between active areas but also the active region itself. However, the mechanical stress problem [21] still remains, and several methods have been proposed [22] to deal with it.

2.2.2.4 Channel Doping

To adjust the threshold voltage of MOSFETs, V_{th}, to that required by a circuit design, the channel doping process is usually required. The doping is carried out by ion implantation usually through a thin

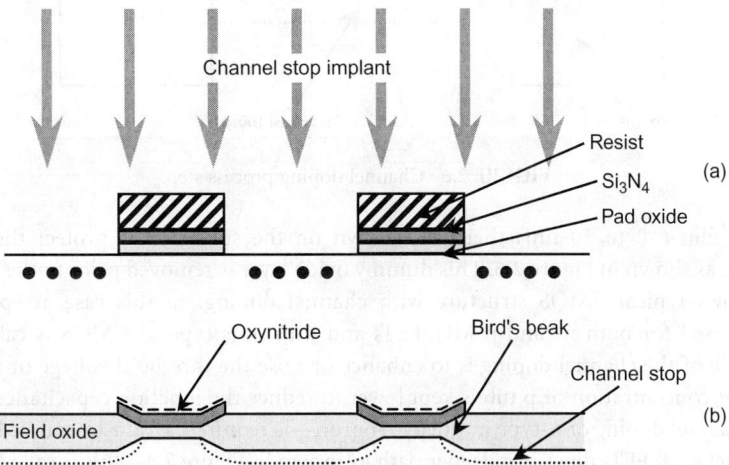

FIGURE 2.4 Process for local oxidation of silicon: (a) after silicon nitride/pad oxide etch and channel-stop implant, (b) after field oxidation, which produces an oxynitride film on nitride.

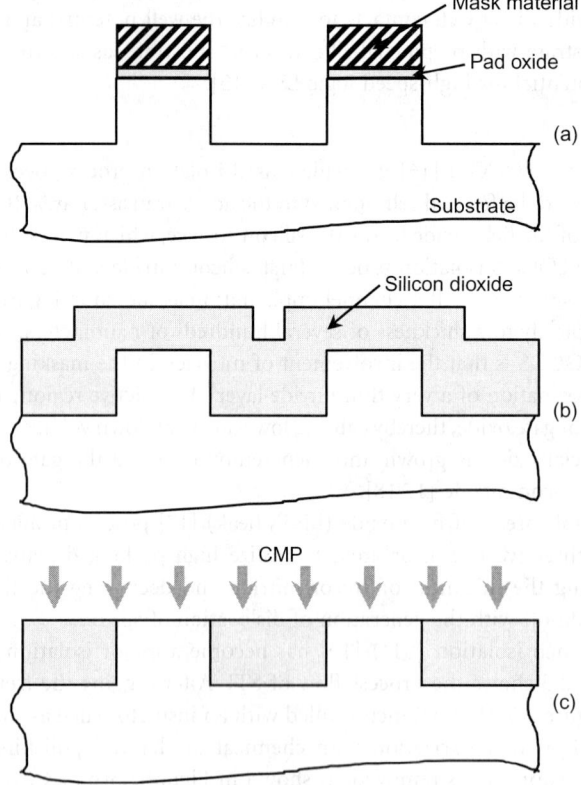

FIGURE 2.5 Process flow of STI. (a) Trenches are formed by RIE. (b) Filling by deposition of SiO2. (c) Planarization by CMP.

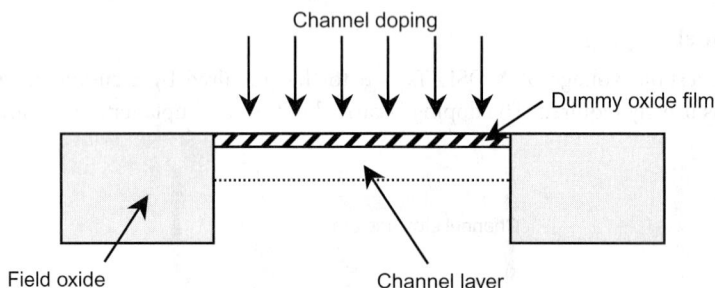

FIGURE 2.6 Channel doping process step.

dummy oxide film (10 to 30 nm) thermally grown on the substrate to protect the surface from contamination, as shown in Figure 2.6. This dummy oxide layer is removed prior to the gate oxidation. Figure 2.7 shows typical CMOS structure with channel doping. In this case, n+ polysilicon gate electrodes are used for both n- and p-MOSFETs and thus, this type of CMOS is called single-gate CMOS. The role of the channel doping is to enhance or raise the threshold voltage of n-MOSFETs. It is desirable that concentration of p tub is kept lower to reduce the junction capacitance of source and drain. Thus, channel doping of p-type impurity—boron—is required. Drain-to-source leakage current in short-channel MOSFETs flows in a deeper path as shown in Figure 2.8—this is called short-channel effects. Thus, heavy doping of the deeper region is effective for suppressing the short-channel effect. This doping is called deep-ion implantation.

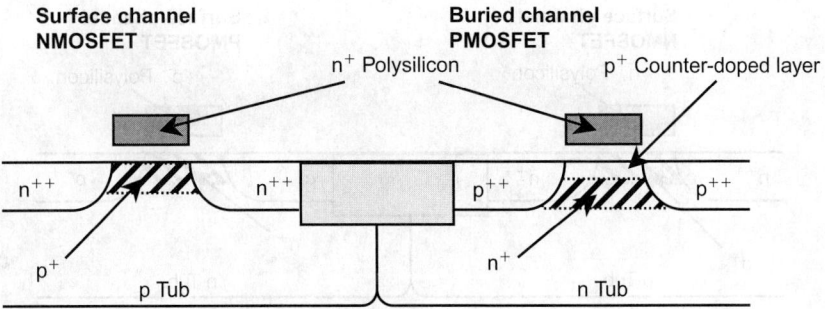

FIGURE 2.7 Schematic cross section of single-gate CMOS structure

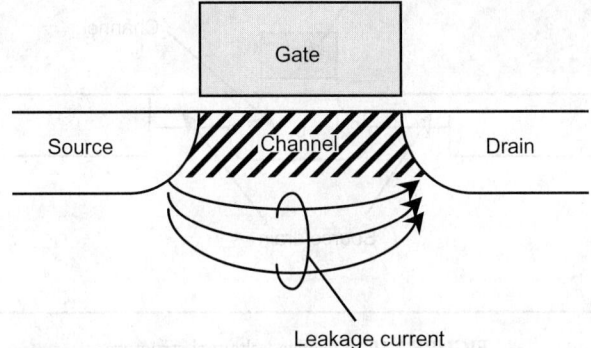

FIGURE 2.8 Leakage current flow in short channel MOSFET.

In the case of p-MOSFET with n^+ polysilicon gate electrode, the threshold voltage becomes too high in the negative direction if there is no channel doping. To adjust the threshold voltage, ultra-shallow p-doped region is formed by the channel implantation of boron. This p-doped layer is often called the counter-doped layer or buried channel layer, and p-MOSFETs with this structure are called buried-channel MOSFETs. (In contrast, MOSFETs without buried channel layer are called surface-channel MOSFETs. n-MOSFETs in this case are the surface-channel MOSFETs.) In the buried-channel case, the short-channel effect is more severe, and thus, deep implantation of n-type impurity such as arsenic or phosphorus is necessary to suppress them.

In deep submicron gate length CMOS, it is difficult to suppress the short-channel effect [23], and thus, p^+ polysilicon electrode is used for p-MOSFETs as shown in Figure 2.9. For n-MOSFETs, n^+ polysilicon electrode is used. Thus, this type of CMOS is called dual-gate CMOS. In the case of p^+ polysilicon p-MOSFET, the threshold voltage becomes close to 0 V because of the difference in work function between n- and p-polysilicon gate electrodes [24–26], and thus, buried layer is not required. Instead, n-type impurity channel doping such as arsenic is required to raise the threshold voltage slightly in the negative direction.

Impurity redistribution during high-temperature LSI manufacturing processes sometimes makes channel profile broader, which causes short-channel effect. To suppress the redistribution, dopant with lower diffusion constant, such as indium, is used instead of boron.

For the purpose of realizing a high-performance transistor, it is important to reduce junction capacitance. To realize lower junction capacitance, a localized diffused channel structure [27,28], as shown in Figure 2.10, is proposed. Since the channel layer exists only around the gate electrode, junction capacitances of source and drain are reduced significantly.

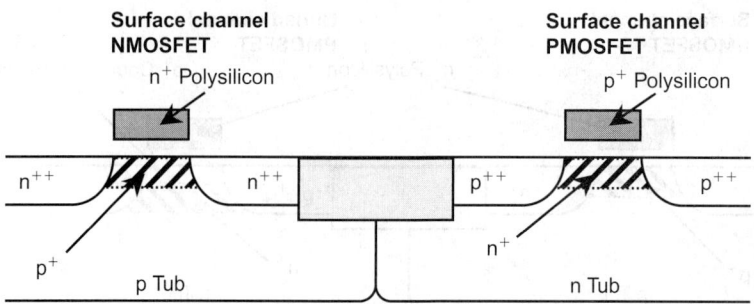

FIGURE 2.9 Schematic cross section of dual-gate CMOS structure.

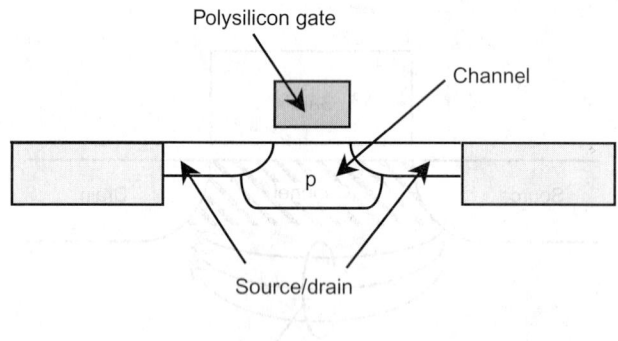

FIGURE 2.10 Localized channel structure.

2.2.2.5 Gate Insulator

Gate dielectric determines several important properties of MOSFETs and thus, uniformity in its thickness, low defect density of the film, low fixed charge and interface state density at the dielectric and silicon interface, small roughness at the interface, high reliability of time-dependent dielectric breakdown (TDDB) and hot-carrier-induced degradation, and high resistivity to boron penetration (explained in this section) are required. As a consequence of the downsizing of MOSFET, the thickness of gate dielectric has become thinner. Generally, the thickness of gate oxide is 7–8 nm for 0.4 μm gate length MOSFETs and 5–6 nm for 0.25 μm gate length MOSFETs.

Silicon dioxide is commonly used for gate dielectrics, which are formed by several methods, such as dry O_2 oxidation and wet or steam (H_2O) oxidation [29]. The steam is produced by the reaction of H_2 and O_2 ambient in the furnace. Recently, H_2O oxidation has been widely used for gate oxidation because of good controllability of oxide thickness and high reliability.

In the case of dual-gate CMOS structure as shown in Figure 2.9, boron penetration from p^+ gate electrode to channel region through gate silicon dioxide, which is described in the following section, is a problem. To prevent this problem, oxynitride has been used as gate dielectric material [30,31]. In general, oxynitride gate dielectric is formed by the annealing process in NH_3, NO (or N_2O) after silicon oxidation, or by direct oxinitridation of silicon in NO (or N_2O) ambient. Figure 2.11 shows the typical nitrogen profile of oxynitride gate dielectric. Recently, remote plasma nitridation [32,33] has been much studied and it is reported that oxynitride gate dielectric grown by the remote plasma method showed better quality and reliability than that grown by the silicon nitridation method.

In the sub-quarter-micron CMOS device regime, gate oxide thickness is close to the limitation of tunneling current flow, around 3 nm thickness. To prevent tunneling current, high dielectric constant, κ, materials, such as Si_3N_4 [34] and Ta_2O_5 [35,36], are proposed instead of silicon dioxide. In these cases, the thickness of gate insulator can be kept at a relatively thick value, because high-κ insulator realizes high gate capacitance, and thus better driving capability.

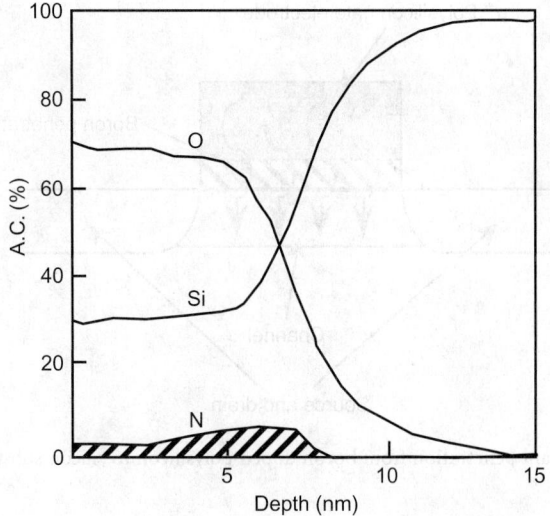

FIGURE 2.11 Oxygen, nitrogen and silicon concentration profile of oxynitride gate dielectric measured by AES.

2.2.2.6 Gate Electrode

Heavily doped polysilicon has been widely used for gate electrode because of its stability to high-temperature LSI fabrication processing. To reduce the resistance of gate electrode which contributes significantly to RC delay time, silicides of refractory metals have been put on the polysilicon electrode [37,38]. Polycide [38], the technique of combining a refractory metal silicide on top of doped polysilicon, has the advantage of preserving the good electric and physical properties at the interface between polysilicon and gate oxide while, at the same time, the sheet resistance of gate electrode is reduced significantly.

For doping the gate polysilicon, ion implantation is usually employed. In the case of heavy doping, dopant penetration from boron-doped polysilicon to the silicon substrate channel region though the gate oxide occurs in the high-temperature LSI fabrication processes as shown in Figure 2.12. (In contrast, usually, penetration of the n-type dopant, such as phosphorus or arsenic does not occur.) When the doping of impurities in the polysilicon is not sufficient, the depletion of gate electrode occurs as shown in Figure 2.13, resulting in significant decrease in the drive capability of transistor as shown in Figure 2.14 [39]. There is a trade-off between the boron penetration and the gate electrode depletion, and so thermal process optimization is required [40]. Recently, polysilicon germanium (SiGe) gate FET is demonstrated to prevent this depletion phenomenon of gate electrode [41].

Gate length is one of most important dimensions defining MOSFET performance, and thus the lithography process for gate electrode patterning requires high-resolution technology.

In the case of light-wave source, g-line (wavelength 436 nm) and i-line (365 nm) of mercury lamp were popular methods. Recently, a higher-resolution process, excimer laser lithography, has been used. In the excimer laser process, KrF (248 nm) [42] and ArF (193 nm) [43] have been proposed and developed. For around 0.25 μm gate length electrode, the KrF excimer laser process is widely used in the production of devices. In addition, electron-beam [44–46] and x-ray [47] lithography techniques are being studied for sub-0.1 μm lithography.

For etching of gate polysilicon, a high-selectivity RIE process is required for selecting polysilicon from SiO_2, because the gate dielectric beneath the polysilicon is a very thin film in the case of recent devices.

2.2.2.7 Source/Drain Formation

Source and drain diffused layers are formed by the ion implantation process. As a consequence of transistor downsizing, at the drain edge (interface of channel region and drain) where reverse-biased pn junctions exist, higher electrical field has been observed. As a result, carriers across these junctions are suddenly accelerated and become hot carriers, which create a serious reliability problem for MOSFET [48].

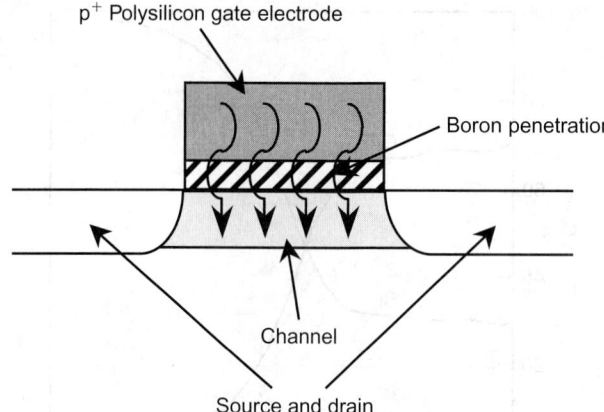

FIGURE 2.12 Dopant penetration from boron doped polysilicon to silicon substrate channel region.

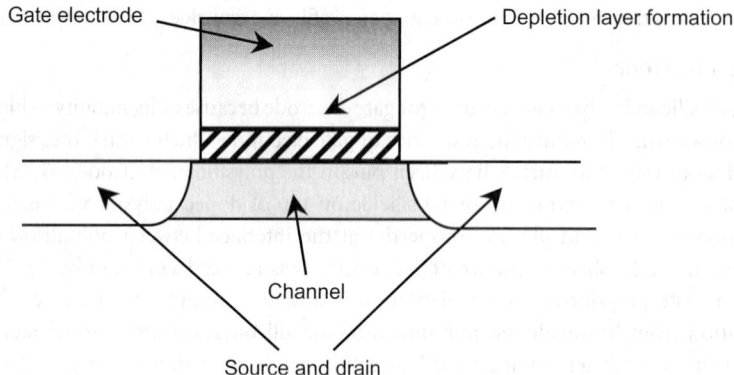

FIGURE 2.13 Depletion of gate electrode in the case that the doping of impurities in the gate electrode is not sufficient.

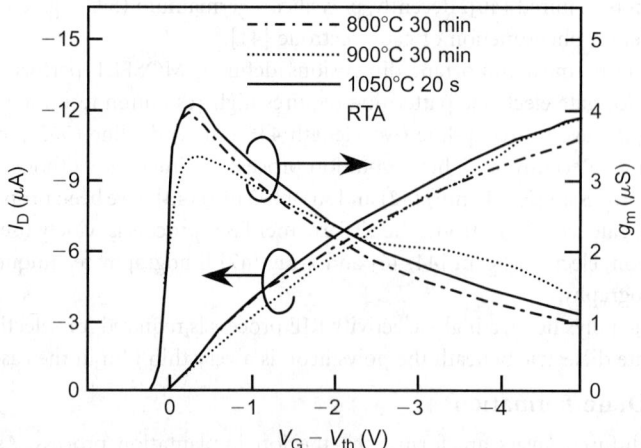

FIGURE 2.14 I_D, g_m – V_G characteristics for various thermal conditions. In the case of 800 °C 30 min, significant decrease of drive capability of transistor occurs because of the depletion of gate electrode.

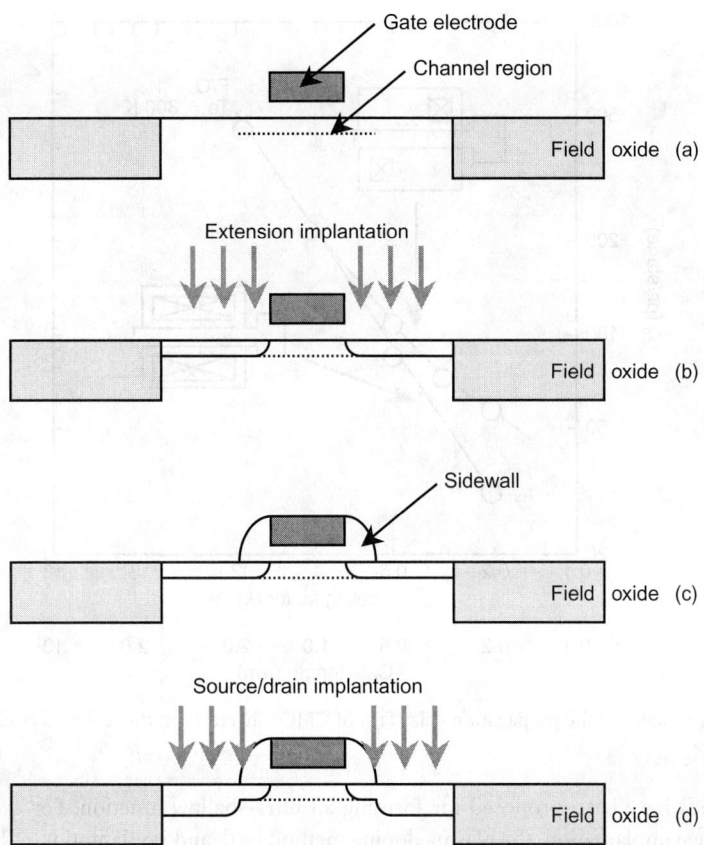

FIGURE 2.15 Process flow of LDD structure. (a) After gate electrode patterning. (b) Extension implantation. (c) Sidewall spacer formation. (d) Source/drain implantation.

To prevent the hot carrier problem, the lightly doped drain (LDD) structure is proposed [49]. The LDD process flow is shown in Figure 2.15. After gate electrode formation, ion implantation is carried out to make extension layers, and the gate electrode plays the role of self-aligned mask which covers channel layer, as shown in Figure 2.15(b). In general, arsenic is doped for n-type extension of NMOS, and BF_2 for p-type extension of PMOS. To prevent the short-channel effect, impurity profile of the extension must be very shallow. Although shallow extension can be realized by ion implantation with low dose, the resistivity of extension layers becomes high, and thus, MOSFET characteristics degrade. Hence, it is very difficult to meet these two requirements. Also, impurities diffusion in this extension affects the short-channel effect significantly. Thus, it is necessary to minimize the thermal process after forming extension.

Insulating film, such as Si_3N_4 or SiO_2, is deposited by the chemical vapor deposition method. Then, etching back RIE treatment is performed on the whole wafer, and, as a result, the insulating film remains only at the gate electrode side as shown in Figure 2.15(c). This remaining film is called a sidewall spacer. This spacer works as a self-aligned mask for deep source/drain n^+ and p^+ doping, as shown in Figure 2.15(d). In general, arsenic is doped for deep source/drain of NMOSFET, and BF_2 for PMOSFET. In the dual-gate CMOS process, gate polysilicon is also doped in this process step to prevent gate electrode depletion.

After that, to make doped impurities activate electrically and to recover from implantation damage, an annealing process, such as rapid thermal annealing (RTA), is carried out.

According to the MOSFET scaling law, when gate length and other dimensions are shrunk by factor k, the diffusion depth also needs to be shrunk by $1/k$. Hence, the diffusion depth of the extension part is required to be especially shallow.

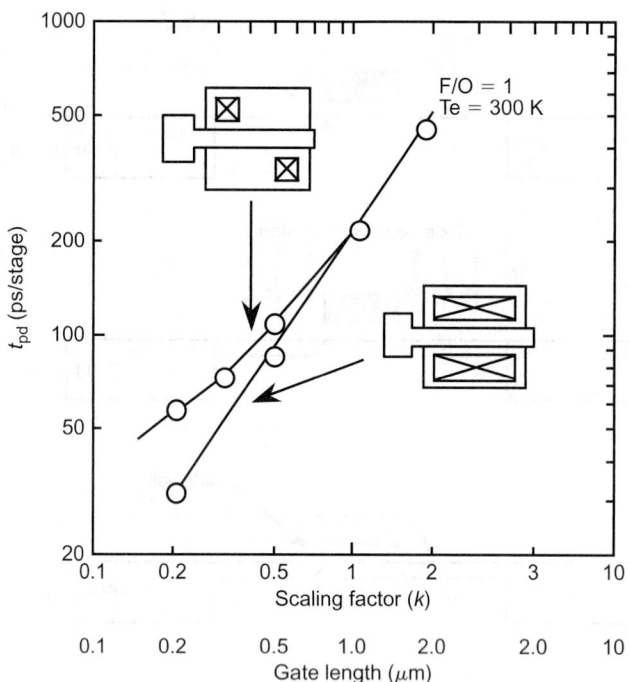

FIGURE 2.16   Dependence of the propagation delay (t_{pd}) of CMOS inverters on the scaling factor, k, or gate length.

Several methods have been proposed for forming an ultra-shallow junction. For example, very low accelerating voltage implantation, the plasma doping method [50], and implantation of heavy molecules, such as $B_{10}H_{14}$ for p-type extension [51], are being studied.

2.2.2.8   Salicide Technique

As the vertical dimension of transistors is reduced with device downscaling, an increase is seen in sheet resistance; both of the diffused layers, such as source and drain, and the polysilicon films, such as the gate electrode. This is becoming a serious problem in the high-speed operation of integrated circuits.

Figure 2.16 shows the dependence of the propagation delay time (t_{pd}) of CMOS inverters on the scaling factor, k, or gate length [52]. These results were obtained by simulations in which two cases were considered. First is the case in which source and drain contacts with the metal line were made at the edge of the diffused layers, as illustrated in the figure inset. In an actual LSI layout, it often happens that the metal contact to the source or drain can be made only to a portion of the diffused layers, since many other signal or power lines cross the diffused layers. The other case is that in which the source and drain contacts cover the entire area of the source and drain layers, thus reducing diffused line resistance. It is clear that without a technique to reduce the diffused line resistance, t_{pd} values cannot keep falling as transistor size is reduced; they will saturate at gate lengths of around a quarter micron.

To prevent this problem—the high resistance of shallow diffused layers and thin polysilicon films—self-aligned silicide (salicide) structures for the source, drain, and gate have been proposed, as shown in Figure 2.17 [53–55].

First, a metal film such as Ti or Co is deposited on the surface of the MOSFET after the formation of the polysilicon gate electrode, gate sidewall, and source and drain diffused layers, as shown in Figure 2.17(b). The film is then annealed by RTA in an inert ambient. During the annealing process, the areas of metal film in direct contact with the silicon layer—that is, the source, drain, and gate electrodes—are selectively converted into the silicide and other areas remain metal, as show in Figure 2.17(c).

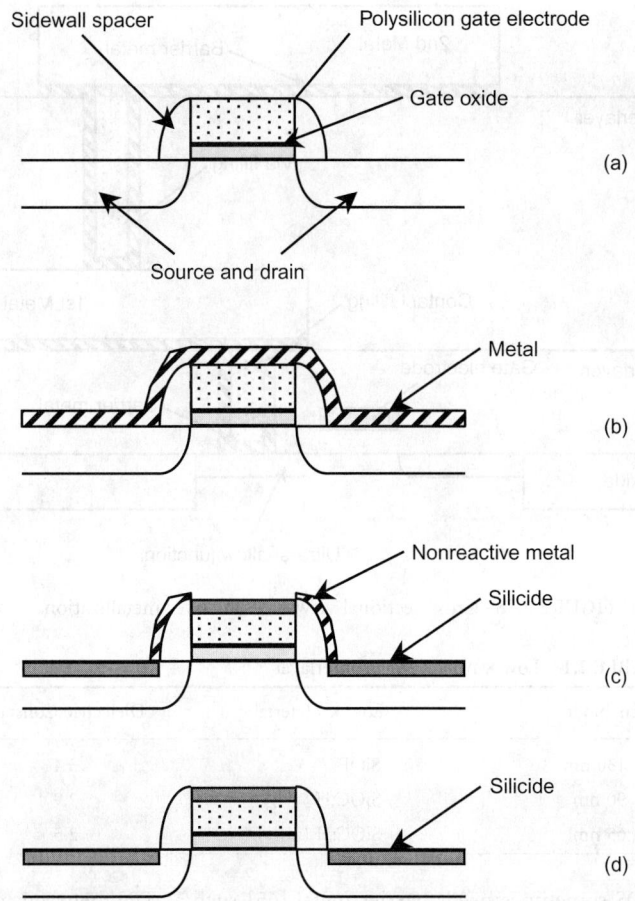

Sidewall spacer Polysilicon gate electrode

Gate oxide

(a)

Source and drain

Metal

(b)

Nonreactive metal

Silicide

(c)

Silicide

(d)

FIGURE 2.17 A typical process flow and schematic cross section of salicide process. (a) MOSFET formation. (b) Metal deposition. (c) Silicidation by thermal annealing. (d) Removal of non-reactive metal.

The remaining metal can be etched off with an acid solution such as $H_2O_2 + H_2SO_4$, leaving the silicide self-aligned with the source, drain, and gate electrode, as shown in Figure 2.17(d).

When the salicide process first came into use, furnace annealing was the most popular heat-treatment process [53–55]; however, RTA [56–58] replaced furnace annealing early on, because it is difficult to prevent small amounts of oxidant entering through the furnace opening, and these degrade the silicide film significantly since silicide metals are easily oxidized. In contrast, RTA reduces this oxidation problem significantly, resulting in reduced deterioration of the film and consequently of its resistance.

For sub-half-micron gate length FETs, $TiSi_2$ [56–58] is widely used as a silicide in LSI applications. However, in the case of ultra-small geometry MOSFETs for VLSI, use of $TiSi_2$ is subject to several problems. When the $TiSi_2$ is made thick, a large amount of silicon is consumed during silicidation, and this should result in problems of junction leakage at the source or drain. On the contrary, if a thin layer of $TiSi_2$ is chosen, agglomeration of the film occurs [59] at higher silicidation temperatures.

However, $CoSi_2$ [60] and NiSi [61] have a large silicidation temperature window for low sheet resistance; hence, it is expected to be widely used as silicidation material for advanced VLSI applications [62].

2.2.2.9 Interconnect and Metallization

Aluminum is widely used as a wiring metal for VLSI. However, in the case of downsized CMOS, electromigration (EM) [63] and stress migration (SM) [64] become serious problems. To prevent these problems, Al–Cu (typically ~0.5 wt% Cu) [65] is a useful wiring material. In addition, ultra-shallow junction

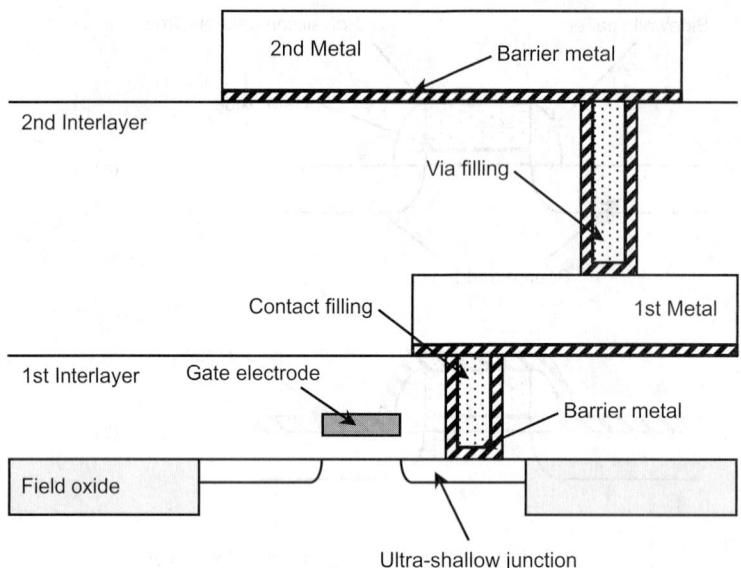

FIGURE 2.18 Cross-sectional view of multi-layer metallization.

TABLE 2.1 Low κ materials for interlayer

Tech. Node	Low-κ Material	Dielectric Constant
130 nm	SiOF	3.4
90 nm	SiOC:H	2.9
65 nm	SiOC:H (pore)	2.5

for downsized CMOS sometimes needs barrier metal [65] such as TiN, between metal and silicon, to prevent junction leakage current.

Figure 2.18 shows a cross-sectional view of a multilayer metallization structure. As a consequence of CMOS downscaling, contact or via aspect ratio becomes larger, and, as a result, filling of contact or via is not sufficient. Hence, new filling techniques, such as W-plug [66,67], are widely used.

In addition, considering both reliability and low resistivity, Cu [68] is a useful wiring material. In the case of Cu interconnects, metal thickness can be reduced for realizing the same interconnect resistance of aluminum case. The reduction of the metal thickness is useful for reducing the capacitance between the dense interconnect wires, resulting in the high-speed operation of the circuit. To reduce RC delay of wire in CMOS LSI, not only the wiring material but also the interlayer material is important. In particular, low dielectric constant, κ, film is widely studied [69]. Table 2.1 shows various low-κ films for ultra-small geometry devices [70]. Also, several low-κ materials are demonstrated for sub-50 nm node processes [71–74].

In the case of Cu wiring, the dual-damascene process [75–77] is being widely used, because it is difficult to realize fine Cu pattern by RIE. Figure 2.19 shows the process flow of Cu dual-damascene metallization. After deposition of dielectric interlayer as shown in Figure 2.19(a), trenches for via contact and wiring area are formed as shown in Figure 2.19(b). By using electroplating method, Cu film was deposited, and then the CMP process [78] is carried out for planarization as shown in Figure 2.19(c). It should be noted that a barrier metal, such as TiN, is essential between Cu and interlayer to prevent Cu diffusion into dielectric layers.

2.2.3 Passive Device for Analog Operation

System-on-chip technology has come into widespread use, and, as a result, the LSI system sometimes requires analog functions. In this case, analog passive devices should be integrated [79,80], as shown in Figure 2.20.

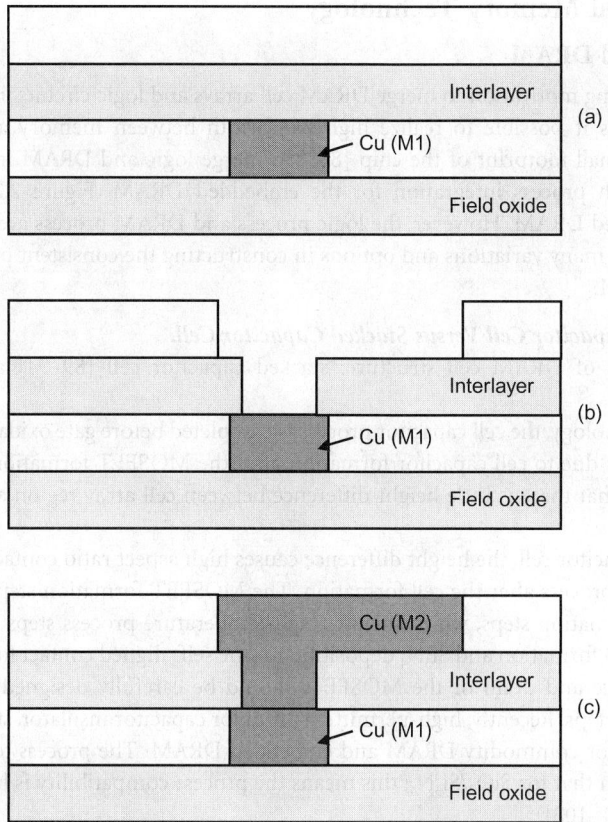

FIGURE 2.19 A typical process flow dual damascene process. (a) Inter layer dielectric film deposition. (b) Via formation. (c) Cu (M2) deposition and planarization.

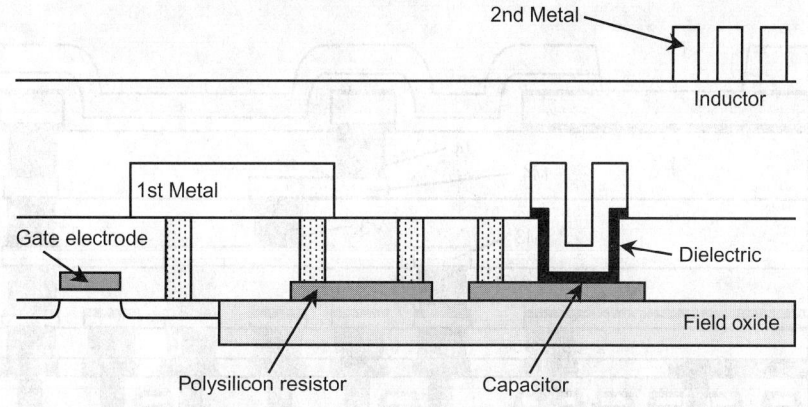

FIGURE 2.20 Various passive devices for analog application.

Resistors and capacitors [81] already have good performance, even for high-frequency applications. However, it is difficult to realize a high-quality inductor on a silicon chip, because of inductance loss in silicon substrate, in which the resistivity is lower than that in the compound semiconductor, such as GaAs, substrate. Relatively higher sheet resistance of aluminum wire used for high-density LSI is another problem. Recently, the quality of inductor has been improved by using thicker Al or Cu wire [82] and by optimizing substrate structure [83–87].

2.2.4 Embedded Memory Technology

2.2.4.1 Embedded DRAM

There has been a strong motivation to merge DRAM cell arrays and logic circuits in a single silicon chip. This approach makes it possible to realize high bandwidth between memory and logic, low power consumption, and small footprint of the chip [88]. To merge logic and DRAM into a single chip, it is necessary to establish process integration for the embedded DRAM. Figure 2.21 shows the typical structure of embedded DRAM. However, the logic process and DRAM process are not compatible with each other. There are many variations and options in constructing the consistent process integration for the embedded DRAM.

2.2.4.1.1 Trench Capacitor Cell Versus Stacked Capacitor Cell.

There are two types of DRAM cell structure: stacked capacitor cell [89–94] and trench capacitor cell [95,96].

In trench cell technology, the cell capacitor process is completed before gate oxidation. Therefore, there is no thermal process due to cell capacitor formation after the MOSFET formation. Another advantage of the trench cell is that there is little height difference between cell array region and peripheral circuit region [97–100].

In the stacked capacitor cell, the height difference causes high aspect ratio contact holes and difficulty in the planarization process after the cell formation. The MOSFET formation steps are followed by the stacked capacitor formation steps, which include high-temperature process steps such as storage node insulator (SiO_2/Si_3N_4) formation and Si_3N_4 deposition for the self-aligned contact formation. The salicide process for the source and drain of the MOSFETs should be carefully designed to endure the high-temperature process steps. Recently, high-permittivity film for capacitor insulator, such as Ta_2O_5 and BST, has been developed for commodity DRAM and embedded DRAM. The process temperature for Ta_2O_5 and BST is lower than that for SiO_2/Si_3N_4; this means the process compatibility is better with such high-permittivity film [101–103].

2.2.4.1.2 MOSFET Structure.

The MOSFET structure in DRAMs is different from that in logic ULSIs. In recent DRAMs, the gate is covered with Si_3N_4 for self-aligned contact process steps in the bit-line contact formation. It is very

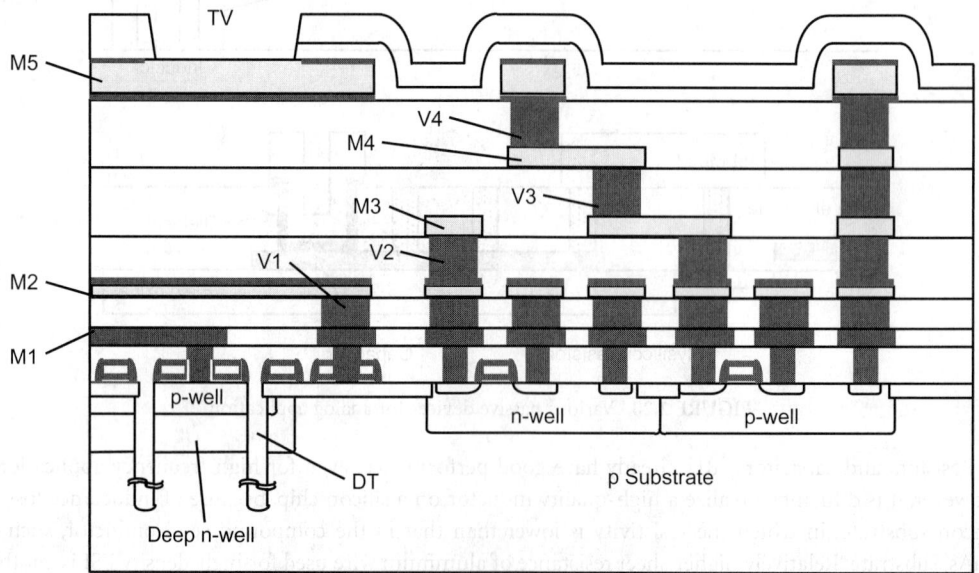

FIGURE 2.21 Schematic cross section of the embedded DRAM including DRAM cells and logic MOSFETs.

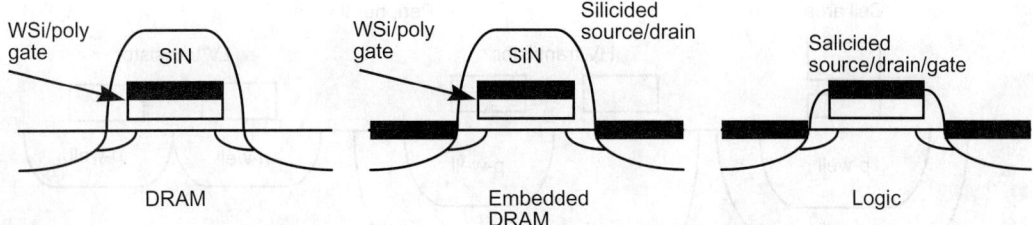

FIGURE 2.22 Typical MOSFET structures for DRAM, embedded DRAM, and logic.

difficult to apply the salicide process to the gate, source, and drain at the same time. A solution to the problem is to apply the salicide process to the source and drain only. A comparison of the MOSFET structures is shown in Figure 2.22. Tsukamoto et al. [89] proposed another approach, namely the use of W-bit line layer as local interconnect in the logic portion.

2.2.4.1.3 Gate Oxide Thickness.

Generally, DRAM gate oxide thickness is greater than that of the logic VLSIs. This is because maximum voltage of the transfer gate in the DRAM cells is higher than V_{CC}, the power supply voltage. In the logic VLSI, the maximum gate voltage is equal to V_{CC} in most cases. To keep up with the MOSFET performance in logic VLSIs, the oxide thickness of the embedded DRAMs needs to be scaled down further than in the DRAM case. To do so, highly reliable gate oxide and new circuit scheme in the word line biasing, such as applying negative voltage to the cell transfer gate, is required.

Another approach is to use thick gate oxide in the DRAM cell and thin gate oxide in the logic [104].

2.2.4.1.4 Fabrication Cost Per Wafer.

The conventional logic VLSIs do not need the process steps for the DRAM cell formation. In contrast, most of the DRAMs use only two layers of aluminum. This raises wafer cost of the embedded DRAMs. Embedded DRAM chips are used only if the market can absorb the additional wafer cost for some reasons: high bandwidth, lower power consumption, small footprint, flexible memory configuration, lower chip assembly cost, etc.

2.2.4.1.5 Next-Generation Embedded DRAM.

Process technology for the embedded DRAM with 90 or 65 nm design rules will include state-of-the-art DRAM cell array and high-performance MOSFETs in the logic circuit [105,106]. The embedded DRAM could be a technology driver because the embedded DRAM contains most of the key process steps for DRAM and logic VLSIs.

2.2.4.2 Embedded Flash Memory Technology [107]

Recently, the importance of embedded flash technology has been increasing and logic chips with non-volatile functions have become indispensable for meeting various market requirements.

Key issues in the selection of an embedded flash cell [108] are (1) tunnel-oxide reliability (damage-less program/erase(P/E) mechanism), (2) process and transistor compatibility with CMOS logic, (3) fast read with low V_{CC}, (4) low power (especially in P/E), (5) simple control circuits, (6) fast program speed, and (7) cell size. This ordering largely depends on target device specification and memory density, and, in general, is different from that of high-density stand-alone memories. NOR-type flash is essential and EEPROM functionality is also required on the same chip [109]. Figure 2.23 shows the typical device structure of NOR-type flash memory with logic device.

2.2.4.2.1 Process Technology [110].

To realize high-performance embedded flash chips, at least three kinds of gate insulators are required beyond the 0.25 μm regime; to form flash tunnel oxide, CMOS gate oxide, high-voltage transistor gate oxide, and I/O transistor gate oxide. Flash cells are usually made by the stacked gate process. Therefore, it is difficult to achieve less than 150% of the cost of pure logic devices.

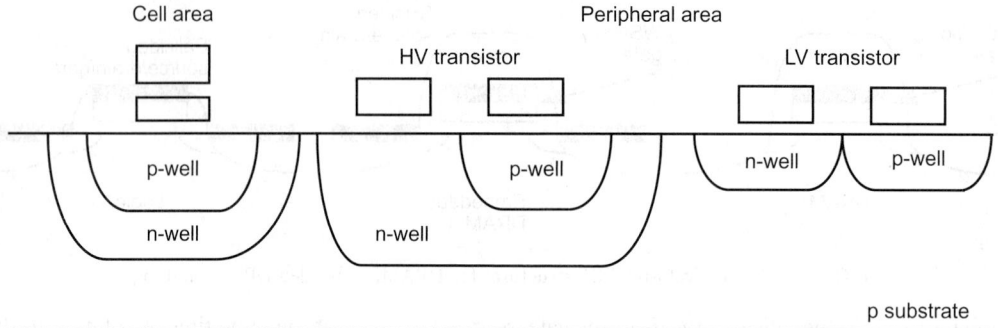

FIGURE 2.23 Device structure schematic view of the NOR flash memories with dual-gate Ti-salicide.

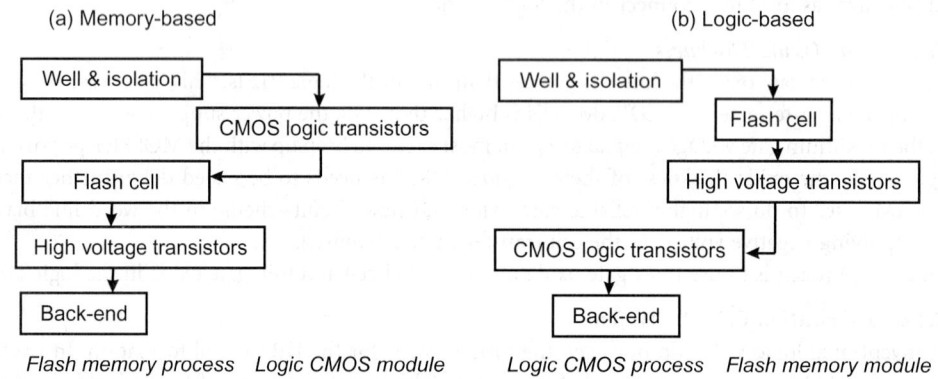

FIGURE 2.24 Process modules.

The two different approaches to realize embedded flash chips are (a) memory-based and (b) logic-based, as shown in Figure 2.24.

(a) is advantageous in that it exploits established flash reliability and yield guaranteed by memory mass production lines, but is disadvantageous for realizing high-performance CMOS transistors due to the additional flash process thermal budget. On the contrary, (b) can use fully CMOS-compatible transistors as they are, but, due to the lack of dedicated mass production lines, great efforts are required to establish flash cell reliability and performance. Historically, (a) has been adopted, but (b) has become more important recently. In general, the number of additional masks required to embed flash cell into logic chips ranges from 4 to 9.

For high-density embedded flash chips, one transistor stack gate cell using channel hot electron programming and channel FN tunneling erasing will be mainstream. For medium- or low-density high-speed embedded flash chips, two transistors will be important in the case of using the low-power P/E method. From the reliability point of view, p-channel cell using band-to-band tunneling induced electron injection [111] and channel FN tunneling ejection are promising since page-programmable EEPROM can also be realized by this mechanism [108].

2.3   BiCMOS Technology

Development of BiCMOS technology began in the early 1980s. In general, bipolar devices are attractive because of their high speed, better gain, better driving capability, and low wide-band noise properties that allow high-quality analog performance. CMOS is particularly attractive for digital applications because of its low power and high packing density. Thus the combination would not only lead to the replacement and improvement of existing ICs but would also provide access to completely new circuits.

Figure 2.25 shows a typical BiCMOS structure [112]. Generally, BiCMOS has vertical npn bipolar junction transistor, lateral pnp transistor, and CMOS on the same chip. Furthermore, if additional mask steps are allowed, passive devices are integrated, as described in the previous section. The main feature of the BiCMOS structure is the existence of a buried layer because bipolar processes require an epitaxial layer grown on a heavily doped n^+ subcollector to reduce collector resistance.

Figure 2.26 shows typical process flow of BiCMOS. This is the simplest arrangement for incorporating bipolar devices and a kind of low-cost BiCMOS. Here, the BiCMOS process is completed with minimum additional process steps required to form the npn bipolar device, transforming the CMOS baseline process to a full BiCMOS technology. For this purpose, many processes are merged. p Tub of n-MOSFET shares an isolation of bipolar devices, n tub of p-MOSFET device is used for the collector, the n^+ source and drain are used for the emitter regions and collector contacts, and also extrinsic base contacts have the p^+ source and drain of PMOS device for common use.

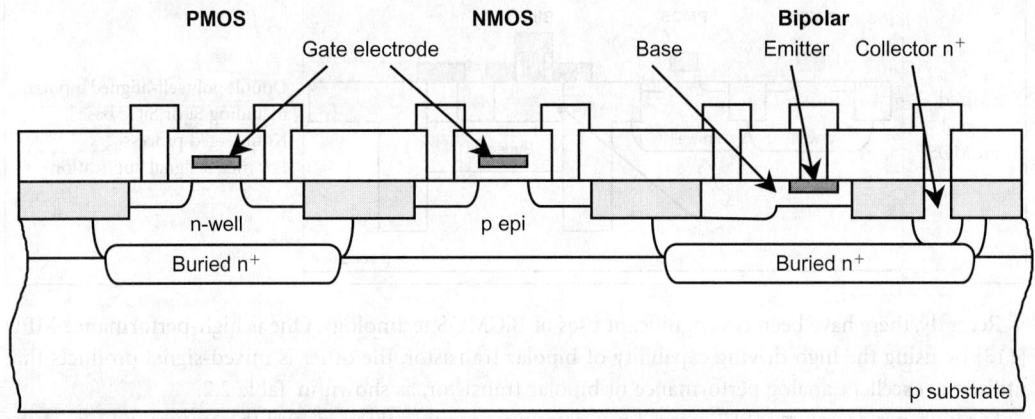

FIGURE 2.25 Cross-sectional view of BiCMOS structure.

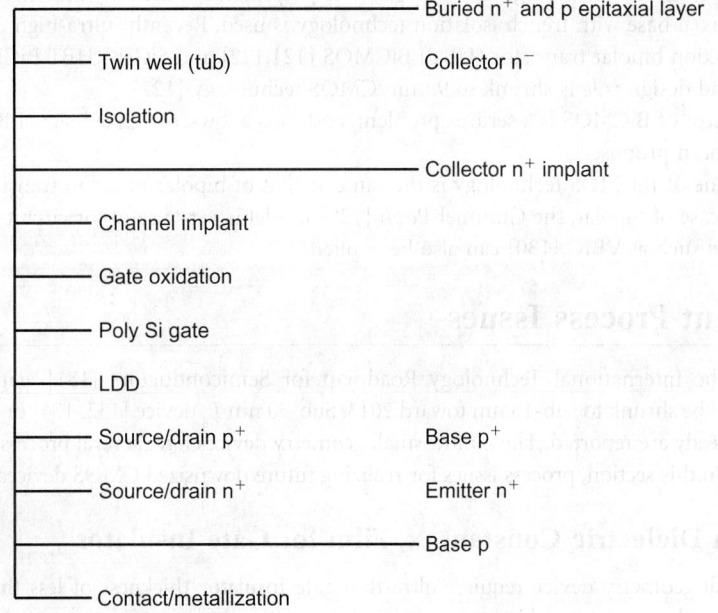

FIGURE 2.26 Typical process flow of BiCMOS device.

TABLE 2.2   Recent BiCMOS structures

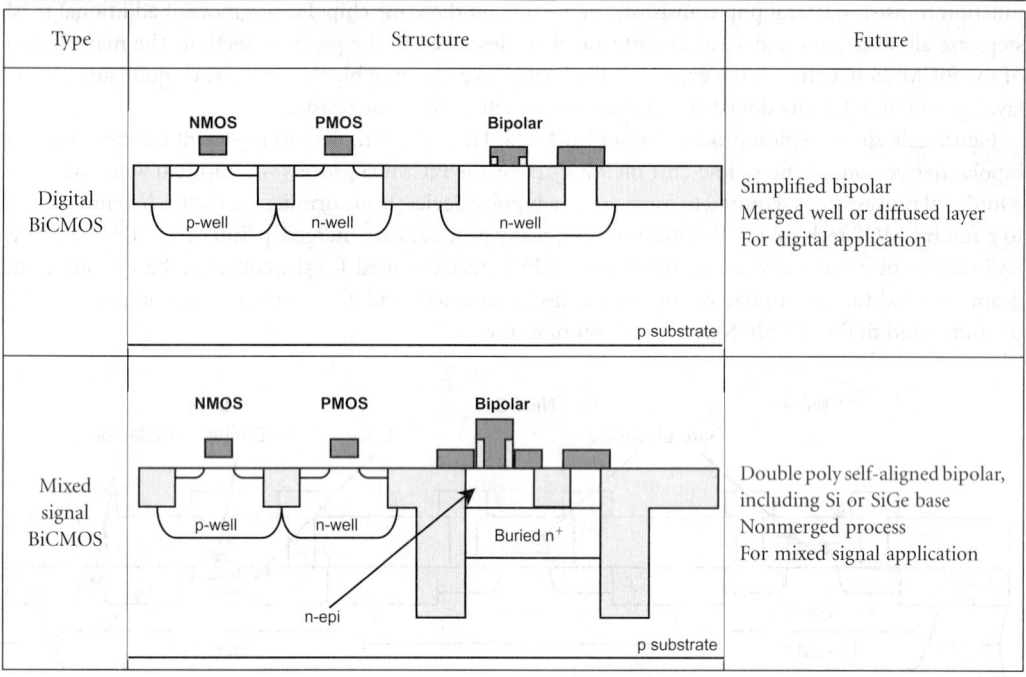

Type	Structure	Future
Digital BiCMOS		Simplified bipolar Merged well or diffused layer For digital application
Mixed signal BiCMOS		Double poly self-aligned bipolar, including Si or SiGe base Nonmerged process For mixed signal application

Recently, there have been two significant uses of BiCMOS technology. One is high-performance MPU [113] by using the high driving capability of bipolar transistor, the other is mixed-signal products that utilize the excellent analog performance of bipolar transistor, as shown in Table 2.2.

For high-performance MPU, merged processes were commonly used, and the mature version of the MPU product has been replaced by CMOS LSI. However, this application is becoming less popular now with reduction in the supply voltage. Mixed-signal BiCMOS requires high performance, especially with respect to f_T, f_{max}, and low-noise figure. Hence, double polysilicon structure with silicon [114] or SiGe [115–120] epitaxial base with trench isolation technology is used. Recently, ultra-high cutoff frequency SiGe heterojunction bipolar transistor (HBT) BiCMOS [121,122] and SiGe:C HBT BiCMOS [123–126] are reported, and design rule is shrunk to 90 nm CMOS technology [127].

Fabrication cost of BiCMOS is a serious problem, and thus a low-cost mixed-signal BiCMOS process [128] has also been proposed.

Modeling issue of BiCMOS technology is the same as that of bipolar junction transistor and CMOS devices. In the case of bipolar, the Gummel-Poon [129] model is very useful for circuit simulation, and advanced model such as VBIC [130] can also be applied.

2.4 Recent Process Issues

According to the International Technology Roadmap for Semiconductors [131], gate length, L_g, of MOSFET would be shrunk to sub-15 nm toward 2013. Sub-50 nm L_g device [132–134] or even sub-15 nm device [135] already are reported. These ultra-small geometry devices have several process issues as shown in Figure 2.27. In this section, process issues for realizing future downsized CMOS devices are introduced.

2.4.1 High Dielectric Constant, κ, Film for Gate Insulator

Since ultra-small geometry device requires ultra-thin gate insulator thickness of less than 3 nm, direct tunneling current through gate oxide is observed. In some cases, this tunnel current might be accepted to realize high driving capability [136–140]. Also, SiON gate insulator with sufficient nitridation can reduce

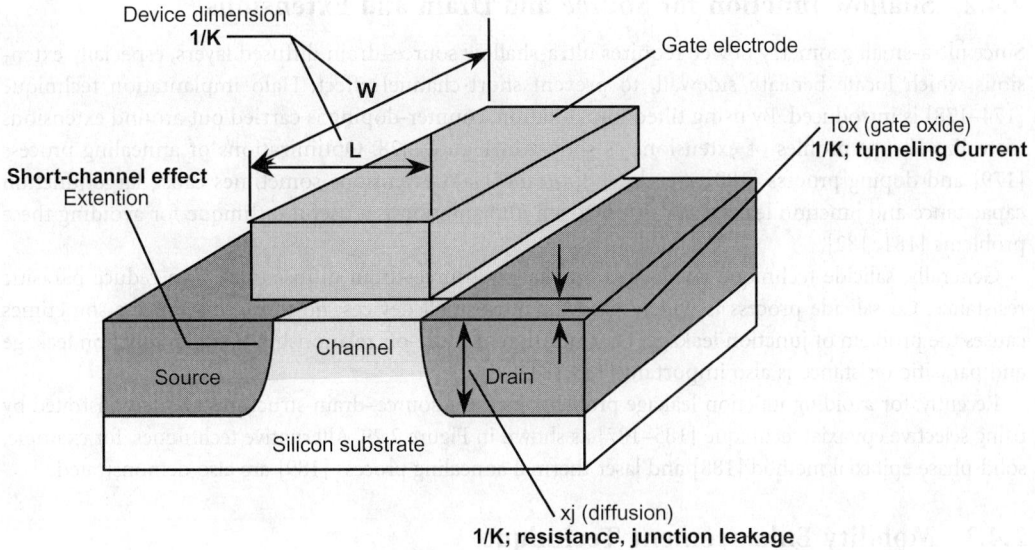

FIGURE 2.27 Process issues for ultra-small geometry transistor.

TABLE 2.3 High κ materials for gate insulator

High-κ Material	Dielectric Constant	Reference
Al_2O_3	10	[144]
Ta_2O_5	20–30	[30,36]
HfO_2	30	[145–147]
HfSiON	10–15	[148,149]
ZrO_2	16–25	[150–153]
$SrTiO_3$	300	[154]
La_2O_3	20–25	[155]

gate leakage current significantly [141,142]. However, in taking account of power consumption of standby mode, especially for mobile applications, these tunneling currents become a severe problem. Because of high dielectric constant, high-κ material of even relatively thick thickness can realize ultra-thin equivalent oxide thickness (EOT) so it can reduce tunneling currents through the gate insulator [143].

Various high-κ films are widely studied as shown in Table 2.3. Although Al_2O_3 [144] and Ta_2O_5 [35,36] were widely noticed in the early stage of high-κ material research, Hf-based material, such as HfO_2 [156, 157] or HfSiON [158,159], have become promising films for sub-50 nm gate length FETs, because of realizing thermal stability and high effective mobility.

For manufacturing high-κ gate insulator CMOS FETs, there are still reliability problems, such as degradation mechanisms [160–163]. Moreover, dielectric constants of gate dielectrics and sidewalls may be very different from each other, resulting in a large discontinuity in dielectric constant; therefore, electric field distribution affected by gate-edge structure should be investigated [164]. In taking account of applying for analog applications, low-frequency behavior is also important [165,166].

For ultra-small geometry FETs with ultra-thin EOT by high-κ gate insulator, metal gate electrode devices are demonstrated to prevent gate electrode resistance problem [167]. Also, these devices can avoid both gate depletion and boron penetration problems. Promising materials as metal gate electrode are TiN [168–170] and TaN [171,172], and dual metal gate structure by using TiN (for p-MOS) and TaSiN (for n-MOS) is also reported [173].

2.4.2   Shallow Junction for Source and Drain and Extensions

Since ultra-small geometry device requires ultra-shallow source–drain diffused layers, especially extensions which locate beneath sidewall, to prevent short-channel effect, Halo implantation technique [174–178] is introduced. By using tilted implantation, counter-doping is carried out around extensions to realize abrupt profiles of extensions as shown in Figure 2.28. Optimizations of annealing process [179] and doping process [180] are also important. Halo extensions sometimes cause large junction capacitance and junction leakage. Tilted nitrogen implantation is a useful technique for avoiding these problems [181, 182].

Generally, salicide technique is adopted on shallow source–drain diffused layers to reduce parasitic resistance. Co salicide process is widely used for ultra-small devices; however, this process sometimes causes the problem of junction leakage. Optimization of trade-off relationship between junction leakage and parasitic resistance is also important [183,184].

Recently, for avoiding junction leakage problem, elevated source–drain structures are demonstrated by using selective epitaxial technique [185–187] as shown in Figure 2.29. Alternative techniques, for example, solid-phase epitaxial method [188] and laser thermal annealing process [189] are also demonstrated.

2.4.3   Mobility Enhancement Technique

Ultra-small geometry device requires low-κ films as gate insulator to reduce gate leakage current; however, interstitial state between low-κ film and silicon causes reduction of mobility [190]. Strained layer by SiGe

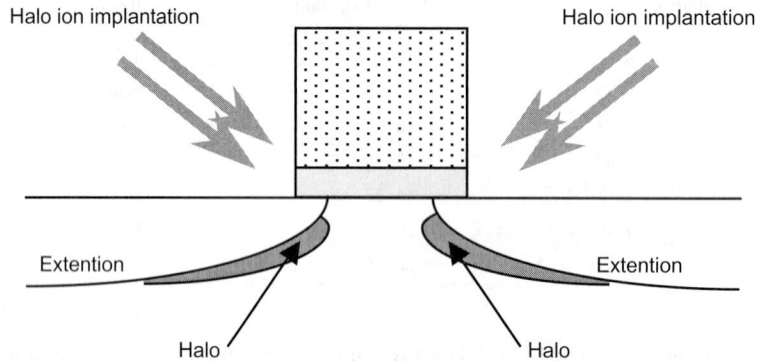

FIGURE 2.28 A schematic of halo implantation

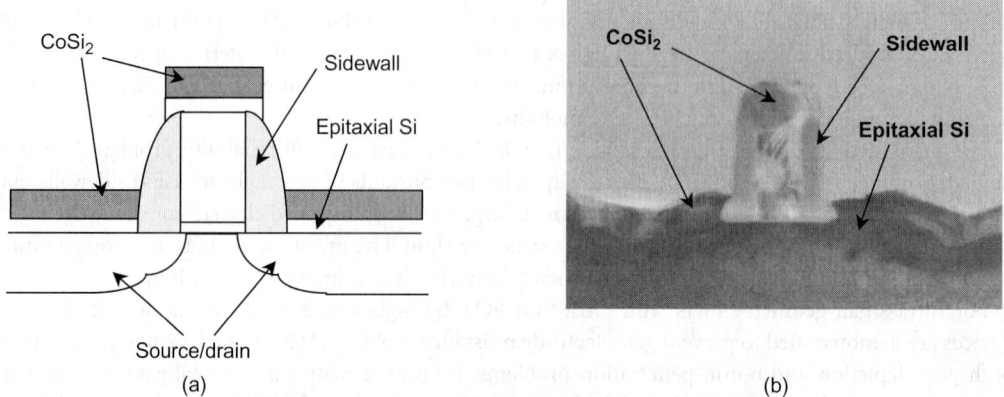

FIGURE 2.29 Structure of raised source/drain/gate FET. (a) Schematic cross section. (b) TEM photograph.

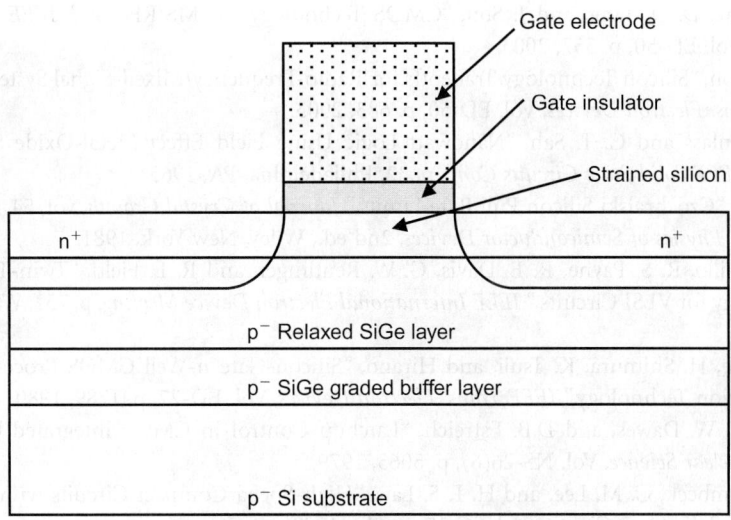

FIGURE 2.30 A schematic of SiGe strained FET.

film can enhance the carrier mobility [191–196] as shown in Figure 2.30. These effects are observed even in ultra-small geometry devices, such as 45 nm gate length FET [197] and 40 nm gate length FET [198].

In contrast, as well known, carrier mobility depends on orientation of silicon wafer. In fact, p-MOSFET realizes high performance on (110) wafer and n-MOSFET on (100) wafer. By using these characteristics, hybrid substrate structures are demonstrated [199–201].

2.4.4 Modeling Issue

As the *de facto* industry standard MOS model, BSIM has been widely used [202]. Latest version BISIM4 can be applied for high-frequency analog circuits [203] or recent process issues [204].

In the case of modeling ultra-small geometry devices, it is necessary to take account of quantum-level phenomena [205] or ballistic effects [206], as a result, physics-based simulation [207] and Monte Carlo simulation [208] are demonstrated. However, pragmatic compact model is important to simulate large-scale level. PSP MOS model [209,210] and HiSIM [211] are useful compact models in the future LSI applications.

2.5 Summary

We described CMOS and BiCMOS technology in this section. CMOS is the most important device structure for realizing the future higher-performance devices required for multimedia and other demanding applications. However, certain problems are preventing the downsizing of device dimensions. To prevent these problems, three-dimensional structures, such as vertical MOSFET [212], FIN-type structure [213,214], and double-gate structure [215] are introduced. In taking account of manufacturing, planer structure is still important [216]. Sub-10 nm CMOS devices have been already reported [217].

BiCMOS technology is also important, especially for mixed-signal applications. However, CMOS device performance has already been demonstrated for RF applications, and thus analog CMOS circuit technology will be very important for realizing the production of analog CMOS.

References

1. S. Kohyama, "SoC Solutions and Technologies for Digital Hypermedia Platform," *IEEE International Electron Device Meeting 1999*, p. 8, Washington, DC, 1999.

2. C. H. Diaz, D. D. Tang, and J. Sun, "CMOS Technology for MS/RF SoC," *IEEE Trans. Electron Devices*, Vol. ED-50, p. 557, 2003.
3. L. E. Larson, "Silicon Technology Tradeoffs for Radio-Frequency/Mixed-Signal System-on-a-Chip," *IEEE Trans. Electron Devices*, Vol. ED-50, p. 683, 2003.
4. F. M. Wanlass and C. T. Sah, "Nanowatt Logic Using Field Effect Metal-Oxide Semiconductor Triode", *IEEE Solid State Circuits Conf.*, p. 32, Philadelphia, PA, 1963.
5. S. N. Rea, "Czochralski Silicon Pull Rate Limits," *Journal of Crystal Growth*, vol. 54, p. 267, 1981
6. S. M. Sze, *Physics of Semiconductor Devices*, 2nd ed., Wiley, New York, 1981.
7. L. C. Parrillo, R. S. Payne, R. E. Davis, G. W. Reutlinger, and R. L. Field, "Twin-Tub CMOS—A Technology for VLSI Circuits," *IEEE International Electron Device Meeting*, p. 752, Washington, DC, 1980.
8. T. Ohzone, H. Shimura, K. Tsuji, and Hirano, "Silicon-Gate n-Well CMOS Process by Full Ion-Implantation Technology," *IEEE Trans. Electron Devices*, Vol. ED-27, p. 1789, 1980.
9. A. Ochoa, W. Dawes, and D.B. Estreich, "Latchup Control in CMOS Integrated Circuits," *IEEE Trans. Nuclear Science*, Vol. NS-26(6), p. 5065, 1979.
10. R. H. Krambeck, C. M. Lee, and H. F. S. Law, "High-Speed Compact Circuits with CMOS," *IEEE Journal of Solid State Circuits*, Vol. SC-17, p. 614, 1982.
11. R. D. Rung, C. J. Dell'Oca, and L. G. Walker, "A Retrograde P-Well for Higher Density CMOS," *IEEE Trans. Electron Devices*, Vol. ED-28, p. 1115, 1981.
12. S. R. Combs, "Scaleable Retrograde P-Well CMOS Technology," *IEEE International Electron Device Meeting*, p.346, Washington, DC, 1981.
13. J. E. Schroeder, A. Ochoa Jr., and P. V. Dressendrfer, "Latch-Up Elimination in Bulk CMOS LSI Circuits," *IEEE Trans. Nuclear Science*, Vol. NS-27, p. 1735, 1980.
14. Y. Sakai, T. Hayashida, N. Hashimoto, O. Mimato, T. Musuhara, K. Nagasawa, T. Yasui, and N. Tanimura, "Advanced Hi-CMOS Device Technology," *IEEE International Electron Device Meeting*, p. 534, Washington, DC, 1981.
15. R. de Werdt, P. van Attekum, H. den Blanken, L. de Bruin, F. op den Buijsch, A. Burgmans, T. Doan, H. Godon, M. Grief, W. Jansen, A. Jonkers, F. Klaassen, M. Pitt, P. van der Plass, A. Stomeijer, R. Verhaar, and J. Weaver, "A 1M SRAM with Full CMOS Cells Fabricated in a 0.7 μm Technology," *IEEE International Electron Device Meeting*, p. 532, Washington, DC, 1987.
16. J. A. Apples, E. Kooi, M. M. Paffen, J. J. H. Schlorje, and W. H. C. G. Verkuylen, "Local Oxidation of Silicon and Its Application in Semiconductor Technology," *Philips Research Report*, Vol. 25, p. 118, 1970.
17. T. A. Shankoff, T. T. Sheng, S. E. Haszko, R. B. Marcus, and T. E. Smith, "Bird's Beak Configuration and Elimination of Gate Oxide Thinning Produced During Selective Oxidation," *Journal of Electrochemical Society*, Vol. 127, p. 216, 1980.
18. S. Nakajima, K. Kikuchi, K. Minegishi, T. Araki, K. Ikuta, and M. Oda, "1 μm 256K RAM Process Technology Using Molybdenum-Polysilicon Gate," *IEEE International Electron Device Meeting*, p. 663, Washington, DC, 1981.
19. G. Fuse, H. Ogawa, K. Tateiwa, I. Nakano, S. Odanaka, M. Fukumoto, H. Iwasaki, and T. Ohzone, "A Practical Trench Isolation Technology with a Novel Planarization Process," *IEEE International Electron Device Meeting*, p. 732, Washington, DC, 1987.
20. K. A. Perry, "Chemical Mechanical Polishing: The Impact of a New Technology on an Industry," *Symposium on VLSI Technology, Digest of Technical Papers*, p. 2, Honolulu, 1998.
21. T. Kuroi, T. Uchida, K. Horita, M. Sakai, Y. Itoh, Y. Inoue, and T. Nishimura, "Stress Analysis of Shallow Trench Isolation for 256MDRAM and Beyond," *IEEE International Electron Device Meeting*, p. 141, San Francisco, CA, 1998.
22. S. Matsuda, T. Sato, H. Yoshimura, A. Sudo, I. Mizushima, Y. Tsumashima, and Y. Toyoshima, "Novel Corner Rounding Process for Shallow Trench Isolation Utilizing MSTS (Micro-Structure Transformation of Silicon)," *IEEE International Electron Device Meeting*, p. 137, San Francisco, CA, 1998.

23. G. J. Hu and R. H. Bruce, "Design Trade-Off between Surface and Buried-Channel FET's," *IEEE Trans. Electron Devices*, Vol. 32, p. 584, 1985.

24. K. M. Cham, D. W. Wenocur, J. Lin, C. K. Lau, and H.-S. Hu, "Submicronmeter Thin Gate Oxide P-Channel Transistors with P+ Poly-Silicon Gates for VLSI Applications," *IEEE Electron Device Letters*, Vol. EDL-7, p. 49, 1986.

25. D. T. Amm, H. Mingam, P. Delpech, and T. T. d'Ouville, "Surface Mobility in P+ and N+ Doped Polysilicon Gate PMOS Transistors," *IEEE Trans. Electron Devices*, Vol. 36, p. 963, 1989.

26. A. Toriumi, T. Mizuno, M. Iwase, M. Takahashi, H. Niiyama, M. Fukumoto, S. Inaba, I. Mori, and M. Yoshimi, "High Speed 0.1 μm CMOS Devices Operating at Room Temperature," *Extended Abstract of International Conference on Solid State Devices and Materials*, p. 487, Tsukuba, Japan, 1992.

27. H. Oyamatsu, M. Kinugawa, and M. Kakumu, "Design Methodology of Deep Submicron CMOS Devices for 1V operation," *Symposium on VLSI Technology, Digest of Technical Papers*, p. 89, Kyoto, Japan, 1993.

28. K. Takeuchi, T. Yamamoto, A. Tanabe, T. Matsuki, T. Kunio, M. Fukuma, K. Nakajima, H. Aizaki, H. Miyamoto, and E. Ikawa, "0.15μm CMOS with High Reliability and Performance," *IEEE International Electron Device Meeting*, p. 883, Washington, DC, 1993.

29. J. R. Ligenza and W. G. Spitzer, "The Mechanism for Silicon Oxidation in Steam and Oxygen," *Journal of Physics and Chemistry of Solids*, Vol. 14. p. 131, 1960.

30. T. Morimoto, H. S. Momose, Y. Ozawa, K. Yamabe, and H. Iwai, "Effects of Boron Penetration and Resultant Limitations in Ultra Thin Pure-Oxide and Nitrided-Oxide, *IEEE International Electron Device Meeting*, p. 429, Washington, DC, 1990.

31. A. Uchiyama, H. Fukuda, T. Hayashi, T. Iwabuchi, and S. Ohno, "High Performance Dual-Gate Sub-Halfmicron CMOSFETs with 6nm-Thick Nitrided SiO_2 Films in an N_2O Ambient," *IEEE International Electron Device Meeting*, p. 425, Washington, DC, 1990.

32. M. Rodder, I.-C. Chen, S. Hattangaly, and J. C. Hu, "Scaling to a 1.0V-1.5V, Sub 0.1μm Gate Length CMOS Technology: Perspective and Challenges", *Extended Abstract of International Conference on Solid State Devices and Materials*, p. 158, Hiroshima, Japan, 1998.

33. M. Rodder, S. Hattangaly, N. Yu, W. Shiau, P. Nicollian, T. Laaksonen, C. P. Chao, M. Mehrotra, C. Lee, S. Murtaza, and A. Aur, "A 1.2V, 0.1 μm Gate Length CMOS Technology: Design and Process Issues," *IEEE International Electron Device Meeting 1998*, p. 623, San Francisco, CA, 1998.

34. M. Khare, X. Guo, X. W. Wang, and T. P. Ma, "Ultra-Thin Silicon Nitride Gate Dielectric for Deep-Sub-Micron CMOS Devices," *Symposium on VLSI Technology, Digest of Technical Papers*, p. 51, Kyoto, Japan, 1997.

35. A. Yagishita, T. Saito, K. Nakajima, S. Inumiya, Y. Akasaka, Y. Ozawa, G. Minamihaba, H. Yano, K. Hieda, K. Suguro, T. Arikado, and K. Okumura, "High Performance Metal Gate MOSFETs Fabricated by CMP for 0.1 μm Regime," *IEEE International Electron Device Meeting*, p. 785, San Francisco, CA, 1998.

36. S. Inumiya, Y. Morozumi, A. Yagishita, T. Saito, D. Gao, D. Choi, K. Hasebe, K. Suguro, Y. Tsunashima, and T. Arikado, "Conformable Formation of High Quality Ultra-Thin Amorphous Ta_2O_5 Gate Dielectric Utilizing Water Assisted Deposition (WAD) for Sub 50 nm Damascene Metal Gate MOSFETs," *IEEE International Electron Device Meeting*, p. 649, San Francisco, CA, 2000.

37. S. P. Murarka, D. B. Fraser, A. K. Shinha, and H. J. Levinstein, "Refractory Silicides of Titanium and Tantalum for Low-Resistivity Gates and Interconnects," *IEEE Trans. Electron Devices*, Vol. ED-27, p. 1409, 1980.

38. H. J. Geipel, Jr., N. Hsieh, M. H. Ishaq, C. W. Koburger, and F. R. White, "Composite Silicide Gate Electrode—Interconnections for VLSI Device Technologies," *IEEE Trans. Electron Devices*, Vol. ED-27, p. 1417, 1980.

39. H. Hayashida, Y. Toyoshima, Y. Suizu, K. Mitsuhashi, H. Iwai, and K. Maeguchi, "Dopant redistribution in dual gate W-polycide CMOS and its improvement by RTA," *Symposium on VLSI Technology, Digest of Technical Papers*, p. 29, Kyoto, Japan, 1989.

40. K. Uwasawa, T. Mogami, T. Kunio, and M. Fukuma, "Scaling Limitations of Gate Oxide in p⁺ Polysilicon Gate MOS Structure for Sub-Quarter Micron CMOS Devices," *IEEE International Electron Device Meeting*, p. 895, Washington, DC, 1993.

41. A. Hokazono, K. Ohuchi, M. Takayanagi, Y. Watanabe, S. Magoshi, Y. Kato, T. Shimizu, S. Mori, H. Oguma, T. Sasaki, H. Yoshimura, K. Miyano, N. Yasutake, H. Suto, K. Adachi, H. Fukui, T. Watanabe, N. Tamaoki, Y. Toyoshima, and H. Ishiuchi, "14 nm Gate Length CMOSFETs Utilizing Low Thermal Budget Process with Poly-SiGe and Ni Salicide," *IEEE International Electron Device Meeting*, p. 639, San Francisco, CA, 2002.

42. T. Ozaki, T. Azuma, M. Itoh, D. Kawamura, S. Tanaka, Y. Ishibashi, S. Shiratake, S. Kyoh, T. Kondoh, S. Inoue, K. Tsuchida, Y. Kohyama, and Y. Onishi, "A 0.15 μm KrF Lithography for 1Gb DRAM Product Using High Printable Patterns and Thin Resist Process," *Symposium on VLSI Technology, Digest of Technical Papers*, p. 84, Honolulu, 1998.

43. S. Hirukawa, K. Matsumoto, and K. Takemasa, "New Projection Optical System for Beyond 150 nm Patterning with KrF and ArF Sources," *Proceedings of 1998 International Symposium on Optical Science, Engineering, and Instrumentation, SPIE's Annual Meeting*, p. 414, 1998.

44. A. Triumi, and M. Iwase, "Lower Submicrometer MOSFETs Fabricated by Direct EB Lithography," *Extended Abstract of the 19th Conference on Solid State Devices and Materials*, p. 347, Tokyo, Japan, 1987.

45. J. A. Liddle and S. D. Berger, "Choice of System Parameters for Projection Electron-Beam Lithography: Accelerating Voltage and Demagnification Factor," *Journal of Vacuum and Science Technology*, Vol. B10(6), p. 2776, 1992.

46. K. Nakajima, H. Yamashita, Y. Kojima, T. Tamura, Y. Yamada, K. Tokunaga, T. Ema, K. Kondoh, N. Onoda, and H. Nozue, "Improved 0.12 μm EB Direct Writing for Gbit DRAM Fabrication," *Symposium on VLSI Technology, Digest of Technical Papers*, p. 34, Honolulu, 1998.

47. K. Deguchi, K. Miyoshi, H. Ban, H. Kyuragi, S. Konaka, and T. Matsuda, "Application of X-Ray Lithography with a Single-Layer Resist Process to Subquartermicron LSI Fabrication," *Journal of Vacuum Science Technology*, Vol. B10(6), p. 3145, 1992.

48. F. Matsuoka, H. Iwai, H. Hayashida, K. Hama, Y. Toyoshima, and K. Maeguchi, "Analysis of Hot Carrier Induced Degradation Mode on PMOSFETs," *IEEE Transactions Electron Devices*, Vol. ED-37, p. 1487, 1990.

49. S. Ogura, P. J. Chang, W. W. Walker, D. L. Critchlow, and J. F. Shepard, "Design and Characteristics of the Lightly-Doped Drain-Source (LDD) Insulated Gate Field Effect Transistor," *IEEE Transactions Electron Devices*, Vol. ED-27, p. 1359, 1980.

50. J. M. Ha, J. W. Park, W. S. Kim, S. P. Kim, W. S. Song, H. S. Kim, H. J. Song, K. Fujihara, M. Y. Lee, S. Felch, U. Jeong, M. Groeckner, K. H. Kim, H. J. Kim, H. T. Cho, Y. K. Kim, D. H. Ko, and G. C. Lee, "High Performance pMOSFET with BF₃ Plasma Doped Gate/Source/Drain and A/D Extension," *IEEE International Electron Device Meeting*, p. 639, San Francisco, CA, 1998.

51. K. Goto, J. Matsuo, T. Sugii, H. Minakata, I. Yamada, and T. Hisatsugu, "Novel Shallow Junction Technology Using Decaborone (B₁₀H₁₄)," *IEEE International Electron Device Meeting*, p. 435, San Francisco, CA, 1996.

52. T. Ohguro, S. Nakamura, M. Saito, M. Ono, H. Harakawa, E. Morifuji, T. Yoshitomi, T. Morimoto, H. S. Momose, Y. Katsumata, and H. Iwai, "Ultra-shallow Junction and Salicide Technique for Advanced CMOS Devices," *Proceedings of the Sixth International Symposium on Ultralarge Scale Integration Science and Technology, Electrochemical Society*, p. 275, May, 1997.

53. C. M. Osburn, M. Y. Tsai, and S. Zirinsky, "Self-Aligned Silicide Conductors in FET Integrated Circuits," *IBM Technical Disclosure Bulletin*, Vol. 24, p. 1970, 1981.

54. T. Shibata, K. Hieda, M. Sato, M. Konaka, R. L. M. Dang, and H. Iizuka, "An Optimally Designed Process for Submicron MOSFET's," *IEEE International Electron Device Meeting*, p. 647, Washington, DC, 1981.

55. C. Y. Ting, S. S. Iyer, C. M. Osburn, G. J. Hu, and A. M. Schweighart, "The Use of TiSi₂ in a Self-Aligned Silicide Technology," *Proceedings of 1st International Symposium on VLSI Science and Technology, Electrochemical Society Meeting*, Vol. 82(7), p. 224, 1982.

56. R. A. Haken, "Application of the Self-Aligned Titanium Silicide Process to Very Large Scale Integrated N-Metal-Oxide-Semiconductor and Complementary Metal-Oxide-Semiconductor Technologies," *Journal of Vacuum Science and Technology*, Vol. B3(6), p. 1657, 1985.

57. N. Kobayashi, N. Hashimoto, K. Ohyu, T. Kaga, and S. Iwata, "Comparison of $TiSi_2$ and WSi_2 for Sub-Micron CMOSs", *Symposium on VLSI Technology, Digest of Technical Papers*, p. 49, 1986.

58. V. Q. Ho and D. Poulin, "Formation of Self-Aligned $TiSi_2$ for VLSI Contacts and Interconnects," *Journal of Vacuum Science and Technology*, Vol. A5, p. 1396, 1987.

59. C. H. Ting, F. M. d'Heurle, S. S. Iyer, and P. M. Fryer, "High Temperature Process Limitation on $TiSi_2$," *Journal of Electrochemical Society*, Vol. 133(12), p. 2621, 1986.

60. C. M. Osburn, M. Y. Tsai, S. Roberts, C. J. Lucchese, and C. Y. Ting, "High Conductivity Diffusions and Gate Regions Using a Self-Aligned Silicide Technology", *Proceedings of 1st International Symposium on VLSI Science and Technology, Electrochemical Society*, Vol. 82–1, p. 213, 1982.

61. T. Morimoto, H. S. Momose, T. Iinuma, I. Kunishima, K. Suguro, H. Okano, I. Katakabe, H. Nakajima, M. Tsuchiaki, M. Ono, Y. Katsumata, and H. Iwai, "A NiSi salicide technology for advanced logic devices," *IEEE International Electron Device Meeting 1991*, p. 653, Washington, DC, 1991.

62. T. Ohguro, M. Saito, E. Morifuji, T. Yoshitomi, T. Morimoto, H. S. Momose, Y. Katsumata, and H. Iwai, "Thermal Stability of $CoSi_2$ Film for CMOS Salicide," *IEEE Trans. Electron Devices*, Vol. ED-47, p. 2208, 2000.

63. T. Kwork, "Effect of Metal Line Geometry on Electromigration Lifetime in Al-Cu Submicron Interconnects," *26th Annual Proceedings of Reliability Physics*, p. 185, 1988.

64. N. Owada, K. Hinode, M. Horiuchi, T. Nishida, K. Nakata, and K. Mukai, "Stress Induced Slit-Like Void Formation in a Fine-Pattern Al-Si Interconnect during Aging Test," *Proceedings of the 2nd International IEEE VLSI Multilevel Interconnection Conference*, p. 173, 1985.

65. T. Kikkawa, H. Aoki, E. Ikawa, and J. M. Drynan, "A Quarter-Micrometer Interconnection Technology Using a TiN/Al-Si-Cu/Al-Si-Cu/TiN/Ti Multilayer Structure," *IEEE Trans. Electron Devices*, Vol. ED-40, p. 296, 1993.

66. F. White, W. Hill, S. Eslinger, E. Payne, W. Cote, B. Chen, and K. Johnson, "Damascene Stud Local Interconnect in CMOS Technology," *IEEE International Electron Device Meeting*, p. 301, San Francisco, CA, 1992.

67. N. Kobayashi, M. Suzuki, and M. Saitou, "Tungsten Plug Technology: Substituting Tungsten for Silicon Using Tungsten Hexafluoride," *Extended Abstract of International Conference on Solid State Devices and Materials*, p. 85, 1988.

68. W. Cote, G. Costrini, D. Eldlstein, C. Osborn, D. Poindexter, V. Sardesai, and G. Bronner, "An Evaluation of Cu Wiring in a Production 64Mb DRAM," *Symposium on VLSI Technology, Digest of Technical Papers*, p. 24, Honolulu, 1998.

69. A. L. S. Loke, J. Wetzel, C. Ryu, W.-J. Lee, and S. S. Wong, "Copper Drift in Low-k Polymer Dielectrics for ULSI Metallization," *Symposium on VLSI Technology, Digest of Technical Papers*, p. 26, Honolulu, 1998.

70. T. Kikkawa, "Current and Future Low-k Dielectrics for Cu Interconnects," *IEEE International Electron Device Meeting*, p. 253, San Francisco, 2000.

71. H. Miyajima, K. Watanabe, K. Fujita, S. Ito, K. Tabuchi, T. Shimayama, K. Akiyama, T. Hachiya, K. Higashi, N. Nakamura, A. Kajita, N. Matsunaga, Y. Enomoto, R. Kanamura, M. Inohara, K. Honda, H. Kamijo, R. Nakata, H. Yano, N. Hayasaka, T. Hasegawa, S. Kadomura, H. Shibata, and T. Yoda, "Challenge of Low-k Materials for 130, 90, 65 nm Node Interconnect Technology and Beyond," *IEEE International Electron Device Meeting*, p. 329, San Francisco, CA, 2004.

72. M. Inohara, I. Tamura, T. Yamaguchi, H. Koike, Y. Enomoto, S. Arakawa, T. Watanabe, E. Ide, S. Kadomura, and S. Sunouchi, "High Performance Copper and Low-k Interconnect Technology Fully Compatible to 90nm-Node SoC Application (CMOS4)," *IEEE International Electron Device Meeting*, p. 77, San Francisco, CA, 2002.

73. T. I. Bao, C. C. Ko, J. Y. Song, L. P. Li, H. H. Lu, Y. C. Lu, Y. H. Chen, S. M. Jang, and M. S. Liang, "90 nm Generation Cu/CVD Low-k (k < 2.5) Interconnect Technology," *IEEE International Electron Device Meeting*, p. 583, San Francisco, CA, 2002.

74. Y. Oku, K. Yamada, T. Goto, Y. Seino, A. Ishikawa, T. Ogata, K. Kohmura, N. Fujii, N. Hata, R. Ichilawa, T. Yoshino, C. Negoro, A. Nakano, Y. Sonoda, S. Takada, H. Miyoshi, S. Oike, H. Tanaka, H. Matsuo, K. Kinoshita, and T. Kikkawa, "Novel Self-Assembled Ultra-Low-κ Porous Silica Films with High Mechanical Strength for 45 nm BEOL Technology," *IEEE International Electron Device Meeting*, p. 139, Washington, DC, 2003.

75. J. Wada, Y. Oikawa, T. Katata, N. Nakamura, and M. B. Anand, "Low Resistance Dual Damascene Process by AL Reflow Using Nb Liner," *Symposium on VLSI Technology, Digest of Technical Papers*, p. 48, Honolulu, 1998.

76. K. Ueno, M. Suzuki, A. Matsumoto, K. Motoyama, T. Tonegawa, N. Ito, K. Arita, Y. Tsuchiya, T. Wake, A. Kubo, K. Sugai, N. Oda, H. Miyamoto, and S. Saito, "A High Reliability Copper Dual-Damascene Interconnection With Direct-Contact Via Structure," *IEEE International Electron Device Meeting*, p. 265, San Francisco, CA, 2000.

77. M. S. Liang, "Challenges in Cu/Low-K Integration," *IEEE International Electron Device Meeting 2004*, p. 313, San Francisco, CA, 2004.

78. S. Kondo, B. U. Yoon, S. Tokitoh, K. Misawa, S. Sone, H. J. Shin, N. Ohashi, and H. Kobayashi, "Low-Pressure CMP for 300-mm Ultra Low-k (k = 1.6–1.8)/Cu Integration," *IEEE International Electron Device Meeting*, p. 151, Washington, DC, 2003.

79. H. S. Momose, R. Fujimoto, S. Ohtaka, E. Morifuji, T. Ohguro, T. Yoshitomi, H. Kimijima, S. Nakamura, T. Morimoto, Y. Katsumata, H. Tanimoto, and H. Iwai, "RF Noise in 1.5 nm Gate Oxide MOSFETs and the Evaluation of the NMOS LNA Circuit Integrated on a Chip," *Symposium on VLSI Technology, Digest of Technical Papers*, p. 96, Honolulu, 1998.

80. C. H. Chen, C. S. Chang, C. P. Chao, J. F. Kuan, C. L. Chang, S. H. Wang, H. M. Hsu, W. Y. Lien, Y. C. Tsai, H. C. Lin, C. C. Wu, C. F. Huang, S. M. Chen, P. M. Tseng, C. W. Chen, C. C. Ku, T. Y. Lin, C. F. Chang, H. J. Lin, M. R. Tsai, S. Chen, C. F. Chen, M. Y. Wei, Y. J. Wang, J. C. H. Lin, W. M. Chen, C. C. Chang, M.C. King, C. M. Huang, C. T. Lin, J. C. Guo, G. J. Chern, D. D. Tang, and J. Y. C. Sun, "A 90nm CMOS MS/RF Based Foundry SOC Technology Comprising Superb 185 GHz f_T RFCMOS and Versatile, High-Q Passive Components for Cost/Performance Optimization," *IEEE International Electron Device Meeting*, p. 39, Washington, DC, 2003.

81. S.-J. Ding, H. Hu, C. Zhu, S. J. Kim, X. Yu, M.-F. Li, B. J. Cho, D. S. H. Chan, M. B. Yu, C. Rustagi, A. Chin, and D.-L. Kwong, "RF, D.C, and Reliability Characteristics of ALD HfO_2-Al_2O_3 Laminate MIM Capacitors for Si RF IC Applications," *IEEE Trans. Electron Devices*, Vol. ED-51, p. 886, 2004.

82. J. N. Burghartz, "Progress in RF Inductors on Silicon—Understanding Substrate Loss," *IEEE International Electron Device Meeting*, p. 523, San Francisco, CA, 1998.

83. T. Yoshitomi, Y. Sugawara, E. Morifuji, T. Ohguro, H. Kimijima, T. Morimoto, H. S. Momose, Y. Katsumata, and H. Iwai, "On-Chip Inductors with Diffused Shield Using Channel-Stop Implant." *IEEE International Electron Device Meeting*, p. 540, San Francisco, CA,1998.

84. Y. Zhuang, M. Vroubel, B. Rejaei, and J. N. Burghartz, "Ferromagnetic RF Inductors and Transformers for Standard CMOS/BiCMOS," *IEEE International Electron Device Meeting*, p. 475, San Francisco, CA, 2002.

85. M.-C. Hsieh, Y.-K. Fang, C.-H. Chen, S.-M. Chen, and W.-K. Yeh, "Design and Fabrication of Deep Submicron CMOS Technology Compatible Suspended High-Q Spiral Inductors," *IEEE Transactions Electron Devices*, Vol. ED-51, p. 324, 2004.

86. C. Liao, C.-W. Liu, and Y.-M. Hsu, "Observation of Explosive Spectral Behaviors in Proton-Enhanced High-Q Inductors and Their Explanations," *IEEE Transactions Electron Devices*, Vol. ED-50, p. 758, 2003.

87. J. N. Burghartz and B. Rejaei, "On the Design of RF Spiral Inductors on Silicon," *IEEE Transactions Electron Devices*, Vol. ED-50, p. 718, 2003.

88. J. Borel, "Technologies for Multimedia Systems on a Chip," *International Solid State Circuit Conference, Digest of Technical Papers,* p. 18, 1997.

89. M. Tsukamoto, H. Kuroda, and Y. Okamoto, "0.25mm W-Polycide Dual Gate and Buried Metal on Diffusion Layer (BMD) Technology for DRAM-Embedded Logic Devices," *Symposium on VLSI Technology, Digest of Technical Papers,* p. 23, 1997.

90. K. Itabashi, S. Tsuboi, H. Nakamura, K. Hashimoto, W. Futoh, K. Fukuda, I. Hanyu, S. Asai, T. Chijimatsu, E. Kawamura, T. Yao, H. Takagi, Y. Ohta, T. Karasawa, H. Iio, M. Onoda, F. Inoue, H. Nomura, Y. Satoh, M. Higashimoto, M. Matsumiya, T. Miyabo, T. Ikeda, T. Yamazaki, M. Miyajima, K. Watanabe, S. Kawamura, and M. Taguchi, "Fully Planarized Stacked Capacitor Cell with Deep and High Aspect Ratio Contact Hole for Giga-bit DRAM," *Symposium on VLSI Technology, Digest of Technical Papers,* p. 21, 1997.

91. K. N. Kim, J. Y. Lee, K. H. Lee, B. H. Noh, S. W. Nam, Y. S. Park, Y. H. Kim, H. S. Kim, J. S. Kim, J. K. Park, K. P. Lee, K. Y. Lee, J. T. Moon, J. S. Choi, J. W. Park, and J.G. Lee, "Highly Manufacturable 1Gb SDRAM," *Symposium on VLSI Technology, Digest of Technical Papers,* p. 10, 1997.

92. Y. Kohyama, T. Ozaki, S. Yoshida, Y. Ishibashi, H. Nitta, S. Inoue, K. Nakamura, T. Aoyama, K. Imai, and N. Hayasaka, "A Fully Printable, Self-aligned and Planarized Stacked Capacitor DRAM Cell Technology for 1Gbit DRAM and Beyond," *Symposium on VLSI Technology, Digest of Technical Papers,* p. 17, 1997.

93. J. M. Drynan, K. Nakajima, T. Akimoto, K. Saito, M. Suzuki, S. Kamiyama, and Y. Takaishi, "Cylindrical Full Metal Capacitor Technology for High-Speed Gigabit DRAMs," *Symposium on VLSI Technology, Digest of Technical Papers,* p. 151, 1997.

94. S. Takehiro, S. Yamauchi, M. Yoshimura, and H. Onoda, "The Simplest Stacked BST Capacitor for the Future DRAMs Using a Novel Low Temperature Growth Enhanced Crystallization," *Symposium on VLSI Technology, Digest of Technical Papers,* p. 153, 1997.

95. L. Nesbit, J. Alsmeier, B. Chen, J. DeBrosse, P. Fahey, M. Gall, J. Gambino, S. Gerhard, H. Ishiuchi, R. Kleinhenz, J. Mandelman, T. Mii, M. Morikado, A. Nitayama, S. Parke, H. Wong, and G. Bronner, "A 0.6 µm² 256Mb Trench DRAM Cell With Self-Aligned BuriEd STrap (BEST)," *IEEE International Electron Device Meeting,* p. 627, Washington, DC, 1993.

96. G. Bronner, H. Aochi, M. Gall, J. Gambino, S. Gernhardt, E. Hammerl, H. Ho, J. Iba, H. Ishiuchi, M. Jaso, R. Kleinhenz, T. Mii, M. Narita, L. Nesbit, W. Neumueller, A. Nitayama, T. Ohiwa, S. Parke, J. Ryan, T. Sato, H. Takato, and S. Yoshikawa, "A Fully Planarized 0.25 µm CMOS Technology for 256Mbit DRAM and Beyond," *Symposium on VLSI Technology, Digest of Technical Papers,* p. 15, 1995.

97. H. Ishiuchi, Y. Yoshida, H. Takato, K. Tomioka, K. Matsuo, H. Momose, S. Sawada, K. Yamazaki, and K. Maeguchi, "Embedded DRAM Technologies," *IEEE International Electron Device Meeting,* p. 33, Washington, DC, 1997.

98. M. Togo, S. Iwao, H. Nobusawa, M. Hamada, K. Yoshida, N. Yasuzato, and T. Tanigawa, "A Salicide-Bridged Trench Capacitor with a Double-Sacrificial-Si3N4-Sidewall (DSS) for High-Performance Logic-Embedded DRAMs," *IEEE International Electron Device Meeting,* p. 37, Washington, DC, 1997.

99. S. Crowder, S. Stiffler, P. Parries, G. Bronner, L. Nesbit, W. Wille, M. Powell, A. Ray, B. Chen, and B. Davari, "Trade-Offs in the Integration of High Performance Devices with Trench Capacitor DRAM," *IEEE International Electron Device Meeting,* p. 45, Washington, DC, 1997.

100. S. Crowder, R. Hannon, H. Ho, D. Sinitsky, S. Wu, K. Winstel, B. Khan, S. R. Stiffler, and S. S. Iyer, "Integration of Trench DRAM into a High-Performance 0.18 µm Logic Technology with Copper BEOL," *IEEE International Electron Device Meeting,* p. 1017, San Francisco, CA, 1998.

101. M. Yoshida, T. Kumauchi, K. Kawakita, N. Ohashi, H. Enomoto, T. Umezawa, N. Yamamoto, I. Asano, and Y. Tadaki, "Low Temperature Metal-Based Cell Integration Technology for Gigabit and Embedded DRAMs," *IEEE International Electron Device Meeting,* p. 41, Washington, DC, 1997.

102. S. Nakamura, M. Kosugi, H. Shido, K. Kosemura, A. Satoh, H. Minakata, H. Tsunoda, M. Kobayashi, T. Kurahashi, A. Hatada, R. Suzuki, M. Fukuda, T. Kimura, M. Nakabayashi, M. Kojima, Y. Nara,

T. Fukano, and N. Sasaki, "Embedded DRAM Technology Compatible to the 0.13 μm High-Speed Logics by Using Ru Pillars in Cell Capacitors and Peripheral Vias," *IEEE International Electron Device Meeting*, p. 1029, San Francisco, CA, 1998.

103. J. M. Drynan, K. Fukui, M. Hamada, K. Inoue, T. Ishigami, S. Kamiyama, A. Matsumoto, H. Nobusawa, K. Sugai, M. Takenaka, H. Yamaguchi, and T. Tanigawa, "Shared Tungsten Structures for FEOL/BEOL Compatibility in Logic-Friendly Merged DRAM," *IEEE International Electron Device Meeting*, p. 849, San Francisco, 1998.

104. M. Togo, K. Noda, and T. Tanigawa, "Multiple-Thickness Gate Oxide and Dual-Gate Technologies for High-Performance Logic-Embedded DRAMs," *IEEE International Electron Device Meeting*, p. 347, San Francisco, CA, 1998.

105. N. Yanagiya, S. Matsuda, S. Inaba, M. Takayanagi, I. Mizushima, K. Ohuchi, K. Okano, K. Takahashi, E. Morifuji, M. Kanda, Y. Matsubara, M. Habu, M. Nishigoori, K. Honda, H. Tsuno, K. Yasumoto, T. Yamamoto, K. Hiyama, K. Kokubun, T. Suzuki, J. Yoshikawa, T. Sakurai, T. Ishizuka, Y. Yoshida, M. Moriuchi, M. Kishida, H. Matsumori, H. Harakawa, H. Oyamatsu, N. Nagashima, S. Yamada, T. Noguchi, H. Okamoto, and M. Kakumu, "65nm CMOS Technology (CMOS5) with High Density Embedded Memories for Broadband Microprocessor Applications," *IEEE International Electron Device Meeting*, p. 57, San Francisco, CA, 2002.

106. Y. Matsubara, M. Habu, S. Matsuda, K. Honda, E. Morifuji, T. Yoshida, K. Kokubun, K. Yasumoto, T. Sakurai, T. Suzuki, J. Yoshikawa, E. Takahashi, K. Hiyama, M. Kanda, R. Ishizuka, M. Moriuchi, H. Koga, Y. Fukuzaki, Y. Sogo, H. Nagashima, Y. Okamoto, S. Yamada, and T. Noguchi, "Fully Compatible Integration of High Density Embedded DRAM with 65nm CMOS Technology (CMOS5)," *IEEE International Electron Device Meeting 2003*, p. 423, Washington, DC, 2003.

107. IEEE Standard #1005.

108. K. Yoshikawa, "Guide-Lines on Flash Memory Cell Selection", *Extended Abstract of International Conference on Solid State Devices and Materials*, p. 138, 1998.

109. H. Watanabe, S. Yamada, M. Tanimoto, M. Mitsui, S. Kitamura, K. Amemiya, T. Tanzawa, E. Sakagami, M. Kurata, K. Isobe, M. Takebuchi, M. Kanda, S. Mori, and T. Watanabe, "Novel 0.44μm² Ti-Salicide STI Cell Technology for High-Density NOR Flash Memories and High Performance Embedded Application," *IEEE International Electron Device Meeting*, p. 975, San Francisco, CA, 1998.

110. C. Kuo, "Embedded Flash Memory Applications, Technology and Design," *1995 IEDM Short Course: NVRAM Technology and Application, IEEE International Electron Device Meeting*, Washington, DC, 1995.

111. T. Ohnakado, K. Mitsunaga, M. Nunoshita, H. Onoda, K. Sakakibara, N. Tsuji, N. Ajika, M. Hatanaka, and H. Miyoshi, "Novel Electron Injection Method Using Band-to-Band Tunneling Induced Hot Electron(BBHE) for Flash Memory with a P-Channel Cell," *IEEE International Electron Device Meeting*, p. 279, Washington, DC, 1995

112. H. Iwai, G. Sasaki, Y. Unno, Y. Niitsu, M. Norishima, Y. Sugimoto, and K. Kannzaki, "0.8μm Bi-CMOS Technology with High f_T Ion-Implanted Emitter Bipolar Transistor," *IEEE International Electron Device Meeting*, p. 28, Washington, DC, 1987.

113. L. T. Clark and G. F. Taylor, "High Fan-In Circuit Design," *Bipolar/BiCMOS Circuits & Technology Meeting*, p. 27, Minneapolis, 1994.

114. H. Nii, C. Yoshino, K. Inoh, N. Itoh, H. Nakajima, H. Sugaya, H. Naruse, Y. Kataumata, and H. Iwai, "0.3 μm BiCMOS Technology for Mixed Analog/Digital Application System," *Bipolar/BiCMOS Circuits & Technology Meeting*, p. 68, Minneapolis, 1997.

115. R. A. Johnson, M. J. Zierak, K. B. Outama, T. C. Bahn, A. J. Joseph, C. N. Cordero, J. Malinowski, K. A. Bard, T. W. Weeks, R. A. Milliken, T. J. Medve, G. A. May, W. Chong, K. M. Walter, S. L. Tempest, B. B. Chau, M. Boenke, M. W. Nelson, and D. L. Harame, "1.8 Million Transistor CMOS ASIC Fabricated in a SiGe BiCMOS Technology," *IEEE International Electron Device Meeting*, p. 217, San Francisco, CA, 1998.

116. M. Carroll, T. Ivanov, S. Kuehne, J. Chu, C. King, M. Frei, M. Mastrapasqua, R. Johnson, K. Ng, S. Moinian, S. Martin, C. Huang, T. Hsu, D. Nguyen, R. Singh, L. Fritzinger, T. Esry, W. Moller, B. Kane, G. Abeln, D. Hwang, D. Orphee, S. Lytle, M. Roby, D. Vitkavage, D. Chesire, R. Ashton, D. Shuttleworth, M. Thoma, S. Lewellen, P. Mason, T. Lai, H. Hsieh, D. Dennis, E. Harris, S. Thomas, R. Gregor, P. Sana, and W. Wu, "COM2 SiGe Modular BiCMOS Technology for Digital, Mixed-Signal, and RF Applications," *IEEE International Electron Device Meeting*, p. 145, San Francisco, CA, 2000.

117. T. Hashimoto, F. Sato, T. Aoyama, H. Suzuki, H. Yoshida, H. Fujii, and T. Yamazaki, "A 73GHz f_T 0.18μm RF-SiGe BiCMOS Technology Thermal Budget Trade-Off and with Reduced Boron-Spike Effect on HBT Characteristics," *IEEE International Electron Device Meeting*, p. 149, San Francisco, CA, 2000.

118. F. Sato, T. Hashimoto, H. Fujii, H. Yoshida, H. Suzuki, and T. Yamazaki, "A 0.18-μm RF SiGe BiCMOS Technology with Collector-Epi-Free Double-Poly Self-Aligned HBTs," *IEEE Trans. Electron Devices*, Vol. ED-50, p. 669, 2003.

119. K. Washio, "SiGe HBT and BiCMOS Technologies," *IEEE International Electron Device Meeting*, p. 113, Washington, DC, 2003.

120. A. T. Tilke, M. Rochel, J. Berkner, S. Rothenhäußer, K. Stahrenberg, J. Wiedermann, C. Wagner, and C. Dahl, "A Low-Cost Fully Self-Aligned SiGe BiCMOS Technology Using Selective Epitaxy and a Lateral Quasi-Single-Poly Integration Concept," *IEEE Trans. Electron Devices*, Vol. ED-51, p. 1101, 2004.

121. K. Washio, E. Ohue, H. Shimamoto, K. Oda, R. Hayami, Y. Kiyota, M. Tanabe, M. Kondo, T. Hashimoto, and T. Harada, "A 0.2-μm 180-GHz-f_{max} 6.7-ps-ECL SOI/HRS Self-Aligned SEG SiGe HBT/CMOS Technology for Microwave and High-Speed Digital Applications," *IEEE Trans. Electron Devices*, Vol. ED-49, p. 271, 2002.

122. T. Hashimoto, Y. Nonaka, T. Tominari, H. Fujiwara, K. Tokunaga, M. Arai, S. Wada, T. Udo, M. Sato, M. Miura, H. Shimamoto, K. Washio, and H. Tomioka, "Direction to Improve SiGe BiCMOS Technology Featuring 200-GHz SiGe HBT and 80-nm Gate CMOS," *IEEE International Electron Device Meeting*, p. 129, Washington, DC, 2003.

123. M. W. Xu, S. Decoutere, A. Sibaja-Hernandez, K. Van Wichelen, L. Witters, R. Loo, E. Kunnen, C. Knorr, A. Sadovnikov, and C. Bulucea, "Ultra Low Power SiGe:C HBT for, 0.18 μm RF-BiCMOS," *IEEE International Electron Device Meeting*, p. 125, Washington, DC, 2003.

124. D. Knoll, K. E. Ehwald, B. Heinemann, A. Fox, K. Blum, H. Rücker, F. Fürnhammer, B. Senapati, R. Barth, U. Haak, W. Höppner, J. Drews, R. Kurps, S. Marschmeyer, H. H. Richter, T. Grobolla, B. Kuck, O. Fursenko, P. Schley, R. Scholz, B. Tillack, Y. Yamamoto, K. Köpke, H. E. Wulf, D. Wolansky, and W. Winkler, "A Flexible, Low-Cost, High Performance SiGe:C BiCMOS Process with a One-Mask HBT Module," *IEEE International Electron Device Meeting*, p. 783, San Francisco, CA, 2002.

125. B. Heinemann, R. Barth, D. Bolze, J. Drews, P. Formanek, O. Fursenko, M. Glante, K. Glowatzki, A. Gregor, U. Haak, W. Höppener, D. Knoll, R. Kurps, S. Marschmeyer, S. Orlowski, H. Rücker, P. Schley, D. Schmidt, R. Scholz, W. Winkler, and Y. Yamamoto, "A Complementary BiCMOS Technology with High Speed npn and pnp SiGe:C HBTs," *IEEE International Electron Device Meeting 2003*, p. 117, Washington, DC, 2003.

126. H. Rücker, B. Heinemann, R. Barth, D. Bolze, J. Drews, U. Haak, W. Höppener, D. Knoll, K. Köpke, S. Marschmeyer, H. H. Richter, P. Schley, D. Schmidt, R. Scholz, P. Tillack, W. Winkler, H.-E. Wulf, and Y. Yamamoto, "SiGe:C CiCMOS Technology with 3.6 ps Gate Delay," *IEEE International Electron Device Meeting*, p. 121, Washington, DC, 2003.

127. K. Kuhn, M. Agostinelli, S. Ahmed, S. Chambers, S. Cea, S. Christensen, P. Fischer, J. Gong, C. Kardas, T. Letson, L. Henning, A. Murthy, H. Muthali, B. Obradovic, P. Packan, S. W. Pae, I. Post, S. Putna, K. Raol, A. Roskowski, R. Soman, T. Thomas, P. Vandervoorn, M. Weiss, and I. Young, "A 90 nm Communication Technology Featuring SiGe HBT Transistors, RF CMOS, Precision R-L-C RF Elements and 1 μm² 6-T SRAM Cell," *IEEE International Electron Device Meeting*, p. 73, San Francisco, CA, 2002.

128. Y.-F. Chyan, T. G. Ivanov, M. S. Carroll, W. J. Nagy, A. S. Chen, and K. H. Lee, "A 50-GHz 0.25-μm High-Energy Implanted BiCMOS (HEIBiC) Technology for Low-Power High-Integration Wireless-Communication System," *Symposium on VLSI Technology, Digest of Technical Papers*, p. 92, Honolulu, 1998.

129. I. Getreu, "Modeling the Bipolar Transistor," Elsevier, New York, 1978.

130. X. Cao, J. MaMacken, K. Stiles, P. Layman, J. J. Liou, A. Ortiz-Conde, S. Moinian, "Comparison of the New VBIC and Conventional Gummel-Poon Bipolar Transistor Models," *IEEE Trans. Electron Devices*, Vol. ED-47, p. 427, 2000.

131. International Technology Roadmap for Semiconductors 2005 (ITRS 2005), http://www.itrs.net/ Common/2005ITRS/Home2005.htm.

132. Q. Xiang, J. Jeon, P. Sachdey, B. Yu, K. C. Saraswat, and M.-R. Lin, "Very High Performance 40nm CMOS with Ultra-Thin Nitride/Oxynitride Stack Gate Dielectric and Pre-Doped Dual Poly-Si Gate Electrodes," *IEEE Trans. Electron Devices*, Vol. ED-47, p. 860, 2000.

133. R. Chau, J. Kavalieros, B. Roberds, R. Schenker, D. Lionberger, D. Barlarge, B. Doyle, R. Arghavani, A. Murthy, and G. Dewey, "30nm Physical Gate Length CMOS Transistors with 1.0 ps n-MOS and 1.7 ps p-MOS Gate Delays," *IEEE Trans. Electron Devices*, Vol. ED-47, p. 45, 2000.

134. E. Morifuji, M. Kanda, N. Yanagiya, S. Matsuda, S. Inaba, K. Okano, K. Takahashi, M. Nishigoori, H. Tsuno, T. Yamamoto, K. Hiyama, M. Takayanagi, H. Oyamatsu, S. Yamada, T. Noguchi, and M. Kakumu, "High Performance 30nm Bulk CMOS for 65nm Technology Node (CMOS5)," *IEEE International Electron Device Meeting*, p. 655, San Francisco, CA, 2002.

135. K. Goto, Y. Tagawa, H. Ohta, H. Morioka, S. Pidin, Y. Momiyama, H. Kokura, S. Inagaki, N. Tamura, M. Hori, T. Mori, M. Kase, K. Hashimoto, M. Kojima, T. Sugii, "High Performance 25 nm Gate CMOSFETs for 65nm Node High Speed MPUs," *IEEE International Electron Device Meeting*, p. 623, Washington, DC, 2003.

136. H. S. Momose, M. Ono, T. Yoshitomi, T. Ohguro, S. Nakamura, M. Saito, and H. Iwai, "Tunneling gate oxide approach to ultra-high current drive in small-geometry MOSFETs," *IEEE International Electron Device Meeting*, p. 593, San Francisco, CA, 1994.

137. H. S. Momose, M. Ono, T. Yoshitomi, T. Ohguro, S. Nakamura, M. Saito, and H. Iwai, "1.5 nm direct-tunneling gate oxide Si MOSFETs," *IEEE Trans. Electron Devices*, Vol. ED-43, p. 1233, 1996.

138. H. S. Momose, M. Ono, T. Yoshitomi, T. Ohguro, S. Nakamura, M. Saito, and H. Iwai, "Prospects for Low-Power, High-Speed MPUs Using 1.5 nm Direct-Tunneling Gate Oxide MOSFETs," *Journal of Solid-State Electronics*, Vol. 41, p. 707, 1997.

139. H. S. Momose, E. Morifuji, T. Yoshitomi, T. Ohguro, M. Saito, T. Morimoto, Y. Katsumata, and H. Iwai, "High-Frequency AC Characteristics of 1.5 nm Gate Oxide MOSFETs," *IEEE International Electron Device Meeting*, p. 105, San Francisco, CA, 1996.

140. H. S. Momose, S. Nakamura, T. Ohguro, T. Yoshitomi, E. Morifuji, T. Morimoto, Y. Katsumata, and H. Iwai, "Study of the Manufacturing Feasibility of 1.5 nm Direct-Tunneling Gate Oxide MOSFETs: Uniformity, Reliability, and Dopant penetration of the Gate Oxide," *IEEE Trans. Electron Devices*, Vol. ED-45, p. 691, 1998.

141. M. Fujiwara, M. Takayanagi, T. Shimizu, and Y. Toyoshima, "Extending Gate Dielectric Scaling Limit by NO Oxynitride: Design and Process Issues for Sub-100 nm Technology," *IEEE International Electron Device Meeting*, p. 227, San Francisco, CA, 2000.

142. S. Inaba, T. Shimizu, S. Mori, K. Sekine, K. Saki, H. Suto, H. Fukui, M. Nagamine, M. Fujiwara, T. Yamamoto, M. Takayanagi, I. Mizushima, K. Okano, S. Matsuda, H. Oyamatsu, Y. Tsunashima, S. Yamada, Y. Toyoshima, and H. Ishiuchi, "Device Performance of Sub-50 nm CMOS with Ultra-Thin Plasma Nitrided Gate Dielectrics," *IEEE International Electron Device Meeting*, p. 651, San Francisco, CA, 2000.

143. Y.-Y. Fan, Q. Xiang, J. An, L. F. Register, and S. K. Banerjee, "Impact of Interfacial Layer and Transition Region on Gate Current Performance of High-K Gate Dielectric Stack: Its Tradeoff With Gate Capacitance," *IEEE Trans. Electron Devices*, Vol. ED-50, p. 433, 2003.

144. D. A. Buchanan, E. P. Gusev, E. Cartier, H. Okorn-Schmidt, K. Rim, M. A. Gribelyuk, A. Mocuta, A. Ajmera, M. Copel, S. Guha, N. Bojarczuk, A. Callegari, C. D'Emic, P. Kozlowski, K. Chan, R. J. Fleming, P. C. Jamison, J. Brown, and R. Arndt, "80 nm Poly-Silicon Gated n-FETs with Ultra-Thin Al_2O_3 Gate Dielectric for ULSI Applications," *IEEE International Electron Device Meeting*, p. 223, San Francisco, CA, 2000.

145. K. Ohnishi, C. S. Kang, R. Choi, H.-J. Cho, Y. H. Kim, S. Krishnan, M. S. Akbar, and J. C. Lee, "Performance of Polysilicon Gate HfO_2 MOSFETs on (100) and (111) Silicon Substrates," *IEEE Electron Device Letters*, Vol. 24, p. 254, 2003.

146. S. J. Lee, H. F. Luan, W. P. Bai, C. H. Lee, T. S. Jeon, Y. Roberts, and D. L. Kwong, "High Quality Ultra Thin CVD HfO_2 Gate Stack with Poly-Si Gate Electrode," *IEEE International Electron Device Meeting*, p. 31, San Francisco, CA, 2000.

147. L. Kang, K. Onishi, Y. Jeon, B. H. Lee, C. Kang, W.-J. Qi, R. Nieh, S. Gopalan, R. Choi, and J. C. Lee, "MOSFET Devices with Polysilicon on Single-Layer HfO_2 High-K Dielectrics," *IEEE International Electron Device Meeting*, p. 35, San Francisco, CA, 2000.

148. M. Koyama, A. Kanako, T. Ino, M. Koike, Y. Kamata, R. Iijima, Y. Kaminuta, A. Takashima, M. Suzuki, C. Hongo, S. Inumiya, M. Takayanagi, and A. Nishiyama, "Effects of Nitrogen in HfSiON Gate Dielectric on the Electrical and Thermal Characteristics," *IEEE International Electron Device Meeting*, p. 849, San Francisco, CA, 2002.

149. K. Sekine, S. Inumiya, M. Sato, A. Kaneko, K. Eguchi, and Y. Tsunashima, "Nitrogen Profile Control by Plasma Nitridation Technique for Poly-Si Gate HfSiON CMOSFET with Excellent Interface Property and Ultra-Low Leakage Current," *IEEE International Electron Device Meeting*, p. 103, Washington, DC, 2003.

150. Y. Ma, Y. Ono, L. Stecker, D. R. Evans, and S. T. Hsu, "Zirconium Oxide Based Gate Dielectrics with Equivalent Oxide Thickness of Less Than 1.0 nm and Performance of Submicron MOSFET Using a Nitride Gate Replacement Process," *IEEE International Electron Device Meeting*, p. 149, Washington, DC, 1999.

151. W.-J. Qi, R. Nieh, B. H. Lee, L. Kang, Y. Jeon, K. Onishi, T. Ngai, S. Banerjee, and J. C. Lee, "MOSCAP and MOSFET Characteristics Using ZrO_2 Gate Dielectric Deposited Directly on Si," *IEEE International Electron Device Meeting*, p. 145, Washington, DC, 1999.

152. C. H. Lee, H. F. Luan, W. P. Bai, S. J. Lee, T. S. Jeon, Y. Senzaki, D. Roberts, and D. L. Kwong, "MOS Characteristics of Thin Rapid Thermal CVD ZrO_2 and Zr Silicate Gate Dielectrics," *IEEE International Electron Device Meeting*, p. 27, San Francisco, CA, 2000.

153. R. E. Nieh, C. S. Kang, H.-J. Cho, K. Onishi, R. Choi, S. Krishnan, J. H. Han, Y.-H. Kim, M. S. Akbar, and J. C. Lee, "Electrical Characterization and Material Evaluation of Zirconium Oxynitride Gate Dielectric in TaN-Gated NMOSFETs with High-Temperature Forming Gas Annealing," *IEEE Trans. Electron Devices*, Vol. ED-50, p. 333, 2003.

154. H.-T. Lue, C.-Y. Liu, and T.-Y. Tseng, "An Improved Two-Frequency Method of Capacitance Measurement for $SrTiO_3$ as High-k Gate Dielectric," *IEEE Electron Device Letters*, Vol. 23, p. 553, 2002.

155. C. Y. Lin, M. W. Ma, A. Chin, Y. C. Yeo, C. Zhu, M. F. Li, and D.-L. Kwong, "Fully Silicided NiSi Gate on La_2O_3 MOSFETs," *IEEE Electron Device Letters*, Vol. 24, p. 348, 2003.

156. B. Tavel, X. Garros, T. Skotnicki, F. Martin, C. Leroux, D. Bensahel, M. N. Séméria, Y. Morand, J. F. Damlencourt, S. Descombes, F. Leverd, Y. Le-Friec, P. Leduc, M. Rivoire, S. Jullian, and R. Pantel, "High Performance 40nm nMOSFETs with HfO_2 Gate Dielectric and Polysilicon Damascene Gate," *IEEE International Electron Device Meeting*, p. 429, San Francisco, CA, 2002.

157. R. Choi, K. Onishi, C. S. Kang, S. Gopalan, R. Nieh, Y. H. Kim, J. H. Han, S. Krishnan, H. Cho, A. Shahriar, and J. C. Lee, "Fabrication of High Quality Ultra-Thin HfO_2 Gate Dielectric MOSFETs Using Deuterium Anneal," *IEEE International Electron Device Meeting*, p. 613, San Francisco, CA, 2002.

158. T. Iwamoto, T. Ogura, M. Terai, H. Watanabe, H. Watanabe, N. Ikarashi, M. Miyamura, T. Tatsumi, M. Saitoh, A. Morioka, K. Watanabe, Y. Saito, Y. Yabe, T. Ikarashi, K. Masuzaki, Y. Mochizuki, and T. Mogami, "A Highly Manufacturable Low Power and High Speed CMOS

FET with Dual Poly-Si Gate Electrodes," *IEEE International Electron Device Meeting*, p. 639, Washington, DC, 2003.

159. T. Aoyama, T. Maeda, K. Torii, K. Yamashita, Y. Kobayashi, S. Kamiyama, T. Miura, H. Kitajima, and T. Arikado, "Proposal of New HfSiON CMOS Fabrication Process (HAMDAMA) for Low Standby Power Devices," *IEEE International Electron Device Meeting*, p. 95, San Francisco, CA, 2004.

160. M. Koyama, H. Satake, M. Koike, T. Ino, M. Suzuki, R. Iijima, Y. Kamimuta, A. Takashima, C. Hongo, and A. Nishiyama, "Degradation Mechanism of HfSiON Gate Insulator and Effect of Nitrogen Composition on the Statistical Distribution of the Breakdown," *IEEE International Electron Device Meeting*, p. 931, Washington, DC, 2003.

161. A. Shanware, M. R. Visokay, J. J. Chambers, A. L. P. Rotondaro, J. McPherson, and L. Colombo, "Characterization and Comparison of the Charge Trapping in HfSiON and HfO_2 Gate Dielectrics," *IEEE International Electron Device Meeting*, p. 939, Washington, DC, 2003.

162. M. Koyama, Y. Kaminuta, T. Ino, A. Kaneko, S. Inumiya, K. Eguchi, M. Takayanagi, and A. Nishiyama, "Careful Examination on the Asymmetric Vfb Shift Poly-Si/HfSiON Gate Stack and its Solution by Hf Concentration Control in the Dielectric Near the Poly-Si Interface with Small EOT Expense," *IEEE International Electron Device Meeting*, p. 499, San Francisco, CA, 2004.

163. T. Watanabe, M. Takayanagi, K. Kojima, K. Sekine, H. Yamasaki, K. Eguchi, K. Ishimaru, and H. Ishiuchi, "Impact of Hf Concentration on Performance and Reliability for HfSiON-CMOSFET," *IEEE International Electron Device Meeting*, p. 507, San Francisco, CA, 2004.

164. M. Ono and A. Nishiyama, "Influence of Structure Around Gate-Edge on High Electric Field Strength in MISFETs With High-*k* Gate Dielectrics," *IEEE Trans. Electron Devices*, Vol. ED-51, p. 68, 2004.

165. B. Min, S. P. Devireddy, Z. Celik-Butler, F. Wang, A. Zlotnicka, H.-H. Tseng, and P. J. Tobin, "Low-Frequency Noise in Submicrometer MOSFETs With HfO_2, HfO_2/Al_2O_3 and $HfAlO_x$ Gate Stacks," *IEEE Trans. Electron Devices*, Vol. ED-51, p. 1315, 2004.

166. E. Simoen, A. Mercha, L. Pantisano, C. Claeys, and E. Young, "Low-Frequency Noise Behavior of SiO_2-HfO_2 Dual-Layer Gate Dielectric nMOSFETs With Different Interfacial Oxide Thickness," *IEEE Trans. Electron Devices*, Vol. ED-51, p. 780, 2004.

167. J. K. Schaeffer, C. Capasso, L. R. C. Fonseca, S. Samavedam, D. C. Gilmer, Y. Liang, S. Kalpat, B. Adetutu, H.-H. Tseng, Y. Shiho, A. Demkov, R. Hegde, W. J. Taylor, R. Gregory, J. Jiang, E. Luckowski, M. V. Raymond, K. Moore, D. Triyoso, D. Roan, B. E. White, Jr., and P. J. Tobin, "Challenge for the Integration of Metal Gate Electrode," *IEEE International Electron Device Meeting*, p. 287, San Francisco, CA, 2004.

168. A. Yagishita, T. Saito, K. Nakajima, S. Inumiya, Y. Akasaka, Y. Ozawa, K. Hieda, Y. Tsunashima, K. Suguro, T. Arikado, and K. Okumura, "High Performance Damascene Metal Gate MOSFET's for 0.1 μm Regime," *IEEE Trans. Electron Devices*, Vol. ED-47, p. 1028, 2000.

169. B. H. Lee, R. Choi, L. Kang, S. Gopalan, R. Nieh, K. Onishi, Y. Jeon, W.-J. Qi, C. Kang, and J. C. Lee, "Characteristics of TiN Gate MOSFET with Ultrathin Hafnium Oxide (8Å-12Å)," *IEEE International Electron Device Meeting*, p. 39, San Francisco, CA, 2000.

170. A. Yagishita, T. Saito, S. Inumiya, K. Matsuo, Y. Tsunashima, K. Suguro, and T. Arikado, "Dynamic Threshold Voltage Damascene Metal Gate MOSFET (DT-DMG-MOS) with Low Threshold Voltage, High Drive Current, and Uniform Electrical Characteristics," *IEEE International Electron Device Meeting*, p. 663, San Francisco, CA, 2000.

171. H. Y. Yu, H. F. Lim, J. H. Chen, M. F. Li, C. Zhu, C. H. Tung, A. Y. Du, W. D. Wang, D. Z. Chi, and D.-L. Kwong, "Physical and Electrical Characteristics of HfN Gate Electrode for Advanced MOS Devices," *IEEE Electron Device Letters*, Vol. 24, p. 230, 2003.

172. C. S. Kang, H.-J. Cho, R. Choi, Y.-H. Kim, C. Y. Kang, S. J. Rhee, C. Choi, M. S. Akbar, and J. C. Lee, "The Electrical and Material Characterization of Hafnium Oxynitride Gate Dielectrics with TaN-Gate Electrode," *IEEE Trans. Electron Devices*, Vol. ED-51, p. 220, 2004.

173. S. B. Samavedam, L. B. La, J. Smith, S. Dakshina-Murthy, E. Luckowski, J. Schaeffer, M. Zavala, R. Martin, V. Dhandapani, D. Triyoso, H. H. Tseng, P. J. Tobin, D. C. Gilmer, C. Hobbs, W. J. Taylor,

J. M. Grant, R. I. Hegde, J. Mogab, C. Thomas, P. Abramowitz, M. Moosa, J. Conner, J. Jiang, V. Arunachalam, M. Sadd, B. Y. Nguyen, and B. White, "Dual-Metal Gate with HfO_2 Gate Dielectric," *IEEE International Electron Device Meeting, 2002*, p. 433, San Francisco, 2002.

174. K. Miyashita, H. Yoshimura, M. Takayanagi, M. Fujiwara, K. Adachi, T. Nakayama, and Y. Toyoshima, "Optimized Halo Structure for 80 nm Physical Gate CMOS Technology with Indium and Antimony Highly Angled Ion Implantation," *IEEE International Electron Device Meeting*, p. 645, Washington, DC, 1999.

175. B. Yu, H. Wang, O. Milic, Q. Xiang, W. Wang, J. X. An, M.-R. Lin, "50nm Gate-Length CMOS Transistor with Super-Halo: Design, Process, and Reliability," *IEEE International Electron Device Meeting*, p. 653, Washington, DC, 1999.

176. H. Wakabayashi, M. Ueki, M. Narihiro, T. Fukai, N. Ikezawa, T. Matsuda, K. Yoshida, K. Takeuchi, Y. Ochiai, T. Mogami, T. Kunio, "45-nm Gate Length CMOS Technology and Beyond Using Steep Halo," *IEEE International Electron Device Meeting*, p. 49, San Francisco, CA, 2000.

177. H. Wakabayashi, M. Ueki, M. Narihiro, T. Fukai, N. Ikezawa, T. Matsuda, K. Yoshida, K. Takeuchi, Y. Ochiai, T. Mogami, T, Kunio, "Sub-50-nm Physical Gate Length CMOS Technology and Beyond Using Steep Halo," *IEEE Trans. Electron Devices*, Vol. ED-49, p. 89, 2002.

178. K. Liu, J. Wu, J. Chen, and A. Jain, "Fluorine-Assisted Super-Halo for Su-50-nm Transistors," *IEEE Electron Device Letters*, Vol. 24, p. 180, 2003.

179. E. Morifuji, A. Ohishi, K. Miyashita, H. Kawashima, T. Nakayama, H. Yoshitomi, and Y. Toyoshima, "An 80 nm Dual-Gate CMOS with Shallow Extensions Formed after Activation Annealing and SALICIDE," *IEEE International Electron Device Meeting*, p. 649, Washington, DC, 1999.

180. S. Inaba, K. Miyano, A. Hokazono, K. Ohuchi, I. Mizushima, H. Oyamatsu, Y. Tsunashima, Y. Toyoshima, and H. Ishiuchi, "Silicon on Depletion Layer FET (SODEL FET) for Sub-50 nm High Performance CMOS Applications: Novel Channel and S/D Profile Engineering Schemes by Selective Si Epitaxial Growth Technology," *IEEE International Electron Device Meeting*, p. 659, San Francisco, CA, 2002.

181. K. Lee, C. Murthy, R. Rengarajan, S. Hegde, and R. Jammy, "Simultaneous Optimization of Short-Channel Effects and Junction Capacitance in pMOSFET Using Large-Angle-Tilt-Implantation of Nitrogen (LATIN)," *IEEE Electron Device Letters*, Vol. 23, p. 547, 2002.

182. Y. Momiyama, K. Okabe, H. Nakao, M. Kase, M. Kojima, and T. Sugii, "Lateral Extension Engineering Using Nitrogen Implantation (N-tub) for High-Performance 40-nm pMOSFETs," *IEEE International Electron Device Meeting*, p. 647, San Francisco, CA, 2002.

183. W.-T. Kang, J.-S. Kim, K.-Y. Lee, Y.-C. Shin, T.-H. Kim, Y.-J. Park, and J.-W. Park, "The Leakage Current Improvement in an Ultrashallow Junction NMOS with Co Sikicided Source and Drain," *IEEE Electron Device Letters*, Vol. 21, p. 9, 2000.

184. K. Goto, A. Fushida, J. Watanabe, T. Sukegawa, Y. Tada, T. Nakamura, T. Yamazaki, and T. Sugii, "A New Leakage Mechanism of Co Salicide and Optimized Process Conditions," *IEEE Trans. Electron Devices*, Vol. ED-46, p. 117, 1999.

185. T. Ohguro, H. Naruse, H. Sugaya, H. Kimijima, E. Morifuji, T. Yoshitomi, T. Morimoto, H. S. Momose, Y. Katsumata, and H. Iwai, "0.12 µm Raised Gate/Source/Drain Epitaxial Channel NMOS Technology," *IEEE International Electron Device Meeting*, p. 927, San Francisco, CA, 1998.

186. A. Hokazono, K. Ohuchi, K. Miyano, I. Mizushima, Y. Tsunashima, and Y. Toyoshima, "Source/Drain Engineering for Sun-100 nm CMOS Using Selective Epitaxial Growth Technique," *IEEE International Electron Device Meeting*, p. 243, San Francisco, CA, 2000.

187. W. Jeamsaksiri, M. Jurczak, L. Grau, D. Linten, E. Augendre, M. De Potter, R. Rooyackers, Piet Wambacq, and G. Badenes, "Gate-Source-Drain Architecture Impact on DC and RF Performance of Sub-100-nm Elevated Source/Drain NMOS Transistors," *IEEE Trans. Electron Devices*, Vol. ED-50, p. 610, 2003.

188. K. Miyano, I. Mizushima, A. Hokazono, K. Ohuchi, and Y. Toyoshima, "Low Thermal Budget Elevated Source/Drain Technology Utilizing Novel Solid Phase Epitaxy and Selective Vapor Phase Etching," *IEEE International Electron Device Meeting*, p. 433, San Francisco, CA, 2000.

189. A. Shima, H. Ashihara, T. Mine, Y. Goto, M. Horiuchi, Y. Wang, S. Talwar, and A. Hiraiwa, "Self-Limiting Laser Thermal Process for Ultra-Shallow Junction Formation of 50-nm Gate CMOS," *IEEE International Electron Device Meeting*, p. 493, Washington, DC, 2003.

190. K. Onishi, C. S. Kang, R. Choi, H.-J. Cho, S. Gopalan, R. E. Nieh, S. A. Krishnan, J. C. Lee, "Improvement of Surface Carrier Mobility of HfO_2 MOSFETs by High-Temperature Forming Gas Annealing," *IEEE Trans. Electron Devices*, Vol. ED-50, p. 384, 2003.

191. J. L. Hoyt, H. M. Nayfeh, S. Eguchi, I. Aberg, G. Xia, T. Drake, E. A. Fitzgerald, and D. A. Antoniadis, "Strained Silicon MOSFET Technology," *IEEE International Electron Device Meeting*, p. 23, San Francisco, CA, 2002.

192. J.-S. Goo, Q. Xiang, Y. Takamura, H. Wang, J. Pan, F. Arasnia, E. N. Paton, P. Besser, M. V. Sidorov, E. Adem, A. Lochtefeld, G. Braithwaite, M. T. Currie, R. Hammond, M. T. Bulsara, and M.-R. Lin, "Scalability of Strained-Si nMOSFETs Down to 25 nm Gate Length," *IEEE Electron Device Letters*, Vol. 24, p. 351, 2003.

193. M. L. Lee and E. A. Fitzgerald, "Optimized Strained Si/Strained Ge Dual-Channel Heterostructures for High Mobility P- and N-MOSFETs," *IEEE International Electron Device Meeting*, p. 429, Washington, DC, 2003.

194. S. Datta, G. Dewey, M. Doczy, B. S. Doyle, B. Jin, J. Kavalieros, R. Kotlyar, N. Metz, N. Zelick, and R. Chau, "High Mobility Si/SiGe Strained Channel MOS Transistors with HfO_2/TiN Gate Stack," *IEEE International Electron Device Meeting*, p. 653, Washington, DC, 2003.

195. S. H. Olsen, A. G. O'Neill, S. Chattopadhyay, L. S. Driscoll, K. S. K. Kwa, D. J. Norris, A. G. Cullis, and D. J. Paul, "Study of Single- and Dual-Channel Designs for High-Performance Strained-Si-SiGe n-MOSFETs," *IEEE Trans. Electron Devices*, Vol. ED-51, p. 1245, 2004.

196. J. Jung, S. Yu, M. L. Lee, J. L. Hoyt, E. A. Fitzgerald, and D. A. Antoniadis, "Mobility Enhancement in Dual-Channel P-MOSFETs," *IEEE Trans. Electron Devices*, Vol. ED-51, p. 1424, 2004.

197. T. Ghani, M. Armstrong, C. Auth, M. Bost, P. Charvat, G. Glass, T. Hoffmann, K. Johnson, C. Kenyon, J. Klaus, B. McIntyre, K. Mistry, A. Murthy, J. Sandford, M. Silberstein, S. Sivakumar, P. Smith, K. Zawadzki, S. Thompson, and M. Bohr, "A 90nm High Volume Manufacturing Logic Featuring Novel 45nm Gate Length Strained Silicon CMOS Transistors," *IEEE International Electron Device Meeting*, p. 978, Washington, DC, 2003.

198. T. Sanuki, A. Oishi, Y. Morimasa, A. Aota, T. Kinoshita, R. Hasumi, Y. Takegawa, K. Isobe, H. Yoshimura, M. Iwai, K. Sunouchi, and T. Noguchi, "Scalability of Strained Silicon CMOSFET and High Drive Current Enhancement in the 40nm Gate Length Technology," *IEEE International Electron Device Meeting*, p. 65, Washington, DC, 2003.

199. M. Yang, M. Ieong, L. Shi, K. Chan, V. Chan, E. Gusev, K. Jenkins, D. Boyd, Y. Ninomiya, D. Pendleton, Y. Surpris, D. Heenan, J. Ott, K. Guarini, C. D'Emic, M. Cobb, P. Mooney, B. To, N. Rovedo, J. Bebedict, R. Mo, and N. Ng, "High Performance CMOS Fabricated on Hybrid Substrate with Different Crystal Orientations," *IEEE International Electron Device Meeting*, p. 453, Washington, DC, 2003.

200. L. Chang, M. Ieong, and M. Yang, "CMOS Circuit Performance Enhancement by Surface Orientation Optimization," *IEEE Trans. Electron Devices*, Vol. ED-51, p. 1621, 2004.

201. T. Komoda, A. Oishi, T. Sanuki, K. Kasai, H. Yoshimura, K. Ohno, M. Iwai, M. Saito, F. Matsuoka, N. Nagashima, and T. Noguchi, "Mobility Improvement for 45nm Node by Combination of Optimized Stress Control and Channel Orientation Design," *IEEE International Electron Device Meeting*, p. 217, San Francisco, CA, 2004.

202. http://www-device.eecs.berkeley.edu/~bsim3/

203. J.-S. Goo, W. Liu, C.-H. Choi, K. R. Green, Z. Yu, T. H. Lee, and R. W. Dutton, "The Equivalence of van der Ziel and BSIM4 Models the Induced Gate Noise of MOSFETs," *IEEE International Electron Device Meeting*, p. 811, San Francisco, CA, 2000.

204. K. M. Cao, W.-C. Lee, W. Liu, X. Jin, P. Su, S. K. H. Fung, J. X. An, B. Yu, and C. Hu, "BISIM4 Gate Leakage Model Including Source-Drain Partition," *IEEE International Electron Device Meeting*, p. 815, San Francisco, CA, 2000.

205. Z. Yu, R. W. Dutton, R. A. Kiehl, "Circuit/Device Modeling at Quantum Level," *IEEE Trans. Electron Devices*, Vol. ED-47, p. 1819, 2000.

206. F. M. Bufler and W. Fichtner, "Scaling of Strained-Si n-MOSFETs into Ballistic Regime and Associated Anisotropic Effects," *IEEE Trans. Electron Devices*, Vol. ED-50, p. 278, 2003.

207. F. Bonani, S. D. Guerrieri, and G. Ghione, "Physical-Based Simulation Technique for Small- and Large-Signal Device Noise Analysis in RF Applications," *IEEE Trans. Electron Devices*, Vol. ED-50, p. 633, 2003.

208. F. M. Bufler, Y. Asahi, H. Yoshimura, C. Zechner, A. Schenk, and W. Fichtner, "Monte Carlo Simulation and Measurement of Nanoscale n-MOSFETs," *IEEE Trans. Electron Devices*, Vol. ED-50, p. 418, 2003.

209. H. Wang and G. Gildenblat, "Scattering Matrix Based Compact MOSFET Model," *IEEE International Electron Device Meeting*, p. 125, San Francisco, CA, 2002.

210. G. Gildenblat, X. Li, H. Wang, W. Wu, R. van Langevelde, A. J. Scholt, G. D. J. Smit, and D. B. M. Klaassen, "Introduction to PSP MOSFET Model," *Technical Proceedings of the Workshop on Compact Modeling*, p. 19, 2005.

211. M. Miura-Mattausch, H. Ueno, M. Tanaka, H. J. Mattausch, S. Kumashiro, T. Yamaguchi, K. Yamashita, and N. Nakayama, "HiSIM: A MOSFET Model for Circuit Simulation Connecting Circuit Performance with Technology," *IEEE International Electron Device Meeting*, p. 109, San Francisco, CA, 2002.

212. H. Liu, J. K. O. Sin, P. Xuan, and J. Bokor, "Characterization of Ultrathin Vertical Channel CMOS Technology," *IEEE Trans. Electron Devices*, Vol. ED-51, p. 106, 2004.

213. B. Yu, L. Chang, S. Ahmed, H. Wang, S. Bell, C.-Y. Yang, C. Tabery, C. Ho, Q. Xiang, T.-J. King, J. Bokor, C. Hu, M.-R. Lin, and D. Kyser, "FinFET Scaling to 10nm Gate Length," *IEEE International Electron Device Meeting*, p. 251, Washington, DC, 2001.

214. E. J. Nowak, B. A. Rainey, D. M. Fried, J. Kedzierski, M. Ieong, W. Leipold, J. Wright, and M. Breitwisch, "A Functional FinFET-DGCMOS SRAM Cell," *IEEE International Electron Device Meeting*, p. 411, Washington, DC, 2001.

215. J.-H. Lee, G. Taraschi, A. Wei, T. A. Langdo, E. A. Fitzgerald, and D. Antoniadis, "Super Self-Aligned Double-Gate (SSDG) MOSFETS Utilizing Oxidation Rate Difference and Selective Epitaxy," *IEEE International Electron Device Meeting 2001*, p. 71, Washington, DC, 2001.

216. H. Iwai, "FSM-CO–," *IEEE International Electron Device Meeting*, p. 11, San Francisco, CA, 2004.

217. H. Wakabayashi, T. Ezaki, M. Hane, T. Ikezawa, T. Sakamoto, H. Kawaura, S. Yamagami, N. Ikarashi, K. Takeuchi, T. Yamamoto, and T. Mogami, "Transport Properties of Sub-10-nm Planar-Bulk- CMOS Devices," *IEEE International Electron Device Meeting*, p. 429, San Francisco, CA, 2004.

3

Silicon-on-Insulator Technology

Sorin Cristoloveanu

Institute of Microelectronics,
Electromagnetism and Photonics

CONTENTS

3.1 Introduction

Silicon-on-insulator (SOI) technology, more specifically silicon-on-sapphire, was originally invented for the niche of radiation-hard circuits. In the last 25 years, a variety of SOI structures have been conceived with the aim of dielectrically separating, using a buried oxide (Fig. 3.1b), the active device volume from the silicon substrate [1,2]. Indeed, in an MOS transistor, only the very top region (0.1 to 0.2-μm thick, i.e., <0.1% of the total thickness) of the silicon wafer is useful for electron transport and device operation, whereas the substrate is responsible for detrimental, parasitic effects (Fig. 3.1a).

More recently, the advent of new SOI materials (Unibond, ITOX) and the explosive growth of portable microelectronic devices have attracted considerable attention on SOI for the fabrication of low-power/ voltage and high-frequency CMOS circuits.

The aim of this chapter is to overview the state-of-the-art of SOI technologies, including the material synthesis (Section 3.2), the key advantages of SOI circuits (Section 3.3), the structure and performance of typical devices (Section 3.4), and the operation modes of fully depleted (FD; Section 3.5) and partially-depleted (PD) SOI MOSFETs (Section 3.6). Sections 3.7 and 3.8 are dedicated to small-geometry effects and innovative transistor architectures. The main challenges that SOI is facing, to successfully surpass bulk-Si in the commercial arena, are critically discussed in Section 3.9.

3.2 Fabrication of SOI Wafers

Many techniques, more or less mature and effective, are available for the synthesis of SOI wafers [1]. However, the overwhelming role is played by the Unibond and Smart-Cut processes.

3.2.1 Silicon on Sapphire

Silicon-on-sapphire (SOS, Fig. 3.2a$_1$) is the initial member of the SOI family. The epitaxial growth of Si films on Al_2O_3 gives rise to small silicon islands that eventually coalesce. The interface transition region contains crystallographic defects owing to the lattice mismatch and Al contamination from the substrate. The electrical properties suffer from lateral stress, in-depth inhomogeneity of SOS films, and defective transition layer [3].

SOS has undergone a significant lifting: larger wafers (6 to 8 in) and thinner films (100 nm) with higher crystal quality and carrier mobility [4]. This improvement is achieved by *solid-phase epitaxial regrowth*. Silicon ions are implanted to amorphize the film and erase the memory of damaged lattice and interface. Annealing allows the epitaxial regrowth of the film, starting from the "seeding" surface toward the Si–Al_2O_3 interface.

Owing to the "infinite" thickness of the insulator, SOS is still attractive for the integration of RF and radiation-hard circuits.

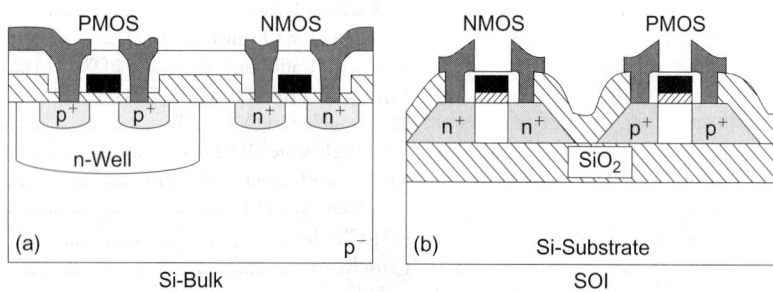

FIGURE 3.1 Basic architecture of MOS transistors in (a) bulk silicon and (b) SOI.

3.2.2 ELO and ZMR

The *epitaxial lateral overgrowth* (ELO) method consists in growing a single-crystal Si film on a seeded and, often, patterned oxide (Fig. 3.2a$_2$). Since the epitaxial growth proceeds in both lateral and vertical directions, the ELO process requires a post-epitaxy thinning of the Si film.

Alternatively, poly-silicon can be deposited directly on SiO$_2$; subsequent *zone melting recrystallization* (ZMR) is achieved by scanning high-energy sources (lasers, lamps, beams, or strip heaters) across the wafer. The ZMR process can be seeded or unseeded; it is basically limited by the lateral extension of single-crystal regions, free from grain subboundaries and associated defects. ELO and ZMR are basic techniques for the integration of vertical and 3-D stacked circuits.

3.2.3 FIPOS

The *full isolation by porous oxidized silicon* (FIPOS) method makes use of the very large surface-to-volume ratio (10^3 cm^2/cm^{-3}) of porous silicon which is, thereafter, subject to selective oxidation (Fig. 3.2a$_3$). The critical step is the conversion of selected p-type regions of the Si wafer into porous silicon, via anodic reaction. From a conceptual viewpoint, FIPOS can combine electroluminescent porous Si devices with fast SOI–CMOS circuits.

3.2.4 SIMOX

In the 1990s, the dominant SOI technology was *separation by implantation of oxygen* (SIMOX). The buried oxide (BOX) is synthesized by internal oxidation during the deep implantation of oxygen ions into a Si wafer. Annealing at high temperature (1320°C for 6 h) is necessary to recover a suitable crystalline quality

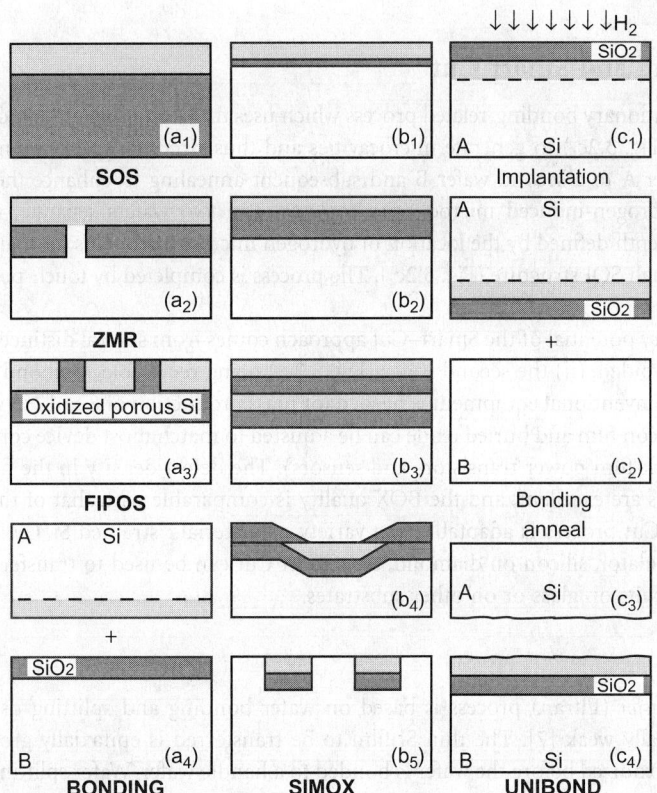

FIGURE 3.2 SOI family: (a) SOS, ZMR, FIPOS, and wafer bonding, (b) SIMOX variants (structure b$_5$ can also be achieved by SON technology), and (c) Unibond processing sequence.

of the film. High current implanters (100 mA) have been conceived to produce 8 in wafers with good thickness uniformity, low defect density (except threading dislocations of 10^4 to 10^6 cm^{-2}), sharp Si–SiO$_2$ interface, robust BOX, and high carrier mobility [5].

The family of SOI structures is presented in Figure 3.2b.

- Thin and thick Si films fabricated by adjusting the implant energy.
- Low-dose SIMOX: a dose of 4×10^{17} O$^+$ cm^{-2} and an additional oxygen-rich anneal for enhanced BOX integrity (ITOX process) yield a 0.1 μm thick BOX (Fig. 3.2b$_1$).
- Standard SIMOX obtained with 1.8×10^{18} O$^+$ cm^{-2} implant dose, at 190 keV and 650°C; the thicknesses of the Si film and BOX are roughly 0.2 and 0.4 μm, respectively (Fig. 3.2b$_2$).
- Double SIMOX (Fig. 3.2b$_3$), where the Si layer sandwiched between the two oxides can serve for interconnects, wave guiding, additional gates, or electric shielding.
- Laterally isolated single-transistor islands (Fig. 3.2b$_4$), formed by implantation through a patterned oxide.
- Interrupted oxides (Fig. 3.2b$_5$), which can be viewed as SOI regions integrated into a bulk Si wafer.

3.2.5   Wafer Bonding

Wafer bonding (WB) and etch-back stand as a more mature SOI technology. An oxidized wafer is mated to another SOI wafer (Fig. 3.2a$_4$). The challenge is to drastically thin down one side of the bonded structure in order to reach the target thickness of the silicon film. Etch-stop layers can be achieved by doping steps (p$^+$/p$^-$, p/n), SiGe, or porous silicon. The advantage of WB is to provide unlimited combinations of BOX and film thicknesses, whereas its weakness comes from the dificulty to produce ultrathin films with good uniformity.

3.2.6   Unibond and Smart Cut

Unibond is a revolutionary bonding-related process which uses the deep implantation of hydrogen into an oxidized Si wafer (Fig. 3.2c$_1$) to generate microcavities and thus circumvent the thinning problem [6,2]. After bonding wafer A to a second wafer B and subsequent annealing to enhance the bonding strength (Fig. 3.2c$_2$), the hydrogen-induced microcavities coalesce. The two wafers separate, not at the bonded interface, but at a depth defined by the location of hydrogen microcavities. This mechanism, named *Smart Cut*, results in a rough SOI structure (Fig. 3.2c$_4$). The process is completed by touch-polishing to erase the surface roughness.

The extraordinary potential of the Smart–Cut approach comes from several distinct advantages: (i) the etch-back step is avoided, (ii) the second wafer (Fig. 3.2c$_3$) being recyclable, Unibond is a "single-wafer" process, (iii) only conventional equipment is needed for mass production of 8 to 12 in wafers, and (iv) the thickness of the silicon film and buried oxide can be adjusted to match most device configurations (ultra-thin CMOS or thick-film power transistors and sensors). The defect density in the film is very low, the electrical properties are excellent, and the BOX quality is comparable with that of the original thermal oxide. The Smart–Cut process is adaptable to a variety of materials: strained Si, Ge, SiGe, SiC or III–V compounds on insulator, silicon on diamond, etc. Smart Cut can be used to transfer already fabricated bulk-Si CMOS circuits on glass or on other substrates.

3.2.7   Eltran

Epitaxial layer transfer (Eltran) process is based on wafer bonding and splitting using porous silicon which is mechanically weak [7]. The thin Si film to be transferred is epitaxially grown on the porous layer and partially oxidized before the wafer is bonded to a handle wafer. Wafer splitting is achieved using a fine water jet. The residual porous silicon is etched away from the SOI wafer and the surface is smoothed by hydrogen annealing. Eltran features the same basic advantages as those listed for Smart Cut, the main difference being the splitting mechanism.

3.2.8 Silicon-On-Nothing (SON)

Silicon-on-nothing (SON) consists of selective epitaxy of sacrificial SiGe regions in a bulk-Si wafer. A silicon film (20 nm or thinner) is epitaxially grown on SiGe. The selective etching of the SiGe layer leaves an empty space (air gap) underneath the film, which can be filled with a dielectric to form a localized SOI structure, integrated in the bulk-Si wafer (as in Fig. 3.2b$_5$) [8]. Alternatively, the suspended Si membrane can be used to fabricate gate-all-around transistors.

3.3 Generic Advantages of SOI

SOI circuits consist of single-device islands dielectrically isolated from each other and from the underlying substrate (Fig. 3.1b). The lateral isolation offers more compact design and simplified technology than in bulk silicon: there is no need of wells or inter-device trenches. In addition, the vertical isolation achieves thin films, and eliminates most of the detrimental substrate effects (latch-up, punch-through, etc).

The source and drain regions extend down to the BOX, thus the junction surface is minimized. This implies reduced leakage currents and junction capacitances which further translates into improved speed, lower power dissipation, and wider temperature range of operation.

The limited extension of drain and source regions allows SOI devices to be less affected by short-channel effects, originated from 'charge sharing' between gate and junctions or from drain-induced barrier lowering. Besides the outstanding tolerance of transient radiation effects, SOI MOSFETs experience a lower electric-field peak than in bulk Si and are potentially more immune to hot-carrier damage.

It is in the highly competitive domain of low-voltage, low-power circuits, operated with one-battery supply (<1.5 V), that SOI can express its entire potential. A small gate voltage gap is suited to switch a transistor from off to on state. SOI offers the possibility to achieve a quasi-ideal subthreshold slope (60 mV/decade at room temperature), hence a threshold voltage below 0.3 V. Low leakage currents limit the *static* power dissipation, as compared with bulk Si, whereas the *dynamic* power dissipation is minimized by the combined effects of low parasitic capacitances and reduced voltage supply.

Two arguments can be given to outline unequivocally the advantage of SOI over bulk Si:

- Operation at similar *voltage* consistently shows about 20 to 30% increase in speed, whereas operation at similar *low-power* dissipation yields as much as 300% performance gain in SOI. It is believed, at least in the SOI community, that SOI circuits of generation (*n*) and bulk-Si circuits from the *next* generation (*n* + 1) perform comparably.
- Bulk Si technology attempts to mimic a number of features that are natural in SOI: the double-gate configuration is reproduced by processing surrounded-gate vertical MOSFETs on bulk Si; full depletion is approached by tailoring a low–high step doping, and the dynamic-threshold operation is borrowed from SOI.

The problem for SOI is that such an enthusiastic list of merits did not perturb the fantastic progress and authority of bulk Si technology. There was no room or need so far for an alternative technology such as SOI, which was utilized only for circuits with high added value. But, in the late 1990s, SOI has been included in the *International Technology Roadmap of Semiconductors* as the best suited CMOS option.

3.4 SOI Devices

3.4.1 CMOS Circuits

High-performance SOI CMOS circuits, compatible with low-power or high-speed ULSI applications have been repeatedly demonstrated on deep submicron devices. High-end microprocessors are being fabricated on SOI by IBM, AMD, Motorola, etc. RF SOI devices also show unchallenged capability in terms of frequency and noise. Extremely low-power circuits for mobile communication, portable processors operated with one-battery supply (0.5 to 1.2 V), and even battery-less watches are currently fabricated on SOI Unibond wafers. The SOI versatility has been taken advantage of for conceiving capacitor-less DRAMs.

SOI is also an ideal substrate for systems-on-chip. FD CMOS SOI circuits operate successfully at temperatures beyond 300°C (for aeronautics and automobiles): the leakage currents are much smaller and the threshold voltage is less temperature sensitive (≈0.5 mV/°C) than in bulk Si [9]. In addition, many SOI circuits are radiation-hard, able to sustain doses above 10 Mrad, for the space industry.

3.4.2 Bipolar Transistors

As a consequence of the small film thickness, most of the bipolar transistors have lateral configuration. The implementation of BiCMOS technology on SOI has resulted in devices with high cutoff frequency. Hybrid MOS-bipolar transistors with increased current drive and transconductance are formed by connecting the gate to the floating body (or base): the MOSFET action governs in strong inversion whereas, in weak inversion, the bipolar current prevails [10].

Vertical bipolar transistors have been processed in thick-film SOI (wafer bonding or epitaxial growth over SIMOX). An elegant solution for thin-film SOI is to replace the buried collector by an inversion layer activated by the back gate [10].

3.4.3 High-Voltage Devices

The outstanding advantage is the dielectric isolation. Power transistors can have lateral (LD–MOSFETs essentially for lightening) or vertical architecture. Lateral double–diffused MOSFETs (LDMOS), with long drift region, were fabricated on SIMOX and showed 90 V–1.3 A capability [11]. Vertical power devices (IGBT, LDMOS, VMOS, etc.) can be accommodated in a thicker wafer-bonding SOI.

SOI offers the possibility to synthesize locally a BOX (SON process or "interrupted" SIMOX, Fig. 3.2b$_5$). Therefore, a vertical power transistor, located in the bulk region of the wafer, can be controlled and rendered "smart" by being located next to a low-power SOI CMOS [12] (Fig. 3.3a). A variant of this concept is the "mezzanine" structure, which served for the fabrication of a 600 V/25 A smart-power

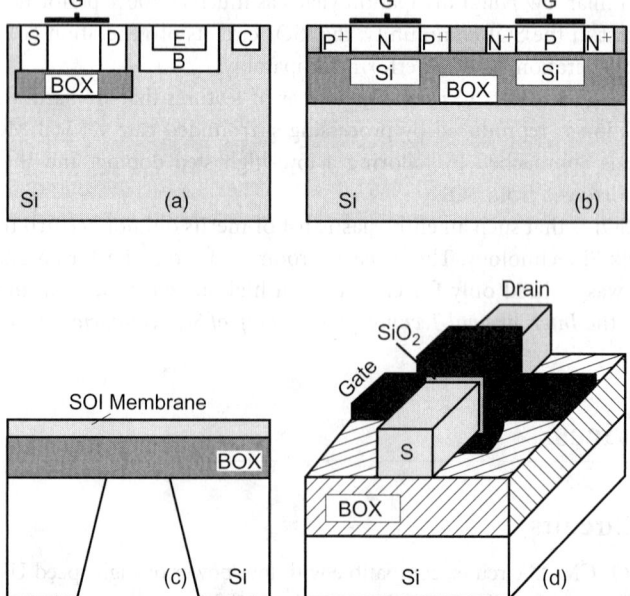

FIGURE 3.3 Examples of innovative SOI devices: (a) combined bipolar (or high power) bulk-Si transistor with low-voltage SOI CMOS circuits, (b) dual-gate transistors, (c) pressure sensor, and (d) gate-all-around (GAA) MOSFET.

device [13]. Double SIMOX (Fig. 3.2b$_3$) has also been used to combine a power MOSFET, with a double-shielded high-voltage lateral CMOS and an intelligent low-voltage CMOS circuit [12].

3.4.4 Innovative Devices

Most innovative devices make use of special SOI features such as the adjustment of the thickness of the Si overlay and BOX, and the implementation of additional layers underneath the BOX (Fig. 3.3b).

SOI is an ideal material for microsensors and MEMS because the Si/BOX interface gives a perfect etch-stop mark, making it possible to fabricate very thin membranes (Fig. 3.3c). Transducers for detection of pressure, acceleration (air bags), gas flow, temperature, radiation, magnetic field, etc. have successfully been integrated on SOI [1,13].

Three-dimensional circuits containing consecutive thin silicon and BOX layers have been demonstrated with the ELO and ZMR methods. For example, an image-signal processor is organized in three levels: photodiode arrays in the upper SOI layer, fast A/D converters in the intermediate SOI layer, and arithmetic units and shift registers in the bottom bulk Si level [14].

The *gate all–around* (GAA) transistor of Figure 3.3d, based on the concept of volume inversion, is fabricated by etching a cavity into the BOX and wrapping the oxidized transistor body into a poly-Si gate [10].

The family of SOI devices also includes optical waveguides and modulators, microwave transistors integrated on high-resistivity SIMOX, twin-gate MOSFETs, and other exotic devices [1,10]. They do not belong to *science fiction:* the devices have already been demonstrated in terms of technology and functionality . . . even if most people still do not believe that they can operate indeed.

Finally, most of the Si nanoelectronic devices (SET, tunneling transistors, quantum dots and wires) have used SOI, either for the ease of processing or for ultrathin film capability [15,16].

3.5 FD SOI Transistors

In SOI MOSFETs (Fig. 3.1b), inversion channels can be activated at both the front Si–SiO$_2$ interface (via gate modulation V_{G_1}) and back Si–BOX interface (via substrate, back-gate bias V_{G_2}).

Full depletion means that the depletion region covers the whole transistor body. The depletion charge is constant and cannot extend according to the gate bias. A better coupling develops between the gate bias and the inversion charge, leading to enhanced drain current. In addition, the front- and back-surface potentials become coupled too. The coupling factor is roughly equal to the thickness ratio between the gate oxide and BOX. The electrical characteristics of one channel vary remarkably with the bias applied to the opposite gate. Owing to *interface coupling*, the front-gate measurements are all reminiscent of the back-gate bias and quality of the BOX and interface.

Totally new $I_D(V_G)$ relations apply to FD SOI–MOSFETs whose complex behavior is controlled by both gate biases. The typical characteristics of the front-channel transistor are schematically illustrated in Figure 3.4, for three distinct bias conditions of the back interface (inversion, depletion, and accumulation), and will be explained next.

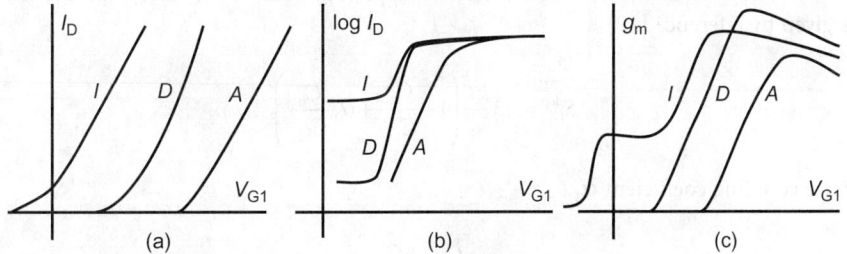

FIGURE 3.4 Generic front-channel characteristics of a FD n-channel SOI MOSFET for accumulation (A), depletion (D), and inversion (I) at the back interface: (a) $I_D(V_{G_1})$ curves in strong inversion, (b) log $I_D(V_{G_1})$ curves in weak inversion, and (c) transconductance $g_m(V_{G_1})$ curves.

3.5.1 Threshold Voltage

The lateral shift of $I_D(V_G)$ curves (Fig. 3.4a) is explained by the linear variation of the front-channel threshold voltage, $V_{T_1}^{dep}$, with back-gate bias. This *potential coupling* causes $V_{T_1}^{dep}$ to decrease linearly, with increasing V_{G_2}, between two plateaus corresponding, respectively, to accumulation and inversion at the back interface [17]

$$V_{T_1}^{dep} = V_{T_1}^{acc} - \frac{C_{si}C_{ox_2}(V_{G_2} - V_{G_2}^{acc})}{C_{ox_1}(C_{ox_2} + C_{si} + C_{it_2})} \tag{3.1}$$

where $V_{T_1}^{acc}$ is the threshold voltage when the back interface is accumulated,

$$V_{T_1}^{acc} = \Phi_{fb_1} + \frac{C_{ox_1} + C_{si} + C_{it_1}}{C_{ox_1}} 2\Phi_F - \frac{Q_{si}}{2C_{ox_1}} \tag{3.2}$$

and $V_{G_2}^{acc}$ is given by

$$V_{G_2}^{acc} = \Phi_{fb_2} - \frac{C_{si}}{C_{ox_2}} 2\Phi_F - \frac{Q_{si}}{2C_{ox_2}} \tag{3.3}$$

In the above equations, C_{si}, C_{ox}, and C_{it} are the capacitances of the FD film, oxide, and interface traps, respectively; Q_{si} is the depletion charge, Φ_F is the Fermi potential, and Φ_{fb} is the flat-band potential. Subscripts 1 and 2 hold for the front- or the back-channel parameters and can be interchanged to account for the variation of the back-channel threshold voltage V_{T_2} with V_{G_1}.

The difference between the two plateaus, $\Delta V_{T_1} = (C_{si}/C_{ox_1})2\Phi_F$, slightly depends on doping, whereas the slope does not. We must insist on the polyvalence of Eqs. (3.1) to (3.3) as compared with the simple case of bulk Si MOSFETs (or partially depleted (PD) MOSFETs), where

$$V_{T_1} = \Phi_{fb_1} + \left(1 + \frac{C_{it_1}}{C_{ox_1}}\right) 2\Phi_F + \frac{\sqrt{4q\epsilon_{si}N_A\Phi_F}}{C_{ox_1}} \tag{3.4}$$

The extension to p-channels or accumulation-mode SOI–MOSFETs is also straightforward [1].

In FD MOSFETs, the threshold voltage decreases in thinner films (i.e., reduced depletion charge), until C_{si} prevails or quantum effects arise and lead to the formation of a 2–D subband system. In ultrathin films ($t_{si} \le 10$ nm), the separation between the ground state and the bottom of the conduction band increases with reducing thickness: a V_T rebound is then observed [18].

3.5.2 Subthreshold Slope

For depletion at the back interface, the subthreshold slope (Fig. 3.4b) is very steep and the subthreshold *swing S* is given by reference [19]

$$S_1^{dep} = 2.3 \frac{kT}{q} \left(1 + \frac{C_{it_1}}{C_{ox_1}} + \alpha_1 \frac{C_{si}}{C_{ox_1}}\right) \tag{3.5}$$

The interface coupling coefficient α_1,

$$\alpha_1 = \frac{C_{ox_2} + C_{it_2}}{C_{si} + C_{ox_2} + C_{it_2}} < 1 \tag{3.6}$$

accounts for the influence of back-interface traps C_{it_2} and BOX thickness C_{ox_2} on the front-channel current [19].

In the ideal case, where $C_{it_{1,2}} \simeq 0$ and the BOX is much thicker than both the film and the gate oxide (i.e. $\alpha_1 \simeq 0$), the swing approaches the theoretical limit $S_1^{dep} \simeq 60$ mV/decade at 300 K. Accumulation at the back interface does decouple the front inversion channel from back-interface defects but, in turn, makes α_1 tend to unity (as in bulk-Si or PD MOSFETs), causing an overall degradation of the swing.

It is worth noting that the above simplified analysis and equations are valid only when the BOX is thick enough so that substrate effects occurring underneath the BOX can be overlooked. The capacitances of the BOX and Si substrate are connected in series. Therefore, the swing may depend, essentially for thin BOXs, on the density of traps and surface charge (accumulation, depletion, or inversion) at the *third* interface: BOX-Si substrate. The general trend is that the subthreshold slope improves for thinner silicon films and thicker BOXs. Film thinning leads to a lower subthreshold swing only in the case of a few states at the silicon layer/BOX interface.

3.5.3 Transconductance

For strong inversion and ohmic region of operation, the front-channel drain current and transconductance are given by

$$I_D = \frac{C_{ox_1} W V_D}{L} \frac{\mu_1}{1 + \theta_1 (V_{G_1} - V_{T_1}) + \theta_2 (V_{G_1} - V_{T_1})^2} (V_{G_1} - V_{T_1}(V_{G_2})) \tag{3.7}$$

$$g_{m_1} = \frac{C_{ox_1} W V_D}{L} \frac{\mu_1 [1 - \theta_2 (V_{G_1} - V_{T_1})^2]}{[1 + \theta_1 (V_{G_1} - V_{T_1}) + \theta_2 (V_{G_1} - V_{T_1})^2]^2} \tag{3.8}$$

where μ_1 is the mobility of front-channel carriers, and $\theta_{1,2}$ are the mobility attenuation coefficients. Coefficient θ_2 reflects the surface roughness scattering and is relevant for ultrathin gate oxides.

The complexity of the transconductance curves in FD MOSFETs (Fig. 3.4c) is explained by the influence of the back-gate bias via $V_{T_1}(V_{G_2})$. The effective mobility and transconductance peak are maximum for depletion at the back interface, owing to combined effects of reduced vertical field and series resistances.

An unusual feature is the distortion of the transconductance (curve I, Fig. 3.4c) which reflects the possible activation of the back channel, before the inversion charge build-up is completed at the front channel [20]. While the front interface is still depleted, increasing V_{G_1} reduces the back threshold voltage and eventually opens the *back* channel. The plateau of the front-channel transconductance (Fig. 3.4c) can be used to derive directly the back-channel mobility.

3.5.4 Volume Inversion

In thin and low-doped films, the simultaneous activation of front and back channels induces by continuity (i.e., charge coupling) the onset of *volume inversion* [21]. Unknown in bulk Si, this effect enables the inversion charge to cover the whole film. Self-consistent solutions of Poisson and Schrödinger equations indicate that the maximum density of the inversion charge can be reached in the middle of the film, away from the interface. For double-gate operation the electric field cancels in the middle of the film enabling the carrier mobility to increase further. This results in higher current drive and transconductance, attenuated influence of interface defects (traps, fixed charges, and roughness), and reduced $1/f$ noise. Note that some degree of volume inversion subsists in single-gate MOSFETs if the film is ultrathin.

Multiple-gate MOSFETs (DELTA, FinFETs, and GAA transistors), designed to take full advantage from volume inversion, also benefit from reduced short-channel effects.

3.5.5 Defect Coupling

In FD MOSFETs, carriers flowing at one interface may sense the presence of defects located at the opposite interface. *Defect coupling* is observed as an apparent degradation of the front-channel properties, which is actually induced by the BOX damage. This unusual mechanism is notorious after back-interface degradation via radiation or hot-carrier injection.

3.6 PD SOI Transistors

In PD SOI MOSFETs, the depletion charge controlled by one or both gates does not extend from an interface to the other. A neutral region subsists and, therefore, the interface coupling effects are disabled. When the body is grounded (via independent body contacts or body-source ties), PD SOI transistors behave very much like bulk–Si MOSFETs and most of the standard $I_D(V_G, V_D)$ equations and design concepts apply. If body contacts are not supplied, the so-called *floating-body* effects arise, leading to detrimental consequences.

3.6.1 Classical Floating-Body Effects

The *kink* effect is due to majority carriers, generated by impact ionization, which collect in the transistor body. The body potential is raised which reduces the threshold voltage. This feedback gives rise to extra drain current (kink) in $I_D(V_D)$ characteristics (Fig. 3.5a), which is annoying in analog circuits.

In weak inversion and for high drain bias, a similar positive feedback (increased inversion charge → more impact ionization → body charging → threshold voltage lowering) is responsible for negative resistance regions, hysteresis in log $I_D(V_G)$ curves, and eventually latch (loss of gate control; Fig. 3.5b).

The floating body may also induce transient effects. A drain current *overshoot* is observed when the gate is turned on (Fig. 3.5c). Majority carriers are expelled from the depletion region and collect in the neutral body increasing the potential. Equilibrium is reached through electron–hole recombination which eliminates the excess majority carriers, making the drain current to decrease gradually with time. A reciprocal *undershoot* occurs when the gate is switched from strong to weak inversion: the current now increases with time (Fig. 3.5d) as the majority carrier generation allows the depletion depth to shrink gradually. In short-channel MOSFETs, the transient times are dramatically reduced because of the additional contribution of source and drain junctions to establish equilibrium.

The high-frequency switching of integrated circuits may prevent the transistor body from reaching equilibrium. The charging and discharging of the body is an iterative process which may cause "history"

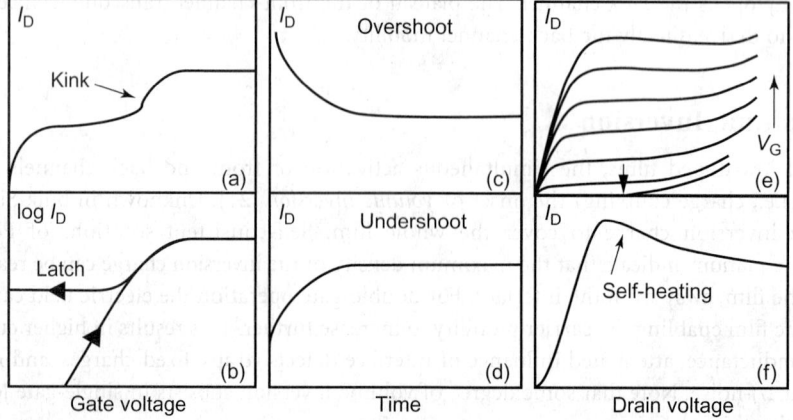

FIGURE 3.5 Parasitic effects in PD SOI MOSFETs: (a) kink in $I_D(V_D)$ curves, (b) latch in $I_D(V_G)$ curves, (c) drain current overshoot, (d) current undershoot, (e) premature breakdown, and (f) self-heating.

effects and dynamic instabilities. In a ring oscillator, the switching delay of an inverter is governed by the amount of available current, which can be higher (overshoot) or lower (undershoot) than at equilibrium. The switching speed depends on the number of previous switches.

An obvious solution to alleviate floating-body effects is to sacrifice chip space for designing body contacts. The problem is that, in ultrathin films with large sheet resistance, the body contacts are far from being ideal. Their intrinsic resistance does not allow the body to be perfectly grounded and may generate additional noise. A floating body is then preferable to a poor body contact.

An exciting PD device is the *dynamic-threshold* DT-MOS transistor. It is simply configured by interconnecting the gate and the body. As the gate voltage increases in weak inversion, the simultaneous raise in body potential causes the threshold voltage to decrease. DT-MOSFETs achieve perfect gate-charge coupling, maximum subthreshold slope, and enhanced current, which are attractive features for low-voltage, low-power circuits.

3.6.2 Gate-Induced Floating-Body Effects

In MOSFETs with ultrathin (<2 nm) gate oxide, the body is charged by the tunneling current giving rise to *gate-induced floating-body effect* (GIFBE). GIFBE occurs even at low drain voltage and is not related to impact ionization. Typical features are a second peak in transconductance (Fig. 3.6) [22,23] and an excess low-frequency noise [24].

The body potential is defined by the balance between the incoming gate tunneling current (body charging) and the outgoing current (body discharging via junction leakage and/or carrier recombination). In PD MOSFETs, the increase in body potential directly lowers the threshold voltage [23], giving rise to a "kink" in the drain current [25] and a second g_m peak (Fig. 3.6a). GIFBE may also occur in FD MOSFETs, in particular when the back interface is biased in accumulation (Fig. 3.6b) [26]: the GIFBE peak gradually distorts and eventually offsets the mobility-related peak. It is clear that the mobility extracted from such a curve is totally meaningless.

GIFBE is a dimensional effect which decreases in shorter MOSFETs because the tunneling current is reduced, whereas the junction leakage is rather constant. GIFBE also decreases in narrower MOSFETs where the carrier lifetime is degraded and the source-body barrier is lower. In addition, GIFBE depends

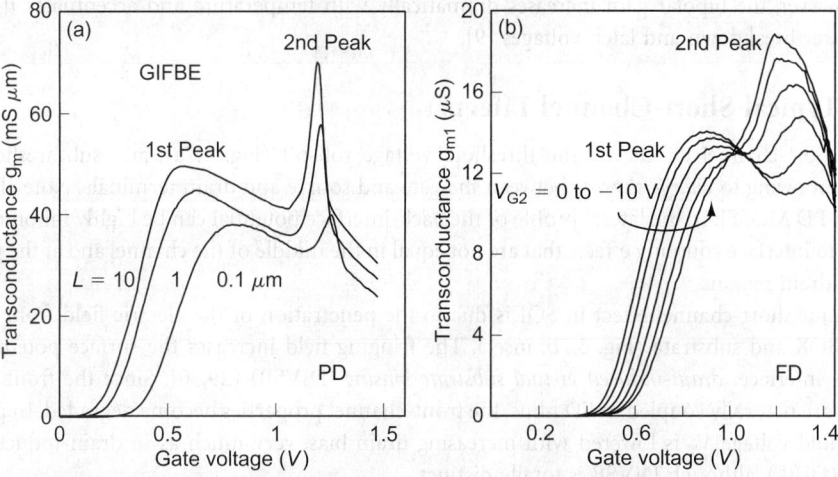

FIGURE 3.6 Transconductance modification by GIFBE in SOI MOSFETs : (a) second peak in normalized transconductance ($g_m \times L$) curves for short and long PD transistors and (b) second peak evolution as a function of back-gate bias in FD SOI MOSFETs ($L = W = 10$ μm, $V_D = 0.1$ V) (from Cassé M. et al., *Solid-State Electron.*, vol. 48, no. 7, pp. 1243–1247, 2004).

on the scanning speed of the gate voltage: for slower measurements, the second peak of the transconductance appears at a lower V_G. The asymmetry between gradual body charging (for increasing V_G) and body discharging (for decreasing V_G) is summarized by a hysteresis in $I_D(V_G)$ curves [22,23].

The transient effects and history effects are dramatically modified because GIFBE enables a faster recovery of the equilibrium body charge.

3.6.3 From Partial to Full Depletion

Partial depletion occurs if the vertical depletion region w_D, controlled by the gate, does not cover the whole body ($w_D < t_{si}$). This old definition does not apply to very short devices, where the lateral depletion regions of the source and drain junctions enhance the overall depletion [27]. The junctions cause a lowering of the *effective* doping seen by the gate, allowing the vertical depletion region to extend deeper. The interesting consequence is that the transition from PD to FD operation is also controlled by the channel length, not only by the doping/thickness ratio. For this reason, PD technology would require excessive doping levels and will be hard to defend for very advanced CMOS nodes (<65 nm).

The discussion above is valid for SOI MOSFETs without pockets. In case of higher doping levels localized near the source and drain (pockets), the transition from PD to PD can show an opposite trend: shorter transistors exhibit a higher effective doping making them more PD [28].

3.7 Small-Geometry Effects

3.7.1 Parasitic Bipolar Transistor

In both FD and PD MOSFETs with submicron length, the source–body junction can easily be turned on. The inherent activation of the lateral bipolar transistor has favorable (extra current flow in the body) or detrimental (premature breakdown, Fig. 3.5e) consequences.

The breakdown voltage is evaluated near the threshold, where the bipolar action prevails. The breakdown voltage is especially lowered for n-channels, shorter devices, thinner films, and higher temperatures. As expected, the impact ionization rate and related floating-body effects are attenuated at high temperature. However, the bipolar gain increases dramatically with temperature and accentuates the bipolar action: lower breakdown and latch voltages [9].

3.7.2 Typical Short-Channel Effects

Familiar short-channel effects are the threshold voltage roll-off (Fig. 3.7a), and subthreshold swing degradation owing to *charge sharing* between the gate and source and drain terminals. Note also that in very short FD MOSFETs, the lateral profile of the back-interface potential can be highly inhomogeneous; this leads to interface coupling effects that are not equal in the middle of the channel and at the proximity of source/drain regions.

The major short-channel effect in SOI is due to the penetration of the electric field from the drain into the BOX and substrate (Fig. 3.7b, inset). The fringing field increases the surface potential at the film–BOX interface: *drain-induced virtual substrate biasing* (DIVSB) [29,30]. Since the front and back interfaces are naturally coupled in FD films, the front-channel properties become degraded. In particular, the threshold voltage V_T is lowered with increasing drain bias, very much as in drain-induced barrier lowering (DIBL), although DIVSB is totally distinct.

The key parameters in SOI are the doping level, film thickness, and BOX thickness [31]. Ultrathin, FD MOSFETs show improved performance in terms of V_T roll-off, DIBL, and DIVSB, as compared with PD SOI or bulk Si transistors (Fig. 3.7) [32]. The worst case occurs when the film thickness corresponds to the transition between full and partial depletion.

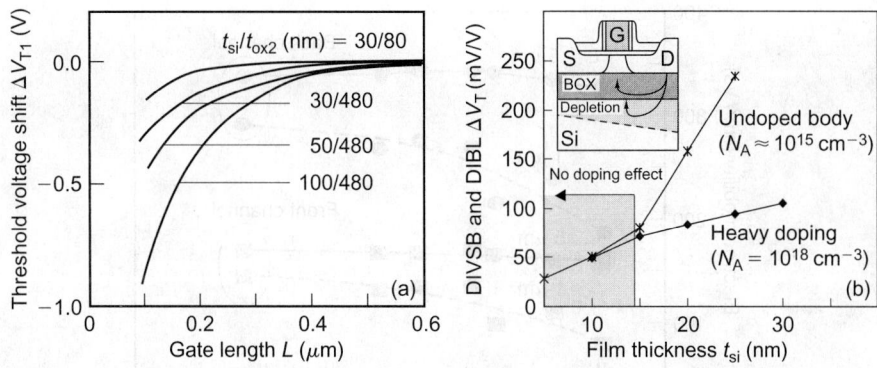

FIGURE 3.7 Typical short-channel effects in FD SOI MOSFETs. (a) Threshold voltage roll-off induced by charge sharing for different thicknesses of film and BOX (from Ohmura, Y., et al., *IEDM Tech. Dig.*, p. 675, 1991). (b) Threshold voltage lowering $\Delta V_T / \Delta V_D$ by DIBL and DIVSB effects versus film thickness and channel doping; the doping effect is cancelled for films thinner than 15 nm (channel length $L = 0.1$ μm).

The transconductance is obviously improved in deep submicron transistors. Velocity saturation occurs as in bulk silicon. The main short-channel limitation of the transconductance comes from series resistance effects.

The lifetime of submicron MOSFETs is affected by hot-carrier injection into the gate oxide(s). The degradation mechanisms are more complex in SOI than in bulk Si, owing to the presence of two oxides, two channels, and related coupling mechanisms [33]. The defects are created at the interface where the carriers flow or at the opposite interface. As a guideline, SOI n–MOSFETs degrade less than bulk Si MOSFETs for $V_G \simeq V_D/2$ (i.e., for maximum substrate current) and more for $V_G \simeq V_T$ (i.e., enhanced hole injection). The device aging is accelerated by accumulating the back interface [33].

3.7.3 Scaling Issues

The scaling strategy in SOI is illustrated in Figure 3.7b. The threshold voltage lowering with drain bias (by DIVSB and DIBL) is compared for highly doped and undoped MOSFETs. A thick undoped film is definitely not suitable. However, film thinning gradually erases the advantage of heavy doping. For 15 nm thick film, ΔV_T becomes reasonable and the doping effect disappears. It is concluded that *undoped and ultrathin* MOSFETs benefit from an unchallenged electrostatic control and are exceptionally robust to short-channel effects. Other advantages are the high carrier mobility and excellent subthreshold slope.

Further solutions for short channels aim at reducing the fringing field penetration in the BOX and substrate: thinner BOX with lower permittivity, double-gate structure, or ground plane (highly doped region or metal layer underneath the BOX) [30,34].

By solving the Poisson equation, it is demonstrated that the minimum channel length is proportional to the film thickness: $L_{min} \simeq 4 t_{si}$ [35]. This guiding rule makes it clear that SOI MOSFETs will break the 10-nm-length barrier as soon as films thinner than 3 nm will be routinely manufactured [36]. As a confirmation, 2.5-nm-long transistors with acceptable characteristics have been simulated for 1-nm-thick SOI films [37]. On the practical side, 6-nm-long SOI MOSFET [38] as well as film thinning down to a few monolayers (1 nm) [39] have been achieved already.

It is worth noting that emerging CMOS technologies such as strained-Si, SiGe, and Ge do not compete with SOI. They *must* be SOI-like: whatever the semiconductor, the electrostatic problems are more or less the same. The semiconductor film should be ultrathin and placed on an insulator. Otherwise, the short-channel effects will become an issue again in extremely small MOSFETs.

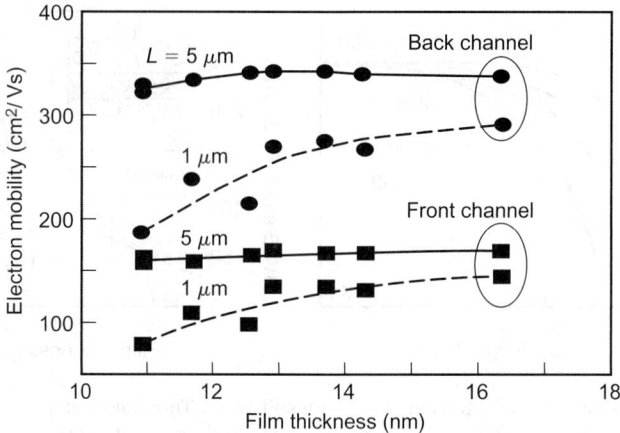

FIGURE 3.8   Front- and back-channel electron mobility versus film thickness in short and long SOI MOSFETs (from Ohata, A. et al., *Proc. ESSDERC 2004*, IEEE, pp. 109–112, 2004).

3.7.4   Ultrathin Channel Effects

The ultrathin body, needed for ultimate scaling, enables very interesting thickness effects. Not only is the interface coupling amplified, but also *supercoupling* may occur [22,40]. Supercoupling reflects the fact that the film tends to behave as a quasi-rectangular well: when the potential at one interface is modified by the gate, the potential of the entire film follows. The notion of front and back channels becomes obsolete and needs to be replaced by the concept of *volume inversion* [21,1] or *volume accumulation*. Since the front and back channels cannot be separated, formulations like "front-channel mobility" should be translated into "mobility seen from the front gate."

In sub-10-nm-thick films, vertical quantum confinement and subband splitting become noticeable, increasing the threshold voltage. Monte Carlo simulations suggest that the carrier mobility is maximum in 3–5-nm-thick films [41]. However, there is no experimental support to date. Early measurements actually indicated the opposite trend, i.e., a mobility degradation in thinner films.

Figure 3.8 shows that in long, 10–20-nm-thick MOSFETs the thickness effect is irrelevant, whereas in short-channel transistors the mobility seems to decrease for thinner films. This difference implies that the apparent mobility degradation in thinner films is merely a series-resistance effect [42]. Other characterization artifacts have been identified (GIFBE, poly depletion, etc.).

3.7.5   Self-Heating and Alternative BOX Layers

Self-heating, induced by the power dissipation, is exacerbated in SOI by the poor thermal conductivity of the surrounding SiO_2 layers. Self-heating is responsible for mobility degradation, threshold voltage shift, and negative differential conductance shown in Figure 3.5f. The temperature raise can exceed 100 to 150°C in SOI [43]. Thin BOXs ($\leq$100 nm) and thicker Si films ($\geq$100 nm) are suitable when self-heating becomes a major issue.

A more revolutionary solution is to modify the generic SOI structure by replacing the standard SiO_2 BOX with Al_2O_3, AlN, SiC, diamond, quartz, etc. (Fig. 3.9a) [44]. These new structures are still SOI, except that the letter *I* is no longer restricted to SiO_2 and recovers the general meaning of buried insulator.

Reducing the self-heating by 50°C represents an immediate gain of more than 25% in mobility ($\mu \sim T^{-1.5}$). This improvement applies simultaneously to electrons and holes and corresponds to the gain in speed expected from the "next" CMOS generation. It follows that the carrier mobility can be engineered not only by using strained silicon and various crystal orientations (<100> and <110>), but also by preventing excessive self-heating. These novel dielectrics are even more attractive for temperature management in high-power SOI devices and for photonic applications.

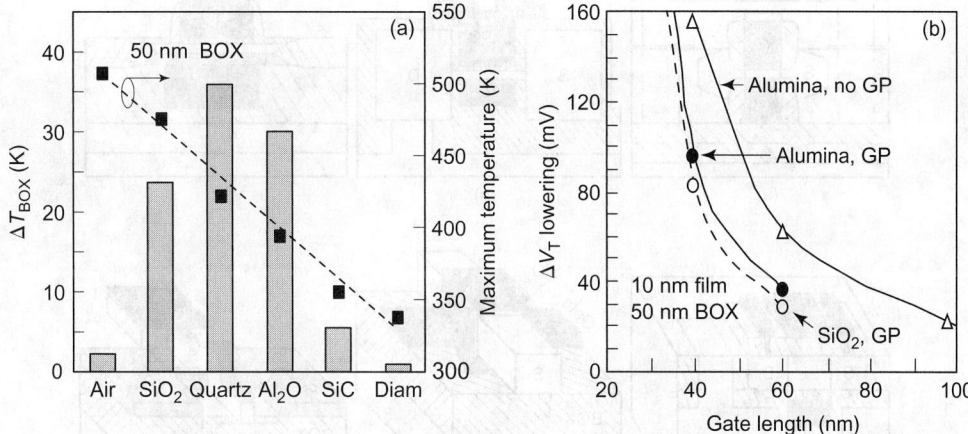

FIGURE 3.9   (a) Temperature difference ΔT_{BOX} (columns) between thick BOX (400 nm) and thin BOX (50 nm) and maximum body temperature (line) for various BOX dielectrics (50-nm-long MOSFETs) (from Bresson, N. et al., *Solid-State Electron.,* vol. 49, no. 9, pp. 1522–1528, 2005). (b) Threshold voltage reduction, induced by DIBL and DIVSB, versus channel length in various FD MOSFETs (SiO_2 or Al_2O_3 BOX, GP or no GP) (from Oshima, K. et al., *Solid-State Electron.,* vol. 48, pp. 907–917, 2004).

A subsequent problem is that the change in the dielectric constant impedes on the 2D distributions of the electric potential in the transistor [44,45]. The classical short-channel effects (charge sharing and DIBL) are marginally degraded for high-K BOX. A 25-nm-long, alumina-BOX MOSFET exhibits only a 25% larger threshold voltage roll-off, which can be further attenuated by thinning the BOX [45]. The control of the fringing fields (DIVSB) is excellent ($\Delta V_T/V_D \simeq 100$ mV/V) for diamond, quartz, SiO_2, and air or modest (250 mV/V) for SiC and Al_2O_3. In the latter case, the device architecture can be optimized by including a ground plane (GP).

Figure 3.9b shows combined solutions: ultrathin Si film (5 to 10 nm), thin BOX (50 nm), and GP. Without GP, DIVSB effect increases exponentially in MOSFETs shorter than 50 nm. A GP is more effective for shorter channels and alumina BOX. The conclusion is twofold: the slight electrostatic disadvantage of alumina BOX is practically erased in GP MOSFETs, and it is minor compared with the huge thermal advantage.

3.8   Multiple-Gate SOI MOSFETs

Innovative transistors with two or more gates are currently being explored for enhanced performance and functionality.

3.8.1   Double-Gate MOSFETs

Double-gate (DG) MOSFETs are ideal devices for electrostatic integrity and ultimate scaling below 10 nm channel length. The formation of front- and back-inversion channels enables volume inversion, which offers enhanced drain current and transconductance. The total inversion charge in DG-mode is roughly twice the inversion charge in single-gate (SG) mode and the subthreshold swing is ideal. The essential aspect is that the minority carriers flow in the middle of the film and experience less surface scattering, hence the mobility [41,46] and radiation hardness [47,10] are improved. Ernst et al. [39] reported an outstanding transconductance increase by more than 200% for 3-nm-thick DG-mode transistor.

The two gates collaborate to provide an excellent electrostatic control, so that short-channel effects (DIBL, DIVSB, punch-through) are reduced. Numerical simulations including quantum effects, band-to-band tunneling, and direct source-to-drain tunneling recommend a body thickness-to-length ratio of roughly 1/2, a condition less stringent than in SG-MOSFETs ($\approx$1/4).

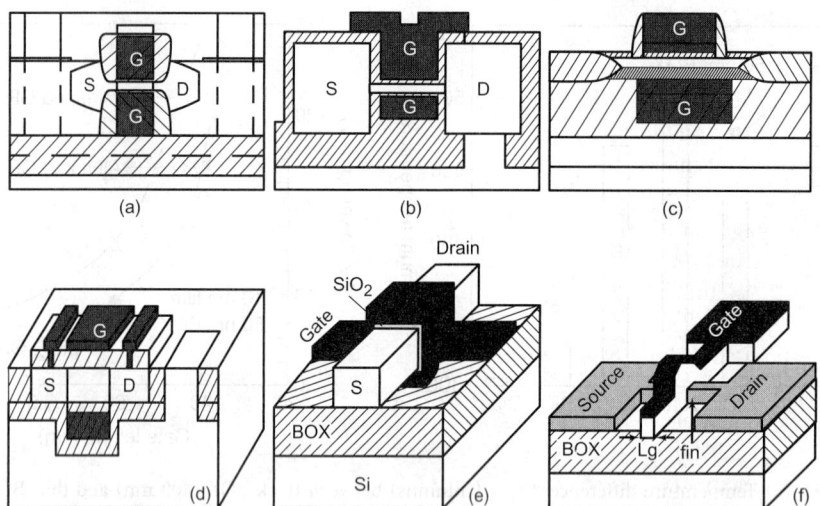

FIGURE 3.10　(a) Technological solutions tested for multiple-gate SOI MOSFETs: (a) self-aligned sacrificial SiGe/Si-body/SiGe stack, where the SiGe layers are subsequently replaced by the gates and the source/drain terminals are formed by selective epitaxy, (b) tunnel epitaxy through the empty space (sacrificial layer) between the gates, (c) wafer bonding, (d) ELO after formation of the bottom gate, (e) GAA, and (f) FinFET.

The main difficulty resides in devising a realistic and pragmatical technology. Several demonstrations are shown in Figure 3.10. Planar process is suitable in many respects, but cannot guarantee the self-alignment of the two gates. The DG technology can be greatly simplified if a reasonable degree of gate misalignment is tolerable. A possibility is to design a longer bottom gate, whereas the channel length is still defined by the source/drain implantation through the shorter top gate (Fig. 3.10c). Surprisingly, the transconductance and drive current may be higher than in "ideal" DG transistor with symmetrical self-aligned gates [48,49]. The reason is the dual action of the longer gate which contributes to volume inversion in the body and simultaneously to accumulation in the source/drain regions. This *field-effect-junction* mechanism contributes to the dynamic lowering of the series resistance. By optimizing the bottom gate length, the gain in transconductance compensates for the parasitic overlapping capacitance.

Such asymmetrical DG transistors have recently been fabricated starting from an SOI wafer and using a wafer-bonding technology. The bottom gate was made on top of the SOI film. This wafer is turned upside-down and bonded to a support Si wafer. After etching the substrate and BOX of the handling SOI wafer, the front gate was formed on the denuded side of the film, roughly aligned to the bottom gate (Fig. 3.10c).

A totally different approach is the vertical DG MOSFET, where the source-body-drain stack and the current flow are perpendicular to the wafer surface. These devices are attractive because the channel length (i.e., body thickness) can be controlled by epitaxy, instead of e-beam lithography. They suffer however from the asymmetry of the source and drain terminals and from the difficulty of achieving tiny pillars with ultrasmall inter-gate distances.

FinFETs are nonplanar DG transistors with relatively easy-to-implement process. In DG-FinFETs (Fig. 3.10f), the gate covers three sides of the body (fin) but the top channel is deactivated by using a thicker dielectric. The FinFET is a semi-vertical device because the current is controlled by the two vertical gates and flows horizontally along the body sidewalls.

A more advanced alternative (MIGFET) is to etch-off the top gate and provide independent contacts to the lateral gates. The advantage is that two gates can play different functions, so reducing the complexity of digital circuits [50].

Although the FinFET performance is promising, two critical scaling issues need to receive attention: (i) the control of the crystal quality and orientation on the sidewalls by wet etching, and (ii) the trimming of the transistor body (inter-gate distance) to keep the short-channel effects under control [51,52].

3.8.2 Triple-Gate MOSFETs

A FinFET with an active top gate is called triple-gate MOSFET (TG-MOSFET). Actually, one single gate governs three different sections of the channel: two vertical and one horizontal (Fig. 3.10f). The gate dielectric should be equally thin on the three sides of the body to avoid multiple threshold voltages. The performance is encouraging. The magnitude of the current can be adjusted via the fin width, but this advantage is debatable because the scaling capability is inferior in wider fins.

By separating the contributions of the different channels, it was found that the carrier mobility is significantly degraded on the fin sidewalls as compared with the top and bottom channels [53]. Process refinements [51,52] and crystal orientation are under investigation to improve FinFET performance.

The coupling effects depend on the fin height, t_{si}, and width, W. A square TG-FinFET ($t_{si} = W = 20$ nm, Fig. 3.11a) features an inhomogeneous vertical variation of the electron concentration and surface potential on the lateral sides [54]. The "measured" threshold voltage corresponds to the lowest position-dependent V_T. Corner effects may also become important, essentially in highly doped bodies.

A narrow and tall fin exhibits two distinct regions [54]. At the bottom of the device, the carrier distribution is inhomogeneous (2D, as in Fig. 3.11a), whereas in the upper region, the carrier profile becomes vertically homogeneous and quasi-1D in the lateral direction. The front channel and the upper regions of the lateral channels are in strong inversion. The electrostatics is controlled by the lateral gates, which can suppress the coupling to the bottom gate (Fig. 3.11b). This coupling coefficient saturates for an aspect ratio of $t_{si}/W \simeq 4$ [54].

For thin and wide fins ($t_{si} \ll W$, as in FD MOSFETs), the lateral gates do not control the body well enough. Instead, the back-gate coupling is strong and modulates the front-channel conduction (Fig. 3.11b).

The geometry optimization aims to reach more current per fin (wider transistors) while avoiding too much sensitivity to substrate-effects. Even for a grounded back gate, a virtual substrate biasing can be induced by radiation, hot-carrier injection, or DIVSB effects. To achieve a low subthreshold swing, both dimensions W and t_{si} should be reasonably small (TG case) or one dimension must be *very* small.

If a very narrow body is manufacturable, the fringing fields are controlled by the lateral gates which define the back-surface potential, blocking the penetration of the fringing field from the drain (DIVSB). This control can be enhanced by allowing the lateral gates to extend vertically into the BOX (π-gate) and laterally underneath the film (Ω-gate) [10]. π-gate and Ω-gate architectures do relax the constraint of ultranarrow fins, but in turn require a thick enough BOX.

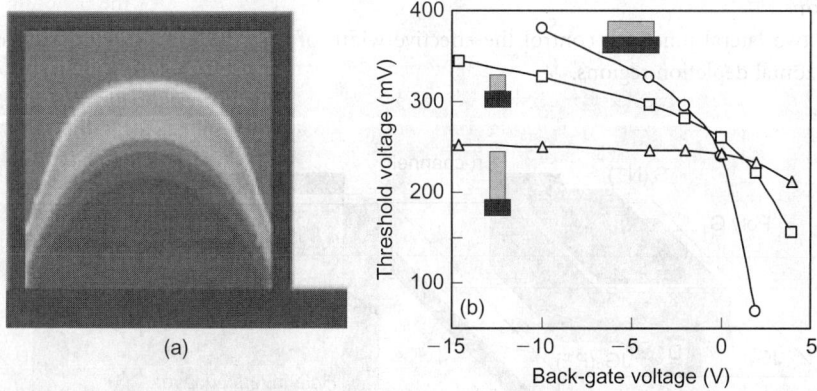

(a)

(b)

FIGURE 3.11   Triple-gate FinFETs: (a) cross-section of the minority carrier distribution in a square (20×20 nm) fin with the gate biased in inversion ($V_{G_1} = +0.5$ V) and the substrate in accumulation ($V_{G_2} = -10$ V); (b) threshold voltage as a function of V_{G_2} in wide, square, and tall fins (aspect ratios: $t_{si}/W = 20/80$, $20/20$, and $80/20$) (from Cristoloveanu et al., *Int. J. High-Speed Electron.*, 2006, in press.)

At this point, it is worth underlining the 3D nature of the coupling effects:

- *Lateral* coupling between the side gates.
- *Vertical* coupling between the top gate and the bottom gate.
- *Longitudinal* coupling between the drain and the body via the fringing fields (DIVSB).

An ultimate and spectacular size effect is related to the transistor *volume*. FinFET technology is capable of releasing devices with all dimensions (thickness, width, and length) in the 10 nm range. A 10^{-18} cm^3 body volume raises fundamental questions. For example, what doping level is induced by one single impurity? Does the impurity position matter? Should atomistic simulations include the silicon atoms one by one?

3.8.3   Gate-All-Around MOSFETs

The GAA technology (Fig. 3.10e), invented by Colinge et al. [10], is complex: (i) formation of a small-size Si membrane, (ii) thermal oxidation, and (iii) wrapping a homogeneous gate. The membrane can be processed in SOI by etching part of the BOX underneath the silicon film [10], or in bulk-Si by SON technology [8]: epitaxy of a sacrificial layer of SiGe, epitaxy of the thin Si film and, finally, removal of the SiGe layer (see Fig. 3.2b$_5$). GAA MOSFETs can also have vertical pillar configuration. The formation of a pillar with small enough diameter is very challenging.

In GAA MOSFETs, the corner regions have a lower threshold voltage and turn on earlier than the main channel. This causes an increase of the leakage current in the off-state and poor subthreshold characteristics. Corner rounding equalizes the minority carrier distribution and suppresses the activation of the parasitic channels. A simpler solution for attenuating the corner effect is to leave the body *undoped* and adjust the threshold voltage with a midgap metal gate.

3.8.4   Four-Gate FET

The four-gate FET (G^4-FET) is a genuine four-gate transistor operated in accumulation or depletion modes [55]. Figure 3.12a shows an inversion-mode, p-channel SOI MOSFET with two N$^+$ body contacts. The same device becomes a G^4-FET when the current is driven by electrons in the perpendicular direction. The majority carriers flow between the body contacts which play the role of source and drain for the G^4-FET (Fig. 3.12a). There are four *independent* gates:

- The usual front and back MOS gates govern the surface accumulation or vertical depletion regions.
- The two lateral junctions control the effective width of the body through the extension of the horizontal depletion regions.

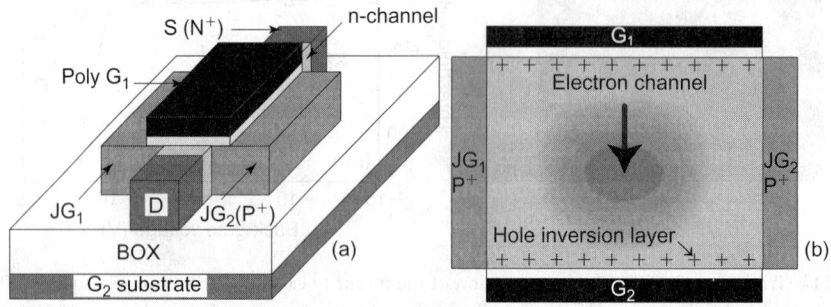

FIGURE 3.12   Basic configuration of the four-gate transistor and cross-section of the carrier distribution for operation in volume mode (depletion-all-around) with inverted interfaces.

The conduction path is modulated by mixed MOS-JFET effects: from a wire-like volume conduction to strongly accumulated front- and back-interface channels. Different models explain the conduction mechanisms in surface accumulation or pure volume modes [56]. The G^4-FET exhibits high current and transconductance and excellent subthreshold swing. Each gate has the capability of switching the transistor on and off. The independent action of the four gates opens promising perspectives for novel applications: mixed-signal circuits, nanoelectronic devices (quantum wires), four-level logic functions with a reduced number of transistors, etc.

Note that the G^4-FET accommodates naturally to scaling. As the gate length of CMOS circuits goes down, the width of the G^4-FET is reduced increasing the junction–gate action. However, the G^4-FET will not compete for minimum size, it will be more suitable for innovative circuit designs.

The most exciting aspect is the depletion-all-around (DAA) mode of operation. The majority carrier channel is surrounded by depletion regions. A quantum wire can be formed (Fig. 3.12b), the dimensions of which are vertically and laterally controlled by the gate bias, not by the lithography. The volume-conduction channel benefits from a double-shielding effect; it is separated from the interface, first by the depletion regions and second by the inversion layers. In the DAA mode, the device features maximum mobility, minimum noise, and unchallenged radiation hardness capability [57]. The G^4-FET structure makes possible the independent cross-conduction of majority carriers (in the volume) and minority carriers (at the interfaces), which is a source for revolutionary devices.

3.9 SOI Challenges

Although SOI is already a mature technology, there are still serious challenges in various domains: fundamental and device physics, technology, device modeling, and circuit design. For example, quantum transport phenomena play an increasing role in ultrathin SOI transistors. It is clear that new physical concepts, ideas, and modeling tools are needed to account for minimum-size mechanisms and to take advantage of them. As far as the technology is concerned, a primary challenge is the fabrication of SOI wafers with ultrathin film, thin BOX, excellent thickness uniformity, low defect content, and reasonable cost.

There is a demand for appropriate characterization techniques, either imported from other semiconductors or entirely conceived for SOI [1]. Such a pure SOI technique is the pseudo-MOS transistor (Ψ–MOSFET, Fig. 3.13) [58]. Ironically, it behaves very much like the MOS device that Shockley attempted to demonstrate 60 years ago but, at that time, he did not have the chance to know about SOI. The inset of Figure 3.13 shows that the Si substrate is biased as a gate and induces a conduction channel (inversion or accumulation) at the film–oxide interface. Source and drain probes are used to measure $I_D(V_G)$ characteristics. The Ψ–MOSFET does not require any processing, hence valuable information is

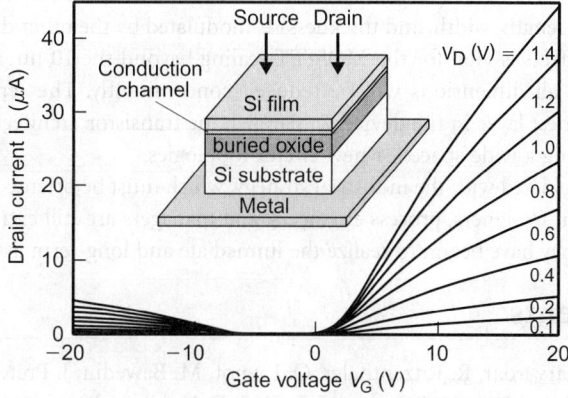

FIGURE 3.13 Pseudo–MOSFET transistor and $I_D(V_G)$ characteristics in SOI.

directly available: quality of the film, interface and oxide, electron/hole mobilities, and lifetime. Contactless optical probing in Ψ–MOSFET configuration would be attractive.

Full CMOS processing must address typical SOI requirements such as the series resistance reduction in ultrathin MOSFETs (via local body oxidation, elevated source and drain structures, etc.), the lowering of the source–body barrier by source engineering (silicidation, SiGe, etc.), the control of the parasitic bipolar transistor, and the limitation of self-heating effects. It is now clear that the best of SOI is certainly not achievable by simply using a very good bulk-Si technology. For example, DG SOI MOSFETs deserve special processing and design.

According to process engineers and circuit designers, PD SOI MOSFETs are more user friendly as they maintain the flavor of bulk-Si technology. In contrast, very thin FD transistors show superior tolerance to short-channel effects. A possible solution is the incorporation of a GP in the BOX.

Advanced physics-based and compact modeling is requested for correct transcription of the transistor behavior, including the transient effects owing to body charging and discharging, floating body mechanisms, bipolar transistor, dual-gate operation, quantum effects, self-heating, and short-channel limitations.

It is obvious that SOI does need SOI-dedicated CAD libraries. This implies a substantial amount of work which, in turn, will guarantee that the advantages and peculiar constraints of SOI devices are properly accounted for. The optimum configuration of memories, microprocessors, DSP, etc. are different in SOI as compared with bulk. Not only can SOI afford to combine FD/PD, low/high power, and DT–MOSFETs into a single chip, but also the basic mechanisms of operation differ.

3.10   Conclusion

SOI offers the opportunity to integrate high-performance and/or innovative devices while expanding the imminent frontiers of the CMOS down-scaling. Since bulk-Si CMOS can hardly continue, the SOI horizon is cleared. Continuous progress in material science and technology is already incorporating novel semiconductors (strained layers, Ge, and SiGe) and dielectrics into SOI structures. These devices are expected to infuse enhanced performance and new functionalities. The short-term prospects of SOI-based microelectronics depend on the penetration rate of high-speed and low-power SOI circuits into the market. Most of the disadvantages of SOI (self-heating, hot-carriers, early breakdown, etc.) tend to disappear for operation at low voltage.

The nanosize SOI MOSFET stands as a perfect device for a smooth transition from microelectronics to nanoelectronics. Recent results for state-of-the-art SOI MOS transistors reveal the mixed flavors of device scaling. Thin tunneling oxides turn on remarkable GIFBE. In nanometer-thick SOI films, the coupling effects are amplified leading to supercoupling and interesting quantum effects. The self-heating issue can be alleviated by thinning the BOX and replacing it with a different dielectrics. A GP or additional gates avoid degrading the electrostatic behavior of the nano-MOSFET.

The family of *size effects* in SOI is very reach because each dimension of the transistor plays a specific role. A given size effect (length, width, and thickness) is modulated by the other dimensions. The control of these 3D coupling effects is vital for the MOSFET scaling beyond the 10 nm channel-length barrier. What is certain is that all dimensions will be reduced concomitantly. The semiconductor body will presumably be the thinnest layer in the device. In parallel, the transistor architecture is rapidly evolving to multiple gates, opening a wide space for new circuit topologies.

A key challenge is associated with the industrial strategy, which must be oriented to overcome the bulk-Si mono-cultural barrier. Designers, process engineers, and managers are still extremely busy loading the bulk-Si machine. But, they have begun to realize the immediate and long-term assets of SOI technology.

Acknowledgments

Special thanks to K. Akarvardar, R. Ritzenthaler, O. Faynot, M. Bawedin, J. Pretet, F. Allibert, M. Cassé, T. Poiroux, A. Ohata, K. Oshima, N. Bresson, H. Iwai, S. Deleonibus, B. Dufrene, B. Blalock, F. Dauge, C. Gallon, A. Vandooren, J-H. Lee, C. Mazuré, S. Eminente, K. Na, P. Gentil, T. Skotnicki, and M. Gri.

References

1. S. Cristoloveanu, and S.S. Li. *Electrical Characterization of SOI Materials and Devices*. Kluwer, Norwell (1995).
2. G.K. Celler, and S. Cristoloveanu. Frontiers of silicon-on-insulator. *J. Appl. Phys.*, vol. 93, pp. 4955–4978 (2003).
3. S. Cristoloveanu. Silicon films on sapphire. *Rep. Prog. Phys.*, vol. 3, p. 327 (1987).
4. R.A. Johnson, P.R. de la Houssey, C.E. Chang, P.-F. Chen, M.E. Wood, G.A. Garcia, I. Lagnado, and P.M. Asbeck. Advanced thin-film silicon-on-sapphire technology: microwave circuit applications. *IEEE Trans. Electron Dev.*, vol. 45, p. 1047 (1998).
5. S. Cristoloveanu. A review of the electrical properties of SIMOX substrates and their impact on device performance. *J. Electrochem. Soc.*, vol. 138, p. 3131 (1991).
6. M. Bruel. Silicon-on-insulator material technology. *Electronics Lett.*, vol. 31, p. 1201 (1995).
7. T. Yonehara. In (S.S. Iyer and A.J. Auberton-Hervé, Eds.) ELTRAN (SOI-Epi wafer) technology. *Silicon Wafer Bonding Technology for VLSI and MEMS Applications*, INSPEC, London, UK, Chap. 4, p. 53 (2002).
8. T. Skotnicki. In *Silicon-On-Insulator Technology and Devices X*, Electrochemical Soc., Pennington, vol. 2001–2003, pp. 391–402 (2001).
9. S. Cristoloveanu, and G. Reichert. Recent advances in SOI materials and device technologies for high temperature. In (I. Golecki, E. Kolawa, B. Gollomp, Eds.) *High Temperature Electronic Materials, Devices and Sensors Conf.*, IEEE Electron Devices Soc., San Diego, USA, pp. 86–93 (1998).
10. J.-P. Colinge. *Silicon-On-Insulator Technology: Materials to VLSI*, Kluwer, Boston, 3rd ed. (2004).
11. J.M. O'Connor, V.K. Luciani, and A.L. Caviglia. High-voltage DMOS power FETs on thin SOI substrates. *IEEE Int. SOI CONF. Proc.*, p. 167 (1990).
12. T. Ohno, S. Matsumoto, and K. Izumi. An intelligent power IC with double buried-oxide layers formed by SIMOX technology. *IEEE Trans. Electron Devices*, vol. 40, p. 2074 (1993).
13. H. Vogt. Advantages and potential of SOI structures for smart sensors. In *SOI Technology and Devices*, Electrochem. Soc., Pennington, p. 430 (1994).
14. T. Nishimura, Y. Inoue, K. Sugahara, S. Kusonoki, T. Kumamoto, S. Nakagawa, M. Nakaya, Y. Horiba, and Y. Akasaka. Three-dimensional IC for high-performance image signal processor. *IEDM Dig.*, p. 111 (1987).
15. A. Zaslavsky, C. Aydin, S. Luryi, S. Cristoloveanu, D. Mariolle, D. Fraboulet, and S. Deleonibus. Ultrathin silicon-on-insulator vertical tunneling transistor. *Appl. Phys. Lett.*, vol. 83, no. 8, pp. 1653–1655 (2003).
16. Y. Ono, Y. Takahashi, K. Yamazaki et al. Si complementary single-electron inverter. In *Technical Digest IEDM*, Piscataway, USA, pp. 367–370 (1999).
17. H-K. Lim, and J.G. Fossum, Threshold voltage of thin-film silicon on insulator (SOI) MOSFETs, *IEEE Trans. Electron Dev.*, vol. 30, p. 1244 (1983).
18. Y. Ohmura, T. Ishiyama, M. Shoji, and K. Izumi. In (P.L.F. Hemment, S. Cristoloveanu, K. Izumi, T. Houston, and S. Wilson, Eds.) Quantum mechanical transport characteristics in ultimately miniaturized MOSFETs/SIMOX. *SOI Technology and Devices*, Electrochem. Soc., Pennington, p. 199 (1996).
19. B. Mazhari, S. Cristoloveanu, D.E. Ioannou, and A.L. Caviglia. Properties of ultrathin wafer-bonded silicon on insulator MOSFETs. *IEEE Trans. Electron Dev.*, vol. ED–38, p. 1289 (1991).
20. T. Ouisse, S. Cristoloveanu, and G. Borel. Influence of series resistances and interface coupling on the transconductance of fully depleted silicon-on-insulator MOSFETs. *Solid-State Electron.*, vol. 35, p. 141 (1992).
21. F. Balestra, S. Cristoloveanu, M. Bénachir, J. Brini, and T. Elewa. Double-gate silicon on insulator transistor with volume inversion: a new device with greatly enhanced performance. *IEEE Electron Device Lett.*, vol. 8, p. 410 (1987).
22. J. Pretet, A. Ohata, F. Dieudonné et al. Scaling issues for advanced SOI devices: gate oxide tunneling, thin buried oxide, and ultrathin films. In (R.E. Sah, M.J. Dean, D. Landheer, K.B. Sundaram, W.D. Brown, and D. Misra, Eds.) *Silicon Nitride and Silicon Dioxide Thin Insulating Films VII*, Electrochem. Soc. Proc., vol. PV-2003-02, Pennington, USA, pp. 476–487 (2003).

23. J. Pretet, T. Matsumoto, T. Poiroux et al. New mechanism of body charging in partially depleted SOI–MOSFETs with ultrathin gate oxide. In *Proc. ESSDERC'02,* University of Bologna, pp. 515–518 (2002).

24. F. Dieudonné, S. Haendler, J. Jomaah, and F. Balestra. Low frequency noise and hot-carrier reliability in advanced SOI MOSFETs. *Solid-State Electron.,* vol. 48, no. 6, pp. 985–997 (2004).

25. A. Mercha, J.M. Rafi, E. Simoen, E. Augendre, and C. Claeys. *IEEE Trans. Electron Dev.,* vol. 50, no. 7, pp. 1675–1682 (2003).

26. M. Cassé, J. Pretet, S. Cristoloveanu et al. Gate-induced floating-body effect in fully depleted SOI MOSFETs with tunneling oxide and back-gate biasing. *Solid-State Electron.,* vol. 48, no. 7, pp. 1243–1247 (2004).

27. F. Allibert, J. Pretet, G. Pananakakis, and S. Cristoloveanu. Transition from partial to full depletion in silicon-on-insulator transistors: impact of channel length. *Appl. Phys. Lett.,* vol. 84, pp. 1192–1194 (2004).

28. S. Zaouia, S. Goktepeli, A.H. Perera, and S. Cristoloveanu. Short-channel, narrow-channel and ultrathin oxide effects in advanced SOI MOSFETs. In (G. Celler, S. Cristoloveanu, J. Fossum, F. Gamiz, and K. Izumi, Eds.) *Silicon-on-Insulator Technology and Devices XII,* Electrochem. Soc. Proc., Pennington, USA, pp. 309–316 (2005).

29. S. Cristoloveanu, T. Ernst, D. Munteanu, and T. Ouisse. Ultimate MOSFETs on SOI: ultra thin, single gate, double gate, or ground plane. *Int. J. High Speed Electron. Syst.,* vol. 10, no. 1, pp. 217–230 (2000).

30. T. Ernst, C. Tinella, and S. Cristoloveanu. Fringing fields in sub–0.1-μm fully depleted SOI MOSFETs: optimization of the device architecture. *Solid-State Electron.,* vol. 46, no. 3, pp. 373–378 (2002).

31. Y. Ohmura, S. Nakashima, K. Izumi, and T. Ishii. 0.1-μm-gate, ultrathin film CMOS device using SIMOX substrate with 80-nm-thick buried oxide layer. *IEDM Tech. Dig.,* p. 675 (1991).

32. F. Balestra, and S. Cristoloveanu. Special mechanisms in thin-film SOI MOSFETs. *Microelectron. Reliab.,* vol. 37, p. 1341 (1997).

33. S. Cristoloveanu. Hot-carrier degradation mechanisms in silicon-on-insulator MOSFETs. *Microelectron. Reliab.,* vol. 37, p. 1003 (1997).

34. H.-S. Wong, D.J. Frank, and P.M. Solomon. Device design considerations for double-gate, ground-plane, and single-gated ultra-thin SOI MOSFET's at the 25 nm channel length generation. *IEDM Tech. Dig.,* p. 407 (1998).

35. R.-H. Yan, A. Ourmazd, and K.F. Lee. Scaling the Si MOSFET: from bulk to SOI to bulk. *IEEE Trans. Electron Dev.,* vol. 39, no. 7, pp. 1704–1710 (1992).

36. D. Franck, S. Laux, and M. Fischetti. Monte Carlo simulation of a 30 nm dual-gate MOSFET: how short can Si go? *IEDM Tech. Dig.,* p. 553 (1992).

37. K.K. Likharev. Electronics below 10 nm. In *Nano and Giga Challenges in Microelectronics,* Elsevier, Amsterdam, pp. 27–68 (2003).

38. B. Doris et al. Device design considerations for ultra-thin SOI MOSFETs. *IEDM Tech. Dig.,* IEEE, Piscataway, pp. 27.3.1–27.3.4 (2003).

39. T. Ernst, S. Cristoloveanu, G. Ghibaudo, T. Ouisse, S. Horiguchi, Y. Ono, Y. Takahashi, and K. Murase. Ultimately thin double-gate SOI MOSFETs. *IEEE Trans. Electron Dev.,* vol. 50, no. 3, pp. 830–838 (2003).

40. A. Ohata, J. Pretet, S. Cristoloveanu, and A. Zaslavsky. Correct biasing rules for virtual DG mode operation in SOI–MOSFETs. *IEEE Trans. Electron Dev.,* vol. 52, no. 1, pp. 124–125 (2005).

41. F. Gamiz, J.B. Roldan, J.A. Lopez-Villanueva et al. Monte Carlo simulation of electron transport in silicon-on-insulator devices. In *Silicon-On-Insulator Technology and Devices X,* Electrochem. Soc. Proc., PV-2001-2003, Pennington, USA, pp. 157–168 (2001).

42. A. Ohata, M. Cassé, S. Cristoloveanu, and T. Poiroux. Mobility issues in ultrathin SOI MOSFETs: thickness variations, GIFBE, and coupling effects. In *Proc. ESSDERC 2004,* IEEE, pp. 109–112 (2004).

43. L.T. Su, K.E. Goodson, D.A. Antoniadis, M.I. Flik, and J.E. Chung. Measurement and modeling of self-heating effects in SOI n–MOSFETs. In *IEDM Tech. Dig.*, p. 357 (1992).

44. N. Bresson, S. Cristoloveanu, C. Mazuré, F. Letertre, and H. Iwai. Integration of buried insulators with high thermal conductivity in SOI MOSFETs: thermal properties and short channel effects. *Solid-State Electron.*, vol. 49, no. 9, pp. 1522–1528 (2005).

45. K. Oshima, S. Cristoloveanu, B. Guillaumot, H. Iwai, and S. Deleonibus. Advanced SOI MOSFETs with buried alumina and ground plane: self-heating and short-channel effects. *Solid-State Electron.*, vol. 48, pp. 907–917 (2004).

46. D. Esseni, M. Mastrapasqua, G.K. Celler et al. An experimental study of mobility enhancement in ultra-thin SOI transistors operated in double-gate mode. *IEEE Trans. Electron Dev.*, vol. 50, no. 3, pp. 802–808 (2003).

47. C.R. Cirba, S. Cristoloveanu, R.D. Schrimpf et al. Total-dose radiation hardness of double-gate ultra-thin SOI MOSFETs. In *Silicon-on-Insulator Technology and Devices XI*, Electrochem. Soc. Proc. vol. 2003–2005, Pennington, USA, pp. 493–498 (2003).

48. F. Allibert, A. Zaslavsky, J. Pretet, and S. Cristoloveanu. Double-gate MOSFETs: is gate alignment mandatory? In *Proc. ESSDERC'2001*, Frontier Group, pp. 267–270 (2001).

49. J. Widiez, F. Daugé, M. Vinet et al. Experimental gate misalignment analysis in double gate SOI MOSFETs. In *Proc. IEEE Int. SOI Conf.*, Charleston, USA (2004).

50. L. Chang, M. Ieong, and M. Yang. CMOS circuit performance enhancement by surface orientation optimization. *IEEE Trans. Electron Dev.*, vol. 51, no. 10, pp. 1621–1627 (2004).

51. Y.X. Liu et al. A highly threshold voltage-controllable 4T FinFET with an 8.5-nm-thick Si-fin channel. *IEEE Electron Dev. Lett.*, vol. 25, pp. 510–512 (2004).

52. W. Xiong, G. Gebara, J. Zaman et al. Improvement of FinFET electrical characteristics by hydrogen annealing. *IEEE Electron Device Lett.*, vol. 25, no. 8, pp. 541–543 (2004).

53. F. Daugé, J. Pretet et al. Coupling effects and channels separation in FinFETs. *Solid-St. Electron.*, vol. 48, pp. 535–542 (2004).

54. S. Cristoloveanu, R. Ritzenthaler, A. Ohata, and O. Faynot. 3D size effects in advanced SOI devices. In *Int. J. High-Speed Electron.* (2006) in press.

55. B.J. Blalock, S. Cristoloveanu, B. Dufrene et al. The multiple-gate MOS–JFET transistor. In *Frontiers in Electronics—Future Chips*, World Scientific, Singapore, vol. 26, pp. 305–314 (2002).

56. K. Akarvardar, B. Dufrene, S. Cristoloveanu et al. Multi-bias dependence of threshold voltage, subthreshold swing, and mobility in G^4-FETs. In *Proc. ESSDERC'03*, Lisbon, pp. 127–130 (2003).

57. K. Akarvardar et al. Total-dose radiation hardness of the SOI 4-gate transistor (G^4-FET). In *Silicon-On-Insulator Technology and Devices XII*, Electrochem. Soc. Proc. Pennington, USA, pp. 99–106 (2005).

58. S. Cristoloveanu, D. Munteanu, and M. Liu. A review of the pseudo–MOS transistor in SOI wafers: operation, parameter extraction, and applications. *IEEE Trans. Electron Dev.*, vol. 47, no. 5, pp. 1018–1027 (2000).

4

SiGe HBT Technology

John D. Cressler

Georgia Institute of Technology

CONTENTS

4.1 Introduction

The concept of "bandgap engineering" has been used for many years in compound semiconductors such as gallium arsenide (GaAs) and indium phosphide (InP) to realize a host of novel electronic devices. A bandgap-engineered transistor is compositionally altered, using more than one type of semiconductor, in a manner which improves a specific device metric of interest (e.g., speed). A transistor designer might choose, for instance, to make a bipolar transistor which has a GaAs base and collector region, but which also has a AlGaAs emitter. Such a "heterostructure" device has electrical properties which are inherently superior to what could be achieved using a single semiconductor. In addition to simply combining two different materials (e.g., AlGaAs and GaAs), bandgap engineering often involves compositional grading of materials from point A to point B within a device. For instance, one might choose to vary the Al content in an AlGaAs/GaAs transistor from a mole fraction of 0.4–0.6 across a given distance within the emitter region.

Device designers have long sought to combine such bandgap engineering techniques enjoyed in compound semiconductor technologies with the fabrication maturity, high integration levels, high yield, and hence low cost associated with conventional silicon (Si)-integrated circuit manufacturing. Epitaxial *silicon–germanium* (*SiGe*) alloys offer considerable advantages for realizing viable bandgap-engineered transistors in the Si material system, either via compressively strained SiGe layers, tensilely strained Si layers, or some combination of the two. Such Si-based bandgap engineering is an exciting development because it allows Si-based electronic devices to achieve performance levels which were once thought impossible, and thus dramatically extends the number of high-performance applications that can be addressed using low-cost Si-based technology. This chapter reviews the recent progress in both SiGe *heterojunction bipolar transistor* (*HBT*) technology.

4.2 SiGe Strained Layer Epitaxy

Si and Ge, being chemically compatible elements, can be intermixed to form a stable alloy. Unfortunately, however, the lattice constant of Si is about 4.2% smaller than that of Ge. The difficulties associated with realizing viable SiGe bandgap-engineered transistors can be traced to the problems encountered in growing high-quality, defect-free epitaxial SiGe alloys in the presence of this lattice mismatch. For electronic applications it is essential to obtain a SiGe film which adopts the same lattice constant as the underlying Si substrate with perfect alignment across the growth interface. In this case, the resultant SiGe alloy is under compressive strain. This strained SiGe film is thermodynamically stable only under a narrow range of conditions which depends on the film thickness and the effective strain (determined by the Ge fraction, and typically expressed in % Ge) [1]. The critical thickness below which the grown film is unconditionally stable depends reciprocally on the effective strain (Figure 4.1). Thus, for practical electronic device applications, SiGe alloys must be thin (typically <100–150 nm) and contain only modest amounts of Ge (typically <20–30%). It is essential for electronic devices that the SiGe films remain thermodynamically stable so that conventional Si fabrication techniques such as high-temperature annealing, oxidation, and ion implantation can be employed without generating defects, thereby maintaining complete compatibility with conventional Si manufacturing.

From an electronic device viewpoint, the property of the strained SiGe alloy that is most often exploited in bipolar transistors is the reduction in bandgap with strain and Ge content (~75 meV per 10% Ge) [2]. This band offset appears mostly in the valence band, which is particularly useful for realizing npn SiGe HBTs and p-channel FETs [3]. While these band offsets are modest compared with those that can be achieved in III–V semiconductors, the Ge content can be compositionally graded to produce local electric fields which aid carrier transport. For instance, in a SiGe HBT the Ge content might be graded from 0 to 15% across distances as short as 50–60 nm, producing built-in drift fields as large as 15–20 kV/cm. Such fields can rapidly accelerate the carriers to scattering limited velocity (1×10^7 cm/s), thereby improving the transistor frequency response. Another benefit of using SiGe strained layers is the enhancement in carrier mobility. This advantage will be exploited in SiGe channel FETs as discussed below.

Epitaxial SiGe strained layers on Si substrates can be successfully grown today by a number of different techniques, including molecular beam epitaxy (MBE), ultra-high-vacuum/chemical vapor deposition (*UHV/CVD*), rapid-thermal CVD (RTCVD), and reduced-pressure CVD (RPCVD). Each growth technique has advantages and disadvantages, but it is generally agreed that UHV/CVD [4], has a number of appealing features for the commercialization of SiGe integrated circuits. These features of UHV/CVD which make it particularly suitable for SiGe manufacturing include (1) batch processing on up to 16 wafers

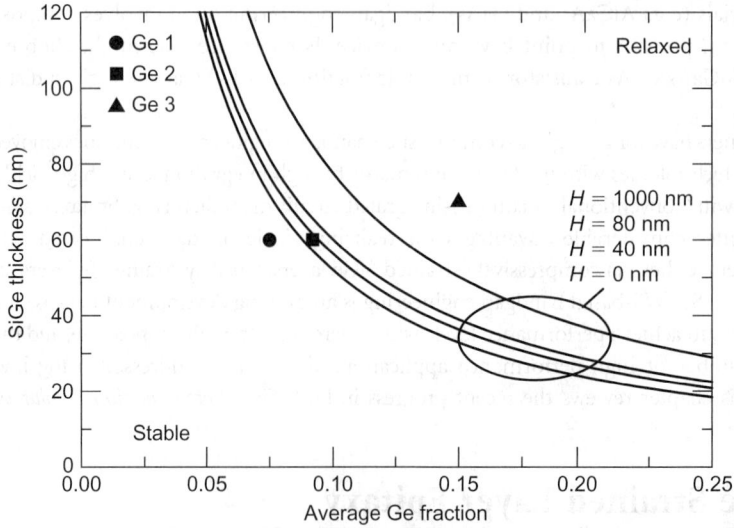

FIGURE 4.1   Silicon–germanium film thickness versus average Ge fraction. The figure shows theoretical stability curves for varying emitter cap thickness (Fischer's theory) and data points for representative stable (Ge 1 and Ge 2) and metastable (Ge 3) profiles. (From J.D. Cressler, *Silicon Heterostructure Handbook: Materials, Fabrication, Devices, Circuits, and Applications of SiGe and Si Strained Layer Epitaxy,* Boca Raton, FL: CRC Press, 2006. With permission.)

simultaneously, (2) excellent doping and thickness control on large (e.g., 200 mm) wafers, (3) very low background oxygen and carbon concentrations, (4) compatibility with patterned wafers and hence conventional Si bipolar and CMOS fabrication techniques, and (5) the ability to compositionally grade the Ge content in a highly controllable manner across short distances. The experimental results presented in this chapter are based on the UHV/CVD growth technique as practiced at IBM Corporation, and are representative of the state of the art in SiGe technology.

4.3 The SiGe Heterojunction Bipolar Transistor

The SiGe HBT is by far the most mature Si-based bandgap-engineered electronic device [5]. The first SiGe HBT was reported in 1987 [6,7], and is in commercial production today worldwide at upwards of 25 different companies. Evolution in transistor performance has been exceptionally rapid (Figure 4.2). Significant steps along the path to manufacturing included the first demonstration of high-frequency (75 GHz) operation of a SiGe HBT in a non-self-aligned structure early 1990 [8]. This result garnered significant attention worldwide since the performance of the SiGe HBT was roughly twice what a state-of-the-art Si BJT(Bipolar Junction Transistor) could achieve. The first fully integrated, self-aligned SiGe HBT technology was demonstrated later in 1990 [9], the first fully integrated 0.5 μ SiGe BiCMOS technology (SiGe HBT + Si CMOS) in 1992 [10], and SiGe HBTs with frequency response above 100 GHz in 1993 and 1994 [11,12]. A number of companies around the world have demonstrated robust SiGe HBT technologies (upwards of 25 companies in 2005) [13–25]. Recent work has focused on practical applications of SiGe HBT circuits for a wide variety of mixed-signal circuit applications [26–28].

Because the intent in SiGe technology is to combine bandgap engineering with conventional Si fabrication techniques, most SiGe HBT technologies appear very similar in structure to conventional Si bipolar technologies. A typical device cross section is shown in Figure 4.3. This SiGe HBT has a planar, self-aligned structure with a polysilicon emitter contact, silicided extrinsic base, and deep- and shallow-trench isolation. A 5–7 level, chemical-mechanical-polishing (CMP) planarized, W-stud, AlCu (or full Cu metallization followed by a final thick Al layer) CMOS-like metalization scheme is used. The extrinsic resistive and capacitive parasitics are intentionally minimized to improve the maximum oscillation frequency (f_{max}) of

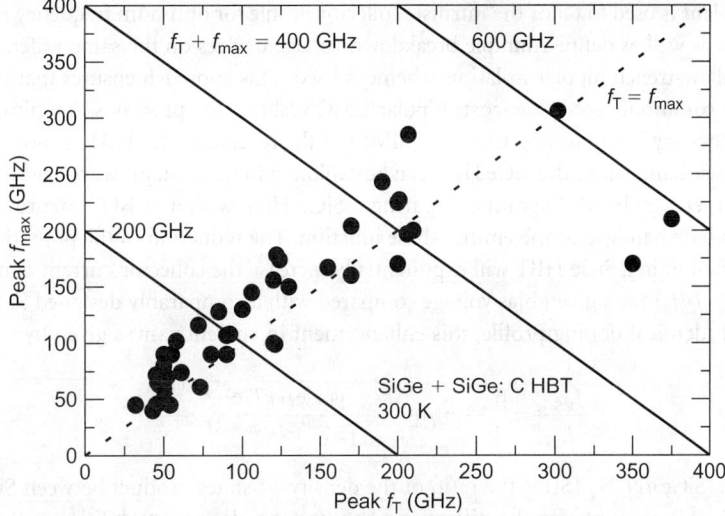

FIGURE 4.2 Transistor peak maximum oscillation frequency (f_{max}) as a function of peak cutoff frequency (f_T) showing the evolutionary path of SiGe HBT performance. (From J.D. Cressler, *Silicon Heterostructure Handbook: Materials, Fabrication, Devices, Circuits, and Applications of SiGe and Si Strained Layer Epitaxy*, Boca Raton, FL: CRC Press, 2006. With permission.)

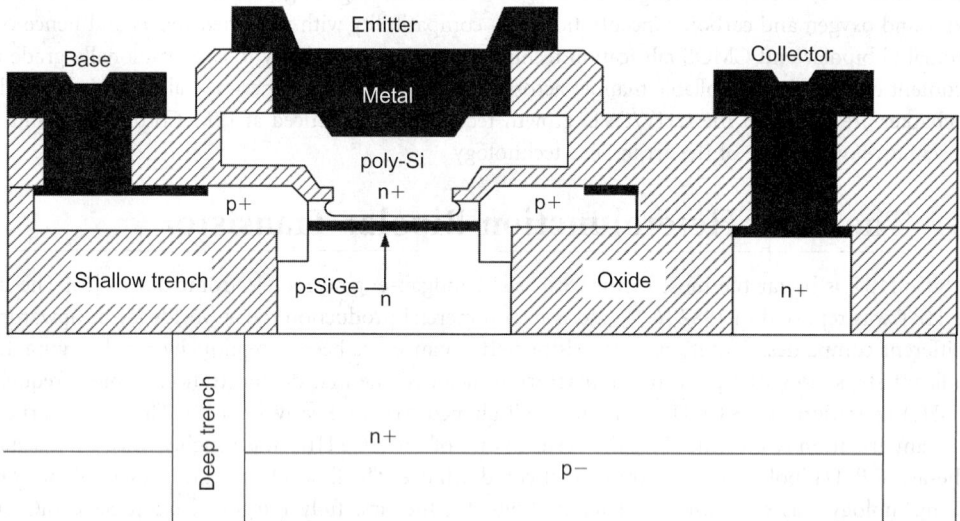

FIGURE 4.3 Schematic cross section of a first-generation self-aligned UHV/CVD SiGe HBT. (From J.D. Cressler, *Silicon Heterostructure Handbook: Materials, Fabrication, Devices, Circuits, and Applications of SiGe and Si Strained Layer Epitaxy,* Boca Raton, FL: CRC Press, 2006. With permission.)

the transistor. Observe that the Ge is introduced only into the thin base region of the transistor, and is deposited with a thickness and Ge content that ensures the film is thermodynamically stable. The in situ boron-doped, graded SiGe base is deposited across the entire wafer using growth techniques such as UHV/ CVD. In areas that are not covered by oxide, the SiGe film consisting of an intrinsic-Si/strained boron-doped SiGe/intrinsic-Si stack is deposited as a perfect single-crystal layer on the Si substrate. Over the oxide, the deposited layer is polycrystalline (poly), and will serve either as the extrinsic base contact of the SiGe HBT, the poly-on-oxide resistor, or the gate electrode of the Si CMOS devices. The metallurgical base and single-crystal emitter widths range from 30 to 90 nm and 25 to 35 nm, respectively. A masked phosphorus implant is used to tailor the intrinsic collector profile for optimum frequency response at high current densities as well as define multiple breakdown voltage devices on the same wafer. A conventional deep-trench/shallow-trench bipolar isolation scheme is used. This approach ensures that the SiGe HBT is compatible with commonly used (low-cost) bipolar/CMOS fabrication processes. A typical doping profile measured by secondary ion mass spectroscopy (SIMS) of the resultant SiGe HBT is shown in Figure 4.4.

The smaller base bandgap of the SiGe HBT can be exploited in three major ways, and is best illustrated by examining an energy band diagram comparing a SiGe HBT with a Si BJT (Figure 4.5). First, note the reduction in base bandgap at the emitter–base junction. The reduction in the potential barrier at the emitter–base junction in a SiGe HBT will exponentially increase the collector current density, and hence current gain ($\beta = J_C/J_B$) for a given bias voltage compared with a comparably designed Si BJT. Compared with a Si BJT of identical doping profile, this enhancement in current gain is given by

$$\frac{J_{C,SiGe}}{J_{C,Si}} = \frac{\beta_{SiGe}}{\beta_{Si}} = \gamma\eta \frac{\Delta E_{g,Ge}(\text{grade})/kT \; e^{\Delta E_{g,Ge}(0)/kT}}{1 - e^{-\Delta E_{g,Ge}(\text{grade})/kT}} \tag{4.1}$$

where $\eta = N_C N_V(\text{SiGe})/N_C N_V (\text{Si})$ is the ratio of the density-of-states product between SiGe and Si, and $\gamma = D_{nb}(\text{SiGe})/D_{nb}(\text{Si})$ accounts for the differences between the electron mobilities in the base between Si and SiGe. The position dependence of the band offset with respect to Si is conveniently expressed as a bandgap grading term $\Delta E_{g,Ge} (\text{grade}) = (\Delta E_{g,Ge} (W_b) - \Delta E_{g,Ge} (0))$. As can be seen in Figure 4.6, which compares the measured Gummel characteristics for two identically constructed SiGe HBTs and Si BJTs, these theoretical expectations are clearly borne out in practice.

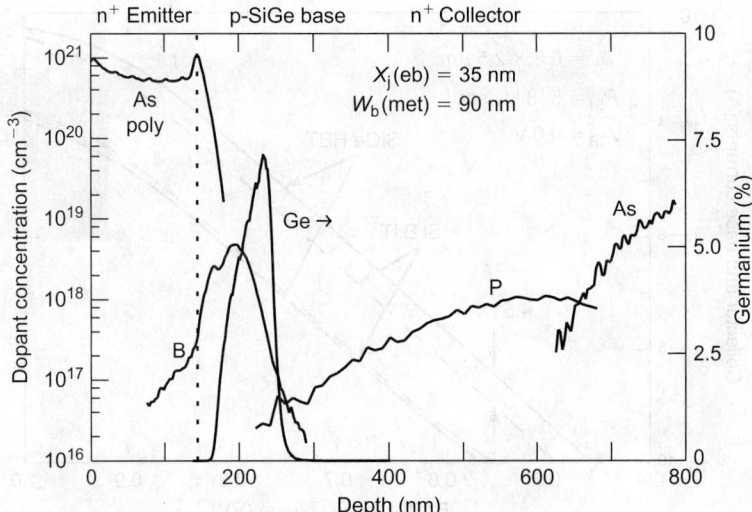

FIGURE 4.4   Secondary ion mass spectroscopy (SIMS) doping profile of a first-generation graded-base SiGe HBT. (From J.D. Cressler, *Silicon Heterostructure Handbook: Materials, Fabrication, Devices, Circuits, and Applications of SiGe and Si Strained Layer Epitaxy*, Boca Raton, FL: CRC Press, 2006. With permission.)

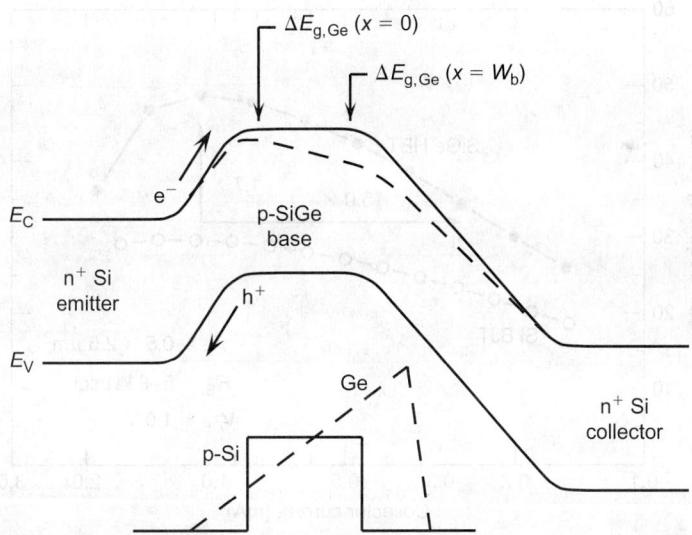

FIGURE 4.5   Energy band diagram for comparably designed Si BJT and graded-base SiGe HBT. (From J.D. Cressler, *Silicon Heterostructure Handbook: Materials, Fabrication, Devices, Circuits, and Applications of SiGe and Si Strained Layer Epitaxy*, Boca Raton, FL: CRC Press, 2006. With permission.)

Second, if the Ge content is graded across the base region of the transistor, the conduction band edge becomes position-dependent (refer to Figure 4.5), inducing an electric field in the base which accelerates the injected electrons. The base transit time is thereby shortened and the frequency response of the transistor is improved according to

$$\frac{\tau_{b,SiGe}}{\tau_{b,Si}} = \frac{f_{T,Si}}{f_{T,SiGe}} = \frac{2}{\eta}\left(\frac{kT}{\Delta E_{g,Ge}(\text{grade})}\right)\left[1 - \frac{1 - e^{-\Delta E_{g,Ge}(\text{grade})/kT}}{\Delta E_{g,Ge}(\text{grade})/kT}\right] \tag{4.2}$$

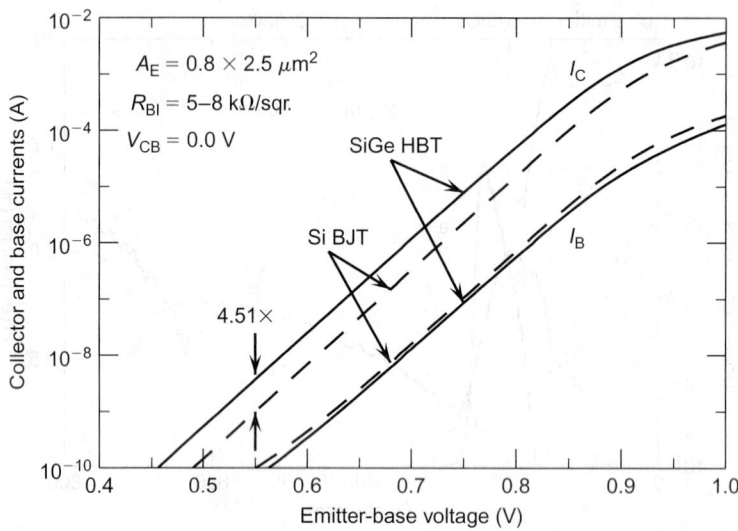

FIGURE 4.6   Measured current–voltage characteristics of both a SiGe HBT and a Si BJT with a comparable doping profile. (From J.D. Cressler, *Silicon Heterostructure Handbook: Materials, Fabrication, Devices, Circuits, and Applications of SiGe and Si Strained Layer Epitaxy*, Boca Raton, FL: CRC Press, 2006. With permission.)

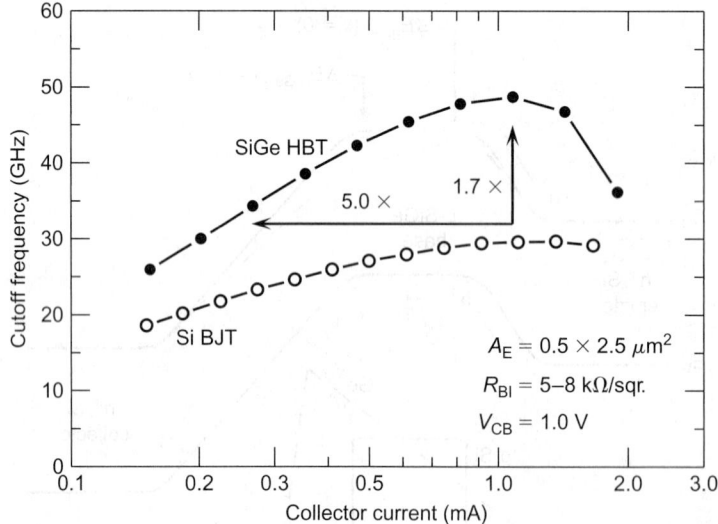

FIGURE 4.7   Measured cutoff frequency as a function of bias current for both a SiGe HBT and a Si BJT with a comparable doping profile. (From J.D. Cressler, *Silicon Heterostructure Handbook: Materials, Fabrication, Devices, Circuits, and Applications of SiGe and Si Strained Layer Epitaxy*, Boca Raton, FL: CRC Press, 2006. With permission.)

Figure 4.7 compares the measured unity gain cutoff frequency (f_T) of a first-generation SiGe HBT and a comparably constructed Si BJT, and shows that an improvement in peak f_T of 1.7× can be obtained with relatively modest Ge profile grading (0 to 7.5% in this case). Aggressive vertical profile scaling has resulted in dramatic improvements in device performance, with peak cutoff frequencies currently in excess of 300 GHz (Figure 4.8).

The final advantage of using a graded Ge profile in a SiGe HBT is the improvement in the output conductance of the transistor, an important analog design metric. For a graded-base SiGe HBT the Early

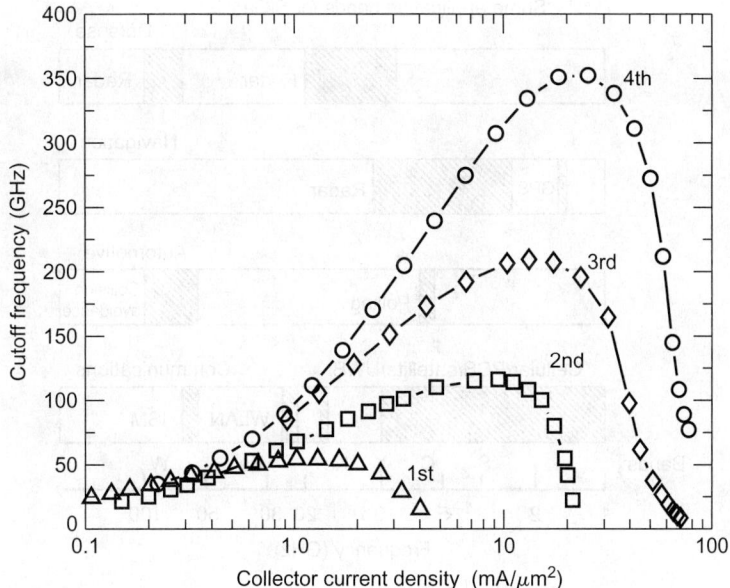

FIGURE 4.8 Cutoff frequency as a function of bias current for four different generations of SiGe HBTs. (From J.D. Cressler, *Silicon Heterostructure Handbook: Materials, Fabrication, Devices, Circuits, and Applications of SiGe and Si Strained Layer Epitaxy*, Boca Raton, FL: CRC Press, 2006. With permission.)

voltage (a measure of output conductance) increases exponentially compared with a Si BJT of comparable doping, according to

$$\frac{V_{A,SiGe}}{V_{A,Si}} = e^{\Delta E_{g,Ge}(\text{grade})/kT} \left[\frac{1 - e^{-\Delta E_{g,Ge}(\text{grade})/kT}}{\Delta E_{g,Ge}(\text{grade})/kT} \right] \qquad (4.3)$$

In essence, the position dependence of the bandgap in the graded-base SiGe HBT weights the base profile toward the collector region, making it harder to deplete the base with collector–base bias, hence yielding a larger Early voltage. A transistor with a high Early voltage has a very flat common-emitter output characteristic, and hence low output conductance. For the device shown in Figure 4.6, the Early voltage increases from 18 V in the Si BJT to 53 V in the SiGe HBT, a 3× improvement.

4.4 Applications and Future Directions

Bandgap-engineered SiGe HBTs have many attractive features which make them ideal candidates for a wide variety of circuit applications. For instance, Si BJTs are well known to have superior low-frequency noise properties compared with compound semiconductor technologies. Low-frequency noise is often a major limitation for RF and microwave systems because it directly limits the spectral purity of the transmitted signal. SiGe HBTs have low-frequency properties as good as or better than Si BJTs, superior to that obtained in AlGaAs/GaAs HBTs and Si CMOS [29,30]. The broadband (RF) noise in SiGe HBTs is competitive with GaAs and InP technology and superior to Si BJTs. In addition, SiGe HBTs have recently been shown to be very robust with respect to ionizing radiation, an important feature for space-based electronic systems [31,32]. Finally, cooling enhances all of the advantages of a SiGe HBT. In striking contrast to a Si BJT, which strongly degrades with cooling, the current gain, Early voltage, cutoff frequency, and maximum oscillation frequency (f_{max}) all improve significantly as the temperature drops [33–35]. This means that the SiGe HBT is well suited for operation in the cryogenic environment (e.g., 77 K),

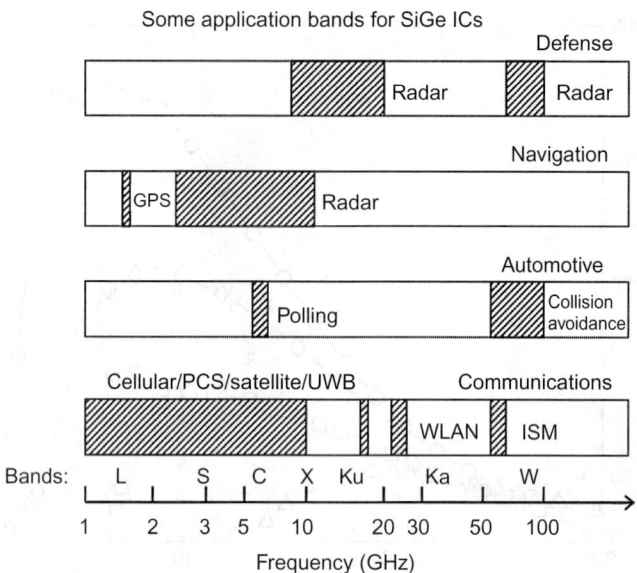

FIGURE 4.9 SiGe HBT application targets for various market sectors, as a function of frequency band. (From J.D. Cressler, *Silicon Heterostructure Handbook: Materials, Fabrication, Devices, Circuits, and Applications of SiGe and Si Strained Layer Epitaxy*, Boca Raton, FL: CRC Press, 2006. With permission.)

historically the exclusive domain of Si CMOS and III–V compound semiconductor technologies. Cryogenic electronics is in growing use in both military and commercial applications such as space-based satellites, high-sensitivity instrumentation, high-T_C superconductors, and future cryogenic computers.

Figure 4.9 summarizes many of the emerging applications of SiGe BiCMOS HBT technology, and span the range of the defense, navigation, automotive, and communications industries, from low RF (900 MHz) to high-mm wave (100 GHz), where it offers an ideal combination of high integration level, high performance, and low cost [36,37]. Figure 4.10 shows a schematic representation of an envisioned single-chip SiGe mm-wave transceiver for very high data rate (>1.0 Gb/s), short-range communications links.

One may logically wonder just how fast SiGe HBTs will be at the end of the day. Transistor-level performance in SiGe HBTs continues to rise at a truly dizzying pace. Both first- and second-generation SiGe HBT BiCMOS technology is widely available and even at the 200 GHz (third-generation) performance level, several companies already have robust commercially available technologies. At present, the most impressive new SiGe HBT result achieves 302 GHz peak f_T and 306 GHz peak f_{max}, a record for any Si-based transistor. This level of performance was achieved at a BV_{CEO} of 1.6 V, a BV_{CBO} of 5.5 V, and a current gain of 660. Noise measurements on these devices yielded NF_{min}/G_{assoc} of 0.45 dB/14 dB and 1.4 dB/8 dB at 10 and 25 GHz, respectively [37]. Measurements of early (unoptimized prototypes) of fourth-generation SiGe HBTs have yielded record values of 375 GHz peak f_T at 300K, and above 500 GHz at 5K. Simulations suggest that THz-level (1000 GHz) intrinsic transistor performance is a realistic goal. It seems likely that we will see SiGe HBTs above-500 GHz peak f_T and f_{max} fully integrated with nanometer-scale (90 nm and below) Si CMOS (possibly strained Si CMOS) within the next 3–5 years.

One might ask, particularly within the confines of ultimate market relevance, why one would even attempt to build 500 GHz SiGe HBTs? If the future "killer app" turns out to be single-chip mm-wave transceiver systems with onboard DSP for broadband multimedia, radar, etc., then the ability of highly scaled, highly integrated, very-high-performance SiGe HBTs to dramatically enlarge the circuit/system design space of the requisite mm-wave building blocks may well prove to be a fruitful (and marketable) path.

Other interesting themes are emerging in the SiGe HBT BiCMOS technology space. One is the the very recent emergence of complementary SiGe (C-SiGe) HBT processes (npn + pnp SiGe HBTs). While very early pnp SiGe HBT prototypes were demonstrated in the early 1990s, only in the last few years have fully

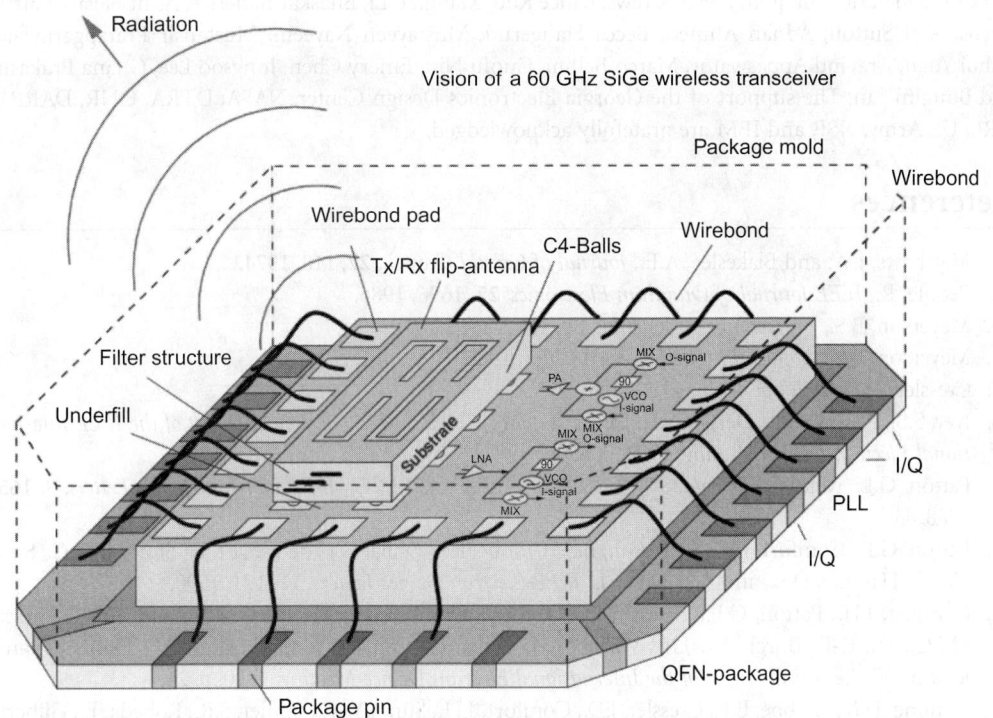

FIGURE 4.10 Vision of a single-chip SiGe mm-wave transceiver. (From J.D. Cressler, *Silicon Heterostructure Handbook: Materials, Fabrication, Devices, Circuits, and Applications of SiGe and Si Strained Layer Epitaxy*, Boca Raton, FL: CRC Press, 2006. With permission.)

complementary SiGe processes been developed, the most mature of which to date has 200 GHz npn SiGe HBTs and 80 GHz pnp SiGe HBTs [37]. Having very-high-speed pnp SiGe HBTs onboard presents a fascinating array of design opportunities aimed particularly at the analog/mixed-signal circuit space. In fact, an additional emerging trend in the SiGe field, particularly for companies with historical pure analog circuit roots, is to target lower peak f_T, but higher breakdown voltages, while simultaneously optimizing the device for core analog applications (e.g., op amps, line drivers, and data converters), designs which might, for instance, target better noise performance, and higher current gain—Early voltage product than mainstream SiGe technologies. One might even choose to put that SiGe HBT platform on top of thick-film SOI for better isolation properties. Another interesting option is the migration of high-speed vertical SiGe HBTs with very-thin-film CMOS-compatible SOI [35]. This technology path would clearly favor the eventual integration of SiGe HBTs with strained Si CMOS, all on SOI, a seemingly natural migratory path. Clearly, SiGe is a highly dynamic field and much is on the horizon. Stay tuned!

Acknowledgments

I would like to thank A. Joseph, D.L. Harame, B.S. Meyerson, D. Ahlgren, and the members of the SiGe team at IBM Corporation as well as the past and present members of my SiGe research team for their contributions, and including David Richey, Alvin Joseph, Bill Ansley, Juan Roldán, Stacey Salmon, Lakshmi Vempati, Jeff Babcock, Suraj Mathew, Kartik Jayanaraynan, Greg Bradford, Usha Gogineni, Gaurab Banerjee, Shiming Zhang, Krish Shivaram, Dave Sheridan, Gang Zhang, Ying Li, Zhenrong Jin, Qingqing Liang, Ram Krithivasan, Yun Luo, Tianbing Chen, Enhai Zhao, Yuan Lu, Chendong Zhu,

Jon Comeau, Jarle Johansen, Joel Andrews, Lance Kuo, Xiangtao Li, Bhaskar Banerjee, Amit Bavasi, Curtis Grens, Akil Sutton, Adnan Ahmed, Becca Haugerud, Mustayeen Nayeem, Mustansir Pratapgarhwala, Jiahui Yuan, Aravind Appaswamy, Marco Bellini, Guofu Niu, Emery Chen, Jongsoo Lee, Gnana Prakash, and Bongim Jun. The support of the Georgia Electronics Design Center, NASA, DTRA, ONR, DARPA, NRL, US Army, NSF, and IBM are gratefully acknowledged.

References

1. Matthews, J.W. and Blakeslee, A.E., *Journal of Crystal Growth*, **27**, 118, 1974.
2. People, R., *IEEE Journal of Quantum Electronics*, **22**, 1696, 1986.
3. Meyerson, B.S., *Proceedings of the IEEE*, **80**, 1592, 1992.
4. Meyerson, B.S., *Applied Physics Letters*, **48**, 797, 1986.
5. Cressler, J.D., *IEEE Spectrum*, 49, March 1995.
6. Iyer, S.S., Patton, G.L., Delage, S.L., Tiwari, S., and Stork, J.M.C., *Technical Digest of the IEEE International Electron Device Meeting*, 1987, p. 874.
7. Patton, G.L., Iyer, S.S., Delage, S.L., Tiwari, S., and Stork, J.M.C., *IEEE Electron Device Letters*, **9**, 165, 1988.
8. Patton, G.L., Comfort, J.H., Meyerson, B.S., Crabbé, E.F., Scilla, G.J., de Fresart, E., Stork, J. M.C., Sun, J.Y.-C., Harame, D.L., and Burghartz, J., *IEEE Electron Device Letters*, **11**, 171, 1990.
9. Comfort, J.H., Patton, G.L., Cressler, J.D., Lee, W., Crabbé, E.F., Meyerson, B.S., Sun, J.Y.-C., Stork, J.M.C., Lu, P.-F., Burghartz, J.N., Warnock, J., Scilla, G., Toh, K.-Y., D'Agostino, M., Stanis, C. and Jenkins, K., *Technical Digest of the International Electron Device Meeting*, 1990, p. 21.
10. Harame, D.L., Crabbé, E.F., Cressler, J.D., Comfort, J.H., Sun, J.Y.-C., Stiffler, S.R., Kobeda, E., Gilbert, M., Malinowski, J., Dally, A.J., Ratanaphanyarat, S., Saccamango, M.J., Rausch, W., Cotte, J., Chu, C., and Stork, J.M.C., *Technical Digest of the International Electron Device Meeting*, 1992, p. 19.
11. Schuppen, A., Gruhle, A., Erben, U., Kibbel, H., and Konig, U., in *Technical Digest of the International Electron Device Meeting*, 1994, p. 377.
12. Crabbé, E.F., Meyerson, B.S., Stork, J.M.C., and Harame, D.L., in *Technical Digest of the International Electron Device Meeting*, 1993, p. 83.
13. Harame, D.L., Stork, J.M.C., Meyerson, B.S., Hsu, K.Y.-J., Cotte, J., Jenkins, K.A., Cressler, J.D., Restle, P., Crabbé, E.F., Subbanna, S., Tice, T.E., Scharf, B.W., and Yasaitis, J.A., *Technical Digest of the International Electron Device Meeting*, 1993, p. 71.
14. Harame, D.L., Schonenberg, K., Gilbert, M., Nguyen-Ngoc, D., Malinowski, J., Jeng, S.-J., Meyerson, B.S., Cressler, J.D., Groves, R., Berg, G., Tallman, K., Stein, K., Hueckel, G., Kermarrec, C., Tice, T., Fitzgibbons, G., Walter, K., Colavito, D., Houghton, T., Greco, N., Kebede, T., Cunningham, B., Subbanna, S., Comfort, J.H., and Crabbé, E.F., *Technical Digest of the International Electron Device Meeting*, 1994, p. 437.
15. Nguyen-Ngoc, D., Harame, D.L., Malinowski, J.C., Jeng, S.-J., Schonenberg, K.T., Gilbert, M.M., Berg, G., Wu, S., Soyuer, M., Tallman, K.A., Stein, K.J., Groves, R.A., Subbanna, S., Colavito, D., Sunderland, D.A., and Meyerson, B.S., *Proceedings of the Bipolar/BiCMOS Circuits and Technology Meeting*, 1995, p. 89.
16. Harame, D.L., Comfort, J.H., Cressler, J.D., Crabbé, E.F., Sun, J.Y.-C., Meyerson, B.S., and Tice, T., *IEEE Transactions on Electron Devices*, **42**, 455, 1995.
17. Harame, D.L., Comfort, J.H., Cressler, J.D., Crabbé, E.F., Sun, J.Y.-C., Meyerson, B.S., and Tice, T., *IEEE Transactions on Electron Devices*, **42**, 469, 1995.
18. Gruhle, A., *Journal of Vacuum Science and Technology B*, **11**, 1186, 1993.
19. Gruhle, A., Kibbel, H., Konig, U., Erben, U., and Kasper, E., *IEEE Electron Device Letters*, **13**, 206, 1992.
20. Schuppen, A., *Technical Digest of the International Electron Device Meeting*, 1995, p. 743.
21. Schuppen, A., Dietrich, H., Gerlach, S., Arndt, J., Seiler, U., Gotzfried, A., Erben, U., and Schumacher, H., *Proceedings of the Bipolar/BiCMOS Circuits and Technology Meeting*, 1996, p. 130.

22. Sato, F., Takemura, H., Tashiro, T., Hirayama, T., Hiroi, M., Koyama, K., and Nakamae, M., in *Technical Digest of the International Electron Device Meeting*, 1992, p. 607.

23. Sato, F., Hashimoto, T., Tatsumi, T., and Tasshiro, T., *IEEE Transactions on Electron Devices*, **42**, 483, 1995.

24. Sato, F., Tezuka, H., Soda, M., Hashimoto, T., Suzaki, T., Tatsumi, T., Morikawa, T., and Tashiro, T., *Proceedings of the Bipolar/BiCMOS Circuits and Technology Meeting*, 1995, p. 158.

25. Meister, T.F., Schafer, H., Franosch, M., Molzer, W., Aufinger, K., Scheler, U., Walz, C., Stolz, M., Boguth, S., and Bock, J., *Technical Digest of the International Electron Device Meeting*, 1995, p. 739.

26. Soyuer, M., Burghartz, J., Ainspan, H., Jenkins, K.A., Xiao, P., Shahani, A., Dolan, M., and Harame, D.L., *Proceedings of the Bipolar/BiCMOS Circuits and Technology Meeting*, 1996, p. 169.

27. Schumacher, H., Erben, U., Gruhle, A., Kibbelk, H., and Konig, U., *Proceedings of the Bipolar/BiCMOS Circuits and Technology Meeting*, 1995, p. 186.

28. Sato, F., Hashimoto, T., Tatsumi, T., Soda, M., Tezuka, H., Suzaki, T., and Tashiro, T., *Proceedings of the Bipolar/BiCMOS Circuits and Technology Meeting*, 1995, p. 82.

29. Cressler, J.D., Vempati, L., Babcock, J.A., Jaeger, R.C., and Harame, D.L., *IEEE Electron Device Letters*, **17**, 13, 1996.

30. Vempati, L., Cressler, J.D., Babcock, J.A., Jaeger, R.C., and Harame, D.L., *IEEE Journal of Solid-State Circuits*, **31**, 1458, 1996.

31. Babcock, J.A., Cressler, J.D., Vempati, L., Clark, S.D., Jaeger, R.C., and Harame, D.L., *IEEE Electron Device Letter*, **16**, 351, 1995.

32. Babcock, J.A., Cressler, J.D., Vempati, L., Jaeger, R.C., and Harame, D.L., *IEEE Transactions on Nuclear Science*, **42**, 1558, 1995.

33. Cressler, J.D., Crabbé, E.F., Comfort, J.H., Sun, J.Y.-C., and Stork, J.M.C., *IEEE Electron Device Letters*, **15**, 472, 1994.

34. Cressler, J.D., Comfort, J.H., Crabbé, E.F., Patton, G.L., Stork, J.M.C., Sun, J.Y.-C., and Meyerson, B.S., *IEEE Transactions on Electron Devices*, **40**, 525, 1993.

35. Cressler, J.D., Crabbé, E.F., Comfort, J.H., Stork, J.M.C., and Sun, J.Y.-C., *IEEE Transactions on Electron Devices*, **40**, 542, 1993.

36. Cressler, J.D. and Niu, G.F., *Silicon-Germanium Heterojunction Bipolar Transistors*, Boston, MA: Artech House, 2003.

37. Cressler, J.D.(Ed.), *Silicon Heterostructure Handbook: Materials, Fabrication, Devices, Circuits, and Applications of SiGe and Si Strained Layer Epitaxy*, Boca Raton, FL: CRC Press, 2006.

5

Silicon Carbide Technology

Philip G. Neudeck
NASA Glenn Research Center

5.1   Introduction

Silicon carbide (SiC)-based semiconductor electronic devices and circuits are presently being developed for use in high-temperature, high-power, and high-radiation conditions under which conventional semiconductors cannot adequately perform. Silicon carbide's ability to function under such extreme conditions is expected to enable significant improvements to a far-ranging variety of applications and systems. These range from greatly improved high-voltage switching for energy savings in public electric power distribution and electric motor drives to more powerful microwave electronics for radar and communications to sensors and controls for cleaner-burning more fuel-efficient jet aircraft and automobile engines [1–7]. In the particular area of power devices, theoretical appraisals have indicated that SiC power MOSFET's and diode rectifiers would operate over higher voltage and temperature ranges, have superior switching characteristics, and yet have die sizes nearly 20 times smaller than correspondingly rated silicon-based devices [8]. However, these tremendous theoretical advantages have yet to be widely realized in commercially available SiC devices, primarily owing to the fact that SiC's relatively immature crystal growth and device fabrication technologies are not yet sufficiently developed to the degree required for reliable incorporation into most electronic systems.

   This chapter briefly surveys the SiC semiconductor electronics technology. In particular, the differences (both good and bad) between SiC electronics technology and the well-known silicon VLSI technology are highlighted. Projected performance benefits of SiC electronics are highlighted for several large-scale applications. Key crystal growth and device-fabrication issues that presently limit the performance and capability of high-temperature and high-power SiC electronics are identified.

5.2   Fundamental SiC Material Properties

5.2.1   SiC Crystallography: Important Polytypes and Definitions

Silicon carbide occurs in many different crystal structures, called polytypes. A more comprehensive introduction to SiC crystallography and polytypism can be found in Reference 9. Despite the fact that all SiC polytypes chemically consist of 50% carbon atoms covalently bonded with 50% silicon atoms, each SiC polytype has its own distinct set of electrical semiconductor properties. While there are over 100 known polytypes of SiC, only a few are commonly grown in a reproducible form acceptable for use as an electronic semiconductor. The most common polytypes of SiC presently being developed for electronics are 3C-SiC, 4H-SiC, and 6H-SiC. The atomic crystal structure of the two most common polytypes is shown in the schematic cross section in Figure 5.1. As discussed much more thoroughly in References 9 and 10, the different polytypes of SiC are actually composed of different stacking sequences of Si–C bilayers (also called Si–C double layers), where each single Si–C bilayer is denoted by the dotted boxes in Figure 5.1. Each atom within a bilayer has three covalent chemical bonds with other atoms in the same (its own) bilayer, and only one bond to an atom in an adjacent bilayer. Figure 5.1a shows the bilayer of the stacking sequence of 4H-SiC polytype, which requires four Si–C bilayers to define the unit cell repeat distance along the c-axis stacking direction (denoted by $<0\,0\,0\,1>$ Miller indices). Similarly, the 6H-SiC polytype illustrated in Figure 5.1b repeats its stacking sequence every six bilayers throughout the crystal along the stacking direction. The $<1\,\overline{1}00>$ direction depicted in Figure 5.1 is often referred to as one of (along with $<11\overline{2}0>$) the a-axis directions. SiC is a polar semiconductor across the c-axis, in that one surface normal to the c-axis is terminated with silicon atoms while the opposite normal c-axis surface is terminated with carbon atoms. As shown in Figure 5.1a, these surfaces are typically referred to as "silicon face" and "carbon face" surfaces, respectively. Atoms along the left-or right-side edge of Figure 5.1a would reside on ($<1\,\overline{1}00>$) "a-face" crystal surface plane normal to the $<1\,\overline{1}00>$ direction. 3C-SiC, also referred to as β-SiC, is the only form of SiC with a cubic crystal lattice structure. The noncubic polytypes of SiC are sometimes ambiguously referred to as α-SiC. 4H-SiC and 6H-SiC are only two of the many

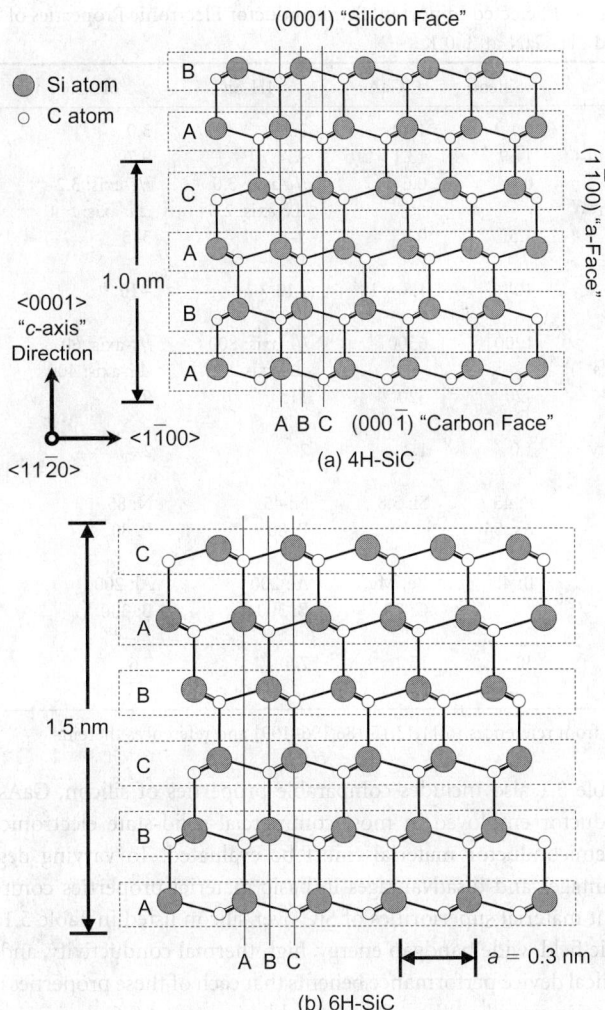

FIGURE 5.1 Schematic cross-sectional depictions of (a) 4H-SiC and (b) 6H-SiC atomic crystal structure, showing important crystallographic directions and surfaces (see text).

possible SiC polytypes with hexagonal crystal structure. Similarly, 15R-SiC is the most common of the many possible SiC polytypes with a rhombohedral crystal structure.

5.2.2 SiC Semiconductor Electrical Properties

Owing to the differing arrangement of Si and C atoms within the SiC crystal lattice, each SiC polytype exhibits unique fundamental electrical and optical properties. Some of the more important semiconductor electrical properties of the 3C, 4H, and 6H SiC polytypes are given in Table 5.1. Much more detailed electrical properties can be found in References 11–13 and references therein. Even within a given polytype, some important electrical properties are nonisotropic, in that they are strong functions of crystallographic direction of current flow and applied electric field (for example, electron mobility for 6H-SiC). Dopant impurities in SiC can incorporate into energetically inequivalent sites. While all dopant ionization energies associated with various dopant incorporation sites should normally be considered for utmost accuracy, Table 5.1 lists only the shallowest reported ionization energies of each impurity.

TABLE 5.1   Comparison of Selected Important Semiconductor Electronic Properties of Major SiC Polytypes with Silicon, GaAs, and 2H-GaN at 300 K

Property	Silicon	GaAs	4H-SiC	6H-SiC	3C-SiC	2H-GaN
Bandgap (eV)	1.1	1.42	3.2	3.0	2.3	3.4
Relative dielectric constant	11.9	13.1	9.7	9.7	9.7	9.5
Breakdown field $N_D = 10^{17}$ cm^{-3} (MVcm^{-1})	0.6	0.6	//c-axis: 3.0 ⊥c-axis: 2.5	//c-axis: 3.2 ⊥c-axis: > 1	1.8	2–3
Thermal Conductivity (W/cm-K)	1.5	0.5	3–5	3–5	3–5	1.3
Intrinsic carrier concentration (cm^{-3})	10^{10}	1.8×10^6	$\sim 10^{-7}$	$\sim 10^{-5}$	~ 10	$\sim 10^{-10}$
Electron mobility at $N_D = 10^{16}$ cm^{-3} (cm^{2}V^{-1}s^{-1})	1200	6500	//c-axis: 800 ⊥c-axis: 800	//c-axis: 60 ⊥c-axis: 400	750	900
Hole mobility at $N_A = 10^{16}$ cm^{-3} (cm^{2}V^{-1}s^{-1})	420	320	115	90	40	200
Saturated electron velocity (10^7 cms^{-1})	1.0	1.2	2	2	2.5	2.5
Donor dopants and shallowest ionization energy (meV)	P: 45 As: 54	Si: 5.8	N: 45 P: 80	N: 85 P: 80	N: 50	Si: 20
Acceptor dopants and shallowest ionization energy (meV)	B: 45	Be, Mg, C: 28	Al: 200 B: 300	Al: 200 B: 300	Al: 270	Mg: 140
2005 Commercial wafer diameter (cm)	30	15	7.6	7.6	15	None

Source: Data compiled from references [6,11,13,15,186,196–199] and references therein.

For comparison, Table 5.1 also includes comparable properties of silicon, GaAs, and GaN. Because silicon is the semiconductor employed in most commercial solid-state electronics, it is the standard against which other semiconductor materials must be evaluated. To varying degrees the major SiC polytypes exhibit advantages and disadvantages in basic material properties compared to silicon. The most beneficial inherent material superiorities of SiC over silicon listed in Table 5.1 are its exceptionally high breakdown electric field, wide bandgap energy, high thermal conductivity, and high carrier saturation velocity. The electrical device performance benefits that each of these properties enables are discussed in the next section, as are system-level benefits enabled by improved SiC devices.

5.3   Applications and Benefits of SiC Electronics

Two of the most beneficial advantages that SiC-based electronics offer are in the areas of high-temperature and high-power device operation. The specific SiC device physics that enables high-temperature and high-power capabilities will be examined first, followed by several examples of revolutionary system-level performance improvements these enhanced capabilities enable.

5.3.1   High-Temperature Device Operation

The wide bandgap energy and low intrinsic carrier concentration of SiC allow SiC to maintain semiconductor behavior at much higher temperatures than silicon, which in turn permits SiC semiconductor device functionality at much higher temperatures than silicon [7]. As discussed in basic semiconductor electronic device physics textbooks [14,15], semiconductor electronic devices function in the temperature range where intrinsic carriers are negligible so that conductivity is controlled by intentionally introduced dopant impurities. Furthermore, the intrinsic carrier concentration n_i is a fundamental prefactor to well-known equations governing undesired junction reverse-bias leakage currents. As temperature increases, intrinsic carriers increase exponentially so that undesired leakage currents grow unacceptably large, and eventually at still higher temperatures, the semiconductor device operation is overcome by uncontrolled conductivity as intrinsic carriers exceed intentional device dopings. Depending upon specific device design, the intrinsic carrier concentration of silicon generally confines silicon device operation to junction temperatures <300°C. SiC's much smaller intrinsic carrier concentration theoretically permits device operation at junction temperatures exceeding 800°C. 600°C SiC device operation has been experimentally demonstrated on a variety of SiC devices (Section 5.6.3).

The ability to place uncooled high-temperature semiconductor electronics directly into hot environments would enable important benefits to automotive, aerospace, and deep-well drilling industries [7,16]. In the case of automotive and aerospace engines, improved electronic telemetry and control from high-temperature engine regions are necessary to more precisely control the combustion process to improve fuel efficiency while reducing polluting emissions. High-temperature capability eliminates performance, reliability, and weight penalties associated with liquid cooling, fans, thermal shielding, and longer wire runs needed to realize similar functionality in engines using conventional silicon semiconductor electronics.

5.3.2 High-Power Device Operation

The high breakdown field and high thermal conductivity of SiC coupled with high operational junction temperatures theoretically permit extremely high-power densities and efficiencies to be realized in SiC devices. The high breakdown field of SiC relative to silicon enables the blocking voltage region of a power device to be roughly 10× thinner and 10× heavier doped, permitting a roughly 100-fold beneficial decrease in the blocking region resistance at the same voltage rating [8]. Significant energy losses in many silicon high-power system circuits, particularly hard-switching motor drive and power conversion circuits, arise from semiconductor switching energy loss [1,3,17]. While the physics of semiconductor device switching loss are discussed in detail elsewhere [18], switching energy loss is often a function of the turn-off time of the semiconductor switching device, generally defined as the time lapse between application of a turn-off bias and the time when the device actually cuts off most of the current flow. In general, the faster a device turns off, the smaller its energy loss in a switched power conversion circuit. For device-topology reasons discussed in References 3,8, and 19–21, SiC's high breakdown field and wide energy bandgap enable much faster power switching than is possible in comparably volt–ampere-rated silicon power-switching devices. The fact that high-voltage operation is achieved with much thinner blocking regions using SiC enables much faster switching (for comparable voltage rating) in both unipolar and bipolar power device structures. Therefore, SiC-based power converters could operate at higher switching frequencies with much greater efficiency (i.e., less switching energy loss) [1,22]. Higher switching frequency in power converters is highly desirable because it permits use of smaller capacitors, inductors, and transformers, which in turn can greatly reduce overall power converter size, weight, and cost [3,22,23].

While SiC's smaller on-resistance and faster switching helps minimize energy loss and heat generation, SiC's higher thermal conductivity enables more efficient removal of waste heat energy from the active device. Because heat energy radiation efficiency increases greatly with increasing temperature difference between the device and the cooling ambient, SiC's ability to operate at high junction temperatures permits much more efficient cooling to take place, so that heat sinks and other device-cooling hardware (i.e., fan cooling, liquid cooling, air conditioning, heat radiators, etc.) typically needed to keep high-power devices from overheating can be made much smaller or even eliminated.

While the preceding discussion focused on high-power switching for power conversion, many of the same arguments can be applied to devices used to generate and amplify RF signals used in radar and communications applications. In particular, the high breakdown voltage and high thermal conductivity coupled with high carrier saturation velocity allow SiC microwave devices to handle much higher power densities than their silicon or GaAs RF counterparts, despite SiC's disadvantage in low-field carrier mobility [5,6,24–26].

5.3.3 System Benefits of High-Power High-Temperature SiC Devices

Uncooled operation of high-temperature and high-power SiC electronics would enable revolutionary improvements to aerospace systems. Replacement of hydraulic controls and auxiliary power units with distributed "smart" electromechanical controls capable of harsh ambient operation will enable substantial jet-aircraft weight savings, reduced maintenance, reduced pollution, higher fuel efficiency, and increased operational reliability [7]. SiC high-power solid-state switches will also enable large efficiency gains in electric power management and control [1,4,27–31]. Performance gains from SiC electronics could enable the public power grid to provide increased consumer electricity demand without building additional generation plants, and improve power quality and operational reliability through "smart" power management. More efficient electric motor drives enabled by SiC will also benefit industrial production systems as well as transportation systems such as diesel-electric railroad locomotives, electric mass-transit systems, nuclear-powered ships, and electric automobiles and buses.

From the above discussions it should be apparent that SiC high-power and high-temperature solid-state electronics promise tremendous advantages that could significantly impact transportation systems and power usage on a global scale. By improving the way in which electricity is distributed and used, improving electric vehicles so that they become more viable replacements for internal combustion-engine vehicles, and improving the fuel efficiency and reducing pollution of the remaining fuel-burning engines and generation plants, SiC electronics promises the potential to better the daily lives of all citizens of planet Earth.

5.4 SiC Semiconductor Crystal Growth

As of this writing, much of the outstanding theoretical promise of SiC electronics highlighted in the previous section has largely gone unrealized. A brief historical examination quickly shows that serious shortcomings in SiC semiconductor material manufacturability and quality have greatly hindered the development of SiC semiconductor electronics. From a simple-minded point of view, SiC electronics development has very much followed the general rule of thumb that a solid-state electronic device can only be as good as the semiconductor material from which it is made.

5.4.1 Historical Lack of SiC Wafers

Reproducible wafers of reasonable consistency, size, quality, and availability are a prerequisite for commercial mass production of semiconductor electronics. Many semiconductor materials can be melted and reproducibly recrystallized into large single crystals with the aid of a seed crystal, such as in the Czochralski method employed in the manufacture of almost all silicon wafers, enabling reasonably large wafers to be mass produced. However, because SiC sublimes instead of melting at reasonably attainable pressures, SiC cannot be grown by conventional melt-growth techniques. Prior to 1980, experimental SiC electronic devices were confined to small (typically ~1 cm²), irregularly shaped SiC crystal platelets grown as a byproduct of the Acheson process for manufacturing industrial abrasives (e.g., sandpaper) [32] or by the Lely process [33]. In the Lely process, SiC sublimed from polycrystalline SiC powder at temperatures near 2500°C are randomly condensed on the walls of a cavity forming small, hexagonally shaped platelets. While these small, nonreproducible crystals permitted some basic SiC electronics research, they were clearly not suitable for semiconductor mass production. As such, silicon became the dominant semiconductor fueling the solid-state technology revolution, while interest in SiC-based microelectronics was limited.

5.4.2 Growth of 3C-SiC on Large-Area (Silicon) Substrates

Despite the absence of SiC substrates, the potential benefits of SiC hostile-environment electronics nevertheless drove modest research efforts aimed at obtaining SiC in a manufacturable wafer form. Toward this end, the heteroepitaxial growth of single-crystal SiC layers on top of large-area silicon substrates was first carried out in 1983 [34], and subsequently followed by a great many others over the years using a variety of growth techniques. Primarily owing to large differences in lattice constant (~20% difference between SiC and Si) and thermal expansion coefficient (~8% difference), heteroepitaxy of SiC using silicon as a substrate always results in growth of 3C-SiC with a very high density of crystallographic structural defects such as stacking faults, microtwins, and inversion domain boundaries [35]. Other large-area wafer materials besides silicon (such as sapphire, silicon-on-insulator, and TiC) have been employed as substrates for heteroepitaxial growth of SiC epilayers, but the resulting films have been of comparably poor quality with high crystallographic defect densities. The most promising 3C-SiC-on-silicon approach to date that has achieved the lowest crystallographic defect density involves the use of undulant silicon substrates [36]. However, even with this highly novel approach, dislocation densities remain very high compared to silicon and bulk hexagonal SiC wafers.

While some limited semiconductor electronic devices and circuits have been implemented in 3C-SiC grown on silicon, the performance of these electronics (as of this writing) can be summarized as severely limited by the high density of crystallographic defects to the degree that almost none of the operational benefits discussed in Section 5.3 has been viably realized. Among other problems, the crystal defects "leak" parasitic current across reverse-biased device junctions where current flow is not desired. Because excessive crystal defects lead to electrical device shortcomings, there are as yet no commercial electronics manufactured in 3C-SiC grown on large-area substrates. Thus, 3C-SiC grown on silicon presently has more potential as a mechanical material in microelectromechanical systems (MEMS) applications (Section 5.6.5) instead of being used purely as a semiconductor in traditional solid-state transistor electronics.

5.4.3 Growth of Hexagonal Polytype SiC Wafers

In the late 1970s, Tairov and Tzvetkov established the basic principles of a modified seeded sublimation growth process for growth of 6H-SiC [37,38]. This process, also referred to as the modified Lely process, was a breakthrough for SiC in that it offered the first possibility of reproducibly growing acceptably large single crystals of SiC that could be cut and polished into mass-produced SiC wafers. The basic growth process is based on heating polycrystalline SiC source material to ~2400°C under conditions, where it sublimes into the vapor phase and subsequently condenses onto a cooler SiC seed crystal [10,37–39]. This produces a somewhat cylindrical boule of single-crystal SiC that grows taller roughly at the rate of a few millimeters per hour. To date, the preferred orientation of the growth in the sublimation process is such that vertical growth of a taller cylindrical boule proceeds along the <0 0 0 1> crystallographic c-axis direction (i.e., vertical direction in Fig. 5.1). Circular "c-axis" wafers with surfaces that lie normal (i.e., perpendicular to within 10°) to the c-axis can be sawed from the roughly cylindrical boule. After years of further development of the sublimation growth process, Cree, Inc., became the first company [40] to sell 2.5 cm diameter semiconductor wafers of c-axis-oriented 6H-SiC in 1989. Correspondingly, the vast majority of SiC semiconductor electronics development and commercialization has taken place since 1990 using c-axis-oriented SiC wafers of the 6H and 4H-SiC polytypes. N-type, p-type, and semi-insulating SiC wafers of various sizes (presently as large as 7.6 cm in diameter) are now commercially available from a variety of vendors [10,39,41]. It is worth noting that attainable substrate conductivities for p-type SiC wafers are more than 10× smaller than for n-type substrates, which is largely due to the difference between donor and acceptor dopant ionization energies in SiC (Table 5.1). More recently, SiC wafers grown with gas sources instead of sublimation of solid sources or a combination of gas and solid sources have also been commercialized [42,43]. Growth of SiC boules and wafers oriented along other crystallographic directions, such as <11$\bar{2}$0> and <1$\bar{1}$00> "a-face" orientations, have also been investigated over the last decade [44]. While these other SiC wafer orientations offer some interesting differences in device properties compared to conventional c-axis-oriented wafers (mentioned briefly in Section 5.5.5), all commercial SiC electronic parts produced (as of this writing) are manufactured using c-axis-oriented wafers.

Wafer size, cost, and quality are all very critical to the manufacturability and process yield of mass-produced semiconductor microelectronics. Compared to commonplace silicon wafer standards, present-day 4H- and 6H-SiC wafers are smaller, more expensive, and generally of inferior quality containing far more crystal imperfections (see Section 5.4.5 below). This disparity is not surprising considering that silicon wafers have undergone nearly five decades of commercial process refinement.

5.4.4   SiC Epilayers

Most SiC electronic devices are not fabricated directly in sublimation-grown wafers, but are instead fabricated in much higher quality epitaxial SiC layers that are grown on top of the initial sublimation-grown wafer. Well-grown SiC epilayers have superior electrical properties and are more controllable and reproducible than bulk sublimation-grown SiC wafer material. Therefore, the controlled growth of high-quality epilayers is highly important in the realization of useful SiC electronics.

5.4.4.1   SiC Epitaxial Growth Processes

An interesting variety of SiC epitaxial growth methodologies, ranging from liquid-phase epitaxy, molecular beam epitaxy, and chemical vapor deposition(CVD) have been investigated [10,45]. The CVD growth technique is generally accepted as the most promising method for attaining epilayer reproducibility, quality, and throughputs required for mass production. In the simplest terms, variations of SiC CVD are carried out by heating SiC substrates in a chamber "reactor" with flowing silicon- and carbon-containing gases that decompose and deposit Si and C onto the wafer allowing an epilayer to grow in a well-ordered single-crystal fashion under well-controlled conditions. Conventional SiC CVD epitaxial growth processes are carried out at substrate growth temperatures between 1400°C and 1600°C at pressures from 0.1 to 1 atm resulting in growth rates of the order of a few micrometers per hour [10,41,46]. Higher temperature (up to 2000°C) SiC CVD growth processes, some using halide-based growth chemistries, are also being pioneered to obtain higher SiC epilayer growth rates of the order of hundreds of micrometers per hour that appear sufficient for growth of bulk SiC boules in addition to very thick epitaxial layers needed for high-voltage devices [42,47,48].

Despite the fact that SiC growth temperatures significantly exceed epitaxial growth temperatures used for most other semiconductors, a variety of SiC CVD epitaxial growth reactor configurations have been developed and commercialized [41,46,49]. For example, some reactors employ horizontal reactant gas flow across the SiC wafer, while others rely on vertical flow of reactant gases; some reactors have wafers surrounded by heated "hot-wall" or "warm-wall" configurations, while other "cold-wall" reactors heat only a susceptor residing directly beneath the SiC wafer. Most reactors used for commercial production of SiC electronics rotate the sample to ensure high uniformity of epilayer parameters across the wafer. SiC CVD systems capable of simultaneously growing epilayers on multiple wafers have enabled higher wafer throughput for SiC electronic device manufacture.

5.4.4.2   SiC Epitaxial Growth Polytype Control

Homoepitaxial growth, whereby the polytype of the SiC epilayer matches the polytype of the SiC substrate, is accomplished by "step-controlled" epitaxy [50–52]. Step-controlled epitaxy is based upon growing epilayers on an SiC wafer polished at an angle (called the "tilt-angle" or "off-axis angle") of typically 3°–8° off the $(0\,0\,0\,1)$ basal plane, resulting in a surface with atomic steps and relatively long, flat terraces between steps. When growth conditions are properly controlled and there is a sufficiently short distance between steps, Si and C adatoms impinging onto the growth surface find their way to step risers, where they bond and incorporate into the crystal. Thus, ordered, lateral "step-flow" growth takes place which enables the polytypic stacking sequence of the substrate to be exactly mirrored in the growing epilayer. SiC wafers cut with nonconventional surface orientations such as $(11\bar{2}0)$ and $(03\bar{3}8)$, provide a favorable surface geometry for epilayers to inherit stacking sequence (i.e., polytype) via step flow from the substrate [53,54].

When growth conditions are not properly controlled when steps are too far apart, as can occur with poorly prepared SiC substrate surfaces that are polished to within <1° of the $(0\,0\,0\,1)$ basal plane, growth adatoms island nucleate and bond in the middle of terraces instead of at the steps. Uncontrolled island nucleation (also referred to as terrace nucleation) on SiC surfaces leads to heteroepitaxial growth of poor-quality 3C-SiC [51,52]. To help prevent spurious terrace nucleation of 3C-SiC during epitaxial growth, most commercial 4H- and 6H-SiC substrates are polished to tilt angles of 8° and 3.5° off the (0 0 0 1) basal plane, respectively. To date, all commercial SiC electronics rely on homoepitaxial layers that are grown on these "off-axis" prepared (0 0 0 1) *c*-axis SiC wafers.

Proper removal of residual surface contamination and defects left over from the SiC wafer cutting and polishing process is also vital to obtaining high-quality SiC epilayers with minimal dislocation defects. Techniques employed to better prepare the SiC wafer surface prior to epitaxial growth range from dry etching to chemical-mechanical polishing (CMP) [55]. As the SiC wafer is heated up in a growth chamber in preparation for initiation of epilayer growth, a high-temperature in-situ pregrowth gaseous etch (typically using H_2 and/or HCl) is usually carried out to further eliminate surface contamination and defects [46,56,57]. It is worth noting that optimized pregrowth processing enables step-flow growth of high-quality homoepilayers even when the substrate tilt angle is reduced to <0.1° off-axis from the (0 0 0 1) basal plane [56]. In this case, axial screw dislocations are required to provide a continual spiral template of steps needed to grow epilayers in the <0 0 0 1> direction while maintaining the hexagonal polytype of the substrate [58].

5.4.4.3 SiC Epilayer Doping

In-situ doping during CVD epitaxial growth is primarily accomplished through the introduction of nitrogen (usually N_2) for n-type and aluminum (usually trimethyl- or triethylaluminum) for p-type epilayers [10,59]. Some alternative dopants such as phosphorus and boron have also been investigated for the n-and p-type epilayers, respectively [59,60]. While some variation in epilayer doping can be carried out strictly by varying the flow of dopant gases, the site-competition doping methodology has enabled a much broader range of SiC doping to be accomplished [59,61]. In addition, site competition has also made moderate epilayer dopings more reliable and repeatable. The site-competition dopant-control technique is based on the fact that many dopants of SiC preferentially incorporate into either Si lattice sites or C lattice sites. As an example, nitrogen preferentially incorporates into lattice sites normally occupied by carbon atoms. By epitaxially growing SiC under carbon-rich conditions, most of the nitrogen present in the CVD system (whether it is a residual contaminant or intentionally introduced) can be excluded from incorporating into the growing SiC crystal. Conversely, by growing in a carbon-deficient environment, the incorporation of nitrogen can be enhanced to form very heavily doped epilayers for ohmic contacts. Aluminum, which is opposite to nitrogen, prefers the Si site of SiC, and other dopants have also been controlled through site competition by properly varying the Si/C ratio during crystal growth. SiC epilayer dopings ranging from 9×10^{14} to 1×10^{19} cm^{-3} are commercially available, and researchers have reported obtaining dopings over a factor of 10 larger and smaller than this range for the n- and p-type dopings [40]. The surface orientation of the wafer also affects the efficiency of doping incorporation during epilayer growth [54]. As of this writing, epilayers available for consumers to specify and purchase to meet their own device application needs have thickness and doping tolerances of ±25% and ±50%, respectively [40]. However, some SiC epilayers used for high-volume device production are far more optimized, exhibiting <5% variation in doping and thickness [41].

5.4.5 SiC Crystal Dislocation Defects

Table 5.2 summarizes the major known dislocation defects found in present-day commercial 4H- and 6H-SiC wafers and epilayers [10,39,41,62,63]. Since the active regions of devices reside in epilayers, the epilayer defect content is clearly of primary importance to SiC device performance. However, as evidenced by Table 5.2, most epilayer defects originate from dislocations found in the underlying SiC substrate prior to epilayer deposition. More details on the electrical impact of some of these defects on specific devices are discussed later in Section 5.6.

The micropipe defect is regarded as the most obvious and damaging "device-killer" defect to SiC electronic devices [64]. A micropipe is an axial screw dislocation with a hollow core (diameter of the order of a micrometer) in the SiC wafer and epilayer that extends roughly parallel to the crystallographic *c*-axis normal to the polished *c*-axis wafer surface [65–67]. These defects impart considerable local strain to the surrounding SiC crystal that can be observed using X-ray topography or optical cross polarizers [39,41,68,69]. Over the course of a decade, substantial efforts by SiC material vendors has succeeded in reducing SiC wafer micropipe densities nearly 100-fold, and some SiC boules completely free of micropipes have been demonstrated [10,41,70]. In addition, epitaxial growth techniques for closing SiC substrate micropipes (effectively dissociating the hollow-core axial dislocation into multiple closed-core dislocations) have been developed [71]. However, this approach has not yet met the demanding electronic reliability requirements for commercial SiC power devices that operate at high electric fields [72].

Even though micropipe "device-killer" defects have been almost eliminated, commercial 4H- and 6H-SiC wafers and epilayers still contain very high densities (>10,000 cm^{-2}, summarized in Table 5.2) of other less-harmful dislocation defects. While these remaining dislocations are not presently specified in SiC material vendor specification sheets, they are nevertheless believed responsible for a variety of nonideal device behaviors that have hindered reproducibility and commercialization of some (particularly high electric field) SiC electronic devices [63,73,74]. Closed-core axial screw dislocation defects are similar in structure and strain properties to micropipes, except that their Burgers vectors are smaller so that the core is solid instead of a hollow void [66,67,75,76]. As shown in Table 5.2, basal plane dislocation defects and threading edge dislocation defects are also plentiful in commercial SiC wafers [39,62].

As discussed later in Section 5.6.4.1.2, 4H-SiC electrical device degradation caused by the expansion of stacking faults initiated from basal plane dislocation defects has hindered commercialization of bipolar power devices [63,74,77,78]. Similar stacking fault expansion has also been reported when doped 4H-SiC epilayers have been subjected to modest (~1150°C) thermal oxidation processing [79,80]. While epitaxial growth techniques to convert basal-plane dislocations into threading-edge dislocations have recently been reported, the electrical impact of threading-edge dislocations on the performance and reliability of high-electric field SiC devices remains to be fully ascertained [39]. It is also important to note that present-day commercial SiC epilayers still contain some undesirable surface morphological features such as "carrot defects" which could affect SiC device processing and performance [40,81,82].

In an exciting initial breakthrough, a Japanese team of researchers reported in 2004 that they achieved a 100-fold reduction in dislocation density in prototype 4H-SiC wafers of up to 3 in. in diameter [83]. While such greatly improved SiC wafer quality offered by this "multiple *a*-face" growth technique should prove highly beneficial to electronic (especially high-power) SiC device capabilities, it remains uncertain as of this writing as to when this significantly more complex (and therefore expensive) growth process will result in commercially viable mass-produced SiC wafers and devices.

5.5   SiC Device Fundamentals

To minimize the development and production costs of SiC electronics, it is important that SiC device fabrication takes advantage of existing silicon and GaAs wafer processing infrastructure as much as possible. As will be discussed in this section, most of the steps necessary to fabricate SiC electronics starting from SiC wafers can be accomplished using somewhat modified commercial silicon electronics processes and fabrication tools.

5.5.1   Choice of Polytype for Devices

As discussed in Section 4, 4H- and 6H-SiC are the far superior forms of semiconductor device quality SiC commercially available in mass-produced wafer form. Therefore, only 4H- and 6H-SiC device processing methods will be explicitly considered in the rest of this section. It should be noted, however, that most of the processing methods discussed in this section are applicable to other polytypes of SiC, except for the case of a 3C-SiC layer still residing on a silicon substrate, where all processing temperatures need to be kept well below the melting temperature of silicon (~1400°C). It is generally accepted that 4H-SiC's substantially higher carrier mobility and shallower dopant ionization energies compared to 6H-SiC (Table 5.1) should make it the polytype of choice for most SiC electronic devices, provided that all other device processing, performance, and cost-related issues play out as being roughly equal between the two polytypes. Further-more, the inherent mobility anisotropy that degrades conduction parallel to the crystallographic *c*-axis in 6H-SiC particularly favors 4H-SiC for vertical power device configurations (Section the 5.6.4). Because the ionization energy of the p-type acceptor dopants is significantly deeper than for the n-type donors, a much higher conductivity can be obtained for the n-type SiC substrates than for the p-type substrates.

5.5.2   SiC-Selective Doping: Ion Implantation

The fact that diffusion coefficients of most SiC dopants are negligibly small (at ≤1800°C) is excellent for maintaining device junction stability, because dopants do not undesirably diffuse as the device is operated long term at high temperatures. Unfortunately, this characteristic also largely (except for B at extreme temperatures [84]) precludes the use of conventional dopant diffusion, a highly useful technique widely employed in silicon microelectronics manufacturing, for patterned doping of SiC.

Laterally patterned doping of SiC is carried out by ion implantation. This somewhat restricts the depth that most dopants can be conventionally implanted to <1 μm using conventional dopants and implan-tation equipment. Compared to silicon processes, SiC ion implantation requires a much higher thermal budget to achieve acceptable dopant implant electrical activation. Summaries of ion implantation pro-cesses for various dopants can be found in [85–96]. Most of these processes are based on carrying out implantation at temperatures ranging from room temperature to 800°C using a patterned (sometimes high-temperature) masking material. The elevated temperature during implantation promotes some lattice self-healing during the implant, so that damage and segregation of displaced silicon and carbon atoms does not become excessive, especially in high-dose implants often employed for ohmic contact formation. Co-implantation of carbon with dopants has been investigated as a means to improve the electrical conductivity of the more heavily doped implanted layers [88,95,97].

Following implantation, the patterning mask is stripped and a higher temperature (~1200 to 1800°C) anneal is carried out to achieve maximum electrical activation of dopant ions. The final annealing conditions are crucial to obtaining desired electrical properties from ion-implanted layers. At higher implant anneal temperature, the SiC surface morphology can seriously degrade [87,98]. Because subli-mation etching is driven primarily by loss of silicon from the crystal surface, annealing in silicon overpressures can be used to reduce surface degradation during high-temperature anneals [99]. Such overpressure can be achieved by close-proximity solid sources such as using an enclosed SiC crucible with SiC lid and/or SiC powder near the wafer, or by annealing in a silane-containing atmosphere. Similarly, robust deposited capping layers such as AlN and graphite, have also proven effective at better preserving SiC surface morphology during high-temperature ion implantation annealing [91,92].

As evidenced by a number of works, the electrical properties and defect structure of 4H-SiC doped by ion implantation and annealing are generally inferior to SiC doped in-situ during epitaxial growth [89,100–103]. Naturally, the damage imposed on the SiC lattice roughly scales with implantation dose. Even though reasonable electrical dopant activations have been achieved, thermal annealing pro-cesses developed to date for SiC have not been able to thoroughly repair all damage imposed on the crystal lattice by higher-dose ion implantations (such as those often used to form heavily doped layers in preparation of ohmic contact formation, Section 5.5.3). The degraded crystal quality of highly implanted SiC layers has been observed to degrade carrier mobilities and minority carrier lifetimes, thereby causing significant degradation to the electrical performance of some devices [90,103]. Until large further improvements to ion-implanted doping of SiC are developed, SiC device designs will have to account for nonideal behavior associated with SiC-implanted layers.

5.5.3   SiC Contacts and Interconnect

All useful semiconductor electronics require conductive signal paths in and out of each device as well as conductive interconnects to carry signals between devices on the same chip and to external circuit elements that reside off-chip. While SiC itself is theoretically capable of fantastic electrical operation under extreme conditions (Section 5.3), such functionality is useless without contacts and interconnects that are also capable of operation under the same conditions. The durability and reliability of metal–semiconductor contacts and interconnects are one of the main factors limiting the operational high-temperature limits of SiC electronics. Similarly, SiC high-power device contacts and metallizations will have to withstand both high temperature and high current density stress never before encountered in silicon power electronics experience.

The subject of metal–semiconductor contact formation is a very important technical field too broad to be discussed in great detail here. For general background discussions on metal–semiconductor contact physics and formation, the reader should consult narratives presented in References 15 and 104. These references primarily discuss ohmic contacts to conventional narrow-bandgap semiconductors such as silicon and GaAs. Specific overviews of SiC metal–semiconductor contact technology can be found in References 105–110.

As discussed in References 105–110, there are both similarities and a few differences between SiC contacts and contacts to conventional narrow-bandgap semiconductors (e.g., silicon, GaAs). The same basic physics and current transport mechanisms that are present in narrow-bandgap contacts such as surface states, Fermi-pinning, thermionic emission, and tunneling, also apply to SiC contacts. A natural consequence of the wider bandgap of SiC is the higher effective Schottky barrier heights. Analogous with narrow-bandgap ohmic contact physics, the microstructural and chemical state of the SiC–metal interface is crucial to contact electrical properties. Therefore, premetal-deposition surface preparation, metal deposition process, choice of metal, and post-deposition annealing can all greatly impact the resulting performance of metal–SiC contacts. Because the chemical nature of the starting SiC surface is strongly dependent on surface polarity, it is not uncommon to obtain significantly different results when the same contact process is applied to the silicon face surface versus the carbon face surface.

5.5.3.1 SiC Ohmic Contacts

Ohmic contacts serve the purpose of carrying electrical current into and out of the semiconductor, ideally with no parasitic resistance. The properties of various ohmic contacts to SiC reported to date are summarized elsewhere [107–110]. While SiC-specific ohmic contact resistances at room temperature are generally higher than in contacts to narrow-bandgap semiconductors, they are nevertheless sufficiently low for most envisioned SiC applications. Lower specific contact resistances are usually obtained to n-type than to p-type 4H- and 6H-SiC. Consistent with narrow-bandgap ohmic contact technology, it is easier to make low-resistance ohmic contacts to heavily doped SiC and thermal annealing is almost always employed to promote favorable interfacial reactions.

Truly enabling harsh-environment SiC electronics will require ohmic contacts that can reliably withstand prolonged harsh-environment operation. Most reported SiC ohmic metallizations appear sufficient for long-term device operation up to 300°C. SiC ohmic contacts that withstand heat soaking under no electrical bias at 500–600°C for hundreds or thousands of hours in nonoxidizing gas or vacuum environments have also been demonstrated [110]. Only recently has successful long-term electrical operation of n-type ohmic contacts in oxidizing 500–600°C air ambients been demonstrated in relatively low-current density devices [111,112]. Further research is needed to obtain similarly durable high-temperature contacts to p-type SiC. Electromigration, oxidation, and other electrochemical reactions driven by high-temperature electrical bias in a reactive oxidizing environment are likely to limit SiC ohmic contact reliability for the most demanding applications that simultaneously require both high temperature and high power (i.e., high current density). The durability and reliability of SiC ohmic contacts is one of the critical factors limiting the practical high-temperature limits of SiC electronics.

5.5.3.2 SiC Schottky Contacts

Rectifying metal–semiconductor Schottky barrier contacts to SiC are useful for a number of devices, including commercialized SiC metal–semiconductor field-effect transistors (MESFETs) and fast-switching rectifiers [40,113]. References 105, 106, 108, and 114 summarize electrical results obtained in a variety of SiC Schottky studies. Owing to the wide bandgap of SiC, almost all unannealed metal contacts to lightly doped 4H- and 6H-SiC are rectifying. Rectifying contacts permit extraction of Schottky barrier heights and diode ideality factors by well-known current–voltage (*I–V*) and capacitance–voltage (*C–V*) electrical measurement techniques [104]. While these measurements show a general trend that Schottky junction barrier height does somewhat depend on metal–semiconductor work function difference, the dependence is weak enough to suggest that surface state charge also plays a significant role in determining the effective barrier height of SiC Schottky junctions. At least some experimental scatter exhibited for identical metals can be attributed to surface cleaning and metal deposition process differences, as well as different barrier height measurement procedures. For example, the work by Teraji et al. [115], in which two different surface-cleaning procedures prior to titanium deposition lead to ohmic behavior in one case and rectifying behavior in the other, clearly shows that the process plays a significant role in determining SiC Schottky contact electrical properties.

It is important to note that nonuniformities in electrical behavior, many of which have been traced to SiC crystal defects (Section 5.4.5) have been documented to exist across the lateral area of most SiC Schottky contacts with areas $>10^{-4}$ cm^2 [116–118]. Furthermore, the reverse current drawn in experimental SiC diodes, while small, is nevertheless larger than expected based on theoretical substitution of SiC parameters into well-known Schottky diode reverse leakage current equations developed for narrow-bandgap semiconductors. Models based on spatially localized Schottky barrier lowering as well as quantum mechanical tunneling owing to higher SiC electric fields, have been proposed to explain the nonideal reverse leakage behavior of SiC Schottky diodes [119–121]. In addition, electric field crowding along the edge of a SiC Schottky contact can also lead to increased reverse-bias leakage current and reduced reverse breakdown voltage [15,18,104]. Edge-termination techniques to relieve electric field edge crowding and improve Schottky rectifier reverse properties are briefly discussed later in Section 5.6.4. The practical operation of rectifying SiC Schottky diodes is usually limited to temperatures below 400°C; above this temperature, reverse-bias thermionic emission leakage current and thermally driven SiC–metal interface degradation (via material intermixing and chemical reactions) tend to become undesirably large.

5.5.4 Patterned Etching of SiC for Device Fabrication

At room temperature, there are no known conventional wet chemicals that etch single-crystal SiC. Most patterned etching of SiC for electronic devices and circuits is accomplished using dry etching techniques. The reader should consult References 122–124 which contain summaries of dry SiC etching results obtained to date. The most commonly employed process involves reactive ion etching (RIE) of SiC in fluorinated plasmas. Sacrificial etch masks (such as aluminum metal) are deposited and photolithographically patterned to protect desired areas from being etched. The SiC RIE process can be implemented using standard silicon RIE hardware and typical 4H- and 6H-SiC RIE etch rates of the order of hundreds of angstroms per minute. Well-optimized SiC RIE processes are typically highly anisotropic with little undercutting of the etch mask, leaving smooth surfaces. One of the keys to achieving smooth surfaces is preventing "micromasking", wherein the masking material is slightly etched and randomly redeposited onto the sample effectively masking very small areas on the sample that were intended for uniform etching. This can result in "grass"-like etch-residue features being formed in the unmasked regions, which is undesirable in most cases.

While RIE etch rates are sufficient for many electronic applications, much higher SiC etch rates are necessary to carve features of the order of tens to hundreds of micrometers deep that are needed to realize advanced sensors, MEMS, and through-wafer holes useful for SiC RF devices. High-density plasma dry-etching techniques such as electron cyclotron resonance and inductively coupled plasma have been developed to meet the need for deep etching of SiC. Residue-free patterned etch rates exceeding a thousand angstroms a minute have been demonstrated [122,123,125–128].

Patterned etching of SiC at very high etch rates has also been demonstrated using photo-assisted and dark electrochemical wet etching [129,130]. By choosing proper etching conditions, this technique has demonstrated a very useful dopant-selective etch-stop capability. However, there are major incompatibilities of the electrochemical process that make it undesirable for VLSI mass production, including extensive preetching and postetching sample preparation, etch isotropy and mask undercutting, and somewhat nonuniform etching across the sample. Laser etching techniques are capable of etching large features, such as via through-wafer holes useful for RF chips [131].

5.5.5   SiC Insulators: Thermal Oxides and MOS Technology

The vast majority of semiconductor-integrated circuit chips in use today rely on silicon metal-oxide–semiconductor field-effect transistors (MOSFETs), whose electronic advantages and operational device physics are summarized in Katsumata's chapter and elsewhere [15,18,132]. Given the extreme usefulness and success of inversion channel MOSFET-based electronics in VLSI silicon (as well as discrete silicon power devices), it is naturally desirable to implement high-performance inversion channel MOSFETs in SiC. Like silicon, SiC forms a thermal SiO_2 when it is sufficiently heated in an oxygen environment. While this enables SiC MOS technology to somewhat follow the highly successful path of silicon MOS technology, there are nevertheless important differences in insulator quality and device processing that are presently preventing SiC MOSFETs from realizing their full beneficial potential. While the following discourse attempts to quickly highlight key issues facing SiC MOSFET development, more detailed insights can be found in References 133–142.

From a purely electrical point of view, there are two prime operational deficiencies of SiC oxides and MOSFETs compared to silicon MOSFETs. First, effective inversion channel mobilities in most SiC MOS-FETs are lower than one would expect based on silicon inversion channel MOSFET carrier mobilities. This seriously reduces the transistor gain and current-carrying capability of SiC MOSFETs, so that SiC MOSFETs are not nearly as advantageous as theoretically predicted. Second, SiC oxides have not proven as reliable and immutable as well-developed silicon oxides, in that SiC MOSFETs are more prone to threshold voltage shifts, gate leakage, and oxide failures than comparably biased silicon MOSFETs. In particular, SiC MOSFET oxide electrical performance deficiencies are attributed to differences between silicon and SiC thermal oxide quality and interface structure that cause the SiC oxide to exhibit undesirably higher levels of interface state densities ($\sim 10^{11}$–10^{13} eV^{-1} cm^{-2}), fixed oxide charges ($\sim 10^{11}$–10^{12} cm^{-2}), charge trapping, carrier oxide tunneling, and lowered mobility of inversion channel carriers.

In highlighting the difficulties facing SiC MOSFET development, it is important to keep in mind that early silicon MOSFETs also faced developmental challenges that took many years of dedicated research efforts to successfully overcome. Indeed, tremendous improvements in 4H-SiC MOS device performance have been achieved in recent years, giving hope that beneficial 4H-SiC power MOSFET devices for operation up to 125°C ambient temperatures might become commercialized within the next few years. For example, 4H-SiC MOSFET inversion channel mobility for conventionally oriented (8° off (0001) c-axis) wafers has improved from <10 to >200 cm^2/V^{-1}s^{-1}, while the density of electrically detrimental SiC–SiO$_2$ interface state defects energetically residing close to the conduction band edge has dropped by an order of magnitude [141,143,144]. Likewise, alternative SiC wafer surface orientations such as (11$\bar{2}$0) and (03$\bar{3}$8) that are obtained by making devices on wafers cut with different crystallographic orientations (Section 5.2.1), have also yielded significantly improved 4H-SiC MOS channel properties [54,139]. One key step to obtaining greatly improved 4H-SiC MOS devices has been the proper introduction of nitrogen-compound gases (in the form of N_2, NO, N_2O, or NH_3) during the oxidation and post-oxidation annealing process [136,137,141,142]. These nitrogen-based anneals have also improved the stability of 4H-SiC oxides to high electric field and high-temperature stressing used to qualify and quantify the reliability of MOSFETs [140]. However, as Agarwal et al. [145] have pointed out, the wide bandgap of SiC reduces the potential barrier impeding tunneling of damaging carriers through oxides grown on 4H-SiC, so that 4H-SiC oxides cannot be expected to attain identical high reliability as thermal oxides on silicon. It is highly probable that alternative gate insulators besides thermally grown SiO_2 will have to be developed for optimized implementation of inversion channel 4H-SiC insulated gate transistors for the most demanding high-temperature and high-power electronic applications. As with silicon MOSFET technology, multilayer dielectric stacks will likely be developed to further enhance SiC MOSFET performance [133,146].

5.5.6 SiC Device Packaging and System Considerations

Hostile-environment SiC semiconductor devices and ICs are of little advantage if they cannot be reliably packaged and connected to form a complete system capable of hostile-environment operation. With proper material selection, modifications of existing IC packaging technologies appear feasible for non-power SiC circuit packaging up to 300°C [147]. Recent work is beginning to address the needs of the most demanding aerospace electronic applications, whose requirements include operation in high-vibration 500–600°C oxidizing-ambient environments, sometimes with very high power [7,148–152]. For example, some prototype electronic packages and circuit boards that can withstand over a thousand hours at 500°C have been demonstrated. Harsh-environment passive components such as inductors, capacitors, and transformers, must also be developed for operation in demanding conditions before the full system-level benefits of SiC electronics discussed in Section 5.3 can be successfully realized.

5.6 SiC Electronic Devices and Circuits

This section briefly summarizes a variety of SiC electronic device designs broken down by major application areas. SiC process and material technology issues limiting the capabilities of various SiC device topologies are highlighted as key issues to be addressed in further SiC technology maturation. Throughout this section, it should become apparent to the reader that the most difficult general challenge preventing SiC electronics from fully attaining beneficial capabilities is attaining long-term high operational reliability, while operating in previously unattained temperature and power density regimes. Because many device reliability limitations can be traced to fundamental material and junction/interface issues already mentioned in Sections 5.4 and 5.5, efforts to enable useful (i.e., reliable) SiC electronics should focus on improvements to these fundamental areas.

5.6.1 SiC Optoelectronic Devices

The wide bandgap of SiC is useful for realizing short-wavelength blue and ultraviolet (UV) optoelectronics. 6H-SiC-based pn junction light-emitting diodes (LEDs) were the first semiconductor devices to cover the blue portion of the visible color spectrum, and became the first SiC-based devices to reach high-volume commercial sales [153]. Because SiC's bandgap is indirect (i.e., the conduction minimum and valence band maximum do not coincide in crystal momentum space), luminescent recombination is inherently inefficient [154]. Therefore, LEDs based on SiC pn junctions were rendered quite obsolete by the emergence of much brighter, much more efficient direct-bandgap Group III-nitride (III-N such as GaN, and InGaN) blue LEDs [155]. However, SiC wafers are still employed as one of the substrates (along with sapphire) for growth of III-N layers used in high-volume manufacture of green and blue nitride-based LEDs.

SiC has proven much more efficient at absorbing short-wavelength light, which has enabled the realization of SiC UV-sensitive photodiodes that serve as excellent flame sensors in turbine-engine combustion monitoring and control [153,156]. The wide bandgap of 6H-SiC is useful for realizing low photodiode dark currents as well as sensors that are blind to undesired near-infrared wavelengths produced by heat and solar radiation. Commercial SiC-based UV flame sensors, again based on epitaxially grown dry-etch mesa-isolated 6H-SiC pn junction diodes, have successfully reduced harmful pollution emissions from gas-fired ground-based turbines used in electrical power generation systems [156]. The low dark-currents of SiC diodes are also useful for X-ray, heavy ion, and neutron detection in nuclear reactor monitoring and enhanced scientific studies of high-energy particle collisions and cosmic radiation [157,158].

5.6.2 SiC RF Devices

The main use of SiC RF devices appears to lie in high-frequency solid-state high-power amplification at frequencies from around 600 MHz (UHF-band) to perhaps as high as a few gigahertz (X-band). As discussed in far greater detail in References 5, 6, 25, 26, 159, and elsewhere, the high breakdown voltage and high thermal conductivity coupled with high carrier saturation velocity allow SiC RF transistors to handle much higher power densities than their silicon or GaAs RF counterparts, despite SiC's disadvantage in low-field carrier mobility (Table 5.1). The higher thermal conductivity of SiC is also crucial in minimizing channel self-heating so that phonon scattering does not seriously degrade carrier velocity. These material advantage RF power arguments apply to a variety of different transistor structures such as MESFETs and static induction transistors (SITs) and other wide bandgap semiconductors (such as Group III-nitrides) besides SiC. The high power density of wide bandgap transistors will prove quite useful in realizing solid-state transmitter applications, where higher power with smaller size and mass are crucial. Fewer transistors capable of operating at higher temperatures reduce matching and cooling requirements, leading to reduced overall size and cost of these systems.

SiC-based high-frequency RF MESFETs are now commercially available [40]. However, it is important to note that this occurred after years of fundamental research tracked down and eliminated poor reliability owing to charge-trapping effects arising from immature semi-insulating substrates, device epilayers, and surface passivation [159]. One key material advancement that enabled reliable operation was the development of "high-purity" semi-insulating SiC substrates (needed to minimize parasitic device capacitances) with far less charge trapping induced than the previously developed vanadium-doped semi-insulating SiC wafers. SiC MESFET devices fabricated on semi-insulating substrates are conceivably less susceptible to adverse yield consequences arising from micropipes than vertical high-power switching devices, primarily because a c-axis micropipe can no longer short together two conducting sides of a high field junction in most areas of the lateral channel MESFET structure.

SiC mixer diodes also show excellent promise for reducing undesired intermodulation interference in RF receivers [160–162]. More than 20 dB dynamic range improvement was demonstrated using non-optimized SiC Schottky diode mixers. Following further development and optimization, SiC-based mixers should improve the interference immunity in situations (such as in aircraft or ships) where receivers and high-power transmitters are closely located.

5.6.3 SiC High-Temperature Signal-Level Devices

Most analog signal conditioning and digital logic circuits are considered "signal level" in that individual transistors in these circuits do not typically require any more than a few milliamperes of current and <20 V to function properly. Commercially available silicon-on-insulator circuits can perform complex digital and analog signal-level functions up to 300°C when high-power output is not required [163]. Besides ICs in which it is advantageous to combine signal-level functions with high-power or unique SiC sensors/MEMS onto a single chip, more expensive SiC circuits solely performing low-power signal-level functions appear largely unjustifiable for low-radiation applications at temperatures below 250–300°C [7].

As of this writing, there are no commercially available semiconductor transistors or integrated circuits (SiC or otherwise) for use in ambient temperatures above 300°C. Even though SiC-based high-temperature laboratory prototypes have improved significantly over the last decade, achieving long-term operational reliability remains the primary challenge of realizing useful 300–600°C devices and circuits. Circuit technologies that have been used to successfully implement VLSI circuits in silicon and GaAs such as CMOS, ECL, BiCMOS, DCFL, etc., are to varying degrees candidates for $T > 300$°C SiC-integrated circuits. High-temperature gate-insulator reliability (Section 5.5.5) is critical to the successful realization of MOSFET-based integrated circuits. Gate-to-channel Schottky diode leakage limits the peak operating temperature of SiC MESFET circuits to around 400°C (Section 5.5.3.2). Therefore, pn junction-based devices such as bipolar junction transistors (BJTs) and junction field effect transistors (JFETs), appear to be stronger (at least in the nearer term) candidate technologies to attain long-duration operation in

300–600°C ambients. Because signal-level circuits are operated at relatively low electric fields well below the electrical failure voltage of most dislocations, micropipes and other SiC dislocations affect signal-level circuit process yields to a much lesser degree than they affect high-field power device yields.

As of this writing, some discrete transistors and small-scale prototype logic and analog amplifier SiC-based ICs have been demonstrated in the laboratory using SiC variations of NMOS, CMOS, JFET, and MESFET device topologies [164–170]. However, none of these prototypes are commercially viable as of this writing, largely owing to their inability to offer prolonged-duration electrically stable operation at ambient temperatures beyond the ~250–300°C realm of silicon-on-insulator technology. As discussed in Section 5.5, a common obstacle to all high-temperature SiC device technologies is reliable long-term operation of contacts, interconnect, passivation, and packaging at $T > 300$°C. By incorporating highly durable high-temperature ohmic contacts and packaging, prolonged continuous electrical operation of a packaged 6H-SiC field effect transistor at 500°C in oxidizing air environment was recently demonstrated [111,112,149].

As further improvements to fundamental SiC device processing technologies (Section 5.5) are made, increasingly durable $T > 300$°C SiC-based transistor technology will evolve for beneficial use in harsh-environment applications. Increasingly complex high-temperature functionality will require robust circuit designs that accommodate large changes in device operating parameters over the much wider temperature ranges (as large as 650°C spread) enabled by SiC. Circuit models need to account for the fact that SiC device epilayers are significantly "frozen-out" owing to deeper donor and acceptor dopant ionization energies, so that nontrivial percentages of device-layer dopants are not ionized to conduct current near room temperature [171]. Because of these carrier freeze-out effects, it will be difficult to realize SiC-based ICs operational at junction temperatures much lower than –55°C (the lower end of U.S. Mil-Spec. temperature range).

5.6.4 SiC High-Power Switching Devices

The inherent material properties and basic physics behind the large theoretical benefits of SiC over silicon for power switching devices were discussed Section 5.3.2. Similarly, it was discussed in Section 5.4.5 that crystallographic defects found in SiC wafers and epilayers are presently a primary factor limiting the commercialization of useful SiC high-power switching devices. This section focuses on the additional developmental aspects of SiC power rectifiers and power switching transistor technologies.

Most SiC power device prototypes employ similar topologies and features as their silicon-based counterparts such as vertical flow of high current through the substrate to maximize device current using minimal wafer area (i.e., maximize current density) [18]. In contrast to silicon, however, the relatively low conductivity of present-day p-type SiC substrates (Section 5.4.3) dictates that all vertical SiC power device structures be implemented using n-type substrates in order to achieve beneficially high vertical current densities. Many of the device design trade-offs roughly parallel well-known silicon power device trade-offs, except for the fact that numbers for current densities, voltages, power densities, and switching speeds are much higher in SiC.

For power devices to successfully function at high voltages, peripheral breakdown owing to edge-related electric field crowding [15,18,104] must be avoided through careful device design and proper choice of insulating/passivating dielectric materials. The peak voltage of many prototype high-voltage SiC devices has often been limited by destructive edge-related breakdown, especially in SiC devices capable of blocking multiple kilovolts. In addition, most testing of many prototype multikilovolt SiC devices has required the device to be immersed in specialized high-dielectric strength fluids or gas atmospheres to minimize damaging electrical arcing and surface flashover at device peripheries. A variety of edge-termination methodologies, many of which were originally pioneered in silicon high-voltage devices, have been applied to prototype SiC power devices with varying degrees of success, including tailored dopant and metal guard rings [172–179]. The higher voltages and higher local electric fields of SiC power devices will place larger stresses on packaging and on wafer insulating materials, so some of the materials used to insulate/passivate silicon high-voltage devices may not prove sufficient for reliable use in SiC high-voltage devices, especially if those devices are to be operated at high temperatures.

5.6.4.1 SiC High-Power Rectifiers

The high-power diode rectifier is a critical building block of power conversion circuits. Recent reviews of experimental SiC rectifier results are given in References 3, 134, 172, 180, and 181. Most important SiC diode rectifier device design trade-offs roughly parallel well-known silicon rectifier trade-offs, except for the fact that current densities, voltages, power densities, and switching speeds are much higher in SiC. For example, semiconductor Schottky diode rectifiers are majority carrier devices that are well known to exhibit very fast switching owing to the absence of minority carrier charge storage that dominates (i.e., slows, adversely resulting in undesired waste power and heat) the switching operation of bipolar pn junction rectifiers. However, the high breakdown field and wide energy bandgap permit operation of SiC metal–semiconductor Schottky diodes at much higher voltages (above 1 kV) than is practical with silicon-based Schottky diodes that are limited to operation below ~200 V owing to much higher reverse-bias thermionic leakage.

5.6.4.1.1 *SiC Schottky Power Rectifiers.*

4H-SiC power Schottky diodes (with rated blocking voltages up to 1200 V and rated on-state currents up to 20 A as of this writing) are now commercially available [40,113]. The basic structure of these unipolar diodes is a patterned metal Schottky anode contact residing on top of a relatively thin (roughly of the order of 10 μm in thickness) lightly n-doped homoepitaxial layer grown on a much thicker (around 200–300 μm) low-resistivity n-type 4H-SiC substrate (8° off axis, as discussed in Section 5.4.4.2) with backside cathode contact metallization [172,182]. Guard ring structures (usually p-type implants) are usually employed to minimize electric field crowding effects around the edges of the anode contact. Die passivation and packaging help prevent arcing/surface flashover harmful to reliable device operation.

The primary application of these devices to date has been switched-mode power supplies, where (consistent with the discussion in Section 5.3.2) the SiC Schottky rectifier's faster switching with less power loss has enabled higher frequency operation and shrinking of capacitors, inductors and the overall power supply size and weight [3,23]. In particular, the effective absence of minority carrier charge storage enables the unipolar SiC Schottky devices to turn off much faster than the silicon rectifiers (which must be pn junction diodes above ~200 V blocking) which must dissipate injected minority carrier charge energy when turned off. Even though the part cost of SiC rectifiers has been higher than competing silicon rectifiers, an overall lower power supply system cost with useful performance benefits is nevertheless achieved. It should be noted, however, that changes in circuit design are sometimes necessary to best enhance circuit capabilities with acceptable reliability when replacing silicon with SiC components.

As discussed in Section 5.4.5, SiC material quality presently limits the current and voltage ratings of SiC Schottky diodes. Under high forward bias, Schottky diode current conduction is primarily limited by the series resistance of the lightly doped blocking layer. The fact that this series resistance increases with temperature (owing to decreased epilayer carrier mobility) drives equilization of high forward currents through each diode when multiple Schottky diodes are paralleled to handle higher on-state current ratings [17].

5.6.4.1.2 *Bipolar and Hybrid Power Rectifiers.*

For higher voltage applications, bipolar minority carrier charge injection (i.e., conductivity modulation) should enable SiC pn diodes to carry higher current densities than unipolar Schottky diodes whose drift regions conduct solely using dopant-atom majority carriers [19–21,172,180]. Consistent with silicon rectifier experience, SiC pn junction generation-related reverse leakage is usually smaller than thermionic-assisted Schottky diode reverse leakage. As with silicon bipolar devices, reproducible control of minority carrier lifetime will be essential in optimizing the switching-speed versus on-state current density performance trade-offs of SiC bipolar devices for specific applications. Carrier lifetime reduction via intentional impurity incorporation and introduction of radiation-induced defects appears feasible. However,

TABLE 5.2 Major Types of Extended Crystal Defects Reported in SiC Wafers and Epilayers

Crystal Defect	Density in Wafer (cm^{-2})	Density in Epilayer (cm^{-2})	Comments
Micropipe (Hollow-core axial screw dislocation)	~10–100 (0)	~10–100 (0)	Known to cause severe reduction in power device breakdown voltage and increase in off-state leakage
Closed–core axial screw dislocation	~10^3–10^4 (~10^2)	~10^3–10^4 (~10^2)	Known to cause reduction in device breakdown voltage, increase in leakage current, reduction in carrier lifetime
Basal plane dislocation	~10^4 (~10^2)	~10^2–10^3 (<10)	Known nucleation source of expanding stacking faults leading to bipolar power device degradation, reduction in carrier lifetime
Threading-edge dislocation	~10^2–10^3 (~10^2)	~10^4 (~10^2)	Impact not well known
Stacking faults (disruption of stacking sequence)	~10–10^4 (0)	~10–10^4 (0)	Faults known to degrade bipolar power devices, reduce carrier lifetime
Carrot defects	N/A	1–10 (0)	Known to causes severe reduction in power device breakdown voltage and increase in off-state leakage
Low-angle grain boundaries	~10^2–10^3 (0)	~10^2–10^3 (0)	Usually more dense near edges of wafers, impact not well known

Note: Numbers in parentheses denote research laboratory "best" results that were not commercially available for use in SiC electronics manufacture in 2005.

the ability to obtain consistently long minority carrier lifetimes (above a microsecond) has proven somewhat elusive as of this writing, indicating that further improvement to SiC material growth processes are needed to enable the full potential of bipolar power rectifiers to be realized [183].

As of this writing, SiC bipolar power rectifiers are not yet commercially available. Poor electrical reliability caused by electrically driven expansion of 4H-SiC epitaxial layer stacking faults initiated from basal plane dislocation defects (Table 5.2) effectively prevented concerted efforts for commercialization of 4H-SiC pn junction diodes in the late 1990s [63,74,184]. In particular, bipolar electron–hole recombination that occurs in forward-biased pn junctions drove the enlargement of stacking disorder in the 4H-SiC blocking layer, forming an enlarging quantum well (based on narrower 3C-SiC bandgap) that effectively degrades transport (diffusion) of minority carriers across the lightly doped junction blocking layer. As a result, the forward voltages of 4H-SiC pn rectifiers required to maintain rated on-state current increase unpredictably and undesirably over time. As discussed in Section 5.4.5, research toward understanding and overcoming this material defect-induced problem has made important progress, so that hopefully SiC bipolar power devices might become commercialized within a few years [39,41].

A drawback of the wide bandgap of SiC is that it requires larger forward-bias voltages to reach the turn-on "knee" of a diode where significant on-state current begins flowing. In turn, the higher knee voltage can lead to an undesirable increase in on-state power dissipation. However, the benefits of 100× decreased drift region resistance and much faster dynamic switching should greatly overcome SiC on-state knee voltage disadvantages in most high-power applications. While the initial turn-on knee of SiC pn junctions is higher (around 3 V) than for SiC Schottky junctions (around 1 V), conductivity modulation enables SiC pn junctions to achieve lower forward voltage drop for higher blocking voltage applications [172,180].

Hybrid Schottky/pn rectifier structures first developed in silicon that combine pn junction reverse blocking with low Schottky forward turn-on should prove extremely useful in realizing application-optimized SiC rectifiers [134,172,180,181]. Similarly, combinations of dual Schottky metal structures and trench pinch rectifier structures can also be used to optimize SiC rectifier forward turn-on and reverse leakage properties [185].

5.6.4.2   SiC High-Power Switching Transistors

Three terminal power switches that use small drive signals to control large voltages and currents (i.e., power transistors) are also critical building blocks of high-power conversion circuits. However, as of this writing, SiC high-power switching transistors are not yet commercially available for beneficial use in power system circuits. As well summarized in References 134, 135, 172, 180, and 186–188, a variety of improving three-terminal SiC power switches have been prototyped in recent years.

The present lack of commercial SiC power switching transistors is largely due to several technological difficulties discussed elsewhere in this chapter. For example, all high-power semiconductor transistors contain high-field junctions responsible for blocking current flow in the off-state. Therefore, performance limitations imposed by SiC crystal defects on diode rectifiers (Sections 5.4.5 and 5.6.4.1) also apply to SiC high-power transistors. Also, the performance and reliability of inversion channel SiC-based MOS field-effect gates (i.e., MOSFETs, IGBTs, etc.) has been limited by poor inversion channel mobilities and questionable gate-insulator reliability discussed in Section 5.5.5. To avoid these problems, SiC device structures that do not rely on high-quality gate insulators, such as the MESFET, JFET, BJT, and depletion-channel MOSFET, have been prototyped toward use as power switching transistors. However, these other device topologies impose non-standard requirements on power system circuit design that make them unattractive compared with the silicon-based inversion-channel MOSFETs and IGBTs. In particular, silicon power MOSFETs and IGBTs are extremely popular in power circuits largely because their MOS gate drives are well insulated from the conducting power channel, require little drive signal power, and the devices are "normally off" in that there is no current flow when the gate is unbiased at 0 V. The fact that the other device topologies lack one or more of these highly circuit-friendly aspects has contributed to the inability of SiC-based devices to beneficially replace silicon-based MOSFETs and IGBTs in power system applications.

As discussed in Section 5.5.5, continued substantial improvements in 4H-SiC MOSFET technology will hopefully soon lead to the commercialization of 4H-SiC MOSFETs. In the meantime, advantageous high-voltage switching by pairing a high-voltage SiC JFET with a lower-voltage silicon power MOSFETs into a single module package appears to be nearing practical commercialization [188]. Numerous designs for SiC doped-channel FETs (with both lateral and vertical channels) have been prototyped, including depletion-channel (i.e., buried or doped channel) MOSFETs, JFETs, and MESFETs [187]. Even though some of these have been designed to be "normally-off" at zero applied gate bias, the operational characteristics of these devices have not (as of this writing) offered sufficient benefits relative to cost to enable commercialization.

Substantial improvements to the gain of prototype 4H-SiC power BJTs have been achieved recently, in large part by changing device design to accommodate for undesired large minority carrier recombination occurring at p-implanted base contact regions [103]. IGBTs, thyristors, Darlington pairs, and other bipolar power device derivatives from silicon have also been prototyped in SiC [134,180,186]. Optical transistor triggering, a technique quite useful in previous high-power silicon device applications, has also been demonstrated for SiC bipolar devices [189]. However, because all bipolar power transistors operate with at least one pn junction injecting minority carriers under forward bias, crystal defect-induced bipolar degradation discussed for pn junction rectifiers (Section 5.6.4.1.2) also applies to the performance of bipolar transistors. Therefore, the effective elimination of basal plane dislocations from 4H-SiC epilayers must be accomplished before any power SiC bipolar transistor devices can become sufficiently reliable for commercialization. SiC MOS oxide problems (Section 5.5.5) will also have to be solved to realize beneficial SiC high-voltage IGBTs. However, relatively poor p-type SiC substrate conductivity may force development of p-IGBTs instead of n-IGBT structures that presently dominate in silicon technology.

As various fundamental SiC power device technology challenges are overcome, a broader array of SiC power transistors tackling increasingly widening voltage, current, and switching speed specification will enable beneficial new power system circuits.

5.6.5   SiC MicroElectromechanical Systems (MEMS) and Sensors

As described in Hesketh's chapter on micromachining in this book, the development and use of silicon-based MEMS continues to expand. While the previous sections of this chapter have centered on the use of SiC for traditional semiconductor electronic devices, SiC is also expected to play a significant role in emerging MEMS applications [124,190]. SiC has excellent mechanical properties that address some shortcomings of silicon-based MEMS such as extreme hardness and low friction reducing mechanical wear-out as well as excellent chemical inertness to corrosive atmospheres. For example, SiCs excellent durability is being examined as enabling for long-duration operation of electric micromotors and micro jet-engine power generation sources where the mechanical properties of silicon appear to be insufficient [191].

Unfortunately, the same properties that make SiC more durable than silicon also make SiC more difficult to micromachine. Approaches to fabricating harsh-environment MEMS structures in SiC and prototype SiC-MEMS results obtained to date are reviewed in References 124 and 190. The inability to perform fine-patterned etching of single-crystal 4H- and 6H-SiC with wet chemicals (Section 5.5.4) makes micromachining of this electronic-grade SiC more difficult. Therefore, the majority of SiC micromachining to date has been implemented in electrically inferior heteroepitaxial 3C-SiC and polycrystalline SiC deposited on silicon wafers. Variations of bulk micromachining, surface micromachining, and micro-molding techniques have been used to fabricate a wide variety of micromechanical structures, including resonators and micromotors. A standardized SiC on silicon wafer micromechanical fabrication process foundry service, which enables users to realize their own application-specific SiC micromachined devices while sharing wafer space and cost with other users, is commercially available [192].

For applications requiring high temperature, low-leakage SiC electronics not possible with SiC layers deposited on silicon (including high-temperature transistors, as discussed in Section 5.6.2), concepts for integrating much more capable electronics with MEMS on 4H/6H SiC wafers with epilayers have also been proposed. For example, pressure sensors being developed for use in higher temperature regions of jet engines are implemented in 6H-SiC, largely owing to the fact that low junction leakage is required to achieve proper sensor operation [152,193]. On-chip 4H/6H integrated transistor electronics that beneficially enable signal conditioning at the high-temperature sensing site are also being developed [112]. With all micromechanical-based sensors, it is vital to package the sensor in a manner that minimizes the imposition of thermomechanical induced stresses (which arise owing to thermal expansion coefficient mismatches over much larger temperature spans enabled by SiC) onto the sensing elements. Therefore (as mentioned previously in Section 5.5.6), advanced packaging is almost as critical as the use of SiC toward usefully expanding the operational envelope of MEMS in harsh environments.

As discussed in Section 5.3.1, a primary application of SiC harsh-environment sensors is to enable active monitoring and control of combustion engine systems to improve fuel efficiency while reducing pollution. Toward this end, SiC's high-temperature capabilities have enabled the realization of catalytic metal–SiC and metal-insulator–SiC prototype gas sensor structures with great promise for emission monitoring applications and fuel system leak detection [194,195]. High-temperature operation of these structures, not possible with silicon, enables rapid detection of changes in hydrogen and hydrocarbon content to sensitivities of parts per million in very small-sized sensors that could easily be placed unobtrusively on an engine without the need for cooling. However, further improvements to the reliability, reproducibility, and cost of SiC-based gas sensors are needed before these systems will be ready for widespread use in consumer automobiles and aircraft. In general, the same can be said for most SiC MEMS, which will not achieve widespread beneficial system insertion until high reliability in harsh environments is assured via further technology development.

5.7 Future of SiC

It can be safely predicted that SiC will never displace silicon as the dominant semiconductor used for the manufacture of the vast majority of the world's electronic chips that are primarily low-voltage digital and analog chips targeted for operation in normal human environments (computers, cell phones, etc.). SiC will only be used where substantial benefits are enabled by SiC's ability to expand the envelope of high-power and high-temperature operational conditions such as the applications described in Section 5.3. Perhaps, the only major existing application area where SiC might substantially displace today's use of silicon is the area of discrete power devices used in power conversion, motor control, and management circuits.

The power device market, along with the automotive sensing market present the largest-volume market opportunity for SiC-based semiconductor components. However, the end consumers in both of these applications demand excruciatingly high reliability (i.e., no operational failures) combined with competitively low overall cost. For SiC electronics technology to have large impact, it must greatly evolve from its present status to meet these demands. There is clearly a very large discrepancy between the revolutionary broad theoretical promise of SiC semiconductor electronics technology (Section 5.3) versus the operational capability of SiC-based components that have actually been deployed in only a few commercial and military applications (Section 5.6). Likewise, a large discrepancy also exists between the capabilities of laboratory SiC devices compared with commercially deployed SiC devices. The inability of many "successful" SiC laboratory prototypes to rapidly transition to commercial product demonstrates both the difficulty and criticality of achieving acceptable reliability and costs.

5.7.1 Future Tied to Material Issues

The previous sections of this chapter have already highlighted major known technical obstacles and immaturities that are largely responsible for hindered SiC device capability. In the most general terms, these obstacles boil down to a handful of key fundamental material issues. The rate at which the most critical of these fundamental issues is solved will greatly impact the availability, capability, and usefulness of SiC semiconductor electronics. Therefore, the future of SiC electronics is linked to investment in basic material research toward solving challenging material-related impediments to SiC device performance, yield, and reliability.

The material challenge that is arguably the biggest key to the future of SiC is the removal of dislocations from SiC wafers. As described previously in this chapter and references therein, the most important SiC power rectifier performance metrics, including device ratings, reliability, and cost are inescapably impacted by high dislocation densities present in commercial SiC wafers and epilayers. If mass-produced SiC wafer quality approached that of silicon wafers (which typically contain less than one dislocation defect per square centimeter), far more capable SiC unipolar and bipolar high-power rectifiers (including devices with kilovolt and kiloampere ratings) would rapidly become widely available for beneficial use in a far larger variety of high-power applications. Similar improvements would also be realized in SiC transistors, paving the way for SiC high-power devices to indeed beneficially displace silicon-based power devices in a tremendously broad and useful array of applications and systems (Section 5.3). This advancement would unlock a much more rapid and broad SiC-enabled power electronic systems "revolution" compared to the relatively slower "evolution" and niche-market insertion that has occurred since SiC wafers were first commercialized roughly 15 years ago. As mentioned in Section 5.4, recent laboratory results [83] indicate that drastic reductions in SiC wafer dislocations are possible using radically new approaches to SiC wafer growth compared to standard boule-growth techniques practiced by all commercial SiC wafer vendors for over a decade. Arguably, the ultimate future of SiC high-power devices may hinge on the development and practical commercialization of low dislocation density SiC growth techniques substantially different from those employed today.

It is important to note that other emerging wide bandgap semiconductors besides SiC theoretically offer similarly large electrical system benefits over silicon semiconductor technology as described in Section 5.3. For example, diamond and some Group III-nitride compound semiconductors (such as GaN; Table 5.1) have high breakdown field and low intrinsic carrier concentration that enables operation at power densities, frequencies, and temperatures comparable to or exceeding SiC. Like SiC, however, electrical devices in these semiconductors are also hindered by a variety of difficult material challenges that must be overcome in order for beneficially high performance to be reliably achieved and commercialized. If SiC electronics capability expansion evolves too slowly compared to other wide bandgap semiconductors, the possibility exists that the latter will capture applications and markets originally envisioned for SiC. However, if SiC succeeds in being the first to offer reliable and cost-effective wide bandgap capability to a particular application, subsequent wide-bandgap technologies would probably need to achieve far better cost/performance metrics in order to displace SiC. It is therefore likely that SiC, to some degree, will continue its evolution toward expanding the operational envelope of semiconductor electronics capability.

5.7.2 Further Recommended Reading

This chapter has presented a brief summary overview of evolving SiC semiconductor device technology. The following publications, which were heavily referenced in this chapter, are highly recommended as supplemental reading to more completely cover SiC electronics technology development in much greater technical detail than possible within this short chapter.

Reference 11 is a collection of invited in-depth papers from recognized leaders in SiC technology development that first appeared in special issues of the journal *Physica Status Solidi* (a 162 (1)) and (b 202, (1)) in 1997. In 2003, the same editors published a follow-on book [12] containing additional invited papers to update readers on new "recent major advances" in SiC since the 1997 book.

One of the best sources of the most up-to-date SiC electronics technology development information is the *International Conference on Silicon Carbide and Related Materials (ICSCRM)*, which is held every 2 years (years ending in odd numbers). To bridge the 24-month gap between international SiC meetings, the *European Conference on Silicon Carbide and Related Materials (ECSCRM)* is held in years ending in even numbers. Since 1999, the proceedings of peer-reviewed papers presented at both the International and European SiC conferences have been published by Trans Tech Publications as volumes in its Materials Science Forum offering, which are available online via paid subscription (http://www.scientific.net). In addition, the meetings of the Materials Research Society (MRS) often hold symposiums and publish proceedings (book and online editions; http://www.mrs.org) dedicated to SiC electronics technology development. Reference 200 is the proceedings of the most recent MRS SiC symposium held in April 2004, and the next such symposium is scheduled for the 2006 MRS spring meeting in San Francisco.

The following technical journal issues contain collections of invited papers from SiC electronics experts that offer more detailed insights than this-chapter, yet are conveniently brief compared to other volumes already mentioned in this section:

1. *Proceedings of the IEEE*, Special Issue on Wide Bandgap Semiconductor Devices, 90 (6), June 2002.
2. *Materials Research Society Bulletin*, Advances in Silicon Carbide Electronics, 30(4), April 2005.

In addition, a variety of internet websites contain useful SiC information and links can be located using widely available internet search engine services. The author of this chapter maintains a website that contains information and links to other useful SiC internet websites at http://www.grc.nasa.gov/WWW/SiC/.

References

1. Baliga, B.J., "Power Semiconductor Devices for Variable-Frequency Drives," *Proceedings of the IEEE* 82(8), 1112, 1994.
2. Baliga, B.J., "Trends in Power Semiconductor Devices," *IEEE Transactions on Electron Devices* 43(10), 1717, 1996.
3. Elasser, A. and Chow, T.P., "Silicon Carbide Benefits and Advantages for Power Electronics Circuits and Systems," *Proceedings of the IEEE* 90(6), 969, 2002.
4. Johnson, C.M., "Clear road ahead?" *Power Engineer* 18(4), 34, 2004.
5. Weitzel, C.E., Palmour, J.W., Carter, C.H., Jr., Moore, K., Nordquist, K.J., Allen, S., Thero, C., and Bhatnagar, M., "Silicon Carbide High Power Devices," *IEEE Transactions on Electron Devices* 43(10), 1732, 1996.

6. Trew, R.J., "SiC and GaN Transistors—Is There One Winner for Microwave Power Applications," *Proceedings of the IEEE* 90(6), 1032, 2002.
7. Neudeck, P.G., Okojie, R.S., and Chen, L.-Y., "High-Temperature Electronics—A Role for Wide-Bandgap Semiconductors," *Proceedings of the IEEE* 90(6), 1065, 2002.
8. Bhatnagar, M. and Baliga, B.J., "Comparison of 6H-SiC, 3C-SiC, and Si for Power Devices," *IEEE Transactions on Electron Devices* 40(3), 645, 1993.
9. Powell, J.A., Pirouz, P., and Choyke, W.J., "Growth and Characterization of Silicon Carbide Polytypes for Electronic Applications," in *Semiconductor Interfaces, Microstructures, and Devices: Properties and Applications*, Ed. Feng, Z.C., Institute of Physics Publishing, Bristol, United Kingdom, 1993, p. 257.
10. Powell, A.R. and Rowland, L.B., "SiC Materials—Progress, Status, and Potential Roadblocks," *Proceedings of the IEEE* 90(6), 942, 2002.
11. Choyke, W.J., Matsunami, H., and Pensl, G., *Silicon Carbide—A Review of Fundamental Questions and Applications to Current Device Technology*, Wiley-VCH, Berlin, 1997.
12. Choyke, W.J., Matsunami, H., and Pensl, G., *Silicon Carbide: Recent Major Advances*, Springer, Berlin, 2003.
13. Ioffe Physico-Technical Institute, http://www.ioffe.ru/SVA/NSM/Semicond/SiC.
14. Pierret, R.F., *Advanced Semiconductor Fundamentals*, Addison-Wesley, Reading, MA, 1987.
15. Sze, S.M., *Physics of Semiconductor Devices, 2nd ed.*, Wiley-Interscience, New York, 1981.
16. Dreike, P.L., Fleetwood, D.M., King, D.B., Sprauer, D.C., and Zipperian, T.E., "An Overview of High-Temperature Electronic Device Technologies and Potential Applications," *IEEE Transactions on Components, Packaging, and Manufacturing Technology* 17(4), 594, 1994.
17. Kodani, K., Matsurnoto, T., Saito, S., Takao, K., Mogi, T., Yatsuo, T., and Arai, K., "Evaluation of parallel and series connection of silicon carbide Schottky barrier diode (SiC-SBD)," in *Power Electronics Specialists Conference, 2004. PESC 04. 2004 IEEE 35th Annual*, 2004, p. 3971.
18. Baliga, B.J., *Modern Power Devices, 1st ed.*, Wiley, New York, 1987.
19. Ruff, M., Mitlehner, H., and Helbig, R., "SiC Devices: Physics and Numerical Simulation," *IEEE Transactions on Electron Devices* 41(6), 1040, 1994.
20. Chow, T.P., Ramungul, N., and Ghezzo, M., "Wide-Bandgap Semiconductor Power Devices," in *Power Semiconductor Materials and Devices*, *Materials Research Society Symposia Proceedings*, vol. 483, Eds. Pearton, S.J., Shul, R.J., Wolfgang, E., Ren, F., and Tenconi, S., Materials Research Society, Warrandale, PA, 1998, p. 89.
21. Bakowski, M., Gustafsson, U., and Lindefelt, U., "Simulation of SiC High-Power Devices," *Physica Status Solidi (a)* 162(1), 421, 1997.
22. Tolbert, L.M., Ozpineci, B., Islam, S.K., and Peng, F.Z., "Impact of SiC Power Electronic Devices for Hybrid Electric Vehicles," in *2002 Future Car Congress Proceedings*, June 3–5, 2002, Arlington, VA, Soceity of Automotive Engineers, Warrendale, PA, 2002.
23. Rupp, R. and Zverev, I., "System Design Considerations for Optimizing the Benefit by Unipolar SiC Power Devices," in Silicon Carbide 2002—Materials, Processing and Devices, *Materials Research Society Symposia Proceedings*, vol. 742, Eds. Saddow, S.E., Larkin, D.J., Saks, N.S., Schoener, A., and Skowronski, M., Materials Research Society, Warrandale, PA, 2003, p. 329.
24. Trew, R.J., "Experimental and Simulated Results of SiC Microwave Power MESFETs," *Physica Status Solidi (a)* 162(1), 409, 1997.
25. Clarke, R.C. and Palmour, J.W., "SiC Microwave Power Technologies," *Proceedings of the IEEE* 90(6), 987, 2002.
26. Morvan, E., Kerlain, A., Dua, C., and Brylinski, C., "Development of SiC Devices for Microwave and RF Power Amplifiers," in *Silicon Carbide: Recent Major Advances*, Eds. Choyke, W.J., Matsunami, H., and Pensl, G., Springer Berlin, 2003, p. 839.
27. Baliga, B.J., "Power ICs In The Saddle," *IEEE Spectrum* 32(7), 34, 1995.
28. Heydt, G.T. and Skromme, B.J., "Applications of High-Power Electronic Switches In The Electric Power Utility Industry and The Needs For High-Power Switching Devices," in Power Semiconductor

Materials and Devices, *Materials Research Society Symposia Proceedings*, vol. 483, Eds. Pearton, S.J., Shul, R.J., Wolfgang, E., Ren, F., and Tenconi, S., Materials Research Society, Warrandale, PA, 1998, p. 3.

29. Ozpineci, B., Tolbert, L.M., Islam, S.K., and Hasanuzzaman, M., "System Impact of Silicon Carbide Power Devices," *International Journal of High Speed Electronics and Systems* 12(2), 439, 2002.

30. Johnson, C.M., "Comparison of Silicon and Silicon Carbide Semiconductors for a 10 kV Switching Application," in *IEEE 35th Annual Power Electronics Specialists Conference*, IEEE, Piscataway, NJ, 2004, p. 572.

31. Ericsen, T., "Future Navy Application of Wide Bandgap Power Semiconductor Devices," *Proceedings of the IEEE* 90(6), 1077, 2002.

32. Acheson, A.G., England Patent 17911, 1892.

33. Lely, J.A., "Darstellung von Einkristallen von Silicium carbid und Beherrschung von Art und Menge der eingebautem Verunreingungen," *Ber. Deut. Keram. Ges.* 32, 229, 1955.

34. Nishino, S., Powell, J.A., and Will, H.A., "Production of Large-Area Single-Crystal Wafers of Cubic SiC for Semiconductor Devices," *Applied Physics Letters* 42(5), 460, 1983.

35. Pirouz, P., Chorey, C.M., and Powell, J.A., "Antiphase Boundaries in Epitaxially Grown Beta-SiC," *Applied Physics Letters* 50(4), 221, 1987.

36. Nagasawa, H., Yagi, K., Kawahara, T., Hatta, N., Pensl, G., Choyke, W.J., Yamada, T., Itoh, K.M., and Schöner, A., "Low-Defect 3C-SiC Grown on Undulant-Si (001) Substrates," in *Silicon Carbide: Recent Major Advances*, Eds. Choyke, W.J., Matsunami, H., and Pensl, G., Springer, Berlin, 2003, p. 207.

37. Tairov, Y.M. and Tsvetkov, V.F., "Investigation of Growth Processes of Ingots of Silicon Carbide Single Crystals," *Journal of Crystal Growth* 43, 209, 1978.

38. Tairov, Y.M. and Tsvetkov, V.F., "General Principles of Growing Large-Size Single Crystals of Various Silicon Carbide Polytypes," *Journal of Crystal Growth* 52, 146, 1981.

39. Sumakeris, J.J., Jenny. J.R., and Powell, A.R., "Bulk Crystal Growth, Epitaxy, and Defect Reduction in Silicon Carbide Materials For Microwave and Power Devices," *MRS Bulletin* 30(4), 280, 2005.

40. Cree, Inc., http://www.cree.com.

41. Powell, A.R., Sumakeris, J.J., Leonard, R.T., Brady, M.F., Muller, S.G., Tsvetkov, V.F., Hobgood, H.M.D., Burk, A.A., Paisley, M.J., and Glass, R.C., "Status of 4H-SiC Substrate and Epitaxial Materials for Commercial Power Applications," in Silicon Carbide 2004—Materials, Processing and Devices, *Materials Research Society Symposium Proceedings*, vol. 815, Eds. Dudley, M., Gouma, P., Kimoto, T., Neudeck, P.G., and Saddow, S.E., Materials Research Society, Warrandale, PA, 2004, p. 3.

42. Ellison, A., Magnusson, B., Sundqvist, B., Pozina, G., Bergman, J.P., Janzen, E., and Vehanen, A., "SiC Crystal Growth by HTCVD," in Silicon Carbide and Related Materials 2003, *Materials Science Forum*, vol. 457–460, Eds. Madar, R., Camassel, J., and Blanquet, E., Trans Tech, Switzerland, 2004, p. 9.

43. Balkrishna, V., Augustine, G., Gaida, W.E., Thomas, R.N., and Hopkins, R.H., "Advanced Physical Vapor Transport Method and Apparatus for Growing High-Purity Single-Crystal Silicon Carbide," US Patent 6,056,820, 2000.

44. Ohtani, N., Katsuno, M., Fujimoto, T., and Yashiro, H., "Defect Formation and Reduction During Bulk SiC Growth," in *Silicon Carbide: Recent Major Advances*, Eds. Choyke, W.J., Matsunami, H., and Pensl, G., Springer, Berlin, 2003, p. 137.

45. Davis, R.F., Kelner, G., Shur, M., Palmour, J.W., and Edmond, J.A., "Thin Film Deposition and Microelectronic and Optoelectronic Device Fabrication and Characterization in Monocrystalline Alpha and Beta Silicon Carbide," *Proceedings of the IEEE* 79(5), 677, 1991.

46. Schöner, A., "New Development in Hot Wall Vapor Phase Epitaxial Growth of Silicon Carbide," in *Silicon Carbide: Recent Major Advances*, Eds. Choyke, W.J., Matsunami, H., and Pensl, G., Springer, Berlin, 2003, p. 229.

47. Kordina, O., Hallin, C., Henry, A., Bergman, J.P., Ivanov, I., Ellison, A., Son, N.T., and Janzen, E., "Growth of SiC by" "Hot-Wall" "CVD and HTCVD," *Physica Status Solidi (b)* 202(1), 321, 1997.

48. Fanton, M., Skowronski, M., Snyder, D., Chung, H.J., Nigam, S., Weiland, B., and Huh, S.W., "Growth of Bulk SiC by Halide Chemical Vapor Deposition," in Silicon Carbide and Related Materials 2003,

Materials Science Forum, vol. 457–460, Eds. Madar, R., Camassel, J., and Blanquet, E., Trans Tech, Switzerland, 2004, p. 87.

49. Rupp, R., Wiedenhofer, A., and Stephani, D., "Epitaxial Growth of SiC in a Single and a Multi Wafer Vertical CVD System: A Comparison," *Materials Science and Engineering B* 61–62, 125, 1999.

50. Kong, H.S., Glass, J.T., and Davis, R.F., "Chemical Vapor Deposition and Characterization of 6H-SiC Thin Films on Off-Axis 6H-SiC Substrates," *Journal of Applied Physics* 64, 2672, 1988.

51. Kimoto, T., Itoh, A., and Matsunami, H., "Step-Controlled Epitaxial Growth of High-Quality SiC Layers," *Physica Status Solidi (b)* 202(1), 247, 1997.

52. Powell, J.A. and Larkin, D.J., "Process-Induced Morphological Defects in Epitaxial CVD Silicon Carbide," *Physica Status Solidi (b)* 202(1), 529, 1997.

53. Burk, A.A., Jr. and Rowland, L.B., "Homoepitaxial VPE Growth of SiC Active Layers," *Physica Status Solidi (b)* 202(1), 263, 1997.

54. Kimoto, T., Yano, H., Negoro, Y., Hashimoto, K., and Matsunami, H., "Epitaxial Growth and Device Processing of SiC on Non-Basal Planes," in *Silicon Carbide: Recent Major Advances*, Eds. Choyke, W.J., Matsunami, H., and Pensl, G., Springer, Berlin, 2003, p. 711.

55. Monnoye, S., Turover, D., and Vincente, P., "Surface Preparation Techniques for SiC Wafers," in *Silicon Carbide: Recent Major Advances*, Eds. Choyke, W.J., Matsunami, H., and Pensl, G., Springer, Berlin, 2003, p. 699.

56. Powell, J.A., Petit, J.B., Edgar, J.H., Jenkins, I.G., Matus, L.G., Yang, J.W., Pirouz, P., Choyke, W.J., Clemen, L., and Yoganathan, M., "Controlled Growth of 3C-SiC and 6H-SiC Films on Low-Tilt-Angle Vicinal (0001) 6H-SiC Wafers," *Applied Physics Letters* 59(3), 333, 1991.

57. Burk, A.A., Jr. and Rowland, L.B., "The Role of Excess Silicon and In Situ Etching on 4H-SiC and 6H-SiC Epitaxial Layer Morphology," *Journal of Crystal Growth* 167(3–4), 586, 1996.

58. Neudeck, P.G. and Powell, J.A., "Homoepitaxial and Heteroepitaxial Growth on Step-Free SiC Mesas," in *Recent Major Advances in SiC*, Eds. Choyke, W.J., Matsunami, H., and Pensl, G., Springer, Berlin, 2003, p. 179.

59. Larkin, D.J., "SiC Dopant Incorporation Control Using Site-Competition CVD," *Physica Status Solidi (b)* 202(1), 305, 1997.

60. Wang, R., Bhat, I.B., and Chow, T.P., "Epitaxial growth of n-type SiC using phosphine and nitrogen as the precursors," *Journal of Applied Physics* 92(12), 7587, 2002.

61. Larkin, D.J., Neudeck, P.G., Powell, J.A., and Matus, L.G., "Site-Competition Epitaxy for Superior Silicon Carbide Electronics," *Applied Physics Letters* 65(13), 1659, 1994.

62. Ha, S., Mieszkowski, P., Skowronski, M., and Rowland, L.B., "Dislocation Conversion in 4H Silicon Carbide Epitaxy," *Journal of Crystal Growth* 244(3–4), 257, 2002.

63. Lendenmann, H., Dahlquist, F., Bergman, J.P., Bleichner, H., and Hallin, C., "High-Power SiC Diodes: Characteristics, Reliability and Relation to Material Defects," in Silicon Carbide and Related Materials 2001, *Materials Science Forum*, vol. 389–393, Eds. Yoshida, S., Nishino, S., Harima, H., and Kimoto, T., Trans Tech Publications, Switzerland, 2002, p. 389.

64. Neudeck, P.G. and Powell, J.A., "Performance Limiting Micropipe Defects in Silicon Carbide Wafers," *IEEE Electron Device Letters* 15(2), 63, 1994.

65. Yang, J.-W., "SiC: Problems in Crystal Growth and Polytypic Transformation," PhD, Case Western Reserve University, 1993.

66. Si, W., Dudley, M., Glass, R., Tsvetkov, V., and Carter, C.H., Jr., "Hollow-Core Screw Dislocations in 6H-SiC Single Crystals: A Test of Frank's Theory," *Journal of Electronic Materials* 26(3), 128, 1997.

67. Si, W. and Dudley, M., "Study of Hollow-Core Screw Dislocations in 6H-SiC and 4H-SiC Single Crystals," in Silicon Carbide, III-Nitrides, and Related Materials, *Materials Science Forum*, vol. 264–268, Eds. Pensl, G., Morkoc, H., Monemar, B., and Janzen, E., Trans Tech Publications, Switzerland, 1998, p. 429.

68. Dudley, M., Huang, X., and Vetter, W.M., "Synchrotron White Beam X-Ray Topography and High-Resolution X-Ray Diffraction Studies of Defects in SiC Sibstrates, Epilayers and Device Structures," in *Silicon Carbide: Recent Major Advances*, Eds. Choyke, W.J., Matsunami, H., and Pensl, G., Springer, Berlin, 2003, p. 629.

69. Ma, X., Dudley, M., Vetter, W., and Sudarshan, T., "Extended SiC Defects: Polarized Light Microscopy Delineation and Synchrotron White-Beam X-Ray Topography Ratification," *Japanese Journal of Applied Physics* 42(9A/B), L1077, 2003.

70. Basceri, C., Khlebnikov, I., Khlebnikov, Y., Sharma, M., Muzykov, P., Stratiy, G., Silan, M., and Balkas, C., "Micropipe-Free Single-Crystal Silicon Carbide (SiC) Ingots via Physical Vapor Transport (PVT)," in Silicon Carbide and Related Materials 2005, Materials Science Forum, vols. 527–529, Eds. Devaty, R.P., Larkin, D.J., and Saddow, S.E., *Trans Tech Publications*, Switzerland, 2006, p. 32.

71. Tsuchida, H., Kamata, I., Izumi, S., Tawara, T., Jikimoto, T., Miyanagi, T., Nakamura, T., and Izumi, K., "Homoepitaxial Growth and Characterization of Thick SiC Layers With a Reduced Micropipe Density," in Materials Research Society Symposium Proceedings, vol. 815, 2004, p. 35.

72. Rupp, R., Treu, M., Turkes, P., Beermann, H., Scherg, T., Preis, H., and Cerva, H., "Influence of Overgrown Micropipes in the Active Area of SiC Schottky Diodes on Long-Term Reliability," in Silicon Carbide and Related Materials 2004, *Materials Science Forum*, vols. 483–485, Eds. Nipoti, R., Poggi, A., and Scorzoni, A., Trans Tech Publicatons, Switzerland, 2005, p. 925.

73. Neudeck, P.G., Huang, W., and Dudley, M., "Study of Bulk and Elementary Screw Dislocation Assisted Reverse Breakdown in Low-Voltage (<250 V) 4H-SiC p$^+$n Junction Diodes—Part 1: DC Properties," *IEEE Transactions on Electron Devices* 46(3), 478, 1999.

74. Ha, S. and Bergman. J.P., "Degradation of SiC High-Voltage pin Diodes," *MRS Bulletin* 30(4), 305, 2005.

75. Wang, S., Dudley, M., Carter, C.H., Jr., Tsvetkov, V.F., and Fazi, C., "Synchrotron White Beam Topography Studies of Screw Dislocations in 6H-SiC Single Crystals," in Applications of Synchrotron Radiation Techniques to Materials Science, *Materials Research Society Symposium Proceedings*, vol. 375, Eds. Terminello, L., Shinn, N., Ice, G., D'Amico, K., and Perry, D., Materials Research Society, Warrandale, PA, 1995, p. 281.

76. Dudley, M., Wang, S., Huang, W., Carter, C.H., Jr., and Fazi, C., "White Beam Synchrotron Topographic Studies of Defects in 6H-SiC Single Crystals," *Journal of Physics D: Applied Physics* 28, A63, 1995.

77. Pirouz, P., Zhang, M., Galeckas, A., and Linnros, J., "Microstructural Aspects and Mechanism of Degradation of 4H-SiC PiN Diodes Under Forward Biasing," in *Materials Research Society Symposium Proceedings*, vol. 815, Eds. Dudley, M., Gouma, P., Kimoto, T., Neudeck, P.G., and Saddow, S.E., Materials Research Society, Warrendale, 2004, p. 91.

78. Stahlbush, R.E., Twigg, M.E., Sumakeris, J.J., Irvine, K.G., and Losee, P.A., "Mechanisms of Stacking Fault Growth in SiC PiN Diodes," in *Materials Research Society Symposium Proceedings*, vol. 815, Eds. Dudley, M., Gouma, P., Kimoto, T., Neudeck, P.G., and Saddow, S.E., Materials Research Society, Warrendale, PA, 2004, p. 103.

79. Okojie, R.S., Xhang, M., Pirouz, P., Tumakha, S., Jessen, G., and Brillson, L.J., "Observation of 4H-SiC to 3C-SiC Polytypic Transformation During Oxidation," *Applied Physics Letters* 79(19), 3056, 2001.

80. Okojie, R.S. and Zhang, M., "Thermoplastic Deformation and Residual Stress Topography of 4H-SiC Wafers," in *Materials Research Society Symposium Proceedings*, vol. 815, Eds. Dudley, M., Gouma, P., Kimoto, T., Neudeck, P.G., and Saddow, S.E., Materials Research Society, Warrendale, PA, 2004, p. 133.

81. Diaz-Guerra, C. and Piqueras, J., "Electron-beam-induced current study of electrically active defects in 4H-SiC," *Journal of Physics: Condensed Matter* 16(2), S217, 2004.

82. Wahab, Q., Ellison, A., Henry, A., Janzén, E., Hallin, C., Persio, J.D., and Martinez, R., "Influence of Epitaxial Growth and Substrate-Induced Defects on the Breakdown of 4H–SiC Schottky Diodes," *Applied Physics Letters* 76(19), 2725, 2000.

83. Nakamura, D., Gunjishima, I., Yamaguchi, S., Ito, T., Okamoto, A., Kondo, H., Onda, S., and Takatori, K., "Ultrahigh-Quality Silicon Carbide Single Crystals," *Nature* 430, 1109, 2004.

84. Soloviev, S., Gao, Y., Wang, X., and Sudarshan, T., "Boron Diffusion into 6H-SiC Through Graphite Mask," *Journal of Electronic Materials* 30(3), 224, 2001.

85. Troffer, T., Schadt, M., Frank, T., Itoh, H., Pensl, G., Heindl, J., Strunk, H.P., and Maier, M., "Doping of SiC by Implantation of Boron and Aluminum," *Physica Status Solidi (a)* 162(1), 277, 1997.

86. Kimoto, T., Itoh, A., Inoue, N., Takemura, O., Yamamoto, T., Nakajima, T., and Matsunami, H., "Conductivity Control of SiC by In-Situ Doping and Ion Implantation" , in Silicon Carbide, III-Nitrides, and Related Materials, *Materials Science Forum*, vols. 264–268, Eds. Pensl, G., Morkoc, H., Monemar, B., and Janzen, E., Trans Tech Publications, Switzerland, 1998, p. 675.

87. Capano, M., Ryu, S., Cooper, J. A. Jr., Melloch, M., Rottner, K., Karlsson, S., Nordell, N., Powell, A., and D. Walker, J., "Surface Roughening in Ion Implanted 4H-Silicon Carbide," *Journal of Electronic Materials* 28(3), 214, 1999.

88. Tone, K. and Zhao, J., "A Comparative Study of C Plus Al Coimplantation and Al Implantation in 4H- and 6H-SiC," *IEEE Transactions on Electron Devices* 46(3), 612, 1999.

89. Saks, N.S., Agarwal, A.K., Mani, S.S., and Hegde, V.S., "Low-Dose Nitrogen Implants in 6H–Silicon Carbide," *Applied Physics Letters* 76(14), 1896, 2000.

90. Saks, N.S., Agarwal, A.K., Ryu, S.-H., and Palmour, J.W., "Low-dose Aluminum and Boron Impants in 4H and 6H Silicon Carbide," *Journal of Applied Physics* 90(6), 2796, 2001.

91. Jones, K.A., Ervin, M.H., Shah, P.B., Derenge, M.A., Vispute, R.D., Venkatesan, T., and Freitas, J.A., "Electrical Activation Processes in Ion Implanted SiC Device Structures," *AIP Conference Proceedings* 680(1), 694, 2003.

92. Negoro, Y., Katsumoto, K., Kimoto, T., and Matsunami, H., "Electronic Behaviors of High-Dose Phosphorus-Ion Implanted 4H-SiC (0001)," *Journal of Applied Physics* 96(1), 224, 2004.

93. Negoro, Y., Kimoto, T., and Matsunami, H., "Aluminum-Ion Implantation into 4H-SiC (11–20) and (0001)" , in *Materials Research Society Symposium Proceedings*, vol. 815, Eds. Dudley, M., Gouma, P., Kimoto, T., Neudeck, P.G., and Saddow, S.E., Materials Research Society, Warrendale, PA, 2004, p. 217.

94. Saks, N.S., Suvorov, A.V., and Capell, D.C., "High-Temperature High-Dose Implantation of Aluminium in 4H-SiC," *Appled Physics Letters* 84(25), 5195, 2004.

95. Schmid, F. and Pensl, G., "Comparison of the Electrical Activation of P+ and N+ Ions Co-Implanted Along With Si+ or C+ Ions Into 4H-SiC," *Applied Physics Letters* 84(16), 3064, 2004.

96. Negoro, Y., Kimoto, T., and Matsunami, H., "Technological Aspects of Ion Implantation in SiC Device Processes", in Silicon Carbide and Related Materials 2004, *Materials Science Forum*, vols. 483–485, Eds. Nipoti, R., Poggi, A., and Scorzoni, A., Trans Tech Publicatons, Switzerland, 2005, p. 599.

97. Zhao, J.H., Tone, K., Weiner, S., R., Caleca, M.A., Du, H., and Withrow, S.P., "Evaluation of Ohmic Contacts to P-Type 6H-SiC Created by C and Al Coimplantation," *IEEE Electron Device Letters* 18(8), 375, 1997.

98. Capano, M.A., Ryu, S., Melloch, M.R., Cooper, J.A., Jr., and Buss, M.R., "Dopant Activation and Surface Morphology of Ion Implanted 4H- and 6H-Silicon Carbide," *Journal of Electronic Materials* 27(4), 370, 1998.

99. Rao, S., Saddow, S.E., Bergamini, F., Nipoti, R., Emirov, Y., and Agrawal, A., "A Robust Process for Ion Implant Annealing of SiC in a Low-Pressure Silane Ambient", in Silicon Carbide 2004—Materials, Processing, and Devices, *Materials Research Society Symposium Proceedings*, vol. 815, Eds. Dudley, M., Gouma, P., Kimoto, T., Neudeck, P.G., and Saddow, S.E., Materials Research Society, Warrendale, PA, 2004, p. 229.

100. Pensl, G., Frank, T., Krieger, M., Laube, M., Reshanov, S., Schmid, F., and Weidner, M., "Implantation-Induced Defects in Silicon Carbide," *Physica B: Condensed Matter* 340–342, 121, 2003.

101. Anwand, W., Brauer, G., Wirth, H., Skorupa, W., and Coleman, P.G., "The Influence of Substrate Temperature On the Evolution of Ion Implantation-Induced Defects in Epitaxial 6H-SiC," *Applied Surface Science* 194(1–4), 127, 2002.

102. Ohno, T., Onose, H., Sugawara, Y., Asano, K., Hayashi, T., and Yatasuo, T., "Electron Microscopic Study on Residual Defects of Al+ or B+ Implanted 4H-SiC," *Journal of Electronic Materials* 28(3), 180, 1999.

103. Huang, C.-F. and Cooper, J.A., Jr., "High Current Gain 4H-SiC npn Bipolar Junction Transistors," *IEEE Electron Device Letters* 24(6), 396, 2003.

104. Rhoderick, E.H. and Williams, R.H., Metal-Semiconductor Contacts Clarendon Press, Oxford, UK, 1988.

105. Porter, L.M. and Davis, R.F., "A Critical Review of Ohmic and Rectifying Contacts for Silicon Carbide," *Materials Science and Engineering B* B34, 83, 1995.

106. Bozack, M.J., "Surface Studies on SiC as Related to Contacts," *Physica Status Solidi (b)* 202(1), 549, 1997.

107. Crofton, J., Porter, L.M., and Williams, J.R., "The Physics of Ohmic Contacts to SiC," *Physica Status Solidi (b)* 202(1), 581, 1997.

108. Saxena, V. and Steckl, A.J., "Building Blocks for SiC Devices: Ohmic Contacts, Schottky Contacts, and p-n Junctions," in *Semiconductors and Semimetals*, vol. 52, Academic Press, New York, 1998, p. 77.

109. Madsen, L., "Formation of Ohmic Conacts to α-SiC and Their Impact on Devices," *Journal of Electronic Materials* 30(10), 1353, 2001.

110. Tanimoto, S., Okushi, H., and Arai, K., "Ohmic Contacts for Power Devices on SiC," in *Silicon Carbide: Recent Major Advances*, Eds. Choyke, W.J., Matsunami, H., and Pensl, G., Springer, Berlin, 2003, p. 651.

111. Okojie, R.S., Lukco, D., Chen, Y.L., and Spry, D.J., "Reliability assessment of Ti/TaSi2/Pt ohmic contacts on SiC after 1000 h at 600 °C," *Journal of Applied Physics* 91(10), 6553, 2002.

112. Spry, D., Neudeck, P., Okojie, R., Chen, L.-Y., Beheim, G., Meredith, R., Mueller, W., and Ferrier, T., "Electrical Operation of 6H-SiC MESFET at 500 °C for 500 h in Air Ambient," in *Proceedings 2004 IMAPS International Conference and Exhibition on High Temperature Electronics (HiTEC 2004)*, May 19–12, 2004, Santa Fe, NM, International Microelectronics and Packaging Society (IMAPS), Washington, DC, 2004.

113. Infineon Technologies, http://www.infineon.com/sic.

114. Itoh, A. and Matsunami, H., "Analysis of Schottky Barrier Heights of Metal/SiC Contacts and Its Possible Application to High-Voltage Rectifying Devices," *Physica Status Solidi (a)* 162(1), 389, 1997.

115. Teraji, T., Hara, S., Okushi, H., and Kajimura, K., "Ideal Ohmic Contact to n-Type 6H-SiC by Reduction of Schottky Barrier Height," *Applied Physics Letters* 71(5), 689, 1997.

116. Schnabel, C.M., Tabib-Azar, M., Neudeck, P.G., Bailey, S.G., Su, H.B., Dudley, M., and Raffaelle, R.P., "Correlation of EBIC and SWBXT Imaged Defects and Epilayer Growth Pits in 6H-SiC Schottky Diodes," in Silicon Carbide and Related Materials 1999, *Materials Science Forum*, vols. 338–342, Eds. Carter, C.H., Jr., Devaty, R.P., and Rohrer, G.S., Trans Tech Publications, Switzerland, 2000, p. 489.

117. Manfredotti, C., Vittone, E., Paolini, C., Olivero, P., Lo Giudice, A., Jaksic, M., and Barrett, R., "Investigation of 4H-SiC Schottky Diodes by Ion and X-ray Micro Beam Induced Charge Collection Techniques," *Diamond and Related Materials* 12(3–7), 667, 2003.

118. Wang, Y., Ali, G.N., Mikhov, M.K., Vaidyanathan, V., Skromme, B.J., Raghothamachar, B., and Dudley, M., "Correlation Between Morphological Defects, Electron Beam-Induced Current Imaging, and the Electrical Properties of 4H-SiC Schottky Diodes," *Journal of Applied Physics* 97(1), 013540, 2005.

119. Bhatnagar, M., Baliga, B.J., Kirk, H.R., and Rozgonyi, G.A., "Effect of Surface Inhomogenities on the Electrical Characteristics of SiC Schottky Contacts," *IEEE Transactions on Electron Devices* 43(1), 150, 1996.

120. Crofton, J. and Sriram, S., "Reverse Leakage Current Calculations for SiC Schottky Contacts," *IEEE Transactions on Electron Devices* 43(12), 2305, 1996.

121. Defives, D., Noblanc, O., Dua, C., Brylinski, C., Barthula, M., Aubry-Fortuna, V., and Meyer, F., "Barrier Inhomogeneities and Electrical Characteristics of Ti/4H-SiC Schottky Rectifiers," *IEEE Transactons on Electron Devices* 46(3), 449, 1999.

122. Yih, P.H., Saxena, V., and Steckl, A.J., "A Review of SiC Reactive Ion Etching in Fluorinated Plasmas," *Physica Status Solidi (b)* 202(1), 605, 1997.

123. Beheim, G., "Deep Reactive Ion Etching of Silicon Carbide," in *The MEMS Handbook*, Ed. Gad-el-Hak, M., CRC Press, Boca Raton, FL, 2002, p. 21-1.

124. Zorman, C.A. and Mehregany, M., "Micromachining of SiC," in *Silicon Carbide: Recent Major Advances*, Eds. Choyke, W.J., Matsunami, H., and Pensl, G., Springer, Berlin, 2003, p. 671.

125. Cao, L., Li, B., and Zhao, J.H., "Inductively Coupled Plasma Etching of SiC for Power Switching Device Fabrication", in Silicon Carbide, III-Nitrides, and Related Materials, *Materials Science Forum*, vols. 264–268, Eds. Pensl, G., Morkoc, H., Monemar, B., and Janzen, E., Trans Tech Publications, Switzerland, 1998, p. 833.

126. McLane, G.F. and Flemish, J.R., "High Etch Rates of SiC in Magnetron Enhanced SF_6 Plasmas," *Applied Physics Letters* 68(26), 3755, 1996.

127. Chabert, P., Proust, N., Perrin, J., and Boswell, R.W., "High rate etching of 4H–SiC using a SF_6/O_2 helicon plasma," *Applied Physics Letters* 76(16), 2310, 2000.

128. Cho, H., Leerungnawarat, P., Hays, D.C., Pearton, S.J., Chu, S.N.G., Strong, R.M., Zetterling, C.-M., Östling, M., and Ren, F., "Ultradeep, low-damage dry etching of SiC," *Applied Physics Letters* 76(6), 739, 2000.

129. Shor, J.S. and Kurtz, A.D., "Photoelectrochemical Etching of 6H-SiC," *Journal of the Electrochemical Society* 141(3), 778, 1994.

130. Shor, J.S., Kurtz, A.D., Grimberg, I., Weiss, B.Z., and Osgood, R.M., "Dopant-Selective Etch Stops in 6H and 3C SiC," *Journal of Applied Physics* 81(3), 1546, 1997.

131. Kim, S., Bang, B.S., Ren, F., D'Entremont, J., Blumenfeld, W., Cordock, T., and Pearton, S.J., "SiC Via Holes by Laser Drilling," *Journal of Electronic Materials* 33(5), 477, 2004.

132. Pierret, R.F., *Field Effect Devices*, Addison-Wesley, Reading, MA, 1983.

133. Afanas'ev, V.V., Ciobanu, F., Pensl, G., and Stesmans, A., "Contributions to the Density of Interface States in SiC MOS Structures," in *Silicon Carbide: Recent Major Advances*, Eds. Choyke, W.J., Matsunami, H., and Pensl, G., Springer, Berlin, 2003, p. 343.

134. Agarwal, A., Das, M., Krishnaswami, S., Palmour, J., Richmond, J., and Ryu, S.H., "SiC Power Devices-An Overview," in Silicon Carbide 2004—Materials, Processing and Devices, *Materials Research Society Symposium Proceedings*, vol. 815, Eds. Dudley, M., Gouma, P., Kimoto, T., Neudeck, P.G., and Saddow, S.E., Materials Research Society, Warrendale, PA, 243, 2004.

135. Agarwal, A., Ryu, S.-H., and Palmour, J., "Power MOSFETs in 4H-SiC: Device Design and Technology," in *Silicon Carbide: Recent Major Advances*, Eds. Choyke, W.J., Matsunami, H., and Pensl, G., Springer, Berlin, 2003, p. 785.

136. Dimitrijev, S., Harrison, H.B., Tanner, P., Cheong, K.Y., and Han, J., "Properties of Nitrided Oxides on SiC," in *Silicon Carbide: Recent Major Advances*, Eds. Choyke, W.J., Matsunami, H., and Pensl, G., Springer, Berlin, 2003, p. 373.

137. Dimitrijev, S. and Jamet, P., "Advances in SiC power MOSFET technology," *Microelectronics Reliability* 43(2), 225, 2003.

138. Saks, N.S., "Hall Effect Studies of Electron Mobility and trapping at the SiC/SiO_2 Interface," in *Silicon Carbide: Related Major Advances*, Eds. Choyke, W.J., Matsunami, H., and Pensl, G., Springer-Berlin, 2003, p. 387.

139. Kimoto, T., Kanzaki, Y., Noborio, M., Kawano, H., and Matsunami, H., "MOS Interface Properties and MOSFET Performance on 4H-SiC and Non-Basal Faces Processed by N_2O Oxidation," in Silicon Carbide 2004—Materials, Processing and Devices, *Materials Research Society Symposium Proceedings*, vol. 815, Eds. Dudley, M., Gouma, P., Kimoto, T., Neudeck, P.G., and Saddow, S.E., Materials Research Society, Warrendale, PA, 2004, p. 199.

140. Krishnaswami, S., Das, M.K., Agarwal, A.K., and Palmour, J.W., "Reliability of Nitrided Oxides in n- and p-Type 4H-SiC MOS Structures," in *Materials Research Society Symposium Proceedings*, vol. 815, Eds. Dudley, M., Gouma, P., Kimoto, T., Neudeck, P.G., and Saddow, S.E., Materials Research Society, Warrendale, PA, 2004, p. 205.

141. Das, M.K., "Recent Advances in (0001) 4H-SiC MOS Device Technology," in Silicon Carbide and Related Materials 2003, *Materials Science Forum*, vol. 457–460, Eds. Madar, R., Camassel, J., and Blanquet, E., Trans Tech, Switzerland, 2004, p. 1275.

142. Dhar, S., Wang. S, Williams, J.R., Pantelides, S.T., Feldman, L.C., "Interface Passivation for Silicon Dioxide Layers on Silicon Carbide," *MRS Bulletin* 30(4), 288, 2005.

143. Gudjonsson, G., Olafsson, H.O., Allerstam, F., Nilsson, P.-A., Sveinbjornsson, E.O., Zirath, H., Rodle, T., and Jos, R., "High Field-Effect Mobility in n-Channel Si Face 4H-SiC MOSFETs With Gate Oxide Grown On Aluminum Ion-Implanted Material," *IEEE Electron Device Letters* 26(2), 96, 2005.

144. Sveinbjornsson, E.O., Olafsson, H.O., Gudjonsson, G., Allerstam, F., Nilsson, P.A., Syvajarvi, M., Yakimova, R., Hallin, C., Rodle, T., and Jos, R., "High Field Effect Mobility in Si Face 4H-SiC MOSFET Made on Sublimation Grown Epitaxial Material," in Silicon Carbide and Related Materials 2004, *Materials Science Forum* vol. 483–485, Eds. Nipoti, R., Poggi, A., and Scorzoni, A., Trans Tech, Switzerland, 841, 2005.

145. Agarwal, A.K., Seshadri, S., and Rowland, L.B., "Temperature Dependence of Fowler-Nordheim Current in 6H- and 4H-SiC MOS Capacitors," *IEEE Electron Device Letters* 18(12), 592, 1997.

146. Tanimoto, S., Tanaka, H., Hayashi, T., Shimoida, Y., Hoshi, M., and Mihara, T., "High-reliability ONO Gate Dielectric for Power MOSFETs", in Silicon Carbide and Related Materials 2004, *Materials Science Forum*, vol. 677–680, Eds. Nipoti, R., Poggi, A., and Scorzoni, A., Trans Tech, Switzerland, 2005, p. 677.

147. Johnson, R.W. and Williams, J., "SiC Power Device Packaging Technologies for 300 to 350 °C Applications," *Materials Science Forum* vol. 483–485, Eds. Nipoti, R., Poggi, A., and Scorzoni, A., Trans Tech, Switzerland, 785, 2005.

148. Salmon, J.S., Johnson, R.W., and Palmer, M., "Thick Film Hybrid Packaging Techniques for 500 °C Operation," in *4th International High Temperature Electronics Conference*, June 14–18, 1998, Albuquerque, NM, IEEE, Piscataway, NJ, 1998, p. 103.

149. Chen, L.-Y., Okojie, R.S., Neudeck, P.G., Hunter, G.W., and Lin, S.-T., "Material System for Packaging 500 °C MicroSystem," in Microelectronic, Optoelectronic, and MEMS Packaging, *Materials Research Society Symposia Proceedings*, vol. 682, Eds. Boudreaux, J.C., Dauskardt, R.H., Last, H.R., and McCluskey, F.P., Materials Research Society, Warrandale, PA, 2001.

150. Savrun, E., "Packaging Considerations for Very High Temperature Microsystems," in *Proceedings 2004 IMAPS International Conference and Exhibition on High Temperature Electronics (HiTEC 2004)*, May 19–12, 2004, Santa Fe, NM, International Microelectronics and Packaging Society (IMAPS), Washington, DC, 2004.

151. Casady, J.B., Dillard, W.C., Johnson, R.W., and Rao, U., "A hybrid 6H-SiC temperature sensor operational from 25 °C to 500 °C," *IEEE Transactions on Components, Packaging, and Manufacturing Technology, Part A* 19(3), 416, 1996.

152. Okojie, R.S., Savrun, E., Phong, N., Vu, N., and Blaha, C., "Reliability Evaluation of Direct Chip Attached Silicon Carbide Pressure Transducers," in *Proceedings of the 2004 IEEE Sensors Conference*, 2004, p. 635.

153. Edmond, J., Kong, H., Suvorov, A., Waltz, D., and Carter, C., Jr., "6H-Silicon Carbide Light-Emitting Diodes and UV Photodiodes," *Physica Status Solidi (a)* 162(1), 481, 1997.

154. Bergh, A.A. and Dean, P.J., *Light-Emitting Diodes*, Clarendon Press, Oxford, 1976.

155. Edmond, J., Abare, A., Bergman, M., Bharathan, J., Lee Bunker, K., Emerson, D., Haberern, K., Ibbetson, J., Leung, M., Russel, P., and Slater, D., "High Efficiency GaN-Based LEDs and Lasers on SiC," *Journal of Crystal Growth* 272(1–4), 242, 2004.

156. Brown, D.M., Downey, E., Kretchmer, J., Michon, G., Shu, E., and Schneider, D., "SiC Flame Sensors for Gas Turbine Control Systems," *Solid-State Electronics* 42(5), 755, 1998.

157. Bertuccio, G., Binetti, S., Caccia, S., Casiraghi, R., Castaldini, A., Cavallini, A., Lanzieri, C., LeDonne, A., Nava, F., and Pizzini, S., "Silicon Carbide for Alpha, Beta, Ion and Soft X-Ray High-Performance Detectors," in *Materials Science Forum* vols. 483–485, Eds. Nipoti, R., Poggi, A., and Scorzoni, A., Trans Tech, Switzerland, 2005, p. 1015.

158. Seshadri, S., Dulloo, A., Ruddy, F., Seidel, J., and Rowland, L., "Demonstration of an SiC Neutron Detector for High-Radiation Environments," *IEEE Transactions on Electron Devices* 46(3), 567, 1999.

159. Sriram, S., Ward. A., Henning, J., and Allen. S.T., "SiC MESFETs for High-Frequency Applications," *MRS Bulletin* 30(4), 308, 2005.

160. Fazi, C. and Neudeck, P., "Use of Wide-Bandgap Semiconductors to Improve Intermodulation Distortion in Electronic Systems," in Silicon Carbide, III-Nitrides, and Related Materials, *Materials Science Forum*, vols. 264–268, Eds. Pensl, G., Morkoc, H., Monemar, B., and Janzen, E., Trans Tech Publications, Switzerland, 1998, p. 913.

161. Eriksson, J., Rorsman, N., and Zirath, H., "4H-Silicon Carbide Schottky Barrier Diodes for Microwave Applications," *IEEE Transactions on Microwave Theory and Techniques* 51(3), 796, 2003.

162. Simons, R.N. and Neudeck, P.G., "Intermodulation-distortion performance of silicon-carbide Schottky-barrier RF mixer diodes," *IEEE Transactions on Microwave Theory and Techniques* 51(2), 669, 2003.

163. Honeywell Solid State Electronics Center, http://www.ssec.honeywell.com.

164. Xie, W., Cooper, J.A., Jr., and Melloch, M.R., "Monolithic NMOS Digital Integrated Circuits in 6H-SiC," *IEEE Electron Device Letters* 15(11), 455, 1994.

165. Brown, D.M., Ghezzo, M., Kretchmer, J., Krishnamurthy, V., Michon, G., and Gati, G., "High-Temperature Silicon Carbide Planar IC Technology and First Monolithic SiC Operational Amplifier IC," in *Transactions Second International High Temperature Electronics Conference*, June 5–10, 1994, Charlotte, NC, Sandia National Laboratories, Albuquerque, NM, 1994, p. XI.

166. Brown, D.M., Downey, E., Ghezzo, M., Kretchmer, J., Krishnamurthy, V., Hennessy, W., and Michon, G., "Silicon Carbide MOSFET Integrated Circuit Technology," *Physica Status Solidi (a)* 162(1), 459, 1997.

167. Ryu, S.H., Kornegay, K.T., Cooper, J.A., Jr., and Melloch, M.R., "Digital CMOS ICs in 6H-SiC Operating on a 5 V Power Supply," *IEEE Transactions on Electron Devices* 45(1), 45, 1998.

168. Diogu, K.K., Harris, G.L., Mahajan, A., Adesida, I., Moeller, D.F., and Bertram, R.A., "Fabrication and Characterization of a 83 MHz High-Temperature β-SiC MESFET Operational Amplifier With an AlN Isolation Layer on (100) 6H-SiC," in *54th Annual IEEE Device Research Conference* IEEE, Santa Barbara, CA, 1996, p. 160.

169. Harris, G.L., Wongchotigul, K., Henry, H., Diogu, K., Taylor, C., and Spencer, M.G., "Beta SiC Schottky Diode FET Inverters Grown on Silicon," in Silicon Carbide and Related Materials: Proceedings of the Fifth International Conference, *Institute of Physics Conference Series*, Eds. Spencer, M.G., Devaty, R.P., Edmond, J.A., Kahn, M.A., Kaplan, R., and Rahman, M., IOP Publishing, Bristol, United Kingdom, 1994, p. 715.

170. Neudeck, P.G., Beheim, G.M., and Salupo, C.S., "600 C Logic Gates Using Silicon Carbide JFET's," in *Government Microcircuit Applications Conference Technical Digest*, March 20–23, 2000, Anaheim, CA, p. 421.

171. Pierret, R.F., *Semiconductor Fundamentals*, Addison-Wesley, Reading, MA, 1983.

172. Cooper, J.A., Jr. and Agarwal, A., "SiC Power-Switching Devices—The Second Electronics Revolution?," *Proceedings of the IEEE* 90(6), 956, 2002.

173. Sochacki, M., Lukasiewicz, R., Rzodkiewicz, W., Werbowy, A., Szmidt, J., and Staryga, E., "Silicon Dioxide and Silicon Nitride as a Passivation and Edge Termination for 4H-SiC Schottky Diodes," *Diamond and Related Materials* 14(3–7), 1138, 2005.

174. Chang, S.-C., Wang, S.-J., Uang, K.-M., and Liou, B.-W., "Design and Fabrication of High Breakdown Voltage 4H-SiC Schottky Barrier Diodes With Floating Metal Ring Edge Terminations," *Solid-State Electronics* 49(3), 437, 2005.

175. Hu, S. and Sheng, K., "A New Edge Termination Technique for SiC Power Devices," *Solid-State Electronics* 48(10–11), 1861, 2004.

176. Beck, A.L., Yang, B., Guo, X., and Campbell, J.C., "Edge Breakdown in 4H-SiC Avalanche Photodiodes," *IEEE Journal of Quantum Electronics* 40(3), 321, 2004.

177. Ayalew, T., Gehring, A., Grasser, T., and Selberherr, S., "Enhancement of Breakdown Voltage for Ni-SiC Schottky Diodes Utilizing Field Plate Edge Termination," *Microelectronics Reliability* 44(9–11), 1473, 2004.

178. Nigam, S., Kim, J., Luo, B., Ren, F., Chung, G.Y., Pearton, S.J., Williams, J.R., Shenai, K., and Neudeck, P., "Influence of Edge Termination Geometry On Performance of 4H-SiC p-i-n Rectifiers," *Solid-State Electronics* 47(1), 61, 2003.
179. Tarplee, M.C., Madangarli, V.P., Zhang, Q., and Sudarshan, T.S., "Design Rules for Field Plate Edge Termination in SiC Schottky Diodes," *IEEE Transactions on Electron Devices* 48(12), 2659, 2001.
180. Chow, T.P., "SiC Bipolar Power Devices," *MRS Bulletin* 30(4), 299, 2005.
181. Sugawara, Y., "High Voltage SiC Devices," in *Silicon Carbide: Recent Major Advances*, Eds. Choyke, W.J., Matsunami, H., and Pensl, G., Springer, Berlin, 2003, p. 769.
182. Rupp, R., Treu, M., Mauder, A., Griebl, E., Werner, W., Bartsch, W., and Stephani, D., "Performance and Reliability Issues of SiC-Schottky Diodes," in Silicon Carbide and Related Materials 1999, *Materials Science Forum*, vols. 338–342, Eds. Carter, C.H., Jr., Devaty, R.P., and Rohrer, G.S., Trans Tech Publications, Switzerland, 2000, p. 1167.
183. Polyakov, A.Y., Li, Q., Huh, S.W., Skowronski, M., Lopatiuk, O., Chernyak, L., and Sanchez, E., "Minority carrier diffusion length measurements in 6H-SiC," *Journal of Applied Physics* 97(5), 053703, 2005.
184. Lendenmann, H., Bergman, J.P., Dahlquist, F., and Hallin, C., "Degradation in SiC Bipolar Devices: Sources and Consequences of Electrically Active Dislocations in SiC," in Silicon Carbide and Related Materials—2002, *Materials Science Forum*, vols. 433–436, Eds. Bergman, P. and Janzen, E., Trans Tech Publications, Switzerland, 2003, p. 901.
185. Schoen, K.J., Henning, J.P., Woodall, J.M., Cooper, J.A., Jr., and Melloch, M.R., "A Dual-Metal-Trench Schottky Pinch-Rectifier in 4H-SiC," *IEEE Electron Device Letters* 19(4), 97, 1998.
186. Chow, T.P., Ramungul, N., Fedison, J., and Tang, Y., "SiC Power Bipolar Transistors and Thyristors," in *Silicon Carbide: Recent Major Advances*, Eds. Choyke, W.J., Matsunami, H., and Pensl, G., Springer, Berlin, 2003, p. 737.
187. Zhao, J.H., "Silicon Carbide Power Field-Effect Transistors," *MRS Bulletin* 30(4), 293, 2005.
188. Friedrichs, P., "Charge Controlled Silicon Carbide Switching Devices," in Silicon Carbide 2004—Materials, Processing and Devices, *Materials Research Society Symposium Proceedings*, vol. 815, Eds. Dudley, M., Gouma, P., Kimoto, T., Neudeck, P.G., and Saddow, S.E., Materials Research Society, Warrendale, PA, 2004, p. 255.
189. Levinshtein, M.E., Ivanov, P.A., Agarwal, A.K., and Palmour, J.W., "Optical Switch-On Of Silicon Carbide Thyristor," *Electronics Letters* 38(12), 592, 2002.
190. Mehregany, M., Zorman, C., Narayanan, N., and Wu, C.H., "Silicon Carbide MEMS for Harsh Environments," *Proceedings of the IEEE* 14(8), 1998.
191. Epstein, A.H., "Millimeter-Scale, Micro-ElectroMechanical Systems Gas Turbine Engines," *ASME Journal of Engineering for Gas Turbines and Power* 126, 205, 2004.
192. FLX Micro, http://www.flxmicro.com.
193. Masheeb, F., Stefanescu, S., Ned, A.A., Kurtz, A.D., and Beheim, G., "Leadless Sensor Packaging For High-Temperature Applications," in *The Fifteenth IEEE International Conference on Micro Electro Mechanical Systems*, January 20–24, 2002, Las Vegas, NV, IEEE, Piscataway, NJ, 2002, p. 392.
194. Hunter, G.W., Neudeck, P.G., Xu, J., Lukco, D., Trunek, A., Artale, M., Lampard, P., Androjna, D., Makel, D., and Ward, B., "Development of SiC-Based Gas Sensors for Aerospace Applications," in Silicon Carbide 2004—Materials, Processing and Devices, *Materials Research Society Symposium Proceedings*, vol. 815, Eds. Dudley, M., Gouma, P., Kimoto, T., Neudeck, P.G., and Saddow, S.E., Materials Research Society, Warrendale, PA, 2004, p. 287.
195. Spetz, A.L. and Savage, S., "Advances in SiC Field Effect Gas Sensors," in *Silicon Carbide: Recent Major Advances*, Eds. Choyke, W.J., Matsunami, H., and Pensl, G., Springer, Berlin, 2003, p. 869.
196. Pensl, G., Morkoc, H., Monemar, B., and Janzen, E., *Silicon Carbide, III-Nitrides, and Related Materials*, Trans Tech Publications, Switzerland, 1998.
197. Harris, G.L., *Properties of SiC*, The Institute of Electrical Engineers, London, 1995.

198. Kemerley, R.T., Wallace, H.B., and Yoder, M.N., "Impact of Wide Bandgap Microwave Devices on DoD Systems," *Proceedings of the IEEE* 90(6), 1059, 2002.

199. Pearton, S.J., Zolper, J.C., Shul, R.J., and Ren, F., "GaN: Processing, defects, and devices," *Journal of Applied Physics* 86(1), 1, 1999.

200. Dudley, M., Gouma, P., Kimoto, T., Neudeck, P.G., and Saddow, S.E., Silicon Carbide 2004—Materials, Processing and Devices, *Materials Research Society Symposium Proceedings*, vol. 815, Materials Research Society, Warrendale, PA, 2004.

6

Passive Components

Ashraf Lotfi

Enpirion, Inc.

CONTENTS

6.1 Magnetic Components

6.1.1 Integration Issues

It is well known and recognized that magnetic components are to be avoided when designing integrated circuits (ICs) due to their lack of integrability. New developments in the field of magnetic component fabrication are promising devices that can be integrated and miniaturized using monolithic fabrication techniques as opposed to today's bulk methods. The driving forces for such developments rest in certain applications that benefit or rely on inductive or magnetically coupled devices using ferromagnetic media. Examples of such applications include tuned RF tanks, matching networks, DC–DC power conversion and regulation, network filters, and line isolators/couplers.

Emerging applications requiring more mobility, lower power dissipation, and smaller component and system sizes have been drivers for the development of highly integrated systems and subsystems. To match these trends, it has become necessary to be able to integrate high-quality magnetic devices (i.e., inductors and transformers) with the systems they operate in as opposed to being stand-alone discrete devices. Not only does their discrete nature prevent further miniaturization, but their very nature also hampers improved performance (e.g., speed).

The main features of a monolithic magnetic device are:

1. High values of inductance compared to air core spirals.
2. Enhanced high-frequency performance.
3. Energy storage, DC bias, and power-handling capabilities.
4. Use of ferromagnetic materials as a magnetic core.
5. Photolithographic fabrication of windings and magnetic core.

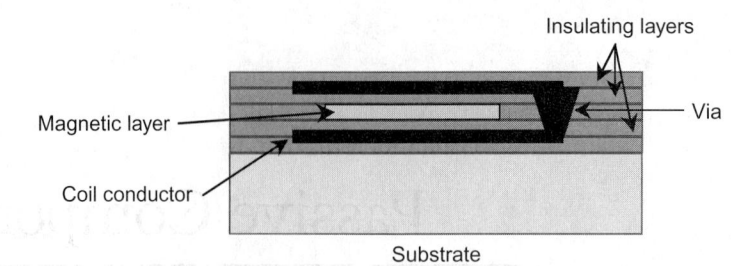

FIGURE 6.1 Cross-section of a monolithic micromagnetic device fabricated using IC methods.

6. Multilayer mask fabrication for complete magnetic device design.
7. Standard or semistandard IC-processing techniques.

Owing to the use of standard photolithography, etching, and patterning methods for their fabrication, monolithic magnetic devices may appear compatible with IC processes. However, two main characteristics make these devices more suitably fabricated off-line from a mainstream IC process:

1. *Coarser design rules.* Usually magnetic device designs do not require submicron geometries as demanded by semiconductor designs. This discrepancy means that an expensive submicron process for these components would unnecessarily raise the device cost.
2. *Use of ferromagnetic core materials.* The use of iron, cobalt, nickel, and their alloys is at the heart of a high-quality magnetic device. Some of these materials are alien and contaminating to semiconductor cleanrooms. As a result, processing sequences and logistics for full integration with semiconductors is still in the development phases.

With these two major differences, integration of magnetics and semiconductors may require the use of multichip modules or single package multidie cases. Full integration into a single monolithic die requires separate processing procedures using the same substrate [1,2,3,5].

The construction of a monolithic micromagnetic device fabricated on a substrate such as silicon or glass is shown in Figure 6.1. In this diagram, the magnetic layer is sandwiched between upper and lower conductor layers that are connected together by means of an electrically conducting "via." This structure is referred to as a *toroidal device* from its discrete counterpart.

Conversely, a dual structure can be made where two magnetic layers sandwich the conductor layer (or layers). This dual structure can be referred to as an *EE device* since it is derived from the standard discrete "EE" core type. In either case, as required by the operation of any magnetically coupled device, the magnetic flux path in the magnetic film and the current flow in the coil conductor are orthogonal in accordance with Ampere's circuital law. Interlayer insulation between conductors and the magnetic layer(s) is necessary, both to reduce capacitive effects and to provide a degree of electrical voltage breakdown. These parameters are affected by the choice of insulator systems used in microfabricated circuits due to the differing values of dielectric constants and breakdown voltages used. Some commonly used insulator systems include silicon dioxide and polyimide, each of which has distinctly different processing methods and physical characteristics. Conductor layers for the coil windings can be fabricated using standard aluminum metallization. In some cases, copper conductors are a better choice due to its higher conductivity and hence lower resistive losses. This is especially important if the device is to handle any significant power. The magnetic film layer is a thin film of a chosen magnetic material typically between 1 and 10 µm in thickness. Such materials can be routinely deposited by standard techniques such as sputtering or electrodeposition. The specific method chosen must yield magnetic films with the desired properties, namely permeability, parallel loss resistance, and maximum flux density. These parameters vary with the deposition conditions and techniques, so significant development and optimization has occurred to produce desirable results.

Since the design may call for energy storage and hence gaps in the core, the fabrication method can be modified to incorporate these features. Figure 6.2 shows the geometry of a planar magnetic core with a gap produced by photolithography. In this case, the gap is formed as a result of the artwork generated

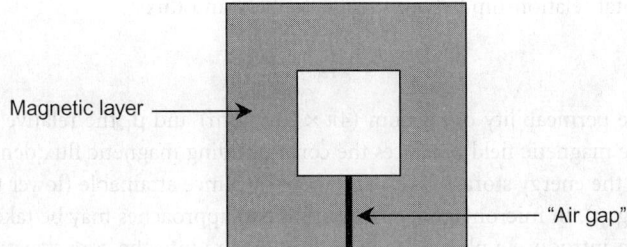

FIGURE 6.2 Planar, single-layer magnetic core configuration for energy storage (top view).

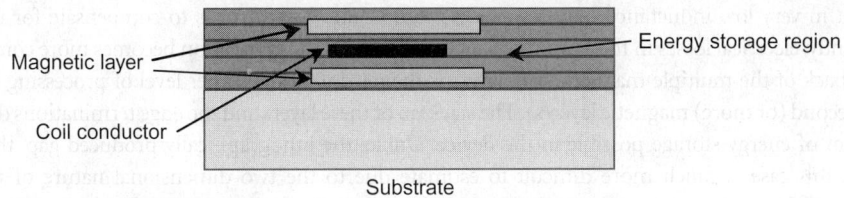

FIGURE 6.3 Multiple magnetic layers for energy storage micromagnetics (cross-sectional view).

for the core design. Figure 6.3 shows the design of a gap using multilayer magnetic films. The energy storage region exists in the insulation between the two magnetic layers.

The fabrication and construction of conductors for the coil usually involves depositing standard interconnect metals (e.g., aluminum) by sputter deposition. The thicknesses are chosen based on the current-carrying capability and the frequency of operation as well as the desired configuration (inductor or transformer). The DC resistance of the conductors must be minimized to reduce DC losses, but the conductor thickness and arrangement must also result in minimal AC losses. This can be accomplished by reduced resistivity (i.e., copper versus aluminum) and by multiple conductor layers to reduce skin and proximity effects at high frequencies.

6.1.2 Designs for Integrated Circuits

Unlike discrete magnetic components, monolithic micromagnetic devices are designed to operate at substantially higher frequencies. Owing to their integrated nature and smaller physical size interconnection and coupling parasitics are lower thus enabling high-frequency response. However, the smaller physical size also places upper limits on characteristics such as inductance, current levels, power levels, and dissipation. With these limits the maximum energy storage, E, is lower. In any inductor, the energy stored due to the current flowing (I) is related to the magnetic fields in the volume of the device by

$$E = \frac{1}{2} L I^2 = \frac{1}{2} \iiint \overline{B} \overline{H} \, dv \qquad (6.1)$$

where L is the transformer or inductor's inductance in Henries and I (A) the maximum current carried by the corresponding winding. This is related to the magnetic flux, $\overline{B}$, and magnetic field, H, present in the volume of the device. So with a small physical volume one can see from Eq. (6.1) that the energy stored is also small. This limited energy storage capability limits these devices to operate in low power circuits. In order to obtain a high B–H product for more energy storage, a combination of high- and low-permeability regions should be fabricated, i.e., a gap in the high-permeability path is introduced. This gap region helps to maintain a high flux density as well as an appreciable field. The highly permeable region however, while being able to maintain high flux density does not support large magnetic fields due to the fundamental relationship between magnetic field and flux:

$$\vec{B} = \mu_0 \mu_r \vec{H} \tag{6.2}$$

In Eq. (6.2), μ_0 is the permeability of vacuum ($4\pi \times 10^{-7}$ H/m) and μ_r the relative permeability of the medium in which the magnetic field produces the corresponding magnetic flux density. The size of this gap both determines the energy storage levels and the inductance attainable (lower than the inductance attainable without a gap). In micromagnetic fabrication two approaches may be taken to create this "air gap" region. One is to introduce a planar lithographical feature into the core structure (Figure 6.2) and the other is to rely on multiple magnetic core layers separated by insulating layers (Figure 6.3). The drawback of lithographical gap is the limits imposed by the design rules. In this case the gap may not be any smaller than the minimum design rule, which as was mentioned could be quite coarse. Excessive gap sizes result in very low inductance requiring an increase in number of turns to compensate for this drop. Consequently, electrical losses in these windings increase and also the fabrication becomes more complicated. The drawback of the multiple magnetic core layers is the need to add another level of processing to obtain at least a second (or more) magnetic layer(s). The stack up of these layers and the edge terminations determine the amount of energy storage possible in the device. Unlike the lithographically produced gap, the energy storage in this case is much more difficult to estimate due to the two-dimensional nature of the edge-termination fields in the gap region surrounding the multilayer magnetic cores. In uniform field cases, the energy stored in the volume of the gap can be obtained from Eqs. (6.1) and (6.2) due to the continuity and uniformity of the flux density vector in both the core and gap regions giving

$$E = \frac{1}{2}LI^2 \approx \frac{1}{2}\frac{B^2 V_{gap}}{\mu_0} \tag{6.3}$$

where V_{gap} is the volume of the gap region in m³. The approximation is valid as long as the gap region carries a uniform flux density and is "magnetically long" compared to the length of the highly permeable core region (i.e., gap length/$\mu_{r\,gap} \gg$ core length/$\mu_{r\,mag}$). Usually this condition can be satisfied with most ferromagnetic materials of choice, but some ferromagnetic materials may have low enough permeabilities to render this approximation invalid. In this event, some energy is stored within the ferromagnetic material and Eq. (6.3) should be modified. Eq. (6.3) is very useful in determining the size of gap needed to support the desired inductance and current levels for the device. For example, if a 250 nH inductor operating at 250 mA of current bias were needed, the gap volume necessary to support these specifications would be about 2×10^{-5} mm³, assuming a material with a maximum flux density of 1.0 T. In the planar device of Figure 6.2 with nominal magnetic film dimensions of 2 μm in the normal direction and 200 μm in the planar direction, the required gap width would be about 5 μm. Since the gap in this case is obtained by photolithography, the minimum feature size for this process would need to be 5 μm. If a different material of lower maximum flux density capability of 0.5 T were used instead, the rated current level of 250 mA would have to be downgraded to 62 mA to prevent saturation of the magnetic material. Conversely, the gap length of 5 μm could be increased to 20 μm while maintaining the same current level assuming that adjustments are made to the turns to maintain the desired inductance. Such trade-offs are common, but are more involved because of the interaction of gap size with inductance level and number of turns.

Another aspect of the design is the conductor for coil windings for an inductor or for primary and secondary windings in the case of a transformer. The number of turns is usually selected based on the desired inductance and turns ratio (for a transformer), which are typically circuit design parameters. As is well known, the number of turns around a magnetic core gives rise to an inductance, L, given by

$$L = \frac{\mu_0 \mu_r N^2 A}{l} \quad \text{(H)} \tag{6.4}$$

In this relation, N is the number of turns around a magnetic core of cross-sectional area, A (m²) and magnetic path length, l (m). The inductance is reduced by the presence of a gap since this will serve to increase the path length. The choice of conductor thickness is always made in light of the AC losses occurring when conductors carry high-frequency currents. The conductors will experience various current redistribution effects due to the presence of eddy currents induced by the high-frequency magnetic fields surrounding the conductors. The well-known skin effect is one of such effects. Current will crowd toward the surface of the conductor and flow mainly in a thickness related to the skin depth, δ,

$$\delta = \frac{1}{\sqrt{\pi f \mu_0 \sigma}} \ \text{(m)} \tag{6.5}$$

For a copper conductor $\delta = 66/\sqrt{f(\text{MHz})}\,\mu\text{m}$. At 10 MHz the skin depth in copper is 20 μm, placing an upper limit on conductor thickness. When the interconnect metallization is aluminum, $\delta = 81/\sqrt{f(\text{MHz})}\,\mu\text{m}$, so the upper limit at 10 MHz becomes 25 μm of metal thickness. Usually the proximity of conductors to one another forces further optimization because of the introduction of losses due to eddy currents induced by neighboring conductors. In this case, conductor thickness should be further adjusted with respect to the skin depth to reduce the induced eddy currents. The increase in conductor resistance due to the combined skin and proximity effects in a simple primary–secondary winding metallization scheme (shown in Figure 6.4) can be calculated as an increase over the DC resistance of the conductor from

$$Rac = Rdc \frac{1}{2} \frac{h}{\delta} \left\{ \frac{\sinh \frac{h}{\delta} + \sin \frac{h}{\delta}}{\cosh \frac{h}{\delta} - \cos \frac{h}{\delta}} + \frac{\sinh \frac{h}{\delta} - \sin \frac{h}{\delta}}{\cosh \frac{h}{\delta} + \cos \frac{h}{\delta}} \right\} \tag{6.6}$$

In this relationship, h is the thickness of the metallization being used and the Rac is obtained once the DC resistance (Rdc, also a function of h) is known. A distinct minimum for Rac can be obtained and yields the lowest possible AC resistance when

$$\frac{h}{\delta} = \frac{\pi}{2} \tag{6.7}$$

with a corresponding minimum value of AC resistance of

$$Rac = \frac{\pi}{2} Rdc \tanh \frac{\pi}{2} = 1.44 \tag{6.8}$$

When the geometry differs from the simple primary to secondary interface of Figure 6.4 to more turns, layers, shapes, etc., a more complicated analysis is necessary [4]. The simple relation of Eq. (6.7) is no longer valid. Nevertheless, this relation provides a very good starting point for many designs. A qualitative explanation of this behavior stems from the fact that a thicker metallization will produce less DC resistance, but provides poor AC utilization due to current crowding near the surface. On the other

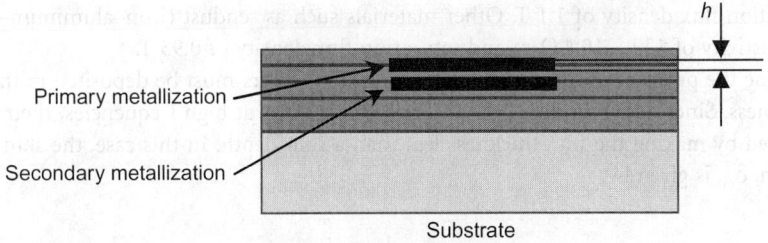

FIGURE 6.4 Configuration of primary and secondary transformer metallization for AC resistance calculation.

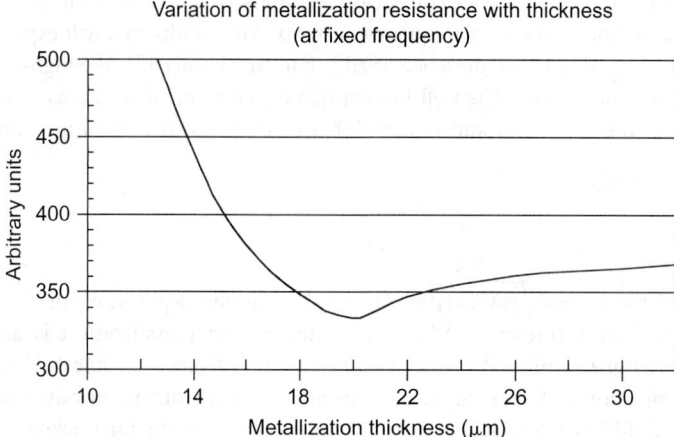

FIGURE 6.5 The variation of high-frequency metallization resistance with metal thickness for the configuration of Figure 6.4. A distinct minimum is observed due to skin and proximity effects.

hand, a thinner metallization increases the DC resistance, while providing better AC conductor utilization. The optimum situation is somewhere in between these two extreme cases as can be seen from Figure 6.5.

The principles presented in this section regarding design issues are at the core of every magnetic component design for ICs. However, many design details especially at elevated frequencies are beyond the scope of this text. It is important to note that many of the limitations on high-frequency designs (100 MHz and higher) are imposed by the properties of the magnetic materials used in the cores of these devices.

6.1.3   Magnetic Core Materials

The most common magnetic materials used for discrete magnetic components operating at higher frequencies are ferrites. This is mainly due to their high resistivity (1–10 Ω m). Despite a low saturation flux density of ~0.3 T, such a high resistivity makes ferrites suitable for applications up to 1 MHz where hysteresis core losses are still limited. When the frequency is raised over 1 MHz the core losses become excessive thus degrading the quality factor and efficiency of the circuit. Moreover, the permeability of all magnetic materials experiences a roll-off beyond a maximum upper frequency. Commonly used ferrites operate up to 1–2 MHz before permeability roll-off occurs. Higher roll-off frequencies are available, but with higher loss factors. Ferrites, however, are not amenable to IC fabrication since they are produced by a high-temperature sintering process. In addition, their low flux saturation levels would not result in the smallest possible device per unit area. A set of more suitable materials for IC fabrication is the magnetic metal alloys usually derived from iron, cobalt, or nickel. These alloys can be deposited as thin films using IC fabrication techniques such as sputtering or electrodeposition and possess saturation flux levels of 0.8 to as high as 2.0 T. Their main drawback due to their metallic nature is much lower resistivity. Permalloy, a common magnetic alloy (~80% nickel and 20% iron) has a resistivity of 20×10^{-8} Ω m, with a saturation flux density of 1.1 T. Other materials such as sendust (iron–aluminum–silicon) have improved resistivity of 120×10^{-8} Ω m and saturation flux density of 0.95 T.

To overcome the problem of low resistivity, the magnetic layers must be deposited in thin films with limited thickness. Since eddy currents flow in the metallic films at high frequencies, their effect can be greatly reduced by making the film thickness less than a skin depth. In this case, the skin depth in the magnetic film, δ_m, is given by

$$\delta_m = \frac{1}{\sqrt{\pi f \mu_0 \mu_r \sigma}} \quad \text{(m)} \tag{6.9}$$

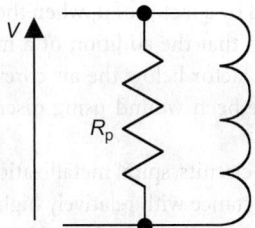

FIGURE 6.6 Magnetic material losses are represented by a parallel shunt resistance.

In a thin film of permalloy ($\mu_r = 2000$), the skin depth at 10 MHz is 3 μm. To limit eddy current losses in the film its thickness must be chosen to be <3 μm. This limitation will conflict with the inductance requirement, since a larger inductance requires a thicker magnetic film (see Eq. (6.4)). Such difficulties can be overcome by depositing the magnetic film in multiple layers insulated from one another to restrict eddy current circulation. Such a structure would still provide the overall thickness needed to achieve the specified inductance while limiting the eddy current loss factor. The use of multilayers also allows the reduction of die size due to the buildup of magnetic core cross section (A in Eq. 6.4) in vertical layers rather than by increasing the planar dimensions. As a result, it can be seen that a trade-off exists between number of layers and die size to yield the most economical die cost.

In addition to eddy current losses due to the low magnetic metal resistivity, hysteresis losses occur as in any magnetic material due to the traversing of the nonlinear B–H loop at the frequency of operation. This is due to the loss of energy needed to rotate magnetic domains within the material. This loss is given by

$$P_{hys} = f \int H \, dB \qquad (6.10)$$

which is the area enclosed by the particular B–H loop demanded by the circuit operation and f the frequency of operation. This loss is expressed in many forms depending on the application. In many cases it is given in the form of a "parallel" or "shunt resistance" and it therefore presents a reduction in impedance to the source as well as a reduction in the overall quality factor of the inductor or transformer. It also represents a finite power loss since this loss is simply V^2/R_p (W), where V is the applied voltage (Fig. 6.6).

Notice that R_p is a nonlinear resistance with both frequency and flux level dependencies. It can be specified at a given frequency and flux level and is usually experimentally measured. It can also be extracted from core loss data usually available in the form

$$P = kf^\alpha B^\beta = \frac{V^2}{R_p} \qquad (6.11)$$

In this relation, k, α, and β are constants for the material on hand. This model is useful for circuit simulation purposes thereby avoiding the nonlinear properties of the magnetic material. However, care should be exercised in using such models since with a large enough excitation, the value of the shunt resistor changes.

6.2   Air Core Inductors

Air core inductors do not use a magnetic core to concentrate the lines of magnetic flux. Instead the flux lines exist in the immediate neighborhood of the coils without tight confinement. As a result, the inductance of an air core coil is considerably lower than one with a magnetic core. In fact, at low frequencies, the quality factor is reduced by a factor of μ_r when the loss factor is small. At high frequencies however, the loss factor may be so high that the addition of a magnetic core and increased inductance actually ends up degrading the quality factor below the air core value. Discrete air core inductors have been used in RF applications and have been wound using discrete magnet or Litz wire onto forming cylinders.

For ICs, especially RF and microwave circuits, spiral metallization deposited on a substrate is a common means to obtain small amounts of inductance with relatively high-quality factors. Inductance values that can be obtained by these techniques are usually in the low nH range (1–20 nH). Estimating inductance values using air cores is much more complicated than in the case of highly permeable cores due to the lack of flux concentration. Formulas have been derived for different spiral air coil shapes (*assuming perfectly insulating substrates*) and are tabulated in several handbooks for inductance calculations [6–9]. An example of a useful inductance formula [9] is

$$L = 4\pi \times 10^{-7} N^2 r_{avg} \left\{ \ln \frac{8r_{avg}}{(r_{out}-r_{in})} + \frac{1}{24} \left(\frac{(r_{out}-r_{in})}{r_{avg}} \right)^2 \left(\ln \frac{8r_{avg}}{(r_{out}-r_{in})} + 3.583 \right) - 0.5 \right\} \quad (6.12)$$

In this formula, r_{out} and r_{in} are the outer and inner radii of the spiral, respectively, and the average radius r_{avg} is

$$r_{avg} = \frac{(r_{out}+r_{in})}{2} \quad (6.13)$$

The formula is an approximation that loses accuracy as the device size becomes large (i.e., large r_{out} with respect to r_{in}).

The loss factors of such devices are strongly influenced by the nonidealities of the substrates and insulators on which they are deposited. For example, aluminum spiral inductors fabricated on silicon with highly doped substrates and epitaxial layers can have significant reduction in quality factor due to the conductivity of the underlying layers. These layers act as ground planes producing the effect of an image of the spiral underneath. This in turn causes a loss in inductance. This can be as much as 30–60% when compared with a spiral over a perfect insulator. In addition, an increase in the loss factor occurs due to circulating eddy currents in the conductive under layers. Increases in the effective resistance of 5–10 times the perfect insulator case is possible, increasing with increased frequency. All these effects can be seen to degrade the performance of these inductors thus requiring design optimization [10–12].

These substrate effects appear in the form of coupling capacitances from the spiral metal to the substrates as well as spreading resistances in the substrate itself. The spreading resistance is frequency-dependent, increasing with higher frequency. The amount of coupling to the substrate depends on the coupling capacitances and hence the separation of the spiral from the substrate. This distance is the dielectric thickness used in the IC process. Only with very large dielectric thicknesses are the substrate effects negligible. In practical cases where it is relatively thin and limited to a few microns, the effects are very large giving an overall quality factor, *Q*, which is significantly lower than the *Q* of the spiral without the substrate. Figure 6.7 shows a typical degradation curve of *Q* on a resistive substrate for "thick" and "thin" separations or dielectric thicknesses. The trends of this curve are also similar if the dielectric thickness variable is replaced by the substrate resistivity as a variable. The exact amount of degradation depends on the separation involved, the dielectric constant, and the resistivity of the substrate. With these quantities known it is possible to construct a circuit model to include these effects and hence solve for the overall quality factor including the substrate effects.

To improve the inductor quality factor on a resistive substrate, some design solutions are possible. One solution to this problem is to increase the substrate resistivity. Another is to design a spiral with

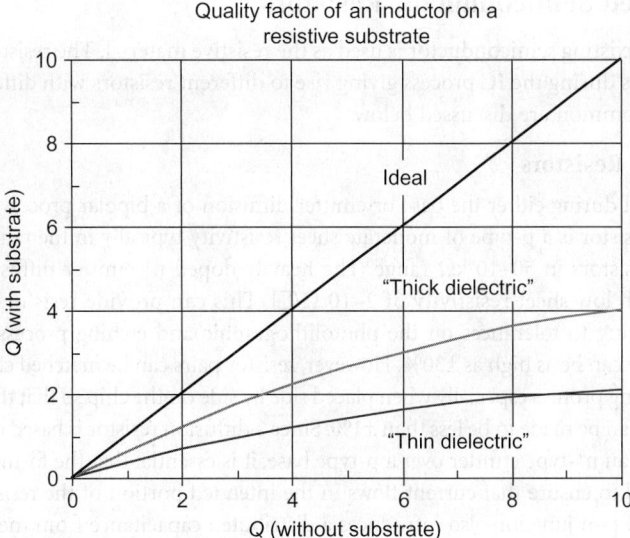

Quality factor of an inductor on a resistive substrate

FIGURE 6.7 Degradation of inductor quality factor by placement on a resistive substrate.

small footprint to reduce coupling to the substrate. In order to offset the increased resistance (which also reduces Q) thicker metallization would be necessary and clearly a trade-off situation arises requiring some design optimization by circuit modeling or more accurately by electromagnetic finite-element analysis.

6.3 Resistors

Resistors have been available for use in ICs for many years. Some of these are made in silicon, so they are directly integrated with the rest of the IC processes. Others, similar to the magnetic device case, are thin-film resistors fabricated in an off-line process that is not necessarily compatible with silicon IC processing. Integrated silicon resistors offer simplicity in fabrication but have less than ideal characteristics with loose tolerances. For this reason, many circuits rely on the ratio of resistor values rather than their absolute values. Thin-film resistors, in contrast, are more superior offering tight tolerances and the ability to trim their absolute value down to very precise values. They also display more stable temperature and frequency dependence.

Usually resistors in ICs are characterized in terms of their sheet resistance rather than their absolute resistance value. Sheet resistance, R_{sheet}, is defined as the resistance of a resistive strip with equal length and width so that

$$R_{sheet} = \frac{\rho}{t} \quad (\Omega/\square) \tag{6.14}$$

where ρ is the material resistivity (Ω m) and t its thickness (m). Once R_{sheet} is given, the resulting resistor value is obtained by multiplying by its length-to-width aspect ratio. To avoid very high aspect ratios, an appropriate sheet resistivity should be used. For example, with $R_{sheet} = 10\ \Omega/\square$ a 10:1 length-to-width ratio would give a 100 Ω resistor. However, to obtain a 1 kΩ resistor it would be better to use a different material, say, $R_{sheet} = 100\ \Omega/\square$ with the same 10:1 ratio instead of using a 100:1 ratio with the low-resistivity material.

6.3.1 Integrated Semiconductor Resistors

In this category the existing semiconductor is used as the resistive material. The resistor may be fabricated at a number of stages during the IC process giving rise to different resistors with different characteristics. Some of the most common are discussed below.

6.3.1.1 Diffused Resistors

These can be formed during either the base or emitter diffusion of a bipolar process. For an npn process the base diffusion resistor is a p-type of moderate sheet resistivity typically in the range of 100–200 $\Omega/\square$. This can provide resistors in 50–10 kΩ range. The heavily doped n$^+$ emitter diffusion will produce an n$^+$-type resistor with low sheet resistivity of 2–10 $\Omega/\square$. This can provide resistors with low values in 1–100 Ω range. Owing to tolerances on the photolithographic and etching processes, the tolerance on the bsolute resistance can be as high as ±30%. However, resistor pairs can be matched closely in temperature coefficients and doping profiles especially when placed side by side on the chip so that the resultant tolerance on the resistor ratio can be made to be less than ±1%. Since a diffusion resistor is based on a p-type base over an n-type epitaxy or an n$^+$-type emitter over a p-type base, it is essential that the formed p–n junctions are always reverse-biased to ensure that current flows in the intended portion of the resistor. The presence of such a reverse-biased p–n junction also introduces a distributed capacitance from the resistor body to the substrate. This will cause high-frequency degradation whereby the resistor value drops from its nominal design value to a lower impedance value due to the shunting capacitance.

6.3.1.2 Pinched Resistors

A variation to the diffused resistor that is used to increase the sheet resistivity of base region is to use the n$^+$-type emitter as a means to reduce the cross-sectional area of the base region, thereby increasing the sheet resistivity. This can increase the sheet resistance to ~1 k$\Omega/\square$. In this case, one end of the n$^+$-type emitter must be tied to one end of the resistor to contain all current flow to the pinched base region.

6.3.1.3 Epitaxial Resistors

High resistor values can be formed using the epitaxial layer since it has higher resistivity than other regions. Epitaxial resistors can have sheet resistances around 5 k$\Omega/\square$. However, epitaxial resistors have even looser tolerances due to the wide tolerances on both epitaxial resistivity and epitaxial layer thickness.

6.3.1.4 MOS Resistors

A MOSFET can be biased to provide a nonlinear resistor. Such a resistor provides much greater values than diffused ones while occupying a much smaller area. With the gate shortened to the drain in a MOSFET, a quadratic relation between current and voltage exists and the device conducts current only when the voltage exceeds the threshold voltage. Under these circumstances, the current flowing in this resistor (i.e., the MOSFET drain current) depends on the width to length ratio of the channel. Hence to increase the resistor value, the aspect ratio of the MOFET should be reduced to give a longer channel length and narrower channel width.

6.3.2 Thin-Film Resistors

As mentioned before in the magnetic core case, a resistive thin-film layer can be deposited (e.g., by sputtering) on the substrate to provide a resistor with very tight absolute value tolerance. In addition, given a large variety of resistor materials a wide range of resistor values can be obtained in small footprints, thereby having very small parasitic capacitances and small temperature coefficients. Some common thin-film resistor materials include tantalum, tantalum nitride, and nickel–chromium. Unlike semiconductor resistors, thin-film resistors can be laser trimmed to adjust their values to very high accuracies of up to 0.01%. Laser trimming can increase the resistor value since the fine beam evaporates a portion of the thin-film material. By its nature, laser trimming is a slow and costly operation that is justified when very high accuracy on absolute values is necessary.

6.4 Capacitors

As in the inductor case, the limitation on integrated capacitors is die size due to the limited capacitance/per unit area available on a die. These limitations are imposed by the dielectrics used with their dielectric constants and breakdown voltages. Most integrated capacitors are either junction capacitors or MOS capacitors [13,14,15,16,17,18].

6.4.1 Junction Capacitors

A junction capacitor is formed when a p–n junction is reversed-biased. This can be formed using the base–emitter, base–collector, or collector–substrate junctions of an npn structure in bipolar ICs. Of course, the particular junction must be maintained in reverse bias to provide the desired capacitance. Since the capacitance arises from the parallel plate effect across the depletion region, whose thickness in turn is voltage-dependent, the capacitance is also voltage-dependent decreasing with increased reverse bias. The capacitance depends on the reverse voltage in the following form:

$$C(V) = \frac{C_0}{(1 + V/\psi_0)^n} \tag{6.15}$$

The built-in potential, ψ_0, depends on the impurity concentrations of the junction being used. For example, $\psi_0 = 0.7$ V for a typical bipolar base–emitter junction. The exponent n depends on the doping profile of the junction. The approximations $n = \frac{1}{2}$ for a step junction and $n = \frac{1}{3}$ for a linearly graded junction are commonly used. The resultant capacitance depends on C_0, the capacitance per unit area with zero bias applied. This depends on the doping level and profile. The base–emitter junction provides the highest capacitance per unit around 1000 pF/mm² with a low breakdown voltage (~5 V). The base–collector junction provides ~100 pF/mm² with a higher breakdown voltage (~40 V).

6.4.2 MOS Capacitors

MOS capacitors are usually formed as parallel plate devices with a top metallization and the high conductivity n⁺ emitter diffusion as the two plates with a thin oxide dielectric sandwiched in between. The oxide is usually a thin layer of SiO_2 with a relative dielectric constant ε_r of 3–4 or Si_3N_4 with ε_r of 5–8. Since the capacitance obtained is $\varepsilon_0 \varepsilon_r A / t_{oxide}$, the oxide thickness, t_{oxide} is critical. The lower limit on the oxide thickness depends on the process yields and tolerances as well as the desired breakdown voltage and reliability. MOS capacitors can provide around 1000 pF/mm² with breakdown voltages up to 100 V. Unlike junction capacitors, MOS capacitors are voltage-independent and can be biased either positively or negatively. Their breakdown, however, is destructive since the oxide fails permanently. Care should be taken to prevent overvoltage conditions.

References

1. N. Saleh and A. Qureshi, "Permalloy thin-film inductors," *Electron. Lett.*, 6, 850–852, 1970.
2. R. Soohoo, "Magnetic film inductors for integrated circuit applications," *IEEE Trans. Magn.*, MAG-15, 1803, 1979.
3. M. Mino, T. Yachi, A. Tago, K. Yanagisawa, K. Sakakakibara, "A new planar microtransformer for use in micro-switching converters," *IEEE Trans. Magn.*, 28, 1969, 1992.
4. J. Vandelac and P. Ziogas, "A novel approach for minimizing high frequency transformer copper loss," *IEEE Trans. Power Electron.*, 3, 266–276, 1988.
5. T. Sato, H. Tomita, A. Sawabe, T. Inoue, T. Mizoguchi, and M. Sahashi, "A magnetic thin film inductor and its application to a MHz switching dc–dc converter," *IEEE Trans. Magn.*, 30, 217–223, 1994.

6. F. Grover, *Inductance Calculations*, Dover, New York, 1946.
7. V. Welsby, *Theory and Design of Inductance Coils*, 2nd ed., MacDonald & Co., London, 1960.
8. C. Walker, *Capacitance, Inductance, and Crosstalk Analysis*, Artech House, Boston, 1990.
9. K.C. Gupta, R. Garg and R. Chadha, *Computer-Aided Design of Microwave Circuits*, Artech House, Dedham, MA, 1981.
10. R. Remke and G. Burdick, "Spiral inductors for hybrid and microwave applications," *Proc. 24th Electron. Components Conf.*, May 1974, pp. 152–161.
11. R. Arnold and J. Pedder, "Microwave characterization of microstrip lines and spiral inductors in MCM-D technology," *IEEE Trans. Comp. Hybrids Manuf. Technol.*, 15, 1038–1045, 1992.
12. N.M. Nguyen and R.G. Meyer, "Si IC-compatible inductors and LC passive filters," *IEEE J. Solid-State Circuits*, 25, 1028–1031, 1990.
13. A. Glaser and G. Subak-Sharpe, *Integrated Circuit Engineering*, Addison-Wesley, Reading, MA, 1977.
14. M.E. Goodge, *Semiconductor Device Technology*, Howard Sams & Co., Inc., Indiana, 1983.
15. A.B. Grebene, *Bipolar and MOS Analog Integrated Circuit Design*, Wiley, New York, 1984.
16. P.R. Gray and R.G. Meyer, *Analysis and Design of Analog Integrated Circuits*, Wiley, New York, 1977.
17. J.E. Sergent and C.A. Harper, *Hybrid Microelectronics Handbook*, McGraw-Hill, New York, 1995.
18. R.A. Levy, *Microelectronic Materials and Processes*, Kluwer Academic Publishers, Dordrecht, The Netherlands, 1989.

7

Power IC Technologies

Akio Nakagawa
Toshiba Corporation

CONTENTS

7.1 Introduction

VLSI technology has advanced so rapidly that gigabit DRAMs have been realized, and the technology faces silicon material limit. Microelectronics mostly advances signal-processing LSIs such as memories and microprocessors. Power systems and the related circuits cannot be outside the influence of VLSI technology [1]. It would be quite strange for power systems alone to still continue to consume a large space while brains become smaller and smaller. In contrast, almost all of the systems require actuators or power devices to control motors, displays, and multimedia equipment. The advance in microelectronics made it possible to integrate large-scale circuits in a small silicon chip, ending up in high system performance and resultant system miniaturization. The system miniaturization inevitably necessitated power IC development. Typical early-developed power ICs were audio power amplifiers, which used bipolar transistors as output devices. pn junction isolation method was well suited to integrate bipolar transistors with control circuits.

Real advancements in intelligent power ICs were triggered by the invention of power DMOSFETs [2] in the 1970s. DMOS transistors have ideal features for output devices of power ICs. No driving DC current is necessary, and large current can be controlled simply by changing the gate voltage. In addition, DMOS switching speed is sufficiently fast.

The on-resistance of vertical DMOSFET has been greatly reduced year after year with the advance in fine lithography in LSI technology. In the mid-1980s, the new concept of Smart Power [3] was introduced. Smart Power integrates bipolar and CMOS devices with vertical DMOS, using a process primarily optimized for polysilicon gate self-aligned DMOS. The main objective is to integrate control and protection circuits with vertical power devices not only to increase device reliability and performance but also to realize easy use of power devices. The concept Smart Power was applied to high-voltage vertical DMOS with drain contact on the backside of the chip because discrete DMOS technology was already well advanced in the early 1980s. Their main application field was the automotive, replacing mechanical relays and eliminating wire harness.

As the technology of microlithography has further advanced, the on-resistance of DMOS, especially low-voltage DMOS, has continuously decreased. In the early 1990s, the on-resistance of low-voltage lateral DMOS became lower than that of bipolar Trs [4]. It was even true that low-voltage lateral DMOS is superior to vertical planar discrete DMOS since fine lithography does not contribute to decrease in on-resistance of vertical DMOS because of the parasitic JFET resistance. Recently, with the introduction of a 0.6 μm design rule, lateral DMOS has become predominant over the wide voltage range from 20 up to 150 V. Mixed technology, called BCD [4], integrating BiCMOS and DMOS, is now widely accepted for low-voltage power ICs.

For high-voltage power ICs, DMOS is not suitable for output devices because of a high on-resistance. Thyristor-like devices, such as GTOs have conventionally been used for high-voltage applications. Integration of thyristor-like devices needs a method of dielectric device isolation (DI). The conventional DI method, called EPIC [5], has been used for high-voltage telecommunication ICs, called SLIC. However, it has problems of high cost and large wafer warpage. In 1985 and 1986, wafer direct-bonding technology was invented [6], and low-cost DI wafers became available. In 1990, it was shown [7] that a high-voltage of 500 V can be realized even in relatively thin SOI by applying a large part of the voltage across the buried oxide. Thin SOI layer can be easily isolated by trenches. This opened up the new field of high-voltage SOI technology. Wafer warpage of directly bonded SOI wafers is very small. This made it possible not only to fabricate large diameter (8 in) SOI wafers but also to apply advanced lithography to SOI power ICs. The chip size of SOI power ICs can be reduced by narrow trench isolation and by use of high-performance lateral IGBTs. The low-cost DI wafers and the chip size reduction have widened the application fields of SOI power ICs, covering the applications of automotive, motor control, and PDP drivers.

7.2 Intelligent Power IC

Figure 7.1 shows typical functions integrated into intelligent power ICs. The most important feature of the intelligent power IC is that a large power can be controlled by logic-level input signals and all the cumbersome circuits such as driving circuits, sense circuits, and protection circuits required for power device control are inside the power ICs.

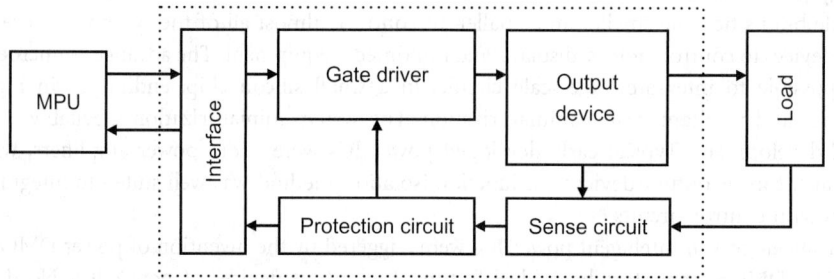

FIGURE 7.1　Typical integrated functions in intelligent power ICs.

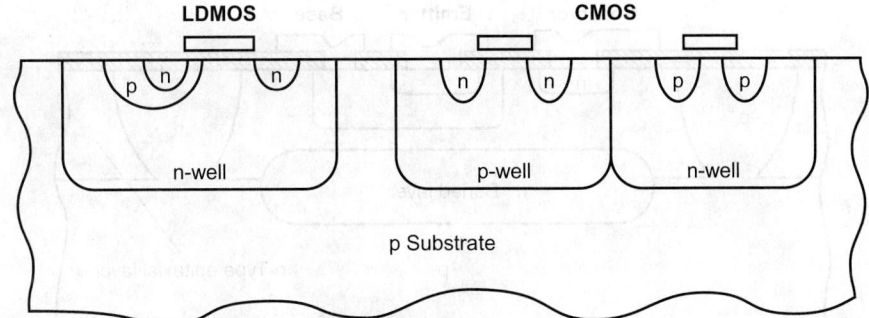

FIGURE 7.2 Self isolation technology.

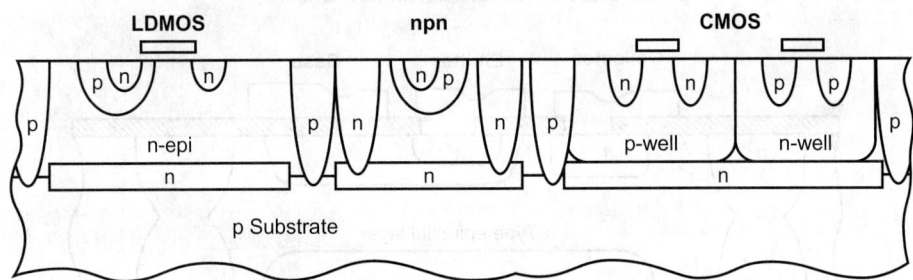

FIGURE 7.3 Junction isolation (JI) technology.

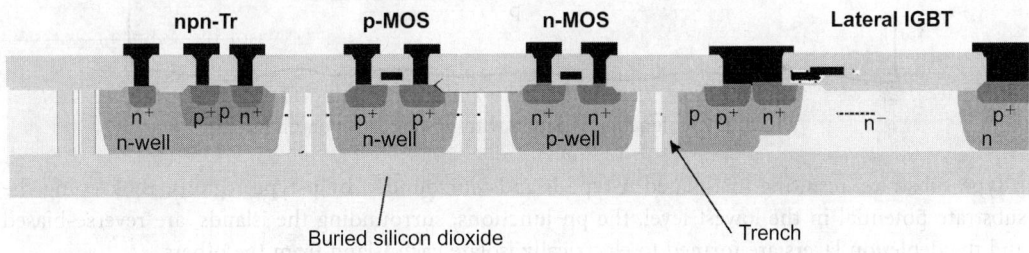

FIGURE 7.4 Dielectric isolation (DI) technology.

Technologies for realizing such power ICs are classified into three categories, as shown in Figures 7.2–7.4. These are self isolation, junction isolation, and dielectric isolation. Self isolation is a method that does not use any special means to isolate each device and each device structure automatically isolates itself from the other.

Junction isolation (JI) is a method that uses reverse-biased junction-depletion layers to isolate each device. JI is the most frequently used method for low-voltage power ICs, using bipolar transistor or DMOS outputs.

Junction isolation is not sufficient to isolate IGBTs or thyristors. Dielectric isolation is a method that uses silicon dioxide film to isolate devices and, thus, offers complete device isolation. Although the EPIC method has conventionally been used, the high cost of wafer fabrication has been a problem. Recently, wafer direct-bonding technology has been invented and bonded SOI wafers are available in a lower price.

7.2.1 pn Junction Isolation

pn junction isolation is the most familiar method and has been used since the beginning of the bipolar IC history. Figure 7.5 shows the cross section of a typical junction isolation structure. First, an n-type epitaxial layer is formed on p-type silicon substrate. p-type diffusion layers are then formed to reach

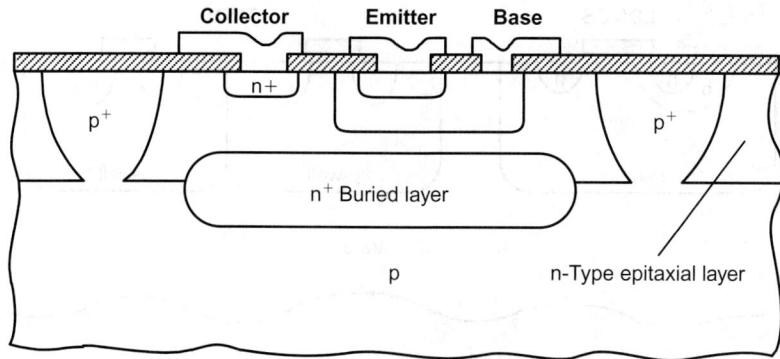

FIGURE 7.5 Junction isolation structure.

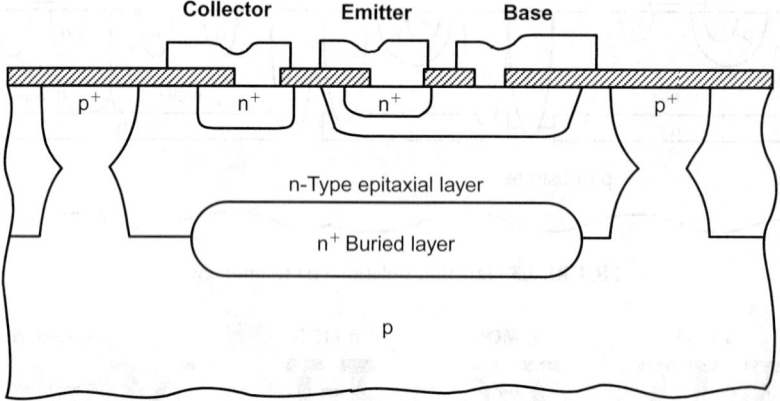

FIGURE 7.6 Junction isolation with upward isolation diffusions.

p-type substrate, resulting in isolated n-type islands surrounded by p-type regions. By keeping the substrate potential in the lowest level, the pn junctions, surrounding the islands, are reverse-biased and the depletion layers are formed to electrically isolate each island from the others.

pn junction isolation is particularly useful for low-voltage power ICs, called BCD technology. BiCMOS circuits and LDMOS devices are integrated in a single chip in BCD technology as seen in Figure 7.3.

If this method is applied to high-voltage power ICs, a thick n-type epitaxial layer is required and deep isolation diffusions are necessary. Deep diffusion accompanies large lateral diffusion, ending up in large isolation area. One solution for this is to use buried p+ diffusion layers for upward isolation diffusions as shown in Figure 7.6. However, 200 V is a practical limit for the conventional pn junction isolation.

A variety of method was proposed to overcome this voltage limit. Figure 7.7 shows a typical example for this [8]. A shallow hole is formed where a high-voltage device is formed before the n-type epitaxial growth. This allows a locally thicker n-type epitaxial layer for high-voltage transistors.

Another distinguished example is shown in Figure 7.8, where n+ substrate is used in place of p-type substrate. p- and n-type epitaxial layers are subsequently formed. This example makes it possible to integrate a vertical DMOSFET with a backside drain contact with junction-isolated BiCMOS control circuits. This structure was proposed as Smart Power in the mid-1980s.

7.2.2 BCD Power ICs and Technology Road Map

Figure 7.9 shows the technology road map for BCD power ICs. Before 1995, power ICs fabrication technology was about 10 years behind that of CMOS. Since 1996, the design rule of power ICs has been rapidly approaching

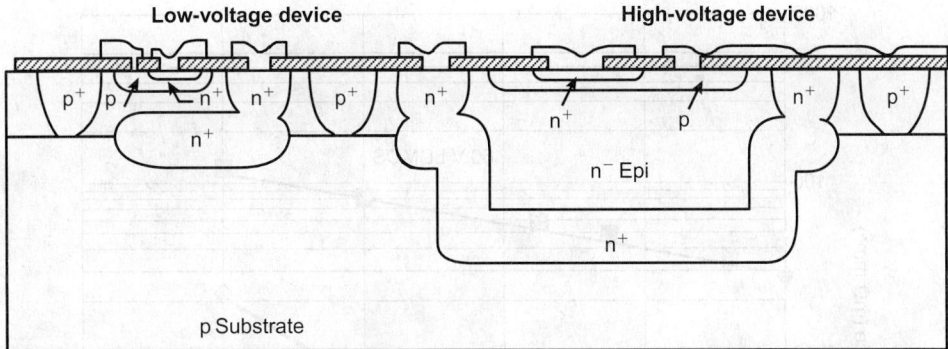

FIGURE 7.7 An example to overcome junction isolation voltage limit.

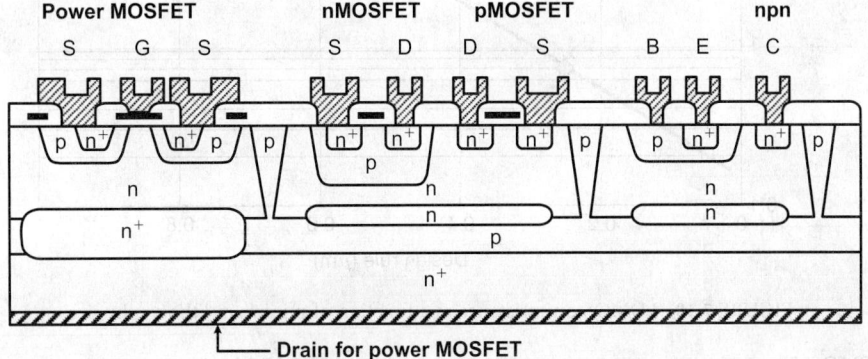

FIGURE 7.8 Junction isolation with VDMOSFET.

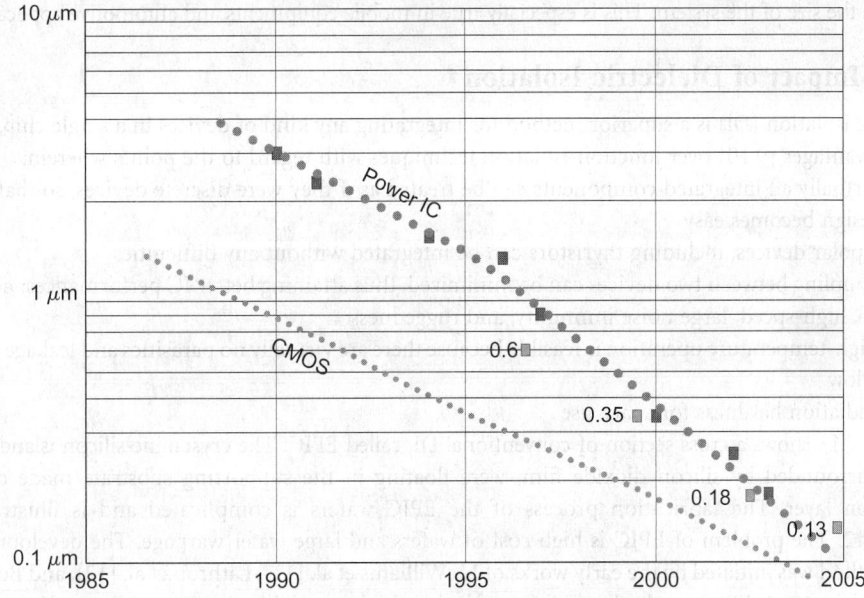

FIGURE 7.9 Process trends of power ICs.

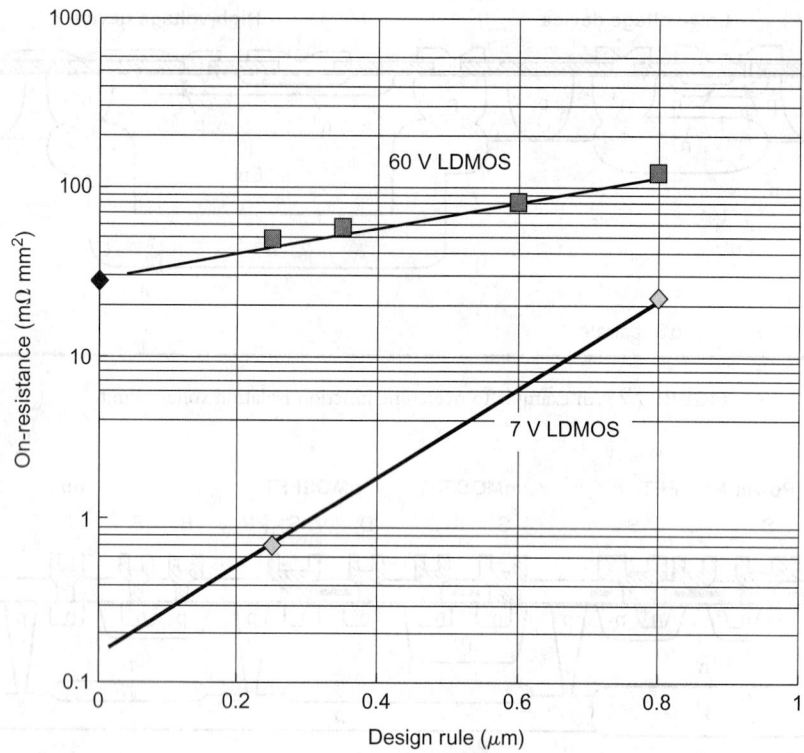

FIGURE 7.10 LDMOS on-resistance as a function of minimum lithography size.

that of CMOS. There are two main reasons. First, lower-voltage LDMOS becomes more important for the application of mobile equipments, such as cell phones. The on-resistance of the low-voltage LDMOS improves significantly by adopting finer design rule as shown in Figure 7.10. The detailed explanation of LDMOS is presented in Section 7.6. The other is the need of system integration. More functionality needs to be integrated to reduce the size of the system. This is especially true in mobile equipments and automotive applications.

7.2.3 Impact of Dielectric Isolation

Dielectric isolation (Dl) is a superior method for integrating any kind of devices in a single chip. DI has many advantages [9,10] over junction-isolation techniques with regard to the points wherein:

1. Virtually all integrated components can be treated as if they were discrete devices, so that circuit design becomes easy
2. Bipolar devices, including thyristors can be integrated without any difficulties
3. Coupling between two devices can be minimized, thus attaining better IC performances: no latch-up, high-speed, large noise immunity, and ruggedness
4. High-temperature operation is feasible because there are virtually no parasitics and leakage current is low
5. Radiation hardness for space use

Figure 7.11 shows a cross section of conventional DI, called EPIC. The crystalline silicon islands, completely surrounded by silicon dioxide film, were floating in the supporting substrate made of thick polysilicon layer. The fabrication process of the EPIC wafers is complicated and is illustrated in Figure 7.12. The problem of EPIC is high cost of wafers and large wafer warpage. The development of EPIC method was initiated by the early works of McWilliams et al. [11], Lathrop et al. [12], and Bouchard et al. [13] in 1964. EPIC method was first applied to high-speed bipolar ICs owing to low parasitic capacitance.

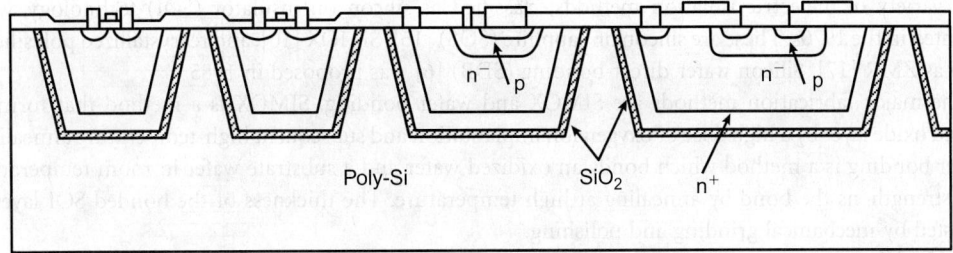

FIGURE 7.11   Dielectric isolation with EPIC technology.

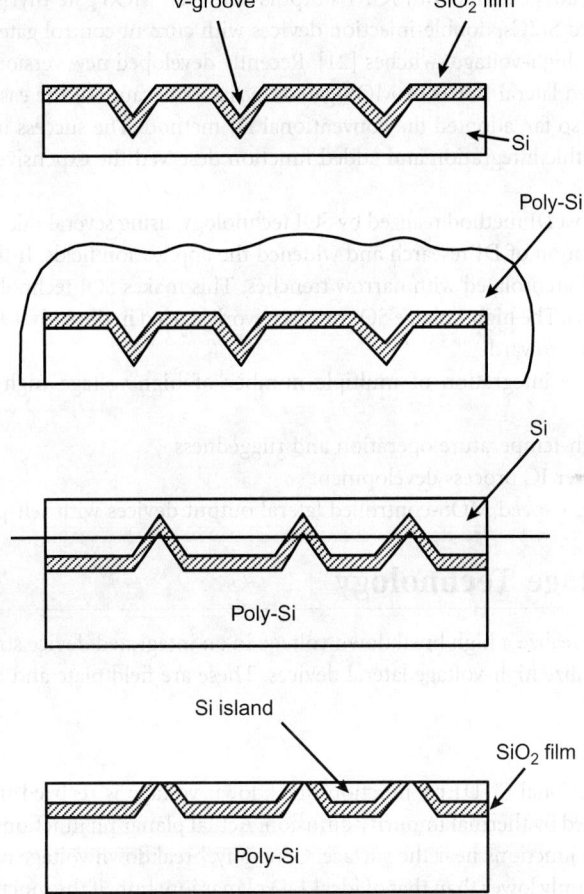

FIGURE 7.12   Fabrication process of EPIC wafers. A very thick polycrystalline silicon layer is deposited on an oxidized single-crystal silicon with a grooved surface. The crystalline silicon is grounded and polished so that the silicon layers are isolated each other by the grooves.

Early work on high-voltage integrated circuits was triggered by the need for display drivers and high-voltage telecommunication circuits. Efforts to achieve high-voltage lateral MOSFETs started in the early 1970s and 800 V lateral MOSFET using RESURF concept and DMOS (DSA [2]) technology was developed for display drivers in 1976, before the RESURF concept was fully established [14].

The need for high-voltage SLICs advanced the EPIC technology because it required electrically floating high-voltage bidirectional switches, which were realized only by the DI technique.

A variety of dielectric isolation methods, classified as silicon on insulator (SOI) technology, were invented in the 1970s. These are silicon on sapphire (SOS) [15], SIMOX [16], and recrystallized polysilicon such as ZMR [17]. Silicon wafer direct-bonding (SDB) [6] was proposed in 1985.

The major fabrication methods are SIMOX and wafer bonding. SIMOX is a method that forms a buried oxide layer by a high dose of oxygen ion implantation and subsequent high-temperature annealing. Wafer bonding is a method which bonds an oxidized wafer and a substrate wafer in room temperature and strengthens the bond by annealing at high temperature. The thickness of the bonded SOI layer is adjusted by mechanical grinding and polishing.

In the late 1980s, the MOS gate power device technology has been greatly improved. Especially, the success in the MOS bipolar composite devices such as IGBTs [18,19] and MCTs [20] made it possible to control a large current by the MOS gate. The large current-handling capability of the MOS gate bipolar devices has accelerated adopting DI with IGBT outputs and even MOS gate thyristors in power ICs.

In the early developed SLICs, double-injection devices with current control gates such as gated diodes and GTOs were used as high-voltage switches [21]. Recently developed new version SLIC (telecommunication ICs) have adopted lateral IGBTs or MOS-gated thyristors because of the ease of gate drive. All the commercialized SLICs, so far, adopted the conventional DI method. The success in SLIC was supported by the fact that monolithic integration and added function deserved the expensive Dl for telecommunication application.

In the 1990s, a low-cost DI method realized by SOI technology, using several micron thick or less silicon layers, changed the situation of DI research and widened the application fields. If the silicon layer is thin, devices in the SOI layer are isolated with narrow trenches. This makes SOI technology very attractive for high-voltage applications. The high-voltage SOI research work started in the early 1990s [1,7]. The research works have been directed toward:

1. Monolithic device integration of multiple number of high-voltage–high-current devices with control circuits
2. ICs allowing high-temperature operation and ruggedness
3. Low-cost Dl power IC process development
4. High-current high-speed MOS-controlled lateral output devices with self-protection functions

7.3 High-Voltage Technology

It is quite important to realize a high breakdown voltage in an integrated device structure. There are two major techniques to realize high-voltage lateral devices. These are field plate and Resurf technique.

7.3.1 Field Plate

It is ideal if one-dimensional (1-D) pn junction breakdown voltage is realized in an actual planar pn junction, which is formed by thermal impurity diffusion. Actual planar pn junctions consist of cylindrical junctions and spherical junctions near the surface. Generally, breakdown voltage of cylindrical or spherical junctions is significantly lower than that of ideal 1-D planar junction, if the junction curvature is small.

A field plate is a simple and frequently used technique to increase the breakdown voltage of actual planar junctions. Figure 7.13 shows an example. Field plates, placed on the thick field oxide, induce depletion layers underneath themselves. The curvature of the formed depletion layers can be reduced with the induced depletion layers, relaxing the curvature effects by the field plate.

7.3.2 Resurf Technology

Resurf technique was originally proposed in 1979 [14] as a method to obtain a high breakdown voltage in conventional JI structure. Figure 7.14 shows a high-voltage structure, where depletion layer develops in the p substrate and n-epitaxial layer. If the epilayer impurity dose, Q_o, is high, premature breakdown occurs before n-epi layer is fully depleted. If the impurity dose in the epilayer is optimized, the epilayer

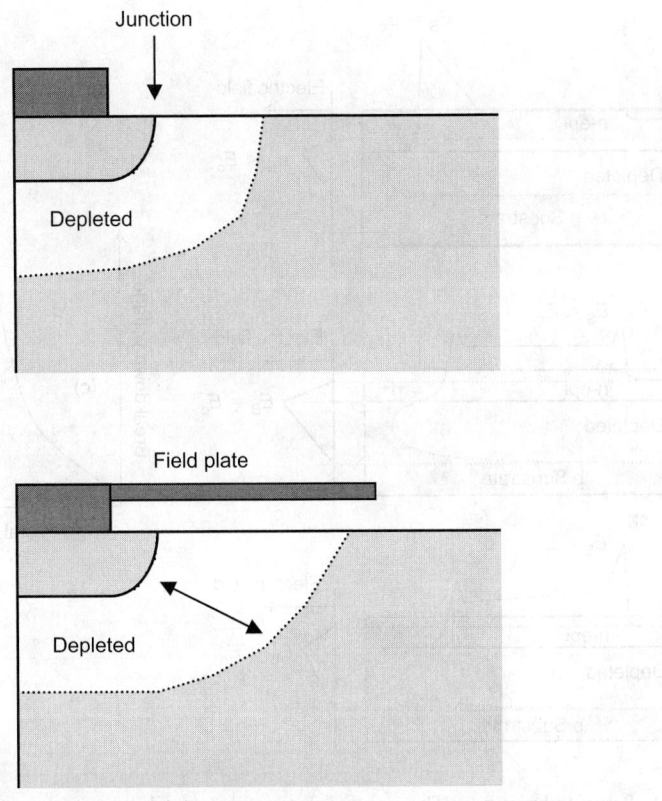

FIGURE 7.13 Field plate structure.

is just completely depleted when breakdown occurs. The achieved breakdown voltage is very high because depletion layer is sufficiently thick both in lateral and vertical direction. The important point is that the total charge, Q_c, in the epilayer is chosen so that the value satisfies the equation

$$Q_c = \varepsilon E_c$$

where E_c denotes critical electric field in silicon (3×10^5 V/cm). This charge can be depleted just when the electric field becomes E_c and the junction breakdown occurs. In other words, the epilayer is completely depleted just when the breakdown occurs, if the total epilayer dose is Q_c/q, which is approximately 2×10^{12} cm^{-2}.

If the n-epi layer dose, Q_c, is excessively low, the epilayer is completely depleted before the vertical electric field reaches the critical field. The premature breakdown again occurs in this case, as the surface electric field exceeds the critical value at the edge of the n$^+$ layer.

7.3.3 High-Voltage Metal Interconnection

In high-voltage power ICs, there must be interconnection metal layers crossing high-voltage junctions. These high-voltage interconnection layers may cause degradation of breakdown voltage of high-voltage devices. These problems are often solved with a thicker insulator layer underneath the interconnection layers. However, special means are required if the breakdown voltage is over 400 V.

Figure 7.15 shows one of the methods to shield the influence of metal interconnection layers on the underlying devices. A spiral-shaped high-resistance polysilicon layer, connecting source and drain electrodes, effectively shields the influence of the interconnection layer on the depletion layer [22]. This is because the potential of high-resistance polysilicon layer is determined by small leakage current.

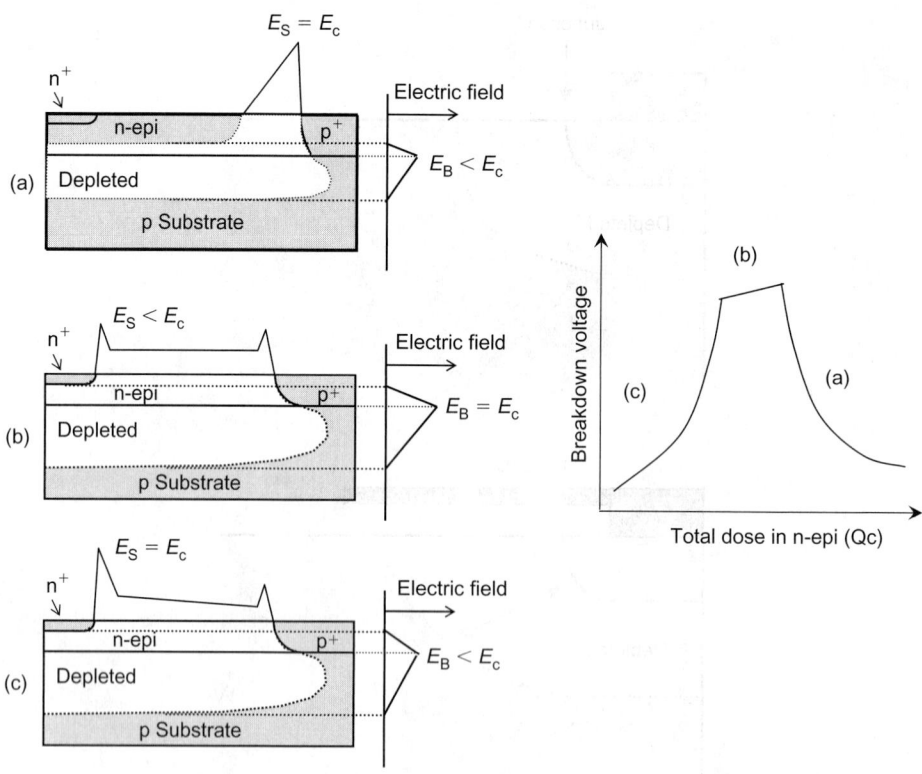

FIGURE 7.14   Resurf technique. (a) The case that shows epilayer total impurity dose, Q_c, is excessively high; (b) the just optimized case; and (c) the case that Q_c is excessively low.

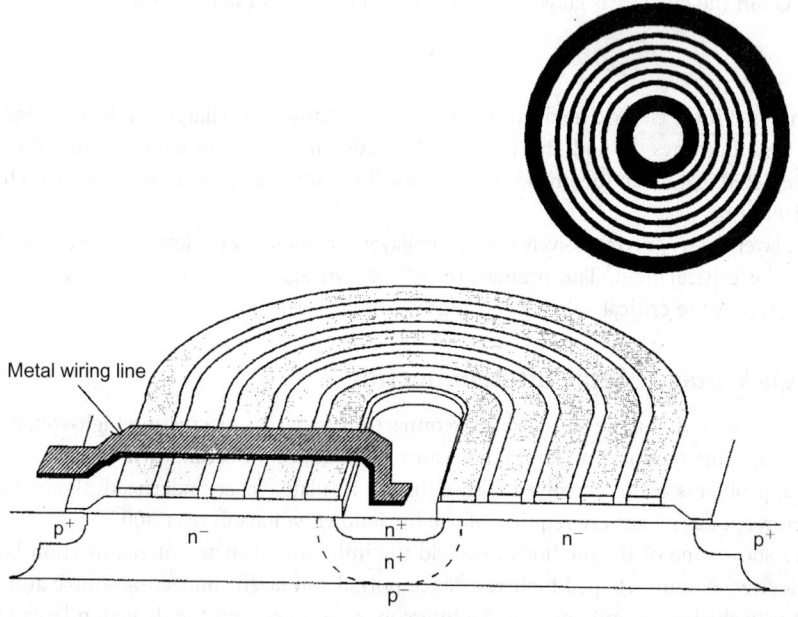

FIGURE 7.15   A method to shield the influence of metal interconnection layer.

Another typical example is multiple floating field plates. The cross-section of the structure is similar to Figure 7.15. The difference is that the polysilicon forms multiple closed rings, which are electrically isolated from each other. Multiple floating field plates also prevent the breakdown voltage reduction due to metal interconnection.

7.4 High-Voltage SOI Technology

SOI power ICs are classified into two categories from the viewpoint of SOI wafer structure. The difference is whether there is a buried n^+ layer on the buried oxide. Figure 7.16 shows a typical device structure, employing an n^+ buried layer on the buried oxide. The breakdown voltage is determined with the thickness of the high-resistivity n layer or the thickness of the depletion layer. The maximum breakdown voltage is limited to below 100 V because of SOI layer thickness or practically available trench depth. For this case, SOI wafers are used as a simple replacement for conventional DI wafers.

Figure 7.17 shows another typical SOI power IC structure, employing a high-voltage lateral IGBT. n^- drift layer is fully depleted by the application of a high-voltage. As the buried oxide and the depletion layer both share the applied voltage, high breakdown voltage is realized in relatively thin SOI [7]. This type of power ICs fully enjoy the features of SOI technology and is described in detail in this section.

There are two big challenges associated with high-voltage devices on SOI. One is how to realize a high breakdown voltage under the influence of substrate ground potential. The other is how to attain a low on-resistance device in the thin silicon layer. In the conventional DI, the wrap-around n^+ region (see Figure 7.16) is used in the DI island to prevent the influence of substrate potential on the device breakdown voltage. However, for thin SOI layers, this method cannot be used. The bottom silicon dioxide layer simply works as an undoped layer as far as Poisson equation is concerned. Thus, a SOI layer on a grounded silicon substrate structure behaves in a way similar to the structure of a doped n-type thin silicon layer on undoped silicon layer (corresponding to silicon dioxide) on grounded p silicon substrate. Thus, the SOI layer works in the same way as a Resurf layer.

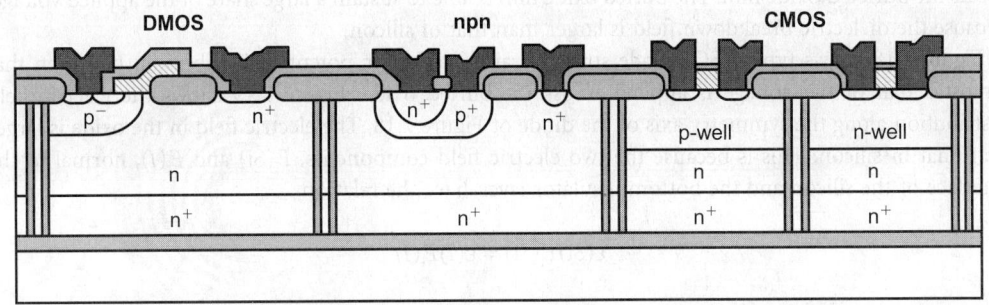

FIGURE 7.16 A high-voltage SOI device structure with n^+ buried layer on the buried oxide.

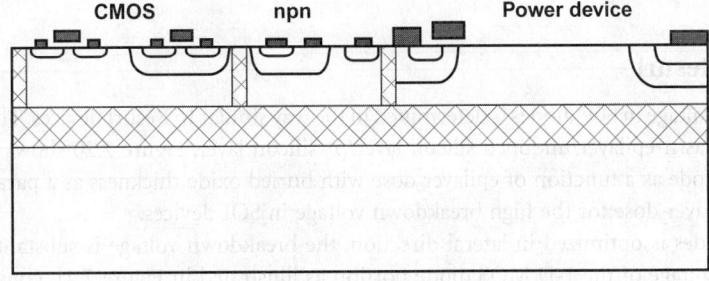

FIGURE 7.17 A high-voltage SOI device structure without n^+ buried layer on the buried oxide.

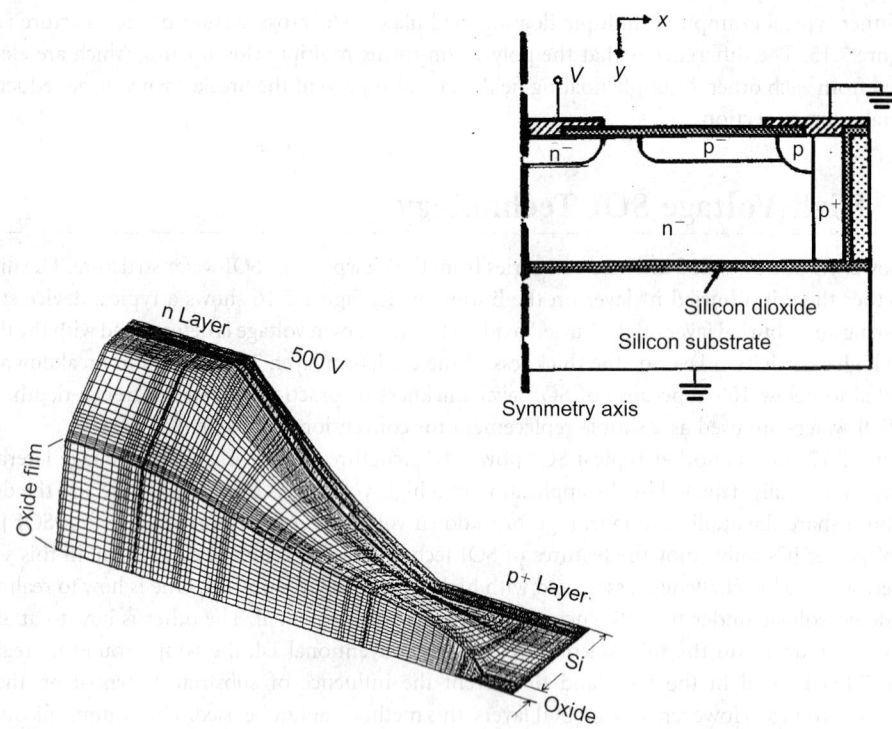

FIGURE 7.18 SOI diode and potential distribution in the diode.

A high breakdown voltage of a SOI layer device can be realized by applying a large part of the voltage across the buried dioxide film. The buried oxide film is able to sustain a large share of the applied voltage, because the dielectric breakdown field is larger than that of silicon.

Figure 7.18 shows typical SOI diode structure and its electric potential distribution. It is seen that almost a half of the voltage is applied across the buried oxide. Figure 7.19 shows the electric field distribution along the symmetry axis of the diode of Figure 7.18. The electric field in the oxide is larger than that in silicon. This is because the two electric field components, $E_t(\text{Si})$ and $E_t(I)$, normal to the interface of the silicon and the bottom insulator layer, have the relation

$$\varepsilon(\text{Si})E_t(\text{Si}) = \varepsilon(I)E_t(I)$$

where $\varepsilon(\text{Si})$ and $\varepsilon(I)$ denote dielectric constants for silicon and silicon dioxide, respectively. Using an insulator film with a lower dielectric constant will increase the device breakdown voltage because the insulator layer sustains a larger share of the applied voltage.

7.4.1 SOI Resurf

The breakdown voltage in SOI diodes is determined in the way similar to Resurf devices. The SOI structure can be regarded as n-epilayer/undoped silicon layer/p⁺ silicon layer. Figure 7.20 shows the breakdown voltage of SOI diode as a function of epilayer dose with buried oxide thickness as a parameter. There is an optimum epilayer dose for the high breakdown voltage in SOI devices.

If the SOI diodes is optimized in lateral direction, the breakdown voltage is substantially limited to the breakdown voltage of the 1-D MOS diode portion as illustrated in Figure 7.21, consisting of n⁺/n⁻/oxide/substrate. Figure 7.22 shows the measured SOI device breakdown voltage as a function of SOI layer

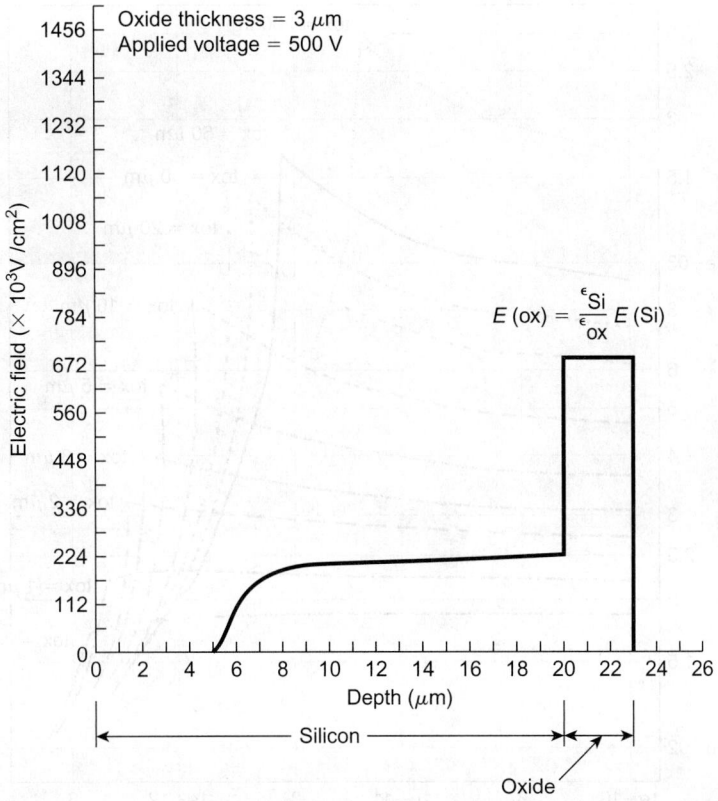

FIGURE 7.19 Electric field distribution along the symmetry axis of diode shown in Figure 7.18.

thickness with buried oxide thickness as a parameter. The calculated breakdown voltage of 1-D MOS diodes are shown together. A 500 V breakdown voltage can be obtained with a 13-μm-thick SOI with 3-μm-thick buried oxide.

It is very difficult to achieve a high breakdown voltage exceeding 600 V in a simple SOI structure, because a thicker buried oxide layer of 4 μm or more is required. The maximum breakdown voltage is substantially limited to the breakdown voltage of the 1-D MOS diode and the actually realized breakdown voltage is lower than this limit. If the influence of the substrate potential can be shielded, it is possible to achieve a higher breakdown voltage in the SOI device.

A new high-voltage SOI device structure, which is free from the above constraints, was proposed in 1991 [1], which realizes 1200 V breakdown voltage even with a thin buried oxide [23].

To improve the breakdown voltage, an SOI structure with shallow n^+ layer diffused from the bottom of SOI layer was proposed [24]. Figure 7.23 shows the structure of an SOI diode with shallow n^+ layer and the electric field strength in the MOS diode portion. The figure also shows the electric field for the structure without shallow n^+ layer. In general, if a larger proportion of the applied voltage is sustained by the bottom oxide layer, a higher breakdown voltage can be achieved. The problem is how to apply higher electric field across the buried oxide without increasing the electric field strength in the SOI layer. This problem can be solved by placing a certain amount of positive charge on the SOI layer–buried oxide interface. The positive charge in the interface shields the high electric field in the buried oxide, so that a voltage across the oxide layer can be increased without applying higher electric field in the SOI layer. The shallow n^+ layer diffused from the bottom is a practical technique to place the positive charge on the SOI layer–buried oxide interface as shown in Figure 7.23. The required dose of the shallow n^+ layer is around 1×10^{12} cm^{-2}.

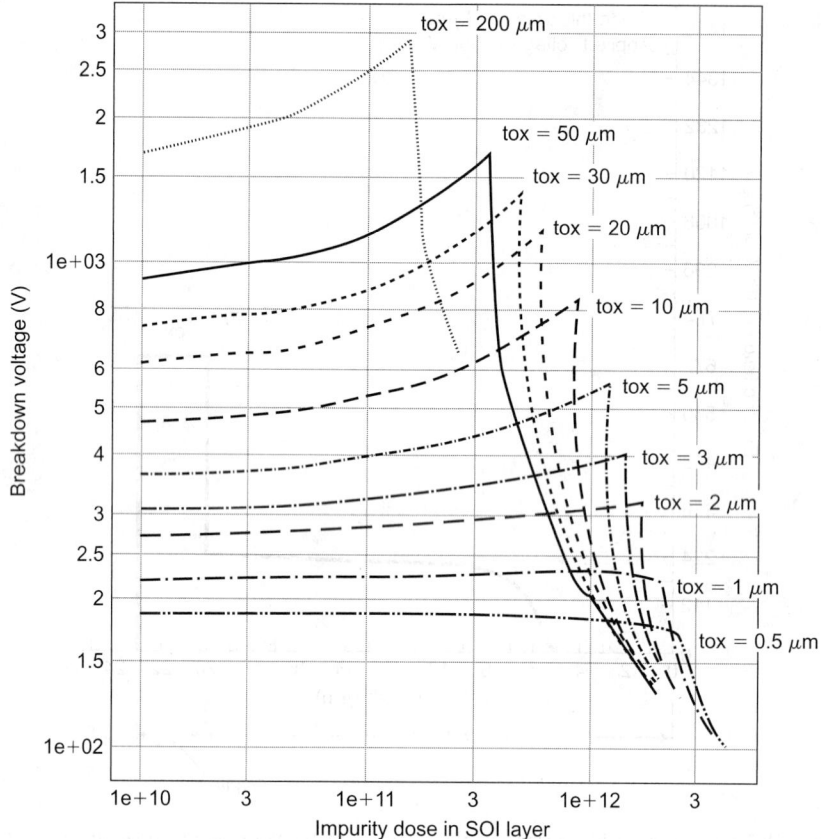

FIGURE 7.20 Calculated breakdown voltage as a function of impurity dose in SOI layer with buried oxide thickness, t_{ox}, as a parameter.

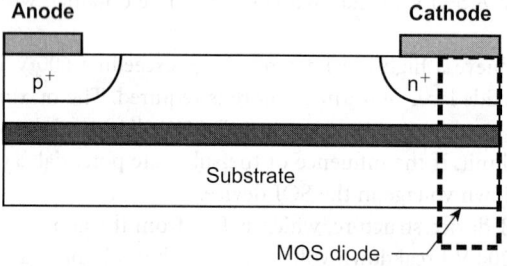

FIGURE 7.21 1-D MOS-diode structure in SOI diode.

7.4.2 Very Thin SOI

Merchant et al. [25] showed that the SOI diode breakdown voltage is significantly enhanced if the SOI layer thickness is very thin, e.g., such as thin as 0.1 μm. As shown in Figure 7.22, reduction of the SOI layer thickness enhanced the breakdown voltage if the thickness is less than 1 μm. This is because the carrier path along the vertical high electric field is as short as the SOI layer thickness, so that the carriers reach the top or bottom surface of the SOI layer before ionizing sufficient amount of carriers for avalanche multiplication along the path. They proposed combination of the very thin SOI layer and a linearly graded impurity profile of the n-type silicon layer for a high-voltage $n^+n^-p^+$ lateral diode. A breakdown voltage of 700 V was realized by this structure.

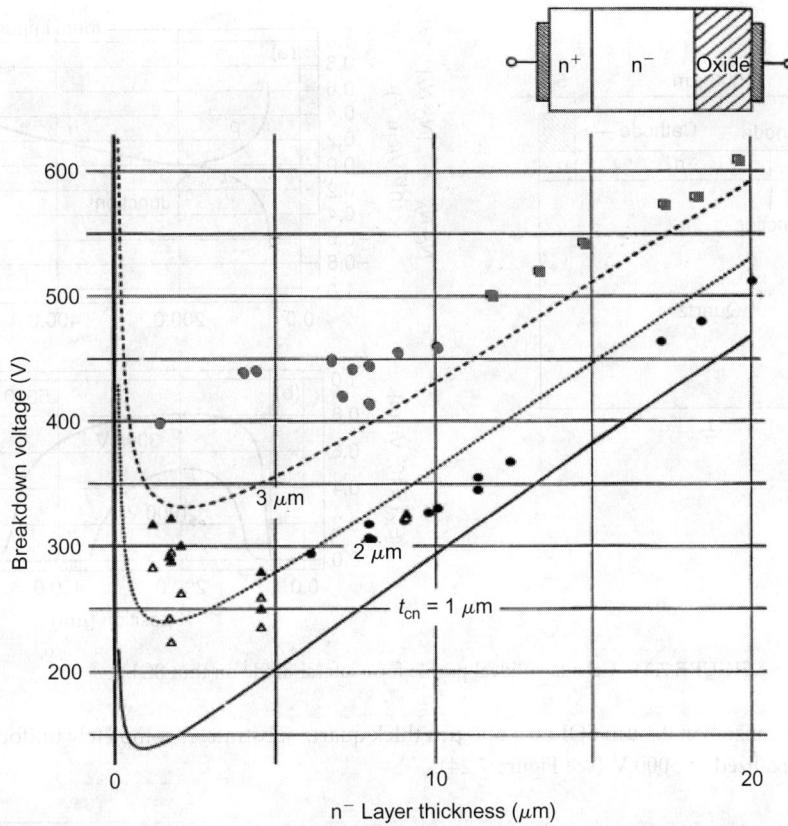

FIGURE 7.22 Measured and calculated SOI device breakdown voltage.

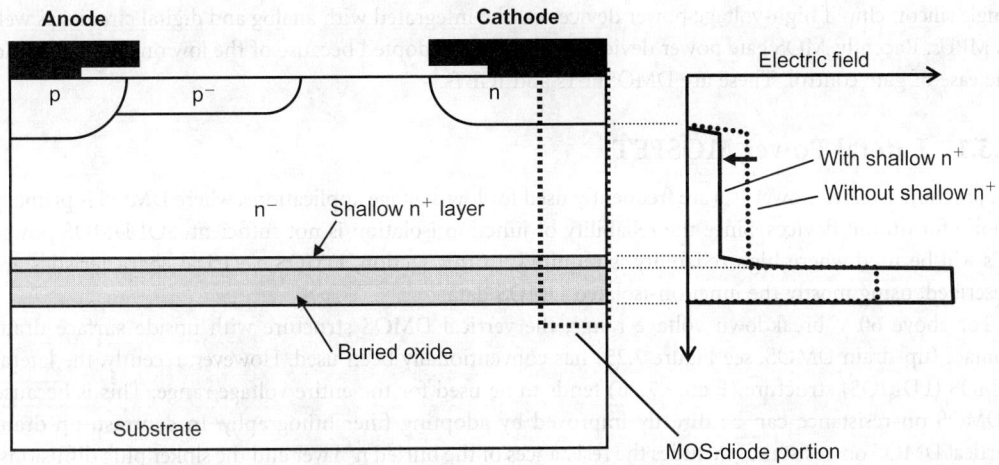

FIGURE 7.23 SOI diode with shallow n⁺ layer diffused from the bottom of SOI layer.

The exact 2-D simulations revealed that the ideal profile for a lateral diode is approximated by a function which is similar to a tangent function, as shown in Figure 7.24. The important point is that the p layer impurity profile should also be graded and that the linearly graded portion is terminated by the exponentially increasing ending portions. By using the proposed profile, a 5000 V lateral diode was

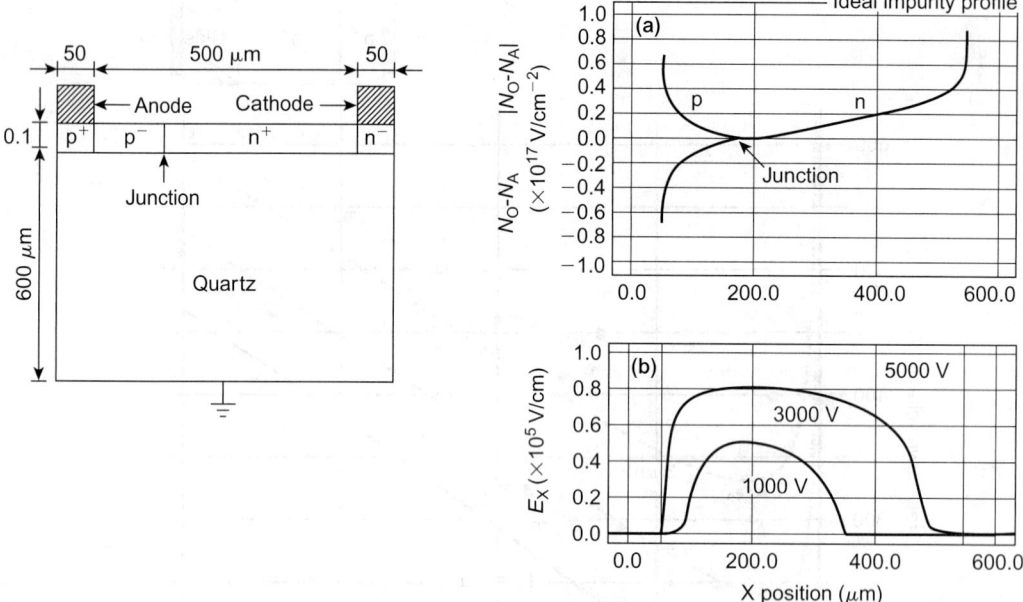

FIGURE 7.24 Calculated ideal profile for a lateral diode on thin SOI.

predicted to be realized on 0.1 μm SOI on a 600-μm thick quartz substrate. A completely uniform lateral electric field is realized at 5000 V (see Figure 7.24).

7.5 Power Output Devices

Output devices are most important part of power ICs. A whole power system can be integrated on a single silicon chip if high-voltage power devices can be integrated with analog and digital circuits as well as MPUs. Recently, MOS gate power devices are primarily adopted because of the low on-resistance and the ease of gate control. These are DMOSFETs and IGBTs.

7.5.1 Lateral Power MOSFET

pn junction-isolated power ICs are frequently used for low-voltage applications, where DMOS is primary choice for output devices. Since the reliability of junction isolation is not sufficient, SOI DMOS power ICs will be used where high reliability is required. In this section, DMOS electrical characteristics are described, using mostly the junction-isolated DMOS data.

For above 60 V breakdown voltage range, the vertical DMOS structure with upside surface drain contact (up-drain DMOS, see Figure 7.25) has conventionally been used. However, recently, the lateral DMOS (LDMOS) structure (Figure 7.26) tends to be used for the entire voltage range. This is because LDMOS on-resistance can be directly improved by adopting finer lithography. In contrast, up-drain vertical DMOS on-resistance includes the resistances of the buried n⁺ layer and the sinker plug diffusions, which are not improved by finer lithography.

Figure 7.27 shows the state-of-the-art DMOS on-resistance as a function of breakdown voltage. The figure also shows the state-of-the-art on-resistance for vertical discrete trench MOSFETs as a comparison. Black squares show lateral DMOS, and triangles show trench MOSFETs. Recently, battery-operated mobile equipment and computer peripherals have opened large application fields, and lateral MOSFETs of less than 60 V are the major output devices. It is an interesting fact that the state-of-the-art on-resistances of lateral DMOS and vertical trench MOSFETs are almost the same. This implies that power

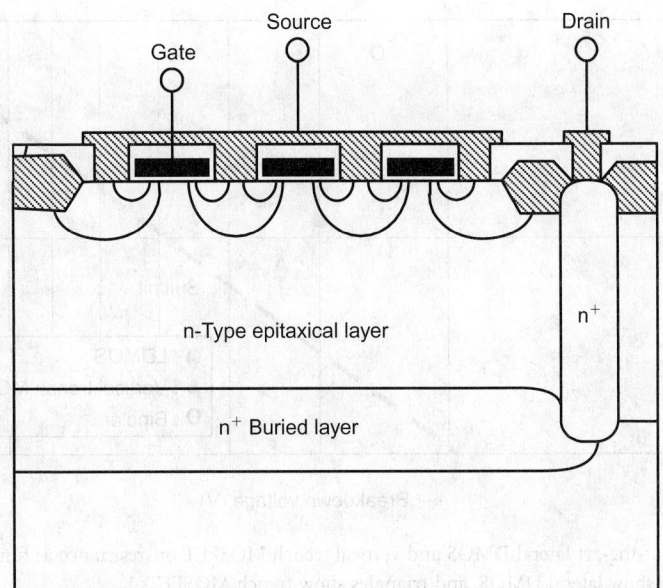

FIGURE 7.25 Vertical DMOS structure with upside surface drain contact.

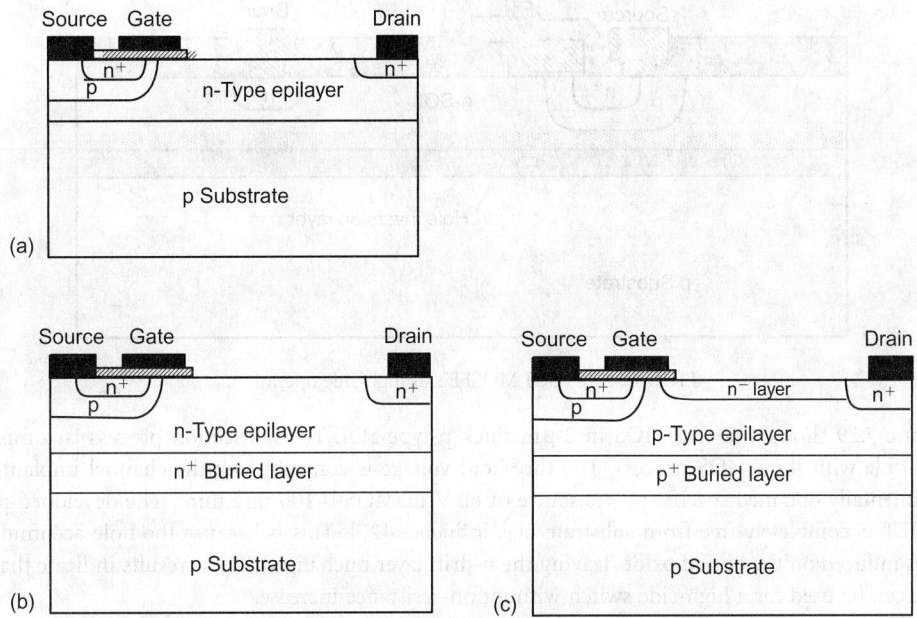

FIGURE 7.26 Three typical lateral DMOS (LDMOS) structures.

ICs with a vertical DMOS output will be replaced by power ICs with a lateral DMOS output, if current capacity is not large—for example, less than 10 A.

High-side switching operation is an important function in automotive applications, especially in case of H-bridges for motor control. The on-resistance of conventional junction-isolated high-voltage MOSFETs, shown in Figure 7.26a, is significantly influenced by the source-to-substrate bias [26], because a large part of the n-type epilayer is depleted. However, in SOI MOSFETs as shown in Figure 7.28, the drift layer is not depleted, but a hole inversion layer is formed on the buried oxide. Thus, the substrate bias influence of SOI LDMOS is small [26].

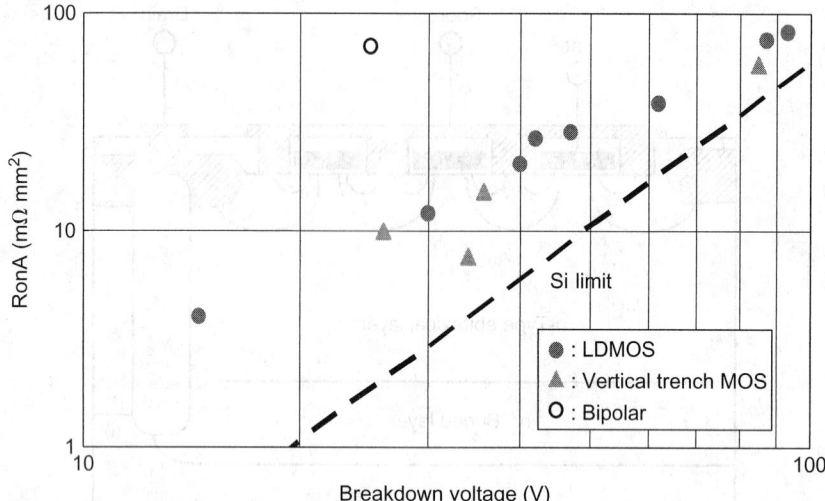

FIGURE 7.27 State-of-the-art lateral DMOS and vertical trench MOSFET on-resistance as function of breakdown voltage. (Black circles show lateral DMOS, and triangles show trench MOSFETs.)

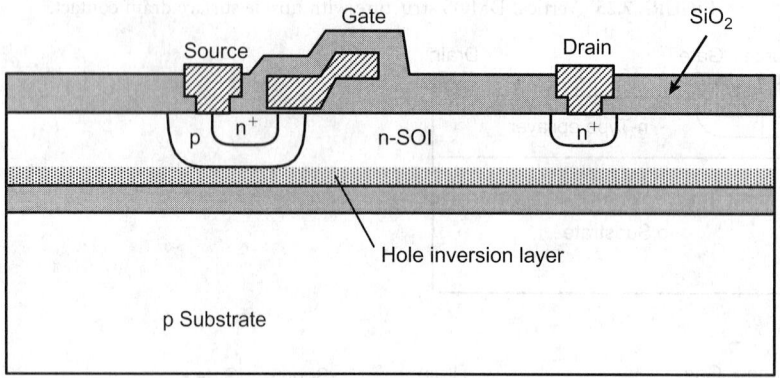

FIGURE 7.28 SOI MOSFET at high-side operation.

Figure 7.29 shows a 60 V DMOS in 2-μm-thick p-type SOI. The fabrication process is completely compatible with the CMOS process. The threshold voltage is controlled by the channel implant. The experimentally obtained specific on-resistance of 60 V LDMOS is 100 mΩ mm². The developed power MOSFET is completely free from substrate bias influence [27]. This is because the hole accumulation layer is induced on the buried oxide, leaving the n-drift layer unchanged. These results indicate that this device can be used for a high-side switch without on-resistance increase.

7.5.2 Adaptive Resurf

Figure 7.30a shows a cross-sectional view of a typical Resurf LDMOS on p-epi layer with n$^+$ buried layer. The Resurf structure is often used to increase the static breakdown voltage and to reduce the on-resistance. Figure 7.31a shows a typical I–V curve of a Resurf LDMOS. The on-resistance is 15.7 mΩ mm² at the gate voltage, V_g, of 5 V. The static breakdown voltage is 30.2 V. The device breakdown voltage decreases to 13.9 V when the device is operating at the gate voltage of 5 V. The breakdown voltage, when the device is in conduction state, is called as on-state breakdown voltage.

The reason why the on-state breakdown voltage is degraded is that the net effective positive charge is reduced by the existence of a large amount of negative electron charges in the Resurf layer due to a large

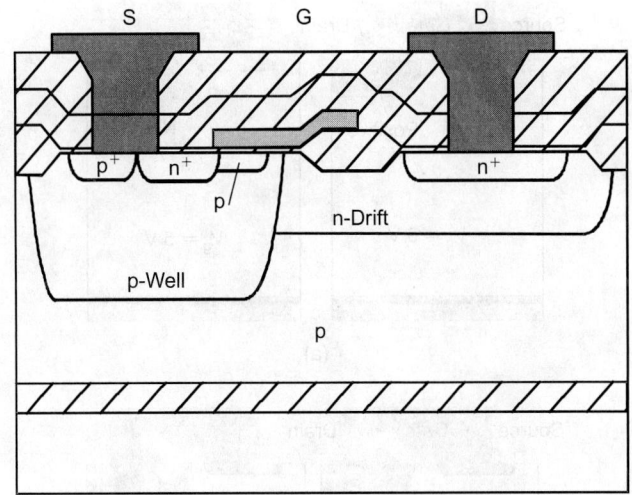

FIGURE 7.29 60 V DMOS in 5-μm-thick p-type SOI.

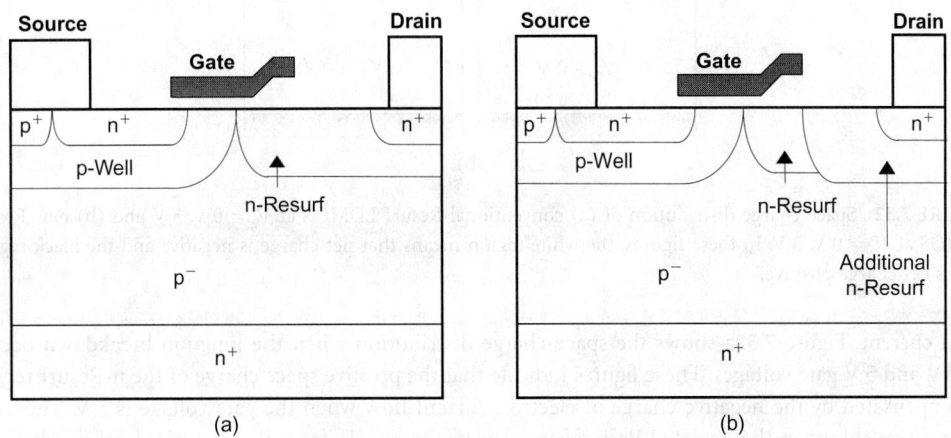

FIGURE 7.30 (a) Conventional resurf LDMOS. (b) Adaptive resurf LDMOS.

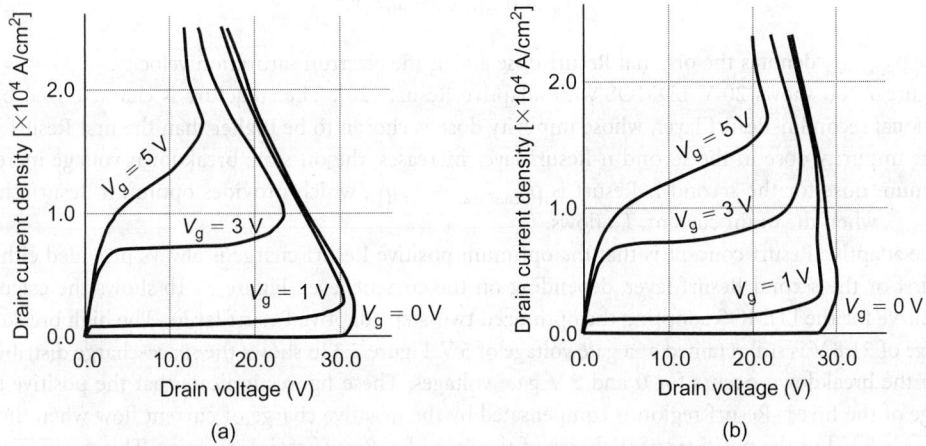

FIGURE 7.31 Current voltage characteristics of (a) conventional Resurf LDMOS, (b) adaptive Resurf LDMOS.

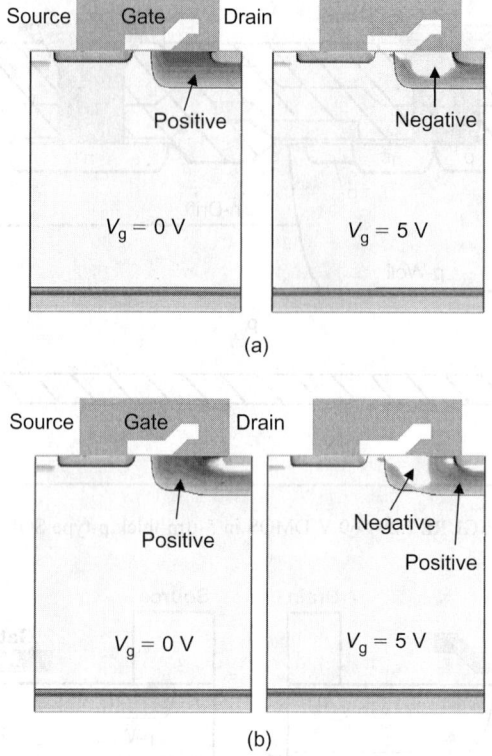

FIGURE 7.32 Space-charge distribution of (a) conventional Resurf LDMOS at V_g = 0 V, 5 V and (b) new Resurf LDMOS at V_g = 0 V, 5 V. In these figures, the white region means that net charge is negative and the black region means net charge positive.

drain current. Figure 7.32a shows the space-charge distribution when the junction breakdown occurs for 0 V and 5 V gate voltages. These figures indicate that the positive space charge of the n-Resurf region is compensated by the negative charge of electron current flow when the gate voltage is 5 V. Thus, the net positive charge in the depleted Resurf layer deviates markedly from the optimized value, when the device is on-state. The net positive Resurf charge ρ_{net} under a drain current of I_D is expressed by

$$\rho_{net} = \rho_{Resurf\ dose} - I_D/qv_s$$

where $\rho_{Resurf\ dose}$ denotes the original Resurf dose and v_s the electron saturation velocity.

Figure 7.30b shows 20 V LDMOS with adaptive Resurf [28]. The structure is characterized by the additional second n-Resurf layer, whose impurity dose is chosen to be higher than the first Resurf layer. As the impurity dose in the second n-Resurf layer increases, the on-state breakdown voltage increases. Optimum dose for the second n-Resurf is $\rho_{Resurf\ dose} + I_D/qv_s$, which provides optimum Resurf charge, $\rho_{Resurf\ dose}$, when the drain current, I_D, flows.

The adaptive Resurf concept is that the optimum positive Resurf charge is always provided either by the first or the second Resurf layer, depending on the current level. Figure 7.31b shows the calculated *I–V* curve for the LDMOS adopting the optimized two-step adaptive Resurf layers. The high breakdown voltage of 21.8 V is still retained at a gate voltage of 5 V. Figure 7.32b shows the space-charge distribution when the breakdown occurs for 0 and 5 V gate voltages. These figures indicate that the positive space charge of the first n-Resurf region is compensated by the negative charge of current flow when the gate voltage is 5 V, but the positive space charge of the second n-Resurf region remains. This positive space-charge region sustains a high on-state breakdown voltage.

7.5.3 Metal Wiring Resistance

The on-resistance of LDMOS itself becomes significantly small as the design rule becomes finer and finer as shown in Figure 7.10. If the device on-resistance becomes comparable to that of the sheet resistance of metal layer, the metal wiring resistance cannot be ignored. For example, the sheet resistance of 1-μm-thick aluminum layer is 28 mΩ/□. The typical aluminum layer thickness of conventional fine CMOS process is less than 1 μm. Figure 7.33 shows calculated LDMOS on-resistance of 1 mm² device area as a function of second metal layer thickness with aluminum layer width, W, as a parameter. The source and drain aluminum layer widths are defined in Figure 7.34. The first aluminum layer thickness is assumed to be 1 μm for all the calculated cases. The total number of the second metal layers are chosen so that the total second metal layer width becomes 1 mm. The overall LDMOS on-resistance depends significantly on the metal thickness and its layout. The pure silicon on-resistance of the calculated device is 40 mΩ mm². It is concluded that the 5 μm or much thicker metal layer is necessary to realize less than 50 mΩ on-resistance devices. Today, a thick copper metal layer is often utilized to realize low on-resistance LDMOS from this viewpoint.

7.5.4 Lateral IGBTs on SOI

IGBTs are suitable for high-voltage medium-current power ICs because of large current capability, as shown in Figure 7.35. IGBT can be recognized as a pnp transistor driven by an n-channel MOSFET for a first-order approximation. Lateral IGBTs can be fabricated by conventional CMOS compatible processes.

The switching speed of bipolar power devices is conventionally controlled by introduction of a lifetime killer. However, lifetime control process is not compatible with conventional CMOS process. There are two ways to control switching speed of power devices. One way is to use thin SOI layers. The switching speed of IGBTs improves as the SOI thickness decreases, because carrier lifetime is effectively decreased by the influence of large carrier recombination at the silicon dioxide interfaces [29]. The other way is to reduce emitter efficiency of the p⁺ drain or collector. The effective methods are (1) emitter short, (2) low-dose emitter [30], (3) high-dose n buffer, and (4) forming an n⁺ layer in the p⁺ emitter [31].

Figure 7.36 shows a cross section of large current lateral IGBTs. Large current capability has been realized by adopting multiple surface channels. Figure 7.37 shows typical current voltage curves of the

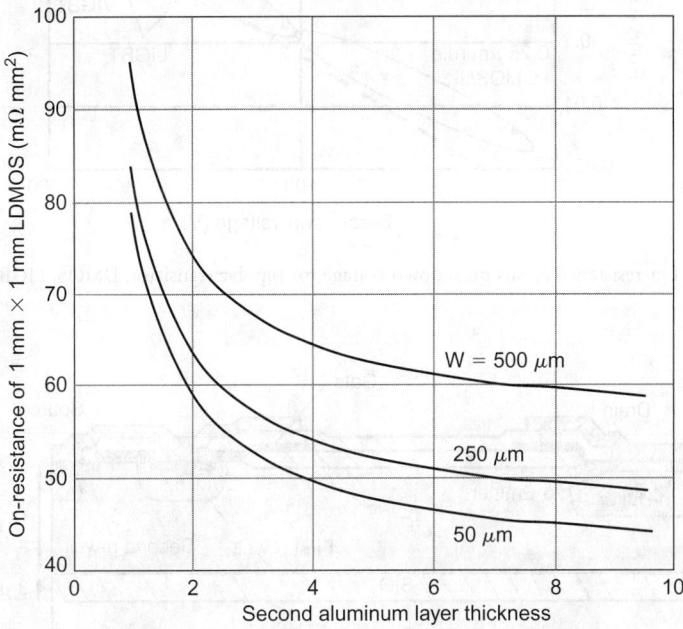

FIGURE 7.33 On-resistance of LDMOS as a function of second metal layer thickness with metal layer width as a parameter.

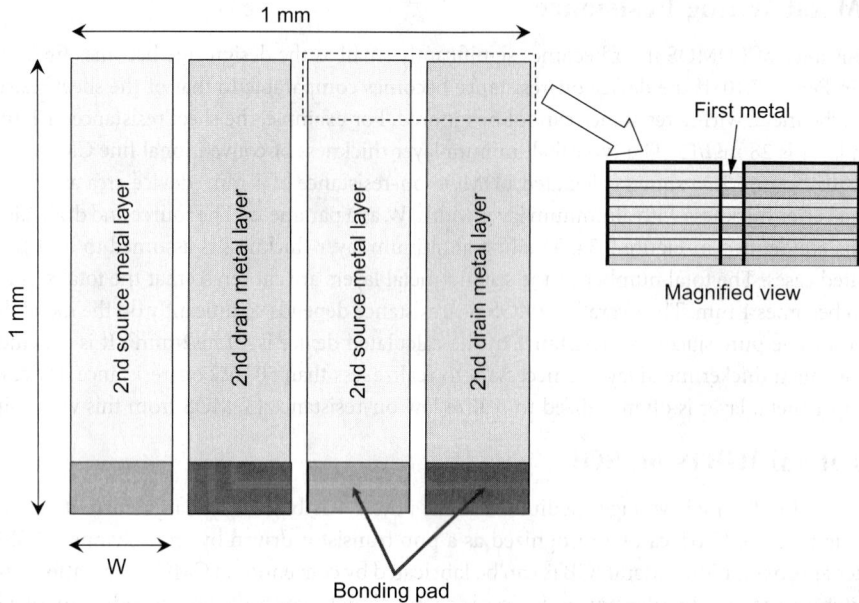

FIGURE 7.34 Metal layer layout of LDMOS in Figure 7.33.

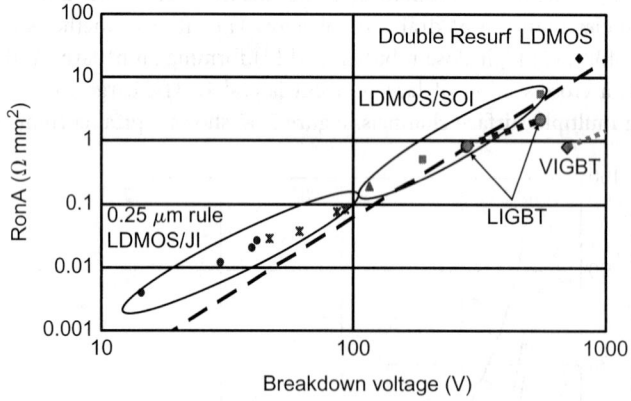

FIGURE 7.35 On-resistance versus breakdown voltage for bipolar transistor, DMOS, LIGBT, and VIGBT.

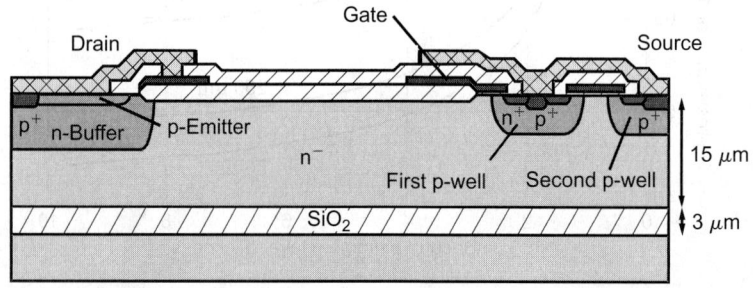

FIGURE 7.36 Cross section of large current lateral IGBT.

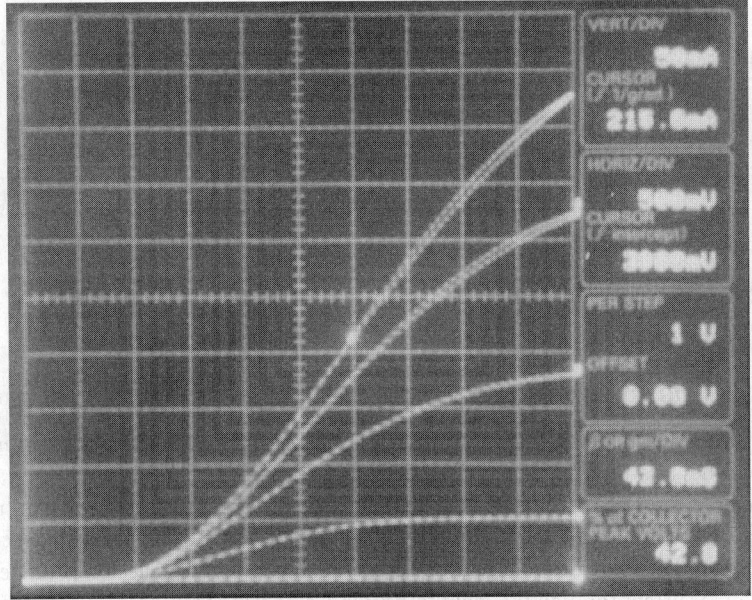

FIGURE 7.37 Typical current–voltage curves of multichannel LIGBT.

multichannel LIGBT. The current is an exponential function of the drain bias (collector bias) for low-voltage range. The current voltage curves seem to have 0.8 V off-set voltage just like a diode. Typical switching speed of the developed LIGBTs is 300 ns.

It is extremely important to increase operating current density of LIGBTs to reduce chip size. This is because output devices occupy most of the chip area and the cost of the power ICs deeply depends on the size of the power devices. The current density of the developed LIGBT is 175 A/cm^2 for 3 V forward voltage.

7.6 Sense and Protection Circuit

Power ICs have a significant advantage over discrete devices on the power switch protection because the protection circuit can be integrated on the same chip. Figure 7.38a shows the over-temperature sense circuit. In this circuit, the junction temperature dependence of diode forward voltage drop is utilized to sense the temperature. The series diodes are located close to the power switches so that the junction temperature of the diodes responds to the temperature change of the power switches. Comparing the

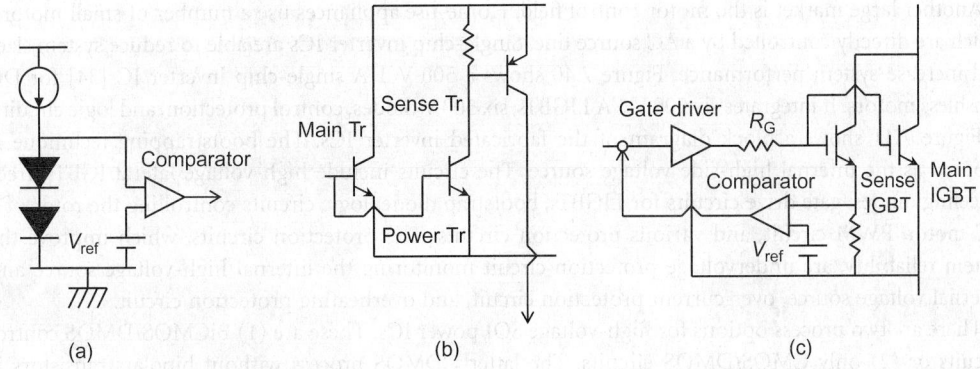

(a)   (b)   (c)

FIGURE 7.38 Sense circuits for over-temperature and over-current. (a) Over-temperature sense circuit. (b) Over-current sense circuit for bipolar Tr. (c) Over-current sense circuit for IGB (MOSFET).

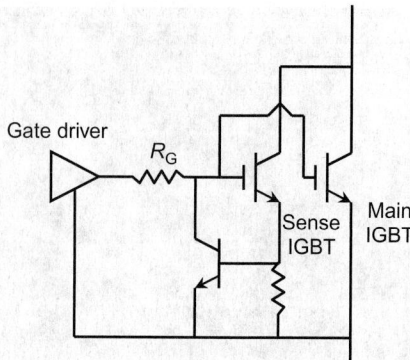

FIGURE 7.39 Typical short-circuit protection method for IGBTs.

forward voltage drop across the diodes with the reference voltage, the circuit senses the over temperature at the power switches and the fault signal is fed back to the gate control logic.

Figure 7.38b shows the over-current sense circuit for bipolar power transistors. The magnitude of current through the main transistor is reflected to the current through the sense transistor with the current mirror circuit configuration. Therefore the voltage drop across the resistor in the collector of the sense transistor is proportion to the current through the main transistor. Comparing the voltage drop with the reference voltage, the circuit detects the over current. Figure 7.38c shows the over-current sense circuit for IGBTs or MOSFETs, which has a similar configuration with Figure 7.38b. In this circuit, the over current is detected by the voltage drop across the resistor in the sense IGBT (MOSFET) emitter (source).

In the short-circuit protection, the collector current should be squeezed or terminated as soon as the short-circuit operation is detected. For this purpose, the protection circuit directly draws down the gate voltage with the short feedback loop. Figure 7.39 shows a typical short-circuit protection circuit for IGBTs. When a high current flows through the IGBT, the voltage drop across the emitter resistor of the sense IGBT directly drives the npn transistor. The transistor draws down the gate voltage so that the collector current is reduced to the level of safe turn-off operation.

7.7 Examples of High-Voltage SOI Power IC

The current main applications of high-voltage SOI power ICs are DC motor control and flat panel display drivers. Recently, technologies for color plasma display panels have been greatly improved, and demands for PDP driver ICs have increased. There are several reports [32,33] on the development of such ICs using SOI wafers. Flat panel display drivers have to integrate a large number of high-voltage devices. Trench isolation and LIGBTs are key techniques for reducing chip size and resultant cost.

Another large market is the motor control field. Home-use appliances use a number of small motors, which are directly controlled by a AC source line. Single-chip inverter ICs are able to reduce system sizes and increase system performance. Figure 7.40 shows a 500 V 1 A single-chip inverter IC [34] for DC brushless motors. It integrates six 500 V 1 A LIGBTs, six 500 V diodes, control protection, and logic circuits.

Figure 7.41 shows a block diagram of the fabricated inverter ICs. The bootstrapping technique is adopted as the internal high-side voltage source. The circuits include high-voltage lateral IGBTs, free-wheeling diodes, gate drive circuits for LIGBTs, bootstrap diode, logic circuits controlling the rotor of a DC motor, PWM circuit, and various protection circuits. The protection circuits, which improve the system reliability, are undervoltage protection circuit monitoring the internal high-voltage source and external voltage source, over-current protection circuit, and overheating protection circuit.

There are two process options for high-voltage SOI power ICs. These are (1) BiCMOS/DMOS control circuits or (2) only CMOS/DMOS circuits. The latter CDMOS process without bipolar transistors is suitable for high-voltage SOI power ICs, because bipolar devices often cause malfunction of the analog circuits in the high-side driver. For example, Figure 7.42a and Figure 7.42b show how the dV/dt current

FIGURE 7.40 Photograph of 500 V 1 A single-chip inverter.

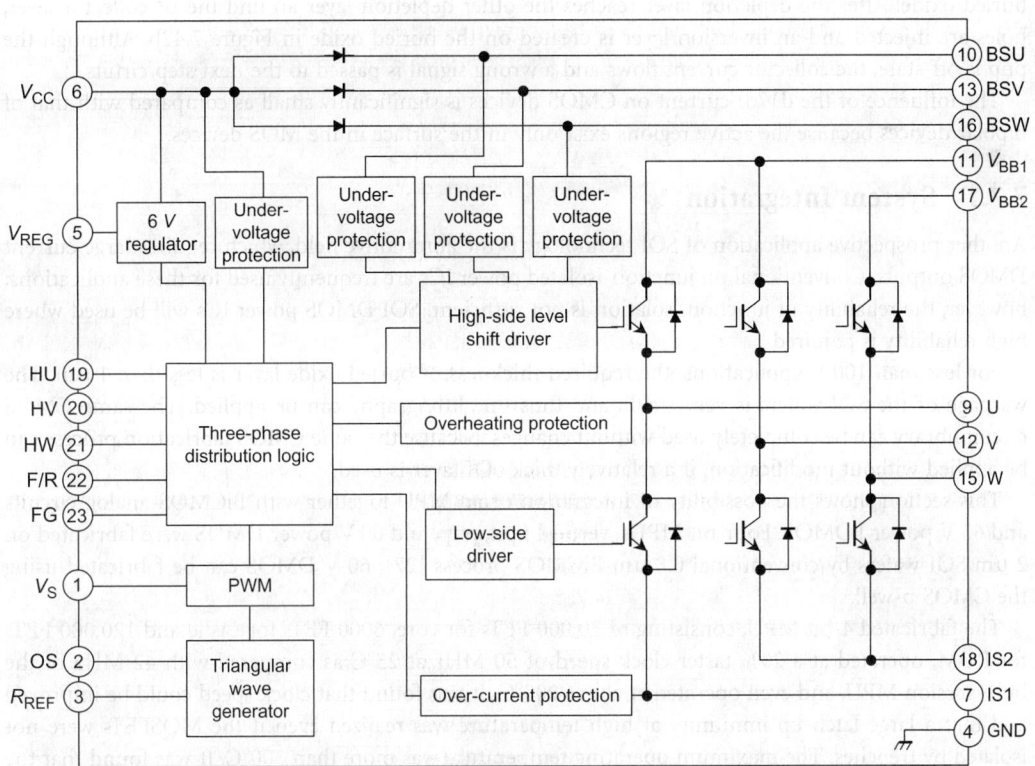

FIGURE 7.41 Circuit diagram of a single-chip inverter IC.

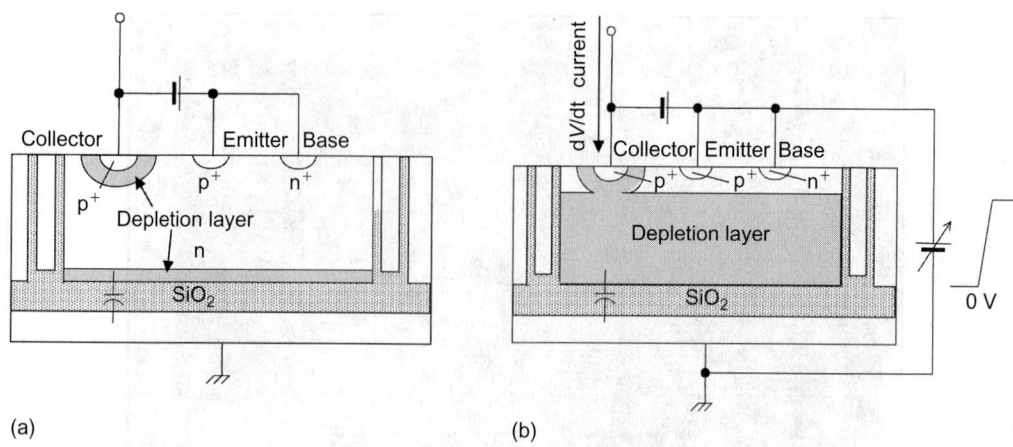

(a)   (b)

FIGURE 7.42 dV/dt current flowing in pnp bipolar transistor in a trench-isolated island. (a) No significant voltage is applied to the SOI island. (b) When a large dV/dt is applied between the SOI island and the substrate, depletion layer develops on the bottom oxide and reaches the p$^+$ collector.

flows and causes malfunction of lateral pnp transistors, if the pnp transistors are used in the high-side driver circuit. Figure 7.42a shows the depletion layer formation when the voltage between the base and the emitter is zero and the pnp is off-state. When the low-side output LIGBT is turned-off, the ground level of the whole high-side driver circuits is elevated and the depletion layer is initially created from the buried oxide. After the depletion layer reaches the other depletion layer around the p$^+$ collector layer, holes are injected and an inversion layer is created on the buried oxide in Figure 7.42b. Although the pnp is off-state, the collector current flows and a wrong signal is passed to the next step ciruits.

The influence of the dV/dt current on CMOS devices is significantly small as compared with that of bipolar devices because the active regions exist only in the surface in the MOS devices.

7.7.1 System Integration

Another prospective application of SOI technology is the automotive field, which requires large current DMOS outputs. Conventional pn junction-isolated power ICs are frequently used for these applications; however, the reliability of junction isolation is not sufficient. SOI DMOS power ICs will be used where high reliability is required.

For less than 100 V applications, the required thickness of buried oxide layer is less than 1 μm. The warpage of the SOI wafers is very small, and thus fine lithography can be applied. The same CMOS circuit library can be completely used without changes, because the same CMOS fabrication process can be applied without modification, if a relatively thick SOI layer is used.

This section shows the possibility of integration of an MPU together with BiCMOS analog circuits and 60 V power LDMOS. Four-bit MPUs, vertical npn, pnp, and 60 V power DMOS were fabricated on 2 μm SOI wafers by conventional 0.8 μm BiCMOS process [27]; 60 V DMOS can be fabricated using the CMOS p-well.

The fabricated 4-bit MPU, consisting of 30,000 FETs for core, 6000 FETs for cashe and 120,000 FETs for ROM, operated at a 20% faster clock speed of 50 MHz at 25°C as compared with 42 MHz of the bulk-version MPU, and even operated at above 200°C. It was found that clock speed could be improved and that a large latch-up immunity at high temperature was realized even if the MOSFETs were not isolated by trenches. The maximum operating temperature was more than 300°C. It was found that the yield of the MPU fabricated on SOI was the same as that on bulk wafers, verifying that the crystal quality of the currently available SOI wafers was sufficiently good. It was also found that both SOI and bulk MPUs could be operated at 300°C if MPUs consist of pure CMOS, although the power consumption of the bulk MPU was larger than that of the SOI MPUs.

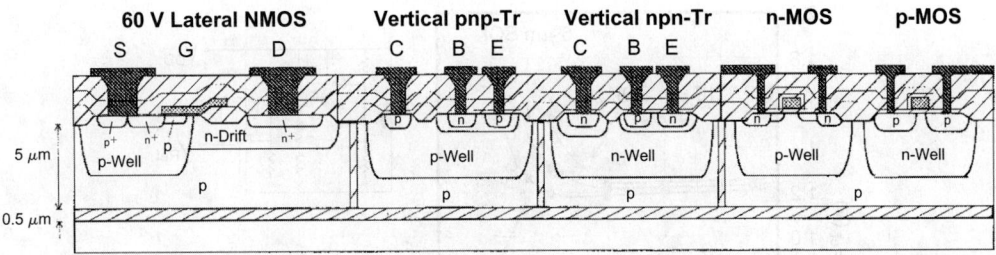

FIGURE 7.43 SOI power IC system integration.

One of the characteristic features of the SOI power IC structure, shown in Figure 7.43, is that there are no buried layers for bipolar transistors. It was found that vertical npn and pnp transistors fabricated on the n- and p-well layers exhibited sufficiently good characteristics, and the typical current gains h_{FE} for the vertical npn and pnp transistors were 80 and 30, respectively.

All these results show that system integration including power LDMOS will be promising.

7.8 High-Temperature Operation of SOI Power ICs

The leakage current of SOI devices simply reduces as the SOI layer becomes thinner, as seen in Figure 7.44 [27]. Small leakage current enables high-temperature operation of SOI power ICs. It was experimentally shown that IGBTs can be operated at a switching frequency of 20 kHz at 200°C, if they are fabricated in thin SOI of less than 5 μm. Maximum operating temperature of analog circuits in SOI increases as the SOI layer becomes thinner. Figure 7.45 shows the output voltage of bandgap reference circuits as a function of temperature with SOI thickness as a parameter. In the circuits, each device was not trench-isolated. If all the devices are trench-isolated, much higher temperature operation can be expected, as shown later.

The operation of 250 V, 0.5 A three-phase single-chip inverter ICs at 200°C, fabricated in 5-μm-thick trench-isolated SOI layer, was demonstrated [35]. CMOS circuits on SOI were found to be capable of operating at 300°C. CMOS-based analog circuits with a minimum number of bipolar transistors were adopted. It was found that the bandgap reference circuit operated at 250°C. A DC brushless motor was successfully operated by the single-chip inverter IC at 200°C. The carrier frequency was 20 kHz.

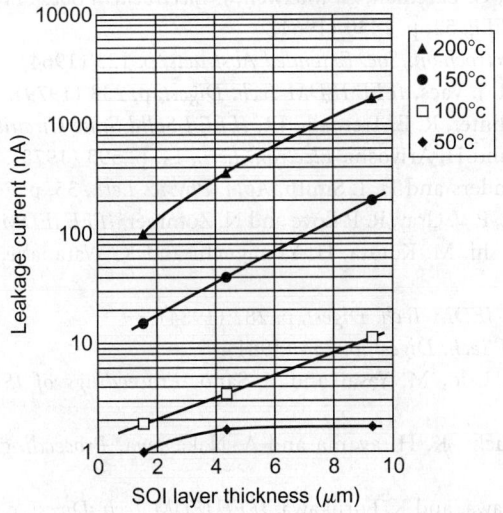

FIGURE 7.44 Leakage current versus SOI layer thickness.

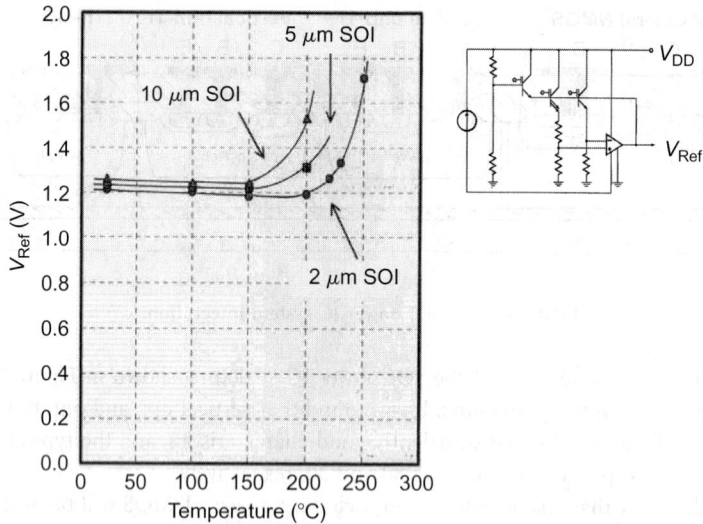

FIGURE 7.45 Output voltage of bandgap reference circuit as function of operation temperature.

References

1. A. Nakagawa, *Proceedings of ISPSD*, Baltimore, Maryland, USA, p. 16 (1991).
2. Y. Tarui, Y. Hayashi and T. Sekigawa, *Proc. of 1ˢᵗ Solid State Devices*, Tokyo, p. 105 (1969).
3. R. S. Wrathall, D. Tam, L. Terry and S. T. Robb, *IEEE IEDM Tech Digest*, p. 408 (1983).
4. B. Murari, F. Bertotti, G. A. Vignola, (Eds), *Smart Power ICs*, Springer-Verlag Berlin Heidelberg (1996).
5. J. D. Beasom, *IEEE IEDM Tech Digest*, p. 41 (1973).
6. M. Shimbo, K. Furukawa, K. Fukuda, and K. Tanzawa, *J. Appl. Phys.* Vol. 60, p. 2987 (1986).
7. A. Nakagawa, N. Yasuhara, and Y. Baba, *Proceedings of ISPSD*, Shinjuku, Tokyo, pp. 97–101 (1990).
8. T. Okabe, K. Sakamoto, and K. Hoya, *Proceedings of ISPSD*, Shinjuku, Tokyo, p. 96 (1988).
9. H. W. Becke, *IEEE IEDM Tech. Digest*, p. 724 (1985).
10. V. Rumennik, *IEEE SPECTRUM*, p. 42, July 1985.
11. D. McWilliams, C. Fa, G. Larchian, O. Maxwell, *J. Electrochem Soc.* 111, p. 153 (1964).
12. J. W. Lathrop, *Proc. IEEE*, 52, p. 1430 (1964).
13. J. Bouchard et al., *Electrochem. Soc. Extended Abstracts*, p. 123 (1964).
14. J. A. Appels and H. M. J. Vaes, *IEEE IEDM Tech. Digest*, p. 238 (1979).
15. R. S. Rosen, M. R. Splinter, R. E. Tremain, JR, *IEEE J Solid State Circuits*, SC-11, p. 431 (1976).
16. K. Izumi, M. Doken and H. Ariyoshi, *Electron. Lett.*, 14, p. 593 (1978).
17. M. W. Geis, D. C. Flanders and H. I. Smith, *Appl. Physics Lett.*, 35, p. 71 (1979).
18. B. J. Baliga, M. S. Adler, P. V. Gray, R. P. Love and N. Zommer, *IEEE IEDM Tech. Digest*, p. 264 (1982).
19. A. Nakagawa, H. Ohashi, M. Kurata, H. Yamaguchi and K. Watanabe, *IEEE IEDM Tech. Digest*, p. 860 (1984).
20. V. A. K. Temple, *IEEE IEDM Tech. Digest*, p. 282 (1984).
21. T. Kamei, *IEEE IEDM Tech. Digest*, p. 254 (1981).
22. K. Endo, Y. Baba, Y. Udo, M. Yasai and Y. Sano, *Proceedings of ISPSD*, Davos, Switzerland, p. 379 (1994).
23. H. Funaki, Y. Yamaguchi, K. Hirayama and A. Nakagawa, *Proceedings of ISPSD*, Kyoto, Japan, p. 25 (1998).
24. N. Yasuhara, A. Nakagawa, and K. Furukawa, *IEEE IEDM Tech. Digest*, p. 141 (1991).

25. S. Merchant, E. Arnold, H. Baumgart, S. Mukherjee, H. Pein and R. Pinker, *Proceedings of ISPSD*, Baltimore, Maryland, USA, p. 31 (1991).

26. E. Arnold, S. Merchant, M. Amato, S. Mukherjee and H. Pein, *Proceedings of ISPSD*, Baltimore, Maryland, USA, p. 242 (1992).

27. H. Funaki, Y. Yamaguchi, Y. Kawaguchi, Y. Terasaki, H. Mochizuki, A. Nakagawa, *IEEE IEDM Tech Digest*, p. 967 (1995).

28. K. Kinoshita, Y. Kawaguchi, and A. Nakagawa, *Proceedings of ISPSD*, Kyoto, pp. 65–68 (1998).

29. I. Omura, N. Yasuhara, A. Nakagawa and Y. Suzuki, *Proceedings of ISPSD*, Monterey, California, p. 248 (1993).

30. H. Funaki, T. Matsudai, A. Nakagawa, N. Yasuhara, Y. Yamaguchi, *Proceedings of ISPSD*, Weimar, Germany, p. 33 (1997).

31. Y. Yamaguchi, A. Nakagawa, N. Yasuhara, K. Watanabe and T. Ogura, *Ext. Abst. of 22nd SSDM*, Sendai, p. 677 (1990).

32. F. Gonzalez, V. Shekhar, Chia-Kuang Chan, B. Choy, N. Chen, *IEEE IEDM Tech digest*, p. 473 (1996).

33. H. Sumida, A. Hirabayashi, H. Shimabukuro, Y. Takazawa and Y. Shigeta, *Proceedings of ISPSD*, Kyoto, p. 137 (1998).

34. A. Nakagawa, H. Funaki, Y. Yamaguchi and F. Suzuki, *Proceedings of ISPSD*, Toronto, Canada, p. 321 (1999).

35. Y. Yamaguchi, N. Yasuhara, T. Matsudai and A. Nakagawa, *Proceedings of PCIM*, Japan, p. 1 (1998).

8

Microelectronics Packaging

Bi-Shiou Chiou

National Chiao Tung University

CONTENTS

8.1 Introduction

An "electronic package" can be defined as the portion of an electronic structure which serves to protect an electronic device from its environments and the environment from the electronic device. Packaging is an art based on the science of establishing interconnections and a suitable operating environment for predominantly electrical circuits. It supplies the chips with wires to distribute signals and power, remove the heat generated by the circuits, and provides them with physical support and environmental protection [1,2]. With continued demand for better performance, the electronics industry has been forcing more and more circuitry onto a silicon chip. As the circuit density increases, the number of electrical connections on a chip increases, and the time it takes for an electrical signal to travel form one chip to another as well as the retention of signal integrity have become important considerations. Besides, power requirements of the chip have increased dramatically and dissipating the heat to keep the chip running at its design temperature is an important requirement. Packaging is rapidly becoming an area of microelectronics technology that can limit the operating speed on an integrated circuit. The evolution of packaging is a response to the need to optimize for cost, performance, and reliability, with the emphasis shifting according to application priorities (see Figure 8.1 and Table 8.1).

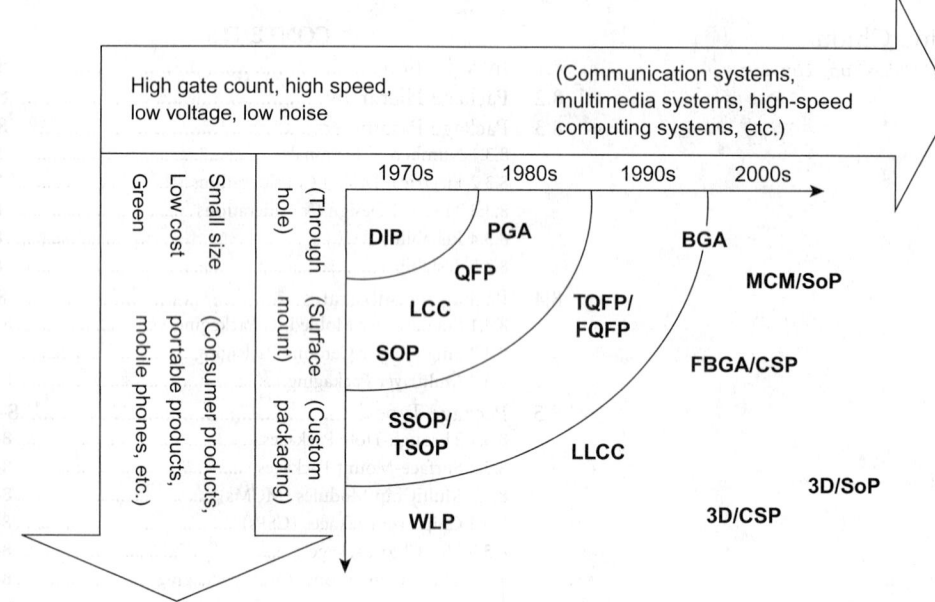

FIGURE 8.1   Evolution of packaging technology shifts according to application priorities.

TABLE 8.1   Electronic Packaging Requirements

Speed	*Size*
Large bandwidth	Compact size
Short inter-chip propagation delay	
Thermal and mechanical	*Test and Reliability*
High heat removal rate	Easy to test
A good match between the thermal coefficients	Easy to modify
of the dice and the chip carrier	Highly reliable
Pin Count and wireability	Low cost
Large I/O count per chip	*Noise*
Large I/O between the first and second level	Low noise coupling among wires
package	Good-quality transmission line
	Good power distribution

Source: From Chen, W.K., *VLSI Handbook,* 1st edition, chapter 11, CRC Press, Boca Raton, FL, 2000.

8.2 Package Hierarchy

Typical electronic systems are made up of several levels of packaging (see Figure 8.2). The zeroth level of packaging is the implementation of gate-to-gate interconnections on a silicon chip (e.g., gold and solder bumps). The semiconductor chip is encapsulated into a package, which constitutes the first level of packaging. A printed circuit board is usually employed (i.e., second level of packaging) because the total circuit and bit count required might exceed that available on a single first-level package. Besides, there may be components that cannot be readily integrated on a chip or first-level package such as capacitors, high-power resistors, and inductors. Therefore, several levels of packaging will be present and each level of packaging has distinctive types of interconnection devices associated with it (see Table 8.2). They are often referred to as a packaging hierarchy. The number of levels within a hierarchy may vary,

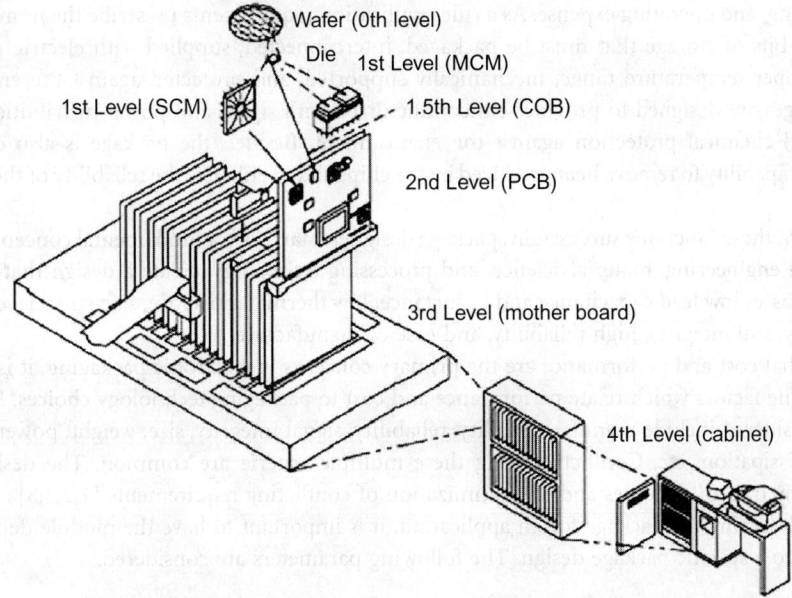

FIGURE 8.2 Packaging hierarchy of a hypothetical digital computer.

TABLE 8.2 Connection Methods for Various Package Levels

Level	Connections	Method	Distance from Center of Chip
0	Gate-to-gate interconnecting and solder bumping on a Si chip	PVD, CVD, and plating	<100 μm
1	Packaging of Si chips into chip carriers	Wire bonding, tape automatic bonding (TAB), flip chip bonding	0.05–0.25 mm
2	Connecting chip module/passive component to PCB	PTH, surface mounting technology (SMT)	0.1–5.0 mm
3	Connecting PCB to mother board	PTH, connector	1.0–5.0 cm
4	Connections between mother boards	Cable, connector	>5.0 cm
5	Connections between subsystems (e.g., local area network)	Cable, optical fibre, electromagnetic wave (wireless)	1–100 m
6	Connections between systems (e.g., communication network)	Cable, optical fibre, electromagnetic wave (wireless)	>600 m

depending on the degree of integration and the totality of packaging needs [1,2]. In the past, the packaging hierarchy contained more levels. As shown in Figure 8.2, dies were mounted on individual chip carriers, which were placed on a printed circuit board. Cards were then plugged into a larger board, and the boards were cabled into a cabinet. Finally, the cabinets were connected to assemble the computer. Today, higher levels of integration make many levels of packaging unnecessary, and this reduces the circuit path length significantly and improves the performance, cost, and reliability of the computers. Ideally, all circuitry one day may be placed on a single piece of semiconductor. Thus, packaging evolution reflects the integrated circuits (ICs) progress [4–6].

8.3 Package Parameters

A successful package design will satisfy all given application requirements at an acceptable design, manufacturing, and operating expense. As a rule, application requirements prescribe the number of logic circuits and bits of storage that must be packaged, interconnected, supplied with electric power, kept within a proper temperature range, mechanically supported, and protected against the environment. Thus, packages are designed to provide semiconductor IC with signal and power distribution, physical support, and chemical protection against the environment. Besides, the package is also designed to provide the capability to remove heat produced by the chips and to enhance the reliability of the packaging structures.

To perform these functions successfully, package designers start with a fundamental concept and, using principles of engineering, material science, and processing technology, create a design that is low cost and encompasses low lead capacitance and inductance, low thermal resistance, safe stress levels, material compatibility, seal integrity, high reliability, and ease of manufacture.

Given that cost and performance are the primary concerns in electronic packaging, it is important to examine the factors which relate performance and cost to packaging technology choices. Factors that must be considered include manufacturability, reliability, signal integrity, size, weight, power consumption, heat dissipation, etc. Conflicts among these multiple criteria are common. The design process involves many tradeoff analyses and the optimization of conflicting requirements [1,2,7,8].

While designing the package for an application, it is important to have the module defined before committing to a specific package design. The following parameters are considered.

8.3.1 Number of Terminals

The establishment of and adherence to good chip design rules are essential to achieving high yields in IC package assembly. The total number of terminals at packaging interfaces is a major cost factor. Signal interconnections and terminals constitute the majority of conducting elements, especially in low-cost packaging. Other conductors supply power and provide ground or other reference voltages.

Circuit partitioning and an appropriate net topology are essential to minimize the reflection noise in the packaging of high-speed circuitry and multichip modules. The number of terminals supporting a group of circuits is strongly dependent on the function of this group. With memory ICs, the smallest pinout can be obtained because the stream of data can be limited to a single bit. Exactly the opposite is the case with groups of logic circuits owing to a random partitioning of a computer. The pinout requirement is one of the key driving parameters for all levels of packaging [1–3,9,10].

8.3.2 Electrical Design Considerations

The two primary electrical functions of an electronic package are to deliver power to the circuits and to carry electrical signals from one circuit to another. Hence, the major electrical design objectives in electronic packaging are to maintain signal fidelity in signal paths and to minimize noise generation in electrical power conductors while minimizing the cost. High-speed systems have unique requirements for packaging technology as a result of the relatively short wavelength of the electromagnetic energy and the circuit components must be considered as distributed elements rather than as lumped elements.

High-speed digital design, in contrast to digital design at low speeds, emphasizes the behavior of passive circuit elements such as wires, circuit boards, and so on. At low speeds, passive circuit elements are just part of a product's packaging. At higher speeds, as a signal propagates through the package, it is degraded owing to reflections and line resistance (see Table 8.3). Controlling the resistance and the inductance

TABLE 8.3   Common Transmission Effect

Waveform	Transmission Line Effect
	Ohmic drop (frequency independent)
	Skin effect (frequency dependent)
	Propagation delay T_d, Rise time degradation ($T_{r,r} > T_{r,s}$)
	Reflection from capacitive load
	Reflection from inductive load
	Spurious noise Switching Oscillation Cross talk Others

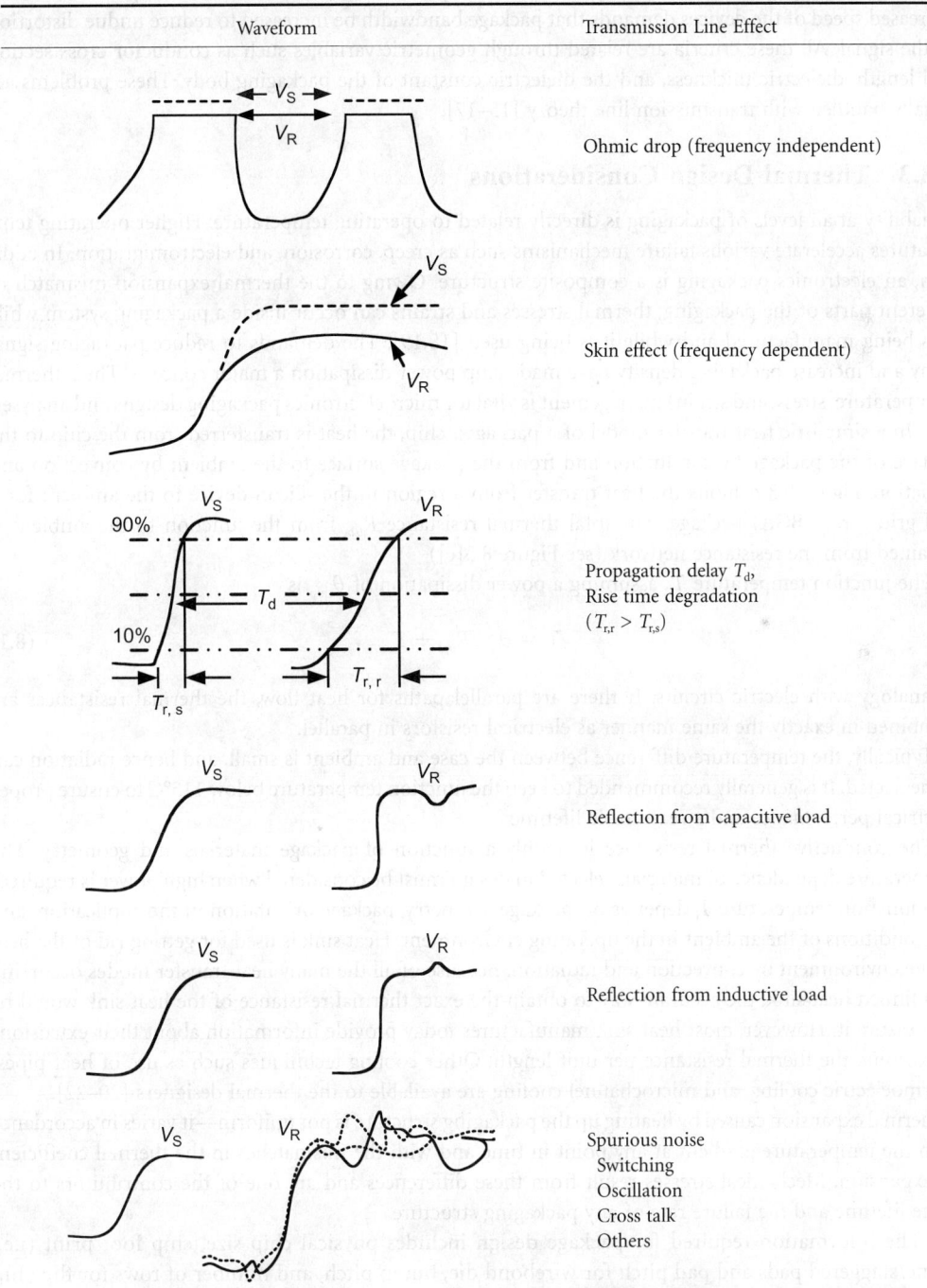

Note: V_S, source signal; V_R, receiver signal.

associated with the power and ground distribution paths to combat ground bounce and the simultaneous switching noise is essential. Besides, controlling the impedance environment of the signal distribution path in the package to mitigate the reflection-related noise is important. Reflections cause an increase in the transition time and may split the signal into two or more pulses with the potential of causing erroneous switching in the subsequent circuit and thus malfunctioning of the system. Controlling the capacitive coupling between signal traces in the signal distribution path to reduce cross talk is also important [11–14]. Increased speed of the devices demands that package bandwidth be increased to reduce undue distortion of the signal. All these criteria are related through geometric variables such as conductor cross section and length, dielectric thickness, and the dielectric constant of the packaging body. These problems are usually handled with transmission line theory [15–17].

8.3.3 Thermal Design Considerations

Reliability at all levels of packaging is directly related to operating temperature. Higher operating temperatures accelerate various failure mechanisms such as creep, corrosion, and electromigration. In addition, an electronics packaging is a composite structure. Owing to the thermal expansion mismatch of different parts of the packaging, thermal stresses and strains can occur inside a packaging system while it is being manufactured and while it is being used [18,19]. The demands to reduce packaging signal delay and increase packaging density have made chip power dissipation a major concern. Thus, thermal (temperature, stress, and strain) management is vital for microelectronics packaging designs and analyses.

In a simplistic heat transfer model of a packaged chip, the heat is transferred from the chip to the surface of the package by conduction and from the package surface to the ambient by convection and radiation. Figure 8.3 exhibits the heat transfer from a region in the silicon device to the ambient for a ball grid array (BGA) package. The total thermal resistance $R_{\theta ja}$ from the junction to the ambient is obtained from the resistance network (see Figure 8.3[c])

The junction temperature T_j, assuming a power dissipation of θ_{chip} is

$$T_j = \theta_{chip} R_{\theta jd} + T_a \tag{8.1}$$

in analogy with electric circuits. If there are parallel paths for heat flow, the thermal resistances are combined in exactly the same manner as electrical resistors in parallel.

Typically, the temperature difference between the case and ambient is small, and hence radiation can be neglected. It is generally recommended to keep the junction temperature below 115°C to ensure proper electrical performance and a reasonable lifetime.

The conductive thermal resistance is mainly a function of package materials and geometry. The temperature dependence of materials selected in design must be considered when high power is required. The junction temperature T_j depends on package geometry, package orientation in the application, and the conditions of the ambient in the operating environment. Heat sink is used for getting rid of the heat of the environment by convection and radiation. Because of all the many heat transfer modes occurring in a finned heat sink, the accurate way to obtain the exact thermal resistance of the heat sink would be to measure it. However, most heat sink manufactures today provide information about their extrusions concerning the thermal resistance per unit length. Other cooling techniques such as use of heat pipes, thermoelectric cooling, and microchannel cooling are available to the thermal designers [20–22].

Thermal expansion caused by heating up the packaging structure is not uniform—it varies in accordance with the temperature gradient at any point in time and with the mismatches in the thermal coefficient of expansion. Mechanical stresses result from these differences and are one of the contributors to the finite lifetime and the failure rate of any packaging structure.

The information required for package design includes physical chip size, chip foot print (i.e., inline/staggered pads and pad pitch for wirebond die, bump pitch, and number of rows for flip chip die), die netlist (signals, power, and ground), type of package (ceramic, laminate, or plastic), package thermal and electrical performance requirements, footprint and pitch for second-level connections

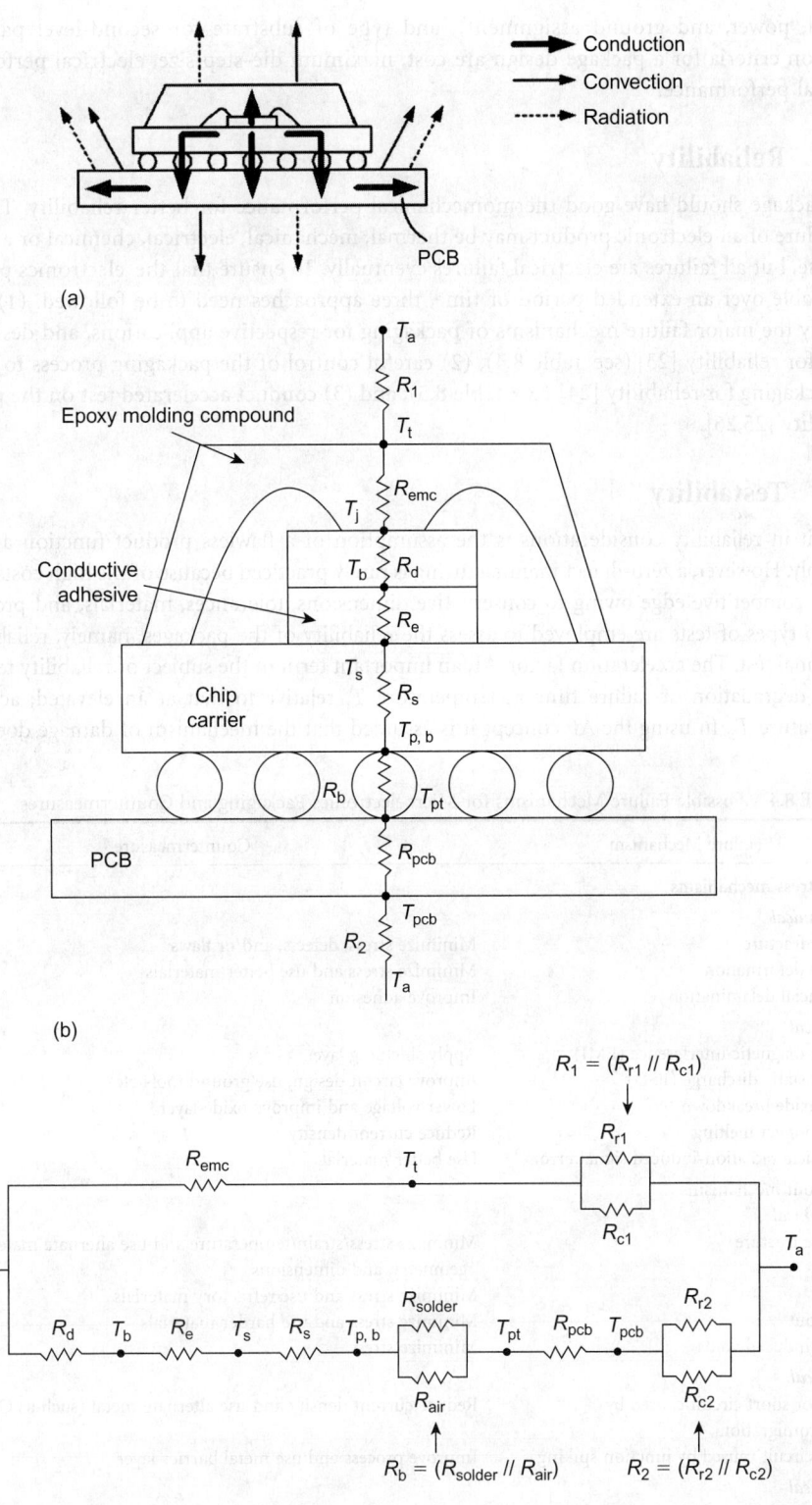

FIGURE 8.3 (a) Path of heat flow; (b) temperature and thermal resistance distribution; and (c) equivalent electrical circuit based on thermal resistance of a BGA package.

(signal, power, and ground assignment), and type of substrate for second-level packaging. The selection criteria for a package design are cost, maximum die-step size, electrical performance, and thermal performance.

8.3.4 Reliability

The package should have good thermomechanical performance for better reliability. The causes for the failure of an electronic product may be thermal, mechanical, electrical, chemical or a combination of these, but all failures are electrical failures eventually. To ensure that the electronics packaging will be reliable over an extended period of time, three approaches need to be followed: (1) understand, identify the major failure mechanisms of packaging for respective applications, and design the packaging for reliability [23] (see Table 8.4); (2) careful control of the packaging process to manufacture the packaging for reliability [24] (see Table 8.5); and (3) conduct accelerated test on the packaging for reliability [25,26].

8.3.5 Testability

Implicit in reliability considerations is the assumption of a flawless product function after its initial assembly. However, a zero-defect manufacturing is rarely practiced because of the high costs and possible loss of competitive edge owing to conservative dimensions, tolerances, materials, and process choices. So, two types of tests are employed to assess the reliability of the packages, namely, reliability test and functional test. The acceleration factor, AF, an important term in the subject of reliability test is the ratio of the degradation or failure time at temperature T_1 relative to that at an elevated, accelerated-test temperature T_2. In using the AF concept it is assumed that the mechanism of damage does not change

TABLE 8.4 Possible Failure Mechanisms for Microelectronics Packaging and Countermeasures

Failure Mechanism	Countermeasure
Overstress mechanisms	
Mechanical	
Brittle fracture	Minimize stress, defects, and/or flaws
Plastic deformation	Minimize stress and use better materials
Interfacial delamination	Improve adhesion
Electrical	
Electromagnetic interference (EMI)	Apply shielding layer
Electrostatic discharge (ESD)	Improve circuit design, use ground tools etc.
Gate oxide breakdown	Lower voltage and improve oxide layer
Interconnect melting	Reduce current density
α particle radiation-induced signal error	Use better material
Wear-out mechanisms	
Mechanical	
Fatigue fracture	Minimize stress/strain/temperature and use alternate materials, geometry, and dimensions
Creep	Minimize stress and use refractory materials
Wear out	Minimize stress and use harder materials
Stress-induced voids	Minimize stress
Electrical	
Open or short circuit caused by electromigration	Reduce current density and use alternate metal (such as Cu)
Short circuit caused by junction spiking	Improve process and use metal barrier layer
Chemical	
Corrosion	Provide sealing and encapsulation
Diffusion	Lower temperature and use diffusion barrier
Dendritic growth	Increase thickness and reduce humidity

TABLE 8.5   Defects Associated with Chip Packaging Process and Countermeasures

Possible Defects	Countermeasure
Die bonding	
Ion contamination	Improve cleaning process
Chip fracture	Use better material and process
Epoxy creep	Use high-temperature epoxy and reduce stress
Excessive electrical resistance, moisture source, and thermal resistance	Use better materials
Incorrect bonding area, thickness, and positioning	Improve personnel training and equipment alignment
Poor adhesion	Use better materials and process
Wire bonding	
Chip scratches, fracture and contaminated	Improve personnel training and equipment alignment
Incorrect wire dress	Use better materials
Incorrect bond placement	Improve personnel training and process
Controlled collapse chip connection	
Bridging, dewetting, and flux residue	
Poor reflow	Use better process and materials
Incorrect solder composition	
Poor wetting	
Plastic encapsulation	
Cracks, voids	
Excessive or inadequate cure	
Inadequate thermal conductivity and inadequate moisture resistance	Use better encapsulant and process
Poor adhesion to lead frame	
Package sealing	
Contamination	Improve cleaning process
Deformed metallization	Use better process and materials
Excessive moisture in package	Use encapsulant with low moisture adsorption
Crack of oxide and passivation in IC	Reduce internal stress
Cleaning	
Contamination and organic residue	Improve personnel training and use better materials and process in cleaning process
Radiation damage to device	
Surface pitting	

in the accelerated-test process. The AF values obtained from accelerated testing under some set of stress-test conditions are incorporated into reliability functions that are valid under use conditions. The stress includes temperature, voltage, current, and humidity, taken singly or in combination. The Arrhenius equation is often used to evaluate the activation energy for degradation. However, the Arrhenius model applies only to thermally activated processes. It is invalid for facture caused by mechanisms such as ESD, mechanical shock, etc. [27,28]. Besides, even for the thermally activated degradation process, it is difficult to isolate and characterize individual damage mechanisms when an admixture or distribution of activation energies may exist [29]. Functional test is used to determine whether a particular function of the board is working properly by toggling the circuitry through its legal states or conditions. Functional test is necessary in circuits to detect all possible faults that could prevent proper circuit operation. Design for test minimizes the cost and maximizes the success and value of the test process. New ESD test methods and equipment are required to comprehend the increasing pincount and shrinking interconnect pitch.

8.4   Packaging Substrates

An IC package falls into three basic categories: single-layer molded IC packages, single-layer ceramic packages, and multilayer packages.

In a single-layer molded IC package, the IC chip is first mechanically bonded to a lead frame and then electrically interconnected with fine wires from the chip bond pads to the corresponding lead-frame fingers. The lead-frame subassembly is imbedded in plastic after molding. In a single-layer ceramic package, the ceramic chip carrier is fabricated using either ceramic green-sheets or dry-pressing processes. For the multilayer type packages, the IC chip is assembled into a prefabricated multilayer substrate made of plastic or ceramic.

8.4.1 Single-Layer Molded IC Packaging

The molding compound used is man-made organic polymer (plastic) which is relatively porous and absorbs or transports water molecules and ions easily. Plastic packages are not very reliable because the aluminum metallization is susceptible to rapid corrosion in the presence of moisture, contaminants, and electric fields. Besides, impurities from the plastic or other materials in the construction of the package can cause threshold shifts or act as catalysts in metal corrosion. Fillers can also affect reliability and thermal performance of the plastic package.

8.4.2 Single-Layer Ceramic Packaging

Pressed ceramic technology packages are used mainly for economically encapsulating ICs and semiconductor devices requiring hermetic seals. Any contaminant present before sealing must be removed to an acceptable level before or during the sealing process. The hermetic package must pass both gross and fine leak tests and also exclude environmental contaminants and moisture for a long period of time.

8.4.3 Multilayer Packaging

The fabrication of a multilayer plastic substrate begins with impregnation of glass cloth with a thermo-setting resin solution to form a stable material termed "prepreg." Several plies of prepreg are sandwiched between sheets of treated copper foil and laminated to form a copper-clad, fully cured epoxy–glass composite core. The cores are then circuitized using photolithographic process. The composite circuit board is then fabricated by interleaving the cores with additional sheets of prepreg and copper foil. Lamination, hole drilling, photolithography, and plating processes are repeated to construct a multilayer printed circuit board.

Multilayered ceramic substrates consist of a cofired stack of ceramic green sheets on which metal wiring is printed and vias are punched for interlayer connections. Ceramic powders of the desired composition are mixed and ground with organic binders, solvents, and plasticizers to form a slurry which is then cast with a doctor's blade to form green sheet. Metallization of the green sheet and viafill are accomplished by thick film screen printing or extrusion filler. A schematic [3] of the whole process of making cofired multilayered substrates is illustrated in Figure 8.4. The tapes shrink during sintering and it is important to obtain uniform and repeatable sintering shrinkage throughout each part. The firing temperature dictates the metal wiring used for the circuitry. The sintering temperature of the powder has to be less than that of the melting temperature of the metal used for conductors. Pure alumina sinters at above 1500°C and hence requires refractory metals such as W or Mo for the wiring. These are generally referred to as high-temperature cofired ceramics (HTCC) and are losing popularity in the industry because of their high-temperature processing. In addition, alumina has a high dielectric constant and high CTE, further limiting its applications.

Glass ceramics can be sintered at relatively lower temperatures (<1000°C) and hence, more conductive metals such as Ag–Pd and Cu are used for the interlayer circuitry. These materials are referred

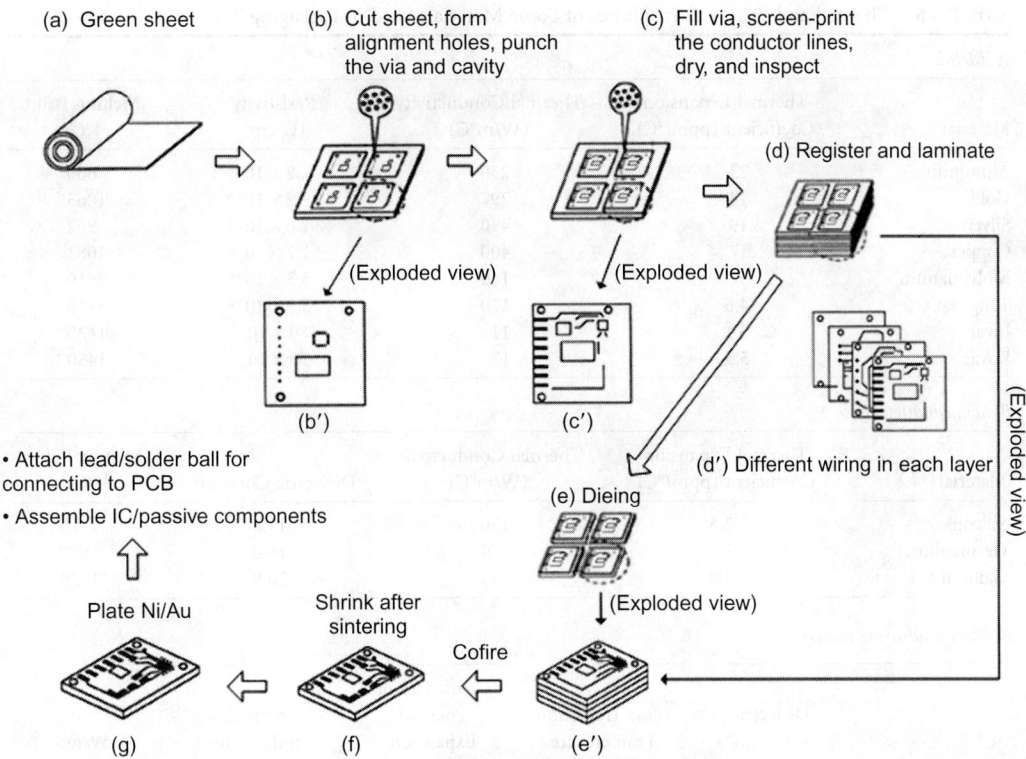

(a) Green sheet

(b) Cut sheet, form alignment holes, punch the via and cavity

(c) Fill via, screen-print the conductor lines, dry, and inspect

(d) Register and laminate

(Exploded view)

(b')

(Exploded view)

(c')

(d') Different wiring in each layer

(Exploded view)

(e) Dieing

(Exploded view)

• Attach lead/solder ball for connecting to PCB

• Assemble IC/passive components

Plate Ni/Au

Shrink after sintering

Cofire

(g)

(f)

(e')

FIGURE 8.4 Process flow for manufacturing cofired multilayer ceramic substrates.

to as low-temperature cofired ceramics (LTCC). Glass ceramics typically consist of glass-forming compounds, mixed with alumina or silica. The glassy phases melt at low temperatures, completely wet the alumina, and aid in sintering. Compositions of glass ceramics can be adjusted to obtain the desired properties such as a thermal expansion coefficient close to that of Si and a low permittivity. The major drawback to glass ceramics in many applications is their very low thermal conductivity. Miniaturization and improved device performance can be achieved by integrating passive components such as resistors, capacitors, and inductors into LTCC substrates. Components made from LTCC are receiving wide attention for high-frequency/RF applications in the telecommunications industry [8].

An ideal substrate material should combine all of the following general characteristics:

- Low dielectric constant and loss factor
- Thermal expansion coefficient close to that of silicon
- High thermal conductivity
- High mechanical strength
- High dielectric strength and resistivity
- Good thermal shock resistance
- Chemical and thermal stability under conditions of processing and use
- Nontoxicity
- Cofirable with Au, Ag, or Cu
- Low cost.

Table 8.6 summarizes the thermal and electrical properties of some materials used in packaging [1,3,30–32].

TABLE 8.6   Thermal and Electrical Properties of Some Materials Used in Packaging

A. Metals

Material	Thermal Expansion Coefficient (ppm/°C)	Thermal Conductivity (W/m°C)	Resistivity (Ω cm)	Melting Point (°C)
Aluminum	23	230	2.8×10^{-6}	660
Gold	14	297	2.2×10^{-6}	1063
Silver	19	430	1.6×10^{-6}	962
Copper	17	400	1.7×10^{-6}	1083
Molybdenum	5	140	5.3×10^{-6}	2610
Tungsten	4.6	170	5.3×10^{-6}	3415
Invar	3.1	11	80×10^{-6}	1425
Kovar	5.3	17	50×10^{-6}	1450

B. Semiconductors

Material	Thermal Expansion Coefficient (ppm/°C)	Thermal Conductivity (W/m°C)	Dielectric Constant	Melting Point (°C)
Silicon	2.5	150	11.8	1415
Germanium	5.7	70	16.0	937
Gallium arsenide	5.8	50	10.9	1238

C. Plastic substrate materials

Material	Dielectric Constant (1MHz)	Glass Transition Temperature T_g (°C)	Coefficient of Thermal Expansion (ppm/°C)	Thermal Conductivity (W/m°C)	Water Absorption (%)
FR-4 + E glass	4.1–4.2	125–135	12–16	0.35	1.1–1.2
BT resin + E glass	3.85–3.95	180–190	—	—	0.8–0.9
Cyanate ester + E glass	3.5–3.6	240–250	—	—	0.6–0.7
PI + E glass	3.95–4.05	>260	11–14	0.35	1.4–1.5
PTFE + E glass	2.45–2.55	327	24	0.26	0.2–0.3

(Continued)

8.5   Package Types

IC packages have been developed over time to meet the requirements of high performance and small size. Figure 8.5 illustrates the size reduction of IC package over time. There are two types of packages for chip module to be connected to PC board: through-hole packages and surface-mount packages [3,7,32].

8.5.1   Through-Hole Packages

Through-hole mounting technology uses precision holes drilled through the board and plated with copper. This copper plating forms the connections between separate layers which consist of thin copper sheets stacked together and insulated by epoxy fiberglass. There are no dedicated via structures to make connections between wiring levels; through holes serve that purpose. The component leads are inserted into plated holes in the board and soldered. The advantage of through-hole package is that it forms a sturdy support for the chip carrier and resists thermal and mechanical stresses caused by the variations in the expansions of components at raised temperatures. Various types of through-hole packages are summarized in Table 8.7.

TABLE 8.6 (Continued)

D. Ceramic substrate materials, Si, Cu, and CVD Diamond

Material	Melting Point (°C)	Dielectric Constant	Dissipation Factor (tan δ)	Resistivity (Ω cm)	Thermal Expansion Coefficient (ppm/°C)	Thermal Conductivity (W/m°C)	Bending Strength (MPa)	Density (kg/cm³)	Signal Propagation Delay[a] (ps/cm)
Al_2O_3	2050	8.5–10	0.0004–0.001	$>10^{14}$	6.5–7.2	22–40	300–385	3.75–4.0	~105
AlN	2677	8.5–10	0.001	$>10^{14}$	2.7–4.6	100–260	280–320	3.2	~100
BeO	2725	6.5–8.9	<0.001	$>10^{15}$	6.3–9.0	260–300	170–240	2.95	~113
BN	3000	4.1–5.4	0.1	$>10^{14}$	2.6–8.6	55–600	110	2.2–3.1	~140
SiC	3100	—	0.05	$>10^{14}$	2.8–4.6	70–270	450	3.0–3.2	
Si_3N_4	1900	5–10	—	$>10^{14}$	2.3–3.2	25–35	255–690	2.4–3.4	~113
SiO_2	1715	3.8	0.004	$>10^{14}$	0.5	1.6	50	2.2	~154
Si	1435	12	—	10^5	2.6	120–150	690	2.33	~122
Glass	—	5.7–7.2	0.006	$>10^{14}$	9.2	2	50	2.9	~140
Cordierite[b]	Decomposes	4.5	0.400	$10^6–10^{12}$	2.5	2.5	70	2.7	~120
Forsterite[b]	1900	6.2	0.50	$10^{10}–10^{12}$	9.8	3.3	170	2.9	~118
Mullite[b]	1850	6.2–6.8	0.02	$>10^{14}$	4.0–4.9	5.0–10	140	3.1	~126
Steatite[c]	—	5.7	0.1	10^{12}	4.2	2.5	170	2.7	~117
Glass ceramics	—	4.5–4.8	0.002	$>10^{13}$	2.5–6.5	0.8–2.3	150–240	—	
Cu	1083			1.6×10^{-6}	18.8	400		8.92	
CVD diamond		5.2		$10^{12}–10^{14}$	0.8–2.0	1300–1500		3.5	

[a]Proportional to the square root of dielectric constant.
[b]Cordierite, $M_2A_2S_5$; forsterite, M_2S; mullite, A_3S; M, MgO; A, Al_2O_3; S, SiO_2.
[c]A multiphase material containing $[Mg_3Si_4O_{10}(OH)_2]_2$.

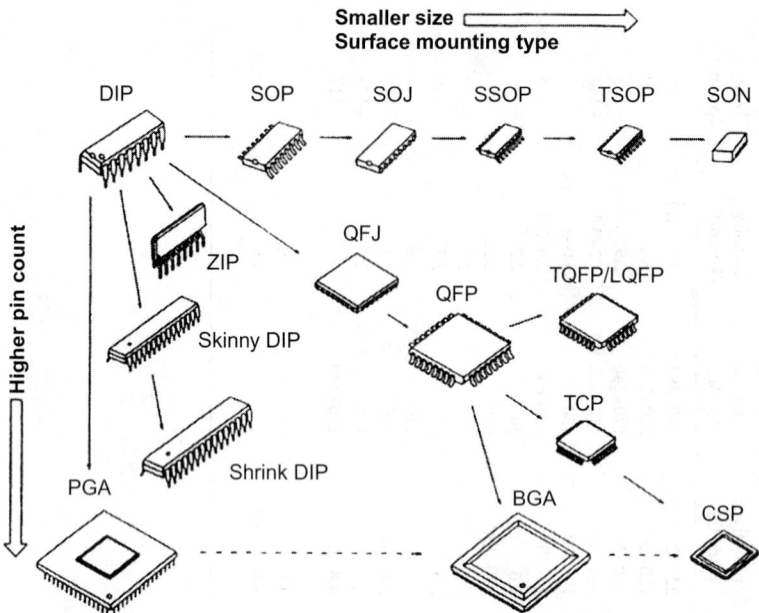

FIGURE 8.5 Packaging trends (from Chen W.K., editor, *VLSI Handbook*, 1st edition, chapter 11, CRC Press, Boca Raton, FL, 2000).

8.5.2 Surface-Mount Packages

In SMT, a chip carrier is soldered to the pads on the surface of a board without requiring any through holes. The advantages of SMT are smaller component sizes, lack of through holes, and the possibility of mounting chips on both sides of the PC board. This reduces package parasitic capacitances and inductances associated with the package pins and board wiring. SMT is the mainstream technology for second-level package. Various types of surface-mount *packages* are summarized in Table 8.8.

8.5.3 MultiChip Modules (MCMs)

In an MCM, several chips are supported on a single package. By eliminating one level of packaging, the inductance and capacitance of the electrical connections among the dice are reduced. There are several advantages of MCMs over single-chip carriers. The MCM minimizes the chip-to-chip spacing and reduces the inductive and capacitive discontinuities between the chips mounted on the substrate by replacing the die-bump-interconnect-bump-die path. There are three primary categories of MCMs: MCM-C (cofire), MCM-D (deposit), and MCM-L (laminate). For MCM-C, modules are constructed on cofired ceramic or glass–ceramic substrates using thick-film screen printing technologies to form the conductor patterns (see Figure 8.4). The term "cofire" implies that the conductors and ceramic are fired at the same time. For MCM-D, modules are formed by the deposition of thin film metals on dielectrics, which may be polymers or inorganic dielectrics. Silicon substrates are used in some MCM-D (silicon-on-silicon packaging) to eliminate the mismatch of thermal expansion. For MCM-L, modules are constructed of plastic laminate-based dielectrics and copper conductors utilizing advanced forms of printed wiring board (PWB) technologies to form the copper interconnects and vias [8,33].

8.5.4 Chip Size Packages (CSPs)

The definition of CSP is that the package area (width) is less than 1.5 (1.2) times that of the chip area (width). CSPs can be divided into two categories: the fan-in type and the fan-out type.

TABLE 8.7 Various Through-Hole Packages

Package Type	Cross Section of Bonding Regions	Description
Dual in-line package (DIP), Shrink DIP(SDIP)		A DIP is a rectangular package with two rows of pins in its two sides. Here, first the die is bonded on the lead frame and in the next step, chip I/O and power/ground pads are wire bonded to the lead frame, and the package is molded in plastic. DIPs are the workhorse of the high-volume and general-purpose logic products. However, the lower available pin counts of DIP is a limiting factor
Single in-line package (SIP)	Chip module, Chip lead, Solder, PCB	
Zigzag in-line package (ZIP)		
Pin grid array (PGA)	Chip module, Pin, PCB	A PGA has leads on its entire bottom surface. It has cavity-up and cavity-down versions. In a cavity-down version, a die is mounted on the same side as the pins facing toward the PC board, and a heat sink can be mounted on its backside to improve the heat flow. When the cavity and the pins are on the same side, the total number of pins is reduced because the area occupied by the cavity is not available for brazed pins. High pin count and larger power dissipation capability of PGAs make them attractive for different types of packaging

Fan-in type CSPs are suitable for memory applications that have relatively low pin counts. Depending on the location of bonding pads on the chip surface, the fan-in type is further divided into two types: the center pad type and the peripheral pad type. The fan-in type CSP keeps all the solder bumps within the chip area by arranging bumps in area array format on the chip surface. The fan-out CSPs are used mainly for logic applications; because of the die size to pin count ratio, the solder bumps cannot be designed within the chip area [34,35].

8.5.5 Flip-Chip Package

Flip-chip technology employs soldering between the integrated circuit die face and the interconnecting substrate. The length of the electrical connections between the chip and the substrate can be

TABLE 8.8   Various Types of Surface Mount Packages

Package Type	Cross-Section of Bonding Regions	Description
Small outline package (SOP) or SOIC, SO thin SOP (TSOP), shrink SOP (SSOP)	The leads bended inward for SOJ	The SOP has gull-wing-shaped leads. It requires less pin spacing than the DIPs and PGAs. SOP packages usually have small lead counts and are used for discrete, analog, and SSI/MSI logic parts
Quad flat-package (QFP), thin QFP (TQFP), plastic leaded chip carrier (PLCC)	L-type lead for QFP and TQFP, PLCC has L lead or J lead	QFP provides pins on all four sides of the rectangular package. Thin QFPs are developed to reduce the weight of the package. PLCC offers higher pin counts than the SOP. J-leaded PLCCs pack denser and are more suitable for automation than the L-leaded PLCCs because their leads do not extend beyond the package
Leadless ceramic chip carrier (LCCC)	The leads are left exposed around the package periphery to provide contacts for surface mounting	LCCCs take advantage of multilayer ceramic technology. Dice in the LCCCs are mounted in cavity-down position, and the back side of the chip faces away from the board, providing a good heat removal path. LCCCs are hermetically sealed
Surface mount PGA (SPGA)	SPGA is similar to PGA, but with flat leads for surface mount	
(BGA)	Similar to PGA, but use solder balls instead of leads or pads for connection	BGA packages transfer heat efficiently. The disadvantages of BGA packages are that solder connections cannot be visually inspected, and removed parts cannot be reused

minimized by placing metallic bumps, usually solder, on the dice, flipping the chips over, aligning them with the contacts pads on the substrate, and reflowing the solder balls in a furnace to establish the bonding between the chips and the package [36,37]. Flip-chip bonding provides electrical connections with minute parasitic inductance and capacitance. Besides, contact pads are distributed over the entire chip surface. This saves silicon area, increases the maximum I/O and power/ground terminals available with a given die size, and provides more efficiently routed signal and power/ground interconnections on the chip.

8.5.6 Three-Dimensional (3-D) packaging

The driving forces behind the development of 3-D packaging technology are similar to the multichip module technology, although the requirements for the 3-D technology are more aggressive. These requirements include the need for significant size and weight reductions, better system performance, higher packaging efficiency, smaller delay, higher reliability, higher number of input/output (I/O) contacts, and smaller operating power. Higher operating speed and lower power consumption can be realized because of the shorter signal paths between circuits that result from 3-D packaging. There are three categories of 3D packaging: (1) package-level 3-D packaging, in which packaged chips are stacked vertically; [38] (2) chip-level 3-D packaging, in which bare chips are stacked and then packaged [39–42]; and (3) wafer-level 3-D packaging, in which wafers are stacked and then diced and packaged (see Figure 8.6). Both package- and chip-level 3-D packagings are under production. Wafer-level 3-D packaging is a future technology and needs long development time. The key technologies for wafer-level

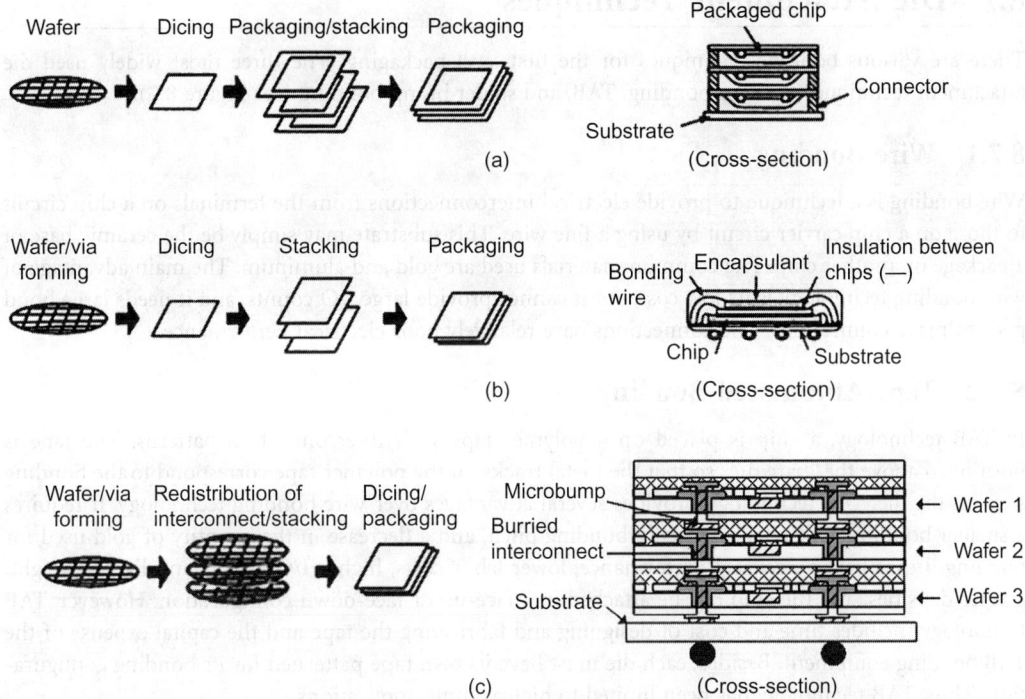

FIGURE 8.6 Process flow and schematic cross-section of 3-D packagings: (a) package-level 3-D packaging; (b) Chip-level 3-D packaging; and (c) wafer-level 3-D packaging.

3-D packaging include: formations of via holes, buried interconnections, and micro-bumps; wafer thinning; wafer alignment and stacking; testing and evaluation of interconnection, micro-bumps and so on [43–46].

8.6   Hermetic Packages

The package environment has two sources: the ambient environment in which the package is placed and the internal environment that results from adsorbed/absorbed moisture and residual ionics from incomplete or inadequate rinsing. In hermetic packages, the packaged cavity is sealed such that external chemical species are prevented from entering into it. However, a finite rate of leakage occurs through diffusion and permeation. Water can dissolve salts and other polar molecules to form an electrolyte, which together with the metal conductors and the potential difference between them, can create leakage paths as well as corrosion problems. Moisture is contributed mainly by the sealing ambient, the absorbed and dissolved water from the sealing materials, lid and the substrate, and the leakage of external moisture through the seal. The permeability to moisture of glasses, ceramics, and metals, is orders of magnitude lower than that of any plastic material. Hence, the hermetic packages are those made of metals, ceramics, and glasses. The common feature of hermetic packages is using a lid or a cap to seal in the semiconductor device mounted on a suitable substrate. The leads entering the package are also hermetically sealed. For microwave circuits, seals to prevent the entry of undesirable liquids and gases as well as to provide EMI protection are required. The type of seal depends on the packaging of the electronics within and the environmental requirements.

8.7   Die Attachment Techniques

There are various bonding techniques for the first-level packaging. The three most widely used die attachment techniques are wire bonding, TAB, and solder bump bonding (see Figure 8.7).

8.7.1   Wire Bonding

Wire bonding is a technique to provide electrical interconnections from the terminals on a chip circuit to those on a chip carrier circuit by using a fine wire. This substrate may simply be the ceramic base of a package or another chip. The common materials used are gold and aluminum. The main advantage of wire-bonding technology is its low cost; but it cannot provide large I/O counts, and it needs large bond pads to make connections. The connections have relatively poor electrical performance.

8.7.2   Tape-Automated Bonding

In TAB technology, a chip is placed on a polymer tape with interconnection patterns. The tape is positioned above the "bare die" so that the metal tracks on the polymer tape correspond to the bonding sites on the die. TAB technology provides several advantages over wire bonding technology. It requires a smaller bonding pad, smaller on-chip bonding pitch, and a decrease in the quantity of gold used for bonding. It has better electrical performance, lower labor costs, higher I/O counts and lighter weight, greater densities, and the chip can be attached in a face-up or face-down configuration. However, TAB technology includes time and cost of designing and fabricating the tape and the capital expense of the TAB bonding equipment. Besides, each die must have its own tape patterned for its bonding configuration. Thus, TAB technology has been limited to high-volume applications.

8.7.3   Solder Bump Bonding

In solder bump-bonding processes, solder bumps are bonded to the pads of the semiconductor chip, the chip is flipped over, the solder bumps are aligned with the contact pads on the substrate and reflowed in a furnace to establish the bonding between the die and the substrate. This technology provides electrical

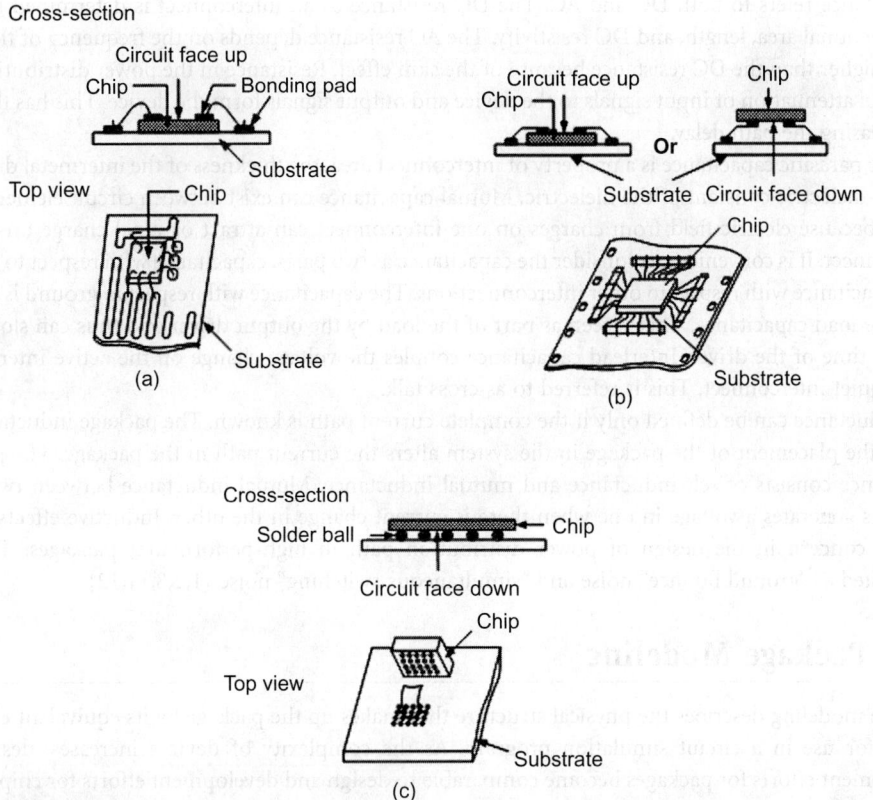

FIGURE 8.7 Various die attachment techniques: (a) wire bonding, (b) TAB, and (c) solder bump bonding.

connections with minute parasitic inductances and capacitances. In addition, the contact pads are distributed over the entire chip surface rather than being confined to the periphery. As a result, the silicon area is used more efficiently, the maximum number of interconnects is increased, and signal interconnections are shortened. But this technique results in poor thermal conduction, difficult inspection of the solder bumps, and possible thermal expansion mismatch between the semiconductor chips and the substrate. Underfill epoxy encapsulant is used to reduce the effect of thermal expansion mismatch between the silicon chip and the organic substrate. The underfill reduces the stresses and strains in the flip-chip solder joints and redistributes the stresses and strains over the entire chip area that would otherwise be increasingly concentrated near the corner solder joints of the chip. Thus, the reliability of the flip-chip solder joints is enhanced [47,48].

8.7.4 Other bonding techniques

There are other bonding techniques such as sea of leads (SoL)-compliant interconnect technique [49], stud bump bonding [37], surface-activated bonding [50], and bonding with conductive adhesives or metal bumps [51,52].

8.8 Package Parasitics

Package parasitics are the electrical resistance, capacitance, and inductance associated with the electrical interconnection of a device package. The parasitics are determined by the geometries and properties of both the interconnect and the intermetal dielectrics.

Resistance refers to both DC and AC. The DC resistance of an interconnect is determined by of its cross-sectional area, length, and DC resistivity. The AC resistance depends on the frequency of the signal and is higher than the DC resistance because of the skin effect. Resistance in the power distribution path results in attenuation of input signals to the device and output signals form the device. This has the effect of increasing the path delay.

The parasitic capacitance is a property of interconnect area, the thickness of the intermetal dielectric, and the dielectric constant of the dielectric. Mutual capacitance can exist between circuit elements. This occurs because electric field from charges on one interconnect can attract or repel charge on another interconnect. It is convenient to consider the capacitance as two parts: capacitance with respect to ground, and capacitance with respect to other interconnections. The capacitance with respect to ground is referred to as the load capacitance. This is seen as part of the load by the output driver and thus can slow down the rise time of the driver. Interlead capacitance couples the voltage change on the active interconnect to the quiet interconnect. This is referred to as cross talk.

Inductance can be defined only if the complete current path is known. The package inductance will vary if the placement of the package in the system alters the current path in the package. The parasitic inductance consists of self-inductance and mutual inductance. Mutual inductance between two interconnects generates a voltage in one when there is current change in the other. Inductive effects are the leading concern in the design of power distribution path in high-performance packages. They are manifested as "ground bounce" noise and "simultaneous switching" noise [1,2,8,11,12].

8.9 Package Modeling

Package modeling describes the physical structure that makes up the package by its equivalent electrical circuit for use in a circuit simulation program. As the complexity of devices increases, design and development efforts for packages become comparable to design and development efforts for chips. Many package design concepts must be simulated to assess their associated performance parameters. Computer-aided design software and test chips are becoming indispensable design tools. Computer-aided design tools are extensively used to analyze the thermal, thermomechanical, mechanical, and electrical parameters of packages and circuit simulation programs are used to evaluate the overall performance of a packaged circuit. Until now, the equivalent electrical circuit extracted from electric modeling incorporated only lumped electrical parameters, but as frequency of operation of the circuits is increasing, the distributed model of the package needs to be developed for high-frequency simulations [1,2,8,11,12,53,54].

8.10 Packaging in Wireless Applications

Wireless applications typically involve RF, high-frequency digital, and mixed-mode circuits. In these cases, circuit performance could be altered by the undesired coupling of a signal from one part of the circuit to another. Hence, wireless packaging requires minimal electrical parasitic effects that need to be well characterized.

Wireless circuitry differs from DC and low-frequency circuits in that the signals carried in both the conductors and the dielectric surrounding the conductors are influenced strongly by the high-frequency electrical properties of the dielectric. Hence the dielectric properties of packages which contain wireless circuits are quite important. Ceramic materials are applied, both as parts of the package as well as for subsystem carrying RF transmission lines. To this end, and to provide electromagnetic shielding, these materials have to be metallized partly. Aluminum nitride, beryllium, aluminum silicon carbide, and CVD diamond show the best thermal conductivity and are therefore applied in high-power applications, while alumina is well known for standard microwave applications (see Table 8.6D).

The trend in wireless applications is to integrate multiple modules on a single chip. So, the thermal management of the whole chip becomes crucial. The IC package must have good thermal properties. Metal as a material shows optimal properties concerning electrical conductivity, thermal conductivity, electromagnetic shielding, mechanical and thermal stability. For thermal expansion, the best match to semiconductor and ceramic material can be achieved with molybdenum, tungsten, or special composites like kovar. At high frequencies, interconnections need to be carefully designed. Microstrip interconnects, coplanar waveguide are mostly used for microwave packaging. Flip-chip packaging has tremendous potential for future RF packaging [55–60]. To improve the performance of the wireless system, it is crucial to reduce the extrinsic parasitics. System integration is key to reduce the parasitics, the size, and cost of a wireless system. Integration of passive components is a major challenge in wireless packages. More and more efforts are being made to integrate passive components and power devices on a chip with the other mixed-signal circuits. The size of the package becomes an issue. Micromachining technology provides a way to make miniature packages that conform to RF circuits, while providing physical and electrical shielding. Conformal packages made by applying micromachining technology provide the capability to isolate individual components between the adjacent circuits.

8.11 Future Trends

Packaging needs are driven as much by market application requirements as by silicon technology. The package cost has to follow the die cost reduction curve. As the complexity of package technology continues to increase, new materials will be needed to meet design, performance, and cost challenges. Efficient thermal management is a must for high speed dies. Significant engineering development will be needed for power increases at each technology generation.

Environmental concerns will continue to impact the selection/design of packaging materials/processes such as materials and surface finishes for lead-free solder assembly and for halogen-free materials [61,62]. For the next-generation substrate, both chip-in-substrate (i.e., embedded active IC devices) packaging technology [63–66] and packaging for flexible applications [67,68] are two important fields. Bumpless area array technologies will be needed to further reduce the package parasitics. System-on-package (SoP) is an emerging concept for highly integrated multifunctional system. SoP may overcome both the computing limitations and integration limitations of system-on-chip (SoC), system-in-package (SiP), MCM, and traditional packaging [61,69–71].

Packaging design and fabrication are increasingly important to system applications. Consideration of factors affecting waveform integrity for both power and signal (i.e., timing, cross talk, and ground bounce) will affect device layout on the chip, chip layout on the package, and interconnect. Package designs no longer can be developed independently of the chip and system; they must be considered concurrently as part of the overall system design. Chip–package–substrate codesign is the trend for high-performance systems.

System-level design capability is a critical issue. An integrated design environment of physical, electrical, thermal, thermo-mechanical, chip, package, and system design needs to be evolved. Coordinated design tools and simulators to address chip, package, and substrate codesign are required to manage the complexity of packaging that is being pushed to its performance limits.

Conventional surface mount packages will dominate in the region of low pin count and low clock frequency. BGA and chip-scale packages will be used for medium pin counts. Bare chip solutions, with high density and good electrical properties, could be very competitive with packaged solutions. However, bare chip solutions have a reliability versus cost tradeoff. Interconnection methods which offer direct pathways between chips to improve the performance of electronic products with minimal disruption to manufacturing infrastructure are needed to meet performance and cost challenges [72].

References

1. Tummala, R.R., Rymaszewski, E.J., and Klopfenstein, A.G., *Microelectronics Packaging Handbook*, 2nd edition, Chapman & Hall, New York, 1997.
2. Seraphim, D.P., Lasky, R.C., and Li, C.Y., editors, *Principles of Electronics Packaging*, "Introduction" McGraw-Hill, New York, 1989.

3. Chen, W.K., editor, *VLSI Handbook*, 1st edition, chapter 11, "Microelectronics Packaging," CRC Press, Boca Raton, FL, 2000.

4. Chang, G.K., Guidotto, D., Huang, Z., Liu, F., Chang, Y.J., Yu, J.J., and Tummala, R.R., "A Digital-Optical System-on-Package Module Architecture for High-Performance Multiprocessors," 17th *IEEE Lasers and Electro-Optics Society Meeting*, vol. 2, p. 581, 2004.

5. Chang, G.K., Guidotti, D., Liu, F., Chang, Y.J., Huang, Z., Sundaram, V., Balaraman, D., Hegde, S., Tummala, R.R., "Chip-to-Chip Optoelectronics SoP on Organic Boards or Packages," *IEEE Trans. Adv. Packaging.*, vol. 27, no. 2, p. 386, 2004.

6. Pendse, R., Marcus, B., Yee, M.H., Yun, J.S., Zahn, B., Jafari, B., Dewey, T., Lau, T., Michael, M., Singh, I., and Starr, O., "MAP (Mobile AGP Processor)—A High-Performance Integrated Graphics Module," *Proc. ECTC*, vol. 1, p. 7, 2004.

7. Blackwell, G. R., editor, *The Electronic Packaging Handbook*, CRC Press, Boca Raton, FL, IEEE Press, Washington, 2000.

8. Brown, W.D., editor, *Advanced Electronic Packaging—with Emphasis on Multichip Modules*, IEEE Press, Washington, 1999.

9. Pham, N., Cases, M., de Araujo, D., Matoglu, E., "Design Methodology for Multiple-Domain Power Distribution Systems," *Proc. ECTC*, vol. 1, p. 542, 2004.

10. Eo, Y., Eisenstadt, W.R., Jin, W.J., Choi, J.W., and Shim, J., "A Compact Multilayer IC Package Model for Efficient Simulation, Analysis, and Design of High-Performance VLSI Circuits," *IEEE Trans. Adv. Packag.*, vol. 26, no. 4, p. 392, 2003.

11. Konsowski, S.G. and Helland, A.R., *Electronic Packaging of High-Speed Circuitry*, McGraw-Hill, New York, 1997.

12. Johnson, H. and Graham, M., *High-Speed Digital Design—A Handbook of Black Magic*, Prentice-Hall, New York, 1993.

13. Choi, S.H. and Roy, K., "Noise Analysis under Capacitive and Inductive Coupling for High-Speed Circuits," Proc. *1st IEEE International Workshop on Electronic Design, Test and Applications (DELTA 02)*, p. 365, 2002.

14. Na, N., Budell, T., Chiu, C., Tremble, E., and Wemple, I., "The Effects of On-chip and Package Decoupling Capacitors and an Efficient ASIC Decoupling Methodology," *Proc. ECTC*, vol. 1, p. 556, 2004.

15. Deutsch, A., Kopcsay, G.V., Restle, P.J., Smith, H.H., Katopis, G., Becker, W.D., Coteus, P.W., Surovic, C.W., Rubin, B.J., Dunne, R.P., Jr., Gallo, T., Jenkins, K.A., Terman, L.M., Dennard, R.H., Sai-Halasz, G.A., Krauter, and B.L., Knebel, D.R., "When are Transmission-Line Effects Important for On-Chip Interconnections?" *IEEE Trans. Microw. Theory Tech.*, vol. 45, no. 10, p. 1836, 1997.

16. Heydari, P. and Pedram, M., "Interconnect Energy Dissipation in High-Speed ULSI Circuits," *Proc. 15th International Conf. on VLSI Design*, p. 132, 2002.

17. Quint, D., Bois, K., and Wang, Y., "A Simplified Cross-Coupling Model for Multiple Balanced Transmission Lines," *Proc. ECTC*, vol. 1, p. 255, 2004.

18. Lau, J. H., editor, *Thermal Stress and Strain in Microelectronics Packaging*, Van Nostrand Reinhold, New York, 1993.

19. Xiu, K. and Ketchen, M., "Thermal Modeling of a Small Extreme Power Density Macro on a High-Power Density Microprocessor Chip in the Presence of Realisic Packaging and Interconnect Structures," *Proc. ECTC*, vol. 1, p. 918, 2004.

20. Peterson, G.P., An Introduction to Heat Pipes, Wiley, New York, 1994.

21. Mallik, A.K., Peterson, G.P., and Weichold, M.H., "On the Use of Micro Heat Pipes as an Integral Part of Semiconductor Devices," *J. Electron. Packag.* vol. 114, p. 436, 1992.

22. Ma, H.B. and Peterson, G.P., "The Minimum Meniscus Radius and Capillary Heat Transport Limit in Micro Heat Pipes," *J. Heat Transfer*, vol. 120, p. 227, 1998.

23. Tummala, R.R., editor, *Fundamentals of Microsystems Packaging*, chapter 5, "Fundamentals of Design for Reliability," McGraw-Hill, New York, 2001.

24. Ohring, M., *Reliability and Failure of Electronic Materials and Devices*, Academic Press, New York, 1998.

25. Yang, D.G., Jansen, K.M.B., Ernst, L.J., Zhang, GQ., van Driel, W.D., Bressers, H.J.L., Fan, X.J., "Prediction of Process-Induced Warpage of IC Packages Encapsulated with Thermosetting Polymers," *Proc. ECTC*, vol. 1, p. 98, 2004.
26. Tunga, K., Kacker, K., Pucha, R.V., Sitaraman, S.K., "Accelerated Thermal Cycling: Is it Different for Lead-free Solder?" *Proc. ECTC*, vol. 2, p. 1579, 2004.
27. Pitarresi, J., Roggeman, B., Chaparala, S., Geng, P., "Mechanical Shock Testing and Modeling of PC Motherboards," *Proc. ECTC*, vol. 1, p. 1047, 2004.
28. Irving, S., Liu, Y., "Free Drop Test Simulation for Portable IC Package by Implicit Transient Dynamics FEM," *Proc. ECTC*, vol. 1, p. 1062, 2004.
29. Pecht, M., Dasgupta, A., Evans, J.W., and Evans, J.Y., *Quality Conformance and Qualification of Microelectronic Packages and Interconnects*, Wiley, New York, 1994.
30. Lau, J.H., Wong, C.P., Prince, J.L., and Nakayama, W., *Electronic Packaging: Design, Materials, Process, and Reliability*, McGraw-Hill, New York, 1998.
31. Gilleo, K., *Area Array Packaging Materials*, McGraw-Hill, New York, 2004.
32. Harper, C. A., *Electronic Packaging and Interconnection Handbook*, 2nd edition, McGraw-Hill, New York, 1997.
33. Doane, D.A. and Franzon, P.D., *Multichip Module Technologies and Alternatives—the Basics*, Van Nostrand Reinhold, New York, 1993.
34. Lau, J.H. and Lee, S.W.R., *Chip Scale Package: Design, Materials, Process, Reliability, and Applications*, McGraw-Hill, New York, 1999.
35. Patwardhan, V., Nguyen, H., Zhang, L., Kelkar, N., Nguyen, L., "Constrained Collapse Solder Joint Formation for Wafer-Level Chip-Scale Packages to Achieve Reliability Improvement," *Proc. ECTC*, vol. 2, p. 1479, 2004.
36. Lau, J.H., *Low-Cost Flip-Chip Technologies*, McGraw-Hill, New York, 2000.
37. Klein, M., Busse, E., Kaschlun, K., Oppermann, H., "Investigation of Cu Stud Bumping for Single-Chip Flip-Chip Assembly," *Proc. ECTC*, vol. 2, p. 1181, 2004.
38. Yano, Y., Sugiyama, T., Ishihara, S., Fukui, Y., Juso, H., Miyata, K., Sota, Y., and Fhjita, K., "Three-Dimensional Very Thin Stacked Packaging Technology for SiP," *Proc. ECTC*, San Diego, p. 1329, 2002.
39. Awad, E., Ding, H., Graf, R.S., and Maloney, J.J., "Stacked-Chip Packaging: Electrical, Mechanical, and Thermal Challenges," *Proc. ECTC*, vol. 2, p. 1608, 2004.
40. Zhang, X. and Tee, T.Y., "Advanced Warpage Prediction Methodology for Matrix Stacked Die BGA during Assembly Processes," *Proc. ECTC*, vol. 1, p. 593, 2004.
41. Watanabe, N., Hasegawa, S., and Asano, T., " Connection Test of Area Bump Using Active-Matrix Switches," *Jpn. J. Appl. Phys.*, vol. 44, no. 4B, p. 2770, 2005.
42. Watanabe, N., Ootani, Y., and Asano, T., "Pyramid Bumps for Fine-Pitch Chip-Stack Interconnection," *Jpn. J. Appl. Phys.*, vol. 44, no. 4B, p. 2751, 2005.
43. Tru-Si Technology, http://www.trusi.com.
44. Pienimaa, S.K., Miettinen, J., and Ristolainen, E., "Stacked Modular Package," *IEEE Trans. Adv. Packaging.*, vol. 27, no. 3, p. 461, 2004.
45. Koyanagi, M., Nakagawa, Y., Lee, K.W., Nakamura, T., Yamada, Y., Inamura, K., Park, K.T., and Kurino, H., "Neuromorphic Vision Chip Fabricated Using Three-Dimensional Integration Technology," *Proc. ISSCC*, p. 270, 2001.
46. Pozder, S., Lu, J.Q., Kwon, Y., Zollner, S., Yu, J., McMahon, J.J., Cale, T.S., Yu, K., Gutmann, R.J., "Back-End Compatibility of Bonding and Thinning Processes for a Wafer-Level 3D Interconnect Technology Platform," *Proc. IITC*, p. 102, 2004.
47. Tsao, P.H., Huang, C., Lii, M.J., Su, B., and Tsai, N.S., "Underfill Characterization for Low-k Dielectric/Cu Interconnect IC Flip-Chip Package Reliability," *Proc. ECTC*, vol. 1, p. 767, 2004.
48. Hannan, N., Kujala, A., Mohan, V., Morganelli, P., and Shah, J., "Investigation of Different Options of Pre-Applied CSP Underfill for Mechanical Reliability Enhancements in Mobil Phones," *Proc. ECTC*, vol. 1, p. 770, 2004.

49. Dang, B., Patel, C., Thacker, H., Bakir, M., Martin, K., and Meindl, J., "Optimal Implementation of Sea of Leads (SoL) Compliant Interconnect Technology," *Proc. IITC*, p. 99, 2004.
50. Suga, T., Kim, T.H., and Howlader, M.M.R., "Combined Process for Wafer Direct Bonding by Means of the Surface Activation Method," *Proc. ECTC*, vol. 1, p. 484, 2004.
51. Yim, M.J., Hwang, J.S., Kim, J.G., Kim, H.J., Kwon, W.S., Jang, K.W., and Paik K.W., "Anisotropic Conductive Adhesives with Enhanced Thermal Conductivity for Flip-Chip Applications," *Proc. ECTC*, vol. 1, p. 159, 2004.
52. Gupta, D., Fria, M., Kalle, F., "A Low-Cost Plated Column Bump Technology for Sub-100 µm and WLP Applications," *Proc. ECTC*, vol. 1, p. 58, 2004.
53. Koukab, A., Banerjee, K., and Declercq, M., "Modeling Techniques and Verification Methodologies for Substrate Coupling Effects in Mixed-Signal System-on-Chip Designs," *IEEE Trans. Comput.-Aided Des. of Integrated Circuits Syst.*, vol. 23, p. 823, 2004.
54. Badaroglu, M., Donnay, S., De Man, H.J., Zinzius, Y.A., Gielen, G.E., Sansen, W., Fondén, T., and Signell, S., "Modeling and Experimental Verification of Substrate Noise Generation in a 220-Kgates WLAN System-on-Chip With Multiple Supplies," *IEEE J. Solid-State Circuits*, vol. 38, no. 7, p. 1250, 2003.
55. Branch, J., Guo, X., Gao, L., Sugavanam, A., Lin, J.J., and O, K.K., "Wireless Communication in a Flip-Chip Package Using Integrated Antennas on Silicon Substrates," *IEEE Electron Dev. Lett.*, vol. 26, no. 2, p. 115, 2005.
56. Wu, J.H., Scholvin, J., and del Alamo, J.A., "A Through-Wafer Interconnect in Silicon for RFICs," *IEEE Trans. Electron Dev.*, vol. 51, no. 11, p. 1765, 2004.
57. Smolders, A.B., Pulsford, N.J., Philippe, P., and van Straten, F.E., "RF SiP: The next Wave for Wireless System Integration," *IEEE Radio Frequency Integrated Circuits Symposium*, MO3D-3, p. 233, 2004.
58. Li, R.L., DeJean, G., Maeng, M., Lim, K., Pinel, S., Tentzeris, M.M., and Laskar, J., "Design of Compact Stacked-Patch Antennas in LTCC Multilayer Packaging Modules for Wireless Applications," *IEEE Trans. Adv. Packag.*, vol. 27, no. 4, p. 581, 2004.
59. Larson, L. and Jessie, D., "Advances in RF Packaging Technologies for Next-Generation Wireless Communications Applications (Invited Paper)," *Proc. IEEE, Custom Integrated Circuits Conference*, p. 323, 2003.
60. Banerji, K., "Impact of 3G Communication IC Packaging," *Proc. IEEE, Electronics Packaging Technol. Conf.*, p. 460, 2000.
61. Semiconductor Industry Associations, *International Technology Roadmap for Semiconductors*; 2004, http://public.itrs.net/Files/2004updateFinal/2004Update.htm.
62. Dunford, S., Canumalla, S., Viswanadharn, P., "Intermetallic Morphology and Damage Evolution under Thermomechanical Fatigue of Lead (Pb)-Free Solder Interconnections," *Proc. ECTC*, vol. 1, p. 726, 2004.
63. Tuominen, R. and Kivilahti, J.K., "A Novel IMB Technology for Integrating Active and Passive Components," *Proc. 4th International Conference on Adhesive Joining and Coating Technology in Electronics Manufacturing*, p. 269, 2000.
64. Aschenbrenner, R., Ostmann, A., Neumann, A., and Reichl, H., "Process Flow and Manufacturing Concept for Embedded Active Devices," *Proc. EPTC*, p. 605, 2004.
65. Chen, Y.H., Lin, J.R., Chen, S.L., Ko, C.T., Kuo, T.Y., Chien, C.W., and Yu, S.P., "Chip-in-Substrate Package, CiSP, Technology," *Proc. EPTC*, p. 595, 2004.
66. Ostmann, A., Neumann, A., Auersperg, J., Ghahremani, C., Sommer, G., Aschenbrenner, R., and Reichl, H., "Integration of Passive and Active Components into Build-Up Layers," *Proc. EPTC*, p. 223, 2002.
67. Huang, J.R., Qian, W., Klauk, H., and Jackson, T.N., "Active-Matrix Pixelized Well Detectors on Polymeric Substrates," *Proc. IEEE, National Aerospace and Electron. Conf.*, p. 476, 2000.
68. Redinger, D., Molesa, S., Yin, S., Farschi, R., and Subramanian, V., "An Ink-Jet-Deposited Passive Component Process for RFID," *IEEE Trans. Electron Dev.*, vol. 51, no. 12, p. 1978, 2004.

69. Tummala, R.R., "SOP: What Is It and Why? A New Microsystem-Integration Technology Paradigm-Moore's Law for System Integration of Miniaturized Convergent Systems of the Next Decade," *IEEE Trans. Adv. Packag.*, vol. 27, no. 2, p. 241, 2004.

70. Tummala, R.R., Swaminathan, M., Tentzeris, M.M., Laskar, J., Chang, G.K., Sitaraman, S., Keezer, D., Guidotti, D., Huang, Z., Lim, K., Wan, L., Bhattacharya, W.K., Sundaram, V., Liu, F., and Raj, P.M., "The SOP for Miniaturized, Mixed-Signal Computing, Communication, and Consumer Systems of the Next Decade," *IEEE Trans. Adv. Packag.*, vol. 27, no. 2, p. 250, 2004.

71. Becker, K.F., Jung, E., Ostmann, A., Braun, T., Neumann, A., Aschenbrenner, R., and Reichl, H., "Stackable System-On-Packages With Integrated Components," *IEEE Trans. Adv. Packag.*, vol. 27, no. 2, p. 268, 2004.

72. Fjelstad, P., "An Alternative PCB Architecture for High-Speed Chip-to-Chip Signal Transmission," *Printed Circuit Design & Manufacture*, p. 34, February, 2005.

9

Multichip Module Technologies

Victor Boyadzhyan
Jet Propulsion Laboratory

John Choma, Jr.
University of Southern California

CONTENTS

9.1   Introduction

From the pioneering days to its current renaissance, the electronics industry has become the largest and most pervasive manufacturing industry in the developed world. Electronic products have the hallmark of innovation, creativity, and cost competitiveness in the world market place. The way the electronics are packaged, in particular, has progressed rapidly in response to customers' demands in general for diverse functions, cost, performances, and robustness of different products. For practicing engineers, there is a need to access the current state of knowledge in design and manufacturing tradeoffs.

Thus arises a need for electronics technology-based knowledge to optimize critical electronic design parameters such as speed, density, and temperature, resulting in performance well beyond PC board design capabilities. By removing discrete component packages and using more densely packed interconnects, electronic circuit speeds increase. The design challenge is to select the appropriate packaging technology, and to manage any resulting thermal problems.

The expanding market for high-density electronic circuit layouts calls for multi-chip modules (MCMs) to be able to meet the requirements of fine track and gap dimensions in signal layers, the retention of accurately defined geometry in multilayers, and high conductivity to minimize losses. Multi-chip module technologies fill this gap very nicely. This chapter provides engineers/scientists with an overview of existing MCM technologies and briefly explains similarities and differences of existing MCM technologies. The text is reinforced with practical pictorial examples, omitting extensive development of theory and details of proofs.

The simplest definition of a multi-chip module (MCM) is that of a single electronic package containing more than one integrated circuit (IC) die [1]. An MCM combines high-performance ICs with a custom-designed common substrate structure that provides mechanical support for the chips and multiple layers of conductors to interconnect them.

One advantage of this arrangement is that it takes better advantage of the performance of the ICs than it does interconnecting individually packaged ICs because the interconnect length is much shorter. The really unique feature of MCMs is the complex substrate structure that is fabricated using multilayer ceramics, polymers, silicon, metals, glass ceramics, laminates, etc. Thus, MCMs are not really new. They have been in existence since the first multi-chip hybrid circuit was fabricated. Conventional PWBs utilizing chip-on-board (COB), a technique where ICs are mounted and wire-bonded directly to the board, have also existed for some time. However, if packaging efficiency (also called silicon density), defined as the percentage of area on an interconnecting substrate that is occupied by silicon ICs, is the guideline used to define an MCM, then many hybrid and COB structures with less than 30% silicon density do not qualify as MCMs. In combination with packaging efficiency, a minimum of four conductive layers and 100 I/O leads has also been suggested as criteria for MCM classification [1].

A formal definition of MCMs has been established by the Institute for Interconnecting and Packaging Electronic Circuits (IPC). They defined three primary categories of MCMs: MCM-L, MCM-C, and MCM-D.

It is important to note that these are simple definitions. Consequently, many IC packaging schemes, which technically do not meet the criteria of any of the three simple definitions, may incorrectly be referred to as MCMs. However, when these simple definitions are combined with the concept of packaging efficiency, chip population, and I/O density, there is less confusion about what really constitutes an MCM. The fundamental (or basic) intent of MCM technology is to provide an extremely dense conductor matrix for the interconnection of bare IC chips. Consequently, some companies have designated their MCM products as high-density interconnect (HDI) modules.

9.2   Multi-Chip Module Technologies

From the above definitions, it should be obvious that MCM-Cs are descended from classical hybrid technology, and MCM-Ls are essentially highly sophisticated printed circuit boards, a technology that has been around for over 40 years. On the other hand, MCM-Ds are the result of manufacturing technologies that draw heavily from the semiconductor industry.

9.2.1   MCM-L

Modules constructed of plastic laminate-based dielectrics and copper conductors utilizing advanced forms of printed wiring board (PWB) technologies to form the interconnects and vias are commonly called "laminate MCMs," or MCM-Ls [2].

Advantages

Economic   Ability to fabricate circuits on large panels with a multiplicity of identical patterns. Reduces manufacturing cost. Quick response to volume orders.

Disadvantages

Technological   More limited in interconnect density relative to advanced MCM-C and MCM-D technologies. Copper slugs and cutouts are used in MCM-Ls for direct heat transfer. This degrades interconnection density.

MCM-L development has involved evolutionary technological advances to shrink the dimensions of interconnect lines and vias. From a cost perspective, it is desirable to use conventional PWB technologies for MCM-L fabrication. This is becoming more difficult as the need for multi-chip modules with higher interconnect density continues.

As MCM technologies are being considered for high-volume consumer products applications, a focus on containing the cost of high-density MCM-Ls is becoming critical.

The most usefull charateristic in assessing the relative potential of MCM-L technology is interconnection density [3,4], which is given by:

$$\text{Packaging efficiency (\%)} = \text{Silicon chip area/Package area} \tag{9.1}$$

The above formula measures how much of the surface of the board can be used for chip mounting pads versus how much must be avoided because of interconnect traces and holes/pads.

9.2.2   MCM-C

These are modules constructed on co-fired ceramic or glass-ceramic substrates using thick-film (screen printing) technologies to form the conductor patterns using fireable metals. The term "co-fired" implies that the conductors and ceramic are heated at the same time. These are also called thick-film MCMs.

Ceramic technology for MCMs can be divided into four major categories

- Thick-film hybrid process
- High-temperature co-fired alumina process (HTCC)
- Low-temperature co-fired ceramic/glass based process (LTCC)
- High T_c aluminum nitride co-fired substrate (AIN)

Thick-film hybrid technology produces by the successive deposition of conductors, dielectric, and/or resistor patterns onto a base substrate [5]. The thick-film material, in the form of a paste, is screenprinted onto the underlying layer, then dried and fired. The metallurgy chosen for a particular hybrid construction depends on a number of factors, including cost sensitivity, conductivity requirements, solderability, wire bondability, and more. A comparative summary of typical ceramic interconnect properties is compiled in Table 9.1.

9.3   Materials for HTCC Aluminum Packages

Metal conductors of tungsten and molybdenum are used for compatibility in co-firing to temperatures of 1600°C. Materials compatibility during co-firing dictates that raw materials of alumina with glass used for densification and any conductor metal powders (W, Mo) must be designed to closely match onset,

TABLE 9.1 A Comparative Summary of Typical Ceramic Interconnect Properties

Item	Thick Film	HTCC	LTCC
Line width (μm)	125	100	100
Via diameter (μm)	250	125	175
Ave. No. conductor layers	1–6	1–75	1–75
Conductor res. (mohm/sq)	2–100	8–12	3–20
ε (dielectric)	5–9	9–10	5–8
CTE	4–7.5	6	3–8
T_c Dielectric (W/mC)	2	15–20	1–2
Relative cost (low volume)	Medium	High	High
Tooling costs	Low	High	High
Capital outlay	Low	High	Medium

rate, and volume shrinkage; promote adhesion; and minimize thermal expansion mismatch between conductor and dielectric.

9.3.1 Processing HTCC Ceramics

The raw materials used in fabrication of aluminum substrates include aluminum oxide, glass, binder, plasticizer, and solvent. Materials specifications are used to control alumina particle size, surface area, impurity, and agglomeration. Glass frit is controlled through composition, glass transition and softening point, particle size, and surface area. Molecular weight, group chemistry, and viscosity controls are used for the binder and plasticizer.

9.3.2 Metal Powder and Paste

A thick-film paste often uses metal powder, glass powder, organic resins, and solvents. Compositions are varied to control screening properties, metal shrinkage, and conductivity. Paste fabrication begins with batch mixing, dispersion, and deagglomeration, which are completed on a three-roll mill.

9.3.3 Thick-Film Metallization

The greensheet is cast, dried, stripped from the carrier film, and blanked into defect-free sheets, typically 200 mm^2. The greensheet is then processed through punching, screening, and inspection operations.

9.3.4 HTCC in Summary

- Electrical performance characteristics include 50-ohm impedance, low conductor resistance, ability to integrate passive components such as capacitors and inductors, the ability to achieve high wiring density (ease of increasing the number of wiring at low cost), the ability to support high-speed simultaneous switching drivers, and the ease of supporting multiple reference voltages.
- Inherent thermal performance characteristics superior to MCM-L and MCM-D.
- Time-demonstrated reliability.

9.4 LTCC Substrates

The use of glass and glass-ceramics in electronic packaging goes back to the invention of semiconductors. Glasses are used for sealing T-O type packages and CERDIPs, as crossover and inter-level dielectrics in hybrid substrates. The success of co-fired alumina substrates spurred the development of the multilayer glass-ceramic substrates. These advantages derive from the higher electrical conductivity lines of copper, silver, or gold; the lower dielectric constant of the glass ceramic; and the closer CTE match of the substrate to silicon.

TABLE 9.2 AlN Properties Comparison

Item	ALN	HTCC	LTCC	BeO	Si	Cu
Thermal conductivity (W/mK)	175	25	2	260	150	394
Density (g/cm)		3.3	3.9	2.6	2.8	8.9
Dielectric constant (Mhz)	8.9	9.5	5.0	6.7		
Dissipation factor	0.0004	0.0004	0.0002	0.0004		
Bending strength (MaP)	320	420	210	220		

Source: From Ref. 2.

Two approaches have been used to obtain glass-ceramic compositions suitable for fabricating self-supporting substrates [6–8]. In the first approach, fine powder of a suitable glass-composition is used that has the ability to sinter well in the glassy state and simultaneously crystallize to become a glass-ceramic. More commonly, mixtures of ceramic powders are used, such as alumna and a suitable glass in nearly equal proportions, to obtain densely sintered substrates [9,10]. Because many glass and ceramic powders can be used to obtain densely sintered glass-ceramic, the actual choice is often made on the basis of other desirable attributes in the resulting glass-ceramic—such as low dielectric constant for lowering the signal propagation delay and coefficient of thermal expansion (CTE) closely matched to the CTE of silicon to improve the reliability of solder interconnections. Unlike the case of a crystallizing glass, the mixed glass and ceramic approach allows for a much wider choice of materials.

9.5   Aluminum Nitride

Aluminum nitride products are used in a variety of commercial and military applications.

Thermal management with solutions such as AlN can provide superior cooling to ensure reliable system operation. AlN packages typically offer a thermal conductivity of 150 to 200 W/mK, a level which can be compared with many metals or other high thermal conductive materials such as berillia (BeO) or silicon carbide (SiC). AlN has a thermal coefficient of expansion of 4.4 ppm, which is better matched to silicon than to alumina or plastics. Table 9.2 provides a comparison of AlN properties.

9.6   Materials for Multi-Layered AlN Packages

Aluminum nitride is a synthetic compound manufactured by two processes: the carbothermal reduction of alumina (Eq. 9.2) and/or direct nitridation of aluminum metal (Eq. 9.3):

$$Al_2O_3 + 3C + N_2 \rightarrow 2AlN + 3CO \qquad (9.2)$$

$$2Al + N_2 \rightarrow 2AlN \qquad (9.3)$$

9.6.1   MCM-D

Modules are formed by the deposition of thin-film metals and dielectrics, which may be polymers or inorganic dielectrics. These are commonly called thin-film MCMs.

Here, the focus will be on materials to fabricate the high-density MCM-D interconnect. The materials of construction can be categorized as the thin-film dielectric, the substrate, and the conductor metallization.

9.7   Thin-Film Dielectrics

Dielectrics for the thin-film packaging are polymeric and inorganic. Here, we will try to be brief and informative about those categories. Thin-film packages have evolved to a much greater extent with polymeric materials. The capability offered by polymers include a lower dielectric constant, the ability to form thicker layers with higher speeds, and lower cost of deposition. Polymer dielectrics have been used as insulating layers in recent microelectronics packaging.

9.8   Carrier Substrates

The thin-film substrate must have a flat and polished surface in order to build upon. The substrate should be inert to the process chemicals, gas atmospheres, and temperatures used during the fabrication of the interconnect. Mechanical properties are particularly important because the substrate must be strong enough to withstand handling, thermal cycling, and shock. The substrate must also meet certain CTE constraints because it is in contact with very large silicon chips on one side and with the package on the other side [11,12]. Thermal conductivity is another important aspect when heat-generating, closely spaced chips need that heat conducting medium. It is informative to state that high-density, large-area processing has generated interest in glass as a carrier material.

Metallic substrates have been used to optimize the thermal and mechanical requirements while minimizing substrate raw material and processing costs. Metallic sandwiches such as Cu/Mo/Cu can be tailored to control CTE and thermal properties. 5%Cu/Mo/5%Cu is reported to have a thermal conductivity (TC) of 135 W/mK, a CTE of 5.1 ppm, an as-manufactured surface finish of 0.813 µm, and a camber of 0.0005 in/in.

9.9   Conductor Metallization

In MCM, fabrication materials chosen will depend on the design, electrical requirements, and process chosen to fabricate the MCM. It is important to note that the most critical requirements for conductor metallization are conductivity and reliability.

9.9.1   Aluminum

Aluminum is a low-cost material that has adequate conductivity and can be deposited and patterned by typical IC techniques. It is resistant to oxidation. It can be sputtered or evaporated, but cannot be electroplated.

9.9.2   Copper

Copper has significant conductivity over aluminum and is more electromigration-resistant. It can be deposited by sputtering, evaporation, electroplating, or electroless plating. Copper rapidly oxidizes, forming a variety of oxides that can have poor adhesion to polymer dielectrics and copper itself.

9.9.3   Gold

Gold is used in thin-film structures to minimize via contact resistance problems caused by oxidation of Cu and Al. Gold can be deposited by sputtering, evaporation, electroplating, or electroless plating. Cost is high with excellent resistivity characteristic. Adhesion is poor and so it requires a layer (50 to 200 nm) of Ti or Ti/W.

9.10   Choosing Substrate Technologies and Assembly Techniques

The MCM designer has the freedom of choosing/identifying substrate and assembly technologies [13–15] from many sources [16,17]. If you are about to start a design and are looking for guidelines, finding a good source of information could be Internet access. In addition to designing to meet specific performance requirements, it is also necessary to consider ease of manufacture. For example, Maxtek publishes a set of design guidelines, that inform the MCM designer of preferred assembly materials and processes, process flows, and layout guidelines.

Substrate Technologies	Assembly Techniques
Chip-on-board	Surface mount
Chip-on-flex	Chip-and-wire
Thick-film ceramic	Mixed technology
Cofired ceramic	Special module
Thin-film ceramic	

Under Substrate Technologies and Assembly techniques it will be very informative to look at some pictorial examples, a primary source of which is Maxtek. The pictorial examples, followed by a brief description of technology or assembly technique shown, will serve as a quick guideline for someone who would like to get a feel of what technologies are viable in the MCM technology domain. Here, it is important to mention that a lot of current space electronic flight projects use MCM technologies for their final deliverables. Project Cassini, for example, used MCM hybrids in telecommunication subassemblies. On this note, take a look at some of the examples of MCM technologies currently on the market.

9.10.1 Chip-on-Board

Chip-on-board substrate technology (Figure 9.1) has low set-up and production costs and utilizes Rigid FR-406, GETEK, BT Epoxy, or other resin boards [3]. Assembly techniques used are direct die attach/ wire bonding techniques, combined with surface-mount technologies.

9.10.2 Chip-on-Flex

Chip-on-Flex substrate technologies [18,19] (Figure 9.2) are excellent for high-frequency, space-constrained circuit implementation. In creating this particular technology, the manufacturer needs Kapton or an equivalent flex-circuit base material with board stiffeners. Here, the die can be wire-bonded, with "glob-top" protection, while other discretes can be surface mounted. Integral inductors can be incorporated (e.g., for load matching).

9.10.3 Thick-Film Ceramic

This technology is the most versatile technology, with low-to-medium production costs. The 1-GHz attenuator above demonstrates the versatility of thick film on ceramic substrate technology (Figure 9.3), which utilizes both standard and custom ICs, printed resistors and capacitors actively trimmed to 0.25% with extremely stable capacitors formed between top plate and ground plane on the other side of substrate. Thick-film thermistor here senses overheating resistor and protects the remainder of the circuit.

FIGURE 9.1 Chip-on-board.

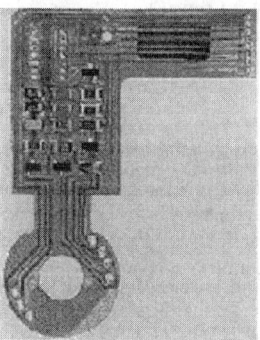

FIGURE 9.2 Chip-on-flex.

FIGURE 9.3 Thick-film ceramic.

FIGURE 9.4 Co-fired ceramic.

9.10.4   Co-Fired Ceramic

Co-fired ceramic MCM substrate technologies (low- or high-temperature co-fired multilayer ceramic) (Figure 9.4) are particularly suited for high-density digital arrays. Despite its high set-up and tooling costs, up to 14 co-fired ceramic layers are available from this particular manufacturer. In this technology, many package styles are available, including DIP, pin-grid array, and flat pack.

9.10.5   Thin-Film Ceramic

For thin-film ceramic technologies (see Figure 9.5), here is the outlined technology features include:

- High-performance MCMs, offset by high set-up and tooling costs

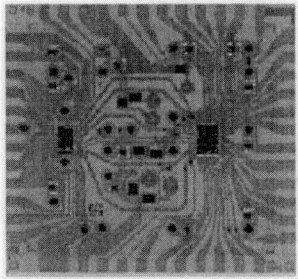

FIGURE 9.5 Thin-film ceramic.

TABLE 9.3 Thick-Film Specifications

Printed Component	Sheet Resistivity per square	Typical Values	Comments
Conductor, Gold	3–5 mohm		Lowest resistivity, 5-mil min. line/space
Etched thick-film	3–5 mohm		2 mil min. line/space
Conductor, Pd-Ag	<50 mohm		Lowest cost, solderable, 10-mil min. line/space
Resistor	3 ohm–1 Mohm		<±100 ppm/°C, laser trimmable to ±0.25%
Capacitor		10, 50, 100, 1500	Untrimmed ±30%, may be actively trimmed dielectric constant
Inductor		4–100 nH	Untrimmed ±10%

Source: Ref. 19.

- Alumina or BeO (as shown here) substrate
- Both sides of substrate can be used, the back side typically being a ground plane, with access through plated-through holes
- Chip-and-wire assembly with epoxied capacitors
- Many packaging styles available, including Kovar or ceramic package and lid
- Primarily used for high-frequency circuits requiring thin-film inductors or controlled impedance lines

For quick reference, some specifications and MCM technology information are summarized in Tables 9.3 and 9.4.

Other thick-film components are available, such as thermistors and spark gaps. In many applications, both sides of a substrate can be used for printed components.

9.11 Assembly Techniques

9.11.1 Surface-Mount Assembly

The surface-mount assembly technique can be categorized under lowest-cost, fastest turnaround assembly method, using pre-packaged components soldered to glass-epoxy board, flex circuit, or thick-film ceramic substrate. Many package styles are available, such as SIP, DIP. Pins may be attached as in-line leads, 90° leads, etc [19].

TABLE 9.4 MCM Substrate Technologies

Technology	Material	Line/Space (mils min.)	Dielectric Constant	Integral R	C	L
Chip-on-board	Glass-epoxy (FR-4), polyimides	4	4.3–5.0 4.0–4.6	Y	Y	
Chip-on-flex	Kapton or equiv. with stiffeners	3 inner, 5 outer	3.2–3.9	Y	Y	
Thick-film ceramic	Alumina	5	9.2–6.3	Y	Y	Y
	BeO	5		Y	Y	
Multilayer ceramic	Lo-fire alumina	5	5.8–9.0	Y	Y	
	Hi-fire alumina	5	9.0–9.6	Y	Y	
Thin-film ceramic	Alumina	1	9.9	Y	Y	Y
	BeO	1	6.3			
Etched thick-film	Alumina	2	9.2	Y	Y	Y

FIGURE 9.6 JPL STRV-2 project: tunneling accelerometer.

9.11.2   Chip-and-Wire Assembly

In order to minimize interconnect electronic circuit parasitics, a high-density layout is one of the assembly techniques recommended as a solution. The highlight of this technique is epoxy attachment/wire bonding of integrated circuits and components (e.g., capacitors) to a glass-epoxy board (chip-on-board). Of course, another way is attachment to a thick- or thin-film ceramic substrate. Currently, many package styles are available and can be listed as follows [19,20]:

- Epoxy seal, using a B-stage ceramic lid epoxies to the substrate (quasi-hermetic seal)
- Encapsulation, in which bare semiconductor die are covered with a "glob top" (low-cost seal)
- Metal (typically Kovar) package with Kovar lid either welded or soldered to the package (hermetic seal)
- Leads are typically plug-in type, SIP, DIP, PGA, etc.

9.11.3   Mixed Technologies

Another category of assembly technique recognized as "mixed technologies" combines chip-and-wire with surface-mount assembly techniques on a single substrate, which may be a glass-epoxy board or ceramic substrate. Heat sink/heat spreaders are available in a variety of materials.

The Maxtek module shown in Figure 9.7 includes a 4-layer, 20-mil-thick glass-epoxy board mounted to a beryllium copper heat spreader.

Selectively gold plated for wire bonding pads and pads at each end for use with elastomeric connectors,
- Three methods of IC die attach
- Epoxied directly to glass-epoxy board

FIGURE 9.7 Example of a mixed-technology assembly technique.

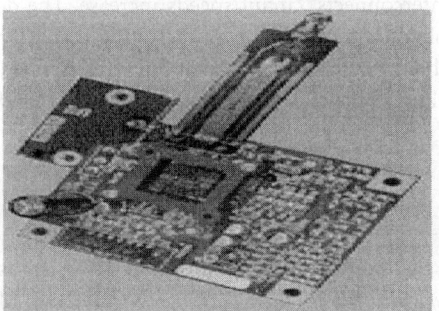

FIGURE 9.8 Example of a special module.

- Epoxied directly to the BeCu heat spreader through a cutout in the board
- Epoxied to the head spreader, through a cutout, via a thermally conductive submount, to electrically isolate the die from the heat spreader
- Solder-mounted resistors and capacitors
- 50-ohm differential signal lines
- IC die may be "glob topped" or covered in either a ceramic or plastic lid for protection

9.11.4 Special Modules

Under special modules we can emphasize and highlight technologies of complex assemblies of mixed-technology substrates, flexible circuits, and/or electromechanical components. (See Figure 9.8.) Complex assemblies of mixed-technology substrates often utilize a B-stage epoxy lid or glob top over chip-and-wire circuitry. These technologies enable designers to provide an integrated solution complex system problems like in a module shown on page 22 which is CRT Driver System capable of displaying 4 million pixels with 1.5-ns rise and fall times. The circuit incorporates a thin-film MCM connected by a special high-frequency connector to an FR-4 board with thick-film ceramic low-inductance load resistor and flex circuit with an integral impedance-matching inductor, connecting directly to the CRT [19,20].

The design engineer of an MCM chip should work with customer to partition the circuit and optimize the design to be implemented in MCM technology. Application of technologies for placement, routing, via minimization, tree searching, and layer estimation will be important to assess at this point. The general function, purpose, and configuration of the active elements, interconnects, and assembly technology should also be assessed, along with key materials and critical properties, representative manufacturing-process flows, potential failure mechanisms, qualification procedures, and design for testability.

Note that two concepts must be carefully examined for successful MCM production. An MCM design is initiated by selecting appropriate technologies from the many options available. The basic choices are for substrate technology and assembly techniques. Design trade-offs are analyzed, and a preliminary specification is completed. Following circuit simulation, prototypes are produced and tested. When the application requires it, an ASIC can be designed to be included in the MCM.

9.12 Summary

In summary, it is customary to give an answer to the fundamental question: What multi-chip modules do for you? . . . and here is the answer . . .

MCMs optimize critical design parameters such as speed, density, and temperature, resulting in performance well beyond PC board design capabilities. By removing discrete component packages and using more densely packed interconnects, circuit speeds increase. The design challenge is to select the appropriate packaging technology, and to manage any resulting thermal problems [20].

MCM technologies found their way and are utilized in the wireless, fiber, and instrumentation markets; space and military programs; and in the real world, they stand in the forefront of best merchant-market technology.

References

1. W.D. Brown, *ELEG 5273 Electronic Packaging*, University of Arkansas.
2. P.E. Garrou and I. Turlik, *Multichip Module Technology Handbook*, McGraw-Hill, 1998.
3. J.H. Reche, "High Density Multichip Interconnect for Advanced Packaging," *Proc. NEPCON West*, 1989, 1308.
4. N.G. Koopman, T.C. Reiley, and P.A. Totta, "Chip and Package Interconnections," in *Microelectronics Packaging Handbook*, Van Nostrand Reinhold, New York, 1989, 361.
5. D. Suranayana et al., "Flip Chip Solder Bump Fatigue Life Enhanced by Polymer Encapsulation," *Proc. 40th ECTC*, 1990, 338
6. J.G. Aday, T.G. Tessier, H. Crews, and J. Rasul, "A Comperative Analysis of High Density PWB Technologies," *Proc. Int. MCM Conference*, Denver, 1996, 239.
7. J.G. Aday, T.G. Tessier, and H. Crews, "Selecting Flip Chip on Board Compatable High Density PWB Technologies," *Int. J. Microcircuits and Electronic Packaging*, vol. 18, No. 4, 1995, 319.
8. Y. Tsukada, S. Tsuchida and Y. Mashimoto, "Surface Laminar Circuitry Packaging," *Proc. 42nd ECTC*, 1992, 22.
9. M. Moser and T.G. Tessier, "High Density PCBs for Enhanced SMT and Bare Chip Assembly Applications," *Proc. Int. MCM Conference*, Denver, 1995, 543.
10. E. Enomoto, M. Assai, Y. Sakaguchi, and C. Ohashi, "High Density Printed Wiring Boards Using Advanced Fully Additive Processing," *Proc. IPC*, 1989, 1.
11. C. Sullivan, R. Funer, R. Rust, and M. Witty, "Low Cost MCM-L for Vehicle Application," *Proc. Int. MCM Conf.*, Denver, 1996, 142.
12. W. Schmidt, "A Revolutionary Answer to Today's and Future Interconnect Challange," *Proc. of Printed Circui World Convention VI*, San Francisco, CA, 1994, T12-1.
13. D.P. Seraphim, D.E. Barr, W.T. Chen, G.P. Schmitt, and R.R. Tummala, "Printed Circuit Board Packaging," in *Microelectronics Packaging Handbook*, Van Nostrand Reinhold, New York, 1989, 853.

14. M.J. Begay, and R. Cantwell, "MCM-L Cost Model and Application Case Study," *Proc. Int. MCM Conference*, Denver, 1994, 332.

15. Compositech Ltd., 120 Ricefield Lane, Hauppauge, NY 11788-2071.

16. H. Holden, "Comparing Costs for Various PWB Build Up Technologies," *Proc. Int. MCM conference*, Denver, 1996, 15.

17. Diekman J. and Mirhej, M., "Nonwoven Aramid Papers: A New PWB Reinforcement Technology," *Proc. IEPS*, 1990, 123.

18. J. Fjelstad, T. DiStefano, and K. Karavakis, "Multilayer Flexiable Circuits with Area Arrray Interconnections for High Performance Electronics," *Proc. 2nd Int. FLEXCON*, 1995, 110.

19. Maxtek, Web site reference: http://www.maxtek.com.

20. James J. Licari and L.R. Enlow, *Hybrid Microcircuit Technology Handbook*, July 1988, 25.

Section II

Devices and their Models

John Choma, Jr.
University of Southern California

10

Bipolar Junction Transistor Circuits

David J. Comer and
Donald T. Comer

Brigham Young University

CONTENTS

10.1 Introduction

The *bipolar junction transistor* (BJT) was the workhorse of the electronics industry from the 1950s through the 1990s. This device was responsible for enabling the computer age as well as the modern era of communications. Although early systems that demonstrated the feasibility of electronic computers used the vacuum tube, this element was too unreliable for dependable, long-lasting computers. The invention of the point-contact transistor in 1947 [1] and the BJT shortly thereafter along with the rapid improvement in this device led to the development of highly reliable electronic computers and modern communication systems.

Integrated circuits (ICs), based on the BJT, became commercially available in the mid-1960s and further improved dependability of the computer and other electronic systems while reducing the size and cost of the overall system. Ultimately, the microprocessor chip was developed in the early 1970s and the age of small, capable, personal computers was ushered in. While the *metal-oxide-semiconductor field-effect transistor* (MOSFET) device is now more prominent than the BJT in the personal computer arena, the BJT is still important in larger high-speed computers, high-frequency communication systems, and power control systems.

Because of the continued improvement in BJT performance and the development of the heterojunction BJT, this device remains very important in the electronics field even as the metal oxide semiconductor (MOS) device becomes more significant.

10.2 Properties of the BJT

The present BJT technology is used to make both discrete component devices as well as IC chips. The basic construction techniques are similar in both cases with primary differences arising in size and packaging. The following description will be provided for the BJT constructed as an IC device on a silicon substrate. These devices are referred to as "junction isolated" devices. The cross-sectional view of a BJT is shown in Figure 10.1 [2]. Lower voltage devices fabricated on IC chips may occupy a surface area of around 100 μm^2. There are three physical regions comprising the BJT. These are the emitter, the base, and the collector. The thickness of the base region between emitter and collector can be a small fraction of a micrometer, while the overall vertical dimension of a device may be a few micrometers.

Thousands of such devices can be fabricated within a silicon wafer. They may be interconnected on the wafer using metal deposition techniques to form a system such as a microprocessor chip or they may be separated into thousands of individual BJTs, each mounted in its own case. The photolithographic methods that make it possible to simultaneously construct thousands of BJTs have led to continually decreasing size and cost of the BJT.

Electronic devices, such as the BJT, are governed by current–voltage relationships that are typically nonlinear and rather complex. In general, it is difficult to analyze devices that obey nonlinear equations, much less developed design methods for circuits that include these devices. The basic concept of modeling an electronic device is to replace the device in the circuit with easy-to-analyze components that approximate the voltage–current characteristics of the device. A model can then be defined as a collection of

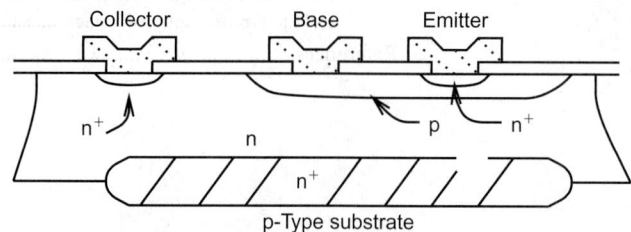

FIGURE 10.1 An integrated npn BJT.

simple components or elements used to represent a more complex electronic device. Once the device is replaced in the circuit by the model, well-known circuit analysis methods can be applied.

There are generally several different models for a given device. One may be more accurate than others, another may be simpler than others, another may model the DC voltage–current characteristics of the device, while still another may model the AC characteristics of the device.

Models are developed to be used for manual analysis or to be used by a computer. In general, the models for manual analysis are simpler and less accurate while the computer models are more complex and more accurate. Essentially, all models for manual analysis and many models for computer analysis include only linear elements. Nonlinear elements are included in computer models, but increase the computation times involved in circuit simulation over the times involved in simulation of linear models.

10.3   Basic Operation of the BJT

The BJT has three possible regions of operation: the *active* region, the *cutoff* region, and the *saturation* region. Amplifier circuits, creating output signals that are larger versions of the input signals, use the active region exclusively. Digital or switching circuits may pass through the active region while making transitions between the saturation and cutoff regions.

To understand the operation of the BJT as an amplifier or a switching circuit, we will discuss the simplified version of the device shown in Figure 10.2. The device shown is an npn device which consists of a p-doped material interfacing on opposite sides to an n-doped material. A pnp device can be created using an n-doped central region with p-doped interfacing regions. Since the npn type of BJT is more popular in present construction processes, the following discussion will center on this device.

The geometry of the device implied in Figure 10.2 is physically more like the earlier alloy transistor. This geometry is also capable of modeling the modern BJT (Figure 10.1) as the theory applies almost equally well to both geometries. Normally, some sort of load would appear in either the collector or emitter circuit, however this is not important to the initial discussion of BJT operation.

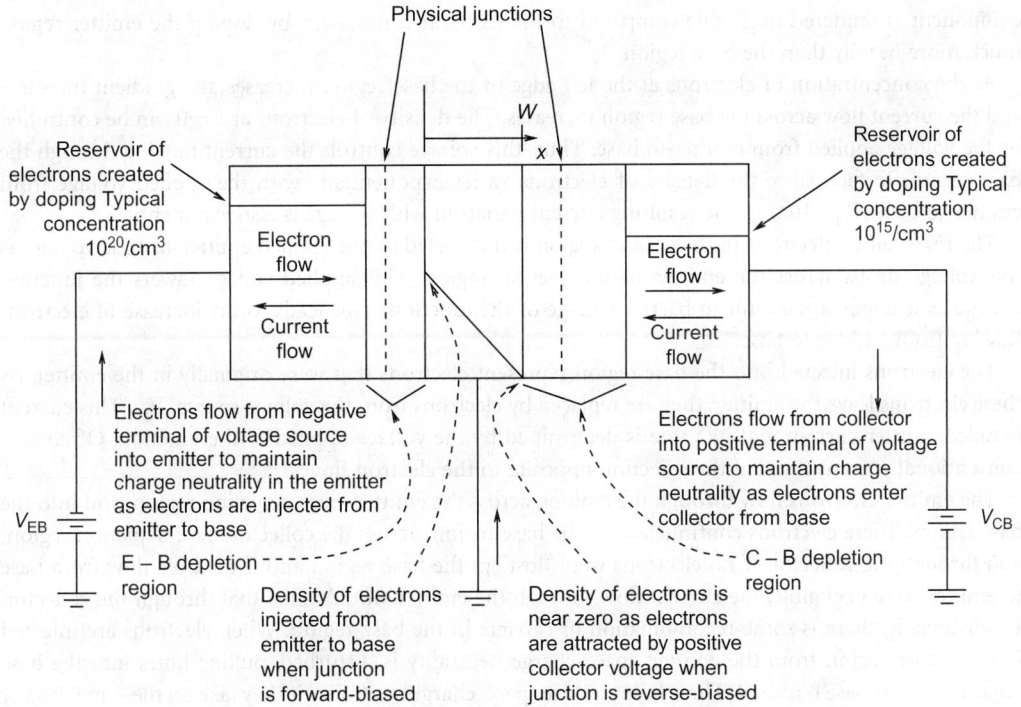

FIGURE 10.2 Distribution of electrons in the active region.

10.3.1   The Active Region

The circuit of Figure 10.2 is in the active region, that is, the emitter–base junction is forward-biased while the collector–base junction is reverse-biased. The current flow is controlled by the profile of electrons in the p-type base region. It is proportional to the slope or gradient of the free electron density in the base region. The well-known diffusion equation can be expressed as [3]

$$I = qD_n A \frac{dn}{dx} = -\frac{qD_n An(0)}{W} \tag{10.1}$$

where q is the electronic charge, D_n the diffusion constant for electrons, A the cross-sectional area of the base region, W the width or thickness of the base region, and $n(0)$ the density of electrons at the left edge of the base region. The negative sign reflects the fact that conventional current flow is opposite to the flow of the electrons.

The concentration of electrons at the left edge of the base region is given by

$$n(0) = n_{bo} e^{qV_{BE}/kT} \tag{10.2}$$

where q is the charge on an electron, k the Boltzmann's constant, T the absolute temperature, and n_{bo} the equilibrium concentration of electrons in the base region. While n_{bo} is a small number, $n(0)$ can be large for values of applied base to emitter voltages of 0.6 to 0.7 V. At room temperature, this equation can be written as

$$n(0) = n_{bo} e^{V_{BE}/0.026} \tag{10.3}$$

In Figure 10.2, the voltage $V_{EB} = -V_{BE}$.

A component of hole current also flows across the base–emitter junction from base to emitter. This component is rendered negligible compared to the electron component by doping the emitter region much more heavily than the base region.

As the concentration of electrons at the left edge of the base region increases, the gradient increases and the current flow across the base region increases. The density of electrons at $x = 0$ can be controlled by the voltage applied from emitter to base. Thus, this voltage controls the current flowing through the base region. In fact, since the density of electrons varies exponentially with the applied voltage from emitter to base (Eq. [10.2]), the resulting current variation with voltage is also exponential.

The reservoir of electrons in the emitter region is unaffected by the applied emitter-to-base voltage as this voltage drops across the emitter–base depletion region. This applied voltage lowers the junction voltage as it opposes the built-in barrier voltage of the junction. This leads to the increase in electrons flowing from emitter to base.

The electrons injected into the base region represent electrons that were originally in the emitter. As these electrons leave the emitter, they are replaced by electrons from the voltage source, V_{EB}. This current is called emitter current and its value is determined by the voltage applied to the junction. Of course, conventional current flows in the direction opposite to the electron flow.

The emitter electrons flow through the emitter, across the emitter–base depletion region, and into the base region. These electrons continue across the base region, across the collector–base depletion region, and through the collector. If no electrons were "lost" in the base region and if the hole flow from base to emitter were negligible, the current flow through the emitter would equal that through the collector. Unfortunately, there is some recombination of carriers in the base region. When electrons are injected into the base region from the emitter, space-charge neutrality is disturbed, pulling holes into the base region from the base terminal. These holes restore space-charge neutrality if they take on the same density throughout the base as the electrons. Some of these holes recombine with the free electrons in the base and the net flow of recombined holes into the base region leads to a small, but finite, value of base current. The electrons that recombine in the base region reduce the total electron flow to the collector. Because the base region is very narrow, only a small percentage of electrons traversing the base region recombine and the emitter current is reduced by a small percentage as it becomes collector current.

In a typical low-power BJT, the collector current might be 0.995 I_E. The current gain from emitter to collector, I_C/I_E, is called α and is a function of the construction process for the BJT. Using Kirchhoff's current law, the base current is found to equal the emitter current minus the collector current. This gives

$$I_B = I_E - I_C = (1 - \alpha)I_E \qquad (10.4)$$

If $\alpha = 0.995$, then $I_B = 0.005 I_E$. Base current is very small compared with emitter or collector current. A parameter β is defined as the ratio of collector current to base current resulting in

$$\beta = \frac{\alpha}{1-\alpha} \qquad (10.5)$$

This parameter represents the current gain from base to collector and can be quite high. For the value of α cited earlier, the value of β is 199.

10.3.2   The Cutoff Region

The cutoff region occurs when both junctions of the BJT are reverse-biased. In this situation, no electrons are injected across the emitter–base junction and no current flows through the base region. All terminals have zero current flow, prompting the name of this region. The BJT now approximates an open circuit from collector to emitter with high voltage drop, but no collector current flow.

10.3.3   The Saturation Region

If the voltage across the base–collector junction becomes forward-biased rather than reverse-biased, and the base–emitter junction is also forward-biased, the device is in the saturation region. This is usually accomplished by lowering the collector voltage to a value less than the base voltage. In this arrangement, the voltage drop from collector to emitter consists of the difference between the drop across the base–emitter junction and the drop across the base–collector junction. This voltage may be only 0.2 V. The BJT now approximates a short circuit from collector to emitter with little voltage drop, but possible high collector current.

10.4   Circuit Applications of the BJT

Two very common applications of the BJT are the linear amplifier and the digital or switching circuit. Many signals of interest are too small to be useful in some desired function. Signals such as the output of a microphone or the output of a radio antenna are in the microvolt to millivolt range. To drive a speaker or to drive a radio receiver, these signals must be amplified to levels of several volts. Thus, amplification is an important function in electronics.

Digital circuits are used to represent binary numbers and to perform logic operations and numerical calculations with binary numbers. The simplest logic circuit must accept two different voltage levels that signify a binary "1" or a binary "0" and produce an appropriate output voltage at either of the two acceptable levels. Digital circuits are often based on the simple BJT switch to be discussed in the following paragraphs.

10.4.1 Voltage Amplification

Since collector current is always a factor of β times the base current when the BJT operates in its active region, current amplification can be quite high if base current is the input current and collector current is the output current. Generally, the quantity of greatest interest to the circuit designer is voltage amplification. Figure 10.3 shows a simple configuration of a BJT voltage amplifier. This circuit is known as the common emitter configuration.

A voltage source is not typically used to forward bias the base–emitter junction in an actual circuit, but we will assume that V_{BB} is used for this purpose. A value of V_{BB} or V_{BE} near 0.6–0.7 V would be appropriate for this situation. The collector supply would be a larger voltage such as 12 V. We will assume that the value of V_{BB} sets the DC emitter current to a value of 1 mA for this circuit. The collector current entering the BJT will be slightly <1 mA, but we will ignore this difference and assume that $I_C = 1$ mA also. With a 4-kΩ collector resistance, a 4-V drop will appear across R_C leading to a DC output voltage of 8 V. The distribution of electrons across the base region for the steady-state or quiescent conditions is shown by the solid line in Figure 10.3.

If a small AC voltage now appears in series with V_{BB}, with a peak value much less than V_{BB}, the injected electron density on the left side of the base region will be modulated. Since this density varies exponentially with the applied voltage (see Eq. [10.2]), a small AC voltage can cause considerable changes in density. The broken lines in Figure 10.3 show the distributions at the positive and negative peak voltages. The collector current may change from its quiescent level of 1 mA to a maximum of 1.1 mA as v_{in} reaches its positive peak and to a minimum of 0.9 mA when v_{in} reaches its negative peak. The output collector voltage will drop to a minimum value of 7.6 V as the collector current peaks at 1.1 mA and will reach a maximum voltage of 8.4 V as the collector current drops to 0.9 mA. The peak-to-peak AC output voltage is then 0.8 V. The peak-to-peak value of v_{in} to cause this change might be 5 mV, giving a voltage gain of $A = -0.8/0.005 = -160$. The negative sign occurs because when v_{in} increases, the collector current increases, but the collector voltage decreases. This represents a phase inversion in the amplifier of Figure 10.3.

In summary, a small change in base-to-emitter voltage causes a large change in emitter current. This current is channeled across the collector and through the load resistance, across which a larger incremental voltage develops. The ratio of incremental output voltage to incremental input voltage is called the amplification factor or, more commonly, the voltage gain of the circuit.

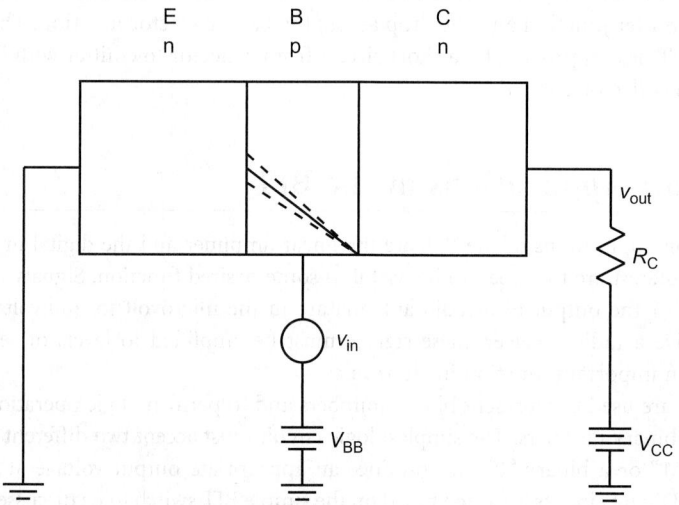

FIGURE 10.3 A BJT amplifier.

10.4.2 The Digital Switch

Figure 10.4 demonstrates a simple switching circuit that functions as a logic level inverter. The two logic levels representing binary 1 and 0 might be +5 and 0 V, respectively. If the lower level is applied to the input, that is, $v_{in} = 0$ V, the base–emitter junction is not forward-biased and the BJT is in its cutoff region. Since no collector current flows through the resistance, no voltage is dropped across this element and the collector or output voltage equals the power supply voltage of 5 V. If the input voltage is raised to 5 V, the base current increases to

$$I_B = \frac{V_{CC} - V_{BEon}}{R_B} \tag{10.6}$$

where V_{BEon} is typically 0.7 V. The collector current increases toward βI_B causing the voltage drop across the load resistance to increase. The collector voltage drops to V_{CEsat}, perhaps 0.2 V, and limits further increase in I_C. This circuit behaves as an inverter. A low input voltage results in a high output voltage and vice-versa. This type of circuit forms the basis of many logic circuits and is discussed further in a later section.

10.5 Nonideal Effects

In addition to the desired effects that lead to amplification, there are some inherent undesired effects that detract from the amplification of the BJT stage. Some of the sources of these undesired effects are ohmic effects, the Early effect, and reactive effects. To accurately analyze the behavior of the BJT, these effects must be included.

10.5.1 Ohmic Effects

The metal connections to the semiconductor regions exhibit some ohmic resistance. The emitter contact resistance and collector contact resistance is often in the ohm range and does not affect the BJT operation in most applications. The base region is very narrow and offers little area for a metal contact. Furthermore, because this region is narrow and only lightly doped compared with the emitter, the ohmic resistance of the base region itself is rather high. The total resistance between the contact and the intrinsic base region may be 20–150 Ω. This resistance can become significant in determining the behavior of the BJT, especially at higher frequencies.

10.5.2 Base-Width Modulation (Early Effect)

The widths of the depletion regions are functions of the applied voltages. The collector voltage generally exhibits the largest voltage change and as this voltage changes, so also does the collector–base depletion region width. As the depletion layer extends further into the base region, the slope of the electron

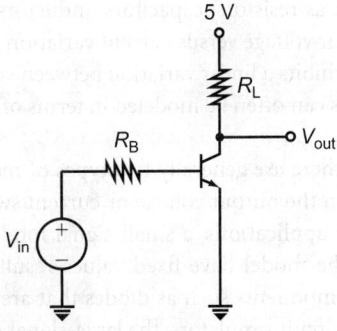

FIGURE 10.4 A switching circuit.

distribution in the base region becomes greater since the width of the base region is decreased. A slightly steeper slope leads to a slightly more collector current. As reverse bias decreases, the base width increases and the current decreases. This effect is called base-width modulation and can be expressed in terms of the Early voltage [2], V_A, by the expression

$$I_C = \beta I_B \left(1 + \frac{V_{CE}}{V_A}\right)$$ (10.7)

The Early voltage will be constant for a given device and is typically in the range 50–100 V for low-power BJTs.

10.5.3 Reactive Effects

Changing the voltages across the depletion regions results in a corresponding change in charge. This leads to an effective capacitance since

$$C = \frac{dQ}{dV}$$ (10.8)

This depletion region capacitance, also called the junction capacitance, is a function of the voltage applied to the junction and can be written as [3]

$$C_{dr} = \frac{C_{J0}}{(\phi - V_{app})^m}$$ (10.9)

where C_{J0} is the junction capacitance at zero bias, ϕ the built-in junction barrier voltage, V_{app} the applied junction voltage, and m a constant. For modern BJTs m is close to 0.33. The applied junction voltage has a positive sign for a forward bias and a negative sign for a reverse bias.

An increase in forward base–emitter voltage results in a higher density of electrons injected into the base region. The charge distribution in the base region changes with this voltage change and this leads to a capacitance called the diffusion capacitance. The diffusion capacitance is a function of the emitter current and can be written as

$$C_D = k_2 I_E$$ (10.10)

where k_2 is a constant for a given device.

10.6 Basic Modeling

Simple electrical components such as resistors, capacitors, inductors, and independent voltage sources can be characterized in terms of their voltage versus current variation or V–I characteristics. For example, an ideal resistor has a model that exhibits a linear variation between voltage and current at all frequencies of operation. More complex devices can often be modeled in terms of these simpler components to allow easier analysis.

For devices such as transistors, there are generally two types of models. A large-signal model is used in most digital applications wherein the output voltage or current swing is quite large, approaching the power supply voltage. In amplifier applications, a small-signal model is used. The small-signal model assumes that all components of the model have fixed values resulting in linear operation, while the large-signal model may include components such as diodes that are highly nonlinear.

Prior to the availability of good circuit simulators, the large-signal models were linearized over certain regions to allow manual analysis to be done. Even earlier simulators were inefficient in solving nonlinear equations and piecewise linear models were often used to save expensive computation time. Later simulators applied more efficient methods of nonlinear equation solution and computing power became much less expensive. At the present time, the most popular simulators, based on Spice (simulation program for integrated circuit emulation), offer linearized models for small-signal circuits and nonlinear models for large-signal circuits.

10.6.1 Large-Signal Models

In some digital or switching applications, the operating point of the transistor may move from the cutoff region through the active region and into the saturation region. A model developed in the early 1950s is the Ebers–Moll model [4]. The Ebers–Moll model for an npn transistor is shown in Figure 10.5. This model, with certain additions, is still used in several modern computer simulators. This model implements the Ebers–Moll equations given by

$$I_E = -I_{ES}(e^{qV_{BE}/kT} - 1) + \alpha_R I_{CS}(e^{qV_{BC}/kT} - 1) \tag{10.11}$$

$$I_C = \alpha_F I_{ES}(e^{qV_{BE}/kT} - 1) - I_{CS}(e^{qV_{BC}/kT} - 1) \tag{10.12}$$

where α_F is the forward current gain from emitter to collector and α_R the current gain from collector to emitter when operating in the inverted mode, that is, with the collector–base junction forward-biased and the emitter–base junction reverse-biased. Because of the geometry and doping of modern transistors, the value of α_R might fall between 0.5 and 0.8 [2]. The parameters I_{ES} and I_{CS} are constants at a given temperature that depend on the geometry and doping of the device.

The form of the equations often used for simulation programs [2] takes advantage of the fact that

$$\alpha_F I_{ES} = \alpha_R I_{CS} = I_S \tag{10.13}$$

where I_S depends on geometry and doping of the base region at a given temperature. This now allows the Ebers–Moll equations to be expressed in terms of I_S without requiring both I_{ES} and I_{CS}.

The nonlinearities involved in the Ebers–Moll model make it very difficult to use in manual analysis. Simple piecewise linear models that change abruptly as the transistor moves into a different operating region are applied to perform approximate manual analysis. Often, when switching between saturation and cutoff takes place, the models of Figure 10.6 can be used to determine appropriate element values to be specified for succeeding simulations [5]. For the cutoff region, Figure 10.6(a) shows an open circuit between all terminals of the BJT. This is based on the Ebers–Moll model of Figure 10.5 assuming that the leakage currents of the reverse-biased junctions are zero. The saturation model of Figure 10.6(b) shows a small voltage source that represents the forward-biased voltage drop from base to emitter

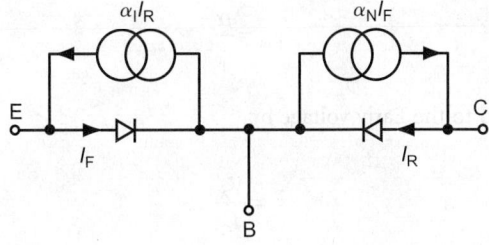

FIGURE 10.5 Ebers–Moll model for a pnp transistor.

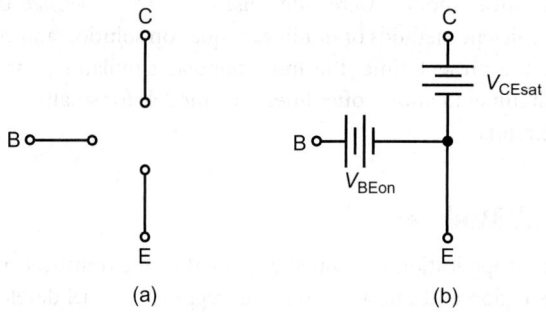

FIGURE 10.6 Approximate models: (a) cutoff; (b) saturation.

(0.6–0.7 V), while an even smaller voltage source represents the saturation voltage from collector to emitter (0.2–0.5 V) of the BJT.

To do more accurate simulations, the earlier computer models evolved to include second-order effects such as base-width modulation (Early effect), changes in β with low and high current levels, base resistance, collector-to-substrate, and other capacitive effects. The Gummel–Poon model, developed in 1970 [6], has been improved over the years and is available in many versions of the Spice simulator. With proper parameter values entered into the simulator, the accuracy of BJT simulation is excellent when using the modern Gummel–Poon model. Although MOSFET modeling has improved dramatically over the last decade, the simulation results for MOSFET circuits are generally not as accurate as those for BJT circuits.

10.6.2 Small-Signal Models

Most amplifier circuits constrain the BJT stages to active region operation. In such cases, the models can be assumed to have constant elements that do not change with signal swing. These models are linear models. The quiescent or bias currents are first found using a nonlinear model to allow the calculation of current-dependent element values for the small-signal model. For example, the dynamic resistance of the base–emitter diode, r_e, is given by

$$r_e = \frac{kT}{qI_E} \tag{10.14}$$

where I_E is the DC emitter current. Once I_E is found from a nonlinear DC equivalent circuit, the value of r_e is calculated and assumed to remain constant as the input signal is amplified.

The equivalent circuit [7] of Figure 10.7 shows the small-signal hybrid-π model that is closely related to the Gummel–Poon model and is used for AC analysis in the Spice program. The capacitance, C_π, accounts for the diffusion capacitance and the emitter–base junction capacitance. The collector–base junction capacitance is designated C_μ. The resistance $r_\pi = (\beta + 1)r_e$. The transconductance, g_m, is given by

$$g_m = \frac{\alpha}{r_d} \tag{10.15}$$

The impedance, r_o, is related to the Early voltage by

$$r_o = \frac{V_A}{I_C} \tag{10.16}$$

R_B, R_E, and R_C are the base, emitter, and collector resistances, respectively.

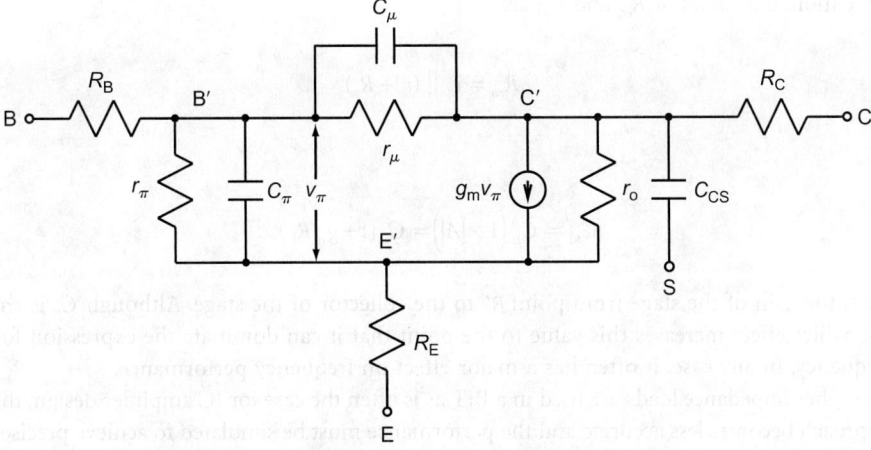

FIGURE 10.7 The hybrid-π small-signal model for the BJT.

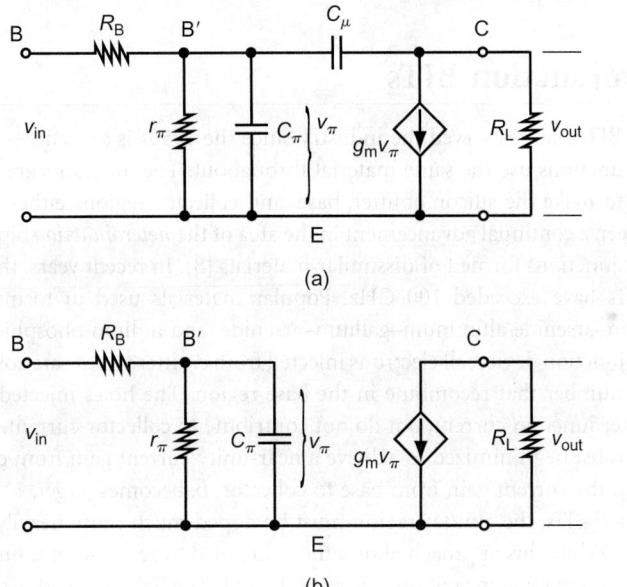

FIGURE 10.8 (a) Approximate equivalent circuit. (b) Unilateral equivalent circuit.

In many high-frequency, discrete amplifier stages and in some high-frequency IC stages, a small value of load resistance will be used. Approximate hand analysis can then be done by neglecting the ohmic resistances, R_E and R_C along with C_{CS}, the collector to substrate capacitance. If the impedance r_o is much larger than the load resistance, R_L, of the high-frequency amplifying stage, it can be considered as an open circuit. The resulting model that allows relatively viable hand analysis is shown in Figure 10.8(a).

For small values of load resistances, the effect of the capacitance can be represented by an input Miller capacitance that appears in parallel with C_π, resulting in a total capacitance of C_T. This simplified circuit now includes two separate loops as shown in Figure 10.8(b) and is referred to as the unilatera equivalent circuit. Typically, the input loop limits the bandwidth of the amplifier stage. When driven with a signal generator having a source resistance of R_s, the upper 3-dB frequency can be approximated as

$$f_{high} = \frac{1}{2\pi R_{eq}(C_\pi + C_M)} \tag{10.17}$$

In this equation, the values of R_{eq} and C_M are

$$R_{eq} = r_\pi \,\|\, (r_b + R_s)$$ (10.18)

and

$$C_M = C_\mu \left(1 + |A|\right) = C_\mu (1 + g_m R_L)$$ (10.19)

where A is the gain of the stage from point B' to the collector of the stage. Although C_μ is small, the resulting Miller effect increases this value to the point that it can dominate the expression for upper 3-dB frequency. In any case, it often has a major effect on frequency performance.

When higher impedance loads are used in a BJT as is often the case for IC amplifier design, the Miller effect approach becomes less accurate and the performance must be simulated to achieve precise results. Silicon BJTs now have current gain bandwidth figures, f_t, that exceed 12 GHz [2]. An amplifier stage constructed with such a device might have an upper corner frequency of 1 GHz.

10.7 Heterojunction BJTs

The classic silicon BJT that has served the industry since the 1950s is sometimes called a *homojunction* device since both junctions use the same material throughout. The junctions are formed by adding the proper impurities to make the silicon emitter, base, and collector regions either n- or p-type. The last two decades have seen a continual advancement in the area of the *heterojunction* bipolar transistor (HBT). These devices have junctions formed of dissimilar materials [8]. In recent years, the values of f_t for some heterojunction BJTs have exceeded 100 GHz. Popular materials used in forming HBTs are silicon/germanium, gallium–arsenide/aluminum–gallium–arsenide, and indium phosphide.

In an npn homojunction device, all electrons injected from emitter to base are collected by the collector, except for a small number that recombine in the base region. The holes injected from base to emitter contribute to emitter junction current, but do not contribute to collector current. This hole component of emitter current must be minimized to achieve a near-unity current gain from emitter to collector. As α approaches unity, the current gain from base to collector, β, becomes larger.

To produce high-β BJTs, the emitter region must be doped much more heavily than the base region as explained earlier. While this approach allows the value of β to reach several hundred, it also leads to some effects that limit the frequency of operation of the BJT. The lightly doped base region causes higher values of base resistance as well as emitter–base junction capacitance. Both of these effects are minimized in the HBT. The Si/Ge HBT uses silicon for the emitter and collector regions and a silicon/germanium mixture for the base region [6]. The difference in energy gap between the silicon emitter material and the silicon/germanium base material results in an asymmetric barrier to current flow across the junction. The barrier for electron injection from emitter to base is smaller than the barrier for hole injection from base to emitter. The base can then be doped more heavily than a conventional BJT to achieve lower base resistance, R_B, but the hole flow across the junction remains negligible due to the higher barrier voltage. The emitter of the HBT can be doped more lightly to lower the junction capacitance, notably C_μ. Large values of β are still possible in the HBT while minimizing frequency limitations.

From the standpoint of analysis, the Spice models for the HBT are structurally similar to those of the BJT. The major differences are in the parameter values. For example, Table 10.1 shows a comparison of typical Spice model parameters for a simple BJT process and a typical modern GaAs HBT process.

Table 10.1 demonstrates the smaller values of forward transit time, ohmic base resistance, and transition capacitance for the HBT compared with the BJT. Because modern HBT devices can operate in tens of GHz range, layout parasitics such as capacitance between substrate and metal interconnect and between

TABLE 10.1 Comparison of BJT and HBT Parameters

Symbol	Parameter Name	BJT	HBT	Unit
IS	Reverse saturation current	2.28E-17	1.591E-26	A
ISE	BE leakage saturation current	0	1E-20	A
ISC	BC leakage saturation current	0	3E-15	A
IKF	Corner for forward beta high-current roll-off	20E-3	0.1	A
IKR	Corner for reverse beta high-current roll-off	∞	0.1	A
IRB	Current where RB falls halfway to its minimum value	∞	0.015	A
ITF	Coefficient for TF collector current dependence	0	7.108	A
VAF	Forward Early voltage	60	1000	V
VAR	Reverse Early voltage	8	1000	V
VTF	Coefficient for TF BC voltage dependence	∞	0.7966	V
VJE	BE built-in potential	0.8043	1.46	V
VJC	BC built-in potential	0.334	1.32	V
VJS	CS built-in potential	0.5364	—	V
RB	Zero-bias base resistance	160	90	Ω
RBM	Minimum base resistance at high currents	160	75.24	Ω
RE	Emitter ohmic resistance	4	32	Ω
RC	Collector ohmic resistance	240	13.28	Ω
CJE	Zero-bias BE transition capacitance	157E-15	9.111E-15	F
CJC	Zero-bias BC transition capacitance	207E-15	11.75E-15	F
CJS	Zero-bias CS transition capacitance	1060E-15	—	F
TF	Ideal forward transit time	157E-12	1.7E-12	s
TR	Ideal reverse transit time	16E-9	1E-9	s
BF	Ideal maximum forward current gain	100	120	
BR	Ideal maximum reverse current gain	1	0.8	
NF	Forward current emission coefficient	1	1.01	
NE	BE leakage emission coefficient	1.5	1.675	
NR	Reverse current emission coefficient	1	1	
NC	BC leakage emission coefficient	2	1.93	
MJE	BE junction grading factor	0.31	0.5	
MJC	BC junction grading factor	0.209	0.5	
MJS	CS junction grading factor	0.270		
XTB	Forward and reverse beta temperature coefficient	0	−0.7	
XTF	Coefficient for TF bias dependence		16.3	
XCJC	Fraction of CJC connected to internal base node	1	0.927	
PTF	Excess phase at $1/(2\pi\,TF)$ Hz	5		deg
EG	Energy gap	1.1	1.424	eV
XT1	Temperature exponent for effect on IS	12.78	3	

different metal interconnects become significant and must be considered in the design and simulation of HBT circuits. Present-day software tools allow the extraction and inclusion of these layout and interconnect parasitics.

Two popular simulators used today for HBT circuits are Agilent's ADS software and a similar CAD tool powered by ANSOFT. These simulators use device models and a syntax similar to Spice. The tools support both the Gummel–Poon models discussed earlier and a more complex model called the vertical bipolar inter-company model (VBIC) [9]. The VBIC model is an extension of the Gummel–Poon model and accounts for additional effects, such as parasitic junction transistor action, avalanche multiplication, and self-heating effects. The self-heating of HBT devices is especially important since the oxide isolation used in the manufacturing process results in a high thermal resistance from the collector to the substrate. Although the VBIC model is thought to be more accurate, current simulators generally run faster using the Gummel–Poon model which tends to be used for initial design verification with final optimization done with VBIC models. Most HBT foundries support both the Gummel–Poon and VBIC models in their design kits for simulation.

10.8 IC Biasing Using Current Mirrors

Differential stages are very important in IC amplifier design. These stages require a constant DC current for proper bias. The diode-biased current sink or current mirror of Figure 10.9 is a popular method of creating a constant current bias for differential stages and other stages requiring a constant current bias.

The concept of the current mirror was developed specifically for analog IC biasing and is a good example of a circuit that takes advantage of the excellent matching characteristics that are possible in ICs. In the circuit of Figure 10.9, the current I_2 is intended to be equal to or "mirror" the value of I_1. Current mirrors can be designed to serve as sinks or sources.

The general function of the current mirror is to reproduce or mirror the input or reference current to the output while allowing the output voltage to assume any value within some specified range. The current mirror can also be designed to generate an output current that equals the input current multiplied by a scale factor K. The output current can be expressed as a function of input current as

$$I_0 = KI_{IN} \tag{10.20}$$

where K can be equal to, less than, or greater than unity. This constant can be established accurately by relative device sizes and will not vary with temperature.

10.8.1 Current Source Operating Voltage Range

Figure 10.10 shows an ideal or theoretical current sink in (a) and a practical sink in (b). The voltage at node A in the theoretical sink can be tied to any potential above or below ground without affecting the value of I. In contrast, in the practical circuit of Figure 10.10(b), it is necessary that the transistor remains in the active region to provide a current of

$$I = \alpha \frac{V_B - V_{BE}}{R} \tag{10.21}$$

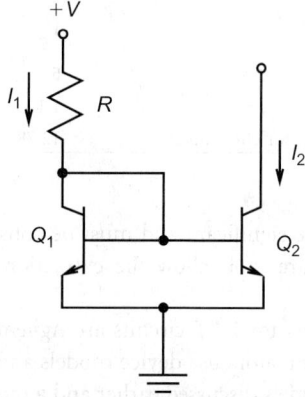

FIGURE 10.9 Current mirror bias stage.

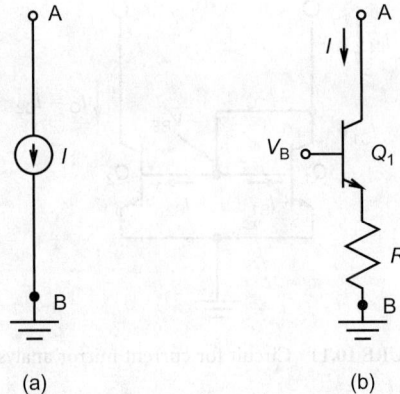

FIGURE 10.10 Current sink circuits: (a) ideal sink; (b) practical sink.

This requires that the collector voltage exceeds the voltage V_B at all times. The upper limit on this voltage is determined by the breakdown voltage of the transistor. The output voltage must then satisfy

$$V_B < V_C < (V_B + BV_{CE}) \tag{10.22}$$

where BV_{CE} is the breakdown voltage from collector to emitter of the transistor. This voltage range over which the current source operates is called the *output voltage compliance range* or the *output compliance*.

10.8.2 Current Mirror Analysis

The current mirror is again shown in Figure 10.11. If devices Q_1 and Q_2 are assumed to be matched devices, we can write

$$I_{E1} = I_{E2} = I_{E0}e^{V_{BE}/V_T} \tag{10.23}$$

where $V_T = kT/q$ and I_{EO} is a constant for a given transistor at a given temperature. The base currents for each device will also be identical and can be expressed as

$$I_{B1} = I_{B2} = \frac{I_{E0}}{\beta+1}e^{V_{BE}/V_T} \tag{10.24}$$

Device Q_1 operates in the active region, but near saturation by virtue of the collector–base connection. This configuration is called a diode-connected transistor. Since the collector to emitter voltage is small, the Early effect is negligible and the collector current for device Q_1 is given by

$$I_{C1} = \beta I_{B1} \approx \frac{\beta}{\beta+1}I_{E0}e^{V_{BE}/V_T} \tag{10.25}$$

The device Q_2 does not have the constraint that $V_{CE} \approx 0.7$ V as device Q_1 has. The collector voltage for Q_2 will be determined by the external circuit that connects to this collector. Thus, the collector current for this device is

$$I_{C2} = \beta I_{B2}\left(1+\frac{V_{C2}}{V_A}\right) \tag{10.26}$$

where V_A is the Early voltage. In effect, the output stage has an output impedance given by Eq. (10.16). The current mirror more closely approximates a current source as the output impedance becomes larger.

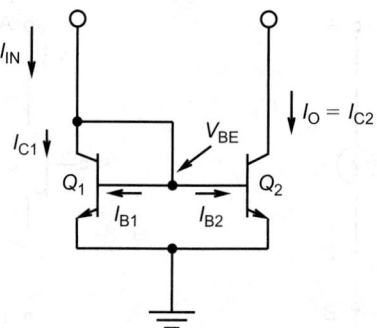

FIGURE 10.11 Circuit for current mirror analysis.

If we limit the voltage V_{C2} to small values relative to the Early voltage, I_{C2} is approximately equal to I_{C1}. For IC designs, the voltage required at the output of the current mirror is generally small, making this approximation valid.

The input current to the mirror is larger than the collector current and is

$$I_{IN} = I_{C1} + 2I_B \tag{10.27}$$

Since $I_{OUT} = I_{C2} = I_{C1} = \beta I_B$, we can write Eq. (10.27) as

$$I_{IN} = \beta I_B + 2I_B = (\beta + 2)I_B \tag{10.28}$$

Relating I_{IN} to I_{OUT} results in

$$I_{OUT} = \frac{\beta}{\beta + 2} I_{IN} = \frac{I_{IN}}{1 + 2/\beta} \tag{10.29}$$

For typical values of β these two currents are essentially equal. Thus, a desired bias current, I_{OUT}, is generated by creating the desired value of I_{IN}.

The current I_{IN} is normally established by connecting a resistance, R_1, to a voltage source V_{CC} to set I_{IN} to

$$I_{IN} = \frac{V_{CC} - V_{BE}}{R_1} \tag{10.30}$$

Control of output bias current or collector current for Q_2 is then accomplished by choosing proper values of V_{CC} and R_1.

Figure 10.12 shows a multiple output current mirror. It can be shown that the output current for each identical device in Figure 10.12 is

$$I_O = \frac{I_{IN}}{1 + (N+1)/\beta} \tag{10.31}$$

where N is the number of output devices.

The current sinks can be turned into current sources by using pnp transistors and a power supply of opposite polarity. The output devices can also be scaled in area to make I_{OUT} larger or smaller than I_{IN}. Figure 10.13 shows a multiple output mirror where output source currents I_2 and I_3 are referenced to input current I_1 as also are sink currents I_4 and I_5. If all device sizes are equal, then $I_3 = 2I_2$ and $I_5 = 2I_4$.

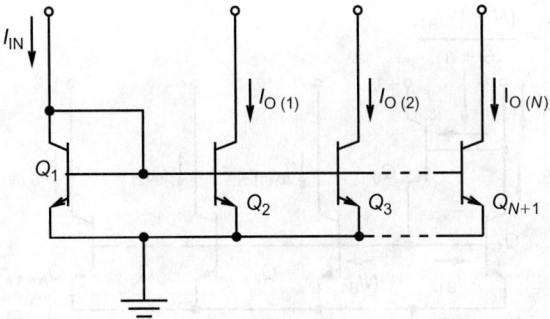

FIGURE 10.12 Multiple output current mirror.

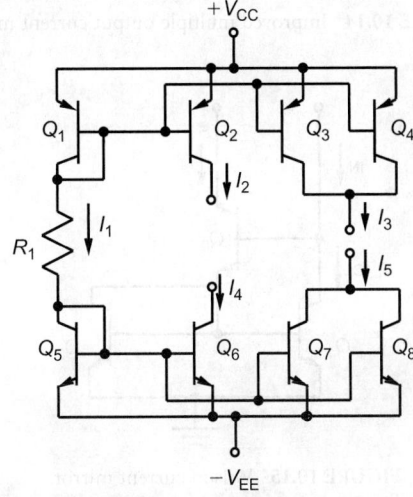

FIGURE 10.13 Multiple output current mirror with sinks and sources.

10.8.3 Current Mirror with Reduced Error

The difference between output current in a multiple output current mirror and the input current can become quite large if N is large. One simple method of avoiding this problem is to use an emitter follower to drive the bases of all devices in the mirror as shown in Figure 10.14.

The emitter follower, Q_0, has a current gain from base to collector of $\beta + 1$ reducing the difference between I_O and I_{IN} to

$$I_{IN} - I_O = \frac{N+1}{\beta+1} I_B \tag{10.32}$$

The output current for each device is

$$I_O = \frac{I_{IN}}{1 + (N+1)/\beta(\beta+1)} \tag{10.33}$$

10.8.4 The Wilson Current Mirror

In the simple current mirrors discussed, it was assumed that the collector voltage of the output stage was small compared with the Early voltage. When this is untrue, the output current will not remain constant,

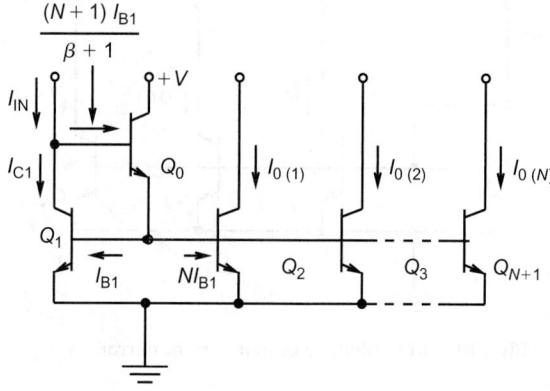

FIGURE 10.14 Improved multiple output current mirror.

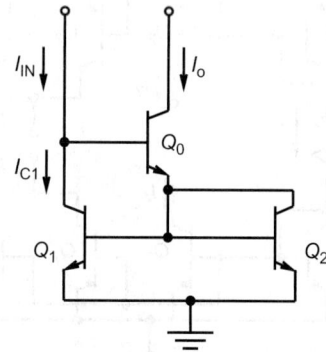

FIGURE 10.15 Wilson current mirror.

but will increase as output voltage (V_{CE}) increases. In other words, the output compliance range is limited with these circuits due to the finite output impedance of the BJT. A modification of the improved output current mirror of Figure 10.14 was proposed by Wilson [10] and is illustrated in Figure 10.15.

The Wilson current mirror is connected such that $V_{CB2} = 0$ and $V_{CB1} = V_{BE0}$. Both Q_1 and Q_2 now operate with a near-zero collector–emitter bias even though the collector of Q_0 might feed into a high voltage point. It can be shown that the output impedance of the Wilson mirror is increased by a factor of $\beta/2$ over the simple mirror. This higher impedance translates into a higher output compliance. This circuit also reduces the difference between input and output current by means of the emitter follower stage.

10.9   The Basic BJT Switch

In digital circuits, the BJT is used as a switch to generate one of only two possible output voltage levels, depending on the input voltage level. Each voltage level is associated with one of the binary digits, for example, the high voltage level may fall between 2.8 and 5 V while the low voltage level may fall between 0 and 0.8 V.

Logic circuits are based on BJT stages that are either in cutoff with both junctions reverse-biased or in a conducting mode with the emitter–base junction forward-biased. When the BJT is "on" or conducting emitter current, it can be in the active regionor the saturation region. If it is in the saturation region, the collector–base region is also forward-biased. The three possible regions of operation are summarized in Table 10.2.

TABLE 10.2   Regions of Operation

Region	Cutoff	Active	Saturation
C–B junction bias	Reverse	Reverse	Forward
E–B junction bias	Reverse	Forward	Forward

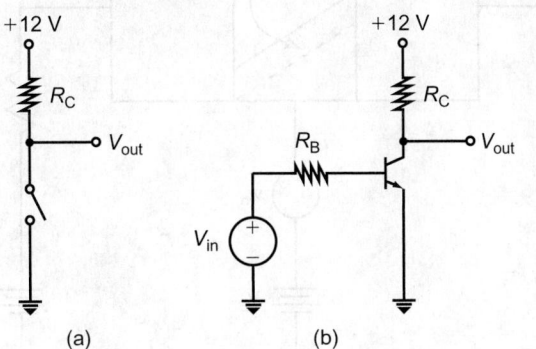

FIGURE 10.16   (a) A switch circuit. (b) The BJT as a switch.

The BJT very closely approximates certain switch configurations. For example, when the switch of Figure 10.16(a) is open, no current flows through the resistor and the output voltage is +12 V. Closing the switch causes the output voltage to drop to 0 V and a current of 12/R flows through the resistance. When the base voltage of the BJT of Figure 10.16(b) is zero or negative, the device is cut off and no collector current flows. The output voltage is +12 V just as in the case of the open switch. If a large enough current is now driven into the base to saturate the BJT, the output voltage becomes very small, ranging from 20 to 500 mV, depending on the BJT used. The saturated state corresponds closely to the closed switch. During the time that the BJT switches from cutoff to saturation, the active region equivalent circuit applies. For high-speed switching of this circuit, appropriate reactive effects must be considered. For low-speed switching these reactive effects can be neglected.

Saturation occurs in the basic switching circuit of Figure 10.16(b) when the entire power supply voltage drops across the load resistance. No voltage, or perhaps a few tenths of volts, then appears from collector to emitter. This occurs when the base current exceeds the value

$$I_{B(sat)} = \frac{V_{CC} - V_{CE(sat)}}{\beta R_L} \tag{10.34}$$

When a transistor switch is driven into saturation, the collector–base junction becomes forward-biased. This situation results in the electron distribution across the base region as shown in Figure 10.17.

The forward bias of the collector–base junction leads to a nonzero concentration of electrons at the right edge of the base region. This results in an excess concentration of electrons in the base that is unnecessary to support the gradient of carriers across this region. When the input signal to the base switches to a lower level to either turn the device off or decrease the current flow, the excess charge must be removed from the base region before the current can begin to decrease.

10.10   High-Speed BJT Switching

There are three major effects that extend switching times in a BJT:

1. The depletion region or junction capacitances are responsible for delay time when the BJT is in the cutoff region.

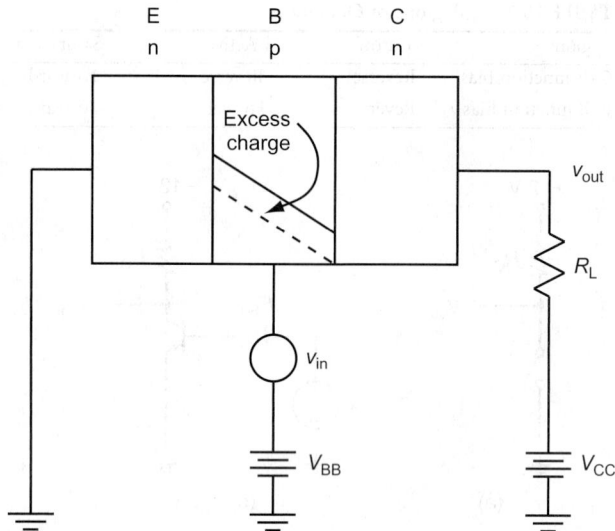

FIGURE 10.17 Electron distribution in the base region of a saturated BJT.

2. The diffusion capacitance and the Miller-effect capacitance are responsible for the rise and fall times of the BJT as it switches through the active region.
3. The storage time constant accounts for the time taken to remove the excess charge from the base region before the BJT can switch from the saturation region to the active region.

There are other second-order effects that are generally negligible compared with the previously listed time lags.

Since the transistor is generally operating as a large-signal device, the parameters such as junction capacitance or diffusion capacitance will vary as the BJT switches. One approach to the evaluation of time constants is to calculate an average value of capacitance over the voltage swing that takes place. Not only is this method used in hand calculations, most computer simulation programs use average values to speed calculations.

10.10.1 Overall Transient Response

Before discussing the individual BJT switching times it is helpful to consider the response of a common-emitter switch to a rectangular waveform[5]. Figure 10.18 shows a typical circuit using an npn transistor.

A rectangular input pulse and the corresponding output are shown in Figure 10.19. In many switching circuits, the BJT must switch from its off state to saturation and later return to the off state. In this case delay time, rise time, saturation storage time, and fall time must be considered in that order to find the overall switching time.

The total waveform is made up of five sections: delay time, rise time, on time, storage time, and fall time. The following list summarizes these points and serves as a guide for future reference:

t'_d = passive delay time; time interval between application of forward base drive and start of collector-current response

t_d = total delay time; time interval between application of forward base drive and the point at which I_C has reached 10% of the final value

t_r = rise time; 10–90% rise time of I_C waveform

t'_s = saturation storage time; time interval between removal of forward base drive and the start of I_C decrease

t_s = total storage time; time interval between removal of forward base drive and the point at which $I_C = 0.9I_{C(sat)}$

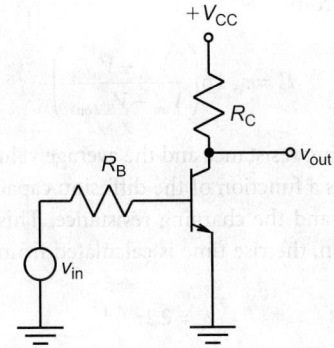

FIGURE 10.18 A simple switching circuit.

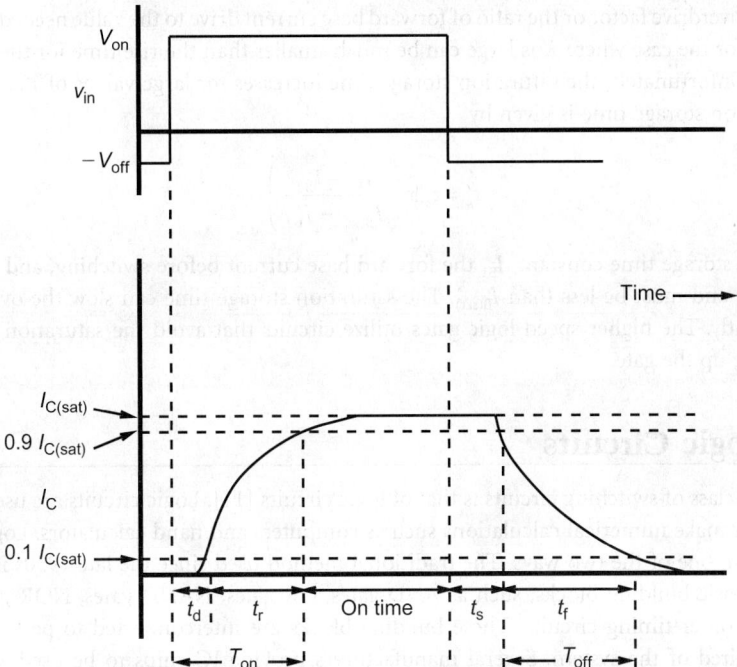

FIGURE 10.19 Input and output waveforms.

t_f = fall time; 90–10% fall time of I_C waveform

T_{on} = total turn-on time; time interval between application of base drive and the point at which I_C has reached 90% of its final value

T_{off} = total turn-off time; time interval between removal of forward base drive and the point at which I_C has dropped to 10% of its value during on time.

Not all applications will require evaluation of each of these switching times. For instance, if the base drive is insufficient to saturate the transistor, t_s will be zero. If the transistor never leaves the active region, delay time will also be zero. The factors involved in calculating the switching times are summarized in the following paragraphs [5].

The passive delay time is found from

$$t'_d = \tau_d \ln\left(\frac{V_{on} + V_{off}}{V_{on} - V_{BE(on)}} \right)$$ (10.35)

where τ_d is the product of the charging resistance and the average value of the two junction capacitances.

The active region time constant is a function of the diffusion capacitance, the collector–base junction capacitance, the transconductance, and the charging resistance. This time constant is denoted by τ. If the transistor never enters saturation, the rise time is calculated from the well-known formula

$$t_r = 2.2\tau$$ (10.36)

If the BJT is driven into saturation, the rise time is found from Ref. [5].

$$t_r = \tau \ln\left(\frac{K - 0.1}{K - 0.9} \right)$$ (10.37)

where K is the overdrive factor or the ratio of forward base current drive to the value needed for saturation. The rise time for the case where K is large can be much smaller than the rise time for the nonsaturating case ($K < 1$). Unfortunately, the saturation storage time increases for large values of K.

The saturation storage time is given by

$$t'_s = \tau_s \ln\left(\frac{I_{B1} - I_{B2}}{I_{B(sat)} - I_{B2}} \right)$$ (10.38)

where τ_s is the storage time constant, I_{B1} the forward base current before switching, and I_{B2} the current after switching and must be less than $I_{B(sat)}$. The saturation storage time can slow the overall switching time significantly. The higher speed logic gates utilize circuits that avoid the saturation region for the BJTs that make up the gate.

10.11   Logic Circuits

The largest subclass of switching circuits is that of logic circuits [11]. Logic circuits are used to construct all systems that make numerical calculations such as computers and hand calculators. Logic systems are now designed in one of the two ways. The traditional method used since the late 1950s is to design the system using basic building blocks, such as AND gates, OR gates, NAND gates, NOR gates, inverters, flip-flops, and other timing circuits. These building blocks are interconnected to perform the overall functions required of the system. Several manufacturers provide IC chips to be used as the building blocks. Little or no circuit design is required to produce the finished system.

The second method is to realize the entire logic system as a system-on-a-chip (SOC). Computer firms design the entire central processing unit along with timing and memory units on a single silicon chip. All internal interconnections between building blocks are done on the chip with external pin connections provided only to interface with peripheral chips. In this case the design of the actual complex IC must be done to produce the logic system.

Building block logic circuits have been used since the 1950s in various forms. The early construction of logic gates was done on printed circuit boards using primarily discrete versions of BJTs, diodes, capacitors, and resistors. In the 1960s, ICs were utilized to create logic gates on a chip. Initially the IC contained two to four gates on a single chip. As the technology for making ICs improved, more gates could be fabricated on a single chip. The fabrication process has progressed through the small-scale integration (SSI) stage to the medium-scale integration (MSI) stage to the large-scale integration (LSI) stage to the very large-scale integration (VLSI) stage to the ultra large-scale integration (ULSI) stage.

Whereas the SSI chip may contain only a few gates requiring six transistors, the complex microcomputer chip might contain over 100 million transistors.

All levels of ICs are available at present. The building block approach uses SSI, MSI, and LSI chips in products that are produced in relatively small quantities. Gates and other logic circuits are generally mounted on printed circuit boards and interconnected to form the desired logic system. For more complex systems or systems that will sell high volumes, VLSI or ULSI technology is used to implement an SOC.

When designing noncritical systems that are not pushing the limits of speed performance or power minimization, it is unnecessary for the logic designer to understand the details of the electronic circuits that make up the gates. The manufacturers have designed these and other logic circuits and specify key parameters that can be used in system design. These circuits are fabricated in several different configurations and from several types of device. Those chips having in common a particular device and configuration are said to belong to a logic family.

Some examples of currently useful families are *transistor-transistor logic* (TTL), *emitter-coupled logic* (ECL), and *complementary MOS* (CMOS) [11]. The TTL and ECL families are based on BJT circuits and are often applied to the building block approach. CMOS logic circuits are provided as building blocks and are also used for most computer SOC implementations. While the following discussion is oriented toward the building block approach, many of the considerations apply to the design of circuits on a single chip.

Within each family of logic are several categories or types of circuit. For example, the TTL family includes the conventional TTL, low-power, Schottky, low-power Schottky, advanced Schottky (AS), and advanced low-power Schottky (ALS) categories. All circuits of a given family must have compatible operating characteristics. The high-level voltage developed at the output of a gate must be sufficient to drive the input of any other gate to the same high level. The low-level output must pull the input of the next stage down to an acceptably low level. Certain current requirements must be met at each voltage level. Each category is compatible to some degree with other categories of the same family. For perfect compatibility, a single category is used for a given design. In general, an entire digital system will use only one or two logic-circuit families. Hundreds to thousands of SSI or MSI logic elements are connected properly to form the required subsystems of the digital system.

Sometimes it is necessary to connect logic elements that are not of the same family. When this is done, an interface between the different elements may be required. An interface consists of circuits that translate the output signals from one family to the input signals required by the other family. Certain families can be combined without interface circuits. A family is said to be compatible with another when both families can be interconnected without requiring interface circuits. Other combinations of families are popular enough that standard interface circuits are provided within the IC families.

10.11.1 Current and Voltage Definitions

To ensure proper operation when logic building blocks are interconnected, certain voltage and current requirements must be met. Some of the key parameters specified by manufacturers are the input/output current/voltages listed below.

V_{IHmin} = minimum input voltage that the logic element is guaranteed to interpret as the high logic level

V_{ILmax} = maximum input voltage that the logic element is guaranteed to interpret as the low logic level

V_{OHmin} = minimum high logic-level voltage appearing at the output terminal of the logic element

V_{OLmax} = maximum low logic-level voltage appearing at the output terminal of the logic element

I_{IHmax} = maximum current that will flow into an input when a specified high logic-level voltage is applied

I_{ILmax} = maximum current that will flow into an input when a specified low logic-level voltage is applied

I_{OH} = current flowing into the output when a specified high-level output voltage is present
I_{OL} = current flowing into an output when a specified low-level output voltage is present
I_{OS} = current flowing into an output when the output is shorted and input conditions are such to establish a high logic-level output.
(Current flowing out of a terminal has a negative value.)

When the output of Gate A is connected to the input of Gate B, the value of V_{OLmax} for Gate A must be less than V_{ILmax} for Gate B and the value of V_{OHmin} for Gate A must be greater than V_{IHmin} for Gate B. In addition, the output currents of Gate A must be compatible with the input current requirements of Gate B for proper operation.

10.11.2 Fan-Out

In a digital system a given gate may drive the inputs to several other gates. The designer must be certain that the driving gate can meet the current requirements of the driven stages at both high and low voltage levels. The number of inputs that can be driven by the gate is referred to as the *fan-out* of the circuit. This figure is expressed in terms of the number of standard inputs that can be driven. Most circuits of a family will require the same input current, but a few may require more. If so, the specs for such a circuit will indicate that the input is equivalent to some multiple of standard loads. For example, a circuit may present an equivalent input of two standard loads. If fan-out of a gate is specified as 10, only five of these circuits could be safely driven. Figure 10.20 shows an AND gate loaded with four inputs, assuming each circuit presents one standard load to the AND gate output.

In several handbooks, the current requirements are given and fan-out can be calculated. For example, one TTL gate has the following current specs:

$$I_{IH} = 40\,\text{mA}; \quad I_{IL} = -1.6\ \text{mA}$$

$$I_{OH} = -400\,\text{mA}; \quad I_{OL} = 16\ \text{mA}$$

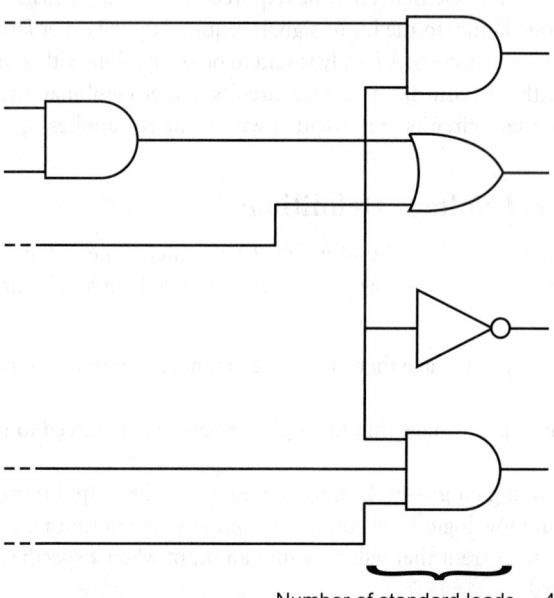

Number of standard loads = 4

FIGURE 10.20 Example of fan-out.

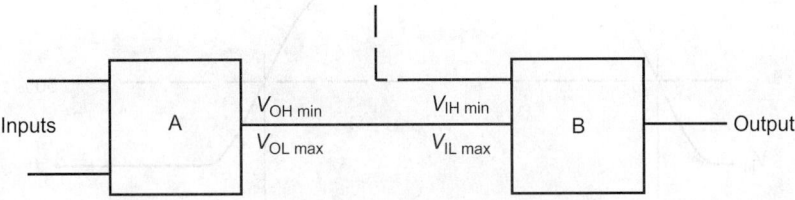

FIGURE 10.21 Stages used to calculate noise margin.

If this gate is to drive several other similar gates, we see that the output current capability of the stage is 10 times that required by the input. We note that the output stage can drive 400 mA into the following stages at the high level and sink 16 mA at the low level. The fan-out of this gate is 10.

10.11.3   Noise Margin

Although current requirement is the major factor in determining fan-out, input capacitance or *noise margin* may further influence this figure. Noise margin specifies the maximum amplitude noise pulse that will not change the state of the driven stage. This assumes that the driving stage presents a worst case logic level to the driven stage. Noise margin can be evaluated from a consideration of the voltage levels V_{IHmin}, V_{ILmax}, V_{OHmin}, and V_{OLmax}. Figure 10.21 shows two logic circuits that are cascaded.

If we assume that $V_{ILmax} = 0.8$ V for circuit B, this means that the input must be <0.8 V to guarantee that circuit B interprets this value as a low level. If circuit A has a value of $V_{OLmax} = 0.4$ V, a noise spike of less than the difference $0.8 - 0.4$ V cannot lead to a level misinterpretation by circuit B. The difference

$$V_{ILmax} - V_{OLmax} \tag{10.39}$$

is called the low-level noise margin. Assuming that $V_{IHmin} = 2$ V for circuit B and $V_{OHmin} = 2.7$ V for circuit A, the high-level margin is $2.7 - 2.0 = 0.7$ V. This high-level noise margin is found from

$$V_{OHmin} - V_{IHmin} \tag{10.40}$$

Since the minimum voltage developed by circuit A at the high level is 2.7 V, while circuit B requires only 2.0 V to interpret the signal as a high level, a negative noise spike of −0.7 V or less will not result in an error.

As we consider the noise margin we recognize that the values calculated in Eq. (10.39) and Eq. (10.40) are worst case values. A particular circuit could have actual noise margins better than those calculated. As more gate inputs are connected to a given output, the voltages generated at both high and low levels are affected as a result of increased current flow. Thus, fan-out is influenced by noise margin.

10.11.4   Switching Times

Another quantity which is used to characterize switching circuits is the speed with which the device responds to input changes. The preceding section discussed the factors that influence transistor switching times. For switching circuits, the graph of Figure 10.22 is useful in defining delay times [11]. This figure assumes an inverting gate.

There is a finite delay between the application of the input pulse and the output response. A quantitative measure of this delay is the difference in time between the point where V_{in} rises to 50% of its final value and the time when V_{out} falls to its 50% point. This quantity is called leading-edge delay t_{pHL}. The trailing-edge delay t_{pLH} is the time difference between 50% points of the trailing edges of the input and output signals. The *propagation delay* is defined as the average of t_{pHL} and t_{pLH}, or

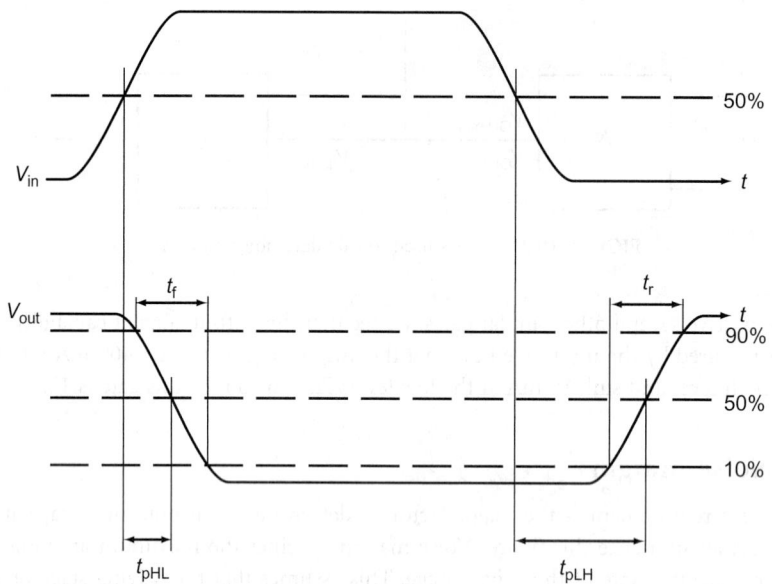

FIGURE 10.22 Definition of switching times.

$$t_{pd} = \frac{t_{pHL} + t_{pLH}}{2} \tag{10.41}$$

Propagation delay time of an IC is a function of passive delay time, rise and fall times, and the saturation storage time of the circuit's individual transistors. Since input and output capacitance will influence the IC switching times, fan-in and fan-out will also affect delay times. Switching times are sometimes specified by graphs showing the various times as functions of the number of standard input loads with specified driving conditions.

An understanding of the definitions given in the preceding paragraphs allows the designer to use logic gates as building blocks in digital systems. The TTL family has been the workhorse for many years in SSI and MSI applications. Fast and versatile, no other line offers as great a variety of circuits. The fabrication of resistors requires more chip volume than transistors do, and TTL chips use several resistors per gate. Consequently, applications in LSI circuits are somewhat limited.

ECL logic is the highest speed BJT family available. It does not offer as wide a variety of circuit types as TTL, but has been popular for use in high-speed applications such as supercomputers.

10.11.5 A TTL Logic Gate

The TTL family was originally based on the multiemitter construction of transistors shown in Figure 10.23 [11]. The operation of the input transistor can be visualized with the help of the circuit of Figure 10.24, which shows the bases of the three transistors connected in parallel, as are the collectors, whereas the emitters are separate

If all emitters are at ground level, the transistors will be saturated by the large base drive. The collector voltage will be only a few tenths of a volt above ground. The base voltage will equal $V_{BE(on)}$, which may be 0.5 V. If one or two of the emitter voltages are raised, the corresponding transistors will shut off. The transistor with an emitter voltage of 0 V will still be saturated, however; and saturation will force the base voltage and collector voltage to remain low. If all three emitters are raised to a higher level, the base and collector voltages will tend to follow this signal.

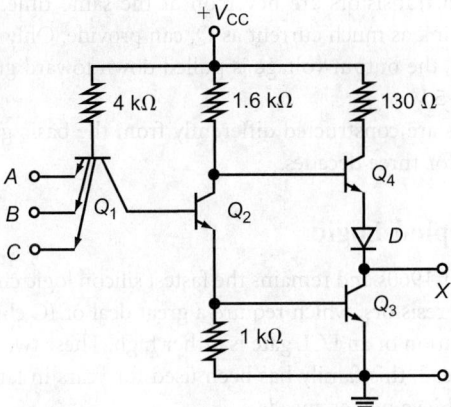

FIGURE 10.23 A basic TTL gate.

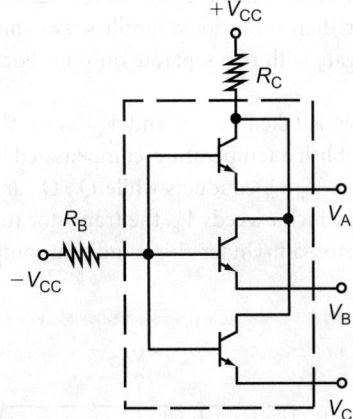

FIGURE 10.24 Equivalent circuit of input gate.

Returning to the basic gate of Figure 10.23, we see that when the low logic level appears at one or more of the inputs, Q_1 will be saturated with a very small voltage appearing at the collector of this stage. Since at least $2V_{BE(on)}$ must appear at the base of Q_2 in order to turn Q_2 and Q_3 on, we can conclude that these transistors are off. When Q_2 is off, the current through the 1.6-kΩ resistance is diverted into the base of Q_4, which then drives the load as an emitter follower.

When all inputs are at the high voltage level, the collector of Q_1 attempts to rise to this level. This turns Q_2 and Q_3 on, which clamps the collector of Q_1 to a voltage of approximately $2V_{BE(on)}$. The base–collector junction of Q_1 appears as a forward-biased diode, whereas in this case the base–emitter junctions are reverse-biased diodes. As Q_2 turns on, the base voltage of Q_4 drops, decreasing the current through the load. The load current tends to decrease even faster than it would if only Q_4 were present, because Q_3 is turning on to divert more current from the load. At the end of the transition, Q_4 is off with Q_2 and Q_3 on. For positive logic, the circuit behaves as a NAND gate.

This arrangement of the output transistors is called a totem pole. In the emitter follower the output impedance is asymmetrical with respect to emitter current. As the emitter follower turns on, the output impedance decreases. Turning the stage off increases the output impedance and can lead to distortion of the load voltage especially for capacitive loads. The totem-pole output stage overcomes this problem.

Transistor Q_3 is called the pull-down transistor and Q_4 is called the pull-up transistor. The circuit is designed such that these two transistors are never on at the same time. If this occurred, Q_3 may be destroyed because it cannot sink as much current as Q_4 can provide. Only one of these stages will be on at any given time. If Q_3 is on, the output voltage is pulled down toward ground; if Q_4 is on, the output voltage is pulled up toward +5 V.

Although newer TTL gates are constructed differently from the basic gate, this configuration played a large role in digital design for three decades.

10.11.6 Emitter-Coupled Logic

ECL was developed in the mid-1960s and remains the fastest silicon logic circuit available. The two major disadvantages of ECL are (1) resistors, which require a great deal of IC chip area, must be used in each gate and (2) the power dissipation of an ECL gate is rather high. These two shortcomings limit the usage of ECL in VLSI systems. Instead, this family has been used for years in larger supercomputers that can afford space and power to achieve higher speeds.

The high speeds obtained with ECL are primarily based on two factors. No device in an ECL gate is ever driven into the saturation region and thus, saturation storage time is never involved as devices switch from one state to another. The second factor is that required voltage swings are not large. Voltage excursions necessary to change an input from the low logic level to the high logic level are minimal. Although noise margins are lower than other logic families, switching times are reduced in this way. Figure 10.25 shows an older ECL gate with two separate outputs. For positive logic, X is the OR output while Y is the NOR output.

Often the positive supply voltage is taken as 0 V and V_{EE} as −5 V due to noise considerations. The diodes and emitter follower Q_5 establish a temperature-compensated base reference for Q_4. When inputs A, B, and C are less than the voltage V_B, Q_4 conducts while Q_1, Q_2, and Q_3 are cut off. If any one of the inputs is switched to the first level, which exceeds V_B, the transistor turns on and pulls the emitter of Q_4 positive enough to cut this transistor off. Under this condition, output Y goes negative while X goes

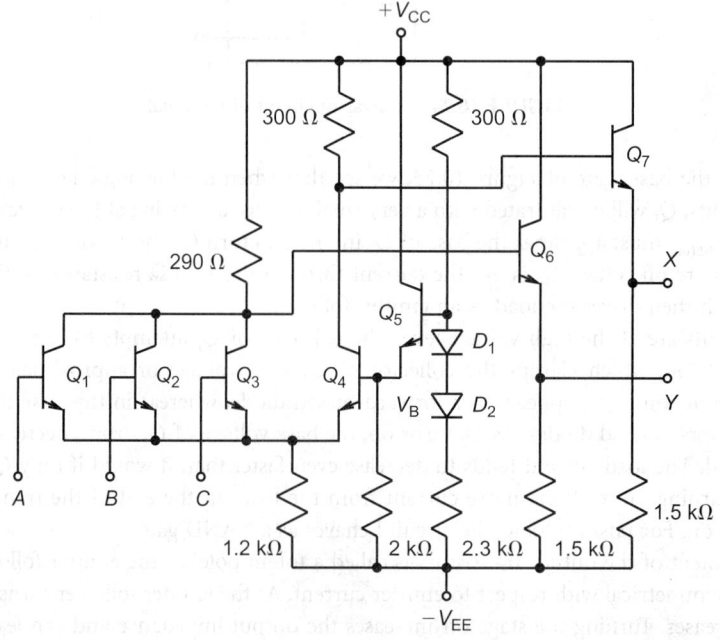

FIGURE 10.25 A basic ECL gate.

positive. The relatively large resistor common to the emitters of Q_1, Q_2, Q_3, and Q_4 prevents these transistors from saturating. In fact, with nominal logic levels of −1.9 and −1.1 V, the current through the emitter resistance is approximately equal before and after switching takes place. Thus, only the current path changes as the circuit switches. This type of operation is sometimes called current mode switching. Although the output stages are emitter followers, they conduct reasonable currents for both logic level outputs and, therefore, minimize the asymmetrical output impedance problem.

In an actual ECL gate, the emitter follower load resistors are not fabricated on the chip. The newer version of the gate replaces the emitter resistance of the differential stage with a current source and replaces the bias voltage circuit with a regulated voltage circuit.

The ECL 100K family has gate propagation delay times around 0.75 ns compared with 2 ns for the older ECL 10K family [12]. The basic ECL architecture has also been extended to faster HBT devices similar to the original ECL logic families. Speed increases correspond to the improvement in gain-bandwidth figures of HBT devices over BJT devices. Propagation delay times of <100 ps are available in logic gates fabricated in InP.

References

1. J.E. Brittain (Ed.), *Turning Points in American Electrical History*, Sec. II-D, IEEE Press, New York, 1977.
2. P.R. Gray, P.J. Hurst, S.H. Lewis, and R.G. Meyer, *Analysis and Design of Analog Integrated Circuits*, 4th ed., Wiley, New York, 2001.
3. D.J. Comer and D.T. Comer, *Fundamentals of Electronic Circuit Design*. Wiley, New York, 2002.
4. J.J. Ebers and J.L. Moll, "Large-signal behavior of junction transistors," *Proceedings of the IRE*, 42, 1954, pp. 1761–72.
5. D.J. Comer, *Modern Electronic Circuit Design*. Addison-Wesley, Reading, MA, 1976.
6. H.K. Gummel and H.C. Poon, "An integral charge control model of bipolar transistors," *Bell Systems Technical Journal*, 49, 1970, pp. 827–52.
7. A. Vladimirescu, *The Spice Book*. Wiley, New York, 1994.
8. B.G. Streetman and S. Banerjee, *Solid State Electronic Devices*, 5th ed., Prentice-Hall, Upper Saddle River, NJ, 2000.
9. C.C. McAndrew, J.A. Seitchik, D.F. Bowers, M. Dunn, M. Foisy I. Getreu, M. McSwain, S. Moninian, J. Parker, D.J. Roulston, M. Schroter, P. van Wijnen, and L.F. Wagner, "VBIC95, the vertical bipolar inter-company model," *IEEE Journal of Solid-State Circuits*, 31, 1996, pp. 1476–83.
10. G.R. Wilson, "A monolithic junction FET–NPN operational amplifier," *IEEE Journal of Solid State Circuits*, SC-3, 1968, pp. 341–9.
11. D.J. Comer, *Digital Logic and State Machine Design*, 3rd ed., Saunders College Publishing, New York, 1995.
12. J.F. Wakerly, *Digital Design Principles and Practices*, 3rd ed., Prentice-Hall, Upper Saddle River, NJ, 2000.

11

RF Passive IC Components

Thomas H. Lee

Maria del Mar Hershenson

Sunderarajan S. Mohan

Kirad Samavati and

C. Patrick Yue
Stanford University

CONTENTS

11.1   Introduction

Passive energy storage elements are widely used in radio-frequency (RF) circuits. Although their impedance behavior often can be mimicked by compact active circuitry, it remains true that passive elements offer the largest dynamic range and the lowest power consumption. Hence, the highest performance will always be obtained with passive inductors and capacitors. Unfortunately, standard integrated circuit technology has not evolved with a focus on providing good passive elements. This chapter describes the limited palette of options available, as well as means to make the most use out of what is available.

11.2   Fractal Capacitors

Of capacitors, the most commonly used are parallel-plate and MOS structures. Because of the thin gate oxides now in use, capacitors made out of MOSFETs have the highest capacitance density of any standard IC option, with a typical value of approximately 7 fF/μm^2 for a gate oxide thickness of 5 nm. A drawback, however, is that the capacitance is voltage dependent. The applied potential must be well in excess of a threshold voltage in order to remain substantially constant. The relatively low breakdown voltage (on the order of 0.5 V/nm of oxide) also imposes an unwelcome constraint on allowable signal amplitudes. An additional drawback is the effective series resistance of such structures, due to the MOS channel resistance. This resistance is particularly objectionable at radio frequencies, since the impedance of the combination may be dominated by this resistive portion.

Capacitors that are free of bias restrictions (and that have much lower series resistance) may be formed out of two (or more) layers of standard interconnect metal. Such parallel-plate capacitors are quite linear and possess high breakdown voltage, but generally offer two orders of magnitude lower capacitance density than the MOSFET structure. This inferior density is the consequence of a conscious and continuing effort by technologists to keep low the capacitance between interconnect layers. Indeed, the vertical spacing between such layers generally does not scale from generation to generation. As a result, the disparity between MOSFET capacitance density and that of the parallel-plate structure continues to grow as technology scales.

A secondary consequence of the low density is an objectionably high capacitance between the bottom plate of the capacitor and the substrate. This bottom-plate capacitance is often a large fraction of the main capacitance. Needless to say, this level of parasitic capacitance is highly undesirable.

In many circuits, capacitors can occupy considerable area, and an area-efficient capacitor is therefore highly desirable. Recently, a high-density capacitor structure using lateral fringing and fractal geometries has been introduced [1]. It requires no additional processing steps, and so it can be built in standard digital processes. The linearity of this structure is similar to that of the conventional parallel-plate capacitor. Furthermore, the bottom-plate parasitic capacitance of the structure is small, which makes it appealing for many circuit applications. In addition, unlike conventional metal-to-metal capacitors, the density of a fractal capacitor increases with scaling.

11.2.1 Lateral Flux Capacitors

Figure 11.1(a) shows a lateral flux capacitor. In this capacitor, the two terminals of the device are built using a single layer of metal, unlike a vertical flux capacitor, where two different metal layers must be used. As process technologies continue to scale, lateral fringing becomes more important. The lateral spacing of the metal layers, s, shrinks with scaling, yet the thickness of the metal layers, t, and the vertical spacing of the metal layers, t_{ox}, stay relatively constant. This means that structures utilizing lateral flux enjoy a significant improvement with process scaling, unlike conventional structures that depend on vertical flux. Figure 11.1(b) shows a scaled lateral flux capacitor. It is obvious that the capacitance of the structure of Figure 11.1(b) is larger than that of Figure 11.1(a).

Lateral flux can be used to increase the total capacitance obtained in a given area. Figure 11.2(a) is a standard parallel-plate capacitor. In Figure 11.2(b), the plates are broken into cross-connected sections [2].

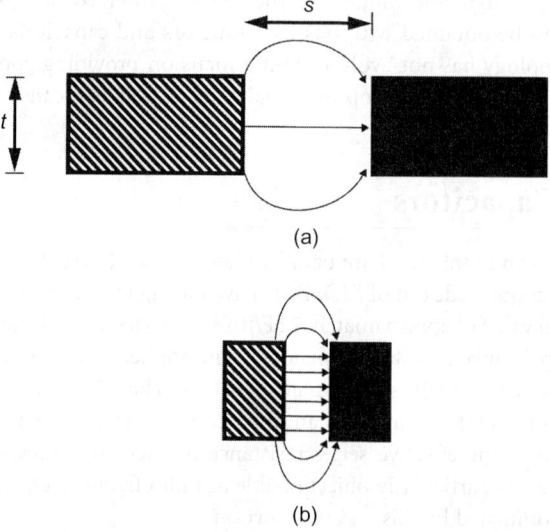

(a)

(b)

FIGURE 11.1 Effect of scaling on lateral flux capacitors: (a) before scaling and (b) after scaling.

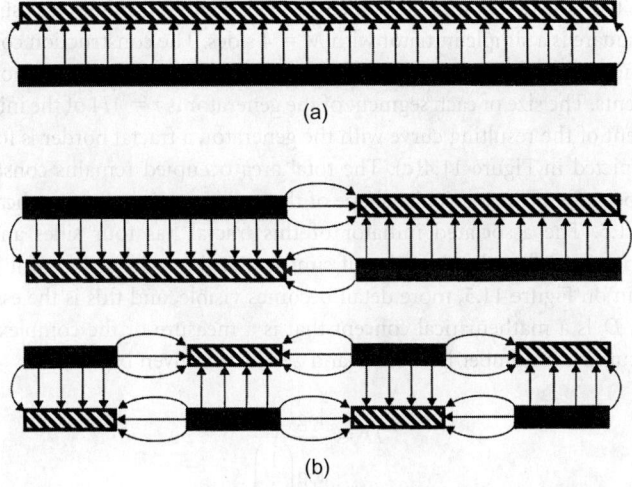

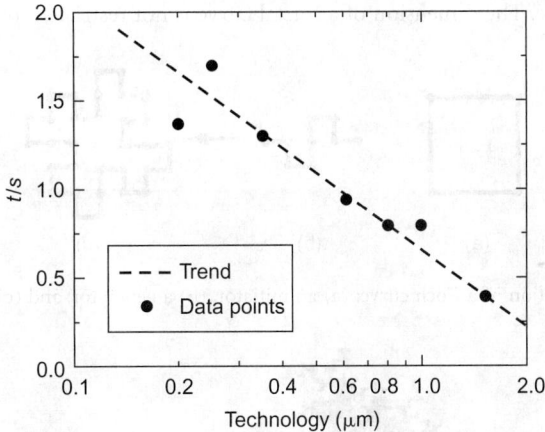

FIGURE 11.2 Vertical flux vs. lateral flux: (a) a standard parallel-plate structure, and (b) cross-connected metal layers.

FIGURE 11.3 Ratio of metal thickness to horizontal metal spacing vs. technology (channel length).

As can be seen, a higher capacitance density can be achieved by using lateral flux as well as vertical flux. To emphasize that the metal layers are cross connected, the two terminals of the capacitors in Figure 11.2(b) are identified with two different shadings. The idea can be extended to multiple metal layers as well.

Figure 11.3 shows the ratio of metal thickness to minimum lateral spacing, t/s, vs. channel length for various technologies [3–5]. The trend suggests that lateral flux will have a crucial role in the design of capacitors in future technologies.

The increase in capacitance due to fringing is proportional to the periphery of the structure; therefore, structures with large periphery per unit area are desirable. Methods for increasing this periphery are the subject of the following sections.

11.2.2 Fractals

A fractal is a mathematical abstract [6]. Some fractals are visualizations of mathematical formulas, while others are the result of the repeated application of an algorithm, or a *rule*, to a *seed*. Many natural phenomena can be described by fractals. Examples include the shapes of mountain ranges, clouds, coastlines, etc.

Some ideal fractals have finite area but infinite perimeter. The concept can be better understood with the help of an example. *Koch islands* are a family of fractals first introduced as a crude model for the shape of a coastline. The construction of a Koch curve begins with an *initiator,* as shown in the example of Figure 11.4(a). A square is a simple initiator with $M = 4$ sides. The construction continues by replacing each segment of the initiator with a curve called a *generator,* an example of which is shown in Figure 11.4(b) that has $N = 8$ segments. The size of each segment of the generator is $r = 1/4$ of the initiator. By recursively replacing each segment of the resulting curve with the generator, a fractal border is formed. The first step of this process is depicted in Figure 11.4(c). The total area occupied remains constant throughout the succession of stages because of the particular shape of the generator. A more complicated Koch island can be seen in Figure 11.5. The associated initiator of this fractal has four sides and its generator has 32 segments. It can be noted that the curve is self similar, that is, each section of it looks like the entire fractal. As we zoom in on Figure 11.5, more detail becomes visible, and this is the essence of a fractal.

Fractal dimension, D, is a mathematical concept that is a measure of the complexity of a fractal. The dimension of a flat curve is a number between 1 and 2, which is given by

$$D = \frac{\log(N)}{\log\left(\dfrac{1}{r}\right)} \tag{11.1}$$

where N is the number of segments of the generator and r is the ratio of the generator segment size to the initiator segment size. The dimension of a fractal curve is not restricted to integer values, hence the

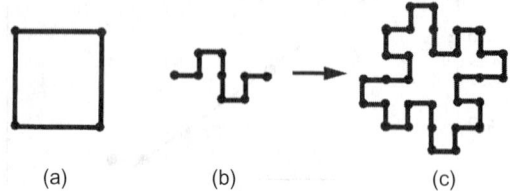

(a)   (b)   (c)

FIGURE 11.4 Construction of a Koch curve: (a) an initiator, (b) a generator, and (c) first step of the process.

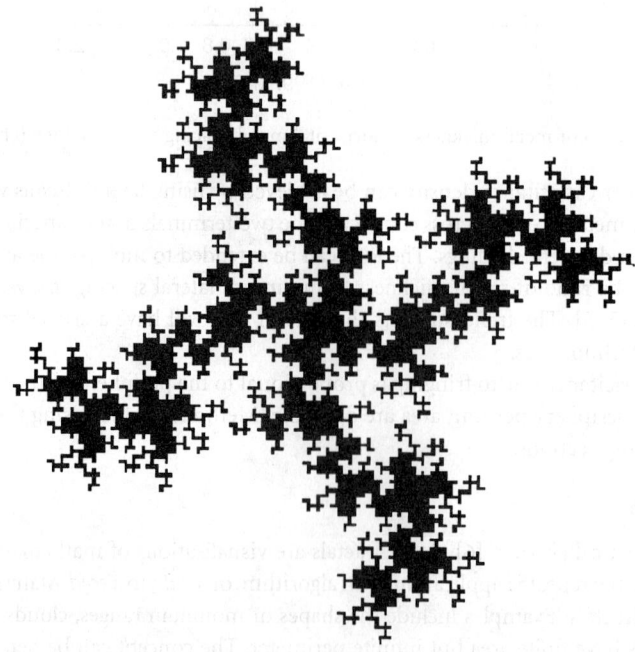

FIGURE 11.5 A Koch island with $M = 4$, $N = 32$, and $r = 1/8$.

term "fractal." In particular, it exceeds 1, which is the intuitive dimension of curves. A curve that has a high degree of complexity, or D, fills out a two-dimensional flat surface more efficiently. The fractal in Figure 11.4(c) has a dimension of 1.5, whereas for the border line of Figure 11.5, $D = 1.667$.

For the general case where the initiator has M sides, the periphery of the initiator is proportional to the square root of the area:

$$P_0 = k \cdot \sqrt{A} \tag{11.2}$$

where k is a proportionality constant that depends on the geometry of the initiator. For example, for a square initiator, $k = 4$; and for an equilateral triangle, $k = 2 \cdot \sqrt[4]{27}$. After n successive applications of the generation rule, the total periphery is

$$P = k\sqrt{A} \cdot (Nr)^n \tag{11.3}$$

and the minimum feature size (the resolution) is

$$l = \frac{k\sqrt{A}}{M} \cdot r^n \tag{11.4}$$

Eliminating n from Eqs. 11.3 and 11.4 and combining the result with Eq. 11.1, we have

$$P = \frac{k^D}{M^{D-1}} \cdot \frac{\left(\sqrt{A}\right)^D}{l^{D-1}} \tag{11.5}$$

Equation 11.5 demonstrates the dependence of the periphery on parameters such as the area and the resolution of the fractal border. It can be seen from Eq. 11.5 that as l tend toward zero, the periphery goes to infinity; therefore, it is possible to generate fractal structures with very large perimeters in any given area. However, the total periphery of a fractal curve is limited by the attainable resolution in practical realizations.

11.2.3   Fractal Capacitor Structures

The final shape of a fractal can be tailored to almost any form. The flexibility arises from the fact that a wide variety of geometries can be used as the initiator and generator. It is also possible to use different generators during each step. This is an advantage for integrated circuits where flexibility in the shape of the layout is desired.

Figure 11.6 is a three-dimensional representation of a fractal capacitor. This capacitor uses only one metal layer with a fractal border. For a better visualization of the overall picture, the terminals of this square-shaped capacitor have been identified using two different shadings. As was discussed before, multiple cross-connected metal layers may be used to improve capacitance density further.

One advantage of using lateral flux capacitors in general, and fractal capacitors in particular, is the reduction of the bottom-plate capacitance. This reduction is due to two reasons. First, the higher density of the fractal capacitor (compared to a standard parallel-plate structure) results in a smaller area. Second, some of the field lines originating from one of the bottom plates terminate on the adjacent plate, instead of the substrate, which further reduces the bottom-plate capacitance as shown in Figure 11.7. Because of this property, some portion of the parasitic bottom-plate capacitor is converted into the more useful plate-to-plate capacitance.

FIGURE 11.6 3-D representation of a fractal capacitor using a single metal layer.

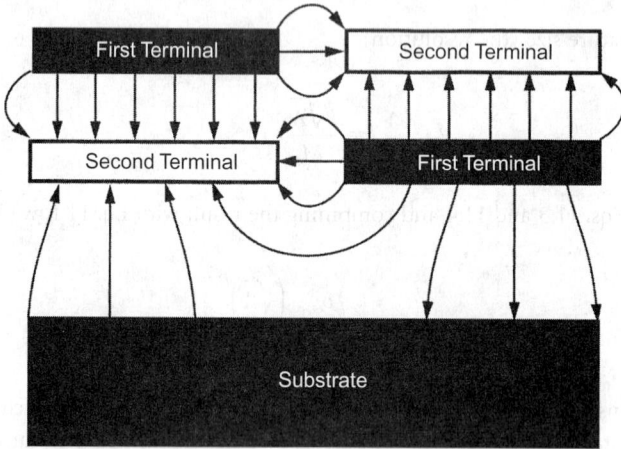

FIGURE 11.7 Reduction of the bottom-plate parasitic capacitance.

The capacitance per unit area of a fractal structure depends on the dimension of the fractal. To improve the density of the layout, fractals with large dimensions should be used. The concept of fractal dimension is demonstrated in Figure 11.8. The structure in Figure 11.8(a) has a lower dimension compared to the one in Figure 11.8(b), so the density (capacitance per unit area) of the latter is higher.

To demonstrate the dependence of capacitance density on dimension and lateral spacing of the metal layers, a first-order electromagnetic simulation was performed on two families of fractal structures. In Figure 11.9, the boost factor is plotted vs. horizontal spacing of the metal layers. The *boost factor* is defined as the ratio of the total capacitance of the fractal structure to the capacitance of a standard parallel-plate structure with the same area. The solid line corresponds to a family of fractals with a moderate fractal dimension of 1.63, while the dashed line represents another family of fractals with $D = 1.80$, which is a relatively large value for the dimension. In this first-order simulation, it is assumed that the vertical spacing and the thickness of the metal layers are kept constant at a 0.8-µm level. As can be seen in Figure 11.9, the amount of boost is a strong function of the fractal dimension as well as scaling.

In addition to the capacitance density, the quality factor, Q, is important in RF applications. Here, the degradation in quality factor is minimal because the fractal structure automatically limits the length of the thin metal sections to a few microns, keeping the series resistance reasonably small. For applications that require low series resistance, lower dimension fractals may be used. Fractals thus add one more degree of freedom to the design of capacitors, allowing the capacitance density to be traded for a lower series resistance.

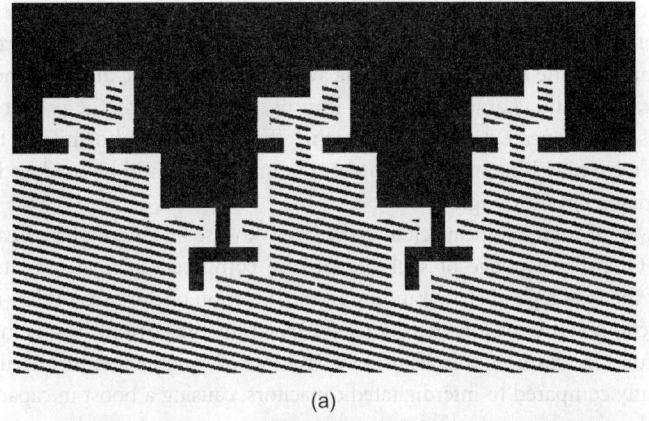

(a)

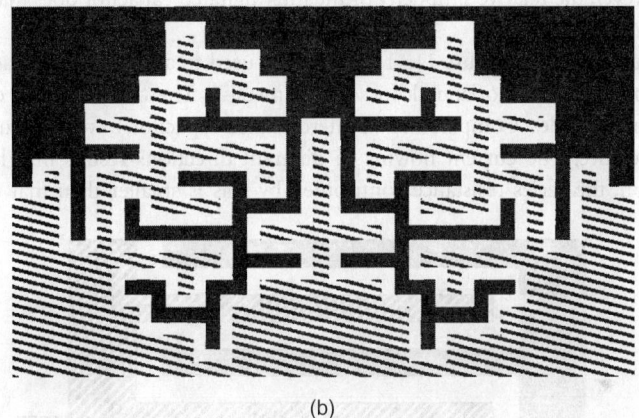

(b)

FIGURE 11.8 Fractal dimension of (a) is smaller than (b).

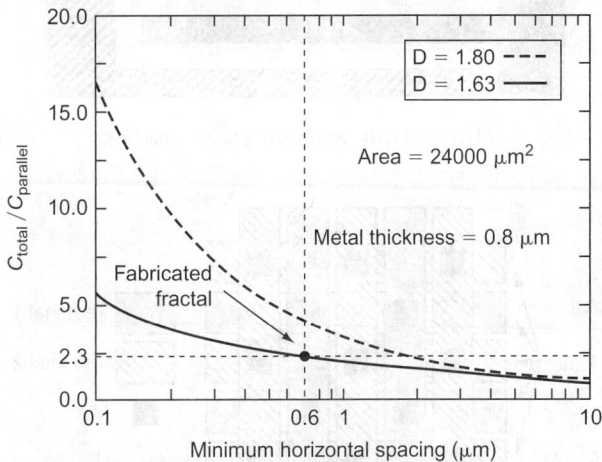

FIGURE 11.9 Boost factor vs. lateral spacing.

In current IC technologies, there is usually tighter control over the lateral spacing of metal layers compared to the vertical thickness of the oxide layers, from wafer to wafer and across the same wafer. Lateral flux capacitors shift the burden of matching away from oxide thickness to lithography. Therefore, by using lateral flux, matching characteristics can improve. Furthermore, the pseudo-random nature of the structure can also compensate, to some extent, the effects of non-uniformity of the etching process. To achieve accurate ratio matching, multiple copies of a unit cell should be used, as is standard practice in high-precision analog circuit design.

Another simple way of increasing capacitance density is to use an interdigitated capacitor depicted in Figure 11.10 [2,7]. One disadvantage of such a structure compared to fractals is its inherent parasitic inductance. Most of the fractal geometries randomize the direction of the current flow and thus reduce the effective series inductance; whereas for interdigitated capacitors, the current flow is in the same direction for all the parallel stubs. In addition, fractals usually have lots of rough edges that accumulate electrostatic energy more efficiently compared to interdigitated capacitors, causing a boost in capacitance (generally of the order of 15%). Furthermore, interdigitated structures are more vulnerable to non-uniformity of the etching process. However, the relative simplicity of the interdigitated capacitor does make it useful in some applications.

The woven structure shown in Figure 11.11 may also be used to achieve high capacitance density. The vertical lines are in metal-2 and horizontal lines are in metal-1. The two terminals of the capacitor are identified using different shades. Compared to an interdigitated capacitor, a woven structure has much less inherent series inductance. The current flowing in different directions results in a higher self-resonant frequency. In addition, the series resistance contributed by vias is smaller than that of an interdigitated

FIGURE 11.10 An interdigitated capacitor.

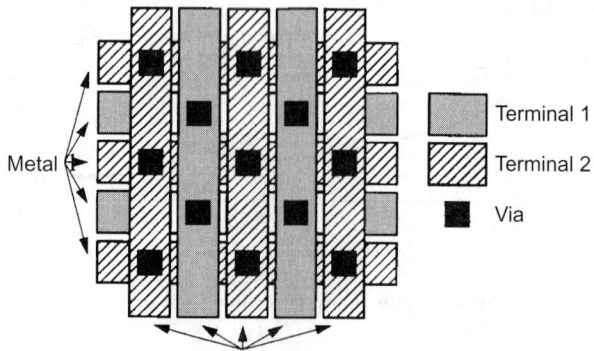

FIGURE 11.11 A woven structure.

capacitor, because cross-connecting the metal layers can be done with greater ease. However, the capacitance density of a woven structure is smaller compared to an interdigitated capacitor with the same metal pitch, because the capacitance contributed by the vertical fields is smaller.

11.3 Spiral Inductors

More than is so with capacitors, on-chip inductor options are particularly limited and unsatisfactory. Nevertheless, it is possible to build practical spiral inductors with values up to perhaps 20 nH and with Q values of approximately 10. For silicon-based RF ICs, Q degrades at high frequencies due to energy dissipation in the semiconducting substrate [8]. Additionally, noise coupling via the substrate at GHz frequencies has been reported [9]. As inductors occupy substantial chip area, they can potentially be the source and receptor of detrimental noise coupling. Furthermore, the physical phenomena underlying the substrate effects are complicated to characterize. Therefore, decoupling the inductor from the substrate can enhance the overall performance by increasing Q, improving isolation, and simplifying modeling.

Some approaches have been proposed to address the substrate issues; however, they are accompanied by drawbacks. Some [10] have suggested the use of high-resistivity (150 to 200 Ω-cm) silicon substrates to mimic the low-loss semi-insulating GaAs substrate, but this is rarely a practical option. Another approach selectively removes the substrate by etching a pit under the inductor [11]. However, the etch adds extra processing cost and is not readily available. Moreover, it raises reliability concerns such as packaging yield and long-term mechanical stability. For low-cost integration of inductors, the solution to substrate problems should avoid increasing process complexity.

In this section, we present the *patterned ground shield* (PGS) [23], which is compatible with standard silicon technologies, and which reduces the unwanted substrate effects. The great improvement provided by the PGS reduces the disparity in quality between spiral inductors made in silicon and GaAs IC technologies.

11.3.1 Understanding Substrate Effects

To understand why the PGS should be effective, consider first the physical model of an ordinary inductor on silicon, with one port and the substrate grounded, as shown in Figure 11.12 [8]. An on-chip inductor is physically a three-port element including the substrate. The one-port connection shown in Figure 11.12 avoids unnecessary complexity in the following discussion and at the same time preserves the inductor characteristics. In the model, the series branch consists of L_s, R_s, and C_s. L_s represents the spiral inductance, which can be computed using the Greenhouse method [12] or well-approximated by simple analytical formulas to be presented later. R_s is the metal series resistance whose behavior at RF is governed by the eddy current effect. This resistance accounts for the energy loss due to the skin effect in the spiral interconnect structure as well as the induced eddy current in any conductive media close to the inductor.

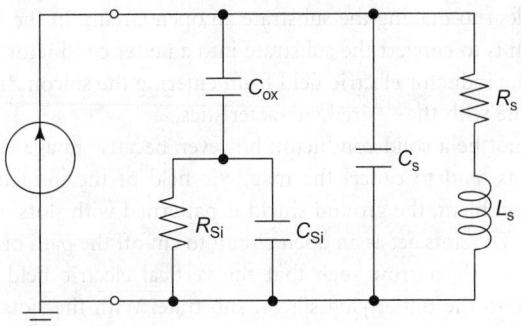

FIGURE 11.12 Lumped physical model of a spiral inductor on silicon.

The series feedforward capacitance, C_s, accounts for the capacitance due to the overlaps between the spiral and the center-tap underpass [13]. The effect of the inter-turn fringing capacitance is usually small because the adjacent turns are almost at equal potentials, and therefore it is neglected in this model. The overlap capacitance is more significant because of the relatively large potential difference between the spiral and the center-tap underpass. The parasitics in the shunt branch are modeled by C_{ox}, C_{Si}, and R_{Si}. C_{ox} represents the oxide capacitance between the spiral and the substrate. The silicon substrate capacitance and resistance are modeled by C_{Si} and R_{Si}, respectively [14,15]. The element R_{Si} accounts for the energy dissipation in the silicon substrate.

Expressions for the model element values are as follows:

$$R_s = \frac{\rho l}{\delta w \left(1 - e^{-\frac{t}{\delta}}\right)} \tag{11.6}$$

$$C_s = nw^2 \cdot \frac{\varepsilon_{ox}}{t_{oxM1-M2}} \tag{11.7}$$

$$C_{ox} = \frac{\varepsilon_{ox}}{2t_{ox}} \cdot l \cdot w \tag{11.8}$$

$$C_{Si} = \frac{1}{2} \cdot l \cdot w \cdot C_{sub} \tag{11.9}$$

$$R_{Si} = \frac{2}{l \cdot w \cdot G_{sub}} \tag{11.10}$$

where ρ is the DC resistivity of the spiral; t is the overall length of the spiral windings; w is the line width; δ is the skin depth; n is the number of crossovers between the spiral and center-tap (and thus $n = N - 1$, where N is the number of turns); $t_{oxM1-M2}$ is the oxide thickness between the spiral and substrate; C_{sub} is the substrate capacitance per unit area; and G_{sub} is the substrate conductance per unit area. In general, one treats C_{sub} and G_{sub} as fitting parameters.

Exploration with the model reveals that the substrate loss stems primarily from the penetration of the electric field into the lossy silicon substrate. As the potential drop in the semiconductor (i.e., across R_{Si} in Figure 11.12) increases with frequency, the energy dissipation in the substrate becomes more severe. It can be seen that increasing R_p to infinity reduces the substrate loss. It can be shown that R_p approaches infinity as R_{Si} goes either to zero or infinity. This observation implies that Q can be improved by making the silicon substrate *either* a perfect insulator or a perfect conductor. Using high-resistivity silicon (or etching it away) is equivalent to making the substrate an open circuit. In the absence of the freedom to do so, the next best option is to convert the substrate into a better conductor. The approach is to insert a ground plane to block the inductor electric field from entering the silicon. In effect, this ground plane becomes a pseudo-substrate with the desired characteristics.

The ground shield cannot be a solid conductor, however, because image currents would be induced in it. These image currents tend to cancel the magnetic field of the inductor proper, decreasing the inductance. To solve this problem, the ground shield is patterned with slots orthogonal to the spiral as illustrated in Figure 11.13. The slots act as an open circuit to cut off the path of the induced loop current. The slots should be sufficiently narrow such that the vertical electric field cannot leak through the patterned ground shield into the underlying silicon substrate. With the slots etched away, the ground strips serve as the termination for the electric field. The ground strips are merged together around the four outer edges of the spiral. The separation between the merged area and the edges is not critical.

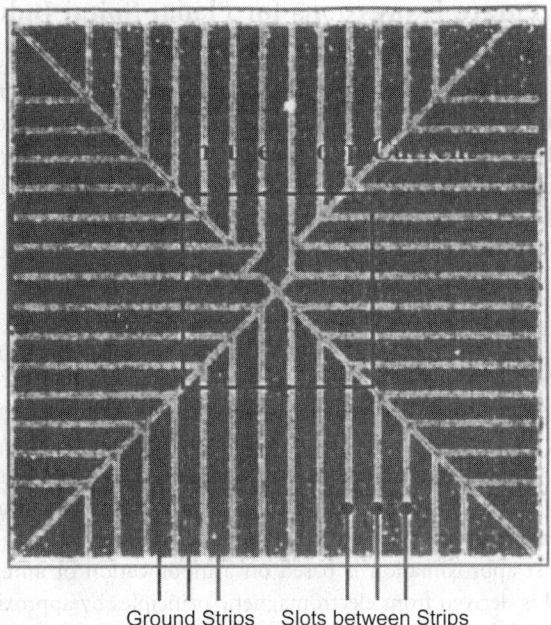

Ground Strips   Slots between Strips

FIGURE 11.13 A close-up photo of the patterned ground shield.

However, it is crucial that the merged area not form a closed ring around the spiral since it can potentially support unwanted loop current. The shield should be strapped with the top layer metal to provide a low-impedance path to ground. The general rule is to prevent negative mutual coupling while minimizing the impedance to ground.

The shield resistance is another critical design parameter. The purpose of the patterned ground shield is to provide a good short to ground for the electric field. Since the finite shield resistance contributes to energy loss of the inductor, it must be kept small. Specifically, by keeping the shield resistance small compared to the reactance of the oxide capacitance, the voltage drop that can develop across the shield resistance is very small. As a result, the energy loss due to the shield resistance is insignificant compared to other losses. A typical on-chip spiral inductor has parasitic oxide capacitance between 0.25 and 1 pF, depending on the size and the oxide thickness. The corresponding reactance due to the oxide capacitance at 1 to 2 GHz is of the order of 100 Ω, and hence a shield resistance of a few ohms is sufficiently small not to cause any noticeable loss.

With the PGS, one can expect typical improvements in Q ranging from 10 to 33%, in the frequency range of 1 to 2 GHz. Note that the inclusion of the ground shields increases C_p, which causes a fast roll-off in Q above the peak-Q frequency and a reduction in the self-resonant frequency. This modest improvement in inductor Q is certainly welcome, but is hardly spectacular by itself. However, a more dramatic improvement is evident when evaluating inductor-capacitor resonant circuits. Such LC tank circuits can absorb the parasitic capacitance of the ground shield. Since the energy stored in such parasitic elements is now part of the circuit, the overall circuit Q is greatly increased. Improvements of factors of approximately two are not unusual, so that tank circuits realized with PGS inductors possess roughly the same Q as those built in GaAs technologies.

As stated earlier, substrate noise coupling can be an issue of great concern owing to the relatively large size of typical inductors. Shielding by the PGS improves isolation by 25 dB or more at GHz frequencies. It should be noted that, as with any other isolation structure (such as a guard ring), the efficacy of the PGS is highly dependent on the integrity of the ground connection. One must often make a tradeoff between the desired isolation level and the chip area that is required to provide a low-impedance ground connection.

11.3.2   Simple, Accurate Expressions for Planar Spiral Inductances

In the previous section, a physically based model for planar spiral inductors was offered, and reference was made to the Greenhouse method as a means for computing the inductance value. This method uses as computational atoms the self- and mutual inductances of parallel current strips. It is relatively straightforward to apply, and yields accurate results. Nevertheless, simpler analytic formulas are generally preferred for design since important insights are usually more readily obtained.

As a specific example, square spirals are popular mainly because of their ease of layout. Other polygonal spirals have also been used to improve performance by more closely approximating a circular spiral. However, a quantitative evaluation of possible improvements is cumbersome without analytical formulas for inductance.

Among alternative shapes, hexagonal and octagonal inductors are used widely. Figures 11.14 through 11.16 and show the layout for square, hexagonal, and octagonal inductors, respectively. For a given shape, an inductor is completely specified by the number of turns n, the turn width w, the turn spacing s, and any one of the following: the outer diameter d_{out}, the inner diameter d_{in}, the average diameter $d_{avg} = 0.5(d_{out} + d_{in})$, or the fill ratio, defined as $\rho = (d_{out} + d_{in})/(d_{out} + d_{in})$. The thickness of the inductor has only a very small effect on inductance and will therefore be ignored here.

We now present three approximate expressions for the inductance of square, hexagonal, and octagonal planar inductors. The first approximation is based on a modification of an expression developed by Wheeler [16]; the second is derived from electromagnetic principles by approximating the sides of the spirals as current sheets; and the third is a monomial expression derived from fitting to a large database of inductors (whose exact inductance values are obtained from a 3-D electromagnetic field solver). All three expressions are accurate, with typical errors of 2 to 3%, and very simple, and are therefore excellent candidates for use in design and optimization.

11.3.2.1   Modified Wheeler Formula

Wheeler [16] presented several formulas for planar spiral inductors, which were intended for discrete inductors. A simple modification of the original Wheeler formula allows us to obtain an expression that is valid for planar spiral integrated inductors:

$$L_{mw} = K_1 \mu_0 \frac{n^2 d_{avg}}{1 + K_2 \rho} \tag{11.11}$$

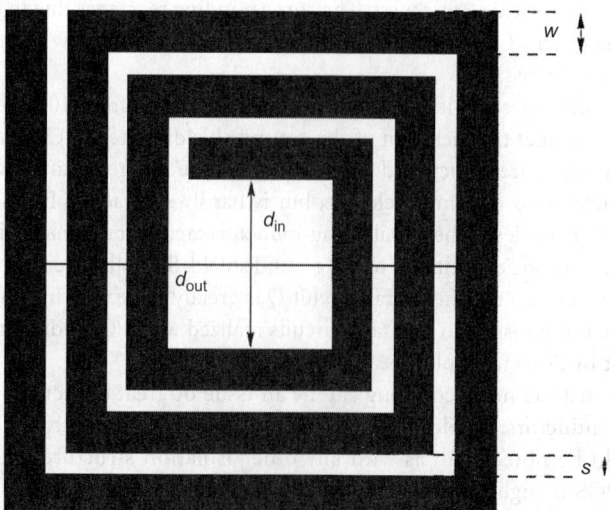

FIGURE 11.14   Square inductor.

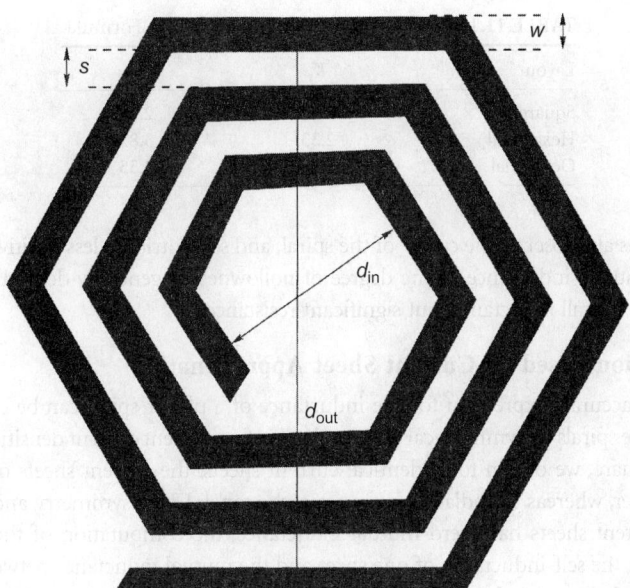

FIGURE 11.15 Hexagonal inductor.

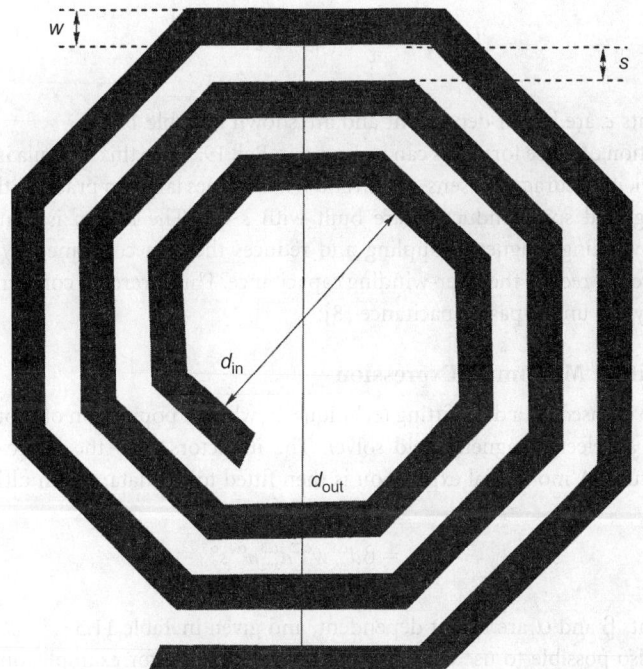

FIGURE 11.16 Octagonal inductor.

where ρ is the fill ratio defined previously. The coefficients K_1 and K_2 are layout dependent and are shown in Table 11.1.

The fill factor ρ represents how hollow the inductor is: for small ρ, we have a hollow inductor ($d_{out} \cong d_{in}$), and for a large ρ we have a filled inductor ($d_{out} \gg d_{in}$). Two inductors with the same average diameter but different fill ratios will, of course, have different inductance values; the filled one has a smaller inductance

TABLE 11.1 Coefficients for Modified Wheeler Formula

Layout	K_1	K_2
Square	2.34	2.75
Hexagonal	2.33	3.82
Octagonal	2.25	3.55

because its inner turns are closer to the center of the spiral, and so contribute less positive mutual inductance and more negative mutual inductance. Some degree of hollowness is generally desired since the innermost turns contribute little overall inductance, but significant resistance.

11.3.2.2 Expression Based on Current Sheet Approximation

Another simple and accurate expression for the inductance of a planar spiral can be obtained by approximating the sides of the spirals by symmetrical current sheets of equivalent current densities [17]. For example, in the case of the square, we obtain four identical current sheets: the current sheets on opposite sides are parallel to one another, whereas the adjacent ones are orthogonal. Using symmetry and the fact that sheets with orthogonal current sheets have zero mutual inductance, the computation of the inductance is now reduced to evaluating the self-inductance of one sheet and the mutual inductance between opposite current sheets. These self- and mutual inductances are evaluated using the concepts of geometric mean distance (GMD) and arithmetic mean distance (AMD) [17,18]. The resulting expression is:

$$L_{\text{gmd}} = \frac{\mu n^2 d_{\text{avg}}}{\pi}\left(c_1(\log c_2/\rho) + c_3\rho\right) \tag{11.12}$$

where the coefficients c_i are layout dependent and are shown in Table 11.2.

A detailed derivation of these formulas can be found in Ref. 19. Since this formula is based on a current sheet approximation, its accuracy worsens as the ratio s/w becomes large. In practice, this is not a problem since practical integrated spiral inductors are built with $s < w$. The reason is that a smaller spacing improves the inter-winding magnetic coupling and reduces the area consumed by the spiral. A large spacing is only desired to reduce the inter-winding capacitance. This is rarely a concern as this capacitance is always dwarfed by the under-pass capacitance [8].

11.3.2.3 Data-Fitted Monomial Expression

Our final expression is based on a data-fitting technique, in which a population of thousands of inductors are simulated with an electromagnetic field solver. The inductors span the entire range of values of relevance to RF circuits. A monomial expression is then fitted to the data, which ultimately yields:

$$L_{\text{mon}} = \beta d_{\text{avg}}^{\alpha1} w^{\alpha2} d_{\text{avg}}^{\alpha3} n^{\alpha4} s^{\alpha5} \tag{11.13}$$

where the coefficients β and α_i are layout dependent, and given in Table 11.3.

Of course, it is also possible to use other data-fitting techniques; for example, one which minimizes the maximum error of the fit, or one in which the coefficients must satisfy given in equalities or bounds. The monomial expression is useful since, like the other expressions, it is very accurate and very simple. Its real value, however, is that it can be used for the optimal design of inductors and circuits containing inductors, using geometric programming, which is a type of optimization method that requires monomial models [20,21].

Figure 11.17 shows the absolute error distributions of these expressions. The plots show that typical errors are in the 1 to 2% range, and most of the errors are below 3%. These expressions for inductance, while quite simple, are thus sufficiently accurate that field solvers are rarely necessary.

TABLE 11.2 Coefficients for Current-Sheet Inductance Formula

Layout	c_1	c_2	c_3
Square	2.00	2.00	0.54
Hexagonal	1.83	1.71	0.45
Octagonal	1.87	1.68	0.60

TABLE 11.3 Coefficients for Monomial Inductance Formula

Layout	b	α_1	α_2	α_3	α_4	α_5
Square	1.66×10^{-3}	−1.33	−0.13	2.50	1.83	−0.022
Hexagonal	1.33×10^{-3}	−1.46	−0.16	2.67	1.80	−0.030
Octagonal	1.34×10^{-3}	−1.35	−0.15	2.56	1.77	−0.032

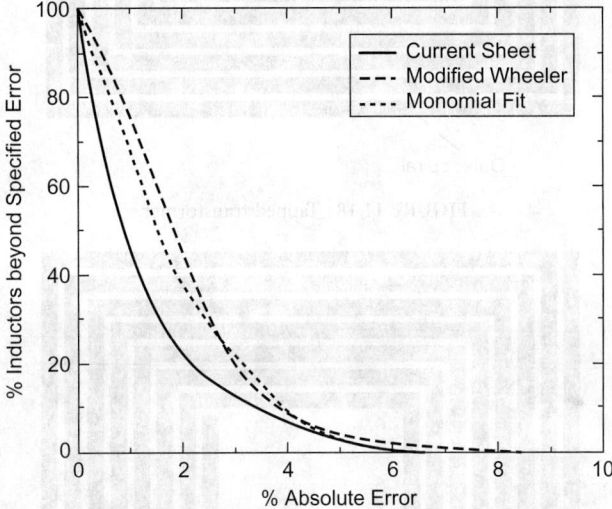

FIGURE 11.17 Error distribution for three formulas, compared to field solver simulations.

These expressions can be included in a physical, scalable lumped-circuit model for spiral inductors where, in addition to providing design insight, they allow efficient optimization schemes to be employed.

11.4 On-Chip Transformers

Transformers are important elements in RF circuits for impedance conversion, impedance matching, and bandwidth enhancement. Here, we present an analytical model for monolithic transformers that is suitable for circuit simulation and design optimization. We also provide simple expressions for calculating the mutual coupling coefficient (k).

We first discuss different on-chip transformers and their advantages and disadvantages. We then present an analytical model along with expressions for the elements in it and the mutual coupling coefficient.

11.4.1 Monolithic Transformer Realizations

Figures 11.18 through 11.23 illustrate common configurations of monolithic transformers. The different realizations offer varying tradeoffs among the self-inductance and series resistance of each port, the mutual coupling coefficient, the port-to-port and port-to-substrate capacitances, resonant frequencies, symmetry, and area. The models and coupling expressions allow these trade-offs to be systematically explored, thereby permitting transformers to be customized for a variety of circuit design requirements.

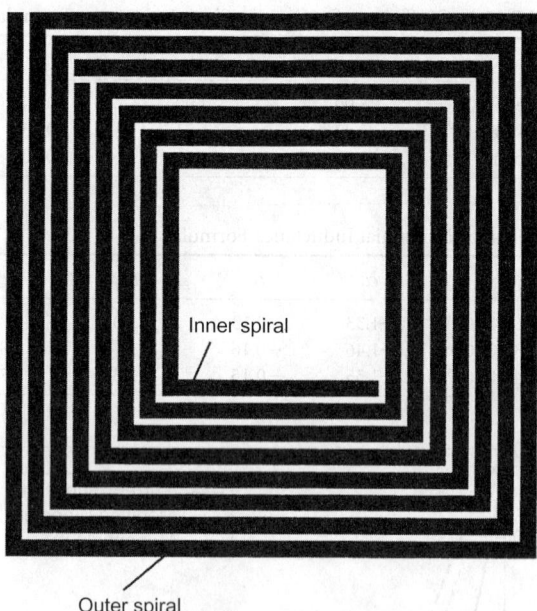

FIGURE 11.18　Tapped transformer.

FIGURE 11.19　Interleaved transformer.

　　The characteristics desired of a transformer are application dependent. Transformers can be configured as three or four-terminal devices. They may be used for narrowband or broadband applications. For example, in single-sided to differential conversion, the transformer might be used as a four-terminal narrowband device. In this case, a high mutual coupling coefficient and high self-inductance are desired, along with low series resistance. On the other hand, for bandwidth extension applications, the transformer is used as a broadband three-terminal device. In this case, a small mutual coupling coefficient and high series resistance are acceptable, while all capacitances need to be minimized [22].

FIGURE 11.20 Stacked transformer with top spiral overlapping the bottom one.

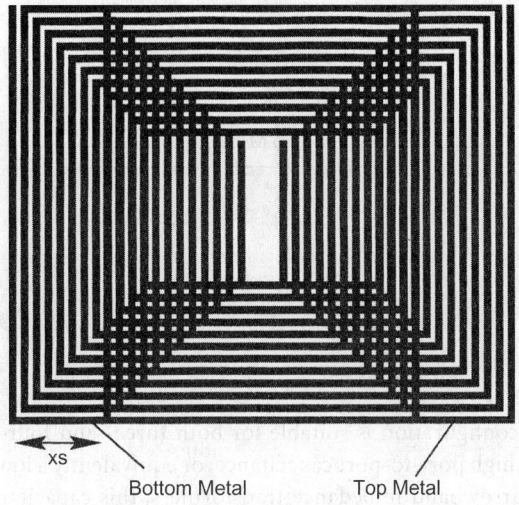

FIGURE 11.21 Stacked transformer with top and bottom spirals laterally shifted.

The tapped transformer (Figure 11.18) is best suited for three-port applications. It permits a variety of tapping ratios to be realized. This transformer relies only on lateral magnetic coupling. All windings can be implemented with the top metal layer, thereby minimizing port-to-substrate capacitances. Since the two inductors occupy separate regions, the self-inductance is maximized while the port-to-port capacitance is minimized. Unfortunately, this spatial separation also leads to low mutual coupling ($k = 0.3$–0.5).

The interleaved transformer (Figure 11.19) is best suited for four-port applications that demand symmetry. Once again, capacitances can be minimized by implementing the spirals with top level metal so that high resonant frequencies may be realized. The interleaving of the two inductances permit moderate coupling ($k = 0.7$) to be achieved at the cost of reduced self-inductance. This coupling may be increased at the cost of higher series resistance by reducing the turn width (w) and spacing (s).

The stacked transformer (Figure 11.20) uses multiple metal layers and exploits both vertical and lateral magnetic coupling to provide the best area efficiency, the highest self-inductance, and highest

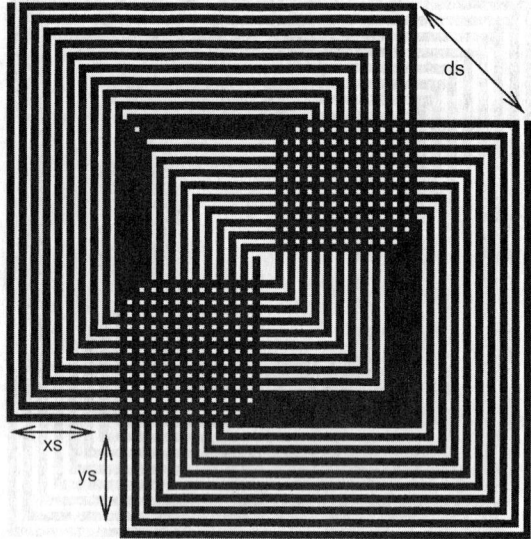

FIGURE 11.22　Stacked transformer with top and bottom spirals diagonally shifted.

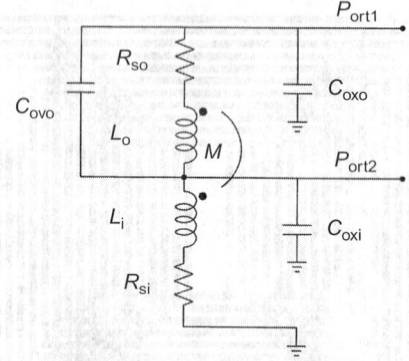

FIGURE 11.23　Tapped transformer model.

coupling ($k = 0.9$). This configuration is suitable for both three- and four-terminal configurations. The main drawback is the high port-to-port capacitance, or equivalently a low self-resonant frequency. In some cases, such as narrowband impedance transformers, this capacitance may be incorporated as part of the resonant circuit. Also, in multi-level processes, the capacitance can be reduced by increasing the oxide thickness between spirals. For example, in a five-metal process, 50 to 70% reductions in port-to-port capacitance can be achieved by implementing the spirals on layers five and three instead of five and four. The increased vertical separation will reduce k by less than 5%. One can also trade off reduced coupling for reduced capacitance by displacing the centers of the stacked inductors (Figures 11.21 and 11.22)

11.4.2　Analytical Transformer Models

Figures 11.23 and 11.24 present the circuit models for tapped and stacked transformers, respectively. The corresponding element values for the tapped transformer model are given by the following equations (subscript o refers to the outer spiral, i to the inner spiral, and T to the whole spiral):

$$L_T = \frac{9.375\mu_0 n_T^2 AD_T^2}{11OD_T - 7AD_T} \tag{11.14}$$

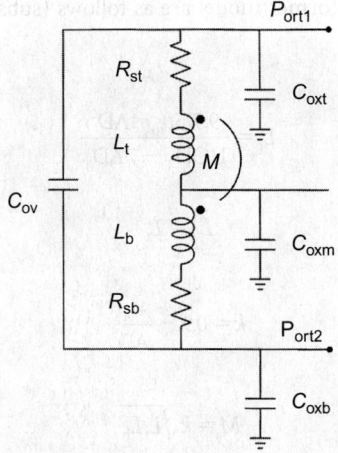

FIGURE 11.24 Stacked transformer model.

$$L_o = \frac{9.375\mu_0 n_o^2 AD_o^2}{11OD_o - 7AD_o} \tag{11.15}$$

$$L_i = \frac{9.375\mu_0 n_i^2 AD_i^2}{11OD_i - 7AD_i} \tag{11.16}$$

$$M = \frac{L_T - L_o - L_i}{2\sqrt{L_o L_i}} \tag{11.17}$$

$$R_{so} = \frac{\rho l_o}{\delta w \left(1 - e^{-\frac{t}{\delta}}\right)} \tag{11.18}$$

$$R_{si} = \frac{\rho l_i}{\delta w \left(1 - e^{-\frac{t}{\delta}}\right)} \tag{11.19}$$

$$C_{ovo} = \frac{\varepsilon_{ox}}{t_{ox,t-b}} \cdot (n_o - 1)w^2 \tag{11.20}$$

$$C_{oxo} = \frac{\varepsilon_{ox}}{2t_{ox}} \cdot l_o w \tag{11.21}$$

$$c_{oxi} = \frac{\varepsilon_{ox}}{2t_{ox}} \cdot (l_o + l_i)w \tag{11.22}$$

where ρ is the DC metal resistivity; δ is the skin depth; $t_{ox,t-b}$ is the oxide thickness from top level metal to bottom metal; n is the number of turns; OD, AD, and ID are the outer, average, and inner diameters, respectively; l is the length of the spiral; w is the turn width; t is the metal thickness; and A is the area.

Expressions for the stacked transformer model are as follows (subscript t refers to the top spiral and b to the bottom spiral):

$$L_t = \frac{9.375\mu_0 n^2 AD^2}{11OD_T - 7AD_T} \tag{11.23}$$

$$L_b = L_t \tag{11.24}$$

$$k = 0.9 - \frac{d_s}{AD} \tag{11.25}$$

$$M = k\sqrt{L_t L_b} \tag{11.26}$$

$$R_{st} = \frac{\rho_t l}{\delta_t w \left(1 - e^{-\frac{t_t}{\delta_t}} \right)} \tag{11.27}$$

$$R_{sb} = \frac{\rho_b l}{\delta_b w \left(1 - e^{-\frac{t_b}{\delta_b}} \right)} \tag{11.28}$$

$$C_{ov} = \frac{\varepsilon_{ox}}{2t_{ox,t-b}} \cdot l \cdot w \cdot \frac{A_{ov}}{A} \tag{11.29}$$

$$C_{oxt} = \frac{\varepsilon_{ox}}{2t_{oxt}} \cdot l \cdot w \cdot \frac{A - A_{ov}}{A} \tag{11.30}$$

$$C_{oxb} = \frac{\varepsilon_{ox}}{2t_{ox}} \cdot l \cdot w \tag{11.31}$$

$$C_{oxm} = C_{oxt} + C_{oxb} \tag{11.32}$$

where t_{oxt} is the oxide thickness from top metal to the substrate; t_{oxb} is the oxide thickness from bottom metal to substrate; k is the coupling coefficient; A_{ov} is the overlap area of the two spirals; and d_s is the center-to-center spiral distance.

The expressions for the series resistances (R_{so}, R_{si}, R_{st}, and R_{sb}), the port-substrate capacitances (C_{oxo}, C_{oxi}, C_{oxt}, C_{oxb}, and C_{oxm}) and the crossover capacitances (C_{ovo}, C_{ovi}, and C_{ov}) are taken from Ref. 8. Note that the model accounts for the increase in series resistance with frequency due to skin effect. Patterned ground shields (PGS) are placed beneath the transformers to isolate them from resistive and capacitive coupling to the substrate [23]. As a result, the substrate parasitics can be neglected.

The inductance expressions in the foregoing are based on the modified Wheeler formula discussed earlier [24]. This formula does not take into account the variation in inductance due to conductor thickness and frequency. However, in practical inductor and transformer realizations, the thickness is small compared to the lateral dimensions of the coil and has only a small impact on the inductance. For typical conductor thickness variations (0.5 to 2.0 μm), the change in inductance is within a few percent for practical inductor geometries. The inductance also changes with frequency due to changes in current distribution within the conductor. However, over the useful frequency range of a spiral, this variation is negligible [23]. When compared to field solver simulations, the inductance expression exhibits a maximum error of 8% over a broad design space (outer diameter OD varying from 100 to 480 μm, L varying from 0.5 to 100 nH, w varying from 2 μm to 0.3OD, s varying from 2 μm to w, and inner diameter ID varying from 0.2 to 0.8OD).

For the tapped transformer, the mutual inductance is determined by first calculating the inductance of the whole spiral (L_T), the inductance of the outer spiral (L_o), the inductance of the inner spiral (L_i), and then using the expression $M = (L_T - L_o - L_i)/2$. For the stacked transformer, the spirals have identical lateral geometries and therefore identical inductances. In this case, the mutual inductance is determined by first calculating the inductance of one spiral (L_T), the coupling coefficient (k) and then using the expression $M = kL_T$. In this last case the coupling coefficient is given by $k = 0.9 - d_s/(AD)$ for $d_s < 0.7AD$, where d_s is the center-to-center spiral distance and AD is the average diameter of the spirals. As d_s increases beyond 0.7AD, the mutual coupling coefficient becomes harder to model. Eventually, k crosses zero and reaches a minimum value of approximately –0.1 at $d_s = $ AD. As d_s increases further, k asymptotically approaches zero. At $d_s = 2AD$, $k = -0.02$, indicating that the magnetic coupling between closely spaced spirals is negligible.

The self-inductances, series resistances, and mutual inductances are independent of whether a transformer is used as a three- or four-terminal device. The only elements that require recomputation are the port-to-port and port-to-substrate capacitances. This situation is analogous to that of a spiral inductor being used as a single- or dual-terminal device.

As with the inductance formulas, the transformer models obviate the need for full field solutions in all but very rare instances, allowing rapid design and optimization.

References

1. Samavati, H. et al., "Fractal capacitors," *1998 IEEE ISSCC Dig. of Tech. Papers*, Feb. 1998.
2. Akcasu, O.E., "High capacitance structures in a semiconductor device," U.S. Patent 5 208 725, May 1993.
3. Bohr, M., "Interconnect scaling — The real limiter to high performance VLSI," *Intl. Electron Devices Meeting Tech. Digest*, pp. 241–244, 1995.
4. Bohr, M. et al., "A high performance 0.25 μm logic technology optimized for 1.8V operation," *Intl. Electron Devices Meeting Tech. Digest*, pp. 847–850, 1996.
5. Venkatesan, S. et al., "A high performance 1.8V, 0.20 μm CMOS technology with copper metallization," *Intl. Electron Devices Meeting Tech. Digest*, pp. 769–772, 1997.
6. Mandelbrot, B.B., *The Fractal Geometry of Nature*, W. H. Freeman, New York, 1983.
7. Pettenpaul, E. et al., "Models of lumped elements on GaAs up to 18 GHz," *IEEE Transactions on Microwave Theory and Techniques*, vol. 36, no. 2, pp. 294–304, Feb. 1988.
8. Yue, C.P., Ryu, C., Lau, J., Lee, T.H., and Wong, S.S., "A physical model for planar spiral inductors on silicon," *International Electron Devices Meeting Technical Digest*, pp. 155–158, Dec. 1996.
9. Pfost, M., Rein, H.-M., and Holzwarth, T., "Modeling substrate effects in the design of high speed Si-bipolar IC's," *IEEE J. Solid-State Circuits*, vol. 31, no. 10, pp. 1493–1501, Oct. 1996.
10. Ashby, K.B., Koullias, I.A., Finley, W.C., Bastek, J.J., and Moinian, S., "High Q inductors for wireless applications in a complementary silicon bipolar process," *IEEE J. Solid-State Circuits*, vol. 31, no. 1, pp. 4–9, Jan. 1996.
11. Chang, J.Y. C., Abidi, A.A., and Gaitan, M., "Large suspended inductors on silicon and their use in a 2-μm CMOS RF amplifier," *IEEE Electron Device Letters*, vol. 14, no. 5, pp. 246–248, May 1993.

12. Greenhouse, H.M., "Design of planar rectangular microelectronic inductors," *IEEE Transactions on Parts, Hybrids, and Packing*, vol. PHP-10, no. 2, pp. 101–109, June 1974.

13. Wiemer, L. and Jansen, R.H., "Determination of coupling capacitance of underpasses, air bridges and crossings in MICs and MMICs," *Electronics Letters*, vol. 23, no. 7, pp. 344–346, Mar. 1987.

14. Ho, I.T. and Mullick, S.K., "Analysis of transmission lines on integrated-circuit chips," *IEEE J. Solid-State Circuits*, vol. SC-2, no. 4, pp. 201–208, Dec. 1967.

15. Hasegawa, H., Furukawa, M., and Yanai, H., "Properties of microstrip line on Si-SiO$_2$ system," *IEEE Transactions on Microwave Theory and Techniques*, vol. MTT-19, no. 11, pp. 869–881, Nov. 1971.

16. Wheeler, H.A., "Simple inductance formulas for radio coils," *Proc. of the IRE*, vol. 16, no. 10, pp. 1398–1400, October 1928.

17. Rosa, E.B., "Calculation of the self-inductances of single-layer coils," *Bull. Bureau of Standards*, vol. 2, no. 2, pp. 161–187, 1906.

18. Maxwell, J.C., *A Treatise on Electricity and Magnetism*, Dover, 3rd ed., 1967.

19. Mohan, S.S., "Formulas for planar spiral inductances," *Tech. Rep., IC Laboratory*, Stanford University, Aug. 1998, http://www-smirc.stanford.edu.

20. Boyd, S. and Vandenberghe, L., "Introduction to convex optimization with engineering applications," Course Notes, 1997, http://www-leland.stanford.edu/class/ee364/.

21. Hershenson, M., Boyd, S.P., and Lee, T.H., "GPCAD: A tool for CMOS op-amp synthesis," in *Digest of Technical Papers, IEEE International Conference on Computer-Aided Design*, Nov. 1998.

22. Lee, T.H., *The Design of CMOS Radio-Frequency Integrated Circuits*, Cambridge University Press, Cambridge, 1998.

23. Yue, C.P. et al., "On-chip spiral inductors with patterned ground shields for Si-based RF ICs," *IEEE J. Solid-State Circuits*, vol. 33, pp. 743–752, May 1998.

24. Wheeler, H.A., "Simple inductance formulas for radio coils," *Proc. of the IRE*, vol. 16, no. 10, pp. 1398–1400, Oct. 1928.

12

CMOS Fabrication

Jeff Jessing
Boise State University

CONTENTS

This chapter* provides a brief overview of CMOS process integration and fabrication unit processes. Starting from virgin silicon wafers, a well-defined collection of semiconductor processes is required to fabricate CMOS integrated circuits (ICs). It is this deliberate combination of processes that is termed *integration*. Circuit design and process integration interact strongly. For instance, the design rule set typically is determined largely by the limitations in the fabrication processes. Therefore, circuit designers, process engineers, and integration engineers must communicate effectively.

    First, we will examine the fundamental processes, called unit processes, which are combined in a deliberate sequence to fabricate CMOS. Our primary focus will be the qualitative understanding of the processes, with limited introduction of quantitative expressions. Usually, the unit processes are repeated many times in a given process sequence. We will discuss a representative modern CMOS process sequence, also called a process flow.

*Figures and text are taken, in large part, from previous work of Jeff Jessing, specifically Chapter 7 in R. J. Baker, *CMOS Circuit Design, Layout, and Simulation*, 2nd ed, Wiley-IEEE Press, 2004, ISBN 047170055X.

12.1   CMOS Unit Processes

In this section we introduce each of the major processes required to fabricate CMOS ICs. First, however, we will discuss wafer preparation, even though it is not a unit process, because it is important to know how manufacturers produce wafers. After that, we will present the unit processes that fabrication facilities use to produce ICs. Grouped by function, they are thermal oxidation, doping processes, photolithography, thin-film removal, and thin-film deposition.

12.1.1   Wafer Manufacture

Silicon is the second-most abundant element in the Earth's crust; however, it occurs exclusively in compounds. Elemental silicon is a man-made material refined from these various compounds, the most common being silica (impure SiO_2). Modern ICs must be fabricated on ultrapure, defect-free slices of single-crystalline silicon, called wafers.

12.1.1.1   Metallurgical-Grade Silicon (MGS)

Wafer production requires three general processes: silicon refinement, crystal growth, and wafer formation. Silicon refinement begins with the reduction of silica in an arc furnace at roughly 2000°C with a carbon source. The carbon effectively "pulls" the oxygen from the SiO_2 molecules, thus chemically reducing the oxide into roughly 98% pure silicon, referred to as MGS. The overall reduction is governed by the following equation:

$$SiO_2(\text{solid}) + 2C(\text{solid}) \rightarrow Si(\text{liquid}) + 2CO(\text{gas}) \qquad (12.1)$$

12.1.1.2   Electrical-Grade Silicon (EGS)

MGS is not pure enough for microelectronic device applications because the electronic properties of a semiconductor such as silicon are extremely sensitive to impurity concentrations. Impurity levels measured at parts per million or less can have dramatic effects on carrier mobilities, lifetimes, etc. It is therefore necessary to further purify the MGS into what is known as EGS. EGS is produced by chlorinating grounded MGS as shown in the following representative, unbalanced equation:

$$MGS(\text{solid}) + HCl(\text{gas}) \rightarrow SiH_4(\text{liquid}) + SiHCl_3(\text{liquid}) \qquad (12.2)$$

Since the reaction by-products are liquids at room temperature, ultrapure EGS can be obtained from fractional distillation and chemical reduction processes. The resultant EGS is in the form of polycrystalline chunks.

12.1.1.3   Czochralski (CZ) Growth

To achieve a single-crystalline form, the EGS must be subjected to a process called CZ growth. A schematic representation of the CZ growth process is shown in Figure 12.1. The polycrystalline EGS is melted in a large quartz crucible. A small seed crystal of known orientation is introduced into the surface of the silicon melt. The seed crystal, rotating in one direction, is slowly pulled from the silicon melt, which is rotating in the opposite direction. Solidification of the silicon onto the seed results in the formation of a growing crystal (called a boule or ingot) that assumes the crystallographic orientation of the seed. In general, the slower the pull rate (typically measured in millimeter per hour), the larger the diameter of the silicon crystal. The silicon boule is then turned down to the appropriate diameter, and flats or notches are ground into the surface of the boule to indicate a precise crystal orientation. Using a special saw, the silicon boule is cut into thin wafers. The wafers are finished by using a chemical-mechanical polishing (CMP) process, as discussed in Section 12.1.5,

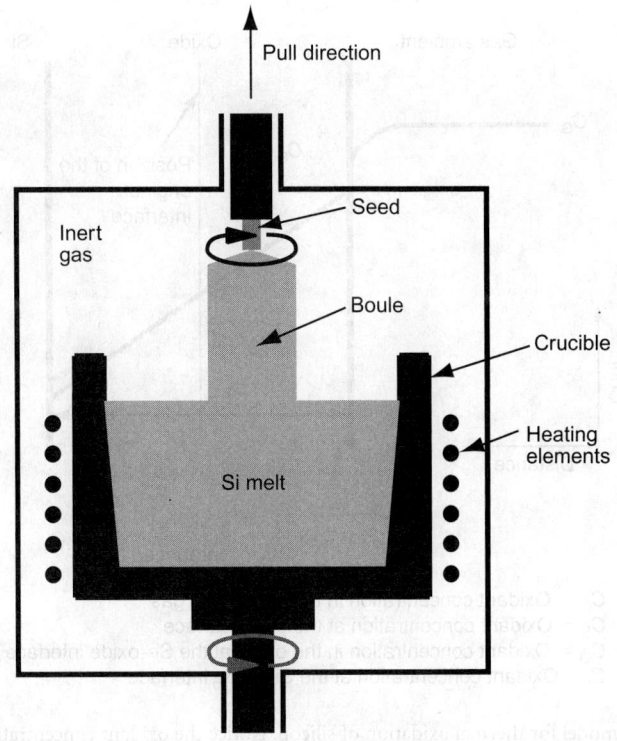

FIGURE 12.1 Simplified diagram showing Czochralski (CZ) crystal growth.

to yield a mirror-like finish on one side of the wafer. Although devices are fabricated entirely within the top 2 μm of the wafer, for adequate mechanical support, the final wafer may be of the order of 1 mm thick (thickness increases with wafer diameter).

12.1.2   Thermal Oxidation

Silicon, when exposed to an oxidant at elevated temperatures, will readily form a thin layer of oxide on all exposed surfaces. Silicon's native oxide is in the form of silicon dioxide (SiO_2). With respect to CMOS fabrication, SiO_2 can serve as a high-quality dielectric in device structures such as gate oxides. During processing, thermally grown oxides can be used as implantation, diffusion, and etch masks. The dominance of silicon as a microelectronic material can be attributed to the existence of this high-quality native oxide and the resultant near-ideal silicon–oxide interface.

Figure 12.2 depicts the basic thermal oxidation process. The silicon wafer is exposed at high temperatures (typically, 900–1200°C) to a gaseous oxidant such as molecular oxygen (O_2) and water vapor (H_2O). For obvious reasons, oxidation in O_2 is called dry oxidation, whereas in H_2O it is called wet oxidation. The gas–solid interface forms a stagnant layer through which the oxidant must diffuse to reach the surface of the wafer. Once at the surface, the oxidant must again diffuse through the oxide layer present. As the oxidant species reaches the silicon–oxide interface, one of the following two reactions occurs:

$$Si + O_2 \rightarrow SiO_2 \quad \text{(dry oxidation)} \tag{12.3}$$

$$Si + 2H_2O \rightarrow SiO_2 + 2H_2 \quad \text{(wet oxidation)} \tag{12.4}$$

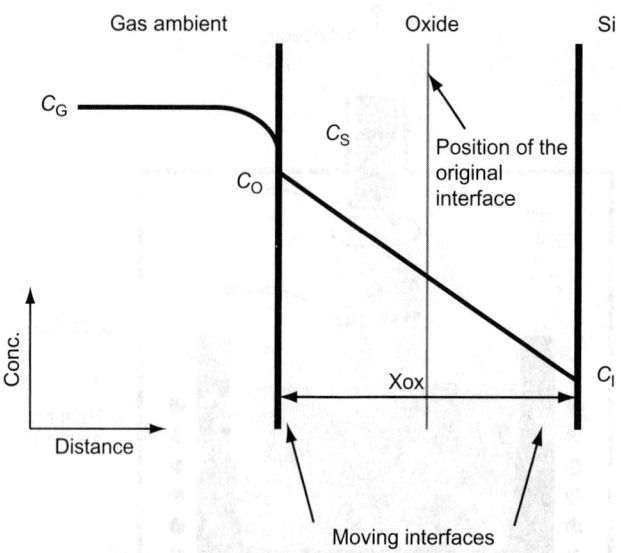

C_G = Oxidant concentration in the bulk of the gas
C_S = Oxidant concentration at the oxide surface
C_O = Oxidant concentration in the oxide at the Si–oxide interface
C_I = Oxidant concentration at the Si–oxide interface

FIGURE 12.2 Simple model for thermal oxidation of silicon. Notice the oxidant concentrations (boundary conditions) in the gas, oxide, and silicon.

It should be emphasized that reactions specified by Eq. (12.3) and Eq. (12.4) occur at the silicon–oxide interface, where silicon is consumed in the reaction. As seen in Figure 12.2, the position of the original interface is at ~50% of the thickness of the oxide.

The rate of thermal oxidation is a function of temperature and rate constants. The rate is directly proportional to temperature. The rate constants are, in turn, a function of gas partial pressures, oxidant type, and silicon wafer characteristics such as doping type, doping concentration, and crystallographic orientation. In general, dry oxidation yields a denser and thus higher quality oxide than does wet oxidation. However, wet oxidation occurs at a much higher rate compared to dry oxidation. Depending on the temperature and existing thickness of oxide present, the overall oxidation rate can be either diffusion limited (e.g., thick oxides at high temperatures) or reaction-rate limited (e.g., thin oxides at low temperatures). Practically speaking, oxide thicknesses are limited to less than a few thousand angstroms and to <1 μm for dry and wet oxidation, respectively.

In a modern fabrication facility, oxidation occurs in either a tube furnace or in a rapid thermal-processing (RTP) tool, as shown schematically in Figure 12.3. A tube furnace consists of a quartz tube surrounded by heating element coils. The wafers are loaded into the heated tube, where oxidants can be introduced through inlets. The function of the RTP is similar, except that the thermal source is heating lamps.

12.1.3 Doping Processes

Controlled introduction of dopant impurities into silicon is necessary to affect majority carrier type, carrier concentration, carrier mobility, carrier lifetime, and internal electric fields. Common n-type dopants in Si are P, As, and Sb. The common p-type dopant in Si is B. The two primary methods of dopant introduction are solid-state diffusion and ion implantation. Historically, solid-state diffusion has been an important doping process; however, ion implantation is the preferred method in modern CMOS fabrication.

12.1.3.1 Ion Implantation

The workhorse method of introducing dopants into the near-surface region of wafers is a process called ion implantation. In ion implantation, dopant atoms (or molecules) are ionized, then accelerated through a large

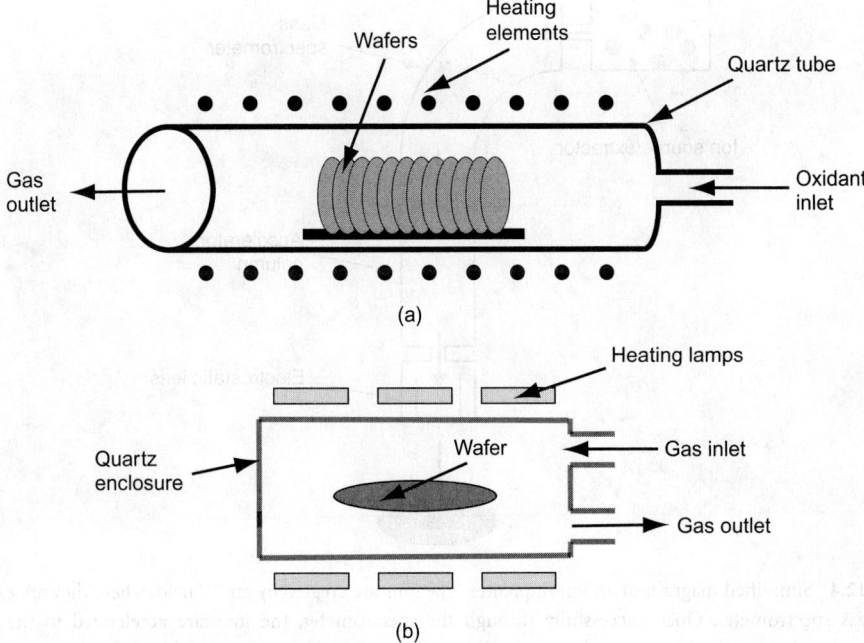

FIGURE 12.3 Simplified schematic representation of (a) an oxidation tube furnace, (b) a rapid thermal anneal tool.

electric potential (a few kilovolts to megavolts) toward a wafer. The highly energetic ions bombard and thus implant into its surface. Obviously, this process leads to a high degree of lattice damage, which is generally repaired by annealing at high temperatures. Because the ions do not necessarily come to rest at a lattice site, an anneal is required to electrically activate the dopant impurities by thermally agitating them into lattice sites.

Figure 12.4 shows a schematic diagram of an ion implanter. The ions are generated by an RF field in the ion source, from which they are subsequently extracted to a mass spectrometer. The spectrometer allows only ions with a user-selected mass to enter the accelerator, where the ions are passed through a large potential field. The ions are then scanned via an electrostatic lens across the surface of the wafer.

A first-order model for an implant doping profile is given by a Gaussian distribution described mathematically as

$$N(x) = N_p \exp[-(x - R_p)^2 / 2\Delta R_p^2] \tag{12.5}$$

where N_p is the peak concentration, R_p the projected range, and ΔR_p the straggle. By inspection, R_p should be identified as the mean distance the ions travel into the silicon and ΔR_p as the associated standard deviation. Figure 12.5 illustrates a typical ion implant profile. Obviously, N_p occurs at a depth of R_p. Moreover, the area under the implant curve corresponds to what is referred to as the implant dose Q_{imp}, given mathematically as

$$Q_{imp} = \int_{\infty}^{0} N(x)\, dx \tag{12.6}$$

Localized implantation is achieved by masking-off regions of the wafer with an appropriately thick material such as oxide, silicon nitride, polysilicon, or photoresist. Since implantation will occur in the masking layer, the thickness must be of sufficient magnitude to stop the ions before they reach the silicon

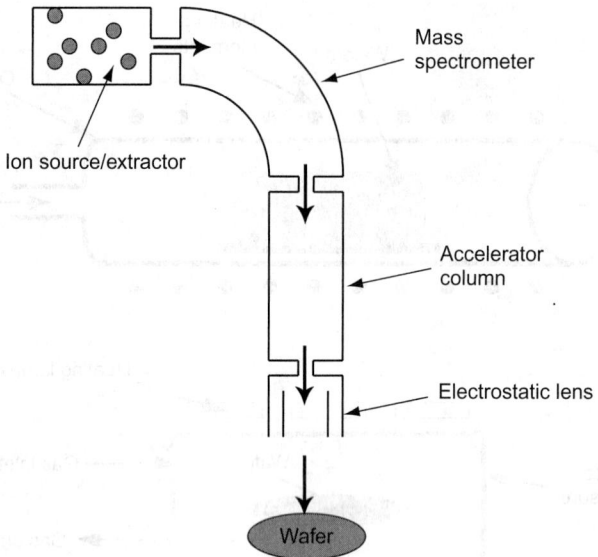

FIGURE 12.4 Simplified diagram of an ion implanter. The ions are created by an RF field, where they are extracted into a mass spectrometer. Once successfully through the spectrometer, the ions are accelerated to the desired magnitude. An electrostatic lens scans the ion beam on the surface of a wafer to achieve the appropriate dose. Electrostatically, the ions can be counted where an integrator can provide the real-time dose.

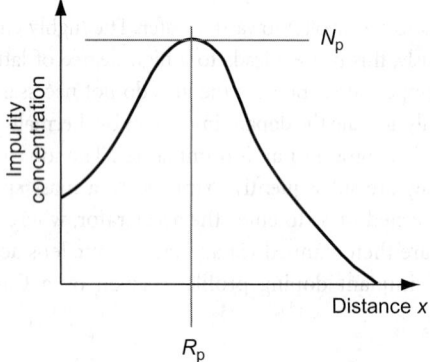

FIGURE 12.5 Ideal implant profile representing Eq. (12.5). The mean depth of the implant is R_p occurring at the N_p. The standard deviation of the distribution is given by ΔR_p. Notice that the peak concentration occurs subsurface and depends on the implant energy.

substrate. Compared with solid-state diffusion, ion implantation has the advantages of being a low-temperature and a highly controlled process.

12.1.3.2 Solid-State Diffusion

Solid-state diffusion is a method of introducing and redistributing dopants. In this section, we will study solid-state diffusion primarily to gain insight into "parasitic" dopant redistribution during thermal processes. In typical CMOS process flows, dopants are introduced via ion implantation into localized regions, which are subsequently processed at high temperatures. Solid-state diffusion inherently occurs in these high-temperature steps, thus spreading out the implant profile in three dimensions. The net effect is to shift the boundary of the implant from its original implant-defined position,

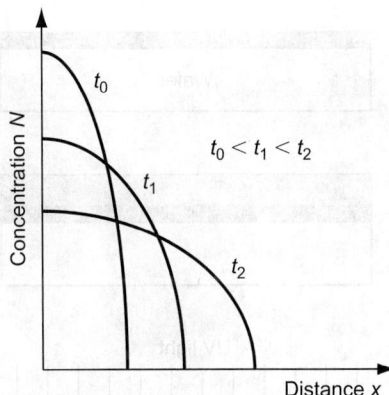

FIGURE 12.6 Idealized limit-source diffusion profile showing the effects of drive-in time on the profile. Notice that $t_0 < t_1 < t_2$, and the peak concentration is at the surface of the substrate ($x = 0$). The area under each curve is equal.

both laterally and vertically. This thermal smearing of the implant profiles must be accounted for during CMOS process flow development. If not, the final device characteristics can differ significantly from those expected.

Solid-state diffusion (or simply diffusion) requires two conditions: a dopant concentration gradient and thermal energy. Diffusion is directly proportional to both. An implanted profile (approximated by a delta function at the surface of the wafer) diffuses to first-order as

$$N(x, t) = [Q_{\mathrm{imp}}]/[(\pi Dt)^{1/2} \exp(-x^2/4Dt)] \qquad (12.7)$$

where Q_{imp} is the implant dose, D the diffusivity of the dopant, and t the diffusion time. Figure 12.6 illustrates this so-called limited-source diffusion of a one-dimensional implant profile. Notice that the areas under the respective curves for a given time are equal.

12.1.4 Photolithography

In the fabrication of CMOS it is necessary to localize processing effects to form a multitude of features simultaneously on the surface of the wafer. The collection of processes that accomplishes this important task—using ultraviolet light, a photomask, and a light-sensitive chemical resistant polymer—is called photolithography. Although many different categories of photolithography exist, they all share the same basic processing steps resulting in micron-to-submicron features generated in the light-sensitive polymer called photoresist. The photoresist patterns can then serve as ion implantation masks and etch masks during subsequent processing steps.

Figure 12.7 outlines the major steps required to implement photolithography patterning of a thermally grown oxide. Photoresist, a viscous liquid polymer, is applied to the top surface of the oxidized wafer. The application typically occurs by dropping (or spraying) a small volume of photoresist onto a rapidly rotating wafer, yielding a uniformly thin film on the surface. Following spinning, the coated wafer is softbaked on a hot plate, which drives out most solvents from the photoresist and improves adhesion to the underlying substrate. Next, the wafers are exposed to ultraviolet light through a mask (or reticle) that contains the layout patterns for a given drawn layer. The three general methods used to expose photoresist are contact, proximity, and projection photolithography.

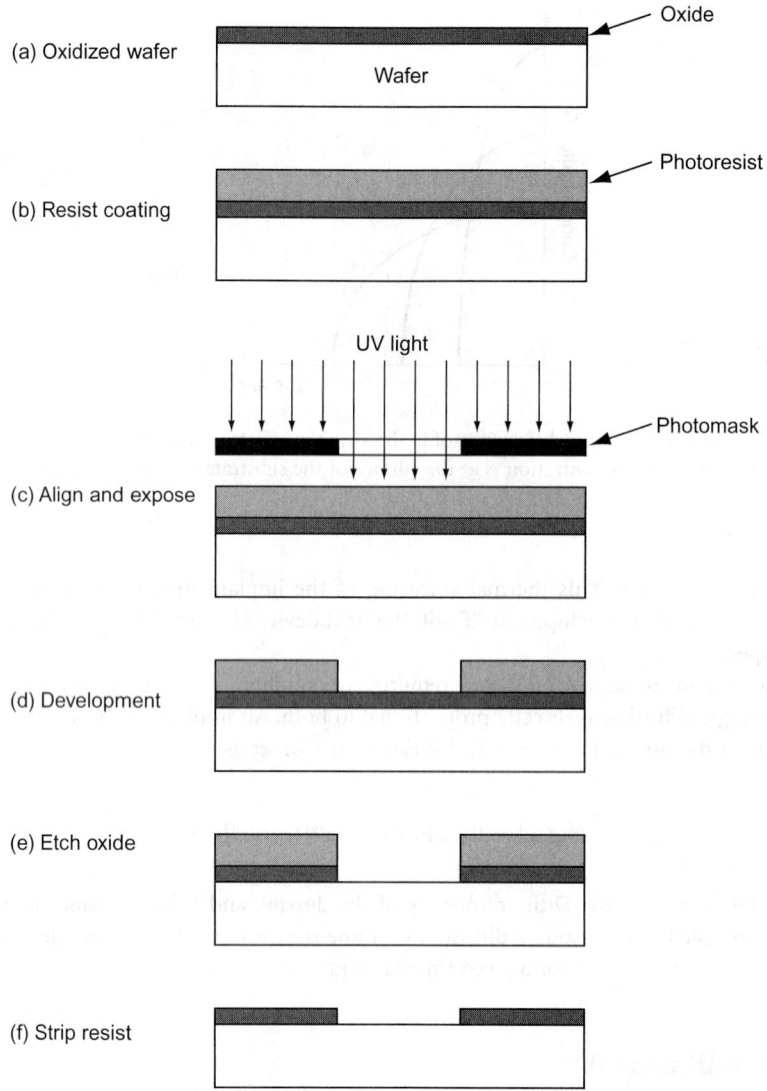

FIGURE 12.7 Simplified representation of the primary steps required for the implementation of photolithography and pattern transfer.

In both contact and proximity photolithography, the mask and the wafers are in contact and in close proximity, respectively, to the surface of the photoresist. Here the mask features are of the same scale as the features to be exposed on the surface. In projection photolithography, which is the dominant type of patterning technology, the mask features are drawn at a larger scale (e.g., 5× or 10×) relative to the features exposed on the surface. This is accomplished with a projection stepper that uses reduction optics to project an image of the mask onto the photoresist surface. For positive-tone photoresist, the ultraviolet light breaks molecular bonds, making the exposed regions more soluble. For negative-tone photoresist, exposure causes polymerization and thus less solubility. The exposed resist-coated wafer is developed in an alkaline solution. Whether a positive image or negative image relative to the mask patterns is generated depends on the formulation of the photoresist. To harden the photoresist for improved etch resistance and to improve adhesion, the newly developed wafers often are hardbaked. At this point, the wafer can be etched to transfer the photoresist pattern into the underlying thin film. Etch processes will be discussed in the next section.

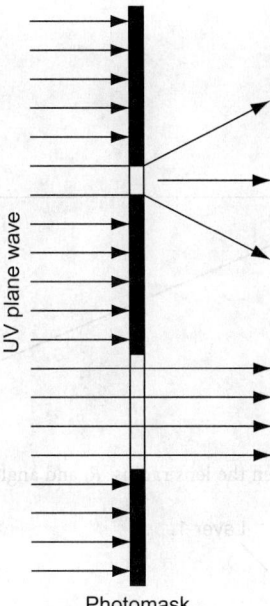

FIGURE 12.8 Diffraction of light. The diffraction effects become significant as the mask feature dimensions approach the wavelength of ultraviolet light. Notice that the diffraction angle is larger for the smaller opening than for the larger opening.

12.1.4.1 Resolution

In general, a given projection stepper has three critical parameters: resolution, depth of focus (DOF), and pattern registration. The diffraction of light caused by the various interfaces in its path limits the minimum printable feature size (Figure 12.8). Resolution is defined as the minimum feature size, M, that can be printed on the surface of the wafer given by

$$M = (c_1 \lambda)/\text{NA} \tag{12.8}$$

where λ is the wavelength of the ultraviolet light source, NA the numerical aperture of the projection lens, and c_1 a constant whose value ranges from 0.5 to 1. The NA of a lens is illustrated in Figure 12.9 and is given mathematically as

$$\text{NA} = n \sin \theta \tag{12.9}$$

where n is the index of refraction of the space between the wafer and the lens and θ the acute angle between the focal point on the surface of the wafer and the edge of the lens radius. Notice that M is directly proportional to wavelength, hence diffraction effects are the primary limitation in printable feature size. To a limit, the NA of the projection lens can be increased to help combat diffraction effects because large NA optics have an increased ability to capture diffracted light.

12.1.4.2 DOF

DOF of the projection optics limits the ability to pattern features at different heights. Mathematically, DOF is given by

$$\text{DOF} = (c_2 \lambda)/(\text{NA}^2) \tag{12.10}$$

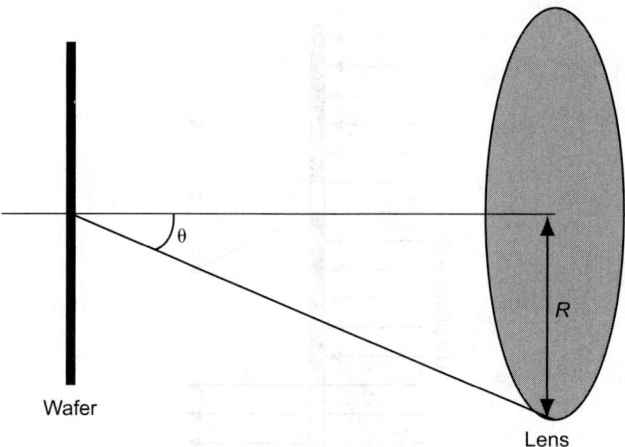

FIGURE 12.9 Relationship between the lens radius, R, and angle θ used for computing the NA.

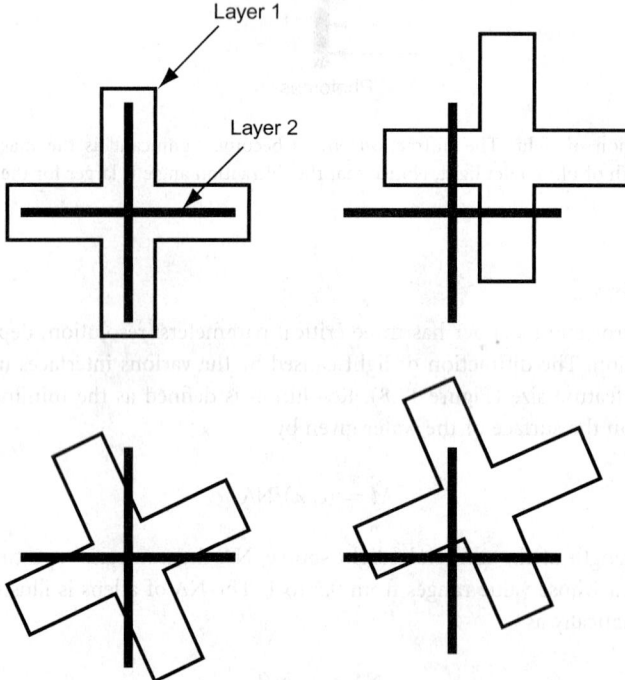

FIGURE 12.10 Simple registration errors that can occur during wafer-to-mask alignment in photolithography: (a) no registration error; (b) x–y registration error; (c) z-rotation registration error; (d) x–y and z-rotation error. Other registration errors exist, but are not discussed here.

where c_2 is a constant ranging in value from 0.5 to 1. As is apparent from Eq. (12.8) and Eq. (12.10), a fundamental trade-off exists between minimum feature size and DOF. In other words, to print the smallest possible features, the surface topography must be minimized (often achieved through the use of CMP as discussed in the next section).

12.1.4.3 Aligning Masks

During CMOS fabrication numerous mask levels (e.g., active, poly, and contacts) are printed on the wafer. Each of these levels must be accurately aligned to one another. Registration is a measure of the level-to-level alignment error. Registration errors occur in x-, y-, and z-rotations, as illustrated in Figure 12.10.

12.1.4.4   Emerging Lithographic Technologies

At first glance, it would seem that the minimum feature size could not be less than the wavelength of the light. However, advanced techniques such as optical proximity correction (OPC) and wavefront engineering of photomasks have been developed to push the resolution limits below the exposure wavelength.

There is significant effort to develop lithographic techniques that can pattern features with dimensions measured in the nanometer regimen. Most of these methods employ the use of small-wavelength light sources in the extreme ultraviolet to x-ray wavelengths. Extreme ultraviolet (EUV) lithography is one of the leading candidates for producing features below 50 nm.

12.1.5   Thin-Film Removal

After photolithography, usually one of the two processes is performed: thin-film etching, which is used to transfer the photoresist patterns to the underlying thin film(s) and ion implantation, which employs the photoresist patterns to block the dopants from selected regions of the wafer surface. Here, we discuss thin-film etching processes based on wet chemical etching and dry-etching techniques. We also discuss the CMP process that is used to remove unpatterned thin films.

12.1.5.1   Thin-Film Etching

After a photoresist pattern is generated, either wet or dry etching is commonly used to transfer patterns into underlying films. Etch rate (thickness removed per unit time), selectivity, and degree of anisotropy are key parameters for both wet and dry etching. Etch rates are typically a strong function of solution (or gas) concentrations and temperature. Selectivity, S is defined as the etch rate ratio of one material to another, given by the selectivity equation

$$S = R_2/R_1 \tag{12.11}$$

where R_2 is the etch rate of the material intended to be removed and R_1 the etch rate of the underlying, masking, or adjacent material that will remain. The degree of anisotropy, A_f, is a measure of how rapidly an etchant will remove material in different directions and is mathematically given by

$$A_f = 1 - (R_l/R_v) \tag{12.12}$$

where R_l is the lateral etch rate and R_v the vertical etch rate. If $A_f = 1$, the etchant is completely anisotropic. However, if $A_f = 0$, the etchant is completely isotropic. In photolithography, the degree of anisotropy is a major factor in the achievable resolution. Figure 12.11 illustrates the effects of etch bias (i.e., $d_{film} - d_{mask}$) on the final feature size. For the submicron features that are required in CMOS, dry-etch techniques are preferred over wet-etch processes because they generally have a higher degree of anisotropy. Both wet and dry etching methods are applied to the removal of metals, semiconductors, and insulators.

12.1.5.2   Wet Etching

In wet etching, a chemical solution is used to remove material. In CMOS fabrication, wet processes are used for cleaning wafers and for thin-film removal. Wet-cleaning processes are repeated numerous times throughout a process flow. Some are targeted toward removal of particulate; others are intended to remove organic and inorganic surface contaminants. Wet etchants can be isotropic (i.e., etch rate is the same in all directions), or anisotropic (i.e., etch rate differs in different directions). However, most of the wet etchants used in CMOS fabrication are isotropic. Wet etchants generally tend to be highly selective compared with the dry-etch processes. To improve the etch uniformity and to aid in the removal of particulate, it is common to ultrasonically vibrate the etchant and use microcontrollers to accurately control the temperature of the bath. Once an etch is completed, the wafers are rinsed in deionized water and then spun dry.

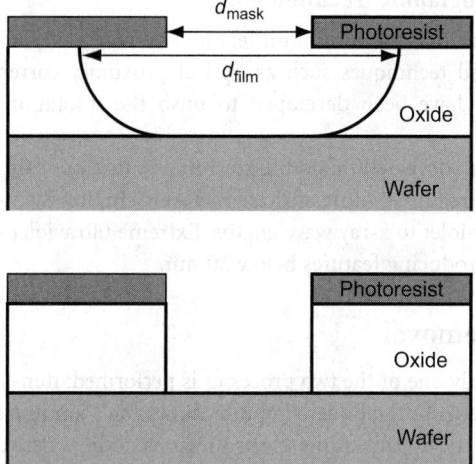

FIGURE 12.11 Diagram showing post-etch profiles: (a) isotropic etch profile; (b) anisotropic etch profile. Notice for (a), that because of isotropy in the etch process the mask opening, d_{mask} does not match the etched opening in the underlying film. The difference between these dimensions is called etch bias.

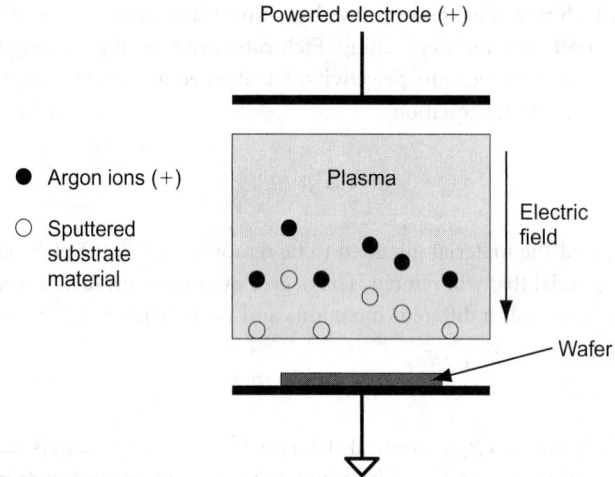

FIGURE 12.12 Simplified schematic diagram of the sputter-etch process. This process is dominated by the physical bombardment of ions on a substrate.

12.1.5.3 Dry Etching

The three general categories of dry-etch techniques in CMOS fabrication are sputter etching, plasma etching, and reactive-ion etching (RIE). Figure 12.12 schematically illustrates a sputter etch process. An inert gas (e.g., argon) is ionized, where the ions are accelerated through an electric field established between two conductive electrodes, called the anode and the cathode. A vacuum in the range of milliTorr must exist between the plates to allow the appropriate ionization and transfer of ions. Under these conditions, a glow discharge or plasma is formed between the electrodes. The plasma consists of positively charged ions and electrons that respond in opposition to the electric field. The wafer that is placed on the cathode is bombarded by positively charged ions that cause material to be ejected off the surface. Essentially, sputter etching is atomic-scale sandblasting. A DC power supply can be used for sputter etching of conductive substrates, while an RF supply is required through capacitive coupling to etch nonconductive substrates. Sputter etching does not tend to be selective, but is very anisotropic.

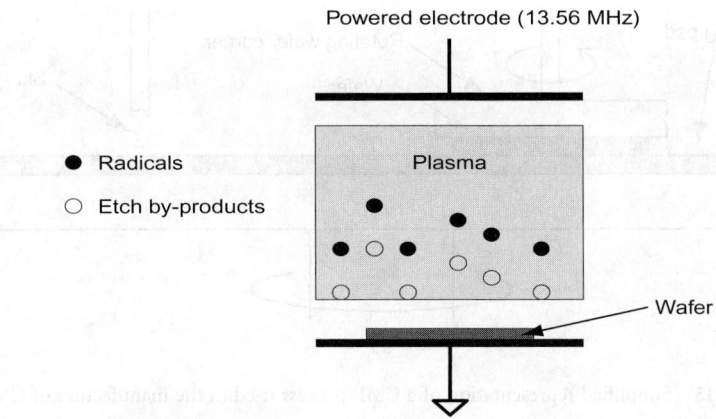

FIGURE 12.13 Simplified schematic diagram of a plasma etch process. This process is dominated by the chemical reactions of radicals at the surface of a substrate.

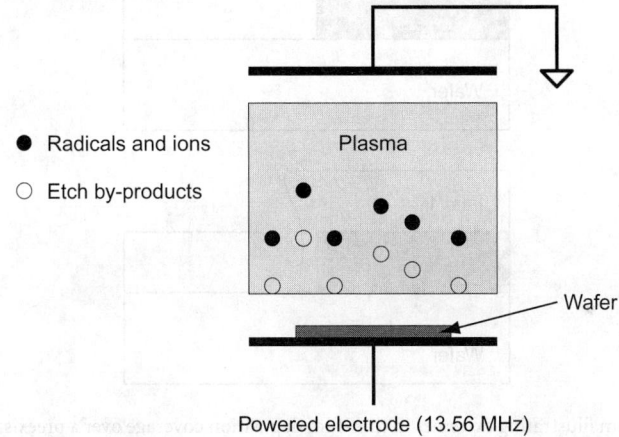

FIGURE 12.14 Simplified schematic diagram of an RIE process. This process has both physical (ion bombardment) and chemical (radicals) components.

A simplified diagram of a plasma etch system is shown in Figure 12.13. A gas or a mixture of gases (e.g., halogens) is ionized, producing reactive species called radicals. A glow discharge or plasma is formed between the electrodes. The radicals chemically react with the surface material, where the reaction products are in the gas phase and are pumped away through a vacuum system. Plasma etching can be very selective, but is typically highly isotropic.

While sputter etching is a purely mechanical process and plasma etching a purely chemical one, RIE is a combination of sputter etching and plasma etching as shown schematically in Figure 12.14. In RIE, a gas or mixture of gases (e.g., fluorocarbons) is ionized, generating radicals and ionized species, both of which interact with the surface of the wafer. In CMOS fabrication, RIE is the dominant etch process because it can provide the benefits of both sputter etching and plasma etching: it can be highly selective *and* highly anisotropic.

12.1.5.4 CMP

Figure 12.15 depicts the key features of CMP. In CMP, an abrasive chemical solution, called a slurry, is introduced between a polishing pad and the wafer. Material on the surface of the wafer is removed by both a mechanical polishing component and a chemical reaction component. In modern CMOS fabrication, CMP is a critical process that planarizes the surface of the wafer prior to photolithography. The planar surface allows the printed feature size to be decreased. CMP can remove metals, semiconductors, and insulators.

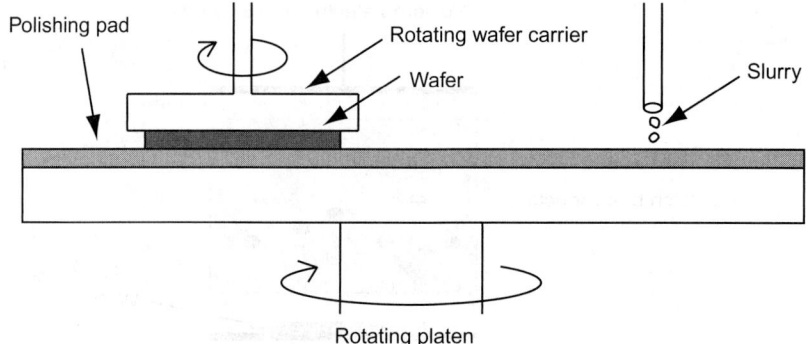

FIGURE 12.15 Simplified representation of a CMP process used in the manufacture of CMOS.

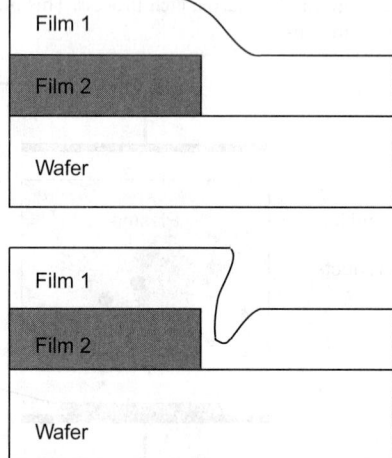

FIGURE 12.16 Diagram illustrating extremes in thin-film deposition coverage over a preexisting film step: (a) good step coverage; (b) poor step coverage.

12.1.6 Thin-Film Deposition

CMOS ICs require insulators, conductors, and semiconductors. Examples of the latter include crystalline silicon for active areas and polycrystalline silicon for gate electrodes or local area interconnects. Gate dielectrics, device isolation, metal-to-substrate isolation, metal-to-metal isolation, passivation, etch masks, implantation masks, diffusion barriers, and sidewall spacers may use insulators such as Si_3N_4, SiO_2, or doped glasses. Conductors—such as aluminum, copper, cobalt, titanium, tungsten, and titanium nitride—are employed for local interconnects, contacts, vias, diffusion barriers, global interconnects, and bond pads.

In this section, we discuss methods of depositing thin films of insulators, conductors, and semiconductors. The two primary categories of thin-film deposition are physical vapor deposition (PVD) and chemical vapor deposition (CVD). Electrodeposition, used to deposit copper for back-end interconnects, is less common and will be briefly introduced.

Deposited films may be characterized by several factors. Of prime importance are inherent film quality as it relates to compositional control, low contamination levels, low defect density, and predictable and stable electrical and mechanical properties. Moreover, film thickness uniformity must be understood and controlled to high levels. To achieve highly uniform CMOS parameters across a wafer, it is common to have to control the film thickness uniformity to $<\pm5$ nm across the surface of the wafer. In addition, film uniformity over topographical features—called step coverage—is of critical importance. As depicted in Figure 12.16, good step coverage results in uniform thickness over all surfaces. In contrast, poor step coverage results in significantly less thickness on vertical surfaces relative to surfaces parallel to the surface of the wafer.

Related to step coverage is what is referred to as "gap fill". This applies to depositing material into a high-aspect ratio opening such as contacts or gaps between adjacent metal lines. Figure 12.17 illustrates a deposition with good gap fill and a deposition that yields a poor one (also called a "keyhole" or void).

12.1.6.1 PVD

In PVD, physical processes produce the constituent atoms (or molecules) that pass through a low-pressure gas phase and subsequently condense on the surface of the substrate. Common PVD processes are evaporation and sputter deposition. Either can be used to deposit a wide range of insulating, conductive, and semi-conductive materials. However, one drawback of PVD is that the resultant films often have poor step coverage.

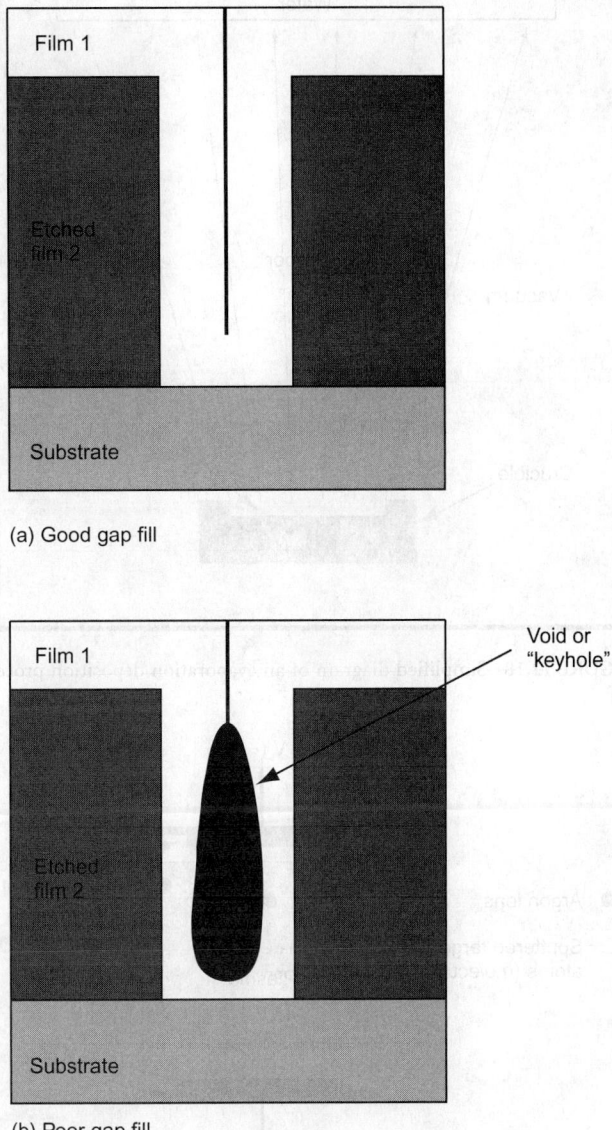

FIGURE 12.17 Diagram illustrating gap-fill profiles (good and bad) of a high-aspect ratio opening filled with a deposited film.

Evaporation is one of the oldest methods of depositing thin films of metals, insulators, and semiconductors. The basic process of evaporation is shown in Figure 12.18. The material to be deposited is heated beyond its melting point in a high vacuum chamber, where the vapor form of the material coats all surfaces exposed within the mean free path of the evaporant. The heat source can be either a heating filament or a focused electron beam.

In simple terms, sputter deposition is similar to sputter etching as discussed in Section 12.1.5, except that the wafer serves as the anode and the cathode is a target material to be deposited. Figure 12.19 outlines a simplified sputter deposition process. An inert gas such as argon is ionized in a low-

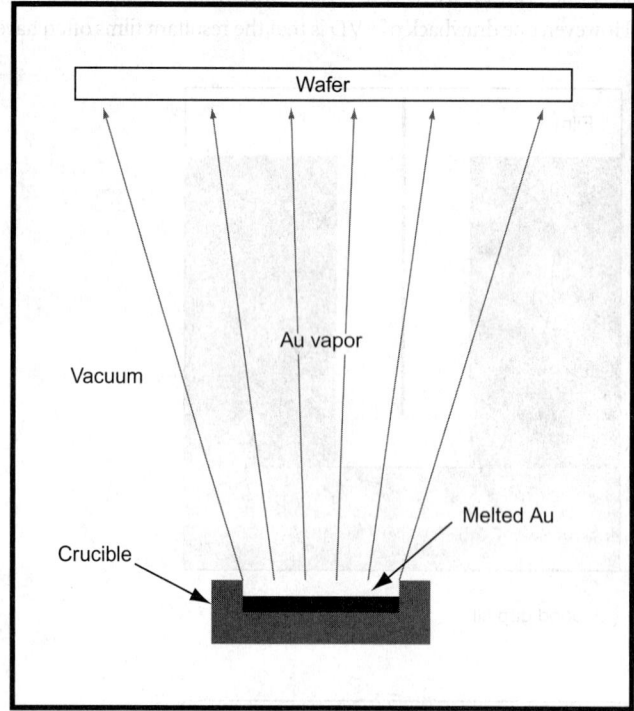

FIGURE 12.18 Simplified diagram of an evaporation deposition process.

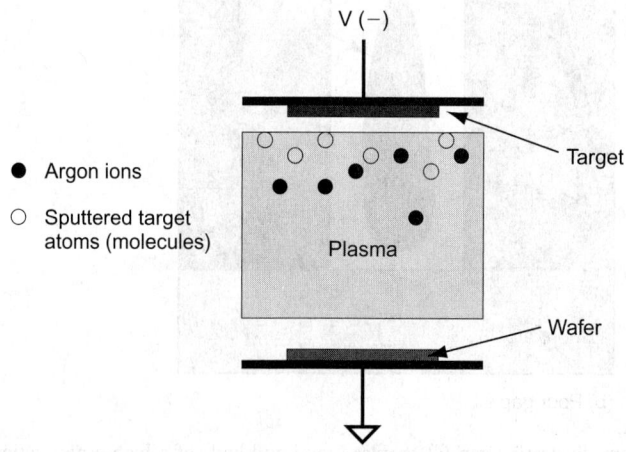

FIGURE 12.19 Simplified diagram of a sputter deposition process.

pressure ambient, where the positively charged ions are accelerated through the electric field toward the target—an ultrahigh-purity disk of material to be deposited. The bombardment of the target with the ions sputters (or ejects) target atoms (or molecules), resulting in their transit to the wafer surface to form a thin film. Sputter deposition, similar to sputter etching, requires a DC supply for conductors; however, a capacitively coupled RF supply must be used to deposit nonconductive materials.

12.1.6.2 CVD

In CVD, reactant gases are introduced into a chamber where chemical reactions between the gases at the surface of the substrate produce a desired film. Common CVD processes are atmospheric pressure (APCVD), low-pressure (LPCVD), and plasma-enhanced (PECVD). Again, a wide variety of insulators, conductors, and semiconductors can be deposited by CVD. Most importantly, the resultant films tend to have better step coverage than those deposited by PVD processes. APCVD occurs at relatively low temperatures in an apparatus similar to an oxidation tube furnace, where an appropriate reactive gas is passed over the wafers (see Figure 12.3). As depicted in Figure 12.20, LPCVD occurs in a reactor in the pressure range of milliTorr to a few Torr. Compared to APCVD, the low-pressure process can yield highly conformal films, but at the expense of a higher deposition temperature. Figure 12.21 shows a

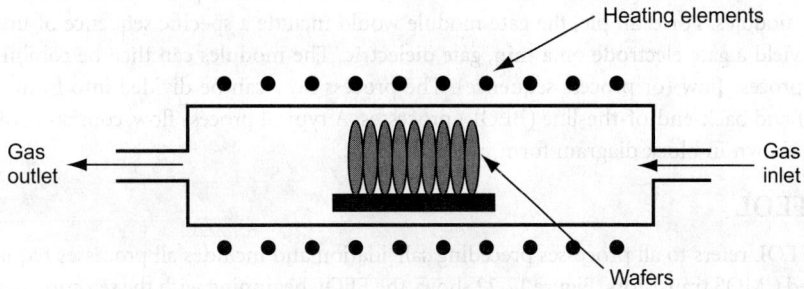

FIGURE 12.20 Simplified schematic diagram of an LPCVD reactor.

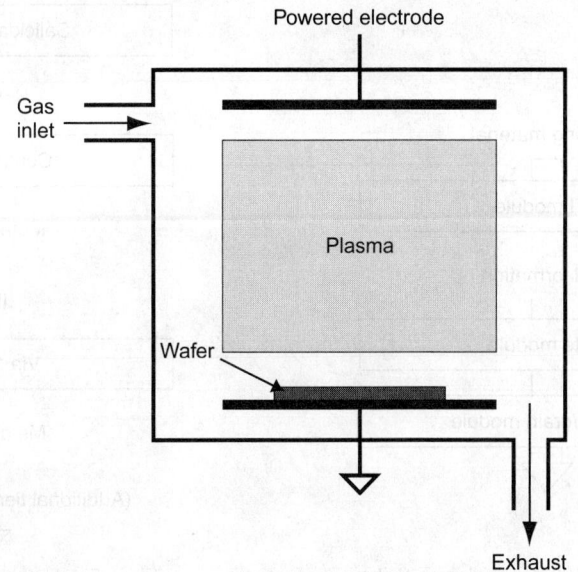

FIGURE 12.21 Simplified schematic diagram of a PECVD reactor.

schematic diagram of a PECVD reactor. In PECVD, a plasma is used to impart energy for the surface reactions, hence allowing for lower temperature deposition. In comparison, PECVD has the advantage of being a low-temperature *and* a highly conformal process. It should be noted that during epitaxial growth of crystalline semiconductors (e.g., Si or $Si_{1-x}Ge_x$) graded layers are often deposited using a CVD process.

12.1.6.3 Electroplating

In modern copper (Cu) backends, electroplating is used to deposit a conformal film of Cu on the conductive regions of the substrate. Electroplating is a method where the wafer is immersed in a bath of an electrolyte (often a salt solution) and an electrical current is driven through the solution from an electrode to the wafer (also serving as an electrode). This process causes a reaction to occur precipitates the metal (e.g., Cu) out of the electrolyte to the surface of the substrate.

12.2 CMOS Process Integration

Process integration is the task of combining a deliberate sequence of unit processes to fabricate integrated microelectronic devices such as PMOS transistors, NMOS transistors, resistors, capacitors, and diodes. A typical CMOS technology consists of a complex arrangement of unit processes in which several hundred steps are required to manufacture ICs on a silicon wafer. Groups of unit processes are combined to form integration modules. For example, the gate module would include a specific sequence of unit processes that would yield a gate electrode on a thin, gate dielectric. The modules can then be combined to yield the overall process flow (or process sequence). The process flow can be divided into front-end-of-the-line (FEOL) and back-end-of-the-line (BEOL) processes. A typical process flow, consisting of numerous modules, is shown in block diagram form in Figure 12.22.

12.2.1 FEOL

Generally, FEOL refers to all processes preceding salicidation and includes all processes required to fully form isolated CMOS transistors. Figure 12.22 shows the FEOL beginning with the selection of the starting

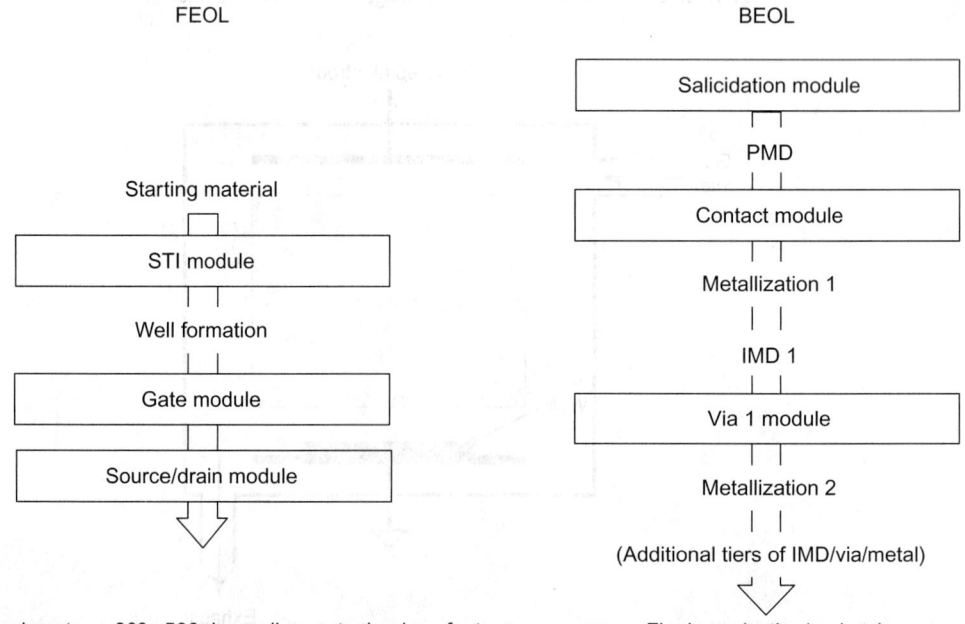

FIGURE 12.22 Typical CMOS process flow, illustrating the difference between FEOL and BEOL processes.

material (i.e., type of silicon wafer to be used). Then the shallow-trench isolation (STI) module forms regions of dielectric between regions of active area. Next, the wells (or tubs) are formed, followed by the gate module, which includes all processes to properly define gate electrodes on a thin oxide. The FEOL concludes with the source/drain module. Its processes form low-doped drain (LDD) extensions and the source/drain regions themselves.

Major differences in the design of the FEOL are often due to the specific technology requirements. For instance, high-speed logic is often implemented with the selection of silicon-on-insulator (SOI) wafers or an SiGe channel structure. The SOI process generally yields devices with lower capacitance than does a comparable bulk technology. Whereas, SiGe technologies are fabricated using an epitaxtial growth of a strained channel that yield higher electron (and hole) mobilities. Regardless of the specific type of FEOL technology, the overall process flow is similar in most cases.

12.2.2 BEOL

All processes subsequent to source/drain formation fall within the BEOL. Hence, BEOL processes "wire" transistors together using multiple layers of dielectrics and metals. The BEOL begins with the salicidation of the polysilicon and source/drain regions. The remaining BEOL processes proceed in repetitive sets of modules to yield lateral and vertical interconnects isolated from one another with dielectrics. Conventionally, aluminum (Al) metallization and tungsten (W) contact and via plugs are used in the BEOL. Higher performance (i.e., lower resistance and lower capacitance) BEOL are implemented using a Cu dual-Damascene/low-k dielectrics for the respective conducting and insulating materials.

It is important to understand that there is a high degree of interrelationship between unit processes within each module and between modules themselves. A seemingly "trivial" change in one unit process in a given module can have dramatic effects on processes in other modules. In other words, there is no such thing as a trivial process change.

12.2.3 CMOS Process Description

Even with a single-device type, process schemes vary with respect to achieving similar structures, so it would be virtually impossible to outline all schemes. However, an example of a generic (but representative), deep-submicron CMOS process flow is presented ("deep" indicates that a deep- or short- wavelength ultraviolet light source is used when patterning the wafers). Our generic CMOS technology will have the following features:

1. FEOL
   (a) STI
   (b) Twin tubs
   (c) Single-level polysilicon
   (d) LDD extensions

2. BEOL
   (a) Fully planarized dielectrics
   (b) Planarized W contacts and via plugs
   (c) Al metallization

Following each major process step, cross sections will be shown. The cross sections were generated with a technology computer-assisted design (TCAD) package called Tsuprem-4 and Taurus-Visual (2D), released by Technology Modeling Associates, Inc. These tools simulate a defined sequence of unit processes, allowing a process to be modeled before actually being implemented in a fabrication facility. For brevity, there are several omissions and consolidations in the process description. These include:

1. Wafer cleaning performed immediately before all thermal processes, metal depositions, and after photoresist removal.

TABLE 12.1   Masks Used in Our Generic CMOS Process

Layer Name	Mask	Aligns to Level	Times Used	Purpose
1 (Active)	Clear	Aligns to notch	1	Defines active areas
2 (p well)	Clear	1	2	Defines NMOS sidewall implants and p-well
3 (n well)	Dark	1	2	Defines PMOS sidewall implants and n-well
4 (poly1)	Clear	1	1	Defines polysilicon
5 (n select)	Dark	1	2	Defines nLDD and n^+
6 (p select)	Dark	1	2	Defines pLDD and P
7 (Contact)	Dark	4	1	Defines contact to poly and active areas
8 (Metal1)	Clear	7	1	Defines metal1
9 (Via1)	Dark	8	1	Defines via1 (connects M1 to M2)
10 (Metal2)	Clear	9	1	Defines metal2
Passivation	Dark	Top-level metal	1	Defines bond pad opening in passivation

2. Individual photolithographic process steps (such as dehydration bake, wafer priming, photoresist application, softbake, alignment, exposure, photoresist development, hardbake, inspection, registration measurement, and critical dimension [CD] measurement).
3. Backside film removal following select CVD processes.
4. Metrology performed to measure particle levels and film thickness, and post-etch CDs.

To implement our CMOS technology we will employ the use of a reticle set as outlined in Table 12.1. The masks are labeled as having either a clear or dark-field. Clear-field masks contain opaque features totaling <50% of the mask area. In contrast, dark-field masks have opaque features accounting for >50% of the total area. Using this mask set, we will assume the exclusive use of positive-tone photoresist processing. When appropriate, representative mask features will be shown that will yield isolated, complementary transistors adjacent to one another.

12.2.4   FEOL Integration

As stated previously, the FEOL encompasses all processing required to fabricate the fully formed, isolated CMOS transistors. This subsection discusses the modules and unit processes required for a representative CMOS process flow.

12.2.4.1   Starting Material

The choice of substrate is strongly influenced by the application and characteristics of the CMOS ICs to be fabricated. Bulk silicon is the least expensive, but may not be the optimal choice in high-performance or harsh-environment CMOS applications. Epitaxial (epi) wafers are heavily doped bulk wafers with a thin, moderately to lightly doped epitaxial silicon layer grown on the surface. The primary advantage of epiwafers is for immunity to latch-up. SOI wafers provide a performance increase and elimination of latch up. However, SOI CMOS is more costly to implement than the bulk or epitechnologies. The three general types of silicon substrates are shown in Figure 12.23.

In current manufacturing, wafer diameters typically range from 150 to 300 mm. Wafer thickness increases with diameter to allow for greater rigidity. The actual CMOS is constructed in the top 1 μm or less of the wafer; the remaining hundreds of micrometers are used solely for mechanical support during device fabrication.

We will use bulk silicon wafers for our CMOS technology in this section. It should be noted that, with relatively minor process and integration adjustments, our technology could be applicable to the epi or

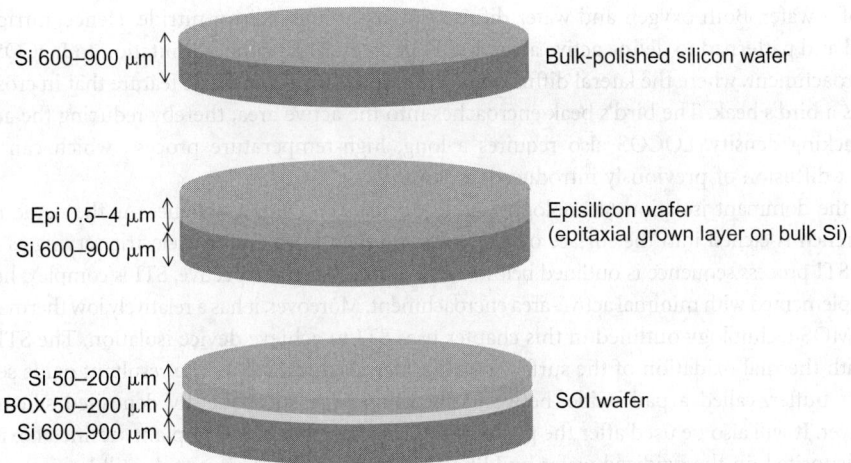

FIGURE 12.23 Three general types of silicon wafers used for CMOS fabrication.

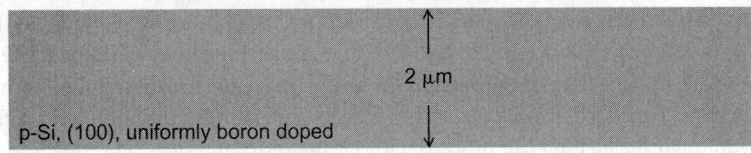

FIGURE 12.24 Simulated cross-sectional view (the top 2 μm) of the bulk wafer in Figure 12.23.

SOI CMOS processes. Figure 12.24, the first of many simulated cross sections, shows only the top 2 μm of silicon. At the beginning of the fabrication process, wafer characteristics such as resistivity, sheet resistance, crystallographic orientation, and bow-and-warp are measured and recorded. Each wafer receives its own number and lot assignment, usually inscribed with a laser.

12.2.4.2 STI-Module

Devices (e.g., PMOS and NMOS) must be electrically isolated from one another. This isolation is of primary importance. It suppresses current leakage between similar as well as dissimilar devices.

One of the simplest isolation methods is to fabricate the CMOS so that a reversed-bias pn junction is formed between the transistors. Oppositely doped regions (e.g., n well adjacent to a p well) can be electrically isolated by tying the n region to the most positive potential in the circuit and the p region to the most negative. As long as the reverse bias is maintained and the breakdown voltage is not exceeded for all operating conditions, only a small diode reverse-saturation current accounts for the leakage current. Because this junction leakage current is directly proportional to the size of the junction area, junction isolation alone is inadequate for the large p and n regions in modern devices.

The second general method of isolation is related to the formation of thick dielectric regions, called field regions, between transistors. The region without the thick dielectric is where the transistors reside. It is called the active area. The relatively thick oxide that forms between active areas is called field oxide (FOX). Polysilicon interconnections formed over the field regions provide localized electrical continuity between transistors. This arrangement inherently leads to the formation of parasitic field-effect transistors. The FOX increases the parasitic transistor's threshold voltage sufficiently so that the device always remains in the off state. The parasitic transistor's threshold voltage can be further enhanced by increasing the surface doping concentration, called channel stops, under the FOX. Two general approaches apply to the formation of FOX regions: local oxidation of silicon (LOCOS) and STI.

LOCOS has been used extensively for 0.5 μm or larger minimum linewidth CMOS technologies. In LOCOS, a diffusion barrier of silicon nitride blocks the thermal oxidation of specific regions on the surface of a wafer. Both oxygen and water diffuse slowly through silicon nitride. Hence, nitride can be deposited and patterned to define active areas and FOX areas. The primary limitation to LOCOS is bird's beak encroachment, where the lateral diffusion of the oxidant forms an oxide feature that in cross section resembles a bird's beak. The bird's beak encroaches into the active area, thereby reducing the achievable circuit-packing density. LOCOS also requires a long, high-temperature process, which can result in significant diffusion of previously introduced dopants.

STI is the dominant isolation technology for sub-0.5 μm CMOS technologies. As the name implies, a shallow trench is etched into the surface of the wafer and then filled with a dielectric serving as the FOX. A typical STI process sequence is outlined below. From a processing perspective, STI is complex; however, it can be implemented with minimal active-area encroachment. Moreover, it has a relatively low thermal budget.

The CMOS technology outlined in this chapter uses STI to achieve device isolation. The STI module begins with thermal oxidation of the surface of the wafer (Figure 12.25). The resultant oxide serves as a film-stress buffer, called a pad oxide, between silicon and the subsequently deposited silicon nitride (Si_3N_4) layer. It will also be used after the post-CMP nitride strip as an ion implant sacrificial oxide. Next, Si_3N_4 is deposited on the oxidized wafer by LPCVD (Figure 12.26). This nitride will later serve as both an implant mask and a CMP stop layer.

Photolithography (Figure 12.27, mask layer 1) is used to produce the appropriate patterns in photoresist to define the active areas. Then endpoint detected RIE transfers the photoresist pattern into the underlying film stack of nitride and oxide. Notice in Figure 12.27 that, under the photoresist, the PMOS and NMOS devices will be fabricated on the left and right side, respectively. The region cleared of photoresist corresponds to the isolation regions. Timed RIE forms 0.4 μm deep silicon trenches with the photoresist softmask present (Figure 12.28). Although the etching can proceed without the resist, the sidewall profile can be tailored to a specific slope with the presence of the polymer during the etch process. Sloped sidewalls can help reduce leakage current from parasitic corner transistors, although at the cost of reduced packing density. Following the silicon etch, O_2 plasma and wet processing strip the photoresist and etch by-products from the surface of the wafer. At this point, the general structural form of the STI is completed.

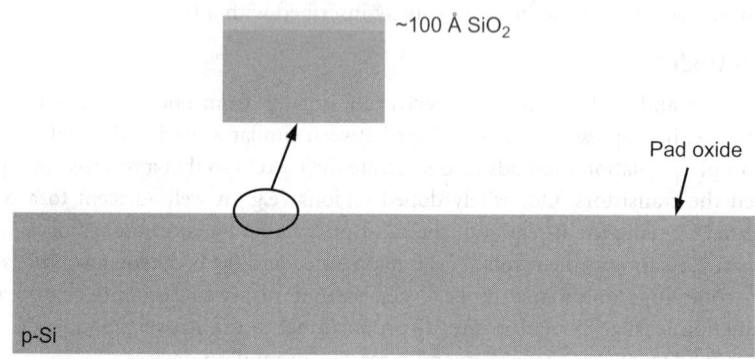

FIGURE 12.25   STI film stack. The oxide is thermally grown at approximately 900°C with dry O_2.

FIGURE 12.26   STI film stack. Silicon nitride deposited by LPCVD at approximately 800°C.

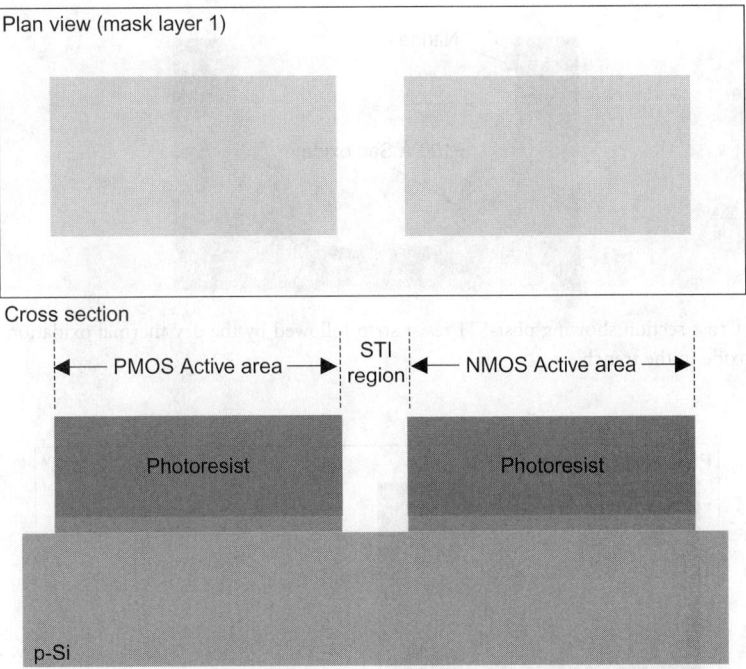

FIGURE 12.27   STI definition, photolithography, and nitride/pad oxide etch with fluorocarbon-based RIE.

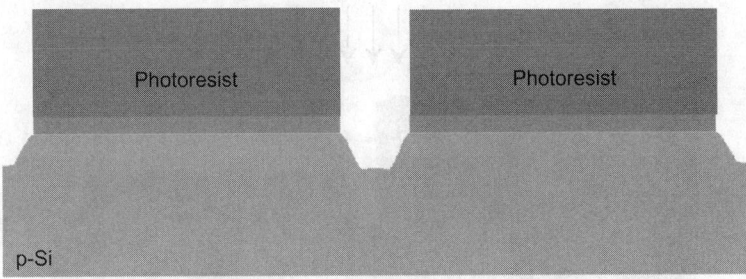

FIGURE 12.28   Timed RIE forming silicon trenches.

The subsequent series of processes improve the STI's effectiveness in suppressing leakage currents. The expanded view of the trench (Figure 12.29) shows a thin oxide that has been thermally grown on the exposed silicon. It should be noted that the nitride provides a barrier to the diffusion of oxygen, hence the oxidation occurs only in the silicon-exposed regions. This oxide, which aids in softening the corner of the trench, will be used as a sacrificial implant layer in the subsequent ion implantation. In general, such implant sacrificial oxides are used to suppress ion channeling in the crystal lattice, minimize lattice damage from the ion bombardment, and protect the silicon surface from contamination. Photolithography (mask layer 2) is used to pattern resist to protect the PMOS sides of the trench during the p-wall implant. A shallow boron fluoride (BF_2) implant is performed to dope what will eventually become the. p-well trench sidewalls (called the p walls) (Figure 12.30). The p-wall implant increases the threshold voltage of the parasitic corner transistor and minimizes leakage under the trench. Then the BF_2-implanted resist is stripped, using O_2 plasma and wet processing (Figure 12.31). Again, photolithography (Figure 12.32, mask layer 3) is used to produce the complementary pattern for the n-wall implant. For the same reasons, this shallow phosphorus implant is introduced into what will become the n-well trench sidewalls. Next, the phosphorus-implanted resist is stripped, using O_2 plasma and wet processing, yielding the structure depicted in Figure 12.33.

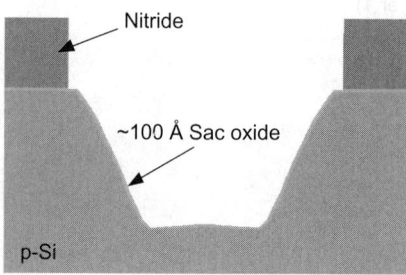

FIGURE 12.29 Cross section showing post-STI resist strip followed by the dry thermal oxidation (at 900°C) of a sacrificial (sac) oxide in the trench.

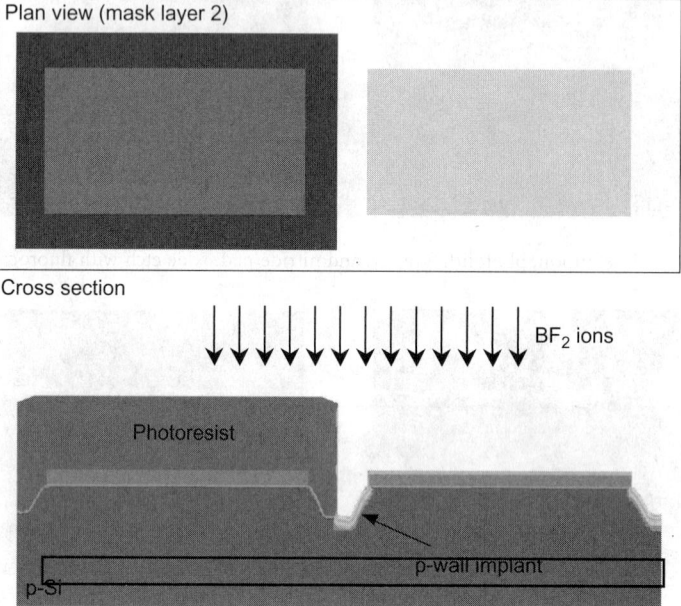

FIGURE 12.30 p-Wall sidewall formation via photolithography and BF_2 implantation.

At this point, the sacrificial oxide has been degraded by the implantations and is likewise stripped, using a buffered hydrofluoric acid (HF) solution. A thin, high-quality thermal oxide is regrown in the trenches to form what is called a trench liner. In general, the liner oxide improves the interface quality between the silicon and the subsequent trench fill, thus suppressing the interface leakage current. Specifically, the formation of the trench liner oxide "cleans" the surface prior to trench fill, anneals sidewall implant damage, and passivates interface states to minimize parasitic leakage paths.

Once the liner oxide is grown, CVD overfills the trenches with a dielectric (Figure 12.34). This trench ill provides the field isolation that is required to increase the threshold voltage of the parasitic field transistors. Further, it serves to block subsequent ion implants. Although, not shown, it is common to use a "block-out" pattern to improve the uniformity of the STI CMP. In Figure 12.35, CMP removes the CVD overfill. The nitride is used as a polish-stop layer. Next, a brief buffered oxide etch removes oxide that may have formed on top of the nitride. Then the nitride is removed from the active areas by using a wet- or dry-etch process (Figure 12.36). Note that the pad oxide remains after this step. At this point, the STI is fully formed.

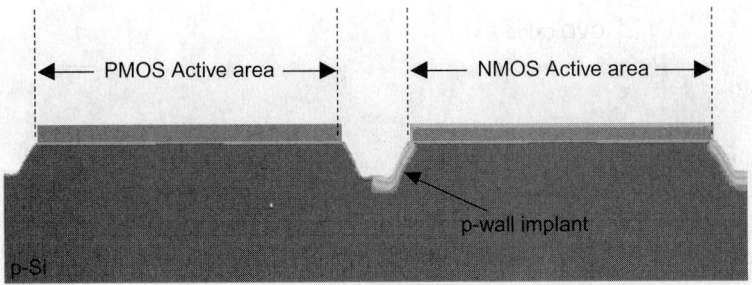

FIGURE 12.31 Post-p-wall photoresist strip using O_2 plasma and wet processing.

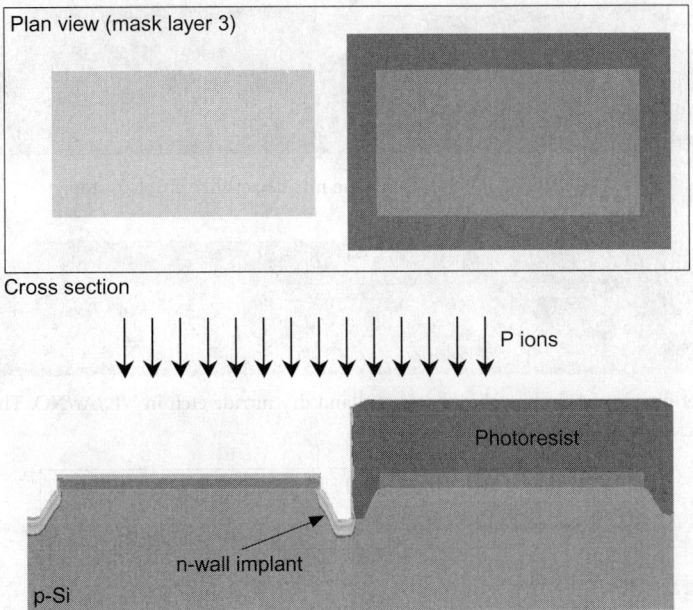

FIGURE 12.32 n-Wall sidewall formation via photolithography and phosphorus implantation.

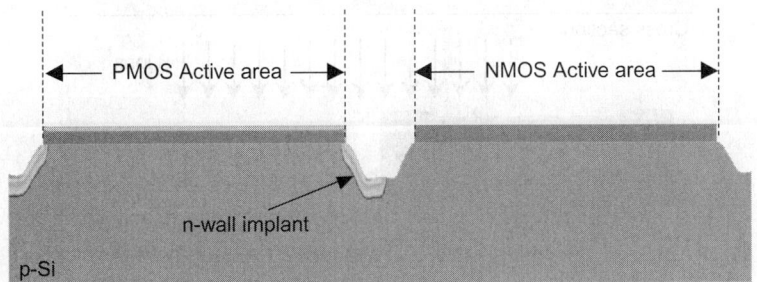

FIGURE 12.33 Post-n-wall photoresist strip using O_2 plasma and wet processing.

12.2.4.3 Twin-Tub Module

CMOS can be implemented in four general forms: n well, p well, twin well (called twin tub), and triple well. The CMOS technology discussed in this chapter uses a twin-well approach. The p well and n well provide the appropriate dopants for the NMOS and PMOS, respectively. Modern wells are implanted with retrograde profiles to maximize transistor performance and reliability.

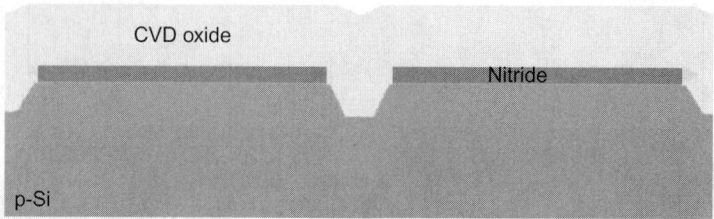

FIGURE 12.34 High-quality, 100-Å-thick liner oxide, thermally grown at 900°C. High-density plasma (HDP) CVD trench fill at room temperature. Notice the trenches are overfilled.

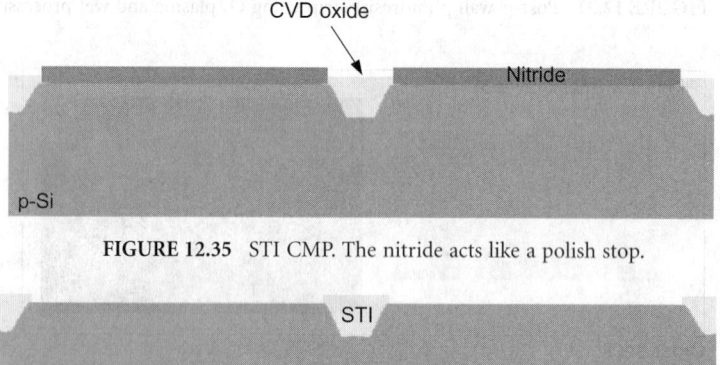

FIGURE 12.35 STI CMP. The nitride acts like a polish stop.

FIGURE 12.36 Wet nitride etch in hot phosphoric acid and dry nitride etch in $NF_3/Ar/NO$. The remaining oxide will be used as a sacrificial oxide for subsequent implants.

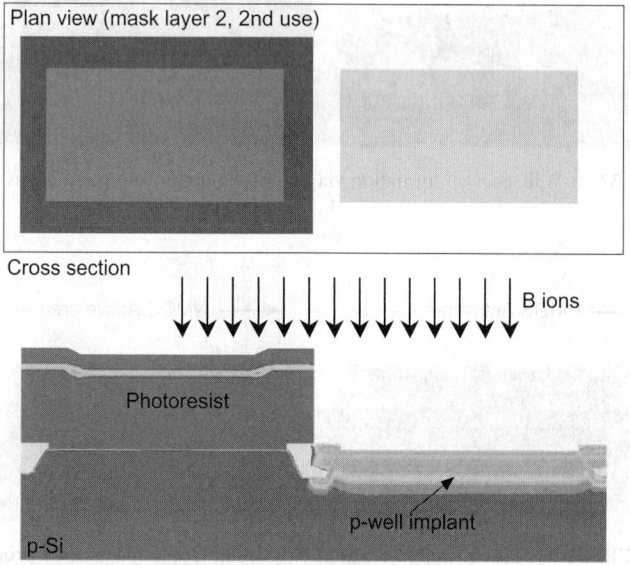

FIGURE 12.37 p-Well formation via photolithography and boron implantation.

Following the STI module, the twin-tub module begins with p-well photolithography (mask layer 2, second use) to generate a resist pattern that covers the PMOS-active regions, but exposes the NMOS-active areas (Figure 12.37). A relatively high-energy boron implant is performed into the NMOS-active areas, but blocked from the PMOS-active area. The pad oxide remaining from the STI module now serves as the sacrificial oxide for the well implants. It should be pointed out that the p well can be formed by a composition of several

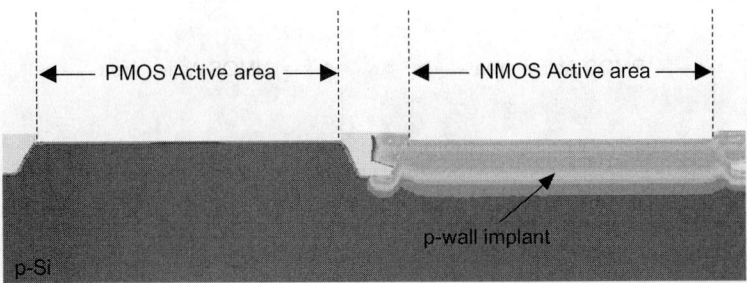

FIGURE 12.38 Post-p-well photoresist strip using O_2 plasma and wet processing.

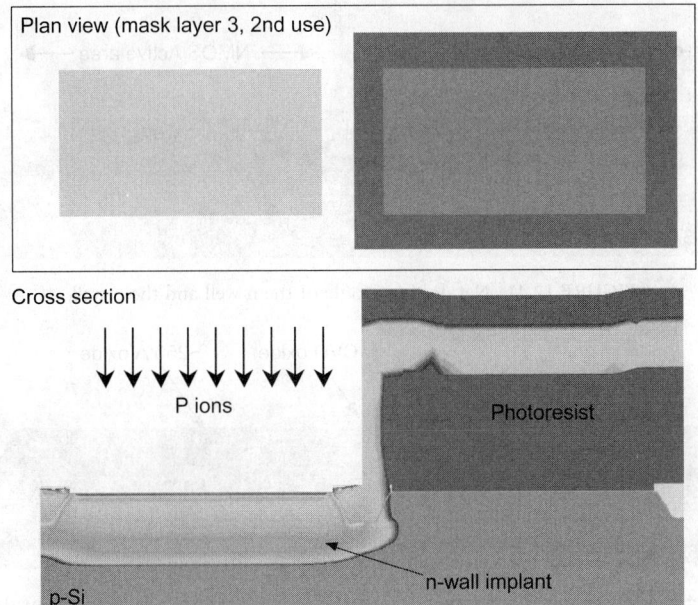

FIGURE 12.39 n-Well formation via photolithography and P implantation.

implants at different doses and energies to achieve the desired retrograde profile. Following the p-well implant, the resist is removed, using O_2 plasma and wet processing, resulting in the structure shown in Figure 12.38.

Next, a complimentary resist pattern is formed using the n-well mask and photolithography (Figure 12.39, mask layer 3, second use). Again, a relatively high-energy implant, this time phosphorus, generates the n well. As with the p well, a multitude of implants may be used to achieve the desired retrograde profile. Following the n-well implant, O_2 plasma and wet processing strip the resist (Figure 12.40). At this point, both the isolation and wells are fully formed. Figure 12.41 shows the cross section of the substrate after the twin-tub module. Notice the net doping profile is given, highlighting both the well and wall implants simultaneously. It should be emphasized that the PMOS are fabricated in the n wells while the NMOS are made in the p wells.

12.2.4.4 Gate Module

The gate module begins (Figure 12.42) with the buffered oxide etching of the remaining thin oxide in the active areas from the twin-tub module. A sacrificial oxide is thermally grown to serve as a threshold-adjust implant oxide and a pregate oxidation "clean-up". Next, a blanket (unpatterned), low-energy BF_2 threshold adjust implant is done (Figure 12.43). It allows for the "tuning" of both the PMOS and NMOS threshold voltages. The single boron implant is common for single work-function gates. However, for dual work-function gates (common in technologies with minimum gate lengths of 250 nm or less),

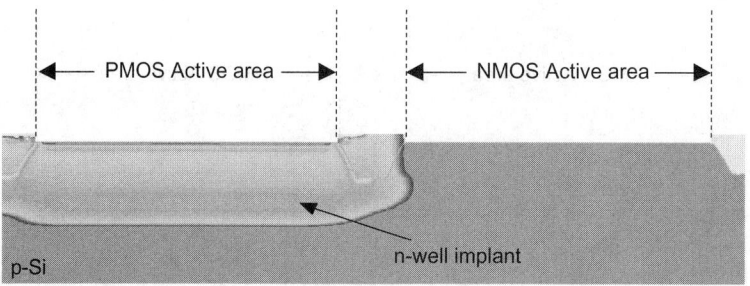

FIGURE 12.40 Post n-well photoresist strip using O_2 plasma and wet processing.

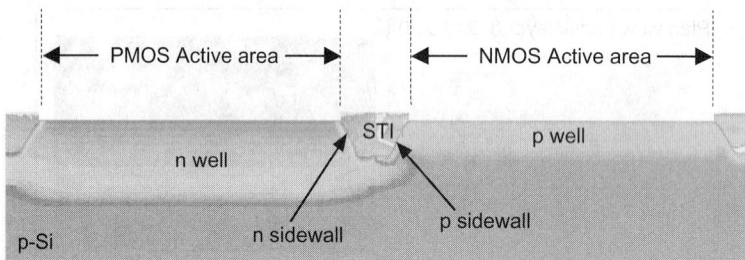

FIGURE 12.41 Net doping profile of the n well and the p well.

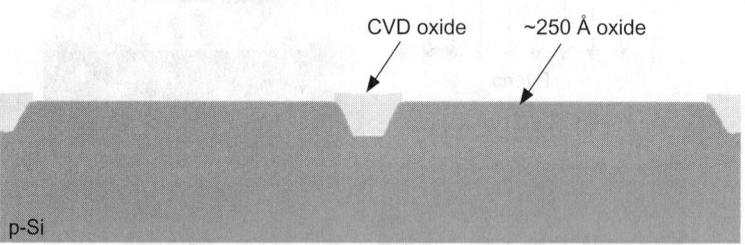

FIGURE 12.42 Wet etch of the remaining trench stack oxide using buffered HF. Sacrificial oxide formation using dry thermal oxidation at approximately 900°C.

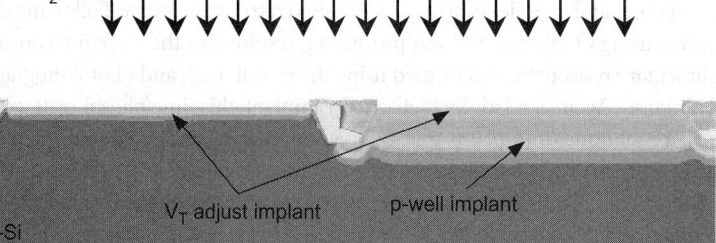

FIGURE 12.43 Blanket low-energy BF_2 implant for NMOS and PMOS threshold voltage V_T adjust.

separate p- and n-type implants are required for the threshold adjustment in the PMOS and NMOS, respectively.

The next set of processes form the "gate stack" (the gate dielectric and polysilicon gate electrode). Of course, the gate stack provides for capacitive coupling to the channel. The sacrificial oxide is stripped by wet processing from the active areas. Then a high-quality, thin oxide is thermally grown to serve as the gate dielectric (Figure 12.44). In modern CMOS, it is common to use nitrided gate oxide by performing the oxidation in O_2 and NO or N_2O. It can be argued that the gate oxidation is the most critical step in the entire process sequence, because the characteristic of the resultant film greatly determines the behavior of the CMOS transistors.

LPCVD polysilicon deposition immediately follows the gate oxidation (Figure 12.45). For single work-function gates, the polysilicon can be doped with phosphorus during polydeposition, or subsequently implanted. For dual work-function gates, the NMOS and PMOS can be doped during the n+ and p+ source/drain implants, respectively.

Once the gate stack is formed, the transistor gates and local interconnects are patterned using photolithography (mask layer 4) to generate the appropriate patterns in photoresist (Figure 12.46). The gate patterning must be precisely controlled, since it determines the gate lengths. Deviations in the resultant

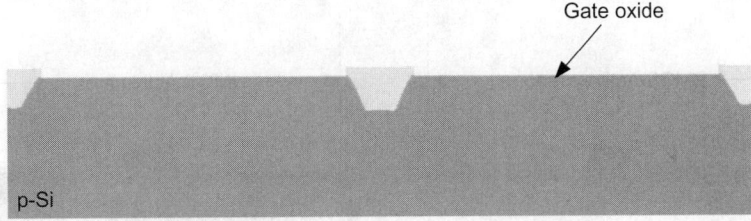

FIGURE 12.44 Removal of sacrificial oxide using buffered HF followed by gate dielectric formation using dry oxidation in an ambient of O_2, NO, and N_2O.

FIGURE 12.45 Polysilicon deposition via LPCVD at ~550°C. Note that the polysilicon deposition must occur immediately following gate oxidation.

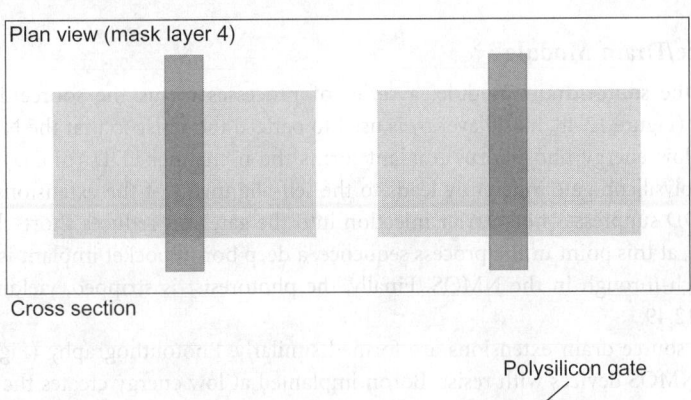

FIGURE 12.46 Gate electrode and local interconnect photolithography and polysilicon RIE.

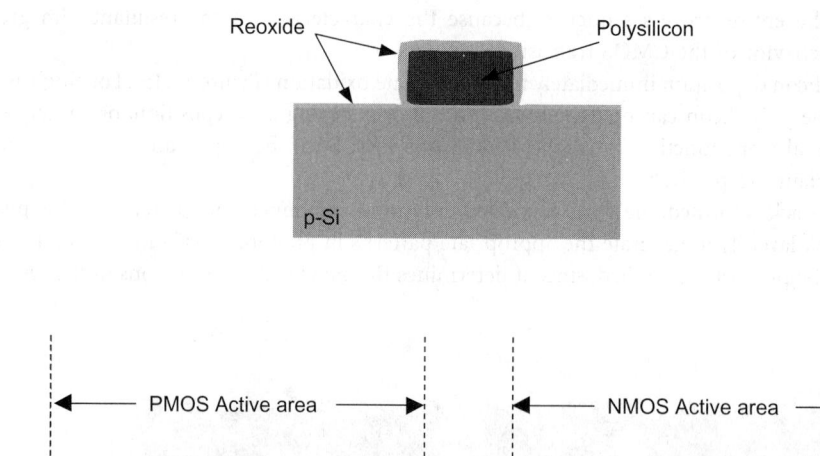

FIGURE 12.47 Polysilicon reoxidation using dry O_2 at approximately 900°C. Notice that the resulting oxide is thicker on the polysilicon than on the active silicon.

physical gate lengths can cause severe performance issues with the CMOS. The ideal gate profiles following the RIE of polysilicon and subsequent stripping of the resist are seen in Figure 12.46.

The gate module concludes with polyreoxidation (Figure 12.47). Thermal oxidation of the polysilicon and active silicon facilitates the growth of a buffer pad oxide for the subsequent nitride spacer deposition and electrically activates the implanted dopants in the polysilicon. Since the polysilicon oxidizes at a faster rate than the crystalline silicon, the resultant oxide thickness is greater on the polysilicon than on the active silicon.

12.2.4.5 Source/Drain Module

At the onset of the source/drain module, a series of processes forms the source/drain extensions. Photolithography (Figure 12.48, mask layer 5) is used to pattern the resist so that the NMOS devices are exposed. Then a low-energy phosphorus implant forms the n-channel LDD (nLDD) extensions. The presence of the polysilicon gate inherently leads to the self-alignment of the extensions with respect to the gate. The nLDD suppresses hot-carrier injection into the gate and reduces short-channel effects in the NMOS. Often, at this point in the process sequence, a deep boron pocket implant is used to prevent source/drain punch-through in the NMOS. Finally, the photoresist is stripped, yielding the structure shown in Figure 12.49.

The p-channel source drain extensions are formed similarly. Photolithography (Figure 12.50, mask layer 6) protects NMOS devices with resist. Boron implanted at low energy creates the p-channel LDD (pLDD) extensions. Again, the polysilicon serves to self-align the implant with respect to the gate electrodes. As was the case with the NMOS, a deep phosphorus pocket implant will suppress punch-through in the PMOS. Finally, the photoresist is stripped (Figure 12.51).

To complete the source/drain extensions, the gate sidewall spacers must be formed prior to the actual source/drain implants. First, conformal silicon nitride (or an oxide) is deposited using LPCVD (Figure 12.52). Following a nitride etch, this nitride will form the LDD sidewall spacers. The spacers function as a mask to the source/drain implants and as a barrier to the subsequent salicide formation. The actual spacers are formed

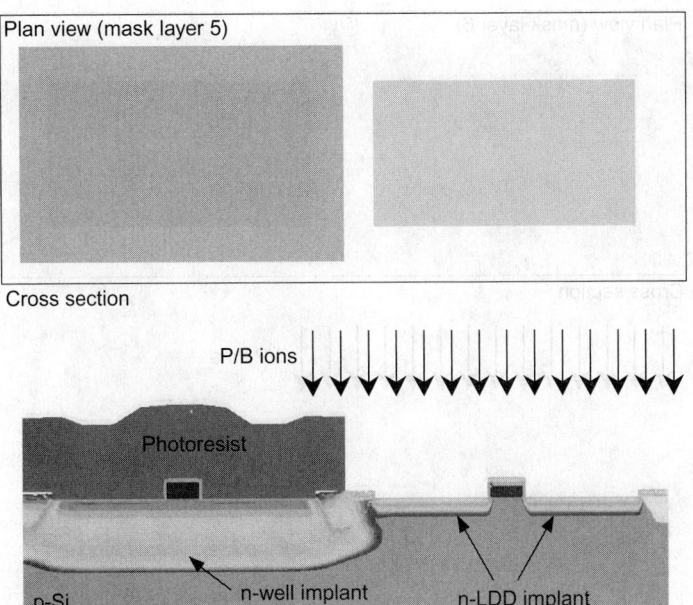

FIGURE 12.48 n-LDD/n-pocket formation using low-energy implantation of P and B, respectively.

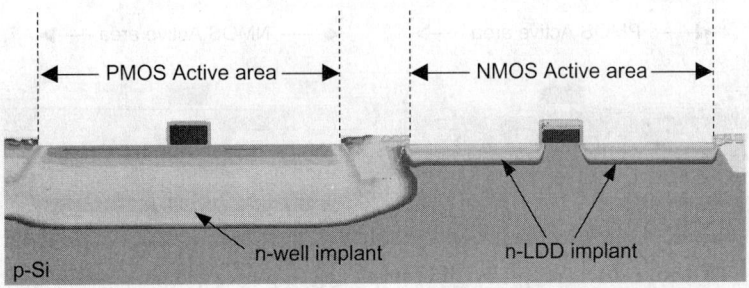

FIGURE 12.49 Post n-LDD resist strip using O_2 plasma and wet processing.

by an unpatterned anisotropic RIE of nitride (Figure 12.53). The spacer etch is end-pointed on the underlying oxide. Because the nitride is the thickest along the polysilicon sidewall, a well-formed insulating region remains on both sides of the polysilicon. This structure is called a spacer.

During source/drain implants, the polysilicon and spacers combine to block the implantation, thus allowing self-alignment for not only the gate but also for the LDD extensions. Stating this, the NMOS source/drains are formed (Figure 12.54). Photolithography (mask layer 5, second use) protects the PMOS with resist while exposing the NMOS. A relatively low-energy, high-dose arsenic implant is performed to form the n^+ regions, and the resist is stripped (Figure 12.55). In addition to source/drain formation, this implant creates the necessary n^+ ohmic contacts.

The PMOS source/drains and p^+ ohmic contacts are formed in a similar manner by photolithography (mask layer 6, second use) and a low-energy, high-dose BF_2 implant (Figure 12.56). Then the resist is stripped (Figure 12.57). The source/drain module concludes with a high-temperature anneal that electrically activates the implants and recrystallizes the damaged silicon. In modern CMOS, the primary reason for choosing polysilicon as the gate electrode material as opposed to metal is that polysilicon can withstand the high temperatures required to activate the source/drain implants.

At this point in our CMOS process sequence, we have fully formed the CMOS transistors and their isolation. This marks the completion of the FEOL. Figure 12.58 summarizes its main features.

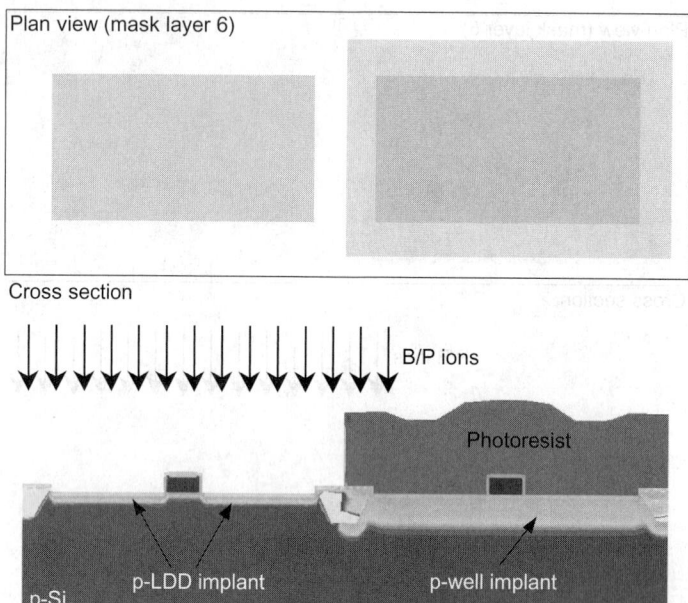

FIGURE 12.50   p-LDD/p-pocket formation using low-energy implantation of B and P, respectively.

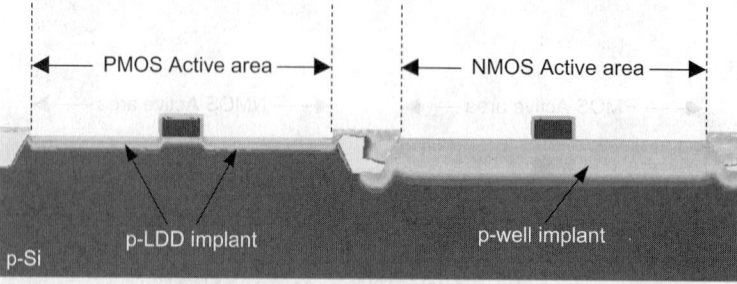

FIGURE 12.51   Post p-LDD resist strip using O_2 plasma and wet processing.

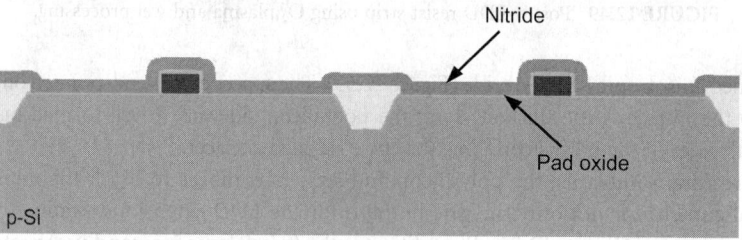

FIGURE 12.52   Sidewall spacer nitride deposition using LPCVD at approximately 800°C.

FIGURE 12.53   Dry, anisotropic, end-pointed RIE of spacer nitrate yielding gate sidewall spacers.

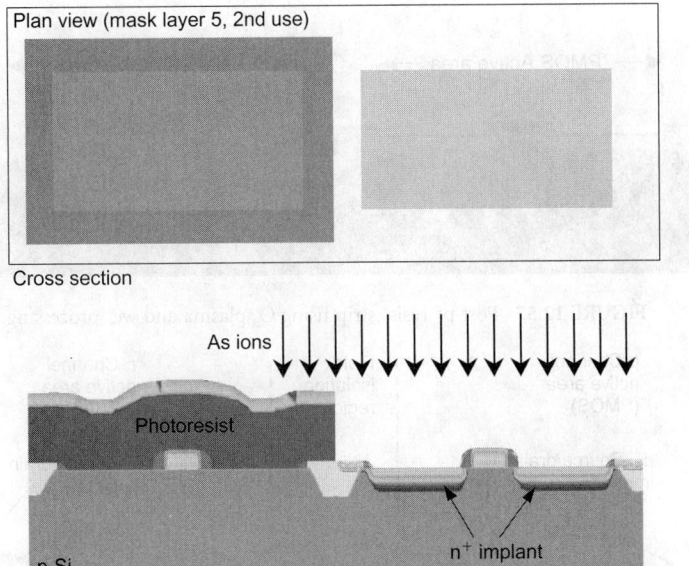

FIGURE 12.54 n⁺ Source/drain formation using a low-energy, high-dose implantation of As.

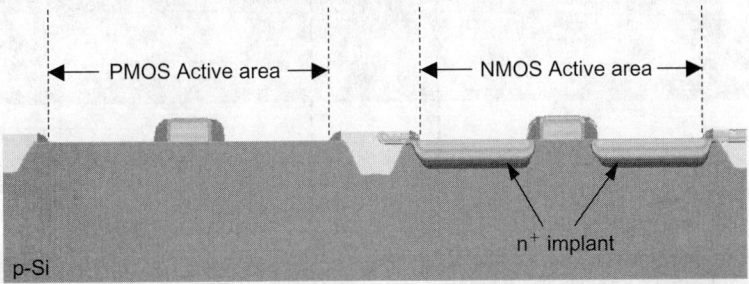

FIGURE 12.55 Post n⁺ resist strip using O_2 plasma and wet processing.

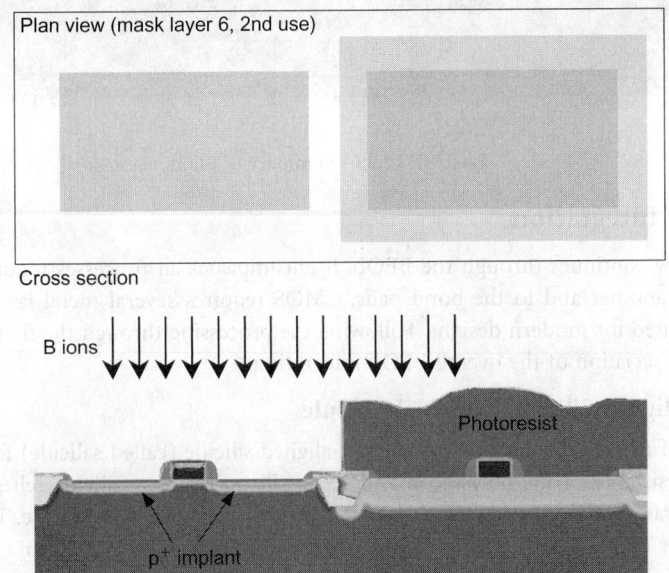

FIGURE 12.56 p⁺ Source/drain formation using a low-energy, high-dose implantation of BF_2.

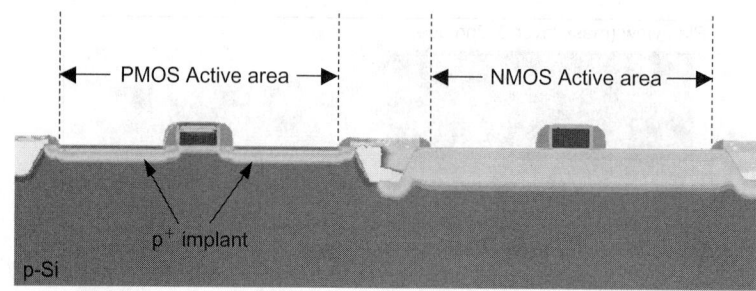

FIGURE 12.57 Post p⁺ resist strip using O_2 plasma and wet processing.

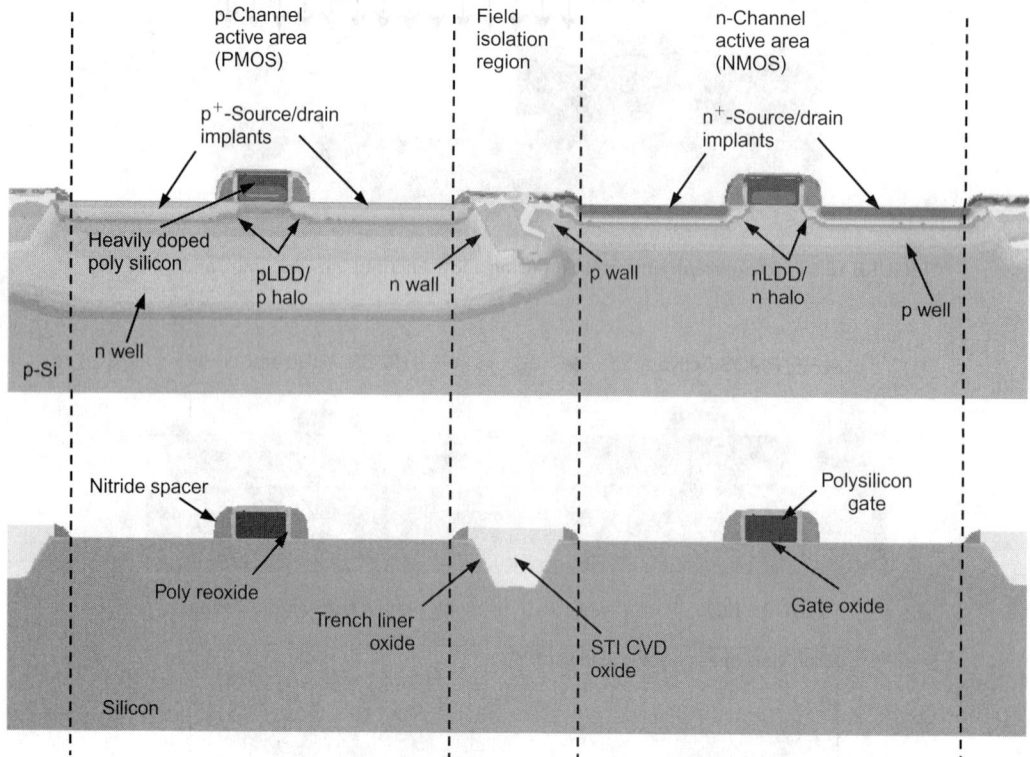

FIGURE 12.58 Summary of FEOL features.

12.2.5 BEOL Integration

CMOS process flow continues through the BEOL. It encompasses all processes required to "wire" the transistors to one another and to the bond pads. CMOS requires several metal layers to achieve the interconnects required for modern designs. Following the processing through the first two metal layers will provide an appreciation of the overall BEOL integration.

12.2.5.1 Self-Aligned Silicide (Salicide) Module

At the boundary of the FEOL and BEOL is the self-aligned silicide (called salicide) formation. Silicide lowers the sheet resistance of the polysilicon and active silicon regions. Salicide relies on the fact that metal silicide generally will not form over dielectric materials such as silicon nitride. Therefore, a metal such as titanium or cobalt can be deposited over the entire surface of the wafer and then annealed to selectively form silicide over exposed polysilicon and silicon. Because of the presence of the trench fill and sidewall spacers, the silicide becomes self-aligned without the need for photopatterning.

The salicide module begins by removing the thin oxide, present from the FEOL, using buffered HF (Figure 12.59). Next, a refractory metal (e.g., titanium or cobalt) is deposited by sputtering (Figure 12.60). To minimize contamination, a thin layer of TiN is deposited as a cap. A relatively low-temperature, nitrogen-ambient, rapid thermal annealing (RTA) is used to react titanium (or cobalt) with the silicon, forming $TiSi_2$ (or $CoSi_x$). The resultant silicide (e.g., C_{49}) is a high-resistivity phase. Also, notice that the underlying nitride and oxide serve to block the formation of the silicide from the sidewalls and trenches, respectively.

To prevent spacer overgrowth of the silicide, the low-resistivity phase is achieved by processing with two separate anneals. The first, described above, forms the high-resistivity phase without the risk of silicide formation on the nitride. The second, described below, occurs after wet chemical etching of the unreacted titanium (or cobalt) using a higher temperature. This causes a phase change (C_{49} to C_{54} for $TiSi_2$) with a much lower resistivity. If one high-temperature anneal was originally performed to achieve the low-resistivity phase, then significant overgrowth can occur, leading to current leakage from the source and drain to the gate of the transistors.

To continue the salicide module, following the first anneal, the unreacted titanium (or cobalt) is wet-etched chemically from the wafer. The second RTA, in argon ambient at a slightly higher temperature, is used to achieve the low-resistivity phase (Figure 12.61).

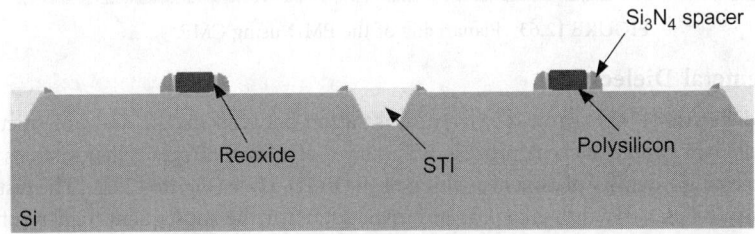

FIGURE 12.59 Removal of the exposed reoxide (present from the FEOL) using buffered HF.

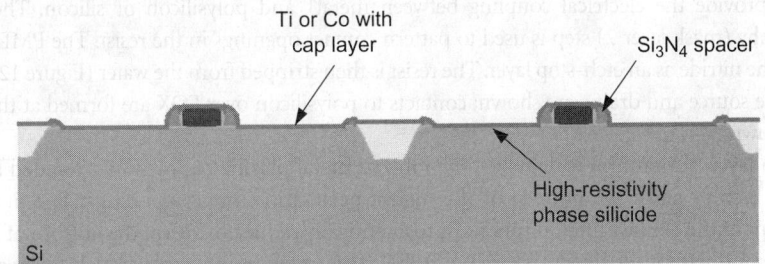

FIGURE 12.60 Titanium or cobalt deposited by PVD followed by the first salicide RTA.

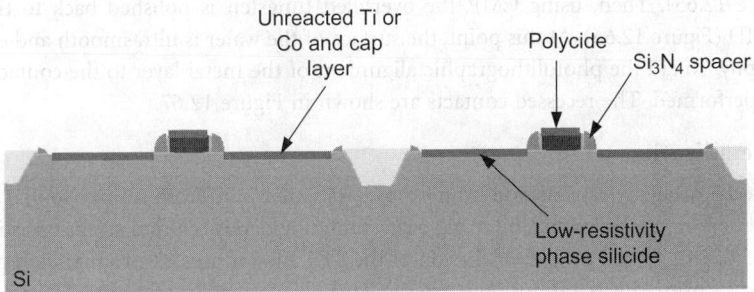

FIGURE 12.61 Wet chemical etch of the unreacted titanium or cobalt followed by the second salicide RTA.

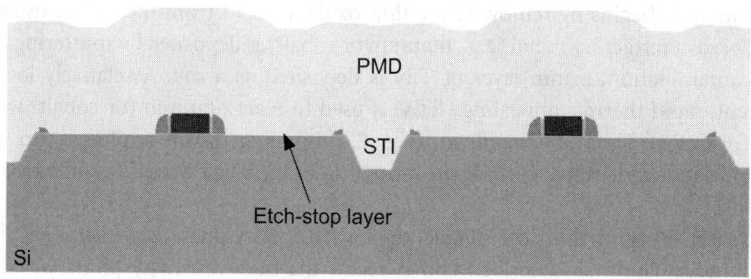

FIGURE 12.62 PMD deposition using high-density plasma CVD.

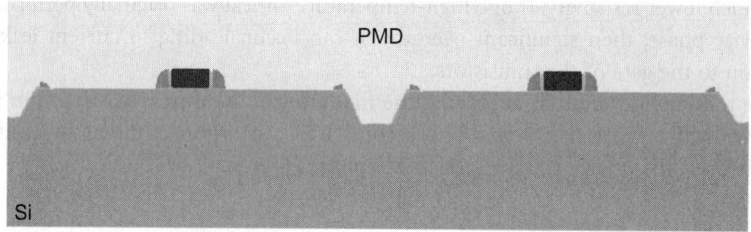

FIGURE 12.63 Planarizing of the PMD using CMP.

12.2.5.2 Premetal Dielectric

The premetal dielectric (PMD) provides electrical isolation between metal1 and polysilicon or silicon. To aid the subsequent contact etching process, a thin layer of silicon nitride is deposited as an etch stop. This is followed by high-density plasma deposition of the PMD oxide (Figure 12.62). The resultant surface of the PMD must be planarized to allow for improved DOF for the subsequent high-resolution photo-patterning of metal1. CMPis used to planarize the PMD (Figure 12.63).

12.2.5.3 Contact Module

The contacts provide the electrical coupling between metal1 and polysilicon or silicon. The first BEOL photolithography (mask layer 7) step is used to pattern contact openings in the resist. The PMD is then dry-etched, using the nitride as an etch-stop layer. The resist is then stripped from the wafer (Figure 12.64). Contact openings to the source and drains are shown; contacts to polysilicon over FOX are formed at the same time, but are not shown.

Next, a thin layer of titanium is deposited by ionized metal plasma deposition, preceded by an in-situ argon sputter etch to clean the bottom of the high-aspect ratio contact openings. The titanium is the first component of the contact liner. It functions to chemically reduce oxides at the bottom of the contacts. The second component of the liner is a thin CVD TiN. This layer's primary purpose is to act as a diffusion barrier to fluorine (which readily etches silicon) that will be used in the subsequent tungsten deposition. The contact openings are filled (actually overfilled) with tungsten using a tungsten fluoride (WF_6) CVD process (Figure 12.65). Then, using CMP, the overfilled tungsten is polished back to the top of the planarized PMD (Figure 12.66). At this point, the surface of the wafer is ultrasmooth and essentially free of all topography. To aid the photolithographic alignment of the metal layer to the contacts, a tungsten recess etch is performed. The recessed contacts are shown in Figure 12.67.

12.2.5.4 Metallization 1

To allow for electrical signal transmission from contact-to-contact and from contact-to-via1, defined metallization must be formed. Following the recess etch, sputtering deposits a film stack consisting of Ti/TiN/Al/TiN (Figure 12.68). The Ti provides adhesion of the TiN and reduces electromigration problems. The bottom TiN serves primarily as a diffusion barrier to $TiAl_3$ formation. The topmost TiN acts as an antireflective coating for the metal photolithography as well as an etch stop for the subsequent via formation.

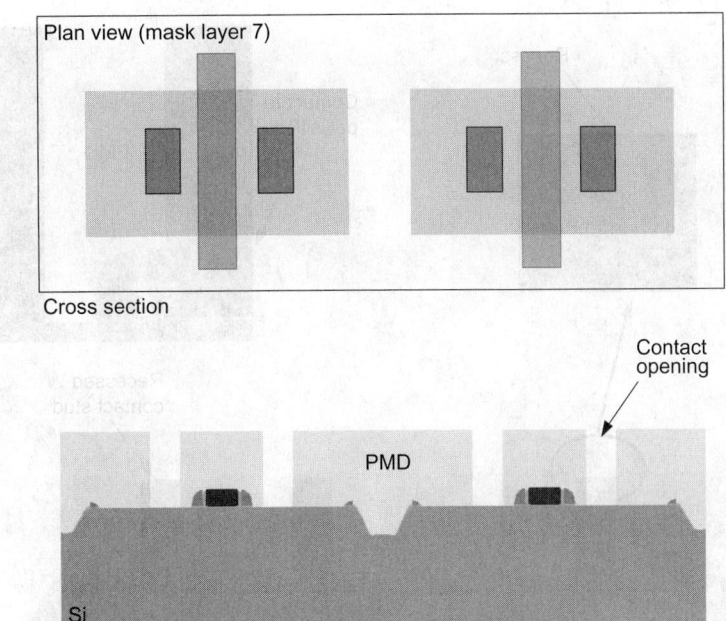

FIGURE 12.64 Contact definition using photolithography and RIE of the PMD. Notice that the contacts to the polysilicon over FOX are formed at the same time, but not shown.

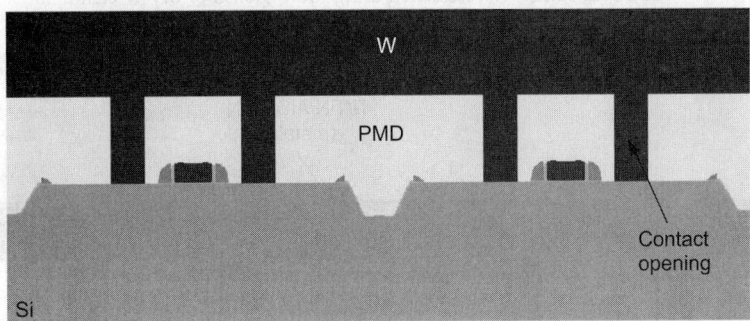

FIGURE 12.65 Ti/TiN liner deposition using IMP and CVD, respectively. W contact fill deposition using WF_6 CVD.

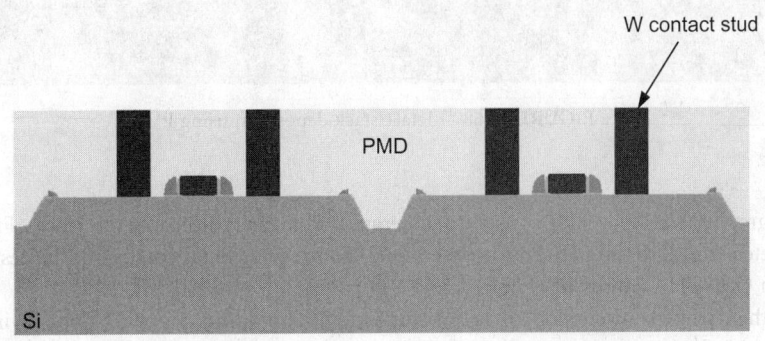

FIGURE 12.66 W CMP forming defined contacts.

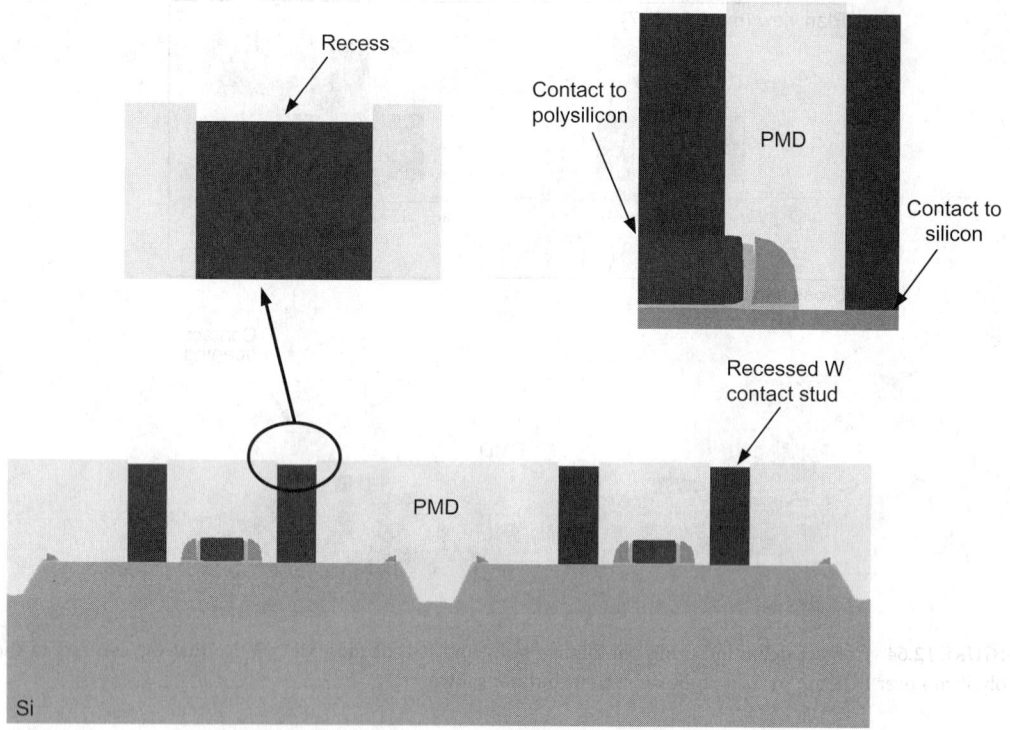

FIGURE 12.67 W-recess etch using "buff" polish or dry W etch.

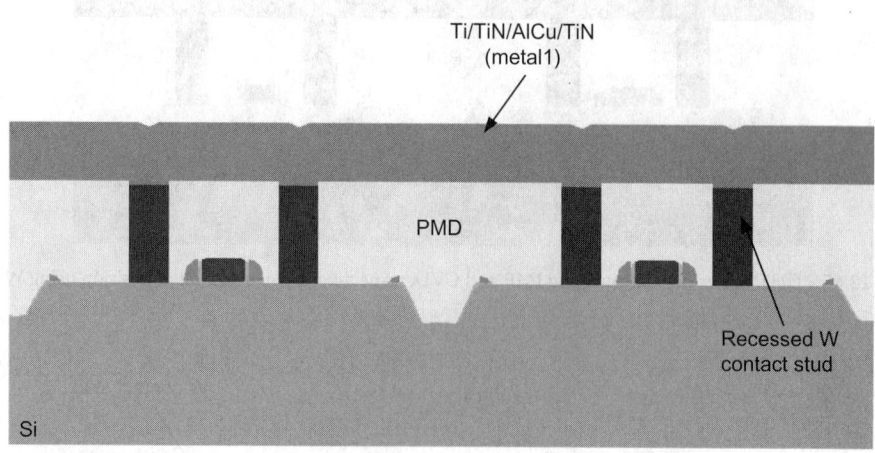

FIGURE 12.68 Metal1 stack deposition using PVD.

Photolithography (mask layer 8) is used to generate the metal1 pattern in the resist (Figure 12.69). A dry-metal etch transfers the pattern into the metal. To prevent metal corrosion, the resist is plasma stripped in an $O_2/N_2/H_2O$ ambient (Figure 12.69).

Note that this is not a discussion of metal implementation using copper. Copper wiring is often implemented with dual-Damascene techniques, where vias and metal layers are simultaneously formed in a series of process steps.

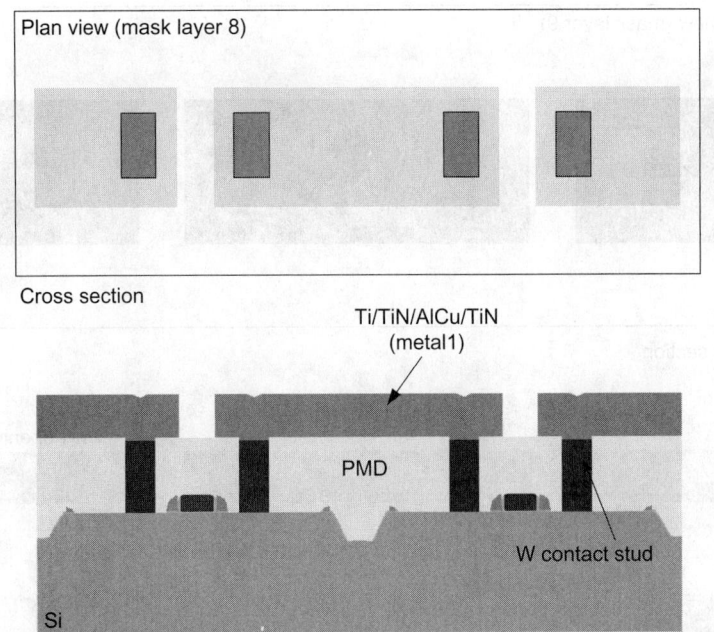

FIGURE 12.69 Metal1 definition using photolithography and dry metal etch.

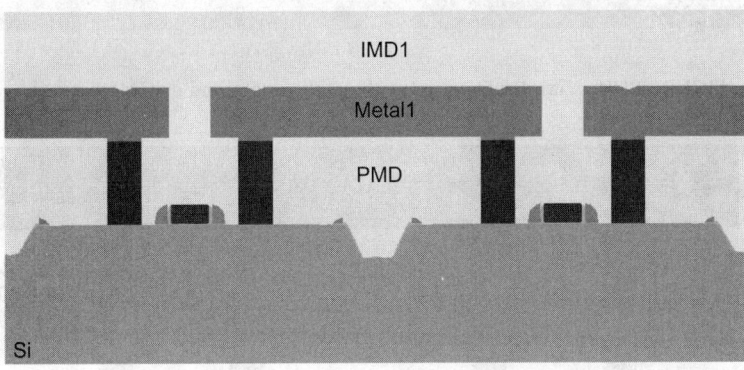

FIGURE 12.70 IMD1 deposition using HDP CVD. This is followed by IMD1 planarization using CMP.

12.2.5.5 Intrametal Dielectric 1 Deposition

The intrametal dielectric 1 (IMD1) provides the electrical isolation between metal1 and metal2. This film usually is deposited using high-density plasma CVD (Figure 12.70). The conformal deposition results in surface topography that must be planarized by CMP to improve the DOF for subsequent photolithographic steps, as in the case of PMD planarization.

12.2.5.6 Via1 Module

Electrical coupling between metal 1 and metal 2 is achieved by the via1 module. The planarized IMD is photolithographically defined (mask layer 9) and an RIE is used to open the vias. The resist is stripped using O_2 plasma and wet processing (Figure 12.71). Next, as with the contact fill, in the step of argon sputter etch followed by deposition of thin layers of IMP titanium and CVD TiN. A WF_6 CVD process is used to deposit (overfill) the vias (Figure 12.72). The excess tungsten is removed by CMP, using the IMD1 as a "polish stop" (Figure 12.73). Again, to provide observable alignment features, a tungsten recess etch is often required (Figure 12.73).

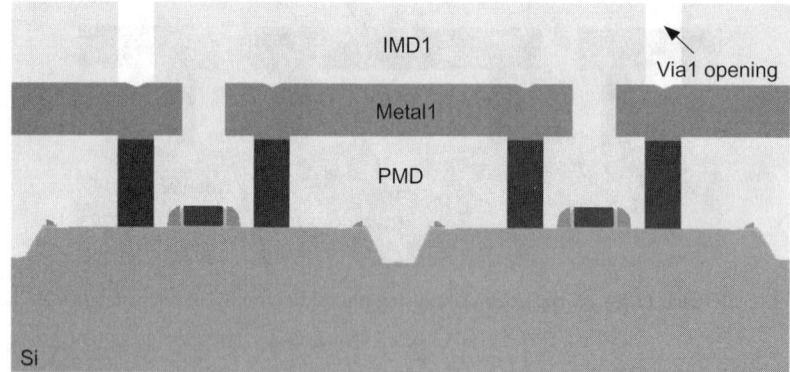

FIGURE 12.71 Via1 definition using photolithography and dry IMD1 etch.

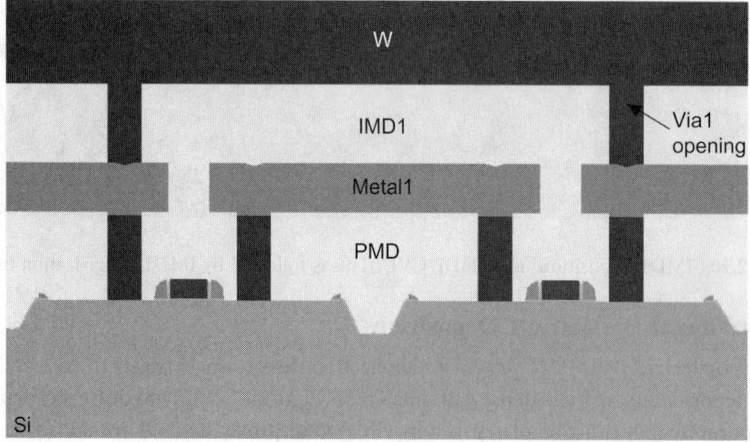

FIGURE 12.72 Ti/TiN liner deposition using IMP and CVD, respectively. W via1 fill deposition using WF_6 CVD.

12.2.5.7 Metallization 2

In a manner similar to the metal1 process, the metal2 stack is deposited (Figure 12.74) and photolithographically defined (Figure 12.75, mask layer 10).

12.2.5.8 Additional Metal/Dielectric Layers

At this point, additional tiers of dielectric/metal layers can be formed by replicating the aforementioned processes. In modern CMOS, the number of metal layers can be greater than eight. It should be noted

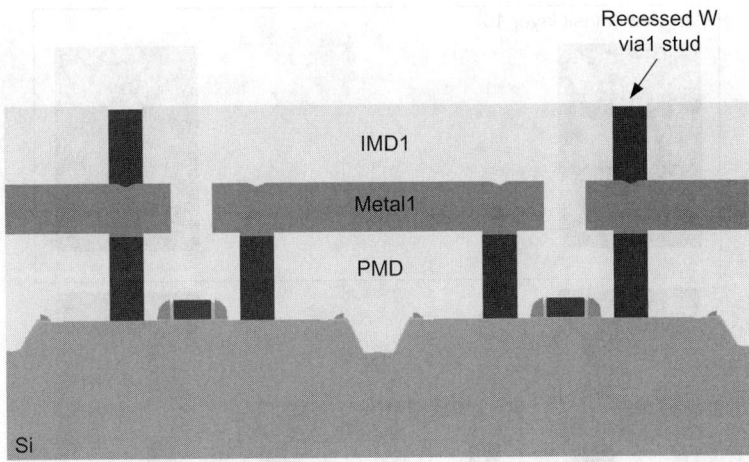

FIGURE 12.73 W CMP forming defined vias. This is followed by W-recess etch using "buff" polish or dry W etch.

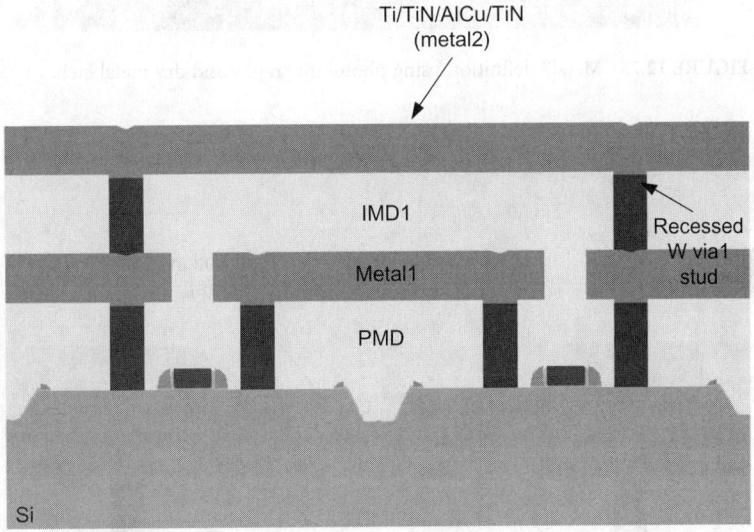

FIGURE 12.74 Metal2 stack deposition using PVD.

that as dielectric/metal layers are added, the cumulative film stresses may cause significant bow/warp in the wafers. Hence, great effort is expended to minimize stresses in the BEOL films.

12.2.5.9 Final Passivation

To protect the CMOS from mechanical abrasion during probe and packaging, and to provide a barrier against contaminants (e.g., H_2O and ionic salts), a final passivation layer must be deposited. The passivation type is determined in large part by the CMOS ICs' packaging. Common passivation layers are doped glass and silicon nitride on deposited oxide. Figure 12.76 shows the resultant cross-section of our CMOS process flow after deposition of the passivation. Finally, the bond pads are opened with photolithography (mask layer *n*) and dry etching the passivation. The CMOS is completed with a resist strip (Figure 12.77).

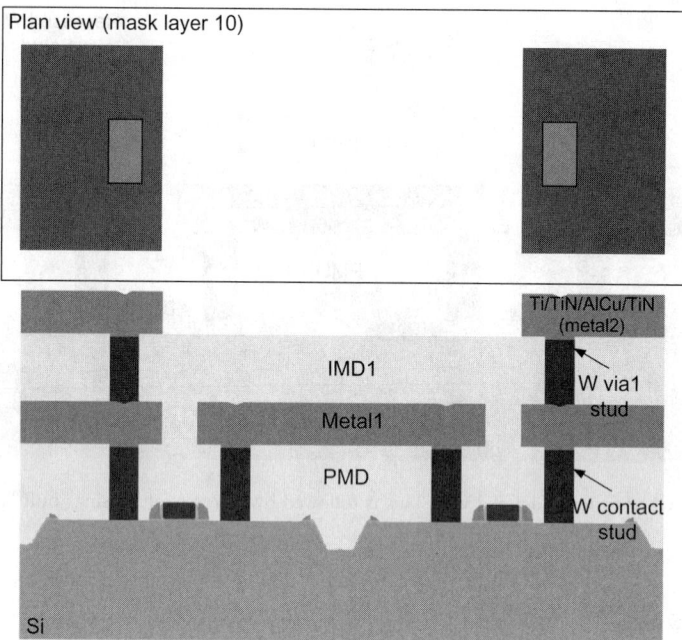

FIGURE 12.75 Metal2 definition using photolithography and dry metal etch.

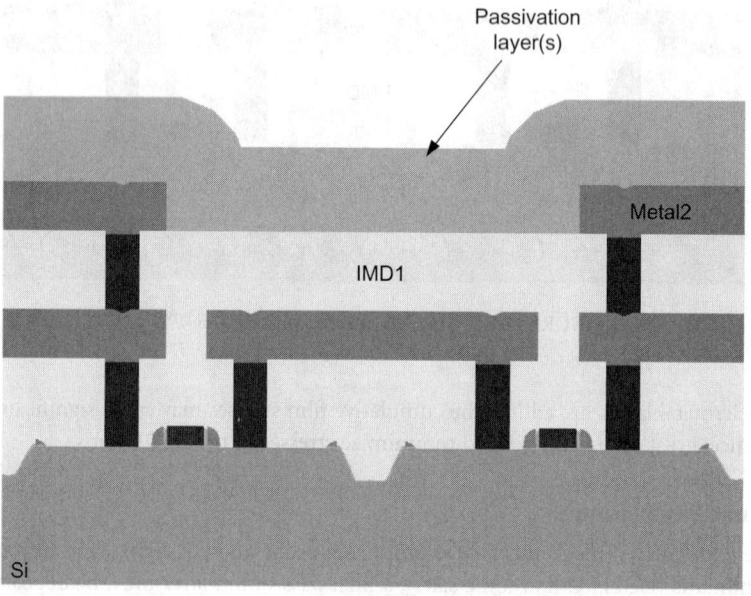

FIGURE 12.76 Deposition of final passivation.

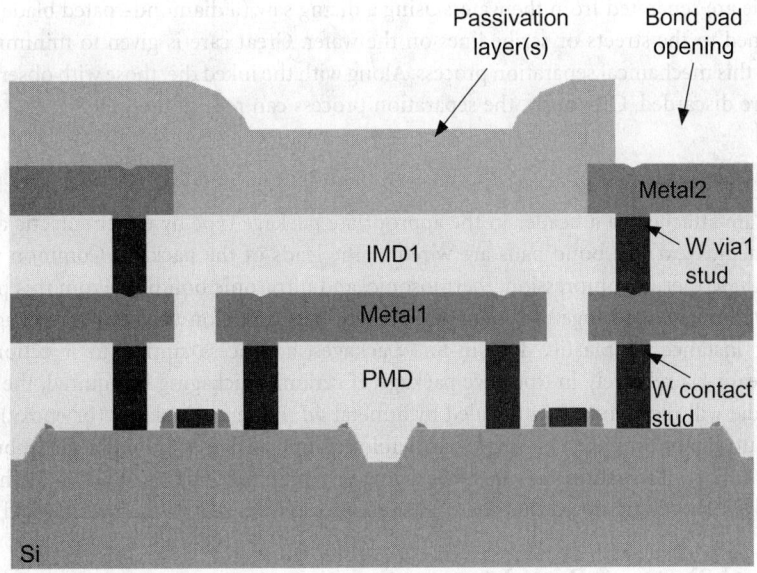

FIGURE 12.77 Bond-pad definition using photolithography and dry etch of passivation.

12.3   Back-End Processes

After final passivation, the wafers are removed from the clean room in preparation for a series of back-end (i.e., post-fab) processes. These processes include wafer probe, die separation, packaging, and final test/burn-in.

12.3.1   Wafer Probe

Generally, dedicated die with parametric structures and devices are stepped into various positions of the wafer. Alternatively, parametric structures are placed in the dividing regions, called streets or scribe lines, between the die. Electrical characterization of these parametric structures and devices is often performed at select points in the fabrication process flow (such as after metal1 patterning). Parameters such as contact resistance, sheet resistance, transistor threshold voltage, saturated drain current, off-current, subthreshold slope, etc., are measured. If problems are observed from the in-line parametric tests, troubleshooting can begin sooner than if testing is only performed after final passivation. Furthermore, wafers that do not meet parametric standards can be removed (or "killed") from the fabrication sequence (thus saving money).

After the completed wafers are removed from the fab, wafer-level probing is performed to check the final device parameters and CMOS IC functionality and performance. Wafer probe is accomplished by using sophisticated testers that can probe individual die (or sets of die) and apply test vectors to determine circuit behavior. Inevitably, a percentage of die will not pass all the vector tests and are thus considered failed. The ratio of good die to total die represents the wafer *yield* given by

$$Y = \text{Number of die passing all tests/total number of die} \tag{12.13}$$

In these cases, the die are marked, often with an ink dot, to indicate a nonfunctional circuit. Since the processing costs of a wafer are fixed, higher yield equates to higher profit.

12.3.2   Die Separation

Before separating individual die, the backs of the wafers are often thinned using a lapping process similar to the CMP discussed in Section 12.1.5. This thinning may be required for specific types of packages. It can also help remove heat from the CMOS circuits.

Next, the die are separated from the wafers using a dicing saw (a diamond-coated blade). The cutting paths are aligned to the streets or scribe lines on the wafer. Great care is given to minimize damage to the die during this mechanical separation process. Along with the inked die, those with observable damage from dicing are discarded. Obviously, the separation process can reduce the yield.

12.3.3   Packaging

The good die are attached to a header in the appropriate package type by either eutectic attachment or epoxy attachment. Next, the bond pads are wired to the leads of the package. Common wire-bonding techniques include thermocompression, thermosonic, and ultrasonic bonding. From this point, packaging is completed by a wide range of different processes, depending on the kind of package the IC will reside in. For instance, in plastic, dual in-line packages, a process similar to injection molding is performed, creating a relatively inexpensive package. If ceramic packaging is required, the attached and wire-bonded die will reside in a cavity sealed by a metal lid. In general, plastic (or epoxy) packages are inexpensive, but do not provide a hermetic seal, while ceramic packages are more costly, but do provide a hermetic seal. For information on other packaging schemes, see Further Reading at the end of this chapter. Finally, it should be noted that the packaging process can add to the overall yield loss.

12.3.4   Final Test and Burn-In

Once packaged, the CMOS parts are tested for final functionality and performance. After this is done, it is common for the parts to go through a burn-in step in which they are operated at extreme temperatures and voltages to weed out infant failures. Additional yield loss can be observed.

12.4   Summary

In this chapter, the fundamental unit processes required for manufacturing CMOS ICs were introduced. These unit processes included thermal oxidation, solid-state diffusion, ion implantation, photolithography, wet chemical etching, dry (plasma) etching, CMP, PVD, and CVD. A brief overview of substrate preparation was also given. Following this foundation, a representative, deep-submicron CMOS process flow was described and the significant issues in both FEOL and BEOL integration were discussed. Finally, an overview of the back-end processes was presented, including wafer probe, die separation, packaging and final test, and burn-in.

Additional Reading

1. S.A. Campbell, *The Science and Engineering of Microelectronic Fabrication*, 2nd ed, Oxford University Press, Oxford, 2001. ISBN 0-19-513605-5.
2. R.C. Jaeger, *Introduction to Microelectronic Fabrication*, 2nd ed, Volume 5 of the Modular Series on Solid State Devices, Prentice-Hall, New York, 2002, ISBN 0-20-144494-1.
3. J.D. Plummer, M.D. Deal, and P.B. Griffin, *Silicon VLSI Technology, Fundamentals, Practice, and Modeling*, Prentice-Hall, New York, 2000, ISBN 0-13-085037-3.
4. G.S. May and S.M. Sze, *Fundamentals of Semiconductor Fabrication*, Wiley, New York, 2003, ISBN 0471232793.

13

Analog Circuit Simulation

J. Gregory Rollins
Antrim Design Systems

CONTENTS

13.1 Introduction

Analog circuit simulation usually means simulation analog circuits or very detailed simulation of digital circuits. The most widely known and used circuit simulation program is SPICE (simulation program with integrated circuit emphasis) of which it is estimated that there are over 100,000 copies in use. SPICE was first written at the University of California at Berkeley in 1975, and was based on the combined work of many reasearchers over a number of years. Research in the area of circuit simulation continues at many universities and industrial sites. Commercial versions of SPICE or related programs are available on a wide variety of computing platforms, from small personal computers to large mainframes. A list of some commercial simulator vendors can be found in the Appendix. The focus of this chapter is the simulators and the theory behind them. Examples are also given to illustrate their use.

13.2 Purpose of Simulation

Computer-aided simulation is a powerful aid during the design or analysis of VLSI circuits. Here, the main emphasis will be on analog circuits; however, the same simulation techniques may be applied to digital circuits, which are, after are, composed of analog circuits. The main limitation will be the size of these circuits because the techniques presented here provide a very detailed analysis of the circuit in question and, therefore, would be too costly in terms of computer resources to analyze a large digital system. However, some of the techniques used to analyze digital systems (like iterated timing analysis or relaxation methods) are closely related to the methods used in SPICE.

It is possible to simulate almost any type of circuit SPICE. The programs have built-in elements for resistors, capacitors, inductors, dependent and independent voltage and current sources, diodes, MOSFETs, JFETs, BJTs, transmission lines, and transformers. Commercial versions have libraries of standard components which have all necessary parameters prefitted to typical specifications. These libraries include items such as discrete transistors, op-amps, phase-locked loops, voltage regulators, logic integrated circuits, and saturating transformer cores. Versions are also available which allow the inclusion of digital models (mixed mode simulation) or behavioral models which allow the easy modeling of mathematical equations and relations.

Computer-aided circuit simulation is now considered an essential step in the design of modern integrated circuits. Without simulation, the number of "trial runs" necessary to produce a working IC would greatly increase the cost of the IC and the critical time to market. Simulation provides other advantages, including:

- The ability to measure "inaccessible" voltages and currents which are buried inside a tiny chip or inside a single transistor.
- No loading problems are associated with placing a voltmeter or oscilloscope in the middle of the circuit, measuring difficult one-shot waveforms or probing a microscopic die.
- Mathematically ideal elements are available. Creating an ideal voltage or current source is trivial with a simulator, but impossible in the laboratory. In addition, all component values are exact and no parasitic elements exist.
- It is easy to change the values of components or the configuration of the circuit.

Unfortunately, computer-aided simulation has it own set of problems, including:

- Real circuits are distributed systems, not the "lumped element models" which are assumed by simulators. Real circuits, therefore, have resistive, capacitive, and inductive parasitic elements present in addition to the intended components. In high-speed circuits, these parasitic elements can be the dominant performance-limiting elements in the circuit, and they must be painstakingly modeled.
- Suitable predefined numerical models have not yet been developed for certain types of devices or electrical phenomena. The software user may be required, therefore, to create his or her own models out of other models which are available in the simulator. (An example is the solid-state thyristor, which may be created from an npn and pnp bipolar transistor).
- The numerical methods used may place constraints on the form of the model equations used. In addition, convergence difficulties can arise, making the simulators difficult to use.
- There are small errors associated with the solution of the equations and other errors in fitting the non-linear models to the transistors which make up the circuit.

13.3   Netlists

Before simulating, a circuit must be coded into a netlist. Figure 13.1 shows the circuit for a simple differential pair. Circuit nodes are formed wherever two or more elements meet. This particular circuit has seven nodes, which are numbered zero to six. The ground or datum node is traditionally numbered as zero. The circuit elements (or branches) connect the nodes.

The netlist provides a description of the topography of a circuit and is simply a list of the branches (or elements) that make up the circuit. Typically, the elements may be entered in any order and each has a unique name, a list of nodes, and either a value or model identifier. For the differential amplifier of Figure 13.1, the netlist is shown in Figure 13.2.

The first line gives the title of the circuit (and is required in many simulators). The next three lines define the three voltage sources. The letter *V* at the beginning tells SPICE that this is a voltage source element. The list of nodes (two in this case) is next followed by the value in volts. The syntax for the resistor is similar to that of the voltage source; the starting letter *R* in the names of the resistors tells SPICE

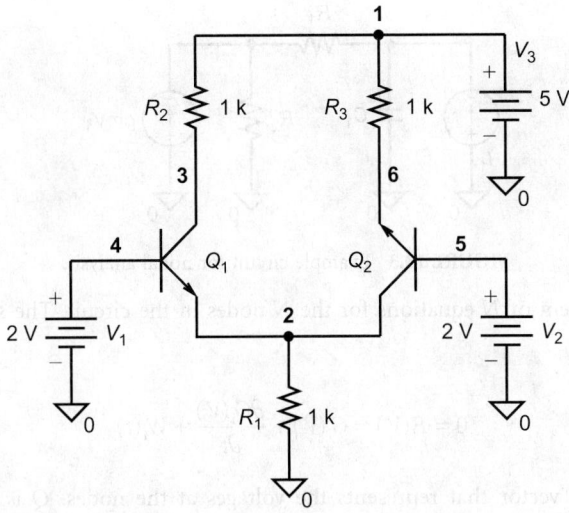

FIGURE 13.1 Circuit for differential pair.

```
Differential pair circuit
V1 4 0 2V
V2 5 0 2V
V3 1 0 5V
R1 2 0 1k
R2 3 1 1k
R3 6 1 1k
Q1 3 4 2 m2n2222
Q2 6 5 2 mq2n2222
.model m2n2222 NPN IS=1e-12 BF=100 BR=5 TF=100pS
```

FIGURE 13.2 Netlist for differential pair.

that these are resistors. SPICE also understands that the abbreviation "k" after a value means 1000. For the two transistors Q_1 and Q_2, the starting letter Q indicates a bipolar transistor. Q_1 and Q_2 each have three nodes and in SPICE, the convention for their ordering is collector, base, emitter. So, for Q_1, the collector is connected to node 3, the base to node 4, and the emitter to node 2. The final entry "m2n2222" is a reference to the model for the bipolar transistor (note that both Q_1 and Q_2 reference the same model). The ".model" statement at the end of the listing defines this model. The model type is npn (for an npn bipolar junction transistor), and a list of "parameter = value" entries follow. These entries define the numerical values of constants in the mathematical models which are used for the bipolar transistor. (Models will be discused in more detail later on.) Most commercial circuit simulation packages come with "schematic capture" software that allows the user to draw the circuit by placing and connecting the elements with the mouse.

13.4   Formulation of the Circuit Equations

In SPICE, the circuits are represented by a system of ordinary differential equations. These equations are then solved using several different numerical techniques. The equations are constructed using Kirchoff's voltage and current laws. The first system of equations pertains to the currents flowing into each node. One equation is written for each node in the circuit (except for the ground node), so the following

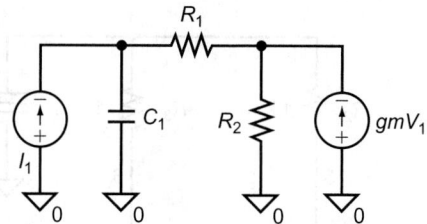

FIGURE 13.3 Example circuit for nodal analysis.

equation is really a system of N equations for the N nodes in the circuit. The subscript i denotes the node index.

$$0 = F_i(V) = G_i(V) + \frac{\partial Q_i(V)}{\partial t} + W_i(t) \tag{13.1}$$

V is an N-dimensional vector that represents the voltages at the nodes. Q is another vector which represents the electrical charge (in Coulombs) at each node. The term W represents any independent current sources that may be attached to the nodes and has units of amperes. The function $G(V)$ represents the currents that flow into the nodes as a result of the voltages V. If the equations are formulated properly, a system of N equations in N unknowns results.

For example, for the circuit of Figure 13.3 which has two nodes, we need to write two equations. At Node 1:

$$0 = (V_1 - V_2)/R_1 + \frac{d(C_1 V_1)}{dt} + I_1 \tag{13.2}$$

We can identify $G(V)$ as $(V_1 - V_2)/R$, the term $Q(V)$ is $C_1 V_1$ and $W(t)$ is simply I_1. Likewise at Node 2:

$$0 = (V_2 - V_1)/R_1 + V_2/R_2 + gmV_1 \tag{13.3}$$

In this example, G and Q are simple linear terms; however, in general, they can be non-linear functions of the voltage vector V.

13.5 Modified Nodal Analysis

Normal nodal analysis that uses only Kirchoff's current law, cannot be used to represent ideal voltage sources or inductors. This is so because the branch current in these elements cannot be expressed as a function of the branch voltage. To resolve this problem, KVL is used to write a loop equation around each inductor or voltage source. Consider Figure 13.4 for an example of this procedure. The unknowns to be solved for are the voltage V_1 at Node 1, V_2 the voltage at Node 2, V_3 the voltage at Node 3, the current flowing through voltage source V_1 which we shall call I_x and the current flowing in the inductor L_1 which we shall call I_1. The system of equations is:

$$0 = V_1/R_1 + I_x$$

$$0 = V_2/R_2 - I_z + I_l$$

$$0 = V_3/R_3 - I_l$$

$$0 = V_1 - V_x + V_2 \tag{13.4}$$

$$0 = V_2 + \frac{d(L_1 I_1)}{dt} - V_3$$

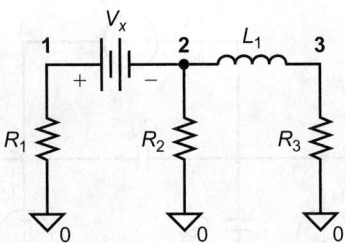

FIGURE 13.4 Circuit for modified nodal analysis.

The use of modified nodal analysis does have the disadvantage of requiring that an additional equation be included for each inductor or voltage source, but has the advantage that ideal voltage sources can be used. The total number of equations to be solved is therefore the number of nodes plus the number of voltages sources and inductors.

13.6 Active Device Models

VLSI circuits contain active devices like transistors or diodes which act as amplifiers. These devices are normally described by a complicated set of non-linear equations. We shall consider a simple model for the bipolar transistor—the Ebers-Moll model. This model is one of the first developed, and while it is too simple for practical application, it is useful for discussion.

A schematic of the Ebers-Moll model is shown in Figure 13.5. The model contains three non-linear voltage-dependent current sources I_c, I_{bf}, and I_{br} and two non-linear capacitances C_{be} and C_{bc}. The current flowing in the three current sources are given by the following equations:

$$I_c = I_s(\exp(V_{be}/V_t)) - \exp(V_{ce}/V_t) \tag{13.5}$$

$$I_{bf} = \frac{I_s}{B_f}(\exp(V_{be}/V_t)-1) \tag{13.6}$$

$$I_{br} = \frac{I_s}{B_r}(\exp(V_{bc}/V_t)-1) \tag{13.7}$$

The voltages V_{be} and V_{bc} are the voltages between base and emitter and the base and collector, respectively. I_s, B_f and B_r are three user-defined parameters which govern the DC operation of the BJT. V_t is the "thermal voltage" or kT/q, which has the numerical value of approximately 0.26 volts at room temperature. Observe that in the normal forward active mode, where $V_{be} > 0$ and $V_{ce} < 0$, I_{br} and the second term in I_c vanish and the current gain of the BJT, which is defined as I_c/I_b becomes numerically equal to B_f. Likewise, in the reverse mode where $V_{ce} > 0$ and $V_{be} < 0$, the reverse gain (I_e/I_b) is equal to B_r.

The two capacitances in Figure 13.5 contribute charge to the emitter, base, and collector, and this charge is given by the following equations:

$$Q_{be} = \tau_f I_s(\exp V_{be}/V_t -1) + C_{je}\int_0^{V_{be}}(1-V/V_{je})^{-m_e} \tag{13.8}$$

$$Q_{bc} = \tau_r I_s(\exp V_{bc}/V_t - 1) + C_{jc}\int_0^{V_{bc}}(1-V/V_{jc})^{-m_c} \tag{13.9}$$

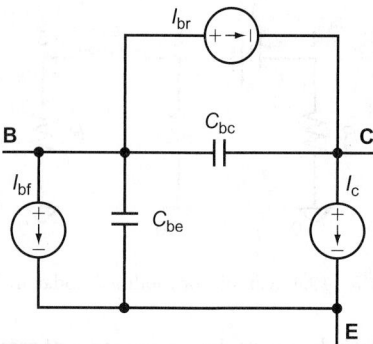

FIGURE 13.5 The Ebers-Moll model for the bipolar transistor.

Q_{be} contributes positive charge to the base and negative charge to the emitter. Q_{bc} contributes positive charge to the base and negative charge to the collector. The first term in each charge expression is due to charge injected into the base from the emitter for Q_{be} and from the collector into the base for Q_{bc}. Observe that the exponential terms in the charge terms are identical to the term in I_c. This is so because the injected charge is proportional to the current flowing into the transistor. The terms τ_f and τ_r are the forward and reverse transit times, respectively, and correspond to the amount of time it takes the electrons (or holes) to cross the base. The second term in the charge expression (the term with the integral) corresponds to the charge in the depletion region of the base–emitter junction for Q_{be} and in the base–collector junction for Q_{bc}. Recall that the depletion width in a pn junction is a function of the applied voltage. The terms V_{je} and V_{jc} are the "built-in" potentials with units of volts for the base–emitter and base–collector junctions. The terms m_c and m_e are the grading coefficients for the two junctions and are related to how rapidly the material changes from n-type to p-type across the junction.

This "simple" model has eleven constants I_s, B_f, B_r, C_{je}, C_{jc}, M_e, M_c, V_{je}, V_{jc}, T_f, and T_r which must be specified by the user. Typically, these constants would be extracted from measured *I-V* and *C-V* data taken from real transistors using a fitting or optimization procedure (typically a non-linear least-squares fitting method is needed). The Ebers-Moll model has a number of shortcomings which are addressed in newer models like Gummel-Poon, Mextram, and VBIC. The Gummel-Poon model has over 40 parameters that must be adjusted to get a good fit to data in all regions of operation.

Models for MOS devices are even more complicated than the bipolar models. Modeling the MOSFET is more difficult than the bipolar transistor because it is often necessary to use a different equation for each of the four regions of operation (off, subthreshold, linear, saturation) and the drain current and capacitance are functions of three voltages (V_{ds}, V_{bs}, and V_{gs}) rather than just two (V_{be} and V_{ce}) as in the case of the BJT. If a Newton-Raphson solver is to be used, the *I-V* characteristics and capacitances must be continuous and it is best if their first derivatives are continuous as well. Furthermore, MOS models contain the width (W) and length (L) of the MOSFET channel as parameters; and for the best utility the model should remain accurate for many values of W and L. This property is referred to as "scalability."

Over the years, literally hundreds of different MOS models have been developed. However, for modern VLSI devices, only three or four are commonly used today. These are the SPICE Level-3 MOS model, the HSPICE Level-28 model (which is a proprietary model developed by Meta Software), the public domain BSIM3 model developed at UC Berkeley, and MOS9 developed at Phillips. These models are supported by many of the "silicon foundries," that is, parameters for the models are provided to chip designers by the foundries. BSIM3 has been observed to provide a good fit to measured data and its *I-V* curves to be smooth and continuous (thereby resulting in good simulator convergence). The main drawback of BSIM3 is that it has over 100 parameters which are related in intricate ways, making extraction of the parameter set a difficult process.

A process known as "binning" is used to provide greater accuracy. When binning is used, a different set of model parameters is used for each range of the channel length and width (L and W). An example

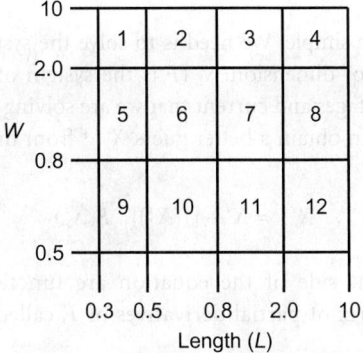

FIGURE 13.6 Binning of MOS parameters.

of this is shown in Figure 13.6. For a given type of MOSFET, 12 complete sets of model parameters are extracted and each is valid for a given range. For example, in Figure 13.6, the set represented by the number "11" would only be valid for channel lengths between 0.8 and 2.0 microns and for channel widths between 0.5 and 0.8 microns. Thus, for a typical BSIM3 model with about 60 parameters, $12 \times 60 = 720$ parameters would need to be extracted in all and this just for one type of device.

Many commercial simulators contain other types of models besides the traditional R, L, C, MOS, and BJT devices. Some simulators contain "behavioral" models which are useful for systems design or integration tasks; examples of these are integrators, multipliers, summation, and LaPlace operator blocks. Some simulators are provided with libraries of prefitted models for commercially available operational amplifiers, logic chips, and discrete devices. Some programs allow "mixed-mode" simulation, which is a combination of logic simulation (which normally allows only a few discrete voltage states) and analog circuit simulation.

13.7 Types of Analysis

For analog circuits, there are three commonly used methods of analysis, these being DC, AC, and transient analysis. DC analysis is used to examine the steady-state operation of a circuit; that is, what the circuit voltages and currents would be if all inputs were held constant for an infinite time. AC analysis (or sinusoidal steady state) examines circuit performance in the frequency domain using phasor analysis. Transient analysis is performed in the time domain and is the most powerful and computationally intensive of the three. For special applications, other methods of analysis are available such as the Harminic-Balance method, which is useful for detailed analysis of non-linear effects in circuits excited by purely periodic signals (like mixers and RF amplifiers).

13.7.1 DC (Steady-State) Analysis

DC analysis calculates the steady-state response of a circuit (with all inductors shorted and capacitors removed). DC analysis is used to determine the operating point (*Q*-point) of a circuit, power consumption, regulation and output voltage of power supplies, transfer functions, noise margin and fanout in logic gates, and many other types of analysis. In addition, a DC solution must be calculated to find the starting point for AC and transient analysis.

To calculate the DC solution, we need to solve Kirchoff's equations formulated earlier. Unfortunately, since the circuit elements will be non-linear in most cases, a system of transcendental equations will normally result and it is impossible to solve this system analytically. The method which has met with the most success is Newton's method or one of its derivatives.

13.7.1.1 Newton's Method

Newton's method is actually quite simple. We need is to solve the system of equations $F(X) = 0$ for X, where both F and X are vectors of dimension N. (F is the system of equations from modified nodal analysis, and X is the vector of voltages and current that we are solving for). Newton's method states that given an initial guess for X^i, we can obtain a better guess X^{i+1} from the equation:

$$X^{i+1} = X^i - [J(X^i)]^{-1} F(X^i) \tag{13.10}$$

Note that all terms on the right side of the equation are functions only of the vector X^i. The term $J(X)$ is a $N \times N$ square matrix of partial derivatives of F, called the Jacobian. Each term in J is given by:

$$J_{i,j} = \frac{\partial F_i(X)}{\partial X_j} \tag{13.11}$$

We assemble the Jacobian matrix for the circuit at the same time that we assemble the circuit equations. Analytic derivatives are used in most simulators.

The −1 in Eq. 13.10 indicates that we need to invert the Jacobian matrix before multiplying by the vector F. Of course, we do not need to actually invert J to solve the problem; we only need to solve the linear problem $F = YJ$ for the vector Y and then calculate $X^{i+1} = X^i - Y$. A direct method such as the LU decomposition is usually employed to solve the linear system.

For the small circuit of Figure 13.3, analyzed in steady state (without the capacitor), the Jacobian entries are:

$$J_{1,1} = 1/R_1 \qquad\qquad J_{1,2} = -1/R_1$$
$$\tag{13.12}$$
$$J_{2,1} = 1/R_1 + gm \qquad\qquad J_{2,2} = 1/R_1 + 1/R_2$$

For a passive circuit (i.e., a circuit without gain), the Jacobian will be symmetric and for any row, the diagonal entry will be greater than the sum of all the other entries.

Newton's method converges quadratically, provided that the initial guess X^i is sufficiently close to the true solution. Quadratically implies that if the distance between X^i and the true solution is d, then the distance between X^{i+1} and the true solution will be d^2. Of course, we are assuming that d is small to start with. Still, programs like SPICE may require 50 or more iterations to achieve convergence. The reason for this is that, often times, the initial guess is poor and quadratic convergence is not obtained until the last few iterations. There are additional complications like the fact that the model equations can become invalid for certain voltages. For example, the BJT model will "explode" if a junction is forward-biased by more than 1 V or so since: $\exp(1/Vt) = 5e16$. Special limiting or damping methods must be used to keep the voltages and currents to within reasonable limits.

13.7.1.2 Example Simulation

Most circuit simulators allow the user to ramp one or more voltage sources and plot the voltage at any node or the current in certain branches. Returning to the differential pair of Figure 13.1, we can perform a DC analysis by simply adding a .DC statement (see Figure 13.7). A plot of the differential output voltage (between the two collectors) and the voltage at the two emitters is shown in Figure 13.8. Observe that the output voltage is zero when the differential pair is "balanced" with 2.0 V on both inputs. The output saturates at both high and low values for V_1, illustrating the non-linear nature of the analysis. This simulation was run using the PSPICE package from MicroSim Corporation. The simulation run is a few seconds on a 486 type PC.

```
Steady state analysis of differential pair.
V1  4  0  2V
V2  5  0  2V
V3  1  0  5V
R1  2  0  1k
R2  3  1  1k
R3  6  1  1k
Q1  3  4  2  m2n2222
Q2  6  5  2  m2n2222
.model m2n2222 NPN IS=1e-12 BF=100 BR=5 TF=100pS
.dc V1 1.0 3.0 0.01
```

FIGURE 13.7 Input file for DC sweep of V_1.

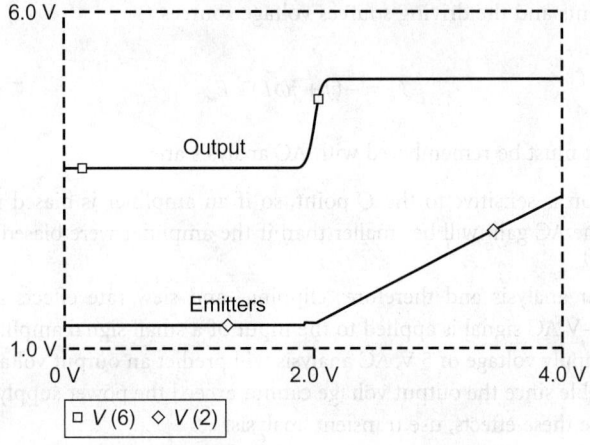

FIGURE 13.8 Output from DC analysis.

13.7.2 AC Analysis

AC analysis is performed in the frequency domain under the assumption that all signals are represented as a DC component V_{dc} plus a small sinusoidal component V_{ac}.

$$V = V_{dc} + V_{ac}\exp(j\omega t)$$ (13.13)

Here, $j = \sqrt{-1}$, ω is the radial frequency $(2\pi f)$, and V_{ac} is a complex number. Expanding (1) about the DC bias point V_{dc} (also referred to as the Q point), we obtain:

$$F(V) = F(V_{dc}) + W_{dc} + W_{ac} + \frac{\partial G(V_{dc})}{\partial V_{dc}}V_{ac} + \frac{\partial}{\partial t}\left(\frac{\partial Q(V_{dc})}{\partial V_{dc}}\right)V_{ac} + \alpha V_{ac}^2 \Lambda$$ (13.14)

The series has an infinite number of terms; however, we assume that if V_{ac} is sufficiently small, all terms above first order can be neglected. The first two terms on the right-hand side are the DC solution and, when taken together, yield zero. The third term W_{ac} is the vector of independent AC current sources which drive the circuit. The partial derivative in the fourth term is the Jacobian element, and the derivative of Q in parentheses is the capacitance at the node. When we substitute the exponential into Eq. 13.14, each term will have an exponential term that can be canceled. The result of all these simplifications is the familiar result:

$$0 = W_{ac} + JV_{ac} + j\omega CV_{ac}$$ (13.15)

This equation contains only linear terms which are equal to the partial derivatives of the original problem evaluated at the Q point. Therefore, before we can solve the AC problem, we must calculate the DC bias point. Rearranging terms slightly, we obtain:

$$V_{ac} = -(J + j\omega C)^{-1} W_{ac} \tag{13.16}$$

The solution at a given frequency can be obtained from a single matrix inversion. The matrix, however, is complex but normally the complex terms share a sparsity pattern similar to the real terms. It is normally possible (in FORTRAN and C++) to create a suitable linear solver by taking the linear solver which is used to calculate the DC solution and substituting "complex" variables for "real" variables. Since there is no non-linear iteration, there are no convergence problems and AC analysis is straightforward and fool-proof.

The same type of analysis can be applied to the equations for modified nodal analysis. The unknowns will of course be currents and the driving sources voltage sources.

$$I_{ac} = -(J + j\omega L)^{-1} E_{ac} \tag{13.17}$$

The only things that must be remembered with AC analysis are:

1. The AC solution is sensitive to the Q point, so if an amplifier is biased near its saturated DC output level, the AC gain will be smaller than if the amplifier were biased near the center of its range.
2. This is a linear analysis and therefore "clipping" and slew rate effects are not modeled. For example, if a 1-V AC signal is applied to the input of a small signal amplifier with a gain of 100 and a power supply voltage of 5 V, AC analysis will predict an output voltage of 100 V. This is of course impossible since the output voltage cannot exceed the power supply voltage of 5 V. If you want to include these effects, use transient analysis.

13.7.2.1 AC Analysis Example

In the following example, we will analyze the differential pair using AC analysis to determine its frequency response. To perform this analysis in SPICE, we need only specify which sources are the AC driving sources (by adding the magnitude of the AC signal at the end) and specify the frequency range on the .AC statement (see Figure 13.9). SPICE lets the user specify the range as linear or "decade," indicating that we desire a logarithmic frequency scale. The first number is the number of frequency points per decade. The second number is the starting frequency, and the third number is the ending frequency.

Figure 13.10 shows the results of the analysis. The gain begins to roll off at about 30 MHz due to the parasitic capacitances within the transistor models. The input impedance(which is plotted in kΩ) begins to roll off

```
AC analysis of differential pair.
V1 4 0 2V AC 1
V2 5 0 2V
V3 1 0 5V
R1 2 0 1k
R2 3 1 1k
R3 6 1 1k
Q1 3 4 2 m2n2222
Q2 6 5 2 mq2n2222
.model m2n2222 NPN IS=1e-12 BF=100 BR=5 TF=100pS
.AC DEC 10 1e3 1e9
```

FIGURE 13.9 Input file for AC analysis.

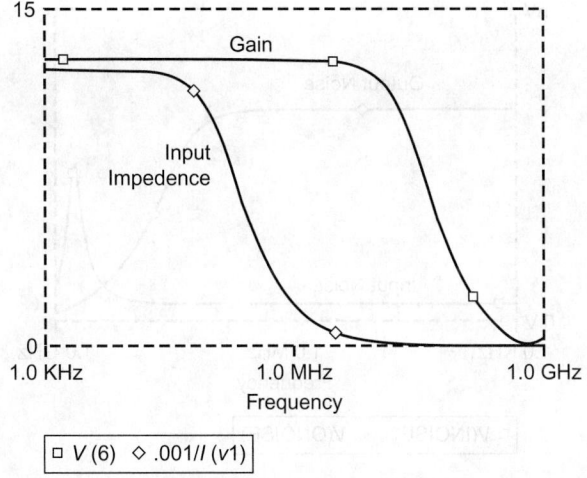

FIGURE 13.10 Gain and input impedance calculated by AC analysis.

at a much lower frequency. The reduction in input impedance is due to the increasing current that flows in the base-emitter capacitance as the current increases. SPICE does not have a method of calculating input impedance, so we have calculated it as $Z = V_{in}/I(V_{in})$, where $V_{in} = 1.0$, using the post-processing capability of PSPICE. This analysis took about 2 seconds on a 486 type PC.

13.7.2.2 Noise Analysis

Noise is a problem primarily in circuits that are designed for the amplification of small signals like the RF and IF amplifiers of a receiver. Noise is the result of random fluctuations in the currents which flow in the circuit and is generated in every circuit element. In circuit simulation, noise analysis, is an extension of AC analysis. During noise analysis, it is assumed that every circuit element contributes some small noise component either as a voltage V_n in series with the element or as a current I_n across the element. Since the noise sources are small in comparison to the DC signal levels, AC small signal analysis is an applicable analysis method.

Different models have been developed for the noise sources. In a resistor, thermal noise is the most important component. Thermal noise is due to the random motion of the electrons:

$$I_n^2 = \frac{4kT\Delta f}{R} \tag{13.18}$$

where T is the temperature, k is Boltzman's constant, and Δf is the bandwidth of the circuit. In a semiconductor diode, shot noise is important. Shot noise is related to the probability that an electron will surmount the semiconductor barrier energy and be transported across the junction:

$$I_n^2 = 2qI_d\Delta f \tag{13.19}$$

There are other types of noise that occur in diodes and transistors; examples are flicker and popcorn noise. Noise sources, in general, are frequency dependent.

Noise signals will be amplified or attenuated as they pass through different circuits. Normally, noise is referenced to some output point called the "summing node." This would normally be the output of the amplifier where we would actually measure the noise. We can call the gain between the summing node and the current flowing in an element j in the circuit $A_j(f)$. Here, f is the analysis frequency since the gain will normally be frequency dependent.

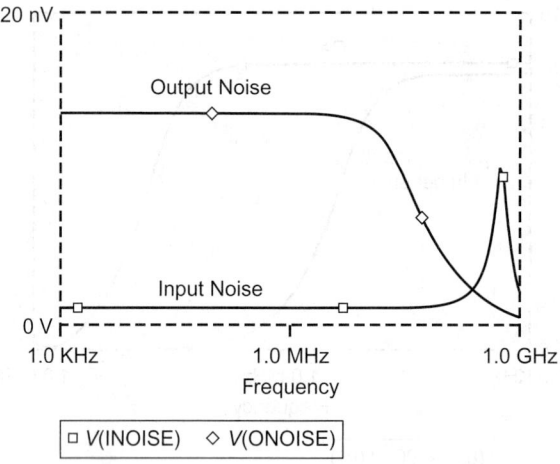

FIGURE 13.11 Noise referenced to output and input.

Noise signals are random and uncorrelated to each other so their magnitudes must be root-mean-squared summed rather than simply summed. Summing all noise sources in a circuit yields:

$$I_n(f) = \sqrt{\sum_j A_j^2(f) I_j^2(f)}$$

(13.20)

It is also common to reference noise back to the amplifier input and this is easily calculated by dividing the above expression by the amplifier gain. Specifying noise analysis in SPICE is simple. All the user needs to do is add a statement specifying the summing node and the input source. Spice then calculates the noise at each as a function of frequency

```
.noise v([6]) V1
```
(13.21)

See Figure 13.11 for example output. Many circuit simulators will also list the noise contributions of each element as part of the output. This is particularly helpful in locating the source of noise problems.

13.7.3 Transient Analysis

Transient analysis is the most powerful analysis capability because the transient response of a circuit is so difficult to calculate analytically. Transient analysis can be used for many types of analysis, such as switching speed, distortion, and checking the operation of circuits like logic gates, oscillators, phase-locked loops, or switching power supplies. Transient analysis is also the most CPU intensive and can require 100 or 1000 times the CPU time of DC or AC analysis.

13.7.3.1 Numerical Method

In transient analysis, time is discretized into intervals called *time steps*. Typically, the time steps are of unequal length, with the smallest steps being taken during intervals where the circuit voltages and currents are changing most rapidly. The following procedure is used to discretize the time-dependent terms in Eq. 13.1.

Time derivatives are replaced by difference operators, the simplest of which is the forward difference operator:

$$\frac{dQ(t_k)}{dt} = \frac{Q(t_{k+1}) - Q(t_k)}{h}$$

(13.22)

where h is the time step given by $h = t_{k+1} - t_k$. We can easily solve for the charge $Q(t_{k+1})$ at the next time point:

$$Q(t_{k+1}) = Q(t_k) - h(G_i(V(t_k)) + W_i(t_k)) \tag{13.23}$$

using only values from past time points. This means that it would be possible to solve the system simply by plugging in the updated values for V each time. This can be done without any matrix assembly or inversion and is very nice. (Note for simple linear capacitors, $V = Q/C$ at each node, so it is easy to get V back from Q.) However, this approach is undesirable for circuit simulation for two reasons. (1) The charge Q, which is a "state variable" of the system, is not a convenient choice since some nodes may not have capacitors (or inductors) attached, in which case they will not have Q values. (2) It turns out that forward (or explicit) time discretization methods like this one are unstable for "stiff" systems, and most circuit problems result in "stiff systems." The term "stiff system" refers to a system that has greatly varying time constants.

To overcome the stiffness problem, we must use implicit time discretization methods which, in essence, means that the G and W terms in the above equations must be evaluated at t_{k+1}. Since G is non-linear, we will need to use Newton's method once again.

The most popular implicit method is the trapezoidal method. The trapezoidal method has the advantage of only requiring information from one past time point and, furthermore, has the smallest error of any method requiring one past time point. The trapezoidal method states that if I is the current in a capacitor, then:

$$I(t_{k+1}) = \frac{dQ}{dt} = 2\frac{Q(V(t_{k+1})) - Q(V(t_k))}{h} - I(t_k) \tag{13.24}$$

Therefore, we need only substitute the above equation into Eq. (13.1) to solve the transient problem. Observe that we are solving for the voltages $V(t_{k+1})$, and all terms involving t_k are constant and will not be included in the Jacobian matrix. An equivalent electrical model for the capacitor is shown in Figure 13.12. Therefore, the solution of the transient problem is in effect a series of DC solutions where the values of some of the elements depend on voltages from the previous time points.

All modern circuit simulators feature automatic time step control. This feature selects small time steps during intervals where changes are occurring rapidly and large time steps in intervals where there is little change. The most commonly used method of time step selection is based on the local truncation error (LTE) for each time step. For the trapezoidal rule, the LTE is given by:

$$\varepsilon = \frac{h^3}{12}\frac{d^3x}{dt^3}(\xi) \tag{13.25}$$

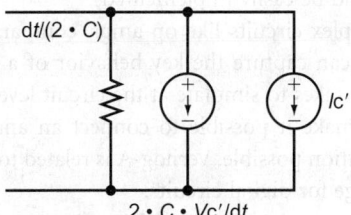

FIGURE 13.12 Electrical model for a capacitor; the two current sources are independent sources. The prime (′) indicates values from a preceding time point.

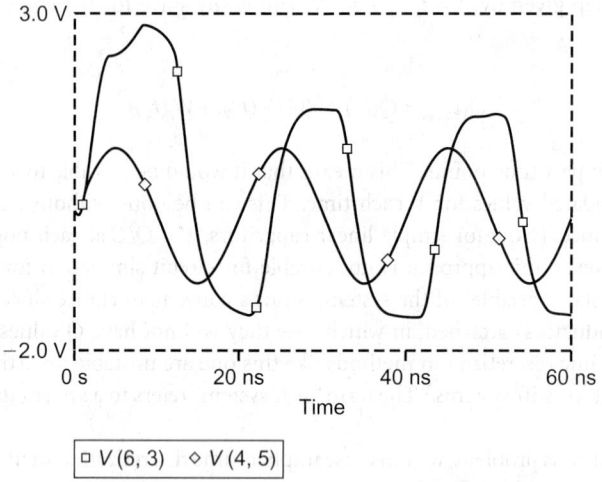

FIGURE 13.13   Transient response $V(6,3)$ of differential amplifier to sinusoidal input at $V(4,5)$.

and represents the maximum error introduced by the trapezoidal method at each time step. If the error (ε) is larger than some preset value, the step size is reduced. If the error is smaller, then the step size is increased. In addition, most simulators select time points so that they coincide with the edges of pulse-type waveforms.

13.7.3.2   Transient Analysis Examples

As a simple example, we return to the differential pair and apply a sine wave differentially to the input. The amplitude (2 V p-p) is selected to drive the amplifier into saturation. In addition, we make the frequency (50 MHz) high enough to see phase shift effects. The output signal is therefore clipped due to the non-linearities and shifted in phase due to the capacitive elements in the transistor models (see Figure 13.13). The first cycle shows extra distortion since it takes time for the "zero-state" response to die out. This simulation, using PSPICE, runs in about one second on a 486 type computer.

13.8   Verilog-A

Verilog-A is a new language designed for simulation of analog circuits at various levels. Mathematical equations can be entered directly as well as normal SPICE-type circuit elements.

Groups of equations and elements can be combined into reusable "modules" that are similar to subcircuits. Special functions are also provided for converting analog signals into digital equivalents, and vice versa. Systems-type elements such as LaPlace operators, integrators, and differentators are also provided. This makes it possible to perform new types of modeling which were not possible in simulators like SPICE:

- Equations can be used to construct new models for electrical devices (for example, the Ebers-Moll model described earlier could be easily implemented).
- Behavioral models for complex circuits like op-amps, comparitors, phase detectors, etc. can be constructed. These models can capture the key behavior of a circuit and yet be simulated in a small fraction of the time it takes to simulate at the circuit level.
- Special interface elements make it possible to connect an analog block to a digital simulator, making mixed-mode simulation possible. Verilog-A is related to and compatible with the popular Verilog-D modeling language for digital circuits.

As an example, consider a phase-locked loop circuit which is designed as an 50X frequency multiplier. A block diagram for the PLL is shown in Figure 13.14 and the Verilog-A input listing is shown in Figures 13.15 and 13.16.

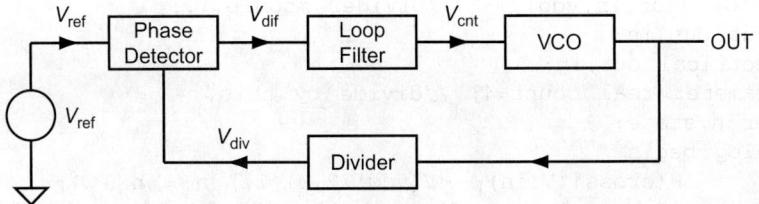

FIGURE 13.14 Block diagram of phase-locked loop.

```
'include "electrical.h"
'timescale 100ns / 10ns

module top;                          //Top level Module
        electrical Vdd, Vref, Vdif, Vcnt, Vout, Vdiv;
        pd #(.gain(15u), .rout(2e6), dir(-1)) p1(Vdif, Vdiv, Vref,Vdd);
        filter f1(Vdif, Vcnt);
        vco #(.gain(2e7), .center(3e7)) v1(Vout, Vcntm Vdd);
        divide #(.count(50)) d1(Vdiv, Vout, Vdd);
        analog begin
                V(Vdd)   <+ 5.0;
                V(Vref)  <+ 2.5+2.5*sin(6.238*1e6*$realtime);
        end
endmodule

module pd (out,vco,ref,vdd);         //Phase detector Module.
        inout out,vco,ref,vdd;
        electrical out,vco,ref,vdd;
        parameter real gain=15u, rout=5e6;  // gain & output impedance
        parameter integer dir=1; //1,-1 pos or neg edge trigger
        integer state;
        analog begin
                @(cross((V(ref) - V(vdd)/2),dir)) state = state -1;
                @(cross((V(vco) - V(vdd)/2),dir)) state = state +1;
                if (state > 1) state = 1;
                if (state < -1) state = -1;
                if (state !=0) I(out) <+ transition(state * gain,0,0,0);
                I(out) <+ V(out)/rout;
        end
endmodule
```

FIGURE 13.15 Part one of Verilog-A listing for PLL.

Simulation of this system at the circuit level is very time consuming due to the extreme difference in frequencies. The phase detector operates at a low frequency of 1.0 MHz, while the VCO operates at close to 50 MHz. However, we need to simulate enough complete cycles at the phase detector output to verify that the circuit correctly locks onto the reference signal.

The circuit is broken up into five blocks or modules: The "top module," VCO, divider, phase detector, and loop filter. The VCO has a simple linear dependence of frequency on the VCO input voltage and produces a sinusoidal output voltage. The VCO frequency is calculated by the simple expression freq = center + gain * ($V_{in} - V_{min}$). Center and gain are parameters which can be passed in when the VCO is created within the top module by the special syntax "#(gain(2e7),.center(3e7)" in the top module. If the parameters are not specified when the module is created,

```
module divide (out,in,vdd);    //Divider module....
       inout out,in;
       electrical out,in;
       parameter real count=4; //divide by this.
    integer n,state;
       analog begin
              @(cross((V(in) - V(vdd)/2.0),1)) n = n + 1;
              if (n >= count/2) begin
                    if (state == 0)state = 1;
                    else state = 0;
                    n = 0;
              end
              V(out) <+ transition(state*5,0,0,0);
       end
endmodule

module vco (vout, vin, vdd);          //VCO module
       inout vin, vout, vdd;
       electrical vin, vout, vdd;
     parameter real gain = 5e6, center = 40e6, vmin=1.8;
     real freq,vinp;
     analog begin
         vinp = V(vin);
         if (vinp < vm) vinp = vmin;
         freq = center + gain*(vinp-vmin);
         V(vout) <+ V(vdd)/2.0*sin(6.28318531*idt(freq,0))+V(vdd)/2.0;
     end
endmodule

'language SPICE
.SUBCKT filter Vin Vout
Rf2 Vin Vout 100
Rfilter Vin VC2 200000
C1 VC2 0 58p
C2 Vin 0 5p
.ends
'endlanguage
```

FIGURE 13.16 Part two of Verilog-A PLL listing.

then the default values specified within the module are used instead. The special V() operator is used to obtain the voltage at a node (in this case V(in) and V(Vdd). The sinusoidal output is created using the SIN and IDT operators. SIN calculates the sin of its argument. I_{dt} calculates the integral of its argument with respect to time. The amplitude of the output is taken from the V_{dd} input, thus making it easy to integrate the VCO block with others. Given that V_{dd} = 5 volts, gain = 2e7 Hz/V center = 3e7 Hz, and in = 1.8, the final expression for the VCO output is:

$$V_{out} = 2.5 + 2.5 \sin\left(3e7 + 2e7 \int 2\pi(V_{in} - 1.8)\,dt\right) \tag{13.26}$$

The phase detector functions as a charge pump which drives current into or out of the loop filter, depending on the phase difference between its two inputs. The @cross(V1,dir) function becomes true whenever signal V_1 crosses zero in the direction specified by dir. This either increments or decrements the variable STATE. The "transition" function is used to convert the STATE signal, which is essentially digital and changes abruptly, into a smoothly changing analog signal which can be applied to the rest of the circuit. The "<+" (or contribution) operator adds the current specified by the equation on the right to the node on the left. Therefore, the phase detector block forces current into the output node whenever the VCO signal leads the reference signal and forcing current out of the output node whenever the reference leads the vco signal. The phase detector also has an output resistance which is specified by parameter ROUT.

The loop filter is a simple SPICE subcircuit composed of two resistors and one capacitor. Of course, this subcircuit could contain other types of elements as well and can even contain other Verilog-A modules. The divider block simply counts zero crossings and, when the count reaches the preset divisor, the output of the divider is toggeled from 0 to 1, or vice versa. The transition function is used to ensure that a smooth, continuous analog output is generated by the divider.

This PLL was simulated using AMS from Antrim Design Systems. The results of the simulations are shown in Figure 13.17. The top of Figure 13.17 shows the output from the loop filter (V_{cnt}). After a few cycles, the PLL has locked onto the reference signal. The DC value of the loop filter output is approximately 2.8 V. Referring back to the VCO model, this gives an output frequency of $2e7*(2.8 - 1.8) + 3e7 = 50$ MHz, which is as expected. The lower portion of Figure 13.17 shows the divider output (V_{div}) and the reference signal (V_{ref}). It can be seen that the two signals are locked in phase. Figure 13.18 shows the VCO output and the divider output. As expected, the VCO frequency is 50 times the divider frequency.

The behavioral models used in this example are extremely simple ones. Typically, more complex models must be used to accurately simulate the operation of an actual PLL. A better model might include effects such as the non-linear dependence of the VCO frequency on the input voltage, the effects on signals introduced through power supply lines, delays in the divider and phase detector, and finite signal rise and fall times. These models can be built up from measurements, or transistor-level simulation of the underlying blocks (a process known as *characterization*). Of course, during the simulation, any of the behavioral blocks could be replaced by detailed transistor level models or complex Verilog-D digital models.

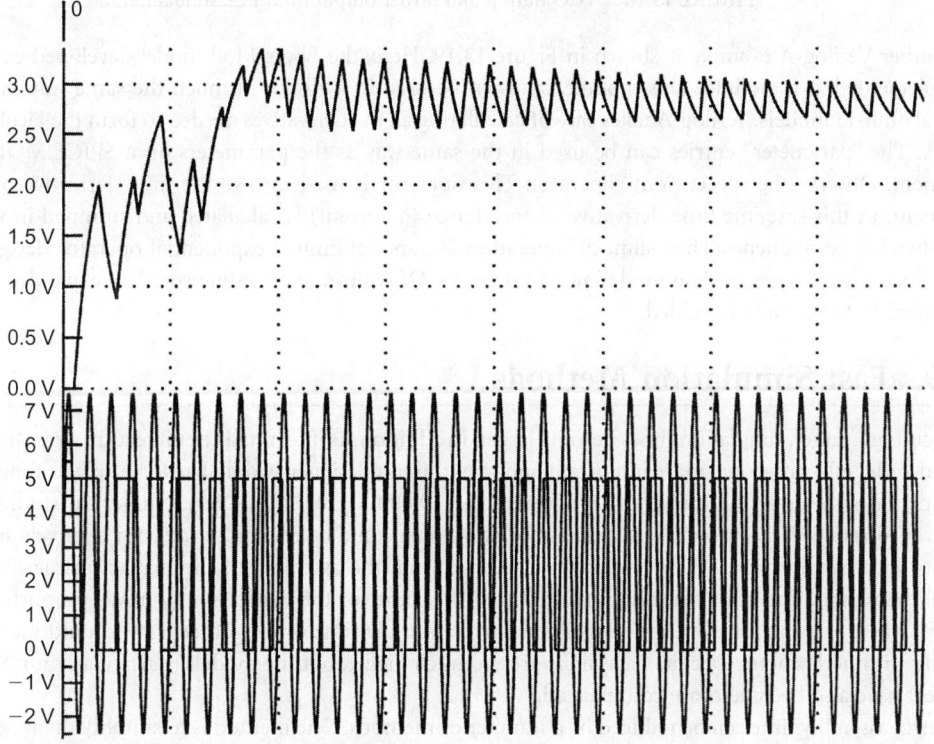

FIGURE 13.17   Loop filter output (top) V_{ref} and divider output (bottom) from PLL simulation.

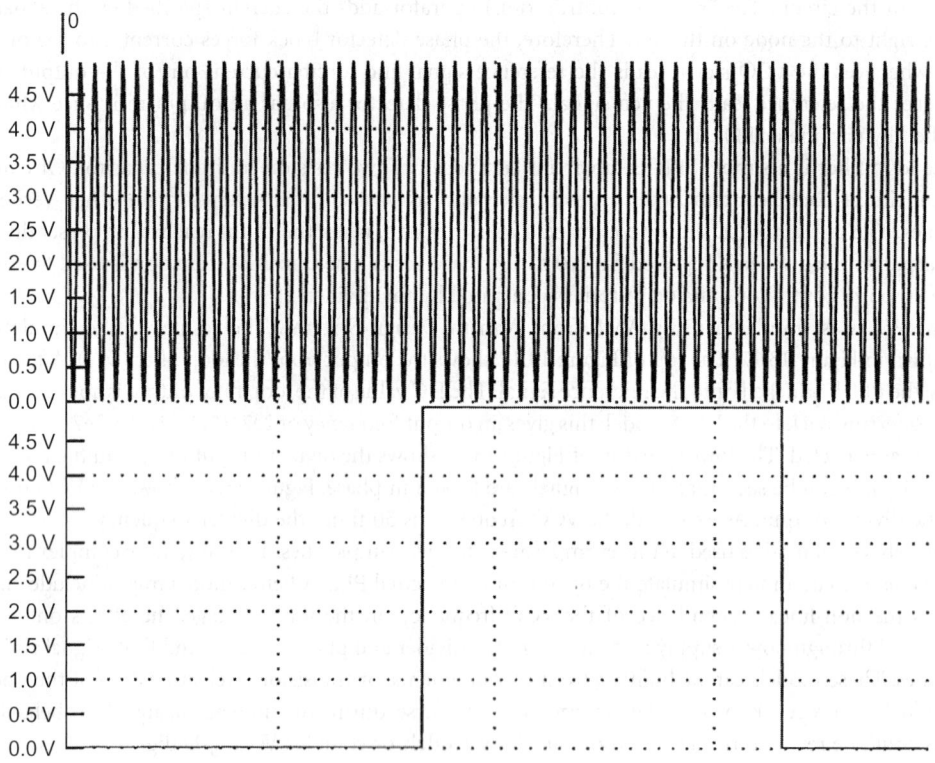

FIGURE 13.18   VCO output and divder output from PLL simulation.

Another Verilog-A example is shown in Figure 13.19. Here, the Ebers-Moll model developed earlier is implemented as a module. This module can then be used in a circuit in much the same way as the normal built-in models. Verilog-A takes care of calculating all the derivatives needed to form the Jacobian matrix. The "parameter" entries can be used in the same way as the parameters on a SPICE.MODEL statement. Observe the special "ddt" operator. This operator is used to take the time derivative of its argument. In this case, the time derivative of the charge (a current) is calculated and summed in with the other DC components. The "$limexp" operation is a special limited exponential operator designed to give better convergence when modeling pn junctions. Of course, this module could be expanded and additional features could be added.

13.9   Fast Simulation Methods

As circuits get larger, simulation times become larger. In addition, as integrated circuit feature sizes shrink, second-order effects become more important and many circuit designers would like to be able to simulate large digital systems at the transistor level (requiring 10,000 to 100,000 nodes). Numerical studies in early versions of SPICE showed that the linear solution time could be reduced to 26% for relatively small circuits with careful coding. The remainder is used during the assembly of the matrix, primarily for model evaluation. The same studies found that the CPU time for the matrix solution was proportional to $n^{1.24}$, where n is the number of nodes. The matrix assembly time on the other hand should increase linearly with node count. Circuits have since grown much bigger, but the models (particularly for MOS devices) have also become more complicated.

Matrix assembly time can be reduced by a number of methods. One method is to simplify the models; however, accuracy will be lost as well. A better way is to precompute the charge and current characteristics for the complicated models and store them into tables. During simulation, the actual current and charges

```
module mybjt (C,B,E);
      inout C,B,E;
      electrical C,B,E;
      parameter real is=1e-16, bf=100, br=10, tf=1n, tr=10n, cje=1p
      cjc=1p,
            vje=0.75, vjc=0.75, mje=0.33 mjc=0.33;
      real ibc, ibe, qbc, qbe, vbe, vbc, cc, cb, x, y;
      analog begin
            vbe=V(B,E);
            vbc=V(B,C);
            ibe=is*($limexp(vbe/$vt)-1);
            ibc=is*($limexp(vbc/$vt)-1);
            cb=ibe/bf + ibc/br;
            cc=ibe - ibc - ibc/br
            if(vbe < 0) begin
                  x=1-vbe/vje;
                  y=exp(-mje*ln(x));
                  qbe=vje*cje*(1-x*y)/(1.0-mje)+tf*ibe;
            end
            else qbe=cje*(vbe+0.5*mje*vbe*vbe/vje)+tf*ibe;
            if(vbc < 0) begin
                  x=1-vbc/vjc;
                  y=exp(-mjc*ln(x));
                  qbc = vjc*cjc*(1-x*y)/(1.0-mjc)+tr*ibc;
            end
            else qbc=cjc*(vbc+0.5*mjc*vbc*vbc/vjc)+tr*ibc;
            ibe=ddt(qbe);
            ibc=ddt(qbc);
            I(C,E) <+ cc - ibc;
            I(B,E) <+ cb + ibe + ibc;
      end
endmodule
```

FIGURE 13.19   Verilog-A implimentation of the Ebers-Moll model.

can be found from table lookup and interpolation, which can be done quickly and efficiently. However, there are some problems:

1. To assure convergence of Newton's method, both the charge and current functions and their derivatives must be continuous. This rules out most simple interpolation schemes and means that something like a cubic spline must be used.

2. The tables can become large. A MOS device has four terminals, which means that all tables will be functions of three independent variables. In addition, the MOSFET requires four separate tables (Id, Qg, Qd, Qb). If we are lucky, we can account for simple parameteric variations (like channel width) by a simple multiplying factor. However, if there are more complex dependencies as is the case with channel length, oxide thickness, temperature, or device type, we will need one complete set of tables for each device.

If the voltages applied to an element do not change from the past iteration to the present iteration, then there is no need to recompute the element currents, charges, and their derivatives. This method is referred to as taking advantage of latency and can result in large CPU time savings in logic circuits, particularly if coupled with a method which only refractors part of the Jacobian matrix. The tricky part is knowing when the changes in voltage can be ignored. Consider, for example, the input to a high-gain op-amp, here ignoring a microvolt change at the input could result in a large error at the output. Use of sophisticated latency-determining methods could also cut into the savings.

Another set of methods are the waveform relaxation techniques which increase efficiency by temporarily ignoring couplings between nodes. The simplest version of the method is as follows. Consider a circuit with n nodes which requires m time points for its solution. The circuit can be represented by the vector equation:

$$F_i(V(t)) + \frac{dQ_i(V(t))}{dt} = 0 \qquad (13.27)$$

Using trapezoidal time integration gives a new function:

$$W_i(V(k)) = F_i(V(k)) + F_i(V(k-1)) + 2[Q_i(V(k)) - Q_i(V(k-1))]/dt = 0 \qquad (13.28)$$

We need to find the $V(k)$ which makes W zero for all k time points at all i nodes. The normal method solves for all n nodes simultaneously at each time point before advancing k. Waveform relaxation solves for all m time points at a single node (calculates the waveform at that node) before advancing to the next node. An outer loop is used to assure that all the individual nodal waveforms are consistent with each other.

Waveform relaxation is extremely efficient as long as the number of outer loops is small. The number of iterations will be small if the equations are solved in the correct order; that is, starting on nodes which are signal sources and following the direction of signal propagation through the circuit. This way, the waveform at node $i + 1$ will depend strongly on the waveform at node i, but the waveform at node i will depend weakly on the signal at node $i + 1$. The method is particularly effective if signal propagation is unidirectional, as is sometimes the case in logic circuits. During practical implementation, the total simulation interval is divided into several subintervals and the subintervals are solved sequentially. This reduces the total number of time points which must be stored. Variants of the method solve small numbers of tightly coupled nodes as a group; such a group might include all the nodes in a TTL gate or in a small feedback loop. Large feedback loops can be handled by making the simulation time for each subinterval less than the time required for a signal to propagate around the loop.

The efficiency of this method can be further improved using different time steps at different nodes, yielding a multi-rate method. This way, during a given interval, small time steps are used at active nodes while large steps are used at inactive nodes (taking advantage of latency).

13.10 Commercially Available Simulators

The simulations in this chapter were performed with the evaluation version of PSPICE from Microsim and AMS from Antrim design systems. The following vendors market circuit simulation software. The different programs have strengths in different areas and most vendors allow you to try their software in-house for an "evaluation period" before you buy.

SPICE2-SPICE3	University of California Berkeley, CA
AMS	Antrim Design Systems, Scotts Valley, CA., www.antrim.com
PSPICE	Orcad Corporation, Irvine, CA, www.orcad.com
HSPICE	Avant! Corporation, Fremont, CA, www.avanticorp.com
ISPICE	Intusoft, SanPedro, CA, www.intusoft.com
SABER	Analogy, Beaverton, OR, www.analogy.com
SPECTRE	Cadence Design Systems, San Jose, CA, www.cadence.com
TIMEMILL	Synopsys Corporation, Sunnyvale, CA, www.synopsys.com
ACCUSIM II	Mentor Graphics, Wilsonville, OR, www.mentorg.com

References

On general circuit simulation:

1. J. Vlach and K. Singhal, *Computer Methods for Circuit Analysis and Design*, New York, Van Nostrand Reinhold, 1983.
2. A.E. Ruehli (Editor), *Circuit Analysis, Simulation and Design, Part 1*, Elsevier Science Publishers, B.V. North Holland, 1981.
3. L. Nagel, SPICE2: A Computer Program to Simulate Semiconductor Circuits, Ph.D. thesis, University of California, Berkeley, 1975.
4. K. Kundert, *The Designers Guide to SPICE and SPECTRE*, Kluwer Academic Publishers, 1995.
5. P.W. Tuinenga, *SPICE, A Guide to Circuit Simulation and Analysis Using PSPICE*, Prentice-Hall, Englewood Cliffs, NJ, 1988.
6. J.A. Connelley and P. Choi, *Macromodeling with SPICE*, Prentice-Hall, Engelwood Cliffs, NJ, 1992.
7. P. Gray and R. Meyer, *Analysis and Design of Analog Integrated Circuits*, Wiley, New York, 1977.
8. A. Vladimiresch, *The SPICE Book*, Wiley, New York, 1994.

On modern techniques:

9. A.E. Ruehli (Editor), *Circuit Analysis, Simulation and Design, Part 2*, Elsevier Science Publishers, B.V. North Holland, 1981.

On device models:

10. P. Antognetti and G. Massobrio, *Semiconductor Modeling with SPICE2*, McGraw-Hill, New York, 2nd ed., 1993.
11. D. Foty, *MOSFET Modeling with SPICE*, Prentice-Hall, Englewood Cliffs, NJ, 1997.
12. BSIM3 Users Guide, www-device.EECS.Berkeley.EDU/~bsim3.
13. MOS-9 models, www-us2.semiconductors.philips.com/Philils_Models.

On Verilog-A:

14. D. Fitzpatrick, *Analog Behavior Modeling with the Verilog-A Language*, Kluwer-Academic Publishers, 1997.

14

Interconnect Modeling and Simulation

Michel S. Nakhla and
Ramachandra Achar

Carleton University

14.1 Introduction

With the rapid developments in VLSI technology, design, and CAD techniques, at both the chip and package level, the central processor cycle times are reaching the vicinity of 1 ns and communication switches are being designed to transmit data that have bit rates faster than 1 Gb/s. The ever-increasing quest for high-speed applications is placing higher demands on interconnect performance and highlights the previously negligible effects of interconnects (Figure 14.1), such as ringing, signal delay, distortion, reflections, and crosstalk [1–33]. In addition, the trend in the VLSI industry toward miniature designs, low power consumption, and increased integration of analog circuits with digital blocks has further complicated the issue of signal integrity analysis. Figure 14.2 describes the effect of scaling of chip on the global interconnect delay. As seen, the global interconnect delay grows as a cubic power of the scaling factor [1]. It is predicted that interconnects will be responsible for nearly 70 to 80% of the signal delay in high-speed systems.

Thousands of engineers, intent on the best design possible, use SPICE [7] on a daily basis for analog simulation and general-purpose circuit analysis. However, the high-speed interconnect problems are not always handled appropriately by the present levels of SPICE. If not considered during the design stage, these interconnect effects can cause logic glitches that render a fabricated digital circuit inoperable, or they can distort an analog signal such that it fails to meet specifications. Since extra iterations in the design cycle are costly, accurate prediction of these effects is a necessity in high-speed designs. Hence, it becomes extremely important for designers to simulate the entire design along with interconnect subcircuits as efficiently as possible while retaining the accuracy of simulation [12,26–65].

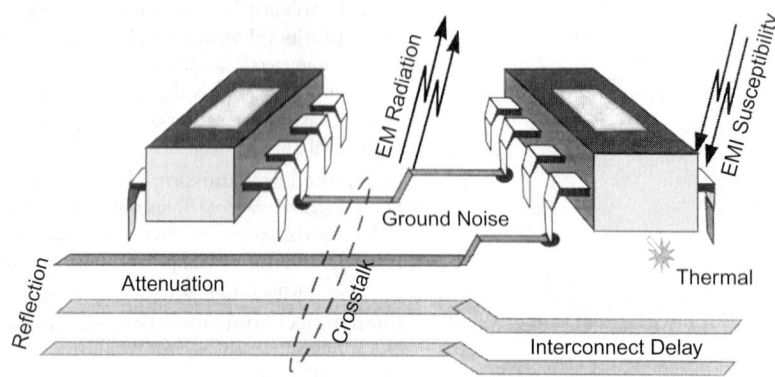

FIGURE 14.1 High-speed interconnect effects.

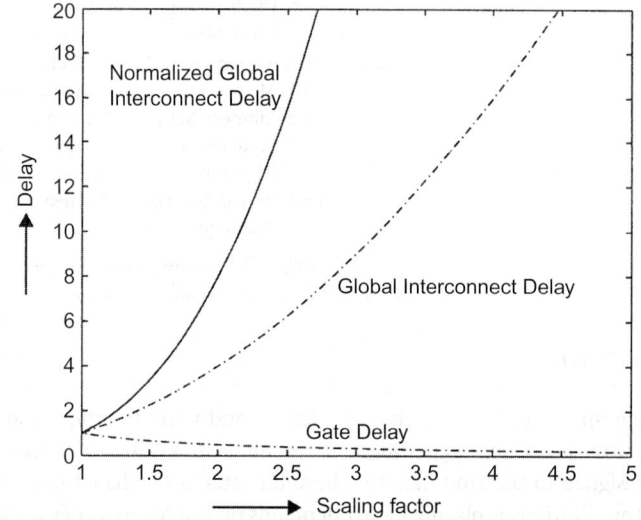

FIGURE 14.2 Impact of scaling on signal delay in high-speed systems.

14.1.1 What Is High Speed?

Speaking on a broader perspective, a "high-speed interconnect" is the one in which the time taken by the propagating signal to travel between its end points cannot be neglected. An obvious factor that influences this definition is the physical extent of the interconnect—the longer the interconnect, the more time the signal takes to travel between its end points. Smoothness of signal propagation suffers once the line becomes long enough for the signal's rise/fall times to roughly match its propagation time through the line. Then, the interconnect electrically isolates the driver from the receivers, which no longer function directly as loads to the driver. Instead, within the time of the signal's transition between its high and low voltage levels, the impedance of the interconnect becomes the load for the driver and also the input impedance to the receivers [1–6]. This leads to various transmission line effects, such as reflections, overshoot, undershoot, and crosstalk, and modeling of these needs the blending of EM and circuit theory.

Alternatively, the term "high-speed" can be defined in terms of the frequency content of the signal. At low frequencies, an ordinary wire (i.e., an interconnect) will effectively short two connected circuits. However, this is not the case at higher frequencies. The same wire, which is so effective at lower frequencies for connection purposes, has too much inductive/capacitive effect to function as a short at higher frequencies. Faster clock speeds and sharper slew rates tend to add more and more high-frequency contents.

An important criterion used for classifying interconnects is the *electrical length* of an interconnect. An interconnect is considered to be "electrically short," if at the highest operating frequency of interest, the interconnect length is physically shorter than one-tenth of the wavelength (i.e., length of the interconnect/$\lambda < 0.1$, $\lambda = v/f$). Otherwise the interconnect is referred as electrically long [1,8]. In most digital applications, the desired highest operating frequency (which corresponds to the minimum wavelength) of interest is governed by the rise/fall time of the propagating signal. For example, the energy spectrum of a trapezoidal pulse is spread over an infinite frequency range; however, most of the signal energy is concentrated near the low-frequency region and decreases rapidly with increase in frequency (this is illustrated in Figure 14.3 for two different instances of rise times: 1 ns and 0.1 ns). Hence, ignoring the high-frequency components of the spectrum above a maximum frequency, f_{max}, will not seriously alter the overall signal shape. Consequently, for all practical purposes, the width of the spectrum can be assumed to be finite. In other words, the signal energy of interest is assumed to be contained in the major lobe of the spectrum, and f_{max} can be defined as corresponding to 3-dB bandwidth point [2,3,25]

$$f_{max} = \frac{0.35}{t_r} \qquad (14.1)$$

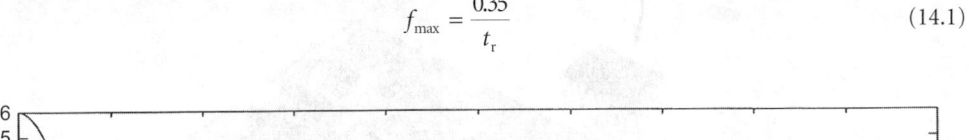

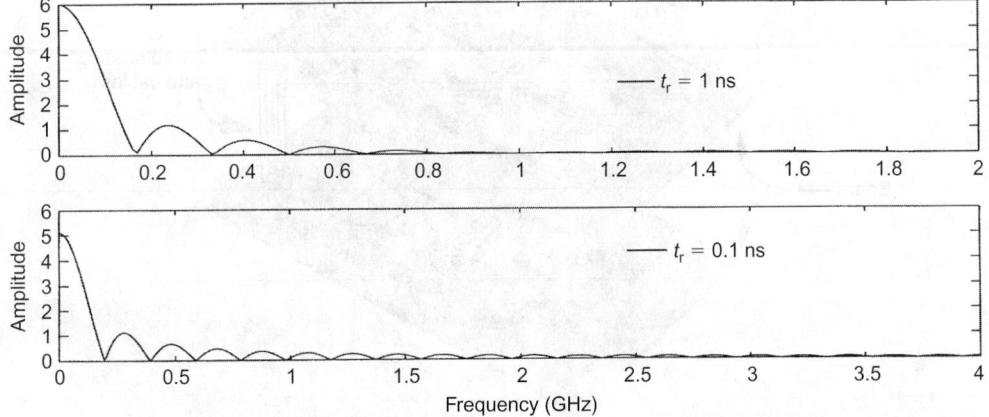

FIGURE 14.3 Frequency spectrum of trapezoidal pulse $t_{pw} = 5$ ns.

where t_r is the rise/fall time of the signal. This implies that, for example, for a rise time of 0.1 ns, the maximum frequency of interest is approximately 3 GHz or the minimum wave-length of interest is 10 cm. In some cases, the limit can be more conservatively set as [25]

$$f_{\text{max}} = \frac{1}{t_r} \qquad (14.2)$$

In summary, the primary factors with regard to high-speed signal distortion effects that should be considered are interconnect length, cross-sectional dimensions, signal slew rate, and clock speed. Other factors that also should be considered are logic levels, dielectric material, and conductor resistance. Electrically short interconnects can be represented by lumped models, whereas electrically long interconnects need distributed or full-wave models.

14.2   Interconnect Models

High-speed system designers are driven by the motivation to have signals with higher clock and slew rates while at the same time to innovate on reducing the wiring cross-section as well as packing the lines together. Reducing the wiring dimensions results in appreciably resistive lines. In addition, all these interconnections may have non-uniform cross-sections caused by discontinuities such as connectors, vias, wire bonds, flip-chip solder balls, redistribution leads, and orthogonal lines. Interconnections can be from various levels of design hierarchy (Figure 14.4) such as on-chip, packaging structures, multi-chip modules, printed circuit boards, and backplanes. On a broader perspective, interconnection technology can be classified into five categories, as shown in Table 14.1 [6], namely, on-chip wiring, thin-film wiring, ceramic carriers, thin-film wiring, printed circuit boards, and shielded cables. For the categories shown in Table 14.1, the wavelength is of the order of 1 to 10 cm. The propagated signal rise times are in the range 100 to 1000 ps. Hence, the line lengths are either comparable or much longer than the signal wavelengths. Depending on the operating frequency, signal rise times, and the nature of the structure, the interconnects can be modeled as lumped (RC or RLC), distributed (frequency-independent/dependent RLCG parameters, lossy, coupled), full-wave models, or measured linear subnetworks.

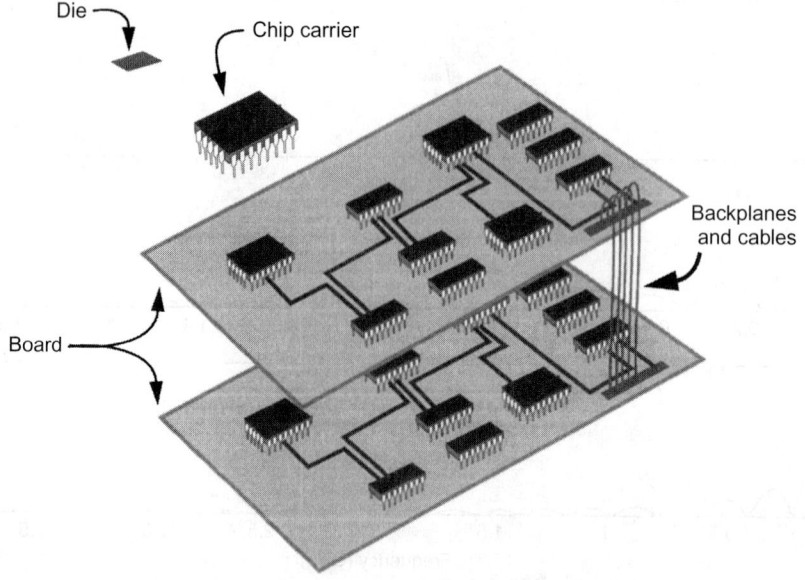

FIGURE 14.4 Interconnect hierarchy.

TABLE 14.1   Interconnect Technologies

Interconnection Type	Line Width (μm)	Line Thickness (μm)	Line Resistance (ohm/cm)	Maximum Length (cm)
On-chip	0.5–2	0.7–2	100–1000	0.3–1.5
Thin-film	10–25	5–8	1.25–4	20–45
Ceramic	75–100	16–25	0.4–0.7	20–50
Printed circuit board	60–100	30–50	0.06–0.08	40–70
Shielded cables	100–450	35–450	0.0013–0.033	150–500

14.2.1 Lumped Models

In the past, interconnect models have been generally restricted to RC tree models. RC trees are RC circuits with capacitors from all nodes to ground, no floating capacitors, no resistors to ground. The signal delay through RC trees were often estimated using a form of the *Elmore delay* [8,34], which provided a dominant time constant approximation for monotonic step responses.

14.2.1.1 Elmore Delay

There are many definitions of delay, given the actual transient response. Elmore delay is defined as the time at which the output transient rises to 50% of its final value. Elmore's expression approximates the mid-point of the *monotonic step response* waveform by the mean of the impulse response as

$$T_D = \int_0^\infty t\dot{v}\mathrm{d}t \tag{14.3}$$

Since $v(t)$ is monotonic, its first derivative (the impulse response) will have the form of a probability density function. The mean of the distribution of the first derivative is a good approximation for the 50% point of the transient portion of $v(t)$. For an RC tree, Elmore's expression can be applied since step responses for these circuits are always monotonic.

However, with increasing signal speeds, and in diverse technologies such as bipolar, BiCMOS, or MCMs, RC tree models are no longer adequate. In bipolar circuits, lumped-element interconnect models may require the use of inductors or grounded resistors, which are not compatible with RC trees. Even for MOS circuits operating at higher frequencies, the effects of coupling capacitances may need to be included in the delay estimate.

RLC circuits with non-equilibrium initial conditions may have responses that are non-monotonic. This typically results in visible signal ringing in the waveform. A single time constant approximation with Elmore delay is not generally sufficient for such circuits. Usually, lumped interconnect circuits extracted from layouts contain large number of nodes, which make the simulation highly CPU intensive. Figure 14.5 shows a general lumped model where $R, L, C,$ and G correspond to the resistance, inductance, capacitance, and conductance of the interconnect, respectively.

14.2.2 Distributed Transmission Line Models

At relatively higher signal speeds, the electrical length of interconnects becomes a significant fraction of the operating wavelength, giving rise to signal-distorting effects that do not exist at lower frequencies. Consequently, the conventional lumped impedance interconnect models become inadequate, and transmission line models based on quasi-TEM assumptions are needed. The *TEM (Transverse Electromagnetic Mode)* approximation represents the ideal case, where both E and H fields are perpendicular to the direction of propagation and it is valid under the condition that the line cross-section is much smaller than the wavelength. However, in practical wiring configurations, the structure has all the inhomogeneities mentioned previously. Such effects

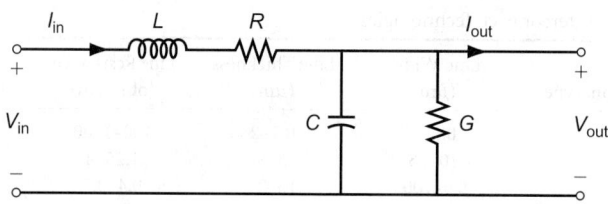

FIGURE 14.5 Lumped-component model.

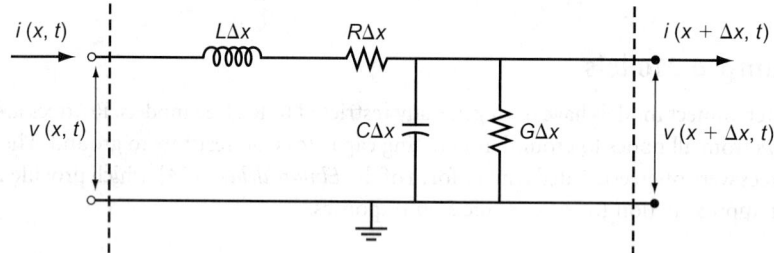

FIGURE 14.6 One segment of distributed transmission line model.

give rise to E or H fields in the direction of propagation. If the line cross-section or the extent of these non-uniformities remain a small fraction of the wavelength in the frequency range of interest, the solution to Maxwell's equations are given by the so-called quasi-TEM modes and are characterized by distributed R, L, C, G per unit length parameters (Figure 14.6).

The basic quasi-TEM model is the simple "delay" line or lossless line ($R = G = 0$). In this case, a signal traveling along a line has the same amplitude at all points, but is shifted in time with a propagation delay per unit length (τ) given by

$$\tau = \sqrt{LC} = \sqrt{\varepsilon\mu} \tag{14.4}$$

where ε and μ are the dielectric permitivity and permeability of the medium, respectively. The characteristic impedance for the lossless case is given by

$$Z = \sqrt{L/C} \tag{14.5}$$

More complicated models include per-unit-length loss (either in the direction of traveling wave or due to dielectric substrate loss) or coupling between adjacent transmission lines, where the coupling may be resistive, inductive, capacitive, or a combination of these [8]. In such cases, the propagation constants (γ) and the characteristic impedances (Z) are given by

$$\gamma = \sqrt{(R+jwL)(G+jwC)} \tag{14.6}$$

$$Z = \sqrt{\frac{R+jwL}{G+jwC}} \tag{14.7}$$

14.2.3 Distributed Models with Frequency-Dependent Parameters

As the operating frequency increases, the per unit length parameters of the transmission line can vary. This is mainly due to varying current distribution in the conductor and ground plane caused by the induced electric field. This phenomenon can be categorized as follows: skin, edge, and proximity effects [8,20,56].

14.2.3.1 Edge and Proximity Effects

The *edge* and *proximity effects* influence the interconnect parameters in the low to medium frequency region. The edge effect causes the current to concentrate near the sharp edges of the conductor, thus raising the resistance. It affects both the signal and ground conductors, but is more pronounced on signal conductors. The proximity effect causes the current to concentrate in the sections of ground plane that are close to the signal conductor. This modifies the magnetic field between the two conductors, which in turn reduces the inductance per unit length. It also raises the resistance per unit length as more current is crowded in the ground plane under the conductor. The proximity effect appears at medium frequencies. While both effects need to be accounted for, the proximity effect seems to have more significance, especially in its effect on the inductance.

14.2.3.2 Skin Effect

The *skin effect* causes the current to concentrate in a thin layer at the conductor surface. It is pronounced mostly at high frequencies on both the signal and ground conductors. The current distribution falls off exponentially as we approach the interior of the conductor. The average depth of current penetration is a function of frequency and is known as *skin depth*, which is given by

$$\delta = \left(\frac{2\rho}{w\mu} \right)^{1/2} \tag{14.8}$$

where w is frequency (rad/s), ρ is the volume resistivity, and μ is the magnetic permeability. This results in the resistance being proportional to the square root of the frequency at very high operating frequencies. The magnetic fields inside the conductors are also reduced due to skin effect. This reduces the internal inductance and therefore the total inductance. At even higher frequencies, as the internal inductance approaches zero, the edge, and proximity effects being fully pronounced, the inductance becomes essentially constant.

14.2.3.3 Typical Behavior of R and L

The frequency plots of R and L of the microstrip in Figure 14.7 are shown in Figure 14.8 and Figure 14.9. These plots present a typical behavior of R and L in general. L starts off as essentially constant until the edge and proximity effects get into effect. The edge effect causes the resistance to increase, and the current crowding under signal conductor (due to the proximity effect) causes the inductance to decrease and resistance to increase. As we go higher in frequency, the edge and proximity effects become fully pronounced and will cause little additional change in R and L, but the skin effect becomes significant. Initially, the inductance is reduced due to skin effect because of the reduction of magnetic fields inside the conductors; but as the contribution of those magnetic fields to the overall inductance becomes insignificant, L becomes essentially constant at very high frequency. The resistance, on the other hand, becomes a direct function of the skin depth and therefore varies with the square root of the frequency.

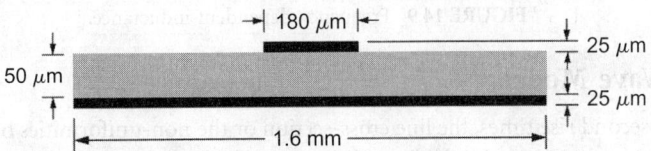

FIGURE 14.7 An interconnect over a ground plane.

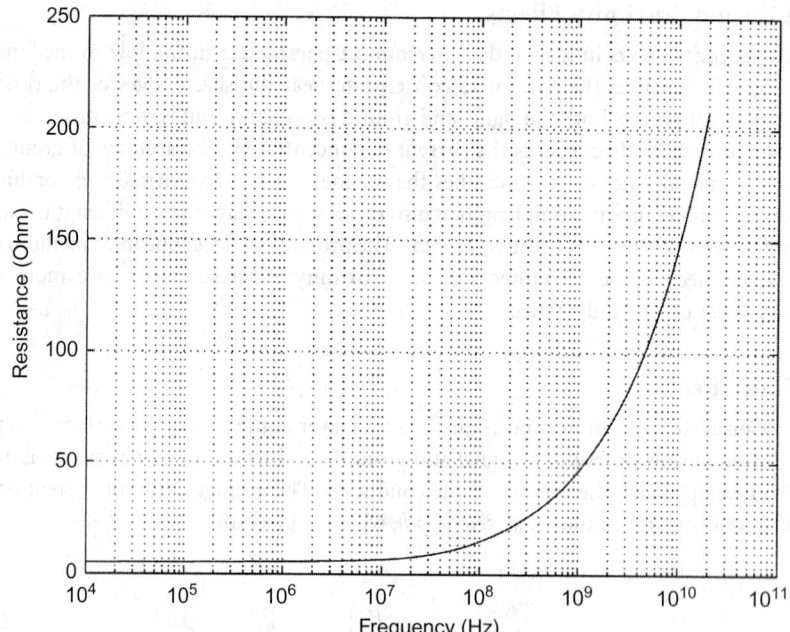

FIGURE 14.8 Frequency-dependent resistance.

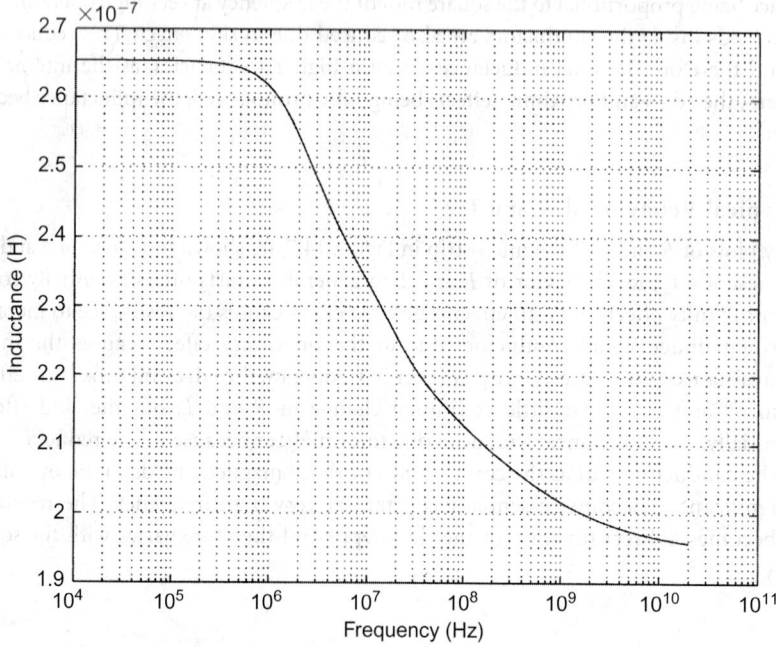

FIGURE 14.9 Frequency-dependent inductance.

14.2.4 Full-Wave Models

At further subnanosecond rise times, the line cross-section or the non-uniformities become a significant fraction of the wavelength and, under these conditions, the field components in the direction of propagation can no longer be neglected (Figure 14.10). Consequently, even the distributed models based on

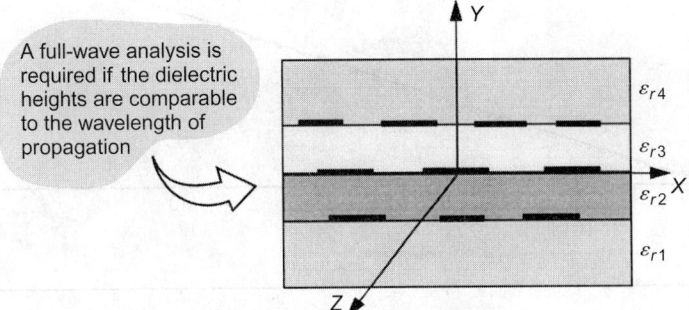

FIGURE 14.10 Cross-sectional view of a multiconductor dispersive system.

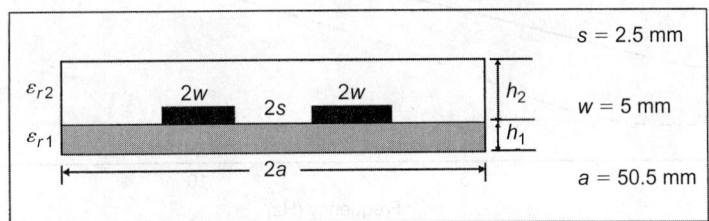

FIGURE 14.11 A coupled interconnect structure.

quasi-TEM approximations become inaccurate in describing the interconnect performance [12–19]. In such situations, full-wave models, which take into account all possible field components and satisfies all boundary conditions, are required to give an accurate estimation of high-frequency effects.

The information that is obtained through a full-wave analysis is in terms of field parameters such as propagation constant, characteristic impedance, etc. A typical behavior of the modal propagation constant and characteristic impedances obtained using the full-wave spectral domain method for the structure shown in Figure 14.11 is given in Figure 14.12. The deviation suffered by the quasi-TEM models with respect to full-wave results is illustrated through a simple test circuit shown in Figure 14.13. As seen from Figure 14.14 for the structure under consideration, quasi-TEM results deviated from full-wave results as early as 400 MHz. The differences in the modeling schemes with respect to the transient responses are illustrated in Figure 14.15 and Figure 14.16. *In general, an increase in the dielectric thickness causes quasi-TEM models to become inaccurate at relatively lower frequencies.* This implies that a full-wave analysis becomes necessary as we move up in the integration hierarchy, from the chip to PCB/system level. It is found that depending on the interconnect structure, when the cross-sectional dimensions approach 1/40 to 1/10 of the effective wavelength, quasi-TEM approximation deviates considerably from full-wave results. (The effective wavelength of a wave propagating in a dielectric medium at a certain frequency is given by

$$\lambda_e = \frac{\text{velocity of the wave}}{\text{frequency}} = \frac{1/(\sqrt{\mu_r E_{\text{effective}}})}{f} \qquad (14.9)$$

Also, a reduction in the separation width between adjacent conductors makes the full-wave analysis more essential, especially for crosstalk evaluation. The same is true with an increase in the dielectric constant values.

However, circuit simulation of full-wave models is highly involved. A circuit simulator requires the information in terms of currents, voltages, and circuit impedances. This demands a generalized method to combine modal results into circuit simulators in terms of a full-wave stencil. Another important issue involved here is the cost of a full-wave analysis associated with each interconnect subnetwork at each frequency point of analysis. For typical high-speed interconnect circuits which need thousands of

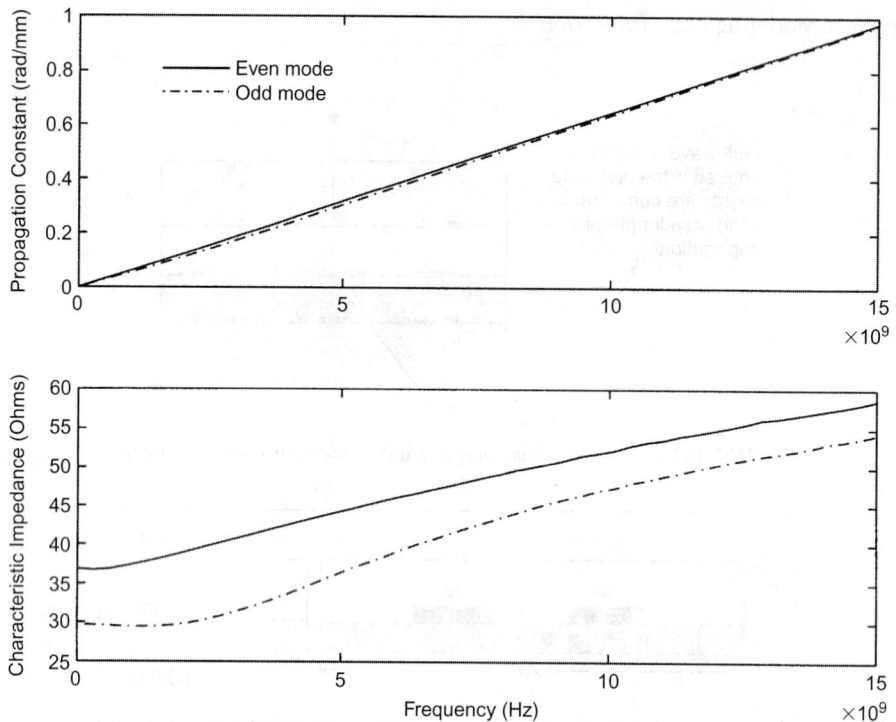

FIGURE 14.12 Modal propagation constans and charactersitic impedances.

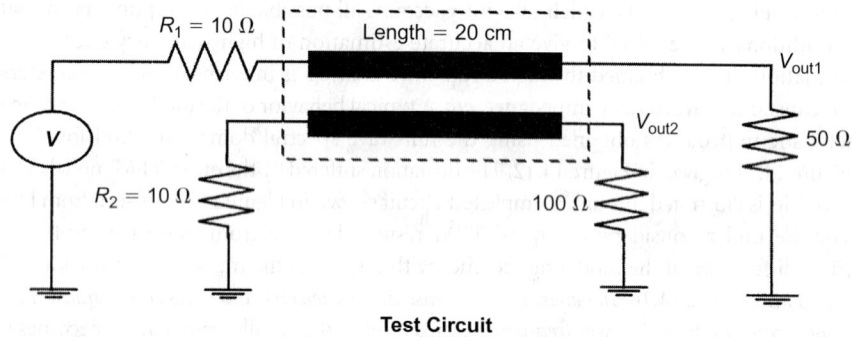

Test Circuit

FIGURE 14.13 Test circuit for simulation of interconnect structure in Figure 18.13.

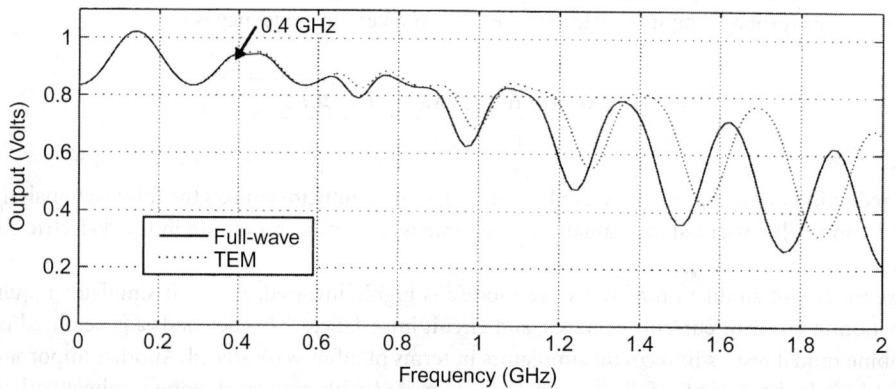

FIGURE 14.14 Comparison of full-wave and quasi-TEM frequency responses.

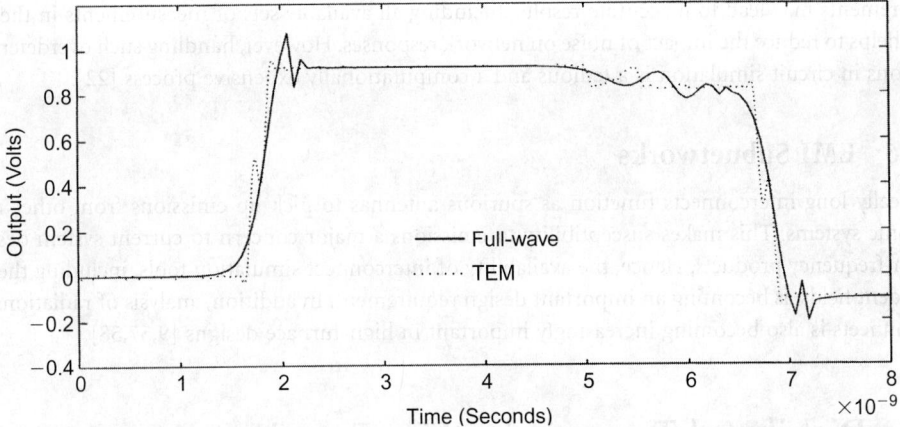

FIGURE 14.15 Comparison of full-wave and quasi-TEM transient responses at node V_{out1}.

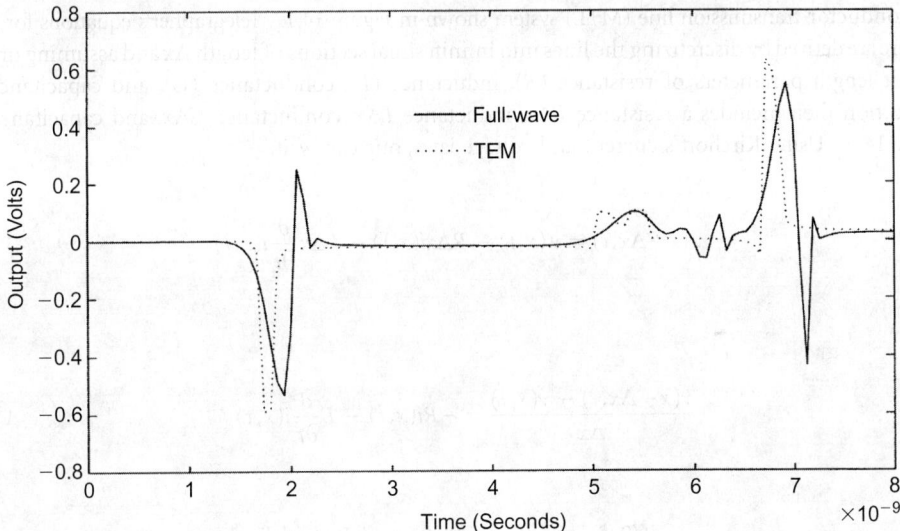

FIGURE 14.16 Comparison of full-wave and quasi-TEM crosstalk responses (node V_{out2}).

frequency point solutions to get accurate responses, it would become prohibitively CPU-expensive to simulate because of the combined cost involved (i.e., evaluation of a full-wave model to obtain modal parameters and computation of circuit response at each point).

14.2.5 Measured Subnetworks

In practice, it may not be possible to obtain accurate analytical models for interconnects because of the geometric inhomogeneity and associated discontinuities. To handle such situations, modeling techniques based on measured data have been proposed in the literature [22,24,59–65]. In general, the behavior of high-speed interconnects can easily be represented by measured frequency-dependent scattering parameters or time-domain terminal measurements. Time-domain data could be obtained by time-domain reflectometry (TDR) measurements [59] or electromagnetic techniques [23,24]. One important factor to note here is that the subnetwork can be characterized by large sets of time-domain data, making the system overdetermined. This kind of measurement data in large number of sets is essential in a practical environment as the measurements are usually contaminated by noise, and the use of only a single set of measurements may lead to inaccurate results. Including all available sets of measurements in the simulation helps to reduce the impact of noise on network responses. However, handling such overdetermined situations in circuit simulation is a tedious and a computationally expensive process [22].

14.2.6 EMI Subnetworks

Electrically long interconnects function as spurious antennas to pick up emissions from other nearby electronic systems. This makes susceptibility to emissions a major concern to current system designers of high-frequency products. Hence, the availability of interconnect simulation tools, including the effect of incident fields, is becoming an important design requirement. In addition, analysis of radiations from interconnects is also becoming increasingly important in high-furnace designs [9,57,58].

14.3   Distributed Transmission Line Equations

Transmission line characteristics are in general described by Telegrapher's equations. Consider the multiconductor transmission line (MTL) system shown in Figure 14.17. Telegrapher's equations for such a structure are derived by discretizing the lines into infinitesimal sections of length Δx and assuming uniform per-unit length parameters of resistance (R), inductance (L), conductance (G), and capacitance (C). Each section then includes a resistance $R\Delta x$, inductance $L\Delta x$, conductance $G\Delta x$, and capacitance $C\Delta x$ (Figure 14.6). Using Kirchoff's current and voltage laws, one can write

$$v(x+\Delta x,t) = v(x,t) - R\Delta x i(x,t) - L\Delta x \frac{\partial}{\partial t} i(x,t) \tag{14.10}$$

or

$$\frac{v(x+\Delta x,t) - v(x,t)}{\Delta x} = -Ri(x,t) - L\frac{\partial}{\partial t}i(x,t) \tag{14.11}$$

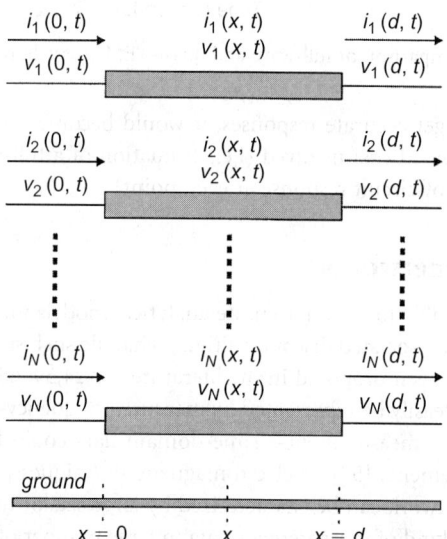

FIGURE 14.17 Multiconductor transmission line system.

Taking the limit $\Delta x \to 0$, one gets

$$\frac{\partial}{\partial x}v(x,t) = -Ri(x,t) - L\frac{\partial}{\partial t}i(x,t) \tag{14.12}$$

Similarly for $i(x,t)$, one can write

$$\frac{\partial}{\partial x}i(x,t) = -Gv(x,t) - C\frac{\partial}{\partial t}v(x,t) \tag{14.13}$$

The equations for a single line also hold good for multiple coupled lines, with a modification that per-unit-length parameters now become matrices (R, L, G, and C) and voltage–current variables become vectors represented by V and I, respectively. Noting this and taking Laplace transform of Eqs. 14.12 and 14.13, one can write

$$\frac{\partial}{\partial x}V(x,s) = -(R+sL)I(x,s) \tag{14.14}$$

$$\frac{\partial}{\partial x}I(x,s) = -(G+sC)V(x,s) \tag{14.15}$$

Equations 14.14 and 14.15 can be written as

$$\frac{\partial}{\partial x}V(x,s) = -Z_pI(x,s) \tag{14.16}$$

$$\frac{\partial}{\partial x}I(x,s) = -Y_pV(x,s) \tag{14.17}$$

where Z_p and Y_p represent the impedance and admittance matrices, given by

$$Z_p = R+sL; \quad Y_p = G+sC \tag{14.18}$$

14.3.1 Eigenvalue-Based Transmission Line Stencil

The two differential equations given in Eqs. 14.16 and 14.17 can be combined into a set of wave equations as

$$\frac{\partial}{\partial x^2}V(x,s) = Z_pY_pV_p(x,s) \tag{14.19}$$

$$\frac{\partial}{\partial x^2}I(x,s) = Y_pZ_pI_p(x,s) \tag{14.20}$$

which will have a solution of the form

$$V_m(x,s) = V_m(0,s)e^{\pm \gamma_m(s)x} \tag{14.21}$$

$$I_m(x,s) = I_m(0,s)e^{\pm \gamma_m(s)x} \tag{14.22}$$

where $\gamma_m(s)$ is the complex propagation constant. Substituting the solution forms of Eqs. 14.21 and 14.22 into wave equation 14.19 yields

$$\det\{\gamma_m^2 U - Z_P Y_P\} V_m(0) = 0 \tag{14.23}$$

where U is an identity matrix. $V_m(0)$ will have nontrivial solutions if γ_m^2 satisfies the eigenvalue problem given by

$$\det\{\gamma_m^2 U - Z_P Y_P\} = 0 \tag{14.24}$$

For inhomogeneous dielectrics, there exist in general m distinct eigenvalues where $m = 1, 2, \ldots, N$. Each eigenvalue has its corresponding eigenvector S_m. Let Γ be a diagonal matrix whose elements are the complex propagation constants $\{\gamma_1, \gamma_2, \ldots, \gamma_N\}$. Let S_v be a matrix with eigenvectors S_m placed in respective columns. The transmission line stencil can now be derived after little manipulations as

$$PV(s) + QI(s) = 0 \tag{14.25}$$

where

$$P = \begin{bmatrix} S_V E_1 S_v^{-1} & -U \\ S_i E_2 S_v^{-1} & 0 \end{bmatrix}; \qquad Q = \begin{bmatrix} S_v E_1 S_v^{-1} & 0 \\ S_i E_2 S_i^{-1} & -U \end{bmatrix} \tag{14.26}$$

$$E_1 = \mathrm{diag}\left\{\frac{e^{(-\gamma_m d)} + e^{(\gamma_m d)}}{2}\right\}; \qquad E_2 = \mathrm{diag}\left\{\frac{e^{(-\gamma_m d)} - e^{(\gamma_m d)}}{2}\right\} \tag{14.27}$$

where d is length of the line and $m = 1, 2, \ldots, N$. S_i is computed as $S_i = Z_P^{-1} S_v \Gamma \cdot V$ and I represent the Laplace-domain terminal voltage and current vectors of multiconductor transmission line, given by

$$V(s) = \begin{bmatrix} V(0) \\ V(d) \end{bmatrix}; \qquad I(s) = \begin{bmatrix} I(0) \\ I(d) \end{bmatrix} \tag{14.28}$$

The MTL stencil described by Eq. 14.25 is widely used in moment matching techniques (MMTs) for transmission line analysis [38]. Another form of MTL stencil is also quite popular and it has the matrix exponential form [44], which is explained below.

14.3.2 Matrix Exponential Stencil

Equations 14.14 and 14.15 can be written in the hybrid form as

$$\frac{\partial}{\partial x}\begin{bmatrix} V(x,s) \\ I(x,s) \end{bmatrix} = (D + sE)\begin{bmatrix} V(x,s) \\ I(x,s) \end{bmatrix} \tag{14.29}$$

where

$$D = \begin{bmatrix} 0 & -R \\ -G & 0 \end{bmatrix}; \quad E = \begin{bmatrix} 0 & -L \\ -C & 0 \end{bmatrix} \tag{14.30}$$

Hybrid transmission line stencil in the exponential form can be written as

$$\begin{bmatrix} V(d,s) \\ I(d,s) \end{bmatrix} = e^{(D+sE)d} \begin{bmatrix} V(0,s) \\ I(0,s) \end{bmatrix} \tag{14.31}$$

Parameters P and Q of the transmission line in Eq. 14.26 can also be computed making use of Eq. 14.31 as follows: Define $T(s)$ as

$$T(s) = \begin{bmatrix} T_{11} & T_{12} \\ T_{21} & T_{22} \end{bmatrix} = e^{(D+sE)d} \tag{14.32}$$

Using Eq. 14.32 and rewriting Eq. 14.31 in the form of Eq. 14.26, we get

$$P = \begin{bmatrix} -T_{11} & U \\ -T_{21} & 0 \end{bmatrix}; \quad Q = \begin{bmatrix} -T_{12} & 0 \\ -T_{22} & U \end{bmatrix} \tag{14.33}$$

14.3.3   Distributed vs. Lumped: Number of Lumped Segments Required

It is often of practical interest to switch between distributed models and lumped representations. In this case, it is necessary to know approximately how many lumped segments are required to approximate a distributed model. For the purpose of illustration, consider LC segments, which can be viewed as low-pass filters. For a reasonable approximation, this filter must pass at least some multiples of the highest frequency content f_{max} of the propagating signal (say, ten times, $f_0 \geq 10f_{max}$). In order to relate these [2,3], we make use of the 3-dB passband of the LC filter given by

$$f_0 = \frac{1}{\pi\sqrt{LdCd}} = \frac{1}{\pi\tau d} \tag{14.34}$$

where d is the length of the line. From Eq. 14.1, we have $f_{max} = 0.35/t_r$ and using Eq. 14.34, we can express the relation $f_0 \geq 10f_{max}$ in terms of the delay of the line and the rise time as

$$\frac{1}{\pi\tau d} \geq 10 \times 0.35/t_r \tag{14.35}$$

or

$$t_r \geq 3.5(\pi\tau d) \approx 10\tau d \tag{14.36}$$

In other words, delay allowed per segment is $0.1t_r$. Hence, the total number of segments (N) required is given by

$$N = \frac{10\tau d}{t_r} \tag{14.37}$$

In the case of RLC segments, in addition to satisfying Eq. 14.36, the series resistance of each segment must also be accounted for. The series resistance Rd representing the ohmic drop should not lead to impedance mismatch, which can result in excessive reflection within the segment [2,3].

Example

Consider a digital signal with rise time of 0.2 ns propagating on a lossless wire of length 10 cm, with a per unit delay of 70.7 ns. This can be represented by a distributed model with per unit length parameters of $L = 5$ nH/cm and $C = 1$ pF/cm. If the same circuit were to be represented by lumped segments, one needs $N = ((10 \times 70.7e^{-12} \times 10)/(0.2e^{-9})) = 35$ sections. It is to be noted that using more sections does not clean up ripples completely, but helps reduce the first overshoot (Gibb's phenomenon). Ripples are reduced when some loss is taken into account.

14.4 Interconnect Simulation Issues

As pointed out earlier, simulation of interconnects is associated with two major bottlenecks: mixed frequency/time problem and the CPU expense.

14.4.1 Mixed Frequency/Time Problem

The major difficulty in simulating these high-frequency models lies in the fact that distributed/full-wave elements, while formulated in terms of partial differential equations, are best described in the frequency domain. Non-linear terminations, on the other hand, can only be given in the time domain. These simultaneous formulations cannot be handled by a traditional ordinary differential equation solver such as SPICE [8,38].

14.4.2 CPU Expense

In general, interconnect networks consist of hundreds or thousands of components, such as resistors, capacitors, inductors, transmission lines, and other levels of interconnect models. At the terminations, there generally exist some nonlinear elements such as drivers and receivers. If only lumped RLC models are considered, ordinary differential equation solvers such as SPICE may be used for simulation purposes. However, the CPU cost may be large, owing to the fact that SPICE is mainly a non-linear simulator and it does not handle large RLC networks efficiently.

14.4.3 Background on Circuit Simulation

Prior to introducing interconnect simulation algorithms, it would be useful to have a glimpse at the basic circuit simulation techniques. Conventional circuit simulators are based on the simultaneous solution of linear equations which are obtained by applying Kirchoff's current law (KCL) to each node in the network. Either for frequency- or time-domain analysis, the first step is to set up the *modified nodal analysis matrix (MNA)* [68]. For example, consider the small circuit in Figure 14.18. Its MNA equations are

$$\left(\begin{bmatrix} G_1 & 0 & 1 & 1 \\ 0 & G_2 & -1 & 0 \\ 1 & -1 & 0 & 0 \\ 1 & -0 & 0 & 0 \end{bmatrix} + s \begin{bmatrix} 0 & 0 & 0 & 0 \\ 0 & C & 0 & 0 \\ 0 & 0 & -L & 0 \\ 0 & 0 & 0 & 0 \end{bmatrix} \right) \begin{bmatrix} V_1 \\ V_2 \\ I_L \\ I_E \end{bmatrix} = \begin{bmatrix} 0 \\ 0 \\ 0 \\ E \end{bmatrix} \tag{14.38}$$

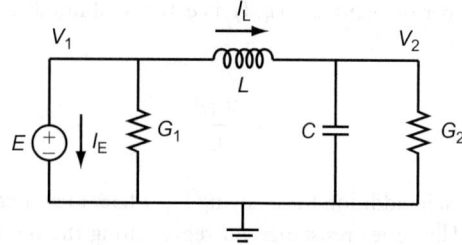

FIGURE 14.18 Example circuit for MNA formulation.

The above equation, representing a simple two-node circuit, has the same form as any other MNA matrix representing a large linear lumped network. Hence, MNA equations in general for lumped linear networks can be written as

$$(G + sW)X(s) - b = 0 \qquad (14.39)$$

or, in the time domain, it can be written as

$$Gx(t) + W\dot{x}(t) - b(t) = 0 \qquad (14.40)$$

14.4.4 Discussion of CPU Cost in Conventional Simulation Techniques

Frequency-domain simulation is conventionally done by solving Eq. 14.39 at each frequency point through LU decomposition and forward-backward substitution. For time-domain simulation, linear multi-step techniques [69] are used. The most common of these integration formulas is the trapezoidal rule, which gives the following difference formula

$$x(t + \Delta t) = x(t) + \Delta t \left(\frac{\dot{x}(t) + \dot{x}(t + \Delta t)}{2} \right) \qquad (14.41)$$

$$\left(G + \frac{2}{\Delta t} W \right) x(t + \Delta t) = \left(\frac{2}{\Delta t} W - G \right) x(t) + (b(t) + b(t + \Delta t)) \qquad (14.42)$$

Note that the trapezoidal rule is an integration formula of order of 2, which is the highest possible order that could ensure absolute stability [69]. Due to such relatively low order, simulators are forced to use small step sizes to ensure the accuracy during transient simulation. Transient response computation requires the LU decomposition of Eq. 14.42 at every time step. It gets further complicated in the presence of nonlinear elements, in which case the Eq. 14.40 gets modified as

$$Gx(t) + W\dot{x}(t) + \phi(x(t)) - b(t) = 0 \qquad (14.43)$$

where $\phi(x(t))$ is a non-linear function of x. The difference equation based on the trapezoidal rule therefore becomes

$$\left(G + \frac{2}{\Delta t} W \right) x(t + \Delta t) + \phi(x(t + \Delta t)) = \left(\frac{2}{\Delta t} W - G \right) x(t) + (b(t) + b(t + \Delta t)) - \phi(x(t)) \qquad (14.44)$$

In the case of nonlinear elements, Newton iterations are used to solve Eq. 14.44, which requires two to three LU decompositions at each time step. This causes (note that W and G matrices for interconnect networks are usually very large) the CPU cost of a time-domain analysis to be very expensive. This led to the development of model-reduction algorithms, which effects a reduction in the order of the linear subnetwork before performing a non-linear analysis so as to yield fast simulation results.

14.4.5 Circuit Equations in the Presence of Distributed Elements

Now consider the general case in which the network ϕ also contains arbitrary linear subnetworks along with lumped components. The arbitrary linear subnetworks may contain lossy coupled transmission lines and also measured subnetworks. Let there be N_t lossy coupled transmission line sets, with n_a

coupled conductors in the linear subnetwork a. Without loss of generality, the modified nodal admittance (MNA) [68,69] matrix for the network ϕ with an impulse input excitation can be formulated as

$$C\frac{\partial}{\partial t}v(t) + Gv(t) + \sum_{a=1} D_a i_a(t) - b\delta(t) = 0, \qquad t \in [0,T] \tag{14.45}$$

where:

- $D_a = [d_{i,j} \in \{0, 1\}]$, where $i \in \{1, 2, ..., N_\phi\}$, $j \in \{1, 2, ..., 2n_a\}$ with a maximum of one non-zero in each row or column, is a selector matrix that maps $i_a(t) \in \mathfrak{R}^{2n_a}$ the vector of terminal currents entering the transmission line subnetwork a, into the node space $\mathfrak{R}^{N}\phi$ of network ϕ.
- N_ϕ is the total number of variables in the MNA formulation.

Using the transmission line stencil given by Eq. 14.25 and taking the Laplace transform of Eq. 14.45 assuming zero initial conditions, one obtains

$$\begin{bmatrix} G+sC & D_1 & \cdots & D_{N_t} \\ P_1 D_1^t & Q_1 & 0 & 0 \\ \cdots & 0 & \cdots & 0 \\ P_{N_t} D_{N_t}^t & 0 & 0 & Q_{N_t} \end{bmatrix} \begin{bmatrix} V(s) \\ I_1(s) \\ \cdots \\ I_{N_t}(s) \end{bmatrix} = \begin{bmatrix} b \\ 0 \\ \cdots \\ 0 \end{bmatrix} \tag{14.46}$$

or

$$Y(s)X(s) = E \tag{14.47}$$

$$X(s) = [V^t(s)I_1^t(s), I_2^t(s), ..., I_{N_t}^t(s)]^t \tag{14.48}$$

$$E = [b^t 0^t ... 0^t] \tag{14.49}$$

Example

To illustrate the above formulation, consider the network shown in Figure 14.19, which is a modified version of the previous example with introduction of a coupled transmission line. The network can be described by Eq. 14.46 as

$$G + sC = \begin{bmatrix} sC_1 + G_1 & 0 & 0 & 0 & 0 & 0 \\ 0 & G_2 & 0 & 0 & 0 & 0 \\ 0 & 0 & G_3 & 0 & -G_3 & 0 \\ 0 & 0 & 0 & sC_2 & 0 & 0 \\ 0 & 0 & -G_3 & 0 & G_3 & 1 \\ 0 & 0 & 0 & 0 & 1 & 0 \end{bmatrix}; \quad D_1 = \begin{bmatrix} 1 & 0 & 0 & 0 \\ 0 & 1 & 0 & 0 \\ 0 & 0 & 1 & 0 \\ 0 & 0 & 0 & 1 \\ 0 & 0 & 0 & 0 \\ 0 & 0 & 0 & 0 \end{bmatrix} \tag{14.50}$$

$$V(s) = [V_1\ V_2\ V_3\ V_4\ V_5 I_e]^t \tag{14.51}$$

$$I_1(s) = [I_a\ I_b\ I_c\ I_d]^t \tag{14.52}$$

$$b = [0\ 0\ 0\ 0\ 0e]^t \tag{14.53}$$

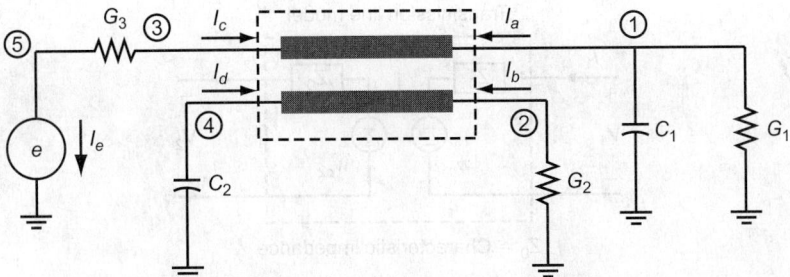

FIGURE 14.19 An example circuit containing transmission line components.

Conventional simulation methods obtain the frequency response of a circuit by solving Eq. 14.46 using LU decomposition and forward-backward substitution at various frequency points (usually thousands of points are required to get an accurate response over a desired frequency range). However, moment-matching techniques such as AWE extract the poles and residues of Eq. 14.46 using one LU decomposition only. The transfer function of the network and its frequent response can then be deduced from poles and residues. In addition to speeding up the simulation, MMTs provide a convenient way to handle mixed frequency/time simulation problem through macromodeling.

14.5 Interconnect Simulation Techniques

The main objective behind the interconnect simulation algorithms is to address the mixed frequency/time problem as well as ability to handle large linear circuits without too much CPU expense. There have been several algorithms proposed for this purpose, and they are discussed below.

14.5.1 Method of Characteristics

The method of characteristics transforms partial differential equations of a transmission line into ordinary differential equations [28,29]. Extensions to this method to allow it to handle lossy transmission lines can also be found in the literature; for example, the iterative waveform relaxation techniques (IWR) [28,29]. The method of characteristics is still used as one of the most practical techniques for simulation of lossless lines.

An analytical solution for Eqs. 14.21 and 14.22 was derived in Reference 28 for two conductor lines which provides the Y parameters of the two-port transmission line network

$$I = YV; \quad \begin{bmatrix} I_1 \\ I_2 \end{bmatrix} = \frac{1}{Z_0(1 - e^{-2\gamma l})} \begin{bmatrix} 1 + e^{-2\gamma l} & -2e^{-\gamma l} \\ -2e^{-\gamma l} & 1 + e^{-2\gamma l} \end{bmatrix} \begin{bmatrix} V_1 \\ V_2 \end{bmatrix} \tag{14.54}$$

where γ is the propagation constant, and Z_0 is the characteristic impedance. V_1 and I_1 are the terminal voltage and current at the near end of the line, V_2 and I_2 are the terminal voltage and current at the far end of the line. The Y parameters of the transmission line are complex functions of s, and in most cases cannot be directly transformed into an ordinary differential equation in the time domain. The method of characteristics succeeded in doing such a transformation, but only for lossless transmission lines. Although this method was originally developed in the time domain using what was referred to as characteristic curves (hence, the name), a short alternative derivation in the frequency domain will be presented here. The frequency-domain approach gives more insight as to the limitations of this technique and possible solutions to those limitations.

By rearranging the terms in Eq. 14.54, we can write

$$V_1 = Z_0 I_1 + W_{c1}$$
$$V_2 = Z_0 I_2 + W_{c2} \tag{14.55}$$

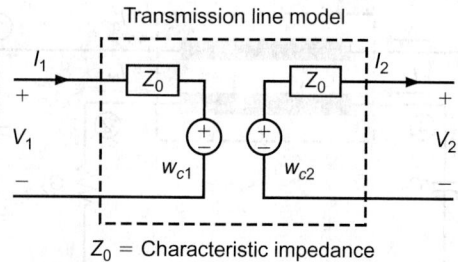

FIGURE 14.20   Macromodel using method of characteristics.

where W_{c1} and W_{c2} are defined as

$$W_{c1} = e^{-\gamma l}[2V_2 - e^{-\gamma l}(Z_0 I_1 + V_1)]$$
$$W_{c2} = e^{-\gamma l}[2V_1 - e^{-\gamma l}(Z_0 I_2 + V_2)] \tag{14.56}$$

Using Eqs. 14.55 and 14.56, a recursive relation for W_{c1} and W_{c2} can be obtained as

$$W_{c1} = e^{-\gamma l}[2V_2 - W_{c2}]$$
$$W_{c2} = e^{-\gamma l}[2V_1 - W_{c1}] \tag{14.57}$$

A lumped model of the transmission line can then be deduced from Eqs. 14.55 and 14.56, as shown in Figure 14.20. If the lines were lossless (in which case the propagation constant is purely imaginary; $\gamma = j\beta$), the frequency-domain expression (Eq. 14.57) can be analytically converted into time domain using inverse Laplace transform as

$$w_{c1}(t + \tau) = 2v_2(t) - w_{c2}(t)$$
$$w_{c2}(t + \tau) = 2v_1(t) - w_{c1}(t) \tag{14.58}$$

where $e^{-j\beta}$ is replaced by a time shift (or delay). Each transmission line can therefore be modeled by two impedances and two voltage-controlled voltage sources with time delay. While this transmission line model is in the time domain and can be easily linked to time-domain circuit simulation, the time shift affects the stability of the integration formula and causes the step size to significantly decrease, therefore increasing the CPU time. For lossy lines, the propagation constant is not purely imaginary and can therefore not be replaced by a pure delay. In that case, analytical expressions for w_{c1} and w_{c2} cannot be found in the time domain, although some numerical techniques were proposed (e.g., the iterative waveform relaxation techniques (IWR)) [28].

Recently, there have been several publications based on approximating the frequency characteristics of transmission-line equations using rational polynomials [27]. Analytical techniques to directly convert partial differential equations into time-domain macromodels based on Padé rational approximations of exponential matrices have also been reported [26].

14.5.2   Moment-Matching Techniques

Interconnect networks generally tend to have large number of poles, spread over a wide frequency range. Although the majority of these poles would normally have very little effect on simulation results, they make the simulation CPU extensive by forcing the simulator to take smaller step sizes.

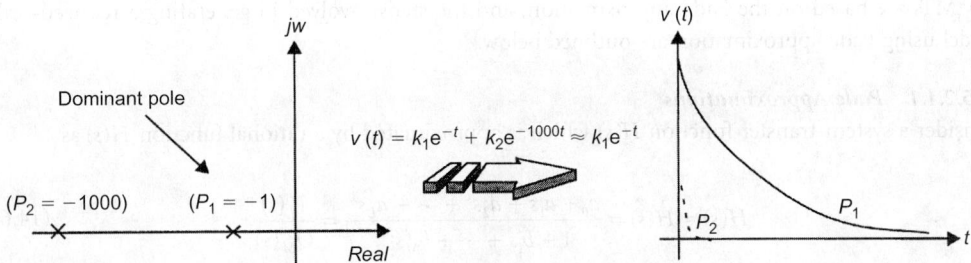

FIGURE 14.21 Summary of the steps involved in the MMT algorithm.

14.5.2.1 Dominant Poles

Dominant poles are those that are close to the imaginary axis and significantly influence the time as well as the frequency characteristics of the system. The moment-matching techniques (MMTs) [34–67] capitalize on the fact that, irrespective of the presence of large number of poles in a system, only the dominant poles are sufficient to accurately characterize a given system. This effect is demonstrated in Figure 14.21, where it is clear that pole P_2 will have little effect on the final transient result.

A brief mathematical description of the underlying concepts of moment-matching techniques is given below. Consider a single input/single output system and let $H(s)$ be the transfer function. $H(s)$ can be represented in a rational form as

$$H(s) = \frac{P(s)}{Q(s)} \tag{14.59}$$

where $P(s)$ and $Q(s)$ are polynomials in s. Equivalently, Eq. 14.59 can be written as

$$H(s) = c + \sum_{i=0}^{N_p} \frac{k_i}{s - p_i} \tag{14.60}$$

where P_i and k_i are the ith pole-residue pair, N_p is the total number of system poles, and c is the direct coupling constant. Next, the time-domain impulse response can be computed in a closed form using inverse Laplace transform as

$$h(t) = c\delta t + \sum_{i=0}^{N_p} k_i e^{p_i t} \tag{14.61}$$

In the case of large networks, N_p, the total number of poles can be of the order of thousands. Generating all the N_p poles will be highly CPU intensive even for a small network; and for large networks, it is completely impractical. MMTs address the above issue by deriving a reduced-order approximation $\hat{H}(s)$ in terms of dominant poles, instead of trying to compute all the poles of a system. Assuming that only L dominant poles were extracted which give a reasonably good approximation to the original system, Eq. 14.59 and the corresponding approximate frequency and time responses can be written as

$$H(s) \approx \hat{H}(s) = \frac{\hat{P}(s)}{\hat{Q}(s)} = \hat{c} + \sum_{i=0}^{L} \frac{\hat{k}_i}{s - \hat{p}_i} \tag{14.62}$$

$$h(t) \approx \hat{h}(t) = \hat{c}\delta t + \sum_{i=0}^{L} \hat{k}_i e^{\hat{p}_i t} \tag{14.63}$$

MMTs are based on the Padé approximation, and the steps involved in generating a reduced-order model using Padé approximation are outlined below.

14.5.2.1.1 Padé Approximations

Consider a system-transfer function $H(s)$ which is approximated by a rational function $\hat{H}(s)$ as

$$H(s) \approx \hat{H}(s) = \frac{a_0 + a_1 s + a_2 s^L + \cdots + a_L s^L}{1 + b_1 s + \cdots + b_M s^M} = \frac{P_L(s)}{Q_M(s)} \tag{14.64}$$

where $a_0, \ldots, a_L, b_1, \ldots, b_M$ are the unknowns (total of $L + M$ variables). Next, expanding $H(s)$ using Taylor series coefficients M_i (moments), we have

$$H(s) \approx \hat{H}(s) = m_0 + m_1 s + m_2 s^2 + \cdots + m_n s^n = \sum_{i=0}^{n} s^i m_i \tag{14.65}$$

The series in Eq. 14.65 is then matched to the lower-order rational polynomial given by Eq. 14.64; hence, the name moment-matching techniques (MMTs), also known as Padé approximation). There are $L + M$ unknowns and hence we need to match only the first $L + M$ moments as

$$\frac{a_0 + a_1 s + a_2 s^L + \cdots + a_L s^L}{1 + b_1 s + \cdots + b_M s^M} = m_0 + m_1 s + m_2 s^2 + \cdots + m_{(L+M)} s^{L+M} \tag{14.66}$$

Cross-multiplying and equating the powers of s starting from s^{L+1} to s^{L+M} on both sides of Eq. 14.66, we can evaluate the denominator polynomial coefficients as

$$\begin{bmatrix} m_{L-M+1} & m_{L-M+2} & \cdots & m_L \\ m_{L-M+2} & \cdots & \cdots & m_{L+1} \\ \cdots & \cdots & \cdots & \cdots \\ m_L & m_{L+1} & \cdots & m_{L+M-1} \end{bmatrix} \begin{bmatrix} b_M \\ b_{M-2} \\ \cdots \\ b_1 \end{bmatrix} = - \begin{bmatrix} m_{L+1} \\ m_{L+2} \\ \cdots \\ m_{L+M} \end{bmatrix} \tag{14.67}$$

The numerator coefficients can then be found by equating the remaining powers of s, starting from s^0 to s^L as

$$a_0 = m_0$$

$$a_1 = m_1 + b_1 m_0$$

$$\cdots \tag{14.68}$$

$$a_L = m_L + \sum_{i=1}^{\min(L, M)} b_i m_{L-i}$$

Equations 14.67 and 14.68 yield an approximate transfer function in terms of rational polynomials. Alternatively, an equivalent pole-residue model can be found as follows. Poles p_i are obtained by applying a root-solving algorithm on denominator polynomial $\hat{Q}(s)$. In order to obtain k_i, expand the approximate transfer function given by Eq. 14.62 using Maclaurin series as

$$\hat{H}(s) = \hat{c} - \sum_{n=0}^{} s^n \left(\sum_{i=0}^{} \frac{\hat{k}_i}{\hat{p}_i^{n+1}} \right) \tag{14.69}$$

Comparing $\hat{H}(s)$ from Eqs. 14.65 and 14.69, we note that

$$m_0 = \hat{c} - \sum_{i=0}^{L} \frac{\hat{k}_i}{\hat{p}_i}$$

$$\ldots \quad (0 < i < 2L)$$

(14.70)

$$m_i = -\sum_{i=0}^{L} \frac{\hat{k}_i}{\hat{p}_i^{i+1}}$$

Residues can be evaluated by writing the equations in Eq. 14.70 in a matrix form as

$$
\begin{bmatrix}
\hat{p}_1^{-1} & \hat{p}_2^{-1} & & \hat{p}_L^{-1} & -1 \\
\hat{p}_1^{-2} & \hat{p}_2^{-2} & & \hat{p}_L^{-2} & 0 \\
 & & \cdots & & \\
 & & \cdots & & \\
\hat{p}_1^{-L-1} & \hat{p}_2^{-L-1} & & \hat{p}_L^{-L-1} & 0
\end{bmatrix}
\begin{bmatrix}
\hat{k}_1 \\
\hat{k}_2 \\
\cdots \\
\hat{k}_L \\
\hat{c}
\end{bmatrix}
= -
\begin{bmatrix}
M_0 \\
M_1 \\
\cdots \\
M_{L-1} \\
M_L
\end{bmatrix}
$$

(14.71)

In the above equations, $\hat{c}$ represents the direct coupling between input and output. There are more exact ways to compute $\hat{c}$ which are not detailed here; interested readers can refer to Reference 35.

14.5.2.2 Computation of Moments

Having outlined the concept of MMTs, to proceed further, we need to evaluate the moments of the system. Consider the circuit equation represented by Eq. 14.46 and expand the response vector $X(s)$ using Taylor series as

$$X(s) = M_0 + M_1 s + M_2 s^2 + \cdots + M_n s^n = \sum_{i=0}^{n} s^i M_i$$

(14.72)

where M_i represents the ith moment-vector. For the purpose of illustration, consider the simple case of network with only lumped models. In this case, the network can be represented in the form Eq. 14.39 as

$$[G + sC][X(s)] = [E]$$

(14.73)

or

$$[G + sC][M_0 + M_1 s + M_2 s^2 + \cdots] = [E]$$

(14.74)

Equating powers of s on both sides of Eq. 14.74, we obtain the following relationships

$$GM_0 = E$$

$$GM_i = -CM_{i-1} \quad i > 0$$

(14.75)

The above equations give a closed form relationship for the computation of moments. The moments of a particular output node of interest (which are represented by m_i in Eqs. 14.65 to 14.71), are picked from moment-vectors M_i. As seen, Eq. 14.75 requires only one LU decomposition and few forward-backward substitutions during the recursive computation of higher-order moments. Since the major cost involved in circuit simulation is due to LU decomposition, MMTs yield very high speed advantage (100 to 1000 times) compared to conventional simulators.

14.5.2.3 Generalized Computation of Moments

In the case of networks containing transmission lines and measured subnetworks, moment computation is not straightforward. A generalized relation for recursive computation of higher-order moments at an expansion point $s = \alpha$ can be derived [44] using Eq. 14.46 as:

$$[\Psi(\alpha)]M_0 = E$$

$$[\Psi(\alpha)]M_n = -\sum_{r=1}^{n} \frac{(\Psi^{(r)}\big|_{s=\alpha})M_{n-r}}{r!} \qquad (14.76)$$

where the superscript r denotes the rth derivative at $s = \alpha$. It can be seen that the coefficient on the left-hand side does not change during higher-order moment computation and, hence, only one LU decomposition would suffice. It is also noted that the lumped networks are a special case of Eq. 18.76 (where $\Psi^{(r)} = 0$ for $r \geq 2$, in which case Eq. 14.76 reduces to the form given by Eq. 14.75).

Having obtained a recursive relationship Eq. 14.76 for higher-order moments, in order to proceed further, we need the derivatives of (Ψ). The derivatives $A_d^{(r)}$ and $B_d^{(r)}$ corresponding to transmission lines can be computed using the matrix exponential-based method [34,38]. Efficient techniques for computation of moments of transmission lines with frequency-dependent parameters [56], full-wave [12], and measured subnetworks [63,65] can also be found in the literature.

14.5.2.4 Computation of Time-Domain Macromodel

Once a pole-residue model describing the interconnect network is obtained, a time-domain realization in the form of state-space equations can be obtained as [39–41,75,76]:

$$\frac{d}{dt}[z_\pi(t)] - [A_\pi][z_\pi(t)] - [B_\pi][i_\pi(t)] = 0$$

$$[v_\pi(t)] - [C_\pi][z_\pi(t)] + [D_\pi][i_\pi(t)] \;\; = 0 \qquad (14.77)$$

where i_π and v_π are the vectors of terminal currents and voltages of the linear subnetwork π. Using standard non-linear solvers or any of the general-purpose circuit simulators, the unified set of differential equations represented by Eq. 14.77 can be solved to yield unified transient solutions for the entire non-linear circuit consisting of high-frequency interconnect subnetworks. For those simulators (such as HSPICE) that do not directly accept the differential equations as input, the macromodel represented by Eq. 14.77 can be converted to an equivalent subcircuit, and is described in the next section. Figure 14.22 summarizes the computational steps involved in the MMT algorithm.

14.5.3 Limitations of Single-Expansion MMT Algorithms

Obtaining a lower-order approximation of the network transfer function using a single Padé expansion is commonly referred as *asymptotic waveform evaluation (AWE)* in the literature. However, due to the

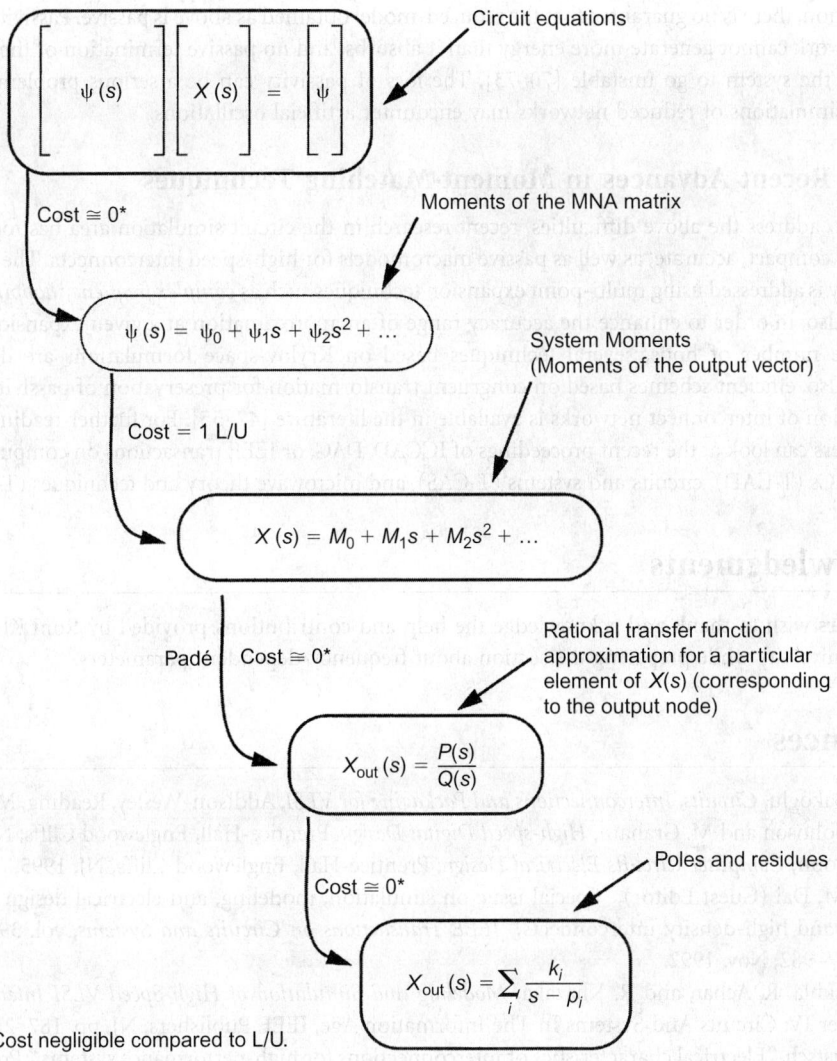

FIGURE 14.22 Summary of the steps involved in the MMT algorithm.

inherent limitations of Padé approximants, MMTs based on single expansion often give inaccurate results. The following is a list of those properties that have the most impact on MMTs.

- The matrix in Eq. 14.67 (which is known as Toeplitz matrix) is ill-conditioned if its size is large. This implies that we can only expect to detect six to eight accurate poles from a single expansion.
- Padé often produces unstable poles on the right-hand side of the complex plane.
- Padé accuracy deteriorates as we move away from the expansion point.
- Padé provides no estimates for error bounds.

Generally, AWE gives accurate results for RC networks, but often fails for non-RC networks. This is due to the fact that the poles of an RC network are all on the real axis and therefore an expansion at the origin could clearly determine which ones are dominant and which ones are not. However, in the case of general RLC networks, it is possible to have some of the dominant poles outside the radius of convergence of the Padé expansion. In systems containing distributed elements, the number of dominant poles will be significantly higher, and it is very difficult to capture all with a single Padé expansion.

In addition, there is no guarantee that the reduced-model obtained as above is passive. Passivity implies that a network cannot generate more energy than it absorbs, and no passive termination of the network will cause the system to go unstable [70–73]. The loss of passivity can be a serious problem because transient simulations of reduced networks may encounter artificial oscillations.

14.5.4 Recent Advances in Moment-Matching Techniques

In order to address the above difficulties, recent research in the circuit simulation area has focused on arriving at compact, accurate, as well as passive macromodels for high-speed interconnects. The problem of accuracy is addressed using multi-point expansion techniques such as *complex frequency hopping (CFH)* [44–46]. Also, in order to enhance the accuracy range of an approximation at a given expansion and to reduce the number of hops, several techniques based on Krylov-space formulations are developed [47–55]. Also, efficient schemes based on congruent transformation for preservation of passivity during the reduction of interconnect networks is available in the literature [47–55]. For further readings, interested readers can look at the recent proceedings of ICCAD, DAC, or IEEE transactions on computer-aided design of ICs (T-CAD), circuits and systems (T-CAS), and microwave theory and techniques (T-MTT).

Acknowledgments

The authors wish to thank and acknowledge the help and contributions provided by Roni Khazaka of Carlton University while preparing the section about frequency-dependent parameters.

References

1. H.B. Bakoglu, *Circuits, Interconnections and Packaging for VLSI*, Addison-Wesley, Reading, MA, 1990.
2. H.W. Johnson and M. Graham, *High-speed Digital Design*, Prentice-Hall, Englewood Cliffs, NJ, 1993.
3. R.K. Poon, *Computer Circuits Electrical Design*, Prentice-Hall, Englewood Cliffs, NJ, 1995.
4. W.W.M. Dai (Guest Editor), "Special issue on simulation, modeling, and electrical design of high-speed and high-density interconnects," *IEEE Transactions on Circuits and Systems*, vol. 39, no. 11, pp. 857–982, Nov. 1992.
5. M. Nakhla, R. Achar, and R. Khazaka, *Modeling and Simulation of High-Speed VLSI Interconnects*, Chapter IV: Circuits And Systems In The Information Age, IEEE Publishers, NJ, pp. 187–215, 1997.
6. A. Deustsch, "Electrical characteristics of interconnections for high-performance systems," *Proceedings of the IEEE*, vol. 86, no. 2, pp. 315–355, Feb. 1998.
7. T.L. Quarles, *The SPICE3 Implementation Guide*, Technical Report, ERL-M89/44, University of California, Berkeley, 1989.
8. C. Paul, *Analysis of Multiconductor Transmission Lines*, John Wiley & Sons, New York, 1994.
9. C. Paul, *Introduction to Electromagnetic Compatibility*, John Wiley & Sons, New York, 92.
10. S. Gao, A. Yang, and S. Kang, "Modeling and simulation of interconnection delays and cross talks in high-speed integrated circuits," *IEEE Trans. on Circuits and Systems*, pp. 1–9, Jan. 90.
11. H. Hasegawa and S. Seki, "Analysis of interconnection delay on very high-speed LSI/VLSI chips using a microstrip line model," *IEEE Trans. Electron Devices*, pp. 1954–1960, Dec. 1984.
12. R. Achar, M. Nakhla, and Q.J. Zhang, "Full-wave analysis of high-speed interconnects using complex frequency hopping," *IEEE Trans. on Computer-Aided Design*, pp. 997–1016, Oct. 98.
13. T. Itoh and R. Mittra, "Spectral domain approach for calculating the dispersion characteristics of microstrip lines," *IEEE Trans. Microwave Theory Tech.*, pp. 496–499, Feb. 1973.
14. D. Mirshekar-Syahkal, *Spectral Domain Method for Microwave Integrated Circuits*, Joinery & Sons, Inc., 1990.
15. R.H. Jansen, "Spectral Domain Approach for microwave integrated circuits," *IEEE Trans. Microwave Theory Tech.*, vol. MTT-33, pp. 1043–1056, Feb. 1985.

16. R. Wang, and O. Wing, "A circuit model of a system of VLSI interconnects for time response computation," *IEEE Trans. Microwave Theory Tech.*, vol. MTT-39, pp. 688–693, April 1991.

17. M.A. Kolbehdari, M. Srinivasan, M. Nakhla, Q.J. Zhang, and R. Achar, "Simultaneous time and frequency domain solution of EM problems using finite element and CFH techniques," *IEEE Trans on Microwave Theory and Techniques*, vol. 44, pp. 1526–1534, Sept. 1996.

18. A.E. Ruehli, "Equivalent circuit models for three dimensional multiconductor systems," *IEEE Trans. Microwave Theory Tech.*, vol. 22, no. 3, pp. 216–224, Mar. 1974.

19. A.E. Ruehli and H. Heeb, "Circuit models for three dimensional geometries including dielectrics," *IEEE Trans. Microwave Theory Tech.*, pp. 1507–1516, Mar. 1992.

20. A.R. Djordjević and T.K. Sarkar, "Closed-form formulas for frequency-dependent resistance and inductance per unit length of microstrip and strip transmission lines," *IEEE Trans. Microwave Theory Tech.*, vol. MTT-42, pp. 241–248, Feb. 1994.

21. A.R. Djordjević, R.F. Harrington, T.K. Sarkar, and M. Bazdar, *Matrix Parameters for Multiconductor Transmission Lines: Software and Users Manual*, Reteach House, Boston, 1989.

22. R. Achar and M. Nakhla, "Efficient transient simulation of embedded subnetworks characterized by S-parameters in the presence of nonlinear elements," *IEEE Transactions on Microwave Theory and Techniques*, vol. 46, pp. 2356–2363, Dec. 1998.

23. Y. Tsuei, A.C. Cangellaris, and J.L. Prince, "Rigorous electromagnetic modeling of chip-to-package (first-level) interconnections," *IEEE Trans. Components Hybrids Manufacturing Technology*, vol. 16, no. 8, pp. 876–883, Aug. 1993.

24. M. Picket-May, A. Taflove, and J. Baron, "FD-TD modeling of digital signal propagation in 3-D circuits with passive and active loads," *IEEE Trans. Microwave Theory Tech.*, vol. 42, no. 8, pp. 1514–1523, Aug. 1994.

25. T. Dhane and D.D. Zutter, "Selection of lumped element models for coupled lossy transmission lines," *IEEE Trans. Computer-Aided Design*, vol. 11, July 1992.

26. X. Li, M. Nakhla, and R. Achar, "A universal closed-loop high-speed interconnect model for general purpose circuit simulators," *Proc. IEEE International Symposium on Circuits and Systems (ISCAS)*, Monterey, CA, pp. 66–69, June, 1998.

27. M. Celik and A.C. Cangellaris, "Simulation of dispersive multiconductor transmission lines by Padé approximation via Lanczos process," *IEEE Trans. MTT*, pp. 2525–2535, Dec. 96.

28. F.Y. Chang, "The generalized method of characteristics for waveform relaxation analysis of lossy coupled transmission lines," *IEEE Trans. Microwave Theory Tech.*, vol. 37, pp. 2028–2038, Dec. 1989.

29. N. Orhanovic, P. Wang, and V.K. Tripathi, "Generalized method of characteristics for time domain simulation of multiconductor lossy transmission lines," *Proceedings IEEE Symposium on Circuits and Systems*, May 1990.

30. E.C. Chang and S.M. Kang, "Computationally efficient simulation of a lossy transmission line with skin effect by using numerical inversion of Laplace transform," *IEEE Transactions on Circuits and Systems*, vol. 39, pp. 861–868, July 1992.

31. R. Griffith and M. Nakhla, "Mixed frequency/time domain analysis on nonlinear circuits," *IEEE Trans. Computer-Aided Design*, vol. 10, no. 8, pp. 1032–1043, Aug. 1992.

32. R. Wang and O. Wing, "Transient analysis of dispersive VLSI interconnects terminated in nonlinear loads," *IEEE Trans. Computer-Aided Design*, vol. 11, pp. 1258–1277, Oct. 1992.

33. S. Lin and E.S. Kuh, "Transient simulation of lossy interconnects based on the recursive convolution formulation," *IEEE Trans. on Circuits and Systems*, vol. 39, no. 11, pp. 879–892.

34. L.T. Pillage and R.A. Rohrer, "Asymptotic waveform evaluation for timing analysis," *IEEE Trans. Computer-Aided Design*, vol. 9, pp. 352–366, Apr. 1990.

35. E. Chiprout and M. Nakhla, *Asymptotic Waveform Evaluation and Moment Matching for Interconnect Analysis*, Kluwer Academic Publishers, Boston, 1993.

36. S. Kumashiro, R.A. Rohrer, and A.J. Strojwas, "Asymptotic waveform evaluation for transient analysis of 3-D interconnect structures," *IEEE Trans. Computer-Aided Design*, vol. 12, no. 7, pp. 988–996, 1993.

37. X. Huang, "Padé Approximation of linear(ized) circuit responses," Ph.D. dissertation, Carnegie Mellon Univ., Nov. 1990.

38. T. Tang and M. Nakhla, "Analysis of high-speed VLSI interconnect using asymptotic waveform evaluation technique," *IEEE Trans. Computer-Aided Design*, pp. 2107–2116, Mar. 92.

39. D. Xie and M. Nakhla, "Delay and crosstalk simulation of high speed VLSI interconnects with nonlinear terminations," *IEEE Trans. Computer-Aided Design*, pp. 1798–1811, Nov. 1993.

40. R. Achar and M. Nakhla, *Minimum Realization of Reduced-Order Models of High-Speed Interconnect Macromodels*, Chapter: *Signal Propagation on Interconnects*, Kluwer Academic Publishers, Boston, 1998.

41. R. Achar and M. Nakhla, "A novel technique for minimum-order macromodel synthesis of high-speed interconnect subnetworks," *Proc. EEE International Symposium on Circuits and Systems (ISCAS)*, Monterey, CA, pp. 70–73, June 1998.

42. R. Griffith, E. Chiprout, Q.J. Zhang, and M. Nakhla, "A CAD framework for simulation and optimization of high-speed VLSI interconnections," *IEEE Transactions on Circuits and Systems*, vol. 39, no. 1, pp. 893–906.

43. J.E. Bracken, V. Raghavan, and R.A. Rohrer, "Interconnect simulation with asymptotic waveform evaluation (AWE)," *IEEE Transactions on Circuits and Systems*, pp. 869–878, Nov. 1992.

44. E. Chiprout and M. Nakhla, "Analysis of interconnect networks using complex frequency hopping," *IEEE Trans. Computer-Aided Design*, vol. 14, pp.186–199, Feb. 1995.

45. R. Sanaie, E. Chiprout, M. Nakhla, and Q.J. Zhang, "A fast method for frequency and time domain simulation of high-speed VLSI interconnects," *IEEE Trans. Microwave Theory Tech.*, vol. 42, no. 12, pp. 2562–2571, Dec. 1994.

46. R. Achar, M. Nakhla and E. Chiprout, "Block CFH: A model-reduction technique for distributed interconnect networks," *Proc. IEEE European Conference on Circuits Theory and Design (ECCTD)*, pp. 396–401, Sept. 1997, Budapest, Hungary.

47. P. Feldmann and R.W. Freund, "Efficient linear circuit analysis by Padé via Lanczos process," *IEEE Transactions on Computer-Aided Design*, vol. 14, pp. 639–649, May 1995.

48. P. Feldmann and R.W. Freund, "Reduced order modeling of large linear subcircuits via a block Lanczos algorithm," in *Proc. Design Automation Conf.*, pp. 474–479, June 1995.

49. M. Silveria, M. Kamon, I. Elfadel, and J. White, "A coordinate-transformed Arnoldi algorithm for generating guaranteed stable reduced-order models of arbitrary RLC circuits," in *Proc. IEEE ICCAD*, Nov. 96.

50. M. Chou and J. White, "Efficient reduced order modeling for the transient simulation of three dimensional interconnect," in *Proc. ICCAD*, pp. 40–44, Nov. 1995.

51. K.J. Kerns and A.T. Yang, "Preservation of passivity during RLC network reduction via split congruence transformations," *IEEE Trans. on Computer-Aided Design*, pp. 582–591, July 1998.

52. A. Odabasioglu, M. Celik and L.T. Pillage, "PRIMA: Passive Reduced-Order Interconnect Macromodeling Algorithm," *Proceedings IEEE ICCAD*, pp. 58–65, Nov. 1997.

53. I.M. Elfadel and D.D. Ling, "A block rational Arnoldi algorithm for multiport passive model-order reduction of multiport RLC networks," *Proc. of ICCAD-97*, pp. 66–71, Nov. 1997.

54. Q. Yu, J.M.L. Wang, and E.S. Kuh, "Multipoint moment-matching model for multiport distributed interconnect networks," *Proc. of ICCAD-98*, pp. 85–90, Nov. 1998.

55. P.K. Gunupudi, M. Nakhla, and R. Achar, "Efficient simulation of high-speed interconnects using Krylov-space techniques," *Proceedings IEEE 7th EPEP*, New York, pp. 292–296, Oct. 1998.

56. R. Khazaka, E. Chiprout, M. Nakhla, and Q.J. Zhang, "Analysis of high-speed interconnects with frequency dependent parameters," *Proc. Intl. Symp. EMC*, pp. 203–208, Zurich, March 1995.

57. R. Khazaka and M. Nakhla, "Analysis of high-speed interconnects in the presence of electromagnetic interference," *IEEE Trans. MTT*, vol. 46, pp. 940–947, July 1998.

58. I. Erdin, R. Khazaka, and M. Nakhla, "Simulation of high-speed interconnects in the presence of incident field," *IEEE Trans. MTT*, vol. 46, pp. 2251–2257, Dec. 1998.

59. S.D. Corey and A.T. Yang, "Interconnect characterization using time-domain reflectometry," *IEEE Trans. Microwave Theory Tech.*, vol. 43, pp. 2151–2156, Sep. 95.

60. B.J. Cooke, J.L. Prince, and A.C. Cangellaris "S-parameter analysis of multiconductor integrated circuit interconnect systems," *IEEE Trans. Computer-Aided Design*, pp. 353–360, Mar. 1992.

61. J.E. Schutt-Aine and R. Mittra, "Scattering parameter transient analysis of transmission lines loaded with nonlinear terminations," *IEEE Trans. Microwave Theory Tech.*, pp. 529–536, 1988.

62. P.C. Cherry and M.F. Iskander, "FDTD analysis of high frequency electronic interconnection effects," *IEEE Trans. Microwave Theory Tech.*, vol. 43, no. 10, pp. 2445–2451, Oct. 1995.

63. M. Celik, A.C. Cangellaris, and A. Deutsch, "A new moment generation technique for interconnects characterized by measured or calculated S-parameters," *IEEE Intl. Microwave Symposium Digest*, pp. 196–201, June 1996.

64. W.T. Beyene and J.E. Schutt-Aine, "Efficient trenasient simulation of high-speed interconnects characterized by sampled data," *IEEE Transactions on CPMT*, Part B, vol. 21, pp. 105–113, Feb. 1998.

65. G. Zheng, Q.J. Zhang, M. Nakhla, and R. Achar, "An efficient approach for simulation of measured subnetworks with complex frequency hopping," *Proceedings IEEE/ACM In. Conf. Computer Aided Design*, San Jose, CA, pp. 23–26, Nov. 1996.

66. G.A. Baker Jr., *Essential of Padé Approximants*, Academic, New York, 1975.

67. J.H. McCabe, "A formal extension of the Padé table to include two point Padé quotients," *J. Inst. Math. Applic.*, vol. 15, pp. 363–372, 1975.

68. C.W. Ho, A.E. Ruehli, and P.A. Brennan, "The modified nodal approach to network analysis," *IEEE Trans. Circuits and Systems*, vol. CAS-22, pp. 504–509, June 1975.

69. J. Vlach and K. Singhal, *Computer Methods for Circuit Analysis and Design*, Van Nostrand Reinhold, New York, 1983.

70. L. Weinberg, *Network Analysis and Synthesis*, McGraw-Hill, New York, 1962.

71. E. Kuh and R. Rohrer, *Theory of Active Linear Networks*, Holden-day, San Francisco, CA, 67.

72. E.A. Guillemin, *Synthesis of Passive Networks*, John Wiley & Sons, New York, 1957.

73. M.E.V. Valkenburg, *Introduction to Modern Network Synthesis*, John Wiley & Sons, New York, 1960.

74. J.W. Demmel, *Applied Numerical Linear Algebra*, SIAM Publishers, Philadelphia, PA, 1997.

75. T. Kailath, *Linear Systems*. Prentice-Hall, Toronto, 1980.

76. C.T. Chen, *Linear System Theory and Design*, Holt, Rinehart and Winston, New York, 1984.

Section
III

Low Power Electronics and Design

Massoud Pedram
University of Southern California

15

System-Level Power Management: An Overview

Ali Iranli
University of Southern California

Massoud Pedram
University of Southern California

CONTENTS

15.1 Introduction

One of the key challenges of computer system design is the management and conservation of energy. This challenge is evident in a number of ways. The goal may be to extend the battery lifetime of a portable, battery-powered device. The processing power, memory, and network bandwidth of such devices are increasing quickly, resulting in an increase in demand for higher power dissipation, while the battery capacity is improving at a much slower pace. Other goals may be to limit the cooling requirements of a computer system or to reduce the financial cost of operating a large computing facility with a high energy bill. This chapter focuses on techniques which dynamically manage electronic systems in order to minimize its energy consumption. Ideally, the problem of managing the energy consumed by electronic systems should be addressed at all levels of design, ranging from low-power circuits and architectures to application and system software capable of adapting to the available energy source. Many research and industrial efforts are currently underway to develop low-power hardware as well as energy-aware application software in the design of energy-efficient computing systems. Our objective in this chapter is to explore what the system software, vis-à-vis the operating system (OS), can do within its own resource management functions to improve the energy efficiency of the computing system without requiring any specialized, low-power hardware or any explicit assistance from application software and compilers. There are two approaches to consider at the OS-level for attacking most of the specific energy-related goals described above. The first is to develop resource management policies that eliminate waste or overhead and allow energy-efficient use of the devices. The second is to change the system workload so as to reduce the amount of work to be done, often by changing the fidelity of objects accessed, in a manner which will be acceptable to the user of the application. This chapter provides a first introduction to these two approaches with appropriate review of related works.

15.2   Background

A system is a collection of components whose combined operation provides a useful service. Typical systems consist of hardware components integrated on single or multiple chips and various software layers. Hardware components are macro-cells that provide information processing, storage, and interfacing. Software components are programs that realize system and application functions. Sometimes, system specifications are required to fit into specific interconnections of selected hardware components (e.g., Pentium processor) with specific system software (e.g., Windows or Linux) called *computational platforms*.

System design consists of realizing a desired functionality while satisfying some design constraints. Broadly speaking, constraints limit the design space and relate the major design trade-off between *quality of service* (QoS) versus cost. QoS is closely related to performance, i.e., system throughput and task latency. QoS relates also to the system *dependability*, i.e., to a class of system metrics such as reliability, availability, and safety that measure the ability of the system to deliver a service correctly, within a given time window and at any time. Design cost relates to design and manufacturing costs (e.g., silicon area and testability) as well as to operation costs (e.g., power consumption and energy consumption per task).

In recent years, the design trade-off of performance versus power consumption has received large attention because of (i) the large number of systems that need to provide services with the energy provided by a battery of limited weight and size, (ii) the limitation on high-performance computation because of heat dissipation issues, and (iii) concerns about dependability of systems operating at high temperatures because of power dissipation. Here we focus on *energy-managed computer* (EMC) systems. These systems are characterized by one or more high-performance processing cores, large on-chip memory cores, various I/O controller cores. The use of these cores will force system designers to treat them as black boxes and abandon the detailed tuning of their performance/energy parameters. On the other hand, various I/O devices are provisioned in the system-level design to maximize the interaction between the user and the system and among different users of the same system.

Dynamic power management (DPM) is a feature of the run-time environment of an EMC system that dynamically reconfigures itself to provide the requested services and performance levels with a minimum number of active components or a minimum activity level on such components. DPM encompasses a set of techniques that achieve energy-efficient computation by selectively turning off (or reducing the performance of) system components when they are idle (or partially unexploited). The fundamental premise for the applicability of DPM is that systems (and their components) experience nonuniform workloads during operation time. Such an assumption is valid for most systems, both when considered in isolation and when internetworked. A second assumption of DPM is that it is possible to predict, with a certain degree of confidence, the fluctuations of workload. In this chapter we present and classify different modeling frameworks and approaches to DPM.

15.3   Modeling Energy-Managed Computers

An EMC models the electronic system as a set of interacting power manageable components (PMCs) controlled by one or more *power managers* (PMs). We model PMCs as black boxes. We are not concerned on how PMCs are designed; instead we focus on how they interact with each other and the operating environment. The purpose of this analysis is to understand what type and how much information should be exchanged between a power manager and system components to implement effective system-wide energy management policies. We consider PMCs in isolation first. Next, we describe DPM for systems with several interacting components.

15.3.1   Power Manageable Components

A PMC is defined to be an atomic block in an electronic system. PMCs can be as complex as a printed circuit board realizing an I/O device, or as simple as a functional unit within a chip. At the system level, a component is typically seen as a black box, i.e., no data is available about its internal architecture. The key attribute of a PMC is the availability of multiple modes of operation, which span the power–performance trade-off curve. Nonpower-manageable components are designed for a given performance target and power dissipation specification. In contrast, with PMCs, it is possible to dynamically switch between high-performance, high-power modes of operation and low-power, low-performance ones so as to provide just enough computational capability to meet a target timing constraint while minimizing the total energy consumption of completing a computational task. In the limit, one can think of a PMC to have a continuous range of operational modes. Clearly, as the number of available operational modes increases the ability to perform fine-grained control of the PMC to minimize the power waste and achieve a certain performance level increases. In practice, the number of modes of operation tends to be small because of the increased design complexity and hardware overhead of supporting multiple power modes.

Another important factor about a PMC is the overhead associated with the PMC transitioning from one mode of operation to next. Typically, this overhead is expressed as transition energy and a delay penalty. If the PMC is not operational during the transition, some performance is lost whenever a transition is initiated. The transition overhead depends on PMC implementation: in some cases the cost may be negligible, but, generally, it is not. Transition overhead plays a significant role in determining the number and type of operational modes enabled by the PMC designer. For example, excessive energy and delay overheads for transitions into and out of a given PMC state may make that state nearly useless because it will be very difficult to recompense the overheads unless the expected duration of contiguous time that the PMC remains in that state is especially long.

Mathematically, one can represent a PMC by a finite-state machine where states denote the operational modes of the PMC and state transitions represent mode transition. Each edge in the state machine has an associated energy and delay cost. In general, low-power states have lower performance and larger transition overhead compared to high-power states. This abstract model is referred to as a *power state machine* (PSM). Many single-chip components such as processors [1], memories [2], and archetypal I/O devices such as disk drives [3], wireless network interfaces [4], and displays [5] can readily be modeled by a PSM.

Example

The StrongARM SA-1100 processor [6] is an example of a PMC. It has three modes of operation: *RUN, STDBY*, and *SLEEP*. The *RUN* mode is the normal operating mode of the SA-1100: every on-chip resource is functional. The chip enters the *RUN* mode after successful power-up and reset. *STDBY* mode allows a software application to stop the CPU when it is not in use, while continuing to monitor interrupt requests on or off chip. In the *STDBY* mode, the CPU can be brought back to the *RUN* mode quickly when an interrupt occurs. *SLEEP* mode offers the greatest power savings and, consequently, the lowest level of available functionality. In the transition from *RUN* or *STDBY*, the SA-1100 performs an orderly shutdown of its on-chip activity. In a transition from *SLEEP* to any other state, the chip steps through a rather complex wake-up sequence before it can resume normal activity. The PSM model of the StrongARM SA-1100 is shown in Figure 15.1. States are marked with power dissipation and performance values, edges are marked with transition times and energy dissipation overheads. The power consumed during transitions is approximately equal to that in the *RUN*

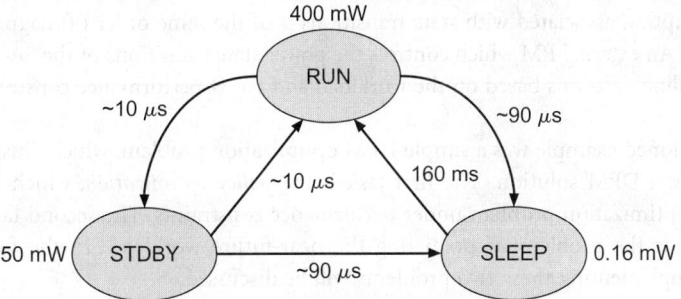

FIGURE 15.1 Power state machine for strong ARM SA1100.

mode. Notice that both *STDBY* and *SLEEP* have null performance, but the time for exiting *SLEEP* is much longer than that for exiting *STDBY* (10 μs versus 160 ms). In contrast, the wake-up time from the *SLEEP* state is much larger, and therefore, it must be carefully compared with the environment's time constants before deciding to shut the processor down. In the limiting case of a workload with no idle periods longer than the time required to enter and exit the *SLEEP* state, a greedy policy which would shut down the processor as soon as an idle period was detected tends to reduce performance without actually saving any power (the ratio of the energy consumption divided by the transition time associated with any of the state transitions is of the same order of power dissipation in the *RUN* state). An external PM that controls the intermode transitions of the SA-1100 processor must observe the workload and make decisions according to a policy whose optimality depends on workload statistics and on predefined performance constraints. Notice that the policy becomes trivial if there are no performance constraints: the PM can keep the processor nearly always in the *SLEEP* state.

15.3.2 Dynamic Power Management Techniques

This section reviews various techniques for controlling the power state of a system and its components. One may consider components as black boxes, whose behavior is abstracted by the PSM model and focus on how to design effective power management policies. Without loss of generality, consider the problem of controlling a single component (or, equivalently, the system as a whole). Furthermore, assume that transitions between different states are instantaneous and the transition energy overhead is nonexistent. In such a system, DPM is a trivial task and the optimum policy is greedy one, i.e., as soon as the system is idle, it can be transitioned to the deepest sleep state available. On the arrival of a request, the system is instantaneously activated. Unfortunately, most PMCs have nonnegligible performance and power costs for state transitions. For instance, if entering a low-power state requires power-supply shutdown, returning from this state to the active state requires a (possibly long) time to (1) turn on and stabilize the power supply and the clock; (2) reinitialize the system; and (3) restore the context. When power state transitions have a cost, finding the optimal DPM policy becomes a difficult optimization problem. In this case, the DPM policy optimization is equivalent to a decision-making problem in which the PM must decide if and when it is worthwhile (from a performance and power dissipation viewpoint) to transition to which low-power state (in case of having multiple low-power states).

Example

Consider the StrongARM SA-1100 processor described in the previous example. Transition times between *RUN* and *STDBY* states are very fast so that the *STDBY* state can be optimally exploited according to a greedy policy possibly implemented by an embedded PM. In contrast, the wake-up time from the *SLEEP* state is much longer and has to be compared with the time constants for the workload variations to determine whether or not the processor should be shut down. In the limiting case of a workload without any idle period longer than the time required to enter and exit the *SLEEP* state, a greedy policy for shutting down the processor (i.e., moving to *SLEEP* state as soon as an *STDBY* period is detected) will result in performance loss, but no power saving. This is because the power consumption associated with state transitions is of the same order of magnitude as that of the *RUN* state. An external PM which controls the power state transitions of the SA-1100 processor must make online decisions based on the workload and target performance constraints.

The aforementioned example was a simple DPM optimization problem, which illustrated the two key steps of designing a DPM solution. The first task is the *policy optimization*, which is the problem of solving a power optimization problem under performance constraints. The second task is the *workload prediction*, which is the problem of predicting the near-future workload. In the following, different approaches for implementing these two problems will be discussed.

The early works on DPM focused on predictive shutdown approaches [7,8], which make use of "timeout"-based policies. A power management approach based on discrete-time Markovian decision processes was

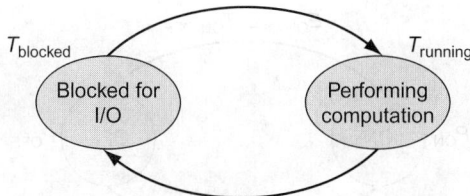

FIGURE 15.2 Event-driven application alternates between blocked and running states.

proposed in Ref. [9]. The discrete-time model requires policy evaluation at periodic time instances and may thereby consume a large amount of power even when no change in the system state has occurred. To surmount this shortcoming, a model based on continuous-time Markovian decision processes (CTMDP) was proposed in Ref. [10]. The policy change under this model is asynchronous and is thus more suitable for implementation as part of a real-time operating system environment. Ref. [11] proposed time-indexed semi-Markovian decision processes for system modeling. Other approaches such as adaptive-learning-based strategies [12], session clustering and prediction strategies [13], online strategies [14,15], and hierarchical system decomposition and modeling [34] have also been utilized to find a DPM policy of EMCs.

In the following sections, we describe various DPM techniques in more detail.

15.3.2.1 Predictive Shutdown Approaches

Applications such as display servers, user-interface functions, and communication interfaces are "event-driven" in nature with intermittent computational activity triggered by external events and separated by periods of inactivity. An obvious way to reduce average power consumption in such applications is to shut the system down during periods of inactivity. This can be accomplished by shutting off the system clock or in certain cases by shutting off the power supply (cf. Figure 15.2).

An event-driven application will alternate between a blocked state where it stalls the CPU waiting for external events and a running state where it executes instructions. Let T_{blocked} and T_{running} denote the average time spent in the blocked and the running states, respectively. One can improve the energy efficiency by as much as a factor of $1 + T_{\text{blocked}}/T_{\text{running}}$ provided that the system is shut down whenever it is in the blocked state.

There are two key questions: (1) how to shut down, and (2) when to shut down. The first question is addressed by developing mechanisms for stopping and restarting the clock or for turning off and on the power supply. The second question is addressed by devising policies such as "shut the system down if the user has been idle for five minutes." Although these two issues are not really independent because the decision about when to shut down depends on the overhead of shutting down the system, the predictive shutdown approaches focus primarily on the question of deciding when to shut down while being aware of the available shutdown mechanisms. Simple shutdown techniques, for example, shutting down after a few seconds of no keyboard or mouse activity, are typically used to reduce power consumption in current notebook computers. However, the event-driven nature of modern applications, together with efficient hardware shutdown mechanisms provided by PMCs, suggests the possibility of a more aggressive shutdown strategy where parts of the system may be shut down for much smaller intervals of time while waiting for events.

In this section we explore a shutdown mechanism where we try to predict the length of idle time based on the computation history, and then shut the processor down if the predicted length of the idle time justifies the cost in terms of both energy and performance overheads of shutting down. The key idea behind the predictive approach can be summarized as follows: "Use history to predict whether T_{blocked} will be long enough to justify a shutdown." Unfortunately, this is a difficult and error-prone task. One therefore has to resort to heuristics to predict T_{blocked} for the near future. Ref. [7,8] present approaches where based on the recent history, a prediction is made as to whether or not the next idle time will be long enough to at least break even with the shutdown overhead. Results demonstrate that for reasonable values of the shutdown overhead, the predictive approaches tend to result in sizeable energy savings compared to the greedy shutdown approach, while the performance degradation remains negligible.

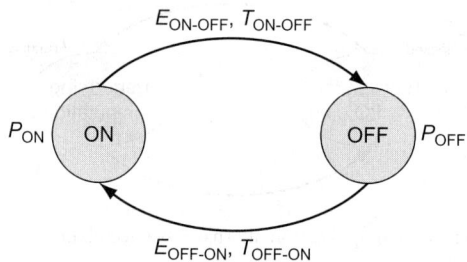

FIGURE 15.3 PSM of a two-state power-manageable component.

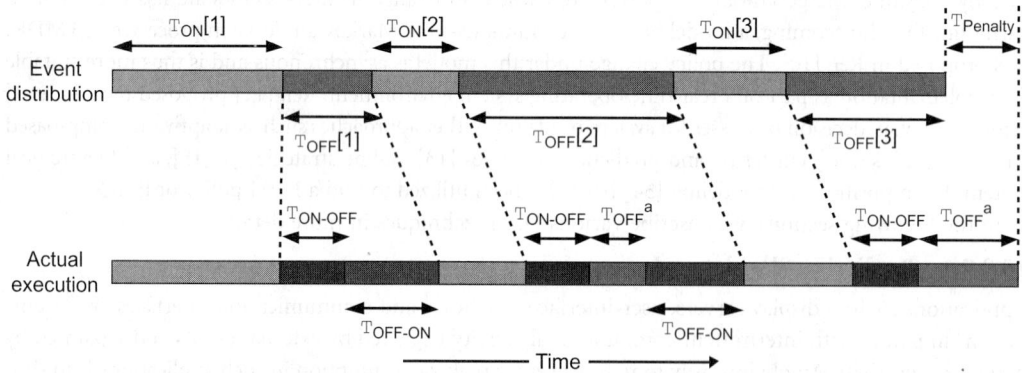

FIGURE 15.4 Graphical illustration of how a simple greedy algorithm can result in a significant delay penalty.

Restricting the analysis to a simple event-driven model of an application program running on the SA-1100 processor and considering internally controlled transfer between RUN and STDBY states and externally controlled transfer between STDBY and SLEEP states of the processor, the system can be modeled by a partially self-power-managed PSM with only two states (cf. Figure 15.3): ON and OFF. The ON state is a macro-state representing the RUN and STDBY states of the processor and the local policy used by the processor itself to move between RUN and STNDBY states depending on the workload. The OFF macro-state is the same as the SLEEP state. The power consumption associated with the ON state is the expected power consumption in this macro-state and is calculated as a function of standby time, local transition probabilities, and energy overhead of the transitions. Transitions between ON and OFF macro-states correspond to transitions between RUN and SLEEP states and their overheads are set accordingly.

The processor starts in the ON state, and makes transitions from ON to OFF back to ON state. Let $T_{ON}[i]$ and $T_{OFF}[i]$ denote the time spent by the application in the ith visit to the ON and the OFF states, respectively. Furthermore, we define T_{ON} as the average of $T_{ON}[i]$ over all i, and similarly T_{OFF} as the average of $T_{OFF}[i]$ over all i. Let $T_{ON\text{-}OFF}$ and $E_{ON\text{-}OFF}$ denote the time and energy dissipation overhead associated with the transfer to OFF state. $T_{OFF\text{-}ON}$ and $E_{OFF\text{-}ON}$ are similarly defined (cf. Figure 15.4). In predictive shutdown approaches the PM predicts the upcoming duration of time for which the system will be idle, $T_{OFF}[i]$, based on the information from the current active period, $T_{ON}[i]$ and previous active and idle periods $T_{ON}[j]$ and $T_{OFF}[j]$ for $j = i-1, i-2, \ldots, 1$. The policy then is to transfer the processor from ON to OFF state if $T_{OFF}[i] \geq T_{BE}$, where T_{BE} is the duration of break-even time. The processor is then turned on (it moves from OFF to ON state) as soon as a new request for data processing comes in.

In Ref. [7], the authors proposed two different approaches for predicting $T_{OFF}[i]$. The first approach uses regression analysis on application traces and calculates $T_{OFF}[i]$ in terms of $T_{OFF}[i-1]$ and $T_{ON}[i]$. Notice that $T_{OFF}[i-1]$ denotes the actual (and not the initially predicted) duration of the OFF time on the $(i-1)$th visit to the OFF state. For their second approach, the authors simplify the analysis based on the observation that long OFF periods are often followed by short ON periods. Therefore, a simple rule is constructed whereby, based on the duration of the current ON period, $T_{ON}[i]$, the PM predicts the duration of the next OFF period, $T_{OFF}[i]$, to be larger or smaller than the break-even time, T_{BE}, and therefore, decides whether it should maintain or change the current power state of the processor.

Ref. [8] improves this approach by using an *exponential average* of the previous OFF periods as the predicted value for the upcoming idle period duration. More precisely,

$$T_{Off}^{est}[i] = aT_{Off}^{act}[i-1] + a(1-a)T_{Off}^{act}[i-2] + a(1-a)^2 T_{Off}^{act}[i-3] + \ldots$$

$$+ a(1-a)^{m-2} T_{Off}^{act}[i-m+1] + (1-a)^m T_{Off}^{est}[i-m]$$

(15.1)

where $0 \leq a \leq 1$ is a weighting coefficient. Parameter a controls the relative weight of recent and past history in the prediction. If $a = 0$, then $T_{OFF}^{est}[i] = T_{OFF}^{act}[i-m]$, i.e., the recent history has no effect on the estimation. In contrast, if $a = 1$, then $T_{OFF}^{est}[i] = T_{OFF}^{act}[i-1]$, i.e., only the immediate past matters in setting the duration of the next idle period. In general, however, this equation favors near-past historical data. For example, for $a = \frac{1}{2}$ $T_{OFF}^{act}[i-1]$ has a weight of $\frac{1}{2}$ whereas $T_{OFF}^{act}[i-3]$ has a weight of $\frac{1}{8}$.

As mentioned before, when the PMC resumes the ON state from the OFF state (i.e., on system wake-up), the PMC suffers a delay penalty of T_{OFF-ON} having to restore the original system state. This delay penalty can have a large negative impact on the PMC's performance. Ref. [8] circumvents this problem by proposing a prewakeup approach before the arrival time of the next event. In this approach, the system starts the activation process immediately after the predicted time interval for the current idle period. Let us consider the case where the predicted idle period is overestimated, i.e., $T_{Off}^{est}[i] = T_{Off}^{act}[i] + d$ for $d > 0$. Two subcases are possible. (1.1) $d \leq T_{OFF-ON}$: the system wakes up after T_{OFF-ON} and the delay penalty is $(T_{OFF-ON} - d)$; (1.2) $d \leq T_{OFF-ON}$: the system is awakened after T_{OFF-ON} time units. Next, consider the case where the predicted idle period is underestimated, i.e., $T_{Off}^{est}[i] = T_{Off}^{act}[i] - d$ for $d \geq 0$. Again we consider two subcases. (2.1) $d \leq T_{OFF-ON}$: the system will wake up after T_{OFF-ON} and immediately starts executing the arrived computational task. There is no energy waste and the delay penalty is $(T_{OFF-ON} - d)$. (2.2) $d > T_{OFF-ON}$: the system will wake up after T_{OFF-ON} and remain ON for a period of $(d - T_{OFF-ON})$ time units ahead of the next required computation. Energy waste is $(d - T_{OFF-ON})P_{ON}$. There is no delay penalty. In summary, the prewakeup policy results in shorter delay penalty in subcases (1.1), (2.1), and (2.2), but it results in energy waste in subcase (2.2).

To alleviate the chances for underestimation of the idle period, Ref. [8] proposes a timeout scheme which periodically examines the PMC to determine whether it is idle but not shut down. If that is the case, then it increases $T_{OFF}^{est}[i]$. The chance of overprediction is reduced by imposing a saturation condition on predictions, i.e., $T_{OFF}^{est}[i] \leq C_{max} T_{OFF}^{est}[i-1]$.

Several other adaptive predictive techniques have been proposed to deal with nonstationary workloads. In the work by Krishnan et al. [16], a set of prediction values for the length of idle period is maintained. In addition, each prediction value is annotated with an indicator to show how successful it would have been if it had been used in the past. The policy then chooses for the length of next idle period the prediction which has the highest indicator value among the set of available ones. Another policy, presented by Helmbold et al. [17], also keeps a list of candidate predictions and assigns a weight to each timeout value based on how well it would have performed for past requests relative to an optimum offline strategy. The actual prediction is then obtained as a weighted average of all candidate predictions. Another approach, introduced by Douglis et al. [18], keeps only one prediction value but adaptively changes the value. In particular, it increases (decreases) the prediction value when this value causes too many (few) shutdowns.

The accuracy of workload prediction can be increased by customizing predictors to a particular class of workloads. This kind of customization restricts its scope of applicability, but also reduces the difficulties of predicting completely general workloads. A recently proposed adaptive technique [19], which is specifically tailored toward hard-disk power management, is based on the observation that disk accesses are clustered in sessions. Sessions are periods of relatively high disk activity separated by long periods of inactivity. Under the assumption that disk accesses are clustered in sessions, adaptation is only used to predict the session length. Prediction of a single parameter is easily accomplished and the reported accuracy is high.

As mentioned earlier, there are periods of unknown duration during which there are no tasks to run and the device can be powered down. These idle periods end with the arrival of a service request. The decision that the online DPM algorithm has to make is when to transition to a lower power state, and which state to transition to. The power-down states are denoted by $s_0, \ldots, s_k$, with associated decreasing power consumptions of $P_0, \ldots, P_k$. At the end of the idle period, the device must return to the highest power state, s_0. There is an associated transition energy e_{ij} and transition time t_{ij} to move from state s_i to s_j. The goal is to minimize the energy dissipation consumed during the idle periods. Online power-down techniques can be evaluated according to a competitive ratio (a ratio of 1 corresponds to the optimal solution). There is a deterministic 2-competitive algorithm for two-state systems, which keeps the service provider in the active state until the total energy consumed is equal to the transition energy. It is recognized that this algorithm is optimally competitive. Furthermore, if the idle period is generated by a known probability distribution, then there is an optimally competitive probability-based algorithm which is $(e/(e-1))$-competitive [20]. For some systems, the energy needed and time spent to go from a higher power state to a lower power state is negligible. Irani et al. show in Ref. [14] that for such systems, the two-state deterministic and probability-based algorithms can be generalized to systems with multiple sleep states so that the same competitive ratios can be achieved. The probability-based algorithm requires information about a probability distribution, which generates the length of the idle period. In Ref. [15], Irani et al. give an efficient heuristic for learning the probability distribution based on recent history.

15.3.2.2 Markovian Decision Process-Based Approaches

The most aggressive predictive power management policies turn off every PMC as soon as it becomes idle. Whenever a component is needed to carry out some task, the component must first be turned on and restored to its fully functional state. As mentioned above, the transition between the inactive and the functional state has latency and energy overheads. As a result, "eager" policies are often unacceptable because they can degrade performance without decreasing power dissipation. The heuristic power management policies are useful in practice although no strong optimality result has been proved for these types of policies. On the other hand, stochastic control based on Markov models has emerged as an effective power management framework. In particular, the stochastic PM techniques have a number of key advantages over predictive techniques. First, they capture a global view of the system, thus allowing the designer to search for a global optimum which can exploit multiple inactive states of multiple interacting resources. Second, they compute the exact solution (in polynomial time) for the performance-constrained power optimization problem. Third, they exploit the vigor and robustness of randomized policies. However, a number of key points must be considered when deciding whether or not to utilize a stochastic DPM technique. First, the performance and power obtained by a policy are expected values, and there is no guarantee that the results will be optimum for a specific workload instance (i.e., a single realization of the corresponding stochastic process). Second, policy optimization requires a priori Markov models of the service provider (SP) and service requester (SR). One can safely assume that the SP model can be precharacterized; however, this assumption may not be true about the SR's model. Third, policy implementation tends to be more involved. An implicit assumption of most DPM techniques is that the power consumption of the PM is negligible. This assumption must be validated on a case-by-case basis, especially for stochastic approaches. Finally, the Markov model for the SR or SP may be only an approximation of a much more complex stochastic process. If the model is not accurate, then the "optimal" policies are also approximate solutions.

In the following, we consider a discrete-time (i.e., slotted time) setting [9]. Time is described by an infinite series of discrete values $t_n = Tn$, where T is the time resolution (or period), and $n \leq N^+$. The EMC is modeled with a single SR (or user) whose requests are en-queued in a single queue, service queue (SQ), and serviced by a single SP. The PM controls over time the behavior of the SP (Figure 15.5).

15.3.2.2.1 Service Requester.

This unit sends requests to the SP. The SR is modeled as a Markov chain whereby the observed variable is the number of requests s_r sent to the SP during time period t_n. The service request process and its relevant parameters are known. Moreover, it is known that in each time period a maximum of S_p requests can be generated.

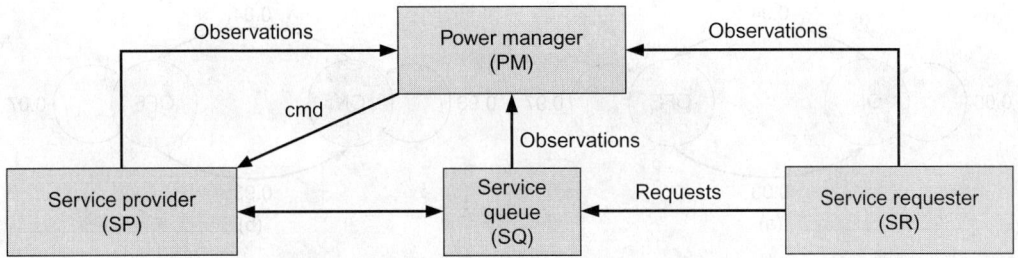

FIGURE 15.5 Illustration of the abstract system model.

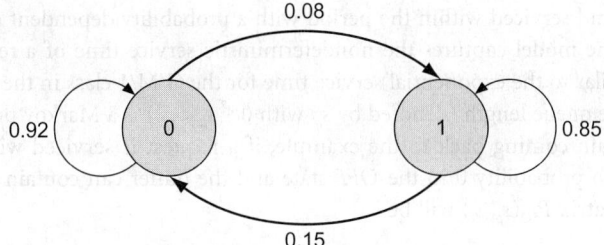

FIGURE 15.6 A Bursty service requester's stochastic model.

Example

Consider a "bursty" workload with a maximum of one request per period, i.e., the SR has two states as depicted in Figure 15.6. Since the workload comes in bursts, a request will be generated at time t_{n+1} with a probability of 0.85 if a request is received at time t_n. On the other hand, if there is no request at time t_n, then with a probability of 0.92 there will be no request at time t_{n+1}. The mean duration of a stream of requests is equal to $1/0.15 = 6.67$ periods.

15.3.2.2.2 Service Provider.

The SP is a PMC which services requests issued by the SR. In each time period, the SP can be in only one state. Each state $s_p \leq \{1, \ldots, S_p\}$ is characterized by a performance level and by a power consumption level. In the simplest example, one could have two states: *ON* and *OFF*. In each period, transitions between power states are controlled by a PM through commands: $cmd \leq CMD = \{1, \ldots, N_C\}$. For example, one can define two simple commands: *Go2Active* and *Go2Sleep*. When a specific command is issued, the SP will move to a new power state with a fixed probability depending on the command *cmd*, and on the departing and arriving states. In other words, when a command is issued by the PM, there is no guarantee that the command is immediately executed. Instead, the command influences the way in which the SP will act in the future. This probabilistic model describes the view where the evolution in time of power states is modeled by a Markov process in which the transition probability matrix is dependent on the commands issued by the PM. In other words, there is one transition probability matrix for each command *cmd*.

Coming back to the SA-1100 example with two states, Figure 15.7 depicts the probabilistic behavior of the device under influence of *Go2Sleep* and *Go2Active* commands.

The transition time from *OFF* to *ON* when *Go2Active* has been issued is a geometric random variable with an average of $1/0.04 = 25$ time periods. Each power state has a specific power consumption rate and performance (e.g., clock speed), which is a function of the state itself. In addition, each transition between two power states in annotated with an energy cost and a latency, representing the overhead of transitioning between the two corresponding states. Such information is usually provided in the data-sheets of the PMCs.

15.3.2.2.3 Service Queue.

When service requests arrive during one period, they are buffered in a queue of length ($S_q \geq 1$). The queue is usually considered to be a FIFO, although other schemes can also be modeled efficiently. The

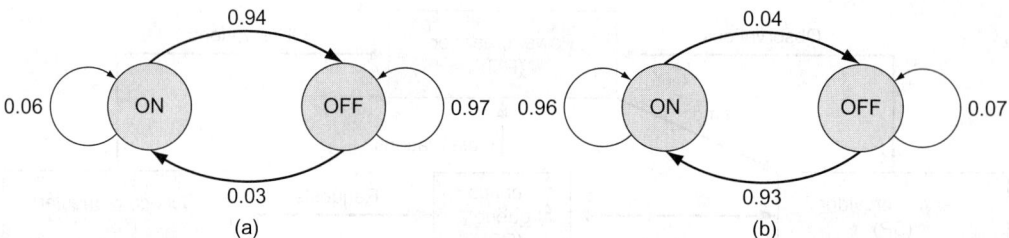

FIGURE 15.7 Stochastic model of the SA-1100. (a) Stochastic model of the PMC when *Go2Sleep* command has been issued. (b) Stochastic model of the PMC when *Go2Active* command has been issued.

request is processed and serviced within the period with a probability dependent on the power state of the SP. In this way, the model captures the nondeterministic service time of a request as a geometric random variable, similar to the exponential service time for the G/M/1 class in the queuing theory [21]. It follows that also the queue length (denoted by s_q, with $0 \le s_q < S_q$) is a Markov process with transition matrix $P_{SQ}(s_p, s_r)$. Again coming back to the example, if a request is serviced with probability 0.9 in the *ON* state and with probability 0 in the *OFF* state and the buffer can contain at most one request, then the transition matrix $P_{SQ}(s_p, s_r)$ will be

$$P_{SP}(ON, 0) = \begin{matrix} & 0 & 1 \\ 0 & 1.0 & 0.0 \\ 1 & 0.9 & 0.1 \end{matrix} \qquad P_{SP}(ON, 1) = \begin{matrix} & 0 & 1 \\ 0 & 0.9 & 0.1 \\ 1 & 0.9 & 0.1 \end{matrix}$$

$$\tag{15.2}$$

$$P_{SP}(OFF, 0) = \begin{matrix} & 0 & 1 \\ 0 & 1.0 & 0.0 \\ 1 & 0.0 & 1.0 \end{matrix} \qquad P_{SP}(OFF, 1) = \begin{matrix} & 0 & 1 \\ 0 & 0.0 & 1.0 \\ 1 & 0.0 & 1.0 \end{matrix}$$

15.3.2.2.4 *Power Manager.*

This component communicates with the SP and attempts to set its state at the beginning of each period by issuing commands from among a finite set, *CMD*. This goal is in turn achieved only in a probabilistic sense, that is, the PM changes the transition matrix of the SP by issuing a particular command. In the aforementioned example, the two possible commands are *Go2Active*, and *Go2Sleep*. The PM has all specifications and collects all relevant information (by observing SR, SQ, and SP) needed for implementing a power management policy. The power consumption of the PM is assumed to be much smaller than that of the PMCs it controls and so it is not a concern. The state of the EMC composed of the SP, the SR, and the queue is then a triplet, $s = (s_p, s_q, s_r)$. Being the composition of three Markov chains, s is a Markov chain (with $S = S_r \times S_p \times S_q$ states), whose transition matrix $P(cmd)$ depends on the command *cmd* issued to the SP by the PM. Hence, the system is fully described by a set of N_C transition matrices, one for each command.

In the above description no mention is made of the energy source (i.e., the battery). In the stochastic approaches the goal is to minimize (or bound) the average power consumption of the SP, and not to maximize the expected battery life. This choice has several advantages: it does not need to consider the details of the power source (rate-dependent energy discharge characteristics, and energy recovery), while it still retains the primary feature of minimizing (or constraining) the power consumption level. However, there have been recent attempts to incorporate the battery behavior in modeling the EMC systems while finding a solution to the DPM problem [22].

At the beginning of each time period t_n, the PM observes the "history" of the system, i.e., the sequence of states and commands up to t_{n-1}. It then controls the SP by making a decision. In *deterministic policies,* the PM makes a single decision to issue specific command on the basis of the history of the system. In contrast, in the much broader set of policies called the *randomized policies,* PM assigns probabilities to every available command and then chooses the command to issue, according to this probability distribution. In this way, even if the same decision is taken in different periods, the actual commands issued by the PM can be different.

Mathematically speaking, let H_n represent history of the EMC, then a decision $\delta(H_n)$ in a randomized policy is a set of probabilities, p_{cmd}, where each p_{cmd} represents the probability of issuing command, *cmd*, given that the history of the system is H_n. A deterministic decision is the special case with $p_{cmd} = 1$ for some command, *cmd*. Over a finite time horizon, the decisions taken by the PM are a finite discrete sequence $\delta_{(1)}, \ldots, \delta_{(n)}$. We call this sequence a policy π. The policy π is the free variable of our optimization problem. If a policy $\pi = (\delta_{(1)}, \ldots, \delta_{(n)})$ is adopted, then we can define $P_n^\pi = \prod_{i=1}^n P_{\delta(i)}$, which is simply the *n*-step transition matrix under policy π.

Example

Consider the example of the previous section. Suppose that the PM observes the following history: s1 = (0, ON, 0), s2 = (1, OFF, 0) (states in periods 1, 2), and *cmd*(1) = *Go2Sleep* (action taken at time 1). A possible decision at time 2 in a randomized policy, when state s2 is observed, consists of setting probabilities for issuing commands *Go2Active* and *Go2Sleep* to $p_{Go2Active} = 0.34$, $p_{Go2Sleep} = 0.66$, respectively. In contrast, in case of a deterministic policy, for example, the PM will decide to issue the *Go2Active* command to the underlying PMC.

15.3.2.2.5 *Policy Optimization.*

The problem is to find the optimal policy (set of state-action pairs) for the PM such that some power-related cost function is minimized subject to a set of performance constraints. Consider that the system is in state $s = (s_p, s_q, s_r)$ at the beginning of time period *n*. A typical example of a cost function is the *expected power consumption* of the SP in that time period, which is denoted as $c(s_p, \delta_{(n)})$ and represents the power consumption when the SP starts in state s_p and the PM takes decision $\delta_{(n)}$. (Note that $c(s_p, \delta_{(n)}) = \sum_{cmd \in \delta(n)} P_{cmd} c(s_p, cmd)$.) A second parameter of interest is the *performance penalty* in that time period, which is denoted by $d(s_q)$ and is typically set to the queue length, s_q. Finally, one can consider the *request loss* in the time period, denoted by $b(s_r, s_q)$. The loss factor is in general set to one when a request arrives ($s_r = 1$) and the queue buffer is full ($s_q = S_q$); otherwise it is set to zero. We are interested in finding the optimum stationary PM policy for the system. This means that decision, δ_n, is only a function of the system state, *s*, and not of the time period at which the decision is made. In other words, the policy sought is one in which the same decision is made for a given global state regardless of time. With this in mind, we can now define a power consumption vector, $\underline{c}(\delta_s)$, a performance penalty vector, $d(\delta_s)$, and a request loss vector, $\underline{b}(\delta_s)$. Each vector has |S| elements. Furthermore, the *s*th element of each vector is the expected value of the corresponding quantity when the system is in global state *s* and decision δ_n is taken. When performing policy optimization, we want to minimize the expected power consumption while keeping the average performance and request loss below some levels specified by the user. Given the probability distribution of the initial system state at the beginning of the first period, p_1, the problem of determining the optimal stationary policy can be formally described as follows:

$$\min_\pi \lim_{N \to \infty} \frac{1}{N} \sum_{n=1}^N p_1 P_{n-1}^\pi \underline{c}(\delta_{(n)})$$

s.t.

$$\lim_{N \to \infty} \frac{1}{N} \sum_{n=1}^N p_1 P_{n-1}^\pi \underline{d}(\delta_{(n)}) \le D_M \tag{15.3}$$

$$\lim_{N \to \infty} \frac{1}{N} \sum_{n=1}^N p_1 P_{n-1}^\pi \underline{b}(\delta_{(n)}) \le B_M$$

where $\underline{c}(\delta_{(n)}) = \underline{c}(\delta_s)$ if $s = s_{(n)}$ and D_M and B_M denote the upper bounds on average required performance penalty and request loss in any time period, respectively. The optimization is carried over the set of all possible policies. Hence, solving the aforementioned optimization appears to be a formidable task. Fortunately, if the delay constraint for the EMC is an active constraint, the optimal power management policy will generally be a randomized policy [23]. The randomized optimal policy can be obtained by solving a linear programming problem as explained in Ref. [9]. This approach offers significant improvements over previous power management techniques in terms of its theoretical framework for modeling and optimizing the EMC.

There are, however, some shortcomings. First, because the EMC is modeled in the discrete-time domain, some assumptions about the PMCs may not hold for real applications, such as the assumption that each event comes at the beginning of a time slice, or the assumption that the transition of the SQ is independent of the transition of the SP, etc. Second, the state transition probability of the system model cannot be obtained accurately. For example, the discrete-time model cannot distinguish the busy state and the idle state because the transitions between these two states are instantaneous. However, the transition probabilities of the SP when it is in these two states are different. Moreover, the PM needs to send control signals to the PMCs in every time slice, which results in heavy signal traffic and a heavy load on the system resources (and, therefore, more power).

Ref. [10] overcomes these shortcomings by introducing a new system model based on CTMDP. As a result of this model, the power management policy becomes asynchronous which is more appropriate for implementation as part of the operating system. The new model considers the correlation between the state of the SQ and the state of the SP, which is more realistic than previous models. Moreover, the service requester model is capable of capturing complex workload characteristics and the overall system model is constructed exactly and efficiently from those of the component models. An analytical-based approach is used to calculate the generator matrix for the joint process of SP–SQ and a tensor sum-based method is utilized to calculate the generator matrix of the joint process of SP–SQ and SR. The policy optimization problem under the CTMDP model can be solved using (exact) linear programming and (heuristic) policy iteration algorithms. Moreover, this work models the service queue as two queues consisting of *low-* and *high-priority* service requests, which furthermore captures the behavior of real-life EMCs.

Because the CTMDP policy is a randomized policy, at times it may not turn off the SP even when there is no request in the SQ. If the stochastic model exactly represents the system behavior, then this policy is optimal. However, in practice, because the stochastic model is not accurate enough, the CTMDP policy may cause unnecessary energy dissipation by not turning off the SP. For example, the real requests pattern on the SP may be quite different from what has been assumed in theory, and the SP idle time may be much longer than one would expect based on the assumption of exponential input interarrival time. In this case, keeping the SP on while it is idle can result in power waste. Qiu et al. [10] thus present an improved CTMDP policy (called CTMDP-Poll) by adding a *polling state*. The functionality of the polling state is very simple. After adding this state, even if the CTMDP policy allows the SQ to stay on when the SQ is empty, the policy will re-evaluate this decision after some random-length period of time. For example, if the SQ is empty and the PM has made a decision (with probability of 0.1) of letting the SQ to stay ON, then after 2s, if there is no change in the SQ, the models will enter the polling state, and the PM will have to re-evaluate its decision. At this time, the probability for it to still let SQ remain on is again 0.1. So as the time goes on, the total probability of the SQ remaining in the ON state reduces in a geometric manner. In this way, one can make sure that the SP will not be idle for too long, resulting in less wasteful energy dissipation.

The timeout policy is an industry standard that has been widely supported by many real systems. A DPM technique based on timeout policies may thus be easier and safer for users to implement. At the same time, it helps them achieve a reasonably good energy-performance trade-off. To implement a more elaborate DPM technique requires the users to directly control the power-down and wake-up sequences of system components, which normally necessitates detailed knowledge of hardware and involves a large amount of low-level programming dealing with the hardware interface and device drivers. Notice also that the various system modules typically interact with each other implying that sudden power-down of a system module may cause the whole system to malfunction or become unstable, i.e., direct control over the state of a system module is a big responsibility that should not be delegated unceremoniously. A DPM technique based on a simple and well-tested timeout policy and incorporated in the operating system will have none of the above concerns. Based on these reasons, Rong and Pedram [24] present a timeout-based DPM technique, which is constructed based on the theory of Markovian processes and is capable of determining the optimal timeout values for an electronic system with multiple power-saving states. More precisely, a Markovian process-based stochastic model is described to capture the power management behavior of an electronic system under the control of a timeout policy. Perturbation analysis is used to construct an offline gradient-based approach to determine the set of optimal timeout values. Finally, online implementation of this approach is also discussed.

15.3.2.3 Petri Net-Based Approaches

The DPM approaches based on Markov decision processes offer significant improvements over heuristic power management policies in terms of the theoretical framework and ability to apply strong mathematical optimization techniques [25–31]. However, previous works based on Markov decision processes only describe modeling and policy optimization techniques for a simple power-managed system. Such a system contains one SP that provides services (e.g., computing and file access), one SQ that buffers the service requests for the SP, and one SR that generates the requests for SP. It is relatively easy to construct the stochastic models of the individual components because their behavior is rather simple. However a significant effort is required to construct the joint model of SP and SQ mostly because of the required synchronization between state transitions of SP and SQ. Furthermore, the size of the Markov process model of the overall system rises rapidly as the number of SPs and SRs is increased.

Generalized stochastic Petri Nets (GSPNs) target more complex power-managed systems as shown in Figure 15.8. The example depicts a typical multiserver (distributed computing) system. Note that this model only captures those system behaviors that are related to the power management. The system contains multiple SPs with their own local SQs (LSQ). There is a SR that generates the tasks (requests) that need to be serviced. The request dispatcher (RD) makes decisions about which SP should service which request. Different SPs may have different power/performance parameters. In real applications, the RD and LSQs can be part of the operating system, while SPs can be multiple processors in a multiprocessor computing system or number of networked computers of a distributed computing system.

The complexity of the modeling problem for the above system is high not only because of the increased number of components, but also because of the complex system behaviors that are present. For example, one needs to consider the synchronization of LSQs and SPs, the synchronization of the SR and LSQs, the dispatch behavior of the RD, and so on. In this situation when complex behaviors must be captured by the system model, the modeling techniques in Ref. [11] become inefficient because they only offer stochastic models for individual components and require that global system behaviors be captured manually. Obviously, we need new DPM modeling techniques for large systems with complex behaviors.

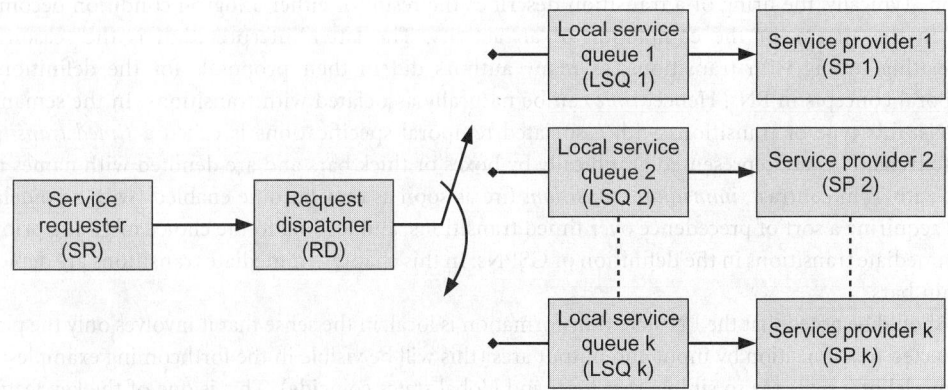

FIGURE 15.8 A multiserver/distributed computing system.

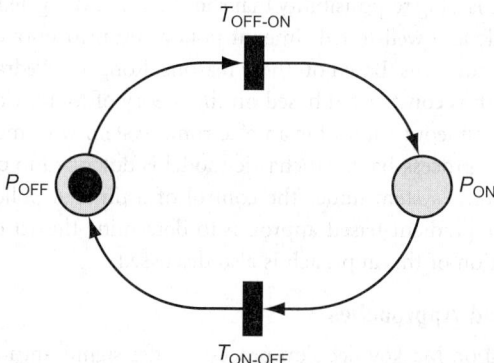

FIGURE 15.9 PN description of a switch.

For a detailed introduction to Petri nets, refer to Ref. [33].

A *Petri net* (PN) model is graphically represented by a directed bipartite graph in which the two types of nodes (places and transitions) are drawn as circles, and either bars or boxes, respectively (cf. Figure 15.9). The edges of the graph are classified (with respect to transitions) as

- Input edges: arrow-headed edges from places to transitions
- Output edges: arrow-headed edges from transitions to places

Multiple (input/output) edges between places and transitions are permitted and annotated with a number specifying their multiplicities.

Places can contain *tokens*, which are drawn as black dots within places. The state of a PN is called *marking*, and is defined by the number of tokens in each place. As in classical automata theory, in PN there is a notion of initial state (*initial marking*). Places are used to describe possible local system states (named conditions or situations). Transitions are used to describe events that may modify the system state. Edges specify the relation between local states and events in two ways: they indicate the local state in which the event can occur, and the local state transformations induced by the event.

The dynamic behavior of the PN is governed by the firing rule. A transition can fire (an event takes place) if all the transition input places (i.e., those places connected to the transition with an arc whose direction is from the place to the transition), contain at least one token. In this case the transition is said to be enabled. The firing of an enabled transition removes one token from all of its input places, and generates one token in each of its output places (i.e., those places connected to the transition with an arc whose direction is from the transition to the place). The firing of a transition is an atomic operation. Tokens are removed from input places, and deposited into output places with one indivisible action. Typically, the firing of a transition describes the result of either a logical condition becoming true in the system, or the completion of an activity. The latter interpretation is the reason for associating timing with transitions, as many authors did in their proposals for the definition of temporal concepts in PNs. Hence, *time* can be naturally associated with transitions. In the semantics of PNs, this type of transitions with associated temporal specifications is called a *timed transition*. These transitions are represented graphically by boxes or thick bars and are denoted with names that start with T. In contrast, *immediate transitions* fire as soon as they become enabled (with zero delay), thus acquiring a sort of precedence over timed transitions, and leading to the choice of giving priority to immediate transitions in the definition of GSPNs. In this chapter, immediate transitions are depicted as thin bars.

It should be noted that the PN state transformation is local, in the sense that it involves only the places connected to a transition by input and output arcs (this will be visible in the forthcoming examples; the PN model of a switch is so simple that local and global states coincide). This is one of the key features of PNs, which allows compact description of distributed systems.

Example

A simple example of a PN model is given in Figure 15.9, where two places P_{ON} and P_{OFF}, and two transitions $T_{ON\text{-}OFF}$ and $T_{OFF\text{-}ON}$ are connected with four arcs. Both places define conditions (i.e., the "ON condition" or the "OFF condition"). The state depicted in Figure 15.9 is such that place P_{OFF} contains one token; thus the "OFF condition" is true; instead, since place P_{ON} is empty, the "ON condition" is false. In this simple example, transition $T_{OFF\text{-}ON}$ is enabled, and it fires, removing one token from P_{OFF} and depositing one token in P_{ON}. The new state is such that the "ON condition" is true and the "OFF condition" is false. In the new state, transition $T_{ON\text{-}OFF}$ is enabled, and it fires restoring the state shown in Figure 15.9. The simple PN model in Figure 15.9 may be interpreted as the PN description of the behavior of a switch.

In a *stochastic petri net* (SPN) model, each timed transition is associated not only with a single transition time but a collection of randomly generated transition times from an exponential distribution. As in case of timed transitions, for the description of SPNs, one can assume that each timed transition possesses a timer. When the transition becomes enabled for the first time after firing, the timer is set to a value that is sampled from the exponential pdf associated with the transition. During all time periods in which the transition is enabled, the timer is decremented. Transitions fire when their timer readout goes down to zero. With this interpretation, each timed transition can be used to model the execution of an activity in a distributed environment; all activities execute in parallel (unless otherwise specified by the PN structure) until they complete. At completion, activities induce a local change of the system state, which is specified with the interconnection of the transition to input and output places. In GSPNs, the exponentially timed and immediate transitions coexist in the same model. In the context of GSPNs, places can be divided into two different classes based on the type of the transitions for which they are inputs, i.e., *vanishing places* and *tangible places*. The place is a vanishing place if it is the only input place of an immediate transition; otherwise the place is called a tangible place.

Let us see how one can use this modeling tool to develop a DPM policy by capturing the exact behavior of a complex EMC system. To capture the energy consumption of the EMC in the GSPN model, the following two definitions are necessary:

1. A GSPN with cost is a GSPN model with the addition of two types of cost: *impulse cost* associate with marking transitions and *rate cost* associated with places. Impulse cost occurs when the GSPN makes a transition from one marking to another. Rate cost is the cost per unit time when the GSPN stays in a certain marking.

2. A controllable GSPN with cost is a GSPN where all or part of the probabilities of timed transitions can be controlled by outside commands.

Example

Consider that a SP in the processor has two power states: {*ON, OFF*}. In the *ON* state, the SP provides service with an average service time of 5 ms. The average time to switch from the *ON* state to *OFF* state is 0.66 ms, and the average time to switch from the *OFF* state to *ON* state is 6 ms. The power consumption of the SP is 2.3 W when it is in the *ON* state and 0.1 W when it is in the *OFF* state. The energy needed to switch from the *ON* state to *OFF* state is 2 mJ, and the energy needed to switch from the *OFF* state to *ON* state is 30 mJ. Assume that the maximum length of the SQ is 3. Figure 15.10 shows the GSPN model of the single-processor system. The input gate $G_{capacity}$ sets the SQ capacity constraint. The place $P_{ON\text{-}OFF}$ denotes the SP status when it is switching from the *ON* state to *OFF* state while the place $P_{OFF\text{-}ON}$ denotes the SP status when it is switching from the *OFF* state to *ON* state. The place $P_{idle(ON)}/P_{idle(OFF)}$ denotes the SP status when it is idle and the power state is *ON/OFF*. The place $P_{work(ON)}$ denotes the SP status when it is working and the power state is *ON*. The SP will have exactly one such status at any time. From the topology of the GSPN, one realizes that the sum of tokens in places $P_{ON\text{-}OFF}$, $P_{idle(ON)}$, $P_{work(ON)}$, $P_{idle(OFF)}$, and $P_{OFF\text{-}ON}$ is 1 at any time.

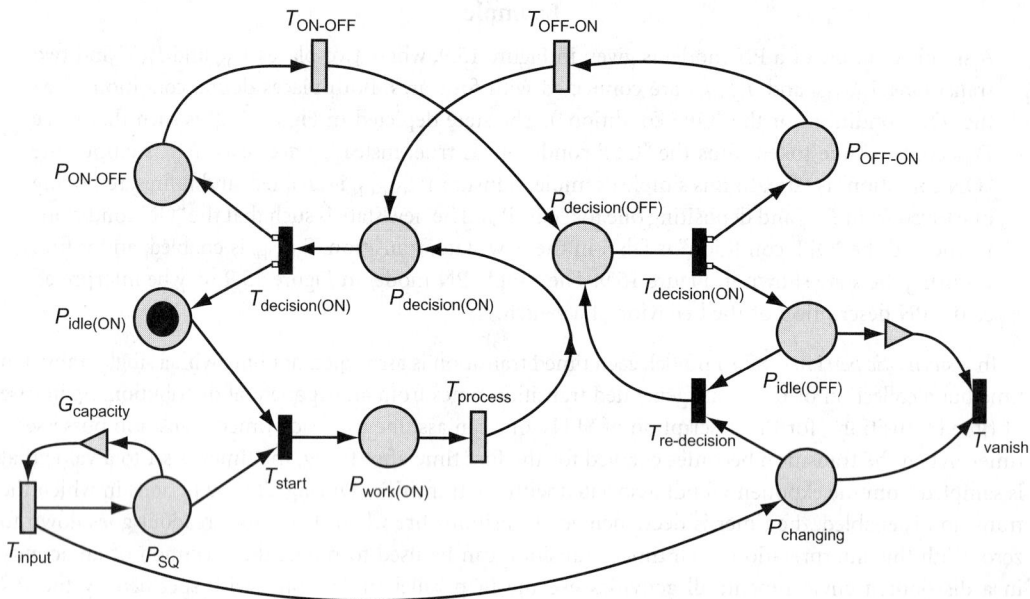

FIGURE 15.10 Example GSPN model of a single requester–single server system.

The number of tokens in P_{SQ} denotes the number of waiting requests in the SQ. The initial marking of $P_{idle(ON)}$ is 1 while the initial marking of the other places is 0, which indicates that the initial state of the SP is idle and the initial state of SQ is empty. The places $P_{decision(ON)}$ and $P_{decision(OFF)}$ are vanishing places. They indicate the very short period of time when the SP is taking command from PM and is in the *ON* or *OFF* state. The place $P_{changing}$ is also a vanishing place. It is an auxiliary place, which indicates that the state of the system is changing so that it is time for the SP to receive the power management command if it is currently idle. T_{ON-OFF} and T_{OFF-ON} are timed activities. They indicate the time needed to switch from the *ON* state to *OFF* state and the time needed to switch from the *OFF* state to *ON* state. $T_{processing}$ is also a timed transition, which indicates the time needed to process one request. T_{input} denotes the time needed to generate the next request. It actually belongs to the GSPN model of the request generation system. $T_{decision(ON)}$ and $T_{decision(OFF)}$ are immediate transitions. They represent the process of randomized action issued by the PM. The two cases in $T_{decision(ON)}$ or $T_{decision(OFF)}$ are mutually exclusive. The case probability equals the action probability, which is marking- and policy-dependent. If the policy is unknown, the GSPN is a controllable GSPN. When the SP is idle and *ON* (a token is in place $P_{idle(ON)}$) and SQ is not empty, the immediate transition T_{start} is completed which indicates that the SP enters the busy state. When the SP is *OFF* (a token is in place $P_{idle(OFF)}$) and the state of SQ is changing (a token is in place $P_{changing}$), the immediate transition $T_{re-decision}$ is completed which indicates that the SP returns to the action taking stage. If the SP is not *OFF* and the state of SQ is changing, the immediate transition T_{vanish} is completed which indicates that the change is ignored.

In the complex system the *request generation system* (RGS) can be very complicated and cumbersome. RGS can generate various types of requests. The generation time of different types of request may be different. Some types of requests can be serviced by several different SPs; whereas some other types of requests can be serviced by only a certain SP. There may exist correlations among the generation of different types of requests. If the SQ is full, the RGS will stop generating request. It will resume request generation when there is vacancy in the SQ.

Example

Assume that there are three types of requests: the first is type A which can only be serviced by SP A, the second is type B which can only be service by SP B, and the third is type AB which can be serviced

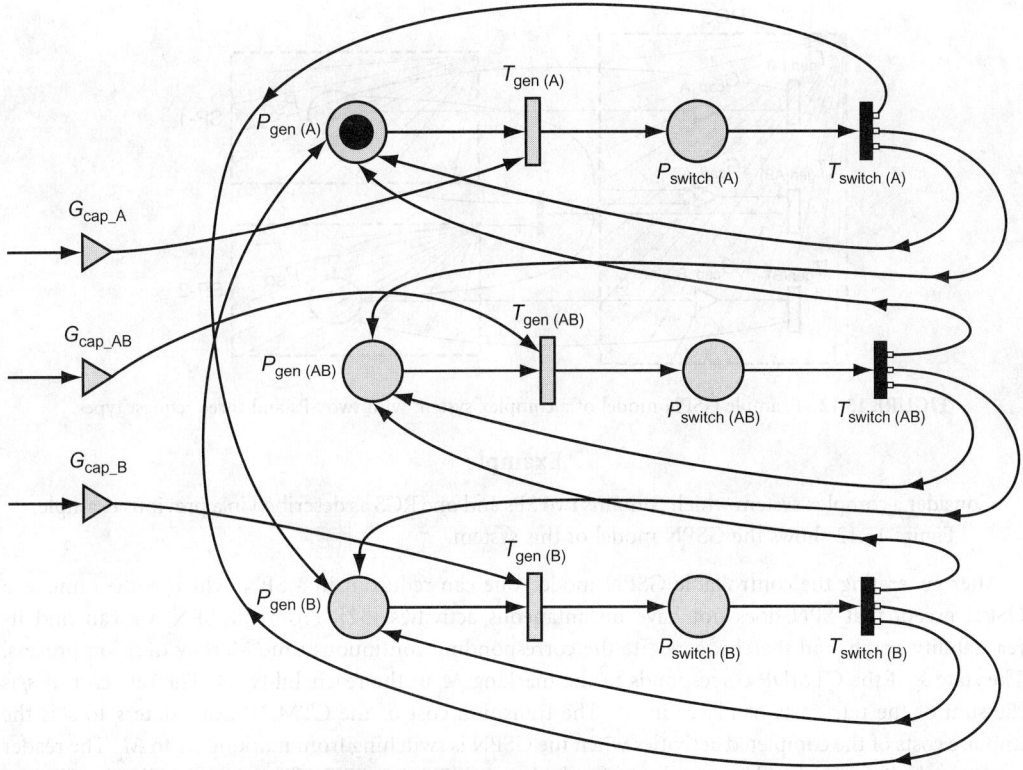

FIGURE 15.11 Example GSPN model for an RGS.

by both SP A and SP B. The correlations among these requests are given by a probabilistic matrix, For example, from the matrix one will know that the probability that a type AB request was issued after a type A request is 0.6. Figure 15.11 shows the GSPN model of this RSG. In this figure, the case probability of activities $T_{switch(A)}$, $T_{switch(B)}$, and $T_{switch(AB)}$ takes value from a probabilistic matrix. The input gate G_{cap_A} represents the condition that the SQ in SP A is not full. The input gate G_{cap_B} represents the condition that the SQ in SP B is not full. The input gate G_{cap_AB} represents the condition that the SQ in SP A or the SQ in SP B is not full. Notice that the input places of these input gates belong to the GSPN model of each SP. These input gates enable or disable the request generation. For example, if the condition given by G_{cap_A} is false, which means that SQ of SP A is full, then the time activity $T_{gen(A)}$ is disabled, which means that request generation procedure of type A request pauses. $T_{gen(A)}$ will be enabled when the condition given by G_{cap_A} becomes true, which means that the request generation procedure resumes when there is a vacancy in the SQ.

To model a complex system composed of several SPs, similar to the one shown in Figure 15.8 with an RGS and interactions among the components, a hierarchical approach can be used. First, each SP is modeled using a single requester–single server model. Next, the requests generated by the RGS are sent to the SP #i with probability p_i through a dispatcher. If the request can only be serviced by SP #i, then p_i is 1. If the request cannot be serviced by SP #i then p_i is 0. In all other cases the probability p_i is controlled by the dispatcher. The optimal dispatch policy can be obtained by solving a Markov decision process. The GSPN model of such complex system contains the following components:

1. The GSPN models of RGS and SPs.
2. A set of input gates $\{G_{cap\_i}\}$. The input place of a G_{cap_i} is the P_{SQ} of all SPs, which can provide service for request type i. The activity of G_{cap_i} is $T_{gen(i)}$. A gate G_{cap_i} indicates the condition that there are free positions in SQ to buffer the request.
3. Arcs from transition $T_{gen(i)}$ in RGS to place P_{SQ} in any SP that can provide service for request i.

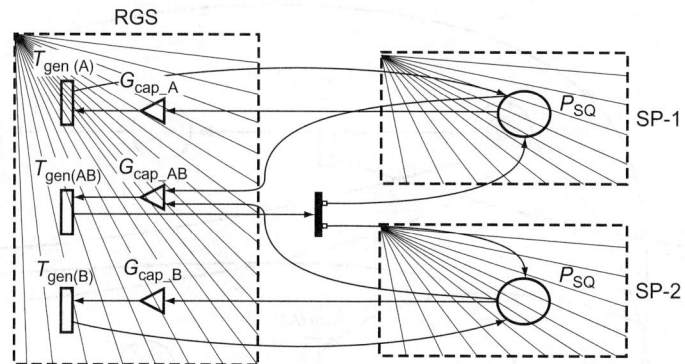

FIGURE 15.12   Example GSPN model of a complex system with two SPs and three request types.

Example

Consider a complex system which contains two SPs and one RGS as described in a previous example. Figure 15.12 shows the GSPN model of this system.

After generating the controllable GSPN model, one can reduce it to a SPN, which is the same as a GSPN except that SPN does not have instantaneous activities [32]. From the SPN, we can find its reachability graph, and thereby, generate the corresponding continuous time Markov decision process. The state s_i of the CTMDP corresponds to the marking M_i in the reachability set. The rate cost of s_i is the sum of the rate costs of places in M_i. The transition cost of the CTMDP from state s_i to s_j is the impulse costs of the completed activities when the GSPN is switching from marking M_i to M_j. The reader may refer to Ref. [32] for the procedure of reducing a GSPN to a SPN. The optimal policy in CTMDP is obtained by solving a set of linear programming. A GSPN can be converted to a CTMDP, hence it can be evaluated efficiently. However, the exponential distribution is not always an appropriate way to represent the transition time. If the transition time has a general distribution, the Markovian property will be destroyed. This problem can be circumvented by using the stage method [31], which approximates the general distributions using the series or parallel combinations of exponential stages.

15.4   Conclusions

This chapter describes various DPM approaches for performing energy-efficient computation: predictive shutdown, Markovian decision process-based, and generalized stochastic Petri net-based approaches. A significant reduction in power consumption can be obtained by employing these DPM techniques. For example, for applications where continuous computation is not being performed, an aggressive shutdown strategy based on an online predictive technique can reduce the power consumption by a large factor compared to the straightforward conventional schemes where the power-down decision in based solely on a predetermined idle time threshold. Moreover, predictive shutdown heuristic may be applied to manage the shutdown of peripherals such as disks. An online algorithm that makes the shutdown decision using a prediction of the time to next disk access can result in higher power reduction compared to more conventional threshold-based policies for disk shutdown.

In contrast, construction of optimal power management policies for low-power system is a critical issue that cannot be addressed by using common sense and heuristic solutions such as those used in predictive shutdown schemes. Stochastic models provide a mathematical framework for the formulation of power-managed devices and workloads. The constrained policy optimization problem can be solved exactly in this modeling framework. Policy optimization can be cast into a linear programming problem and solved in polynomial time by efficient interior point algorithms. Moreover, trade-off curves of power versus performance can be computed. Furthermore, adaptive algorithms can compute optimal policies in systems where workloads are highly nonstationary and the service provider model changes over time.

CTMDP-based techniques introduce a new and more complete model of the system components as well as the model of the whole system. This mathematical framework captures the characteristics of the real applications more accurately which is mainly because the problem is solved in continuous-time domain while previous approaches solve the problem in discrete-time domain.

A shortcoming of DTMDP or CTMDP-based techniques is that it is very difficult to use these modeling frameworks when attempting to represent complex systems, which in turn consist of multiple closely inter-acting SPs and must cope with complicated synchronization schemes. In this case, GSPN and the correspond-ing modeling techniques based on the theory of GSPN have proven to be quite effective. The constructed GSPN model can be automatically converted to an isomorphic continuous-time Markov decision process. From the corresponding Markov decision process, one can calculate the optimal DPM policy, which achieves minimum power consumption for given delay constraints. In real applications, the interarrival time of service requests may not follow an exponential distribution, for example, they could have heavy-tail distributions such as Pareto distribution. This problem can be solved by using the "stage method" (i.e., approximating the given source of requests by a series–parallel connection of exponentially distributed sources).

Acknowledgment

This work was sponsored in part by DARPA and NSF CSN program office.

References

1. S. Gary and P. Ippolito, "PowerPC 603, a microprocessor for portable computers," *IEEE Design Test Comput.*, vol. 11, pp. 14–23, 1994.
2. "Advanced micro devices," in AM29SLxxx Low-Voltage Flash Memories, 1998.
3. E. Harris, S.W. Depp, W.E. Pence, S. Kirkpatrick, M. Sri-Jayantha, and R.R. Troutman, "Technology directions for portable computers," *Proc. IEEE*, vol. 83, p. 636–657, 1996.
4. M. Stemm and R. Katz, "Measuring and reducing energy consumption of network interfaces in hand-held devices," *IEICE Trans. Commun.*, vol. E80-B, pp. 1125–1131, 1997.
5. H. Shim, N. Chang, and M. Pedram, "A backlight power management framework for the battery-operated multi-media systems," *IEEE Design Test Comput.*, vol. 21, pp. 388–396, 2004.
6. SA-1100 Microprocessor Technical Reference Manual, Intel, 1998.
7. M. Srivastava, A. Chandrakasan, and R. Brodersen, "Predictive system shutdown and other architectural techniques for energy efficient programmable computation," *IEEE Trans. VLSI Syst.*, vol. 4, pp. 42–55, 1996.
8. C.-H. Hwang and A. Wu, "A predictive system shutdown method for energy saving of event-driven computation," In *Proceedings of International Conference on Computer-Aided Design*, pp. 28–32, November 1997.
9. L. Benini, G. Paleologo, A. Bogliolo, and G. De Micheli, "Policy optimization for dynamic power management," *IEEE Trans. Computer-Aided Design*, vol. 18, pp. 813–833, 1999.
10. Q. Qiu, Q. Wu, and M. Pedram, "Stochastic modeling of a power managed system-construction and optimization," *IEEE Trans. Computer-Aided Design*, vol. 20, pp. 1200–1217, 2001.
11. T. Simunic, L. Benini, P. Glynn, and G. De Micheli, "Event-driven power management," *IEEE Trans. Computer-Aided Design*, vol. 20, pp. 840–857, 2001.
12. E.-Y. Chung, L. Benini, and G.D. Micheli, "Dynamic power management using adaptive learning trees," In *Proceedings of ICCAD*, 1999.
13. Y. Lu and G. DeMicheli, "Adaptive hard disk power management on personal computers," In *Proceedings of the Great Lakes Symposium on VLSI*, 1999.
14. S. Irani, R. Gupta, and S. Shukla, "Competitive analysis of dynamic power management strategies for systems with multiple power savings states," In *IEEE Conference on Design, Automation and Test in Europe*, 2002.

15. S. Irani, S. Shukla, and R. Gupta, "Online strategies for dynamic power management in systems with multiple power saving states," In *IEEE Transactions on Embedded Computing Systems*, 2003.
16. P. Krishnan, P. Long, and J. Vitter, "Adaptive disk spin-down via optimal rent-to-buy in probabilistic environments," In *International Conference on Machine Learning*, pp. 322–330, July 1995.
17. D. Helmbold, D. Long, and E. Sherrod, "Dynamic disk spin-down technique for mobile computing," In *IEEE Conference on Mobile Computing*, pp. 130–142, November 1996.
18. F. Douglis, P. Krishnan, and B. Bershad, "Adaptive disk spin-down policies for mobile computers," In *Second USENIX Symposium on Mobile and Location-Independent Computing*, pp. 121–137, April 1995.
19. Y. Lu and G. De Micheli, "Adaptive hard disk power management on personal computers," In *Great Lakes Symposium on VLSI*, pp. 50–53, February 1999.
20. A. Karlin, M. Manasse, L. McGeoch, and S. Owicki, "Randomized competitive algorithms for non-uniform problems," *ACM-SIAM Symposium on Discrete Algorithms*, pp. 301–309, 1990.
21. D. Gross and C.M. Harris, *Fundamentals of Queuing Theory*, Wiley, Indianapolis, U.S.A. 1985.
22. P. Rong and M. Pedram, "Battery-aware power management based on Markovian decision processes," In *Proceedings of the International Conference on Computer Aided Design*, pp. 712–717, November 2002.
23. E.V. Denardo, "On linear programming in a Markov decision problem," *Manage. Sci.*, vol. 16, pp. 281–288, 1970.
24. P. Rong and M. Pedram, "Determining the optimal timeout values for a power-managed system based on the theory of Markovian processes: offline and online algorithms," In *Proceedings of Design Automation and Test in Europe*, 2006.
25. U. Narayan Bhat, "*Elements of Applied Stochastic Processes*," Wiley, Indianapolis, U.S.A. 1984.
26. B. Miller, "Finite state continuous time Markov decision processes with an finite planning horizon." *SIAM J. Control*, vol. 5, pp. 266–281, 1968.
27. B. Miller, "Finite state continuous time Markov decision processes with an infinite planning horizon." *J. Math. Anal. Appl.*, 22, pp. 552–569, 1968.
28. R.A. Howard, *Dynamic Programming and Markov Processes*, Wiley, New York, 1960.
29. D.P. Heyman and M.J. Sobel, *Stochastic Models in Operations Research*, McGraw-Hill, New York, U.S.A. 1982.
30. G. Bolch, S. Greiner, H.D. Meer, and K.S. Trivedi, *Queuing Networks and Markov Chains*, Wiley, Indianapolis, U.S.A. 1998.
31. L. Kleinrock, *Queuing Systems. Volume I: Theory*, Wiley-Interscience, New York, 1981.
32. M.A. Marsan, G. Balbo, G. Conte, S. Donatelli, and G. Franceschinis, *Modeling with Generalized Stochastic Petri Nets*, Wiley, New York, 1995.
33. J. Wang, *Timed Petri Nets: Theory and Application*, Springer—Mathematics, New York, U.S.A. 1998.
34. Z. Ren, B.H. Krogh, and R. Marculescu, "Hierarchical adaptive dynamic power management," In *Proceedings of the Conference on Design, Automation and Test in Europe*–vol. 1, February 2004.

16

Communication-Based Design for Nanoscale SoCs

Umit Y. Ogras

Radu Marculescu
Carnegie Mellon University

CONTENTS

16.1 Introduction

Systems-on-chip (SoCs) designed at nanoscale domain will soon contain billions of transistors [1]. This makes it possible to integrate large amounts of embedded memory and hundreds of IP cores running multiple concurrent processes on a single chip. This chapter focuses on the communication-centric SoC design, while the computational aspects are discussed in Chapter 17 [2]. The design of complex SoCs faces a number of design challenges. First, the richness of computational resources places tremendous demands on the communication resources and, consequently, the entire design methodology changes from computation-based design to communication-based design. Second, global interconnects cause severe on-chip synchronization errors, unpredictable delays, and high power consumption. Finally, increasing complexity, costs, and tight time-to-market constraints require new design methodologies that favor design reuse at all levels of abstraction [3–5]. As a result, novel on-chip communication architectures that can effectively address all these problems are highly needed.

Owing to their limited bandwidth, legacy bus-based architectures fail to solve the above-mentioned problems. Moreover, a global bus with large capacitive load causes long delays and high-power consumption. While point-to-point communication architectures may provide the required performance, the

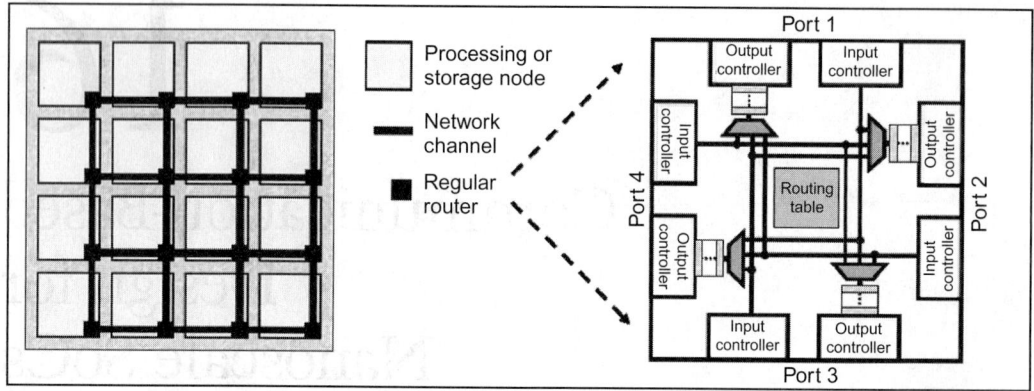

FIGURE 16.1 Regular tile-based networks-on-chip architecture and a simple on-chip router with four ports are shown.

design area and lack of scalability become two major issues for implementing complex applications; this is mainly due to the dedicated channels between all the communicating IP pairs [6]. In contrast to these traditional methods, the recently proposed networks-on-chip (NoC) architecture provides a large bandwidth with moderate area overhead [5–9].

16.1.1 Advantages of NoC Architectures

NoC architectures consist of a number of heterogeneous cores such as CPU or DSP modules, video processors, embedded memory blocks, and application-specific components, as shown in Figure 16.1. Each core has an embedded router which connects the core to other nodes in the network via an interconnection network. The advantages of NoCs include scalability, reusability, and predictability as detailed below.

- *Scalability.* Since the communication between different nodes is achieved by routing packets instead of wires, a large number of cores can be connected without using long global wires. Moreover, as the number of cores in the design increases, the network bandwidth increases accordingly. As a result, the NoC paradigm provides a highly scalable communication architecture.
- *Design reuse.* The modularity of the NoC approach offers a great potential for reusing the network routers and other IP cores. The routers, interconnect, and lower-level communication protocols can be designed, optimized, and verified only once and reused subsequently in a large number of products. Likewise, the IP cores complying with the network interface can be reused across many different applications.
- *Predictability.* The structured nature of the global wires facilitates well-controlled and optimized electrical parameters. In turn, these controlled parameters make possible the use of aggressive signaling circuits which can reduce the power dissipation and propagation delay significantly [5].

16.1.2 NoC Design Space

From a design methodology standpoint, designing on-chip networks differs significantly from designing large-scale interconnection networks [10], mainly due to the distinctive cost functions and constraints imposed by the on-chip functionality. For example, the energy consumption is one of the major constraints in NoC design, while it is hardly a consideration for large-scale networks. Moreover, unlike large-scale networks, NoCs are highly application-specific. Therefore, an ample portion of the design effort will likely go into optimizing the network for a specific application or class of applications.

NoC design problems can be conceived as representing a three dimensional (3D) design space [11]. The first dimension of this space is represented by the *communication infrastructure* (e.g., network topology and width of the channel links). This dimension defines how nodes are interconnected to each other and reflects the fundamental properties of the underlying network. The second dimension, namely the *communication paradigm,* captures the dynamics of transferring messages (e.g., *XY* versus adaptive routing) inside the network. Finally, the third dimension, *application mapping,* defines how different tasks that implement the target application are mapped to the network nodes (e.g., how various traffic sources and sinks are distributed across the network). In the design-automation community, design space exploration along each dimension has been considered to some extent without explicitly referring to such a classification.

The simplicity of regular grid architectures, as opposed to the complexity of the fully customized topologies, favors the design approaches where a modular network topology is chosen a priori. Once the topology is fixed, the main design problem becomes mapping the target application to the given topology, while optimizing one or more design variables, such as performance or energy [12–14]. As shown in Figure 16.2, fixing the topology and routing function a priori, requires the least design effort but, at the same time, limits severely the quality of the design one can achieve. The design quality can be improved by evaluating a set of standard topologies (e.g., torus or other higher dimensional meshes) and choosing the best alternative for a particular application; this is because the selection of the network topology has a considerable impact on area, performance, and system power consumption [15,16]. It is also possible to partially customize the network architecture to match the target application. For example, one can optimize the buffer allocation to the network nodes, rather than simply using the uniform buffer allocation [17]. Indeed, the total buffer budget affects the routers area significantly, while its distribution over the routers has a huge impact on performance.

The standard network topologies (e.g., star, tree, torus, hypercube, etc.) are suited for scenarios where we have limited or no knowledge about the traffic in the network. However, when the performance, area, and power that can be harvested by the standard topologies do not satisfy the design constraints, the information about the application-specific communication requirements can be exploited for optimization purposes [18–20]. By synthesizing fully customized topologies, a much larger portion of the design space can be explored compared to simply limiting the design to standard topologies.

Obviously, the actual level of customization, illustrated in Figure 16.2, will be determined by factors such as design cost, product volume, and market conditions. Nevertheless, effective design automation tools targeting different customization levels, similar to those discussed in the following sections, should be available to designers to address a wide range of design problems.

In the remaining part of this chapter we first overview the *engineering* aspects of NoC design. Toward this end, concrete results ranging from algorithms for application mapping [12–14], all the way down to silicon implementation [21–23] will illustrate the NoC paradigm in action. As the network paradigm moves

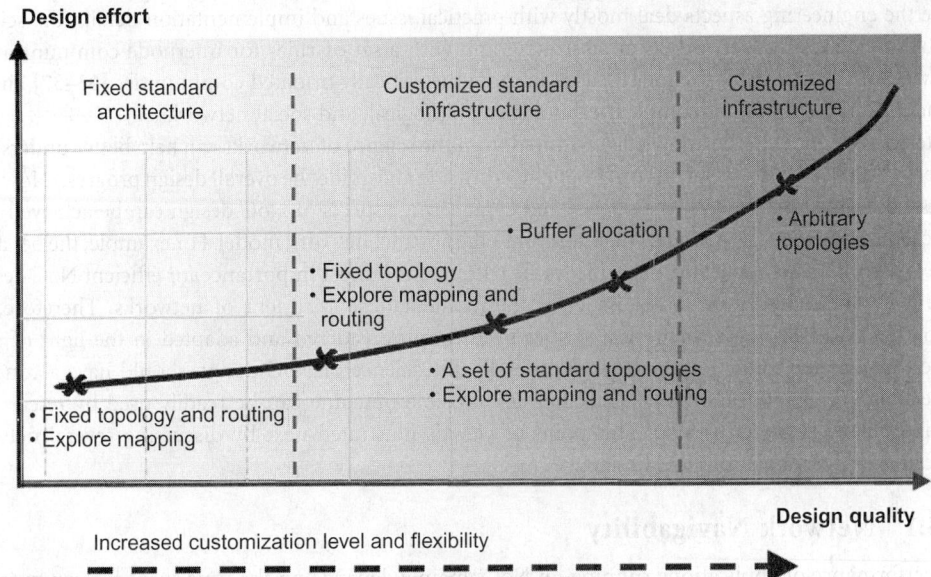

FIGURE 16.2 Design space exploration for NoC architectures.

closer to silicon, designing NoCs becomes increasingly based on concrete metrics for energy and performance. Consequently, the NoC design becomes more of a *science* on its own. Consequently, the science of NoC design is overviewed in Section 16.3, while its interplay with the engineering aspects is discussed in Section 16.4.

16.2   Engineering Issues in NoC Design

The engineering aspects of NoC design are mostly related to practical issues ranging from clocking and signaling strategies at physical level, all the way up to operating system design at software level [7]. Finding solutions to problems into this category is very important for bringing the NoC paradigm closer to practice. A typical example of work dealing primarily with the engineering aspects of NoC design is the aSoC communication architecture, which provides one of the first examples of an NoC implementation [21]. Similarly, the SPIN interconnect architecture implements a 32-port network architecture supporting best-effort traffic in a 0.13 mm process [22]. A more recent implementation considers a highly optimized hierarchical star topology [23]. Bertozzi et al. [24] present a design flow starting from a high-level application specification that incorporates topology mapping and selection to derive an optimized NoC configuration with respect to different design objectives.

Some applications require guarantees on packets delivery time. Generating NoC architectures with guaranteed services have been investigated [25–27]. In Refs. [25,26], the Quality of Service (QoS) is ensured by allocating the necessary bandwidth under a time division multiplexing scheme and using routers which support both guaranteed and best-effort services. In [27], the authors an asynchronous NoC with wormhole routing and provide service guarantees through the use of virtual channels. Finally, the approach in Ref. [28] provides QoS based on different traffic classes.

Efficient estimation and optimization of power consumption is also an important problem in NoC design. Wang et al. [29] present a power model for routers and a power-performance simulator for interconnection networks. Ye et al. [30] analyze the power consumption in switch fabrics and propose a system-level energy model for NoCs. Finally, a power-performance evaluator for mesh-based NoCs is presented in Ref. [31].

16.3   The Science of NoC Design

While the engineering aspects deal mostly with practical issues and implementation details, the science of NoCs perceives the network as an abstract graph with a set of rules for internode communication. This view of the network is frequently used in more theoretically oriented communities [33–37] aiming to understand the nature of complex technological, biological, and social networks.

Interestingly enough, the knowledge acquired from the science of networks can help better understand the very nature of design problems on the engineering side and guide the overall design progress. However, successful interaction between the science and engineering aspects of NoC design can be achieved only by incorporating the NoC-specific constraints into the abstract network model. For example, the behavior of the network under application-specific traffic pattern has utmost importance for efficient NoC design, while it is not necessarily a major issue for the mathematical treatment of networks. Therefore, the approaches based on these theoretical studies need to be considered and adapted in the light of such application-specific traffic patterns. More generally, the science of NoC design should have a concrete connection to the application side such that any possible outcome can be readily used by researchers working on the engineering side. This point of view is illustrated next by discussing several network properties in the context of NoC design.

16.3.1   Network Navigability

The performance of applications running on NoCs mainly depends on the time spent for computation and communication. While the computation time is primarily determined by the design of the IP cores implementing the application, the communication time depends heavily on the ability of the network to move packets around in an efficient manner. Indeed, if the network enables fast and efficient packet navigation, then the communication time can be reduced significantly.

Some technological networks such as WWW, consist of more than 1 billion nodes but are clearly much easier to navigate compared to other networks (e.g., a 2D mesh) of equal size. Indeed, it is possible to reach any web page of interest with just a few clicks due to the small diameter of the WWW; as suggested recently, the longest path between any two nodes in such a network is around 19 hops [37]. For comparison, the diameter of a 2D mesh network with same number of nodes would be more than 50,000.

In general, the standard network architectures adopted from parallel computer networks [10] are only scalable in the sense that new nodes can be easily added to an already existing network, without distorting the structure and reducing the available bandwidth per node. However, the *diameter* (D) and the *average internode distance** of a $n \times n$ *2D mesh* network are proportional to n. Therefore, navigating in these networks becomes easily a major problem.

For application-specific NoCs, the detailed understanding of the communication workload can be exploited for optimization purposes [18–20]. Customized NoC topologies do not only improve the network performance, but also result in a better resource utilization. In contrast, the implementation complexity of such customized topologies, lack of standardization, etc. may represent a major handicap for the fully customized architectures. Therefore, the improvement in area and performance comes at the cost of losing the network regularity.

Fortunately, there exists a huge class of networks, commonly referred to as *small-world* networks [33], where one can search for network topologies with superior properties compared to standard topologies considered so far. Indeed, small-world networks are characterized by small internode distances and high clustering coefficients. Recent theoretical studies show that regular networks can be transformed into small-world networks by introducing long-range connections between remotely located nodes [33,34]. Inspired by this idea, a good compromise between the well-structured networks and fully customized topologies can be obtained by customizing the grid-like NoC architectures with application-specific long-range links [32]. In Section 16.4, we address this very issue and show the subtle interplay between the network science and engineering aspects.

16.3.2 Clustering Effects

An important characteristic of the small-world networks is their dense structure which can be measured by the clustering coefficient [33]. To compute the clustering coefficient of node i (denoted by C_i), we need to focus on its neighbors. A node with n_i neighbors can have at most $n_i(n_i - 1)/2$ links among them. If we denote by l_i the number of actual links that do exist between the neighbors, then C_i is expressed as

$$C_i = \frac{2l_i}{n_i(n_i - 1)} \tag{16.1}$$

In other words, C_i is the ratio between the number of links that connect the neighbors of node i and the maximum number of links that can possibly exist among all its neighbors. Hence, C_i measures how tightly the neighbors of node i are connected to each other. The clustering coefficient of the network (C_N) is then found by averaging the clustering coefficients over all the nodes (i.e., $C_N = mean(C_i)$).

Implementing an application across a network requires mapping the application tasks to the network nodes. In many applications, a subset of tasks tend to communicate more frequently with each other compared to the remaining nodes. Intuitively, keeping the nodes that communicate more frequently close to each other is desirable both for performance and energy considerations. However, the nodes in the network can have only a limited number of immediate neighbors, depending on the degree of the node (i.e., the number of input/output links to the node) and other practical considerations. For this reason,

---

*Under the uniform traffic assumption.

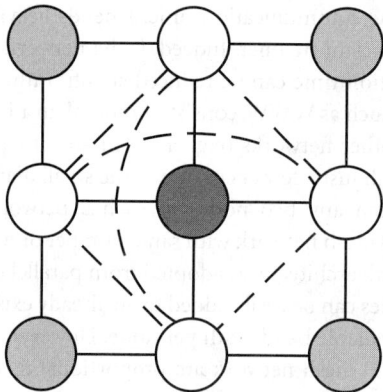

FIGURE 16.3 Illustration of computing the clustering coefficient.

the applications can be mapped more efficiently to those network topologies where the network clustering closely follows the inherent clustering inside the application.

The clustering coefficient for most standard topologies, such as mesh networks and hypercubes, is equal to zero, because none of the immediate neighbors of any given node are directly connected to each other. For example, in Figure 16.3, the central node is connected to four immediate neighbors via solid lines. There are six possible connections (shown with dashed lines) between the four neighbors. However, since none of these connections does exist in reality, the clustering coefficient is zero. Similar arguments hold for the remaining nodes in a 2D network. As a result, the application-specific customization of the network topology can be also beneficial for improving the clustering of the network to match the application characteristics. Indeed, the NoC architecture customization via long-range link insertion improves the network clustering, when a long-range link of size 2, similar to the dashed lines in Figure 16.3, is inserted to the network. Hence, combined with the reduced internode distances due to the long-range links, improved clustering moves the initial topology closer to a small-world network.

16.3.3 Network Performance under Application-Specific Traffic

Traditionally, the performance of NoCs is measured in terms of the average packet latency and maximum throughput of the network. It is possible, however, to define a unifying performance metric which provides valuable information about the average packet latency *and* network throughput [32]. At low traffic loads, the average packet latency exhibits a weak dependence on the traffic injection rate. However, when the traffic injection rate exceeds a critical value (λ_c), the packet delivery times rise abruptly and the network throughput starts to collapse (see Figure 16.4). The state of the network before congestion (i.e., the region on left-hand side of the critical value) is the *congestion-free* state, while the state beyond the critical value (right-hand side) is said to be the *congested* state. Finally, the transition from the free to the congested state is known as the *phase transition* region. Since larger values for λ_c imply higher sustainable throughput values (and the latency beyond λ_c increases abruptly), λ_c can be used as a convenient performance metric. For this reason, finding an analytical expression for λ_c is desirable for fast performance evaluation which can be used in an optimization loop.

16.3.3.1 Analytical Evaluation of λ_c

Let us denote the total number of messages in the network, at time *t*, by $N(t)$ and the aggregated packet injection rate by λ. In the *free state* regime (i.e., $\lambda < \lambda_c$), the network is in the steady state, so the packet injection rate equals the packet ejection rate. As a result, we can equate the injection and ejection rates

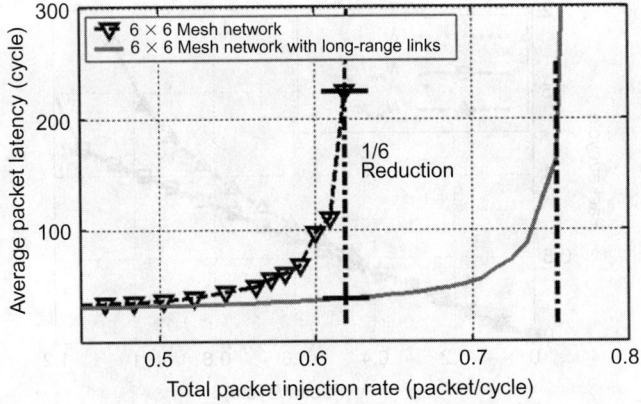

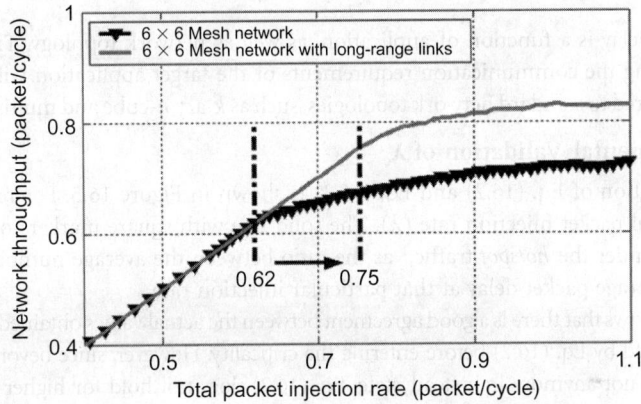

FIGURE 16.4 The evolution of the average packet latency and network throughput as the traffic load increases shown a 6×6 standard mesh before and after insertion of long-range links (Section 16.4). Larger critical traffic load implies higher achievable throughput and smaller latency.

to obtain the following approximation:

$$\lambda = \frac{N_{\text{ave}}}{\tau_{\text{ave}}} \tag{16.2}$$

where τ_{ave} is the *average time* each packet spends in the network, and $N_{\text{ave}} = \langle N(t) \rangle$ the average number of packets in the network. The exact value of τ_{ave} is a function of the traffic injection rate, topology, routing strategy, etc. We observe that τ_{ave} shows a weak dependence on the traffic injection rate when the network is in the free state. Hence the *free packet delay* (τ_0), which can be computed analytically, can be used to approximate τ_{ave}. If we denote the average number of packets in the network, at the onset of the criticality, by $N_{\text{ave}}^{\text{c}}$ we can write the following relation:

$$\lambda_{\text{c}} \approx \frac{N_{\text{ave}}^{\text{c}}}{\tau_0} \tag{16.3}$$

This approximation acts also as an *upper bound* for the critical load λ_{c}, since $\tau_0 \leq \tau_{\text{ave}}(\tau_{\text{c}})$. We note that this relation can be also found using *mean field* [35] and *distance* models [36], where $N_{\text{ave}}^{\text{c}}$ is approximated by the number of nodes in the network, under the assumption that the utilization of the routers is close to unity at the onset of the criticality.

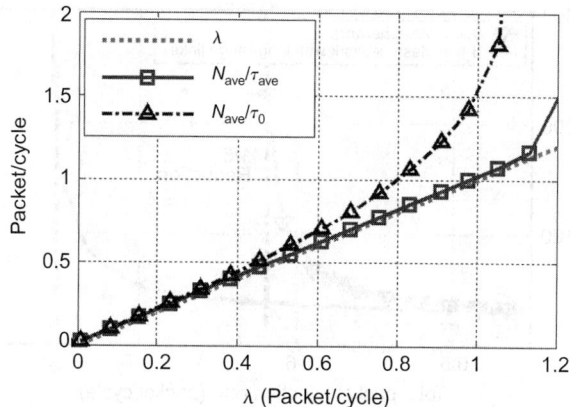

FIGURE 16.5 Experimental verification of Eqs. (16.2) and (16.3) for a 8×8 mesh network.

The free packet delay is a function of application, as well as network topology. Therefore, a network architecture matching the communication requirements of the target application will deliver a superior performance compared to standard network topologies, such as k-ary n-cube and multistage networks [10].

16.3.3.2 Experimental Validation of λ_c

Experimental validation of Eq. (16.2) and Eq. (16.3) is shown in Figure 16.5. For reference, the dotted line shows the actual packet injection rate (λ). The solid line with square markers on it is obtained for an 8×8 network under the *hotspot* traffic,[†] as the ratio between the average number of packets in the network and the average packet delay at that particular injection rate.

This plot clearly shows that there is a good agreement between the actual values obtained through simulation and the ones predicted by Eq. (16.2) before entering the criticality. However, since beyond the critical traffic value the network is not anymore in a steadystate, Eq. (16.2) does not hold for higher-injection rates. The dashed line with triangular markers on it (Figure 16.5) illustrates the upper bound in Eq. (16.3). We observe that this analytical expression provides a good approximation and holds the upper bound property true.

16.4   Bridging the Science and Engineering of NoC Design

This section addresses the customization of regular NoC architectures through application-specific long-range link insertion as a way to illustrate the interaction between the science and engineering of network design. The basic idea of inducing small-world effects into regular networks via long-range links comes from physics [33]. We explore this idea in the context of application-specific NoC architecture customization by introducing NoC-specific constraints into problem formulation. Specifically, the application-specific long-range link insertion is first formulated as a constraint optimization problem and then a practical solution is presented. Finally, the impact of the presented approach on performance, energy consumption, and area is thoroughly evaluated.

16.4.1   Long-Range Link Insertion as a Constrained Optimization Problem

For NoCs, the long-range links should be inserted in a smart manner rather than randomly (as in Refs. [33,34]) due to the following reasons:

- Long-range links have a measurable impact on the network performance.
- The information about the communication workload can be effectively used for optimization purposes.

[†]Three nodes are selected arbitrarily to act as hotspot nodes. Each node in the network sends packets to these hotspot nodes with a higher probability compared to the remaining nodes.

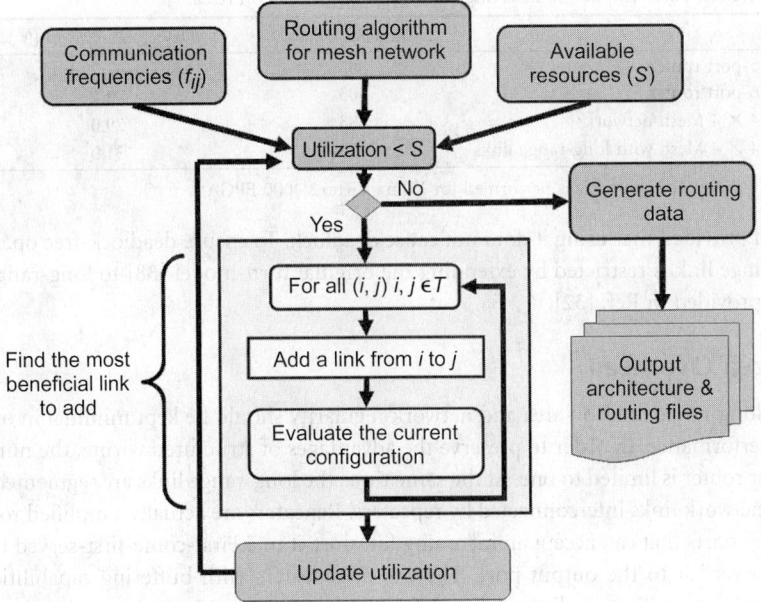

FIGURE 16.6 The flow of the long-range link insertion algorithm.

- Inserting long-range links has an associated cost in terms of area and wiring resources, so there exists a constraint on the proportion of long-range links that can be utilized.

Consequently, application-specific long-range link insertion is modeled as a constraint-optimization problem [32]. Figure 16.6 outlines long-range link insertion algorithm which inserts long-range links with the objective of maximizing the critical traffic load λ_c, subject to the available resources. Given a network architecture and a target application characterized by different communication frequencies (f_{ij}) among the network tiles, the algorithm first estimates the critical traffic load, λ_c. Then, the improvement in λ_c after the insertion of each long-range link is evaluated, and the long link which delivers the largest gain is permanently inserted to the network. Since this step requires a special mechanism for packet routing, a routing strategy is also developed, as described in Section 16.4.2. Finally, the link-insertion procedure repeats until all available resources are used up.

16.4.2 Deadlock-free Routing Algorithm

There are a number of practical issues regarding packet routing that are critical for NoCs:

- First, while theoretical studies [35,36] assume infinite buffering resources, the amount of limited on-chip resource should be taken into account when dealing with NoCs.
- Second, inserting long-range links can cause cyclic dependencies. Combined with finite buffers, the arbitrary use of long-range links may cause deadlock states.
- Finally, without a customized mechanism in place, the newly added long-range links cannot be utilized by the default routing strategy.

To address these issues, a routing algorithm is developed for the use of long-range connections. More precisely, the routers without a long-range connection use the default routing strategy. For all other routers, the distance to the destination, with and without the long-range link, is computed. Since the routing decision is made locally, only the long-range link connected to the current router is taken into account. If using the long-range link results in a shorter distance to the destination, then the long-range

TABLE 16.1 Impact of Inserting of Long-Range Links on Area

	Number of Slices	Device Utilization (%)
5-port router	397	1.7
6-port router	503	2.2
4 × 4 Mesh network	6683	29.0
4 × 4 Mesh with long-range links	7152	31.0

Note: The synthesis is performed for Xilinx Virtex2 4000 FPGA.

link is utilized provided that using it does not cause deadlock. To ensure deadlock-free operation the use of the long-range link is restricted by extending the original turn-model [38] to long-range links. More discussion is provided in Ref. [32].

16.4.3 Area Overhead

The effect of long-range links on area and network regularity should be kept minimal in order to justify the gains in performance. In order to preserve the advantages of structured wiring, the number of long-range links per router is limited to one. At the same time, the long-range links are segmented into regular, fixed-length, network links interconnected by repeaters. Repeaters are actually simplified routers consisting of only two ports that can accept an incoming flit, store it in a First-come-first-served (FIFO) buffer, and finally forward it to the output port. The use of repeaters with buffering capabilities guarantees latency-insensitive operation, as discussed in Ref. [39].

The feasibility of the proposed methodology and realistic measurements on area overhead are demonstrated by using an FPGA prototype consisting of a 4M gate Xilinx Virtex2 FPGA. As shown in Table 16.1, a router with five ports designed for a 2D mesh network utilizes 397 slices, while a router with six ports utilizes 503 slices of the target device.

In addition to this, a pure mesh network, and a mesh network with four long-range links consisting of 12 regular link segments in total were synthesized. It has been observed that the extra links induce about 7% area overhead. This overhead has to be taken into account, while computing the maximum number of long-range links that can be added to a regular mesh network.

16.4.4 Experimental Study

Demonstrating the potential of such a theoretical approach is a crucial step toward its widespread use in practice. For this reason, the impact of long-range links on the performance and energy consumption is evaluated using an FPGA prototype and a cycle accurate C++-based NoC simulator. Figure 16.7 shows some FPGA measurements for a standard mesh network before and after inserting the long-range links. The evaluations are performed using realistic benchmarks retrieved from the E3S benchmark suite [41] using hotspot traffic. We observe that by inserting long-range links, the critical traffic load increases significantly resulting in an improvement in the average packet latency and network throughput. Similar improvements are obtained from the C++ simulations [32].

16.4.4.1 Comparison against Torus and Higher Dimensional Networks

On-chip implementation of higher dimensional mesh and torus networks looks similar to implementing customized topologies using long-range links. However, there is a fundamental difference in the sense that the application-specific customization approach finds the optimal links to be inserted based on a rigorous analysis rather than by inserting them based on a fixed rule. In fact, the topologies synthesized using long-range links reduce to the standard higher dimensional networks, if we replace the optimal link-insertion algorithm with a static rule for links insertion.

Owing to the optimization process, a network architecture obtained by application-specific long-range link insertion can achieve better performance compared to a standard higher dimensional network, although it utilizes less resources. To demonstrate this fact, a 4×4 2D torus network with folded links [5], and a mesh network with eight unidirectional links found using our proposed technique are compared. Long-range

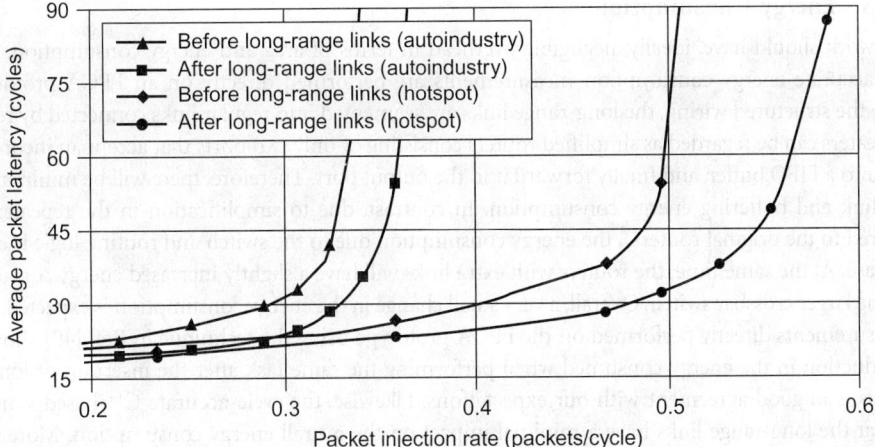

FIGURE 16.7 The comparison of the average packet latency for *hotspot* and *autoindustry* benchmarks before and after the insertion of long-range links.

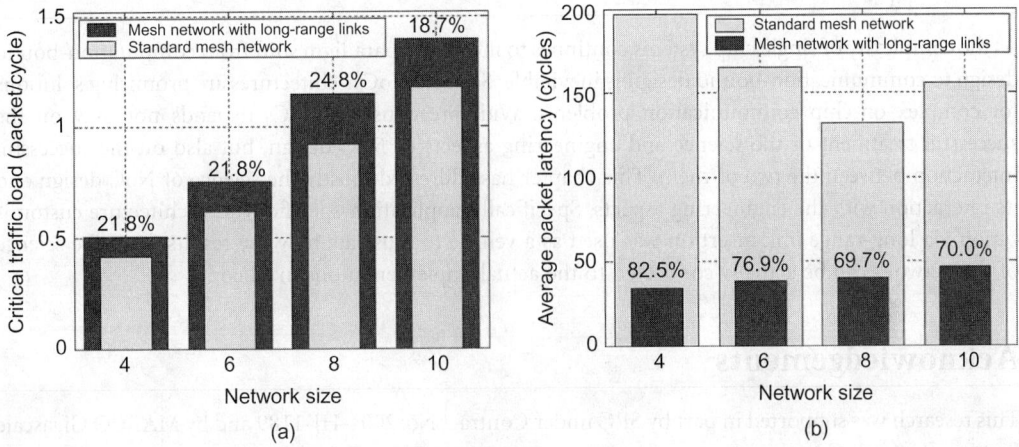

FIGURE 16.8 Scalability results. Performance of the proposed technique for larger network sizes.

links inserted on top of the mesh network consist of 12 regular link segments. This amounts to half of the regular links required to convert the mesh network to a torus. The simulations show 4% improvement in the critical traffic load compared to the torus network. Likewise, the average packet latency, at 0.48 packet/cycle injection rate which is close to the critical load of the torus network, drops from 77.0 to 34.4 cycles with the use of application-specific long-range links. This significant gain is obtained by utilizing only half of resources since inserting the most beneficial links for a given traffic pattern make more sense than blindly adding wrap-around channels all over the network as is the case for the folded torus.

16.4.4.2 Scalability Analysis

The scalability of the long-range link insertion technique is evaluated using networks of sizes ranging from 4 × 4 to 10 ×10. Figure 16.8 shows that consistent improvements are obtained when the network size scales up. For example, by inserting only six long-range links (which is equivalent to 32 regular links total) to a 10 ×10 network, the critical load of the network under hotspot traffic shifts from 1.18 to 1.40 packet/cycle giving a 18.7% improvement. This result is similar to the gain obtained for smaller networks. Figure 16.8(a) also reveals that the critical traffic load grows with the network size due to the increase in the total available bandwidth. Likewise, we observe a consistent reduction in the average packet latency across different network sizes, as shown in Figure 16.8(b).

16.4.4.3 Energy Consumption

The network should have, ideally, negligible overhead in terms of area and energy consumption. For this reason, accurate energy consumption measurements are performed directly on an FPGA prototype. To preserve the structured wiring, the long-range links are segmented into regular links connected by repeaters. The repeaters can be regarded as simplified routers consisting of only two ports that accept an incoming flit, store it into a FIFO buffer, and finally forward it to the output port. Therefore, there will be minimal impact on the link and buffering energy consumption. In contrast, due to simplification in the repeater design (compared to the original routers), the energy consumption due to the switch and routing logic is expected to decrease. At the same time, the routers with extra links will have a slightly increased energy consumption due to the larger crossbar switch. Overall, a very small change in the energy consumption is expected. Indeed, the measurements directly performed on the FPGA prototype using the technique in Ref. [40] show about 2.2% reduction in the energy consumed when performing the same task after the insertion of long-range links; this is in good agreement with our expectations. Likewise, the cycle-accurate C^{++}-based simulations show that the long-range links have a minimal impact on the overall energy consumption. More detailed evaluation of energy consumption can be found in [42].

16.5 Conclusion

As the complexity of single chip systems continues to increase, a paradigm shift from computation-bound design to communication-bound design is inevitable. Scalable NoC architectures are promising solutions for complex on-chip communication problems. Widespread use of NoCs depends not only on the successful treatment of the science and engineering aspects of NoC design, but also on the successful interaction between the two of them. This chapter has addressed mostly the science of NoC design and its interaction with the engineering aspects. Specifically, application-specific NoC architecture customization via long-range link insertion was used as a vehicle to illustrate how the results from the science of the network can be actually connected to the actual implementations in silicon.

Acknowledgements

This research was supported in part by SRC under Contract No. 2004-HJ-1189 and by MARCO Gigascale Systems Research Center.

References

1. Semiconductor Association, *The International Technology Roadmap for Semiconductors* (ITRS), The Semiconductor Industry Association, San Jose, CA, 2004.
2. R.P. Dick, L. Shang, N.K. Jha, *Power-Aware Architectural Synthesis*, VLSI Handbook, Wai-Kai Chen (ed.), Second Edition, CRC Book Press, 2006, Chapter 17.
3. K. Keutzer, S. Malik, A.R. Newton, J. Rabaey, A. Sangiovanni-Vincentelli, "System-level design: Orthogonalization of concerns and platform-based design," *IEEE Transactions on Computer-Aided Design of Integrated Circuits and Systems*, 19, pp. 1523–1543, 2000.
4. R. Ho, K. Mai, and M. Horowitz, "The future of wires," *Proceedings of the IEEE*, 2001, Vol. 89, No. 4 pp. 490–504.
5. W. Dally, B. Towles, "Route packets, not wires: On-chip interconnection networks," *Proceedings of DAC*, Las Vegas, NV, June 2001.
6. E. Bolotin, I. Cidon, R. Ginosar, A. Kolodny, "Cost considerations in network on chip," *Integration, the VLSI Journal*, 38, 19–42, 2004.
7. A. Jantsch, H. Tenhunen (Eds.), *Networks-on-Chip*. Kluwer, Dordrecht, 2003.

8. A. Hemani, A. Jantsch, S. Kumar, A. Postula, J. Oberg, M. Millberg and D. Lindqvist, "Network on a chip: an architecture for billion transistor era," *Proceedings of IEEE NorChip Conference*, Turku, Finland, 2000.

9. L. Benini, G. De Micheli, "Networks on chips: a new SoC paradigm," *IEEE Computer*, 35, 2002.

10. J. Duato, S. Yalamanchili, L.M. Ni, *Interconnection Networks: An Engineering Approach.* Morgan Kaufmann, Los Altos, CA, 2002.

11. U.Y. Ogras, J. Hu, R. Marculescu, "Key research problems in NoC design: A holistic perspective," *Proceedings of CODES+ISSS*, Jersey City, NJ, September 2005.

12. J. Hu, R. Marculescu, "Energy- and performance-aware mapping for regular NoC architectures," *IEEE Transactions on CAD of Integrated Circuits and Systems*, 24, pp. 551–562, 2005.

13. S. Murali, G. De Micheli, "Bandwidth-constrained mapping of cores onto NoC architectures," *Proceedings of DATE*, Paris, France, February, 2004.

14. G. Ascia, V. Catania, M. Palesi, "Multi-objective mapping for mesh-based NoC architectures," *Proceedings ISSS+CODES*, Stockholm, Sweden, September, 2004.

15. S. Murali, G. De Micheli, "SUNMAP: A tool for automatic topology selection and generation for NoCs," *Proceedings of DAC*, San Diego, CA, June 2004.

16. M. Kreutz, L. Carro, C.A. Zeferino, A.A. Susin, "Communication architectures for system-on-chip," *Proceedings of Symposium on Integrated Circuits and Systems Design*, Pirenopolis, Brazil, September 2001.

17. J. Hu, R. Marculescu, "Application-specific buffer space allocation for networks-on-chip router design," *Proceedings of ICCAD*, San Jose, CA, November 2004.

18. A. Pinto, L.P. Carloni, A.L. Sangiovanni-Vincentelli, "Efficient synthesis of networks on chip," *Proceedings of ICCD*, San Jose, CA, October 2003.

19. K. Srinivasan, K.S. Chatha, G. Konjevod, "Linear programming based techniques for synthesis of network-on-chip architectures," *Proceedings of ICCD*, San Jose, CA, October, 2004.

20. U.Y. Ogras, R. Marculescu, "Energy- and performance-driven customized architecture synthesis using a decomposition approach," *Proceedings of DATE*, Munich, Germany, March 2005.

21. J. Liang, S. Swaminathan, R. Tessier, "aSoC: a scalable, single-chip communications architecture," *Proceedings of International Conference on PACT*, Philadelphia, PA, October 2000.

22. A. Adriahantenaina, A. Greiner, "Micro-network for SoC: Implementation of a 32-Port SPIN network," *Proceedings of DATE*, Munich, Germany, March 2003.

23. Kangmin Lee, S. Lee, S. Kim, H. Choi, D. Kim, M. Lee, H. Yoo, "A 51 mW 1.6 GHz on-chip network for low-power heterogeneous SoC platform," *Proceedings of International Solid-State Circuits Conference* (ISSCC), San Francisco, Feburary 2004.

24. D. Bertozzi, A. Jalabert, S. Murali, R. Tamhankar, S. Stergiou, L. Benini, G. De Micheli, "NoC synthesis flow for customized domain specific multiprocessor systems-on-chip," *IEEE Transactions on Parallel and Distributed Systems*, 16, 113–129, 2005.

25. K. Goossens, J. Dielissen, O.P. Gangwal, S.G. Pestana, A. Radulescu, E. Rijpkema, "A design flow for application-specific networks-on-chip with guaranteed performance to accelerate SOC design and verification," *Proceedings of DATE*, Munich, Germany, March 2005.

26. M. Millberg, E. Nilsson, R. Thid, A. Jantsch, "Guaranteed bandwidth using looped containers in temporally disjoint networks within the Nostrum Network-on-Chip," *Proceedings of DATE*, Paris, France, February 2004.

27. T. Bjerregaard, J. Sparso, "A Router architecture for connection-oriented service guarantees in the MANGO clockless Network-on-Chip," *Proceedings of DATE*, Munich, Germany, March 2005.

28. E. Bolotin, I. Cidon, R. Ginosar, A. Kolodny, "QoS architecture and design process for networks-on-chip," *Journal of Systems Architecture*, Special issue on Network on Chip, 50, pp. 105–128, 2004.

29. H. Wang, X. Zhu, L. Peh, S. Malik, "Orion: A power-performance simulator for interconnection networks," *Proceedings of MICRO 35*, Istanbul, Turkey, November 2002.

30. T.T. Ye, L. Benini, G. De Micheli, "Analysis of power consumption on switch fabrics in network routers," *Proceedings of DAC*, Anaheim, CA, June 2003.

31. N. Banerjee, P. Vellank, K.S. Chatha, "A power and performance model for network-on-chip archi-tectures," *Proceedings of DATE*, Paris, France, February 2004.

32. U.Y. Ogras, R. Marculescu, ' "It's a small world after all": NoC Performance Optimization via Long Link Insertion,' in IEEE Trans. on Very Large Scale Integration Systems, Special Section on Hardware/ Software Codesign and System Synthesis, 14(7), pp. 693–706, July 2006.

33. D.J. Watts, S.H. Strogatz, "Collective dynamics of small-world networks," *Nature*, 393, 440–442, 1998.

34. M.E.J. Newman, D.J. Watts, "Scaling and percolation in the small-world network model," *Physical Review E*, 60, 7332–7342, 1999.

35. H. Fuks, A. Lawniczak, "Performance of data networks with random links," *Mathematics and Computers in Simulation*, 51, pp. 101–117, 1999.

36. M. Woolf, D. Arrowsmith, R.J. Mondragon, J.M. Pitts, "Optimization and phase transitions in a chaotic model of data traffic," *Physical Review E*, 66, pp. 046–106, 2002.

37. R. Albert, A. Barabasi, "Statistical mechanics of complex networks," *Reviews of Modern Physics*, 74, pp. 47–97, 2002.

38. C.J. Glass, L.M. Ni, "The turn model for adaptive routing," *Proceedings of ISCA*, Australia, May 1992.

39. L.P. Carloni, K.L. McMillan, A.L. Sangiovanni-Vincentelli, "Theory of latency-insensitive design," *IEEE Transactions on CAD of Integrated Circuits and Systems*, pp. 1059–1076, 20, 2001.

40. H.G. Lee, K. Lee, Y. Choi, N. Chang, "Cycle-accurate energy measurement and characterization of FPGAs," *Analog IC and Signal Processing*, 42, pp. 146–154, 2005.

41. R.P. Dick, D.L. Rhodes, W. Wolf, "TGFF: Task graphs for free," *Proceedings of International Workshop on Hardware/Software Codesign*, Seattle, Washington, March 1998.

42. U.Y. Ogras, R. Marculescu, H.G. Lee, N. Chang, "Communication Architecture Optimization: Making the Shortest Path Shorter in Regular Networks-on-Chip," *Proceedings of DATE*, Munich, Germany, March 2006.

17

Power-Aware Architectural Synthesis

Robert P. Dick
Northwestern University

Li Shang
Queen's University

Niraj K. Jha
Princeton University

CONTENTS

17.1 Introduction

Power consumption is one of the most important parameters of modern electronic systems. It impacts the performance, cooling costs, packaging costs, and reliability of integrated circuits (ICs), as well as the life spans of battery-powered electronics. Therefore, it is important to consider power consumption during design.

The complexity and scope of automatic design continues to increase. It is now possible to automatically design, i.e., synthesize, complex ICs and systems from high-level specifications without designer intervention. Architectural synthesis has been an active research area for more than a decade. Since addressing power consumption at higher levels of the design process increases the potential for improvement, researchers have developed a wide range of power optimization and management techniques to address IC power consumption issues during architectural synthesis. In this chapter, we present techniques for the synthesis of low-power ICs and systems. In particular, we focus on power-aware behavioral synthesis and system synthesis.

The rest of this chapter is organized as follows. In Section 17.1.1, we introduce and define behavioral synthesis and system synthesis. In Section 17.2, we describe the contributors to IC and system power consumption and discuss a number of techniques to improve power and thermal characteristics. Many of these techniques will prove useful in both behavioral synthesis and system synthesis. In Sections 17.3 and 17.4, we provide details on behavioral synthesis and system synthesis and indicate areas of active research. Section 17.5 points out a few commercial architectural synthesis products. We conclude in Section 17.6.

17.1.1   Architectural Synthesis Overview

In 1958 and 1959, Jack Kilby and Robert Noyce independently built the first ICs. Although the simple applications of early ICs enabled fully manual design, within 10 years, engineers were designing large-scale integration (LSI) ICs containing tens of thousands of transistors. In the late 1960s and early 1970s, fully manual design became impractical and engineers began automating the design process. Table 17.1 gives a chronology of areas of active research and development in electronic design automation. Note that the first research in an area may have appeared before the area was of wide interest, e.g., some researchers had already made great progress in behavioral synthesis before the 1990s. As indicated in Table 17.1, tasks that consist of simple actions repeatedly applied were the most straightforward and the first to be automated. However, as design complexity continued to increase, engineers found it necessary to automate increasingly complicated and creative tasks that had once required the efforts of skilled designers. In recent years, two trends are apparent: higher levels of the design process have been automated and power consumption has become a first-order design characteristic. In the past 5 years, these trends have converged; research on power-aware architectural synthesis has proceeded at a rapid pace.

Figure 17.1 illustrates the conventional levels or stages of digital system design. Physical design, i.e., deciding on the physical locations and shapes of transistors, functional units, and processors, as well as communication, clock distribution, and power distribution networks, was largely automated in the

TABLE 17.1 Chronology of Active Research and Development Topics in Electronic Design Automation

1958–1965	Manual design
1965–1975	Schematic capture, automated mask production Circuit simulation Automatic routing
1975–1985	Automated placement Design rule checking Layout vs. schematic checking
1985–1990	Hardware description languages Logic synthesis Static timing analysis
1990–1995	Behavioral synthesis Formal verification
1995–2000	Hardware–software cosynthesis SoC synthesis **Low-power design becomes critical**
2000–2005	NoC synthesis, platform-based design Synthesis for new processes, e.g., microfluidics and MEMS **Thermal and reliability issues become critical**

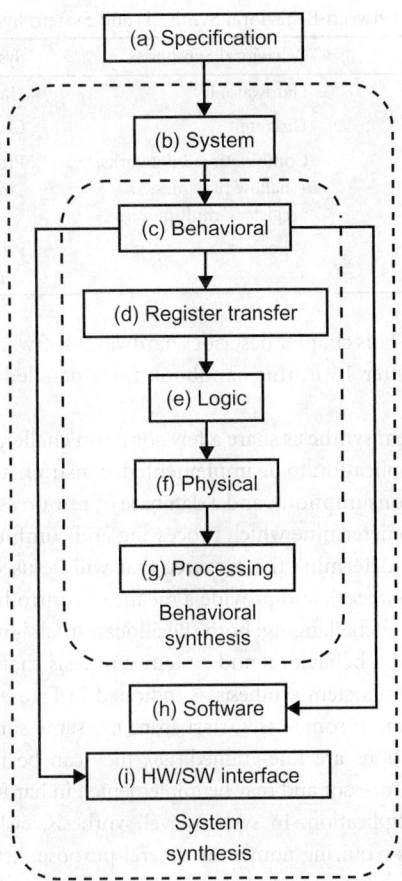

FIGURE 17.1 Digital system design levels.

1960s and 1970s. However, this area remains open, with continued improvement over past work and new algorithms to deal with changes brought about by process scaling. Combinational logic synthesis, the efficient design of combinational networks that implement Boolean expressions, advanced rapidly in the 1970s and 1980s. Register-transfer level (RTL) optimizations, such as retiming, underwent substantial advances in the 1990s.

This trend of automating increasingly high levels of the design process continues to this day. Sophisticated algorithms are now used to automatically design, or synthesize, very large-scale integration (VLSI) circuits and hardware–software systems, starting from high-level descriptions of application behavior. These synthesis algorithms automatically make design decisions at many levels, ranging from architectural level to physical level, to optimize performance, energy consumption, thermal characteristics, price, and reliability. Power consumption is now a critical cost for synthesized architectures. It influences packaging and cooling costs, performance, reliability, and battery life span. Moreover, optimizing power and thermal characteristics greatly increases the complexity of synthesis. This chapter provides a taxonomy of synthesis problems, describes how state-of-the-art synthesis algorithms solve these problems, and indicates trends that will influence future work in the field.

Architectural synthesis may be broken into two main areas: behavioral synthesis and system synthesis. This chapter describes these areas and explains methods of reducing power consumption during synthesis. However, each area is broad; they cannot be exhaustively covered here. System synthesis has its roots in hardware–software cosynthesis, with much current activity in system-on-chip (SoC) synthesis

TABLE 17.2 Differences between Behavioral Synthesis and System Synthesis

	Behavioral Synthesis	System Synthesis
Implementation	Hardware, IC	Hardware–software system
Timing model	Discrete	Continuous
Processing element model	Combinational networks, shallow pipelines, registers, multiplexers	Processors, protocol translators, memories
Communication resource model	Wires	Protocols over buses, network-on-chip, or wires

and network-on-chip synthesis. This chapter describes hardware–software cosynthesis and SoC synthesis but defers to Marculescu's Chapter 16 in this handbook for a detailed treatment of network-on-chip synthesis.

Behavioral synthesis and system synthesis share a few common challenges. In both cases, starting from an abstract description of the application to be implemented, constraints on the costs of the system (e.g., price, performance, and power consumption), and a database of resources that may be used to implement the application, it is necessary to determine which processing and communication resources will be used in the final design (*allocation*[*]), determine the resource that will be used for each particular operation and communication event (*assignment*), and provide a means of controlling the times at which all events occur (*scheduling*). These tasks are challenging; both the allocation/assignment and scheduling problems are NP-complete [1].[†] In summary, behavioral and system synthesis share a number of hard problems.

Behavioral synthesis differs from system synthesis as indicated in Figure 17.1 and Table 17.2. Behavioral synthesis and system synthesis can, in some cases, start from the same sorts of specifications. However, in behavioral synthesis, most operations are fine-grained, i.e., they can be represented by short instruction sequences for a general-purpose processor and may be implemented in hardware as a combinational network or a shallow pipeline, e.g., multiplication. In system-level synthesis, tasks are generally coarse-grained. They may be complex procedures requiring numerous general-purpose instructions or sequential hardware implementations, e.g., fast Fourier transform. In behavioral synthesis, it is generally assumed that the entire specification is implemented in synthesized hardware. In system synthesis, hardware and software are both used in the implementation. Differences in task granularity and implementation style (hardware-only or hardware–software) lead, in turn, to other differences between behavioral synthesis and system synthesis.

The simplicity of components, e.g., functional units and wires, in behavioral synthesis simplifies clocking, scheduling, and interface problems; it is usually possible to assume a globally synchronous system, a discrete time schedule in which all operations take small integer numbers of clock cycles, and straightforward interfacing between components. However, the simplicity of individual components is offset by their quantity. Quickly determining the impact of architectural decisions on the floorplans[‡] and thermal profiles of designs containing hundreds or thousands of frequently parallel operations is extremely challenging.

In system synthesis, the number of components is limited. However, they are generally more complicated than arithmetic functional units, e.g., instruction processors or protocol translators. The system synthesis algorithm may not have control over the implementation of each complex component. Therefore, providing for global synchronization and communication is more challenging. Interface synthesis, i.e., synthesizing the interfaces between hardware components as well as software and hardware, is an

---

[*]Some behavioral synthesis researchers define *allocation* to be the assignment of tasks and communication events to resources as well as the selection of resources.

[†]Garey and Johnson [1] provide an introduction to the theory of NP-completeness. For the purpose of this chapter, the implications can be summarized as follows: nobody has ever developed and reported an algorithm that can quickly produce optimal solutions to large instances of these problems and there is strong evidence (but no proof) that such an algorithm cannot be implemented using conventional, deterministic computers.

[‡]A floorplan indicates the positions of all architectural components in an IC.

active area of research in system synthesis. Unlike behavioral synthesis, tasks in system synthesis need not take a small integer number of clock cycles: time values are modeled as reals, not integers. Moreover, some operations may have dramatically higher execution times than others. Therefore, a number of discrete time domain scheduling algorithms that are promising in behavioral synthesis are not directly applicable in system synthesis.

17.2 Challenges of Low-Power Synchronous System Synthesis and Design

This section introduces the fundamentals necessary to understand the sources of power consumption in synchronous digital systems (Section 17.2.1). It then describes a number of techniques that may be used during behavioral synthesis and system synthesis to improve power and thermal characteristics (Sections 17.2.2–17.2.5).

17.2.1 Power Overview

With increasing system integration, as well as aggressive technology scaling, power consumption has become a major challenge in digital system design. In high-performance computer systems, power and thermal issues are key design concerns. Power management and optimization techniques are essential for minimizing system temperature to permit reliable operation. For portable devices, prolonging battery lifetimes and minimizing packaging costs are primary design challenges. Power also interacts with other design metrics, such as performance, cost, and reliability, thereby further increasing design complexity. For example, the failure rate of electronic devices is a strong function of system temperature, which is in turn controlled by system power dissipation. Therefore, increasing power consumption results in the need for more complicated cooling and packaging solutions to sustain system reliability, which in turn increases costs. As projected by International Technology Roadmap for Semiconductors (ITRS) [2], power will continue to be a limiting factor in future technologies. There is an increasing need to address power issues in a systematic way at all levels of the design process.

In digital CMOS circuits, power dissipation is the sum of dynamic power, $P_{dynamic}$, and static power, P_{static}. Dynamic power, $P_{dynamic}$, results from charging and discharging of the capacitance of CMOS gates and interconnect during circuit switching, P_{switch}, and the power during transient short-circuits when inputs are in transition, $P_{short\ circuit}$. For synchronous CMOS designs, switching power is one of the dominant sources of power consumption. It is a function of physical capacitance, C, switching activity, s,§ clock frequency, f, and supply voltage, V_{dd}:

$$P_{switch} = \frac{1}{2} sCV_{dd}^2 f \tag{17.1}$$

In CMOS, the other major source of power consumption, static power, P_{static}, results from leakage current. Leakage current has five basic components: reverse-biased PN junction current, subthreshold leakage, gate leakage, punch-through current and gate tunneling current. Of these five components, subthreshold and gate leakage will remain dominant during the next few years. The subthreshold leakage power is given by

$$P_{subthreshold} = I_{sub} \frac{W}{L} V_{dd} e^{-V_{th}/nV_T} \tag{17.2}$$

---

§ The product of physical capacitance, C, and switching activity, s, is also called switched capacitance.

where I_{sub} and n are technology parameters, W and L device geometries, V_{th} is the threshold voltage, and V_T the thermal voltage constant [3]. Gate leakage is the current between the gate terminal and any of the other three terminals (drain, source, or body). As a result of technology scaling, gate leakage increases exponentially due to decreasing gate oxide thickness.

From Eqs. (17.1) and (17.2), it can be seen that total power consumption may be reduced by attacking operating voltage, capacitance, switching activity, threshold voltage, transistor size, and temperature. In real designs, the variables upon which total power consumption depends are often closely related: reducing one may increase another. In addition, reducing power consumption may have a negative impact on other design metrics. A synthesis algorithm must simultaneously consider, and trade off, these design metrics.

17.2.2 Operating Voltage-Oriented Techniques

Reducing operating voltage, V_{dd}, is one of the most promising techniques for reducing dynamic power consumption. As indicated by Eq. (17.1), $P_{dynamic}$ is quadratically related to V_{dd}. All other things being equal, halving V_{dd} reduces $P_{dynamic}$ to one-fourth of its initial value. However, this reduction has a negative impact on circuit performance [4]:

$$f = \frac{k(V_{dd} - V_{th})^{\alpha}}{V_{dd}} \qquad (17.3)$$

where k is a design-specific constant and α a process-specific constant ranging from 1 to 2. As a result, for low values of V_{th} and $\alpha \simeq 2$, and all other things being equal, halving V_{dd} implies halving clock frequency, f. Some of the following sections describe techniques for reducing operating voltage without degrading performance.

17.2.2.1 Multiple Simultaneous Operating Voltages

ICs contain timing-critical and -noncritical combinational logic paths between memory elements (latches and flip-flops). It is possible to selectively decrease the operating voltage(s) of gates on the noncritical paths, thereby reducing $P_{dynamic}$ without reducing performance. Multiple voltage techniques may be used within architectural synthesis. Although it is not essential, processing elements[¶] sharing the same voltage are often placed in contiguous regions called voltage islands to simplify power distribution. Communication between different voltage regions relies on level converters. This physical requirement for contiguous regions dramatically changes the IC floorplan, thereby changing communication power consumption, wire delays, and thermal properties. These changes, in turn, impact the original design properties, e.g., combinational path criticality, optimal clock frequency, and operation cycle times. It is necessary to consider the consequences of using multiple voltages at different design levels, i.e., architectural and physical. Recent multiple voltage behavioral [5] and system [6] synthesis techniques allow solution of the voltage-level assignment problem concurrently with one or more of the other following problems: processing element selection, assignment of tasks to processors, scheduling, and floorplanning.

17.2.2.2 Dynamic Voltage (and Frequency) Scaling

In addition to varying the operating voltages of subcircuits by position, it is possible to vary operating voltages in time. Dynamic voltage scaling is generally carried out in conjunction with frequency scaling to prevent timing violations. It allows an IC to adaptively adjust operating voltage to minimize power consumption without violating timing constraints. Dynamic voltage and frequency scaling (DVFS) interacts closely with scheduling: some schedules allow timing slack to be used for power minimization without the violation of deadlines while others leave little opportunity for power minimization. Synthesis algorithms have been developed for both offline DVFS and online DVFS [7], for which predictions of future system behavior are used to amortize the cost of voltage and frequency changes over longer low-power periods.

[¶]When used in a general context, the term processing element will be used to refer to both functional units, e.g., multipliers and adders, as well as system-level processing elements, e.g., microprocessor cores.

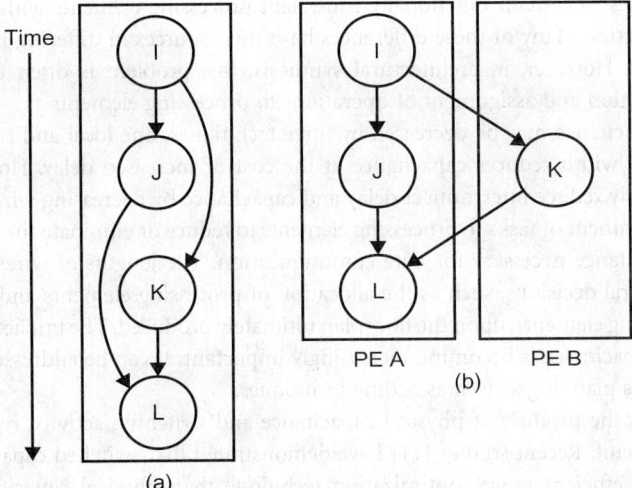

FIGURE 17.2 Series and parallel implementations of a dataflow graph. (a) Serial implementation. (b) Parallel implementation.

17.2.2.3 Scheduling and Timing

Scheduling is the process of selecting the orders and execution start times of operations and communication events. In some cases, a system's original schedule may not permit the reduction of operating voltage(s) without performance degradation. For example, some operations may be immediately followed by other operations, leaving little spare time for voltage reduction. Changing operation start times and orders can open up opportunities for greater reductions in power consumption via operating voltage reduction. It is also possible to change the number of clock cycles and frequency for an operation, thereby allowing a decrease in power consumption without degrading computational throughput.

17.2.2.4 Power for Performance and Area Techniques

Even if it seems that attempts to reduce the operating voltage for tasks on critical timing paths will result in performance degradation, it is sometimes possible to buy back the lost performance at a cost in area. Consider the example in Figure 17.2. In the serial implementation shown on the left, the processing element must operate at a high voltage at all times to meet the performance requirements. By adding another processing element and parallelizing the operations, as shown by the parallel implementation to the right, it is possible to finish execution early, thereby providing enough timing slack to permit operating voltage to be reduced to 3/4 its initial value, thereby reducing dynamic power consumption to 9/16 its initial value. This general technique of buying voltage reduction at the cost of performance and gaining back the lost performance through increased area and design complexity forms the basis of a number of techniques in low-power architectural synthesis [8–10].

17.2.3 Switched Capacitance-Oriented Techniques

It is possible to reduce both active device and interconnect capacitance via a number of synthesis techniques. Reducing a CMOS gate in size reduces the capacitance driven by the previous gate. However, this also increases resistance to the power and ground rails, increasing the delay of the subsequent gate. In many cases, several devices are not on the critical timing path of the system. Their sizes may be reduced to reduce driven capacitance. This technique shares properties with operating voltage reduction. However, the potential for improvement to dynamic power consumption is generally smaller because the relationship between power and capacitance-dependent delay is subquadratic, i.e., reducing operating voltage is generally a better choice than reducing capacitance. However, reducing capacitance does come with two additional advantages: area efficiency and no need for multi-voltage support. During architectural synthesis, it is common for libraries to contain functionally equivalent processing elements with different power and performance properties. Many of these differences have their sources in differing internal gate and wire capacitance values. However, in architectural synthesis, this problem is often encompassed by processing element selection and assignment of operations to processing elements.

Interconnect self-capacitance may be decreased by three techniques, one local and two architectural. Decreasing interconnect width reduces capacitance at the cost of increased delay. However, it is also possible to simultaneously reduce interconnect delay and capacitance by decreasing wire length. Finally, one can change the assignment of tasks to processing elements to reduce or eliminate the inter-processing element-switched capacitance necessary for data communication. The lengths of wires are decided by the impact of architectural decisions, such as the allocation of processing elements and the assignment of operations to processing elements, upon the floorplan ultimately produced. The impact of interconnect coupling on effective capacitance is becoming increasingly important. It can be addressed during architectural synthesis via bus planning as well as coding techniques.

Switched capacitance, the product of physical capacitance and switching activity, reflects the actual run-time load of the circuit. Recent studies [11] have demonstrated that switched capacitance minimization is a much more efficient power optimization technique than physical capacitance reduction. Switched capacitance reduction techniques have been developed at all levels of the design hierarchy. Architecture-level techniques [12], such as power management, data encoding, glitch suppression, and architectural transformation, are widely used in low-power behavioral and system synthesis.

17.2.4 Leakage Power Techniques

Most work in low-power synthesis explicitly targets dynamic power consumption. This is not surprising. Even at the 90-nm process node, dynamic power accounts for over 90% of total power in modern processors. However, research indicates that a half or more of the total power consumption will result from leakage at the 25-nm process node [2]. Subthreshold leakage is an exponential function of chip temperature. As a result, increasing temperature from 25°C to 100°C can result in subthreshold leakage being the dominant source of power consumption.

As indicated in Eq. (17.3), it is necessary to reduce V_{th} in unison with V_{dd} to maintain good performance. However, reduction in threshold voltage increases subthreshold leakage. This problem may be addressed by using multiple threshold voltages [13], such as multi-V_{th}, adaptive body biasing, etc. During synthesis, high threshold voltages can be assigned to functional units along noncritical timing paths to reduce subthreshold leakage while functional units on critical paths operate at lower threshold voltages to maintain performance.

Power gating reduces subthreshold leakage power consumption by inserting sleep transistors in series with pull-up or pull-down paths of functional units to control their leakage power dependent on the sleep transistor inputs [13]. NMOS transistors with high threshold voltages are typically used as sleep transistors. This circuit topology is known as MTCMOS. Other techniques, e.g., exploiting the transistor stack effect, transistor sizing, and supply voltage scaling, may also be used to minimize subthreshold leakage power consumption.

17.2.5 Temperature-Oriented Techniques

All other things being equal, increasing IC power consumption increases temperature. Using temperature-aware techniques in architectural synthesis is a complex task. IC temperature is affected by many factors, including IC dynamic and leakage power profile, interconnect power profile, as well as the packaging and cooling solution. Many of these power profiles are only available after physical design, i.e., floorplanning. Although power optimization techniques can reduce average chip temperature, local thermal hotspots due to unbalanced chip power profiles may result in thermal emergencies, e.g., reliability problems due to electromigration. In addition, subthreshold leakage power consumption has an exponential relationship with chip temperature. Without temperature optimization, leakage power

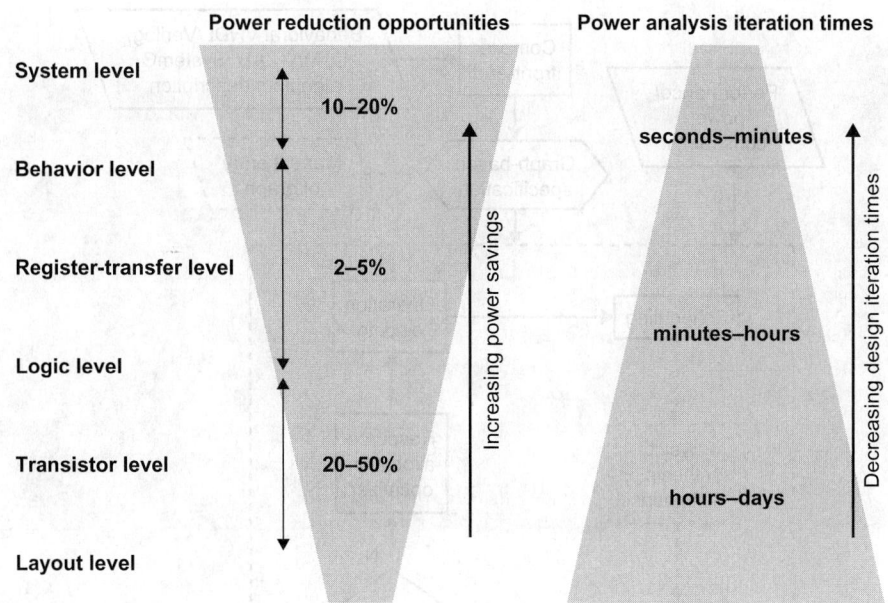

FIGURE 17.3 Benefits of high-level power analysis and optimization. (From A. Raghunathan et al., *High-Level Power Analysis and Optimization.* Kluwer Academic Publishers, Boston, 1997.)

can dominate power consumption. To address IC thermal problems, it is critical to integrate architectural synthesis with physical synthesis and thermal analysis to form a complete thermal optimization flow [5]. Thermal modeling and analysis also need to be incorporated into the inner optimization loop to guide IC synthesis. However, detailed thermal characterization requires 3D full chip-package thermal analysis, which may have high computational complexity. Thermal analysis may easily become the performance bottleneck for thermal-aware synthesis.

17.2.6 Potential of Power Optimization at Different Design Levels

Although power minimization techniques were first developed at the device level, postponing power optimization until this stage of the design process neglects opportunities at higher levels. As indicated in Figure 17.3, considering power minimization at earlier stages of the synthesis or design process has a number of advantages. It yields greater potential for improvement. Moreover, it indirectly improves solution quality because many candidate designs may be considered at higher levels of synthesis due to the use of more abstract (hierarchical) system modeling.

17.3 Low-Power Behavioral Synthesis

Behavioral synthesis, or high-level synthesis, is the automatic design of an IC starting from an implementation-independent description of the design's behavior, a description of the functional units and communication resources available, and constraints on performance and power.

Figure 17.4 gives an overview of a behavioral synthesis optimization flow. Note that, although this flow is representative, other high-level meta-algorithms exist. For example, it would be possible to use a mixed integer linear program (MILP) solver on a unified behavioral synthesis problem formulation, in which case there would be no allocation, binding, and scheduling optimization loop.

Although the input to a behavioral synthesis system can take many forms, the most common are software language, hardware description language, or graph-based specifications. An example C input

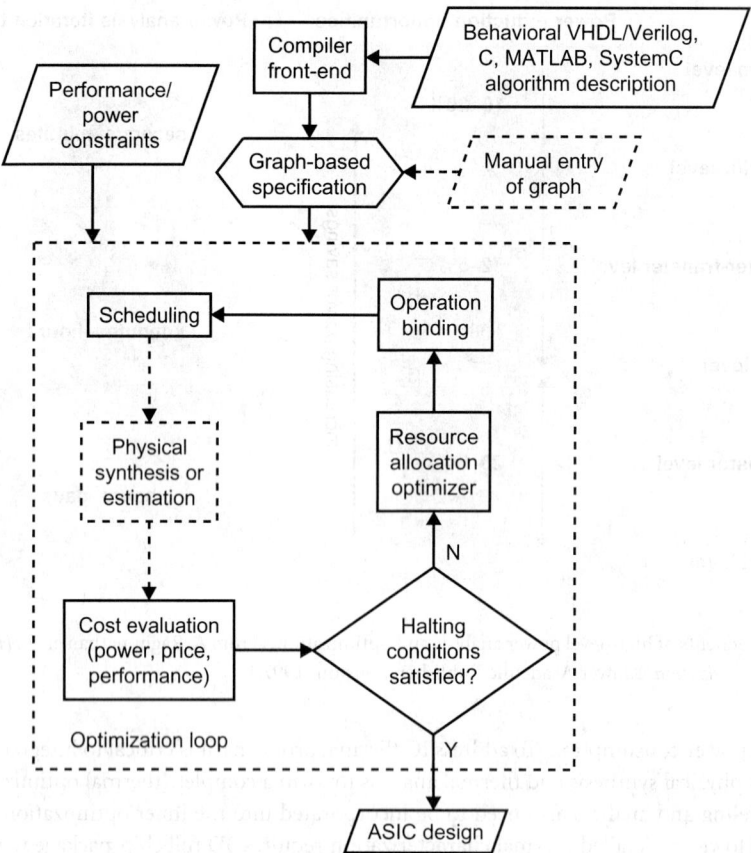

FIGURE 17.4 Behavioral synthesis algorithm.

```
void FIR_filter(int n, int order, int * a, int * x, int * y) {
  int i, j;
  for (i = order - 1; i < n; ++i) {
    y[i] = 0;
    for (j = 0; j < order; ++j) {
      y[i] += a[j] * x[i - j];
    }
  }
}
```

FIGURE 17.5 Finite impulse response filter code.

file for a finite impulse response filtering algorithm is shown in Figure 17.5. As shown in Figure 17.4, regardless of the starting point, behavioral synthesis systems use compilers [14,15] to convert specifications into (possibly synchronous) data flow graphs (DFGs) or control-data flow graphs (CDFGs) for further optimization. Translation and performance optimization of the code in Figure 17.5 results in the

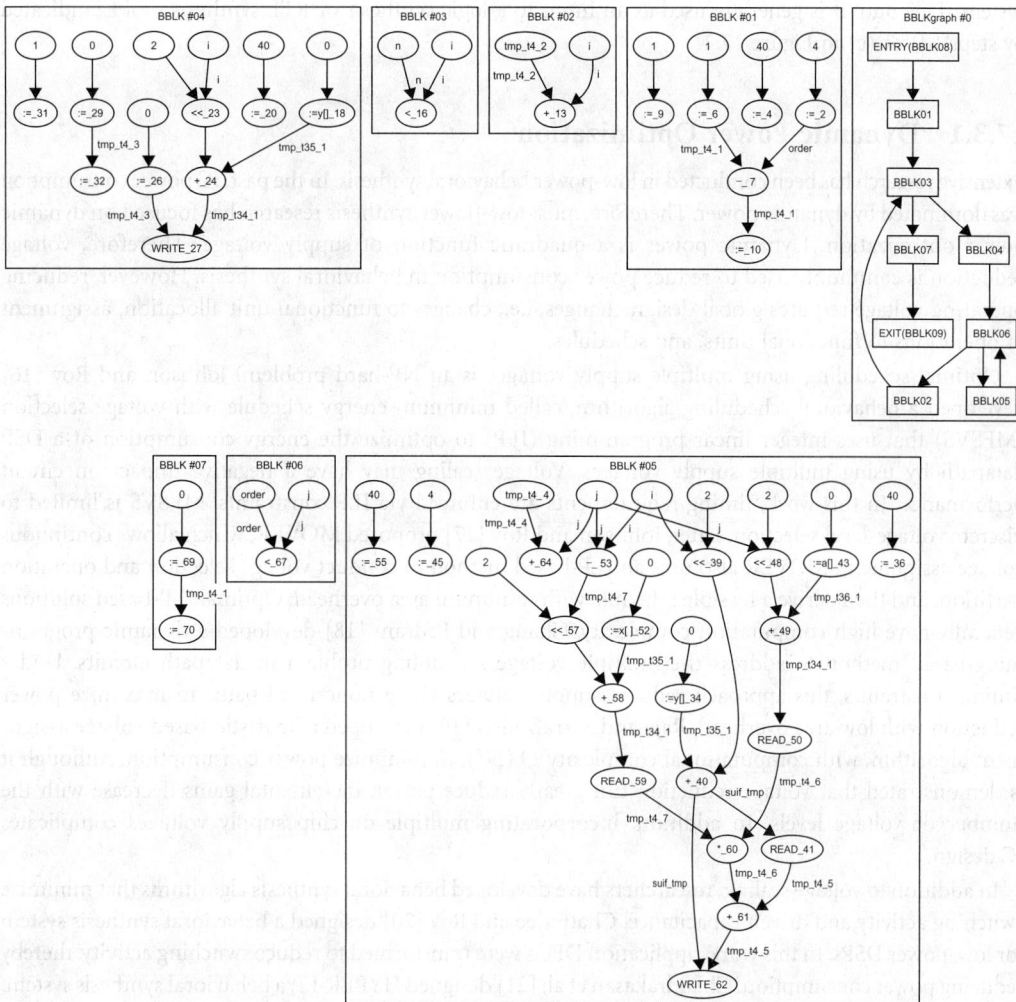

FIGURE 17.6 Finite impulse response filter CDFG. (Courtesy of Dr. Rajarshi Mukherjee at Synopsys and Dr. David Zaretsky at Binachip, Inc.)

control-data flow graph shown in Figure 17.6. In this figure, the graph at the upper-right shows the flow of control among the basic blocks, i.e., straight-line sequences of code that may be represented with DFGs. Within each basic block, the nodes without incoming edges represent variables or constants and the other nodes represent operations on the data arriving on the incoming arcs. In addition to a description of the algorithm to be implemented, behavioral synthesis tools require models for the hardware resources that may be used in the implementation. For example, the user may provide a library of performance and power models for the available functional units, e.g., adders, multipliers, and registers. These models may be provided as part of the resource library or automatically generated by commercial timing and power analysis tools.

A behavioral synthesis algorithm does functional unit allocation, operation binding, and scheduling to optimize performance, IC area, and possibly power. Power optimization, e.g., minimizing switched wire capacitance, may require physical information and, therefore, floorplanning block placement within behavioral synthesis. The product of behavioral synthesis is a complete RTL description of the synthesized system. This output is generally used as an input to a logic synthesis or RTL synthesis tool as indicated by steps (d) or (e) in Figure 17.1.

17.3.1 Dynamic Power Optimization

Extensive research has been conducted in low-power behavioral synthesis. In the past, IC power consumption was dominated by dynamic power. Therefore, most low-power synthesis research has focused on dynamic power optimization. Dynamic power is a quadratic function of supply voltage. Therefore, voltage reduction is commonly used to reduce power consumption in behavioral synthesis. However, reducing operating voltage requires global design changes, i.e., changes to functional unit allocation, assignment of operations to functional units, and schedules.

Optimal scheduling using multiple supply voltages is an NP-hard problem. Johnson and Roy [16] developed a behavioral scheduling algorithm, called minimum energy schedule with voltage selection (MESVS) that uses integer linear programming (ILP) to optimize the energy consumption of a DSP datapath by using multiple supply voltages. Voltage scaling may have a negative impact on circuit performance. In this work, timing requirements are enforced via ILP constraints. MESVS is limited to discrete voltage-level selection. Later, Johnson and Roy [17] proposed MOVER, which allows continuous voltage assignment. MOVER also uses an ILP-based method to conduct voltage selection and operation partition, and then derive a feasible schedule with minimum area overhead. Optimal ILP-based solutions generally have high computation complexity. Chang and Pedram [18] developed a dynamic programming-based method to address the multiple voltage scheduling problem in datapath circuits. Under timing constraints, this approach reduces supply voltages along noncritical paths to maximize power reduction with low area overhead. Raje and Sarrafzadeh [19] developed a heuristic-based voltage assignment algorithm, with computational complexity $O(N^2)$, to minimize power consumption. Although it is demonstrated that voltage reduction can greatly reduce power, incremental gains decrease with the number of voltage levels. In addition, incorporating multiple on-chip supply voltages complicates IC design.

In addition to voltage scaling, researchers have developed behavioral synthesis algorithms that minimize switching activity and driven capacitance. Chatterjee and Roy [20] designed a behavioral synthesis system for low-power DSPs. In this work, application DFGs were transformed to reduce switching activity, thereby reducing power consumption. Chandrakasan et al. [21] designed HYPER-LP, a behavioral synthesis system. HYPER-LP uses algorithmic transformations enable voltage scaling and effective capacitance reduction. Kumar et al. [22] developed a profile-driven behavioral synthesis algorithm, using profiling to characterize the run-time activities of DFG-based system models. Low-power behavioral synthesis is then conducted to minimize estimated system switching activity. Chang and Pedram [23] proposed an allocation and binding technique to minimize the switching activity in registers. In this work, statistical methods are used to characterize the switching activities of registers. A max-cost flow algorithm was then proposed to conduct power-optimal register assignment. Chang and Pedram [24] also proposed a low-power binding technique to minimize the power consumption of datapath functional units, in which power optimization is formulated as a max-cost multicommodity flow problem. Dasgupta and Karri [25] proposed simultaneous binding and scheduling techniques to reduce switching activity, and hence the power consumption, of buses. Mehra et al. [26] proposed behavioral synthesis techniques for low-power real-time applications. By preserving locality and regularity in input behavior during resource assignment, this technique reduces the need for global buses, thereby reducing power consumption. Ercegovac et al. [27] proposed a behavioral synthesis system that uses multiple precision arithmetic units to support low-power ASIC synthesis. In this work, system resource allocation is conducted through multigradient search and task assignment is based on a modified Karmarkar–Karp's number partitioning heuristic.

A few researchers have developed high-level synthesis algorithms that combine numerous power optimization techniques. Musoll and Cortadella [28] proposed several high-level power-optimization techniques, including loop interchange, operand reordering, operand sharing, idle units, and operand correlation, for reducing the activities of functional units. Raghunathan and Jha [29] designed SCALP, an iterative-improvement-based behavioral synthesis system, for low-power data-intensive applications. SCALP provides a rich set of behavioral optimization techniques, including architectural transformation, scheduling, clock selection, module selection, and hardware allocation and assignment. Khouri et al. [30] showed how to perform low-power behavioral synthesis for control-flow intensive algorithms. This work uses an iterative improvement framework to perform design space exploration. Behavioral power optimization techniques, including loop unrolling, module selection, resource sharing, and multiplexer network restructuring, are done concurrently.

17.3.2   Physical-Aware Power Optimization

In conventional behavioral synthesis, physical implementation details were generally ignored when making architectural decisions. Continued process scaling has required fundamental changes to IC synthesis. At present, physical design details must be considered during all stages of IC synthesis. Many of the techniques use physical information, e.g., floorplan block placements, to optimize switched capacitance better [31–34], as explained in Section 17.2.3. Although they do not use a floorplan, Lyuh et al. [35] optimize assignment of communication events to interconnect buses, and the order of (capacitively coupled) wires within buses, to reduce effective switched capacitance. Prabhakaran and Banerjee [36] proposed a simultaneous scheduling, binding, and floorplanning algorithm to address the power consumption of interconnect during behavioral synthesis. Zhong and Jha [37] presented an interconnect-aware low-power behavioral synthesis algorithm, called ISCALP, that minimizes power consumption in interconnects through interconnect-aware binding. Recently, Gu et al. [38] designed a fast, high-quality incremental floorplanning and behavioral synthesis system that concurrently optimizes performance, power, and area.

17.3.3   Leakage Power Optimization

As a result of technology scaling, leakage power consumption is becoming increasingly significant in digital CMOS circuits. Khouri and Jha [39] were the first to propose a method of reducing leakage power consumption during behavioral synthesis. They proposed an iterative algorithm using dual-V_{th} technology. Through each iteration, a greedy prioritization approach is used to identify the functional unit with maximum leakage power reduction potential, and then replace it with a higher-V_{th} functional unit. Gopalakrishnan and Katkoori [40] proposed KnapBind, a leakage-aware resource allocation and binding algorithm to minimize datapath leakage power consumption. This work maximizes the idle time of datapath modules. MTCMOS functional modules with large idle time slots are placed into sleep mode when they are idle. Tang et al. [41] proposed a heuristic to minimize leakage power consumption during behavioral synthesis. The synthesis problem is formulated as the maximum weight-independent set problem. Datapath components with maximum or near-maximum leakage-saving potentials are identified and replaced with low-leakage library modules. Leakage power is a strong function of chip temperature. Mukherjee et al. [42] proposed a temperature-aware resource-binding technique to minimize leakage power consumption during behavioral synthesis. The proposed iterative resource-binding technique minimizes chip peak temperature by balancing the chip power profile, thereby reducing leakage power.

17.3.4   Thermal Optimization

Increasing performance requirements and system integration are dramatically increasing IC power density, and hence chip temperature. Thermal effects are becoming increasingly important during IC design. Mukherjee et al. [43] addressed thermal issues during behavioral synthesis. They proposed temperature-aware resource allocation and binding algorithms to minimize chip peak temperature. Gu et al. [5] designed TAPHS, a thermal-aware unified physical and behavioral synthesis system. TAPHS incorporates a complete set of integrated behavioral and physical thermal optimization techniques, including voltage assignment, voltage island generation, and thermal-aware floorplanning, to jointly optimize chip temperature, power, performance, and area. Thermal-aware behavioral synthesis algorithms must determine the temperature profiles of a tremendous number of candidate designs. Recently, researchers have developed and publicly released fast and accurate thermal analysis tools specifically for this purpose [44].

17.4   Low-Power System Synthesis

System synthesis has its roots in hardware–software cosynthesis. Early hardware–software cosynthesis algorithms took, as input, a high-level description of the application's required functionality, descriptions of available hardware, e.g., instruction processors and application-specific integrated circuits (ASICs), as well as performance and power requirements. The hardware–software cosynthesis algorithm automatically produced a design for the desired application, often consisting of application-specific and general-purpose processors mounted on a printed circuit board. The main focus of most hardware–software cosynthesis algorithms is partitioning applications between instruction processors and application-specific cores/ICs.

   SoC synthesis algorithms target hardware–software systems implemented on single ICs. Although their functionality overlaps with hardware–software cosynthesis algorithms, SoC synthesis algorithms also place great weight on synthesizing (heterogeneous) communication buses or networks. In addition, some consider the interaction between architectural and physical design to solve the entire SoC synthesis problem better.

   Figure 17.7 illustrates a system synthesis optimization flow. Although this flow is representative, some flows, e.g., those using constructive algorithms, may differ. Initially, a description of the algorithm to be implemented is provided in a high-level language such as MATLAB, C, or SystemC. This description is

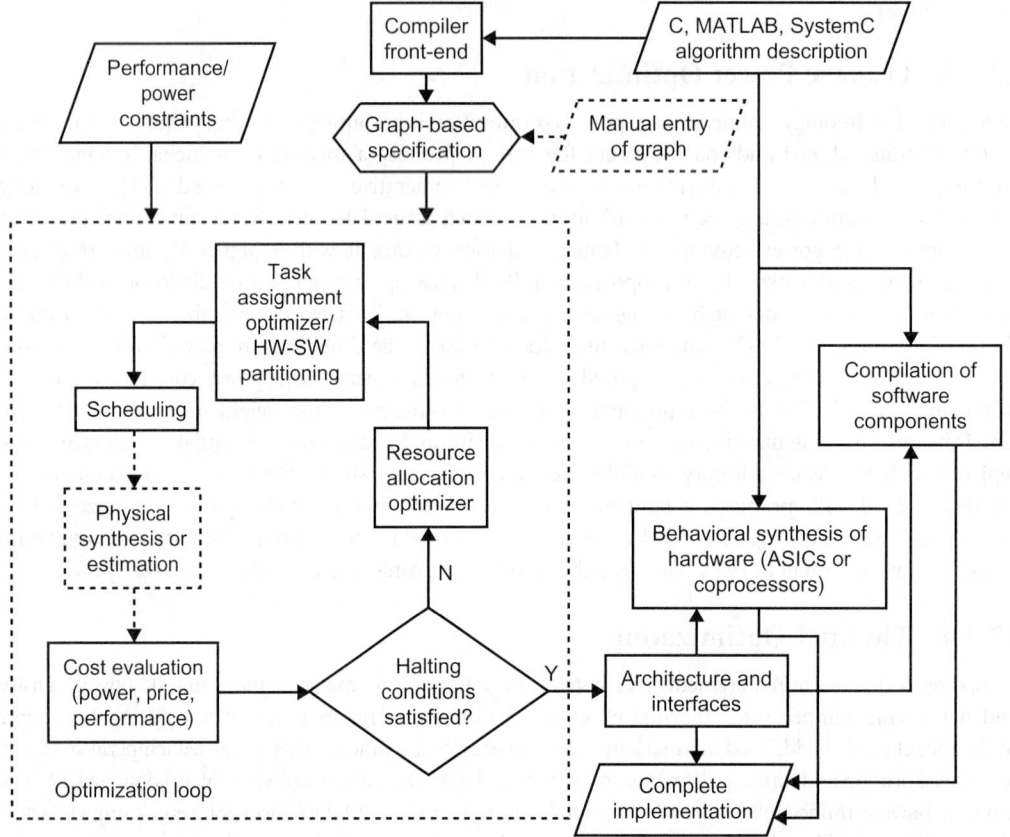

FIGURE 17.7   System synthesis algorithm.

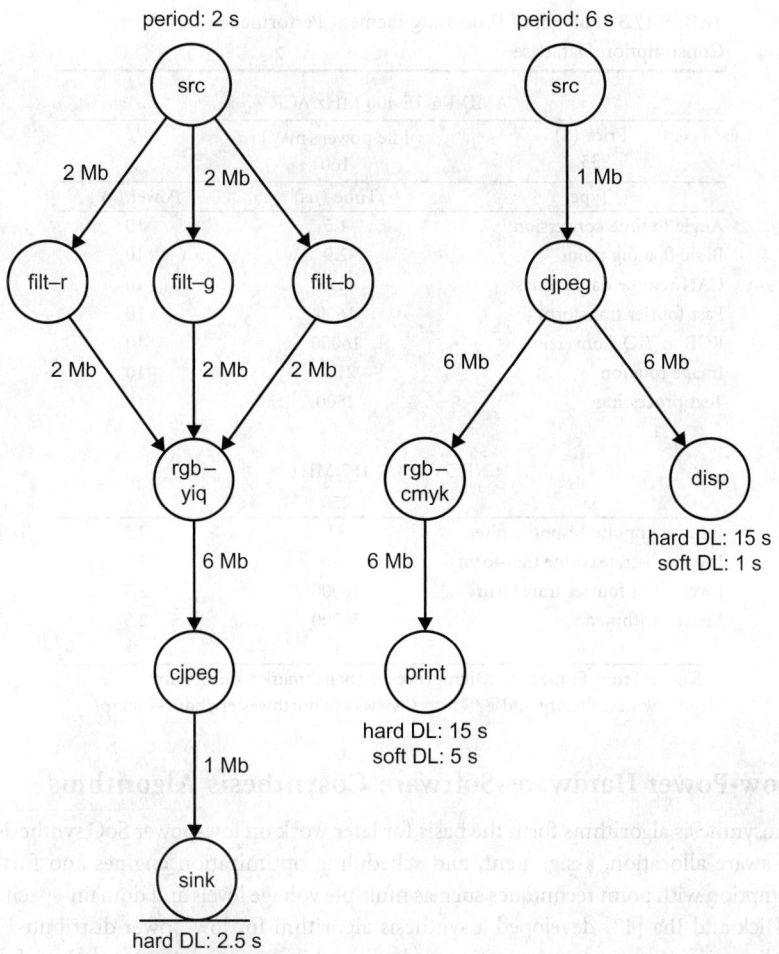

FIGURE 17.8 Example image processing application specification. (From E3S: The Embedded System Synthesis Benchmarks Suite. http://www.eecs.northwestern.edu.)

then translated into a graph representation by a compiler front-end. Note that these first stages may be omitted if a graph-based specification is available. One such graph format, shown in Figure 17.8, is a task set composed of multiple directed acyclic graphs in which nodes represent tasks and edges represent data dependencies. Timing constraints may be expressed as deadlines (DL) on nodes. Different tasks may be invoked periodically with different periods.

In addition to the required functionality, a database containing price, power consumption, execution time, and other characteristics of processing elements and communication resources is also provided. A portion of one such database is shown in Table 17.3.

Potential architectures consisting of processing element allocations, assignments of tasks to processing elements, and a schedule of all tasks and communication events are then optimized. Costs such as price, power, and execution time are then evaluated. The process repeats until acceptable solutions are produced. The resulting architectures are then completed by using behavioral synthesis to generate application-specific cores or FPGA configurations for the hardware-implemented tasks and using a compiler to generate executable code for the software-implemented tasks. Note that many existing system synthesis algorithms only solve subsets of the entire system synthesis problem.

TABLE 17.3 Portion of Processing Element Performance and Power
Consumption Database

AMD K6-2E 400 MHz/ACR		
Price ($)	Idle power (mW)	
33	160	
Type	Time (μs)	Power (W)
Angle to time conversion	1.5	10
Basic floating point	2.9	10
CAN remote data request	0.35	10
Fast fourier transform	1600	10
RGB to YIQ Conversion	16000	10
Image rotation	2100	10
Text processing	2800	10
⋮	⋮	⋮
NEC VR5432 167 MHz		
33	250	
Infinite impulse response filter	83	2.5
Inverse discrete cosine transform	840	2.5
Inverse fast fourier transform	16000	2.5
Matrix arithmetic	36000	2.5
⋮	⋮	⋮

Source: From Embedded Microprocessor Benchmark Consortium.
http://www.eembc.org and E3S http://www.eecs.northwestern.edu/~dickrp/.

17.4.1 Low-Power Hardware–Software Cosynthesis Algorithms

Low-power cosynthesis algorithms form the basis for later work on low-power SoC synthesis. They build upon power-aware allocation, assignment, and scheduling optimization engines and further improve power consumption with point techniques such as multiple voltage levels and domain-specific scheduling algorithms. Dick and Jha [47] developed a synthesis algorithm for low-power distributed systems that simultaneously optimizes power consumption and price while honoring hard real-time deadlines. Dave et al. [48] developed a constructive algorithm to solve the low-power multirate distributed system cosynthesis problem. Shang and Jha [49] presented a method of synthesizing low-power systems containing dynamically reconfigurable FPGAs.

Much of the early work in low-power hardware–software cosynthesis was based on the assumption that processing elements are off-the-shelf parts with strict constraints on operating voltages. Later work relaxed this assumption, considering multiple operating voltages and DVFS (described in Section 17.2). Gruian and Kuchcinski [50] developed a dual-voltage task-scheduling algorithm for reducing power consumption. Kirovski and Potkonjak [51] developed an integrated DVFS and system synthesis algorithm for independent tasks mapped to a bus-based multiprocessor. Schmitz and Al-Hashimi [52] developed a genetic algorithm to incorporate DVFS into an energy minimization technique for distributed embedded systems. It takes the power variations of tasks into account while performing DVFS. An offline voltage-scaling heuristic is proposed that is fast enough for use in system synthesis, starting from real-time periodic task graphs. Yan et al. [53] proposed a scheduling algorithm that uses DVFS and adaptive body biasing to optimize both dynamic and leakage power consumption. Analytical solutions are derived to determine the optimal supply voltage and bias voltage. Then, the optimal energy consumption is determined under real-time constraints.

DVS can also be applied to communication links. Naturally, performing simultaneous DVS in the processors and communication links in a distributed system can yield greater power savings than performing DVS in the processor alone. Luo et al. [54] presented such a method. In addition to honoring real-time constraints, their scheduling algorithm also efficiently distributes timing slack among tasks and multihop communication events.

Quality of service (QoS) is an important consideration in designing systems for real-time multimedia and wireless communication applications. Qu and Potkonjak [55] proposed a technique for partitioning a set of applications among multiple processors and determining a DVFS schedule to minimize energy consumption under constraints on QoS. The applications are assumed to be independent, have the same arrival times and no deadline constraints.

17.4.2 Low-Power System-on-Chip Synthesis Algorithms

The low-power SoC problem combines elements of hardware–software cosynthesis and behavioral synthesis. Like hardware–software cosynthesis, tasks may be implemented with general-purpose instruction processors or application-specific hardware accelerators. However, the synthesis algorithm potentially has greater control over the details of hardware implementation, opening new options for power optimization.

Methods of estimating SoC power consumption are essential to enable design exploration and synthesis. Bergamaschi et al. [56] developed an SoC analysis tool that estimates power and may be used within a system synthesis flow. Lajolo et al. [57] described a number of ASIC and instruction processor power estimation techniques that may be used in system synthesis. Based on these power estimation algorithms, synthesis algorithms may select and optimize SoC designs.

Power estimation techniques can be used to guide the search for high-quality solutions during the synthesis of low-power or low-temperature SoCs. Givargis et al. [58] developed a method of pruning the set of SoC candidate architectures to efficiently arrive at low-power designs. They determine which elements of the solution are independent from each other, thereby decomposing the problem into small, independent problems. Fei and Jha [59] describe a functional partitioning method for synthesizing low-power real-time distributed embedded systems whose constituent nodes are SoCs. The input specification, given as a set of task graphs, is partitioned and each portion is implemented as an SoC. Hung et al. [6] give a method of using voltage islands and thermal analysis within SoC synthesis to minimize peak temperature. Hong et al. [60] presented an algorithm to select a processor core and instruction/data cache configuration to best enable DVFS.

Communication networks have a large impact on the power consumption, performance, and feasibility of SoC designs. As a result, a number of researchers have worked on low-power, communication-centric SoC synthesis. Dick and Jha [61] developed a low-power SoC synthesis algorithm that optimizes power consumption, performance, and area. It uses floorplanning block placement to estimate communication delay, power consumption, and wire congestion. Lyonnard et al. [62] developed a low-power SoC synthesis algorithm that gives great attention to communication network synthesis. Instead of estimating physical characteristics via floorplanning, this work focuses on logical bus structure and communication protocol modeling. Hu et al. [63] optimize SoC bus bit-width under a fixed processing element allocation, task assignment, and schedule. Results for a seven-core H.263 encoder are presented. Thepayasuwan et al. [64] used simulated annealing to design bus topologies and demonstrated results for a JPEG SoC design. They did parasitic extraction for performance estimation and reduced power consumption by minimizing bus length. They proposed using the algorithm as a synthesis postprocessing step. Hu et al. [65] presented a method of using voltage islands in SoC designs that minimizes power consumption, area overhead, and number of voltage islands. Pasricha et al. [66] developed an algorithm for floorplan-aware synthesis of bus topologies that meet combinational delay constraints imposed by bus cycle times. This work assumes a fixed IP core allocation and task assignment. It minimizes bus count and bus width under explicit communication throughput constraints.

Conventional SoC designs typically contain a limited number of modules connected by on-chip buses or point-to-point links. However, as the number of on-chip modules grows in the coming years, bus or point-to-point link communication will face serious problems due to increasing global wire delay. To address these issues, in SoC designs buses are gradually being replaced by more sophisticated on-chip communication networks [67]. An on-chip network may consume a significant portion of an SoC power budget [68]. Therefore, power and power-related design problems, such as thermal problems [69], are of great concern in network-on-chip designs. The design and synthesis of on-chip networks supporting multihop routing has grown into an active and broad research area. Readers may refer to Marculescu's Chapter 16 in this handbook for a detailed treatment of this area.

17.5 Commercial Products

Although complete and general low-power system synthesis tools are not yet available, a number of supporting tools have been released. As described in Section 17.3, the performance and power consumption of functional units can be automatically determined via logic synthesis and analysis tools such as PrimeTime and PrimePower from Synopsys, Encounter from Cadence, Blast Power and Blast Fusion QT from Magma Design Automation, as well as Synplify from Synplicity.

Behavioral synthesis has reached a level of maturity at which a number of commercial products are available. Cynthesizer from Forte Design Systems synthesizes an RTL description from a SystemC algorithm. The Get2Chip synthesis tool, now owned by Cadence, translates Superlog to RTL. A number of synthesis tools target FPGAs. The DSP Synthesis tool from AccelChip starts from MATLAB, CoDeveloper from Impulse Accelerated Technologies starts from C, Mitrion's virtual processor starts from a C-like language, and BINACHIP's FREEDOM compiler starts from (digital signal processing) instruction processor executables.

17.6 Conclusions

As indicated in Section 17.5, behavioral synthesis is a commercially supported alternative to RTL design. A number of companies offer solutions to portions of the system synthesis problem. Both areas remain open with active research on new application domains, new synthesis algorithms, and new implementation technologies. Power and thermal optimization techniques in behavioral synthesis and system synthesis are necessary to improve performance, battery life, reliability, product size, and cooling costs. During the next 5 years, we can expect behavioral and system synthesis to continue to displace and supplement manual architectural design for high-complexity products that are produced in limited volumes, e.g., application-specific embedded systems. In addition, we can expect continued research on power and thermal aware synthesis and the industrial application of mature techniques.

References

1. M.R. Garey and D.S. Johnson, *Computers and Intractability: A Guide to the Theory of NP-Completeness*. W.H. Freeman, NY, 1979.
2. International Technology Roadmap for Semiconductors. http://public.itrs.net.
3. S.M. Martin, K. Flautner, T. Mudge, and D. Blaauw, "Combined dynamic voltage scaling and adaptive body biasing for lower power microprocessors under dynamic workloads," in *Proc. Int. Conf. Computer-Aided Design*, pp. 721–725, November 2002.
4. K.A. Bowman, B.L. Austin, J.C. Eble, X. Tang, and J.D. Meindl, "A physical alpha-power law MOSFET model," *J. Solid-State Circuits*, vol. 34, pp. 1410–1414, 1999.
5. Z.P. Gu, Y. Yang, C. Zhu, J. Wang, R.P. Dick, and L. Shang, "TAPHS: Thermal-aware unified physical-level and high-level synthesis," in *Proc. Asia & South Pacific Design Automation Conf.*, January 2006.
6. W. Hung, G. Link, Y. Xie, N. Vijaykrishnan, N. Dhanwada, and J. Conner, "Temperature-aware voltage islands architecting in system-on-chip design," in *Proc. Int. Conf. Computer Design*, October 2005.
7. N.K. Jha, "Low power system scheduling and synthesis," in *Proc. Int. Conf. Computer-Aided Design*, pp. 259–263, November 2001.
8. L. Goodby, A. Orailoglu, and P.M. Chau, "Microarchitecture synthesis of performance-constrained, low-power VLSI designs," in *Proc. Int. Conf. Computer Design*, October 1994.

9. A. Raghunathan and N.K. Jha, "An iterative improvement algorithm for low power data path synthesis," in *Proc. Int. Conf. Computer-Aided Design*, pp. 597–602, November 1995.
10. R.S. Martin and J.P. Knight, "Power profiler: Optimizing ASICs power consumption at the behavioral level," in *Proc. Design Automation Conf.*, June 1995.
11. A. Raghunathan, N.K. Jha, and S. Dey, *High-Level Power Analysis and Optimization*. Kluwer Academic Publishers, Boston, 1997.
12. J. Rabaey and M. Pedram, eds., *Low Power Design Methodologies*. Kluwer Academic Publishers, Boston, 1996.
13. A. Agarwal, C.H. Kim, S. Mukhopadhyay, and K. Roy, "Leakage in nano-scale technologies: Mechanisms, impact and design considerations," in *Proc. Design Automation Conf.*, pp. 6–11, June 2004.
14. M.W. Hall, J.M. Anderson, S.P. Amarasinghe, B.R. Murphy, S.-W. Liao, E. Bugnion, and M. S. Lam, "Maximizing multiprocessor performance with the SUIF compiler," *IEEE Trans. Computers*, December 1996.
15. P.P. Chang, S.A. Mahlke, W.Y. Chen, N.J. Water, and W. mei W. Hwu, "IMPACT: An architectural framework for multiple-instruction-issue processors," in *Proc. Int. Symp. Computer Architecture*, May 1991.
16. M. Johnson and R.K. Roy, "Optimal selection of supply voltages and level conversion during datapath scheduling under resource constraints," in *Proc. Int. Conf. Computer Design*, pp. 72–77, October 1996.
17. M.C. Johnson and K. Roy, "Datapath scheduling with multiple supply voltages and level converters," *ACM Trans. Design Automation Electronic Systems*, vol. 2, no. 3, pp. 227–248, 1997.
18. J. Chang and M. Pedram, "Energy minimization using multiple supply voltages," in *Proc. Int. Symp. Low Power Electronics & Design*, pp. 157–162, August 1996.
19. S. Raje and M. Sarrafzadeh, "Variable voltage scheduling," in *Proc. Int. Symp. Low Power Electronics & Design*, pp. 9–14, August 1995.
20. A. Chatterjee and R.K. Roy, "Synthesis of low power linear DSP circuits using activity metrics," in *Proc. Int. Conf. VLSI Design*, pp. 261–264, January 1994.
21. A.P. Chandrakasan, M. Potkonjak, R. Mehra, J. Rabaey, and R. Brodersen, "Optimizing power using transformations," *IEEE Trans. Computer-Aided Design of Integrated Circuits and Systems*, vol. 14, pp. 12–51, 1997.
22. N. Kumar, S. Katkoori, L. Rader, and R. Vemuri, "Profile-driven behavioral synthesis for low power VLSI systems," *IEEE Design & Test of Computers*, vol. 13, pp. 70–84, 1995.
23. J.M. Chang and M. Pedram, "Register allocation and binding for low power," in *Proc. Design Automation Conf.*, June 1995.
24. J. Chang and M. Pedram, "Module assignment for low power," in *Proc. European Design Automation Conf.*, pp. 376–381, September 1996.
25. A. Dasgupta and R. Karri, "Simultaneous scheduling and binding for power minimization during microarchitecture synthesis," in *Proc. Int. Symp. Low-Power Design*, April 1994.
26. R. Mehra, L. M. Guerra, and J. M. Rabaey, "Low Power Architecture Synthesis and Impact of Exploiting Locality," *J. {VLSI} Signal Processing 8*, vol. 13, pp. 877–888, August 1996.
27. M. Ercegovac, D. Kirovski, and M. Potkonjak, "Low-power behavioral synthesis optimization using multiple precision arithmetic," in *Proc. Design Automation Conf.*, pp. 568–573, June 1999.
28. E. Musoll and J. Cortadella, "High-level synthesis techniques for reducing the activity of functional units," in *Proc. Int. Symp. Low Power Electronics & Design*, pp. 99–104, August 1995.
29. A. Raghunathan and N.K. Jha, "SCALP: An iterative-improvement-based low-power data path synthesis system," *IEEE Trans. Computer-Aided Design of Integrated Circuits and Systems*, vol. 16, pp. 1260–1277, 1997.
30. K.S. Khouri, G. Lakshminarayana, and N.K. Jha, "High-level synthesis of low power control-flow intensive circuits," *IEEE Trans. Computer-Aided Design of Integrated Circuits and Systems*, vol. 18, pp. 1715–1729, 1999.

31. D.W. Knapp, "Fasolt: A program for feedback-driven data-path optimization," *IEEE Trans. Computer-Aided Design of Integrated Circuits and Systems,* vol. 11, pp. 677–695, 1992.

32. J.P. Weng and A.C. Parker, "3D scheduling: High-level synthesis with floorplanning," in *Proc. Design Automation Conf.,* June 1992.

33. Y.M. Fang and D.F. Wong, "Simultaneous functional-unit binding and floorplanning," in *Proc. Int. Conf. Computer-Aided Design,* November 1994.

34. W.E. Dougherty and D.E. Thomas, "Unifying behavioral synthesis and physical design," in *Proc. Design Automation Conf.,* June 2000.

35. C.-G. Lyuh, T. Kim, and K.-W. Kim, "Coupling-aware high-level interconnect synthesis," *IEEE Trans. Computer-Aided Design of Integrated Circuits and Systems,* vol. 23, pp. 157–164, 2004.

36. P. Prabhakaran and P. Banerjee, "Parallel Algorithms for Simultaneous Scheduling, Binding and Floorplanning in High-Level Synthesis," *Proc. Int. Symp. Circuits & Systems,* pp. 372–376, May 1998.

37. L. Zhong and N.K. Jha, "Interconnect-aware low power high-level synthesis," *IEEE Trans. Computer-Aided Design of Integrated Circuits and Systems,* vol. 24, pp. 336–351, 2005.

38. Z.P. Gu, J. Wang, R.P. Dick, and H. Zhou, "Incremental exploration of the combined physical and behavioral design space," in *Proc. Design Automation Conf.,* pp. 208–213, June 2005.

39. K.S. Khouri and N.K. Jha, "Leakage power analysis and reduction during behavioral synthesis," *IEEE Trans. Computer-Aided Design of Integrated Circuits and Systems,* vol. 10, pp. 876–885, 2002.

40. C. Gopalakrishnan and S. Katkoori, "KnapBind: An area-efficient binding algorithm for low-leakage datapaths," in *Proc. Int. Conf. Computer Design,* pp. 430–435, October 2003.

41. X. Tang, H. Zhou, and P. Banerjee, "Leakage power optimization with dual-Vth library in high-level synthesis," in *Proc. Design Automation Conf.,* pp. 202–207, June 2005.

42. R. Mukherjee, S.O. Memik, and G. Memik, "Peak temperature control and leakage reduction during binding in high level synthesis," in *Proc. Int. Symp. Low Power Electronics & Design,* pp. 251–256, August 2005.

43. R. Mukherjee, S.O. Memik, and G. Memik, "Temperature-aware resource allocation and binding in high-level synthesis," in *Proc. Design Automation Conf.,* June 2005.

44. Yonghong Yang, Zhenyu (Peter) Gu, Changyun Zhu, Robert P. Dick, and Li Shang, "ISAC: Integrated Space and Time Adaptive Chip-Package Thermal Analysis," *IEEE Trans. on Computer-Aided Design of Intergrated Circuits and Systems.* To appear.

45. "Embedded microprocessor benchmark consortium." http://www.eembc.org.

46. "E3S: The embedded system synthesis benchmarks suite." E3S link at http://www.ece.northwestern.edu/~dickrp.

47. R.P. Dick and N.K. Jha, "MOGAC: A multiobjective genetic algorithm for hardware-software co-synthesis of distributed embedded systems," *IEEE Trans. Computer-Aided Design of Integrated Circuits and Systems,* vol. 17, pp. 920–935, 1998.

48. B.P. Dave, G. Lakshminarayana, and N.K. Jha, "COSYN: Hardware-software co-synthesis of heterogeneous distributed embedded systems," *IEEE Trans. VLSI Systems,* vol. 7, pp. 92–104, 1999.

49. L. Shang and N.K. Jha, "Hardware-software co-synthesis of low power real-time distributed embedded systems with dynamically reconfigurable FPGAs," in *Proc. Int. Conf. VLSI Design,* pp. 345–352, January 2002.

50. F. Gruian and K. Kuchcinski, "LEneS: Task scheduling for low-energy systems using variable supply voltage processors," in *Proc. Asia & South Pacific Design Automation Conf.,* pp. 449–455, January 2001.

51. D. Kirovski and M. Potkonjak, "System-level synthesis of low-power hard real-time systems," in *Proc. Design Automation Conf.,* pp. 697–702, June 1997.

52. M. Schmitz and B.M. Al-Hashimi, "Considering power variations of DVS processing elements for energy minimization in distributed systems," in *Proc. Int. Symp. System Synthesis,* November 2001.

53. L. Yan, J. Luo, and N.K. Jha, "Combined dynamic voltage scaling and adaptive body biasing for heterogeneous distributed real-time embedded systems," in *Proc. Int. Conf. Computer-Aided Design,* pp. 30–37, November 2003.

54. J. Luo, L.-S. Peh, and N.K. Jha, "Simultaneous dynamic voltage scaling of processors and communication links in real-time distributed embedded systems," in *Proc. Design, Automation & Test in Europe Conf.,* March 2003.

55. G. Qu and M. Potkonjak, "Energy minimization with quality of service," in *Proc. Int. Symp. Low Power Electronics & Design,* pp. 43–49, August 2000.

56. R.A. Begamaschi, Y. Shin, N. Dhanwada, S. Bhattacharya, W.E. Dougherty, I. Nair, J. Darringer, and S. Paliwal, "SEAS: A system for early analysis of SoCs," in *Proc. Int. Conf. Hardware/Software Codesign and System Synthesis,* pp. 150–155, October 2003.

57. M. Lajolo, A. Raghunathan, S. Dey, L. Lavagno, and A. Sangiovanni-Vincentelli, "Efficient power estimation techniques for HW/SW systems," in *Proc. Alessandro Volta Memorial Wkshp. Low-Power Design,* March 1999.

58. T. Givargis, F. Vahid, and J. Henkel, "System-level exploration for Pareto-optimal configurations in parameterized systems-on-a-chip," in *Proc. Int. Conf. Computer-Aided Design,* pp. 25–30, November 2001.

59. Y. Fei and N.K. Jha, "Functional partitioning for low power distributed systems of systems-on-a-chip," in *Proc. Int. Conf. VLSI Design,* January 2002.

60. I. Hong, D. Kirovski, G. Qu, M. Potkonjak, and M.B. Srivastava, "Power optimization of variable voltage core-based systems," *IEEE Trans. Computer-Aided Design of Integrated Circuits and Systems,* vol. 18, pp. 1702–1714, 1999.

61. R.P. Dick and N.K. Jha, "MOCSYN: Multiobjective core-based single-chip system synthesis," in *Proc. Design, Automation & Test in Europe Conf.,* pp. 263–270, March 1999.

62. D. Lyonnard, S. Yoo, A. Baghdadi, and A.A. Jerraya, "Automatic generation of application-specific architectures for hetereogeneous multiprocessor system-on-chip," in *Proc. Design Automation Conf.,* pp. 518–523, June 2001.

63. J. Hu, Y. Deng, and R. Marculescu, "System-level point-to-point communication synthesis using floorplanning information," in *Proc. Int. Conf. VLSI Design,* January 2002.

64. N. Thepayasuwan, V. Damle, and A. Doboli, "Bus architecture synthesis for hardware-software co-design of deep submicron systems on chip," in *Proc. Int. Conf. Computer Design,* January 2003.

65. J. Hu, Y. Shin, N. Dhanwada, and R. Marculescu, "Architecting voltage islands in core-based system-on-a-chip designs," in *Proc. Int. Symp. Low Power Electronics & Design,* pp. 180–185, August 2004.

66. S. Pasricha, N. Dutt, and E. Bozorgzadeh, "Floorplan-aware automated synthesis of bus-based communication architectures," in *Proc. Design Automation Conf.,* June 2005.

67. W.J. Dally and B. Towles, "Route packets, not wires: On-chip interconnection networks," in *Proc. Design Automation Conf.,* pp. 684–689, June 2001.

68. H.-S. Wang, X. Zhu, L.-S. Peh, and S. Malik, "Orion: A power-performance simulator for interconnection networks," in *Proc. Int. Symp. Microarchitecture,* pp. 294–305, November 2002.

69. L. Shang, L.-S. Peh, A. Kumar, and N.K. Jha, "Thermal modeling, characterization and management of on-chip networks," in *Proc. Int. Symp. Microarchitecture,* pp. 67–80, December 2004.

18

Dynamic Voltage Scaling for Low-Power Hard Real-Time Systems

Jihong Kim
Seoul National University

Flavius Gruian
Lund University

Dongkun Shin
Samsung Electronics Co., LTD.

CONTENTS

18.1 Introduction

Energy consumption has become a dominant design constraint for modern VLSI systems, especially for mobile-embedded systems that operate with a limited energy source such as batteries. For these systems, the battery lifetime is often a key product differentiating factor, making the reduction of energy consumption an important optimization goal. Even for nonportable VLSI systems such as high-performance microprocessors, the energy consumption is still an important design constraint, because large heat dissipations in high-end microprocessors often result in the device thermal degradation, system malfunction, or in some cases, nonrecoverable crash. To solve these problems effectively, power-aware design techniques are necessary over a wide range of hardware and software design abstractions, including circuit, logic, architecture, compiler, operating system, and application levels.

Dynamic voltage scaling (DVS) [1], which can be applied in both hardware and software desgin abstractions, is one of the most effective design techniques in minimizing the energy consumption of VLSI systems. Since the energy consumption E of complementary metal oxide semiconductors (CMOS) circuits has a quadratic dependency on the supply voltage, lowering the supply voltage reduces the energy consumption significantly. When a given application does not require the peak performance of a VLSI system, the clock speed (and its corresponding supply voltage) can be dynamically adjusted to the lowest level that still satisfies the performance requirement, saving the energy consumed without perceivable performance degradations. This is the key principle of a DVS technique.

For example, consider a task with a deadline of 25 ms, running on a processor with the 50 MHz clock speed and 5.0 V supply voltage. If the task requires 5×10^5 cycles for its execution, the processor executes the task in 10 ms and becomes idle for the remaining 15 ms. (We call this type of an idle interval the *slack* time.) However, if the clock speed and the supply voltage are lowered to 20 MHz and 2.0 V, it finishes the task at its deadline (=25 ms), resulting in 84% energy reduction.

Since lowering the supply voltage also decreases the maximum achievable clock speed [2], various DVS algorithms for real-time systems have the goal of reducing supply voltage dynamically to the lowest possible level while satisfying the tasks' timing constraints. For real-time systems where timing constraints must be strictly satisfied, a fundamental energy-delay tradeoff makes it more challenging to dynamically adjust the supply voltage so that the energy consumption is minimized while not violating the timing requirements. In this paper, we focus on DVS algorithms for hard real-time systems.

For hard real-time systems, there are two types of voltage-scheduling approaches depending on the voltage scaling granularity: intratask DVS (IntraDVS) and intertask DVS (InterDVS). The intratask DVS algorithms [3,4] adjust the voltage within an individual task boundary, while the intertask DVS algorithms determine the voltage on a task-by-task basis at each scheduling point. The main difference between the two approaches is whether the slack times are used for the current task or for the tasks that follow. InterDVS algorithms distribute the slack times from the current task for the following tasks, while IntraDVS algorithms use the slack times from the current task for the current task itself.

The effectiveness of a DVS algorithm largely depends on two steps: *slack identification* and *slack distribution*. In the slack distribution step, the goal is to identify idle intervals as much as possible in advance. Identified slack intervals make it possible to scale the supply voltage for energy minimization. Most of existing techniques take advantages of a priori knowledge on programs or task sets (e.g., program structure, task set specification) as well as dynamic workload variations during run time. The goal of the slack distribution step is to assign the most appropariate amount of the slack for the next code segment/task to be executed. Since the appropriate slack amount for the next segment/task is determined by many factors including future execution behaviors, optimally distributing the identified slack is a challenging problem. Existing DVS algorithms are mostly different in these two steps.

Our main purpose of this paper is to survey and present representative DVS techniques proposed for both IntraDVS and InterDVS in a unified fashion. We present taxonomies of IntraDVS algorithms and InterDVS algorithms, respectively. Within each category, we describe key techniques for the slack identification and slack distribution steps. Two-layered introduction of DVS algorithms would help readers understand the overview of DVS as well as important details.

The rest of the paper is organized as follows. Before IntraDVS and InterDVS algorithms are explained, we briefly review power-related background concepts and describe the characteristics of variable-voltage processors in Section 18.2. We present intra- and intertask voltage-scheduling algorithms, respectively, in Sections 18.3 and 18.4. We conclude with a summary in Section 18.5.

18.2 Background

18.2.1 Energy-Delay Relationship

Modern microprocessors are implemented using CMOS circuits. To examine the tradeoff between energy and performance in variable voltage processors, we first describe the physical characteristics of CMOS circuits, especially in terms of energy consumption and circuit delay.

The power P_{CMOS} dissipated on a CMOS logic can be decomposed into two types, static power P_{static} and dynamic power $P_{dynamic}$ [5]. In the ideal case, CMOS circuits do not dissipate static power, since in steady state there is no open path from source to ground. However, there are always leakage currents and short-circuit currents which lead to a static power consumption. In the past, the static power took only a tiny fraction of the total power consumption. However, the leakage power is expected to exceed dynamic power consumption in future technology as the minimum feature size is dropped below 65 nm [5]. In this paper, we focus on dynamic power dissipation which is still a major power consumer in current VLSI systems.

The dynamic power of CMOS circuits is dissipated when the output capacitance is charged or discharged, and is given by $P_{dynamic} = \alpha C_L V_{dd}^2 f_{clk}$, where α is the switching activity (the average number of high-to-low transitions per cycle), C_L the load capacitance, V_{dd} the supply voltage, and f_{clk} the clock frequency. Since we will be focusing on dynamic power, the total power dissipation can be approximated as $P_{CMOS} \approx P_{dynamic} = \alpha C_L V_{dd}^2 f_{clk}$, and the energy consumption during the time interval $[0,T]$ is given by $E = \int_0^T P(t)\,dt \propto V_{dd}^2 f_{clk} T = V_{dd}^2 N_{cycle}$, where $P(t)$ is the power dissipation at t and N_{cycle} the number of clock cycles during the interval $[0,T]$. These equations indicate that a significant energy saving can be achieved by reducing the supply voltage V_{dd}; a decrease in the supply voltage by a factor of 2 yields a decrease in the energy consumption by a factor of 4.

Unfortunately, the supply voltage cannot be reduced for free. The circuit delay t_{delay}, which determines the possible maximum clock frequency, is dependent on the supply voltage [2]: $f_{clk}^{-1} \propto t_{delay} \propto V_{dd}/(V_{dd} - V_{th})^\gamma$, where V_{th} is the threshold voltage and γ the velocity saturation index ($1.3 \leq \gamma \leq 2$). Assuming that V_{th} is sufficiently small, the relation between the clock frequency and the supply voltage is given by $f_{clk} \propto V_{dd}^{(\gamma-1)}$. Hence, the clock frequency should be scaled along with the supply voltage. As a consequence, reducing the supply voltage yields energy saving but leads to performance degradation. Real-time scheduling and energy minimization are therefore tightly coupled problems, and should be tackled in conjunction for better results.

18.2.2 Variable-Voltage Processor Models

The ideal model of a variable speed processor is able to run at a continuous range of clock frequencies and voltages. Moreover, since the goal is to consume as little energy as possible, for a given clock frequency there is a unique optimal supply voltage. This is the lowest voltage for which the circuit delay still permits the given clock frequency. The supply voltage and the energy consumption per cycle are therefore uniquely determined by the clock frequency. In the following, instead of using the absolute clock frequency as a basis to describe the processor clock and supply settings, the term processor speed will be used. The processor speed is the relative value of the clock frequency f compared to a reference clock frequency f_{ref}, which is usually also the maximum clock frequency $s_f = f/f_{ref}$. A processor running at half speed will thus have the clock frequency half the reference frequency, with all the resultant consequences in terms of supply voltage, power, and energy consumption.

Using the equations introduced in Section 18.2.1, the voltage and power dissipation at frequency f can be written in terms of their reference values $V_f = V_{ref} s_f^{1/(\gamma-1)}$ and $P_f = P_{ref} s_f^{1+2/(\gamma-1)}$, where the velocity saturation index γ is approximated by 2.0 in the classical MOSFET model. More accurate models [5] show that γ is closer to 1.3, yet this does not affect the power dissipation is a convex function of the processor speed. In fact, since tasks execute clock cycles, it makes more sense to talk about the energy consumption per clock cycle for a certain frequency, e_f, than to talk about the power dissipation P_f

$$e_f = P_f(1/f) = P_{ref} s_f^{2/(\gamma-1)} s_f(1/f) = P_{ref} s_f^{2/(\gamma-1)}(1/f_{ref}) = e_{ref} s_f^{2/(\gamma-1)}$$

For $\gamma = 2$, e_f depends quadratically on the processor speed s_f. This is the commonly used model in voltage-scheduling research. Finally, the energy of a task that executes a certain number cycles N_f at each frequency $f \in F$ can be computed as $E_{ideal} = \sum_{f \in F} N_f e_f = e_{ref} \sum_{f \in F} N_f s_f^{1+2/(\gamma-1)}$. In the equation, the energy is denoted by E_{ideal} since it does not consider the effects of speed switching on energy. The ideal model of the variable speed processor can switch between clock frequencies and supply voltages without any time or energy overhead. A more realistic model of a variable speed processor has to address two real problems: first, the range of available processor speeds is limited and discrete. Second, there are speed transition overheads both in time and energy. We will now look at these problems in more detail.

In practice, the range of available processor speeds can only be discrete. This comes from the fact that the core clock frequency is generated internally by a phase-locked loop (PLL) or delay loop logic (DLL) using an external, fixed frequency clock. The internally generated frequency is a multiple of the external frequency. The supply voltage follows then the steps imposed by the available clock frequency steps. However, even on a discrete range of speeds, one can simulate a continuous range of speeds. The virtual clock frequency can be obtained by running different parts of a given task at different real clock frequencies. To simulate a desired frequency f_v, it is enough to use two real frequencies, one higher frequency $f_H > f_v$ and one lower frequency $f_L < f_v$. A task requiring N clock cycles will then run N_H clock cycles at f_H, and N_L ($= N - N_H$) clock cycles at f_L. To determine N_H and N_L, it is enough to check that the time covered by running N clock cycles at f_v is equal to the time covered by running N_H cycles at f_H plus N_L at f_L: $N/f_v = N_H/f_H + (N - N_H)/f_L$. Finally, if we take into account the fact that the number of clock cycles has to be an integer, we obtain the following solution: $N_H = \lceil N(f_v^{-1} - f_L^{-1})/(f_H^{-1} - f_L^{-1}) \rceil$ and $N_L = N - N_H$.

Note that the virtual frequency obtained using this splitting may be slightly higher than the desired virtual frequency. This difference is negligible for tasks using a sufficiently large number of clock cycles. If switching between the two frequencies takes a relatively important interval of time $t_{H \rightarrow L}$, one may take this into account: $N/f_v = N_H/f_H + (N - N_H)/f_L + t_{H \rightarrow L}$. From the viewpoint of the energy consumption, it is optimal if one chooses f_H and f_L to be the closest bounding frequencies for f_v [6,7]. Using this execution model, for a discrete range of speeds, the real energy function becomes then a piecewise-linear convex function [7]. Between any two adjacent real frequencies, the energy varies in a linear manner.

In modern processors, the clock signal accounts for a large part of energy consumption. To reduce jitter, noise, and energy consumption, the high speed core clock signals are today generated on-chip, using PLL or DLL. An external slow, and thus low energy clock signal is used by the on-chip PLL/DLL to generate the fast core clock. Changing the frequency of the PLL output signal has certain latency, since the loop has to adjust to the new frequency. This means that during the time in which the PLL relocks, the processor has to stall. So there is a certain time overhead when switching between speeds. The voltage supply design may also contribute to the speed switching overhead. This happens for the architectures where the processor must stall until the supply voltage stabilizes. Of course, if both the supply voltage and clock frequency change simultaneously, only the slowest of the two operations will affect the switch latency. However, many processors are designed such that they can keep executing instructions at constant rate while the voltage switches between two levels, with the working clock frequency determined by the lowest voltage. Moreover, depending on the number of speed switches relative to the performed tasks, the time overhead may be small enough to be considered negligible.

18.2.3 Variable Voltage Processor Examples

We briefly describe five examples of variable voltage processors. The first one is the result of an academic research project at UC Berkeley, and the rest are industry developments by Transmeta, AMD, and Intel. The Transmeta and AMD approaches include both hardware features and software managers for power efficiency. This makes them rather transparent to the software developer. The Intel and Berkeley solutions are focused on the hardware support, offering full control to the software developer.

UC Berkeley's lpARM [1].   The lpARM processor, developed at UC Berkeley, is a low-power, ARM core-based architecture, capable of run-time voltage and clock frequency changes. The prototype described in Ref. [1] (0.6 μ technology) is, reportedly, able to run at clock frequencies in the 5 to 80 MHz range, with 5 MHz increments. The supply voltage is adjustable in the 1.2–3.8 V range.

Transmeta Crusoe's LongRun [8].   Crusoe is a Transmeta processor family (TM5x00), with a VLIW (Very Long Instruction Word) core and x86 Code Morphing software that provides x86-compatibility. Besides four power management states, these processors support run-time voltage and clock frequency hopping. Frequency can change in steps of 33 MHz and the supply voltage in steps of 25 mV, within the hardware's operating range. The number of available speeds depends thus on the model. The TM5600 model, for example, operates in normal mode between 300–667 MHz and 1.2–1.6 V, meaning 11 different speed settings. The corresponding power consumption varies between 1.5 and 5.5 W. The speed is decided using feedback from the Code Morphing algorithm, which reports the utilization. The LongRun manager employs this feedback to compute and control the optimal clock speed and voltage. Note that this is a fine grain control, transparent to the programmer. The algorithms we present in this paper require direct control over the processor speed, and would substitute or augment LongRun. Nevertheless, the Crusoe architecture is a successful example of a variable voltage processor, widely used in low-power systems. A comparison with a conventional mobile x86 processor using Intel SpeedStep, running a software DVD player shows the TM5600 to consume almost one third of the power of the mobile x86 (6 W for TM5600 versus 17 W for the mobile x86).

AMD's PowerNow! [9].   AMD introduced PowerNow!, a technology for on-the-fly independent control of voltage and frequency. Their embedded processors from the AMD-K6-2E+ and AMD-K6-IIIE+ families are all implementing PowerNow!. AMD PowerNow! is able to support 32 different core voltage settings ranging from 0.925 to 2.00 V with voltage steps of 25 or 50 mV. Clock frequency can change in steps of 33 or 50 MHz, from an absolute low of 133 or 200 MHz, respectively. The voltage and frequency changes are controlled through a special block, the enhanced power management (EPM) block. At a speed change, an EPM timer ensures stable voltage and PLL frequency, operation which can take at most 200 μs. During this time, instruction processing stops. A comparison with a Pentium III 600+ using Intel SpeedStep shows that the AMD's processor with PowerNow! consumes around 50% less power than the Pentium with SpeedStep (3 W for AMD-K6-2E+ versus 7 W for Pentium III 600+).

Intel's SpeedStep [10].   Intel's SpeedStep is probably the earliest solution from the ones presented here, and consequently the weakest one. Besides normal operation, SpeedStep defines the following low-power states: Sleep, Deep Sleep, and Deeper Sleep. It only specifies two speeds, orthogonal with the power states, a Maximum Performance Mode (fast clock, high voltage, high power) and a Battery Optimized Mode (slower clock, lower voltage, power efficient). For instance, Mobile Intel Pentium 4-M Processor [10] uses 1.3 and 1.2 V for the two speeds, while the clock frequencies are 1.8 (or as low as 1.4 GHz depending on the model) and 1.2 GHz, respectively. The power consumption of the Mobile Pentium 4 is anywhere between 30 (Maximum Performance 1.8 GHz) and 2.9 W (in Deeper Sleep, 1 V). Switching between speeds requires going to Deep Sleep, change the voltage and frequency, and wake up again, procedure which requires at least 40 μs.

Intel's XScale [11].   Intel has recently come out with XScale, an ARM core-based architecture that supports on-the-fly clock frequency and supply voltage changes. The frequency can be changed directly, by writing values in a register, while the voltage has to be provided from and controlled via an off-chip source. The XScale core specification allows 16 different clock settings, and four different power modes (one ACTIVE and three other). The actual meaning of these settings are dependent on the application specific standard product (ASSP). For instance, the 80,200 processor supports clock frequencies up to 733 MHz, adjustable in steps of 33–66 MHz. The core voltage can vary between 0.95 and 1.55 V. Switching between speeds takes around 30 μs, and the power consumption for the 80,200 (core plus pin power) is anywhere between 1 W (at maximum speed) and a few μW (in sleep mode).

These examples show that variable speed processors become more and more common. They usually have a discrete range of voltages and clock frequencies, and exhibit latency when switching between speeds. Voltage-scheduling algorithms targeting energy efficiency have to take into account these characteristics of real processors. The scheduling algorithms presented in this paper make good use of the hardware capabilities of such processors, especially in hard real-time environments.

18.3   IntraTask Voltage Scaling for Hard Real-Time Systems

18.3.1   A Taxonomy of IntraDVS

The main feature of IntraDVS algorithm is how to select the program points where the voltage and clock will be scaled. Depending on the selection mechanism, we can classify IntraDVS algorithms into five categories as shown in Figure 18.1.

Segment-based IntraDVS techniques partition a task into several segments [12,13]. After executing a segment, they adjust the clock speed and supply voltage exploiting the slack times from the executed segments of a program. The segment-based IntraDVS approach was improved into *collaborative* IntraDVS techniques [14]. In the collaborative IntraDVS, OS and compiler work together to scale the supply voltage within a task. At the off-line stage, the compiler partitions an application into several segments and instruments the application with temporal hints. During run time, the operating system periodically changes the processor's clock frequency and voltage based on actual execution behaviors (combined with the inserted temporal hints).

Path-based IntraDVS techniques use the control flow information to find slack times [3,15,16]. They select all the program locations where the remaining workload is changed, and insert voltage scaling codes into the identified program locations at compile time. The inserted voltage scaling code modifies the supply voltage when executed, exploiting slack times coming from run-time variations of different execution paths.

In designing a path-based IntraDVS algorithm, two key issues exist. The first issue is how to predict the remaining execution cycles. Depending on the prediction method, several kinds of IntraDVS algorithms have been proposed, using the remaining worst-case execution path (*RWEP-IntraDVS*) [3], using the remaining average-case execution path (*RAEP-IntraDVS*) [15], and using the remaining optimal case execution path (*ROEP-IntraDVS*) [16]. The RAEP-IntraDVS and the ROEP-IntraDVS generate more energy-efficient schedules over the RWEP-IntraDVS, because they exploit the profile information of a task execution. While the remaining execution cycles are predicted based on the most frequent execution path in the RAEP-IntraDVS [15], the ROEP-IntraDVS uses the optimal predicted execution cycles based on the profile information [16].

The second issue of path-based IntraDVS is how to determine the voltage scaling points in the program code. The optimal points are the earliest points where we can detect the run-time slack. While the original

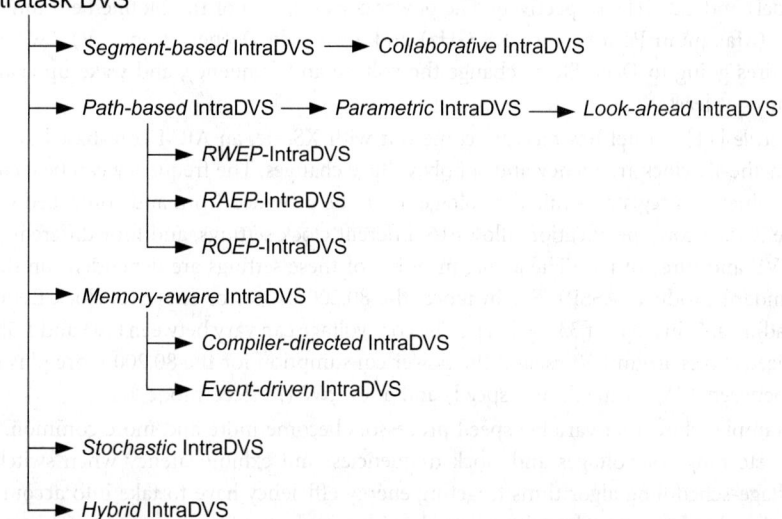

FIGURE 18.1   A taxonomy of dynamic voltage-scheduling techniques for real-time systems.

path-based IntraDVS techniques find the voltage scaling points using the control flow information of a program, the *parametric* IntraDVS and *look-ahead* IntraDVS techniques identify the voltage scaling points using the data flow information of the program as well as the control flow information to find earlier voltage scaling points [17,18].

Memory-aware IntraDVS utilizes the CPU idle times due to external memory stalls. While the *compiler-driven* IntraDVS [19] identifies the program regions where the CPU is mostly idle due to memory stalls at compile time, the *event-driven* IntraDVS [20,21] uses several performance-monitoring events to capture the CPU idle time at run time.

Stochastic IntraDVS uses the stochastic information on the program's execution time [4,22]. This technique is motivated by the idea that, from the energy consumption perspective, it is usually better to "start at low speed and accelerate execution later when needed" than to "start at high speed and reduce the speed later when the slack time is found" in the program execution. It finds a speed schedule that minimizes the expected energy consumption while still meeting the deadline. A task starts executing at a low speed and then gradually accelerates its speed to meet the deadline. Since an execution of a task might not follow the worst-case execution path (WCEP), it can happen that high-speed regions are avoided.

The main limitation of these IntraDVS techniques is that they have no global view of the task set in multitask environments. On the basis of an observation that a cooperation between IntraDVS and InterDVS could result in more energy-efficient systems, the *hybrid IntraDVS* technique selects either the *intra mode* or the *inter mode* when slack times are available during the execution of a task [23]. At the inter mode, the slack time identified during the execution of a task is transferred to the following other tasks. Therefore, the speed of the current task is not changed by the slack time produced by itself. At the intra mode, the slack time is used for the current task, reducing its own execution speed.

18.3.2 Segment-Based IntraDVS

The segment-based IntraDVS technique partitions a task into several segments and considers them as sequentially executed tasks [12]. The voltage-scheduling algorithm of segment-based IntraDVS can be summarized as follows:

1. After a task is divided into N segments, the following parameters are computed by static analysis techniques [24] or directly measured:
   - C_{WC} and C_{WC}^i: the worst-case execution cycles (WCEC) of the whole task and the i-th segment, respectively
   - C_{RW}^i: the remaining WCEC of the ith segment, i.e., $\sum_{j=i}^N C_{\mathrm{WC}}^i$

2. Additionally, the following parameters are determined:
   - D: the relative deadline of the whole task
   - f_{ref}: the clock speed at which the workload C_{WC} can be completed before the deadline D, i.e., $f_{\mathrm{ref}} = C_{\mathrm{WC}}/D$
   - T_{WC}^i: the worst case execution time (WCET) of the ith segment under the clock frequency f_{ref}, i.e., $T_{\mathrm{WC}}^i = C_{\mathrm{WC}}^i/f_{\mathrm{ref}}$
   - T_{RW}^i: the remaining WCET of the ith segment under the clock frequency f_{ref}, i.e., $T_{\mathrm{RW}}^i = C_{\mathrm{RW}}^i/f_{\mathrm{ref}}$

3. Before the ith segment is executed, the slack time T_{slack} is determined as $D - T_{\mathrm{ACC}}^i - T_{\mathrm{RW}}^i$, where T_{ACC}^i is the accumulated execution time from the first segment to the $(i-1)$th segment. The target clock frequency f_{tar} is computed efficiently to exploit the slack time T_{slack}.

Figure 18.2 illustrates the parameters defined above, assuming that the execution of the second segment was just completed. Steps 1 and 2 are performed at compile time, while step 3 is performed at run time. Using this technique, each segment can utilize the slack times generated from the previously executed segments.

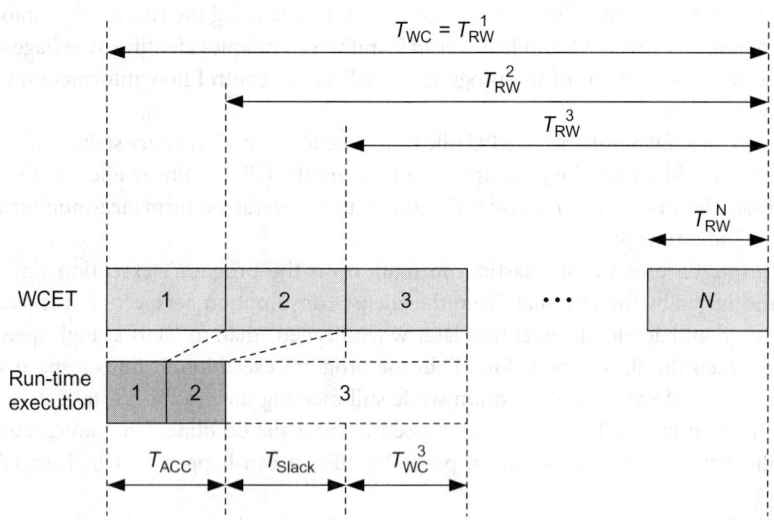

FIGURE 18.2 IntraDVS using a fixed-length segment.

In step 3, we should determine the target clock frequency f_{tar} using T_{slack}, that is, we should distribute the slack time available. Different slack distribution policies are possible and some of the representative approaches are as follows [13]:

- Give the identified slack to the immediately following segment. The target clock frequency f_{tar} is determined as follows:

$$f_{tar} = \frac{C^i_{WC}}{T_{slack} + T^i_{WC}} \qquad (18.1)$$

- Distribute the slack time evenly to all remaining segments:

$$f_{tar} = \frac{C^i_{RW}}{T_{slack} + T^i_{RW}} \qquad (18.2)$$

- Combine the identified slack with the estimated future slacks (which may come from the following segments' early completions):

$$f_{tar} = \frac{C^i_{WC}}{T_{slack} + T^i_{WC} + \sum^N_{j=i+1}(T^j_{WC} - T^j_{AC})} \qquad (18.3)$$

where T^j_{AC} is the average case execution time of the jth segment under the clock frequency f_{ref}.

Although the last approach usually works better than two other approaches, it should be used only when the processor provides the clock frequency under which the remaining worst-case workload can be completed before the deadline. Therefore, the following condition should be satisfied:

$$\frac{C^{i+1}_{RW}}{D - T^i_{ACC}} \leq f_{max} \qquad (18.4)$$

where f_{max} is the maximum clock speed provided by the variable voltage processor.

A key problem of the segment-based IntraDVS is how to divide an application into segments. Automatically partitioning an application code is not trivial. For example, consider the problem of determining the granularity of speed changes. Ideally, the more frequently the voltage is changed, the more efficiently the application can exploit dynamic slacks, saving more energy. However, there is energy and time overhead associated with each speed adjustment. Therefore, we should determine how far apart any two voltage scaling points should be. Since the distance between two consecutive voltage scaling points varies depending on the execution path taken at run time, it is difficult to determine the length of voltage scaling intervals statically.

One solution is to use both the compiler and the operating system to adapt performance and reduce energy consumption of the processor. The collaborative IntraDVS [14] uses such an approach and provides a systematic methodology to partition a program into segments considering branch, loop, and procedure call. The compiler does not insert voltage scaling codes between segments but annotates the application program with so-called power management hints (PMH) based on program structure and estimated worst-case performance. A PMH conveys path-specific run-time information about a program's progress to the operating system. It is very low cost instrumentation that collects path-specific information for the operating system about how the program is behaving relative to the worst-case performance. The operating system periodically invokes a power management point (PMP) to change the processor's performance based on the timing information from the PMH. This collaborative approach has the advantage that the lightweight hints can collect accurate timing information for the operating system without actually changing the performance. Further, the periodicity of performance/energy adaptation can be controlled independently of PMH to better balance the high overhead of adaptation.

We can also partition a program based on the workload type. For example, the required decoding time for each frame in an MPEG decoder can be separated into two parts [25]: a frame-dependent (FD) part and a frame-independent (FI) part. The FD part varies greatly according to the type of the incoming frame, whereas the FI part remains constant regardless of the frame type. The computational workload for an incoming frame's FD part (W_{FD}^P) can be predicted by using a frame-based history, i.e., maintaining a moving-average of the FD time for each frame type. The FI time is not predicted since it is constant for a given video sequence (W_{FI}). Because the total predicted workload is ($W_{FD}^P + W_{FI}$), given a deadline D, the program starts with the cock speed f_{FD} as follows:

$$f_{FD} = \frac{W_{FD}^P + W_{FI}}{D} \qquad (18.5)$$

For the MPEG decoder, since the FD part (such as the IDCT and motion compensation steps) is executed before the FI part (such as the frame dithering and frame display steps), the FD time prediction error is recovered inside that frame itself, i.e., during the FI part, so that the decoding time of each frame can be maintained satisfying the given frame rate. This is possible because the workload of the FI part is constant for a given video stream and easily obtained after decoding the first frame. When a misprediction occurs (which can be detected by comparing the predicted FD time (W_{FD}^P) with the actual FD time (W_{FD}^A)), an appropriate action must be taken during the FI part to compensate for the misprediction.

If the actual FD time was smaller than the predicted value, there will be an idle interval before the deadline. Hence, we can scale down the voltage level during the FI part's processing. On the other hand, if the actual FD time was larger than the predicted value, a corrective action must be taken to meet the deadline. This is accomplished by scaling up the voltage and frequency during the FI part so as to make up for the lost time. Since the elapsed time consumed during the FD part is W_{FD}^A / f_{FD}, the FI part should start with the following cock speed f_{FI}:

$$f_{FI} = \frac{W_{FI}}{D - (W_{FD}^A / f_{FD})} \qquad (18.6)$$

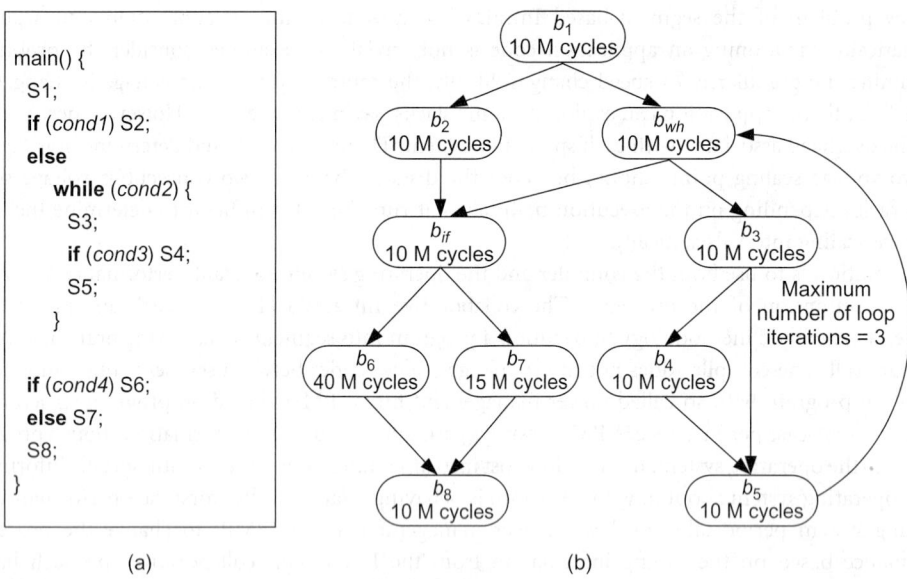

```
main() {
  S1;
  if (cond1) S2;
  else
    while (cond2) {
      S3;
      if (cond3) S4;
      S5;
    }
  if (cond4) S6;
  else S7;
  S8;
}
```

(a)   (b)

FIGURE 18.3 An example program *P*: (a) an example real-time program with the 2 s deadline and (b) its CFG representation G_P.

18.3.3 Path-Based IntraDVS

At a specific program point, two kinds of slack times can be identified: backward slack and forward slack. While the backward slack is generated from the early completion of the executed program segments, the forward slack is generated when the change of remaining workload is estimated. Though the segment-based IntraDVS utilizes the backward slack times, the path-based IntraDVS exploits the forward slack times based on the program's control flow.

Consider a hard real-time program *P* with the deadline of 2 s shown in Figure 18.3(a). The control flow graph (CFG) G_P of the program *P* is shown in Figure 18.3(b). In G_P, each node represents a basic block of *P* and each edge indicates the control dependency between basic blocks. The number within each node indicates the number of execution cycles of the corresponding basic block. The back edge from b_5 to b_{wh} models the **while** loop of the program *P*. The WCEC of this program is 2×10^8 cycles.

Path-based IntraDVS [3] adjusts the clock speed within the task depending on the control flow. For example, when the program control flow follows the execution path $\pi_1 = (b_1, b_2, b_{if}, b_6, b_8)$ of Figure 18.3(b), the clock speed is initially determined to complete the WCEP before the deadline, i.e., 100 MHz. However, we can reduce the clock speed at the edge (b_1, b_2) because we know this control flow does not follow the WCEP. In the path-based IntraDVS algorithm, we identify appropriate program locations where the clock speed should be adjusted, and inserts clock and voltage scaling codes to the selected program locations at compile time. The branching edges of the CFG, i.e., branch or loop statements, are the candidate locations for inserting voltage scaling codes because the remaining execution cycles are changed at those locations.

The path-based IntraDVS consists of two key steps: (1) one to predict the execution path of application program at compile time and (2) the other to adjust the clock speed depending on the real execution path taken at run time. In the first step, using the predicted execution path, we calculate the remaining predicted execution cycles (RPEC) $\delta(b_i)$ at a basic block b_i which is a branching node in G_P as follows:

$$\delta(b_i) = c(b_i) + \mathcal{P}(\delta(b_j), \delta(b_k)) \qquad (18.7)$$

where $c(b_i)$ is the execution cycles for the basic block b_i and $\mathcal{P}$ the prediction function. The basic blocks b_j and b_k are the immediate successor nodes of b_i in the CFG. Depending on the prediction method for execution path, the function $\mathcal{P}$ is determined. For example, if we take the WCEP as an execution path to be taken at run time, $\mathcal{P}(\alpha, \beta)$ will be equal to $\max(\alpha, \beta)$. With the predicted value of $\delta(b_i)$, we set the initial clock frequency and its corresponding voltage assuming that the task execution will follow the predicted execution path. We call the predicted execution path as the *reference path*, because the clock speed is determined based on the execution path.

For a loop, we use the following equation to predict the remaining execution cycles for the loop L:

$$\delta(L) = c(H_L) + (c(H_L) + c(B_L))N_{pred}(L) + \delta(post_L) \tag{18.8}$$

where $c(H_L)$ and $c(B_L)$ mean the execution cycles of the header and the body of the loop L, respectively.[*] $N_{pred}(L)$ is the predicted number of loop iterations and $post_L$ denotes the successor node of the loop, which is executed just after the loop termination.

At run time, if the actual execution deviates from the (predicted) reference path (say, by a branch instruction), the clock speed can be adjusted depending on the difference between the remaining execution cycles of the reference path and that of the newly deviated execution path. If the new execution path takes significantly longer to complete its execution than the reference execution path, the clock speed should be *raised* to meet the deadline constraint. In contrast, if the new execution path can finish its execution earlier than the reference execution path, the clock speed can be *lowered* to save energy.

For run-time clock speed adjustment, voltage scaling codes are inserted into the selected program locations at compile time. The branching edges of the CFG, i.e., branch or loop statements, are the candidate locations for inserting voltage scaling codes. They are called *voltage scaling points* (VSPs), because the clock speed and voltage are adjusted at these points. There are two types of VSPs, the B-type VSP and L-type VSP. The B-type VSP corresponds to a branch statement while the L-type VSP maps into a loop statement. VSPs can be also categorized into Up-VSPs or Down-VSPs, where the clock speed is raised or lowered, respectively. At each VSP (b_i, b_j), the clock speed is determined using $\delta(b_i)$ and $\delta(b_j)$ as follows:

$$f(b_j) = \frac{\delta(b_j)}{T_j} = f(b_i)\frac{\delta(b_j)}{\delta(b_i)-c(b_i)} = f(b_i)r(b_i,b_j) \tag{18.9}$$

where $f(b_i)$ and $f(b_j)$ are the clock speeds at the basic blocks b_i and b_j, respectively. T_j is the remaining time until the deadline from the basic block b_j and $r(b_i, b_j)$ the *speed update ratio* of the edge (b_i, b_j).

For a loop, if the actual number of loop iterations is N_{actual}, the clock speed is changed after the loop as follows:

$$f(post_L) = f(pre_L)\frac{\max((c(H_L)+c(B_L))(N_{actual}(L)-N_{pred}(L)),0) + \delta(post_L)}{\max((c(H_L)+c(B_L))(N_{pred}(L)-N_{actual}(L)),0) + \delta(post_L)} \tag{18.10}$$

where $f(pre_L)$ is the clock speed before executing the loop L. If N_{actual} is larger (smaller) than N_{pred}, the clock speed is increased (decreased).

18.3.3.1 IntraDVS Using Worst-Case Timing Information

Among the prediction policies for the remaining execution cycles, the simplest and most conservative one is to use the remaining worst-case execution cycles (RWEC) for RPEC. This technique is called as RWEP-IntraDVS [3]. For the prediction function $\mathcal{P}$, the following function is used to calculate the RWEC δ_w:

$$\delta_w(b_i) = c(b_i) + \max(\delta_w(b_j), \delta_w(b_k)) \tag{18.11}$$

---

[*] B_L can be composed of several basic blocks.

For example, in Figure 18.3(b), $\delta(b_{if}) = c(b_{if}) + \max(\delta(b_6),\delta(b_7)) = 6 \times 10^7$ cycles. The predicted total execution cycles calculated by this equation is same as WCEC and there are only Down-VSPs in a CFG. So the speed is dropped at all VSPs and there is no possibility of a deadline miss.

For the L-type VSP, we should use the maximum loop iteration number N_w as follows:

$$\delta_w(L) = c(H_L) + (c(H_L) + c(B_L)) N_w + \delta(\text{post}_L) \tag{18.12}$$

For the CFG in Figure 18.3(b), the clock speed is lowered at the B-type VSPs, (b_1, b_2), (b_3, b_5), (b_{if}, b_7) and the L-type VSP, (b_{wh}, b_{if}). At the VSP (b_{if}, b_7), the clock speed is reduced as follows:

$$f(b_7) = f(b_{if}) \frac{\delta(b_7)}{\delta(b_{if}) - c(b_{if})} = f(b_{if})0.5 \tag{18.13}$$

18.3.3.2 IntraDVS Using Profile Information

Although the RWEP-IntraDVS reduces the energy consumption significantly while guaranteeing the deadline, this is a pessimistic approach because it always predicts that the longest path will be executed. If we use the average-case execution path (ACEP) as a reference path, a more efficient voltage schedule can be generated. To find the ACEP, we should utilize the profile information on the program execution.

18.3.3.2.1 RAEP-IntraDVS

The ACEP is an execution path that will be most likely to be executed. The average-case remaining execution cycles δ_a are defined by the following equation when b_j and b_k are the immediate successor nodes of a branching node b_i:

$$\delta_a(b_i) = c(b_i) + \begin{cases} \delta_a(b_j) & \text{if } p_j \geq p_k \\ \delta_a(b_k) & \text{otherwise} \end{cases} \tag{18.14}$$

p_j and p_k are the probabilities of b_j and b_k being executed after b_i at run time, respectively.

For L-type VSPs, we use the average number N_{avg} of loop iterations:

$$\delta_a(L) = c(H_L) + c(H_L) + c(B_L))N_{avg}(L) + \delta_a(\text{post}_L) \tag{18.15}$$

It can be easily shown that using the ACEP instead of the WCEP is generally more energy-efficient. For a typical program, about 80% of the program execution occurs in only 20% of its code, which is called the hot paths [26]. To achieve high-energy efficiency, an IntraDVS algorithm should be optimized so that these hot paths are energy-efficient. If we use one of the hot paths as a reference path, the speed change graph for the hot paths will be a near flat curve with little changes in the clock speed, which gives the best energy efficiency under a given amount of work.

However, it is not always energy-efficient to use the ACEP as a reference path. This is because we consider only the probability of the execution path to decide the reference path. Especially, when the average-case execution cycle is significantly smaller than the WCEC, the energy consumption of RAEP-IntraDVS can be larger than that of RWEP-IntraDVS if the WCEP is actually executed at run time. If we modify the definition of the average-case remaining execution cycle as follows, we can consider both the probability and the execution cycles of the remaining execution path:

$$\delta_a(b_i) = c(b_i) + \begin{cases} \delta_a(b_j) & \text{if } \delta_a(b_j)p_j \geq \delta_a(b_k)p_k \\ \delta_a(b_k) & \text{otherwise} \end{cases} \tag{18.16}$$

18.3.3.2.2 *ROEP-IntraDVS*

Using the profile information, we can compute the optimal voltage schedule. Consider a simple CFG with only one branching node b_i. Basic blocks b_j and b_k are the immediate successor nodes of b_i. Following the execution of b_i, basic blocks b_j or b_k is executed with the probabilities p_j and p_k, respectively. Then, we can represent the average energy consumption as follows:[†]

$$\bar{E} \propto c(b_i)f(b_i)^2 + p_j\delta(b_j)f(b_j)^2 + p_k\delta(b_k)f(b_k)^2$$

$$= c(b_i)f(b_i)^2 + p_j\delta(b_j)\left(f(b_i)\frac{\delta(b_j)}{\delta(b_i) - c(b_i)} \right)^2 + p_k\delta(b_k)\left(f(b_i)\frac{\delta(b_k)}{\delta(b_i) - c(b_i)} \right)^2$$

$$\bar{E} \propto c(b_i) + p_j\frac{\delta(b_j)^3}{(\delta(b_i) - c(b_i))^2} + p_k\frac{\delta(b_k)^3}{(\delta(b_i) - c(b_i))^2} \tag{18.17}$$

The right-hand side of Eq. (18.17) is minimized when $\delta(b_i)$ has the following value [16]:

$$\delta(b_i) = c(b_i) + \sqrt[3]{\delta(b_j)^3 p_j + \delta(b_k)^3 p_k} \tag{18.18}$$

We call $\delta(b_i)$ of Eq. (18.18) the length of the ROEP starting from b_i. If we set the speed of each basic block b_i to $\delta(b_i)/T_i$, where T_i is the remaining time to the deadline from b_i, we can get the optimal voltage schedule. (The detailed proof is found in Ref. [16].) Since this technique uses an virtual optimal execution path as a reference path, it is called ROEP-IntraDVS. One special feature of ROEP-IntraDVS is that none of feasible execution paths is a reference path. Therefore, each branching edge becomes a voltage scaling point and it is necessary to select the final VSPs from the candidate VSPs considering the voltage scaling overhead.

18.3.3.3 IntraDVS Using Data Flow Analysis

The original path-based IntraDVS techniques select the voltage scaling points using the control flow information (i.e., branch and loop) of a target program. For example, in Figure 18.4(a), the path-based IntraDVS algorithm finds a B-type VSP where the remaining RWEC are reduced, because the basic block $b4$ is not executed. However, we can infer the direction of the branch earlier than when the basic block $b2$ is executed because the values of x and y are not changed after the basic block $b1$. Figure 18.4(b) shows the modified program which adjusts the clock speed and the supply voltage between $b1$ and $b2$. The program in Figure 18.4(b) consumes less energy than the one in Figure 18.4(a), because the clock speed is more uniform across multiple basic blocks if $x + y \leq 0$. This example shows that we can improve the energy performance of IntraDVS further if we can move voltage scaling points to the earlier part of the program.

18.3.3.3.1 *Parametric IntraDVS*

The parametric IntraDVS technique [17] scales voltage/frequency based upon the parameterization of the RWEC of a task. The parametric RWEC formula of the task can be determined by a static analysis of a program. For example, in Figure 18.5(a), the original IntraDVS technique uses the variable l to keep track of the number of loop iterations. At the L-type VSP, the clock speed is reduced if l is smaller than the maximum number M of loop iterations. However, if we analyze the loop using a parametric WCET analysis technique such as in Ref. [27], we can know that the basic block $b2$ is executed by $[c/k]$ times. Therefore, if $[c/k] < M$, the voltage can be lowered. If we know the values of c and k in advance, we can reduce the clock speed before the loop is executed. The parametric IntraDVS technique can insert the P-type VSP before the loop as shown in Figure 18.5(b).

[†]We assume that the clock frequency is proportional to the supply voltage.

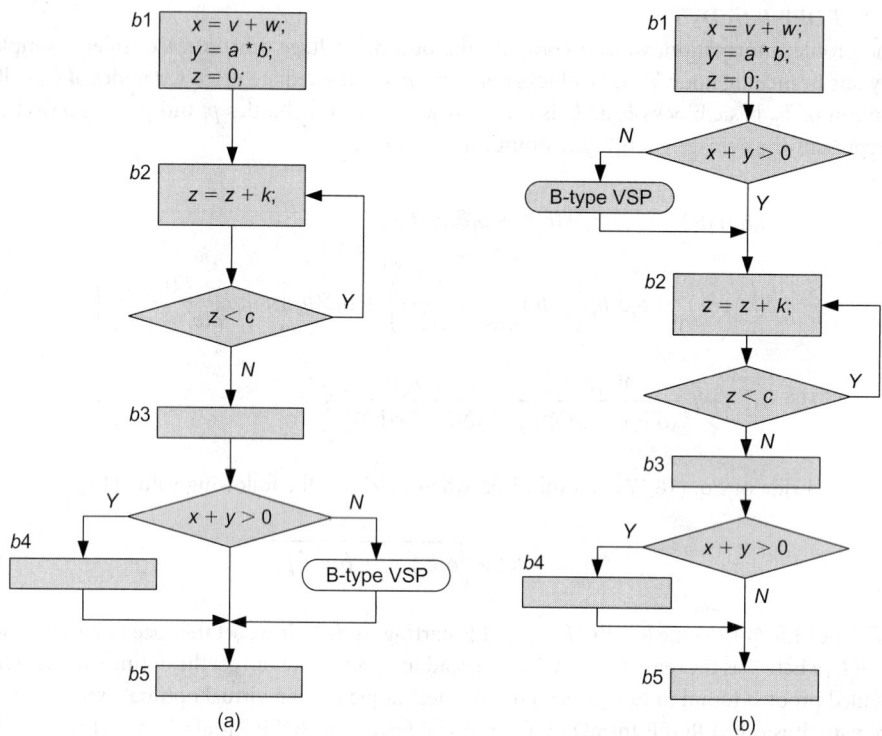

FIGURE 18.4 An example program for: (a) original IntraDVS; (b) look-ahead IntraDVS.

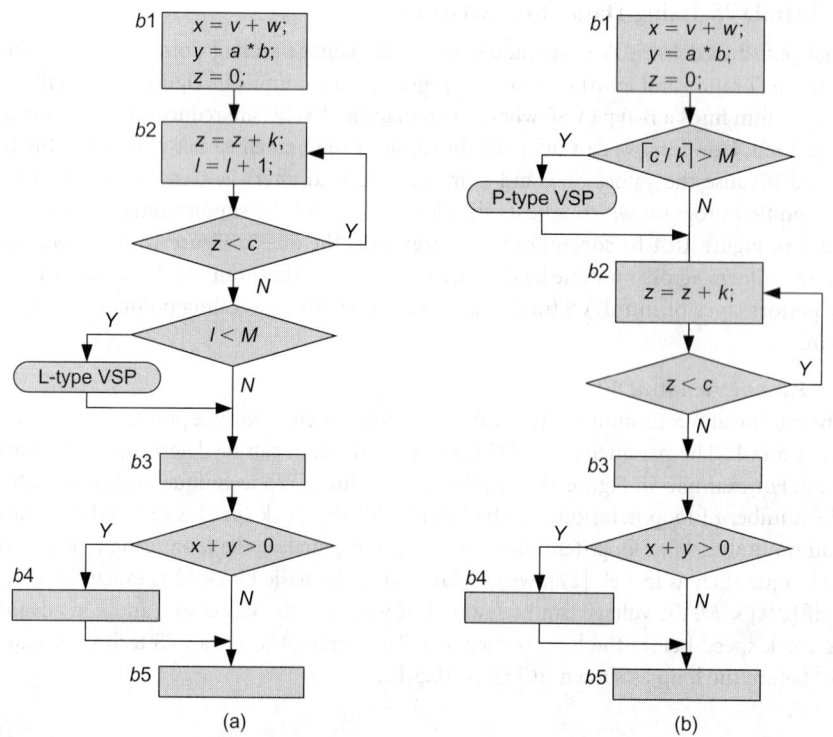

FIGURE 18.5 An example program for an L-type VSP: (a) original IntraDVS; (b) parametric IntraDVS.

18.3.3.3.2 Look-ahead IntraDVS

It is not trivial to derive the number of loop iterations from a program in the parametric IntraDVS. The *look-ahead IntraDVS* (LaIntraDVS) [18] technique uses the data flow analysis technique to identify earlier voltage scaling points. Using data flow analysis, we can decide program locations where each variable is defined and used. For LaIntraDVS, we need the following post-processing steps after the voltage scaling points are selected by the original IntraDVS algorithm:

1. Given an original voltage scaling point s, we identify the branch condition $C(s)$ which is the necessary condition for s to be executed at run time. Using the variables in the expression of $C(s)$, we compose a set of condition variables $V(s)$. For example, in Figure 18.4(a), the branch condition for the B-type voltage scaling point is $C(s) = (x + y > 0)$. The variables in $C(s)$ are x and y (i.e., $V(s) = \{x, y\}$).
2. The look-ahead point set $\mathbb{L}(s, v_i)$ is identified for each variable v_i in $V(s)$ using a data flow analysis technique. The look-ahead points are the earliest program points where we can get the value of v_i that will not be changed until s (i.e., $\mathbb{L}(s, x) = \{b1\}$ and $\mathbb{L}(s, y) = \{b1\}$).
3. Using the look-ahead point sets, the look-ahead voltage scaling points, after which the value of branch condition $C(s)$ does not change, are identified.
4. We insert the voltage scaling codes at the look-ahead voltage scaling points.

In Figure 18.4(a), the clock speed is adjusted at the B-type VSP as follows:

$$f_{\text{tar}} = f_{b3} \frac{\delta(b5)}{\delta(b4)} \tag{18.19}$$

where f_{b3} means the clock speed during the execution of the basic block $b3$. However, in Figure 18.4(b), the clock speed is adjusted at the B-type VSP as follows:

$$f_{\text{tar}} = f_{b1} \frac{\delta(b2) - c(b4)}{\delta(b2)} \tag{18.20}$$

If the condition variables are expressed with other variables in the identified look-ahead point, we can recursively examine the look-ahead point further using the multistep LaIntraDVS, which tries to find earlier look-ahead points using the data flow of the variables. In this case, we need to consider the overhead of the compensation code which should be inserted at the final look-ahead VSP to reflect the instructions to be executed at the intermediate look-ahead points. (The look-ahead IntraDVS technique can be considered as a more general version of Walsh's parametric IntraDVS technique [17], because the LaIntraDVS technique handles both the branch structure and the loop structure, while the parametric IntraDVS technique considered only the loop structure. In addition, LaIntraDVS uses the multistep approach.)

18.3.4 Memory-Aware IntraDVS

The memory-aware IntraDVS differs from the path-based IntraDVS in the type of CPU slacks being exploited. While the path-based IntraDVS takes advantage of the difference between the predicted execution path and the real execution path of applications, the memory-aware IntraDVS exploits slacks from the memory stalls. The idea is to identify the program regions in which the CPU is mostly idle due to memory stalls, and slow them down for energy reduction. If the system architecture supports the overlapped execution of the CPU and memory operations, such a CPU slow down will not result in a serious system performance degradation, hiding the slow CPU speed behind the memory hierarchy accesses which are on the critical path. There are two kinds of approaches to identify the memory-bound regions: analyzing a program at compile time and monitoring run-time hardware events.

18.3.4.1 Compiler-Directed IntraDVS

We can divide the total program execution time T into three portions:

$$T = W_c + W_m + W_b \tag{18.21}$$

where W_c, W_m, and W_b are defined as follows:

- W_c: the total amount of times when the CPU is busy while the memory is idle.
- W_m: the total amount of times when the memory is busy while the CPU is stalled (waiting for data from the memory).
- W_b: the total amount of times when both the CPU and memory are active at the same time.

When the CPU clock speed is reduced, W_m does not change, while W_c and W_b increase. Assume that the program in the reduced CPU clock speed behaves exactly the same for every program step as the program in the normal speed, but only executed in "slow motion".[‡] Then, if we scale down the clock speed by the factor of η, the execution time will be increased as follows:

$$T_{new}(\eta) = \eta W_c + \max(W_m + W_b, \eta W_b) \tag{18.22}$$

Defining that the performance penalty θ is $(T_{new}(\eta) - T)/T$, we can determine the CPU slowdown factor η with respect to the performance penalty θ as follows [19, 28]:

$$1 \leq \eta \leq 1 + \min(\theta / W_c, W_m / W_b) \tag{18.23}$$

The compiler-directed IntraDVS partitions a program into multiple program regions. It assigns different slowdown factors to different selected regions so as to maximize the overall energy savings without violating the global performance penalty constraint. The application program is partitioned not to introduce too much overhead due to switches between different voltages/frequencies. That is, the granularity of the region needs to be large enough to compensate for the overhead of voltage and frequency adjustments.

18.3.4.2 Event-Driven IntraDVS

It is very difficult to calculate the exact CPU idle time of a program in a static manner such as during the compilation time because on/off-chip latencies are severely affected by dynamic behavior, such as cache statistics and different access overheads for different external devices. So, these unpredictable dynamic behaviors should be captured at run time.

Event-driven IntraDVS [21] makes use of run-time information about the external memory access statistics to perform CPU voltage and frequency scaling with the goal of minimizing the energy consumption while controlling the performance penalty. The technique relies on dynamically constructed regression models that allow the CPU to calculate the expected workload and slack time for the next time slot. This is achieved by estimating and exploiting the ratio of the total off-chip access time to the total on-chip computation time. To capture the CPU idle time, several performance monitoring events (such as ones collected by the performance-monitoring unit [PMU] of the XScale processor) can be used. Using the performance-monitoring events, we can count the number of instructions being executed and the number of external memory accesses at run time.

[‡]This may not be the case in practice, for instance, due to out-of-order instruction executions.

18.3.5 Stochastic IntraDVS

Stochastic IntraDVS can be completely implemented inside the operating systems and does not require special compiler support. Consequently, it interferes less with the actual task than the compiler-based methods. This also means that tasks do not need to be re-compiled when the architecture changes. In principle, this approach computes a voltage schedule only once, when the task starts executing. During task execution, no re-scheduling is done, but the supply voltage is changed at well established intervals.

Stochastic IntraDVS is based on an observation that tasks with variable execution time usually finish before their WCET. Therefore, it makes sense to execute first at a low voltage and accelerate the execution, instead of executing at high voltage first and decelerate. In this manner, if a task instance does not take the WCET, high-voltage regions can be skipped altogether. This approach uses stochastic data to build a multiple voltage schedule, in which the processor speed increases towards the deadline. The purpose for using stochastic data is to minimize the average case energy consumption.

The stochastic voltage schedule for a task is obtained using the probability distribution function of the task execution time. This probability distribution function can be obtained off-line, via simulations, or built and improved at run time. Let us denote by X the random variable associated with the number of clock cycles used by a task instance. We will use the cumulative density function, cdf_x, of the random variable X (i.e., $cdf_x = P(X \leq x)$). This function reflects the probability that a task instance finishes before a certain number of clock cycles. If W is the worst-case number of clock cycles, $cdf_W = 1$. Deciding a voltage schedule for the task means that for every clock cycle up to W, we decide a specific voltage level (and the corresponding processor speed). Each cycle y, depending on the voltage level used, will consume a specific energy, e_y. But each of these cycles are executed with a certain probability, so the average energy consumed by cycle y can be computed as $(1 - cdf_y)e_y$. To obtain the average energy for the whole task, we have to consider all the cycles up to W:

$$\bar{E} = \sum_{0 < y \leq W} (1 - cdf_y)e_y \tag{18.24}$$

This is the value we want to minimize by choosing appropriate voltage levels for each cycle.

A task has to complete its execution within the relative deadline, D. If we denote the clock cycle time associated to the clock cycle y by λ_y, this constraint can be written as

$$\sum_{0 < y \leq W} \lambda_y \leq D \tag{18.25}$$

From these two equations, we can determine the clock cycle time as follows:

$$\lambda_y = D \frac{\sqrt[3]{1 - cdf_y}}{\sum_{0 < y \leq W} \sqrt[3]{1 - cdf_y}} \tag{18.26}$$

Though we can get the optimal schedule using the clock cycle time of each clock cycle, it is impractical to implement because variable voltage processors provide finite clock/voltage levels. Moreover, OS must issue a voltage scaling command each time it wants to change the speed even if the processor has continuous clock levels. Therefore, the virtual clock cycle times should be converted to available values by distributing the virtual clock cycle time between two existing clock frequencies. Using the speed dithering technique, the stochastic IntraDVS technique can be supported in real platforms.

18.4   InterTask Voltage Scaling for Hard Real-Time Systems

Where intratask dynamic voltage scheduling targets energy reduction by scheduling speeds inside a task, without employing knowledge about other tasks that might run in the system, intertask scheduling addresses speed selection at task group level. The advantage of using a group level speed scheduling comes from the possibility of distributing the workload more evenly over tasks and execution time. As a result, more time is spent at lower speeds, leading to lower energy consumption.

As detailed in the previous chapter, intratask DVS means, in principle, selecting the right processor speed for specific task sections. At that level, timing is strictly determined by the speed to section assignment, and therefore intratask DVS scheduling and speed selection at task level refer basically to the same notion. However, moving up to groups of tasks, intertask DVS refers not only to assigning speeds to tasks, but also to deciding a start time for each task. Alternatively, one can view this as assigning certain time intervals for each task, within which these have to start and complete their execution in an optimal manner from the energy point of view. Thus, in general the intertask DVS problem is at least as hard as the classic scheduling problem. Occasionally, research in this area addresses intertask DVS in two steps, first deciding on the order of task execution (*scheduling*) followed by assigning speeds to individual tasks (*speed selection*). However, we will use *dynamic voltage scheduling* or *speed scheduling* to refer to this process as a whole.

Typically, at task group level one is not interested in how exactly each individual task makes use of its assigned time interval. In general, it is assumed that tasks can optimally select their execution speed inside the assigned interval. In this sense, intratask and intertask DVS are orthogonal and can and should be used to complement each other. Nevertheless, it may not always pay off to use the best, and usually most complex, techniques at both individual task and task group level. For once, intertask strategies may compensate for poor choices of intratask decisions, making complex task-level approaches obsolete. Secondly, the energy optimization version of the *law of diminishing returns*§ says that less and less energy is gained for the same added effort. Consequently, the scheduling overhead might end up consuming more energy than it is actually gained.

Given the extensive research and results in the area of hard real-time scheduling, it is understandable that the majority of the intertask DVS strategies build on top of classic hard real-time scheduling approaches. In fact, because of the strong dependency between energy, processor speed, and timing, using DVS to reduce energy consumption while keeping hard deadlines is more of a challenge in these kind of real-time systems rather than in any other. Although DVS techniques designed for soft real-time, Quality of Service, and user perceived latency oriented systems do exist, this chapter focuses on hard real-time approaches, as these are still applicable for less constrained systems.

From the architectural point of view, multiprocessor systems appear to be at least as efficient in terms of energy as uniprocessor systems. DVS strategies especially designed for multiprocessor systems do exist [29–43]. However, given the restricted space, only techniques designed for uniprocessor systems are presented in the following. Nevertheless, these can be extended for multiprocessor architectures in the same way classic real-time algorithms can be derived for multiprocessor systems.

18.4.1   A Taxonomy of InterDVS

A classification of the existing intertask DVS hard real-time approaches is by no means trivial. Other than that, there are a large number of InterDVS algorithms proposed in recent years (perhaps, more than necessary in practice!), most InterDVS algorithms are a blend of various techniques and methods, which make it difficult to classify them. Very few of these belong exclusively to one class, and therefore we will

---

§Economic law stating that if one input used in the manufacture of a product is increased while all other inputs remain fixed, a point will eventually be reached at which the input yields progressively smaller increases in output (*Encyclopaedia Britannica* Online).

TABLE 18.1 An Inter-DVS Taxonomy

Occurence		Foundation	References
Static (off-line)	Fully	EDF	[7,46–52]
		Fixed priorities	[50,53–55]
	Mostly	EDF	[56–58]
		RM	[59–61]
		Others	[62,63]
Dynamic (run-time)		EDF	[64–67]
		RM	[59,61,65,66,68]

rather classify scheduling decisions than complete approaches. One can group intertask DVS scheduling methods according to:

- Their *occurrence*, or the moment they are employed. Approaches range from fully static, when all the decisions regarding scheduling and speed selection are taken off-line, before the system becomes functional to mostly dynamic, where run-time speed management and scheduling are employed.
- Their *complexity*, or the overhead required by the scheduling strategy. Since, generally inter-task DVS, just as classic scheduling, is a hard problem, optimal algorithms are expensive. Heuristics with lower overhead may be employed at the expense of efficiency.
- Their *foundation*, or classic scheduling algorithm they build upon. To guarantee hard real-time requirements, most approaches employ a classic scheduling algorithm, such as rate-monotonic (RM) [44] or earliest deadline first (EDF) [45], which is then extended with various off-line and run-time speed selection strategies.
- Their *flexibility*, or the ability to accommodate and employ run-time task execution variations. Some approaches use only the WCET of tasks to take scheduling decisions while others may employ profiling, statistic information, and even execution history, to adapt to run-time variations.

Notice that the criteria identified above are not necessarily independent, as for example, high overhead, but accurate methods are more likely to be employed off-line than at run time. Furthermore, depending on the foundation, intertask DVS methods may be more or less flexible, employing predominantly run time or off-line decisions. In this paper, we classify existing techniques mainly according to occurrence and foundation, as shown in Table 18.1.

18.4.2 Models and Assumptions for Task Groups

In this section, we introduce the models and assumptions commonly used in hard real-time inter-task DVS. In particular, of interest are those pertaining task groups, while in regard to the processor models, the same models and assumptions used for intra-DVS remain valid.

To capture tasks and their interactions (communications), *task graphs* are preferred, especially in multi-processor architectures, which require explicit modeling of communication channels. On single processors, shared memory is used for intertask communications, which lends itself to simpler models. Often, tasks are even considered to be independent. In this cases, modeling systems using *task sets* will often suffice. In fact, tasks sets are the preferred model in classic real-time scheduling. From a real-time perspective, the most important parameters of a task τ_i in a set $T = \{\tau_i\}_{i=1,\ldots,N}$ are:

1. *Its rate or period*, T_i. To simplify the problem, certain approaches assume that all tasks have the same period. Nevertheless, in general tasks are assumed to have different periods. For each period, a different task *instance* (job, ϕ) needs to execute. The point in time when a job becomes ready to execute is referred to as *arrival* or *release* time, A_i. A task set hyper-period H_i is the time interval after which the pattern of tasks arrival repeats itself. Some approaches focus on *aperiodic* tasks, where the rates are unknown.
2. *Its deadline*, D_i. In hard real-time systems keeping deadlines is vital for the safety of the system. Occasionally, tasks are assumed to have deadlines equals to periods $D_i = T_i$ or smaller than their periods $D_i < T_i$. The case when $D_i > T_i$ is rather uncommon, and hard to handle even in classic real time.

3. *Its execution pattern.* Differently from the classic real time, this parameter is not measured in time, but rather in clock cycles, as the same number of clock cycles takes different time depending on the processor speed. Apart from that, as in classic real-time scheduling, a task execution pattern can be:

(a) fixed, when it does not vary from instance to instance C_i, or

(b) variable from instance to instance (C_{ij} in instance j), between a best case BCE_i and a worst case WCE_i, usually modeled by a random variable X_i associated to a certain probability distribution η_i.

4. *Its switched capacitance, SW_i.* The switched capacitance is an important parameter affecting the power, and thus energy consumption. Often tasks are assumed to employ roughly the same processor resources, meaning that the switched capacitance remains constant from task to task and instance to instance. In this case, the switched capacitance can actually be omitted, as it does not have any effect on the processor-level scheduling decisions. Nevertheless, tasks and even instances can vary in the way they use the processor. A data-processing task may actually employ more resources (large switched capacitance) than a control task.

Additional parameters are occasionally used to model tasks in some intertask DVS approaches to model tasks, mainly to integrate the classic real-time scheduling and current intertask DVS scheduling more closely. However, these additional parameters do not bring anything new to the basic inter-task DVS principles. With the measures introduced above, a number of others taken from the classic real-time theory will prove useful in dealing with DVS scheduling. The most important in this is sense processor utilization.

Definition 18.1   The processor utilization for a give task set $\{\tau_i\}_{i=1,...,N}$ is computed as $U = \sum_{i=1,...,N} C_i / T_i$, where C_i is the time needed to execute C_i clock cycles at the maximum (reference) clock frequency.

For tasks with variable execution pattern, this measure usually refers to the *worst-case utilization*, for which C_i is replaced by WCE_i.

18.4.3   Static Scheduling

In static intertask DVS, the schedule and task speeds are fixed off-line, before the system becomes operational. Normally, all the parameters required to compute an exact schedule need to be available. This is in fact one drawback of employing static scheduling, since for tasks with variable execution time one has to assume the worst case for all instances. For highly dynamic systems, where task execution can vary a lot compared to the worst case, fully static schedules are not very effective. Additionally, they usually need to cover a full hyperperiod, which could include an extremely large number of instances. Consequently, the amount of data these methods need to provide to the run-time system may be extremely large, as the speeds and exact start times for each instance have to be stored for run-time use. In contrast, off-line techniques may be computationally intensive, as they are applied before the system becomes operational when the designer can afford more time to find a good schedule. Furthermore, the run-time overhead for static scheduling is minimal, as no computations need to take place, since task speeds and start times are already decided.

18.4.3.1   Overview of Techniques

One of the earliest publications on DVS [46] presents an optimal algorithm for static speed scheduling of independent task sets in an EDF-like manner. Refs. [53,54] present similar algorithms for sets of tasks with fixed priorities. Starting from the same initial work, Ref. [51] presents a heuristic that accounts for speed transition overheads and discrete voltage levels. Ref. [7] improves upon these, by presenting an optimal[¶] EDF DVS algorithm that accounts for discrete voltage levels and tasks with nonuniform load (switched capacitances).

A mixed-integer linear programming (MILP) approach for off-line EDF DVS with switching overheads on a dual-speed processor is presented in Ref. [49], along with a simpler but fast heuristic. The same

---

[¶]The optimality claim holds for sets of tasks that can be preempted at any time, even within a clock cycle (*inifinitesimal preemption*).

authors adopt a generalized network flow model in Ref. [52] to solve the problem optimally[||] for discrete voltage levels.

In Ref. [47], a nonpreemptive intertask DVS is presented, for sets of independent tasks with known arrival times, deadlines, and execution times. Their heuristic has a two-phase scheduling algorithm at its core. In the first phase a feasible schedule is found, assuming that all tasks run at the nominal voltage (or maximal speed). A priority function based on interval work demand is used to schedule the most constrained tasks first. The second phase attempts to iteratively lower each task speed until no energy reduction occurs. As the objective function is randomized with a certain offset, the best solution out of a series of schedules is selected. The algorithm is claimed to have an $O(n^3)$ complexity, for sets of n tasks. A similar algorithm, including switching activity as a task parameter is presented in Ref. [48].

In Refs. [50,55], several off-line speed-scheduling techniques for minimising power and energy consumption are presented. All start from valid real-time schedules (RM, EDF) and extend them with speed assignment. Gradually the speed of each task is decreased to a minimum beyond which deadlines start to be violated. The algorithmic complexity of these methods depends on the number of iterations taken, thus on the number of available speeds.

18.4.3.2 A Classic Preemptive Static Approach

The off-line scheduling algorithm presented in Ref. [46] has inspired many of the recent approaches, and therefore it can be considered to be a classic static speed-selection technique. The method finds optimal speeds for a set of jobs scheduled with the preemptive earliest deadline first strategy, by using a processor utilization-like measure. With the notations starting from 18.2, a few necessary definitions adapted from Ref. [45] are given below.

> **Definition 18.2** Given a set of real-time jobs running on a variable speed processor, the processor demand of the job set on the interval $[t_1,t_2)$ is $h_{[t_1,t_2)} = \sum_{[A_k,D_k)\subseteq[t_1,t_2)} C_k$.

In other words, it is the amount of computation in clock cycles, required by the jobs that can execute after t_1 and have to finish before moment t_2.

> **Definition 18.3** Given a set of real-time jobs running on a variable speed processor, the loading factor on the interval $[t_1,t_2)$ is the fraction of the interval needed to execute its job, that is, $u_{[t_1,t_2)} = h_{[t_1,t_2)}/(t_2 - t_1)$ (interval intensity in Ref. [46]).

Note that for any interval, the loading factor is in fact the clock frequency needed to carry out the jobs associated with that interval. Furthermore, these are minimal values, since if a lower clock speed is used, the jobs cannot be completed inside that interval.

> **Definition 18.4** An interval I for which $u_I = \sup_{0 \le t_1 < t_2)} u[t_1,t_2)$, meaning that it maximizes $u[t_1,t_2)$ is called critical interval. The associated job set $\{\phi_j\}_{[A_j,D_j)\subseteq I}$ is called critical group.

Given the above definitions and observations, the speed scheduling algorithm from Ref. [46] iteratively assigns speeds to critical groups and eliminates them from the job set, while also extracting their associated interval from the scheduling interval. A pseudocode description of the algorithm is given in Box 18.1. Until all jobs are scheduled, the algorithm finds the critical interval and associated jobs, assigns them a unique speed based on processor load and extracts them from the unscheduled set of jobs.

The algorithmic complexity for this algorithm is $O(N^2)$, with N being the number of jobs. With more specialized data structures, this can be reduced to $O(N \log^2 N)$ according to Ref. [46]. Nevertheless, the number of jobs to consider for scheduling depends very much on the rates of a periodic task set. For the set of only five tasks from Example 18.2, the number of jobs in a hyperperiod is around 1,50,000, yielding a very long run time for the algorithm. This is in fact a drawback present in all the approaches that have to unroll a whole hyperperiod to carry out the scheduling.

[||]Again, given an infinitesimal task preemption capability.

BOX 18.1   Optimal Static DVS for preemptive EDF jobs

```
let J = {φᵢ}ᵢ₌₁,...,ₙ // the initial set of jobs
let I = [A,D) // the scheduling interval
repeat
        *find [a,b) for which u₍ₐ,ᵦ₎ = sup₍ₜ₁,ₜ₂₎⊆I u₍ₜ₁,ₜ₂₎ // critical interval
        let J₍ₐ,ᵦ₎ = {φi|[Aᵢ, Dᵢ) ⊆ [a,b)}ᵢ∈ⱼ // critical group
        forall φ ∈ J₍ₐ,ᵦ₎ do //same speed for all critical job
                let sφ = u₍ₐ,ᵦ₎
        let I = [A,D = D - (b - a)) //reduced scheduling interval
        let J = J - J₍ₐ,ᵦ₎ //discard scheduled jobs
        // update remaining jobs' parameters to reflect
        // the changes in the scheduling interval
        forall φₖ ∈ J do
                if Aₖ ≥ a then
                        if Aₖ > b then let Aₖ = Aₖ - (b - a)
                                else let Aₖ = a
        *similar adjustment for Dₖ
until J = φ //all jobs scheduled
```

Example 18.1

Obtaining a Static DVS Schedule

Consider the following set S of four jobs modeled by their arrival, deadline, and execution time, respectively:

$$S = \{\phi_1(0,10,1), \phi_2(1,7,1), \phi_3(2,5,2), \phi_4(4,6,1)\} \tag{18.27}$$

To find the critical interval and set of jobs, the following intervals are interesting. [0,10) containing all jobs, yielding a loading factor of $(1 + 1 + 2 + 1)/(10 - 0) = 0.5$; [1,7) containing jobs ϕ_2, ϕ_3, and ϕ_4, with a loading factor of $(1 + 2 + 1)/(7 - 1) = 0.66$; and [2,6), containing the overlapping jobs ϕ_3 and ϕ_4, with the largest loading factor $(2 + 1)/(6 - 2) = 0.75$. Consequently, the critical jobs are ϕ_3 and ϕ_4, which are the first to be scheduled at a speed of 0.75 of the maximal speed. Having dealt with the interval [2,6) and the corresponding jobs, they must be excluded from the current problem, to obtain the reduced set of jobs $\{\phi'_1(0,6,1), \phi'_2(1,3,1)\}$. Now the step of finding critical jobs for this reduced problem can be started. The interesting intervals this time are [0,6) for all jobs, with a loading factor of $(1 + 1)/(6 - 0) = 0.33$ and [1,3) for job ϕ_2, with a loading factor of $1/(3 - 1) = 0.5$. The critical job is in this case ϕ_2, and will be scheduled to execute at speed 0.5. Reducing the problem further, the only remaining job is $\phi''_1(0, 4, 1)$, yielding a loading factor of 0.25. Thus, the last job to be scheduled ϕ_1 will execute at speed 0.25. The final schedule, together with the power consumption at any time, is depicted in Figure 18.6. The energy consumption for this arrangement turns out to be 60% of the maximum speed energy.

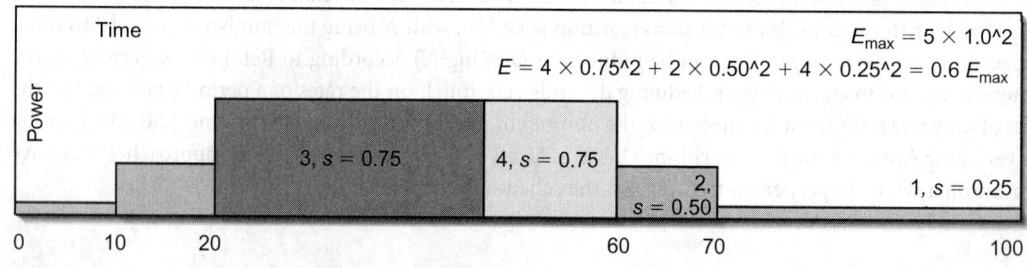

FIGURE 18.6   The static schedule for the four jobs in Example 18.1.

18.4.4 Off-Line Decisions for Run-Time Scheduling

Static scheduling is not flexible enough to accommodate or take advantage of run-time variations appearing from instance to instance. Run-Time speed management is necessary in this case. Nevertheless, all run-time techniques do employ some sort of off-line phase in order to benefit from the a priori knowledge of certain task parameters. Such off-line decisions include besides classic real-time scheduling choices (i.e. fixing task priorities), determining upper bounds on processor speeds for each task (maximum required speeds, MRS) or even forcing preemption points and altering task priorities, all in order to make the run-time speed management more efficient.

However, it is common that typical static approaches are used in an off-line phase even for run-time strategies, making the distinction between static scheduling and the off-line phase of run-time approaches less clear. In general, static DVS approaches tend to provide a complete solution to the scheduling problem, while off-line decisions are intended to complement and help run-time DVS scheduling, without loading the system with possibly unnecessary data. The large amount of data regarding job speeds and start times that most static DVS techniques provide is becoming obsolete, as run-time speed management takes over.

18.4.4.1 Maximum Required Speeds for RM Scheduling

The maximum required speed (MRS) method presented in this section is an example of using static task data to improve the run-time schedule via off-line decisions. The main idea of this approach rests on reaching a close to 100% processor utilization by identifying tasks that can run slower. More precisely, MRS is not a scheduling algorithm in itself, but a procedure to compute the smallest possible processing speeds for which a task set is still schedulable with a specific on-line algorithm. At run time, the task set is still scheduled according to the on-line algorithm, but the processor speeds are set using the off-line computed values. Note that the speed choice must not affect the feasibility of the schedule. For tasks with fixed execution time, the only modification of the non-DVS run-time algorithms resides in adding a speed switching sequence. The MRS technique can be applied to a number of classic real-time strategies, including fixed priority scheduling.

For RM scheduling, one may choose to use as a limit the utilization imposed by the condition proposed by Liu and Layland in Ref. [44]. With this approach the MRS is unique for all tasks and equal to

$$S_{RM-MRS} = \frac{U}{N(2^{1/N}-1)} \tag{18.28}$$

At a closer look, the schedule feasibility condition proposed in Ref. [44] is a sufficient one and covers the worst possible case for the task group characteristics. An exact analysis as proposed in Ref. [69] may reveal further possibilities for stretching tasks while still meeting the deadlines. On the basis of this, Ref. [59] describes a method to compute the maximum required frequency (speed) for a task set. We go even further, and instead of computing a single common maximum required speed for the whole task set $\{\tau_i\}_{i=1,...,N}$, as in Ref. [59], we compute individual speeds for each task τ_i. Note that the speed required by a task is inverse proportional to the task stretching factor. Finding the MRS is in fact equivalent to finding the minimal stretching factors $\{\alpha_i\}_{i=1,...,N}$. We focus on computing the α factors.

As introduced in Section 18.4.2, a task model includes its deadline D_i and period T_i. Since MRS is a static method, the third relevant task parameter in our case is the WCET. The task WCET, denoted by C_i in the following, refers to the time required by the worst case execution pattern (WCE) at the reference (and fastest) frequency f_{ref}: $C_i = WCE_i/f_{ref}$. Note that for a task with a unique execution pattern, where BCE = WCE = C, WCET can also be written as $C_i = C_i/f_{ref}$. Furthermore, we consider that the tasks in the group are indexed according to their priority, computed as in RM scheduling ($1/T_i$).

We compute the stretching factors in an iterative manner, starting from the highest priority tasks and continuing with lower priority tasks. Consider that index q points to the latest task which has been assigned a stretching factor. Initially, $q = 0$. Each of the tasks $\{\tau_i\}_{q < i \leq N}$ has to be executed before one of its scheduling points S_i as defined in Ref. [69]

$$S_i = \left\{ kT_j \mid 1 \leq j \leq i \wedge 1 \leq k \leq \left\lfloor \frac{T_i}{T_j} \right\rfloor \right\} \tag{18.29}$$

The above equation defines all the scheduling points when the deadlines are equal to task periods, $T_i = D_i$. For task sets where $T_i \neq D_i$, we need to change the set of scheduling points according to:

$$S_i' = \{t \mid t \in S_i \wedge t < D_i\} \cup \{D_i\} \tag{18.30}$$

Task τ_i exactly meets its deadline if there exists a scheduling point $S_{ij} \in S_i$ for which the following relation holds:

$$\sum_{1 \leq r \leq q} \alpha_r C_r \left\lceil \frac{S_{ij}}{T_r} \right\rceil + \alpha_{ij} \sum_{q < p \leq i} C_p \left\lceil \frac{S_{ij}}{T_p} \right\rceil = S_{ij} \tag{18.31}$$

Note that for the tasks which already have assigned a stretching factor we used that one, α_r, while for the rest of the tasks we assumed they will all use the same and yet to be computed stretching factor, α_{ij}, which is dependent on the scheduling point. For task τ_i, the best scheduling choice, from the energy point of view, is the largest of its α_{ij}. At the same time, from Eq. (18.5), this is equal for all tasks $\{\tau_i\}_{q < i \leq N}$. In fact, there is a task with index m for which its best stretching factor is the smallest among all other tasks:

$$\exists m, q < m \leq N, \quad \text{such that} \quad \max_{q < j \leq m} (\alpha_{mj}) = \min_{q < i \leq n} \left(\max_{q < j \leq i} (\alpha_{ij}) \right) \tag{18.32}$$

Note that this does not necessarily correspond to the last task, τ_N. If $q = 0$, this task sets the minimal clock frequency as computed in Ref. [59]. Having found the index m, all tasks between q and m can be at most stretched (equally) by the stretching factor of m. Thus, we assign the stretching factors as

$$\alpha_r = \max_{q < j \leq m} (\alpha_{mj}), \quad r = q + 1, \ldots, m \tag{18.33}$$

With this, an iteration of the algorithm for finding the stretching factors is complete. The next iteration then proceeds for $q = m$. The process ends when q reaches N, meaning all tasks have been given their own off-line stretching factors. Finally, the maximum required processor speed for each task is given by the inverse of its off-line stretching factor

$$s_i = 1/\alpha_i, \quad i = 1, \ldots, N \tag{18.34}$$

For a concrete example on how to compute the MRS for a set five tasks, refer Example 18.2.

18.4.4.2   Maximum Required Speeds for EDF Scheduling

EDF strategy, as a dynamic priority scheduling method, is more successful than RM, being able to utilize the processor up to 100%. Intuitively, allowing tasks to run slower until processor utilization reaches 100% should be the preferred method. Nevertheless, determining MRS for off-line tasks is only easy for particular cases, that can be successfully handled by classic real-time analysis. For a generic hybrid task set, Theorem 3.11 from Ref. [45] states a sufficient for EDF schedule feasibility:

$$\sum_{i = 1, \ldots, N} \frac{C_i}{\min\{D_i, T_i\}} \leq 1 \tag{18.35}$$

This result can be used to determine processor speeds for some particular situations:

(a) $D_i \geq T_i$. In this case, the above condition is also necessary, the left side becoming the classic expression for processor utilization. The recommended (and optimal) unique speed for all tasks is then $s_{EDF-MRS} = U$, or the worst-case processor utilization at the highest clock frequency. Since this technique is straightforward, it is commonly adopted in EDF with DVS scheduling [56,66].

(b) $D_i < T_i$. Unfortunately, in this case the feasibility analysis becomes intractable [70,71]. Consequently, finding optimal speeds—which is equivalent to examining families of task sets—is an even more difficult problem. Yet, upper bounds for maximum required speeds can be found using the left side of the sufficiency condition. Nevertheless, it is possible that this upper bound computes to a value >1, even if the task set can be scheduled under EDF. For these situations, other methods have to be employed for finding better speed bounds.

(c) $D_i \leq T_i$, $T_i = T$. For sets of tasks with the same period, the EDF schedule feasibility problem is solvable in polynomial time (sequencing with release times and deadlines, SS1, with preemption from Ref. [72]). For tasks with different arrival times (*asynchronous*), the method presented in Section 18.4.3.2 can be used. A similar but simpler strategy can be applied to sets of tasks with the same arrival times (*synchronous*) [62].

Example 18.2

RM–MRS and EDF–MRS for a Set of Five Tasks

This example contains the results of MRS for the set of five tasks described in Table 18.2. For this set, the task deadlines are equal to the task periods. For the RM-scheduling method, the stretching factors are computed individually. Note that tasks 3 and 4 can be stretched off-line more than 1 and 2, while 5 has the largest stretching factor. The processor utilization changes from 0.687 to 0.994. Observe also that the stretching factors for the lower priority tasks require more iterations to compute. For the EDF scheduling, there is a single stretching factor, common to all tasks, equal to 1/0.687. The maximum required processor speeds relative to the reference speed are obtained by inverting the α factors.

If we consider that the tasks have fixed execution pattern, we can easily compute the energy consumptions for the RM–MRS and EDF–MRS. For this we found out the number of executed instances of each task over the task set hyperperiod, computed as the least common multiplier (lcm) of the task periods. For our example, lcm(5,11,45,130,370) is 476190. Next, assuming that the energy consumed during a clock cycle is dependent on the square of the processor speed, the energy of a task instance is proportional to $C_i s_i^2$. Finally, we can sum up the energy consumption after the number of instances for each task. The numerical results are detailed in Table 18.3. Note that, for this example, we assumed that no power is consumed during idle and speed switching. Also, the processor is ideal in the sense that it can run at any speed under the reference speed. It is interesting to note that, for this case, the energy consumed using RM–MRS is very close to that using EFD–MRS. Both approaches manage to save about 52% of the energy consumed by using only the classic RM and EDF employing the same reference speed for all tasks. Moreover, the EDF–MRS energy is minimal, as all tasks execute at the same speed, fully using the processor.

TABLE 18.2 A Numerical Example of MRS

| | | | MRS α Factors (and Speeds) | | |
| | Task | | RM | | EDF |
No.	WCET (C)	Period (T)	α (Speeds)	Iterations	α (Speed)
1	1	5	1.428 (0.700)	1	—
2	5	11	1.428 (0.700)	1	—
3	1	45	1.785 (0.560)	2	1.4556 (0.687)
4	1	130	1.785 (0.560)	2	—
5	1	370	2.357 (0.424)	3	—

TABLE 18.3 RM–MRS and EFD–MRS Energy Consumption for the Task set in Table 18.2

Task No.	Instance Energy		Instances per Hyperperiod	All Instances Energy	
	RM–MRS	EDF–MRS		RM–MRS	EDF-MRS
1	0.4904	0.4720	95238	46704.0	44952.3
2	2.4520	2.3599	43290	106147.0	102160.1
3	0.3139	0.4720	10582	3321.6	4994.7
4	0.3139	0.4720	3663	1149.8	1728.9
5	0.1800	0.4720	1287	231.7	607.5
Total energy consumption				157554.1	154443.5
% from max speed energy (327220.0)				48.15%	47.20%

18.4.4.3 Extensions

The approaches mentioned before employed a rather simplistic view of the system, which is often sufficient. Nevertheless, more realistic models can be used along with more refined methods, as briefly reviewed below.

Task-Specific Power Characteristics. For tasks with different power characteristics (different switching capacitance), the techniques presented above are not optimal, as tasks consuming more power should be run slower than tasks that consume little power. In Ref. [57] an optimal approach is described similar to the RM–MRS (Section 18.4.4.1) for EDF-scheduled tasks with different power characteristics.

Scheduling with Synchronization. Task synchronization and blocking due to accessing common resources is yet another case for which using the same speed for all tasks is not optimal. Again, based on a real-time analysis similar to the one in Section 18.4.4.1, the authors of Refs. [58,60,73] describe a method for determining off-line maximum required speeds for tasks with synchronization, scheduled with RM and EDF.

Variable Execution Patterns. All the approaches described earlier, work either on tasks with fixed execution patterns or use the worst-case execution exclusively. Nevertheless, for tasks with variable execution pattern, knowing something about the stochastic behavior of the tasks can help to derive better off-line schedules. One approach that employs the expected execution time for this purpose is the uncertainty-based ordering (UBO) [62,63].

UBO is based on the observations that scheduling short tasks or tasks with large variations (uncertainties) in execution early, allows the run-time speed manager to select a low processing speed early. In contrast, allowing for uncertainties to persist, forces the run-time manager to keep selecting high speeds, in order to accommodate even the worst-case behavior.

In this context, Ref. [63] defines a new priority function for sets of independent tasks with variable execution pattern, unique period, and deadline. This priority function sets in fact an order among those tasks whose order does not affect the real-time behavior of the task set. Going further on this idea, Ref. [62] presents a UBO method for EDF-scheduled jobs with synchronous release times. Moreover, by examining the ready tasks between two successive deadlines, the approach splits certain tasks by introducing preemption points, if the execution order can be improved in this manner.

18.4.5 Run-Time Scheduling

Off-Line DVS techniques are not able to adapt to run-time variations in task execution times. To take advantage of idle times appearing from such variations, run-time scheduling techniques must be employed. Nevertheless, run-time DVS is most often used in conjunction with static techniques, since many of the system parameters are already known at design time. In this context, run-time methods often perform small speed adjustments on top of an already existing schedule or scheduling strategy. However, for sets of tasks with large run-time variations in execution pattern, such techniques are essential for further reducing the energy consumption.

In general, run-time DVS decisions may be taken at any time while the system is in operation. Nevertheless, the time points that are most likely to introduce variations to the off-line decisions taken for a certain task set, are those in which task instances finish their execution. These are the moments when, if tasks finish early, new idle times are introduced into the system. Consequently, in most run-time DVS techniques, scheduling points, namely task arrivals and task completion times, become thus also the points where speed-related decisions are taken. However, not all scheduling points need to include a speed decision. Furthermore, the moments used to take speed-related decisions and scheduling points may be totally decoupled, having the scheduler only giving suggestions (hints) to a speed manager that changes processor speed at its own pace, as in the intratask approach from Ref. [74]. Regardless, the efficiency of the run-time DVS is highly dependent on basically two mechanisms. First, the idle time detection/estimation is essential to determining the available slack time, used to slow down tasks. Second, the method of distributing this slack to instances determines the time spent at low speeds, thus the efficiency of the DVS strategy. In the following, the two mechanisms will be identified as *slack estimation* and *slack distribution.*

Normally, for systems where variations occur frequently, DVS techniques should be employed often. In this case, the overhead introduced by the scheduling decisions should be small enough to be worthwhile. In contrast, more expensive DVS techniques may be used infrequently, if run-time variations are also infrequent. As the overhead of run-time DVS methods must be rather low, such run-time decisions must be simple, taken usually through fast heuristics. Additionally, they only affect a small number of task instances, often just those waiting to execute. Finally, for hard real-time systems, such decisions should not negatively affect the real-time properties of the system exposed by the off-line analyses. All deadlines must still be met.

18.4.5.1 Overview of Techniques

One of the simplest run-time DVS mechanisms can be found in the *lppsRM/EDF* strategies, introduced in Refs. [59,65]. Basically, DVS is in effect only when there is a single instance ready to run on the processor. In those cases, the instance may occupy the processor until the nearest arrival of another task instance. The execution speed for the instance ready to run is then chosen in such a manner that the execution will complete right before the next task arrival (NTA). For this reason, this method will be referred to as *stretch-to-NTA*. Note that the only time the processor runs at a lower speed is for isolated instances now and then, employing a greedy slack distribution mechanism. Furthermore, although slack might be produced early during the schedule, it will only be used by the last instance in the ready queue. This leads to a rather inefficient use of the slack, as most instances run at the maximal speed, while occasionally an instance will run at a very low speed.

A rather similar, but more efficient approach, is the cycle-conserving RM (*ccRM*) from Ref. [66]. Again, the slack produced by tasks finishing early is used to lower the speed of the instances ready to execute, which can be more than one this time. This method employs a better slack distribution strategy, as more than one isolated instance can use lower speeds. Along with *ccRM*, in Ref. [66] the same authors describe two EDF-based techniques. Cycle-conserving EDF (*ccEDF*) distributes newly produced slack more evenly over a larger number of instances than *ccRM*. In particular, in each scheduling point the processor speed is adjusted according to the current processor utilization, as in situation (a) in Section 18.4.4.2. The current utilization is computed as a sum of the actual utilization for tasks that completed their instances in the current period and the worst-case utilization for the tasks waiting to execute. Techniques that recompute the processor utilization in order to determine a new speed are often referred to as *utilization updating* methods. An extension of the *ccEDF*, the more aggressive look-ahead EDF (*laEDF*), is also presented in Ref. [66]. The processor speed is further reduced by deferring as much work as possible beyond the current deadline, while still making sure that the deadlines can be met even in the worst case.

The two approaches presented in Ref. [64], designed on top of EDF policy, handle the available slack in a more explicit manner. The mechanism used by the dynamic reclaiming algorithm (DRA) distributes the newly produced slack in a greedy manner to the next highest priority instance from the ready queue. Assigning slack according to task priorities is essential, as the real-time behavior of the system is conserved since lower priority task do not interfere with higher priority tasks. The second approach presented in

Ref. [64] is a more aggressive speed reduction (AGR) technique, that speculatively assumes that future instances will probably exhibit a computational demand lower than worst case. In this approach, slack is distributed to tasks according to their expected computational demands.

A similar mechanism for explicitly managing slack is taken in the EDF targeted *lpSEH* [23]. Slack is considered to be produced by both higher and lower priority instances finished before the currently executing instance. More slack can thus be detected and used to lower the speed of the current task. Nevertheless, all slack is used to compute the speed of the task about to execute, making the slack distribution strategy more greedy than AGR or the utilization-based *laEDF*.

Explicit slack management can also be successfully employed in RM-based DVS policies. The *lpWDA* [68] approach uses a work demand-based approach to estimate the available slack for the current task instance. All the time that is not demanded by other tasks, higher priority that are allowed to execute or lower priority that got the chance to execute before the current task is considered to be slack and used by the current instance. Since exact knowledge of the work demand is costly to obtain on-line, safe heuristics are employed to estimate the demand. A more detailed comparison between different existing run-time strategies, including experimental evaluations can be found in Refs. [23] and [75].

18.4.5.2 A Slack Management Policy for Fixed Priority Tasks

In this section, we give an example of a slack distribution strategy, introduced first in Ref. [61], builing on top of RM scheduling. This method is designed to work on processors with an arbitrary number of speeds and it has a low computational complexity, independent of the characteristics of the task sets. Briefly, in this policy, an early finishing task may pass on its unused processor time to any of the tasks executing next. But this slack cannot be used by any task at any time since deadlines have to be met. This problem is solved by considering several levels of slacks, with different priorities, similar to the slack stealing algorithm [76].

For a task set $\{\tau_i\}_{1 \le i \le N}$ that exhibits M different priorities, M levels of real-time slack $\{S_j\}_{1 \le j \le M}$ are used. Without great loss of generality consider that the tasks have different priorities, or $M = N$. Also consider that the task set and slack levels are already ordered by priority, where level 1 corresponds to the highest priority. The slack in each level is a cumulative value, the sum of the unused processor times remaining from the tasks with higher priority. Initially, all level slacks S_j are set to 0. At run time, the slack levels are managed as follows:

- Whenever an instance k of a task τ_i with priority i *starts executing*, it can use an arbitrary part ΔC_i^k of the slack available at level i, S_i. So the allowed execution time, for instance, k of task τ_i will be

$$A_i^k = C_i + \Delta C_i^k \tag{18.36}$$

  where C_i is task τ_i WCET for the maximum (reference) processor speed, equal to WCE_i/f_{ref}. The remaining slack from level i cannot be used again on the same level. Therefore the slack level i is reset to 0. This can also be seen as a degradation of the slack from level i into level $i + 1$ slack. To summarize, each level of slack will be updated according to

$$S_j' = \begin{cases} 0, & j \le i \\ S_j - \Delta C_i^k, & j > i \end{cases} \tag{18.37}$$

- Whenever a task instance *finishes its execution*, it will generate some slack if it finishes before its allowed time. If X_i^k is the actual execution time of instance k of task τ_i, the generated slack is

$$\Delta A_i^k = A_i^k - X_i^k \tag{18.38}$$

This slack can be used by the lower priority tasks. In this case, the slack levels are updated according to

$$S_j'' = \begin{cases} S_j, & j \le i \\ S_j + \Delta A_i^k, & j > i \end{cases} \tag{18.39}$$

- *Idle processor times* are subtracted for all slacks. This ensures that the critical instance from the classic RM analysis remains the same.

The computational complexity required by the on-line method is, thus, linearly dependent to the number of slack levels: $O(M)$.

Note that task instances can only use slack generated from higher priority tasks and produce lower priority slack. Whenever the lowest priority task starts executing, all slack levels are reset. Note also that not necessarily all slacks at one level is used by a single task. Various strategies can be employed, but we mention here only two:

- *Greedy*. The task gets all the slack available for its level:

$$\Delta C_i^k = S_i \tag{18.40}$$

- *Mean proportional*. Given the mean execution time $\bar{X}_i$ for each task instance waiting to execute, the slack is proportionally distributed according to

$$\Delta C_i^k = S_i \frac{\bar{X}_i}{\sum_{\tau j \in \text{ReadyQ}} \bar{X}_j} \tag{18.41}$$

Experimentally [61], it turns out that the *mean proportional* policy performs marginally better than the *greedy* one. However, depending on the processor, the overhead needed to maintain the information required by the *mean proportional* method might cancel its gains compared to the simpler *greedy* strategy.

It is important to notice that the scheduling policy described above can be easily combined with other mechanisms, run-time, off-line, or intratask. The *stretch-to-NTA* run-time strategy can be additionally employed. The off-line minimum required speed computation for RM may also be used in a preparation phase. Furthermore, at task level, intra-DVS can be additionally employed. Nevertheless, as mentioned before, at some point the overhead needed by the additional mechanism will eventually cancel out its gain. Therefore, it is important to carry out an evaluation and tuning of the methods to be used, and select only those that give the largest energy reductions. A simulator such as *SimDVS* [23] is essential in this sense.

18.5 Concluding Remarks

We have described representative DVS techniques proposed for hard real-time systems. Taking advantage of workload variations within a single task execution as well as workload fluctuations from running multiple tasks, DVS provides effective low-power solutions by adjusting the supply voltage, which is a dominant factor in power consumption of embedded systems. In this paper, we focused on two main steps of a DVS algorithm, namely the slack identification step and slack distribution step. We reviewed these steps in two types of voltage scheduling techniques, intra- and intertask DVS.

Intratask DVS techniques identify the slack times due to the early completion of real-time applications and exploit them within a task boundary by lowering the operating clock frequency and supply voltage. Depending on how to identify the slack times and how to select the voltage scaling points, different kinds of IntraDVS techniques have been introduced. The intratask slack times can be identified by observing the execution path or the run-time architectural events (such as memory stall). The energy efficiency of

IntraDVS techniques is affected significantly depending on how accurately the techniques can predict the future workload. Using the profile information of task executions, for example, may give a more efficient IntraDVS algorithm. Since IntraDVS changes the supply voltage in a fine granularity, IntraDVS techniques require to consider the voltage scaling overhead. In addition, the available voltage levels should be taken into account in scaling the voltage level.

Intertask DVS addresses energy reduction at the task group level. Hard real-time systems can benefit from intertask DVS, while maintaining their real-time behavior. The majority of hard real-time interDVS techniques build upon classic real-time scheduling strategies in order to fulfill their timing requirements. Such techniques, ranging from fully off line to run time, are based on accurately estimating the slack time and efficiently distributing it to instances. Intertask DVS mechanisms are orthogonal to intratask techniques, and are often used in conjunction with the latter. Nevertheless, overly complex DVS approaches may introduce overheads that cancel out the energy gains, depending on the processor, application, and the mechanisms involved.

Acknowledgments

Jihong Kim was supported in part by the MIC (Ministry of Information and Communication), Korea, under the ITRC (Information Technology Research Center) support program supervised by the IITA (Institute of Information Technology Assessment) (IITA-2005-C1090-0502-0031), and by MIC & IITA through IT Leading R&D Support Project. All the correspondence should be sent to a corresponding author, Jihong Kim (jihong@davinci.snu.ac.kr).

References

1. T.D. Burd, T. A. Pering, A.J. Stratakos, and R.W. Brodersen, A dynamic voltage scaled microprocessor system. *IEEE Journal of Solid State Circuits*, 35(11), pp. 1571–1580, Nov. 2000.
2. T. Sakurai and A. Newton. Alpha-power law MOSFET model and Its application to CMOS inverter delay and other formulars. *IEEE Journal of Solid State Circuits*, 25, 584–594, 1990.
3. D. Shin, J. Kim, and S. Lee, Intra-task voltage scheduling for low-energy hard real-time applications. *IEEE Design and Test of Computers*, 18, 20–30, 2001.
4. F. Gruian, Hard real-time scheduling using stochastic data and dvs processors. *Proceedings of International Symposium on Low Power Electronics and Design*, pp. 46–51, 2001.
5. J.M. Rabaey and M. Pedram, *Low Power Design Methodologies*. Kluwer Academic Publishers, Dordrecht, 1996.
6. T. Ishihara and H. Yasuura, Voltage scheduling problem for dynamically variable voltage processors. *Proceedings of International Symposium On Low Power Electronics and Design*, pp. 197–202, 1998.
7. W.-C. Kwon and T. Kim, Optimal voltage allocation techniques for dynamically variable voltage processors. *Proceedings of Design Automation Conference*, pp. 125–130, 2003.
8. Transmeta Corporation, *Crusoe Processor. http://www.transmeta.com.* 2000.
9. AMD Corporation, *PowerNow! Technology. http://www.amd.com.* 2000.
10. Intel Corporation, *Mobile Intel Pentium 4 Processor-M with 512KB L2 Cache on .13 Micron Process at 1.6 GHz and 1.7 GHz Datasheet*, 2002.
11. Intel Corporation, *Intel XScale Technology. http://developer.intel.com/design/intelxscale.* 2001.
12. S. Lee and T. Sakurai, Run-time voltage hopping for low-power real-time systems. *Proceedings of Design Automation Conference*, pp. 806–809, 2000.
13. D. Mosse, H. Aydin, B. Childers, and R. Melhem, Compiler-assisted dynamic power-aware scheduling for real-time applications. *Proceedings of Workshop on Compiler and OS for Low Power*, 2000.
14. N. AbouGhazaleh, B. Childers, D. Mosse, R. Melhem, and M. Craven, Energy management for real-time embedded applications with compiler support. *Proceedings of Conference on Language, Compiler, and Tool Support for Embedded Systems*, San Diego, CA, USA, pp. 238–246, 2002.

15. D. Shin and J. Kim, A profile-based energy-efficient intra-task voltage scheduling algorithm for hard real-time applications. *Proceedings of International Symposium on Low Power Electronics and Design*, pp. 271–274, 2001.

16. J. Seo, T. Kim, and K.-S. Chung, Profile-based optimal intra-task voltage scheduling for hard real-time applications. *Proceedings of Design Automation Conference*, pp. 87–92, 2004.

17. B. Walsh, R. van Engelen, K. Gallivan, J. Birch, and Y. Shou, Parametric intra-task dynamic voltage scheduling. *Proceedings of Workshop on Compilers and Operating Systems for Low Power*, 2003.

18. D. Shin and J. Kim, Optimizing intra-task voltage scheduling using data flow analysis. *Proceedings of Asia and South Pacific Design Automation Conference*, pp. 703–708, 2005.

19. C-H. Hsu and U. Kremer, The design, implementation, and evaluation of a compiler algorithm for CPU energy reduction. *Proceedings of ACM SIGPLAN Conference on Programming Languages, Design, and Implementation*, pp. 38–48, 2003.

20. A. Weissel and F. Bellosa, Process cruise control: Event-driven clock scaling for dynamic power management. *Proceedings of International Conference on Compilers, Architecture, and Synthesis for Embedded Systems*, pp. 238–246, 2002.

21. K. Choi, R. Soma, and M. Pedram, Fine-grained dynamic voltage and frequency scaling for precise energy and performance trade-off based on the ratio of off-chip access to on-chip computation times. *IEEE Transactions on Computer Aided Design*, 24, 18–28, 2005.

22. J.R. Lorch and A.J. Smith, Improving dynamic voltage scaling algorithms with pace. *Proceedings of ACM SIGMETRICS Conference*, pp. 50–61, 2001.

23. D. Shin, W. Kim, J. Jeon, and J. Kim, Simdvs: An integrated simulation environment for performance evaluation of dynamic voltage scaling algorithms. *Lecture Notes in Computer Science*, 2325, 141–156, 2003.

24. S.-S. Lim, Y.H. Bae, G.T. Jang, B.-D. Rhee, S.L. Min, C.Y. Park, H. Shin, K. Park, and C.S. Kim, An accurate worst case timing analysis for risc processors. *IEEE Transactions on Software Engineering*, 21, 593–604, 1995.

25. K. Choi, W-C. Cheng, and M. Pedram, Frame-based dynamic voltage and frequency scaling for an mpeg player. *Journal of Low Power Electronics*, 1, 27–43, 2005.

26. T. Ball and J.R. Larus, Using paths to measure, explain, and enhance program behavior. *IEEE Computer*, 33, 57–65, 2000.

27. B. Lisper. Fully automatic, parametric worst-case execution time analysis. *Proceedings of ACM SIGMETRICS Conference*, pp. 85–88, 2003.

28. C-H. Hsu, U. Kremer, and M. Hsiao, Compiler-directed dynamic frequency and voltage scheduling. *Proceedings of Workshop on Power-Aware Computer Systems*, 2000.

29. F. Gruian, System-level design methods for low-energy architectures containing variable voltage processors. *Lecture Notes in Computer Science*, 2008, 1–12, 2000.

30. L. Yan, J. Luo, and N.K. Jha, Combined dynamic voltage scaling and adaptive body biasing for heterogeneous distributed real-time embedded systems. *Proceedings of International Conference on Computer Aided Design*, pp. 30–37, 2003.

31. J. Luo and N.K. Jha, Battery-aware static scheduling for distributed real-time embedded systems. In *Proceedings of Design Automation Conference*, pp. 444–449, 2001.

32. J. Luo and N.K. Jha, Static and dynamic variable voltage scheduling algorithms for real-time heterogeneous distributed embedded systems. *Proceedings of International Conference on VLSI Design*, pp. 719–726, 2002.

33. J. Luo and N.K. Jha, Power-conscious joint scheduling of periodic task graphs and aperiodic tasks in distributed real-time systems. *Proceedings of International Conference on Computer Aided Design*, pp. 357–364, 2000.

34. N.K. Bambha, S.S. Bhattacharyya, J. Teich, and E. Zitzler, Hybrid global/local search strategies for dynamic voltage scaling in embedded multiprocessors. *Proceedings of International Symposium on Hardware/Software Codesign*, pp. 243–248, 2001.

35. M.T. Schmitz, B.M. Al-Hashimi, and P. Eles, Energy-efficient mapping and scheduling for DVS enabled distributed embedded systems. *Proceedings of Design, Automation and Test in Europe Conference and Exhibition*, pp. 514–521, 2002.

36. M.T. Schmitz, B.M. Al-Hashimi, and P. Eles, Synthesizing energy-efficient embedded systems with LOPOCOS. *Design Automation for Embedded Systems*, 6, 401–424, 2002.

37. M.T. Schmitz and B.M. Al-Hashimi, Considering power variations of dvs processing elements for energy minimization in distributed systems. *Proceedings of International Symposium on System Synthesis*, pp. 250–255, 2001.

38. M.T. Schmitz, B.M. Al-Hashimi, and P. Eles, Iterative schedule optimization for voltage scalable distributed embedded systems. *ACM Transactions on Embedded Computing Systems*, Vol. 3, Issue 1, pp. 182–217, Feb. 2004.

39. A. Andrei, M. Schmitz, P. Eles, Z. Peng, and B.M. Al-Hashimi, Overhead-conscious voltage selection for dynamic and leakage energy reduction of time-constrained systems. *Proceedings of Design, Automation and Test in Europe Conference and Exhibition*, 2004.

40. R. Mishra, N. Rastogi, D. Zhu, D. Mossé, and R. Melhem, Energy aware scheduling for distributed real-time systems. *Proceedings of International Parallel and Distributed Processing Symposium*, 2003.

41. G. Varatkar and R. Marculescu, Communication-aware task scheduling and voltage selection for total systems energy minimization. *Proceedings of International Conference on Computer Aided Design*, pp. 510–517, 2003.

42. Y. Zhang, X. Hu, and Z. Chen, Energy minimization of real-time tasks on variable voltage processors with transition energy overhead. *Proceedings of Asia and South Pacific Design Automation Conference*, pp. 65–70, 2003.

43. Y. Zhang, X. Hu, and Z. Chen, Task scheduling and voltage selection for energy minimization. *Proceedings of Design Automation Conference*, pp. 183–188, 2002.

44. C.L. Liu and J.W. Layland. Scheduling algorithms for multiprogramming in a hard real time environment. *Journal of ACM*, 20, 46–61, 1973.

45. J.A. Stankovic, M. Spuri, K. Ramamritham, and G.C. Buttazzo, *Deadline Scheduling For Real-Time Systems: EDF and Related Algorithms*. Kluwer Academic Publishers, Dordrecht, 1998.

46. F. Yao, A. Demers, and S. Shenker, A scheduling model for reduced CPU energy. *Proceedings of IEEE Annual Foundations of Computer Science*, pp. 374–382, 1995.

47. I. Hong, D. Kirovski, G. Qu, M. Potkonjak, and M.B. Srivastava, Power optimization of variable voltage core-based systems. *Proceedings of Design Automation Conference*, pp. 176–181, 1998.

48. A. Manzak and C. Chakrabarti, Variable voltage task scheduling for minimizing energy or minimizing power. *Proceedings of IEEE International Conference on Acoustics, Speech, and Signal Processing*, Vol. 6, pp. 3239–3242, 2000.

49. V. Swaminathan and K. Chakrabarty. Investigating the effect of voltage-switching on low-energy task scheduling in hard real-time systems. *Proceedings of Asia and South Pacific Design Automation Conference*, pp. 251–254, 2001.

50. A. Manzak and C. Chakrabarti, Variable voltage task scheduling algorithms for minimizing energy. *Proceedings of International Symposium on Low Power Electronics and Design*, pp. 279–282, 2001.

51. B. Mochocki, X. Hu, and G. Quan, A realistc variable voltage scheduling model for real-time applications. *Proceedings of International Conference on Computer Aided Design*, pp. 726–731, 2002.

52. V. Swaminathan and K. Chakrabarty, Generalized network flow techniques for dynamic voltage scaling in hard real-time systems. *Proceedings of International Conference on Computer Aided Design*, 2003.

53. G. Quan and X. Hu, Energy efficient fixed-priority scheduling for real-time systems on variable voltage processors. *Proceedings of Design Automation Conference*, pp. 828–833, 2001.

54. G. Quan and X. Hu, Minimum energy fixed-priority scheduling for variable voltage processors. *Proceedings of Design, Automation and Test in Europe Conference and Exhibition*, pp. 782–787, 2002.

55. A. Manzak and C. Chakrabarti, Variable voltage task scheduling algorithms for minimizing energy/power. *IEEE Transactions on Very Large Scale Integration Systems*, 11, 270–276, 2003.

56. A. Sinha and A.P. Chandrakasan, Energy efficient real-time scheduling [microprocessors]. *Proceedings of International Conference on Computer Aided Design*, pp. 458–463, 2001.

57. H. Aydin, R. Melhem, D. Mossé, and P. Mejía-Alvarez, Determining optimal processor speeds for periodic real-time tasks with different power characteristics. *Proceedings of Euromicro Conference on Real-Time Systems*, pp. 225–232, 2001.

58. R. Jejurikar and R. Gupta, Energy aware EDF scheduling with task synchronization for embedded real-time systems. Technical Report CECS 02-24, Department of Information and Computer Science, University of California Irvine, 2002.

59. Y. Shin and W. Choi, Power conscious fixed priority scheduling for real-time systems. *Proceedings of Design Automation Conference*, pp. 134–139, 1999.

60. R. Jejurikar and R. Gupta, Energy aware task scheduling with task synchronization for embedded real time systems. *Proceedings of International Conference on Compilers, Architectures, and Synthesis for Embedded Systems*, pp. 164–169, 2002.

61. F. Gruian, Hard real-time scheduling for low-energy using stochastic data and DVS processors. *Proceedings of International Symposium on Low Power Electronics and Design*, pp. 46–51, 2001.

62. F. Gruian, Energy-Centric Scheduling for Real-Time Systems. Doctoral dissertation, Department of Computer Science, Lund Institute of Technology, 2002.

63. F. Gruian and K. Kuchcinski, Uncertainty-based scheduling: Energy-efficient ordering for tasks with variable execution time. *Proceedings of International Symposium on Low Power Electronics and Design*, pp. 465–468, 2003.

64. H. Aydin, P. Mejía-Alvarez, D. Mossé, and R. Melhem, Dynamic and aggressive scheduling techniques for power-aware real-time systems. *Proceedings of IEEE Real-Time Systems Symposium*, pp. 95, 2001.

65. Y. Shin, K. Choi, and T. Sakurai, Power optimization of real-time embedded systems on variable speed processors. *Proceedings of International Conference on Computer Aided Design*, pp. 365–368, 2000.

66. P. Pillai and K.G. Shin, Real-time dynamic voltage scaling for low-power embedded operating systems. *Proceedings of Symposium on Operating System Principles*, pp. 89–102, 2001.

67. W. Kim, J. Kim, and S.L. Min, A dynamic voltage scaling algorithm for dynamic-priority hard real-time systems using slack time analysis. *Proceedings of Design, Automation and Test in Europe Conference and Exhibition*, pp. 788–794, 2002.

68. W. Kim, J. Kim, and S.L. Min, Dynamic voltage scaling algorithm for fixed-priority real-time systems using work-demand analysis. *Proceedings of International Symposium on Low Power Electronics and Design*, pp. 396–401, 2003.

69. J. Lehoczky, L. Sha, and Y. Ding, The rate monotonic scheduling algorithm: exact characterization and average case behavior. *Proceedings of Real Time Systems Symposium*, pages 166–171, 1989.

70. J.Y.-T. Leung and M.L. Merrill, A note on preemptive scheduling of periodic, real-time tasks. *Information Processing Letters*, pp. 115–118, Nov. 1980.

71. Baruah, S., Mok, A., Rosier, L., "Algorithms and complexity concerning the preemptively scheduling of periodic, real-time tasks on one processor." *Real-Time Systems Journal*, Vol. 2, pp. 301–324, 1990.

72. M.R. Garey and D.S. Johnson, *Computers and Intractability: A Guide to the Theory of NP-Completeness*. W.H. Freeman and Company, New York, 1979.

73. R. Jejurikar and R. Gupta, Efficiency and optimality of static slowdown for periodic tasks in real-time embedded systems. Technical Report CECS 02-03, Department of Information and Computer Science, University of California Irvine, 2002.

74. N. AbouGhazaleh, D. Mossé, B. Childers, R. Melhem, and M. Craven, Collaborative operating system and compiler power management for real-time applications. *Proceedings of IEEE Real-Time and Embedded Technology and Applications Symposium*, 2003.

75. W. Kim, D. Shin, H.S. Yun, J. Kim, and S.L. Min, Performance comparison of dynamic voltage scaling algorithms for hard real-time systems. *Proceedings of IEEE Real-Time and Embedded Technology and Applications Symposium*, 2002.

76. J. Lehoczky and S. Ramos-Thuel, An optimal algorithm for scheduling soft-aperiodic tasks in fixed-priority preemptive systems. *Proceedings of Real Time Systems Symposium*, pp. 110–123, 1992.

19

Low Power Microarchitecture Techniques

Emil Talpes
Advanced Micro Devices, Inc.
One AMD Place
P.O. Box 3453
Sunnyvale, California 94088-3453

Diana Marculescu
Carnegie Mellon University

CONTENTS

19.1 Introduction

Very few domains have grown as fast as computer architecture. In the last 25 years, integrated microprocessors have evolved from small oddities to become the virtually single source of computing power. They now form the core of almost every electronic device, ranging from small, battery-operated devices to massively parallel computer systems. Their size, performance, and price make them ideal for an extremely wide range of applications.

Two decades ago, microprocessors were doing exactly what the Instruction Set Architecture (ISA) specified: read one instruction, decode and execute it, and then repeat the same steps for the next one. Each instruction required several clock cycles, and the execution time for the entire program was equal to the sum of the execution times for its instructions. Speeding up the processor meant reducing the cycle time, while sometimes reducing the number of cycles per instruction as well. As the number of transistors that could be crammed on a single chip increased, they started exceeding what is typically necessary for such a processor. Thus, the idea of executing more than one instruction at a time has emerged. Designers started using temporal parallelism, overlapping instructions in different phases of execution. Such pipelined processors were able to achieve a much higher throughput than their predecessors, close to one instruction per clock cycle.

The next step in microarchitecture evolution was the superscalar processing paradigm, essentially consisting of multiple pipelines connected together. These processors fetch two or more adjacent instructions, decode them, and then try to execute them in parallel if no data dependency is detected. Finally, out-of-order capabilities have been developed, allowing the processor more freedom in picking the instructions to be executed in parallel. Such processors can achieve a throughput significantly higher than one instruction per cycle, while also being able to work at very high clock speeds.

The current state-of-the-art high-end microprocessors are based on this superscalar, out-of-order microarchitecture. These processors can fetch multiple instructions during each clock cycle, dynamically predicting which will be needed next, well in advance of the actual execution. Instructions are decoded in parallel, and data dependencies are verified. To eliminate any possible false dependencies, the registers are renamed inside a register file that is much larger than the one specified by the ISA. Instructions are reordered according to their data dependencies trying to maximize throughput and, in the end, they are executed by a parallel execution core.

However, in spite of the growing internal complexity, microprocessors must maintain their relative ease of use. They can still be treated as simple "black boxes" which sequentially execute the instructions of a program, and, in most cases, they can still execute programs written two decades ago. All these new mechanisms and capabilities are encapsulated inside a layer that maintains backward compatibility with applications developed for much simpler microprocessors. As long as the programmer obeys the conventions of the ISA, it is not necessary for him to know any of the implementation details, allowing him to choose among a multitude of compatible processors. While this compatibility layer often complicates the microarchitecture even further, it also allows the designers to push the envelope and include features that are not part of the original ISA.

19.2   Power Consumption

The evolution of the microprocessor architecture over the last couple of decades has been facilitated by tremendous improvements in silicon process technologies. As predicted in 1965 by Intel's cofounder Gordon Moore, the number of transistors placed on a single chip has indeed doubled roughly every couple of years. Each new process technology brings smaller and faster devices, allowing designers to create more complex architectures working at faster clock speeds.

With a reduction in size, a single transistor consumes less power with each new process technology. Using state-of-the-art technology to implement an older microarchitecture is a simple method for lowering the power consumption of a processor and it is sometimes used in mid-life product updates. However, the performance level achieved by the resulting processor would be unacceptable in most situations. Most of the time, a new process technology is accompanied by either a completely new microarchitecture or an updated one that is optimized for higher clock speeds.

While the size of a single transistor is typically halved with each new process technology, the die sizes have remained fairly constant over time and are dictated primarily by economic factors. Today's microarchitectures use the inexpensive silicon real estate by emphasizing execution parallelism, both temporal and spatial. This translates into more and more transistors switching during every clock cycle. While 15 years ago Intel's i486 processor had approximately 1.2M transistors, the latest dual core Pentium 4 uses about 230M transistors (Figure 19.1).

In addition, clock frequencies have gone up as well, driven by both longer instruction pipelines and advances in process technology. Intel's i486 was released in 1989 at 33 MHz, but Pentium 4 today reaches almost 4 GHz. Since power consumption is directly proportional to both the clock frequency and the number of devices, it is easy to understand why power consumption has evolved into a major problem. The latest Pentium 4 processor dissipates more than 100 W for a silicon area of ~1 cm² [1], starting to show the limitations of our cooling and power delivery capabilities.

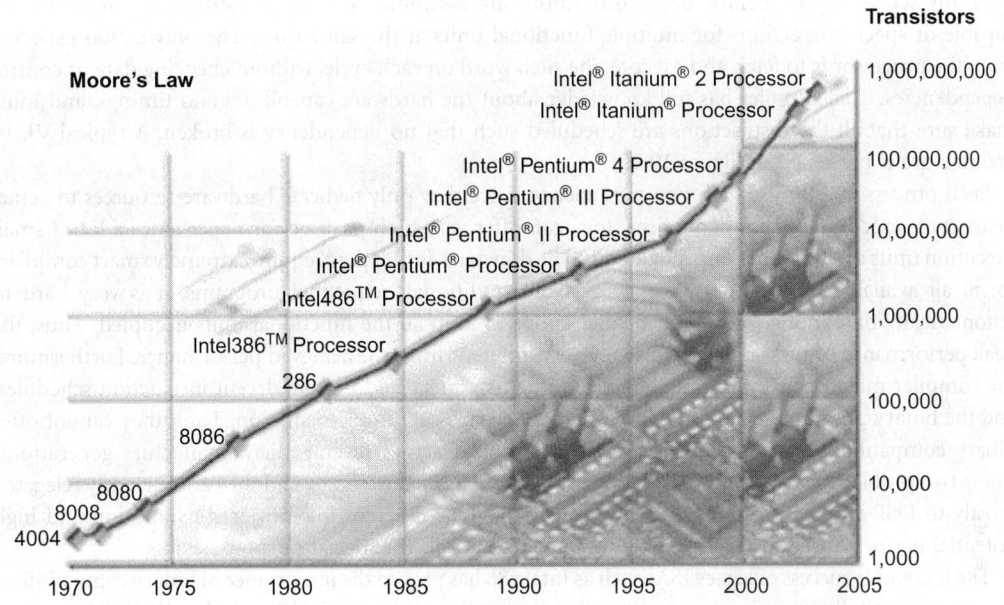

FIGURE 19.1 On-chip transistor count increase for the Intel processors (*Source*: Intel).

19.3 Techniques for Low-Power Processor Design

Traditionally, performance concerns have taken priority over energy costs or power consumption. Power efficiency has been addressed mainly at the technology level, through lower supply voltages, smaller transistors, silicon-on-insulator (SOI) technology, better packaging, etc. Nevertheless though, power dissipation has become one of the primary design constraints for modern processors, and thus microarchitecture designers must now take it into consideration as well.

At the microarchitectural level, the methods to reduce the dynamic power consumption fall under two fundamental categories: **capacitance reduction** and **dynamic voltage/frequency scaling** (DVS). These address specific aspects of the dynamic power consumption equation

$$P = \tfrac{1}{2}V_{dd}^2 f_{clk} C_{eff} \tag{19.1}$$

19.3.1 Limiting the Switched Capacitance

The first class of methods aims at reducing the work performed by the processor during each clock cycle, and, in consequence, the switched capacitance C_{eff}. Most of these techniques come somewhat against the current industry trend which is to add more hardware and perform more speculative work. All this additional hardware is frequently blamed for the inefficiency of modern processors, since it consumes a lot of power, while offering limited performance benefit.

The first and most obvious method to create a low-power processor is to return to basics and limit the amount of on-chip logic to only useful program computation. Such processors do not include any hardware for scheduling instructions or for data/control speculation. Many of them are single-issue processors, and, depending on the performance point targeted by these designs, they can be pipelined (e.g., various ARM implementations [2]) or nonpipelined (e.g., various i8051 compatible micro-controllers from various companies [3]).

A related category is composed of multiple-issue processors built under the very long instruction word (VLIW) paradigm. Such designs are capable of executing several instructions in parallel, but they do not offer any scheduling capability. Basic instructions are assembled into long instruction words that are capable of specifying actions for multiple functional units at the same time. The only action expected from the processor is to fetch and execute one such word on each cycle, without checking data or control dependencies. The compiler has full knowledge about the hardware capabilities and timings, and must make sure that all the instructions are scheduled such that no dependency is broken. A typical VLIW architecture is presented in Figure 19.2.

Such processors promise high power efficiency, since they only dedicate hardware resources to actual instruction execution. At the same time, they can offer very high peak performance since a lot of small execution units can be placed on a single chip. The drawback is that they require extremely smart compilers to fill all available execution slots. Starting from traditional, sequential programs, it is very hard to automatically find enough independent operations to keep all the functional units occupied. Thus, the peak performance of these devices is usually very different from the achieved performance. Furthermore, the compiler must know everything about the microarchitecture to create decent instruction schedules, and the binary program becomes tied to a particular hardware implementation. Thus, they cannot offer binary compatibility across a larger processor family or across several microarchitecture generations. These two problems have severely limited the spread of VLIW processors, and they are currently relegated mostly to DSP-style applications. However, owing to their inherent low-power consumption and high potential performance, they can provide very good solutions for special-purpose devices.

The incredible success of legacy ISAs such as Intel x86 has proven the importance of binary compatibility. Binary compatibility has also proven more important than the benefits brought by RISC ISAs such as Alpha [4] or PA-RISC [5]. Processors using these instruction sets do not require complicated hardware wrappers and decoders, being able to dedicate more resources to actual program execution. For economical reason though, these additional logic blocks are maintained at the expense of power and complexity.

An attempt to improve the power efficiency by getting rid of these hardware wrappers has been made by Transmeta [6], customizing a low-power VLIW processor for general-purpose computation. Crusoe and Efficeon are essentially VLIW processors surrounded by a software layer that does just-in-time compilation for x86 ISA programs. While these processors are able to run legacy x86 binaries and offer low-power consumption, their performance varies greatly with the application behavior. They take a steep penalty whenever translation has to be performed, and work well only if the code locality is very good.

Even though statically scheduled (in-order) processors consume less power per clock cycle, their dynamically scheduled (out-of-order) counterparts are currently much more popular. This is mainly caused by performance reasons: out-of-order processors are typically faster, and they are also less demanding about compiler quality. They also tend to perform better as clock frequencies increase, since they can schedule around long latency memory instructions. Thus, several techniques have been proposed for reducing the power consumption in large superscalar, out-of-order processors.

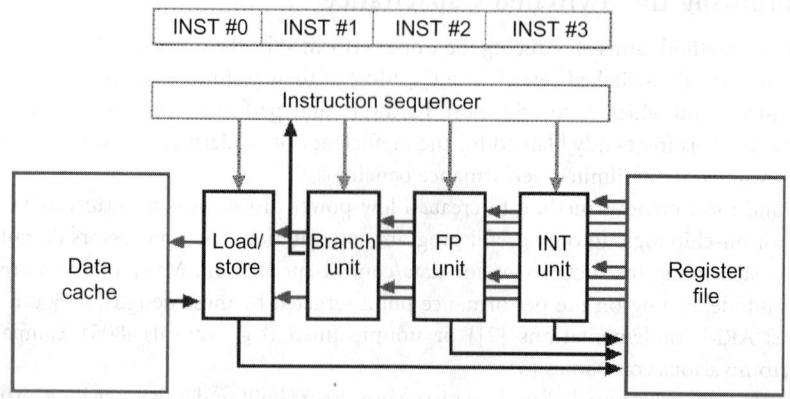

FIGURE 19.2   Typical VLIW architecture.

Traditionally, a superscalar processor contains a large number of resources of various types so that it can accommodate different instruction mixes. For example, to achieve a sustained throughput of three instructions per clock cycle, the execution cores of Pentium 4 and Opteron include many more execution units (9 and 10, respectively). Even though all these resources are infrequently used, they still consume power and this inefficiency can be targeted by various methods.

Guarded evaluation [7] has been proposed as a static technique for reducing the power required by a combinational circuit when some of its input values are not necessary through successive time steps. This method works by stopping the propagation of signal transitions through unused circuits, effectively limiting the dynamic power consumption (Figure 19.3). A slightly more general version of this technique is clock gating, proposed in Ref. [8] for saving the power wasted by units that are temporarily not used. Clock gating also targets the propagation of signal transitions and works by shutting off the clock signal in the latches placed in front of the target circuit. In various forms, these techniques are now used widely to limit the power consumed in areas of the design that are not exercised during specific computations.

A different research direction targets the speculative nature of these superscalar, out-of-order engines. While speculation can dramatically increase overall performance, the ultimate result is dependent on the accuracy of these predictions. A prediction which ultimately proves to be wrong is very costly in terms of power consumption and it does not help at all in terms of performance (it may or may not impact the performance depending upon the actual implementation). Thus, it has been proposed to use confidence estimators and limit the amount of speculation to only the cases when the probability of success is very high. These techniques decrease the overall power consumption by reducing the amount of work that is ultimately thrown away when a prediction proves to be wrong. Such mechanisms have been proposed for both data and control speculation [9,10].

Another technique which targets the assumption that performance can be increased by throwing more hardware at the problem is resource scaling [11]. This method targets highly complex processor implementations, which use a number of resources to extract a high level of instruction-level parallelism (ILP). While some applications might exercise the entire set of resources (usually scientific or media applications); in other cases the inherent level of ILP is limited and most of the resources remain unused (control-bound applications). Such applications would maintain a relatively similar execution pattern on much simpler processors. The resource scaling technique takes advantage of this fact and turns off (through clock gating) some of the units available, saving power while maintaining a comparable performance.

All the ideas mentioned so far attempt to increase efficiency by reducing the amount of power spent on *useless computation*. However, even under the idealized assumption that everything works perfectly and no useless work is performed, the dynamically scheduled processor still requires more power per instruction than the simpler VLIW implementation. Such processors use long pipelines, performing operations like branch predictions, register renaming, reordering, etc. The concept of work reuse [12–14] tries to bring the superscalar, out-of-order processor closer to the efficiency of a VLIW.

In Pentium 4, a special trace-cache has been placed in the pipeline, after the x86 decoding stages. By storing decoded instructions (uops) in the trace-cache, the whole decode stage can be shut down for

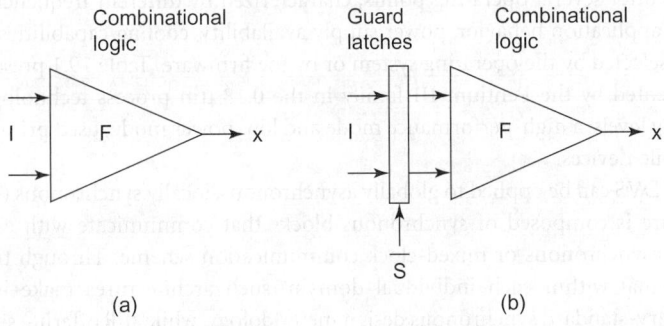

FIGURE 19.3   Combinational circuit with unguarded input I (a) and with guard latches (b) [7]. S is the guard signal.

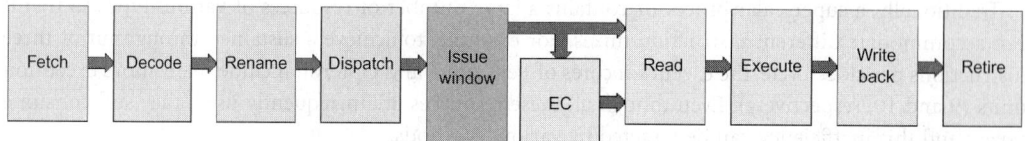

FIGURE 19.4   EC-based processor microarchitecture.

significant periods of time while the rest of the execution engine continues working. When a hit in the trace-cache occurs, instructions do not need to be decoded again and can be fed into the pipeline directly from the trace-cache.

An extension of this technique [14] uses an **execution cache** (**EC**) placed deep in the pipeline to shorten the critical instruction path (Figure 19.4). If the cache is placed after the issue stage, instructions that are fetched, decoded, and have already had registers renamed can be stored in *issue order* (instead of in *program order*) in the EC. Assuming that the EC efficiency is very good, most of the time instructions are executed out of this cache. Instructions are retrieved from the EC in issue order, and they can be sent directly to the execution engine in a VLIW-like fashion. Using clock gating, the front-end of the pipeline is shut off, reducing the effective work that needs to be performed for each retired instruction. However, when instructions are not found in the EC, the front-end resumes its role as a scheduler, creating traces for further reuse. This technique will be described in detail in Section 19.4.

19.3.2   DVS

The second class of methods relies on the observation that users need top performance only in infrequent cases. Modern processors offer more than adequate performance on most applications, being overdesigned for the few situations when more power is required. This holds especially true for processors that go into personal devices such as PDAs, cell phones, even laptop and desktop computers. Many of the applications run on these devices are bound by user input, so the performance of the underlying processor is largely irrelevant. Another class of applications that has become very popular contains media applications. In such cases, the optimal performance is actually predefined based on the media type and a faster processor will not improve the user experience.

A related, but slightly different case, is represented by mobile devices. Even though more performance would be desirable in some cases, most of the time they are limited by battery capacity. Thus, a compromise can be struck by giving up on some of the potential performance for a longer battery life.

All processors that are currently intended for mobile applications support techniques such as frequency and voltage scaling [6,15,16] to reduce their power requirements on those applications where high performance is not required. DVS is a very effective technique for reducing power consumption since significant reductions can be obtained at the expense of relatively small performance drops. While the performance is only impacted by the reduction in clock frequency, the power also goes down quadratically with the voltage supply. As can be seen in Table 19.1, power consumption decreases much faster than the overall performance.

Such processors offer several operating points, characterized by different frequencies and V_{dd} values. Depending on the application behavior, power supply availability, cooling capabilities, etc., one of these available points is selected by the operating system or by the firmware. Table 19.1 presents the SpeedStep capability implemented by the Pentium III family in the 0.18 μm process technology. This processor offers two operating levels: a high-performance mode and low-power mode, used primarily for extending battery life in mobile devices.

A special case of DVS can be applied to globally asynchronous, locally synchronous (GALS) processors. Such an architecture is composed of synchronous blocks that communicate with each other only on demand, using an asynchronous or mixed-clock communication scheme. Through the use of a locally generated clock signal within each individual domain, such architectures make it possible to take advantage of industry-standard synchronous design methodology, while still offering some of the benefits of asynchronous circuits. Thus, they do not require a global clock distribution network and deskewing

TABLE 19.1   SpeedStep Implementation in the Coppermine Pentium III Processor [16]

Maximum Performance Mode			Battery-Optimized Mode		
Frequency (MHz)	Voltage (V)	Maximum Power Consumption (W)	Frequency (MHz)	Voltage (V)	Maximum Power Consumption (W)
600	1.6	20.0	500	1.35	12.2
650	1.6	21.5	500	1.35	12.2
700	1.6	23.0	550	1.35	13.2
750	1.6	24.6	550	1.35	13.2
800	1.6	25.9	650	1.35	15.1
850	1.6	27.5	700	1.35	16.1
900	1.7	30.7	700	1.35	16.1
1000	1.7	34.0	700	1.35	16.1

circuitry. Also, they allow the supply voltages and clock frequencies to be scaled per domain, offering significantly more flexibility. However, the overhead introduced by communicating data across clock domain boundaries may become a fundamental drawback, limiting the performance of these systems. Thus, the choice of granularity for these synchronous blocks or islands must be very carefully done in order to prevent the interdomain communication from becoming a significant bottleneck. At the same time, the choice of the interdomain communication scheme as well as of the on-the-fly mechanisms for per domain DVS become critical when analyzing overall power–performance trends. A processor microarchitecture using the GALS+DVS techniques is presented in Section 19.5.

19.3.3   Reducing the Static Power Consumption

A third class of methods targets a different aspect of the power consumption in modern microprocessors: the static power. While dynamic power is consumed when a transistor switches between "on" and "off", static power is leaked through the transistor junctions even when it does not perform any useful work.

Traditionally, the static power has been several orders of magnitude smaller than the dynamic power consumption, thus being largely ignored by processor designers. However, as transistors become smaller with each new process technology, the isolator layers shrink and leakage currents increase. Furthermore, the number of such leaky transistors placed on a chip increases with each microarchitecture generation. As a result, the static power consumption has become a first-class concern, equally important when compared to the dynamic power consumption for modern-processor microarchitecture design.

All techniques proposed for limiting the static power target the leakage current. One such solution is to stack multiple "off" transistors on any path between V_{dd} and ground, exponentially reducing the current that leaks on the path. Such an example is presented in Figure 19.5.

Power gating [17] relies on the presence of these sleep transistors, placed besides the target logic modules. Depending on whether the sleep transistors are placed toward the V_{dd} or the ground rails, the method uses a gated V_{dd} or a gated ground, respectively.

While this method works very well for combinational circuits, special care must be taken when it is applied to sequential logic. In this case the modified voltage levels can interfere with the memory elements, destroying the state of the circuit when the sleep transistor is turned off. This problem is discussed in Ref. [18], where a special design called DRI cache is proposed for the L1 caches of a microprocessor. The cache stores additional information, allowing it to detect timeout conditions and turn-off sections which are not exercised by the current program. Furthermore, the DRAM cells use sleep transistors and can be gated to reduce the static power consumption. The design uses the gated ground methodology to retain the cache content while different sections are turned off.

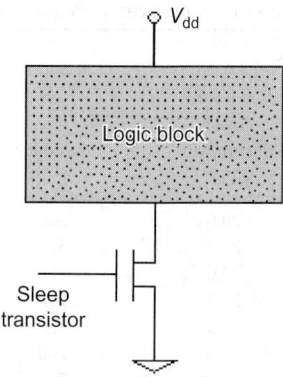

FIGURE 19.5   Sleep transistor usage.

A related method for reducing the static power consumption is the drowsy cache [19]. The microarchitectural decisions are taken in a fashion similar to the DRI-cache design, the hardware deciding when the cache lines can be turned off without impacting the overall performance. However, the circuit implementation is more complex: it relies on adaptive body biasing, a technique that dynamically modifies the effective V_t of the memory cells. The higher V_t translates into a lower leakage current, reducing the static power for cache lines that are turned off.

19.4   Case Study 1: EC-Based Microarchitectures

As mentioned in the previous section, this technique reduces the power consumption of a superscalar, out-of-order microprocessor through dynamic work reuse. To achieve this, when an instruction is reexecuted, the goal is to reuse as much as possible from the computations performed during previous executions. Obviously, the work performed in the fetch and decode stages is identical each time a specific trace is executed. Using a special register file structure, the work performed by the rename and issue stages can also be reused.

To reuse the work and reduce the branch misprediction path, an EC is inserted deep in the pipeline, after the issue stage. To avoid the performance penalty incurred by an extra pipeline stage between the issue and the execution stages, instructions are issued in parallel to both the execution stage and the EC. The conceptual microarchitecture is illustrated in Figure 19.6.

Normally, instructions are fetched from the I-Cache through the fetch stage and then decoded. In the next stage, physical registers are assigned for each logical register, avoiding potential false dependencies. The resulting instructions are placed in the issue window for dependency checking. A number of independent instructions are issued to the execution stage and, in parallel, added to a *fill buffer* to create program traces. When enough instructions are placed in this fill buffer, the entire program sequence is stored in the EC in the issue order, for later potential reuse.

In this setting, the branch mispredict path can be significantly shortened by feeding the execution units directly from the EC whenever possible. Initially, when the EC is empty, instructions are launched from the Issue Window, while a trace is built in parallel in the EC (*trace-segment build* phase). Upon a mispredict (or a trace completion condition), a search is performed to identify a possible next trace starting at that point, and should a hit occur, the instructions continue to be executed from the EC, on the alternative execution path. When operating on this alternative execution path, the processor behaves essentially like a VLIW core with instructions being fetched from the execution cache and sent directly to the execution engine. If a miss is encountered on a trace search, the pipeline front-end must be launched again and a new trace is built.

19.4.1   EC Internal Structure

Similar to the conventional trace-cache implementations, this microarchitecture divides the program into traces of instructions that are stored in a order different from the one given by their original addresses.

This design allows to implicitly encode information about the reordering work done in the fetch and issue stages through the actual order in which instructions are stored. However, the cache chosen in the proposed architecture is structurally different from the trace-cache typically used for increasing the fetch bandwidth.

When stored in issue order, instructions lose their original, logical order and they can be retrieved only on a sequential basis. However, in order to allow for traces to be reused, the start address of each trace needs to correspond to a physical address in the memory space. Instructions from two consecutive traces cannot be interleaved, so at each change of trace the processor must restart trace execution in order. Specifically, at each trace end, a *trace look-up* step must be performed. While most of the time the performance penalty associated with this look-up can be hidden (the look-up being started in advance), there are certain conditions when this is not possible. Together with the need for an in-order start of each trace, this leads to some performance penalty associated with each trace change.

To minimize the overall performance impact of this design, the created traces must be as long as possible. While most trace-cache designs proposed in the literature limit the traces to at the most three basic blocks, it is desirable to include as many instructions as possible in a trace if there is no mispredict encountered. However, as they get longer, the number of traces that can be accommodated in a reasonably sized cache decreases. This leads to a decrease in the hit rate at trace look-up, so a higher utilization of the front-end pipeline is achieved. While this does not significantly impact the overall performance, it increases the power consumption since both the front-end of the pipeline and the EC (in trace-build mode) are simultaneously utilized. The block architecture of the EC is presented in Figure 19.7.

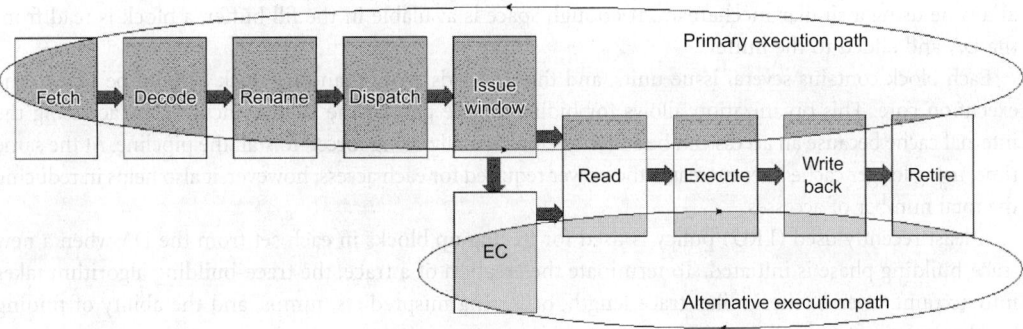

FIGURE 19.6   Superscalar microarchitecture using an EC for reusing scheduled instruction streams.

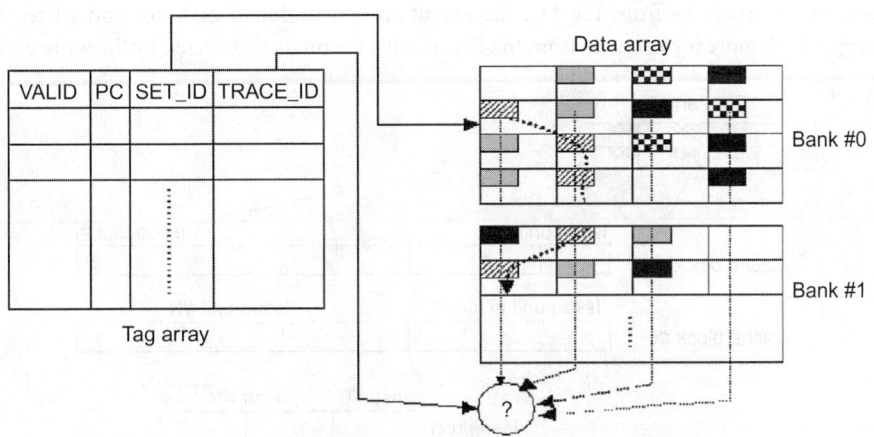

FIGURE 19.7   EC structure.

The EC structure consists of a *tag array* (TA) and a corresponding *data array* (DA). The TA is an associative cache, addressed using the translated program counter. It is used for trace look-up and it should be as fast as possible to reduce the performance overhead associated with searching for a new trace. The *SET_ID* value obtained from the TA points to the set in the DA that contains the trace start. The DA is a multiway set associative cache composed of multiple memory banks. A comparison with the *TRACE_ID* is performed for each block in the set to identify the correct starting block for the next trace. The next chunk of instructions is located in one of the blocks of the following set, and so on (see Figure 19.7). A special end-of-trace marker identifies the end of the trace.

By knowing beforehand which set will be accessed next, a new look-up for each subsequent access can be avoided. Furthermore, knowing the next set allows the use of multiple memory banks to implement the DA. While one of the banks is used, the others can be turned off, resulting in further energy savings. Thus, the entire array is used only when accessing the first instructions in a trace. On all subsequent accesses, the energy consumed by the line decoder and by the unused banks can be used. Depending on the application, the relative number of accesses made to only one bank with respect to the total number of accesses can vary between 1:2 (e.g., *gcc*) and 3:4 (e.g., for floating-point benchmarks).

Inside each block, an arbitrary number of issue units is stored (Figure 19.8). An issue unit consists of independent instructions that can be issued in parallel to the functional units. Since issue units are recorded during the trace-building phase (when the front end of the pipeline is used) and then reused in all subsequent executions of the trace, the processor will take the same optimizing decisions each time it executes the code.

All instructions coming from the issue window are first assembled into traces using the fill buffer and then recorded in the EC. The fill buffer can accommodate two DA blocks, and when enough instructions are available to fill a block, they are written to the EC. When reading from the EC, one issue unit is issued at a time using a similar mechanism. If enough space is available in the fill buffer, a block is read from the DA and added to the buffer.

Each block contains several issue units, and thus it needs more than one clock cycle to be sent to the execution core. This organization allows for hiding a large part of the latency incurred by accessing the internal cache because an access can be started before actually being forced to stall the pipeline. At the same time, using longer cache lines increases the power required for each access; however, it also helps in reducing the total number of accesses.

A least recently used (LRU) policy is used for freeing up blocks in each set from the DA when a new trace-building phase is initiated. To terminate the creation of a trace, the trace-building algorithm takes into account several criteria like: trace length, occurring mispredicts, jumps, and the ability of finding another existing trace starting at the current point.

When creating a new trace, instructions are added until the trace grows beyond a certain maximum length or when a branch mispredict occurs and the execution must resume from a different address. When fetching instructions from the EC, the execution is abandoned on trace end (detected when attempting to fetch more instructions from the DA) or on a mispredict (detected by the write back stage).

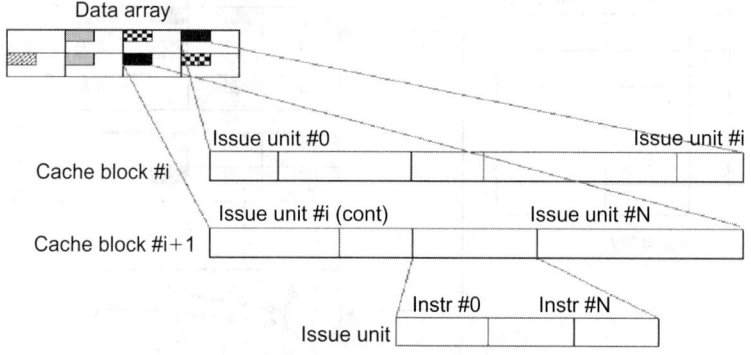

FIGURE 19.8   EC block organization.

A trace look-up is then performed and, if this search misses, the front-end of the pipeline is restarted. Should a hit occur, instructions are issued on the alternative execution path directly from the EC.

19.4.2 The Register File

Placing the above-described EC deep in the pipeline, after the issue stage, allows for reusing the work done by all the units belonging to the front-end stages. This also implies that register renaming is not performed on the instructions issued directly from the EC. Being different at each trace run, the values held by the register file cannot be stored in the EC and reused. However, the register renaming operation is only performed in trace-creation mode. This operating mode assumes that the virtual-to-architected register mapping is the same at the beginning of each trace. Some architectural changes need to be made to the register pool and control unit to ensure that this condition can be satisfied. The logic structure needed for implementing each register is presented in Figure 19.9.

This structure employs a special pool of physical registers (organized as a circular buffer) for renaming every logical register of the ISA. Unlike in a typical register file, where an architected register can be renamed to any physical register, an architected register can be renamed by using only the physical registers of the corresponding pool. However, each subsequent "write" goes to a different physical register of the pool using a deterministic algorithm. This approach solves the problem of potential false data dependencies and can be used for implementing register renaming. Because of its very predictable behavior, this mechanism is similar to what was proposed for implementing modulo-scheduling [20].

When going through the rename stage, each instruction is allocated a physical register as destination, other than the one holding the last known value for the corresponding architected register. Having different physical destinations, instructions can write the result as soon as it is available, setting the V (valid) bit. This bit is used to detect the register file locations, which hold valid data. If all the source registers for an instruction have this bit set, the instruction can be issued to the execution unit. The bit is cleared when the physical register is allocated as a destination in the rename stage and it is set when the value is written in the write back stage. The actual value in the physical register (register 0: N-1) is only accessed when we have to read or write the value from the architected register.

The speculated (S) bits mark the committed values and they are cleared only after the instruction is retired. Whenever a rollback condition occurs, the index is reverted to the "last committed" value for the architected register. A physical register cannot be assigned as a destination for a new instruction (in the register renaming stage) if the associated S bit is set. If this happens, there are not enough physical registers to perform renaming at this moment and the rename stage stalls.

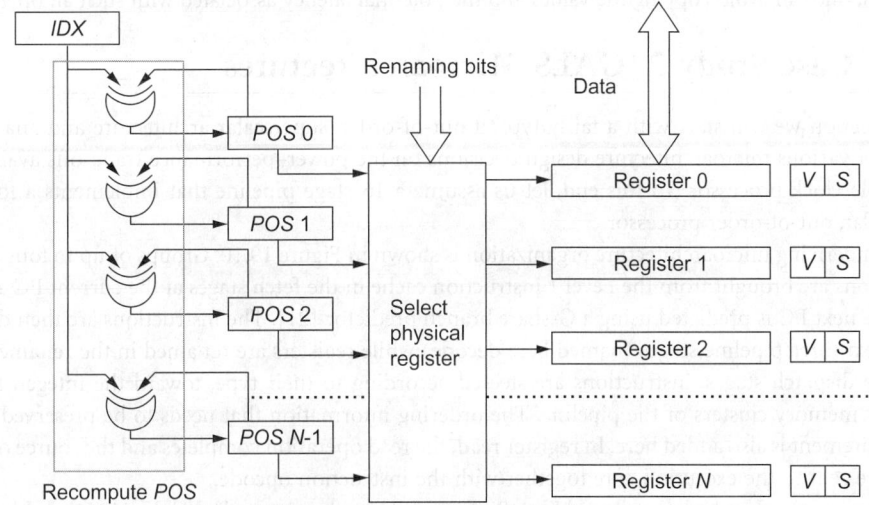

FIGURE 19.9   Architected register structure.

The N indices (values $POS\ 0 - POS\ N\text{-}1$) are initialized with consecutive values $(0, 1, 2, \ldots, N\text{-}1)$ and represent the logical order of the registers in the circular queue. IDX is a pointer in this queue, representing the most recent register used for writing. All accesses (to the value field or to the status bits V and S) are associative, comparing the renaming information against the POS tags. The physical registers with tag 0 ($POS = 0$) hold the actual value when a trace starts execution and the POS tags remain constant until the trace ends, representing the logical order (and the allocation order) of the circular queue.

19.4.3   Register Renaming

When the instruction reaches the renaming stage, physical registers must be assigned as its source and destination registers. The IDX value is the index of the physical register holding the last value written to that architected register. It is incremented (modulo N) for each "write" operation so successive writes to the same architected register will actually use different physical registers. For each "read", the IDX value is read and assigned to the instruction as a physical register indicator.

The S bit is checked for the corresponding physical register and, if it is found set, the pipeline is stalled. Otherwise, S is set and the V bit is reset to mark the value as not yet available. V will be set when the result is computed and written back to the register, while S will be deleted later, when the instruction is retired.

Each trace generation is done with an initial value of $IDX = 0$, meaning that the correct value for the register is stored in the location marked by $POS = 0$. If this condition is respected, all the subsequent executions of a trace can be done without further renaming the registers. The caveat is that this requires a *checkpoint* to be performed when a trace execution ends: the POS values must be recomputed for the circular buffer, so each time it starts with the latest value for that architected register. This can be done by subtracting the IDX value from the POS, but this would require a separate adder for each physical register. However, since the physical order of the registers is not relevant and it does not have to match the logical one, the same effect can be obtained by XOR-ing the IDX with each POS value. By doing so, all registers have different tags ranging between 0 and $N - 1$ and the register holding the last value receives $POS = 0$. An example of register renaming is presented in Appendix A.

As can be seen, all renaming information is already present. Since the execution starts with all valid values for the architected registers in the physical register zero and the first "writes" uses register one as a destination, they will not destroy the old value. Of course, when reexecuting the trace, physical register zero in each pool may be different from that of the last time. The associative scheme ensures that all the traces will see the same register configuration each time they are executed. Another valid option here is to physically copy the values toward the origin of each register queue when a trace ends. However, by using the associative approach, one can avoid copying the values and the potential latency associated with such an operation.

19.5   Case Study 2: GALS Microarchitectures

In this section we will start with a fairly typical out-of-order, superscalar architecture and analyze the impact of various microarchitecture design decisions on the power–performance trade-offs available in a multiple clock processor. To this end, let us assume a 16-stage pipeline that implements a four-way superscalar, out-of-order processor.

The underlying microarchitecture organization is shown in Figure 19.10. Groups of up to four aligned instructions are brought from the Level 1 instruction cache in the **fetch** stages at the current PC address, while the next PC is predicted using a G-share branch predictor [21]. The instructions are then decoded in the next three pipeline stages (named here **decode**) while registers are renamed in the **rename** stages. After the **dispatch** stages, instructions are steered according to their type, toward the integer, floating point, or memory clusters of the pipeline. The ordering information that needs to be preserved for in-order retirement is also added here. In **register read**, the read operation completes and the source operand values are sent to the execution core together with the instruction opcode.

Instructions are placed in a distributed Issue Buffer (similar to the one used by Alpha 21264) and reordered according to their data dependencies. Independent instructions are sent in parallel to the out-of-order

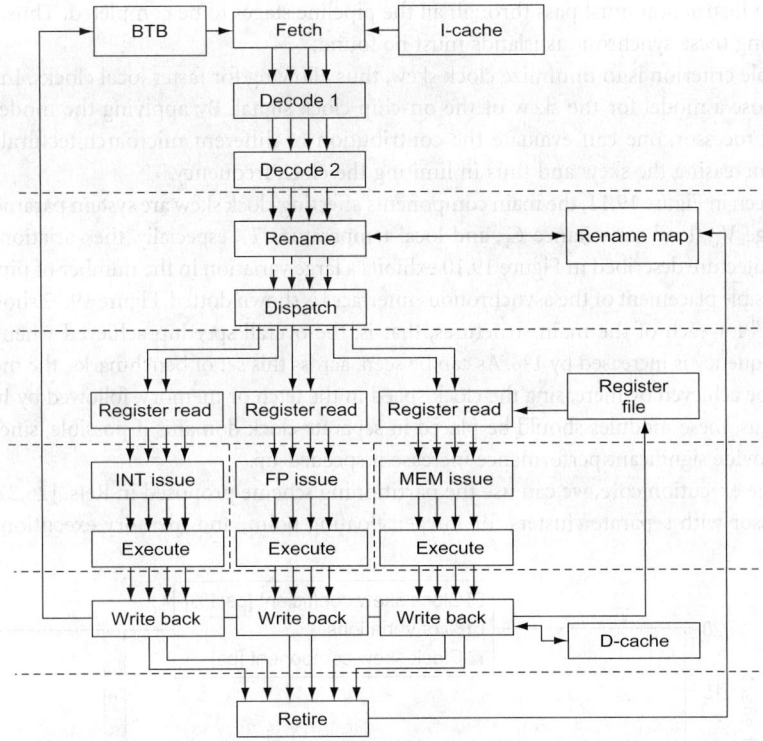

FIGURE 19.10   The baseline microarchitecture.

execution core. The execution can take one or more clock cycles (depending on the type of functional unit that executes the instruction) and the results are written back to the register file in the **write back** stages. Finally, the instructions are reordered for in-order retirement, according to the tags received during **dispatch**.

Of extreme importance for our design exploration is the choice of various design knobs that impact the overall power–performance trade-offs in GALS processors. Since the primary focus is on the microarchitecture level, we chose to omit several lower-level issues in this study.

- *Local clock generation.* Each clock domain in a GALS system needs its own local clock generator; ring oscillators have been proposed as a viable clock generation scheme [22,23].
- *Failure modeling.* A system with multiple clock domains is prone to synchronization failures; we do not attempt to model these since their probabilities are rather small for the communication mechanisms considered (but nonzero) [23,24] and this type of microarchitecture does not target mission-critical systems.

Instead, we are focusing on the following microarchitecture design knobs:

- The choice of the **communication** scheme among frequency islands.
- The granularity chosen for the **frequency islands**.
- The **dynamic control strategy** for adjusting voltage/speed of clock domains so as to achieve better power efficiency.

19.5.1   The Choice of Clock Domain Granularity

To assess the impact of introducing a mixed-clock interface on the overall performance of the baseline pipeline, let us assume that the pipeline is broken into several synchronous blocks. The natural approach—minimize the communication over the synchronous blocks' boundaries—does not necessarily work here. An instruction must pass through all the pipeline stages to be completed. Thus, other criteria for determining these synchronous islands must be found.

One possible criterion is to minimize clock skew, thus allowing for faster local clocks. In Ref. [25], the authors propose a model for the skew of the on-chip clock signal. By applying the model to an Alpha 21264 microprocessor, one can evaluate the contribution of different microarchitectural and physical elements in increasing the skew and thus in limiting the clock frequency.

As can be seen in Figure 19.11, the main components affecting clock skew are system parameter variations (supply voltage V_{dd}, load capacitance C_L, and local temperature T), especially, the variations in C_L. Since the microarchitecture described in Figure 19.10 exhibits a large variation in the number of pipeline registers clocked, a possible placement of the asynchronous interfaces is shown dotted. Figure 19.12 shows the "speed-up coefficient" for each of the main structures, that is, the overall speedup achieved when the domain's local clock frequency is increased by 1%. As can be seen, across this set of benchmarks, the most significant speedup can be achieved by increasing the clock speed in the fetch or memory, followed by Integer and FP partitions. Thus, these modules should be placed in separate clock domains if possible, since individually they could provide significant performance increase if speeded-up.

To break the execution core, we can use the partitioning scheme proposed in Refs. [26,27]. By starting with a processor with separate clusters for integer, floating point, and memory execution units (much

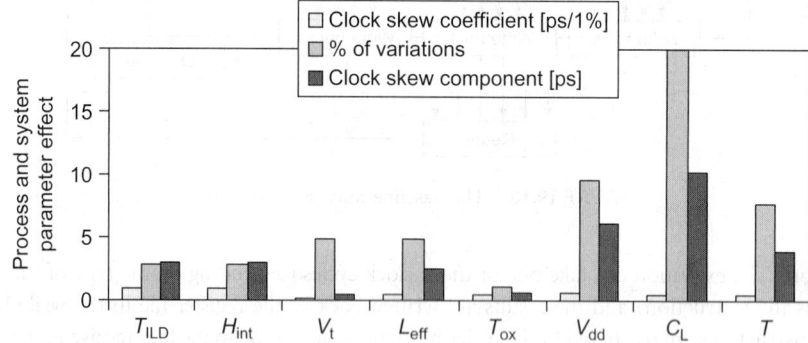

FIGURE 19.11 Total clock skew for Alpha 21264 [22] as function of interconnect parameters T_{ILD} (interlevel dielectric thickness), and H_{int} (interconnect thickness); technology parameters V_t (threshold voltage), L_{eff} (transistor channel length), and T_{ox} (oxide thickness); and system parameters V_{dd} (supply voltage), C_L (load capacitance), and T (temperature).

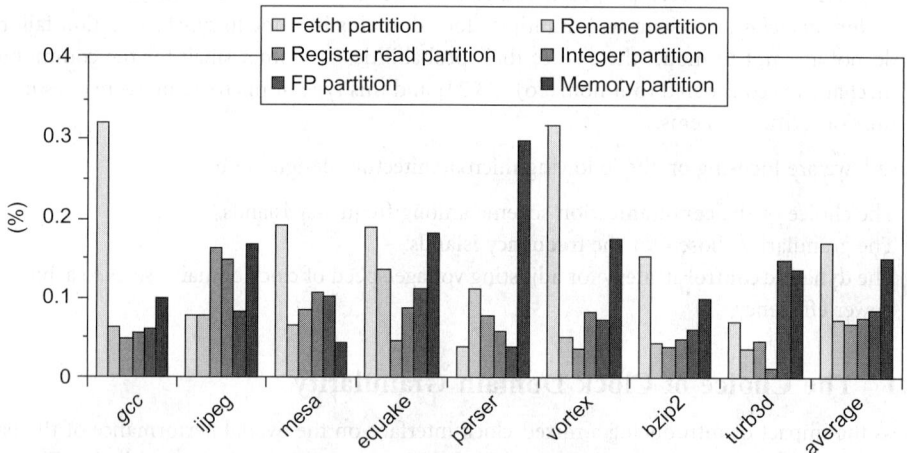

FIGURE 19.12 Performance increase for 1% per-domain speed-up.

like the Alpha 21264 design) we can naturally separate these clusters into three synchronous modules. The drawback of this scheme is that it increases significantly the latency of forwarding a result across the asynchronous interface toward another functional unit. This effect can be seen mainly in load-use operations that are executed in separate clock domains, imposing a significant penalty on the overall performance in some programs.

To limit the latency of reading or writing data from the registers, the register read and the write back stages must be placed together, in the same synchronous partition as the register file. Following the same rationale, the rename and retire stages both need to access the rename table, so they must be placed in the same partition.

Following these design choices, we can now split the pipeline into at least four clock regions. The first one is composed of the fetch stage, together with all the branch prediction and instruction cache logic. The two decode stages can be included either in the first clocking region or in the second one—all the instructions that pass from fetch to decode will be passed down the pipeline to rename.

To limit the load capacitance variations and also considering the bitwidth increase of the pipeline after Decode, we can introduce an asynchronous boundary here. The second clocking region will be organized around the renaming mechanism and it will also contain the reorder buffer and the retire logic. Given the variation in the register width for the rest of the pipeline, an asynchronous boundary can also be introduced after dispatch. The third clocking region must be organized around the register file, including the register read and write back stages. Finally, the out-of-order part of the pipeline (the Issue logic and the execution units) is split into separate clusters that amount to three different clock regions. The forwarding paths can thus be internal—toward a unit with the same type and placed in the same clock region—or external—toward other clock regions.

19.5.2   Choice of Interdomain Communication Scheme

One of the most important aspects of implementing a GALS microprocessor is choosing an asynchronous communication protocol. For high-performance processors, the bandwidth and latency of the internal communication are both important and a trade-off is harder to identify. Several mechanisms have been proposed for asynchronous data communication between synchronous modules in a larger design [23].

The conventional scheme to tackle such problems is the extensive use of synchronizers—a double-latching mechanism that conservatively delays a potential read, waiting for data signals to stabilize as shown in Figure 19.13(a). Even though data are produced before time step 2, the synchronizer enforces its availability at the consumer only at time step 4. This makes classical synchronizers rather unattractive, as their use decreases performance and the probability of failure for the whole system rises with the number of synchronized signals.

Pausable clocks (Figure 19.13[b]) have been proposed as a scheme that relies on stretching the clock periods on the two communicating blocks until the data are available or the receiver is ready to accept them [28]. If T is greater than an arbitrary threshold, then the read can proceed, otherwise the active

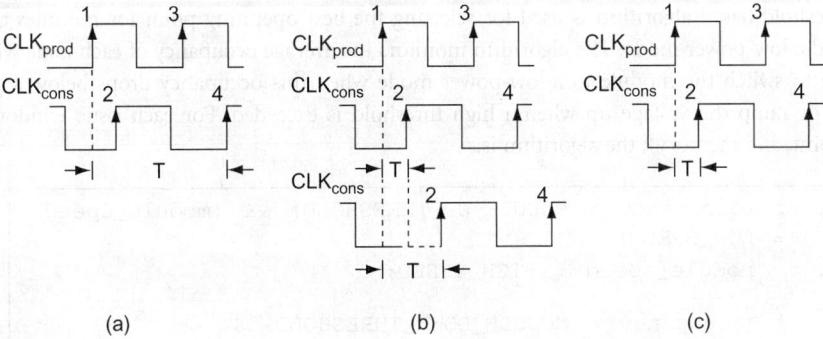

         (a)   (b)   (c)

FIGURE 19.13   Timing diagram—synchronizers (a), pausable clocks (b), and arbiters (c), CLK_{prod} and CLK_{cons} are the producer and consumer clock signals, respectively.

edge 2 of the consumer clock is delayed. While the latency is better in this approach, it assumes that asynchronous communication is infrequent. Stretching the clock is reflected in the performance of each synchronous block and thus, it is most effective when the two blocks use a similar clock frequency. It can also be an effective mechanism when the whole block must wait anyway until data are available.

Another approach is to use arbiters for detecting any timing violation condition—Figure 19.13(c). In this case, data produced at time step 1 may be available at time step 2 if T is larger than a certain threshold. While the mechanism is conceptually similar to that of synchronizers, it offers a smaller latency.

Asynchronous FIFO queues have been proposed in Ref. [29], using either synchronizers or arbiters. Such an approach works well under the assumption that the FIFO is neither completely full, nor completely empty. The scheme retains the extra latency introduced by the use of synchronizers, but improves the bandwidth through pipelining. For the nominal operation of this structure (when the FIFO is neither empty, nor full), a potential read is serviced using a cell different from the one handling the next write, so both can be performed without synchronization.

All these mechanisms reduce the error probability to very low levels, but they cannot ensure that metastability will never occur. However, as Ginosar [24] showed recently, the error rate can be reduced as much as it is desired. Typically, the mean time to failure is of the order of hundreds of years, at least an order of magnitude higher than the time between soft error occurrences [30] or the expected life of the product.

19.5.3   Choice of Dynamic Control Strategy

One of the main advantages offered by the GALS approach is the ability to run each synchronous module at a different clock frequency. If the original pipeline stages are not perfectly balanced, the synchronous blocks that we obtain after the partitioning can naturally be clocked at different frequencies. For example, if the longest signal path belongs to the Register Renaming module, in the GALS approach, we could potentially run the Execution Core at a higher clock frequency than the fully synchronous design.

Furthermore, even if we start with a perfectly balanced design (or we resize transistors to speed up longer signal paths), we can slow down synchronous blocks that are off the critical path, while keeping the others running at nominal speed. The slower clock domains could also operate at a lower supply voltage, thus producing additional power savings. Since energy consumption is quadratically dependent on V_{dd}, reducing it can lead to significant energy benefits, while latency (D) is increased accordingly.

$$D \propto \frac{V_{dd}}{(V_{dd} - V_t)^\alpha} \qquad (19.2)$$

where α is a technology-dependent factor, which is 1.2–1.6 for current technologies [31] and V_t the threshold voltage.

To exploit nonuniform program profiles and noncriticality of various workloads, different schemes have been proposed for selecting the optimal frequency and voltage supply in a GALS processor. In Ref. [26], a simple threshold-based algorithm is used for selecting the best operating point for modules that have a normal and a low-power mode. The algorithm monitors the average occupancy of each issue window and can decide to switch the module to a low-power mode when this occupancy drops below a predefined threshold, or ramp the voltage up when a high threshold is exceeded. For each issue window (integer, floating point, and memory), the algorithm is:

```
1. if (occupancy > MODULE_UP_THRESHOLD) && (module_speed
   == LOW_SPEED)
2.    module_speed = HIGH_SPEED;

3. if (occupancy < MODULE_DOWN_THRESHOLD) &&
   (module_speed == HIGH_SPEED)
4.    module_speed = LOW_SPEED;
```

A more complex model is proposed in Ref. [27]. Here, an attack-decay algorithm is assumed for selecting the best-operating point for processors that offer a wide range of frequencies and supply voltages. The algorithm monitors the instruction window occupancy and, based on its *variation*, decides whether the frequency should be increased or decreased. Any significant variation triggers a rapid change of the clock frequency to counter it. For small or no variations, the clock frequency is decayed continuously, while monitoring performance (Appendix B).

The instruction window occupancy is not the only significant aspect that can be considered for deciding a switch. Even though an instruction window could have high occupancy, this could be due to a bottleneck in another cluster. If load operations are delayed, it is very likely that instructions will accumulate in the integer cluster as well. However, speeding up the clock in the integer domain will not improve the performance. In this case, taking decisions based only on local issue queue occupancy will not help and the number of *interdomain data dependencies* (that is, the number of pending dependencies to or from another clock domain) may be more significant than the issue window occupancy.

Furthermore, both Refs. [26,27] allow DVS just for the execution core, assuming that the clock speed of the front-end is critical for the overall performance, and thus should not be reduced. However, there are large sequences of code, where the usable parallelism (defined here in terms of IPC—instructions committed per clock cycle) is significantly smaller than the theoretical pipeline throughput. In these cases, it makes sense to reduce the speed of the front-end since it produces more instructions than can be processed by the back-end.

With these observations, we can modify the previously described methods to include both information about the number of *interdomain dependencies* and DVS algorithm for the *front end* of the pipeline. Thus, we obtain:

```
1. if ((inter_domain_dependencies + occupancy) >
       MODULE_UP_THRESHOLD) && (module_speed == LOW_SPEED)
2.     module_speed = HIGH_SPEED;

3. if ((inter_domain_dependencies + occupancy) <
       MODULE_DOWN_THRESHOLD) && (module_speed == HIGH_SPEED)
4. module_speed = LOW_SPEED;
```

For the front-end clock domain, the algorithm is:

```
1. if (front_end_throughput / back_end_throughput)
       < FRONT_END_UP_THRESHOLD
2.   module_speed = HIGH_SPEED;

3. if (front_end_throughput / back_end_throughput
       > FRONT_END_DOWN_THRESHOLD)
4.   module_speed = LOW_SPEED;
```

A similarly modified algorithm can be derived from the attack-decay approach, using the same combined metric and allowing for variations in the front-end frequency.

19.6   Conclusions and Future Research Directions

When it comes to performance, superscalar processor designers have always been the last to accept a possible compromise. As intended for applications, where raw performance is the primary target, the last bit of potential efficiency is usually squeezed from each architectural design. Traditionally, power dissipation issues have been addressed mainly at the technology level, through lower supply voltages, smaller transistors, SOI technology, better packaging, etc. Nevertheless though, power dissipation has lately become one of the design constraints for modern processors, and thus the microarchitecture designer must now take power requirements into consideration as well. We have already reached the limits of our capabilities for economical cooling and power delivery, and future microarchitecture designs will have to fit within a roughly similar power envelope.

This reality makes power consumption a first rate design challenge, and future microprocessors will have to spend less power per instruction to become economically viable. While various DVS techniques can lower power consumption in cases where high performance is not required, this is primarily a solution for portable devices. In general, processors need to become more power efficient, performing less-speculative work, which is eventually thrown out and utilizing less logic real estate for performing useful work. In this respect, techniques like the EC help in reducing the switched capacitance per committed instruction. By concentrating the activity in the back-end portion of the pipeline, this method allows the entire front-end of the processor to be gated off to limit the dynamic power consumption.

A different challenge comes from the increased variability associated with smaller process technologies. Since it becomes harder to make chip-wide decisions without sacrificing performance and power efficiency, local solutions will start to gain importance. The GALS design methodology fits very well into this trend, allowing logic islands to communicate asynchronously on a per-need basis. Each such island can be designed and optimized separately, limiting the impact of process variability on the entire chip.

Finally, the decreasing feature sizes translate into smaller transistor junctions and thinner isolator layers. Such transistors leak current even when they do not switch state, and this leakage power has already become a problem for the existing process technologies. While dynamic power consumption can be limited by clock gating, this static power is consumed by all transistors of the design, even when they do not perform any work at all. Depending on the type of logic block, there are several solutions for dealing with this problem (see Section 19.3.3), but all of them have drawbacks and tend to affect performance. Since the leakage current increases exponentially with each new process technology, static power consumption is gradually gaining the same importance as the dynamic power.

References

1. INTEL Corp, "Intel Pentium 4 Processors 560, 550, 540 and 520 Supporting Hyper-Threading Technology on 90 nm Process Technology", Datasheet, Document 302351-002, www.intel.com.
2. INTEL Corp, "Intel XScale Core Developer's Manual", Manual, Document 27347302, www.intel.com.
3. Philips Corp, "80C51 8-bit microcontroller family", Datasheet, Document 8XC51_8XC52_6, www.semiconductors.philips.com.
4. R.E. Kessler, E.J. McLellan, and D.A. Webb, "The Alpha 21264 Microprocessor Architecture", in *IEEE Micro*, 19(2), pp. 24–36, March 1999.
5. Hewlet Packard Corp, "*PA-RISC 1.1. Architecture and Instruction Set Reference Manual*", Hewlett Packard, third edition, 1994, HP Part Number 09740-90039.
6. A. Klaiber, "*The Technology Behind Crusoe Processors*", http://www.transmeta.com.
7. V. Tiwari, S. Malik, and P. Ashar, "Guarded Evaluation: Pushing Power Management to Logic Synthesis/Design", in *International Symposium on Low Power Design*, April 1995.
8. F. Theeuwen and E. Seelen, "Power Reduction Through Clock Gating by Symbolic Manipulation", in *Workshop on Logic and Architecture Synthesis*, 1996.
9. R. Moreno, L. Pinuel, S. del Pino, and F. Tirado, "Power-Efficient Value Speculation for High-Performance Microprocessors", in *Proceedings of the Euromicro*, p. 1292, 2000.
10. J.L. Aragón, J. González, and A. González, "Power-Aware Control Speculation through Selective Throttling", in *Proceedings of the High-Performance Computer Architecture*, 2003.
11. A. Iyer and D. Marculescu, "Power Aware Microarchitecture Resource Scaling", in *Proceeding of the IEEE Design, Automation and Test in Europe Conference (DATE)*, Munich, Germany, March 2001.

12. INTEL Corp, "Trace based instruction caching", US Patent US6170038.

13. B. Solomon, A. Mendelson, D. Orenstein, Y. Almog, and R. Ronen, "Micro-Operation Cache: A Power Aware Frontend for Variable Instruction Length ISA" in *Proceedings of the International Symposium on Low-Power Electronics and Design*, pp. 4–9, August 2001.

14. E. Talpes and D. Marculescu, "Reusing Scheduled Instructions for Improving the Power Efficiency of Superscalar Processors", in *IEEE Transactions on VLSI*, 13, May 2005.

15. L.T. Clark, "Circuit Design of Xscale™ Microprocessors" in *2001 Symposium on VLSI Circuits, Short Course on Physical Design for Low-Power and High-Performance Microprocessor Circuits*, June 2001.

16. INTEL Corp, "Mobile Intel Pentium III Processors, Intel SpeedStep Technology", Datasheet, Document CS-007509, www.intel.com.

17. H. Zhigang, A. Buyuktosunoglu, V. Srinivasan, V. Zyuban, H. Jacobson, and P. Bose, "Microarchitectural Techniques for Power Gating of Execution Units", in *Proceedings of the International Symposium on Low-Power Electronics and Design*, 2004.

18. S. Yang, B. Falsafi, M. D. Powell, K. Roy, and V.J. Vijaykumar, "An Integrated Circuit/Architecture Approach to Reducing Leakage in Deep-Submicron High-Performance I-Caches", in *Proceedings of the International Symposium on High-Performance Computer Architecture*, 2001.

19. K. Flautner, N. Kim, S. Martin, D. Blaauw, and T. Mudge, "Drowsy Caches: Simple Techniques for Reducing Leakage Power", in *Proceedings of the International Symposium on Computer Architecture*, 2002.

20. B.R. Rau, M.S. Schlansker, and P.P. Tirumalai, "Code Generation Schema for Modulo Scheduled Loops," in *Proceedings of the 25th Annual International Symposium on Microarchitecture*, December 1992.

21. S. McFarling, "Combining branch predictors", *Technical Report DEC WRL Technical Note TN-36*, DEC Western Research Laboratory, 1993.

22. R. Ronen, A. Mendelson, K. Lai, Shih-Lien Lu, F. Pollack, and J.P. Shen, "Coming Challenges in Microarchitecture and Architecture", in *Proceedings of the IEEE*, 89(3), March 2001.

23. J. Muttersbach, T. Vilinger, and W. Fichtner, "Practical Design of Globally-Asynchronous Locally Synchronous Systems", in *Proceedings of the 6th International Symposium on Advanced Research in Asynchronous Circuits and Systems*, Israel, April 2000.

24. R. Ginosar, "Fourteen Ways to Fool Your Synchronizer", in *Proceedings of the International Symposium on Asynchronous Circuits and Systems*, 2003.

25. P. Zarkesh-Ha, T. Mule', and J.D. Meindl, "Characterization and Modeling of Clock Skew with Process Variations", in *IEEE Custom Integrated Circuit Conference*, pp. 441–444, May 1999.

26. A. Iyer and D. Marculescu, "Power Efficiency of Multiple Clock, Multiple VoltageCores", in *Proceedings of the IEEE/ACM International Conference on Computer-Aided Design (ICCAD)*, San Jose, CA, Nov. 2002.

27. G. Semeraro, G. Magklis, R. Balasubramonian, D. Albonesi, S. Dwarkadas, and M. L. Scott, "Energy-Efficient Processor Design Using Multiple Clock Domains with Dynamic Voltage and Frequency Scaling", in *Proceedings of the Symposium on High-Performance Computer Architecture*, February 2002.

28. J. Muttersbach, T. Villiger, H. Kaeslin, N. Felber, and W. Fichtner, "Globally Asynchronous Locally Synchronous Architectures to Simplify the Design of On-Chip Systems", in *Proceedings of the 12th IEEE International ASIC/SOC Conference*, September 1999.

29. T. Chelcea and S. Nowick, "Robust Interfaces for Mixed Systems with Application to Latency-Insensitive Protocols", in *Proceedings of the Design Automation Conference*, June 2001.

30. P. Shivakumar, M. Kistler, S. Keckler, D. Burger, and L. Alvisi, "Modeling the Effect of Technology Trends on the Soft Error Rate of Combinational Logic", in *Proceedings of the International Conference on Dependable Systems and Networks*, 2002.

31. K. Chen and C. Hu, "Performance and Vdd Scaling in Deep Submicrometer CMOS", in *IEEE Journal of Solid State Circuits (JSSC)*, vol. 33, no. 10, pp. 1586–1589, October 1998.

Appendix A. Example of Register Renaming

For an easier understanding of the renaming mechanism, let us assume a simple test trace. Assume a code is executed in a loop, with each iteration consisting of the following sequence:

```
Addr0:
Mov r1, #5          ;r1 <- 5
Mov r2, r0          ;r2 <- r0
Add r3, r1, r0      ;r3 <- r1 + r0
Sub r2, r3, r1      ;r2 <- r3 - r1
Xor r1, r1, r2      ;r1 <- r1 ^ r2
Jmp Addr0           ;go back to the beginning
```

Let us assume that the trace will be created unrolling only two successive iterations of this loop (this can hold or the trace can be longer, but it does not make a difference for the register renaming algorithm). For this example, we also assume the use of four physical registers for each architected register. After unrolling, the trace becomes:

```
 0. Mov r1, #5          ;r1 <- 5
 1. Mov r2, r0          ;r2 <- r0
 2. Add r3, r1, r0      ;r3 <- r1 + r0
 3. Sub r2, r3, r1      ;r2 <- r3 - r1
 4. Xor r1, r1, r2      ;r1 <- r1 ^ r2
 5. Jmp Addr0           ;go back to the beginning (will continue with
                        ;6 after the loop is unrolled in the EC)
 6. Mov r1, #5          ;r1 <- 5
 7. Mov r2, r0          ;r2 <- r0
 8. Add r3, r1, r0      ;r3 <- r1 + r0
 9. Sub r2, r3, r1      ;r2 <- r3 - r1
10. Xor r1, r1, r2      ;r1 <- r1 ^ r2
11. Jmp Addr0           ;go back to the beginning (restart the same
                        trace)
```

When this trace is executed for the first time, the front-end of the pipeline, issue and register renaming is performed. At the beginning of the trace, the *IDX* field is zero for all the four registers used here. The registers are renamed as follows:

```
 0. Mov r1, #5          ;r1.1 <- 5 (IDX is incremented for r1)
 1. Mov r2, r0          ;r2.1 <- r0.0(IDX is incremented for r2, IDX
                        ;is found 0 for r0)
 2. Add r3, r1, r0      ;;r3.1 <- r1.1 + r0.0
 3. Sub r2, r3, r1      ;r2.2 <- r3.1 - r1.1 (IDX is again
                        incremented for r2)
 4. Xor r1, r1, r2      ;r1.2 <- r1.1 ^ r2.2
 5. Jmp Addr0           ;continue with instruction 6
 6. Mov r1, #5          ;r1.3 <- 5
 7. Mov r2, r0          ;r2.3 <- r0.0
 8. Add r3, r1, r0      ;r3.2 <- r1.3 + r0.0
 9. Sub r2, r3, r1      ;r2.0 <- r3.2 - r1.3
10. Xor r1, r1, r2      ;r1.0 <- r1.3 ^ r2.0
11. Jmp Addr0           ;go back to the beginning
```

In representing the above trace, we have used the notation *ri.j* for physical register *j* in the pool associated with the architected register *i*.

When the trace ends, the physical queues for each architected register need to be reorganized. For register zero, *IDX* remains set at zero, so all *POS* fields will be the same. For registers *r1* and *r2*, *IDX* will be zero again (in our example, we have four "writes" per trace for these registers), so the relative register ordering in the circular queue remains the same. For *r3*, *IDX* is two at the end of the trace, so the *XOR* will actually modify the *POS* values. Physical register two will become zero and the rest will receive new positions in the circular queue. When retrieving the same trace from the EC, we execute directly:

```
 0.  Mov  r1.1,  #5
 1.  Mov  r2.1,  r0.0
 2.  Add  r3.1,  r1.1,  r0.0
 3.  Sub  r2.2,  r3.1,  r1.1
 4.  Xor  r1.2,  r1.1,  r2.2
 5.  Jmp  Addr0
 6.  Mov  r1.3,  #5
 7.  Mov  r2.3,  r0.0
 8.  Add  r3.2,  r1.3,  r0.0
 9.  Sub  r2.0,  r3.2,  r1.3
10.  Xor  r1.0,  r1.3,  r2.0
11.  Jmp  Addr0
```

Appendix B. Attack-Decay DVS Algorithm

```
1.  if ((prev_occupancy-occupancy > THRESHOLD) && (old_IPC-IPC <
        THRESHOLD))
2.      module_speed -= ATTACK;
3. else

4. if (occupancy-prev_occupancy > THRESHOLD)
5.      module_speed += ATTACK;
6. else

7. if  (module_speed == HIGH_SPEED) &&
        (counter > MAX_LIMIT)
8.      module_speed -= ATTACK;
9. else

10.    if (module_speed == LOW_SPEED) &&
       (counter > MAX_LIMIT)
11.    module_speed += ATTACK;
12.    else

13.    module_speed -= DECAY;

14.    if (module_speed <= LOW_SPEED) {
15.    module_speed = LOW_SPEED;
16.    counter++;
17.    }
```

(continued)

```
18.      if (module_speed >= HIGH_SPEED) {
19.      module_speed = HIGH_SPEED;
20.      counter++;
21.  }

22.   prev_occupancy = occupancy
```

Architecture and Design Flow Optimizations for Power-Aware FPGAs

Aman Gayasen and
N. Vijaykrishnan
Pennsylvania State University

CONTENTS

Field programmable gate arrays (FPGAs) constitute one of the fastest growing market segments in the semiconductor industry. At the time they were introduced, FPGAs were viewed as devices to implement glue logic, but the modern FPGA has grown to be capable of implementing entire systems. This, coupled with the traditional advantages of using FPGAs, namely, short time to market, ease of use, and reprogrammability, has brought FPGAs to the forefront of today's technology choices.

Even with all the progress FPGAs have made, their power efficiency remains significantly inferior to that of custom ICs (but better than that of processors). For example, a JPEG 2000 implementation on an FPGA consumed three times more power than a JPEG 2000 ASIC [1]. This disadvantage has kept FPGAs away from mobile applications, where low energy consumption is crucial. Moreover, as technology scales, power consumption is becoming an issue not only in mobile applications but also in tethered ones. Therefore, power optimization is crucial to sustain the growth of FPGAs.

Both industry and academia are working toward reducing the power consumption in FPGAs. An example is the announcement of standby mode in some of the upcoming Lattice FPGAs [2]. Altera's Stratix-II FPGAs also boast of an architecture tailored to consume less power. Meanwhile Xilinx uses transistors with different oxide thicknesses and threshold voltages to reduce leakage in the transistors that are not performance-critical. Academic researchers have proposed power reduction techniques at most of the design steps involved in using FPGAs—from technology mapping [3] to placement and routing [4] to bit generation [5,6]. Several architectural and circuit-level studies have also been conducted [7–10].

This chapter first analyzes the power consumption in an FPGA, and then presents methods to reduce it. The power-saving techniques are categorized into architecture-, CAD-, circuit-, or multi-level techniques. All these techniques will focus on SRAM-based FPGAs, such as those offered by Xilinx and Altera.

20.1 FPGA Architectures

We will start with the traditional view of an island-style FPGA (see Figure 20.1). It consists of a two-dimensional (2-D) array of configurable logic blocks (CLBs) in a sea of routing wires. These routing wires connect among themselves through programmable switches, forming *switch blocks*. Similarly, these wires also connect to the CLBs, forming *connection boxes*. We will use the term *segment* or *wire* to refer to a routing wire. A routing *channel* consists of multiple such wires. A *net*, in contrast, refers to a logical signal in the user design, which typically will be routed using a number of routing wires. The term *channel width* refers to the number of wires in the routing channel.

The modern FPGA has grown to be more complex than the one shown in Figure 20.1. Figure 20.2 shows the Virtex-2 FPGA architecture, which represents the state of the art. It stores the configuration

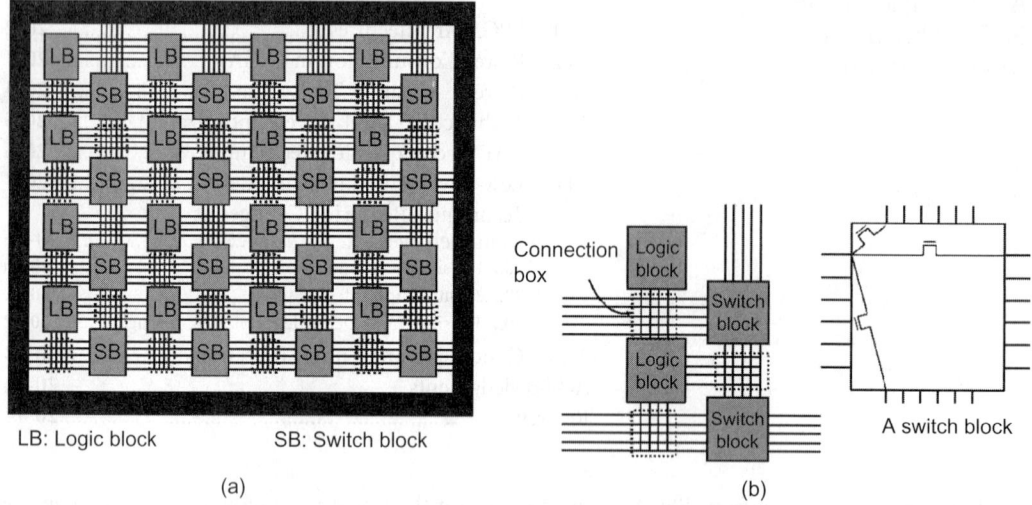

FIGURE 20.1 Traditional FPGA architecture.

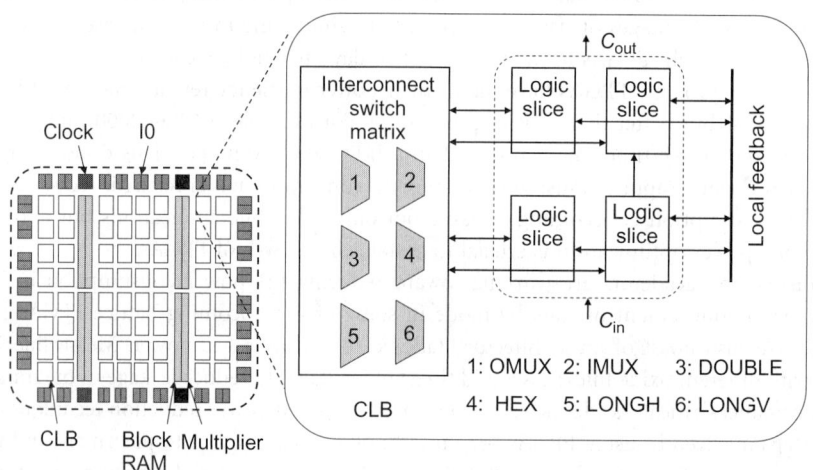

FIGURE 20.2 Virtex-2 FPGA architecture.

information in SRAM cells, each of which consists of six transistors. The basic logic element in Virtex-2 is called a slice. A slice consists of two look up tables (LUTs), two flip-flops (FFs), fast carry logic, and some wide Muxes [11]. A CLB in turn consists of four slices and an interconnect switch matrix. The interconnect switch matrix consists of large multiplexers (as large as 32-to-1) controlled by configuration SRAM cells. Note that Figure 20.2 is not drawn to scale, and in reality the interconnect switches account for nearly 70% of the CLB area. The FPGA contains an array of such CLBs along with block RAMs (BRAMs), multipliers, and IO blocks. Altera's FPGAs are also similar in technology to Virtex-2.

Antifuse-based FPGAs, offered by Actel, constitute a different category of FPGAs, which can be programmed only once. Actel and Lattice also manufacture some flash-based FPGAs. In this chapter, we will limit the discussion to SRAM-based FPGAs.

20.2   Power Consumption in FPGAs

Power in FPGAs can be divided into dynamic and static powers. Dynamic power is expended only when there is some activity in the circuit, and consists of switching power as well as short-circuit power. Static power consists of leakage currents associated with all transistors in the FPGA, and auxiliary power that is consumed in some analog circuits that use a static current as reference. These analog circuits are used for internal clock generation, and for some IO standards. To summarize,

$$P_{total} = P_{dynamic} + P_{static}$$
$$P_{dynamic} = P_{switching} + P_{short-circuit}$$
$$P_{switching} = \sum_{i=0}^{N_{nets}} 0.5 C(i) V^2 f$$
$$P_{static} = P_{leakage} + P_{aux}$$

where N_{nets} is the total number of nets, $C(i)$ the load capacitance of net i, V the voltage swing of the net, and f the toggle rate.

Switching power dominates the dynamic power, and is much larger than that in ASICs. This is mainly due to a larger interconnect power. Most wires in an FPGA have got programmable switches connected to them, which are configured to implement the required connections. These switches add parasitic capacitances to the wires, which in turn increase the load capacitance of the net that uses the wire. Consequently, the switching power for most nets in an FPGA is much larger than that in ASICs. However, this is not the only reason for the large switching power. Because of limited number of wires available in the FPGA's routing channel, not all nets are routed using the shortest possible path. Instead, some nets take a longer route because the wires needed to implement the shortest path have been assigned to another net. For example, in a timing-driven router, nets which are on the critical path follow the shortest path, whereas those on noncritical paths may take more circuitous routes.

Figure 20.3 shows the breakdown of dynamic power in a Virtex-2 FPGA when random input vectors are applied [12]. Note that the routing fabric consumes most of the power. This is not surprising, because

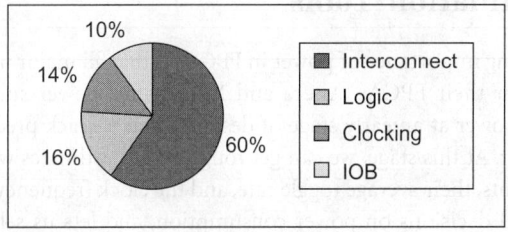

FIGURE 20.3   Dynamic power breakdown for Virtex-2 FPGA. (From L. Shang et al., Dynamic power consumption in Virtex™-II FPGA family, In *Proceedings of ACM/SIGDA International Symposium on Field-Programmable Gate Arrays*, 2002.)

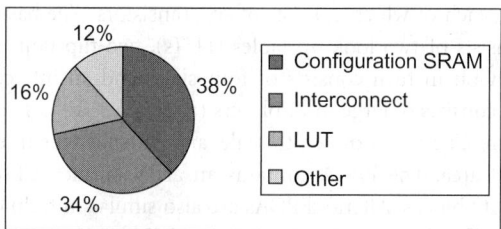

FIGURE 20.4   Leakage power breakdown for Spartan-3 FPGA. (From T. Tuan and B. Lai, Leakage Power analysis of a 90nm FPGA, In *Custom Integrated Circuits Conference*, 2003.)

connections in an FPGA usually go through multiple switches, each of them adding some load capacitance as well as resistance to the net's path. Even if a net does not connect through switch boxes, it is usually loaded by parasitic capacitances of the unused switches that could potentially be utilized for providing connections.

Leakage power dominates the static power, and can be further broken into active leakage—that consumed in the used portion of FPGA—and inactive leakage—that consumed in the transistors that are not used by the design. Inactive leakage is a big portion of total leakage, because for most FPGA designs, a large number of transistors remain unused. In fact, for many designs, inactive leakage is the dominant portion of total leakage.

Figure 20.4 shows the breakdown of leakage power for a Spartan-3 FPGA, which uses 90-nm process technology. Note the large contribution of the configuration SRAMs. These transistors can easily be optimized for leakage, because they do not affect the run-time performance of the FPGA. In fact, more recent FPGAs, such as Virtex-4, claim a much lower leakage power in their configuration SRAM. Since the routing fabric takes up most of the area in the FPGA, it is not surprising that the other major portion of leakage comes from the muxes used to implement programmable interconnect.

Manufacturers usually design FPGAs to support a large variety of customers, with varying demands for resources. This makes power reduction in FPGAs a very challenging task. For example, the routing channel width in a commercial FPGA is kept near maximum, so that almost all designs remain routable. Consequently, even if some users do not need such a large channel width, they pay in terms of power consumption. Recently, some FPGA manufacturers have started offering domain-specific FPGAs, in which the FPGA is optimized for a certain class of applications. Currently, this is done at a larger granularity, for example, by varying the number of DSP blocks, or the embedded memory, based on the application requirements. Such custom embedded blocks affect the performance as well as power of the FPGA. In general, a design, if implemented using custom hard-wired blocks, dissipates much less power (as less as 5–10% [13]) than if implemented using the configurable logic. Therefore, a signal-processing application will benefit both in energy and performance by using the DSP blocks available in some FPGAs.

20.3   Power Estimation Tools

An indication of the growing importance of power in FPGAs is that all major manufacturers now provide power estimation tools for their FPGAs. Altera and Xilinx offer power spreadsheets and web-based tools [14,15] to estimate power at an early stage of design, when we lack precise information about the design's switching behavior. At this stage, we can get rough power estimates with as little information as the number of logic elements, their average toggle rate, and the clock frequency. This helps us understand the impact of architectural decisions on power consumption, and lets us select a design that suits the needs of the application.

In addition to the spreadsheets, both Xilinx and Altera provide detailed power estimators that can read postroute simulation results to improve the accuracy of power estimation. The power estimation relies on capacitance models, which are used to calculate the load capacitances of all nets. For example, the Xilinx power tool, called XPower, reads the postroute simulation results and provides the power of every net and slice in the FPGA.

We also have a power estimator available from the University of British Columbia [16]. This is integrated into VPR, an academic place and route tool for FPGAs [17], and is limited to the architectures that VPR can support. This tool calculates the switching activity of all internal nodes by propagating the input toggle rates, assuming that the inputs are uncorrelated. Since this tool does not take simulation results as an input, its accuracy may be low for some designs. Another, more accurate, power estimator is available from UCLA [8]. This tool back-annotates capacitance and delay values from the routed design, and then performs a cycle-accurate power simulation.

Until now we have discussed the characteristics of power consumption in FPGAs and some tools that estimate this power. The rest of this chapter will focus on power optimization techniques. We have categorized these techniques into architecture-, CAD-, and circuit level, and those that encompass multiple levels.

20.4   Architecture-Level Power Optimization

An FPGA can be broadly divided into logic and routing portions. Since different parameters characterize the architectures of these two portions, we will look at these separately.

The FPGA logic architecture is defined by the LUT size, the cluster size, and by any nonstandard logic components used. While all these architectural factors influence the power consumption, most studies have focused on LUT and cluster sizes.

LUT size affects the FPGA power in several ways. We observe that the power consumed in a single LUT increases with LUT size—for the same function, a six-input LUT dissipates more power than a four-input LUT. However, the number of LUTs needed to implement a design decreases as the LUT size increases. Consequently, there is an optimal LUT size that gives the lowest logic power for the entire FPGA. Note that LUT size influences the number of nets in the routing fabric, and therefore, indirectly, affects the interconnect power too. This implies that deciding the best LUT size demands estimating the total power and not just the logic power. Such a detailed exploration has shown that a LUT size of 4 inputs consumes the least power as well as energy [18]. Interestingly, a size-4 LUT was earlier shown to be the most area efficient as well [19].

The next logic architecture parameter we will discuss is cluster size. Clustering of LUTs has several advantages. Since the intracluster connections are much faster than the intercluster ones, a timing-driven clustering algorithm can pack the LUTs on the critical path into the same cluster. Furthermore, clustering reduces the problem size for the placer, and therefore, improves its run time as well as performance. Clustering can also help to reduce the load capacitance, and consequently power, for short connections that get absorbed within the cluster. Li et al. [8] explored different clusters consisting of 4, 8, or 12 LUTs, and concluded that a 12-LUT cluster gives the lowest power-delay product. This happens because the larger cluster size helps reduce the interconnect resources between logic blocks, and therefore, reduces interconnect power.

The FPGA routing architecture is described by the lengths of the segments, the connection flexibilities, and the types of routing switches used. For example, a Virtex-2 FPGA uses segments of lengths 1, 2, 6, and long (spanning entire row or column), with all of the routing switches being buffered. The connection flexibility for a switch block is defined as the number of connections per segment in the switch block. In Virtex-2, this number varies depending on the segment, and is not a constant for the switch box.

Segment length affects both delay and power of the nets. Short segments lead to more hops in the routing of a net. The switches connecting these segments add resistance and capacitance to the net's route, and therefore, increase the net's delay and power. In contrast, a very long segment may force the net to use a longer route, which again increases the capacitance of the net, and consequently the delay as well as power. An optimal segment length would minimize the routing energy, and most likely, an architecture containing multiple segment lengths would be better than one with all segments of the same length. We refer the reader to Vassiliadis et al. [20] for such an exploratory study.

Another parameter defining a routing architecture is the type of routing switches used to connect segments to each other. Li et al. [8] explored three kinds of routing architectures. In the first architecture, 50% of the routing switches used buffers, and the other 50% used pass transistors. In the second, all the routing switches used pass transistors, and in the third, all of them used buffers. The first architecture gave the lowest power and power-delay product.

The reader should be warned that these architectural results depend very strongly on the technology and circuits used to implement the FPGA. Quite possibly, the energy-optimal points will change when the implementation technology is changed, or a different circuit style is adopted.

20.5   CAD Techniques for Reducing Power

A typical FPGA CAD flow begins with technology mapping, followed by clustering, placement, routing, and finally a step that generates the configuration bits for download on the FPGA (see Figure 20.5). While current commercial tools usually optimize for either area or speed, each of these steps can be optimized for power as well. We visit some such techniques in this section.

Technology mapping converts a netlist composed of logic gates into a netlist of LUTs. A performance-driven mapper minimizes the depth of the combinational network, which has previously been done optimally using tools such as FlowMap [21] and CutMap [22]. These tools model each logic gate in the design as a node in a graph, with all the connections as edges. To map into K-input LUTs, nodes are merged or decomposed to create groups with the number of incoming edges equal to or less than K. To produce a depth-optimal mapping, some of the nodes in the graph are duplicated. However, this increases the number of nodes, and thereby the number of connections in the mapped netlist, which increases the amount of power consumed by the design. A simple technique to make the mapping power-aware would be to discourage the duplication of nodes [3]. Another approach to reduce power in the mapper is to include the switching activities of nets in the mapping algorithm. This technique tries to construct LUTs that absorb high-activity nets, thereby removing them from the netlist. The combination of the above two techniques has been estimated to reduce energy by 7.6% compared to CutMap, and by 16.9% when compared to FlowMap [4]. Another algorithm was proposed by Li et al. [23], which used an efficient

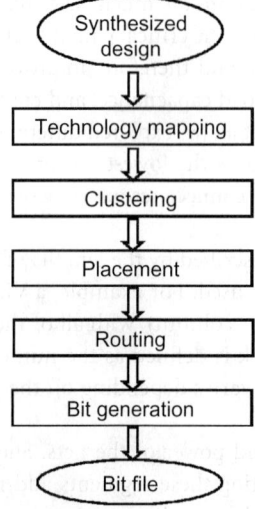

FIGURE 20.5   FPGA CAD flow.

network-flow computation method to perform the low-power technology mapping. They report 14% power reduction compared to CutMap.

The next step in the CAD flow involves clustering (packing) LUTs and FFs into logic blocks. Modern FPGAs use clustered logic blocks, consisting of multiple LUTs, FFs, and other logic elements. The connections within a cluster are faster and consume less power than intercluster connections. Clustering can be viewed as a preplacement step where logic elements are packed together to reduce the complexity of the final placement. The goals of packing could be to minimize area—in which case, it will try to pack all clusters fully, or delay—in which case, it will try to pack logic elements on the critical path together, or routability—in which case, it will minimize the number of distinct nets connected to a cluster. Clustering can also be made power-driven by clustering logic elements such that high-activity nets use the intracluster connections. Since intracluster connections have lower capacitance than the intercluster connections, the total power consumption is reduced if high-activity nets use the former. This technique is estimated to reduce the energy by 12.6% for clusters of size 4 [4]. Another technique, which uses Rent's rule to minimize routing area at the clustering phase, reduces FPGA power by approximately 13% because of a decrease in the active wire length [24].

Clustering is followed by placement, in which these packed clusters are assigned coordinates on the FPGA. VPR uses simulated annealing for placement, with the following cost functions:

$$Cost_{wiring} = \sum_{i=1}^{N_{nets}} q(i)[bb_x(i) + bb_y(i)]$$

$$Cost_{timing} = \sum_{\forall i,j \in circuit} Delay(i,j)\, Criticality(i,j)^{CE}$$

where N_{nets} is the number of nets, $bb_x(i)$ and $bb_y(i)$ are the x and y dimensions of the bounding box of net i, $q(i)$ is a scaling term, $Delay(i,j)$ the estimated delay of the connection from source i to sink j, CE a constant, and $Criticality(i,j)$ an indication of how close to the critical path the connection is [25]. The overall change in the cost, ΔC is calculated as follows.

$$\Delta C = \lambda \frac{\Delta Cost_{timing}}{PreviousCost_{timing}} + (1-\lambda)\frac{\Delta Cost_{wiring}}{PreviousCost_{wiring}}$$

where $PreviousCost_{timing}$ and $PreviousCost_{wiring}$ are normalizing factors updated once every temperature, and λ is a user-defined constant that decides the relative importance of the cost components.

With cost terms to reduce wiring and critical-path delay, the simulated annealing placer, in the current form, does not optimize power. However, the simple structure of the algorithm allows us easily to make it power-aware by adding another cost term that increases the cost for high-activity nets as follows [4]:

$$Cost_{power} = \sum_{i=1}^{N_{nets}} q(i)[bb_x(i) + bb_y(i)]Activity(i)$$

Lamoureux and Wilton [4] showed that the above cost function reduces FPGA power by 6.7%, but simultaneously increases the critical-path delay by 4.0%. Therefore, the power-delay product (energy) remains almost the same.

The next CAD step involves routing the nets. Routing in FPGAs is constrained by the number of tracks and switches available in the fabric. The timing-driven router of VPR uses a pathfinder-based algorithm with the following cost function to evaluate a routing track n while forming a connection from source i to sink j.

$$Cost(n) = Crit(i,j)\,delay(n) + (1 - Crit(i,j))\,congestion(n)$$

where $Crit(i, j)$ is the same as in the placement, $delay(n)$ the Elmore delay of node n, and $congestion(n)$ a cost denoting the congestion at the node n that is updated as routing progresses. Therefore, the router favors a track with lower delay and congestion. Furthermore, it gives priority to delay for nets on the critical path, and to congestion for the other nets.

To make the router power-aware, the cost function can be modified by adding a cost based on activity of the net as follows [4]:

$$Cost(n) = Crit(i, j)\, delay(n) + (1 - Crit(i, j))[ActCrit(i)cap(n) + (1 - ActCrit(i))congestion(n)]$$

where $cap(n)$ is the capacitance associated with routing resource node n and $ActCrit(i)$ the activity criticality of net i:

$$ActCrit(i) = min\left(\frac{Activity(i)}{MaxActivity}, MaxActCrit \right)$$

where $Activity(i)$ is the switching activity of net i, MaxActivity the maximum switching activity of all the nets, and MaxActCrit the maximum activity criticality that any net can have. Using the above cost function reduced the energy by 2.6% on the average.

The final CAD step generates a bit file that stores the FPGA's configuration information. Some power optimizations can be done even at this stage. Anderson et al. [5] observed that a routing mux leaks less when its output is kept at the logic value of 1. Therefore, changing the configuration bits of the LUTs to maximize the probability that their outputs are at logic 1, can reduce FPGA leakage. In fact, this technique reduces the active leakage (leakage in used portion of the FPGA) by 25% on average [5]. Reconfiguring the LUTs also influences their dynamic power. However, finding the power-optimal configurations for all the LUTs in a design is a difficult problem. Therefore, a technique that processes clusters of LUTs and reduces the power in such local clusters was proposed [6]. This reduced the dynamic power by 20.6%.

An important point to remember is that all these techniques are not completely independent, and their individual power savings do not accumulate if all of them are applied. An example is that power optimization in the mapper reduces the power by 7.6% and power-aware clustering reduces it by 12.6%, but when both of them are applied to a design, the total power reduction is only 17.6%, and not the 20.2% it would be if they were perfectly cumulative [4].

20.6 Low-Power Circuit Techniques

Now let us look at some circuit techniques for reducing power in FPGAs. An obvious approach would be to resize all transistors for optimal power. Since this approach applies equally well to any integrated circuit, we do not discuss this further. Instead, in this section, we describe some of the techniques proposed specifically for FPGAs.

One technique to reduce dynamic power consumption in the routing is to use a low-swing interconnect circuit, which uses a lower voltage in the routing fabric. A drawback of most conventional low-swing techniques is the slow speed of the receiver circuit, and the short-circuit current at the receiver end. George et al. [7] mitigated this problem by employing cascode circuitry and differential circuits at the receiver (see Figure 20.6). They assumed pass-transistor switches in the switch blocks, and modified the drivers and receivers for connections with the CLB output and input pins, respectively. This low-swing circuit reduced the energy by a factor of 2 over full-swing interconnect when V_{DDH} of 1.5 V and V_{DDL} of 0.8 V were used.

Another circuit technique uses redundant memory cells to reduce leakage energy (see Figure 20.7). Routing multiplexers in FPGAs are usually implemented in multiple stages to reduce parasitic capacitance in intermediate or output nodes and to minimize the number of programmable memory cells. For example, a two-stage implementation of a pass-transistor-based multiplexer is shown in Figure 20.7(a). It is composed of several smaller multiplexers, and to reduce the total number of SRAM cells, the same

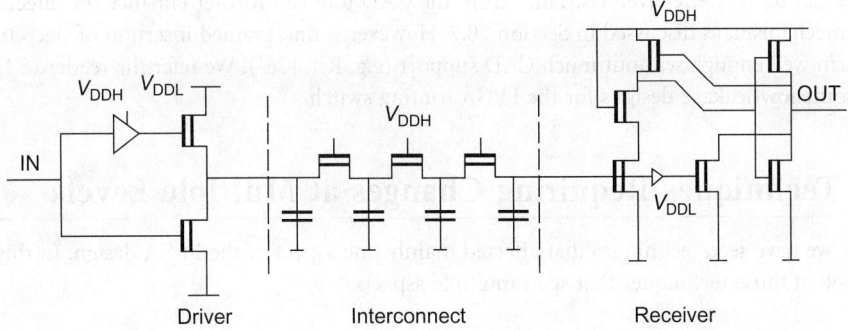

FIGURE 20.6   Low-swing interconnect circuit. (From V. George et al., The design of a low energy FPGA, In *Proceedings of International Symposium on Low Power Electronics and Design*, 1999.)

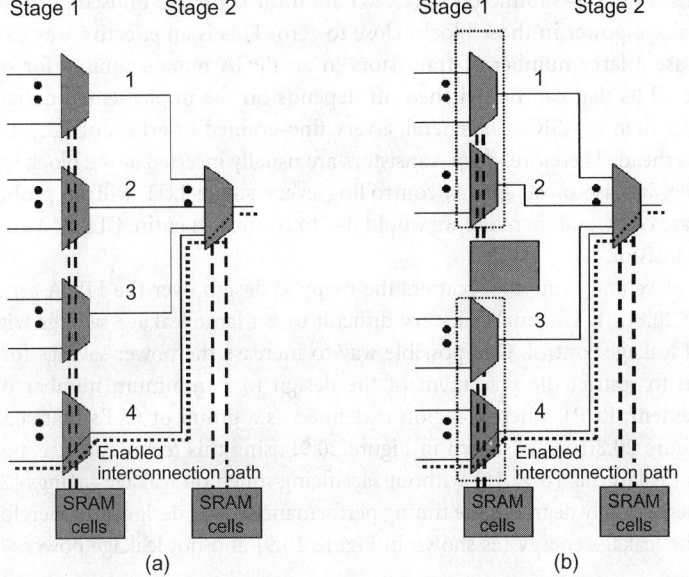

FIGURE 20.7   (a) A two-stage implementation of a pass-transistor-based mux. (b) A two-stage mux with redundant memory cells. (From A. Rahman and V. Polavarapuv, Evaluation of low-leakage design techniques for field programmable gate arrays, In *Proceedings of ACM/SIGDA International Symposium on Field-Programmable Gate Arrays, 2004.*)

SRAM cell configures one pass transistor from each multiplexer in stage 1. When this mux is used, one of the paths from the inputs to the output is activated. However, since all small muxes in stage 1 share their select signals, each of them pass one of their inputs to their outputs. This drives the intermediate nodes 1, 2, 3, and 4 to V_{DD} or V_{SS} (depending on the value of the input passed by the corresponding state 1 mux), which in turn keeps the drain-to-source voltage, V_{DS}, of all disabled pass transistors as V_{DD} or V_{SS}. Under these conditions, the disabled pass transistors with $V_{DS} = V_{DD}$ contribute to leakage power. Figure 20.7(b) shows a low-leakage technique where the pass transistors that are not included in an enabled interconnection path are turned off and the subthreshold leakage current through a series of connected NMOS devices determines intermediate node voltage [9]. Although such an implementation requires additional SRAM cells for granular controllability, they reduce the leakage current of pass transistors in disabled input-to-output paths. Since these SRAM cells can be easily optimized for leakage by high-V_t devices, the impact on total leakage power due to integration of additional SRAM cells is minimal.

Leakage can also be reduced by inserting sleep transistors to cut off the supply to unused portions in an FPGA. Since the FPGA contains a large number of unused transistors for any given user design, sleep transistors can be very effective. Assistance from the CAD tool can further enhance the effectiveness of the sleep mechanism, as discussed in Section 20.7. However, a fine-grained insertion of sleep transistors can perform well enough without much CAD support (e.g. Ref. [26]). We refer the reader to Lodi et al. [10] for some low-leakage designs for the FPGA routing switch.

20.7 Techniques Requiring Changes at Multiple Levels

Until now we have seen techniques that affected mainly one aspect of the FPGA design. In this section, we will look at those techniques that span multiple aspects.

20.7.1 Using Sleep Transistors

Sleep transistors are low-leakage transistors used to cut off the supply to an unused or idle circuit block. They are widely used in deep-submicron ASICs to cut off the supply to unused portions of a design, and thus make the leakage power in those blocks close to zero. This is an effective way to reduce leakage in FPGAs too, because a large number of transistors in an FPGA remain unused for most user designs. Since the specific CLBs that can be switched off depends on the implemented design, sleep transistor insertion is trickier than in ASICs. In general, a very fine-grained insertion of sleep transistors incurs a very large area overhead. Therefore, sleep transistors are usually inserted at the block level instead of gate level. For example, in case of an FPGA, controlling every single LUT will be prohibitive due to the implementation area cost, and therefore we would like to control an entire CLB or a group of CLBs using the same sleep transistor.

Normally, the place and route tool scatters the mapped design over the FPGA array to optimize for speed (see Figure 20.8(a)). This makes it very difficult to get large leakage savings without resorting to very fine-grained leakage control. One possible way to increase the power savings for a coarse-grained leakage control is to restrict the placement of the design to a minimum number of regions (region constrained placement, RCP), where a region is defined as a group of CLBs that share the same sleep transistor (see Figure 20.8(b)). As shown in Figure 20.9, using this technique, it is possible to use very large region sizes (as large as 16 × 16) without sacrificing much on leakage savings [27]. Note that this change in placement usually degrades the timing performance of the design, and therefore, it is important that we look at the leakage energy (as shown in Figure 20.9) and not leakage power.

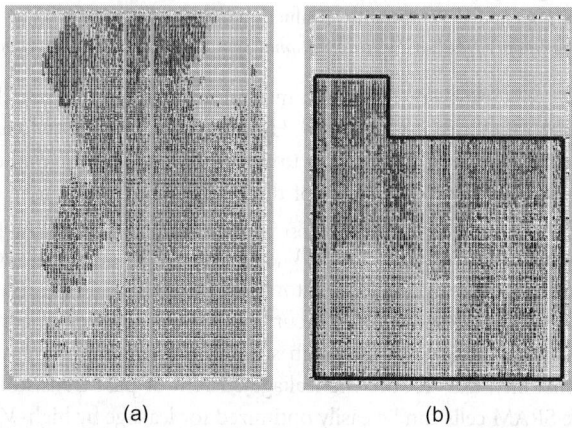

(a)   (b)

FIGURE 20.8   Different placements for an example design. (a) Normal placement. (b) RCP. (From A. Gayasen et al., Reducing leakage energy in FPGAs using region-constrained placement, In *Proceedings of International Symposium on Field-Programmable Gate Arrays*, 2004.)

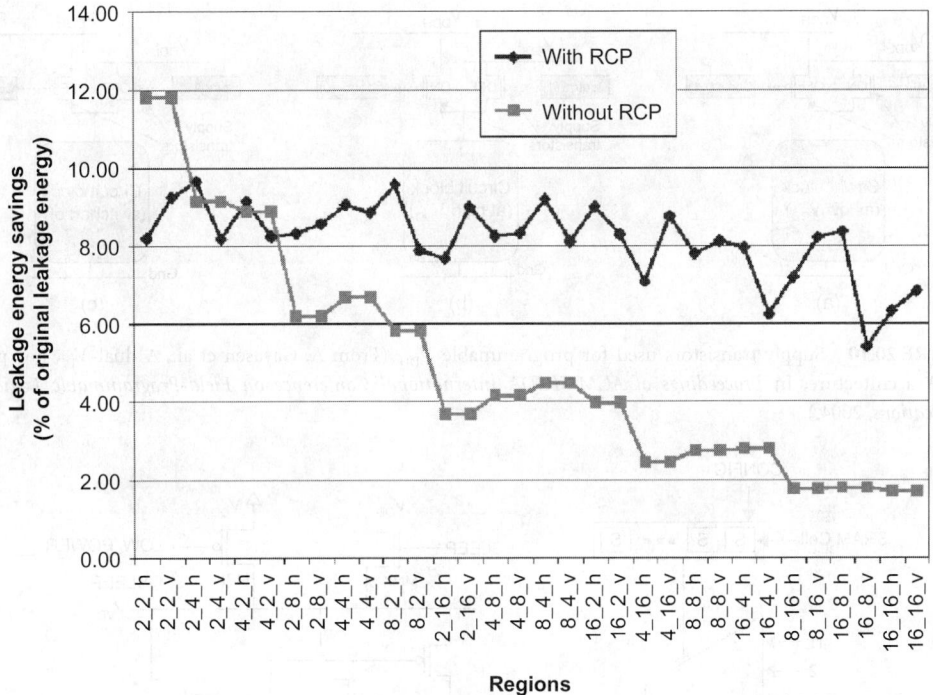

FIGURE 20.9   Average leakage energy savings for RCP and normal placement. *X*-axis shows the *x* and *y* dimensions of regions in the format *x*–*y*. (From A. Gayasen et al., Reducing leakage energy in FPGAs using region-constrained placement, In *Proceedings of International Symposium on Field-Programmable Gate Arrays*, 2004.)

20.7.2   Dual-Supply Techniques

Most user designs have only few combinational paths that are timing-critical and which need to be run at maximum performance. The other noncritical paths can be run at a reduced performance and save power, while maintaining the overall performance of the design. One way to save power in noncritical paths is by reducing the power supply voltage to those paths. It becomes more challenging for FPGAs because the critical paths depend on the user design that is implemented on the FPGA. This prohibits the FPGA manufacturers from fixing a lower supply for portions of the device. There are two ways to circumvent this problem. The first technique fixes the supply voltages for individual CLBs at the time of fabrication of the FPGA, thus creating a mixed pool of fast and slow CLBs [28]. The complexity in this case lies in the CAD tool since it has to make sure that the timing-critical blocks in a user design are mapped to the fast CLBs. Furthermore, there may not be enough fast CLBs in the FPGA to obtain optimal performance for some designs. These difficulties hinder the use of this technique for power reduction.

The second technique uses a CLB-level configurable supply voltage (see Figure 20.10) to lower the supply for a noncritical block [28–31]. Using the circuit shown in Figure 20.10, it is possible to set the supply voltage of every CLB to either high (V_{DDH}) or low (V_{DDL}). The value of the supply is controlled by using two SRAM cells. Although this incurs a large area overhead (since the supply transistors must be sized large for performance), it reduces power by more than 50% compared to a single-supply FPGA. An additional advantage of this technique is that both the supply transistors can be switched off to cut off the supply completely for unused CLBs, and therefore save leakage power. This method needs CAD support to decide which CLBs can be run on a lower supply. Another complication is the need for level converters whenever a low supply block drives a high supply block. A level converter scales a V_{DDL} value to V_{DDH}. Without the level conversion, the PMOS in the V_{DDH} buffer will not be completely switched off, leading to a large static current. However, level converters add area, power, and delay overheads to the

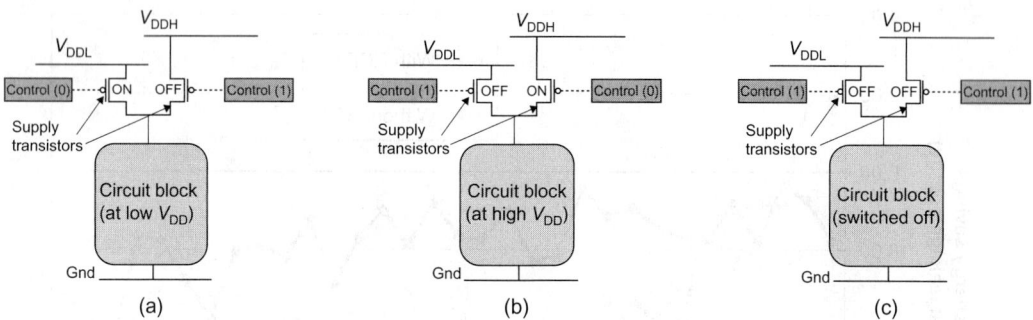

FIGURE 20.10 Supply transistors used for programmable V_{DD}. (From A. Gayasen et al., A dual-V_{DD} low power FPGA architecture, In *Proceedings of ACM/SIGDA International Conference on Field-Programmable Logic and Applications*, 2004.)

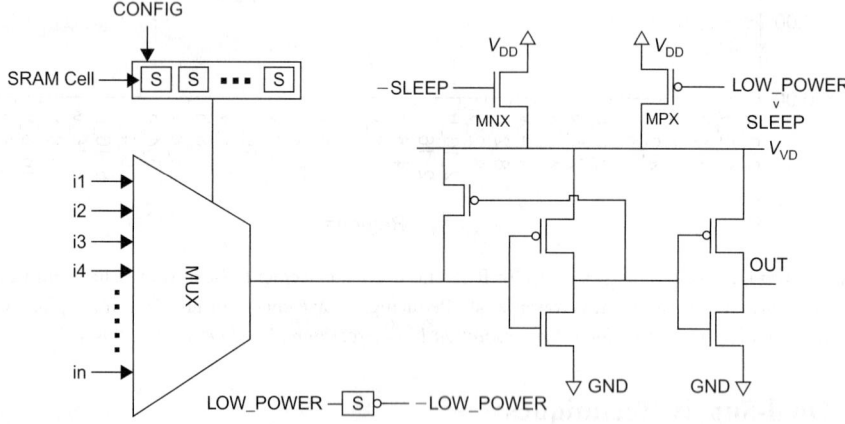

FIGURE 20.11 Low-power routing switch. (From J.H. Anderson and F. Najm, Low power programmable routing circuitry for FPGAs, In *Proceedings of IEEE International Conference on Computer-Aided Design*, 2004.)

implemented design. We can reduce these overheads by minimizing the number of level converters. For example, Gayasen et al. [30] experimented with level converters only at CLB pins.

To reduce the area overhead of a dual-supply FPGA, Anderson and Najm [32] proposed a routing switch that generates two voltage levels from a single supply by using the threshold drop across an NMOS transistor (MNX in Figure 20.11). This eliminates the area penalty associated with routing two power grids. In Figure 20.11, when MPX is switched off, but MNX is kept on, the routing switch functions in this low supply mode. When both MNX and MPX are switched off, then the routing switch goes into the sleep mode. The states of MPX and MNX are controlled using configuration SRAMs.

20.7.3   Using Input Dependence of Mux Leakage

The above technique of using two supply voltages incurs an area penalty of the order of 50% [33], which indicates that it should be used only for applications that can tolerate some increase in cost. Since for most applications this is not the case, we need to explore ways to reduce power with less area penalty. By limiting the V_{DD} configurability to logic blocks, Lin et al. [18] reduced the area overhead to 17%. We have already seen in Section 20.5 that optimizing the CAD-flow for low power reduced the consumption by 22.6%. We have also seen that inverting the logic implemented in some LUTs can possibly reduce leakage in the used portion of the design. This was possible because the mux leakage strongly depends on the values of its inputs.

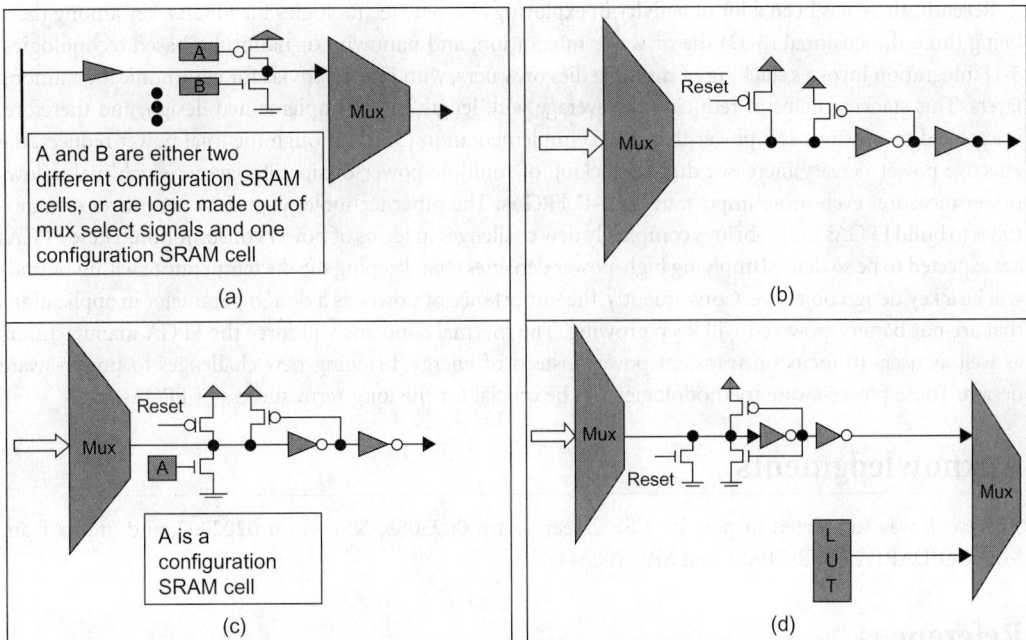

FIGURE 20.12 Circuit modifications to enable input control for routing muxes. (From S. Srinivasan, Leakage control in FPGA routing circuit, In *Proceedings of Asia and South Pacific Design Automation Conference*, 2005.)

Now, we look at another technique that requires small changes in the FPGA circuits to allow the unused inputs of routing muxes to be set at desired values, and hence, reduce leakage. Figure 20.12 shows the modified circuits for the routing muxes [34]. Figure 20.12(a) shows an obvious way to control the inputs of a mux. It uses two configuration bits at every input, which select between "0" and "1." When both are not selected, the user signal is used as input. Although this circuit provides the maximum flexibility, its large area overhead makes it impractical to be used in a real FPGA. This area penalty can be reduced by observing that since most of the inputs of the multiplexers are driven by other muxes, it is sufficient to set the mux *outputs* to desired values. Figure 20.12(b) shows an area-efficient implementation of such a method, where the reset mechanism ensures that all the undriven signals are pulled up to a logic state "1." Figure 20.12(d) shows a similar circuit where the mux output is set to "0" instead of "1." Finally, Figure 20.12(c) shows a circuit to set the mux output at either "0" or "1," depending on what is desired for least leakage. With this capability to set unused segments at desired values, the CAD tool can decide on the optimal configurations of muxes and unused LUTs to reduce leakage.

20.8 Conclusion and Future Directions

Reducing the power consumption in FPGAs is an active research area. With efforts from various research groups both in academia and in the industry, the goal of developing a low-power FPGA that can be used in battery-powered devices looks feasible. Using some simple CAD techniques, we can reduce FPGA power by up to 20%, while more exotic techniques like dual-supply FPGAs can reduce the power by more than 50%.

In this chapter, we discussed the techniques to reduce the power consumption of the FPGA core, consisting of an array of CLBs. The modern FPGA contains other components like DSP blocks, high-speed transceivers, and embedded memory. Although low-power techniques for these components may have been proposed in other contexts, applying them to FPGAs will be crucial for a commercially successful low-power FPGA.

Recently, there has been a lot of activity in exploring alternate technologies for FPGAs, key among them being three-dimensional (3-D) die or wafer integration, and nanowire or nanotube-based technologies. 3-D integration involves stacking of multiple dies or wafers, with interlayer vias for communication among layers. This stacking helps in reducing the average wire-length in the implemented design, and therefore, is expected to consume less power than a 2-D implementation [35]. Although the total power reduces, the effective power density increases due to stacking of multiple power-dissipating layers, which makes low-power measures even more important for 3-D FPGAs. The other technology—use of nanowires or nano-tubes to build FPGAs [36]—brings completely new challenges in terms of power consumption. These FPGAs are expected to be so dense (implying high-power densities) that keeping the die temperature within bounds will be a key design objective. Consequently, the importance of power as a design parameter in applications that are not battery-powered will keep growing. The thermal concerns will force the FPGA manufacturers as well as users to focus on transient power instead of energy, bringing new challenges to power-aware design. These power-aware methodologies will be crucial for the long-term success of FPGAs.

Acknowledgments

This work was supported in part by NSF career award 0093085, NSF grant 0202007, and grants from MARCO/DARPA (GSRC:PAS) and SRC (00541).

References

1. P.R. Schumacher. "An efficient optimized JPEG 2000 tier-1 coder hardware implementation". *Proc. SPIE*, vol. 5150, 2003.
2. "Lattice Reduces Standby Current for LatticeXP FPGA Family by a Factor of 1000". Press article August 8, 2005. http://www.latticesemi.com/corporate/press/product/2005/pr080805.cfm.
3. J.H. Anderson and F. Najm. "Power-aware technology mapping for LUT-based FPGAs". In *Proceedings of IEEE International Conference on Field-Programmable Technology*, 2002.
4. J. Lamoureux and S. Wilton. "On the interaction between power-aware FPGA CAD algorithms". In *Proceedings of ACM/SIGDA International Symposium on Field-Programmable Gate Arrays*, 2003.
5. J. H. Anderson, F. Najm, and T. Tuan. "Active leakage power optimization for FPGAs". In *Proceedings of ACM/SIGDA International Symposium on Field-Programmable Gate Arrays*, 2004.
6. B. Kumthekar, L. Benini, E. Macii, and F. Somenzi. "In-place power optimization for LUT-based FPGAs". In *Proceedings of ACM/IEEE Design Automation Conference*, 1998.
7. V. George, H. Zhang, and J. Rabaey. "The design of a low energy FPGA". In *Proceedings of International Symposium on Low Power Electronics and Design*, 1999.
8. F. Li, D. Chen, L. He, and J. Cong. "Architecture evaluation for power-efficient FPGAs". In *Proceedings of ACM/SIGDA International Symposium on Field-Programmable Gate Aarrays*, 2003.
9. A. Rahman and V. Polavarapuv. "Evaluation of low-leakage design techniques for field programmable gate arrays". In *Proceedings of ACM/SIGDA International Symposium on Field-Programmable Gate Arrays*, 2004.
10. A. Lodi, L. Ciccarelli, and R. Giansante. "Combining low-leakage techniques for FPGA routing design". In *Proceedings of ACM/SIGDA International Symposium on Field-Programmable Gate Arrays*, 2005.
11. Xilinx product datasheets. http://www.xilinx.com/literature.
12. L. Shang, A.S. Kaviani, and K. Bathala. "Dynamic power consumption in Virtex[tm]-II FPGA family". In *Proceedings of ACM/SIGDA International Symposium on Field-Programmable Gate Arrays*, 2002.
13. A. Telikepalli. "Performance vs. power: Getting the best of both worlds". *Xcell J.*, 54, 2005.
14. Altera power tools. http://www.altera.com/support/devices/estimator/powpowerplay.html.
15. Xilinx power tools. http://www.xilinx.com/power.
16. K. Poon, A. Yan, and S. Wilton. "A flexible power model for FPGAs". In *Proceedings of International Conference on Field Programmable Logic and Applications*, 2002.

17. V. Betz and J. Rose. "VPR: A new packing, placement and routing tool for FPGA research". In *International Workshop on Field-Programmable Logic and Applications*, 1997.

18. Y. Lin, F. Li, and L. He. "Circuits and architectures for field programmable gate array with configurable supply voltage". *IEEE Trans. VLSI Systems*, vol. 13, no. 9, pp. 1035–1047, 2005.

19. J. Rose, R. Francis, D. Lewis, and P. Chow. "Architecture of field-programmable gate arrays: The effect of logic block functionality on area efficiency". *J. Solid-State Circuits*, vol. 25, no. 5, pp. 1217–1225, 1990.

20. N. Vassiliadis, S. Nikolaidis, S. Siskos, and D. Soudris. "The effect of the interconnection architecture on the FPGA performance and energy consumption". In *Proceedings of the 12th IEEE Mediterranean Electrotechnical Conference*, 2004.

21. J. Cong and Y. Ding. "FlowMap: An optimal technology mapping algorithm for delay optimization in lookup-table based FPGA designs". *IEEE Trans. Computer-Aided Design*, vol. 13, no. 1, pp. 1–12, 1994.

22. J. Cong, C. Wu, and E. Ding. "Cut ranking and pruning: Enabling a general and efficient FPGA mapping solution". In *Proceedings of ACM/SIGDA International Symposium on Field-Programmable Gate Arrays*, 2003.

23. H. Li, S. Katkoori, and W-K. Mak. "Power minimization algorithms for LUT-based FPGA technology mapping". *ACM Trans. Design Automation of Electronic Systems*, vol. 9, no. 1, pp. 33–51, 2004.

24. A. Singh and M. Marek-Sadowska. "Efficient circuit clustering for area and power reduction in FPGAs". In *Proceedings of ACM/SIGDA International Symposium on Field-Programmable Gate Arrays*, 2002.

25. A. Marquardt, V. Betz, and J. Rose. "Timing-driven placement for FPGAs". In *Proceedings of ACM/SIGDA International Symposium on Field-Programmable Gate Arrays*, 2000.

26. B. Calhoun, F. Honore, and A. Chandrakasan. "Design methodology for fine-grained leakage control in MTCMOS". In *Proceedings of International Symposium on Low Power Electronics and Design*, 2003.

27. A. Gayasen, Y. Tsai, N. Vijaykrishnan, M. Kandemir, M.J. Irwin, and T. Tuan. "Reducing leakage energy in FPGAs using region-constrained placement". In *Proceedings of International Symposium on Field-Programmable Gate Arrays*, 2004.

28. F. Li, Y. Lin, L. He, and J. Cong. "Low-power FPGA using pre-defined dual-Vdd/dual-Vt fabrics". In *Proceedings of ACM/SIGDA International Symposium on Field-Programmable Gate Arrays*, 2004.

29. F. Li, Y. Lin, and L. He. "FPGA power reduction using configurable dual-Vdd". In *Proceedings of ACM/IEEE Design Automation Conference*, 2004.

30. A. Gayasen, K. Lee, N. Vijaykrishnan, M. Kandemir, M.J. Irwin, and T. Tuan. "A dual-Vdd low power FPGA architecture". In *Proceedings of ACM/SIGDA International Conference on Field-Programmable Logic and Applications*, 2004.

31. F. Li, Y. Lin, and L. He. "Vdd programmability to reduce FPGA interconnect power". In *Proceedings of IEEE International Conference on Computer-Aided Design*, 2004.

32. J.H. Anderson and F. Najm. "Low power programmable routing circuitry for FPGAs". In *Proceedings of IEEE International Conference on Computer-Aided Design*, 2004.

33. Y. Lin and L. He. "Leakage efficient chip-level dual-vdd assignment with time slack allocation for FPGA power reduction". In *Proceedings of ACM/IEEE Design Automation Conference*, 2005.

34. S. Srinivasan, A. Gayasen, N. Vijaykrishnan, M.J. Irwin, and T. Tuan. "Leakage control in FPGA routing fabric". In *Proceedings of Asia and South Pacific Design Automation Conference*, 2005.

35. A. Rahman, S. Das, A. Chandrakasan, and R. Reif. "Wiring requirement and three-dimensional integration of field-programmable gate arrays". In *Proceedings of ACM/IEEE Workshop on System Level Interconnect Prediction*, 2001.

36. A. Dehon. "Design of programmable interconnect for sublithographic programmable logic arrays". In *Proceedings of International Symposium on Field Programmable Gate Arrays*, 2005.

37. T. Tuan and B. Lai. "Leakage power analysis of a 90 nm FPGA". In *Custom Integrated Circuits Conference*, 2003.

21

Technology Scaling and Low-Power Circuit Design

Ali Keshavarzi

Intel Corporation

CONTENTS

21.1 Introduction

Technologists and designers are encountering several challenges in maintaining historical rates of performance improvement and energy reduction with CMOS technology scaling as we entered the sub-100 nm technology generations, where we are introducing 65 nm technology to high-volume manufacturing. Researchers have identified barriers to continued scaling of technology and circuit power supply voltage [1]. For a microprocessor design to achieve high-performance and low-power, fundamental device physics issues such as transistor short-channel effects, device parameter variations, excessive subthreshold leakage, junction leakage, and gate oxide leakage show up at aggressively scaled technologies. Furthermore, some of the key bottlenecks are related to reducing device parasitics such as source/drain resistances and gate overlap capacitances. Excessive subthreshold and gate oxide leakage are emerging as serious problems. Functionality of special circuits such as wide fan-in domino circuits, SRAM cell stability, bitline delay scaling, and clock and interconnect power consumptions are issues that designers are facing. In addition, energy efficiency of the microarchitecture of general-purpose microprocessors is starting to play a more critical role in the performance versus power and area trade-offs. Potential solutions to the device technology scaling challenges at gate lengths approaching 10 nm are discussed in Section 21.2. Section 21.3 describes some promising circuit and design techniques to control leakage power. Energy-efficient microarchitecture trends are elucidated in Section 21.4.

21.2 Technology Scaling Challenges

21.2.1 Device Performance and Energy Scaling

Figure 21.1 plots technology nodes and transistor physical gate length in micrometers on the *y*-axis as a function of time on the *x*-axis. This plot shows that the actual transistor physical gate length dimension is more aggressive than the dimension of the technology node of interest. Nevertheless, we are continuing to scale the dimensions. In the sub-180 nm technology generations, it is difficult to maintain traditional constant electric-field supply voltage scaling (Figure 21.2 and Figure 21.3). Constant-field voltage scaling requires scaling of supply voltage as we scale the transistor gate length. If we scale the supply voltage, then we have to

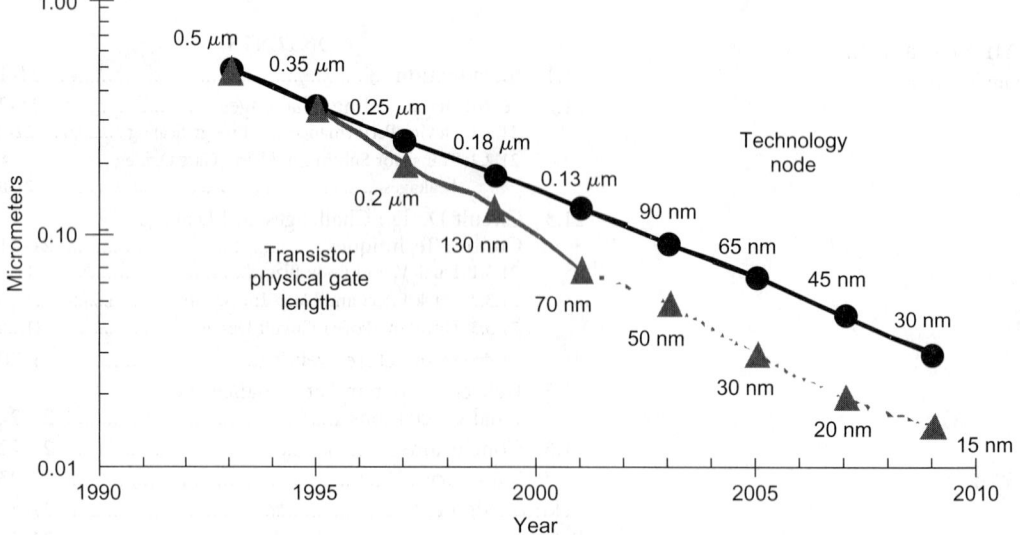

FIGURE 21.1 Physical gate length scaling trend.

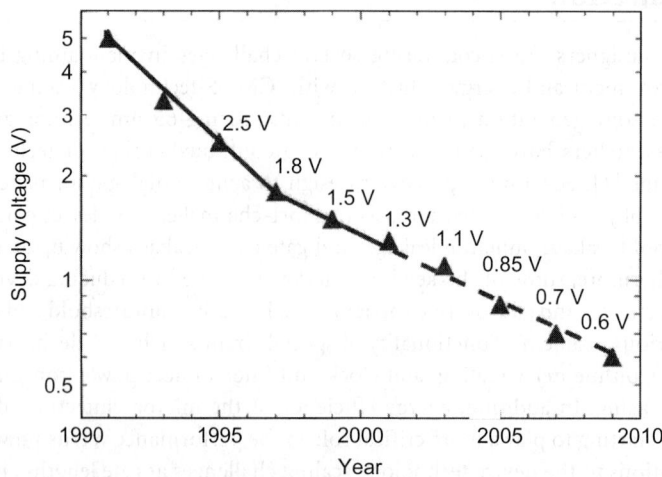

FIGURE 21.2 Supply voltage scaling trend.

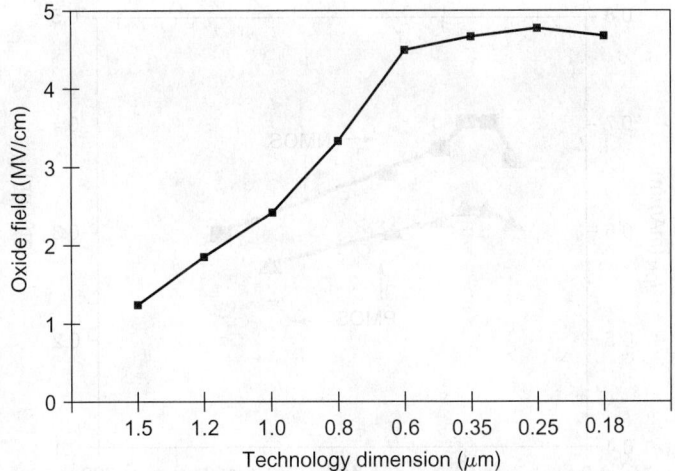

FIGURE 21.3 Electric field across the gate oxide.

scale the transistor threshold voltage similarly to sustain the required gate overdrive to maintain transistor drive current and performance. Figure 21.3 shows that we followed constant-field scaling from 0.6 μm technology node to 180 nm node. Owing to the nonscalability of threshold voltage from excessive leakage current considerations, threshold voltage has not been scaled aggressively and supply voltage has not scaled aggressively (aggressive means scaling by ~30% and less aggressive means scaling by ~15% which is less than the scaling factor of 30%). Consequently, we have not fulfilled true constant-field voltage scaling post 180 nm technology node. Essentially, the electric field across the gate dielectric has been increasing by 10% per generation because we have not reduced the supply voltage with the same scaling factor as we have scaled the gate oxide thickness. This has been made possible by the superior long-term reliability offered by physically thinner gate oxides. In order to improve the delay of driving constant capacitance loads such as those posed by interconnects in high-performance microprocessor designs, the transistor saturation current per unit width must remain constant or increase from one technology generation to the next. This performance gain has been accomplished by reducing the rate of supply voltage scaling (from 30% per generation to 15% per generation), clearly at the expense of increasing switching power density. Low-power circuit operation relies on aggressive scaling of supply voltage which has been challenged to maintain high-performance operation.

Extrinsic source/drain resistance, gate overlap capacitance, and junction capacitance do not scale in a desired fashion. This limits the circuit delay improvements achievable from large intrinsic device saturation currents. Reducing the depth of the source/drain junction extension or tip (shallow junctions) improves short-channel effects and thus allows a shorter gate length. However, tip depth of <40 nm causes the tip resistance to become so large that the drive current becomes smaller. Figure 21.4 shows improvement in NMOS as well as PMOS transistor drive current as we reduce source/drain junction depths owing to improving the transistor's electrostatics and its short-channel effects. However, transistor drive current has an optimum at approximately 40 nm. Below junction depths of 40 nm, one can observe that the transistor drive current becomes less owing to an increase in transistor source/drain resistance (what device engineers refer to as $R_{external}$). Increasing the tip doping concentration beyond the solid solubility limit can help alleviate this problem to some extent. However, this is very difficult to achieve. Furthermore, abruptness of the doping profiles, in vertical as well as lateral directions, must be increased to help out this limitation.

Spreading resistance from the inversion layer to the source/drain extension region also limits the drive current. This makes it difficult to scale the gate overlap length below 10 nm. Overlap is the region where source/drain junction extends under the gate. Overlap is required for transistor operation. However, its distance should be minimized to reduce parasitic gate overlap capacitance running across the width of the transistor at the source/drain to gate edge. Reducing this overlap distance below 10 nm impacts spreading resistance negatively and lowers transistor drive current and device performance.

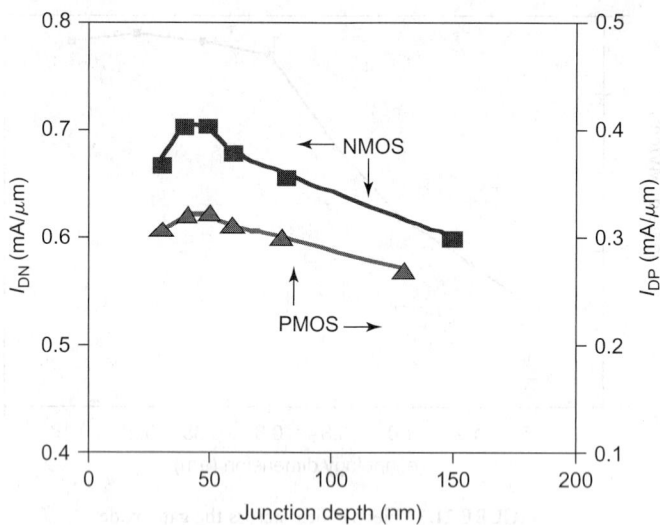

FIGURE 21.4 Optimal source/drain extension depth.

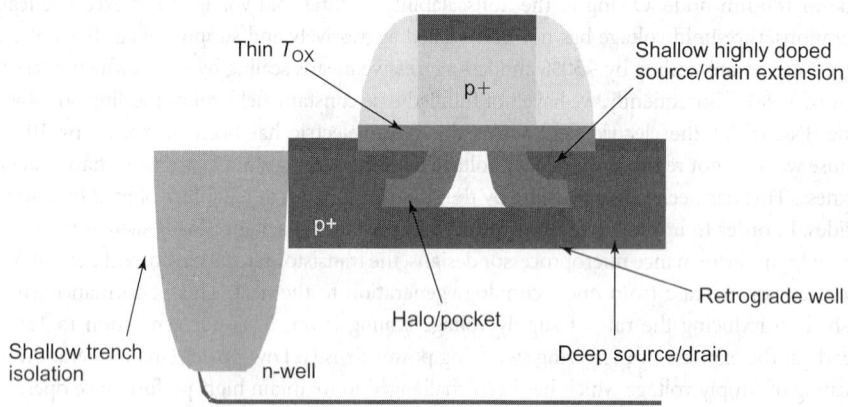

FIGURE 21.5 Halo and retrograde well—features of advanced high-performance bulk CMOS transistor.

Advanced bulk CMOS transistors use sophisticated channel implants including halo implant and retrograde well to assist designing a device that does not suffer from excessive short-channel effects (Figure 21.5). By changing the doping profile in the channel region, the distribution of the electric field and potential contours can be changed. The objective is to optimize the channel doping profile to minimize transistor I_{OFF} while maximizing transistor drive current, I_{ON}. Steep retrograde wells and halo implants have been extensively used in transistor scaling to achieve the above-mentioned objective [2–8]. The trend is in the direction of increasing channel doping, demanded by subsurface punch-through control and short-channel effect reduction. However, the increased doping causes capacitance of the gate-edge junction sidewall to increase with technology scaling. This degrades delays of wide-OR circuits such as bitlines in the cache. In spite of all these limitations, device and transistor delays represented by CV/I metric are reducing and delays well beyond a terahertz is achievable at sub-1 V supply voltages using a traditional planar bulk CMOS device structure in the 15–30 nm gate length regime (Figure 21.6 and Figure 21.7) [9–11].

A fully depleted SOI device structure, referred to as the depleted substrate transistor (DST) and ultrathin body MOSFETs are promising for alleviating many of the challenges discussed before

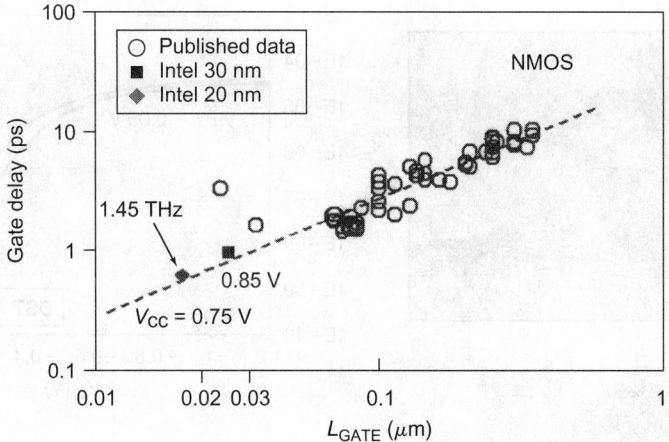

FIGURE 21.6 *CV/I* delay scaling trends for bulk CMOS.

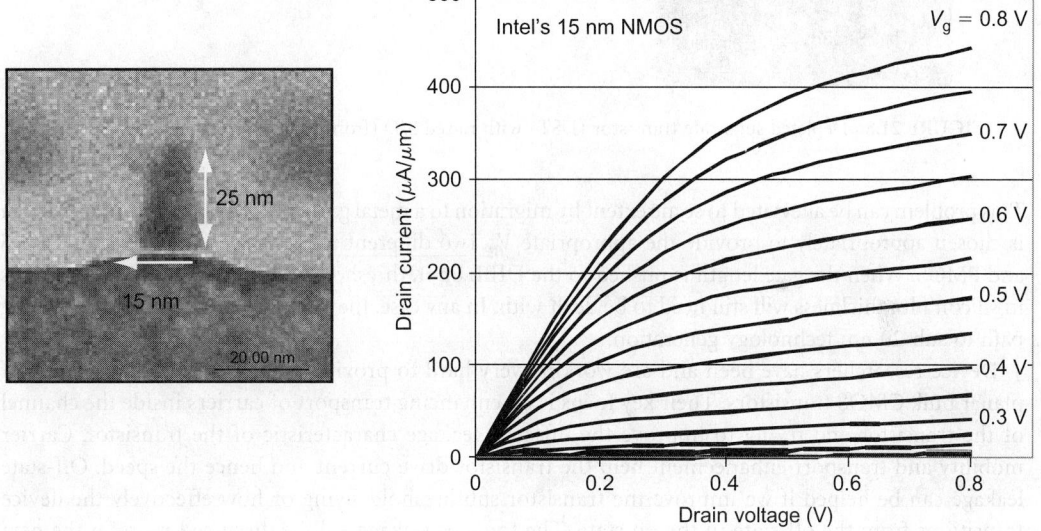

FIGURE 21.7 Intel's 15 nm bulk NMOS transistor.

(Figure 21.8). The subthreshold swing is much steeper in the DST compared to either bulk or partially depleted SOI devices when the silicon film thickness is below 30 nm, resulting in a fully depleted channel. This allows V_t to be reduced for a specific leakage target and boosts the drive current. Furthermore, the oxide layer below the silicon channel completely eliminates subsurface punch-through and junction leakage currents. Therefore, channel doping can be reduced. This reduces the gate-edge junction sidewall capacitance dramatically.

The source/drain extension depth in the DST can be scaled by simply scaling the silicon thickness to improve short-channel effects. The buried oxide layer also serves as a diffusion stopper and creates more abrupt vertical doping profiles in the source/drain region. When combined with a raised source/drain structure, the drive current improvement owing to lower parasitic resistance is as much as 30%. The main challenge associated with further development of the DST with conventional polysilicon gates is achieving a sufficiently tight control of the silicon film thickness since the threshold voltage is quite sensitive to the film thickness. Control of this thickness is rather challenging, costly, and a key barrier before considering the DST as a viable transistor-architecture option in high-volume manufacturing.

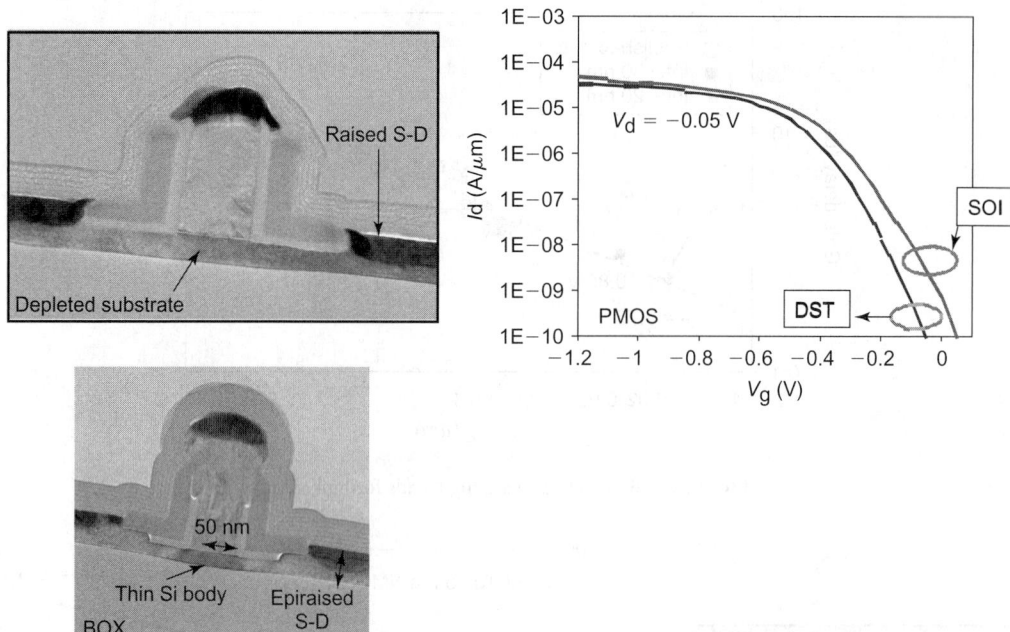

FIGURE 21.8 Depleted substrate transistor (DST) with raised S/D (from Chau R., et al., *IEDM*, 2001).

This problem can be alleviated to some extent by migration to a metal gate electrode, whose work function is chosen appropriately to provide the appropriate V_t. Two different metals may be needed for NMOS and PMOS. When the gate length is pushed to the DIBL limit, threshold voltage sensitivity to variations in silicon film thickness will still need to be dealt with. In any case, the DST provides a promising scaling path to sub-20 nm technology generation.

Device researchers have been and are working very hard to provide paths for extending scaling of planar bulk CMOS transistors. Their key focus is on enhancing transport of carriers inside the channel of the transistor and trying to improve the off-state leakage characteristic of the transistor. Carrier mobility and transport enhancement help the transistor drive current and hence the speed. Off-state leakage can be helped if we improve the transistor subthreshold swing or how effectively the device transitions from the off state to the on state. The topic of leakage will be discussed more in the next section. Enhancing carrier channel transport properties has been achieved by strained Si [12–14] and more research is being conducted by new materials and structures that enable high mobility inside the transistor channel. Intel improved their transistor drive currents by optimization and enhancements of strain silicon techniques [12–14]. For PMOS, uniaxial strain was applied from raised source/drain regions by using epitaxial SiGe film. It was shown that channel strain was significantly improved by increasing Ge content in the SiGe film and optimizing the source–drain recess geometry. For NMOS, strain is applied through sacrificial films from the use of tensile cap films (Figure 21.9 and Figure 21.10). Research toward improving carrier mobility inside transistor channel has continued to such an extent that recently, compound semiconductors are being considered for channel of NMOS transistors [15]. These researchers used indium antimonide (InSb) that has the highest electron mobility and saturation velocity among all known semiconductors to fabricate very fast switching field-effect transistors. Enhancing carrier mobility for NMOS is important as carrier mobility for PMOS can further get enhanced by strain. For future technologies, carbon nanotube transistors (CNTs) have shown that they have very high intrinsic carrier mobility [16,17]. But many technological limitations and research questions remain unanswered for these new material systems before they become a viable solution for advanced development.

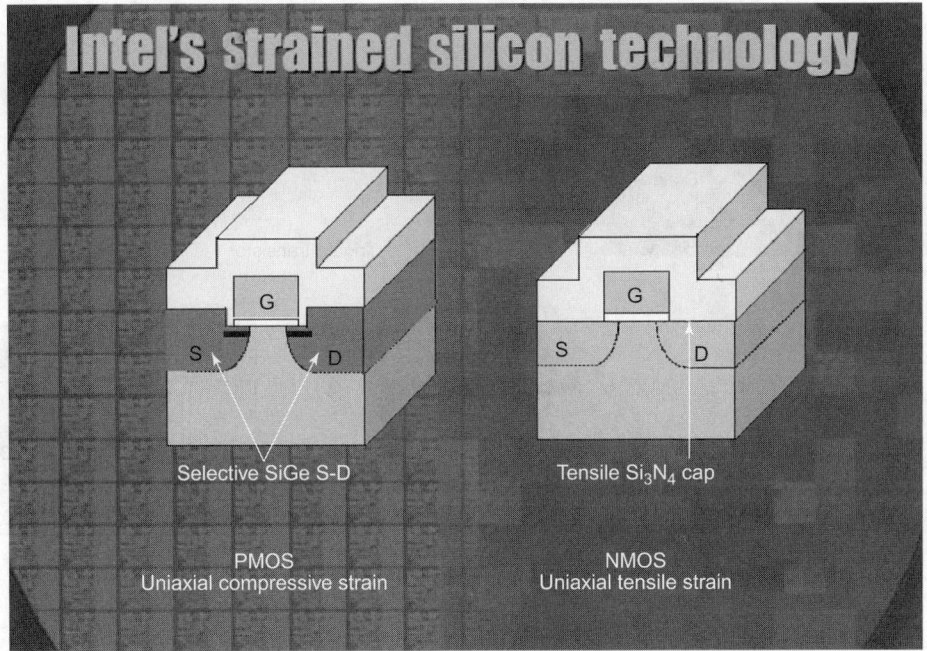

FIGURE 21.9 Strained Si technology for improving carrier mobility and transport.

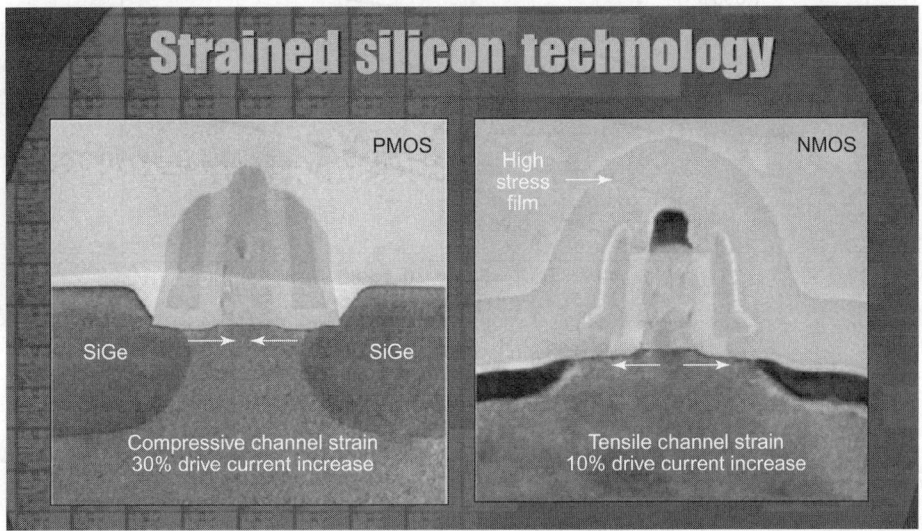

FIGURE 21.10 TEM picture of strained Si technology for improving carrier mobility and transport.

Nonplanar multigate structures and ultrathin body MOSFETs such as FinFETs [18] and trigate transistors [19] are an extension of fully depleted CMOS devices with better electrostatics and subthreshold slope that was discussed for DST and SOI technology earlier (Figure 21.11). DST can be viewed as single gate [20], FinFET is a double-gate structure, and trigate as its name applies has three gates that makes this device almost a surround gate structure with very good electrostatics. These structures provide better transistor scalability by enhancing device electrostatics. However, many integration challenges remain before these structures can be fabricated and make it into high-volume manufacturing. Furthermore, we need to make sure nonplanar devices provide more current in a given area footprint. Therefore, layout of these nonplanar devices should in effect provide the benefits of area scaling in accordance with Moore's

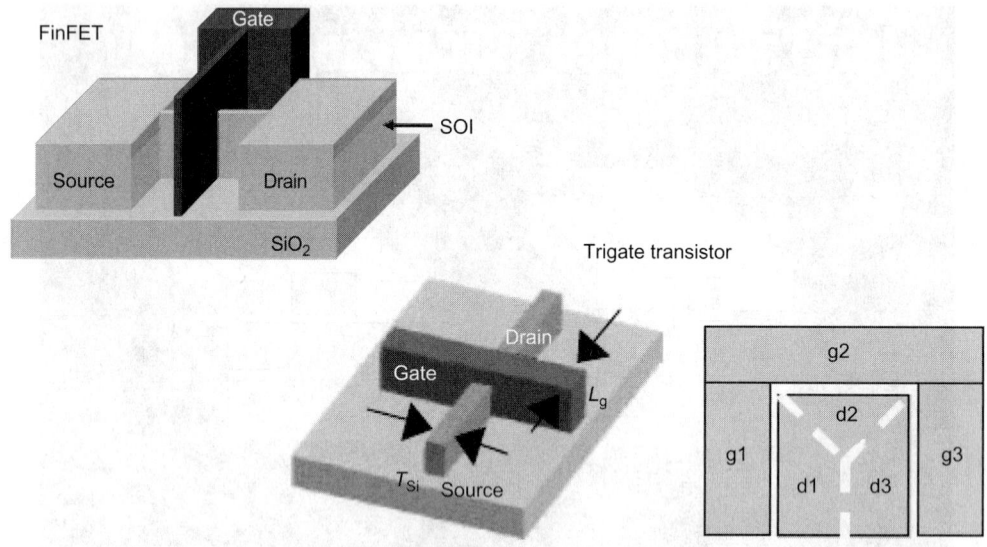

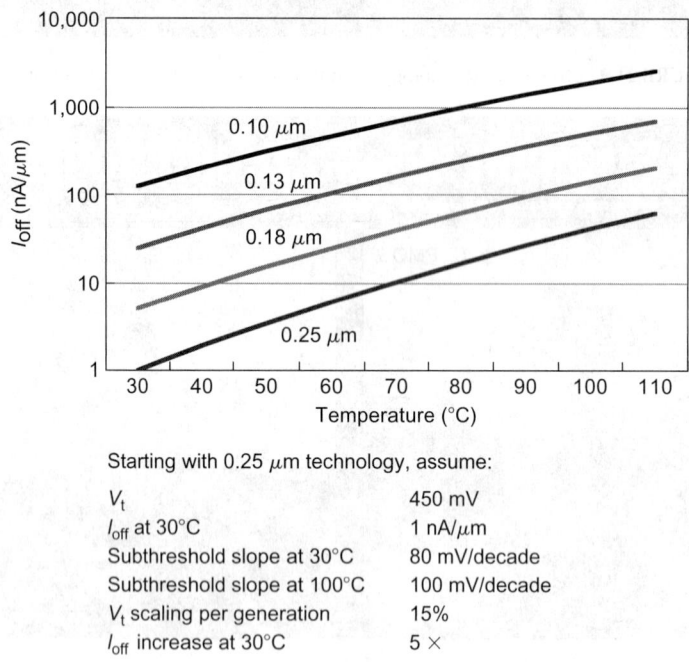

FIGURE 21.11 Multigate nonplanar transistor structures.

Starting with 0.25 μm technology, assume:

V_t	450 mV
I_{off} at 30°C	1 nA/μm
Subthreshold slope at 30°C	80 mV/decade
Subthreshold slope at 100°C	100 mV/decade
V_t scaling per generation	15%
I_{off} increase at 30°C	5 ×

FIGURE 21.12 I_{off} versus temperature versus technology.

Law. Managing parasitics and fabricating these devices are also not trivial. We will discuss some of these challenges here for the sake of completeness.

21.2.2 Transistor Subthreshold and Gate Oxide Leakages

Subthreshold leakage current of a transistor is increasing by ~5× per generation (Figure 21.12 and Figure 21.13). At high temperature, it exceeds 1000 nA/μm in sub-100 nm technology nodes (Figure 21.14). As the physical gate oxide thickness approaches sub-10 Å regime, gate oxide leakage becomes <100 A/cm² (Figure 21.15) owing to direct band-to-band tunneling. Although gate oxide

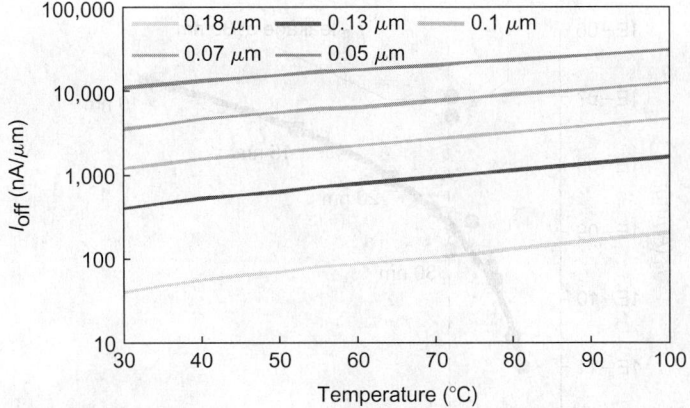

FIGURE 21.13 I_{off} versus temperature for more scaled technologies.

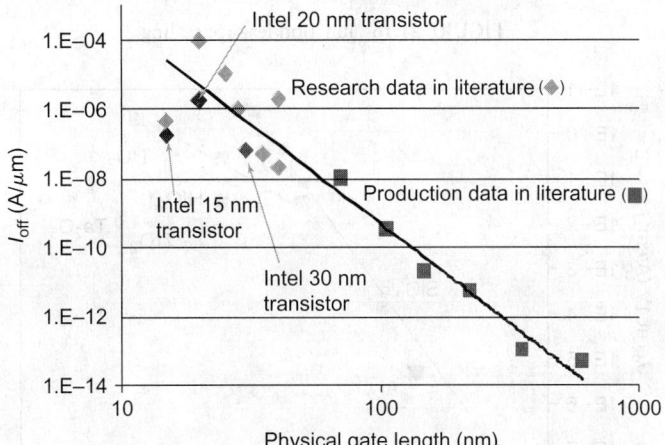

FIGURE 21.14 Subthreshold leakage scaling trend.

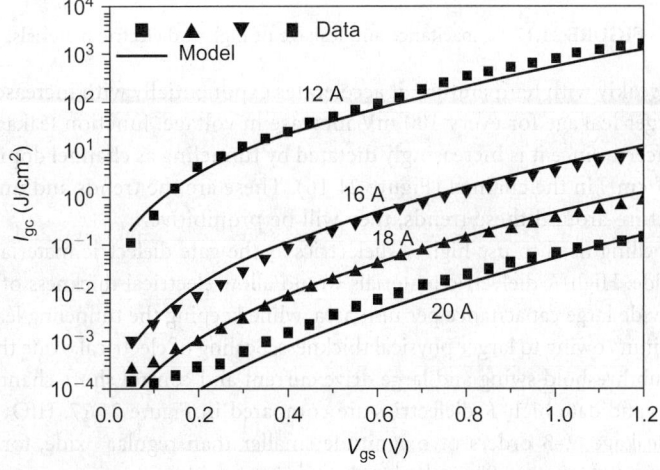

FIGURE 21.15 Gate oxide leakage scaling with thickness and voltage.

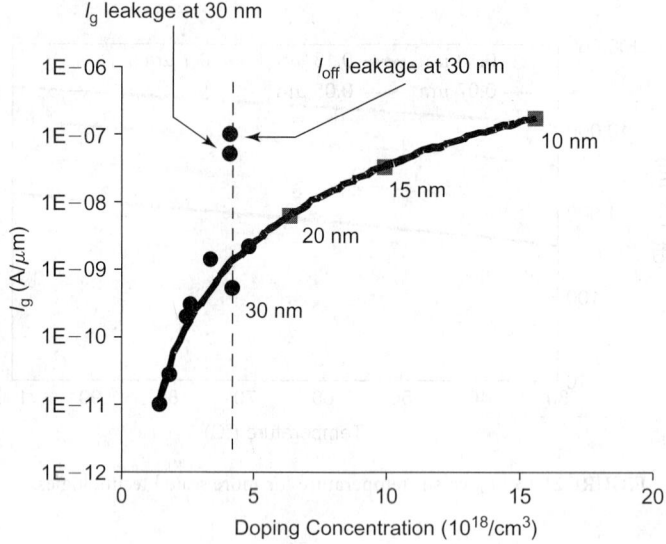

FIGURE 21.16 Junction leakage scaling.

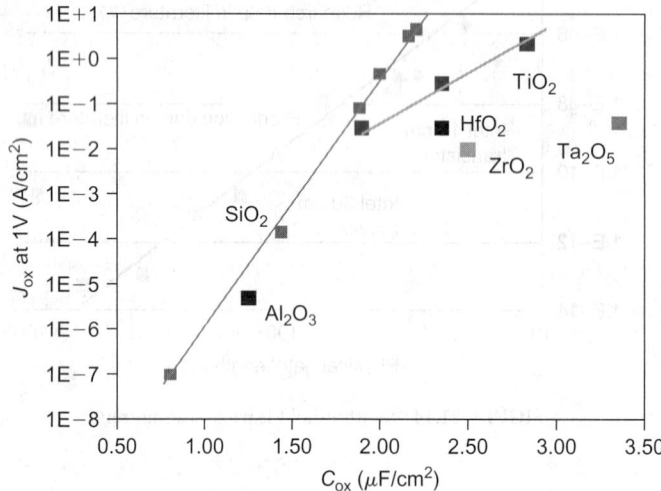

FIGURE 21.17 Capacitance and leakage of high-K dielectric materials.

leakage increases weakly with temperature, it accelerates exponentially with increase in supply voltage at a rate of 2× larger leakage for every 100 mV increase in voltage. Junction leakage is an additional component of concern, since it is increasingly dictated by tunneling as channel doping concentrations approach 5×10^{18} cm^{-3} in the channel (Figure 21.16). These are the trends and unless we design the transistor architecture around these trends, they will be prohibitive.

There is a compelling need to use high-K dielectrics as the gate dielectric material to replace silicon dioxide or oxynitride. High-K dielectric materials would allow electrical thickness of the gate dielectric to be scaled to provide large capacitance per unit area, while keeping the tunneling leakage per unit area within acceptable limits owing to larger physical thickness. Scaling of electrical oxide thickness is essential to provide sharp subthreshold swing and large drive current and control short-channel effects. Characteristics of several candidate high-K dielectrics are compared in Figure 21.17. HfO$_2$ and ZrO$_2$ provide the smallest gate leakage, 2–3 orders of magnitude smaller than regular oxide, for a target electrical thickness in the sub-10 Å regime. Since the bandgap reduces with increasing permittivity, gate leakage owing to thermal emission dominates for materials with very high K values. Thus, Ta$_2$O$_5$ that has a very

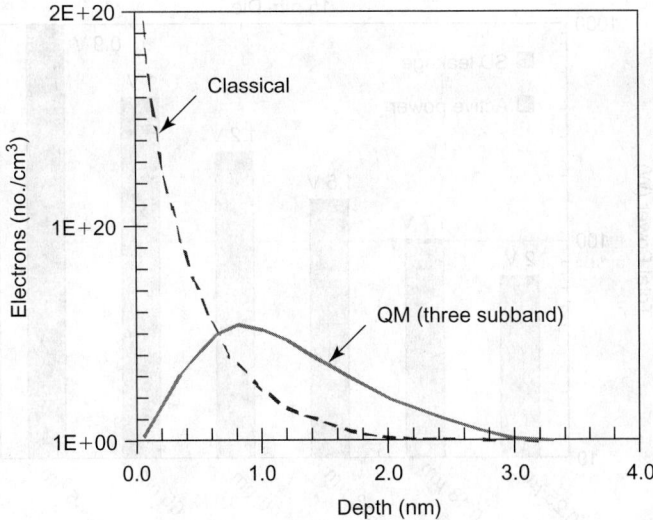

FIGURE 21.18 Increase in electrical oxide thickness by QM effects.

high dielectric constant value is not as attractive for gate dielectric application. Of course, many process integration challenges need to be resolved and silicon-dielectric interface quality needs to be improved (for better reliability and higher carrier mobility) for these new dielectric materials to provide improvements to CMOS circuit delay.

The scaling of electrical gate oxide thickness is limited by poly depletion and separation of inversion layer charge from the oxide–silicon interface at high vertical fields owing to quantum-mechanical (QM) effects (Figure 21.18). Each of these effects adds approximately 5 Å to the effective electrical oxide thickness at the highest gate voltage. Even if we reduce the physical gate oxide thickness to almost zero in a polysilicon gate electrode system, we still see an effective electrical gate oxide thickness (owing to polydepletion and QM effects). These effects become more significant as the gate voltage increases. Thus, when averaged over the entire gate voltage range from V_t to the maximum supply voltage, their impacts on drive current are less severe. Nevertheless, to maximize the benefit of migrating to a high-K gate dielectric, polydepletion should be reduced or eliminated. Polydepletion has historically been reduced by doping the polysilicon gate electrode heavily. Increasing polydoping beyond the solid solubility limit is desirable. Transition to a metal gate fully eliminates polydepletion. But metal gates with appropriate work functions for NMOS and PMOS must be identified and process integration issues must be resolved. Combining metal gates with high-K dielectrics (in ultrathin fully depleted devices) to set the threshold voltage by work function engineering is a promising approach that also addresses the polydepletion problem. However, additional process complexities owing to two different gate electrode metals, one for NMOS and one for PMOS, will be incurred.

The trend in future transistor development will be toward using high-K dielectric with metal gates. Carrier mobility and transport will continue to improve and we may go toward nonplanar structures (ultrathin body, trigate, etc.) to enhance transistor electrostatics and short-channel effects. Advanced development of whatever solution scaling provides for us requires a tedious focus on reducing device parasitics. Transistors' parasitic capacitances and series resistances are critical for efficient and cost-effective device integration before transistors are fabricated in high-volume manufacturing.

21.3   Circuit Design Challenges and Leakage Control Techniques

We have established that in technology scaling as supply voltage scales down, transistor threshold voltage needs to reduce to maintain high performance, resulting in excessive subthreshold leakage power. In this section, circuit techniques for leakage avoidance, control, and tolerance to mitigate the subthreshold leakage will be discussed.

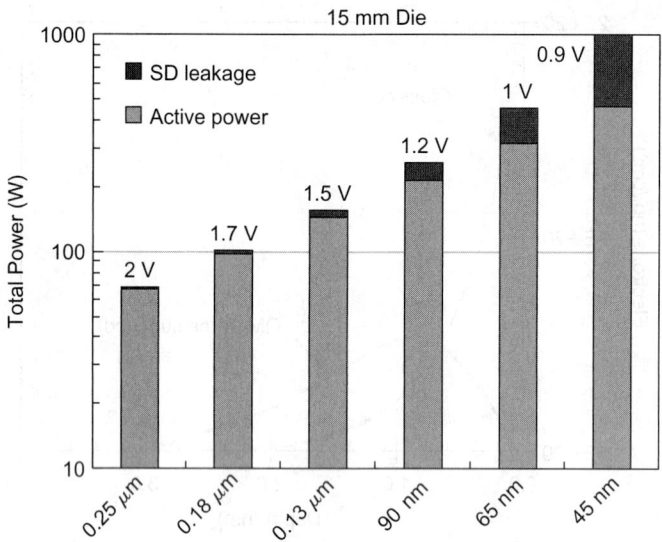

FIGURE 21.19 Switching and leakage power scaling trends.

Leakage power has become a larger fraction of the total active power of microprocessors (Figure 21.19). This poses serious challenges for heat removal and power delivery in high-performance processors. Excessive leakage power can also cause thermal runaway during burn-in, and impact the burn-in cost. Subthreshold leakage dominates at high temperature and gate oxide leakage is a significant contributor to the burn-in leakage power owing to the higher voltage used. Of course, standby leakage power at room temperature also needs to be kept sufficiently small for battery-operated systems. Consequently, today's designers are aware of power and managing leakage and power are part of design objectives. Researchers have suggested various means for designs to be power aware that will be discussed in the following sections.

21.3.1 Dual-V_t and Body Bias

Dual-V_t designs can reduce leakage power during active operation, burn-in, and standby. Two V_t's are provided by the process technology for each transistor. Performance-critical transistors are made low-V_t to provide the target chip performance. The rest of the transistors are made high-V_t to minimize leakage power [21]. Since the full-chip frequency is dictated by only a fraction of transistors in the critical paths, this selective V_t assignment is possible without degrading the overall chip performance achievable by using a single low-V_t transistor everywhere. Figure 21.20 shows an example circuit block, where all low V_t design provides 24% delay improvement over all high V_t design. Notice that as you start inserting low V_t devices (y-axis), the delay improves (x-axis). Only 34% of the total transistor width needs to be low-V_t in this example, to get the same frequency as using low-V_t everywhere. Typically, low-V_t device leakage is 10× higher than high-V_t. Thus, by carefully employing low-V_t up to 34% of the total width, 24% delay improvement is possible with ~3× increase in leakage, compared to all high-V_t design.

Another technique to reduce leakage power during burn-in and standby is to apply reverse-body bias (RBB) to the transistors to increase V_t since high performance is not required during these modes. There is an optimal RBB value that minimizes leakage power as shown in Figure 21.21 [22]. Using RBB values larger than this value causes the junction leakage current to increase and overall leakage power to go up. In sub-100 nm technology generation, approximately 500 mV RBB is optimal. Two to three times reduction in leakage current is achievable. However, effectiveness of RBB reduces as channel lengths become smaller or V_t values are lowered (Figure 21.22) [23]. Essentially, the V_t-modulation capability by RBB weakens as short-channel effects become worse or body effect diminishes owing to lower channel doping.

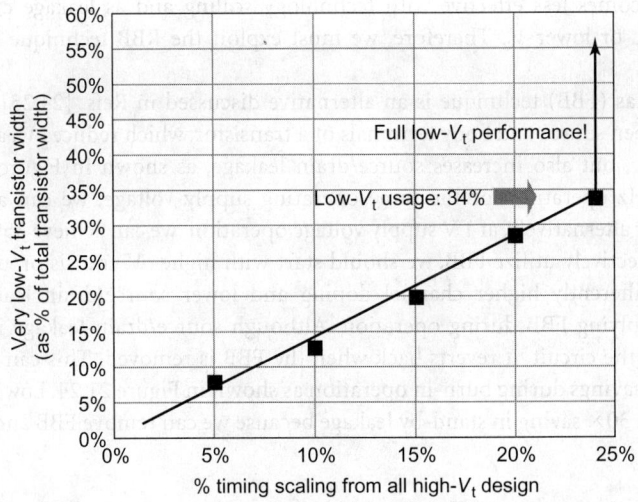

FIGURE 21.20　Performance versus leakage in dual-V_t designs.

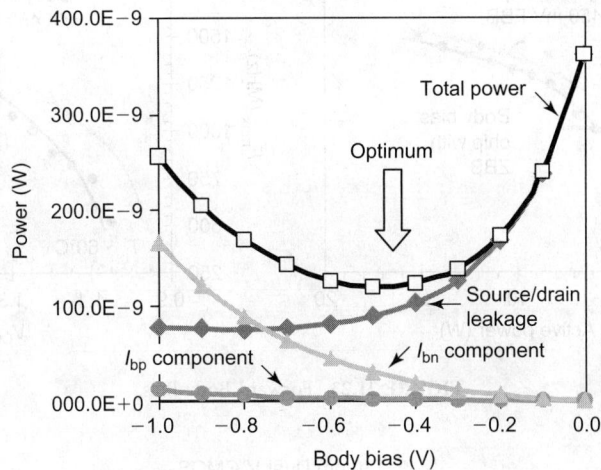

FIGURE 21.21　Optimum Reverse Body Bias.

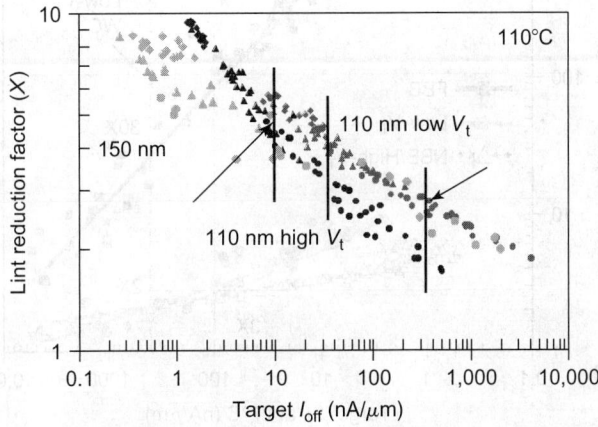

FIGURE 21.22　Subthreshold leakage reduction by Reverse Body Bias.

Therefore, RBB becomes less effective with technology scaling and as leakage currents are pushed higher by shorter L or lower V_t. Therefore, we must exploit the RBB technique for leakage control while it lasts.

Forward-body bias (FBB) technique is an alternative discussed in Refs. [24,25]. In this technique, we apply FBB between source and body terminals of a transistor, which reduces V_t, and hence improves circuit performance, but also increases source/drain leakage, as shown in Figure 21.23. This figure shows that at 1 GHz operation for the same operating supply voltage, we can achieve 25% power saving with FBB. Or alternatively, at 1 V supply voltage operation, we can achieve 35% higher frequency of operation. To effectively utilize FBB, we should start with higher V_t transistor, provided by process technology with inherently higher channel doping and lower source/drain leakage, and improve performance by applying FBB during operation. Although source/drain leakage is increased during active operation of the circuit, it reverts back when the FBB is removed. This can be used to provide substantial leakage savings during burn-in operation as shown in Figure 21.24. Low V_t devices achieved by FBB can result in 30× saving in stand-by leakage because we can remove FBB and apply RBB during stand-by.

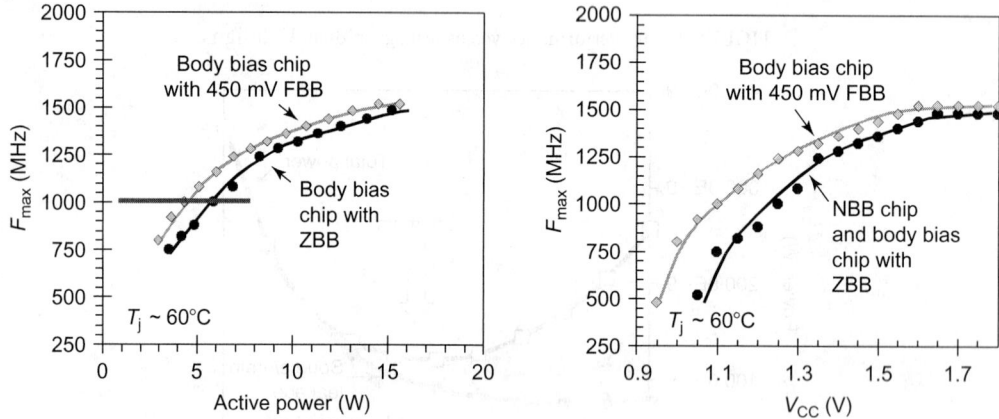

FIGURE 21.23 Forward Body Bias.

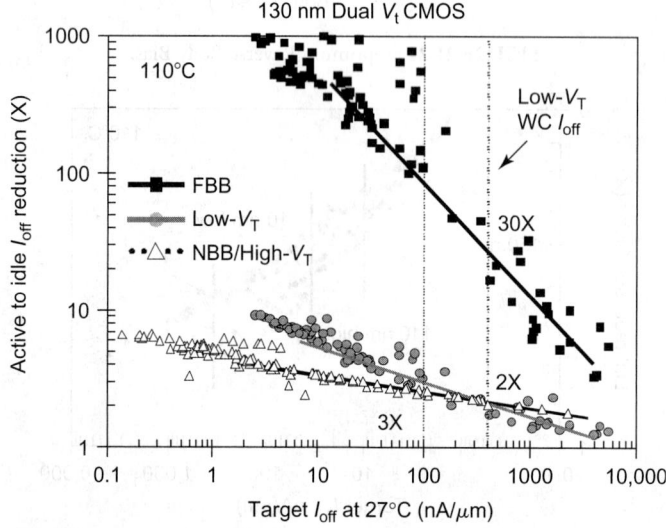

FIGURE 21.24 Stand-by leakage control by Forward Body Bias.

A combination of RBB and FBB together can be applied adaptively in a technique called adaptive body biasing (ABB) to reduce source/drain leakage with reduced performance or improve performance with higher source/drain leakage [26]. Figure 21.25 and Figure 21.26 show a test chip and resulting bin split by employing the ABB technique. In this experiment, a test chip is subdivided into 23 subsites, and each site has circuitry to apply FBB and RBB in small increments. PMOS transistors in each site may be biased individually; however, all NMOS transistors of the test chip have the same body bias. During the test, slower dies are applied with FBB to improve performance and faster, leaky dies are applied RBB to reduce source/drain leakage. The experiment shows that 100% yield can be achieved with significant boost to the high frequency bin by applying ABB at the full die level (ignoring subsites). Furthermore, by applying selective body bias to PMOS transistors in the different subsites, the high-frequency bin grows even further to 97% [27].

ABB has also some merit in dealing with increased circuit leakage and frequency variability that will not be discussed here. However, Figure 21.27 shows how leakage, power, and frequency spreads can be modulated by applying body bias. ABB can deal with both die-to-die and within-die parameter variations.

Although the effectiveness of dual-V_t designs for low-power applications is shown, there are several challenges in implementing it. One of the issues is the variation that exists in transistor threshold voltage. Each target V_t has a variation around it. Consideration of this variation and desired targets to set V_t values has created challenges in design. Making two precise V_t values has become daunting considering

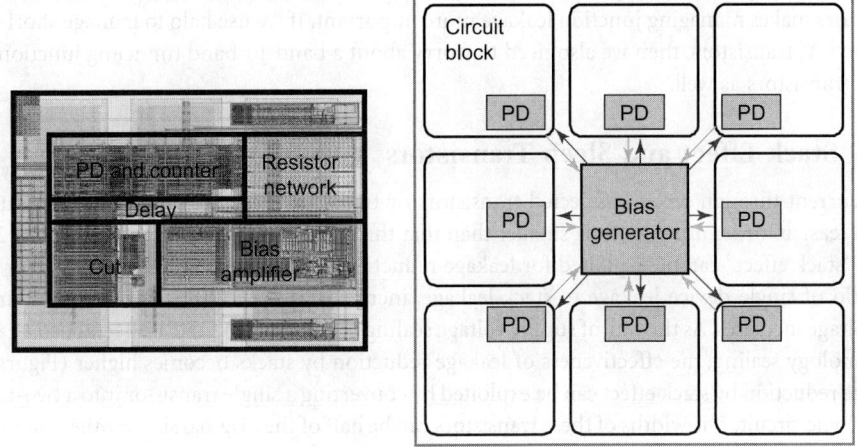

FIGURE 21.25 Testchip to study Adaptive Body Bias.

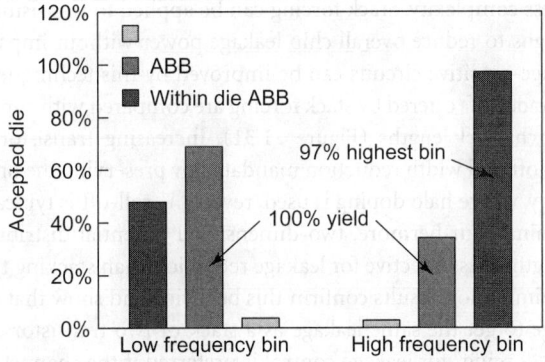

FIGURE 21.26 Improvement of bin split by Adaptive Body Bias.

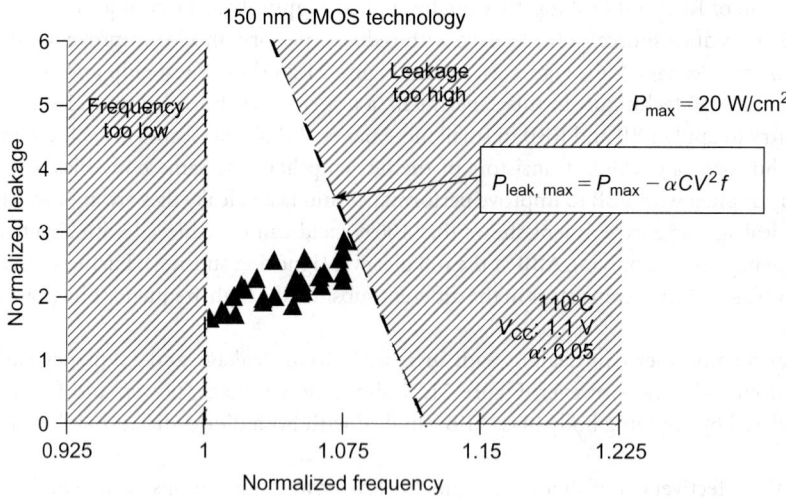

FIGURE 21.27 ABB and parameter variation.

variation, thereby reducing the effectiveness of dual-V_t designs [28]. From transistor design point of view, targeting a low V_t transistor competes with managing device short-channel effects and architecting high V_t transistors makes managing junction leakage more important. If we use halo to manage short-channel effect of low V_t transistors, then we also need to worry about a band-to-band tunneling junction leakage in low V_t transistors as well.

21.3.2 Stack Effect and Sleep Transistors

Leakage current through series-connected transistors or transistor "stacks," with more than one device "off," is at least an order of magnitude smaller than that through a single device (Figure 21.28) [29]. This so-called "stack effect" can be exploited for leakage reduction in circuits. The stack effect factor, defined as the ratio of single device leakage to stack leakage, increases as the DIBL factor becomes larger and supply voltage increases. As the rate of supply voltage scaling diminishes and DIBL effects become stronger with technology scaling, the effectiveness of leakage reduction by stacks becomes higher (Figure 21.29).

Leakage reduction by stack effect can be exploited by converting a single transistor into a two-transistor stack in a logic circuit. The widths of these transistors can be half of the original size or other combinations can be chosen to preserve the same input capacitance load as the original single device. Leakage versus delay trade-off provided by this "stack forcing" technique applied to both high-V_t and low-V_t devices is illustrated in Figure 21.30. Clearly, stack forcing can be used to emulate additional higher V_t devices without increasing process complexity. Stack forcing can be applied to transistors in noncritical paths in single-V_t or dual-V_t designs to reduce overall chip leakage power without impacting chip performance. Also, robustness of leakage-sensitive circuits can be improved by this technique.

Leakage versus delay trade-offs offered by stack forcing are compared with similar trade-offs achievable by increasing transistor channel lengths (Figure 21.31). Increasing transistor length reduces leakage because of threshold roll-off and width reduction mandated by preserving the original input capacitance. In sub-100 nm technology, where halo doping is used, reverse V_t roll-off is typically observed for channel lengths higher than nominal. Furthermore, two-dimensional potential distribution effects dictate that doubling the channel length is less effective for leakage reduction than stacking two transistors, especially when the DIBL is high. Simulation results confirm this behavior and show that channel length has to be made three times as large to get the same leakage as a stack of two transistors, resulting in 60% worse delay. Clearly, then "stack forcing" for leakage control is preferred if the channel length needs to be more than doubled to achieve the target low leakage.

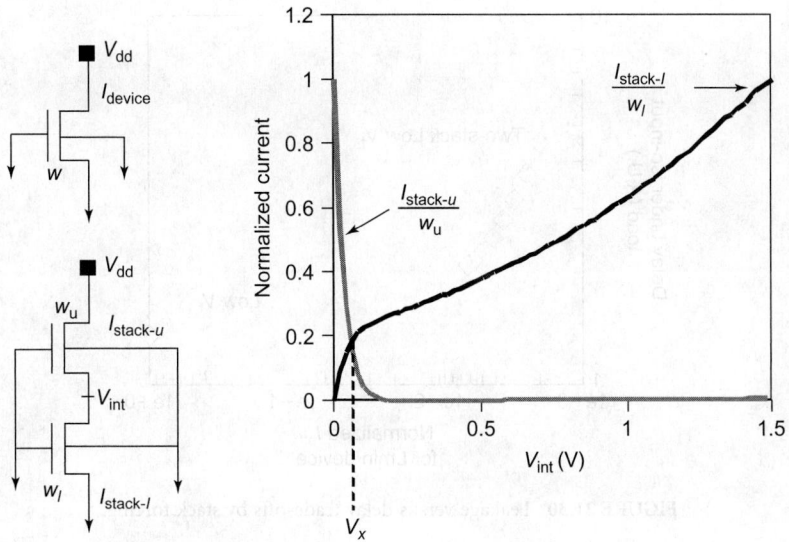

FIGURE 21.28 Leakage current of transistor stacks—stack effect.

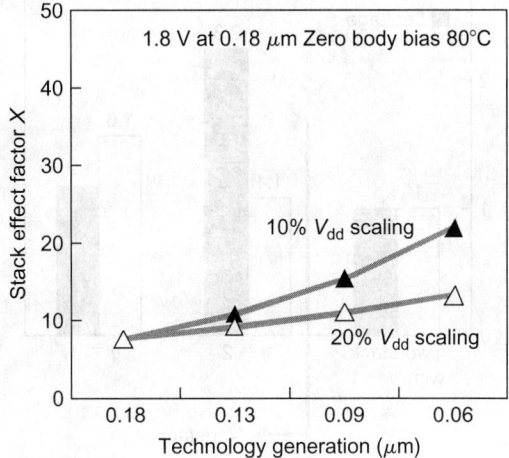

FIGURE 21.29 Scaling of stack effect factor.

Typically, large circuit blocks contain some series-connected devices in complex logic gates. These so-called "natural stacks" can be exploited to reduce standby leakage [30]. Leakage power of a large circuit block such as a 32-bit static CMOS Kogge-Stone adder depends strongly on the primary input vector (Figure 21.32). The total "off" device width and the number of transistor stacks with two or more "off" devices change as primary input vectors change. This causes the leakage power to vary with input vector. When a circuit block is "idle," one can store the input vector that provides the least amount of leakage at the primary input flops. This can reduce the standby leakage power by 2×. There is no performance overhead since this predetermined input vector can be encoded in the feedback path of the input flip-flop. The minimum time required in standby mode, so that the energy overhead for entry and exit into this mode is <10% of the leakage energy saved, is tens of microseconds. This time reduces further with technology scaling as leakage levels increase, making this technique more attractive. Of course, EDA tools will be needed to identify this "lowest leakage" input vector efficiently during design phase for each circuit block.

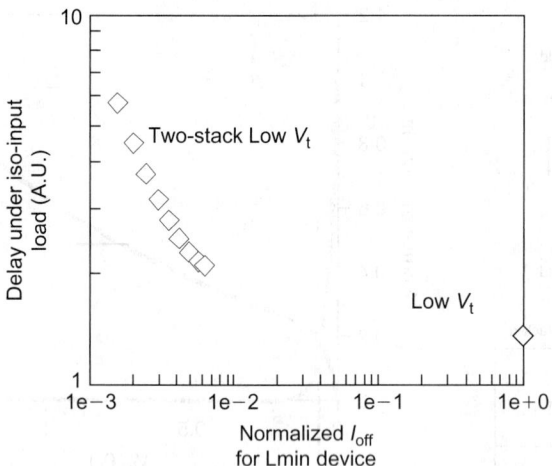

FIGURE 21.30   Leakage versus delay trade-offs by stack forcing.

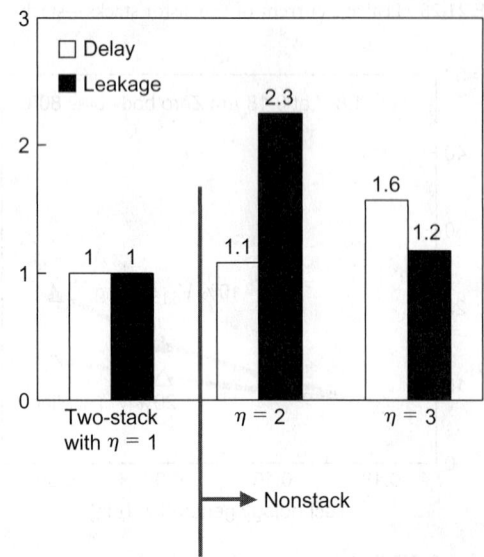

FIGURE 21.31   Stack forcing versus longer channel length.

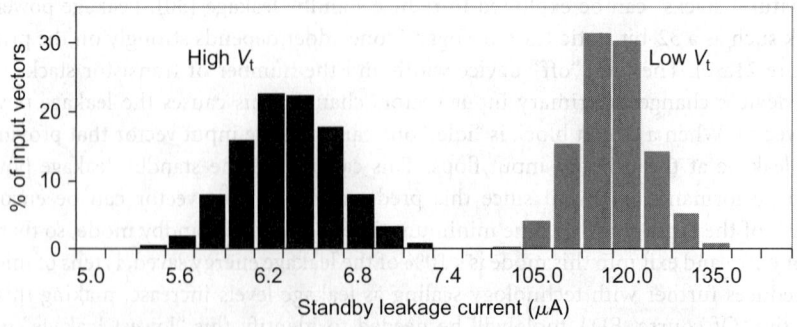

FIGURE 21.32   Leakage control by natural stacks.

Another promising leakage control technique is to employ sleep transistors [31], which act like brute-force switches between logic and power rails, which are turned off when the logic is not in use to reduce source/drain leakage, as shown in Figure 21.33. The sleep transistors may be high or low V_t, and may be simply turned off ($V_{gs} = 0$) or underdriven ($V_{gs} < 0$), and the leakage reduction varies accordingly. Switching sleep transistors on and off consumes energy, and this must be taken into account in evaluating the overall leakage reduction benefit. Figure 21.34 shows the potential benefits of sleep transistors when applied to an ALU, and compares it to body bias [32,33]. For the same performance (frequency), the sleep transistor-based ALU must operate at higher supply voltage, resulting in higher active power when compared to body bias. Yet, the leakage savings owing to sleep transistor are higher, and the total power of the ALU is the lowest with sleep transistors for the same performance.

The use of sleep transistors has not been limited to logic design only, but has been extended to static RAMs or SRAMs [34–36]. Since leakage power is a significant fraction of the total cache power, it is imperative to use such low-leakage techniques in conventional memory designs. This however, comes at

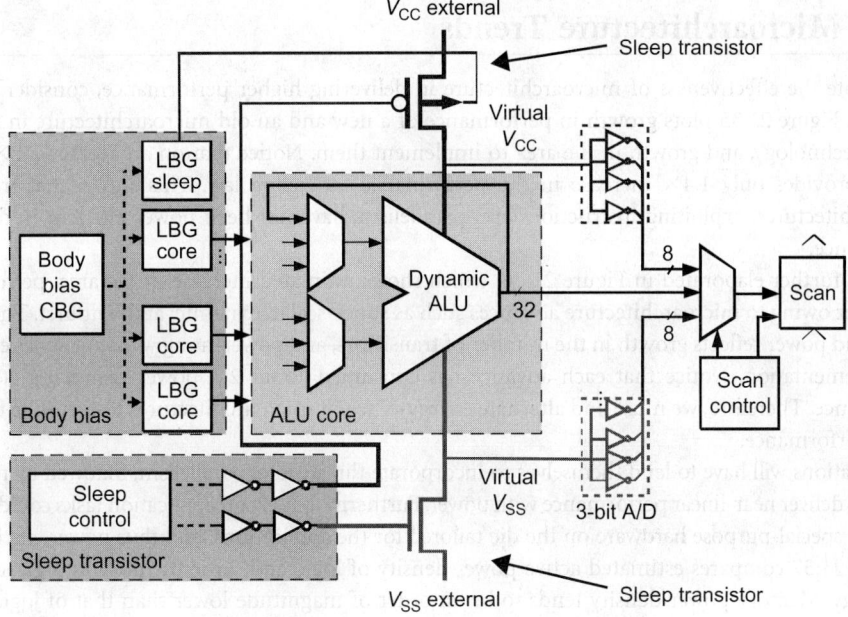

FIGURE 21.33　ALU with sleep transistors for reducing source/drain leakage.

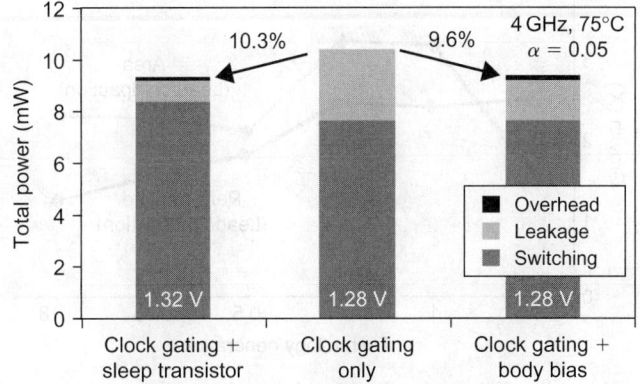

FIGURE 21.34　Reduction by sleep transistors of power including the overhead by 10%.

the cost of memory access speed and cell stability. Proper design techniques are necessary to lower the leakage in memories, without affecting the data retention capability or read/write stability.

21.3.3   UltraLow-Power Circuit Design

In recent years, the demand for power-sensitive, battery-operated, and hand-held devices has increased substantially, thereby necessitating research in an area of ultralow power circuits capable of working in a few tens to a few hundreds of kilohertz. Subthreshold logic (where the supply voltage, V_{DD} is below the threshold voltage, V_t) has emerged as a popular choice for realizing extremely low-power digital systems [37,38]. The advantage in power reduction comes not only from the reduced V_{DD}, but also from the reduced gate capacitance in the subthreshold region. Considerable work has already been done in realizing digital signal processors in subthreshold domain [39] and optimizing the design techniques for the same [40]. Parameter variation poses a challenge for designing circuits operating in subthreshold region.

21.4   Microarchitecture Trends

To evaluate the effectiveness of microarchitecture in delivering higher performance, consider Pollack's rule [41]. Figure 21.35 plots growth in performance of a new and an old microarchitecture in the same process technology, and growth in the area to implement them. Notice that on an average a 2× growth in area provides only 1.4× increase in the performance—a square law. This shows that traditional microarchitectures, exploiting instruction level parallelism, have not been power efficient in delivering performance.

This is further elaborated in Figure 21.36, which shows estimated increase in die area, performance, and power owing to microarchitecture advances such as super-scalar, dynamic, and netburst. The growth in area and power reflects growth in the number of transistors, and power-hungry circuit styles employed for implementation. Notice that each advance has consumed about 2× power delivering 40% more performance. Therefore, we must find alternate energy-efficient microarchitectures to continue to deliver higher performance.

Applications will have to lend themselves to incorporate thread-level parallelism, followed by multiprocessing to deliver near-linear performance with power. Furthermore, certain application tasks could be easily served by special-purpose hardware on the die tailored for the applications, and thus power efficient.

Figure 21.37 compares estimated active power density of logic and static memory in a given process technology. Memory power density tends to be an order of magnitude lower than that of logic. This is because only a part of the memory is accessed at any given time. Also, memory transistors can withstand

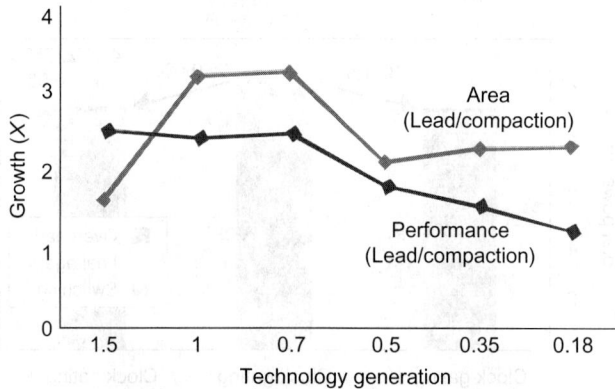

FIGURE 21.35   Microarchitecture efficiency trends. (Note that performance has been measured using SpecINT and SpecFP.)

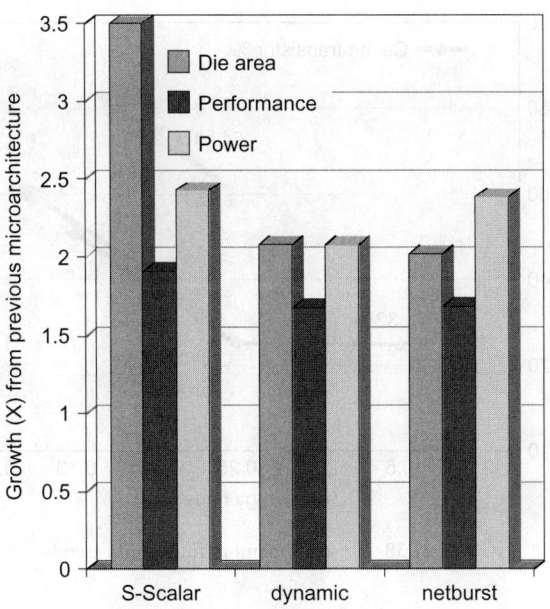

FIGURE 21.36 Performance, power, and area trade-offs of general purpose microarchitectures.

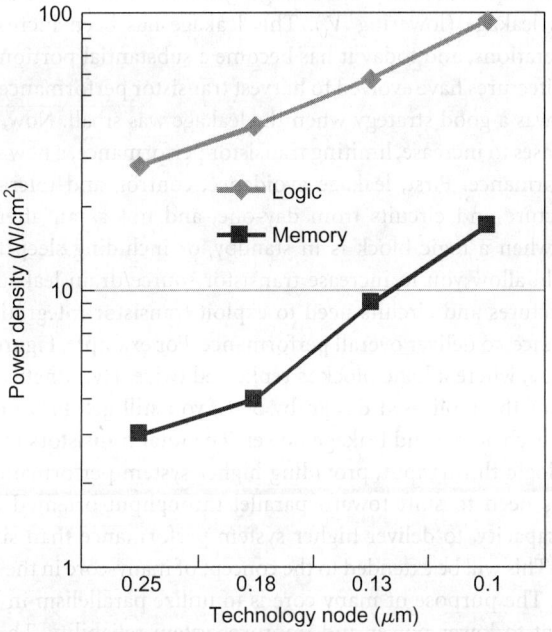

FIGURE 21.37 Power density of memory versus logic.

relatively higher threshold voltages, reducing the leakage power compared to logic. To make up for the loss of transistor performance, memory operations can be pipelined, with modest increase in latency.

Therefore, future microarchitectures could exploit lower power density of memory to stay on the performance trend, and yet lower active and leakage power. The trend is already evident as shown in Figure 21.38, which plots cache memory transistors in microprocessors in several technology generations. Future microarchitectures will use even bigger caches to continue to deliver higher performance [42].

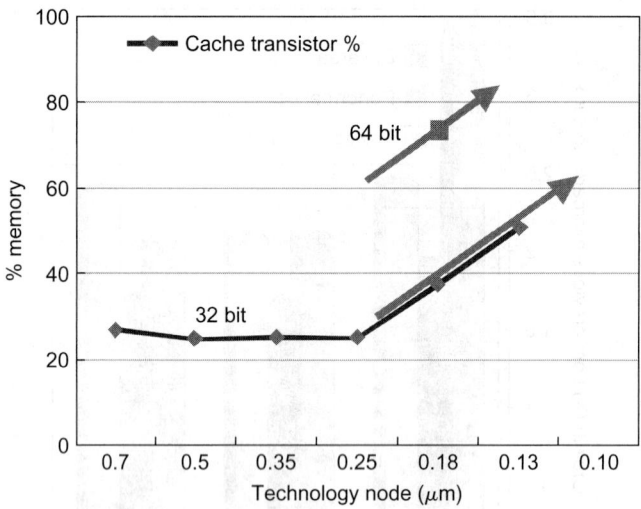

FIGURE 21.38 On-chip memory integration trends.

21.5 Delivering System Performance and Final Discussions

Transistor performance improves every technology generation and part of it is due to increased transistor source/drain leakage (lowering V_t). This leakage has been increasing exponentially over several technology generations, and today it has become a substantial portion of the total power [27]. Circuits and microarchitectures have evolved to harvest transistor performance to deliver overall system performance, and this was a good strategy when the leakage was small. Now, and in the future, when source/drain leakage ceases to increase, limiting transistor performance, a new strategy must be adopted to deliver system performance. First, leakage avoidance, control, and tolerance should be compre- hended in the architecture and circuits from day-one, and not as an after-thought. For example, activating stack effect when a logic block is in standby, or including sleep transistor control in the microarchitecture would allow you to increase transistor source/drain leakage and hence the perfor- mance. Second, architectures and circuits need to exploit transistor integration capacity, rather than raw transistor performance, to deliver overall performance. For example, Figure 21.39 shows a through- put-oriented architecture, where a logic block is replicated twice. Hypothetically, if you reduce supply voltage and frequency of the replicated design by 30%, you still get 40% more throughput, though with 30% reduction in each active and leakage power. The total transistors in the design doubled, but resulted in 40% more logic throughput, providing higher system performance. That is why microar- chitectures and circuits need to shift toward parallel throughput-oriented architectures, exploiting transistor integration capacity, to deliver higher system performance than simply depending on raw transistor performance. This will be extended to the concept of many-core in the future high-performance microprocessor design. The purpose of many core is to utilize parallelism in hardware not to achieve higher performance, but to lower power and improve system reliability. This approach may help in dealing with the challenge of higher variability.

21.6 Conclusions

We described CMOS scaling challenges for gate lengths approaching 10 nm and potential solutions in circuits and microarchitecture. These solutions may appear difficult, but are more mature and less risky than other proposed alternatives for CMOS. That is why, CMOS is, for now, and for the foreseeable future.

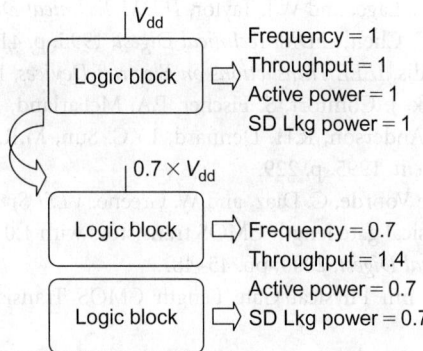

FIGURE 21.39 A throughout-oriented design—higher logic throughput with lower power.

Interactions among technologists, circuit designers, and architects have now become even more important to optimize the entire system particularly for low-power designs. Technology, circuits, and micro-architecture need to make concerted efforts to solve leakage, power and variation issues, and limitations.

21.7 Summary

Scaling of CMOS technology continues in spite of tremendous technology development barriers, design challenges, and prohibitive costs. Today, the 65 nm CMOS technology node is moving from development to high-volume manufacturing, while research and development continues on future technology nodes including 45 nm, 30 nm, and beyond.

However, the design of ICs in these scaled technologies faces growing limitations. It is increasingly difficult to sustain supply and threshold voltage scaling to provide the required performance increase, limit energy consumption, control power dissipation, and maintain reliability. These requirements pose several difficulties across a range of disciplines spanning technology, fabrication, circuits, systems, design, and architecture.

On the technology front, the question arises whether we can continue to scale CMOS technology or whether we are close to the end of the ITRS road map. Should we continue along the traditional CMOS scaling path—reduce effective oxide thickness, improve channel mobility, and minimize parasitics—or consider a more radical departure from planar CMOS to nonplanar device structures such as tri-gate and FinFET thin body transistors? Can we translate the device electrostatic improvement to performance gain?

On the design front, while researchers are exploring various circuit design techniques to deal with leakage and process variation, they have also started studying how to optimize circuits and systems with nonplanar CMOS devices.

Acknowledgments

The author wishes to acknowledge his colleagues, Shekhar Borkar, Vivek De, Robert Chau, and Suman Datta for valuable discussions.

References

1. V. De and S. Borkar, "Technology and design challenges for low power and high performance microprocessors," *International Symposium on Low Power Electronics and Design*, 1999, pp. 163–168.
2. S.E. Thompson, P.A. Packan, and M.Y. Bohr, "MOS Scaling: Transistor Challenges for the 21ˢᵗ Century," *Intel Technology Journal*, 1998.
3. S.E. Thompson, P.A. Packan, and M.Y. Bohr, *VLSI Technology Symposium Digest*, 1996, p. 154.

4. S. Venkatesan, J.W. Lutze, C. Lage, and W.J. Taylor, *IEDM Technical Digest*, 1995, p. 419.
5. M. Rodder, S. Aur, and I.-C. Chen, *IEDM Technical Digest*, 1995, p. 415.
6. J.B. Jacobs and D. Antoniadis, *IEEE Transactions on Electron Devices*, 1995, p. 870.
7. G.G. Shahidi, J.D. Warnock, J. Comfort, S. Fischer, P.A. McFarland, A. Acovic, T.I. Chappell, B.A. Chappell, T.H. Ning, C.J. Anderson, R.H. Dennard, J.Y.C. Sun, M.R. Polcari, and B. Davari, *IBM Journal Research Development*, 1995, p. 229.
8. M. Cao, P. Griffin, P. Vande Voorde, C. Diaz, and W. Greene, *VLSI Symposium Digest*, 1997, p. 85.
9. R. Chau et al., "30 nm physical gate length CMOS transistors with 1.0 ps n-MOS and 1.7 ps p-MOS gate delays," *IEDM Technical Digest*, 2000, pp. 45–48.
10. R. Chau, "30 nm and 20 nm Physical Gate Length CMOS Transistors," *Silicon Nanotechnology Workshop*, 2001.
11. R. Chau et al., "A 50 nm Depleted-Substrate CMOS Transistor (DST)," *IEDM Technical Digest*, 2001, pp. 621–624.
12. K. Mistry et al., *Symposium on VLSI Technology Digest*, 2004.
13. T. Ghani et al., *IEDM Tech. Dig.*, 2003, pp. 197–200.
14. P. Bai et al., *IEDM Tech. Dig.*, 2004, pp. 657–660.
15. N. Lindert et al., *IEEE Electron Device Letters*, **22**, pp. 487–489, 2001.
16. R. Chau, B. Doyle, J. Kavalieros, D. Barlage, A. Murthy, M. Doczy, R. Rios, T. Linton, R. Arghavani, B. Jin, S. Datta, and S. Hareland, "Advanced Depleted-Substrate Transistors: Single-Gate, Double-Gate and Tri-Gate," *2002 International Conference on Solid State Devices and Materials (SSDM 2002)*, Nagoya, Japan.
17. R. Chau et al., *IEDM Tech. Digest*, 2001, pp. 621–624.
18. L. Wei et al., VLSI 2000, *13th International Conference on VLSI Design*, 2000.
19. A. Keshavarzi, S. Narendra, S. Borkar, C. Hawkins, K. Roy, and V. De, "Technology scaling behavior of optimum reverse body bias for standby leakage power reduction in CMOS IC's," *International Symposium on Low Power Electronics and Design*, 1999, pp. 252–254.
20. A Keshavarzi, S. Ma, S. Narendra, B. Bloechel, K. Mistry, T. Ghani, S. Borkar, and V. De, "Effectiveness of reverse body bias for leakage control in scaled dual Vt CMOS ICs," *International Symposium on Low Power Electronics and Design*, 2001, pp. 207–212.
21. A Keshavarzi, S. Narendra, B. Bloechel, S. Borkar, and V. De, "Forward body bias for microprocessors in 130 nm technology generation and beyond," pp. 312–315.
22. S. Narendra, M. Haycock, V. Govindarajulu, V. Erraguntla, H. Wilson, S. Vangal, A. Pangal, E. Seligman, R. Nair, A. Keshavarzi, B. Bloechel, G. Dermer, R. Mooney, N. Borkar, S. Borkar, and V. Vivek De, "1.1V 1GHz communications router with on-chip body bias in 150nm CMOS." *IEEE International Solid-State Circuits Conference (ISSCC), 2002*, Digest of Technical Papers, Vol. 2, pp. 218–482.
23. J. Tschanz, J. Kao, S. Narendra, R. Nair, D. Antoniadis, A. Chandrakasan, and De Vivek, "Adaptive body bias for reducing impacts of die-to-die and within-die parameter variations on microprocessor frequency and leakage." *IEEE International Solid-State Circuits Conference, 2002*. Digest of Technical Papers. Feb. 2002, Vol. 1, pp. 422–478.
24. S. Borkar, "Circuit techniques for subthreshold leakage avoidance, control and tolerance," *IEEE International Electron Devices Meeting, 2004*. Digest of Technical Papers, pp. 421–424.
25. A. Srivastava, D. Sylvester, and D. Blaauw, "Statistical Optimization of Leakage Power Considering Process Variations using Dual-V_{th} and Sizing," *2004 Proceedings of Design Automation Conference (DAC 2004)*, pp. 773–778.
26. S. Narendra, S. Borkar, V. De, D. Antoniadis, and A. Chandrakasan, "Scaling of stack effect and its application for leakage reduction," *International Symposium on Low Power Electronics and Design, 2001*, Aug., pp. 195–200.
27. Y. Ye, S. Borkar, and V. De, "A new technique for standby leakage reduction in high-performance circuits," *Symposium on VLSI Circuits, 1998*, Digest of Technical Papers. 11–13, June 1998, pp. 40–41.

28. J. Tschanz, S. Narendra, Ye Yibin, B. Bloechel, S. Borkar, and V. De, "Dynamic-sleep transistor and body bias for active leakage power control of microprocessors," *IEEE International Solid-State Circuits Conference, 2003*. Digest of Technical Papers. vol. 1, pp. 102–481.

29. A. Alvandpour, R. Krishnamurthy, K. Soumyanath, and S. Borkar, "A conditional keeper technique for sub-0.13μ wide dynamic gates," *Symposium on VLSI Circuits, 2001*. Digest of Technical Papers. 14–16, June 2001, pp. 29–30.

30. R. Krishnamurthy, A. Alvandpour, G. Balamurugan, N. Shanbhagh, K. Soumyanath, and S. Borkar, "A 0.13 μm 6 GHz 256 × 32b leakage-tolerant register files," *Symposium on VLSI Circuits, 2001*. Digest of Technical Papers. 14–16, June 2001, pp. 25–26.

31. C.H. Kim, J. Kim, I. Chang, and K. Roy, "PVT-Aware Leakage Reduction for On-die Caches with Improved Read Stability," *IEEE International Solid-State Circuits Conference*, San Francisco, CA, 2005, p. 482.

32. K. Zhang, U. Bhattacharya, Z. Chen, F. Hamzaoglu, D. Murray, N. Vallepalli, Y. Wang, B. Zheng, and M. Bohr, "SRAM Design on 65-nm CMOS technology with dynamic sleep transistor for leakage reduction," *IEEE Journal of Solid-State Circuits*, vol. 40, no. 4, 2005, pp. 895–901.

33. K. Zhang, U. Bhattacharya, Z. Chen, F. Hamzaoglu, D. Murray, N. Vallepalli, Y. Wang, B. Zheng, and M. Bohr, "SRAM design on 65nm CMOS technology with integrated leakage reduction scheme," Digest of Technical Papers, *2004 Symposium on VLSI Circuits*, June 2004, pp. 294–295.

34. H. Soleman, K. Roy, and B.C. Paul, "Robust Subthreshold Logic for Ultra-Low Power Operation," *IEEE Transactions on VLSI Systems*, vol. 9, no. 1, pp. 90–99, 2001.

35. A. P. Chandrakasan, S. Sheng, and R.W. Broderson, "Low-Power CMOS Digital Design," *IEEE Journal of Solid-State Circuits*, April 1992, vol. 27, pp. 473–484.

36. A. Wang and A.P. Chandrakasan, "A 180-mV Subthreshold FFT Processor Using a Minimum Energy Design Methodology," *IEEE Journal of Solid-State Circuits*, 2005, vol. 40, no. 1, pp. 310–319.

37. A. Raychowdhury, B.C. Paul, S. Bhunia, and K. Roy, "Computing with Subthreshold Leakage: Device/Circuit/Architecture Co-design for Ultralow Power Subthreshold Operation," *IEEE Transactions on Very Large Scale Integration Systems*, 2005, vol. 13, no. 11, pp. 1213–1224.

38. F. Pollack, "New Microarchitecture Challenges in the Coming Generations of CMOS Process Technologies," *Micro32*, 1999.

39. G. Sery, S. Borkar, and V. De, *DAC*, 2002, pp. 78–83.

40. S. Datta, T. Ashley, R. Chau, K. Hilton, R. Jefferies, T. Martin, and T. Phillips, "85nm Gate Length Enhancement and Depletion mode InSb Quantum Well Transistors for Ultra High Speed and Very Low Power Digital Logic Applications," *Digest of International Electron Device Meeting*, Dec. 2005.

41. A. Javey, J. Guo, D. Farmer, Q. Wang, E. Yenilmez, R. Gordon, M. Lundstrom, and H. Dai, "Self-Aligned Ballistic Molecular Transistors and Electrically Parallel Nanotube Arrays," *Nanoletter*, vol. 4, 2004, pp. 1319–1322.

42. A. Javey, J. Guo, Q. Wang, M. Lundstrom and H. Dai, "Ballistic Carbon Nanotube Field-effect Transistors," *Nature*, vol. 427, 2003, pp. 654–657.

Section IV

Amplifiers

Rolf Schaumann
Portland State University

0-8493-XXXX-X/04/$0.00+$1.50
© 2006 by CRC Press LLC

Section IV

Amplifiers

Rolf Schaumann

22

CMOS Amplifier Design

Harry W. Li
Formerly with the University of Idaho at Boise

R. Jacob Baker
University of Idaho at Boise

Donald C. Thelen
American Microsystems, Inc.

CONTENTS

22.1 Introduction

This chapter discusses the design, operation, and layout of CMOS analog amplifiers and subcircuits (current mirrors, biasing circuits, etc.). To make this discussion meaningful and clear, we need to define some important variables related to the DC operation of MOSFETs (Figure 22.1). Figure 22.1(a) shows the simplified schematic representations of n- and p-channel MOSFETs. We say simplified because, when these symbols are used, it is *assumed* that the fourth terminal of the MOSFET (i.e., the body connection) is connected to either the lowest potential on the chip (V_{SS} or ground for the NMOS) or the highest potential (V_{DD} for the PMOS). Figure 22.1(b) shows the more general schematic representation of n- and p-channel MOSFETs. We are assuming that, although the drain and source of the MOSFETs are interchangeable, drain current flows from the top of the device to the bottom. Because of the assumed direction of current flow, the drain terminal of the n-channel is on the top of the symbol, while the drain terminal of the p-channel is on the bottom of the schematic symbol. The following are short descriptions of some important characteristics of MOSFETs that will be useful in the following discussion.

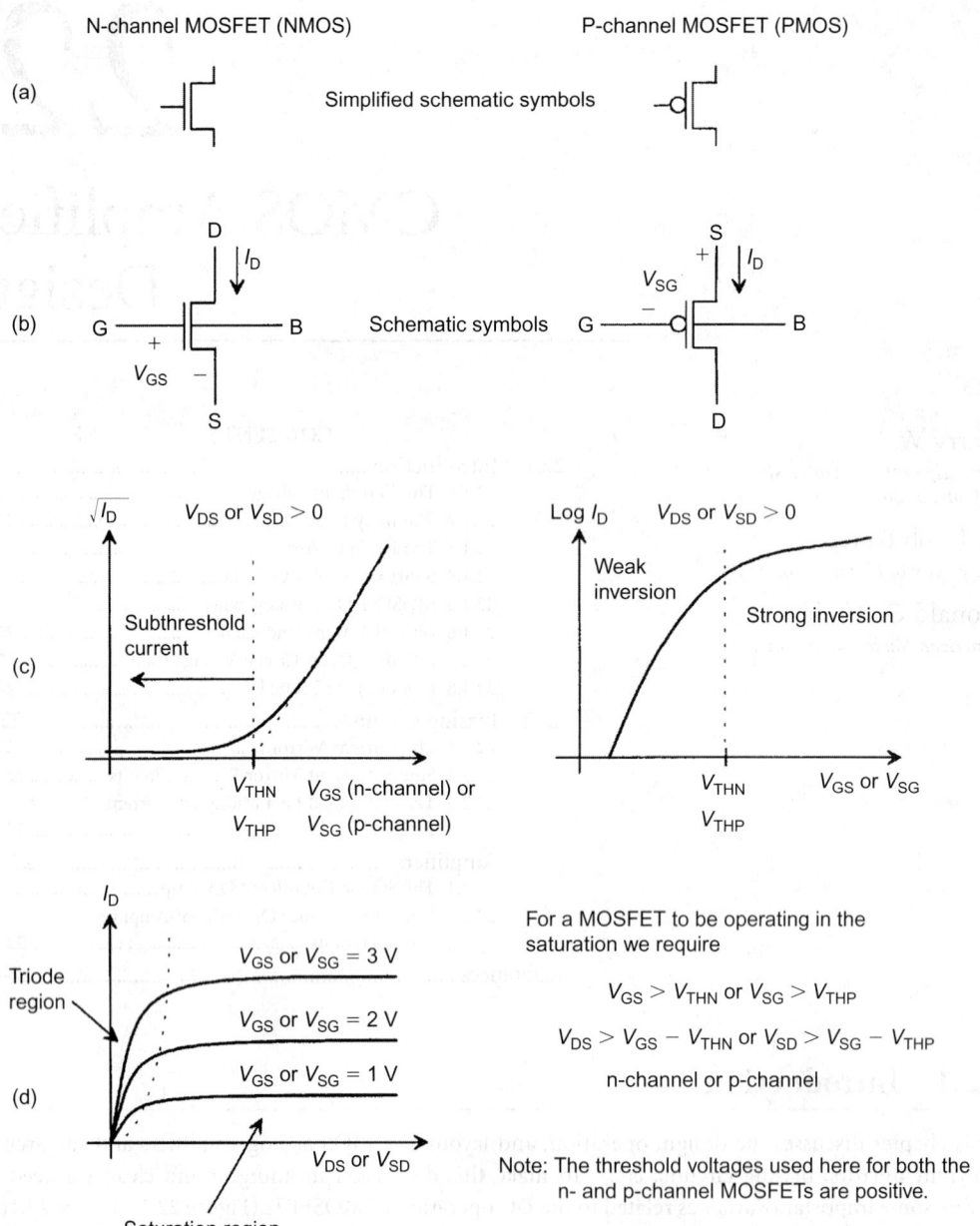

FIGURE 22.1 MOSFET device characteristics.

22.1.1 The Threshold Voltage

Loosely defined, the threshold voltage, V_{THN} or V_{THP}, is the minimum gate-to-source voltage (V_{GS} for the n-channel or V_{SG} for the p-channel) that causes a current to flow when a voltage is applied between the drain and source of the MOSFET. As shown in Figure 22.1(c) the threshold voltage is estimated by plotting the square root of the drain current against the gate-source voltage of the MOSFET and looking at the intersection of the line tangent with this plot with the x-axis (V_{GS} for the n-channel). As seen in the figure, a current does flow below the threshold voltage of the device. This current is termed, for obvious reasons, the *subthreshold current*. The subthreshold current is characterized by plotting the log of the drain current against the gate-source voltage. The slope of the curve in the subthreshold region (sometimes also called the *weak inversion region*) is used to specify how the drain current changes with V_{GS}. A typical value for the reciprocal of the slope of this curve is 100 mV/dec. An equation relating the drain current of an n-channel MOSFET operating in the subthreshold region to V_{GS} is (assuming $V_{DS} > 100$ mV):

$$I_D = I_{D0} \cdot \frac{W}{L} \cdot \exp\left[\frac{q(V_{GS} - V_{THN})}{kT \cdot N}\right] \tag{22.1}$$

where W and L are the width and length of the MOSFET, I_{D0} is a measured constant, k is Boltzmann's constant (1.38×10^{-23} J/K), T is temperature in Kelvin, q is the electronic charge (1.609×10^{-23} C), and N is the slope parameter. Note that the slope of the log I_D vs. V_{GS} curve in the subthreshold region is

$$\frac{\Delta \log I_D}{V_{GS}} = \frac{q \cdot \log e}{N \cdot kT} \tag{22.2}$$

22.1.2   The Body Effect

The threshold voltage of a MOSFET is dependent on the potential between the source of the MOSFET and its body. Consider Figure 22.2, showing the situation when the body of an n-channel MOSFET is connected to ground and the source of the MOSFET is held V_{SB} above ground. As V_{SB} is increased (i.e., the potential on the source of the MOSFET increases relative to ground), the minimum V_{GS} needed to cause appreciable current to flow increases (V_{THN} increases as V_{SB} increases). We can relate V_{THN} to V_{SB} using the body-effect coefficient, γ, by

$$V_{THN} = V_{THN0} + \gamma\sqrt{|2\phi_F| + V_{SB}} - \sqrt{|2\phi_F|} \tag{22.3}$$

where V_{THN0} is the zero-bias threshold voltage when $V_{SB} = 0$ and ϕ_F is the surface electrostatic potential [1] with a typical value of 300 mV. An important thing to notice here is that the threshold voltage tends to change less with increasing source-to-substrate (body) potential (increasing V_{SB}).

22.1.3   The Drain Current

In the following discussion, we will assume that the gate-source voltage of a MOSFET is greater than the threshold voltage so that a reasonably sized drain current can flow ($V_{GS} > V_{THN}$ or $V_{SG} > V_{THP}$). If this is the case, the MOSFET operates in either the triode region or the saturation region [Figure 22.1(d)]. The

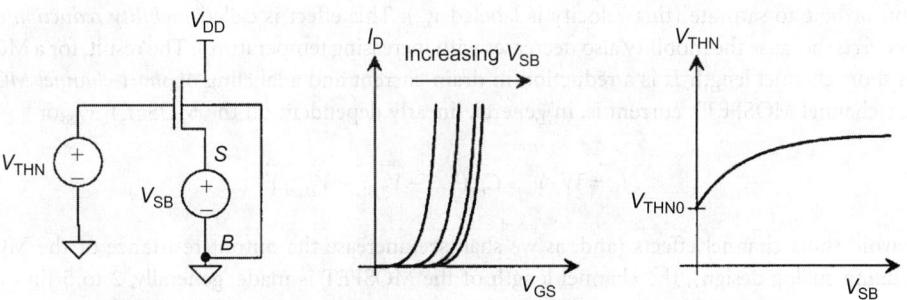

FIGURE 22.2   Illustration of the threshold voltage dependence on body-effect.

drain current of a *long L* n-channel MOSFET operating in the *triode region*, is given by

$$I_D = \beta\left[(V_{GS} - V_{THN})V_{DS} - \frac{V_{DS}^2}{2}\right] \tag{22.4}$$

assuming long *L* with $V_{DS} \leq V_{GS} - V_{THN}$. Note that the MOSFET behaves like a voltage-controlled resistor when operating in the deep triode region with a channel resistance (the resistance is measured between the drain and source of the MOSFET) approximated by

$$R_{CH} = \frac{1}{\beta(V_{GS} - V_{THN}) - \beta V_{DS}} \quad \text{or} \quad \frac{1}{\beta(V_{GS} - V_{THN})} \quad \text{if} \quad V_{DS} \ll V_{GS} - V_{THN} \tag{22.5}$$

When doing analog design, it is often useful to implement a resistor whose value is dependent on a controlling voltage (a voltage-controlled resistor). When the NMOS device is operating in the saturation region, the drain current is given by

$$I_D = \beta(V_{GS} - V_{THN})^2[1 + \lambda(V_{DS} - V_{DS,sat})] \tag{22.6}$$

assuming long *L* with $V_{DS} \geq V_{GS} - V_{THN}$, where $V_{DS,sat} = V_{GS} V_{THN}$.

The *transconductance parameter*, β, is given by

$$\beta = KP \cdot \frac{W}{L} = \mu_{n,p} C_{ox} \cdot \frac{W}{L} \tag{22.7}$$

where $\mu_{n,p}$ is the mobility of either the electron or hole and C_{ox} is the oxide capacitance per unit area [ε_{ox}/t_{ox}, the dieletric constant of the gate oxide (35.4 aF/μm) divided by the gate oxide thickness]. Typical values for KP_n, KP_p, and C_{ox} for a 0.5-μm process are 150 μA/V², 50 μA/V², and 4 fF/μm², respectively. Also, an important thing to note in these equations for an n-channel MOSFET, is that V_{GS}, V_{DS}, and V_{THN} can be directly replaced with V_{SG}, V_{SD}, and V_{THP}, respectively, to obtain the operating equations for the p-channel MOSFET (keeping in mind that all quantities under normal conditions for operation are positive.) Also note that the saturation slope parameter λ (also known as the channel length/mobility modulation parameter) determines how changes in the drain-to-source voltage affect the MOSFET drain current and thus the MOSFET output resistance.

22.1.4 Short-Channel MOSFETs

As the channel length of a MOSFET is reduced, the electron and hole mobilities, μ_n and μ_p, start to get smaller. The mobility is simply a ratio of the electron or hole velocity to the applied electric field. Reducing the channel length increases the applied electric field while at the same time causing the velocity of the electron or hole to saturate (this velocity is labeled v_{sat}). This effect is called *mobility reduction* or *hot-carrier effects* (because the mobility also decreases with increasing temperature). The result, for a MOSFET with a short channel length *L*, is a reduction in drain current and a labeling of *short-channel MOSFET*. A short-channel MOSFET's current is, in general, linearly dependent on the MOSFET V_{GS} or

$$I_D = W \cdot v_{sat} \cdot C_{ox}(V_{GS} - V_{THN} - V_{DS,sat}) \tag{22.8}$$

To avoid short-channel effects (and, as we shall see, increase the output resistance of the MOSFET when doing analog design), the channel length of the MOSFET is made, generally, 2 to 5 times larger than the minimum allowable *L*. For a 0.5-μm CMOS process, this means we make the channel length of the MOSFETs 1.0 to 2.5 μm.

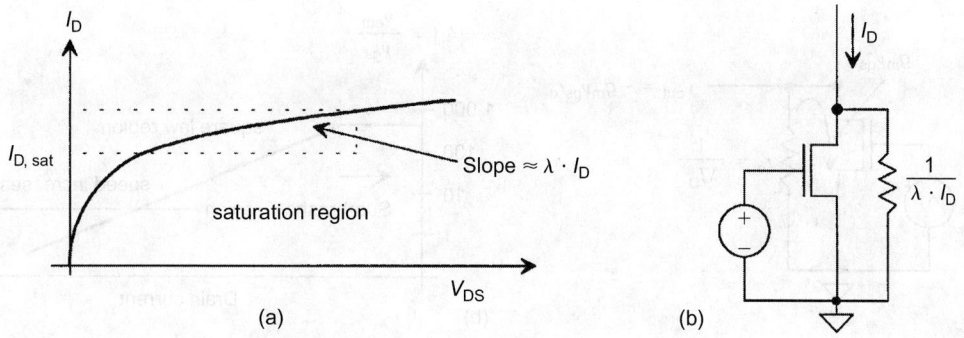

FIGURE 22.3 Output resistance of a MOSFET.

22.1.5 MOSFET Output Resistance

An important parameter of a MOSFET in analog applications is its output resistance. Consider the portion of a MOSFET's I-V characteristics shown in Figure 22.3(a). When the MOSFET is operating in the saturation region, the slope of I_D, because of changes in the drain current with changes in the drain-source voltage, is relatively small. If this change were zero, the drain current would be a fixed value independent of the voltage between the drain and source of the MOSFET (in other words, the MOSFET would act like an ideal current source.) Even with the small change in current with V_{DS}, we can think of the MOSFET as a current source. To model the changes in current with V_{DS}, we can simply place a resistor across the drain and source of the MOSFET [Figure 22.3(b)]. The value of this resistor is

$$r_O \approx \frac{1}{\lambda I_D} \tag{22.9}$$

where λ is in the range of 0.1 to 0.01 V^{-1}.

At this point, several practical comments should be made: (1) in general, to increase λ, the channel length is made 2 to 5 times the minimum allowable channel L (this was also necessary to reduce the short-channel effects discussed above); (2) the value of λ is normally determined empirically; trying to determine λ from an equation is, in general, not too useful; (3) the output resistance of a MOSFET is a function of the MOSFET's drain current. The exact value of this current is not important when estimating the output resistance. Whether $I_{D,sat}$ (the drain current at $V_{D,sat}$) or the actual operating point current, I_D, is used is not practically important when determining r_o.

22.1.6 MOSFET Transconductance

It is useful to determine how a change in the gate-source voltage changes the drain current of a MOSFET operating in the saturation region. We can relate the change in drain current, i_d, to the change in gate-source voltage, v_{gs}, using the MOSFET transconductance, g_m, or

$$g_m = \frac{\Delta i_D}{\Delta v_{GS}} \Rightarrow i_d = g_m v_{gs} \tag{22.10}$$

Neglecting the output resistance of a MOSFET, we can write the sum of the DC and AC (or changing) components of the drain current and gate-source voltage using

$$i_D = i_d(AC) + I_D(DC) = \frac{\beta}{2}(V_{GS}(DC) + v_{gs}(AC) - V_{THN})^2 \tag{22.11}$$

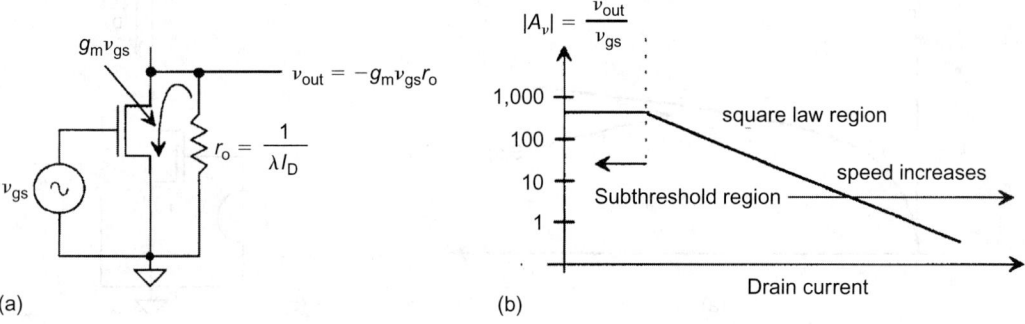

FIGURE 22.4   Open-circuit voltage gain (DC biasing not shown).

If we hold the DC values constant and assume they are large compared to the AC components, then by simply taking the derivative of i_D with respect to v_{gs}, we can determine g_m. Doing this results in

$$g_m = \beta(V_{GS} - V_{THN}) = \sqrt{2\beta I_D} \tag{22.12}$$

Following this same procedure for the MOSFET operating in the subthreshold region results in

$$g_m = \frac{I_D \cdot q}{kT} \tag{22.13}$$

Notice how the change in transconductance is linear with drain current when the device is operating in the subthreshold region. The larger incremental increase in g_m is due to the exponential relationship between I_D and V_{GS} when the device is operating in the weak inversion region (as compared to the square law relationship when the device is operating in the strong inversion region).

22.1.7   MOSFET Open-Circuit Voltage Gain

At this point, we can ask, "What's the largest possible voltage gain I can get from a MOSFET under ideal biasing conditions?" Consider the schematic diagram shown in Figure 22.4(a) without biasing circuitry shown and with the effects of finite MOSFET output resistance modeled by an external resistor. The open-circuit voltage gain can be written as

$$|A| = \frac{v_{out}}{v_{gs}} = g_m r_o = \frac{\sqrt{2\beta I_D}}{\lambda I_D} = \frac{\sqrt{2\beta}}{\lambda \sqrt{I_D}} \text{ (for normal operation)} \tag{22.14}$$

which increases as the DC drain biasing current is reduced (and the MOSFET intrinsic speed decreases) until the MOSFET enters the subthreshold region. Once in the subthreshold region, the voltage gain flattens out and becomes

$$A_v = g_m r_o = \frac{I_D \cdot q}{kT} \cdot \frac{1}{\lambda I_D} = \frac{q}{\lambda \cdot kT} \text{ (in the subthreshold region)} \tag{22.15}$$

22.1.8   Layout of the MOSFET

Figure 22.5 shows the layout of both n-channel and p-channel devices. In this layout, we are assuming that a p-type substrate is used for the body of the NMOS (the body connection for the n-channel MOSFET is made through the p+ diffusion on the left in the layout). An n-well is used for the body

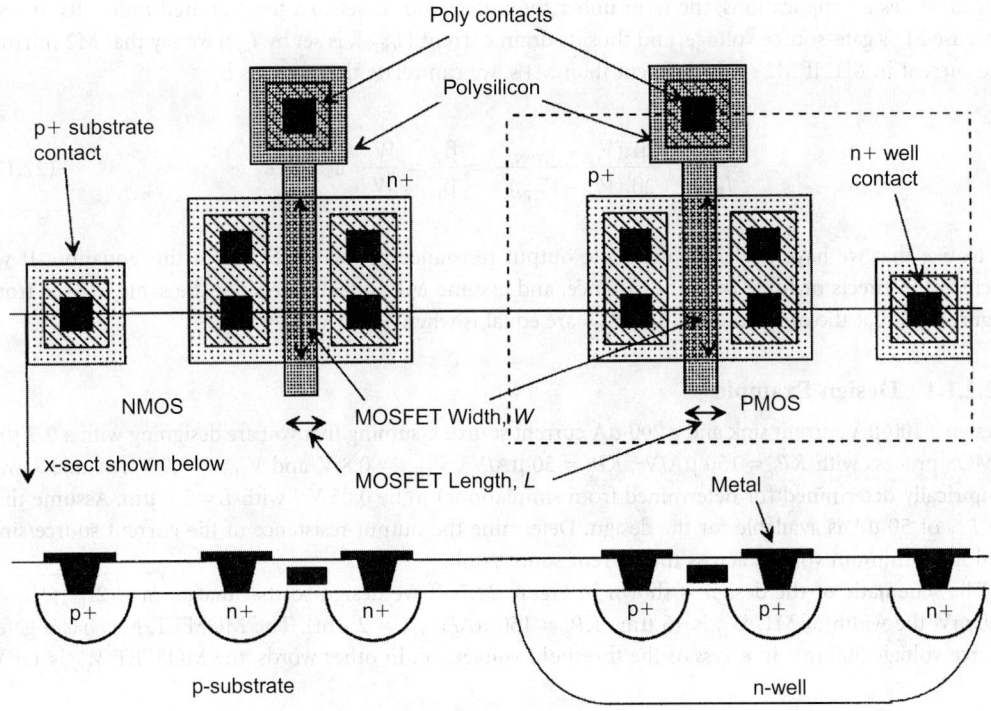

FIGURE 22.5 Layout and cross-sectional views of NMOS and PMOS devices.

of the PMOS devices (the connection to the n-well is made through the n$^+$ diffusion on the right in the layout). An important thing to notice from this layout is that the intersection of polysilicon (poly for short) and n$^+$ (in the p-substrate) or p$^+$ (in the n-well) forms a MOSFET. The length and width of the MOSFET, as seen in Figure 22.5, is determined by the size of this overlap. Also note that the four metal connections to the terminals of each MOSFET, in this layout, are floating; that is, not connected to anything but the MOSFETs themselves. It is important to understand, since the p-substrate is common to all MOSFETs on the chip, that this would require the body of the n-channel MOSFET (again the p$^+$ diffusion on the left in the layout) be connected to V_{SS} (ground for most digital applications).

22.2 Biasing Circuits

A fundamental component of any CMOS amplifier is a biasing circuit. This section presents important design topologies and considerations used in the design of CMOS biasing circuits. We begin this section with a discussion of current mirrors.

22.2.1 The Current Mirror

The basic CMOS current mirror is shown in Figure 22.6. For the moment, we will not concern ourselves with the implementation of I_{REF}. By tying M1's gate to its drain, we set V_{GS} at a value given by

$$V_{GS} = V_{THN} + \sqrt{\frac{2I_{REF}}{\beta_1}}$$

(22.16)

For most design applications, the term under the square root is set to a few hundred millivolts or less. Because M2's gate-source voltage, and thus its drain current (I_{OUT}), is set by I_{REF}, we say that M2 mirrors the current in M1. If M2's β is different than M1's, we can relate the currents by

$$\frac{I_{OUT}}{I_{REF}} = \frac{2\beta_2(V_{GS}-V_{THN})^2}{2\beta_1(V_{GS}-V_{THN})^2} = \frac{\beta_2}{\beta_1} = \frac{W_2}{W_1} \quad \text{if } L_1 = L_2 \tag{22.17}$$

Notice that we have neglected the finite output resistance of the MOSFETs in this equation. If we include the effects of finite output resistance, and assume M1 and M2 are sized the same, we see from Figure 22.6 that the only time I_{REF} and I_{OUT} are equal is when $V_{GS} = V_{OUT} = V_{DS2}$.

22.2.1.1 Design Example

Design a 100-μA current sink and a 200-μA current source assuming that you are designing with a 0.5-μm CMOS process with $KP_n = 150\ \mu\text{A/V}^2$, $KP_p = 50\ \mu\text{A/V}^2$, $V_{THN} = 0.8$ V, and $V_{THP} = 0.9$ V. Assume λ was empirically determined (or determined from simulations) to be 0.05 V^{-1} with $L = 2.0\ \mu$m. Assume that an I_{REF} of 50 μA is available for the design. Determine the output resistance of the current source/sink and the minimum voltage across the current source/sink.

The schematic of the design is shown in Figure 22.7. If we design so that that term $\sqrt{(2I_{REF})/\beta_1}$ is 300 mV, the width of M1, W_1, is 15 μm ($KP_n = 150\ \mu\text{A/V}^2$, $L = 2\ \mu$m). The MOSFET, M1, has a gate-source voltage 300 mV in *excess* of the threshold voltage, or, in other words, the MOSFET V_{GS} is 1.1 V.

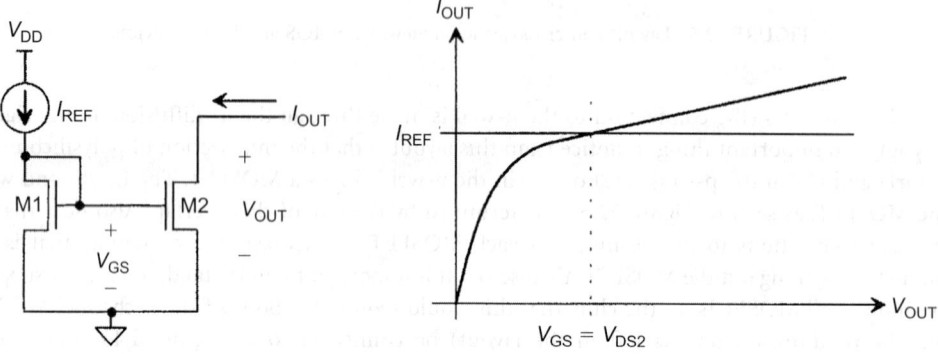

FIGURE 22.6 Basic CMOS current mirror.

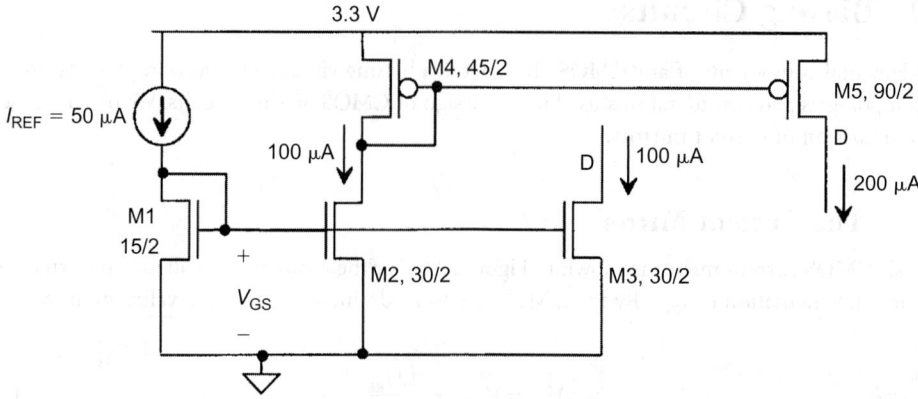

FIGURE 22.7 Design example.

For M2 and M3 to sink 100 μA (twice the current in M1), we simply increase their widths to 30 μm for the same V_{GS} bias supplied by M1. Note the minimum voltage required across M3, V_{DS3}, in order to keep M3 out of the triode region (V_{DS3} $V_{GS} - V_{THN}$), is simply the *excess gate voltage*, $\sqrt{(2I_{REF})/\beta_1}$, or, for this example, 300 mV. (Note: simply put, 300 mV is the minimum voltage on the drain of M3 required to keep it in the saturation region.) Increasing the widths of the MOSFETs lowers the minimum voltage required across the MOSFETs (lowers the excess gate voltage) so they remain in saturation at the price of increased layout area. Also, differences in MOSFET threshold voltage become more significant, affecting the matching between devices. Note that the output resistance of M3 is simply.

The purpose of M2 should be obvious at this point; it provides a 100-μA bias for the p-channel current mirror M4/M5. Again, if we set M4's excess gate voltage to 300 mV (so that the V_{SG} of the p-channel MOSFET is 1.2 V), the width of M4 can be calculated (assuming that L is 2 μm and $KP_p = 50$ μA/V²) to be 45 μm (or a factor of 3 times the n-channel width due to the differences in the transconductance parameters of the MOSFETs). Since the design required a current source of 200 μA, we increase the width of the p-channel, M5, to 90 μm (so that it mirrors twice the current that flows in M4). The output resistance of the current source, M5, is 100 kΩ, while the maximum voltage on the drain of M5 is 3 V (in order to keep M5 in saturation.)

22.2.1.2　Layout of Current Mirrors

In order to get the best matching between devices, we need to lay the MOSFETs out so that differences in the mirrored MOSFETs' widths and lengths are minimized. Figure 22.8 shows the layout of the M1 (15/2) and M2 (30/2) MOSFETs of Figure 22.7. Notice how, instead of laying M2 out in a fashion similar to M1 (i.e., a single poly over active strip), we split M2 into two separate MOSFETs that have the same shape as M1.

The matching between current mirrors can also be improved using a *common-centroid* layout (Figure 22.9). Parameters such as the threshold voltage and *KP* in practice can have a somewhat linear change with position on the wafer. This may be the result of varying temperature on the top of the wafer during processing, or fluctuations in implant dose with position. If we lay MOSFETs M2 and M1 out as shown in Figure 22.9(a) (which is the same way they were laid out in Figure 22.8), M1 has a "weight" of 1 while M2 has a weight of 5. These numbers may correspond to the threshold voltages of three individual MOSFETs with numerical values 0.81, 0.82, and 0.83 V. By using the layout shown in Figure 22.9(b), M1's or M2's average weight is 2. In other words, using the threshold voltages as an example, M1's threshold voltage is 0.82 V while the average of M2's threshold voltage is also 0.82 V. Similar discussions

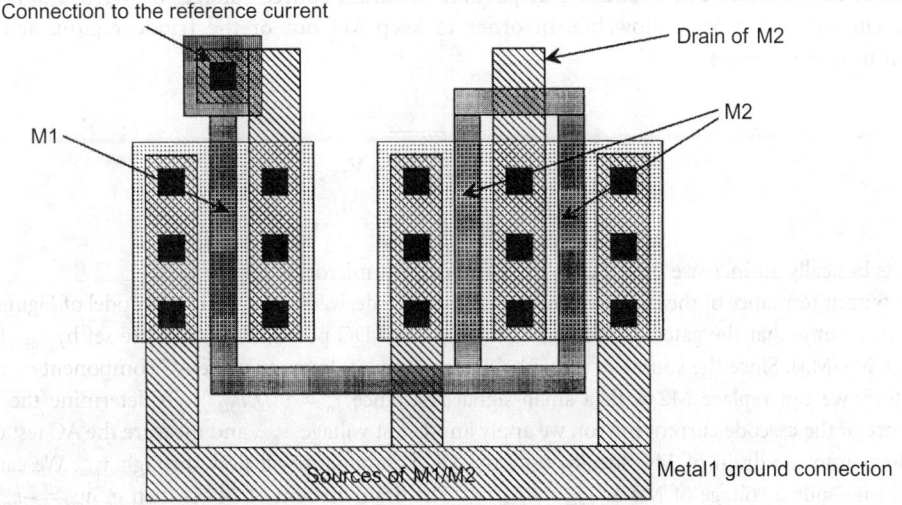

FIGURE 22.8　Layout of MOSFET mirror M1/M2 in Figure 22.7.

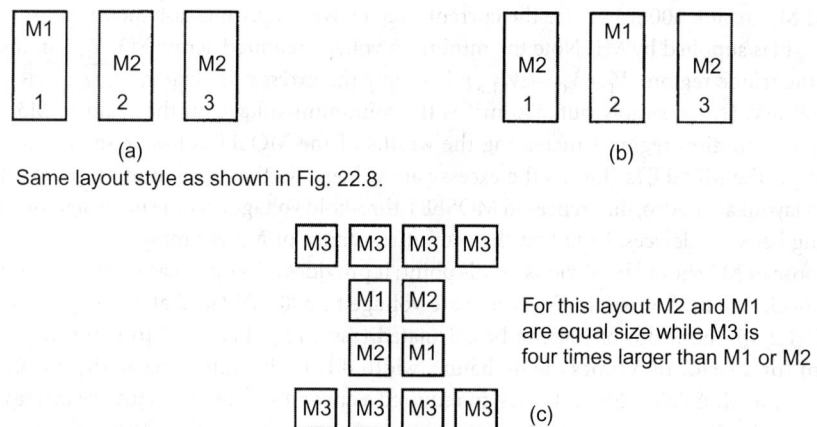

(a)   (b)

Same layout style as shown in Fig. 22.8.

For this layout M2 and M1
are equal size while M3 is
four times larger than M1 or M2

(c)

FIGURE 22.9 Common-centroid layout used to improve matching.

can be made if the transconductance parameters vary with position. Figure 22.9(c) shows how three devices can be matched using a common-centroid layout (M2 and M1 are the same size while M3 is 4 times their size.) A good exercise at this point is to modify the layout of Figure 22.9(c) so that M2 is twice the size of M1 and one half the size of M3.

22.2.1.3 The Cascode Current Mirror

We saw that in order to improve the matching between the two currents in the basic current mirror of Figure 22.6, we needed to force the drain-source voltage of M2 to be the same as the drain-source voltage of M1 (which is also V_{GS} in Figure 22.6). This can be accomplished using the cascode connection of MOSFETs shown in Figure 22.10. The term "cascode" comes from the days of vacuum tube design where a cascade of a common-cathode amplifier driving a common-grid amplifier was used to increase the speed and gain of an overall amplifier design.

Using the cascode configuration results in higher output resistance and thus better matching. For the following discussion, we will assume that M1 and M3 are the same size, as are M2 and M4. Again, remember that M1 and M2 form a current mirror and operate in the same way as previously discussed. The addition of M3 and M4 helps force the drain-source voltages of M1/M2 to the same value. The minimum V_{OUT} allowable, in order to keep M4 out of the triode region, across the current mirror increases to

$$V_{OUT,min} = 2\sqrt{\frac{2I_{REF}}{\beta_{1,3}}} + V_{THN} \tag{22.18}$$

which is basically an increase of V_{GS} over the basic current mirror of Figure 22.6.

The output resistance of the cascode configuration can be derived with the circuit model of Figure 22.11. Here, we assume that the gates of M4 and M2 are at fixed DC potentials (which are set by I_{REF} flowing through M1/M3). Since the source of M2 is held at ground, we know that the AC component of v_{gs2} is 0. Therefore, we can replace M2 with a small-signal resistance $r_o = (1/\lambda I_{OUT})$. To determine the output resistance of the cascode current mirror, we apply an AC test voltage, v_{test}, and measure the AC test current that flows into the drain of M4. Ideally, only the DC component will flow through v_{test}. We can write the AC gate-source voltage of M4 as $v_{gs4} = -i_{test} \cdot r_o$. The drain current of M4 is then $g_{m4}v_{gs4} = -i_{test} \cdot g_{m4}r_o$ while the current through the small-signal output resistance of M4, r_o, is $(v_{test} - (-i_{test}r_o))/r_o$. Combining

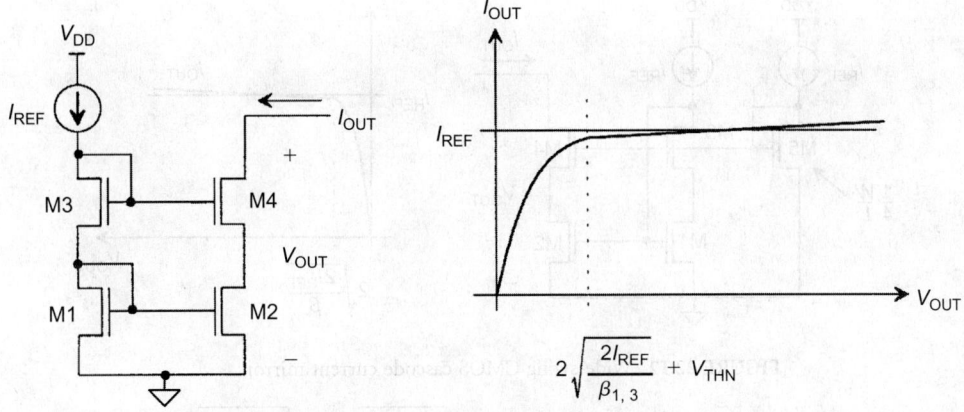

FIGURE 22.10 Basic cascode CMOS current mirror.

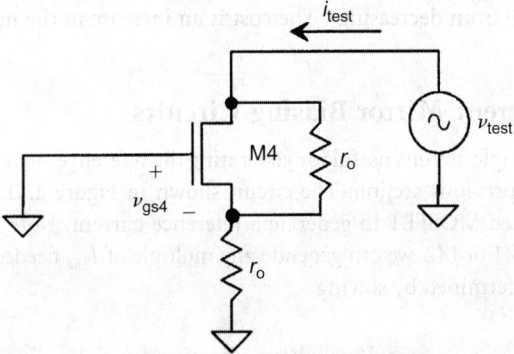

FIGURE 22.11 Determining the small-signal output resistance of the cascode current mirror.

these equations yields the cascode output resistance of

$$R_{\text{out, cascode}} = \frac{v_{\text{test}}}{i_{\text{test}}} = r_o(1 + g_m r_o) + r_o \approx g_m r_o^2 \tag{22.19}$$

The cascode current source output resistance is $g_m r_o$ (the open-circuit voltage gain of a MOSFET) times larger than r_o (the simple current mirror output resistance.) The main drawback of the cascode configuration is the increase in the minimum required voltage across the current sink in order for all MOSFETs to remain in the saturation region of operation.

22.2.1.4 Low-Voltage Cascode Current Mirror

If we look at the cascode current mirror of Figure 22.10, we see that the drain of M2 is held at the same potential as the drain of M1, that is, VGS or $\sqrt{(2I_{\text{REF}})/\beta} + V_{\text{THN}}$. We know that the voltage on the drain of M2 can be as low as $\sqrt{(2I_{\text{REF}})/\beta}$ before it starts to enter the triode region. Knowing this, consider the *wide-swing current mirror* shown in Figure 22.12. Here, "wide-swing" means the minimum voltage across the current mirror is $2\sqrt{(2I_{\text{REF}})/\beta}$, the sum of the excess gate voltages of M2 and M4. To understand the operation of this circuit, assume that M1 through M4 have the same *W/L* ratio (their βs are all equal). We know that the V_{GS} of M1 and M2 is $\sqrt{(2I_{\text{REF}})/\beta} + V_{\text{THN}}$. It is desirable to keep M2's drain at $\sqrt{(2I_{\text{REF}})/\beta}$ for wide-swing operation. This means, since M3/M4 are the same size

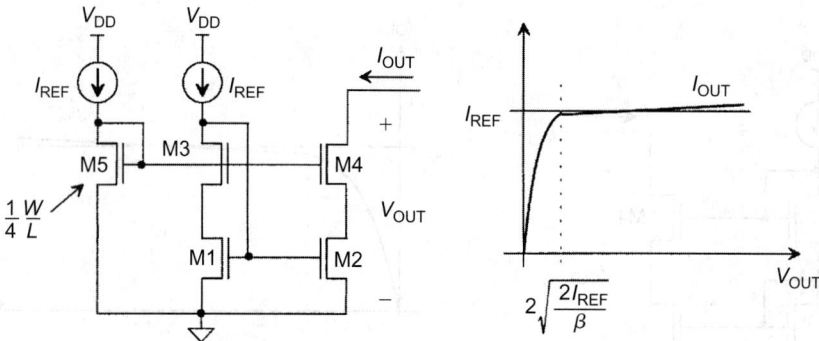

FIGURE 22.12 Wide-swing CMOS cascode current mirror.

as M1/M2, the gate voltage of M3/M4 must be $V_{GS} + \sqrt{(2I_{REF})/\beta}$ or $2\sqrt{(2I_{REF})/\beta} + V_{THN}$. By sizing M5's channel width so that it is one fourth of the size of the other transistor widths and forcing I_{REF} through the diode connected M5, we can generate this voltage. We should point out that the size (its W/L ratio) of M5 can be further decreased, say to 1/5, in order to keep M1/M2 from entering the triode region (and the output resistance from decreasing). The cost is an increase in the minimum allowable voltage across M2/M4, that is, V_{OUT}.

22.2.2 Simple Current Mirror Biasing Circuits

Figure 22.13 shows two simple circuits useful for generating the reference current, I_{REF}, used in the current mirrors discussed in the previous section. The circuit shown in Figure 22.13(a) uses a simple resistor with a gate-drain connected MOSFET to generate a reference current. Note how, by adding MOSFETs mirroring the current in M1 or M2, we can generate any multiple of I_{REF} needed. The reference current, of Figure 22.13(a), can be determined by solving

$$I_{REF} = \frac{V_{DD} - V_{GS}}{R} = \frac{\beta}{2}(V_{GS} - V_{THN})^2 \tag{22.20}$$

Figure 22.13(b) shows a MOSFET-only bias circuit. Since the same current flows in M1 and M2, we can mirror off of either MOSFET to generate our bias currents. The current flowing in M1/M2 is designed using

$$I_{REF} = \frac{\beta_1}{2}(V_{GS} - V_{THN})^2 = \frac{\beta_2}{2}(V_{DD} - V_{GS} - V_{THN})^2 \tag{22.21}$$

Notice in both equations above that the reference current is a function of the power supply voltage. Fluctuations, as a result of power supply noise, in V_{DD} directly affect the bias currents. In the next section, we will present a method for generating currents that reduces the currents' sensitivity to changes in *VDD*.

22.2.2.1 Temperature Dependence of Resistors and MOSFETS

Figure 22.14 shows how a resistor changes with temperature, assuming a linear dependence. The temperature coefficient is used to relate the value of a resistor at room temperature, or some known temperature T_0, to the value at a different temperature. This relationship can be written as

$$R(T) = R(T_0) \cdot [1 + \text{TCR}(T - T_0)] \tag{22.22}$$

where TCR is the temperature coefficient of the resistor ppm/°C (parts per million, a multiplier of 10^{-6}, per degree C). Typical values for TCRs for n-well, n+, p+, and poly resistors are 2000, 500, 750, and 100 ppm/°C, respectively.

Figure 22.15 shows how the drain current of a MOSFET changes with temperature. At low gate-source voltages, the drain current increases with increasing temperature. This is a result of the threshold voltage decreasing with increasing temperature which dominates the *I-V* characteristics of the MOSFET. The temperature coefficient of the threshold voltage (NMOS or PMOS), TCV_{TH}, is generally around -3000 ppm/°C. We can relate the threshold voltage to temperature using

$$V_{\text{TH}}(T) = V_{\text{TH}}(T_0)[1 + \text{TCV}_{\text{TH}}(T - T_0)]\qquad(22.23)$$

At larger gate-source voltages, the drain current decreases with increasing temperature as a result of the electron or hole mobility decreasing with increasing temperature. In other words, at low gate-source voltages, the temperature changing the threshold voltage dominates the *I-V* characteristics of the MOSFET; while at larger gate-source voltages, the mobility changing with temperature dominates. Note that at

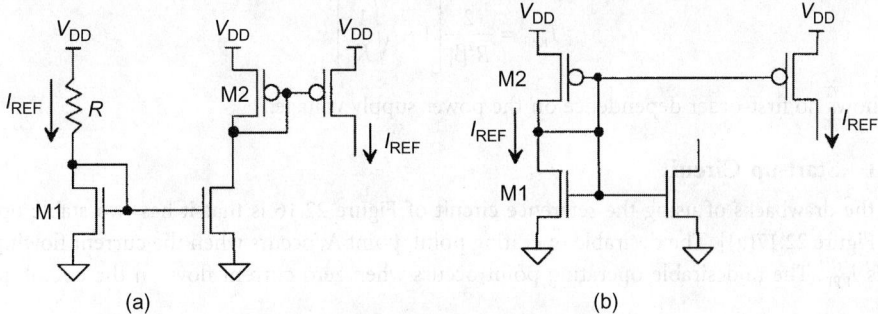

(a)   (b)

FIGURE 22.13 Simple biasing circuits.

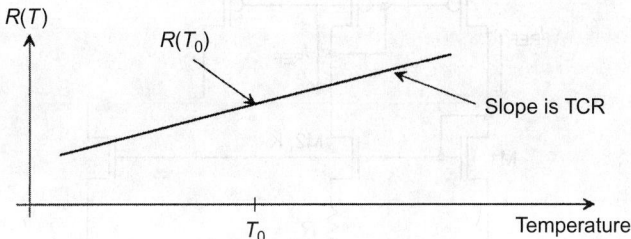

FIGURE 22.14 Variation of a resistor with temperature.

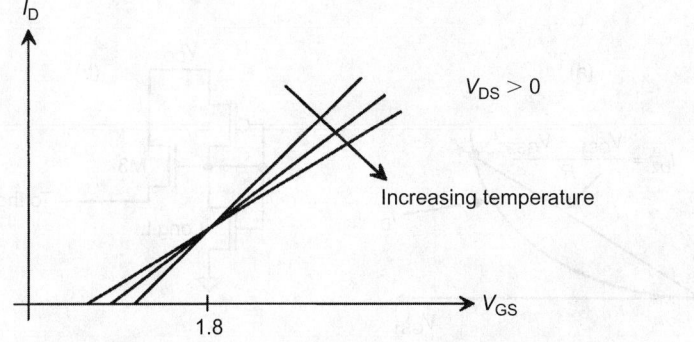

FIGURE 22.15 Temperature characteristics of a MOSFET.

around 1.8 V, for a typical CMOS process, the drain current does not change with temperature. The mobility can be related to temperature by

$$\mu(T) = \mu(T_0)\left(\frac{T}{T_0}\right)^{-1.5} \tag{22.24}$$

22.2.3 The Self-Biased Beta Multiplier Current Reference

Figure 22.16 shows the self-biased beta multiplier current reference. This circuit employs positive feedback, with a gain less than one, to reduce the sensitivity of the reference current to power supply changes. MOSFET M2 is made K times wider than MOSFET M1 (in other words, $\beta_2 = K\beta_1$; hence, the name beta multiplier). We know from this figure that, $V_{GS1} = V_{GS2} + I_{REF}R$, where $V_{GS1} = \sqrt{(2I_{REF})/\beta_1} + V_{THN}$ and $V_{GS2} = \sqrt{(2I_{REF})/K\beta_1} + V_{THN}$; therefore, we can write the current in the circuit as

$$I_{REF} = \frac{2}{R^2\beta_1}\left[1 - \sqrt{\frac{1}{K}}\right]^2 \tag{22.25}$$

which shows no first-order dependence on the power supply voltage.

22.2.3.1 Start-up Circuit

One of the drawbacks of using the reference circuit of Figure 22.16 is that it has two stable operating points [Figure 22.17(a)]. The desirable operating point, point A, occurs when the current flowing in the circuit is I_{REF}. The undesirable operating point occurs when zero current flows in the circuit, point B.

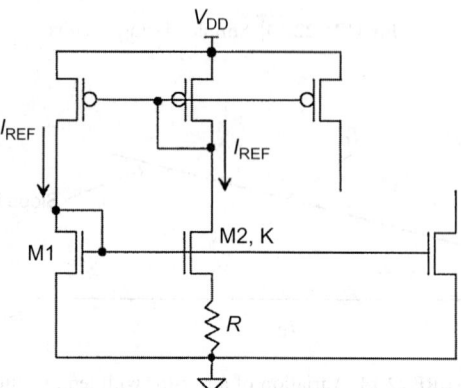

FIGURE 22.16 Beta multiplier current reference.

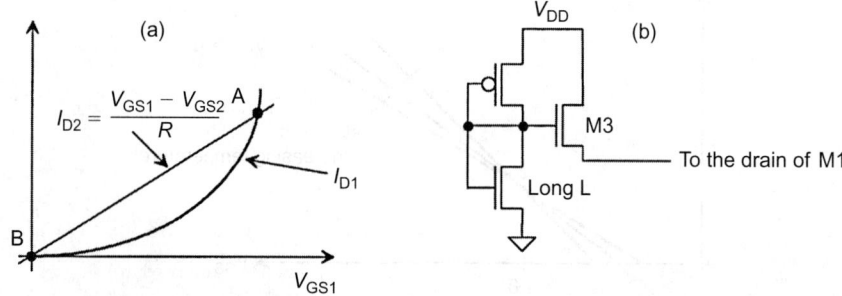

FIGURE 22.17 Start-up circuit for the beta multiplier circuit shown in Figure 22.16.

Because of the possibility of zero current flowing in the reference, a start-up circuit should always be used when using the beta multiplier. The purpose of the start-up circuit is to ensure that point B is avoided. When designed properly, the start-up circuit [Figure 22.17(b)] does not affect the beta multiplier operation when I_{REF} is non-zero (M3 is off when operating at point A).

22.2.3.2 A Comment About Stability

Since the beta multiplier employs positive feedback, it is possible that the circuit can become unstable and oscillate. However, with the inclusion of the resistor in series with the source of M2, the gain around the loop, from the gate of M2 to the drain/gate of M1, with the loop broken between the gates of M1 and M2, is less than one, keeping the reference stable. Adding a large capacitance across R, however, can increase the loop gain to the point of instability. This situation could easily occur if R is bonded out to externally set the current.

22.3 Amplifiers

Now that we have introduced biasing circuits and MOS device characteristics, we will immediately dive into the design of operational amplifiers. Since space is very limited, we will employ a top-down approach in which a series of increasingly complex circuits are dissected stage by stage and individual blocks analyzed.

Operational amplifiers typically are composed of either two or three stages consisting of a differential amplifier, a gain stage and an output stage as seen in Figure 22.18. In some applications, the gain stage and the output stage are one and the same if the load is purely capacitive. However, if the output is to drive a resistive load or a large resistive load, then a high current gain buffer amplifier is used at the output. Each stage plays an important role in the performance of the amplifier.

The differential amplifier offers a variety of advantages and is always used as the input to the overall amplifier. Since it provides common-mode rejection, it eliminates noise common on both inputs, while at the same time amplifying any differences between the inputs. The limit for which this common mode rejection occurs is called *common-mode range* and signifies the upper and lower common mode signal values for which the devices in the diff-amp are saturated. The differential amplifier also provides gain. The gain stage is typically a common-source or cascode type amplifier. So that the amplifier is stable, a compensation network is used to intentionally lower the gain at higher frequencies. The output stage provides high current driving capability for either driving large capacitive or resistive loads. The output stage typically will have a low output impedance and high signal swing characteristics. In some cases, it may be advantageous to add bipolar devices to improve the performance of the circuitry. These will be presented as the block level circuits are analyzed.

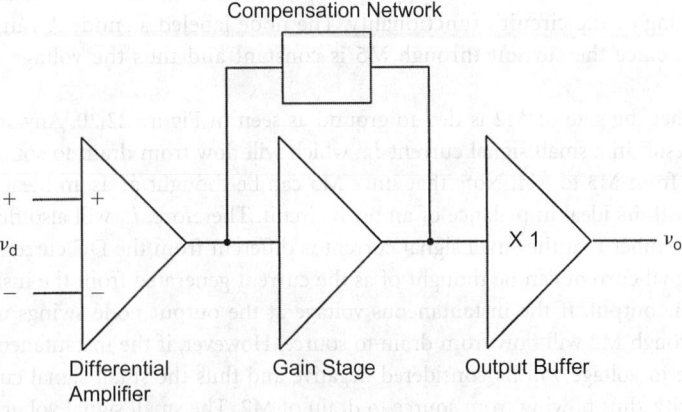

FIGURE 22.18 Block diagram for a generic op-amp.

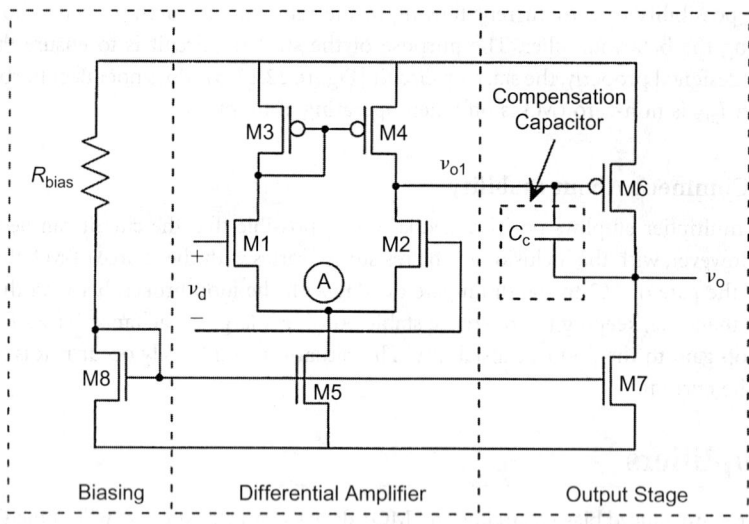

FIGURE 22.19 Basic two-stage op-amp.

22.3.1   The Simple Unbuffered Op-Amp

Examine the simple operational amplifier shown in Figure 22.19. Here, the amplifier can be segregated into a biasing block, a differential input stage, an output stage, and a compensation capacitor.

The biasing circuit is a simple current mirror driver, consisting of the resistor R_{bias} and the transistor M8. The current through M8 is mirrored through both M5 and M7. Thus, the current through the entire circuit is set by the value of R_{bias} and the relative W/Ls of M8, M5, and M7. The actual values of this current will be discussed a little bit later on. When designing with R_{bias}, one must be careful not to ignore the effect of temperature on R_{bias}, and thus the values of the currents through the circuit. We will see later on how the bias current greatly affects the performance of the amplifier. For the commercial temperature range of 0°C to 125°C, the current through M8 should be simulated with R_{bias} at values of ±30% of its nominal value. Other, more sophisticated voltage references (as discussed earlier) can be used in place of the resistor reference, and will be presented as we progress.

22.3.1.1   The Differential Amplifier

The differential amplifier is composed of M1, M2, M3, M4, and M5, with M1 matching M2 and M3 matching M4. The transistor M5 can be replaced by a current source in the ideal case to enhance one's understanding of the circuit's functionality. The node labeled as node A can be thought of as a virtual ground, since the current through M5 is constant and thus the voltage at node A is also constant.

Now assume that the gate of M2 is tied to ground as seen in Figure 22.20. Any small signal on the gate of M1 will result in a small signal current i_{d1}, which will flow from drain to source of M1 and will also be mirrored from M3 to M4. Note that since M5 can be thought of as an ideal current source, it can be replaced with its ideal impedance of an open circuit. Therefore, i_{d1} will also flow from source to drain of M2. Remember that the small signal current is different from the DC current flowing through M2. The small signal current can be thought of as the current generated from the instantaneous (AC + DC) voltage at the output. If the instantaneous voltage at the output node swings up, then the small signal current through M2 will flow from drain to source. However, if the instantaneous voltage swings down, the change in voltage will be considered negative and thus the small signal current will assume an identical polarity, thus flowing from source to drain of M2. The small signal voltage produced at the output node will then be $2i_{d1}$ times the output impedance of the amplifier. In this case, R_{out} is simply

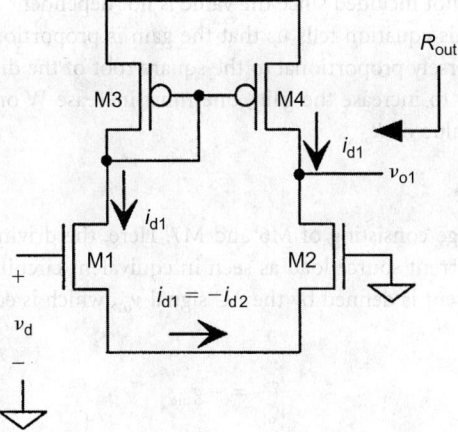

FIGURE 22.20 Pseudo-AC circuit showing small signal currents.

$r_{o2} \| r_{o4}$. The value of the small signal current, i_{d1}, is simply

$$i_{d1} = g_{m1} v_{gs1} \tag{22.26}$$

and the differential voltage, $v_d = v_{gs1} + v_{gs2}$. Therefore, since $v_{gs1} = v_{gs2}$, then,

$$\frac{i_{d1}}{v_d} = \frac{g_{m1}}{2} \tag{22.27}$$

the small signal output voltage is simply

$$v_{o1} = 2 \cdot i_{d1} \cdot r_{o2} \| r_{o4} \tag{22.28}$$

and the small signal voltage gain can then be easily derived as

$$\frac{v_{o1}}{v_d} = g_{m1,2}(r_{o2} \| r_{o4}) \tag{22.29}$$

Now let us examine this equation more carefully. Substituting the expressions for g_m and r_o, the previous equation becomes

$$\frac{v_{o1}}{v_d} \approx \sqrt{2\beta_{1,2} I_{D1,2}} \cdot \frac{1}{2\lambda I_{D1,2}} = K' \cdot \sqrt{\frac{W_{1,2}}{L_{1,2} I_{D1,2}}} \cdot \frac{1}{\lambda} \tag{22.30}$$

where K' is a constant, which is uncontrollable by the designer. When designing analog circuits, it is just as important to understand the effects of the controllable variables on the specification as it is to know the absolute value of the gain using hand calculations. This is because the hand analysis and computer simulations will vary a great deal because of the complex modeling used in today's CAD tools. Examining Eq. (22.30), and knowing that the effect of λ on the gain diminishes as L increases such that $1/\lambda$ is directly proportional to channel length. Then, a proportionality can be established between $W/L_{1,2}$ and the drain current versus the small signal gain such that

$$\frac{v_{o1}}{v_d} \propto \sqrt{\frac{W_{1,2} \cdot L_{1,2}}{I_{D1,2}}} \tag{22.31}$$

Notice that the constant was not included since the value is not dependent on anything the designer can adjust. The importance of this equation tells us that the gain is proportional to the square root of the product of W and L and inversely proportional to the square root of the drain current through M1 and M2, which is also 1/2 I_{D6}. So to increase the gain, one must increase W or L or decrease the value I_{D6}, which is dependent on the value of R_{bias}.

22.3.1.2 The Gain Stage

Now examine the output stage consisting of M6 and M7. Here, the driving transistor, M6, is a simple inverting amplifier with a current source load as seen in equivalent circuit shown in Figure 22.21. The value of the small signal current is defined by the AC signal v_{o1}, which is equal to v_{sg6}. Therefore,

$$\frac{i_{d6}}{v_{o1}} = -g_{m6} \tag{22.32}$$

and the gain of the stage is simply

$$\frac{i_{d6}}{v_{o1}} \cdot \frac{v_o}{i_{d6}} = \frac{v_o}{v_{o1}} = -g^{m6} \cdot r_{o6} \parallel r_{o7} \tag{22.33}$$

Again, we can write the gain in terms of the designer's variables, so that the gain of the amplifier can be expressed as a proportion of

$$\frac{v_o}{v_{o1}} \propto \sqrt{\frac{W_6 \cdot L_{6,7}}{I_{D6,7}}} \tag{22.34}$$

Therefore, overall, the gain of the entire amplifier is

$$\frac{v_o}{v_d} = g_{m1,2}(r_{o2} \parallel r_{o4}) \cdot -g^{m6}(r_{o6} \parallel r_{o7}) \tag{22.35}$$

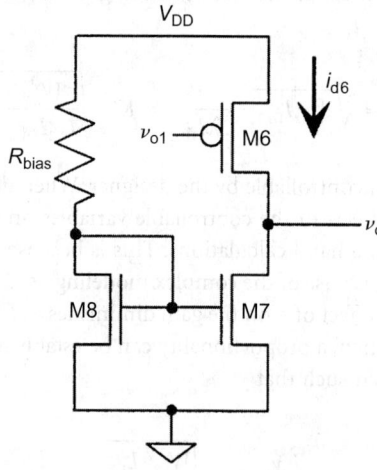

FIGURE 22.21 Output stage circuit.

or as the proportionalities

$$\frac{v_o}{v_d} \propto \sqrt{\frac{W_{1,2} \cdot L_{1,2}}{I_{D1,2}}} \cdot \sqrt{\frac{W_6 \cdot L_{6,7}}{I_{D6,7}}}$$ (22.36)

So, the key variables for adjusting gain are the drain currents (the smaller the better) and the W and L ratios of M1, M2, and M6 (the larger the better). Of course, there are lower and upper limits to the drain currents and the W/Ls, respectively, that we will examine as we analyze other specifications that will ultimately determine the bounds of adjustability.

22.3.1.3 Frequency Response

Now examine the amplifier circuit shown in Figure 22.22. In this particular circuit, the compensation capacitor, C_C, is removed and will be re-added shortly. The capacitors C_1 and C_2 represent the total lumped capacitance from each ground. Since the output nodes associated with each output is a high impedance, these nodes will be the dominant frequency-dependent nodes in the circuit. Since each node in a circuit contributes a high-frequency pole, the frequency response will be dominated by the high-impedance nodes.

An additional capacitor could have been included at the gate of M4 to ground. However, the equivalent impedance from the gate of M4 to ground is approximately equal to the impedance of a gate-drain connected device or $1/g_{m3}$ as seen in Figure 22.23(a)–(c). If a controlled source has the controlling voltage directly across its terminals [Figure 22.23(b)], then the effective resistance is simply the controlling voltage (in this case, v_{gs}) divided by the controlled current ($g_m v_{gs}$), which is $1/g_m \| r_o$ or approximately $1/g_m$ if r_o is much greater than $1/g_m$ [Figure 22.23(c)]. Therefore, the impedance seen from the gate of M4 to ground

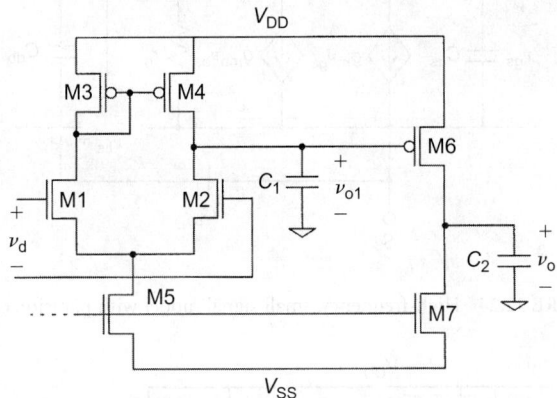

FIGURE 22.22 Two-stage op-amp with lumped parasitic capacitors.

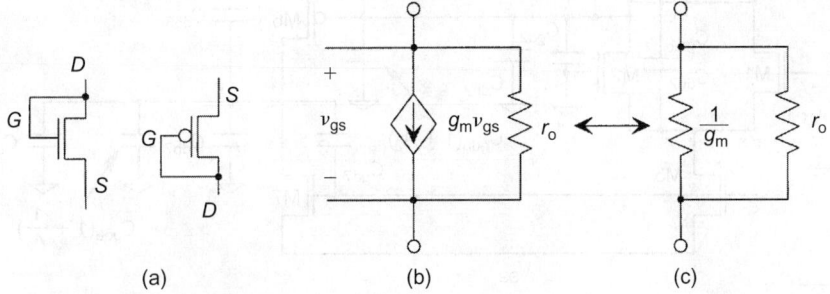

FIGURE 22.23 Equivalent resistance for a gate-drain connected MOSFET device.

is low, and the pole associated with the node will be at a much higher frequency than the high-impedance nodes. The same holds true with the node associated with the drain of M5. That node is considered an AC ground, so it has no effect on the frequency response of the small signal circuit. The only remaining node is the node which defines the current mirrors (the gate of M8). Since this node is not in the small signal path, and is a DC bias voltage, it can be also be considered to be an AC ground.

One can approximate the frequency response of the amplifier by examining both the effective impedance and parasitic capacitance at the output of each stage. The parasitic capacitances can be seen in the small signal model of a MOSFET in Figure 22.24. The capacitors C_{gb}, C_{sb}, and C_{db} represent the bulk depletion capacitors of the transistors, while C_{gd} and C_{gs} represent the overlap capacitances from gate to drain and gate to source, respectively. Referring to Figure 22.25, which shows the parasitic capacitors explicitly, C_1 can now be written as

$$C_1 = C_{db4} + C_{gd4} + C_{db2} + C_{gd2} + C_{gs6} + C_{gd6}(1 + A_2) \qquad (22.37)$$

Note that C_{gd4} is included as capacitor to ground since the low impedance caused by the gate-drain device of M3 can be considered equivalent to an AC ground. The capacitor C_{db2} is also connected to AC ground at the source-coupled node consisting of M1 and M2.

Miller's theorem was used to determine the effect of the bridging capacitor C_{db6}, connected from the gate to the drain of M6. Miller's theorem approximates the effects of the gate-drain capacitor by replacing

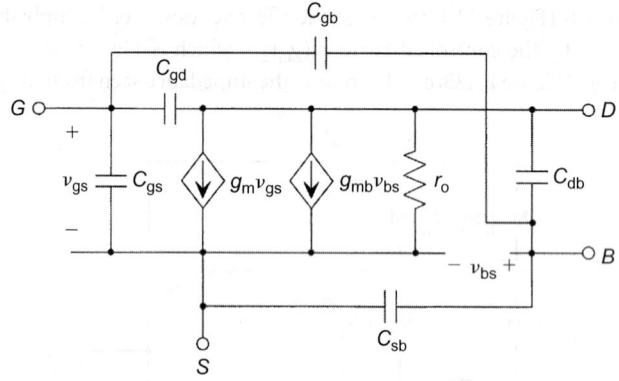

FIGURE 22.24 High-frequency, small signal model with parasitic capacitors.

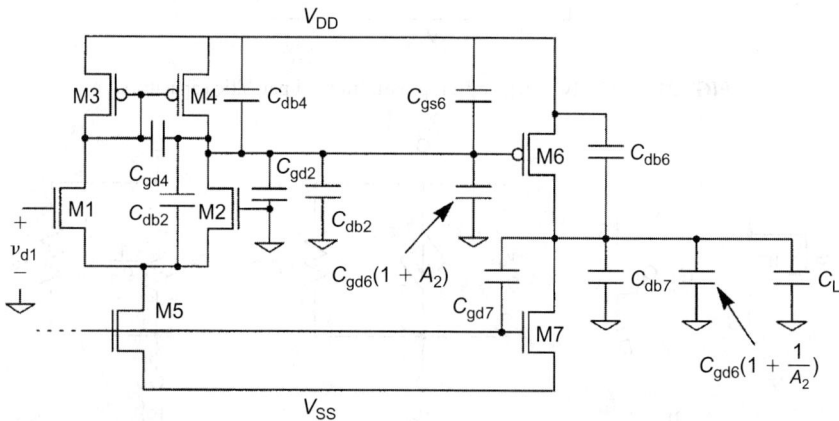

FIGURE 22.25 Two-stage op-amp with parasitics shown explicitly.

the bridging capacitor with an equivalent input capacitor of value $C_{db6} \cdot (1 - A_2)$ and an equivalent output capacitor with a value of $C_{db6} \cdot (1 - 1/A_2)$. The term A_2 is the gain across the original bridging capacitor and is $-g_{m6} \cdot r_{o6} \| r_{o7}$. The reader should consult Ref. 2 for a proof of Miller's theorem.

The capacitor C_2 can also be determined by examining Figure 22.25,

$$C_2 = C_{db6} + C_{db7} + C_{gd7} + C_{gd6} \cdot \left(1 + \frac{1}{A_2}\right) + C_L \tag{22.38}$$

Now assume that C_1 is greater than C_2. This means that the pole associated with the diff-amp output will be lower in frequency than the pole associated with the output of the output stage. Thus, a good model for the op-amp can be seen in Figure 22.26. The transfer function, ignoring C_C, will then become

$$\frac{v_o(s)}{v_{d1}(s)} = -g_{m1}(r_{o2}\|r_{o4}) \cdot g_{m6}(r_{o6}\|r_{o7}) \cdot \frac{1}{\left(\dfrac{f}{jf_{p1}}+1\right)\left(\dfrac{f}{f_{p2}}+1\right)} \tag{22.39}$$

where

$$f_{p1} = \frac{1}{2\pi R_{out1}C_1}, f_{p2} = \frac{1}{2\pi R_{out}C_2} \tag{22.40}$$

The plot of the transfer function can be seen in Figure 22.27. Note that in examining the frequency response, that the phase margin is virtually zero. Remember that phase margin is the difference between phase at the frequency at which the magnitude plot reaches 0 dB (also known as the gain-bandwidth product) and the phase at the frequency at which the phase has shifted $-180°$. It is recommended for stability reasons, that the phase margin of any amplifier be at least 45° (60° is recommended). A phase margin below 45° will result in long settling times and increased propagation delays. The system can also be thought of as a simple second-order linear controls system with the phase margin directly affecting the transient response of the system.

22.3.1.4   Compensation

Now we will include C_C in the circuit. If C_C is much greater than C_{gd6}, then the C_C will dominate the value of C_1 [especially since it is multiplied by $(1 - A_2)$] and will cause the pole, f_{p1}, to roll-off much earlier than without C_C to a new location, f_{p1}. One could solve, using circuit analysis, the circuit shown in Figure 22.26 with C_C included to also prove that the second pole, f_{p2}, moves further out [3] to a higher frequency, f_{p2}. Ideally, the addition of C_C will cause an equivalent single pole roll-off of the overall

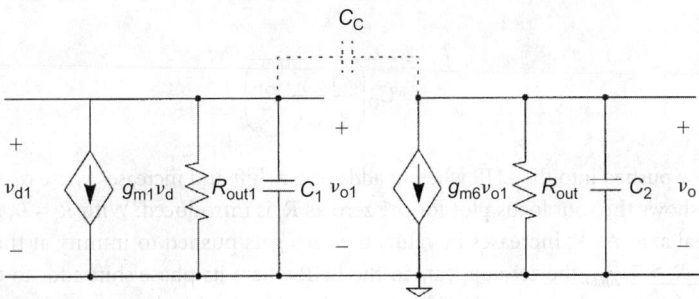

FIGURE 22.26 Model used to determine the frequency response of the two-stage op-amp.

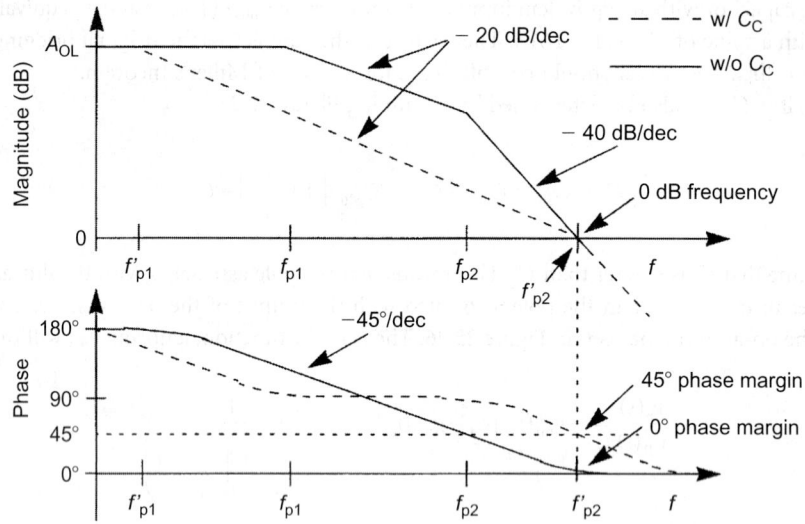

FIGURE 22.27 Magnitude and phase of the two-stage op-amp with and without compensation.

frequency response. The second pole should not begin to affect the frequency response until after the magnitude response is below 0 dB. The new values of the poles are

$$f'_{p1} \approx \frac{1}{2\pi(g_{m6}R_{out})C_C R_{out1}}, f'_{p2} \approx \frac{g_{m6}C_C}{2\pi \cdot (C_2 C_1 + C_2 C_C + C_C C_1)} \approx \frac{g_{m6}}{2\pi \cdot C_2} \tag{22.41}$$

If the previously discussed analysis was performed, one would also see that by using Miller's theorem, we are neglecting a right-hand plane (RHP) zero that could have negative consequences on our phase margin, since the phase of an RHP zero is similar to the phase of a left-hand plane (LHP) zero. An RHP zero behaves similarly to an LHP zero when examining the magnitude response; however, the phase response will cause the phase plot to shift −180° more quickly. The RHP zero is at a value of

$$f_{z1} = \frac{g_{m6}}{C_C} \tag{22.42}$$

To avoid effects of the RHP zero, one must try to move the zero out well beyond the point at which the magnitude plot reaches 0 dB (suggested rule of thumb: factor of 10 greater). The comparison of the frequency response of the two-stage op-amp with and without C_C can be seen in Figure 22.27.

One remedy to the zero problem is to add a resistor, R_z, in series with compensation capacitor as seen in Figure 22.28. The expression of the zero after adding the resistor becomes

$$f_{z1} = \frac{1}{C_C\left(\frac{1}{g_{m6}} - R_z\right)} \tag{22.43}$$

and the zero can be pushed into the LHP where it adds phase shift and increases phase margin if $R_z > 1/g_{m6}$. Figure 22.29 [4] shows the root locus plot for the zero as R_z is introduced. With $R_z = 0$, the zero location is on the RHP real axis. As R_z increases in value, the zero gets pushed to infinity at the point at which $R_z = 1/g_{m6}$. Once $R_z > 1/g_{m6}$, the zero appears in the LHP where its phase shift adds to the overall phase response, thus improving phase margin. This type of compensation is commonly referred to as *lead compensation*, and is commonly used as a simple method for improving the phase margin. One should

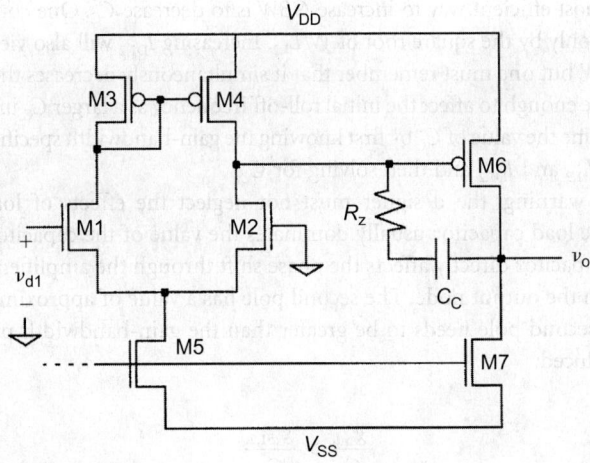

FIGURE 22.28 Compensation including a nulling resistor.

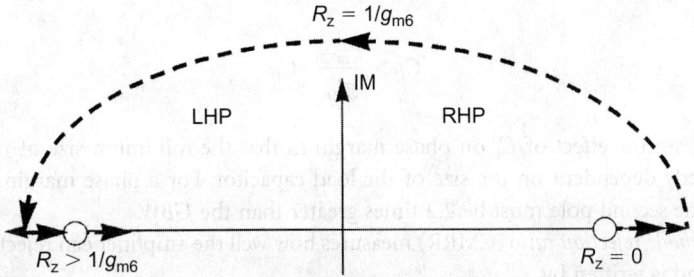

FIGURE 22.29 Root locus plot of how the zero shifts from the RHP to the LHP as R_z increases in value.

be careful about using R_z, since the absolute values of the resistors are not well predicted. The value of the resistor should be simulated over its maximum and minimum values to ensure that no matter if the zero is pushed into the LHP or the RHP, that the value of the zero is always 10 times greater than the gain-bandwidth product.

22.3.1.5 Other AC Specifications

Referring back to the equations for gain and the first pole location, we can now see adjustable parameters for affecting frequency response by plugging in the proportionalities for their corresponding factors. Since the value of the resistors are directly proportional to the length of L (the longer the L, the higher the resistance, since longer channel lengths diminish the effects of channel length modulation, λ).

The *gain-bandwidth product* for the compensated op-amp is the open-loop gain multiplied by the bandwidth of the amplifier (as set by f_{p1}). Therefore, the gain-bandwidth product is

$$GBW = (g_{m1} \cdot r_{o2} \| r_{o4})(g_{m6} \cdot r_{o6} \| r_{o7})\left(\frac{1}{2\pi(g_{m6}R_{out})C_C R_{out1}}\right) = \frac{g_{m1}}{2\pi C_C} \tag{22.44}$$

Or we can write the expression as

$$GBW \propto \frac{\sqrt{\dfrac{W}{L_{1,2}}I_{D1,2}}}{C_C} \tag{22.45}$$

which shows that the most efficient way to increase *GBW* is to decrease C_C. One could increase the $W/L_{1,2}$ ratio, but its increase is only by the square root of $W/L_{1,2}$. Increasing $I_{D1,2}$ will also yield an increase (also to the square root) in *GBW*, but one must remember that it simultaneously decreases the open-loop gain. The value of C_C must be large enough to affect the initial roll-off frequency as a larger C_C improves phase margin. One could easily determine the value of C_C by first knowing the gain-bandwidth specification, then iteratively choosing values for $W/L_{1,2}$ and $I_{D1,2}$ and then solving for C_C.

One other word of warning: the designer must not neglect the effects of loading on the circuit. Practically speaking, the load capacitor usually dominates the value of the capacitor, C_2 in Figure 22.22. The value of the load capacitor directly affects the phase shift through the amplifier by changing the pole location associated with the output node. The second pole has a value of approximately $g_{m6}/2\pi C_L$ as was seen earlier. Since the second pole needs to be greater than the gain-bandwidth product, the following relationship can be deduced:

$$\frac{g_{m6}}{C_L} > \frac{g_{m1,2}}{C_C} \tag{22.46}$$

or

$$C_C > \frac{g_{m1,2}}{g_{m6}} \cdot C_L \tag{22.47}$$

Thus, one can see the effect of C_L on phase margin in that the minimum size of the compensation capacitor is directly dependent on the size of the load capacitor. For a phase margin of 60°, it can be shown [3] that the second pole must be 2.2 times greater than the *GBW*.

The *common-mode rejection ratio* (CMRR) measures how well the amplifier can reject signals common to both inputs and is written by

$$\text{CMRR} = 20\log\left|\frac{A_d}{A_{cm}}\right| = 20\log\left|\frac{v_o/v_d}{v_o/v_{cm}}\right| \tag{22.48}$$

where V_{cm} is a common mode input signal, which in this case is composed of an AC and a DC component (the higher the value of CMRR, the better the rejection). The circuit for calculating common mode gain can be seen in Figure 22.30. The gain through the output stage will be the same for both the differential gain, A_d, and the common-mode gain, A_{cm}. The last stage will cancel itself out in the expression for CMRR;

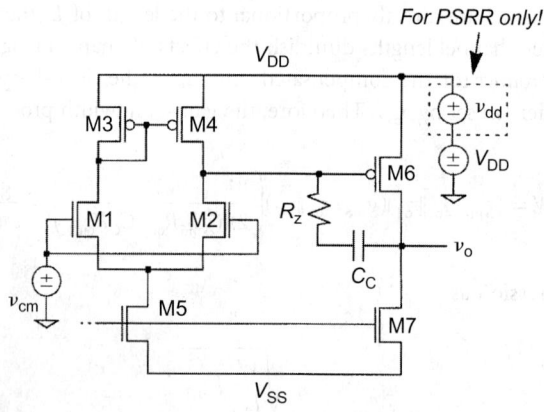

FIGURE 22.30 Circuit used to determine CMRR and PSRR.

thus, the differential amplifier will determine how well the entire amplifier rejects common mode signals. This rejection is one of the most advantageous reasons for using a diff-amp as an input stage. If the inputs are subjected to the same noise source, the diff-amp has the ability to reject the noise signal and only amplify the difference in the inputs. For the diff-amp used in this example, the common-mode gain is

$$\frac{v_{o1}}{v_{cm}} = -\frac{1}{2g_{m4}r_{o5}}$$
(22.49)

This equation bears some explanation. Since the inputs are tied together, the source-coupled node can no longer be considered an AC ground. Therefore, the resistance of the tail current device, M5, must be considered in the analysis. The AC currents flowing through both M1 and M2 are equal and $v_{gs3} = v_{gs4}$. The output impedance of the diff-amp will drop considerably when using a common-mode signal due to the feedback loop consisting of M1, M3, M4, and M2. As a result, the common-mode signal appearing on the drains of M3 and M4 will be identical.

One can determine this gain by using half-circuit analysis (Figure 22.31). This is equivalent to a common source amplifier with a large source resistance of value $2r_{o5}$. When using half-circuit analysis, the current source resistance doubles because the current through the driving device M1 is one half of the original tail current. Therefore, the gain of this circuit can be approximately determined as the negative ratio between the resistance attached to the drain of M1 divided by the resistance attached to the source of M1, or

$$\frac{v_{o1}}{v_{cm}} = -\frac{1/g_{m3,4}}{2r_{o5}}$$
(22.50)

Adding the expression for the differential gain of the diff-amp, v_{o1}/v_d, we can write the expression for the CMRR as

$$\text{CMRR} = 20\log(2g_{m1,2}g_{m3,4}(r_{o2}\|r_{o4})r_{o5})$$
(22.51)

or as a proportion,

$$\text{CMRR} \propto 20\log\left(\sqrt{\frac{W_{1,2}\cdot L_{1,2}}{I_{D1,2}}}\cdot\sqrt{\frac{(W/L)_{3,4}}{2I_{D5}}}\cdot L_5\right)$$
(22.52)

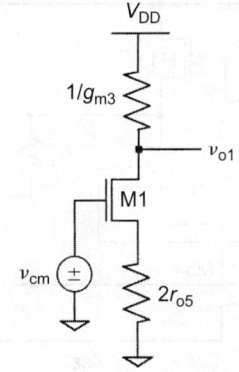

FIGURE 22.31 Half-circuit used to determine the common-mode gain.

So, it can be seen that the most efficient manner in which to increase the CMRR of this amplifier is to increase the channel length of M5 (the tail current device). This, too, has a large signal implication that we will discuss later in this section.

The *power supply rejection ratio* (PSRR) measures how well the amplifier can reject changes in the power supply. This is also a critical specification because it would be desirable to reject noise on the power supply outright. PSRR from the positive supply is defined as:

$$\text{PSRR}^{v_{dd}} = \frac{v_o/v_d}{v_o/v_{dd}} \tag{22.53}$$

where the gain v_o/v_{dd} is the small signal gain from v_{dd} (refer back to Figure 22.30) to the output of the amplifier with the input signal, v_d, equal to zero. PSRR can also be measured from V_{ss} by inserting a small signal source in series with ground. One should be careful, however, when simulating this specification to make sure that the inputs are properly biased so as to ensure that they are in saturation before simulating. It is best to use a fully differential (differential input and differential output) to most effectively reject power supply noise (to be discussed later).

22.3.1.6 Large Signal Considerations

Other considerations that must be discussed are the large signal tradeoffs. One cannot ignore the effects of adjusting the small signal specifications on the large signal characteristics. The large signal characteristics that are important include the common-mode range, slew rate, and output signal swing.

Slew rate is defined as the maximum rate of change of the output voltage due to a change in the input voltage. For this particular amplifier, the maximum output voltage is ultimately limited by how fast the tail current device (M5) can charge and discharge the compensation capacitor. The slew rate can then be approximated as

$$SR = \frac{dV_O}{dt} \approx \frac{I_{D5}}{C_C} \tag{22.54}$$

Typically, the diff-amp is the major limitation when considering slew rate. However, the tradeoff issues again come into play. If I_{D5} is increased too much, the gain of the diff-amp may decrease below a satisfactory amount. If C_C is made too small, then the phase margin may decrease below an acceptable amount.

The *common-mode range* is defined as the range between the maximum and minimum common-mode voltages for which the amplifier behaves linearly. Referring to Figure 22.32, suppose that the common-mode

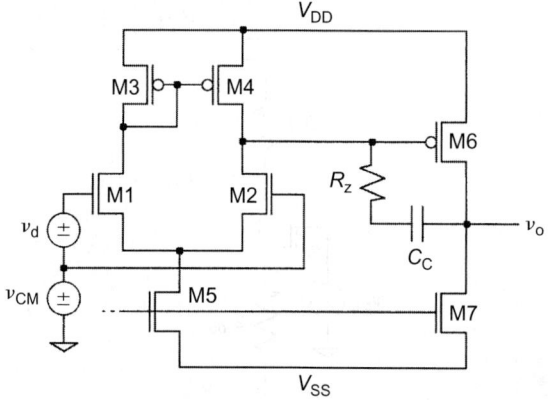

FIGURE 22.32 Determining the CMR for the two-stage op-amp.

voltage is DC value and that the differential signal as is also shown. If the common-mode voltage is swept from ground to V_{DD}, there will be a range for which the amplifier will behave normally and where the gain of the amplifier is relatively constant. Above or below that range, the gain drops considerably because the common-mode voltage forces one or more devices into the triode region.

The maximum common-mode voltage is limited by both M1 and M2 going into triode. This point can be defined by a borderline equation in which $V_{DS1,2} = V_{GS1,2} - V_{THN}$ or, in this case, $V_{D1,2} = V_{G1,2} - V_{THN}$. Substituting $V_{DD} - V_{SG3}$ for $V_{D1,2}$, and solving for $V_{G1,2}$, which now represents the maximum common-mode voltage, the expression becomes

$$V_{G1,2(max)} = V_{DD} - \left[\sqrt{\frac{I_{D5}}{\beta_3}} + V_{THP} \right] + V_{THN} \tag{22.55}$$

where the value of V_{SG3} is written in terms of its drain current using the saturation equation. If the threshold voltages are assumed to be approximately the same value, then the equation can be written as

$$V_{G1,2(max)} = V_{DD} - \sqrt{\frac{I_{D5}}{\beta_3}} = V_{DD} - \sqrt{\frac{L_3 \cdot I_{D5}}{W_3 \cdot K_3}} \tag{22.56}$$

The minimum voltage is limited by M5 being driven into nonsaturation by the common-mode voltage source. The borderline equation ($V_{D5} = V_{G5} - V_{THN}$) for this transistor can then be used with $V_{D5} = V_{G1,2} - V_{GS1,2}$ and writing both V_{G5} and $V_{GS1,2}$ in terms of its drain current yields

$$V_{G1,2(min)} = V_{SS} + \sqrt{\frac{2I_{D5}}{\beta_5}} + \sqrt{\frac{I_{D5}}{\beta_{1,2}}} = V_{SS} + \sqrt{\frac{2L_5 \cdot I_{D5}}{W_5 \cdot K_5}} + \sqrt{\frac{L_{1,2} \cdot I_{D5}}{W_{1,2} \cdot K_{1,2}}} \tag{22.57}$$

Now notice the influencing factors for improving the common-mode range ($V_{G1,2(max)}$ is increased and $V_{G1,2(min)}$ is decreased). Assume that V_{DD} and V_{SS} are defined by the circuit application and are not adjustable. To make $V_{G1,2(max)}$ as large as possible, I_{D5} and L_3 should be made as small as possible while W_3 is made as large as possible. And to make $V_{G1,2(min)}$ as small as possible, L_5, I_{D5}, and $L_{1,2}$ should be made as small as possible while increasing W_5 and $W_{1,2}$ as large as possible. Making the drain current as small as possible is in direct conflict with the slew rate. Decreasing L_5 will also degrade the common-mode rejection ratio, and increasing W_3 will affect the pole location of the output node associated with the diff-amp, thus altering the phase margin. All these tradeoffs must be considered as the designer chooses a circuit topology and begins the process of iterating to a final design.

The output swing of the amplifier is defined as the maximum and minimum values that can appear on the output of the amplifier. In analog applications, we are concerned with producing the largest swing possible while keeping the output driver, M6, in saturation. This can be determined by inspecting the output stage. If the output voltage exceeds the gate voltage of M6 by more than a threshold voltage, then M6 will go into triode. Thus, since the gate voltage of M6 is defined as $V_{DD} - V_{SG3,4}$, the maximum output voltage will be determined by the size of M3 and M4. The larger the channel width of M3 and M4, the smaller the value of $V_{SG3,4}$ and the higher the output can swing. However, again the tradeoff of making M3 and M4 too large is the reduction in bandwidth due to the increased parasitic capacitance associated with the output of the diff-amp.

The minimum value of the output swing is limited by the gate voltage of M7, which is defined by biasing circuitry. Again using the borderline equation, the drain of M7 may not go below the gate of M7 by more than a V_{THN}. Thus, to improve the swing, the value of V_{GS7} must be made small, which implies that the value of V_{GS8} and V_{GS5} also be made small, resulting in large values of M5, M8, and M7. This is not a very wise option because increasing M5 causes all the PMOS devices to increase by the same factor, resulting in large devices for the entire circuit. By carefully designing the bias device M8, one can design V_{GS8} to be around 0.3 V above VTHN. Thus, the output can swing to within 0.3 V of V_{SS}.

22.3.1.7 Tradeoff Example

When designing amplifiers, the tradeoff issues that occur are many. For example, there are many effects that occur just by increasing the drain current through the diff-amp: the open-loop gain goes down (by the square root of I_D) while the bandwidth increases (by I_D) due to the fact that the resistors r_{o2} and r_{o4} decrease by $1/I_D$. The overall effect is an increase (by the square root of I_D) in GBW, as predicted earlier. A table summarizing the various tradeoffs that occur from attempting to increase the DC gain can be seen in Table 22.1. It is assumed that if a designer only takes the one action listed that the following secondary effects will occur. In fact, the designer should understand the secondary effects enough to where a counteraction is taken to offset the secondary effects.

The key for the entire circuit design is the size of M5; the remaining transistors can be written as factors of W_5. The minimum amount of current flowing through M5 is determined by the slew rate. Since M3 and M4 carry half the current of M5, then the widths of M3 and M4 can be determined by assuming that $V_{SG3} = V_{SG4} = V_{GS5}$

$$\frac{2I_{D3,4}}{I_{D5}} \approx \frac{2K_p \cdot (W/L)_{3,4} \cdot (V_{SG3,4} - |V_{THP}|)^2}{K_n \cdot (W/L)_5 \cdot (V_{GS5} - V_{THN})^2} \approx \frac{2K_p \cdot (W/L)_{3,4}}{K_n \cdot (W/L)_5} \tag{22.58}$$

and since $L_3 = L_5$, and $K_n = 3K_p$, then that leads to the conclusion that $W_{3,4} = 1.5 \cdot W_5$. If the nulling resistor is used in the compensation network, the values for M6 and M7 are determined by the amount of load capacitance attached to the output. If a large capacitance is present, the widths of M6 and M7 will need to be large so as to provide enough sinking and sourcing current to and from the load capacitor. Suppose it was decided that the amount of current needed for M6 and M7 was twice that of M5. Then, W_7 would be twice as large as W_5, and W_6 would be six times larger than W_5 to account for the differing K values. Alternatively, if everything is saturated, then $I_{D3} = I_{D4}$, and the drain voltage at the output of the diff-amp is identical to the drain voltage of M3. This implies that under saturation conditions, the gate of M6 is at the same potential as the gate of M4; thus, again it must be emphasized that we are talking about quiescent conditions here, and the current through M6 will be defined by the ratio of M6 to M4. Therefore, since M6 is carrying four times as much current as M3, then $W_6 = 4W_{3,4}$, which is six times the value of W_5. Thus, every device except M1 and M2 is written in terms of M5.

TABLE 22.1   Tradeoff Issues for Increasing the Gain of the Two-Stage Op-Amp

Desire	Action(s)	Secondary effects
Increase DC gain	Increase W/L1,2	Decreases phase margin
		Increases GBW
		Increases CMRR
		Decreases CMR
	Decrease ID5	Decreases SR
		Increases CMR
		Increases CMRR
		Increases phase margin
	Increase W/L6	Increases phase margin
		Increases output swing
	decrease ID6	Decreases output current drive
		Decreases phase margin

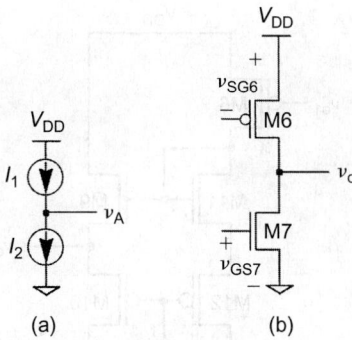

FIGURE 22.33 Illustration of a high-impedance node.

The sizes of M1 and M2 are the most critical of the amplifier. If $W_{1,2}$ are made too large, then C_C will be big due to its relationship with C_L and the ratio of g_{m1} and g_{m6} [Eq. (22.48)]. However, if $W_{1,2}$ is made too small, the gain may not meet the requirements needed.

One word about high-impedance nodes. If two current sources are in a series as shown in Figure 22.33(a), then the value of the voltage between them is difficult to predict. In the ideal case, both current sources have infinite impedances, so any slight mismatch between I_1 and I_2 will result in large swings in v_A. The same holds true for Figure 22.33(b). Since the two devices form a high impedance at the output, and each device can be considered a current source, any mismatches in the currents defined by $\beta_2(V_{SG2} - V_{THP})^2$ and $\beta_1(V_{GS1} - V_{THN})^2$ will result in large voltage offsets at the output, with the device with the larger defined current being driven into triode. Thus, the smaller of the two defined currents will be the one flowing through both devices. Another way to visualize this condition is to place a large resistor representing the output impedance from v_o to ground. Any difference between the two transistor currents will flow into or out of the resistor, creating a large voltage offset. Feedback is typically used around the op-amp to stabilize the DC output value.

22.3.1.8 A Word about Circuit Simulation

Circuit simulators have become powerful design tools for analysis of complicated analog circuits. However, the designer must be very careful about the role of the simulator in the design. When simulating high-gain amplifier circuits, it is important to understand the trends and inner working of the circuit before simulations begin. One should always *interpret* rather than blindly trust the simulation results (the latter is a guaranteed recipe for disaster!). For example, the previously mentioned high-impedance nodes should always be given careful consideration when simulating the circuit. Because these nodes are highly dependent on λ, predicting the actual DC value of the nodes either through simulation or hand analysis is a near impossibility. Before any AC analysis is performed, check the values of the DC points in the circuit to ensure that every device is in saturation. Failure to do so will result in very wrong answers.

22.3.1.9 Other Output Stages

With the preceding design, the output stage was a high-impedance driver, capable of handling only capacitive loads. If a resistive load is present, an additional stage should be added that has a low output impedance and high current drive capability. An example of the output stage can be seen in Figure 22.34. Here, the output impedance is simply $1/g_{m9}||1/g_{m10}$. Since we do not wish to have a large output impedance, the values for L_9 and L_{10} should be made as small as possible. The transistors M11 and M12 are used to help bias the output devices such that M9 and M10 are just barely on under quiescent conditions. This kind of amplifier is known as a class AB output stage and has limitations in CMOS due to the body effect.

In some cases, it is advantageous to use a bipolar output driver as seen in Figure 22.35. Since most BiCMOS processes provide only one flavor of BJT (an npn), the transistor Q1 can be used for extra current drive. This results in a dual-sloped transfer curve characteristic as the output stage goes from

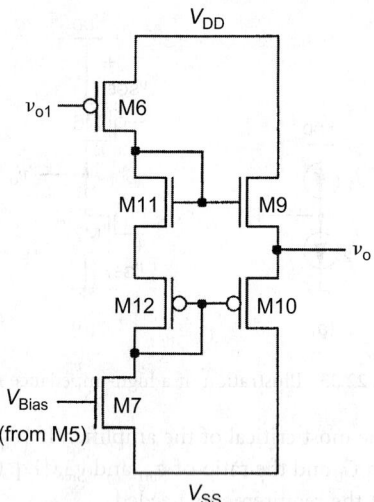

FIGURE 22.34 A low-impedance output stage.

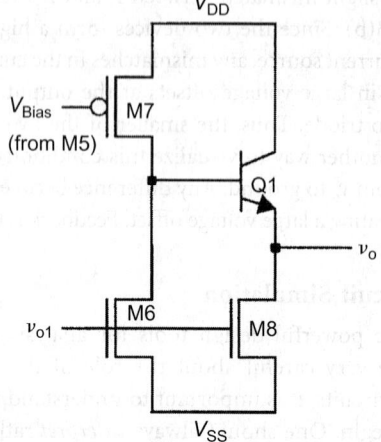

FIGURE 22.35 Using an NPN BJT as an output driver.

sourcing to sinking. It should be noted that one could use this output stage with the complementary version of the two-stage amplifier previously discussed.

Another BiCMOS output stage can be seen in Figure 22.36.[5] This is known as a "pseudo-push-pull" output stage. M6 and M7 can be output of the previously discussed two-stage op-amp. With the new pseudo-push-pull output attached, the amplifier is able to achieve high output swing with large current drive capability using very little silicon area. The transistor MQ1 is for level shifting purposes only. When the output signal needs to swing in the positive direction, Q2 behaves just like the BJT output driver shown in Figure 22.35. When the output swings in the negative direction, Q1 drives the gate of M9 down, thus increasing I_{D9}. The increase in current is mirrored via M10 to M11, which is able to sink a large amount of current. The output voltage at its lowest voltage is approximately the same voltage as the emitter voltage of Q2. The transistor Q2 provides the needed low output impedance.

Another advantage of using BJT devices in the design is to provide large g_ms as compared to the MOS counterpart. BiCMOS circuits can be constructed, which offer high input impedance and gain bandwidth products. If both npn and pnp devices are available in the BiCMOS process, then the output stage seen in Figure 22.37 can be used. This circuit functions as a buffer circuit with very low output impedance.

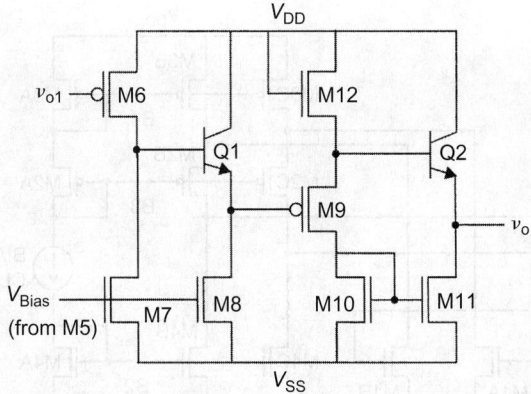

FIGURE 22.36 A "pseudo" push-pull npn only output driver.

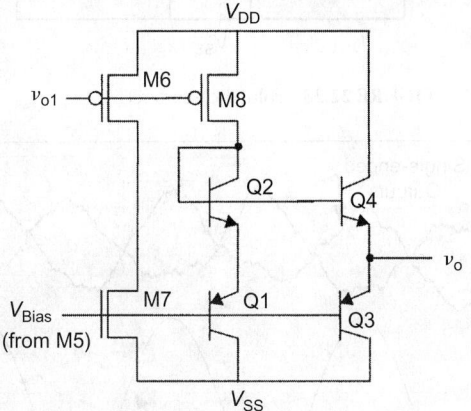

FIGURE 22.37 A low-impedance output stage using both npn and pnp devices.

22.3.2 High-Performance Operational-Amplifier Considerations

In the commercial world, it seems there are not many applications that require simple, easy to design op-amps. High bit rate communication systems and over-sampled data converters push the bandwidth and slew rate capabilities of CMOS op-amps, while battery-powered systems are required to squeeze just enough performance out of micro-amps of current, and a volt or two of power supply. Sensor interfaces and audio systems demand low noise and distortion. To further complicate things, CMOS analog circuits are often integrated with large digital circuits, making isolation from switching noise a major concern. This section will present solutions to some of the problems faced by op-amp designers who, because of budget constraints or digital compatibility, do not have the option to use bipolar junction transistors in their design. We will then give some hints on where the designer might use bipolar transistors if they are available.

22.3.2.1 Power Supply Rejection

Fully differential circuits like the OTA shown in Figure 22.38 are used in mixed signal circuits because they provide good rejection of substrate noise and power supply noise. As long as the noise coupled from the substrate or power supply is equal for both outputs, the difference between the two signals is noise-free (differential component of the noise is zero). This is illustrated in Figure 22.39. The top two traces in Figure 22.39 are a differential signal corrupted with common-mode noise. The bottom trace is the difference between these two noisy signals. If the next circuit in the path has good common-mode

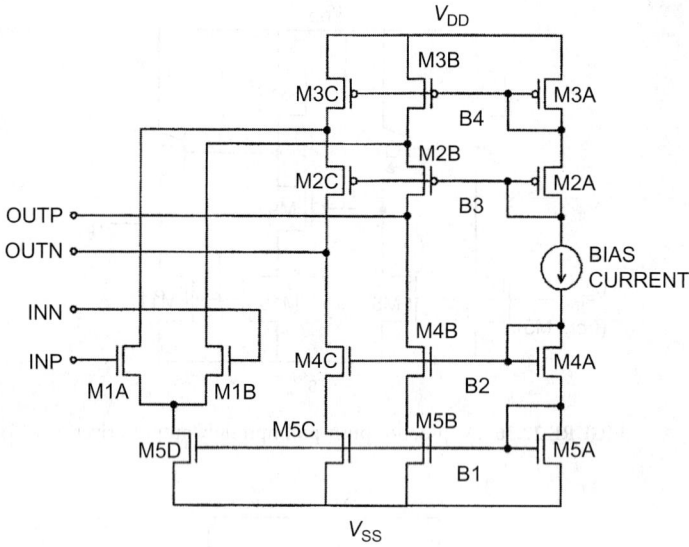

FIGURE 22.38 Folded cascode OTA.

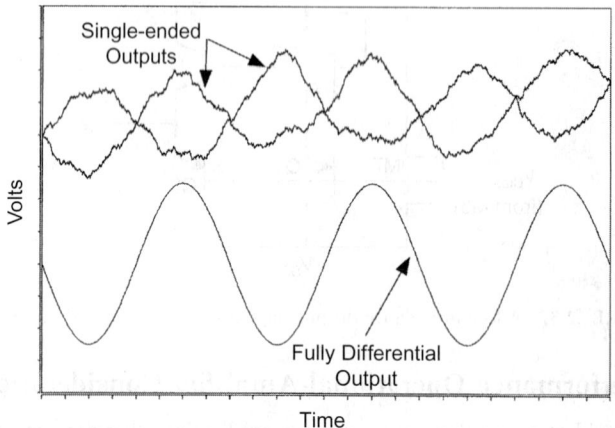

FIGURE 22.39 Simulation output illustrating the difference between single-ended and fully differential signals.

rejection, the substrate and power supply noise will be ignored. In practical circuits, mismatches between the transistors of symmetrical halves of the differential circuit will lead to imperfect matching of noise on the outputs, and therefore reduced rejection of power supply noise. Common centroid layouts and large geometry transistors are necessary to minimize mismatches. Differential circuits are capable of twice the signal swing of single-ended circuits, making them especially welcome in low-voltage and low-noise applications.

Single-stage or multiple-stage op-amps can be made differential, but each stage requires a common-mode feedback circuit to give the differential output a common-mode reference. Consider the folded cascode OTA shown in Figure 22.38. If the threshold voltage of M5A is slightly larger than the threshold voltage of M5B and M5C, the pull-down currents will be larger than the pull-up currents. This small current difference, in combination with the very high output impedance of the cascode current mirrors, will cause the output voltages to be pegged at the negative power supply. This common-mode error cannot be corrected by applying feedback to the differential pair. A common-mode feedback circuit is needed to find the average of the output voltages, and control the pull-up or pull-down current in the outputs to maintain this average at the desired reference. A center-tapped resistor between the outputs could be used to sense the common-mode voltage if the outputs were buffered to drive such a load. Since a folded cascode OTA cannot drive resistors, a switched capacitor would be a better choice to sense the common-mode voltage as shown in Figure 22.40. The PH1 and PH2 clock signals must be non-overlapping. When the PH1 switches are closed, C_{1A} and C_{1B} are discharged to zero, while C_{2A} and C_{2B} provide feedback to the common-mode amplifier. The PH1 switches are then opened, and a moment later, the PH2 switches closed. The charge transfer that takes place moves the center tap between C_{2A} and C_{2B} toward the average of the two output voltages. After many clock cycles, the input to the common-mode feedback amplifier will be the average of the two voltages. C_{1A} and C_{1B} can be precharged to a bias voltage to provide a level shift. This allows direct feedback from the common-mode circuit to the pull-up bias as shown in Figure 22.41.

A BiCMOS version of the folded cascode amplifier can be seen in Figure 22.42 [6]. Here, it is again assumed that only npn devices are available. Note that to best utilize the npn devices, the folded cascoded uses a diff-amp with P-channel input devices. The amplifier also uses the high swing current mirror presented in Section 22.2.

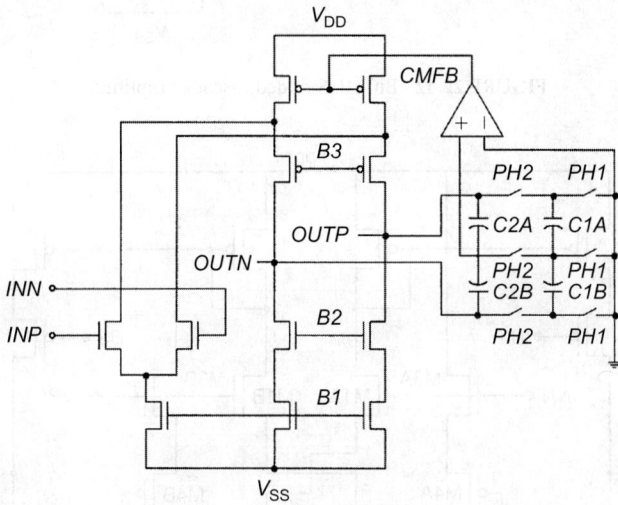

FIGURE 22.40 Folded cascode OTA using switched capacitor common mode feedback.

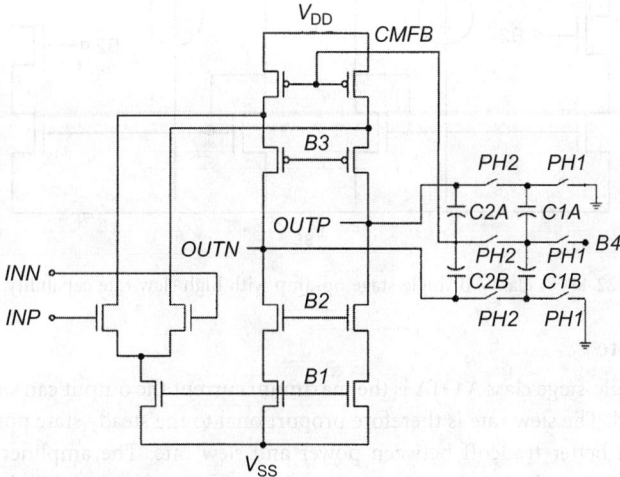

FIGURE 22.41 Using a bias voltage to precharge the switched capacitor common mode feedback capacitors.

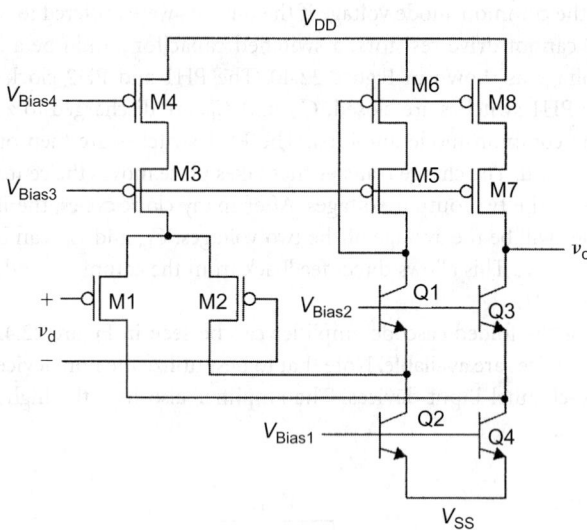

FIGURE 22.42 BiCMOS folded cascode amplifier.

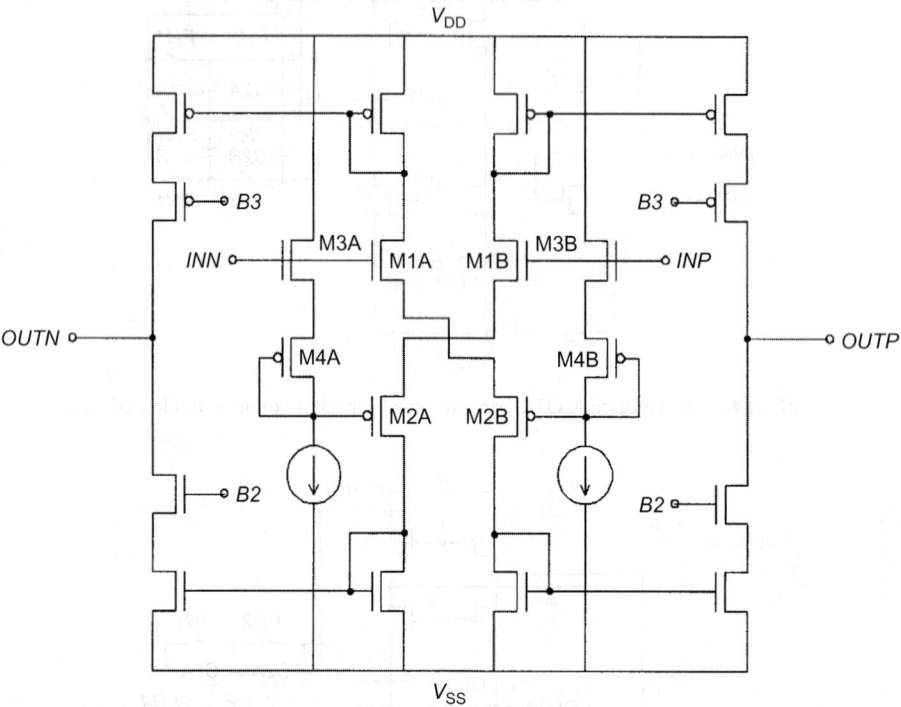

FIGURE 22.43 A class AB single stage op-amp with high slew rate capability.

22.3.2.2 Slew Rate

The slew rate of a single-stage class A OTA is the maximum current the output can source or sink, divided by the capacitive load. The slew rate is therefore proportional to the steady state power consumed. Class AB amplifiers give a better tradeoff between power and slew rate. The amplifier's maximum output current is not the same as the steady-state current. An example of a class AB single-stage op-amp is shown in Figure 22.43. The differential pair is replaced by M1A, M1B, M2A, and M2B. A level shifter consisting of M3A, M3B, M4A, and M4B couples the input signal to M2A and M2B, and sets up the zero input bias current. If the width of M2A and M2B are three times the width of M1A and M1B, the small signal voltage on the nodes between M1 and M2 will be approximately zero. The current available from one of the input transistors is approximately

$$I_{out} = KP\frac{W}{L}(V_{in} + V_{bias} + V_{THN})^2 \tag{22.59}$$

The differential current from the input stage is

$$I_{OUTP} - I_{OUTN} = 2 \cdot \beta(V_{in} + V_{BIAS} + V_{THN}) \tag{22.60}$$

It is interesting to note that the non-linearities cancel. The output current becomes non-linear again as soon as one of the input transistors turns off. The maximum current available from the input transistors is not limited by a current source as is a differential pair. It should be noted that it is impossible to keep all transistors saturated for low power supplies and large input common-mode swings. A similar BiCMOS circuit with high slew rate can be seen in Figure 22.44 [7].

Adaptive bias is another method to reduce the ratio of supply current to slew rate. Adaptive bias senses the current in each side of a conventional differential pair, and increases the bias current when the current on one side falls below a preset value. This guarantees that neither side of the differential pair will turn off. The side that is using most of the current must therefore be supplied with much more than the zero input amount. The differential pair current can be sensed by measuring the gate-to-source voltages as shown in Figure 22.45, or by measuring the current in the load as shown in Figure 22.46. Both of these adaptive bias schemes depend on a feedback loop to control the current in the differential pair. These circuits improve settling time for low power, low bandwidth circuits, but the delay in the feedback path is a problem for high-speed circuits.

A second form of adaptive bias can be used when it is known that increased output current is needed just after clock edges, such as in switched capacitor filters. This type of adaptive bias can be realized using

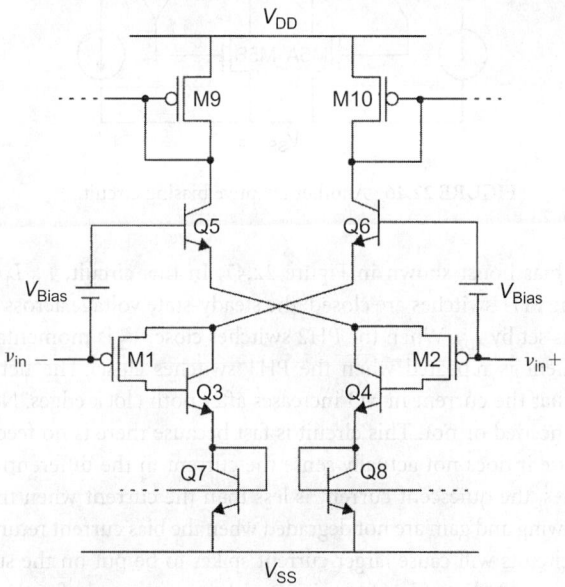

FIGURE 22.44 A BiCMOS class AB input stage.

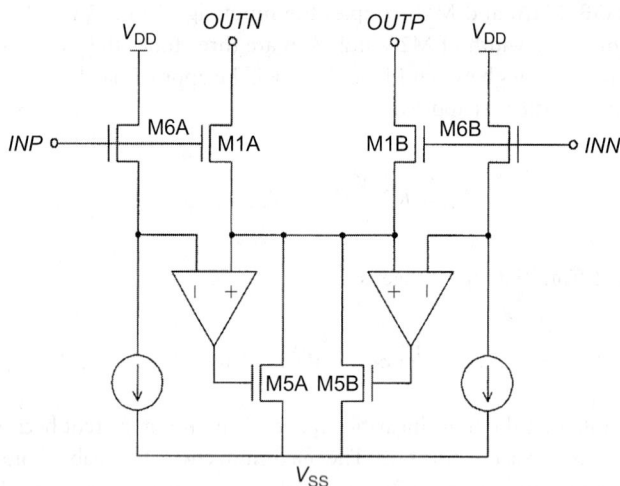

FIGURE 22.45 Adaptive biasing scheme for improved slew rate performance.

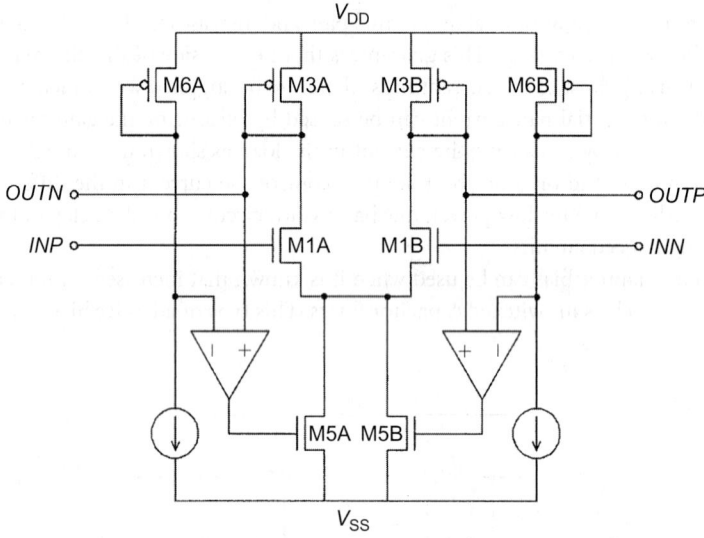

FIGURE 22.46 Another adaptive biasing circuit.

the switched capacitor bias boost shown in Figure 22.47. In this circuit, $I_{D1} \cdot L_1/W_1 > I_{D2} \cdot L_2/W_2$, which makes $V_1 > V_2$. When the PH1 switches are closed, the steady-state voltage across the capacitor is $V_1 - V_2$, and the current in M3 is set by I_{D2}. When the PH2 switches close, V_2 is momentarily increased, while V_1 is decreased. The transient is repeated when the PH1 switches close. The net effect of the switched capacitor bias boost is that the current in M3 increases after both clock edges. Notice that the current is increased, whether it is needed or not. This circuit is fast because there is no feedback loop, but it is not the most efficient because it does not actually sense the current in the differential pair.

In all three approaches, the quiescent current is less than the current when the output is required to slew. Therefore, output swing and gain are not degraded when the bias current returns to its quiescent value. All three adaptive bias circuits will cause larger current spikes to be put on the supplies by the op-amps. The width of the power supply lines should be increased to compensate for the increased *IR* drop and crosstalk. Enhanced slew rate circuits are not linear time invariant systems because the transconductance

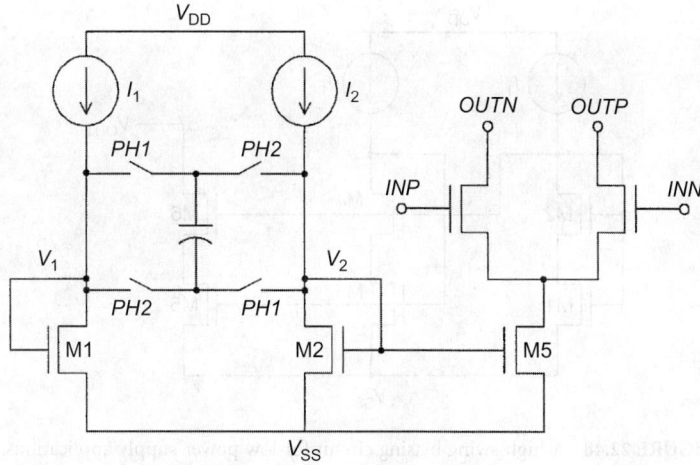

FIGURE 22.47   Adaptive biasing using a switched capacitor circuit.

and output impedance of the transistors are bias current dependent, and the bias is time varying. A transient analysis is the most dependable way to evaluate settling time in this case.

22.3.2.3   Output Swing

Decreasing power supply voltages put an uncomfortable squeeze on the design engineer. To maintain the desired signal-to-noise ratio with a smaller signal swing, circuit impedances must decrease, which often cancels any power savings that may be gained by a lower supply voltage. To get the best signal swing, differential circuits are used. If the output stages use cascode current mirrors, bias voltages must be generated, which keep both the mirror and cascode transistors in saturation with the minimum voltage on the output. Another example of a high swing bias circuit (refer also to Section 22.2) and a current mirror is shown in Figure 22.48. First, let the W/L ratio of M2, M4, and M6 be equal, and the W/L ratio of M3 and M5 be equal. Now recall that to keep a MOSFET in saturation

$$V_{DS} \geq V_{GS} - V_{THN} \tag{22.61}$$

The minimum output voltage that will keep both M5 and M6 in saturation with proper biasing is

$$V_{out} \geq V_{GS6} - 2V_{THN} + V_{GS5} \tag{22.62}$$

ignoring the bulk effects. For a given current, the minimum drain voltage can be rewritten as

$$V_{DS} \geq \sqrt{\frac{I_D \cdot L}{K \cdot W}} \tag{22.63}$$

The equation for the minimum V_{OUT} can be rewritten as

$$V_{OUT} \geq \sqrt{\frac{I_{OUT} \cdot L_6}{K_6 \cdot W_6}} + \sqrt{\frac{I_{OUT} \cdot L_5}{K_5 \cdot W_5}} \tag{22.64}$$

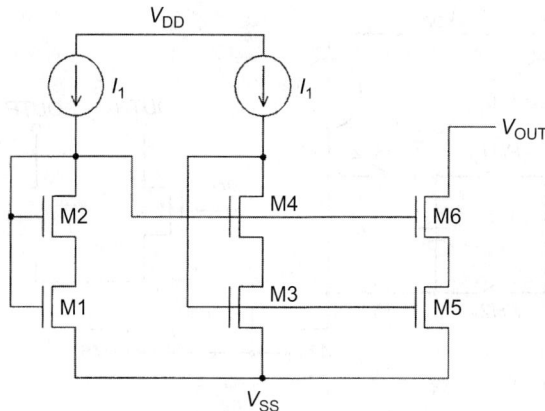

FIGURE 22.48 A high swing biasing circuit for low power supply applications.

The trick to making this bias generator work is setting V_{DS} of M1 equal to the minimum V_{DS} required by M3 and M5. M1 is biased in the linear region, while we wish to keep M3 and M5 saturated. It is a good idea to set $L_1 = L_3 = L_5$ to match etching tolerances. A second trick is to make sure M3 and M4 stay saturated. M4 is inserted between the gate and drain connections of M3 to make $V_{DS3} = V_{DS5}$. If the W/L ratio of M3 is too small, M3 will be forced out of saturation by the source of M4. If the W/L_3 is too large, the gate voltage of M3 will not be large enough to keep M4 in saturation.

22.3.2.4 DC Gain

Start by calculating the DC gain of the folded cascode OTA in Figure 22.38. If we assume the output impedance of the n-channel cascode current mirror is much greater than the output impedance of the individual transistors, then the DC gain is approximately

$$\frac{v_o}{v_{in}} = -g_{m1} \cdot r_{o1} \| r_{o3} \cdot g_{m2} \cdot r_{o2} \tag{22.65}$$

If we assume that the current from the M3 splits equally to M1 and M2, the gain can be written as

$$\frac{v_o}{v_{in}} \propto \frac{\sqrt{W_1 L_1 W_2 L_2}}{I_{D1}} \cdot \frac{L_3}{2L_1 + L_3} \tag{22.66}$$

We can see that the gate area of the differential pair and cascode transistors must both double each time current is doubled to maintain the same gain. We also note that it is desirable to make $L_3 > L_1$. If the current in the amplifier were raised to increase gain-bandwidth, or slew rate, it would be desirable to increase the widths of the transistors by the same factor to maintain output swing.

Regulated gate cascode outputs increase the gain of the OTA by effectively multiplying the g_m of M4 by the gain of the RGC amplifier. The stability of the RGC amplifier loop must be considered. An example of a gain boosted output stage is shown in Figure 22.49.

22.3.2.5 Gain Bandwidth and Phase Margin

Again, start with the transfer function for the folded cascode OTA of Figure 22.38. If we assume the output impedances of the n-channel cascode current mirrors are very large, and the gain of M2 is much

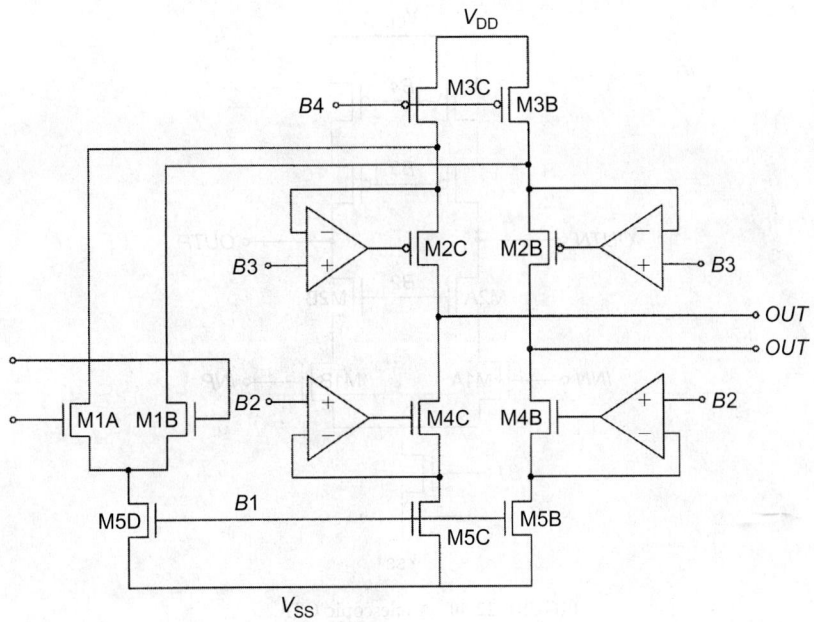

FIGURE 22.49 OTA with regulated gate cascode output.

greater than one, we have

$$\frac{v_o}{v_{in}} = \frac{g_{m1} r_{o1} g_{m2} r_{o2}}{(C_1 \cdot C_{out} \cdot r_{o1} \cdot r_{o2})\left(s^2 + \left(\dfrac{g_{m2}}{C_1} + \dfrac{1}{C_1 \cdot r_{o1}} + \dfrac{1}{C_1 \cdot r_{o2}} + \dfrac{1}{C_{out} \cdot r_{o2}}\right)s + \dfrac{1}{C_1 \cdot C_{out} \cdot r_{o1} \cdot r_{o2}}\right)} \qquad (22.67)$$

where r_{o1} is now the parallel combination of r_{o1} and r_{o3}. If we further assume that the poles are spaced far apart, and that g_{m2} is much larger than $1/r_{o1}$ and $1/r_{o2}$, then the gain-bandwidth product is g_{m1}/C_{load}. The second pole, which will determine phase margin, is approximately

$$\omega = \frac{g_{m2}}{C_1} + \frac{1}{C_{out} \cdot r_{o2}}$$

The depletion capacitance of the drains of M1 and M3 will also add to this capacitance. As a first cut, let $C_1 = K_c \cdot W_2 \cdot L_2$. Now the equation for the second pole boils down to

$$\omega \approx \frac{\sqrt{K \cdot I_{D2}}}{K_C \cdot L_2 \cdot \sqrt{W_2 \cdot L_2}} + \frac{I_{D2}}{L_2 \cdot C_{out}}$$

To get maximum phase margin, we clearly want to use as short a channel length as the gain specification will allow. The folded cascode OTA and two-stage OTA both have n-channel and p-channel transistors in the signal path. Since holes have lower mobility than electrons, it is necessary to make a silicon p-channel transistor about three times wider than an n-channel transistor of the same length to get the same transconductance. The added parasitic capacitance of the wider p-channel transistor is a hindrance for high-speed design. The telescopic OTA shown in Figure 22.50 has only n-channel transistors in the

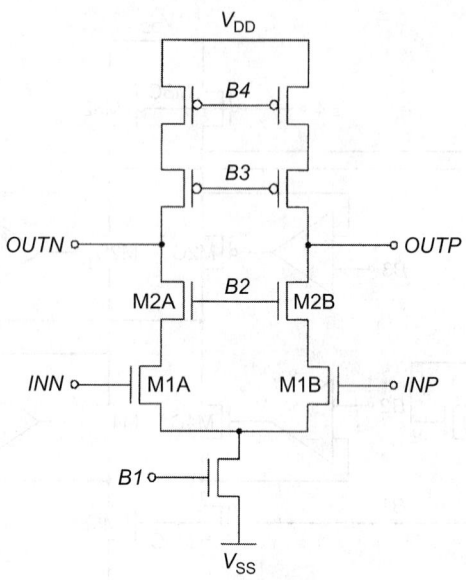

FIGURE 22.50 A telescopic OTA.

signal path, and can therefore achieve very high bandwidths with acceptable phase margin. Its main drawback is that the output common-mode voltage must be more positive than the input common-mode voltage. This amplifier can achieve even wider bandwidth with acceptable phase margin if M2 is replaced by an npn bipolar transistor.

References

1. R.J. Baker, H.W. Li, and D.E. Boyce, *CMOS: Circuit Design, Layout, and Simulation*, IEEE Press, 1998.
2. A.S. Sedra and K.C. Smith, *Microelectronic Circuits, fourth edition*, Oxford University Press, London, 1998.
3. P.E. Allen and D.R. Holberg, *CMOS Analog Circuit Design*, Saunders College Publishing, Philadelphia, 1987.
4. P.R. Gray, *Basic MOS Operational Amplifier Design—An Overview, Analog MOS Integrated Circuits*, IEEE Press, 1980.
5. H. Qiuting, A CMOS power amplifier with a novel output structure, *IEEE J. of Solid-State Circuits*, vol. 27, no. 2, pp. 203–207, Feb. 1992.
6. M. Ismail, and T. Fiez, *Analog VLSI: Signal and Information Processing*, McGraw-Hill, Inc., New York, 1994.
7. S. Sen and B. Leung, A class-AB high-speed low-power operational amplifier in BiCMOS technology, *IEEE J. of Solid-State Circuits*, vol. 31, no. 9, pp. 1325–1330, Sept. 1996.

23

Bipolar Junction
Transistor Amplifiers

David J. Comer and
Donald T. Comer

Brigham Young University

CONTENTS

23.1   Introduction

23.1.1   The Need for Amplification

The field of electronics includes many applications wherein important information is generated in the form of a voltage or current waveform. Often this voltage is too small to perform the function for which it is intended. Examples of this situation are the audio microphone in a sound amplification signal, the infrared diode detector of a heat-seeking missile, the output of the light intensity sensor of a CD player, the output of a receiving antenna for an amplitude modulation (AM) or frequency modulation (FM) signal, the cell phone signal received by the base station before retransmission, and the output of a receiving dish antenna in a satellite TV system. Each of these signals have peak voltages in the range of hundreds of microvolts to hundreds of millivolts. To transmit a signal over phone lines or through the atmosphere and then activate a speaker, or to illuminate a cathode-ray tube or other type of TV display may require a signal of several volts rather than millivolts.

An audio amplification system may require peak voltages of tens of volts to drive the speaker. The speakers in a CD system or in radio receivers may also require several volts to produce appropriate levels of volume. The video information for a TV signal may need to be amplified to a value of thousands of volts to drive the picture tube or display. Even in computer systems, the signals recovered from the floppy or hard disk drives must be amplified to logic levels to become useful. Amplification of each of these signals is imperative to make the signal become useful in its intended application. Consequently, amplifiers make up a part of almost every electronic system designed today.

At the present time, the bipolar junction transistor (BJT) and the metal-oxide semiconductor field-effect transistor (MOSFET) are the major devices used in the amplification of electronic signals.

23.1.2   A Brief History of Amplifying Devices

Amplification is very important that the so-called era of electronics was ushered in only after the development of the triode vacuum tube that made amplification possible in 1906 [1]. This device enabled public address systems and extended distances over which telephone communications could take place. Later, this element allowed the small signal generated by a receiving antenna located miles away from a transmitter to be amplified and demodulated by a radio receiver circuit. Without amplifiers, communications over long distances would be difficult if not impossible.

Improvement of the vacuum tube continued for several years with the tetrode and pentode tubes emerging in the late 1920s. These devices enabled the rapid development of AM radio and television. Vacuum tubes dominated the electronics field well into the 1950s serving as the basic element in radios, television, instrumentation, and communication circuits.

The forerunner of the BJT, the point-contact transistor, was invented in 1947 [2]. The BJT followed shortly and grew to dominance in amplifier applications by the mid-1960s. From then until the 1990s, this device had no peer in the electronics field. In the 1990s, the MOSFET became important as an amplifying device, however, the silicon BJT along with the newer heterojunction BJT (HBT), continue to be used in many amplifier applications.

23.2   Amplifier Types and Applications

Amplifiers can be classified in various ways. The input and output variables can be used to describe the amplifier. The frequency range of the output variable or the power delivered to a load can also describe a major characteristic of an amplifier. In the case of an operational amplifier (op amp), the mathematical operations that can be performed by the circuit suggest the name of the amplifier.

23.2.1 Input and Output Variables

Electronic transducers generally produce either voltage or current as the quantity to be amplified. After amplification, the signal variable of interest could again be voltage or current, depending on the circuit or device driven. A given amplifier can then have either current or voltage as the input signal and current or voltage as the output signal. As the input variable changes, the amplifier produces a corresponding output variable change. Gain of an amplifier is defined as the ratio of output variable change to the input variable change as shown in Figure 23.1.

There are four possible ratios for the gain. These are summarized in Table 23.1 along with the corresponding amplifier type.

The voltage amplifier is used more often than the others, but each has specific applications. Some photo detectors generate output current proportional to the light intensity and this current is the input variable to the amplifier. The output variable may be voltage for this application. There are other occasions when an amplifier must generate an output current that is proportional to the input variable, thus, a given application may require any one of the four possible types listed in Table 23.1.

23.2.2 Frequency Range

The frequency range over which an amplifier exhibits useful gain is an important specification. An audio amplifier may have a relatively constant gain from tens of Hz up to nearly 20 kHz. The magnitude of this constant gain is often called the midband gain. Those frequencies at which the magnitude of gain falls by 3 dB from the midband value are called 3-dB frequencies. For a high-fidelity audio amplifier, the lower 3-dB frequency may be 20 Hz while the upper 3-dB frequency may be 20 kHz. The bandwidth of this amplifier extends from 20 Hz to 20 kHz. In spite of the relatively small absolute bandwidth, this type of amplifier is referred to as a *wideband amplifier*. The ratio of the upper 3-dB frequency to the lower 3-dB frequency is 20,000/20, which is 1000 in this case.

Another important type of amplifier is that used in radio receiver circuits. The receiver for an AM signal consists of several stages that amplify over a frequency range from about 450 to about 460 kHz. The gain may be maximum at 455 kHz, but drops very rapidly toward zero when the frequency is below 450 or above 460 kHz. This type of amplifier is called a *narrowband amplifier* because the ratio of upper 3-dB frequency to lower 3-dB frequency has a near-unity value of 460/450 in this case.

23.2.3 Output Power

Some audio speakers may require hundreds of watts of power to operate at the desired level. An light emitting diode (LED) display may only require milliwatts of power. The amplifiers that drive these devices

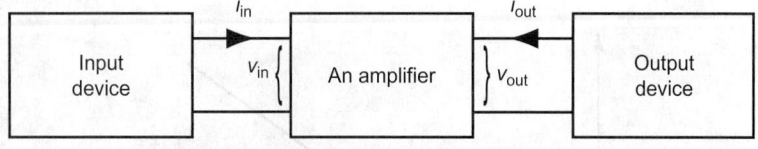

FIGURE 23.1 An amplifier circuit.

TABLE 23.1 Amplifiers based on Output/Input Ratio

Input Variable	Output Variable	Gain	Amplifier Type
Voltage	Voltage	v_{out}/v_{in}	Voltage amplifier
Current	Current	i_{out}/i_{in}	Current amplifier
Voltage	Current	i_{out}/v_{in}	Transadmittance amplifier
Current	Voltage	v_{out}/i_{in}	Transimpedance amplifier

can be classified as high power or low power, respectively. While there are no fixed power values to differentiate, amplifiers that generate outputs in the milliwatt range are called low-power amplifiers or simply amplifiers. Amplifiers that produce outputs in the watt to several watt range are called power amplifiers or high-power amplifiers. The bulk of integrated circuit (IC) amplifiers are used in low-power applications although pulse-width modulation (PWM) IC amplifier stages can deliver several watts of power to a load.

23.2.4 Distortion in Power Stages

There are two major causes of nonlinear distortion in BJT stages [3]. The first is the change of current gain β with changes in I_C or V_{CE}. If an undistorted base current enters the device, the output current is distorted as a result of the change in current gain as the output quantities vary. This effect can also be explained for a common-emitter stage in terms of the output characteristics which show unequal spacing as I_C or V_{CE} is changed. Obviously, if the output signal amplitude is limited, less distortion occurs. In power stages, very large output-current changes may be prevalent, and higher distortion levels are to be expected.

The second important cause of distortion arises from the nonlinearity of the base–emitter characteristics of the transistor. When a voltage source drives a common-emitter stage with little series resistance, the base current can be quite distorted. This current varies with voltage in the same way that diode current varies with diode voltage. It is possible to use this input distortion to partially offset the output distortion; however, feedback techniques are most often used in distortion reduction. A perfect, distortion-free amplifier would have transfer characteristics that form a straight line as shown in Figure 23.2.

The BJT with no emitter degeneration has a collector current that relates to base–emitter voltage given by

$$I_C = \beta K_1 e^{q V_{BE}/kT}$$

where k is the Boltzmann's constant. This function is highly nonlinear, and only small variations of V_{BE} can be applied to approximate linear behavior. An emitter resistance or emitter degeneration can be added to make the circuit more linear at the expense of reduced gain. This amounts to feedback to improve the nonlinear distortion. Another method of reducing distortion is to use several cascaded stages to achieve a very high gain; then feedback is applied around the amplifier to decrease the distortion.

23.2.4.1 Total Harmonic Distortion

In an amplifier circuit, the output signal should ideally contain only those frequency components that appear in the input signal. The nonlinear nature of amplifying devices introduces extraneous frequencies in the output signal that are not contained in the input signal. These unwanted signals are referred

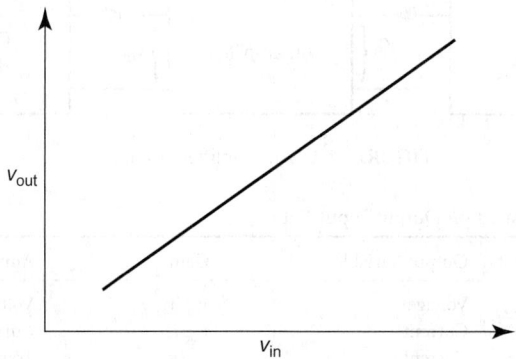

FIGURE 23.2 Transfer characteristics of a distortion-free amplifier.

to as harmonic distortion. One measure of the amplifier's performance is called the *total harmonic distortion* (THD).

If a sinusoidal signal is applied to the amplifier input, the output will also contain a major component of this signal at the fundamental frequency, with amplitude designated v_f. In addition, there will be smaller harmonic components in the output signal. These amplitude values will be designated $v_2, v_3, v_4, \ldots, v_n$. Fortunately, in engineering applications, the harmonics decrease in amplitude as frequency increases. Thus, perhaps only the second or third harmonic is large enough to affect the THD. The THD can be defined as a percentage by

$$\text{THD} = \frac{\sqrt{v_2^2 + v_3^2 + v_4^2 + \cdots}}{v_f} \times 100 \tag{23.1}$$

In dB, this parameter becomes

$$\text{THD} = 10 \log\left(\frac{v_2^2 + v_3^2 + v_4^2 + \cdots}{v_f^2}\right) \tag{23.2}$$

The THD of an amplifier is a major specification used to characterize the amplifier performance.

23.2.4.2 Intercept Points

The unwanted harmonic distortion of an amplifier is a function of output signal size. As the signal becomes larger, the linear output will increase, and so will the distortion. In fact, unwanted components due to harmonic distortion may increase more rapidly than the linear component.

Typically, an amplifier with a small output signal, compared with the size of the active region, will have a large linear output component or fundamental component relative to the harmonic components. As signal size increases, the size of the second or third harmonic may become as large as the fundamental component. Generally, this will not occur before the output signal reaches the extremes of the active region, consequently it would be impossible to measure the size of the output signal that results in a harmonic component that is equal in amplitude to the fundamental component.

The concept of the *intercept point* allows this output to be approximated. A plot, often in dB, is made of the input power versus the output power of the fundamental component. This plot is a straight line with a slope of unity for smaller output signals. As the output approaches the edges of the active region, the output power increases much more slowly than does the input power. Finally, the output power is limited to some fixed value as the output signal size is limited by the active region boundaries. Figure 23.3 shows this variation.

For small signals, each harmonic component is small. The third harmonic is plotted in Figure 23.3 in addition to the fundamental component. As input power is increased, the third harmonic component may increase at three times the rate of increase in the fundamental component. Before the input power is high enough to significantly limit the fundamental power, a straight-line extension is added to the fundamental power plot and the third harmonic power plot. The point where these two extensions intersect is called the *third-order intercept* point (TOI). Note that this point often falls beyond the actual output power that can be achieved by the amplifier. The TOI is specified in terms of the input power required at this point. For example, in Figure 23.3, the TOI is −10 dB. The higher the intercept point, the more linear is the amplifier.

A second-order intercept can also be defined in much the same way as the TOI is defined. For communication amplifiers, the TOI is generally more important than the second-order intercept because the third-order term can result in a frequency that falls within the desired passband of the amplifier. The TOI is an oft-used specification for power stages as well as high-frequency amplifiers.

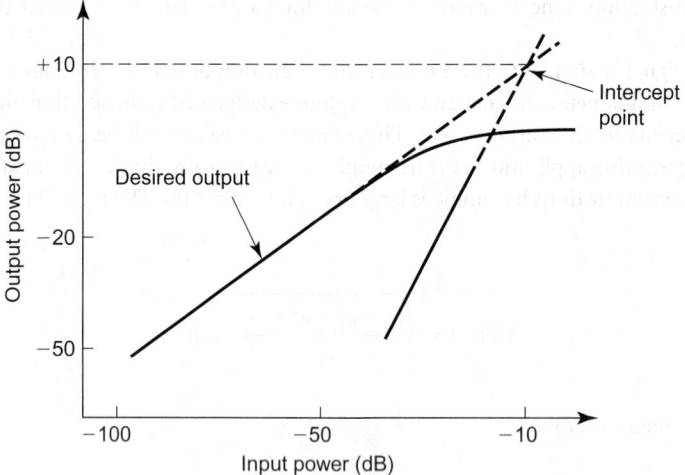

FIGURE 23.3   Amplifier output power versus input power.

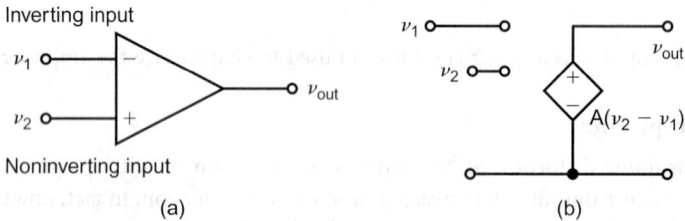

FIGURE 23.4   (a) Symbol for the op amp. (b) Ideal equivalent circuit.

23.2.5   Operational Amplifiers

The op amp was developed long before the IC was invented. It was used in analog computers that were designed to solve differential equations by simulation. This amplifier formed the basis of circuits that performed mathematical operations such as weighting, summing, subtracting, and integrating electrical signals. The op amp has two features that allow these operations to be accomplished. The first feature is a differential input to allow subtraction. The second feature is that a virtual ground can be created when a feedback resistor is connected between the amplifier output and the inverting or negative input terminal. This virtual ground allows perfect summation of currents into the inverting terminal and also allows perfect integration of an input signal.

Before the IC op amp was developed, the discrete circuit op amp was quite expensive and large, perhaps costing $200 and occupying a volume of 250 cm³. With its low cost and small size, the IC op amp has now made a rarely used component, one of the most popular IC chips in the electronics field. The major limitation of the op amp is its frequency response that limits its use in very high-frequency applications.

The near-ideal op amp has a very high gain, a high input impedance, and a low output impedance. The symbol for this device is shown in Figure 23.4(a) along with the ideal equivalent circuit of Figure 23.4(b).

23.3   Differences between Discrete and IC Amplifier Design

Prior to the early 1960s, electronic circuits were constructed from discrete component circuits. Discrete resistors, capacitors, inductors, and transistors are packaged individually to be connected with wires or printed circuit board (PCB) conductors to create a functioning electronic circuit. The discrete component circuit is still required in certain applications in the 21st century, but the IC is now the dominant form of implementation for an electronic system. There are major differences in the design principles used for the IC and the discrete circuit as the following section will explain [4].

23.3.1 Component Differences

Discrete resistors, capacitors, and inductors are available in a very large range of values. Although many amplifier circuits are designed with element values of 5–10% tolerances, very precise values can be obtained at a higher cost. The range of resistor values extends from a few Ω to many MΩ and capacitor values range from fF to hundreds of μF. Metal core or air-core inductors are available over a wide range of values as also are transformers with a wide choice of turns ratios.

For most standard fabrication processes, the IC chip becomes too large to be useful when the total resistance of resistors on the chip exceed some value such as 100 kΩ. The same applies if the total capacitance exceeds perhaps 50–100 pF. It is only in the last decade that inductors could be fabricated on a chip. Even now, integrated inductors with limited values in the nH range are possible and losses may limit Q values to the range 2 to 8 [5].

A major problem with IC components is the lack of precise control of values. Resistors or capacitors are typically fabricated with an absolute accuracy of around 20%. It is possible, however, to create resistive ratios or capacitive ratios that approach a 0.1% accuracy, even though absolute value control is very poor in standard processes.

The cost of a discrete circuit is determined by different variables than that of the IC. This results in a design philosophy for ICs that differs from that of discrete component circuits. Before ICs were available, a designer attempted to minimize the number of transistors in a circuit. In fact, some companies estimated the overall component cost of a circuit by multiplying the number of transistors in the circuit by some predetermined cost per stage. This cost per stage was determined mainly by the transistor cost plus a small amount added to account for the passive component cost. The least expensive item in the discrete circuit is generally the resistor which may cost a few cents.

In an IC, the greatest cost is often associated with the component that requires the most space. The BJT typically requires much less space than resistors or capacitors and is the least expensive component on the chip. An IC designer's attempt to minimize space or simplify the fabrication process for a given circuit generally results in much larger numbers of transistors and smaller numbers of resistors and capacitors than the discrete circuit version would contain. For example, a discrete circuit bistable flip-flop of 1960 vintage contained two transistors, six resistors, and three capacitors. A corresponding IC design had 18 transistors, two resistors, and no capacitors.

Matching of components in IC design can be considerably better than discrete design, but also presents unique problems as well. Components that are made from identical photographic masks can vary in size and value because of uneven etching rates influenced by nearby structures or by asymmetrical processes. Dummy components must often be included to achieve similar physical environments for all matched components. Certain processes must be modified to result in symmetrical results. Metal conductors deposited on the IC chip can introduce parasitics or modify performance of the circuit if placed in certain areas of the chip, thus care must be taken in placing the metal conductors. Of course, these kinds of considerations are unnecessary in discrete circuit design.

While discrete circuits can use matched devices such as differential stages, often single-ended stages can be designed to meet relatively demanding specifications. Discrete components can be produced with very tight tolerances to lead to circuit performance that falls within the accuracy of the specifications. Circuit design can be based on the absolute accuracy of key components such as resistors or capacitors. Although it may add to the price of a finished circuit, simple component-tuning methods can be incorporated into the production of critical discrete circuits.

Since the absolute values of resistors and capacitors created by standard processes for ICs cannot be determined with great accuracy, matching of components and devices is used to achieve acceptable performance. Differential stages are very common in IC design since matched transistors and components are easy to create even if absolute values cannot be controlled accurately.

Some analog circuits do not require high component densities while others may pack a great deal of circuitry into a small chip space. For low-density circuits, more resistors and small capacitors may be used, but for high-density circuits, these elements must be minimized. The next section considers circuits that replace resistors and capacitors with additional BJTs in IC amplifier building blocks. The BJT current mirror, which is quite popular in the biasing of IC amplifiers has been discussed in Chapter 10 of this handbook. This circuit is also used to replace the load resistance of a conventional amplifier to make it easier to integrate on a chip.

23.3.2 Parasitic Differences

One advantage of an IC is the smaller internal capacitance parasitics of each device. A proper layout of an IC leads to small parasitic device capacitances that can be much less than corresponding values for discrete devices. Although the discrete and IC device may be fabricated with similar processes, the collector and emitter of the IC device are typically much smaller and usually connect to fine pitch external leads/ connectors. The discrete device must make external connections to the base, collector, and emitter and these external connection can add a few picofarads of capacitance. An on-chip IC device generally connects to another device adding minimal parasitic capacitance. However, external connects from the chip to the package must be carefully designed to minimize parasitic capacitance as signals are brought off chip.

Another difference relates to the parasitic capacitance associated with IC resistors and inductors. These elements have capacitance as a result of the oxide dielectric between metal layers and the chip substrate. The frequency response of IC resistors and inductors must be considered in critical designs.

At higher frequencies (above a ~1 GHz) other considerations influence the IC design. To maintain the advantage of low cost in IC circuits over discrete realizations, the devices on the chip are made much smaller than their discrete counterparts. Current and power levels must be limited in the smaller devices to avoid gain roll-off due to high-level injection effects or overheating. The lower currents necessitate higher impedance levels in most bipolar chip designs.

For many analog circuits, terminations of 50 Ω are used at I/O ports. This requires large current magnitudes to reach the specified voltage levels. For example, a 20 mA peak current would be required to develop a peak voltage of 1 V. This is not a difficult problem in discrete circuit design using PCBs. The limitation on current and power in the chip design leads to less flexibility with higher impedance interconnects to develop acceptable voltages. High-frequency chip designs must also deal with bond wire or other package inductance in the range 0.1–0.5 nH and the package capacitance that may reach 5 pF. A buffer is often needed to generate the off-chip signals with sufficient current to develop the specified voltages, but IC chip heating due to power dissipation of the buffer stage must be carefully considered. Common thermal centroid techniques are used to prevent unbalanced heating of matched devices.

23.4 Building Blocks for IC Amplifiers

This section discusses several single-stage amplifying circuits that may be combined with other stages to create a high-performance amplifier.

23.4.1 Small-Signal Models

Most amplifier circuits constrain the BJT stages to active region operation. In such cases, the models can be assumed to have constant elements that do not change with signal swing. These models are linear models. The quiescent or bias currents are first found using a nonlinear model to allow the calculation of current-dependent element values for the small-signal model. For example, the dynamic resistance of the base-emitter diode, r_e, is given by

$$r_e = \frac{kT}{qI_E} \tag{23.3}$$

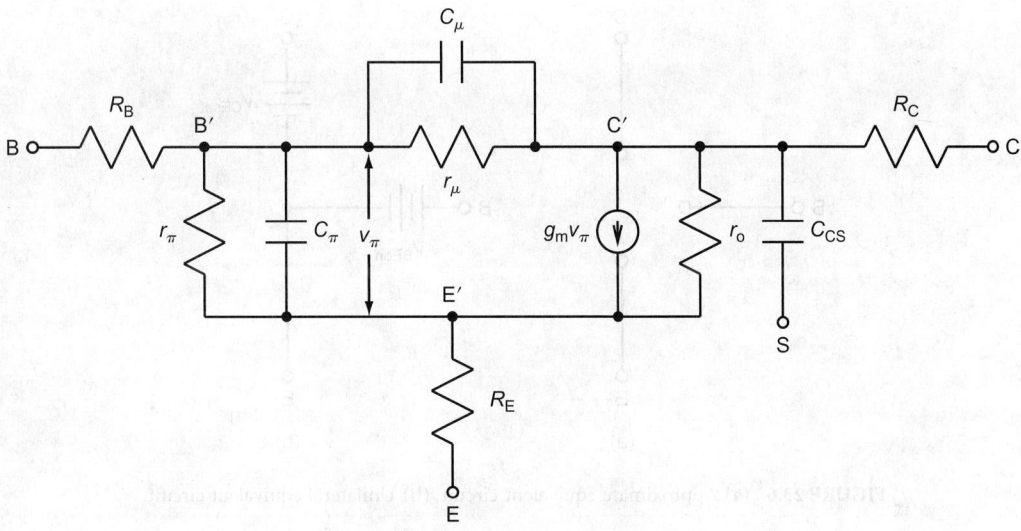

FIGURE 23.5   The hybrid-π small-signal model for the BJT.

where I_E is the DC emitter current. Once I_E is found from a nonlinear DC equivalent circuit, the value of r_e is calculated and assumed to remain constant as the input signal is amplified.

The equivalent circuit [6] of Figure 23.5 shows the small-signal hybrid-π model that is closely related to the Gummel–Poon model and is used for AC analysis in the Spice program.

The capacitance, C_π, accounts for the diffusion capacitance and the emitter–base junction capacitance. The collector–base junction capacitance is designated C_μ. The resistance, r_π, is equal to $(\beta + 1)r_e$. The transconductance, g_m, is given by

$$g_m = \frac{\alpha}{r_d} \tag{23.4}$$

The impedance, r_o, is related to V_A, the Early voltage, and I_C, collector current at $V_{CE} = 0$ by

$$r_o = \frac{V_A}{I_C} \tag{23.5}$$

R_B, R_E, and R_C are the base, emitter, and collector resistances, respectively.

In many high-frequency, discrete amplifier stages and in some high-frequency IC stages, a small value of load resistance will be used. Approximate hand analysis can then be done by neglecting the ohmic resistances, R_E and R_C along with C_{CS}, the collector to substrate capacitance. If the impedance r_o is much larger than the load resistance, R_L, of the high-frequency amplifying stage it can be considered an open circuit. The resulting model that allows relatively viable hand analysis is shown in Figure 23.6(a).

For small values of load resistances, the effect of the capacitance can be represented by an input Miller capacitance that appears in parallel with C_π. This simplified circuit now includes two separate loops as shown in Figure 23.6(b) and is referred to as the unilateral equivalent circuit. Typically, the input loop limits the bandwidth of the amplifier stage. When driven with a signal generator having a source resistance of R_s, the upper 3-dB frequency can be approximated as

$$f_{high} = \frac{1}{2\pi R_{eq}(C_\pi + C_M)} \tag{23.6}$$

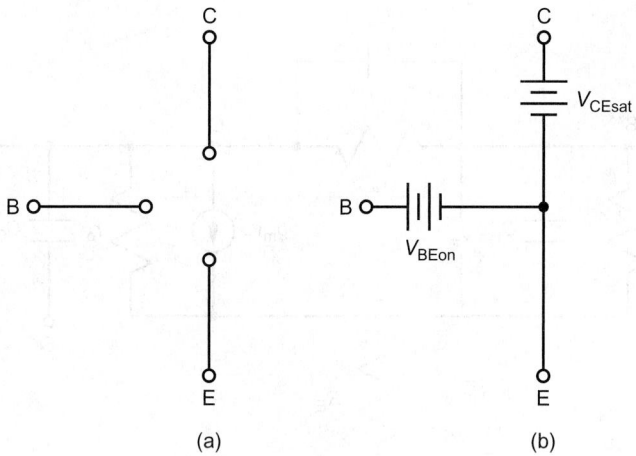

FIGURE 23.6 (a) Approximate equivalent circuit. (b) Unilateral equivalent circuit.

In this equation, the values of R_{eq} and C_M are

$$R_{eq} = r_\pi \parallel (r_b + R_s) \tag{23.7}$$

and

$$C_M = C_\mu \left(1 + |A| \right) = C_\mu (1 + g_m R_L) \tag{23.8}$$

where A is the gain of the stage from point B′ to the collector of the stage. Although C_μ is small, the resulting Miller effect increases this value to the point that it can dominate the expression for upper 3-dB frequency. In any case, it often has a major effect on frequency performance.

When higher impedance loads are used in a BJT as is often the case for IC amplifier design, the Miller effect approach becomes less accurate and the performance must be simulated to achieve more accurate results. However, the unilateral circuit that includes C_μ and C_{CS} from the collector to ground after reflecting the effects of C_μ to the input loop can be used to approximate the amplifier stage performance. Manually calculated results are generally within 5–10% of simulation results.

23.4.2 Active Loads

To achieve high voltage gains and eliminate load resistors, active loads are used in BJT IC amplifier stages [4]. In a conventional common-emitter stage, the gain is limited by the size of the collector resistance. The midband voltage gain from base to collector of a common-emitter stage is given by

$$A_V = -\frac{\alpha R_C}{(r_e + R_E)}$$

It would be possible to increase this voltage gain by increasing R_C, however, making R_C large can lead to some serious problems. A large collector load requires a low quiescent collector current to result in proper bias. This may lead to lower values of β, since current gain in a silicon transistor typically falls at low levels of emitter current. To achieve a voltage gain of 1000 V/V, a collector load of perhaps 100–200 kΩ might be required. The low collector current needed for proper bias, a few microamps, would lead to a low value of β and a very high value of r_e. The desired high voltage gain may not be achievable under these conditions.

It would be desirable if the collector load presented a low resistance to DC signals, but presented a high incremental resistance. This combination of impedances would result in a stable operating point along with a high gain. A perfect current source with infinite incremental resistance along with a finite DC current flow satisfies the requirements, but is not a practical solution. In contrast, circuits that approximate current sources are relatively easy to construct. A good approximation to the current source is obtained by using another transistor for the collector load of an amplifying transistor. Not only can this device present a low DC and high incremental impedance, it is a simple element to implement on a chip. This transistor that replaces the resistive load is referred to as an active load.

The circuit of Figure 23.7 demonstrates one type of BJT active load. The transistor Q1 is the amplifying element with Q2 acting as the load. Transistor Q1 looks into the collector of Q2. The incremental output impedance at the collector of a transistor having an emitter resistance in the low kΩ range can easily exceed 500 kΩ. With such a high impedance, Q2 approximates a current source.

The DC collector currents of both transistors are equal in magnitude. This magnitude can be set to a value that leads to a reasonable value of β. Since Q2 has a very high output impedance, the midband voltage gain will be determined primarily by the collector-to-emitter resistance of Q1 and can be calculated from

$$A_{MB} = \frac{-\beta_1 r_{ce1}}{R_g + r_{\pi 1}} \qquad (23.9)$$

where r_{ce1} is the output impedance of Q1. If the generator resistance, R_g is negligible, this equation reduces to

$$A_{MB} = -\frac{r_{ce1}}{r_{e1}} = -\frac{(V_A + V_{CQ1})/I_C}{V_T/I_E} \approx -\frac{V_A + V_{CQ1}}{V_T} \qquad (23.10)$$

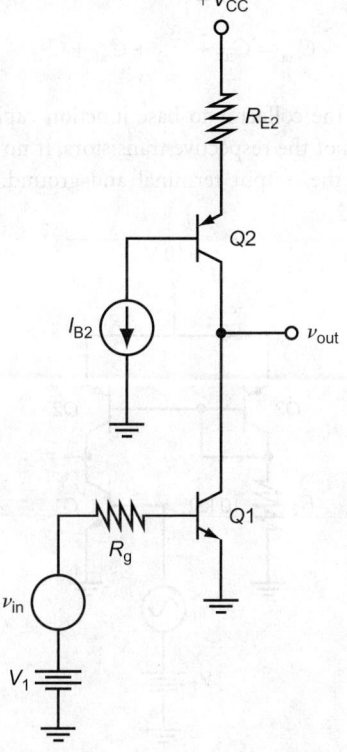

FIGURE 23.7 An active load amplifier.

For an Early voltage of $V_A = 80$ and $V_T = 0.026$ V, a small-signal voltage gain exceeding -3000 V/V could result. In a normal application, this stage would drive a second stage. The input impedance of the second stage will load the output impedance of the first stage, further lowering the gain. Depending on the input impedance of the second stage and the impedance of the active load stage, the gain magnitude may still exceed 1000 V/V.

The concept of an active load that presents a large incremental resistance while allowing a large DC quiescent current is important in IC design. In addition to the current source load just considered, the current mirror stage can also be used to provide the active load of a differential stage as discussed in the next subsection.

23.4.3   The Common Emitter Stage

A simple configuration for an IC amplifying stage is shown in Figure 23.8. In this stage, the output impedance of the current mirror is not large enough to be considered an open circuit as it was in the circuit of Figure 23.7. Thus, the analysis will have to account for this element.

The high-frequency response of discrete BJT stages is often determined by the input circuit, including the Miller effect capacitance. The collector load resistance in a discrete stage is usually small enough that the output circuit does not affect the upper corner frequency. In the circuit of Figure 23.8, as in many IC amplifier stages, the output impedance is very high compared with the discrete stage. For this circuit, the output impedance of the amplifier consists of the output impedance of Q2 in parallel with that of Q1. This value will generally be several tens of kΩ.

The equivalent circuit of the amplifier of Figure 23.8 is indicated in Figure 23.9. The value of R_{out} is

$$R_{\text{out}} = r_{o1} \parallel r_{o2} = r_{ce1} \parallel r_{ce2} \tag{23.11}$$

The capacitance in parallel with R_{out} is approximately

$$C_{\text{out}} = C_{\mu 1} + C_{\mu 2} + C_{cs1} + C_{cs2} \tag{23.12}$$

In this equation, $C_{\mu 1}$ and $C_{\mu 2}$ are the collector-to-base junction capacitances and C_{cs1} and C_{cs2} are the collector-to-substrate capacitances of the respective transistors. If no generator resistance is present, $C_{\mu 1}$ will also appear in parallel with the output terminal and ground. When R_g is present, we will still

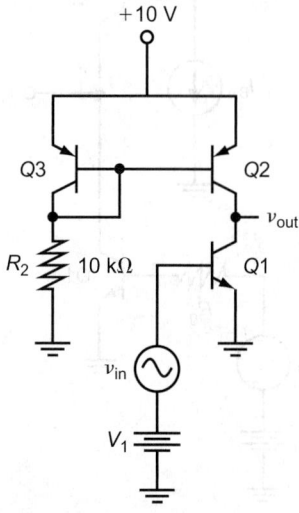

FIGURE 23.8   A common-emitter amplifier with current mirror active load.

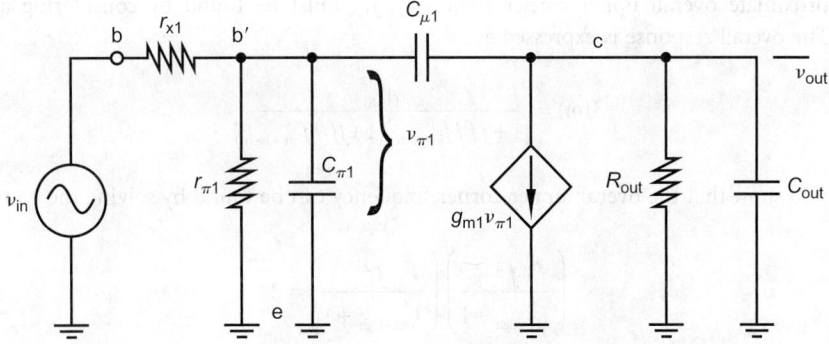

FIGURE 23.9 Equivalent circuit for the amplifier of Figure 23.8.

approximate the output capacitance with the same equation, although feedback effects between the output and the bases of Q1 and Q2 actually modify the value slightly.

The midband gain is easy to evaluate as

$$A_{MB} = \frac{-\beta_1 R_{out}}{R_g + r_{x1} + r_{\pi1}}$$ (23.13)

The upper corner frequency is now more difficult to evaluate than that of the discrete circuit with its low value of collector load resistance. In the discrete circuit, the input loop generally determines the overall upper corner frequency of the circuit. Although the Miller effect will be much larger in the IC stage, lowering the upper corner frequency of the input loop, the corner frequency of the output loop will also be smaller due to the large value of R_{out}. Both frequencies may influence the overall upper corner frequency of the amplifier.

As mentioned previously, it is difficult to manually calculate the upper corner frequency of the stage and an accurate value is generally found by simulation. An approximation of the upper corner frequency can be manually found by reflecting the bridging capacitance, C_μ, to both the input and the output. The value reflected to the input side, across terminals b′ and e, is

$$(1 + |A_{b'c1}|)C_{\mu1}$$ (23.14)

as in the discrete circuit amplifier. Thus, the total input capacitance in parallel with $r_{\pi1}$ is

$$C_{in} = C_{b'e1} + (1 + |A_{b'c1}|)C_{\mu1}$$ (23.15)

The major component of $C_{b'e1}$ is the diffusion capacitance, C_π, of Q1.

The upper corner frequency resulting from the input circuit of this stage is

$$f_{in-high} = \frac{1}{2\pi C_{in} R_{eq}}$$ (23.16)

where $R_{eq} = (R_g + r_{x1}) \| r_{\pi1}$.

The upper corner frequency resulting from the output side of the stage is

$$f_{out-high} = \frac{1}{2\pi C_{out} R_{out}}$$ (23.17)

The approximate overall upper corner frequency, f_{2o}, must be found by considering a two-pole response. The overall response is expressed as

$$A(\omega) = \frac{A_{MB}}{\left(1 + jf/f_{in-high}\right)\left(1 + ff/f_{out-high}\right)}$$ (23.18)

It is easy to show that the overall upper corner frequency can be found by solving the equation

$$\left(\frac{f^2}{f_{in-high}^2 + 1}\right)\left(\frac{f^2}{f_{out-high}^2 + 1}\right) = 2$$ (23.19)

Although this method is not as accurate as the simulation, results will often be within 5–10%.

23.4.4   The Common-Base Stage

The common-base stage shown in Figure 23.10 has an advantage and a disadvantage when compared with the common-emitter stage. The advantage is that the Miller effect, that is, the multiplication of apparent capacitance at the input, is essentially eliminated. The noninverting nature of the gain does not lead to an increased input capacitance. Furthermore, the capacitance between input (emitter) and output (collector) is generally negligible. The upper corner frequency is then higher than that of a comparable common-emitter stage.

The disadvantage is the low current and power gain of the common-base stage compared with the common-emitter stage. The low-frequency current gain for the common-base stage equal α and is slightly less than unity. In the common-emitter stage, the current gain equals β and may be over 200. Since voltage gain is similar for the two stages, the power gain is also much lower for the common-base stage. The input resistance at low frequencies is also much lower than that of the common-emitter stage and can load the previous stage.

The equation for voltage gain is

$$A_{MB} = \frac{\alpha_1 R_{out}}{r_{x1}/(\beta+1) + r_{e1} + R_g}$$ (23.20)

where R_{out} is the parallel combination of r_{out1} and r_{ce2}. The output impedance of $Q1$ depends on the generator resistance, R_g, and ranges from r_{ce1} for $R_g = 0$ to βr_{ce1} when R_g approaches infinity. The midband voltage gain is similar to that of the common-emitter stage.

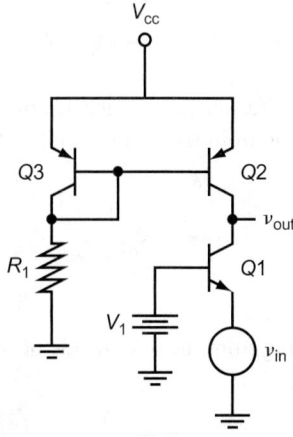

FIGURE 23.10   Common-base amplifier.

The expression for voltage gain as a function of frequency is

$$A = \frac{A_{\text{MB}}}{\left(1 + j(f/f_{\text{in–high}})\right)\left(1 + j(f/f_{\text{out–high}})\right)} \tag{23.21}$$

where $f_{\text{in–high}}$ is the corner frequency of the input circuit and $f_{\text{out–high}}$ the corner frequency of the output circuit. These values are

$$f_{\text{in–high}} = \frac{1}{2\pi C_{\pi 1}(r_{\text{e1}} \| R_{\text{g}})} \tag{23.22}$$

and

$$f_{\text{out–high}} = \frac{1}{2\pi C_{\text{out}} r_{\text{out}}} \tag{23.23}$$

The output capacitance consists of the collector to base capacitances and the collector to substrate capacitances of both Q1 and Q2, that is,

$$C_{\text{out}} = C_{\mu 1} + C_{\mu 2} + C_{\text{cs1}} + C_{\text{cs2}} \tag{23.24}$$

Generally, the corner frequency of the input circuit is considerably higher than the corner frequency of the output circuit for high-gain IC stages and the overall upper corner frequency is approximated by $f_{\text{out–high}}$.

The midband voltage gain of the common-base and common-emitter stages decreases with increasing generator resistance. In multistage amplifiers, the generator resistance of one stage is the output resistance of the previous stage. If this value is large, the gain of the following stage may be small. Furthermore, the upper corner frequency of the common-emitter stage is affected by the generator resistance. Large values of R_{g} can lead to small values of upper corner frequency.

23.4.5   The Emitter Follower

A stage that can be used to minimize the adverse effect on frequency response caused by a generator resistance is the emitter follower. Although this stage has a voltage gain near unity, it can be driven by a higher voltage gain stage while the emitter follower can drive a low impedance load. A typical stage is shown in Figure 23.11.

The output stage of the npn current mirror, Q2, serves as a high impedance load for the emitter follower, Q1. An equivalent circuit that represents the emitter follower of Figure 23.11 is indicated in Figure 23.12. For this circuit, $g_{\text{m1}} = \alpha_1/r_{\text{e}} \approx 1/r_{\text{e}}$.

This circuit can be analyzed to result in a voltage gain of

$$A = \frac{C_{\pi 1}/(\Pi C(R_{\text{g}} + r_{\text{x1}}))(j\omega + (g_{\text{m1}} r_{\pi 1} + 1)/r_{\pi 1} C_{\pi 1})}{-\omega^2 + bj\omega + d} \tag{23.25}$$

where

$$\Pi C = C_{\text{out2}} C_{\mu 1} + C_{\text{out2}} C_{\pi 1} + C_{\pi 1} C_{\mu 1} \tag{23.26}$$

$$b = \frac{1}{\Pi C}\left(\frac{C_{\pi 1} + C_{\text{out2}}}{R_{\text{g}} + r_{\text{x1}}} + \frac{C_{\pi 1} + C_{\mu 1}(1 + g_{\text{m1}} r_{\text{ce2}})}{r_{\text{ce2}}} + \frac{C_{\text{out2}} + C_{\mu 1}}{r_{\pi 1}}\right) \tag{23.27}$$

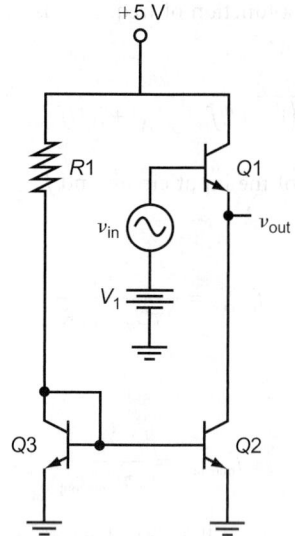

FIGURE 23.11 The emitter follower.

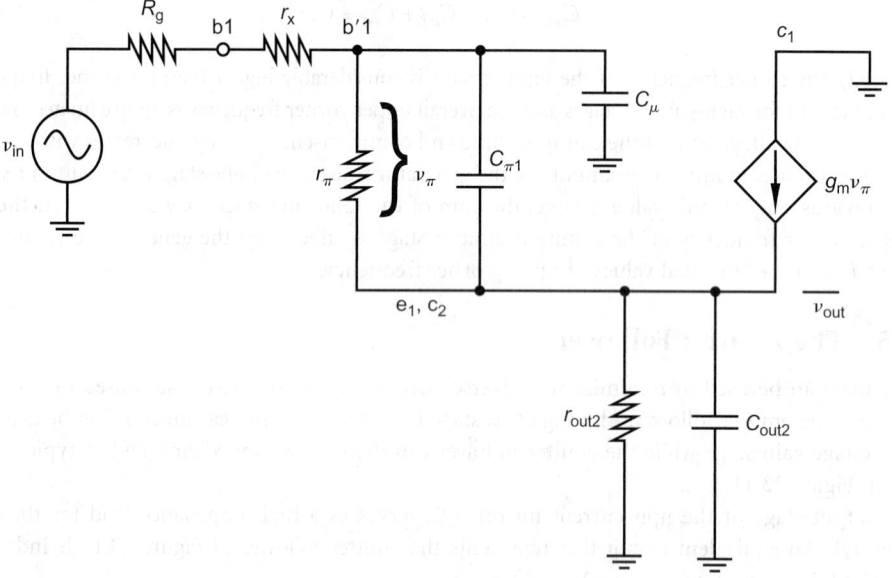

FIGURE 23.12 Equivalent circuit for the emitter follower.

and

$$d = \frac{R_g + r_{x1} + r_{\pi1} + r_{ce2}(g_{m1}r_{\pi1} + 1)}{\prod C(R_g + r_{x1})r_{\pi1}r_{ce2}} \qquad (23.28)$$

Note that the output capacitance of Q2 can be approximated as the sum of $C_{\mu2}$ and C_{cs2}.
The midband voltage gain is found from Eq. (23.25) by letting $\omega \to 0$. This value is

$$A_{MB} = \frac{(1 + g_{m1}r_{\pi1})r_{ce2}}{R_g + r_{x1} + r_{\pi1} + (1 + g_{m1}r_{\pi1})r_{ce2}} \qquad (23.29)$$

This gain is very near unity for typical element values.

The bandwidth is more difficult to calculate since the response has one zero and two poles. The zero for the circuit of Figure 23.12 is typically larger than the lowest frequency pole. If these frequencies were canceled, the larger pole would determine the corner frequency. Since they do not cancel, the overall upper corner frequency is expected to be smaller than the larger pole frequency. An accurate calculation can be made from Eq. (23.25) when the parameters are known.

In high-frequency design, it must be recognized that the output impedance of the emitter follower can become inductive [7].

23.5 Compound Stage Building Blocks for IC Amplifiers

Several of the single stages of the previous section can be combined to construct high-performance two-stage building blocks.

23.5.1 The Cascode Amplifier Stage

One of the problems with the common-emitter stage using an active load is the Miller effect. This stage has a high voltage gain from base to collector. The circuit of Figure 23.8 has an inverting voltage gain, A_{MB}, that can be quite high, perhaps −700 V/V. The base–collector junction capacitance is multiplied by $(1 + |A_{MB}|)$ and reflected to the input loop. This capacitance adds to the diffusion capacitance from point b′ to point e and can lower the upper corner frequency to a relatively small value.

In contrast, the common-base stage of Figure 23.10 has essentially no Miller effect, but has a low input impedance and can load the impedance of the driving source. The cascode stage combines the best features of both the common-emitter and common-base stages. The cascode amplifier stage of Figure 23.13 minimizes the capacitance reflected to the input. The input impedance to the circuit is that of the common-emitter stage and is two orders of magnitude higher than the common-base stage. In this circuit, the input capacitance is primarily composed of the diffusion capacitance of Q1. The voltage gain from base-to-collector of Q1 is quite low since the collector load of this device consists

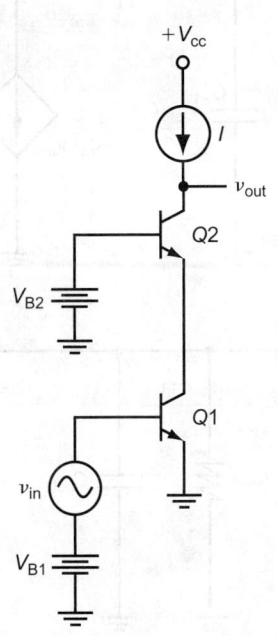

FIGURE 23.13 The cascode amplifier.

of the impedance looking into the emitter of Q2. This impedance is approximately equal to the base-emitter diode resistance of Q2 which is

$$r_{e2} = \frac{26}{I_{E2}}$$

With low voltage gain, the Miller effect of the first stage is minimized. The upper device passes the incremental signal current of Q1 to its collector and develops a large voltage across the current source impedance. There is no Miller multiplication of capacitance from the input of Q2 (emitter) to the output (collector) since the gain is noninverting and negligible capacitance exists between emitter and collector. Thus, the cascode stage essentially eliminates Miller effect capacitance and its resulting effect on upper corner frequency.

A high-frequency equivalent circuit of this stage is shown in Figure 23.14. The resistance R includes any generator resistance and the base resistance, r_{x1} of Q1. The resistance r_{cs} is the output resistance of the current source. The output capacitance is the sum of $C_{\mu2}$, C_{cs2}, and any capacitance at the current source output. The resistance r_{out2} can be quite large since Q2 sees a large emitter resistance looking into the collector of Q1. This emitter load leads to negative feedback that increases the output resistance of Q2.

The midband voltage gain is calculated from the equivalent circuit of Figure 23.14 after eliminating the capacitors. This gain is found rather easily by noting that the input current to Q1 is

$$i_{b1} = \frac{v_{in}}{R + r_{\pi1}} \tag{23.30}$$

This current will be multiplied by β_1 to become collector current in Q1. This current also equals the emitter current of Q2. The emitter current of Q2 is multiplied by α_2 become collector current of Q2. The output voltage is then

$$v_{out} = i_{c2}R_3 \tag{23.31}$$

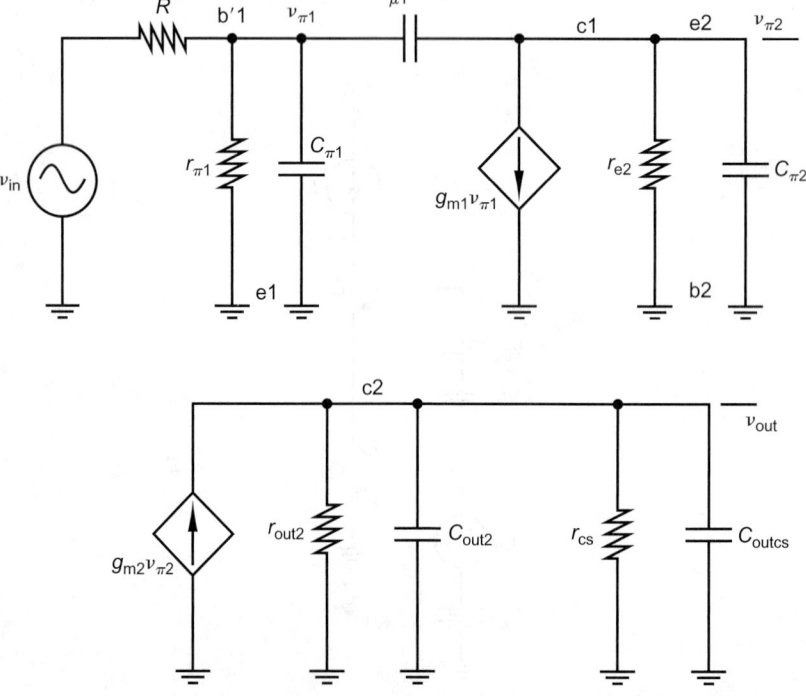

FIGURE 23.14 Equivalent circuit for the cascode amplifier.

where $R_3 = r_{out2} \parallel r_{cs}$. This resistance could be very large if the current source resistance, r_{cs}, is large. The value of r_{out2} will be high since the emitter of Q2 sees a resistance of r_{ce1}. Combining this information results in a midband voltage gain of

$$A_{MB} = \frac{-\beta_1 \alpha_2 R_3}{R + r_{\pi 1}} \tag{23.32}$$

If no generator resistance is present and if $R_3 = 100 \text{ k}\Omega$, this midband voltage gain might exceed 5000 V/V. The gain as a function of frequency can be found as

$$A = A_{MB} \frac{1}{[1 + j\omega C_{\pi 1}(r_{\pi 1} \parallel R)]} \frac{1}{[1 + j\omega r_{e2} C_{\pi 2}]} \frac{1}{[1 + j\omega R_3 C_{out}]} \tag{23.33}$$

Typically, the corner frequency of the second frequency term in Eq. (23.33), that is,

$$f_2 = \frac{1}{2\pi r_{e2} C_{\pi 2}}$$

is much higher than that of the first term,

$$f_{in-high} = \frac{1}{2\pi (r_{\pi 1} \parallel R) C_{\pi 1}}$$

This is especially true if R is large compared with $r_{\pi 1}$. In this case, since $r_{e1} \approx r_{e2}$ as a result of equal emitter currents, then $r_{e2} \ll (\beta + 1) r_{e1}$. For hand analysis of the cascode circuit, the second term in the expression for gain is often neglected.

The gain can then be written as

$$A = A_{MB} \frac{1}{(1 + j(f/f_{in-high})} \frac{1}{1 + j (f/f_{out-high})} \tag{23.34}$$

where

$$f_{out-high} = \frac{1}{2\pi C_{out} R_3} \tag{23.35}$$

and $f_{in-high}$ was defined previously.

The capacitance C_{out} is the sum of the current source output capacitance and the output capacitance of Q2 giving

$$C_{out} = C_{out2} + C_{outcs}$$

If a current mirror with output stage Q3 generates the collector bias current for Q1 and Q2, the output capacitance is

$$C_{out} = C_{\mu 2} + C_{\mu 3} + C_{cs2} + C_{cs3}$$

23.5.2 The Differential Stage

The differential pair or differential stage is very important in the electronic field. Virtually every op amp chip includes a differential pair as the input stage. Some advantages of the differential stage are its relative

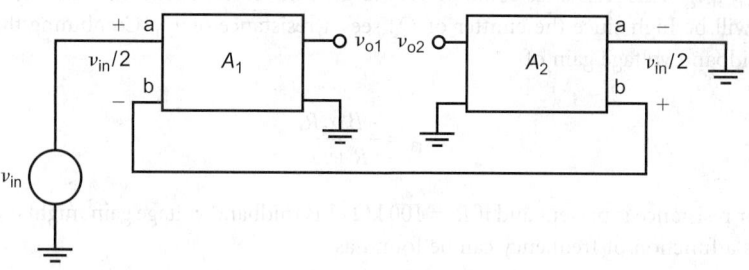

FIGURE 23.15   A differential pair.

immunity to temperature effects and power supply voltage changes, and its ability to amplify DC signals. The differential amplifier uses a pair of identical stages connected in a configuration that allows the temperature drift of one stage to cancel that of the other stage. The basic configuration of a differential stage is shown in Figure 23.15. The two devices can be connected and operated in several different configurations. The mode of operation most often used with the differential amplifier is the *differential input–double-ended output* mode. Differential input refers to a situation wherein the voltage appearing across the input of one stage is equal in magnitude, but opposite in polarity to the voltage appearing across the input of the other stage. Double-ended output refers to the fact that the output voltage is taken as the difference in voltage between the output voltage of each stage. In single-stage amplifiers, the output voltage appears between the circuit output and ground. This is called a *single-ended* output. In Figure 23.15, the double-ended output voltage is

$$V_{\text{out}} = V_{o1} - V_{o2}$$

or

$$V_{\text{out}} = V_{o2} - V_{o1}$$

depending on the choice of output terminals. If this differential pair used a single-ended output, it could be taken between the output of stage 1 and ground or the output of stage 2 and ground.

A second input mode that could be used is the *common mode*. If the same signal is applied to both inputs, the circuit is said to operate in common mode. If the amplifier stages have exactly equal gains, the signals V_{o1} and V_{o2} will be equal. The double-ended output signal would then be zero in the ideal differential stage operating in common mode. In practice the two amplifier gains will not be identical; thus, the common mode output signal will have a small value. The common-mode gain A_{CM} is defined as

$$A_{\text{CM}} = \frac{V_{o1} - V_{o2}}{V_{\text{CM}}} = A_1 - A_2 \tag{23.36}$$

where V_{CM} is the common-mode voltage applied to both inputs and A_1 and A_2 are the voltage gains of the two stages.

The differential gain applies when the input signals v_1 and v_2 are not equal and the gain in this case is

$$A_{\text{D}} = \frac{V_{o1} - V_{o2}}{v_1 - v_2} = \frac{A_1 v_1 - A_2 v_2}{v_1 - v_2} = \frac{V_{\text{out}}}{v_1 - v_2} \tag{23.37}$$

where A_{D} is the differential voltage gain of the amplifier with a double-ended output. If the double-ended output had been defined as $v_{o2} - v_{o1}$ rather than $v_{o1} - v_{o2}$, then only the algebraic sign of A_{D} would change.

In the general case, both common-mode and differential signals will be applied to the amplifier. This situation arises, for example, when the differential inputs are driven by preceding stages that have an output consisting of a DC bias voltage and an incremental signal voltage.

Using superposition, the output voltage for this case can be calculated from

$$v_{\text{out}} = A_D(v_1 - v_2) + A_{CM}v_{CM} \tag{23.38}$$

The common-mode input voltage can be found from

$$v_{CM} = \frac{v_1 + v_2}{2} \tag{23.39}$$

In the ideal situation with perfectly symmetrical stages, the common-mode input would lead to zero output, thus a measure of the asymmetry of the differential pair is the common-mode rejection ratio defined as

$$\text{CMRR} = 20\log\frac{|A_D|}{|A_{CM}|}\ \text{dB} \tag{23.40}$$

If $|A_D| = 100$ V/V and $|A_{CM}| = 0.01$ V/V, this value would be

$$\text{CMRR} = 20\log\frac{100}{0.01} = 80\ \text{dB}$$

The larger the CMRR, the smaller is the effect of A_{CM} on the output voltage compared with A_D.

A big advantage of the differential stage is in the cancellation of drift at the double-ended output. Temperature drifts in each stage are often common-mode signals. For example, the change in forward voltage across the base–emitter junction with constant current is about −2 mV/°C. As the temperature changes, each junction voltage changes by the same amount. These changes can be represented as equal voltage signals applied to the two inputs. If the stages are closely matched, very little output drift will be noted. The integrated differential amplifier can perform considerably better than its discrete counterpart since component matching is more accurate and a relatively uniform temperature prevails throughout the chip. Power supply noise is also a common-mode signal and has little effect on the output signal if the common-mode gain is low.

As previously mentioned, the mode of operation most often used with the differential amplifier is the differential input, double-ended output mode. In this configuration, the input voltage applied to one stage should be equal in magnitude, but opposite in polarity to the voltage applied to the other stage. One method of obtaining these equal magnitude, opposite polarity signals using only a single input source is shown in Figure 23.16.

If the input resistances of both stages are equal, half of v_{in} will drop across each stage. While the voltage from terminal a to terminal b of stage 1 represents a voltage drop, the voltage across the corresponding terminals of stage 2 represents a rise in voltage. We can write

$$v_1 = \frac{v_{\text{in}}}{2}$$

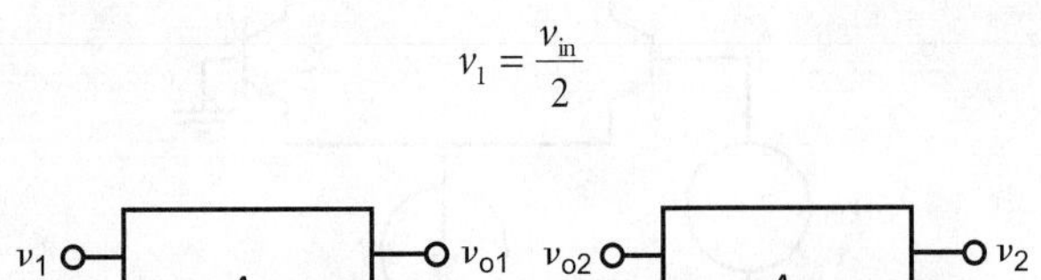

FIGURE 23.16 A differential input obtained from a single input source.

and

$$v_2 = -\frac{v_{in}}{2}$$

Each stage has the same magnitude of input voltage, but is opposite in polarity to that of the other stage. From Eq. (23.37) we can write the differential gain as

$$A_D = \frac{v_{out}}{v_{in}} = \frac{v_{o2} - v_{o1}}{v_1 - v_2} = \frac{v_{o2} - v_{o1}}{v_{in}/2 - (-v_{in}/2)}$$

$$= \frac{A_1 \dfrac{v_{in}}{2} + A_2 \dfrac{v_{in}}{2}}{\dfrac{v_{in}}{2} + \dfrac{v_{in}}{2}} = \frac{A_1 + A_2}{2} \tag{23.41}$$

23.5.3   The BJT Differential Pair

A BJT differential pair is shown in Figure 23.17.

23.5.3.1   Small-Signal Voltage Gain

Figure 23.18 represents a simple equivalent circuit for the BJT differential pair. The steps in calculating the single-ended or double-ended voltage gain of a differential pair are

1. Calculate the input current as $i_{in} = v_{in}/R_{in}$.
2. Note that the base currents are related to i_{in} by $i_{b1} = i_{in}$ and $i_{b2} = -i_{in}$.
3. Calculate collector currents as $i_{c1} = \beta_1 i_{b1}$ and $i_{c2} = \beta_2 i_{b2}$.
4. Calculate collector voltages from $v_{o1} = -i_c R_{Ceq1}$ and $v_{o2} = -i_2 R_{Ceq2}$, where

$$R_{Ceq} = R_C \| r_{out} \tag{23.42}$$

With these values, the single-ended or double-ended voltage gain can be found. In the normal situation, we assume perfectly symmetrical pairs with $\beta_1 = \beta_2 = \beta$ and $R_{C1} = R_{C2} = R_C$. We also assume that the bias current I splits equally between the two stages giving $I_{E1} = I_{E2} = I/2$.

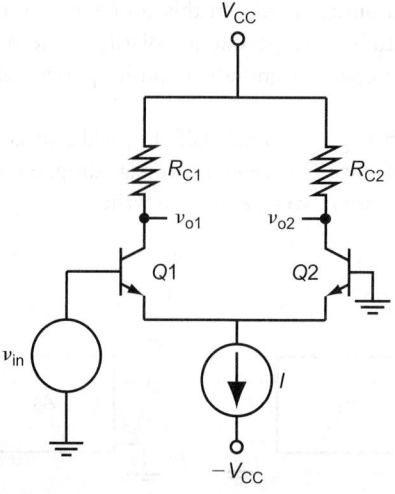

FIGURE 23.17   The BJT differential pair.

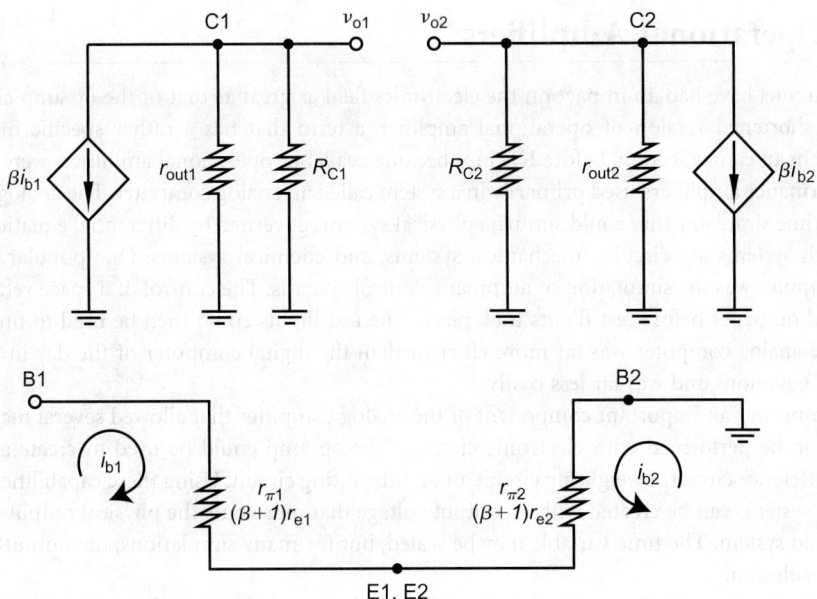

FIGURE 23.18 Equivalent circuit of the differential pair.

The single-ended gain of the first stage of Figure 23.17 is found to be

$$A_{S1} = \frac{v_{o1}}{v_{in}} = \frac{-\beta R_{Ceq}}{2(\beta+1)r_e} \tag{23.43}$$

The single-ended gain of the second stage equals this value in magnitude, but shows no phase inversion and therefore, has a positive algebraic sign.

The double-ended differential gain is

$$A_D = \frac{v_{o2} - v_{o1}}{v_{in}} = \frac{(\beta v_{in}/2(\beta+1)r_e)R_{Ceq} - ((-\beta v_{in}/2(\beta+1)r_e)R_{Ceq})}{v_{in}} = \frac{\beta R_{Ceq}}{(\beta+1)r_e} \tag{23.44}$$

This differential double-ended voltage gain is equal in magnitude to the voltage gain of a single transistor amplifier with a load of R_C and no external emitter resistance. The advantages of decreased temperature drift and good power supply rejection in critical applications, and a DC input reference of 0 V are typically far more important than the necessity of using an extra device, especially for IC chips.

The previous calculations for the differential pair assume there is no load across the output terminals. If this circuit drives a succeeding stage, the input impedance to this following stage will load the differential pair. One approach to this problem results from noting that when the collector of one BJT is driven positive by the input signal, the other collector moves an equal voltage in the opposite direction. Each end of the load resistor, R_L, is driven in equal but opposite directions. The midpoint of the resistor is always at 0 V, for incremental signals. To calculate the loaded voltage gain, a resistance of $R_L/2$ can be placed in parallel with each collector resistance to give an equivalent collector resistance of

$$R_{Ceq} = R_C \| R_L/2 \tag{23.45}$$

Eq. (23.44) can now be used to calculate the loaded differential gain.

23.6 Operational Amplifiers

Very few circuits have had an impact on the electronics field as great as that of the op amp circuit. This name is a shortened version of operational amplifier, a term that has a rather specific meaning. As mentioned in an earlier section, before IC chips became available, operational amplifiers were expensive, high-performance amplifiers used primarily in a system called an analog computer. The analog computer was a real-time simulator that could simulate physical systems governed by differential equations. Examples of such systems are circuits, mechanical systems, and chemical systems. One popular use of the analog computer was the simulation of automatic control systems. The control of a space vehicle had to be designed on paper before test flights took place. The test flights could then be used to fine-tune the design. The analog computer was far more efficient than the digital computer of the day in simulating differential equations and was far less costly.

The op amp was an important component of the analog computer that allowed several mathematical operations to be performed with electronic circuits. The op amp could be used to create a summing circuit, a difference circuit, a weighting circuit, or an integrating circuit. Using these capabilities, complex differential systems can be created with an output voltage that represents the physical output variable of the simulated system. The time variable may be scaled, but for many simulations, the output represents a real-time solution.

One of the key features of an op amp is the differential input to the amplifier. This allows differences to be formed, and also allows the creation of a virtual ground or virtual short across the input terminals. This virtual short is used in summing several current signals into a node without affecting the other input current signals. These signals are then summed and easily converted into an output voltage.

The virtual ground also allows the formation of rather accurate integrals of the op amp using an additional resistor and capacitor. This feature is essential in the simulation of differential equations.

Other amplifiers are sometimes mistakenly referred to as op amps, but unless these amplifiers possess the capability to create a virtual ground and do mathematical operations, this is a misnomer.

The first IC op amp was introduced by Fairchild Semiconductor Corporation in 1964. This chip, designated as μA702 or 702, was followed shortly by 709 which was the first analog IC product to receive wide industry acceptance. The National LM101 and Fairchild 741 advanced the field and eliminated the external compensation capacitors required for the earlier models. The ensuing popularity and low price of the 741 allowed op amps to be treated as a component rather than a subcircuit that must be designed.

Today, op amps are available in bipolar (BJT), complementary metal oxide semiconductor (CMOS), and BiCMOS (BJT/CMOS) technologies and designers have the option of including dozens of op amps and other circuits on a single chip. Modern op amps generally use the same architecture developed in the LM101/741 circuits with performance improvements resulting from improved processing techniques.

The configuration of an op amp offers one important advantage for IC amplifiers. This advantage is the op amp's ability to allow the input signal to be referenced to any DC voltage, including ground, within the allowed input range. This eliminates the need for a large coupling capacitor to isolate a DC input transducer from an amplifier input.

23.6.1 A Classical IC Op Amp Architecture

While the technology used to implement op amps has changed considerably over the years, the basic architecture has remained remarkably constant [4]. This section will first discuss that architecture and some approaches to its implementation. After this topic is considered, the subject will turn to methods of specifying amplifier performance.

The architecture of many op amps appears as shown in Figure 23.19. The first section is a differential amplifier required by all op amps to allow a virtual ground and the implementation of mathematical operations. This stage is generally designed to have a very high differential voltage gain, perhaps a few thousand. The bandwidth is generally rather low as a result of the high voltage gain. In some cases, this

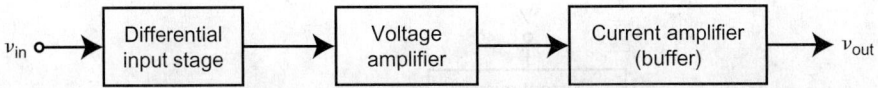

FIGURE 23.19 Classical op amp architecture.

differential amplifier will use an active load that will also convert the double-ended output of the differential stage into a single-ended output that will drive the second voltage amplifier.

The second stage of Figure 23.19 is a single-ended voltage amplifier with a relatively high voltage gain. This gain may reach values of 500 V/V. Often this stage is used to compensate the op amp, a topic that will be discussed later in this chapter. For now it is sufficient to say that this stage will probably be an inverting stage that can multiply the apparent value of some capacitance placed between the input and output of the stage. The large capacitive load presented to the output of the first stage due to the Miller effect will lead to a very low upper corner frequency, perhaps 10–100 Hz, for the first stage.

The last stage is the output stage. It may be nothing more than an emitter follower that has a large current amplification along with a voltage gain that is near unity. This stage will have a very high upper corner frequency.

23.6.2 A High-Gain Differential Stage

The normal way to achieve high voltage gains is to use an active load for the amplifying stages. The incremental resistance presented to the amplifying stage is very high, while the DC voltage across the active load is small. One popular choice for an active differential stage load is the current mirror. This load provides very high voltage gain and also converts the double-ended output signal into a single-ended signal referenced to ground. Figure 23.20 shows a block diagram of such an arrangement.

With no input signal applied to the differential stage, the tail current splits equally between I_{diff1} and I_{diff2}. The input current to the mirror equals this value of $I_{tail}/2$. This value is also mirrored to the output of the mirror giving $I_{out} = I_{tail}/2$. We will assume that the voltage between the current mirror output and the second differential stage is approximately zero, although this assumption is unnecessary to achieve the correct result.

When a signal is applied to the differential input, it may increase the current I_{diff1} by a peak value of ΔI. The input current to the mirror now becomes

$$I_{in} = \frac{I_{tail}}{2} + \Delta I$$

The output current from the mirror also equals this value. However, the input signal to the differential stage will decrease I_{diff2} by the same amount that I_{diff1} increases. Thus, we can write

$$I_{diff2} = \frac{I_{tail}}{2} - \Delta I$$

The current to the resistance R increases from its quiescent value of zero to

$$I_R = I_{out} - I_{diff2} = 2\Delta I$$

The incremental output voltage resulting is then

$$v_{out} = 2R\Delta I \tag{23.46}$$

When an incremental input signal is applied to the differential pair, half of this voltage will drop across each base–emitter junction of the pair. This results in equal incremental differential stage currents in the

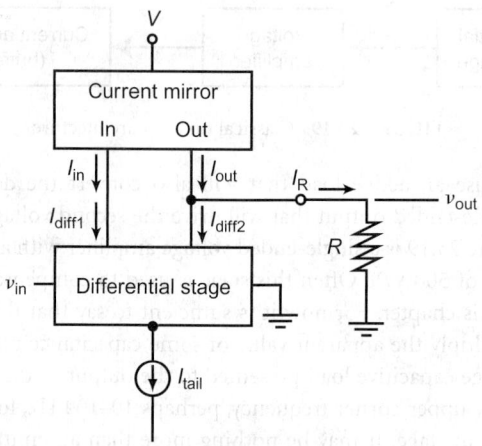

FIGURE 23.20 Differential stage with current mirror load.

two output devices, but they will be in opposite directions. An incremental input signal, v_{in}, will produce incremental currents of

$$i_{diff1} = \Delta I = \frac{g_m v_{in}}{2} \tag{23.47}$$

and

$$i_{diff2} = -\Delta I = -\frac{g_m v_{in}}{2} \tag{23.48}$$

where g_m is the transconductance of devices 1 and 2. Assuming negligibly large output resistances of the current mirror and the differential stage, the incremental output voltage becomes

$$v_{out} = 2i_{diff1}R = g_m R v_{in}$$

with a resulting midband voltage gain of

$$A_{MB} = g_m R \tag{23.49}$$

If the output resistances of the mirror and differential stage are significant, the voltage gain can be found by combining these resistances in parallel with the load resistance to form $R_{eff} = R_{out} \parallel R$. This resistance then replaces R in Eq. (23.49). The load resistance may, in fact, be the incremental input resistance of the following stage. Very large values of voltage gain can result from this configuration. For the BJT, the transconductance is given by $g_m = \alpha/r_e \approx 1/r_e$.

While this expression for voltage gain is the same as that for the differential gain of a resistive load stage, given by Eq. (23.44), two significant points should be made. First of all, the impedance R can be much greater than any resistive load that can be used in a differential stage. Large values of R in the differential stage would cause saturation of the stages for reasonable values of tail currents. In addition, large values of R are more difficult to fabricate on an IC chip. The current mirror solves this problem. The second point is that the output voltage of the differential pair with a current mirror load is a single-ended output which can be applied to a following simple amplifier stage. However, the rejection of common-mode variables caused by temperature change or power supply voltage changes is still in effect with the current mirror stage. If a resistive load differential stage must provide a single-ended output, the gain drops by a factor of 2, compared with the double-ended output, and common-mode rejection no longer takes place.

23.6.3 The Second Amplifier Stage

Before the purpose of the second amplifier stage can be fully understood, a discussion on circuit stability must take place.

23.6.3.1 Feedback and Stability of the Op Amp

The op amp is used in a feedback configuration for essentially all amplifying applications. Because of nonideal or parasitic effects, it is possible for the feedback amplifier to exhibit unstable behavior. Oscillations at the output of the amplifier can exist, having no relationship to the applied input signal. This, of course, negates the desired linear operation of the amplifier. It is necessary to eliminate any undesired oscillation signal from the amplifier.

The open-loop voltage gain of a three-stage amplifier such as that of Figure 23.19 may be represented by a gain function of

$$A = \frac{A_{MB}}{\left(1 + j\frac{\omega}{P_1}\right)\left(1 + j\frac{\omega}{P_2}\right)\left(1 + j\frac{\omega}{P_3}\right)} \tag{23.50}$$

where P_1, P_2, and P_3 are the dominant poles of gain stages 1, 2, and 3. The quantity A_{MB} is the low-frequency or midband gain of the amplifier.

When this op amp is used in a negative feedback configuration such as that shown in Figure 23.21, the loop gain must be analyzed to see if the conditions for oscillation occur. The feedback factor for this circuit is

$$F = \frac{R_2}{R_2 + R_F} \tag{23.51}$$

To check the stability of this circuit, a zero volt input signal (short circuit) can be applied to the noninverting input terminal. The loop gain from inverting terminal to output and back to inverting terminal can then be found with the noninverting terminal shorted to ground. When this is done, it is found [4] that the amplifier is unstable for most practical values of values of R_F and R_2. The worst case condition occurs as R_2 approaches infinity and R_F approaches zero as in the case of a unity-gain voltage buffer.

One measurement of stability is referred to as the *phase margin*. This quantity is defined in terms of the phase shift of AF that exists when the magnitude of AF has dropped to unity. The number of degrees

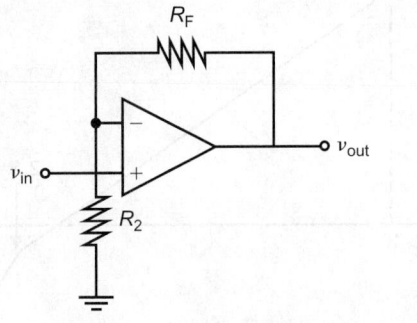

FIGURE 23.21 A noninverting op amp stage.

less than −180° at this frequency is called the phase margin. Most amplifiers target a phase margin of 45° or more.

One possibility to achieve a reasonable phase margin is to intentionally modify one of the three upper corner frequencies of the op amp. For example, if the pole associated with the first stage is lowered by several factors, the amplifier can become stable, even when used as a unity-gain voltage buffer.

The frequency response of this gain is sketched in Figure 23.22 for a midband gain of 300,000. We note that the magnitude of this gain has fallen below a value of unity (0 dB) before the second upper corner frequency of 2×10^6 is reached. The phase shift of AF at the frequency where the magnitude has dropped to unity may be −130° resulting in a phase margin of 50°. Normally, engineers do not destroy bandwidth of a stage intentionally, however in this case it is necessary to stabilize the operation of the op amp in the feedback configuration.

Lowering the upper corner frequency of one stage is not a trivial matter. While it may not need to be reduced to 10 rad/s in a practical op amp, it is often lowered to 10–100 Hz. A capacitor must be added to the appropriate point in the stage to drop this frequency, but a relatively large capacitor is required. In the early days of the IC op amp, the two terminals between which a capacitor was to be added were connected to two external pins of the chip. A discrete capacitor of sufficient value was then added externally.

In 1967, the capacitor was added to the IC chip using the Miller effect to multiply the capacitor value. Returning to Figure 23.19, it is seen that a capacitor can be added between input and output of the second stage. This is typically a capacitor of value 30 pF that, due to the Miller effect, is multiplied by a factor of $(A_2 + 1)$ where $-A_2$ is the gain of the second stage. The Miller capacitance is reflected to the input of the second stage that loads the output of the first stage. This large effective capacitance, driven by the large output impedance of the first stage produces a very low upper corner frequency for the first stage.

In a 741 op amp design, the gain A_2 is approximately −400 V/V. With a 30 pF capacitance bridging input to output, the effective capacitance at the input is ~0.012 μF. This creates a bandwidth for the op amp of 10 Hz or 62.8 rad/s.

This method of solving the instability problem is referred to as *dominant pole compensation*. The lower value of pole frequency is seen from Figure 23.22 to dominate the amplifier performance up to frequencies above the useful range of gain.

It should be mentioned that op amps for specific applications may not need to be stabilized for the unity gain configuration. The minimum gain required by the amplifier may be 2 or 3 V/V or some other value that is easier to compensate. In such cases, the dominant pole need not be decreased to the level of the general purpose op amp. The resulting frequency performance of the op amp can then be higher than that of the general purpose stage.

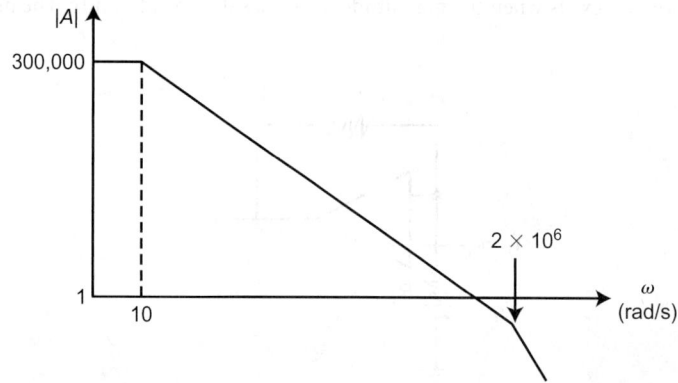

FIGURE 23.22 Op amp frequency response.

23.6.4 Op Amp Specifications

There are several nonideal effects in the op amp that detract from its overall performance. Mismatching of devices in the input differential pair, required bias currents, and junction voltage temperature effects can degrade the performance. These effects are described in terms of the following specifications. The diagram of Figure 23.23 is used to define various terms used in these definitions.

Input offset voltage (V_{OS}) Mismatch of the transistors in the differential input stage leads to a finite output DC voltage when both inputs are shorted to ground. This finite output voltage is called the *output offset voltage*. A slight voltage mismatch in the differential pair is amplified by succeeding stages to create a larger voltage at the output. Inaccurate biasing of later stages also contribute to the output offset. Inaccuracies in later stages are amplified by smaller factors than are early stage inaccuracies.

The *input offset voltage* is the voltage that must be applied across the differential input terminals to cause the output voltage of the op amp to equal zero. Theoretically, this voltage could be found by measuring the output voltage when the input terminals are shorted, then dividing this value by the gain of the op amp. In practice, this may not be possible as the gain may not be known or the output offset may exceed the size of the active region. The value of V_{OS} is typically a few millivolts for monolithic or IC op amps.

Input offset voltage drift (TCV_{OS}): Temperature changes affect certain parameters of the transistors, leading to a drift in the output DC offset voltage with temperature. In BJT devices, the voltage across the base-to-emitter junction for a constant emitter current drops by ~2 mV/°C. Small drifts of voltage in early stages will be amplified by following stages to produce relatively large drifts in output voltage. Because the output DC signal may not exist within the active region of the op amp, the drift is again referred to the input.

The *input offset voltage drift* is defined as the change in V_{OS} for a 1°C change in temperature (near room temperature). A value of 10–20 μV/°C is typical for IC op amps.

Input bias current (I_B). A BJT differential stage will require a finite amount of base current for biasing purposes. This is true even if both inputs are grounded. The *input bias current* of an op amp is defined as the average value of bias current into each input with the output driven to zero. The two bias currents are generally slightly different so I_B is

$$I_B = \frac{(I_{B1} + I_{B2})}{2} \tag{23.52}$$

Input offset current (I_{OS}). The *input offset current* is the difference between the two input bias currents when the output is at 0 V. This parameter is

$$I_{OS} = |I_{B1} - I_{B2}| \tag{23.53}$$

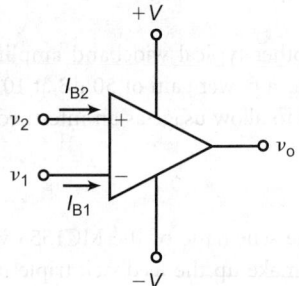

FIGURE 23.23 An op amp.

Common-mode input voltage range (CMVR). The voltage range over which the inputs can be simultaneously driven without causing deterioration of op amp performance is called the *common-mode voltage range* or CMVR. In most op amps, the CMVR is a few volts less than the rail-to-rail value of the power supplies. In many applications, both inputs are forced to move together due to the virtual short between the input terminals when negative feedback is used.

Common-mode rejection ratio (CMRR). The ratio of input common-mode voltage to change in input offset voltage is called the *common-mode rejection range* or CMRR. An equivalent definition is the ratio of differential voltage gain to common-mode voltage gain. IC op amps range from 80 to 100 dB for the CMRR. This parameter was mentioned earlier in the chapter and is a measure of the mismatch of incremental gain from each of the two inputs to output. If the incremental gains from each input to output were equal, the CMRR would be infinite.

Power supply rejection ratio (PSRR). The *power supply rejection ratio* or PSRR is the ratio of change in the input offset voltage to a unit change in one of the power supply voltages. An op amp with two power supplies requires that a PSRR be specified for each power supply.

23.7 Wideband Amplifiers

The development of wideband amplifiers with discrete circuits progressed rapidly during World War II with vacuum tubes and into the 1960s with BJT circuits. Although many important techniques were perfected between 1940 and 1960, these methods are not generally applicable to wideband IC design. Many of these approaches to wideband amplifier design required relatively large inductors for shunt peaking and relatively large coupling capacitors to AC couple individual stages. While small inductors are now available on IC chips, the range of values limit their usage. Likewise, the limitation on IC capacitor sizes preclude the use of coupling capacitors for IC amplifiers. These difficulties are mitigated in IC design by using additional transistors, however, the benefit of past theoretical developments is then largely unused. Furthermore, the absence of coupling capacitors in IC amplifiers leads to a much stronger interaction between DC and AC design of each stage than is present in discrete design. Two of the more popular methods of IC wideband amplifier design that overcome the limitations on IC circuits are based on composite stages or feedback cascades [7,8].

23.7.1 Composite or Compound Stages

The wideband amplifier of Figure 23.24 is a classic architecture on which many present amplifiers are based. This schematic is for the RCA CA3040 [8]. The input is a buffered, differential, cascode pair. The transistors T_1 and T_4 are emitter follower stages, while the pair of devices T_2 and T_3 forms a cascode stage as also does the pair of devices T_5 and T_6. Transistor T_9 with its emitter degeneration resistance forms a high output impedance current source. Transistors T_7 and T_8 buffer the output signals to allow relatively high output currents.

Since the input is differential, the input signal can be referenced to ground eliminating the need for a coupling capacitor. The frequency response of the gain extends from DC up to ~55 MHz with a constant gain of 30 dB.

The Motorola MC1490 [8] is another typical wideband amplifier chip that can be used for radio frequency or audio applications. It has a power gain of 50 dB at 10 MHz and 35 dB at 100 MHz. It also has a built-in automatic gain control to allow usage as an intermediate frequency amplifier.

23.7.2 Feedback Cascades

The circuit of Figure 23.25 shows the schematic of the MC1553 wideband amplifier using a feedback triple [7]. Transistors T_1, T_2, and T_3 make up the feedback triple using resistor R_F to create a feedback path. This path establishes the incremental voltage gain and also provides DC feedback to keep T_1, T_2, and T_3 in their active regions. There is a second feedback path from the current mirror output transistor

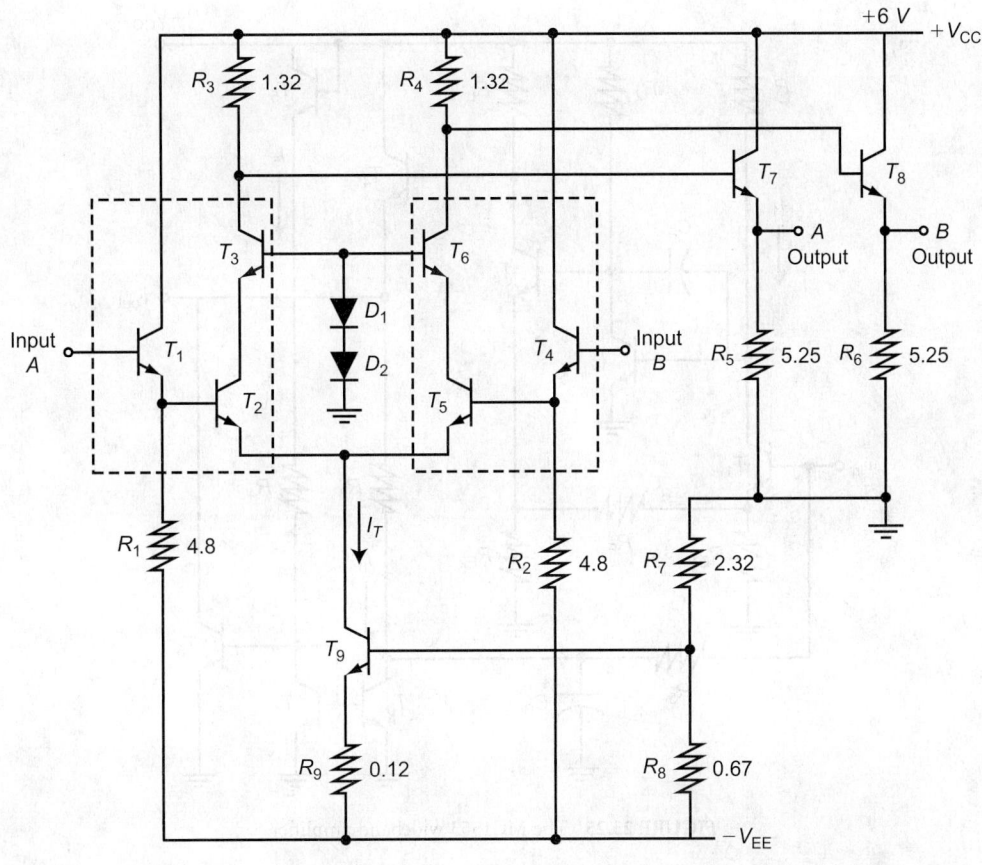

FIGURE 23.24 The RCA CA3040 wideband amplifier.

T_6, but this is primarily DC feedback. The capacitor C_B is a large, external bypass capacitor that must be added to the circuit to decouple the AC feedback to the input stage. This IC amplifier has a voltage gain of 50 V/V with a bandwidth of 50 MHz.

Although the capacitor C_B is required in some applications, if the amplifier is driven by a low impedance source, it is unnecessary to add this element.

In recent years, both wide- and narrowband amplifiers in the 1–6 GHz range have become more significant. Many companies that provide amplifiers in this frequency range use HBT designs with SiGe, GaAs, GaN, or other material to achieve the necessary frequency performance.

23.8 IC Power Output Stages

If an IC power amplifier is to occupy a relatively small volume, the output power will be limited. This is due to the limitation imposed on the thermal conductivity of a small-area power device. A discrete transistor mounted on a large heat sink will exhibit a much higher thermal conductivity than that of the smaller IC chip. This limits the power that can be dissipated by the output device or devices of the IC as the junction temperature will be higher with the lower thermal conductivity of the IC chip. Typically, this limitation leads to an IC power output stage [3] that is implemented in a high-efficiency configuration. Generally, the output stage for power outputs in the range of a few watts will use a class-B configuration, while those with tens or hundreds of watts will use a class-D configuration.

One of the most significant limitations on dissipation is the junction temperature. As the temperature rises, several potentially dangerous effects may occur. First, the solder or alloys used in the transistor can

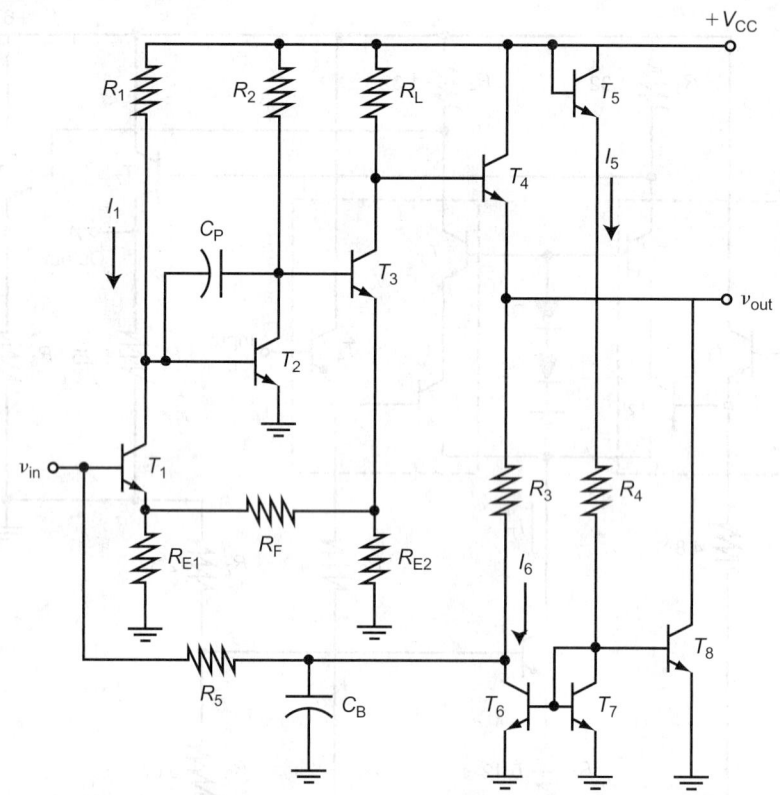

FIGURE 23.25   The MC1553 wideband amplifier.

be softened or even melted. Second, the impurity profiles in the doped regions can be affected if elevated temperatures exist for long periods of time. A third result of higher temperatures is the increase in collector leakage current. In power transistors, this current doubles for an incremental increase in temperature of 7–10°C. The leakage current, I_{co}, at temperature T_2 can be related to I_{co} at temperature T_1 by

$$I_{co}(T_2) = I_{co}(T_1)\, 2^{(T_2 - T_1)/T_K}$$

where T_K can range from 7 to 10°C. For a temperature increase of 100°C, the minimum factor of increase in I_{co} is 1024 ($2^{100/10}$), whereas the maximum factor is ~20,000($2^{100/7}$). This marked increase in I_{co} can lead to increased power dissipation which leads to an increased temperature followed by a further increase in I_{co}. In some cases, this feedback effect is large enough to cause thermal runaway and destroy the transistor. This effect is minimized by placing a resistance in the emitter to decrease the forward bias on the base–emitter junction as current increases.

In other situations, the leakage current can approach the value of the quiescent collector current. Since leakage current is not controlled by the applied input signal, its effects can severely limit the amplifying properties of the stage. For these reasons, the manufacturer places a limit on the maximum allowable junction temperature of the device.

The maximum junction temperature is therefore an important quantity that limits the power a transistor can deliver. The junction temperature will be determined by the power being dissipated by the transistor, the thermal conductivity of the transistor case, and the heat sink that is being used. The collector junction is the point at which most power is dissipated; hence, it is this junction that concerns us here.

Basically, manufacturers specify the allowable dissipation of a transistor in two ways. One way is to specify the maximum junction temperature along with the thermal resistance between the collector junction and the exterior of the case. Although this method is very straightforward, it incorrectly implies that the allowable power increases indefinitely as the transistor is cooled to lower temperatures. Actually, there is a maximum limit on the allowable dissipation of the transistor which is reflected by the second method of specification. This method shows a plot of allowable power dissipation versus the temperature of the mounting base. Quite often this plot shows power dissipation versus ambient temperature, where an infinite heat sink is used. However, if the transistor could be mounted on an infinite heat sink, the ambient temperature would equal the mounting base temperature; thus both plots convey the same information. This second method indicates the maximum allowable power dissipation, in addition to the maximum junction temperature.

The maximum limit on power dissipation ensures that chip temperature differentials do not become excessive as excess power is dissipated in collector regions. It also minimizes the possibility of excessive collector currents in typical applications.

23.8.1   Thermal Resistance

The thermal resistance of a material is defined as the ratio of the temperature difference across the material to the power flow through the material. This assumes that the temperature gradient is linear throughout. The symbol θ is used for thermal resistance, and

$$\theta = \frac{\Delta T}{P} = \frac{\text{Temperature difference across conductor}}{\text{Power flowing through conductor}} \tag{23.54}$$

The diagram of Figure 23.26 represents the thermal circuit of an IC chip, including the output transistor, surrounded by free air. Here, θ_{JM} is the thermal resistance from collector to mounting base and θ_{A} the thermal resistance of that portion of air in contact with the mounting base; T_{A} the temperature of air far away from the transistor; T_{M} the mounting base temperature; and T_{J} the collector junction temperature. The power P will be determined by the electrical circuit that includes the output transistor; P, in turn, will determine the temperatures T_{J} and T_{M}. The temperature T_{J} can be written as

$$T_{\text{J}} = T_{\text{A}} + P(\theta_{\text{JM}} + \theta_{\text{A}}) \tag{23.55}$$

This equation shows that as more power is dissipated by the output transistor, T_{J} must rise.

For high-power IC chips mounted in a TO-3 package, θ_{A} is usually many times greater than θ_{JM}. If θ_{JM} were 1°C/W, then θ_{A} might be 5–20°C/W. Of course, θ_{A} depends on the area of the IC package.

The diagram of Figure 23.27 represents the thermal circuit of the IC chip mounted on a heat sink. The power that can be dissipated without exceeding the maximum junction temperature is found from solving the preceding equation to result in

$$P_{\text{max}} = \frac{T_{\text{J}} - T_{\text{A}}}{\theta_{\text{JM}} + \theta_{\text{HS}}} \tag{23.56}$$

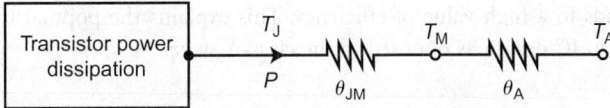

FIGURE 23.26   Transistor dissipating power in free air.

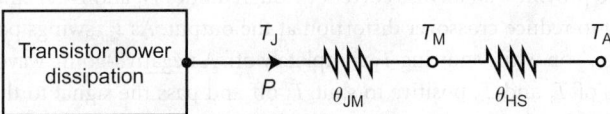

FIGURE 23.27   Transistor mounted on heat sink.

The values of θ_{HS} presented between the chip case and free air might range from 0.4°C/W for air-cooled systems to 2°C/W for flat vertical-finned aluminum heat sinks to 8°C/W for cylindrical heat sinks that slide over the chip package. For each thermal circuit, the amount of allowable power dissipation is fixed.

23.8.2   Circuit or Conversion Efficiency

The efficiency of a power output stage is a measure of its effectiveness in converting DC power into AC output power. It is defined as

$$\eta = \frac{P_{out}}{P_S} \tag{23.57}$$

where P_{out} is the average output power and P_S the power supplied by the DC power supply.

A useful relationship between the allowable dissipation of the chip and the maximum output power can be found by assuming that the power delivered from the DC source is dissipated by the output transistor and the load. This assumption would be very inaccurate for a class-A, resistive load stage, in which the resistor dissipates significant DC power. For many class-B or class-D stages, the assumption is not unreasonable. In equation form, this assumption is expressed by

$$P_S = P_{out} + P_T \tag{23.58}$$

where P_T is the actual dissipation of the output transistor or transistors.

In terms of the circuit efficiency, the transistor dissipation can be written as

$$P_T = (1 - \eta)P_S = (1 - \eta)\frac{P_{out}}{\eta} \tag{23.59}$$

Solving for P_{out} in terms of P_T leads to

$$P_{out} = \frac{\eta}{1 - \eta} P_T \tag{23.60}$$

The effect of circuit efficiency on output power can be demonstrated by assuming that the maximum allowable output stage dissipation is 5 W. If the circuit efficiency is 50% or $\eta = 0.5$, the maximum output power calculated from Eq. (23.60) is also 5 W. If the efficiency is increased to 78.5%, the maximum efficiency of an ideal class-B stage, the maximum output power is found to be 18.26 W. Increasing the circuit efficiency to 98% a figure approached in a near-ideal class-D stage, the maximum output power becomes 245 W. In this case, an output stage that can dissipate 5 W delivers 245 W to the load. Although these calculations are based on some idealizations, it clearly shows the importance of using a circuit configuration that leads to a high value of efficiency. This explains the popularity of the class-B stage and the class-D stage in IC design as opposed to the class-A stage.

23.8.3   Class-B Output Stage

A class-B stage that can be integrated is shown in Figure 23.28. The current source I_1 consists of an IC current-source stage to provide a small bias current through diodes D_1 and D_2. A small quiescent collector current I_2 is necessary to reduce crossover distortion at the output. As v_{in} swings positive the input to T_2 goes negative, turning T_2 on while shutting T_1 completely off. A negative-going waveform at the amplifier input drives the bases of T_1 and T_2 positive to shut T_2 off and pass the signal to the output through T_1.

The amplifier can be made *short-circuit proof* by limiting the output current that can flow if the output terminal is accidentally shorted to one of the supplies. Figure 23.29 indicates the additional

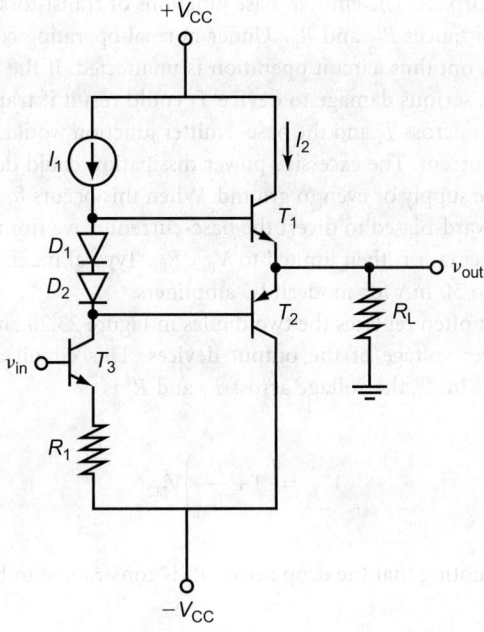

FIGURE 23.28   A class-B IC stage.

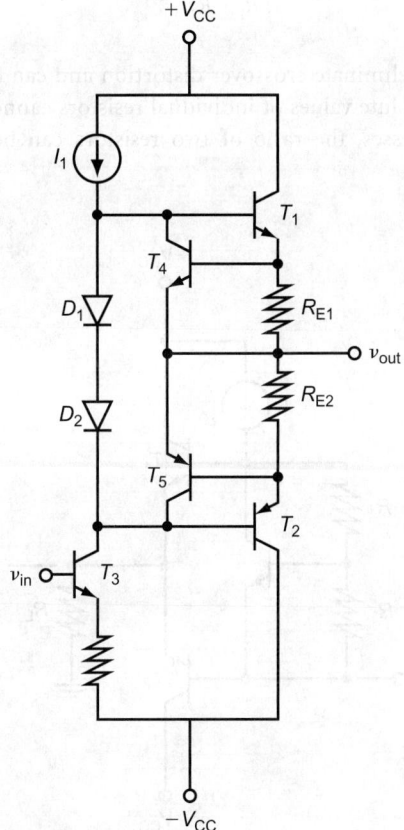

FIGURE 23.29   Class-B stage with short-circuit protection.

circuitry required for this purpose. The emitter–base junctions of transistors T_4 and T_5 are driven by the voltage drops across the resistances R_{E1} and R_{E2}. Under normal operating conditions, these voltages are too small to turn T_4 and T_5 on; thus, circuit operation is unaffected. If the output is short-circuited to the negative supply voltage, serious damage to device T_1 could result if transistor T_4 were not present. A large voltage would appear across T_1 and the base–emitter junction would be forward-biased resulting in a high value of emitter current. The excessive power dissipation could destroy T_1 if the output were to be shorted to the negative supply or even to ground. When this occurs for the circuit of Figure 23.29, T_4 becomes sufficiently forward-biased to divert the base-current drive from T_1. The maximum current that can flow in the output circuit is then limited to V_{BE4}/R_{E1}. Typical maximum currents for the short-circuit case range from 10 to 50 mA for modern IC amplifiers.

The V_{BE} multiplier circuit often replaces the two diodes in Figure 23.28 and Figure 23.29 to get better cancellation of the crossover voltage of the output devices. This circuit appears in Figure 23.30. If negligible base current flows in T_3, the voltage across R_1 and R_2 is

$$V_{CE3} = \left(1 + \frac{R_1}{R_2}\right) V_{BE3} \tag{23.61}$$

This voltage is found by noting that the drop across R_2 is constrained to be V_{BE3}, and this value must also equal

$$\frac{R_2}{R_1 + R_2} V_{CE3}$$

The voltage V_{CE3} is used to eliminate crossover distortion and can be adjusted by the ratio of the two resistors. Whereas the absolute values of individual resistors cannot be accurately determined in standard IC fabrication processes, the ratio of two resistors can be determined to the required accuracy.

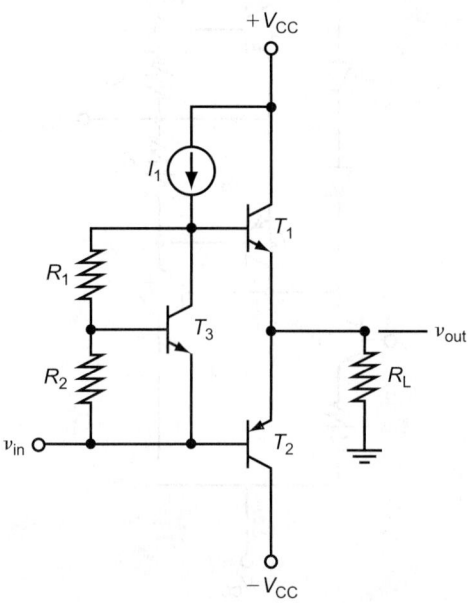

FIGURE 23.30   Bias circuit using V_{BE} multiplier.

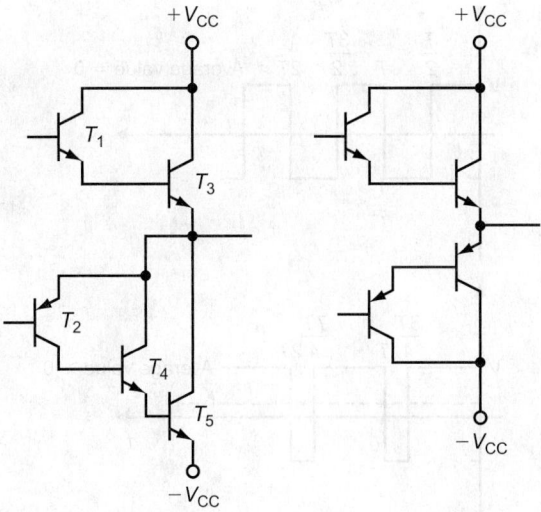

FIGURE 23.31 Modified Darlington and Darlington output stages.

There are several other output configurations based on the complementary emitter follower. A popular one is the Darlington output stage modified for IC amplifiers as shown in Figure 23.31. The current gain of this stage is very high, approximately equal to β^2. If high-gain pnp devices are available, a Darlington pair similar to the upper npn pair can replace devices T_2, T_4, and T_5. For the Darlington pairs, a larger difference of input bias voltage must be provided to the inputs of the respective pairs due to the larger voltage drop between each input and output which is now $2V_{BE(on)}$ instead of just $V_{BE(on)}$. The V_{BE} multiplier circuit can be designed to generate this increased bias voltage.

23.8.4 Class-D Output Stages

IC class-D amplifiers using PWM have been reported with efficiencies of 90% or more at 10-W output and a frequency response from 20 Hz to 20 kHz. Many IC chips are available that drive power BJTs or MOSFETs, delivering 30–50 W to a load. Larger discrete circuits report audio amplifiers based on the class-D stage that deliver 600 W per stereo channel [9]. This type of amplifier is often used in *low-end* car radios.

PWM is used to reduce the power dissipated by the transistor while delivering a large power to the load. The varying load signal is applied by means of output devices that switch between on and off states. The resulting load voltage has a rectangular waveform that contains an average or DC value dependent on the duty cycle. In addition, the load voltage would also contain several AC components, however, these unwanted components can be filtered before reaching the load.

Any periodic waveform can be represented by a Fourier series consisting of a DC component (if present), a fundamental frequency component, and higher harmonics of the fundamental frequency. A rectangular wave switching between $+V$ and $-V$ that remains positive for t^+ seconds is said to have a *duty cycle* of

$$d = \frac{t^+}{T} \tag{23.62}$$

where T is the repetition period of the waveform. A rectangular waveform with a 50% duty cycle contains no DC or average value, but a change of duty cycle will give the waveform an average value as shown in Figure 23.32. If the period, T, remains constant as the duty cycle is varied, the average value of a rectangular waveform of amplitude V is

$$V_{av} = V(2d - 1) \tag{23.63}$$

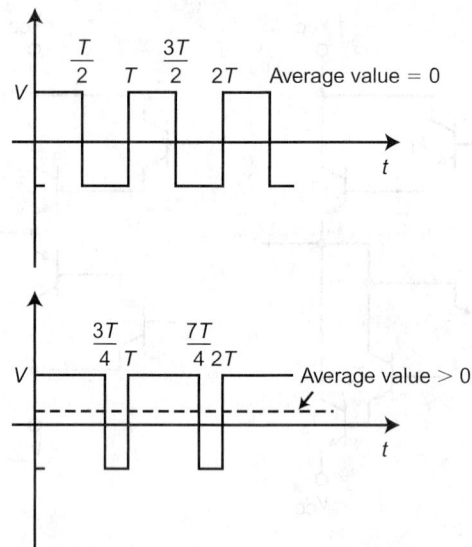

FIGURE 23.32 Rectangular waveforms with different duty cycles.

The average value varies directly with d. The Fourier coefficients of the AC components also vary as d is changed and new frequencies may be introduced, but in general, these components are of little interest to us as they can be easily eliminated. If, for example, the repetition frequency is 200 kHz, all AC components of the waveform will be greater than this value and far out of the audio range. Now if we vary the duty cycle sinusoidally at some low frequency, the average value will also vary sinusoidally. Mathematically, we can express this by saying that if $d = 0.5 + k\sin \omega t$, then the average value is

$$V_{av} = 2kV\sin \omega t \qquad (23.64)$$

The waveform with variable duty cycle can be filtered by a low-pass filter to eliminate all frequencies above ω. The result is a low-frequency output sinusoid with an amplitude that varies in the same manner as the duty cycle. A block diagram of PWM amplification is shown in Figure 23.33.

The voltage control circuit must have a 50% duty cycle when the input signal is at 0 V. As v_{in} becomes nonzero, the duty cycle varies proportionally. The high-power switching stage amplifies this rectangular wave and applies the output to a low-pass circuit that allows only the changes in average value to pass to the load. The output signal is then proportional to the input signal, but can be at a much higher power level than the input signal. The power output stage may dissipate only 5–10% of the power delivered to the load.

The major advantage of PWM is that the output transistors need to operate in only two states to produce the rectangular waveform: either fully on or fully off. In saturation, we know that the very small voltage drop across a transistor leads to very low-power dissipation. A very small dissipation is also present when the transistor is cut off. If switching times were negligible, no device power loss would occur during the transition between states. Actually, there is a finite switching time and this leads to an increased total dissipation of the output stages. Still, the efficiency figures for the class-D amplifier are very high, as reported earlier. This leads to higher possible power outputs and smaller chip areas for integrated PWM amplifiers. The stages can be direct-coupled to the load, which eliminates the necessity of capacitors. Nonlinear distortion can be less than that of class-B stages, and matching of transistors is unnecessary.

In contrast, the disadvantages of this amplifier ultimately dictate the limits of usefulness of the PWM scheme. The upper frequency response is limited to a small fraction of the switching frequency. The operating frequency of power transistors generally decreases with higher power ratings. It follows that the upper corner frequency of the amplifier may be lower for higher power transistors. Furthermore, a low-pass filter may be required to eliminate the unwanted frequency components of the waveform. The generation of radio frequencies or electromagnetic interference by the switching circuits can also present problems in certain applications.

In addition to compound emitter followers, the power output stages can be designed in several arrangements. Figure 23.34 shows two possible configurations. The diodes appearing across the output transistors are present to protect the transistors against inductive voltage surges. If the filter is inductive, the current reversals that occur over short switching times generate very large voltage spikes, unless the protective diodes are used.

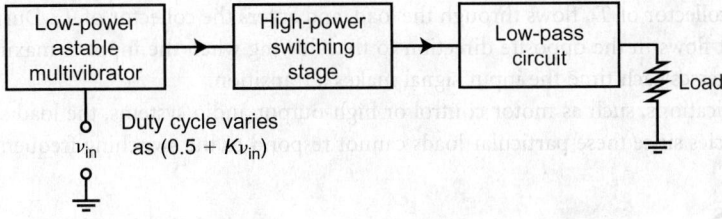

FIGURE 23.33 Architecture of a PWM amplifier.

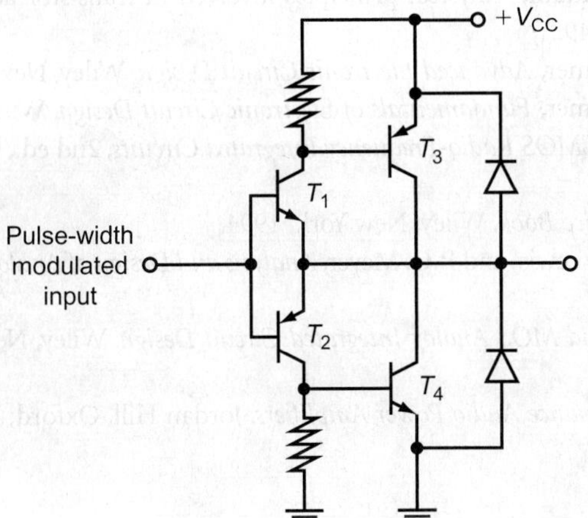

FIGURE 23.34 Class-D output stages.

In the push–pull circuit, the low-power, pulse-width modulated input turns T_1 and T_3 on when the signal is at its maximum value. Transistors T_2 and T_4 are off at this time, and current is forced through the load. When the signal switches to the minimum value, T_1 and T_3 go off, while T_2 and T_4 turn on to pull current through the load.

Figure 23.34(b) shows a bridge circuit that can drive a floating load with a single power supply. When the input signal reaches its maximum value, T_1 and T_3 are on while T_2 and T_4 are held off. The input signal is inverted and applied to the bases of T_5 and T_6. This inverted signal is at its minimum value during the time when the normal input is maximum; thus, T_5 and T_7 will be off while T_6 and T_8 are on. Current will leave the collector of T_3, flow through the load, and enter the collector of T_8. When the input assumes the most negative value, T_1, T_3, T_6 and T_8 turn off while T_2, T_4, T_5, and T_7 turn on. Current now leaves the collector of T_7, flows through the load, and enters the collector of T_4. During this period, the load current flows in the opposite direction to that flowing when the input is maximum. The load current then reverses each time the input signal makes a transition.

In some applications, such as motor control or high-output audio systems, the load serves as a filter of high frequencies since these particular loads cannot respond to the switching frequencies.

References

1. T. Lewis, *Empire of the Air*, Harper Perennial, New York, 1991.
2. J. Bardeen and W.H. Brittain, "Physical principles involved in transistor action," *Bell System Technical Journal*, 28, 239–247, 1949.
3. D.J. Comer and D.T. Comer, *Advanced Electronic Circuit Design*. Wiley, New York, 2002.
4. D.J. Comer and D.T. Comer, *Fundamentals of Electronic Circuit Design*. Wiley, New York, 2002.
5. T.H. Lee, *The Design of CMOS Radio-Frequency Integrated Circuits*, 2nd ed., Cambridge University Press, New York, 2004.
6. A. Vladimirescu, *The Spice Book*. Wiley, New York, 1994.
7. P.R. Gray, P.J. Hurst, S.H. Lewis, and R.G. Meyer, *Analysis and Design of Analog Integrated Circuits*, 4th ed., Wiley, New York, 2001.
8. A.B. Grebene, *Bipolar and MOS Analog Integrated Circuit Design*. Wiley, New York, 2003 (reference for new book).
9. B. Duncan, *High Performance Audio Power Amplifiers*. Jordan Hill, Oxford; Newnes London, UK, 1996.

24

High-Frequency Amplifiers

Chris Toumazou
University of London

Alison Burdett
Toumaz Technology Limited

CONTENTS

24.1   Introduction

As the operating frequency of communication channels for both video and wireless increases, there is an ever-increasing demand for high-frequency amplifiers. Furthermore, the quest for single-chip integration has led to a whole new generation of amplifiers predominantly geared toward CMOS very large-scale integration (VLSI). In this chapter we will focus on the design of high-frequency amplifiers for potential applications in the front-end of video, optical, and RF systems. Figure 24.1 shows for example, the architecture of a typical mobile phone transceiver front-end. With channel frequencies approaching the 2 GHz range coupled with demands for reduced chip size and power consumption, there is an increasing quest for VLSI at microwave frequencies. The shrinking feature size of CMOS has facilitated the design of complex analog circuits and systems in the 1–2 GHz range, where more traditional low-frequency lumped circuit techniques are now becoming feasible. Since the amplifier is the core component in such systems, there has been an abundance of circuit design methodologies for high-speed, low-voltage, low-noise, and low-distortion operation.

   This chapter will present various amplifier designs that aim to satisfy these demanding requirements. In particular we will review, and in some cases present new ideas for power amps, low-noise amplifiers (LNAs), and transconductance cells, which form core building blocks for systems such as in Figure 24.1. Section 24.2 begins by reviewing the concept of current feedback, and shows how this concept can be employed in the development of low-voltage, high-speed, constant-bandwidth CMOS amplifiers. The next two sections of the chapter focus on amplifiers for wireless receiver applications, investigating performance requirements and design strategies for optical receiver amplifiers (Section 24.3) and high-frequency LNAs (Section 24.4). Section 24.5 considers the design of amplifiers for the transmitter side, and in particular the design and feasibility of class E power amps is discussed. Finally, Section 24.6 reviews a very recent low-distortion amplifier design strategy termed "log-domain", which has shown enormous potential for high-frequency, low-distortion tunable filters.

24.2   Current Feedback OP-AMP (CFOA)

24.2.1   CFOA Basics

The operational amplifier (op-amp) is one of the most fundamental building blocks of analog circuit design [1,2]. High-performance signal-processing functions such as amplifiers, filters, and oscillators can be readily implemented with the availability of high-speed, low-distortion op-amps. In the last decade, the development of complementary bipolar technology has enabled the implementation of single-chip video op-amps [3–7]. The emergence of op-amps with nontraditional topologies such as the CFOA has improved the speed of these devices even further [8–11]. CFOA structures are well known for their ability to overcome (to a first-order approximation) the gain-bandwidth trade-off and slew rate limitation that characterizes traditional voltage feedback op-amps [12].

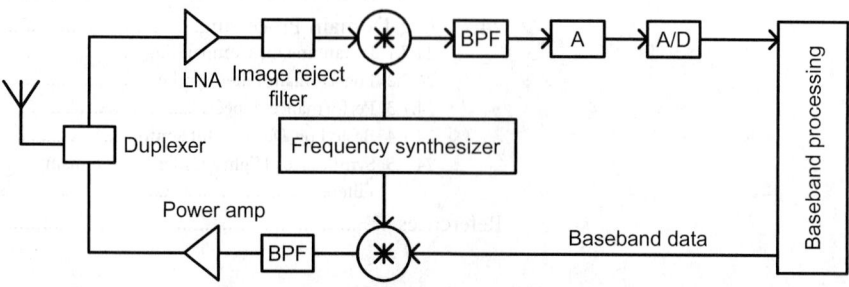

FIGURE 24.1   Generic wireless transceiver architecture.

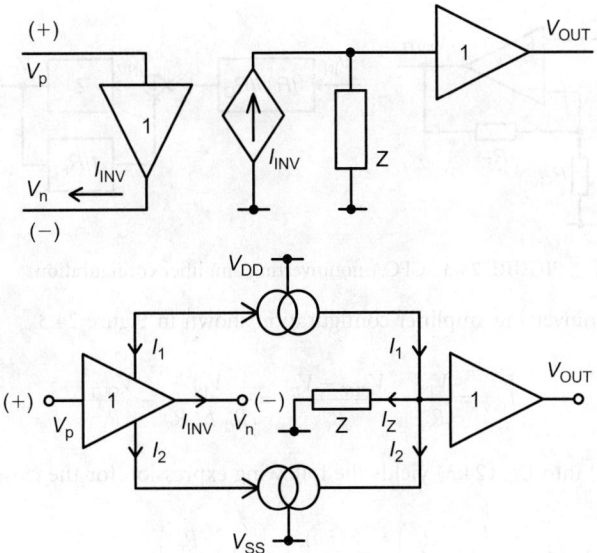

FIGURE 24.2 CFOA macromodel.

Figure 24.2 shows a simple macromodel of a CFOA, along with a simplified circuit diagram of the basic architecture. The topology of the CFOA differs from the conventional voltage feedback op-amp (VOA) in two respects. First, the input stage of a CFOA is a unity-gain voltage buffer connected between the inputs of the op-amp. Its function is to force V_n to follow V_p, very much like a conventional VOA does via negative feedback. In the case of the CFOA, because of the low-output impedance of the buffer, current can flow in or out of the inverting input, although in normal operation (with negative feedback) this current is extremely small. Second, a CFOA provides a high open-loop transimpedance gain $Z(j\omega)$, rather than open-loop voltage gain as with a VOA. This is shown in Figure 24.2, where a current-controlled current source senses the current I_{INV} delivered by the buffer to the external feedback network, and copies this current to a high-impedance $Z(j\omega)$. The voltage conveyed to the output is given by

$$V_{OUT} = Z(j\omega)I_{INV} \Rightarrow \frac{V_{OUT}}{I_{INV}}(j\omega) = Z(j\omega) \tag{24.1}$$

When the negative feedback loop is closed, any voltage imbalance between the two inputs owing to some external agent, will cause the input voltage buffer to deliver an error current I_{INV} to the external network. This error current $I_{INV} = I_1 - I_2 = I_z$ is then conveyed by the current mirrors to the impedance $Z(j\omega)$, resulting in an ouput voltage as given by Eq. (24.1). The application of negative feedback ensures that V_{OUT} will move in the direction that reduces the error current I_{INV} and equalizes the input voltages.

We can approximate the open-loop dynamics of the CFOA as a single pole response. Assuming that the total impedance $Z(j\omega)$ at the gain node is the combination of the output resistance of the current mirrors R_o in parallel with a compensation capacitor C we can write

$$Z(j\omega) = \frac{R_o}{1 + j\omega R_o C} = \frac{R_o}{1 + j(\omega/\omega_o)} \tag{24.2}$$

where $\omega_o = 1/R_o C$ represents the frequency, where the open-loop transimpedance gain is 3 dB down from its low frequency value R_o. In general R_o is designed to be very high in value.

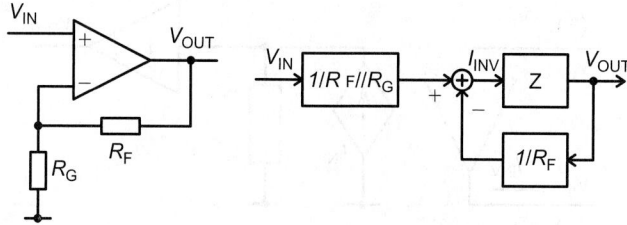

FIGURE 24.3 CFOA noninverting amplifier configuration.

Referring to the noninverting amplifier configuration shown in Figure 24.3

$$I_{INV} = \frac{V_{IN}}{R_G} - \frac{V_{OUT} - V_{IN}}{R_F} = \frac{V_{IN}}{R_G // R_F} - \frac{V_{OUT}}{R_F} \tag{24.3}$$

Substituting Eq. (24.1) into Eq. (24.3) yields the following expression for the closed-loop gain:

$$A_{CL}(j\omega) = \left(1 + \frac{R_F}{R_G}\right)\frac{Z(j\omega)}{R_F + Z(j\omega)} = \left(1 + \frac{R_F}{R_G}\right)\frac{1}{1 + R_F/Z(j\omega)} \tag{24.4}$$

Combining Eq. (24.2) and Eq. (24.4), and assuming that the low-frequency value of the open-loop transimpedance is much higher than the feedback resistor ($R_o \gg R_F$) gives

$$A_{CL}(j\omega) = \left(1 + \frac{R_F}{R_G}\right)\frac{1}{1 + j(R_F\omega/R_o\omega_o)} = \frac{A_{Vo}}{1 + j(\omega/\omega_\alpha)} \tag{24.5}$$

Referring to Eq. (24.5), the closed-loop gain $A_{Vo} = 1 + R_F/R_G$, while the closed-loop -3 dB frequency ω_α is given by

$$\omega_\alpha = \frac{R_o}{R_F}\omega_o \tag{24.6}$$

Eq. (24.6) indicates that the closed-loop bandwidth does not depend on the closed-loop gain as in the case of a conventional VOA, but is determined by the feedback resistor R_F. Explaining this intuitively, the current available to charge the compensation capacitor at the gain node is determined by the value of the feedback resistor R_F and not R_o, provided that $R_o \gg R_F$. So once the bandwidth of the amplifier is set via R_F, the gain can be independently varied by changing R_G. The ability to control the gain independently of bandwidth constitutes a major advantage of CFOA over conventional VOA.

The other major advantage of the CFOA compared to the VFOA is the inherent absent of slew rate limiting. For the circuit of Figure 24.2, assume that the input buffer is very fast and thus a change in voltage at the noninverting input is instantaneously converted in to the inverting input. When a step ΔV_{IN} is applied to the noninverting input, the buffer output current can be derived as

$$I_{INV} = \frac{V_{IN} - V_{OUT}}{R_F} + \frac{V_{IN}}{R_G} \tag{24.7}$$

Eq. (24.7) indicates that the current available to charge/discharge the compensation capacitor is proportional to the input step regardless of its size, i.e., there is no upper limit. The rate of change of the output voltage is thus

$$\frac{dV_{OUT}}{dt} = \frac{I_{INV}}{C} \Rightarrow V_{OUT}(t) = \Delta V_{IN}\left(1 + \frac{R_F}{R_G}\right)\left(1 - e^{-t/R_FC}\right) \tag{24.8}$$

Eq. (24.8) indicates an exponential output transition with time constant $\tau = R_F C$. Similar to the small-signal frequency response, the large-signal transient response is governed by R_F alone regardless of the magnitude of the closed-loop gain. The absence of slew rate limiting allows for faster settling times and eliminates slew rate-related nonlinearities.

In most practical bipolar realizations, Darlington-pair transistors are used in the input stage to reduce input bias currents, which makes the op-amp somewhat noisier and increases the input offset voltage. This is not necessary in CMOS realizations owing to the inherently high MOSFET input impedance. However, in a closed-loop CFOA, R_G should be much larger than the output impedance of the buffer. In bipolar realizations it is fairly simple to obtain a buffer with low output resistance, but this becomes more of a problem in CMOS owing to the inherently lower gain of MOSFET devices. As a result, R_G typically needs to be higher in a CMOS CFOA than in a bipolar realization, and consequently R_F needs to be increased above the value required for optimum high-frequency performance. Additionally, the fact that the input buffer is not in the feedback loop imposes linearity limitations on the structure, especially if the impedance at the gain node is not very high. Regardless of these problems, CFOAs exhibit excellent high-frequency characteristics and are increasingly popular in video and communications applications [13].

The following sections outline the development of a novel low-output impedance CMOS buffer, which is then employed in a CMOS CFOA to reduce the minimum allowable value of R_G.

24.2.2 CMOS Compound Device

A simple PMOS source follower is shown in Figure 24.4. The output impedance seen looking into the source of M1 is approximately $Z_{out} = 1/g_m$, where g_m is the small-signal transconductance of M1. To increase g_m the drain current of M1 could be increased, which leads to an increased power dissipation. Alternatively, the dimensions of M1 can be increased, resulting in additional parasitic capacitance and hence an inferior frequency response. Figure 24.5 shows a configuration, which achieves a higher transconductance than the simple follower of Figure 24.4 for the same bias current [11]. The current of M2 is fed back to M1 through the a:1 current mirror. This configuration can be viewed as a compound transistor, whose gate is the gate of M1 and whose source is the source of M2. The impedance looking into the compound source can be approximated as $Z_{out} = (g_{m1} - ag_{m2})/(g_{m1}g_{m2})$, where g_{m1} and g_{m2} represent the small-signal transconductance of M1 and M2, respectively. The output impedance can be made small by setting the current mirror transfer ratio $a = g_{m1}/g_{m2}$.

The p-compound device is practically implemented as in Figure 24.6. To obtain a linear voltage transfer function from node 1 to 2, the gate-source voltages of M1 and M3 must cancel. The current mirror (M4–M2) acts as NMOS–PMOS gate-source voltage matching circuit [14] and compensates for the difference in the gate-source voltages of M1 and M3, which would normally appear as an output offset. DC analysis, assuming a square law model for the MOSFETs, shows that the output voltage exactly follows the input voltage. However, in practice, channel length modulation and body effects preclude exact cancellation [15].

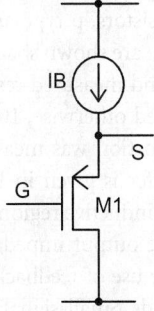

FIGURE 24.4 Simple PMOS source follower.

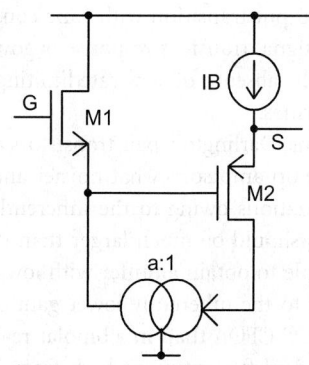

FIGURE 24.5 Compound MOS device.

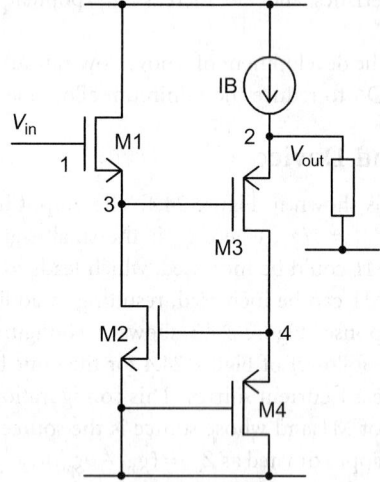

FIGURE 24.6 Actual p-compound device implementation.

24.2.3 Buffer and CFOA Implementation

The CFOA shown in Figure 24.7 has been implemented in a single-well 0.6 μm digital CMOS process [11]; the corresponding layout plot is shown in Figure 24.8. The chip has an area of 280 μm × 330 μm and a power dissipation of 12 mW. The amplifier comprises two voltage followers (input and output) connected by cascoded current mirrors to enhance the gain node impedance. A compensation capacitor ($C_c = 0.5$ pF) at the gain node ensures adequate phase margin and thus closed-loop stability. The voltage followers have been implemented with two compound transistors, p type and n type, in a push–pull arrangement. Two such compound transistors in the output stage are shown shaded in Figure 24.7. The input voltage follower of the CFOA was initially tested open loop, and measured results are summarized in Table 24.1. The load is set to 10 kΩ//10 pF except where mentioned otherwise, 10 kΩ being a limit imposed by overall power dissipation of the chip. Intermodulation distortion was measured with two tones separated by 200 kHz. The measured output impedance of the buffer is given in Figure 24.9. It remains below 80 Ω up to a frequency of about 60 MHz, when it enters an inductive region. A maximum impedance of 140 Ω is reached around 160 MHz. Beyond this frequency, the output impedance is dominated by parasitic capacitances. The inductive behavior is characteristic of the use of feedback to reduce output impedance and can cause stability problems when driving capacitive loads. Small-signal analysis (summarized in Table 24.2) predicts a double zero in the output impedance [15].

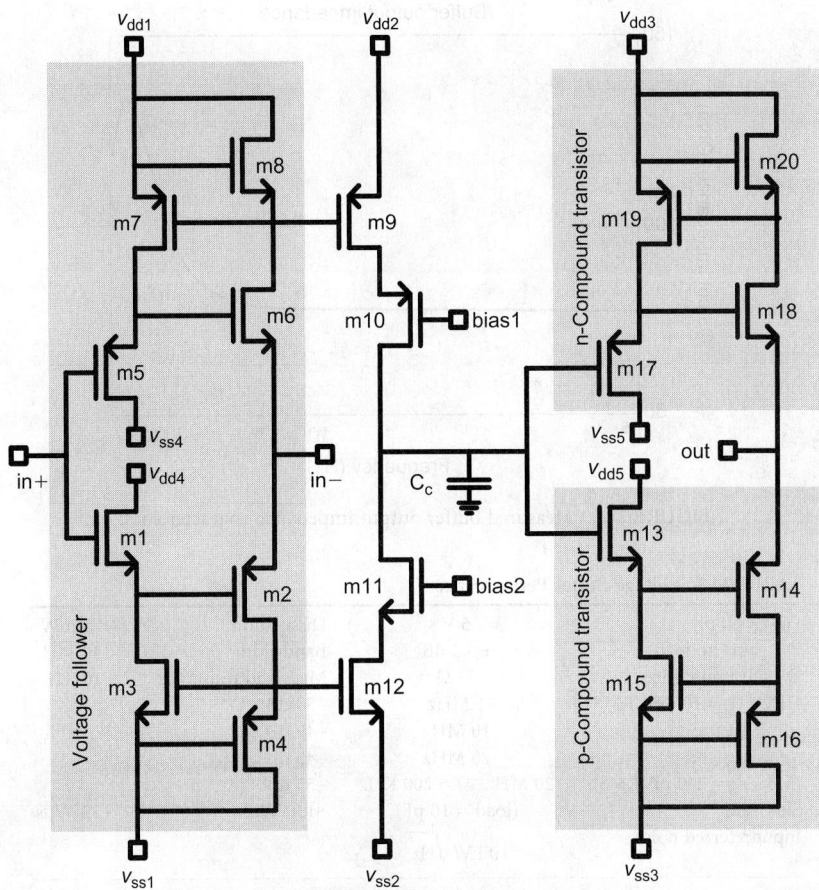

FIGURE 24.7 CFOA schematic.

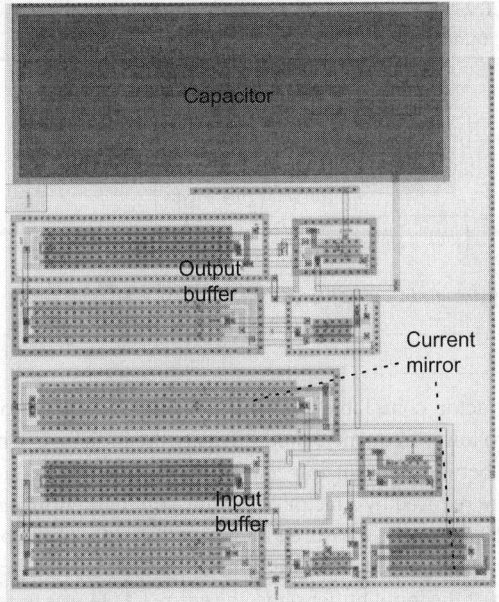

FIGURE 24.8 CFOA layout plot.

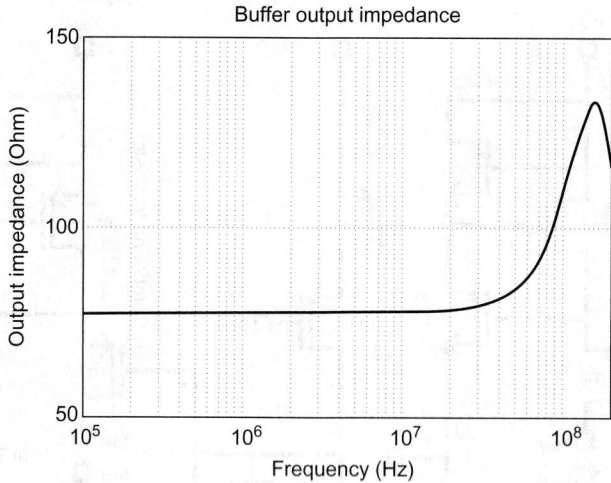

FIGURE 24.9   Measured buffer output impedance characteristics.

TABLE 24.1   Voltage Buffer Performance

Power supply	5 V	Dissipation	5 mW
DC gain (no load)	−3.3 dB	Bandwidth	140 MHz
Output impedance	75 Ω	Min. load resistance	10 KΩ
HD2 (V_{in} = 200 mVRMS)	1 MHz	−50 dB	
	10 MHz	−49 dB	
	20 MHz	−45 dB	
IM3 (V_{in} = 200 mVRMS)	20 MHz, Δf = 200 KHz	−53 dB	
Slew rate	(load = 10 pF)	+130 V/μs	−72 V/μs
Input referred noise			
	10 nV/$\sqrt{Hz}$		

Note: load=10 kΩ//10pF, except for slew rate measurement.

TABLE 24.2   Voltage Transfer Function and Output Impedance of Compound Device

$$Z_{out} = \frac{G}{\left(gm1 + gds1 + gds2\right)\left(gm3 + gds3\right)\left(gm4 + gds4\right)}$$

$$\frac{V_{out}}{V_{in}} = \frac{gm1\,gm3\left(gm4 + gds4\right)}{\left(gm1 + gds1 + gds2\right)\left(gm3 + gds3\right)\left(gm4 + gds4\right) + g_L G}$$

$$G = \left(gm1 + gds1 + gds2\right)\left(gm4 + gds4 + gds3\right) - gm2\,gm3$$

Decreasing the value of factor G in Table 24.2 will not only reduce the output impedance, but will also move the double zero to lower frequencies and intensify the inductive behavior. The principal trade-off in this configuration is between output impedance magnitude and inductive behavior. In practice, the output impedance can be reduced by a factor of 3 while still maintaining good stability when driving capacitive loads. Figure 24.10 shows the measured frequency response of the buffer. Given the low-power dissipation, excellent slew rates have been achieved (Table 24.2).

After the characterization of the input buffer stage, the entire CFOA was tested to confirm the suitability of the compound transistors for the implementation of more complex building blocks. Open-loop

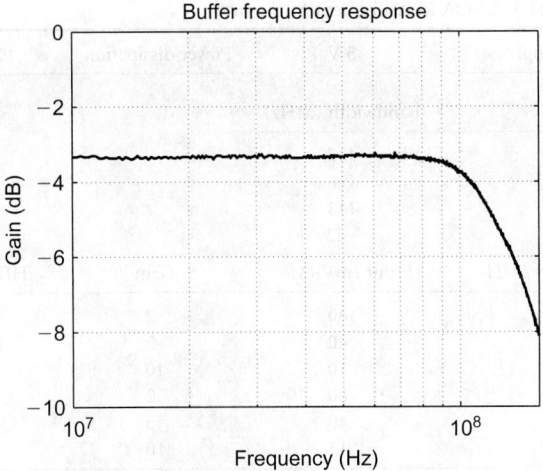

FIGURE 24.10 Measured buffer frequency response.

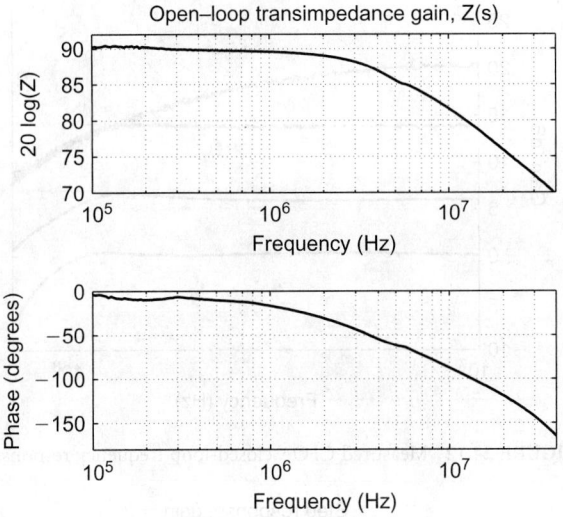

FIGURE 24.11 Measured CFOA open-loop transimpedance gain.

transimpedance measurements are shown in Figure 24.11. The bandwidth of the amplifier was measured at gain settings of 1, 2, 5, and 10 in a noninverting configuration, and the feedback resistor was trimmed to achieve maximum bandwidth at each gain setting separately. CFOA measurements are summarized in Table 24.3, loading conditions are again 10 kΩ//10 pF.

Figure 24.12 shows the measured frequency response for various gain settings. The bandwidth remains constant at 110 MHz for gains of 1, 2, and 5 consistent with the expected behavior of a CFOA. The bandwidth falls to 42 MHz for a gain of 10 owing to the finite output impedance of the input buffer stage which serves as the CFOA inverting input. Figure 24.13 illustrates the step response of the CFOA driving a 10 kΩ//10 pF load at a voltage gain of 2. It can be seen that the inductive behavior of the buffers has little effect on the step response. Finally, distortion measurements were carried out for the entire CFOA for gain settings 2, 5, and 10 and are summarized in Table 24.3. HD2 levels can be further improved by employing a double-balanced topology. A distortion spectrum is shown in Figure 24.14; the onset of HD3 is owing to clipping at the test conditions.

TABLE 24.3 CFOA Measurement Summary

Power supply	5 V	Power dissipation	12 mW
Gain	Bandwidth (MHz)		
1	117		
2	118		
5	113		
10	42		
Frequency (MHz)	Input (mVRMS)	Gain	HD2 (dB)
1	140	2	−51
	40	5	−50
	10	10	−49
10	80	2	−42
	40	5	−42
	13	10	−43

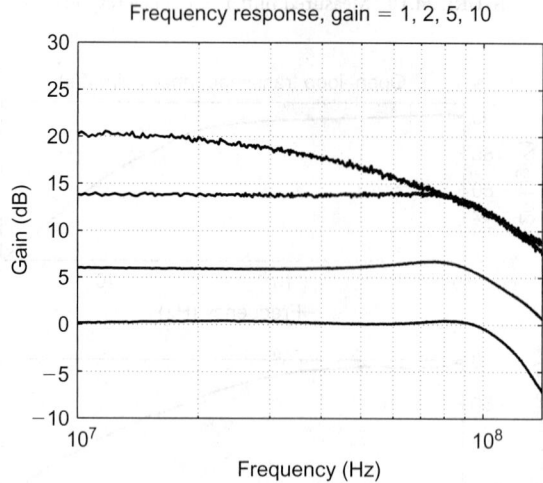

FIGURE 24.12 Measured CFOA closed-loop frequency response.

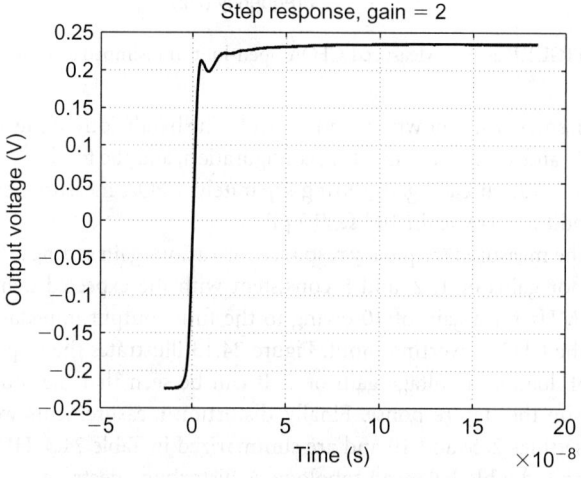

FIGURE 24.13 Measured CFOA step response.

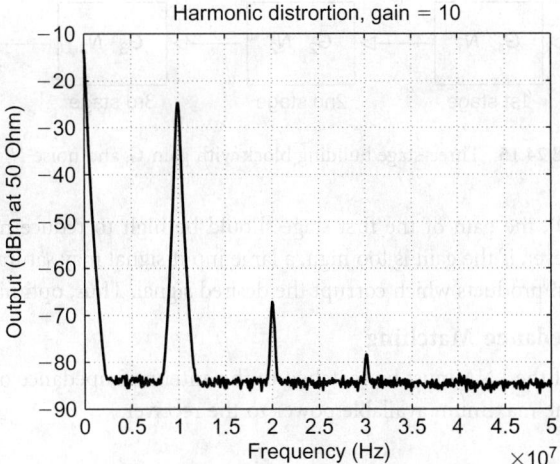

FIGURE 24.14 CFOA harmonic distortion measurements.

24.3 RF LNAs

This section reviews the important performance criteria demanded of the front-end amplifier in a wireless communication receiver. The design of CMOS LNAs for front-end wireless communication receiver applications is then addressed. The following section considers the related topic of LNAs for optical receiver front-ends.

24.3.1 Specifications

The front-end amplifier in a wireless receiver must satisfy demanding requirements in terms of noise, gain, impedance matching, and linearity.

24.3.1.1 Noise

Since the incoming signal is usually weak, the front-end circuits of the receiver must possess very low noise characteristics so that the original signal can be recovered. Provided that the gain of the front-end amplifier is sufficient so as to suppress noise from the subsequent stages, the receiver noise performance is determined predominantly by the front-end amplifier. Hence the front-end amplifier should be an LNA.

24.3.1.2 Gain

The voltage gain of the LNA must be high enough to ensure that noise contributions from the following stages can be safely neglected. As an example, Figure 24.15 shows the first three stages in a generic front-end receiver, where the gain and output-referred noise of each stage are represented by G_i and N_i ($i = 1, 2, 3$), respectively. The total noise at the third-stage output is given by

$$N_{out} = N_{in} G_1 G_2 G_3 + N_1 G_2 G_3 + N_2 G_3 + N_3 \qquad (24.9)$$

This output noise (N_{out}) can be referred to the input to derive an equivalent input noise (N_{eq}):

$$N_{eq} = \frac{N_{out}}{Gain} = \frac{N_{out}}{G_1 G_2 G_3} = N_{in} + \frac{N_1}{G_1} + \frac{N_2}{G_1 G_2} + \frac{N_3}{G_1 G_2 G_3} \qquad (24.10)$$

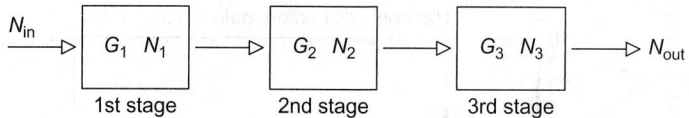

FIGURE 24.15 Three-stage building block with gain G_i and noise N_i per stage.

According to Eq. (24.10), the gain of the first stage should be high to reduce noise contributions from subsequent stages. However, if the gain is too high, a large input signal may saturate the subsequent stages yielding intermodulation products which corrupt the desired signal. Thus, optimization is inevitable.

24.3.1.3 Input Impedance Matching

The input impedance of the LNA must be matched to the antenna impedance over the frequency range of interest to transfer the maximum available power to the receiver.

24.3.1.4 Linearity

Unwanted signals at frequencies fairly near the frequency band of interest may reach the LNA with signal strengths many times higher than that of the desired signal. The LNA must be sufficiently linear to prevent these out-of-band signals from generating intermodulation products within the desired frequency band, and thus degrading the reception of the desired signal. Since third-order mixing products are usually dominant, the linearity of the LNA is related to the "third-order intercept point" (IP3), which is defined as the input power level that results in equal power levels for the output fundamental frequency component and the third-order intermodulation components. The dynamic range of a wireless receiver is limited by noise at the lower and nonlinearity at the upper band.

24.3.2 CMOS Common-Source (CS) LNA: Simplified Analysis

24.3.2.1 Input Impedance Matching by Source Degeneration

For maximum power transfer, the input impedance of the LNA must be matched to the source resistance which is normally 50 Ω. Impedance matching circuits consist of reactive components and therefore are (ideally) lossless and noiseless. Figure 24.16 shows the small-signal equivalent circuit of a CS LNA input stage with impedance matching circuit, where the gate-drain capacitance C_{gd} is assumed to have negligible effect and is thus neglected [16,17]. The input impedance of this CS input stage is given by

$$Z_{in} = j\omega(L_g + L_s) + \frac{1}{j\omega C_{gs}} + \frac{g_m}{C_{gs}}L_s = \frac{g_m}{C_{gs}}L_s = R_s \qquad (24.11)$$

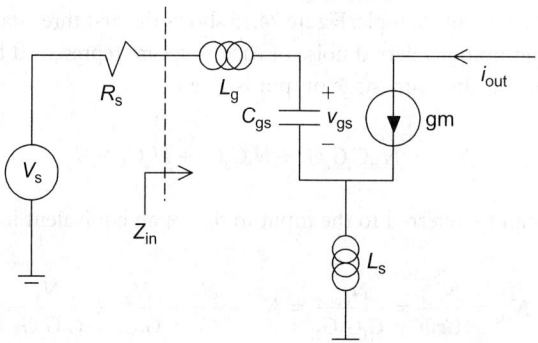

FIGURE 24.16 Simplified small-signal equivalent circuit of the CS stage.

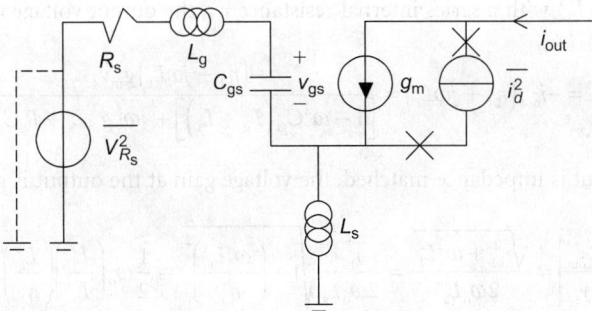

FIGURE 24.17 Simplified noise equivalent circuit of the CS stage.

Thus for matching, the following two conditions must be satisfied:

$$\omega_o^2 = \frac{1}{(L_g + L_s)C_{gs}} \quad \text{and} \quad \frac{g_m}{C_{gs}}L_s = R_s \tag{24.12}$$

24.3.2.2 Noise Figure of CS Input Stage

Two main noise sources exist in a CS input stage as shown in Figure 24.17; thermal noise from the source resistor R_s (denoted $\overline{v_{R_s}^2}$) and channel thermal noise from the input transistor (denoted $\overline{i_d^2}$). The output noise current due to $\overline{v_{R_s}^2}$ can be determined from Figure 24.17 as

$$\overline{i_{nout1}^2} = \frac{g_m^2 \ \overline{v_{R_s}^2}}{\omega^2(g_m L_s + R_s C_{gs})^2} = \frac{g_m^2}{4\omega^2 R_s^2 C_{gs}^2} \overline{v_{R_s}^2} \tag{24.13}$$

while the output noise current due to $\overline{i_d^2}$ can be evaluated as

$$\overline{i_{nout2}} = \frac{\overline{i_d}}{\left(1 + \dfrac{g_m L_s}{R_s C_{gs}}\right)} = \frac{1}{2}\overline{i_d} \quad \therefore \ \overline{i_{out2}^2} = \frac{1}{4}\overline{i_d^2} \tag{24.14}$$

From Eq. (24.13) and Eq. (24.14) the noise figure of the CS input stage is determined as

$$NF = 1 + \frac{\overline{i_{nout2}^2}}{\overline{i_{nout1}^2}} = 1 + \Gamma\left(\frac{\omega_o^2 R_s C_{gs}^2}{g_m}\right) = 1 + \Gamma\left(\frac{L_s}{L_s + L_g}\right) \tag{24.15}$$

In practice, any inductor (especially a fully integrated inductor) has an associated resistance which will contribute thermal noise, degrading the noise figure in Eq. (24.15).

24.3.2.3 Voltage Amplifier with Inductive Load

Referring to Figure 24.15, the small-signal current output is given by

$$i_{out} = \frac{g_m v_s}{\left[1 - \omega^2 C_{gs}\left(L_g + L_s\right)\right] + j\omega\left(g_m L_s + R_s C_{gs}\right)} \tag{24.16}$$

For an inductive load (L_1) with a series internal resistance r_{L1}, the output voltage is thus

$$v_{out} = -i_{out}\left(r_{L1} + j\omega L_1\right) = \frac{-\left(r_{L1} + j\omega L_1\right)g_m v_s}{\left[1 - \omega^2 C_{gs}\left(L_g + L_s\right)\right] + j\omega\left(g_m L_s + R_s C_{gs}\right)} \tag{24.17}$$

Assuming that the input is impedance matched, the voltage gain at the output is given by

$$\left|\frac{v_{out}}{v_s}\right| = \frac{\sqrt{r_{L1}^2 + \omega_o^2 L_1^2}}{2\omega_o L_s} = \frac{r_{L1}}{2\omega_o L_s}\sqrt{1 + \left(\frac{\omega_o L_1}{r_{L1}}\right)^2} \cong \frac{1}{2}\omega_o\left(\frac{L_1}{L_s}\right)\left(\frac{L_1}{r_{L1}}\right) \tag{24.18}$$

24.3.3 CMOS CS LNA: Effect of C_{gd}

In the analysis so far, the gate-drain capacitance (C_{gd}) has been assumed to be negligible. However, at very high frequencies, this component cannot be neglected. Figure 24.18 shows the modified input stage of a CS LNA including C_{gd}, and where an input AC-coupling capacitance C_{in} has also been included. Small-signal analysis shows that the input impedance is now given by

$$Z_{in} = \frac{g_m L_s}{C_{gs} + C_{gd}\left[j\omega L_s g_m + \dfrac{g_m\left(1 - \omega^2 L_s C_{gd}\right)}{\dfrac{1}{Z_L} + j\omega C_{gd}}\right]} \tag{24.19}$$

Eq. (24.19) exhibits resonance frequencies which occur when

$$1 - \omega^2 L_s C_{gs} = 0 \quad \text{and} \quad 1 - \omega^2 L_g C_{in} = 0 \tag{24.20}$$

Eq. (24.19) indicates that the input impedance matching is degraded by the load Z_L when C_{gd} is included in the analysis.

24.3.3.1 Input Impedance with Capacitive Load

If the load Z_L is purely capacitive, that is,

$$Z_L = \frac{1}{j\omega C_L} \tag{24.21}$$

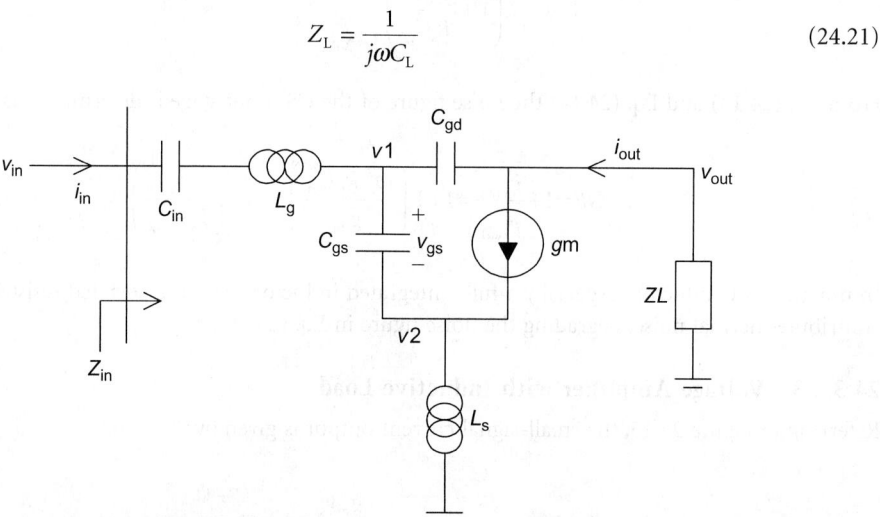

FIGURE 24.18 Noise-equivalent circuit of the CS stage, including effects of C_{gd}.

then the input impedance can be easily matched to the source resistor R_s. Substituting Eq. (24.21) for Z_L, the bracketed term in the denominator of Eq. (24.19) becomes

$$d_1 = j\omega L_s g_m + \frac{g_m\left(1 - \omega^2 L_s C_{gd}\right)}{j\omega\left(C_{gd} + C_L\right)} = 0 \tag{24.22}$$

under the condition that

$$1 - \omega^2 L_s\left(2C_{gd} + C_L\right) = 0 \tag{24.23}$$

The three conditions in Eq. (24.20) and Eq. (24.23) should be met to ensure input impedance matching. However, in practice, we are unlikely to be in the situation of using a load capacitor.

24.3.3.2 Input Impedance with Inductive Load

If $Z_L = j\omega L_L$, the CS LNA input impedance is given by

$$Z_{in} = \frac{g_m L_s}{C_{gs} + j\omega C_{gd} g_m\left[L_s + L_L\left(\dfrac{1 - \omega^2 L_s C_{gd}}{1 - \omega^2 L_L C_{gd}}\right)\right]} \tag{24.24}$$

To match a purely resistive input, the value of the reactive term in Eq. (24.24) must be negligible; this is difficult to achieve.

24.3.4 Cascode CS LNA

24.3.4.1 Input Matching

As outlined in the above section, the gate-drain capacitance (C_{gd}) degrades the input impedance matching and therefore reduces the power transfer efficiency. In order to reduce the effect of C_{gd}, a cascoded structure can be used [18–20]. Figure 24.19 shows a cascode CS LNA. Since the voltage gain from the gate to the drain of M1 is unity, the gate-drain capacitance (C_{gd1}) no longer sees the full input–output voltage swing which greatly improves the input–output isolation. The input impedance can be approximated by Eq. (24.11), thus allowing a simple matching circuit to be employed [18].

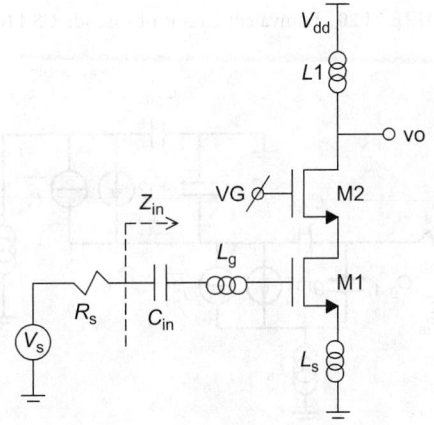

FIGURE 24.19 Cascode CS LNA.

24.3.4.2 Voltage Gain

Figure 24.20 shows the small-signal equivalent circuit of the cascode CS LNA. Assuming that the input is fully matched to the source, the voltage gain of the amplifier is given by

$$\frac{v_{out}}{v_s} = -\frac{1}{2}\left(\frac{j\omega L_1}{1-\omega^2 L_1 C_{gd2}}\right)\left(\frac{g_{m2}}{g_{m2}+j\omega C_{gs2}}\right)\left[\frac{g_{m1}}{\left(1-\omega^2 L_s C_{gs1}\right)+j\omega L_s g_{m1}}\right] \tag{24.25}$$

At the resonant frequency, the voltage gain is given by

$$\frac{v_{out}}{v_s}(\omega_o) = -\frac{1}{2}\left(\frac{L_1}{L_s}\right)\left(\frac{1}{1-\omega_o^2 L_1 C_{gd2}}\right)\frac{1}{1+j\omega_o\left(C_{gs1}/g_{m2}\right)} \approx -\frac{1}{2}\left(\frac{L_1}{L_s}\right)\frac{1}{1+j\left(\omega_o/\omega_T\right)} \tag{24.26}$$

From Eq. (24.26), the voltage gain is dependent on the ratio of the load and source inductance values. Therefore, high gain accuracy can be achieved since this ratio is largely process independent.

24.3.4.3 Noise Figure

Figure 24.21 shows an equivalent circuit of the cascode CS LNA for noise calculations. Three main noise sources can be identified; the thermal noise voltage from R_s, and the channel thermal noise currents from

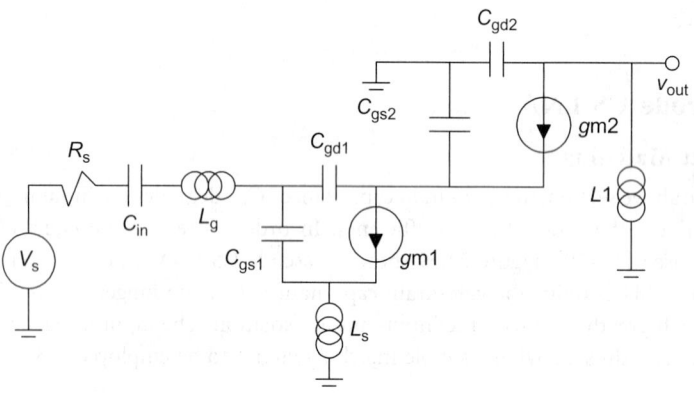

FIGURE 24.20 Equivalent circuit of cascode CS LNA.

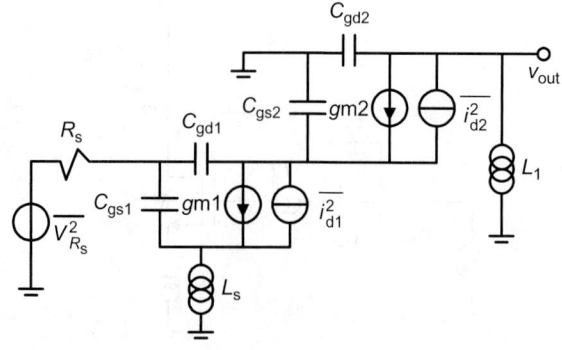

FIGURE 24.21 Noise-equivalent circuit of cascode CS LNA.

M1 and M2. Assuming that the input impedance is matched to the sources, the output noise current due to $\overline{v_{RS}^2}$ can be derived as

$$\overline{i_{out1}} = \frac{1}{2j\omega_o L_s\left(1 - \omega_o^2 L_1 C_{gd2}\right)}\left(\frac{g_{m2}}{g_{m2} + j\omega_o C_{gs2}}\right)\overline{v_{RS}} \tag{24.27}$$

The output noise current contribution due to i_{d1}^2 of M1 is given by

$$\overline{i_{out2}} = \frac{1}{2\left(1 - \omega_o^2 L_1 C_{gd2}\right)}\left(\frac{g_{m2}}{g_{m2} + j\omega_o C_{gs2}}\right)\overline{i_{d1}} \tag{24.28}$$

The output noise current due to i_{d2}^2 of M2 is given by

$$\overline{i_{out3}} = \frac{j\omega_o C_{gs2}}{\left(1 - \omega_o^2 L_1 C_{gd2}\right)\left(g_{m2} + j\omega_o C_{gs2}\right)}\overline{i_{d2}} \tag{24.29}$$

The noise figure of the cascode CS LNA can thus be derived as

$$NF = 1 + \frac{\overline{i_{out2}^2}}{\overline{i_{out1}^2}} + \frac{\overline{i_{out3}^2}}{\overline{i_{out1}^2}} = 1 + \Gamma\left(1 + \frac{4\omega_o^2 C_{gs2}^2}{g_{m1}g_{m2}}\right) \tag{24.30}$$

To improve the noise figure, the transconductance values (g_m) of M1 and M2 should be increased. Since the gate-source capacitance (C_{gs2}) of M2 is directly proportional to the gate width, the gate-width of M2 cannot be enlarged to increase the transconductance. Instead, this increase should be realized by increasing the gate-bias voltage.

24.4 Optical Low-Noise Preamplifiers

Figure 24.22 shows a simple schematic diagram of an optical receiver, consisting of a photodetector, a preamplifier, a wide-band voltage amplifier, and a predetection filter. Since the front-end transimpedance preamplifier is critical in determining the overall receiver performance, it should possess a wide bandwidth so as not to distort the received signal, high gain to reject noise from subsequent stages, low noise to achieve high sensitivity, wide dynamic range, and low intersymbol interference (ISI).

24.4.1 Front-End Noise Sources

Receiver noise is dominated by two main noise sources: the detector (PIN photodiode) noise and the amplifier noise. Figure 24.22 illustrates the noise-equivalent circuit of the optical receiver.

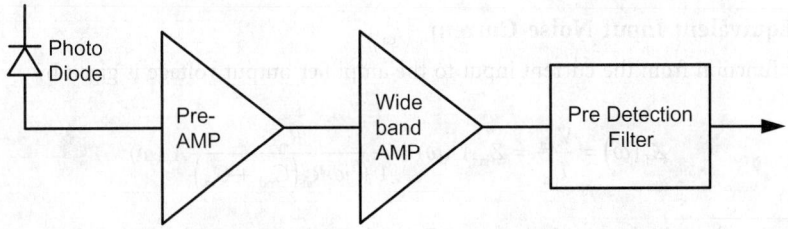

FIGURE 24.22 Optical receiver front end.

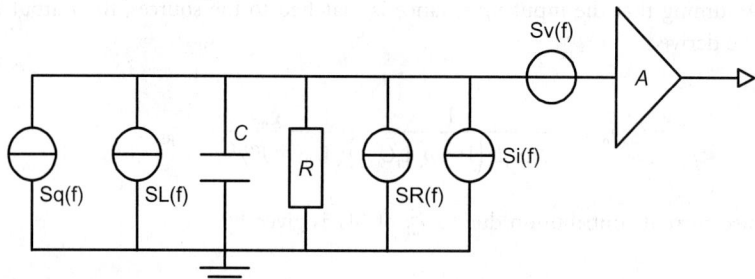

FIGURE 24.23 Noise-equivalent circuit of the front-end optical receiver.

24.4.1.1 PIN Photodiode Noise

The noise generated by a PIN photodiode arises mainly from three shot noises contributions: quantum noise $S_q(f)$, thermally generated dark-current shot noise $S_D(f)$ and surface leakage-current shot noise $S_L(f)$. Other noise sources in a PIN photodiode such as series resistor noise are negligible in comparison. The quantum noise $S_q(f)$, also called signal-dependent shot noise, is produced by the light-generating nature of photonic detection and has a spectral density $S_q(f) = 2qI_{pd}\Delta f$, where I_{pd} is the mean signal current arising from the Poisson statistics. The dark-current shot noise $S_D(f)$ arises in the photodiode bulk material. Even when there is no incident optical power, a small reverse leakage current still flows resulting in shot noise with a spectral density $S_D(f) = 2qI_{DB}\Delta f$, where I_{DB} is the mean thermally generated dark current. The leakage shot noise $S_L(f)$ occurs because of surface effects around the active region, and is described by $S_L(f) = 2qI_{SL}\Delta f$, where I_{SL} is the mean surface leakage current.

24.4.1.2 Amplifier Noise

For a simple noise analysis, the pre- and postamplifiers in Figure 24.22 are merged to a single amplifier with a transfer function of $A_v(\omega)$. The input impedance of the amplifier is modeled as a parallel combination of R_{in} and C_{in}.

If the photodiode noise is negligibly small, the amplifier noise will dominate the whole receiver noise performance as can be inferred from Figure 24.23. The equivalent noise current and voltage spectral densities of the amplifier are represented as $S_i(A^2/Hz)$ and $S_v(V^2/Hz)$, respectively.

24.4.1.3 Resistor Noise

The thermal noise generated by a resistor is directly proportional to the absolute temperature T and is represented by a series noise voltage generator or by a shunt noise current generator [21] of value

$$\overline{v_R^2} = 4kTR\Delta f \quad \text{or} \quad \overline{i_R^2} = 4kT\frac{1}{R}\Delta f \tag{24.31}$$

where k is the Boltzmann's constant and R the resistance.

24.4.2 Receiver Performance Criteria

24.4.2.1 Equivalent Input Noise Current $\langle \overline{i_{eq}^2} \rangle$

The transfer function from the current input to the amplifier output voltage is given by

$$Z_T(\omega) = \frac{V_{out}}{I_{pd}} = Z_{in}A_v(\omega) = \frac{R_{in}}{1 + j\omega R_{in}\left(C_{pd} + C_{in}\right)}A_v(\omega) \tag{24.32}$$

where C_{pd} is the photodiode capacitance, and R_{in} and C_{in} the input resistance and capacitance of the amplifier, respectively. Assuming that the photodiode noise contributions are negligible and that the amplifier noise sources are uncorrelated, the equivalent input noise current spectral density can be derived from Figure 24.23 as

$$S_{eq}(f) = S_i + \frac{S_v}{\left[Z_{in}\right]^2} = S_i + S_v\left[\frac{1}{R_{in}^2} + (2\pi f)^2(C_{pd} + C_{in})^2\right] \qquad (24.33)$$

The total mean-square noise output voltage $\langle v_{no}^2 \rangle$ is calculated by combining Eq. (24.32) and Eq. (24.33) as follows:

$$\langle v_{no}^2 \rangle = \int_0^\infty S_{eq}(f)\left|Z_T(f)\right|^2 df \qquad (24.34)$$

This total noise voltage can be referred to the input of the amplifier by dividing it by the squared DC gain $\left|Z_T(0)\right|^2$ of the receiver to give an equivalent input mean-square noise current

$$\langle i_{eq}^2 \rangle = \frac{\langle v_{no}^2 \rangle}{\left|Z_T(0)\right|^2} = \left(S_i + \frac{S_v}{R_{in}^2}\right)\int_0^\infty \frac{\left|Z_T(f)\right|^2}{\left|Z_T(0)\right|^2} df + S_v\left[2\pi(C_{pd} + C_{in})\right]^2 \int_0^\infty f^2 \frac{\left|Z_T(f)\right|^2}{\left|Z_T(0)\right|^2} df$$

$$= \left(S_i + \frac{S_v}{R_{in}^2}\right)I_2 B + \left[2\pi(C_{pd} + C_{in})\right]^2 I_3 B^3 S_v \qquad (24.35)$$

where B is the operating bit-rate and I_2 ($= 0.56$) and I_3 ($= 0.083$) the Personick second and third integrals, respectively, as given in Ref. [22].

According to Morikoni et al. [23], the Personick integral in Eq. (24.35) is correct only if a receiver produces a raised-cosine output response from a rectangular input signal at the cutoff bit rate above which the frequency response of the receiver is zero. However, the Personick integration method is generally preferred when comparing the noise (or sensitivity) performance of different amplifiers.

24.4.2.2 Optical Sensitivity

Optical sensitivity is defined as the minimum received optical power incident on a perfectly efficient photodiode connected to the amplifier, such that the presence of the amplifier noise corrupts on average only one bit per 10^9 bits of incoming data. Therefore, a detected power greater than the sensitivity level guarantees system operation at the desired performance. The optical sensitivity is predicted theoretically by calculating the equivalent input noise spectral density of the receiver [24], and is calculated as

$$S = 10\log_{10}\left(Q\frac{hc}{q\lambda}\sqrt{\langle i_{eq}^2 \rangle}\frac{1}{1\text{ mW}}\right) \text{ (dBm)} \qquad (24.36)$$

where h is the Planck's constant, c the speed of light, q the electronic charge, and λ the wavelength of light (in micrometer) in an optical fibre. $Q(= \sqrt{SNR})$, where SNR represents the required signal-to-noise ratio (SNR). The value of Q should be 6 for a bit error rate (BER) of 10^{-9} and 7.04 for a BER of 10^{-12}. The relation between Q and BER is given by

$$BER = \frac{\exp(-Q^2/2)}{\sqrt{2\pi}Q} \qquad (24.37)$$

Since the number of photogenerated electrons in a single bit is very large (more than 10^4) for optoelectronic-integrated receivers [25], Gaussian statistics of the above BER equation can be used to describe the detection probability in PIN photodiodes.

24.4.2.3 SNR at the Photodiode Terminal

Among the photodiode noise sources, quantum noise is generally dominant and can be estimated as

$$\left\langle \overline{i_n^2} \right\rangle_q = 2qI_{pd}B_{eq} \tag{24.38}$$

where I_{pd} is the mean signal current and B_{eq} is the equivalent noise bandwidth. The SNR referred to the photodiode terminal [22] is thus given by

$$SNR = \frac{I_{pd}^2}{\left\langle \overline{i_n^2} \right\rangle_{pd} + \dfrac{4kTB_{eq}}{R_B} + \left\langle \overline{i_{eq}^2} \right\rangle_{amp}} \tag{24.39}$$

where all noise contributions owing to the amplifier are represented by the equivalent noise current $\left\langle \overline{i_{eq}^2} \right\rangle_{amp}$. It is often convenient to combine the noise contributions from the amplifier and the photodiode with the thermal noise from the bias resistor by defining a noise figure NF

$$\left\langle \overline{i_n^2} \right\rangle_{pd} + \frac{4kTB_{eq}}{R_B} + \left\langle \overline{i_{eq}^2} \right\rangle_{amp} = \frac{4kTB_{eq}NF}{R_B} \tag{24.40}$$

The SNR at the photodiode input is thus given by

$$SNR \cong \frac{I_{pd}^2 R_B}{4kTB_{eq}NF} \tag{24.41}$$

24.4.2.4 ISI

When a pulse passes through a band-limited channel, it gradually disperses. When the channel bandwidth is close to the signal bandwidth, the expanded rise and fall times of the pulse signal will cause successive pulses to overlap, deteriorating the system performance and giving higher error rates. This pulse overlapping is known as ISI. Even with raised-signal power levels, the error performance cannot be improved [26].

In digital optical communication systems, sampling at the output must occur at the point of maximum signal to achieve the minimum error rate. The output pulse shape should therefore be chosen to maximize the pulse amplitude at the sampling instant and should give a zero at other sampling points, i.e., at multiples of $1/B$, where B is the data rate. Although the best choice for this purpose is the sinc-function pulse, in practice, a raised-cosine spectrum pulse is used instead. This is because the sinc-function pulse is very sensitive to changes in the input pulse shape and variations in component values, and because it is impossible to generate an ideal sinc function.

24.4.2.5 Dynamic Range

The dynamic range of an optical receiver quantifies the range of detected power levels within which correct system operation is guaranteed. Dynamic range is conventionally defined as the difference between the minimum input power (which determines sensitivity) and the maximum input power (limited by overload level). Above the overload level, the BER rises owing to the distortion of the received signal.

24.4.3   Transimpedance (TZ) Amplifiers

High-impedance (HZ) amplifiers are effectively open-loop architectures, and exhibit a high gain, but a relatively low bandwidth. The frequency response is similar to that of an integrator, and thus HZ amplifiers require an output equalizer to extend their frequency capabilities. In contrast, the transimpedance (TZ) configuration exploits resistive negative feedback, providing an inherently wider bandwidth and eliminating the need for an output equalizer. In addition, the use of negative feedback provides a relatively low input resistance and thus the architecture is less sensitive to the photodiode parameters. In a TZ amplifier, the photodiode bias resistor R_B can be omitted, since bias current is now supplied through the feedback resistor.

In addition to wider bandwidth, TZ amplifiers offer a larger dynamic range because the transimpedance gain is determined by a linear feedback resistor, and not by a nonlinear open-loop amplifier, as is the case for HZ amplifiers. The dynamic range of TZ amplifiers is set by the maximum voltage swing available at the amplifier output, provided no integration of the received signal occurs at the front end. Since the TZ output stage is a voltage buffer, the voltage swing at the output can be increased with high current operation. The improvement in dynamic range in comparison to the HZ architecture is approximately equal to the ratio of open-loop to closed-loop gain [27]. Conclusively, the TZ configuration offers the better performance compromise compared to the HZ topology, and hence this architecture is preferred in optical receiver applications.

A schematic diagram of a TZ amplifier with PIN photodiode is shown in Figure 24.24. With an open-loop high-gain amplifier and a feedback resistor, the closed-loop transfer function of the TZ amplifier is given by

$$Z_T(s) = \frac{-R_f}{\left(\dfrac{1+A}{A}\right) + sR_f\left[\dfrac{C_{in} + (1+A)C_f}{A}\right]} \cong \frac{R_f}{1 + sR_f\left(\dfrac{C_{in}}{A} + C_f\right)} \tag{24.42}$$

where A is the open-loop mid-band gain of the amplifier that is assumed to be greater than unity, R_f the feedback resistance, C_{in} the total input capacitance of the amplifier including the photodiode and the parasitic capacitance and C_f represents the stray feedback capacitance. The −3 dB bandwidth of TZ amplifier is approximately given by

$$f_{-3dB} = \frac{(1+A)}{2\pi R_f C_T} \tag{24.43}$$

where C_T is the total input capacitance including the photodiode capacitance. The TZ amplifier can thus have wider bandwidth by increasing the open-loop gain, although the open-loop gain cannot be increased indefinitely without stability problems.

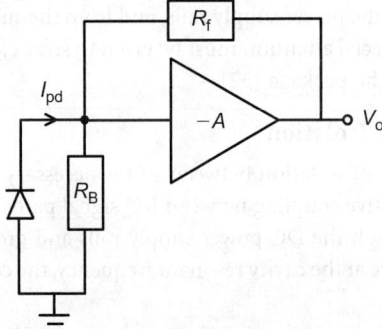

FIGURE 24.24 Schematic diagram of a transimpedance amplifier with photodiode.

However, a trade-off between low noise and wide bandwidth exists, since the equivalent input noise current spectral density of the TZ amplifier is given by

$$S_{eq}(f) = \frac{4kT}{R_f} + \frac{4kT}{R_B} + S_i(f)S_v(f)\left[\left(\frac{1}{R_f} + \frac{1}{R_B}\right)^2 + \left(2\pi f\right)^2\left(C_{pd} + C_{in}\right)^2\right] \quad (24.44)$$

where C_{in} is the input capacitance of the input transistor. Increasing the value of R_f not only reduces the noise current in Eq. (24.44) but also shrinks the bandwidth in Eq. (24.43). This conflict may be mitigated by making A in Eq. (24.43) as large as the closed-loop stability allows [28]. However, the feedback resistance R_f cannot be increased indefinitely owing to the dynamic range requirements of the amplifier, since too large a feedback resistance causes the amplifier to be overloaded at high signal levels. This overloading can be avoided by using automatic gain control circuitry which automatically reduces the transimpedance gain in discrete steps to keep the peak output signal constant [27].

The upper limit of R_f is set by the peak amplitude of the input signal. Since the DC transimpedance gain is approximately equal to the feedback resistance R_f, the output voltage is given by $I_{pd}R_f$, where I_{pd} is the signal photocurrent. If this output voltage exceeds the maximum voltage swing at the output, the amplifier will be saturated and the output will be distorted, yielding bit errors. The minimum value of R_f is determined by the output signal level at which the performance of the receiver is degraded owing to noise and offsets. For typical fiber-optic communication systems, the input signal power is unknown, and may vary from just above the noise floor to a value great enough to generate 0.5 mA at the detector diode [29].

The TZ configuration has some disadvantages over the HZ amplifiers. The power consumption is fairly high, partly owing to the broadband operation provided by the negative feedback. A propagation delay exists in the closed loop of the feedback amplifier which may reduce the phase margin of the amplifier and cause peaking in the frequency response. Additionally, any stray feedback capacitance C_f will further deteriorate the AC performance.

Among the three types of TZ configuration in CMOS technology (CS, common-drain and common-gate TZ amplifiers), the common-gate configuration has potentially the highest bandwidth owing to its inherently lower input resistance. Using a common-gate input configuration, the resulting amplifier bandwidth can be made independent of the photodiode capacitance (which is usually the limiting factor toward achieving gigahertz preamplifier designs). Recently, a novel common-gate TZ amplifier has been demonstrated which shows superior performance compared with various other configurations [30,31].

24.4.4 Layout for HF Operation

Wide-band high-gain amplifiers have isolation problems irrespective of the choice of technology. Coupling from output to input, from the power supply rails, and from the substrate are all possible. Therefore careful layout is necessary, and special attention must be given to stray capacitance both on the integrated circuit (IC) and associated with the package [32].

24.4.4.1 Input/Output (I/O) Isolation

For stable operation, a high level of isolation between I/O is necessary. Three main factors degrade the I/O isolation [33,34]; first, capacitive coupling between I/O signal paths through the air and through the substrate, second, feedback through the DC power supply rails and ground-line inductance, and third, the package cavity resonance, since at the cavity resonant frequency, the coupling between I/O can become very large.

In order to reduce the unwanted coupling (or to provide good isolation, typically more than 60 dB) between I/O, the I/O pads should be laid out to be diagonally opposite each other on the chip with a thin "left-to-right" geometry between the I/O. The small input signal enters on the left-hand side of the chip, while the large output signal exits on the far right-hand side. This helps to isolate the sensitive input stages from the larger signal output stages [35,36].

Using fine line widths and shielding are effective techniques to reduce coupling through the air. Substrate coupling can be reduced by shielding and by using a thin and low-dielectric substrate. Akazawa et al. [33] suggest a structure for effective isolation: a coaxial-like signal-line for high shielding, and a very thin dielectric DC feed-line structure for low characteristic impedance.

24.4.4.2   Reduction of Feedback Through the Power Supply Rails

Careful attention should be given to the layout of power supply rails for stable operation and gain flatness. Power lines are generally inductive, thus on-chip capacitive decoupling is necessary to reduce the high-frequency power line impedance. However, a resonance between these inductive and capacitive components may occur at frequencies as low as several hundreds of megahertz, causing a serious dip in the gain-frequency response and an upward peaking in the isolation-frequency characteristics. One way to reduce this resonance is to add a series damping resistor to the power supply line, making the Q factor of the LC (inductor-capacitor) resonance small. Additionally, the power supply line should be widened to reduce the characteristic impedance/inductance. In practice, if the characteristic impedance is as small as several ohms, the dip and peaking do not occur even without resistive termination [33].

Resonance also occurs between the IC pad capacitance (C_{pad}) and the bond-wire inductance (L_{bond}). This resonance frequency is typically above 2 GHz in miniature RF packages. Also in layout, the power supply rails of each IC chip stage should be split from the other stages to reduce the parasitic feedback (or coupling effect through wire-bonding inductance) which causes oscillation [34]. This helps to minimise crosstalk through power supply rail. The IC is powered through several pads and each pad is individually bonded to the power supply line.

24.4.4.3   I/O Pads

The bond pads on the critical signal path (e.g., input pad and output pads) should be made as small as possible to minimize the pad-to-substrate capacitance [35]. A floating n-well placed underneath the pad will further reduce the pad since the well capacitance will appear in series with the pad capacitance. This floating well also prevents the pad metal from spiking into the substrate.

24.4.4.4   High-Frequency (HF) Ground

The best possible HF grounds to the sources of the driver devices (and hence the minimization of interstage cross talk) can be obtained by separate bonding of each source pad of the driver MOSFETs to the ground plane that is very close to the chip [36]. A typical bond wire has a self-inductance of a few nanohertz, which can cause serious peaking within the bandwidth of amplifiers or even instability. By using multiple bond wires in parallel, the ground line inductance can be reduced to <1 nH.

24.4.4.5   Flip-chip Connection

In noisy environments, the noise-insensitive benefits of optical fibers may be lost at the receiver connection between the photodiode and the preamplifier. Therefore proper shielding or the integration of both components onto the same substrate is necessary to prevent this problem. However, proper shielding is costly, while integration restricts the design to GaAs technologies.

As an alternative, the flip-chip interconnection technique using solder-bumps has been used [37,38]. Small solder bumps minimize the parasitics owing to the short interconnection lengths and avoid damages by mechanical stress. Moreover, the technique needs relatively low-temperature bonding and hence further reduces damage to the devices. Easy alignment and precise positioning of the bonding can be obtained by a self-alignment effect. Loose chip alignment is sufficient because the surface tension of the molten solder during reflow produces precise self-alignment of the pads [34]. Solder bumps are fabricated onto the photodiode junction area to reduce parasitic inductance between the photodiode and the preamplifier.

24.5   Fundamentals of RF Power Amplifier Design

24.5.1   Power Amplifier (PA) Requirements

An important functional block in wireless communication transceivers is the PA. The transceiver PA takes as input the modulated signal to be transmitted, and amplifies this to the power level required to drive the antenna. Because the levels of power required to transmit the signal reliably are often fairly high, the PA is one of the major sources of power consumption in the transceiver. In many systems, power consumption may not be a major concern, as long as the signal can be transmitted with adequate power. For battery-powered systems, however, the limited amount of available energy means that the power consumed by all devices must be minimized so as to extend the transmit time. Therefore, power efficiency is one of the most important factors when evaluating the performance of a wireless system.

The basic requirement for a PA is the ability to work at low supply voltages as well as high operating frequencies, and the design becomes especially difficult owing to the trade-offs between supply voltage, output power, distortion, and power efficiency which can be made. Moreover, since the PA deals with large signals, small-signal analysis methods cannot be applied directly. As a result, both the analysis and the design of PAs are challenging tasks.

This section will first present a study of various configurations employed in the design of state-of-the-art nonlinear RF PAs. Practical considerations toward achieving full integration of PAs in CMOS technology will also be highlighted.

24.5.2   PA Classification

PAs currently employed for wireless communication applications can be classified into two categories: linear power amplifiers and nonlinear power amplifiers. For linear PAs, the output signal is controlled by the amplitude, frequency, and phase of the input signal. Conversely for nonlinear PAs, the output signal is only controlled by the frequency of input signal.

Conventionally, linear PAs can be classified as Class A, Class B, or Class AB. These PAs produce a magnified replica of the input signal voltage or current waveform, and are typically used where accurate reproduction of both the envelope and the phase of the signal is required. However, either poor power efficiency or large distortion prevents them from being extensively employed in wireless communications.

Many applications do not require linear RF amplification. Gaussian Minimum Shift Keying (GMSK) [39], the modulation scheme used in the European standard for mobile communications (GSM), is an example of constant envelope modulation. In this case, the system can make use of the greater efficiency and simplicity offered by nonlinear PAs. The increased efficiency of nonlinear PAs such as Class C, Class D, and Class E, results from techniques that reduce the average collector voltage–current product (i.e., power dissipation) in the switching device. Theoretically, these switching-mode PAs have 100% power efficiency since ideally there is no power loss in the switching device.

24.5.2.1   Linear PAs

24.5.2.1.1   *Class A*

The basic structure of the class A PA is shown in Figure 24.25 [40]. For Class A amplification, the conduction angle of the device is 360°, that is the transistor is in its active region for the entire input cycle. The serious shortcoming with Class A PAs is their inherently poor power efficiency, since the transistor is always dissipating power. The efficiency of a single-ended Class A PA is ideally limited to 50%. However, in practice, few designs can reach this ideal efficiency owing to additional power loss in the passive components. In an inductorless configuration, the efficiency is only about 25% [41].

24.5.2.1.2   *Class B*

A PA is defined as Class B when the conduction angle for each transistor of a push–pull pair is 180° during any one cycle. Figure 24.26 shows an inductorless Class B PA. Since each transistor only conducts

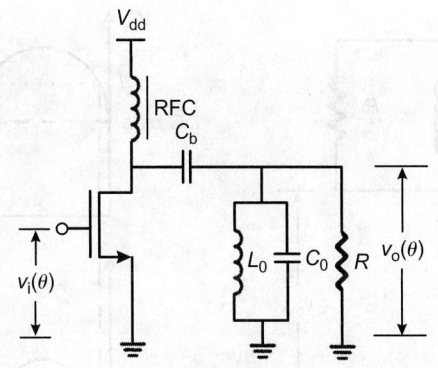

FIGURE 24.25 Single-ended Class A PA.

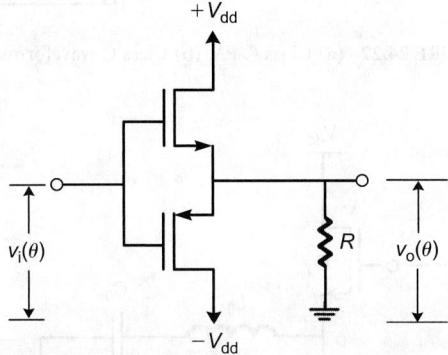

FIGURE 24.26 Inductorless Class B PA.

for half of the cycle, the output suffers crossover distortion owing to the finite threshold voltage of each transistor. When no signal is applied, there is no current flowing, and as a result any current through either device flows directly to the load, thereby maximazing the efficiency. The ideal efficiency can reach 78% [41] allowing this architecture to be of use in applications where linearity is not the main concern.

24.5.2.1.3 *Class AB*
The basic idea of Class AB amplification is to preserve the Class B push–pull configuration while improving the linearity by biasing each device slightly above the threshold. The implementation of Class AB PAs is similar to that of Class B configurations. By allowing the two devices to conduct current for a short period, the output voltage waveform during the crossover period can be smoothed which thus reduces the crossover distortion of the output signal.

24.5.2.2   Nonlinear PAs

24.5.2.2.1 *Class C*
A Class C PA is the most popular nonlinear PA used in the RF band. The conduction angle is less than 180° since the switching transistor is biased on the verge of conduction. A portion of the input signal will make the transistor operate in the amplifying region, and thus the drain current of the transistor is a pulsed signal. Figures 24.27(a) and (b) show the basic configuration of a Class C PA and its corresponding waveforms; clearly the input and output voltages are not linearly related.

The efficiency of an ideal Class C amplifier is 100% since at any point in time either the voltage or the current waveform is zero. In practice this ideal situation cannot be achieved, and the power efficiency

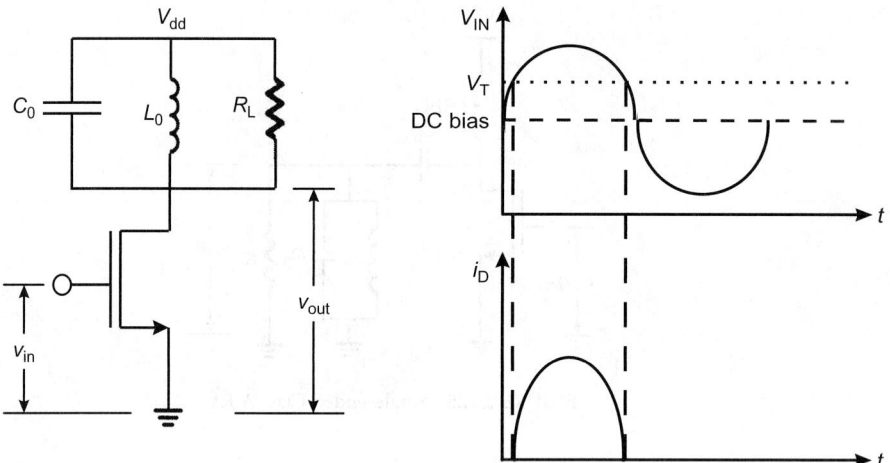

FIGURE 24.27 (a) Class C PA; (b) Class C waveforms.

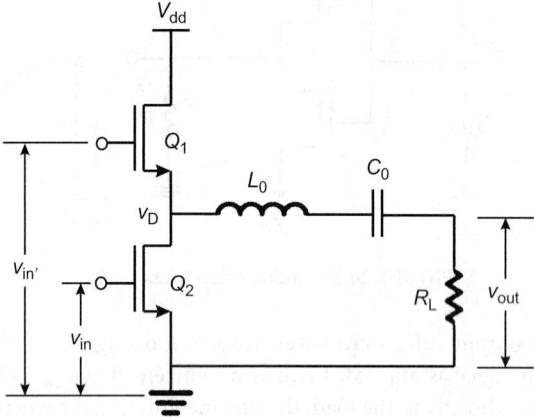

FIGURE 24.28 Class D PA.

should be maximized by reducing the power loss in the transistor. That is, minimize the current through the transistor when the voltage across the output is high, and minimize the voltage across the output when the current flows through the device.

24.5.2.2.2 *Class D*

A Class D amplifier employs a pair of transistors and a tuned output circuit, where the transistors are driven to act as a two-pole switch and the output circuit is tuned to the switching frequency. The theoretical power efficiency is 100%. Figure 24.28 shows the voltage-switching configuration of a Class D amplifier. The input signals of transistors Q_1 and Q_2 are out of phase, and consequently, when Q_1 is on, Q_2 is off, and vice versa. Since the load network is a tuned circuit, we can assume that it provides little impedance to the operating frequency of the voltage v_d and high impedance to other harmonics. Since v_d is a square wave, its Fourier expansion is given by

$$v_d(\omega t) = V_{DC}\left[\frac{1}{2} + \frac{2}{\pi}\sin(\omega t) + \frac{2}{3\pi}\sin(3\omega t) + \dots\right] \tag{24.45}$$

The impedance of the RLC series load at resonance is equal to R_L, and thus the current is given by

$$i_L(\omega t) = \frac{2V_{DC}}{\pi R_L}\sin(\omega t) \qquad (24.46)$$

Each of the devices carries the current during one half of the switching cycle. Therefore, the output power is given by

$$P_o = \frac{2}{\pi^2}\frac{V_{DC}^2}{R_L} \qquad (24.47)$$

Design efforts should focus on reducing the switching loss of both transistors as well as generating the input driving signals.

24.5.2.2.3 *Class E*

The idea behind the Class E PA is to employ nonoverlapping output voltage and output current waveforms. Several criteria for optimizing the performance can be found in [42]. Following these guidelines, Class E PAs have high power efficiency, simplicity, and relatively high tolerance to circuit variations [43]. Since there is no power loss in the transistor as well as in the other passive components, the ideal power efficiency is 100%. Figure 24.29 shows a class E PA and the corresponding waveforms are given in Figure 24.30.

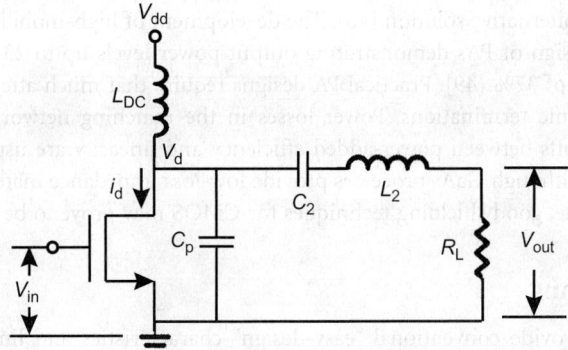

FIGURE 24.29 Class E PA.

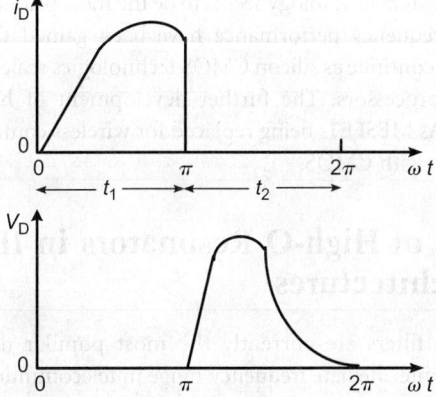

FIGURE 24.30 Waveforms of Class E operation.

The Class E waveforms indicate that the transistor should be completely off before the voltage across it changes, and that the device should be completely on before it starts to allow current to flow through it. References [44,45] demonstrate practical Class E operation at RF frequencies using a GaAs process.

24.5.3 Practical Considerations for RF Power Amplifiers

More recently, single-chip solutions for RF transceivers have become a goal for modern wireless communications owing to potential savings in power, size and cost. CMOS must clearly be the technology of choice for a single-chip transceiver owing to the large amount of digital baseband processing required. However, the PA design presents a bottleneck toward full integration, since CMOS PAs are still not available. The requirements of low supply voltage, gigahertz-band operation, and high output power make the implementation of CMOS PAs very demanding. The proposal of "microcell" communications may lead to a relaxed demand for output power levels which can be met by designs such as that described in Reference [46], where a CMOS Class C PA has demonstrated up to 50% power efficiency with 20 mW output power.

Nonlinear PAs seem to be popular for modern wireless communications owing to their inherent high power efficiency. Since significant power losses occur in the passive inductors as well as the switching devices, the availability of on-chip low-loss passive inductors is important. The implementation of CMOS on-chip spiral inductors has therefore become an active research topic [47].

Owing to the poor spectral efficiency of a constant envelope modulation scheme, the high-power efficiency benefit of nonlinear PAs is eliminated. A recently proposed linear transmitter using a nonlinear PA may prove to be an alternative solution [48]. The development of high-mobility devices such as SiGe HBTs has led to the design of PAs demonstrating output power levels up to 23 dBm at 1.9 GHz with power-added efficiency of 37% [49]. Practical PA designs require that much attention be paid to issues of package and harmonic terminations. Power losses in the matching networks must be absolutely minimized, and trade-offs between power-added efficiency and linearity are usually achieved through impedance matching. Although GaAs processes provide low-loss impedance matching structures on the semi-insulating substrate, good shielding techniques for CMOS may prove to be another alternative.

24.5.4 Conclusions

Although linear PAs provide conventional "easy-design" characteristics and linearity for modulation schemes such as $\pi/4$-DQPSK, modern wireless transceivers are more likely to employ nonlinear PAs owing to their much higher power efficiency. As the development of high-quality on-chip passive components makes progress, the trend toward full integration of the PA is becoming increasingly plausible.

The rapid development of CMOS technology seems to be the most promising choice for PA integration, and vast improvements in frequency performance have been gained through device scaling. These improvements are expected to continue as silicon CMOS technologies scale further, driven by the demand for high-performance microprocessors. The further development of high-mobility devices such as SiGe HBTs may finally see GaAs MESFETs being replaced for wireless communication applications, since SiGe technology is compatible with CMOS.

24.6 Applications of High-Q Resonators in IF-Sampling Receiver Architectures

Transconductance-C (gm-C) filters are currently the most popular design approach for realizing continuous-time filters in the intermediate frequency range in telecommunications systems. This section will consider the special application area of high-Q resonators for receiver architectures employing IF-sampling.

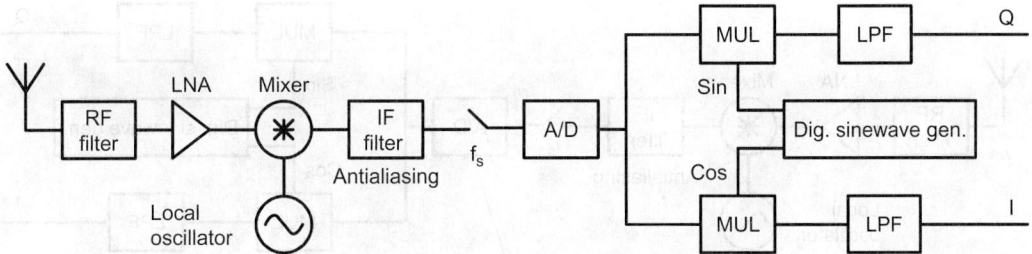

FIGURE 24.31   IF-sampling receiver.

24.6.1   IF Sampling

A design approach for contemporary receiver architectures that is currently gaining popularity is IF digitization, whereby low-frequency operations such as second mixing and filtering can be performed more efficiently in the digital domain. A typical architecture is shown in Figure 24.31. The IF signal is digitized, mixed with the quadrature phases of a digital sinusoid, and lowpass filtered to yield the quadrature baseband signals. Since processing takes place in the digital domain, I/Q mismatch problems are eliminated. The principal issue in this approach, however, is the performance required from the A/D converter. Noise referred to the input of the A/D must be very low so that selectivity remains high. At the same time, the linearity of the A/D must be high to minimize corruption of the desired signal through intermodulation effects. Both the above requirements should be achieved at an input bandwidth commensurate with the value of the IF frequency, and at an acceptable power budget.

Oversampling has become popular in recent years because it avoids many of the difficulties encountered with conventional methods for A/D and D/A conversion. Conventional converters are often difficult to implement in fine-line VLSI technology, because they require precise analog components and are very sensitive to noise and interference. In contrast, oversampling converters trade off resolution in time for resolution in amplitude, in such a way that the imprecise nature of the analog circuits can be tolerated. At the same time, they make extensive use of digital signal-processing power, taking advantage of the fact that fine-line VLSI is better suited for providing fast digital circuits than for providing precise analog circuits. Therefore, IF-digitization techniques utilizing oversampling Sigma–Delta modulators are very well suited to modern submicron CMOS technologies, and their potential has made them the subject of active research.

Most Delta–Sigma modulators are implemented with discrete-time circuits, switched-capacitor (SC) implementations being by far the most common. This is mainly due to the ease with which monolithic SC filters can be designed as well as the high linearity which they offer. The demand for high speed $\Sigma\Delta$ oversampling A/D converters, especially for converting bandpass signals, makes it necessary to look for a technique that is faster than SC. This demand has stimulated researchers to develop a method for designing continuous-time $\Delta\Sigma$ A/Ds. Although continuous-time modulators are not easy to integrate, they possess a key advantage over their discrete-time counterparts. The sampling operation takes place inside the modulator loop, making it is possible to "noise-shape" the errors introduced by sampling, and provide a certain amount of anti-aliasing filtering at no cost. On the other hand, they are sensitive to memory effects in the D/As and are very sensitive to jitter. They must also process continuous-time signals with high linearity. In communications applications, meeting the latter requirement is complicated by the fact that the signals are located at very high frequencies.

As shown in Figure 24.32, integrated bandpass implementations of continuous-time modulators require integrated continuous-time resonators to provide the noise shaping function. The gm-C approach of realizing continuous-time resonators offers advantages of complete system integration and total design freedom. However, the design of CMOS high-Q high-linearity resonators at the tens of megahertz is very challenging. Since the linearity of the modulator is limited by the linearity of the resonators utilized, the continuous-time resonator is considered to be the most demanding analog subblock of a bandpass

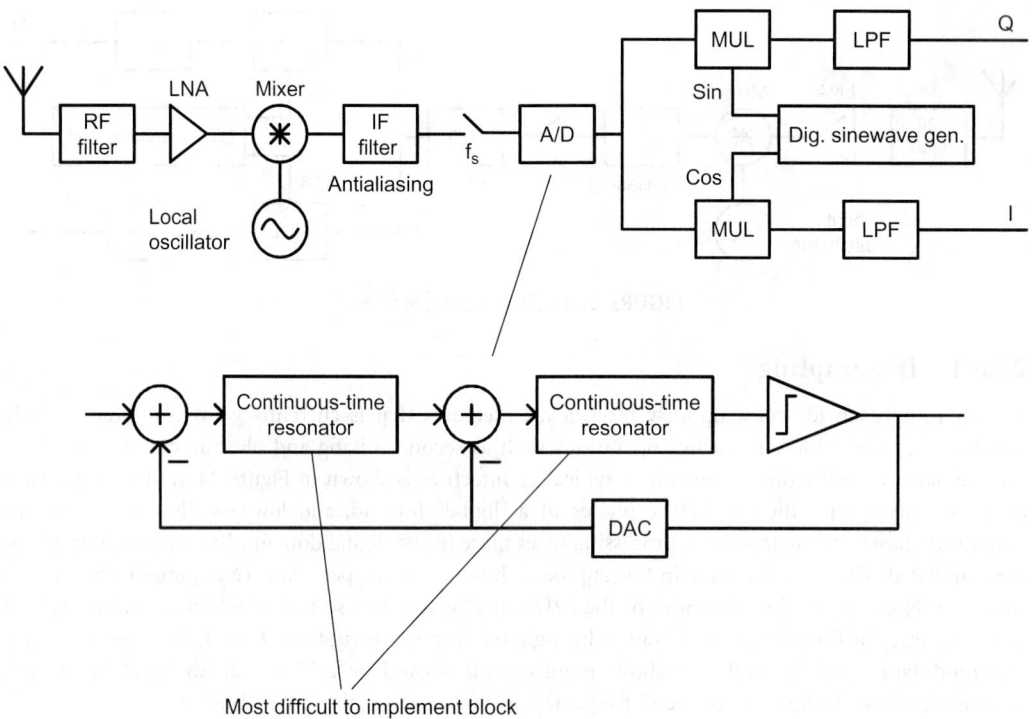

Most difficult to implement block

FIGURE 24.32 Continuous-time ΣΔ A/D in IF-sampling receiver.

TABLE 24.4 Fully Integrated Continuous-Time
Resonator Specifications

Resonator specifications	
Centre frequency	50 MHz
Quality factor	50
Spurious-Free Dynamic Range	>30 dB
Power dissipation	Minimal

continuous-time Sigma–Delta modulator. Typical specifications for a gm-C resonator used to provide the noise-shaping function in a ΣΔ modulator in a mobile receiver (see Figure 24.32) are summarized in Table 24.4.

24.6.2   Linear Region Transconductor Implementation

The implementations of fully integrated high-selectivity filters operating at tens to hundreds of megahertz provides benefits for wireless transceiver design, including chip area economy and cost reduction. The main disadvantages of on-chip active filter implementations when compared to off-chip passives include increased power dissipation, deterioration in the available dynamic range with increasing Q, and Q and resonant frequency integrity (because of process variations, temperature drifts, and aging, automatic tuning is often unavoidable especially in high-Q applications). The transconductor-capacitor (Gm-C) technique is a popular technique for implementing high-speed continuous-time filters and is widely used in many industrial applications [50]. Because Gm-C filters are based on integrators built from an open-loop transconductance amplifier driving a capacitor, they are typically very fast, but have limited linear dynamic range. Linearization techniques which reduce distortion levels can be used, but often lead to a compromise between speed, dynamic range, and power consumption.

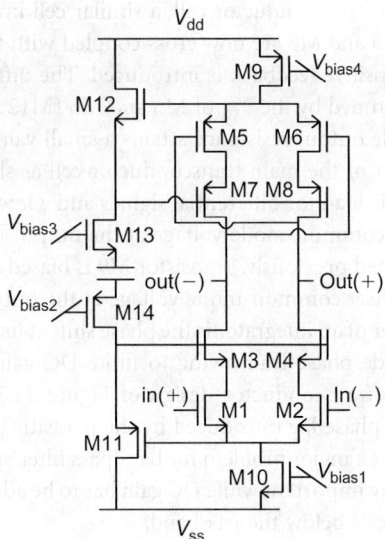

FIGURE 24.33 Triode region transconductor.

As an example of the trade-offs in design, consider the transconductor shown in Figure 24.33. This design consists of a main transconductor cell (M1, M2, M3, M4, M10, M11, and M14) with a negative resistance load (M5, M6, M7, M8, M9, M12, and M13). Transistors M1 and M2 are biased in the triode region of operation using cascode devices M3 and M4 and determine the transconductance gain of the cell. In the triode region of operation the drain current versus terminal voltage relation can be approximated (for simple hand calculations) as $I_D = K\left[2\left(V_{GS} - V_T\right)V_{GS} - V_{DS}^2\right]$, where K and V_T are the transconductance parameter and the threshold voltage, respectively. Assuming that V_{DS} is constant for both M1 and M2, both the differential-mode and the common-mode transconductance gains can be derived as $G_{DM} = G_{CM} = 2KV_{DS}$, which can thus be tuned by varying V_{DS}.

The high value of common-mode transconductance is undesirable since it may result in regenerative feedback loops in high-order filters. To improve the CMRR transistor and avoid the formation of such loops, M10 is used to bias the transconductor thus transforming it from a pseudodifferential to a fully differential transconductor [51]. Transistors M11 and M14 constitute a floating voltage source, thus maintaining a constant drain-source voltage for M1 and M2.

The nonlinearities in the voltage to current transfer of this stage are mainly due to three effects. The first is the finite impedance levels at the sources of the cascode devices, which cause a signal-dependent variation of the corresponding drain-source voltages of M1 and M2. A fast-floating voltage source and large cascode transistors therefore need to be used to minimize this nonlinearity. The second cause of nonlinearity is the variation of carrier mobility μ of the input devices M1 and M2 with $V_{GS} - V_T$, which becomes more apparent when short-channel devices are used ($K = \mu C_{ox} W/2L$). A simple first-order model for transverse-field mobility degradation is given by $\mu = \mu_0/(1 + \theta(V_{GS} - V_T))$, where μ_0 and θ are the zero field mobility and the mobility reduction parameter, respectively. Using this model the third-order distortion can be determined by a Maclaurin series expansion as $\theta^2/4(1 + \theta(V_{CM} - V_T))$ [52]. This expression cannot be regarded as exact, although is useful to obtain insight. Furthermore, it is valid only at low frequencies, where reactive effects can be ignored and the coefficients of the Maclaurin series expansion are frequency independent. At high frequencies or when very low values of distortion are predicted by the Maclaurin series method, a generalized power series method (Volterra series) must be employed [53,54]. Finally, a further cause of nonlinearity is mismatch between M1 and M2 which can be minimized by good layout. A detailed linearity analysis of this transconductance stage is presented in Ref. [55].

To provide a load for the main transconductor cell, a similar cell implemented by p-devices is used. The gates of the linear devices M5 and M6 are now cross coupled with the drains of the cascode devices M7 and M8. In this way, weak positive feedback is introduced. The differential-mode output resistance can now become negative and is tuned by the V_{DS} of M5 and M6 (M12 and M13 form a floating voltage source), while the common-mode output resistance attains a small value.

When connected to the output of the main transconductor cell as shown in Figure 24.33, the cross-coupled p cell forms a high-ohmic load for differential signals and a low-ohmic load for common-mode signals, resulting in a controlled common-mode voltage at the output [52,56]. CMRR can be increased even further using M10 as described previously. Transistor M9 is biased in the triode region of operation and is used to compensate the offset common-mode voltage at the output.

The key performance parameter of an integrator is the phase shift at its unity gain frequency. Deviations from the ideal $-90°$ phase include phase lead owing to finite DC gain and phase lag owing to high-frequency parasitic poles. In the transconductor design of Figure 24.33, DC gain is traded for phase accuracy, thus compensating the phase lag introduced by the parasitic poles. The reduction in DC gain for increased phase accuracy is not a major problem for bandpass filter applications, since phase accuracy at the center frequency is extremely important while DC gain has to be adequate to ensure that attenuation specifications are met at frequencies below the passband.

From simulation results using parameters from a 0.8 μm CMOS process, with the transconductor unity gain frequency set at 50 MHz, third-order intermodulation components were observed at -78 dB with respect to the fundamental signals (two input signals at 49.9 and 50.1 MHz were applied, each at 50 mV$_{pp}$).

24.6.3 A Gm-C Bandpass Biquad

24.6.3.1 Filter Implementation

The implementation of on-chip high-Q resonant circuits presents a difficult challenge. Integrated passive inductors have generally poor quality factors which limits the Q of any resonant network in which they are employed. For applications in the hundreds of megahertz to a few gigahertz, one approach is to implement the resonant circuit using low-Q passive on-chip inductors with additional Q-enhancing circuitry. However, for lower frequencies (tens of megahertz), on-chip inductors occupy a huge area and this approach is not attractive.

As disscussed above, an alternative method is to use active circuitry to eliminate the need for inductors. Gm-C based implementations are attractive owing to their high-speed potential and good tuneability. A bandpass biquadratic section based upon the transconductor of Figure 24.33 is shown in Figure 24.34. The transfer function of Figure 24.34 is given by:

$$\frac{V_o}{V_i} = \frac{g_{mi}R_o}{\left(R_oC\right)^2} \cdot \frac{\left(1 + sR_oC\right)}{\left\{s^2 + s\dfrac{2R_o + g_m^2R_o^2R}{R_o^2C} + \dfrac{1 + g_m^2R_o^2}{R_o^2C^2}\right\}} \tag{24.48}$$

R_o represents the total resistance at the nodes owing to the finite output resistance of the transconductors. R represents the effective resistance of the linear region transistors in the transconductor (see Figure 24.33), and is used here to introduce damping and control the Q. From Eq. (24.48) it can be shown that $\omega_o \approx g_m/C$, $Q \approx g_m R_o/(2 + R_o R g_m^2)$, $Q_{max} = Q|_{R=0} = (g_m R_o)/2$ and $A_o = g_m Q$. Thus, g_m is used to set the central frequency, R to control the Q, and g_{mi} to control the bandpass gain A_o. A dummy g_{mi} is used to provide symmetry and thus better stability owing to process variations, temperature, and aging.

One of the main problems when implementing high-Q high-frequency resonators is maintaining stability of the center frequency ω_o and the quality factor Q. This problem calls for very careful layout

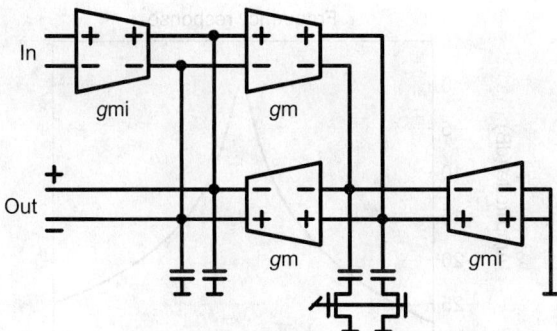

FIGURE 24.34 Biquad bandpass.

and the implementation of an automatic tuning system. Another fundamental limitation associated with available dynamic range occurs, namely that the dynamic range of high-Q Gm-C filters has been found to be inversely proportional to the filter Q [56]. The maximum dynamic range is given by

$$DR = \frac{V_{max}^2}{V_{noise}^2} = \frac{V_{max}^2 \cdot C}{4kT\xi Q} \tag{24.49}$$

where V_{max} is the maximum root mean square (RMS) voltage across the filter capacitors, C the total capacitance, k the Boltzman's constant, T the absolute temperature, and ξ the noise factor of the active circuitry ($\xi = 1$ corresponds to output noise equal to the thermal noise of a resistor of value $R = 1/g_m$, where g_m is the transconductor value used in the filter).

In practice the dynamic range achieved will be less than this maximum value owing to the amplification of both noise and intermodulation components around the resonant frequency. This is a fundamental limitation, and the only solution is to design the transconductors for low noise and high linearity. The linearity performance in narrowband systems is characterized by the spurious-free dynamic range (SFDR). SFDR is defined as the signal to noise ratio (SNR) when the power of the third-order intermodulation products equals the noise power. As shown in [55], the SFDR of the resonator in Figure 24.34 is given by

$$SFDR = \frac{1}{2(kT)^{2/3}} \left(\frac{3V_{o,peak}^2 C}{4\xi IM_{3,int}} \right)^{2/3} \frac{1}{Q^2} \tag{24.50}$$

where $IM_{3,int}$ is the third-order intermodulation point of the integrator used to implement the resonator. The SFDR of the resonator thus deteriorates by 6 dB if the quality factor is doubled, assuming that the output swing remains the same. In contrast, implementing a resonant circuit using low-Q passive on-chip inductors with additional Q-enhancing circuitry leads to a dynamic range amplified by a factor Q_o, where Q_o is the quality factor of the on-chip inductor itself [57]. However, as stated above, for frequencies in the tens of megahertz, on-chip inductors occupy a huge area and thus the Q_o improvement in dynamic range is not high enough to justify the area increase.

24.6.3.2 Simulation Results

To confirm operation, the filter shown in Figure 24.34 has been simulated in HSPICE using process parameters from a commercial 0.8 μm CMOS process. Figure 24.35 shows the simulated frequency and phase response of the filter for a center frequency of 50 MHz and a quality factor of 50. Figure 24.36

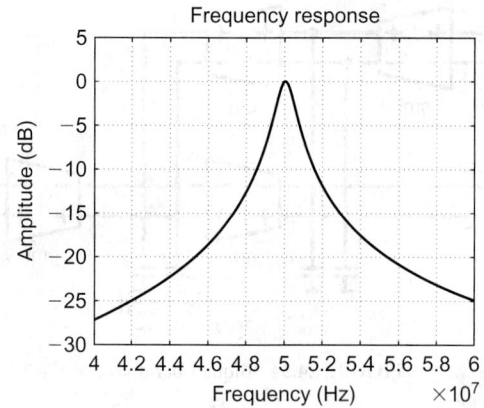

FIGURE 24.35 Simulated bandpass frequency response.

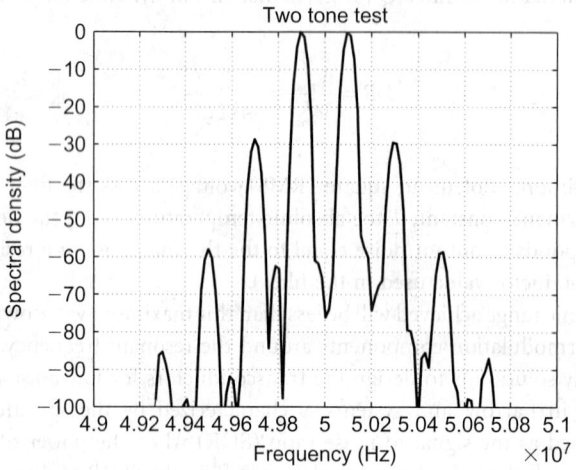

FIGURE 24.36 Simulated two-tone intermodulation test.

TABLE 24.5 Bandpass Biquad Filter Simulation Results

Power dissipation (Supply voltage = 5 V)	12.5 mW
Common-mode output offset	<1 mV
Center frequency	50 MHz
Quality factor	50
Output noise voltage (integrated over the band from 40 to 60 MHz with $Q = 50$)	500 μVRMS
Output signal voltage (so that intermodulation components are at the same level as the noise, $Q = 50$)	25.2 mVRMS
SFDR ($Q = 50$)	34 dB

shows the simulated output of the filter when the input consists of two tones at 49.9 and 50.1 MHz, respectively, each at 40 mV$_{pp}$. At this level of input signal, the third-order intermodulation components were found to be at the same level as the noise. Thus, the predicted SFDR is about 34 dB with $Q = 50$. Table 24.5 summarizes the simulation results.

24.7 Log-Domain Processing

24.7.1 Instantaneous Companding

The concept of "instantaneous companding" is an emerging area of interest within the field of analog IC design. Currently, the main area of application for this technique is the implementation of continuous-time, fully integrated filters with wide dynamic range, high-frequency potential, and wide tuneability.

With the drive toward lower supply voltages and higher operating frequencies, traditional analog IC design methodologies are proving inadequate. Conventional techniques to linearize inherently nonlinear devices require an overhead in terms of increased power consumption or reduced operating speed. Recently, the use of companding, originally developed for audio transmission, has been proposed as an elegant solution to the problem of maintaining dynamic range and high-frequency operation under low supply voltage [58]. In this approach, nonlinear amplifiers are used to compress the dynamic range of the input signal to ensure, for example, that signal levels always remain above a certain noise threshold, but below levels that may cause overload. The overall system must thus adapt its operation to ensure that input–output linearity is maintained as the instantaneous signal level alters. Although the system is input–output linear, signals internal to the system are inherently nonlinear. Companding has traditionally been realized in two ways, syllabic and instantaneous, depending on how the system adapts in response to the changing input signal level. In syllabic companding, the level of compression (and expansion) is a nonlinear function of a slowly varying property of the signal, e.g., envelope or power. In contrast, instantaneous companding adapts the compression and expansion ratios instantaneously with the changing input signal amplitude.

Perhaps, the best-known recent example of this approach is the log-domain technique, where the exponential *I–V* characteristics of bipolar junction transistors (BJTs) are directly exploited to implement input–output linear filters. Since the large signal device equations are utilized, there is no need for small signal operation or local linearization techniques, and the resulting circuits have the potential for wide dynamic range and high-frequency operation under low power supply voltages. Log-domain circuits are generally implemented using BJTs and capacitors only and thus are inherently suitable for integration. The filter parameters are easily tuneable, which makes them robust. In addition, the design procedure is systematic, which suggests that these desirable features can be reproduced for different system implementations. The following section provides an introduction to the synthesis of log-domain filters and highlights various performance issues. Interested readers are advised to consult Refs. [59–74] for a more detailed treatment of the subject.

24.7.2 Log-Domain Filter Synthesis

The synthesis of log-domain filters using state–space transformations was originally proposed by Frey [60]. As an example of this methodology, consider a biquad filter with the following transfer function:

$$Y(s) = \frac{s\omega_o u_1(s) + \omega_o^2 u_2(s)}{s^2 + (\omega_o/Q)s + \omega_o^2} \tag{24.51}$$

Thus, using u_1 as input gives a bandpass response at the output y, while using u_2 as input gives a lowpass response. This system can also be described by the following state–space equations

$$\begin{aligned}
\dot{x}_1 &= -(\omega_o/Q)x_1 - \omega_o x_2 + \omega_o u_1 \\
\dot{x}_2 &= -\omega_o x_1 - \omega_o u_2 \\
y &= x_1
\end{aligned} \tag{24.52}$$

where a "dot" denotes differentiation in time. To transform these linear state equations into non-linear nodal equations that can be directly implemented using bipolar transistors, the following exponential transformations are defined:

$$x_1 = I_S \exp\left(\frac{V_1}{V_t}\right), \qquad u_1 = \left(\frac{I_S^2}{I_o}\right) \exp\left(\frac{V_{in1}}{V_t}\right)$$

$$x_2 = I_o \exp\left(\frac{V_2}{V_t}\right), \qquad u_2 = I_S \exp\left(\frac{V_{in2}}{V_t}\right) \tag{24.53}$$

These transformations map the linear state variables to currents flowing through BJTs biased in the active region. I_S represents the BJT reverse saturation current, while V_t is the thermal voltage. Substituting these transformations into in the state equations 24.52 gives:

$$C\dot{V}_1 = -\frac{I_o}{Q} - \frac{I_o^2}{I_S} \exp\left(\frac{V_2 - V_1}{V_t}\right) + I_S \exp\left(\frac{V_{o1} - V_1}{V_t}\right)$$

$$C\dot{V}_2 = I_S \exp\left(\frac{V_1 - V_2}{V_t}\right) - I_S \exp\left(\frac{V_{o2} - V_2}{V_t}\right) \tag{24.54}$$

$$y = I_S \exp\left(\frac{V_1}{V_t}\right)$$

In Eq. (24.54) a tuning current $I_o = C\omega_o V_t$ is defined, where C is a scaling factor which represents a capacitance. The linear state-space equations have thus been transformed into a set of nonlinear nodal equations, and the task for the designer is now to realize a circuit architecture that will implement these nonlinear equations. Considering the first two equations in Eq. (24.54), the terms on the LHS can be implemented as currents flowing through grounded capacitors of value C connected at nodes V_1 and V_2, respectively. The expressions on the RHS can be implemented by constant current sources in conjunction with appropriately biased bipolar transistors to realize the exponential terms. The third equation in Eq. (24.54) indicates that the output y can be obtained as the collector current of a bipolar transistor biased with a base-emitter voltage V_1.

Figure 24.37 shows a possible circuit implementation, which is derived using Gilbert's translinear circuit principle [61]. The detailed circuit implementation is described in more detail in Ref. [60]. The center frequency of this filter is given by $\omega_o = (I_o/CV_t)$ and can thus be tuned by varying the value of the bias current I_o.

24.7.3 Performance Aspects

24.7.3.1 Tuning Range

Both the quiescent bias current I_o and capacitance value C can be varied to alter the filter response. However, the allowable capacitor values are generally limited to within a certain range. On the lower side, C must not become smaller than the parasitic device capacitance. The base-emitter diffusion capacitance of the transistors is particularly important, since this is generally the largest device capacitance (up to the picofaraday range) and is also nonlinear. In addition, as the value of C decreases, it becomes more difficult to match capacitance values. The silicon area available limits the maximum value of C; for example in a typical technology, a 50 pF poly-poly capacitor consumes 100 μm $\times$ 1000 μm of silicon area.

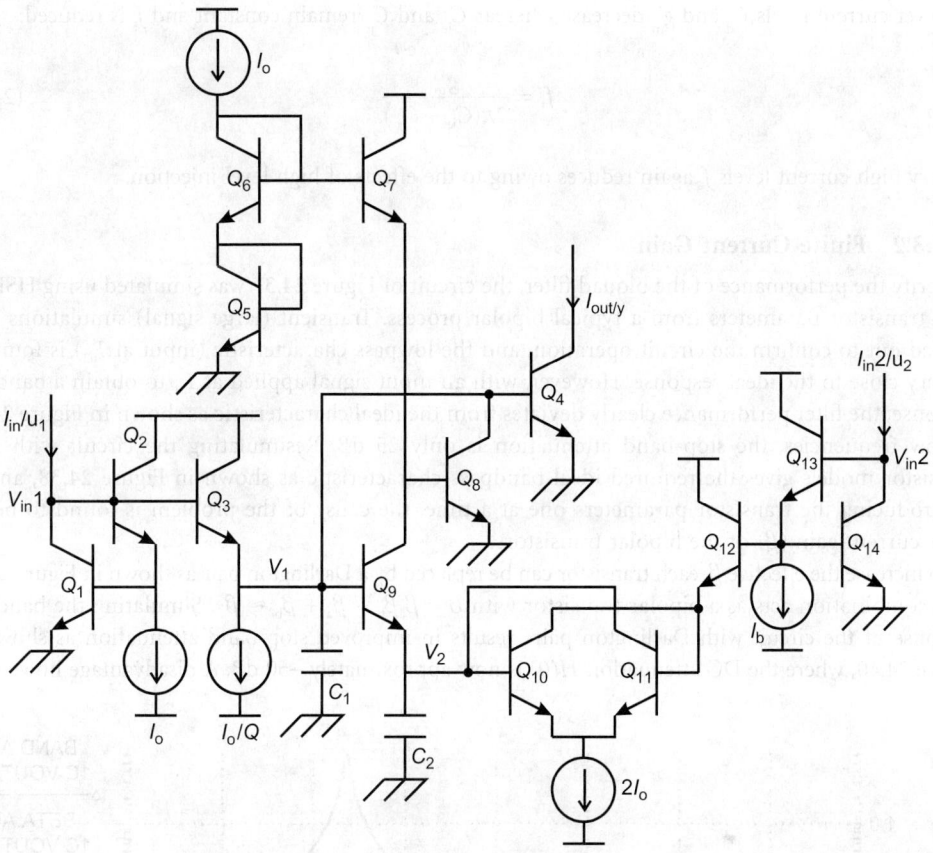

FIGURE 24.37 A log-domain biquadratic filter. A lowpass response is obtained by applying a signal at I_{u_2} and keeping I_{u_1} constant, while a bandpass response is obtained by applying a signal at I_{u_1} and keeping I_{u_2} constant.

The range of allowable currents in a modern BJT is in principle fairly large (several decades); however, at very low and very high current levels, the current gain (β) is degraded. For high-frequency operation particular attention must be paid to the actual f_t of the transistors, which is given by Ref. [62].

$$f_t = \frac{g_m}{2\pi(C_\pi + C_\mu)} = \frac{g_m}{2\pi(C_d + C_{je} + C_{jc})} \tag{24.55}$$

C_{je} and C_{jc} represent the emitter and collector junction capacitance respectively, and are only a weak function of the collector bias current. C_d represents the base-emitter diffusion capacitance and is proportional to the instantaneous collector current; $C_d = (\tau_f I_c / V_t)$ where τ_f is the effective base transit time. g_m represents the device transconductance which is again dependent on the collector current; $g_m = (I_c / V_t)$. At high current levels the diffusion capacitance is much larger than the junction capacitance, and thus

$$f_t = \frac{g_m}{2\pi C_d} = \frac{1}{2\pi \tau_f} \tag{24.56}$$

At lower current levels C_d and g_m decrease whereas C_{je} and C_{jc} remain constant and f_t is reduced:

$$f_t = \frac{g_m}{2\pi(C_{je} + C_{jc})}$$ (24.57)

At very high current levels f_t again reduces owing to the effects of high level injection.

24.7.3.2 Finite Current Gain

To verify the performance of the biquad filter, the circuit of Figure 24.37 was simulated using HSPICE with transistor parameters from a typical bipolar process. Transient (large signal) simulations were carried out to confirm the circuit operation, and the lowpass characteristic (input at I_{u2}) is found to be very close to the ideal response. However, with an input signal applied at I_{u1} to obtain a bandpass response, the filter performance clearly deviates from the ideal characteristic as shown in Figure 24.38. At low frequencies, the stop-band attenuation is only 25 dB. Resimulating the circuit with ideal transistor models gives the required ideal bandpass characteristic as shown in Figure 24.38, and by reintroducing the transistor parameters one at a time, the cause of the problem is found to be the finite current gain (β) of the bipolar transistors.

To increase the effective β, each transistor can be replaced by a Darlington pair as shown in Figure 24.39. This combination acts as a bipolar transistor with $\beta = \beta_a\beta_b + \beta_a + \beta_b \approx \beta^2$. Simulating the bandpass response of the circuit with Darlington pairs results in improved stop-band attenuation as shown in Figure 24.40, where the DC attenuation $H(0)$ is now approximately −50 dB. A disadvantage however is

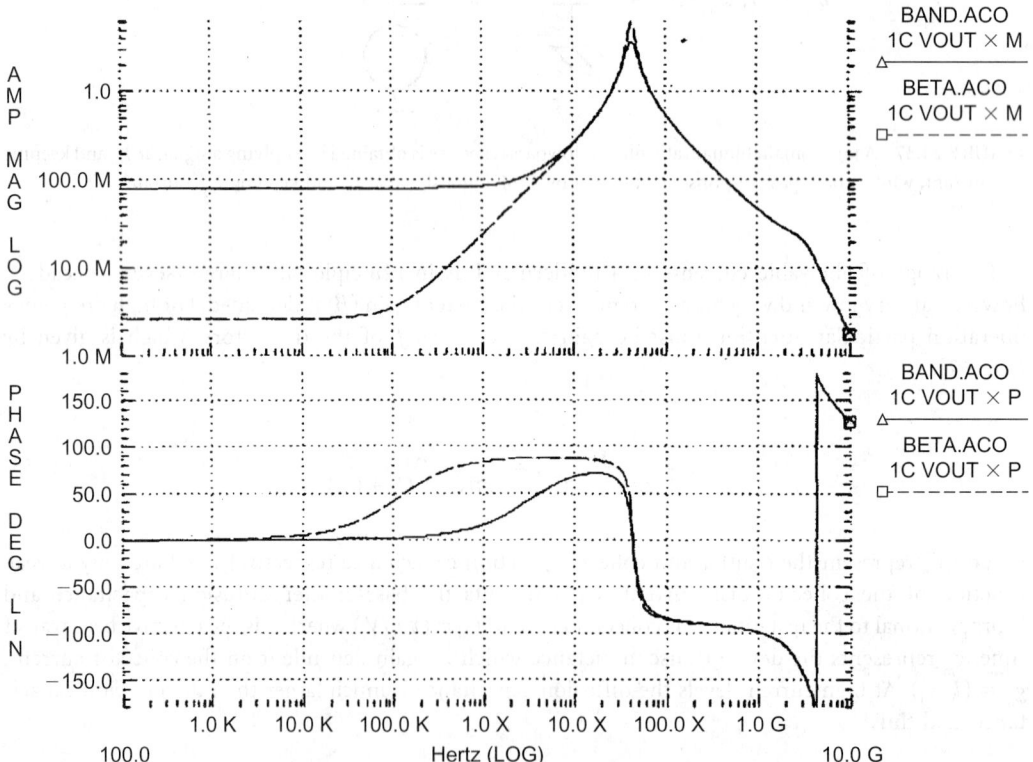

FIGURE 24.38 Bandpass response of the biquad of Figure. Notice the poor low-frequency stop-band attenuation (~25dB) if real transistor models are used (solid line), versus the much better stop-band attenuation with β set to 1000 (dashed line).

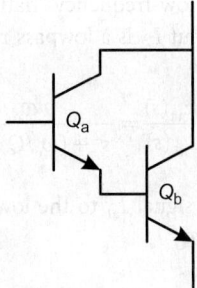

FIGURE 24.39 Darlington pair.

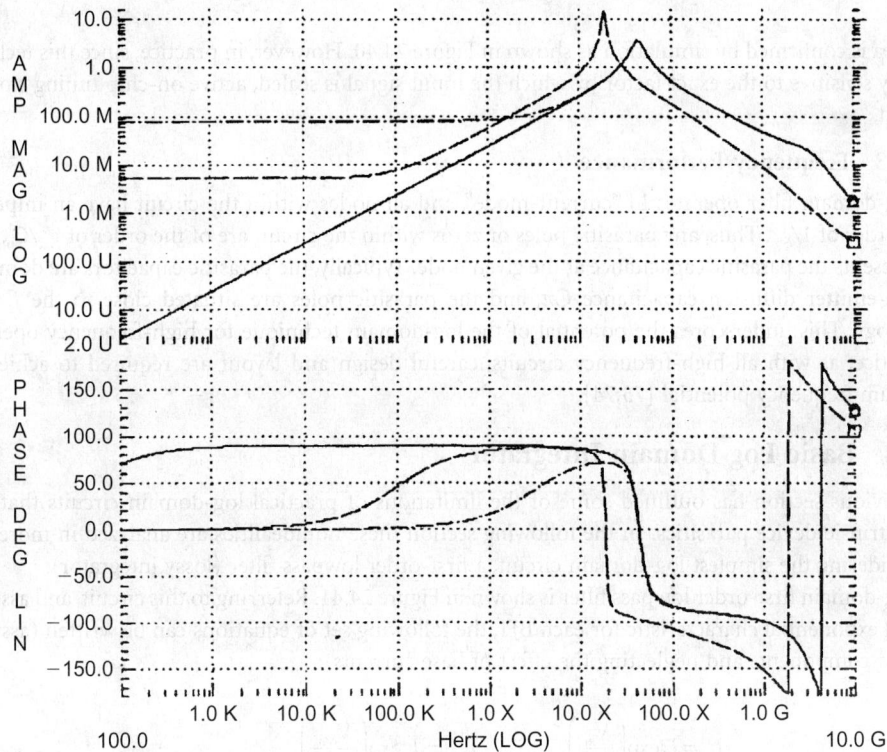

FIGURE 24.40 Possible solutions for the effect of finite β. The "dashed-dotted" line is the original response with a stop-band attenuation of only 25 dB. The dashed line (−50 dB) is the result of replacing all BJTs by Darlington pairs. The solid line is the result of feeding a fraction of the input signal to the second (i.e., lowpass) input.

that a higher supply voltage is now required, since the effective base-emitter voltage V_{be} of the Darlington pair is double that of a single device. In addition, the current in device Q_a of Figure 24.39 is now β times smaller than the design current I_o, resulting in a much lower f_t for this device.

An alternative method to improve the bandpass characteristic is described below. The (nonideal) transfer function from the bandpass input I_{u1} is described by

$$H_1(s) = \frac{I_{out}(s)}{I_{u1}(s)} = \frac{\omega_o(s + \omega_z)}{s^2 + (\omega_o/Q)s + \omega_o^2} \tag{24.58}$$

ω_Z is a parasitic zero, which describes the low-frequency "flattening out" of the bandpass characteristic. The transfer function from the second input I_{u2} is a lowpass response

$$H_2(s) = \frac{I_{out}(s)}{I_{u2}(s)} = \frac{\omega_o^2}{s^2 + (\omega_o/Q)s + \omega_o^2} \tag{24.59}$$

By applying a scaled version of the input signal I_{u1} to the lowpass input I_{u2}, the unwanted zero ω_z can thus be compensated. Setting $I_{u2} = -(\omega_Z/\omega_O)I_{u1}$

$$I_{out}(s) = \frac{\omega_o(s + \omega_z)I_{u1}(s) + \omega_o^2\left(-\dfrac{\omega_z}{\omega_o}I_{u1}(s)\right)}{s^2 + (\omega_o/Q)s + \omega_o^2} = \frac{\omega_o s}{s^2 + (\omega_o/Q)s + \omega_o^2}I_{u1(s)} \tag{24.60}$$

This idea is confirmed by simulation as shown in Figure 24.40. However, in practice, since this technique it is very sensitive to the exact factor by which the input signal is scaled, active on-chip tuning would be required.

24.7.3.3   Frequency Performance

The log-domain filter operates in "current-mode", and all nodes within the circuit have an impedance of the order of $1/g_m$. Thus, any parasitic poles or zeros within the circuit are of the order of g_m/C_p, where C_p represents the parasitic capacitance at the given node. Typically, the parasitic capacitors are dominated by base-emitter diffusion capacitance $C\pi$, and the parasitic poles are situated close to the f_t of the technology. This underscores the potential of the log-domain technique for high-frequency operation. In practice, as with all high-frequency circuits, careful design and layout are required to achieve the maximum frequency potential [73,74].

24.7.4   Basic Log-Domain Integrator

The previous section has outlined some of the limitations of practical log-domain circuits that result from intrinsic device parasitics. In the following section these nonidealities are analyzed in more detail by considering the simplest log-domain circuit, a first-order lowpass filter (lossy integrator).

A log-domain first-order lowpass filter is shown in Figure 24.41. Referring to this circuit, and assuming an ideal exponential characteristic for each BJT, the following set of equations can be written (assuming matched components and neglecting the effect of base currents):

$$I_{in} = I_S \exp\left(\frac{V_{in}}{V_t}\right), \qquad I_{out} = I_S \exp\left(\frac{V_{out}}{V_t}\right)$$

$$V_{out} = V_c + V_t \ln\left(\frac{I_o}{I_S}\right), \qquad I_o + C\frac{dV_c}{dt} = I_S \exp\left(\frac{V_{in} - V_c}{V_t}\right) \tag{24.61}$$

Combining these equations results in the following linear first-order differential equation:

$$\left(\frac{CV_t}{I_o}\right)\frac{dI_{out}}{dt} + I_{out} = I_{in} \tag{24.62}$$

Taking the Laplace transform produces the following linear transfer function:

$$\frac{I_{out}(s)}{I_{in}(s)} = \frac{I_o}{I_o + sCV_t} \tag{24.63}$$

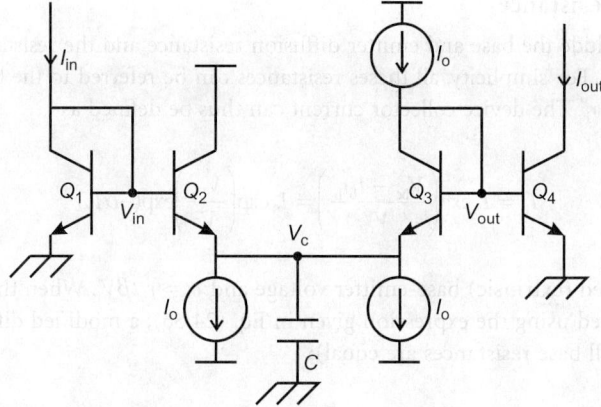

FIGURE 24.41 A first-order lowpass filter (lossy integrator).

Eq. (24.63) describes a first-order lowpass circuit with −3 dB frequency $f_c = I_o/2\pi CV_t$. This transfer function has been derived assuming an ideal exponential BJT characteristic; however, in practice, device nonidealities will introduce deviations from this ideal exponential characteristic, resulting in transfer function errors and distortion. This is similar to the case of "conventional" linear system design, where any deviation from the ideal linear building-block response will contribute to output distortion. A brief discussion of the performance limitations of log-domain filters owing to device nonidealities is given below; further discussion of distortion and performance limitations can be found in Refs. [63,64].

24.7.4.1 Effects of Finite Current Gain

Assuming all transistors have equal β, Eq. (24.62) is modified to the following nonlinear differential equation (neglecting terms with β^2 or higher in the denominator):

$$\left(\frac{CV_t}{I_o}\right)\frac{dI_{out}}{dt}\left(1 + \frac{I_o}{I_{out}(1+\beta)}\right) + I_{out}\left(1 + \frac{I_{out}}{\beta I_o}\right) + \frac{2I_o}{1+\beta} = I_{in}\left(1 + \frac{I_{in}}{\beta}\right) \tag{24.64}$$

An analytical solution to Eq. (24.64) is difficult to obtain, thus a qualitative discussion is given. At low frequencies, neglecting the differential terms, finite β causes a DC gain error and quadratic (even-order) distortion. At high frequencies, the differential term is modified depending on the values of I_{out} and β, therefore scalar error and modulation of the 3 dB frequency are expected. In practice, the effects of finite β are further complicated owing to its dependence on frequency and device collector current.

24.7.4.2 Component Mismatch

Emitter area mismatches cause variations in the saturation current I_S between transistors. Taking the emitter area into account, Eq. (24.62) can be rewritten as

$$\left(\frac{CV_t}{I_o}\right)\frac{dI_{out}}{dt} + I_{out} = \lambda I_{in} \tag{24.65}$$

where $\lambda = (I_{S3}I_{S4}/I_{S1}I_{S2}) = (A_3A_4/A_1A_2)$. It is clear from Eq. (24.65) that area mismatches introduce only a change in the proportionality constant or DC gain of the integrator, and do not have any effect on the linearity or time constant of the integrator. The gain error can be compensated by easily adjusting one of the DC bias currents, thus I_S mismatches do not seem a significant problem.

24.7.4.3 Ohmic Resistance

Ohmic resistances include the base and emitter diffusion resistance and the resistance of interconnects and contact interfaces. For simplicity, all theses resistances can be referred to the base as an equivalent ohmic base resistance r_b. The device collector current can thus be defined as

$$I_c = I_S \exp\left(\frac{V_{be} - I_b r_b}{V_t}\right) = I_S \exp\left(\frac{V_{be}}{V_t}\right)\exp(-\alpha I_c) \tag{24.66}$$

where V_{be} is the applied (extrinsic) base-emitter voltage and $\alpha = r_b/\beta V_t$. When the first-order filter of Figure 24.41 is analyzed using the expression given in Eq. (24.66), a modified differential equation is obtained (assuming all base resistances are equal):

$$\left(\frac{CV_t}{I_o}\right)\frac{d[I_{out}\exp(\alpha I_{out})]}{dt} + I_{out}\exp(\alpha I_{out}) = \left(\exp(\partial\alpha I_o)\right)I_{in}\exp(\alpha I_{in})\exp\left[-\alpha\frac{CV_t}{I_o}\frac{dI_{out}}{dt}\left(1 + \alpha I_{out}\right)\right] \tag{24.67}$$

The term $\partial\alpha$ represents the difference between two α values and will be close to zero if all base resistances are assumed equal, thus can be neglected. At frequencies well below ω_o the time derivative in the exponent can also be neglected to give

$$\left(\frac{CV_t}{I_o}\right)\frac{d\left[I_{out}\exp(\alpha I_{out})\right]}{dt} + I_{out}\exp(\alpha I_{out}) = I_{in}\exp(\alpha I_{in}) \tag{24.68}$$

Expanding the differential term in Eq. (24.68)

$$I_{out}\exp(\alpha I_{out}) + \tau\frac{dI_{out}}{dt}\exp(\alpha I_{out}) + \tau\alpha I_{out}\frac{dI_{out}}{dt}\exp(\alpha I_{out}) = I_{in}\exp(\alpha I_{in}) \tag{24.69}$$

Eq. (24.69) is clearly no longer linear, thus distortion products will be present at the output. To quantify this distortion, we apply as an input signal $I_{in} = A\exp(j\omega t)$, and expect as output

$$I_{out} = B\exp j(\omega w + \theta) + c\exp j(2\omega t + \phi) + D\exp j(3\omega t + \psi) \tag{24.70}$$

Expanding the exponential terms in Eq. (24.69) by the first few terms of their series expansion

$$\exp(x) = 1 + x + \frac{x^2}{2} + \frac{x^3}{6} \tag{24.71}$$

and identifying the terms in $\exp(j\omega t)$, $\exp(2j\omega t)$, etc. results in

$$B\exp j\theta = \frac{A}{1 + s\tau}, \quad C\exp j\phi = \alpha A^2 p(\tau), \quad D\exp j\psi = \alpha^2 A^3 q(s\tau) \tag{24.72}$$

The expression for B is (as expected) the first-order transfer function of the system. The expressions for C and D give the second and third harmonic terms

$$HD_2 = \frac{|C\exp j\phi|}{|A|} \qquad HD_3 = \frac{|D\exp j\psi|}{|A|} \tag{24.73}$$

$p(s\tau)$ and $q(s\tau)$ are rational functions in $s\tau$, thus the output distortion is frequency dependent. Distortion is low at low frequencies and peaks around the cutoff frequency ($s\tau = 1$) where $p = 0.25$ and $q = 2$. The maximum distortion levels can be approximated as

$$\text{HD}_{2(\text{max})} = 0.25\alpha A = 0.25\frac{r_b}{\beta V_t}I_{\text{in}}$$

$$\text{HD}_{3(\text{max})} = 2(\alpha A)^2 = 2\left(\frac{r_b}{\beta V_t}\right)^2 I_{\text{in}}^2$$

(24.74)

These expressions demonstrate that when the voltage drop across the base resistance becomes comparable to V_t, there is significant distortion. This analysis also predicts that the distortion is larger at frequencies closer to the cutoff frequency, which is confirmed by circuit simulation.

In practice, the base resistance can be minimized by good layout (multiple base contacts), or by connecting several transistors in parallel. The latter, however, decreases the current through each transistor, which may lead to a decrease in f_t. It should be noted that from a technological point of view, a high f_t and a low r_b are opposing goals. To achieve a high f_t, the base should be as shallow as possible, reducing base transit time. At the same time, however, a shallow base increases the silicon (intrinsic) base resistance.

24.7.4.4 Early Effect

The early effect (base-width modulation) causes the collector current to vary with the collector-emitter and base-collector voltages. Considering the variation of collector-emitter voltage, the collector current can be written as $I_c = (1 + V_{ce}/V_A)\exp(V_{be}/V_t)$, where V_A is the forward-biased early voltage. An analysis of the circuit of Figure 24.41 shows that the early effect introduces a scalar error to the DC gain as in the case of emitter area (I_S) mismatch. In practice, the base-width modulation of the devices also introduces distortion because V_{ce} and V_{bc} are signal dependent. However, since voltage swings in current-mode circuits are minimized, this is not believed to be a major source of distortion.

24.7.4.5 Frequency Limitations

Each bipolar transistor has various intrinsic capacitors, the most important being C_μ (base-collector junction capacitance), C_{cs} (collector-substrate junction capacitance), and C_π (the sum of the base-emitter junction capacitance and the base-emitter diffusion capacitance). The junction capacitors depend only slightly on the operating point, while the base-emitter diffusion capacitance is proportional to the bias current, as given by $C_d = \tau_f I_C/V_t$. To determine the position of the parasitic poles and zeros, Figure 24.41 should be analyzed using large-signal device models. Unfortunately, the complexity of the resulting expressions renders this approach impractical even for the simplest log-domain circuits. To gain an intuitive understanding of the high-frequency limitations of the circuit, a small-signal analysis can be performed. Although log-domain circuits are capable of large-signal operation, this small-signal approach is justified to some extent since small signals can be considered as a "special case" of large signals. If the circuit fails to operate correctly for small signals, then it will almost certainly fail in large-signal operation (unfortunately, the opposite does not hold; if a circuit operates correctly for small signals, it does not necessarily work well for large signals). Analyzing the circuit of Figure 24.41, replacing each transistor by a small-signal (hybrid-π) equivalent model (comprising g_m, r_b, r_π, C_π, C_μ) [62], results in the following expression:

$$\frac{I_{\text{out}}}{I_{\text{in}}} = \left(\frac{g_{m4}}{g_{m1}}\right)\frac{1 + \tau_z s}{(1 + \tau_{p1}s)(1 + \tau_{p2}s)}$$

(24.75)

τ_{p1} is the time constant of the first (dominant) pole, given by

$$\tau_{p1} = \frac{C}{g_{m2}} + \left(C_{\mu4} + C_{\pi4}\right)\left(r_{b4} + \frac{1}{g_{m2}} + \frac{1}{g_{m3}}\right)$$

$$+ \frac{C_{\pi1}}{g_{m1}} + \frac{C_{\pi2}}{g_{m2}} + \frac{C_{\pi3}}{g_{m3}} + C_{\mu1}r_{b1} + C_{\mu2}r_{b2} + C_{\mu3}r_{b3} \tag{24.76}$$

Ideally, this pole location should depend only on the design capacitance C, and not on the device parasitics. Therefore $C_{\mu}r_b$, $C_{\pi}/g_m \ll C/g_{m2}$. Since $1/C_{\mu}r_b$ is typically much greater than f_t, this first constraint does not form a limit. The value of C_{π}/g_m depends on the operating point. For large currents, C_{π} is dominated by diffusion capacitance C_d so that $C_{\pi}/g_m = 1/f_t$. For smaller currents, g_m decreases while C_{π} becomes dominated by junction capacitance C_{je}, so that $C_{\pi}/g_m \gg 1/f_t$. Thus, it would seem that the usable cutoff frequency of the basic log-domain first-order filter is limited by the actual f_t of the transistors. The second pole time constant τ_{p2} (assuming that $\tau_{p1} = C/g_{m2}$), is

$$\tau_{p2} = \left(C_{\mu4} + C_{\pi4}\right)\left(r_{b4} + \frac{1}{g_{m3}}\right) + \left(C_{\mu2} + C_{\pi2}\right)\left(r_{b2} + \frac{1}{g_{m1}}\right)$$

$$+ \frac{C_{\pi1} + C_{cs1}}{g_{m1}} + \frac{C_{\pi13} + C_{cs3}}{g_{m3}} + C_{\mu1}r_{b1} + C_{\mu4}r_{b4} \tag{24.77}$$

This corresponds approximately to the f_t of the transistors, though Eq. (24.77) shows that the collector-substrate capacitance also contributes toward limiting the maximum operating frequency. The zero time constant τ_Z is given by

$$\tau_z = \left(C_{\pi1} + C_{\mu1}\right)r_{b1} + \frac{C_{\pi2}}{g_{m2}} + \frac{C_{\pi4}}{g_{m4}} + C_{\mu4}r_{b4} \tag{24.78}$$

This is of the same order of magnitude as the second pole. This means that the first zero and the second pole will be close together, and will compensate to a certain degree. However, in reality, there are more poles and zeros than Eq. (24.75) would suggest, and it is likely that others will also occur around the actual f_t of the transistors.

24.7.4.6 Noise

Noise in companding and log-domain circuits is discussed in some detail in Refs. [66–68], and a complete treatment is beyond the scope of this discussion. For linear (noncompanding) circuits, noise is generally assumed to be independent of signal level, and SNR will increase with increasing input signal level. This is not true for log-domain systems. At small input signal levels, the noise value can be assumed to be approximately constant, and an increase in signal level will give an increase in SNR. At high signal levels, the instantaneous value of noise will increase, and thus the SNR levels out at a constant value. This can be considered as an intermodulation of signal and noise power. For the class A circuits which have been discussed earlier, the peak signal level is limited by the DC bias current. In this case the large-signal noise is found to be of the same order of magnitude as the quiescent noise level, and thus a linear approximation is generally acceptable (this is not the case for class AB circuits).

24.7.5 Synthesis of Higher-Order Log-Domain Filters

The state-space synthesis technique outlined above proves difficult if the implementation of high-order filters is required, since it becomes difficult to define and manipulate a large set of state equations. One solution is to use the signal flow graph (SFG) synthesis method proposed by Perry and Roberts [69] to simulate LC ladder filters using log domain building blocks. The interested reader is also referred to references [70–72], which present modular and transistor-level synthesis techniques that can be easily extended to higher order filters.

References

1. A. Sedra and K. Smith, *Microelectronic Circuits*, Oxford Publishers, Oxford, 1998.
2. P.R. Gray and R.G. Meyer, *Analysis and Design of Analog Integrated Circuits*, Wiley, New York, 1993.
3. W.H. Gross, "New High Speed Amplifier Designs, Design Techniques and Layout Problems", in *Analog Circuit Design*, Eds. J.S. Huijsing, R.J. van der Plassche, and W. Sansen, Kluwer Academic Publishers, Dordrecht, 1993.
4. D.F. Bowers, "The Impact of New Architectures on the Ubiquitous Operational Amplifier", in *Analog Circuit Design*, Eds. J.S. Huijsing, R.J. van der Plassche, and W. Sansen, Kluwer Academic Publishers, Dordrecht, 1993.
5. J. Fonderie and J.H. Huijsing, "Design of Low-Voltage Bipolar Opamps", in *Analog Circuit Design*, Eds. J.S. Huijsing, R.J. van der Plassche, and W. Sansen, Kluwer Academic Publishers, Dordrecht, 1993.
6. M. Steyaert and W. Sansen, "Opamp Design towards Maximum Gain-Bandwidth", in *Analog Circuit Design*, Eds. J.S. Huijsing, R.J. van der Plassche, and W. Sansen, Kluwer Academic Publishers, Dordrecht, 1993.
7. K. Bult and G. Geelen, "The CMOS Gain-Boosting Technique", in *Analog Circuit Design*, Eds. J.S. Huijsing, R.J. van der Plassche, and W. Sansen, Kluwer Academic Publishers, Dordrecht, 1993.
8. J. Bales, "A Low-Power, High-Speed, Current Feedback Opamp with a Novel Class AB High-Current Output Stage", *IEEE J. Solid-State Circuits*, vol. 32, no. 9, p. 1470, 1997.
9. C. Toumazou, "Analogue Signal Processing: The 'Current' Way of Thinking", *Int. Journal of High-Speed Electronics*, vol. 32, nos. 3 and 4, 1992, p. 297.
10. K. Manetakis and C. Toumazou, "A new CMOS CFOA suitable for VLSI Technology", *Electron. Letters*, vol. 32, no. 12, pp. 1090–1092, 1996.
11. K. Manetakis, C. Toumazou and C. Papavassiliou, "A 120 MHz, 12 mW CMOS Current Feedback Opamp", *Proc. of IEEE Custom Int. Circuits Conf.*, 1998, p. 365.
12. D.A. Johns and K. Martin, *Analog Integrated Circuit Design*, Wiley, New York, 1997.
13. C. Toumazou, J. Lidgey and A. Payne, "Emerging Techniques for High-frequency BJT Amplifier Design: A current-mode perspective", *Int. Conf. on Electron. Circuits Syst.*, Parchment Press, Cairo, 1994.
14. M.C.H. Cheng and C. Toumazou, "3 V MOS Current Conveyor Cell for VLSI Technology", *Electron. Lett.*, vol. 29, 1993, p. 317.
15. K. Manetakis, "Intermediate Frequency CMOS Analogue Cells for Wireless Communications", PhD Thesis, Imperial College London, 1998.
16. R.A. Johnson, C.E. Chang, P. R. de la Houssaye, M.E. Wood, G.A. Garcia, P.M. Asbeck and I. Lagnado, "A 2.4 GHz Silicon-on-Sapphire CMOS Low-Noise Amplifier", *IEEE Microwave and Guided Wave Lett.*, vol. 7, no. 10, pp. 350–352, 1997.
17. A.N. Karanicolas, "A 2.7 V 900 MHz CMOS LNA and Mixer", *IEEE Digest of I.S.S.C.C.*, pp. 50–51, 1996.
18. D.K. Shaffer and T.H. Lee, "A 1.5-V, 1.5-GHz CMOS Low Noise Amplifier", *IEEE J.S.S.C.*, vol. 32, no. 5, pp. 745–759, 1997.
19. J.C. Rudell, J.-J. Ou, T.B. Cho, G. Chien, F. Brianti, J.A. Weldon and P.R. Gray, "A 1.9 GHz Wide-Band IF Double Conversion CMOS Integrated Receiver for Cordless Telephone Applications", *Digest of IEEE I.S.S.C.C.*, pp. 304–305, 1997.
20. E. Abou-Allam et al., "CMOS Front End RF Amplifier with On-Chip Tuning", *Proc. of IEEE ISCAS'96*, pp. 148–151, 1996.
21. P.R. Gray and R.G. Meyer, "Analysis and Design of Analogue Integrated Circuits and Systems", Chapter 11, 3rd ed., Wiley, New York, 1993.

22. M.J.N. Sibley, "Optical Communications", chapters 4–6, Macmillan, New York, 1995.

23. J.J. Morikuni, A. Dharchoudhury, Y. Leblebici and S.M. Kang, "Improvements to the Standard Theory for Photoreceiver Noise", *J. Lightwave Tech.*, vol. 12, no. 4, pp. 1174–1184, 1994.

24. A.A. Abidi, "Gigahertz Transresistance Amplifiers in Fine Line NMOS", *IEEE J.S.S.C.*, vol. SC-19, no. 6, pp. 986–994, 1984.

25. M.B. Das, J. Chen and E. John, "Designing Optoelectronic Integrated Circuit (OEIC) Receivers for High Sensitivity and Maximally Flat Frequency Response", *J. of Lightwave Tech.*, vol. 13, no. 9, pp. 1876–1884, 1995.

26. B. Sklar, *Digital Communication: Fundamentals and Applications*, Prentice-Hall, New York, 1988.

27. S.D. Personick, "Receiver Design for Optical Fiber Systems", *IEEE Proc.*, vol. 65, no. 12, pp. 1670–1678, 1977.

28. J.M. Senior, *Optical Fiber Communications: Principles and Practice*, chapters 8–10, PHI, Hertfordshire, UK, 1985.

29. N. Scheinberg, R.J. Bayruns and T.M. Laverick, "Monolithic GaAs Transimpedance Amplifiers for Fiber-Optic Receivers", *IEEE J.S.C.C.*, vol. 26, no. 12, pp. 1834–1839, 1991.

30. C. Toumazou and S.M. Park, "Wide-band low noise CMOS transimpedance amplifier for gigahertz operation", *Electron. Lett.*, vol. 32, no. 13, pp. 1194–1196, 1996.

31. S.M. Park and C. Toumazou, "Giga-hertz Low Noise CMOS Transimpedance Amplifier", *Proc. IEEE ISCAS*, vol. 1, pp. 209–212, June 1997.

32. D.M. Pietruszynski, J.M. Steininger and E.J. Swanson, "A 50-Mbit/s CMOS Monolithic Optical Receiver", *IEEE J.S.S.C.*, vol. 23, no. 6, pp. 1426–1432, 1988.

33. Y. Akazawa, N. Ishihara, T. Wakimoto, K. Kawarada and S. Konaka, "A Design and Packaging Technique for a High-Gain, Gigahertz-Band Single-Chip Amplifier", *IEEE J.S.S.C.*, vol. SC-21, no. 3, pp. 417–423, 1986.

34. N. Ishihara, E. Sano, Y. Imai, H. Kikuchi and Y. Yamane, "A Design Technique for a High-Gain, 10-GHz Class-Bandwidth GaAs MESFET Amplifier IC Module", *IEEE J.S.S.C.*, vol. 27, no. 4, pp. 554–561, 1992.

35. M. Lee and M.A. Brooke, "Design, Fabrication, and Test of a 125 Mb/s Transimpedance Amplifier Using MOSIS 1.2 *μ*m Standard Digital CMOS Process", *Proc. 37th Midwest Sym., Cir. and Sys.*, vol. 1, pp. 155–157, August 1994.

36. R.P. Jindal, "Gigahertz-Band High-Gain Low-Noise AGC Amplifiers in Fine-Line NMOS", *IEEE J.S.S.C.*, vol. SC-22, no. 4, pp. 512–520, 1987.

37. N. Takachio, K. Iwashita, S. Hata, K. Onodera, K. Katsura and H. Kikuchi, "A 10 Gb/s Optical Heterodyne Detection Experiment Using a 23 GHz Bandwidth Balanced Receiver", *IEEE Trans. M.T.T.*, vol. 38, no. 12, pp. 1900–1904, 1990.

38. K. Katsura, T. Hayashi, F. Ohira, S. Hata and K. Iwashita, "A Novel Flip-Chip Interconnection Technique Using Solder Bumps for High-Speed Photoreceivers", *J. Lightwave Tech.*, vol. 8, no. 9, pp. 1323–1326, 1990.

39. K. Murota and K. Hirade, "GMSK Modulation for Digital Mobile Radio Telephony", *IEEE Trans. Commun.*, vol. 29, pp. 1044–1050, 1981.

40. H. Krauss, C.W. Bostian, and F.H. Raab, *Solid State Radio Engineering*, Wiley, New York, 1980.

41. A.S. Sedra and K.C. Smith, *Microelectronic Circuits*, 4th ed. Oxford University Press, New York, 1998.

42. N.O. Sokal and A.D. Sokal, "Class E, a New Class of High-Efficiency Tuned Single-Ended Switching Power Amplifiers", *IEEE Journal of Solid-State Circuits*, vol. SC-10, pp. 168–176, 1975.

43. F.H. Raab, "Effects of Circuit Variations on the Class E Tuned Power Amplifier", *IEEE Journal of Solid-State Circuits*, vol. SC-13, pp. 239–247, 1978.

44. T. Sowlati, C.A.T. Salama, J. Sitch, G. Robjohn and D. Smith, "Low-Voltage, High-Efficiency Class E GaAs Power Amplifiers for Mobile Communications", in *IEEE GaAs IC Symp. Tech. Dig.*, pp. 171–174, 1994.

45. T. Sowlati, C.A.T. Salama, J. Sitch, G. Robjohn and D. Smith, "Low-Voltage, High-Efficiency GaAs Class E Power Amplifiers for Wireless Transmitters", *IEEE Journal of Solid-State Circuits*, vol. SC-13, no. 10, pp. 1074–1080, 1995.

46. A. Rofougaran, G. Chang. J.J. Rael, J.Y.-C. Chang, M. Rofougaran, P.J. Chang, M. Djafari, M.K. Ku, J. Min, E.W. Roth, A.A. Abidi and H. Samueli, "A Single-Chip 900 MHz Spread-Spectrum Wireless Transceiver in 1-μm CMOS—Part I: Architecture and Transmitter Design", *IEEE Journal of Solid-State Circuits*, vol. SC-33, no. 4, pp. 515–534, 1998.

47. J. Chang, A.A. Abidi and M. Gaitan, "Large Suspended Inductors on Silicon and their Use in a 2-μm CMOS RF Amplifier", *IEEE Electron Device Letters*, vol. 14, no. 5, pp. 246–248, 1993.

48. T. Sowlati, Y. Greshishchev, C.A.T. Salama, G. Rabjohn and J. Sitch, "Linearized High-Efficiency Class E Power Amplifier for Wireless Communications", *IEEE Custom Integrated Circuits Conf. Proc.*, pp. 201–204, 1996.

49. G.N. Henderson, M.F. O'Keefe, T.E. Boless and P. Noonan, "SiGe Bipolar Junction Transistors for Microwave Power Applications", *IEEE MTT-S Int. Microwave Symp. Dig.*, pp. 1299–1302, 1997.

50. Y. Tsividis, "Integrated Continuous-Time Filter Design – An Overview", *IEEE Journal of Solid-State Circuits*, vol. 29, no. 3, 1994.

51. F. Rezzi, A. Baschirotto and R. Castello, 'A 3 V 12–55 MHz BiCMOS Pseudo-Differential Continuous-Time Filter', *IEEE Trans. on Circuits and Systems-I*, vol. 42, no. 11, 1995.

52. B. Nauta, *Analog CMOS Filters for Very High Frequencies*, Kluwer Academic Publishers, Dordrecht, 1993.

53. C. Toumazou, F. Lidgey and D. Haigh, *Analogue IC Design: The Current-Mode Approach*, Peter Peregrinus Ltd. for IEEE Press, 1990.

54. S. Szczepanski and R. Schauman, "Nonlinearity-Induced Distortion of the Transfer Function Shape in High-Order Filters", *Kluwer Journal of Analog Int. Circuits and Signal Processing*, vol. 3, pp. 143–151, 1993.

55. K. Manetakis, "Intermediate Frequency CMOS Analogue Cells for Wireless Communications", PhD Thesis, Imperial College London, 1998.

56. S. Szczepanski, "VHF Fully-Differential Linearized CMOS Transconductance Element and Its Applications", *Proc. IEEE Int. Symp. Circuits Syst. (ISCAS)*, London, 1994.

57. S. Pipilos and Y. Tsividis, "RLC Active Filters with Electronically Tunable Center Frequency and Quality Factor", *Electron. Letters*, vol. 30, no. 6, 1994.

58. O. Shoaei and W.M. Snelgrove, "A Wide-range Tunable 25–110 MHz BiCMOS Continuous-Time Filter", *Proc. IEEE Int. Symp. Circuits Syst.* (ISCAS), Atlanta, 1996.

59. Y. Tsividis, "Externally Linear Time Invariant Systems and Their Application to Companding Signal Processors", *IEEE Trans. CAS-II*, vol. 44 no. 2, pp. 65–85, 1997.

60. D. Frey, "Log-domain filtering: an approach to current-mode filtering", *IEE Proc. G*, vol. 140, pp. 406–416, 1993.

61. B. Gilbert, "Translinear circuits: a proposed classification", *Electron. Lett.*, vol. 11, no. 1, pp. 14–16, 1975.

62. P. Grey and R. Meyer, *Analysis and Design of Analog Integrated Circuits*, 3rd ed., Wiley, New York, 1993.

63. E.M. Drakakis, A. Payne and C. Toumazou, "Log-domain state-space: a systematic transistor-level approach for log-domain filtering", accepted for publication in *IEEE Trans. CAS-II*, vol. 46, Issue 3, pp. 290–305, 1999.

64. V. Leung, M. El-Gamal and G. Roberts, "Effects of transistor non-idealities on log-domain filters", *Proc. IEEE Int. Symp. Circuits Syst.*, Hong Kong, pp. 109–112, 1997.

65. D. Perry and G. Roberts, "Log domain filters based on LC-ladder synthesis", *Proc. 1997 IEEE Int. Symp. on Circuits and Syst. (ISCAS)*, Seattle, pp. 311–314, 1995.

66. J. Mulder, M. Kouwenhoven and A. van Roermund, "Signal × noise intermodulation in translinear filters", *Electron. Lett*, vol. 33, no. 14, pp. 1205–1207, 1997.

67. M. Punzenberger and C. Enz, "Noise in instantaneous companding filters", *Proc. 1997 IEEE Int. Symp. Circuits Syst.*, Hong Kong, pp. 337–340, June 1997.

68. M. Punzenberger and C. Enz, "A 1.2 V low-power BiCMOS class-AB log-domain filter", *IEEE J. Solid-State Circuits*, vol. SC-32, no. 12, pp. 1968–1978, 1997.

69. D. Perry and G. Roberts, "Log-domain filters based on LC ladder synthesis", *Proc. 1995 IEEE Int. Symp. Circuits Syst.*, Seattle, pp. 311–314, 1995.

70. E. Drakakis, A. Payne and C. Toumazou, "Bernoulli operator: a low-level approach to log-domain processing", *Electron. Lett.*, vol. 33, no. 12, pp. 1008–1009, 1997.
71. F. Yang, C. Enz and G. Ruymbeke "Design of low-power and low-voltage log-domain filters", *Proc. 1996 IEEE Int. Symp. Circuits Syst.*, Atlanta, pp. 125–128, 1996.
72. J. Mahattanakul and C. Toumazou, "Modular log-domain filters", *Electron. Lett.*, vol. 33, no. 12, pp. 1130–1131, 1997.
73. D. Frey, "A 3.3 V electronically tuneable active filter useable to beyond 1 GHz", *Proc. 1994 IEEE Int. Symp. Circuits Syst.*, London, pp. 493–496, 1994.
74. M. El-Gamal, V. Leung, and G. Roberts, "Balanced log-domain filters for VHF applications", *Proc. 1997 IEEE Int. Symp. Circuits Syst.*, Monterey, pp. 493–496, 1997.

25

Operational Transconductance Amplifiers

Mohammed Ismail
Ohio State University

Seok-Bae Park
Ohio State University

Ayman A. Fayed
Texas Instruments, Inc.

R.F. Wassenaar
University of Twente

CONTENTS

25.1 Introduction

In many analog or mixed analog–digital VLSI applications, an operational amplifier may not be appropriate to use for an active element. For example, when designing integrated high-frequency active filter circuitry, a much simpler building block, called an operational transconductance amplifier (OTA), is often used [1]. This type of amplifier is characterized as a voltage-driven current source and in its simplest form is a combination of a differential input pair with a current mirror as shown in Figure 25.1. It is a simple circuit with a relatively small chip area. Further, it has a high bandwidth and also a good common-mode rejection ratio up to very high frequencies. The small signal transconductance, $g_m = \partial I_{out}/\partial V_{in}$, can be controlled by the tail current. This chapter discusses CMOS OTA design for modern VLSI applications. We begin the chapter with a brief study of noise in OTAs, followed by OTA design techniques.

25.2 Noise Behavior of the OTA

The noise behavior of the OTA is discussed here. Attention will be paid to thermal and flicker noise and to the fact that, for minimal noise, some voltage gain, from the input of the differential pair to the input of the current mirror, is required. Then, only the noise of the input pair becomes dominant and the other noise sources can be neglected to first order. The noise behavior of a single MOS transistor is modeled by a single noise voltage source. This noise voltage source is placed in series with the input (gate) of a "noiseless" transistor. Figure 25.2(a) shows the simple OTA, including the noise sources, while Figure 25.2(b) shows the same circuit with all the noise referred to the input of the stage.

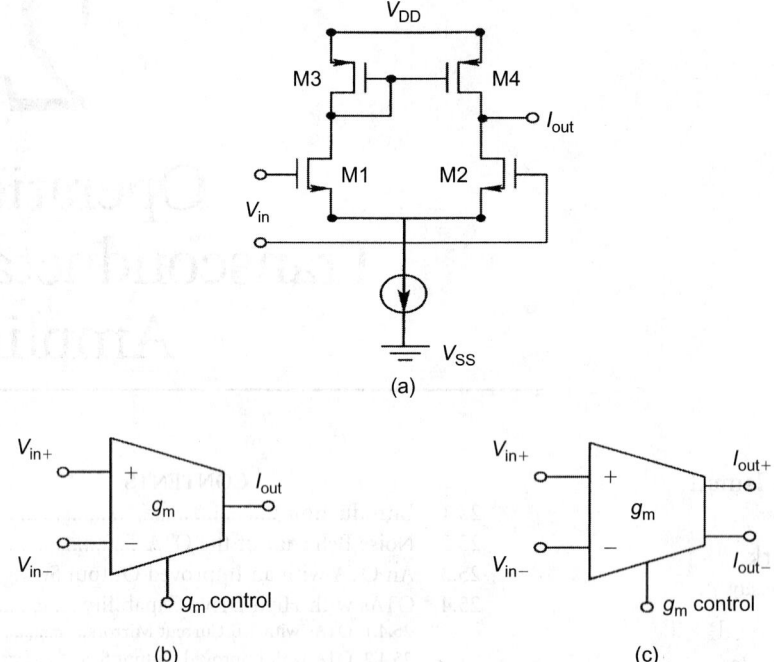

FIGURE 25.1 (a) An NMOS differential pair with a PMOS current mirror forming an OTA; (b) the symbol for a single-ended OTA; and (c) the symbol for a fully differential OTA.

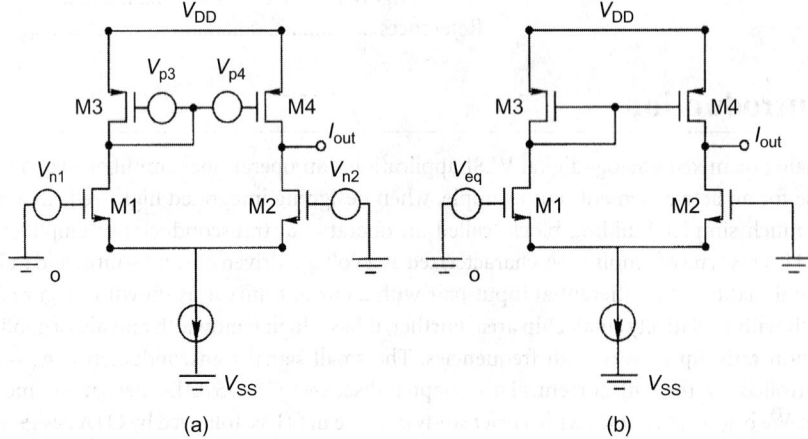

FIGURE 25.2 (a) The OTA with its noise voltage sources; and (b) the same circuit with the noise voltage sources referred to one of the input nodes.

All the noise sources indicated in Figure 25.2(a) are converted into equivalent input noise voltages, which are then added to form a single noise source at the input (Figure 25.2(b)). As a result, we obtain (assuming $g_{m1} = g_{m2}$ and $g_{m3} = g_{m4}$) the following mean-square input referred noise voltage:

$$\overline{V_{eq}^2} = \overline{V_{n1}^2} + \overline{V_{n2}^2} + \left(\frac{g_{m3}}{g_{m1}}\right)^2 (\overline{V_{p3}^2} + \overline{V_{p4}^2}) \tag{25.1}$$

The thermal noise contribution of one transistor, over a band Δf, is written as

$$\overline{V_{th}^2} = \frac{2}{3} 4kT \frac{1}{g_m} \Delta f \qquad (25.2)$$

where k is the Boltzman constant and T the absolute temperature.

The equivalent noise voltage $\overline{V_{th_{eq}}^2}$ becomes

$$\overline{V_{th_{eq}}^2} = \frac{2}{3} 4kT \left(\frac{1}{g_{m1}} + \frac{1}{g_{m2}} + \left(\frac{g_{m3}}{g_{m1}} \right)^2 \left(\frac{1}{g_{m3}} + \frac{1}{g_{m4}} \right) \right) \Delta f \qquad (25.3)$$

and because $g_{m1} = g_{m2}$ and $g_{m3} = g_{m4}$, $\overline{V_{th_{eq}}^2}$ becomes

$$\overline{V_{th_{eq}}^2} = \frac{16}{3} kT \left(\frac{1}{g_{m1}} + \left(\frac{g_{m3}}{g_{m1}} \right)^2 \frac{1}{g_{m3}} \right) \Delta f \qquad (25.4)$$

or

$$\overline{V_{th_{eq}}^2} = \frac{16}{3} \frac{kT}{g_{m1}} \left(1 + \frac{g_{m3}}{g_{m1}} \right) \Delta f \qquad (25.5)$$

Expressing g_m in physical parameters results in

$$\overline{V_{th_{eq}}^2} = \frac{16kT}{3\sqrt{\mu_n C_{ox} (W/L)_1 I_0}} \left(1 + \sqrt{\frac{\mu_p (W/L)_3}{\mu_n (W/L)_1}} \right) \Delta f \qquad (25.6)$$

where I_0 represents the tail current of the differential pair. Note that the term between brackets represents the relative noise contribution of the current mirror. This term can be neglected if M3 and M4 are chosen relatively long and narrow in comparison to M1 and M2.

It should be mentioned that the thermal noise of an N-MOS transistor and a P-MOS transistor with equal transconductance is the same. In most standard IC processes, a 3–10 times lower $1/f$ noise is observed for P-MOS transistors in comparison to N-MOS transistors of the same size. However, in modern processes, the $1/f$ noise contribution of N- and P-MOS transistors tends to be equal.

For the $1/f$ noise, it is usually assumed for standard IC processes that

$$\overline{V_{1/f}^2} = \frac{K'}{WLC_{ox}f} \Delta f \qquad (25.7)$$

where K' is the flicker noise coefficient in the range of 10^{-24} J for N-MOS transistors and in the range of 3×10^{-25} to 10^{-25} J for P-MOS transistors. The equivalent $1/f$ input noise source of the OTA in Figure 25.2(b) yields

$$\overline{V_{eq(1/f)}^2} = \frac{2K_n' \Delta f}{W_1 L_1 C_{ox} f} \left(1 + \frac{K_p' \mu_p L_1^2}{K_n' \mu_n L_3^2} \right) \qquad (25.8)$$

Here, the noise contributions of the current mirror (M3, M4) will be negligible if L_3 is chosen much larger than L_1.

The offset voltage of a differential pair is lowest when the transistors are in the weak-inversion mode; but on the contrary, the mismatch in the current transfer of a current mirror is lowest when the transistors are deep in strong inversion. Hence, the conditions that have to be fulfilled for both minimal equivalent input noise and minimal offset are easy to combine.

25.3   An OTA with an Improved Output Swing

A CMOS OTA with an output swing much higher than that in Figure 25.1(a) is shown in Figure 25.3. This configuration needs two extra current mirrors and consumes more current, but the output voltage "window" is, in the case when common-mode input voltage is zero, about doubled. The rules discussed earlier for sizing the input transistors and current-mirror transistors to reduce noise and offset still apply. However, there is still a tradeoff. On the one hand, a high voltage gain from the input nodes to the current mirror is good for reducing noise and mismatch effects; in contrast, too much gain also reduces the upper limit of the common-mode input voltage range and the phase margin needed to ensure stability (this will be discussed later) [2]. A voltage gain on the order of 3–10 is advised. The frequency behavior of the OTA in Figure 25.3 is rather complex since there are two different signal paths in parallel, as shown in Figure 25.4. In this scheme, r_p represents the parallel value of the output resistance of the stage ($r_{o6} \parallel r_{o8}$) and the load resistance (R_L); therefore,

$$r_p = r_{o6} \parallel r_{o8} \parallel R_L \tag{25.9}$$

The capacitor C_p represents the sum of the parasitic output capacitance and the load capacitance $C_p = C_o + C_L$. Using the half-circuit principle for the differential pair, a fast signal path can be seen from M2 via current mirror M7, M8 to the output. This signal path contributes an extra high-frequency pole. The other signal path leads from transistor M1 via both current mirrors M3, M4 and M5, M6 to the output. In this path, two extra poles are added. The transfer of both signal paths and their combination are shown in the plots in Figure 25.5, assuming equal pole positions of all

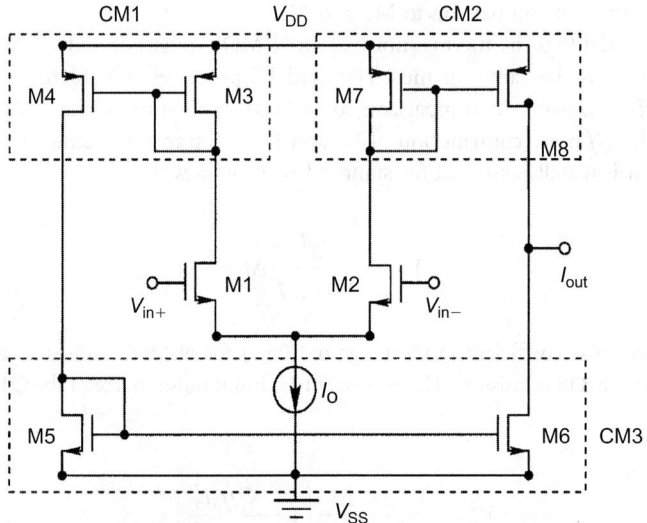

FIGURE 25.3 An OTA with an improved output window.

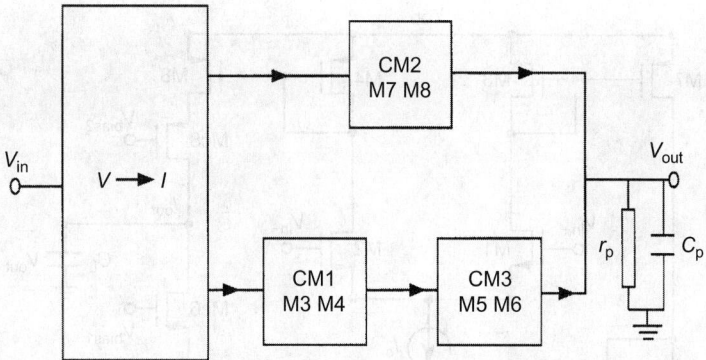

FIGURE 25.4 The signal paths of the OTA in Figure 25.3.

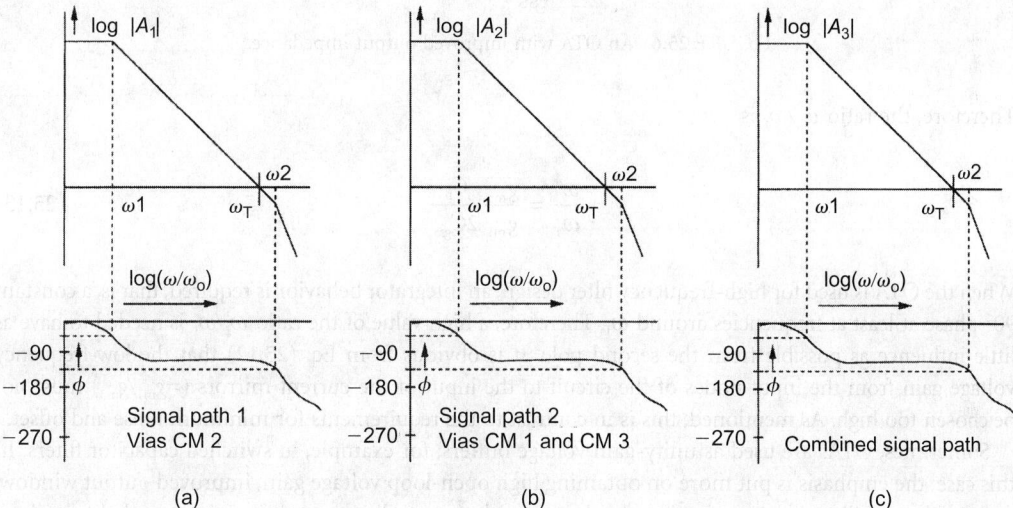

FIGURE 25.5 (a) The Bode plot belonging to signal path 1 in the OTA in Figure 25.3 and 25.4, (b) signal path 2, and (c) the combined signal path.

three current mirrors. Note that the first (dominant) pole (ω_1) is determined by r_p and C_p:

$$\omega_1 = \frac{1}{r_p C_p} \qquad (25.10)$$

The second pole (ω_2) is determined by the transconductance of M3 and the sum of the gate–source capacitance of M3 and M4. If M3 and M4 are equal, the second pole is located at

$$\omega_2 = \frac{g_{m3}}{2C_{gs3}} \qquad (25.11)$$

The unity-gain corner frequency ω_T of the loaded OTA is at

$$\omega_T = \frac{g_{m1}}{C_p} \qquad (25.12)$$

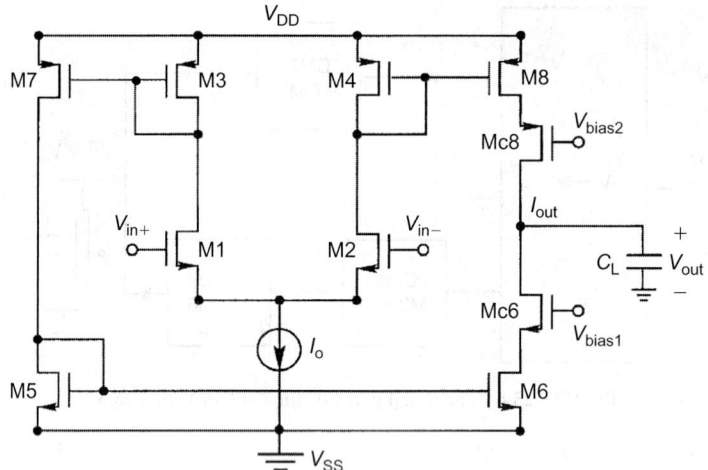

FIGURE 25.6 An OTA with improved output impedance.

Therefore, the ratio ω_2/ω_T is

$$\frac{\omega_2}{\omega_T} = \frac{g_{m3}}{g_{m1}}\frac{C_p}{2C_{gs3}} \tag{25.13}$$

When the OTA is used for high-frequency filter design, an integrator behavior is required, that is, a constant 90° phase at least at frequencies around ω_T. Therefore, a high value of the ratio ω_2/ω_T is needed to have as little influence as possible from the second pole. It is obvious from Eq. (25.12) that the low-frequency voltage gain from the input nodes of the circuit to the input of the current mirrors ($=g_{m1}/g_{m3}$) must not be chosen too high. As mentioned, this is in contrast to the requirements for minimum noise and offset.

Sometimes, OTAs are used as unity-gain voltage buffers; for example, in switched capacitor filters. In this case, the emphasis is put more on obtaining high open-loop voltage gain, improved output window, and good capability to drive capacitive loads efficiently (or small resistors); its integrator behavior is of less importance.

To increase the unloaded voltage gain, cascode transistors can be added in the output stage. This greatly increases the output impedance of the OTA and hardly decreases the phase margin. The penalty that has to be paid is an additional pole in the signal path and some reduction of the maximum possible output swing. This reduction can be very small if the cascode transistors are biased on the weak-inversion mode. The open-loop voltage gain can be on the order of 40–60 dB. A possible realization of such a configuration is shown in Figure 25.6 [3].

25.4   OTAs with High Drive Capability

For driving capacitive loads (or small resistors), a large available output current is necessary. In the OTAs shown so far, the amount of output current available is equal to twice the quiescent current (i.e., the tail current I_0). In some situations, this current can be too small. There are several ways to increase the available current in an efficient way. To achieve this, four design principles will be discussed here.

1. Increasing the quiescent current by using current mirrors with a current transfer ratio greater than 1.
2. Using a two-transistor level structure to drive the output transistors.
3. Adaptive biasing techniques.
4. Class AB techniques.

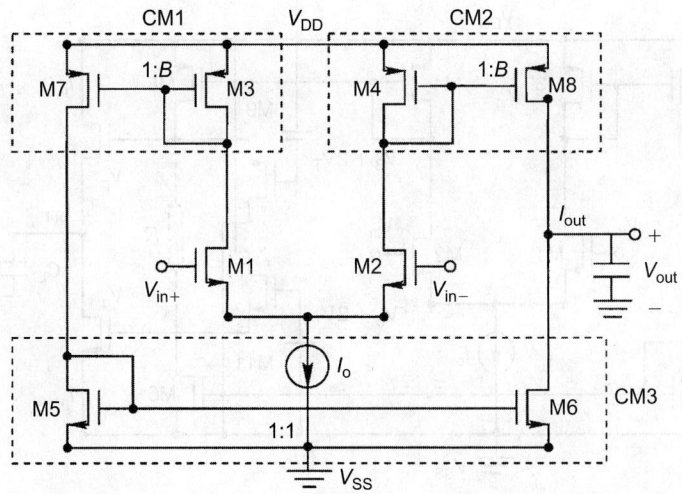

FIGURE 25.7 An OTA with improved load current using 1:B current mirrors.

25.4.1 OTAs with 1:B Current Mirrors

One way to increase available output current is to increase the transfer ratio of the current mirrors CM1 and CM2 by a factor *B*, as indicated in Figure 25.7 [4]. The amount of available output current and also the overall transconductance increase by the same factor. Unfortunately, the –3 dB frequency of the CM1–CM2 current mirrors will be reduced by a factor $(B + 1)/2$ due to the larger gate–source capacitance of the mirror output transistors. Moreover, ω_T will increase, ω_2 will decrease, and the ratio ω_2/ω_T will be strongly deteriorated. The amount of available output current though is *B* times the tail current. It is also possible to increase the current transfer ratio of current mirror CM3 instead of CM1. A better current efficiency then results, but at the expense of more asymmetry in the two signal paths. Although the amount of the maximum available output current is *B* times the tail current in both situations, the ratio between the maximum available current and quiescent current of the output stage remains equal to two, just as in the OTAs discussed previously.

25.4.2 OTA with Improved Output Stage

Another way of increasing the maximal available output current is illustrated in Figure 25.8 [5]. It improves upon the factor 2 relationship between quiescent and maximal available current. Assuming equal *K* factors for all transistors shown in the circuit leads to the conclusion that the effective gate–source voltage of transistor M11 ($= V_{GS11} - V_{T11}$) equals that of transistor M1 ($= V_{GS1} - V_{T1}$), since they carry the same current, assuming that transistors M1, M4, and M6 are in saturation. Because the current drawn through transistor M9 is equal to the current in transistor M2, their effective source–gate voltages must also be equal assuming equal *K* factor for M2 and M9. Since the sum of the effective gate–source voltages transistors M11 and M12, and also of M9 and M10, is fixed and equal to V_B, a situation exists which is equivalent to the two transistor level structure described in Ref. [6].

The ratio between the maximum available output current and the quiescent current of the output stage can be chosen by the designer. It is equal to $(V_B/(V_B - V_{GS0}))^2$, where V_{GS0} is the quiescent gate–source voltage of transistor M11.

If the OTA is used in an overdrive situation $|V_{in}| > \sqrt{(2I_0)/K}$, then either M6 or M5 will be cut off, while the other transistor carries its maximum current. As a result, one of the output transistors (M10 or M12) carries its maximum current, while the other transistor is in a low-current standby situation. The maximum current that one of the output transistors carries is therefore proportional to V_B^2. With the high ohmic resistor *R* (indicated in Figure 25.8 with dotted lines), this maximum current corresponds

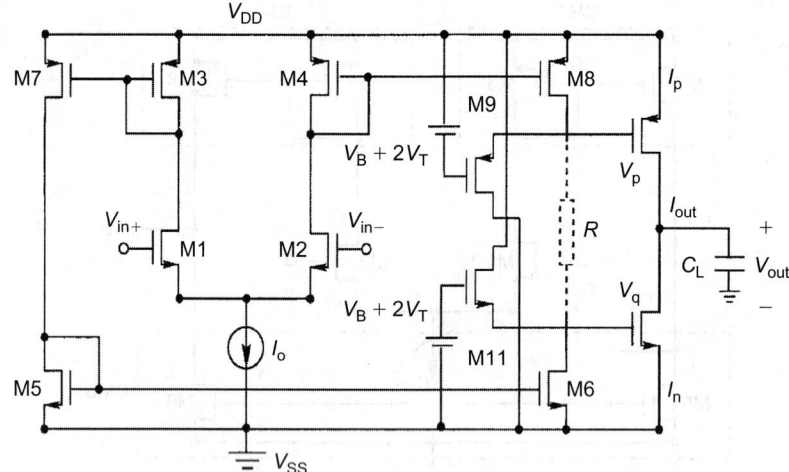

FIGURE 25.8 An OTA with an improved ratio between the maximum available current and the quiescent current of the output stage.

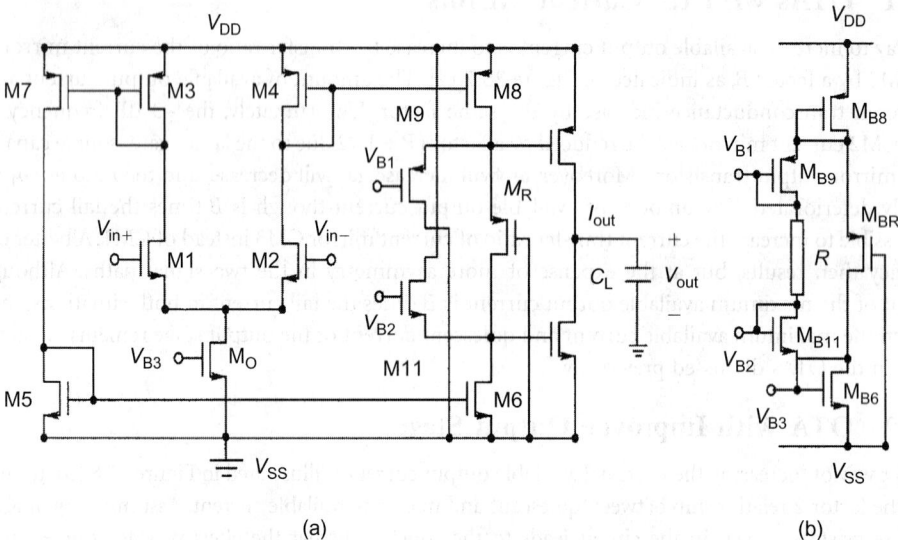

FIGURE 25.9 (a) The complete OTA, (b) and its bias stage.

to either $(V_P - V_{SS} - V_{TN})^2$ or $(V_{DD} - V_Q - V_{TP})^2$, because in that situation no current flows through the resistor. Hence, with the extra resistor, it becomes possible to increase the maximum current in overdrive situations and therefore reduce the slewing time. Because resistor R is chosen to be high, it does not disturb the behavior of the circuit discussed previously. In practice, resistor R is replaced by transistor M_R working in the triode region, as shown in Figure 25.9(a). Figure 25.9(b) shows the circuit which was used in Ref. [6] for biasing the gates of transistors M0, M9, and M11. It is much like the so-called replica biasing. The current in the circuit is strongly determined by the voltage across R (and its value) and is therefore very sensitive to variations in the supply voltage.

25.4.3 Adaptively Biased OTAs

Another combination of high available output current with low standby current can be realized by making the tail current of the differential input pair signal-dependent. Figure 25.10 shows the basic idea of such

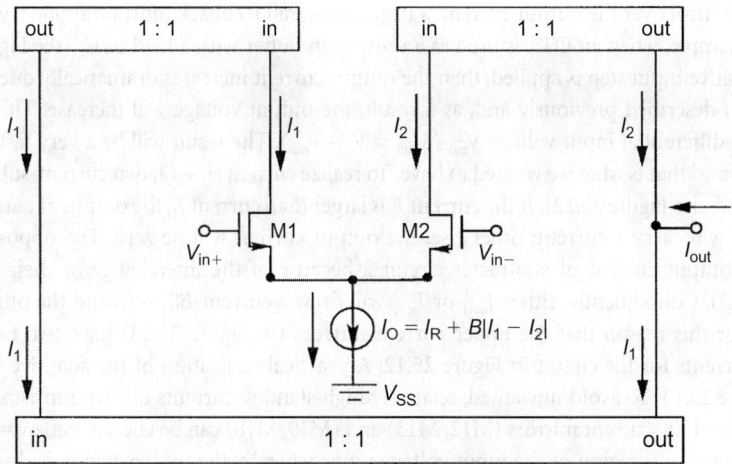

FIGURE 25.10 An OTA with an input-dependent tail current.

an OTA with adaptive biasing [7]. The tail current I_0 of the differential pair is the sum of a fixed value I_R and an additional current equal to the absolute value of the difference between the drain currents multiplied by the current feedback factor B ($I_0 = I_R + B|I_1 - I_2|$). Therefore, with zero differential input voltage, only a low bias current I_R flows through the input pair. A differential input voltage, V_{ind}, will cause a difference in the drain currents which will increase the tail current. This, in turn, again gives rise to a greater difference in the drain current, and so on. This is the kind of positive feedback that can bring the differential input pair from the weak-inversion mode into the strong-inversion mode, depending on the input voltage and the chosen current feedback factor B.

Normally, when $V_{ind} = V_{in+} - V_{in-}$ is small, the input transistors are in weak inversion. The differential output current ($I_1 - I_2$) of a differential pair operating in weak inversion equals the tail current times $\tanh((qV_{ind})/(2AkT))$. This leads to the following equation:

$$(I_1 - I_2) = I_R + B\left|I_1 - I_2\right|\tanh\left(\frac{qV_{ind}}{2AkT}\right) \tag{25.14}$$

or

$$(I_1 - I_2) = \frac{\tanh\left(qV_{ind}/2AkT\right)}{1 - B\left|\tanh\left(qV_{ind}/2AkT\right)\right|}I_R \tag{25.15}$$

and because $I_{out} = (I_1 - I_2)$

$$I_{out} = \frac{\tanh\left(qV_{ind}/2AkT\right)}{1 - B\left|\tanh\left(qV_{ind}/2AkT\right)\right|}I_R \tag{25.16}$$

However, in the case of large currents, this expression will no longer be valid since M1 – M2 will leave the weak-inversion domain and enter the strong-inversion region. If that is the case, the output current becomes

$$I_{out} = \begin{cases} \dfrac{k}{2}V_{ind}\sqrt{\dfrac{4I_R}{K} - (1 - B^2)V_{ind}^2} + B\dfrac{K}{2}V_{ind}^2 & \text{for } V_{ind} > 0 \\[4mm] \dfrac{k}{2}V_{ind}\sqrt{\dfrac{4I_R}{K} - (1 - B^2)V_{ind}^2} - B\dfrac{K}{2}V_{ind}^2 & \text{for } V_{ind} < 0 \end{cases} \tag{25.17}$$

To keep some control over the output current, a negative overall feedback must be applied, which is usually the case. For example, when an OTA is used as a unity-gain buffer with a load of C_L (see Figure 25.11) and assuming a positive input step is applied, then the output current increases dramatically due to the positive feedback action described previously and, as a result, the output voltage will increase. This will lead to a decrease of the differential input voltage V_{ind} ($V_{ind} = V_s - V_{out}$). The result will be a very fast settling of the output voltage, and that is what we wanted to have. To realize current $|I_1 - I_2|$, two current-subtracter circuits can be combined (see Figure 25.12). If the current I_2 is larger than current I_1, the output of current-subtracter circuit 1 (I_{out1}) will carry a current; otherwise, the output current will be zero. The opposite situation is found for the output current of subtracter circuit 2 because of the interchange of their input currents ($I_{out2} = B(I_1 - I_2)$). Consequently, either I_{out1} or I_{out2} will draw a current $B|I_1 - I_2|$ and the other current will be zero. It is for this reason that the upper current mirrors (in Figure 25.13) have two extra outputs to support the currents for the circuit in Figure 25.12. A practical realization of the adaptive-biasing OTA is shown in Figure 25.13. To avoid unwanted, relatively high standby currents due to transistors mismatches, the transfer ratio of the current mirrors (M12, M13) and (M19, M18) can be chosen somewhat larger than 1. This ensures an inactive region of the input voltage range whereby the feedback loop is deactivated.

Another example of an adaptive tail current circuit is shown in Figure 25.14 [8]. It has a normal OTA structure except that the input pair is realized in twofold, and the tail current transistor is used in a feedback loop. This feedback loop includes the inner differential pair and tail current transistor M0 as well as a minimum current selector, the current source I_U, transistor M_R, and a current sink I_L. The minimum current selector [9] delivers an output current equal to the lowest value of its input currents ($I'_{out} = Min(I'_1, I'_2)$). The feedback loop ensures that the output current of the minimum current selector is equal to the difference in currents between the upper and lower current sources. Assume that the upper current carries a current

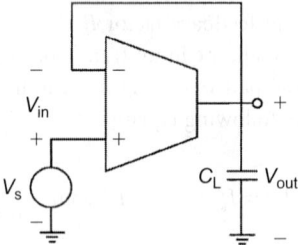

FIGURE 25.11 An OTA used as a unity-gain buffer.

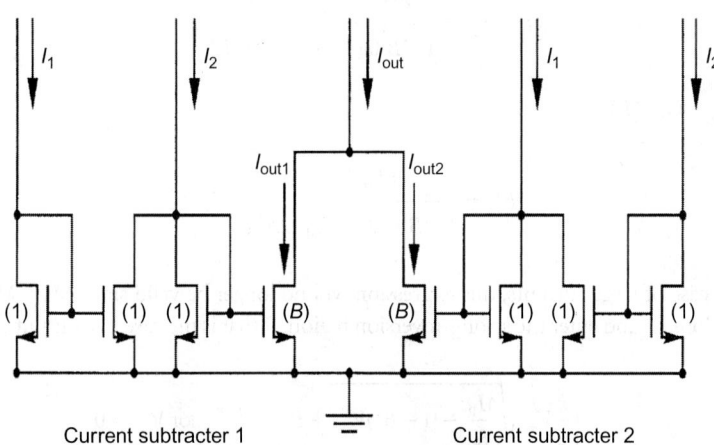

FIGURE 25.12 A combination of two current subtracters for realizing the adaptive biasing current for the circuit in Figure 25.10.

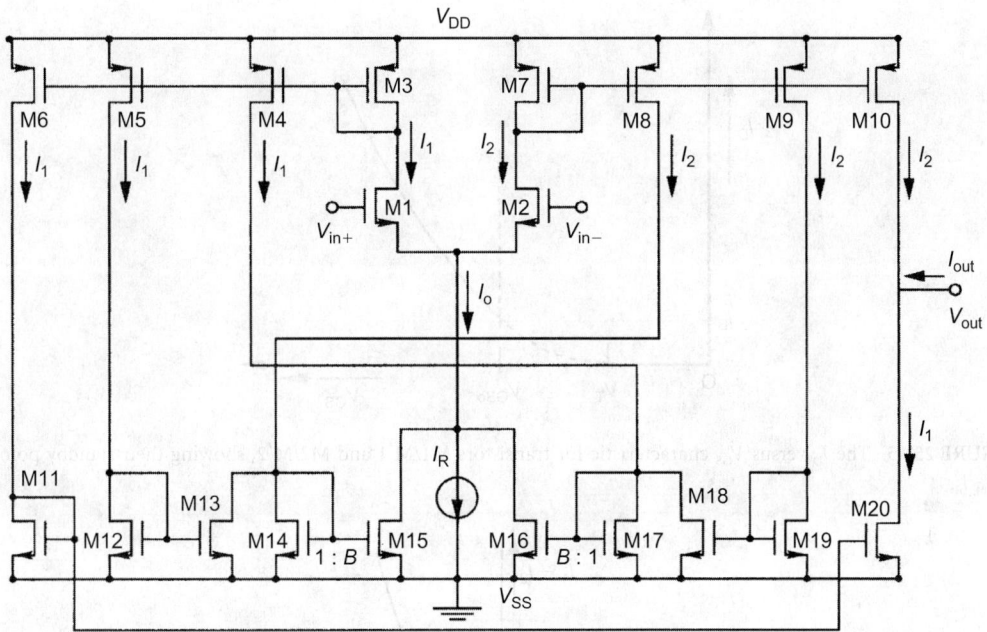

FIGURE 25.13 A practical realization of OTA with an adaptive biasing of its tail current.

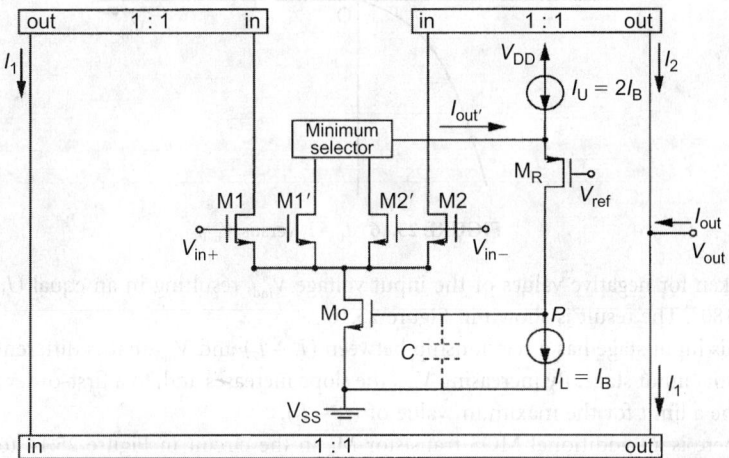

FIGURE 25.14 An OTA using a minimum selector for adapting the tail current.

$2I_B$ and the lower current source carries I_B, then the feedback loop will bias the tail current in such a way that either I'_1 or I'_2 becomes equal to I_B; for positive values of V_{ind}, that will be I'_2. It should be realized that at $V_{ind} = 0$, all four input transistors are biased at the same gate–source voltage (V_{GS0}), corresponding to a drain current I_B. In the case of positive input voltages, the gate–source voltage of M2/M'2 will not change.

Therefore, all the input voltage will be added to the bias voltage of M1/M'1, that is,

$$V_{GS1} = V_{GS0} + V_{ind} \qquad (25.18)$$

Figure 25.15 shows the I_D versus V_{GS} characteristic for both transistors M1/M'1 and M2/M'2. Accordingly, the relationship between $(I_1 - I_2)$ versus V_{ind} (for $V_{ind} > 0$) follows the right side of the $I_D - V_{GS}$ curve of M1, starting from the standby point (V_{GS0}, I_B) as indicated by the solid curve in Figure 25.15. A similar

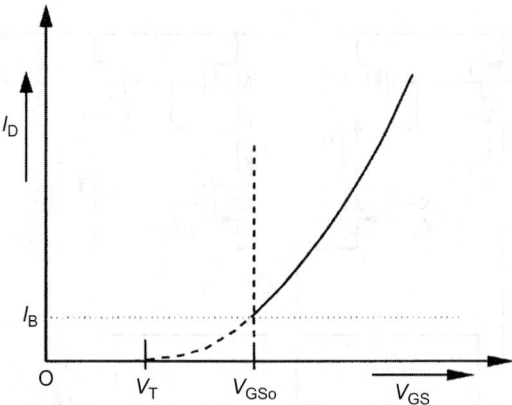

FIGURE 25.15 The I_D versus V_{GS} characteristic for transistors M1/M$'$1 and M2/M$'$2, showing their standby point V_{GS0}, I_B.

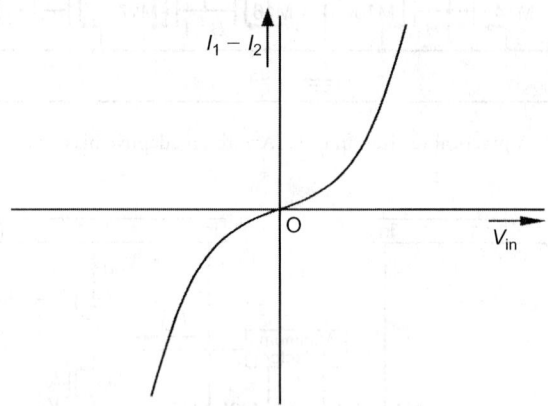

FIGURE 25.16 $I_1 - I_2$ versus V_{ind}.

view can be taken for negative values of the input voltage V_{ind}, resulting in an equal $(I_1 - I_2)$ versus V_{in} curve rotated 180°. The result is shown in Figure 25.16.

Note that this input stage has a relationship between $(I_1 - I_2)$ and V_{ind} that is different from that of a simple differential input stage. By increasing V_{ind}, the slope increases and, to a first-order approximation, there will not be a limit for the maximum value of $(I_1 - I_2)$.

Note that there is an additional MOS transistor M_R in the circuit in Figure 25.14 to fix the output voltage of the minimum current selector circuit. The lower current source I_L is necessary to be able to discharge the gate–source capacitor C of M0 (indicated in Figure 25.14 with dotted lines). The OTA in Figure 25.14 is simpler than that in Figure 25.13. However, its bandwidth is lower due to the high impedance of node P in the feedback loop.

25.4.4 Class AB OTAs

Another possibility to design an OTA with a good current efficiency is to use an input stage exhibiting class AB characteristics [10]. The input stage in Figure 25.17 contains two CMOS pairs [11] connected as Class AB input transistors. They are driven by four source followers. By applying a differential input voltage, the current through one of the input pairs will increase while the current through the other will decrease. The maximum current that can flow through the CMOS pair is, to first order, unlimited. In practice, it is limited by the supply voltage, the K_{eq} factor, the mobility reduction factor, and the series resistance. The currents are delivered to the output with the help of two current mirrors. In the OTA shown in Figure 25.17, only one of the two outputs of each CMOS pair is used. The other output currents flow directly to the supply rails. Instead of wasting the other output currents, they can be used to supply an extra output. So with the addition of two current mirrors, an OTA with complementary outputs as shown in Figure 25.18 can be achieved [12]. An improvement of the output impedance and low-frequency voltage gain can be obtained by cascoding the output transistors of the current mirrors (Figure 25.19). Usually, this reduces the output window. The function of transistors M41–M44 is to control the DC output voltages. They form a part of a common-mode feedback system, which will be discussed next.

The relationship between the differential input voltage V_{in} and one of the output currents I_{out} is shown in Figure 25.20. There is a linear relationship between V_{ind} and I_{out} for small to moderate values of V_{ind}.

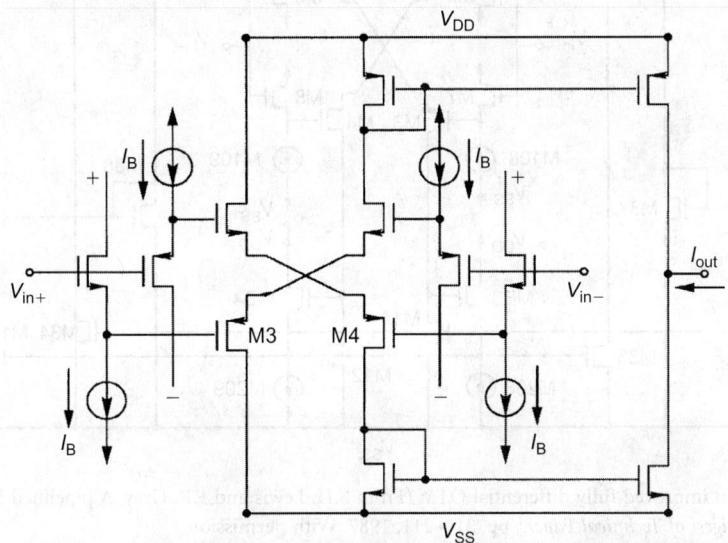

FIGURE 25.17 An OTA having a class AB input stage.

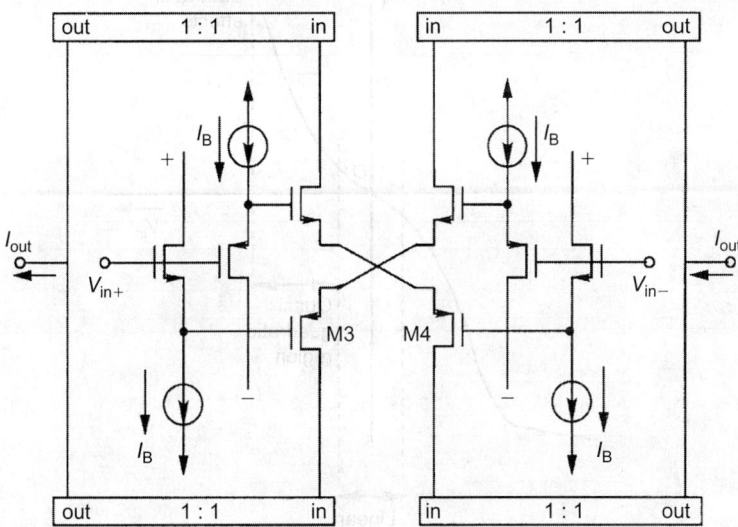

FIGURE 25.18 An OTA having a class AB input stage and two complementary outputs.

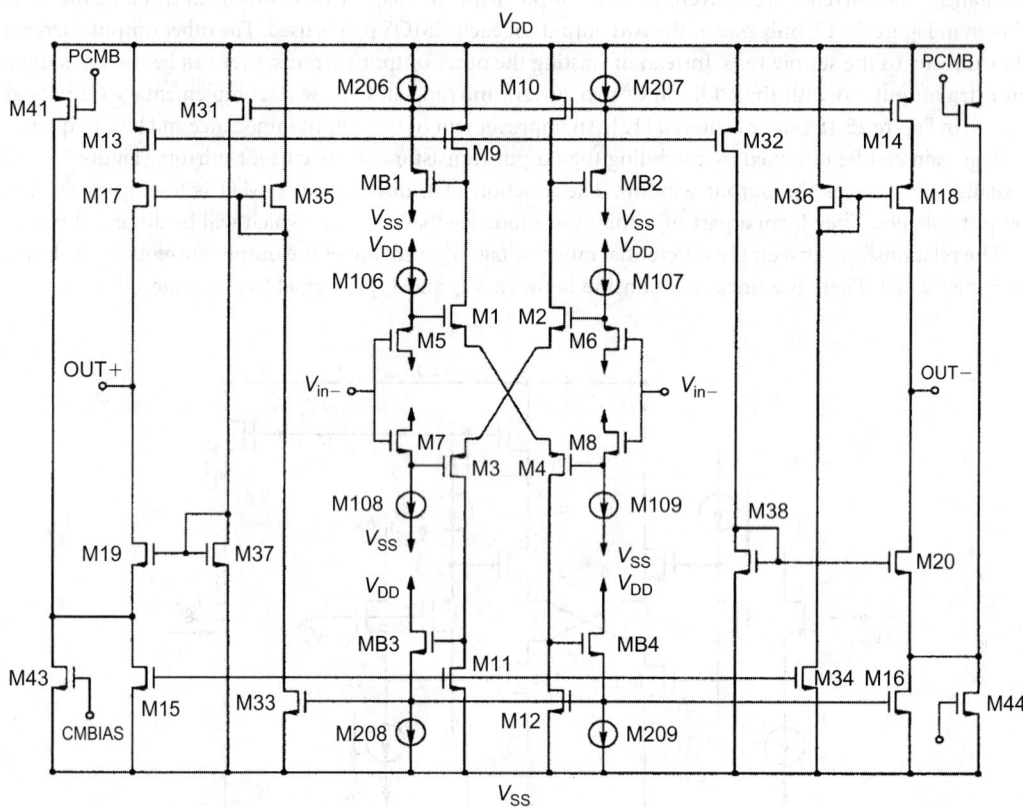

FIGURE 25.19 An improved fully differential OTA. (From S.H. Lewis and P.R. Gray, A pipelined 5MHz 9b ADC, *IEEE ISSCC' 87 Digest of Technical Papers*, pp. 210–211, 1987. With permission.)

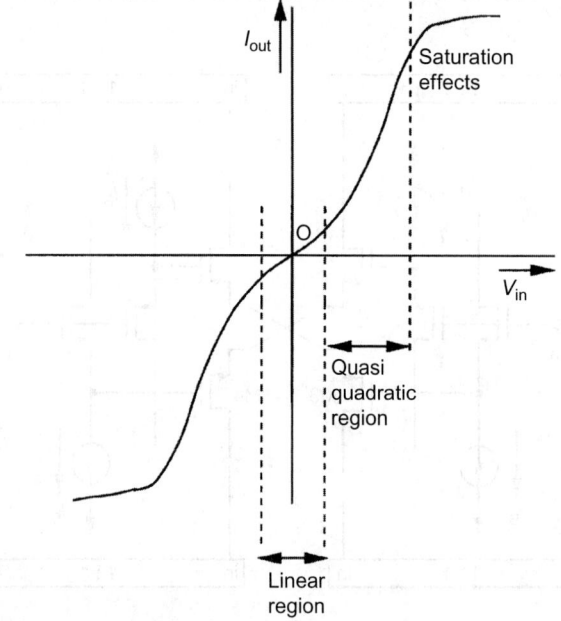

FIGURE 25.20 The $V_{in} \rightarrow I_{out}$ characteristic of the OTA in Figure 25.19.

In the case of larger values of V_{ind}, one of the CMOS pairs becomes cut off, resulting in a quasi-quadratic relationship. At a further increase of V_{ind}, the output current will be somewhat saturated due to mobility reduction and to the fact that one of the transistors of the CMOS pair leaves saturation mode. The latter effect is, of course, also strongly dependent on the common input voltage.

25.5 Common-Mode Feedback

A fully differential OTA circuit, as in Figure 25.19, has many advantages compared with its single-ended counterpart. It is a basic building block in filter design. A fully differential approach, in general, leads to a more efficient current use, doubling of the maximum output-voltage swing, and an improvement of the power-supply rejection ratio (PSRR). It also leads to a significant reduction of the total harmonic distortion, since all even harmonics are canceled out due to the symmetrical structure. Even when there is a small imperfection in the symmetry, the reduction in distortion will be significant.

However, this type of symmetrical circuit needs an extra feedback loop. The feedback around a single-ended OTA usually only provides a differential-mode feedback and is ineffective for common-mode signals.

So, in the case of the fully differential OTA, a common-mode feedback (CMFB) circuit is needed to control the common output voltage. Without a CMFB, the common-mode output voltage of the OTA is not defined and it may drift out of its high-gain region. The general structure of a simple OTA circuit with a differential output and a CMFB circuit is shown in Figure 25.21. The need for a CMFB circuit is a drawback since it counters many of the advantages of the fully differential approach. The CMFB circuit requires chip area and power, introduces noise, and limits the output-voltage swing.

Figure 25.22(b) shows a simple implementation of a CMFB circuit. A differential pair (M1, M2) is used to sense the common-mode output voltage. So, the voltage at the common source of this differential pair (V_s) is used. Its voltage provides, with a level shift of one V_{GS}, the common-mode output voltage of the OTA. The voltage at this node is the first-order insensitive to the differential input voltage. The relationship between the differential input voltage V_{in} of the differential pair, superimposed on a common-mode input voltage V_{CM}, and its common-source voltage V_s is shown in Figure 25.22(a). The common-mode output voltage of the OTA is determined by the V_{GS} of M1/M2 and M9/M10 and can be controlled by the voltage source V_0. There might be an offset in the DC value of the two output voltages due to a mismatch in transistors M9 and M10.

If the amplitude of the differential output voltage increases, the common-mode voltage will not remain constant, but will be slightly modulated by the differential output voltage, with a modulation frequency

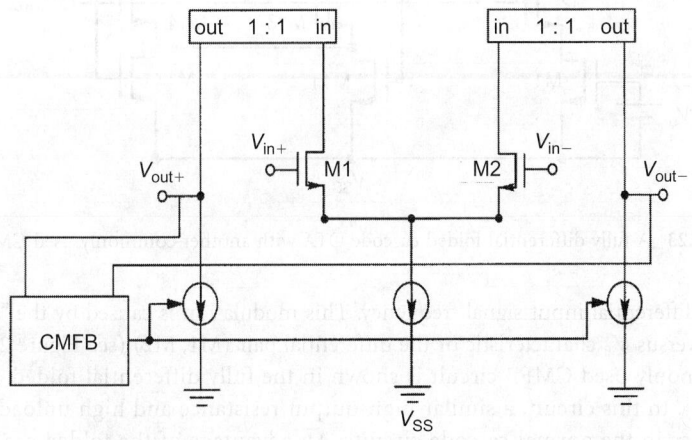

FIGURE 25.21 The general structure of a simple OTA circuit having a differential output and the required CMFB.

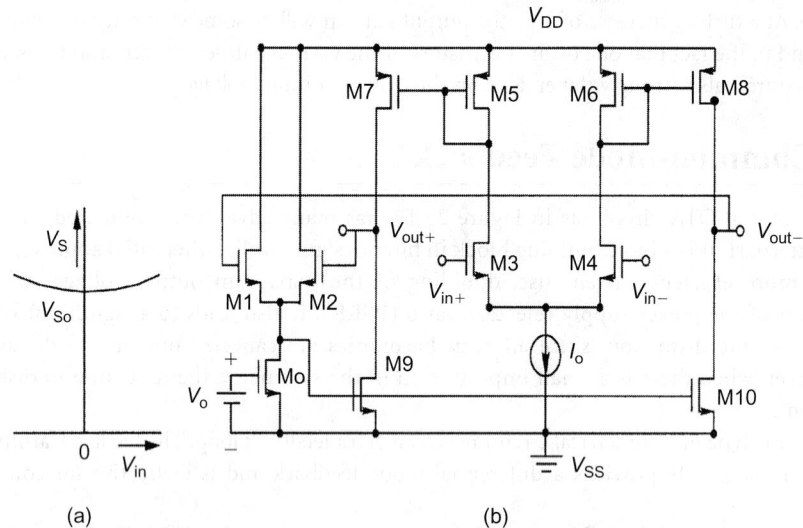

FIGURE 25.22 (a) The relationship between the differential input voltage, superimposed on a common-mode voltage V_{CM} of a differential pair (M1, M2) and its common-source voltage V_s. (b) a fully differential OTA with the differential pair (M1, M2) for providing a common-mode feedback.

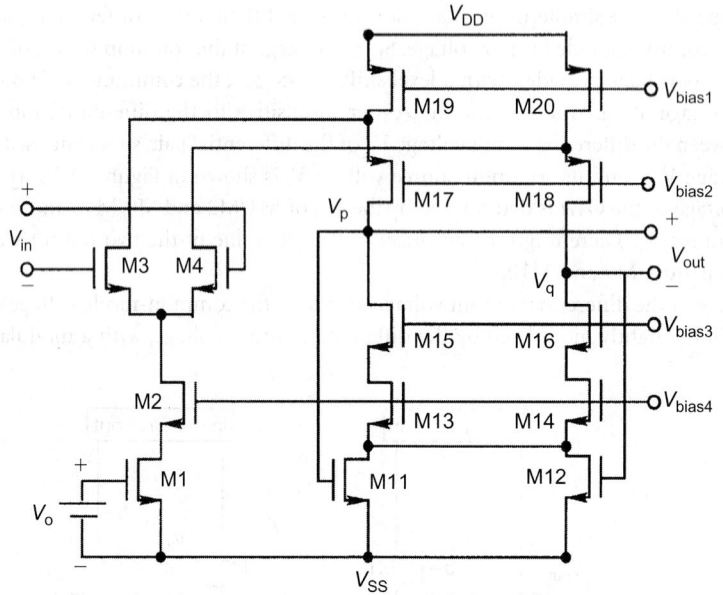

FIGURE 25.23 A fully differential folded cascode OTA with another commonly used CMFB circuit.

that is twice the differential input signal frequency. This modulation is caused by the "nonflat" characteristic of the V_s versus V_{in} characteristic of the differential pair (M1, M2) (see Figure 25.22(a)).

Another commonly used CMFB circuit is shown in the fully differential folded cascode OTA in Figure 25.23 [13]. In this circuit, a similar high-output resistance and high unloaded voltage gain can be achieved as in the normal cascode circuits. An advantage of the folded cascode technique, however, is a higher accuracy in the signal-current transfer because current mirrors are avoided.

In Figure 25.23, all transistors are in saturation, with the exception of M1, M11, and M12, which are in the triode region. The CMFB is provided with the help of M11 and M12. These two transistors sense the output voltages V_P and V_Q. Since they operate in the triode region, their sum-current is insensitive to the differential output voltage ($V_P - V_Q$) and depends only on the common output voltage (($V_P + V_Q$)/2). Because the current that flows through M17 and M18 forces the value of the above-mentioned sum-current, they also determine, together with V_{bias4}, the common-mode output voltage. By choosing V_{bias1} in such a way that I_{M19} is twice I_{M17}, and making the width of transistor M1 twice that of M11 (=M12), the nominal common-mode output voltage will be equal to the gate voltage of M1.

25.6 Filter Applications with Low-Voltage Highly Linear OTAs

Usually, G_m-C filters are considered suitable candidates for high-speed and low-power applications. Compared with the SC op-amp and RC op-amp techniques, the applicability of the G_m-C filter is limited by the low dynamic range and medium, even poor, linearity. A low-voltage, highly linear voltage-controlled transconductor [14, 15] shown in Figure 25.24 is used to implement a low-voltage, G_m-C all-pass adaptive forward equalizer (FE) stage operating at 125 Mbps. The adaptive forward equalizer is a part of a larger repeater/transceiver that enables IEEE 1394b transceivers to communicate over variable length of UTP-5 cables for up to 100 m. The repeater receives the data from one IEEE 1394b transceiver and retransmits it over the UTP-5 cable. On the other end of the cable, another repeater is used to receive the data (using the equalizer), and sends it to another IEEE 1394b transceiver. The equalizer is a cascade of two stages of the G_m-C stage shown in Figure 25.25. The control voltage is used to adapt the G_m of the transconductors, and consequently change the location of the poles in the forward equalizer. The control voltage is generated using an adaptive loop that estimates the attenuation introduced by the cable (which is a function of the cable length) and adjusts the response of the forward equalizer accordingly. To meet the jitter specification required for adequate performance of transceivers, the transconductor used in the forward equalizer stage has to have a THD of 30 dB or better at an input voltage range from 0 to 500 mV and data rate of 125 Mbps. Figure 25.26 shows the frequency response of the forward

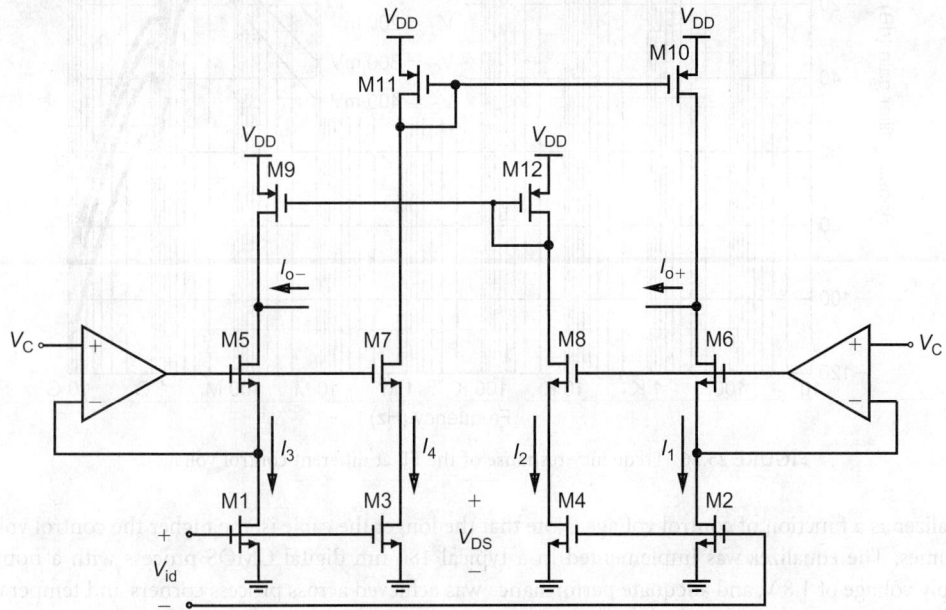

FIGURE 25.24 A low-voltage, highly linear voltage-controlled transconductor.

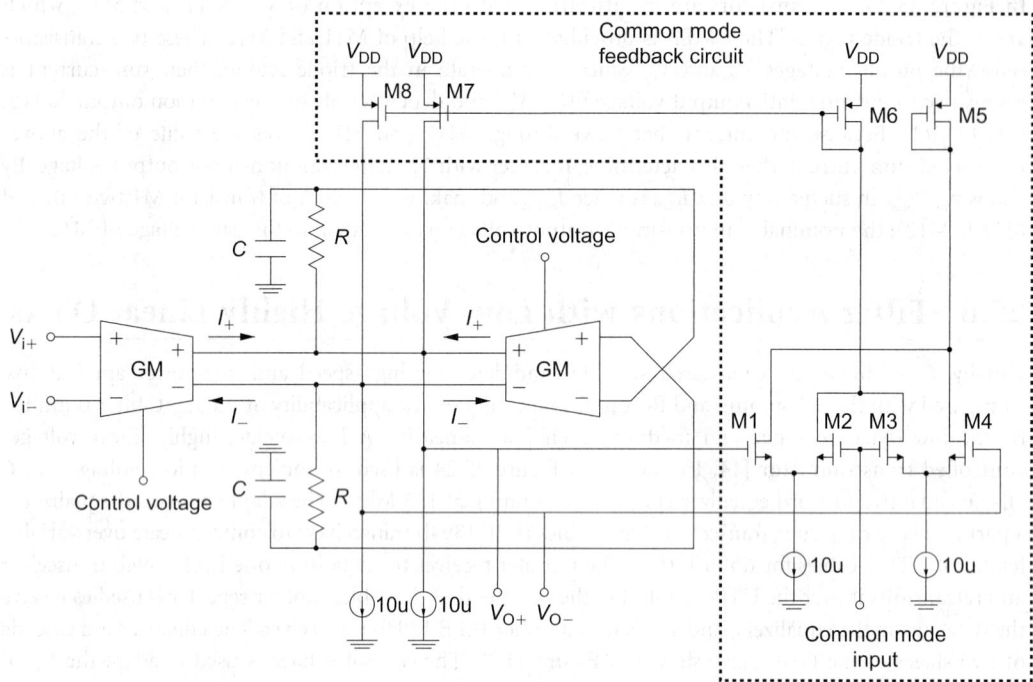

FIGURE 25.25   A G_m-C filter implemented using the transconductor in Figure 25.24.

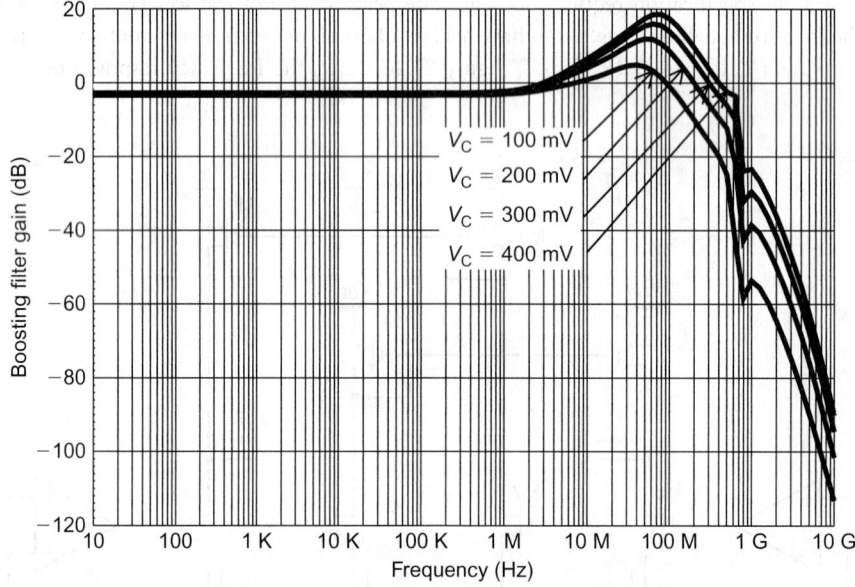

FIGURE 25.26   Frequency response of the FE at different control voltages.

equalizer as a function of control voltage. Note that the longer the cable is, the higher the control voltage becomes. The equalizer was implemented in a typical 180-nm digital CMOS process with a nominal supply voltage of 1.8 V, and adequate performance was achieved across process corners and temperature range of –40 to 125°C at supply voltage as low as 1.6 V. The transconductor occupied an area of 1945 μm^2, and consumed an average power of 418 μW at 125 Mbps.

References

1. M. Ismail and T. Fiez, *Analog VLSI Signal and Information Processing*, McGraw-Hill, New York, 1994.
2. E.A. Vittoz, The design of high-performance analog circuits on digital CMOS chips, *IEEE J. Solid-State Circuits*, vol. SC-20, pp. 657–665, 1985.
3. F. Krummenacher, High voltage gain CMOS OTA for micro-power SC filters, *Electron. Lett.*, vol. 17, pp. 160–162, 1981.
4. M.S.J. Steyaert, W. Bijker, P. Vorenkamp, and J. Sevenhans, ECL-CMOS and CMOS-ECL interface in 1.27mm CMOS for 150 MHz digital ECL data transmission systems, *IEEE J. Solid-State Circuits*, vol. SC-26, pp. 15–24, 1991.
5. S.L. Wong and C.A.T. Salama, An efficient CMOS buffer for driving large capacitive loads, *IEEE J. Solid-State Circuits*, vol. SC-21, pp. 464–469, 1986.
6. R.F. Wassenaar, Analysis of Analog C-MOS Circuits, Ph.D. thesis, University of Twente, The Netherlands, 1996.
7. M.G. Degrauwe, J. Rijmenants, E.A. Vittoz, and H.J. DeMan, Adative biasing CMOS amplifiers, *IEEE J. Solid-State Circuits*, vol. SC-17, pp. 522–528, 1982.
8. E. Seevinck, R.F. Wassenaar, and W. de Jager, Universal adaptive biasing principle for micro-power amplifiers, *Proceedings of ESSCIRC'84*, pp. 59–62, September 1984.
9. R.F. Wassenaar, Current-mode minimax circuit, *IEEE Circuits Devices*, vol. 8, p. 47, 1992.
10. S. Dupuie and M. Ismail, High frequency CMOS transconductors, Chap. 5, in *Analogue IC Design: The Current-Mode Approach*, C. Toumazou, F.J. Lidgey, and D.G. Haight (Eds.), Peter Peregrinus Ltd., London, 1990.
11. E. Seevinck and R.F. Wassenaar, A versatile CMOS linear transconductor/square-law function circuit, *IEEE J. Solid-State Circuts*, vol. SC-22, pp. 366–377, 1987.
12. S.H. Lewis and P.R. Gray, A pipelined 5MHz 9b ADC, *IEEE ISSCC'87 Digest of Technical Papers*, pp. 210–211, 1987.
13. T.C. Choi, R.T. Kaneshiro, R. Brodersen, and P.R. Gray, High-frequency CMOS switched capacitor filters for communication applications, *IEEE ISSCC'83 Digest of Technical Papers*, pp. 246–247, 314, 1983.
14. A.A. Fayed and M. Ismail, A low-voltage, highly-linear voltage-controlled transconductor, *IEEE Trans. Circuits Systems II*, vol. 52, no. 12, pp. 831–835, Dec. 2005.
15. A.A. Fayed and M. Ismail, *Adaptive Techniques for Analog and Mixed Signal ICs*, Springer, Boston, to appear.

Section V

Logic Circuits

Saburo Muroga
University of Illinois at Urbana-Champaign

0-8493-XXXX-X/04/$0.00+$1.50
© 2006 by CRC Press LLC

26
Expressions of Logic Functions

Saburo Muroga

University of Illinois
at Urbana-Champaign

CONTENTS

26.1 Introduction to Basic Logic Operations

In a contemporary digital computer, logic operations for computational tasks are usually done with signals that take values of 0 or 1. These logic operations are performed by many logic networks which constitute the computer. Each logic network has input variables $x_1, x_2, \ldots, x_n$ and output functions $f_1, f_2, \ldots, f_m$. Each of the input variables and output functions take only binary value, 0 or 1. Now let us consider one of these output functions, f. Any **logic function** f can be expressed by a **combination table** (also called a **truth table**) exemplified in Table 26.1.

26.1.1 Basic Logic Expressions

Any logic function can be expressed with three basic logic operations: OR, AND, and NOT. It can also be expressed with other logic operations, as explained in a later section.

TABLE 26.1   Combination Table

x	y	z	f
0	0	0	0
0	0	1	0
0	1	0	1
0	1	1	1
1	0	0	0
1	0	1	0
1	1	0	0
1	1	1	1

The **OR operation** of n variables $x_1, x_2, \ldots, x_n$ yields the value 1 whenever at least one of the variables is 1, and 0 otherwise, where each of $x_1, x_2, \ldots, x_n$ assumes the value 0 or 1. This is denoted by $x_1 \vee x_2 \vee \ldots \vee x_n$. The OR operation defined above is sometimes called **logical sum**, or **disjunction**. Also, some authors use "+", but throughout Section V, we use $\vee$, and + is used to mean an arithmetic addition.

The **AND operation** of n variables yields the value 1 if and only if all variables $x_1, x_2, \ldots, x_n$ are simultaneously 1. This is denoted by $x_1 \cdot x_2 \cdot x_3 \ldots x_n$. These dots are usually omitted: $x_1 x_2 x_3 \ldots x_n$. The AND operation is sometimes called **conjunction**, or **logical product**.

The **NOT operation** of a variable x yields the value 1 if $x = 0$, and 0 if $x = 1$. This is denoted by $\bar{x}$ or x'. The NOT operation is sometimes called **complement** or **inversion**.

Using these operations, AND, OR, and NOT, a logic function, such as the one shown in Table 26.1 can be expressed in the following formula:

$$f = \bar{x}y\bar{z} \vee \bar{x}yz \vee xyz \tag{26.1}$$

26.1.2 Logic Expressions

Expressions with logic operations with AND, OR, and NOT, such as Eq. 26.1, are called **switching expressions** or **logic expressions**. Variables x, y, and z are sometimes called **switching variables** or **logic variables**, and they assume only binary values 0 and 1. In logic expressions such as Eq. 26.1, each variable, x_i, appears with or without the NOT operation, that is, as $\bar{x}_i$ or x_i. Henceforth, $\bar{x}_i$ and x_i are called the **literals** of a variable x_i.

26.1.3 Logic Expressions with Cubes

Logic expressions such as $f = x\bar{y} \vee yz \vee \bar{x}y\bar{z}$ can be expressed alternatively as a set, {(10-), (-11), (010)}, using components in a vector expression such that the first, second, and third components of the vector represent x, y, and z, respectively, where the value "1" represents x_i, "0" represents $\bar{x}_i$, and "-" represents the lack of the variable. For example, (10-) represents $x\bar{y}$. These vectors are called **cubes**. Logic expressions with cubes are used often because of their convenience for processing by a computer.

26.2 Truth Tables

The value of a function f for different combinations of values of variables can be shown in a table, as exemplified in Tables 26.1 and 26.2. The table for n variables has 2^n rows. Thus the table size increases rapidly as n increases.

Under certain circumstances, some of the combinations of input variable values never occur, or even if they occur, we do not care what values f assumes. These combinations are called **don't-care conditions**, or simply **don't-cares**, and are denoted by "d" or "*", as shown in Table 26.2.

TABLE 26.2 Truth Table with Don't-Care Conditions

Decimal Number of Row	Variables			Function
	x	y	z	f
0	0	0	0	0
1	0	0	1	1
2	0	1	0	d
3	0	1	1	0
4	1	0	0	d
5	1	0	1	1
6	1	1	0	1
7	1	1	1	0

26.2.1   Decimal Specifications

A concise means of expressing the truth table is to list only rows with $f = 1$ and d, identifying these rows with their decimal numbers in the following **decimal specifications**. For example, the truth table of Table 26.2 can be expressed, using Σ, as

$$f(x, y, z) = \sum (1, 5, 6) + d(2, 4)$$

If only rows with $f = 0$ and d are considered, the truth table in Table 23.2 can be expressed, using Π, as

$$f(x, y, z) = \Pi(0, 3, 7) + d(2, 4)$$

26.3   Karnaugh Maps

Logic functions can be visually expressed using a Karnaugh map, which is simply a different way of representing a truth table, as exemplified for four variables in Figure 26.1(a). For the case of four variables, for example, a Karnaugh map consists of 16 cells; that is, 16 small squares as shown in Figure 26.1(a). Here, two-bit numbers along the horizontal line above the squares show the values of x_1 and x_2, and two-bit binary numbers along the vertical line on the left of the squares show the values of x_3 and x_4. The top left cell in Figure 26.1(a) has 1 inside for $x_1 = x_2 = x_3 = x_4 = 0$. Also, the cell in the second row and the second column from the left has d inside. This means $f = d$ (i.e., don't-care) for $x_1 = 0$, $x_2 = 1$, $x_3 = 0$, and $x_4 = 1$. The binary numbers that express variables are arranged in such a way that binary numbers for any two cells that are horizontally or vertically adjacent differ in only one bit position. Also, the two numbers in each row in the first and last columns differ in only one bit position and are interpreted to be adjacent. Also, the two numbers in each column in the top and bottom rows are similarly interpreted to be adjacent. Thus, the four cells in the top row are interpreted to be adjacent to the four cells in the bottom row in each column. The four cells in the first column are interpreted to be adjacent to the four cells in the last column in each row. With this arrangement of cells and this interpretation, a Karnaugh map is more than a concise representation of a truth table; it can express many important algebraic concepts, as we will see later. A Karnaugh map is a two-dimensional representation of the 16 cells on the surface of a torus, as shown in Figure 26.1(b), where the two ends of the map are connected vertically and horizontally.

Figure 26.2 shows the correspondence between the cells in the map in Figure 26.2(a) and the rows in the truth table in Figure 26.2(b). Notice that the rows in the truth table are not shown in consecutive order in the Karnaugh map. The Karnaugh map labeled with variable letters, instead of with binary numbers, shown in Figure 26.2(c), is also often used. Although a 1 or 0 shows the function's value corresponding to a particular cell, 0 is often not shown in each cell. Cells that contain 1's are called **1-cells** (similarly, **0-cells**).

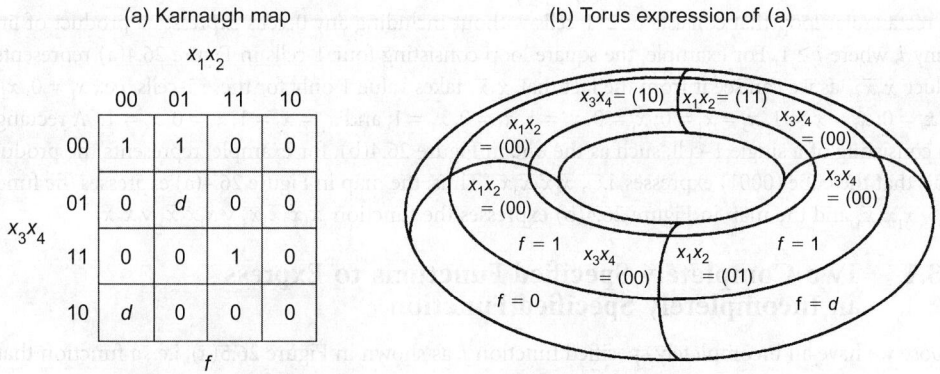

FIGURE 26.1   Karnaugh map for four variables.

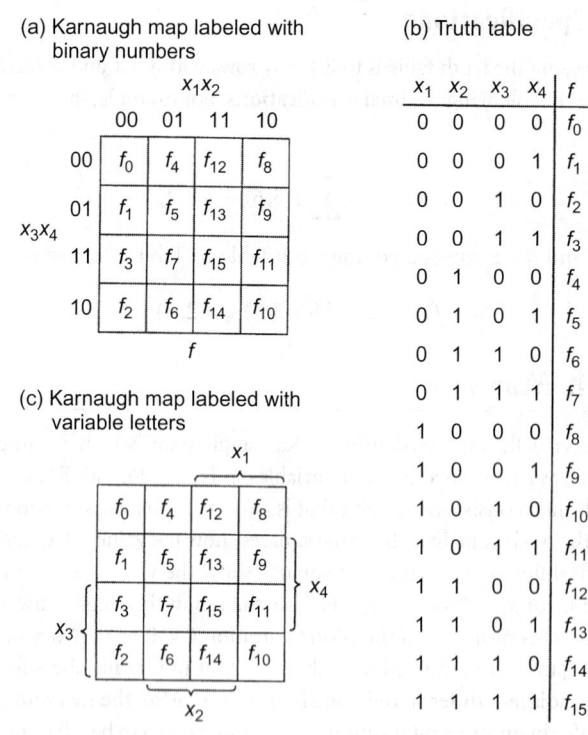

FIGURE 26.2 Correspondence between the cells in a Karanaugh map for four variables and the rows in a truth table.

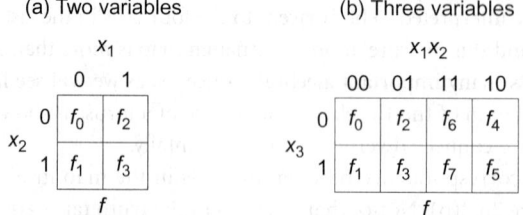

FIGURE 26.3 Karanaugh map for two and three variables.

Patterns of Karnaugh maps for two and three variables are shown in Figures 26.3(a) and (b), respectively. As we extend this treatment to the cases of 5 or more variables, the maps, which will be explained in a later subsection, become increasingly complicated.

A rectangular loop that consists of 2^i 1-cells without including any 0-cells expresses a product of literals for any i, where $i \geq 1$. For example, the square loop consisting four 1-cells in Figure 26.4(a) represents the product $x_2\bar{x}_3$, as we can see it from the fact that $x_2\bar{x}_3$ takes value 1 only for these 1-cells, i.e., $x_1 = 0$, $x_2 = 1$, $x_3 = x_4 = 0$; $x_1 = x_2 = 1$, $x_3 = x_4 = 0$; $x_1 = 0$, $x_2 = 1$, $x_3 = 0$, $x_4 = 1$; and $x_1 = x_2 = 1$, $x_3 = 0$, $x_4 = 1$. A rectangular loop consisting of a single 1-cell, such as the one in Figure 26.4(b), for example, represents the product of literals that the cube (0001) expresses, i.e., $\bar{x}_1\bar{x}_2\bar{x}_3x_4$. Thus, the map in Figure 26.4(a) expresses the function $x_2\bar{x}_3 \vee x_1x_3\bar{x}_4$ and the map in Figure 26.4(b) expresses the function $\bar{x}_1\bar{x}_2\bar{x}_3x_4 \vee x_1x_2\bar{x}_3 \vee x_2\bar{x}_4$.

26.3.1 Two Completely Specified Functions to Express an Incompletely Specified Function

Suppose we have an incompletely specified function f, as shown in Figure 26.5(a), i.e., a function that has some don't-cares. This incompletely specified function f can be expressed alternatively with two completely specified functions, f^{ON} and f^{OFF}, shown in Figure 26.5(b) and (c), respectively. f^{ON} is called **ON-set**

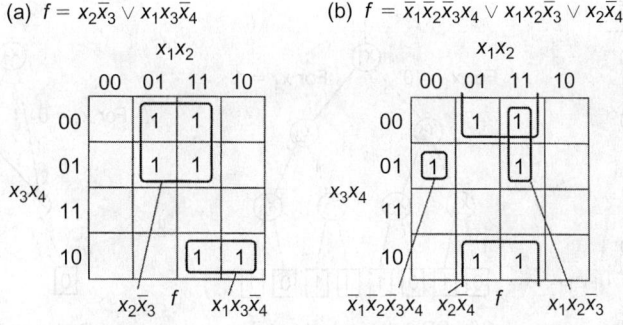

FIGURE 26.4   Products of literals expressed on Karnaugh maps.

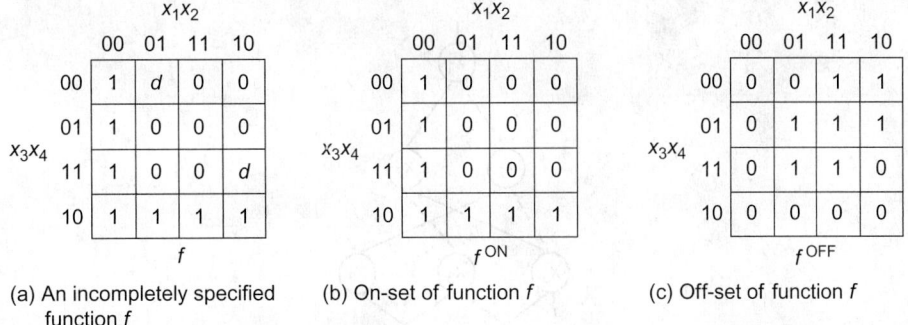

FIGURE 26.5   Expression of an incompletely specified function f with two completely specified functions.

of f and is the function whose value is 1 for $f = 1$ and 0 for $f = 0$ and d. f^{OFF} is called **OFF-set** of f and is 1 for $f = 0$ and 0 for $f = 1$ and d. **Don't-care set** of f can be derived as

$$f^{DC} = \overline{f^{ON} \vee f^{OFF}}$$

26.4   Binary Decision Diagrams

Truth tables or logic expressions can be expressed with **binary decision diagrams**, which are usually abbreviated as **BDDs**. Compared with logic expressions or truth tables, BDDs have unique features, such as unique concise representation, processing speed, and the memory space, as discussed in a later section.

   Let us consider a logic expression, $x_1\bar{x}_2 \vee x_3$, for example. This can be expressed as the truth table shown in Figure 26.6(a). Then, this can be expressed as the BDD shown in Figure 26.6(b). It is easy to see why Figure 26.6(b) represents the truth table in Figure 26.6(a). Let us consider the row, $(x_1 x_2 x_3) = (011)$, for example, in Figure 26.6(a). In Figure 26.6(b), starting from the top node which represents x_1, we go down to the left node which represents x_2, following the dotted line corresponding to $x_1 = 0$. From this node, we go down to the second node from the left which represents x_3, following the solid line corresponding to $x_2 = 1$. From this node, we go down to the fourth rectangle from the left which represents, $f = 1$, following the solid line corresponding to $x_3 = 1$. Thus, we have the value of f that is shown for the row, (x_1, x_2, x_3) $= (011)$ in the truth table in Figure 26.6(a). Similarly, for any row in the truth table, we reach the rectangle that shows the value of f identical to that in the truth table in Figure 26.6(a), by following a solid or dotted line corresponding to 1 or 0 for each of x_1, x_2, x_3, respectively. BDD in Figure 26.6(b) can be simplified to

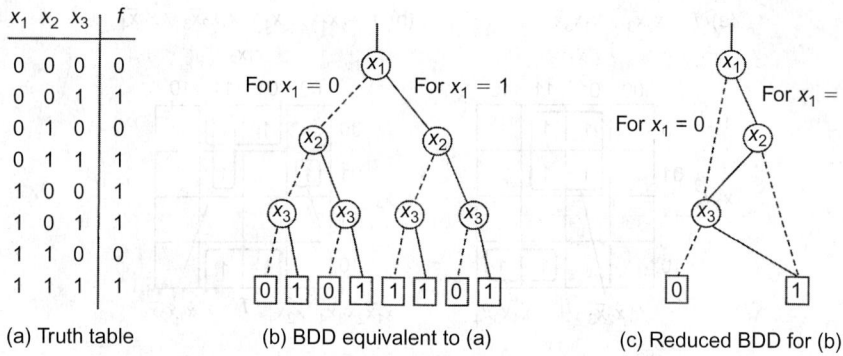

x_1 x_2 x_3	f
0 0 0	0
0 0 1	1
0 1 0	0
0 1 1	1
1 0 0	1
1 0 1	1
1 1 0	0
1 1 1	1

(a) Truth table   (b) BDD equivalent to (a)   (c) Reduced BDD for (b)

FIGURE 26.6 Binary decision diagram.

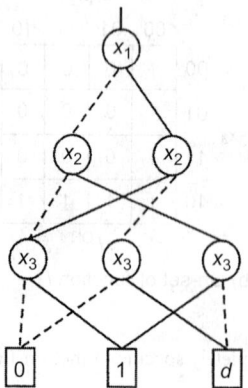

FIGURE 26.7 Reduced BDD with don't-cares.

the BDD shown in Figure 26.6(c), which is called the reduced BDD, by the reduction to be described in a later section.

When a function has don't-cares, d's, we can treat it in the same manner by considering a rectangle for d's, as shown in Figure 26.7.

27

Basic Theory of Logic Functions

Saburo Muroga

University of Illinois
at Urbana-Champaign

CONTENTS

27.1   Basic Theorems

Theory on logic functions where the values of variables and their functions are 0 or 1 only is called **switching theory.** Here, let us discuss the basics of switching theory.

Let us denote the set of input variables by the vector expression $(x_1, x_2, \ldots, x_n)$. There are 2^n different input vectors when each of these n variables assume the value 1 or 0. An input vector $(x_1, x_2, \ldots, x_n)$ such that $f(x_1, x_2, \ldots, x_n) = 1$ or 0 is called a **true (input) vector**, or a **false (input) vector** of f, respectively. Vectors with n components are often called **n-dimensional vectors** if we want to emphasize that there are n components. When the value of a logic function f is specified for each of the 2^n vectors (i.e., for every combination of the values of $x_1, x_2, \ldots, x_n$), f is said to be **completely specified**. Otherwise, f is said to be **incompletely specified**; that is, the value of f is specified for fewer than 2^n vectors. Input vectors for which the value of f is not specified are called **don't-care conditions** usually denoted by "d" or "*," as described in Chapter 23. These input vectors are never applied to a network whose output realizes f, or the values of f for these input vectors are not important. Thus, the corresponding values of f need not be considered.

If there exists a pair of input vectors $(x_1, \ldots, x_{i-1}, 0, x_{i+1}, \ldots, x_n)$ and $(x_1, \ldots, x_{i-1}, 1, x_{i+1}, \ldots, x_n)$ that differ only in a particular variable x_i, such that the values of f for these two vectors differ, the logic function $f(x_1, x_2, \ldots, x_n)$ is said to be **dependent on** x_i. Otherwise, it is said to be **independent of** x_i. In this case, f can be expressed without the x_i in the logic expression of f. If f is independent of x_i, x_i is called a **dummy variable**. If $f(x_1, x_2, \ldots, x_n)$ depends on all its variables, it is said to be **non-degenerate**; otherwise, **degenerate**. For example, $x_1 \vee x_2 x_3 \vee x_2 \bar{x}_3$ can be expressed as $x_1 \vee x_2$ without dummy variable x_3.

27.1.1   Logic Expressions and Expansions

Given variable x_i, x_i and $\bar{x}_i$ are called **literals** of x_i, as already explained.

Definition 27.1: (1) A conjunction (i.e., a logical product) of literals where a literal for each variable appears at most once is called a **term** (or a **product**). A term may consist of a single literal.

A disjunction (i.e., logical sum) of terms is called a **disjunctive form** (or a sum of products). (2) Similarly, a disjunction (i.e., a logical sum) of literals where a literal for each variable appears at most once is called an **alterm**. An alterm may consist of a single literal. A conjunction of alterms is called a **conjunctive form** (or a product of sums).

For example, $x_1\bar{x}_2x_3$ is a term, and $x_1 \vee x_2 \vee \bar{x}_3$ is an alterm. Also, $x_1x_2 \vee x_1 \vee x_2$ and $x_1 \vee \bar{x}_1x_2$ are disjunctive forms that are equivalent to the logic function $x_1 \vee x_2$, but $x_1 \vee x_2(\bar{x}_1 \vee \bar{x}_2)$ is not a disjunctive form, although it expresses the same function. A disjunctive form does not contain products of literals that are identically 0 (e.g., $x_1\bar{x}_2x_2$) from the first sentence of (1). Similarly, a conjunctive form does not contain disjunctions of literals that are identically 1 (e.g., $x_1 \vee \bar{x}_2 \vee x_2 \vee x_3$).

The following expressions of a logic function are important special cases of a disjunctive form and a conjunctive form.

Definition 27.2: Assume that n variables, $x_1, x_2, \ldots, x_n$, are under consideration. (1) A **minterm** is defined as the conjunction of exactly n literals, where exactly one literal for each variable (x_i and $\bar{x}_i$ are two literals of a variable x_i) appears. When a logic function f of n variables is expressed as a disjunction of minterms without repetition, it is called the **minterm expansion** of f. (2) A **maxterm** is defined as a disjunction of exactly n literals, where exactly one literal for each variable appears. When f is expressed as a conjunction of maxterms without repetition, it is called the **maxterm expansion** of f.

For example, when three variables, x_1, x_2, and x_3, are considered, there exist $2^3 = 8$ minterms: $\bar{x}_1\bar{x}_2\bar{x}_3, \bar{x}_1\bar{x}_2x_3, \bar{x}_1x_2\bar{x}_3, \bar{x}_1x_2x_3, x_1\bar{x}_2\bar{x}_3, x_1\bar{x}_2x_3, x_1x_2\bar{x}_3$, and $x_1x_2x_3$. For the given function $x_1 \vee x_2x_3$, the minterm expansion is $\bar{x}_1x_2x_3 \vee x_1\bar{x}_2\bar{x}_3 \vee x_1\bar{x}_2x_3 \vee x_1x_2\bar{x}_3 \vee x_1x_2x_3$ and the maxterm expansion is $(x_1 \vee x_2 \vee x_3)(x_1 \vee x_2 \vee \bar{x}_3)(x_1 \vee \bar{x}_2 \vee x_3)$, as explained in the following.

Also notice that the row for each true vector in a truth table and also its corresponding 1-cell in the Karnaugh map correspond to a minterm. If $f = 1$ for $x_1 = x_2 = 1$ and $x_3 = 0$, then this row in the truth table and its corresponding 1-cell in the Karnaugh map corresponds to a minterm $x_1x_2\bar{x}_3$. Also, as will be described in a later section, the row for each false vector in a truth table and also its corresponding 0-cell in the Karnaugh map corresponds to a maxterm. For example, if $f = 0$ for $x_1 = x_2 = 1$ and $x_3 = 0$, then this row in the truth table and its corresponding 0-cell in the Karnaugh map corresponds to a maxterm $(x_1 \vee x_2 \vee x_3)$.

Theorem 27.1: Any logic function can be uniquely expanded with minterms and also with maxterms.

For example, $f(x_1, x_2) = x_1 \vee x_2\bar{x}_3$ can be uniquely expanded with the minterms as

$$x_1 \vee x_2\bar{x}_3 = \bar{x}_1x_2\bar{x}_3 \vee x_1\bar{x}_2\bar{x}_3 \vee x_1\bar{x}_2x_3 \vee x_1x_2\bar{x}_3 \vee x_1x_2x_3$$

and also can be uniquely expressed with maxterms as

$$x_1 \vee x_2\bar{x}_3 = (x_1 \vee x_2 \vee x_3)(x_1 \vee x_2 \vee \bar{x}_3)(x_1 \vee \bar{x}_2 \vee \bar{x}_3).$$

These expansions have different expressions but both express the same function $x_1 \vee x_2\bar{x}_3$.

The following expansions, called **Shannon's expansions**, are often useful.

Any function $f(x_1, x_2, \ldots, x_n)$ can be expanded into the following expression with respect x_1:

$$f(x_1, x_2, \ldots, x_n) = x_1 f(1, x_2, x_3, \ldots, x_n) \vee \bar{x}_1 f(0, x_2, x_3, \ldots, x_n)$$

where $f(1, x_2, x_3, \ldots, x_n)$ and $f(0, x_2, x_3, \ldots, x_n)$, which are $f(x_1, x_2, \ldots, x_n)$ with x_1 set to 1 and 0, respectively, are called **cofactors**. By further expanding each of $f(1, x_2, x_3, \ldots, x_n)$ and $f(0, x_2, x_3, \ldots, x_n)$ with respect to x_2, we have

$$f(x_1, x_2, \ldots, x_n) = x_1x_2 f(1, x_2, x_3, \ldots, x_n) \vee x_1 \bar{x}_2 f(1, 0, x_3, \ldots, x_n)$$

$$\vee \bar{x}_1x_2 f(0, 1, x_3, \ldots, x_n) \vee \bar{x}_1 \bar{x}_2 f(0, 0, x_3, \ldots, x_n)$$

Then we can further expand with respect to x_3. And so on.

Also, similarly

$$f(x_1, x_2, \ldots, x_n) = (x_1 \vee f(0, x_2, x_3, \ldots, x_n))(\overline{x}_1 \vee f(1, x_2, x_3, \ldots, x_n))$$

$$f(x_1, x_2, \ldots, x_n) = (x_1 \vee x_2 \vee f(0, 0, x_3, \ldots, x_n))(x_1 \vee \overline{x}_2 \vee f(0, 1, x_3, \ldots, x_n))$$

$$(\overline{x}_1 \vee x_2 \vee f(1, 0, x_3, \ldots, x_n))(\overline{x}_1 \vee \overline{x}_2 \vee f(1, 1, x_3, \ldots, x_n))$$

And so on.

These expansions can be extended to the case with m variables factored out, where $m \leq n$, although the only expansions for $m = 1$ (i.e., x_1) and 2 (i.e., x_1 and x_2) are shown above. Of course, when $m = n$, the expansions become the minterm and maxterm expansions.

Theorem 27.2: De Morgan's Theorem —

$$\overline{(x_1 \vee x_2 \vee \ldots \vee x_n)} = \overline{x}_1 \overline{x}_2 \ldots \overline{x}_n \quad \text{and} \quad \overline{(x_1 x_2 \ldots x_n)} = \overline{x}_1 \vee \overline{x}_2 \vee \ldots \vee \overline{x}_n$$

A logic function is realized by a **logic network** that consists of **logic gates**, where logic gates are realized with hardware, such as transistor circuits.

De Morgan's Theorem 27.2 has many applications. For example, it asserts that a NOR gate, i.e., a logic gate whose output expresses the complement of the OR operation on its inputs, with noncomplemented variable input $x_1, x_2, \ldots, x_n$ is interchangeable with an AND gate, i.e., a logic gate whose output expresses the AND operation on its complemented variable inputs $\overline{x}_1, \overline{x}_2, \ldots, \overline{x}_n$, since the outputs of both gates express the same function. This is illustrated in Figure 27.1 for $n = 2$.

Definition 27.3: The **dual** of a logic function $f(x_1, x_2, \ldots, x_n)$ is defined as $\overline{f}(\overline{x}_1, \overline{x}_2, \ldots, \overline{x}_n)$ where $\overline{f}$ denotes the complement of the entire function $f(x_1, x_2, \ldots, x_n)$ [in order to denote the complement of a function $f(x_1, x_2, \ldots, x_n)$, the notation $\overline{f(x_1, x_2, \ldots, x_n)}$ might be used instead of $\overline{f}$]. Let it be denoted by $f^d(x_1, x_2, \ldots, x_n)$. In particular, if $f(x_1, x_2, \ldots, x_n) = f^d(x_1, x_2, \ldots, x_n)$, then $f(x_1, x_2, \ldots, x_n)$ is called a **self-dual function**.

For example, when $f(x_1, x_2) = x_1 \vee x_2$ is given, we have $f^d(x_1, x_2) = \overline{f}(\overline{x}_1, \overline{x}_2) = \overline{x}_1 \vee \overline{x}_2$. This is equal to $x_1 x_2$ by the first identity of Theorem 27.2. In other words, $f^d(x_1, x_2) = x_1 x_2$. The function $x_1 x_2 \vee x_2 x_3 \vee x_1 x_3$ is self-dual, as can be seen by applying the two identities of Theorem 27.2.

Notice that, if f^d is the dual of f, the dual of f^d is f.

The concept of a dual function has many important applications. For example, it is useful in the conversion of networks with different types of gates, as in Figure 27.2, where the replacement of the AND and OR gates in Figure 27.2(a) by OR and AND gates, respectively, yields the output function f^d in Figure 27.2(b), which is dual to the output f of Figure 27.2(a).

As will be explained in a later chapter, a logic gate in CMOS is another important application example of the concept of "dual," where **CMOS** stands for complementary MOS and is a **logic family**, i.e., a type of transistor circuit for realizing a logic gate. Duality is utilized for reducing the power consumption of a logic gate in CMOS.

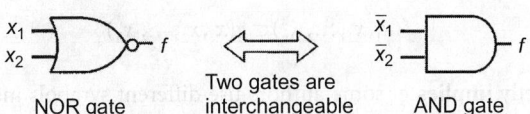

FIGURE 27.1 Application of De Morgan's theorem.

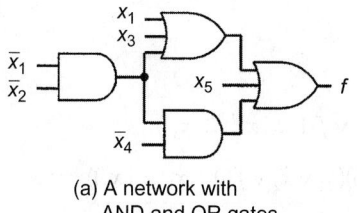
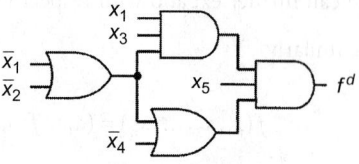

(a) A network with   (b) A network with AND
    AND and OR gates   and OR gates interchanged

FIGURE 27.2 Duality relation between two networks.

The following theorem shows a more convenient method of computing the dual of a function than direct use of Definition 27.3.

Theorem 27.3: Generalized De Morgan's Theorem—Let $f(x_1, x_2, \ldots, x_n)$ be a function expressed by "$\vee$" and "$\cdot$" (and possibly also by parentheses and the constants 0 and 1). Let $g(x_1, x_2, \ldots, x_n)$ be a function that is obtained by replacing every "$\vee$" and "$\cdot$" by "$\cdot$" and "$\vee$", respectively, throughout the logic expression of $f(x_1, x_2, \ldots, x_n)$ (and also, if 0 or 1 is contained in the original expression, by replacing 0 and 1 by 1 and 0, respectively). Then,

$$f^d(x_1, x_2, \ldots, x_n) = g(x_1, x_2, \ldots, x_n)$$

For example, when $f(x_1, x_2) = x_1 \vee x_2 \cdot \bar{x}_3$ is given, $f^d = x_1 \cdot (x_2 \vee \bar{x}_3)$ is obtained by this theorem. Here, notice that in f, the calculation of $\cdot$ precedes that of $\vee$; and in f^d, the $\vee$ must correspondingly be calculated first [thus, the parentheses are placed as $(x_2 \vee \bar{x}_3)$]. When $f = x_1 \vee 0 \cdot x_2 \cdot \bar{x}_3$ is given, $f^d = x_1 \cdot (1 \vee x_2 \vee \bar{x}_3)$ results by this theorem. When $f = x_1 \vee 1 \cdot x_2 \cdot \bar{x}_3$ is given, $f^d = x_1 \cdot (0 \vee x_2 \vee \bar{x}_3)$ results.

For example, the dual of $x_1 \vee x_2 x_3$ is $\bar{x}_1 \vee \bar{x}_2 \bar{x}_3$ according to Definition 27.3, which is a somewhat complicated expression. But by using the generalized De Morgan's theorem, we can immediately obtain the expression without bars, $x_1 \cdot (x_2 \vee x_3) = x_1 x_2 \vee x_1 x_3$.

27.2 Implication Relations and Prime Implicants

In this section, we discuss the algebraic manipulation of logic expressions; that is, how to convert a given logic expression into others. This is very useful for simplification of logic expression. Although simplification of a logic expression based on a Karnaugh map, which will be discussed in Chapter 28, is convenient in many cases, algebraic manipulation is more convenient in many other situations [3].

Definition 27.4: Let two logic functions be $f(x_1, x_2, \ldots, x_n)$ and $g(x_1, x_2, \ldots, x_n)$. If every vector $(x_1, x_2, \ldots, x_n)$ satisfying $f(x_1, x_2, \ldots, x_n) = 1$ satisfies also $g(x_1, x_2, \ldots, x_n) = 1$ but the converse does not necessarily hold, we write

$$f(x_1, x_2, \ldots, x_n) \subseteq g(x_1, x_2, \ldots, x_n) \tag{27.1}$$

and we say that f **implies** g. In addition, if there exists a certain vector $(x_1, x_2, \ldots, x_n)$ satisfying simultaneously $f(x_1, x_2, \ldots, x_n) = 0$ and $g(x_1, x_2, \ldots, x_n) = 1$, we write

$$f(x_1, x_2, \ldots, x_n) \subset g(x_1, x_2, \ldots, x_n) \tag{27.2}$$

and we say that f **strictly implies** g (some authors use different symbols instead of $\subset$). Therefore, Eq. 27.1 means $f(x_1, x_2, \ldots, x_n) \subset g(x_1, x_2, \ldots, x_n)$ or $f(x_1, x_2, \ldots, x_n) = g(x_1, x_2, \ldots, x_n)$. These relations are called **implication relations**. The left- and right-hand sides of Eq. (27.1) or (27.2) are called **antecedent** and **consequent**, respectively. If an implication relation holds between f and g. That is, if $f \subseteq g$ or $f \supseteq g$ holds, f and g are said to be **comparable** (more precisely, "$\subseteq$-comparable" or "implication-comparable"). Otherwise, they are **incomparable**.

When two functions, f and g, are given, we can find by the following methods at least whether or not there exists an implication relation between f and g; for example, using a truth table for f and g, directly based on Definition 27.4. If and only if there is no row in which $f = 1$ and $g = 0$, the implication relation $f \subseteq g$ holds. Furthermore, if there is at least one row in which $f = 0$ and $g = 1$, the relation is tightened to $f \subset g$. Table 27.1 shows the truth table for $f = x_1 \bar{x}_3 \vee \bar{x}_1 x_2 x_3$ and $g = x_1 \vee x_2$. There is no row with $f = 1$ and $g = 0$, so $f \subseteq g$ holds. Furthermore, there is a row with $f = 0$ and $g = 1$, so the relation is actually $f \subset g$.

Although "g implies f" means "if $g = 1$, then $f = 1$," it is to be noticed that "g does not imply f" does not necessarily mean "f implies g" but does mean either "f implies g" or "g and f are incomparable." In other words, it does mean "if $g = 1$, then f becomes a function other than the constant function which is identically equal to 1." (As a special case, f could be identically equal to 0.) Notice that "g does not imply f" does not necessarily mean "if $g = 0$, then $f = 0$."

Definition 27.5: An **implicant** of a logic function f is a term that implies f.

For example, x_1, x_2, $x_1 x_2$, and $x_1 x_3 \bar{x}_2$ are examples of implicants of the function $x_1 \vee x_2$. But $\bar{x}_1 \bar{x}_2$ is not. Notice that $x_1 x_3$ is an implicant of $x_1 \vee x_2$ even though $x_1 \vee x_2$ is independent of x_3. (Notice that every product of an implicant of f with any dummy variables is also an implicant of f. Thus, f has an infinite number of implicants.) But $x_1 x_2$ is not an implicant of $f = x_1 x_3 \vee \bar{x}_2$ because $x_1 x_2$ does not imply f. (When $x_1 x_2 = 1$, we have $f = x_3$, which can be 0. Therefore, even if $x_1 x_2 = 1$, f may become 0.) Some implicants are not obvious from a given expression of a function. For example, $x_1 x_2 \vee \bar{x}_1 x_2$ has implicants $x_2 x_3$ and $x_2 x_3 x_4$. Also, $x_1 \bar{x}_2 \vee x_2 x_3 \vee \bar{x}_1 x_3$ has an implicant x_3 because, if $x_3 = 1$, $x_1 \bar{x}_2 \vee x_2 x_3 \vee \bar{x}_1 x_3$ becomes $x_1 \bar{x}_2 \vee x_2 \vee \bar{x}_1 = x_1 \vee x_2 \vee \bar{x}_1 = (x_1 \vee \bar{x}_1) \vee x_2 = 1 \vee x_2$, which is equal to 1.

Definition 27.6: A term P is said to **subsume** another term Q if all the literals of Q are contained among those of P. If a term P which subsumes another term Q contains literals that Q does not have, P is said to **strictly subsume** Q.

For example, term $x_1 \bar{x}_2 x_3 \bar{x}_4$ subsumes $x_1 x_3 \bar{x}_4$ and also itself. More precisely speaking, $x_1 \bar{x}_2 x_3 \bar{x}_4$ strictly subsumes $x_1 x_3 \bar{x}_4$. Notice that Definition 27.6 can be equivalently stated as follows: "A term P is said to subsume another term Q if P implies Q; that is, $P \subseteq Q$. Term P strictly subsumes another term Q if $P \subset Q$."

Notice that when we have terms P and Q, we can say, "P implies Q" or, equivalently, "P subsumes Q." But the word "subsume" is ordinarily not used in other cases, except for comparing two alterms (as we will see in Section 28.4). For example, when we have functions f and g that are not in single terms, we usually do not say "f subsumes g."

On a Karnaugh map, if the loop representing a term P (always a single rectangular loop consisting of 2^i 1-cells because P is a product of literals) is part of the 1-cells representing function f, or is contained in the loop representing a term Q, P implies f or subsumes Q, respectively. Figure 27.3 illustrates this.

TABLE 27.1 Example for $f \subseteq g$

x_1	x_2	x_3	f	g
0	0	0	0	0
0	0	1	0	1
0	1	0	0	1
0	1	1	1	1
1	0	0	1	1
1	0	1	0	1
1	1	0	1	1
1	1	1	0	1

Conversely, it is easy to see that, if a term P, which does not contain any dummy variables of f, implies f, the loop for P must consist of some 1-cells of f, and if a term P, which does not contain any dummy variables of another term Q, implies Q, the loop for P must be inside the loop for Q.

The following concept of "prime implicant" is useful for deriving a simplest disjunctive form for a given function f (recall that logic expressions for f are not unique) and consequently for deriving a simplest logic network realizing f.

> **Definition 27.7:** A **prime implicant** of a given function f is defined as an implicant of f such that no other term subsumed by it is an implicant of f.

For example, when $f = x_1 x_2 \vee \bar{x}_1 x_3 \vee x_1 x_2 x_3 \vee \bar{x}_1 x_2 x_3$ is given, $x_1 x_2$, $\bar{x}_1 x_3$, and $x_2 x_3$ are prime implicants. But $x_1 x_2 x_3$ and $\bar{x}_1 x_2 x_3$ are not prime implicants, although they are implicants (i.e., if any of them is 1, then $f = 1$). Prime implicants of a function f can be obtained from other implicants of f by stripping off unnecessary literals until further stripping makes the remainder no longer imply f. Thus, $x_2 x_3$ is a prime implicant of $x_1 x_2 \vee \bar{x}_1 x_3$, and $x_2 x_3 x_4$ is an implicant of this function but not a prime implicant. As seen from this example, some implicants, such as $x_2 x_3$, and accordingly some prime implicants are not obvious from a given expression of a function. Notice that, unlike implicants, **a prime implicant cannot contain a literal of any dummy variable of a function.**

On a Karnaugh map, all prime implicants of a given function f of at least up to four variables can be easily found. As is readily seen from Definition 27.7, each rectangular loop that consists of 2^i 1-cells, with i chosen as large as possible, is a prime implicant of f. If we find all such loops, we will have found all prime implicants of f. Suppose that a function f is given as shown in Figure 27.4(a). Then, the prime implicants are shown in Figure 27.4(b). In this figure, we cannot make the size of the rectangular loops

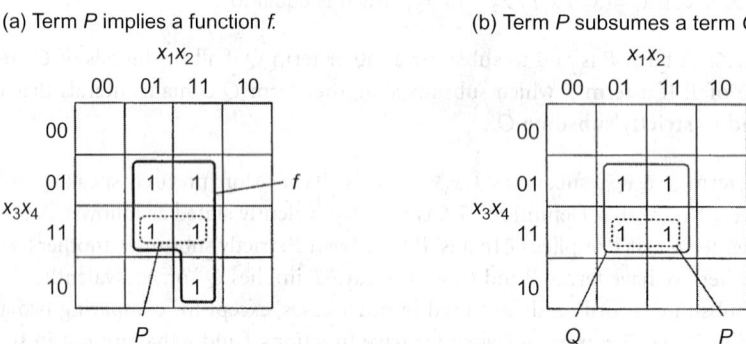

FIGURE 27.3   Comparison of "imply" and "subsume."

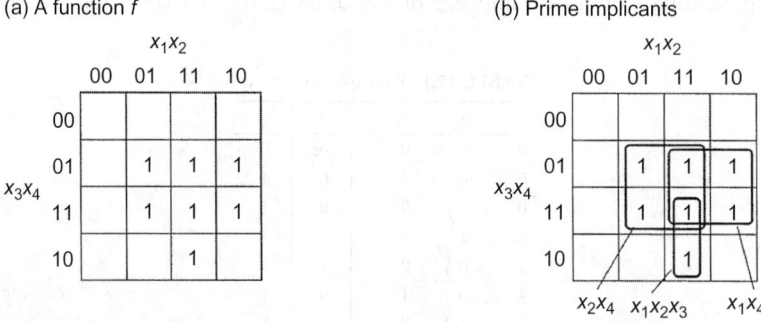

FIGURE 27.4   Expression of prime implicants on Karnaugh maps.

any bigger. (If we increase the size of any one of these loops, the new rectangular loop will contain a number of 1-cells that is not 2^i for any i, or will include one or more 0-cells.)

27.2.1 Consensus

Next, let us systematically find all prime implicants, including those not obvious, for a given logic function. To facilitate our discussion, let us define a consensus.

> **Definition 27.8:** Assume that two terms, P and Q, are given. If there is exactly one variable, say x, appearing noncomplemented in one term and complemented in the other—in other words, if $P = xP'$ and $Q = \bar{x}Q'$ (no other variables appear complemented in either P' or Q', and noncomplemented in the other)—then the product of all literals except the literals of x, that is, $P'Q'$ with duplicates of literals deleted, is called the **consensus** of P and Q.

For example, if we have two terms, $x_1\bar{x}_2x_3$ and $\bar{x}_1\bar{x}_2x_4\bar{x}_5$, the consensus is $\bar{x}_2x_3x_4\bar{x}_5$. But $x_1\bar{x}_2x_3$ and $\bar{x}_1x_2x_4\bar{x}_5$ do not have a consensus because two variables, x_1 and x_2, appear noncomplemented and complemented in these two terms.

A consensus can easily be shown on a Karnaugh map. For example, Figure 27.5 shows a function $f = x_1x_2 \vee \bar{x}_1x_4$. In addition to the two loops shown in Figure 27.5(a), which corresponds to the two prime implicants, x_1x_2 and $\bar{x}_1x_4$, of f, this f can have another rectangular loop, which consists of 2^i 1-cells with i chosen as large as possible, as shown in Figure 27.5(b). This third loop, which represents x_2x_4, the consensus of x_1x_2 and $\bar{x}_1x_4$, intersects the two loops in Figure 27.5(a) and is contained within the 1-cells that represent x_1x_2 and $\bar{x}_1x_4$. This is an important characteristic of a loop representing a consensus. Notice that these three terms, x_2x_4, x_1x_2, and $\bar{x}_1x_4$, are prime implicants of f. **When rectangular loops of 2^i 1-cells are adjacent (not necessarily exactly in the same row or column), the consensus is a rectangular loop of 2^i 1-cells, with i chosen as large as possible, that intersects and is contained within these loops. Therefore, if we obtain all largest possible rectangular loops of 2^i 1-cells, we can obtain all prime implicants, including consensuses, which intersect and are contained within other pairs of loops.** Sometimes, a consensus term can be obtained from a pair consisting of another consensus and a term, or a pair of other consensuses that do not appear in a given expression. For example, $x_1\bar{x}_2$ and x_2x_3 of $x_1\bar{x}_2 \vee x_2x_3 \vee \bar{x}_1x_3$ yield consensus x_1x_3, which in turn yields consensus x_3 with $\bar{x}_1x_3$. Each such consensus is also obtained among the above largest possible rectangular loops.

As we can easily prove, **every consensus that is obtained from terms of a given function f implies f.** In other words, every consensus generated is an implicant of f, although not necessarily a prime implicant.

27.2.2 Derivation of All Prime Implicants from a Disjunctive Form

The derivation of all prime implicants of a given function f is easy, using a Karnaugh map. If, however, the function has five or more variables, the derivation becomes increasingly complicated on a Karnaugh map. Therefore, let us discuss an algebraic method, which is convenient for implementation in a computer

(a) A function f

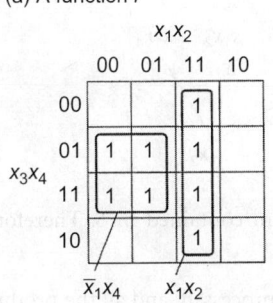

(b) Consensus as a prime implicant

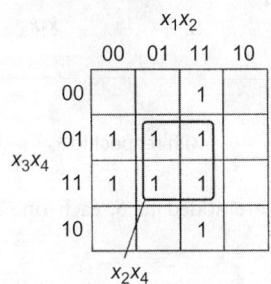

FIGURE 27.5 Expression of a consensus on Karnaugh maps.

program, although for functions of many variables even algebraic methods are too time consuming and we need to resort to heuristic methods.

The following algebraic method to find all prime implicants of a given function, which Tison [4,5] devised, is more efficient than the previously known **iterated-consensus method**, which was proposed for the first time by Blake in 1937 [2].

Definition 27.9: Suppose that p products, $A_1, \ldots, A_p$, are given. A variable such that only one of its literals appears throughout $A_1, \ldots, A_p$ is called a **unate variable**. A variable such that both of its literals appear in $A_1, \ldots, A_p$ is called a **binate variable**.

For example, when $x_1\bar{x}_2$, $\bar{x}_1x_3$, $x_3\bar{x}_4$ and $\bar{x}_2x_4$ are given, x_1 and x_4 are binate variables, since x_1 and x_4 as well as their complements, $\bar{x}_1$ and $\bar{x}_4$ appear, and x_2 and x_3 are unate variables.

Procedure 27.1: The Tison Method—Derivation of All Prime Implicants of a Given Function

Assume that a function f is given in a disjunctive from $f = P \vee Q \vee \ldots \vee T$, where $P, Q, \ldots, T$ are terms, and that we want to find all prime implicants of f. Let S denote the set $\{P, Q, \ldots, T\}$.

1. Among $P, Q, \ldots, T$ in set S, first delete every term subsuming any other term. Among all binate variables, choose one of them.

For example, when $f = x_1x_2x_4 \vee x_1x_3 \vee x_2\bar{x}_3 \vee x_2x_4 \vee x_3\bar{x}_4$ is given, delete $x_1x_2x_4$, which subsumes x_2x_4. The binate variables are x_3 and x_4. Let us choose x_3 first.

2. For each pair of terms, that is, one with the complemented literal of the chosen variable and the other with the noncomplemented literal of that variable, generate the consensus. Then add the generated consensus to S. From S, delete every term that subsumes another.

For our example, x_3 is chosen as the first binate variable. Thus we get consensus x_1x_2 from pair x_1x_3 and $x_2\bar{x}_3$, and consensus $x_1\bar{x}_4$ from pair x_1x_3 and $\bar{x}_3\bar{x}_4$. None subsumes another. Thus, S, the set of prime implicants, becomes x_1x_3, $x_2\bar{x}_3$, x_2x_4, $\bar{x}_3\bar{x}_4$, x_1x_2, and $x_1\bar{x}_4$.

S:

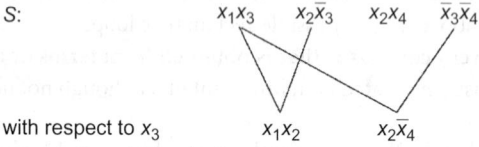

with respect to x_3

3. Choose another binate variable in the current S. Then go to Step 2. If all binate variables are tried, go to Step 4.

For the above example, for the second iteration of Step 2, we choose x_4 as the second binate variable and generate two new consensuses as follows:

S:

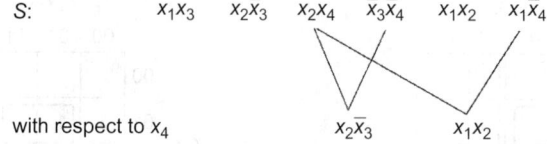

with respect to x_4

But when they are added to S, each one subsumes some term contained in S. Therefore, they are eliminated.

4. The procedure terminates because all binate variables are processed, and all the products in S are desired prime implicants.

The last expression is called the **complete sum** or **the all-prime-implicant disjunction**. The complete sum is the first important step in deriving the most concise expressions for a given function. □

Generation of prime implicants for an incompletely specified function, which is more general than the case of completely specified function described in Procedure 27.1, is significantly speeded up with the use of BDD (described in Chapters 27.1 and 27.4) by Coudert and Madre [1].

References

1. Coudert, O. and J.C. Madre, "Implicit and incremental computation of primes and essential primes of Boolean functions," *Design Automation Conf.*, pp. 36–39, 1992.
2. Brown, F.M., "The origin of the iterated consensus," *IEEE Tr. Comput.*, p. 802, Aug. 1968.
3. Muroga, S., *Logic Design and Switching Theory*, John Wiley & Sons, 1979 (now available from Krieger Publishing Co.).
4. Tison, P., Ph.D. dissertation, Faculty of Science, University of Grenoble, France, 1965.
5. Tison, P., "Generalization of consensus theory and application to the minimization of Boolean functions," *IEEE Tr. Electron. Comput.*, pp. 446–456, Aug. 1967.

28

Simplification of Logic Expressions

Saburo Muroga

*University of Illinois
at Urbana-Champaign*

CONTENTS

28.1 Minimal Sums

In this chapter, using only AND and OR gates, we will synthesize a two-level logic network. This is the fastest network, if we assume that every gate has the same delay, since the number of levels cannot be reduced further unless a given function can be realized with a single AND or OR gate. If there is more than one such network, we will derive the simplest network. Such a network has a close relation to important concepts in switching algebra, that is, irredundant disjunctive forms and minimal disjunctive forms (or minimal sums), as we discuss in the following.

Now let us explore basic properties of logic functions.

For many functions, some terms in their complete sums are redundant. In other words, even if we eliminate some terms from a complete sum, the remaining expression may still represent the original function for which the complete sum was obtained. Thus, we have the following concept.

Definition 28.1: An **irredundant disjunctive form** for *f* (sometimes called an irredundant sum-of-products expression or an irredundant sum) is a disjunction of prime implicants such that removal of any of the prime implicants makes the remaining expression not express the original *f*.

An irredundant disjunctive form for a function is not necessarily unique.

Definition 28.2: Prime implicants that appear in every irredundant disjunctive form for *f* are called **essential prime implicants** of *f*. Prime implicants that do not appear in any irredundant disjunctive form for *f* are called **absolutely eliminable prime implicants** of *f*. Prime implicants that appear in some irredundant disjunctive forms for *f* but not in all are called **conditionally eliminable prime implicants** of *f*.

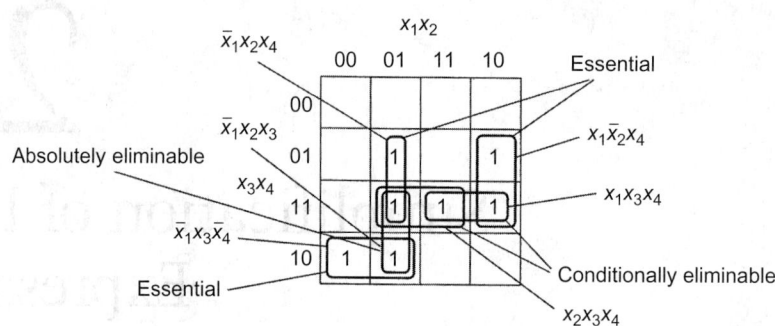

FIGURE 28.1 Different types of prime implicants.

Different types of prime implicants are shown in the Karnaugh map in Figure 28.1 for a function $f = \bar{x}_1 x_3 \bar{x}_4 \vee \bar{x}_1 x_2 x_3 \vee \bar{x}_1 x_2 x_4 \vee x_2 x_3 x_4 \vee x_1 x_3 x_4 \vee x_1 \bar{x}_2 x_4$. Here, $\bar{x}_1 x_2 \bar{x}_4$, $\bar{x}_1 x_2 x_4$, and $x_1 \bar{x}_2 x_4$ are essential prime implicants, $x_2 x_3 x_4$ and $x_1 x_3 x_4$ are conditionally eliminable prime implicants and $\bar{x}_1 x_2 x_3$ is absolutely eliminable prime implicant.

The concepts defined in the following play an important role in switching theory.

Definition 28.3: Among all irredundant disjunctive forms of f, those with a minimum number of prime implicants are called **minimal sums** (some authors call them as "sloppy minimal sum"). Among minimal sums, those with a minimum number of literals are called **absolute minimal sums** for f.

Irredundant disjunctive forms for a given function f can be obtained by deleting prime implicants one by one from the complete sum in all possible ways, after obtaining the complete sum by the Tison method discussed in Chapter 27. Then the minimal sums can be found among the irredundant disjunctive forms. Usually, however, this approach is excessively time-consuming because a function has many prime implicants when the number of variables is very large. When the number of variables is too large, derivation of a minimal sum is practically impossible.

Later, we will discuss efficient methods to derive minimal sums within reasonable computation time when the number of variables is few.

28.2 Derivation of Minimal Sums by Karnaugh Map

Because of its pictorial nature, a Karnaugh map is a very powerful tool for deriving manually all prime implicants, irredundant disjunctive forms, minimal sums, and also absolute minimal sums. Algebraic concepts such as prime implicants and consensuses can be better understood on a map.

One can derive all prime implicants, irredundant disjunctive forms, and minimal sums by the following procedures on Karnaugh maps, when the number of variables is small enough for the map to be manageable.

Procedure 28.1: Derivation of Minimal Sums on a Karnaugh Map

This procedure consists of three steps:

1. On a Karnaugh map, encircle all the 1-cells with rectangles (also squares as a special case), each of which consists of 2^i 1-cells, choosing i as large as possible, where i is a non-negative integer. Let us call these loops **prime implicant loops**, since they correspond to prime implicants in the case of the Karnaugh map for four or fewer variables. (In the case of a five- or six-variable map, the correspondence is more complex, as will be explained later.) Examples are shown in Figure 28.2(a).

2. Cover all the 1-cells with prime-implicant loops so that removal of any loops leaves some 1-cells uncovered. These sets of loops represent **irredundant disjunctive forms**. Figures 28.2(b) through 28.2(e) represent four irredundant disjunctive forms, obtained by choosing the loops in Figure 28.2(a) in four different ways. For example, if the prime-implicant loop $x_1 \bar{x}_3 x_4$ is omitted in Figure 28.2(b), the 1-cells for $(x_1, x_2, x_3, x_4) = (1\ 1\ 0\ 1)$ and $(1\ 0\ 0\ 1)$ are not covered by any loops.

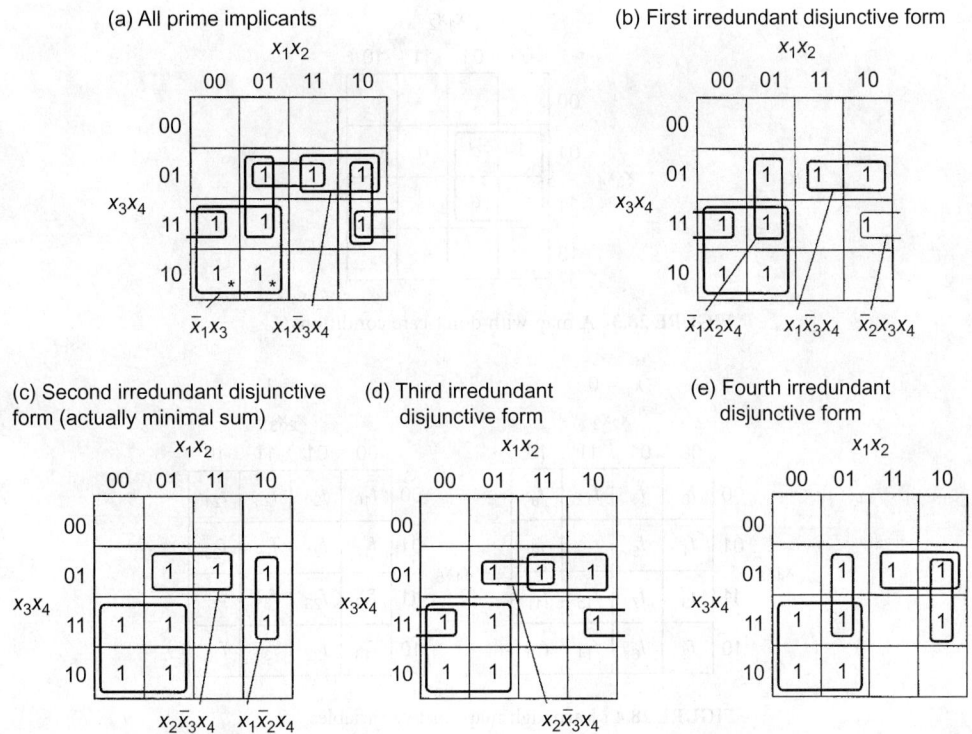

FIGURE 28.2 Irredundunt disjunctive forms and a minimal sum.

3. From the sets of prime implicant loops formed in Step 2 for irredundant disjunctive forms, choose the sets with a minimum number of loops. Among these sets, the sets that contain as many of the largest loops as possible (a larger loop represents a product of fewer literals) represent minimal sums. Figure 28.2(c) expresses the unique minimal sum for this function, since it contains one less loop than Figure 28.2(b), (d), or (e).

It is easy to see, from the definitions of prime implicants, irredundant disjunctive forms, and minimal sums, why Procedure 28.1 works.

When we derive irredundant disjunctive forms or minimal sums by Procedure 28.1, the following property is useful. When we find all prime-implicant loops by Step 1, some 1-cells may be contained in only one loop. Such 1-cells are called **distinguished 1-cells** and are labeled with asterisks. (The 1-cells shown with asterisks in Figure 28.2(a) are distinguished 1-cells.) A prime implicant loop that contains distinguished 1-cells is called an **essential prime implicant loop**. The corresponding prime implicant is an essential prime implicant, as already defined. In every irredundant disjunctive form and every minimal sum to be found in Step 2 and 3, respectively, essential prime implicants must be included, since each 1-cell on the map must be contained in at least one prime implicant loop and distinguished 1-cells can be contained only in essential prime implicant loops. Hence, if essential prime implicant loops are first identified and chosen, Procedure 28.1 is quickly processed.

Even if the don't-care condition d is contained in some cells, prime implicants can be formed in the same manner, by simply regarding d as being 1 or 0 whenever necessary to draw a greater prime implicant loop. For example, in Figure 28.3, we can draw a greater rectangular loop by regarding two d's as being 1. One d is left outside and is regarded as being 0. We need not consider loops consisting of d's only.

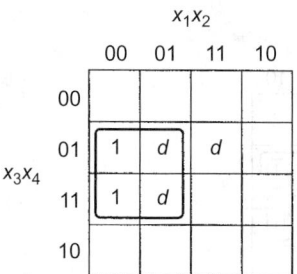

FIGURE 28.3 A map with don't-care conditions.

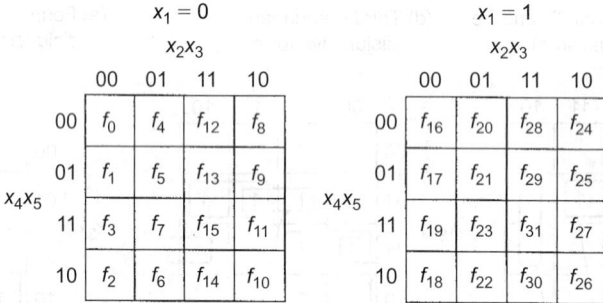

FIGURE 28.4 Karnaugh map for five variables.

28.2.1 Maps for Five and Six Variables

The Karnaugh map is most useful for functions of four or fewer variables, but it is often useful also for functions of five or six variables. A map for five variables consists of two four-variable maps, as shown in Figure 28.4, one for each value of the first variable. A map for six variables consists of four four-variable maps, as shown in Figure 28.5, one for each combination of values of the first two variables. Note that **the four maps in Figure 28.5 are arranged so that binary numbers represented by x_1 and x_2 differ in only one bit horizontally and vertically** (the map for $x_1 = x_2 = 1$ goes to the bottom right-hand side).

In a five-variable map, 1-cells are combined in the same way as in the four-variable case, with the additional feature that rectangular loops of 2^i 1-cells that are on different four-variable maps can be combined to form a greater loop replacing the two original loops only if they occupy the same relative position on their respective four-variable maps. Notice that these loops may be inside other loops in each four-variable map. For example, if f_{15} and f_{31} are 1, they can be combined; but even if $f_{15} = f_{29} = 1$, f_{15} and f_{29} cannot. In a six-variable map, only 1-cells in two maps that are horizontally or vertically adjacent can be combined if they are in the same relative positions. In Figure 28.5, for example, if f_5 and f_{37} are 1, they can be combined; but even if f_5 and f_{53} are 1, f_5 and f_{53} cannot. Also, four 1-cells that occupy the same relative positions in all four-variable maps can be combined as representing a single product. For example, f_5, f_{21}, f_{37}, and f_{53} can be combined if they are 1.

In the case of a five-variable map, we can find prime implicant loops as follows.

Procedure 28.2: Derivation of Minimal Sums on a Five-Variable Map

1. Unlike Step 1 of Procedure 28.1 for a function of four variables, this step requires the following two substeps to form prime implicant loops:
   a. On each four-variable map, encircle all the 1-cells with rectangles, each of which consists of 2^i 1-cells, choosing the number of 1-cells contained in each rectangle as large as possible. Unlike the case of Procedure 28.1, **these loops are not necessarily prime implicant loops** because they may not represent prime implicant, depending on the outcome of substep b.

In Figures 28.6 and 28.7, for example, loops formed in this manner are shown with solid lines.

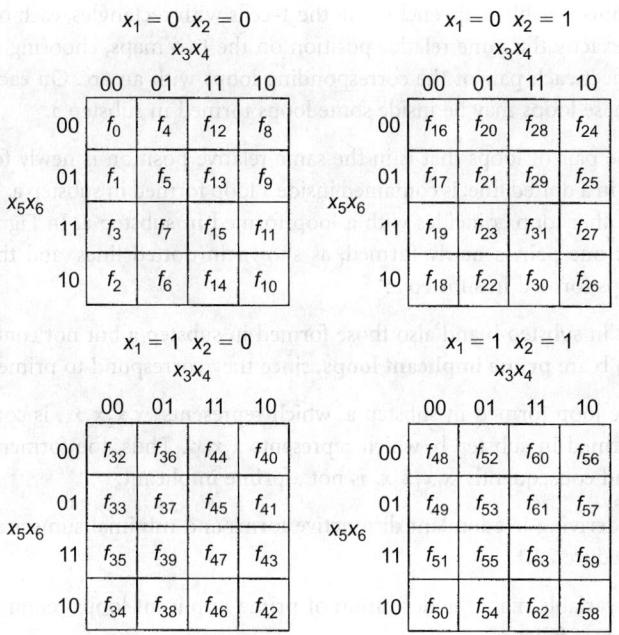

FIGURE 28.5 Karnaugh map for six variables.

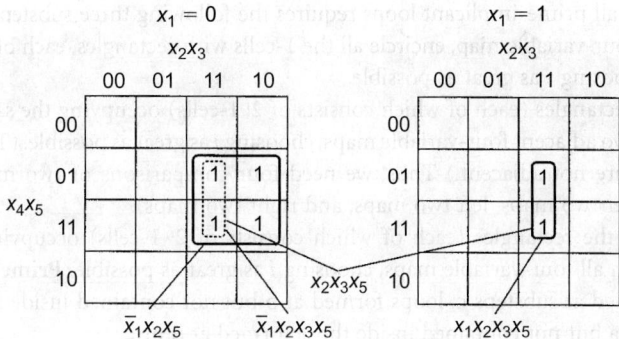

The prime implicants of this function are $\bar{x}_1 x_2 x_3$ and $x_2 x_3 x_5$.

FIGURE 28.6 Prime implicant loops on a map for five variables.

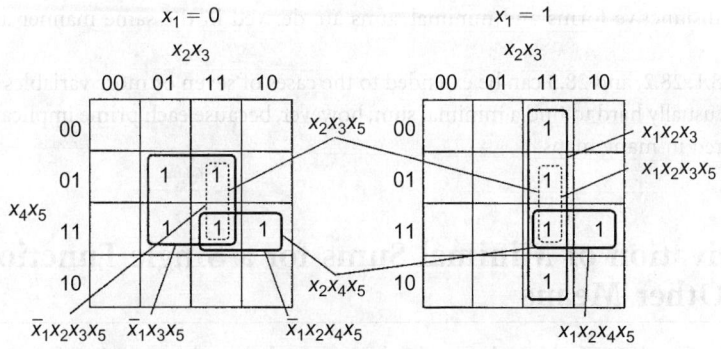

The prime implicants of this function are $\bar{x}_1 x_3 x_5$, $x_2 x_3 x_5$, $x_2 x_4 x_5$, and $x_1 x_2 x_3$.

FIGURE 28.7 Prime implicant loops on a map for five variables.

b. On each four-variable map, encircle all the 1-cells with rectangles, each of which consists of 2^i 1-cells in exactly the same relative position on the two maps, choosing i as great as possible. Then connect each pair of the corresponding loops with an arc. On each four-variable map, some of these loops may be inside some loops formed in substep a.

In Figure 28.6, one pair of loops that is in the same relative position is newly formed. One member of the pair, shown in a dotted line, is contained inside a loop formed in substep a. The pair is connected with an arc. The other loop coincides with a loop formed in substep a. In Figure 28.7, two pairs of loops are formed: one pair is newly formed, as shown in dotted lines, and the second pair is the connection of loops formed in substep a.

The loops formed in substep b and also those formed in substep a but not contained in any loop formed in substep b are **prime implicant loops,** since they correspond to prime implicants.

In Figure 28.6, the loop formed in substep a, which represents $\bar{x}_1 x_2 x_3 x_5$, is contained in the prime implicant loop formed in substep b, which represents $x_2 x_3 x_5$. Thus, the former loop is not a prime implicant loop, and consequently $\bar{x}_1 x_2 x_3 x_5$ is not a prime implicant.

2. Processes for deriving irredundant disjunctive forms and minimal sums are the same as Steps 2 and 3 of Procedure 28.1. ☐

In the case of six-variable map, the derivation of prime implicant loops requires more comparisons of four-variable maps, as follows.

Procedure 28.3: Derivation of Minimal Sums on a Six-Variable Map

1. Derivation of all prime-implicant loops requires the following three substeps:
   a. On each four-variable map, encircle all the 1-cells with rectangles, each of which consists of 2^i 1-cells, choosing i as great as possible.
   b. Find the rectangles (each of which consists of 2^i 1-cells) occupying the same relative position on every two adjacent four-variable maps, choosing i as great as possible. (Two maps in diagonal positions are not adjacent.) Thus, we need four comparisons of two maps (i.e., upper two maps, lower two maps, left two maps, and right two maps).
   c. Then find the rectangles (each of which consists of 2^i 1-cells) occupying the same relative position on all four-variable maps, choosing i as great as possible. **Prime implicant loops** are loops formed at substeps c, loops formed at b but not contained inside those at c, and loops formed at a but not contained inside those formed at b or c.
2. Processes for deriving irredundant disjunctive forms and minimal sums are the same as Steps 2 and 3 of Procedure 28.1. ☐

An example is shown in Figure 28.8.

Irredundant disjunctive forms and minimal sums are derived in the same manner as in the case of four variables.

Procedures 28.1, 28.2, and 28.3 can be extended to the cases of seven or more variables with increasing complexity. It is usually hard to find a minimal sum, however, because each prime implicant loop consists of 1-cells scattered in many maps.

28.3 Derivation of Minimal Sums for a Single Function by Other Means

Derivation of a minimal sum for a single function by Karnaugh maps is convenient for manual processing because designers can know the nature of logic networks better than automated minimization, to be described in Chapter 30, but its usefulness is limited to functions of four or five variables.

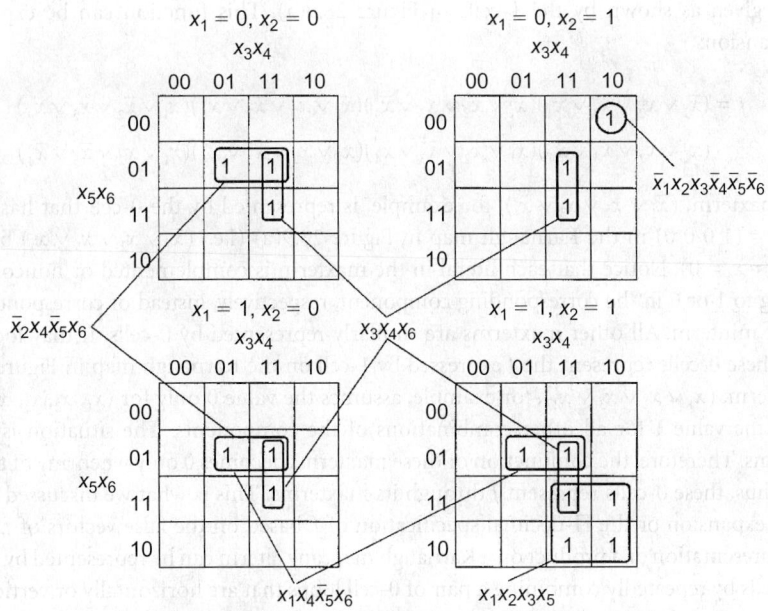

FIGURE 28.8 Prime-implicant loops on a map for six variables.

There are several methods for derivation of a minimal sum for a single function. A minimal sum can be found by forming a so-called a covering table where the minterms of a given function f are listed on the horizontal coordinate and all the prime implicants of f are listed on the vertical coordinate. A minimal sum can be described as a solution, that is, a minimal set of prime implicants that covers all the minterms [2]. The feasibility of this approach based on the covering table is limited by the number of minterms and prime implicants rather than the number of variables, which is the limiting factor for the feasibility of the derivation of minimal sums based on Karnaugh maps. This approach based on the table can be converted to an algebraic method, with prime implicants derived by the Tison method, as described by Procedure 4.6.1 in Ref. 3. This is much faster than the approach based on the table. Another approach is generation of irredundant disjunctive forms and then derive a minimal sum among them [4,5]. The feasibility of this approach is limited by the number of consensuses rather than the number of variables, minterms, or prime implicants.

As the number of variables, minterms, or prime implicants increases, the derivation of absolute minimal sums or even prime sums becomes too time-consuming, although the enhancement of the feasibility has been explored [1]. When too time-consuming, we need to resort to heuristic minimization, as will be described in Chapter 30 and Chapter 45, Section 45.3.

28.4   Prime Implicates, Irredundant Conjunctive Forms, and Minimal Products

The "implicates," "irredundant conjunctive forms," and "minimal products," which can be defined all based on the concept of conjunctive form, will be useful for deriving a minimal network that has OR gates in the first level and one AND gate in the second level.

First, let us represent the maxterm expansion of a given function f on a Karnaugh map. Unlike the map representation of the minterm expansion of f, where each minterm contained in the expansion is represented by a 1-cell on a Karnaugh map, each maxterm is represented by a 0-cell. Suppose that a function f is given as shown by the 1-cells in Figure 28.9(a). This function can be expressed in the maxterm expansion:

$$f = (\bar{x}_1 \vee x_2 \vee x_3 \vee x_4)(\bar{x}_1 \vee x_2 \vee x_3 \vee \bar{x}_4)(\bar{x}_1 \vee \bar{x}_2 \vee \bar{x}_3 \vee x_4)(x_1 \vee \bar{x}_2 \vee x_3 \vee x_4)$$

$$(x_1 \vee \bar{x}_2 \vee x_3 \vee \bar{x}_4)(x_1 \vee \bar{x}_2 \vee \bar{x}_3 \vee \bar{x}_4)(x_1 \vee x_2 \vee x_3 \vee \bar{x}_4)(x_1 \vee x_2 \vee \bar{x}_3 \vee \bar{x}_4)$$

The first maxterm, $(\bar{x}_1 \vee x_2 \vee x_3 \vee x_4)$, for example, is represented by the 0-cell that has components $(x_1, x_2, x_3, x_4) = (1\ 0\ 0\ 0)$ in the Karnaugh map in Figure 28.9(a) (i.e., $(\bar{x}_1 \vee x_2 \vee x_3 \vee x_4)$ becomes 0 for $x_1 = 1, x_2 = x_3 = x_4 = 0$). Notice that each literal in the maxterm is complemented or noncomplemented, corresponding to 1 or 0 in the corresponding component, respectively, instead of corresponding to 0 or 1 in the case of minterm. All other maxterms are similarly represented by 0-cells. It may look somewhat strange that these 0-cells represent the f expressed by 1-cell on the Karnaugh map in Figure 28.9(a). But the first maxterm, $(\bar{x}_1 \vee x_2 \vee x_3 \vee x_4)$, for example, assumes the value 0 only for $(x_1, x_2, x_3, x_4) = (1\ 0\ 0\ 0)$ and assumes the value 1 for all other combinations of the components. The situation is similar with other maxterms. Therefore, the conjunction of these maxterms becomes 0 only when any of the maxterms becomes 0. Thus, these 0-cells represent f through its maxterms. This is what we discussed earlier about the maxterm expansion or the Π-decimal specification of f, based on the false vectors of f.

Like the representation of a product on a Karnaugh map, any alterm can be represented by a rectangular loop of 2^i 0-cells by repeatedly combining a pair of 0-cell loops that are horizontally or vertically adjacent in the same rows or columns, where i is a non-negative integer. For example, the two adjacent 0-calls in the same column representing maxterms $(\bar{x}_1 \vee x_2 \vee x_3 \vee x_4)$ and $(\bar{x}_1 \vee x_2 \vee x_3 \vee \bar{x}_4)$ can be combined to form alterm $\bar{x}_1 \vee x_2 \vee x_3$, by deleting literals x_4 and $\bar{x}_4$, as shown in a solid-line loop in Figure 28.9(b).

The function f in Figure 28.9 can be expressed in the conjunctive form $f = (\bar{x}_1 \vee x_2 \vee x_3)(x_1 \vee \bar{x}_2 \vee x_3)$ $(x_1 \vee \bar{x}_4)(\bar{x}_1 \vee \bar{x}_2 \vee \bar{x}_3 \vee x_4)$, using a minimum number of such loops. The alterms in this expansion are represented by the loops shown in Figure 28.9(b).

Definition 28.4: An implicate of f is an alterm implied by a function f.

Notice that an implicate's relationship with f is opposite to that of an implicant that implies f.

For example, $(x_1 \vee x_2 \vee x_3)$ and $(x_1 \vee x_2 \vee \bar{x}_3)$ are implicates of function $f = x_1 \vee x_2$, since whenever $f = 1$, both $(x_1 \vee x_2 \vee x_3)$ and $(x_1 \vee x_2 \vee \bar{x}_3)$ become 1. The implication relationship between an alterm P and a function f can sometimes be more conveniently found, however, by using the property "f implies P if $f = 0$ whenever $P = 0$," which is a restatement of the property "f implies P if $P = 1$ whenever $f = 1$" defined earlier (we can easily see this by writing a truth table for f and P). Thus, $(x_1 \vee x_3)$, $(x_1 \vee \bar{x}_2)$, and $(x_1 \vee x_2 \vee x_3)$ are implicates of $f = (x_1 \vee \bar{x}_2)(\bar{x}_2 \vee x_3)$, because when $(x_1 \vee x_3)$, $(x_1 \vee \bar{x}_2)$, or $(x_1 \vee x_2 \vee x_3)$ is 0, f is 0. [For example, when $(x_1 \vee x_3)$ is 0, $x_1 = x_3 = 0$ must hold. Thus, $f = (\bar{x}_2)(x_2) = 0$. Consequently, $(x_1 \vee x_3)$ is an alterm implied by f, although this is not obvious from the given expressions of f and

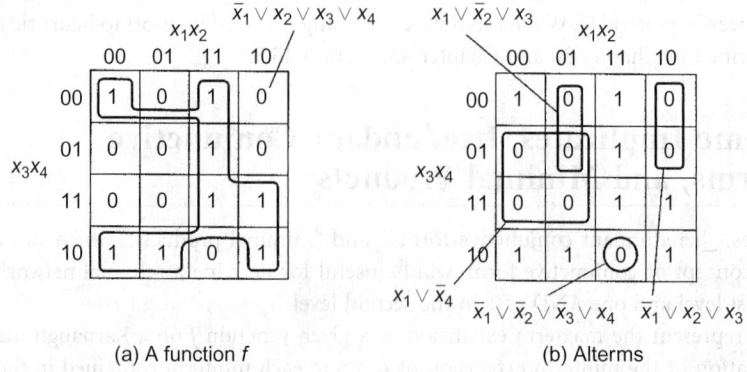

(a) A function f (b) Alterms

FIGURE 28.9 Representation of a function f on a map with alterms.

$(x_1 \vee x_3).$] Also, $(x_1 \vee \bar{x}_2 \vee x_4)$ and $(x_1 \vee x_2 \vee x_3 \vee x_4)$, which contain the literals of a dummy variable x_4 of this f, are implicates of f.

Definition 28.5: An alterm P is said to **subsume** another alterm Q if all the literals in Q are among the literals of P.

The alterm $(x_1 \vee \bar{x}_2 \vee x_3 \vee x_4)$ subsumes $(\bar{x}_2 \vee x_3)$. Summarizing the two definitions of "subsume" for "alterms" in Definition 28.5 and "terms" in Definition 27.6, we have that **"P subsumes Q" simply means "P contains all the literals in Q," regardless of whether P and Q are terms or alterms**. But the relationships between "subsume" and "imply" in the two cases are opposite. If an alterm P subsumes another alterm Q, then $Q \subseteq P$ holds; whereas if a term P subsumes another term Q, then $P \subseteq Q$ holds.

Definition 28.6: A **prime implicate** of a function f is defined as an implicate of f such that no other alterm subsumed by it is an implicate of f.

In other words, if deletion of any literal from an implicate of f makes the remainder not an implicate of f, the implicate is a prime implicate. For example, $(x_2 \vee x_3)$ and $(x_1 \vee x_3)$ are prime implicates of $f = (x_1 \vee \bar{x}_2)(x_2 \vee x_3)(\bar{x}_1 \vee x_2 \vee x_3)$, but $x_1 \vee x_3 \vee x_4$ is not a prime implicate of f, although it is still an implicate of this f. As seen from these examples, some implicates are not obvious from a given conjunctive form of f. Such an example is $x_1 \vee x_3$ in the above example. As a matter of fact, $x_1 \vee x_3$ can be obtained as the consensus by the following Definition 28.7 of two alterms, $(x_1 \vee \bar{x}_2)$ and $(x_2 \vee x_3)$, of f. Also notice that, unlike implicates, prime implicates cannot contain a literal of any dummy variable of f.

On a Karnaugh map, **a loop for an alterm P that subsumes another alterm Q is contained in the loop for Q**. For example, in Figure 28.10, the loop for alterm $(x_1 \vee \bar{x}_3 \vee x_4)$, which subsumes alterm $(x_1 \vee \bar{x}_3)$, is contained in the loop for $x_1 \vee \bar{x}_3$. Thus, **a rectangular loop that consists of 2^i 0-cells, with i as large as possible, represents a prime implicate of f**.

The consensus of two **alterms**, V and W, is defined in a manner similar to the consensus of two terms.

Definition 28.7: If there is exactly one variable, say x, appearing noncomplemented in one alterm and complemented in the other (i.e., if two alterms, V and W, can be written as $V = x \vee V'$ and $W = \bar{x} \vee W'$, where V' and W' are alterms free of literals of x), the disjunction of all literals except those of x (i.e., $V' \vee W'$ with duplicate literals deleted) is called the **consensus** of the two alterms V and W.

For example, when $V = x \vee y \vee \bar{z} \vee u$ and $W = \bar{x} \vee y \vee u \vee \bar{v}$ are given, their consensus is $y \vee \bar{z} \vee u \vee \bar{v}$.

On a Karnaugh map, a consensus is represented by the largest rectangular loop of 2^i 0-cells that intersects, and is contained within, two adjacent loops of 0-cells that represent two alterms. For example, in Figure 28.11 the dotted-line loop represents the consensus of two alterms, $(\bar{x}_2 \vee \bar{x}_4)$ and $(\bar{x}_1 \vee x_2 \vee \bar{x}_3)$, which are represented by the two adjacent loops.

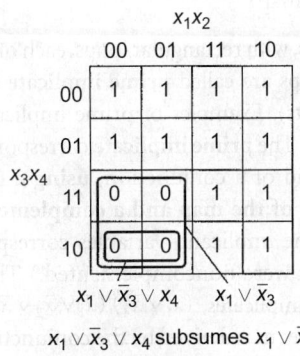

$x_1 \vee \bar{x}_3 \vee x_4$ subsumes $x_1 \vee \bar{x}_3$
$(x_1 \vee \bar{x}_3)$ implies $(x_1 \vee \bar{x}_3 \vee x_4)$

FIGURE 28.10 An alterm that subsumes another alterm.

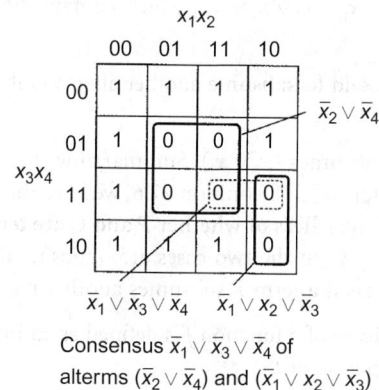

FIGURE 28.11   Consensus of two adjacent alterms.

All prime implicates of a function f can be algebraically obtained from a conjunctive form for f by modifying the Tison method discussed in Procedure 27.1; that is, by using dual operations in the method.

We can define the following concepts, which are dual to irredundant disjunctive forms, minimal sums, absolute minimal sums, essential prime implicants, complete sums, and others.

Definition 28.8: An **irredundant conjunctive form** for a function f is a conjunction of prime implicates such that removal of any of them makes the remainder not express f. The **minimal products** are irredundant conjunctive forms for f with a minimum number of prime implicates. The **absolute minimal products** are minimal products with a minimum number of literals. Prime implicates that appear in every irredundant conjunctive form for f are called **essential prime implicates** of f. **Conditionally eliminable prime implicates** are prime implicates that appear in some irredundant conjunctive forms for f, but not in others. **Absolutely eliminable prime implicates** are prime implicates that do not appear in any irredundant conjunctive form for f. The **complete product** for a function f is the product of all prime implicates of f.

28.5   Derivation of Minimal Products by Karnaugh Map

Minimal products can be derived by the following procedure, based on a Karnaugh map.

Procedure 28.4: Derivation of Minimal Products on a Karnaugh Map

Consider the case of a map for four or fewer variables (cases for five or more variables are similar, using more than one four-variable map).

1. Encircle **0-cells**, instead of 1-cells, with rectangular loops, each of which consists of 2^i 0-cells, choosing i as large as possible. These loops are called **prime implicate loops** because they represent prime implicates (not prime implicants). Examples of prime implicate loops are shown in Figure 28.12 (including the dotted-line loop). The **prime implicate** corresponding to a loop is formed by making a **disjunction** of literals, instead of a conjunction, **using a noncomplemented variable corresponding to 0 of a coordinate of the map and a complemented variable corresponding to 1.** (Recall that, in forming a prime implicant, variables corresponding to 0's were complemented and those corresponding to 1's were noncomplemented.) Thus, corresponding to the loops of Figure 28.12, we get the prime implicates, $(x_1 \vee \bar{x}_4), (\bar{x}_1 \vee \bar{x}_2 \vee \bar{x}_3)$ and $(\bar{x}_2 \vee \bar{x}_3 \vee \bar{x}_4)$.

2. Each **irredundant conjunctive form** is derived by the conjunction of prime implicates corresponding to a set of loops, so that removal of any loop leaves some 0-cells uncovered by loops. An example is the set of two solid-line loops in Figure 28.12, from which the irredundant conjunctive form $(x_1 \vee \bar{x}_4)(\bar{x}_1 \vee \bar{x}_2 \vee \bar{x}_3)$ is derived.

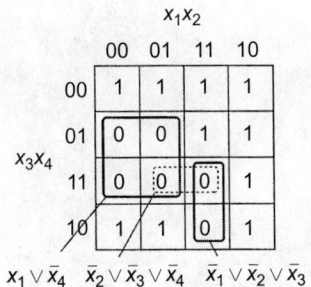

FIGURE 28.12 Prime-implicate loops and the corresponding prime implicates.

3. Among sets of a minimum number of prime implicate loops, the sets that contain as many of the largest loops as possible yield **minimal products**.

The don't-care conditions are dealt with in the same manner as in case of minimal sums. In other words, whenever possible, we can form a larger prime implicate loop by interpreting some d's as 0-cells. Any prime-implicate loops consisting of only d-cells need not be formed.   ☐

Procedure 28.4 can be extended to five or more variables in the same manner as Procedures 28.2 and 28.3, although the procedure will be increasingly complex.

References

1. Coudert, O., "Two-Level Logic Minimization: An Overview," *Integration*, vol. 17, pp. 97–140, Oct. 1994.
2. McCluskey, E.J., "Minimization of Boolean functions," *Bell System Tech. J.*, vol. 35, no. 5, pp. 1417–1444, Nov. 1956.
3. Muroga, S., *Logic Design and Switching Theory*, John Wiley & Sons, 1979 (now available from Krieger Publishing Co.).
4. Tison, P., Ph.D. dissertation, Faculty of Science, University of Grenoble, France, 1965.
5. Tison, P., "Generalization of consensus theory and application to the minimization of Boolean functions," *IEEE Tr. Electron. Comput.*, pp. 446–456, Aug. 1967.

FIGURE 28.17 Prime implicate map and the corresponding prime implicates.

References

1. Coudert, O., "Two-Level Logic Minimization: An Overview," *Integration*, Vol. 17, pp. 97-140, October 1994.

2. McCluskey, E. J., "Minimization of Boolean Functions," *Bell System Tech. J.*, Vol. 35, pp. 1417-1444, November 1956.

3. Mano, M., *Computer Design and Architecture*, 2nd Edition, John Wiley & Sons, 1978.

4. Brand, D., Ph.D. dissertation, Faculty of Science, University of Grenoble, France, 1985.

5. Tison, P., "Generalization of consensus theory and application to the minimization of Boolean functions," *IEEE Trans. on Computers*, pp. 446-456, Aug. 1967.

29

Binary Decision Diagrams

Shin-ichi Minato
NTT Network Innovation
Laboratories

Saburo Muroga
University of Illinois
at Urbana-Champaign

CONTENTS

29.1 Basic Concepts

Binary decision diagrams (BDDs), which were introduced in Chapter 26, Section 26.1, are a powerful means for computer processing of logic functions because in many cases, with BDDs, smaller memory space is required for storing logic functions and values of functions can be calculated faster than with truth tables or logic expressions. As logic design has been done in recent years with computers, BDDs are extensively used because of these features. BDDs are used in computer programs for automation of logic design, verification (i.e., identifying whether two logic networks represent the identical logic functions), diagnosis of logic networks, simplification of transistor circuits (such as ECL and MOSFET circuits, as explained in Chapters 38, 39, and 40), and other areas, including those not related to logic design.

In this chapter, we discuss the basic data structures and algorithms for manipulating BDDs. Then we describe the variable ordering problem, which is important for the effective use of BDDs. The concept of BDD was devised by Lee in 1959 [15]. Binary decision programs that Lee discussed are essentially binary decision diagrams. Then, in 1978, its usefulness for expressing logic functions was shown by Akers [3,4]. But since Bryant [6] developed a reduction method in 1986, it has been extensively used for design automation for logic design and related areas.

From a truth table, we can easily derive the corresponding binary decision diagram. For example, the truth table shown in Figure 29.1(a) can be converted into the BDD in Figure 29.1(b). But there are generally many BDDs for a given truth table, that is, the logic function expressed by this truth table. For example, all BDDs shown in Figure 29.1(c) through (e) represent the logic function that the truth table in Figure 29.1(a) expresses. Here, note that in each of Figures 29.1 (b), (c), and (d), the variables appear in the same order and none of them appears more than once in every path from the top node. But in (e), they appear in different orders in different paths. BDDs in Figures 29.1(b), (c), and (d) are called **ordered BDDs** (or **OBDDs**). But the BDD in Figure 29.1(e) is called an **unordered BDD**. These BDDs can be reduced into a simple BDD by the following procedure.

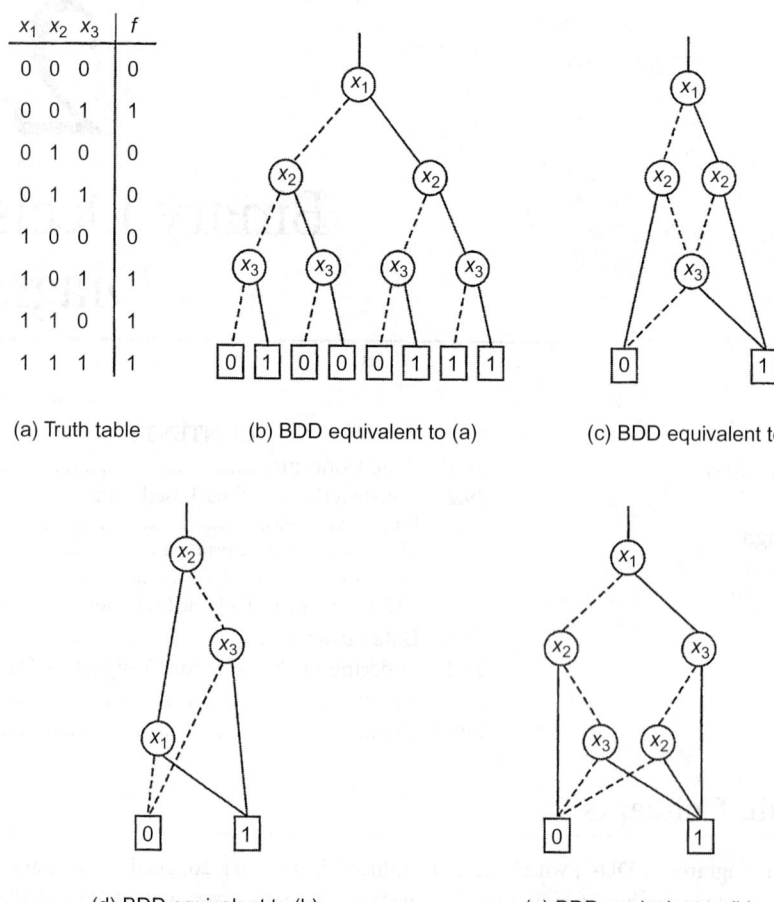

x_1 x_2 x_3	f
0 0 0	0
0 0 1	1
0 1 0	0
0 1 1	0
1 0 0	0
1 0 1	1
1 1 0	1
1 1 1	1

(a) Truth table

(b) BDD equivalent to (a)

(c) BDD equivalent to (b)

(d) BDD equivalent to (b)

(e) BDD equivalent to (b)

FIGURE 29.1 Truth table and BDDs for $x_1 x_2 \vee \bar{x}_2 x_3$.

In a BDD, the top node is called the **root** that represents the given function $f(x_1, x_2, \ldots, x_n)$. Rectangles that have 1 or 0 inside are called **terminal nodes**. They are also called **1-terminal** and **0-terminal**. Other nodes are called **non-terminal nodes** denoted by circles with variables inside. They are also called simply **nodes**, differentiating themselves from the 0- and 1-terminals. From a node with x_i inside, two lines go down. The solid line is called **1-edge**, representing $x_i = 1$; and the dotted line is called **0-edge**, representing $x_i = 0$.

In an OBDD, the value of the function f can be evaluated by following a path of edges from the root node to one of the terminal nodes. If the nodes in every path from the root node to a terminal node are assigned with variables, $x_1, x_2, x_3, \ldots$, and x_n, in this order, then f can be expressed as follows, according to the Shannon expansion (described in Chapter 27). By branching with respect to x_1 from the root node, $f(x_1, x_2, \ldots, x_n)$ can be expanded as follows, where $f(0, x_2, \ldots, x_n)$ and $f(1, x_2, \ldots, x_n)$ are functions that the nodes at the low ends of 0-edge and 1-edge from the root represent, respectively:

$$f(x_1, x_2, \ldots, x_n) = \bar{x}_1 f(0, x_2, \ldots, x_n) \vee x_1 f(1, x_2, \ldots, x_n)$$

Then, by branching with respect to x_2 from each of these two nodes that $f(0, x_2, \ldots, x_n)$ and $f(1, x_2, \ldots, x_n)$ represent, each $f(0, x_2, \ldots, x_n)$ and $f(1, x_2, \ldots, x_n)$ can be expanded as follows:

$$f(0, x_2, \ldots, x_n) = \bar{x}_2 f(0, 0, x_3, \ldots, x_n) \vee x_2 f(0, 1, x_3, \ldots, x_n)$$

and

$$f(1, x_2, \ldots, x_n) = \bar{x}_2 f(1, 0, x_3, \ldots, x_n) \lor x_2 f(1, 1, x_3, \ldots, x_n)$$

And so on. As we go down from the root node toward the 0- or 1-terminal, more variables of f are set to 0 or 1. Each term excluding x_i or $\bar{x}_i$ in each of these expansions [i.e., $f(0, x_2, \ldots, x_n)$ and $f(1, x_2, \ldots, x_n)$ in the first expansion, $f(0, 0, \ldots, x_n)$ and $f(0, 1, \ldots, x_n)$ in the second expansion, etc.], are called **cofactors**. Each node at the low ends of 0-edge and 1-edge from a node in an OBDD represents cofactors of the **Shannon expansion** of the logic function at the node, from which these 0-edge and 1-edge come down.

Procedure 29.1: Reduction of a BDD

1. For the given BDD, apply the following steps in any order.
   a. If two nodes, v_a and v_b, that represent the same variable x_i, branch to the same nodes in a lower level for each of $x_i = 0$ and $x_i = 1$, then combine them into one node that still represents variable x_i.

In Figure 29.2(a), two nodes, v_a and v_b, that represent variable x_i, go to the same node v_u for $x_i = 0$ and the same node v_v for $x_i = 1$. Then, according to Step a, two nodes, v_a and v_b, are merged into one node v_{ab}, as shown in Figure 29.2(b). The node v_{ab} represents variable x_i.

   b. If a node that represents a variable x_i branches to the same node in a lower level for both $x_i = 0$ and $x_i = 1$, then that node is deleted, and the 0- and 1-edges that come down to the former are extended to the latter.

In Figure 29.3(a), node v_a that represents variable x_i branches to the same node v_b for both $x_i = 0$ and $x_i = 1$. Then, according to Step b, node v_a is deleted, as shown in Figure 29.3(b), where the node v_a is a don't-care for x_i, and the edges that come down to v_a are extended to v_u.

   c. Terminals nodes with the same value, 0 or 1, are merged into the terminal node with the same original value.

All the 0-terminals in Figure 29.1(b) are combined into one 0-terminal in each of Figures 29.1(c), (d), and (e). All the 1-terminals are similarly combined.

2. When we cannot apply any step after repeatedly applying these steps (a), (b), and (c), the **reduced ordered BDD** (i.e., **ROBDD**) or simply called the **reduced BDD** (i.e., **RBDD**), is obtained for the given function.

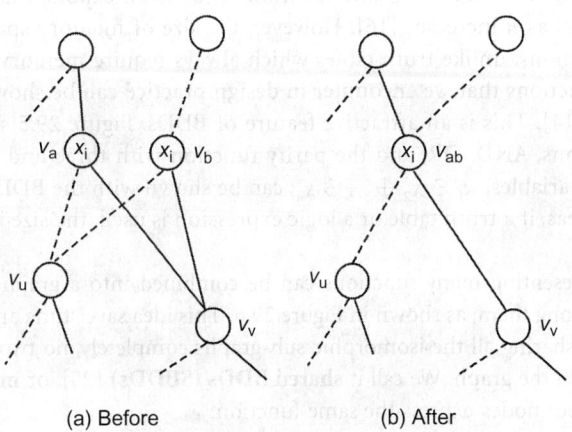

(a) Before   (b) After

FIGURE 29.2 Step 1(a) of Procedure 29.1.

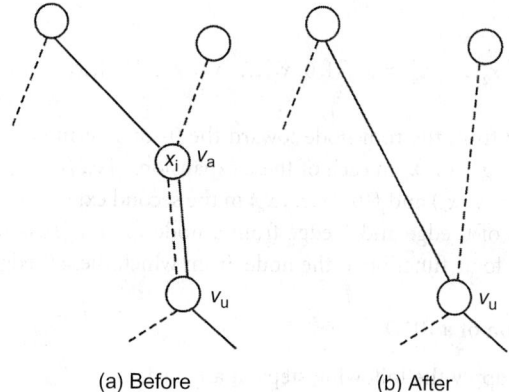

(a) Before   (b) After

FIGURE 29.3   Step 1(b) of Procedure 29.1.

The following process is faster than the repeated application of Step 1(a) of Procedure 29.1. Suppose a BDD contains two nodes, v_a and v_b, both of which represent x_i, such that a sub-BDD, B_1, that stretches downward from v_a is completely identical to another sub-BDD, B_2, that stretches downward from v_b. In this case, each sub-BDD is said to be **isomorphic** to the other. Then the two sub-BDDs can be merged. For example, the two sub-BDDs, B_1 and B_2, shown in the two dotted rectangles in Figure 29.4(a), both representing x_4, can be merged into one, B_3, as shown in Figure 29.4(b).

> **Theorem 29.1:** Any completely or incompletely specified function has a unique reduced ordered BDD and any other ordered BDD for the function in the same order of variables (i.e., not reduced) has more nodes.

According to this theorem, the ROBDD is unique for a given logic function when the order of the variables is fixed. (A BDD that has don't-cares can be expressed with addition of the d-terminal or two BDDs for the ON-set and OFF-set completely specified functions, as explained with Figure 23.5. For details, see Ref. 24.) Thus, ROBDDs give canonical forms for logic functions. This property is very important to practical applications, as we can easily check the equivalence of two logic functions by only checking the isomorphism of their ROBDDs. **Henceforth in this chapter, ROBDDs will be referred to as BDDs for the sake of simplicity.**

It is known that a BDD for an n-variable function requires an exponentially increasing memory space in the worst case, as n increases [16]. However, the size of memory space for the BDD varies with the types of functions, unlike truth tables which always require memory space proportional to 2^n. But many logic functions that we encounter in design practice can be shown with BDDs without large memory space [14]. This is an attractive feature of BDDs. Figure 29.5 shows the BDDs representing typical functions, AND, OR, and the parity function with three and n input variables. The parity function for n variables, $x_1 \oplus x_2 \oplus \ldots \oplus x_n$, can be shown with the BDD of $2(n-1) + 1$ nodes and 2 terminals; whereas, if a truth table or a logic expression is used, the size increases exponentially as n increases.

A set of BDDs representing many functions can be combined into a graph that consists of BDDs sharing sub-graphs among them, as shown in Figure 29.6. This idea saves time and space for duplicating isomorphic BDDs. By sharing all the isomorphic sub-graphs completely, no two nodes that express the same function coexist in the graph. We call it **shared BDDs (SBDDs)** [25], or **multi-rooted BDDs**. In a shared BDD, no two root nodes express the same function.

Shared BDDs are now widely used and those algorithms are more concise than ordinary BDDs. In the remainder of this chapter, we deal with shared BDD only.

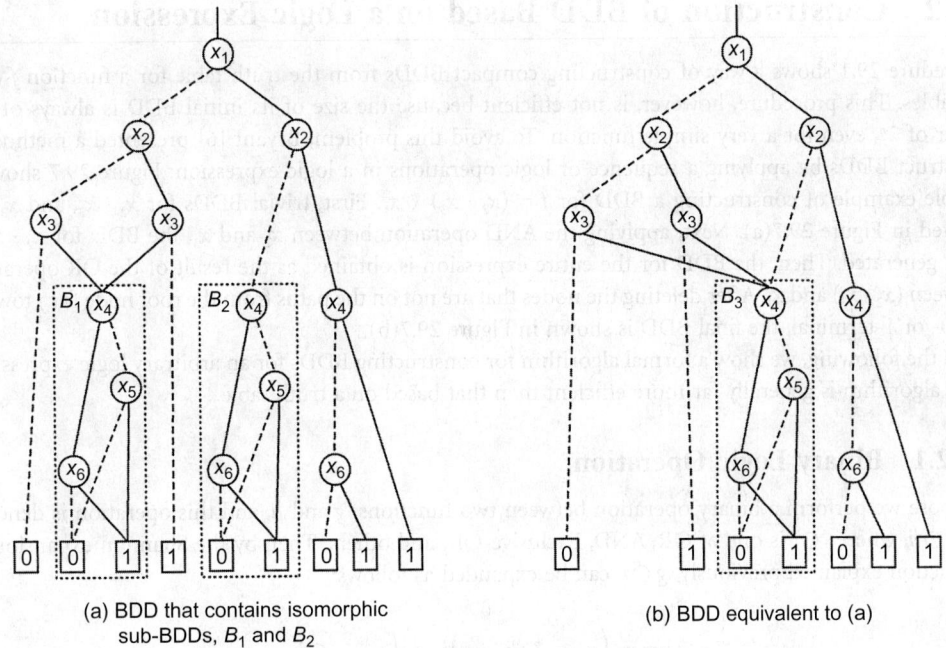

(a) BDD that contains isomorphic   (b) BDD equivalent to (a)
sub-BDDs, B_1 and B_2

FIGURE 29.4 Merger of two isomorphic sub-BDDs.

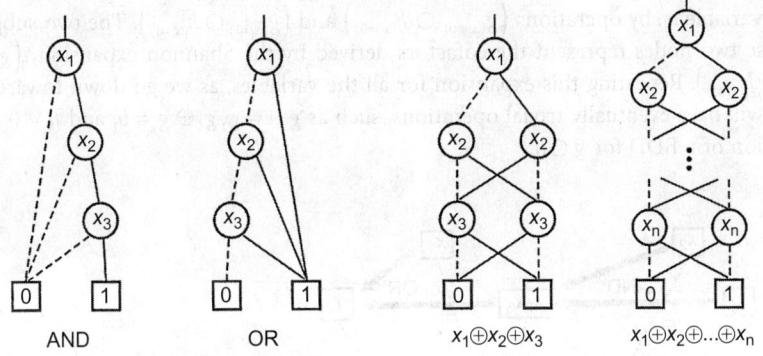

AND   OR $x_1 \oplus x_2 \oplus x_3$ $x_1 \oplus x_2 \oplus \ldots \oplus x_n$

FIGURE 29.5 BDDs for typical logic functions.

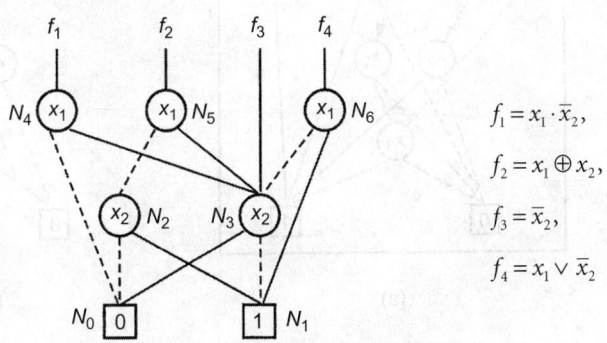

$$f_1 = x_1 \cdot \bar{x}_2,$$
$$f_2 = x_1 \oplus x_2,$$
$$f_3 = \bar{x}_2,$$
$$f_4 = x_1 \vee \bar{x}_2$$

FIGURE 29.6 A shared BDD.

29.2 Construction of BDD Based on a Logic Expression

Procedure 29.1 shows a way of constructing compact BDDs from the truth table for a function f of n variables. This procedure, however, is not efficient because the size of its initial BDD is always of the order of 2^n, even for a very simple function. To avoid this problem, Bryant [6] presented a method to construct BDDs by applying a sequence of logic operations in a logic expression. Figure 29.7 shows a simple example of constructing a BDD for $f = (x_2 \cdot x_3) \vee x_1$. First, trivial BDDs for $x_2 \cdot x_3$, and x_3 are created in Figure 29.7(a). Next, applying the AND operation between x_2 and x_3, the BDD for $x_2 \cdot x_3$ is then generated. Then, the BDD for the entire expression is obtained as the result of the OR operation between $(x_2 \cdot x_3)$ and x_1. After deleting the nodes that are not on the paths from the root node for f toward the 0- or 1-terminal, the final BDD is shown in Figure 29.7(b).

In the following, we show a formal algorithm for constructing BDDs for an arbitrary logic expression. This algorithm is generally far more efficient than that based on a truth table.

29.2.1 Binary Logic Operation

Suppose we perform a binary operation between two functions, g and h, and this operation is denoted by $g \bigcirc h$, where "$\bigcirc$" is one of OR, AND, Exclusive-OR, and others. Then by the Shannon expansion of a function explained previously, $g \bigcirc h$ can be expanded as follows:

$$g \bigcirc h = \bar{x}_i \cdot \left(g_{(x_i=0)} \bigcirc h_{(x_i=0)} \right) \vee x_i \cdot \left(g_{(x_i=1)} \bigcirc h_{(x_i=1)} \right)$$

with respect to variable x_i, which is the variable of the node that is in the highest level among all the nodes in g and h. This expansion creates the 0- and 1-edges which go to the next two nodes from the node for the variable x_i, by operations $\left(g_{(x_i=0)} \bigcirc h_{(x_i=0)} \right)$ and $\left(g_{(x_i=1)} \bigcirc h_{(x_i=1)} \right)$. The two subgraphs whose roots are these two nodes represent the cofactors derived by the Shannon expansion, $\left(g_{(x_i=0)} \bigcirc h_{(x_i=0)} \right)$ and $\left(g_{(x_i=1)} \bigcirc h_{(x_i=1)} \right)$. Repeating this expansion for all the variables, as we go down toward the 0- or 1-terminal, we will have eventually trivial operations, such as $g \cdot 1 = g, g \oplus g = 0$, and $h \vee 0 = h$, finishing the construction of a BDD for $g \bigcirc h$.

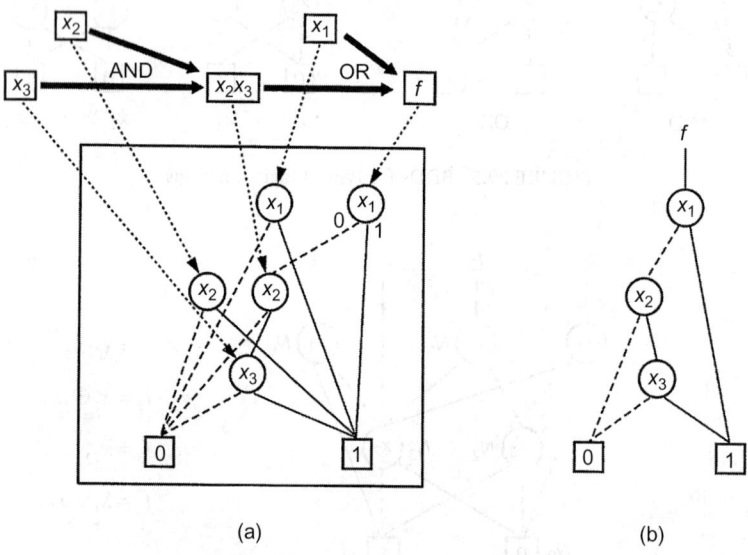

(a) (b)

FIGURE 29.7 Generation of BDDs for $f = (x_2 \cdot x_3) \vee x_1$.

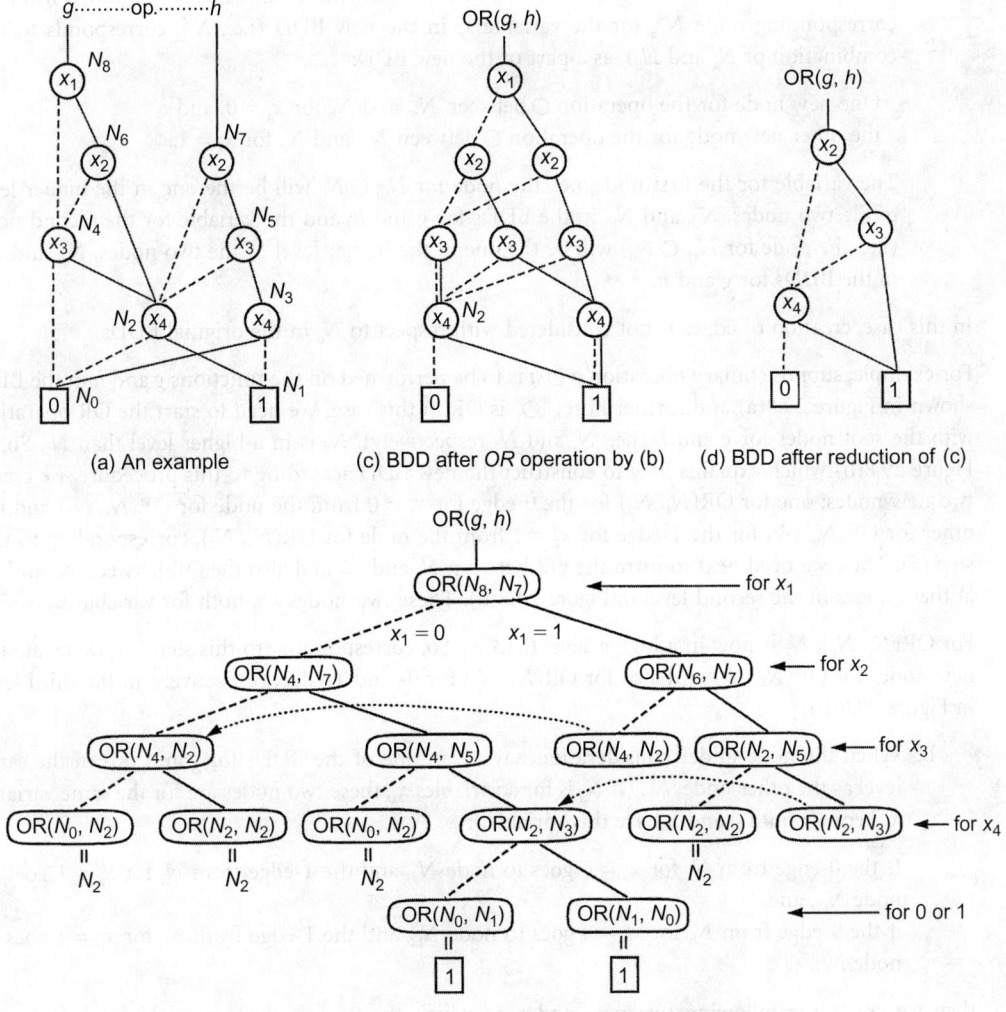

(a) An example   (c) BDD after *OR* operation by (b)   (d) BDD after reduction of (c)

(b) Procedural structure for forming the new BDD

FIGURE 29.8 Procedure of binary operation.

When BDDs for functions g and h are given, we can derive a new BDD for function $f = g \bigcirc h$ by the following procedure.

Procedure 29.2: Construction of a BDD for Function $f \bigcirc h$, Given BDDs for g and h.

Given BDDs for functions g and h (e.g., Figure 29.8(a)), let us construct a new BDD (e.g., Figure 29.8(b)) with respect to the specified logic operation $g \bigcirc h$ and then reduce it by the previous Procedure 29.1 (e.g., Figure 29.8(d)).

1. Starting with the root nodes for g and h and going down toward to the 0- or 1-terminal, apply repeatedly steps a and b in the following, considering steps c and d.

   a. When the node under consideration, say N_a, in one of the two BDDs for g and h is in a higher level than the other node, N_b:

   If N_a and N_b are for variables x_a and x_b, respectively, and if the 0-edge from N_a for $x_a = 0$ goes to node N_{a0} and the 1-edge from N_a for $x_a = 1$ goes to node N_{a1}, then we create the following two new nodes, to which the 0- and 1-edges go down from the corresponding node N'_a for the variable x_a in the new BDD (i.e., N'_a corresponds to this combination of N_a and N_b), as a part of the new BDD:

> One new node for the operation $\bigcirc$ between N_{a0} and N_b for $x_a = 0$, and
> the other new node for the operation $\bigcirc$ between N_{a1} and N_b for $x_a = 1$.

The variable for the first node, i.e., the node for $N_{a0} \bigcirc N_b$ will be the one in the higher level of the two nodes, N_{a0} and N_b, in the BDDs for g and h; and the variable for the second node (i.e., the node for $N_{a1} \bigcirc N_b$) will be the one in the higher level of the two nodes, N_{a1} and N_b, in the BDDs for g and h.

In this case, creation of edges is not considered with respect to N_b in the original BDDs.

For example, suppose binary operation $g \bigcirc h$ is to be performed on the functions g and h in the BDD shown in Figure 29.8(a) and furthermore, "$\bigcirc$" is OR in this case. We need to start the OR operation with the root nodes for g and h (i.e., N_8 and N_7 respectively). N_8 is in a higher level than N_7. So, in Figure 29.8(b) which explains how to construct the new BDD according to this procedure, we create two new nodes; one for $\mathrm{OR}(N_4, N_7)$ for the 0-edge for $x_1 = 0$ from the node for $\mathrm{OR}(N_8, N_7)$ and the other for $\mathrm{OR}(N_6, N_7)$ for the 1-edge for $x_1 = 1$ from the node for $\mathrm{OR}(N_8, N_7)$, corresponding to this step (a). Thus, we need next to form the OR between N_4 and N_7 and also the OR between N_6 and N_7 at these nodes in the second level in Figure 29.8(b). These two nodes are both for variable x_2.

For $\mathrm{OR}(N_4, N_7)$, N_7 is now in a higher level than N_4. So, corresponding to this step (a), we create the new nodes for $\mathrm{OR}(N_4, N_2)$ and also for $\mathrm{OR}(N_4, N_5)$ for 0- and 1-edge, respectively, in the third level in Figure 29.8(b).

b. When the node under consideration, say N_a, in one of the BDDs for g and h is in the same level as the other node, N_b. (If N_a is for a variables x_a, these two nodes are for the same variable x_a because both g and h have this variable.):

> If the 0-edge from N_a for $x_a = 0$ goes to node N_{a0} and the 1-edge from N_a for $x_a = 1$ goes to node N_{a1}, and
> if the 0-edge from N_b for $x_b = 0$ goes to node N_{b0} and the 1-edge from N_b for $x_b = 1$ goes to node N_{b1},

then we create the following two new nodes, to which the 0- and 1-edges come down from the corresponding node N'_a for the variable x_a in the new BDD (i.e., N'_a corresponds to this combination of N_a and N_b), as a part of the new BDD:

> one new node for the operation $\bigcirc$ between N_{a0} and N_{b0} for $x_a = 0$, and
> the other new node for the operation $\bigcirc$ between N_{a1} and N_{b1} for $x_a = 1$.

The variable for the first node, i.e., the node for $N_{a0} \bigcirc N_{b0}$ will be the one in the higher level of the two nodes, N_{a0} and N_{b0}, in the BDDs for g and h; and the variable for the second node, i.e., the node for $N_{a1} \bigcirc N_{b1}$ will be the one in the higher level of the two nodes, N_{a1} and N_{b1}, in the BDDs for g and h.

For the example in Figure 29.8(b), we need to create two new nodes for the operations, $\mathrm{OR}(N_4, N_2)$ and $\mathrm{OR}(N_2, N_5)$ to which the 0- and 1-edges go for $x_2 = 0$ and $x_2 = 1$, respectively, from the node $\mathrm{OR}(N_6, N_7)$ because N_6 and N_7 are in the same level for variable x_2 in Figure 29.8(a). These two nodes are both for variable x_3.

c. In the new BDD for the operation $g \bigcirc h$, we need not have more than one subgraph that represent the same logic function. So, during the construction of the new BDD, whenever a new node (i.e., a root node of a subgraph) is for the same operation as an existing node, we need not continue creation of succeeding nodes from that new node.

For the example, in Figure 29.8(b), the node for operation OR(N_4, N_2) appears twice, as shown with a dotted arc, and we do not need to continue the creation of new nodes from the second and succeeding nodes for OR(N_4, N_2). OR(N_2, N_3) also appears twice, as shown with a dotted arc.

    d. In the new BDD for the operation $g \bigcirc h$, if one of N_a and N_b is the 0- or 1-terminal, or N_a and N_b are the same, i.e., $N_a = N_b$, then we can apply the logic operation that $g \bigcirc h$ defines. If $g \bigcirc h$ is for AND operation, for example, the node for AND(N_1, N_a) represents N_a because N_1 is the 1-terminal in Figure 29.8(a), and furthermore, if N_a represents function g, this node represents g.

Also, it is important to notice that only this step (d) is relevant to the logic operation defined by $g \bigcirc h$, whereas all other steps from (a) through (c) are irrelevant of the logic operation $g \bigcirc h$.

For the example in Figure 29.8(b), the node for operation OR(N_0, N_2) represents N_2, which is for variable x_4. Also, the node for operation OR(N_0, N_1) represents constant value 1 because N_0 and N_1 are the 0- and 1-terminals, respectively, and consequently, $0 \vee 1$.

By using binary operation $f \oplus 1$, $\overline{f}$ can be performed and its processing time is linearly proportional to the size of a BDD. However, it is improved to a constant time by using the **complement edge**, which is discussed in the following. This technique is now commonly used.

  2. Convert each node in the new BDD that is constructed in Step 1 to a node with the corresponding variable inside. Then derive the reduced BDD by Procedure 29.1.

For the example, each node for operation OR(N_a, N_b) in Figure 29.8(b) is converted to a node with the corresponding variable inside in Figure 29.8(c), which is reduced to the BDD in Figure 29.8(d).

29.2.2 Complement Edges

Complement edge is a technique to reduce computation time and memory requirement of BDDs by using edges that indicates to complement the function of the subgraph pointed to by the edge, as shown in Figure 29.9(a). This idea was first shown by Akers [3] and later discussed by Madre and Billon [18]. The use of complement edges brings the following outstanding merits.

- The BDD size is reduced by up to a half
- Negation can be performed in constant time
- Logic operations are sped by applying the rules, such as $f \cdot \overline{f} = 0$, $f \vee \overline{f} = 1$, and $f \oplus \overline{f} = 1$

Use of complement edges may break the uniqueness of BDDs. Therefore, we have to adopt the two rules, as illustrated in Figure 29.9(b):

  1. Using the 0-terminal only.
  2. Not using a complement edge as the 0-edge of any node (i.e., use it on 1-edge only). If necessary, the complement edges can be carried over to higher nodes.

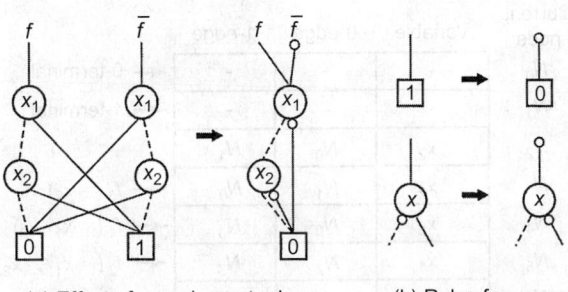

(a) Effect of complement edges   (b) Rules for usage

FIGURE 29.9 Complement edges.

29.2.3 Derivation of a Logic Expression from a BDD

A logic expression for f can be easily derived from a BDD for f. A path from the root node for f to the 1-terminal in a BDD is called **1-path**, where the values of the variables on the edges on this path make f equal 1. For each 1-path, a product of literals is formed by choosing x_i for $x_i = 1$ or its complement, $\bar{x}_i$ for $x_i = 0$. The disjunction of such products for all 1-paths yields a logic expression for f. For example, in Figure 29.8(a), the sequence of nodes, N_8-N_6-N_4-N_2-N_1, is a 1-path and the values of variables on this path, $x_1 = 1$, $x_2 = 0$, $x_3 = 1$, and $x_4 = 1$, make the value of f (g for this example) equal 1. Finding all 1-paths, a logic expression for f is derived as $f = x_1\bar{x}_2x_3x_4 \vee x_1x_2x_4 \vee \bar{x}_1x_3x_4$. It is important to notice that logic expressions that are derived from all 1-paths are usually not minimum sums for the given functions.

29.3 Data Structure

In a typical realization of a BDD manipulator, all the nodes are stored in a single table in the main memory of the computer. Figure 29.10 is a simple example of realization for the BDD shown in Figure 29.6. Each node has three basic attributes: input variable and the next nodes accessed by the 0- and 1-edges. Also, 0- and 1-terminals are first allocated in the table as special nodes. The other non-terminal nodes are created one by one during the execution of logic operations.

Before creating a new node, we check the reduction rules of Procedure 29.1. If the 0- and 1-edges go to the same next node (Step 1(b) of Procedure 29.1) or if an equivalent node already exists (Step 1(a)), then we do not create a new node but simply copy that node as the next node. To find an equivalent node, we check a table which displays all the existing nodes. The hash table technique is very effective to accelerate this checking. (It can be done in a constant time for any large-scale BDDs, unless the table overflows in main memories.)

When generating BDDs for logic expressions, such as Procedure 29.2, many intermediate BDDs are temporarily generated. It is important for memory efficiency to delete such unnecessary BDDs. In order to determine the necessity of the nodes, a **reference counter** is attached to each node, which shows the number of incoming edges to the node.

In a typical implementation, the BDD manipulator consumes 20 to 30 bytes of memory for each node. Today, there are personal computers and workstations with more than 100 Mbytes of memory, and those facilitate us to generate BDDs containing as many as millions of nodes. However, the BDDs still grow beyond the memory capacity in some practical applications.

29.4 Ordering of Variables for Compact BDDs

BDDs are a canonical representation of logic functions under a fixed order of input variables. A change of the order of variable, however, may yield different BDDs of significantly different sizes for the same function. The effect of variable ordering depends on logic functions, changing sometimes dramatically the size of BDDs. Variable ordering is an important problem in using BDDs.

Current node	Variable	0-edge	1-edge	
N_0	-	-	-	← 0-terminal
N_1	-	-	-	← 1-terminal
N_2	x_2	N_0	N_1	
N_3	x_2	N_1	N_0	← $f_3 (= \bar{x}_2)$
N_4	x_1	N_0	N_3	← $f_1 (= x_1 \cdot \bar{x}_2)$
N_5	x_1	N_2	N_3	← $f_2 (= x_1 \oplus x_2)$
N_6	x_1	N_3	N_1	← $f_4 (= x_1 \vee \bar{x}_2)$

FIGURE 29.10 Table-based realization of a shared BDD.

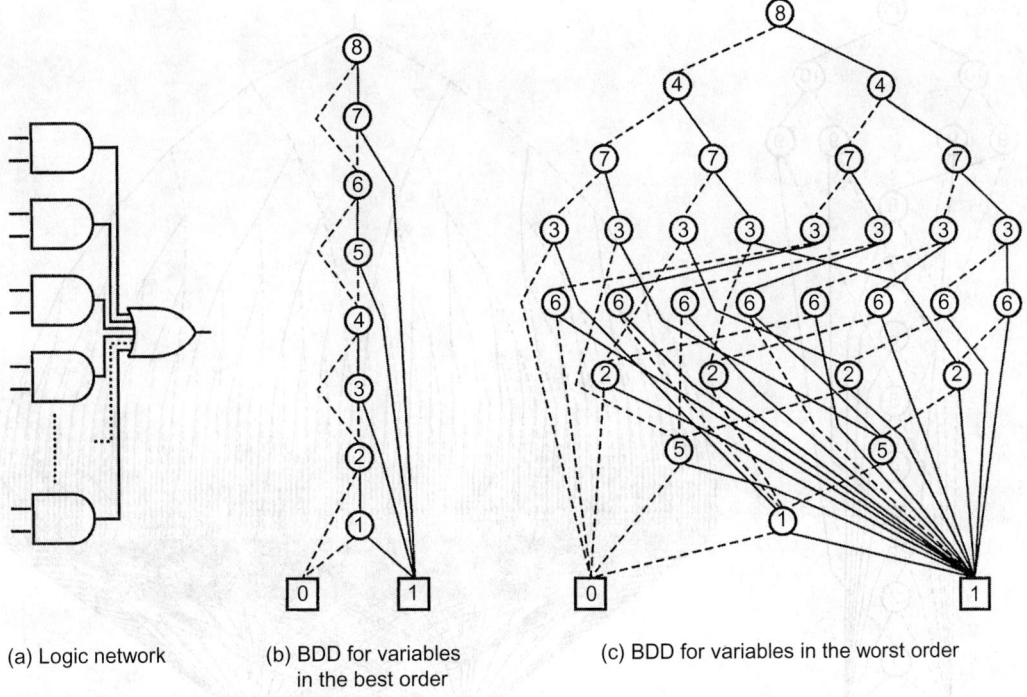

(a) Logic network

(b) BDD for variables in the best order

(c) BDD for variables in the worst order

FIGURE 29.11 BDDs for 2-level logic network with AND/OR gates.

It is generally very time consuming to find the best order [30]. Currently known algorithms are limited to run on the small size of BDDs with up to about 17 variables [13]. It is difficult to find the best order for larger problems in reasonably short processing time. However, if we can find a fairly good order, it is useful for practical applications. There are many research works on heuristic methods of variable ordering.

Empirically, the following properties are known on the variable ordering.

1. Closely-related variables:

   Variables that are in close relationship in a logic expression should be close in variable order (e.g., x_1 in $x_1 \cdot x_2 \vee x_3 \cdot x_4$ is in closer relationship with x_2 than x_3). The logic network of AND-OR gates with $2n$ inputs in 2-level shown in Figure 29.11(a) has $2n$ nodes in the best order with the expression $(x_1 \cdot x_2) \vee (x_3 \cdot x_4) \vee \dots \vee (x_{2n-1} \cdot x_{2n})$ as shown for $n = 2$ in Figure 29.11(b), while it needs $(2^{n+1} - 2)$ nodes in the worst order, as shown in Figure 29.11(c). If the same order of variables as the one in Figure 29.11(b) is kept on the BDD, Figure 29.11(c) represents the function $(x_1 \cdot x_{n+1}) \vee (x_2 \cdot x_{n+2}) \vee \dots \vee (x_n \cdot x_{2n})$.

2. Influential variables:

   The variables that greatly influence the nature of a function should be at higher position. For example, the **8-to-1 selector** shown in Figure 29.12(a) can be represented by a linear size of BDD when the three control inputs are ordered high, but when the order is reversed, it becomes of exponentially increasing size as the number of variables (i.e., the total number of data inputs and control inputs) increases, as shown in Figure 29.12(b).

Based on empirical rules like this, Fujita et al [9]. and Malik et al [20]. presented methods; in these methods, an output of the given logic networks is reached, traversing in a depth-first manner, then an input variable that can be reached by going back toward the inputs of the network is placed at highest position in variable ordering. Minato[25]devised another heuristic method in which each output function of the given logic networks is weighted and then input variables are ordered by the weight of each input

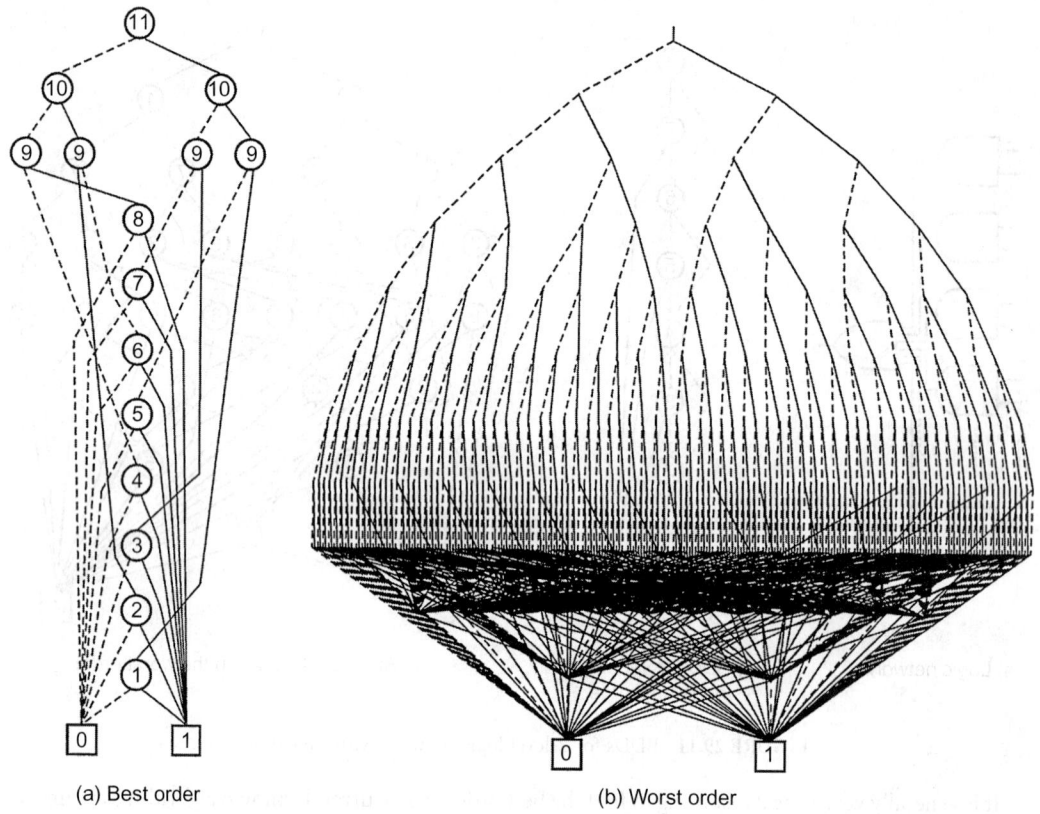

(a) Best order   (b) Worst order

FIGURE 29.12   BDDs for 8-to-1 selector.

variable determined by how each input can be reached from an output of the network. Butler et al [8]. proposed another heuristic based on a measure which uses not only the connection configuration of the network, but also the output functions of the network. These methods probably find a good order before generating BDDs. They find good orders in many cases, but there is no method that is always effective to a given network.

Another approach reduces the size of BDDs by reordering input variables. A greedy local exchange (swapping adjacent variables) method was developed by Fujita et al [10]. Minato [22] presented another reordering method which measures the width of BDDs as a cost function. In many cases, these methods find a fairly good order using no additional information. A drawback of the approach of these methods is that they cannot start if an initial BDD is too large.

One remarkable work is **dynamic variable ordering,** presented by Rudell [27]. In this technique, the BDD package itself determines and maintains the variable order. Every time the BDDs grow to a certain size, the reordering process is invoked automatically. This method is very effective in terms of the reduction of BDD size, although it sometimes takes a long computation time.

Table 29.1 shows experimental results on the effect of variable ordering. The logic network "sel8" is an 8-bit data selector, and "enc8" is an 8-bit encoder. "add8" and "mult6" are an 8-bit adder and a 6-bit multiplier. The rest is chosen from the benchmark networks in MCNC' 90 [1]. Table 29.1 compares four different orders: the original order, a random order, and two heuristic orders. The results show that the heuristic ordering methods are very effective except for a few cases which are insensitive to the order.

Unfortunately, there are some hard examples where variable ordering is powerless. For example, Bryant [6] proved that an n-bit multiplier function requires an exponentially increasing number of BDD nodes for any variable order, as n increases. However, for many other practical functions, the variable ordering methods are useful for generating compact BDDs in a reasonably short time.

TABLE 29.1 Effect of Variable Ordering

	Network Feature			BDD Size (with complement edges)			
Name	No. of Inputs	No. of Outputs	No. of Inputs to All Gates	Original	Random	Heur-1	Heur-2
sel8	12	2	29	16	88	23	19
enc8	9	4	31	28	29	28	27
add8	17	9	65	83	885	41	41
mult6	12	12	187	2183	2927	2471	2281
vg2	25	8	97	117	842	97	86
c432	36	7	203	3986	(>500k)	27302	1361
c499	41	32	275	115654	(>500k)	52369	40288
c880	60	26	464	(>500k)	(>500k)	23364	9114

Note: Heuristic-1: Heuristic order based on connection configuration [5].
Heuristic-2: BDD reduction by exchanging variables [14].

29.5 Remarks

Several groups developed BDD packages, and some of them are open to the public. For example, the CUDD package [12] is well-known to BDD researchers in the U.S. Many other BDD packages may be found on the Internet. BDD packages, in general, are based on the quick search of hash tables and linked-list data structures. They greatly benefit from the property of the **random access machine model** [2], where any data in main memory can be accessed in constant time.

Presently, considerable research is in progress. Detection of total or partial symmetry of a logic function with respect to variables has been a very time-consuming problem, but now it can be done in a short time by BDDs [26,29]. Also, decomposition of a logic function, which used to be very difficult, can be quickly solved with BDD [5,21]. A number of new types of BDDs have been proposed in recent years. For example, the Zero-Suppressed BDD (ZBDD) [23] is useful for solving covering problems, which are used in deriving a minimal sum and other combinatorial problems. The Binary Moment Diagram (BMD) [7] is another type of BDD that is used for representing logic functions for arithmetic operations. For those who are interested in more detailed techniques related to BDDs, several good surveys [11,24,28] are available.

References

1. *ACM/SIGDA Benchmark Newsletter*, DAC '93 Edition, June 1993.
2. Aho, A.V., J.E. Hopcroft, and J.D. Ullman, *The Design and Analysis of Computer Algorithms*, Addison-Wesley, Reading, MA, 1974.
3. Akers, S.B., "Functional testing with binary decision diagrams," *Proc. 8th Ann. IEEE Conf. Fault-Tolerant Comput.*, pp. 75–82, 1978.
4. Akers, S.B., "Binary decision diagrams," *IEEE Trans. on Computers*, vol. C–27, no. 6, pp. 509–516, June 1978.
5. Bertacco, V. and M. Damiani, "The disjunctive decomposition of logic functions," *ICCAD '97*, pp. 78–82, Nov. 1997.
6. Bryant, R.E., "Graph-based algorithms for Boolean function manipulation," *IEEE Trans. on Computers*, vol. C-35, no. 8, pp. 677–691, Aug. 1986.
7. Bryant, R.E. and Y.-A. Chen, "Verification of arithmetic functions with binary moment diagrams," *Proc. 32nd ACM/IEEE DAC*, pp. 535–541, June 1995.
8. Butler, K.M., D.E. Ross, R. Kapur, and M.R. Mercer, "Heuristics to compute variable orderings for efficient manipulation of ordered binary decision diagrams," *Proc. of 28th ACM/IEEE DAC*, pp. 417–420, June 1991.

9. Fujita, M., H. Fujisawa, and N. Kawato, "Evaluation and improvement of Boolean comparison method based on binary decision diagrams," *Proc. IEEE/ACM ICCAD '88,* pp. 2–5, Nov. 1988.
10. Fujita, M., Y. Matsunaga, and T. Kakuda, "On variable ordering of binary decision diagrams for the application of multi-level logic synthesis," *Proc. IEEE EDAC '91,* pp. 50–54, Feb. 1991.
11. Hachtel, G. and F. Somenzi, *Logic Synthesis and Verification Algorithms,* Kluwer Academic Publishers, 1996.
12. http://vlsi.colorado.edu/software.html.
13. Ishiura, N., H. Sawada, and S. Yajima, "Minimization of binary decision diagrams based on exchanges of variables," *Proc. IEEE/ACM ICCAD '91,* pp. 472–475, Nov. 1991.
14. Ishiura, N. and S. Yajima, "A class of logic functions expressible by a polynomial-size binary decision diagrams," in *Proc. Synthesis and Simulation Meeting and International Interchange (SASIMII '90. Japan),* pp. 48–54, Oct. 1990.
15. Lee, C.Y., "Representation of switching circuits by binary-decision programs," *Bell Sys. Tech. Jour.,* vol. 38, pp. 985–999, July 1959.
16. Liaw, H.-T. and C.-S. Lin, "On the OBDD-representation of general Boolean functions," *IEEE Trans. on Computers,* vol. C-41, no. 6, pp. 661–664, June 1992.
17. Lin, B. and F. Somenzi, "Minimization of symbolic relations," *Proc. IEEE/ACM ICCAD '90,* pp. 88–91, Nov. 1990.
18. Madre, J.C. and J. P. Billon, "Proving circuit correctness using formal comparison between expected and extracted behaviour," *Proc. 25th ACM/IEEE DAC,* pp. 205–210, June 1988.
19. Madre, J.C. and O. Coudert, "A logically complete reasoning maintenance system based on a logical constraint solver," *Proc. Int'l Joint Conf. Artificial Intelligence (IJCAI'91),* pp. 294–299, Aug. 1991.
20. Malik, S., A.R. Wang, R.K. Brayton, and A.L. Sangiovanni-Vincentelli, "Logic verification using binary decision diagrams in a logic synthesis environment," *Proc. IEEE/ACM ICCAD '88,* pp. 6–9, Nov. 1988.
21. Matsunaga, Y., "An exact and efficient algorithm for disjunctive decomposition," *SASIMI '98,* pp. 44–50, Oct. 1998.
22. Minato, S., "Minimum-width method of variable ordering for binary decision diagrams," *IEICE Trans. Fundamentals,* vol. E75-A, no. 3, pp. 392–399, Mar. 1992.
23. Minato, S., "Zero-suppressed BDDs for set manipulation in combinatorial problems," *Proc. 30th ACM/IEEE DAC,* pp. 272–277, June 1993.
24. Minato, S., *Binary Decision Diagrams and Applications for VLSI CAD,* Kluwer Academic Publishers, 1995.
25. Minato, S., N. Ishiura, and S. Yajima, "Shared binary decision diagram with attributed edges for efficient Boolean function manipulation," *Proc. 27th IEEE/ACM DAC,* pp. 52–57, June 1990.
26. Möller, D., J. Mohnke, and M. Weber, "Detection of symmetry of Boolean functions represented by ROBDDs," *Proc. IEEE/ACM ICCAD '93,* pp. 680–684, Nov. 1993.
27. Rudell, R., "Dynamic variable ordering for ordered binary decision diagrams," *Proc. IEEE/ACM ICCAD '93,* pp. 42–47, Nov. 1993.
28. Sasao, T., Ed., *Representation of Discrete Functions,* Kluwer Academic Publishers, 1996.
29. Sawada, H., S. Yamashita, and A. Nagoya, "Restricted simple disjunctive decompositions based on grouping symmetric variables," *Proc. IEEE Great Lakes Symp. on VLSI,* pp. 39–44, Mar. 1997.
30. Tani, S., K. Hamaguchi, and S. Yajima, "The complexity of the optimal variable ordering problems of shared binary decision diagrams," *The 4th Int'l Symp. Algorithms and Computation, Lecture Notes in Computer Science,* vol. 762, Springer, 1993, pp. 389–398.

30

Logic Synthesis with AND and OR Gates in Two Levels

Saburo Muroga

University of Illinois
at Urbana-Champaign

CONTENTS

30.1 Introduction

When logic networks are realized in transistor circuits on an integrated circuit chip, each gate in logic networks usually realizes more complex functions than AND or OR gates. But handy design methods are not available for designing logic networks with such complex gates under very diversified complex constraints such as delay time and layout rules. Thus, designers often design logic networks with AND, OR, and NOT gates as a starting point for design with more complex gates under complex constraints. AND, OR, and NOT gates are much easier for human minds to deal with. Logic networks with AND, OR, and NOT gates, after designing such networks, are often converted into transistor circuits. This conversion process illustrated in Figure 30.1 (the transistor circuit in this figure will be explained in later chapters) is called **technology mapping**. As can be seen, technology mapping is complex because logic gates and the corresponding transistor logic gates usually do not correspond one to one before and after technology mapping and layout has to be considered for speed and area.

Also, logic gates are realized by different types of transistor circuits, depending on design objectives. They are called **logic families**. There are several logic families, such as ECL, nMOS circuits, static CMOS, and dynamic CMOS, as discussed in Chapters 33, 38, and 39. Technology mapping is different for different logic families.

First, let us describe the design of logic networks with AND and OR gates in two levels, because handy methods are not available for designing logic networks with AND and OR gates in more than two levels.

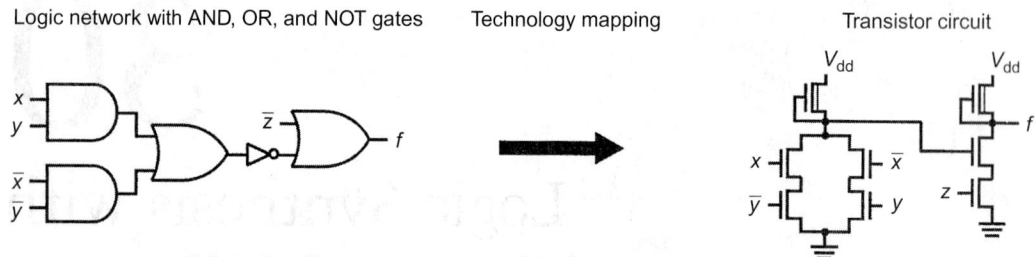

FIGURE 30.1 Conversion of a logic network with simple gates by technology mapping.

Although logic networks with AND and OR gates in two levels may not be directly useful for designing transistor circuits to be laid out on an integrated circuit chip, there are some cases where they are directly usable, such as programmable logic arrays to be described in a later chapter.

30.2 Design of Single-Output Minimal Networks with AND and OR Gates in Two Levels

Suppose that we want to obtain a two-level network with a minimum number of gates and, then, as a secondary objective, a minimum number of connections under Assumptions 30.1 in the following, regardless of whether we have AND or OR gates, respectively, in the first and second levels, or in the second and first levels. In this case, we have to design a network based on the minimal sum and another based on the minimal product, and then choose the better network. Suppose that we want to design a two-level AND/OR network for the function shown in Figure 30.2(a). This function has only one minimal sum, as shown with loops in Figure 30.2(a). Also, it has only one minimal product, as shown in Figure 30.3(a). The network in Figure 30.3(b), based on this minimal product, requires one less gate, despite more loops, than the network based on the minimal sum in Figure 30.2(b), and consequently the network in Figure 30.3(b) is preferable.

The above design procedure of a minimal network based on a minimal sum is meaningful under the following assumptions.

Assumptions 30.1: (1) The number of levels is at most two; (2) Only AND gates and OR gates are used in one level and second the other level; (3) Complemented variables $\bar{x}_i$'s as well as noncomplemented x_i's for each i are available as the network inputs; (4) No maximum fan-in restriction is imposed on any gate in a network to be designed; (5) Among networks realizable in two levels, we will choose networks that have a minimum number of gates. Then, from those with the minimum number of gates, we will choose a network that has a minimum number of connections.

If any of these is violated, the number of logic gates as the primary objective and the number of connections as the secondary objective are not minimized. If we do not have the restriction "at most two levels," we can generally have a network of fewer gates.

Karnaugh maps have been widely used because of convenience when the number of variables is small. But when the number of variables is many, maps are increasingly complex and processing with them become tedious, as discussed in Chapter 28. Furthermore, the corresponding logic networks with AND and OR gates in two levels are not useful because of excessive fan-ins and fan-outs.

30.3 Design of Multiple-Output Networks with AND and OR Gates in Two Levels

So far, we have discussed the synthesis of a two-level network with a single output. In many cases in practice, however, we need a two-level network with multiple outputs; so here let us discuss the synthesis of such a network, which is more complex than a network with a single output.

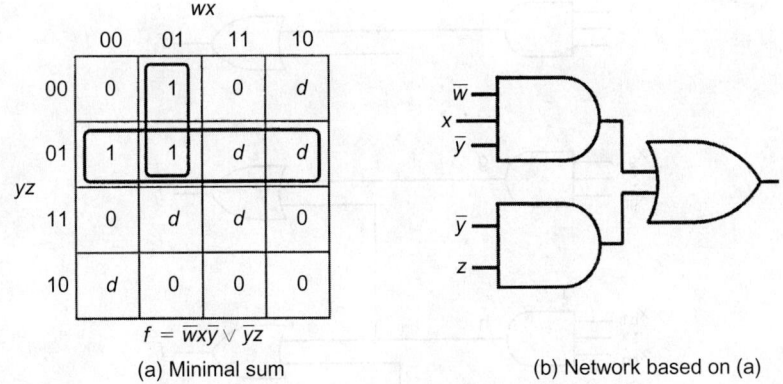

$$f = \overline{w}x\overline{y} \vee \overline{y}z$$

(a) Minimal sum

(b) Network based on (a)

FIGURE 30.2 Minimal sum and the corresponding network.

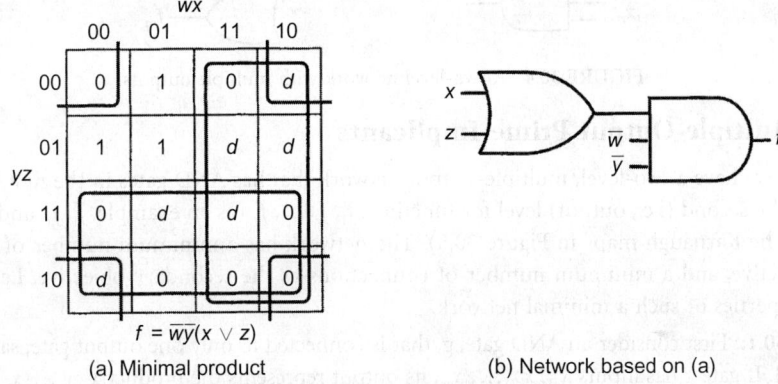

$$f = \overline{w}\,\overline{y}(x \vee z)$$

(a) Minimal product

(b) Network based on (a)

FIGURE 30.3 Minimal product and the corresponding network.

An obvious approach is to design a network for each output function separately. But this approach usually will not yield a more compact network than will synthesis of the functions collectively, because, for example, in Figure 30.4, the AND gate h, which can be shared by two output gates for f_i and f_j, must be repeated in separate networks for f_i and f_j by this approach.

Before discussing a design procedure, let us study the properties of a minimal two-level network that has only AND gates in the first level and only OR gates for given output functions $f_1, f_2, \ldots, f_m$ in the second level, as shown in Figure 30.4, where "a minimal network" means that the network has a minimum number of gates as the primary objective and a minimum number of connections as the secondary objective. The number of OR gates required for this network is at most m. Actually, when some functions are expressed as single products of literals, the number of OR gates can be less than m because these functions can be realized directly at the outputs of the AND gates without the use of OR gates. Also, when a function f_i has a prime implicant consisting of a single literal, that literal can be directly connected to the OR gate for f_i without the use of an AND gate. (These special cases can be treated easily by modifying the synthesis under the following assumption.) However, for simplicity, let us **assume that every variable input is connected only to AND gates in the first level and every function f_i, $1 \leq i \leq m$, is realized at the outputs of OR gates in the second level.** This is actually required with some electronic realizations of a network (e.g., when every variable input needs to have the same delay to the network outputs, or when PLAs which will be described in Chapter 45 are used to realize two-level networks).

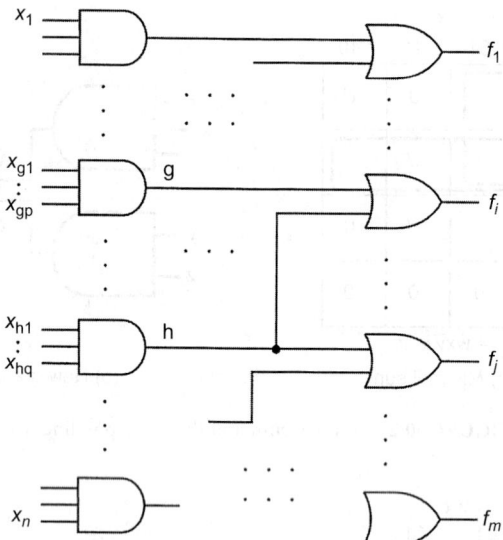

FIGURE 30.4 A two-level network with multiple outputs.

30.3.1 Multiple-Output Prime Implicants

Suppose that we have a two-level, multiple-output network that has AND gates in the first level, and m OR gates in the second (i.e., output) level for functions $f_1, f_2, \ldots, f_m$ (as an example, f_1, f_2, and f_3 are given as shown in the Karnaugh maps in Figure 30.5). The network has a minimum number of gates as the primary objective, and a minimum number of connections as the secondary objective. Let us explore the basic properties of such a minimal network.

Property 30.1: First consider an AND gate, g, that is connected to only one output gate, say for f_i, in Figure 30.4. If gate g has inputs $x_{g1}, x_{g2}, \ldots, x_{gp}$, its output represents the product $x_{g1}x_{g2}\ldots x_{gp}$. Then, if the product assumes the value 1, f_i becomes 1. Thus, the product $x_{g1}x_{g2}\ldots x_{gp}$ is an implicant of f_i. Since the network is minimal, gate g has no unnecessary inputs, and the removal of any input from gate g will make the OR gate for f_i express a different function (i.e., in Figure 30.5, the loop that the product represents becomes larger, containing some 0-cells, by deleting any variables from the product and then the new product is not an implicant of f_i). Thus, $x_{g1}, x_{g2}, \ldots, x_{gp}$ is a prime implicant of f_i.

Property 30.2: Next consider an AND gate, h, that is connected to two output gates for f_i and f_j in Figure 30.4. This time, the situation is more complicated. If the product $x_{h1}x_{h2}\ldots x_{hq}$ realized at the output of gate h assumes the value 1, both f_i and f_j become 1 and, consequently, the product f_if_j of the two functions also becomes 1 (thus, if a Karnaugh map is drawn for f_if_j as shown in Figure 30.5, the map has a loop consisting of only 1-cells for $x_{h1}x_{h2}\ldots x_{hq}$). Thus $x_{h1}x_{h2}\ldots x_{hq}$ is an implicant of product f_if_j. The network is minimal in realizing each of functions $f_1, f_2, \ldots, f_m$; so, if any input is removed from AND gate h, at least one of f_i and f_j will be a different function and $x_{h1}x_{h2}\ldots x_{hq}$ will no longer be an implicant of f_if_j. (That is, if any input is removed from AND gate h, a loop in the map for at least one of f_i and f_j will become larger. For example, if x_1 is deleted from M: $x_1\bar{x}_2x_3$ for f_1f_2, the loop for $\bar{x}_2x_3$, product of the remaining literals, appears in the maps for f_1 and also for f_2 in Figure 30.5. The loop in the map for f_2 contains 0-cell. Thus, if a loop is formed in the map for f_1f_2 as a product of these loops in the maps for f_1 and f_2, the new loop contains 0-cell in the map for f_1f_2. This means that if any variable is removed from $x_{h1}x_{h2}\ldots x_{hq}$, the remainder is not an implicant of f_if_j.) Thus, $x_{h1}x_{h2}\ldots x_{hq}$ is a prime implicant of the product f_if_j. (But notice that the product $x_{h1}x_{h2}\ldots x_{hq}$ is not necessarily a prime implicant of each single function f_i or f_j, as can be seen in Figure 30.5. For example, M: $x_1\bar{x}_2x_3$ is a prime implicant of f_1f_2 in Figure 30.5. Although this product, $x_1\bar{x}_2x_3$, is a prime implicant of f_2, it is not a prime implicant of f_1 in Figure 30.5.)

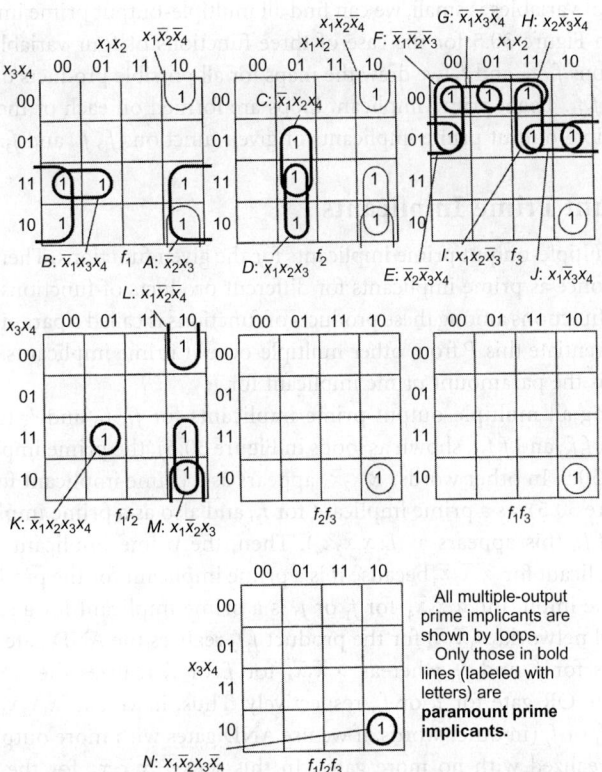

FIGURE 30.5 Multiple-output prime implicants on Karnaugh maps.

Generalizing this, we obtain the following conclusion. Suppose that the output of an AND gate is connected to r OR gates for functions $f_{i1}, \ldots, f_{ir}$. Since the network is minimal, this gate has no unnecessary inputs. Then the product of input variables realized at the output of this AND gate is a prime implicant of the product $f_{i1} \ldots f_{ir}$ (but is not necessarily a prime implicant of any product of $r-1$ or fewer, of these $f_{i1}, \ldots, f_{ir}$).

As in the synthesis of a single-output network, we need to find all prime implicants and then develop disjunctive forms by choosing an appropriate set of prime implicants. (Notice that each of these disjunctive forms is not necessarily a minimal sum for one of functions, f_i.) But, unlike the synthesis of a single-output network, in this case we must consider all prime implicants not only for each of the given functions $f_1, \ldots, f_m$, but also for all possible products of them, $f_1 f_2, f_1 f_3, \ldots, f_{m-1} f_m; f_1 f_2 f_3, f_1 f_2 f_4, \ldots; \ldots; \ldots; f_1 f_2 \ldots f_{m-1} f_m$.

Definition 30.1: Suppose that m functions $f_1, \ldots, f_m$ are given. All prime implicants for each of these m functions, and also all prime implicants for every possible product of these functions, that is,

$$f_1, f_2, \ldots, f_m$$

$$f_1 f_2, f_1 f_3, \ldots, f_{m-1} f_m,$$

$$f_1 f_2 f_3, f_1 f_2 f_4, \ldots, f_{m-2} f_{m-1} f_m$$

$$\ldots, \ldots, \ldots,$$

$$\ldots, \ldots, \ldots,$$

$$f_1 f_2 \ldots f_{m-1}, f_1 f_2 \ldots f_{m-2} f_m, \ldots, f_2 f_3 \ldots f_m$$

$$f_1 f_2 f_3 \ldots f_{m-1} f_m$$

are called the **multiple-output prime implicants** of $f_1, \ldots, f_m$.

When the number of variables is small, we can find all multiple-output prime implicants on Karnaugh maps, as illustrated in Figure 30.5 for the case of three functions of four variables. In addition to the maps for given functions f_1, f_2, and f_3, we draw the maps for all possible products of these functions; that is, $f_1 f_2$, $f_2 f_3$, $f_1 f_3$, and $f_1 f_2 f_3$. Then, prime-implicant loops are formed on each of these maps. These loops represent all the multiple-output prime implicants of given functions f_1, f_2, and f_3.

30.3.2   Paramount Prime Implicants

Suppose we find all multiple-output prime implicants for the given functions. Then, if a prime implicant P appears more than once as prime implicants for different products of functions, P for the product of the largest number of functions among these products of functions is called a **paramount prime implicant for P** in order to differentiate this P from other multiple-output prime implicants. As a special case, if P appears only once, it is the paramount prime implicant for P.

For example, among all multiple-output prime implicants for f_1, f_2, and f_3 (i.e., among all prime implicants for $f_1 f_2$, $f_2 f_3$, $f_1 f_3$, and $f_1 f_2 f_3$ shown as loops in Figure 30.5), the prime implicant $x_1 \bar{x}_2 \bar{x}_4$ appears three times in Figure 30.5. In other words, $x_1 \bar{x}_2 \bar{x}_4$ appears as a prime implicant for the function f_1 (see the map for f_1 in Figure 30.5), as a prime implicant for f_2, and also as a prime implicant for the product $f_1 f_2$ (in the map for $f_1 f_2$, this appears as $L: x_1 \bar{x}_2 \bar{x}_4$). Then, the prime implicant $x_1 \bar{x}_2 \bar{x}_4$ for $f_1 f_2$ is the paramount prime implicant for $x_1 \bar{x}_2 \bar{x}_4$ because it is a prime implicant for the product of two functions, f_1 and f_2, but the prime implicant $x_1 \bar{x}_2 \bar{x}_4$ for f_1 or f_2 is a prime implicant for a single function f_1 or f_2 alone. In the two-level network, $x_1 \bar{x}_2 \bar{x}_4$ for the product $f_1 f_2$ realizes the AND gate with output connections to the OR gates for f_1 and f_2, whereas $x_1 \bar{x}_2 \bar{x}_4$ for f_1 or f_2 realizes the AND gate with output connection to only the OR gate for f_1 or f_2, respectively. Thus, if we use $x_1 \bar{x}_2 \bar{x}_4$ for the product $f_1 f_2$ instead of $x_1 \bar{x}_2 \bar{x}_4$ for f_1 or f_2 (in other words, if we use AND gates with more output connections) then the network will be realized with no more gates. In this sense, $x_1 \bar{x}_2 \bar{x}_4$ for the product $f_1 f_2$ is more desirable than $x_1 \bar{x}_2 \bar{x}_4$ for the function f_1 or f_2. Prime implicant $x_1 \bar{x}_2 \bar{x}_4$ for the product $f_1 f_2$ is called a paramount prime implicant, as formally defined in the following, and is shown with label L in a bold line in the map for $f_1 f_2$ in Figure 30.5; whereas $x_1 \bar{x}_2 \bar{x}_4$ for f_1 or f_2 alone is not labeled and is also not in a bold line in the map for f_1 or f_2. Thus, **when we provide an AND gate whose output realizes the prime implicant $x_1 \bar{x}_2 \bar{x}_4$, we can connect its output connection in three different ways, i.e., to f_1, f_2, or both f_1 and f_2, realizing the same output functions, f_1, f_2, and f_3. In this case, the paramount prime implicant means that we can connect the largest number of OR gates from this AND gate; in other words, this AND gate has the largest coverage, although some connections may turn out to be redundant later.**

> **Definition 30.2:** Suppose that when all the multiple-output prime implicants of $f_1, \ldots, f_m$ are considered, a product of some literals, P, is a prime implicant for the product of k functions f_{p1}, $f_{p2}, \ldots, f_{pk}$ (possibly also prime implicants for products of $k - 1$ or fewer of these functions), but is not a prime implicant for any product of more functions that includes all these functions f_{p1}, $f_{p2}, \ldots, f_{pk}$. (For the above example, $x_1 \bar{x}_2 \bar{x}_4$ is a prime implicant for the product of two functions, f_1 and f_2, and also a prime implicant for a single function, f_1 or f_2. But $x_1 \bar{x}_2 \bar{x}_4$ is not a prime implicant for the product of more functions, including f_1 and f_2, that is, for the product $f_1 f_2 f_3$.) Then, P for the product of $f_{p1}, f_{p2}, \ldots, f_{pk}$ is called the **paramount prime implicant** for this prime implicant ($x_1 \bar{x}_2 \bar{x}_4$ for $f_1 f_2$ is a paramount prime implicant, but $x_1 \bar{x}_2 \bar{x}_4$ for f_1 or f_2 is not). As a special case, if P is a prime implicant for only one function but not a prime implicant for any product of more than one function, it is a paramount prime implicant (B: $\bar{x}_1 x_3 x_4$ for f_1 is such an example in Figure 30.5).

In Figure 30.5, only labeled loops shown in bold lines represent paramount prime implicants.

If a two-level network is first designed only with the AND gates that correspond to the paramount prime implicants, we can minimize the number of gates. This does not necessarily minimize the number of connections as the secondary objective. But in some important electronic realizations of two-level networks, such as PLAs (which will be explained later) this is not important. Thus, let us consider only the minimization of the number of gates in the following for the sake of simplicity.

30.3.3 Design of a Two-Level Network with a Minimum Number of AND and OR Gates

In the following procedure, we will derive a two-level network with a minimum number of gates (but without minimizing the number of connections as the secondary objective) by finding a minimal number of paramount prime implicants to represent the given functions.

Procedure 30.1: Design of a Multiple-Output Two-Level Network That Has a Minimum Number of Gates Without Minimizing the Number of Connections

We want to design a two-level network that has AND gates in the first level and OR gates in the second level. We shall assume that variable inputs can be connected to AND gates only (not to OR gates), and that the given output functions $f_1, f_2, \ldots, f_m$ are to be realized only at the outputs of OR gates.

Suppose we have already found all the paramount prime implicants for the given functions and their products.

1. Find a set of a smallest number of paramount prime implicant loops that covers all 1-cells in Karnaugh maps for the given functions $f_1, f_2, \ldots, f_m$. In this case, maps for their products, such as $f_1 f_2$, need not be considered. If there is more than one such set, choose a set that has as large loops as possible (i.e., choose a set of loops such that the total number of inputs to the AND gates that correspond to these loops is the smallest).

For example, suppose that the three functions of four variables, f_1, f_2, f_3, shown in the Karnaugh maps in Figure 30.5 are given. Then, using only the bold-lined loops labeled with letters (i.e., paramount prime implicants), try to cover all 1-cells in the maps for f_1, f_2, and f_3 only (i.e., using the only top three maps in Figure 30.5). Then, we find that more than one set of loops have the same number of loops with the same sizes. *AKLCDMNFHJ*, one of these sets, covers all functions f_1, f_2, and f_3, as illustrated in Figure 30.6.

2. Construct a network corresponding to the chosen set of paramount prime implicant loops.

Then, the network of 13 gates shown in Figure 30.7 has been uniquely obtained. Letter N (i.e., $x_1 \bar{x}_2 x_3 \bar{x}_4$), for example, is a paramount prime implicant for the product $f_1 f_2 f_3$, so the output of an AND gate with inputs, x_1, $\bar{x}_2$, x_3, and $\bar{x}_4$, is connected to the OR gates for f_1, f_2, and f_3.

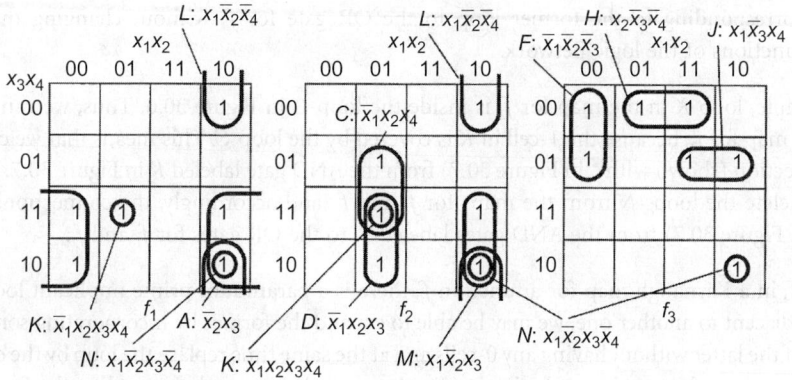

FIGURE 30.6 Covering f_1, f_2, and f_3 with the minimum number of paramount prime implicants.

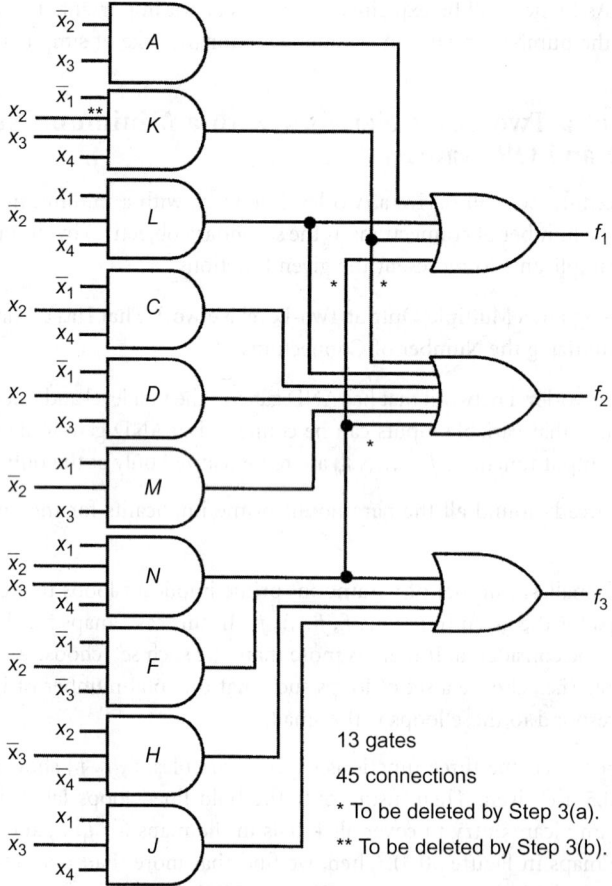

FIGURE 30.7 The network corresponding to *AKLCDMNFHJ*.

3. Then delete unnecessary connections, or replace some logic gates by new ones, by the following steps from the logic networks derived in Step 2, by considering whether some Karnaugh maps have a paramount prime implicant loop that is inside another one, or adjacent to another one.

   a. If, in a Karnaugh map for a function f, there is a paramount prime implicant loop that is inside another one, then the former is not necessary, because the 1-cells contained in this loop are covered by the latter loop. Thus, we can delete the connection from the AND gate corresponding to the former loop to the OR gate for f, without changing the output functions of the logic network.

   For example, loop K in the map for f_2 is inside the loop C in Figure 30.6. Thus, we can delete K from the map for f_2, because the 1-cell in K is covered by the loop C. This means that we can delete the connection (shown with * in Figure 30.7) from the AND gate labeled K in Figure 30.7. Similarly we can delete the loops N from the maps for f_1 and f_2, and accordingly, the connections (shown with * in Figure 30.7) from the AND gates labeled N to the OR gates for f_1 and f_2.

   b. If, in a Karnaugh map for a function f, there is a paramount prime implicant loop that is adjacent to another one, we may be able to expand the former by incorporating some 1-cells of the latter without having any 0-cells and at the same time replace the loop by the expanded loop (i.e., the number of logic gates unchanged). If we can do so, replace the former loop by the expanded loop. The expanded loop represents a product of fewer literals. Thus, we can delete the connection to the AND gate corresponding to the expanded loop without changing the output functions of the logic network.

For example, loop K in the map for f_1 has an adjacent loop A. Then we can replace the loop K by a larger loop (i.e., loop B in Figure 30.5) by incorporating the 1-cell on its left, which represents the product $\bar{x}_1 x_3 x_4$. Also, K appears only in the map for f_2, beside K in the map for f_1. K in the map for f_2 is concluded to be eliminable, so the AND gate for K in Figure 30.7 can be replaced by the new gate for B, keeping the number of logic gates unchanged. This means that we can delete the connection of input x_2 (shown with ** in Figure 30.7) to the AND gate labeled K in Figure 30.7 (i.e., this is essentially replacement of K by B). Thus we can delete in totally, 4 connections from the logic network shown in Figure 30.7, ending up with a simpler network with 13 gates and 41 connections.

When the number of paramount prime implicants is very small, we can find, directly on the maps, a minimum number of paramount prime implicants that cover all 1-cells in the maps for $f_1, f_2, \ldots, f_m$ (not their products) and derive a network with a minimum number of gates, using Procedure 30.1. But when the number of paramount prime implicants is many, the algebraic procedure of Section 4.6 in Ref. 5 is more efficient.

Procedure 30.1 with the deletion of unnecessary connections, however, may not necessarily yield a minimum number of connections as the secondary objective, although the number of gates is minimized as the primary objective. If we want to have a network with a minimum number of connections as the secondary objective, although the network has the same minimum number of gates as the primary objective, then we need to modify Procedure 30.1 and then delete unnecessary connections, as described as Procedure 5.2.1 in Ref. 5. But this procedure is more tedious and time-consuming.

30.3.4 Networks That Cannot be Designed by the Preceding Procedure

Notice that the design procedures in this section yield only a network that has all AND gates in the first level and all OR gates in the second level. (If 0-cells on Karnaugh maps are worked on instead of 1-cells in Procedure 30.1, we have a network with all OR gates in the first level and all AND gates in the second level.) If AND and OR gates are mixed in each level, or the network need not be in two levels, Procedure 30.1 does not guarantee the minimality of the number of logic gates [6].

These two problems can be solved by the integer programming logical design method (to be mentioned in Chapter 34, Section 34.5), which is complex and can be applied to only networks of a small number of gates.

30.3.5 Applications of Procedure 30.1

Procedure 30.1 has important applications, such as PLAs, it but cannot be applied for designing large PLAs. For multiple-output functions with many variables, absolute minimization is increasingly time-consuming. Using BDD (described in Chapter 29), Coudert, Madre, and Lin extended the feasibility of absolute minimization [2,3]. But as it is becoming too time-consuming, we have to give up absolute minimization, resorting to heuristic minimization, such as a powerful method which is called MINI [4] and was later improved as ESPRESSO [1].

References

1. Brayton, R., G. Hachtel, C. McMullen, and A. Sangiovanni-Vincentelli, *Logic Minimization Algorithms for VLSI Synthesis*, Kluwer Academic Publishers, 1984.
2. Lin, B., O. Coudert, and J.C. Madre, "Symbolic prime generation for multiple-valued functions," *DAC* 1992, pp. 40–44, 1992.
3. Coudert, O., "On Solving Covering Problems," *DAC*, pp. 197–202, 1996.

4. Hong, S.J., R.G. Cain and D.L. Ostapko, "MINI: a heuristic approach for logic minimization," *IBM Jour. Res. Dev.*, pp. 443–458, Sept. 1974.
5. Muroga, S., *Logic Design and Switching Theory*, John Wiley & Sons, 1979 (Now available from Krieger Publishing Co.).
6. Weiner, P. and T.F. Dwyer, "Discussions of some flaws in the classical theory of two level minimizations of multiple-output switching networks," *IEEE Tr. Comput.*, pp. 184–186, Feb. 1968.

31

Sequential Networks

Saburo Muroga
University of Illinois
at Urbana-Champaign

CONTENTS

31.1 Introduction

A logic network is called a **sequential network** when the values of its outputs depend not only on the current values of inputs but also on some of the past values, whereas a logic network is called a **combinational network** when the values of its outputs depend only on the current values of inputs but not on any past values. Analysis and synthesis of sequential networks are far more complex than combinational networks. When reliable operation of the networks is very important, the operations of logic gates are often synchronized by clocks. Such clocked networks, whether they are combinational or sequential networks, are called **synchronous networks**, and networks without clock are called **asynchronous networks**.

31.2   Flip-Flops and Latches

Because in a sequential network the outputs assume values depending not only on the current values but also on some past values of the inputs, a sequential network must remember information about the past values of its inputs. Simple networks called **flip-flops** that are realized with logic gates are usually used as memories for this purpose. Semiconductor memories can also serve as memories for sequential networks, but flip-flops are used if higher speed is necessary to match the speed of logic gates. Let us explain the simplest flip-flops, which are called **latches**.

31.2.1   *S-R* Latches

The network in Figure 31.1(a) which is called an **S-R latch**, consists of two NOR gates. Assume that the values at terminals S and R are 0, and the value at terminal Q is 0 (i.e., $S = R = 0$ and $Q = 0$). Since gate 1 has inputs of 0 and 0, the value at terminal $\bar{Q}$ is 1 (i.e., $\bar{Q} = 1$). Since gate 2 in the network has two inputs, 0 and 1, its output is $Q = 0$. Thus, signals 0 and 1 are maintained at terminals Q and $\bar{Q}$, respectively, as long as S and R remain 0. Now let us change the value at S to 1. Then, $\bar{Q}$ is changed to 0, and Q becomes 1 after a short time delay. Even if $Q = 1$ and $\bar{Q} = 0$ were their original values, the change of the value at S to 1 still yields $Q = 1$ and $\bar{Q} = 0$. In other words, Q is set to 1 by supplying 1 to S, no matter whether we originally had $Q = 0$, $\bar{Q} = 1$, or $Q = 1$, $\bar{Q} = 0$. Similarly, when 1 is supplied to R with S remaining at 0, Q and $\bar{Q}$ are set to 0 and 1, respectively, after a short time delay, no matter what values they had before. Thus, we get the first three combinations of the values of S and R shown in the table in Figure 31.1(b). In other words, as long as $S = R = 0$, the values of Q and $\bar{Q}$ are not changed. If $S = 1$, Q is set to 1. If $R = 1$, Q is set to 0. Thus, S and R are called **set** and **reset terminals**, respectively. In order to let the latch work properly, the value 1 at S or R must be maintained until new values of Q and $\bar{Q}$ are established. The S-R latch is usually denoted as in Figure 31.1(c). An S-R latch can also be realized with NAND gates, as shown in Figure 31.1(d). Latches and flip-flops have a direct reset-input terminal and a direct set-input terminal, although these input terminals are omitted in Figure 31.1 and in the figures for other flip-flops for the sake of simplicity. These input terminals are convenient for initial setting to $Q = 1$, or resetting to $Q = 0$.

When $S = R = 1$ occurs, the outputs Q and $\bar{Q}$ are both 0. If S and R simultaneously return to 0, these two outputs cannot maintain 0. Actually, a simultaneous change of S and R to 0 or 1 is physically impossible, often causing the network to malfunction, unless we make the network more sophisticated, such as synchronization of logic gates by a clock, as will be explained later. If S returns to 0 from 1 before R does, we have $Q = 0$ and $\bar{Q} = 1$. If R returns to 0 from 1 before S does, we have $Q = 1$ and $\bar{Q} = 0$. Thus, it is not possible to predict what values we will have at the outputs after having $S = R = 1$. The output of this network is not defined for $S = R = 1$, as this combination is not used. **For simplicity, let us assume that only one of the inputs to any network changes at a time, unless otherwise noted. This is a reasonable and important assumption.**

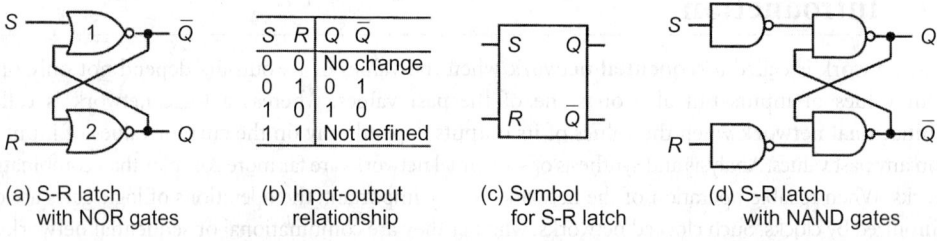

S	R	Q Q̄
0	0	No change
0	1	0 1
1	0	1 0
1	1	Not defined

(a) S-R latch with NOR gates

(b) Input-output relationship

(c) Symbol for S-R latch

(d) S-R latch with NAND gates

FIGURE 31.1   S-R latches.

31.2.2 Flip-Flops

Usually, *S-R* latches are used in designing asynchronous sequential networks, although sequential networks can be designed without them. For example, the memory function can be realized with longer loops of gates than the loop in the latch, and also more sophisticated flip-flops than *S-R* latches can be used. For example, a loop consisting of a pair of inverters and a special gate called a transmission gate is used in CMOS networks, as we will see in Chapter 39, Section 39.2. But for synchronous networks, **raceless flip-flops** (described later in this chapter) are particularly important.

31.3 Sequential Networks in Fundamental Mode

In a sequential network, the value of the network output depends on both the current input values and some of past input values stored in the network. We can find what value the output has for each individual combination of input values, depending on what signal values are stored in the network. This essentially means interpreting the signals stored inside the network as new input variables called **internal variables** and then interpreting the entire network as a combinational network of the new input variables, plus the original inputs, which are **external input variables**.

Let us consider the sequential network in Figure 31.2. Assume that the inputs are never changed unless the network is in a stable condition, that is, unless none of the internal variables is changing. Also assume that, whenever the inputs change, only one input changes at a time. Let y_1, $\bar{y}_1$, y_2, and $\bar{y}_2$ denote the outputs of the two *S-R* latches.

Let us assume that $y_1 = y_2 = 0$ (accordingly, $\bar{y}_1 = \bar{y}_2 = 1$). $x_1 = 0$, and $x_2 = 1$. Then, as can be easily found, we have $z_1 = 1$ and $z_2 = 0$ for this combination of values. Because of $x_1 = 0$ and $x_2 = 1$, the inputs of the latches have values $R_2 = 1$ (accordingly, $y_2 = 0$) and $S_1 = S_2 = R_1 = 0$. Then y_1 and y_2 remain 0. As long as x_1, y_1, and y_2 remain 0 and x_2 remains 1, none of the signal values in this network changes and, consequently, this combination of values of x_1, x_2, y_1, and y_2 is called a **stable state**.

Next let us assume that x_1 is changed to 1, keeping $x_2 = 1$ and $y_1 = y_2 = 0$. For this new combination of input values, we get $z_1 = 0$ and $z_2 = 0$ after a time delay of τ, where τ is the switching time (delay time) of each gate, assuming for the sake of simplicity that each gate has the same τ (this is not necessarily true in practice). The two latches have new inputs $S_1 = 1$ and $S_2 = R_1 = R_2 = 0$ after a time delay of τ. Then, they have new output values $y_1 = 1$ (previously 0), $\bar{y}_1 = 0$, $y_2 = 0$, and $\bar{y}_2 = 1$ after a delay due to the response time of the latches. Outputs z_1 and z_2 both change from 0 to 1. After this change, the network does not change any further. Summarizing the above, we can say that, when the network has the combination $x_1 = x_2 = 1$, $z_1 = z_2 = y_1 = y_2 = 0$, the network does not remain in this combination, but changes into the new combination $x_1 = x_2 = z_1 = z_2 = y_1 = 1$, $y_2 = 0$. Also, outputs z_1 and z_2 change into $z_1 = z_2 = 1$, after assuming the values $z_1 = z_2 = 0$ temporarily. After the network changes into the

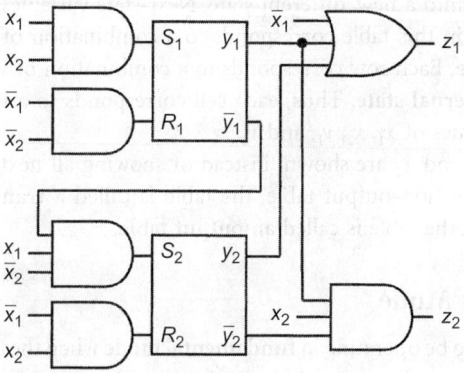

FIGURE 31.2 A sequential network with S-R latches.

TABLE 31.1 Transition-Output Table in Fundamental Mode

combination $x_1 = x_2 = y_1 = 1$, $y_2 = 0$, and $z_1 = z_2 = 1$, it remains there. The combination $x_1 = x_2 = 1$ and $z_1 = z_2 = y_1 = y_2 = 0$ is called an **unstable state.** The transition from an unstable to a stable state, such as the above transition to the stable state, is the key to the analysis of a sequential network.

31.3.1 Transition Tables

This and other transitions for other combinations of values of x_1, x_2, y_1, y_2, z_1, and z_2 can be shown on the map in Table 31.1, which is called a **transition-output table,** showing the next values of y_1 and y_2 as Y_1 and Y_2, respectively. The entry in each cell in Table 31.1 is next states Y_1, Y_2 and outputs z_1 and z_2. The above transition from $x_1 = 0$, $x_2 = 1$, $y_1 = y_2 = 0$ to $x_1 = x_2 = y_1 = 1$, $y_2 = 0$ is shown by the line with an arrow in Table 31.1. For this transition, the network is initially in the cell labeled with * in Table 31.1; and when x_1 changes to 1, the network moves to the next cell labeled with the dot with this cell's entry, $Y_1 = 1$, $Y_2 = 0$, and $z_1 = z_2 = 0$, during the transient period. Here, it is important to notice that in this cell, next values of internal variables y_1 and y_2, i.e., Y_1 and Y_2 are shown but the network actually has current values of y_1 and y_2, that is, $y_1 = y_2 = 0$ during this transition period corresponding to this cell. Y_1 and Y_2 in this cell, which is in an unstable state, indicates what values y_1 and y_2 should take after the transition. Then, the network moves to the bottom cell where the values of y_1 and y_2 are identical to Y_1 and Y_2, respectively, as indicated by the line with an arrow, because Y_1 and Y_2 will be current y_1 and y_2 after the transition. The present values of the internal variables are shown with y_1 and y_2 in small letters and their next values are with Y_1 and Y_2 in capital letters. More specifically, variables y_1 and y_2 are called **present-state (internal) variables**, and Y_1 and Y_2 are called **next-state (internal) variables.** As can easily be seen, when the values of Y_1 and Y_2 shown inside a cell are identical to those of y_1 and y_2 shown on the left of the table, respectively, a state containing these values of Y_1, Y_2, y_1, and y_2 is stable, because there is no transition of y_1 and y_2 into a new, different state. Next-state variables in stable states are encircled in Table 31.1. Each column in this table corresponds to a combination of values of network inputs x_1 and x_2, that is, an **input state.** Each row corresponds to a combination of values of internal variables y_1 and y_2, that is, a present **internal state.** Thus, each cell corresponds to a **total state**, or simply a **state**; that is, a combination of values of x_1, x_2, y_1, and y_2.

When only next states Y_1 and Y_2 are shown, instead of showing all next states Y_1, Y_2, and outputs z_1 and z_2 in each cell of a transition-output table, the table is called a **transition table**, and when only outputs z_1 and z_2 are shown, the table is called an **output table.**

31.3.2 Fundamental Mode

A sequential network is said to be operating in **fundamental mode** when the transition occurs horizontally in the transition table corresponding to each network input change and, then, unless the new state in the same row is stable, the transition continues vertically (not diagonally), settling in a new stable state, such as the transition shown with the line with an arrow in Table 31.1. The transitions of the network take place under the assumption that only one of the network inputs changes at a time, only when the network is not in transition; that is, only when the network is settled in a stable state.

By using a transition-output table, the output sequence [i.e., the sequence of output values corresponding to any input sequence, (i.e., the sequence of input values)] can be easily obtained. In other words, the output sequence can be obtained by choosing columns corresponding to the input sequence and then moving to stable states whenever states.

31.4 Malfunctions of Asynchronous Sequential Networks

An asynchronous sequential network does not work reliably unless appropriate binary numbers are assigned to internal variables for each input change.

31.4.1 Racing Problem of Sequential Networks

A difference in time delays of signal propagation along different paths may cause a malfunction of an asynchronous sequential network that is called a **race**. This is caused by a difference in the delay times of gates.

Let us consider the transition-output table of a certain asynchronous sequential network shown in Table 31.2.

Suppose that the network is in the stable state $(x_1, x_2, y_1, y_2) = (1110)$. If the inputs change from $(x_1, x_2) = (11)$ to (10), the network changes into the unstable state $(x_1, x_2, y_1, y_2) = (1010)$, marked with * in Table 31.2. Because of $y_1 = 1$, $y_2 = 0$, $Y_1 = 0$ and $Y_2 = 1$, two logic gates whose outputs represent y_1 and y_2 in the network must change their output values simultaneously. However, it is extremely difficult for the two gates to finish their changes at exactly the same time, because no two paths that reach these gates have identical time delays. Actually, one of these two logic gates finishes its change before the other does. In other words, we have one of the following two cases:

1. y_2 changes first, making the transition of (y_1, y_2) from (10) to (11) and leading the network into stable state $(x_1, x_2, y_1, y_2) = (1011)$.
2. y_1 changes first, making the transition of (y_1, y_2) from (10) to (00) and reaching the unstable state $(x_1, x_2, y_1, y_2) = (1000)$, and then y_2 changes, making the further transition of (y_1, y_2) from (00) to (01), settling in the stable state $(x_1, x_2, y_1, y_2) = (1001)$.

Thus, the network will go to either state $(y_1, y_2) = (11)$, in case 1, or state $(y_1, y_2) = (00)$, in case 2, instead of going directly to $(y_1, y_2) = (01)$. If $(y_1, y_2) = (11)$ is reached, state (11) is stable, and the network stops here. If (00) is reached, this is an unstable state, and another transition to the next stable state, (01),

TABLE 31.2 Transition-Output Table in Fundamental Mode

$y_1 y_2$	$x_1 x_2$			
	00	01	11	10
00	(00), 0	(00), 0	11, 0**	01, 0
01	(01), 0	00, 0	11, 1	(01), 1
11	(11), 0	10, 0	10, 1	(11), 1
10	—	00, 0	(10), 0	01, 0*

$Y_1 Y_2$, z

occurs. State (01) is the **destination stable state** (i.e., desired stable state), but (11) is not. Thus, depending on which path of gates works faster, the network may malfunction. This situation is called a **race**. The network may or may not malfunction, depending on which case actually occurs.

Next, suppose that the network is in the stable state $(x_1, x_2, y_1, y_2) = (0100)$ and that inputs $(x_1, x_2) = (01)$ change to (11). The cell labeled with ** in Table 31.2 has $(Y_1 Y_2) = (11)$. Thus, y_1 and y_2 must have simultaneous changes. Depending on which one of the two logic gates whose outputs represent y_1 and y_2 changes its output faster, there are two possible transitions. But in this case, both end up in the same stable state, $(y_1, y_2) = (10)$. This is another race, but the network does not have a malfunction depending on the order of change of internal variables. Hence, this is called a **noncritical race**, whereas the previous race for cases 1 and 2 is termed a **critical race** because the performance is unpredictable (it is hard to predict which path of gates will have the signal change faster), possibly causing the network to malfunction. We may have more complex critical racing. In other words, if we have a network such that the output of the gate whose output represents y_1 feeds back to the input of a gate on a path that reaches the gate whose output represents y_1, the y_1 may continue to change its value from 0 to 1 (or from 1 to 0), then change back to 0 by feedback of new value 1, and so on. This **oscillatory race** may continue for a long time.

31.4.2 Remedies for the Racing Problem

Whenever a network has critical races, we must eliminate them for reliable operation. One approach is to make the paths of gates have definitely different time delays. This approach, however, may not be most desirable for the following reasons. First, the speed may be sacrificed by adding gates for delay. Second, there are cases where this approach is impossible if a network contains more than one critical race. By eliminating a critical race in one column in the transition table by using a path of different time delay, a critical race in another column may occur or may be aggravated.

A better approach is to choose some entries in the transition table so that no critical races occur. Then, on the basis of this new table, we synthesize a network with the desired performance, according to the synthesis method to be described later. A change of entries in some unstable states without changing destination stable states, such that only one internal variable changes its value at a time, is one method for eliminating critical races. The critical race discussed above can be eliminated from Table 31.2 by replacing the entry marked with * by (00), as shown in Table 31.3, where only Y_1 and Y_2 are shown without the network output z. If every entry that causes the network to malfunction can be changed in this manner, a reliable network with the performance desired in the original is produced, because every destination stable state to which the network should go is not changed, and the output value for the destination stable state is also not changed. (We need not worry about noncritical races, since they cause no malfunctions.) However, sometimes, there are entries for some unstable states that cannot be changed

TABLE 31.3 Transition Table
in Fundamental Mode, Derived
by Modifying Table 31.2

$y_1 y_2$	$x_1 x_2$ 00	01	11	10
00	(00)	(00)	11	01
01	(01)	00	11	(01)
11	(11)	10	10	(11)
10	—	00	(10)	00

$Y_1 Y_2$

TABLE 31.4 Transition Table
in Fundamental Mode

y_1y_2	x_1x_2 00	01	11	10
00	⓪⓪			
01	⓪①			
11	①①			
10	01	①⓪		

Y_1Y_2

in this manner. For example, consider Table 31.4 (some states are not shown, for the sake of simplicity). The entry (01) for $(x_1, x_2, y_1, y_2) = (0010)$ is such an entry and causes a critical race. [State $(x_1, x_2, y_1, y_2) = (0010)$ in Table 31.2 may have the same property, but actually, the network never enters this state because of the assumption that inputs x_1 and x_2 do not change simultaneously.] Since this entry requires simultaneous transitions of two internal variables, y_1 and y_2, we need to change it to (00) or (11). Both lead the network to stable states different from the destination stable state (01).

When a change of entries in unstable states does not work, we need to redesign the network completely by adding more states (e.g., 8 states with 3 internal variables, y_1, y_2, and y_3, instead of 4 states with 2 internal variables, y_1 and y_2, for Table 31.4) without changing transitions among stable states, as we will see later in this chapter. This redesign may include the reassignment of binary numbers to states, or the addition of intermediate unstable states through which the network goes from one stable state to another. The addition of more states, reassignment of binary numbers to states, and addition of intermediate unstable states for this redesign, however, is usually cumbersome. So, designers usually prefer the use of synchronous sequential networks because design procedures are simpler and the racing problem, including oscillatory races due to the existence of feedback loops in a sequential network, is completely eliminated.

31.5   Different Tables for the Description of Transitions of Sequential Networks

When we look at a given sequential network from the outside, we usually cannot observe the values (i.e., binary numbers) of the internal variables; that is, the inside of the network (e.g., if the network is implemented in an IC package, no internal variables may appear at its pins). Also, binary numbers are not very convenient to use. Therefore, binary numbers for the internal states of the network may be replaced in a transition-output table by arbitrary letters or decimal numbers. The table that results is called a **state-output table**. For example, Table 31.5 is the state-output table obtained from the transition-output table in Table 31.2, where y_1y_2 is replaced by s and Y_1Y_2 is replaced by S. A present state of the internal state is denoted by s and its next state by S.

In some state tables, networks go from one stable state to another through more than one unstable state, instead of exactly one unstable state. For example, in the state-output table in Table 31.5, when inputs (x_1, x_2) change from (00) to (01), the network goes from stable state $(x_1, x_2, s) = (00C)$ to another stable state, $(01A)$, by first going to unstable state $(01C)$, then to intermediate unstable state $(01D)$, and finally to stable state $(01A)$. Here, $(x_1, x_2, s) = (01D)$ is called an **intermediate unstable state**. Such a multiple transition from one stable state to another in a state-output table cannot be observed well from outside the network, and is not important as far as the external performance of the network is concerned.

TABLE 31.5 State-Output Table
Derived from Table 31.2

s	00	01	11	10
A	(A), 0	(A), 0	C, 0	B, 0
B	(B), 0	A, 0	C, 1	(B), 1
C	(C), 0	D, 0	D, 1	(C), 1
D	—	A, 0	(D), 0	B, 0

$x_1 x_2$ (column group); S, z

TABLE 31.6 Flow-Output Table
Derived from Table 31.5

s	00	01	11	10
A	(A), 0	(A), 0	D, 0	B, 0
B	(B), 0	A, 0	D, 1	(B), 1
C	(C), 0	A, 0	(D), 1	C, 1
D	—	A, 0	(D), 0	B, 0

$x_1 x_2$ (column group); S, z

Even if each intermediate unstable state occurring during multiple transitions is replaced by the corresponding ultimate stable state, it does not make any difference if the network performance is observed from outside. Such a table is called a **flow-output table**. The flow-output table corresponding to the state-output table in Table 31.5 is shown in Table 31.6, where D in the intermediate unstable state $(x_1, x_2, s) = (01C)$, for example, in Table 31.5 is replaced by A in Table 31.6.

31.6 Steps for the Synthesis of Sequential Networks

Let us first introduce the general model for sequential networks and then describe a sequence of steps for designing a sequential network.

31.6.1 General Model of Sequential Networks

A sequential network may be generally expressed in the schematic form shown in Figure 31.3. A large block represents a loopless network of logic gates only (without any flip-flops). All loops with and without flip-flops (i.e., both loops that do not contain flip-flops and loops that contain flip-flops) are drawn outside the large block.

This loopless network has external input variables $x_1, \ldots, x_n$, and internal variables $y_1, \ldots, y_p$, as its inputs. It also has external output variables $z_1, \ldots, z_m$, and **excitation variables** $e_1, \ldots, e_q$, as its outputs. Some of the excitation variables $e_1, \ldots, e_q$ are inputs to the flip-flops, serving to excite the flip-flops. The remainder of the excitation variables are starting points of the loops without flip-flops, which end up at some of the internal variables. For loops without flip-flops, $e_i = Y_i$ holds for each i. For example, the network in Figure 31.2 can be redrawn in the format of Figure 31.3, placing the latches outside the loopless network.

31.6.2   Synthesis as a Reversal of Network Analysis

The outputs $e_1, \ldots, e_q, z_1, \ldots, z_m$ of the combinational network inside the sequential network in Figure 31.3 express logic functions that have $y_1, \ldots, y_p, x_1, \ldots, x_n$ as their variables. Thus, if these logic functions are given, the loopless network can be designed by the methods discussed in earlier chapters. This means that we have designed the sequential network, since the rest of the general model in Figure 31.3 is simply loops with or without flip-flops to be placed outside this combinational network. But we cannot derive these logic functions $e_1, \ldots, e_q, z_1, \ldots, z_m$ directly from the given design problem, so let us find, in the following, what gap to fill in.

When loops have no flip-flops, $e_i = Y_i$ holds for each i, where Y is the next value of y, and the binary-value relationship between the inputs $y_1, \ldots, y_p, x_1, \ldots, x_n$ and the outputs $Y_1, \ldots, Y_q, z_1, \ldots, z_m$ of the combinational network inside the sequential network in Figure 31.3 is expressed by a transition-output table. But when loops have flip-flops, we need the relationship between $e_1, \ldots, e_q, z_1, \ldots, z_m$ and $y_1, \ldots, y_p, x_1, \ldots, x_n$, where $e_1, \ldots, e_q$ are the inputs $S_1, R_1, S_2, R_2 \ldots$ of the latches. A table that shows this relationship is called an **excitation-output table**. For this purpose, we need to derive the output-input relation of the S-R latch, as shown in Table 31.8, by reversing the input-output relationship in Table 31.7, which shows the output y of a latch and its next value Y for every feasible combination of the values of S and R.

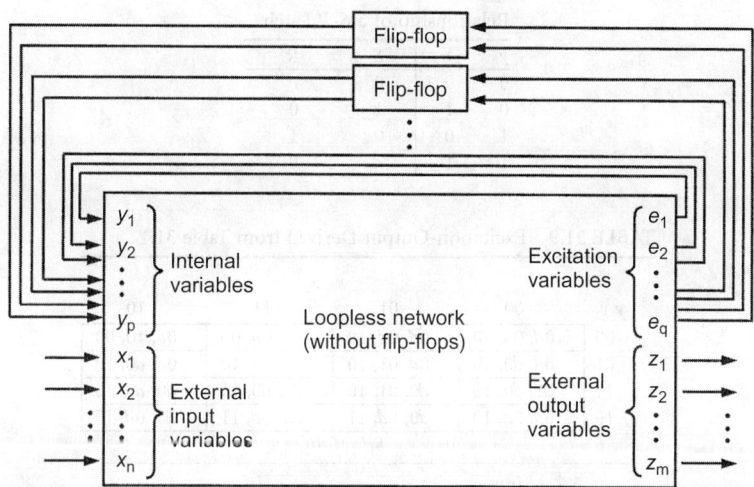

FIGURE 31.3   A general model of a sequential network.

TABLE 31.7   Input-Output Relationship of a S-R Latch

S	R	y	Y
0	0	0	0
		1	1
0	1	0	0
		1	0
1	0	0	1
		1	1

In Table 31.8, d's mean don't-care conditions. For example, $y = Y = 0$ in Table 31.8 results from $S = R = 0$ and also from $S = 0$, $R = 1$ in Table 31.7. Therefore, the first row in Table 31.8, $y = Y = 0$, $S = 0$, $R = d$, is obtained, because $y = Y = 0$ results from $S = 0$ only, but R can be 0 or 1; that is, don't-care, d. By using Table 31.8, the transition-output table in Table 31.1, for example, is converted into the excitation-output table in Table 31.9. For example, corresponding to $y_1 = Y_1 = 0$ in the first row and the first column in Table 31.1, $S_1 = 0$, $R_1 = d$ is obtained as the first $0d$ in the cell in the first row and the first column of Table 31.9, because the first row in Table 31.8 corresponds to this case. Of course, when a network, for example, Figure 31.2 is given, we can derive the excitation-output table directly from the network in Figure 31.2 rather than the transition-output table in Table 31.1 (which was derived for Figure 31.2). But when we are going to synthesize a sequential network from a transition-output table, we do not have the network yet and we need to construct an excitation-output table from a transition-output table, using Table 31.8.

31.6.3 Design Steps for Synthesis of Sequential Networks

A sequential network can be designed in the sequence of steps shown in Figure 31.4. The required performance of a network to be designed for the given problem is first converted into a flow-output table. Then this table is converted into a state-output table, and next into a transition-output table, by choosing an appropriate assignment of binary numbers to all states. Then, the transition-output table is converted into an excitation-output table if the loops contain flip-flops. If the loops contain no flip-flops, the excitation-output table need not be prepared, since it is identical to the transition-output table. Finally, a network is designed, using the logic design procedures discussed in the preceding chapters.

TABLE 31.8   Output-Input
Relationship of a S-R Latch

y	Y	S	R
0	0	0	d
0	1	1	0
1	0	0	1
1	1	d	0

TABLE 31.9   Excitation-Output Derived from Table 31.1

	\multicolumn{4}{c}{x_1, x_2}			
$y_1 y_2$	00	01	11	10
------	------	------	------	------
00	$0d$, $0d$, 10	$0d$, $0d$, 10	10, $0d$, 00	$0d$, 10, 00
01	$0d$, $d0$, 10	$0d$, 01, 10	10, $d0$, 10	$0d$, $d0$, 10
11	01, $d0$, 10	$d0$, 01, 10	$d0$, $d0$, 10	$d0$, $d0$, 10
10	$0d$, $0d$, 10	$d0$, $0d$, 11	$d0$, $0d$, 11	$d0$, $0d$, 10

$$S_1 R_1, \ S_2 R_2, \ z_1 z_2$$

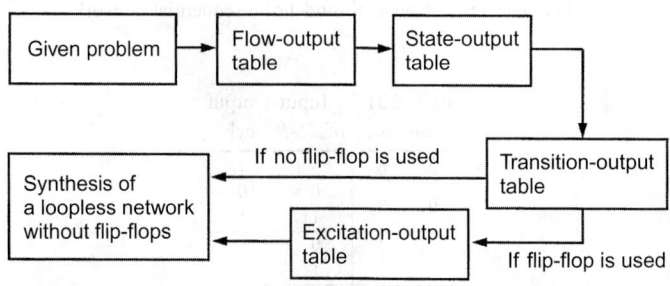

FIGURE 31.4   Steps for designing a sequential network.

31.7 Synthesis of Sequential Networks

Since the use of clocks has many advantages, synchronous sequential networks with clocks are in many cases preferred to asynchronous sequential networks. In addition to the ease of design and elimination of hazards including racing problems, the speed of logic networks can sometimes (e.g., Domino CMOS) be improved by synchronizing the operations of logic gates by clocks, and waveforms of signals can be reshaped into clean ones.

For synchronous sequential networks, more sophisticated flip-flops than latches are usually used, along with clocks. With these flip-flops, which are called raceless flip-flops, network malfunctioning due to hazards can be completely eliminated. Design of sequential networks with raceless flip-flops and clocks is much simpler than that of networks in fundamental mode, since we can use simpler flow-output and state-output tables, which are said to be in skew mode, and multiple changes of internal variables need not be avoided in assigning binary numbers to states.

31.7.1 Raceless Flip-Flops

Raceless flip-flops are flip-flops that have complex structures but eliminate network malfunctions due to races. Most widely used raceless flip-flops are master-slave flip-flops and edge-triggered flip-flops. A **master-slave flip-flop** (or simply, an MS flip-flop) consists of a pair of flip-flops called a master flip-flop and a slave flip-flop. Let us explain the features of master-slave flip-flops based on the ***J-K* master-slave flip-flop** shown in Figure 31.5. For the sake of simplicity, all the gates in these flip-flops are assumed to have equal delay times, although in actual electronic implementations, this may not be true. Its symbol is shown in Figure 31.6, where the letters *MS* are shown inside the rectangle. Each action of the master-slave flip-flop is controlled by the leading and trailing edges of a clock pulse, as explained in the following.

The *J-K* master-slave flip-flop in Figure 31.5 works with the clock pulse, as illustrated in Figure 31.7, where the rise and fall of the pulse are exaggerated for illustration. When the clock has the value 0 (i.e., $c = 0$) NAND gates 1 and 2 in Figure 31.5 have output values 1, and then the flip-flop consisting of gates 3 and 4 does not change its state. As long as the clock stays at 0, gates 5 and 6 force the flip-flop consisting of gates 7 and 8 to assume the same output values as those of the flip-flop consisting of gates 3 and 4 (i.e., the former is slaved to the latter, which is the master). Each of the master and slave is a slight modification of the latch in Figure 31.1(d).

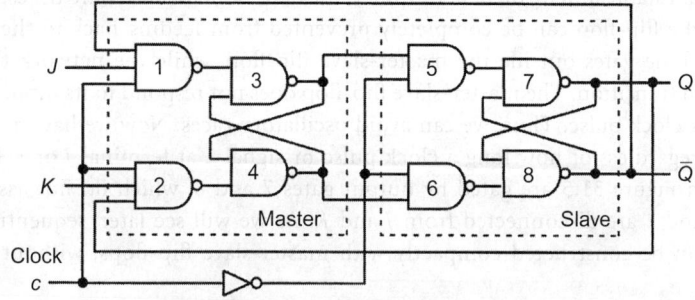

FIGURE 31.5 *J-K* master-slave flip-flop.

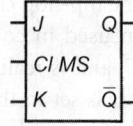

FIGURE 31.6 Symbol for *J-K* master-slave flip-flop.

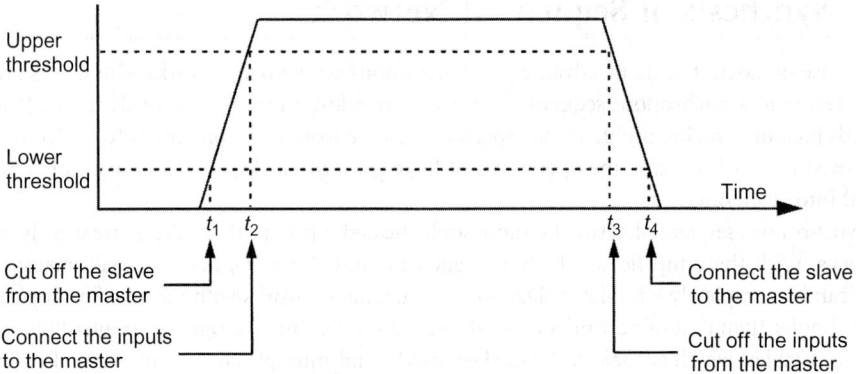

FIGURE 31.7 Clock pulse waveform.

When the clock pulse starts to rise to the lower threshold at time t_1 in Figure 31.7, gates 5 and 6 are disabled: in other words, the slave is cut off from the master. (The lower threshold value of the clock pulse still presents the logic value 0 to gates 1 and 2. The clock waveform is inverted to gates 5 and 6 through the inverter. The inputs to gates 5 and 6 from the inverter present the logic value 0 also to gates 5 and 6, because the inverted waveform is still close to logic value 1 but is not large enough to let gates 5 and 6 work. The inverter is actually implemented by a diode which is forward-biased, so that the network works in this manner.) When the clock pulse reaches the upper threshold at t_2, (i.e., $c = 1$), gates 1 and 2 are enabled, and the information at J or K is read into the master flip-flop through gate 1 or 2. Since the slave is cut off from the master by disabled gates 5 and 6, the slave does not change its state, maintaining the previous output values of Q and $\bar{Q}$. When the clock pulse falls to the upper threshold at t_3 after its peak, gates 1 and 2 are disabled, cutting off J and K from the master. In other words, the outputs of 1 and 2 become 1, and the master maintains the current output values. When the clock pulse falls further to the lower threshold at t_4, gates 5 and 6 are enabled and the information stored at the master is transferred to the slave, gates 7 and 8.

The important feature of the master-slave flip-flop is that the reading of information into the flip-flop and the establishment of new output values are done at different times; in other words, the outputs of the flip-flop can be completely prevented from feeding back to the inputs, possibly going through some gates outside the master-slave flip-flop, while the network that contains the flip-flop is still in transition. The master-slave flip-flop does not respond to its input until the leading edge of the next clock pulse. Thus, we can avoid oscillatory races. Now we have a J-K flip-flop that works reliably, regardless of how long a clock pulse or signal 1 at terminal J or K lasts, since input gates 1 and 2 in Figure 31.5 are gated by output gates 7 and 8, which do not assume new values before gates 1 and 2 are disconnected from J and K. As we will see later, sequential networks that work reliably can be constructed compactly with master-slave flip-flops, without worrying about hazards.

Other types of master-slave flip-flops are also used. The **T (type) master-slave flip-flop** (this is also called toggle flip-flop, trigger flip-flop, or T flip-flop) has only a single input, labeled T, as shown in Figure 31.8(a), which is the J-K master-slave flip-flop with J and K tied together as T. Whenever we have $T = 1$ during the clock pulse, the outputs of the flip-flop change, as shown in Figure 31.8(b). The T-type flip-flop, denoted as Figure 31.8(c), is often used in counters. The **D (type) master-slave flip-flop** (D implies "delay") has only a single input D, and is realized, as shown in Figure 31.9(a). As shown in Figure 31.9(b), no matter what value Q has, Q is set to the value of D that is present during the clock pulse. The D-type flip-flop is used for delay of a signal or data storage and is denoted by the symbol shown in Figure 31.9(c).

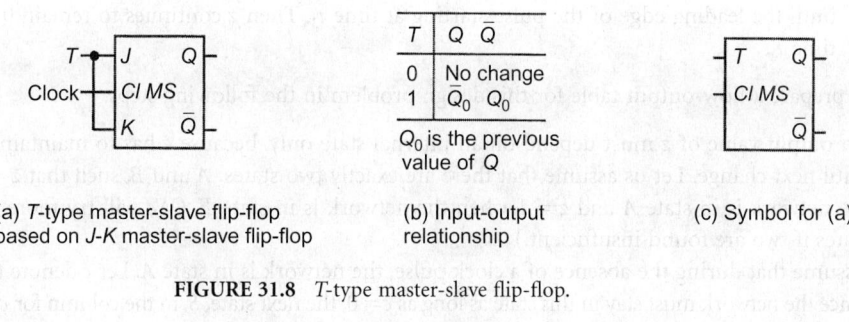

(a) *T*-type master-slave flip-flop based on *J-K* master-slave flip-flop

(b) Input-output relationship

(c) Symbol for (a)

FIGURE 31.8 *T*-type master-slave flip-flop.

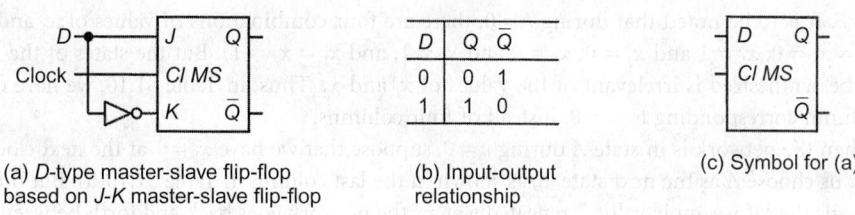

(a) *D*-type master-slave flip-flop based on *J-K* master-slave flip-flop

(b) Input-output relationship

(c) Symbol for (a)

FIGURE 31.9 *D*-type master-slave flip-flop.

Edge triggered flip-flops are another type of raceless flip-flop. Either the leading or the trailing edge of a clock pulse (not both) causes a flip-flop to respond to an input, and then the input is immediately disengaged from the flip-flop. Edge-triggered flip-flops are mostly used in the same manner as master-slave flip-flops.

31.7.2 Example of Design of a Synchronous Sequential Network

Let us now explain synthesis of a synchronous sequential network with an example.

Specification for the Design Example

Suppose we want to synthesize a clocked network with two inputs, x_1 and x_2, and single output z under the following specifications:

1. Inputs do not change during the presence of clock pulses, as illustrated in Figure 31.10. Inputs x_1 and x_2 cannot assume value 1 simultaneously during the presence of clock pulses. During clock pulses, an input signal of value 1 appears at exactly one of two inputs, x_1 and x_2, of the network, or does not appear at all.
2. The value of z changes as follows.
   a. The value of z becomes 1 when the value 1 appears during the clock pulse at the same input at which the last value 1 appeared. Once z becomes 1, z remains 1 regardless of the presence or absence of clock pulses, until signal 1 starts to appear at the other input during clock pulses. (This includes the following case. Suppose that we have $z = 0$, when signal 1 appears at one of the inputs during a clock pulse. Then, signal 0 follows at both x_1 and x_2 during the succeeding clock pulses. If signal 1 comes back to the same input, z becomes 1.) As illustrated in Figure 31.10, z becomes 1 at time t_1 at the leading edge of the second pulse because signal 1 is repeated at input x_1. Then z continues to remain 1 even though signal 1 appears at neither input at the third pulse starting t_2.
   b. The value of z becomes 0 when the value 1 appears at the other input during the clock pulse. Once z becomes 0, z remains 0 regardless of the presence or absence of clock pulses, until signal 1 starts to appear at the same input during clock pulses. In Figure 31.10, z continues to be 1

until the leading edge of the pulse starting at time t_3. Then z continues to remain 0 until the time t_4.

Let us prepare a flow-output table for this design problem in the following steps.

1. An output value of z must depend on an internal state only, because z has to maintain its value until next change. Let us assume that there are exactly two states, A and B, such that $z = 0$ when the network is in state A and $z = 1$ when the network is in state B. (We will try more than two states if two are found insufficient.)

2. Assume that during the absence of a clock pulse, the network is in state A. Let c denote the clock. Since the network must stay in this state as long as $c = 0$, the next state, S, in the column for $c = 0$ must be A, as shown in Table 31.10(a). Similarly, the next state for the second row in the column $c = 0$ must be B. It is to be noted that during $c = 0$, there are four combinations of values of x_1 and x_2 (i.e., $x_1 = x_2 = 0$; $x_1 = 1$ and $x_2 = 0$; $x_1 = 0$ and $x_2 = 1$; and $x_1 = x_2 = 1$). But the states of the network to be synthesized is irrelevant of the values of x_1 and x_2. Thus, in Table 31.10, we have only one column corresponding to $c = 0$, instead of four columns.

3. When the network is in state A during $c = 0$, suppose that we have $x_1 = 1$ at the next clock pulse. Let us choose A as the next state, S, as shown in the last column in Table 31.10(a). But this choice means that, if we apply value 1 repeatedly at x_1, the network goes back and forth between the two states in the first and last columns in the first row in Table 31.10(a). Then $z = 1$ must result from the specification of the network performance; but since the network stays in the first row, we must have $z = 0$. This is a contradiction. Thus, the next state for $(x_1, x_2, s) = (10A)$ cannot be A.

4. Next assume that the next state for $(x_1, x_2, s) = (10A)$ is B, as shown in Table 31.10(b). Suppose that the value 1 has been alternating between x_1 and x_2. If we had the last 1 at x_2, the network must be currently in A because of $z = 0$ for alternating 1's. When the next 1 appears at x_1, $z = 0$

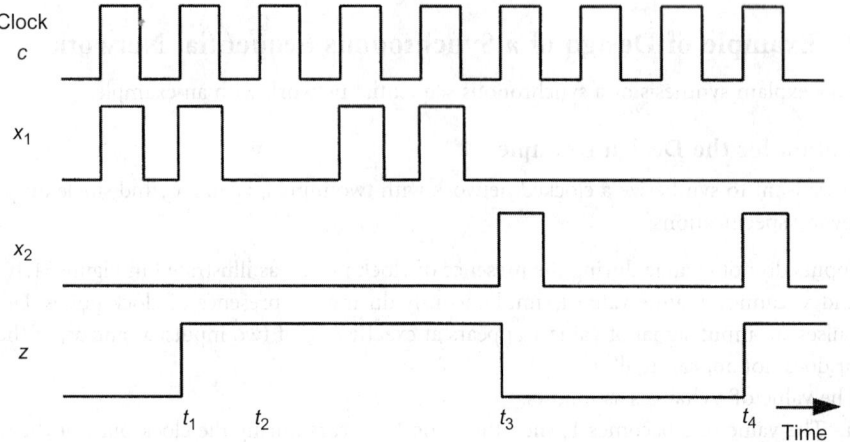

FIGURE 31.10 Waveform for the design example.

TABLE 31.10   Two States Are Not Enough

z	s	$c = 0$	00	01	11	10		z	s	$c = 0$	00	01	11	10
				$c = 1$ x_1, x_2								$c = 1$ x_1, x_2		
0	A	$A, 0$			—	A		0	A	$A, 0$			—	B
1	B	$B, 1$			—			1	B	$B, 1$			—	
				S, z								S, z		

(a) Entering A is not correct.   (b) Entering B is not correct.

must still hold because the value 1 is still alternating. But the network will produce $z = 1$ for state (10A), because B is assumed to correspond to $z = 1$. This is a contradiction, and the choice of B is also wrong.

5. In conclusion, two states are not sufficient, so we try again with more states. For this example, considering only two states corresponding to $z = 0$ and 1 is not appropriate.

6. In the above, we had contradictions by assuming only two states, A and B, corresponding to $z = 0$ and 1; we did not know which input's 1 led to each state. Thus, in addition to the values of z, let us consider which input had the last 1. In other words, we have four states corresponding to the combinations of z and the last 1, as shown in Table 31.11. At this stage, we do not know whether or not three states are sufficient. But let us assume four states for the moment. As a matter of fact, both three states and four states require two internal variables. Hence, in terms of the number of internal variables, it does not matter whether we have three or four states, although the two cases may lead to different networks.

7. **Derivation of a flow-output table in skew mode**: Let us form a flow-output table, assuming the use of J-K master-slave flip-flops. In the column of $c = 0$ in Table 31.11, the network must stay in each state during the absence of a clock pulse. Thus, all the next states S in all the cells in the column of $c = 0$ must be identical to s in each row. Suppose that the network is in state $(c, x_1, x_2, s) = (000A)$ with output $z = 0$ after having the last 1 at x_1. When the value 1 appears at x_1 and c becomes 1, the next state S must be B for the following reason. When the current clock pulse disappears, this 1 at x_1 will be "the last 1" at x_1 (so S must be A or B) and z will have to be 1 because of the repeated occurrence of 1's at x_1. (This contradicts $z = 0$, which we will have if A is entered as S.) Hence, the possibility of S being A is ruled out, and S must be B. Since value 1 is repeated at x_1, we have $z = 1$. The next states and the values of output z in all other cells in Table 31.11 can be found in a similar manner.

Let us analyze how a transition among stable states occurs in this table. According to the problem specification, inputs x_1 and x_2 can change only during the absence of clock pulses.

Suppose that during the absence of a clock pulse, the network is in state $(c, x_1, x_2, s) = (000A)$ in Table 31.11. Suppose that inputs (x_1, x_2) change from (00) to (10) sometime during the absence of a clock pulse and $(x_1, x_2) = (10)$ lasts at least until the trailing edge of the clock pulse. If this transition is interpreted on Table 31.11, the network moves from $(c, x_1, x_2, s) = (000A)$ to (110A), as shown by the dotted-line with an arrow, at the leading edge of the clock pulse. Then the network must stay in this state, (110A), until the trailing edge of the clock pulse, because the outputs of the J-K master-slave flip-flops keep the current values during the clock pulse, changing to its new values only at the trailing edge of the clock pulse and thus the internal state s does not change yet to its new state S. (This is different from the fundamental mode, in which the network does not stay in this state unless the state is a stable one, and vertically moves to a stable state in a different row in the same column.) Then the network moves from $(c, x_1, x_2, s) = (110A)$ to (010B), as shown by the solid-line with an arrow in Table 31.11, when s assumes the new state S at the trailing edge of the clock pulse. Thus, the transition occurs horizontally and then **diagonally** in

TABLE 31.11 Flow-Output Table in Skew-Mode

Last 1 at	z s	$c = 0$	$c = 1$ x_1, x_2 00	01	11	10
x_1	0 A	A, 0	A, 0	C, 0	—	B, 0
x_1	1 B	B, 1	B, 1	C, 0	—	B, 1
x_2	0 C	C, 0	C, 0	D, 1	—	A, 0
x_2	1 D	D, 1	D, 1	D, 1	—	A, 0

S, z

Table 31.11. This type of transition is called **skew mode**, in order to differentiate it from the fundamental mode.

Notice, however, that in the new state (110A) at the leading edge of the clock pulse in the above transition from $(c, x_1, x_2, s) = (000A)$, the network z assumes the new output value. Consequently, in this new stable state during $c = 1$ in Table 31.11, the new current value of the network output, z, is shown, while the value of the internal state in this new stable state shows the next state S, though the network is actually in s. In the fundamental mode, when the network moves horizontally to a new unstable state in the same row in the state-output table, the current value of the network output is shown and the internal state represents the next state, S (in this sense the situation is not different), but the network output lasts only during a short, transient period, unless the new state is stable. In contrast, in skew mode, the output value for the new state is not transient (because the network stays in this state during the clock pulse) and is essential in the description of the network performance.

Let us synthesize a network for this flow-output table in skew mode later.

8. **Derivation of a flow-output table in fundamental mode**: Next let us try to interpret this table in fundamental mode. Table 31.11 shows that, when the network placed in state $(c, x_1, x_2, s) = (000A)$ receives $x_2 = 1$ before the appearance of the next pulse, the next state will be C at the leading edge of the next pulse. But the network goes to unstable state $(c, x_1, x_2, s) = (101C)$ that has entry D, since if we assume fundamental mode, it must move vertically, instead of the diagonal transition in skew mode. Hence, the network must go further to stable state (101D), without settling in the desired stable state, C, if the network still keeps $x_2 = 1$ and $c = 1$. Therefore, the network cannot be in fundamental mode. (Recall that the next state entries in a flow-output table, unlike a state-output table, do not show intermediate unstable states but do show the destination stable states.) The above difficulty can be avoided by adding two new rows, E and F, as shown in Table 31.12. When the network placed in state $(c, x_1, x_2, s) = (000A)$ receives $x_2 = 1$, the network goes to the new stable state F in column $(x_1, x_2) = (01)$ and in row F. For this state F, $z = 0$, without causing contradiction. When the clock pulse disappears, the network goes to stable state (000C) after passing through unstable state (000F). The problem with the other states is similarly eliminated. All stable states are encircled as stable states.

The values of z for stable states are easily entered. The values of z for unstable states can be entered with certain freedom. For example, suppose that the network placed in state $(c, x_1, x_2, s) = (000A)$ receives $x_1 = 1$. Then the next state S is B. In this case, we may enter 0 or 1 as z for (110A) for the following reason. We have $z = 0$ for the initial stable state A during $c = 0$ and $z = 1$ for the destination stable state B during $c = 1$ and correspondingly $z = 0$ or 1. This does not make much difference as far as the external behavior of the network is concerned, because the network stays

TABLE 31.12   Flow-Output Table in Fundamental Mode

| | | | | $c = 1$
x_1, x_2 | | |
Last 1 at	$z\ s$	$c = 0$	00	01	11	10
x_1	0 A	Ⓐ, 0	Ⓐ, 0	F, 0	—	B, d
x_1	0 E	A, 0	—	—	—	Ⓔ, 0
x_1	1 B	Ⓑ, 1	Ⓑ, 1	F, d	—	Ⓑ, 1
x_2	0 C	Ⓒ, 0	Ⓒ, 0	D, d	—	E, 0
x_2	0 F	C, 0	—	Ⓕ, 0	—	—
x_2	1 D	Ⓓ, 1	Ⓓ, 1	Ⓓ, 1	—	E, d

S, z

in this unstable state (110A) for the short transient period and it simply means that $z = 1$ appears a little bit earlier or later. The network that results, however, may be different. Accordingly, $z = d$ (d denotes "don't-care") would be the best assignment, since this gives flexibility in designing the network later.

31.7.3   Design of Synchronous Sequential Networks in Skew Mode

Now let us design a synchronous sequential network based on a flow-output table in skew mode, using J-K master-slave flip-flops.

As pointed out previously, when master-slave flip-flops are used, the network does not have racing hazards even if internal variables make multiple changes, because the flip-flops do not respond to any input changes during clock pulses. Thus, we need not worry about hazards due to multiple changes of internal variables and consequently appropriate assignment of binary numbers to states in forming a transition-output table from a state-output table or a flow-output table in the design steps in Figure 31.4. But the number of gates, connections, or levels in a network to be designed can differ, depending on how binary numbers are assigned to states (it is of secondary importance compared with the hazard problem, which makes networks useless if they malfunction). Making state assignments without considering multiple changes of internal variables is much easier than having to take these changes into account.

Let us derive the transition-output table shown in Table 31.13 from Table 31.11, using a state assignment as shown.

Reversing the inputs and outputs relationship of J-K master-slave flip-flops shown in Table 31.14(a), we have the output-input relationship shown in Table 31.14(b) (other master-slave flip-flop types can be treated in a similar manner). Table 31.14(b) shows what values inputs J and K must take for each change of internal variable y to its next value Y. (In order to have $y = Y = 0$, $J = K = 0$ or $J = 0$ and $K = 1$ must hold, as we can see in Table 31.14(a) and thus we have $S = 0$ and $R = d$ in Table 31.14(b).) Using

TABLE 31.13   Transition-Output Table in Skew Mode

				$c = 1$		
				x_1, x_2		
s	$y_1 y_2$	$c = 0$	00	01	11	10
A	00	00, 0	00, 0	11, 0	dd, d	01, 0
B	01	01, 1	01,1	11, 0	dd, d	01, 1
C	11	11, 0	11, 0	10, 1	dd, d	00, 0
D	10	10, 1	10, 1	10, 1	dd, d	00, 0
				$Y_1 Y_2, z$		

TABLE 31.14   Input-Output Relationship and Output-Input Relationships of J-K Master-Slave Flip-Flop

(a) Input-output relationship					(b) Output-input relationship			
Inputs		Outputs			Outputs		Inputs	
J	K	y	Y		y	Y	S	R
0	0	0	0		0	0	0	d
		1	1		0	1	1	d
0	1	0	0		1	0	d	1
		1	0		1	1	d	0
1	0	0	1					
		1	1					
1	1	0	1					
		1	0					

TABLE 31.15   Excitation-Output Derived from Table 31.13

| | | $c=1$ | | | |
| | | x_1, x_2 | | | |
$y_1 y_2$	$c=0$	00	01	11	10
00	0d, 0d, 0	0d, 0d, 0	1d, 1d, 0	dd, d, 0	0d, 1d, 0
01	0d, d0, 0	0d, d0, 0	1d d0, 0	dd, d, 0	0d, d0, 0
11	d0, d0, 0	d0, d0, 0	d0, d1, 0	dd, d, 0	d1, d1, 0
10	d0, 0d, 0	d0, 0d, 0	d0, 0d, 0	dd, d, 0	d1, 0d, 0

$$J_1 K_1, J_2 K_2, z$$

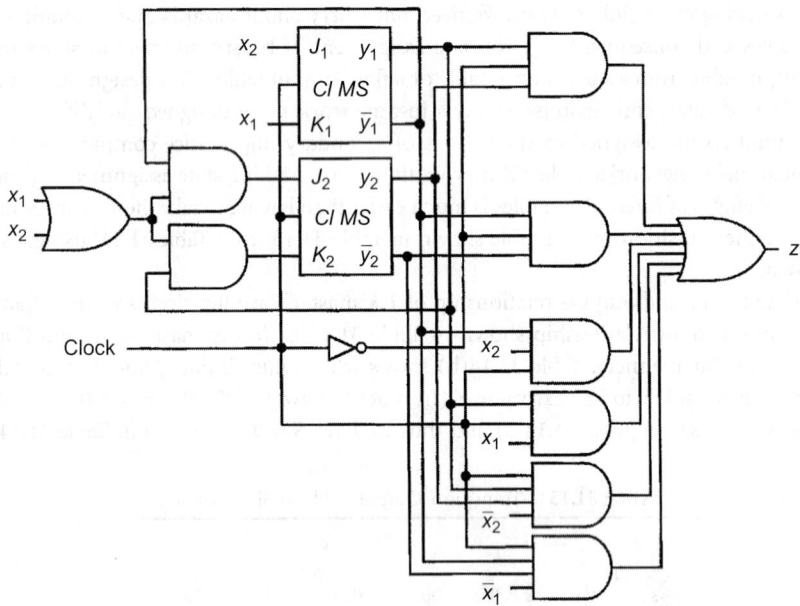

FIGURE 31.11   Synthesized network based on Table 31.15.

Table 31.14(b), we form the excitation table in Table 31.15. Decomposing Table 31.15 into five Karnaugh maps for J_1, K_1, J_2, K_2 and z, we can find a minimal sum for each of J_1, K_1, J_2, K_2, and z. On the basis of these logic expressions, we design the loopless network inside the general model of sequential networks in Figure 31.3. Then, placing two *J-K* master-slave flip-flops outside this loopless network, we have designed the sequential network shown in Figure 31.11.

Master-slave flip-flops do not respond to changes in their inputs when and after their outputs change until the leading edges of next clock pulses. Thus, no network malfunction due to races occurs, and no post-analysis of whether or not the designed networks malfunction due to this is necessary. This is the advantage of clocked networks with raceless flip-flops.

31.7.4   Design of Asynchronous Sequential Networks in Fundamental Mode

Now let us design an asynchronous sequential network based on a flow-output table in fundamental mode.

According to the design steps of Figure 31.4, we have to derive a transition-output table from the flow-output table shown in Table 31.12, by deriving a state-output table by assigning appropriate binary numbers to states such that multiple changes do not occur for every transition from one binary number to another, possibly changing some intermediate unstable states to others. Then, if the designers want to use *S-R* latches, we can derive an excitation table by finding the output-input relationship of the *S-R* latch, as illustrated with Tables 31.7, 31.8, and 31.9. Then, we can design the loopless network inside the general model of Figure 31.3 by deriving minimal sums from the Karnaugh maps decomposed from the excitation-output table. If the designers do not want to use *S-R* latches, we can design the loopless network inside the general model of Figure 31.3 by deriving minimal sums from the Karnaugh maps decomposed from the transition-output table without deriving an excitation-output table.

In the case of an asynchronous sequential network, we need post-analysis of whether the designed network works reliably.

31.7.5 Advantages of Skew Mode

The advantages of skew-mode operation with raceless flip-flops, such as master-slave or edge-triggered flip-flops, can be summarized as follows:

1. We can use no more complex and often simpler flow-output tables (or state-output tables) in skew mode than are required in fundamental mode, making design easier (because we need not consider both unstable and stable states for each input change, and need not consider adding extra states, or changing intermediate unstable states, which are to avoid multiple changes of internal variables).

2. State assignments are greatly simplified because we need not worry about hazard due to multiple changes of internal variables. (If we want to minimize the number of gates, connections, or levels, we need to try different state assignments. This is less important than the reliable operations of the networks to be synthesized.)

3. Networks synthesized in skew mode usually require fewer internal variables than those in fundamental mode.

4. After the network synthesis, we do not need to check whether the networks contain racing hazards or not. This is probably the greatest of all the advantages of skew mode, since checking hazards and finding remedies is usually very cumbersome and time-consuming.

References

1. Kohavi, Z., *Switching and Automata Theory*, 2nd ed., McGraw-Hill, 1978.
2. McCluskey, E.J., *Logic Design Principles: With Emphasis on Testable Semicustom Circuits*, Prentice-Hall, 1986.
3. Miller, R., *Switching Theory*, vol. 2, John Wiley & Sons, 1965.
4. Muroga, S., *Logic Design and Switching Theory*, John Wiley & Sons (now available from Krieger Publishing Co.), 1979.
5. Roth, C.H. Jr., *Fundamentals of Logic Design*, 4th ed., West Publishing Co., 1992.
6. Unger, S.H., *The Essence of Logic Circuits*, 2nd ed., IEEE Press, 1997.

32

Logic Synthesis with AND and OR Gates in Multi-Levels

Yuichi Nakamura
NEC Corporation

Saburo Muroga
*University of Illinois
at Urbana-Champaign*

CONTENTS

32.1 Logic Networks with AND and OR Gates in Multi-Levels

In logic networks, the number of levels is defined as the number of gates in the longest path from external inputs to external outputs. When we design logic networks with AND and OR gates, those in multi-levels can be designed with no more gates than those in two levels. Logic networks in multi-levels have more levels than those in two levels, but this does not necessarily mean that those in multi-levels have greater delay time than those in two levels because a logic gate that has many fan-out connections generally has greater delay time than gates that have fewer fan-out connections (Remark 32.1). Also, a logic gate that has many fan-in connections from other logic gates tends to have larger area in the chip and longer delay time than other gates that have fewer fan-in connections. Thus, if we want to design a logic network with a small delay time and small area, we need to design a logic network in many levels, keeping maximum fan-out and fan-in of each gate under a reasonably small limit.

> **Remark 32.1:** When the line width in an IC chip is large, the delay time of logic gates is larger than those over connections and, once a logic network is designed, it can be laid out on the chip usually without further modifications. But when the line width becomes very short, under 0.25 μm, long connections add more delay due to parasitic capacitance and resistance than the delay of gates. But length of connections cannot be known until making layout on an IC chip after finishing logic design. So at the time of logic design, prior to layout, designers know only the number of fan-out connections from each gate, and this is only partial information on delay estimation. Thus, when the line width becomes very short, it is difficult to estimate precisely the delay of a logic network at the time of logic design. We need to modify a logic network interactively, as we lay it out on the chip.

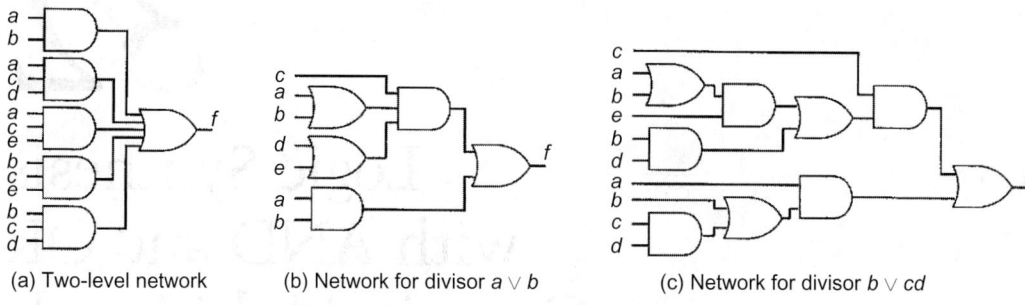

(a) Two-level network　　(b) Network for divisor $a \vee b$　　(c) Network for divisor $b \vee cd$

FIGURE 32.1　Networks for $f = ab \vee acd \vee ace \vee bce \vee bcd$.

Such a multi-level logic network can be derived by rewriting a logic expression with parentheses. For example, a logic expression

$$f = ab \vee acd \vee ace \vee bce \vee bcd \qquad (32.1)$$

can be realized with five AND gates and one OR gate in two levels, as illustrated in Figure 32.1(a). However,

$$f = c(a \vee b)(d \vee e) \vee ab \qquad (32.2)$$

can be obtained by rewriting the expression with parentheses, as explained in the following. This logic expression can be realized with three OR gates and two AND gates in three levels, as illustrated in Figure 32.1(b). The network in Figure 32.1(b) would have a smaller area and smaller delay than the one in Figure 32.1(a) because of fewer logic gates and a smaller maximum fan-in (a logic gate with five fan-ins, for example, has more than twice the area and delay of a gate with two fan-ins).

A logic network in two levels with the fewest gates can be derived by minimal sums or minimal products, which can be derived by reasonably simple algorithms (described in Chapter 30). But if we try to derive multi-level logic networks, only few reasonable algorithms are known. One of them is the weak division described in the following, although the minimality is not guaranteed and its execution is not straightforward. Another algorithm is the special case (i.e., AND and OR gates) of the map-factoring method (described in Chapter 34).

32.2　General Division

Rewriting of a logic expression using parentheses can be done by the following division. The division is based on the use of sub-expressions that can be found in the given logic expression. The given logic expression in a sum-of-products can be rewritten with parentheses if it has common sub-expressions. For example, the logic expression in Eq. 32.1 can be converted to the following expression, using a sub-expression $(a \vee b)$:

$$f = cd(a \vee b) \vee ce(a \vee b) \vee ab$$

This can be further rewritten into the following, by sharing the common sub-expression $(a \vee b)$:

$$f = c(a \vee b)(d \vee e) \vee ab \qquad (32.3)$$

This rewriting can be regarded symbolically as division. Rewriting of the expression in Eq. 32.1 into the one in Eq. 32.3 may be regarded as division with the divisor $x = a \vee b$, the quotient $q = cd \vee ce$ and the remainder $r = ab$. Then, the expression f can be represented as follows:

$f = xq \vee ab$, with divisor $x = a \vee b$, quotient $q = cd \vee ce = c(d \vee e)$, and remainder $r = ab$.

The division is symbolically denoted as f/x. Generally, the quotient should not be 0, but the remainder may be 0.

The division, however, may yield many different results because there are many possibilities in choosing a divisor and also the given logic function can be written in many different logic expressions, as explained in the following.

Division can be repeated on one given logic expression. Suppose $f = \bar{a}b \vee \bar{a}c \vee \bar{b}a \vee \bar{b}c \vee \bar{c}a \vee \bar{c}b$ is given. Repeating division three times, choosing successively $b \vee c$, $a \vee c$, and $a \vee b$ as divisors, the following result is derived:

$$f = x_1\bar{a} \vee x_2\bar{b} \vee x_3\bar{c} \text{ with divisors } x_1 = b \vee c, \ x_2 = a \vee c, \text{ and } x_3 = a \vee b.$$

32.3 Selection of Divisors

Among the divisions, those with a certain type of divisor to be described in the following are called **weak divisions**. The objective of the weak division is the derivation of a logic network with a logic expression having a minimal total number of literals, repeatedly applying the weak division to the given logic expression until the division cannot apply to the logic expression any further. In this case, the total number of literals is intended to be an approximation of the total number of inputs to all the logic gates, although not exactly, as explained later. Thus, we should start with a logic expression in a minimal sum of products by using two-level logic minimization [1,2] before weak division in order to obtain a good result.

In the weak division, the divisor selection is the most important problem to produce a compact network because there are many divisor candidates to be found in the given logic expression and the result of the weak division depends on divisor selection. For example, the expression $f = ab \vee acd \vee ace \vee bce \vee bcd$ in Eq. 32.1 with 14 literals has many divisor candidates, a, b, c, $a \vee b$, $b \vee cd$, and others. When $a \vee b$ is first selected as the divisor, the resultant network illustrated in Figure 32.1(b) for $f = c(a \vee b)(d \vee e) \vee ab$ with 7 literals is obtained. On the other hand, if $b \vee cd$ is first selected as the divisor, the resultant network for $f = c(e(a \vee b) \vee bd) \vee a(b \vee cd)$ with 10 literals illustrated in Figure 32.1(c) is obtained, which is larger than the network illustrated in Figure 32.1(b).

Divisors can be derived by finding sub-expressions called kernels. All the kernels for the given logic expression can be found as follows.

First, all the subsets of products in the given logic expression are enumerated. Next, for each subset of products, the product of the largest number of literals that is common with all the products in this subset is found. This product of the largest number of literals is called a **co-kernel**. Then the sum of products, from each of which this co-kernel (i.e., the product of the largest number of literals) is eliminated is obtained as a **kernel**. For example, the sum of products $abc \vee abd$ has the co-kernel ab as the product of the largest number of literals that is common to all the products, abc and abd. The kernel of $abc \vee abd$ is $c \vee d$. However, $ab \vee ac \vee d$ has no kernels, because it has no common literals for all products. The kernel $b \vee c$ with co-kernel a is found when the subset of products, $ab \vee ac$, is considered.

Certainly, by trying all divisor candidates and selecting the best one, we can derive a network as small as possible by the weak division. However, such an exhaustive search is too time-consuming, requiring a huge memory space. The branch-and-bound method is generally more efficient than the exhaustive search [3].

Thus, the heuristic method that the type of divisor candidates is restricted to sum of products with specific feature is proposed [2]. This method is called **kernel decomposition,** reducing the number of divisor candidates.

A kernel of an expression for f is the sum with at least two products such that all the products in the sum contain no common literals (e.g., $ab \vee c$ is a kernel, but $ab \vee ac$ and abc are not kernels, because all products, ab and ac, in $ab \vee ac$ contain a, and abc is a single product), especially, a kernel whose subsets contain no other kernels is called a **level-0 kernel** (e.g., $a \vee b$ is a level-0 kernel, but $ab \vee ac \vee d$ is not a level-0 kernel because sub-expression $ab \vee ac$ contains kernel $b \vee c$). The level of kernels is defined recursively as a **level-K kernel** contains at the next lower level-$(K-1)$ kernel (e.g., $ab \lor ac \lor d$ is a level-1 kernel because it contains level-0 kernel $b \lor c$).

Usually, a level-K kernel with $K \geq 1$ is not used as a divisor to save processing time, because the results obtained by all level kernels as divisors are the almost same as those obtained by using only the level-0 kernels. Thus, all the kernels that are not level-0 kernels are excluded from divisor candidates.

For example, the logic expression illustrated in Figure 32.1, $f = ab \lor acd \lor ace \lor bce \lor bcd$, has 16 kernels as shown in Table 32.1. By eliminating all the level-1 kernels, $ad \lor ae \lor bd \lor be$, $ea \lor eb \lor bd$ and others from Table 32.1, we have the divisor candidates in Table 32.2.

The next step is the selection of one divisor from all candidates. A candidate that decreases the largest number of literals in the expression by the weak division is selected as a divisor. If there is a tie, choose one of them. The difference in the number of literals before and after the weak division by the kernel is called the **weight of the kernel.** In the above example, the result of the weak division by the kernel $b \lor cd$ is $f = a(b \lor cd) \lor ace \lor bce \lor bcd$. The weight of the kernel $b \lor cd$ is 1, because the number of literals is reduced from 14 to 13 by this division. The weight of the kernels can be easily calculated by the number of literals in kernels and co-kernels without execution of weak division, because the quotient of the division by a kernel is a product of a co-kernel and other sub-expressions. The weight of the kernels for this example is shown in Table 32.2.

Then the kernel $a \lor b$ is selected as a divisor with the largest weight. The expression $f = ab \lor acd \lor ace \lor bce \lor bcd$ is divided by $a \lor b$. In the next division, $d \lor e$ is selected and divides the expression after division by $a \lor b$. Finally, the network $f = c(a \lor b)(d \lor e) \lor ab$ illustrated in Figure 32.1(b) is obtained.

These operations, enumeration of all the kernels, calculation of the weights of all kernels, selection of the largest one, and division are applied repeatedly until no kernel can be found. In this case, we can choose a different sequence of divisors, deriving a different result. But often we do not have a different result, so it may not be worthwhile to try many different sequences of divisors.

Instead of the kernel decomposition method, a faster method is proposed [4]. In this method, the divisor candidates are restricted to only 0-level kernels with two products, along with introduction of complement-sharing that when both $\bar{a}b \lor a\bar{b}$ and $\bar{a}\bar{b} \lor ab$ are among divisor candidates, one is realized with a logic gate while realizing the other by complementing it by an inverter ($\overline{\bar{a}b \lor a\bar{b}} = (a \lor \bar{b})(\bar{a} \lor b) = \bar{a}\bar{b} \lor ab$ for this example). Although the restriction is stronger than the kernel decomposition method, the method produces smaller networks and can run faster than the kernel decomposition in many cases.

TABLE 32.1   Kernals for $f = ab \lor acd \lor ace \lor bce \lor bcd$

Kernel	Co-kernel	Level
$ad \lor ae \lor bd \lor be$	c	1
$ea \lor eb \lor bd$	c	1
$eb \lor ed \lor cd$	c	1
$ab \lor ae \lor bd$	c	1
$ad \lor ae \lor be$	c	1
$a \lor ce \lor cd$	b	1
$b \lor cd$	a	0
$b \lor ce$	a	0
$a \lor be$	b	0
$a \lor cd$	b	0
$d \lor e$	ac	0
$ad \lor be$	c	0
$a \lor b$	cd	0
$ae \lor bd$	c	0
$a \lor b$	ce	0
$d \lor e$	bc	0

TABLE 32.2 0-level Kernals and Co-kernels
Derived from Table 32.1

Kernel	Co-kernel	Weight of Kernels
$b \vee cd$	a	1
$b \vee ce$	a	1
$a \vee be$	b	1
$a \vee cd$	b	1
$ad \vee be$	c	1
$ae \vee bd$	c	1
$a \vee b$	ce, cd	6
$d \vee e$	ac, bc	6

32.4 Limitation of Weak Division

Although simpler networks can be easily derived by the weak division, the weak division cannot derive certain types of logic networks because of its restrictions in its rewriting of logic expressions with parentheses. Rewriting of logic expressions without such restrictions is called **strong division**. For example, complements of sub-expressions, such as $f = (a\bar{b} \vee a\bar{c}) \vee \bar{b}a \vee c$, is not used in the weak division. Also, the two literals, x and $\bar{x}$, for each variable x are regarded as different variables without using identities such as $ab \vee a = a$, $a\bar{a} = 0$, and $a \vee \bar{a} = 1$ in the weak division.

Suppose the sum-of-products $f = a\bar{b} \vee a\bar{c} \vee \bar{b}a \vee b\bar{c} \vee \bar{c}a \vee c\bar{b}$ is given. This can be rewritten in the following two different logic expressions:

$$f_1 = a\bar{b} \vee a\bar{c} \vee \bar{b}a \vee b\bar{c} \vee \bar{c}a \vee c\bar{b}$$

and

$$f_2 = a\bar{a} \vee a\bar{b} \vee a\bar{c} \vee \bar{b}a \vee b\bar{b} \vee b\bar{c} \vee \bar{c}a \vee c\bar{b} \vee c\bar{c},$$

using the identities $a\bar{a} = 0$, $b\bar{b} = 0$, and $c\bar{c} = 0$

They can be further rewritten as follows:

$$f_1 = a(\bar{b} \vee \bar{c}) \vee b(\bar{a} \vee \bar{c}) \vee c(\bar{a} \vee \bar{c})$$

and

$$f_2 = (\bar{a} \vee \bar{b} \vee \bar{c})(a \vee b \vee c)$$

Both of these expressions can be written in the following expressions, using the divisors, quotients, and remainders, which are 0 in this particular example:

$$f_1 = x_{11}q_{11} \vee x_{12}q_{12} \vee x_{13}q_{13} \vee r_1$$

with divisors $x_{11} = \bar{b} \vee \bar{c}$, $x_{12} = \bar{a} \vee \bar{c}$, $x_{13} = \bar{a} \vee \bar{b}$, quotients $q_{11} = a$, $q_{12} = b$, $q_{13} = c$, and remainder $r_1 = 0$

$$f_2 = x_{21}q_{21} \vee r_2$$

with divisor $x_{21} = a \vee b \vee c$, quotient $q_{21} = \bar{a} \vee \bar{b} \vee \bar{c}$, and remainder $r_2 = 0$.

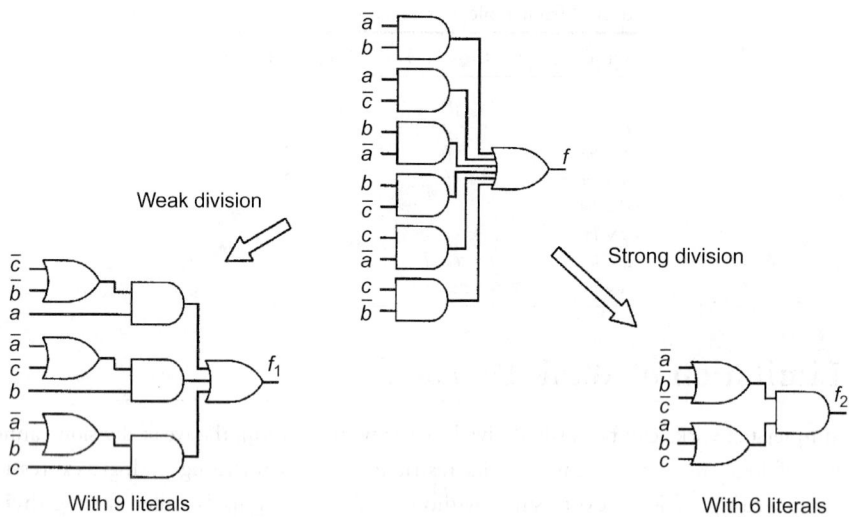

FIGURE 32.2 Strong and weak divisions.

Corresponding to these expressions, we have two different logic networks, as shown in Figure 32.2. The function f_2 is derived by division by a divisor above but actually cannot be obtained by the weak division. Thus, the logic network for f_2 is labeled as the result by strong division in Figure 32.2. Strong division is rewriting of logic expressions using any form, including the complement of a sub-expression and accordingly has far greater possibilities than the division explained so far.

The results of division are evaluated by the number of literals contained in each of the obtained expressions. This number is an approximation of the total number of fan-ins of gates in networks in the following sense: inputs to a logic gate from other gates are not counted as literals. In Figure 32.2, the number of literals of f_1 is 9, and the number of literals of f_2 is 6. But in the logic network for f_1, an input to each of three AND gates in Figure 32.2, for example, is not counted as a literal. Counting them, the total number of fan-ins of all logic gates in the logic network for f_1, which is 15 in Figure 32.2, is larger than the total number of fan-ins of all gates in the logic network for f_2, 8.

References

1. R.K. Brayton, G.D. Hachtel, C. McMullen, and A. Sangiovanni-Vincentelli, *Logic Minimization Algorithms for VLSI Synthesis*, Kluwer Academic Publishers, Boston, 1984.
2. R.K. Brayton, A. Sangiovanni-Vincentelli, and A. Wang, "MIS: A Multiple-Level Logic Optimization System," *IEEE Transaction on CAD*, CAD-6(6), pp. 1062–1081, July 1987.
3. G. De Micheli, A. Sangiovanni-Vincentelli, and P. Antognetti, *Design System for VLSI Circuits: Logic Synthesis and Silicon Compilation*, Martinus Nijhoff Publishers, pp. 197–248, 1987.
4. J. Rajski, and J. Vasudevamurthy, "Testability Preserving Transformations in Multi-level Logic Synthesis," *IEEE ITC*, pp. 265–273, 1990.

33

Logic Properties of Transistor Circuits

Saburo Muroga

University of Illinois
at Urbana-Champaign

CONTENTS

33.1 Basic Properties of Connecting Relays

Relays are probably the oldest means to realize logic operations. Relays, which are electromechanical devices, and their solid-state equivalents (i.e., transistors) are extensively used in many industrial products, such as computers. Relays are conceptually simple and appropriate for introducing physical realization of logic operations. More importantly, the connection configuration of a relay contact network is the same as that of transistors inside a logic gate realized with transistors, in particular MOSFETs (which stands for metal-oxide semiconductor field effect transistors).

A **relay** consists of an armature, a magnet, and a metal contact. An armature is a metal spring made of magnetic material with a metal contact on it. There are two different types of relays: a make-contact relay and a break-contact relay.

A **make-contact relay** is a relay such that, when there is no current through the magnet winding, the contact is open. When a direct current is supplied through the magnet winding, the armature is attracted to the magnet and, after a short time delay, the contact is closed. This type of relay contact is called a "make-contact" and is usually denoted by a lower-case x. The current through the magnet is denoted by a capital letter X, as shown in Figure 33.1.

A **break-contact relay** is a relay such that when there is no current through the magnet winding, the contact closes. When a direct current is supplied, the armature is attracted to the magnet and, after a short time delay, the contact opens. This type of relay contact is called a "break-contact" and is usually denoted by $\bar{x}$. The current through the magnet is again denoted by X, as shown in Figure 33.2.

In either case of a make-contact relay or a break-contact relay, no current in a magnet is represented by $X = 0$, and the flow of a current is represented by $X = 1$. Then $x = X$, no matter whether $X = 0$ or 1. But the contact of a make-contact relay is open or closed according as $X = 0$ or 1, because the contact is expressed by x, whereas the contact of a break-contact relay is closed or open according as $X = 0$ or 1, because the contact is expressed by $\bar{x}$.

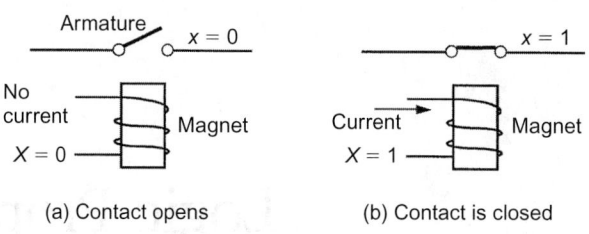

FIGURE 33.1 A make-contact relay.

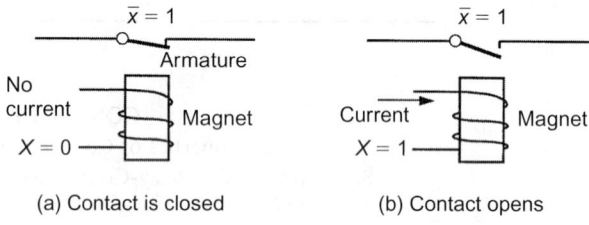

FIGURE 33.2 A break-contact relay.

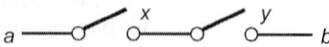

FIGURE 33.3 Series connection of relay contacts.

TABLE 33.1 Combinations of States for the Series Connection in Figure 33.3

(a)				(b)		
x	y	Entire Path Between a and b		x	y	f
Open	Open	Open		0	0	0
Open	Closed	Open		0	1	0
Closed	Open	Open		1	0	0
Closed	Closed	Closed		1	1	1

Let us connect two make-contacts x and y in series, as shown in Figure 33.3. Since X and x assume identical values at any time, the magnet, along with its symbol X, will henceforth be omitted in figures unless it is needed for some reason. Then we have the combinations of states shown in Table 33.1(a), which has only two states, "open" and "closed." Let f denote the state of the entire path between terminals a and b, where f is called the **transmission** of the network. Since "open" and "closed" of a make-contact are represented by $x = 0$ and $x = 1$, respectively, Table 33.1(a) may be rewritten as shown in Table 33.1(b). This table shows the AND of x and y, defined in Chapter 26 and denoted by $f = xy$. Thus the network of a series connection of make-contacts realizes the AND operation of x and y.

Let us connect two make-contacts x and y in parallel as shown in Figure 33.4. Then we have the combinations of states shown in Table 33.2(a). Replacing "open" and "closed" by 0 and 1, respectively, we may rewrite Table 33.2(a) as shown in Table 33.2(b). This table shows the OR of x and y, defined in Chapter 27 and denoted by $f = x \vee y$.

33.2 Analysis of Relay-Contact Networks

Let us analyze a relay contact network. "Analysis of a network" means the description of the logic performance of the network in terms of a logic expression [1].

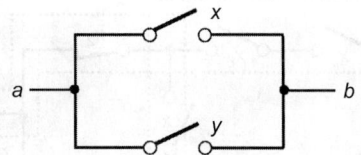

FIGURE 33.4 Parallel connection of relay contacts.

TABLE 33.2 Combinations of States for the Parallel Connection in Figure 33.4

(a)			(b)		
x	y	Entire Path Between a and b	x	y	f
Open	Open	Open	0	0	0
Open	Closed	Closed	0	1	1
Closed	Open	Closed	1	0	1
Closed	Closed	Closed	1	1	1

33.2.1 Transmission of Relay-Contact Networks

We now discuss general procedures to calculate the transmission of a network in which relay contacts are connected in a more complex manner. The first general procedure is based on the concept of tie sets, defined as follows.

Definition 33.1: Consider a path that connects two external terminals, a and b, and no part of which forms a loop. Then the literals that represent the contacts on this path are called a **tie set** of this network.

An example of a tie set is the contacts $x_1, x_6, \bar{x}_2$, and x_5 on the path numbered 1 in Figure 33.5.

Procedure 33.1: Derivation of the Transmission of a Relay-Contact Network by Tie Sets

Find all the tie sets of the network. Form the product of all literals in each tie set. Then the disjunction of all these products yields the transmission of the given network. $\qquad\square$

These tie sets represent all the shortest paths that connect terminals a and b. As an example, the network of Figure 33.5 has the following tie sets:

For path 1: $x_1, x_6, \bar{x}_2, x_5$
For path 2: x_2, x_4, x_5
For path 3: x_1, x_6, x_3, x_4, x_5
For path 4: $x_2, x_3, \bar{x}_2, x_5$

Then, we get the transmission of the network:

$$f = x_1 x_6 \bar{x}_2 x_5 \lor x_2 x_4 x_5 \lor x_1 x_6 x_3 x_4 x_5 \lor x_2 \bar{x}_2 x_3 x_5 \qquad (33.1)$$

where the last term, $x_2 \bar{x}_2 x_3 x_5$, may be eliminated, since it is identically equal to 0 for any value of x_2.

Procedure 33.1 yields the transmission of the given network because all the tie sets correspond to all the possibilities for making f equal 1. For example, the first term, $x_1 x_6 \bar{x}_2 x_5$, in Eq. 33.1 becomes 1 for the combination of variables $x_1 = x_6 = x_5 = 1$ and $x_2 = 0$. Correspondingly, the two terminals a and b of the network in Figure 33.5 are connected for this combination.

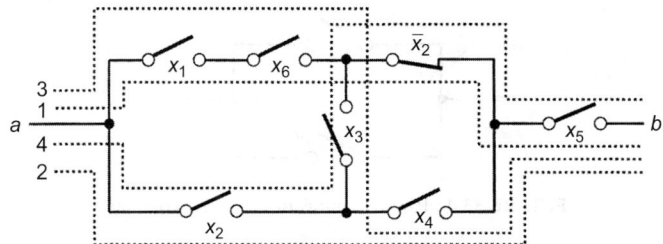

FIGURE 33.5 Tie sets of non-series-parallel network.

The second general procedure is given after the following definition.

Definition 33.2: Consider a set of contacts that satisfy the following conditions:

1. If all of the contacts in this set are opened simultaneously (ignoring functional relationship among contacts; in other words, even if two contacts, x and $\bar{x}$, are included in this set, it is assumed that contacts x and $\bar{x}$ can be opened simultaneously), the entire network is split into exactly two isolated subnetworks, one containing terminal a and the other containing b.

2. If any of the contacts are closed, the two subnetworks can be connected.

Then, the literals that represent these contacts are called a **cut set** of this network. ☐

As an example, let us find all the cut sets of the network in Figure 33.5, which is reproduced in Figure 33.6. First, let us open contacts x_1 and x_2 simultaneously, as shown in Figure 33.6(a). Then terminals a and b are completely disconnected (thus condition 1 of Definition 33.2 is satisfied). If either of contacts x_1 and x_2 is closed, the two terminals a and b can be connected by closing the remaining contacts (thus, condition 2 of Definition 33.2 is satisfied). We have all the cut sets shown in the following list and also in Figure 33.6(b):

For cut 1: x_1, x_2
For cut 2: x_6, x_2
For cut 3: x_1, x_3, x_4
For cut 4: x_6, x_3, x_4
For cut 5: $\bar{x}_2, x_3, x_2$
For cut 6: $\bar{x}_2, x_4$
For cut 7: x_5

Procedure 33.2: Derivation of the Transmission of Relay-Contact Network by Cut Sets

Find all the cut sets of a network. Form the disjunction of all literals in each cut set. Then the product of all these disjunctions yields the transmission of the given network. ☐

On the basis of the cut sets in the network of Figure 33.5 derived above, we get

$$f = (x_1 \vee x_2)(x_6 \vee x_2)(x_1 \vee x_3 \vee x_4)(x_6 \vee x_3 \vee x_4)(\bar{x}_2 \vee x_3 \vee x_2)(\bar{x}_2 \vee x_4)(x_5) \tag{33.2}$$

(a) Cut set 1   (b) All cut sets

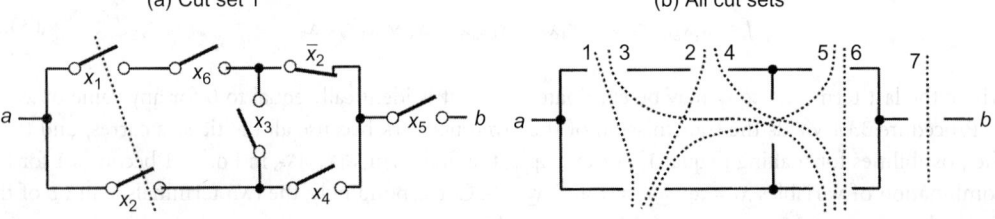

FIGURE 33.6 Cut sets of the network in Figure 33.5.

This expression looks different from Eq. 33.1, but they are equivalent, since we can get identical truth tables for both expressions.

Procedure 33.2 yields the transmission of a relay contact network because all the cut sets correspond to all possible ways to disconnect two terminals, *a* and *b*, of a network; that is, all possibilities of making *f* equal 0. Any way to disconnect *a* and *b* which is not a cut set constitutes some cut set plus additional unnecessary open contacts, as can easily be seen.

The disjunction inside each pair of parentheses in Eq. 33.2 corresponds to a different cut set. A disjunction that contains the two different literals of any variable (e.g., $(\bar{x}_2 \vee x_3 \vee x_2)$ in Eq. 33.2 contains two literals, $\bar{x}_2$ and x_2, of the variable x_2) is identically equal to 1 and is insignificant in multiplying out *f*. Therefore, every cut set that contains the two literals of some variable need not be considered in Procedure 33.2.

33.3   Transistor Circuits

Bipolar transistor and MOSFET are currently the two most important types of transistors for integrated circuit chips, although MOSFET is becoming increasingly popular [2]. A transistor is made of pure silicon that contains a trace of impurities (i.e., n-type silicon or p-type silicon). When a larger amount of impurity than standard is added, we have n^+- and p^+-type silicon. When less, we have n^-- and p^--type silicon. A bipolar transistor has a structure of **n-type region** (or simply n-region) consisting of n-type silicon and **p-type region** (or simply p-region) consisting of p-type silicon, as illustrated in Figure 33.7, different from that of MOSFET illustrated in Figure 33.10.

33.3.1   Bipolar Transistors

An implementation example of an n-p-n bipolar transistor, which has three electrodes (i.e., an emitter, a base, and a collector) is shown in Figure 33.7(a) along with its symbol in Figure 33.7(b). A p-n-p transistor has the same structure except p-type regions and n-type regions exchanged (n^+- and p^+-type regions also exchanged).

Suppose that an n-p-n transistor is connected to a power supply with a resistor and to the ground, as shown in Figure 33.8(a). When the input voltage v_i increases, the collector current i_c gradually increases, as shown in Figure 33.8(b). (Actually, i_c is 0 until v_i reaches about 0.6 V. Then i_c gradually increases and then starts to saturate.) As i_c increases, the output voltage v_0 decreases from 5 V to 0.3 V or less because of the voltage difference across the resistor *R*, as shown in Figure 33.8(c). Therefore, when the input v_i is a high voltage (about 5 V), the output v_0 is a low voltage (about 0.3 V), and when v_i is a low voltage (about 0.3 V),

(a) Structure of n-p-n transistor

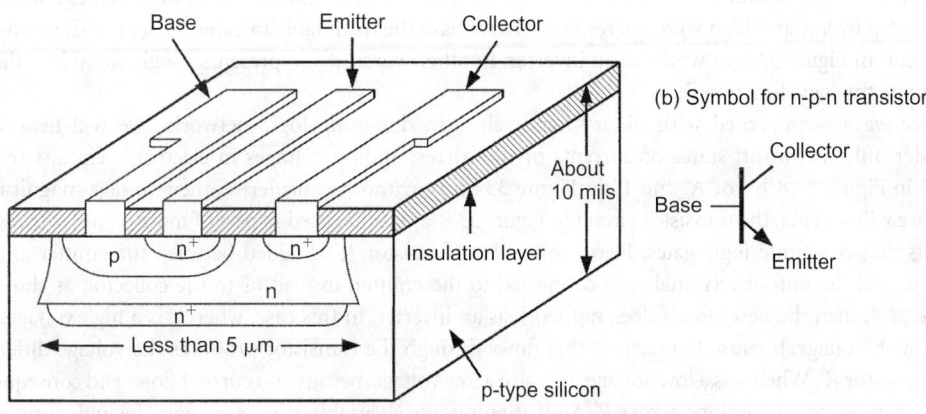

(b) Symbol for n-p-n transistor

FIGURE 33.7   n-p-n transistor.

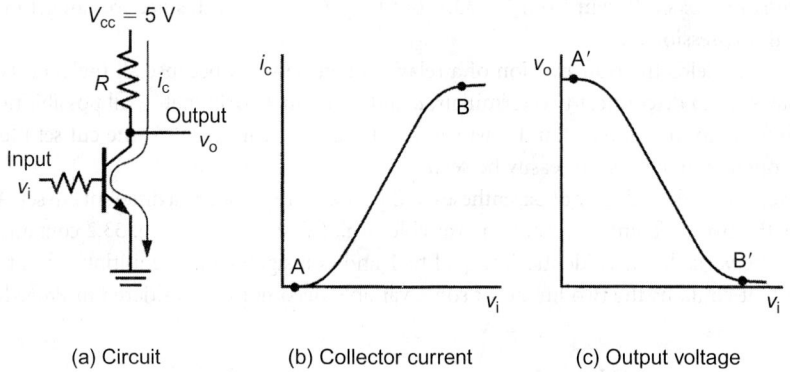

(a) Circuit   (b) Collector current   (c) Output voltage

FIGURE 33.8 Inverter circuit.

TABLE 33.3 Input-Output Relations of the Inverter in Figure 33.8(a)

(a) Voltage Required		(b) Truth Table	
Input v_i	Output v_0	v_i	v_0
Low voltage	High voltage	0	1
High voltage	Low voltage	1	0

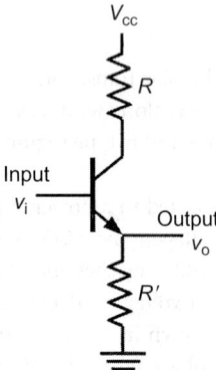

FIGURE 33.9 Emitter follower.

v_0 is a high voltage (about 5 V). This is illustrated in Table 33.3(a). Thus, if binary logic values 0 and 1 are represented by low and high voltages, respectively, we have the truth table in Table 33.3(b). This means that the circuit in Figure 33.8(a) works as an **inverter**. In other words, if v_i represents a logic variable x, then v_0 represents the logic function $\bar{x}$.

Since we are concerned with binary logic values in designing logic networks, we will henceforth consider only the on-off states of currents or the corresponding voltages in electronic circuits (e.g., A and B in Figure 33.8(b), or A′ and B′ in Figure 33.8(c)), without considering their voltage magnitudes.

As we will see later, the transistor circuit in Figure 33.8(a) is often used as part of more complex transistor circuits that constitute logic gates. Here, notice that if resistor R' is added between the emitter and the ground, and the output terminal v_0 is connected to the emitter, instead of to the collector, as shown in Figure 33.9, then the new circuit does not work as an inverter. In this case, when v_i is a high voltage, v_0 is also a high voltage, because the current that flows through the transistor produces the voltage difference across resistor R'. When v_i is a low voltage, v_0 is also a low voltage, because no current flows and consequently no voltage difference develops across R'. So if v_i represents a variable x, v_0 represents the logic function x, and no logic operation is performed. The transistor circuit in Figure 33.9, which is often called an **emitter**

follower, works as a current amplifier. This circuit is used often as part of other circuits to supply a large output current by connecting the collector of the transistor directly to the V_{cc} without R.

A logic gate based on bipolar transistors generally consists of many transistors which are connected in a more complex manner than Figure 33.8 or 33.9, and realizes a more complex logic function than $\bar{x}$. (See Chapter 38 on ECL.)

33.3.2   MOSFET (Metal-Oxide Semiconductor Field Effect Transistor)

In integrated circuit chips, two types of MOSFETs are usually used; that is, **n-channel enhancement-mode MOSFET** (or abbreviated as n-channel enhancement MOS, or enhancement nMOS) and **p-channel enhancement-mode MOSFET** (or abbreviated as p-channel enhancement MOS, or enhancement pMOS). The structure of the former is illustrated in Figure 33.10(a). They are expressed by the symbols shown in Figure 33.10 (b) and (c), respectively. Each of them has three terminals: **gate, source,** and **drain.** In Figure 33.10(a), the gate realized with metal is shown for the sake of simplicity, but a more complex structure, called **silicon-gate MOSFET,** is now far more widely used. **The "gate" in Figure 33.10 should not be confused with "logic gates."** The thin area underneath the gate between the source and drain regions in Figure 33.10(a) is called a **channel,** where a current flows whenever conductive.

Suppose that the source of an n-channel enhancement-mode MOSFET is grounded and the drain is connected to the power supply of 3.3 V through resistor R, as illustrated in Figure 33.11(a). When the

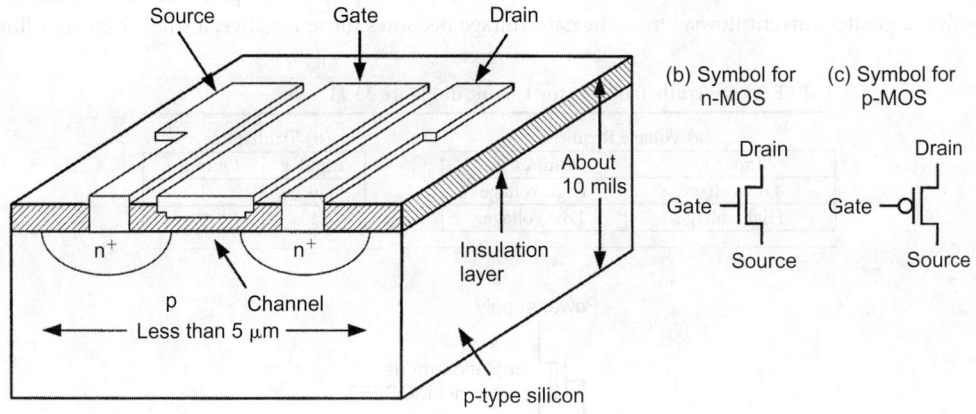

FIGURE 33.10   MOSFET.

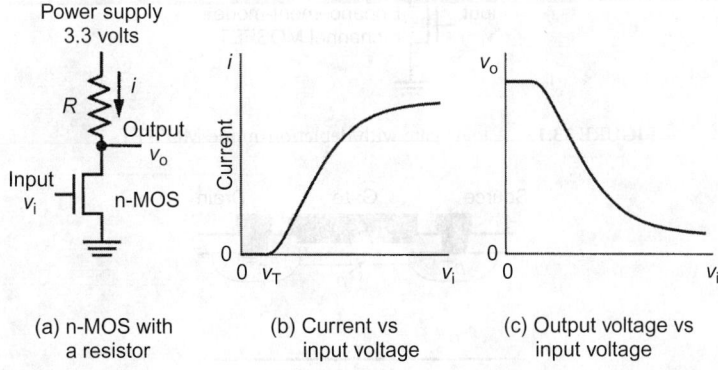

FIGURE 33.11   Inverter.

input voltage v_i increases from 0 V, the current i, which flows from the power supply to the ground through R and the MOSFET, increases as shown in Figure 33.11(b), but for v_i smaller than the threshold voltage V_T, essentially no current flows. Then because of the voltage drop across R, the output voltage v_0, decreases, as shown in Figure 33.11(c). Since we use binary logic, we need to use only two different voltage values, say 0.2 and 3.3 V. If v_i is 0.2 V, no current flows from the power supply to the ground through the MOSFET and v_0 is 3.3 V. If v_i is 3.3 V, the MOSFET becomes conductive and a current flows. V_0 is 0.2 V because of the voltage drop across R. Thus, if v_i is a low voltage (0.2 V), v_0 is a high voltage (3.3 V); and if v_i is a high voltage, v_0 is a low voltage, as shown in Table 33.4(a). If low and high voltages represent logic values 0 and 1, respectively, in other words, if we use **positive logic**, Table 33.4(a) is converted to the truth table in Table 33.4(b). (When low and high voltages represent 1 and 0, respectively, this is said to be in **negative logic**.) Thus, if v_i represents logic variable x, output v_0 represents function $\bar{x}$. This means that the electronic circuit in Figure 33.11(a) works as an **inverter**.

Resistor R shown in Figure 33.11 occupies a large area, so it is usually replaced by a MOSFET, called an n-channel depletion-mode MOSFET, as illustrated in Figure 33.12, where the depletion-mode MOSFET is denoted by the MOSFET symbol with double lines. Notice that **the gate of this depletion-mode MOSFET is connected to the output terminal instead of the power supply**, the logic gate with depletion-mode MOSFET replacing the resistor work in the same manner as before. Logic gates with depletion-mode MOSFETs work faster and are more immune to noise.

The n-channel **depletion-mode MOSFET** is different from the n-channel enhancement-mode MOSFET in having a thin n-type silicon layer embedded underneath the gate, as illustrated in Figure 33.13. When a positive voltage is applied at the drain against the source, a current flows through this thin n-type silicon layer even if the voltage at the gate is 0 V (against the source). As the gate voltage becomes more positive, a greater current flows. Or, as the gate voltage becomes more negative, a smaller current flows.

TABLE 33.4   Truth Table for the Circuit in Figure 33.11

(a) Voltage Required		(b) Truth Table	
Input v_i	Output v_0	Input x	Output f
Low voltage	High voltage	0	1
High voltage	Low voltage	1	0

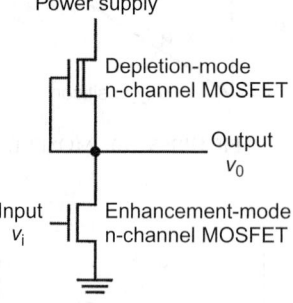

FIGURE 33.12   A logic gate with depletion-mode MOSFET.

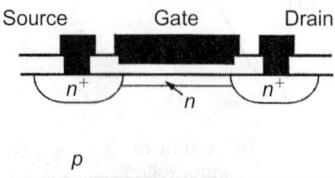

FIGURE 33.13   *n*-Channel depletion-mode MOSFET.

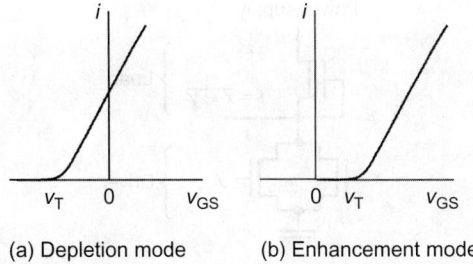

(a) Depletion mode   **(b) Enhancement mode**

FIGURE 33.14 Shift of threshold voltage V_T (V_{GS} is a voltabe between gate and source).

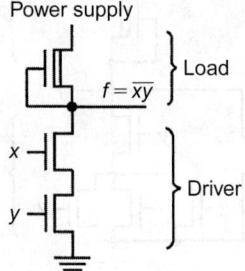

FIGURE 33.15 Logic gate for the NAND function.

TABLE 33.5 Truth Table for the Circuit in Figure 33.15

(a) Voltage Relation			(b) Truth Table		
x	y	f	x	y	f
Low	Low	High	0	0	1
Low	High	High	0	1	1
High	Low	High	1	0	1
High	High	Low	1	1	0

If the gate voltage decreases beyond threshold voltage V_T, no current flows. This relationship, called a **transfer curve**, between the gate voltage V_{GS} (against the source) and the current i is shown in Figure 33.14(a), as compared with that for the n-channel enhancement-mode MOSFET shown in Figure 33.14(b).

By connecting many n-channel enhancement MOSFETs, we can realize any **negative function**, i.e., the complement of a sum-of-products where only non-complemented literals are used ($\overline{x \vee yz}$ is an example of negative function). For example, if we connect three MOSFETs in series, including the one for resistor replacement, as shown in Figure 33.15, the output f realizes the NAND function of variables x and y. Only when both inputs x and y have high voltages, two MOSFETs for x and y become conductive and a current flows through them. Then the output voltage is low. Otherwise, at least one of them is non-conductive and no current flows. Then the output voltage is high. This relationship is shown in Table 33.5(a). In positive logic, this is converted to the truth table 33.5(b), concluding that the circuit represents $\overline{xy}$ which is called the NAND function. Figure 33.16 shows a logic circuit for $\overline{x \vee y}$ (called the NOR function) by connecting MOSFETs in parallel. A more complex example is shown in Figure 33.17.

The MOSFET that is connected between the power supply and the output terminal is called a **load** or **load MOSFET** in each of Figure 33.15 through 33.17. Other MOSFETs that are directly involved in logic operations are called a **driver** or **driver MOSFETs** in each of these circuits.

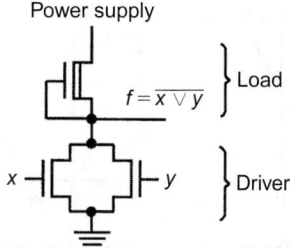

FIGURE 33.16   Logic gate for the NOR function.

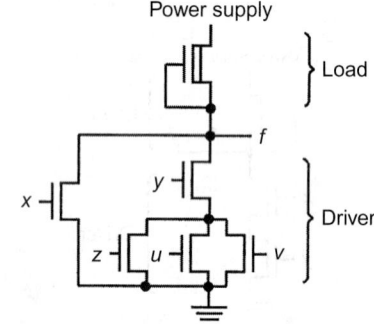

FIGURE 33.17   A logic gate with many MOSFETs.

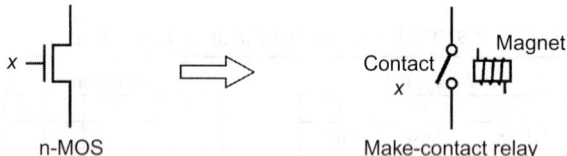

FIGURE 33.18   Analogy between n-MOS and a make-contact relay.

Procedure 8.3: Calculation of the Logic Function of a MOS Logic Gate

The logic function f for the output of each of these MOS logic gates can be obtained as follows.

1. Calculate the transmission of the driver, regarding each n-channel MOSFET as a make-contact of relay, as illustrated in Figure 33.18. When $x = 1$, a current flows through the magnet in the make-contact relay in Figure 33.18 and the contact x is closed and becomes conductive, whereas n-MOS becomes conductive when $x = 1$, i.e., input x of nMOS is a high voltage.
2. Complement it.   □

For example, the transmission of the driver in Figure 33.16 is $x \vee y$. Then by complementing it, we have the output function $f = \overline{x \vee y}$. The output function of a more complex logic gate, such as Figure 33.18, can be calculated in the same manner. Thus, **a MOS circuit expresses a negative function with respect to input variables connected to driver MOSFETs.**

33.3.3   Difference in the Behavior of n-MOS and p-MOS Logic Gates

As illustrated in Figure 33.19, an n-MOS logic gate behaves differently from a p-MOS logic gate. In the case of an n-MOS logic gate which consists of all nMOSFETs, illustrated in Figure 33.19(a), the power supply of the n-MOS logic gate must be positive voltage, say +3.3 V, whereas the power supply of the p-MOS logic gate which consists of all pMOSFETs, illustrated in Figure 33.19(b), must be a negative

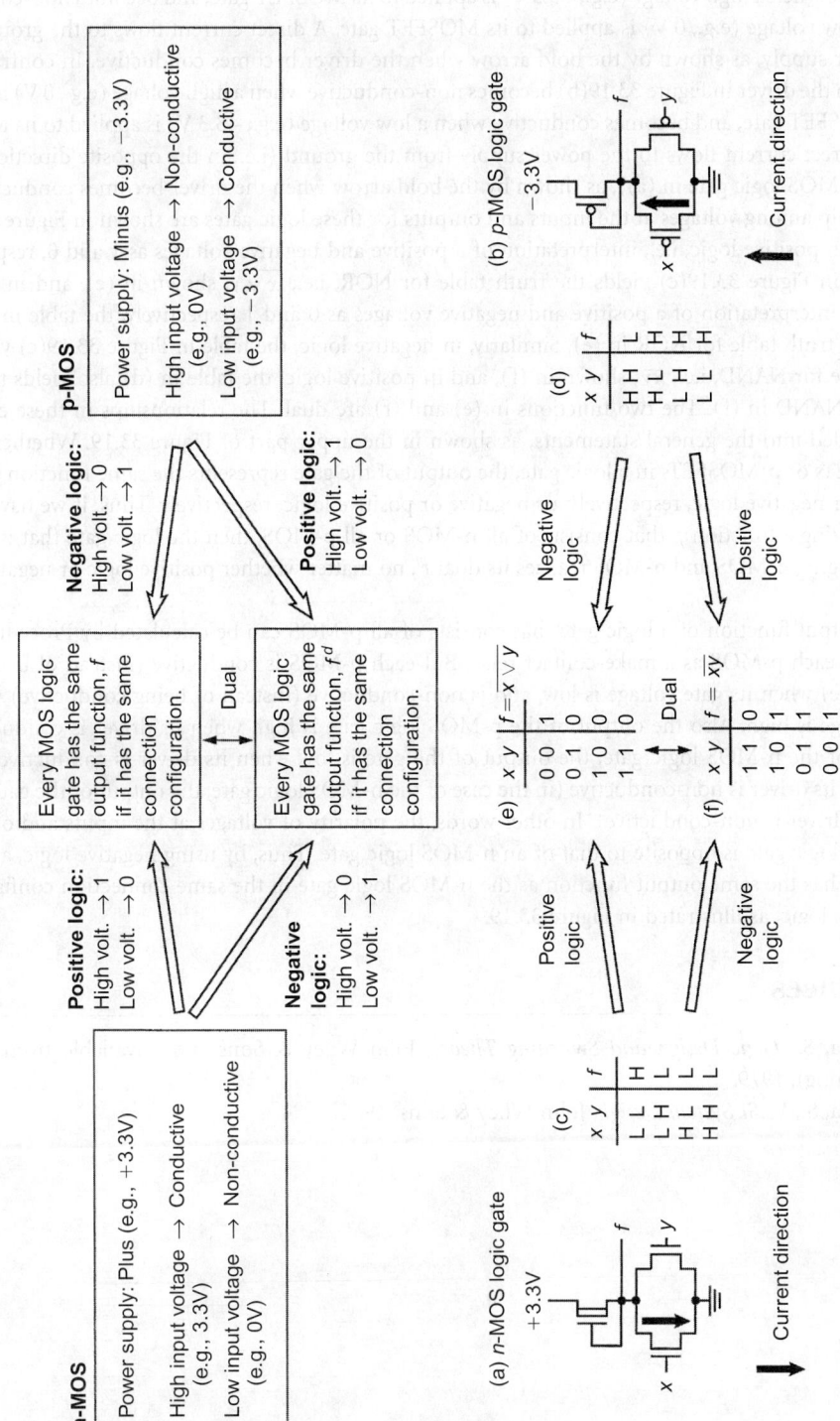

FIGURE 33.19 Behavior of logic gates with n-MOS and p-MOS.

n-MOS

Power supply: Plus (e.g., +3.3V)

High input voltage → Conductive (e.g., 3.3V)

Low input voltage → Non-conductive (e.g., 0V)

Positive logic:
High volt. → 1
Low volt. → 0

Negative logic:
High volt. → 0
Low volt. → 1

Every MOS logic gate has the same output function, f if it has the same connection configuration.

Dual

Every MOS logic gate has the same output function, f^d if it has the same connection configuration.

p-MOS

Power supply: Minus (e.g., −3.3V)

High input voltage → Non-conductive (e.g., 0V)

Low input voltage → Conductive (e.g., −3.3V)

Negative logic:
High volt. → 0
Low volt. → 1

Positive logic:
High volt. → 1
Low volt. → 0

(a) n-MOS logic gate

+3.3V

Current direction

(b) p-MOS logic gate

−3.3V

Current direction

(c)

x y	f
L L	H
L H	L
H L	L
H H	L

(d)

x y	f
H H	L
H L	H
L H	H
L L	H

(e) $f = \overline{x \lor y}$

x y	f
0 0	1
0 1	0
1 0	0
1 1	0

Dual

(f) $f = \overline{xy}$

x y	f
1 1	0
1 0	1
0 1	1
0 0	1

Positive logic

Negative logic

Negative logic

Positive logic

voltage. Otherwise, these logic gates do not work. Each n-MOS in the driver in Figure 33.19(a) becomes conductive when a high voltage (e.g., +3.3 V) is applied to its MOSFET gate, and becomes non-conductive when a low voltage (e.g., 0 V) is applied to its MOSFET gate. A direct current flows to the ground from the power supply, as shown by the bold arrow when the driver becomes conductive. In contrast, each p-MOS in the driver in Figure 33.19(b) becomes non-conductive when a high voltage (e.g., 0 V) is applied to its MOSFET gate, and becomes conductive when a low voltage (e.g., −3.3 V) is applied to its MOSFET gate. A direct current flows to the power supply from the ground (i.e., in the opposite direction to the case of n-MOS logic gate in (a)), as shown by the bold arrow when the driver becomes conductive. The relationship among voltages at the inputs and outputs for these logic gates are shown in Figure 33.19(c) and (d). In positive logic, i.e., interpretation of a positive and negative voltages as 1 and 0, respectively, the table in Figure 33.19(c) yields the truth table for NOR, i.e., $\overline{x \vee y}$, shown in (e), and in negative logic, i.e., interpretation of a positive and negative voltages as 0 and 1 respectively, the table in (d) also yields the truth table for NOR in (e). Similarly, in negative logic, the table in Figure 33.19(c) yields the truth table for NAND, i.e., $\overline{xy}$, shown in (f), and in positive logic, the table in (d) also yields the truth table for NAND in (f). The two functions in (e) and (f) are dual. The relationships in these examples are extended into the general statements, as shown in the upper part of Figure 33.19. Whether we use n-MOSFETs or p-MOSFETs in a logic gate, the output of the gate represents the same function by using positive or negative logic, respectively, or negative or positive logic, respectively. Thus, if we have a logic gate, realizing a function f, that consists of all n-MOS or all p-MOS, then the logic gate that is derived by exchanging n-MOS and p-MOS realizes its dual f^d, no matter whether positive logic or negative logic is used.

The output function of a logic gate that consists of all p-MOS can be calculated by Procedure 33.3, regarding each p-MOS as a make-contact relay. But each p-MOS is conductive (instead of being non-conductive) when its gate voltage is low, and is non-conductive (instead of being conductive) when its gate voltage is high. Also the output of the p-MOS logic gate is high when its driver is conductive (in the case of the n-MOS logic gate, the output of the gate is low when its driver is conductive) and is low when its driver is non-conductive (in the case of the n-MOS logic gate, the output of the gate is high when its driver is non-conductive). In other words, the polarity of voltages at the inputs and output of a p-MOS logic gate is opposite to that of an n-MOS logic gate. Thus, by using negative logic, a p-MOS logic gate has the same output function as the n-MOS logic gate in the same connection configuration in positive logic, as illustrated in Figure 33.19.

References

1. Muroga, S., *Logic Design and Switching Theory*, John Wiley & Sons (now available from Krieger Publishing), 1979.
2. Muroga, S., *VLSI System Design*, John Wiley & Sons, 1982.

34

Logic Synthesis with NAND (or NOR) Gates in Multi-Levels

Saburo Muroga

*University of Illinois
at Urbana-Champaign*

CONTENTS

34.1 Logic Synthesis with NAND (or NOR) Gates

In the previous sections, we have discussed the design of a two-level network with AND and OR gates in **double-rail input logic** (i.e., both x_i and $\bar{x}_i$ for each x_i are available as network inputs) that has a minimum number of gates as the primary objective and a minimum number of connections as the secondary objective. If a network need not be in two levels, we may be able to further reduce the number of gates or connections, but there is no known simple systematic design procedure for this purpose, whether tabular, algebraic, or graphical, that guarantees the minimality of the network. (The integer programming logic design method [4,6] can do this but is complex, requiring long processing time.) But when multi-level minimal networks with NAND gates only (or NOR gates only) are to be designed, there is a method called the **map-factoring method** to design a logic network based on a Karnaugh map.

In designing a logic network in **single-rail input logic** (i.e., only one of x_i and $\bar{x}_i$ for each x_i is available as a network input) with the map-factoring method, it is less easy to see the minimality of the number of gates, although when two-level minimal networks with NAND gates in double-rail input logic are to be designed, it is as easy to see the minimality as two-level minimal networks with AND and OR gates in double-rail input logic on Karnaugh maps. By using designer's intuition based on the pictorial nature of a Karnaugh map, at least reasonably good networks in multi-levels can be derived after trial-and-error efforts. As a matter of fact, minimal networks can sometimes be obtained, although the method does not enable us to prove their minimality. However, if we are satisfied with reasonably good networks, the map-factoring method is useful for manual design. It is actually an extension of the Karnaugh map method for minimal two-level networks with AND and OR gates discussed so far, with far greater flexibility: by the map-factoring method, we can design not only in two-levels but also in multi-levels in single-rail or double-rail input logic, and also two-level minimal networks with NAND gates in double-rail input logic that can be designed by the map-factoring method are essentially two-level minimal networks with AND and OR gates, as will be discussed later. The map-factoring method, which was first described in Chapter 6 of Ref. [3] for single-rail input logic as discussed later in this chapter, is extended here with minor modification [5].

Logic networks with NOR gates only or NAND gates only are useful in some cases, such as gate arrays (to be described in Chapter 46) because the simple connection configuration of MOSFETs in each of these gates makes the area small with high speed.

In designing a logic network in multi-levels, we usually minimize the number of logic gates as the primary objective and then the number of connections as the secondary objective. This is because the design with the minimization of the number of logic gates as the primary objective and then the number of connections as the secondary objective is easier than the minimization of the number of connections as the primary objective and then the number of logic gates as the secondary objective, although the minimization of the number of connections, or the lengths of connections (which is far more difficult to minimize), is important because connections occupy significantly large areas on an integrated circuit chip. But judging the results by an experiment by the integer programming logic design method in limited scale, we have the same or nearly same minimal logic networks by either approach [7].

34.2 Design of NAND (or NOR) Networks in Double-Rail Input Logic by the Map-Factoring Method

NAND gates and NOR gates, which are realized with MOSFETs, are often used in realizing integrated circuit chips, although logic gates that express negative functions which are more complex than NAND or NOR are generally used. (A "negative function" is the complement of a disjunctive form of non-complemented variables. An example is $\overline{xy \vee z}$.) NAND gates are probably more often used than other types of logic gates realizing negative functions because a NAND gate in CMOS has a simple connection configuration of MOSFETs and is fast.

Let us consider representing a NAND gate on a Karnaugh map. The output of the NAND gate with inputs x, $\overline{y}$, and z shown in Figure 34.1(a) can be expressed as the loop for $x\overline{y}z$ consisting of only 0-cells on the map in Figure 34.1(b). (Recall that this loop represents product $x\overline{y}z$ on an ordinary Karnaugh map in the previous chapters.) In this case, it is important to note that only 0-cells are contained inside the loop

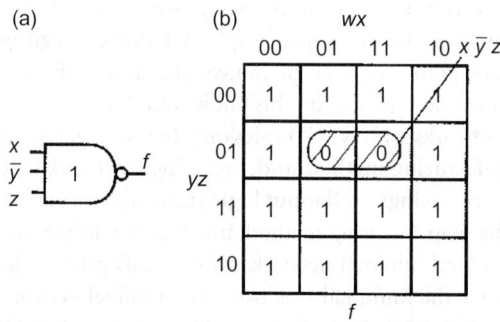

FIGURE 34.1 Representation of a NAND gate on a map.

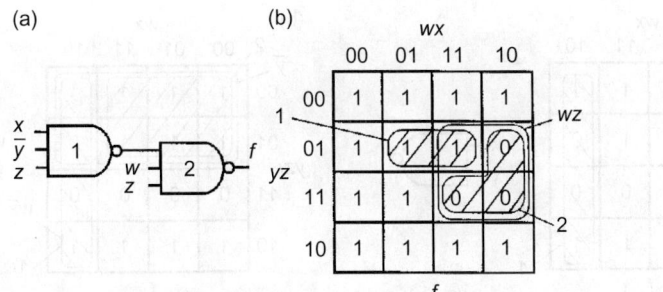

FIGURE 34.2 Representation of a network of two NAND gates.

and all the 1-cells are outside. This is because the output of this NAND gate is the complement of the AND operation of inputs, x, $\bar{y}$, and z (i.e., $\overline{x\bar{y}z}$). Thus, if all the 0-cells in the map can be encircled by a single rectangular loop representing a product of some literals (i.e., the number of 0-cells constituting this loop is 2^i where i is a non-negative integer), the map represents the output of a NAND gate whose inputs are the literals in the product represented by the loop. The value of f is 0 for $x = \bar{y} = z = 1$ (i.e., $x = z = 1$ and $y = 0$) in both Figures 34.1(a) and (b). The value of f is 1 if at least one of x, $\bar{y}$, and z is 0.

Next, connect a new NAND gate (numbered gate 2) to the output of the above NAND gate (i.e., gate 1) and also connect inputs w and z to the new gate, as shown in Figure 34.2(a). Unlike the case of gate 1 explained in Figure 34.1, the output f of this new NAND gate is not expressed by the entire loop for wz in Figure 34.2(b), but is expressed by the portion of the loop (i.e., only 0-cells inside the loop for wz) because the input of gate 2 from gate 1 becomes 0 for the combination, $w = x = z = 1$ and $y = 0$, and consequently f becomes 1 for this combination (if there were no connection from gate 1, the rectangular loop representing wz contains only 0-cells, like gate 1 explained in Figure 34.1). This may be interpreted as the rectangular loop representing wz for gate 2 being **inhibited** by the loop for gate 1, as shown in Figure 34.2(b). The remainder of the loop (i.e., all 0-cells, which is actually all the 0-cells throughout the map) is encircled by a loop and is shaded. This shaded loop represents the output of gate 2 and is said to be **associated** with the output of gate 2. In other words, the loop (labeled wz) which represents wz denotes NAND gate 2, whereas the shaded loop (labeled 2) inside this loop in Figure 34.2(b) denotes the output function of gate 2. Notice that the entire loop for gate 1 is shaded to represent the output function of gate 1, because gate 1 has no inputs from other gates (i.e., gate 1 is inhibited by no other shaded loops) and consequently the loop representing gate 1 coincides with the shaded loop representing the output of gate 1 (i.e., the shaded loop associated with the output of gate 1).

Now let us state a formal procedure to design a NAND network on a Karnaugh map.

Procedure 34.1: The Map-Factoring Method: Design of a Network in Double-Rail Input Logic with as few NAND Gates as Possible

1. Make the first rectangular loop of 2^i cells. This loop may contain 1-cells, 0-cells, d-cells, or a mixture. Draw a NAND gate corresponding to this loop. As inputs to this gate, connect all the literals in the product that this loop represents. Shade the entirety of this loop.

For example, let us synthesize a network for $f = \overline{wy} \vee x\bar{y} \vee \bar{z}$ shown in the Karnaugh map in Figure 34.3(a). Let us make the first rectangular loop as shown in Figure 34.3(a). (Of course, the first loop can be chosen elsewhere.) This loop represents product $w\bar{x}$. Draw gate 1, corresponding to this loop and connect inputs w and $\bar{x}$ to this gate. Shade the entirety of this loop because this gate has no input from another gate and consequently the entirety of this loop is associated with the output function of gate 1.

2. Make a rectangular loop consisting of 2^i cells, encircling 1-cells, 0-cells, d-cells, or a mixture. Draw a NAND gate corresponding to this loop. To this gate, connect literals in the product that this loop represents.

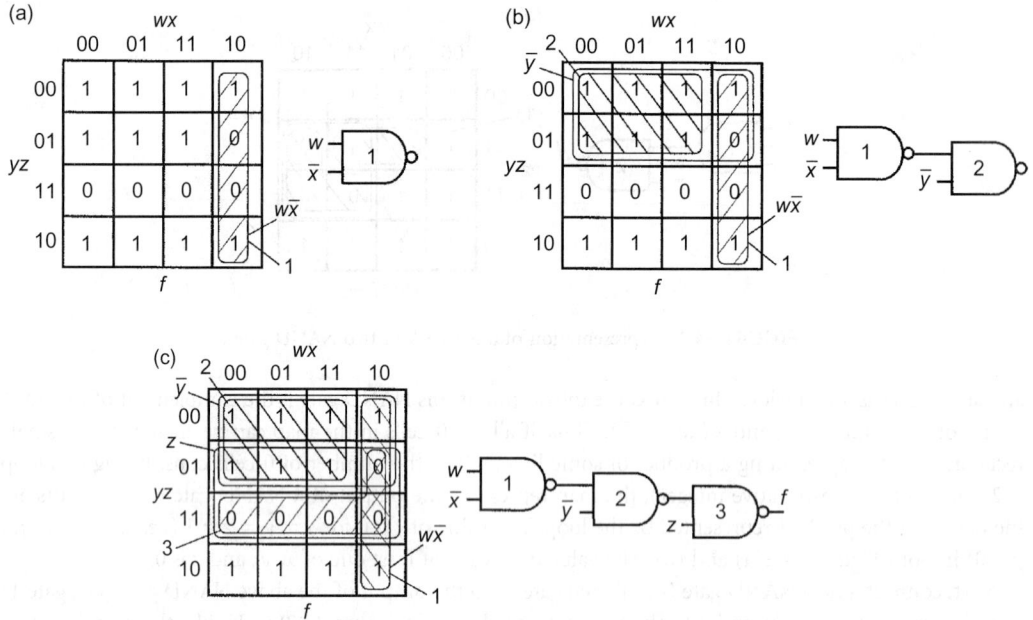

FIGURE 34.3   Example for Procedure 34.1.

Up to this point, this step is identical to Step 1. Now, to this new gate, we further connect the outputs of some or all of the gates already drawn, if we choose to do so. There are the following possibilities:

   a. If we choose not to connect any previous gate to the new gate, the new loop is entirely shaded.
   b. If we choose to connect some or all of the previously drawn gates to the new gate, encircle and shade the area inside the new loop, excluding the shaded loops of the previously drawn gates connected to the new gate. The shaded loop thus formed is associated with the output of this new gate.

Let us continue our example of Figure 34.3. Let us make the loop labeled $\bar{y}$ shown in Figure 34.3(b) as a next rectangular loop consisting of 2^i cells. Draw the corresponding gate 2. Connect input $\bar{y}$ to gate 2 because this loop represents $\bar{y}$. If we choose to connect the output of gate 1 also to gate 2 (i.e., by choosing the case b above), the shaded loop labeled 2 in Figure 34.3(b) represents the output of gate 2.

   3. Repeat Step 2 until the following condition is satisfied:

Termination condition: When a new loop and the corresponding new gate are introduced, all the 0-cells on the entire map and possibly some d-cells constitute the shaded loop associated with the output of the new gate.

Continuing our example, let us make the loop labeled z as a next rectangular loop consisting of 2^i, as shown in Figure 34.3(c). Draw the corresponding gate 3 with input z connected. Choosing the case b in Step 2, connect the output of gate 2 as input of gate 3. (In this case b, we have three choices; that is, connection of the output of gate 1 only, connection of the output of gate 2 only, and connection of both outputs of gates 1 and 2. Let us take the second choice now.) Then, the output of gate 3 is expressed by the shaded loop labeled 3 in Figure 34.3(c). Now, the termination condition is satisfied: all the 0-cells on the entire map constitute the shaded loop associated with the output of new gate 3. Thus, a network for the given function f has been obtained in Figure 34.3(c).

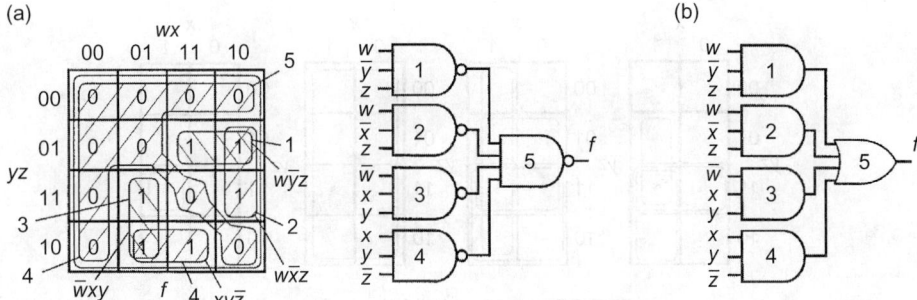

FIGURE 34.4 Network in two levels.

For the sake of simplicity, the example does not contain d-cells. Even when a map contains d-cells, that is, cells for don't-care conditions, the map-factoring method, Procedure 34.1, can be easily used by appropriately interpreting each d-cell as a 0-cell or a 1-cell only when we examine the termination condition in Step 3. $\square$

Notice that we can choose different loops (including the first one in Step 1) in each step, leading to different final networks.

34.2.1 Networks with AND and OR Gates in Two Levels, as a Special Case

If we circle only 1-cells possibly along with some d-cells but without any 0-cells in each step, we can derive a logic network with NAND gates in two levels, as illustrated in Figure 34.4(a). This can be easily converted to the network with AND and OR gates in two levels shown in Figure 34.4(b).

34.2.2 Consideration of Restrictions Such as Maximum Fan-in

If the restriction of maximum fan-in or fan-out is imposed, loops and connections must be chosen so as not to violate it. With the map-factoring method, it is easy to take such a restriction into consideration. Also, the maximum number of levels in a network can be easily controlled.

34.2.3 The Map-Factoring Method for NOR Network

A minimal network of NOR gates for a given function f can be designed by the following approach.

Use the map-factoring method to derive a minimal network of NAND gates for f^d, the dual of the given function f. Then replace NAND gates in the network with NOR gates. The result will be a minimal network of NOR gates for f.

34.3 Design of NAND (or NOR) Networks in Single-Rail Input Logic

In Section 34.2, we discussed the design of a multi-level NAND networks in double-rail input logic, using the map-factoring method. Now let us discuss the design of a multi-level NAND network in single-rail input logic (i.e., no complemented variables are available as inputs to the network) using the map-factoring method.

First let us define **permissible loops** on a Karnaugh map. A permissible loop is a rectangle consisting of cells that contains the cell whose coordinates are variable values of all 1's (i.e., the cell marked with the asterisk in each map in Figure 34.5), where i is one of 0, 1, 2, ... In other words, **a permissible loop must contain the particular cell denoted by the asterisk in each map in Figure 34.5, where this permissible loop consists of 1-cells, 0-cells, d-cells (i.e., don't-care cells), or a mixture**. All permissible loops for three variables are shown in Figure 34.5.

In the following, let us describe the map-factoring method. The procedure is the same as Procedure 34.1 except the use of permissible loop, instead of rectangular loop of 2^i-cells at any place on the map, for representing a gate.

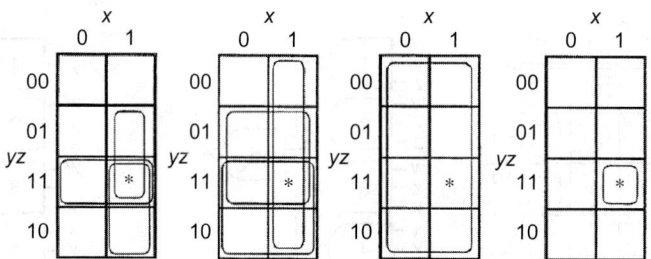

FIGURE 34.5 Permissible loops for three variables.

Procedure 34.2:   The Map-Factoring Method: Design of a Network in Single-Rail Input with as Few NAND Gates as Possible.

We want to design a network with as few NAND gates as possible, under **the assumption that non-complemented variables but no complemented variables are available as network inputs.**

1. Make the first permissible loop, encircling 1-cells, 0-cells, d-cells, or a mixture of them. Draw a NAND gate corresponding to this loop. As inputs to this gate, connect all the literals in the product that this loop represents. Shade the entirety of this loop.

For example, when a function $f = x \vee y \vee \overline{z}$ is given, let us choose the first permissible loop, as shown in Figure 34.6(a), although there is no reason why we should choose this particular loop. (There is no guiding principle for finding which loop should be the first permissible loop. Another loop could be better, but we cannot guess at this moment which loop is the best choice. Another possibility in choosing the first permissible loop will be shown later in Figure 34.7(a).) The loop we have chosen is labeled 1 and is entirely shaded. Then, NAND gate 1 is drawn. Since the loop represents x, we connect x to this gate.

2. Make a permissible loop, encircling 1-cells, 0-cells, d-cells, or mixture of them. Draw a NAND gate corresponding to this loop. To this new gate, connect literals in the product that this loop represents.

In the above example, a new permissible loop is chosen as shown in Figure 34.6(b), although there is no strong reason why we should choose this particular permissible loop. This loop represents product yz, so y and z are connected to the new gate, which is labeled 2.

To this new NAND gate, we further connect the outputs of some gates already drawn, if we choose to do so. There are the following possibilities.

   a. If we choose not to connect any previous gate to the new gate, the new permissible loop is entirely shaded.

If we prefer not to connect gate 1 to gate 2 in the above example, we get the network in Figure 34.6(b). Ignoring the shaded loop for gate 1, the new permissible loop is entirely shaded and is labeled 2, as shown in Figure 34.6(b).

   b. If we choose to connect some previously drawn gates to the new gate, encircle and shade the area inside the new permissible loop, excluding the shaded loops of the previously drawn gates which are connected to the new gate. In other words, the new permissible loop, except its portion inhibited by the shaded areas associated with the outputs of the connected previously drawn gates, is shaded.

In the above example, if we choose to connect gate 1 to gate 2, we obtain the network in Figure 34.6(b'). The portion of the new permissible loop that is not covered by the shaded loop associated with gate 1 is shaded and labeled 2.

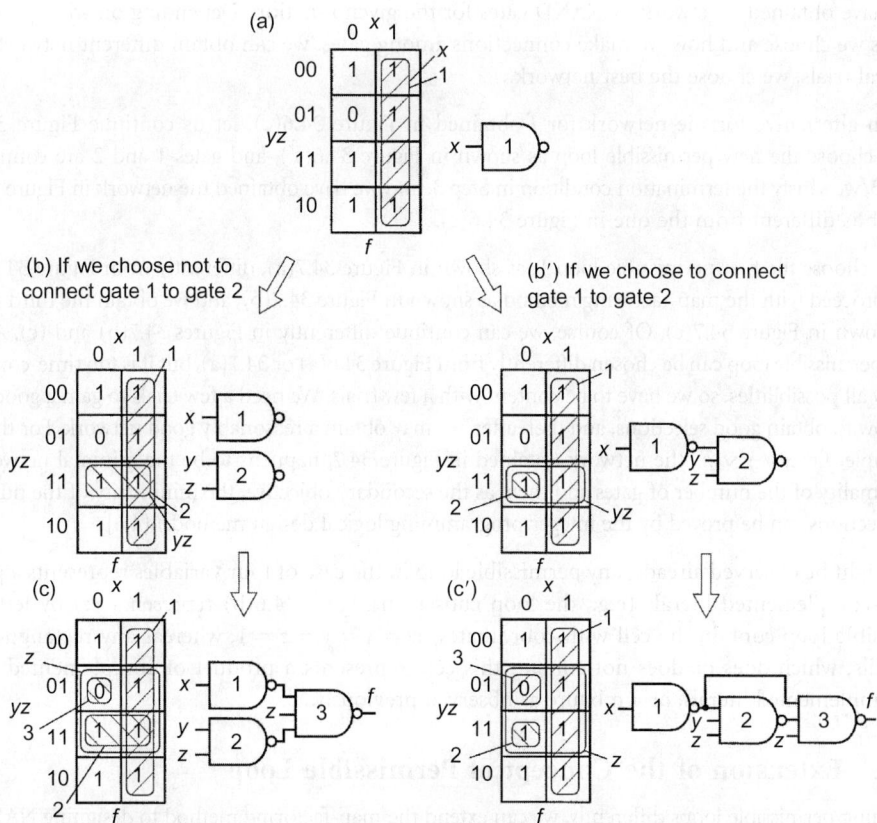

FIGURE 34.6 Map-factoring method applied for $f = x \vee y \vee \bar{z}$.

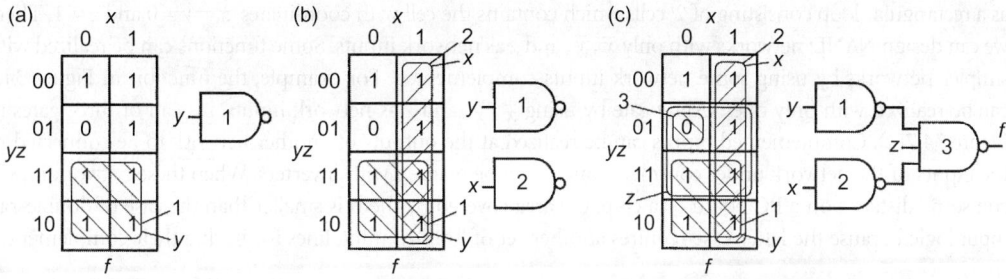

FIGURE 34.7 Network obtained by choosing permissible loops different from Figure 34.6.

3. Repeat Step 2 until the following condition is satisfied:

Termination condition: When a new permissible loop and the corresponding new gate are introduced, all the 0-cells on the entire map and possibly some d-cells constitute the shaded loop associated with the output of the new gate (i.e., the shaded loop associated with the output of the new gate contains all the 0-cells on the entire map, but no 1-cell).

Let us continue Figure 34.6(b). If we choose the new permissible loop as shown in Figure 34.6(c), and if we choose to connect gates 1 and 2 to the new gate 3, the above termination condition has been satisfied; in other words, all the 0-cells on the entire map constitute the shaded loop associated with the output of the new gate 3.

We have obtained a network of NAND gates for the given function. Depending on what permissible loops we choose and how we make connections among gates, we can obtain different networks. After several trials, we choose the best network.

As an alternative for the network for f obtained in Figure 34.6(c), let us continue Figure 34.6(b′). If we choose the new permissible loop as shown in Figure 34.6(c′), and gates 1 and 2 are connected to gate 3, we satisfy the termination condition in Step 3, and we have obtained the network in Figure 34.6(c′), which is different from the one in Figure 34.6(c).

If we choose the first permissible loop 1, as shown in Figure 34.7(a), differently from Figure 34.6(a), we can proceed with the map-factoring method as shown in Figure 34.7(b), and we obtain the third network as shown in Figure 34.7(c). Of course, we can continue differently in Figures 34.7(b) and (c). Also, the first permissible loop can be chosen differently from Figure 34.6(a) or 34.7(a), but it is too time-consuming to try all possibilities, so we have to be content with a few trials. We need a few trials to gain a good feeling of how to obtain good selections, and thereafter, we may obtain a reasonably good network. For the above example, $f = x \vee y \vee \bar{z}$, the network obtained in Figure 34.7, happens to be the minimal network (the minimality of the number of gates and then, as the secondary objective, the minimality of the number of connections can be proved by the integer programming logical design method.) [4,6].   □

As might be observed already, any permissible loop in the case of four variables represents a product of non-complemented literals (e.g., the loop chosen in Figure 34.6(b) represents yz) by letting the permissible loop contain the cell with coordinates, $w = x = y = z = 1$, whereas any rectangular loop of 2^i cells, which does or does not contain this cell, represents a product of complemented literals, non-complemented literals, or a mixture as observed previously.

34.3.1 Extension of the Concept of Permissible Loop

By defining permissible loops differently, we can extend the map-factoring method to designing NAND gate networks with some variables complemented and others non-complemented as network inputs. This is useful when it is not convenient to have other combinations of complemented variables and non-complemented variables as network inputs (e.g., long connections are needed). For example, let us define a permissible loop as a rectangular loop consisting of 2^i cells which contains the cell with coordinates, $x = y = 0$ and $z = 1$. Then, we can design NAND networks with only $\bar{x}$, $\bar{y}$, and z as network inputs. Some functions can be realized with simpler networks by using some network inputs complemented. For example, the function in Figure 34.7 can be realized with only one NAND gate by using $\bar{x}$, $\bar{y}$, and z as network inputs, instead of three gates in Figure 34.7(c). Complemented inputs can be realized at the outputs of another network to be connected to the inputs of the network under consideration, or can be realized with inverters. When these network inputs run some distance on a PC board or a chip, the area covered by them is smaller than the case of double-rail input logic because the latter case requires another set of long-running lines for having their complements.

34.4 Features of the Map-Factoring Method

Very often in design practice, we need to design small networks. For example, we need to modify large networks by adding small networks, designing large networks by assembling small networks, or designing frequently used networks which are to be stored as cells in a cell library (these will be discussed in Chapter 48). Manual design is still useful in these cases because designers can understand very well the functional relationships among inputs, outputs, and gates and also complex constraints.

The map-factoring method has unique features that other design methods based on Karnaugh maps with AND and OR gates described in the previous sections do not have. The map-factoring method can synthesize NAND (or NOR) gate networks in single-rail input or double-rail input logic, in not only in two-levels but also in multi-levels. Also, constraints such as maximum fan-in, maximum fan-out, and the number of levels can be easily taken into account. In contrast, methods for designing logic networks with AND and OR gates on Karnaugh maps described in the previous sections can yield networks under several strong restrictions stated previously, such as only two levels and no maximum fan-in restriction. But networks in two levels in double-rail input logic with NAND (or NOR) gates derived by the map-factoring method are essentially networks in two-level of AND and OR gates in double rail-input logic. The usefulness of the map-factoring method is due to the use of a single gate type, NAND gate (or NOR gate type). Although the map-factoring method is intuitive, derivation of minimal networks with NAND gates in multi-levels is often not easy because there is no good guide line for each step (i.e., where we choose a loop) and in this sense, the method is heuristic.

The map-factoring method can be extended to a design problem where some inputs to a network are independent variables, $x_1, x_2, \ldots, x_n$ and other inputs are logic functions of these variable inputs.

34.5 Other Design Methods of Multi-Level Networks with a Minimum Number of Gates

If we are not content with heuristic design methods, such as the map-factoring method, that do not guarantee the minimality of networks and we want to design a network with a minimum number of NAND (or NOR) gates (or a mixture) under arbitrary restrictions, we cannot do so within the framework of switching algebra, unlike the case of minimal two-level AND/OR gate networks (which can be designed based on minimal sums or minimal products), and the **integer programming logic design method** is currently the only method available [4,6]. This method is not appropriate for hand processing, but can design minimal networks for up to about 10 gates within reasonable processing time by computer. The method can design multiple-output networks also, regardless of whether functions are completely or incompletely specified. Also, networks with a mixture of different gate types (such as NAND, NOR, AND, or OR gates), with gates having double outputs, or with wired-OR can be designed, although the processing time and the complexity of programs increase correspondingly. Also, the number of connections can be minimized as the primary, instead of as the secondary, objective.

Although the primary purpose of the integer programming logic design method is the design of minimal networks with a small number of variables, minimal networks for some important functions of an arbitrary number of variables (i.e., no matter how large n is for a function of n variables), such as adders [1,7] and parity functions [2], have been derived by analyzing the intrinsic properties of minimal networks for these networks.

References

1. Lai, H.C. and S. Muroga, "Minimum binary adders with NOR (NAND) gates," *IEEE Tr. Computers*, vol. C-28, pp. 648–659, Sept. 1979.
2. Lai, H.C. and S. Muroga, "Logic networks with a minimum number of NOR (NAND) gates for parity functions of n variables," *IEEE Tr. Computers*, vol. C-36, pp. 157–166, Feb. 1987.
3. Maley, G.A., and J. Earle, *The Logical Design of Transistor Digital Computers*, Prentice-Hall, Englewood Cliffs, NJ, 1963.
4. Muroga, S. "Logic design of optimal digital networks by integer programming," *Advances in Information Systems Science*, vol. 3, ed. by J.T. Tou, Plenum Press, pp. 283–348, 1970.
5. Muroga, S., *Logic Design and Switching Theory*, John Wiley & Sons (now available from Krieger Publishing Co.), 1979.
6. Muroga, S., "Computer-aided logic synthesis for VLSI chips," *Advances in Computers*, vol. 32, ed. by M.C. Yovits, Academic Press, pp. 1–103, 1991.
7. Muroga, S. and H.C. Lai, "Minimization of logic networks under a generalized cost function," *IEEE Tr. Computers*, vol. C-25, pp. 893–907, Sept. 1976.
8. Sakurai, A. and S. Muroga, "Parallel binary adders with a minimum number of connections," *IEEE Tr. Computers*, vol. C-32, pp. 969–976, Oct. 1983. (In Figure 7, labels a_0 and $\bar{c}_0$ should be exchanged.)

35

Logic Synthesis with a Minimum Number of Negative Gates

Saburo Muroga

University of Illinois
at Urbana-Champaign

CONTENTS

35.1 Logic Design of MOS Networks

A MOS logic gate can express a negative function and it is not directly associated with a simple logic expression such as a minimal sum. So, it is not a simple task to design a network with MOS logic gates so that the logic capability of each MOS logic gate to express a negative function is utilized to the fullest extent. Here let us describe one of few such design procedures that design transistor circuits directly from the given logic functions [7,8].

A logic gate whose output represents a negative function is called a **negative gate**. A MOS logic gate is a negative gate. We now design a network with a minimum number of negative gates. The **feed-forward network** shown in Figure 35.1 (the output of each gate feeds forward to the gates in the succeeding levels) can express any loopless network. Let us use Figure 35.1 as the general model of a loopless network.

The following procedure designs a logic network with a minimum number of negative gates, assuming that only non-complemented input variables (i.e., $x_1, x_2, \ldots, x_n$) are available as network inputs (i.e., single-rail input logic), based on this model [5,6,9].

> **Procedure 35.1: Design of Logic Networks with a Minimum Number of Negative Gates in Single-Rail Input Logic**
>
> We want to design a MOS network with a minimum number of MOS logic gates (i.e., negative gates) for the given function $f(x_1, x_2, \ldots, x_n)$. (The number of interconnections among logic gates is not necessarily minimized.) It is assumed that only non-complemented variables are available as network inputs. The network is supposed to consist of MOS logic gates g_i's whose outputs are denoted by u_i's, as shown in Figure 35.1.

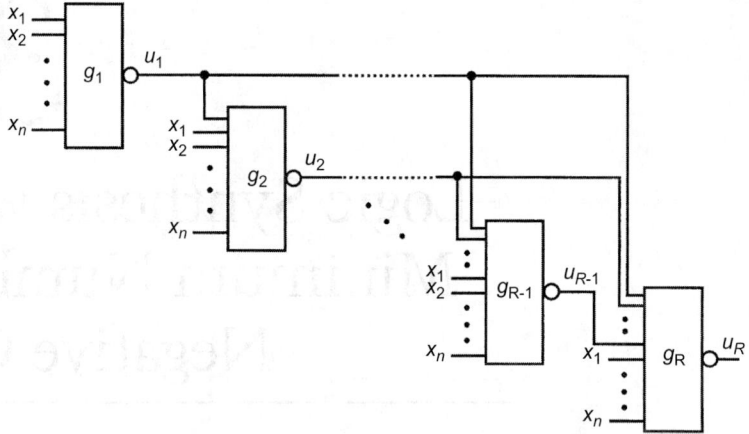

FIGURE 35.1 A feed-forward network of negative gates.

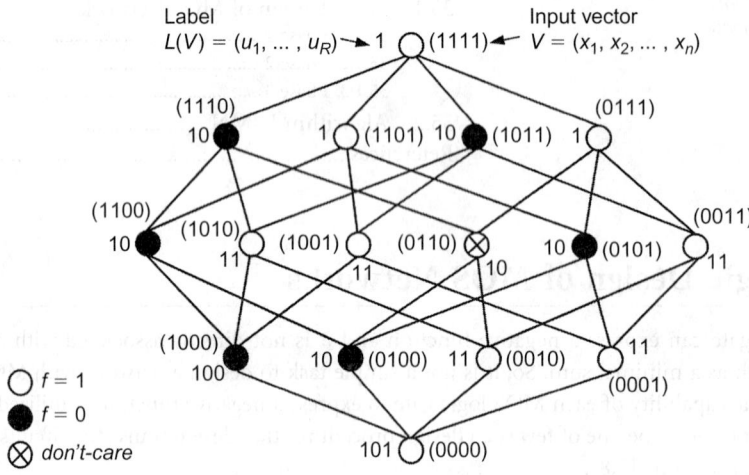

FIGURE 35.2 Lattice example for Procedure 35.1.

35.1.1 Phase 1

1. Arrange all input vectors $V = (x_1, x_2, \ldots, x_n)$ in a lattice, as shown in Figure 35.2, where the nodes denote the corresponding input vectors shown in parentheses. White nodes, black nodes, and nodes with a cross in a circle, ⊗, denote true vectors, false vectors, and don't-care vectors, respectively. The number of 1's contained in each vector V is defined as the **weight** of the vector. All vectors with the same weight are on the same level, placing vectors with greater weights in higher levels, and every pair of vectors that differ only in one bit position is connected by a short line.

Figure 35.2 is a lattice example for an incompletely specified function of four variables.

2. We assign the label $L(V)$ to each vector $V = (x_1, x_2, \ldots, x_n)$ in the lattice in Steps 2 and 3. Henceforth, $L(V)$ is shown without parentheses in the lattice.

First assign the value of f to the vector $(11 \ldots 1)$ of weight n at the top node as $L(11 \ldots 1)$. If f for the top node is "don't-care," assign 0.

In the example in Figure 35.2, we have $L(11 \ldots 1) = 1$ because the value of f for the top node is 1, as shown by the white node.

3. When we finish the assignment of $L(V)$ to each vector of weight w, where $0 < w \le n$, assign $L(V')$ to each vector V' of weight $w-1$, the **smallest binary number** satisfying the following conditions:

If $f(V')$ is not "don't-care,"

    a. The least significant bit of $L(V')$ is $f(V')$ (i.e., the least significant bit of $L(V')$ is 0 or 1, according to whether f is 0 or 1), and

    b. The other bits of $L(V')$ must be determined such that $L(V') \ge L(V)$ holds for every vector V of weight w that differs from V' in only one bit position. In other words, the binary number represented by $L(V')$ is not smaller than the binary number represented by $L(V)$.

If $f(V')$ is "don't-care," ignore (a), but consider (b) only. [Consequently, the least significant bit of $L(V')$ is determined such that (b) is met.]

For the example we get a label $L(1110) = 10$ for the node for vector $V' = (1110)$ because the last bit must be $0 = f(1110)$ by (a), and the number must be equal to or greater than the label 1 for the top node for vector (1111) by (b). Also, for vector $V' = (1000)$, we get a label $L(1000) = 100$ because the last bit must be $0 = f(1000)$ by (a), and the label $L(1000)$ as a binary number must be equal to or greater than each of the labels, 10, 11, and 11, already assigned to the three nodes (1100), (1010), and (1001), respectively.

4. Repeat Step 3 until a label $L(00\ldots0)$ is assigned to the bottom vector $(00\ldots0)$. Then the bit length of $L(00\ldots0)$ is the minimum number of MOS logic gates required to realize f. Denote it by R.

Then make all previously obtained $L(V)$ into binary numbers of the same length as $L(00\ldots0)$ by adding 0's in front of them such that every label $L(V)$ has exactly R bits. For the example, we have $R = 3$, so the label $L(11\ldots1) = 1$ obtained for the top node is changed to 001 by adding two 0's.

35.1.2 Phase 2

Now let us derive MOS logic gates from the $L(V)$'s found in Phase 1.

1. Denote each $L(V)$ obtained in Phase 1 as $(u_1,\ldots, u_i, u_{i+1}, \ldots, u_R)$. As will be seen later, $u_1,\ldots,u_i$, $u_{i+1},\ldots, u_R$ are log functions realized at the outputs of logic gates, $g_1,\ldots, g_i, g_{i+1},\ldots, g_R$, respectively, as shown in Figure 35.1.

2. For each $L(V) = (u_1,\ldots, u_i, u_{i+1},\ldots,u_R)$ that has $u_{i+1} = 0$, make a new vector $(V,u_1,\ldots, u_i)$ (i.e., $(x_1,\ldots, x_n, u_1,\ldots, u_i)$) which does not include $u_{i+1},\ldots, u_R$.

This means the following. For each i, find all labels $L(V)$'s whose $(i + 1)$-th bit is 0 and then for each of these labels, we need to create a new vector $(x_1,\ldots, x_n, u_1,\ldots, u_i)$ by containing only the first i bits of the label $L(V)$. For example, for $u_1 = 0$ (i.e., by setting i of "$u_{i+1} = 0$" to 0), the top node of the example lattice in Figure 35.2 has the label $(u_1, u_2, u_3) = (001)$ which has $u_1 = 0$ (the label 1 which was labeled in Step 3 of Phase 1 was changed to 001 in Step 4). For this label, we need to create a new vector $(x_1, x_2, x_3, x_4, u_1,\ldots, u_i)$, but the last bit u_i becomes u_0 because we set $i = 0$ for $u_{i+1} = 0$. There is no u_0, so the new vector is $(x_1, x_2, x_3, x_4) = (1111)$, excluding $u_1,\ldots, u_i$. (In this sense, the case of $u_1 = 0$ is special, unlike the other cases of $u_2 = 0$ and $u_3 = 0$ in the following.) For other nodes with labels having $u_1 = 0$, we create new vectors in the same manner.

Next, for $u_2 = 0$ (i.e., by setting i of "$u_{i+1} = 0$" to 1), the top node has the label $(u_1, u_2, u_3) = (001)$ which has $u_2 = 0$. For this label, we need to create a new vector $(x_1, x_2, x_3, x_4, u_1)$, including u_1 this time. So, the new vector $(x_1, x_2, x_3, x_4, u_1) = (11110)$ results as the label $L(1111) = 001$ for the top node. Also, a new vector (11010) results for $L(1101) = 001$, and so on.

3. Find all the minimal vectors from the set of all the vectors found in Step 2, where the **minimal vectors** are defined as follows. When $a_k \ge b_k$ holds for every k for a pair of distinct vectors $A = (a_1,\ldots, a_m)$ and

$B = (b_1,\ldots, b_m)$, then the relation is denoted by

$$A \succ B$$

and B is said to be smaller than A. In other words, A and B are compared bit-wise. If no vector in the set is smaller than B, B is called a minimal vector of the set. For example, $(10111) \succ (10101)$ because a bit (i.e., 1 or 0) in every bit position of (10111) is greater than or equal to the corresponding bit in (10101). But (10110) and (10101) are **incomparable** because a bit in each of the first four bit positions of (10110) is greater than or equal to the corresponding bit of (10101), but the fifth bit of (10110) (i.e., 0) is smaller than the corresponding bit (i.e., 1) of (10101).

For the example, for $u_1 = 0$, the minimal vectors are (0100), (0010), (0001). Then, for $u_2 = 0$, we get (11110), (11010), (01110), (10001), and (00001) by Step 2. Then the minimal vectors are (11010), (01110), and (00001). Here, notice that (11110), for example, cannot be a minimal vector because $(11110) \succ (11010)$. (11010) cannot be compared with other two, (01110) and (0001), with respect to $\succ$.

4. For every minimal vector, make the product of the variables that have 1's in the components of the vector, where the components of the vector $(V, u_1,\ldots, u_i)$ denote variables $x_1, x_2,\ldots, x_n, u_1,\ldots, u_p$, in this order. For example, we form $x_1x_2x_4$ for vector (11010). Then make a disjunction of all these products and denote it by $\bar{u}_{i+1}$.

For the example, we get

$$\bar{u}_2 = x_1x_2x_4 \vee x_2x_3x_4 \vee u_1$$

from the minimal vectors (11010), (01110), and (00001) for $u_2 = 0$.

5. Repeat Steps 2 through 4 for each of $i = 1, 2,\ldots, R-1$.

For the example, we get

$$\bar{u}_1 = x_2 \vee x_3 \vee x_4 \text{ and } \bar{u}_3 = x_1x_3x_4u_2 \vee x_1u_1 \vee x_2u_2$$

6. Arrange R MOS logic gates in a line along with their output functions, $u_1,\ldots, u_R$. Then construct each MOS logic gate according to the disjunctive forms obtained in Steps 4 and 5, and make connections from other logic gates and input variables (e.g., MOS logic gate g_2, whose output function is u_2, has connections from x_1, x_2, x_3, x_4, and u_1 to the corresponding MOSFETs in g_2, according to disjunctive form $\bar{u}_2 = x_1x_2x_4 \vee x_2x_3x_4 \vee u_1$). The network output, u_3 (i.e., $u_3 = \overline{x_1x_3x_4u_2 \vee x_1u_1 \vee x_2u_2}$, which can be rewritten as $\bar{x}_1\bar{x}_2 \vee x_1\bar{x}_3x_4 \vee x_2x_3x_4 \vee \bar{x}_2x_3\bar{x}_4$) realizes the given function f.

For the example, we get the network shown in Figure 35.3 (the network shown here is with logic gates in n-MOS only but it is easy to convert this into CMOS, as we will see in Chapter 39).

35.1.3 Phase 3

The bit length R in label $L(00\ldots0)$ for the bottom node shows the number of MOS logic gates in the network given at the end of Phase 2. Thus, if we do not necessarily choose the smallest binary number in Step 3 of Phase 1, but choose a binary number still satisfying the other conditions (i.e., (a) and (b)) in Step 3 of Phase 1, then we can still obtain a MOS network of the same minimum number of MOS logic gates as long as the bit length R of $L(00\ldots0)$ is kept the same. (For the top node also, we do not need to choose the smallest binary number as $L(11\ldots)$, no matter whether f for the node is don't-care.)

This freedom may change the structure of the network, although the number of logic gates is still the minimum. Among all the networks obtained, there is a network that has a minimum number of logic

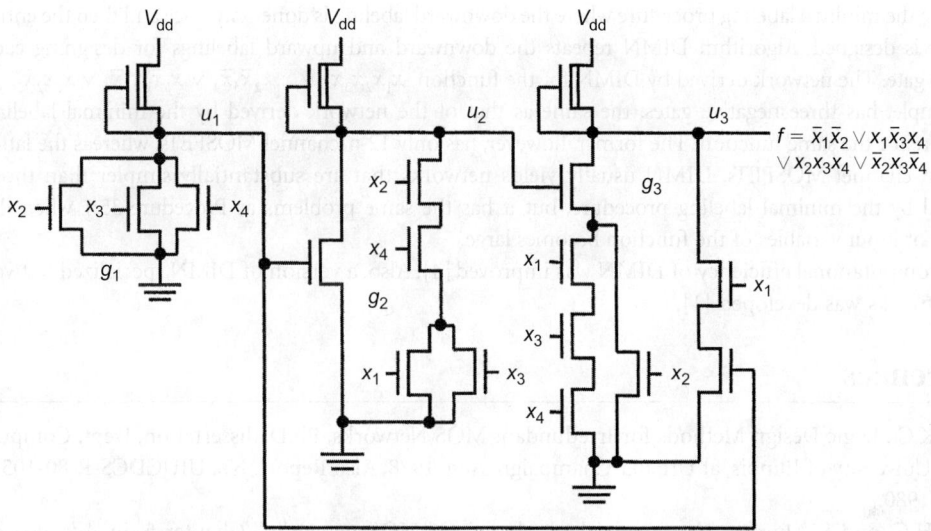

FIGURE 35.3 MOS network based on the labels $L(V)$ in Figure 35.2.

gates as its primary objective, and a minimum number of interconnections as its secondary objective. (Generally, it is not easy to find such a network because there are too many possibilities.) ☐

Although the number of MOS logic gates in the network designed by Procedure 35.1 is always minimized, the networks designed by Procedure 35.1 may have the following problems: (1) the number of interconnections is not always minimized; (2) some logic gates may become very complex so that these logic gates may not work properly with reasonably small gate delay times. If so, we need to split these logic gates into a greater number of reasonably simple logic gates, giving up the minimality of the number of logic gates. Also, after designing several networks according to Phase 3, we may be able to find a satisfactory network.

35.2 Algorithm DIMN

Compared with the problem (1) of Procedure 35.1, problem (2) presents a far more serious difficulty. Thus, Algorithm DIMN (an acronym for Design of Irredundant Minimal Network) was developed to design a MOS network in single-rail input logic such that the number of gates is minimized and every connection among cells is irredundant (i.e., if any connection among logic gates is removed, the network output will be changed) [2,3]. Algorithm DIMN is powerful but is far more complex than the minimal labeling procedure (i.e., Phases 1 and 2 of Procedure 35.1). So, let us only outline it.

Procedure 35.2: Outline of Algorithm DIMN

1. All the nodes of a lattice are labeled by the minimal labeling procedure (i.e., Phases 1 and 2 of Procedure 35.1), starting with the top node and moving downward. Let the number of bits of each label be R. Then all the nodes are labeled by a procedure similar to the minimal labeling procedure, starting with the bottom node which is now labeled with the largest binary number of R bits, and moving upward on the lattice. Then, the first negative gate with irredundant MOSFETs is designed after finding as many don't-cares as possible by comparing two labels at each node which are derived by these downward and upward labelings.

2. The second negative gate with irredundant MOSFETs is designed after downward and upward labelings to find as many don't-cares as possible. This process is repeated to design each gate until the network output gate with irredundant MOSFETs is designed. ☐

Unlike the minimal labeling procedure where the downward labeling is done only once and then the entire network is designed, Algorithm DIMN repeats the downward and upward labelings for designing each negative gate. The network derived by DIMN for the function $x_1x_2x_3x_4 \vee x_1x_2\bar{x}_3\bar{x}_4 \vee x_1\bar{x}_2x_3\bar{x}_4 \vee x_1\bar{x}_2\bar{x}_3x_4$, for example, has three negative gates, the same as that of the network derived by the minimal labeling procedure for the same function. The former, however, has only 12 n-channel MOSFETs, whereas the latter has 20 n-channel MOSFETs. DIMN usually yields networks that are substantially simpler than those designed by the minimal labeling procedure, but it has the same problems as Procedure 35.1 when the number of input variables of the function becomes large.

The computational efficiency of DIMN was improved [4]. Also, a version of DIMN specialized to two-level networks was developed [1].

References

1. Hu, K.C., Logic Design Methods for Irredundant MOS Networks, Ph.D. dissertation, Dept. Comput. Sci., University of Illinois, at Urbana-Champaign, Aug. 1978, Also Report. No. UIUCDCS-R-80-1053, 317, 1980.
2. Lai, H.C. and S. Muroga, "Automated logic design Of MOS networks," Chapter 5, in *Advances in Information Systems Science*, vol. 9, ed. by J. Tou, Plenum Press, New York, pp. 287–336, 1985.
3. Lai, H.C. and S. Muroga, "Design of MOS networks in single-rail input logic for incompletely specified functions," *IIEEE Tr. CAD*, 7, pp. 339–345, March 1988.
4. Limqueco, J.C., Algorithms for the Design of Irredundant MOS Networks, Master thesis, Dept. Comput. Sci., University of Illinois, Urbana, IL, 87, 1998.
5. Liu, T.K., "Synthesis of multilevel feed-forward MOS networks," *IEEE TC*, pp. 581–588, June 1977.
6. Liu, T.K., "Synthesis of feed-forward MOS network with cells of similar complexities," *IEEE TC*, pp. 826–831, Aug. 1977.
7. Muroga, S., *VLSI System Design*, John Wily & Sons, 1982.
8. Muroga, S., *Computer-Aided Logic Synthesis for VLSI Chips*, ed. by M. C. Yovits, vol. 32, Academic Press, pp. 1–103, 1991.
9. Nakamura, K., N. Tokura, and T. Kasami, "Minimal negative gate networks," *IEEE TC*, pp. 5–11, Jan. 1972.

36

Ko Yoshikawa
NEC Corporation

Saburo Muroga
*University of Illinois
at Urbana-Champaign*

Logic Synthesizer with Optimizations in Two Phases

Logic networks and then their corresponding transistor circuits to be laid out on integrated chips have been traditionally designed manually, spending long time and repeatedly making mistakes and correcting them. As the cost and size of transistor circuits continue to decline, logic networks that are realized by transistor circuits have been designed with an increasingly large number of logic gates. Manual design of such large logic networks is becoming too time-consuming and prone to design mistakes, thus necessitating automated design. Logic synthesizers are automated logic synthesis systems used for this purpose and transform the given logic functions into technology-dependent logic circuits that can be easily realized as transistor circuits. The quality of technology-dependent logic circuits derived by logic synthesizers is not necessarily better than manually designed ones, at least for those with a small number of logic gates, but technology-dependent logic circuits for those with millions of logic gates cannot be designed manually for reasonably short times.

Automated design of logic networks has been attempted since the early 1960s [3]. Since the beginning of the 1960s, IBM has pushed research of design automation in logic design. In the 1970s, a different type of algorithm, called the Transduction method, was devised for automated design of logic networks, as described in Chapter 37. Since the beginning of the 1980s, the integration size of integrated chips has tremendously increased, research and development of automated logic synthesis has become very active at several places [1,2,4,9] and powerful logic synthesizers have become commercially available [6].

There are different types of logic synthesizers. But here, let us describe a logic synthesizer that derives a technology-dependent logic circuit in two phases [7], although there are logic synthesizers where, unlike the following **synthesizer in two phases**, only technology-dependent optimization is used throughout the transformation of the given logic functions into technology-dependent logic circuits that are easily realizable as transistor circuits:

In the first phase, an optimum logic network that is not necessarily realizable as a transistor circuit is designed by Boolean algebraic approaches which are easy to use. This phase is called **technology-independent optimization**.

In the second phase, the logic network derived in the first phase is converted into a technology-dependent logic circuit that is easily realizable as a transistor circuit. (Note that all logic synthesizers, including the logic synthesizer in two phases, must eventually have technology-dependent logic circuits that are easily realizable, as transistor circuits.) This phase is called **technology-dependent optimization**.

The logic synthesizer in two phases has variations, depending on what optimization algorithms are used, and some of them have the advantage of short processing time.

Suppose a logic network is designed for the given functions by some means, including design methods described in the previous chapters. Figure 36.1 is an example of such logic networks. Then the logic network is to be processed by many logic optimization algorithms in sequence, as shown in Figure 36.2. As illustrated in Figure 36.1, a subnetwork (i.e., several logic gates that constitute a small logic network by themselves) that has a single output logic function is called a **node,** and its output function is

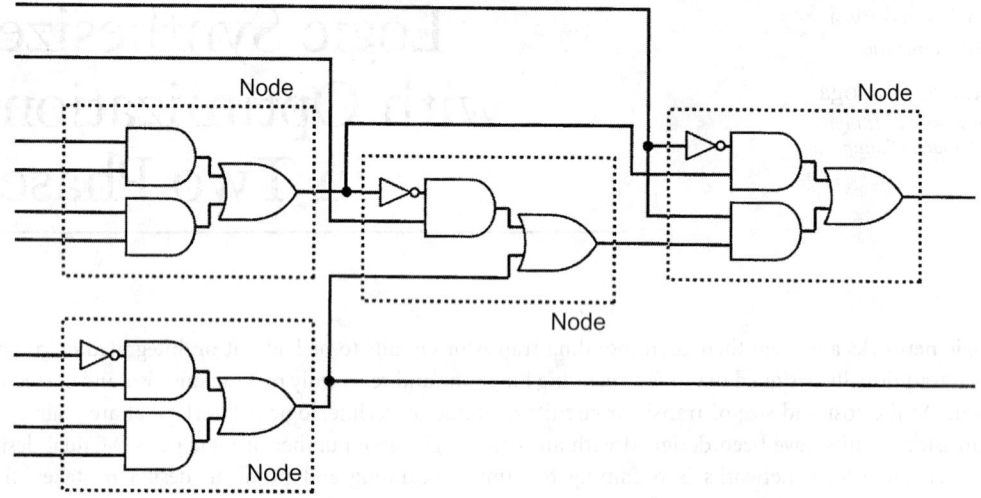

FIGURE 36.1 Logic network to be processed by a logic synthesizer.

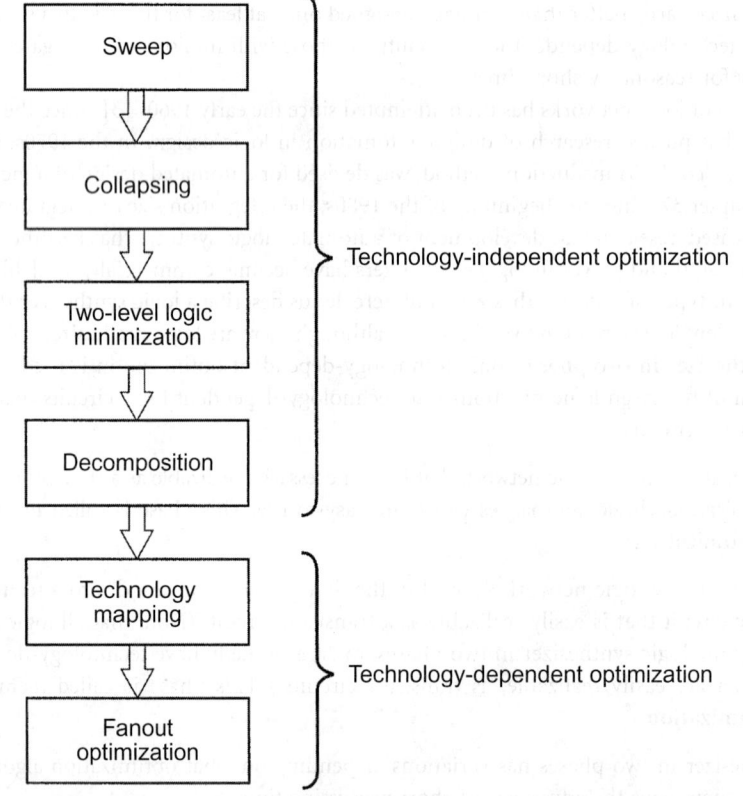

FIGURE 36.2 A logic optimization flow in the logic synthesizer in two phases.

<footer>

</footer>

expressed as a sum-of-products form. Thus, the entire logic network is expressed as a network of the nodes. During the first phase of technology-independent optimization, each node does not correspond to a transistor circuit. But the second phase of technology dependent-optimization, as illustrated in Figure 36.3, converts the logic network derived by the first phase into technology-dependent circuits, in which each node corresponds to a cell [i.e., a small transistor circuit (shown in each dot-lined loop in Figure 36.3)] in a library of cells. The cell library consists of a few hundred cells where a cell is a small transistor circuit with good layout. These cells are designed beforehand and can be repeatedly used for logic synthesis.

A typical flow of logic optimization in the logic synthesis in two phases works as follows and is illustrated in Figure 36.2, although there may be variations:

1. Sweep, illustrated in Figure 36.4: Nodes that do not have any fan-outs are deleted. A node that consists of only a buffer or an inverter node is merged into next nodes. The following rules optimize nodes that have inputs of constant values, 0 or 1:

$$A \cdot 1 = A, A \cdot 0 = 0, \ A \vee 1 = 1, A \vee 0 = A$$

2. Collapsing (also often called flattening), illustrated in the upper part of Figure 36.5: More than one node is merged into one node, so that it has a more complicated logic function than before.

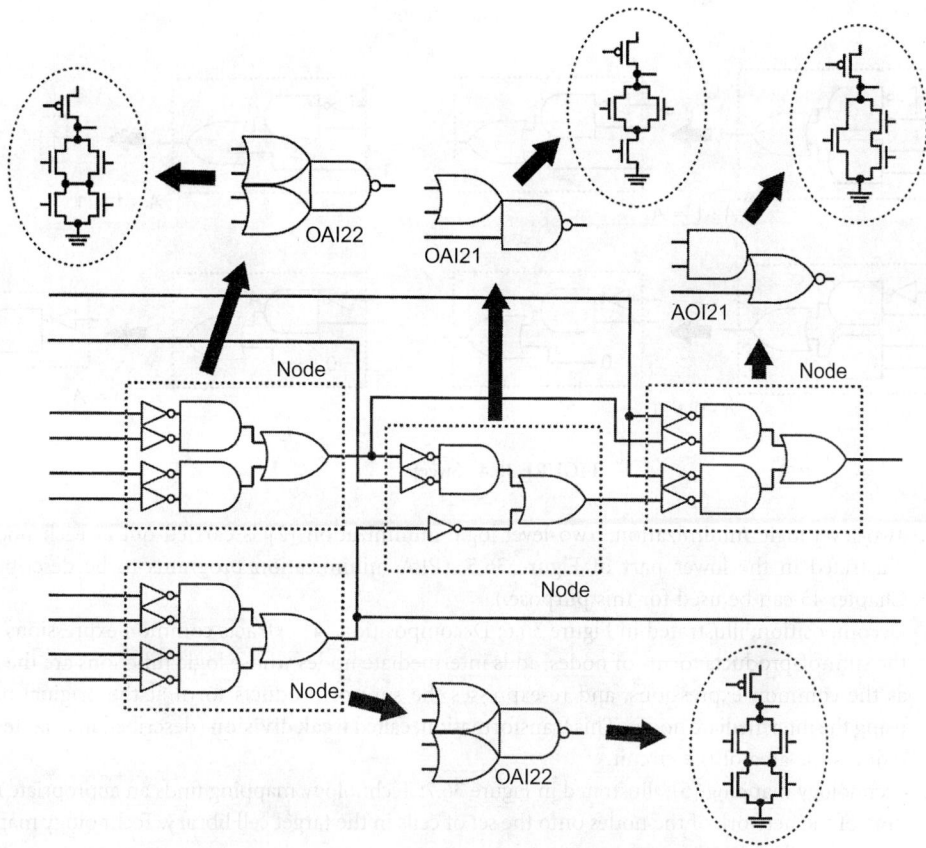

OAI22, OAI21, and AOI21 are the cells in the cell library.

FIGURE 36.3 Technology-dependent logic circuit.

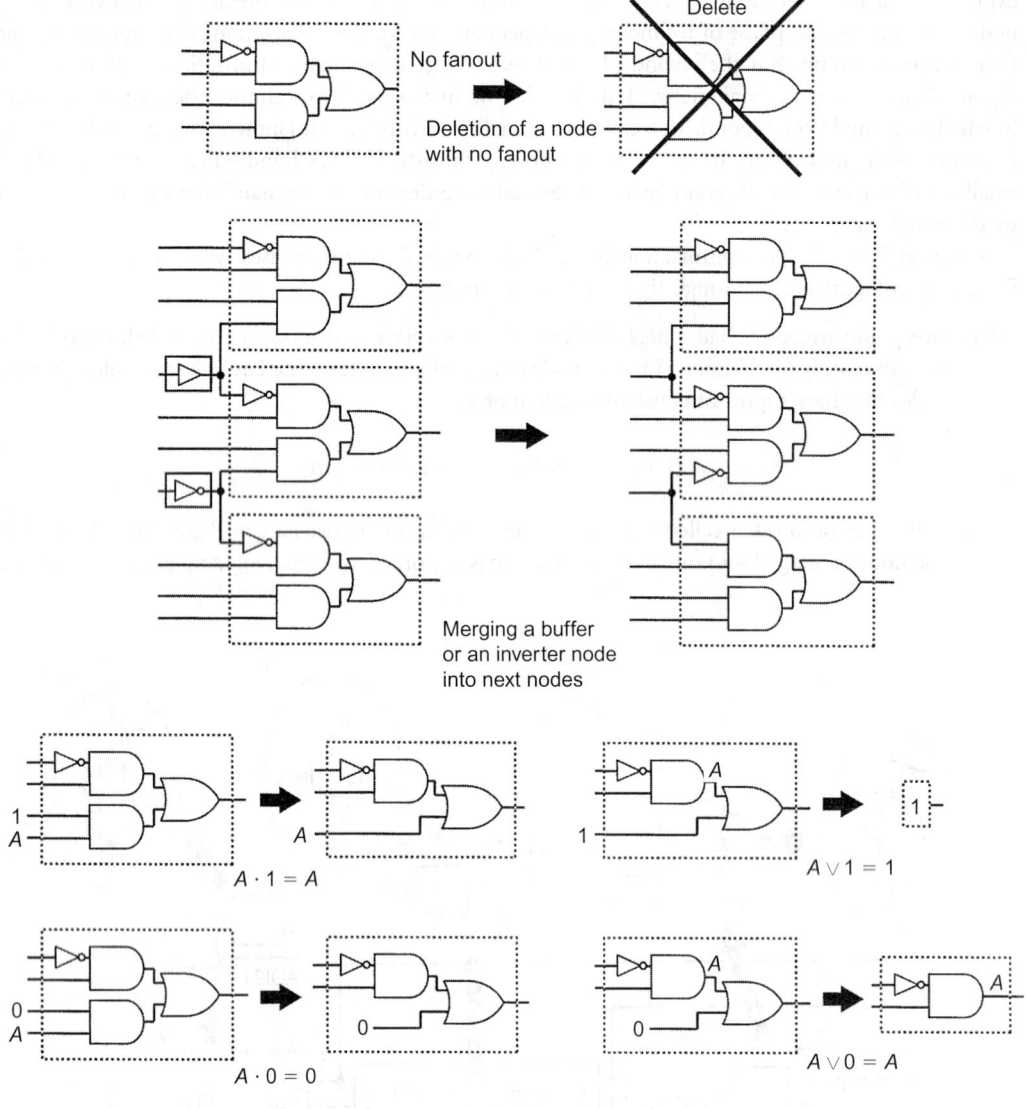

FIGURE 36.4 Sweep.

3. Two-level logic minimization: Two-level logic minimization [2] is carried out at each node, as illustrated in the lower part of Figure 36.5. (PLA minimization programs to be described in Chapter 45 can be used for this purpose.)

4. Decomposition, illustrated in Figure 36.6: Decomposition [4] extracts common expressions from the sum-of-products forms of nodes, adds intermediate nodes whose logic functions are the same as the common expressions, and re-expresses the sum-of-products form at the original nodes, using the intermediate nodes. This transformation, called **weak division** (described in Chapter 32), reduces the area of the circuit.

5. Technology mapping [5], illustrated in Figure 36.7: Technology mapping finds an appropriate mapping of the network of the nodes onto the set of cells in the target cell library. Technology mapping works as follows:

   a. Select a tree structure: a tree structure where all nodes have one fan-out node is extracted from the circuit.

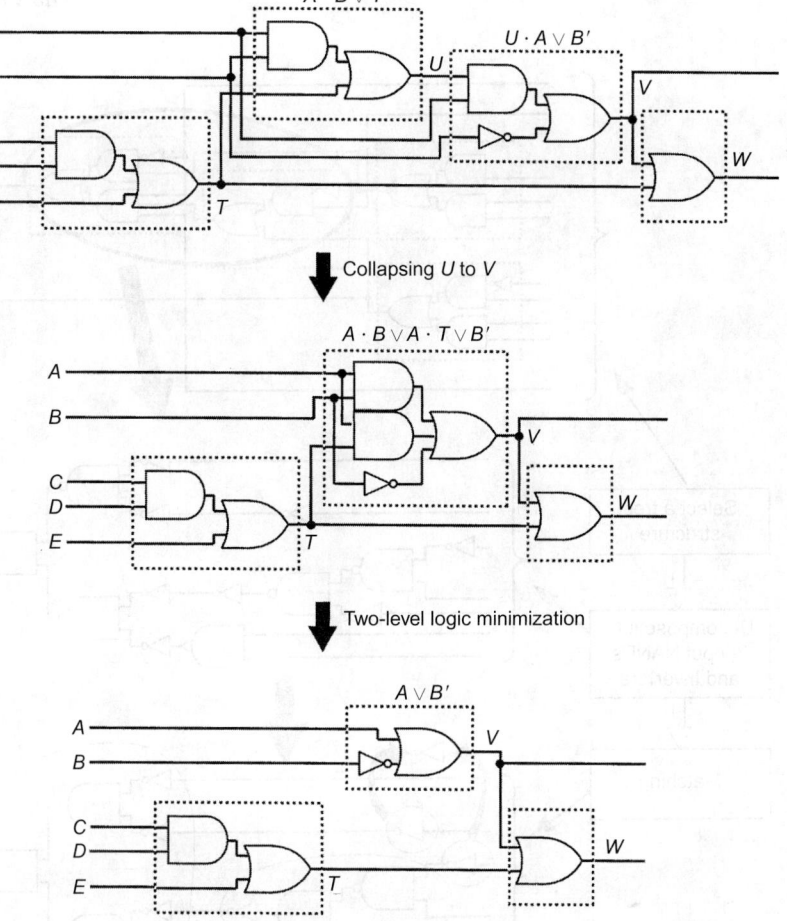

FIGURE 36.5 Collapsing and two-level logic minimization.

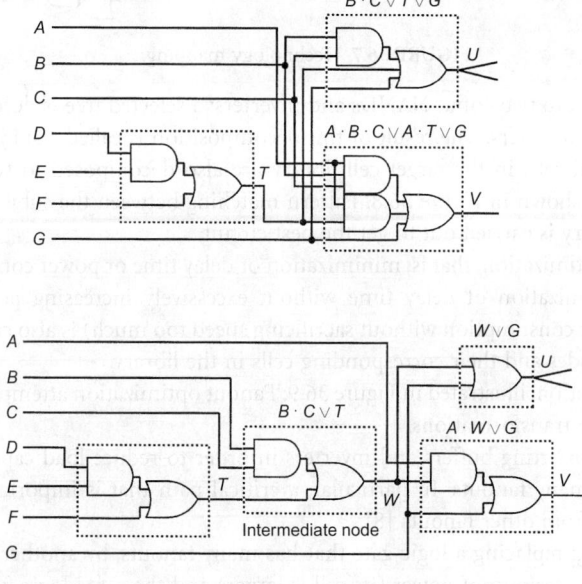

FIGURE 36.6 Decomposition.

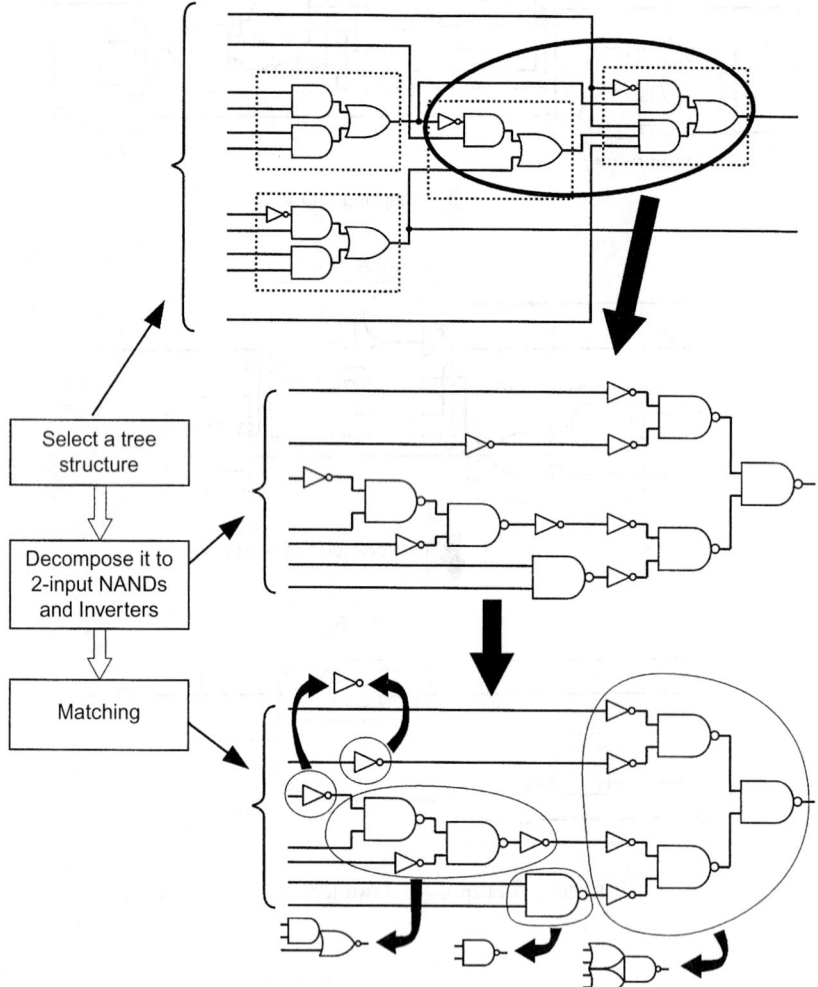

FIGURE 36.7 Technology mapping.

b. Decompose it to two-input NANDs and inverters: a selected tree is decomposed to two-input NANDs and inverters. The result of this decomposition is called a subject tree.

c. Matching: all cells in the target cell library are also decomposed to two-input NANDs and inverters, as shown in Figure 36.8. Pattern matching between the subject tree and the cells in the cell library is carried out to get the best circuit.

Timing or power optimization, that is, minimization of delay time or power consumption under other conditions (e.g., minimization of delay time without excessively increasing power consumption, or minimization of power consumption without sacrificing speed too much) is also carried out by changing the structure of the nodes and their corresponding cells in the library.

6. Fanout optimization, illustrated in Figure 36.9: Fanout optimization attempts to reduce delay time by the following transformations.

a. Buffering: inserting buffers and inverters in order to reduce load capacitance of logic gates that have many fanouts. In particular, a critical path that is important for performance is separated from other fanouts [8].

b. Repowering: replacing a logic gate that has many fanouts, by another cell in the cell library that has greater output power (its cell is larger) and the same logic function as the original one. This does not change the structure of the circuit [8].

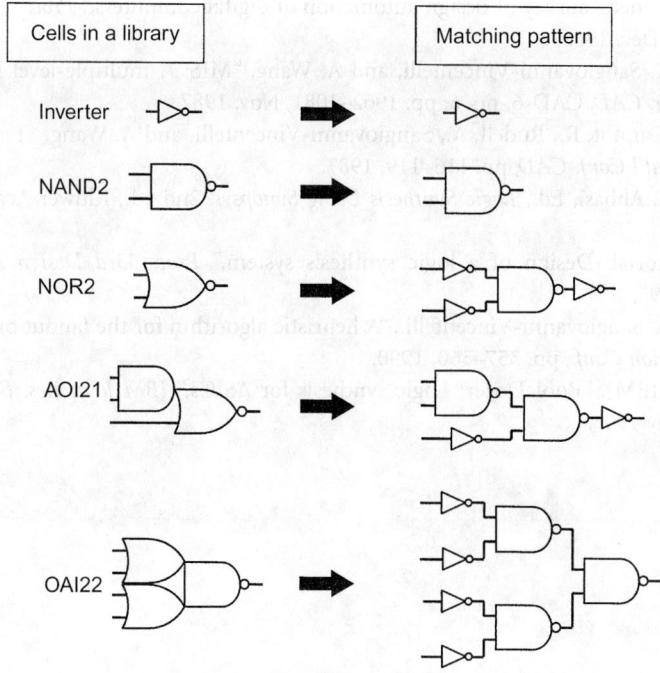

FIGURE 36.8 Pattern trees for a library.

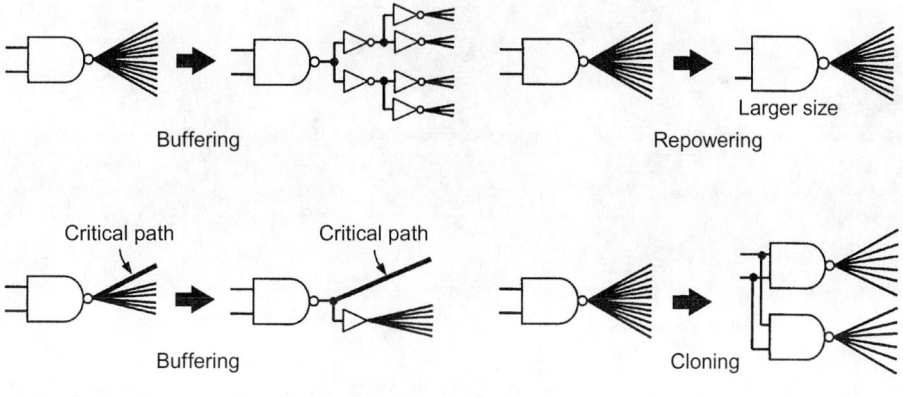

FIGURE 36.9 Fan-out optimization.

    c. Cloning: creating a clone node for a logic gate that has many fanouts and distributing the fanouts among these nodes to reduce the load capacitance.

This is a typical logic optimization flow. Many other optimization algorithms can be incorporated within this flow to synthesize better circuits.

References

1. Bergamaschi, R.A. et al., "High-level synthesis in an industrial environment," *IBM Jour. Res. and Dev.*, vol. 39, pp. 131–148, Jan./March 1995.
2. Brayton R.K., G.D. Hachtel, C. McMullen, and A. Sangiovanni-Vincentelli, *Logic Minimization Algorithms for VLSI Synthesis*, Kluwer Academic Publishers, Boston, 1984.

3. Breuer, M.A., "General survey of design automation of digital computers," *Proc. IEEE*, vol. 54, no. 12, pp. 1708–1721, Dec. 1966.
4. Brayton R.K., A. Sangiovanni-Vincentelli, and A. Wang, "MIS: A multiple-level logic optimization system," *IEEE Tr. CAD*, CAD-6, no. 6, pp. 1062–1081, Nov. 1987.
5. Detjens E., G., Gannot, R., Rudell, A., Sangiovanni-Vincentelli, and A. Wang, "Technology mapping in MIS," *Proc. Int'l Conf. CAD*, pp. 116–119, 1987.
6. Kurup, P. and T. Abbasi, Ed., *Logic Synthesis Using Synopsys*, 2nd ed., Kluwer Academic Publishers, 322, 1997.
7. Rudell R., "Tutorial: Design of a logic synthesis system," *Proc. 33rd Design Automation Conf.*, pp. 191–196, 1996.
8. Singh K.J. and A. Sangiovanni-Vincentelli., "A heuristic algorithm for the fanout problem," *Proc. 27th Design Automation Conf.*, pp. 357–360, 1990.
9. Stok, L., et al. (IBM), "BooleDozer: Logic synthesis for ASICs," *IBM Jour. Res. Dev.*, vol. 40, no. 4, pp. 407–430, July 1996.

37

Logic Synthesizer by the Transduction Method

Saburo Muroga

University of Illinois
at Urbana-Champaign

CONTENTS

37.1 Technology-Dependent Logic Optimization

In this chapter, a logic optimization method called the transduction method, which was developed in the early 1970s, is described. Unlike the logic synthesizer in two phases described in Chapter 36, we can have logic synthesizers with totally technology-dependent optimization, based on the transduction method, and also can have transistor circuits with better quality, although it is more time-consuming to execute the method. Some underlying ideas in the method came from analyzing the minimal logic networks derived by the integer programming logic design method mentioned in Chapter 34, Section 34.5. The integer programming logic design method is very time-consuming, although it can design absolutely minimal networks. The processing time almost exponentially increases as the number of logic gates increases, and it appears to be excessively time consuming to design minimal logic networks with over about 10 logic gates. Unlike the integer programming logic design method, logic synthesizers based on the transduction method are heuristic, just like any other logic synthesizer for synthesizing large circuits, and the minimality of the transistor circuits derived is not guaranteed, but one can design circuits with far greater numbers of transistors.

37.2 Transduction Method for the Design of NOR Logic Networks

The transduction method can be applied on any types of gates, including negative gates (i.e., MOS logic gates), AND gates, OR gates, or logic gates for more complex logic operations; **but in the following, the transduction method is described with logic networks of NOR gates for the sake of simplicity.** The basic concept of the transduction method is "permissible functions," which will be explained later. This method, which is drastically different from any previously known logic design method, is called the **transduction method (trans**formation and re**duction**) because it repeats the transformation and reduction as follows:

Procedure 37.1: Outline of the transduction method

Step 1. Design an initial NOR network by any known design method. Or, any network to be simplified can be used as an **initial network**.

Step 2. Remove the redundant part of the network, by use of permissible functions. (This is called the **pruning procedure**, as defined later).

Step 3. Perform **transformation** (which is local or global) of the current network, using permissible functions.

Step 4. Repeat Steps 2 and 3 until no further improvement is possible. □

Let us illustrate how Procedure 37.1 simplifies the initial network shown in Figure 37.1(a), as an example. By a transformation (more precisely speaking, transformation called "connectable condition," which is explained later), we have the new network shown in Figure 37.1(b), by connecting the output of gate v_6 to the gate v_8 (shown by a bold line). By the pruning procedure, the connection from gate v_6 to gate v_4 (shown

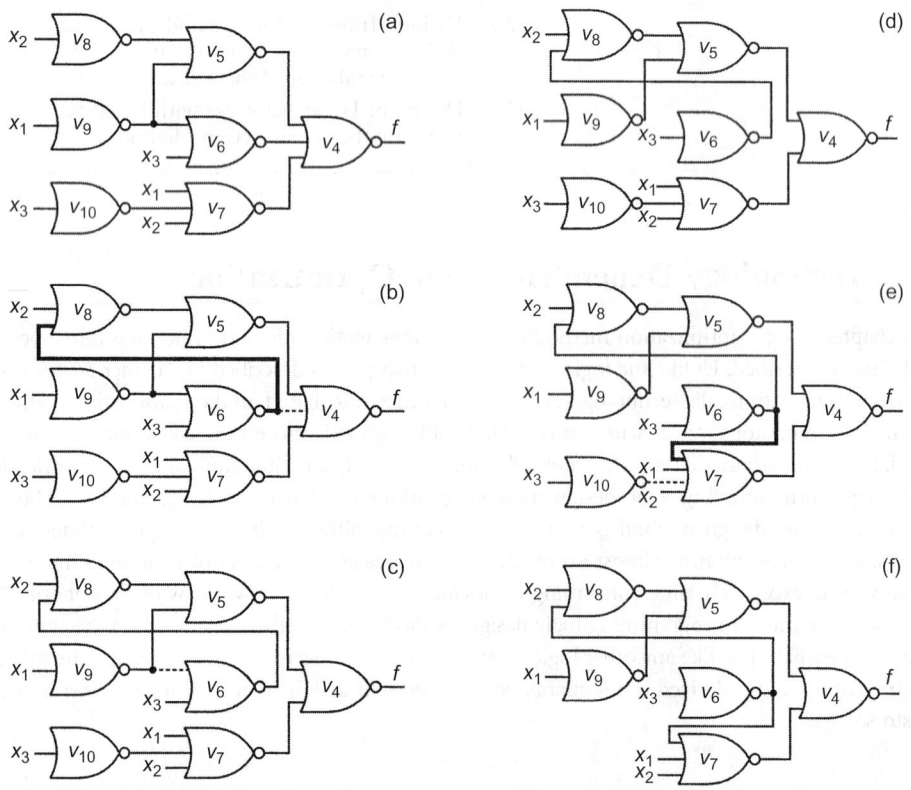

FIGURE 37.1 An example for the simplification of a logic network by the transduction method.

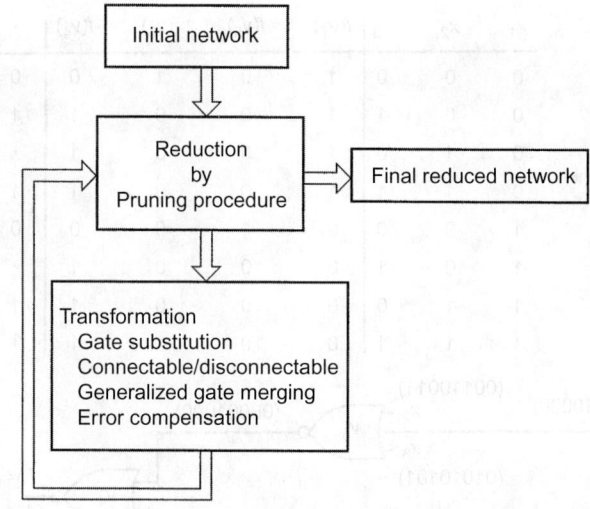

FIGURE 37.2 Basic structure of the transduction method.

by a dotted line) is deleted, deriving the new network shown in Figure 37.1(c), and then the connection from gate v_9 to gate v_6 is deleted, deriving the new network shown in Figure 37.1(d). Then, by a transformation ("connectable condition" again), we have the new network shown in Figure 37.1(e), by connecting the output of gate v_6 to the gate v_7. By the pruning procedure, we can delete the output connection from gate v_{10} to gate v_7, and then we can delete gate v_{10}. During all these transformations and prunings, the network output f is not changed. Thus, the initial network with 7 gates and 13 connections, shown in Figure 37.1(a), has been simplified to the one with 6 gates and 11 connections, shown in Figure 37.1(f), by the transduction method.

Many transformation procedures for Step 3 were devised to develop efficient versions of the transduction method. Reduction of a network is done in Step 2 and also sometimes in Step 3.

The initial network stated in Step 1 of Procedure 37.1 can be designed by any known method and need not be the best network. The transduction method is essentially an improvement of a given logic network, that is, the initial network by repeated transformation and reduction, as illustrated in the flow diagram in Figure 37.2. In contrast, traditional logic design is the design of a logic network (i.e., the initial network in the case of the Transduction method) for given logic functions and once derived, no improvement is attempted. This was natural because it was not easy to find where to delete gates or connections in logic networks. For example, can we simplify the network in Figure 37.1(a), when it is given? The transduction method shows how to do it.

The transduction method can reduce a network of many NOR gates into a new network with as few gates as possible. Also, for a given function, the transduction method can be used to generate networks of different configurations, among which we can choose those suitable for the most compact layouts on chips, since chip areas are greatly dependent on connection configurations of networks.

The transduction method was developed in the early 1970s with many reports and summarized in Refs. 9 and 10.

37.2.1 Permissible Functions

First, let us discuss the concept of permissible functions, which is the core concept for the transduction method.

Definition 37.1: If no output function of a network of NOR gates changes by replacing the function realized at a gate (or an input terminal) v_j, or a connection c_{ij}, with a function g, then function g is called a **permissible function** for that v_j, or c_{ij}, respectively. (Note that in this definition, changes in don't-care values, *, of the network output functions do not matter.)

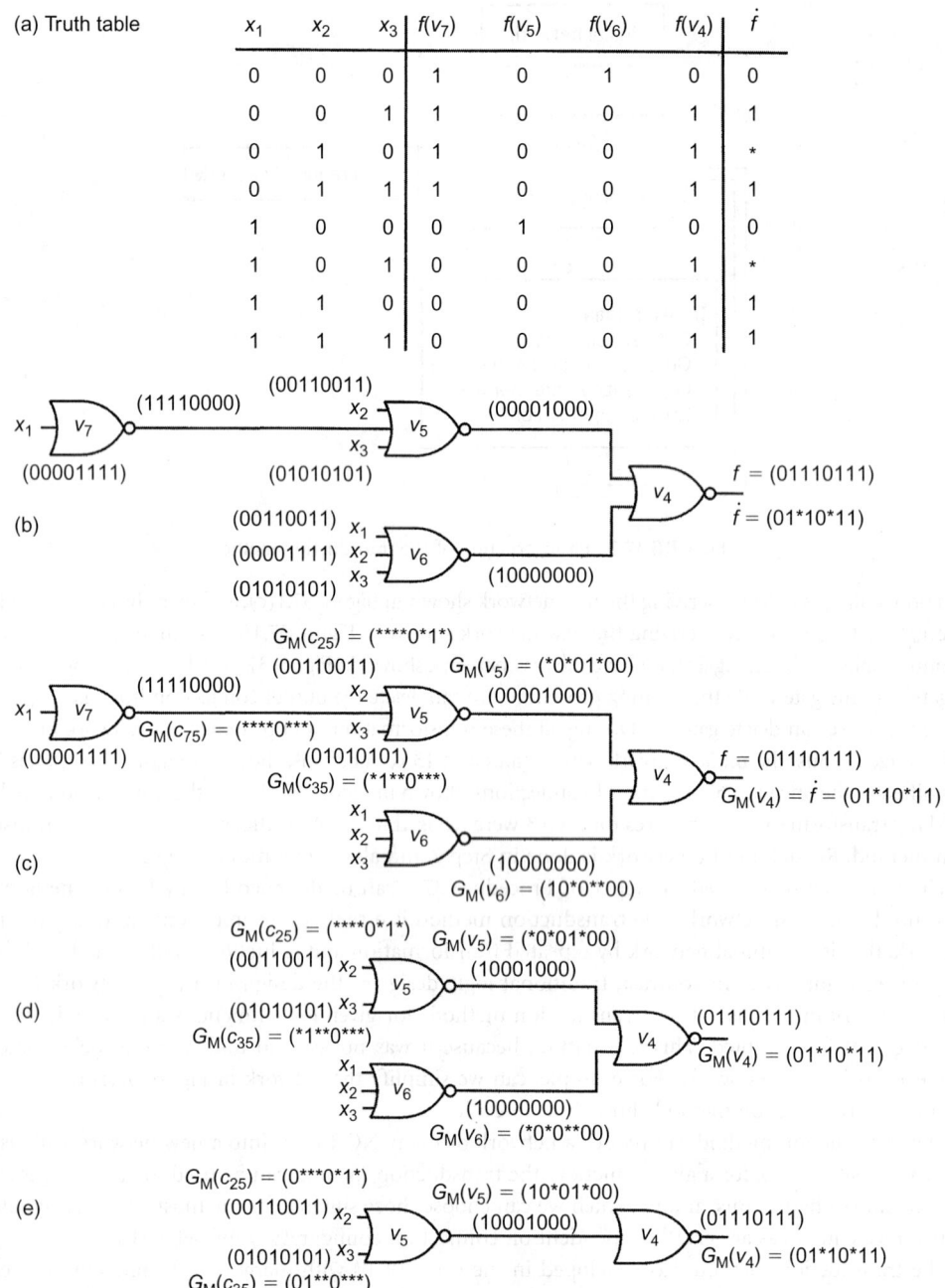

(a) Truth table

x_1	x_2	x_3	$f(v_7)$	$f(v_5)$	$f(v_6)$	$f(v_4)$	$\dot{f}$
0	0	0	1	0	1	0	0
0	0	1	1	0	0	1	1
0	1	0	1	0	0	1	*
0	1	1	1	0	0	1	1
1	0	0	0	1	0	0	0
1	0	1	0	0	0	1	*
1	1	0	0	0	0	1	1
1	1	1	0	0	0	1	1

FIGURE 37.3 Permissible functions of a network.

For example, we want to design a network for a function $\dot{f}(x_1, x_2, x_3)$ shown in the truth table in Figure 37.3(a). Suppose the network in Figure 37.3(b) is designed by some means to realize function $\dot{f}(x_1, x_2, x_3) = (01*10*11)$. The output function realized at each gate v_i in Figure 37.3(b) is denoted by $f(v_i)$ in Figure 37.3(a). The columns in Figure 37.3(a) are shown horizontally as vectors in Figure 37.3(b). For example, the output function of gate v_5, $f(v_5)$, is shown by vector (00001000) placed just outside gate v_5 in Figure 37.3(b). (Henceforth, columns in a truth table will be shown by vectors in figures for logic networks.) Notice that the function $\dot{f}(x_1, x_2, x_3) = (01{*}10{*}11)$ (denoted by f with a dot on top of it) is a function to be realized by a network and generally is incompletely specified, containing don't cares (i.e., *'s), whereas the output function $f(v_4)$ (denoted by f without a dot on top of it) of gate v_4, for example, in the actually realized network in Figure 37.3(b) is completely specified, containing no * because the value of the function $f(v_4)$ realized at the output of gate v_4 is 0 or 1 for each combination of values of input variables $x_1 = (00001111)$, $x_2 = (00110011)$, and $x_3 = (01010101)$ (assuming x_i contains no *'s in its vector for each i).

Let us find permissible functions at the output of gate v_5. Because the first component of the output $\dot{f}$ of gate v_4 is 0, the first component of the output of gate v_6 is 1, and v_4 is a NOR gate, the first component of a permissible function at the output of gate v_5 can be 0 or 1 (in other words, *). Because the second component of the output $\dot{f}$ of v_4 is 1 and v_4 is a NOR gate, the second component of a permissible function at the output of v_5 must be 0. Because the third component of $\dot{f}$ at v_4 is *, the third component of a permissible function at the output of gate v_5 can be 0 or 1 (i.e., *). Calculating every component in this manner, we will find that (00001000), (10001000), ... are all permissible functions at the output of gate v_5, including the original vector (00001000) shown at v_5 in Figure 37.3(b). In other words, even if we replace the function $f(v_5)$ realized at gate v_5 by any of these permissible functions, the network output at v_4 still realizes the given function $\dot{f}(x_1, x_2, x_3) = (01{*}10{*}11)$.

Notice that the value of the signal at a connection c_{ij} from gate v_i to v_j is always identical to the output value of v_i, but permissible functions for a connection c_{ij} are separately defined from those for a gate v_i. As we shall see later, this is important when gate v_i has more than one output connection. When v_i has only one output connection c_{ij}, permissible functions for v_i are identical to those for c_{ij} (the networks in Figure 37.3 are such cases and the permissible functions for each c_{ij} are identical to those for v_i); but when v_i has more than one output connection, they are not necessarily identical, as we will see later.

Usually, there is more than one permissible function for each of v_i and c_{ij}. But when we find a set of permissible functions for a gate or connection, all the vectors representing these permissible functions can be expressed by one vector, as discussed in the following. This is convenient for processing.

For example, suppose the output function $f(v_i)$ of a gate v_i in a network has the values shown in the column labeled $f(v_i)$ in the truth table in Table 37.1, where the network has input variables $x_1, x_2, \ldots, x_n$. Let us write this column as a vector, $f(v_i) = (0011\ldots)$. Suppose any outputs of the network do not change even if this function $f(v_i)$ in Table 37.1 is replaced by $g_1 = (0101\ldots)$, $g_2 = (0001\ldots), \ldots$, or $g_h = (0..1\ldots)$ shown in Table 37.1. In other words, $g_1, \ldots, g_h$ are permissible functions (they are not necessarily all the permissible functions at gate v_i). Then, permissible functions, g_1 through g_h, can be expressed by a single vector $G(v_i) = (0{*}{*}1\ldots)$ for the following reasons:

1. Suppose permissible functions, g_1 and g_2, in Table 37.1 differ only in their second components. In other words, the second component is 1 in g_1 and 0 in g_2. Consequently, even if the second component of the output function of gate v_i is don't care (i.e., *), no network output will change. Thus, $(0{*}01\ldots)$ (i.e., the vector $g_1 = (0101\ldots)$ with the second component replaced by *) is another permissible function at gate v_i. This can be interpreted as follows. Original permissible functions, g_1 and g_2, can be replaced by the new permissible function, interpreting * to mean 0 or 1; in other words, $(0{*}01\ldots)$ means $g_1 = (0101\ldots)$ and $g_2 = (0001\ldots)$.

TABLE 37.1 Truth Table

x_1	$\cdots$	x_{n-1}	x_n	$f(v_i)$	g_1	g_2	g_3	$\cdots$	g_h	$G(v_i)$
0	$\cdots$	0	0	0	0	0	0	$\cdots$	0	0
0	$\cdots$	0	1	0	1	0	1	$\cdots$	$\cdots$	*
0	$\cdots$	1	0	1	0	0	1	$\cdots$	$\cdots$	*
0	$\cdots$	1	1	1	1	1	1	$\cdots$	1	1
$\cdots$	$\cdots$			$\cdots$						$\cdots$
$\cdots$	$\cdots$			$\cdots$				$\cdots$		$\cdots$
$\cdots$	$\cdots$			$\cdots$						$\cdots$

When there is a permissible function g_j with * in the second component, any other permissible function, g_k, can be replaced by g_k itself with its second component replaced by *, even if other components of g_j may not be identical to those of g_k. For example, a permissible function, $g_3 =$ (0111 ...), in Table 37.1 can be replaced by another permissible function (0*11 ...). This is because the value of each component of the output function is independent of any other component (in other words, the value of the output function at each gate for a combination of values of $x_1, x_2, \ldots, x_n$ is completely independent of those for other combinations). This means that if the second component of one permissible function, g_j, is *, then the second components of all other permissible functions can be *, where $1 \leq j \leq h$.

2. Suppose the first and fourth components are 0 and 1, respectively, in permissible functions, g_1 through g_h.

3. Then, the set of permissible functions, $g_1, g_2, \ldots, g_h$, for gate v_i can be expressed as a single vector, $G(v_i) = (0{**}1\ldots)$, as shown in the last column in Table 37.1.

37.2.2 Pruning Procedure

From the basic properties of permissible functions just observed, we have the following.

Theorem 37.1: The set of all permissible functions at a gate (or an input terminal) or connection, or any subset of it, can be expressed by a single incompletely specified function.

Henceforth, let G denote the vector that this single incompletely specified function expresses. $G(v_i) = (0{**}1\ldots)$ in Table 37.1 is such an example.

When an arbitrary NOR gate network, such as Figure 37.3(b), is given, it is generally very difficult to find which connections and/or gates are unnecessary (i.e., redundant) and can be deleted from the network without changing any output functions of the network. But, it is easy to do so by the following fundamental property based on the concept of permissible function:

If a set of permissible functions at a gate, a connection, or an input terminal in a given NOR gate network has no 1-component, in other words, consists of components, 0's, *'s, or a mixture of 0's and *'s, then that gate (and its output connections), connection, or input terminal, is redundant. In other words, it can be removed from the network without changing the network outputs, because the function on this gate, connection, or input terminal, can be the vector consisting of all 0-components (by changing every * to 0); and even if all the output connections of it are removed from the NOR gates whose inputs have these connections, the output functions of the network do not change. (This is because an input to a NOR gate that has more than one input is 0, and it will not affect the output of that NOR gate and hence is removable.)

Based on this property, we can simplify the network by the following procedure:

Procedure 37.2: Pruning procedure

We calculate a set of permissible functions for every gate, connection, and input terminal in a network of NOR gates, starting with the connections connected to output terminals of the network and moving toward the input terminals of the network. Whenever the single vector, G, that this set of permissible functions expresses, is found to consist of only 0 or * components, without 1-components during the calculation, we can delete that gate, connection, or input terminal without changing any network output.   □

In this case, it is important to notice that we generally can prune only one of the gates, connections, and input terminals whose G's consist of only 0-components and *-components, as we will see later. If we want to prune more than one of them, we generally need to recalculate permissible functions of all gates, connections, and input terminals throughout the network after pruning one.

37.2.3 Calculation of Sets of Permissible Functions

Let us introduce simple terminology for later convenience. If there is a connection c_{ij} from the output of v_i to the input of v_j, v_i is called an **immediate predecessor** of v_j, and v_j is called an **immediate successor** of v_i. $IP(v_i)$ and $IS(v_i)$ denote the set of all immediate predecessors of v_i and the set of all immediate successors of v_i, respectively. If there is a path of gates and connections from the output of gate v_i to the input of v_j, then v_i is called a **predecessor** of v_j and v_j is called a **successor** of v_i. $P(v_i)$ and $S(v_i)$ denote the set of all predecessors of v_i and the set of all successors of v_i, respectively.

Thus far, we have considered only an arbitrary set of permissible functions at a gate, connection, or input terminal, v_i (or c_{ij}). Now let us consider the set of all permissible functions at v_i (or c_{ij}).

> **Definition 37.2:** The set of all permissible functions at any gate, connection, or input terminal is called the **maximum set of permissible functions**, abbreviated as **MSPF**. This set can be expressed by a single vector, as already shown.

The MSPF of a gate, connection, or input terminal can be found by examining which functions change the network outputs. But MSPFs throughout the network can be far more efficiently calculated by the following procedure.

Procedure 37.3: Calculation of MSPFs in a network

MSPFs of every gate, connection, and input terminal in a network can be calculated, starting from the outputs of the network and moving toward the input terminals, as follows:

1. Suppose we want to realize a network that has input variables $x_1, x_2, \ldots, x_n$ and output functions $f_1, f_2, \ldots, f_m$, that may contain don't-care components (i.e., *'s). Then, suppose we have actually realized a network with R NOR gates for these functions by some means and the output gates of this network realize $f_1, f_2, \ldots, f_m$, respectively. ($v_1, v_2, \ldots, v_n$ are regarded as input terminals where $x_1, x_2, \ldots, x_n$ are provided, respectively.) Notice that the output functions, $f_1, f_2, \ldots, f_m$, of this network do not contain any don't-care components because each network output function is 1 or 0 for each combination of values of $x_1, x_2, \ldots, x_n$. Calculate output function $f(v_j)$ at each gate v_j in this network.
   Then, the MSPF, $G_M(v_t)$, of each gate, v_t, whose output function is one of the network outputs, f_k, is set to $\dot{f}_k$ (not f_k, which is a completely specified function at v_t of the actually realized network); that is, $G_M(v_t) = \dot{f}_k$, where $k = 1, 2, \ldots, m$.

2. Suppose a gate v_j in the network has MSPF, $G_M(v_j)$. Then, the MSPF of each connection c_{ij} to the input of this gate can be calculated as follows:
   If a component of $G_M(v_j)$ is 0 and the disjunction (i.e., logical sum, or OR) of the values of the corresponding components of the input connections other than c_{ij} is 0, then the MSPF of input connection c_{ij} is 1. If they are 1 and 0, respectively, the MSPF of c_{ij} is 0. And so on. This operation, denoted by $\square$, is summarized in Table 37.2, where the first operand is a component of $G_M(v_j)$ and the second operand is the disjunction of the component values of the input connections other than c_{ij} and where "–" denotes "undefined" because v_j is a NOR gate and consequently these cases do not occur. Calculate each component of $G_M(v_j)$ in this manner.
   If this gate v_j has only one input connection c_{ij} (i.e., v_j becomes an inverter), the MSPF of c_{ij} is 1, 0, or * according as $G_M(v_j)$ is 0, 1, or *.
   This calculation can be formally stated by the following formula:

$$G_M(c_{ij}) = G_M(v_j) \, \square \, (\vee f(v)) \qquad (37.1)$$

$$v \in IP(v_j)$$

$$v \neq v_i$$

TABLE 37.2   Definition of □

	Second operand, $(\vee f(v))$ i.e., $v \in IP(v_j)$ $v \neq v_i$		
□	0	1	*
First operand, i.e., $G_M(v_i)$ 0	1	*	–
1	0	–	–
*	*	*	–

– undefined

where if $IP(v_i) = \{v_i\}$ (in other words, gate v_j has only one input connection from v_i), then

$$(\vee \; f(v))$$
$$v \in IP(v_j)$$
$$v \neq v_i$$

is regarded as the function that is constantly 0.

For example, let us calculate MSPFs of the network shown in Figure 37.3(b) which realizes the function $\dot{f}(x_1, x_2, x_3) = (01\!*\!10\!*\!11)$ given in Figure 37.3(a). Because gate v_4 is the network output, this $\dot{f}$ is the MSPF of gate v_4 according to Step 1; that is, $G_M(v_4) = (01\!*\!10\!*\!11)$ as shown in Figure 37.3(c). Gate v_4 has input connections realizing functions $f(c_{54}) = (00001000)$ and $f(c_{64}) = (10000000)$. Then, $G_M(c_{54})$, MSPF of c_{54}, can be calculated as follows. The first component of $G_M(c_{54})$ is *, using Table 37.2, because the first component of $G_M(v_4)$ is 0 and the first component of $f(c_{64})$ (which, in this case, is the disjunction of the first components of functions realized at connections other than c_{54} because c_{64} is the only connection other than c_{54}) is 1. Proceeding with the calculation based on Table 37.2, we have $G_M(c_{54}) = (*0\!*\!01\!*\!00)$. ($G_M(c_{54})$, and other $G_M(c_{ij})$'s are not shown in Figure 37.3.)

3. The MSPF of a gate or input terminal, v_i, can be calculated from $G_M(c_{ij})$ as follows:
   If a gate or input terminal, v_i, has only one output connection, c_{ij}, whose MSPF is $G_M(c_{ij})$, then $G_M(v_i)$, MSPF of v_i, is given by:

$$G_M(v_i) = G_M(c_{ij}) \tag{37.2}$$

Thus, we have $G_M(v_5) = G_M(c_{54})$ and this is shown in Figure 37.3(c).

If v_i has more than one output connection to gates v_j's, then $G_M(v_i)$ is not necessarily identical to $G_M(c_{ij})$. In this case, the MSPF for any gate or input terminal, v_i, in the network is given by the following H:

$$H = (H^{(1)}, H^{(2)}, \ldots, H^{(2^n)}) \tag{37.3}$$

whose w-th component is given by:
$H^{(w)} = *$ if $f_k^{(w)} \supseteq f_k'^{(w)}$ for every k such that $1 \leq k \leq m$, and
$H^{(w)} = f^{(w)}(v_i)$ if $f_k^{(w)} \supseteq f_k'^{(w)}$ does not hold for some k such that $1 \leq k \leq m$,
where $f_k' = (f_k'^{(1)}, f_k'^{(2)}, \ldots, f_k'^{(2^n)})$ is the new k-th output function of the network, to which the k-th output of the original network, f_k, changes by the following reconfiguration: insert an inverter between gate or input terminal, v_i, and its immediate successor gates, v_j's (so the immediate successor gates receive $\overline{f}(v_i)$ instead of the original function $f(v_i)$ at v_i). Here, "$\supseteq$,"

which means ⊃ or =, is used as an ordinary set inclusion, i.e., $^* \supseteq 0$, $^* \supseteq 1$, $1 \supseteq 1$, and $0 \supseteq 0$, interpreting * as the set of 1 and 0. $f^{(w)}(v_i)$ is the w-th component of $f(v_i)$. Notice that the calculation of the MSPF for v_i, based on Eq. 37.3, is done by finding out the new values of the network outputs, $f_1, f_2, \ldots, f_m$, for the preceding reconfiguration, without using $G_M(c_{ij})$'s. Thus, this calculation is complex and time-consuming.

4. Repeat Steps 2 and 3 until finishing the calculation of MSPFs throughout the network.

Let us continue the example in Figure 37.3(c), where input variables, x_1, x_2, and x_3, are at input terminals v_1, v_2, and v_3, respectively, and then for the input variables, $x_1 = f(v_1) = (0\ 0\ 0\ 0\ 1\ 1\ 1\ 1)$, $x_2 = f(v_2) = (0\ 0\ 1\ 1\ 0\ 0\ 1\ 1)$, and $x_3 = f(v_3) = (0\ 1\ 0\ 1\ 0\ 1\ 0\ 1)$, the outputs of gates are $f(v_4) = (0\ 1\ 1\ 1\ 0\ 1\ 1\ 1)$, $f(v_5) = (0\ 0\ 0\ 0\ 1\ 0\ 0\ 0)$, $f(v_6) = (1\ 0\ 0\ 0\ 0\ 0\ 0\ 0)$, and $f(v_7) = (1\ 1\ 1\ 1\ 0\ 0\ 0\ 0)$. Because gate v_5 has only one output connection, $G_M(v_5) = G_M(c_{54})$, according to Step 3. The first component of $G_M(c_{25})$ (i.e., the MSPF of connection, c_{25}) from input terminal v_2 (which actually has input variable x_2), is * by Table 37.2 because the first component of $G_M(v_5)$ is *. The second component of $G_M(v_2)$ is * by Table 37.2 because the second component of $G_M(v_5)$ is 0 and the disjunction of the second components of $f(c_{75})$ and $f(c_{35})$ is $1 \vee 1 = 1$. And so on. □

37.2.4 Derivation of an Irredundant Network Using the MSPF

The following procedure provides a quick means to identify redundant gates, connections, or input terminals in a network. By repeating the calculation of MSPF at all gates, connections, and input terminals throughout the network and the deletion of some of them by the pruning procedure (Procedure 37.2), we can derive an irredundant logic network as follows, where "**irredundant network**" means that if any connection, gate, or input terminal is deleted, some network outputs will change; in other words, every connection, gate, or input terminal is not redundant.

Procedure 37.4: Derivation of an irredundant network using the MSPF

1. Calculate the MSPFs for all gates, connections, and input terminals throughout the network by Procedure 37.3, starting from each output terminal and moving toward input terminals.
2. During Step 1, we can remove a connection, gate, or input terminal, without changing any output function of the network, as follows:
   a. If there exists any input connection of a gate whose MSPF consists of 0's and *'s only, remove it. We can remove only one connection because if two or more connections are simultaneously removed, some network outputs may change.
   b. As a result of removing any connection, there may be a gate without any output connections. Then remove such a gate.
3. If a connection and possibly a gate are removed in Step 2, return to Step 1 with the new network. Otherwise, go to Step 4.
4. Terminate the procedure and we have obtained an irredundant network. □

This procedure does not necessarily yield a network with a minimum number of gates or connections, because an irredundant network does not necessarily have a minimum number of gates or connections. But the procedure is useful in many cases because every network with a minimum number of gates or connections must be irredundant and an obtained network may be minimal.

Example 37.1: Let us apply Procedure 37.4 to the network shown in Figure 37.3(c). Because each gate has only one output connection, $G_M(v_i) = G_M(c_{ij})$ holds for every v_i. Thus, using Eq. 37.1 and $G_M(v_4) = \dot{f}(x_1, x_2, x_3)$, MSPFs are obtained as follows:

$$G_M(c_{54}) = G_M(v_5) = G_M(v_4) \ \square \ f(v_6) = (^*\ 0\ ^*\ 0\ 1\ ^*\ 0\ 0),$$

$$G_M(c_{64}) = G_M(v_6) = G_M(v_4) \ \square \ f(v_5) = (1\ 0\ ^*\ 0\ ^*\ ^*\ 0\ 0), \text{ and}$$

$$G_M(c_{75}) = G_M(v_7) = G_M(v_5) \ \square \ (x_2 \vee x_3) = (^*\ ^*\ ^*\ ^*\ 0\ ^*\ ^*\ ^*)$$

Then we can remove connection c_{75} and consequently gate v_7 by the pruning procedure, as shown in Figure 37.3(d). For this new network, we obtain the following:

$$f(v_4) = (0\ 1\ 1\ 1\ 0\ 1\ 1\ 1), f(v_5) = (1\ 0\ 0\ 0\ 1\ 0\ 0\ 0),$$

$$f(v_6) = (1\ 0\ 0\ 0\ 0\ 0\ 0\ 0),$$

$$G_M(v_4) = \dot{f}(x_1, x_2, x_3) = (0\ 1\ ^*\ 1\ 0\ ^*\ 1\ 1),$$

$$G_M(c_{54}) = G_M(v_5) = (^*\ 0\ ^*\ 0\ 1\ ^*\ 0\ 0), \text{ and}$$

$$G_M(c_{64}) = G_M(v_6) = G_M(v_4)\ \square\ f(v_5) = (^*\ 0\ ^*\ 0\ ^*\ ^*\ 0\ 0)$$

Then we can remove connection c_{64} and then gate v_6. Thus, the original network of four gates and nine connections is reduced to the network of two gates with three connections in Figure 37.3(e), where G_M's are recalculated and no gate can be removed by the pruning procedure.   □

Whenever we delete any one connection, gate, or input terminal by the pruning procedure, we need to calculate MSPFs of the new network from scratch. If we cannot apply the pruning procedure, the final logic network is irredundant. This means that if any gate, connection, or input terminal is deleted from the network, some of the network outputs change. It is important to notice that an irredundant network is completely testable; that is, we can detect, by providing appropriate values to the input terminals, whether or not the network contains faulty gates or connections. (A redundant network may contain gates or connections such that it is not possible to detect whether they are faulty.)

Use of MSPFs is time-consuming because the calculation of MSPFs by Eq. 37.3 in Step 3 of Procedure 37.3 is time-consuming and also because we must recalculate MSPFs whenever we delete a connection, gate, or input terminal by the pruning procedure.

37.2.5   Calculation of Compatible Sets of Permissible Functions

Reduction of calculation time is very important for developing a practical procedure. For this purpose, we introduce the concept of a compatible set of permissible functions which is a subset of an MSPF.

Definition 37.3: Let V be the set of all the gates and connections in a network, and e be a gate (or input terminal) v_i or a connection c_{ij} in this set. All the sets of permissible functions for all e's in V, denoted by $G_C(e)$'s, are called **compatible sets of permissible functions** (abbreviated as **CSPFs**), if the following properties hold with every subset T of V:

     a. Replacement of the function at each gate or connection, t, in set T by a function $f(t) \in G_C(t)$ results in a new network where each gate or connection, e, such that $e \notin T$ and $e \in V$, realizes function $f(e) \in G_C(e)$.

     b. The condition a holds, no matter which function $f(t)$ in $G_C(t)$ is chosen for each t.   □

Consequently, even if the function of a gate or connection is replaced by $f(v_t) \in G_C(v_t)$, the function $f(v_u)$ at any other gate or connection still belongs to the original $G_c(v_u)$.

CSPFs at all gates and connections in a network can be calculated as follows.

Procedure 37.5: Calculation of CSPFs in a logic network

CSPFs of all gates, connections, and input terminals in a network can be calculated, starting from the network outputs and moving toward the input terminals, as follows:

     1. Suppose we want to realize a network that has inputs $x_1, x_2 \ldots x_n$ and output functions $\dot{f}_1, \dot{f}_2, \ldots,$ $\dot{f}_m$, which may contain don't-care components (i.e., *'s). Then, suppose we have actually realized a network with R NOR gates for these functions by some means.

Calculate output function $f(v_j)$ at each gate v_j in this network.

Then, CSPF of each gate, v_p whose output function is one of the network outputs, f_k, is

$G_C(v_t) = \dot{f}_k$ (not f_k), where $k = 1, 2, \ldots, m$.

2. Unlike MSPFs, the CSPF at each gate, connection, or input terminal is highly dependent on the order of processing gates, connections, and input terminals in a network. So, before starting the calculation of CSPFs, we need to determine an appropriate order of processing. Selection of a good processing **order**, denoted by r, is important for the development of an efficient transduction method. Let $r(v_i)$ denote the ordinal number of gates or connections, v_i in this order.

Suppose a gate v_j in the network has CSPF, $G_C(v_j)$. Then, CSPF of each input connection c_{ij}, $G_C(c_{ij})$, of this gate can be calculated by the following formula which is somewhat different from Eq. 37.1:

$$G_C(c_{ij}) = \{G_C(v_j) \square (\vee f(v))\} \# f(v_i)$$

$$v \in IP(v_j)$$

$$r(v) > r(v_i) \tag{37.4}$$

by calculating the terms on the right-hand side in the following steps:

a. Calculate the disjunction (i.e., component-wise disjunction) of all the functions $f(v_t)$'s of immediate predecessor gates, v_t's, of the gate v_j, whose ordinal numbers $r(v_t)$'s are greater than the ordinal number $r(v_i)$ of the gate v_i. If there is no such gate v_p the disjunction is 0.

b. Calculate the expression inside { } in Eq. 37.4 by using Table 37.2 (the definition of $\square$). In this table, $G_C(v_j)$ is used as the first operand, and

$$(\vee f(v))$$

$$v \in IP(v_j)$$

$$r(v) > r(v_i)$$

(i.e., the disjunction calculated in Step a) is used as the second operand.

c. Calculate the right-hand side of Eq. 37.4 using Table 37.3, with the value calculated in Step b as the first operand and $f(v_i)$ as the second operand.

For example, suppose gate v_j in Figure 37.4 has CSPF, $G_C(v_j) = (010^*)$, and input connections c_{ij}, c_{gj}, and c_{hj}, from gates v_i, v_g, and v_h, whose functions are $f(v_i) = (1001)$, $f(v_g) = (0011)$, and $f(v_h) = (0010)$. (Notice that $f(v_a) = f(c_{ab})$ always holds for any gate or input terminal, v_a, no matter whether v_a has more than one output connection.) Suppose we choose an order, $r(v_i) < r(v_g) < r(v_h)$. Then the first component of $G_C(c_{ij})$ is $\{ 0 \square (0 \vee 0) \}$#1 because the first components of $G_C(v_j)$, $f(v_g)$, $f(v_h)$, and $f(v_i)$, which appear in this order in Eq. 37.4, are 0, 0, 0, and 1, respectively. Using Tables 37.2 and 37.3, this becomes $\{ 0 \square (0 \vee 0) \}$#1 $=\{ 0 \square 0 \}$#1 $= 1$#1 $= 1$. Calculating

TABLE 37.3 Definition of #

	#	Second operand, i.e., $f(v_i)$		
		0	1	*
First operand, i.e.,	0	0	*	*
$G_C(v_j) \square (\vee f(v))$	1	*	1	*
	*	*	*	*
$v \in IP(v_j)$				
$r(v) > r(v_i)$				

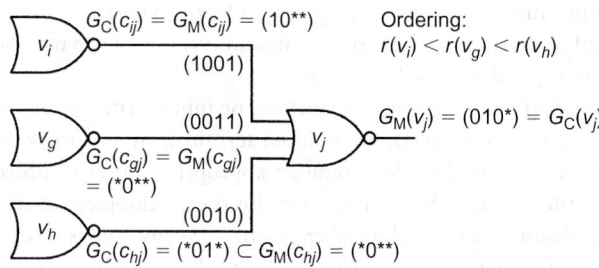

FIGURE 37.4 An example of the calculation of MSPFs and CSPFs.

other components similarly, we have $G_C(c_{ij}) = (10^{**})$. We have $G_C(c_{gj}) = (^*0^{**})$ because the fourth component of $G_C(c_{gj})$, for example, is $\{^* \square\ 0\}\#1$ since the fourth components of $G_C(v_j), f(v_h)$, and $f(v_g)$ are *, 0, and 1, respectively. In this case, the value of $f(v_i)$ is not considered unlike the calculation of MSPF, $G_M(c_{gj})$, because $f(v_i)$ is not included in Eq. 37.4 due to the order, $r(v_i) < r(v_g)$. Also, we have $G_C(c_{hj}) = (^*01^*)$ because the fourth component of $G_C(c_{hj})$, for example, becomes $\{^* \square\ 0\}\#0$ since the fourth components of $G_C(v_j)$,

$$(\vee f(v))$$

$$v \in IP(v_j),$$

$$r(v) > r(v_i)$$

and $f(v_h)$ are *, 0 (because no gate v such that $r(v) > r(v_h)$), and 0, respectively. In this case, $f(v_i)$ and $f(v_g)$ are not considered.

For comparison, let us also calculate MSPFs by Procedure 37.3. The MSPFs of connections c_{ij}, c_{gj}, and c_{hj} can be easily calculated as $G_M(c_{ij}) = (10^{**})$, $G_M(c_{gj}) = (^*0^{**})$, and $G_M(c_{hj}) = (^*0^{**})$, respectively, as shown in Figure 37.4. Comparing with the CSPFs, we can find $G_C(c_{ij}) = G_M(c_{ij})$ and $G_C(c_{gj}) = G_M(c_{gj})$. But $G_C(c_{hj})$ is a subset of $G_M(c_{hj})$ (denoted as $G_C(c_{hj}) \subset G_M(c_{hj})$) because the third component of $G_C(c_{hj}) = (^*01^*)$ is 1, which is a subset of the third component, * (i.e., 0 or 1), of $G_M(c_{hj}) = (^*0^{**})$, while other components are identical.

The CSPF and MSPF of a gate, connection, or input terminal can be identical. For the gate with $G_C(v_j) = G_M(v_j) = (010^*)$ shown in Figure 37.5, for example, we have $G_C(c_{ij}) = G_M(c_{ij}) = (10^{**})$, $G_C(c_{gj}) = G_M(c_{gj}) = (^*0^{**})$, and $G_C(c_{hj}) = G_M(c_{hj}) = (^*01^*)$ with the ordering $r(v_i) < r(v_g) < r(v_h)$. As can be seen in the third components of G_M's in the example in Figure 37.4, when gate v_j in a network has more than one input connection whose w-th component is 1 and we have $G_M^{(w)}(v_j) = 0$, the w-th components of MSPFs for all these input connections are *'s, as seen in the third components of $G_M(c_{ij})$, $G_M(c_{gj})$, and $G_M(c_{hj})$ in the example in Figure 37.4. But the w-th components of CSPFs, however, are *'s except for one input connection whose value is required to be 1, as seen in the third components of $G_C(c_{hj})$ in the example in Figure 37.4. Which input connection is such an input connection depends upon order r. Intuitively, an input connection to the gate v_j from an immediate predecessor gate that has a smaller ordinal number in order r will probably tend to have more *'s in its CSPF and, consequently, have a greater probability for this input connection to be removed.

3. The CSPF of a gate or input terminal, v_i, can be calculated from $G_C(c_{ij})$ as follows.
   If a gate or input terminal, v_i, has only one output connection, c_{ij}, whose CSPF is $G_C(c_{ij})$, then $G_C(v_i)$, CSPF of v_i, is given by

$$G_C(v_i) = G_C(c_{ij}) \tag{37.5}$$

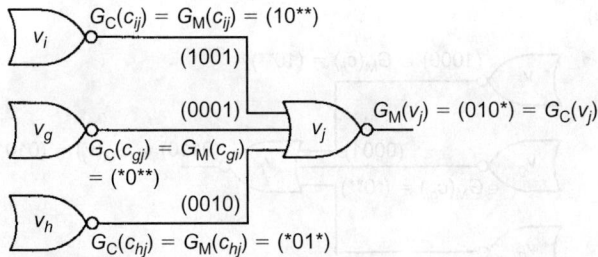

FIGURE 37.5 An example of the calculation of MSPFs and CSPFs.

If v_i has more than one output connection, then $G_C(v_i)$ is not necessarily identical to $G_C(c_{ij})$. In this case, the CSPF for any gate or input terminal, v_i, in a network is given by the following:

$$G_C(v_i) = \bigcap_{v_j \in IS(v_i)} G_C(c_{ij}) \tag{37.6}$$

where the right-hand side of Eq. 37.6 is the intersection of $G_C(c_{ij})$'s of output connections, c_{ij}'s, of gate v_i.

Unlike Eq. 37.3 for the case of MSPFs, Eq. 37.6 is simple and can be calculated in a short time.

4. Repeat Steps 2 and 3 until the calculation of CSPFs throughout the network is finished. □

A gate or connection may have different CSPFs if the order r of processing is changed. On the other hand, each gate or connection has a unique MSPF, independent of the order of processing.

37.2.6 Comparison of MSPF and CSPF

It is important to notice the difference in the ways of defining MSPF and CSPF. Any function $f(v_i)$ (or $f(c_{ij})$), that belongs to the MSPF of a gate, connection, or input terminal, v_i (or c_{ij}), can replace the original function realized at this v_i (or c_{ij}) without changing any network output, keeping the functions at all other gates, connections, or input terminals intact. If functions at more than one gate, connection, and/or input terminal are simultaneously replaced by permissible functions in their respective MSPFs, some network outputs may change. In the case of CSPF, simultaneous replacement of the functions at any number of gates, connections, and input terminals by permissible functions in their respective CSPFs does not change any network output.

> **Example 37.2:** This example illustrates that if functions realized at more than one gate, connection, or input terminal are simultaneously replaced by permissible functions in their respective MSPFs, some network outputs may change, whereas simultaneous replacement by permissible functions in their respective CSPFs does not change any network outputs. Let us consider the network in Figure 37.6(a) where all the gates have the same MSPFs as those in Figure 37.4. In Figure 37.6(a), let us simultaneously replace functions (1001), (0011), and (0010) realized at the inputs of gate v_j in Figure 37.4 by (1000), (0001), and (1001), respectively, such that $(1000) \in G_M(c_{ij}) = (10^{**})$, $(0001) \in G_M(c_{gj}) = (^*0^{**})$, and $(1001) \in G_M(c_{hj}) = (^*0^{**})$ hold. Then the output function of gate v_j in Figure 37.6(a) becomes (0110). But we have $(0110) \notin G_M(v_j)$ because the third component 1 is different from the third component 0 of $G_M(v_j)$. So, (0110) is not a permissible function in MSPF, $G_M(v_j)$. But if we replace the function at only one input to gate v_j by a permissible function of that input, the output function of v_j is still a permissible function in MSPF, $G_M(v_j)$. For example, if only (0011) at the second input of gate v_j in Figure 37.4

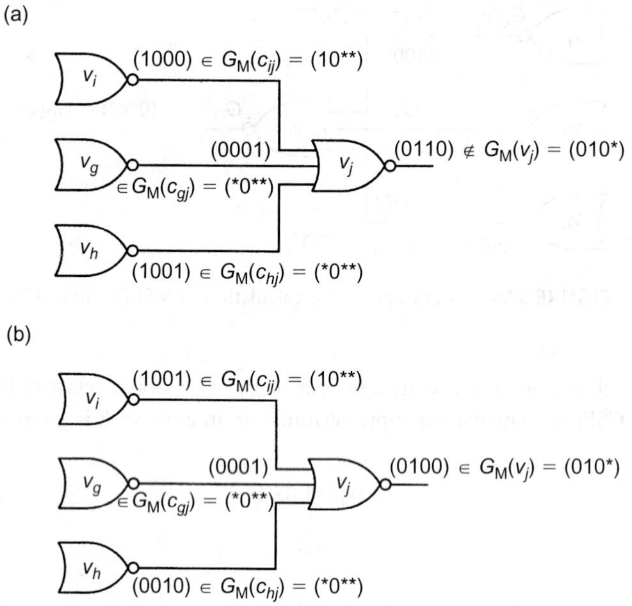

(a)

(1000) ∈ $G_M(c_{ij})$ = (10**)

v_i

(0001)
∈ $G_M(c_{gj})$ = (*0**)

v_g

(0110) ∉ $G_M(v_j)$ = (010*)

v_j

v_h

(1001) ∈ $G_M(c_{hj})$ = (*0**)

(b)

(1001) ∈ $G_M(c_{ij})$ = (10**)

v_i

(0001)
∈ $G_M(c_{gj})$ = (*0**)

v_g

(0100) ∈ $G_M(v_j)$ = (010*)

v_j

v_h

(0010) ∈ $G_M(c_{hj})$ = (*0**)

FIGURE 37.6 MSPFs.

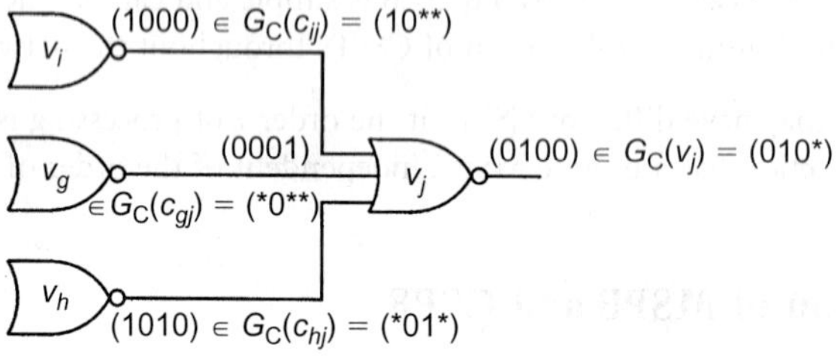

(1000) ∈ $G_C(c_{ij})$ = (10**)

v_i

(0001)
∈ $G_C(c_{gj})$ = (*0**)

v_g

(0100) ∈ $G_C(v_j)$ = (010*)

v_j

v_h

(1010) ∈ $G_C(c_{hj})$ = (*01*)

FIGURE 37.7 CSPFs.

is replaced by (0001), the output function of v_j becomes (0100), which is still a permissible function of $G_M(v_j)$, as shown in Figure 37.6(b).

If we use CSPF, we can replace more than one function. For example, let us consider the network in Figure 37.7, where gate v_j has $G_C(v_j)$, the same as $G_C(v_j)$ in Figure 37.4. The functions at the inputs of gate v_j in Figure 37.7 belong to CSPFs calculated in Figure 37.4; in other words, (1000) ∈ $G_C(c_{ij})$ = (10**), (0001) ∈ $G_C(c_{gj})$ = (*0**), and (1010) ∈ $G_C(c_{hj})$ = (*01*). Even if all functions (1001), (0011), and (0010) in Figure 37.4 are simultaneously replaced by these functions, function (0100) realized at the output of gate v_j is still a permissible function in $G_C(v_j)$. □

Procedures based on CSPFs have the following advantages and disadvantages:

1. For the calculation of CSPFs, we need not use Eq. 37.3 which is time-consuming.
2. Even if a redundant connection is removed, we need not recalculate CSPFs for the new network. In other words, CSPFs at different locations in the network are independent of one another, whereas MSPFs at these locations may not be. Thus, using CSPFs, we can simultaneously remove more than one connection; whereas using MSPFs, we need to recalculate MSPFs throughout the network, whenever one connection is removed.

  If, however, we use CSPFs instead of MSPFs, we may not be able to remove some redundant connections by the pruning procedure because each CSPF is a subset of its respective MSPF and depends on processing order r. Because gates with smaller ordinal number in order, r, tend to have more *-components, the probabilities of removing these gates (or their output connections) are greater. Thus, if a gate is known to be irredundant, or hard to remove, we can assign a larger ordinal number in order r to this gate and this will help giving *-components to the CSPFs of other gates.

3. The property (2) is useful for developing network transformation procedures based on CSPFs, which will be discussed later.

4. Because each CSPF is a subset of a MSPF, the network obtained by the use of CSPFs is not necessarily irredundant. But if we use MSPFs for pruning after repeated pruning based on CSPFs, then the final network will be irredundant.

For these reasons, there is a tradeoff between the processing time and the effectiveness of procedures.

37.2.7 Transformations

We can delete redundant connections and gates from a network by repeatedly applying the pruning procedure (in other words, by repeating only Step 2, without using Step 3, in Procedure 37.1, the outline of the transduction method). In this case, if MSPF is used, as described in Procedure 37.4, the network that results is irredundant. However, to have greater reduction capability, we developed several transformations of a network. By alternatively repeating the pruning procedure (Step 2 in Procedure 37.1) and transformations (Step 3), we can reduce networks far more than by the use of only one of them.

The following gate substitution procedure is one of these transformations.

Procedure 37.6: Gate substitution procedure

If there exist two gates (or input terminals), v_i and v_j, satisfying the following conditions, all the output connections of v_j can be replaced by the output connections of v_i without changing network outputs. Thus, v_j is removable.

1. $f(v_i) \in G(v_j)$, where $G(v_j)$ is a set of permissible functions of v_j which can be an MSPF or a CSPF.
2. v_i is not a successor of v_j (no loop will be formed by this transformation). ☐

This is illustrated in Figure 37.8. The use of MSPFs may give a better chance for a removal than their subsets such as CSPFs, although the calculation of MSPFs is normally time-consuming.

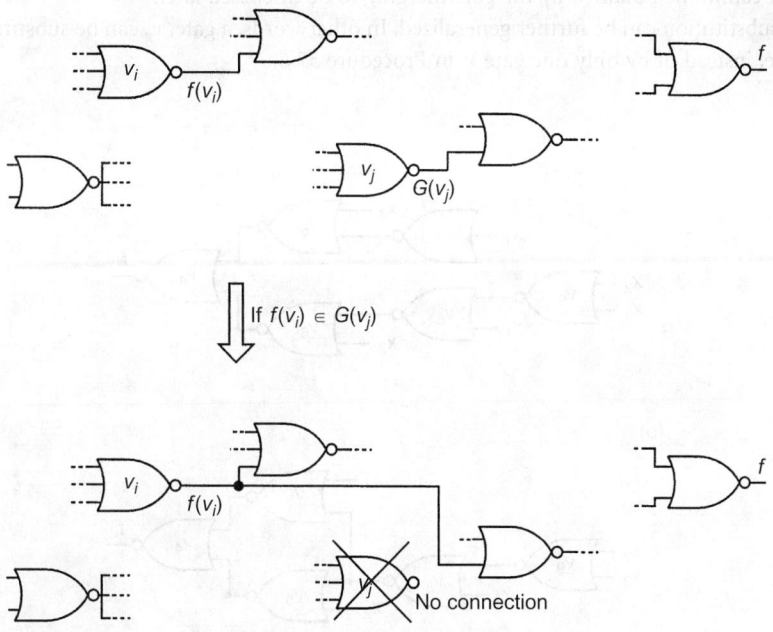

FIGURE 37.8 Gate substitution.

Example 37.3: Let us apply Procedure 37.6 to the network shown in Figure 37.9(a), which realizes the function $f = \bar{x}_1 x_2 \vee \bar{x}_1 x_3 \vee x_2 x_3$. Functions realized at input terminals and gates are as follows:

$$x_1 = f(v_1) = (0\ 0\ 0\ 0\ 1\ 1\ 1\ 1),\ x_2 = f(v_2) = (0\ 0\ 1\ 1\ 0\ 0\ 1\ 1),$$
$$x_3 = f(v_3) = (0\ 1\ 0\ 1\ 0\ 1\ 0\ 1),$$
$$f(v_4) = (0\ 1\ 1\ 1\ 0\ 0\ 0\ 1),\ f(v_5) = (0\ 0\ 0\ 0\ 1\ 0\ 1\ 0),$$
$$f(v_6) = (1\ 0\ 0\ 0\ 1\ 1\ 0\ 0),\ f(v_7) = (1\ 1\ 1\ 1\ 0\ 0\ 0\ 0),$$
$$f(v_8) = (0\ 1\ 1\ 1\ 0\ 0\ 0\ 0),\ \text{and}\ f(v_9) = (1\ 0\ 0\ 0\ 1\ 0\ 0\ 0)$$

Let us consider the following two different approaches.

1. Transformation by CSPFs: CSPFs for the gates are calculated as follows if at each gate, the input in the lower position has higher processing order than the input in the upper position:

$$G_C(v_4) = (0\ 1\ 1\ 1\ 0\ 0\ 0\ 1),\ G_C(v_5) = (*\ 0\ 0\ 0\ *\ *\ 1\ 0),$$
$$G_C(v_6) = (1\ 0\ 0\ 0\ 1\ 1\ *\ 0),\ G_C(v_7) = (*\ *\ 1\ *\ *\ *\ 0\ *),$$
$$G_C(v_8) = (0\ 1\ *\ *\ 0\ 0\ *\ *),\ \text{and}\ G_C(v_9) = (1\ 0\ *\ *\ *\ *\ *\ *)$$

Because $f(v_8) \in G_C(v_7)$, the network in Figure 37.9(b) is obtained by substituting connection c_{86} for c_{75}. Then, gate v_7 is removed, yielding a simpler network.

2. Transformation by MSPFs: In this case, the calculation of MSPFs is very easy because each gate in Figure 37.9(a) has only one output connection. MSPFs for gates are as follows:

$$G_M(v_4) = (0\ 1\ 1\ 1\ 0\ 0\ 0\ 1),\ G_M(v_5) = (*\ 0\ 0\ 0\ *\ *\ 1\ 0),$$
$$G_M(v_6) = (1\ 0\ 0\ 0\ *\ 1\ *\ 0),\ G_M(v_7) = (*\ *\ 1\ *\ *\ *\ 0\ *),$$
$$G_M(v_8) = (0\ 1\ *\ *\ *\ 0\ *\ *),\ \text{and}\ G_M(v_9) = (1\ 0\ *\ *\ *\ *\ *\ *)$$

Here, $G_M(v_7) = G_C(v_7)$, and we get the same result in Figure 37.9(b).

This result cannot be obtained by the gate merging to be discussed later. □

This gate substitution can be further generalized. In other words, a gate, v_p, can be substituted by more than one gate, instead of by only one gate v_i in Procedure 37.6.

(a)

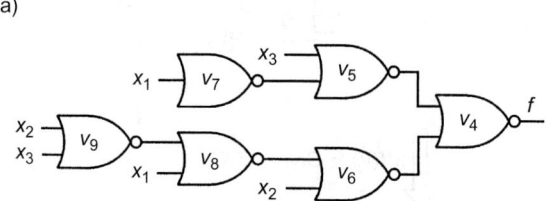

(b)

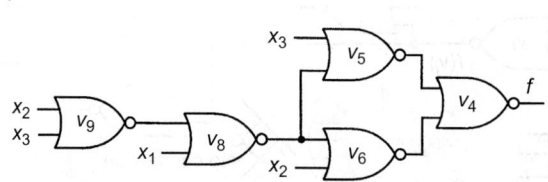

FIGURE 37.9 An example of gate substitution.

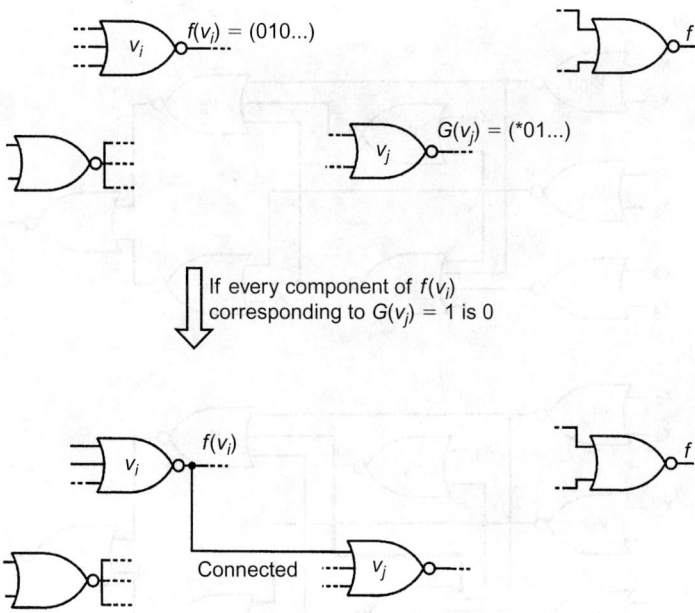

FIGURE 37.10 Connectable condition.

If we connect a new input to a gate or disconnect an existing input from a gate, the output of the gate may be modified. But if the new output is still contained in the set of permissible functions at this gate, the modification does not change the network outputs. By applying this procedure, we can change the network configuration and possibly can remove connections and/or gates. Even if we cannot reduce the network, a modification of the network is useful for further applications of other transformations. We have such a procedure if the connectable condition stated in the following or the disconnectable condition stated after the following is satisfied.

Procedure 37.7: Connectable condition

Let $G(v_j)$ be a set of permissible functions for gate v_j which can be an MSPF or a CSPF. We can add a connection from input terminal or gate, v_i, to v_j without changing network outputs, if the following conditions are satisfied:

 1. $f^{(w)}(v_i) = 0$ for all w's such that $G^{(w)}(v_j) = 1$.
 2. v_i is not a successor of v_j (no loop will be formed by this transformation).

This is illustrated in Figure 37.10.

This transformation procedure based on the connectable condition can be extended into the forms that can be used to change the configuration of a network. When we cannot apply any transformations to a given network, we may be able to apply those transformations after these extended transformations.

Example 37.4: If we add two connections to the network in Figure 37.11(a), as shown in bold lines in Figure 37.11(b), then the output connection (shown in a dotted line) of gate v_{12} becomes disconnectable and v_{12} can be removed. Then we have the network shown in Figure 37.11(c). ☐

Procedure 37.8: Disconnectable condition

If we can find a set of inputs of gate v_k such that the disjunction of the w-th component of the remaining inputs of v_k is 1 for every w satisfying $G^{(w)}(v_k) = 0$, then this set of inputs can be deleted, as being redundant, without changing network outputs. ☐

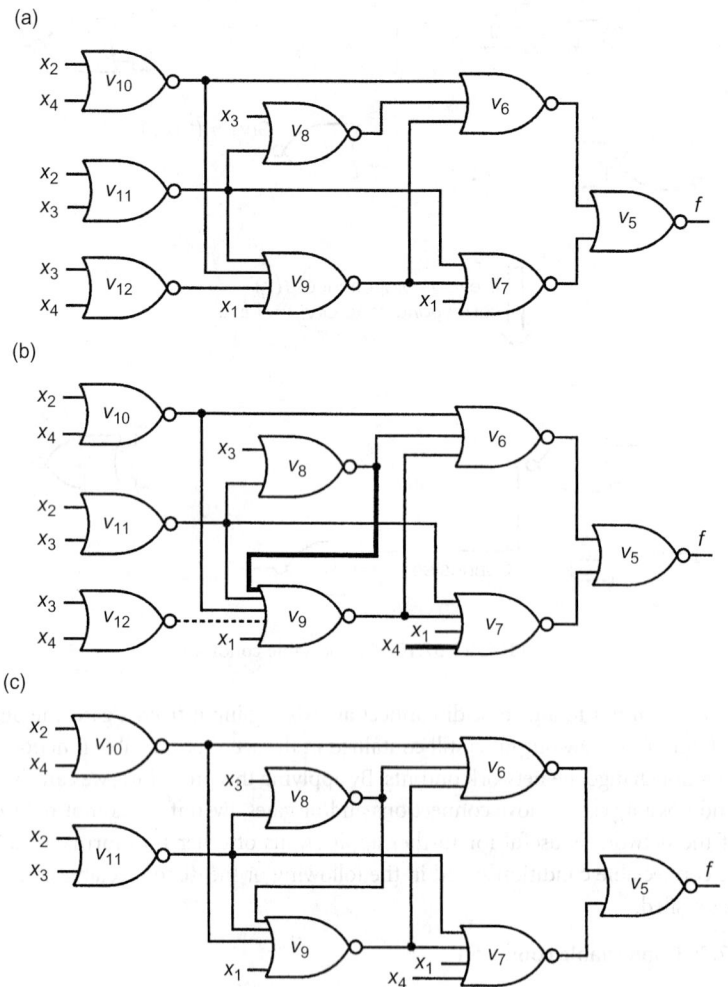

FIGURE 37.11 An example of the connectable/disconnectable conditions.

Procedures 37.7 and 37.8 will be collectively referred as the **connectable/disconnectable conditions** (or **procedures**).

In the network in Figure 37.12(a), x_2 is connectable to gate v_6, and x_1 is connectable to gate v_7. After adding these two connections, the outputs of v_6 and v_7 become identical, so v_7 can be removed as shown in Figure 37.12(b). This transformation is called **gate merging**. This can be generalized, based on the concept of permissible functions, as follows.

Procedure 37.9: Generalized gate merging

1. Find two gates, v_i and v_j, such that the intersection, $G_C(v_i)G_C(v_j)$, of their CSPFs is not empty, as illustrated in Figure 37.13.
2. Consider an imaginary gate, v_{ij}, whose CSPF is to be $G_C(v_i)G_C(v_j)$.
3. Connect all the inputs of gate v_i and v_j to v_{ij}. If v_{ij} actually realizes a function in $G_C(v_i)G_C(v_j)$, then v_{ij} can be regarded as a merged gate of v_i and v_j. Otherwise, v_i and v_j cannot be merged without changing network outputs in this generalized sense.
4. If v_{ij} can replace both v_i and v_j, then remove redundant inputs of v_{ij}.   □

Next, let us outline another transformation, called the **error compensation procedure**. In order to enhance the gate removal capability of the transduction method, the concept of permissible function is

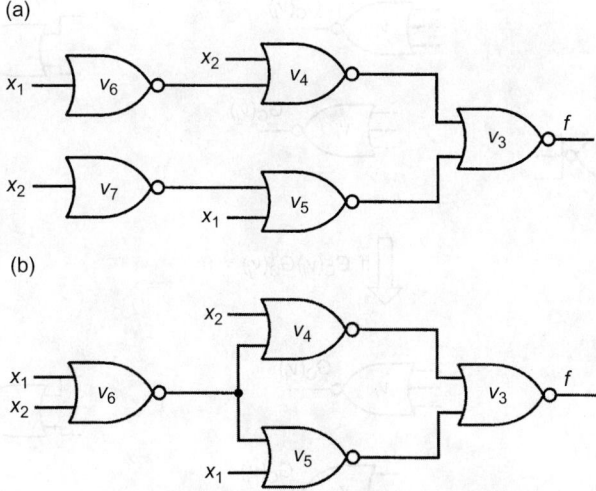

FIGURE 37.12 An example of gate merging.

generalized to "a permissible function with errors" [5]. Because the transformation procedures based on the error compensation are rather complicated, we outline the basic idea of these procedures along with an example, as follows:

1. Remove an appropriate gate or connection from the network.
2. Calculate errors in components of functions at gates or connections that are caused by the removal of the gate or connection, and then calculate permissible functions with errors throughout the network. These permissible functions with errors represent functions with erroneous components (i.e., components whose values are erroneous) as well as ordinary permissible functions that have no erroneous components.
3. Try to compensate for the errors by changing the connection configuration of the network. In order to handle the errors, the procedures based on ordinary permissible functions are modified.

Example 37.5: Figure 37.14(a) shows a network whose output at gate v_5 realizes

$$(1\ 0\ 0\ 0\ 0\ 0\ 0\ 0\ 0\ 0\ 1\ 0\ 1\ 1\ 0\ 1)$$

In order to reduce the network, let us remove gate v_8, having the network in Figure 37.14(b) whose output at gate v_5 is $(1\ 0\ 0\ 0\ 0\ \underline{1}\ 0\ 0\ 0\ 0\ 1\ 0\ 1\ 1\ 0\ 1)$.

Note that the outputs of the two networks differ only in the 6-th components (underlined). We want to compensate for the value of the erroneous component of the latter network by adding connections. Functions realized at the input terminals v_1 through v_4 and gates v_5 through v_{12} in the original network in Figure 37.14(a) are as follows:

$$x_1 = f(v_1) = (0\ 0\ 0\ 0\ 0\ 0\ 0\ 0\ 1\ 1\ 1\ 1\ 1\ 1\ 1\ 1)$$
$$x_2 = f(v_2) = (0\ 0\ 0\ 0\ 1\ 1\ 1\ 1\ 0\ 0\ 0\ 0\ 1\ 1\ 1\ 1)$$
$$x_3 = f(v_3) = (0\ 0\ 1\ 1\ 0\ 0\ 1\ 1\ 0\ 0\ 1\ 1\ 0\ 0\ 1\ 1)$$
$$x_4 = f(v_4) = (0\ 1\ 0\ 1\ 0\ 1\ 0\ 1\ 0\ 1\ 0\ 1\ 0\ 1\ 0\ 1)$$
$$f(v_5) = (1\ 0\ 0\ 0\ 0\ 0\ 0\ 0\ 0\ 0\ 1\ 0\ 1\ 1\ 0\ 1)$$
$$f(v_6) = (0\ 0\ 0\ 0\ 1\ 0\ 1\ 0\ 0\ 0\ 0\ 0\ 0\ 0\ 1\ 0)$$
$$f(v_7) = (0\ 1\ 0\ 1\ 0\ 0\ 0\ 0\ 1\ 1\ 0\ 1\ 0\ 0\ 0\ 0)$$
$$f(v_8) = (0\ 1\ 0\ 1\ 1\ 1\ 1\ 1\ 0\ 0\ 0\ 0\ 0\ 0\ 0\ 0)$$

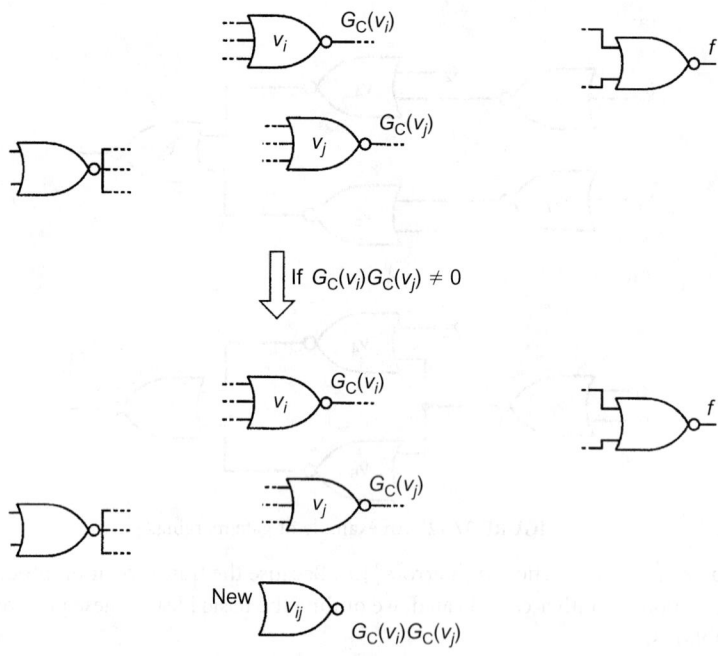

If the output function of gate v_{ij} realizes a permissible function in $G_C(v_i)G_C(v_j)$ by connecting all the inputs of gates v_i and v_j to the input of gate v_{ij}

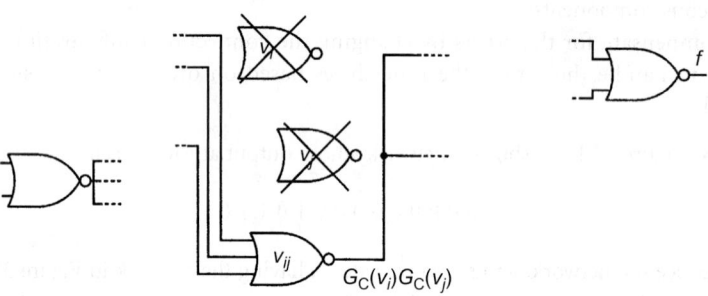

FIGURE 37.13   Generalized gate merging.

$$f(v_9) = (0\ 0\ 1\ 1\ 0\ 0\ 1\ 1\ 0\ 0\ 0\ 0\ 0\ 0\ 0\ 0)$$
$$f(v_{10}) = (1\ 0\ 1\ 0\ 0\ 0\ 0\ 0\ 0\ 1\ 0\ 0\ 0\ 0\ 0)$$
$$f(v_{11}) = (0\ 0\ 0\ 0\ 0\ 0\ 0\ 1\ 1\ 0\ 0\ 1\ 1\ 0\ 0)$$
$$f(v_{12}) = (1\ 1\ 0\ 0\ 1\ 1\ 0\ 0\ 0\ 0\ 0\ 0\ 0\ 0\ 0)$$

The values of the 6-th component (i.e., the values corresponding to the input combination $x_1 = x_3 = 0$ and $x_2 = x_4 = 1$) are shown in Figures 37.14(a) and (b). Components of vectors representing CSPFs can be calculated independently, so we calculate CSPFs for all components except the 6-th component (shown by "–") of the network in Figure 37.14(b), as follows:

$$G_C(v_5) = (1\ 0\ 0\ 0\ 0 - 0\ 0\ 0\ 0\ 1\ 0\ 1\ 1\ 0\ 1)$$
$$G_C(v_6) = (0\ *\ *\ *\ 1 - 1\ *\ *\ *\ 0\ *\ 0\ 0\ 1\ 0)$$

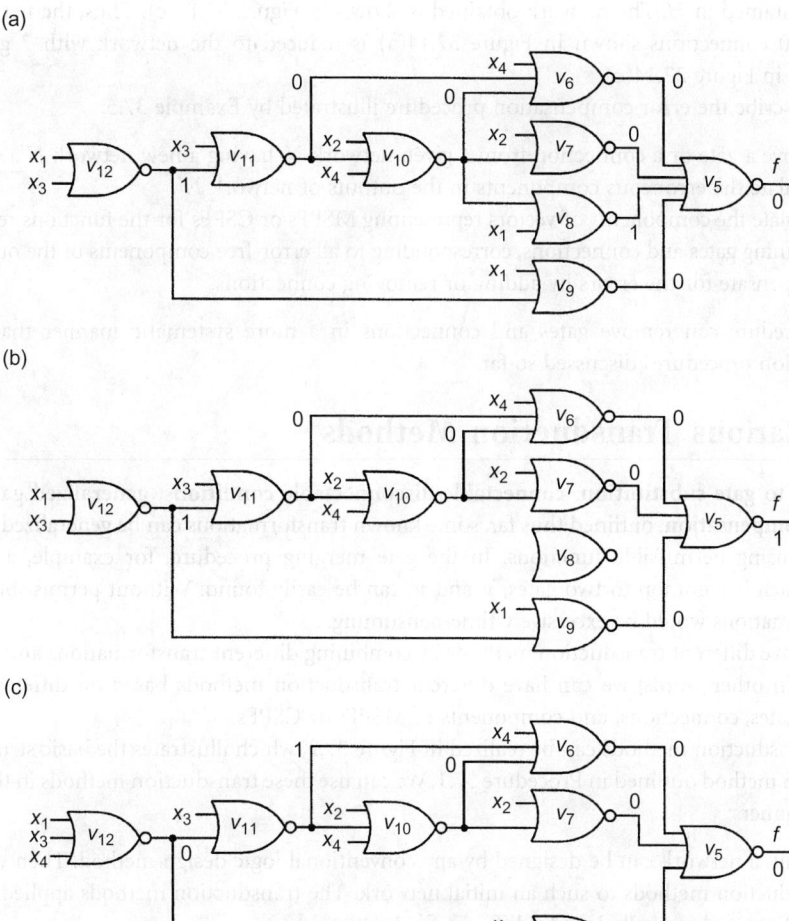

FIGURE 37.14 An example of error compensation.

$$G_C(v_7) = (0\ 1\ *\ 1\ *-*\ *\ 1\ 1\ 0\ 1\ 0\ 0\ *\ 0)$$
$$G_C(v_9) = (0\ *\ 1\ *\ *-*\ 1\ *\ *\ 0\ *\ 0\ 0\ *\ 0)$$
$$G_C(v_{10}) = (1\ 0\ *\ 0\ 0-0\ *\ 0\ 0\ 1\ 0\ *\ *\ 0\ *)$$
$$G_C(v_{11}) = (0\ *\ *\ *\ 0-0\ *\ 1\ *\ 0\ *\ 1\ *\ 0\ *)$$
$$G_C(v_{12}) = (1\ *\ 0\ *\ 1-*\ 0\ 0\ *\ *\ *\ 0\ *\ *\ *)$$

If we can change the 6-th component of any of $f(v_6)$, $f(v_7)$, or $f(v_9)$ (i.e., immediate predecessors of gate v_5) from 0 to 1, the error in the network output can be compensated, as can be seen in Figure 37.14(b) where v_8 is removed. The value 0 in the 6-th component of the output at gate, v_6, v_7, or v_9, is due to $x_4 = 1$, $x_2 = 1$, or $f(v_{12}) = 1$, respectively. If we want to change the output of v_9 from 0 to 1, the 6-th component of $f(v_{12})$ must be 0. If we can change the output of v_{12} to any function in the set of permissible functions

$$H = (1\ *\ 0\ *\ 1\ 0\ *\ 0\ 0\ *\ *\ *\ 0\ *\ *\ *)$$

that is $G_C(v_{12})$ except the 6-th component specified to 0, the error will be compensated. We can generate such a function by connecting x_4 to gate v_{12} and consequently by changing the output of v_{12} into

$$(1\ 0\ 0\ 0\ 1\ 0\ 0\ 0\ 0\ 0\ 0\ 0\ 0\ 0\ 0\ 0)$$

which is contained in *H*. The network obtained is shown in Figure 37.14(c). Thus, the network with 8 gates and 20 connections shown in Figure 37.14(a) is reduced to the network with 7 gates and 18 connections in Figure 37.14(c).   □

Let us describe the error compensation procedure illustrated by Example 37.5:

1. Remove a gate or a connection from a given network *N*, having a new network *N′*.
2. Calculate the erroneous components in the outputs of network *N′*.
3. Calculate the components of vectors representing MSPFs or CSPFs for the functions realized at the remaining gates and connections, corresponding to all error-free components of the outputs of *N′*.
4. Compensate for the errors by adding or removing connections.

This procedure can remove gates and connections in a more systematic manner than the other transformation procedures discussed so far.

37.3 Various Transduction Methods

In addition to **gate substitution**, **connectable/disconnectable conditions**, **generalized gate merging**, and **error compensation**, outlined thus far, some known transformations can be generalized for efficient processing, using permissible functions. In the gate merging procedure, for example, a permissible function which is common to two gates, v_i and v_p, can be easily found. Without permissible functions, the transformations would be excessively time-consuming.

We can have different transduction methods by combining different transformations and the pruning procedure. In other words, we can have different transduction methods based on different orders in processing gates, connections, and components of MSPFs or CSPFs.

These transduction methods can be realized in Figure 37.2, which illustrates the basic structure of the transduction method outlined in Procedure 37.1. We can use these transduction methods in the following different manners:

1. An initial network can be designed by any conventional logic design method. Then we apply the transduction methods to such an initial network. The transduction methods applied to different initial networks usually lead to different final networks.
2. Instead of applying a transduction method only once, we can apply different transduction methods to an initial network in sequence. In each sequence, different or identical transduction methods can be applied in different orders. This usually leads to many different final networks.

Thus, if we want to explore the maximum potential of the transduction methods, we need to use them in many different ways, as explained in 1 and 2 [3,4].

37.3.1 Computational Example of the Transduction Method

Let us show a computational example of the transduction method. Suppose the initial network, which realizes a four-variable function, given as illustrated in Figure 37.15(a), and this function has a minimal network shown in Figure 37.15(b) (its minimality is proved by the integer programming logic design method). Beginning with the initial network of 12 gates shown in Figure 37.15(a), the transduction method with error-compensation transformation (this version was called NETTRA-E3) produced the tree of solutions shown in Figure 37.16. The size of the tree can be limited by the program parameter, NEPMAX [6]. (In Figure 37.16, NEPMAX was set to 2. If set to 8, we will have a tree of 81 networks). The notation "*a/b:c*" in Figure 37.16 means a network numbered *a* (numbered according to the order of generation), consisting of *b* gates and *c* connections, and a line connecting a larger network with a smaller one means that the smaller is derived, treating the larger as an initial network. In Figure 37.16, it is important to notice that while some paths lead to terminal nodes representing minimal networks, others lead to terminal nodes representing networks not very close to the minimal. By comparing the numbers of gates and connections in the networks derived at the terminal nodes of this solution tree, a best solution can be found.

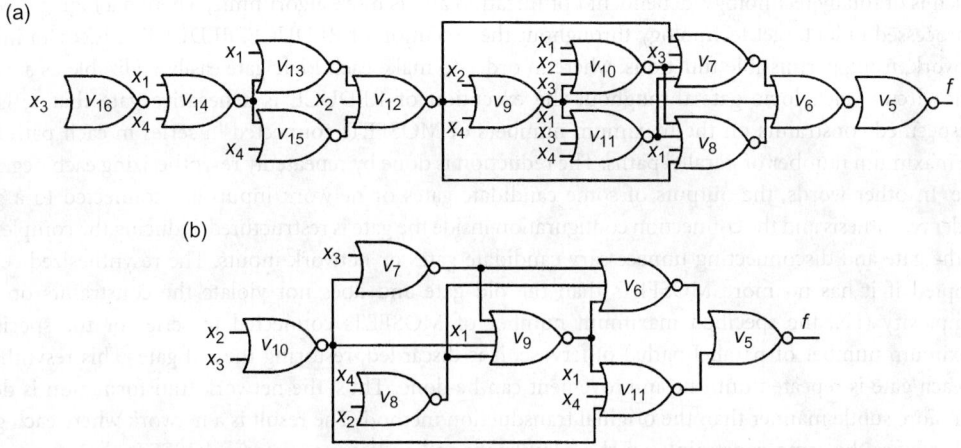

FIGURE 37.15 Initial and final networks for Figure 37.6.

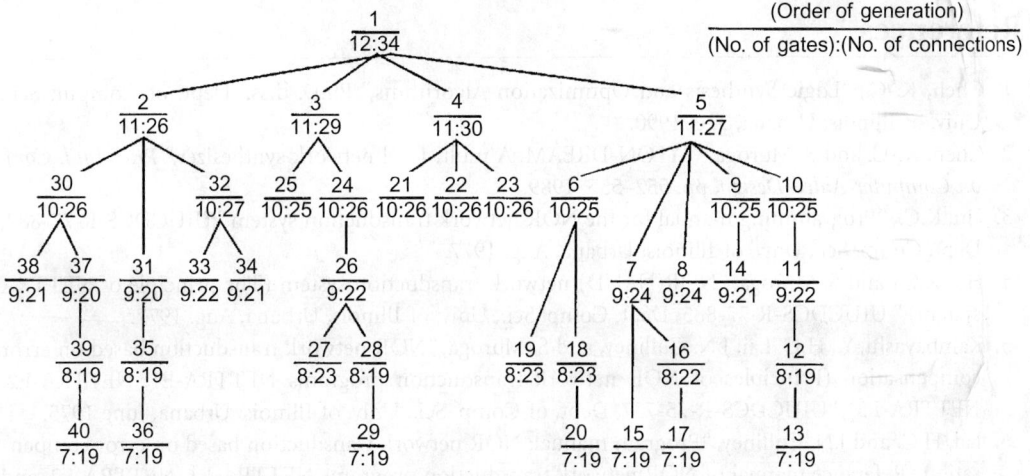

FIGURE 37.16 A tree of solutions generated by the transduction method based on error compensation.

Intermediate solutions are logic networks with different connection configurations of negative gates, so some of them may be more appropriate for layout than others.

37.4 Design of Logic Networks with Negative Gates by the Transduction Method

The transduction method has been described for the logic networks of NOR gates for the sake of simplicity, but it can be applied to logic networks of other types of gates, such as MOS logic gates and a mixture of AND gates and OR gates, tailoring its basic concepts (i.e., permissible functions and transformation). In this sense, it is important to understand what features different types of logic gates and consequently corresponding transistor circuits have in terms of logic operations and network transformations. In order to test the feasibility of design of logic networks with negative gates (MOS logic gates are negative gates) by the transduction method, a few synthesizers, called **SYLON** (an acronym for SYnthesis of LOgic Networks), were developed by modifying the transduction method [1,2,7,8,11,12].

Some SYLON logic synthesizers consist of a mixture of technology-dependent optimization and technology-independent optimization. Here, let us outline **SYLON-REDUCE** [7], a logic synthesizer which is of totally technology-dependent optimization and is more algorithmic, wherein a logic network is processed in its target technology throughout the execution of REDUCE. REDUCE reduces an initial network, using permissible functions, where in order to make each logic gate easily realizable as a MOS logic circuit, each logic gate throughout the execution of REDUCE is a negative gate that satisfies prespecified constraints on the maximum numbers of MOSFETs connected in series in each path and the maximum number of parallel paths. The reduction is done by repeatedly resynthesizing each negative gate. In other words, the outputs of some candidate gates or network inputs are connected to a gate under resynthesis and the connection configuration inside the gate is restructured, reducing the complexity of the gate and disconnecting unnecessary candidate gates or network inputs. The resynthesized cell is adopted if it has no more MOSFETs than the old gate and does not violate the constraints on the complexity (i.e., the specified maximum number of MOSFETs connected in series or the specified maximum number of parallel paths) otherwise, it is discarded, restoring the old gate. This resynthesis of each gate is repeated until no improvement can be done. Thus, the network transformation is done in a more subtle manner than the original transduction method. The result is a network where each gate still satisfies the same constraints on the complexity and contains no more MOSFETs than the corresponding gate in the original network and the connection configuration of the network may be changed.

References

1. Chen, K.-C., "Logic Synthesis and Optimization Algorithms," Ph.D. diss., Dept. of Comput. Sci., Univ. of Illinois, Urbana, 320, 1990.
2. Chen, K.-C. and S. Muroga, "SYLON-DREAM: A multi-level network synthesizer," *Proc. Int'l. Conf. on Computer-Aided Design*, pp. 552–555, 1989.
3. Hu, K.C., "Programming manual for the NOR network transduction system," UIUCDCS-R-77-887, Dept. Comp. Sci., Univ. of Illinois, Urbana, Aug. 1977.
4. Hu, K.C., and S. Muroga, "NOR(NAND) network transduction system (The principle of NETTRA system)," UIUCDCS-R-77-885, Dept. Comp. Sci., Univ. of Illinois, Urbana, Aug. 1977.
5. Kambayashi, Y., H.C. Lai, J.N. Culliney, and S. Muroga, "NOR network transduction based on error compensation (Principles of NOR network transduction programs NETTRA-E1, NETTRA-E2, NETTRA-E3)," UIUCDCS-R-75-737, Dept. of Comp. Sci., Univ. of Illinois, Urbana, June 1975.
6. Lai, H.C. and J.N. Culliney, "Program manual: NOR network transduction based on error compensation (Reference manual of NOR network transduction programs NETTRA-E1, NETTRA-E2, and NETTRA-E3)," UIUCDCS-R-75-732, Dept. Comp. Sci., Univ. of Illinois, Urbana, June 1975.
7. Limqueco, J.C. and S. Muroga, "SYLON-REDUCE: A MOS network optimization algorithm using permissible functions," *Proc. Int'l. Conf. on Computer Design*, Cambridge, MA, pp. 282–285, Sept. 1990.
8. Limqueco, J.C., "Logic Optimization of MOS Networks," Ph.D. thesis, Dept. of Comput. Sci., University of Illinois, Urbana, 250, 1992.
9. Muroga, S., "Computer-aided logic synthesis for VLSI chips," *Advances in Computers*, vol. 32, Ed. by M.C. Yovits, Academic Press, pp. 1–103, 1991.
10. Muroga, S., Y. Kambayashi, H.C. Lai, and J.N. Culliney, "The transduction method—Design of logic networks based on permissible functions," *IEEE TC*, 38, 1404–1424, Oct. 1989.
11. Xiang, X.Q., "Multilevel Logic Network Synthesis System, SYLON-XTRANS, and Read-Only Memory Minimization Procedure, MINROM," Ph.D. diss., Dept. of Comput. Sci., Univ. of Illinois, Urbana, 286, 1990.
12. Xiang, X.Q. and S. Muroga, "Synthesis of multilevel networks with simple gates," *Int'l. Workshop on Logic Synthesis*, Microelectronic Center of North Carolina, Research Triangle Park, NC, May 1989.

38

Emitter-Coupled Logic

Saburo Muroga

University of Illinois
at Urbana-Champaign

CONTENTS

38.1 Introduction

ECL, which stands for emitter-coupled logic, is based on bipolar transistors and is currently the logic family with the highest speed and a high output power capability, although power consumption is also the highest. ECL is more complicated to fabricate, covers a larger chip area, and is more expensive than any other logic family. As the speed of CMOS improves, ECL is less often used but is still useful for the cases where high speed, along with large output power, are necessary, such as high-speed transmission over communication lines. ECL has three types of transistor circuits: standard ECL logic gates, their modification with wired logic, and ECL series-gating [4,6,10,13].

38.2 Standard ECL Logic Gates

A standard ECL logic gate is a stand-alone logic gate, and logic networks can be designed using many as building blocks. Its basic circuit is shown in Figure 38.1. ECL has unique logic capability, as explained in the following.

The logic operation of the ECL gate shown in Figure 38.1 is analyzed in Figure 38.2, where the input z in Figure 38.1 is eliminated for simplicity. The resistors connected to the bases of transistors, T_x and T_y, are for protecting these transistors from possible damage due to heavy currents through a transistor, but not for logic operations, and can be eliminated if there is no possibility for an excessively heavy current to flow. When input x has a high voltage representing logic value 1 and y has a low voltage representing logic value 0, transistor T_x becomes conductive, T_y becomes non-conductive, and a current flows through T_x and resistor R_1, as illustrated in Figure 38.2(a). In this case the voltage at the emitter of transistor T_r becomes higher than -4 V due to the current through resistor R_p, as shown. Consequently, the voltage at this emitter becomes higher and the voltage at its base becomes not sufficiently high against its emitter to make T_r conductive, so there is no current through T_r and resistor R_2. Consequently, transistor T_f has a high voltage at its base, which makes T_f conductive, and output f has a high voltage representing logic value 1. On the other hand, transistor T_g is almost non-conductive (actually a small

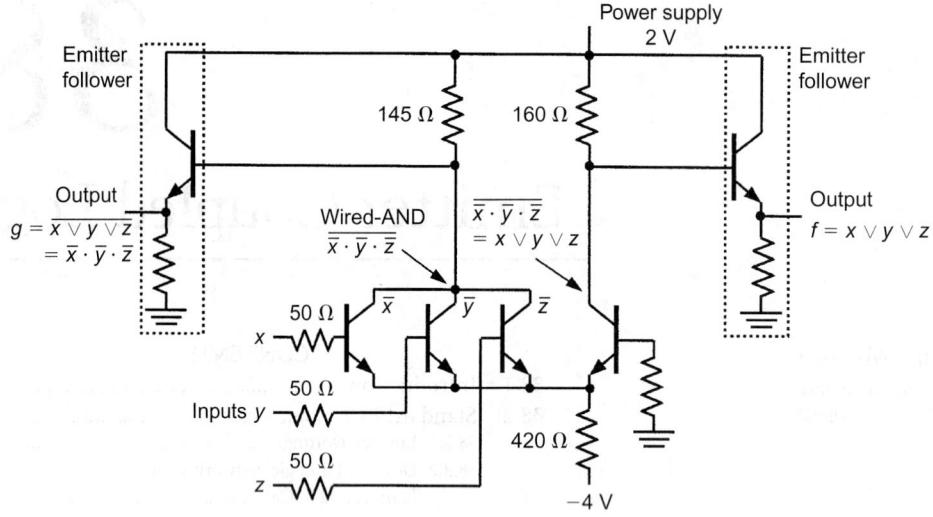

FIGURE 38.1 Basic circuit of standard ECL logic gate.

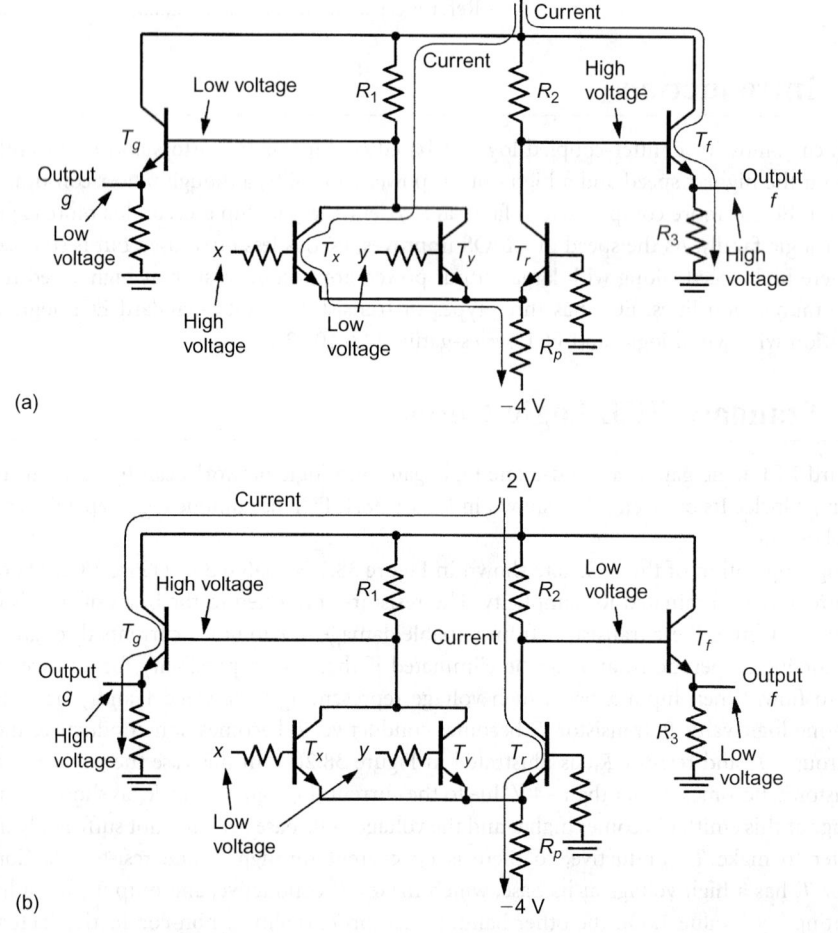

FIGURE 38.2 Logic operation of ECL gate in Figure 38.1.

current flows, but let us ignore it for simplicity), since its base has a low voltage due to the voltage drop developed across resistor R_1 by the current shown. Thus, output g has a low voltage representing logic value 0.

Even when y (instead of x), or both x and y, has a high voltage, the above situation is not changed, except for the current through T_y.

Next, suppose that both inputs x and y have low voltages, as shown in Figure 38.2(b). Then there is no current through resistor R_1. Since the base of transistor T_r has a higher voltage than its emitter (0 V at the base and −0.8 V at the emitter), a current flows through R_2 and T_r, as illustrated in Figure 38.2(b). Thus, T_f has a low voltage at its base and becomes almost non-conductive (more precisely speaking, less conductive). Output f has, consequently, a low voltage, representing logic value 0. Transistor T_g has a high voltage at its base and becomes conductive. Thus, output g has a high voltage, representing logic value 1.

Therefore, a current flows through only one of R_1 and R_2, switching quickly between these two paths. Notice that resistor R_p in Figure 38.2 which is connected to a power supply of minus voltage is essential for this current steering because the voltage at the top end of R_p determines whether T_r becomes conductive or not. The emitter followers (shown in the dot-lined rectangles in Figure 38.1) can deliver heavy output currents because an output current flows only through either transistor T_f or T_g and the on-resistance of the transistor is low.

The above analysis of Figure 38.2 leads to the truth table in Table 38.1. From this table, the network in Figure 38.2 has two outputs: $f = x \vee y$ and $g = \overline{x \vee y}$.

In a similar manner, we can find that the ECL gate in Figure 38.1 has two outputs: $f = x \vee y \vee z$ and $g = \overline{x \vee y \vee z}$.

The gate is denoted by the symbol shown in Figure 38.3. The simultaneous availability of OR and NOR as the double-rail output logic, with few extra components, is the unique feature of the ECL gate, making its logic capability powerful.

38.2.1 Emitter-Dotting

Suppose we have the emitter follower circuit shown in Figure 38.4(a) (also shown in the dot-lined rectangles in Figure 38.1), as part of an ECL logic gate. Its output function, f, at the emitter of the bipolar transistor is 1 when the voltage at the emitter is high (i.e., the bipolar transistor in (a) is conductive), and f is 0 when the voltage at the output terminal is low (i.e., the bipolar transistor in (a) is non-conductive). Then, suppose there is another alike circuit whose output function at the emitter is g. If the emitters of these two circuits are tied together as shown in Figure 38.4(b), the new output function at the tied point is $h = f \vee g$, replacing the original functions, f and g. This connection is called **emitter-dotting**, realizing **Wired-OR**. The tied point represents the new function $f \vee g$ because if both transistors, T_1 and T_2, are non-conductive, the voltage at the tied point is low; otherwise (i.e., if one of T_1 and T_2, or both, is conductive), the voltage at the tied point is high.

TABLE 38.1   Truth Table for Figure 38.2

Inputs		Outputs	
x	y	f	g
0	0	0	1
0	1	1	0
1	0	1	0
1	1	1	0

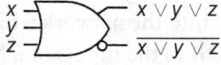

FIGURE 38.3   Symbol for the standard ECL logic gate of Figure 38.1.

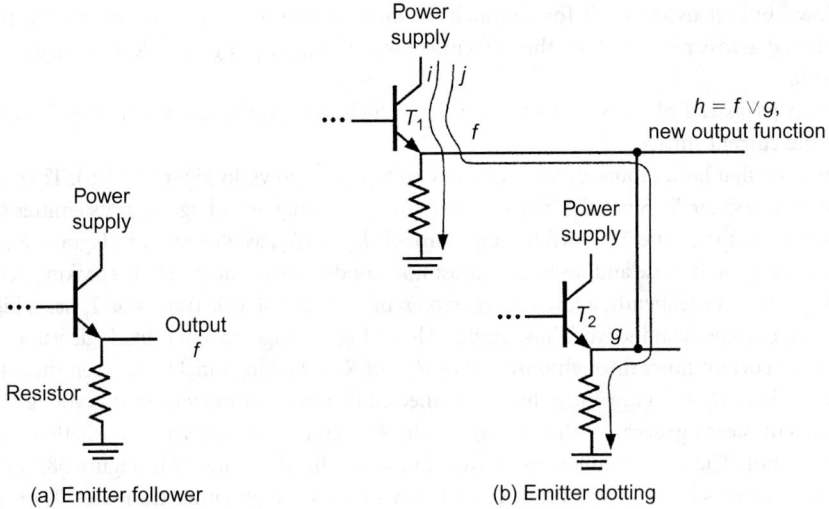

FIGURE 38.4 Emitter-dotting.

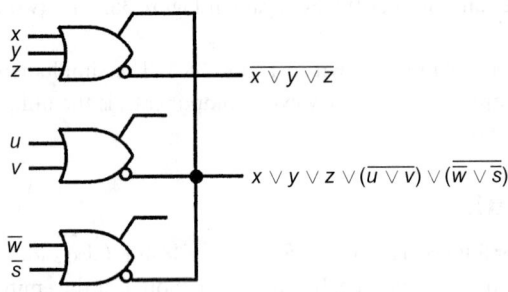

FIGURE 38.5 Wired-OR of ECL gates.

Wired-OR is an important feature of the ECL gate. The ECL gate in Figure 38.1 has two emitter followers: one for f and the other for g. As shown in Figure 38.5, the OR of the outputs can be realized without using an extra gate, simply by tying together these outputs. This is very convenient in logic design. **If one output is Wired-ORed with another ECL gate, it does not express the original function. And it cannot be further Wired-ORed to other gates if we want to realize Wired-OR with the original function.** But if the same output is repeated by adding an emitter follower inside the gate, as shown in Figure 38.6 (i.e., the emitter follower inside a dot-lined rectangle in Figure 38.1), then the new output can be Wired-ORed with another gate output or connected without Wired-OR to the succeeding gates. In the ECL gate at the top position in Figure 38.6, for example, the first output $f = x \vee y \vee z$ is connected to gates in the next level, while the same f in the second output is used to produce the output $\overline{u \vee v} \vee x \vee y \vee z$ by Wired-ORing with the output of the second ECL gate.

38.2.2 Design of a Logic Network with Standard ECL Gates

An ECL gate network can be designed, starting with a network of only NOR gates, for the following reason. Consider a logic network of ECL logic gates shown in Figure 38.7(a), where Wired-ORs are included. This network can be converted into the network without Wired-OR shown in (b) by directly connecting connections in each Wired-OR to the inputs of a NOR gate without changing the outputs at gates 4 and 5, possibly sacrificing the maximum fan-in restriction. Then, two NOR outputs of

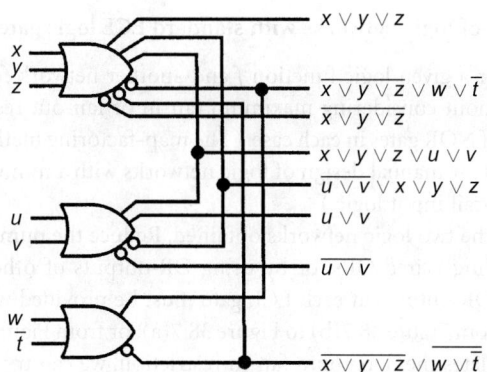

FIGURE 38.6 Multiple-output ECL gates for Wired-OR.

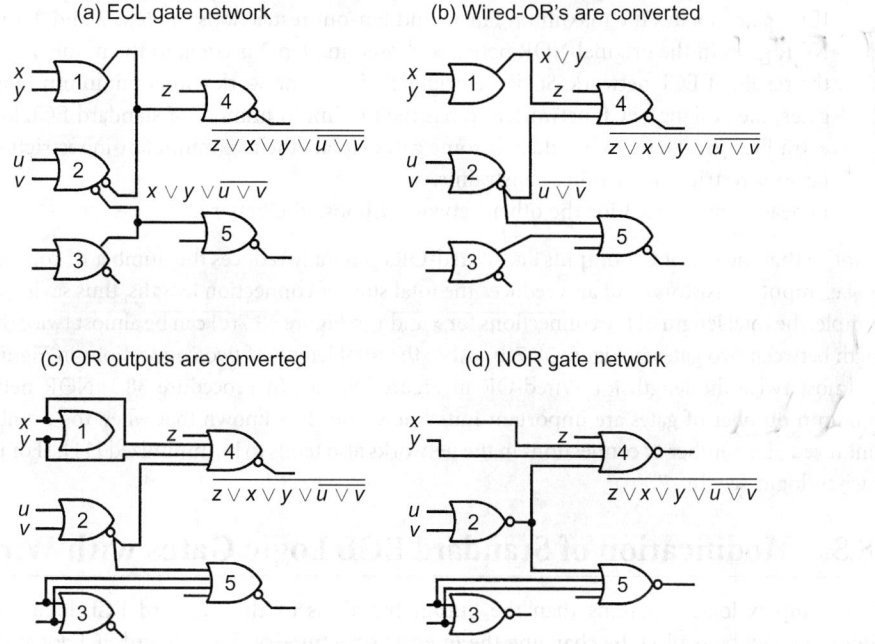

FIGURE 38.7 Conversion of an ECL gate network into a NOR gate network.

gate 2 in (a), for example, can be combined into one in (b). Then, this network can be converted into the network shown in (c) by eliminating OR outputs of all gates 1, 2, and 3 in (b), and connecting inputs of these gates directly to gates 4 and 5. Thus, the network in (c) that expresses the same outputs as the network in (a) consists of NOR gates only (i.e., the outputs of gates 4 and 5 in (c) are the same as those in (a)), possibly further sacrificing the maximum fan-in restriction at some gates. Notice that in this conversion, the number of gates does not change or decreases (if an ECL gate, like gate 1 in (c), has no outputs used, it can be deleted from (c)). Thus, from the given network of standard ECL gates with Wired-ORs, we can derive a NOR network of the same or fewer number of gates, possibly with greater fan-in at some gates, as shown in Figure 38.7(d). Even if each gate has many NOR outputs or OR outputs, the situation does not change.

When we want to design a minimal standard ECL gate network for given functions f and $\bar{f}$, it can be designed by reversing the preceding conversion, as follows.

Procedure 38.1: Design of logic networks with standard ECL logic gates

1. Design a network for a given logic function f and another network for its complement, $\bar{f}$ using NOR gates only without considering maximum fan-in or fan-out restriction at each gate. Use a minimum number of NOR gates in each case. (The map-factoring method described in Chapter 31 is usually convenient for manual design of logic networks with a minimum number of NOR gates in single- or double-rail input logic.)

2. Choose one among the two logic networks obtained. Reduce the number of input connections to each gate, by providing Wired-ORs, or by using OR-outputs of other gates, if possible. In this case, extra NOR or OR outputs at each ECL gate must be provided whenever necessary (like the reverse conversion from Figure 38.7(b) to Figure 38.7(a), or from Figure 38.7(c) to Figure 38.7(b)). Thus, if any gate violates the maximum fan-in restriction, we can try to avoid it by using Wired-ORs or OR outputs.

3. This generally reduces fan-out of gates also; but if any gate still violates the maximum fan-out restriction, try to avoid it by using extra ECL gates (no simple good methods are known for doing this). The output ECL gate of this network presents f and $\bar{f}$.

   If no gate violates the maximum fan-in and fan-out restrictions in Steps 2 and 3, the number of NOR gates in the original NOR network chosen in Step 2 is equal to the number of ECL gates in the resultant ECL network. So, if we originally have a network with a minimum number of NOR gates, the designed ECL network also has the minimum number of standard ECL logic gates. But extra ECL gates have to be added if some gates violate the maximum fan-in restriction, maximum fan-out restriction, or other constraints.

4. Repeat Steps 2 and 3 for the other network. Choose the better one.   □

   Notice that the use of OR outputs and Wired-ORs generally reduces the number of connections or fan-ins (i.e., input transistors) and also reduces the total sum of connection lengths, thus saving chip area. For example, the total length of the connections for x and y in Figure 38.7(c) can be almost twice the connection length between two gates in Figure 38.7(b). Also, the total length of two connections in Figure 38.7(b) can be almost twice the length for Wired-OR in Figure 38.7(a). In Procedure 38.1, NOR networks with a minimum number of gates are important initial networks. It is known that when the number of gates is minimized, the number of connections in the networks also tends to be minimized [11]. (For the properties of wired logic, see Ref. 7.)

38.3   Modification of Standard ECL Logic Gates with Wired Logic

More complex logic functions than the output functions of the standard ECL logic gate shown in Figure 38.1 can be realized by changing the internal structures of the standard ECL logic gates. In other words, if we connect points inside one ECL gate to some points of another ECL gate, we can realize a complex logic function with a simpler electronic circuit configuration. In other words, we can realize logic functions by freely connecting transistors, resistors, and diodes, instead of regarding the fixed connection configuration of transistors, resistors, and diodes as logic gates whose structure cannot be changed. This approach could be called **transistor-level logic design**. Wired logic is a powerful means for this, and collector-dotting and emitter-dotting are the basic techniques of wired logic.

38.3.1   Collector-Dotting

Collector-dotting is commonly used in bipolar transistor circuitry to realize the **Wired-AND operation**. Suppose we have the inverter circuit shown in Figure 38.8(a) as part of an ECL logic gate. Its output function, f, at the collector of the bipolar transistor is 1 when the voltage at the collector is high, and f is 0 when the voltage at the collector is low. Then, suppose there is another like circuit whose output function at the collector is g. If the collectors of these two circuits, instead of the emitters for emitter-dotting in Figure 38.4(b), are tied together as shown in Figure 38.8(b), the new output function at the tied

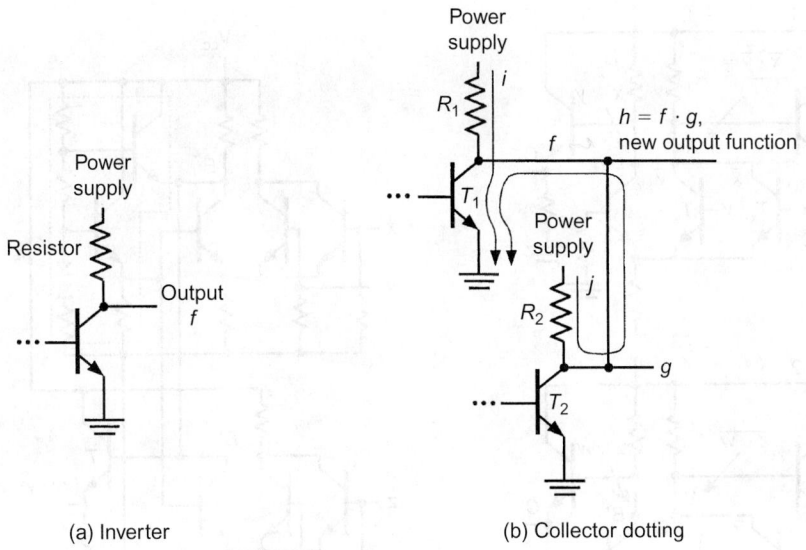

FIGURE 38.8   Collector-dotting.

point is $h = f \cdot g$, replacing the original functions, f and g. This connection is called **collector-dotting**, realizing **Wired-AND**. The tied point represents the new function $f \cdot g$ because if one of T_1 and T_2, or both is conductive in Figure 38.8(b), the voltage at the tied point is low; otherwise (i.e., only when both transistors, T_1 and T_2, are non-conductive), the voltage at the tied point can be high.

In Figure 38.9(a), Figure 38.2 is repeated as gate 1 and gate 2. Transistor T_x has input x at its base, and its collector represents function $\bar{x}$ if T_y does not exist because when its base has a low voltage (i.e., logic value 0), its collector has a high voltage (i.e., logic value 1) and vice versa. Similarly, the collector of transistor T_y represents $\bar{y}$, if T_x does not exist. Then by tying together these collectors (i.e., collector-dotting), the tied point (i.e., point A) represents $\bar{x} \cdot \bar{y} = \overline{x \vee z}$, as already explained with respect to Figure 38.2. Notice that the collector of T_x and the collector of T_y do not represent the original functions $\bar{x}$ and $\bar{y}$ respectively, after collector-dotting. Since the voltage level at B is always opposite to that at A, point B represents $x \vee y$.

We can use collector-dotting more freely. Point A in gate 1 in Figure 38.9(a) can be connected to point A' or B' in gate 2. Point B can also be connected to point A' or B'. Such connections realize Wired-AND or collector-dotting. By collector-dotting points B and B' as shown in Figure 38.9(b), point B 7 (also B') represents new function $(x \vee y) \cdot (z \vee w)$, which also appears at the emitter of transistor T. After this collector-dotting, points B and B' do not represent the original functions $x \vee y$ and $z \vee w$, respectively, anymore. Also, note that the function at any point that is not collector-dotted, such as A and A', is unchanged by collector-dotting of B and B'. In Figure 38.9(b), two transistors, two diodes, and resistors (shown in the dotted line) are added for adjustment of voltage and current. But they have nothing to do with logic operations.

Another example is the parity function $\bar{x}y \vee x\bar{y}$ realized by connecting two ECL gates as shown in Figure 38.10. The parity function requires four ECL gates if designed with the standard ECL logic gates as shown in Figure 38.10(c), but can be realized by the much simpler electronic circuit of Figure 38.10(b). In other words, $\bar{x}y$ and $x\bar{y}$ are realized by Wired-AND, then these two products are Wired-ORed in order to realize $\bar{x}y \vee x\bar{y}$. In Figure 38.10 as well as Figure 38.9, some resistors or transistors may be necessary for electronic performance improvement (since resistors R_1 and R_2 draw too much current in gate 1 in Figure 38.10(a), new resistors are added in Figure 38.10(b) in order to clamp the currents), and unnecessary resistors or transistors may be deleted, although such an addition or elimination of resistors or transistors has nothing to do with logic operations.

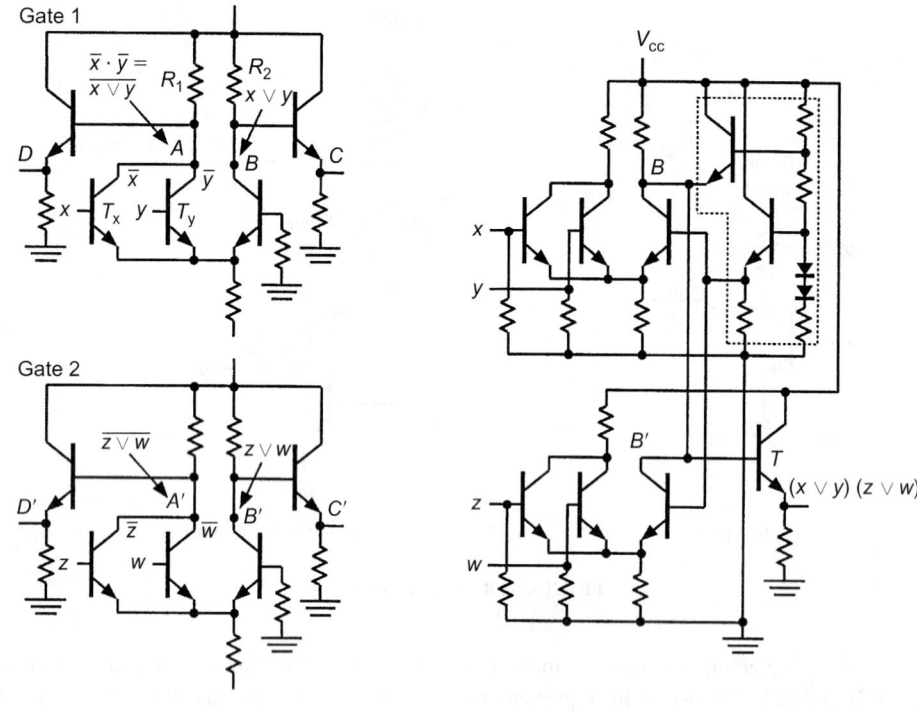

(a) Two standard ECL gates

(b) Complex function realized by Wired-AND

FIGURE 38.9 Example of Wired-AND.

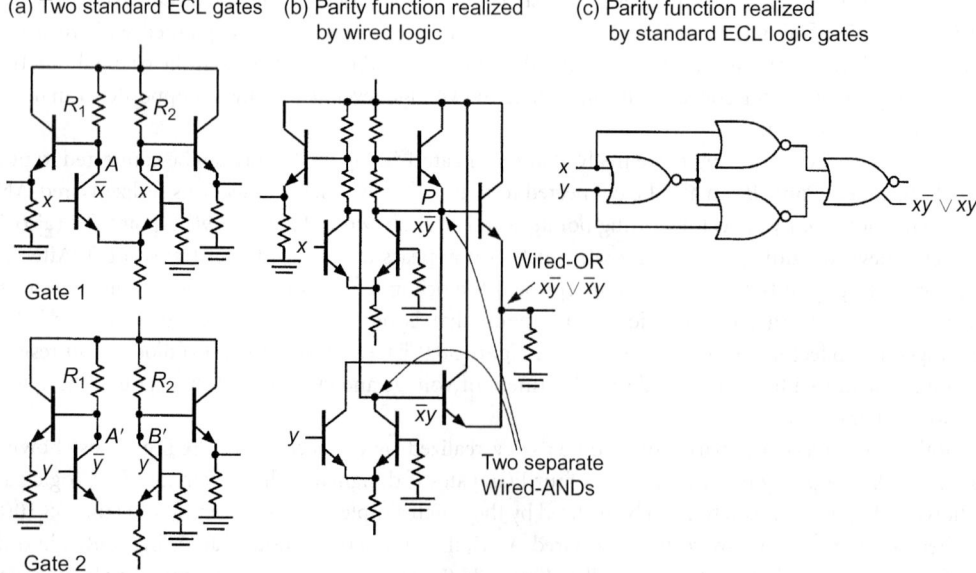

FIGURE 38.10 Parity function realized by ECL gate with Wired logic.

38.4   ECL Series-Gating Circuits

Figure 38.11(a) shows the basic pair of bipolar transistors in **series-gating ECL**, where A is connected to a power supply and B is connected to another power supply of minus voltage through a resistor. (Notice that this pair is T_x and T_r in Figure 38.2, from which T_y is eliminated.) Transistor T_1 has an input x connected to its base and the other transistor T_2 has a constant voltage v_{ref} at its base. As illustrated in Figure 38.2, v_{ref} is grounded through a resistor (i.e., $v_{ref} = 0$), where this resistor is for protection of transistor T_r from damage by a heavy current, and v_{ref} works as a reference voltage against changes of x. (The voltage at v_{ref} can be provided by a subcircuit consisting of resistors, diodes, and transistors, like the one in Figure 38.9(b).) The collector of T_1 represents a logic function $\bar{x}$ because the collector of T_1 can have a high voltage (i.e., logic value 1) only when T_1 is non-conductive, that is, the input is a low voltage ($x = 0$).

The collector of T_2 represents function x because it becomes a high voltage only when the input x is a high voltage (i.e., when the input is a high voltage, T_1 becomes conductive and T_2 becomes non-conductive because a current flows at any time through exactly one of two transistors, T_1 and T_2. Thus, the collector of T_2 becomes a high voltage).

In Figure 38.11(b), we have two pairs of transistors. In other words, we have the pair with input y, in addition to the pair with input x shown in (a). Then let us connect them in series without R_{py}, R_1 and the power supply for R_1, as shown in (c). The voltage at the collector of T_3 is low only when T_3 and T_1 are both conductive and, consequently, a current i flows through T_3 and T_1. The voltage at the collector of T_3 is high when either T_3 or T_1 is non-conductive (i.e., $x = 0$ or $y = 0$) and consequently no current (i.e., i) flows through T_3 and T_1. Thus, the collector of T_3 represents the function $\overline{xy}$, replacing the original function $\bar{y}$ shown in (b). This can be rewritten as $\overline{xy} = \bar{x} \vee \bar{y}$, so **series-gating can be regarded as the OR operation**.

Many of the basic pair of transistors shown in Figure 38.11(a) are connected in a tree structure, as shown in Figure 38.12, where inputs x, y, and z, as well as reference voltages, $v_{ref\text{-}1}$, $v_{ref\text{-}2}$, and $v_{ref\text{-}3}$, need to be at appropriate voltage levels. Then the complement of all minterms can be realized at the collectors of transistors in the top level of the series connections. Two of these complemented minterms (i.e., $\overline{\bar{x} \vee \bar{y} \vee z}$ and $\overline{\bar{x} \vee \bar{y} \vee \bar{z}}$) are shown with emitter followers, as examples at the far right end of Figure 38.12.

Some of these collectors of transistors in the top level can be collector-dotted to realize the desired logic functions, as illustrated in Figure 38.13. Notice that **once collectors are collector-dotted, these collectors do not express their respective original functions**.

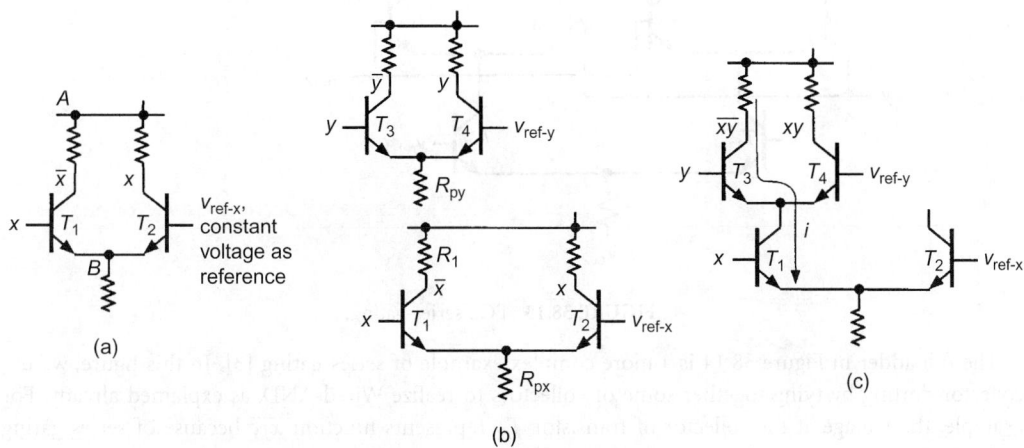

FIGURE 38.11   Series-gating.

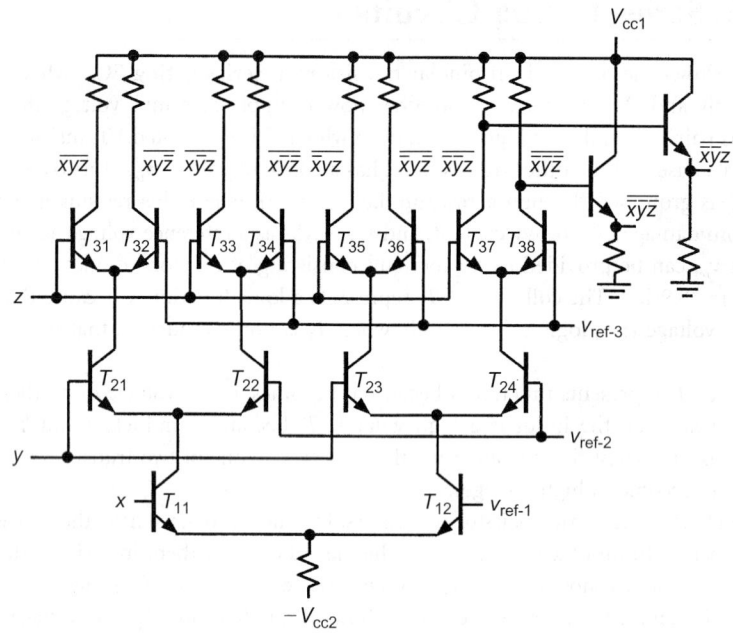

FIGURE 38.12 ECL series-gating.

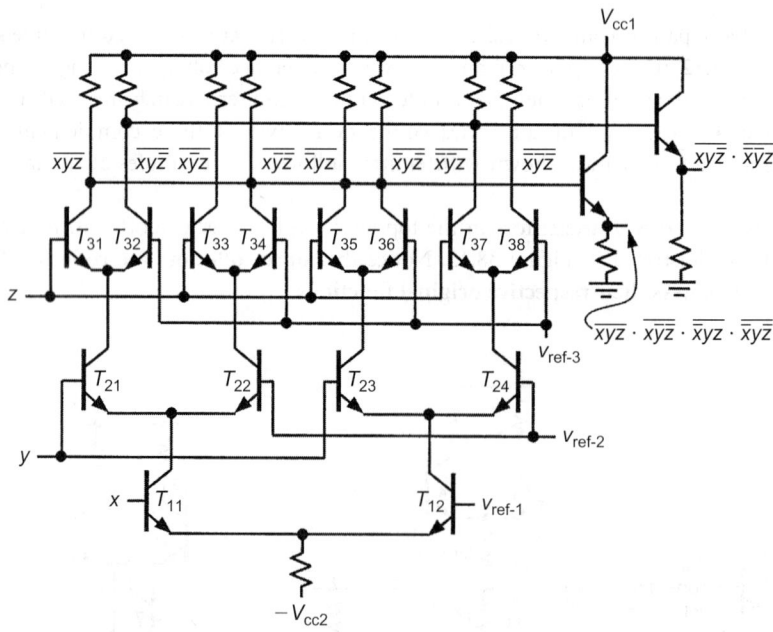

FIGURE 38.13 ECL series-gating.

The full adder in Figure 38.14 is a more complex example of series-gating [3]. In this figure, we use collector-dotting by tying together some of collectors to realize Wired-AND, as explained already. For example, the voltage at the collector of transistor T_{31} represents function $\overline{xyz}$ because of series-gating with T_{11}, T_{21}, and T_{31}. Usually, at most, three transistors are connected in series (the two transistors in the bottom level, T_{01} and T_{02}, in Figure 38.14 are for controlling the amount of current as part of the

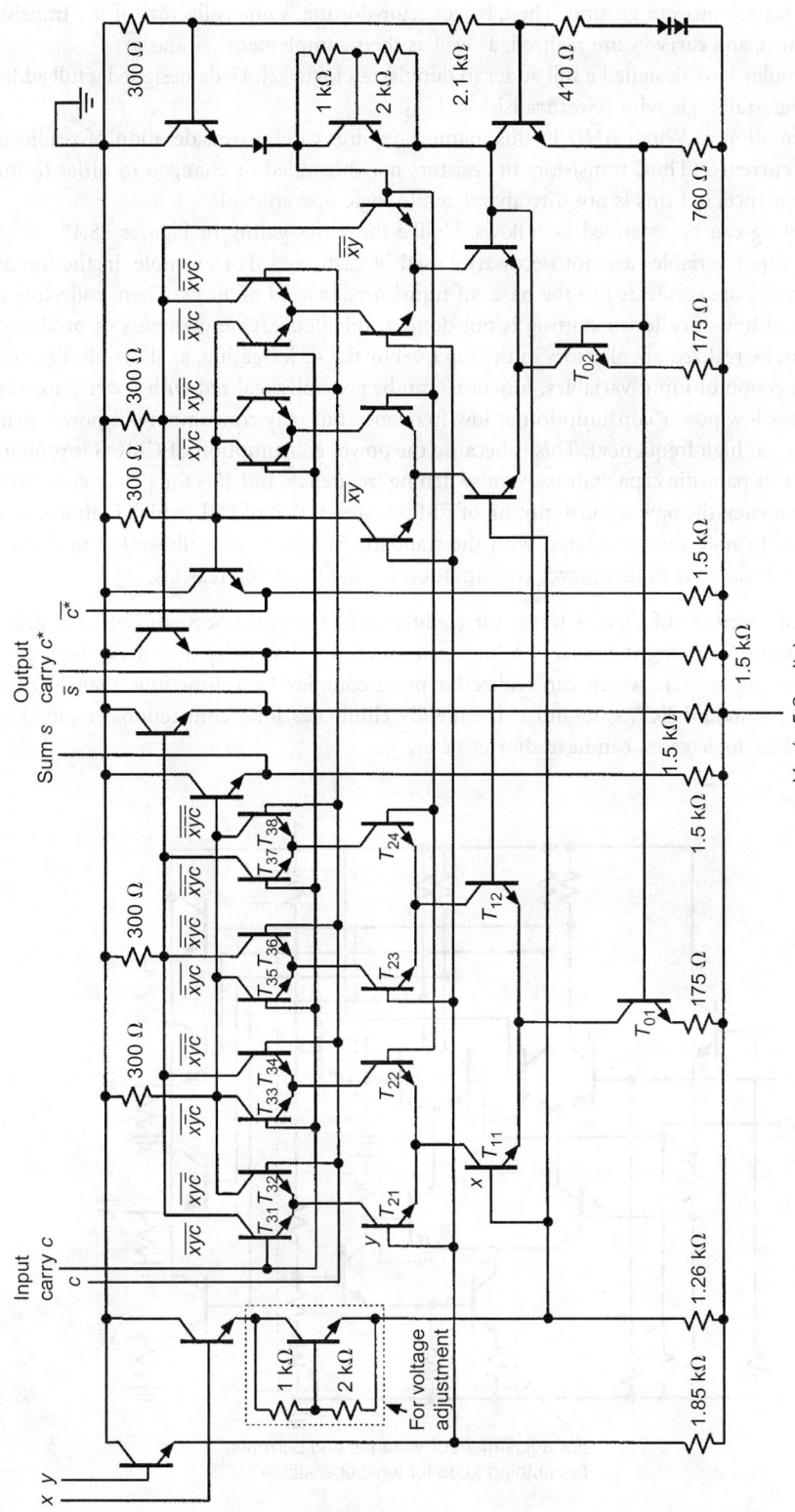

FIGURE 38.14 ECL full adder with series-gating. (From Ref. 3.)

power supply). This is because too many transistors in series tend to slow down the speed of the gate due to parasitic capacitances to ground. Then, by collector-dotting, some collectors of the transistors in the top level, sum s, and carry $c*$ are realized, as well as their complements, $\bar{s}$ and $\bar{c}*$.

Baugh and Wooley have designed a full adder in double-rail logic [2]. Ueda designed a full adder with ECL gates in single-rail logic with fewer transistors [16].

The implementation of Wired-AND in this manner requires careful consideration of readjustments of voltages and currents. (Thus, transistors or resistors may be added or changed in order to improve electronic performance, but this is not directly related to logic operations.)

ECL series-gating can be extended as follows. Unlike the series-gating in Figures 38.12, 38.13, and 38.14, the same input variables are not necessarily used in each level. For example, in the top level in Figure 38.15, y and z are connected to the bases of transistors, instead of all y's. Then, collectors can be collector-dotted, although collector-dotting is not done in this figure. Complements of products, such as $\overline{xy}$ and $\overline{xz}$, can be realized at collectors in the top level by the series-gating, as shown in Figure 38.15. By this free connection of input variables, functions can be generally realized with fewer transistors.

CMOS has very low power consumption at low frequency but may consume more power than ECL at high speed (i.e., at high frequency). This is because the power consumption of CMOS is proportional to CFV^2, where C is parasitic capacitance, F is a switching frequency, and V is the power supply voltage. Thus, at high frequency, the power consumption of CMOS exceeds that of ECL, which is almost constant.

It is important to note that compared with the standard ECL logic gate illustrated in Figure 38.1, series-gating ECL is faster with low power consumption for the following reasons:

- Because of speed-up of bipolar transistor (reduction of base thickness, and others), delay time over connections among standard ECL logic gates is greater than delay time inside logic gates and then series-gating ECL, which can realize far more complex logic functions than NOR or OR realized by standard ECL gate and consequently eliminates long connections required among standard ECL logic gates, can have shorter delay.

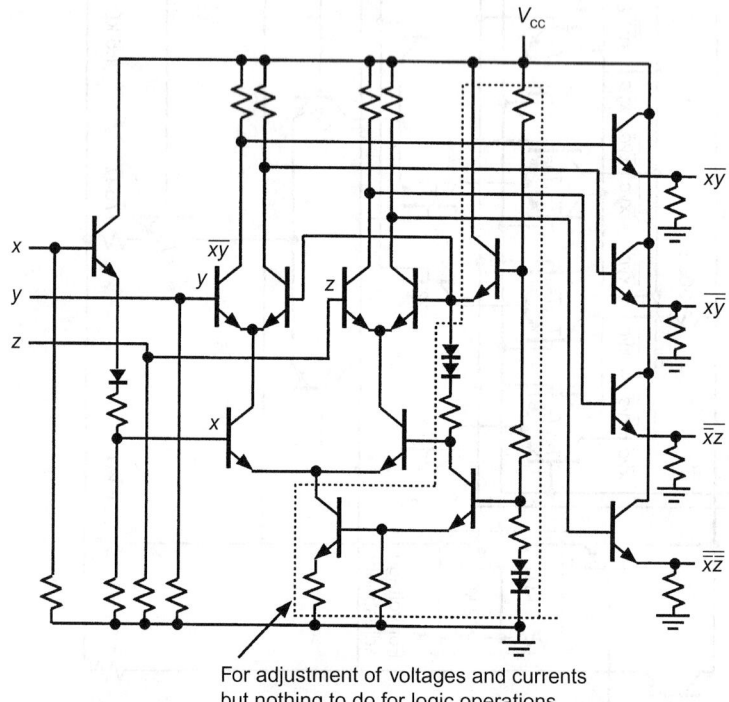

For adjustment of voltages and currents
but nothing to do for logic operations

FIGURE 38.15 Series-gating.

- A series-gating ECL circuit has lower power consumption than a logic network with standard ECL logic gates because power supplies to all standard ECL logic gates are combined into one for the series-gating ECL circuit and a current flows in only one path at any time [1,5,9,12].
- Then, in recent years, the power consumption of series-gating ECL is reduced with improved circuits [8].
- The power consumption of series-gating ECL can also be reduced by active pull-down of some emitters [14,15].

References

1. Abe, S., Y. Watanabe, M. Watanabe, and A. Yamaoka, "M parallel series computer for the changing market," *Hitachi Review*, vol. 45, no. 5, pp. 249–254, 1996.
2. Baugh, C.R. and B.A. Wooley, "One bit full adder," U.S. Patent 3,978,329, August 31, 1976.
3. Garret, L.S., "Integrated-circuit digital logic families III—ECL and MOS devices," *IEEE Spectrum*, pp. 30–42, Dec. 1970.
4. Gopalan, K.G., *Introduction to Digital Microelectronic Circuits*, McGraw-Hill, 1996.
5. Higeta, K. et al., "A soft-error-immune 0.9-ns 1.15-Mb ECL-CMOS SRAM with 30-ps 120 k logic gates and on-chip test circuitry," *IEEE Jour. of Solid-State Circuits*, vol. 31, no. 10, pp. 1443–1450, Oct. 1996.
6. Jager, R.C., *Microelectronics Circuit Design*, McGraw-Hill, 1997.
7. Kambayashi, Y. and S. Muroga, "Properties of wired logic," *IEEE TC*, vol. C-35, pp. 550–563, 1986.
8. Kuroda, T., et al., "Capacitor-free level-sensitive active pull-down ECL circuit with self-adjusting driving capability," *Symp. VLSI Circuits*, pp. 29–30, 1993.
9. Mair, C.A., et al., "A 533-MHz BiCMOS superscaler RISC microprocessor," *IEEE JSSC*, pp. 1625–1634, Nov. 1997.
10. Muroga, S., *VLSI System Design*, John Wiley and Sons, 1982.
11. Muroga, S. and H.-C. Lai, "Minimization of logic networks under a generalized cost function," *IEEE TC*, pp. 893–907, Sept. 1976.
12. Nambu, H., et al., "A 0.65-ns, 72-kb ECL-CMOS RAM macro for a 1-Mb SRAM," *IEEE Jour. of Solid-State Circuits*, vol. 30, no. 4, pp. 491–499, April 1995.
13. Sedra, A.S. and K.C. Smith, *Microelectronic Circuits*, 4th ed., Oxford University Press, 1998.
14. Shin, H.J., "Self-biased feedback-controlled pull-down emitter follower for high-speed low-power bipolar logic circuits," *Symp. VLSI Circuits*, pp. 27–28, 1993.
15. Toh, K.-Y. et al., "A 23-ps/2.1-mW ECL gate with an AC-coupled active pull-down emitter-follower stage," *Jour. SSC*, pp. 1301–1306, Oct. 1989.
16. Ueda, T., Japanese Patent Sho 51-22779, 1976.

39

CMOS

Saburo Muroga
University of Illinois
at Urbana-Champaign

CONTENTS

39.1 CMOS (Complementary MOS)

A CMOS logic gate consists of a pair of subcircuits, one consisting of nMOSFETs and the other pMOSFETs, where all MOSFETs are of enhancement mode described in Chapter 33, Section 33.3. **CMOS**, which stands for complementary MOS [6,8–10], means that the nMOS and pMOS subcircuits are complementary. As a simple example, let us explain CMOS with the inverter shown in Figure 39.1. A p-channel MOSFET is connected between the power supply of positive voltage V_{dd} and the output terminal, and an n-channel MOSFET is connected between the output terminal and the negative side, V_{ss}, of the above power supply, which is usually grounded. When input x is a high voltage, pMOS becomes non-conductive and nMOS becomes conductive. When x is a low voltage, pMOS becomes conductive and nMOS becomes non-conductive. This is the property of pMOS and nMOS when the voltages of the input and the power supply are properly chosen, as explained with Figure 33.19. In other words, when either pMOS or nMOS is conductive, the other is non-conductive. When x is a low voltage (logic value 0), pMOS is conductive, with non-conductive nMOS, and the output voltage is a high voltage (logic value 1), which is close to V_{dd}. When x is a high voltage, nMOS is conductive, with non-conductive pMOS, and the output voltage is a low voltage. Thus, the CMOS logic gate in Figure 39.1 works as an inverter. The pMOS subcircuit in this figure essentially works as a variable load.

When x stays at either 0 or 1, one of pMOS and nMOS subcircuits in Figure 39.1 is always non-conductive, and consequently no current flows from V_{dd} to V_{ss} through these MOSFETs. In other words, when no input changes, the power consumption is simply the product of the power supply voltage V (if V_{ss} is grounded, V is equal to V_{dd}) and a very small current of a non-conductive MOSFET. (Ideally, there should be no current flowing through a non-conductive MOSFET, but actually a very small current which is less than 10 nA flows. Such an undesired, very small current is called a **leakage current**.) This is called the **quiescent power consumption**. Since the leakage current is typically a few nanoamperes,

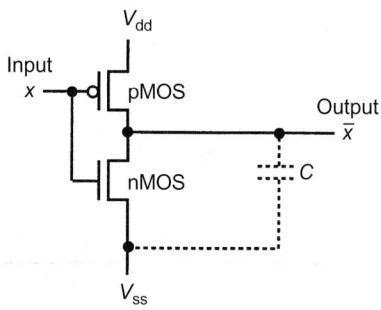

FIGURE 39.1   CMOS inverter.

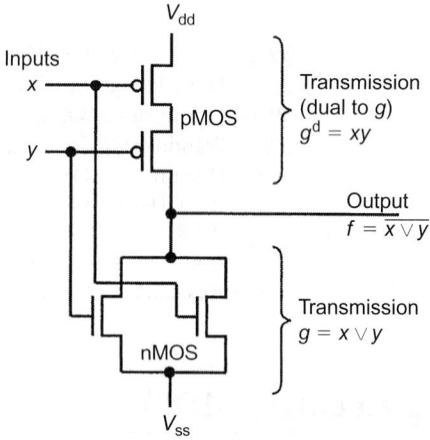

FIGURE 39.2   CMOS NOR gate.

the quiescent power consumption of CMOS is less than tens of nW, which is very small compared with those for other logic families.

Whenever the input x of this CMOS logic gate changes to a low voltage (i.e., logic value 0), the parasitic capacitance C at the output terminal (including parasitic capacitances at the inputs of the succeeding CMOS logic gates, to which the output of this logic gate is connected) must be charged up to a high voltage through the conductive pMOS. (A current can be as large as 0.3 milliamperes or more.) Then, when the input x changes to a high voltage (i.e., logic value 1) at the next input change, the electric charge stored in the parasitic capacitance must be discharged through the conductive nMOS. Therefore, much larger power consumption than the quiescent power consumption occurs whenever the input changes. This dynamic power consumption due to the current during this transition period is given by CFV^2, where C is the parasitic capacitance, V is the power supply voltage, and F is the switching frequency of the input. Thus the power consumption of CMOS is a function of frequency. CMOS consumes very little power at low frequency, but it consumes more than ECL as the frequency increases. As the integration size increases, CMOS is being almost exclusively used in VLSI because of low power consumption. But even CMOS has difficulty in dissipation of the heat generated in the chip when switching frequency increases. In order to alleviate this difficulty, valiants, such as dynamic CMOS, have been used which will be described later.

39.1.1   Output Logic Function of a CMOS Logic Gate

Let us consider a CMOS logic gate in which many MOSFETs of the enhancement mode are connected in each of the pMOS and nMOS subcircuits (e.g., the CMOS logic gate in Figure 39.2). By regarding the pMOS subcircuit as a variable load, the output function f can be calculated in the same manner as the one of an nMOS logic gate:

1. Calculate the transmission between the output terminal and V_{ss} (or the ground), considering nMOSFETs as make-contacts of relays (transmission and relays are described in Chapter 33).
2. Then, its complement is the output function of this CMOS logic gate.

Thus, the CMOS logic gate in Figure 39.2, for example, has the output function $f = \overline{x \vee y}$.

We can prove that if the pMOS subcircuit of any CMOS logic gate has the transmission between V_{dd} and the output terminal, calculated by regarding each pMOS as a make-contact relay, and this transmission is the dual of the transmission of the nMOS subcircuit, then one of the pMOS and nMOS subcircuits is always non-conductive, with the other conductive, for any combination of input values. (Note that regarding each pMOS as a make-contact relay, as we do each nMOS as a make-contact relay, means finding a relationship of connection configuration between nMOS subcircuit and pMOS subcircuit.) In the CMOS logic gate of Figure 39.2, the pMOS subcircuit has transmission $g^d = xy$, which is dual to the transmission $g = x \vee y$ of the nMOS subcircuit, where the superscript d on g means "dual." Thus, any CMOS logic gate has the unique features of unusually low quiescent power consumption and dynamic power consumption CV^2F.

The input resistance of a CMOS logic gate is extremely high and at least 10^{14} Ω. This permits large fan-outs from a CMOS logic gate. Thus, if inputs do not change, CMOS has almost no maximum fan-out restriction. The practical maximum fan-out is 30 or more, which is very large compared with other logic families. If the number of fan-out connections from a CMOS logic gate is too many, the waveform of a signal becomes distorted. Also, fan-out increases the parasitic capacitance and consequently reduces the speed, so fan-out is limited to a few when high speed is required.

In addition to extremely low power consumption, CMOS has the unique feature that CMOS logic networks work reliably even if power supply voltage fluctuates, temperature changes over a wide range, or there is plenty of noise interference. This makes use of CMOS appropriate in rugged environments, such as for automobile, in factories, and weapons.

39.1.2 Problem of Transfer Curve Shift

Unfortunately, when a CMOS logic gate has many inputs, its transfer curve (which shows the relationship between input voltage and output voltage of a CMOS logic gate) shifts, depending on how many of the inputs change values. For example, in the two-input NAND gate shown in Figure 39.3(a), the transfer curve for the simultaneous change of the two inputs (1 and 2) is different from that for the change of only input 1, with input 2 kept at a high voltage. This is different from an nMOS logic gate (or a pMOS logic gate) discussed in the previous sections, where every driver MOSFET in its conductive state must have a much lower resistance than the load MOSFET in order to have a sufficiently large voltage swing. But if only input 1 in Figure 39.3(a), for example, changes, the resistance of the pMOS subcircuit is twice as large as that for the simultaneous change of the two inputs 1 and 2; so parasitic capacitance, C, is charged in a shorter time in the latter case. Other examples are shown in (b) and (c) in Figure 39.3. Because of this problem of transfer curve shift, the number of inputs to a CMOS logic gate is practically limited to four if we want to maintain good noise immunity. If we need not worry about noise immunity, the number of inputs to a CMOS logic gate can be greater.

39.2 Logic Design of CMOS Networks

The logic design of CMOS networks can be done in the same manner as that of nMOS logic networks, because the nMOS subcircuit in each CMOS logic gate, with the pMOS subcircuit regarded as a variable load, essentially performs the logic operation, as seen from Figure 39.2. The design procedures discussed for nMOS networks in Chapter 33, Section 33.3 can be used more effectively than in the case of nMOS

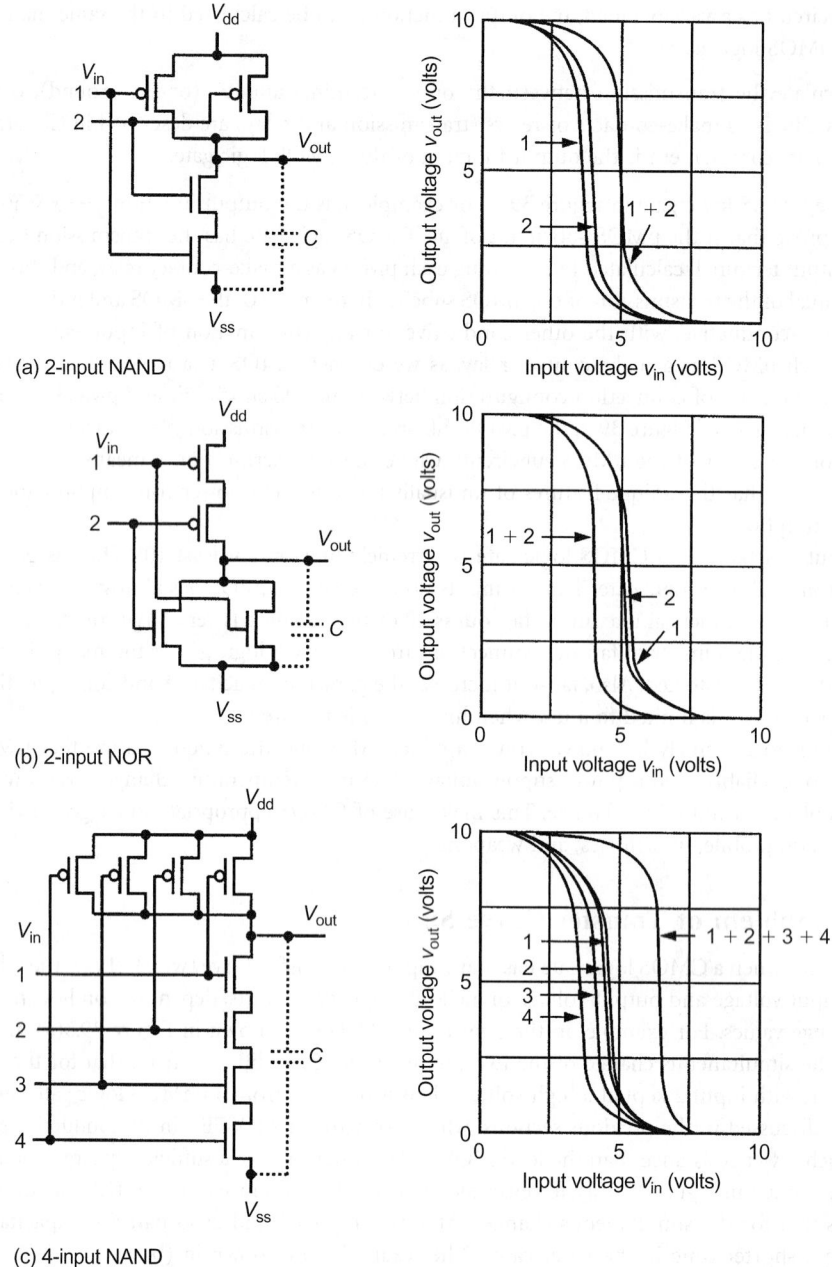

(a) 2-input NAND

(b) 2-input NOR

(c) 4-input NAND

FIGURE 39.3   CMOS transfer curves for different numbers of inputs.

networks, because more than four MOSFETs can be in series inside each logic gate, unless we are concerned about the transfer-curve shift problem or high-speed operation. Also, an appropriate use of transmission gates, discussed in the next paragraph, often simplifies networks.

The **transmission gate** shown in Figure 39.4 is a counterpart of the pass transistor (i.e., transfer gate) of nMOS, and is often used in CMOS network design. It consists of a pair of p-channel and n-channel MOSFETs. The control voltage d is applied to the gate of the n-channel MOSFET, and its complement $\bar{d}$ is applied to the gate of the p-channel MOSFET. If d is a high voltage, both MOSFETs become

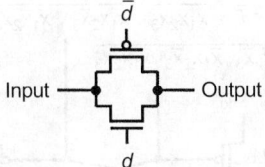

FIGURE 39.4 Transmission gate.

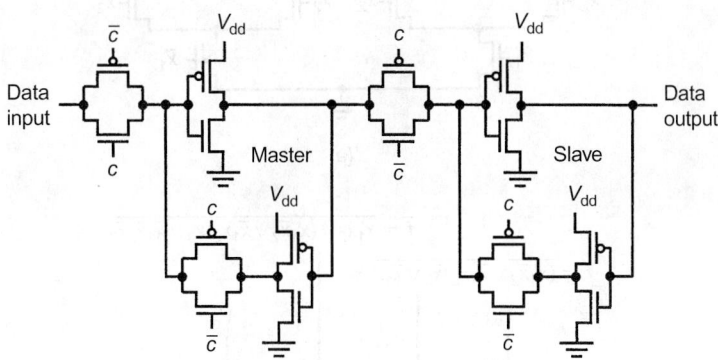

FIGURE 39.5 D-type flip-flop (c is a clock).

conductive, and the input is connected to the output. (Unlike a pass transistor with nMOS whose output voltage is somewhat lower than the input voltage, the output voltage of the transmission gate in CMOS is the same as the input voltage after the transition period.) If d is a low voltage, both MOSFETs become non-conductive, and the output is disconnected from the input, keeping the output voltage (which gradually becomes low because of current leakage) at its parasitic capacitance, as it was before the disconnection. Since the input and output are interchangeable, the transmission gate is bidirectional. A D-type flip-flop is shown in Figure 39.5, as an example of CMOS circuits designed with transmission gates.

A full adder in CMOS with transmission gates is shown in Chapter 41. Also, pass transistors realized in nMOS can been used mixed with CMOS logic gates to reduce area or power consumption, as will be discussed in Chapter 40.

39.3 Logic Design in Differential CMOS Logic

Differential CMOS logic is a logic gate that works very differently from the CMOS logic gate discussed so far. It has two outputs, f and its complement, $\bar{f}$, and works like a flip-flop such that when one output is a high voltage, it always makes the other output have a low voltage.

A logic gate in **cascode voltage switch logic**, which is abbreviated as **CVSL**, is illustrated in Figure 39.6. CVSL is sometimes called **differential logic** because CVSL is a CMOS logic gate that realizes both an output function, f, and its complement, $\bar{f}$, switching their values quickly. The CVSL gate shown in Figure 39.6(a) has a driver that is a tree consisting of nMOSFETs, where each pair of nMOSFETs in one level has input x_i and its complement $\bar{x}_i$. The top end of each path in the tree expresses the complement of a minterm (just like series-gating ECL described in Chapter 38). Then by connecting some of these top ends to both the gate of one pMOSFET (i.e., P1), and the drain of the other pMOSFET (i.e., P2), we can realize the complement of a sum-of-products. The connection of the remaining top ends to both the gate of P2 and the drain of P1 realizes its complement. The outputs in Figure 39.6(a) realize

$$f = x_1 \bar{x}_2 \bar{x}_3 \vee \bar{x}_1 x_2 \bar{x}_3 \vee \bar{x}_1 \bar{x}_2 x_3$$

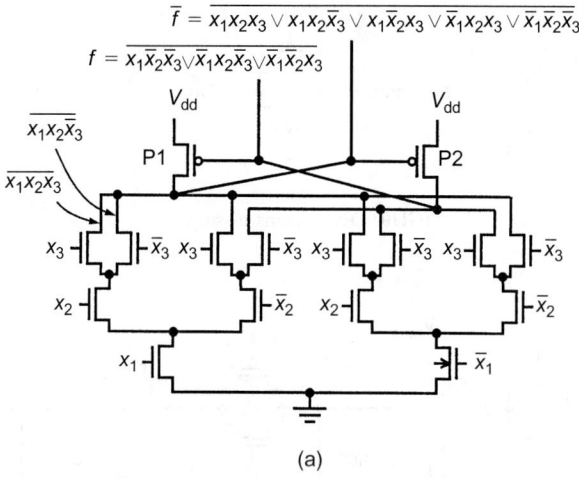

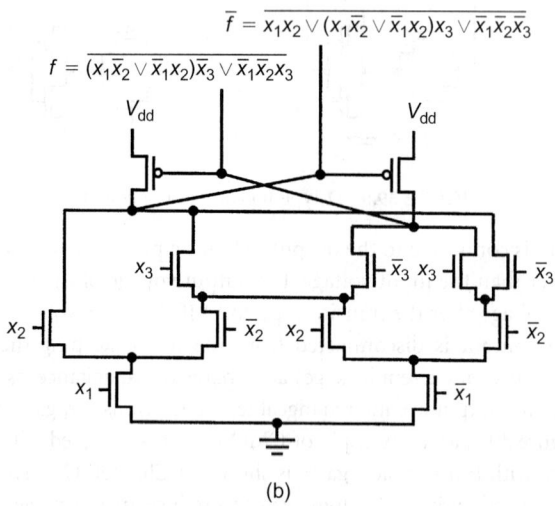

FIGURE 39.6   (a) Static CVSL; (b) Static CVSL for the same functions as in (a).

and

$$\overline{f} = \overline{x_1x_2x_3 \vee x_1x_2\overline{x}_3 \vee x_1\overline{x}_2x_3 \vee \overline{x}_1x_2x_3 \vee \overline{x}_1\overline{x}_2\overline{x}_3}$$

When one path in the tree is conductive, the output of it, say the output f, has a lower voltage, and P1 (i.e., the pMOSFET whose gate is connected to f) becomes conductive. Then, the other output $\overline{f}$ has a high voltage. The driver tree in Figure 39.6(a), which resembles series-gating ECL, can be simplified as shown in (b). (The tree structure in CSVL in (b) can be obtained from an ordered reduced binary decision diagram described in Chapters 26 and 29.) Notice that P1 and P2 in (a) are cross-connected for fast switching.

CVSL can be realized without tree structure. Figure 39.7 shows such a CSVL gate, for $f = \overline{xy}$ and its complement. The connection configuration of nMOSFETs connected to one output terminal is dual to that connected to the other output terminal (in Figure 39.7, nMOSFETs for the output terminal for f are connected in series, while nMOSFETs for $\overline{f}$ are connected in parallel).

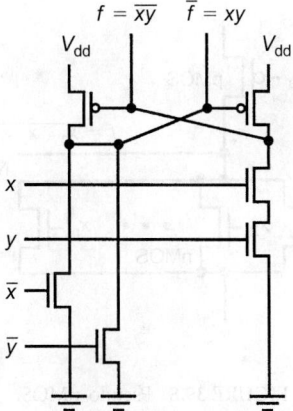

FIGURE 39.7 Static CVSL.

CVSL has a variant called dynamic CVSL, to be described later. In order to differentiate from this, CVSL here is usually called **static CVSL**.

39.4 Layout of CMOS

Layout of CMOS logic networks is more complex than logic networks where only nMOSFETs or pMOSFETs are used because in CMOS networks, nMOSFETs and pMOSFETs need to be fabricated with different materials. This makes layout of CMOS complicated. We often need to consider appropriate connection configuration of MOSFETs inside each logic gate and a logic network such that speed or area is optimized.

When we want to raise the speed, the switching speed of the pMOS subcircuit should be comparable to that of the nMOS subcircuit. Since the mobility of electrons is roughly 2.5 times higher than that of holes, the channel width W of p-channel MOSFETs must be much wider (often 1.5 to 2 times because the width can be made smaller than 2.5 times due to different doping levels) than that of n-channel MOSFETs (which work based on electrons) in order to compensate for the low speed of p-channel MOSFETs (which work based on holes) if the same channel length is used in each. Thus, for high-speed applications, NAND logic gates such as Figure 39.3(a) are preferred to NOR logic gates, because the channel width of p-channel MOSFETs in series in a NOR logic gate such as Figure 39.3(b) must be further increased such that the parasitic capacitance C can be charged up in a shorter time with low series resistance of these p-channel MOSFETs.

When designers are satisfied with low speed, the same channel width is chosen for every p-channel MOSFET as for n-channel MOSFETs, since a CMOS logic gate requires already more area than a pMOS or nMOS logic gate and a further increase by increasing the channel width is not desirable unless really necessary.

39.5 Pseudo-nMOS

In the case of the CMOS discussed so far, each CMOS logic gate consists of a pair of pMOS and nMOS subcircuits which realize dual transmission functions, as explained with Figure 39.2. In design practice, however, some variations are often used. For example, Figure 39.8 realizes NOR. The pMOS subcircuit consists of only a single MOSFET. Thus, the chip area is reduced and the speed is faster than static CMOS discussed so far because of a smaller parasitic capacitance. But more power is dissipated for some combinations of input values [1] because the pMOS subcircuit in (a) is always conductive [9,10].

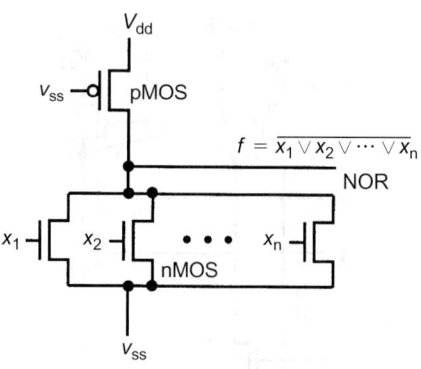

FIGURE 39.8 Pseudo-nMOS.

39.6 Dynamic CMOS

Dynamic CMOS [4] is a CMOS logic gate that works in a manner very different from the CMOS logic gates discussed so far, which are usually called static CMOS. Using clock pulse, a parasitic capacitance is precharged, and then it is evaluated whether the parasitic capacitance is discharged or not, depending on the values of input variables. Dynamic CMOS has been used often for high speed because parasitic capacitance can be made small due to unique connection configuration of MOSFETs inside a logic gate, although good layout is not easy. Power consumption is not necessarily small. (In static CMOS, a current flows from the power supply to the ground during transition period. But in dynamic CMOS, there is no such current. But this does not mean that dynamic CMOS consumes less power because when input variables do not change their values, static CMOS does not consume power at all, whereas dynamic CMOS may consume power by repeating precharging. Dynamic CMOS and static CMOS have completely different power consumption mechanisms.)

39.6.1 Domino CMOS

Domino CMOS, illustrated in Figure 39.9, consists of pairs of a CMOS logic gate and an inverter CMOS logic gate [4]. The first CMOS logic gate in each pair (such as logic gates 1, 3, and 5) has the pMOS subcircuit, consisting of a single pMOSFET with clock, and the nMOS subcircuit, consisting of many nMOSFETs with logic inputs and a single nMOSFET with a clock. The first CMOS logic gate is followed by an inverter CMOS logic gate (such as logic gates 2, 4, and 6). When a clock pulse is absent at all terminals labeled c, all parasitic capacitances (shown by dotted lines) are charged to value 1 (i.e., a high voltage) because all pMOSFETs are conductive. This process is called **precharging**. Thus, the outputs of all inverters become value 0. Suppose that $x = v = 1$ (i.e., a high voltage) and $y = z = u = 0$ (i.e., a low voltage). When a clock pulse appears, that is, $c = 1$, all pMOSFETs become non-conductive but the nMOS subcircuit in each of logic gates 1 and 5 becomes conductive, discharging parasitic capacitance. Then the outputs of logic gates 1, 2, 3, 4, 5, and 6 become 0, 1, 1, 0, 0, and 1, respectively. Notice that the output of logic gate 3 remains precharged because its nMOSFET for u remains non-conductive. Domino CMOS has the following advantages:

- It has a small area because the pMOS subcircuit in each logic gate consists of a single pMOSFET.
- It is faster (about twice) than the static CMOS discussed so far because parasitic capacitances are reduced by using a single pMOS in each logic gate and the first logic gate is buffered by an inverter. Also, an inverter, such as logic gate 2, has smaller parasitic capacitance at its output because it connects to only nMOSFET in logic gate 3, for example, compared to static CMOS where it connects to both pMOSFET and nMOSFET in each of next static CMOS logic gates, to which the output of this static CMOS is connected. This also makes domino CMOS faster.

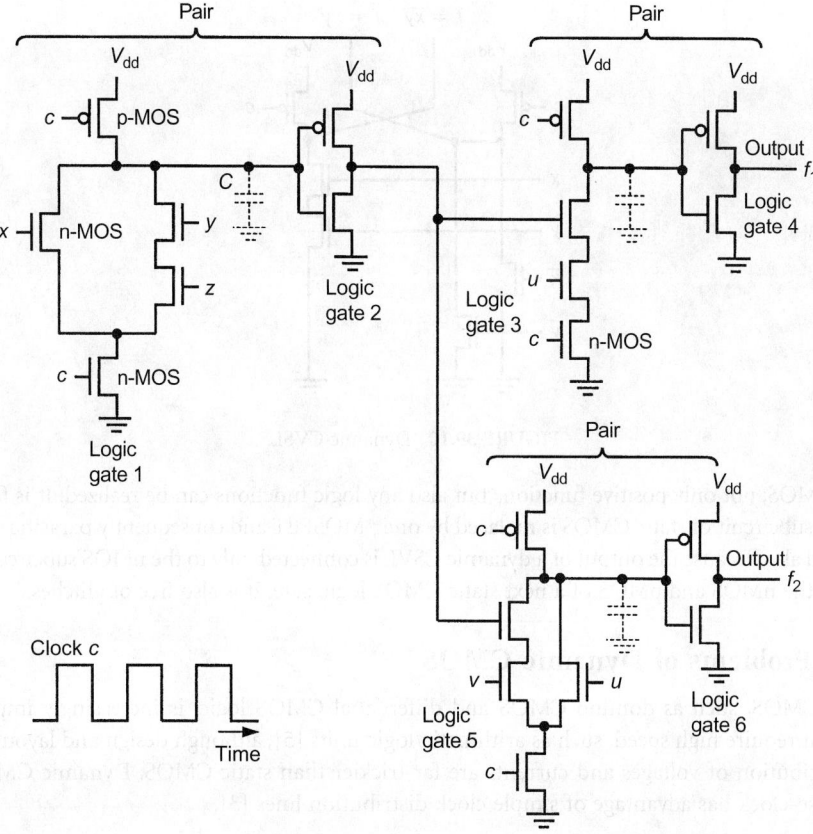

FIGURE 39.9 Domino CMOS.

- It is free of glitches (i.e., transition is smooth) because at the output of each logic gate, a high voltage remains or decreases, but no voltage increases from low to high [7].

Domino CMOS has the following disadvantage:

- Only positive functions with respect to input variables can be realized. (If both x_i and $\bar{x}_i$ for each x_i is available as network inputs, the network can realize any function. But if only one of them, say x_i, is available, functions that are dependent $\bar{x}_i$ on cannot be realized by a domino CMOS logic network.)

So we have to have domino CMOS networks in double-rail input logic (e.g., Ref. 2), or to add inverters, whenever necessary. Thus, although the number of MOSFETs in domino CMOS networks in single-rail input logic, such as Figure 39.9, is almost half of static CMOS networks, the number of MOSFETs in such domino CMOS networks to realize any logic functions may become comparable to the number of MOSFETs in static CMOS networks in single-rail input logic.

39.6.2 Dynamic CVSL

Static CVSL, which is previously described, can be easily converted into **dynamic CVSL** which is faster, as illustrated in Figure 39.10. The parasitic capacitance of two output terminals are precharged through each of the two pMOSFETs during the absence of a clock pulse. Dynamic CVSL works in a similar manner to domino CMOS. Notice that two pMOSFETs are not cross-connected like static CVSL and we have essentially two independent logic gates. Dynamic CSVL with two outputs, f and $\bar{f}$, is in double-rail logic, so unlike

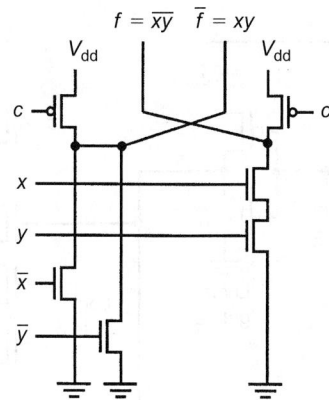

FIGURE 39.10 Dynamic CVSL.

domino CMOS, not only positive functions but also any logic functions can be realized. It is fast because the pMOS subcircuit of static CMOS is replaced by one pMOSFET and consequently parasitic capacitance is small and also because the output of a dynamic CSVL is connected only to the nMOS subcircuits, instead of to both the nMOS and pMOS of a next static CMOS logic gate. It is also free of glitches.

39.6.3 Problems of Dynamic CMOS

Dynamic CMOS, such as domino CMOS and differential CMOS logic, is increasingly important for circuits that require high speed, such as arithmetic/logic units [5], although design and layout of appropriate distribution of voltages and currents are far trickier than static CMOS. Dynamic CMOS with a single-phase-clock has advantage of simple clock distribution lines [3].

References

1. Cooper, J.A., J.A. Copland, and R.H. Krambeck, "A CMOS microprocessor for telecommunications applications," *ISSCC '77*, pp. 137–138.
2. Heikes, C., "A 4.5 mm^2 multiplier array for a 200MFLOP pipelined coprocessor," *ISSCC '94*, pp. 290–291. (Double-rail domino CMOS.)
3. Ji-Ren, Y., I. Karlsson, and C. Svensson, "A true single-phase-clock dynamic CMOS circuit technique," *IEEE JSSC*, pp. 899–901, Oct. 1987.
4. Krambeck, R.H., C.M. Lee, and H.-F.S. Law, "High-speed compact circuits with CMOS," *IEEE JSSC*, pp. 614–619, June 1982. (First paper on domino CMOS)
5. Lu, F. and H. Samueli, "A 200-MHz CMOS pipelined multiplier-accumulator using quasi-domino dynamic full-adder cell design," *IEEE JSSC*, pp. 123–132, Feb. 1993.
6. Muroga, S., *VLSI System Design*, John Wiley & Sons, 1982.
7. Murphy, et al., "A CMOS 32b single chip microprocessor," *ISSCC '81*, pp. 230, 231, 276.
8. Shoji, M., *CMOS Digital Circuit Technology*, Prentice-Hall, 1988.
9. Vyemura, J.P., *CMOS Logic Circuit Design*, Kluwer Academic Publishers, 1999.
10. Weste, N.H.E. and K. Eshraghian, *Principles of CMOS VLSI Design*, 2nd ed., Addison Wesley, 1993.

40

Pass Transistors

Kazuo Yano
Hitachi Ltd.

Saburo Muroga
*University of Illinois
at Urbana-Champaign*

CONTENTS

40.1 Introduction

A MOSFET is usually used such that a voltage at the gate terminal of the MOSFET controls a current between its source and drain. When the voltages at the gate terminals of MOSFETs that constitute a logic gate are regarded as input variables x, y, and so on, the voltage at the output terminal represents the output function f. Here, $x = 1$ means a high voltage and $x = 0$ a low voltage. For example, MOSFETs in the logic gate in CMOS shown in Figure 40.1(a) are used in this manner and the output function f represents $\overline{x \vee y}$. But a MOSFET can be used such that an input variable x applied at one of source and drain of the MOSFET is delivered to the other or not, depending on whether a voltage applied at the gate terminal of the MOSFET is high or low. For example, the n-channel MOSFET shown in Figure 40.1(b) works in this manner and the MOSFET used in this manner is called a **transfer gate** or, more often, a **pass transistor**. When the control voltage c is high, the MOSFET becomes conductive and the input voltage x appears at the output terminal f (henceforth, let letters, f, x, and others represent terminals as well as voltages or signal values), no matter whether the voltage x is high or low. When the control voltage c is low, the MOSFET becomes non-conductive and the input voltage at x does not appear at the output terminal f.

Pass transistors have been used for simplification of transistor circuits. Logic functions can be realized with fewer MOSFETs than logic networks of logic gates where MOSFETs are used like those in Figure 40.1(a). The circuit in Figure 40.2, for example, realizes the even-parity function $\overline{x \oplus y}$ [2]. This circuit works in the following manner. When x and y are both low voltages, n-channel MOSFETs, 1 and 2, are non-conductive and consequently no current flows from the power supply V_{dd} to the terminal x or y. Thus, the output voltage f is high because it is the same as the power supply voltage at V_{dd}. When x is a low voltage and y is a high voltage, MOSFET 1 becomes conductive and 2 is non-conductive. Consequently, a current flows from the power supply to x because x is a low voltage. Thus, the output voltage at f is low. Continuing the analysis, we have the truth table shown in Figure 40.3(a). Then we can derive the truth table for function $f = \overline{x \oplus y}$ in Figure 40.3(b) by regarding a low and high voltages as 0 and 1, respectively. A logic gate for this function in CMOS requires 8 MOSFETs, as shown in Figure 40.4, whereas the circuit realized with pass transistors in Figure 40.2 requires only three MOSFETs. Notice that inputs x and y are connected to the sources of MOSFETs 1 and 2, unlike MOSFET in ordinary logic gates. Signal x at the source of MOSFET 1 is either sent to the drain or not, according to whether or not its MOSFET gate has a high voltage.

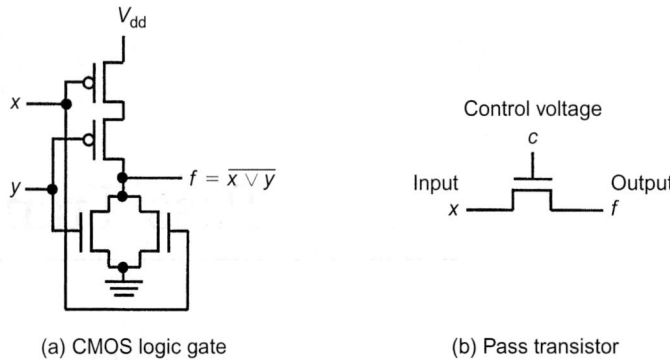

(a) CMOS logic gate   (b) Pass transistor

FIGURE 40.1 CMOS logic gate and pass-transistor circuit.

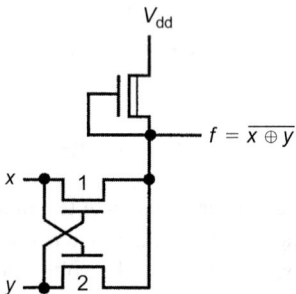

FIGURE 40.2 Circuit with pass transistors for the even-parity function.

(a) Voltage relationship			(b) Logic function		
x	y	f	x	y	f
Low volt.	Low volt.	High volt.	0	0	1
Low volt.	High volt.	Low volt.	0	1	0
High volt.	Low volt.	Low volt.	1	0	0
High volt.	High volt.	High volt.	1	1	1

FIGURE 40.3 Behavior of the circuit in Figure 40.2.

Pass transistors, however, are sometimes used inside an ordinary logic gate, mixed with ordinary MOSFETs. MOSFETs 1, 2, and 3 in Figure 40.5 are such pass transistors. (Actually, the pair of 1 and 2 is a transmission gate to be described in the following and also in Chapter 39, Section 39.2) Logic networks in CMOS where MOSFETs are used in the same way as those in Figure 40.1(a) can sometimes be simplified by an appropriate use of pass transistors, possibly with speed-up.

The wide use of a pass transistor is found in the DRAM memory cell, which consists of one capacitor and one pass transistor. The control of charging to or discharging from the memory capacitor is done through the pass transistor. This shows the impact of this technique in area reduction, power consumption reduction, and possibly also in speed-up. Pass transistors are also used in the arithmetic-logic unit of a computer, which requires speed and small area, such as fast adders (actually the logic gate in Figure 40.5 is part of such an adder), multipliers and multiplexers [3,7,11,12].

The circuit in Figure 40.6(a) in double-rail input logic (i.e., x, $\bar{x}$, y, and $\bar{y}$ are available as inputs) realizes the odd-parity function $x \oplus y$. A circuit for the inverter shown by the triangle with a circle in Figure 40.6(a) is shown in (b).

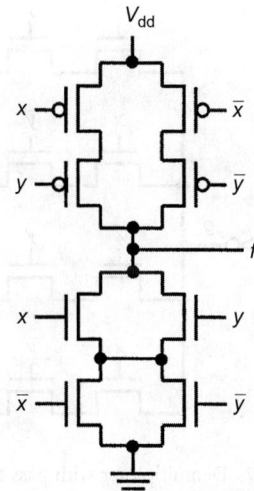

FIGURE 40.4 CMOS logic gate for function $f = \overline{x \oplus y}$.

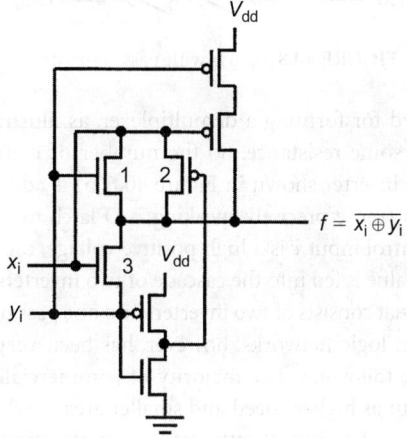

FIGURE 40.5 Pass transistors in a logic gate.

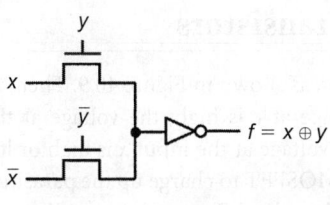

(a) Circuit for the parity function

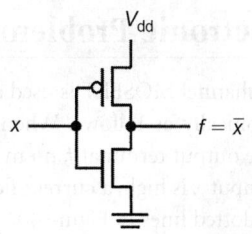

(b) Circuit for the inverter in (a)

FIGURE 40.6 Parity function realized with pass transistors.

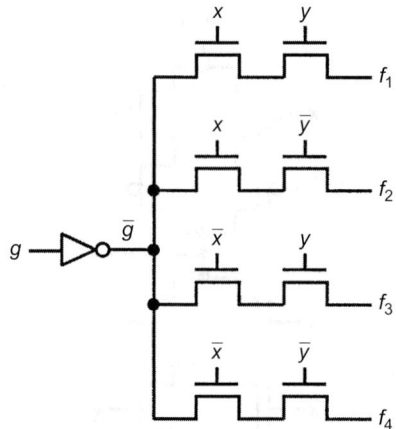

FIGURE 40.7 Demultiplexer with pass transistors.

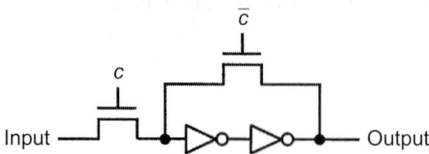

FIGURE 40.8 Latch with pass transistors.

Pass transistors are often used for forming a demultiplexer, as illustrated in Figure 40.7. Series connection of pass transistors has some resistance. So the number of control variables (here, x and y) is limited to at most four and the inverter shown in Figure 40.6(b) is added after input g as a buffer.

Using pass transistors, a latch (more precisely speaking, a D latch to store data) can be constructed as shown in Figure 40.8. When control input c is a high positive voltage, the feedback loop is cut by the pass transistor with $\bar{c}$, and the input value is fed into the cascade of two inverters. When c becomes a low voltage, the input is cut off and the loop that consists of two inverters and one pass transistor retains the information.

The use of pass transistors in logic networks, however, has been very limited because of electronics problems to be discussed in the following. The majority of commercially available logic networks have been in static CMOS circuits. But as higher speed and smaller area are desired, this situation is gradually changing. The pass-transistor logic has recently attracted much attention under these circumstances and is anticipated to be widely used in the near future for its area/power saving and high-performance benefits.

Beside pass transistors, there are many other unconventional MOS networks. All these networks are useful for simplification of electronic realization of logic networks or for improvement of performance, although complex adjustments of voltages or currents are often required.

40.2 Electronic Problems of Pass Transistors

Suppose an n-channel MOSFET is used as a pass transistor, as shown in Figure 40.9. Then, the MOSFET behaves electronically as follows. When the control voltage at c is high, the voltage at the input x is delivered to the output terminal f, no matter whether the voltage at the input x is high or low. But if the voltage at the input x is high, a current flows through the MOSFET to charge up the parasitic capacitance (shown in the dotted lines in Figure 40.9) at f and stops when the difference between the voltage at f and the voltage at c reaches the threshold voltage, making the voltage at f somewhat lower than the voltage at x. When the voltage at c becomes low, the pass transistor becomes non-conductive and the electric charge stored on the parasitic capacitance gradually leaks to the ground through the parasitic resistor.

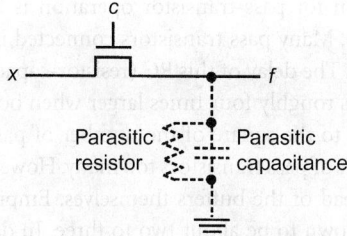

FIGURE 40.9 Electronic behavior of a pass transistor.

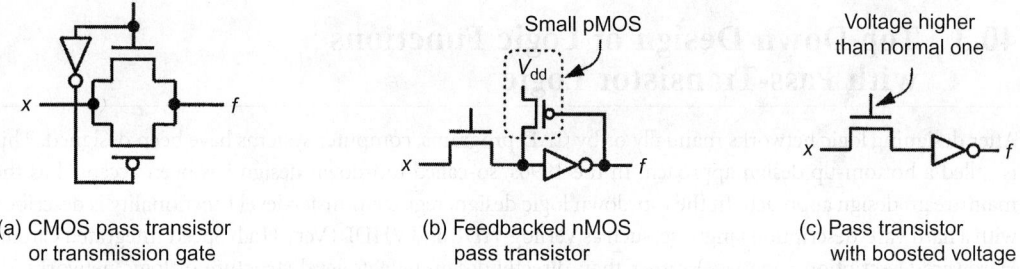

FIGURE 40.10 Techniques to avoid the influence of the low output voltage.

If the voltage at the input x is low when the control voltage at c is high, the electric charge stored on the parasitic capacitance flows to the ground through the pass transistor and input terminal x also if it has not completely leaked to the ground yet.

This complex electronic behavior of the pass transistor makes a circuit with pass transistors unreliable. The intermediate value of the voltage at f, which is lower by the threshold voltage than the voltage at x or partially leaked voltage, causes unpredictable operations of the logic network when the voltage at f is fed to ordinary CMOS logic gates in the next stage. Moreover, it degrades the switching speed of the CMOS logic gates. In the worst case, the circuit loses noise margin or it does not operate properly.

There are three techniques to avoid this drawback. The first one is to combine an nMOS pass transistor and a pMOS pass transistor in parallel, as shown in Figure 40.10(a). With this technique, when the pass transistor is conductive, the output voltage at f reaches exactly the same value as the input voltage at x, no matter whether the input voltage is high or low. This pair of nMOS and pMOS pass transistors is sometimes called a **transmission gate**. Although this has better stability over the pass transistor circuit in Figure 40.9, it consumes roughly twice as large an area.

The second approach is to use a pMOS feedback circuit at the output of the nMOS pass transistor, as shown in Figure 40.10(b). The gate of a p-channel MOSFET is driven by the CMOS inverter (shown as the triangle with a small circle), which works as an amplifier. When the CMOS inverter discharges the electric charge at the output, it also turns on the feedback pMOS to raise the pass transistor output to the power supply voltage, eliminating the unreliable operation. One limitation of this approach is that it does not solve the degradation of switching speed due to low voltage, because the speed is determined by the initial voltage swing before the pMOS turns on. Area increase with this approach is smaller than the transmission gate in Figure 40.10(a).

The third approach is to raise the gate voltage swing up to the normal voltage plus threshold voltage, which is used in DRAM and referred to as "word boost," as shown in Figure 40.10(c). This approach requires a boost driver every time the gate signal is generated, which is difficult to use in logic functions in general. In addition, a voltage that is higher than the power supply voltage is applied to the gate of a MOSFET, which requires special care against breakdown and reliability problems (these need to be solved by increasing the thickness of gate insulation).

Another important consideration for pass-transistor operation is how many pass transistors can be connected in series without buffers. Many pass transistors connected in series can be treated as a serially connected resistor-capacitor circuit. The delay of this RC (resistor-capacitor) circuit, which is proportional to the product of R and C, becomes roughly four times larger when both R and C are doubled. Thus, the delay of this circuit is proportional to the square of the number of pass transistors. This means that it is not beneficial to increase the number of pass transistors too many. However, short-pitch insertion of CMOS inverters increases the delay overhead of the buffers themselves. Empirically, the optimal pass-transistor stages for delay minimization is known to be about two to three. In design practice, the number of pass transistors cannot be arbitrarily chosen because designers want to have a certain number of fan-out connections and a buffer cannot support too many fan-outs and pass transistors. Also, the structure of a logic network cannot be arbitrarily chosen because of area size and power consumption.

40.3 Top-Down Design of Logic Functions with Pass-Transistor Logic

After designing logic networks manually or by CAD programs, computer systems have been designed. This is called a bottom-up design approach. In the 1990s, so-called top-down design has been accepted as the mainstream design approach. In the top-down logic design, register-transfer-level functionality is described with a hardware-description language, such as Verilog-HDL and VHDL (Very High Speed Integrated Circuit Hardware Description Language) rather than directly designing gate-level structure of logic networks, or "netlist." And then, this is converted to logic networks of logic gates using a logic synthesizer (i.e., CAD programs for automated design of logic networks). This process resembles the compilation process of the software construction and it is sometimes referred to as "compile." Based on this netlist, placement and routing of transistors are done automatically on an IC chip. By using this top-down approach, a logic designer can focus on the functional aspect of the logic rather than the in-depth structural aspect. This enhances the productivity. Also, this enables one to easily port one design in one technology to another.

Automated design of logic networks with pass transistors has been difficult to realize because of complex electronic behavior. So, conventionally, pass-transistor logic has been manually designed, particularly in arithmetic modules as shown in this section. But as reduction of power consumption, speedup or area reduction is strongly desired, this is changing. Logic design based on selectors with pass-transistors can be done in this top-down manner [10]. Pass transistors have been used often as a selector by combing two pass transistors, as shown in Figure 40.11(a). A selector is also called a multiplexer. The output f of the selector becomes input x when $c = 1$ and input y when $c = 0$. Figure 40.11(b) shows a selector realized in a logic gate and also in pass transistors. Compared with the selector in a CMOS logic gate shown on the left side of Figure 40.11(b) which consists of ten MOSFETs, the selector in pass transistors shown on the right side of Figure 40.11(b) consists of only four MOSFETs, reducing the number of MOSFETs to less than half, and consequently the area. A selector is known to be a universal logic element because it can be used as an AND, an OR, and an XOR (i.e., Exclusive-OR) by changing its inputs, as shown in Figure 40.11(c). This property is also useful in the top-down design approach discussed in the following. The speed of a logic network with pass transistors is sometimes improved up to 2 times better than a conventional CMOS logic network, depending on logic functions.

One limitation of this pass-transistor selector is that it suffers a relatively slow switching speed when the control signal arrives later than selected input signals. This is because an inverter is needed for the selector to have a complementary signal applied to the gate of a pass transistor.

To circumvent this limitation, CPL (which stands for the complementary pass-transistor logic) has been conceived [11,12]. In CPL, complementary signals are used for both inputs and outputs, eliminating the need for the inverter. The circuits that require complementary signals like CPL are sometimes categorized as dual-rail logics. Because of the need for complementary signal, CPL is sometimes twice as large as CMOS, but is sometimes surprisingly small if a designer succeeds in fully utilizing the functionality of a pass-transistor circuit. A very fast and compact CPL full adder, a multiplier, and a

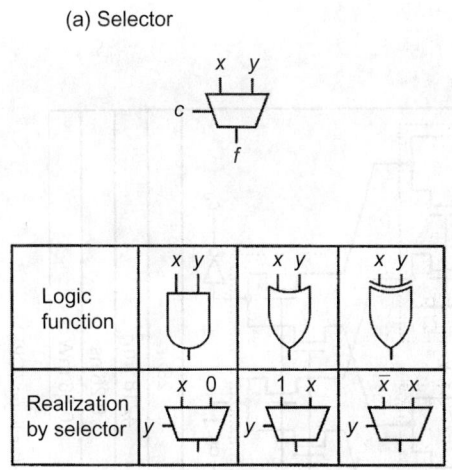

(a) Selector

(c) Logic functions realized by selectors

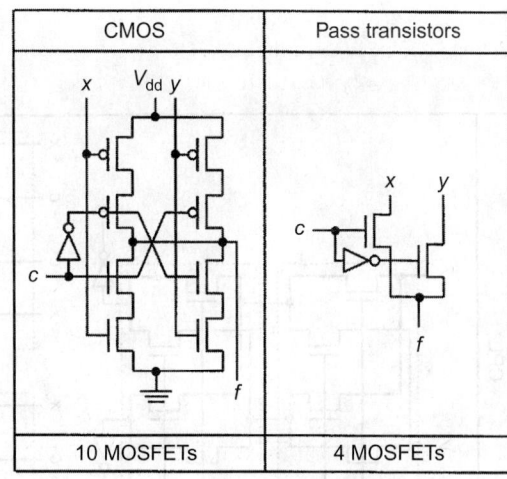

(b) Selectors realized in a CMOS logic gate and pass transistors

FIGURE 40.11 Selectors and various logic functions realized by pass transistors.

carry-propagate-chain circuit have been reported. A full adder realized with CMOS logic gates is compared with a full adder realized with selectors in pass transistors in Figure 40.12. Speed and power consumption are significantly improved.

Variants of CPL have been also reported, including DPL [8], SRPL [5], and SAPL [4].

However, conventional switching theory, on which widely used logic synthesizers are based, cannot be conveniently used for this purpose because it is very difficult to derive convenient logic expressions based on the output function $xc \vee y\bar{c}$ of the selector. Instead, BDD (i.e., binary decision diagrams) are used, as follows.

A simple approach to use a selector as the basic logic element is to build a binary tree of pass-transistor selectors, as shown in Figure 40.13. The truth table shown in Figure 40.13(a) is directly mapped into the tree structure shown in Figure 40.13(b). When $x = 1$, $y = 0$, and $z = 1$, for example, the third 1 from the left in the top of Figure 40.13(b) is connected to the output as $f = 1$. This original tree generally has redundancy, so it should be reduced to an irredundant form as shown in Figure 40.13(c). This approach is simple and effective when the number of input variables is less than 5 or so. However, this does not work for functions with more input variables, because of the explosive increase of the tree size.

To solve this, a binary decision diagram (i.e., BDD), has been utilized [10]. Basic design flow of BDD-based pass-transistor circuit synthesis is shown in Figure 40.14. The logic expressions for functions f_1 and f_2 shown in Figure 40.14(a) are converted to the BDD in (b). Then, buffers (shown as triangles) are inserted in Figure 40.14(c). In this case, only locations where the buffers should be inserted in Figure 40.14(d) are specified and the nature of the BDD in Figure 40.14(c) is not changed. In both Figures 40.14(b) and (c), each solid line denotes the value 1 of a variable and each dotted line the value 0. For example, the downward solid line from the right-hand circle with w inside denotes $w = 1$. From f_1, if we follow dotted lines three times and then the solid line once in each of (b) and (c), we reach the 0 inside the rectangle in the bottom. This means that $f_1 = 0$ for $w = x = y = 0$ and $z = 1$.

Preparation of an appropriate cell library based on selectors is required, as shown in Figure 40.15, which consists of a simple two-input selector (Cell 1) and its variants (Cells 2 and 3). The inverters shown with a dot inside the triangle in Figure 40.15, which is different from the simple inverter shown in Figure 40.6(b), is to keep the electric charge on the parasitic capacitance at its input. In Figure 40.14(d), the inverters of this kind have to be inserted corresponding to the buffers shown in (c). But in this case, the insertion has to be done such that the outputs f_1 and f_2 have the same polarity in both (c) and (d) because the inverters change signal values from 1 to 0 or from 0 to 1.

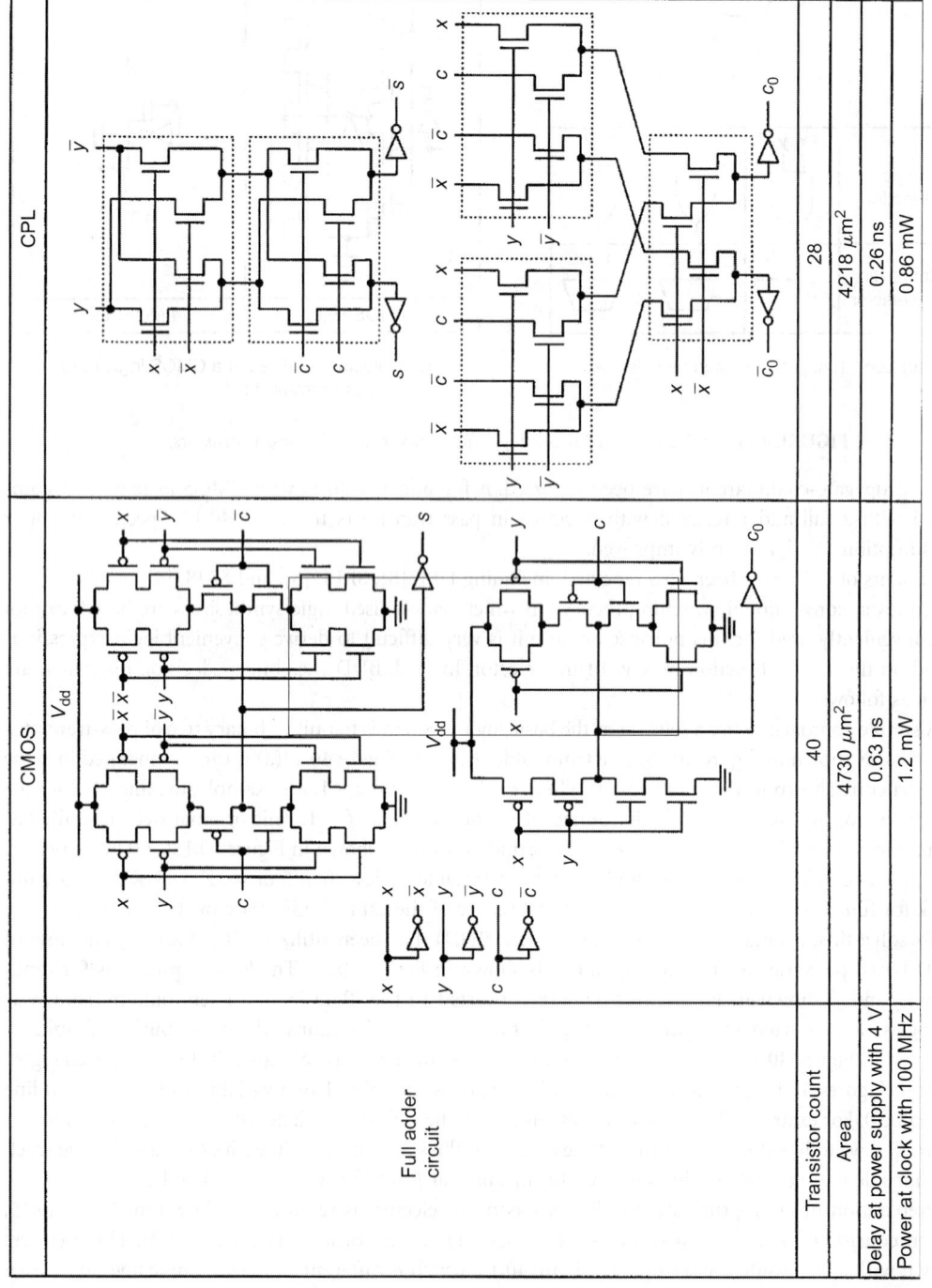

FIGURE 40.12 Full adder realized in CMOS logic gates and complementary pass-transistor logic (CPL).

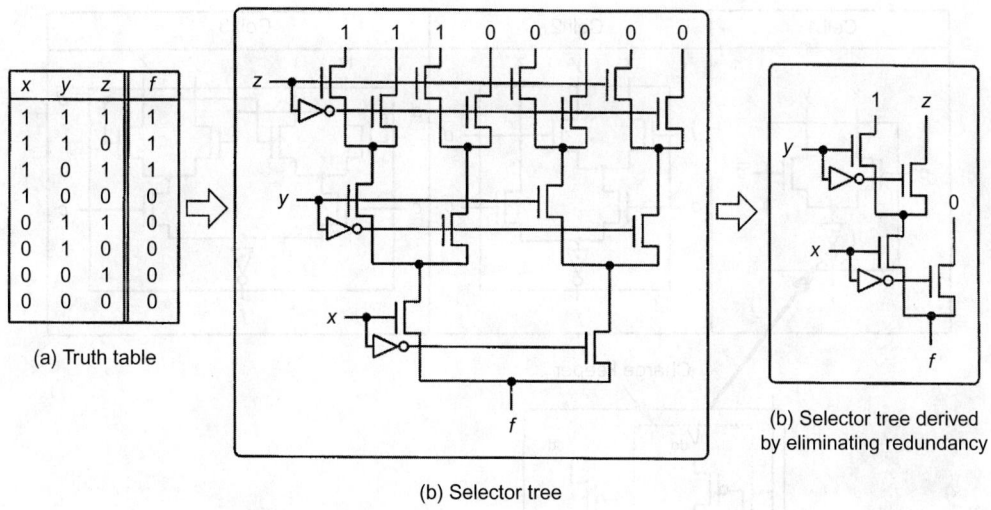

x	*y*	*z*	*f*
1	1	1	1
1	1	0	1
1	0	1	1
1	0	0	0
0	1	1	0
0	1	0	0
0	0	1	0
0	0	0	0

(a) Truth table

1 1 1 0 0 0 0
z
y
x
f
(b) Selector tree
1
z
y
0
x
f
(b) Selector tree derived by eliminating redundancy

FIGURE 40.13 Binary-tree-based simple construction method of pass-transistor logic.

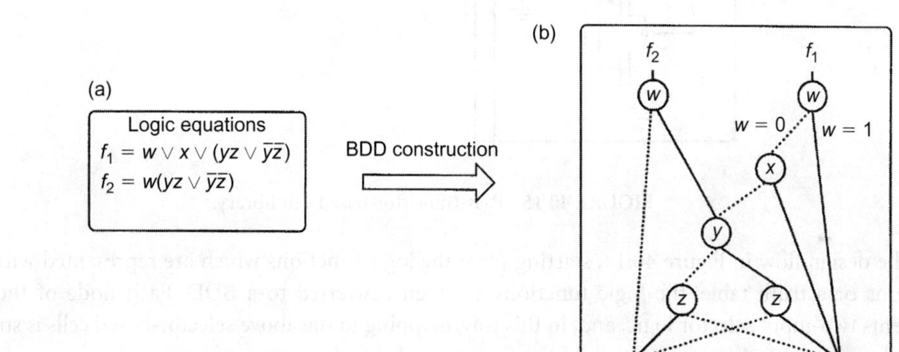

(a)
Logic equations
f1 = w ∨ x ∨ (yz ∨ ȳz̄)
f2 = w(yz ∨ ȳz̄)
BDD construction
(b)
f2
f1
w
w
w = 0
w = 1
x
y
z
z
0
1
Buffer insertion
(c)
f2
f1
w
w
x
y
z
z
0
1
Cell mapping
(d)
f2
f1
Cell 1
Cell 2
w
Vdd
x
Cell 1
y
z

FIGURE 40.14 Design flow of BDD-based pass-transistor logic synthesis.

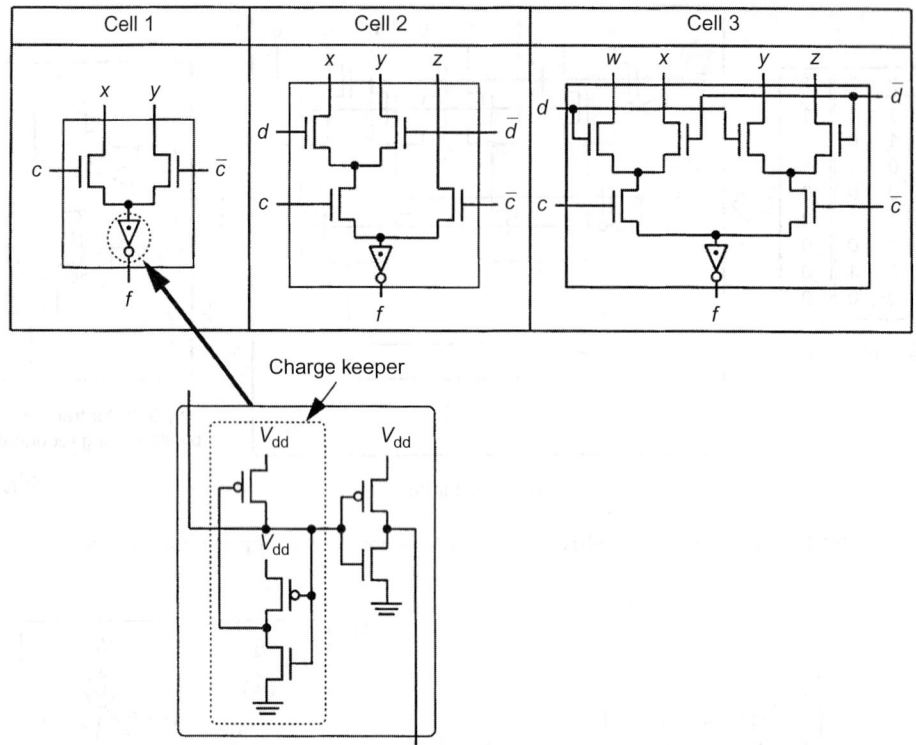

FIGURE 40.15　Pass-transistor-based cell library.

In the design flow in Figure 40.14, starting from the logic functions which are represented with logic equations or a truth table, the logic functions are then converted to a BDD. Each node of the BDD represents two-input selector logic, and, in this way, mapping to the above selector-based cells is straight-forward, requiring only consideration of the fan-out and signal polarity.

One difficulty of this approach with BDD is optimization of the logic depth, that is, the number of pass transistors from an input to an output. One important desired capability of a logic synthesizer for this approach is the control of the logic depth, for example, so that a designer can limit the delay time from an input to an output. It is difficult to incorporate this requirement in the framework of a BDD. Another difficulty of the BDD-based synthesis is that the number of pass transistors connected in series increases linearly as the number of inputs increases and this number may become excessive.

To solve these difficulties, MPL (which stands for multi-level pass-transistor logic) and its representation, multi-level BDD, have been proposed [6]. In the above simple BDD-based approach, the output of a pass-transistor selector is connected only to the source-drain path of another pass-transistor selector. This causes the above difficulty. In MPL, the output of a pass-transistor selector is flexibly connected to either a source-drain path or the gate of another MOSFET. Because of this freedom, the delay of the circuit can be flexibly controlled. It is known empirically that the delay, especially of a logic network having a large number of input variables, is reduced by a factor of 2, compared to the simple BDD approach.

Another important extension of pass-transistor logic is to incorporate CMOS circuits in a logic network [9]. Logic networks based on pass transistors are not always smaller than CMOS logic networks in area, delay, and power consumption. They are effective when selectors fit well to the target logic functions. Otherwise, conventional CMOS logic networks are a better choice. For example, a simple NAND function implemented in CMOS logic network has better delay, area, and power consumption than its pass-transistor-based counterpart. Combining pass-transistor logic and CMOS logic gives the best solution.

Pass-transistor logic synthesis is still not as well developed as CMOS-based logic synthesis. However, even at its current level of development, it has shown generally positive results. In other words, 10 to 30% power reduction is possible, as compared with pure CMOS [9], showing enough potential [1] to be further exploited in future research.

References

1. A.P. Chandrakasan, S. Sheng, and R.W. Brodersen, "Low-power CMOS digital design," *IEEE J. Solid-State Circuits*, SC-27, pp. 473–484, 1992.
2. Frei, A.H., W.K. Hoffman, and K. Shepard, "Minimum area parity circuit building block," *ICCC 80*, pp. 680–684, 1980.
3. K. Kikuchi, Y. Nukada, Y. Aoki, T. Kanou, Y. Endo, and T. Nishitani, "A single-chip 16-bit 25ns realtime video/image signal processor," *1989 IEEE International Solid-State Circuits Conference*, pp. 170–171, 1989.
4. M. Matsui, H. Hara, Y. Uetani, L-S. Kim, T. Nagamatsu, Y. Watanabe, A. Chiba, K. Matsuda, and T. Sakurai, "A 200 MHz 13 mm^2 2-D DCT macrocell using sense-amplifying pipeline flip-flop scheme," *IEEE J. Solid-State Circuits*, vol. 29, pp. 1482–1489, 1994.
5. A. Parameswar, H. Hara, and T. Sakurai, "A high speed, low power, swing restored pass-transistor logic based multiply and accurate circuit for multimedia applications," *IEEE 1994 Custom Integrated Circuits Conference*, pp. 278–281, 1994.
6. Y. Sasaki, K. Yano, S. Yamashita, H. Chikata, K. Rikino, K. Uchiyama, and K. Seki, "Multi-level pass-transistor logic for low-power ULSIs," *IEEE Symp. Low Power Electronics*, pp. 14–15, 1995.
7. Y. Shimazu, T. Kengaku, T. Fujiyama, E. Teraoka, T. Ohno, T. Tokuda, O. Tomisawa, and S. Tsujimichi, "A 50MHz 24b floating-point DSP," *1989 IEEE International Solid-State Conference*, pp. 44–45, 1989.
8. M. Suzuki, N. Ohkubo, T. Yamanaka, A. Shimizu, and K. Sasaki, "A 1.5 ns 32 b CMOS ALU in double pass-transistor logic," *1993 IEEE International Solid-State Circuits Conference Digest of Technical Papers*, pp. 90, 91, 267, 1993.
9. S. Yamashita, K. Yano, Y. Sasaki, Y. Akita, H. Chikata, K. Rikino, and K. Seki, "Pass-transistor/CMOS collaborated logic: The best of both worlds," *1997 Symp. VLSI Circuits Digest of Technical Papers*, pp. 31–32, 1997.
10. K. Yano, Y. Sasaki, K. Rikino, and K. Seki, "Top-down pass-transistor logic design," *IEEE J. Solid-State Circuits*, vol. 31, pp. 792–803, 1996.
11. K. Yano, T. Yamanaka, T. Nishida, M. Saito, K. Shimohigashi, and A. Shimizu, "A 3.8ns CMOS 16×16 multiplier using complementary pass-transistor logic," IEEE *1989 Custom Integrated Circuits Conference*, 10.4.1-4, 1989.
12. K. Yano, T. Yamanaka, T. Nishida, M. Saito, K. Shimohigashi, and A. Shimizu, "A 3.8-ns CMOS 16×16-b multiplier using complementary pass-transistor logic," *IEEE J. Solid-State Circuits*, SC-25, pp. 388–395, 1990.

41
Adders

Naofumi Takagi
Nagoya University

Haruyuki Tago
Toshiba Semiconductor Company

Charles R. Baugh
C. R. Baugh and Associates

Saburo Muroga
*University of Illinois
at Urbana-Champaign*

CONTENTS

41.1 Introduction

Adders are the most common arithmetic circuits in digital systems. Adders are used to do subtraction and also are key components of multipliers and dividers, as described in Chapters 42 and 43. There are various types of adders with different speeds, areas, and configurations. We can select an appropriate one which satisfies given requirements. For the details of adders and addition methods, see Refs. 2, 4–6, 12, 15, 17, and 20.

41.2 Addition in the Binary Number System

Before considering adders, let us take a look at addition in the binary number system.

In digital systems, numbers are usually represented in the binary number representation, although the most familiar number representation to us is the decimal number representation. The binary number representation is with the radix (base) 2 and the digit set $\{0, 1\}$, while the decimal number representation is with the radix 10 and the digit set $\{0, 1, 2, \ldots, 9\}$. For example, a binary number (i.e., a number in the binary number representation) $[1101]$ represents $1 \cdot 2^3 + 1 \cdot 2^2 + 0 \cdot 2^1 + 1 \cdot 2^0 = 13$, whereas a decimal number (i.e., a number in the decimal number representation) $[8093]$ for example, represents $8 \cdot 10^3 + 0 \cdot 10^2 + 9 \cdot 10^1 + 3 \cdot 10^0 = 8093$.

In the binary number representation, an integer is represented as $[x_{n-1}x_{n-2}\ldots x_0]$ where each binary digit, called a bit, x_i, is one of the elements of the digit set $\{0, 1\}$. The binary representation $[x_{n-1}x_{n-2}\ldots x_0]$ represents the integer $\sum_{i=0}^{n-1} x_i \cdot 2^i$.

By the binary number representation, we can represent not only an integer, but also a number that has a fractional part as well as an integral part, as by the decimal number representation. The binary representation $[x_{n-1}x_{n-2}\ldots x_0.x_{-1}x_{-2}\ldots x_{-m}]$ represents the number $\sum_{i=-m}^{n-1} x_i \cdot 2^i$. For example, $[1101.101]$ represents 13.625. By a binary representation with n-bit integral part and m-bit fractional part, we can represent 2^{n+m} numbers in the range from 0 to $2^n - 2^{-m}$.

TABLE 41.1   Truth Table for One-bit Addition

x_i	y_i	c_i	c_{i+1}	s_i
0	0	0	0	0
0	0	1	0	1
0	1	0	0	1
0	1	1	1	0
1	0	0	0	1
1	0	1	1	0
1	1	0	1	0
1	1	1	1	1

Let us consider addition of two binary numbers, $X = [x_{n-1}x_{n-2} \ldots x_0.x_{-1}x_{-2} \ldots x_{-m}]$ and $Y = [y_{n-1}y_{n-2} \ldots y_0.y_{-1}y_{-2} \ldots y_{-m}]$. We can perform addition by calculating the sum at the i-th position, s_i and the carry to the next higher position c_{i+1} from $x_i, y_i,$ and the carry from the lower position c_i according to the truth table shown in Table 41.1, successively from the least significant bit to the most significant one, that is, from $i = -m$ to $n - 1$, where $c_{-m} = 0$. Then, we have a sum $S = [s_n s_{n-1}s_{n-2} \ldots s_0.s_{-1}s_{-2} \ldots s_{-m}]$, where $s_n = c_n$. (i.e., s_n is actually a carry c_n from the $(n - 1)$-th position).

There are two major methods for representing negative numbers in the binary number representation. One is **sign and magnitude representation**, and the other is **two's complement representation**.

In the sign and magnitude representation, the sign and the magnitude are represented separately, as in the usual decimal representation. The first bit is the sign bit and the remaining bits represent the magnitude. The sign bit is normally selected to be 0 for positive numbers and 1 for negative numbers. For example, 13.625 is represented as [01101.101] and −13.625 is represented as [11101.101]. The sign and magnitude binary representation $[x_{n-1}x_{n-2} \ldots x_0.x_{-1}x_{-2} \ldots x_{-m}]$ represents the number $(-1)^{x_{n-1}} \cdot \sum_{i=-m}^{n-2} x_i \cdot 2^i$. By the sign and magnitude representation with n-bit integral part (including a sign bit) and m-bit fractional part, we can represent $2^{n+m} - 1$ numbers in the range from $-2^{n-1} + 2^{-m}$ to $2^{n-1} - 2^{-m}$.

In the two's complement representation, a positive number is represented in exactly the same manner as in the sign and magnitude representation. On the other hand, a negative number $-X$, where X is a positive number, is represented as $2^n - X$. For example, −13.625 is represented as [100000.000] − [01101.101], i.e., [10010.011]. The first bit of the integral part (the most significant bit) is 1 for negative numbers, indicating the sign of the number. The binary representation of $2^n - X$ is called the **two's complement of X**. We can obtain the representation of $-X$, i.e., $2^n - X$ by complementing the representation of X bitwise and adding [0.00...001] (i.e., 1 in the position of 2^{-m} but 0 in all other positions). It is because when $X = [x_{n-1}x_{n-2} \ldots x_0.x_{-1}x_{-2} \ldots x_{-m}] = \sum_{i=-m}^{n-1} x_i \cdot 2^i$, the negation of X, i.e., $-X$, becomes $2^n - X = (\sum_{i=-m}^{n-1} 2^i + 2^{-m}) - X = \sum_{i=-m}^{n-1}(1-x_i) \cdot 2^i + 2^{-m}$, where $1 - x_i$ is 1 or 0, according to whether x_i is 0 or 1, i.e., $1 - x_i$ is the complement of x_i. For example, given a binary number [01101.101], we can obtain its negation [10010.011] by complementing [01101.101] bitwise and adding [0.001] to it, i.e., by [10010.010] + [0.001]. By a two's complement representation with n-bit integral part (including a sign bit) and m-bit fractional part, we can represent 2^{n+m} numbers in the range from -2^n to $2^n - 2^{-m}$.

Each of all the binary number representations described so far (i.e., the positive number representation, sign and magnitude representation, and two's complement representation) can express essentially the same numbers, that is, 2^{n+m} numbers, although the second case expresses $2^{n+m} - 1$ numbers (i.e., one number less) when a number is expressed with $n + m$ bits. Thus, these representations essentially do not lose the precision, no matter whether or not one of the $n + m$-bits is used as a sign bit, although the range of the numbers is different in each case.

When we add two numbers represented in the sign and magnitude representation, we calculate the sign and the magnitude separately. When the operands (the augend and the addend) are with the same sign, the sign of the sum is the same as that of the operands, and the magnitude of the sum is the sum of those of the operands. A carry, 1, from the most significant position of the magnitude part indicates overflow. On the other hand, when the signs of the operands are different, the sign of the sum is the same as that of the operand with larger magnitude, and the magnitude of the sum is the difference of those of the operands.

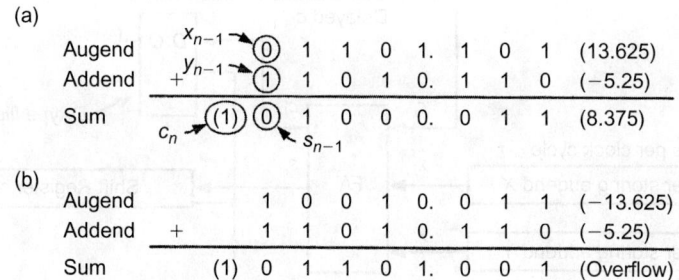

FIGURE 41.1   Examples of addition of numbers in two's complement representation.

The addition of two numbers represented in the two's complement representation, $X + Y$, for $X = [x_{n-1}x_{n-2}\ldots x_0 . x_{-1} x_{-2}\ldots x_{-m}]$ and $Y = [y_{n-1}y_{n-2}\ldots y_0 . y_{-1}y_{-2}\ldots y_{-m}]$ where x_{n-1} and y_{n-1} are their sign bits, can be done as follows, no matter whether each of X and Y is a positive or negative number:

1. The sign bits are added in the same manner as the two bits, x_i and y_i, in any other bit position, that is, according to Table 41.1. As illustrated in Figure 41.1(a), the sign bits, x_{n-1} and y_{n-1}, and the carry, c_{n-1}, from the $(n-2)$-th position are added, producing the sum bit s_{n-1} (in the sign bit position) and the carry c_n. Always, c_n is ignored, no matter whether it is 1 or 0.
2. When $x_{n-1} = y_{n-1}$ (i.e., X and Y have the same sign bit), an overflow occurs if s_{n-1} is different from x_{n-1} and y_{n-1} (i.e., $c_n \oplus c_{n-1} = 1$ means an overflow, as can be seen from Table 41.1). This case is illustrated in Figure 41.1(b) because we have $s_{n-1} = 0$ while $x_{n-1} = y_{n-1} = 1$ and hence s_{n-1} is not equal to x_{n-1} or y_{n-1}.

Next let us consider the subtraction of two numbers represented in the two's complement representation, $X - Y$, that is, subtraction of Y from X, where each of X and Y is a positive or negative number. This can be done as addition as explained in the previous paragraph after taking the two's complement of Y (i.e., deriving $2^n - Y$), no matter whether Y is a negative or positive number. Actually, the subtraction, $X - Y$, can be realized by the addition of X and the bitwise complement of Y with a carry input of 1 to the least significant position. This is convenient for realizing a subtracter circuit, whether it is a serial or parallel adder (to be described later).

Henceforth, let us consider addition of n-bit positive binary integers (without the sign bit) for the sake of simplicity. Let the augend, addend, and sum be $X = [x_{n-1}x_{n-2}\ldots x_0]$, $Y = [y_{n-1}y_{n-2}\ldots y_0]$, and $S = [s_n s_{n-1} s_{n-2}\ldots s_0]$ with $s_n = c_n$, respectively, where each of x_i, y_i, and s_i assumes a value of 0 or 1.

41.3   Serial Adder

A serial adder operates similarly to manual addition. The serial adder, at each step, calculates the sum and carry at one bit position. It starts at the least significant bit position (i.e., $i = 0$) and each successive next step it sequentially moves to the next more significant bit position where it calculates the sum and carry. At the n-th step, it calculates the sum and carry at the most significant bit position (i.e., $i = n-1$). In other words, the serial adder serially adds augend X and addend Y by adding x_i, y_i, and c_i at the i-th bit position from $i = 0$ to $n-1$. From the truth table shown in Table 41.1, we have sum bit $s_i = x_i \oplus y_i \oplus c_i$ and carry to the next higher bit position $c_{i+1} = x_i \cdot y_i \lor c_i \cdot (x_i \lor y_i)$ (also $c_{i+1} = x_i \cdot y_i \lor c_i \cdot (x_i \oplus y_i)$), where "$\cdot$" is AND, "$\lor$" is OR, and "$\oplus$" is XOR, and henceforth, "$\cdot$" will be omitted. This serial addition can be realized by the logic network, called a **serial adder**, or **bit-serial adder**, shown in Figure 41.2, where its operation is synchronized by a clock. The addition of each i-th bit is done at a rate of one bit per cycle of clock, producing sum bits, s_i's, at the same rate, from the least significant bit to the most significant one. In each cycle, s_i and c_{i+1}, are calculated from x_i, y_i, and the carry from the previous cycle, c_i. The core logic network, shown in the rectangle in Figure 41.2, for this one-bit addition for the i-th bit position is called

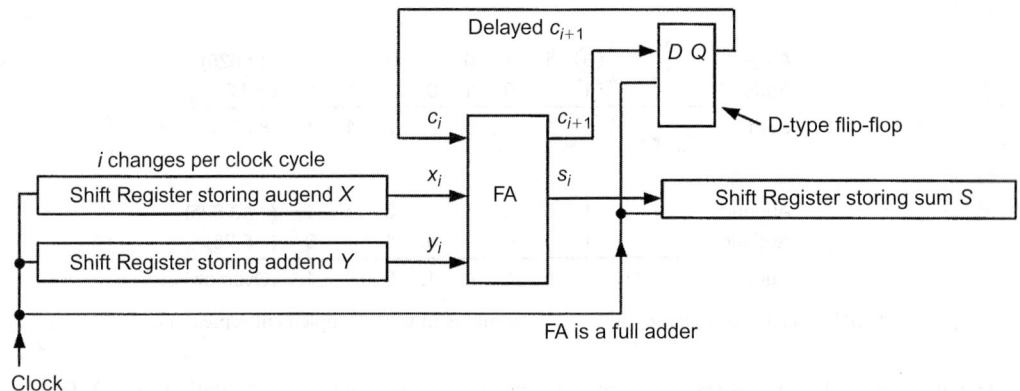

FIGURE 41.2 A serial adder.

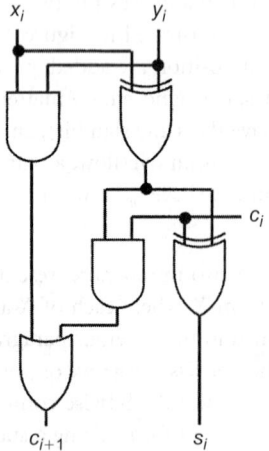

FIGURE 41.3 A full adder.

a **full adder** (abbreviated as **FA**). We obtain a logic network for an FA shown in Figure 41.3 using AND, OR, and XOR gates. A D-type flip-flop may be used as a delay element which stores the carry for a cycle. Full adders realized in ECL (emitter-coupled logic) are described in Chapter 38. FAs with a minimum number of logic gates are known for different types of logic gates [10].

A serial subtracter can be constructed with a minor modification of a serial adder, as explained in the last paragraph of Section 41.2.

41.4 Ripple Carry Adder

A parallel adder performs addition at all bit positions simultaneously, so it is faster than serial adders.

The simplest parallel adder is a **ripple carry adder**. An n-bit ripple carry adder is constructed by cascading n full adders, as shown in Figure 41.4. The carry output of each FA is connected to the carry input of the FA of the next higher position. The amount of its hardware is proportional to n. Its worst-case delay is proportional to n because of ripple carry propagation. In designing an FA for a fast ripple carry adder, it is critical to minimize the delay from the carry-in, c_i, to the carry-out, c_{i+1}.

An FA can be realized with logic gates, such as AND gates, OR gates, and XOR gates, as exemplified in Figure 41.3, and also can be realized with MOSFETs, including pass transistors [18,23], such that a carry c_i goes through logic gates which have some delays. But the speed of adders is very important for

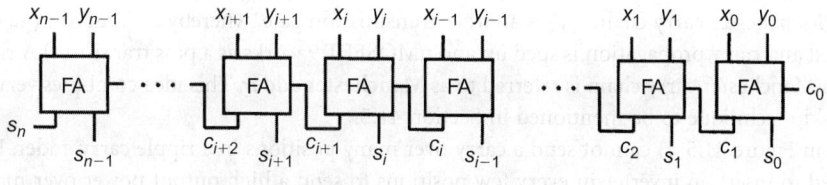

FA is a full adder.

FIGURE 41.4 A ripple carry adder.

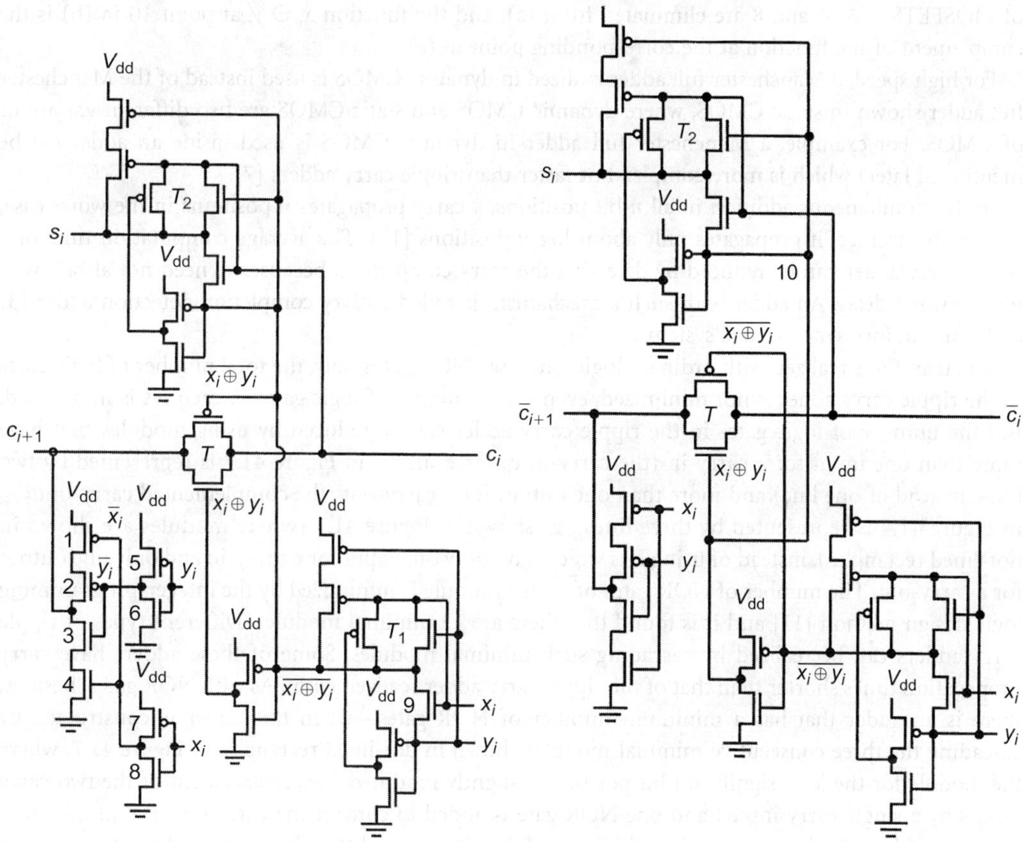

(a) Full adder with non-complemented carries.

(b) Full adder with complemented carries.

FIGURE 41.5 A Manchester-type full adder in CMOS.

the speed of the entire computer. So, FAs are usually realized with more sophisticated transistor circuits using MOSFETs such that a carry c_i can propagate fast to higher bit positions through pass transistors [22]. An example of such an adder is shown in Figure 41.5, being realized in CMOS. In Figure 41.5(a), a carry propagates through transmission gate, T (described in Chapter 39, Section 39.2). When we have $x_i = y_i = 0$, T becomes non-conductive and nMOSFETs, 3 and 4, become conductive. Then, the carry-out, c_{i+1}, becomes 0 because the carry-out terminal c_{i+1} is connected to the ground through 3 and 4. When $x_i = y_i = 1$, T becomes non-conductive and pMOSFETs, 1 and 2, become conductive. Then, the carry-out, c_{i+1}, becomes 1 because the carry-out terminal c_{i+1} is connected to the power supply V_{dd} through 1 and 2. When $x_i = 0$ and $y_i = 1$, or $x_i = 1$ and $y_i = 0$, T becomes conductive and the carry-out terminal is connected to neither V_{dd} nor the ground, so a carry-in c_i is sent to c_{i+1} as a carry-out. Thus, we have the values of c_{i+1} for different combinations of values of x_i, y_i, and c_i, as shown in Table 41.1. This carry-path is called a Manchester carry chain. (T_1 is another transmission gate, whereby a circuit on the carry path is simplified and carry propagation is sped up and nMOSFET 9 works as a pass transistor.) A ripple carry adder with Manchester carry chain is referred to as **Manchester adder**. This idea combines very well with the carry skip technique to be mentioned in Section 41.5.

The FA in Figure 41.5(a) cannot send a carry over many positions in a ripple carry adder. For speed-up, we need to insert an inverter in every few positions to send a high output power over many higher bit positions. In order to reduce the number of inverters which have delays in themselves, we can use the FA shown in Figure 41.5(b) which works with complemented carries. An example with insertion of an inverter at the end of every four cascaded FAs is shown in Figure 41.6, where a block of four of Figure 41.5(a) and a block of four of Figure 41.5(b) are alternated. In Figure 41.5(b), inverters consisting of MOSFETS 5, 6, 7, and 8 are eliminated from (a), and the function $x_i \oplus y_i$ at point 10 in (b) is the complement of the function at the corresponding point in (a).

For high speed, a Manchester full adder realized in dynamic CMOS is used instead of the Manchester full adder shown in static CMOS, where dynamic CMOS and static CMOS are two different variations of CMOS. For example, a Manchester full adder in dynamic CMOS is used inside an adder (to be mentioned later) which is more complex but faster than ripple carry adders [7].

In the simultaneous addition in all n-bit positions, a carry propagates n positions in the worst case, but on the average, it propagates only about $\log_2 n$ positions [13]. The average computation time of a ripple carry adder can be reduced by detecting the carry completion, because we need not always wait for the worst delay. An adder with such a mechanism is called a carry completion detection adder [3] and is useful for asynchronous systems.

When an FA is realized with ordinary logic gates, say NOR gates only, the total number of logic gates in the ripple carry adder is not minimized, even if the number of logic gates in each FA is minimized. But the number of logic gates in the ripple carry adder can be reduced by using modules that have more than one input for a carry-in (the carry-in c_i, for example in Figure 41.7 is represented by two lines, instead of one line) and more than one output for a carry-out (the complemented carry-out $\bar{c}_{i+1}$ in Figure 41.7 is represented by three lines), as shown in Figure 41.7 (where modules are shown in dot-lined rectangles), instead of using FAs which have only one input for a carry-in and only one output for a carry-out. The number of NOR gates of such a module is minimized by the integer-programming logic design method [11] and it is found that there are 13 minimal modules. Different types of ripple carry adders can be realized by cascading such minimal modules. Some of these adders have carry propagation times shorter than that of the ripple carry adder realized with FAs with NOR gates. Besides, there is an adder that has a minimum number of NOR gates—when this adder is constructed by cascading the three consecutive minimal modules shown in dot-lined rectangles in Figure 41.7, where the module for the least significant bit position is slightly modified (i.e., replacement of the two carry inputs by a single carry input) and one NOR gate is added to convert the carry out in multiple-lines from the adder into the carry out in a single line. Then it is proved that the total number of NOR gates in this ripple carry adder is minimum for any value of n [8]. Also, there is a ripple adder such that the number of connections, instead of the number of logic gates, is minimized [14]. Related adders are referred to in Section 2.4.3 of Ref. 11.

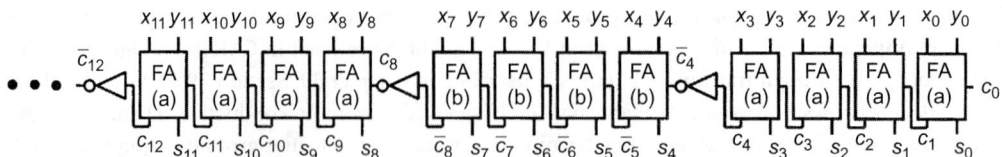

FA (a) and FA (b) here are Figs. 41.5 (a) and (b), respectively.

FIGURE 41.6 A ripple carry adder realized by connecting Figures 41.5(a) and (b).

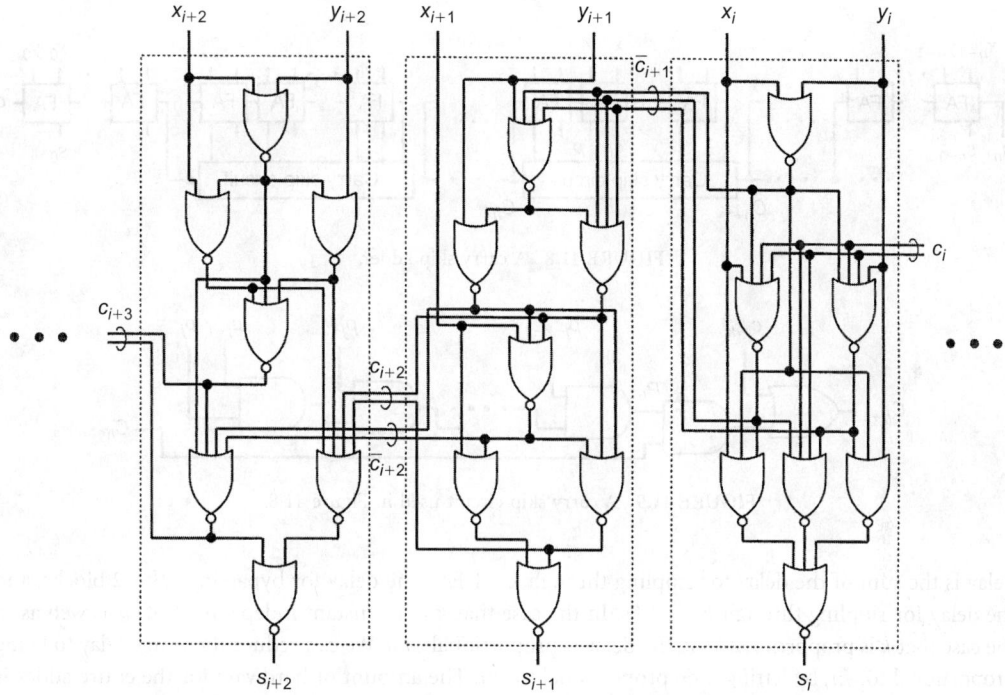

A ripple adder that has the minimal number of NOR gates for any arbitrary bit length can be realized by cascading these three consecutive minimal modules.

FIGURE 41.7 Three consecutive minimal modules for a ripple carry adder that has the minimal number of NOR gates for any arbitrary bit length.

41.5 Carry Skip Adder

In a ripple carry adder, a carry propagates through the i-th FA when $x_i \neq y_i$, i.e., $x_i \oplus y_i = 1$. Henceforth, we denote $x_i \oplus y_i$ as p_i. A carry propagates through a block of consecutive FAs, when all p_i's in the block are 1. This condition (i.e., all p_i's are 1) is called the carry propagation condition of the block.

A **carry skip adder** is a ripple carry adder that is partitioned into several blocks of FAs, attaching a carry skip circuit to each block, as shown in Figure 41.8 [9]. A carry skip circuit detects the carry propagation condition of the block and lets the carry from the next lower block bypass the block when the condition holds. In Figure 41.8, carry skip circuits are not attached to the blocks at the most and least significant few positions because the attachment does not speed up the carry propagation much.

In the carry skip circuit included in Figure 41.8, the carry output, C_{h+1}, from block h that consists of k FAs starting from j-th position is calculated as:

$$C_{h+1} = c_{j+k} \vee P_h C_h$$

where c_{j+k} is the carry from the FA at the most significant position of the block,

$$P_h = p_{j+k-1} p_{j+k-2} \cdots p_j$$

is the formula for the carry propagation condition of the block, C_h is the carry from the next lower block, and p_i's are calculated in FAs. An example of carry skip circuit is shown in Figure 41.9.

A carry may ripple through FAs in the block where it is generated, bypass the blocks where the carry propagation condition holds, and then, ripple through FAs in the block where the carry propagation condition does not hold. When all blocks are of the same size, k FAs, the worst case occurs when a carry is generated at the least significant position and propagate to the most significant position. The worst

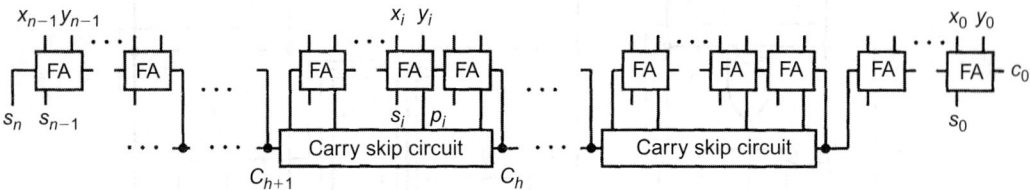

FIGURE 41.8 A carry skip adder.

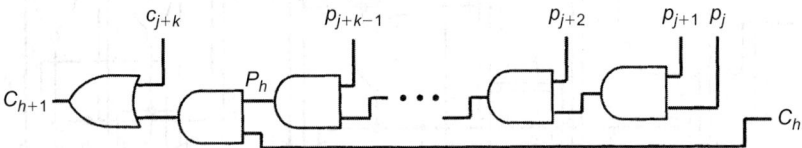

FIGURE 41.9 A carry skip circuit used in Figure 41.8.

delay is the sum of the delay for rippling through $k - 1$ FAs, the delay for bypassing $n/k - 2$ blocks, and the delay for rippling through $k - 1$ FAs. In the case that k is a constant independent of n, as well as in the case that k is proportional to n, the delay is proportional to n. We can reduce the worst delay to being proportional to $\sqrt{n}$, by letting k be proportional to $\sqrt{n}$. The amount of hardware for the entire adder is proportional to n in any case.

Applying the principle used to develop the carry skip adder borrowed from the ripple carry adder, we have a two-level carry skip adder from the basic carry skip adder, for further improvements. Recursive application of the principle yields a multi-level carry skip adder [12].

41.6 Carry Look-Ahead Adder

As previously stated, the carry, c_{i+1}, produced at the i-th position is calculated as $c_{i+1} = x_i \cdot y_i \vee c_i \cdot (x_i \oplus y_i)$. This means that a carry is generated if both x_i and y_i are 1, or an incoming carry is propagated if one of x_i and y_i is 1 and the other is 0. Therefore, letting g_i denote $x_i y_i$, we have $c_{i+1} = g_i \vee c_i p_i$, where $p_i = x_i \oplus y_i$. Here, g_i is the formula for the carry generation condition at the i-th position, i.e., when g_i is 1, a carry is generated at this position. Substituting $g_{i-1} \vee p_{i-1} c_{i-1}$ for c_i, we get

$$c_{i+1} = g_i \vee p_i g_{i-1} \vee p_i p_{i-1} c_{i-1}$$

Recursive substitution yields

$$c_{i+1} = g_i \vee p_i g_{i-1} \vee p_i p_{i-1} g_{i-2} \vee \cdots \vee p_i p_{i-1} \cdots p_0 c_0$$

A **carry look-ahead adder** can be realized according to this expression, as illustrated in Figure 41.10 for the case of four bits [21]. According to this expression, c_{i+1}'s are calculated at all positions in parallel.

It is hard to realize an n-bit carry look-ahead adder precisely according to this expression, unless n is small, because maximum fan-in and fan-out restriction is violated at higher positions. Large fan-out causes large delay, so the maximum fan-out is restricted. Also, the maximum fan-in is usually limited to 5 or less. There are some ways to alleviate this difficulty, as follows.

One way is the partition of carry look-ahead adders into several blocks such that each block consists of k positions, starting from the j-th one. In a block h, the carry at each position, c_{i+1}, where $j \leq i \leq j + k - 1$, is calculated as

$$c_{i+1} = g_i \vee p_i g_{i-1} \vee \cdots \vee p_i p_{i-1} \cdots p_{j+1} g_j \vee p_i p_{i-1} \cdots p_{j+1} p_j C_h$$

where C_h, i.e., c_j is the carry from the next lower block. The carry from the next lower block, C_h, goes to only the positions of the block h, so the fan-outs and fan-ins do not increase beyond a certain value. Therefore, we can form an n-bit adder by cascading n/k k-bit carry look-ahead adders, where k is a small constant independent of n, often 4. The worst delay of this type of adder and also the hardware amount are proportional to n.

Another way of alleviating the difficulty is recursively applying the principle of carry look-ahead to groups of blocks. The carry output of block h, C_{h+1} ($= c_{j+k}$), is calculated as

$$C_{h+1} = g_{j+k-1} \vee p_{j+k-1} g_{j+k-2} \vee \cdots \vee p_{j+k-1} p_{j+k-2} \cdots p_{j+1} g_j \vee p_{j+k-1} p_{j+k-2} \cdots p_{j+1} p_j C_h$$

This means that in the block, a carry is generated if

$$G_h = g_{j+k-1} \vee p_{j+k-1} g_{j+k-2} \vee \cdots \vee p_{j+k-1} p_{j+k-2} \cdots p_{j+1} g_j$$

is 1, and an incoming carry is propagated if $P_h = p_{j+k-1} p_{j+k-2} \cdots p_{j+1} p_j$ is 1. G_h is the formula for the carry generation condition of the block and P_h is the formula for the carry propagation condition of the block. (They are shown as P and G in Figure 41.10) Let us consider a super-block, that is, a group of several consecutive blocks. The carry generation and the carry propagation condition of a super-block are detected from those of the blocks, in the same way that G_h and P_h are detected from g_i's and p_i's. Once the carry input to a super-block is given, carry outputs from the blocks in the super-block are calculated immediately. Consequently, we obtain a fast adder in which small carry look-ahead circuits which include carry calculation circuits are connected in a tree form. Figure 41.11 shows a 4-bit carry look-ahead circuit and Figure 41.12 shows a 16-bit carry look-ahead adder using the 4-bit carry look-ahead circuits, where carry look-ahead circuits are shown as CLA in Figure 41.12. The worst delay of this type of adder is proportional to log n. The number of logic gates is proportional to n.

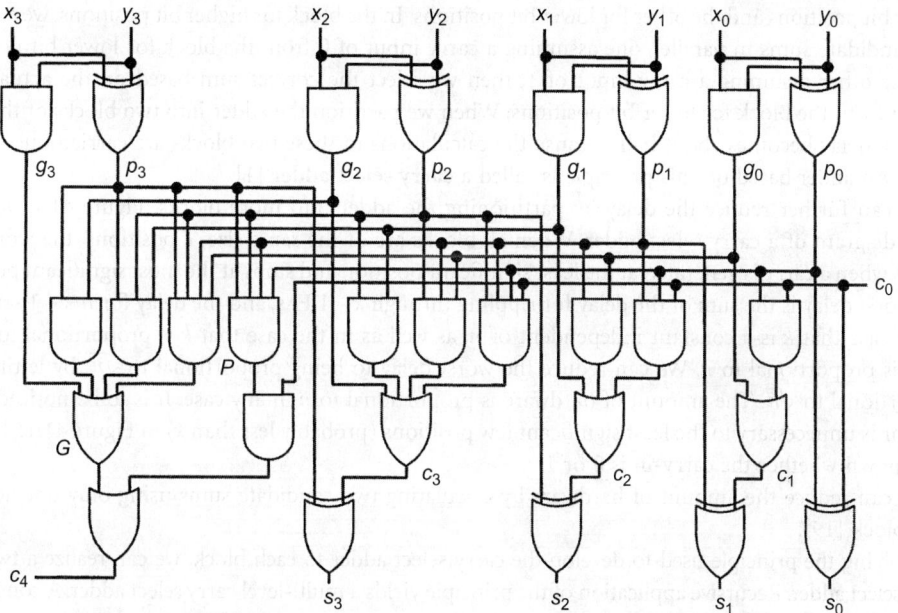

FIGURE 41.10 A 4-bit carry look-ahead adder.

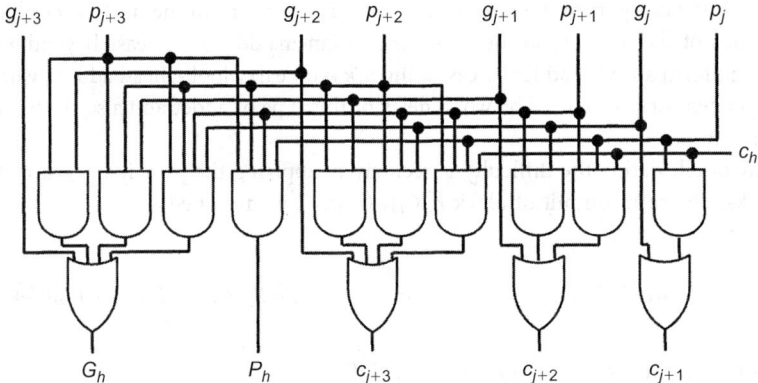

FIGURE 41.11   A 4-bit carry look-ahead circuit.

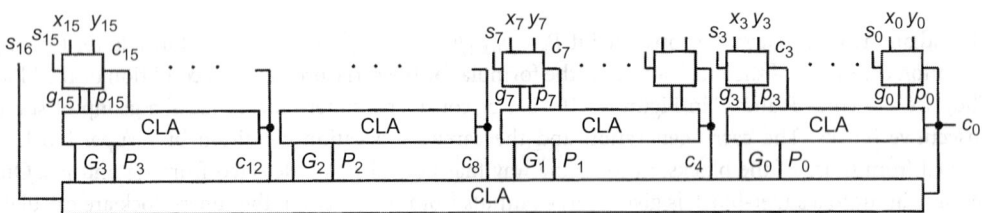

FIGURE 41.12   A 16-bit carry look-ahead adder. CLA stands for a carry look-ahead circuit which includes the carry calculation circuit shown in Figure 41.10.

41.7   Carry Select Adder

We can reduce the worst delay of a ripple carry adder by partitioning the adder into two blocks: one for higher bit positions and the other for lower bit positions. In the block for higher bit positions, we calculate two candidate sums in parallel, one assuming a carry input of 0 from the block for lower bit positions and the other assuming a carry input of 1, then we select the correct sum based on the actual carry output from the block for lower bit positions. When we partition the adder into two blocks of the same size, the delay becomes about half because the calculations in these two blocks are carried out concurrently. An adder based on this principle is called a **carry select adder** [1].

We can further reduce the delay by partitioning the adder into more blocks. Figure 41.13 shows a block diagram of a carry select adder. When all blocks are of the same size, k positions, the worst case occurs when a carry is generated at the least significant position and stops at the most significant position. The worst delay is the sum of the delay for rippling through $k - 1$ FAs, and the delay for $n/k - 1$ selectors. In the case that k is a constant independent of n, as well as in the case that k is proportional to n, the delay is proportional to n. We can reduce the worst delay to being proportional to $\sqrt{n}$, by letting k be proportional to $\sqrt{n}$. The amount of hardware is proportional to n in any case. It is to be noticed that a selector is unnecessary to the least significant few positions (probably less than k) in Figure 41.13 because it is known whether the carry-in is 0 or 1.

We can reduce the amount of hardware by calculating two candidate sums using only one adder in each block [19].

Applying the principle used to develop the carry select adder to each block, we can realize a two-level carry select adder. Recursive application of the principle yields a multi-level carry select adder. A conditional sum adder [16] can be regarded as the extreme case.

A carry select adder is used in a microprocessor with high performance [7].

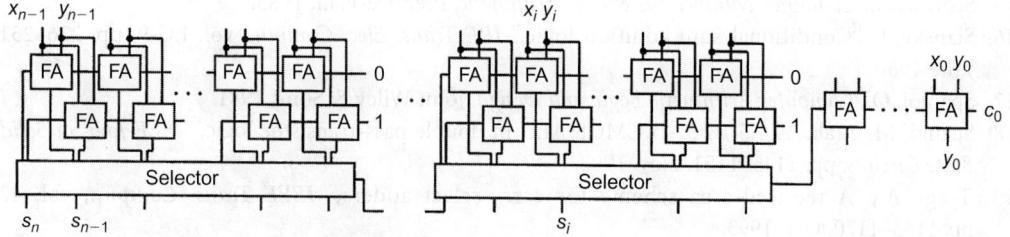

FIGURE 41.13 A carry select adder.

41.8 Carry Save Adder

When we add up several numbers sequentially, it is not necessary to propagate the carries during each addition. Instead, the carries generated during an addition may be saved as partial carries and added with the next operand during the next addition. Namely, we can accelerate each addition by postponing the carry propagation. This leads to the concept of carry save addition. We may add up numbers by a series of carry save additions, followed by a carry propagate addition. Namely, for multiple-operand addition, only one carry propagate addition is required.

An adder for carry save addition is referred to as a **carry save adder**, while the adders mentioned in the previous section are called **carry propagate adders**. A carry save adder sums up a partial sum and a partial carry from the previous stage as well as an operand and produces a new partial sum and partial carry. An *n*-bit carry save adder consists of just *n* full adders without interconnections among them.

References

1. Bedrij, O.J., "Carry-select adder," *IRE Trans. Elec. Comput.*, vol. EC-11, pp. 340–346, June 1962.
2. Cavanagh, J.J.F., *Digital Computer Arithmetic—Design and Implementation*, McGraw-Hill, 1984.
3. Gilchrist, B., J.H. Pomerene, and S.Y. Wong, "Fast carry logic for digital computers," *IRE Trans. Elec. Comput.*, vol. EC-4, pp. 133–136, 1955.
4. Hennessy, J.L. and D.A. Patterson, *Computer Architecture—A Quantitative Approach*, Appendix A, Morgan Kaufmann Publishers, 1990.
5. Hwang, K., *Computer Arithmetic—Principles, Architecture, and Design*, John Wiley & Sons, 1979.
6. Koren, I., *Computer Arithmetic Algorithms*, Prentice Hall, 1993.
7. Kowaleski, J.A. Jr. et al., "A dual-execution pipelined floating-point CMOS processor," *ISSCC Digest of Technical Papers*, pp. 358–359, Feb. 1996.
8. Lai, H.-C. and S. Muroga, "Minimum parallel binary adders with NOR (NAND) gates," *IEEE TC*, pp. 648–659, Sept. 1979.
9. Lehman, M. and N. Burla, "Skip techniques for high-speed carry propagation in binary arithmetic units," *IRE Trans. Elec. Comput.*, vol. EC-10, pp. 691–698, Dec. 1961.
10. Liu, T.-K., K. Hohulin, L.-E., Shiau, and S. Muroga, "Optimal one-bit full adders with different types of gates," *IEEE TC*, pp. 63–70, Jan. 1974.
11. Muroga, S., "Computer-aided logic synthesis for VLSI chips," pp. 1–103, *Advances in Computers*, vol. 32, Ed. by M.C. Yovits, Academic Press, 1991.
12. Omondi, A.R., *Computer Arithmetic Systems—Algorithms, Architecture and Implementations*, Prentice-Hall, 1994.
13. Reitwiesner, G.W., "The determination of carry propagation length for binary addition," *IRE Trans. Elec. Comput.*, vol. EC-9, pp. 35–38, 1960.
14. Sakurai, A. and S. Muroga, "Parallel binary adders with a minimum number of connections," *IEEE Trans. Comput.*, C-32, pp. 969–976, Oct. 1983. (Correction: In Figure 7, labels, a_0 and $\bar{c}_0$, should be interchanged.)

15. Scott, N.R., *Computer Number Systems & Arithmetic*, Prentice-Hall, 1985.

16. Slansky, J., "Conditional sum addition logic," *IRE Trans. Elec. Comput.*, vol. EC-9, pp. 226–231, June 1960.

17. Spaniol, O., *Computer Arithmetic Logic and Design*, John Wiley & Sons, 1981.

18. Suzuki, M. et al., "A 1.5-ns 32-b CMOS ALU in double pass-transistor logic," *IEEE Jour. of Solid-State Circuits*, pp. 1145–1151, Nov. 1993.

19. Tyagi, A., "A reduced-area scheme for carry-select adders," *IEEE Trans. Comput.*, vol. 42, pp. 1163–1170, Oct. 1993.

20. Waser, S. and M.J. Flynn, *Introduction to Arithmetic for Digital Systems Designers*, Holt, Rinehart and Winston, 1982.

21. Weinberger, A. and J.L. Smith, "A one-microsecond adder using one-megacycle circuitry," *IRE Trans. Elec. Comput.*, vol. EC-5, pp. 65–73, 1956.

22. Weste, N.H.E. and K. Eshraghian, *Principles of CMOS VLSI Design*, 2nd ed., Addison Wesley, 1993.

23. Zhuang, N. and H. Wu, "A new design of the CMOS full adder," *IEEE JSSC*, pp. 840–844, May 1992. (Full adder with transfer gates.)

42

Multipliers

Naofumi Takagi
Nagoya University

Charles R. Baugh
C.R. Baugh and Associates

Saburo Muroga
*University of Illinois
at Urbana-Champaign*

CONTENTS

42.1 Introduction

Many microprocessors and digital signal processors now have fast multipliers in them. There are several types of multipliers with different speeds, areas, and configurations. For the details of multipliers and multiplication methods, see Refs. 3–5, 7–10, and 14.

Here, let us consider multiplication of n-bit positive binary integers (without the sign bit) where a multiplicand $X = [x_{n-1}x_{n-2}\ldots x_0]$ is multiplied by a multiplier $Y = [y_{n-1}y_{n-2}\ldots y_0]$ to derive the product $P = [p_{2n-1}p_{2n-2}\ldots p_0]$, where each of x_i, y_i, and p_i takes a value of 0 or 1.

42.2 Sequential Multiplier

A **sequential multiplier** works in a manner similar to manual multiplication of two decimal numbers, although two binary numbers are multiplied in this case. A multiplicand $X = [x_{n-1}x_{n-2}\ldots x_0]$ is multiplied by each bit of a multiplier $Y = [y_{n-1}y_{n-2}\ldots y_0]$, forming the multiplicand-multiple $Z = [z_{n-1}z_{n-2}\ldots z_0]$, where $z_i = x_iy_j$ for each $i = 0, \ldots, n-1$. Then, Z is shifted left by j bit positions and is added, in all digit positions in parallel, to the partial product P_{j-1} which has been formed by the previous steps, to generate the partial product P_j. Repeating this step for $j = 0$ to $n-1$, the product $P = [p_{2n-1}p_{2n-2}\ldots p_0]$ of $2n$ bits is derived. The only difference of this sequential multiplier from the manual multiplication is the repeated addition of each multiplicand-multiple, instead of one-time addition of all multiplicand-multiples at the end.

This sequential multiplier is realized, as shown in Figure 42.1, which consists of a Multiplicand Register of n-bits for storing multiplicand X, a Shift Register of $2n$-bits for storing multiplier Y and partial product P_{j-1}, a Multiplicand-Multiple Generator (denoted as MM Generator) for generating a multiplicand-multiple $y_j \cdot X$, and a Parallel Adder of n-bits. Initially, X is stored in the Multiplicand Register and Y is stored in the lower half (i.e., the least significant bit positions) of the Shift Register where the upper half of the Shift Register stores 0. This sequential multiplier performs one iteration step described above in each clock cycle. In other words, in each clock cycle, a multiplier bit y_j is read from the right-most position of Shift Register. A multiplicand-multiple $y_j \cdot X$ is produced by Multiplicand-Multiple Generator, which is X or 0 based on whether y_j is 1 or 0, and is fed to Parallel Adder. The upper n-bit of the partial product is read from the upper half of Shift Register and also fed to Parallel Adder. The content of Shift Register

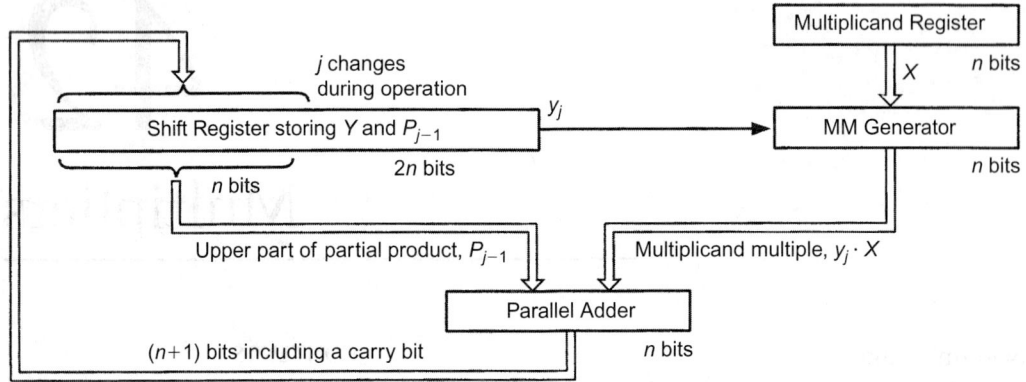

FIGURE 42.1 A sequential multiplier.

is shifted one position to the right. The $(n + 1)$-bit output of Parallel Adder including the carry output, which is the upper part of the updated partial product, is stored into the upper $(n + 1)$ positions of Shift Register. After n cycles, Shift Register holds the $2n$-bit product, P.

We can use any adder described in Chapter 38. The faster the adder, the shorter the clock cycle and hence the faster the multiplier. When we use a carry save adder as Parallel Adder, we have to modify Shift Register so that its upper half stores two binary numbers (i.e., a partial sum and a partial carry). Besides the carry save adder, we need a carry propagate adder for summing up the final partial sum and carry.

We can accelerate sequential multiplication by processing multiple-bits of the multiplier per clock cycle. When k-bits of multiplier Y are processed per cycle, an n-bit multiplication is performed through n/k clock cycles. There are two methods for processing multiple-bits of the multiplier per cycle. One is generating several candidate multiplicand-multiples and then choosing an appropriate one among them. The other is generating k multiplicand-multiples and summing them up in one cycle. Also, these two methods can be combined.

In the first method, Multiplicand-Multiple Generator generates 2^k different multiplicand-multiples, 0, X, $2X$, $3X$, ..., $(2^k - 1)X$. For example, when $k = 2$, Multiplicand-Multiple Generator generates $2X$ and $3X$, as well as X and 0. Multiplicand-Multiple Generator consists of a look-up table containing the necessary multiples of the multiplicand and a selector for selecting an appropriate multiple. The look-up table need not hold all multiples, because several of them can be generated from others by shifting whenever necessary. For example, when $k = 2$, only $3X$ must be pre-computed and be held. **Extended Booth's method** [13] is useful for reducing the number of pre-computed multiples. When k-bits are processed per cycle, k-bit Booth's method is applied, which recodes multiplier Y into radix-2^k redundant signed-digit representation with the digit set $\{-2^{k-1}, -2^{k-1} + 1, ..., 0, ..., 2^{k-1} - 1, 2^{k-1}\}$, where each digit in radix 2^k takes a value among those in this digit set. Y is recoded into $\hat{Y}$ by considering $k + 1$ bits of Y, i.e., $y_{kj+k-1}, y_{kj+k-2}, ..., y_{kj+1}, y_{kj}, y_{kj-1}$, (i.e., $k + 1$ bits among $y_{n-1}, y_{n-2}, ..., y_0$ of Y) per cycle, instead of only a single bit of Y (say, y_j) per cycle, as illustrated in Figure 42.2(a). More specifically, the j-th digit $\hat{y}_j$ of the recoded multiplier, where $j = 0, 1, ..., \lceil (n+1)/k \rceil - 1$, is calculated as $\hat{y}_j = -2^{k-1} \cdot y_{kj+k-1} + 2^{k-2} \cdot y_{kj+k-2} + \cdots + 2 \cdot y_{kj+1} + y_{kj} + y_{kj-1}$. In this case, since all components of $Y = [y_{n-1} y_{n-2} \cdots y_0]$ are recoded for every k components at a time, the recoded number becomes $\hat{Y} = [\hat{y}_{\lceil (n+1)/k \rceil - 1} \hat{y}_{\lceil (n+1)/k \rceil - 2} \cdots \hat{y}_0]$ with $\lceil (n + 1)/k \rceil$ components, in contrast to multiplier $Y = [y_{n-1} y_{n-2} \cdots y_0]$ which has n components. Then we have a multiplicand-multiple $\hat{y}_j \cdot X$. Since the negation of a multiple can be produced by complementing it bitwise and adding 1 at the least significant position (this 1 is treated as a carry into the least significant position of Parallel Adder), the number of multiples to be held is reduced. For example, when $k = 2$, the multiplier is recoded to radix-4 redundant signed-digit representation with the digit set $\{-2, -1, 0, 1, 2\}$ by means of the 2-bit Booth's method (i.e., the extended Booth's method with $k = 2$) as $\hat{y}_j = -2y_{2j+1} + y_{2j} + y_{2j-1}$ and all multiples are produced from X by shift and/or complementation. In 2-bit Booth recoding, multiplier Y is partitioned into 2-bit blocks, and then at the j-th block, the recoded multiplier digit $\hat{y}_j$ is calculated from the two bits of the block, i.e., y_{2j+1} and y_{2j}, and the higher bit of the next lower block, i.e., y_{2j-1}, according to the rule shown in Table 42.1. For example, 11110001010 is recoded to $2\ 0\ \bar{2}\ 1\ \bar{1}\ \bar{2}$, where $\bar{1}$ and $\bar{2}$ denote -1 and -2, respectively, as illustrated in Figure 42.2(b). (In this case, whenever a bit is not available in Y, such as the next lower bit of the least significant bit, it is regarded as 0.) When the extended Booth's method is employed, Parallel Adder is modified such that negative numbers can be processed.

In the second method for processing multiple-bits of the multiplier per clock cycle, Parallel Adder sums up $k + 1$ numbers, using k adders. Any adder can be used, but usually carry save adders are used for the sake of speed and cost. Carry save adders can be connected either in series or in tree form. Of course, the latter is faster, but the structure of Parallel Adder becomes more complicated because of somewhat irregular wire connections. By letting k be n, we have a parallel multiplier, processing the whole multiplier-bits in one clock cycle, as will be mentioned in the following section.

(a) Recoding for k-bit Booth's method

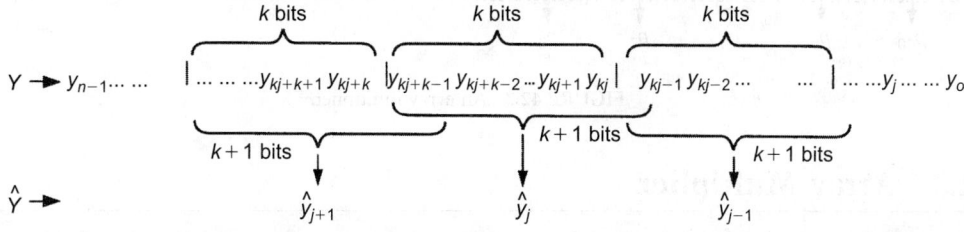

(b) An example of the 2-bit Booth's method based on Table 39.1

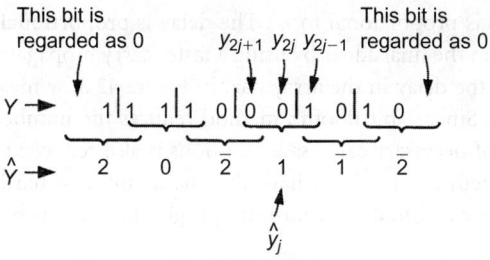

FIGURE 42.2 Recoding in the extended Booth's method.

TABLE 42.1 The Recording Rule of 2-bit Booth's Method

y_{2j+1}	y_{2j}	y_{2j-1}	$\hat{y}_j$
0	0	0	0
0	0	1	1
0	1	0	1
0	1	1	2
1	0	0	-2
1	0	1	-1
1	1	0	-1
1	1	1	0

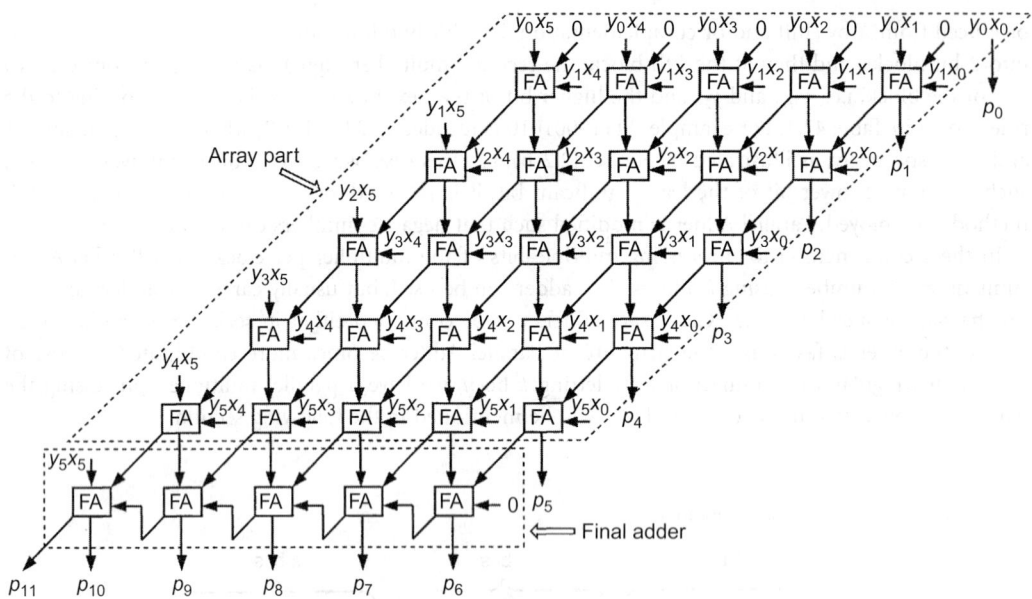

FIGURE 42.3 An array multiplier.

42.3 Array Multiplier

The simplest parallel multiplier is an **array multiplier** [2] in which the multiplicand-multiples (i.e., $(y_j \cdot X)$'s) are summed up one by one by means of a series of carry save adders. It has a two-dimensional array structure of full adders as shown in Figure 42.3. Each row of full adders except the bottom one forms a carry save adder. The bottom row forms a ripple carry adder for the final carry propagate addition. An array multiplier is suited for VLSI realization because of its regular cellular array structure. The number of logic gates is proportional to n^2. The delay is proportional to n.

We can reduce the delay in the final adder by using a faster carry propagate adder such as a carry select adder. Also, we can reduce the delay in the array part in Figure 42.3 by means of 2-bit Booth's method mentioned in Section 42.2. Since 2-bit Booth's method reduces the number of multiplicand-multiples to about half, the number of necessary carry save additions is also reduced to about half, and hence, the delay in the array part is reduced to about half. But the amount of hardware is not reduced much because a 2-bit Booth recoder and multiplicand-multiple generators, which essentially work as selectors, are required.

Another method to reduce the delay in the array part is to double the accumulation stream [6]. Namely, we divide the multiplicand-multiples into two groups, sum up the members of each group by a series of carry save adders independently of the other group, and then sum up the two accumulations into one. The delay in the array part is reduced to about half. The 2-bit Booth's method can be combined with this method. We can further reduce the delay by increasing the number of accumulation streams, although it complicates the circuit structure.

42.4 Multiplier Based on Wallace Tree

In the **multiplier based on Wallace tree** [13], the multiplicand-multiples are summed up in parallel by means of a tree of carry save adders. A carry save adder sums up three binary numbers and produces two binary numbers (i.e., a partial sum and a partial carry). Therefore, using $n/3$ carry save adders in parallel, we can reduce the number of multiplicand-multiples from n to about $2n/3$. Then, using about

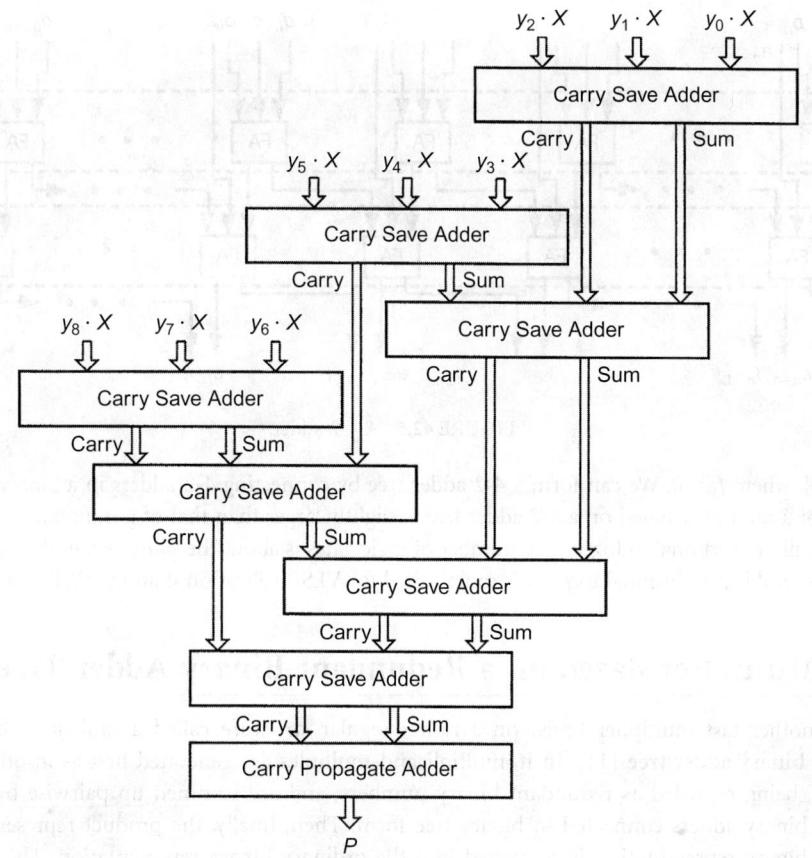

FIGURE 42.4 A multiplier with Wallace tree.

$2n/9$ carry save adders, we can further reduce it to $4n/9$. Applying this principle about $\log_{3/2} n$ times, the number of multiplicand-multiples can be reduced to only two. Finally, we sum up these two multiplicand-multiples by means of a fast carry propagate adder.

Figure 42.4 illustrates a block diagram of a multiplier based on Wallace tree. This consists of full adders, just like the array multiplier described previously. (Recall that a carry save adder consists of full adders.) The delay is small. It is proportional to $\log n$ when a fast carry propagate adder with $O(\log n)$ delay is used for the final addition. The number of logic gates is about the same as that of an array multiplier and is proportional to n^2. However, its circuit structure is not suited for VLSI realization because of the complexity.

The 2-bit Booth's method can be also applied to this type of multiplier. However, it is not as effective as in the array multiplier, because the height of the Wallace tree decreases by only one or two, even though the number of the multiplicand-multiples is reduced to about half.

A full adder, which is used as the basic cell in a multiplier based on Wallace tree, can be regarded as a counter which counts up 1's in the three-input bits and outputs the result as a 2-bit binary number. Namely, a full adder can be regarded as a 3-2 counter. We can also use larger counters, such as 7-3 counters and 15-4 counters, as the basic cells, instead of full adders.

We can increase the regularity in the circuit structure by replacing Wallace tree with a 4-2 adder tree [12], where a 4-2 adder is formed by connecting two carry save adders, shown in the dot-lined rectangles, in series, as shown in Figure 42.5. A 4-2 adder is an adder that sums up four binary numbers, $A = [a_{n-1} \ldots a_0]$, $B = [b_{n-1} \ldots b_0]$, $C = [c_{n-1} \ldots c_0]$, and $D = [d_{n-1} \ldots d_0]$, and produces two binary numbers, $E = [e_n \ldots e_0]$ and

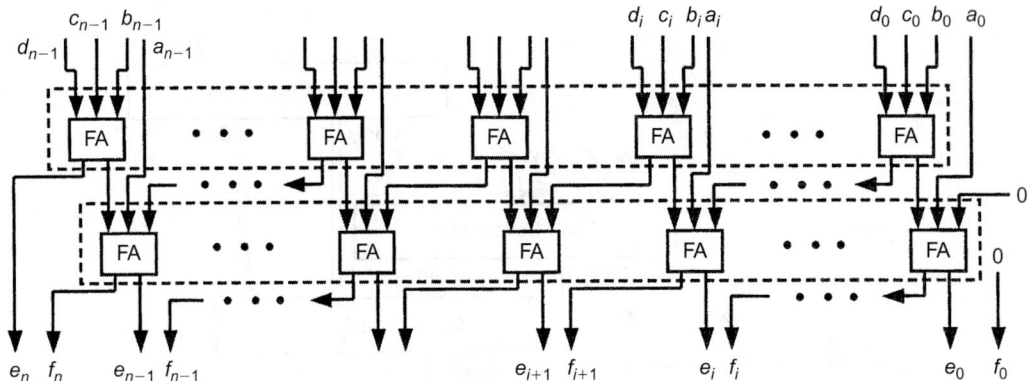

FIGURE 42.5 A 4-2 adder.

$F = [f_n \ldots f_0]$, where $f_0 = 0$. We can form a 4-2 adder tree by connecting 4-2 adders in a binary tree form. The delay of a multiplier based on a 4-2 adder tree is slightly larger than that of a multiplier with Wallace tree but is still proportional to $\log n$. The number of logic gates is about the same as a multiplier based on Wallace tree, and is proportional to n^2. It is more suited for VLSI realization than the Wallace tree.

42.5   Multiplier Based on a Redundant Binary Adder Tree

There is another fast multiplier based on a rather regular structure called a **multiplier based on a redundant binary adder tree** [11]. In it, multiplicand-multiples are generated first as in other parallel multipliers, being regarded as redundant binary numbers, and are summed up pairwise by means of redundant binary adders connected in binary tree form. Then, finally, the product represented in the redundant binary representation is converted into the ordinary binary representation. The redundant binary representation, also called the binary signed-digit representation, is a binary representation with a digit set $\{\bar{1}, 0, 1\}$, where $\bar{1}$ denotes -1 [1]. An n-digit redundant binary number $A = [a_{n-1}a_{n-2}\ldots a_0]$ has the value $\sum_{i=0}^{n-1} a_i \cdot 2^i$, where a_i takes a value $\bar{1}$, 0, or 1. There may be several redundant binary numbers which have the same value. For example, $[0101]$, $[011\bar{1}]$, $[1\bar{1}01]$, $[1\bar{1}1\bar{1}]$, and $[10\bar{1}\bar{1}]$ all represent 5. Because of this redundancy, we can add two redundant binary numbers without carry propagation.

Let us consider the addition of two redundant binary numbers, that is, the augend, $A = [a_{n-1}a_{n-2}\ldots a_0]$, and the addend, $B = [b_{n-1}b_{n-2}\ldots b_0]$ to derive the sum, $S = [s_n s_{n-1}\ldots s_0]$, where each of a_i, b_i, and s_i takes a value -1, 0, or 1. The addition without carry propagation is done in two steps. In the first step, an intermediate carry c_{i+1}, and intermediate sum d_i in the i-th position are determined such that $a_i + b_i = 2c_{i+1} + d_i$ is satisfied (the 2 of $2c_{i+1}$ means shifting c_{i+1} to the next higher digit position as a carry), where each of c_{i+1} and d_i is -1, 0, or 1. In this case, c_{i+1} and d_i are determined such that a new carry will not be generated in the second step. In the second step, in each digit position, sum s_i is determined by adding intermediate sum d_i and intermediate carry c_i from the next lower position, where c_i takes a value, -1, 0, or 1.

Suppose one of addend digit a_i and addend digit b_i is 1 and the other is 0. If $c_{i+1} = 0$ and $d_i = 1$ in the first step, a new carry will be generated for $c_i = 1$ from the next lower digit position in the second step. So, if there is a possibility of $c_i = 1$, we choose $c_{i+1} = 1$ and $d_i = \bar{1}$. On the other hand, if there is a possibility of $c_i = \bar{1}$, we choose $c_{i+1} = 0$ and $d_i = 1$. This makes use of the fact that 1 can be expressed by $[01]$ and $[1\bar{1}]$ in the redundant binary number representation. Whether c_i becomes 1 or $\bar{1}$ can be detected by examining a_{i-1} and b_{i-1} in the next lower digit position but not further lower digit positions. For other combinations of the values of a_i, b_i, and c_i, c_{i+1} and d_i can be similarly determined.

In the second step, s_i is determined by adding only two digits, c_i and d_i. Suppose $c_i = 1$. Then s_i is 0 or 1, based on whether d_i is $\bar{1}$ or 0. Notice that two combinations, $c_i = d_i = 1$ and $c_i = d_i = \bar{1}$, never occur. For other combinations of the values of c_i and d_i, s_i is similarly determined. Consequently, sum digit s_i at each digit position can be determined by the three digit positions of the angend and addend, a_i, a_{i-1} and a_{i-2}, and b_i, b_{i-1}, and b_{i-2}.

A binary number is a redundant binary number as it is, so there is no need for converting it to the redundant binary number representation. But conversion of a redundant binary number to a binary number requires an ordinary binary subtraction. Namely, we subtract the binary number that is derived by replacing every 1 by 0 and $\bar{1}$ by 1 in the redundant binary number, from the binary number that is derived by replacing $\bar{1}$ by 0. For example, $[1\bar{1}0\bar{1}1]$ is converted to $[00111]$ by the subtraction $[10001] - [01010]$. We need borrow propagate subtraction for this conversion. This conversion in a multiplier based on a redundant binary adder tree corresponds to the final addition in the ordinary multipliers.

References

1. Avizienis, A., "Signed-digit number representations for fast parallel arithmetic," *IRE Trans. Elec. Comput.*, vol. EC-10, pp. 389–400, Sep. 1961.
2. Baugh, C.R. and B.A. Wooly, "A two's complement parallel array multiplier," *IEEE Trans. Computers*, vol. C-22, no. 12, pp. 1045–1047, Dec. 1973.
3. Cavanagh, J.J.F., *Digital Computer Arithmetic—Design and Implementation*, McGraw-Hill, 1984.
4. Hennessy, J.L. and D.A. Patterson, *Computer Architecture—A Quantitative Approach, Appendix A*, Morgan Kaufmann Publishers, 1990.
5. Hwang, K., *Computer Arithmetic—Principles, Architecture, and Design*, John Wiley & Sons, 1979.
6. Iwamura, J., et al., "A 16-bit CMOS/SOS multiplier-accumulator," *Proc. IEEE Intnl. Conf. on Circuits and Computers*, 12.3, Sept. 1982.
7. Koren, I., *Computer Arithmetic Algorithms*, Prentice-Hall, 1993.
8. Omondi, A.R., *Computer Arithmetic Systems—Algorithms, Architecture and Implementations*, Prentice-Hall, 1994.
9. Scott, N.R., *Computer Number Systems & Arithmetic*, Prentice-Hall, 1985.
10. Spaniol, O., *Computer Arithmetic Logic and Design*, John Wiley & Sons, 1981.
11. Takagi, N., H. Yasuura, and S. Yajima, "High-speed VLSI multiplication algorithm with a redundant binary addition tree," *IEEE Trans. Comput.*, vol. C-34, no. 9, pp. 789–796, Sep. 1985.
12. Vuillemin, J.E., "A very fast multiplication algorithm for VLSI implementation," *Integration, VLSI Journal*, vol. 1, no. 1, pp. 39–52, Apr. 1983.
13. Wallace, C.S., "A suggestion for a fast multiplier," *IEEE Trans. Elec. Comput.*, vol. EC-13, no. 1, pp. 14–17, Feb. 1964.
14. Waser, S. and M.J. Flynn, *Introduction to Arithmetic for Digital Systems Designers*, Holt, Rinehart and Winston, 1982.

43

Dividers

Naofumi Takagi
Nagoya University

Saburo Muroga
*University of Illinois
at Urbana-Champaign*

CONTENTS

43.1 Introduction

There are two major classes of division methods: subtract-and-shift methods and multiplicative methods. For details of division methods and dividers, see Refs. 2, 5–11, and also Ref. 3 for subtract-and-shift division.

Suppose two binary numbers, X and Y, are normalized in the range equal to or greater than 1/2 but smaller than 1 and also $X < Y$ holds for the sake of simplicity. Then, a dividend $X = [0.1x_2 \ldots x_n]$, which is an n-bit normalized binary fraction, is to be divided by a divisor, $Y = [0.1y_2 \ldots y_n]$, which is another n-bit normalized binary fraction, where each of x_i and y_i takes a value of 0 or 1. We assume to calculate the quotient $Z = [0.1z_2 \ldots z_n]$ which satisfies $|X/Y - Z| < 2^{-n}$, where z_i takes a value of 0 or 1.

43.2 Subtract-and-Shift Dividers

The **subtract-and-shift divider** works in a manner similar to manual division of one decimal number by another, using paper and pencil. In the case of manual division of decimal numbers, each x_i of dividend X and y_i of divisor Y is selected from $\{0, 1, \ldots, 9\}$ (i.e., the radix is 10). Each z_i of quotient Z is selected from $\{0, 1, \ldots, 9\}$. In the following case of dividers for a digital system, dividend X and divisor Y are binary numbers, so each x_i of X and y_i of Y is selected from $\{0, 1\}$, but the quotient (which is not necessarily represented as a binary number) is expressed in radix r. The radix r, is usually chosen to be 2^k (i.e., 2, 4, 8, and so on.) So, a quotient is denoted with $Q = [0.q_1q_2 \ldots q_{\lceil n/k \rceil}]$, to be differentiated from $Z = [0.1z_2 \ldots z_n]$ expressed in binary numbers, although both Q and Z will henceforth be called quotients.

The subtract-and-shift method with a radix r iterates the recurrence step of replacing R_j by $r \cdot R_{j-1} - q_j \cdot Y$, where q_j is the j-th quotient digit and R_j is the partial remainder after the determination of q_j. Initially, R_{j-1} for $j - 1 = 0$; that is, R_0 is X. Each recurrence step consists of the following four substeps. Suppose that $r = 2^k$.

1. Shift of the partial remainder R_{j-1} to the left by k bit positions to produce $r \cdot R_{j-1}$.
2. Determination of the quotient digit q_j by quotient-digit selection.

3. Generation of the divisor multiple $q_j \cdot Y$.
4. Subtraction of $q_j \cdot Y$ from $r \cdot R_{j-1}$ to calculate R_j.

The dividers for a digital system have many variations, depending on the methods in choosing the radix, the quotient-digit set from which q_j is chosen (q_j is not necessarily 0 or 1, even if it is in radix 2), and the representation of the partial remainder. The simplest cases are with a radix of 2 and the partial remainder represented in the non-redundant form. When $r = 2$, the recurrence is $R_j = 2R_{j-1} - q_j \cdot Y$. There are three methods: the restoring, the non-restoring, and the SRT methods.

43.2.1 Restoring Method

In the radix-2 restoring method, a quotient digit, q_j, is chosen from the quotient-digit set $\{0, 1\}$. When $2R_{j-1} - Y \geq 0$ ($2R_{j-1}$ means shift of R_{j-1} by one bit position to the left), 1 is selected, and otherwise 0 is selected. Namely, $R_j' = 2R_{j-1} - Y$ is calculated first, and then, when $R_j' \geq 0$ holds, we set $q_j = 1$ and $R_j = R_j'$, and otherwise $q_j = 0$ and $R_j = R_j' + Y$ (i.e., Y is added back to R_j'). For every j, R_j is kept in the range, $0 \leq R_j < Y$. The j-th bit of the quotient in binary number $Z = [0.1z_2 \ldots z_n]$, z_j, is equal to q_j (i.e., z_j is the same as q_j in radix 2 in this case of the restoring method). This method is called the **restoring method**, because Y is added back to R_j' when $R_j' < 0$. For speed-up, we can use $2R_{j-1}$ as R_j by keeping R_{j-1}, instead of adding back Y to R_j', when $R_j' < 0$. Figure 43.1 shows an example of radix-2 restoring division.

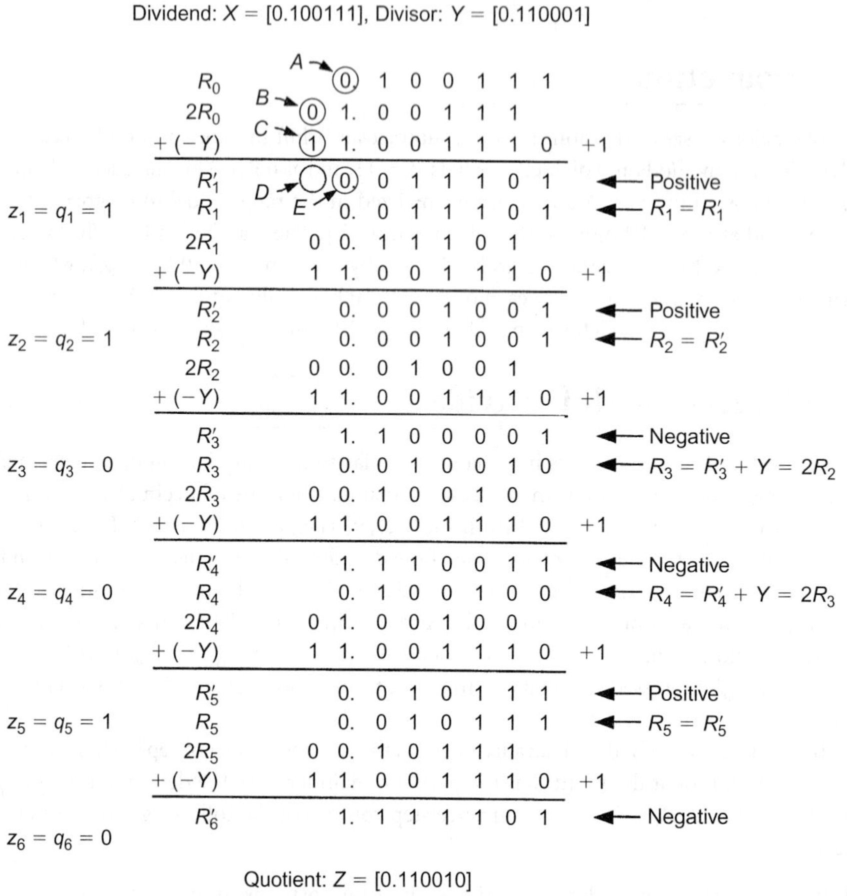

FIGURE 43.1 An example of radix-2 restoring division. (Each of A, B, and C is a sign bit. Notice that D is always equal to E and is ignored.)

43.2.2 Non-Restoring Method

In the radix-2 non-restoring method, a quotient digit, q_j is chosen from the quotient-digit set $\{-1, 1\}$. Quotient digit q_j is chosen according to the sign of R_{j-1}. In other words, we set $q_j = 1$ when $R_{j-1} \geq 0$ and, otherwise, we set $q_j = -1$, abbreviated as $q_j = \bar{1}$. Then, $R_j = 2R_{j-1} - q_j \cdot Y$ is calculated. Even if R_j is negative, Y is not added back in this method, so this method is called the **non-restoring method**. Note that since we have $R_0 = X > 0$ and $(1/2) \leq X < Y < 1$ by the assumption, we always have $q_1 = 1$ and $R_1 = 2R_0 - Y = 2X - Y > 0$, and hence, $q_2 = 1$. For every j, R_j is kept in the range, $-Y \leq R_j < Y$. The j-th bit of the quotient in binary number $Z = [0.1z_2 \ldots z_n]$, z_j, is 0 or 1, based on whether q_{j+1} is -1 or 1. And we have always $z_n = 1$. For example, when $Q = [0.11\bar{1}1\bar{1}1]$ where $\bar{1}$ denotes -1, we have $Z = [0.110011]$. (The given number $[0.11\bar{1}1\bar{1}1]$ is calculated as $[0.111001] - [0.000110] = [0.110011]$. In other words, the number derived by replacing all 1's by 0's and all $\bar{1}$'s by 1's in the given number is subtracted from the number derived by replacing all $\bar{1}$'s by 0's in the given number. This turns out to be a simple conversion between z_j and q_{j+1}, as stated above, without requiring the subtraction.) Combining this conversion with the recurrence on R_j yields the method in which $R_1 = 2X - Y$, and for $j \geq 2$, when $R_{j-1} \geq 0$, $z_{j-1} = 1$ and $R_j = 2R_{j-1} - Y$ and, otherwise, $z_{j-1} = 0$ and $R_j = 2R_{j-1} + Y$. Figure 43.2 shows an example of radix-2 non-restoring division. Note that the remainder for the restoring method is always negative, whereas the remainder for the non-storing method (also the SRT method to be described in the following) can be negative, and consequently when the remainder is negative, the quotient of the latter is greater by 2^{-n} than the former. (This explains the difference between the quotients in Figures 43.1 and 43.2, where $R_6 = 1$ in Figure 43.2 indicates that the remainder is negative.)

Figure 43.3 shows a radix-2 non-restoring divider that performs one recurrence step in each clock cycle. Register For Partial Remainder R_{j-1} initially stores the dividend X and then, during the division, stores a partial remainder R_{j-1} which may become negative. Divisor, Y, in Register For Divisor Y is added to or subtracted from the twice of the R_{j-1} stored in Register For Partial Remainder R_{j-1} by Adder/Subtracter, based on whether the left-most bit of Register For Partial Remainder R_{j-1} (i.e., the sign bit of R_{j-1}) is 1 or 0, and then the result (i.e., R_j) is stored back into Register For Partial Remainder R_{j-1}. Concurrently, the complement of the sign bit of R_{j-1} (i.e., z_{j-1}) is fed to Shift Register For Z_{j-2} (which stores the partial quotient $Z_{j-2} = [0.1z_2 \ldots z_{j-2}]$) from the right end, and the partial quotient stored in Shift Register For Z_{j-2} is shifted

Dividend: $X = [0.100111]$, Divisor: $Y = [0.110001]$

```
                                                          Each of A, B and C is a sign bit.
                        R0    A →  ⓪  1 0 0 1 1 1          Notice that D is always equal to
                       2R0    B → ⓪ 1. 0 0 1 1 1          E and is ignored.
       q1 = 1       + (−Y)    C → ① 1. 0 0 1 1 1 0   +1
                        R1    D →  ⓪⓪ 0 1 1 1 0 1    ← Positive
                       2R1    E → ⓪ 0. 1 1 1 0 1
   z1 = 1  q2 = 1    + (−Y)      1 1. 0 0 1 1 1 0   +1
                        R2      0. 0 0 1 0 0 1        ← Positive
                       2R2     0 0. 0 1 0 0 1
   z2 = 1  q3 = 1    + (−Y)      1 1. 0 0 1 1 1 0   +1
                        R3      1. 1 0 0 0 0 1        ← Negative
                       2R3     1 1. 0 0 0 0 1
   z3 = 0  q4 = 1̄    +    Y     0 0. 1 1 0 0 0 1
                        R4      1. 1 1 0 0 1 1        ← Negative
                       2R4     1 1. 1 0 0 1 1
   z4 = 0  q5 = 1̄    +    Y     0 0. 1 1 0 0 0 1
                        R5      0. 0 1 0 1 1 1        ← Positive
                       2R5     0 0. 1 0 1 1 1
   z5 = 1  q6 = 1    + (−Y)    1 1. 0 0 1 1 1 0   +1
                        R6      1. 1 1 1 1 0 1        ← Negative
```

Quotient: $Z = [0.110011]$

FIGURE 43.2 An example of radix-2 non-restoring division.

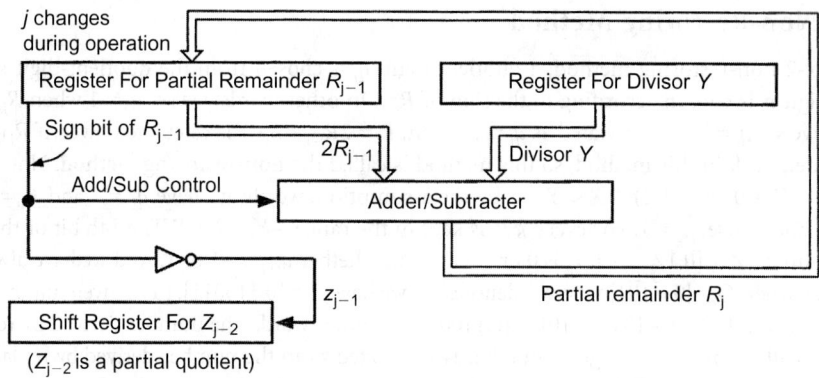

FIGURE 43.3 A radix-2 non-restoring divider.

Dividend: X = [0.100111], Divisor: Y = [0.110001]

```
                A
        R₀    B ⌐ ⓪. 1  0  0  1  1  1
       2R₀      ⓪ 1. 0  0  1  1  1              ⟵ 1/2 ≤ 2R₀ (actually, 1 < 2R₀)
q₁ = 1 + (−Y)  C ① 1. 0  0  1  1  1  0   + 1
        R₁    D → ⓪. 0  1  1  1  0  1
       2R₁    E   0  0. 1  1  1  0  1              ⟵ 1/2 ≤ 2R₁
q₂ = 1 + (−Y)     1  1. 0  0  1  1  1  0   + 1
        R₂        0. 0  0  1  0  0  1
       2R₂        0  0. 0  1  0  0  1              ⟵ −1/2 ≤ 2R₂ < 1/2
q₃ = 0                                             ⟵ No addition
       ─────────────────────────────────
        R₃        0. 0  1  0  0  1  0
       2R₃        0  0. 1  0  0  1  0              ⟵ 1/2 ≤ 2R₃
q₄ = 1 + (−Y)     1  1. 0  0  1  1  1  0   + 1
        R₄        1. 1  1  0  0  1  1
       2R₄        1  1. 1  0  0  1  1              ⟵ −1/2 ≤ 2R₄ < 1/2
q₅ = 0                                             ⟵ No addition
       ─────────────────────────────────
        R₅        1. 1  0  0  1  1  0
       2R₅        1  1. 0  0  1  1  0              ⟵ 2R₅ < −1/2
q₆ = 1̄ +   Y     0  0. 1  1  0  0  0  1
        R₆        1. 1  1  1  1  0  1
```

Quotient: Q = [0.11010$\bar{1}$] → Z = [0.110011]

FIGURE 43.4 An example of radix-2 SRT division. (Each of A, B, and C is a sign bit. Notice that D is always equal to E and is ignored.)

one position to the left. The divider performs one recurrence step in each clock cycle. After n cycles, Shift Register For Z_{j-2} holds the quotient Z. We can use any carry propagate adder (subtracter) as the Adder/ Subtracter. The faster the adder, the shorter the clock cycle, and hence, the faster the divider.

43.2.3 SRT Method

In the radix 2 SRT method, the quotient-digit set is $\{-1, 0, 1\}$. In this case, −1 or 0 or 1 is selected as q_j, based on whether $2R_{j-1} < -(1/2)$ or $-(1/2) \leq 2R_{j-1} < (1/2)$ or $(1/2) \leq 2R_{j-1}$. When 0 is selected as q_j, no addition or subtraction is performed for the calculation of R_j, and hence, the computation time may be shortened. Quotient digit q_j is determined from the values of the three most significant bits (shown in each of the dot-lined rectangles in Figure 43.4) of $2R_{j-1}$, as the radix-2 SRT division is exemplified in

Figure 43.4. R_{j-1} satisfies $-Y \le R_{j-1} < Y$ and is represented in a two's complement representation with 1-bit integral part (and n-bit fractional part).

In the above three methods, carry (or borrow) propagates in the calculation of R_j in each recurrence step because the partial remainder is represented in the non-redundant form. We can accelerate the SRT method by using a redundant form for representing the partial remainder, and performing the calculation of R_j without carry (or borrow) propagation. Let us consider the use of the carry save form. R_j, which satisfies $-Y \le R_j < Y$, is represented in a two's complement carry save form with 1-bit integral part (and n-bit fractional part.) (In the two's complement carry save form, both the partial carry and the partial sum are represented in the two's complement representation.) In this case, -1 or 0 or 1 is selected as q_j, based on whether $2\hat{R}_{j-1} \le -1$ or $2\hat{R}_{j-1} = -(1/2)$, or $2\hat{R}_{j-1} \ge 0$, where $2\hat{R}_{j-1}$ denotes the value of the three most significant digits (shown in each of the dot-lined rectangles in Figure 43.5), i.e., down to the

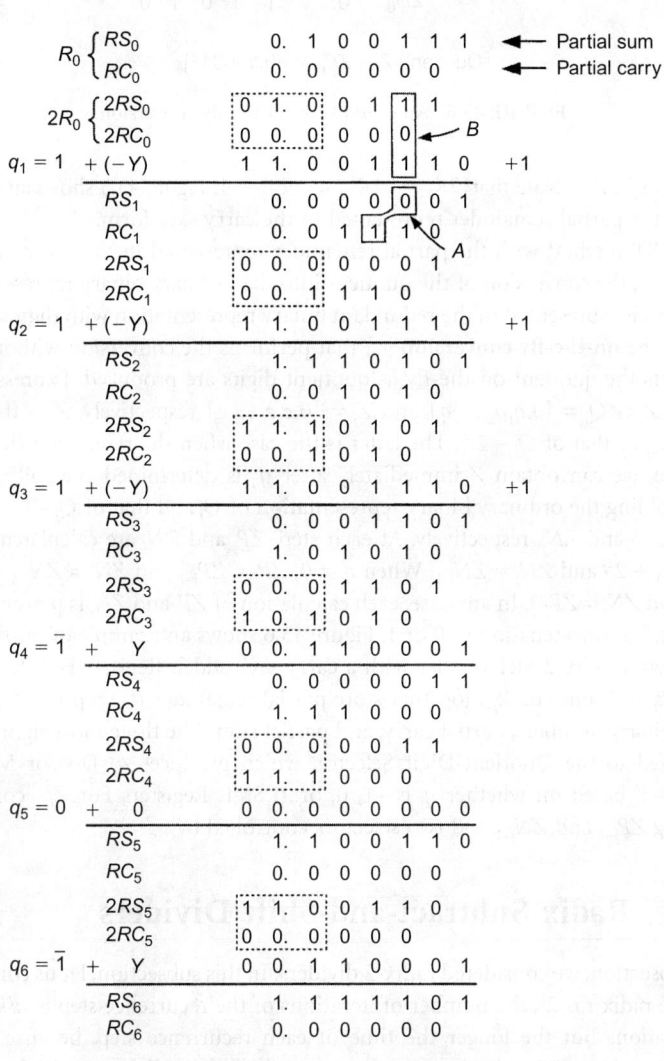

FIGURE 43.5 An example of radix-2 SRT division with partial remainder represented in the carry save form. (The bits shown inside A are calculated from the bits shown inside B.)

$$
\begin{array}{lll}
q_1 = 1 & ZP_1 & 0.\ 1 \\
& ZN_1 & 0.\ 0 \\[4pt]
q_2 = 1 & ZP_2 & 0.\ 1\ 1 \\
& ZN_2 & 0.\ 1\ 0 \\[4pt]
q_3 = 1 & ZP_3 & 0.\ 1\ 1\ 1 \\
& ZN_3 & 0.\ 1\ 1\ 0 \\[4pt]
q_4 = \bar{1} & ZP_4 & 0.\ 1\ 1\ 0\ 1 \\
& ZN_4 & 0.\ 1\ 1\ 0\ 0 \\[4pt]
q_5 = 0 & ZP_5 & 0.\ 1\ 1\ 0\ 1\ 0 \\
& ZN_5 & 0.\ 1\ 1\ 0\ 0\ 1 \\[4pt]
q_6 = \bar{1} & ZP_6 & 0.\ 1\ 1\ 0\ 0\ 1\ 1 \\
& ZN_6 & 0.\ 1\ 1\ 0\ 0\ 1\ 0
\end{array}
$$

Quotient: $Z = ZP_6 = [0.110011]$

FIGURE 43.6   An example of on-the-fly conversion.

first binary position, of $2R_{j-1}$. Note that $2\hat{R}_{j-1} \leq 2R_{j-1} < 2\hat{R}_{j-1} + 1$. Figure 43.5 shows an example of radix-2 SRT division with the partial remainder represented in the carry save form.

In the radix-2 SRT method with the partial remainder represented in the carry save form, as in the original SRT method, the conversion of the quotient into the ordinary binary representation is required because the quotient is represented in the redundant binary representation with digit set $\{-1, 0, 1\}$. There is a method called the **on-the-fly conversion** [4] that performs the conversion without carry propagate addition. It converts the quotient on-the-fly as quotient digits are produced. Expressing the bits up to the j-th of Q and Z as $Q_j = [0.q_1q_2 \ldots q_j]$ and $Z_j = [0.z_1z_2 \ldots z_j]$ respectively, Z_j is the ordinary binary representation of Q_j or that of $Q_j - 2^{-j}$. The latter is the case when the remaining (lower) part of Q_n is negative. Therefore, we can obtain Z immediately after q_n is determined (i.e., all quotient digits are determined), by holding the ordinary binary representation of Q_j and that of $Q_j - 2^{-j}$ at each recurrence step. Let them be ZP_j and ZN_j, respectively. At each step, ZP_j and ZN_j are calculated as follows. When $q_j = -1$, $ZP_j = ZN_{j-1} + 2^{-j}$ and $ZN_j = ZN_{j-1}$. When $q_j = 0$, $ZP_j = ZP_{j-1}$ and $ZN_j = ZN_{j-1} + 2^{-j}$. When $q_j = 1$, $ZP_j = ZP_{j-1} + 2^{-j}$ and $ZN_j = ZP_{j-1}$. In any case, each calculation of ZP_j and ZN_j is performed by a selection of ZP_{j-1} or ZN_{j-1} and a concatenation of 0 or 1. Figure 43.6 shows an example of on-the-fly conversion.

Figure 43.7 shows a radix-2 SRT divider with a carry save adder. Register For Partial Carry For R_{j-1} and Register For Partial Sum For R_{j-1} together store partial remainder R_{j-1} represented in the carry save form, i.e., by two binary numbers, partial carry, and partial sum. The three most significant bits of these two registers are fed to the Quotient-Digit Selector, which produces q_j. Divisor Multiple Generator generates Y, 0, or $-Y$, based on whether q_j is -1, 0, or 1. Shift Registers For Z_{j-1} consists of two shift registers for storing ZP_{j-1} and ZN_{j-1} and two selectors controlled by q_j.

43.3   Higher Radix Subtract-and-Shift Dividers

In the previous subsection, we considered radix-2 dividers. In this subsection, let us consider higher radix dividers. When the radix r is 2^k, the number of iterations of the recurrence step is n/k. The larger the k, the fewer the iterations but the longer the time of each recurrence step, because of the additional complexity in the quotient-digit selection and the generation of the divisor multiples.

A redundant digit set, especially a redundant symmetric signed-digit set $\{-a, -a + 1, \ldots, -1, 0, 1, \ldots, a - 1, a\}$, where $a \geq r/2$, is often used to obtain fast algorithms. A larger a reduces complexity of the quotient-digit selection but increases the complexity of the generation of the divisor multiples. The use of the signed-digit set makes the conversion of the quotient into the ordinary binary representation necessary. The partial remainder can be represented in either non-redundant form, or redundant form (e.g., carry save form). By the use of a redundant form, we can perform addition/subtraction without carry/borrow propagation, and hence fast. However, it slightly complicates the quotient-digit selection and doubles the number of register bits required for storing the partial remainder.

Among the radix-4 methods, the method with the quotient-digit set of $\{-2,-1,0,1,2\}$ and the redundant partial remainder is the most popular, because the generation of the divisor multiples is easy and carry save addition can be used for speed-up. In this method, R_j, which satisfies $-(2/3)Y \le R_j < (2/3)Y$, is represented in the carry save form (or redundant binary representation with the digit set $\{-1,0,1\}$) with 1-bit integral part (and n-bit fractional part). ($R_0 = X$ and $X < (2/3)Y$ must hold.) Quotient digit q_j is determined from the seven most significant digits of R_{j-1} and the five most significant bits of Y. (Actually not five but four bits of Y are required, since its most significant bit is always 1.) The on-the-fly conversion of the quotient can be extended to radix-4. The essential part of a divider based on this method is very similar to that shown in Figure 43.7. The seven most significant bits of the content of Register For Partial Carry For R_{j-1} and Register For Partial Sum For R_{j-1}, as well as the five (actually four) significant bits of the content of Register For Divisor Y are fed to Quotient Digit Selector. Divisor Multiple Generator generates $-2Y, -Y, 0, Y,$ and $2Y$. This type of divider is used in floating point arithmetic units of several microprocessors.

An efficient radix-8 method is with the quotient-digit set of $\{-7, -6, \dots, -1, 0, 1, \dots, 6, 7\}$ and the redundant partial remainder. Quotient digit q_j is determined from the eight most significant digits of R_{j-1} and the four most significant bits of Y. Quotient digit q_j is decomposed into two components, $qh_j \in \{-8,-4,0,4,8\}$ and $ql_j \in \{-2,-1,0,1,2\}$, and R_j is calculated through two carry save additions.

As the radix increases, the quotient-digit selection becomes more complex. This complexity can be reduced by restricting the divisor to a range close to 1, by prescaling the divisor. We can preserve the value of the quotient by prescaling the dividend with the same factor as the divisor. For example, we can design an efficient radix-16 method with the quotient-digit set of $\{-10, -9, \dots, -1, 0, 1, \dots, 9, 10\}$, where q_j is determined from the ten most significant digits of R_{j-1}, by scaling the divisor into $(8107/8192) \le Y \le (8288/8192)$. In this method, q_j is decomposed into two components, $qh_j \in \{-8, -4, 0, 4, 8\}$ and $ql_j \in \{-2, -1, 0, 1, 2\}$, and R_j is calculated through two carry save additions.

Since the recurrence step becomes more complex as the radix increases, a higher radix division step is implemented by several lower radix (i.e., radix of 2 or 4) stages. We can design a more efficient divider

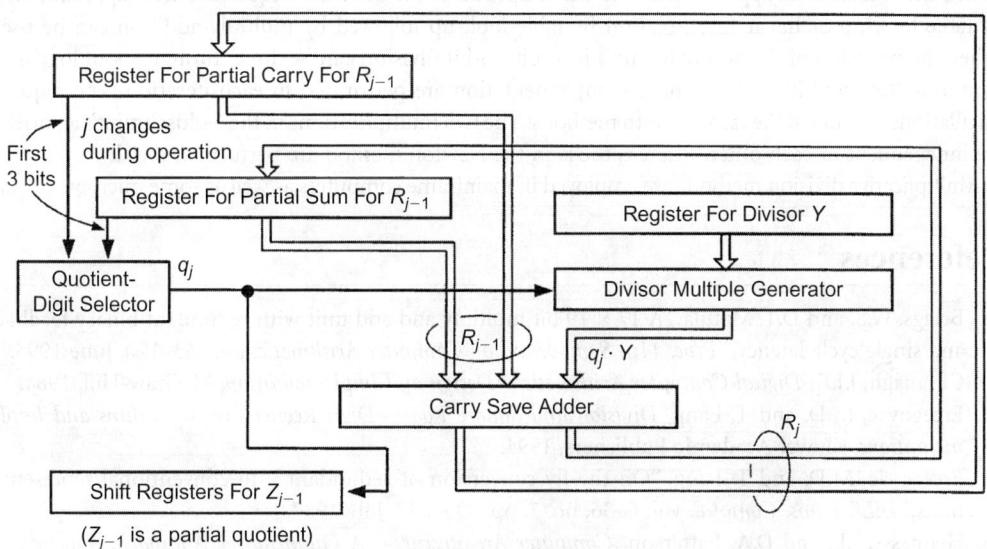

FIGURE 43.7 A radix-2 SRT divider with carry save adder.

by not merely concatenating lower radix stages but overlapping them so that the delay and the cost of the hardware for the recurrence step are reduced.

43.4   Even Higher Radix Dividers with a Multiplier

Division with even higher radix, say radix-2^8, based on the iteration of calculating R_j by $r \cdot R_{j-1} - q_j \cdot Y$, requires a multiplier for the generation of the divisor multiple $q_j \cdot Y$. This multiplier is also used for prescaling the divisor very close to 1, so that the quotient digit is produced by truncation or rounding of the shifted partial remainder to the left of the radix point.

Many variations of the method are feasible, depending on the choice of the radix, the quotient-digit set, the representation of the partial remainder (non-redundant or redundant), quotient-digit selection (by truncation or rounding), calculation of the scaling factor, and scaling. The scaling factor can be calculated by any method that produces an approximation of the reciprocal of the divisor. Direct approximation by table look-up and linear interpolation by table look-up followed by multiply-addition are examples. There are three ways of performing scaling: scaling the divisor and the dividend, scaling the divisor and the quotient, and scaling the partial remainder in each iteration. A version of this method is realized in a math-coprocessor [1].

43.5   Multiplicative Dividers

Multiplicative methods perform division through iterative multiplications and are employed in systems that have fast multipliers. The Newton method and Goldschmidt's algorithm are well-known.

The Newton method calculates the reciprocal of the divisor Y by the Newton-Raphson method. Beginning with an approximation to the reciprocal of the divisor, U_0, $1/Y$ is obtained through the iterative calculation of $U_{i+1} = U_i \cdot (2 - U_i \cdot Y)$. Multiplying the dividend X with the obtained $1/Y$, we can obtain the quotient.

Goldschmidt's algorithm is based on the fact that the value of a fraction is unchanged by multiplying both the numerator and the denominator by the same number. Namely, $X/Y = (X \cdot D_0 \cdot D_1 \cdot D_2 \ldots)/(Y \cdot D_0 \cdot D_1 \cdot D_2 \ldots)$ holds. When D_i's are selected so that $Y \cdot D_0 \cdot D_1 \cdot D_2 \ldots$ approaches to 1, $X \cdot D_0 \cdot D_1 \cdot D_2 \ldots$ approaches X/Y. We can obtain the quotient through the iterative calculations of $D_i = 2 - Y_i$, $Y_{i+1} = Y_i \cdot D_i$, and $X_{i+1} = X_i \cdot D_i$. We use an approximation to $1/Y$ as D_0.

In either method, an approximation to the reciprocal of the divisor is required. Direct approximation by table look-up or linear interpolation by table look-up followed by multiply-addition can be used. When the precision of the approximation is m-bits, n-bit division can be done through about $\log_2(n/m)$ iterations. Two multiplications and a complementation are performed in each iteration. The required calculations are almost the same in both methods. The two multiplications in the Goldschmidt's algorithm are independent of each other, whereas those in the Newton method are performed serially.

Multiplicative division methods are employed in mainframe computers as well as some microprocessors.

References

1. Briggs, W.S. and D.T. Matula, "A 17×19 bit multiply and add unit with redundant binary feedback and single cycle latency," *Proc. 11th Symposium on Computer Arithmetic*, pp. 163–170, June 1993.
2. Cavanagh, J.J.F., *Digital Computer Arithmetic—Design and Implementation*, McGraw-Hill, 1984.
3. Ercegovac, M.D. and T. Lang, *Division and Square Root—Digit-Recurrence Algorithms and Implementations*, Kluwer Academic Publishers, 1994.
4. Ercegovac, M.D. and T. Lang, "On-the-fly conversion of redundant into conventional representations," *IEEE Trans. Comput.*, vol. C-36, no. 7, pp. 895–897, July 1987.
5. Hennessy, J.L. and D.A. Patterson, *Computer Architecture—A Quantitative Approach*, Appendix A, Morgan Kaufmann Publishers, 1990.

6. Hwang, K., *Computer Arithmetic—Principles, Architecture, and Design*, John Wiley & Sons, 1979.
7. Koren, I., *Computer Arithmetic Algorithms*, Prentice-Hall, 1993.
8. Omondi, A.R., *Computer Arithmetic Systems—Algorithms, Architecture and Implementations*, Prentice-Hall, 1994.
9. Scott, N.R., *Computer Number Systems & Arithmetic*, Prentice-Hall, 1985.
10. Spaniol, O., *Computer Arithmetic Logic and Design*, John Wiley & Sons, 1981.
11. Waser, S. and M.J. Flynn, *Introduction to Arithmetic for Digital Systems Designers*, Holt, Rinehart and Winston, 1982.

44

Full-Custom and Semi-Custom Design

Saburo Muroga

University of Illinois
at Urbana-Champaign

44.1 Introduction

As integrated circuits become more inexpensive and compact, many new types of products, such as digital cameras, digital camcorders, and digital television [2], are being introduced, based on digital systems. Consequently, logic design must be done under many different motivations. Since each case is different, we have different design problems. For example, we have to choose an appropriate IC (integrated circuit) logic family, since these cases have different performance requirements (scientific computers require high speed, but wristwatches require very low power consumption), although in recent years, CMOS has been more widely used than other IC logic families, such as ECL, which has been used for fast computers.

Logic functions that are frequently used by many designers, such as a full adder, are commercially available as off-the-shelf IC packages. (A package means an IC chip or a discrete component encased in a container.) Logic networks that realize such logic networks are often called **standard (logic) networks**. A single component, such as a resistor and a capacitor, is also commercially available as an off-the-shelf discrete component package. Logic networks can be assembled with these off-the-shelf packages. In many cases, not only performance requirements but also compactness and low cost are very important for products such as digital cameras. So, digital systems must accordingly be realized in IC packages that are designed, being tailored to specific objectives, rather than assembling many of these off-the-shelf packages on pc-boards, although assembling with these off-the-shelf packages has the advantage of ease of partial design changes.

Here, however, let us consider two important cases of designing an IC chip inside such an IC package, which is not off-the-shelf, that leads to two sharply contrasting logic design approaches: quick design and high-performance design. Quick design of IC chips is called **semi-custom design** (recently called **ASIC design**, abbreviating Application Specific Integrated Circuit design), whereas deliberate design for high performance is called **full-custom design** because full-custom design is fully customized to high performance. Full-custom design is discussed in this chapter, and different approaches of semi-custom design will be discussed in the succeeding chapters.

44.1.1   Semi-Custom Design

When manufacturers introduce new products or computers, it is ideal to introduce them with the highest performance in the shortest time. But it is usually very difficult to attain both, so one of them must be emphasized, based on the firm's marketing strategy against competitors. Often, quick introduction of new computers or new merchandise with digital systems is very important for a manufacturer in terms of profits. (In some cases, introduction of a new product one year earlier than a competitor's generates more than twice the total income that the competitor gets [1].) This is because the firm that introduces the product can capture all initial potential customers at highest prices, and latecomers are left with only the remaining fewer customers, selling at lower prices. This difference in timing often means a big difference in profits. In other words, the profits due to faster introduction of new products on a market often far exceed the profits due to careful design. The use of off-the-shelf IC packages, including off-the-shelf microprocessors, is used to be a common practice for shortening design time. But recent progress enables us to design digital systems in an IC chip more compactly with higher performance than before by curtailing time-consuming layout of transistor circuits on chips and by extensively using CAD programs. Thus, in the case of small volume production, the design cost, part of the product cost, is reduced. This makes semi-custom design appropriate for debugging or prototyping of new design. But extensive use of CAD programs tends to sacrifice the performance or compactness of the semi-custom-designed IC chips. Semi-custom design, or ASIC design, has several different approaches, as will be discussed in later chapters [3,4]. Design of logic networks with the highest performance requires deliberate design of logic networks, design of transistor circuits, layout of these transistor circuits most compactly, and manufacturing of them. Such logic networks are called **random-logic gate networks** and are realized by full-custom design. In contrast to full-custom design, semi-custom design simplifies design and layout of transistor circuits to save expenses and design time. Depending on how design and layout of transistor circuits are simplified (e.g., repetition of small transistor subcircuit, or not so compact layout) and even how logic design is simplified, we have variants of semi-custom design.

44.1.2   Full-Custom Design

Full-custom design is logic design to attain the highest performance or smallest size, utilizing the most advanced technology. Designers usually try to improve the economic aspect, that is, performance per cost, at the same time. Full-custom design with the most advanced technology usually takes many years to achieve final products, because new technology must often be explored at the same time. Hence, this is the other extreme to the above quick design in terms of design time. Every design stage is carefully done for the maximum performance, and transistor circuits are deliberately laid out on chips most compactly, spending months by many draftpeople and engineers. CAD programs are used but not extensively as in the case of semi-custom design. When CAD programs for high performance are not available, for example, for the most compact layout of transistor circuits to which is required for high performance—manual design is used, possibly mixed with the use of some CAD programs. Also, once mistakes sneak into some stages in the long sequence of design, designers have to repeat at least part of the long sequence of design stages to correct them. So, every stage is deliberately tested with substantial effort.

44.1.3   Motivation for Semi-Custom Design

It should be noticed that the cost of a digital system highly depends on the production volume of a chip. The cost of an IC package can be approximately calculated by the following formula:

$$\begin{bmatrix} \text{Total cost} \\ \text{of an IC} \\ \text{package} \end{bmatrix} = \frac{[\text{Design expenses}]}{[\text{Production volume}]} + \begin{bmatrix} \text{Manufacturing} \\ \text{cost per IC} \\ \text{package} \end{bmatrix} \qquad (44.1)$$

The second term on the right-hand side of Eq. 44.1, [Manufacturing cost per IC package], is fairly proportional to the size of each chip when the complexity of manufacturing is determined, being usually on the order of dollars, or tens of dollars in the case of commercial chips. In the case of full-custom design, chips are deliberately designed by many designers spending many months. So, [Design expenses], the first term on the right-hand side of Eq. 44.1 is very high and can easily be on the order of tens of millions of dollars. Thus, the first term is far greater than the second term, making [Total cost of an IC package] very expensive, unless [Production volume] is very large, being on the order of more than tens of millions. Many digital systems that use IC chips are produced in low volume and [Design expenses] must be very low. Semi-custom design is for this purpose and CAD programs need to be used extensively for shortening design time and manpower in order to reduce [Design expenses]. In this case, [Manufacturing cost per IC chip] is higher than that in the case of full-custom design because the size of each chip is larger.

Thus, we can see the following from the formula in Eq. 44.1: chips by semi-custom design are cheaper in small production volume than those by full-custom design, but more expensive in high production volume. But chips by full-custom design are cheaper in the case of high volume production, and are expensive for low volume production.

44.2 Full-Custom Design Sequence of a Digital System

Full-custom design flow of a digital system follows a long sequence of different design stages, as follows.

First, the architecture of a digital system is designed by a few people. The performance or cost of the entire system is predominantly determined by architectural design, which must be done based on good knowledge of all other aspects of the system, including logic design and also software to be run. If an inappropriate architecture is chosen, the best performance or lowest cost of the system cannot be achieved, even if logic networks, or other aspects like software, are designed to yield the best results. For example, if microprogramming is chosen for the control logic of a microcomputer based on ROM, it occupies too much of the precious chip area, sacrificing performance and cost, although we have the advantages of short design time and design flexibility. Thus, if performance or manufacturing cost is important, realization of control logic by logic networks (i.e., hard-wired control logic) is preferred. Actually, every design stage is important for the performance of the entire system. **Logic design is also one of key factors for computer performance, such as architecture design, transistor circuit design, layout design, compilers, and application programs. Even if other factors are the same, computer speed can be significantly improved by deliberate logic design**.

Next, appropriate IC logic families and the corresponding transistor circuit technology are chosen for each segment of the system. Other aspects such as memories are simultaneously determined in greater detail. We do not use expensive, high-speed IC logic families where speed is not required.

Architecture and transistor circuits are outside the scope of this handbook, so they are not discussed here further.

The next stage in the design sequence is the design of logic networks, considering cost reduction and the highest performance, realizing functions for different segments of the digital system. Logic design requires many engineers for a fairly long time.

Then, logic networks are converted into transistor circuits. This conversion is called **technology mapping**. It is difficult to realize the functions of the digital system with transistor circuits directly, skipping logic design, although experienced engineers can design logic networks and technology mapping at the same time, at least partly. Logic design with AND, OR, and NOT gates, using conventional switching theory, is convenient for human minds because AND, OR, and NOT gates in logic networks directly correspond, respectively, to basic logic operations, AND, OR, and NOT in logic expressions. Thus, logic design with AND, OR, and NOT gates is usually favored for manual design by designers and then followed by technology mapping. For example, the logic network with AND and OR gates shown in Figure 44.1(a) is technology-mapped into the MOS circuit shown in Figure 44.1(c). A variety of IC logic families, such

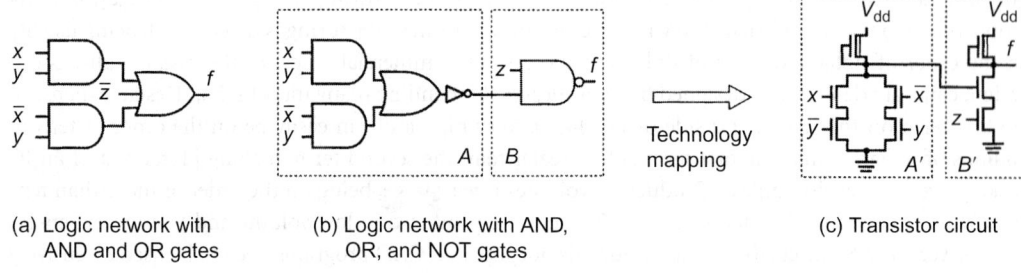

(a) Logic network with   (b) Logic network with AND,   (c) Transistor circuit
AND and OR gates   OR, and NOT gates

FIGURE 44.1 Technology mapping.

as static MOS circuits and dynamic MOS circuits, are now used to realize logic gates. Thus, the relationships between logic networks with AND, OR, and NOT gates and those in transistor circuits are complex because logic gates realized in transistor circuits do not have one-to-one correspondence with AND, OR, and NOT gates, as illustrated in Figure 44.1. The function $f = x\bar{y} \vee \bar{x}y \vee \bar{z}$ in Figure 44.1 can be realized with two AND gates realizing $x\bar{y}$ and $\bar{x}y$, and then with an OR gate which has inputs from the AND gates and $\bar{z}$, as shown in Figure 44.1(a). But if the function is realized with AND, OR, and NOT gates, as shown in Figure 44.1(b), then conversion of it into the MOS circuit shown in Figure 44.1(c) is easier because correspondence between subnetworks in (b) and transistor logic gates in (c) is clear, where dot-lined rectangles, *A* and *B*, in (b) correspond *A'* and *B'* in (c), respectively.

Then, after technology mapping, these transistor circuits are laid out on a chip. Layout is also a painstaking endeavor for many draftpersons. The above design stages are highly interactive and iterative because, if bad design is made in a certain stage, good design in other stages cannot compensate for it, thus yielding poor performance or cost increase of the entire chip. In particular, logic network design and layout design are highly interactive.

In this case, it is important to notice the difference of delay time of signal propagation in logic networks from that in transistor circuits laid out on a chip. In previous chapters, we have assumed for the sake of simplicity that signal propagation has delay time only on gates (for the sake of simplicity, equal delay time is assumed on every gate) but not on connections. But in the case of transistor circuits, signal propagation on each connection has significant delay time, which can be greater than delay time of gates. The longer the connection, the greater the delay time on that connection. The larger the number of connections (i.e., fan-out connections) from the output of a gate, the greater the delay time of each connection. Also, each logic gate realized in transistor circuit may have a different delay time. The greater the fan-in of a gate, the greater the delay time. Restrictions on maximum fan-out and maximum fan-in are very important for the performance of logic networks. Thus, if we want to have fast transistor circuits, we need to consider these relationships in designing logic networks with AND and OR gates. Consideration of only the number of levels is not sufficient.

Then, IC chips are manufactured and are assembled with pc-boards into a digital system.

In the case of the high-performance design discussed above, every effort is made to realize digital systems with the best performance (usually speed), while simultaneously considering the reduction of cost.

When we want to develop digital systems of high performance, using the most advanced technology, much greater manpower and design time are required than that needed for semi-custom design approaches. The actual design time requirement depends on how ambitious the designers are. High-performance microprocessor chips are usually designed by full-custom design, typically taking 3 to 5 years with a large number of people, perhaps several dozen engineers. If the digital system is not drastically different from previous models, design time can be shorter with fewer people; but if the system is based on many new ideas, it may be longer.

As we become able to pack an increasingly large number of networks in a single IC chip every year, the full-custom design of VLSI chips (including microcomputers) of high performance with the most advanced technology is beginning to require far greater design effort and longer time. Thus, more extensive use of improved CAD programs is inevitable. This is because a new generation of microprocessor chips has been introduced every 4 years, having a few times as many transistors on a chip. Compared with systems of 10 years ago, contemporary systems consist of two or three order more transistors, although the physical size of these systems are far smaller. For example, IBM's first personal computer, introduced in 1981, was installed with only 16 kilobytes RAM (expandable to 64 kilobytes) and Intel's microprocessor 8080, which consists of 4800 transistors. But Intel's microprocessor, Pentium III with 500 MHz, introduced in 1999 consists of about 9,500,000 transistors (an approximate number of logic gates can be obtained by dividing the number of transistors by a number between 3 and 5).

In addition to the use of transistor circuits as logic gates, memories are becoming widely used to implement logic networks, being mixed with gates. Also, software is often implemented with ROMs (read-only memories) as firmware, since ROMs are cheaper and smaller than RAMs (random-access memories). Because of these developments, we have complex problems in designing logic networks with a mixture of gates, software, and memories. Essentially, boundaries among logic design, transistor circuits, software, and architecture have disappeared. The number of transistors, or logic gates, used in digital systems is increasing all the time. In designing such gigantic digital systems, it is becoming extremely important to design without errors, necessitating extensive testing in every design stage. To cope with these complex problems, CAD programs with new logic design methods have been developed in recent years. For example, recent CAD programs for logic design can synthesize far larger logic networks than manual design can, and appropriate logic expressions can be derived for functions with a large number of variables by using BDDs.

References

1. Davidow, W., "How microprocessors boost profits," *Electronics,* pp. 105–108, July 11, 1974; p. 92, Jan. 23, 1975.
2. Jurgen, R.K., *Digital Consumer Electronics Handbook,* McGraw-Hill, 1997.
3. Muroga, S., *VLSI System Design,* John Wiley & Sons, 1982.
4. Weste, N.H.E. and K. Eschraghian, *Principles of CMOS VLSI Design: A Systems Perspective,* 2nd ed., Addison-Wesley, 1993.

45

Programmable Logic Devices

Saburo Muroga

University of Illinois
at Urbana-Champaign

CONTENTS

45.1 Introduction

Hardware realization of logic networks is generally very time-consuming and expensive. Also, once logic functions are realized in hardware, it is difficult to change them. In some cases, we need logic networks that are easily changeable. One such case is logic networks whose output functions need to be changed frequently, such as control logic in microprocessors, or logic networks whose outputs need to be flexible, such as additional functions in wrist watches and calculators. Another case is logic networks that need to be debugged before finalizing. **Programmable logic devices** (i.e., **PLDs**) are for this purpose. On these PLDs, all transistor circuits are laid out on IC chips prior to designers' use, considering all anticipated cases. With PLDs, designers can realize logic networks on an IC chip, by only deriving concise logic expressions such as minimal sums or minimal products, and then making connections among pre-laid logic gates on the chip. So, designers can realize their own logic networks quickly and inexpensively using these pre-laid chips, because they need not design logic networks, transistor circuits, and layout for each design problem. Thus, designers can skip substantial time of months for hardware design. CAD programs for deriving minimal sums or minimal products are well developed [1], so logic functions can be realized very easily and quickly as hardware, using these CAD programs. The ease in changing logic functions without changing hardware is just like programming in software, so the hardware in this case is regarded as "programmable." Programmable logic arrays (i.e., PLAs) and FPGAs are typical programmable logic devices.

PLDs consists of mask-programmable PLDs and field-programmable PLDs. **Mask-programmable PLDs** (i.e., **MPLDs**) can be made only by semiconductor manufacturers because connections are made by custom masks. Manufacturers need to make few masks for connections out of all of more than 20 masks, according to customer's specification on what logic functions are to be realized. Unlike mask-programmable PLDs, **field-programmable PLDs** (i.e., **FPLDs**) can be programmed by users and are economical only for small production volume, whereas MPLDs are economical for high production volume. Logic functions can be realized quicker on FPLDs than on MPLDs, saving payment of charges for custom masks for connections to semiconductor manufacturers, but they are larger, more expensive, and slower because of addition of electronic circuits for programmability.

Classification of PLDs is somewhat confusing in publications. Gate arrays to be described in Chapter 46 are regarded as PLDs in a broad sense in some publications, as shown in the following table, though PLDs may not include **field-programmable gate array**s (i.e., **FPGA**s) in some publications. Field-programmable PLDs and FPGAs, are regarded as FPGAs in a broad sense. FPGAs in some cases have arrays of PLDs inside and are sometimes called **complex PLD**s to differentiate them from PLDs which are simpler.

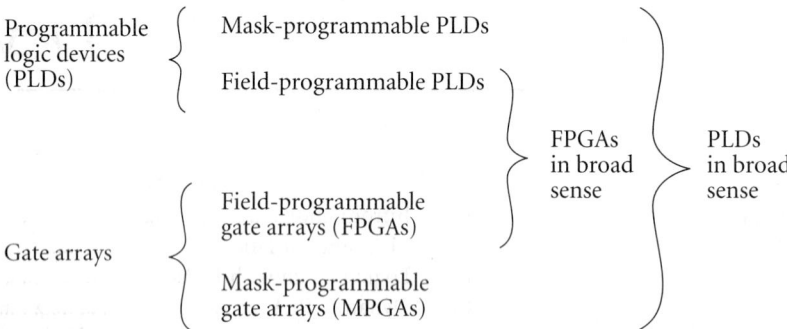

45.2   PLAs and Variations

Programmable logic arrays (i.e., PLAs) are one of programmable logic devices. Logic functions realized in PLAs can be as easily changed as software is.

A **programmable logic array** (abbreviated as **PLA**), which was devised by Proebsting [9], is a special type of ROM (which stands for **Read-Only Memory**), although its usage is completely different from that of ROMs. MOSFETS are arranged in a matrix on a chip, as illustrated in Figure 45.1(a). A PLA consists of an **AND array** and an **OR array**. In order to store logic expressions, connections between the MOSFET gates and the vertical lines in the AND array and also connections between the MOSFET gates and the horizontal lines in the OR array are set up by semiconductor manufacturers during fabrication according to customer' specifications. Since for these connections only one mask out of many necessary masks needs to be custom-made, PLAs are inexpensive when the production volume is high enough to make the custom preparation cost of the connection mask negligibly small. Because of low cost and design flexibility, PLAs are extensively used in VLSI chips, such as microprocessor chips for general computation and microcontroller chips for home appliances, toys, and watches.

When MOSFET gates are connected, as denoted by the large dots in Figure 45.1(a), we have, $\overline{x\bar{y}\bar{z}}, \overline{\bar{x}z}, \overline{xyz}$, at the outputs P_1, P_2, P_3 of the AND array, respectively, since $P_1, P_2,$ and P_3 represent the outputs of NAND gates, if negative logic is used with n-MOS. Here, **negative logic, where a high voltage and a low voltage are regarded as signal 0 and 1, respectively, is used for the sake of the convenience in deriving sums-of-products, i.e., disjunctive forms for the output functions $f_1, f_2,$ and f_3.** (If positive logic is used, where a high voltage and a low voltage are regarded as signal 1 and 0, respectively, then $P_1, P_2,$ and P_3 represent the outputs of NOR gates, and $f_1, f_2,$ and f_3 are expressed in the forms of product-of-sums, i.e., in conjunctive forms. Since most people prefer disjunctive forms, negative logic is usually used in the case of PLAs.) Then, the outputs $f_1, f_2,$ and f_3 of the OR-array also represent the outputs of NAND gates with P_1, P_2, P_3 as their inputs. Thus,

$$f_1 = \overline{P_1 P_3} = \bar{P_1} \vee \bar{P_3} = x\bar{y}\bar{z} \vee xyz$$

$$f_2 = \bar{P_2} = \bar{x}z$$

$$f_3 = \overline{P_1 P_2} = \bar{P_1} \vee \bar{P_2} = x\bar{y}\bar{z} \vee \bar{x}z$$

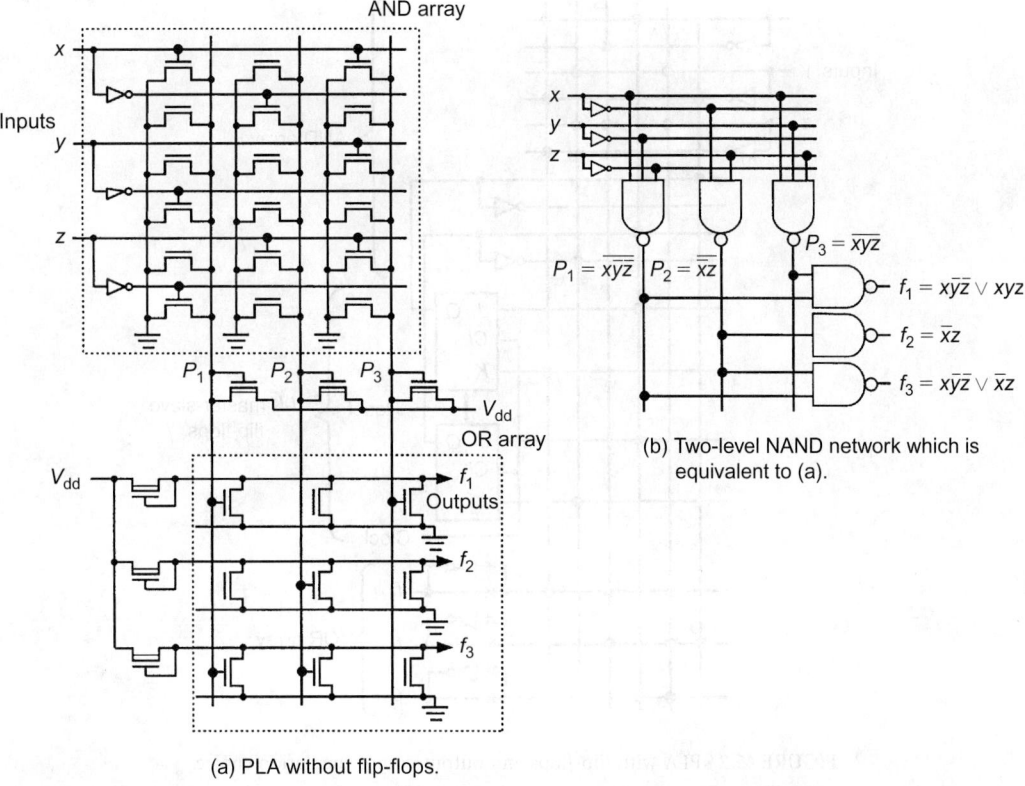

(a) PLA without flip-flops.

(b) Two-level NAND network which is equivalent to (a).

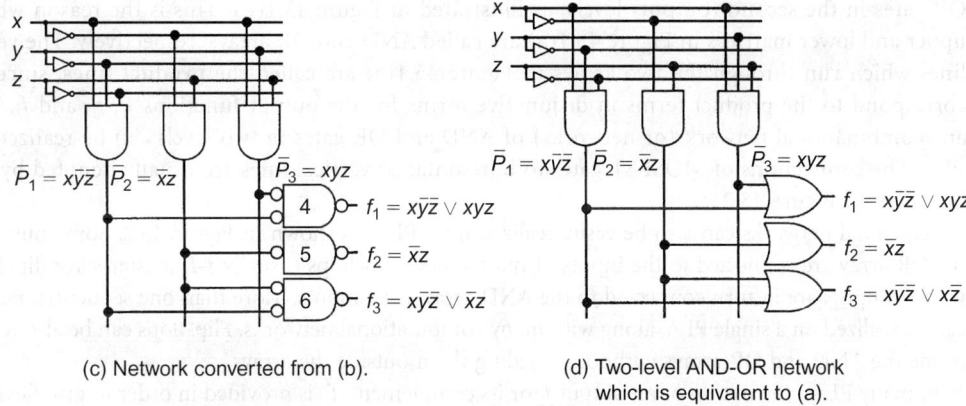

(c) Network converted from (b).

(d) Two-level AND-OR network which is equivalent to (a).

FIGURE 45.1 A PLA and equivalent logic networks.

(It is to be noted that in the AND array, the inputs x, y, and z have their complements by inverters.) Therefore, the two arrays in Figure 45.1(a) represent a network of NAND gates in two levels, as illustrated in Figure 45.1(b). This can be redrawn in Figure 45.1(c) by moving the bubbles (i.e., inverters) to the inputs of the NAND gates 4, 5, and 6 without changing the outputs f_1, f_2, and f_3. Then, gate 4, for example, can be replaced by an OR gate because f_1 is

$$f_1 = \overline{\overline{P_1}\overline{P_3}} = \overline{P_1} \vee \overline{P_3}$$

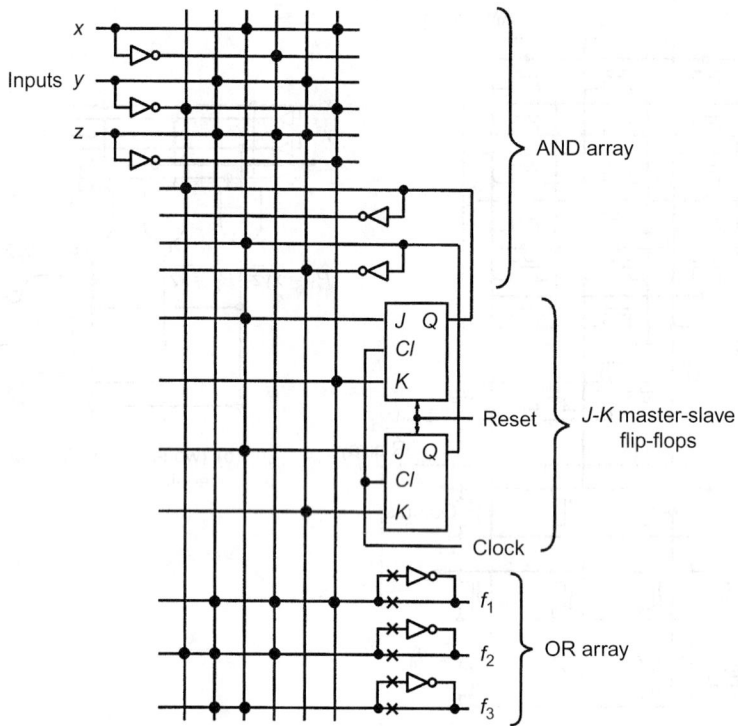

FIGURE 45.2   PLA with flip-flops and output-complementation choice.

by De Morgan's theorem. Thus, this is interpreted as a network of AND gates in the first level and OR gates in the second (output) levels, as illustrated in Figure 45.1(d). This is the reason why the upper and lower matrices in Figure 45.1(a) are called AND and OR arrays, respectively. The vertical lines which run through the two arrays in Figure 45.1(a) are called the **product lines**, since they correspond to the product terms in disjunctive forms for the output functions f_1, f_2, and f_3. Thus, any combinational network (or networks) of AND and OR gates in two levels can be realized by a PLA. The connections of MOSFET gates to horizontal or vertical lines are usually denoted by dots, as shown in Figure 45.2.

   Sequential networks can also be easily realized on a PLA, as shown in Figure 45.2. Some outputs of the OR array are connected to the inputs of master-slave flip-flops (usually J-K master-slave flip-flops), whose outputs are in turn connected to the AND array as its inputs. More than one sequential network can be realized on a single PLA, along with many combinational networks. Flip-flops can be also realized inside the AND and OR arrays without providing them outside the arrays.

   In many PLAs, the option of an output f_i or its complement $\overline{f_i}$ is provided in order to give flexibility, as illustrated in the lower right-hand corner of Figure 45.2. By disconnecting one of the two ×'s at each output, we can have either f_i or $\overline{f_i}$ as output, as illustrated in Figure 45.3. When f_i has too many products in its disjunctive form and cannot be realized on a PLA, its complement $\overline{f_i}$ may have a sufficiently small number of terms to be realizable on the PLA, or vice versa.

   If the number of product lines in a PLA is too many, each horizontal line gets too long with a significant increase in parasitic capacitance. Then, if the majority of the MOSFET gates provided are connected to this horizontal line, the input or its inverter has too many fan-out connections on this horizontal line. Similarly, the total number of horizontal lines cannot be too large. In other words, the array size of a PLA is limited because of speed considerations. In contrast, the size of a ROM can be much larger, since we can use more than one decoder, or use a complex decoding scheme.

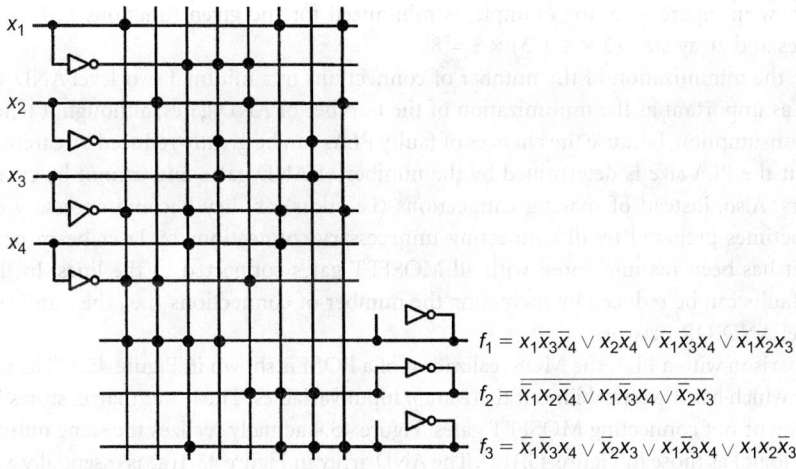

$f_1 = x_1\bar{x}_3\bar{x}_4 \vee x_2\bar{x}_4 \vee \bar{x}_1\bar{x}_3x_4 \vee \bar{x}_1x_2x_3$

$f_2 = \overline{\bar{x}_1x_2\bar{x}_4} \vee x_1\bar{x}_3x_4 \vee \bar{x}_2x_3$

$f_3 = \bar{x}_1\bar{x}_3x_4 \vee \bar{x}_2x_3 \vee x_1\bar{x}_3x_4 \vee x_1x_2\bar{x}_3$

FIGURE 45.3 A PLA minimized for the given functions, f_1, f_2, and f_3.

The PLAs whose connections of some MOSFETs to lines are made by custom masks by semiconductor manufacturers are called **mask-programmable PLAs** (abbreviated as **MPLAs**).

A **field-programmable PLA** (abbreviated as **FPLA**) is also available, using fuses or anti-fuses, where unlike fuses, anti-fuses are initially not connected but can be connected by applying a voltage that is higher than a normal voltage. In an FPLA, a user can set up a dot pattern by blowing fuses or connecting anti-fuses connected to some MOSFETs by temporarily feeding excessively high voltages. In this realization, a special electronic circuit to blow fuses or to connect anti-fuses must be provided in addition to the PLA arrays, and this adds extra area to the entire area. In large-volume production, FPLAs are more expensive due to this extra size than PLAs, but when users need a small number of PLAs, FPLAs are much cheaper and convenient, since users can program FPLAs by themselves, inexpensively and quickly, instead of waiting weeks for the delivery of MPLAs from semiconductor manufacturers.

In contrast to the above FPLAs based on fuses, FPLAs whose undesired connections are disconnected by laser beam are also available. In this case, the chip size is smaller than that of the above FPLAs, since the electronic circuits to blow fuses are not necessary, but special laser equipment is required.

FPLAs are less expensive than MPLAs for small production volumes, although for high production volumes, MPLAs are much less expensive. In particular, when designers want to use MPLAs but their design is not completely debugged, they should try their design ideas with FPLAs and then switch to MPLAs only after debugging is complete, because if a semiconductor manufacturer is already working on MPLAs, sudden interruption of the work due to the discovery of design mistakes is unprofitable for both the manufacturer and the customer.

45.3 Logic Design with PLAs

Minimization techniques for multiple-output logic functions discussed in Chapter 30 can be used to minimize the size of a PLA. If the number of AND gates in a two-level AND-OR network (i.e., the number of distinct multiple-output prime implicants in disjunctive forms) for the given output functions is minimized, we can minimize the number of product lines, t. Thus, the array size $(2n + m)t$ of a PLA is minimized when the PLA has n inputs, m outputs, and t product lines, where n and m are given. Also, if the total number of connections in a two-level AND-OR network is minimized as the secondary objective, as we do in the minimization of a multiple-output logic function, then the number of dots (i.e., connected intersections of the product lines and the horizontal lines) in the PLA is minimized. Therefore, the derivation of a minimal two-level network with AND and OR gates by the minimization techniques known in switching theory is very important for the minimal and reliable design of PLAs.

The PLA show in Figure 45.3, for example, is minimized for the given functions f_1, f_2, and f_3, with 8 product lines and array size, $(2 \times 4 + 3) \times 8 = 88$.

However, the minimization of the number of connections in a minimal two-level AND-OR network may not be as important as the minimization of the number of AND gates, although it tends to reduce the power consumption, because the chances of faulty PLAs can be greatly reduced by careful fabrication of chips. But the PLA size is determined by the number of AND gates and cannot be changed by any other factors. Also, instead of making connections (i.e., dots) as they become necessary on a PLA, a PLA is sometimes prepared by disconnecting unnecessary connections by laser beam or by blowing fuses after it has been manufactured with all MOSFET gates connected to the lines. In this case, the chances of faults can be reduced by increasing the number of connections (i.e., the number of dots) in the two-level AND-OR network.

For comparison with a PLA, the MOS realization of a ROM is shown in Figure 45.4. The upper matrix is a decoder which has 2^n vertical lines if there are n input variables. The lower matrix stores information by connecting or not connecting MOSFET gates. Figure 45.4 actually realizes the same output functions (in negative logic) as those in Figure 45.1(a). The AND array in Figure 45.1(a) is essentially a counterpart of the decoder in Figure 45.4, or the decoder may be regarded as a fixed AND array with 2^n product lines, which is the maximum number of the product lines in a PLA. The AND array in Figure 45.1(a) has only three vertical lines, whereas the decoder in Figure 45.4 has eight fixed vertical lines. This indicates the compact information packing capability of PLAs. PLAs are smaller than ROMs, although the packing advantage of PLAs varies, depending on functions. For example, if we construct a ROM that realizes the functions of the PLA of Figure 45.3, in a manner similar to Figure 45.4, the decoder consists of 8 horizontal lines and 16 vertical lines, and the lower matrix for information storage consists of 16 vertical lines and 3 horizontal lines. Thus, the ROM requires the array size of $16 \times (8 + 3) = 176$, compared with 88 in Figure 45.3.

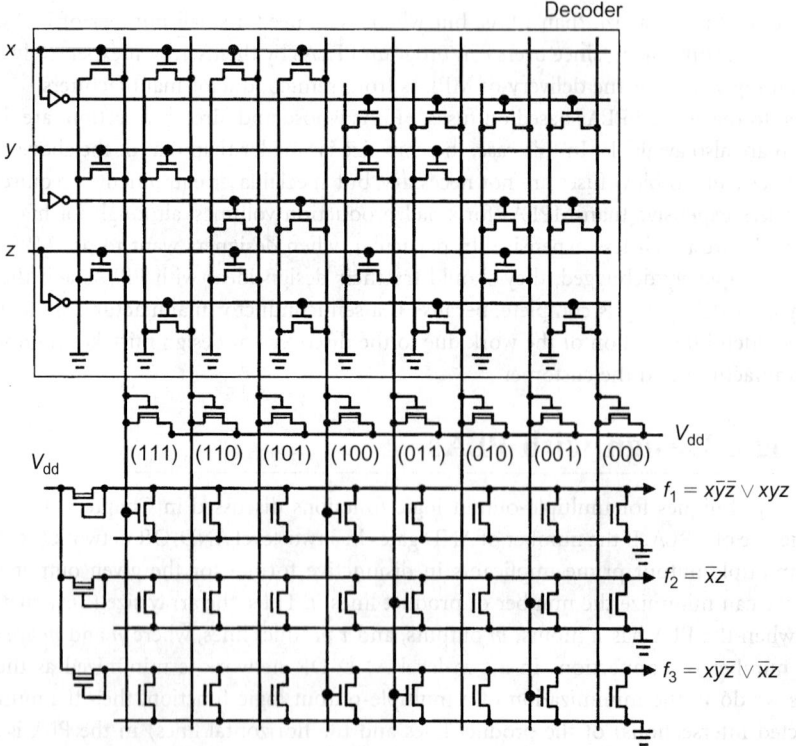

FIGURE 45.4 ROM that corresponds to the PLA in Figure 45.1.

Generally, the size difference between PLAs and ROMs sharply increases as the number of input variables increases.

A PLA, however, cannot store some functions, such as $x_1 \oplus x_2 \oplus \cdots \oplus x_n$ if n is large, because 2^{n-1} product lines are required and the number of these lines is excessively large for a PLA. (The horizontal lines become too long with excessive fan-out and parasitic capacitance.) However, we can store these functions in a ROM with an appropriate decoding scheme.

Of course, in the case of ROMs, storing a truth table without worrying about conversion of given logic functions into a minimal sum is convenient, although it makes the ROM size bigger than the PLA size.

Minimal two-level networks of AND and OR gates for the absolute minimization of the PLA size can be derived by the minimization methods discussed in earlier chapters, if a function to be minimized has either at most several variables, or many more variables but with a simple relationship among its prime implicants [8]. But otherwise, we have to be content with near-minimal networks instead of minimal networks. In many cases, efforts to reduce the PLA size, even without reaching an absolute minimum, result in significant size reduction. Also, CAD programs have been developed with heuristic minimization methods [12,13], such as the one by Hong et al. [7], which was the first powerful heuristic procedure drastically different from conventional minimization procedures. MINI, PLA minimization program of Hong, et al., was later improved to ESPRESSO by Rudell, Brayton, et al. [1,10,11]. Recently, however, Coudert and Madre [2–6] developed a new method for absolute minimization by implicitly expressing prime implicants and minterms using BDDs described in Chapter 29. By this method, absolute minimization of functions with greater numbers of variables is more feasible than before, although it is still time-consuming.

45.4 Dynamic PLA

If we want to realize a PLA in CMOS, instead of static nMOS circuit that has been discussed in Chapter 33, Section 33.3, in order to save power consumption, then a PLA in CMOS requires a large area because we need pMOS and nMOS subcircuits. Thus, instead of static CMOS, the dynamic CMOS illustrated in Figure 45.5(a) is usually used. During the absence of a clock pulse of the first- and second-phase clocks, ϕ_1 and ϕ_2 (i.e., during $\phi_1 = \phi_2 = 0$ (low voltage, using positive logic)) shown in Figure 45.5(b), pMOSFETs, T_1, T_2, and T_3, become conductive and nMOSFETs, T_4, T_5, and T_6 become non-conductive precharging vertical lines, P_1, P_2, and P_3. When a clock pulse of the first-phase clock, ϕ_1, appears but a clock-pulse of the second-phase clock, ϕ_2, does not appear yet, i.e., when $\phi_1 = 1$ (high voltage) and $\phi_2 = 0$, pMOSFETs, T_1, T_2, and T_3, become non-conductive and nMOSFETs, T_4, T_5, and T_6, become conductive. Then, depending on the values of x, y, and z, some verticle lines, P_1, P_2, and P_3 are discharged through some of the nMOSFETs in the AND array. (For example, if $y = 0$ (low voltage), P_1 is discharged through nMOSFETs A.) A clock pulse of the second-phase clock, ϕ_2, is still absent (i.e., $\phi_2 = 0$), so pMOSFETs, T_7, T_8, and T_9, become conductive and nMOSFETs T_{10}, T_{11}, and T_{12}, become non-conductive, precharging horizontal lines, f_1, f_2, and f_3. When a clock pulse of the first-phase clock, ϕ_1, is still present, and a clock pulse of the second-phase clock, ϕ_2, appears, i.e., when $\phi_1 = \phi_2 = 1$, pMOSFETs, T_7, T_8, and T_9, become non-conductive and nMOSFETs, T_{10}, T_{11}, and T_{12}, become conductive. Then, some of horizontal lines, f_1, f_2, and f_3, are discharged through some of the nMOSFETs in the OR array, depending on which of the vertical lines, P_1, P_2, and P_3, are still charged.

45.5 Advantages and Disadvantages of PLAs

PLAs, like ROMs which are more general, have the following advantages over random-logic gate networks, where random-logic gate networks are those that are compactly laid out on an IC chip:

1. There is no neeed for the time-consuming logic design of random-logic gate networks and even more time-consuming layout.
2. Design checking is easy, and design change is also easy.

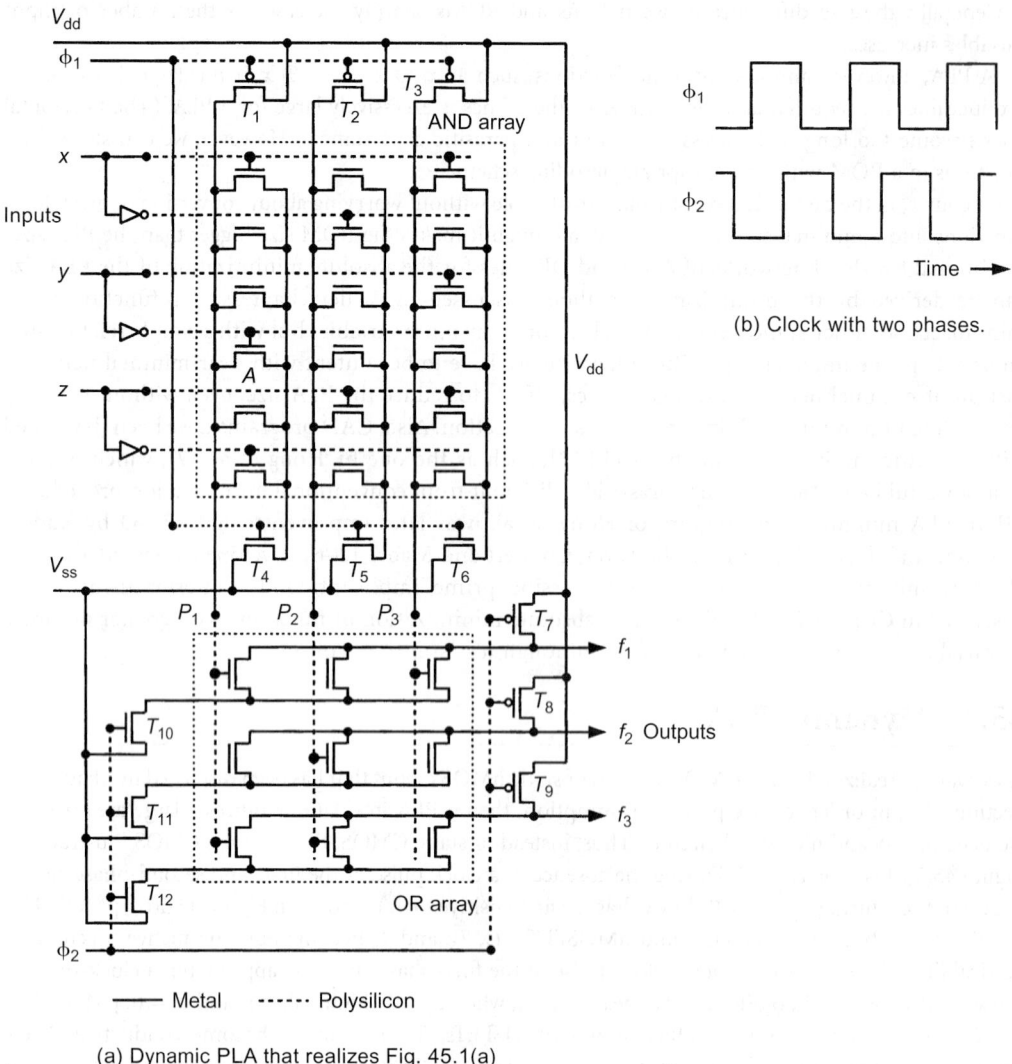

(b) Clock with two phases.

(a) Dynamic PLA that realizes Fig. 45.1(a)

FIGURE 45.5 Dynamic PLA.

3. Layout is far simpler than that for random-logic gate networks, and thus is far less time-consuming.
4. When new IC fabrication technology is introduced, we can use previous design information with ease but without change, making adoption of the new technology quick and easy.
5. Only the connection mask needs to be custom-made.
6. Considering all these, PLA is a very inexpensive approach, greatly shortening desing time.

PLAs have the following disadvantages compared with random-logic gate networks:

1. Random-logic gate networks have higher speed than PLAs or ROMs.
2. Random-logic gate networks occupy smaller chip areas than PLAs or ROMs, although the logic design and the layout of random-logic gate networks are far more tedious and time-consuming.
3. Also, with large production volumes, random-logic gate networks are cheaper than PLAs or ROMs.

PLAs have the following advantage and disadvantage, compared with ROMs:

- For storing the same functions or tasks, PLAs can be smaller than ROMs; generally, the size difference sharply increases as the number of input variables increases.
- The small size advantages of PLAs diminishes as the number of terms in a disjunctive form increases. Thus, PLAs cannot store complex functions, i.e., functions whose disjunctive forms consist of many product terms.

45.5.1 Applications of PLAs

Considering the above advantages and disadvantages, PLAs have numerous unique applications. A microprocessor chip uses many PLAs because of easy of design change and check. In particular, PLAs are used in its control logic, which is complex and requires many changes, even during its design. Also, PLAs are used for code conversions, microprogram address conversions, decision tables, bus priority resolvers, and memory overlay.

When a new product is to be manufactured in small volume or test-marketed, PLAs is a choice. When the new product is well received in the market and does not need further changes, PLAs can be replaced by random-logic gate networks for low cost for high volume production and high speed. Also, a full-custom design approach is very time-consuming, probably taking months or years, but if PLAs are used in the control logic, a number of different custom-design chips with high performance can be made quickly by changing only one connection mask for the PLAs, although these chips cannot have drastically different performance and functions.

45.6 Programmable Array Logic

A **programmable array logic** (**PAL**) is a special type of a PLA where the OR array is not programmable. In other words, in a PAL, the AND array is programmable but the OR array is fixed; whereas in a PLA, both arrays are programmable. The advantage of PALs is the elimination of fuses in the OR array in Figure 45.1(a) and special electronic circuits to blow these fuses. Since these special electronic circuits and programmable OR array occupy a very large area, the area is significantly reduced in PAL. Since single-output, two-level networks (i.e., many AND gates in the first level and one OR gate as the network output) are needed most often in desing practice, many single-output two-level networks which are mutually unconnected are placed in some PAL packages.

In digital systems, many non-standard networks are still used because designers want to differentiate their computers from competitors'. But logic functions that designers want to have are too diverse to be standardized by semiconductor manufacturers. When off-the-shelf IC packages for standard networks, including microprocessors and their peripheral networks, are assembled on pc boards, many non-standard networks are usually required for interfacing them to other key networks or for minor modifications. So, they require many discrete components and IC packages, each of which has a smaller number of transistors, in addition to a microprocessor package with millions of gates, occupying a significant share of the areas on pc boards. Now, we can make connections inside PALs, instead of custom-making pc boards. Custom-made pc boards are expensive and time-consuming because connection patterns on pc boards need to be designed, these pc boards need to be manufactured and then the holes of pc boards have to be soldered to the pins of IC packages. The replacement by PAL packages can substantially reduce the area, time, and cost. If we consider related factors such as reductions of cabinet size, power consumption, and fans, the significance of this reduction is further appreciated.

There are mask-programmable PALs and field-programmable PALs (i.e., FPALs). When logic design is not finalized and needs to be changed often, FPAL packages can reduce expense and time for repeatedly redesigning and remaking pc boards.

References

1. Brayton, R.K., G.D. Hachtel, C.T. McMullen, and A.L. Sangiovanni-Vincentelli, *Logic Minimization Algorithms for VLSI Synthesis*, Kluwer Academic Publishers, 1984.
2. Coudert, O., "Two-level logic minimization: an overview," *Integration, The VLSI Jour.*, pp. 97–140, Oct. 1994.
3. Coudert, O., "Doing two-level logic minimization 100 times faster," *Symposium on Discrete Algorithms (SODA)*, San Francisco, pp. 112–121, Jan. 1995.
4. Coudert, O., "On solving covering problems," *33rd DAC*, pp. 197–202, June 1996.
5. Coudert, O., J.C. Madre, H. Fraisse, and H. Touati, "Implicit prime cover computation: An overview," *Proc. SASIMI'93 (Synthesis and Simulation Meeting and Int'l Interchange)*, Nara, Japan, Oct. 1993.
6. Coudert, O. and J.C. Madre, "New ideas for solving covering problems," *32nd DAC*, San Francisco, pp. 641–646, June 1995.
7. Hong, S.J., R.G. Cain, and D.L. Ostapko, "Mini: A heuristic approach for logic minimization," *IBM JRD*, pp. 443–458, Sept. 1974.
8. Muroga, S., *Logic Design and Switching Theory*, John Wiley & Sons, 1979. (Now available from Krieger Publishing Co.)
9. Proebsting, R., *Electronics*, p. 82, Oct. 28, 1976.
10. Rudell, R.L. and A.L. Sangiovanni-Vincentelli, "Multiple valued minimization for PLA optimization," *IEEE Transactions on CAD*, vol. 6, no. 5, pp. 727–750, Sept. 1987.
11. Sasao, T., "An application of multiple-valued logic to a design of programmable logic arrays," *Proc. of Int'l Symposium on Multiple-Valued Logic*, 1978.
12. Smith, M.J.S., *Application-Specific Integrated Circuits*, Addison-Wesley, 1997.
13. Venkateswaran, R. and P. Mazumder, "A survey of DA techniques for PLD and FPGA based systems," *Integration, the VLSI Jour.*, vol. 17, no. 3, pp. 191–240, Nov. 1994.

46

Gate Arrays

Saburo Muroga
University of Illinois
at Urbana-Champaign

CONTENTS

46.1   Mask-Programmable Gate Arrays

Among all ASIC chips, gate arrays are most widely used because with gate arrays, we can easily realize logic functions that require a far more logic gates than with PLAs. If PLAs are used, such logic functions would require far larger area and delay time. Also, design time with gate arrays is shorter than with the standard cell design approach to be described in Chapter 48.

A gate array is an IC chip on which gates are placed in matrix form without connections among the gates, as illustrated in Figure 46.1(a). By connecting gates, we can realize logic networks, as exemplified in Figure 46.1(b). But actually, logic gates are not realized in a gate array. Instead of gates, cells, each of which consists of unconnected components, are arranged in matrix form, and each cell can realize one of a few different types of gates by connecting these components. Then, by connecting these gates as illustrated in Figure 46.1(b), networks can be realized. Only two or three masks for connections and contact windows (i.e., small metallic areas between MOSFETs and connections) have to be custom-made, instead of all two dozen masks required for full-custom design. Also, because only the connection layout, along with the placement of gates, needs to be considered, CAD can be effectively used, thereby greatly reducing the layout time. Thus, the design with gate arrays is very inexpensive and quick, compared with full-custom design. Gate arrays of CMOS have been extensively used in many computers [1–3].

In Figure 46.1(b), connections among gates are run in narrow strips of space between columns or rows of gates. These strips of space are called **routing channels**. In gate arrays that were commercially available for the first time, routing channels were provided between columns or rows of logic gates. Now, gate arrays without routing channels are also available, running connections over gates, as it becomes easy to do so because many metal layers are available. Such gate arrays are called **sea-of-gate arrays**. Gate arrays with a large number of gates are usually sea-of-gates without routing channels. But gate arrays with routing channels are still used for those with a small number of gates. A relatively large number of pads are necessary, even for such small gate arrays. Then, the number of pads determines the area size of gate arrays and it does not matter whether or not routing channels are provided. If routing channels are provided, then two metal layers are sufficient for a higher yield. Gate arrays in sea-of-gates require three or more metal layers, so they are expensive.

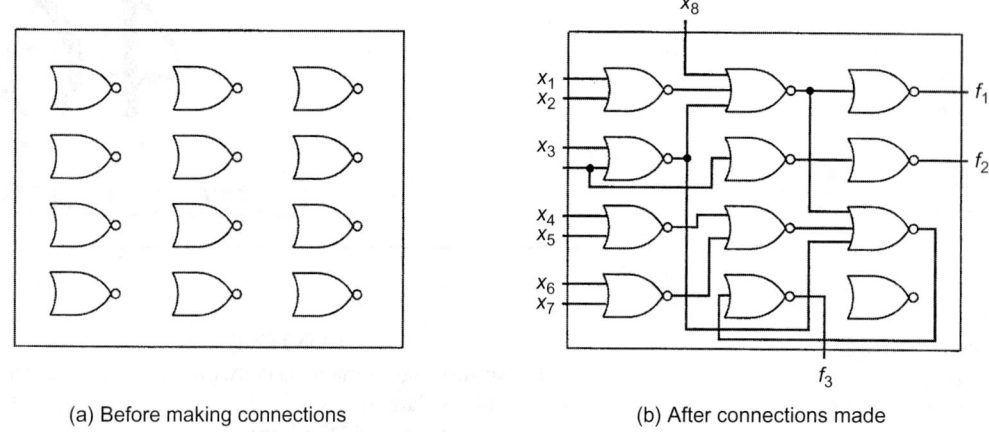

(a) Before making connections   (b) After connections made

FIGURE 46.1   Gate array.

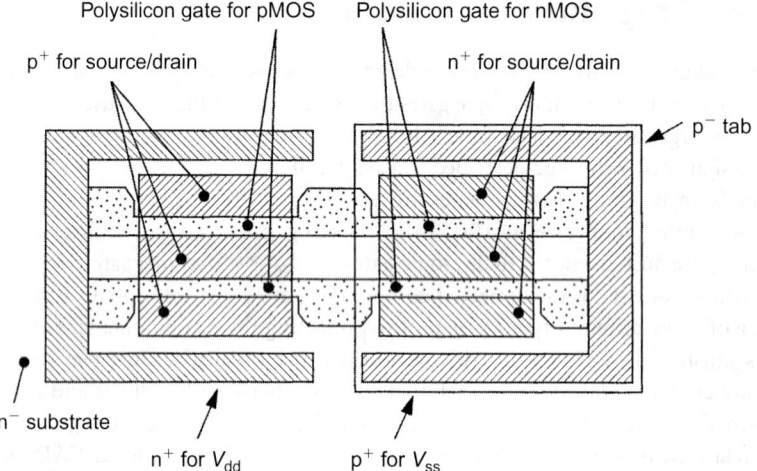

FIGURE 46.2   A cell of CMOS gate array. (Courtesy of Fujitsu Ltd. With permission.)

46.2   CMOS Gate Arrays

CMOS gate arrays are commercially available from many manufacturers in slightly different layout forms. As an example, Figure 46.2 shows a cell of a CMOS gate array, where a pair of pMOSFETs and a pair of nMOSFETs are placed on the left and right, respectively, without connections between them. The NAND gate shown in Figure 46.3(a) can be realized by connecting the components shown in Figure 46.2 by two metal layers as shown in Figure 46.3(b). These two metal layers are formed by forming the first metal layer shown in Figure 46.3(c), the insulation layer (not shown), and then the second metal layer shown in (d). The inverter shown in Figure 46.4(a) can be realized by connections as shown in Figure 46.4(b).

Many different patterns other than that in Figure 46.2 are available for the components of a cell.

(a) NAND gate.

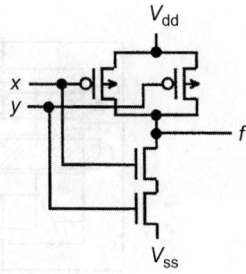

(b) The connections consist of two metal layers shown in (c) and (d).

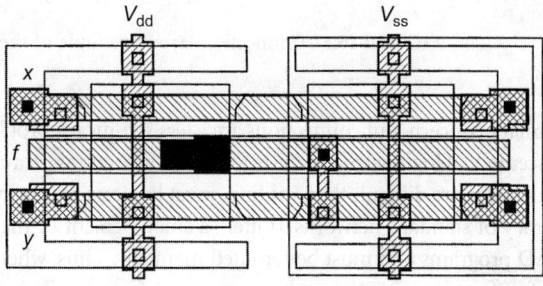

(c) First metal connection layer.

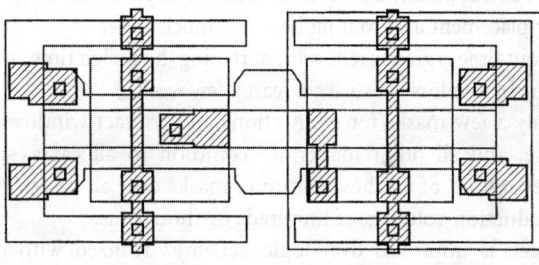

(d) Second metal connection layer.

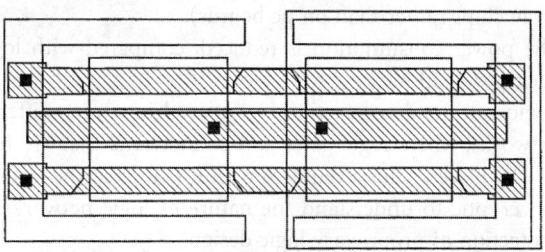

FIGURE 46.3 Connection layout example of the cell in Figure 46.2. (Courtesy of Fijitsu Ltd. With permission.)

46.3 Advantages and Disadvantages of Gate Arrays

Gate arrays have the following advantages:

1. Even when fabrication technology or electronic circuitry changes, gate arrays can be designed in a short time. In other words, only one cell of transistors (or few different cells) needs to be carefully designed and laid out, and this layout can be repeated on a chip.
2. After designers design logic gate networks (although this is still very time-consuming, the minimization of the number of gates or delays, under constraints such as maximum fan-out, would be a designer's

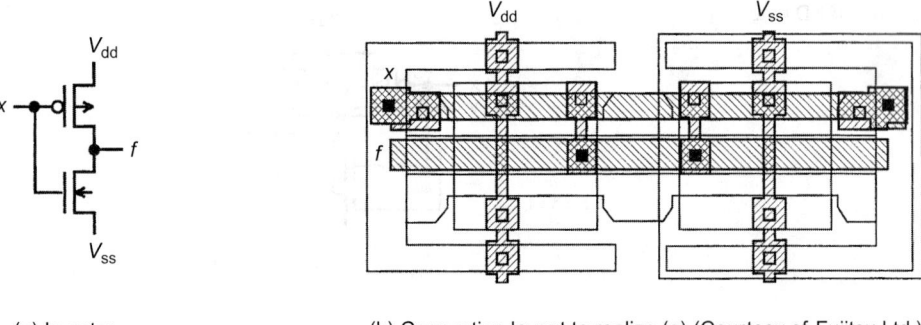

(a) Inverter.   (b) Connection layout to realize (a) (Courtesy of Fujitsu Ltd.)

FIGURE 46.4   Connection layout example of the cell in Figure 46.2.

primary concern, but connections are a less significant problem), CAD programs automatically do the placement of logic gates and the routing of connections on a chip, although in complex cases, placement, which is more difficult than routing, must be done manually by designers, possibly using a placement library of standard networks. (Often, a small percent of the connections cannot be processed well by CAD programs and must be rerouted manually. Thus, when the number of connections is very large, even a few percent means a large number of connections need to be rerouted. So, it is important to reduce this percentage.) It is to be noted that because the gate positions are prefixed on the array, CAD for placement and routing becomes much easier to use than other cases. For the above reasons, the layout time is greatly reduced, shortening the design time, and consequently design expenses. (Delivery time by vendors is usually at least a few weeks.)

3. Only a few masks for connections and contact windows must be custom-made for each design case, and all other masks are common to all cases, spreading the initial investment for the preparation of all these common masks over all cases. Thus, gate arrays are cost-effective in low production volumes of hundreds or thousands.

4. Speed is improved over logic networks realized with off-the-shelf packages or PLAs because interconnections among gates are shortened on the average (most interconnections are inside gate array chips rather than on pc boards).

5. The power consumption is reduced, compared with logic networks realized with off-the-shelf packages or PLAs.

6. Logic gates and connections are placed in a very different manner from full-custom-designed networks, where logic gates and connections that are functionally closely related are placed in nearby places. Thus, even if competitors look at the layout of the gate array chips, the layouts are too cryptic to understand the nature of logic networks. In this sense, gate arrays are good for protection of proprietary logic design.

On the other hand, gate arrays have the following disadvantages.

1. Chip size is large because logic gates are not compactly laid out. For example, each pair of nMOSFETs is placed in an individual p$^-$-tab without sharing p$^-$-tab with many other nMOSFETs, as can be seen in Figure 46.2.

2. The percentage of unused gates is possibly high, for the following reasons. Depending on the types of networks or which parts of a large network are placed in a gate array chip, all spacings provided for connections can be used up (by taking a detour if the shortest paths are filled up by other connections), or all the pins of the chip can be used up by incoming and outgoing connections. In either case, many gates may not be used at all, and fewer than half the gates on a chip are used in some cases. Because of these disadvantages, the average size of a gate array chip is easily four or five times as large as that of a full-custom-designed chip, or it can be even greater, for the same

logic networks. The cost difference would be greater (the cost is not necessarily linearly proportional to chip size) for the same production volume.

3. It is difficult to keep gate delays uniform. As the number of fan-outs and the length of fan-out connections increase, delays increase dramatically. (If delay times of gates are not uniform, the network tends to generate spurious output signals.) In the case of full-custom design, the increase of gate delay by long or many-output connections of a gate can be reduced by redesigning the transistor circuit (e.g., increasing transistor size for delivering greater output power and accordingly reducing the delay). But such a precise adjustment is not possible in the case of gate arrays.

Responding to a variety of different user needs in terms of speed, power consumption, cost, design time, ease of change, and possibly others, a large number of different gate arrays are commercially available from semiconductor manufacturers or are used in-house by computer manufacturers. Different numbers of gates are placed on a chip, with different configuration capabilities. Some gate arrays contain memories, for example.

References

1. Okabe, M. et al., "A 400k-transistor CMOS sea-of-gate array with continuous track allocation," *IEEE J. Solid-State Circuits*, pp. 1280–1286, Oct. 1989.
2. Muroga, S., *VLSI System Design*, John Wiley & Sons, 1982.
3. Price, J.E., "VLSI chip architecture for large computers," in *Hardware and Software Concepts in VLSI*, Edited by G. Rabbat, Van Nostrand Reinhold Co., pp. 95–115, 1983.

47

Field-Programmable Gate Arrays

Saburo Muroga

University of Illinois
at Urbana-Champaign

CONTENTS

47.1 Introduction

Field-programmable gate arrays or **FPGAs** are programmable by users. In other words, users easily and inexpensively realize their own logic networks in hardware, using FPGAs. The change of the logic networks is as easy as software is. FPGAs, however, have a greater variety of hardware and architecture than PLAs or gate arrays. If fuses or anti-fuses are used for hardware programmability, logic functions realized on an FPGA cannot be changed, once realized. But the addition of random-access memory (RAM) to hardware of FPGAs by Freeman [6,7] such that logic functions can be easily changed by changing information stored in the RAM substantially has enhanced the usefulness of FPGAs. By storing information into RAMs of FPGAs, logic functions can be rewritten in a short time as frequently as we need. In this sense, FPGAs with RAMs are not a straightforward extension of the gate arrays in Chapter 46, which are mask-programmable, and are very different from the gate arrays. With the changeability of the logic functions on an FPGA by changing information in its RAMs, the nature of programmability of an FPGA is essentially the same as that of software which is stored in the RAM of a general-purpose computer. With FPGAs, debugging or prototyping of new design can be done as easily and quickly as software. But an FPGA performs much faster than software on computer.

FPGAs have other types of hardware, in addition to those with RAMs. Thus, hardware of FPGAs consists of PLDs, logic gates, random-access memory, and often other types of components such as non-volatile memory. FPGAs from different manufacturers have different organization of PLDs, logic gates, random-access memory, and other types of components. In other words, different manufacturers have FPGAs in different architectures, but all of them have the same common feature: that the layout of a unit is repeated in matrix form. In this case, the unit is a circuit consisting of PLDs, logic gates, random-access memory, and other types of components that is far more complex than "gates" which are repeated in matrix in gate arrays. Logic networks realized in FPGAs are slower by two or three orders of magnitude than those realized in full-custom design, but are much faster by several orders than simulation of logic functions by software. Even application programs can be run on FPGAs and perform much faster than on general-purpose computer in many cases.

As the price of FPGAs goes down with higher speed, FPGAs are replacing other semi-custom design approaches in many applications.

47.2   Basic Structures of FPGAs

In the case of mask-programmable gate arrays, designers have to wait a few weeks for delivery of finished gate arrays from semiconductor manufacturers because the semiconductor manufacturers must prepare custom masks (although the number of custom masks for gate arrays is fewer than the case of the standard-cell library approach described in Chapter 48). With FPGAs, designers can realize their design on FPGA chips by themselves in minutes. Thus, FPGAs are becoming popular [1,2,8–10].

Several different types of structures for FPGAs are available commercially. All of them have a basic structures that consists of many logic blocks or logic cells, accompanied by a large number of pre-laid lines for connecting these logic blocks. So, some manufacturers call FPGAs **logic block arrays** (**LBAs**). One has a structure similar to a gate array with routing channels where each logic cell in a gate array is replaced with a logic block, as shown in Figure 47.1. Another one is similar to sea-of-gate array, as shown in Figure 47.2 illustrated with 16 logic blocks. Also, there is a structure similar to standard cells (to be discussed in the next chapter) where there are routing channels between a pair of rows of logic blocks, as shown in Figure 47.3. There is a structure where outputs of logic blocks are connected to the inputs of other logic blocks through bus lines, as shown in Figure 47.4.

The internal structure of logic blocks or logic cells differs, depending on the manufacturer. A logic block consists of SRAMs (used as look-up tables), PALs, NAND gates, along with multiplexers, flip-flops, and others. Lines are pre-laid horizontally and vertically and are connected to the inputs and outputs of logic blocks by **programmable switches**. Various programmable switches, such as fuses, anti-fuses, RAMs, and non-volatile memories, are provided by different manufacturers. Each line actually consists of many short line segments and only necessary line segments are connected in order not to add unnecessary delay due to parasitic capacitance by using an excessive number of line segments. Line segments are also connected by programmable switches.

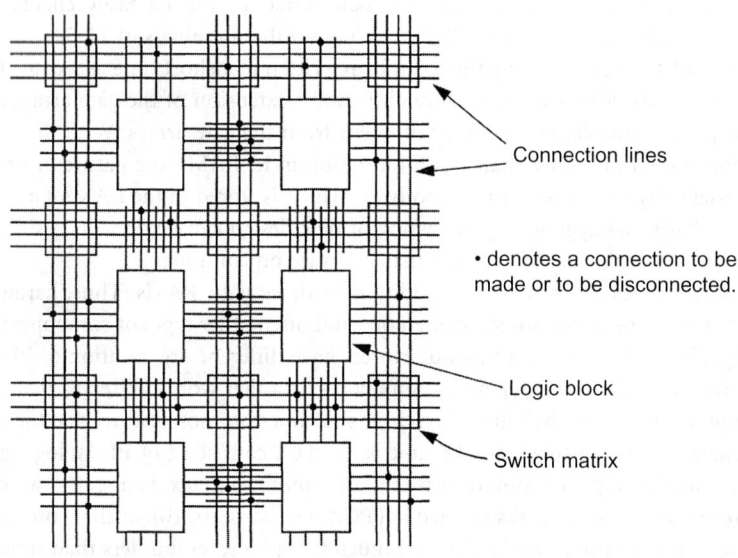

FIGURE 47.1 FPGA type of gate array with routing channels.

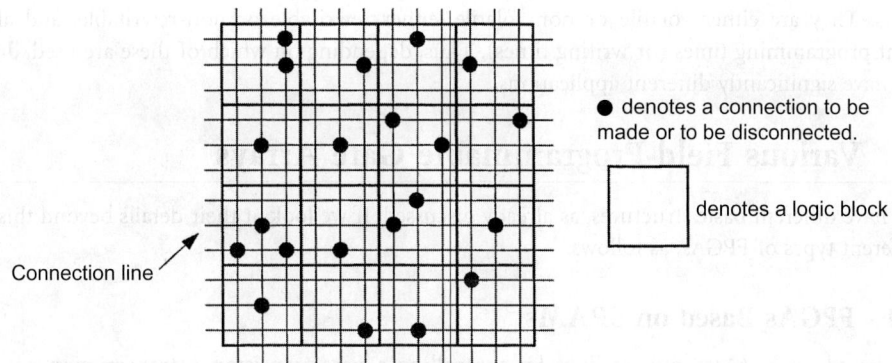

FIGURE 47.2 FPGA type of sea-of-gate array.

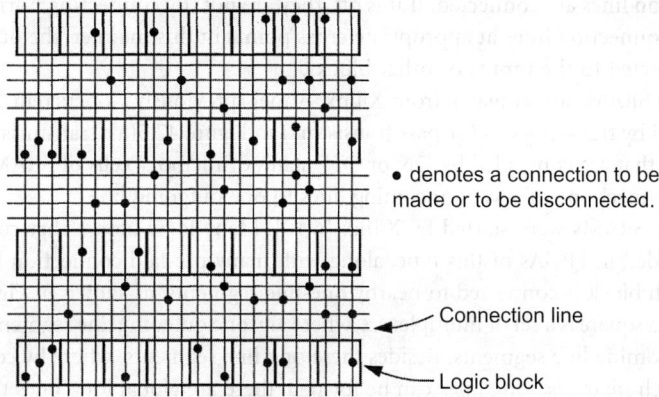

FIGURE 47.3 FPGA type of row-based array.

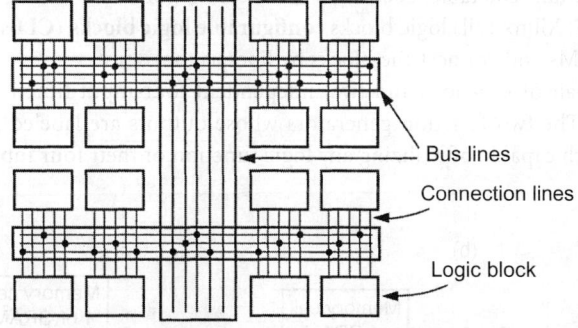

FIGURE 47.4 FPGA type of bus-based array.

In addition to these logic blocks and pre-laid lines, there are different types of input/output control blocks, that is, blocks for inputs for receiving signals from the outside of the FPGA and for outputs for sending signals to the outside.

Each programmable switch consists of many switching elements. A typical FPGA contains hundreds of thousands or more switching elements and how these elements are realized is essential for performance, costs, and size of the FPGA. Fuses, anti-fuses, SRAMs, and non-volatile memories are used as switching elements. They are either volatile or non-volatile, either rewritable or non-rewritable, and all have different programming times (or writing times). Thus, depending on which of these are used, different FPGAs have significantly different applications.

47.3 Various Field-Programmable Gate Arrays

FPGAs have different basic structures, as already discussed. If we look at their details beyond this, there are different types of FPGAs, as follows.

47.3.1 FPGAs Based on SRAMs

FPGAs based on SRAMs connect lines by controlling a pass transistor, a transmission gate, or a multiplexer, each of which is controlled by a flip-flop (i.e., one memory cell of SRAM) as shown in Figure 47.5. Any horizontal line and a vertical line shown in Figure 47.1 can be connected at some of their cross points (shown with dots in Figure 47.1). Suppose a flip-flop is used. If the flip-flop is on, these two connection lines are connected; if it is off, they are not. By connecting horizontal connection lines and vertical connection lines at appropriate cross points in this manner, the outputs of one logic block can be connected to the inputs of other blocks.

FPGAs based on SRAMs are available from Xilinx, Atmel (previously Concurrent Logic), and others. Lines are connected by transfer gates (or pass transistors) in Figure 47.5(a), transmission gates in (b), or multiplexers in (c) that are controlled by ON or OFF states of memory cells of SRAM. A multiplexer is typically used to connect one of several incoming lines to one outgoing line.

FPGAs based on SRAMs were started by Xilinx [6,10]. Many logic blocks that contain SRAMs and flip-flops are provided in FPGAs of this type, along with many pre-laid connection lines, as illustrated in Figure 47.1. Each block is connected to nearby one-line segments, as shown in Figure 47.6. A switch matrix shown with a square is a set of multiplexers, where any one outgoing line segment can be connected to any of many incoming line segments. Besides these one-line segments, where by connecting many of them through switch matrices, long lines can be formed, there are global long lines that can be formed by connecting fewer segments, each of which has longer segment length. (Note that a long line formed by connecting many one-line segments has a long delay because delays of many switch matrices are added to the delay of the lines due to parasitic capacitance.) In this sense, delay times on lines that consist of many line segments are unpredictable. Each logic block consists of a RAM along with several flip-flops, as shown in Figure 47.7. Xilinx calls logic blocks **configurable logic blocks (CLBs)**. Users can store logic functions in these SRAMs and connect the blocks by lines as they want.

Each CLB packs a pair of flip-flops and two independent four-input function generators, as illustrated in Figure 47.7. The two function generators whose outputs are labeled F′ and G′ are realized with RAMs and are each capable of realizing any logic function of their four inputs, offering designers

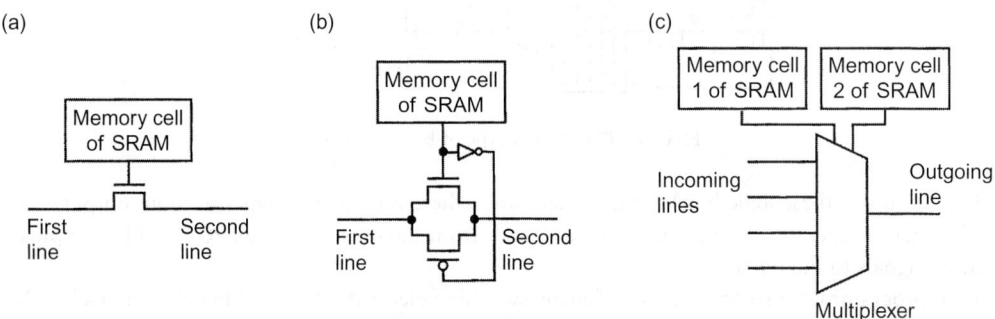

FIGURE 47.5 Connection of lines controlled by SRAM.

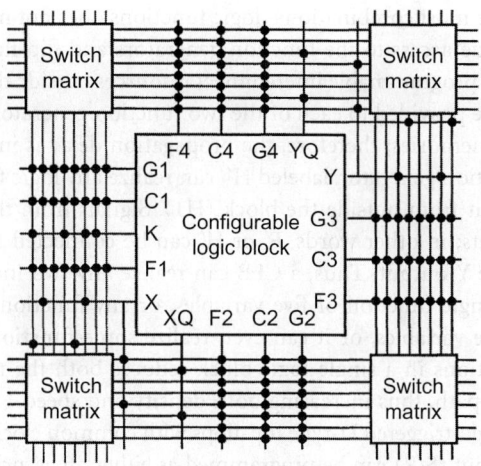

FIGURE 47.6 Typical connection of a configurable logic block to single line sections.

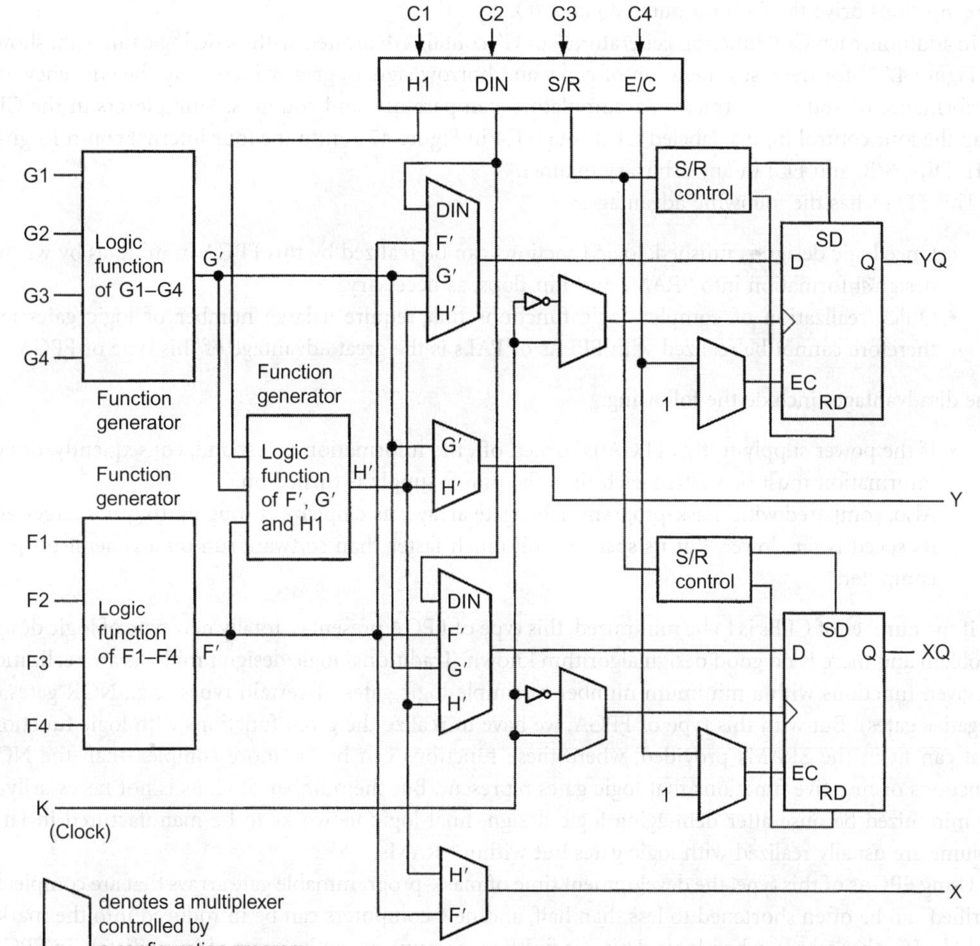

C1   C2   C3   C4

H1   DIN   S/R   E/C

G1
G2   Logic
        function   G'
G3   of G1–G4
G4
        Function
        generator

        Function   Logic
        generator   function   H'
                      of F', G'
                      and H1
F1
F2   Logic
        function   F'
F3   of F1–F4
F4

K
(Clock)

DIN
F'
G'
H'

G'
H'

DIN
F'
G'
H'

H'
F'

S/R
control

SD
D   Q — YQ

EC
RD

Y

S/R
control

SD
D   Q — XQ

EC
RD

X

denotes a multiplexer
controlled by
a configuration program

FIGURE 47.7 Simplified block diagram of configurable logic block of XC4000, FPGA of Xilinx, Inc.

sufficient flexibility, because most combinational logic functions does not need more than four inputs. Thirteen CLB inputs provide access to the function generators and flip-flops. These inputs and four CLB outputs connect to the programmable interconnect resources outside the block. Four independent inputs, F1–F4 or G1–G4, are provided to each of the two function generators. The function generators are used as look-up table memories; therefore, the propagation delay is independent of the function being realized. A third function generator, labeled H′, can realize any logic function of its three inputs, F′ and G′, and a third input from outside the block (H1). Signals from the function generators can exit the CLB on two outputs; in other words, F′ or H′ can be connected to the X output, and G′ or H′ can be connected to the Y output. Thus, a CLB can realize any two independent functions of up to four variables, or any single function of five variables, or any function of four variables together with some functions of five variables, or it can even realize some functions of up to nine variables. Realizing a variety of functions in a single logic block reduces both the number of blocks required and the delay in the signal path, thus increasing both density and speed.

The CLB contains also edge-triggered D-type flip-flops with common clock (K) and clock enable (EC) inputs. A third common input (S/R) can be programmed as either an asynchronous set- or reset-signal independently for each of the two flip-flops; this input can also be disabled for either flip-flop. Each flip-flop can be triggered on either the rising or falling clock edge. The source of a flip-flop data input is programmable; it is driven either by the functions F′, G′, and H′, or the Direct In (DIN) block input. The flip-flops drive the CLB outputs, XQ and YQ.

In addition, each CLB function generator, F′ or G′, contains dedicated arithmetic/logic unit (not shown in Figure 47.7) for the fast generation of carry and borrow signals, greatly increasing the efficiency and performance of adders, subtracters, accumulators, comparators, and counters. Multiplexers in the CLB map the four control inputs, labeled C1 through C4 in Figure 47.7, into the four internal control signals (H1, DIN, S/R, and EC) in any arbitrary manner.

This FPGA has the following advantages:

- Once logic design is finished, logic functions can be realized by this FPGA in minutes by writing design information into SRAMs and flip-flops, as necessary.
- Quick realization of complex logic functions that require a large number of logic gates and therefore cannot be realized with FPLAs or PALs is the great advantage of this type of FPGA.

The disadvantages include the following:

- If the power supply to this FPGA is turned off, the information is lost and, consequently, design information must be written each time the power supply is turned on.
- Also, compared with mask-programmable gate arrays, its chip size is roughly 10 times larger and its speed is far slower. But its speed is still much faster than software run on a general-purpose computer.

If the number of CLBs is to be minimized, this type of FPGA presents a totally new type of logic design problem and there is no good design algorithm known. Traditional logic design problems are realization of given functions with a minimum number of simple logic gates of certain types (e.g., NOR gates or negative gates). But with this type of FPGA, we have to realize the given functions with logic functions that can fit in the SRAMs provided, where these functions can be far more complex than the NOR functions or negative functions that logic gates represent. But the number of CLBs is not necessarily to be minimized because after debugging logic design, final logic networks to be manufactured in large volume are usually realized with logic gates but without RAMs.

Using FPGAs of this type, the development time of mask-programmable gate arrays that are completely verified can be often shortened to less than half, and new computers can be introduced into the market quickly. In other words, when logic design is finished, computers can be shipped immediately in FPGAs to major customers for testing. After these customers find bugs, the designers fix them and the production of mask-programmable gate arrays can be started. Without FPGAs, the computers have to be shipped to customers for testing after manufacturing mask-programmable gate arrays—which takes several weeks. Then, when the customers find bugs, the designers start the production of the second mask-programmable gate arrays. Thus, when mask-programmable gate arrays are used, a new gate array must be manufactured, spending a few weeks, for each correction of bugs and testing by customers. The completion of the computer will be significantly delayed and with far greater expense. If the design is complex, the designers need repetition of tests by the customers. In contrast to this, if FPGAs are used, the designers can send the debugged design to the customers online, and customers can start the second test instantly by revising the contents of the FPGAs.

Atmel's FPGA uses SRAMs, but logic blocks and architecture are similar to Actel's, unlike Xilinx's whose logic blocks contain SRAMs which are used as look-up tables. FPGAs of Plessey are based on SRAMs, but the architecture looks like the sea-of-gate array shown in Figure 47.2. Altera also has FPGAs based on SRAMs, using the architecture with non-volatile memory [10]. FPGAs of Algotronix use SRAMs, but logic blocks are mostly multiplexers. AT&T Microelectronics also has FPGAs based on SRAMs similar to Xilinx's.

47.3.2   FPGAs Based on ROMs with Anti-Fuses

There are FPGAs based on ROMs that are programmable with anti-fuses. These FPGAs are available from Actel, QuickLogic, and Crosspoint Solutions. Connection of lines in this type of FPGA can be made by the use of anti-fuses. Once an anti-fuse becomes conductive, it cannot be non-conductive again, unlike FPGAs based on SRAMs. Typical applications utilizing 85% of the available logic gates, however, need to use only 2 to 3% of these anti-fuses (i.e., to be changed to conductive state). Note that if fuses, instead of antifuses, are used, 97% or 98% of fuses need to be disconnected, requiring disconnection of a huge number of fuses.

Actel's FPGAs are of row-based structure, as illustrated in Figure 47.3 [3–5,10]. The actual logic block array of Actel is shown in Figure 47.8. A logic block (Actel calls it a logic module) for a simple model of Actel's FPGA consists of three multiplexers and one OR gate, as shown in Figure 47.9, where some of their inputs and outputs can be complemented. This logic module, which has eight inputs and a single output, can realize many different logic gates with some inputs inverted, as follows: four basic logic gates

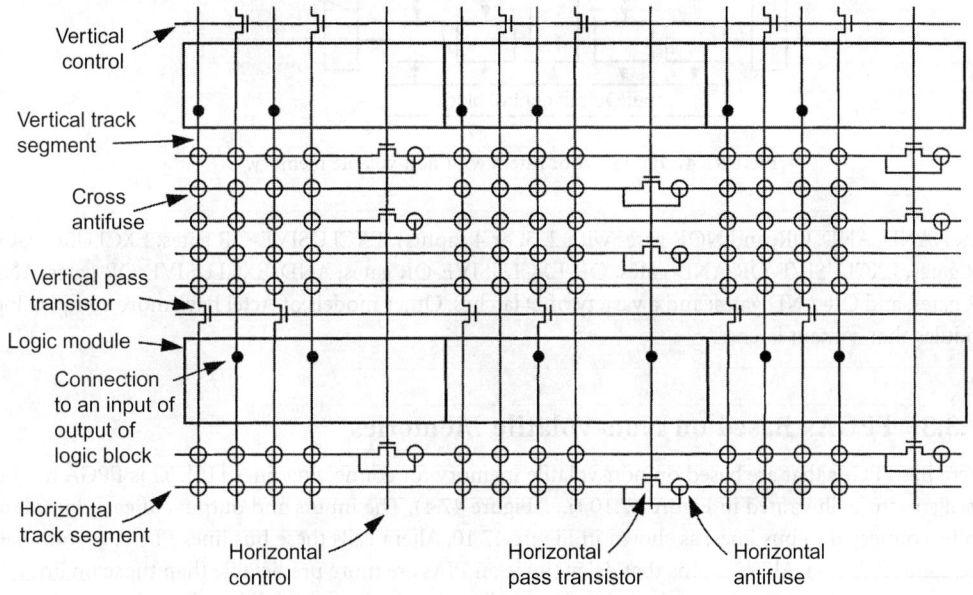

FIGURE 47.8 Logic module array of Actel.

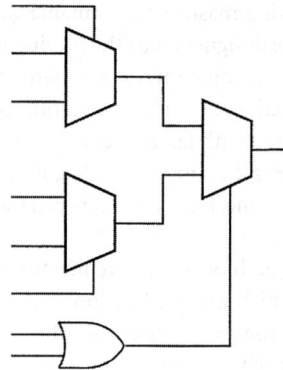

FIGURE 47.9 Logic block of Actel.

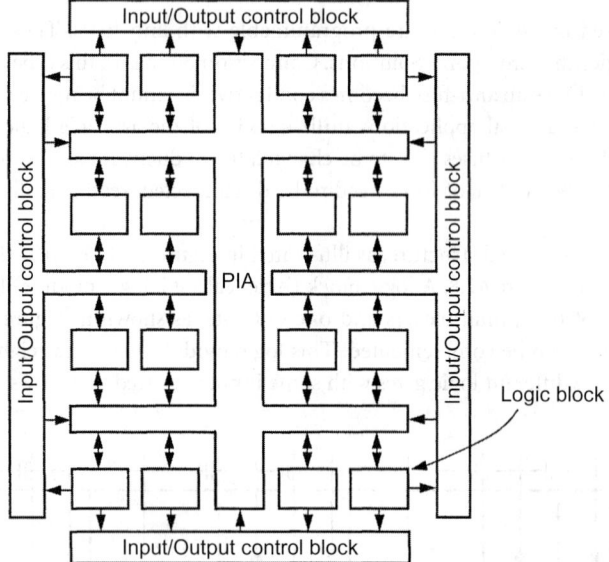

FIGURE 47.10 FPGA of Altera with non-volatile memory.

(i.e., NAND, AND, OR, and NOR gates with 2, 3, or 4 inputs); EXCLUSIVE-OR gates, EXCLUSIVE-OR-OR gates, EXCLUSIVE-OR-AND gates, OR-EXCLUSIVE-OR gates, AND-EXCLUSIVE-OR gates, AND-OR gates, and OR-AND gates; and a variety of D latches. Other models of Actel have more complex logic modules that contain latches.

47.3.3 FPGAs Based on Non-Volatile Memories

Altera has FPGAs that are based on non-volatile memory for connecting lines [10]. This FPGA has bus-based structure illustrated in Figure 47.10 (i.e., Figure 47.4). The inputs and outputs of each logic block can be connected to bus lines, as shown in Figure 47.10. Altera calls these bus lines PIA (Programmable Interconnect Array). Altera claims that delay times on PIAs are more predictable than those on lines that consist of many line segments. A logic block contains of 16 logic macrocells, such as the one shown in Figure 47.11.

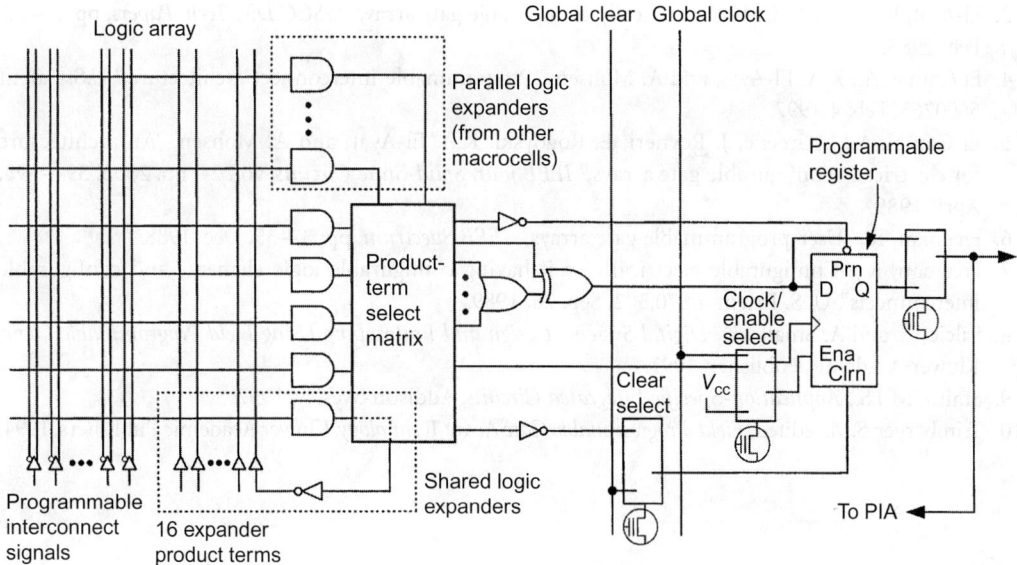

FIGURE 47.11 Macrocell of FPGA of Altera.

FPGAs of Advanced Micro Devices and Lattice Semiconductor Corp. are based on a different type of non-volatile memory, using PALs as logic blocks that can be connected with switch matrixes similar to Altera's.

47.4 Features of FPGAs

If we order semiconductor manufacturers to make mask-programmable gate arrays, we have to wait several weeks and pay twenty-thousand to hundreds of thousands of dollars. But with FPGAs, we can program FPGAs in minutes by ourselves and need to pay in the tens of dollars. But a FPGA can pack only about one-tenth the number of logic gates in a mask-programmable gate array because devices for user programmability, such as SRAMs, non-volatile memory, and anti-fuses, take up large areas. Mask-programmable gate arrays are faster by two orders of magnitude and far cheaper for large production volume. Thus, for debugging or verifying logic design that needs to be done quickly, FPGAs are used, and then mask-programmable gate arrays are used for large volume production after completing debugging or verification.

As seen already, FPGAs are classified into two types, depending on the types of devices used for programmability: rewritable FPGAs and non-rewritable FPGAs. Then, they are accordingly used for completely different purposes. Rewritable FPGAs, such as Xilinx's based on SRAMs and Altera's based on non-volatile memory, can be repeatedly rewritten in minutes. Non-rewritable FPGAs cannot be changed once programmed, but still have the advantages of realizing inexpensive logic chips with faster speed.

The area size of different devices for programmability also gives different advantages and disadvantages. A non-volatile memory cell is roughly four to five times larger than anti-fuse; a memory cell of SRAM is two times larger than a non-volatile memory cell. Anti-fuses are much smaller. So, because of smaller parasitic capacitance, FPGAs based on anti-fuses tend to be faster than those based on non-volatile memory or SRAMs.

References

1. Brown, S.D., et al., *Field-Programmable Gate Arrays*, Kluwer Academic Publishers, 1992.
2. Chan, P.K. and S. Mourad, *Digital Design Using Field Programmable Gate Arrays*, Prentice-Hall, 1994.

3. El-Ayat, K., et al., "A CMOS electrically configurable gate array," *ISSCC Dig. Tech. Papers*, pp. 76–77, Feb. 1988.

4. El Gamal, A., K.A. El-Ayat, and A. Mohsen. "Programmable interconnect architecture," U.S. patent 5600265, Feb. 4 1997.

5. El Gamal, A., J. Greene, J. Reyneri, E. Rogoyski, K.A. El-Ayat, and A. Mohsen, "An architecture for electrically configurable gate arrays," *IEEE Jour. Solid-State Circuits*, vol. 24, no. 2, pp. 394–398, April 1989.

6. Freeman, R., "User-programmable gate arrays," *IEEE Spectrum*, pp. 32–35, Dec. 1988.

7. Freeman, R., "Configurable electrical circuit having configurable logic elements and configurable interconnects", U. S. Patent 4,870,302, Sep. 26, 1989.

8. Salcic, Z. and A. Smailagic, *Digital Systems Design and Prototyping Using Field Programmable Logic*, Kluwer Academic Publisher, 1997.

9. Smith, M.J.S., *Application-Specific Integrated Circuits*, Addison-Wesley, 1997.

10. Trimberger, S.M., edited, *Field-Programmable Gate Array Technology*, Kluwer Academic Publishers, 1994.

48

Cell-Library Design Approach

Saburo Muroga
*University of Illinois
at Urbana-Champaign*

CONTENTS

48.1 Introduction

Compact layouts for basic logic gates (such as NOR gates) and small networks that are frequently used (such as flip-flops and a full adder) are designed by professional designers with time-consuming efforts and are stored in computer memories (i.e., magnetic disks). Each layout is called a **cell** and a collection of these layouts is called a **cell library**. Once a cell library is ready, any other designer can call up specific cells on the monitor of a computer. By arranging them and adding connections among them by the use of a mouse, the designer can make a layout of the entire logic network. When the layout is complete, photomasks are automatically prepared by computer. Such a design approach is called the **cell-library design approach** [1–3].

48.2 Polycell Design Approach

When the layout of every cell has the same height, although its width may be different, this approach is called the **polycell design approach** (or **standard-cell design approach**). It is often called the cell-library design approach, but it is also called the polycell or standard-cell design approach in order to avoid confusion with the words "cell" and "library" from those which are used in the gate arrays. The height of all cells are chosen the same, even though area is wasted inside a cell, such that many cells are connected into rows, as exemplified in Figure 48.1, which shows an example with three rows of cells.

Routing (i.e., connections among cells) is mainly done in space between adjacent rows of cells, as shown in Figure 48.1. This space for routing is called a **routing channel**. Routing, in addition to design of logic networks, is the major task in the polycell design approach and is greatly facilitated by CAD programs.

Connections from one routing channel to others can be done through narrow passages, which are provided between cells in each row. In some cases, connections are allowed to go through cells horizontally or vertically (without going through the narrow passages provided) in order to reduce wiring areas. In this case, detailed routing rules on where in each cell connections are allowed to go through must be incorporated in the CAD programs.

The preparation of a cell library takes time (probably more than a dozen man-months for a small library), but layout time of chips, using a cell library, is greatly reduced by the polycell design approach

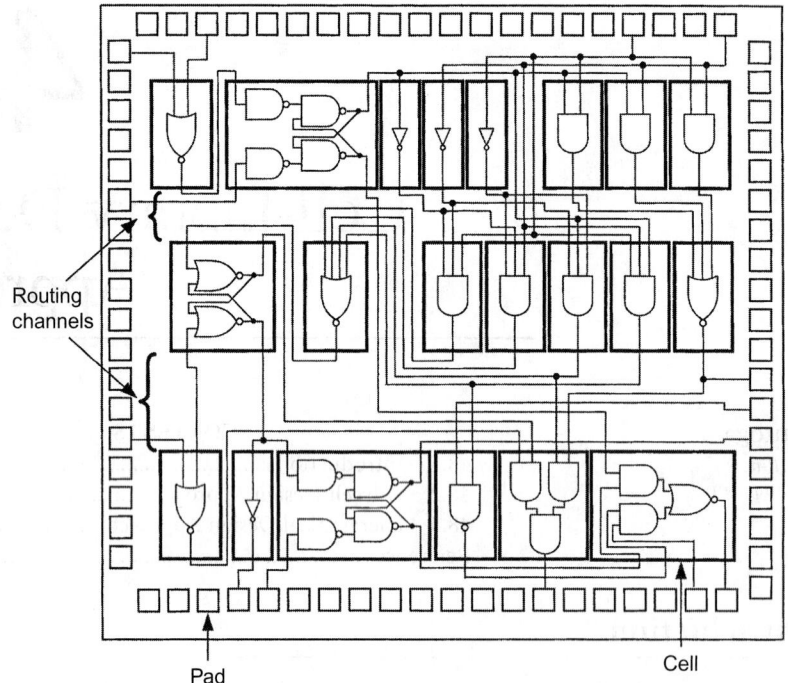

FIGURE 48.1   Polycell layout.

(probably by an order of magnitude), compared with full-custom design which requires the most compact layout of gates and connections. This is a great advantage. When new fabrication technology is to be adopted, only cells have to be newly laid out; whereas in the case of full-custom design, the entire chip has to be laid out compactly with new technology. Thus, the cell library approach can adapt to a new technology in a shorter time than the full-custom design, but chip size and performance are sacrificed. Compared with the full-custom layout of random-logic gate networks, chip size is many times larger, although it is much smaller than the size realized with a gate array. In order to facilitate routing among cells laid out, all cells are prepared with equal height, keeping their inputs and outputs at the sides that face routing channels. Then, cells for simple functions become very thin and those for complex functions become squat. Thus, the area in each cell is not efficiently utilized. Also, interconnections among cells take up a significant share of the chip area.

Unlike gate arrays where only masks for connections and contact windows are to be custom-made, here all masks must be custom-made, so the polycell design approach needs expensive initial investment, despite its advantage of smaller chip size than gate arrays. However, since the layout is highly automated with interactive CAD programs for placement and routing, the layout of the entire chip is not as time-consuming as those by full-custom design, and consequently the polycell design approach is favored in many cases. In this sense, the polycell design approach is not as purely semi-custom design as the gate array design approach. But the approach is not purely full-custom design either, since the chips are not as compact as those by the full-custom design. Thus, it might be called a pseudo-full-custom design approach, although some people call it a full-custom design approach.

The polycell design approach is cost-effective when the production volume is in hundreds of thousands of dollars, since the initial investment costs $20,000 or more, which is much lower than tens of millions dollars for the full-custom design. The polycell design approaches have been extensively used when the chip compactness attained by the full-custom design approach is not required.

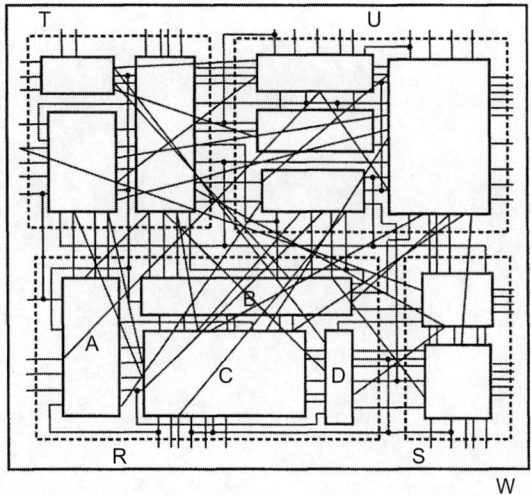

FIGURE 48.2 Hierarchical design approach.

48.3 Hierarchical Design Approach

The cell library design approaches, using cells of different shapes and sizes, can reduce the chip size more than the polycell design approach, because by keeping the same height, a large portion of the area of each cell is wasted, and by keeping all connections among cells in routing channels, the connection area may not be minimized. Moreover, by using a **hierarchical approach** based on cells of different shapes and sizes—in other words, by treating many cells as a building block in a higher level, and many such building blocks as a building block in a next higher level, and so on—we can further reduce the chip area, as illustrated in Figure 48.2, because global area minimization can be treated better, even though this is done on the monitor. In other words, cells A, B, C, and D are assembled into a block R (shown in a dot-lined rectangle), as shown in Figure 48.2. Then, such blocks, R, S, T and U, shown in dot-lined rectangles are assembled into a bigger block W, which is a block in a higher level than blocks R, S, T, and U, as shown in Figure 48.2. But this is much more time-consuming than the polycell design approach, and the development of efficient CAD programs is harder. It appears to be difficult to make the difference of chip area from full-custom designed chips within about 20%, although the areas of full-custom designed chips vary greatly with designers and, accordingly, comparison is not simple.

References

1. Lauther, U., "Cell based VLSI design system," in *Hardware and Software Concepts in VLSI*, Ed. by G. Rabbat, Van Nostrand Reinhold, pp. 480–494, 1983.
2. Kick, B. et al. "Standard-cell-based design methodology for high-performance support chips," *IBM Jour. Res. Dev.*, pp. 505–514, July/Sept. 1997.
3. Muroga, S., *VLSI System Design*, John Wiley & Sons, 1982.

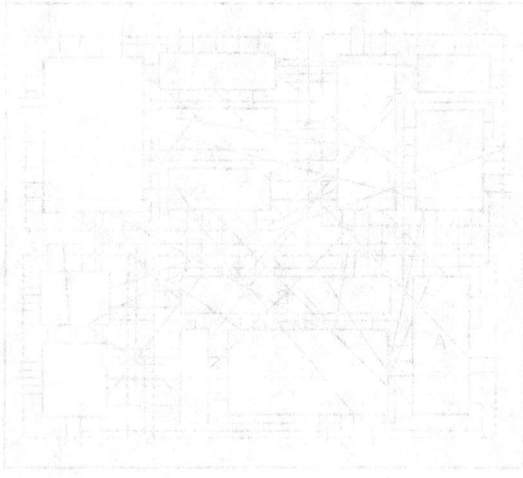

49

Comparison of Different Design Approaches

Saburo Muroga

University of Illinois
at Urbana-Champaign

49.1 Introduction

As discussed so far, there is an almost continuous spectrum of different design approaches from design approaches with off-the-shelf packages, semi-custom design approaches, to the full-custom design approach, depending on the degree of the regularity of transistor arrangement in the layout and the simplification of realization of logic functions (i.e., whether logic functions are realized by a minimal sum or a logic network). Naturally, comparison of these approaches by many different criteria such as performance, design time, and chip size is very complex. Also, different design approaches are often mixed on the same chip, further complicating the comparison. Reliable comparison data are rarely available. But here we will try to give some idea of the advantages and disadvantages of the different design approaches [1–4].

49.2 Design Approaches with Off-the-Shelf Packages

Here, let us compare different design approaches from the viewpoint of design of logic networks or digital systems. This comparison includes off-the-shelf packages (where a package means a component or an IC chip encased in a container). But among all off-the-shelf packages, let us compare only off-the-shelf discrete component packages and off-the-shelf IC packages (many commonly used logic functions are commercially available as off-the-shelf IC packages), excluding off-the-shelf processors that are fully programmable. We exclude off-the-shelf fully programmable processor packages, such as microprocessor or microcontroller packages, because these off-the-shelf packages are essentially general-purpose computers that are used by writing appropriate software and accordingly do not require design of logic networks.

The major advantage of the off-the-shelf packages is the ease of partial changes of the entire computer or replacement of faulty parts. But there are many disadvantages, such as low reliability (due to connections outside these packages), high cost, and bulkiness. When off-the-shelf packages are assembled on pc boards and further into cabinets, the overall system costs make a substantial difference because of additional costs. In the case of custom-designed IC packages of large integration size (the integration size of an integrated-circuit chip means the number of transistors packed in a chip), all additional costs, such as those for pc boards and fans, become zero or insignificant, but we still have to consider test costs, which are even higher because of more stringent test requirements. Thus, custom-designed IC packages of large integration size are cost-effective for high volume production, which justifies high initial investment.

49.3   Full- and Semi-Custom Design Approaches

An approximate comparison of different design approaches in terms of design time and chip area is given in Figure 49.1 in logarithmic scale. For each design situation, designers must choose the most appropriate approach, considering tradeoffs between design time (which is closely related to design cost) and chip area (which is related to manufacturing cost and performance). In this comparison, a design approach in a higher position in Figure 49.1 takes less time to finish the design, but the finished chip is larger and slower in speed than those in lower positions.

Logic functions can be realized in ROMs (read-only-memory) as a truth table. As the number of variables increases, the required memory size exponentially increases. So, its use is limited.

Among all custom (semi- and full-custom) design approaches, PLAs are the quickest to realize logic functions because we can do so simply by deriving minimum sums or minimum products with the use of CAD (computer-aided design) programs, skipping designing logic networks, and also layout and customization of many masks (where manufacturing cost is affected by how many masks need to be customized). But performance, size, and cost are sacrificed. PLAs require only one custom-made mask for connections, while the other custom design approaches require two or more custom-made masks among all of about two dozen required masks. Field-programmable PLAs have larger chip areas than mask-programmable ones, although no custom masks are required.

In using gate arrays, users need to design logic networks but can realize logic networks by the layout of connections among logic gates with CAD programs. Layout and customization of many masks can be skipped. Performance is much higher than PLAs. Gate arrays require two custom-made masks for connections among all of about two dozens masks. Field-programmable gate arrays have a variety. Among them, non-rewritable FPGAs have larger chip areas than mask-programmable ones, although no custom masks are required. Rewritable FPGAs (e.g., those with RAMs) are so different from the other custom design approaches that they cannot be properly compared in Figure 49.1.

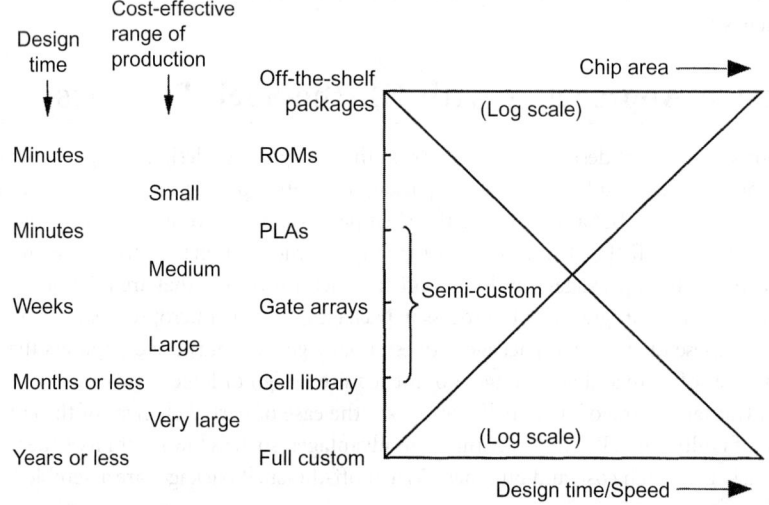

FIGURE 49.1   Chip area versus design time.

The cell-library design approach requires logic design and layout of connections with CAD, using a cell library which requires a one-time design effort. But the layout of logic gates and connections inside cells and layout of transistor circuits can be skipped, though all masks need to be customized. All of about two dozen masks need to be custom-made, making the cell-library design approach more expensive than gate arrays and PLAs.

The full-custom design approach requires deliberate design of all aspects of a digital system, although semi-custom design approaches along with appropriate CAD programs are used wherever speed is not critical for the speed of the entire system or frequent changes are expected, in the entire system. For example, PLAs are used in the control unit of the full-custom design system. All masks need to be custom-made. The full-custom design approach is the most expensive, taking the longest design time, but the approach yields chips with the highest performance and the lowest cost for high volume production.

A crude estimation of design time is of the order of minutes for ROMs and PLAs, weeks for gate arrays, months for cell-library approach, and years for full-custom design approaches, with the appropriate number of engineers assigned in each case, as shown in Figure 49.1. A crude estimation of the cost-effective range of production, as shown in Figure 49.1, is small volume production for the upper range of the cost-effective production of ROMs (i.e., the lower range of the cost-effective production of PLAs), and so on. And the full-custom design approach cannot be more cost-effective than the cell-library approach, unless the production volume is very high.

Rewritable FPGAs have a unique feature that is very different from the other custom design approaches: the ease of changing information in the memories of the FPGAs. With this feature, a designer can send new information to customers over communication lines, instead of sending hardware. So, the customers can change the information on their FPGAs instantly and very inexpensively. Thus, rewritable FPGAs are replacing other custom design approaches in many applications.

An approximate relationship between cost per package and production volume is illustrated in Figure 49.2, although this may change depending on many factors, such as fabrication technology, logic families, system size, and performance. For each production volume, there is the most economical design approach. But the comparison is difficult and has to be done carefully in each case, because each approach

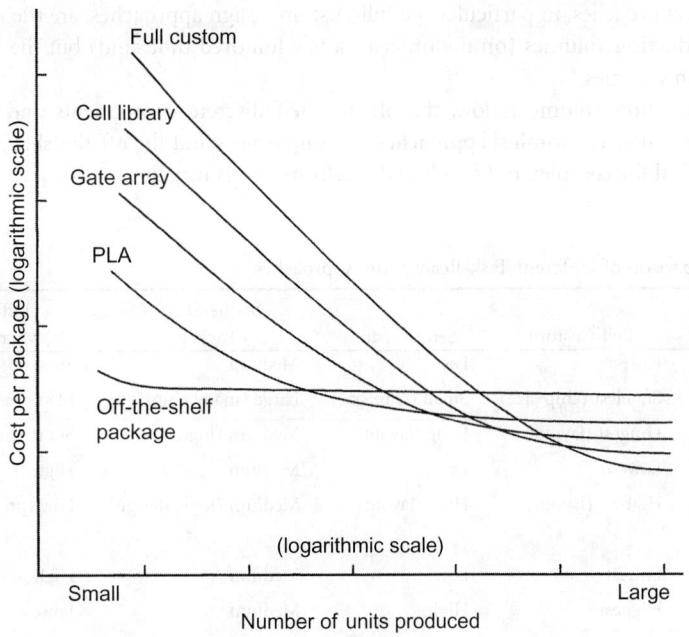

FIGURE 49.2 Package cost versus production volume.

has variations and it makes a difference whether or not libraries of cells or macrocells are prepared from scratch. (Notice that in Figure 49.2, design approaches are shown in thin-line curves for the sake of simplicity, but actually they should be represented in very broad lines.) The cost per package for the off-the-shelf package design approach is fairly uniform over the entire range, but it increases for low production volumes because the development cost becomes significant as initial investment in the overall package cost. The relationship shown in this figure will change as the integration size of an IC chip increases, because the dependence on CAD will inevitably increase.

49.4 Comparison of All Different Design Approaches

As discussed so far, we have a very wide spectrum of different design approaches, from full-custom design approaches to the design approaches with off-the-shelf packages, as illustrated in Table 49.1. Digital systems can be designed by combining them. Depending upon different criteria imposed by different design motivations, such as speed, power consumption, size, design time, ease of changes, and reliability, designers can use the following approaches:

1. Custom-design full- and semi-custom approaches
2. Off-the-shelf discrete components and off-the-shelf IC packages, along with memory packages
3. Off-the-shelf microcomputers along with off-the-shelf IC packages

The full-custom design approaches give us the highest performance and reliability or the smallest chip size, although they are most time-consuming. (Even in the case of microcomputers, the full-custom designed microcomputers have better performance and smaller size than off-the-shelf microcomputers, by being tailored to the users' specific needs.) This is one end of the wide spectrum of different design approaches. At the other end, the off-the-shelf microcomputers give us a design approach where the development time is shortest, by programming rather than by chip design (including logic design), and the design changes are the easiest. The off-the-shelf discrete components and off-the-shelf IC packages give us logic networks tailored to specific needs with less programming than the off-the-shelf microcomputers.

Custom design approaches, in particular the full-custom design approaches, are the most economical for very high production volumes (on the order of a few hundred thousand) but the least economical for low production volumes.

When the production volume is low, the off-the-shelf discrete components and off-the-shelf IC packages give us the most economical approaches for simple tasks, but the off-the-shelf microcomputers are more economical for complex tasks, although performance is usually sacrificed.

TABLE 49.1   Comparison of Different Task-Realization Approaches

	Full-Custom	Semi-Custom	Off-the-Shelf IC Package	Off-the-Shelf Microcomputer
Speed	Fastest	Fast	Medium	Slowest
Size	Smallest (chip size)	Small (chip size)	Large (many chips)	Medium (many chips)
Development time	Longest (layout)	Long (layout)	Medium (logic design)	Short (programming)
Flexibility	Lowest	Low	Medium	High
Initial investment	Highest (layout)	High (layout)	Medium (logic design)	Low (programming)
Unit Cost				
High volume	Lowest	Low	Medium	Highest
Low volume	Highest	High	Medium	Lowest
Reliability	Highest	High	Low	Medium

References

1. Fey, C.F. and D.E. Paraskevopoulos, "A techno-economic assessment of application-specific integrated circuits: current and future trends," *Proc. IEEE*, pp. 829–841, June 1987.
2. Fey, C.F. and D.E. Paraskevopoulos, "Economic aspects of technology selection: Level of integration, design productivity, and development schedules," in *VLSI Handbook*, Ed. by J. Di Jacomo, McGraw-Hill, pp. 25.3–25.27, 1989.
3. Fey, C.F. and D.E. Paraskevopoulos, "Economic aspects of technology selection: Costs and risks," in *VLSI Handbook*, Ed. by J. Di Jacomo, McGraw-Hill, pp. 26.1–26.21, 1989.
4. Muroga, S., *VLSI System Design*, John Wiley & Sons, 1982.

Section VI

Memory, Registers and System Timing

Bing J. Sheu
Taiwan Semiconductor Manufacturing Company

0-8493-XXXX-X/04/$0.00+$1.50
© 2006 by CRC Press LLC

50

System Timing

Baris Taskin
Drexel University

Ivan S. Kourtev
University of Pittsburgh

Eby G. Friedman
University of Rochester

CONTENTS

50.1 Introduction

The concept of *data* or *information* processing arises in a variety of fields. Understanding the principles behind this concept is fundamental to computer design, communications, manufacturing process control, biomedical engineering, and an increasingly large number of other areas of technology and science. It is impossible to imagine modern life without computers for generating, analyzing, and retrieving large amounts of information, as well as for communicating information to end users regardless of their location.

Technologies for designing and building microelectronics-based computational equipment have been steadily advancing ever since the first commercial *discrete integrated circuits* were introduced in the late 1950s [1].* As predicted by *Moore's law* in the 1960s [2], integrated circuit (IC) densities have been doubling

---

*Monolithic ICs were introduced in the 1960s.

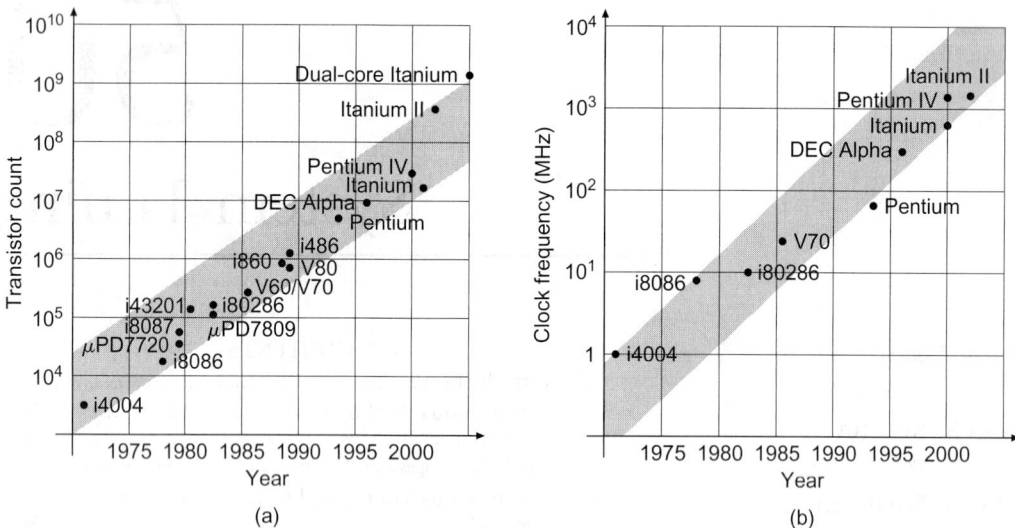

FIGURE 50.1 Moore's law—exponential increase in circuit integration and clock frequency. (a) Evolution of the number of transistors per IC. (b) Evolution of clock frequency.

approximately every 18 months, and this doubling in size has been accompanied by a similar exponential increase in circuit speed (or more precisely, clock frequency). These trends of steadily increasing
circuit size and clock frequency are illustrated in Figure 50.1(a) and Figure 50.1(b), respectively. As a
result of this revolution in semiconductor technology, it is not unusual for modern integrated circuits
to contain hundreds of millions of switching elements (i.e., transistors) packed into a chip area as
large as 500 mm² [3–6]. Such technological capability is due to advances in both design methodologies
and physical manufacturing technologies. Research and experience demonstrate that this trend of
exponentially increasing integrated circuit-based computational power will continue into the foreseeable future.

Integrated circuit performance is typically characterized [7] by the *speed of operation,* the available
circuit functionality, and the *power consumption,* and there are multiple factors which directly affect these
performance characteristics. While each of these factors is significant, on the technological side, increased
circuit performance has been largely achieved by the following approaches:

- Reduction in feature size (technology scaling), that is, the capability of manufacturing physically
  smaller and faster device structures
- Increase in chip area, permitting a larger number of circuits and therefore greater on-chip
  functionality
- Advances in packaging technology, permitting the increasing volume of data traffic between an
  integrated circuit and its environment as well as the efficient removal of heat generated during
  circuit operation

The most complex integrated circuits are referred to as very large scale integration (VLSI) circuits. This
term describes the complexity of modern integrated circuits consisting of hundreds of thousands to many
millions of active transistor elements. Presently, the leading integrated circuit manufacturers have a technological capability for the mass production of VLSI circuits with feature sizes as small as 65 nm [8]. These
technologies are identified with the terms *nanometer* or *very deep submicrometer* (VDSM) technologies.

As these dramatic advances in fabrication technologies take place, integrated circuit performance is
often limited by effects closely related to the very reasons behind these advances such as small geometry
interconnect structures. Circuit performance becomes strongly dependent and limited by electrical issues
that are particularly significant in deep submicrometer integrated circuits. *Signal delay* and related

waveform effects are among those phenomena that have great impact on high performance integrated circuit design methodologies and the resulting system implementation. In the case of fully synchronous VLSI systems, these effects have the potential to create catastrophic failures due to the limited time available for signal propagation between logic gates.

Specifically, in Section 50.2, general timing and operational properties of synchronous circuits are presented. In Section 50.3, the modeling of synchronous circuit components (suitable for computer manipulation) is presented. Also in Section 50.3, the impact of the clock distribution network on circuit timing is described. In Section 50.4, system timing is analyzed. First, system timing properties of edge-triggered and level-sensitive circuits are analyzed. Then, clock skew scheduling methodologies for both types of circuit structures are described. Last, the limitations to improvements in circuit performance achievable through clock skew scheduling are presented. The section is finalized with an appendix containing a glossary of the many terms used throughout this chapter.

50.2   Synchronous VLSI Systems

Owing to the relative simplicity in the design process, the analysis and optimization of VLSI circuits are generally based on logic components operating under a fully synchronous synchronization scheme. In the following sections, these design concepts are briefly reviewed and related fundamental properties are identified. The operational components of VLSI systems relevant to system timing are highlighted.

50.2.1   General Overview

Typically, a digital VLSI system performs a complex computational algorithm, such as a fast Fourier transform or a RISC[†] architecture microprocessor. Although modern VLSI systems contain a large number of components, these systems normally employ only a limited number of different kinds of *logic elements* or *logic gates*. Each logic element accepts certain input signals and computes an output signal for use by other logic elements. At the logic level of abstraction, a VLSI system is a *network* of hundreds of thousands or more logic gates whose terminals are *interconnected* by wires to implement a target algorithm.

The switching variables acting as inputs and outputs of a logic gate in a VLSI system are represented by tangible physical quantities,[‡] while a number of these devices are interconnected to yield the desired function of each logic gate. The specific physical characteristics are collectively summarized with the term *technology*, encompassing details such as the type and behavior of the devices that can be built, the number and sequence of manufacturing steps, and the impedance of the different interconnect materials. Today, several technologies make possible the implementation of high-performance VLSI systems—these technologies are best exemplified by CMOS, bipolar, BiCMOS, and gallium arsenide [9,10]. CMOS technology, in particular, exhibits many desirable performance characteristics, such as low power consumption, high density, ease of design, and moderate to high speed. Owing to these excellent performance characteristics, CMOS technology has become the dominant VLSI technology used today.

The design of a digital VLSI system requires a great deal of effort to consider a broad range of architectural and logical issues; that is, choosing the appropriate gates and interconnections among these gates to achieve the required circuit function. No design is complete, however, without considering the *dynamic* (or transient) characteristics of the signal propagation, or, alternatively, the changing behavior of signals with *time*. Every computation performed by a switching circuit involves multiple signal transitions between logic states and requires a *finite* amount of time to complete. The voltage at every circuit node must reach a specific value for the computation to be completed. State-of-the-art integrated circuit design is therefore largely centered around the difficult task of predicting and properly interpreting signal waveform shapes at various points in a circuit.

[†]RISC, reduced instruction set computer.

[‡]Quantities such as the *electrical voltages* and *currents* in electronic devices.

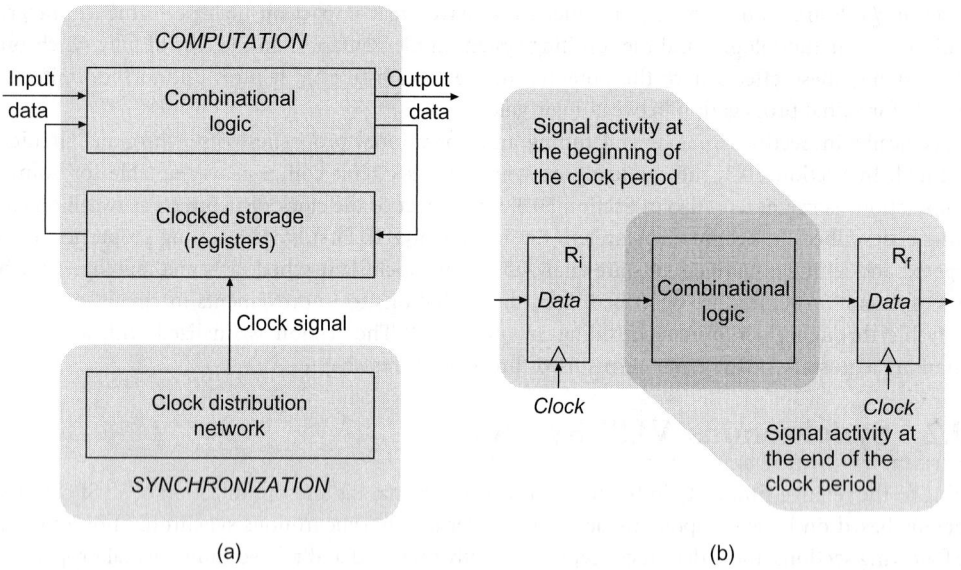

FIGURE 50.2 A synchronous system. (a) Finite-state machine model of a synchronous system. (b) A local data path.

In a typical VLSI system, millions of signal transitions determine the individual gate delays and the overall speed of the system. Some of these signal transitions can be executed *concurrently* while others must be executed in a strict *sequential* order [11]. The sequential occurrence of the latter operations—or signal transition *events*—must be properly coordinated in time such that logically correct system operation is guaranteed and the results are reliable (in the sense that these results can be repeated). This coordination is known as *synchronization* and is critical to ensuring that any pair of logical operations in a circuit with a precedence relationship proceed in the proper order. In modern digital integrated circuits, synchronization is achieved at all stages of the system design process and operation by a variety of techniques, known as a *timing discipline* or *timing scheme* [9,12–14]. With some exceptions, these circuits are based on a *fully synchronous* timing scheme, specifically developed to cope with the finite speed required by the physical signals to propagate through the system.

An example of a *fully synchronous* system is shown in Figure 50.2(a). As illustrated in Figure 50.2(a), there are three recognizable components in this system. The first component—the logic gates, collectively referred to as the *combinational logic*—provides the range of operations that a system executes. The second component—the *clocked storage* elements or simply the *registers*—are elements that store the results of the logical operations. Together, the combinational logic and registers constitute the *computational* portion of the synchronous system and are interconnected in a way that implements the required system function. The third component of the synchronous system—known as the *clock distribution network*—is a highly specialized circuit structure which does not perform a computational process but rather provides an important control capability. The clock generation and distribution network controls the overall synchronization of the circuit by *generating* a time reference and properly *distributes* this time reference to every register.

The normal operation of a synchronous system, such as the example finite-state machine shown in Figure 50.2(a), consists of the iterative execution of computations in the combinational logic followed by the storage of the processed results in the registers. The actual process of storage is temporally controlled by the clock signal and occurs once the signal transients in the logic gate outputs are completed and the outputs have settled to a valid state. At the beginning of each computational cycle, the inputs of the system together with the data stored in the registers initiate a new switching process. As time proceeds, the signals propagate through the logic, generating results at the logic output. By the end of the clock period, these results are stored in the registers. During the following clock cycle, the stored data values start propagating through the logic, progressing toward the system outputs.

The operation of a digital system can therefore be thought of as the sequential execution of a large set of simple computations that occur concurrently in the combinational logic portion of the system. The concept of a *local data path* is a useful abstraction for each of these simple operations and is shown in Figure 50.2(b). The magnitude of the delay of the combinational logic is bound by the requirement of storing data in a register within a clock period. The initial register R_i is the storage element at the beginning of the local data path and provides some or all of the input signals for the combinational logic at the beginning of the computational cycle (defined by the beginning of the clock period). The *combinational path* ends with the data successfully latching within the final register R_f where the results are stored at the end of the computational cycle. Registers act as *sources* and *sinks* for the data between the clock cycles.

50.2.2   Advantages and Drawbacks of Synchronous Systems

The behavior of a fully synchronous system is well defined and controllable as long as the *time window* provided by the clock period is sufficiently long to allow every signal in the circuit to propagate through the required logic gates and interconnect wires and successfully latch within the final register. In designing the system and choosing the proper clock period, however, two contradictory requirements must be satisfied. First, the smaller the clock period, the more computational cycles can be performed by the circuit in a given amount of time. Alternatively, the time window defined by the clock period must be sufficiently long such that the slowest signals reach the destination registers before the current clock cycle is concluded and the following clock cycle is initiated.

Such an organization of computation has certain clear advantages that propel a fully synchronous timing scheme to remain as the primary choice for digital VLSI systems:

- The properties and variations are simple and well understood.
- The scheme eliminates the nondeterministic behavior of the propagation delay in the combinational logic (due to environmental and process fluctuations and unknown input signal patterns) such that the system exhibits a deterministic behavior corresponding to the implemented algorithm.
- The circuit design does *not* need to be concerned with glitches in the combinational logic outputs, so the only relevant dynamic characteristic of the logic is the *propagation delay*.
- The state of the system is completely defined within the storage elements—this fact greatly simplifies certain aspects of the design, debug, and test phases in developing a large system.

A synchronous paradigm, however, also has certain limitations that make the design of synchronous VLSI systems increasingly challenging:

- This synchronous approach has a serious drawback in that the timing scheme requires the overall circuit to operate as slow as the *slowest* register-to-register path. Thus, the global speed of a fully synchronous system depends upon those paths in the combinational logic with the largest delays—these paths are also known as the *worst-case* or *critical* paths. In a typical VLSI system, the propagation delays in the combinational paths are distributed unevenly so there may be many paths with delays much smaller than the clock period. Although these paths could operate correctly at a lower clock period—higher clock frequency—it is those paths with the largest delays that bound the clock period, thereby imposing a limit on the overall system speed. This imbalance in propagation delays is sometimes so dramatic that the system speed is dictated by only a handful of very slow paths.
- The clock signal has to be distributed to tens of thousands of storage registers scattered throughout the system. A significant portion of the system area and dissipated power is therefore devoted to the clock distribution network (reviewed in Section 50.3)—a circuit structure that does not perform any computational function.

- The reliable operation of the system depends upon the assumptions concerning the value of the propagation delays, which, if not satisfied, can lead to catastrophic timing violations and render the system unusable.

50.3 Synchronous Timing and Clock Distribution Networks

The timing of a synchronous VLSI system is characteristically analyzed at the level of its synchronous building blocks, the local data paths. In the following sections, the simple and effective modeling of synchronous building blocks is described. The effects of clock distribution networks on circuit operation are also presented.

50.3.1 Background

As described in Section 50.2, most high-performance digital integrated circuits implement data-processing algorithms based on the iterative execution of basic operations. Typically, these algorithms are highly *parallelized* and *pipelined* by inserting clocked registers at specific locations throughout the circuit. The synchronization strategy for these clocked registers in the vast majority of VLSI-based digital systems is a fully synchronous approach. It is not uncommon for the computational process in these systems to be spread over hundreds of thousands of functional logic elements and tens of thousands of registers.

For such synchronous digital systems to function properly, the vast number of switching events require a strict temporal ordering. This strict ordering is enforced by a global synchronization signal known as the *clock signal*. For a fully synchronous system to operate correctly, the clock signal must be delivered to every register at a precise *relative* time. The delivery function is accomplished by a circuit and interconnect structure known as a *clock distribution network* [15].

Multiple factors affect the propagation delay of the data signals through the combinational logic gates and the interconnect. Since the clock distribution network is composed of logic gates and interconnection wires, the signals in the clock distribution network are also delayed. Moreover, the dependence of the correct operation of a system on the signal delay in the clock distribution network is far greater than on the delay of the logic gates. Recall that by delivering the clock signal to registers at precise times, the clock distribution network essentially quantizes the time of a synchronous system (into clock periods), thereby permitting the simultaneous execution of operations.

The nature of the on-chip clock signal has become a primary factor in limiting circuit performance, causing the clock distribution network to become a performance bottleneck for high-speed VLSI systems. The primary source of load for the clock distribution network has shifted from the logic gates to the *interconnect*, thereby changing the physical nature of the load from a lumped capacitance (C) to a distributed *resistive-capacitive* (RC) load and eventually a distributed *resistive-capacitive-inductive* (RLC) load [7,16,17]. These interconnect impedances degrade the on-chip signal waveform shapes and increase the path delay. Furthermore, uncertainty is introduced into the signal timing due to statistical variations in the parameters characterizing the circuit elements along the clock and data signal paths, caused by the imperfect control of the manufacturing process and the environment. These changes in circuit behavior have a profound impact on both the choice of synchronous design methodology and on the overall circuit performance. Among the most important consequences are increased power dissipated by the clock distribution network as well as the increasingly challenging timing constraints that must be satisfied to avoid timing violations [3–6,15,18–20].

50.3.2 Definitions and Notation

A synchronous digital system is a network of logic gates and registers whose input and output terminals are interconnected by wires. A sequence of connected logic gates (no registers) is called a signal path. Signal paths bounded by registers are called sequentially adjacent paths.

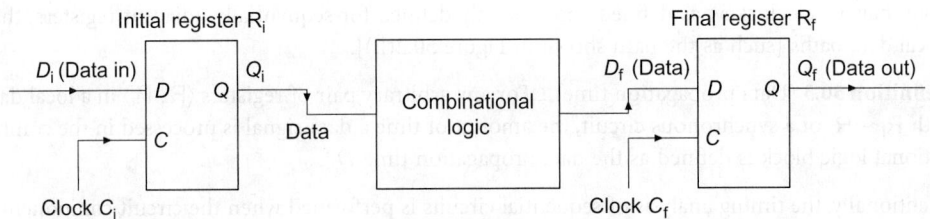

FIGURE 50.3 A local data path.

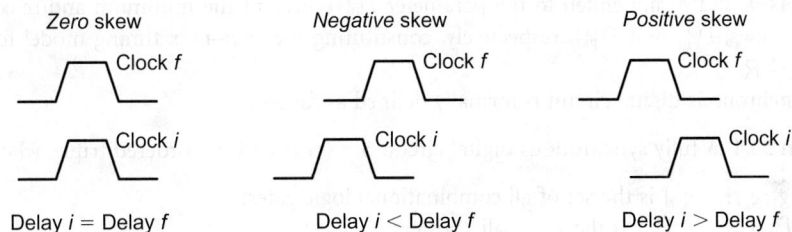

FIGURE 50.4 Lead/lag relationships causing clock skew to be zero, negative, or positive.

Definition 50.1 (Sequentially adjacent pair of registers). For an arbitrary *ordered* pair of registers $\langle R_i, R_f \rangle$ in a synchronous circuit, one of the following two situations can be observed. Either there exists at least one signal path that connects some output of R_i to some input of R_f or any input of R_f *cannot* be reached from any output of R_i by propagating through a sequence of logic elements *only*. In the former case—denoted by $R_1 \rightsquigarrow R_2$—the pair of registers $\langle R_i, R_f \rangle$ is called a *sequentially adjacent pair of registers* and switching events at the output of R_i can possibly affect the input of R_f during the same clock period. A sequentially adjacent pair of registers is also referred to as a *local data path* [15].

A sample local data path with a register (a flip-flop or a latch) is shown in Figure 50.3.[8] In Figure 50.3, the clock signals C_i and C_f driving the initial register R_i and the final register R_f, respectively, of the local data path are shown.

Definition 50.2 For any ordered pair of registers $\langle R_i, R_j \rangle$ in a fully synchronous circuit driven by the clock signals C_i and C_j, respectively, the clock skew $T_{\text{Skew}}(i, j)$ is defined as the difference:

$$T_{\text{Skew}}(i, j) = t_{cd}^{i} - t_{cd}^{j} \qquad (50.1)$$

where t_{cd}^{i} and t_{cd}^{j} are the clock delays of the clock signals C_i and C_j, respectively.

In Definition 50.2, the clock delays t_{cd}^{i} and t_{cd}^{j} are with respect to some reference point. A commonly used reference point is the source of the clock distribution network. Note that the clock skew $T_{\text{Skew}}(i, j)$ as defined in Definition 50.2 obeys the antisymmetric property,

$$T_{\text{Skew}}(i, j) = -T_{\text{Skew}}(j, i) \qquad (50.2)$$

Depending on the values of t_{cd}^{i} and t_{cd}^{f}, the skew can be *zero* ($t_{cd}^{i} = t_{cd}^{f}$), *negative* ($t_{cd}^{i} < t_{cd}^{f}$), or *positive* ($t_{cd}^{i} > t_{cd}^{f}$). This behavior is illustrated in Figure 50.4.

---

[8]Examples of local data paths with flip-flops and latches are shown in Figure 50.17 and Figure 50.21, respectively.

Note that the clock skew as defined above is only defined for sequentially-adjacent registers, that is, for local data paths [such as the path shown in Figure 50.2(b)].

Definition 50.3 (Data propagation time). For any arbitrary pair of registers $\langle R_i, R_f \rangle$ in a local data path $R_i \rightsquigarrow R_f$ of a synchronous circuit, the amount of time a data signal is processed in the combinational logic block is defined as the data propagation time $D_P^{i,f}$.

Conventionally, the timing analysis of sequential circuits is performed when the circuit components are modeled with min–max timing models. In the min–max timing model, the delay information of a circuit component is represented with two quantities; the minimum corresponding to the delay of the component under best-case operation conditions and the maximum for the worst-case operation conditions. The subscripts m and M appended to the parameter $D_P^{i,f}$ represent the minimum and maximum data propagation times, $D_{Pm}^{i,f}$ and $D_{PM}^{i,f}$, respectively, constituting the min–max timing model for the local data path $R_i \rightsquigarrow R_f$.

A fully synchronous digital circuit is formally defined as follows:

Definition 50.4 A fully synchronous digital circuit $S = \langle G, R, C \rangle$ is an ordered triple, where

- $G = \{g_1, g_2, \ldots, g_M\}$ is the set of all combinational logic gates,
- $R = \{R_1, R_2, \ldots, R_N\}$ is the set of all registers, and
- $C = \|c_{i \times j}\|_{N \times N}$ is a matrix describing the connectivity of G where for every element $c_{i,j}$ of C

$$c_{i,j} = \begin{cases} 0, & \text{if } R_i \rightsquigarrow R_j \\ 1, & \text{if } R_i \not\rightsquigarrow R_j \end{cases}$$

Note that in a fully synchronous digital system there are no purely combinational signal *cycles*, that is, the input of any logic gate G_k cannot be reached by starting at the same gate and propagating through a sequence of combinational logic gates only [15,21].

50.3.3   Graph Model of a Fully Synchronous Digital Circuit

Certain properties of a synchronous digital circuit may be better understood by analyzing a graph model of a circuit. A synchronous digital circuit can be modeled as a *directed graph* [22,23] G with a *vertex set* $V = \{v_1, \ldots, v_N\}$ and an *edge set* $E = \{e_1, \ldots, e_{N_P}\} \subseteq V \times V$. An example of a circuit graph G is illustrated in Figure 50.5(a). The number of registers in the circuit is $|V| = N$ where the vertex v_k corresponds to the register R_k. The number of local data paths in the circuit is $|E| = N_P = 11$ for the example shown in Figure 50.5. An edge is directed from v_i to v_j iff $R_i \rightsquigarrow R_j$. In the case where multiple paths between a sequentially adjacent pair of registers $R_i \rightsquigarrow R_j$ exist, only *one* edge connects v_i to v_j. The *underlying* graph

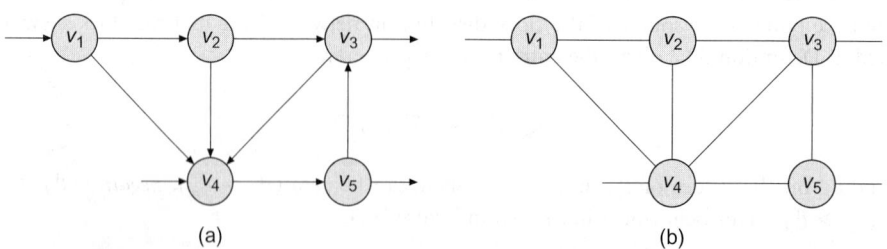

(a)   (b)

FIGURE 50.5 Graphs G and its underlying graph G_u of a circuit with $N = 5$ registers. (a) The directed graph G. (b) The underlying graph G_u of G.

G_u of the graph G is a *nondirected* graph that has the same vertex set V, where the directions have been removed from the edges. The underlying graph G_u of the graph G depicted in Figure 50.5(a) is shown in Figure 50.5(b). Furthermore, an input or an output of the circuit is indicated in Figure 50.5 by an edge incident to only one vertex.

50.3.4 Clock Skew Scheduling

The majority of the approaches used to design a clock distribution network simplify the performance goals by targeting minimal or zero global clock skew [24–26], which can be achieved by different routing strategies [27–30], buffered clock tree synthesis, symmetric *n*-ary trees [31] (most notably H-trees), or a distributed series of buffers connected as a mesh [15,32]. A zero clock skew scheme is established by distributing the clock signal to all synchronous components of a circuit with identical clock delays. In other words, the clock skew evaluates to zero on all of the local data paths of a zero clock skew circuit:

$$\forall \langle \mathsf{R}_i, \mathsf{R}_f \rangle : t_{cd}^i = t_{cd}^f \implies T_{\text{Skew}}(i, f) = 0 \qquad (50.3)$$

For zero clock skew systems, the clock period T_{CP} is limited by the largest maximum data propagation time $D_{PM}^{i,f}$ on the circuit:

$$\min T_{CP} = \max_{\forall \langle \mathsf{R}_i, \mathsf{R}_f \rangle} (D_{PM}^{i,f}) \qquad (50.4)$$

If the circuit operates at any clock period less than the largest maximum data propagation time, a *timing hazard* occurs. For any clock period greater than this value, the circuit is fully functional (no timing hazards occur). Finding a clock period T_{CP} for which a zero clock skew circuit is fully functional (equal to or greater than the largest data propagation time $D_{PM}^{i,f}$), is always possible, making it convenient to design zero clock skew systems. Consequently, the application of zero clock skew schemes has been central to the design of fully synchronous digital circuits for decades [15,33].

The vector column of clock delays $T_{CD} = [t_{cd}^1, t_{cd}^2, \ldots]^T$ is called a *clock schedule* [15,34]. A clock schedule that satisfies Eq. (50.3) is called a *trivial* clock schedule. Note that a trivial clock schedule T_{CD} implies global *zero* clock skew since for any i and f, $t_{cd}^i = t_{cd}^f$, thus, $T_{\text{Skew}}(i, f) = 0$. If T_{CD} is chosen such that the timing constraints of a circuit are satisfied for every local data path $\mathsf{R}_i \leadsto \mathsf{R}_f$, T_{CD} is called a *consistent* clock schedule.

The goal of nonzero clock skew scheduling is to compute a consistent clock schedule that is not trivial, while improving the circuit performance. It has been shown in Refs. [15,24–26,35–37] that by adopting a nonzero clock skew synchronization scheme, synchronous circuits can operate at clock periods less than the largest maximum data propagation time of the circuit. In nonzero clock skew systems, the clock signal delays t_{cd} at certain registers are intentionally delayed to provide additional data-processing time on slower local data paths. Mathematically, the nonzero clock skew values (also called *useful skew*) evaluate to $T_{\text{Skew}}(i, f) \neq 0$ for some (or *all*) local data paths $\mathsf{R}_i \leadsto \mathsf{R}_f$ of the circuit.

The process of determining a consistent clock schedule T_{CD} can be considered as the mathematical problem of optimizing the circuit performance under the timing constraints of a circuit. However, there are important practical issues to consider before a clock schedule can be properly implemented. A clock distribution network must be synthesized such that the clock signal is delivered to each register with the proper delay so as to satisfy the clock skew schedule T_{CD}. Furthermore, this clock distribution network must be constructed so as to minimize the deleterious effects of *interconnect impedances* and *process parameter variations* on the implemented clock schedule. Synthesizing the clock distribution network typically consists of determining a *topology* for the network, together with the circuit design and physical layout of the *buffers* and *interconnect* within the clock distribution network [15].

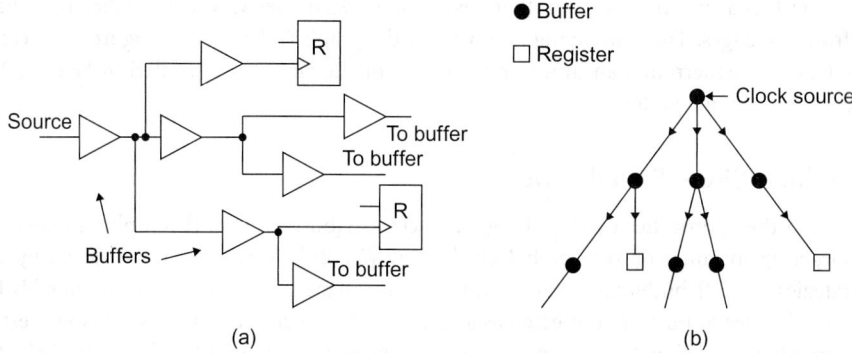

FIGURE 50.6 Tree structure of a clock distribution network. (a) Circuit structure of the clock distribution network. (b) Equivalent graph of a clock tree structure that corresponds to the circuit shown in (a).

50.3.5   Structure of a Clock Distribution Network

The clock distribution network is frequently organized as a rooted tree structure [15,22,24], as illustrated in Figure 50.6, and is often called a *clock tree* [15]. A circuit schematic of a clock distribution network is shown in Figure 50.6(a). An abstract graphical representation of the tree structure depicted in Figure 50.6(a) is shown in Figure 50.6(b). The unique source of the clock signal is at the root of the tree. This signal is distributed from the source to every register in the circuit through a sequence of buffers and interconnect. Typically, a buffer in the network drives a combination of other buffers and registers in the VLSI circuit. An interconnection network of wires connects the output of the driving buffer to the inputs of these driven buffers and registers. An *internal node* of the tree corresponds to a buffer and a *leaf node* of the tree corresponds to a register. There are N leaves[¶] in the clock tree labeled F_1 through F_N where leaf F_j corresponds to register R_j. A clock tree topology that implements a given clock schedule T_{CD} must enforce a clock skew $T_{Skew}(i,f)$ for each local data path $R_i \rightsquigarrow R_f$ of the circuit to ensure that the timing constraints of the circuit are satisfied. This topology, however, can be affected by three important issues relating to the operation of a fully synchronous digital system.

50.3.5.1   Linear Dependency of the Clock Skews

An important corollary related to the *conservation property* [15] of clock skew is that there exists a *linear dependency* among the clock skews of a global data path that form a cycle in the underlying graph of the circuit. Specifically, if $v_0, e_1, v_1 (\neq v_0), \dots, v_{k-1}, e_k, v_k \equiv v_0$ is a cycle in the underlying graph of the circuit,

$$0 = [t_{cd}^0 - t_{cd}^1] + [t_{cd}^1 - t_{cd}^2] + \cdots = \sum_{i=0}^{k-1} T_{Skew}(i,i+1) \tag{50.5}$$

The property described by Eq. (50.5) is illustrated in Figure 50.5 for the undirected cycle v_1,v_4,v_3,v_2,v_1. Note that

$$0 = (t_{cd}^1 - t_{cd}^4) + (t_{cd}^4 - t_{cd}^3) + (t_{cd}^3 - t_{cd}^2) + (t_{cd}^2 - t_{cd}^1)$$

$$= T_{Skew}(1,4) + T_{Skew}(4,3) + T_{Skew}(3,2) + T_{Skew}(2,1) \tag{50.6}$$

[¶]The number of registers N in the circuit.

The importance of this property is that Eq. (50.5) describes the inherent correlation among certain clock skews within a circuit. These correlated clock skews therefore *cannot* be independently optimized. Returning to Figure 50.5, note that it is not necessary that a directed cycle exists in the directed graph G of a circuit for Eq. (50.5) to hold. For example, v_2, v_3, v_4 is not a cycle in the directed circuit graph G in Figure 50.5(a) but v_2, v_3, v_4 is a cycle in the undirected circuit graph G_u in Figure 50.5(b). In addition, $T_{Skew}(2,3) + T_{Skew}(3,4) + T_{Skew}(4,2) = 0$, that is, the skews $T_{Skew}(2,3)$, $T_{Skew}(3,4)$, and $T_{Skew}(4,2)$ are linearly dependent. A maximum of $(|V| - 1) = (N - 1)$ clock skews can be chosen independently of each other in a circuit, which is easily proven by considering a spanning tree of the underlying circuit graph G_u [22,23]. Any spanning tree of G_u will contain $(N - 1)$ edges—each edge corresponding to a local data path—and the addition of any other edge of G_u will form a cycle such that Eq. (50.5) holds for this cycle. Note, for example, that for the circuit modeled by the graph shown in Figure 50.5, *four* independent clock skews can be chosen such that the remaining three clock skews can be expressed in terms of the independent clock skews.

The interdependency of the clock skew values makes the analysis of clock skew scheduling methods a difficult problem. Owing to this interdependency characteristic, a clock skew value cannot be determined independent of the remaining clock skews, thus a typical clock-skew scheduling method must simultaneously encompass the analysis of all local data paths. Such simultaneous analysis of all local data paths in a given synchronous circuit is typically structured by including the timing constraints of every local data path in a single optimization problem.

50.3.5.2 Differential Character of the Clock Tree

In a given circuit, the clock signal delay t_{cd}^j from the clock source to the register R_j is equal to the sum of the propagation delays of the buffers on the unique path that exists between the root of the clock tree and the leaf F_j corresponding to the jth register. Furthermore, if $R_i \rightsquigarrow R_f$ is a sequentially adjacent pair of registers, there is a portion of the two paths—denoted P_{if}^*—between the root of the clock tree and R_i and R_f, respectively, that is common to both paths. This concept is illustrated in Figure 50.7. A portion of a clock tree is shown in Figure 50.7 where each of the vertices 1 through 9 corresponds to a buffer in the clock tree. The vertices 4, 5, and 9 are leaves of the tree and correspond to the registers R_4, R_5, and

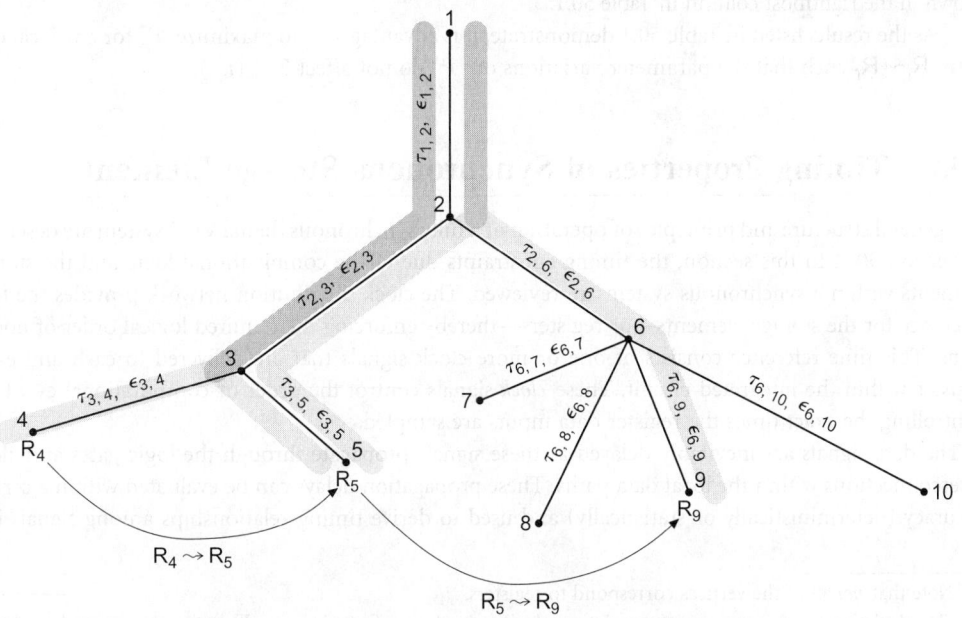

FIGURE 50.7 Illustration of the differential nature of the clock tree.

TABLE 50.1 Target and Actual Values of the Clock Skews for the Local Data Paths $R_4 \rightsquigarrow R_5$ and $R_5 \rightsquigarrow R_9$ Shown in Figure 50.7

	Target Skew	Actual Skew Bounds
$T_{\text{Skew}}(4,5)$	$\tau_{3,4} - \tau_{3,5}$	$\tau_{3,4} - \tau_{3,4} \pm (\epsilon_{3,4} + \epsilon_{3,4})$
$T_{\text{Skew}}(5,9)$	$\tau_{2,3} + \tau_{3,5} - \tau_{2,6} - \tau_{6,9}$	$\tau_{2,3} + \tau_{3,5} - \tau_{2,6} - \tau_{6,9} \pm (\epsilon_{2,3} + \epsilon_{3,5} + \epsilon_{2,6} + \epsilon_{6,9})$

R_9, respectively.[||] The local data paths $R_4 \rightsquigarrow R_5$ and $R_5 \rightsquigarrow R_9$ are indicated in Figure 50.7 with arrows while the paths of the clock signals to each of the registers R_4, R_5, and R_9 are shown in Figure 50.7 as lightly shaded. The portion of the clock signal paths common to both registers of a local data path is shaded darker in Figure 50.7—note the segments $1 \rightarrow 2 \rightarrow 3$ for $R_4 \rightsquigarrow R_5$ and $1 \rightarrow 2$ for $R_5 \rightsquigarrow R_9$.

Similarly, there is a portion of the clock signal path to any of the registers R_i and R_f in a sequentially adjacent pair of registers $R_i \rightsquigarrow R_f$, denoted by P_{if}^i and P_{if}^f, respectively, that is unique to this register. Returning to Figure 50.7, the segments $3 \rightarrow 4$ and $3 \rightarrow 5$ are unique to the clock signal paths to the registers R_4 and R_5 while the segments $2 \rightarrow 3 \rightarrow 5$ and $2 \rightarrow 6 \rightarrow 9$ are unique to the clock signal paths to the registers R_5 and R_9, respectively.

Note that the clock skew $T_{\text{Skew}}(i,f)$ between the sequentially adjacent pair of registers $R_i \rightsquigarrow R_f$ is equal to the difference between the accumulated buffer propagation delays between P_{if}^i and P_{if}^f, that is, $T_{\text{Skew}}(i,f) = \text{Delay}(P_{if}^i) - \text{Delay}(P_{if}^f)$. Therefore, any variation of circuit parameters over P_{if}^* will not affect the value of the clock skew $T_{\text{Skew}}(i,f)$. For the example shown in Figure 50.7, $T_{\text{Skew}}(4,5) = \text{Delay}(P_{4,5}^4) - \text{Delay}(P_{4,5}^5)$ and $T_{\text{Skew}}(5,9) = \text{Delay}(P_{5,9}^5) - \text{Delay}(P_{5,9}^9)$.

This differential feature of the clock tree suggests an approach for minimizing the effects of process parameter variations on the correct operation of the circuit. To illustrate this approach, each branch $p \rightarrow q$ of the clock tree shown in Figure 50.7 is labeled with two numbers—$\tau_{p,q} > 0$ is the intended delay of the branch and $\epsilon_{p,q} \geq 0$ is the maximum error (deviation) of this delay.[**] In other words, the actual delay of the branch $p \rightarrow q$ is in the interval $[\tau_{p,q} - \epsilon_{p,q}, \tau_{p,q} + \epsilon_{p,q}]$. With this notation, the *target* clock skew values for the local data paths $R_4 \rightsquigarrow R_5$ and $R_5 \rightsquigarrow R_9$ are shown in the middle column in Table 50.1. The bounds of the actual clock skew values for the local data paths $R_4 \rightsquigarrow R_5$ and $R_5 \rightsquigarrow R_9$ (considering the ϵ variations) are shown in the rightmost column in Table 50.1.

As the results listed in Table 50.1 demonstrate, it is advantageous to maximize P_{if}^* for any local data path $R_i \rightsquigarrow R_f$ such that the parameter variations on P_{if}^* do not affect $T_{\text{Skew}}(i,f)$.

50.4 Timing Properties of Synchronous Storage Elements

The general structure and principles of operation of a fully synchronous digital VLSI system are described in Section 50.2. In this section, the timing constraints due to the combinational logic and the storage elements within a synchronous system are reviewed. The clock distribution network provides the time reference for the storage elements—or registers—thereby enforcing the required logical order of operations. This time reference consists of one or more clock signals that are delivered to each and every register within the integrated circuit. These *clock* signals control the order of computational events by controlling the exact times the register data inputs are sampled.

The data signals are inevitably delayed as these signals propagate through the logic gates and along interconnections within the local data paths. These propagation delays can be evaluated within a certain accuracy (deterministically or statistically) and used to derive timing relationships among signals in a

---

[||]Note that *not* all of the vertices correspond to registers.

[**]The deviation ε is due to parameter variations during ciruit manufacturing as well as to environmental conditions during operation of the circuit.

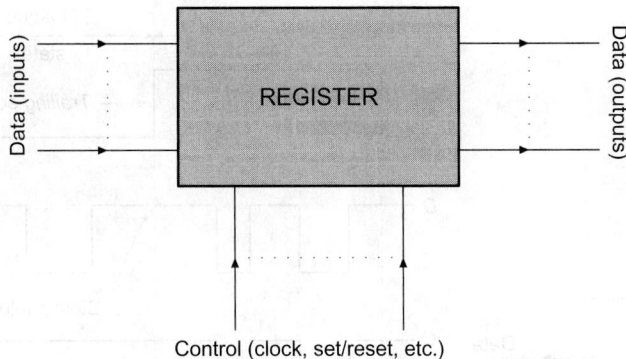

FIGURE 50.8 A general view of a register.

circuit. In this section, the properties of commonly used types of registers and their local timing relationships for different types of local data paths are described. After discussing registers in general in Section 50.4.1, the properties of level-sensitive registers (latches) and the significant timing parameters of these registers are reviewed in Sections 50.4.2 and 50.4.3, respectively. Edge-triggered registers (flip-flops) and related timing parameters are analyzed in Sections 50.4.4 and 50.4.5, respectively. Properties and definitions related to the clock distribution network are reviewed in Section 50.4.6. Finally, the mathematical foundation for analyzing timing violations in flip-flops and latches are discussed in Sections 50.4.7 and 50.4.8, respectively.

50.4.1 Storage Elements

The storage elements (registers) encountered throughout VLSI systems vary widely in function and temporal relationships. Independent of these differences, however, all storage elements share a common feature—the existence of two groups of signals with largely different purpose. A generalized view of a register is depicted in Figure 50.8. The I/O signals of a register can be divided into two groups as shown in Figure 50.8. One group of signals—called the *data* signals—consists of input and output signals of the storage element. These input and output signals are connected to the *data* signal terminals of other storage elements as well as to the terminals of ordinary logic gates. Another group of signals—identified by the name *control* signals—are those signals that control the storage of the data signals in the registers but do not participate in the logical computation process.

Certain control signals enable the storage of a data signal in a register independently of the values of any data signals. These control signals are typically used to initialize the data in a register to a specific well-known value. Other control signals—such as a *clock* signal—control the process of storing a data signal within a register. In a synchronous circuit, each register has at least one clock (or control) signal input.

The two major groups of storage elements (registers) are considered in the following sections based on the type of relationship that exists between the data and clock signals of these elements. In *latches*, it is the specific value or level of a control signal[*†] that determines the data storage process. Therefore, latches are also called *level-sensitive registers*. In contrast to latches, a data signal is stored in *flip-flops*, controlled by an *edge* of a control signal. For that reason, flip-flops are also called *edge-triggered registers*. The timing properties of latches and flip-flops are described in the following sections.

50.4.2 Latches

A *latch* is a register whose behavior depends upon the value or level of the clock signal [9,38–44]. A latch is therefore often referred to as a *transparent* latch, a *level-sensitive* register, or a *polarity hold* latch. A simple

---

[*†]This signal is most frequently the clock signal.

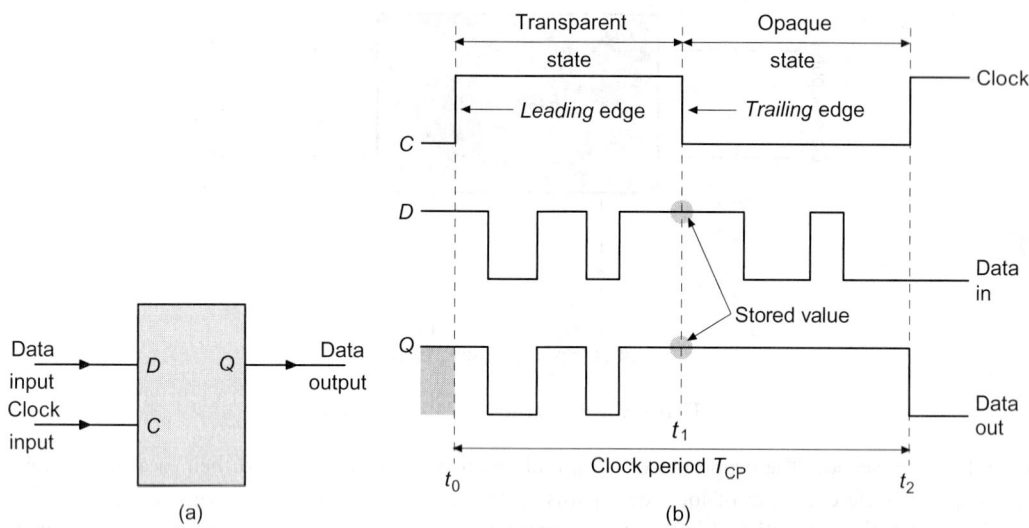

FIGURE 50.9 Schematic representation and principle of operation of a level-sensitive register (latch). (a) A level-sensitive register or latch. (b) Idealized operation of the latch shown in (a).

TABLE 50.2   Operation of the Positive-Polarity D Latch

Clock	Output	State
High	Passes input	Transparent
Low	Maintains output	Opaque

type of latch with a clock signal C and an input signal D is depicted in Figure 50.9(a)—the output of the latch is typically labeled Q. This type of latch is also known as a D latch. The operation of a D latch is illustrated in Figure 50.9(b).

The register illustrated in Figure 50.9 is a *positive-polarity*[**‡] latch since the register is transparent during that portion of the clock period for which C is high. The operation of this positive latch is summarized in Table 50.2.

As described in Table 50.2 and illustrated in Figure 50.9(b), the output signal of the latch follows the data input signal while the clock signal remains high, i.e., $C = 1 \Rightarrow Q = D$. Therefore, the latch is said to be in a *transparent* state during the interval $t_0 < t < t_1$ as shown in Figure 50.9(b). When the clock signal C changes from 1 to 0, the current value of D is stored in the register and the output Q remains fixed to that value regardless of whether the data input D changes. The latch does *not* pass the input data signal to the output, but rather holds onto the last value of the data signal when the clock signal made the high-to-low transition. By analogy with the term *transparent* introduced above, this state of the latch is called *opaque* and corresponds to the interval $t_1 < t < t_2$ as shown in Figure 50.9(b) where the input data signal is isolated from the output port. As shown in Figure 50.9(b), the clock period is $T_{CP} = t_2 - t_0$.

The edge of the clock signal that causes the latch to switch to the transparent state is identified as the *leading* edge of the clock pulse. In the case of the positive latch shown in Figure 50.9(a), the leading edge of the clock signal occurs at time t_0. The opposite direction edge of the clock signal is identified as the *trailing* edge—the falling edge at time t_1 shown in Figure 50.9(b). Note that for a negative latch, the leading edge is a high-to-low transition and the trailing edge is a low-to-high transition.

[**‡]Or simply a *positive* latch.

50.4.3 Parameters of Latches

Registers such as the *D* latch illustrated in Figure 50.9 and the flip-flops described in Sections 50.4.4 and 50.4.5 are built of discrete transistors. The exact relationships among the signals on the terminals of a register can be presented and evaluated in analytic form [45–47]. In this section, however, registers are considered at a higher level of abstraction to hide the details of the specific electrical implementation. The latch parameters are briefly introduced next.

Note that the remaining portion of this section uses an extensive notation for various parameters of signals and storage elements. A glossary of terms used throughout this section is listed in the appendix.

50.4.3.1 Minimum Width of the Clock Pulse

The *minimum width of the clock pulse* C^L_{Wm} is the *minimum* permissible width of this portion of the clock signal during which the latch is transparent. In other words, C^L_{Wm} is the length of the time interval between the leading and the trailing edge of the clock signal such that the latch will operate properly. The minimum width of the clock pulse is determined by multiple factors, including the technological limitations of the manufacturing process and the clock signal generation circuit. Further increasing the value of C^L_{Wm} will *not* affect the values of D^L_{DQ}, δ^L_S, and δ^L_H (defined in Sections 50.4.3.3, 50.4.3.4, and 50.4.3.5, respectively). The minimum width of the clock pulse, $C^L_{Wm} = t_6 - t_1$, is illustrated in Figure 50.10. The clock period is $T_{CP} = t_8 - t_1$.

50.4.3.2 Latch Clock-to-Output Delay

The *clock-to-output delay* D^L_{CQ} (typically called the clock-to-Q delay) is the propagation delay of the latch from the *clock* signal terminal to the output terminal. The value of $D^L_{CQ} = t_2 - t_1$ is depicted in Figure 50.10 and is defined assuming that the *data* input signal has settled sufficiently early to a stable value. Setting the data input signal earlier with respect to the leading clock edge will not affect the value of D^L_{CQ}.

50.4.3.3 Latch Data-to-Output Delay

The *data-to-output delay* D^L_{DQ} (typically called the data-to-Q delay) is the propagation delay of the latch from the *data* signal terminal to the output terminal. The value of D^L_{DQ} is determined assuming that the clock signal has set the latch to the transparent state sufficiently early. Making the leading edge of the

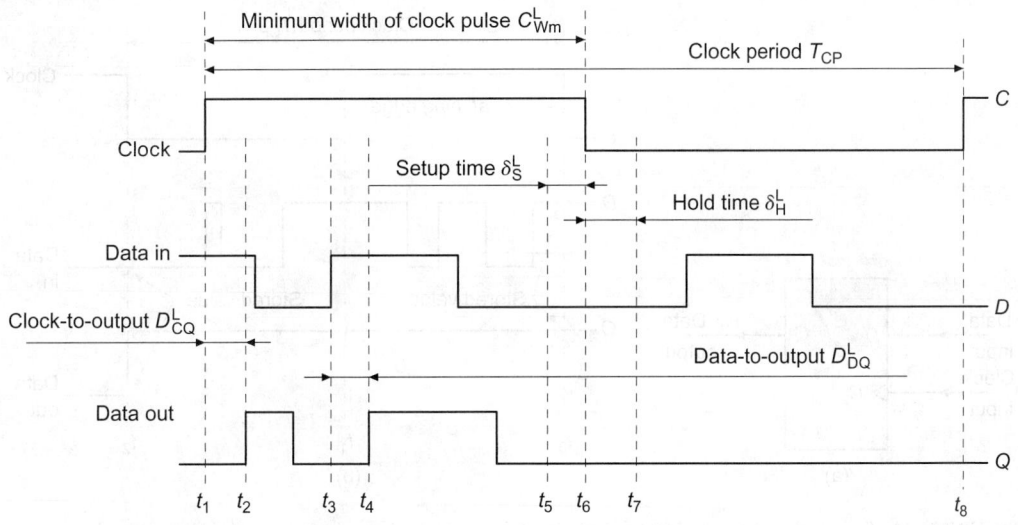

FIGURE 50.10 Parameters of a level-sensitive register.

clock signal occur earlier will not change the value of D_{DQ}^L. The data-to-output delay $D_{DQ}^L = t_4 - t_3$ is illustrated in Figure 50.10.

50.4.3.4　Latch Setup Time

The *latch setup time* $\delta_S^L = t_6 - t_5$, shown in Figure 50.10, is the *minimum* time between a change in the data signal and the trailing edge of the clock signal such that the new value of D would propagate to the output Q of the latch and be stored within the latch during the opaque state.

50.4.3.5　Latch Hold Time

The *latch hold time* δ_H^L is the minimum time after the trailing clock edge that the data signal must remain constant so that this value of D is successfully stored in the latch during the opaque state. This definition of δ_H^L assumes that the last change of the value of D has occurred no later than δ_S^L before the trailing edge of the clock signal. The term $\delta_H^L = t_7 - t_6$ is shown in Figure 50.10.

Sections 50.4.3.1 through 50.4.3.5 are used to refer to any latch in general or, to a specific instance of a latch when this instance can be unambiguously identified. To explicitly refer to a specific instance R_i of a latch, the parameters are additionally shown with a superscript. For example, D_{CQ}^{Li} refers to the clock-to-output delay of latch R_i. Also, adding m and M to the subscript of D_{CQ}^L and D_{DQ}^L may be used to refer to the *minimum* and *maximum* values of D_{CQ}^L and D_{DQ}^L, respectively.

50.4.4　Flip-Flops

An *edge-triggered register* or *flip-flop* is a type of register which, unlike the latches described in Sections 50.4.2 and 50.4.3, is never transparent with respect to the input data signal [9,38–44]. The output of a flip-flop normally does not follow the input data signal at any time during the register operation but rather holds onto a previously stored data value until a new data signal is stored in the flip-flop. A simple type of flip-flop with a clock signal C and an input signal D is shown in Figure 50.11(a)—similar to latches, the output of a flip-flop is usually labeled Q. This specific type of register, shown in Figure 50.11(a), is called a D flip-flop. The operation of a D flip-flop is illustrated in Figure 50.11(b).

In typical flip-flops, data is stored either on the rising edge (low-to-high transition) or on the falling edge (high-to-low transition) of the clock signal. The flip-flops are known as *positive-edge-triggered* and *negative-edge-triggered* flip-flops, respectively. The term *latching, storing*, or *positive* edge is used to identify the edge of the clock signal on which storage in the flip-flop occurs. For clarity, the latching edge of the

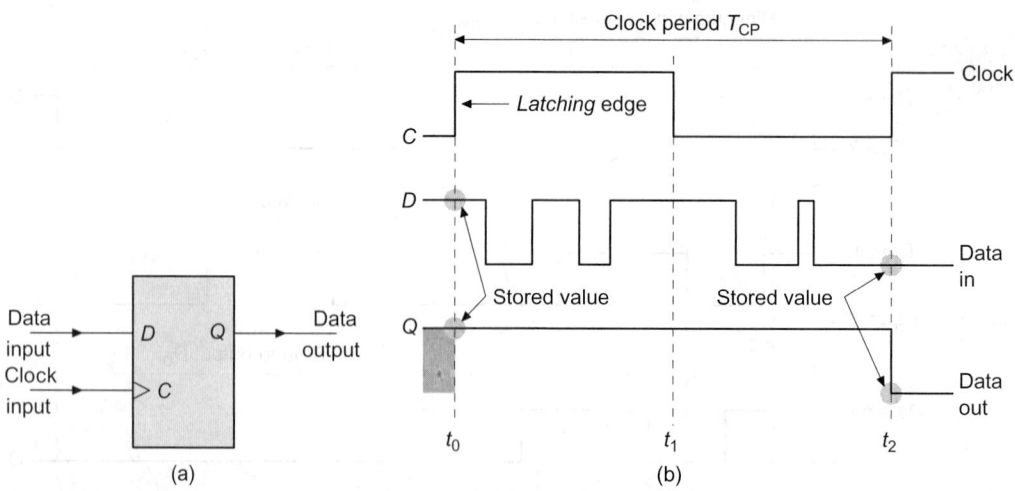

FIGURE 50.11　Schematic representation and principle of operation of an edge-triggered register (flip-flop). (a) An edge-triggered register or a flip-flip. (b) Idealized operation of the flip-flop shown in (a).

clock signal for flip-flops will also be called the leading edge (compare to the discussion of latches in Sections 50.4.2 and 50.4.3). Also note that certain flip-flops—known as *double-edge-triggered* (DET) flip-flops [48–52]—can store data at either edge of the clock signal. The complexity of these flip-flops, however, is significantly higher and the use of DET flip-flops is not very common.

As shown in the timing diagram in Figure 50.11(b), the output of the flip-flop remains unchanged most of the time regardless of the transitions in the data signal. Only values of the data signal in the vicinity of the storing edge of the clock signal can affect the output of the flip-flop. Therefore, changes in the output will only be observed when the currently stored data has a logic value x and the storing edge of the clock signal occurs while the input data signal has a logic value of $\bar{x}$.

50.4.5 Parameters of Flip-Flops

The significant timing parameters of an edge-triggered register are similar to those of latches (recall Section 50.4.3) and are presented next. These parameters are illustrated in Figure 50.12.

50.4.5.1 Minimum Width of the Clock Pulse

The minimum width of the clock pulse C_{Wm}^F is the *minimum* permissible width of the time interval between the latching edge and *non*latching edge of the clock signal. The minimum width of the clock pulse $C_{Wm}^F = t_6 - t_3$ is shown in Figure 50.12 and is the minimum interval between the latching and nonlatching edges of the clock pulse such that the flip-flop will operate correctly. Further increasing C_{Wm}^F will *not* affect the values of the setup time δ_S^F and hold time δ_H^F (defined in Sections 50.4.5.3 and 50.4.5.4, respectively). The clock period $T_{CP} = t_6 - t_1$ is also shown in Figure 50.12.

50.4.5.2 Flip-Flop Clock-to-Output Delay

As shown in Figure 50.12, the *clock-to-output delay* D_{CQ}^F of the flip-flop is $D_{CQ}^F = t_5 - t_3$. This propagation delay parameter—typically called the clock-to-Q delay—is the propagation delay from the clock signal terminal to the output terminal. The value of D_{CQ}^F is defined assuming that the data input signal has settled to a stable value sufficiently early. Setting the data input any earlier with respect to the latching clock edge will not affect the value of D_{CQ}^F.

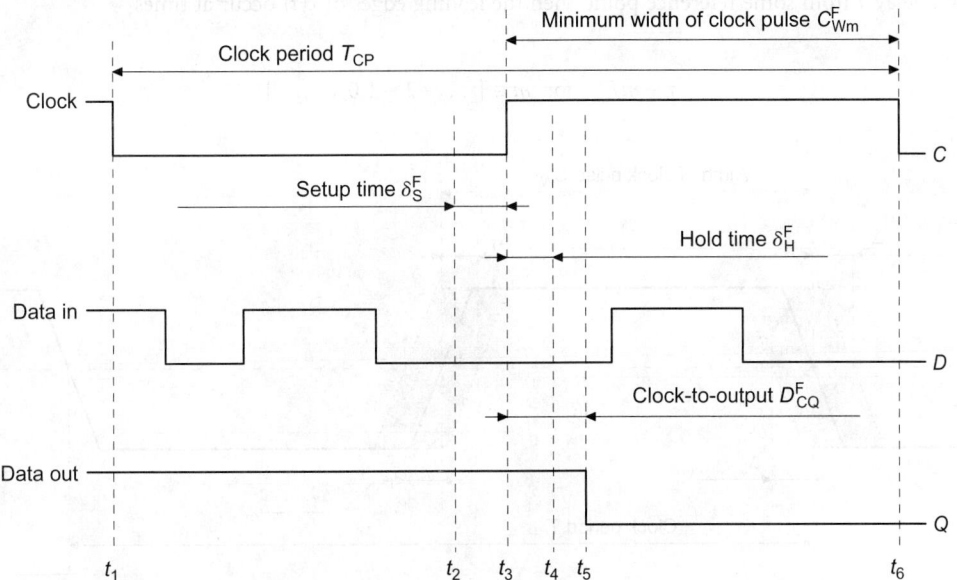

FIGURE 50.12 Parameters of an edge-triggered register.

50.4.5.3 Flip-Flop Setup Time

The flip-flop *setup time* δ_S^F is shown in Figure 50.12 as $\delta_S^F = t_3 - t_2$. The parameter δ_S^F is defined as the *minimum* time between a change in the data signal and the latching edge of the clock signal such that the new value of D propagates to the output Q of the flip-flop and is successfully latched within the flip-flop.

50.4.5.4 Flip-Flop Hold Time

The flip-flop *hold time* δ_H^F is the minimum time after the arrival of the latching clock edge in which the data signal must remain constant to successfully store the D signal within the flip-flop. The hold time $\delta_H^F = t_4 - t_3$ is illustrated in Figure 50.12. This definition of the hold time assumes that the last change of D has occurred no later than δ_S^F before the arrival of the latching edge of the clock signal.

Note that similar to latches, the parameters of these edge-triggered registers refer to any flip-flop in general, or to a specific instance of a flip-flop when this instance is uniquely identified. To explicitly refer to a specific instance i of a flip-flop, the flip-flop parameters are additionally shown with a superscript. For example, δ_S^{Fi} refers to the setup time parameter flip-flop i. Also, adding m and M to the subscript of D_{CQ}^F may be used to refer to the *minimum* and *maximum* values of D_{CQ}^F, respectively.

50.4.6 The Clock Signal

The clock signal is typically delivered to each storage element within a circuit. This signal is crucial to the correct operation of a fully synchronous digital system. The storage elements serve to establish the relative sequence of events within a system such that those operations that cannot be executed concurrently operate sequentially on the proper data signals.

A typical clock signal $c(t)$ in a synchronous digital system is shown in Figure 50.13. The *clock period* T_{CP} of $c(t)$ is indicated in Figure 50.13. To provide the highest possible clock frequency, the objective is for T_{CP} to have the smallest value such that

$$\forall n: \quad c(t) = c(t + nT_{CP}) \tag{50.7}$$

where n is an integer. The width of the clock pulse C_W, shown in Figure 50.13, is explained in Sections 50.4.3.1 and 50.4.5.1.

Typically, the period of the clock signal T_{CP} is a constant, that is, $\partial T_{CP}/\partial t = 0$. If the clock signal $c(t)$ has a delay τ from some reference point, then the leading edges of $c(t)$ occur at times

$$\tau + mT_{CP} \quad \text{for} \quad m \in \{\ldots, -2, -1, 0, 1, 2, \ldots\} \tag{50.8}$$

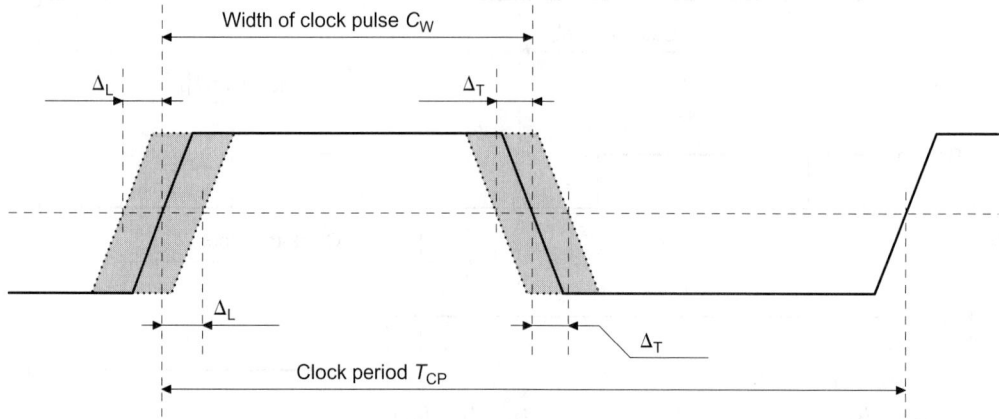

FIGURE 50.13 A typical clock signal.

and the trailing edges of $c(t)$ occur at times

$$\tau + C_W + mT_{CP} \quad \text{for} \quad m \in \{\ldots, -2, -1, 0, 1, 2, \ldots\} \tag{50.9}$$

In practice, however, it is possible for the edges of a clock signal to fluctuate in time, that is, *not* to occur precisely at the times described by Eq. (50.8) and Eq. (50.9) for the leading and trailing edges, respectively. This phenomenon is known as *clock jitter* and may be due to various causes such as variations in the manufacturing process, ambient temperature, power supply noise, and oscillator characteristics.

To account for this clock jitter, the following parameters are introduced:

- The maximum deviation Δ_L of the leading edge of the clock signal such that the leading edge occurs anywhere in an interval $(\tau + kT_{CP} - \Delta_L, \tau + kT_{CP} + \Delta_L)$

- The maximum deviation Δ_T of the trailing edge of the clock signal such that the leading edge occurs anywhere in the interval $(\tau + C_W + kT_{CP} - \Delta_T, \tau + C_W + kT_{CP} + \Delta_T)$

50.4.6.1 Synchronization Schemes

Traditionally, a single-phase clock signal such as the waveform shown in Figure 50.14 is used for synchronization. This relatively easy to implement and analyze single-phase clocking scheme (built with a single-phase clock signal) has several shortcomings in satisfying the timing requirements of circuits manufactured in nanoscale technologies. Under a single-phase clocking scheme, for instance, it becomes infeasible for signals to propagate across the entire area of an integrated circuit within a single clock cycle. To satisfy the increasingly complex timing requirements of integrated circuits, advanced synchronization methodologies such as *multiclock domains* and *multiphase clock signals* are used. These two concepts are briefly reviewed and the operational characteristics are described in this section.

Multiclock domains consist of two or more nonidentical clock signals delivered within a circuit for synchronization. It is possible to use well-tuned clocking schemes within each clock domain, improving the overall operational characteristics of the circuit. The clock signals are typically generated by separate oscillators, thus the frequency and phase information of the clock signals can be independent. The availability of multiple clock signals enables relatively independent synchronization of different domains within a circuit. Communication between (and among) multiple clock domains (the timing of local data paths between multiple clock domains) are managed by simultaneously considering the properties of the multiple clock domains simultaneously.

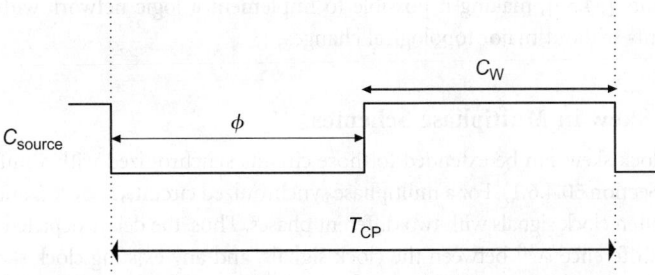

FIGURE 50.14 A single-phase synchronization clock signal.

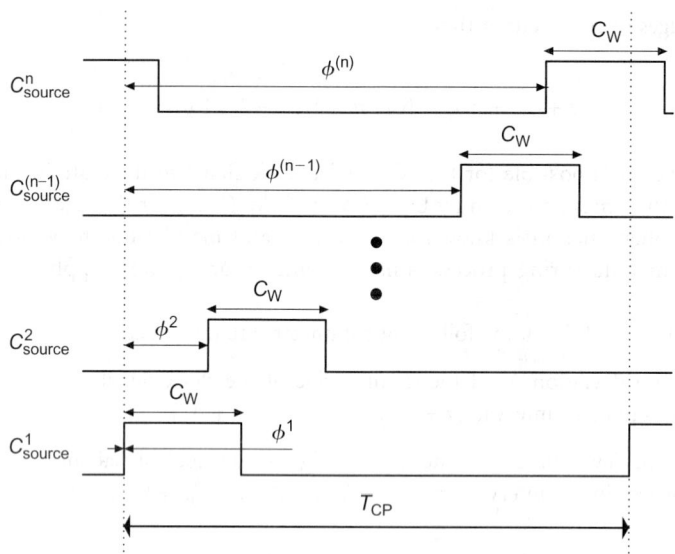

FIGURE 50.15 A generic multiphase synchronization clock.

Multiphase clock signals are generated and used to provide alternate phases of the clock signal for synchronization. Multiphase synchronization clock signals are not as customizable as the clock signals in a multiclock domain application. Multiphase synchronization, however, incurs a smaller design and architecture overhead. Clock waveforms for a multiphase synchronization scheme are shown in Figure 50.15. In Figure 50.15, the set of clock signals $C^{\text{global}} = \{C^1, \ldots, C^m\}$ constitutes the n-phase clocking scheme. The subscripts denote the location of the clock signals in the circuit. For instance, C^1_{source} denotes the clock signal at the clock source of the clock phase C^1. When this clock signal is delivered to an arbitrary register R_k, the (delayed) signal is represented by C^1_k. The start time ϕ^{P_i} of the clock signal phase C^{P_i} is defined with respect to a common reference clock cycle. The *phase shift operator* $\phi^{P_iP_f}$ [53] is used to transform variables among different clock phases. The phase shift operator $\phi^{P_iP_f}$ is defined as the algebraic difference $\phi^{P_iP_f} = \phi^{P_i} - \phi^{P_f} + kT$, where k is the number of clock cycles occurring between phases C^{P_i} and C^{P_f}. Note that for a single-phase clocking scheme, the phase shift operator evaluates to $\phi^{if} = T$.

Dual-phase, nonoverlapping clock signals are easier to generate in practice as compared to generating overlapping phases or arbitrary n-phase clock signals. In the implementation of a dual-phase synchronization scheme, a single-phase clock is typically distributed throughout the circuit and inverted locally to generate the nonoverlapping second phase. Consequently, such dual-phase schemes are more popular as compared to more complex multiphase synchronization schemes. The dual-phase synchronization scheme is particularly popular in level-sensitive circuits. Two-phase, level-sensitive circuits operate similarly to single-phase edge-triggered circuits [54,55], making it possible to implement a logic network with either flip-flop or latch storage elements without major topological changes.

50.4.6.2 Clock Skew in Multiphase Schemes

The definition of clock skew can be extended to those circuits synchronized with a multiphase synchronization scheme (see Section 50.4.6.1). For a multiphase synchronized circuit, Clock i and Clock f (shown in Figure 50.4) are often clock signals with two different phases. Thus, the delays depicted in Figure 50.4 must consider the *phase* difference $\phi^{P_iP_f}$ between the clock signals, and any existing clock skew. Clock skew in a system synchronized with a multiphase synchronization scheme is illustrated in Figure 50.16. In Figure 50.16, clock skew is shown for the generic multiphase synchronization scheme described in Figure 50.15. The

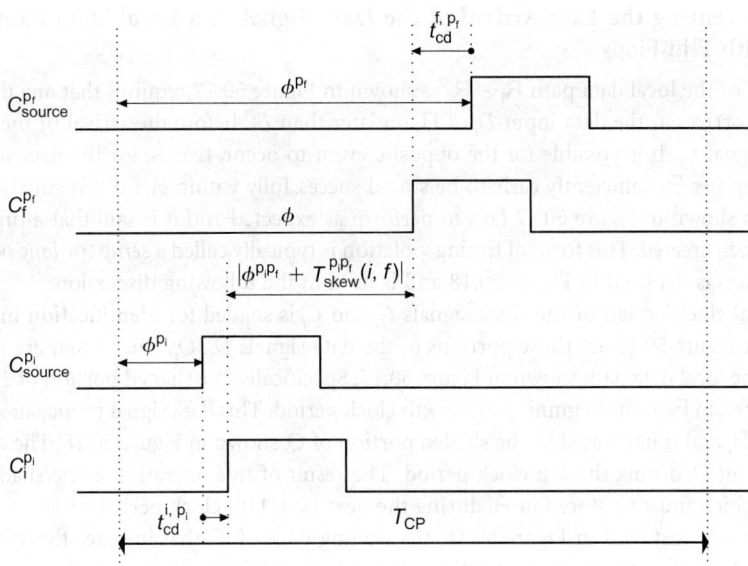

FIGURE 50.16 Multiphase clock skew.

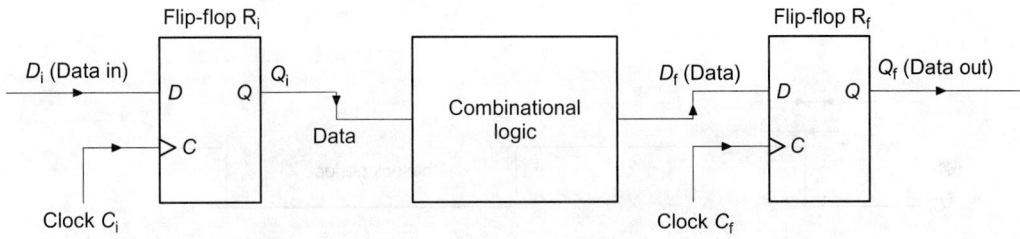

FIGURE 50.17 A single-phase local data path with flip-flops.

parameters t_{cd}^{i,p_i} and t_{cd}^{f,p_f} are the delay of the clock signals $C_i^{p_i}$ and $C_f^{p_f}$ from the clock source to the registers R_i and R_f, respectively. Using this notation, multiphase clock skew is represented as

$$T_{skew}^{p_i p_f}(i, f) = t_{cd}^{i,p_i} - t_{cd}^{f,p_f}$$

In Sections 50.4.7 and 50.4.8, the timing behavior of edge-triggered and level-sensitive circuits, respectively, are analyzed considering a single synchronization clock. The analysis can be systematically extended to circuits synchronized by a multiphase clock, using the multiphase clock skew definition presented above.

50.4.7 Analysis of a Single-Phase Local Data Path with Flip-Flops

A local data path composed of two flip-flops and combinational logic between the flip-flops is shown in Figure 50.17.

An analysis of the timing properties of the local data path shown in Figure 50.17 is presented in the following sections. First, the timing relationships to prevent the late arrival of data signals to R_f are examined in Section 50.4.7.1. The timing relationships to prevent the early arrival of signals to the register R_f are described in Section 50.4.7.2. The analyses presented in Sections 50.4.7.1 and 50.4.7.2 borrow some of the notation from Refs. [13,14]. Similar analyses of synchronous circuits from a timing perspective can be found in Refs. [53,56–59].

50.4.7.1 Preventing the Late Arrival of the Data Signal in a Local Data Path with Flip-Flops

The operation of the local data path $R_i \leadsto R_f$ as shown in Figure 50.17 requires that any data signal that is stored in R_f arrives at the data input D_f of R_f no later than δ_S^{Ff} before the arrival of the latching edge of the clock signal C_f. It is possible for the opposite event to occur, that is, for the data signal D_f *not* to arrive at the register R_f sufficiently early to be stored successfully within R_f. If this situation occurs, the local data path shown in Figure 50.17 fails to perform as expected and it is said that a timing *failure* or *violation* has been created. This form of timing violation is typically called a *setup* (or *long path*) violation. A setup violation is depicted in Figure 50.18 and is used in the following discussion.

The identical clock period of the clock signals C_i and C_f is shaded for identification in Figure 50.18. Also shaded in Figure 50.18 are those portions of the data signals D_i, Q_i, and D_f that are relevant to the operation of the local data path shown in Figure 50.17. Specifically, the shaded portion of D_i corresponds to the data stored in R_i at the beginning of the kth clock period. This data signal propagates to the output of the register R_i and is illustrated by the shaded portion of Q_i shown in Figure 50.18. The combinational logic operates on Q_i during the kth clock period. The result of this operation is the shaded portion of the signal D_f which must be stored in R_f during the next $(k + 1)$th clock period.

Observe that as illustrated in Figure 50.18, the leading edge of C_i that initiates the kth clock period occurs at time $t_{cd}^i + kT_{CP}$. Similarly, the leading edge of C_f that initiates the $(k + 1)$th clock period occurs at time $t_{cd}^f + (k + 1)T_{CP}$. Therefore, the *latest arrival time* t_{AM}^{Ff} of D_f at R_f must satisfy

$$t_{AM}^{Ff} \leq [t_{cd}^f + (k+1)\,T_{CP} - \Delta_L^F] - \delta_S^{Ff} \qquad (50.10)$$

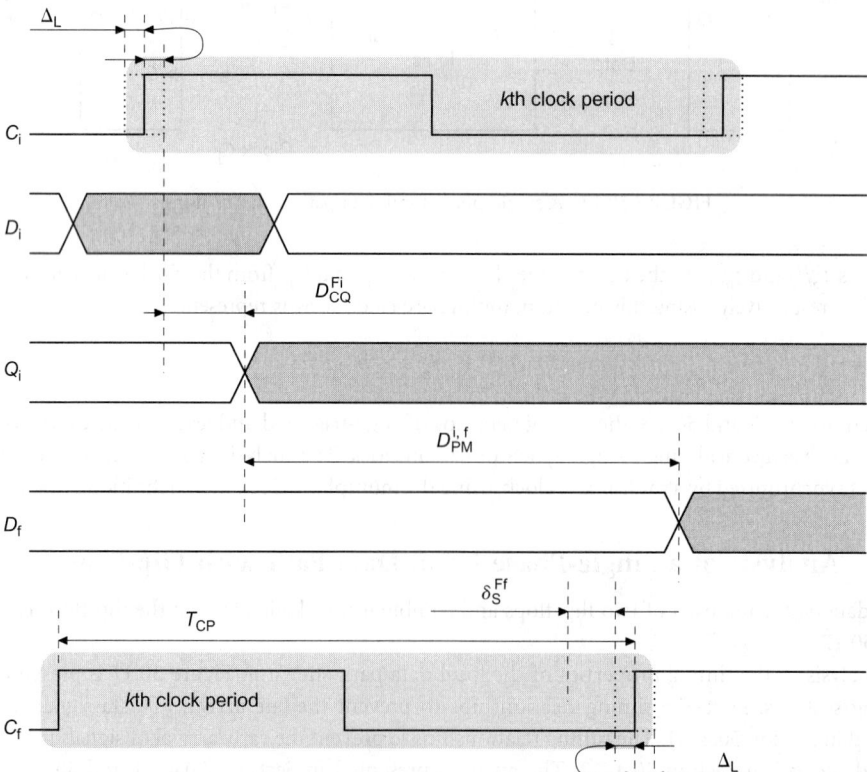

FIGURE 50.18 Timing diagram of a local data path with flip-flops with violation of the setup constraint.

The term $[t_{cd}^f + (k+1)T_{CP} - \Delta_L^F]$ on the right-hand side of Eq. (50.10) corresponds to the critical situation of the leading edge of C_f arriving earlier by the maximum possible deviation Δ_L^F. The $-\delta_S^{Ff}$ term on the right-hand side of Eq. (50.10) accounts for the setup time of R_f (recall the definition of δ_S^F from Section 50.4.5.3). Note that the value of t_{AM}^{Ff} in Eq. (50.10) consists of two components:

1. The latest arrival time t_{QM}^{Fi} that a valid data signal Q_i appears at the output of R_i, that is, the sum $t_{QM}^{Fi} = t_{cd}^i + kT_{CP} + \Delta_L^F + D_{CQM}^{Fi}$ of the latest possible arrival time of the leading edge of C_i and the maximum clock-to-Q delay of R_i.

2. The maximum propagation delay $D_{PM}^{i,f}$ of the data signals through the combinational logic block L_{if} and interconnect along the path $R_i \rightsquigarrow R_f$.

Therefore, t_{AM}^{Ff} can be described as

$$t_{AM}^{Ff} = t_{QM}^{Fi} + D_{PM}^{i,f} = \left(t_{cd}^i + kT_{CP} + \Delta_L^F + D_{CQM}^{Fi}\right) + D_{PM}^{i,f} \tag{50.11}$$

By substituting Eq. (50.11) into Eq. (50.10), the timing condition guaranteeing correct signal arrival at the data input D of R_f is

$$\left(t_{cd}^i + kT_{CP} + \Delta_L^F + D_{CQM}^{Fi}\right) + D_{PM}^{i,f} \leq [t_{cd}^f + (k+1)T_{CP} - \Delta_L^F] - \delta_S^{Ff} \tag{50.12}$$

The above inequality can be transformed by subtracting the kT_{CP} terms from both sides of Eq. (50.12). Furthermore, certain terms in Eq. (50.12) can be grouped together and, by noting that $t_{cd}^i - t_{cd}^f = T_{Skew}(i,f)$ is the clock skew between the registers R_i and R_f,

$$T_{Skew}(i,f) + 2\Delta_L^F \leq T_{CP} - \left(D_{CQM}^{Fi} + D_{PM}^{i,f} + \delta_S^{Ff}\right) \tag{50.13}$$

Note that a violation of Eq. (50.13) is illustrated in Figure 50.18.

The timing relationship Eq. (50.13) represents three important results describing the late arrival of the signal D_f at the data input of the final register R_f in a local data path $R_i \rightsquigarrow R_f$:

1. Given *any* value of $T_{Skew}(i,f), \Delta_L^F, D_{PM}^{i,f}, \delta_S^{Ff}$, and D_{CQM}^{Fi}, the late arrival of the data signal at R_f can be prevented by controlling the value of the clock period T_{CP}. A sufficiently large value of T_{CP} can always be chosen to relax Eq. (50.13) by increasing the upper bound described by the right-hand side of Eq. (50.13).

2. For correct operation, the clock period T_{CP} does *not* necessarily have to be greater than the term $(D_{CQM}^{Fi} + D_{PM}^{i,f} + \delta_S^{Ff})$. If the clock skew $T_{Skew}(i,f)$ is properly controlled, choosing a particular negative value*§ for the clock skew will relax the left side of Eq. (50.13), thereby permitting (13) to be satisfied despite $T_{CP} - (D_{CQM}^{Fi} + D_{PM}^{i,f} + \delta_S^{Ff}) < 0$.

3. Both the term $2\Delta_L^F$ and the term $(D_{CQM}^{Fi} + D_{PM}^{i,f} + \delta_S^{Ff})$ are harmful in the sense that these terms impose a *lower bound* on the clock period T_{CP} (as expected). Although negative skew can be used to relax the inequality Eq. (50.13), these two terms work against relaxing the values of T_{CP} and $T_{Skew}(i,f)$.

Finally, the relationship Eq. (50.13) may be rewritten in a form that clarifies the upper bound on the clock skew $T_{Skew}(i,f)$ imposed by Eq. (50.13):

$$T_{Skew}(i,f) \leq T_{CP} - (D_{CQM}^{Fi} + D_{PM}^{i,f} + \delta_S^{Ff}) - 2\Delta_L^F \tag{50.14}$$

---

*§ More precisely, any negative value within a specified range.

50.4.7.2 Preventing the Early Arrival of the Data Signal in a Local Data Path with Flip-Flops

Late arrival of the signal D_f at the data input of R_f (see Figure 50.17) is analyzed in Section 50.4.7.1. In this section, the analysis of the timing relationships of the local data path $R_i \leadsto R_f$ to prevent early data arrival of D_f is presented. To this end, recall from the discussion in Section 50.4.5.4 that any data signal D_f stored in R_f must lag the arrival of the leading edge of C_f by at least δ_H^{Ff}. It is possible for the opposite event to occur, that is, for a new data D_f^{new} to overwrite the value of D_f and be stored within the register R_f. If this situation occurs, the local data path shown in Figure 50.17 will not perform as desired because of a catastrophic timing violation known as a *hold* (or *short path*) violation.

In this section, hold timing violations are analyzed. It is shown that a hold violation is more dangerous than a setup violation since a hold violation cannot be removed by simply adjusting the clock period T_{CP} (unlike the case of a data signal arriving late where T_{CP} can be increased to satisfy Eq. [50.13]). A hold violation is depicted in Figure 50.19.

Note that in Figure 50.18, a data signal stored in R_i during the kth clock period arrives too late to be stored in R_f during the $(k + 1)$th clock period. In Figure 50.19, however, the data stored in R_i during the kth clock period arrives at R_f too early and *destroys* the data stored in R_f during the same kth clock period. To clarify this concept, certain portions of the data signals in Figure 50.19 are shaded for easy identification. The data D_i stored in R_i at the beginning of the kth clock period is shaded. This data signal propagates to the output of the register R_i and is illustrated by the shaded portion of Q_i shown

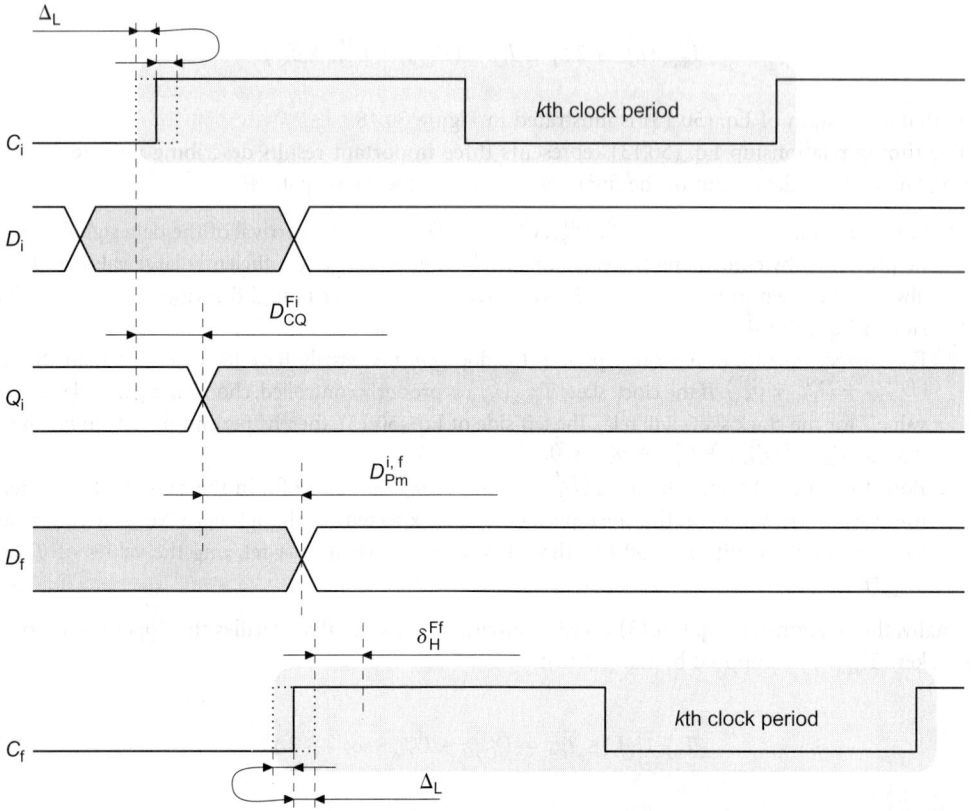

FIGURE 50.19 Timing diagram of a local data path with flip-flops featuring a hold time violation.

in Figure 50.19. The output of the logic (left unshaded part in Figure 50.19) is stored within the register R_f at the beginning of the $(k + 1)$th clock period. Finally, the shaded portion of D_f corresponds to the data that must be stored in R_f at the beginning of the kth clock period.

Note that, as illustrated in Figure 50.19, the leading (or latching) edge of C_i that initiates the kth clock period occurs at time $t_{cd}^i + kT_{CP}$. Similarly, the leading (or latching) edge of C_f that initiates the kth clock period occurs at time $t_{cd}^f + kT_{CP}$. Therefore, the earliest arrival time t_{Am}^{Ff} of the data signal D_f at the register R_f must satisfy the following condition:

$$t_{Am}^{Ff} \geq (t_{cd}^f + kT_{CP} + \Delta_L^F) + \delta_H^{Ff} \tag{50.15}$$

The term $(t_{cd}^f + kT_{CP} + \Delta_L^F)$ on the right-hand side of Eq. (50.15) corresponds to the critical situation of the leading edge of the kth clock period of C_f arriving late by the maximum possible deviation Δ_L^F. Note that the value of t_{Am}^{Ff} in Eq. (50.15) has two components.

1. The earliest arrival time t_{Qm}^{Fi} that a valid data signal Q_i appears at the output of R_i, that is, the sum $t_{Qm}^{Fi} = t_{cd}^i + kT_{CP} - \Delta_L^F + D_{CQm}^{Fi}$ of the earliest arrival time of the leading edge of C_i and the minimum clock-to-Q delay of R_i
2. The minimum propagation delay D_{Pm}^{if} of the signals through the combinational logic block L_{if} and interconnect wires along the path $R_i \leadsto R_f$.

Therefore, t_{Am}^{Ff} can be described as

$$t_{Am}^{Ff} = t_{Qm}^{Fi} + D_{Pm}^{if} = (t_{cd}^i + kT_{CP} - \Delta_L^F + D_{CQm}^{Fi}) + D_{Pm}^{if} \tag{50.16}$$

By substituting Eq. (50.16) into Eq. (50.15), the timing condition that guarantees that D_f does *not* arrive too early at R_f is

$$(t_{cd}^i + kT_{CP} - \Delta_L^F + D_{CQm}^{Fi}) + D_{Pm}^{if} \geq (t_{cd}^f + kT_{CP} - \Delta_L^F) + \delta_H^{Ff} \tag{50.17}$$

The inequality Eq. (50.17) can be further simplified by regrouping terms and noting that $t_{cd}^i - t_{cd}^f = T_{\text{Skew}}(i, f)$ is the clock skew between the registers R_i and R_f.

$$T_{\text{Skew}}(i, f) - 2\Delta_L^F \geq -(D_{CQm}^{Fi} + D_{Pm}^{if}) + \delta_H^{Ff} \tag{50.18}$$

Recall that a violation of Eq. (50.18) is illustrated in Figure 50.19.

The timing relationship described by Eq. (50.18) provides certain important characteristics describing the early arrival of the signal D_f at the data input of the final register R_f of a local data path:

1. Unlike Eq. (50.13), the inequality Eq. (50.18) does not depend on the clock period T_{CP}. Therefore, a violation of Eq. (50.18) *cannot* be corrected by simply manipulating the value of T_{CP}. A synchronous digital system with hold violations is nonfunctional, while a system with setup violations will still operate correctly at a reduced speed.[*][¶] Owing to this behavior, hold violations result in catastrophic timing failure and are considered significantly more dangerous than the setup violations described in Section 50.4.7.1. In conventional zero-clock skew systems, hold violations are typically fixed by inserting delays along the local data paths. For systems where clock skew is manipulated for improved circuit performance (nonzero clock skew circuits), this method is not appropriate.[‖][*]

---

[*][¶]Increasing the clock period T_{CP} to satisfy Eq. (50.13) is equivalent to reducing the frequency of the clock signal.

[*][‖]A delay insertion method can be used on nonzero clock skew circuits as will be discussed in Section 50.4.9.3, but the timing characteristics are different.

---

2. The relationship Eq. (50.18) can be satisfied with a sufficiently large value of clock skew $T_{\text{Skew}}(i, f)$. However, both of the terms $2\Delta_L^F$ and δ_H^{Ff} are harmful in the sense that these terms impose a lower bound on the clock skew $T_{\text{Skew}}(i, f)$. Although positive skew may be used to relax Eq. (50.18), these two terms work against relaxing the values of $T_{\text{Skew}}(i, f)$ and $(D_{CQm}^{Fi} + D_{Pm}^{i,f})$.

Finally, the relationship Eq. (50.18) can be rewritten to stress the lower bound imposed on the clock skew $T_{\text{Skew}}(i, f)$ by Eq. (50.18):

$$T_{\text{Skew}}(i, f) \geq -(D_{Pm}^{i,f} + D_{CQm}^{Fi}) + \delta_H^{Ff} + 2\Delta_L^F \qquad (50.19)$$

50.4.7.3 Clock Skew Scheduling of Edge-Triggered Circuits

Previous research [20,26,37] has indicated that tight control over the clock skew rather than the clock delays is necessary for the circuit to operate reliably. Relationships Eq. (50.14) and Eq. (50.19) are used in Ref. [37] to determine a *permissible range* of the allowed clock skew for each local data path of a circuit with edge-triggered flip-flops. The concept of a permissible range for the clock skew $T_{\text{Skew}}(i, f)$ of a local data path $R_i \rightsquigarrow R_f$ is illustrated in Figure 50.20. For simplicity, the tolerances Δ_L^F and Δ_T^F of the clock signal are ignored in the formulation. When $T_{\text{Skew}}(i, f) \in [(-D_{Pm}^{if} - D_{CQm}^{Fi} + \delta_H^{Ff}), (T_{CP} - D_{PM}^{if} - D_{CQM}^{Fi} + \delta_S^F)]$—as shown in Figure 50.20—Eq. (50.14) and Eq. (50.19) are satisfied. The clock skew $T_{\text{Skew}}(i,f)$ is not *permitted* to be in the interval $(-\infty, -D_{Pm}^{if} - D_{CQm}^{Fi} + \delta_H^{Ff})$ because a race condition will be created. The clock skew is not *permitted* to be in the interval $(T_{CP} - D_{PM}^{if} - D_{CQM}^{Fi} - \delta_S^{Ff}, +\infty)$, either, because in that case, the minimum clock period will be limited.

Also, note that the reliability of a circuit is related to how well the circuit is protected against potential timing violations. Therefore, the reliability of any local data path $R_i \rightsquigarrow R_f$ of a circuit (and therefore of the entire circuit) is increased in two ways:

1. By choosing the clock skew $T_{\text{Skew}}(i, f)$ for a local data path as far as possible from the borders of the permissible range interval, that is, by (ideally) positioning the clock skew $T_{\text{Skew}}(i, f)$ in the middle of the permissible range:

$$T_{\text{Skew}}(i, f) = \frac{1}{2}[T_{CP} - (D_{PM}^{if} + D_{Pm}^{if}) - (D_{CQm}^{Fi} + D_{CQM}^{Fi} + \delta_S^{Ff} - \delta_H^{Ff})]$$

2. By increasing the width of the permissible range of the local data path $R_i \rightsquigarrow R_f$.

Owing to the linear dependence of the clock skews shown in Section 50.3.4.1, however, it may *not* be possible to build a typical circuit such that for each local data path $R_i \rightsquigarrow R_f$, the clock skew $T_{\text{Skew}}(i, f)$ is in the middle of the permissible range. Alternative methods are possible to solve this problem as close as possible to the *ideal* solution [20].

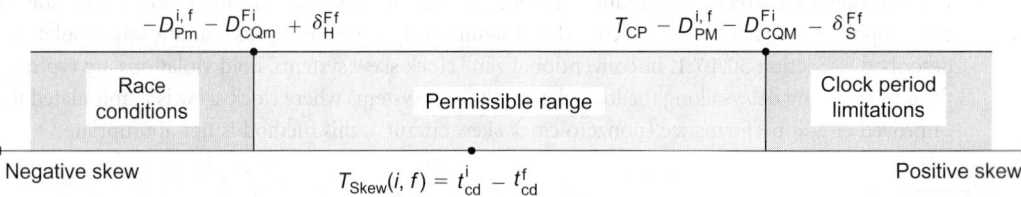

FIGURE 50.20 The permissible range of the clock skew of a local data path $R_i \rightsquigarrow R_f$. A timing violation exists if $T_{\text{Skew}}(i, f) \notin [(-D_{Pm}^{if} - D_{CQm}^{Fi} + \delta_H^{Ff}), (T_{CP} - D_{PM}^{if} - D_{CQM}^{Fi} - \delta_S^{Ff})]$.

TABLE 50.3 LP Model Clock Skew Scheduling of Edge-Triggered Circuits Targeting Minimum Clock Period

$$\begin{aligned}
\min \quad & T_{CP} \\
\text{s.t.} \quad & T_{Skew}(i,f) \leq T_{CP} - D_{PM}^{i,f} - D_{CQM}^{Fi} - \delta_S^{Ff} \\
& T_{Skew}(i,f) \geq -D_{Pm}^{i,f} - D_{CQm}^{Fi} + \delta_H^{Ff}
\end{aligned}$$

TABLE 50.4 QP Model Clock Skew Scheduling of Edge-Triggered Circuits Targeting Safety Against Process Parameter Variations

$$\begin{aligned}
\min \quad & \epsilon = (s-g)^2 = \sum_{k=1}^{p}(s_k - g_k)^2 \\
\text{s.t.} \quad & Bs = 0 \\
& l_k \leq s_k \leq u_k \text{ for } k \in \{1 \cdots p\}
\end{aligned}$$

Fishburn first demonstrated in Ref. [34] how linear programming techniques can be used to solve for a nontrivial clock schedule T_{CD} so as to satisfy Eq. (50.14) and Eq. (50.19) while minimizing the clock period T_{CP}. This linear programming formulation is presented in Table 50.3. Moreover, Fishburn used his LP framework to formulate the clock skew scheduling problem such that the timing reliability of a circuit is improved.

Since Fishburn's pioneering studies, the clock skew scheduling problem of edge-triggered circuits has been extensively addressed by linear and quadratic programming approaches [20,34,60–62]. Most effective of these approaches reported to date is the quadratic programming approach described in Ref. [20], which is given in Table 50.4. In Table 50.4, s are g are vectors of actual and target skews for the local data paths. From a reliability perspective, the target skew g_k for each local data path p_k can be identified as the midpoint of the path dependent permissible ranges. Mathematically, the midpoint is computed $g_k = (l_k + u_k)/2$, where l_k and u_k are the lower and upper bounds of the permissible range illustrated in Figure 50.20. The objective of this QP formulation is to minimize the least square error (LSE) of the actual skew values over the target skew values. The problem constraints are a reiteration of the permissible range requirements illustrated in Figure 50.20 and the linear dependency properties of the clock skew values discussed in Section 50.3.4.1.

Overall, the aggregate of experimental results reported in [20,34,60–63] suggest that clock skew scheduling of edge-triggered circuits permits *approximately* 30% shorter clock periods on average as compared to conventional, zero clock skew, edge-triggered circuits.[†] Most popular solutions exhibit run times comparable to the run times of conventional timing analysis methods of zero clock skew circuits. The scalability of these clock skew scheduling methods, however, are not highly comparable to those of conventional timing analysis methods because of the necessity to simultaneously analyze all (nonzero clock skew) timing paths in a typical clock skew scheduling application.

50.4.8 Analysis of a Single-Phase Local Data Path with Latches

A local data path consisting of two level-sensitive registers (or latches) and the combinational logic between these registers (or latches) is shown in Figure 50.21. Similar to the analysis in Section 50.4.7, a single-phase synchronization clock is selected. An analysis of the timing properties of the local data path shown in Figure 50.21 is offered in the following sections. The timing relationships to prevent the late arrival of the data signal at the latch R_f are examined in Section 50.4.8.1. The timing relationships to prevent the early arrival of the data signal at the latch R_f are examined in Section 50.4.8.2.

[†] The experimental results are reported for clock period minimization of the ISCAS'89 suite of benchmark circuits.

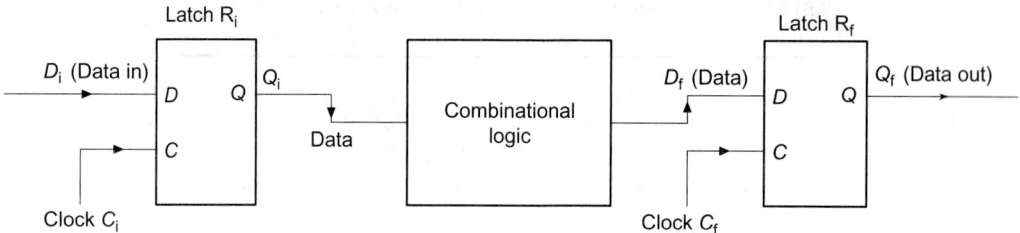

FIGURE 50.21 A single-phase local data path with latches.

The analyses presented in this section build on assumptions regarding the timing relationships among the signals of a latch similar to those assumptions used in Section 50.4.7. Specifically, it is guaranteed that every data signal arrives at the data input of a latch no later than δ_S^L time before the trailing clock edge. Also, this data signal must remain stable at least δ_H^L time after the trailing edge, that is, no new data signal should arrive at a latch δ_H^L time after the latch has become opaque.

Note that these operational properties of latches are not identical to flip-flops. In flip-flops, the setup and hold requirements described above are relative to the *leading*—not to the trailing—edge of the clock signal. This behavior is due to the transparency of the latches during the active level of the clock signal. This transparency permits clock periods smaller than the largest data propagation time, even on a zero clock skew circuit (remember Eq. [50.4]), by arriving after the leading edge of the clock signal and before the trailing edge:

$$\min T_{CP} \leq \max_{\forall \langle R_i, R_f \rangle} (D_{PM}^{if})$$

(50.20)

This operational property of latches is called *time borrowing* [53] (or *cycle stealing* [56]), as the propagation on one local data path *borrows* (or *steals*) time from the propagation on the next local data path by arriving after the leading edge. A nonzero clock skew synchronous circuit with latches can benefit both from the positive impact of clock skew scheduling and the inherent advantageous property of time borrowing.

50.4.8.1 Preventing the Late Arrival of the Data Signal in a Local Data Path with Latches

A data signal propagation scenario similar to the example illustrated in Figure 50.18 is assumed in the following discussion. A data signal D_i is stored in the latch R_i during the kth clock period. The data Q_i stored in R_i propagates through the combinational logic L_{if} and the interconnect along the path $R_i \rightsquigarrow R_f$. In the $(k+1)$th clock period, the result D_f of the computation in L_{if} is stored within the latch R_f. The signal D_f must arrive at least δ_S^L time before the trailing edge of C_f in the $(k+1)$th clock period.

Similar to the discussion presented in Section 50.4.7.1, the *latest arrival time* t_{AM}^{Lf} of D_f at the D input of R_f must satisfy

$$t_{AM}^{Lf} \leq [t_{cd}^f + (k+1)T_{CP} + C_{Wm}^L - \Delta_T^L] - \delta_S^{Lf}$$

(50.21)

Note the difference between Eq. (50.21) and Eq. (50.10). In Eq. (50.10), the first term on the right-hand side is $[t_{cd}^f + (k+1)T_{CP} - \Delta_L^F]$, while in Eq. (50.21), the first term on the right-hand side has an additional term C_{Wm}^L. The addition of C_{Wm}^L is due to the characteristic that unlike flip-flops, a data signal is stored in the latches, shown in Figure 50.21, at the trailing edge of the clock signal (the C_{Wm}^L term). Similar to the case of flip-flops in Section 50.4.7.1, the term $[t_{cd}^f + (k+1)T_{CP} + C_{Wm}^L - \Delta_T^L]$ in the right-hand side of

Eq. (50.21) corresponds to the critical situation of the trailing edge of the clock signal C_f arriving earlier by the maximum possible deviation Δ_T^L.

Observe that the value of t_{AM}^{Lf} in Eq. (50.21) consists of two components:

1. The latest arrival time t_{QM}^{Li} when a valid data signal Q_i appears at the output of the latch R_i
2. The maximum signal propagation delay through the combinational logic block L_{if} and the interconnect along the path $\mathsf{R}_i \rightsquigarrow \mathsf{R}_f$

The arrival time t_{AM}^{Lf} therefore can be defined as

$$t_{AM}^{Lf} = D_{PM}^{i,f} + t_{QM}^{Li} \qquad (50.22)$$

However, unlike the situation of flip-flops discussed in Section 50.4.7.1, the term t_{QM}^{Li} on the right-hand side of Eq. (50.22) is not the sum of the delays through the register R_i. This characteristic occurs because the value of t_{QM}^{Li} depends upon whether the signal D_i arrived *before* or *during* the transparent state of R_i in the kth clock period. The value of t_{QM}^{Li} in Eq. (50.22) is therefore the greater of the following two quantities:

$$t_{QM}^{Li} = \max[(t_{AM}^{Li} + D_{DQM}^{Li}), (t_{cd}^{i} + kT_{CP} + \Delta_L^L + D_{CQM}^{Li})] \qquad (50.23)$$

There are two terms in the right-hand side of Eq. (50.23):

1. The term $(t_{AM}^{Li} + D_{DQM}^{Li})$ corresponds to the situation in which D_i arrives at R_i after the leading edge of the kth clock period
2. The term $(t_{cd}^{i} + kT_{CP} + \Delta_L^L + D_{CQM}^{Li})$ corresponds to the situation in which D_i arrives at R_i before the leading edge of the kth clock pulse arrives

By substituting Eq. (50.23) into Eq. (50.22), the latest time of arrival t_{AM}^{Lf} is

$$t_{AM}^{Lf} = D_{PM}^{i,f} + \max[(t_{AM}^{Li} + D_{DQM}^{Li}), (t_{cd}^{i} + kT_{CP} + \Delta_L^L + D_{CQM}^{Li})] \qquad (50.24)$$

which is substituted into Eq. (50.21) to obtain

$$D_{PM}^{i,f} + \max[(t_{AM}^{Li} + D_{DQM}^{Li}), (t_{cd}^{i} + kT_{CP} + \Delta_L^L + D_{CQ}^{Li})] \le [t_{cd}^{f} + (k+1)T_{CP} + C_{Wm}^L - \Delta_T^L] - \delta_S^{Lf} \qquad (50.25)$$

Eq. (50.25) is an expression for the inequality that must be satisfied to prevent the late arrival of a data signal at the data input D of the register R_f. By satisfying Eq. (50.25), setup violations in the local data path with latches shown in Figure 50.21 are avoided. For a circuit to operate correctly, Eq. (50.25) must be enforced for any local data path $\mathsf{R}_i \rightsquigarrow \mathsf{R}_f$ consisting of the latches R_i and R_f.

The max operation in Eq. (50.25) creates a mathematically difficult situation because it is unknown which of the quantities under the max operation is greater. To overcome this obstacle, this max operation may be split into two conditions:

$$D_{PM}^{i,f} + (t_{AM}^{Li} + D_{DQM}^{Li}) \le [t_{cd}^{f} + (k+1)T_{CP} + C_{Wm}^L + \Delta_T^L] - \delta_S^{Lf} \qquad (50.26)$$

$$D_{PM}^{i,f} + (t_{cd}^{i} + kT_{CP} + \Delta_L^L + D_{CQM}^{Li}) \le [t_{cd}^{f} + (k+1)T_{CP} + C_{Wm}^L - \Delta_T^L] - \delta_S^{Lf} \qquad (50.27)$$

Noting that the clock skew $T_{Skew}(i,f) = t_{cd}^{i} - t_{cd}^{f}$, Eq. (50.26) and Eq. (50.27) can be rewritten as

$$D_{PM}^{i,f} + (t_{AM}^{Li} + D_{DQM}^{Li}) \le [t_{cd}^{f} + (k+1)T_{CP} + C_{Wm}^L - \Delta_T^L] - \delta_S^{Lf} \qquad (50.28)$$

$$T_{\text{Skew}}(i,f) + \left(\Delta_L^L + \Delta_T^L\right) \leq \left(T_{CP} + C_{Wm}^L\right) - \left(D_{CQM}^{Li} + D_{PM}^{if} + \delta_S^{Lf}\right) \tag{50.29}$$

Similar to Sections 50.4.7.1 and 50.4.7.2, Eq. (50.29) can be rewritten in a form that clarifies the upper bound on the clock skew $T_{\text{Skew}}(i,f)$ imposed by Eq. (50.29):

$$D_{PM}^{if} + \left(t_{AM}^{Li} + D_{DQM}^{Li}\right) \leq \left[t_{cd}^f + (k+1)T_{CP} + C_{Wm}^L - \Delta_T^L\right] - \delta_S^{Lf} \tag{50.30}$$

$$T_{\text{Skew}}(i,f) \leq \left(T_{CP} + C_{Wm}^L - \Delta_L^L - \Delta_T^L\right) - \left(D_{CQM}^{Li} + D_{PM}^{if} + \delta_S^{Lf}\right) \tag{50.31}$$

50.4.8.2 Preventing the Early Arrival of the Data Signal in a Local Data Path with Latches

A data signal propagation scenario similar to the example illustrated in Figure 50.19 is assumed in the following discussion. Recall the difference between the late arrival of a data signal at R_f and the early arrival of a data signal at R_f (see Section 50.4.7.2). In the former case, the data signal stored in the latch R_i during the kth clock period arrives too late to be stored in the latch R_f during the $(k+1)$th clock period. In the latter case, the data signal stored in the latch R_i during the kth clock period propagates to the latch R_f too early and overwrites the data signal that was already stored in the latch R_f during the same kth clock period. These constraints hold true for synchronous circuits with latches as well, with some changes due to the subtle differences in operation between flip-flops and latches.

In order for the proper data signal to be successfully latched within R_f during the kth clock period, there should not be any changes in the signal D_f until at least the hold time after the arrival of the storing (trailing) edge of the clock signal C_f. The earliest arrival time t_{Am}^{Lf} of the data signal D_f at the register R_f must therefore satisfy the following condition:

$$t_{Am}^{Lf} \geq \left(t_{cd}^f + kT_{CP} + C_{Wm}^L + \Delta_T^L\right) + \delta_H^{Lf} \tag{50.32}$$

The term $\left(t_{cd}^f + kT_{CP} + C_{Wm}^L + \Delta_T^L\right)$ on the right-hand side of Eq. (50.32) corresponds to the critical situation of the trailing edge of the kth clock period of the clock signal C_f arriving late by the maximum possible deviation Δ_T^L. Note that the value of t_{Am}^{Lf} in Eq. (50.32) consists of two components:

1. The earliest arrival time t_{QM}^{Li} that a valid data signal Q_i appears at the output of the latch R_i, that is, the sum $t_{Qm}^{Li} = t_{cd}^i + kT_{CP} - \Delta_L^L + D_{CQm}^{Li}$ of the earliest arrival time of the leading edge of the clock signal C_i and the minimum clock-to-Q delay D_{CQm}^{Li} of R_f.
2. The minimum propagation delay D_{Pm}^{if} of the signal through the combinational logic L_{if} and the interconnect along the path $R_i \rightsquigarrow R_f$.

Therefore, t_{Am}^{Lf} can be described as

$$t_{Am}^{Lf} = t_{Qm}^{Li} + D_{Pm}^{if} = \left(t_{cd}^i + kT_{CP} - \Delta_L^L + D_{CQm}^{Li}\right) + D_{Pm}^{if} \tag{50.33}$$

By substituting Eq. (50.33) into Eq. (50.32), the timing condition guaranteeing that D_f does *not* arrive too early at the latch R_f is

$$\left(t_{cd}^i + kT_{CP} - \Delta_L^L + D_{CQm}^{Li}\right) + D_{Pm}^{if} \geq \left(t_{cd}^f + kT_{CP} + C_{Wm}^L + \Delta_T^L\right) + \delta_H^{Lf} \tag{50.34}$$

The inequality shown in Eq. (50.34) can be further simplified by reorganizing the terms and noting that $t_{cd}^i - t_{cd}^f = T_{\text{Skew}}(i,f)$ is the clock skew between the registers R_i and R_f:

$$T_{\text{Skew}}(i,f) - \left(\Delta_L^L + \Delta_T^L\right) \geq -\left(D_{CQm}^{Li} + D_{Pm}^{if}\right) + \delta_H^{Lf} \tag{50.35}$$

The timing relationship described by Eq. (50.35) represents two important results describing the early arrival of the signal D_f at the data input of the final latch R_f of a local data path:

1. The relationship in Eq. (50.35) does not depend on the value of the clock period T_{CP}. Therefore, if a hold timing violation in a synchronous system has occurred,[††] this timing violation is catastrophic.
2. The relationship in Eq. (50.35) can be satisfied with a sufficiently large value of the clock skew $T_{Skew}(i,f)$. Furthermore, both the term $(\Delta_L^L + \Delta_T^L)$ and the term δ_H^{Lf} are harmful in the sense that these terms impose a lower bound on the clock skew $T_{Skew}(i,f)$. Although positive skew $T_{Skew}(i,f) > 0$ can be used to relax Eq. (50.35), these two terms make it difficult to satisfy the inequality for specific values of $T_{Skew}(i,f)$ and $(D_{CQm}^{Li} - D_{Pm}^{i,f})$.

Furthermore, the relationship can be rewritten to emphasize the lower bound on the clock skew $T_{Skew}(i,f)$ imposed by Eq. (50.35):

$$T_{Skew}(i,f) \geq (\Delta_L^L + \Delta_T^L) - (D_{CQm}^{Li} + D_{Pm}^{i,f}) + \delta_H^{Lf} \tag{50.36}$$

50.4.8.3 Clock Skew Scheduling of Level-Sensitive Circuits

Level-sensitive circuits are gaining popularity in state-of-the-art integrated circuits due to the smaller size, lower power consumption, and higher speed operation [64–66]. A timing analysis of level-sensitive circuits, however, is difficult, as outlined in Sections 50.4.8.1 and 50.4.8.2. In particular, the transparency property of latches imposes nonlinear timing constraints such as Eq. (50.25).

In conventional timing analysis (without clock skew scheduling), the nonlinearity of the timing constraints are solved with iterative approaches [59,67,68]. The application of clock skew scheduling, however, requires a more sophisticated framework than these types of iterative processes. In Ref. [69], a linear programming approach to solve the clock skew scheduling problem of level-sensitive circuits is proposed. This LP solution, presented in Table 50.5, proposes the mechanics to linearize the originally nonlinear timing constraints. In Table 50.5, the term $FI(j)$ represents the fanin of a register R_j. The tolerances Δ_L^L and Δ_T^L of the clock signal, for simplicity, are ignored in this formulation. This LP problem formulation is used as a framework to address various timing analysis problems of synchronous circuits with latches.

As shown in Refs. [69,70], nonzero clock skew, level-sensitive circuits *might* permit improved circuit performance as compared to nonzero clock skew, edge-triggered circuits. The experimental results reported for nonzero clock skew, level-sensitive circuits indicate that the minimum clock period for this type of circuit

TABLE 50.5 LP Model of Clock Skew Scheduling of Level-Sensitive Circuits Targeting the Minimum Clock Period

$$\min T_{CP} + M[\sum_{\forall j}(t_{Qm}^{Lj} + t_{QM}^{Lj}) + \sum_{\forall j: FI(j) \geq 1}(t_{AM}^{Lj} + t_{Am}^{Lj})]$$

$$\begin{aligned}
s.t. \quad & t_{Am}^{Lf} \geq \delta_H^{Lf} \\
& t_{AM}^{Lf} \leq T_{CP} - \delta_S^{Lf} \\
& t_{Qm}^{Li} \geq t_{Am}^{Li} + D_{DQm}^{Li} \\
& t_{Qm}^{Li} \geq T_{CP} - C_W^L + D_{CQm}^{Li} \\
& t_{QM}^{Li} \geq t_{AM}^{Li} + D_{DQM}^{Li} \\
& t_{QM}^{Li} \geq T_{CP} - C_W^L + D_{DQM}^{Li} \\
& \forall n: t_{Am}^{Lf} \leq t_{Qm}^{Li_n} + D_{Pm}^{i_n,f} + T_{skew}(i_n,f) - T_{CP} \\
& \forall n: t_{AM}^{Lf} \geq t_{QM}^{Li_n} + D_{PM}^{i_n,f} + T_{skew}(i_n,f) - T_{CP} \\
& t_{AM}^{Lf} \geq t_{Am}^{Lf} \\
& t_{QM}^{Lf} \geq t_{Qm}^{Lf}
\end{aligned}$$

[††]As described by the inequality (35) not being satisfied.

is comparable to the minimum clock period observed in nonzero clock skew, edge-triggered circuits (Section 50.4.7.3)—approximately 30% shorter clock periods on average are reported for both types of circuits. However, for some circuits (from the experimental benchmark circuits), level-sensitive circuits are shown to be superior. The improved operational characteristics of these circuits are due to simultaneously considering time borrowing (due to the inherent transparency property of latches) and clock skew scheduling.

50.4.9 Limitations in System Timing

Both zero clock skew and nonzero clock skew circuits are subject to limitations in the minimum clock period at which these circuits are fully operational. Remember from Section 50.3.3 that the limit for a zero clock skew circuit is the slowest local data path of the circuit (the path with the largest data propagation time $D_{PM}^{i,f}$). Consequently, a timing analysis of zero clock skew circuits is centered around identifying the N slowest local data paths of a circuit and ensuring that there are no timing hazards on any of the local data paths for a given clock schedule and a given clock period. Typically, this type of timing analysis is performed with the goal of satisfying all setup time constraints on the N selected paths. As mentioned in Sections 50.4.7 and 50.4.8, this objective can be achieved by lowering the clock frequency until all setup time constraints of the form of Eq. (50.14) [where $T_{Skew}(i, f) = 0$] are satisfied. Any remaining hold time violations can then be removed by inserting delay elements—a procedure called *delay padding* [71].

The limitations on nonzero clock skew circuits are more complicated. These limitations are caused by various circuit topologies and, unlike zero clock skew circuits, both setup and hold time violations are hard to remove. The limitations on the minimum clock period of a nonzero clock skew circuits are caused by the following three factors:

1. Uncertainty of the data propagation time along the local data paths [34]
2. The total data propagation time of the data path cycles [55]
3. The difference between the total data propagation time on reconvergent paths [72]

The first of these three limitations occurs on every single local data path of a synchronous circuit while the second and third limitations only occur on those circuits where the topology of the circuit graph includes cycles and reconvergent paths, respectively. A circuit with all three limitations will ultimately be affected from the most dominant limitation. In this section, these limitations are described for edge-triggered circuits—equivalent limitations on level-sensitive circuits can be similarly derived. To simplify the presentation, it is assumed that the type of limitation that is being discussed is the most dominant.

50.4.9.1 Uncertainty of Data Propagation Times

The uncertainty of the data propagation times is modeled by the min–max timing delay models (Definition 50.3.3) in timing analysis. The algebraic difference between the maximum data propagation time $D_{PM}^{i,f}$ and the minimum data propagation time $D_{Pm}^{i,f}$ on a local data path $R_i \rightsquigarrow R_f$ constitutes the delay uncertainty. For a critical local data path, the trailing edge of the *previous* clock cycle is the hold time before the earliest arrival of the data signal D_f at register R_f. The trailing edge of the *current* clock cycle is the setup time after the latest arrival of the data signal D_f at register R_f. This situation is depicted on an example edge-triggered local data path in Figure 50.22. Note that in Figure 50.22, the tolerance of the clock signals are ignored for the sake of simplicity. For such a critical timing path, the setup and hold time constraints [inequalities Eq. (50.10) and Eq. (50.15), respectively] satisfy the equality conditions.[‡‡] Owing to this limitation, the clock period cannot be minimized any further than

$$\min T_{CP} = \max_{\forall R_i \rightsquigarrow R_f} [D_{PM}^{i,f} + \delta_S^{Ff} - (D_{Pm}^{i,f} + \delta_H^{Ff})] \tag{50.37}$$

The shaded region in Figure 50.22 illustrates the timing criticality, causing the limitation on T_{CP}.

[‡‡]These constraints have no available slack for improvement.

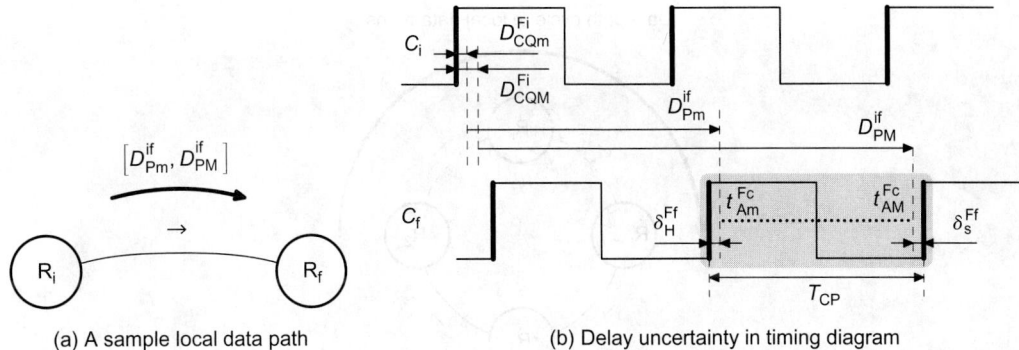

(a) A sample local data path   (b) Delay uncertainty in timing diagram

FIGURE 50.22 Limitation on the minimum clock period T_{CP} caused by the delay uncertainty of a local data path. (a) A sample local data path. (b) Delay uncertainty in timing diagram.

50.4.9.2 Data Path Cycles

Limitations due to data path cycles occur due to the accumulation of the timing relationships over a cycle of local data paths. In a zero clock skew circuit, the circuit topology is almost irrelevant in the timing analysis because each local data path is analyzed independent of any neighbors. The timing of local data paths of a nonzero clock skew circuit, however, is interdependent. For a cycle of local data paths, this interdependency regains the form described in Section 50.3.4.1. In this linear dependency form, the minimum clock period is further limited by the criticality of the local data paths along the cycle (in addition to the limitations caused by the delay uncertainty of each local data path along the cycle). This limitation is illustrated for a sample local data path cycle in Figure 50.23.

The cyclic traveling path for the data signal over a data path cycle, such as the example circuit shown in Figure 50.23, leads to stringent operating conditions under nonzero clock skew. The local data paths along the cycle operate without any slack time, because any existing slack on these local data paths is distributed over the paths through the mechanics of the clock skew scheduling process. In such circuits where a data path cycle is critical, the minimum clock period depends on two factors. The first factor is the number of local data paths n along the cycle. For n local data paths on the cycle, n clock cycles must have passed after each completion of the cycle on a register. The second factor is the total delay of the data signal over the local data paths along the cycle. This total delay time includes the setup time δ_S^{Ff} and maximum clock-to-output time D_{CQM}^{Fi} of each register along the cycle, the maximum data propagation time $D_{PM}^{i,f}$ of each local data path along the cycle, and the tolerances of the clock signal (which are ignored for simplicity). The limitation on the minimum clock period by the data path cycles is given by

$$\min T_{CP} = \frac{\sum_{\forall R_i \to R_f \text{ on cycle}} (D_{CQM}^{Fi} + D_{PM}^{i,f} + \delta_S^{Ff})}{n} \tag{50.38}$$

The shaded region in Figure 50.23 illustrates the timing criticality, causing the limitation on T_{CP}.

50.4.9.3 Reconvergent Paths

A *reconvergent path* is composed of a series of two or more local data paths with a common source register (*divergent* register) and a common sink register (*convergent* register). A *reconvergent system* is composed of at least two parallel reconvergent paths. The interdependency of the timing of local data paths in a nonzero clock skew system occurs explicitly in a reconvergent system because of reconvergent fanout. A data signal that is initially stored in the divergent register starts propagating simultaneously through all of the reconvergent paths but arrives at the convergent register at (possibly) *different* times. In the case of nonidentical numbers of registers in two reconvergent paths, the data signals may arrive at the

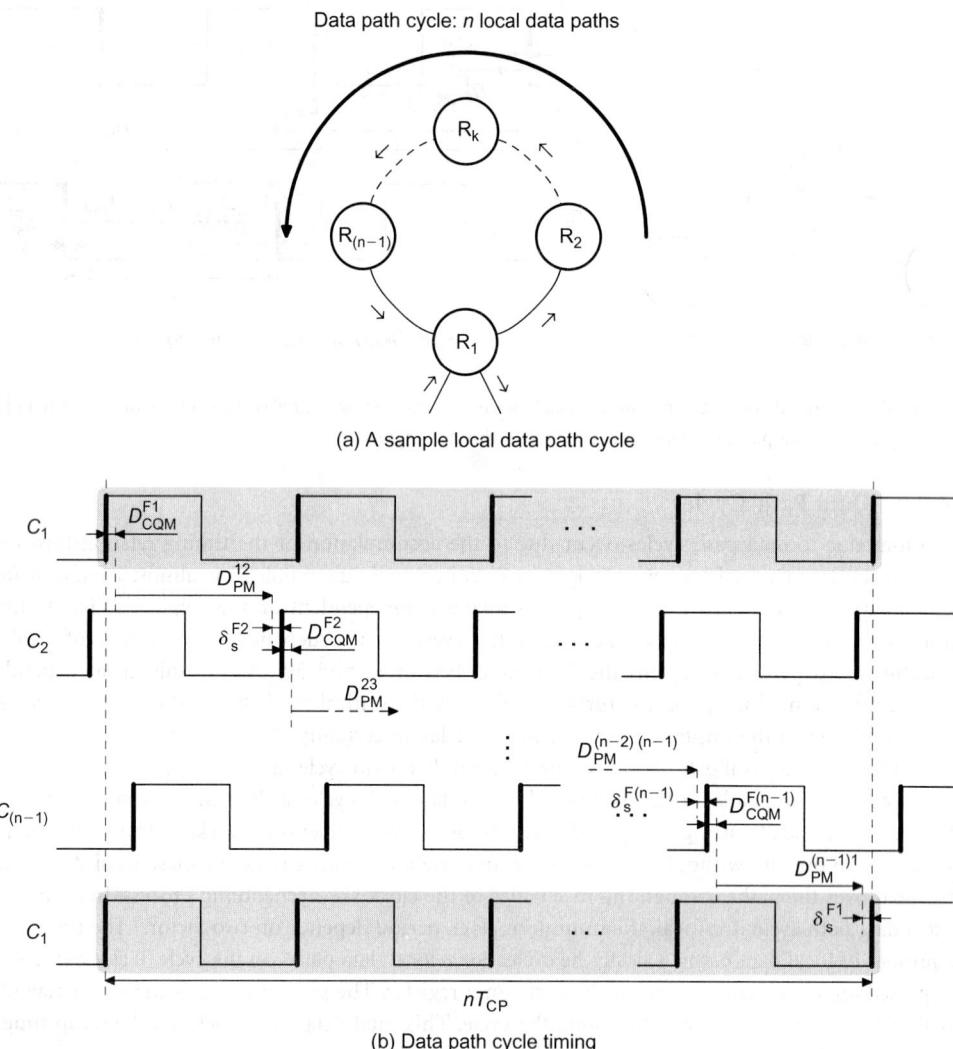

(a) A sample local data path cycle

(b) Data path cycle timing

FIGURE 50.23 Limitation on the minimum clock period T_{CP} caused by data path cycles. (a) A sample local data path cycle. (b) Data path cycle timing.

convergent register during different clock cycles. The timing of all reconvergent paths is satisfied by collectively analyzing the arrival time of the data signals at the convergent register over a duration of (possibly) multiple clock cycles. In Figure 50.24, the limitation such a reconvergent system imposes on the minimum clock period of a nonzero clock skew circuit is illustrated.

In Figure 50.24, two reconvergent paths with m and n registers (excluding divergent and convergent registers) respectively, are considered. The total propagation time of the data signal on the two reconvergent paths are shown. Let the propagation time on the reconvergent paths with m and n registers be the longest and shortest total propagation times, respectively. After propagating along these two paths $(m+1)$ and $(n+1)$ clock cycles must have elapsed, respectively, by the time the data signals arrive at the convergent register. When critical, the reconvergent path with n registers is matched with the trailing edge of the nth clock cycle, while the reconvergent path with m registers is matched with the trailing edge of the $(m+1)$th clock cycle. Thus, the algebraic difference between the two total data propagation

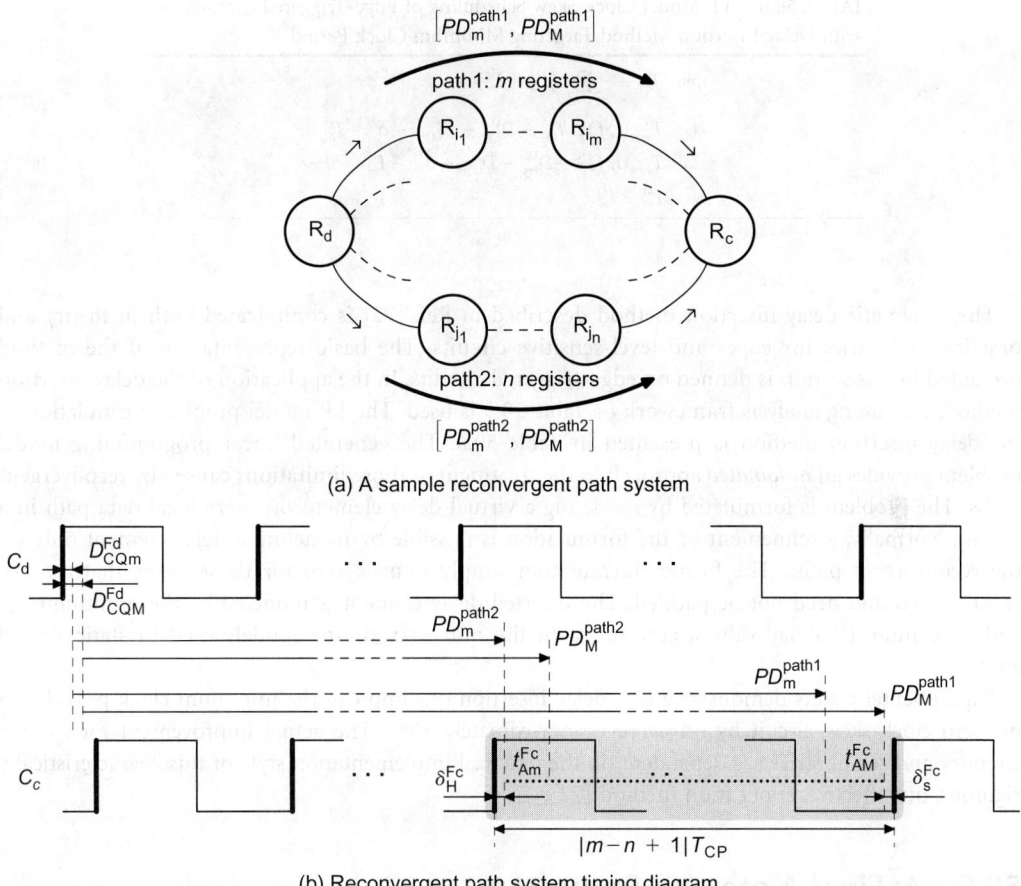

(a) A sample reconvergent path system

(b) Reconvergent path system timing diagram

FIGURE 50.24 Limitation on the minimum clock period T_{CP} caused by reconvergent paths. (a) A sample reconvergent path system. (b) Reconvergent path system timing diagram.

times along the reconvergent paths limits the minimum clock period. Mathematically, the limitation of the reconvergent paths on the minimum clock period of nonzero clock skew circuits is given by

$$\min T_{CP} = \frac{PD_M^{\text{path1}} - PD_m^{\text{path2}} + \delta_S^{Fconvergent} + \delta_H^{Fconvergent}}{|m - n + 1|} \tag{50.39}$$

where PD_M^{pathp} and PD_m^{pathp} represent the maximum and minimum total data propagation times between the divergent and convergent registers over path 1 and path 2, respectively.

Unlike the limitations caused by the delay uncertainty of the local data paths and the total data propagation times along the data path cycles, the limitations caused by reconvergent paths can be mitigated. The mitigation procedure offered in Ref. [72] involves *systematic delay insertion* on one or more of the reconvergent paths to decrease the algebraic difference $(PD_M^{\text{path1}} - PD_m^{\text{path2}})$ of Eq. (50.39), which consequently improves the minimum clock period T_{CP}. Note that it is possible to increase the path delay PD_m^{path2} without increasing PD_M^{path1} because both paths are determined by two different series of local data paths.[†§]

[†§] The minimum and maximum total data propagation times along a reconvergent system may be observed on the same reconvergent path. In such a case, delay insertion is not beneficial.

TABLE 50.6 LP Model Clock Skew Scheduling of Edge-Triggered Circuits with Delay Insertion Method Targeting Minimum Clock Period

$$\min \ T_{CP} + \sum_{\forall \overset{?}{i} \sim \overset{?}{f}} (I_M^{i,f} + I_m^{i,f})$$

$$\text{s.t.} \quad T_{skew}(i,f) \le T_{CP} - D_{PM}^{i,f} - D_{CQM}^{Fi} - \delta_S^{Ff} - I_M^{i,f}$$

$$T_{skew}(i,f) \ge -D_{Pm}^{i,f} - D_{CQm}^{Fi} + \delta_H^{Ff} - I_m^{i,f}$$

$$I_M^{i,f} \ge I_m^{i,f}$$

The systematic delay insertion method described in Ref. [72] is complicated both in theory and practice, and varies for edge- and level-sensitive circuits. The basic representation of the method presented in this section is defined on edge-triggered circuits. In the application of the delay insertion method, the timing analysis framework of Table 50.3 is used. The LP model problem formulation of the delay insertion method is presented in Table 50.6. The generated linear programming model problem provides an *automated* approach to the treatment of those limitations caused by reconvergent paths. The problem is formulated by modeling a virtual delay element on every local data path in a circuit. Normally, a refinement of the formulation is possible by modeling a delay element only on the reconvergent paths. The former formulation simply returns zero for those paths that are not reconvergent and need not be padded. The inserted delay element is modeled by the minimum $I_m^{i,f}$ and maximum $I_M^{i,f}$ delay values, agreeing with the min–max timing models used in static timing analysis.

Experimental results demonstrate that delay insertion can improve the minimum clock period of a nonzero clock skew circuit by on average approximately 10%. The actual improvement for specific circuitry and cell libraries are dependent on the practical implementation style of this characteristically rigorous, but effective, application method [72].

50.5 A Final Note and Summary

In this chapter, the general properties of system timing for synchronous circuits are outlined. The timing properties of registers and local data paths as applicable to overall system timing are analyzed. The timing hazards of synchronous circuits are defined for circuits built with both edge-triggered flip-flops and level-sensitive latches. The benefits of clock skew scheduling in improving circuit performances while eliminating timing hazards are described.

Note that in a fully synchronous digital VLSI system it is possible to encounter types of local data paths different from those circuits analyzed in this section. For example, a local data path may begin with a *positive*-polarity, edge-triggered register R_i and end with a *negative*-polarity, edge-triggered register R_f. It is also possible that different types of registers are used, e.g., a register with more than one data input, or a pulsed latch [73]. In each particular case, the analyses described in this section illustrate a general methodology to determine the proper timing relationships specific to that system. Similar reasoning can be applied to the treatment of other specific timing problems, such as clock period verification [53,59].

References

1. J.S. Kilby, Invention of the integrated circuit, *IEEE Trans. Electron Devices*, vol. ED-23, pp. 648–654, 1976.
2. G.E. Moore, Cramming more components onto integrated circuits, *Electronics*, vol. 38, 1965.
3. T. Takayanagi, J. Shin, B. Petrick, J. Su, H. Levy, H. Pham, J. Son, N. Moon, D. Bistry, M. Singh, V. Mathur, and A. Leon, A dual-core 64 b ultraSPARC microprocessor for dense server applications, *Proceedings of the IEEE International Solid-State Circuits Conference*, vol. 2, pp. 58–513, February 2004.

4. J. Clabes, J. Friedrich, M. Sweet, J. Dilullo, S. Chu, D. Plass, J. Dawson, P. Muench, L. Powell, M. Floyd, B. Sinharoy, M. Lee, M. Goulet, J. Wagoner, N. Schwartz, S. Runyon, G. Gorman, P. Restle, R. Kalla, J. McGill, and S. Dodson, Design and implementation of the POWER5^{TM} microprocessor, *Proceedings of the IEEE International Solid-State Circuits Conference*, vol. 1, pp. 56–57, February 2004.

5. S. Naffziger, B. Stackhouse, and T. Grutkowski, The implementation of a 2-core multi-threaded Itanium-family processor, *Proceedings of the IEEE International Solid-State Circuits Conference*, pp. 182–184, February 2005.

6. D. Pham, S. Asano, M. Bolliger, M. Day, H. Hofstee, C. Johns, J. Kahle, A. Kameyama, J. Keaty, Y. Masubuchi, M. Riley, D. Shippy, D. Stasiak, M. Suzuoki, M. Wang, J. Warnock, S. Weitzel, D. Wendel, T. Yamazaki, and K. Yazawa, The design and implementation of a first-generation CELL processor, *Proceedings of the IEEE International Solid-State Circuits Conference*, pp. 184–186, February 2005.

7. H.B. Bakoglu, *Circuits, Interconnections, and Packaging for VLSI.* Addison-Wesley, 1990.

8. P. Bai, C. Auth, S. Balakrishnan, M. Bost, R. Brain, V. Chikarmane, R. Heussner, M. Hussein, J. Hwang, D. Ingerly, R. James, J. Jeong, C. Kenyon, E. Lee, S.-H. Lee, N. Lindert, M. Liu, Z. Ma, T. Marieb, A. Murthy, R. Nagisetty, S. Natarajan, J. Neirynck, A. Ott, C. Parker, J. Sebastian, R. Shaheed, S. Sivakumar, J. Steigerwald, S. Tyagi, C. Weber, B. Woolery, A. Yeoh, K. Zhang, and M. Bohr, A 65nm logic technology featuring 35nm gate lengths, enhanced channel strain, 8 Cu interconnect layers, low-k ILD and 0.57 μm^2 SRAM cell, *Proceedings of the IEEE International Electron Devices Meeting*, pp. 657–660, December 2004.

9. N.W. Weste and K. Eshraghian, *Principles of CMOS VLSI Design: A Systems Perspective.* Addison-Wesley, 2nd ed., 1992.

10. J.P. Uyemura, *Introduction to VLSI Circuits and Systems.* Wiley, 2002.

11. C. Mead and L. Conway, *Introduction to VLSI Systems.* Addison-Wesley, 1980.

12. F. Anceau, A synchronous approach for clocking VLSI systems, *IEEE J. Solid-State Circuits*, vol. SC-17, pp. 51–56, 1982.

13. M. Afghani and C. Svensson, A unified clocking scheme for VLSI systems, *IEEE J. Solid State Circuits*, vol. SC-25, pp. 225–233, 1990.

14. S.H. Unger and C.-J. Tan, Clocking schemes for high-speed digital systems, *IEEE Trans. Comput.*, vol. C-35, pp. 880–895, 1986.

15. E.G. Friedman, *Clock Distribution Networks in VLSI Circuits and Systems.* IEEE Press, 1995.

16. S. Bothra, B. Rogers, M. Kellam, and C.M. Osburn, Analysis of the effects of scaling on interconnect delay in ULSI circuits, *IEEE Trans. Electron Devices*, vol. ED-40, pp. 591–597, 1993.

17. Y.I. Ismail and E.G. Friedman, Effects of inductance on the propagation delay and repeater insertion in VLSI circuits, *IEEE Trans. VLSI Syst.*, vol. 8, pp. 195–206, April 2000.

18. M. Saint-Laurent, M. Swaminathan, and J. Meindl, On the micro-architectural impact of clock distribution using multiple PLLs, *Proceedings of the IEEE International Conference on Computer Design*, pp. 214–220, September 2001.

19. J. Wood, T. Edwards, and S. Lipa, Rotary traveling-wave oscillator arrays: A new clock technology, *IEEE J. Solid-State Circuits*, vol. 36, pp. 1654–1665, 2001.

20. I.S. Kourtev and E.G. Friedman, *Timing Optimization Through Clock Skew Scheduling.* Kluwer Academic, 2000.

21. C.E. Leiserson and J.B. Saxe, A mixed-integer linear programming problem which is efficiently solvable, *J. Algorithms*, vol. 9, pp. 114–128, 1988.

22. T.H. Cormen, C.E. Leiserson, and R.L. Rivest, *Introduction to Algorithms.* MIT Press, 1989.

23. D.B. West, *Introduction to Graph Theory.* Prentice-Hall, 1996.

24. J.L. Neves and E.G. Friedman, Topological design of clock distribution networks based on non-zero clock skew specification, *Proceedings of the IEEE Midwest Symposium on Circuits and Systems*, pp. 468–471, August 1993.

25. J.G. Xi and W.W.-M. Dai, Useful-skew clock routing with gate sizing for low power design, *Proceedings of the ACM/IEEE Design Automation Conference*, pp. 383–388, June 1996.

26. J.L. Neves and E.G. Friedman, Design methodology for synthesizing clock distribution networks exploiting non–zero localized clock skew, *IEEE Trans. VLSI Syst.*, vol. VLSI-4, pp. 286–291, 1996.

27. M.A.B. Jackson, A. Srinivasan, and E.S. Kuh, Clock routing for high-performance ICs, *Proceedings of the ACM/IEEE Design Automation Conference*, pp. 573–579, June 1990.

28. R.-S. Tsay, An exact zero-skew clock routing algorithm, *IEEE Trans. Computer-Aided Design Integrated Circuits Syst.*, vol. CAD-12, pp. 242–249, 1993.

29. N.-C. Chou and C.-K. Cheng, On general zero-skew clock net construction, *IEEE Trans. VLSI Systems*, vol. VLSI-3, pp. 141–146, 1995.

30. N. Ito, H. Sugiyama, and T. Konno, ChipPRISM: Clock routing and timing analysis for high-performance CMOS VLSI chips, *Fujitsu Sci. Tech. J.*, vol. 31, pp. 180–187, 1995.

31. N. Gaddis and J. Lotz, A 64-b quad-issue CMOS RISC microprocessor, *IEEE J. Solid-State Circuits*, vol. SC-31, pp. 1697–1702, 1996.

32. W.J. Bowhill et al., Circuit implementation of a 300-MHz 64-bit second-generation CMOS alpha CPU, *Digital Tech. J.*, vol. 7, pp. 100–118, 1995.

33. T.-C. Lee and J. Kong, The new line in IC design, *IEEE Spectrum*, pp. 52–58, March 1997.

34. J.P. Fishburn, Clock skew optimization, *IEEE Trans. Comput.*, vol. C–39, pp. 945–951, 1990.

35. E.G. Friedman, The application of localized clock distribution design to improving the performance of retimed sequential circuits, *Proceedings of the IEEE Asia-Pacific Conference on Circuits and Systems*, pp. 12–17, December 1992.

36. I.S. Kourtev and E.G. Friedman, Simultaneous clock scheduling and buffered clock tree synthesis, *Proceedings of the IEEE International Symposium on Circuits and Systems*, pp. 1812–1815, June 1997.

37. J.L. Neves and E.G. Friedman, Optimal clock skew scheduling tolerant to process variations, *Proceedings of the ACM/IEEE Design Automation Conference*, pp. 623–628, June 1996.

38. L.A. Glasser and D.W. Dobberpuhl, *The Design and Analysis of VLSI Circuits*. Addison-Wesley, 1985.

39. J.P. Uyemura, *Circuit Design for CMOS VLSI*. Kluwer Academic, 1992.

40. S.-M. Kang and Y. Leblebici, *CMOS Digital Integrated Circuits: Analysis and Design*. McGraw-Hill, 1996.

41. A.S. Sedra and K.C. Smith, *Microelectronic Circuits*. Oxford University Press, 4th ed., 1997.

42. Z. Kohavi, *Switching and Finite Automata Theory*. McGraw-Hill, 2nd ed., 1978.

43. M.M. Mano and C.R. Kime, *Logic and Computer Design Fundamentals*. Prentice-Hall, Inc., 1997.

44. W. Wolf, *Modern VLSI Design: A Systems Approach*. Prentice-Hall, Inc., 1994.

45. T. Kacprzak and A. Albicki, Analysis of metastable operation in RS CMOS flip-flops, *IEEE J. Solid-State Circuits*, vol. SC-22, pp. 57–64, 1987.

46. T.A. Jackson and A. Albicki, Analysis of metastable operation in D latches, *IEEE Trans. Circuits Syst. I Fundamental Theory Appl.*, vol. CAS I–36, pp. 1392–1404, 1989.

47. E.G. Friedman, Latching characteristics of a CMOS bistable register, *IEEE Trans. Circuits Syst. I Fundamental Theory Appl.*, vol. CAS I–40, pp. 902–908, 1993.

48. S.H. Unger, Double-edge-triggered flip-flops, *IEEE Trans. Comput.*, vol. C-30, pp. 447–451, 1981.

49. S.-L. Lu, A novel CMOS implementation of double-edge-triggered D-flip-flops, *IEEE J. Solid State Circuits*, vol. SC-25, pp. 1008–1010, 1990.

50. M. Afghani and J. Yuan, Double-edge-triggered D-flip-flops for high-speed CMOS circuits, *IEEE J. Solid State Circuits*, vol. SC-26, pp. 1168–1170, 1991.

51. R. Hossain, L. Wronski, and A. Albicki, Double edge triggered devices: Speed and power constraints, *Proceedings of the IEEE International Symposium on Circuits and Systems*, vol. 3, pp. 1491–1494, May 1993.

52. G.M. Blair, Low-power double-edge triggered flipflop, *Electron. Lett.*, vol. 33, pp. 845–847, 1997.

53. M.R. Dagenais and N.C. Rumin, On the calculation of optimal clocking parameters in synchronous circuits with level-sensitive latches, *IEEE Trans. Computer-Aided Design*, vol. CAD-8, pp. 268–278, 1989.

54. A. Ishii, C. Leiserson, and M. Papaefthymiou, Optimizing two-phase, level-clocked circuitry, *Proceedings of the Brown/MIT Conference: Advanced Research on VLSI Parallel Systems*, pp. 245–264, March 1992.

55. M.C. Papaefthymiou and K. Randall, Edge-triggering vs. two-phase level-clocking, *Proceedings of the Symposium on Research in Integrated Systems*, pp. 201–218, March 1993.

56. I. Lin, J.A. Ludwig, and K. Eng, Analyzing cycle stealing on synchronous circuits with level-sensitive latches, *Proceedings of the ACM/IEEE Design Automation Conference*, pp. 393–398, June 1992.
57. J. Lee, D.T. Tang, and C.K. Wong, A timing analysis algorithm for circuits with level-sensitive latches, *IEEE Trans. Computer-Aided Design*, vol. CAD-15, pp. 535–543, 1996.
58. T.G. Szymanski, Computing optimal clock schedules, *Proceedings of the ACM/IEEE Design Automation Conference*, pp. 399–404, June 1992.
59. K.A. Sakallah, T.N. Mudge, and O.A. Olukotun, $checkT_c$ and $minT_c$: Timing verification and optimal clocking of synchronous digital circuits, *Proceedings of the IEEE/ACM International Conference on Computer-Aided Design*, pp. 552–555, November 1990.
60. C. Albrecht, B. Korte, J. Schietke, and J. Vygen, Cycle time and slack optimization for VLSI-chips, *Proceedings of the IEEE/ACM International Conference on Computer-Aided Design*, pp. 232–238, November 1999.
61. S. Held, B. Korte, J. Massberg, M. Ringe, and J. Vygen, Clock scheduling and clocktree construction for high performance ASICs, *Proceedings of the International Conference on Computer-Aided Design*, pp. 232–239, November 2003.
62. K. Ravindran, A. Kuehlmann, and E. Sentovich, Multi-domain clock skew scheduling, *Proceedings of the International Conference on Computer-Aided Design*, pp. 801–808, November 2003.
63. R. Mader, E.G. Friedman, A. Litman, and I.S. Kourtev, Large scale clock skew scheduling techniques for improved reliability of digital synchronous circuits, *Proceedings of the IEEE International Symposium on Circuits and Systems*, vol. 1, pp. 357–360, May 2002.
64. S. Naffziger, G. Colon-Bonet, T. Fischer, R. Riedlinger, T. Sullivan, and T. Grutkowski, The implementation of the Itanium 2 microprocessor, *IEEE J. Solid-State Circuits*, vol. 37, pp. 1448–1460, 2002.
65. J. Warnock, Circuit design issues for the POWER4 chip, *Proceedings of the International Symposium on VLSI Technology, Systems, and Applications*, pp. 125–128, October 2003.
66. C. Webb, C. Anderson, L. Sigal, K. Shepard, J. Liptay, J.D. Warnock, B. Curran, B. Krumm, M. Mayo, P. Camporese, E. Schwarz, M. Farrell, P. Restle, R. Averill III, T. Slegel, W. Houtt, Y. Chan, B. Wile, T. Nguyen, P. Emma, D. Beece, C. Ching-Te, and C. Price, A 400-MHz S/390 microprocessor, *IEEE J. Solid-State Circuits*, vol. 32, pp. 1665–1675, 1997.
67. T.M. Burks, K.A. Sakallah, and T.N. Mudge, Critical paths in circuits with level-sensitive latches, *IEEE Trans. VLSI Syst.*, vol. 3, pp. 273–291, 1995.
68. T.G. Szymanski and N. Shenoy, Verifying clock schedules, *Proceedings of the IEEE/ACM International Conference on Computer-Aided Design*, pp. 124–131, November 1992.
69. B. Taskin and I.S. Kourtev, Linear timing analysis of SOC synchronous circuits with level-sensitive latches, *Proceedings of the IEEE ASIC/SOC Conference*, pp. 358–362, September 2002.
70. B. Taskin and I.S. Kourtev, Linearization of the timing analysis and optimization of level-sensitive digital synchronous circuits, *IEEE Trans. VLSI Syst.*, vol. 12, pp. 12–27, 2004.
71. N. Shenoy, R.K. Brayton, and A.L. Sangiovanni-Vincentelli, Minimum padding to satisfy short path constraints, *Proceedings of the IEEE/ACM International Conference on Computer-Aided Design*, pp. 156–161, November 1993.
72. B. Taskin and I.S. Kourtev, Delay insertion in clock skew scheduling, *Proceedings of the ACM International Symposium on Physical Design*, pp. 47–54, April 2005.
73. D. Harris, *Skew-Tolerant Circuit Design*. Morgan Kaufmann, 2001.

Appendix

Glossary of Terms

The following notations are used in this section:

1. *Clock Signal Parameters*

T_{CP}: the clock period of a circuit

Δ_L: the tolerance of the leading edge of any clock signal

Δ_T: the tolerance of the trailing edge of any clock signal

Δ_L^L: the tolerance of the leading edge of a clock signal driving a latch

Δ_T^L: the tolerance of the trailing edge of a clock signal driving a latch

Δ_L^F: the tolerance of the leading edge of a clock signal driving a flip-flop

Δ_T^F: the tolerance of the trailing edge of a clock signal driving a flip-flop

C_{Wm}^L: the minimum width of the clock signal in a circuit with latches

C_{Wm}^F: the minimum width of the clock signal in a circuit with flip-flops

C^{p_i}: the clock signal phase p_i

ϕ^{p_i}: the delay of clock signal C^{p_i} with respect to common clock cycle

t_{cd}^i: the delay of single-phase clock signal phase at register R_i with respect to common time reference

t_{cd}^{i,p_i}: the delay of clock signal phase p_i at register R_i with respect to common time reference

2. *Latch Parameters*

D_{CQ}^L: the clock-to-output delay of a latch

D_{CQ}^{Li}: the clock-to-output delay of the latch R_i

D_{CQm}^L: the minimum clock-to-output delay of a latch

D_{CQm}^{Li}: the minimum clock-to-output delay of the latch R_i

D_{CQM}^L: the maximum clock-to-output delay of a latch

D_{CQM}^{Li}: the maximum clock-to-output delay of the latch R_i

D_{DQ}^L: the data-to-output delay of a latch

D_{DQ}^{Li}: the data-to-output delay of the latch R_i

D_{DQm}^L: the minimum data-to-output delay of a latch

D_{DQm}^{Li}: the maximum data-to-output delay of the latch R_i

D_{DQM}^L: the minimum data-to-output delay of a latch

D_{DQM}^{Li}: the maximum data-to-output delay of the latch R_i

δ_S^L: the setup time of a latch

δ_S^{Li}: the setup time of the latch R_i

δ_H^L: the hold time of a latch

δ_H^{Li}: the hold time of the latch R_i

t_{AM}^L: the latest arrival time of the data signal at the data input of a latch

t_{AM}^{Li}:   the latest arrival time of the data signal at the data input of the latch R_i

t_{Am}^{L}:   the earliest arrival time of the data signal at the data input of a latch

t_{Am}^{Li}:   the earliest arrival time of the data signal at the data input of the latch R_i

t_{QM}^{L}:   the latest arrival time of the data signal at the data output of the latch

t_{QM}^{Li}:   the latest arrival time of the data signal at the data output of the latch R_i

t_{Qm}^{L}:   the earliest arrival time of the data signal at the data output of a latch

t_{Qm}^{Li}:   the earliest arrival time of the data signal at the data output of the latch R_i

3. *Flip-Flop Parameters*

D_{CQ}^{F}:   the clock-to-output delay of a latch

D_{CQ}^{Fi}:   the clock-to-output delay of the latch R_i

D_{CQm}^{F}:   the minimum clock-to-output delay of a latch

D_{CQm}^{Fi}:   the minimum clock-to-output delay of the latch R_i

D_{CQM}^{F}:   the maximum clock-to-output delay of a latch

D_{CQM}^{Fi}:   the maximum clock-to-output delay of the latch R_i

δ_{S}^{F}:   the setup time of a latch

δ_{S}^{Fi}:   the setup time of the latch R_i

δ_{H}^{F}:   the hold time of a latch

δ_{H}^{Fi}:   the hold time of the latch R_i

t_{AM}^{F}:   the latest arrival time of the data signal at the data input of a latch

t_{AM}^{Fi}:   the latest arrival time of the data signal at the data input of the latch R_i

t_{Am}^{F}:   the earliest arrival time of the data signal at the data input of a latch

t_{Am}^{Fi}:   the earliest arrival time of the data signal at the data input of the latch R_i

t_{QM}^{F}:   the latest arrival time of the data signal at the data output of a latch

t_{QM}^{Fi}:   the latest arrival time of the data signal at the data output of the latch R_i

t_{Qm}^{F}:   the earliest arrival time of the data signal at the data output of a latch

t_{Qm}^{Fi}:   the earliest arrival time of the data signal at the data output of the latch R_i

4. *Local Data Path Parameters*

$R_i \rightsquigarrow R_f$:   a local data path from register R_i to register R_f exists

$R_i \not\rightsquigarrow R_f$:   a local data path from register R_i to register R_f does not exist

51

ROM/PROM/EPROM

Jen-Sheng Hwang
National Science Council, Taiwan

CONTENTS

51.1 Introduction

Read-only memory (ROM) is the densest form of semiconductor memory, which is used for the applications such as video game software, laser printer fonts, dictionary data in word processors, and sound-source data in electronic musical instruments.

The ROM market segment grew well through the first half of the 1990s, closely coinciding with a jump in personal computer (PC) sales and other consumer-oriented electronic systems, as shown in Figure 51.1 [1]. Because a very large ROM application base (video games) moved toward compact disc ROM-based systems (CD-ROM), the ROM market segment declined. However, greater functionality memory products have become relatively cost-competitive with ROM. It is believed that the ROM market will continue to grow moderately through the year 2003.

51.2 ROM

Read-only memories (ROMs) consist of an array of core cells whose contents or state is preprogrammed by using the presence or absence of a single transistor as the storage mechanism during the fabrication process. The contents of the memory are therefore maintained indefinitely regardless of the previous history of the device and/or the previous state of the power supply.

51.2.1 Core Cells

A binary core cell stores binary information through the presence or absence of a single transistor at the intersection of the wordline and bitline. ROM core cells can be connected in two possible ways: a parallel NOR array of cells or a series NAND array of cells each requiring one transistor per storage cell. In this case, either connecting or disconnecting the drain connection from the bitline programs the ROM cell.

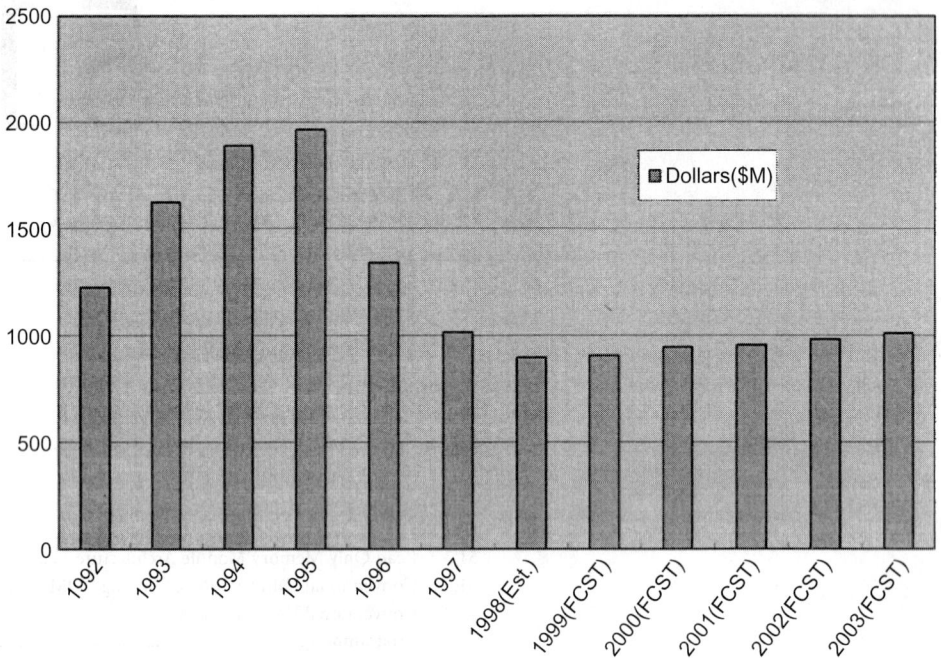

FIGURE 51.1 The ROM market growth and forecast.

The NOR array is larger as there is potentially one drain contact per transistor (or per cell) made to each bitline. Potentially, the NOR array is faster as there are no serially connected transistors as in the NAND array approach. However, the NAND array is much more compact as no contacts are required within the array itself. However, the serially connected pull-down transistors that comprise the bitline are potentially very slow [2].

Encoding multiple-valued data in the memory array involves a one-to-one mapping of logic value to transistor characteristics at each memory location and can be implemented in two ways:

(i) adjust the width-to-length (W/L) ratios of the transistors in the core cells of the memory array; or
(ii) adjust the threshold voltage of the transistors in the core cells of the memory array [3].

The first technique works on the principle that W/L ratio of a transistor determines the amount of current that can flow through the device (i.e., the transconductance). This current can be measured to determine the size of the device at the selected location and hence the logic value stored at this location. In order to store 2 bits per cell, one would use one of four discrete transistor sizes. Intel Corp. used this technique in the early 1980s to implement high-density look-up tables in its i8087 math co-processor. Motorola Inc. also introduced a four-state ROM cell with an unusual transistor geometry that had variable W/L devices. The conceptual electrical schematic of the memory cell along with the surrounding peripheral circuitry is shown in Figure 51.2 [2].

51.2.2 Peripheral Circuitry

The four states in a two-bit per cell ROM are four distinct current levels. There are two primary techniques to determine which of the four possible current levels an addressed cell generates. One technique compares the current generated by a selected memory cell against three reference cells using three separate sense amplifiers. The reference cells are transistors with W/L ratios that fall in between the four possible standard transistor sizes found in the memory array as illustrated in Figure 51.3 [2].

the approach is similar. A low-resolution sense-amp device (CAD converter) can distinguish between reading a two-bit per cell device is possible since the different states for linearly rising voltages to match the input voltages of the cell. This time signal can then be mapped to the equivalent two-bit binary code. In this operation, the memory is accessed.

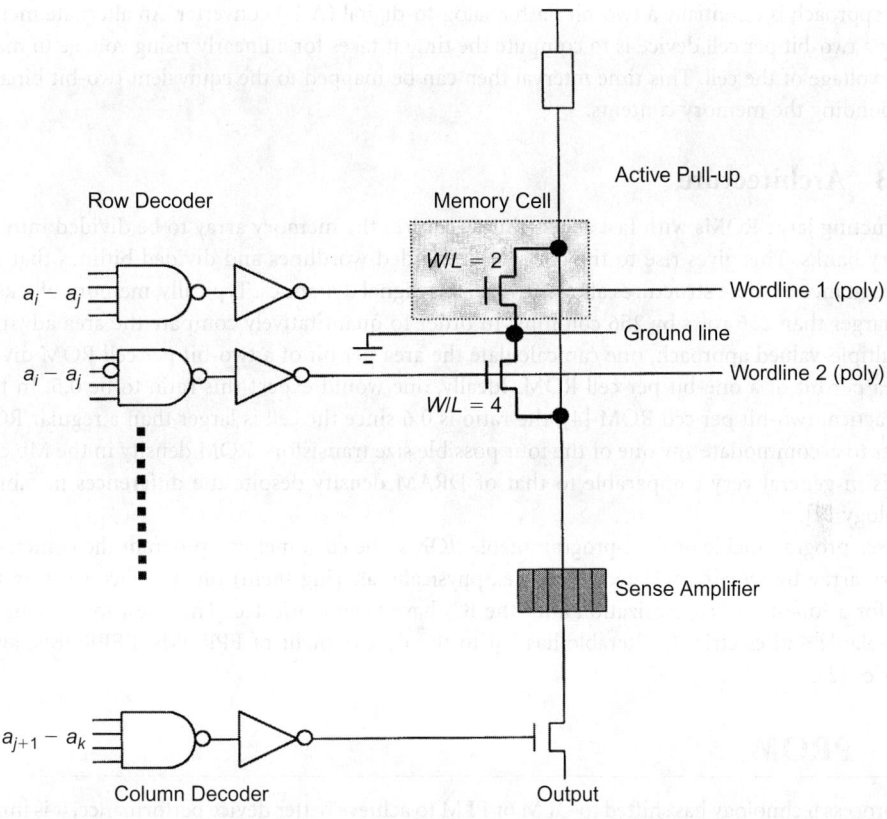

FIGURE 51.2 Geometry-variable multiple-valued NOR ROM.

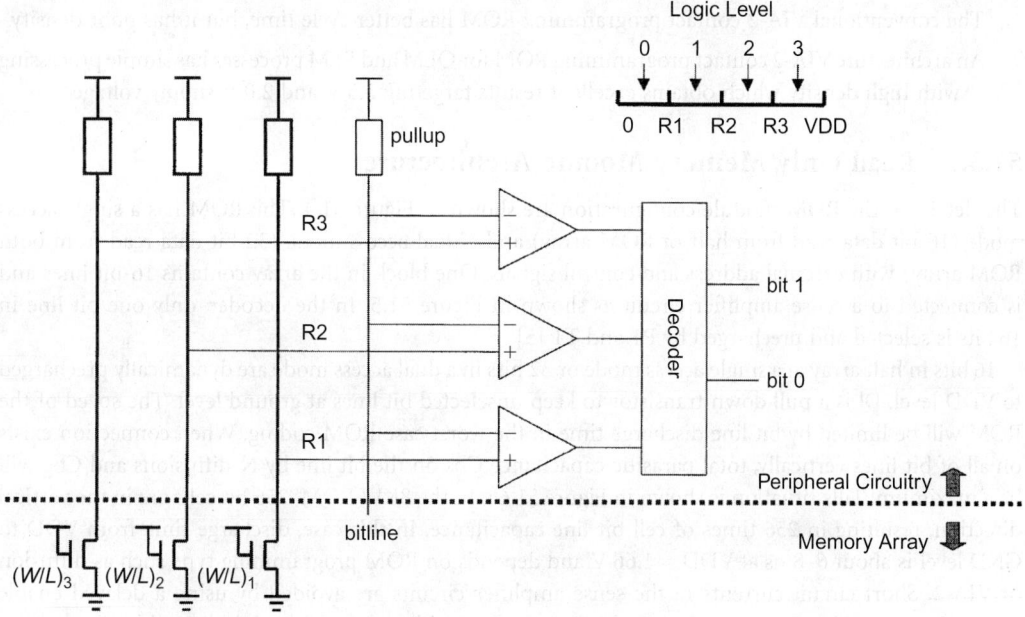

FIGURE 51.3 ROM sense amplifier.

The approach is essentially a two-bit flash analog-to-digital (A/D) converter. An alternate method for reading a two-bit per cell device is to compute the time it takes for a linearly rising voltage to match the output voltage of the cell. This time interval then can be mapped to the equivalent two-bit binary code corresponding the memory contents.

51.2.3   Architecture

Constructing large ROMs with fast access times requires the memory array to be divided into smaller memory banks. This gives rise to the concept of divided wordlines and divided bitlines that reduces the capacitance of these structures allowing for faster signal dynamics. Typically, memory blocks would be no larger than 256 rows by 256 columns. In order to quantitatively compare the area advantage of the multiple-valued approach, one can calculate the area per bit of a two-bit per cell ROM divided by the area per bit of a one-bit per cell ROM. Ideally, one would expect this ratio to be 0.5. In the case of a practical two-bit per cell ROM [4], the ratio is 0.6 since the cell is larger than a regular ROM cell in order to accommodate any one of the four possible size transistors. ROM density in the Mb capacity range is in general very comparable to that of DRAM density despite the differences in fabrication technology [2].

In user-programmable or field-programmable ROMs, the customer can program the contents of the memory array by blowing selected fuses (i.e., physically altering them) on the silicon substrate. This allows for a "one-time" customization after the ICs have been fabricated. The quest for a memory that is nonvolatile and electrically alterable has led to the development of EPROMs, EEPROMs, and flash memories [2].

51.3   PROM

Since process technology has shifted to QLM or PLM to achieve better device performance, it is important to develop a ROM technology that offers short TAT, high density, high speed, and low power. There are many types of ROM each with merits and demerits [5]:

The diffusion programming ROM has excellent density but has a very long process cycle time.

The conventional VIA-2 contact programming ROM has better cycle time, but it has poor density.

An architecture VIA-2 contact programming ROM for QLM and PLM processes has simple processing with high density which obtains excellent results targeting 2.5 V and 2.0 V supply voltage.

51.3.1   Read Only Memory Module Architecture

The details of the ROM module configuration are shown in Figure 51.4. This ROM has a single access mode (16-bit data read from half of ROM array) and a dual access mode (32-bit data read from both ROM array) with external address and control signals. One block in the array contains 16-bit lines and is connected to a sense amplifier circuit as shown in Figure 51.5. In the decoder, only one bit line in 16 bits is selected and precharged by P1 and T1 [5].

16 bits in half array at a single access mode or 32 bits in a dual access mode are dynamically precharged to VDD level. Dl is a pull down transistor to keep unselected bit lines at ground level. The speed of the ROM will be limited by bit line discharge time in the worst case ROM coding. When connection exists on all of bit lines vertically, total parasitic capacitance Cbs on the bit line by N-diffusions and Cbg will be a maximum. Tills situation is shown in Figure 51.6a. In the 8KW ROM, 256 bit cells are in the vertical direction, resulting in 256 times of cell bit line capacitance. In this case, discharge time from VDD to GND level is about 6–8 ns at VDD = 1.66 V and depends on ROM programming type such as diffusion or VIA-2. Short circuit currents in the sense amplifier circuits are avoided by using a delayed enable signal (Sense Enable). There are dummy bit lines on both sides of the array as indicated in Figure 51.4. This line contains "0" s on all 256 cells and has the longest discharge time. It is used to generate timing

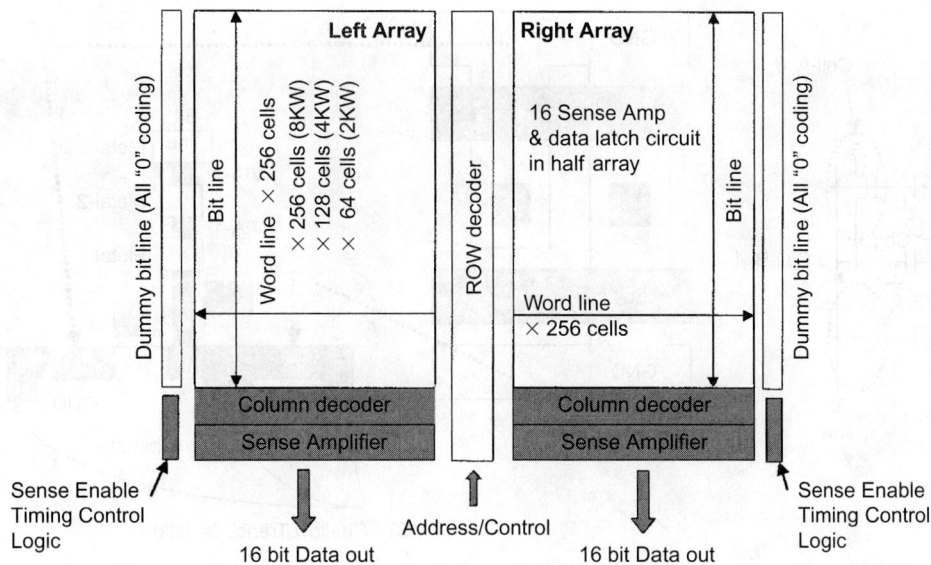

FIGURE 51.4 ROM module array configuration.

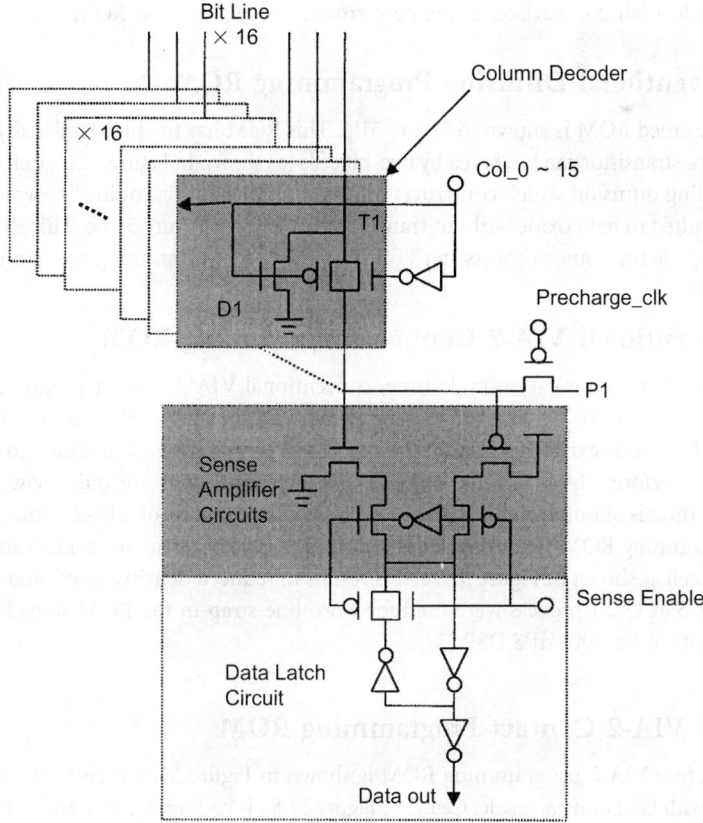

FIGURE 51.5 Detail of low power selective bit line precharge and sense amplifier circuits.

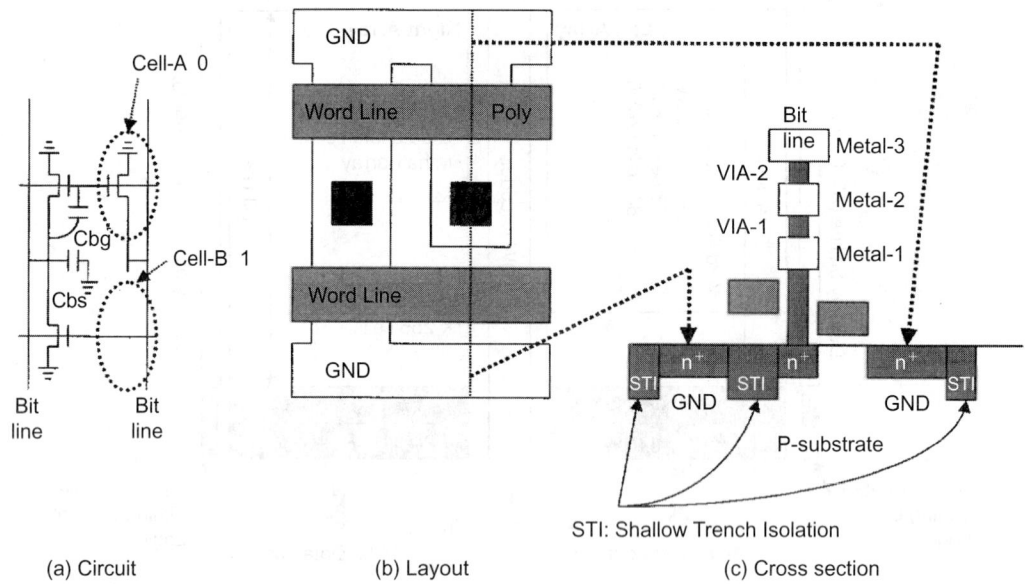

FIGURE 51.6 Diffusion programming ROM.

for a delayed enable signal that activates the sense amplifier circuits. These circuits were used for all types of ROM to provide a fair comparison of the performance of each type of ROM [5].

51.3.2 Conventional Diffusion Programming ROM

Diffusion programmed ROM is shown in Figure 51.6. This ROM has the highest density because bit line contact to discharge transistor can be shared by two-bit cells (as shown in Figure 51.6). Cell-A in Figure 51.6a is coding "0" adding diffusion which constructs transistor, but Cell-B is coding "1" which does not have diffusion and resulted in field oxide without transistor as shown in Figure 51.6c. This ROM requires very long fabrication cycle time since process steps for the diffusion programming are required [5].

51.3.3 Conventional VIA-2 Contact Programming ROM

In order to obtain better fabrication cycle time, conventional VIA-2 contact programming ROM was used as shown in Figure 51.7, Cell-C in Figure 51.7a is coding "1" Cell-D is coding "1". There are determined by VIA-2 code existence on bit cells. The VIA-2 is final stage of process and base process can be completed just before VIA-2 etching and remaining process steps are quite few. So, VIA-2 ROM fabrication cycle time is about 1/5 of the diffusion ROM. The demerit of VIA-2 contact and other type of contact programming ROM was poor density. Because diffusion area and contact must be separated in each ROM bit cell as shown in Figure 51.7c, this results in reduced density, speed, and increased power. Metal-4 and VIA-3 at QLM process were used for word line strap in the ROM since RC delay time on these nobles is critical for 100MIPS DSP [5].

51.3.4 New VIA-2 Contact Programming ROM

The new architecture VIA-2 programming ROM is shown in Figure 51.8. A complex matrix constructs each 8-bit block with GND on each side. Cell-E in Figure 51.8a is coding "0". Bit4 and N4 are connected by VIA-2. Cell-F is coding "1" since Bit5 and N5 are disconnected. Coding other bit lines (Bit 0, 1, 2, 3, 5, 6, and 7) follow the same procedure. This is one of the coding examples to discuss worst case operating speed.

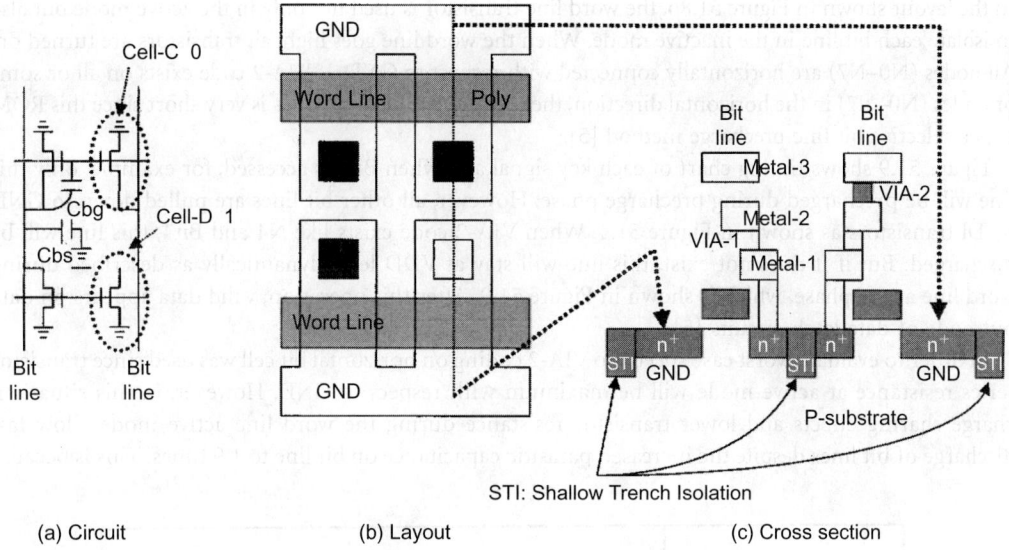

(a) Circuit (b) Layout (c) Cross section

FIGURE 51.7 Conventional VIA-2 programming ROM.

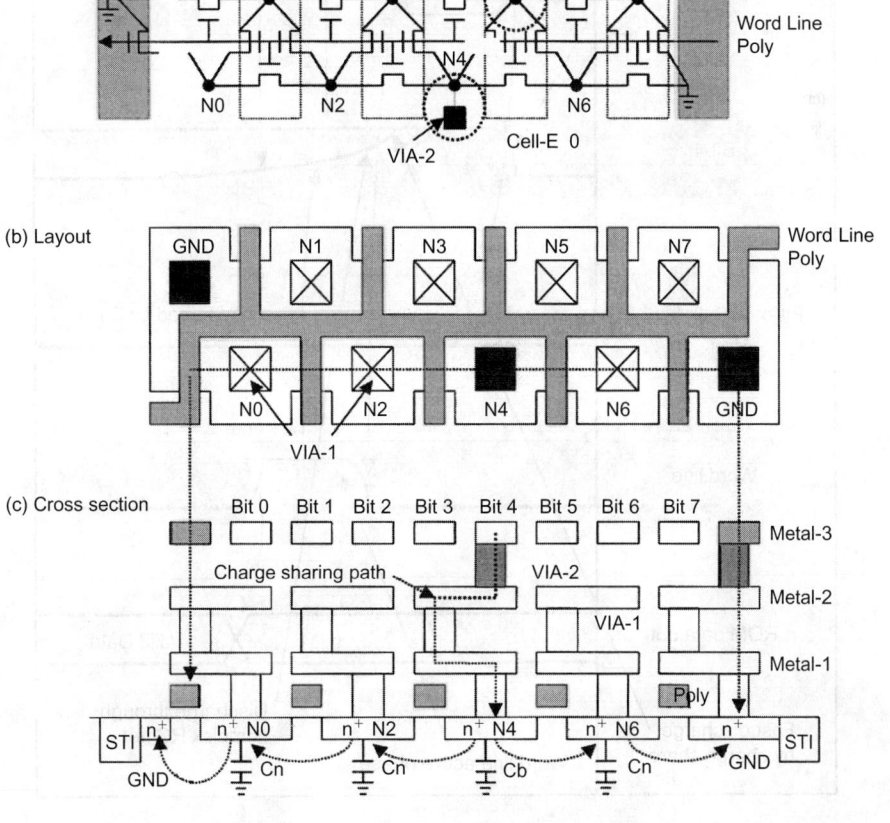

FIGURE 51.8 New VIA-2 programming ROM.

In the layout shown in Figure 51.8b, the word line transistor is used not only in the active mode but also to isolate each bit line in the inactive mode. When the word line goes high, all transistors are turned on. All nodes (N0–N7) are horizontally connected with respect to GND. If VIA-2 code exists on all or some of nodes (N0–N7) in the horizontal direction, the discharge time of bit lines is very short since this ROM uses a selective bit fine precharge method [5].

Figure 51.9 shows timing chart of each key signal and when Bit4 is accessed, for example, only this line will be precharged during precharge phase. However, all other bit lines are pulled down to GND by Dl transistors as shown in Figure 51.4. When VIA-2 code exists like N4 and Bit4, this line will be discharged. But if it does not exist, this line will stay at VDD level dynamically as described during word line active phase, which is shown in Figure 51.9. After this operation, valid data appears on data out node of data latch circuits [5].

In order to evaluate worst case speed, no VIA-2 coding on horizontal bit cell was used since transistor series resistance at active mode will be maximum with respect to GND. However, in this situation, charge sharing effects and lower transistor resistance during the word line active mode allow fast discharge of bit lines despite the increased parasitic capacitance on bit line to 1.9 times. This is because

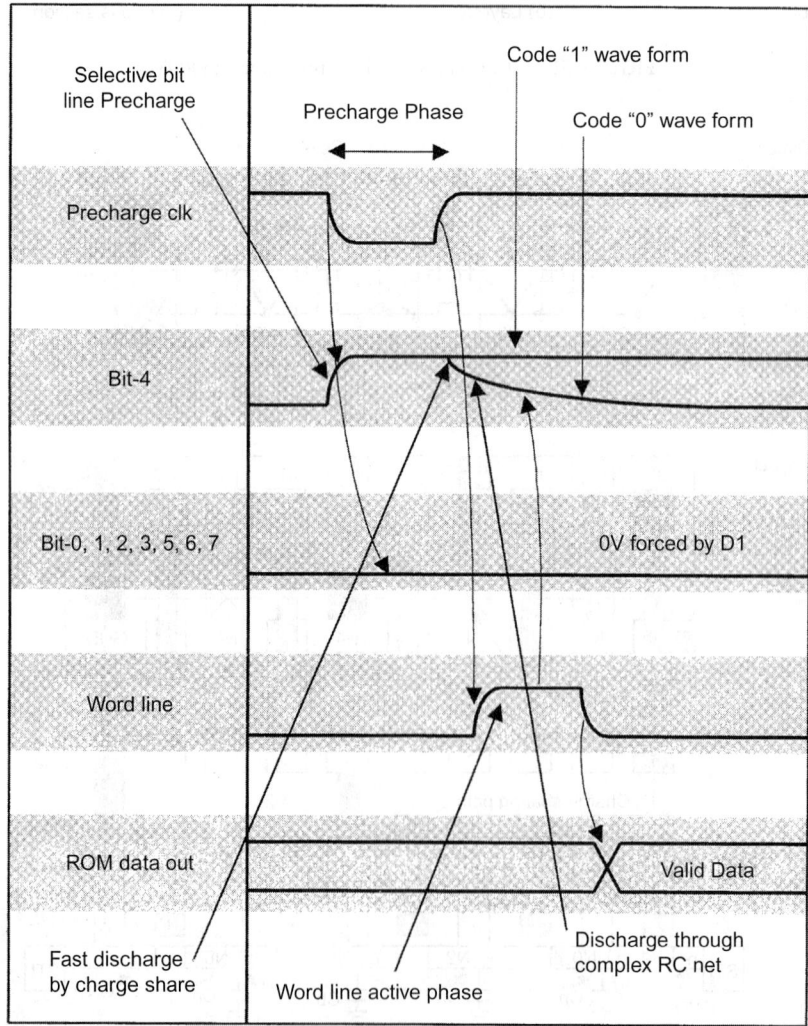

FIGURE 51.9 Timing chart of new VIA-2 programming ROM.

all other nodes (N0–N7) will stay at GND dynamically. The capacitance ratio between bit line (Cb) and all nodes except N4 (Cn) was about 20:1. Fast voltage drop could be obtained by charge sharing at the initial stage of bit line discharging. About five voltage drop could be obtained on 8KW configuration through the charge-sharing path shown in Figure 51.8c. With this phenomenon, the full level discharging was mainly determined by complex transistor RC network connected to GND as shown in Figure 51.8a. This new ROM has much wider transistor width than conventional ROMs and much smaller speed degradation due to process deviations, because conventional ROMs typically use the minimum allowable transistor size to achieve higher density and are more sensitive due to process variations [5].

51.3.5 Comparison of ROM Performance

The performance comparison of each type of ROM are listed in Table 51.1. 8KW ROM module area ratio was indicated using same array configuration, and peripheral circuits with layout optimization to achieve fair comparison. The conventional VIA-2 ROM was 20% bigger than diffusion ROM, but new VIA-2 ROM was only 4% bigger. TAT ratio (days for processing) was reduced to 0.2 due to final stage of process steps. SPICE simulations were performed to evaluate each ROM performance considering low voltage applications. The DSP targets 2.5 V and 2.0 V supply voltage as chip specification with low voltage comer at 2.3 V and 1.8 V, respectively. However, a lower voltage was used in SPICE simulations for speed evaluation to account for the expected 7.5 supply voltage reduction due to the IR drop from the external supply voltage on the DSP chip. Based on this assumption, VDD = 2.13 V and VDD = 1.66 V were used for speed evaluation. The speed of new VIA-2 ROM was optimized at 1.66 V to get over 100 MHz and demonstrated 106 MHz operation at VDD = 1.66 V, 125dc, (based on typical process models). Additionally, 149 MHz at VDD = 2.13 V, 125dc was demonstrated with the typical model and 123 MHz using the slow model. This is a relatively small deviation induced by changes in process parameters such as width reduction of the transistors. By using the fast model, operation at 294 MHz was demonstrated without any timing problems. This means the new ROM has very high productivity with even three sigma of process deviation and wide range of voltages and temperatures [5].

TABLE 51.1 Comparison of ROM Performance

Comparison Item	Diffusion ROM	Conventional VIA-2 ROM	New VIA-2 ROM
8KW (Area ratio)	1.0	1.2	1.04
TAT (Day ratio)	1.0	0.2	0.2
Speed @ 2.13 V, 125dc. Weak.	83 MHz	86 MHz	123 MHz
Speed @ 2.13 V, 125dc. Typical.	166 MHz	98 MHz	149 MHz
Speed @ 2.81 V, –40dc. Strong.	277 MHz	179 MHz	294 MHz
Speed @ 1.66 V. 125dc, Typical.	103 MHz	75 MHz	106 MHz
Power@2.81 V, –40dc. Strong. 100 MHz. (16-bit single access)	15.6 mW	19.3 mW	2 UrnW
Power@2.81 V@40dc. Strong. 100 MHz. (32-bit dual access)	29.6 mW	37.1 mW	401 mW

Performance was measured with worst coding (all coding "1").

References

1. Karls, J., *Status 1999: A Report On The Integrated Circuit Industry*, Integrated Circuit Engineering Corporation, 1999.
2. Gulak, P.G., A review of multiple-valued memory technology, IEEE International Symposium on Multi-valued Logic, 1998
3. Rich, D.A., A Survey of Multi valued memories, *IEEE Trans. On Comput.*, vol. C-35, no. 2, pp. 99–106, Feb. 1986.
4. Prince, B., *Semiconductor Memories*, 2nd ed., John Wiley & Sons Ltd., New York, 1991.
5. Takahashi, H., Muramatsu, S., and Itoigawa, M., A new contact programming ROM architecture for digital signal processor, Symposium on VLSI Circuits, 1998.

52

SRAM

Yuh-Kuang Tseng

*Industrial Research
and Technology Institute, Taiwan*

CONTENTS

52.1 Read/Write Operation

Figure 52.1 shows a simplified readout circuit for an SRAM. The circuit has static bit-line loads composed of pull-up PMOS devices M1 and M2. The bit-lines are pulled up to VDD by bit-line load transistors M1 and M2. During the read cycle, one word-line is selected. The bit line BL is discharged to a level determined by the bit-line load transistor M1, the accessed transistor N1, and the driver transistor N2 as shown in Figure 52.1(b). At this time, all selected memory cells consume a dc column current flowing through the bit-line load transistors, accessed transistors, and driver transistors. This current flow increases the operating power and decreases the access speed of the memory.

Figure 52.2 shows a simplified circuit diagram for SRAM write operation. During the write cycle, the input data and its complement are placed on the bit-lines. Then the word-line is activated. This will force the memory cell to flip into the state represented on the bit-lines, whereas the new data is stored in the memory cell. The write operation can be described as follows. Consider a high voltage level and a low voltage level are stored in both node 1 and node 2, respectively. If the data is to be written into the cell, then node 1 becomes low and node 2 becomes high. During this write cycle, a dc current will flow from VDD through bit-line load transistor M1 and write circuits to ground. This extra dc current flow in write cycle increases the power consumption and degrades the write speed performance. Moreover, in the tail portion of write cycle, if data 0 has been written into node 1 as shown in Figure 52.2, the turn-on word-line transistor N1 and driver transistor N2 form a discharge circuit path to discharge the bit-line voltage. Thus, the write recovery time is increased. In high-speed SRAM, write recovery time is an important component of the write cycle time. It is defined as the time necessary to recover from the write cycle to the read state after the WE signal is disabled [1]. During the write recovery period, the selected cell is in the quasi-read condition [2], which consumes dc current as in the case of read cycle.

Based on the above discussion, the dc current problems that occur in the read and write cycles should be overcome to reduce power dissipation and improve speed performance. Some solutions for the dc current problems of conventional SRAM will be described. During the active mode (read cycle or write cycle), the word-line is activated, and all selected columns consume a dc current. Thus, the word-line activation duration should be shortened to reduce the power consumption and improve speed performance during the active mode. This is possible by using the Address Transition Detection (ATD) technique [3] to

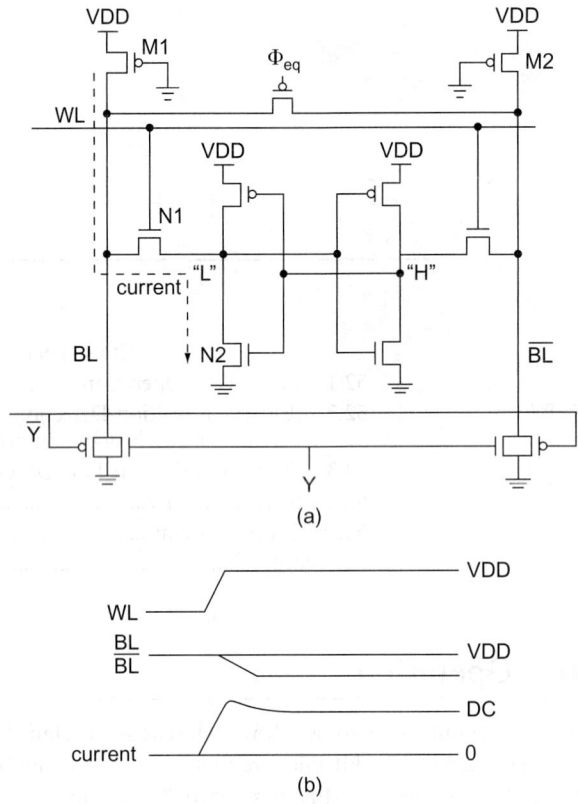

FIGURE 52.1 (a) Simplified readout circuit for an SRAM, (b) signal waveform.

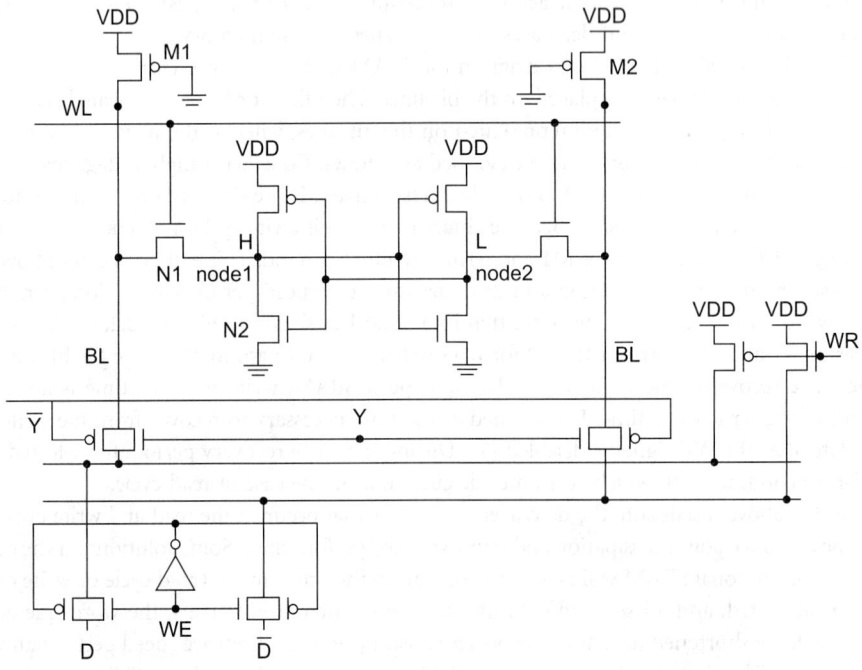

FIGURE 52.2 Simplified circuit diagram for SRAM write operation.

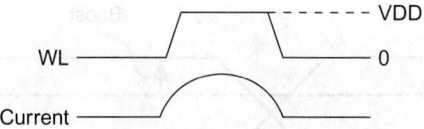

FIGURE 52.3 Word-line signal and current reduction by pulsing the word line.

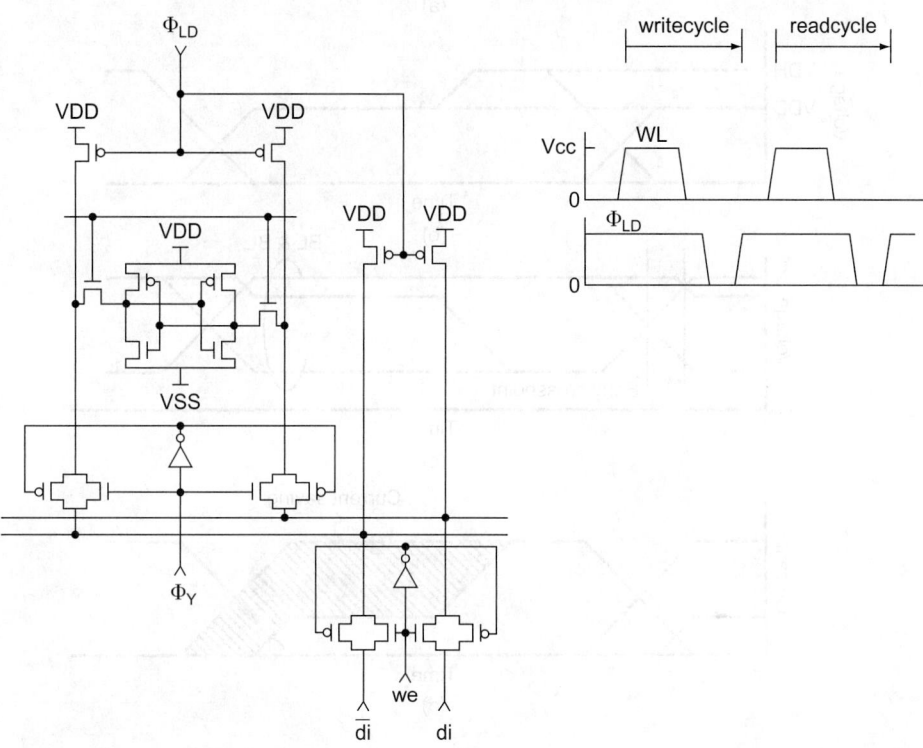

FIGURE 52.4 Simplified circuit configuration and time diagram for read and write operation.

generate the pulsed word-line signal with enough time to achieve the read and write operation, as shown in Figure 52.3.

However, the memory cells asserted by the pulsed word-line signal still consume dc current from VDD through bit-line load transistors, accessed transistors, and driver transistors or write circuits to the ground during the word-line activation period. A dynamic bit-line loads circuit technique [2,4–6] can be used to eliminate the dc power consumption during operation period.

Figure 52.4 shows a simplified circuit configuration and time diagram for read and write operation. In the read cycle, the bit-line load transistors are turned off because the Φ_{LD} signal is in the high state. The bit-line load consists of only the stray capacitance. Therefore, the selected memory cell can rapidly drive the bit-line load, resulting in a fast access time. Moreover, the dc column current consumed by the other activated memory cells can be eliminated. Similarly, the dc current consumption in the write cycle can be eliminated.

A memory cell's readout current I_{cell} depends on the channel conductance of the transfer gates in a memory cell. As the supply voltage is scaled down, the speed performance of SRAM is decreased, significantly, due to small cell's readout current. To increase the channel conductance, widening the channel width and/or boosting word-line voltage are used. For low-voltage operation, boosting the word-line voltage is effective in shortening the delay time, in contrast to widening the channel width. However,

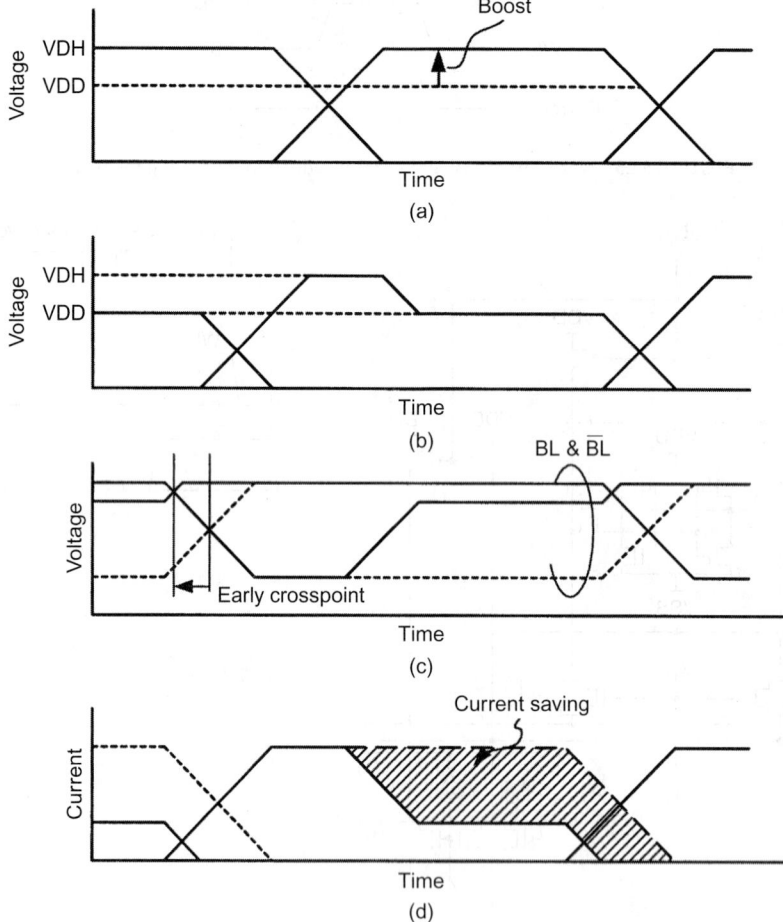

FIGURE 52.5 Step-down boosted-word-line scheme: (a) conventional boosted word-line, (b) step-down boosted word-line, (c) bit-line transition, and (d) current consumption of a selected memory cell. (From Ref. 7.)

this causes an increased power dissipation and a large transition time due to enhanced bit-line swing. To solve these problems, a step-down boosted-word-line scheme that shortens the readout time with little power dissipation penalty was reported by Morimura and Shibata in 1998 [7].

The concept of this scheme is shown in Figure 52.5(b), in contrast to the conventional full-boosted-word-line scheme in Figure 52.5(a). The step-down boosted-word-line scheme also boosts the selected word-line, but the boosted period is restricted only at the beginning of memory cell access. This enables the sensing operation to start early, by fast bit-line transition. During the sensing period of bit-line signals, the word-line potential is stepped down to the supply voltage to suppress the power dissipation; the reduced bit-line signals are sufficient to read out data by current sensing, and the reduced bit-line swing is effective in shortening the bit-line transition time in the next read cycle (Figure 52.5(c)). As a result, fast readout is accomplished with little dissipation penalty (Figure 52.5(d)).

The step-down boosted-word-line scheme is also used in data writing. In the writing cycle, the proposed scheme is just as effective in reducing the memory-cell current because the memory cells unselected by column-address signals consume the same power as in the read cycle. The boosted word-line voltage shortens the time for writing data because it increases the channel conductance of the access transistor in the selected memory cells. The writing recovery operation starts after the word-line voltage is stepped

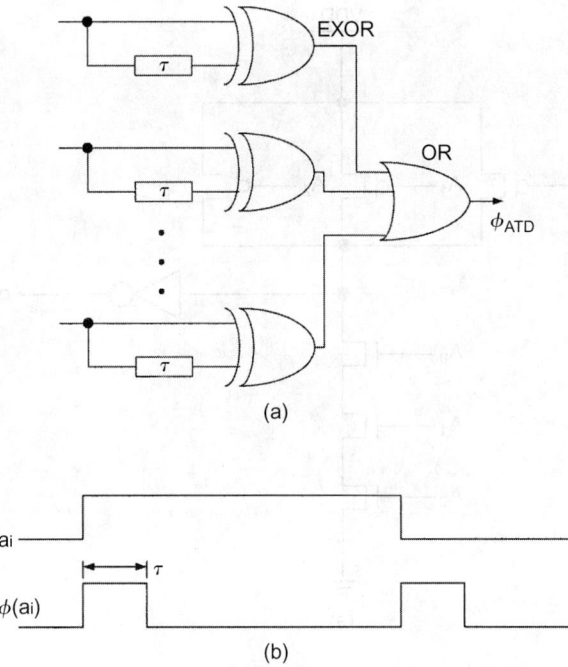

FIGURE 52.6 (a) Summation circuit of all ATD pulses generated from all address transitions (b) ATD pulse waveform. (From Ref. 10.)

down. Reducing the memory cell's current accelerates the recovery operation of lower bit-lines. So, a shorter recovery time than that of the conventional full-boosted-word-line scheme is obtained.

Other circuit techniques for dc column current reduction, such as divided word-line (DWL) [8] and hierarchical word decoding (HWD) [9] structures will be described in the following sections.

52.2 Address Transition Detection (ATD) Circuit for Synchronous Internal Operation [1,10]

The address transition detection (ATD) circuit plays an important role in achieving internal synchronization of operation in SRAM. ATD pulses can be used to generate the different time signals for pulsing word-lines, sensing amplifier, and bit-line equalization. The ATD pulse activating $\phi_{(ai)}$ is generated with XOR circuits by detecting "L" to "H" or "H" to "L" transitions of any input address signal a_i, as shown in Figure 52.6. All the ATD pulses generated from all the address input transitions are summed up to one pulse, ϕ_{ATD} as shown in Figure 52.6. The pulse width of ϕ_{ATD}, is controlled by the delay element τ. The pulse width is usually stretched out with a delay circuit and used to reduce or speed up signal propagation in the SRAM.

52.3 Decoder and Word-Line Decoding Circuit [10–13]

Two kinds of decoders are used in SRAM: the row decoder and the column decoder. Row decoders are needed to select one row of word-lines out of a set of rows in the array. A fast decoder can be implemented by using AND/NAND and OR/NOR gates. Figure 52.7 shows the schematic diagrams of static and dynamic AND gate decoders. The static NAND-type structure is chosen due to its low power consumption, that is, only the decoded row transitions. The dynamic structure is chosen due to its speed and power improvement over conventional static NAND gates.

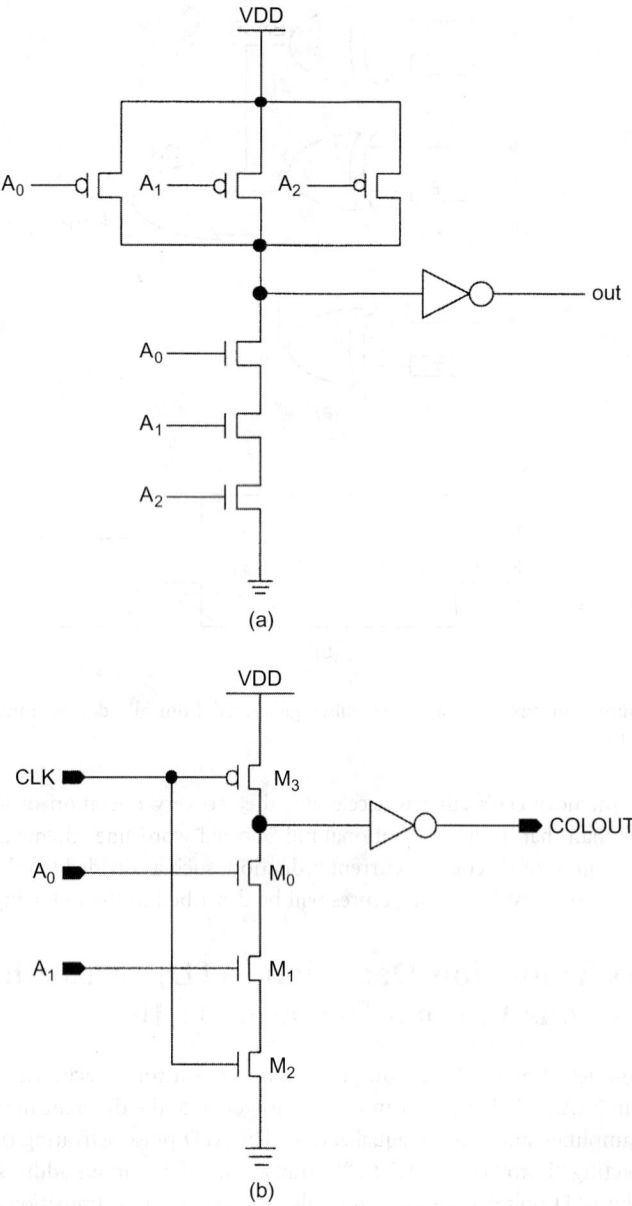

FIGURE 52.7 Circuit diagrams of a three-input AND gate: (a) static CMOS, (b) dynamic CMOS.

From a low-voltage operation standpoint, a dynamic NOR-base decoding would provide lower delay times through the decoder due to the limited amount of stacking of devices. Figure 52.8 shows circuit diagrams of dynamic NOR gates. The dynamic CMOS gate as shown in Figure 52.8(a) consists of input-NMOSs whose drain nodes are precharged to a high level by a PMOS when a clock signal Φ is at a low level, and conditionally discharged by the input-NMOSs when a clock signal Φ is at a high level. The delay time of the dynamic NOR/OR gate does not increase when the number of input signals increases. This is because only one PMOS and two NMOSs are connected in series, even if the number of input signals is large. However, the output of the OR signal is slower than that of the NOR signal because the OR signal is generated from the inverter driven by the NOR signal.

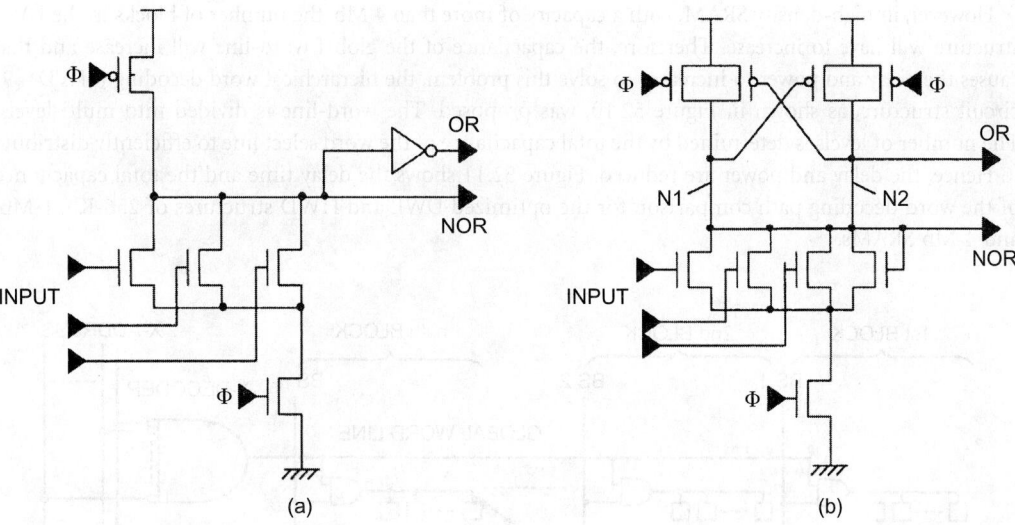

FIGURE 52.8 Circuit diagrams of three-input NOR/OR gates: (a) dynamic CMOS, (b) SCL> (From Ref. 11)

Figure 52.8(b) shows the source-coupled-logic (SCL) [11] NOR/OR circuit. When a clock signal Φ is at a low level, the drain nodes of the NMOS (N1, N2) are precharged to a high level in the circuit. If at least one of input signals of the circuit is at a high level and the clock Φ then turns to a high level, node N1 is discharged to a low level and node N2 remains at a high level. On the other hand, if all the input signals are at a low level and Φ then turns to a high level, node N2 is discharged and node N1 remains at a high level. The SCL circuit can produce an OR signal and a NOR signal simultaneously. Thus, the SCL circuit is suitable for predecoders that have a large number of input signals and for address buffers that need to produce OR and NOR signals simultaneously.

Column decoders select the desired bit pairs out of the sets of bit pairs in the selected row. A typical dynamic AND gate decoder as shown in Figure 52.7(b) can be used for column decoding because the AND structure meets the delay requirements (column decode is not in the worst-case delay path) and does so at a much lower power consumption.

A highly integrated SRAM adopts a multi-divided memory cell array structure to achieve high-speed word decoding and reduce column power dissipation. For this purpose, many high-speed word-decoding circuit architectures have been proposed, such as divided word-line (DWL) [8] and hierarchical word decoding (HWD) [9] structures. The multi-stage decoder circuit technique is adopted in both word-decoding circuit structures to achieve high-speed and low-power operation. The multi-stage decoder circuit has advantages over the one-stage decoder in reducing the number of transistors and fanin. Also, it reduces the loading on the address input buffers. Figure 52.9 shows the decoder structure for a typical partitioned memory array with divided word-line (DWL). The cell array is divided into N_B blocks. If the SRAM has N_C columns, each block contains N_C/N_B columns. The divided word-line in each block is activated by the global word-line and the vertical block select line. Consequently, only the memory cells connected to one divided word-line within a selected block are accessed in a cycle. Hence, the column current is reduced because only the selected columns switch. Moreover, the word-line selection delay, which is the sum of the global word-line delay and the divided word-line delay, is reduced. This is because the total capacitance of the global word-line is smaller than that of a conventional word-line. The delay time of each divided word-line is small due to the short length. In the block decoder, an additional signal Φ, which is generated from an ATD pulse generator, can be adopted to enable the decoder and ensure the pulse activated word-line.

However, in high-density SRAM, with a capacity of more than 4 Mb, the number of blocks in the DWL structure will have to increase. Therefore, the capacitance of the global word-line will increase and that causes the delay and power to increase. To solve this problem, the hierarchical word decoding (HWD) [9] circuit structure, as shown in Figure 52.10, was proposed. The word-line is divided into multi-levels. The number of levels is determined by the total capacitance of the word select line to efficiently distribute it. Hence, the delay and power are reduced. Figure 52.11 shows the delay time and the total capacitance of the word decoding path comparison for the optimized DWL and HWD structures of 256-Kb, 1-Mb, and 4-Mb SRAMs.

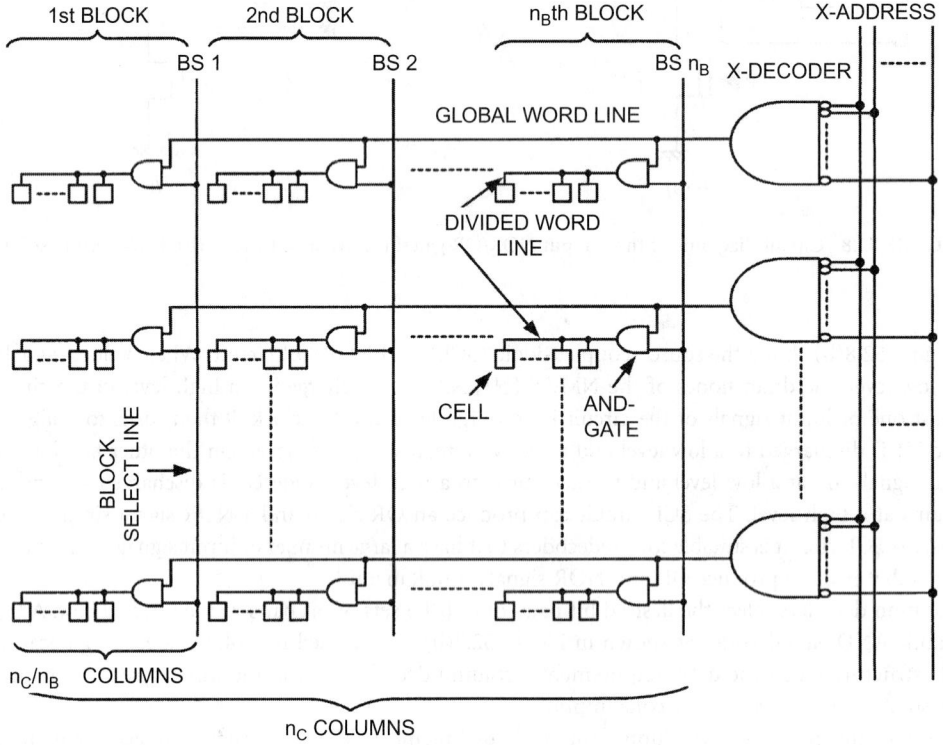

FIGURE 52.9 Divided word-line (DWL) structure. (From Ref. 8.)

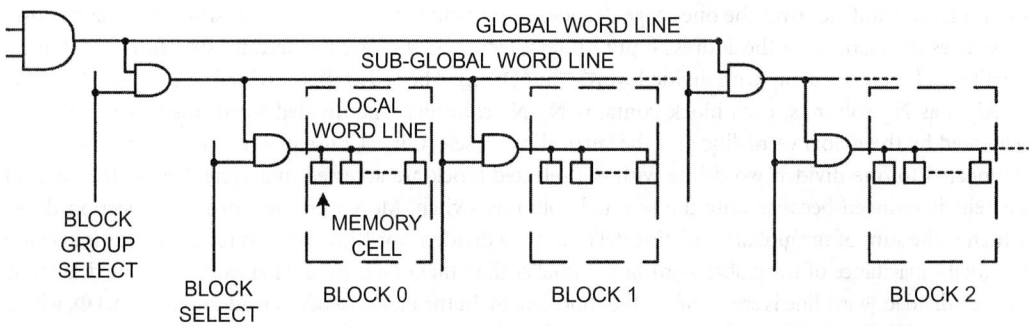

FIGURE 52.10 Hierarchical word decoding structure.

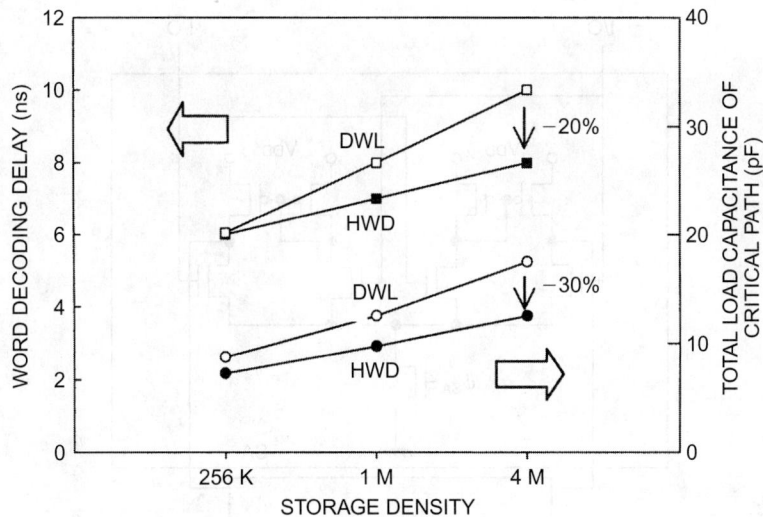

FIGURE 52.11 Comparison of DWL and HWD. (From Ref. 9. With permission.)

52.4 Sense Amplifier [10]

During the read cycle, the bit-lines are initially precharged by bit-line load transistors. When the selected word-line is activated, one of the two bit-lines is pulled low by driver transistor, while the other stays high. The bit-line pull-down speed is very slow due to the small cell size and large bit-line load capacitance. Differential sense amplifiers are used for speed purposes because they can detect and amplify a very small level difference between two bit-lines. Thus, a fast sense amplifier is an important factor in realizing fast access time.

Figure 52.12 shows a switching scheme of well-known current-mirror sense amplifiers [14]. Two amplifiers are serially connected to obtain a full supply voltage swing output because one stage of the amplifier does not provide enough gain for a full swing. The signal Φ_{SA} is generated with an ATD pulse. It is asserted for a period of time, enough to amplify the small difference on data lines; then it is deactivated and the amplified output is latched. Hence, the switch reduces the power consumption, especially at relatively low frequencies.

A latch-type sense amplifier such as a PMOS cross-coupled amplifier [15], as shown in Figure 52.13, greatly reduces the dc current after amplification and latching, because the amplifier provides a nearly full supply voltage swing with positive feedback of outputs to PMOSFETs. As a result, the current in the PMOS cross-coupled sense amplifier is less than one fifth of that in a current-mirror amplifier. Moreover, this positive feedback effect gives much faster sensing speed than the conventional amplifier. To obtain correct and fast operation, the equalization element EQL is connected between the output terminals and are turned on with pulse signals Φ_S and its complement during the transition period of the input signals.

However, the latch-type sense amplifier has a large dependence on the input voltage swing, especially at low current operation conditions. An NMOS source-controlled latched sense amplifier [16] as shown in Figure 52.14 is able to quickly amplify an input voltage swing as small as 10 mV. The sense amplifier consists of two PMOS loads, two NMOS drivers, and two feedback inverters. The sense amplifier control (SAC) signal is driven by the CS input buffer, and Φ_S is a sense-amplifier equalizing pulse generated by the ATD pulse. The gate terminal of the NMOS driver is connected to the local data bus (LD1 and LD2), and the source terminal of the NMOS driver is controlled by the feedback inverter

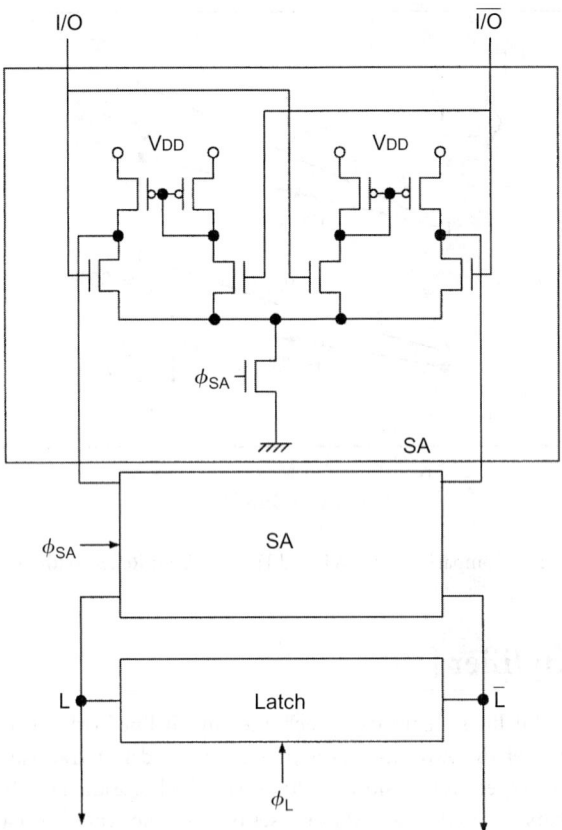

FIGURE 52.12 Two-stage current-mirror sense amplifier. (From Refs. 10 and 14. With permission.)

connected to the opposite output node of sense amplifier. Thus, the NMOS driver connected to the high-going output node turns off immediately. Therefore, the charge-up time of that node can be reduced because no current is wasted in the NMOS driver.

A bidirectional sense amplifier, called a bidirectional read/write shared sense amplifier (BSA) [17], is shown in Figure 52.15. The BSA plays three roles. It functions as a sense amplifier for read operations, and it serves as a write circuit and a data input buffer for write operations. It consists of an 8-to-1 column selector and bit-line precharger, a CMOS dynamic sense amplifier, an SR flip-flop, and an I/O circuit.

Eight bit-line pairs are connected to a CMOS dynamic sense amplifier through CMOS transfer gates. The BLSW signal is used to select a column and to precharge bit-lines. When the BLSW signal is high, one of eight bit-line pairs is connected to the sense amplifier. When the BLSW signal is low, all bit-line pairs are precharged to VDD level. The SAEQB signal controls the sense amplifier equalization. When the SAEQB signal is low, sense nodes D and DB are equalized and precharged to the VDD level. The SENB signal activates the CMOS dynamic sense amplifier. The SR flip-flop holds the result. The output circuit consists of four p-channel transistors. If the result is high, I/O is connected to VDD (3.3 V) and IOB is connected to VDD (3 V) through p-channel devices. VDDL is a 3-V power supply provided externally. The I/O pair is connected to the sense amplifier through p-channel transfer gates controlled by ISWB. During write operations, ISWB falls to connect the I/O pair to the sense amplifier.

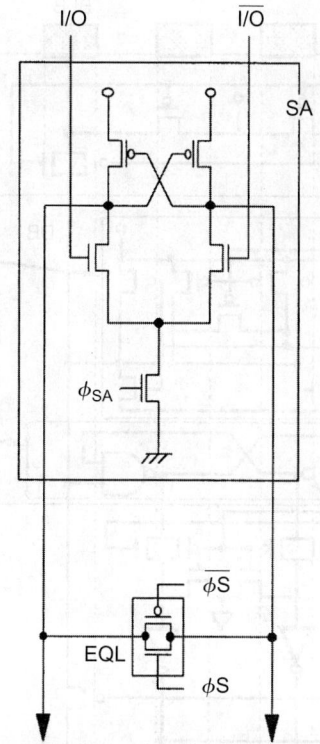

FIGURE 52.13 PMOS cross-coupled amplifier. (From Ref. 15. With permission.)

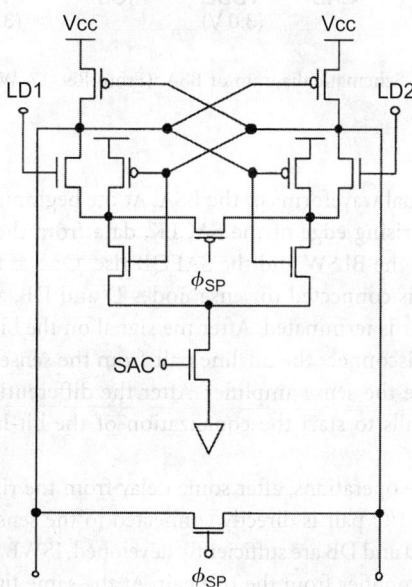

FIGURE 52.14 NMOS source-controlled latched sense amplifier. (From Ref. 16. With permission.)

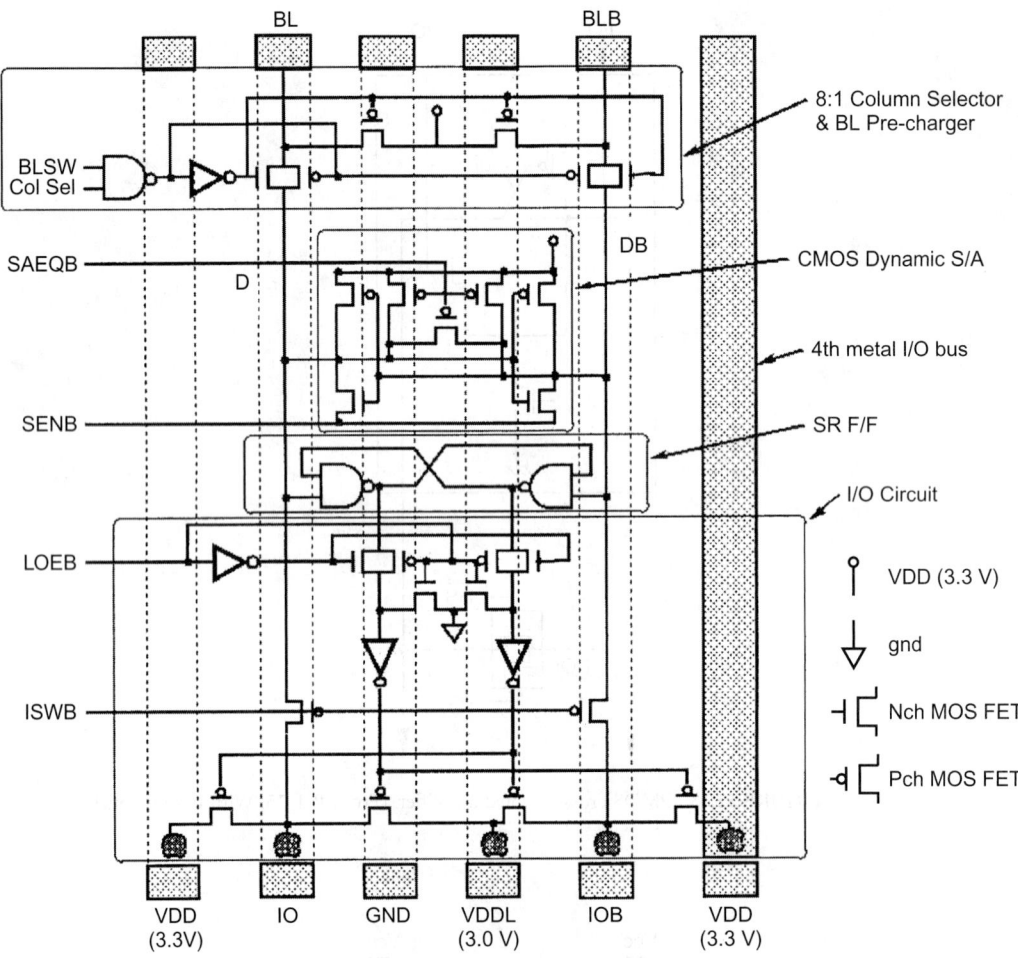

FIGURE 52.15   Schematic diagram of BSA. (From Ref. 17. With permission.)

Figure 52.16 shows operational waveforms of the BSA. At the beginning of the read operations, after some intrinsic delay from the rising edge of the SACLK, data from the selected cell is read onto the bit-line pair. At the same time, the BLSW and the SAEQB rise. One of the eight CMOS transfer gates is turned on, the bit-line pair is connected to sense nodes D and DB, and precharging of the CMOS sense amplifier and bit-line pair is terminated. After the signal on the bit-line pair signal is sufficiently developed, the BLSW falls to disconnect the bit-line pair from the sense nodes D and DB. At the same time, the SENB falls to activate the sense amplifier. After the differential output data is latched onto the SR flip-flop, the SAEQB falls to start the equalization of the bit-line pair and the CMOS sense amplifier.

At the beginning of the write operations, after some delay from the rising edge of SACLK, the ISWB signal falls, and the differential I/O pair is directly connected to the sense amplifier through p-channel transfer gates. After the signals D and DB are sufficiently developed, ISWB turns off the p-channel transfer gates to disconnect the sense amplifier from the I/O pair. At the same time, the SENB falls to sense the data, and BLSW rise to connect the sense amplifier to the bit-line pair. After the data is written into

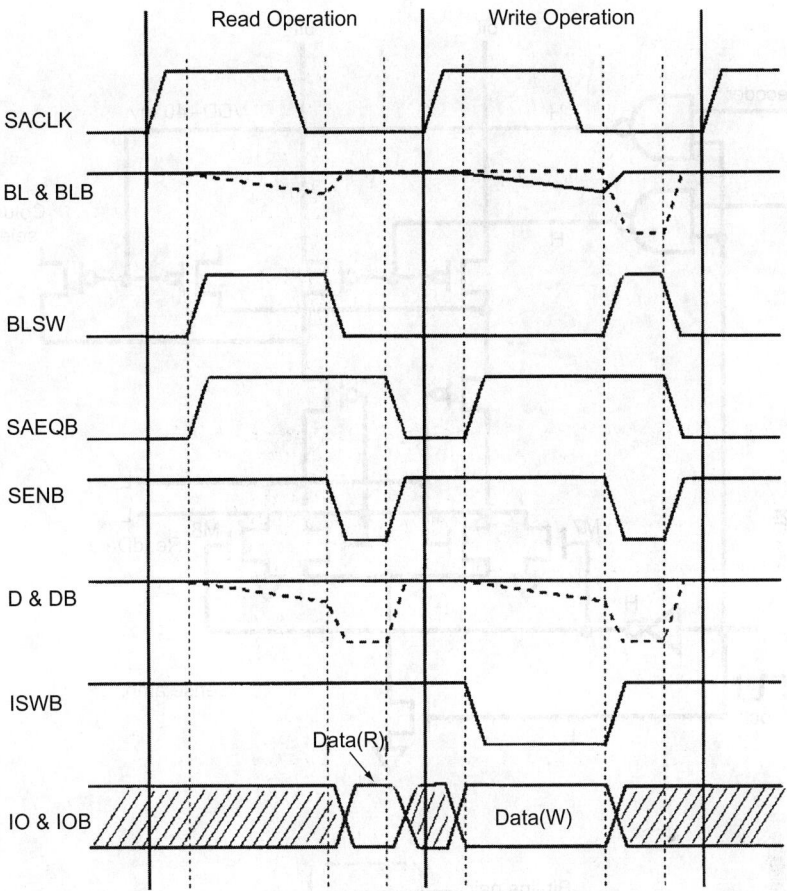

FIGURE 52.16 Operational waveforms of the BSA. (From Ref. 17. With permission.)

the selected memory cell, SAEQB and BLSW fall to start equalization of the bit-line pair and the CMOS sense amplifier.

Conventional sense amplifiers operate incorrectly when threshold voltage deviation is larger than bit-line swing, a current-sensing sense amplifier proposed by Izumikawa et al. in 1997 can continue to operate normally [18]. Figure 52.17 illustrates the sense amplifier operations. Bit-lines are always charged up to VDD through load PMOSFETs. When memory-cells are selected with a word-line, the voltage difference in a bit-line pair appears (Figure 52.17(a)). During this period, all column-select PMOSFETs are off, and no dc current flows in the sense amplifier. The sense amplifier differential outputs, referred to as ReadData, are equalized at ground level through pull-down NMOSFETs M7 and M8.

After a 40-mV difference appears in a bit-line pair, power switch M9 of the sense amplifier and one column-select pair of PMOSFETs are set to on (Figure 52.17(b)). The difference in bit-line voltages causes a current difference between the differential pair PMOS in the sense amplifier, which appears as an output voltage difference. This voltage difference is amplified, and the read operation is accomplished. The current is automatically cut off because of the CMOS inverter. Consequently, the small bit-line swing is sensed without dc current consumption.

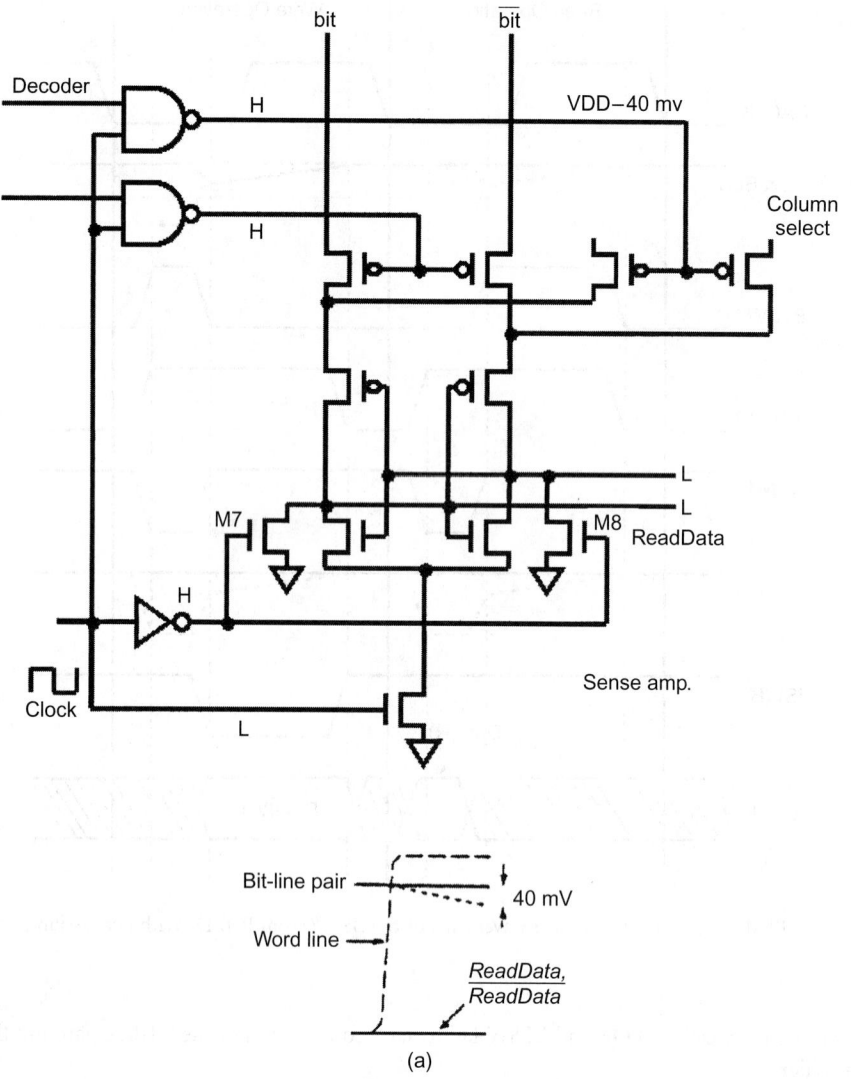

FIGURE 52.17(a) Sense amplifier operation: (a) before sensing. (From Ref. 18. With permission.)

52.5 Output Circuit [4]

The key issue for designing the high-speed SRAM with byte-wide organization is noise reduction. There are two kinds of noise: VDD noise and GND noise. In the high-speed SRAM with byte-wide organization, when the output transistors drive a large load capacitance, the noise is generated and multiplied by 8 because eight outputs may change simultaneously. It is a fundamentally serious problem for the data zero output. That is to say, when the output NMOS transistor drives the large load capacitance, the GND potential of the chip goes up because of the peak current and the parasitic inductance of the GND line. Therefore, the address buffer and the ATD circuit are influenced by the GND bounce, and unnecessary signals are generated.

Figure 52.18 shows a noise-reduction output circuit. The waveforms of the noise-reduction output circuit and conventional output circuit are shown in Figure 52.19. In the conventional circuit, nodes

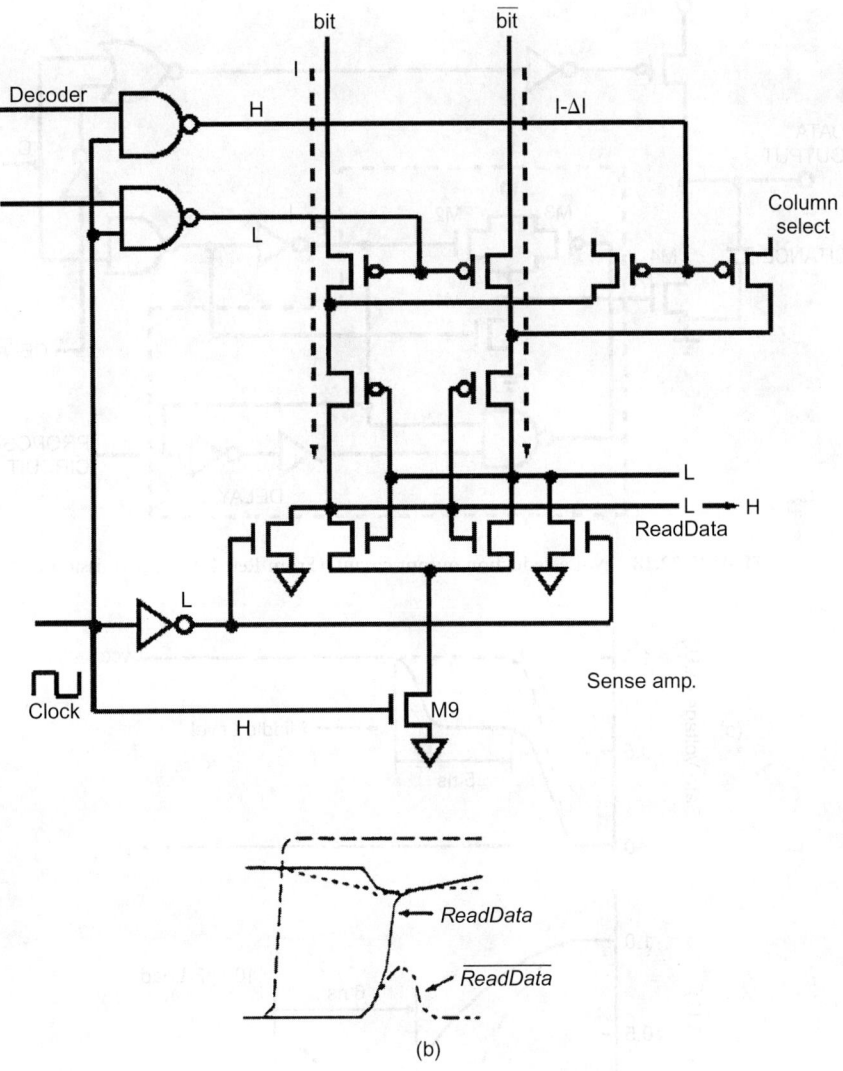

FIGURE 52.17(b) Sense amplifier operation: (b) sensing. (From Ref. 18. With permission.)

A and B are connected directly as shown in Figure 52.18. Its operation and characteristics are shown by the dotted lines in Figure 52.18. Due to the high-speed driving of transistor M4, the GND potential goes up, and the valid data are delayed by the output ringing. A new noise-reduction output circuit consists of one PMOS transistor, two NMOS transistors, one NAND gate, and the delay part (its characteristics are shown by the solid lines in Figure 52.19). The operation of this circuit is explained as follows. The control signals CE and OE are at high level and signal WE is at low level in the read operation. When the data zero output of logical high level is transferred to node C, transistor M1 is cut off, and M2 raises node A to the middle level. Therefore, the peak current that flows into the GND line through transistor M4 is reduced to less than one half that of the conventional circuit because M4 is driven by the middle level. After a 5-ns delay from the beginning of the middle level, transistor M3 raises node A to the VDD level. As a result, the conductance of M4 becomes maximum, but the peak current is small because of the low output voltage. Therefore, the increase of GND potential is small, and the output ringing does not appear.

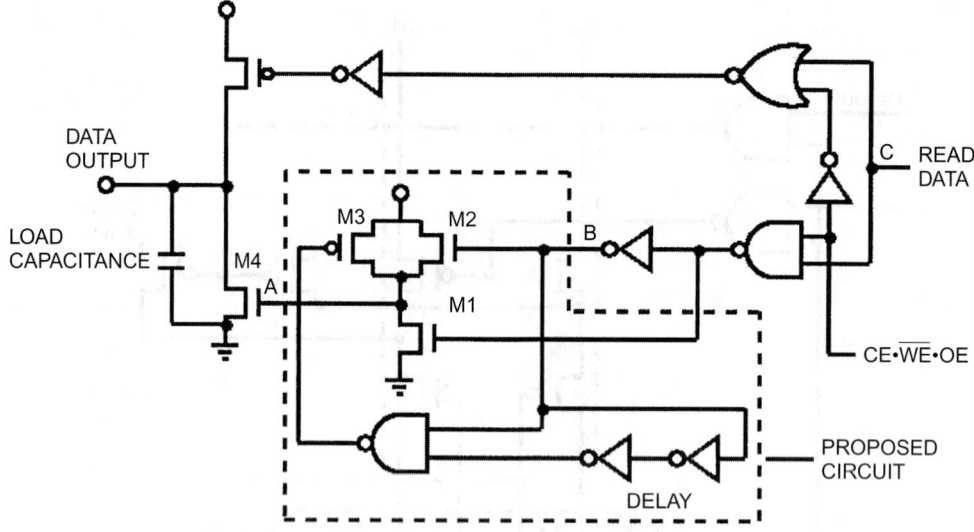

FIGURE 52.18   Noise-reduction output circuit. (From Ref. 4. With permission.)

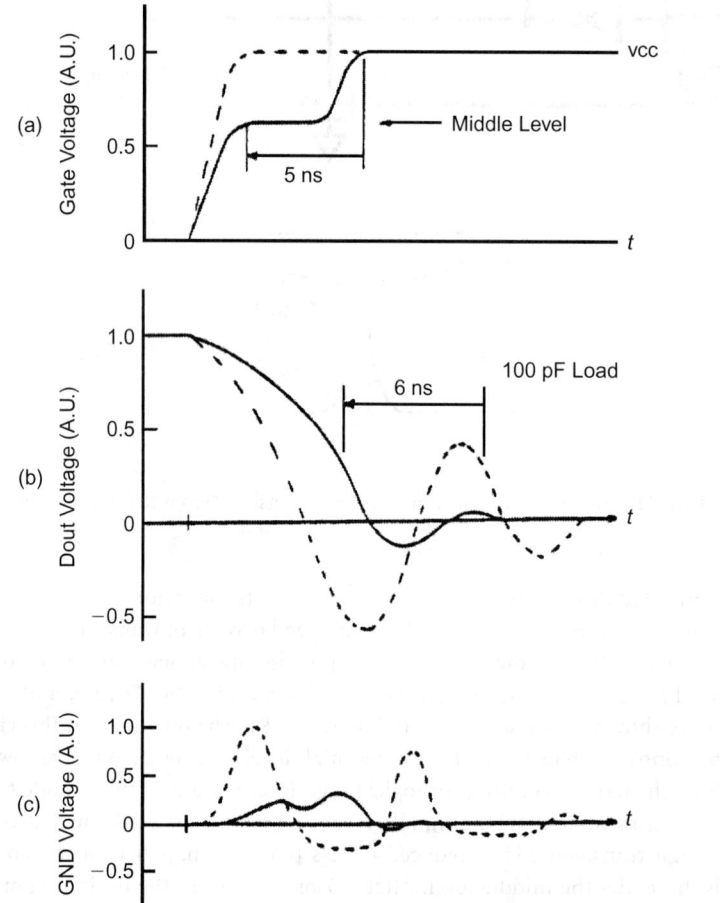

FIGURE 52.19 Waveforms of noise-reduction output circuit (solid line) and conventional output circuit: (a) gate bias, (b) data output, and (c) GND bounce. (From Ref. 4. With permission.)

References

1. Bellaouar, A. and Elmasry, M.I., *Low-Power Digital VLSI Design Circuit and Systems*, Kluwer Academic Publishers, 1995.
2. Ishibashi, K. et al., "A 1-V TFT-Load SRAM Using a Two-Step Word-Voltage Method," *IEEE J. Solid-State Circuits*, vol. 27, no. 11, pp. 1519–1524, Nov. 1992.
3. Chen, C.-W. et al., "A Fast 32KX8 CMOS Static RAM with Address Transition Detection," *IEEE J. Solid-State Circuits*, vol. SC-22, no. 4, pp. 533–537, Aug. 1987.
4. Miyaji, F. et al., "A 25-ns 4-Mbit CMOS SRAM with Dynamic Bit-Line Loads," *IEEE J. Solid-State Circuits*, vol. 24, no. 5, pp. 1213–1217, Oct. 1989.
5. Matsumiya, M. et al., "A 15-ns 16-Mb CMOS SRAM with Interdigitated Bit-Line Architecture," *IEEE J. Solid-State Circuits*, vol. 27, no. 11, pp. 1497–1502, Nov. 1992.
6. Mizuno, H. and Nagano, T., "Driving Source-Line Cell Architecture for Sub-1V High-Speed Low-Power Applications," *IEEE J. Solid-State Circuits*, no. 4, pp. 552–557, Apr. 1996.
7. Morimura, H. and Shibata, N., "A Step-Down Boosted-Wordline Scheme for 1-V Battery-Operated Fast SRAM's," *IEEE J. Solid-State Circuits*, no. 8, pp. 1220–1227, Aug. 1998.
8. Yoshimito, M. et al., "A Divided Word-Line Structure in the Static RAM and Its Application to a 64 K Full CMOS RAM," *IEEE J. Solid-State Circuits*, vol. SC-18, no. 5, pp. 479–485, Oct. 1983.
9. Hirose, T. et al., "A 20-ns 4-Mb CMOS SRAM with Hierarchical Word Decoding Architecture," *IEEE J. Solid-State Circuits*, vol. 25, no. 5, pp. 1068–1074, Oct. 1990.
10. Itoh, K., Sasaki, K., and Nakagome, Y., "Trends in Low-Power RAM Circuit Technologies," *Proceedings of the IEEE*, pp. 524–543, Apr. 1995.
11. Nambu, H. et al., "A 1.8-ns Access, 550-MHz, 4.5-Mb CMOS SRAM," *IEEE J. Solid-State Circuits*, vol. 33, no. 11, pp. 1650–1657, Nov. 1998.
12. Cararella, J.S., "A Low Voltage SRAM For Embedded Applications," *IEEE J. Solid-State Circuits*, vol. 32, no. 3, pp. 428–432, Mar. 1997.
13. Prince, B., *Semiconductor Memories: A Handbook of Design, Manufacture, and Application,* 2nd edition, John Wiley & Sons, 1991.
14. Minato, O. et al., "A 20-ns 64 K CMOS RAM," in *ISSCC Dig. Tech. Papers*, pp. 222–223, Feb. 1984.
15. Sasaki, K., et al., "A 9-ns 1-Mbit CMOS SRAM," *IEEE J. Solid-State Circuits*, vol. 24, no. 5, pp. 1219–1224, Oct. 1989.
16. Seki, T. et al., "A 6-ns 1-Mb CMOS SRAM with Latched Sense Amplifier," *IEEE J. Solid-State Circuits*, vol. 28, no. 4, pp. 478–482, Apr. 1993.
17. Kushiyama, N. et al., "An Experimental 295 MHz CMOS 4K X 256 SRAM Using Bidirectional Read/Write Shared Sense Amps and Self-Timed Pulse Word-Line Drivers," *IEEE J. Solid-State Circuits*, vol. 30, no. 11, pp. 1286–1290, Nov. 1995.
18. Izumikawa, M. et al., "A 0.25-μm CMOS 0.9-V 100M-Hz DSP Core," *IEEE J. Solid-State Circuits*, vol. 32, no. 1, pp. 52–60, Jan. 1997.

53

Embedded Memory

Chung-Yu Wu
National Chiao Tung University

CONTENTS

53.1 Introduction

As CMOS technology progresses rapidly toward the deep submicron regime, the integration level, performance, and fabrication cost increase tremendously. Thus, low-integration low-performance small circuits or systems chips designed using deep submicron CMOS technology are not cost-effective. Only high-performance system chips that integrate CPU (central processing unit), DSP (digital signal processing) processors or multimedia processors, memories, logic circuits, analog circuits, etc. can afford the deep submicron technology. Such system chips are called system-on-a-chip (SOC) or system-on-silicon (SOS) [1,2]. A typical example of SOC chips is shown in Figure 53.1.

Embedded memory has become a key component of SOC and more practical than ever for at least two reasons [3]:

1. Deep submicron CMOS technology affords a reasonable tradeoff for large memory integration in other circuits. It can afford ULSI (ultra large-scale integration) chips with over 10^9 elements on a single chip. This scale of integration is large enough to build an SOC system. This size of circuitry inevitably contains different kinds of circuits and technologies. Data processing and storage are the most primitive and basic components of digital circuits, so that the memory implementation on logic chip has the highest priority. Currently in quarter-micron CMOS technology, chips with up to 128 Mbits of DRAM and 500 Kgates of logic circuit, or 64 Mbits of DRAM and 1 Mgates of logic circuit, are feasible.

2. Memory bandwidth is now one of the most serious bottlenecks to system performance. The memory bandwidth is one of the performance determinants of current von Neuman-type MPU

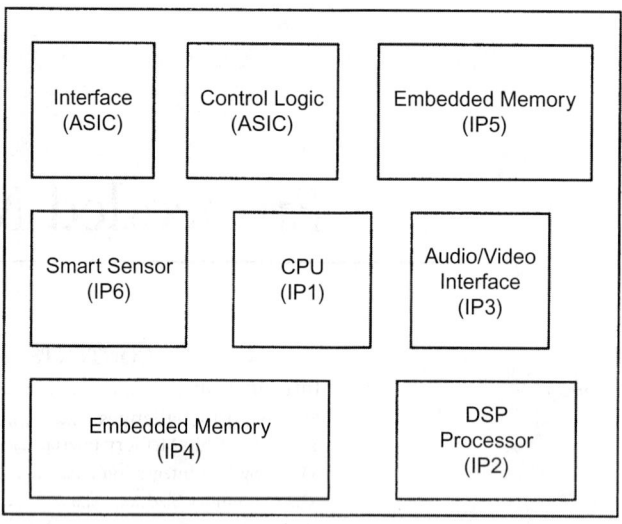

FIGURE 53.1 An example of system-on-a-chip (SOC).

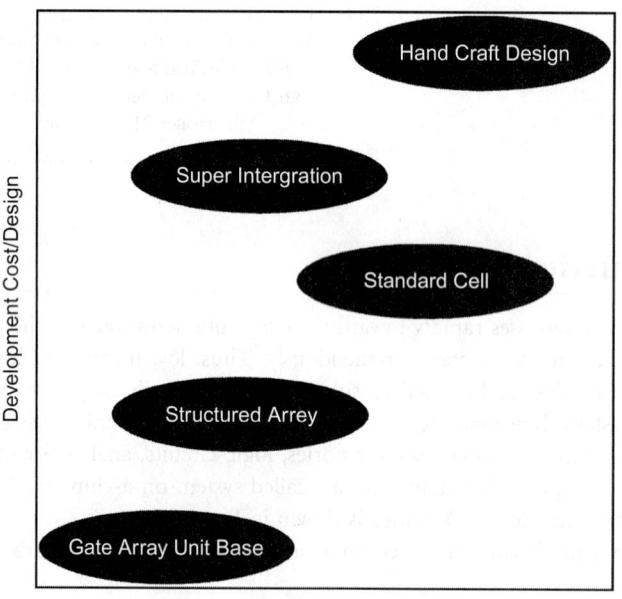

FIGURE 53.2 Various design styles for embedded memories. (From Ref. 3.)

(microprocessing unit) systems. The speed gap between MPUs and memory devices has been increased in the past decade. As shown in Figure 53.1, the MPU speed has improved by a factor of 4 to 20 in the past decade. On the other hand, in spite of exponential progress in storage capacity, minimum access times for each quadrupled storage capacity have improved only by a factor of two, as shown in Figure 53.2. This is partly due to the I/O speed limitation and to the fact that major efforts in semiconductor memory development have focused on density and bit cost improvements. This speed gap creates a strong demand for memory integration with MPU on the same chip. In fact, many MPUs with cycle times better than 60 ns have on-chip memories.

The new trend in MPUs, (i.e., RISC architecture) is another driving force for embedded memory, especially for cache applications [4]. RISC architecture is strongly dependent on memory bandwidth, so that high-performance, non-ECL-based RISC MPUs with more than 25 to 50 MHz operation must be equipped with embedded cache on the chip.

53.2 Merits and Challenges

The main characteristics of embedded memories can be summarized as follows [5].

53.2.1 On-Chip Memory Interface

Advantages include:

1. Replacing off-chip drivers with smaller on-chip drivers can reduce power consumption significantly, as large board wire capacitive loads are avoided. For instance, consider a system which needs a 4-Gbyte/s bandwidth and a bus width of 256 bits. A memory system built with discrete SDRAMs (16-bit interface at 100 MHz) would require about 10 times the power of an embedded DRAM with an internal 256-bit interface.
2. Embedded memories can achieve much higher fill frequencies [6], which is defined as the bandwidth (in Mbit/s) divided by the memory size in Mbit (i.e., the fill frequency is the number of times per second a given memory can be completely filled with new data), than discrete memories. This is because the on-chip interface can be up to 512 bits wide, whereas discrete memories are limited to 16 to 64 bits. Continuing the above example, it is possible to make a 4-Mbit embedded DRAM with a 256-bit interface. In contrast, it would take 16 discrete 4-Mbit chips (256 K × 16) to achieve the same width, so the granularity of such a discrete system is 64 Mbits. But the application may only call for, say, 8 Mbits of memory.
3. As interface wire lengths can be optimized for application in embedded memories, lower propagation times and thus higher speeds are possible. In addition, noise immunity is enhanced.

Challenges and disadvantages include:

1. Although the power consumption per system decreases, the power consumption per chip may increase. Therefore, junction temperature may increase and memory retention time may decrease. However, it should be noted that memories are usually low-power devices.
2. Some sort of minimal external interface is still needed in order to test the embedded memory. The hybrid chip is neither a memory nor a logic chip. Should it be tested on a memory or logic tester, or on both?

53.2.2 System Integration

Advantages include:

1. Higher system integration saves board space, packages, and pins, and yields better form factors.
2. Pad-limited design may be transformed into non-pad-limited by choosing an embedded solution.
3. Better speed scalability, along with CMOS technology scaling.

Challenges and disadvantages include:

1. More expensive packages may be needed. Also, memories and logic circuits require different power supplies. Currently, the DRAM power supply (2.5 V) is less than the logic power supply (3.3 V); but this situation will reverse in the future due to the back-biasing problem in DRAMs.
2. The embedded memory process adds another technology for which libraries must be developed and characterized, macros must be ported, and design flows must be tuned.

3. Memory transistors are optimized for low leakage currents, yielding low transistor performance, whereas logic transistors are optimized for high saturation currents, yielding high leakage currents. If a compromise is not acceptable, expensive extra manufacturing steps must be added.
4. Memory processes have fewer layers of metal than do logic circuit processes. Layers can be added at the expense of fabrication cost.
5. Memory fabs are optimized for large-volume production of identical products, for high capacity utilization and for high yield. Logic fabs, while sharing these goals, are slanted toward lower batch sizes and faster turnaround time.

53.2.3 Memory Size

The advantage is that:

1. Memory size can be customized and memory architecture can be optimized for dedicated applications.

Challenges and disadvantages include:

1. On the other hand, the system designer must know the exact memory requirement at the time of design. Later extensions are not possible, as there is no external memory interface. From the customer's point of view, the memory component goes from a commodity to a highly specialized part that may command premium pricing. As memory fabrication processes are quite different, second-sourcing problems abound.

53.3 Technology Integration and Applications [3,5]

The memory technologies for embedded memories have a wide variation—from ROM to RAM—as listed in Table 53.1 [3]. In choosing these technologies, one of the most important figure of merits is the compatibility to logic process.

1. Embedded ROM: ROM technology has the highest compatibility to logic process. However, its application is rather limited. PLA, or ROM-based logic design, is a well-used but rather special case of embedded ROM category. Other applications are limited to storage for microcode or well-debugged control code. A large size ROM for tables or dictionary applications may be implemented in generic ROM chips with lower bit cost.
2. Embedded EPROM/E^{2}ROM: EPROM/E^{2}PROM technology includes high-voltage devices and/or thin tunneling insulators, which require two to three additional mask steps and processing steps to logic process. Due to its unique functionality, PROM-embedded MPUs [7] are well used. To minimize process overhead, single poly E^{2}PROM cell has been developed [8]. Counterparts to this approach are piggy-back packaged EPROM/MPUs or battery-backed SRAM/MPUs. However, considering process technology innovation, on-chip PROM implementation is winning the game.
3. Embedded SRAM is one of the most frequently used memory embedded in logic chips. Major applications are high-speed on-chip buffers such as TLB, cache, register file, etc. Table 53.2 gives a comparison of some approaches for SRAM integration. A six-transistor cell approach may be the most highly compatible process, unless any special structures used in standard 6-Tr SRAMs are employed. The bit density is not very high. Polysilicon resistor load 4-Tr cells provide higher bit density with the cost of process complexity associated with additional polysilicon-layer resistors. The process complexity and storage density may be compromised to some extent using a single layer of polysilicon. In the case of a polysilicon resistor load SRAM, which may have relaxed specifications with respect to data holding current, the requirement for substrate structure to achieve good soft error immunity is more relaxed as compared to low stand-by generic SRAMs. Therefore, the TFT (thin-film transistor) load cell may not be required for several generations due to its complexity.

TABLE 53.1 Embedded Memory Technologies and Applications

Embedded Memory Technology	Compatibility to Logic Process	Applications
ROM	Diffusion, Vt, Contact programming High compatibility to logic process	Microcode, program storage PAL, ROM-based logic
E/E²prom	High-voltage device, tunneling insulator required	Program, parameter storage, sequencer, learning machine
SRAM	6-Tr/4-Tr single/double poly load cells. Wide range of compatibility	High-speed buffers, cache memory
DRAM	Gate capacitor/4-T/planar/stacked/ trench cells. Wide range of compatibility	High-density, high bit rate storage

Source: From Ref. 3.

TABLE 53.2 Embedded SRAM Options

SRAM Cell Type	Features
CMOS 6-Tr cell	No extra process steps to logic Lower bit density (Cell size, A_{cell} = 2.0 a.u.) Wide operational margin Low data-load current
NMOS 4-Tr Polysilicon Load Cell -Single Poly:	1 additional step to logic process Higher density (A_{cell} = 1.25 a.u.)
-Double Poly:	3 addititional steps to logic process Higher density (A_{cell} = 1 a.u.)

Source: From Ref. 3.

TABLE 53.3 Embedded DRAM Technology Options

Technology	Features
Standard DRAM Trench/Stacked Cell	High density (cell size A_{cell} = 1 a.u.) Large process overhead, >45% additional to logic
Planar C-plate poly-Si Cell	High density (A_{cell} = 1.3 a.u.) Process overhead >35% additional to logic
Gate capacitor +	Relatively high density (A_{cell} = 2.5 a.u.)
1-Tr Cell	No additional process to logic
4-Tr Cell	High speed, short cycle time Density is equivalent to 2-poly SRAM cell (equiv. to SRAM excpt refresh. A_{cell} = 5 a.u.)

Source: From Ref. 3.

4. Embedded DRAM (eDRAM) is not as widely used as SRAMs. Its high density features, however, are very attractive. Several different embedded DRAM approaches are listed in Table 53.3. A trench or stacked cell used in commodity DRAMs has the highest density, but the complexity is also high. The cost is seldom attractive when compared to a multi-chip approach using standard DRAM, which is the ultimate in achieving low bit cost. This type of cell is well suited for ASM (application specific memory), which will be described in the next section. A planar cell with multiple (double) polysilicon tructures is also suitable for memory-rich applications [9]. A gate capacitor storage

cell approach can be fully compatible to logic process providing relatively high density [10]. The four-Tr cell (4-Tr SRAM cell minus resistive load) provides the same speed and density as SRAM, but full compatibility to logic process and requires refresh operation [11].

53.4 Design Methodology and Design Space [3,5]

53.4.1 Design Methodology

The design style of embedded memory should be selected according to applications. This choice is critically important for the best performance and cost balancing. Figure 53.2 shows the various design styles to implement embedded memories.

The most primitive semi-custom design style is based on unit the memory cell. It provides high flexibility in memory architecture and short design TAT (turn around time). However, the memory density is the lowest among various approaches.

The structured array is a kind of gate array that has a dedicated memory array region in the master chip that is configurable to several variations of memory organizations by metal layer customization. Therefore, it provides relatively high density and short TAT. Configurability and fixed maximum memory area are the limitations to this approach.

The standard cell design has high flexibility to the extent that the cell library has a variety of embedded memory designs. But in many cases, new system design requires new memory architectures. The memory performance and density is high, but the mask-to-chip TAT tends to be long.

Super integration is an approach that integrates existing chip design, including I/O pads, so the design TAT is short and proven designs can be used. However, availability of memory architecture is limited and the mask-to-chip TAT is long.

Hand-craft design (does not necessarily mean the literal use of human hands, but heavy interactive design) provides the most flexibility, high performance, and high density; but design TAT is the longest. Thus, design cost is the highest so that the applications are limited to high-volume and/or high-end systems. Standard memories, well-defined ASMs, such as video memories [12], integrated cache memories [13], and high-performance MPU-embedded memories, are good examples.

An eDRAM (embedded DRAM) designer faces a design space that contains a number of dimensions not found in standard ASICs, some of which we will subsequently review. The designer has to choose from a wide variety of memory cell technologies which differ in the number of transistors and in performance.

Also, both DRAM technology and logic technology can serve as a starting point for embedding DRAM. Choosing a DRAM technology as the base technology will result in high memory densities but suboptimal logic performance. On the other hand, starting with logic technology will result in poor memory densities, but fast logic circuits. To some extent, one can therefore trade logic speed against logic area. Finally, it is also possible to develop a process that gives the best of both worlds—most likely at higher expense. Furthermore, the designer can trade logic area for memory area in a way heretofore impossible.

Large memories can be organized in very different ways. Free parameters include the number of memory banks, which allow the opening of different pages at the same time, the length of a single page, the word width, and the interface organization. Since eDRAM allows one to integrate SRAMs and DRAMs, the decision between on/off-chip DRAM-and SRAM/DRAM-partitioning must be made.

In particular, the following problems must be solved at the system level:

Optimizing the memory allocation
Optimizing the mapping of the data into memory such that the sustainable memory bandwidth
   approaches the peak bandwidth
Optimizing the access scheme to minimize the latency for the memory clients and thus minimize the
   necessary FIFO depth

The goals are to some extent independent of whether or not the memory is embedded. However, the number of free parameters available to the system designer is much larger in an embedded solution, and the possibility of approaching the optimal solution is thus correspondingly greater. On the other hand, the complexity is also increased. It is therefore incumbent upon eDRAM suppliers to make the tradeoffs transparent and to quantize the design space into a set of understandable if slightly suboptimal solutions.

53.5 Testing and Yield [3,5]

Although embedded memory occupies a minor portion of the total chip area, the device density in the embedded memory area is generally overwhelming. Failure distribution is naturally localized at memory areas. In other words, embedded memory is a determinant of total chip yield to the extent that the memory portion has higher device density weighted by its silicon area.

For a large memory-embedded VLSI, memory redundancy is helpful to enhance the chip yield. Therefore, the embedded-memory testing, combined with the redundancy scheme, is an important issue. The implementation of means for direct measurement of embedded memory on wafer as well as in assembled samples is necessary.

In addition to off-chip measurement, on-chip measurement circuitry is essential for accurate AC evaluation and debugging. Testing DRAMs is very different from testing logic. In the following, the main points of notice are discussed.

The fault models of DRAMs explicitly tested for are much richer. They include bit-line and word-line failures, crosstalk, retention time failures, etc.

The test patterns and test equipment are highly specialized and complex. As DRAM test programs include a lot of waiting, DRAM test times are quite high, and test costs are a significant fraction of total cost.

As DRAMs include redundancy, the order of testing is: (1) pre-fuse testing, (2) fuse blowing, (3) post-fuse testing. There are thus two wafer-level tests.

The implication on eDRAMs is that a high degree of parallelism is required in order to reduce test costs. This necessitates on-chip manipulation and compression of test data in order to reduce the off-chip interface width. For instance, Siemens Corp. offers a synthesizable test controller supporting algorithmic test pattern generation (ATPG) and expected-value comparison [partial built-in self test (BIST)].

Another important aspect of eDRAM testing is the target quality and reliability. If eDRAM is used for graphics applications, occasional "soft" problems, such as too short retention time of a few cells, are much more acceptable than if eDRAM is used for program data. The test concept should take this cost-reduction potential into account, ideally in conjunction with the redundancy concept.

A final aspect is that a number of business models are common in eDRAM, from foundry business to ASIC-type business. The test concept should thus support testing the memory, either from a logic tester or a memory tester, so that the customer can do memory testing on his logic tester if required.

53.6 Design Examples

Three examples of embedded memory designs are described. The first one is a flexible embedded DRAM design from Siemens Corp. [5]. The second one is the embedded memories in MPEG environment from Toshiba Corp. [14]. The last one is the embedded memory design for a 64-bit superscaler RISC microprocessor from Toshiba Corp. and Silicon Graphics, Inc. [15].

53.6.1 A Flexible Embedded DRAM Design [5]

There is an increasing gap between processor and DRAM speed: processor performance increases by 60% per year in contrast to only a 10% improvement in the DRAM core. Deep cache structures are used to alleviate this problem, albeit at the cost of increased latency, which limits the performance of many applications. Merging a microprocessor with DRAM can reduce the latency by a factor of 5 to 10, increase the bandwidth by a factor of 50 to 100, and improve the energy efficiency by a factor of 2 to 4 [16].

Developing memory is a time-consuming task and cannot be compared with a high-level based logic design methodology which allows fast design cycles. Thus, a flexible memory concept is a prerequisite for a successful application of eDRAM. Its purpose is to allow fast construction of application-specific memory blocks that are customized in terms of bandwidth, word width, memory size, and the number of memory banks, while guaranteeing first-time-right designs accompanied by all views, test programs, etc.

A powerful eDRAM approach that permits fast and safe development of embedded memory modules is described. The concept, developed by Siemens Corp. for its customers, uses a 0.24-μm technology based on its 64/256 Mbit SDRAM process [5]. Key features of the approach include:

Two building-block sizes, 256 Kbit and 1 Mbit; memory modules with these granularities can be constructed

Large memory modules, from 8 to 16 Mbit upwards, achieving an area efficiency of about 1 Mbit/mm^2

Embedded memory sizes up to at least 128 Mbits

Interface widths ranging from 16 to 512 bits per module

Flexibility in the number of banks as well as the page length

Different redundancy levels, in order to optimize the yield of the memory module to the specific chip

Cycle times better than 7 ns, corresponding to clock frequencies better than 143 MHz

A maximum bandwidth per module of about 9 Gbyte/s

A small, synthesizable BIST controller for the memory (see next section)

Test programs, generated in a modular fashion

Siemens Corp. has made eDRAM since 1989 and has a number of possible applications of its eDRAM approach in the pipeline, including TV scan-rate converters, TV picture-in-picture chips, modems, speech-processing chips, hard-disk drive controllers, graphics controllers, and networking switches. These applications cover the full range of memory sizes (from a few Mbits to 128 Mbits), interface widths (from 32 to 512 bits), and clock frequencies (from 50 to 150 MHz), which demonstrates the versatility of the concept.

53.6.2 Embedded Memories in MPEG Environment [14]

Recently, multimedia LSIs, including MPEG decoders, have been drawing attention. The key requirements in realizing multimedia LSIs are their low-power and low-cost features. This example presents embedded memory-related techniques to achieve these requirements, which can be considered as a review of the state-of-the-art embedded memory macro techniques applicable to other logic LSIs.

Figure 53.3 shows embedded memory macros associated with the MPEG2 decoder. Most of the functional blocks use their own dedicated memory blocks and, consequently, memory macros are rather small and distributed on a chip. Memory blocks are also connected to a central address/data bus for implementing direct test mode.

An input buffer for the IDCT is shown in Figure 53.4. Eight 16-bit data from D0 to D7 come from the inverse quantization block sequentially. The stored data should then be read out as 4-bit chunks orthogonal to the input sequence. The 4-bit data is used to address a ROM in the IDCT to realize a distributed arithmetic algorithm.

The circuit diagram of an orthogonal memory whose circuit diagram is shown in Figure 53.5. It realizes the above-mentioned functionality with 50% of the area and the power that would be needed if the IDCT input buffer were built with flip-flops. In the orthogonal memory, word-lines and bit-lines run both vertically and horizontally to achieve the functionality. The macro size of the orthogonal memory is 420 μm $\times$ 760 μm, with a memory cell size of 10.8 μm $\times$ 32.0 μm.

FIFOs and other dual-port memories are designed using a single-port RAM operated twice in one clock cycle to reduce area, as shown in Figure 53.6. A dual-port memory cell is twice as large as a single-port memory cell.

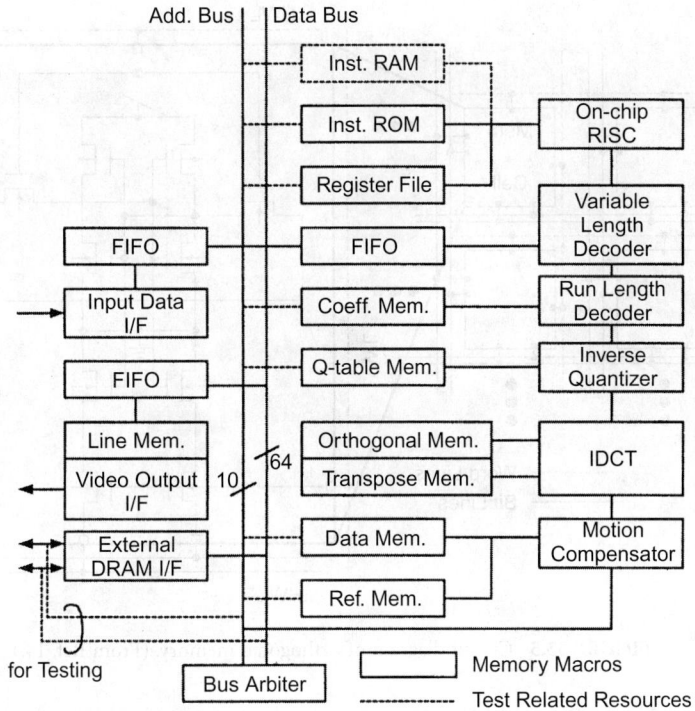

FIGURE 53.3 Block diagram of MPEG2 decoder LSI. (From Ref. 14.)

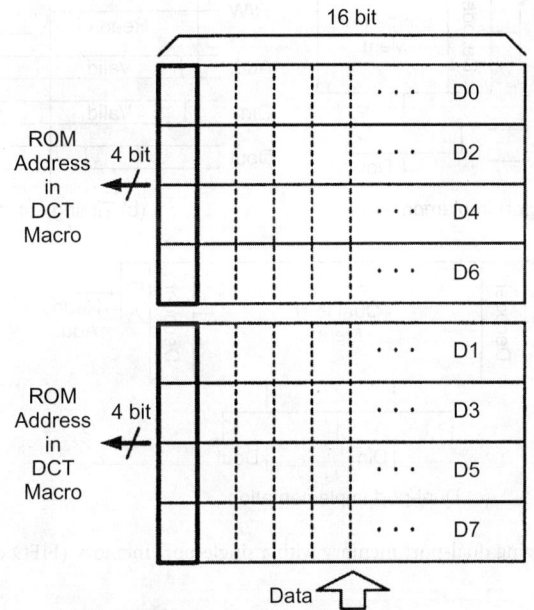

FIGURE 53.4 Input buffer structure for IDCT. (From Ref. 14.)

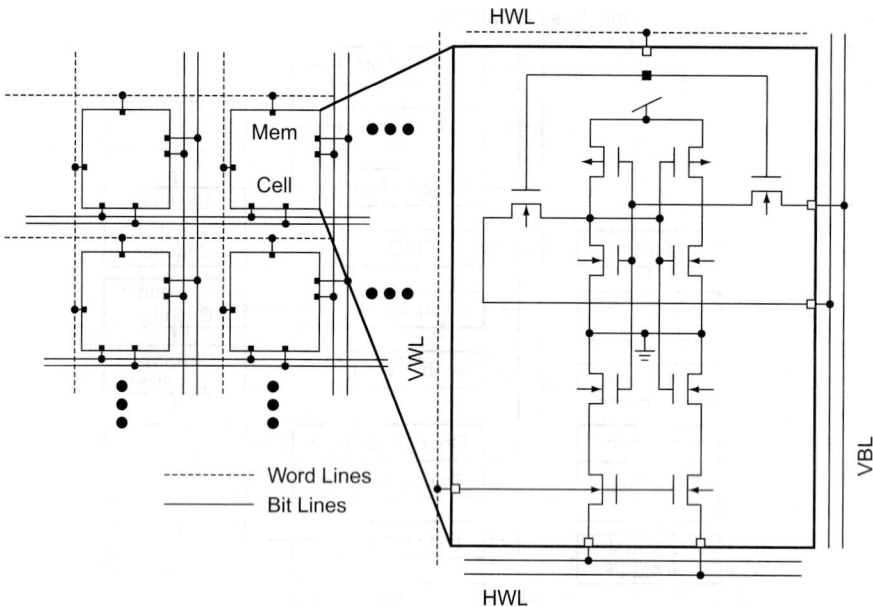

FIGURE 53.5 Circuit diagram of orthogonal memory. (From Ref. 14.)

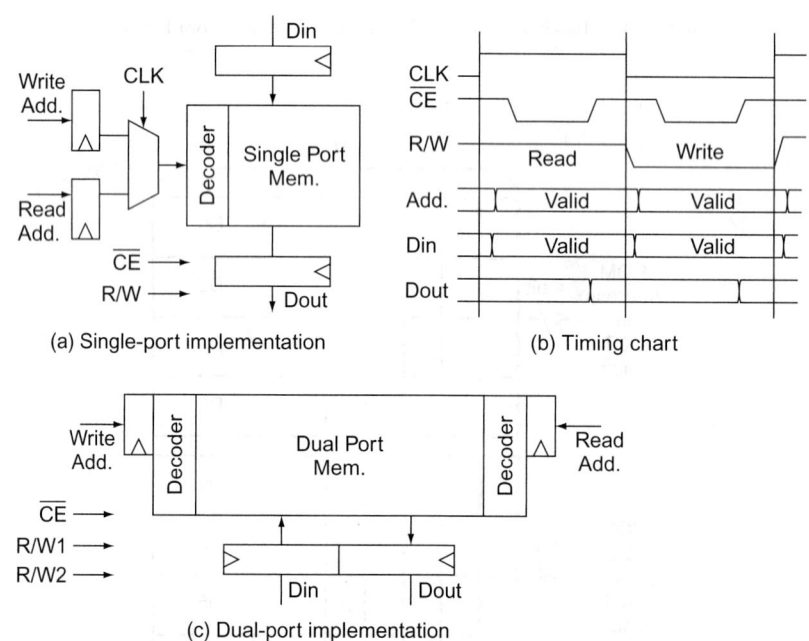

FIGURE 53.6 Realizing dual-port memory with a single-port memory (FIFO case). (From Ref. 14.)

All memory blocks are synchronous self-timed macros and contain address pipeline latches. Otherwise, the timing design needs more time, since the lengths of the interconnections between latches and a decoder vary from bit to bit. Memory power management is carried out using a Memory Macro Enable signal when a memory macro is not accessed, which reduces the total memory power to 60%.

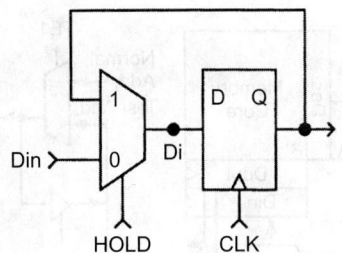

FIGURE 53.7 Optimized flip-flop. (From Ref. 14.)

TABLE 53.4 Comparison of Various Memory Test Strategies

Items	Direct	Scan	BIST
Area	○	Δ	X
Test time	○	X	○
Pattern control	○	○	X
Bus capacitance	Δ	○	○
At-speed test	○	X	○

○: Good   Δ: Fair   X: Poor
Source: Ref. 14.

Flip-flop (F/F) is one of the memory elements in logic LSIs. Since digital video LSIs tend to employ several thousand F/Fs on a chip, the design of the F/F is crucial for small area and low power. The optimized F/F with hold capability is shown in Figure 53.7. Due to the optimized smaller transistor sizes, especially for clock input transistors, and a minimized layout accomodating a multiplexer and a D-F/F in one cell, 40% smaller power and area are realized compared with a normal ASIC F/F.

Establishing full testability of on-chip memories without much overhead is another important issue. Table 53.4 compares three on-chip memory test strategies: a BIST (Built-In Self Test), a scan test, and a direct test. The direct test mode, where all memories can be directly accessed from outside in a test mode, is implemented because of its inherent small area. In a test mode, DRAM interface pads are turned into test pins and can access to each memory block through internal buses, as shown in Figures 53.3 and 53.8.

The present MPEG2 decoder contains a RISC whose firmware is stored in an on-chip ROM. In order to make the debugging easy and extensive, an instruction RAM is put outside the pads in parallel to the instruction ROM and activated by an Al-masterslice in an initial debugging stage as shown in Figure 53.9. For a sample chip mounted in a plastic package, the instruction RAM is cut out by a scribe line. This scheme enables extensive debugging and early sampling at the same time for firmware-ROM embedded LSIs.

53.6.3 Embedded Memory Design for a 64-Bit Superscaler RISC Microprocessor [15]

High-performance embedded memory is a key component in VLSI systems because of the high-speed and wide bus width capability eliminating inter-chip communication. In addition, multi-ported buffer memories are often demanded on a chip. Furthermore, a dedicated memory architecture that meets the special constraint of the system can neatly reduce the system critical path.

On the other hand, there are several issues in embedded RAM implementation. The specialty or variety of the memories could increase design cost and chip cost. Reading very wide data causes large power dissipation. Test time of the chip could be increased because of the large memory. Therefore, design efficiency, careful power bus design, and careful design for testability are necessary.

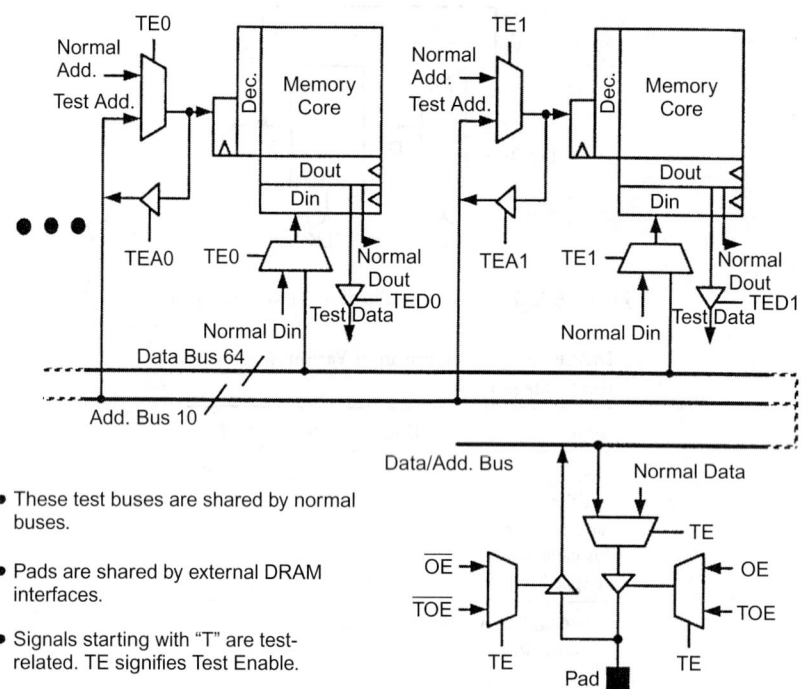

FIGURE 53.8 Direct test architecture for embedded memories. (From Ref. 14.)

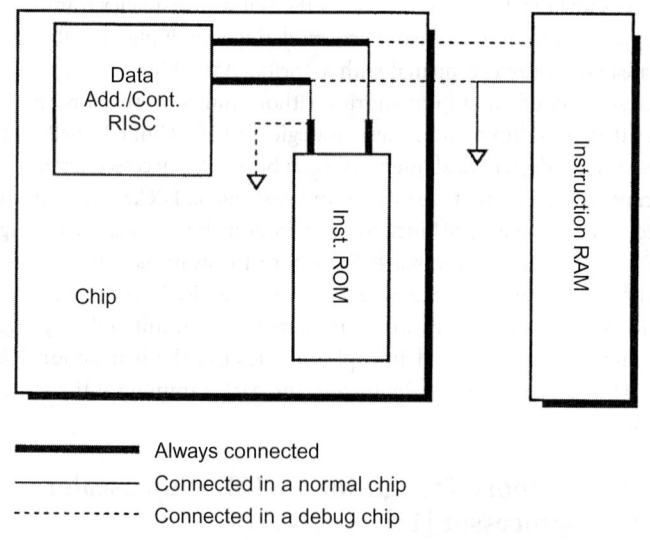

FIGURE 53.9 Instruction RAM masterslice for code debugging. (From Ref. 14.)

TFP is a high-speed and highly concurrent 64-bit superscaler RISC microprocessor, which can issue up to four instructions per cycle [17,18]. Very wide bandwidth of on-chip caches is vital in this architecture. The design of the embedded RAMs, especially on caches and TLB, is reported.

The TFP integer unit (IU) chip implements two integer ALU pipelines and two load/store pipelines. The block diagram is shown in Figure 53.10. A five-stage pipeline is shown in Figure 53.11. In the TFP IU chip, RAM blocks occupy a dominant part of the real estate. The die size is 17.3 mm × 17.3 mm.

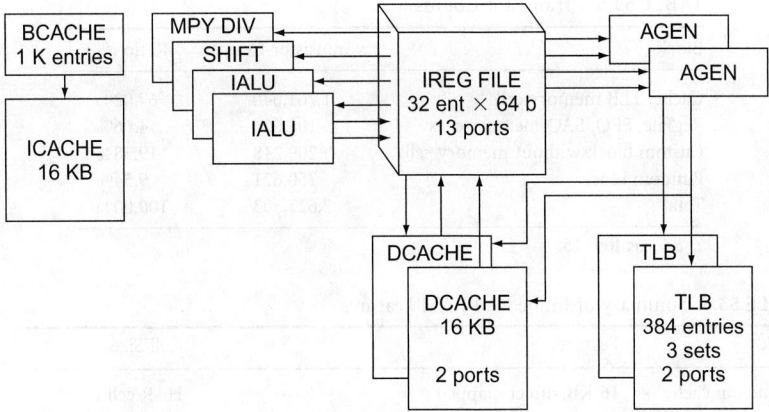

FIGURE 53.10   Block diagram of TFP IU. (From Ref. 15.)

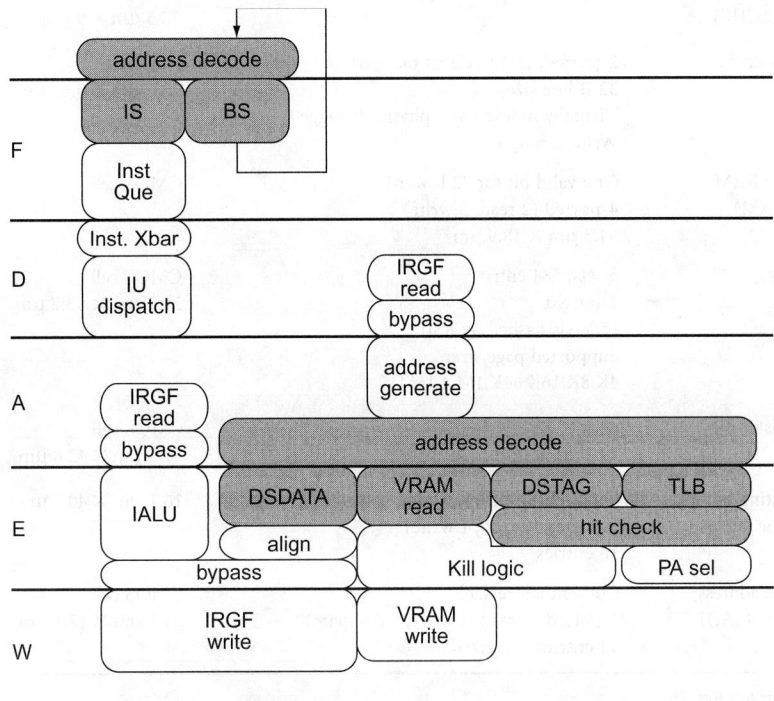

FIGURE 53.11   TFP IU pipelining. (From Ref. 15.)

In addition to other caches, TLB, and register file, the chip also includes two buffer queues: SAQ (store address queue) and FPQ (floating point queue). Seventy-one percent of all overall 2.6 million transistors are used for memory cells. Transistor counts of each block are listed in Table 53.5.

The first generation of TFP chip was fabricated using Toshiba's high-speed 0.8 μm CMOS technology: double poly-Si, triple metal, and triple well. A deep n-well was used in PLL and cache cell arrays in order to decouple these circuits from the noisy substrate or power line of the CMOS logic part. The chip operates up to 75 MHz at 3.1 V and 70°C, and the peak performance reaches 300 MIPS.

TABLE 53.5 Transistor Counts

Block	Transistor Count	Ratio (%)
Cache, TLB memory cell	1,761,040	67.02%
RegFile, FPQ, SAQ memory cells	106,624	4.06%
Custom block without memory cell	209,218	19.38%
Random blocks	250,621	9.54%
Total	2,627,503	100.00%

Source: Ref. 15.

TABLE 53.6 Summary of Embedded RAM Features

Block	Feature	Cell Size
Instruction cache	16 KB, direct mapped	Hi-R cell
(ICACHE)	32 B line size Vitually addressed 4 instructions per cycle	6.75 μm $\times$ 9 μm
Branch Cache (BCACHE)	1 K entries, direct mapped	Hi-R cell 6.75 μm $\times$ 9 μm
Data cache	2-ported, 16 KB, direct mapped 32 B line size Virtually indexed and physically tagged Write through	Hi-R cell 12.6 μm $\times$ 9.45 μm
Valid RAM (VRAM)	One valid bit for 32 b word 4-ported (2 read, 2 write) 34.3 μm $\times$ 18.9 μm	CMOS cell
TLB	3 sets, 384 entries 2-ported Index is hashed by ASID Supported page size: 4K,8K,16K,64K,1M,4,16M	CMOS cell 21.2 μm $\times$ 13.7 μm
Register file	64 b $\times$ 32 entries 13-ported (9 read, 4 write)	CMOS cell 59.5 μm $\times$ 42.8 μm
Floating point queue (FPQ)	Dispatches 4 floating-point instructions per cycle 3-ported (2 read, 1 write) 16 entries	16.1 μm $\times$ 40.7 μm
Store address queue (SAQ)	Content addressable 3-ported (1 read, 1 write, 1 compare) 32 entries, 2 banked	CMOS cell 35.1 μm $\times$ 17.1 μm

Source: Ref. 15.

Features of each embedded memory are summarized in Table 53.6. Instruction, branch, and data caches are direct mapped because of the faster access time. High-resistive poly-Si load cells are used for these caches since the packing density is crucial for the performance.

Instruction cache (ICACHE) is 16 KB of virtual address memory. It provides four instructions (128 bit wide) per cycle. Branch cache (BCACHE) contains branch target address with one flag bit to indicate a predicted branch. BCACHE contains 1-K entries and is virtually indexed in parallel with ICACHE.

Data cache (DCACHE) is 16 KB, dual ported, and supports two independent memory instructions (two loads, or one load and one store) per cycle. Total memory bandwidth of ICACHE and DCACHE reaches 2.4 GB/s at 75 MHz. Floating point load/store data bypass DCACHE and go directly to bigger external global cache [17,19]. DCACHE is virtually indexed and physically tagged.

TLB is dual ported, three-set-associative memory containing 384 entries. A unique address comparison scheme is employed here, which will be described in the following section. It supports several different page sizes, ranging from 4 KB to 16 MB. TLB is indexed by low-order 7 bits of virtual page number (VPN). The index is hashed by exclusive-OR with a low-order ASID (address space identifier) so that many processes can co-exist in TLB at one time.

Since several different RAMs are used in TFP chip, the design efficiency is important. Consistent circuit schemes are used for each of the caches and TLB RAMs. Layout is started from the block that has the tightest area restriction, and the created layout modules are exported to other blocks with small modification.

The basic block diagram of cache blocks is shown in Figure 53.12, and timing diagram is shown in Figure 53.13. Unlike a register file or other smaller queue buffers, these blocks employ dual-railed bit-lines. To achieve 75-MHz operation in the worst-case condition, it should operate at 110 MHz under typical conditions. In this targeted 9-ns cycle time, address generation is done about 3 ns before the end of the cycle, as shown in Figure 53.11. To take advantage of this big address set-up time, address is received by transparent latch: TLAT_N (transparent while clock is low) instead of flip-flop. Thus, decode is started as soon as address generation is done and is finished before the end of the cycle. Another transparent latch—TLAT_P (transparent while clock is high)—is placed after the sense amplifier and it holds read data while the clock is low.

Word-line (WL) is enabled while clock is high. Since the decode is already finished, WL can be driven to "high" as fast as possible. The sense amplifier is enabled (SAE) with a certain delay after the word-line. The paired current-mirror sense amplifier is chosen since it provides good performance without overly strict SAE timing. Bit-line is precharged and equalized while the clock is low. The clock-to-data delay of DCACHE, which is the biggest array, is 3.7 ns under typical conditions: clock-to-WL is 0.9 ns and WL-to-data is 2.8 ns. Since on-chip PLL provides 50% duty clock, timing pulses such as SAE or WE (write enable) are created from system clock by delaying the positive edge and negative edge appropriately.

As both word-line and sense amplifier are enabled in just half the time of one cycle, the current dissipation is reduced by half. However, the power dissipation and current spike are still an issue because

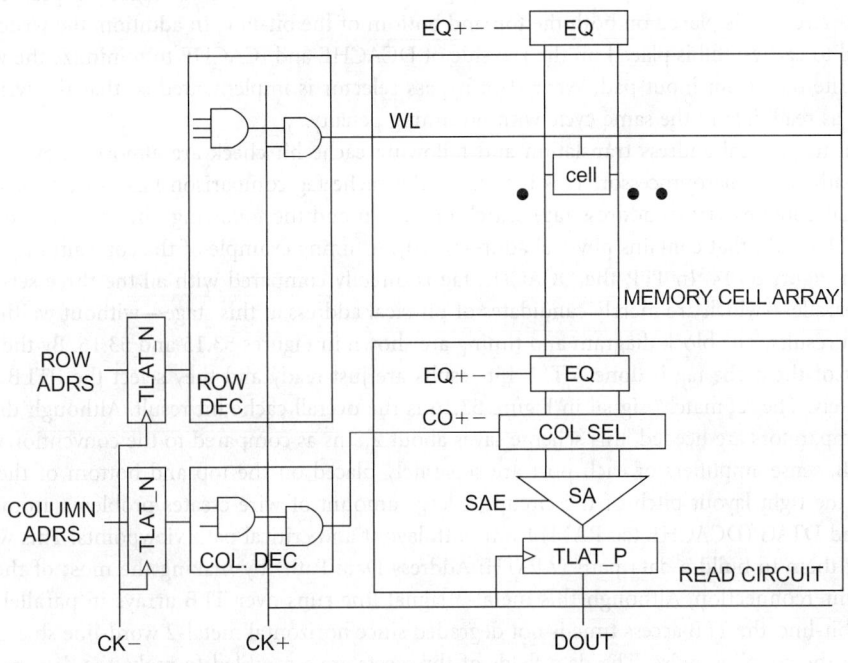

FIGURE 53.12 Basic RAM block diagram. (From Ref. 15)

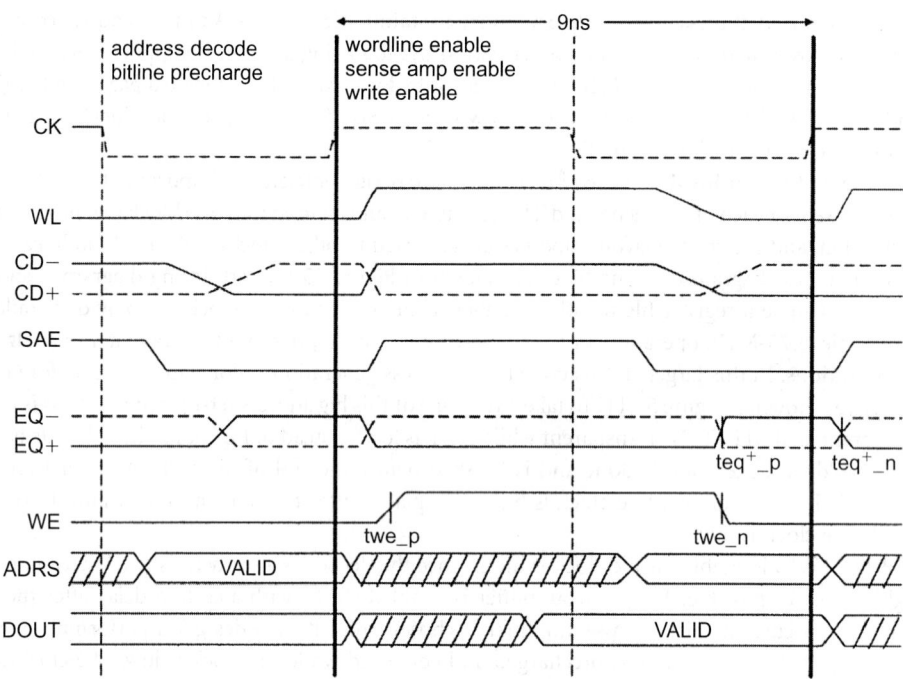

FIGURE 53.13 RAM timing diagram. (From Ref. 15)

the read/write data width is extremely large. Robust power bus matrix is applied in the cache and TLB blocks so that the dc voltage drop at the worst place is limited to 60 mV inside the block.

From a minimum cycle time viewpoint, write is more critical than read because write needs bigger bit-line swing, and the bit-line must be precharged before the next read. To speed up precharge time, precharge circuitry is placed on both the top and bottom of the bit-line. In addition, the write circuitry dedicated to cache-refill is placed on the top side of DCACHE and ICACHE to minimize the wire delay of the write data from input pad. Write data bypass selector is implemented so that the write data is available as read data in the same cycle with no timing penalty.

Virtual to physical address translation and following cache hit check are almost always one of the critical paths in a microprocessor. This is because the cache tag comparison has to wait for the VTLB (RAM that contains virtual address tag) search operation and the following physical address selection from PTLB (RAM that contains physical address) [20]. A timing example of the conventional scheme is shown in Figure 53.14. In TFP, the DCACHE tag is directly compared with all the three sets of PTLB data in parallel—which are merely candidates of physical address at this stage—without waiting for the VTLB hit results. The block diagram and timing are shown in Figures 53.15 and 53.16. By the time this hit check of the cache tag is done, VTLB hit results are just ready and they select the PTLB hit result immediately. The "ePmatch" signal in Figure 53.16 is the overall cache hit result. Although three times more comparators are needed, this scheme saves about 2.8 ns as compared to the conventional one.

In TLB, sense amplifiers of each port are separately placed on the top and bottom of the array to mitigate the tight layout pitch of the circuit. A large amount of wire creates problems around VTLB, PTLB, and DTAG (DCACHE tag RAM) from both layout and critical path viewpoints. This was solved by piling them to build a data path (APATH: Address Data Path) by making the most of the metal-3 vertical interconnection. Although this metal-3 signal line runs over TLB arrays in parallel with the metal-1 bit-line, the TLB access time is not degraded since horizontal metal-2 word-line shields the bit-line from the coupling noise. The data fields of three sets are scrambled to make the data path design tidy; 39-bit (in VTLB) and 28-bit (in PTLB) comparators of each set consist of optimized AND-tree.

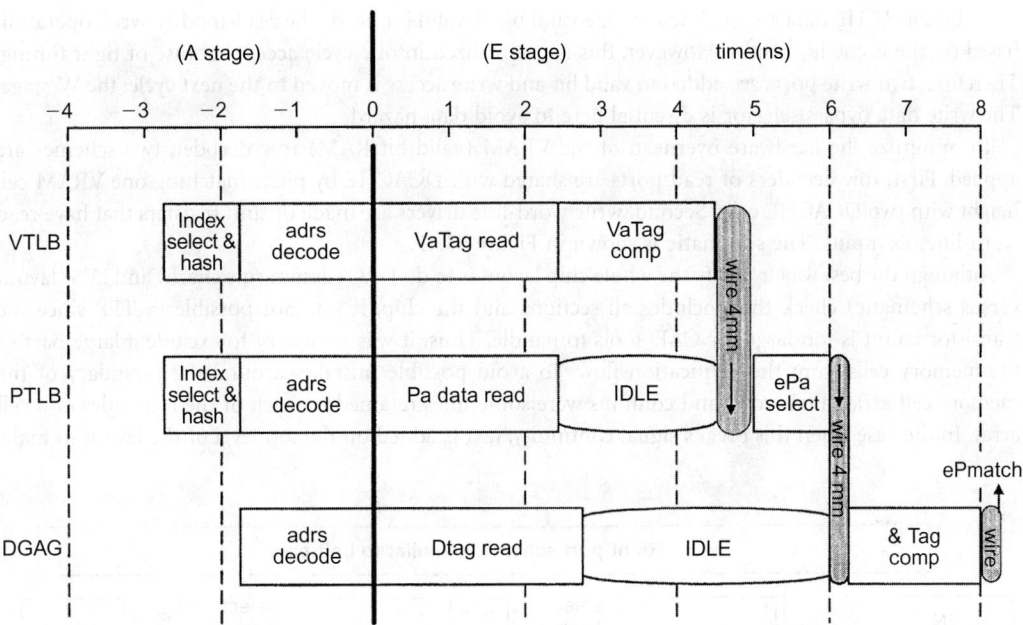

FIGURE 53.14 Conventional physical cache hit check. (From Ref. 15.)

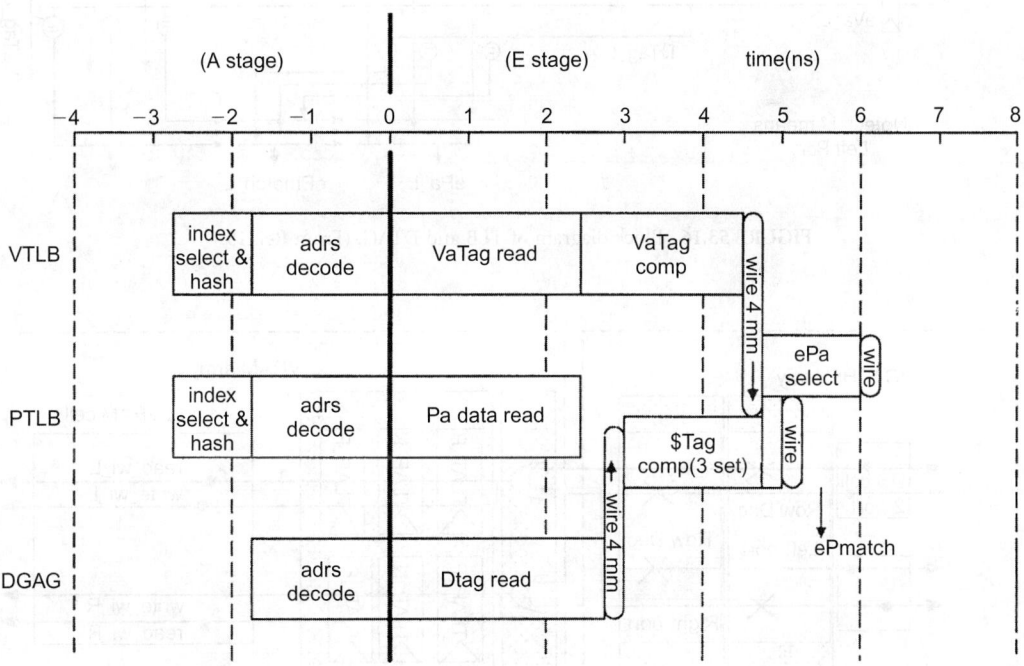

FIGURE 53.15 TFP physical cache hit check. (From Ref. 15.)

Wired-OR type comparators are rejected because a longer wired-OR node in this array configuration would have a speed penalty.

As TFP supports different page sizes, VPN and PFN (page frame number) fields change, depending on the page size. The index and comparison field of TLB are thus made selectable by control signals.

32-bit DCACHE data are qualified by one valid bit. A valid bit needs the read-modify-write operation based on the cache hit results. However, this is not realized in one cycle access because of tight timing. Therefore, two write ports are added to valid bit and write access is moved to the next cycle: the W-stage. The write data bypass selector is essential here to avoid data hazard.

To minimize the hardware overhead of the VRAM (valid bit RAM) row decoder, two schemes are applied. First, row decoders of read ports are shared with DCACHE by pitch-matching one VRAM cell height with two DCACHE cells. Second, write word-line drivers are made of shift registers that have read word-lines as inputs. The schematic is shown in Figure 53.17.

Although the best way to verify the whole chip layout is to do DRC (design rule check) and LVS (layout versus schematic) check that includes all sections and the chip, it was not possible in TFP since the transistor count is too large for CAD tools to handle. Thus, it was necessary to exclude a large part of the memory cells from the verification flow. To avoid possible mistakes around the boundary of the memory cell array, a few rows and columns were sometimes retained on each of the four sides of a cell array. In the case when this breaks signal continuity, text is added on the top level of the layout to make

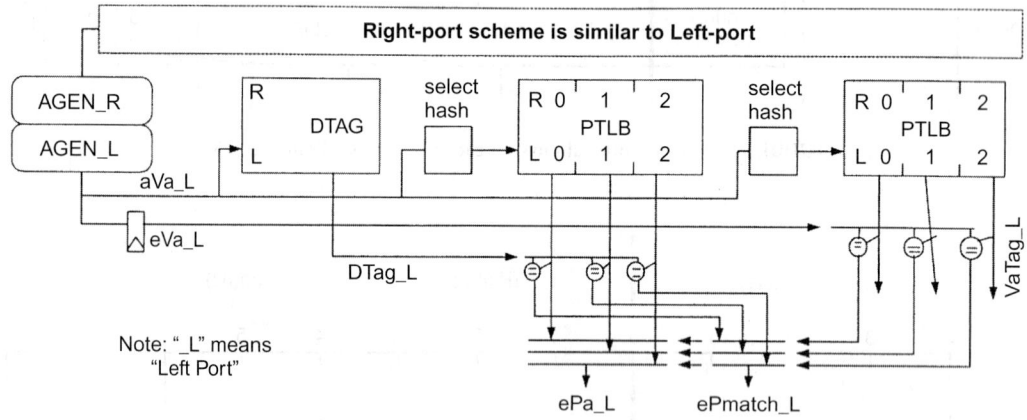

FIGURE 53.16 Block diagram of TLB and DTAG. (From Ref. 15.)

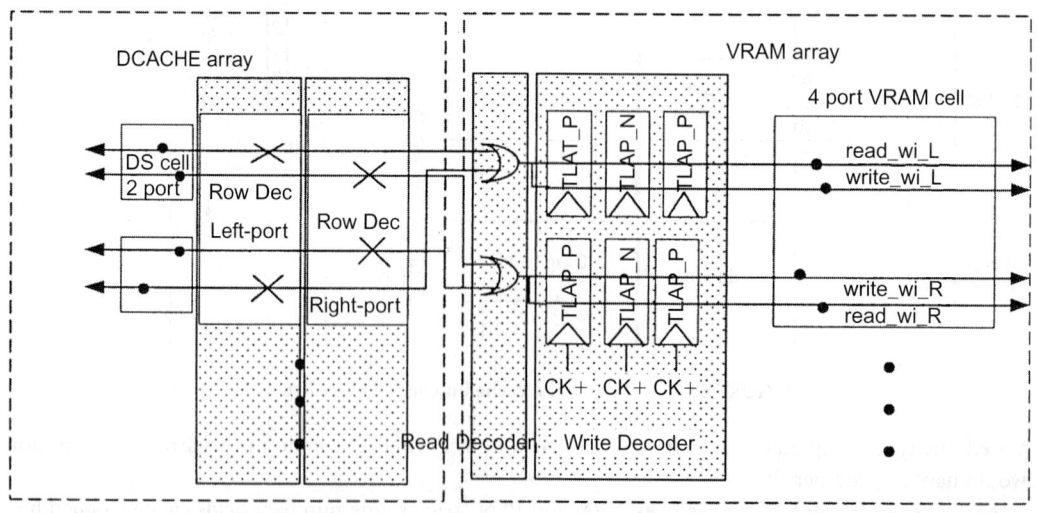

FIGURE 53.17 VRAM row decoder. (From Ref. 15.)

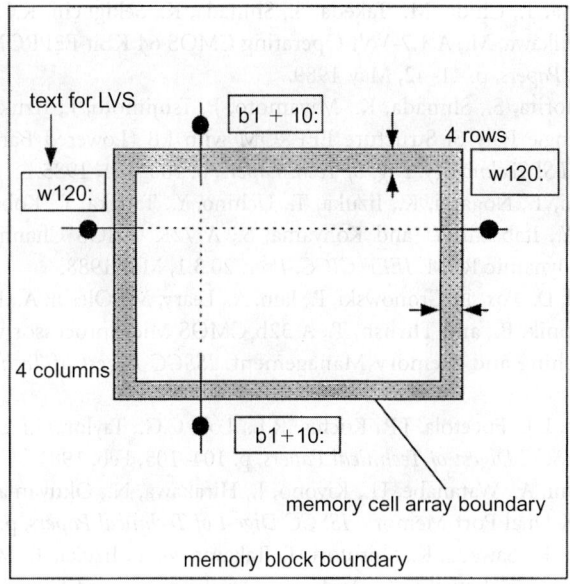

text for LVS

b1 + 10:

4 rows

w120:

w120:

4 columns

b1 + 10:

memory cell array boundary

memory block boundary

FIGURE 53.18   RAM layout verification. (From Ref. 15.)

a virtual connection, as shown in Figure 53.18. These works are basically handled by CAD software plus small programming without editing the layout by hand.

Direct testing of large on-chip memory is highly preferable in VLSI because of faster test time and complete test coverage. TFP IU defines cache direct test in JTAG test mode, in which cache address, data, write enable, and select signals are directly controlled from the outside. Thus, very straightforward evaluation is possible. Utilizing 64-bit, general-purpose bus that runs across the chip, the additional hardware for the data transfer is minimized.

Since defect density is a function of device density and device area, large on-chip memory can be a determinant of total chip yield. Raising embedded memory yield can directly lead to the rise of the chip yield. Failure symptoms of the caches have been analyzed by making a fail-bit-map, and this has been fed back to the fabrication process.

References

1. Borel, J., Technologies for Multimedia Systems on a Chip. In *1997 International Solid State Circuits Conference, Digest of Technical Papers*, 40, 18–21, Feb. 1997.
2. De Man, H., Education for the Deep Submicron Age: Business as Usual? In *Proceedings of the 34th Design Automation Conference*, p. 307–312, June 1997.
3. Iizuka, T., Embedded Memory: A Key to High Performance System VLSIs. *Proceedings of 1990 Symposium on VLSI Circuits*, p. 1–4, June 1990.
4. Horowitz, M., Hennessy, J., Chow, P., Gulak, P., Acken, J., Agrawal, A., Chu, C., McFarling, S., Przybylski, S., Richardson, S., Salz, A., Simoni, R., Stark, D., Steenkiste, P., Tjiang, S., and Wing, M., A 32b Microprocessor with On-chip 2K-Byte Instruction Cache. *ISSCC Dig. of Tech. Papers*, p. 30–31, Feb. 1987.
5. Wehn, N. and Hein, S., Embedded DRAM architectural trade-offs. *Proceedings of Design, Automation and Test in Europe*, p. 704–708, 1998.
6. Przybylski, S.A., New DRAM Technologies: A Comprehensive Analysis of the New Architectures. Report, 1996.

7. Wada, Y., Maruyama, T., Chida, M., Takeda, S., Shinada, K., Sekiguchi, K., Suzuki, Y., Kanzaki, K., Wada, M., and Yoshikawa, M., A 1.7-Volt Operating CMOS 64 KBit E2PROM. *Symp. on VLSI Circ., Kyoto, Dig. of Tech. Papers*, p. 41–42, May 1989.

8. Matsukawa, M., Morita, S., Shinada, K., Miyamoto, J., Tsujimoto, J., Iizuka, T., and Nozawa, H., A High Density Single Poly Si Structure EEPROM with LB (Lowered Barrier Height) Oxide for VLSI's. *Symp. on VLSI Technology, Dig. of Tech. Papers*, p. 100–101, 1985.

9. Sawada, K., Sakurai, T., Nogami, K., Iizuka, T., Uchino, Y., Tanaka, Y., Kobayashi, T., Kawagai, K., Ban, E., Shiotari, Y., Itabashi, Y., and Kohyama, S., A 72K CMOS Channelless Gate Array with Embedded 1Mbit Dynamic RAM. *IEEE CICC, Proc.* 20.3.1, May 1988.

10. Archer, D., Deverell, D., Fox, F., Gronowski, P., Jain, A., Leary, M., Olesin, A., Persels, S., Rubinfeld, P., Schmacher, D., Supnik, B., and Thrush, T., A 32b CMOS Microprocessor with On-Chip Instruction and Data Caching and Memory Management. *ISSCC Digest of Technical Papers*, p. 32–33; Feb. 1987.

11. Beyers, J.W., Dohse, L.J., Fucetola, J.P., Kochis, R.L., Lob, C.G., Taylor, G.L., and Zeller, E.R., A 32b VLSI CPU Chip. *ISSCC Digest of Technical Papers*, p. 104–105, Feb. 1981.

12. Ishimoto, S., Nagami, A., Watanabe, H., Kiyono, J., Hirakawa, N., Okuyama, Y., Hosokawa, F., and Tokushige, K., 256K Dual Port Memory. *ISSCC Digest of Technical Papers*, p. 38–39, Feb. 1985.

13. Sakurai, T., Nogami, K., Sawada, K., Shirotori, T., Takayanagi, T., Iizuka, T., Maeda, T., Matsunaga, J., Fuji, H., Maeguchi, K., Kobayashi, K., Ando, T., Hayakashi, Y., and Sato, K., A Circuit Design of 32Kbyte Integrated Cache Memory. *1988 Symp. on VLSI Circuits*, p. 45–46, Aug. 1988.

14. Otomo, G., Hara, H., Oto, T., Seta, K., Kitagaki, K., Ishiwata, S., Michinaka, S., Shimazawa, T., Matsui, M., Demura, T., Koyama, M., Watanabe, Y., Sano, F., Chiba, A., Matsuda, K., and Sakurai, T., Special Memory and Embedded Memory Macros in MPEG Environment. *Proceedings of IEEE 1995 Custom Integrated Circuits Conference*, p. 139–142, 1995.

15. Takayanagi, T., Sawada, K., Sakurai, T., Parameswar, Y., Tanaka, S., Ikumi, N., Nagamatsu, M., Kondo, Y., Minagawa, K., Brennan, J., Hsu, P., Rodman, P., Bratt, J., Scanlon, J., Tang, M., Joshi, C., and Nofal, M., Embedded Memory Design for a Four Issue Superscaler RISC Microprocessor. *Proceedings of IEEE 1994 Custom Integrated Circuits Conference*, p. 585–590, 1994.

16. Patterson, D. et al., Intelligent RAM (IRAM): Chips that Remember and Compute. In *1997 International Solid State Circuits Conference, Digest of Technical Papers*, 40, 224–225, February 1997.

17. Hsu, P., Silicon Graphics TFP Micro-Supercomputer Chip Set. *Hot Chips V Symposium Record*, p. 8.3.1–8.3.9, Aug. 1993.

18. Ikumi, N. et al., A 300 MIPS, 300 MFLOPS Four-Issue CMOS Superscaler Microprocessor. *ISSCC 94 Digest of Technical Papers*, Feb. 1994.

19. Unekawa, Y. et al., A 110 MHz/1Mbit Synchronous TagRAM. *1993 Symposium on VLSI Circuits Digest of Technical Papers*, p. 15–16, May 1993.

20. Takayanagi, T. et al., 2.6 Gbyte/sec Cache/TLB Macro for High-Performance RISC Processor. *Proceedings of CICC'91*, p. 10.21.1–10.2.4, May 1991.

54

Flash Memories

Rick Shih-Jye Shen
Frank Ruei-Ling Lin
Amy Hsiu-Fen Chou
Evans Ching-Song Yang and
Charles Ching-Hsiang Hsu
National Tsing-Hua University

CONTENTS

54.1 Introduction

In past decades, owing to process simplicity, stacked-gate memory devices have become the mainstream in the non-volatile memory market. This chapter is divided into seven sections to review the evolution of stacked-gate memory, device operation, device structures, memory array architectures, and flash memory system. In Section 54.2, a short historical review of stacked-gate memory device and the current flash device are described. Following this, the current–voltage characteristics, charge injection/ejection mechanisms, and the write/erase configurations are mentioned in detail. Based on the descriptions of device operation, some modifications in the memory device structure to improve performance are addressed in Section 54.4. Following the introductions of single memory device cells, descriptions of the memory array architectures are employed in Section 54.6 to facilitate the understanding of device operation. In Section 54.7, a table lists the history of flash memory development over the past decade. Finally, Section 54.8 is dedicated to the issues related to implementation of a flash memory system.

54.2   Review of Stacked-Gate Non-Volatile Memory

The concept of a memory device with a floating gate was first proposed by Kahng and Sze in 1967 [1]. The suggested device structure was started from a basic MOS structure. As shown in Figure 54.1, the insulator in the conventional MOS structure was replaced with a thin oxide layer (I1), an isolated metal layer (M1), and a thick oxide layer (I2). These stacked oxide and metal layers led to the so-called MIMIS structure. In this device structure, the first insulator layer I1 had to be thin enough to allow electrons injected into the floating gate M1. Besides, the second insulator layer I2 is required to be thick enough to avoid the loss of stored charge during charge injection operation. During electron injection operation, a high electric field (~10 MV/cm) enables the electron tunneling through I1 directly, and the injected electrons are captured in the floating gate and thus change the I–V characteristics. On the other hand, a negative voltage is applied at the external gate to remove the stored electrons during the discharge operation by the same direct tunneling mechanism. Owing to the very thin oxide layer I1, the defects in the oxide and the back tunneling phenomena lead to a poor charge retention capability. However, this MIMIS structure demonstrated, for the first time, the possibility of implementation of non-volatile memory device based on the MOS structure.

After MIMIS was invented, several improvements were proposed to enhance the performance of MIMIS. One was the utilization of dielectric material with a large amount of electron-trapping centers as a replacement of the floating metal gate [2,3]. The injected electrons would be trapped in the bulk and also at the interface traps in the dielectric material, such as silicon nitride (Si_3N_4), Al_2O_3, Ta_2O_5. The device structure with these insulating layers as electron storage node was referred as a *charge trapping device*. Another solution to improve the oxide quality and charge retention capability was the increase of the thickness of the tunnel dielectric I1. This device structure based on the MIMIS structure but with a thicker insulating layer was also referred as *floating gate device*.

In the initial development period, the charge trapping devices had several advantages compared with floating gate devices. They allowed high density, good write/erase endurance capability, and fast programming/erase time. However, the main obstacle for the wide application in charge trapping devices was the poorer charge retention capability than in floating gate devices. On the other hand, the floating gate devices showed a major drawback of not being electrically erasable. Therefore, the erase operation had to be preceded by the time-consuming UV-irradiation process. However, the floating gate devices had been

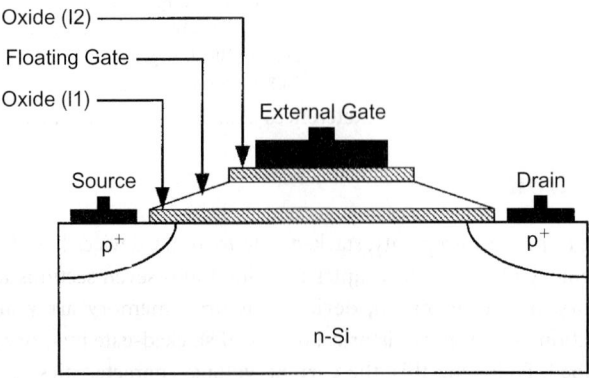

FIGURE 54.1   Schematic cross-section of MIMIS structure.

applied successfully because of the following advantages and improvements. First, the floating gate devices were compatible with the standard double polysilicon NMOS process and then became compatible with CMOS process after minor modification. Second, an excellent charge retention capability was obtained because of the thicker gate oxide. Besides, the thicker oxide leads to a relieved gate disturbance issue. Furthermore, the development of electrical erase operation technique during the 1980s made the write/erase operation easier and more efficient. Based on these reasons, most commercial non-volatile memory companies focused their research efforts on the floating gate devices. Therefore, floating gate devices have become the mainstream product in the non-volatile market.

A high operation voltage is unavoidable when the thickness of oxide I1 increases in MIMIS structure. Thus, another way to achieve electron injection was necessary to make the injection operation more efficient. In 1971, the introduction of a memory element with avalanche injection scheme was demonstrated [4]. This first operating floating gate device—named Floating gate Avalanche injection MOS (FAMOS), as shown in Figure 54.2—was a p-channel MOSFET in which no electrical contact was made to the silicon gate. The injection operation of the FAMOS memory structure is initiated by avalanche phenomena in the drain region underneath the gate. The electron-hole pair generation is caused by applying a high reversed bias at the drain/substrate junction. Some of generated electrons drift toward the floating gate by the positive oxide field which is induced by the capacitive coupling between floating gate and drain. However, the inefficient injection process was the major drawback in this device structure.

In order to improve the injection efficiency, the Stacked-gate Avalanche injection MOS (SAMOS) with an external gate was proposed, as shown in Figure 54.3. Owing to the additional gate bias, the programming speed was improved by an increased drift velocity of electrons in the oxide and the field induced energy barrier lowering at the Si–SiO$_2$, interface. Besides, by employing this control gate, the electrical erase operation became possible by building up a high electric field across the inter-polysilicon dielectric.

All the stacked-gate devices mentioned above are p-channel devices, which utilize avalanche injection scheme. However, if a smaller access time is required for the read operation, n-channel devices are necessary because of higher channel carrier mobility. Since the avalanche injection in an n-channel device is based on the hole injection, other injection mechanisms are required for n-channel stacked-gate memory cells. There are two major injection schemes for the n-channel memory cell. One is the channel hot electron injection (CHEI) and the other one is high electric field (Fowler-Nordheim, FN) tunneling mechanism. These two operation schemes lead to different device structures. The memory devices using the CHEI scheme allow a thicker gate oxide, whereas the memory devices using FN tunneling scheme require thinner oxide. In 1980, researches at Intel Corp. proposed the FLOTOX (FLOating gate Tunnel OXide) device, as shown in Figure 54.4, in which the electrons are injected into and ejected from the floating gate through a high-quality thin oxide

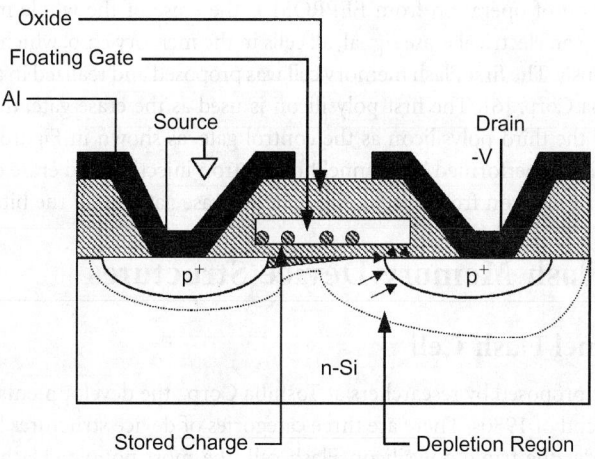

FIGURE 54.2 Schematic cross-section of FAMOS structure.

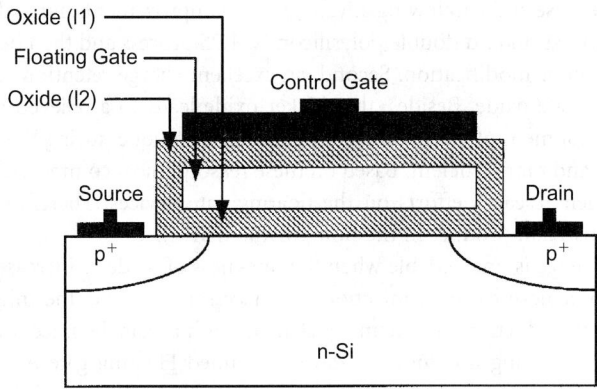

FIGURE 54.3   Schematic cross-section of p-channel SAMOS structure.

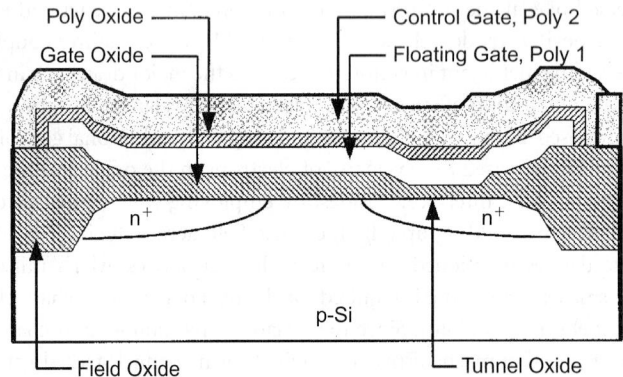

FIGURE 54.4   Schematic cross-section of FLOTOX structure.

region outside the channel region [5]. The FLOTOX cell must be isolated by a select transistor to avoid the over-erase issue and therefore it consists of two transistors. Although this limits the density of such memory in comparison with EPROM and the Flash cell, it enables the byte-by-byte erase and reprogramming operation without having to erase the entire chip or sector. Based on this, the FLOTOX cell is suitable for the applications in which low density, high reliability, and non-volatile memory are required.

Another modification of operation from EEPROM is the erase of the whole memory chip instead of erasing a byte. By using an electrical erase signal, all cells in the memory chip, which is called a Flash device, are erased simultaneously. The first Flash memory cell was proposed and realized in a three-layer polysilicon technology by Toshiba Corp. [6]. The first polysilicon is used as the erase gate, the second polysilicon as the floating gate, and the third polysilicon as the control gate, as shown in Figure 54.5(c). In this device, programming operation is performed by channel hot electron injection and erase operation is carried out by extracting the stored electron from the floating gate to erase gate for all the bits at the same time.

54.3   Basic Flash Memory Device Structures

54.3.1   n-Channel Flash Cell

Based on the concept proposed by researchers at Toshiba Corp., the developments in Flash memory have burgeoned since the end of 1980s. There are three categories of device structures based on the n-channel MOS structure. Besides the triple polysilicon Flash cell, the most popular Flash cell structures are the ETOX cell and the split-gate cell.

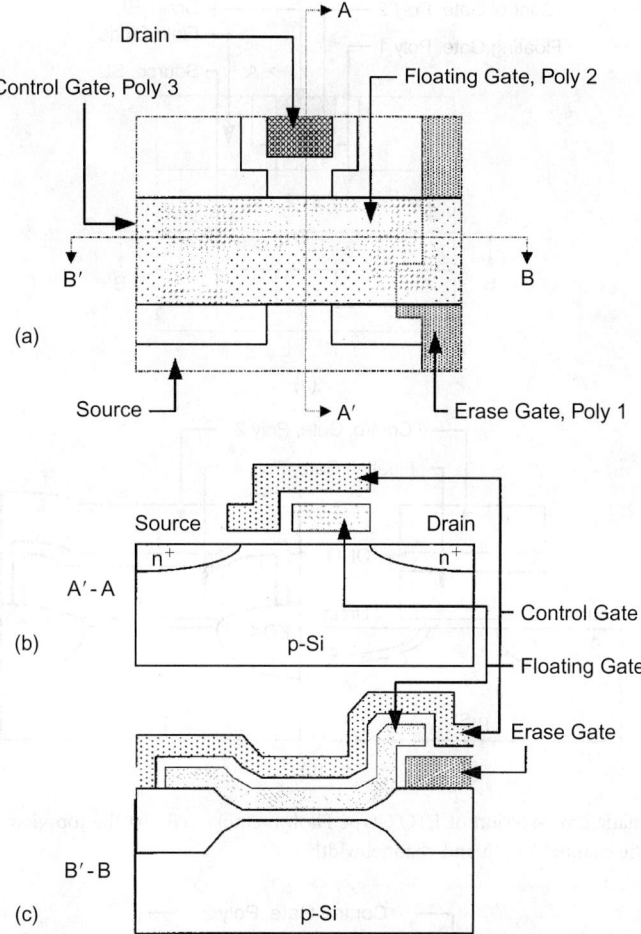

FIGURE 54.5 Tripe-gate Flash memory structure proposed by Toshiba: (a) layout of the cell; (b) cross-section along the channel length, and (c) cross-section along the channel width.

In 1985, Mukherjee et al. [7,9] proposed a source-erase Flash cell called the ETOX (EPROM with Tunnel OXide). This cell structure is the same as that of the UV-EPROM, as shown in Figure 54.6, but with a thin tunnel oxide layer. The cell is programmed by CHEI and erased by applying a high voltage at the source terminal.

A split-gate memory cell was proposed by Samachisa et al. in 1987 [8]. This split-gate Flash cell with a drain-erase type has two polysilicon layers, as shown in Figure 54.7. The cell can be regarded as two transistors in series. One is a floating gate memory, which is similar to an EPROM cell; the other, which is used as a select transistor, is an enhancement transistor controlled by the control gate.

54.3.2 p-Channel Flash Cell

The p-channel Flash memory cell was first proposed by Hsu et al. in 1992 [9]. Recently, several studies have been done on this device structure [10–13]. This Flash cell structure is similar to the ETOX cell but with p-channel. The erase mechanism is still by FN tunneling. As to the electron injection, there are two injection schemes that can be employed: CHEI and BBHE (Band-to-Band tunneling induced Hot Electron injection) [11]. The p-channel Flash cell features high electron injection efficiency, scalability, immunity to the hot hole injection and reduced oxide field during programming. Based on these advantages, the p-channel Flash memory cell seems to reveal a high potential for future low-power Flash applications.

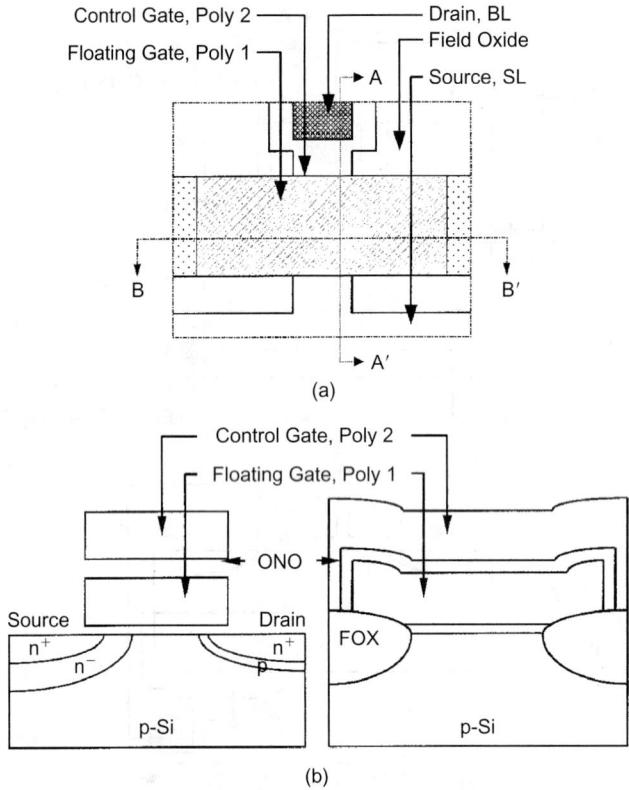

(a)

(b)

FIGURE 54.6 Schematic cross-section of ETOX-type Flash memory cell: (a) the top view of the cell, and (b) the cross-section along the channel length and channel width.

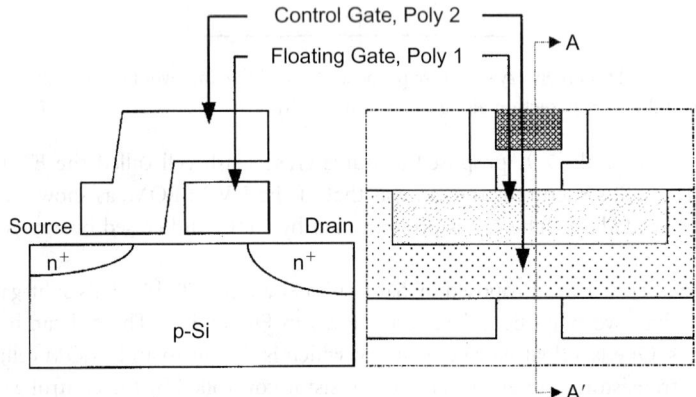

FIGURE 54.7 Schematic cross-section of split-gate Flash memory cell.

54.4 Device Operations

54.4.1 Device Characteristics

54.4.1.1 Capacitive Coupling Effects and Coupling Ratios

The I–V characteristics of stacked gate can be derived from the MOSFET characteristics accompanying with the capacitive-coupling factors. For a stacked-gate device, the device structure can be depicted as

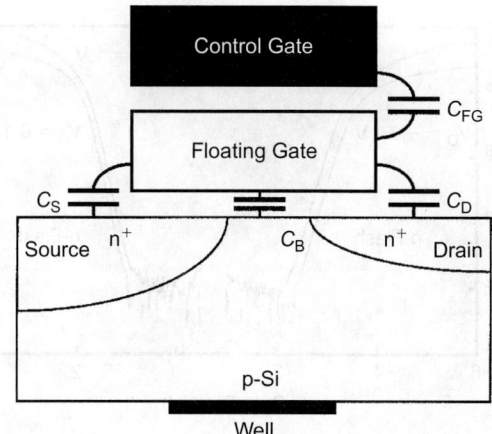

FIGURE 54.8 Schematic cross-section of stacked-gate device and its equivalent capacitive model.

an equivalent capacitive circuit, as shown in Figure 54.8. Owing to being isolated from other terminals, the potential of floating gate, V_{FG}, can be expressed as not only the total contributions from four terminals of the device, but also from the contribution of the stored charge in the floating gate:

$$V_{FG} = \frac{C_{FG}}{C_{TOTAL}} V_G + \frac{C_B}{C_{TOTAL}} V_{WELL} + \frac{C_D}{C_{TOTAL}} V_D + \frac{C_S}{C_{TOTAL}} V_S - \frac{Q}{C_{TOTAL}} \qquad (54.1)$$

$$C_{TOTAL} = C_{FG} + C_B + C_D + C_S \qquad (54.2)$$

and

$$\alpha_{FG} = \frac{C_{FG}}{C_{TOTAL}}, \alpha_B = \frac{C_B}{C_{TOTAL}}, \alpha_D = \frac{C_D}{C_{TOTAL}}, \alpha_S = \frac{C_S}{C_{TOTAL}} \qquad (54.3)$$

where C_{FG}, C_B, C_D, and C_S are the capacitances between floating gate and control gate, well terminal, drain terminal, and source terminal, respectively. Q is the charge stored on the floating gate, and α_{FG}, α_B, α_D, α_S are the gate, well, drain, and source coupling ratios, respectively.

54.4.1.2 Current–Voltage Characteristics

The current–voltage relationship in a stacked-gate device has been studied and modeled in detail [14,15]. By employing Eq. 54.1 for general I–V characteristics in MOSFETs, a simplified I-V relationship in stacked gate devices can be obtained:

$$V_{FG} = \frac{C_{FG}}{C_{TOTAL}} V_G + \frac{C_D}{C_{TOTAL}} V_D - \frac{Q}{C_{TOTAL}}$$

$$= \alpha_{FG} \left(V_G + \frac{C_D}{C_{FG}} V_D - \frac{Q}{C_{FG}} \right) \qquad (54.4)$$

for $V_S = V_{WELL} = 0\,V$

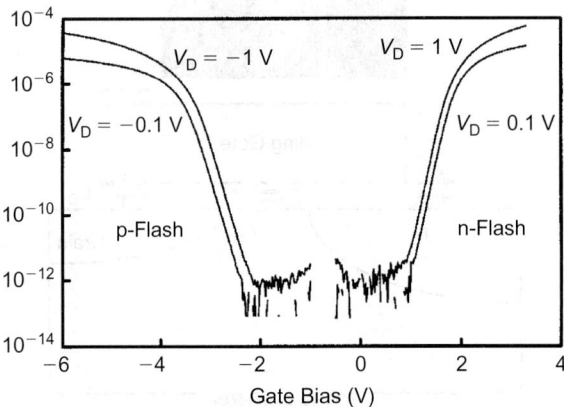

FIGURE 54.9 The subthreshold characteristics of n- and p-channel Flash memory cells.

In the linear region,

$$
I_D = \frac{\mu n \cdot C_{ox} \cdot W}{L}\left(V_{FG} - V_{TH} - \frac{V_D}{2}\right)\cdot V_D
$$

$$
= \frac{\alpha_{FG}\cdot \mu n \cdot C_{ox}\cdot W}{L}\left[V_G + \left(\frac{C_D}{C_{FG}} - \frac{1}{2}\right)V_D - \frac{Q}{C_{FG}} - \frac{V_{TH}}{\alpha_{FG}}\right]V_D
$$

(54.5)

And also in saturation region,

$$
I_D = \frac{\mu n \cdot C_{ox}\cdot W}{2L}(V_{FG} - V_{TH})^2
$$

$$
= \frac{\alpha_{FG}^2 \cdot \mu n \cdot C_{ox}\cdot W}{2L}\left(V_G + \frac{C_D}{C_{FG}}V_D - \frac{Q}{C_{FG}} - \frac{V_{TH}}{\alpha_{FG}}\right)^2
$$

(54.6)

From Eqs. 54.5 and 54.6, it is clearly demonstrated that the stacked-gate device suffers from drain bias coupling during operation. An increase of drain current can be observed, both in output characteristics and transfer characteristics. Figure 54.9 shows the subthreshold characteristics of both the n-channel and p-channel Flash devices. An obvious increase of the subthreshold current can be observed while the drain bias increases. In addition, the increased drain current characteristics in the saturation region are shown in Figure 54.10.

54.4.1.3   Threshold Voltage of Flash Memory Devices

Threshold voltage is defined as the minimum voltage needed to turn on the device. For a stacked-gate device, the threshold voltage measured from the control gate is an indicator of charge storage condition. From Eq. 54.4, we can obtain

$$
V_{FGTH} = \alpha_{FG}\left(V_{GTH} + \frac{C_D}{C_{FG}}V_D - \frac{Q}{C_{FG}}\right)
$$

(54.7)

According to this equation, there exists a linear relationship between threshold voltage measured from floating gate and control gate, drain bias, and stored charge amount. The threshold voltage measured from the floating gate is only determined by the process procedures and device structures. Therefore, the

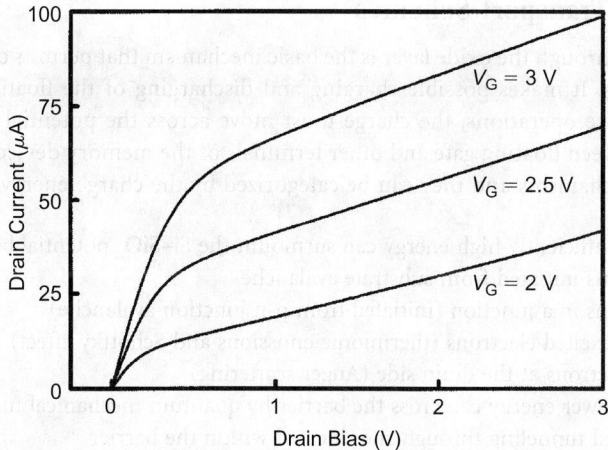

FIGURE 54.10 The output characteristics of stacked-gate memory cells.

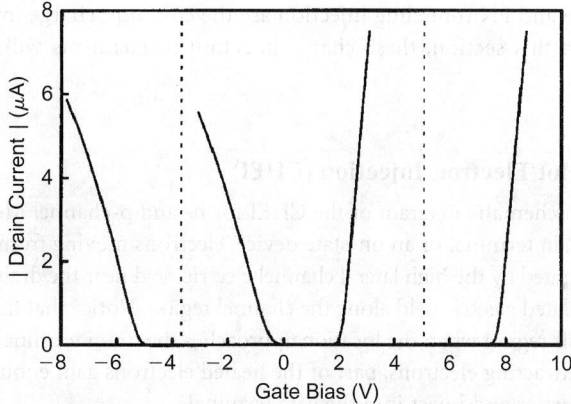

FIGURE 54.11 The transfer characteristics of n- and p-channel Flash memory cells.

change of the threshold voltage measured from control gate linearly depends on the change of the stored charge amount under a fixed drain bias in a specific stacked-gate device. Thus, this can be expressed as

$$\Delta V_{\text{GTH}} = \frac{\Delta Q}{C_{\text{FG}}} \tag{54.8}$$

Based on this relationship, the amount of charge storage in stacked-gate memory cell can be monitored by the measured threshold voltage. As shown in Figure 54.11, the transfer characteristic shifts toward a higher gate bias region, while the increasing amount of electrons are stored in the floating gate for both n- and p-channel Flash memory cells. Thus, device conduction during read operation determines the stored information of the stacked-gate devices. At a specific gate bias condition for reading, as shown in Figure 54.11, the memory with/without stored charge would lead to different amounts of drain current. The stored electron in the floating gate leads no current flow through the channel at the "READ" bias in the n-channel Flash cell, whereas the channel would conduct at the read operation for the p-channel cell with the electron stored in the floating gate. The sense amplifier in the peripheral circuit can detect the drain current and provide the stored information for external applications.

54.4.2 Carrier Transport Schemes

Transport of charge through the oxide layer is the basic mechanism that permits operation of stacked-gate memory devices. It makes possible charging and discharging of the floating gate. In order to achieve the write/erase operations, the charge must move across the potential barrier built by the insulating layers between floating gate and other terminals of the memory device. There are different charge transport mechanisms and they can be categorized by the charge energy [16]:

1. Charges with sufficiently high energy can surmount the Si–SiO$_2$ potential barrier, including:
   a. Hot electrons initiated from substrate avalanche
   b. Hot electrons in a junction (initiated from p-n junction avalanche)
   c. Thermally excited electrons (thermionic emissions and Schottky effect)
   d. "Lucky" electrons at the drain side (Auger scattering)
2. Charges with lower energy can cross the barrier by quantum mechanical tunneling effects:
   a. Trap-assisted tunneling through sites located within the barrier
   b. Direct tunneling when the tunneling distance is equal to the thickness of the oxide
   c. Fowler-Nordheim (FN) tunneling

Hot carrier injection and FN tunneling injection are the common charge injection mechanisms in Flash memory cells. In this section, these charge injection mechanisms will be described in more detail.

54.4.2.1 Channel Hot Electron Injection (CHEI)

Figure 54.12 shows the schematic diagram of the CHEI for n- and p-channel MOSFET. When applying a high voltage at the drain terminal of an on-state device, electrons moving from the source terminal to the drain side are accelerated by the high lateral channel electric field near the drain terminal. Figure 54.13 shows the plots of simulated electric field along the channel region. Notice that the electric field increases abruptly in the pinch-off region when the location approaches the drain terminal. Under the oxide field, which is favorable for attracting electrons, part of the heated electrons gain enough energy to surmount the Si–SiO$_2$ potential barrier and inject into the gate terminal.

Figure 54.14 shows the qualitative plot of gate current characteristic for n-channel MOSFETs. For the gate bias in the region "I", a quite small gate current can be characterized. In this subthreshold region, the carrier injection mainly originates from the avalanche injection, which will be discussed in the next section. In region II, the channel conducts and the channel current increases as the gate bias increases and thus the gate current induced by CHEI increases. As the gate bias increases further, the gate current peaks at a high gate bias. Following the peak value of the gate current, the decreasing gate current is mainly caused by the decrease of the lateral electric field, as illustrated in region III.

On the other hand, the measured gate current characteristic in p-channel MOSFETs is shown in Figure 54.15. Owing to the large potential barrier and short mean free path, the hot hole generated and accelerated in the channel cannot gain enough energy to surmount the oxide barrier. Thus, electron current initiated by channel hot electrons is still the dominant component of gate current in the p-channel MOSFET [17,18]. Besides, the gate current peaks at a lower gate bias in a p-channel MOSFET and has a larger peak value than that in an n-channel MOSFET. In larger gate bias regions, the gate current is dominated by hole injection, which may be caused by the oxide field favoring the injection of the conducting holes into the gate terminal [19].

In the 1980s, there were several approaches to describe the channel hot electron injection into the gate terminal. Takeda, et al. [20] modeled the gate current in n-channel MOSFETs as thermionic emission from the heated electron gas over the Si–SiO$_2$ potential barrier. This thermionic gate current model, referred as the "effective electron temperature model," assumes that the heated electrons become an electron gas with a Maxwellian distribution with an effective temperature $T_e(x)$.

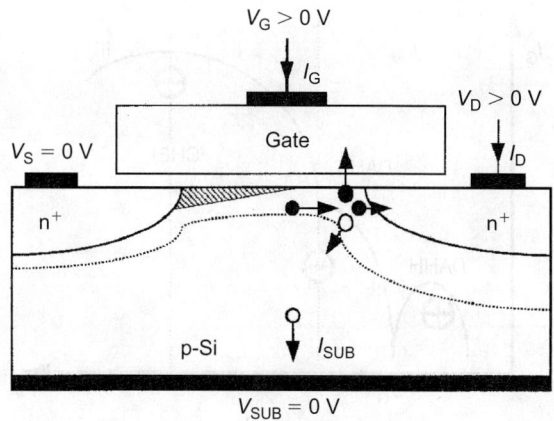

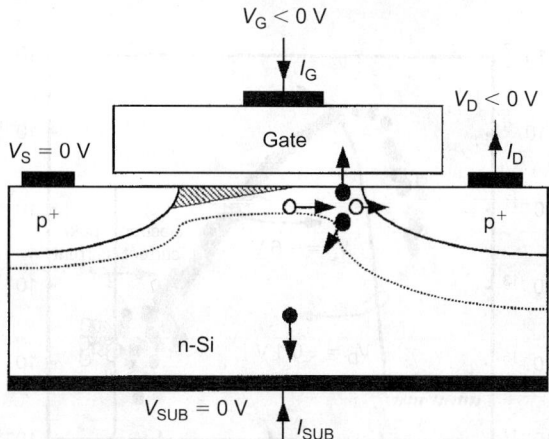

FIGURE 54.12 Schematic illustration of the channel hot carrier effect in (a) n-channel MOSFET, and (b) p-channel MOSFET.

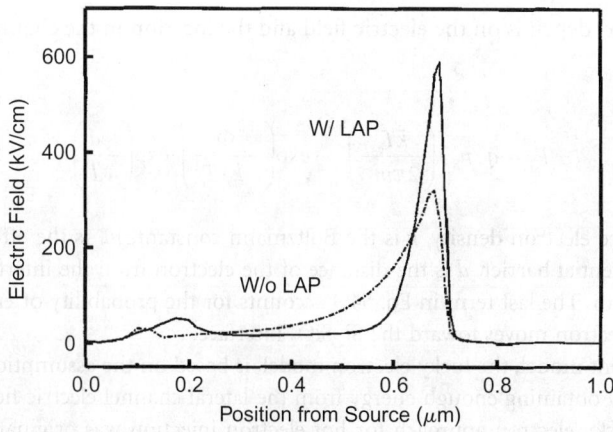

FIGURE 54.13 Simulated electric field along the channel in the n-channel MOSFET.

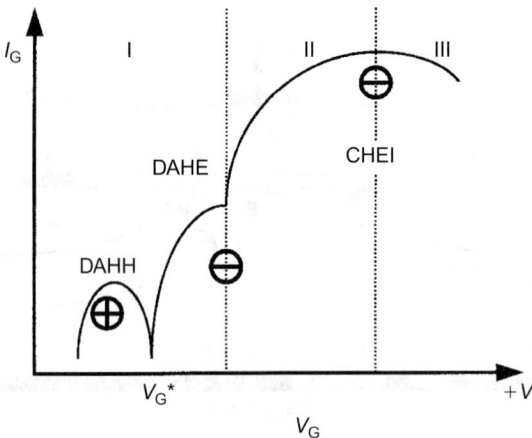

FIGURE 54.14 Schematic gate current behavior in n-channel MOSFET.

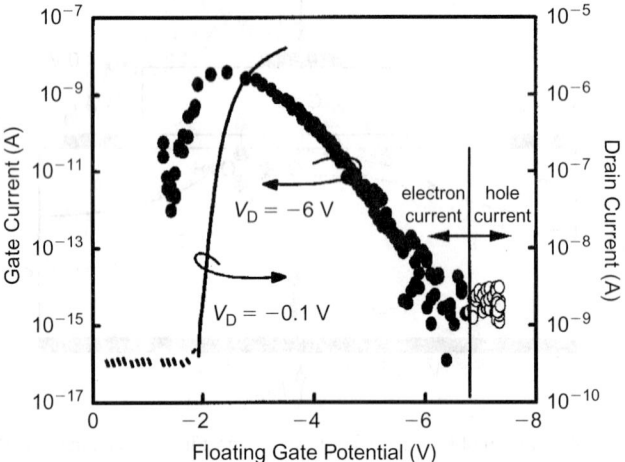

FIGURE 54.15 The gate current behavior of p-channel MOSFET measured from the threshold voltage shift of the stacked-gate structure.

The temperature $T_e(x)$ depends on the electric field and the location in the channel. The gate current is given by

$$J_G = q \cdot n_s \cdot \left(\frac{kT_e}{2\pi m^*} \right)^{1/2} \cdot \exp\left(-\frac{\Phi_B}{k \cdot T_e} \right) \cdot \exp\left(-\frac{d}{l} \right) \tag{54.9}$$

where n_s is the surface electron density, k is the Boltzmann constant, m^* is the effective electron mass, Φ_B is the Si–SiO$_2$ potential barrier, d is the distance of the electron from the interface at $T_e(x)$, and the λ is the mean free path. The last term in Eq. 54.9 accounts for the probability of energy loss due to the collision while the electron moves toward the Si–SiO$_2$ interface.

Another gate current model, the lucky electron model, is based on the assumption that an electron is injected into oxide by obtaining enough energy from the lateral channel electric field without suffering any collision. The lucky electron approach for hot electron injection was originated by Shockley [21] and Verway et al. [22], who applied it in the study of substrate hot electron injection in MOSFETs and subsequently refined and verified by Ning et al. [23]. Hu modified the substrate lucky electron injection model and applied it to CHEI in MOSFETs [24]. In this model, there are three probabilities to describe the physical mechanism responsible for CHEI gate current [25]. They are (1) the probability of a hot electron to gain enough kinetic energy and normal momentum, (2) the probability of not suffering any inelastic collision during transport to the Si–SiO$_2$ interface, and (3) the probability of not suffering collision in oxide image-potential well. Thus, the gate current originated from CHEI is given by

$$I_G = \int_0^L I_D \frac{(P_1 \cdot P_2 \cdot P_3)}{\lambda_r} dx \tag{54.10}$$

where I_D is the channel current, L is the channel length, and λ_r is the redirection scattering mean free path. P_1 is the probability that an electron can gain the energy equals the energy barrier under the channel electric field E without suffering optical phonon scattering and can be expressed as

$$P_1 = \exp\left(-\frac{\Phi_B}{E\lambda}\right) \tag{54.11}$$

where λ is the mean free path for optical phonon scattering. P_2 is the probability of not suffering any inelastic collision during transport to the Si–SiO$_2$ interface and can be expressed as

$$P_2 = \frac{\int_{y=0}^{\infty} n(y) \cdot \exp\left(-\frac{y}{\lambda}\right) dy}{\int_{y=0}^{\infty} n(y) dy} \tag{54.12}$$

The last probability factor is the scattering in the oxide image-potential well. P_3 can be expressed as [26]:

$$P_3 = \exp\left(-\frac{y_o}{\lambda_{ox}}\right) \tag{54.13}$$

Ong et al. modified the lucky electron model to analyze the hot electron injection effects in p-channel MOSFETs [27,28]. Based on Eq. 54.10 and substituting substrate current (I_{SUB}) for drain current (I_D), the gate current in p-channel MOSFETs can be expressed as:

$$I_G = \int_{y=0}^{y=L} I_{SUB} \frac{(P_1 \cdot P_2 \cdot P_3)}{\lambda_r} dy \tag{54.14}$$

After describing the channel hot electron injection mechanisms, the charge injection characteristics based on the CHEI scheme are discussed. First, the output characteristics ($I_D - V_D$) of a memory cell are taken into account. The output characteristic of a stacked-gate device can be regarded as an injection indicator to examine the effects of channel hot electron injection under different device operation conditions and device structures. The output characteristics of the n-channel Flash memory under a high gate bias are shown in Figure 54.16(a). The drain current rolls off at a lower drain bias as the channel length of the device decreases. This indicates obviously that the channel length reduction results in the increase of the lateral channel electric field and therefore the enhancement of hot electron injection.

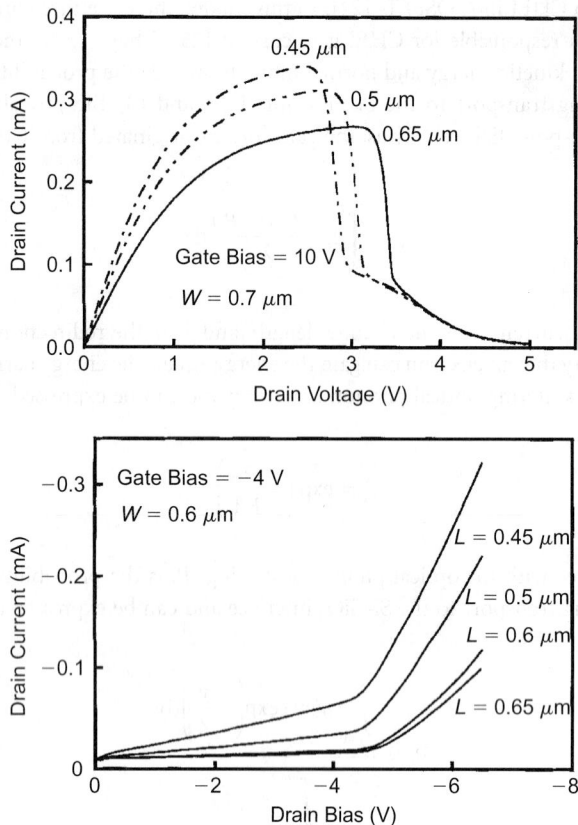

FIGURE 54.16　(a) The output characteristics of the n-channel Flash memory at high gate bias, and (b) the output characteristics of the p-channel Flash memory at high gate bias.

As the electron injection initiates, the stored electrons retard the conduction of the channel and the device is gradually turned off owing to the continuous electron injection. On the contrary, the output characteristics in the p-channel Flash memory, as shown in Figure 54.16(b), reveal a quite different I–V behavior after electron injection. Owing to the reduction of threshold voltage after electron injection, the enhancement of further channel conduction can be observed as the drain bias increases.

Second, the programming characteristics of the n- and p-channel Flash memory are demonstrated. Figure 54.17(a) shows the gate bias effects on the CHEI programming characteristics in an n-channel Flash memory cell. The threshold voltage increases as the electron injection process prolongs and then saturates at different values for different gate biases. On the other hand, Figure 54.17(b) shows the CHEI programming characteristics in a p-channel Flash memory cell. Compared with the n-channel cell, the programming characteristic in the p-channel Flash cell reveals a large dependence on the gate bias condition. This is mainly caused by the CHEI that distributes within a narrower gate bias condition. The gate current in the p-MOSFET peaks at lower gate bias and decreases steeply when the gate bias becomes more negative. Therefore, the injected electrons during programming accompanied by the control gate bias lead to a more negative floating gate potential and the programming behavior is quite different at different gate bias conditions.

54.4.2.2　Drain Avalanche Hot Carrier (DAHC) Injection

As shown in the region I of Figure 54.14, the characteristic of the gate current is still a function of the gate voltage in n-channel MOSFETs. When V_G is smaller than V_G^*, drain avalanche hot hole (DAHH) is the dominant carrier injected into the gate. On the other hand, when V_G is larger than V_G^*, drain avalanche

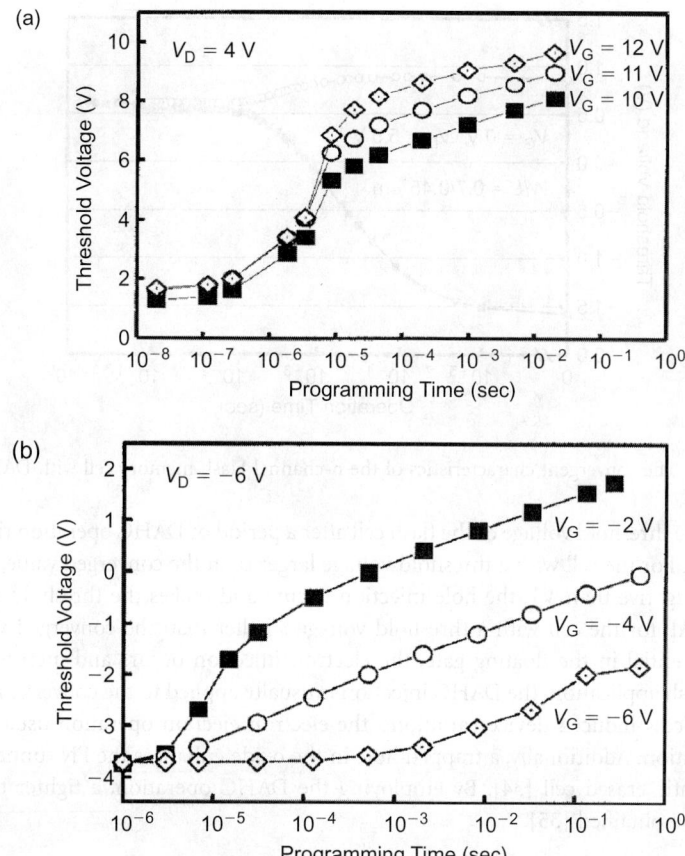

FIGURE 54.17 (a) The programming characteristics of the n-channel Flash memory using channel hot electron injection scheme; (b) the programming characteristics of the p-channel Flash memory using channel hot electron injection.

hot electron (DAHE) is the dominant carrier injected into the gate terminal. V_G^* is the point at which the amounts of the injected hot hole and injected hot electron are in balance. At this gate bias condition, the gate current is not observed.

Conceptually, the existence of hot hole injection seems questionable because of the high barrier (3.8 eV) for hole injection at the Si–SiO$_2$ interface. However, hot hole gate currents have been experimentally identified and modeled [29,32]. Hofmann et al. [30] employed the effective electron temperature model [20] and the concept of oxide scattering effects [25] based on the two-dimensional distribution of electric field, charge carrier, and current density calculated by computer simulator. The hot hole injection and hot electron injection initiated by the avalanche generation were manifested qualitatively. Saks et al. [32] proposed a modified floating gate technique to characterize these extremely small gate currents. It showed that a small positive gate current exists for gate bias near the threshold voltage. They also suggested that the hole current increases with increasing drain bias and decreasing effective channel length, which is analogous to the dependencies for channel hot electron injection. Comparison of hot hole and hot electron gate current as a function of the effective channel length also suggested that the lateral electric field near the drain plays an important role in the hole injection.

In the stacked-gate devices, in the DAHH region, holes are injected into the floating gate, which increases the floating gate voltage gradually, and finally the floating gate voltage reaches the point V_G^*. On the contrary, in the DAHE region, electrons are injected into the floating gate, which decreases the floating gate, and the floating gate voltage also reaches the point V_G^*. Thus, the threshold voltage of the stacked-gate device would distribute at a specific value after the DAHC injection operation. As shown

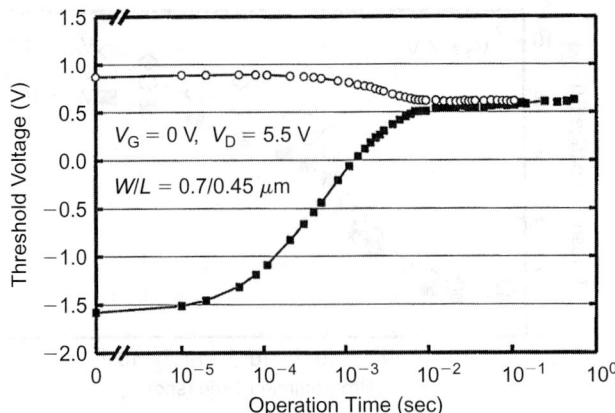

FIGURE 54.18 The convergent characteristics of the n-channel Flash memory cell with DAHC operation.

in Figure 54.18, the threshold voltage of the flash cell after a period of DAHC operation time can converge to a specific value. For the cell with a threshold voltage larger than the converged value, the floating gate voltage is more negative than V_G^*, the hole injection occurs and makes the threshold voltage decrease. On the other hand, for the cell with a threshold voltage smaller than the converged value, it reveals a more positive potential in the floating gate, the electron injection occurs and increases the threshold voltage. In the Flash application, the DAHC injection is usually applied to the convergent operation [33]. Owing to the process-induced device variations, the electron ejection operation usually causes a wide threshold distribution. Additionally, a trapped hole in the oxide enhances the FN tunneling current and generates the erratic erased cell [34]. By employing the DAHC operation, a tighter threshold voltage distribution can be obtained [35].

54.4.2.3 Band-to-Band Tunneling Induced Hot Carrier Injection (BBHC)

Carrier injection initiated by band-to-band tunneling accompanied by lateral junction electric field is also an important charge transport mechanism in Flash memory. As shown in Figure 54.19, the BBHC operation conditions for n- and p-channel lead to different charge injection behaviors. For n-channel MOSFETs, the negative gate bias and positive drain bias lead to the possible hole injection toward the gate terminal. For p-channel MOSFETs, the operation conditions lead to the possible electron injection toward the gate terminal. The initiation of the BBHC injection can be divided into two procedures. One is the band-to-band tunneling, and the other is the acceleration due to lateral electric field and injection due to favorable oxide field.

The band-to-band tunneling phenomenon is usually referred as gate-induced drain leakage current [36]. When a high drain voltage is applied with a grounded gate terminal, a deep depletion region is formed underneath the gate-to-drain overlap region. Electron-hole pairs are generated by the tunneling of valence band electrons into the conduction band and then collected by the drain and substrate terminals, separately. Since the minority carriers (hole in n-MOSFET and electron in p-MOSFET) generated by band-to-band tunneling in the drain region flow to the substrate due to the lateral electric field, the deep depletion region is always present and the band-to-band tunneling process proceeds without forming an inversion layer. The band-to-band tunneling characteristic can be estimated by the calculation of electric field distribution and the tunneling probability [37,38]. Based on the depletion approximation and the assumption of uniform impurity distribution, the electric field $E(x)$ in the depletion region is given by

$$E(x) = \frac{Q \cdot N_o}{\varepsilon_{si}} \sqrt{\frac{2 \cdot \varepsilon_{si} \cdot V_{bend}}{q \cdot N_o}} \left(1 - x\sqrt{\frac{q \cdot N_o}{2 \cdot \varepsilon_{si} \cdot V_{bend}}}\right) \qquad (54.15)$$

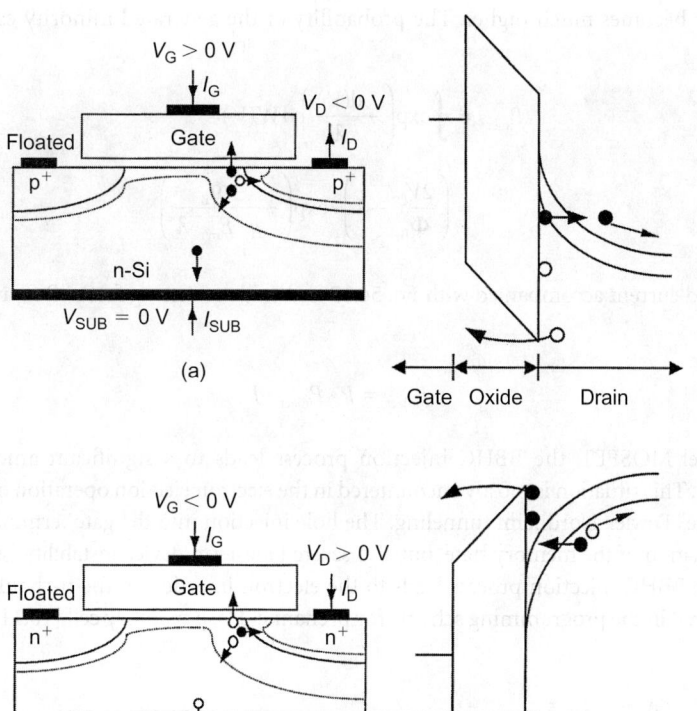

FIGURE 54.19 The schematic illustration for BBHC injection for: (a) n-channel MOSFET, and (b) p-channel MOSFET.

where V_{bend} is the value of the band bending, N_o is the impurity density, and x is the coordinate normal to the Si–SiO$_2$ interface. The continuity equation at the Si–SiO$_2$ interface can be expressed as

$$\varepsilon_{si} \cdot E(x = 0) = \varepsilon_{ox} \cdot E_{ox} = \varepsilon_{ox} \frac{V_D - V_{bend}}{T_{ox}} \qquad (54.16)$$

The tunneling characteristics are usually approximated by the relationship derived from the reverse biased p-n junction tunnel diode [39]:

$$J = B_1 \cdot E^2 \exp\left(-\frac{B_2}{E}\right) \qquad (54.17)$$

where B_1 and B_2 are physical constants. Most of the generated minority carriers are drained away from the substrate terminal. However, owing to the sufficient lateral electric field across the depletion region, these hot carriers may encounter Auger scattering and generate another electron-hole pair [40]. When the drain bias is higher than Si–SiO$_2$ barrier, the top barrier position seen by the cold generated minority carriers is lower at the depletion edge in the channel. Thus, the injection probability of the minority carrier becomes much higher. The probability of the generated minority carrier injection is given by [41].

$$P_{\text{inject}} = \int \exp\left(-\frac{d(V)}{\lambda}\right) dW(V)$$

$$\approx \left(\frac{2V_{\text{D}}}{\Phi_{\text{B}}} - 1\right) \cdot \exp\left(-\frac{\Phi_{\text{B}}}{q \cdot E_{\text{m}} \cdot \lambda}\right) \tag{54.18}$$

Thus, the injected current accompanied with Eq. 54.17 and oxide scattering factor P expressed in Eq. 54.13 can be given by

$$J_{\text{inject}} = P \cdot P_{\text{inject}} \cdot J \tag{54.19}$$

In the n-channel MOSFET, the BBHC injection process leads to a significant amount of hot hole injection [42,43]. This situation is mostly encountered in the electron ejection operation of a Flash memory device with "edge" Fowler-Nordheim tunneling. The hole injection into the gate terminal would result in not only the deviation of the memory state, but also severe long-term device instability issues. However, on the contrary, the BBHC injection process leads to the electron injection in the p-channel MOSFET and has been employed in the programming scheme for p-channel Flash memory cell [10,11]. Figure 54.20(a)

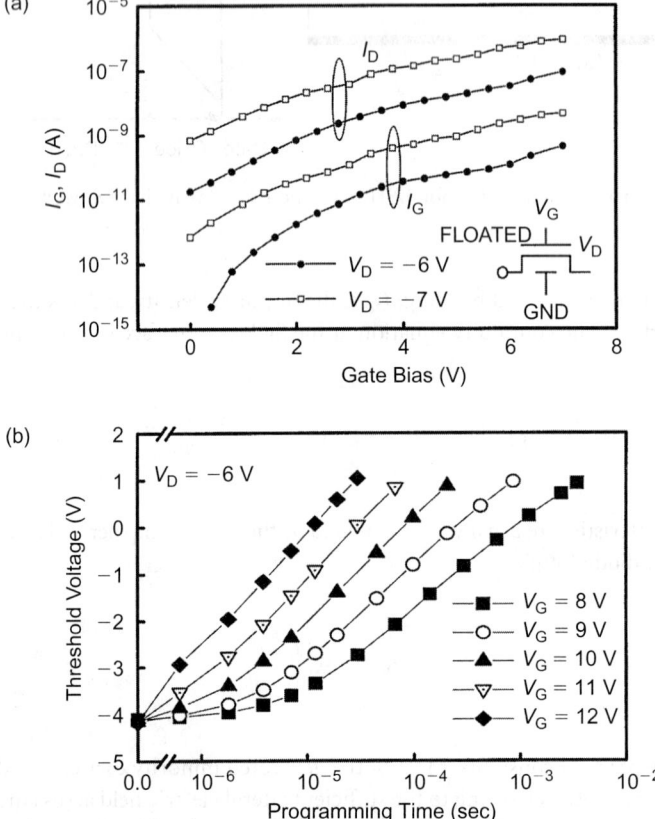

FIGURE 54.20   (a) The BBHE behavior in p-channel MOSFET with different bias conditions; and (b) the programming characteristics in p-channel Flash memory cell with BBHE injection scheme.

shows the BBHE characteristics of the p-channel MOSFET. The drain and gate currents monotonically increase with respect to the gate bias because of the increase of the band-to-band tunneling efficiency and the more favorable oxide field for electron injection. Owing to operating in the off state, the electron injection efficiency of the BBHE scheme is much larger than that in the CHEI operation. The BBHE injection reveals a rather high injection efficiency (I_G/I_D) up to 10^{-2}, which provides a quite efficient programming operation for the p-channel Flash cell [10]. Figure 54.20(b) shows the programming characteristics based on the BBHE injection mechanism. The programming time is greatly shortened as the control gate voltage increases. As compared with the CHEI scheme shown in Figure 54.17(b), the BBHE approach indeed reveals a faster programming speed.

54.4.2.4 Fowler-Nordheim (FN) Tunneling

The FN tunneling formula proposed by Fowler and Nordheim in 1928 can be described as

$$J_{tunnel} = C_o \cdot E^2 \cdot \exp\left(-\frac{4\sqrt{2m^* \cdot \Phi_B^3}}{3 \cdot q \cdot \hbar \cdot E}\right) \tag{54.20}$$

where J_{tunnel} and E are the tunneling current density and electric field across the oxide layer, respectively. Besides, C_o is a material-dependent constant and m^* is the carrier effective mass. The tunneling theory is developed using the semi-classical independent electron model. For a carrier with energy qU_o, the general expression for the transmission coefficient T_c through on energy barrier depends on the barrier shape $U(x)$, as shown in Figure 54.21. The value of T_c is derived using the WKB (Wentzel-Kramers-Brillouin) approximation [44,46].

$$\ln T_c = -\sqrt{\frac{8 \cdot m^* \cdot q}{\hbar}} \cdot \int_0^{X_{tunnel}} \sqrt{U(x) - U_o}\, dx \tag{54.21}$$

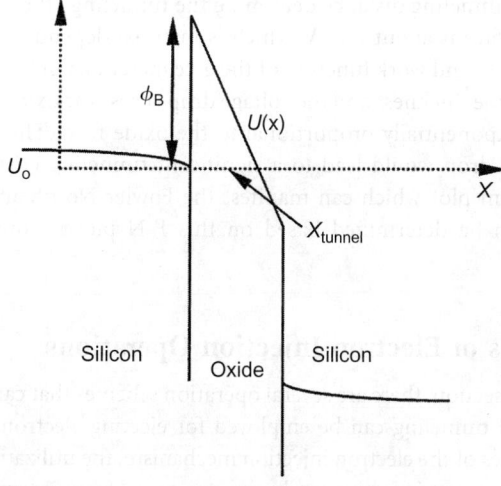

FIGURE 54.21 Schematic diagram of the potential barrier in the polysilicon-oxide-silicon system under applied high voltage.

The tunneling current is obtained by integrating the product of the density of states $N_c(W)$ and the transmission coefficient from lowest occupied energy W_G to infinity,

$$J_{tunnel} = \int_{W_G}^{\infty} N_c(W)Tc(W)\,dW \tag{54.22}$$

This expression is valid for any barrier shape. Under a strong oxide field E, the effective barrier is triangular and the coefficient can be obtained by integrating,

$$U(x) = \phi_B - E \cdot x \tag{54.23}$$

$$\ln T_c = \frac{-4\sqrt{2 \cdot m^* \cdot \Phi_B^3}}{3 \cdot \hbar \cdot q \cdot |E|} \tag{54.24}$$

where Φ_B is the barrier height, $\Phi_B = q\phi_B$.

Solving Eqs. 54.22 and 54.24 with the assumption that only electrons at the Fermi level contribute to the current yields the Fowler-Nordheim formula for the tunneling current density J_{tunnel} at high electric field:

$$J_{tunnel} = \frac{q^3 \cdot E^2}{16 \cdot \pi^2 \cdot \hbar \cdot \Phi_B} \cdot \exp\left(-\frac{4\sqrt{2 \cdot m^* \cdot \Phi_B^3}}{3 \cdot \hbar \cdot q \cdot E}\right) \tag{54.25}$$

This equation can also be expressed as

$$J_{tunnel} = \alpha \cdot E^2 \exp\left(-\frac{\beta}{E}\right) \tag{54.26}$$

where α and β are Fowler-Nordheim constants. The value of α is in the range of 4.7×10^{-5} to 6.32×10^{-7} A/V^2 and β is in the range of 2.2×10^8 to 3.2×10^8 V/cm [47]. The barrier height and tunneling distance determine the tunneling efficiency. Generally, the barrier height at the Si–SiO$_2$ interface is about 3.1 eV, which is material dependent. This parameter is determined by the electron affinity and work function of the gate material. On the other hand, the tunneling distance depends on the oxide thickness and the voltage drop across the oxide. As indicated in Eq. 54.26, the tunneling current is exponentially proportional to the oxide field. Thus, a small variation in the oxide thickness or voltage drop would lead to a significant tunneling current change. Figure 54.22 shows the Fowler-Nordheim plot which can manifest the Fowler-Nordheim constants α and β. The Si–SiO$_2$ barrier height can be determined based on this F-N plot by quantum-mechanical (QM) modeling [48].

54.4.3   Comparisons of Electron Injection Operations

As mentioned in the above section, there are several operation schemes that can be employed for electron injection, whereas only FN tunneling can be employed for ejecting electrons out of the floating gate. Owing to the specific features of the electron injection mechanism, the utilization of an electron injection scheme thereby determines the device structure design, process technology, and circuit design. The main features of CHEI and FN tunneling for n-channel Flash memory cell and also CHEI and BBHE injection for p-channel Flash memory cell are compared in Tables 54.1 and 54.2.

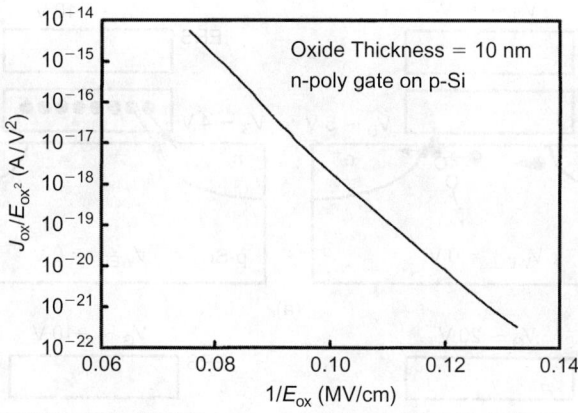

FIGURE 54.22 Fowler-Nordheim plot of the thin oxide.

TABLE 54.1 Comparisons of Fowler-Nordheim Tunneling and Channel Hot Electron Injection as Programming Scheme for Stacked-Gate Devices

FN Tunneling Injection Scheme	CHEI Scheme
Low power consumption	High power consumption
• Single external power supply	• Complicated circuitry technique
High oxide field	Low oxide field
• Thinner oxide thickness required	• Oxide can be thicker
• Higher trap generation rate	• Higher oxide integrity
• More severe read disturbance issue	• Low read disturbance issue
• Highly technological problem	
Slower programming speed	Faster programming speed

TABLE 54.2 Comparisons of Band-to-Band Tunneling Induced Hot Electron Injection and Channel Hot Electron Injection as Programming Scheme for Stacked-Gate Devices

	BBHE Injection Scheme	CHEI Scheme
Power consumption	Lower	Higher
Injection efficiency	Higher	Lower
Programming speed	Faster	Slower
Electron injection window	Wider	Narrower
Oxide field	Higher	Lower

54.4.4 List of Operation Modes

The employment of different electron transport mechanisms to achieve the programming and erase operations can lead to different device operation modes. Typically, in commercial applications, there are three different operation modes for n-channel Flash cells and two different operation modes for p-channel Flash cells. In the n-channel cell, as shown in Figure 54.23, the write/erase operation modes include: (1) programming operation with CHEI and erase operation with FN tunneling ejection at source or drain side [6–8,49–61], as shown in Figure 54.23(a), usually referred as NOR-type operation mode; (2) programming operation with FN tunneling ejection at drain side and erase operation with FN tunneling injection through channel region [62–70], as shown in Figure 54.23(b), usually referred as AND-type operation mode; and (3) programming and erase operations with FN tunneling injection/ejection through channel region [71–78], usually referred as NAND-type operation mode. As to the p-channel cell, as shown in Figure 54.24, the write/erase operation modes include: (1) programming operation with CHEI at drain side and erase operation with FN tunneling ejection through channel region [9], as shown in Figure 54.24(a); (2) programming operation with BBHE at drain side and erase operation with FN tunneling injection through channel region [10,11], as shown in Figure 54.24(b).

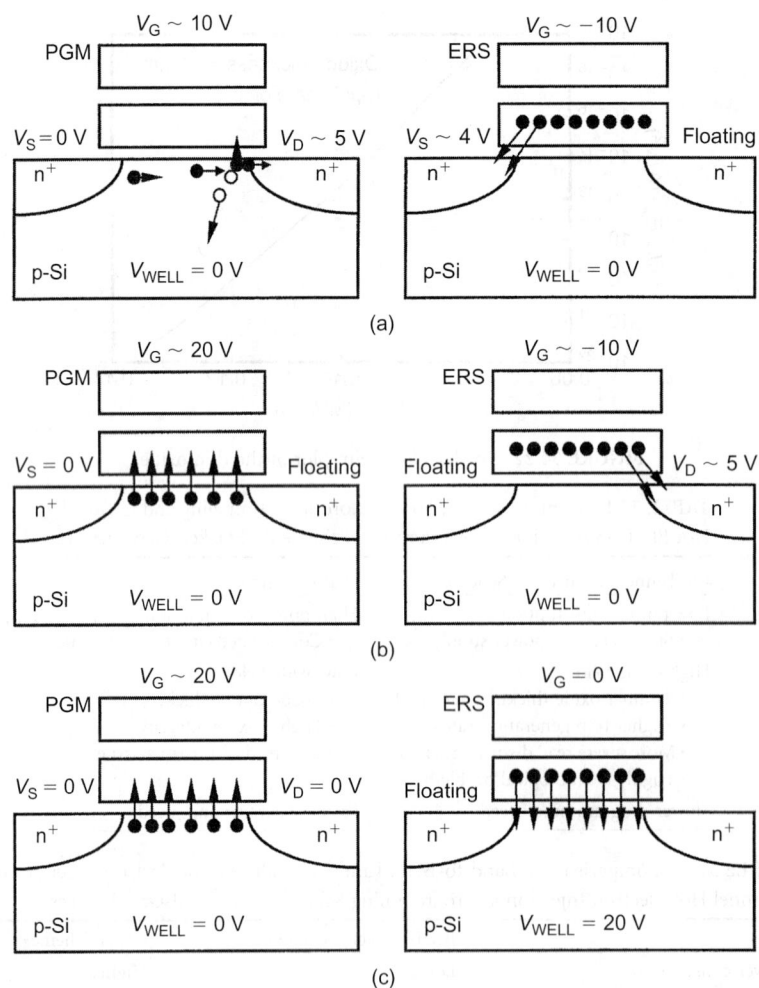

FIGURE 54.23 Different n-channel Flash write/erase operations: (a) programming operation with CHEI at drain side and erase operation with FN tunneling ejection at source side; (b) programming operation with FN tunneling ejection at drain side and erase operation with tunneling injection through channel region; and (c) programming and erase operations with FN tunneling injection/ejection through channel region.

These operation modes not only lead to different device structures but also different memory array architectures. The main purpose of utilizing various device structures for different operation modes is based on the consideration of the operation efficiency, reliability requirements, and fabrication procedures. In addition, the operation modes and device structures determine, and also are determined by, the memory array architectures. In the following sections, the general improvements of the Flash device structures and the array architectures for specific operation modes are described.

54.5 Variations of Device Structure

54.5.1 CHEI Enhancement

As mentioned above, alternative operation modes are proposed to achieve pervasive purposes and various features, which are approached either by CHEI or FN tunneling injection. Furthermore, it is indicated that the over 90% of the Flash memory product ever shipped is the CHEI-based Flash memory device [79].

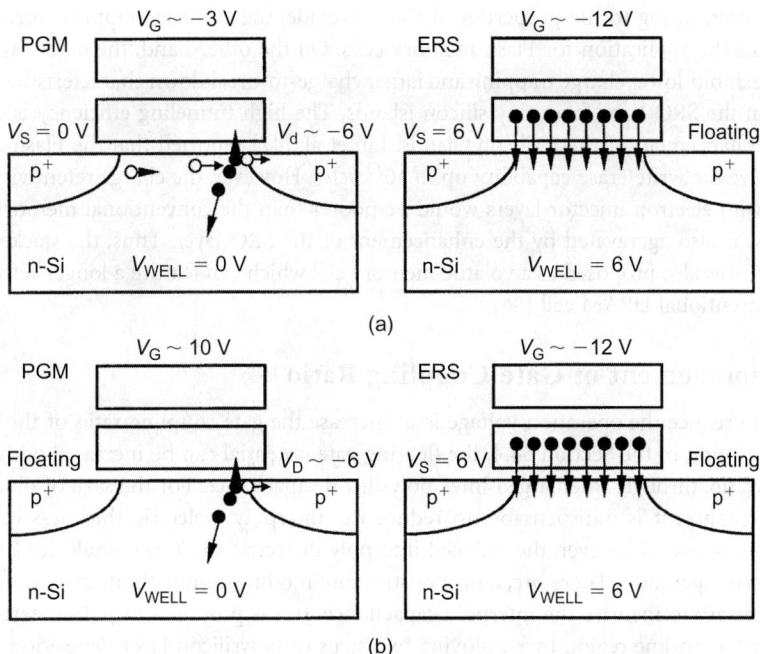

FIGURE 54.24 Different p-channel Flash write/erase operations: (a) programming operation with CHEI at drain side and erase operation with FN tunneling ejection through channel region; and (b) programming operation with BBHE at drain side and erase operation with FN tunneling injection through channel region.

With the major manufacturers' competition, many innovations and efforts are dedicated to improve the performance and reliability of CHEI schemes [50,53,56,57,61,80–83]. As described in Eq. 54.11, an increase in the electric field can enhance the probability of the electrons gaining enough energy. Therefore, the major approach to improve the channel hot electron injection efficiency is to enhance the electric field near the drain side. One of the structure modifications is utilizing the large-angle implanted p-pocket (LAP) around the drain to improve the programming speed [56,57,60,83]. LAP has also been used to enhance the punch-through immunity for scaling down capability [50,53]. As demonstrated in Figure 54.13, the device with LAP has a twofold maximum electric field of that in the device without LAP structure. According to our previous report [83], additionally, the LAP cell with proper process design can satisfy the cell performance requirements such as read current and punch-through resistance and also reliable long-term charge retention. Besides, the utilization of the p-pocket implantation can achieve the low-voltage operation and feasible scaling down capability simultaneously.

54.5.2 FN Tunneling Enhancement

From the standpoint of power consumption, the programming/erase operation based on the FN tunneling mechanism is unavoidable because of the low current during operation. As the dimension of Flash memory continues scaling down, in order to lower the operation voltage, a thinner tunnel oxide is needed. However, it is difficult to scale down the oxide thickness further due to reliability concerns. There are two ways to overcome this issue. One method is to raise the tunneling efficiency by employing a layer of electron injector on top of the tunnel oxide. Another method is to improve the gate coupling ratio of the memory cell without changing the properties of the insulator between the floating gate and well.

The electron injectors on the top of the tunnel oxide enhance the electric field locally and thus the tunneling efficiency is improved. Therefore, the onset of tunneling behavior takes place at a lower operation voltage. There are two materials used as electron injectors: polyoxide layer [84] and silicon-rich oxide (SRO) layer [85]. The surface roughness of the polyoxide is the main feature for electron injectors. However, owing to the properties of the polyoxide, the electron trapping during write/erase operation limits the application for Flash memory cells. On the other hand, the oxide layer containing excess silicon exhibits lower charge trapping and larger charge-to-breakdown characteristics. These silicon components in the SRO layer form tiny silicon islands. The high tunneling efficiency is caused by the electric field enhancement of these silicon islands. Lin et al. [47] reported that the Flash cell with SRO layer can achieve the write/erase capability up to 10^6 cycles. However, the charge retentivity of the Flash memory cell with electron injector layers would be poorer than the conventional memory cell because the charge loss is also aggravated by the enhancement of the SRO layer. Thus, the stacked-gate device with SRO layer was also proposed as a volatile memory cell which can feature a longer refresh time than that in the conventional DRAM cell [86].

54.5.3   Improvement of Gate Coupling Ratio

Another way to reduce the operation voltage is to increase the gate coupling ratio of the memory cell. From the description in the Section 54.4, the floating gate potential can be increased with an increased gate coupling ratio, through an enlarged inter-polysilicon capacitance. For the sake of obtaining a large interpoly capacitance, it is indispensable to reduce the interpoly dielectric thickness or increase the interpoly capacitor area. However, the reduced interpoly dielectric thickness would lead to charge loss during long-term operation. Therefore, a proper structure modification without increasing the effective cell size is necessary to increase the interpoly capacitance. It was proposed to put an extended floating gate layer over the bit-line region by employing two steps of polysilicon layer deposition [68,87]. Such device structure with memory array modifications would achieve a smaller effective cell size and a high coupling ratio (up to 0.8). Shirai et al. [88] proposed the process modification the increase to effective area on the top surface of the floating gate layer. This modified process, which forms a hemispherical-grained (HSG) polysilicon layer, can achieve a high capacitive coupling ratio (up to 0.8). However, the charge retentivity would be a major concern in considering the material as the electric injector.

54.6   Flash Memory Array Structures

54.6.1   NOR Type Array

In general, most of the Flash memory array, as shown in Figure 54.25(a), is the NOR-type array [49–61]. In this array structure, two neighboring memory cells share a bit-line contact and a common source line. Therefore, a half the drain contact size and half the source line width is occupied in the unit memory cell. Since the memory cell is connected to the bit-line directly, the NOR-type array features random access and lower series resistance characteristics. The NOR-type array can be operated in a larger read current and thus a faster read operation speed. However, the drawback of the NOR-type array is the large cell area per unit cell. In order to maintain the advantages in NOR-type array and also reduce the cell size, there were several efforts to improve the array architectures. The major improvement in the NOR-type array is the elimination of bit-line contacts—the employment of buried bit-line configuration [52]. This concept evolves from the contactless EPROM proposed by Texas Instruments Inc. in 1986 [89]. By using this contactless bit-line concept, the memory cell has a 34% size reduction.

54.6.2   AND Type Families

Another modification of the NOR-type array accompanied by a different operation mode is the AND-type array. In the NOR-type array, the CHEI is used as the electron injection scheme. However, owing to the considerations of power consumption and series resistance contributed by the buried bit-line/ source, both the programming and erase operations utilize FN tunneling to eliminate the above concerns. Some improvements and modifications based on the NOR-type array have been proposed, including DIvided-bitline NOR (DINOR) proposed by Mitsubishi Corp. [65,68], Contactless NOR (AND)

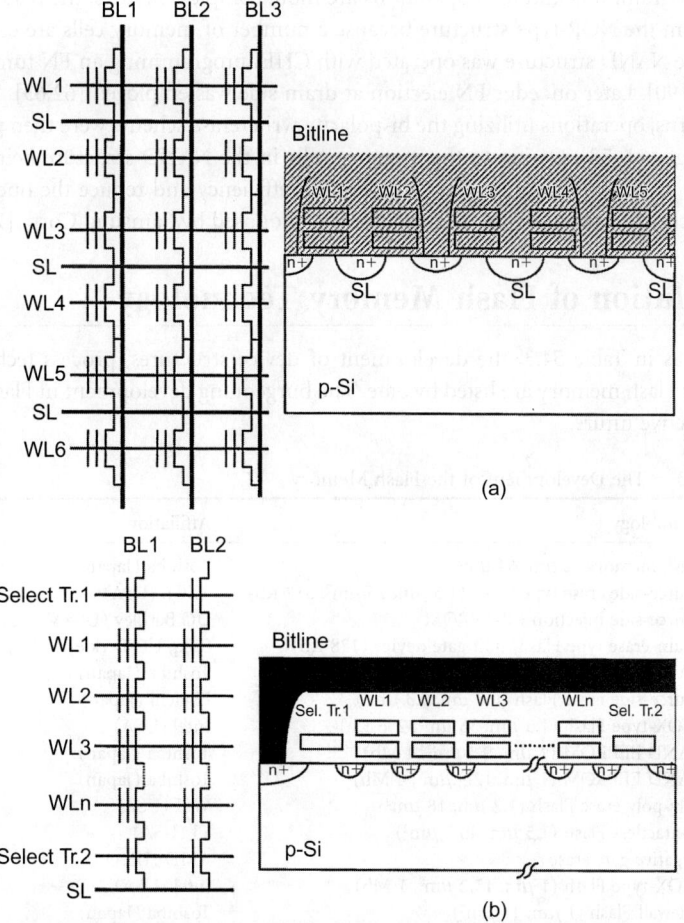

FIGURE 54.25 (a) Schematic top view and cross-section of the NOR-type Flash memory array; and (b) schematic top view and cross-section of the NAND-type Flash memory array.

proposed by Hitachi Corp. [64,66], Asymmetrical Contactless Transistor (ACT) cell by Sharp Corp.[69], and Dual String NOR (DuSNOR) by Samsung Corp. [70] and Macronix, Inc. [67]. The DINOR architecture employs the main bit-line and sub-bit-line configuration to reduce the disturbance issue during FN programming. The AND and DuSNOR structures consist of strings of memory cells with n+ buried source and bit-lines. String-select and ground-select transistors are attached to the bit and source line, respectively. In DuSNOR structure, a smaller cell size can be realized because every two adjacent cell strings share a source line. Although a smaller cell size can be obtained utilizing the buried bit-line and source line, the resistance of the buried diffusion line would degrade the read performance. The read operation consideration will be the dominant factor in determining the size of a memory string in the AND and DuSNOR structures.

54.6.3 NAND Type Array

In order to realize a smaller Flash memory cell, the NAND structure was proposed in 1987 [90]. As shown in Figure 54.25(b), the memory cells are arranged in series. It was reported that the cell size of the NAND structure is only 44% of that in the NOR-type array under the same design rules. The operation mechanisms of a single memory cell in the NAND architecture is the same as NOR and AND architectures.

However, the programming and read operations are more complex. Besides, the read operation speed is lower than that in the NOR-type structure because a number of memory cells are connected in series.

Originally, the NAND structure was operated with CHEI programming an FN tunneling through the channel region [90]. Later on, edge FN ejection at drain side was employed [62,63]. However, owing to reliability concerns, operations utilizing the bi-polarity write/erase scheme were then proposed to reduce the oxide damage [71–78]. Owing to the memory cells in the NAND structure being operated by FN write and erase, in order to improve the FN operation efficiency and reduce the operation voltage, the booster plate technology on the NAND structure was proposed by Samsung Corp. [77]

54.7 Evolution of Flash Memory Technology

In this section, as in Table 54.3, the development of device structures, process technology, and array architectures for Flash memory are listed by date. The burgeoning development in Flash memory devices reveals a prospective future.

TABLE 54.3   The Development of the Flash Memory

Year	Technology	Affiliation	Ref.
1984	Flash memory (2 μm, 64 μm^2)	Toshiba (Japan)	6
1985	Source-side erase type Flash (1.5 μm, 25 μm^2, 512 Kb)	EXCL (USA)	7
1986	Source-side injection (SI-EPROM)	UC Berkley (USA)	49
1987	Drain-erase type Flash, split gate device (128 Kb)	Seeq, UC Berkley (USA)	8
1987	NAND structure EEPROM (1 μm, 6.43 μm^2, 512 Kb)	Toshiba (Japan)	90
1987	Source-side erase Flash (0.8 μm, 9.3 μm^2)	Hitachi (Japan)	50
1988	ETOX-type Flash (1.5 μm, 36 μm^2, 256 Kb)	Intel (USA)	91
1988	NAND EEPROM (1 μm, 9.3 μm^2, 4 Mb)	Toshiba (Japan)	62
1988	NAND EEPROM (1 μm, 12.9 μm^2, 4 Mb)	Toshiba (Japan)	63
1988	Poly-poly erase Flash (1.2 μm, 18 μm^2)	WSI (USA)	92
1988	Contactless Flash (1.5 μm, 40.5 μm^2)	TI (USA)	93
1989	Negative gate erase	AMD (USA)	94
1989	ETOX-type Flash (1 μm, 15.2 μm^2, 1 Mb)	Intel (USA)	95
1989	Sidewall Flash (1 μm, 14 μm^2)	Toshiba (Japan)	51
1989	Punch-through-erase	Toshiba (Japan)	96
1990	Well-erase, bi-polarity W/E operation	Toshiba (Japan)	71, 72
1990	NAND, new self-aligned patterning (0.6 μm, 2.3 μm^2)	Toshiba (Japan)	97
1990	Contactless Flash, ACEE (0.8 μm, 8.6 μm^2, 4 Mb)	TI (USA)	98
1990	FACE cell (0.8 μm, 4.48 μm^2)	Intel (USA)	52
1990	Negative gate erase (0.6 μm, 3.6 μm^2, 16 Mb)	Mitsubishi (Japan)	54
1990	Tunnel diode-based contactless Flash	TI (USA)	99
1990	p-Pocket EPROM cell (0.6 μm, 16 Mb)	Toshiba (Japan)	53
1991	SAS process	Intel (USA)	100
1991	PB-FACE cell (0.8 μm, 4.16 μm^2)	Intel (USA)	101
1991	Burst-pulse erase (0.6 μm, 3.6 μm^2)	NEC (Japan)	56
1991	SSW-DSA cell (0.4 μm, 1.5 μm^2, 64 Mb)	NEC (Japan)	57
1991	Sector erase (0.6 μm, 3.42 μm^2, 16 Mb)	Hitachi (Japan)	64
1991	Self-convergence erase	Toshiba (Japan)	33, 35
1991	Virtual ground, auxiliary gate (0.5 μm, 2.59 μm^2)	Sharp (Japan)	59
1992	AND cell (0.4 μm, 1.28 μm^2, 64 Mb)	Hitachi (Japan)	66
1992	DINOR array (0.5 μm, 2.88 μm^2, 16 Mb)	Mitsubishi (Japan)	65
1992	2-Step erase method	NEC (Japan)	102
1992	Buried source side injection	TI (USA)	60
1992	p-Channel Flash Cell with SRO layer	IBM (USA)	9
1993	HiCR cell (0.4 μm, 1.5 μm^2, 64 Mb)	NEC (Japan)	87
1993	3-D sidewall Flash	Philip, Stanford (USA)	103
1993	Asymmetrical offset S/D DINOR (0.5 μm, 1.0 μm^2)	Mitsubishi (Japan)	68
1993	NAND EEPROM (0.4 μm, 1.13 μm^2, 64 Mb)	Toshiba (Japan)	74

(Continued)

TABLE 54.3 *(Continued)* The Development of the Flash Memory

Year	Technology	Affiliation	Ref.
1994	Self-convergent method	Motorola (USA)	104
1994	Substrate hot electron (SHE) erase	Mitsubishi (Japan)	105
1994	Dual-bit Split-Gate (DSG) cell (multi-level cell)	Hyundai (Korea)	106
1994	SA-STI NAND EEPROM (0.35 μm, 0.67 μm^2, 256 Mb)	Toshiba (Japan)	75
1994	SST cell	SST (USA)	124
1994	AND cell (0.25 μm, 0.4 μm^2, 256 Mb)	Hitachi (Japan)	107
1995	Multi-level NAND EEPROM	Toshiba (Japan)	108
1995	Convergence erase scheme	UT, AMD (USA)	109
1995	DuSNOR array (0.5 μm, 1.6 μm^2)	Samsung (Korea)	70
1995	CISEI programming scheme	AT&T, Lucent (USA)	110
1995	SAHF cell (0.3 μm, 0.54 μm^2, 256 Mb)	NEC (Japan)	88
1995	P-Flash with BBHE scheme (0.4 μm)	Mitsubishi (Japan)	10
1995	ACT cell (0.3 μm, 0.39 μm^2)	Sharp (Japan)	69
1995	Multi-level with self-convergence scheme	National (USA)	111
1995	Multi-level SWATT NAND cell (0.35 μm, 0.67 μm^2)	Toshiba (Japan)	112
1995	SCIHE injection scheme	AMD (USA)	113
1995	Alternating word-line voltage pulse	NKK (Japan)	114
1996	Self-limiting programming p-Flash	Mitsubishi (Japan)	11
1996	High speed NAND (HS-NAND) (2 μm^2, 16 Mb)	Samsung (Korea)	76
1996	Booster plate NAND (0.5 μm, 32 Mb)	Samsung (Korea)	77
1996	Shared bitline NAND (256 Mb)	Samsung (Korea)	115
1997	Φ-Cell	SGS-Thomson (France)	116
1997	NAND with STI (256 Mb)	Toshiba (Japan)	117
1997	Shallow groove isolation (SGI)	Hitachi (Japan)	118
1997	Word-line self-boosting NAND	Samsung (Korea)	119
1997	SPIN cell	Motorola (USA)	120
1997	Booster line technology for NAND	Samsung (Korea)	121
1997	AMG array	WSI (USA)	122
1997	High k interpoly dielectric	Lucent (USA)	123
1997	Self-convergent operation for p-Flash	NTHU (ROC)	12

54.8 Flash Memory System

54.8.1 Applications and Configurations

Flash memory is a single-transistor memory with floating gate for storing charges. Since 1985, the mass production of Flash memory has shared the market of non-volatile memory. The advantages of high density and electrical erasable operation make Flash memory an indispensable memory in the applications of programmable systems, such as network hubs, modems, PC BIOS, microprocessor-based systems, etc. Recently, image cameras and voice recorders have adopted Flash memory as the storage media. These applications require battery operation, which cannot afford large power consumption. Flash memory, a true non-volatile memory, is very suitable for these portable applications because stand-by power is not necessary.

In the interest of portable systems, the specification requirements of Flash memory include some special features that other memories (e.g., DRAM, SRAM) do not have. For example, multiple internal voltages with single external power supply, power-down during stand-by, direct execution, simultaneous erase of multiple blocks, simultaneous re-program/erase of different blocks, precise regulation of internal voltage, embedded program/erase algorithms to control threshold voltage. Since 1995, an emerging need of Flash memory is to increase the density by doubling the number of bits per cell. The charge stored in the floating gate is controlled precisely to provide multi-level threshold voltage. The information stored in each cell can be 00, 01, 10, or 11. Using multi-level storage can decrease the cost per bit tremendously. The multi-level Flash memories have two additional requirements: (1) fast sensing of multi-level information, and (2) high-speed multi-level programming. Since the memory cell characteristics would be degraded after cycling, which leads to fluctuation of programmed states, fast sensing and fast programming are challenged by the variation of threshold voltage in each level.

Another development is analog storage of Flash memory, which is feasible for image storage and voice record. The threshold voltage can be varied continuously between the maximum and minimum values to meet the analog requirements. Analog storage is suitable for recording the information that can tolerate distortion between the storing information and the restored information (e.g., image and speech data).

Before exploring the system design of Flash memory, the major differences between Flash memory and other digital memory, such as SRAM and DRAM, should be clarified. First, multiple sets of voltages are required in Flash memory for programming, erase, and read operations. The high-voltage related circuit is a unique feature that differs from other memories (e.g., DRAM, SRAM). Second, the characteristics of Flash memory cell are degrading because of stress by programming and erasing. The controlling of an accurate threshold voltage by an internal finite state machine is the special function that Flash memory must have. In addition to the mentioned features, address decoding, sense amplifier, and I/O driver are all required in Flash memory. The system of Flash memory, as a result, can be regarded as a simplified mixed-signal product that employs digital and analog design concepts.

Figure 54.26 shows the block diagram of Flash memory. The word-line driver, bit-line driver, and source-line driver control the memory array. The word-line driver is high-voltage circuitry, which includes a logic X-decoder and level shifter. The interface between the bit-line driver and the memory array is the Y-gating. Along the bit-line direction, the sense amplifier and data input/output buffer are in charge of reading and temporary storage of data. The high-voltage parts include charge-pumping and voltage regulation circuitry. The generated high voltage is used to proceed programming and erasing operations. Behind the X-decoder, the address buffer catches the address. Finally, a finite state machine, which executes the operation code, dictates the operations of the system. The heart of the finite state machine is the clocking circuit, which also feeds the clock to a two-phase generator for charge-pumping circuits. In the following sections, the functions of each block will be discussed in detail.

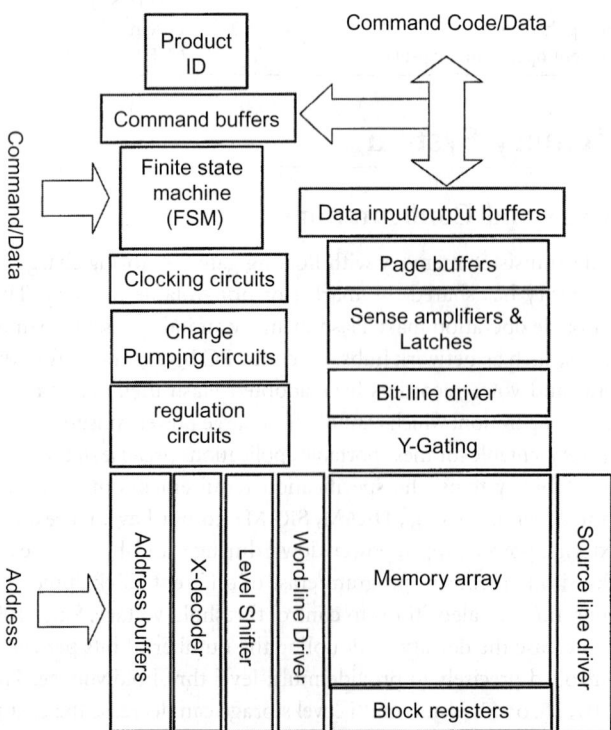

FIGURE 54.26 The block diagram of Flash memory system.

54.8.2 Finite State Machine

A finite state machine (FSM) is a control unit that processes commands and operation algorithms. Figure 54.27(a) demonstrates an example of an FSM. Figure 54.27(b) shows the details of an FSM. The command logic unit is an AND-OR-based logic unit that generates next state codes, while the state register latches the current state. The current state is related to the previous state and input state. State transitions follow the designated state diagram or state table that describe the functionality to translate state codes into controlling signals that are required by other circuits in the memory. The tendency to develop Flash memories goes in the direction of simultaneous program, erase, and read in different blocks. The global FSM takes charge of command distribution, address transition detection (ATD), and data input/output. The address command and data are queued when the selected FSM is busy. The local FSM deals with operations, including read, program, and erase, within the local block. The local FSM is activated and completes an operation independently when a command is issued. The global FSM manages the tasks distributing among local FSMs according to the address. The hierarchical local and global FSM can provide parallel processing; for instance, one block is being programmed while the other block is being erased. This feature of simultaneous read/write reduces the system overhead and speeds up the Flash memory. One example of the algorithm used in the FSM is shown in Figure 54.28. The global FSM loads operating code (OP code) first, then the address transition detection (ATD) enables latch of the address when a different but valid address is observed. The status of the selected block is checked if the command can be executed right away, whereas the command, address, and/or data input are stored in the queues. The queue will be read when the local FSM is ready for executing the next command. The operation code and address are decoded. Sense amplifiers are activated if a read command is issued. Charge-pumping circuits are back to work if a write command is issued. After all preparations are made, the process routine begins, which will be explained later. Following the completion of the process routine, the FSM checks its queues. If there is any command queued for delayed operation, the local FSM reads the queued data and continues the described procedures. Since these operations are invisible to the external systems, the system overhead is reduced.

The process routine is shown in Figure 54.29. The read procedure waits for the completion signal of the sense amplifier, and then the valid data is sent immediately. The programming and erasing operations require a verification procedure to ascertain completion of the operation. The iteration of program-verification and erase-verification proceeds to fine-tune the threshold voltage. However, if the verification

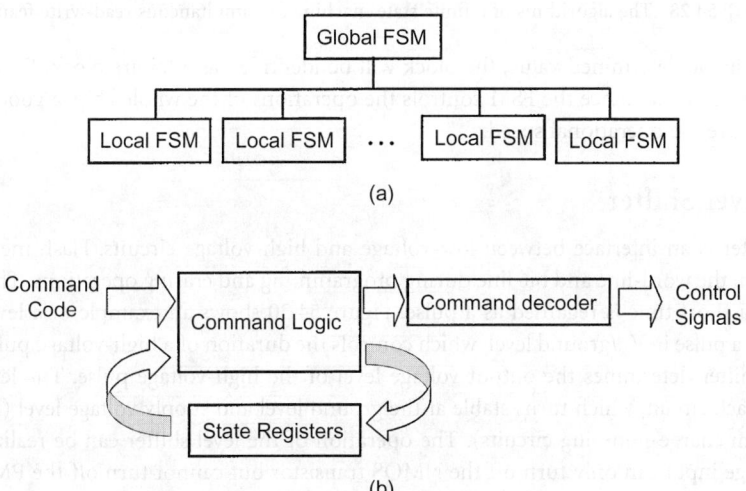

FIGURE 54.27 (a) The hierarchical architecture of a finite state machine; and (b) the block diagram of a finite state machine.

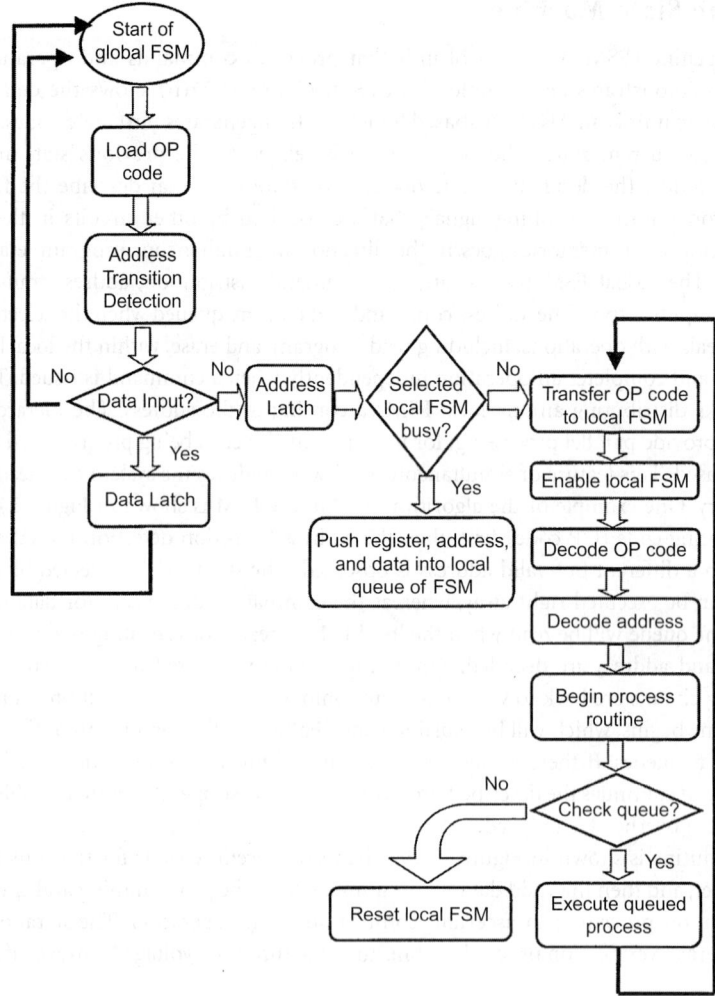

FIGURE 54.28 The algorithms of a finite state machine for simultaneous read-write feature.

time exceeds the predetermined value, the block will be identified as a failure block. Further operation to this block is inhibited. Since the FSM controls the operations of the whole chip, a good design of the FSM can improve the operational speed.

54.8.3 Level Shifter

The level shifter is an interface between low-voltage and high-voltage circuits. Flash memory requires high voltage on the word-line and bit-line during programming and erasing operations. The high voltage appearing in a short time is regarded as a pulse. Figure 54.30 shows an example of a level shifter. The input signal is a pulse in V_{cc}/ground level, which controls the duration of a high-voltage pulse. The supply of the level shifter determines the output voltage level of the high-voltage pulse. The level shifter is a positive feedback circuit, which turns stable at the ground level and supply voltage level (high voltage is generated from charge-pumping circuits). The operation of the level shifter can be realized as follows. The low-voltage input can only turn off the NMOS transistor but cannot turn off the PMOS parts. On the other hand, high voltage can only turn off the PMOS transistor. Therefore, generation of two mutually inverted signals can turn off the individual loading path and provide no leakage current during stand-by. The challenges of the design are the transition power consumption and the possibility of latch-up.

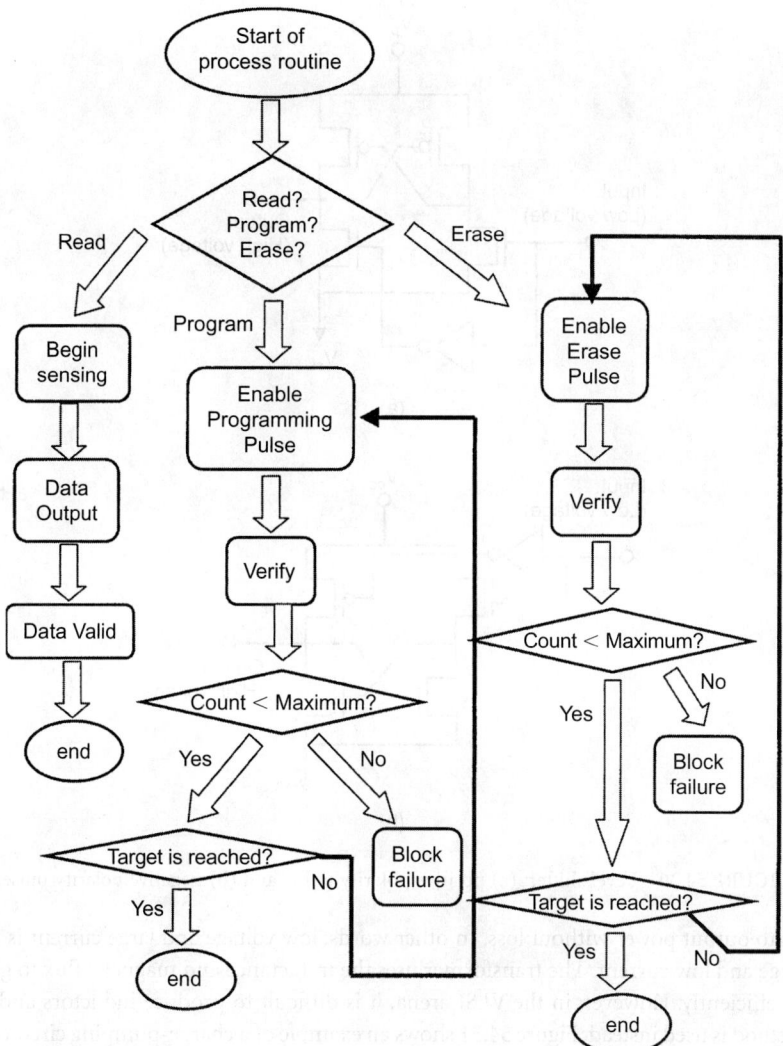

FIGURE 54.29 The algorithm of the process routine in Figure 54.28.

The delay of the feedback loop will result in large leakage current flowing from the high-voltage supply to ground. The leakage current is similar to the transition current of conventional CMOS circuits, but larger due to the delay of the feedback loop. As the large leakage current occurs due to generated substrate current by hot carriers, the level shifter is susceptible to latch-up. The design of the level shifter should focus on speeding up the feedback loop and employing a latch-up-free apparatus. More sophisticated level shifters should be designed to provide tradeoff between switching power and the switching speed.

The level shifter is used in the word-line driver and the bit-line driver if the bit-line requires a voltage larger than the external power supply. The driver is expected to be small because the word-line pitch is nearly minimum feature size. Thus, the major challenges are to simplify the level shifter and to provide a high-performance switch.

54.8.4 Charge-Pumping Circuit

The charge-pumping circuit is a high-voltage generator that supplies high voltage for programming and erasing operations. This kind of circuit is well-known in power equipment, such as power supplies, high-voltage switches, etc. A conventional voltage generator requires a power transformer, which transforms

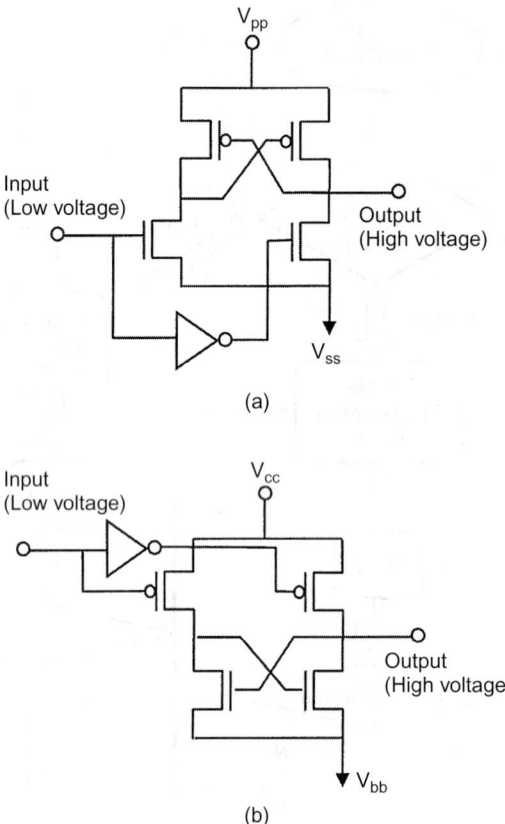

(a)

(b)

FIGURE 54.30   Level shifter: (a) positive polarity pulse, and (b) negative polarity pulse.

input power to output power without loss. In other words, low voltage and large current is transformed to high voltage and low current. The transformer uses the inductance and magnetic flux to generate high voltage very efficiently. However, in the VLSI arena, it is difficult to produce inductors and the charge-pumping method is used instead. Figure 54.31 shows an example of a charge-pumping circuit that consists of multiple-stage pumping units. Each unit is composed of a one-way switch and a capacitor. The one-way switch is a high-voltage switch that does not allow charge to flow back to the input. The capacitor stores the transferred charge and gradually produces high voltage. No two consecutive stages operate at the same time. In other words, when one stage is transferring the charge, the next stage and the previous stage should serve as an isolation switch, which eliminates charge loss. Therefore, a two-phase clocking signal is required to proceed with the charge-pumping operation, producing no voltage drop between the input and output of the switch and large current drivability of the output. In addition, the voltage level must be higher than the previous stage. Therefore, the two-phase clocking signal must be level-shifted to individual high voltages to turn on and off the one-way switch in each pumping unit. A smaller charge-pumping or a more sophisticated level-shift circuit can be employed as self-boosted parts. The generated high voltage, in most cases, is higher than the required voltage. A regulation circuit, which can generate stable voltage and is immune to the fluctuation of external supply voltage and the operating temperature, is used to regulate the voltage and will be described later.

54.8.5   Sense Amplifier

The sense amplifier is an analog circuit that amplifies small voltage differences. Many circuits can be employed—from the simplest two-transistor cross-coupled latches to the complicated cascaded current-mirrors

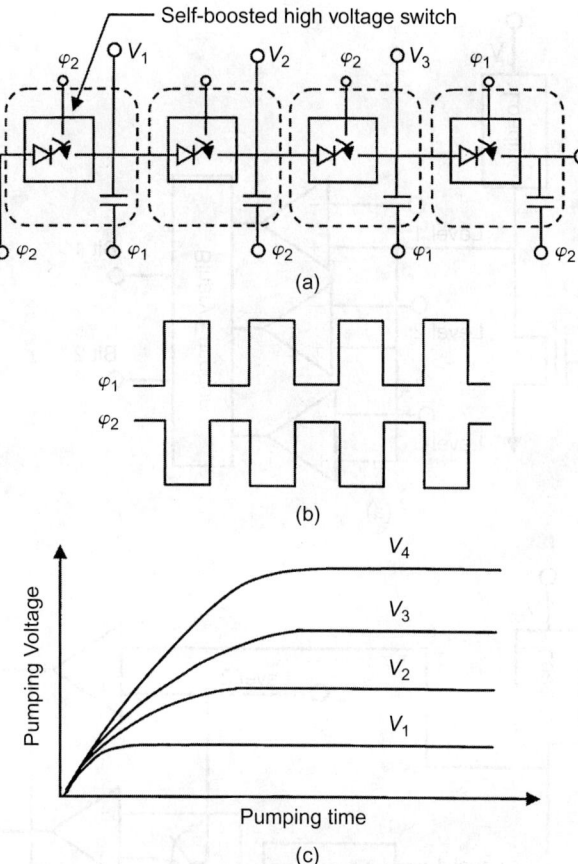

FIGURE 54.31 (a) Charge-pumping circuit; (b) two-phase clock; and (c) pumping voltage.

sense amplifiers. Here, a symbolic diagram is used to represent the sense amplifier in the following discussion. The focus of the sensing circuit is on multi-level sensing, which is currently the engineering issue in Flash memory. Figures 54.32(a) and (b) show the schemes of parallel sensing and consecutive sensing, respectively. These two schemes are based on analog-to-digital conversion (ADC). Information stored in the Flash memory can be read simultaneously with multiple comparators working at the same time. The outputs of the comparators are encoded into N digits for 2^N levels. Figure 54.32(b) shows the consecutive sensing scheme. The sensing time will be N times longer than the parallel sensing for 2^N levels. The sensing algorithm is a conventional binary search that compares the middle values in the consecutive range of interest. Only one sense amplifier is required for a cell. In the example, the additional sense amplifier is used for speeding up the sensing process. The second-stage sense amplifier can be pre-charged and prepared while the first-stage sense amplifier is amplifying the signal. Thus, the sensing time overhead is reduced.

When a multi-level scheme is used, the threshold voltage should be as tight as possible for each level. The depletion of unselected cells is strictly inhibited because the leakage current from unselected cells will destroy the true signal, which leads to error during sensing. Another challenge in multi-level sensing is the generation of reference voltages. Since the reference voltages are generated from the power supply, the leakage along the voltage divider path is unavoidable. Besides, the generated voltages are susceptible to the temperature variation and process-related resistance variation. If the variation of reference voltages cannot be minimized to a certain value, the ambiguous decision would be made for multi-level sensing due to unavoidable threshold spread for each level. Therefore, to provide high-sensitivity sense amplifier and to generate precise and robust reference voltages are the major developing goals for more than four-level Flash memory.

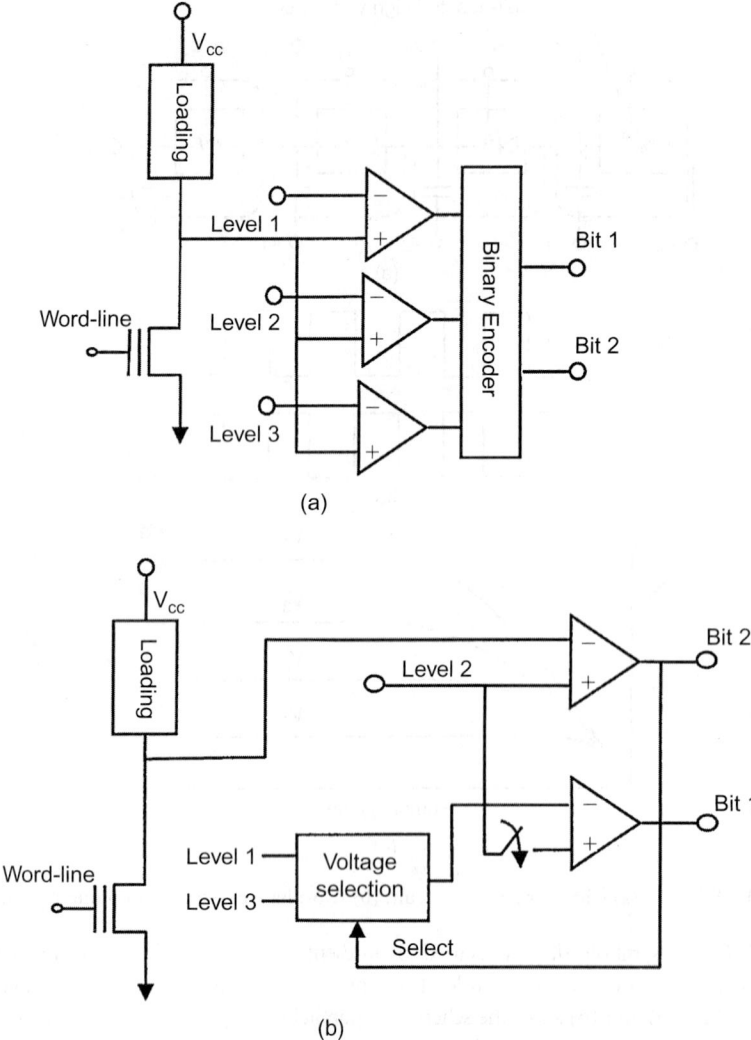

(a)

(b)

FIGURE 54.32   (a) Parallel sensing scheme, and (b) consecutive sensing scheme.

54.8.6   Voltage Regulator

A voltage regulator is an accurate voltage generator that is immune to temperature variation, process-related variation, and parasitic component effects. The concept of voltage regulation arises from the temperature-compensated device and the negative feedback circuits. Semiconductor carrier concentration and mobility are all dependent on the ambient temperature. Some devices have positive temperature coefficients, while others have negative coefficients. We can use both kinds of devices to produce a composite device for complete compensation. Figure 54.33 shows two back-to-back connected diodes that can be insensitive to the temperature over the temperature range of interest, if the doping concentration is properly designed. The forward-bias diode is negatively sensitive to temperature: the higher the temperature, the lower the cut-in voltage. On the other hand, the reverse-bias diode shows a reverse characteristic in the breakdown voltage. When connecting the two diodes and optimizing the diode characteristics, the regulated voltage can be insensitive to temperature. Nevertheless, the generated voltage is usually not what we want. A feedback loop, as shown in Figure 54.34, is needed to generate precise programming and erasing voltage. The charge-pumping output voltage and drivability are functions of

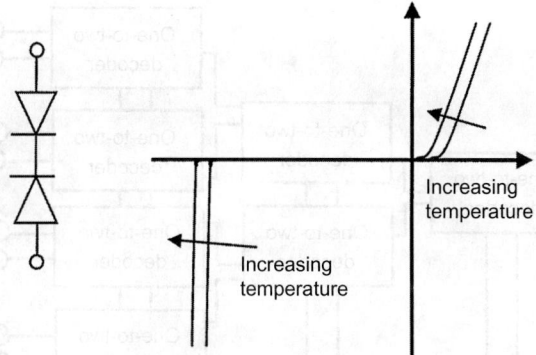

FIGURE 54.33 (a) Back-to-back connected temperature-compensated dual diodes; and (b) the characteristics of a diode as a function of temperature.

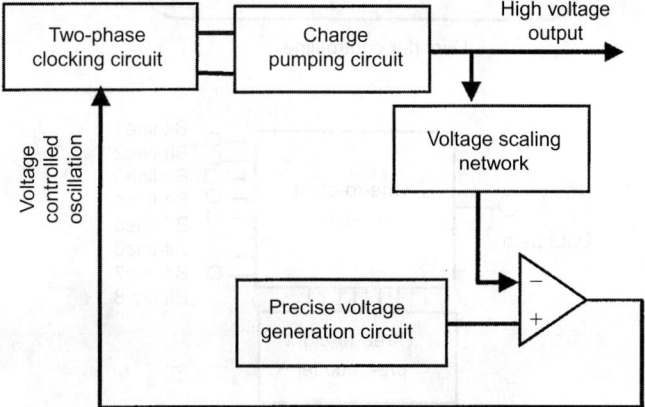

FIGURE 54.34 Voltage regulation block diagram.

the two-phase clocking frequency. The pumping voltage can be scaled to be compared with the precise voltage generator to provide a feedback signal for the clocking circuit whose frequency can be varied. With the feedback loop, the generated voltage can be insensitive to temperature. Whatever the desired output voltage is, the structure can be applied in general to produce temperature-insensitive voltage.

54.8.7 Y-Gating

Y-gating is the decoding path of bit-lines. The bit-line pitch is as small as the minimum feature size. One register and one sense amplifier per bit-line is difficult to achieve. Y-gating serves as a switch that makes multiple bit-lines share one latch and one sense amplifier. Two approaches—indirect decoding and direct decoding—used as the Y-gating are shown in Figures 54.35(a) and (b), respectively. Regarding the indirect decoding, if 2^N bit-lines are decoded using one-to-two decoding unit, the cascaded stages are required with N decoding control lines. However, when the direct decoding schemes is used, 2^N bit-lines require 2^N decoding lines to establish a one-to-$2N$ decoding network, and the pre-decoder is required to generate the decoding signal. The area penalty of indirect decoding is reduced but the voltage drop along the decoding path is of concern. To avoid the voltage drop, a boosted decoding line should be used to overcome the threshold voltage of the passing transistor. Another approach to eliminate voltage drop is the employment of a CMOS transfer gate. However, the area penalty arises again due to well-to-well isolation. Since Flash memory is very sensitive to the drain voltage, the boosted decoding control lines, together with the indirect decoding scheme, are suggested.

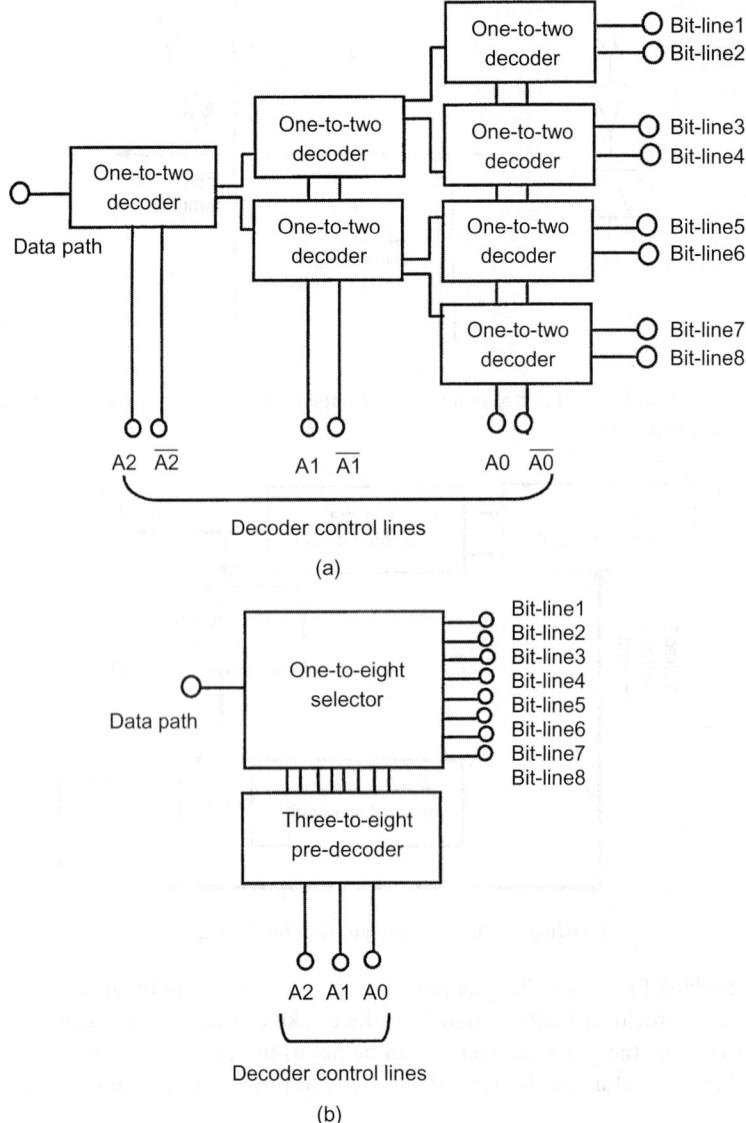

FIGURE 54.35 (a) Indirect decoding, and (b) direct decoding.

54.8.8 Page Buffer

A page buffer is static memory (SRAM-like memory) that serves as a temporary storage of input data. The page buffer also serves as temporary storage of read data. With the page buffer, Flash memory can increase its throughput or bandwidth during programming and read, because external devices can talk to the page buffer in a very short time without waiting for the slow programming of Flash memory. After the input data is transferred to the page buffer, the Flash memory begins programming and external devices can do other tasks. The page size should be carefully designed according to the applications. The larger the page size, the more data can be transferred into Flash memory without having to wait for the completion of programming. However, the area penalty limits the page size. There exists a proper design of page buffer for the application of interest.

54.8.9 Block Register

The block register stores the information about the individual block. The information includes failure of the block, write inhibit, read inhibit, executable operation, etc., according to the applications of interest. Some blocks, especially the boot block, are write-inhibited after first programming. This prevents virus injection in some applications, such as PC BIOS. The block registers are also Flash memory cells for storing block information, which will not disappear after power-off. When the local FSM is working on a certain block, the first thing is to check the status of the block by reading the register. If the block is identified as a failure block, no further operation can be made in this block.

54.8.10 Summary

Flash memory is a system with mixed analog and digital systems. The analog circuits include voltage-generation circuits, analog-to-digital converter circuits, sense amplifier circuits, and level-shifter circuits. These circuits require excellent functionality but small area consumption. The complicated analog designs in the pure-analog circuit do not meet the requirements of Flash memory, which requires large array efficiency, large memory density, and large storage volume. Therefore, the design of these analog circuits tends toward reduced design and qualified function. On the other hand, the digital parts of Flash memory are not as complicated as those digital circuits used in pure digital signal process circuits. Therefore, the mixed analog and digital Flash memory system can be implemented in a simplified way. Furthermore, Flash memory is a memory cell-based system. All the functions of the circuits are designed according to the characteristics of the memory cell. Once the cell structure of a memory differs, it will result in a completely different system design.

References

1. Kahng, D. and Sze, S.M., A floating gate and its application to memory devices, *Bell Syst. Tech. J.*, vol. 46, p. 1283, 1967.
2. Frohman-Bentchlowsky, D., An integrated metal-nitride-oxide-silicon (MNOS) memory, *IEDM Tech. Dig.*, 1968.
3. Pao, H.C and O'Connel, M., *Appl. Phys. Lett.* no. 12, p. 260, 1968.
4. Frohman-Bentchlowsky, D., A fully decoded 2048-bit electrically programmable FAMOS read only memory, *IEEE J. Solid-State Circuits*, vol. SC-6, no. 5, p. 301, 1971.
5. Johnson, W., Perlegos, G., Renninger, A., Kuhn, G., and Ranganath, T., A 16k bit electrically erasable non-volatile memory, *Tech. Dig. IEEE ISSCC*, p. 152, 1980.
6. Masuoka, F., Asano, M., Iwahashi, H., Komuro, T., and Tanaka, S., A new Flash EEPROM cell using triple polysilicon technology, *IEDM Tech. Dig.*, p. 464, 1984.
7. Mukherjee, S., Chang, T., Pang, R., Knecht, M., and Hu, D., A single transistor EEPROM cell and its implementation in a 512K CMOS EEPROM, *IEDM Tech. Dig.*, p. 616, 1985.
8. Samachisa, G., Su, C.-S., Kao, Y.-S., Smarandoiu, G., Wang, C.Y.-M., Wong, T., and Hu, C., A 128K Flash EEPROM using double-polysilicon technology, *IEEE J. Solid-State Circuits*, vol. SC-22, no. 5, p. 676, 1987.
9. Hsu, C.C.-H., Acovic, A., Dori, L., Wu, B., Lii, T., Quinlan, D., DiMaria, D., Taur, Y., Wordeman, M., and Ning, T., A high speed, low power p-channel Flash EEPROM using silicon rich oxide as tunneling dielectric, *Ext. Abstract of 1992 SSDM*, p. 140, 1992.
10. Ohnakado, T., Mitsunaga, K., Nunoshita, M., Onoda, H., Sakakibara, K., Tsuji, N., Ajika, N., Hatanaka, M., and Miyoshi, H., Novel electron injection method using band-to-band tunneling induced hot electron (BBHE) for Flash memory with p-channel cell, *IEDM Tech. Dig.*, p. 279, 1995.
11. Ohnakado, T., Takada, H., Hayashi, K., Sugahara, K., Satoh, S., and Abe, H., Novel self-limiting program scheme utilizing n-channel select transistors in p-channel DINOR Flash memory, *IEDM Tech. Dig.*, 1996.

12. Shen, S.-J., Yang, C.-S., Wang, Y.-S., and Hsu, C.C.-H., Novel self-convergent programming scheme for multi-level p-channel Flash memory, *IEDM Tech. Dig.*, p. 287, 1997.

13. Chung, S.S., Kuo, S.N., Yih, C.M., and Chao, T.S., Performance and reliability evaluations of p-channel Flash memories with different programming schemes, *IEDM Tech. Dig.*, 1997.

14. Wang, S.T., On the I-V characteristics of floating gate MOS transistors, *IEEE Trans. Electron Devices*, vol. ED-26, no. 9, p. 1292, 1979.

15. Liong, L.C. and Liu, P.-C., A theoretical model for the current-voltage characteristics of a floating gate EEPROM cell, *IEEE Trans. Electron Devices*, vol. ED-40, no. 1, p. 146, 1993.

16. Manthey, J.T., Degradation of Thin Silicon Dioxide Films and EEPROM Cells, Ph.D. dissertation, 1990.

17. Ng, K.K. and Taylor, G.W., Effects of hot-carrier trapping in n and p channel MOSFETs, *IEEE Trans. Electron Devices*, vol. ED-30, p. 871, 1983.

18. Selmi, L., Sangiorgi, E., Bez, R., and Ricco, B., Measurement of the hot hole injection probability from Si into SiO_2 in p-MOSFETs, *IEDM Tech. Dig.*, p. 333, 1993.

19. Tang, Y., Kim, D.M., Lee, Y.-H., and Sabi, B., Unified characterization of two-region gate bias stress in submicronmeter p-channel MOSFET's, *IEEE Electron Device Lett.*, vol. EDL-11, no. 5, p. 203, 1990.

20. Takeda, E., Kume, H., Toyabe, T., and Asai, S., Submicrometer MOSFET structure for minimizing hot carrier generation, *IEEE Trans. Electron Devices*, vol. ED-29, p. 611, 1982.

21. Shockley, W., Problems related to p-n junction in silicon, *Solid-State Electron.*, vol. 2, p. 35, 1961.

22. Verwey, J.F., Kramer, R.P., and de Maagt B.J., Mean free path of hot electrons at the surface of boron-doped silicon, *J. Appl. Phys.*, vol. 46, p. 2612, 1975.

23. Ning, T.H., Osburn, C.M., and Yu, H.N., Emission probability of hot electrons from silicon into silicon dioxide, *J. Appl. Phys.*, vol. 48, p. 286, 1977.

24. Hu, C., Lucky-electron model of hot-electron emission, *IEDM Tech. Dig.*, p. 22, 1979.

25. Tam, S., Ko, P.-K., and Hu, C., Lucky-electron model of channel hot electron injection in MOS-FET's, *IEEE Trans. Electron Devices*, vol. ED-31, p. 1116, 1984.

26. Berglung, C.N. and Powell, R.J., Photoinjection into SiO2. Electron scattering in the image force potential well, *J. Appl. Phys.*, vol. 42, p. 573, 1971.

27. Ong, T.-C., Ko, P.K., and Hu, C., Modeling of substrate current in p-MOSFET's, *IEEE Electron Device Lett.*, vol. EDL-8, no. 9, p. 413, 1987.

28. Ong, T.-C., Seki, K., Ko, P.K., and Hu, C., P-MOSFET gate current and device degradation, *Proc. IEEE/IRPS*, p. 178, 1989.

29. Takeda, E., Suzuki, N., and Hagiwara, T., Device performance degradation due to hot carrier injection at energies below the $Si\text{-}SiO_2$ energy barrier, *IEDM Tech. Dig.*, p. 396, 1983.

30. Hofmann, K.R., Werner, C., Weber, W., and Dorda, G., Hot-electron and hole emission effects in short n-channel MOSFET's, *IEEE Trans. Electron Devices*, vol. ED-32, no. 3, p. 691, 1985.

31. Nissan-Cohen, Y., A novel floating-gate method for measurement of ultra-low hole and electron gate currents in MOS transistors, *IEEE Electron Device Lett.*, vol. EDL-7, no. 10, p. 561, 1986.

32. Sak, N.S., Hereans, P.L., Hove, L.V.D., Maes, H.E., DeKeersmaecker, R.F., and Declerck, G.J., Observation of hot-hole injection in NMOS transistors using a modified floating gate technique, *IEEE Trans. Electron Devices*, vol. ED-33, no. 10, p. 1529, 1986.

33. Yamada, S., Suzuki, T., Obi, E., Oshikiri, M., Naruke, K., and Wada, M., A self-convergence erasing scheme for a simple stacked gate Flash EEPROM, *IEDM Tech. Dig.*, p. 307, 1991.

34. Ong, T.C., Fazio, A., Mielke, N., Pan, S., Righos, N., Atwood, G., and Lai, S., Erratic erase in ETOX Flash memory array, *Proc. Symp. on VLSI Technology*, p. 83, 1993.

35. Yamada, S., Yamane, T., Amemiya, K., and Naruke, K., A self-convergence erase for NOR Flash EEPROM using avalanche hot carrier injection, *IEEE Trans. Electron Devices*, vol. ED-43, no. 11, p. 1937, 1996.

36. Chen, J., Chan, T.Y., Chen, I.C., Ko, P.K., and Hu, C., Subbreakdown drain leakage current in MOSFET, *IEEE Electron Device Lett.*, vol. EDL.-8, no. 11, p. 515, 1987.

37. Chan, T.Y., Chen, J., Ko, P.K., and Hu, C., The impact of gate-induced drain leakage on MOSFET scaling, *IEDM Tech. Dig.*, p. 718, 1987.

38. Shrota, R., Endoh, T., Momodomi, M., Nakayama, R., Inoue, S., Kirisawa, R., and Masuoka, F., An accurate model of sub-breakdown due to band-to-band tunneling and its application, *IEDM Tech. Dig.*, p. 26, 1988.

39. Chang, C. and Lien, J., Corner-field induced drain leakage in thin oxide MOSFET's, *IEDM Tech. Dig.*, p. 714, 1987.

40. Chen, I.-C., Coleman, D.J., and Teng, C.W., Gate current injection initiated by electron band-to-band tunneling in MOS devices, *IEEE Electron Device Lett.*, vol. EDL-10, no. 7, p. 297, 1989.

41. Yoshikawa, K., Mori, S., Sakagami, E., Ohshima, Y., Kaneko, Y., and Arai, N., Lucky-hole injection induced by band-to-band tunneling leakage in stacked gate transistor, *IEDM Tech. Dig.*, p. 577, 1990.

42. Haddad, S., Chang, C., Swanminathan, B., and Lien, J., Degradation due to hole trapping in Flash memory cells, *IEEE Electron Device Lett.*, vol. EDL-10, no. 3, p. 117, 1989.

43. Igura, Y., Matsuoka, H., and Takeda, E., New device degradation due to Cold carrier created by band-to-band tunneling, *IEEE Electron Device Lett.*, vol. 10, no. 5, p. 227, 1989.

44. Lenzlinger, M. and Snow, E.H., Fowler-Nordheim tunneling into thermally grown SiO_2, *J. Appl. Phys.*, vol. 40, no. 1, p. 278, 1969.

45. Weinberg, Z.A., On tunneling in MOS structure, *J. Appl. Phys.*, vol. 53, p. 5052, 1982.

46. Ricco, B. and Fischetti, M.V., Temperature dependence of the currents in silicon dioxide in the high field tunneling regime, *J. Appl. Phys.*, vol. 55, p. 4322, 1984.

47. Lin, C.J., Enhanced Tunneling Model and Characteristics of Silicon Rich Oxide Flash Memory, Ph.D. dissertation, 1996.

48. Olivo, P., Sune, J., and Ricco, B., Determination of the Si-SiO_2 barrier height from the Fowler-Nordheim plot, *IEEE Electron Device Lett.*, vol. EDL-12, no. 11, p. 620, 1991.

49. Wu, A.T., Chan, T.Y., Ko, P.K., and Hu, C., A source-side injection erasable programmable read-only-memory (SI-EPROM) device, *IEEE Electron Device Lett.*, vol. EDL-7, no. 9, p. 540, 1986.

50. Kume, H., Yamamoto, H., Adachi, T., Hagiwara, T., Komori, K., Nishimoto, T., Koike, A., Meguro, S., Hayashida, T., and Tsukada, T., A Flash-erase EEPROM cell with an asymmetric source and drain structure, *IEDM Tech. Dig.*, p. 560, 1987.

51. Naruke, K., Yamada, S., Obi, E., Taguchi, S., and Wada, M., A new Flash-erase EEPROM cell with a side-wall select-gate on its source side, *IEDM Tech. Dig.*, p. 603, 1989.

52. Woo, B.J., Ong, T.C., Fazio, A., Park, C., Atwood, D., Holler, M., Tam, S., and Lai, S., A novel memory cell using Flash array contact-less EPROM (FACE) technology, *IEDM Tech. Dig.*, p. 91, 1990.

53. Ohshima, Y., Mori, S., Kaneko, Y., Sakagami, E., Arai, N., Hosokawa, N., and Yoshikawa, K., Process and device technologies for 16M bit EPROM's with large-tilt-angle implanted p-pocket cell, *IEDM Tech. Dig.*, p. 95, 1990.

54. Ajika, N., Obi, M., Arima, H., Matsukawa, T., and Tsubouchi, N., A 5 volt only 16M bit Flash EEPROM cell with a simple stacked gate structure, *IEDM Tech. Dig.*, p. 115, 1990.

55. Manos, P. and Hart, C., A self-aligned EPROM structure with superior data retention, *IEEE Electron Device Lett.*, vol. EDL-11, no. 7, p. 309, 1990.

56. Kodama, N., Saitoh, K., Shirai, H., Okazawa, T., and Hokari, Y., A 5V only 16M bit Flash EEPROM cell using highly reliable write/erase technologies, *Proc. Symp. on VLSI Technology*, p. 75, 1991.

57. Kodama, N., Oyama, K., Shirai, H., Saitoh, K., Okazawa, T., and Hokari, Y., A symmetrical side wall (SSW)-DSA cell for a 64-M bit Flash memory, *IEDM Tech. Dig.*, p. 303, 1991.

58. Liu, D.K.Y., Kaya, C., Wong, M., Paterson, J., and Shah, P., Optimization of a source-side-injection FAMOS cell for Flash EPROM application, *IEDM Tech. Dig.*, p. 315, 1991.

59. Yamauchi, Y., Tanaka, K., Shibayama, H., and Miyake, R., A 5V-only virtual ground Flash cell with an auxiliary gate for high density and high speed application, *IEDM Tech. Dig.*, p. 319, 1991.

60. Kaya, C., Liu, D.K.Y., Paterson, J., and Shah, P., Buried source-side injection (BSSI) for Flash EPROM programming, *IEEE Electron Device Lett.*, vol. EDL-13, no. 9, p. 465, 1992.

61. Yoshikawa, K., Sakagami, E., Mori, S., Arai, N., Narita, K., Yamaguchi, Y., Ohshima, Y., and Naruke, K., A 3.3V operation nonvolatile memory cell technology, *Proc. Symp. on VLSI Technology*, p. 40, 1992.

62. Shirota, R., Itoh, Y., Nakayama, R., Momodomi, M., Inoue, S., Kirisawa, R., et al., A new NAND cell for ultra high density 5V-only EEPROM's, *Proc. Symp. on VLSI Technology*, p. 33, 1988.

63. Momodomi, M., Kirisawa, R., Nakayama, R., Aritome, S., Endoh, T., Itoh, T., et al., New device technologies for 5V- only 4Mb EEPROM with NAND structure cell, *IEDM Tech. Dig.*, p. 412, 1988.

64. Kume, H., Tanaka, T., Adachi, T., Miyamoto, N., Saeki, S., Ohji, Y., et al., A 3.42 μm^2 Flash memory cell technology conformable to a sector erase, *Proc. Symp. on VLSI Technology*, p. 77, 1991.

65. Onoda, H., Kunori, Y., Kobayashi, S., Ohi, M., Fukumoto, A., Ajika, N., and Miyoshi, H., A novel cell structure suitable for a 3 volt operation, sector erase Flash memory, *IEDM Tech. Dig.*, p. 599, 1992.

66. Kume, H., Kato, M., Adachi, T., Tanaka, T., Sasaki, T., and Okazaki, T., A 1.28 μm^2 contactless memory cell technology for a 3V-only 64M bit EEPROM, *IEDM Tech. Dig.*, p. 991, 1992.

67. Method for manufacturing a contact-less floating gate transistor, U.S. Patent 5453391, 1993.

68. Ohi, M., Fukumoto, A., Kunori, Y., Onoda, H., Ajika, N., Hatanaka, M., and Miyoshi, H., An asymmetrical offset source/drain structure for virtual ground array Flash memory with DINOR operation, *Proc. Symp. on VLSI Technology*, p. 57, 1993.

69. Yamauchi, Y., Yoshimi, M., Sato, S., Tabuchi, H., Takenaka, N., and Sakiyam, K., A new cell structure for sub-quarter micron high density Flash memory, *IEDM Tech. Dig.*, p. 267, 1995.

70. Kim, K.S., Kim, J.Y., Yoo, J.W., Choi, Y.B., Kim, M.K., Nam, B.Y., et al. A novel dual string NOR (DuSNOR) memory cell technology scalable to the 256M bit and 1G bit Flash memory, *IEDM Tech. Dig.*, p. 263, 1995.

71. Kirisawa, R., Aritome, S., Nakayama, R., Endoh, T., Shirota, R., and Masuoka, F., A NAND structures cell with a new programming technology for highly reliable 5V-only Flash EEPROM, *Proc. Symp. on VLSI Technology*, p. 129, 1990.

72. Aritome, S., Kirisawa, R., Endoh, T., Nakayama, R., Shirota, R., Sakui, K., Ohuchi, K., and Masuoka, F., Extended data retention characteristics after more than 10^4 write and erase cycles in EEPROM's, *Proc. IEEE/IRPS*, p. 259, 1990.

73. Endoh, T., Iizuka, H., Aritome, S., Shirota, R., and Masuoka, F., New write/erase operation technology for Flash EEPROM cells to improve the read disturb characteristics, *IEDM Tech. Dig.*, p. 603, 1992.

74. Aritome, S., Hatakeyama, K., Endoh, T., Yamaguchi, T., Shuto, S., Iizuka, H., et al., A 1.13 μm^2 memory cell technology for reliable 3.3V 64M NAND EEPROM's, *Ext. Abstract of 1993 SSDM*, p. 446, 1993.

75. Aritome, S., Satoh, S., Maruyama, T., Watanabe, H., Shuto, S., Hermink, G.J., Shirota, R., Watanabe, S., and Masuoka, F., A 0.67 μm^2 self-aligned shallow trench isolation cell (SA-STI cell) for 3V-only 256M bit NAND EEPROM's, *IEDM Tech. Dig.*, p. 61, 1994.

76. Kim, D.J., Choi, J.D., Kim, J. Oh, H.K., and Ahn, S.T., and Kwon, O.H., Process integration for the high speed NAND Flash memory cell, *Proc. Symp. on VLSI Technology*, p. 236, 1996.

77. Choi, J.D., Kim, D.J., Jang, D.S., Kim, J., Kim, H.S., Shin, W.C., Ahn, S.T., and Kwon, O.H., A novel booster plate technology in high density NAND Flash memories for voltage scaling down and zero program disturbance, *Proc. Symp. on VLSI Technology*, p. 238, 1996.

78. Entoh, T., Shimizu, K., Iizuka, H., and Masuoka, F., A new write/erase method to improve the read disturb characteristics based on the decay phenomena of the stress induced leakage current for Flash memories, *IEEE Trans. Electron Device*, vol. ED-45, no. 1, p. 98, 1998.

79. Lai, S.K., NVRAM technology, NOR Flash design and multi-level Flash, *IEDM NVRAM Technology and Application Short Course*, 1995.

80. Yamada, S., Hiura, Y., Yamane, T., Amemiya, K., Ohshima, Y., and Yoshikawa, K., Degradation mechanism of Flash EEPROM programming after programming/erase cycles, *IEDM Tech. Dig.*, p. 23, 1993.

81. Cappelletti, P., Bez, R., Cantarelli, D., and Fratin, L., Failure mechanisms of Flash cell in program/erase cycling, *IEDM Tech. Dig.*, p. 291, 1994.

82. Liu, Y.C., Guo, J.-C., Chang, K.L., Huang, C.I., Wang, W.T., Chang, A., and Shone, F., Bitline stress effects on Flash EPROM cells after program/erase cycling, *IEEE Nonvolatile Semiconductor Memory Workshop*, 1997.

83. Shen, S.-J., Chen, H.-M., Lin, C.-J., Chen, H.-H., Hong, G., and Hsu, C.C.-H., Performance and reliability trade-off of large-tilted-angle implant p-pocket (LAP) on stacked-gate memory devices, *Japan. J. Appl. Phys.*, vol. 36, part 1, no. 7A, p. 4289, 1997.

84. DiMaria, D.J., Dong, D.W., Pesavento, F.L., Lam, C., and Brorson, B.D., Enhanced conduction and minimized charge trapping in electrically alterable read-only memories using off-stoichiometric silicon dioxide films, *J. Appl. Phys.*, vol. 55, p. 300, 1984.

85. Lin, C.-J., Hsu, C.C.-H., Chen, H.-H., Hong, G., and Lu, L.S., Enhanced tunneling characteristics of PECVD silicon-rich-oxide (SRO) for the application in low voltage Flash EEPROM, *IEEE Trans. Electron Device*, vol. ED-43, no. 11, p. 2021, 1996.

86. Shen, S.-J., Lin C.-J., and Hsu, C.C.-H, Ultra fast write speed, long refresh time, low FN power operated volatile memory cell with stacked nanocrystalline Si film, *IEDM Tech. Dig.*, p. 515, 1996.

87. Hisamune, Y.S., Kanamori, K., Kubota, T., Suzuki, Y., Tsukiji, M., Hasegawa, E., et al., A high capacitive-coupling ratio (HiCR) cell for 3V-only 64 M bit and future Flash memories, *IEDM Tech. Dig.*, p. 19, 1993.

88. Shirai, H., Kubota, T., Honma, I., Watanabe, H., Ono, H., and Okazawa, T., A 0.54 μm^2 self-aligned, HSG floating gate cell (SAHF cell) for 256M bit Flash memories, *IEDM Tech. Dig.*, p. 653, 1995.

89. Esquivel, J., Mitchel, A., Paterson, J., Riemenschnieder, B., Tieglaar, H., et al., High density contactless, self aligned EPROM cell array technology, *IEDM Tech. Dig.*, p. 592, 1986.

90. Masuoka, F., Momodomi, M., Iwata, Y., and Shirota, R., New ultra high density EPROM and Flash EEPROM with NAND structure cell, *IEDM Tech. Dig.*, p. 552, 1987.

91. Kynett, V.N., Baker, A., Fandrich, M.L., Hoekstra, G.P., Jungroth, O., Hreifels, J.A., et al., An in-system re-programmable 32K × 8 CMOS Flash memory, *IEEE J. Solid Stat.*, vol. SC-23, no. 5, p. 1157, 1988.

92. Kazerounian, R., Ali, S., Ma, Y., and Eitan, B., A 5 volt high density poly-poly erase Flash EPROM cell, *IEDM Tech. Dig.*, p. 436, 1988.

93. Gill, M., Cleavelin, R., Lin, S., D'Arrigo, I., Santin, G., Shah, P., et al., A 5-volt contactless 256K bit Flash EEPROM technology, *IEDM Tech. Dig.*, p. 428, 1988.

94. Flash EEPROM array with negative gate voltage erase operation, U.S. Patent 5077691, filed:1989.

95. Kynett, V.N., Fandrich, M.L., Anderson, J., Dix, P., Jungroth, O., Hreifels, J.A., et al., A 90ns one-million erase/program cycle 1Mbit Flash memory, *IEEE J. Solid-State Circuits.*, vol. SC-24, no. 5, p. 1259, 1989.

96. Endoh, T., Shirota, R., Tanaka, Y., Nakayama, R., Kirisawa, R., Aritome, S., and Masuoka, F., New design technology for EEPROM memory cells with 10 million write/erase cycling endurance, *IEDM Tech. Dig.*, p. 599, 1989.

97. Shirota, R., Nakayama, R., Kirisawa, R., Momodomi, M., Sakui, K., Itoh, Y., et al., A 2.3 μm^2 memory cell structure for 16M bit NAND EEPROM's, *IEDM Tech. Dig.*, p. 103, 1990.

98. Riemenschneider, B., Esquivel, A.L., Paterson, J., Gill, M., Lin, S., Schreck, J., et al., A process technology for a 5-volt only 4M bit Flash EEPROM with an 8.6 μm^2 cell, *Proc. Symp. on VLSI Technology*, p. 125, 1990.

99. Gill, M., Cleavelin, R., Lin, S., Middendorf, M., Nguyen, A., Wong, J., et al., A novel sub-lithographic tunnel diode based 5V-only Flash memory, *IEDM Tech. Dig.*, p. 119, 1990.

100. Self-aligned source process and apparatus, U.S. Patent 5103274, filed:1991.

101. Woo, B.J., Ong, T.C., and Lai, S., A poly-buffered FACE technology for high density Flash memories, *Proc. Symp. on VLSI Technology*, p. 73, 1991.

102. Oyama, K., Shirai, H., Kodama, N., Kanamori, K., Saitoh, K., et al., A novel erasing technology for 3.3V Flash memory with 64 Mb capacity and beyond, *IEDM Tech. Dig.*, p. 607, 1992.

103. Pein, H. and Plummer, J.D., A 3-D side-wall Flash EPROM cell and memory array, *IEEE Electron Device Lett.*, vol. EDL-14, no. 8, p. 415, 1993.

104. Dhum, D.P., Swift, C.T., Higman, J.M., Taylor, W.J., Chang, K.T., Chang, K.M., and Yeargain, J.R., A novel band-to-band tunneling induced convergence mechanism for low current, high density Flash EEPROM applications, *IEDM Tech. Dig.*, p. 41, 1994.

105. Tsuji, N., Ajika, N., Yuzuriha, K., Kunori, Y., Hatanaka, M., and Miyoshi, H., New erase scheme for DINOR Flash memory enhancing erase/write cycling endurance characteristics, *IEDM Tech. Dig.*, p. 53, 1994.

106. Ma. Y., Pang, C.S., Chang, K.T., Tsao, S.C., Frayer, J.E., Kim, T., Jo, K., Kim, J., Choi, I., and Park, H., A dual-bit split-gate EEPROM (DSG) cell in contactless array for single Vcc high density Flash memories, *IEDM Tech. Dig.*, p. 57, 1994.

107. Kato, M., Adachi, T., Tanaka, T., Sato, A., Kobayashi, T., Sudo, Y., et al., A 0.4 µm self-aligned contactless memory cell technology suitable for 256M bit Flash memory, *IEDM Tech. Dig.*, p. 921, 1994.

108. Hemink, G.J., Tanaka, T., Endoh, T., Aritome, S., and Shirota, R., Fast and accurate programming method for multi-level NAND EEPROM's, *Proc. Symp. on VLSI Technology*, p. 129, 1995.

109. Hu, C.-Y., Kencke, D.L., Banerjee, S.K., Richart, R., Bandyopadhyay, B., Moore, B., Ibok, E., and Garg, S., A convergence scheme for over-erased Flash EEPROM's using substrate-bias-enhanced hot electron injection, *IEEE Electron Device Lett.*, vol. EDL-16, no. 11, p. 500, 1995.

110. Bude, J.D., Frommer, A., Pinto, M.R., and Weber, G.R., EEPROM/Flash sub 3.0V drain-source bias hot carrier writing, *IEDM Tech. Dig.*, p. 989, 1995.

111. Chi, M.H and Bergemont, A., Multi-level Flash/EPROM memories: new self-convergent programming methods for low-voltage applications, *IEDM Tech. Dig.*, p. 271, 1995.

112. Aritome, S., Takeuchi, Y., Sato, S., Watanabe, H., Shimizu, K., Hemink, G., and Shirota, R., A novel side-wall transistor cell (SWATT cell) for multi-level NAND EEPROMs, *IEDM Tech. Dig.*, p. 275, 1995.

113. Hu, C.-Y., Kencke, D.L., Banerjee, S.K., Richart, R., Bandyopadhyay, B., Moore, B., Ibok, E., and Garg, S., Substrate-current-induced hot electron (SCIHE) injection: a new convergence scheme for Flash memory, *IEDM Tech. Dig.*, p. 283, 1995.

114. Gotou, H., New operation mode for stacked gate Flash memory cell, *IEEE Electron Device Lett.*, vol. EDL-16, no. 3, p. 121, 1995.

115. Shin, W.C., Choi, J.D., Kim, D.J., Kim, J., Kim, H.S., Mang, K.M., et al., A new shared bit line NAND cell technology for the 256Mb Flash memory with 12V programming, *IEDM Tech. Dig.*, p. 173, 1996.

116. Papadas, C., Guillaumot, B., and Cialdella, B., A novel pseudo-floating-gate Flash EEPROM device (-cell), *IEEE Electron Device Lett.*, vol. EDL-18, no. 7, p. 319, 1997.

117. Shimizu, K., Narita, K., Watanabe, H., Kamiya, E., Takeuchi, Y., Yaegashi, T., Aritome, S., and Watanabe, T., A novel high-density $5F^2$ NAND STI cell technology suitable for 256Mbit and 1Gbit Flash memories, *IEDM Tech. Dig.*, p. 271, 1997.

118. Kobayashi, T., Matsuzaki, N., Sato, A., Katayama, A., Kurata, H., Miura, A., Mine, T., Goto, Y., et al., A 0.24 $µm^2$ cell process with 0.18 µm width isolation and 3-D interpoly dielectric films for 1Gb Flash memories, *IEDM Tech. Dig.*, p. 275, 1997.

119. Choi, J.D., Lee, D.G., Kim, D.J., Cho, S.S., Kim, H.S., Shin, C.H., and Ahn, S.T., A triple polysilicon stacked Flash memory cell with wordline self-boosting programming, *IEDM Tech. Dig.*, p. 283, 1997.

120. Chen, W.-M., Swift, C., Roberts, D., Forbes, K., Higman, J., Maiti, B., Paulson, W., and Chang, K.-T., A novel flash memory device with split gate source side injection and ONO charge storage stack (SPIN), *Proc. Symp. on VLSI Technology*, p. 63, 1997.

121. Kim, H.S., Choi, J.D., Kim, J., Shin, W.C., Kim, D.J., Mang, K.M., and Ahn, S.T., Fast parallel programming of multi-level NAND Flash memory cells using the booster-line technology, *Proc. Symp. on VLSI Technology*, p. 65, 1997.

122. Roy, A., Kazerounian, R., Irani, R., Prabhakar, V., Nguyen, S., Slezak, Y., et al., A new Flash architecture with a 5.8l2 scalable AMG Flash cell, *Proc. Symp. on VLSI Technology*, p. 67, 1997.

123. Lee, W.-H., Clemens, J.T., Keller, R.C., and Manchanda, L., A novel high K interpoly dielectric (IPD) Al_2O_3 for low voltage/high speed Flash memories: erasing in msec at 3.3V, *Proc. Symp. on VLSI Technology*, p. 117, 1997.

124. Kianian, S., et al., A novel 3-volt-only, small sector erase, high density Flash EEPROM, *Proc. Symp. on VLSI Tech.*, p. 71, 1994.

55

Dynamic Random Access Memory

Kuo-Hsing Cheng

Tamkang University

CONTENTS

55.1   Introduction

The first dynamic RAM (DRAM) was proposed in 1970 with a capacity of 1 Kb. Since then, DRAMs have been the major driving force behind VLSI technology development. The density and performance of DRAMs have increased at a very fast pace. In fact, the densities of DRAMs have quadrupled about every three years.

The first experimental Gb DRAM was proposed in 1995 [1,2] and remains commercially available in 2000. However, multi-level storage DRAM techniques are used to improve the chip density and to reduce the defect-sensitive area on a DRAM chip [3,4]. The developments in VLSI technology have produced DRAMs that realize a cheaper cost per bit compared with other types of memories.

55.2   Basic DRAM Architecture

The basic block diagram of a standard DRAM architecture is shown in Figure 55.1. Unlike SRAM, the addresses on the standard DRAM memory are multiplexed into two groups to reduce the address input pin counts and to improve the cost-effectiveness of packaging. Although the number of address input pin counts can be reduced by half using the multiplexed address scheme on the standard DRAM memory, the timing control of the standard DRAM memory becomes more complex and the operation speed is

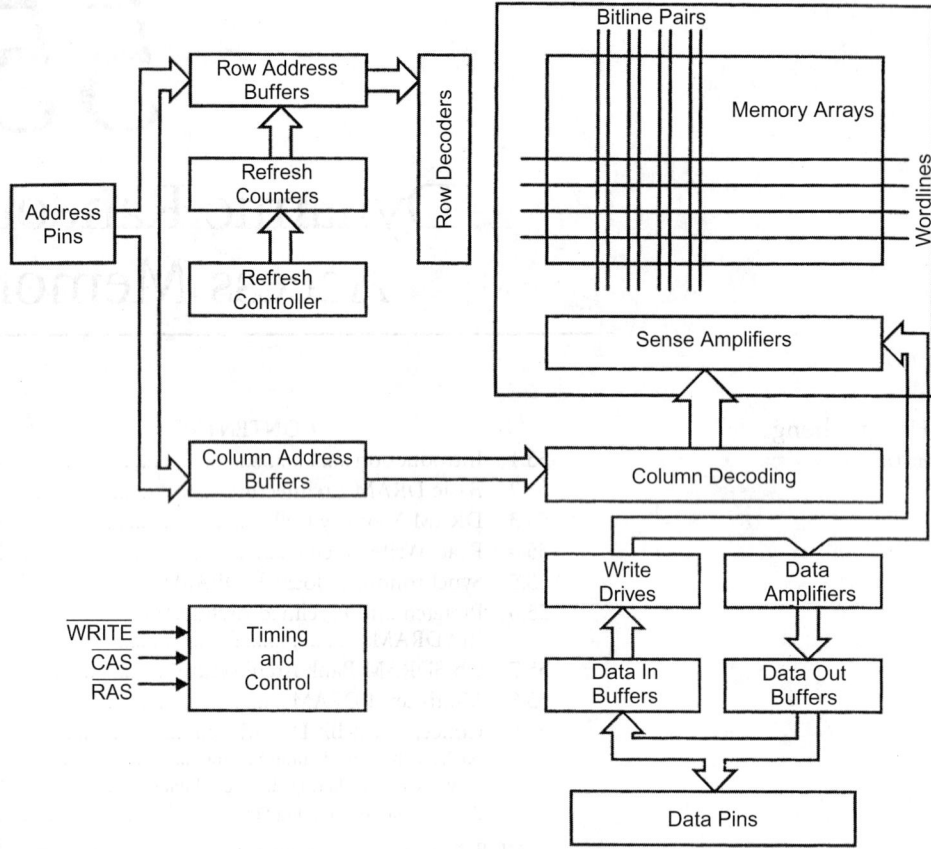

FIGURE 55.1 Basic block diagram of a standard DRAM architecture.

reduced. For high-speed DRAM applications, separate address input pins can be used to reduce the timing control complexity and to improve the operation speed.

In general, the address transition detector (ATD) circuit is not needed in a DRAM memory. DRAM controller provides $\underline{R}$ow $\underline{A}$ddress $\underline{S}$trobe ($\overline{RAS}$) and $\underline{C}$olumn $\underline{A}$ddress $\underline{S}$trobe ($\overline{CAS}$) to latch in the row addresses and the column addresses. As shown in Figure 55.1, the pins of a standard DRAM are:

- Address: which are multiplexed in time into two groups, the row addresses and the column addresses
- Address control signals: the Row Address Strobe $\overline{RAS}$ and the Column Address Strobe $\overline{CAS}$
- Write enable signal: $\overline{WRITE}$
- Input/output data pins
- Power-supply pins

An example of address-multiplexed DRAM timing during basic READ mode is shown in Figure 55.2. The row-falling edge of the address strobe ($\overline{RAS}$) samples the address and starts the READ operation mode. The row addresses are supplied into the address pins and then comes the row address strobe ($\overline{RAS}$) signal. Column addresses are not required until the row addresses are sent in and latched. The column addresses are applied into address pins and then latched in by the column address strobe ($\overline{CAS}$) signal. The access time t_{RAS} is the minimum time for the $\overline{RAS}$ signal to be low and t_{RC} is the minimum READ cycle time. Notice that the multiplexed address arrangement penalizes the access time of the standard DRAM memory.

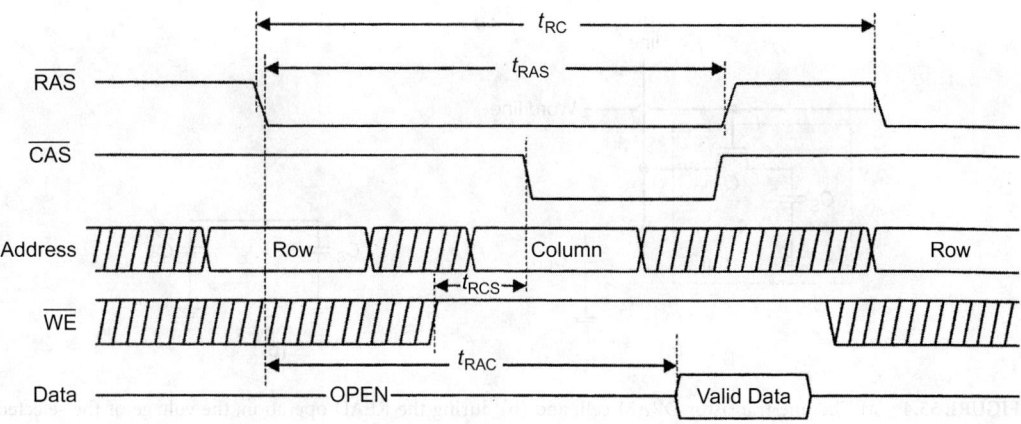

FIGURE 55.2   Read timing diagram for 4M × 1 DRAM.

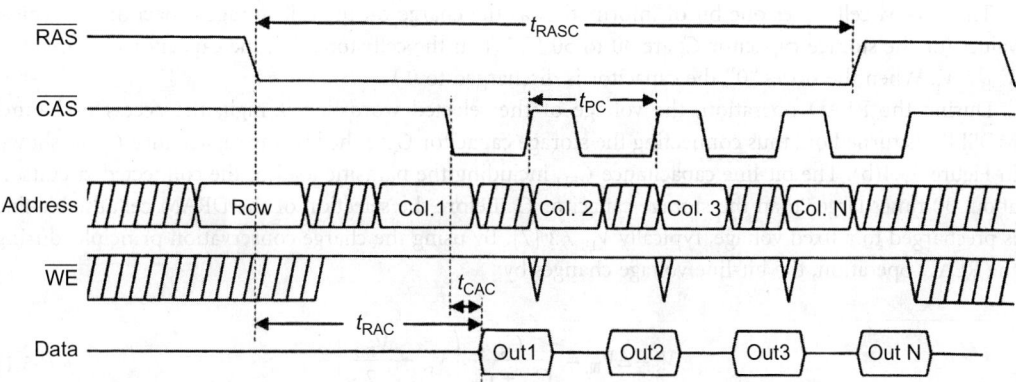

FIGURE 55.3   Fast page mode read timing diagram.

The CMOS DRAMs have several rapid access modes in addition to the basic modes. Figure 55.3 shows an example of the rapid access modes. The timing waveform shown in Figure 55.3 for DRAM operation is the page mode operation. In this mode, the row addresses are applied to the address pins and then clocked by the row address strobe $\overline{RAS}$ signal, and the column addresses are latched into the DRAM chip on the falling edge of $\overline{CAS}$ signal as in a basic READ mode. Along a selected row, the individual column bit can be rapidly accessed, and readout is randomly controlled by the column address and the column address strobe $\overline{CAS}$. By using the page mode, the access time per bit is reduced.

55.3   DRAM Memory Cell

In early CMOS DRAM storage cell design, three-transistor and four-transistor cells were used in 1-Kb and 4-Kb generations. Later, a particular one-transistor cell, as shown in Figure 55.4(a), became the industry standard [5,6]. The one-transistor (1T) cell achieves smaller cell size and low cost. The cell consists of an n-channel MOSFET and a storage capacitor C_s. The charge is stored in the capacitor C_s and the n-channel MOSFET functions as the access transistor. The gate of the n-channel MOSFET is

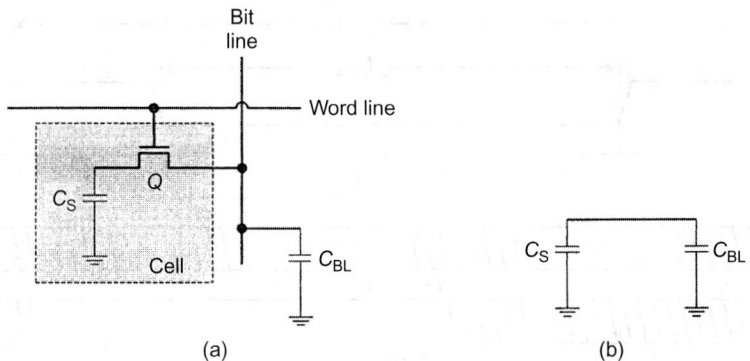

FIGURE 55.4 (a) The one-transistor DRAM cell; and (b) during the READ operation, the voltage of the selected word-line is high, thus connecting the storage capacitor C_s to the bit-line capacitance C_{BL}.

connected to the word-line WL and its source/drain is connected to the bit-line. The bit-line has a capacity C_{BL}, including the parasitic load of the connected circuits.

The DRAM cell stores one bit of information as the charge on the cell storage capacitor C_s. Typical values for the storage capacitor C_s are 30 to 50 fF. When the cell stores "1", the capacitor is charged to $V_{DD} - V_t$. When the stores "0", the capacitor is discharged to 0 V.

During the READ operation, the voltage of the selected word-line is high; the access n-channel MOSFET is turned on, thus connecting the storage capacitor C_s to the bit-line capacitance C_{BL} as shown in Figure 55.4(b). The bit-line capacitance C_{BL}, including the parasitic load of the connected circuits, is about 30 times larger than the storage capacitor C_s. Before the selection of the DRAM cell, the bit-line is precharged to a fixed voltage, typically $V_{DD}/2$ [7]. By using the charge conservation principle, during the READ operation, the bit-line voltage changes by

$$V_s = \Delta V_{BL} = \frac{C_S}{C_{BL} + C_S}\left(V_{cs} - \frac{V_{DD}}{2}\right) \tag{55.1}$$

Here, V_{cs} is the storage voltage on the DRAM cell capacitor C_s. A ratio $R = C_{BL}/C_s$ is important for the read sensing operation. If the cell stores "1" with a voltage $V_{cs} = V_{DD} - V_t$, we have the small bit-line sense signal

$$\Delta V(1) = \frac{1}{1+R}\left(\frac{V_{DD}}{2} - V_t\right) \tag{55.2}$$

If the cell stores "0" with a voltage $V_{cs} = 0$, we have the small bit-line sense signal

$$\Delta V(0) = \frac{1}{1+R}\left(\frac{V_{DD}}{2}\right) \tag{55.3}$$

Since ratio $R = C_{BL}/C_s$ is large, these readout bit-line sense signals $\Delta V(1)$ and $\Delta V(0)$ are very small. Typical values for the sense signal are about 100 mv.

For low-voltage operation, the supply voltage V_{DD} is reduced. Thus, a lower R ratio is required to maintain the sense signals to have enough margin against noise. The main approach is to use a large

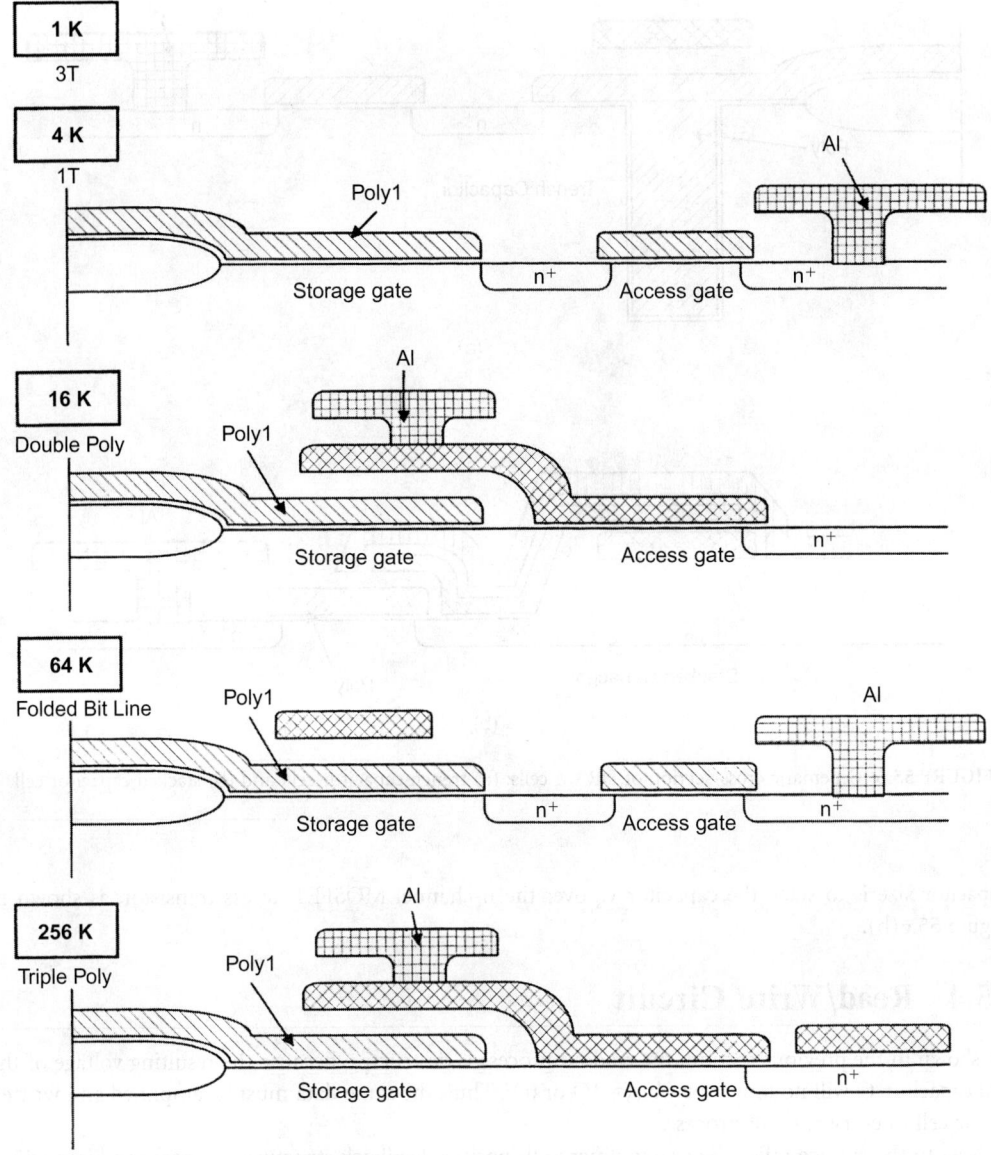

FIGURE 55.5 Structural innovations of planar DRAM cells.

cell storage capacitor C_s. As shown in Figure 55.5, a conventional C_s was implemented by a simple planar-type capacitor. The charge storage in the cell takes place on both the poly-1 gate oxide and the depletion capacitances. The planar DRAM cells have been used in the 1-T DRAMs from the 16 kb to the 1 Mb. The limits of the planar DRAM cell for retaining sufficient capacitance were reached in the mid-1980s in the 1-Mb DRAM. With the increased density higher than 1 Mb, smaller horizontal geometry on the surface of the wafer can be achieved by making increased use of the vertical dimension [8]. One approach is to use a trench capacitor, as shown in Figure 55.6(a) [9]. It is folded vertically into the surface of the silicon in the form of a trench. Another approach for reducing horizontal

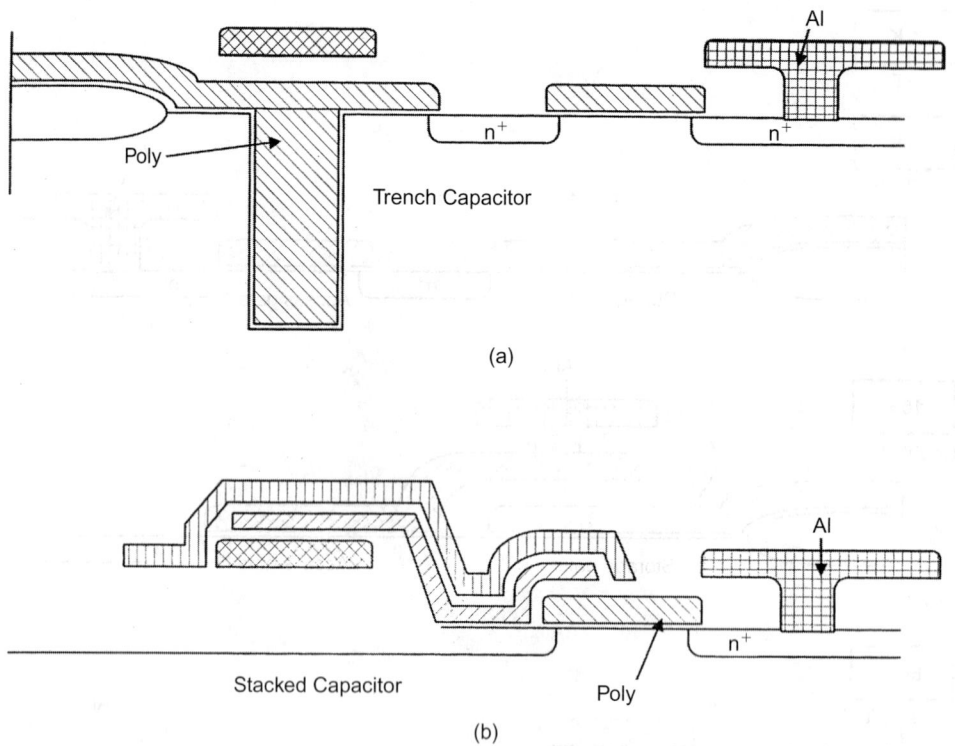

FIGURE 55.6 Schematic cross-section of DRAM cells: (a) trench capacitor cell, and (b) stacked capacitor cell.

capacitor size is to stack the capacitor C_s over the n-channel MOSFET access transistor, as shown in Figure 55.6(b).

55.4 Read/Write Circuit

As shown in the previous section, the readout process is destructive because the resulting voltage of the cell capacitor C_s will no longer be $(V_{DD} - V_t)$ or 0 V. Thus, the same data must be amplified and written to the cell in every readout process.

Next to the storage cells, a sense amplifier with positive feedback structure, as shown in Figure 55.7, is the most important component in a memory chip to amplify the small readout signal in the readout process. The input and output nodes of the differential positive feedback sense amplifier are connected to the bit-lines BL and $\overline{BL}$. The small readout signal appearing between BL and $\overline{BL}$ is detected by the differential sense amplifier and amplified to a full-voltage swing at BL and $\overline{BL}$. For example, if the DRAM memory cell in BL has a stored data "1", then a small positive voltage $\Delta V(1)$ will be generated and added to the bit-line BL voltage after the readout process. The voltage in the bit-line BL will be $\Delta V(1) + V_{DD}/2$. In the same time, the bit-line $\overline{BL}$ will keep its previous precharged voltage level, which is precharged to $V_{DD}/2$. Thus, the small positive voltage $\Delta V(1)$ appears between BL and $\overline{BL}$, with V_{BL} higher than $V_{\overline{BL}}$, immediately after the readout process. It is amplified by the differential sense amplifier. The waveforms of V_B before and after activating the sense amplifier are shown in Figure 55.8. After the sensing and restoring operations, the voltage V_{BL} rises to V_{DD}, and the voltage $V_{\overline{BL}}$ falls to 0 V. The output at BL is then sent to the DRAM output pin.

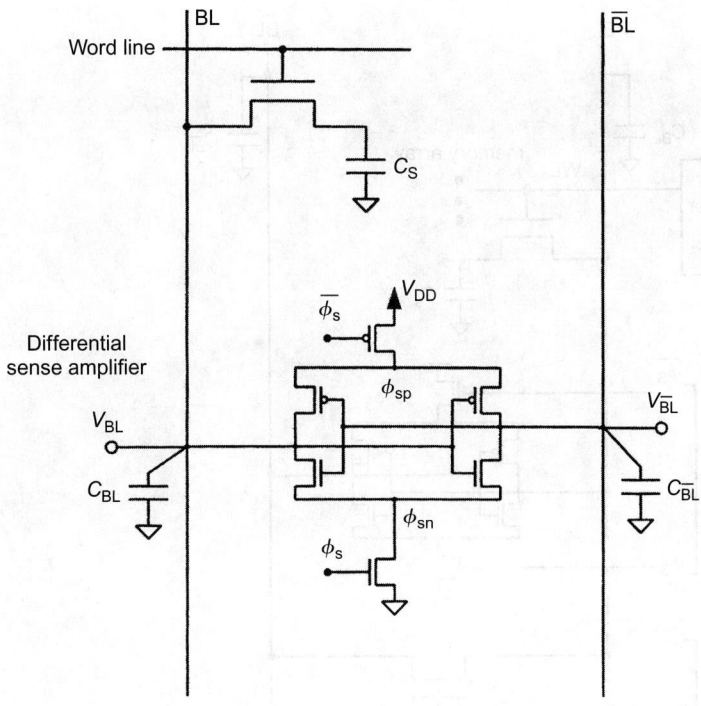

FIGURE 55.7 A differential sense amplifier connected to the bit-line.

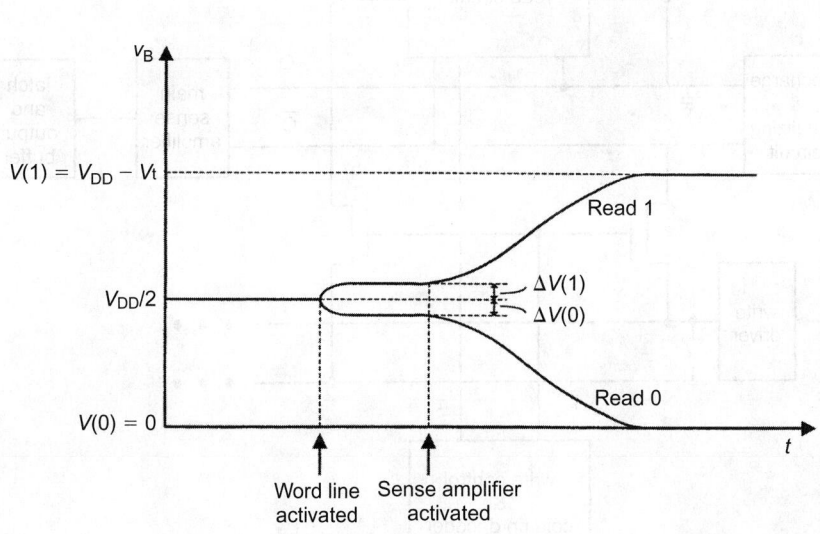

FIGURE 55.8 Timing waveform of V_B.

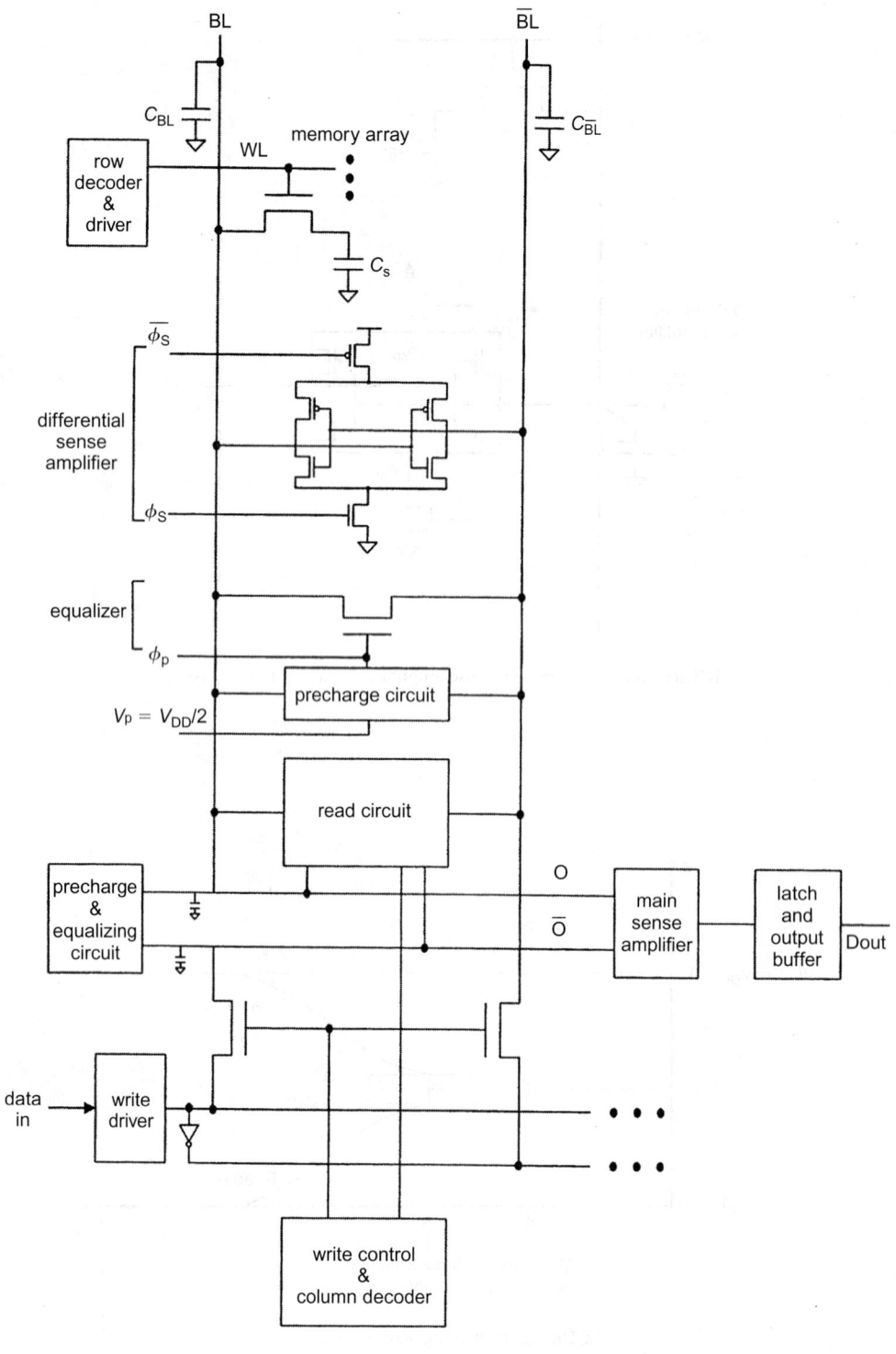

FIGURE 55.9(a) Schematic circuit diagram of DRAM.

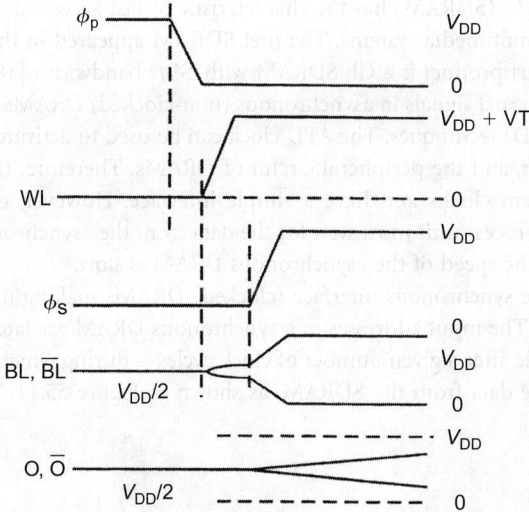

FIGURE 55.9(b) READ operation waveforms.

The various circuits for read, write precharge, and equalization function are shown in Figure 55.9. The sequence of the read operation is performed as follows.

1. Initially, both the bit-lines BL and $\overline{\text{BL}}$ are precharged to $V_{DD}/2$ and equalized before the data readout process. The precharge and equalizer circuits are activated by rising the control signal Φ_p. This will cause the bit-lines BL and $\overline{\text{BL}}$ to be at equal voltage. The control signal Φ_p goes low after the precharge and equalization.

2. The signal WL is selected by the row decoder. It goes up to connect the storage cell to the bit-lines BL and $\overline{\text{BL}}$. A small voltage difference then appears between the bit-lines. The voltage level of the word-line signal WL can be greater than V_{DD} to overcome the threshold voltage drop of the n-channel MOSFET transistor. Thus, the stored voltage level of data "1" at the memory cell can be raised to V_{DD}.

3. Once a small voltage difference is generated between the bit-lines BL and $\overline{\text{BL}}$ by the storage cell, the differential sense amplifier is turned on by pulsing the sense control signal Φ_s high and the sense control signal $\overline{\Phi_s}$ low. Then, the small voltage difference is amplified by the differential sense amplifier. The voltage levels in BL and $\overline{\text{BL}}$ will quickly move to V_{DD} or 0 V by the regenerative action of the positive feedback operation in the differential sense amplifier.

4. After the readout sensing and restoring operations, the voltage levels of the bit-lines have a full voltage swing. Then the differential voltage levels at the bit-lines are read out to the differential output lines O and $\overline{O}$, through a read circuit. A main sense amplifier is used to read and to amplify the output-lines. After these processes, the output data is selected and transferred to the output buffer.

In the write mode, the write control signal $\overline{WRITE}$ is activated. Selected bit-lines BL and $\overline{\text{BL}}$ are connected to a pair of input data controlled by the write control and write driver. The write circuit drives the voltage levels at the bit-lines to V_{DD} or 0 V, and the data are transferred to the DRAM cell when access transistor is turned on.

55.5 Synchronous (Clocked) DRAMs

The application of multimedia is a very hot topic nowadays, and the multimedia systems require high speed and large memory capacity to improve the quality of data processing. Under this trend, high density, high bandwidth, and fast access time are the key requirements of future DRAMs.

The synchronous DRAM (SDRAM) has the characteristic of fast access speed, and is widely used for memory application in multimedia systems. The first SDRAM appeared in the 16-Mb generation, and the current state-of-the-art product is a Gb SDRAM with GB/s bandwidth [10–14].

Conventionally, the internal signals in asynchronous (non-clocked) DRAMs are generated by "address transition detection" (ATD) techniques. The ATD clock can be used to activate the address decoder and driver, the sense amplifier, and the peripheral circuit of DRAMs. Therefore, the asynchronous DRAMs require no external system clocks and have a simple interface. However, during the asynchronous DRAM access cycle, the process unit must wait for the data from the asynchronous DRAM, as shown in Figure 55.10. Therefore, the speed of the asynchronous DRAM is slow.

On the other hand, the synchronous interface (clocked) DRAMs making it under the control of the edge of the system clock. The input addresses of a synchronous DRAM are latched into the DRAM, and the output data is available after a given number of clock cycles—during which the processor unit is free and does not wait for the data from the SDRAM, as shown in Figure 55.11. The block diagram of an

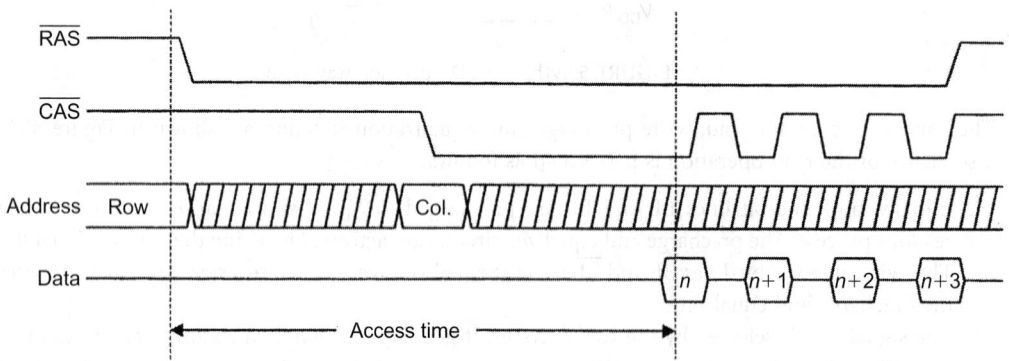

FIGURE 55.10   Read cycle timing diagram for asynchronous DRAM.

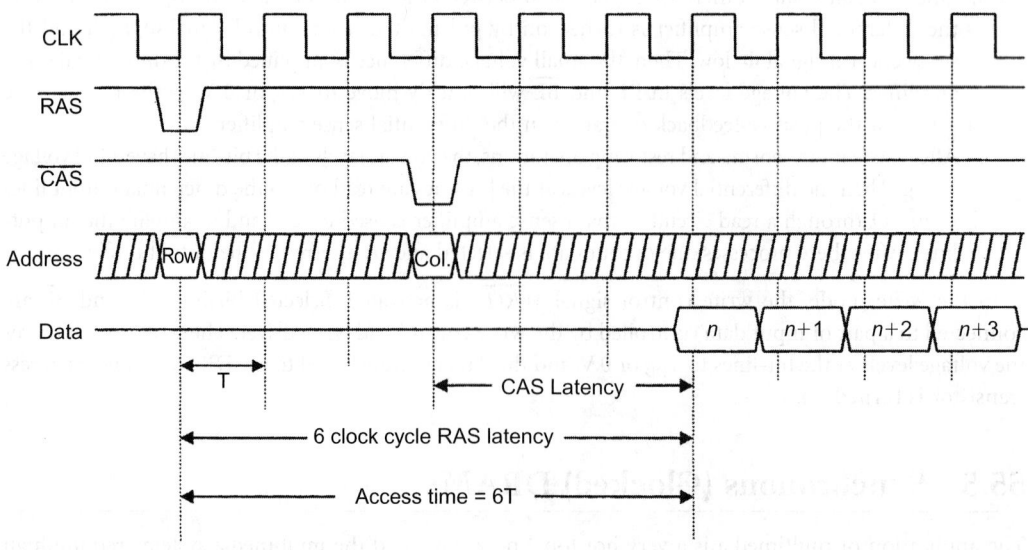

FIGURE 55.11   Read cycle timing diagram for synchronous DRAM.

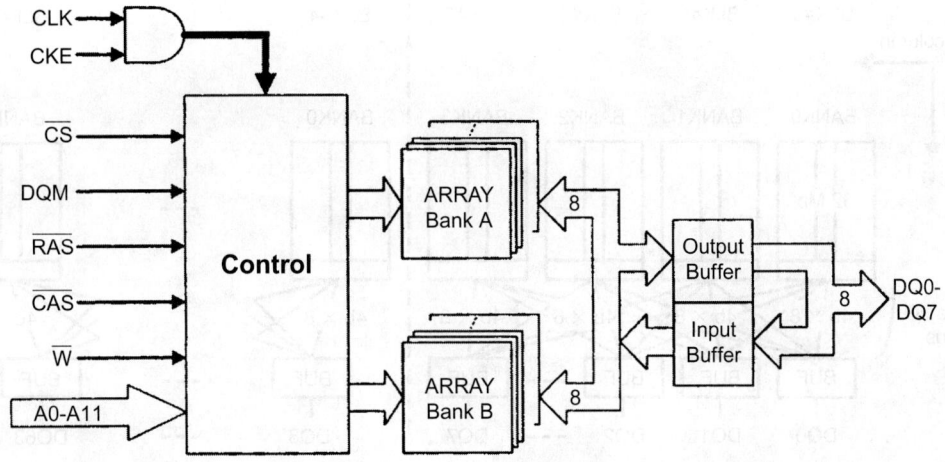

FIGURE 55.12 Block diagrams of a synchronous DRAM.

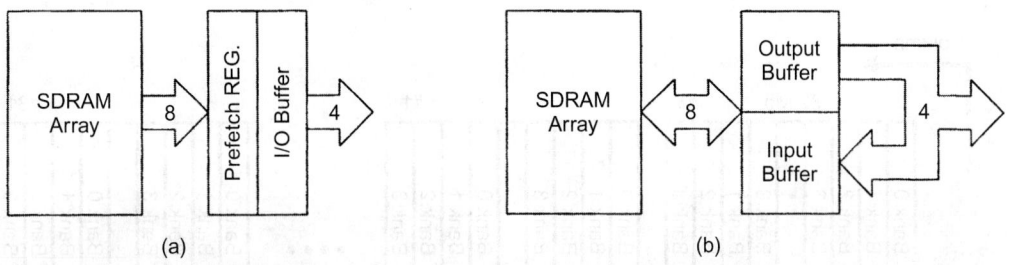

FIGURE 55.13 Block diagrams of two types of synchronous DRAM output: (a) prefetch (b) pipelined.

SDRAM is shown in Figure 55.12. With the synchronous interface scheme, the effective operation speed of a given system is improved.

55.6 Prefetch and Pipelined Architecture in SDRAMs

The system clock activates the SDRAM architecture. In order to speed up the average access time, it is possible to use the system clock to store the next address in the input latch or to be sequentially clocked out for each address access output from the output buffer, as shown in Figure 55.13 [15].

During the read cycle of the prefetch SDRAM, more than one data word is fetched from the memory array and sent to the output buffer. Using the system clock to control the prefetch register and buffer, multiple words of data can be sequentially clocked out for each address access. As shown in Figure 55.13, the SDRAM has a 6-clock-cycle RAS latency to prefetch 4-bit data.

55.7 Gb SDRAM Bank Architecture

To consider the Gb SDRAM realization, the chip layout and bank/data bus architecture is important for data access. Figure 55.14 shows the conventional bank/data bus architecture of 1-Gb SDRAM [16]. It contains 64 DQ pins, 32×32-Mb SDRAM blocks, and four banks; and they all prefetch 4 bits. During the read cycle,

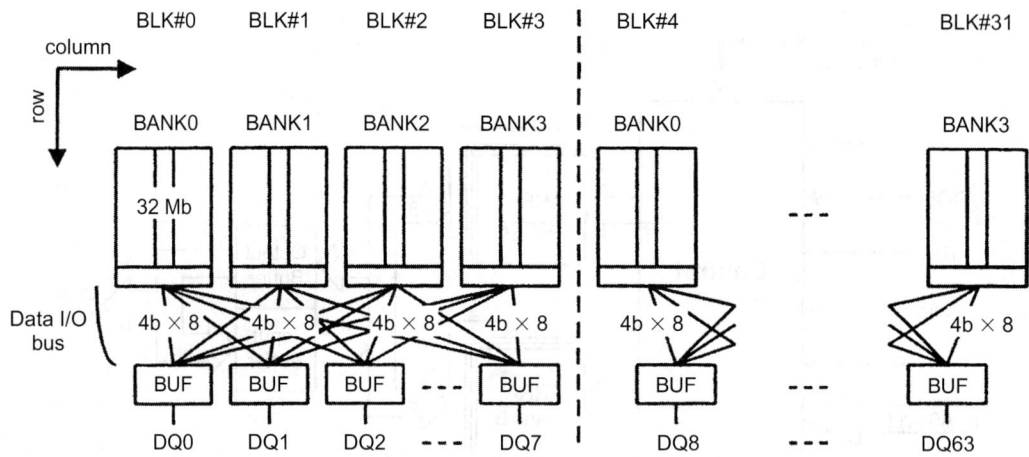

FIGURE 55.14 1-Gb SDRAM bank/data bus architecture.

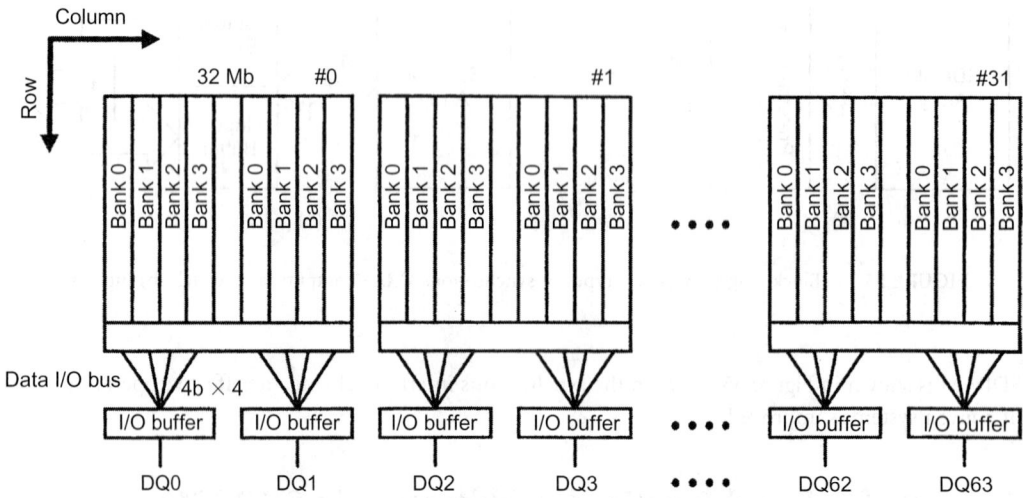

FIGURE 55.15 1-Gb SDRAM D-bank architecture.

the eight 32-Mb DRAM blocks of one bank are accessed simultaneously. The 256-bit data is accessed to the 64 DQ pins and 4 bits are prefetched. In an activated 32-Mb array block, 32-bit data is accessed and associated with eight specific DQ pins. Therefore, it requires a data I/O bus switching circuit between the 32-Mb SDRAM bank and the eight DQ pins. It makes the data I/O bus more complex, and the access time is slower.

In order to simplify the bus structure, the distributed bank (D-bank) architecture is proposed as shown in Figure 55.15. The 1-Gb SDRAM is implemented by 32 × 32-Mb distributed banks. A 32-Mb distributed bank contains two 16-Mb memory arrays as shown in Figure 55.16. The divided word-line technique is used to activate the segment along the column direction. Using this scheme, each of the eight 2-Mb segments is selectively activated; sense amplifiers of one of the eight segments are activated; and all the 16-K sense amplifiers are activated simultaneously. As compared with the conventional architecture, the distributed bank architecture has a much simplified data I/O bus structure.

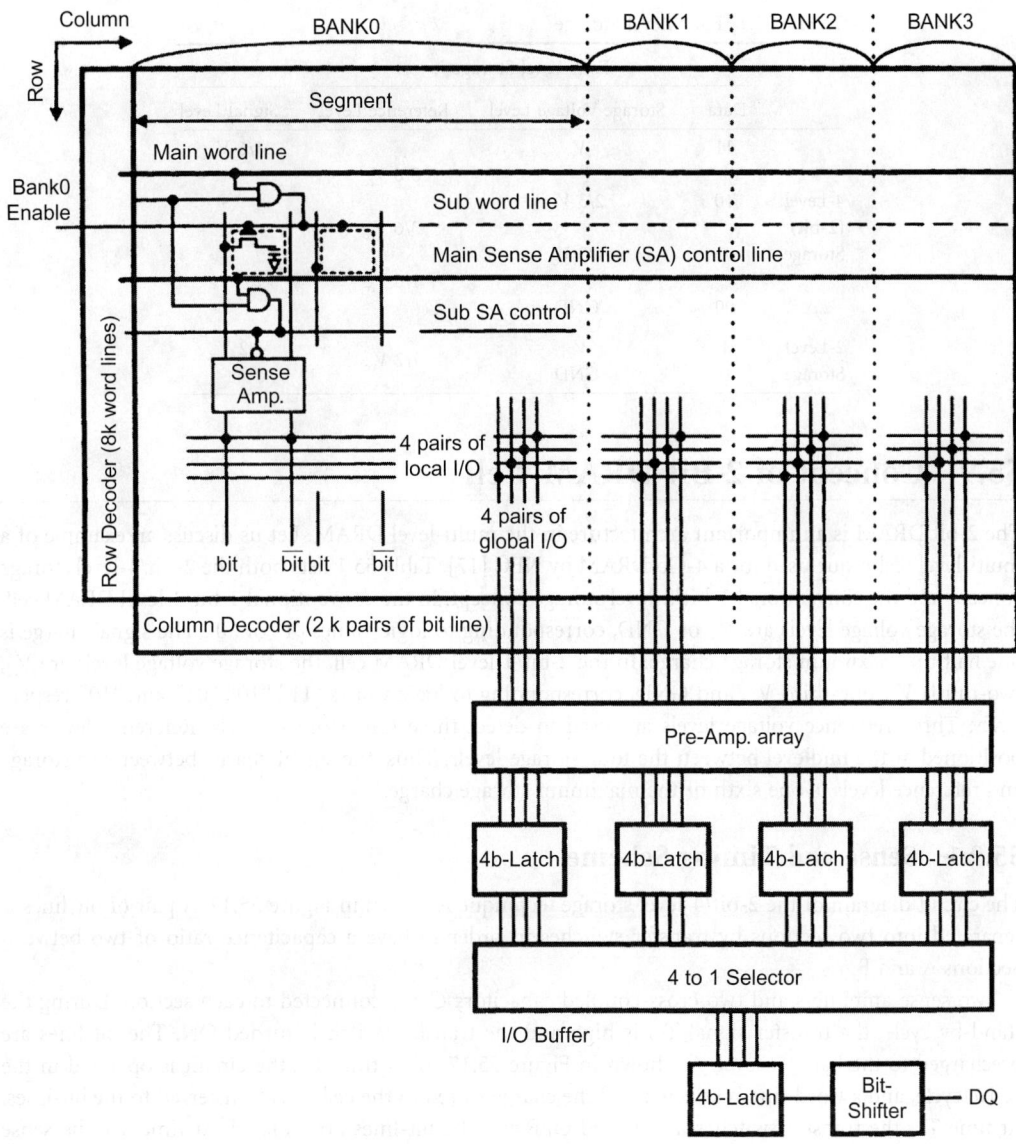

FIGURE 55.16 16-Mb memory array for architecture D-bank.

55.8 Multi-Level DRAM

In modern application-specific IC (ASIC) memory designs, there are some important items—memory capacity, fabrication yield, and access speed—that need to be considered. The memory capacity required for ASIC application has been increasing very rapidly, and the bit-cost reduction is one of the most important issues for file application DRAMs. In order to achieve high yield, it is important to reduce the defect-sensitive area on a chip.

The multi-level storage DRAM technique is one of the circuit technologies that can reduce the effective cell size. It can store multiple voltage levels in a single DRAM cell. For example, in a four-level system, each DRAM cell corresponds to 2-bit data of "11", "10", "01", and "00". Thus, the multi-level storage technique can improve the chip density and reduce the defect-sensitive area on a DRAM chip, and it is one of the solutions to the "density and yield" problem.

TABLE 55.1 Four-Level Storage

		Four-Level Storage		
	Data	Storage Voltage Level	Reference Level	Signal Level
4-Level (2-bit) Storage	11	V_{cc}		$1/6\ V_{cc}$
			$5/6\ V_{cc}$	
	10	$2/3\ V_{cc}$		
			$3/6\ V_{cc}$	
	01	$1/3\ V_{cc}$		
			$1/6\ V_{cc}$	
	00	GND		
2-Level Storage	1	V_{cc}	$1/2\ V_{cc}$	$1/2\ V_{cc}$
	0	GND		

55.9   Concept of 2-Bit DRAM Cell

The 2-bit DRAM is an important architecture in the multi-level DRAM. Let us discuss an example of a multi-level technique used for a 4-Gb DRAM by NEC [17]. Table 55.1 lists both the 2-bit/4-level storage concept and the conventional 1-bit/2-level storage concept. In the conventional 1-bit/2-level DRAM cell, the storage voltage levels are V_{cc} or GND, corresponding to logic values "1" or "0". The signal charge is one half the maximum storage charge. In the 2-bit/4-level DRAM cell, the storage voltage levels are V_{cc}, two-thirds V_{cc}, one-third V_{cc}, and GND, corresponding to logic values "11", "10", "01", and "10", respectively. Three reference voltage levels are used to detect these four storage levels. Reference levels are positioned at the midlevel between the four storage levels. Thus, the signal charge between the storage and reference levels is one sixth of the maximum storage charge.

55.9.1   Sense and Timing Scheme

The circuit diagram of the 2-bit/4-level storage technique is shown in Figure 55.17. A pair of bit-lines is separated into two sections by transfer switches in order to have a capacitance ratio of two between Sections A and B.

Two sense amplifiers and two cross-coupled capacitors C_c are connected to each section. During the stand-by cycle, the transfer signal TG is high and the transfer switch is turned ON. The bit-lines are precharged to the half-V_{cc} level. As shown in Figure 55.17(b), at time T1, the circuit is operated in the active cycle, and a word-line is selected and the charge stored in the cell C_s is transferred to the bit-lines. At time T2, the transfer switches are turned OFF and the bit-lines are isolated. At time T3, the sense amplifier in Section A is activated and the bit-lines in Section A are driven to V_{cc} and GND, depending on the stored data. The amplified data in Section A is the most significant bit (MSB) of the stored data because the reference level is half-V_{cc}.

At the same time interval, the MSB is transferred to the bit-lines in Section B through a cross-coupled capacitor C_c. It can change the bit-line level in Section B for subsequent least significant bit (LSB) sensing. At time T4, the sense amplifier in section B is activated and the LSB is sensed. At time T5, the transfer switch is turned ON, the charge on each bit-line is shared, and the read-out data is restored to the memory cell.

55.9.2   Charge-Sharing Restore Scheme

Table 55.2 lists the restored level generated by the charge-sharing restore scheme. The MSB is latched in Section A, and the LSB is latched in Section B. The capacitance ratio between Sections A and B is 2. The charge of the MSB and the charge of the LSB are combined on the bit-line, and the restore level $V_{restore}$ is generated.

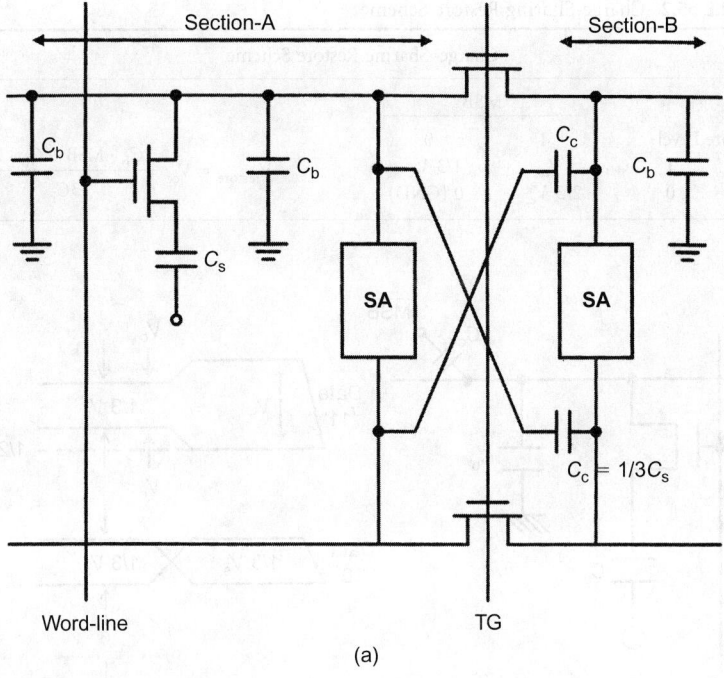

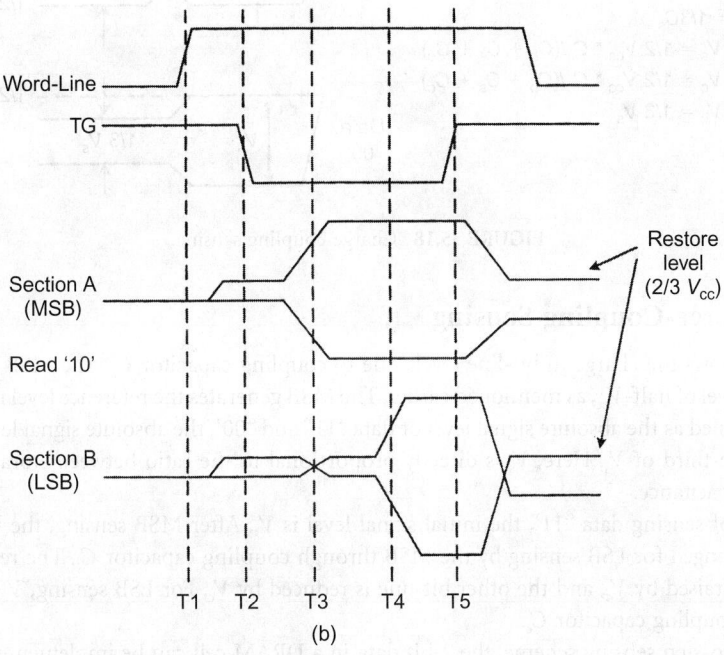

FIGURE 55.17 Principle of sense and restore: (a) circuit diagram, and (b) timing diagram.

TABLE 55.2 Charge-Sharing Restore Scheme

		Charge–Sharing Restore Scheme		
		MSB		
Restore Level		1	0	
LS	1	V_{cc}	$1/3\ V_{cc}$	$V_{restore} = V_{cc} \dfrac{2C_b \cdot MSB + C_b \cdot LSB}{3C_b}$
B	0	$2/3\ V_{cc}$	0 (GND)	

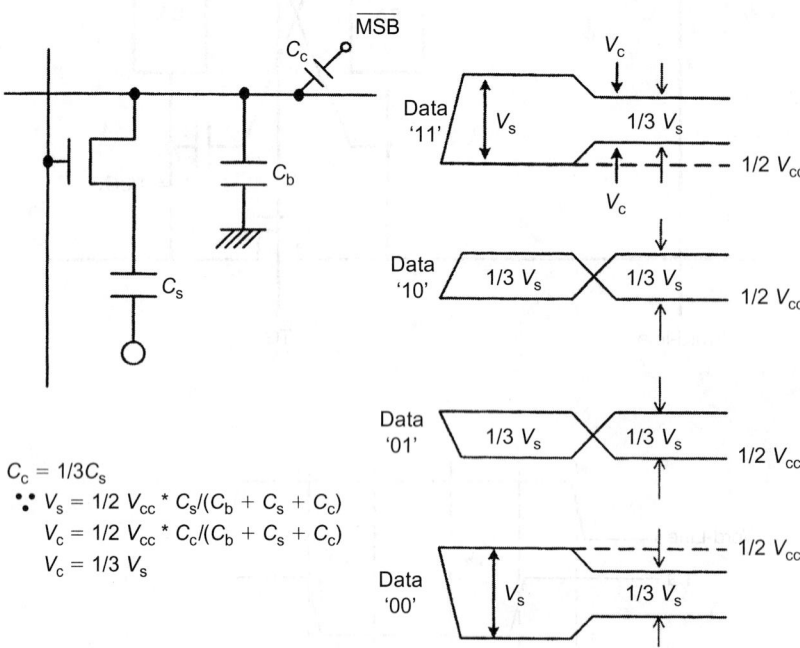

$$C_c = 1/3C_s$$
$$\therefore\ V_s = 1/2\ V_{cc} * C_s/(C_b + C_s + C_c)$$
$$V_c = 1/2\ V_{cc} * C_c/(C_b + C_s + C_c)$$
$$V_c = 1/3\ V_s$$

FIGURE 55.18 Charge-coupling sensing.

55.9.3 Charge-Coupling Sensing

Figure 55.18 shows the charge in bit-line levels due to coupling capacitor C_c. The MSB is sensed using the reference level of half-V_{cc}, as mentioned earlier. The MSB generates the reference level for LSB sensing. When V_s is defined as the absolute signal level of data "11" and "00", the absolute signal level of data "10" and "01" is one-third of V_s. Here, V_s is directly proportional to the ratio between storage capacitor C_s and bit-line capacitance.

In the case of sensing data "11", the initial signal level is V_s. After MSB sensing, the bit-line level in Section B is changed for LSB sensing by the MSB through coupling capacitor C_c. The reference bit-line in Section B is raised by V_c, and the other bit-line is reduced by V_c. For LSB sensing, V_c is one-third of V_s due to the coupling capacitor C_c.

Using the two-step sensing scheme, the 2-bit data in a DRAM cell can be implemented.

References

1. Sekiguchi., T. et al., "An Experimental 220MHz 1Gb DRAM," *ISSCC Dig. Tech. Papers*, pp. 252–253, Feb. 1995.
2. Sugibayashi, T. et al., "A 1Gb DRAM for File Applications," *ISSCC Dig. Tech. Papers*, pp. 254–255, Feb. 1995.

3. Murotani, T. et al., "A 4-Level Storage 4Gb DRAM," *ISSCC Dig. Tech. Papers*, pp. 74–75, Feb. 1997.

4. Furuyama, T. et al., "An Experimental 2-bit/Cell Storage DRAM for Macrocell or Memory-on-Logic Application," *IEEE J. Solid-State Circuits*, vol. 24, no. 2, pp. 388–393, April 1989.

5. Ahlquist, C.N. et al., "A 16k 384-bit Dynamic RAM," *IEEE J. Solid-State Circuits*, vol. SC-11, no. 3, Oct. 1976.

6. El-Mansy, Y. et al., "Design Parameters of the Hi-C SRAM cell," *IEEE J. Solid-State Circuits*, vol. SC-17, no. 5, Oct. 1982.

7. Lu, N.C.C., "Half-V_{DD} Bit-Line Sensing Scheme in CMOS DRAM's," *IEEE J. Solid-State Circuits*, vol. SC-19, no. 4, Aug. 1984.

8. Lu, N.C.C., "Advanced Cell Structures for Dynamic RAMs," *IEEE Circuits and Devices Magazine*, pp. 27–36, Jan. 1989.

9. Mashiko, K. et al., "A 4-Mbit DRAM with Folded-Bit-Line Adaptive Sidewall-Isolated Capacitor (FASIC) Cell," *IEEE J. Solid-State Circuits*, vol. SC-22, no. 5, Oct. 1987.

10. Prince, B., et al., "Synchronous Dynamic RAM," *IEEE Spectrum*, p. 44, Oct. 1992.

11. Yoo, J.-H. et al., "A 32-Bank 1Gb DRAM with 1GB/s Bandwidth," *ISSCC Dig. Tech. Papers*, pp. 378–379, Feb. 1996.

12. Nitta, Y. et al., "A 1.6GB/s Data-Rate 1Gb Synchronous DRAM with Hierarchical Square-Shaped Memory Block and Distributed Bank Architecture," *ISSCC Dig. Tech. Papers*, pp. 376–377, Feb. 1996.

13. Yoo, J.-H. et al., "A 32-Bank 1 Gb Self-Strobing Synchronous DRAM with 1 Gbyte/s Bandwidth," *IEEE J. Solid-State Circuits*, vol. 31, no. 11, pp. 1635–1644, Nov. 1996.

14. Saeki, T. et al., "A 2.5-ns Clock Access, 250-MHz, 256-Mb SDRAM with Synchronous Mirror Delay," *IEEE J. Solid-State Circuits*, vol. 31, no. 11, pp. 1656–1668, Nov. 1996.

15. Choi, Y. et al., "16Mb synchronous DRAM with 125Mbyte/s data rate," *IEEE J. Solid-State Circuits*, vol. 29, no. 4, April 1994.

16. Sakashita, N. et al., "A 1.6GB/s Data-Rate 1-Gb Synchronous DRAM with Hierarchical Square Memory Block and Distributed Bank Architecture," *IEEE J. Solid-State Circuits*, vol. 31, no. 11, pp. 1645–1655, Nov. 1996.

17. Okuda, T. et al., "A Four-Level Storage 4-Gb DRAM," *IEEE J. Solid-State Circuits*, vol. 32, no. 11, pp. 1743–1747, Nov. 1997.

18. Prince, B., *Semiconductor Memories*, 2nd edition, John Wiley & Sons, 1993.

19. Prince, B., *High Performance Memories New Architecture DRAMs and SRAMs Evolution and Function*, 1st edition, Betty Prince, 1996.

20. *Toshiba Applications Specific DRAM Databook*, D-20, 1994.

56

Content-Addressable Memory

Chi-Sheng Lin and
Bin-Da Liu
National Cheng Kung University

CONTENTS

56.1 Introduction

In the ordinary memory circuit designs, such as DRAMs and SRAMs, the memory devices utilize write cycle to store data and read cycle to retrieve the stored data by addressing specific memory location, called an address. In other words, each read cycle and write cycle can access only one specific memory location which is indicated by an address. As a result, the data access of ordinary memory devices is operated in a sequential manner. In the high-speed data search applications, for instance, Internet routers [1–5], image processing [6–8], and pattern recognitions [9–11], the time required for finding data stored in memory array is as short as possible to achieve high-speed data search performance. Because of the sequential data access manner, the data search performance of the ordinary memory devices relies on fast memory bandwidth that cannot be well applied to search-intensive applications. If a memory device can provide useful function, called search function, that compares desired search data with all the data stored in the memory array simultaneously, then the data search performance of the memory device will be improved greatly. To realize the parallel data comparison operation, the data stored in the memory array have to be identified for access by the contents of the stored data themselves rather than by their addresses. Memory that is accessed in this way is called content-addressable memory (CAM).

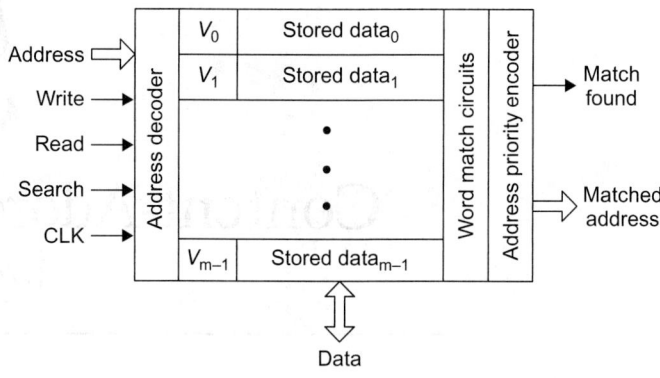

FIGURE 56.1   Functional block diagram of a CAM.

56.2   CAM Architecture

A general CAM architecture usually consists of the data memory with valid bit field, the address decoder, the word match circuit, and the address priority encoder, as shown in Figure 56.1 [12]. The memory organization of the CAM is composed of the data memory and the valid bit field, where the valid bit field indicates the availability of data stored in the related memory location. The address decoder is used to indicate the specific memory location for the read and write operations. To realize the data search operation, the word match circuit is designed to compare the search data with the data stored in the related memory location. In the word match circuit, the data matching is performed by comparing every bit in the stored data with the search data. If every bit in the stored data matches with every corresponding bit in the search data, then the stored data matches the search data, else the stored data mismatches the search data. In the CAM architecture, all word match circuits are operated in parallel to speed up data search operation, and the parallel data comparison may search out more than one matched data stored in CAM array during each data search operation. To output the address of the best-matched stored data, the address priority encoder is used to choose the highest priority address among those matched stored data during each data search operation. In addition, one additional 'Match Found' output signal will be required, since it is possible that none of the stored data matches the search data.

Generally, CAMs have three operation modes: read, write, and search modes. The read and write modes access and manipulate data in the CAM array in the same way as an ordinary memory. The benefit of the CAM is realized in search operation mode to achieve powerful data search performance. In this operation, a desired search data is sent into CAM to compare with all valid data stored in CAM array simultaneously, and the highest priority address among those matched stored data is sent to the output port "Matched Address." Based on the parallel data comparison architecture, the CAM circuit performs large amounts of comparison operation during each data search operation for searching out the address of the best matched data stored in CAM array, the high-speed data search feature makes the CAM to be an appealing memory architecture for search-intensive applications.

56.3   CAM Cell Circuits

In the CAM architecture, the majority parts of this circuit are constructed by large amounts of CAM cell, each CAM cell is implemented by bit storage along with bit comparison circuit. In the CAM cell designs, the bit storage is utilized to store input data, and the bit comparison circuit is applied to compare a desired search data with the stored data. Two kinds of CAM cells are widely used in CAMs: binary CAM (BCAM) cell and ternary CAM (TCAM) cell. The conventional BCAM cell requires nine [12–14] or ten transistors [15], while the conventional TCAM cell requires seventeen transistors [16] in standard CMOS circuit design. Since more transistors result in a higher die area, the CAM cells have relatively low-memory density compared with standard memory cells, such as DRAMs and SRAMs. The low-memory density limits the circuit capacity for CAM applications. In most CAM applications, the required CAM size is smaller than that of ordinary memory devices, as a result of the required search tables in these CAM applications are quite small. In the following sections, both BCAM and TCAM cell designs are introduced.

56.3.1 BCAM Cells

In the CAM circuit design, the BCAM cell is typically constructed of nine- or 10-transistor structures as shown in Figure 56.2. The nine-transistor BCAM cell circuit as shown in Figure 56.2(a) consists of an ordinary six-transistor SRAM cell to store the input data and an XOR-type comparison circuit M1 and M2 with a resultant transistor M3 to drive a match-line (ML). During the write cycle, the input data BL and its complement BLB are placed on the bit-lines. Then the word-line (WL) is driven forcing the stored data Q to be replaced by the input data BL. To read out the stored data Q, the word-line WL is driven placing the stored data Q and its complement QB on the bit-lines BL and BLB, respectively. Then a read sense amplifier is used to detect and amplify the small level difference between two bit-lines BL and BLB. As the result, the read and write operations of the BCAM cell are similar to that of ordinary memory cells. In most CAM applications, since the required CAM size is quite small, the parasitic capacitance of each bit-line is smaller than that of ordinary memory devices. For this reason, the read and write operations of CAM are faster than that of ordinary memory devices.

In the nine-transistor BCAM cell design, the comparison circuits M1 and M2 are designed as an XOR logic circuit to realize bit comparison operation. During search cycle, the search data BL is sent into this cell to compare with the stored data Q using the comparison circuit M1 and M2. In this cycle, if the search data BL is equivalent to the stored data Q, then the comparison circuit outputs a logic 0 to turn off the resultant transistor M3, and the match-line ML is floating indicating that the data comparison is matched. Otherwise, if the search data BL is different from the stored data Q, then the comparison circuit outputs a logic 1 to turn on the resultant transistor M3, and the match-line ML is falling down to VSS indicating that the data comparison is mismatched. Table 56.1 shows the truth table of the nine-transistor BCAM cell. As summarized in this table, if the data comparison is mismatched then the match-line ML is VSS (the resultant transistor M3 is turned on). Otherwise, the match-line ML is floating (the resultant transistor M3 is turned off). In this BCAM cell design, since the comparison circuit (M1 and M2) is implemented using pass transistor logic (PTL) XOR circuit, the output potential of the comparison circuit M1 and M2 is less than or equal to (VDD − VSS − Vtn), where Vtn is the threshold voltage of NMOS transistor [17]. Therefore, the operating voltage (VDD − VSS) must be greater than 2Vtn for turning on

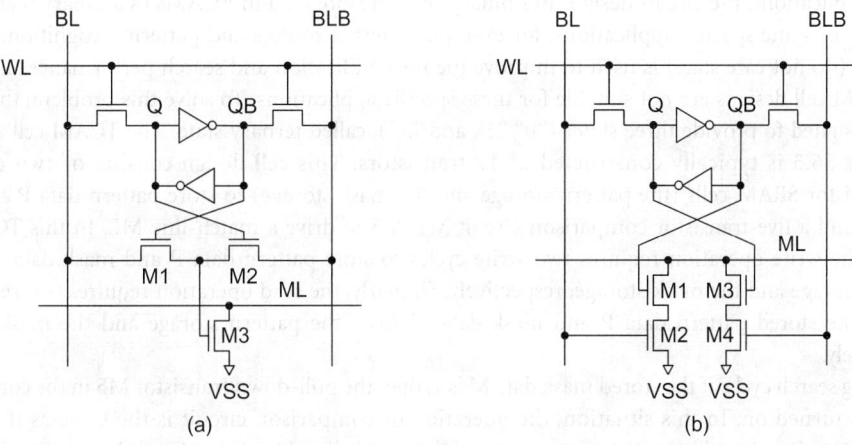

FIGURE 56.2 BCAM cells: (a) nine-transistor BCAM cell; (b) 10-transistor BCAM cell.

TABLE 56.1 Truth Table of the Nine-Transistor BCAM Cell

Search Data (BL)	Stored Data (Q)	Resultant Transistor (M3)	Match-Line (ML)	Comparison Result
0	0	OFF	Floating	Match
0	1	ON	VSS	Mismatch
1	0	ON	VSS	Mismatch
1	1	OFF	Floating	Match

the resultant transistor M3. Based on the nine-transistor BCAM cell design, if the circuit operates in high operating voltage condition, then the falling time of the match-line ML can be very short, as a result the parasitic capacitance of the match-line ML consists only of the drain capacitance of the resultant transistor M3, and the longest pull-down path from the match-line ML to ground node VSS contains only one transistor M3. For this reason, the cell design achieves high-speed data search performance. However, if the circuit operates in low operating voltage condition, then the match-line ML requires a long falling time, since the weak turn-on transistor M3 limits the discharge current from the match-line ML to the ground node VSS. Therefore, the cell design has low data search performance. As in the above-mentioned circuit operation, the operating voltage of nine-transistor BCAM cell design cannot be reduced efficiently for high-speed data search applications.

To reduce the required operating voltage in BCAM cell design, the 10-transistor BCAM cell as shown in Figure 56.2(b) is a better choice than the nine-transistor BCAM cell. In this cell design, the function is the same as that of the nine-transistor BCAM cell. To reduce the operating voltage, the comparison circuit with resultant transistor in the 10-transistor BCAM cell is implemented using the NMOS part of CMOS XNOR gate (M1–M4) with inputs BL and Q as shown in this figure. Since the required operating voltage of CMOS logic circuit is lower than that of PTL logic circuit, the 10-transistor BCAM cell design has low operating voltage feature. However, in this cell design, since the parasitic capacitance of the match-line ML contains at least two drain capacitances of M1 and M3 transistors, the large parasitic capacitance not only reduces the switching speed of the match-line ML but also consumes large amounts of dynamic power dissipation during data search operation. For this reason, the 10-transistor BCAM cell design is not suitable for high-speed and low-power data search applications.

56.3.2 TCAM Cells

In the BCAM cell designs, the stored data has two states ("0" and "1") called binary state. In most data search applications, the circuit design uses binary state to store data in BCAMs as a binary search table. However, in some specific applications, for example, Internet routers and pattern recognition, an extra state "X" (do not care state) is used to improve memory utilization and search performance. Therefore, the BCAM cell designs are not suitable for these specific applications. To solve this problem, the TCAM cell is designed to provide three states ("0", "1", and "X"), called ternary state. The TCAM cell as shown in Figure 56.3 is typically constructed of 17 transistors. This cell design consists of two ordinary six-transistor SRAM cells (the pattern storage and the mask storage) to store pattern data P and mask data M, and a five-transistor comparison circuit M1–M5 to drive a match-line ML. In this TCAM cell design, the write operation requires two write cycles to store pattern data P and mask data M in the pattern storage and the mask storage, respectively. Similarly, the read operation requires two read cycles to read the stored pattern data P and mask data M from the pattern storage and the mask storage, respectively.

During search cycle, if the stored mask data M is 1, then the pull-down transistor M5 in the comparison circuit is turned on. In this situation, the operation of comparison circuit is the same as that of the 10-transistor BCAM cell, and the output state of the match-line ML depends on the comparison result of both the search data BL and the stored data P. Otherwise, if the stored mask data M is "0," then the

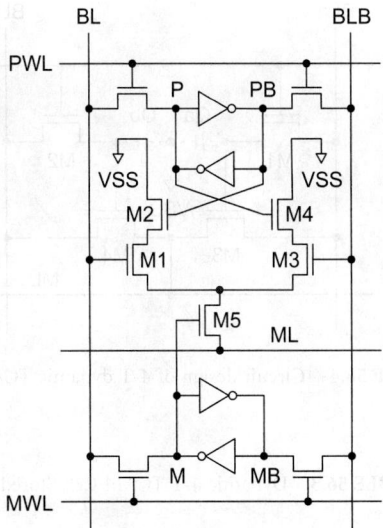

FIGURE 56.3 Seventeen-transistor TCAM cell.

TABLE 56.2 Truth Table of the 17-Transistor TCAM Cell

Search Data (BL)	Stored State	Stored Data (M, P)	Match-Line (ML)	Comparison Result
0 or 1	X	0, 0 or 1	Floating	Match
0	0	1, 0	Floating	Match
0	1	1, 1	VSS	Mismatch
1	0	1, 0	VSS	Mismatch
1	1	1, 1	Floating	Match

pull-down transistor M5 in the comparison circuit is turned off. In this situation, the output state of the match-line ML is floating, no matter what the search data BL and the stored data P are. In other words, the data comparison is always matched when the stored mask M data is "0." Table 56.2 shows the truth table of the 17-transistor TCAM cell. As the summarized in this table, the stored state of TCAM cell has "0", "1", and "X" states, where the "X" state makes the stored data always matched with the search data, no matter what the search data BL and the stored data P are.

Since more transistors result in more hardware cost, the 17-transistor TCAM cell is not suitable for low-cost applications. To reduce hardware cost, a dynamic four-transistor (4-T) TCAM cell design is provided [18]. The four-transistor dynamic TCAM cell, as shown in Figure 56.4, consists of a dynamic storage M1 and M2, and a comparison circuit M3 and M4. During the write cycle, the input data BLa and BLb are, respectively, stored in the nodes Qa and Qb by driving the word-line WL. In this cycle, the transistors M1 and M2 are turned on, and the input data BLa and BLb are stored as the charge on their related node capacitances. When the stored data is "1", the related node capacitance is charged to VDD − VSS − Vthn. When the stored data is "0", the related node capacitance is discharged to VSS. During search cycle, the transistors M3 and M4 are arranged in a PTL-type XOR logic circuit to perform a comparison circuit. Table 56.3 shows how the three states are stored in the dynamic 4-T TCAM cell. In the dynamic 4-T TCAM cell design, although the circuit has only four transistors that reduce hardware cost and power dissipation, the dynamic circuit operation requires complex circuit controller (e.g., dynamic refresh circuit) and careful circuit design to prevent some dynamic circuit design problems. Therefore, the dynamic 4-T TCAM cell design is only provided for some cost-sensitive CAM applications.

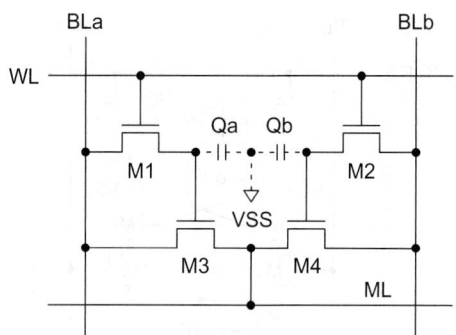

FIGURE 56.4   Circuit design of 4-T dynamic TCAM cell.

TABLE 56.3   Dynamic 4-T TCAM Cell Stored States

Qa	Qb	Stored State
0	0	X
0	1	0
1	0	1
1	1	Not allowed

56.4   CAM Word Circuits

In the general CAM architecture, the word match circuit is designed using dynamic CMOS circuit design to improve data match performance and hardware cost. In the dynamic CMOS circuit design as shown in Figure 56.5, the circuit operation can be separated into two phases: precharge phase (CLK = VSS) and evaluation phase (CLK = VDD). In the precharge phase, owing to the input signal CLK being VSS, the pull-up transistor M1 is turned on and the pull-down transistor M2 is turned off. In this case, a charge current Ip is created by the pull-up transistor M1 to charge the parasitic capacitance CL, and the output node OUT is rising up to VDD. In the evaluation phase, as a result of the input signal CLK being VDD, the pull-up transistor M1 is turned off and the pull-down transistor M2 is turned on. In this case, the output node OUT is conditionally falling down to VSS depending on the function of the N-logic block. If the N-logic block contains a short path from the output node OUT to the pull-down transistor M2 during evaluation phase, then a discharge current Ie is created by the N-logic block and the pull-down transistor M2 to discharge the parasitic capacitance CL, and the output node OUT is falling down to VSS. Otherwise, the parasitic capacitance CL is floating and the output node OUT remains at VDD.

Based on the dynamic CMOS circuit design, the dynamic word match circuit as illustrated in Figure 56.6 is used to realize data match operation. With *n*-bit stored data, the dynamic word match circuit usually consists of the number of *n* CAM cells, the valid bit circuit (VBC), the dynamic circuit controller (DCC), and the match line sense amplifier (MLSA). The VBC is used to indicate the availability of data stored in the word match circuit and the DCC is utilized to control dynamic circuit operation. If the stored state in the VBC is invalid, then the match-line ML is falling down to VSS indicating that the stored data is invalid, and the stored data is always mismatched with the search data. Otherwise, the word match circuit performs a data match function. In the output stage, the MLSA is adopted to amplify the voltage swing of the match-line ML for improving the circuit speed. To identify the match result of the word match circuit, the comparison results of all CAM cells in each word match circuit are collected, and the

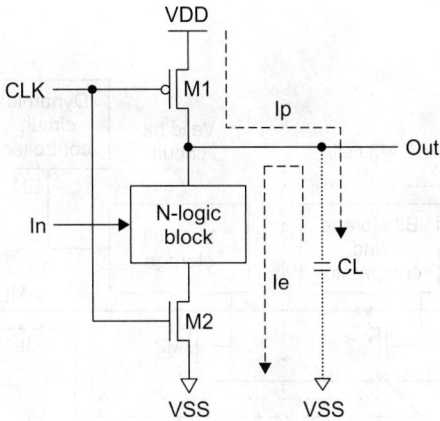

FIGURE 56.5 Block diagram of dynamic CMOS circuit.

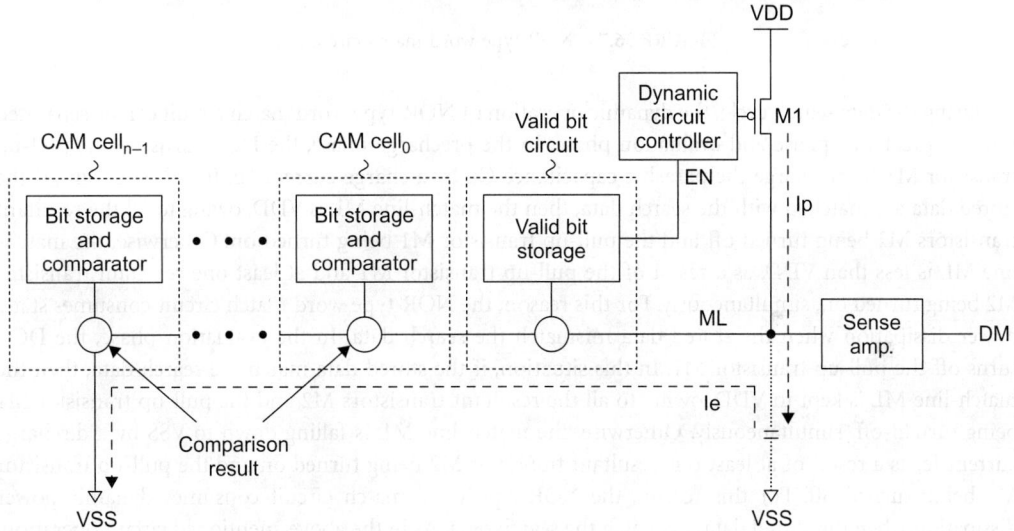

FIGURE 56.6 Dynamic word match circuit.

different collection approaches result in different circuit features. In the following three sections, some popular word match circuit designs are described.

56.4.1 NOR-Type Word Match Circuit

In the NOR-type word match circuit as shown in Figure 56.7 [12], this design collects the resultant transistors M2 of all CAM cells in parallel connection. In this circuit, if the stored data match the search data, then every bit in the search data matches with every corresponding bit in the stored data, else at least one bit in the stored data is mismatched with the corresponding bit in the search data. In addition, referring to the truth table of the nine-transistor BCAM cell shown in Table 56.1, if the comparison result is matched, then the resultant transistor M3 is turned on, else the resultant transistor M3 is turned off. According to the results mentioned above, if the search data match with the data stored in the NOR-type word match circuit, then all the resultant transistors M2 are turned off, else at least one resultant transistor M2 is turned on.

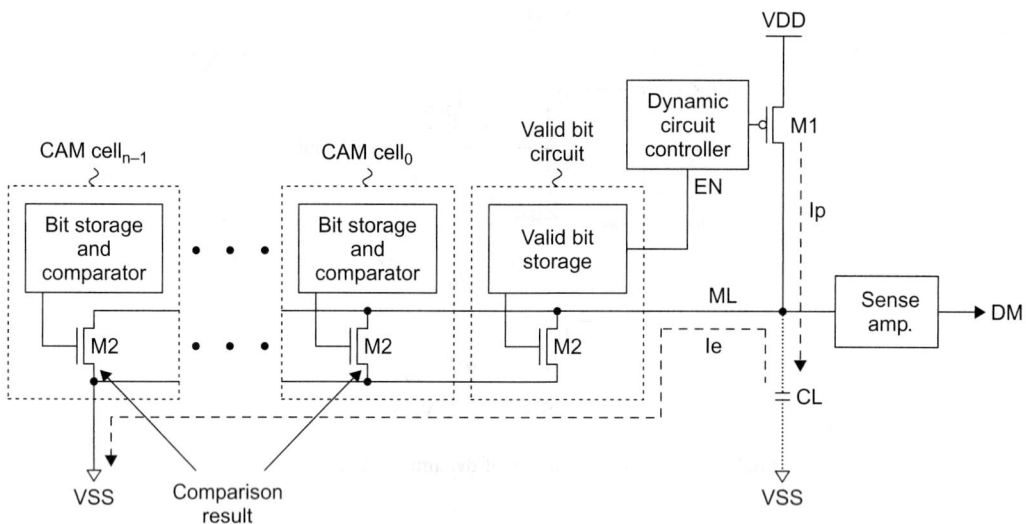

FIGURE 56.7   NOR-type word match circuit.

During the data search cycle, the dynamic operation of NOR-type word match circuit can be separated into the precharge phase and evaluation phase. In the precharge phase, the DCC turns on the pull-up transistor M1 to precharge the parasitic capacitance CL by a charge current Ip. In this situation, if the stored data are matched with the search data, then the match-line ML is VDD, owing to all the resultant transistors M2 being turned off and the pull-up transistor M1 being turned on. Otherwise, the match-line ML is less than VDD, as a result of the pull-up transistor M1 and at least one resultant transistor M2 being turned on, simultaneously. For this reason, the NOR-type word match circuit consumes static power dissipation when the stored data mismatch the search data. In the evaluation phase, the DCC turns off the pull-up transistor M1. In this situation, if the stored data match the search data, then the match-line ML is kept to VDD, owing to all the resultant transistors M2 and the pull-up transistor M1 being turned-off, simultaneously. Otherwise, the match-line ML is falling down to VSS by a discharge current Ie, as a result of at least one resultant transistor M2 being turned on and the pull-up transistor M1 being turned off. For this reason, the NOR-type word match circuit consumes dynamic power dissipation when the stored data mismatch the search data. As in the above-mentioned circuit operation, the NOR-type word match circuit consumes static and dynamic power dissipations simultaneously when the stored data mismatch the search data. Generally, in each data search cycle, almost all data stored in the CAM array are mismatched with the search data. Therefore, the NOR-type word match circuit consumes large amounts of static and dynamic search power dissipations.

In the NOR-type word match circuit, since the design collects the resultant transistors M2 of all CAM cells in parallel connection, the longest pull-down path from the match-line ML to the ground node VSS contains only one transistor M2 that reduces the discharge time of the match-line ML during the evaluation phase. Therefore, the NOR-type word match circuit achieves high-speed data search performance. However, the parallel connection of transistors M2 results in a large parasitic capacitance CL that consumes large dynamic search power dissipation when the stored data mismatch the search data.

56.4.2   NAND-Type Word Match Circuit

One of the most popular techniques for reducing search power dissipation in the word match circuit is the one using the NAND-type structure as shown in Figure 56.8 [14]. In the NAND-type word match circuit, this design collects the resultant transistors M2 of all the CAM cells in series connection. Moreover, the circuit operation of each CAM cell in the NAND-type word match circuit is opposite to that of the

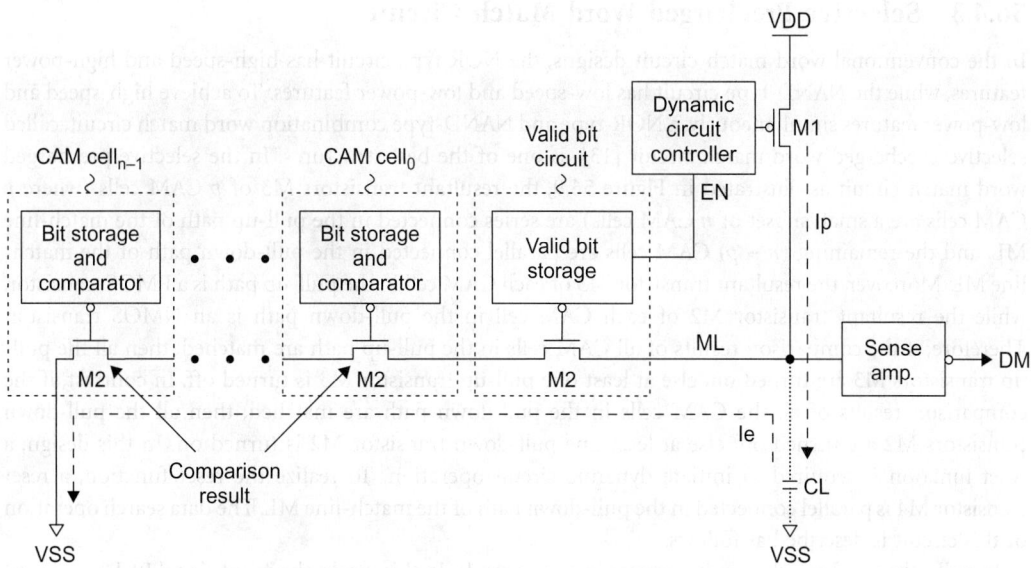

FIGURE 56.8 NAND-type word match circuit.

NOR-type word match circuit. In the NAND-type word match circuit, if the comparison result is matched, then the resultant transistor M2 in the CAM cell is turned off, else the resultant transistor M2 in the CAM cell is turned on. Based on the NAND-type word match circuit design, if the stored data match the search data, then the resultant transistors M2 in all the CAM cells are turned on, else at least one resultant transistor M2 in the CAM cell is turned off.

During the data search cycle, the dynamic operation of NAND-type word match circuit can also be separated into the precharge phase and evaluation phase. In the precharge phase, the DCC turns on the pull-up transistor M1 to precharge the parasitic capacitance CL by a charge current Ip. In this situation, if stored data are mismatched with the search data, then the match-line ML is VDD, owing to at least one resultant transistor M2 being turned off and the pull-up transistor M1 being turned on. Otherwise, the match-line ML is less than VDD, as a result of the pull-up transistor M1 and all the resultant transistors M2 are turned on, simultaneously. For this reason, the NAND-type word match circuit consumes static power dissipation when the stored data match the search data. In the evaluation phase, the DCC turns off the pull-up transistor M1. In this situation, if the stored data mismatch the search data, then the match-line ML is kept to VDD, owing to the pull-up transistor M1 and at least one resultant transistor M2 being turned off, simultaneously. Otherwise, the match-line ML is falling down to VSS by a discharge current Ie, as a result of all the resultant transistors M2 being turned on and the pull-up transistor M1 being turned off. Therefore, the NAND-type word match circuit consumes dynamic power dissipation when the stored data match the search data. As in the above-mentioned circuit operation, the NAND-type word match circuit consumes static and dynamic power dissipations simultaneously when the stored data match the search data. Since only a few data stored in the CAM array are matched with the search data during each data search operation, the NAND-type word match circuit saves large amounts of static and dynamic search power dissipations compared to the NOR-type word match circuit. In this circuit, since the design collects the resultant transistors M2 of all CAM cells in series connection and the longest pull-down path from the match-line ML to the ground node VSS contains all the transistors M2, the long pull-down path structure increases the discharge time of the match-line ML during the evaluation phase. Therefore, the operation speed of the NAND-type word match circuit is slower than that of the NOR-type word match circuit.

56.4.3 Selective Precharged Word Match Circuit

In the conventional word match circuit designs, the NOR-type circuit has high-speed and high-power features, while the NAND-type circuit has low-speed and low-power features. To achieve high-speed and low-power features simultaneously, a NOR-type and NAND-type combination word match circuit, called selective precharged word match circuit [13], is one of the best structures. In the selective precharged word match circuit as illustrated in Figure 56.9, the resultant transistors M3 of p CAM cells (where p CAM cells are a small subset of n CAM cells) are series connected in the pull-up path of the match-line ML, and the remaining $(n - p)$ CAM cells are parallel connected in the pull-down path of the match-line ML. Moreover, the resultant transistor M3 of each CAM cell in the pull-up path is a PMOS transistor, while the resultant transistor M2 of each CAM cell in the pull-down path is an NMOS transistor. Therefore, if the comparison results of all CAM cells in the pull-up path are matched, then all the pull-up transistors M3 are turned on, else at least one pull-up transistor M3 is turned off. In contrast, if the comparison results of all the CAM cells in the pull-down path are matched, then all the pull-down transistors M2 are turned off, else at least one pull-down transistor M2 is turned on. In this design, a reset function is required to initiate dynamic circuit operation. To realize the reset function, a reset transistor M4 is parallel connected in the pull-down path of the match-line ML. The data search operation of this circuit is described as follows.

Initially, the word match circuit operates in a reset mode. In this mode, the input signal RST is assigned to VDD to turn on the reset transistor M4 and a discharge current IRST is generated by the turn-on transistor M4 to discharge the parasitic capacitance CL. As a result, the potential of the match-line ML is falling down to VSS. After that, the input signal RST is assigned to VSS to turn off the reset transistor M4 and the circuit is operated in dynamic operation. In the precharge phase, the pull-up transistor M1 is turned on by the DCC. In this phase, if the comparison results of all the CAM cells in the pull-up path are matched, then a charge current Ip is created by the turn-on transistors M1 and M3 to charge the

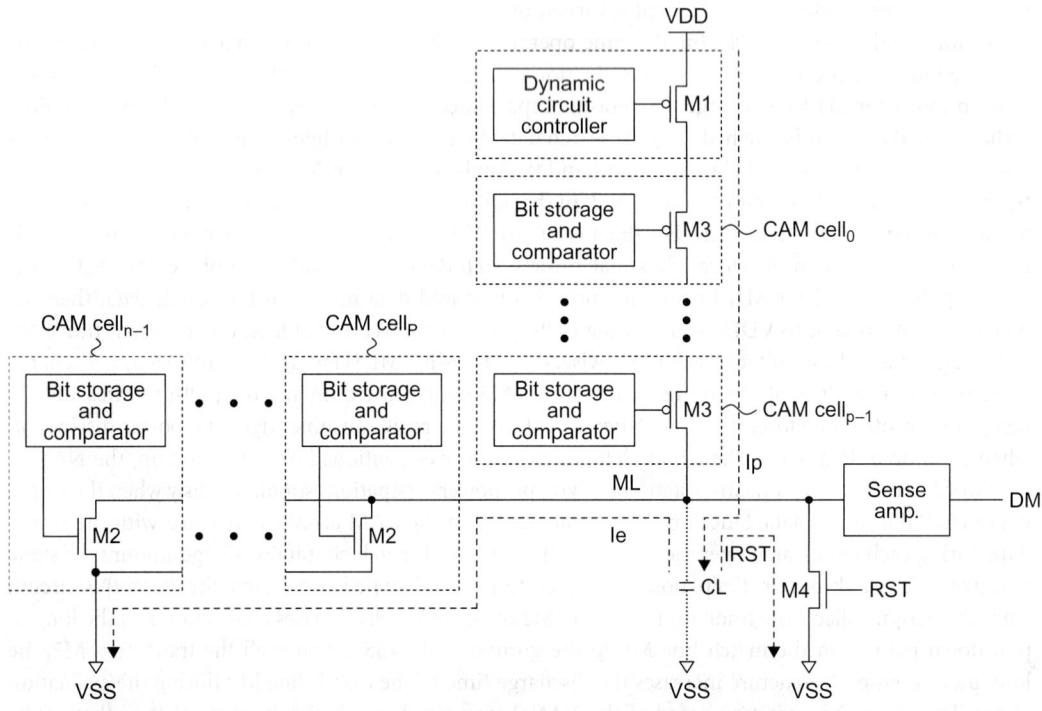

FIGURE 56.9 Selective precharged word match circuit.

parasitic capacitance CL, and the potential of the match-line ML is rising up to VDD. Otherwise, the match-line ML is kept to VSS. In the evaluation phase, if the comparison results of all the CAM cells in the pull-down path are matched, then the potential of the match-line ML is fixed. Otherwise, a discharge current Ie is created by at least one turn-on transistor M2 to discharge the parasitic capacitance CL, and the potential of the match-line ML is falling down to VSS. To summarize the dynamic circuit operation addressed above, if the stored data match the search data (the comparison results of all the CAM cells are matched), then the potential of the match-line ML is VDD. Otherwise (the comparison result of at least one CAM cell is mismatched), the potential of the match-line ML is VSS.

In the selective precharged word match circuit, the search speed is dominated by the number of series-connected transistors M3 (in this case, the number of series connected transistors M3 is p) in the pull-up path of the match-line ML, since the charge time of the parasitic capacitance CL is proportional to the number of series-connected transistors M3 in the pull-up path of the match-line ML. Therefore, the p is as small as possible to improve search performance. In addition, the power dissipation of the selective precharged word match circuit is dominated by the number of parallel-connected transistors M2 (in this case, the number of parallel-connected transistors M2 is $n - p$) in the pull-down path of the match-line ML, since the discharge probability of the parasitic capacitance CL is proportional to the number of parallel-connected transistors M2 in the pull-down path of the match-line ML. For this reason, the p is as large as possible to reduce power dissipation. To summarize the above-mentioned circuit operation, the search speed and the power dissipation of the selective precharged word match circuit are between the NOR-type and NAND-type word match circuits. If p is set to 0, then the selective precharged word match circuit is similar to the NOR-type word match circuit. If p is set to n, then the selective precharged word match circuit is similar to the NAND-type word match circuit. Therefore, both the NOR-type and the NAND-type word match circuits are the special cases of the selective precharged word match circuit, and the selective precharged word match circuit can choose a suitable p value for diverse CAM applications.

56.5 A Low-Power Precomputation-Based CAM Design

In the conventional CAM design, the word match circuit adopts dynamic operation to improve search speed and hardware cost. However, the dynamic circuit has some design issues [17] such as clock skew, low noise margin, and charge sharing. To avoid these drawbacks in the dynamic CAM design, a static CAM design, called precomputation-based CAM (PB-CAM) [19] is one of the best approaches for high-speed and low-power BCAM applications. In the following section, the low-power PB-CAM design is described.

56.5.1 PB-CAM Architecture

The functional block diagram of the PB-CAM architecture, as shown in Figure 56.10, consists of the data memory with parameter field, the address decoder, the PB-CAM word match circuit, and the address priority encoder. Compared to the conventional CAM architecture shown in Figure 56.1, the PB-CAM architecture does not use CLK signal, which is required by dynamic circuit to perform data search operation, as a result of which the PB-CAM word match circuit adopts static pseudo-NMOS logic structure to realize data match operation. In addition, the PB-CAM architecture uses the parameter field to replace the valid bit field of conventional CAM design. The parameter field is utilized to perform precomputation skill and valid bit function simultaneously.

To address the low-power PB-CAM architecture, the design concept for this architecture is introduced. The memory organization of the PB-CAM architecture, as shown in Figure 56.11, is composed of the data memory, the parameter memory, and the parameter extractor. In the write operation, the parameter extractor extracts the parameter of the input data, and then stores the input data and its parameter into the data memory and the parameter memory, respectively. The functional definition of parameter in the PB-CAM architecture is that if data A is the same as data B, then the parameter of data A is the same as the parameter of data B. Based on the parameter definition, if the parameter of data A mismatches the parameter of data B, then

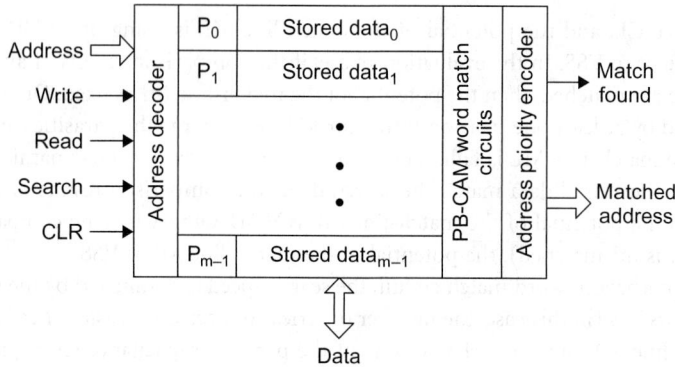

FIGURE 56.10 Functional block diagram of the PB-CAM architecture.

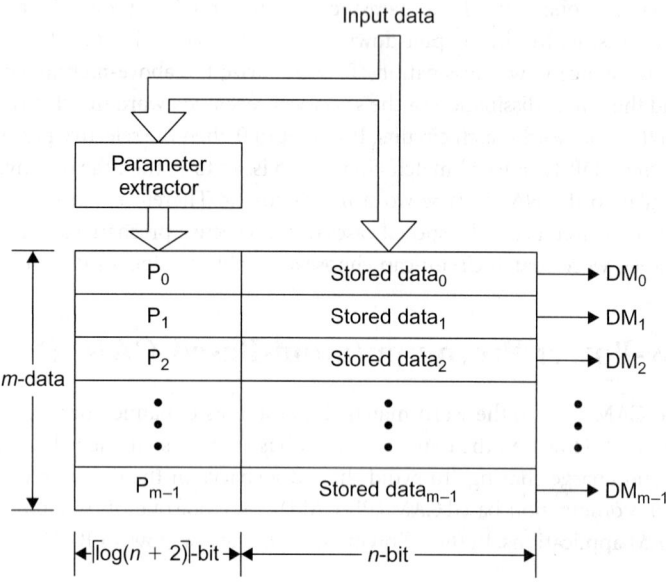

FIGURE 56.11 Memory organization of the PB-CAM architecture.

data A mismatches data B (by the $P \Rightarrow Q \Leftrightarrow \sim Q \Rightarrow \sim P$ theory). Using the parameter comparison approach, the major parts of mismatched data that reduce most of data comparison operations can be identified.

During search operation, in order to reduce large amounts of comparison operation, the operation of the PB-CAM is separated into two comparison processes. In the first comparison process, the parameter extractor extracts the parameter of a desired search data and the parameter comparison circuits then compare the parameter of the search data with all parameters stored in parameter memory in parallel. Recalling the parameter comparison as mentioned above, the data related to this stored parameter concurrently mismatch the search data, if the stored parameter mismatches the parameter of the search data. Otherwise, the data related to this stored parameter has yet to be identified (unidentified). Using the results of the first comparison process, the search data is only compared with those unidentified data to identify any match in the second comparison process. Based on the two comparison processes, if major parts of stored parameter mismatch the parameter of the search data, then most of the comparisons in the second comparison process are largely reduced. The function of the parameter comparison process is just like a filter, the major parts of mismatched data in the first comparison process are filtered to reduce most of the comparisons in the second comparison process. In the PB-CAM design, the parameter comparison process is also known as a precomputation process. Although the data search operation uses two comparison processes to identify any match, both the comparison processes are performed in parallel to improve the data searching speed.

56.5.2 Parameter Extraction Circuit

In the PB-CAM architecture, the parameter extractor dominates most parts of comparison power, since this circuit decides the number of unidentified data remaining after the parameter comparison process. In addition, both parameter memory and parameter comparison circuit of the PB-CAM architecture require extra hardware cost and power dissipation compared with conventional CAM architecture. Therefore, the design concept of the parameter extractor is to filter as many mismatched data as possible in the parameter comparison process with the probable shortest bit length of the parameter. Some functions can be used to realize the parameter extraction in the PB-CAM architecture, such as 1's count function, parity function, and remainder function. In the PB-CAM architecture, the design adopts 1's count function to perform the parameter extraction, because the 1's count function not only filters large amounts of mismatched data with few bit length, but also reduces the transistor count of the PB-CAM cell to seven-transistor. With an n-bit data length, there are $n + 1$ kinds of 1's count (from zero 1's count to n 1's count). Furthermore, it is necessary to add an extra kind of 1's count to indicate the availability of stored data. Based on the parameter extraction function, the minimal bit length of parameter is $\lceil \log(n+2) \rceil$. The required bit length of the parameter is shown in Figure 56.11. In the PB-CAM design, with m words by n bits CAM size, the average number of data comparison in the second comparison process is $m/(n + 1)$, since there are $n + 1$ kinds of 1's count, and only one kind of 1's count (matches with the 1's count of the search data) is unidentified in the parameter comparison process. For example, with a 128 words by 30 bits CAM size, if the search data is 01234567_{16}, then the parameter of the search data is 12. Therefore, the stored data mismatches the search data when its parameter is not 12. Since the range of parameter value is from 0 to 30 and only one parameter value is unidentified in which the parameter value is 12, the average number of data comparison of the second comparison process is $128/31 \approx 4$.

56.5.3 PB-CAM Word Circuit

According to the conventional CAM architecture, the circuit design of CAM word structure adopts dynamic CMOS circuit to improve overall system performance and hardware cost. However, there are some drawbacks to perform CAM word function with dynamic circuit. (1) The dynamic circuit needs an extra precharge time for each data searching operation. (2) The dynamic circuit has some problems such as charge sharing and noise problems. (3) A clock signal is necessary to handle the circuit operation. (4) The noise margin of dynamic circuit is less than V_{tn}.

To eliminate these drawbacks in the conventional CAM word structure, a static pseudo-NMOS word structure shown in Figure 56.12 is one of best structures for realizing word match function. However, the main problem associated with the pseudo-NMOS circuit is its static power dissipation that occurs whenever the pull-down transistor M3 is turned on. Based on the static pseudo-NMOS word circuit design, the static power dissipation occurs when the stored data mismatch the search data, as a result of at least one pull-down transistor M3 being turned on. In general, with m words CAM size, there are $(m - 1)$ stored data mismatched with the search data per data search operation. For this reason, the static power dissipation becomes one of the critical issues in the static pseudo-NMOS word match circuit.

To reduce the static power dissipation in the static pseudo-NMOS CAM word circuit, a novel static pseudo-NMOS CAM word circuit based on the PB-CAM architecture is shown in Figure 56.13. In the PB-CAM word circuit, the parameter comparison circuit is used to control the pull-up transistor M1. Recalling the design concept of the PB-CAM architecture, with m words by n bits CAM size, the average number of data comparison in the second comparison process equals to $m/(n + 1)$. Using the proposed

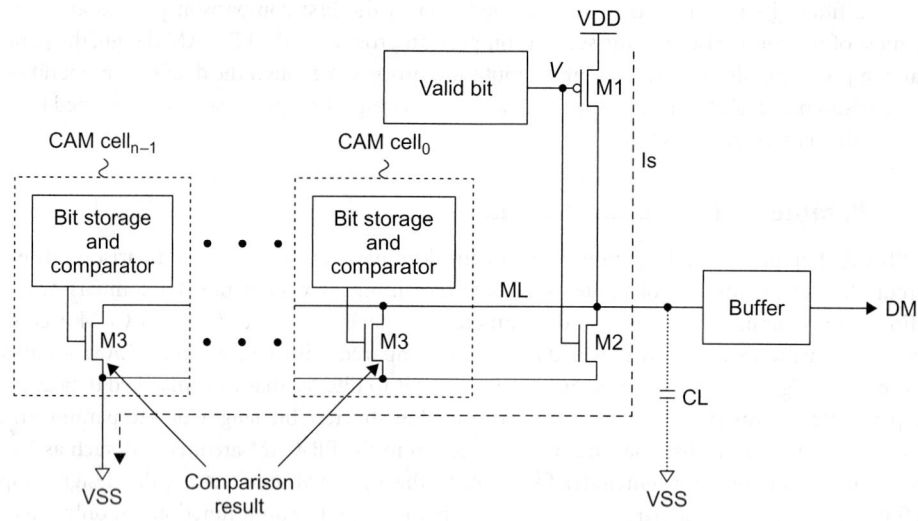

FIGURE 56.12   Static pseudo-NMOS CAM word circuit.

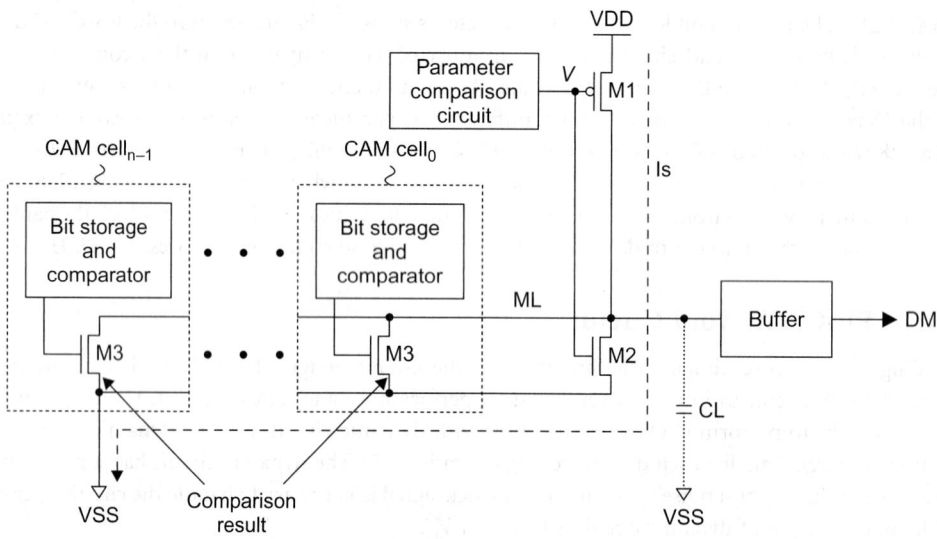

FIGURE 56.13   Static pseudo-NMOS CAM word circuit using the precomputation skill.

precomputation skill, only $m/(n + 1)$ static pseudo-NMOS PB-CAM word circuits turn on its pull-up transistor M1 by its parameter comparison circuit. In addition, one stored data among those $m/(n + 1)$ static pseudo-NMOS PB-CAM word circuits matches the search data. Therefore, the number of PB-CAM word circuits that consume static power is reduced to $(m/(n + 1)) - 1$. For example, with 128 words by 30 bits CAM size, the average number of PB-CAM word circuits consuming static power is equal to $(128/31) - 1 \approx 3$, the static pseudo-NMOS PB-CAM word circuits reduce much of static power dissipation.

56.5.4   PB-CAM Cell

In the previous BCAM circuit design, the BCAM cell is typically constructed by nine-transistor structure as shown in Figure 56.2(a). There are some drawbacks in the conventional BCAM cell circuit.

(1) The BCAM cell requires nine transistors that consume large hardware cost. (2) The BCAM cell design uses the PTL-type XOR gate in the bit comparison circuit, the operating voltage of the BCAM cell circuit cannot be reduced efficiently. (3) The BCAM cell adopts XOR gate to perform bit comparison operation, the XOR gate demands more power consumption than the other standard gates such as NAND gate and NOR gate.

Unlike the conventional BCAM cell design, the PB-CAM cell is a seven-transistor cell structure as shown in Figure 56.14. This cell incorporates a standard five-transistor D-latch device to store the input bit and a NAND-type bit comparison circuit containing two transistors M2 and M3 to drive a word match-line ML. To achieve low-voltage operation, the feedback inverter (INV 2) is a weak-driving design allowing the input data (BL) to be stored in the D-latch device easily. In the conventional BCAM cell design as shown in Figure 56.2, if the search data BL mismatches the stored data Q, then the word match line is VSS, else the word match line is floating. However, in the PB-CAM cell design, the data comparison function is different from that of the conventional BCAM cell design. The truth table of both conventional BCAM cell and PB-CAM cell are shown in Table 56.1 and Table 56.4, respectively. According to both truth tables, the comparison result of the PB-CAM cell does not meet the requirement of the conventional BCAM cell design when the search data BL is 0 and the stored data Q is 1. In this situation, the search data BL mismatches the stored data Q; however, the data comparison result of the PB-CAM cell is matched (since the match-line ML is floating).

Although the PB-CAM cell has an input condition that gives rise to a different result compared with conventional BCAM cell in the search operation, it can be ignored based on the PB-CAM word structure as shown in Figure 56.15. The PB-CAM word circuit has three cases for the search operation. In the first case, the search data BL equals the stored data Q. Since $BL_i = Q_i$, for all i, the output of PB-CAM cell equals the output of conventional BCAM cell. In the second case, the parameter of the search data is not equal to the parameter of the stored data ($V = 1$). Since the pull-up transistor M1 is turned off and the pull-down transistor M2 is turned on by signal V, the match line ML is VSS disregarding the comparison results of PB-CAM cells. The last case is that the search data BL is not equal to the stored data Q, but the parameter of the search data equals the parameter of stored data ($V = 0$). As a result of BL ≠ Q and

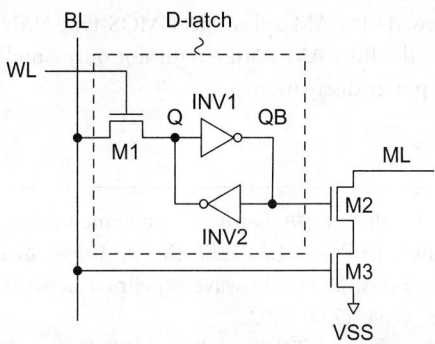

FIGURE 56.14 Proposed seven-transistor PB-CAM cell.

TABLE 56.4 Truth Table of the BCAM Cell and PB-CAM Cell

Search Data (BL)	Stored Data (Q)	Match-line (ML)	Comparison Result
0	0	Floating	Match
0	1	Floating	Match[1]
1	0	VSS	Mismatch
1	1	Floating	Match

[1]The comparison result is different to the conventional BCAM cell in Table 56.1.

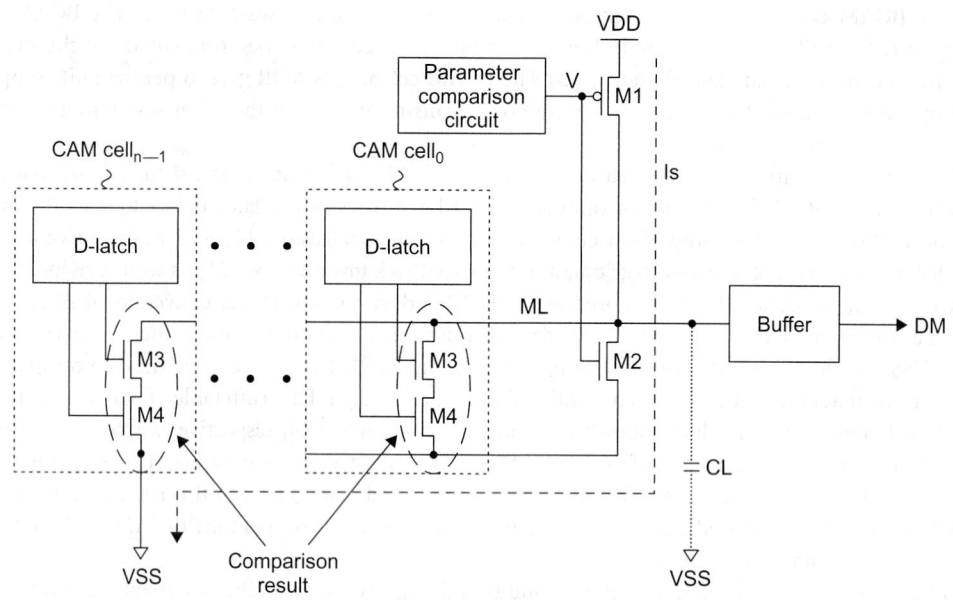

FIGURE 56.15   Proposed PB-CAM word structure with seven-transistor PB-CAM cell.

$V = 0$, this condition exists at least two bit positions i, j, where $n - 1 \leq i, j \leq 0$, such that $(BL_i, Q_i) = (1, 0)$ and $(BL_j, Q_j) = (0, 1)$, respectively. Recalling the results of Table 56.4, the mismatched pattern $(BL_i, Q_i) = (1, 0)$ is detected by the PB-CAM cell, and the comparison result is mismatched. For this reason, another mismatched pattern $(BL_i, Q_i) = (0, 1)$ (the input condition that results in different comparison result between the PB-CAM cell and the conventional BCAM cell) is ignored. To summarize the PB-CAM word circuit with the seven-transistor PB-CAM cell, the data comparison result of the PB-CAM word circuit is the same as that of the conventional BCAM word circuit with the BCAM cell. In addition, the bit comparison circuit in the proposed PB-CAM cell adopts CMOS-type NAND gate to replace conventional PTL-type XOR gate. Therefore, the PB-CAM word circuit not only simplifies hardware design, but also reduces operating voltage and power dissipation.

References

1. M. Sato, K. Kubota, and T. Ohtsuki, "A hardware implementation of gridless routing based on content-addressable memory," in *Proc. ACM/IEEE Design Automat Conf.*, June 1990, pp. 24–28.
2. J. Nyathi and J.G. Delgado-Frias, "A hybrid wave pipelined network router," *IEEE Trans. Circuits Syst. –I*, vol. 49, no. 12, pp. 1764–1772, 2002.
3. D.E. Taylor, J.S. Turner, J.W. Lockwood, T.S. Sproull, and D.B. Parlour, "Scalable IP lookup for Internet routers," *IEEE J. Selected Areas Commun.*, vol. 21, no. 4, pp. 522–534, 2003.
4. Z. Liang, J. Wu, and K. Xu, "A TCAM-based IP lookup scheme for multi-nexthop routing," in *Proc. Int. Comput. Networks Mobile Comput.*, October 2003, pp. 128–135.
5. F. Zane, N. Girija, and A. Basu, "Coolcams: power-efficient TCAMs for forwarding engines," in *Proc. IEEE INFOCOM'03*, March 2003, vol. 1, pp. 42–52.
6. Y.C. Shin, R. Sridha, V. Demjanenko, P.W. Palumbo, and S.N. Srihari, "A special-purpose content-addressable memory chip for real-time image processing," *IEEE J. Solid-State Circuits*, vol. 27, no. 5, pp. 737–744, 1992.
7. T. Ogura, M. Nakanishi, T. Baba, Y. Nakabayshi, and R. Kasai, "A 336-k-bit content-addressable memory for highly parallel image processing," in *Proc. IEEE Cust. Int. Circuits Conf.*, May 1996, pp. 13.4.1–13.4.4.

8. T. Ikenaga and T. Ogura, "A fully-parallel 1 Mb CAM LSI for real-time pixel-parallel image processing," *IEEE J. Solid-State Circuits*, vol. 35, no. 4, pp. 536–544, 2000.

9. H. Yamada, M. Hirata, H. Nagai, and K. Takahashi, "A high-speed string-search engine," *IEEE J. Solid-State Circuits*, vol. 22, no. 5, pp. 829–834, 1987.

10. M. Verleysen, B. Sirletti, A.M. Vandemeulebroecke, and P.G.A. Jespers, "Neural networks for high-storage content-addressable memory: VLSI circuit and learning algorithm," *IEEE J. Solid-State Circuits*, vol. 24, no. 3, pp. 562–569, 1989.

11. M. Meribout, T. Ogura, and M. Nakanishi, "On using the CAM concept for parametric curve extraction," *IEEE Trans. Image Processing*, vol. 9, no. 12, pp. 2126–2130, 2000.

12. H. Miyatake, M. Tanaka, and Y. Mori, "A design for high-speed low-power CMOS fully parallel content-addressable memory macros," *IEEE J. Solid-State Circuits*, vol. 36, no. 11, pp. 956–968, 2001.

13. C.A. Zukowski, and S.Y. Wang, "Use of selective precharge for low-power content-addressable memories," in *Proc. IEEE Int. Symp. Circuits Syst.*, June. 1997, vol. 3, pp. 1788–1791.

14. F. Shafai, K.J. Schultz, G.F.R. Gibson, A.G. Bluschke, and D.E. Somppi, "Fully parallel 30 MHz, 2.5 Mb CAM," *IEEE J. Solid-State Circuits*, vol. 33, pp. 1690–1696, 1998.

15. P.F. Lin and J.B. Kuo, "A 1 V 128 kb four-way set-associative CMOS cache memory using wordline-oriented tag-compare (WOTC) structure with the content-addressable-memory (CAM) 10-transistor tag cell," *IEEE J. Solid-State Circuits*, vol. 36, no. 4, pp. 666–675, 2001.

16. M. Kobayashi, T. Murase, and A. Kuriyama, "A longest prefix match search engine for multi-gigabit IP processing," in *Proc. IEEE Int. Conf. Commn.*, June 2000, vol. 3, pp. 18–22.

17. N.H.E. Weste and K. Eshraghian, *Principles of CMOS VLSI Design: A Systems Perspective*. Reading, MA: Addison-Wesley, 1985.

18. J.G. Delgado-Frias, A. Yu, and J. Nyathi, "A dynamic content-addressable memory using a 4-transistor cell," in *Proc. Int. Workshop Design Mixed-Mode Int. Circuits Applicat.*, July 1999, pp. 110–113.

19. C.S. Lin, J.C. Chang, and B.D. Liu, "A low-power precomputation-based fully parallel content-addressable memory," *IEEE J. Solid-State Circuits*, vol. 38, no. 4, pp. 654–662, 2003.

57

Low-Power Memory Circuits

Martin Margala
University of Massachusetts

CONTENTS

57.1 Introduction

In recent years, a rapid development in VLSI fabrication has led to decreased device geometries and increased transistor densities of integrated circuits (ICs), and circuits with high complexities and very high frequencies have started to emerge. Such circuits consume an excessive amount of power and generate an increased amount of heat. Circuits with excessive power dissipation are more susceptible to run time failures and present serious reliability problems. Increased temperature from high-power processors tends to exacerbate several silicon failure mechanisms. Every 10°C increase in operating temperature approximately doubles a component's failure rate. Increasingly expensive packaging and cooling strategies are required as a chip power increases [1,2]. Owing to these concerns, circuit designers are realizing the importance of limiting power consumption and improving energy efficiency at all levels of the design. The second driving force behind the low-power design phenomenon is a growing class of personal computing devices, such as portable desktops, digital pens, audio- and video-based multimedia products, and wireless communications and imaging systems, such as personal digital assistants, personal communicators, and smart cards. These devices and systems demand high-speed, high-throughput computations, complex functionalities and often real-time processing capabilities [3,4]. The performance of these devices is limited by the size, weight, and lifetime of batteries. *Serious reliability problems, increased design costs and battery-operated applications prompted the IC design community to look more aggressively for new approaches and methodologies that produce more power-efficient designs, which means significant reductions in power consumption for the same level of performance.* Memory circuits form an integral part of every system design as dynamic random access memories (RAMs), static RAMs, Ferroelectric RAMs, MRAMs, ROMs or flash memories, significantly contributing to the system-level power consumption. Two examples of recently presented reduced-power processors show that 43 and 50.3%, respectively, of the total system power consumption is attributed to memory circuits [5,6]. Therefore, reducing the power dissipation in memories can significantly improve the system power-efficiency, performance, reliability, and overall costs.

In this section, all sources of power consumption in different types of memories will be identified, several low-power techniques will be presented and the latest developments in low-power memories will be analyzed.

57.2   Read-Only Memory

Read-only memories (ROMs) are widely used in a variety of applications (permanent code storage for microprocessors or data look-up tables in multimedia processors) for a fixed long-term data storage. The high area density and new submicron technologies with multiple metal layers increase the popularity of ROMs for a low-voltage low-power environment. In the following section, sources of power dissipation in ROMs and applicable efficient low-power techniques are examined.

57.2.1   Sources of Power Dissipation

A basic block diagram of an ROM architecture is presented in Figure 57.1 [7,8]. It consists of an address decoder, a memory controller, a column multiplexer/driver, and a cell array. Table 57.1 shows an example of a power dissipation in a 2K × 18 ROM designed in 0.6 μm CMOS technology at 3.3 V and clocked at 10 MHz [8]. The cell array dissipates 89% of the total ROM power and 11% is dissipated in the decoder, control logic, and the drivers. The majority of the power consumed in the cell array is due to the precharging of large capacitive BLs. During the read and write cycles more than 18 BLs are switched per access, because the wordline selects more BLs than is necessary. Example in Figure 57.2 shows a

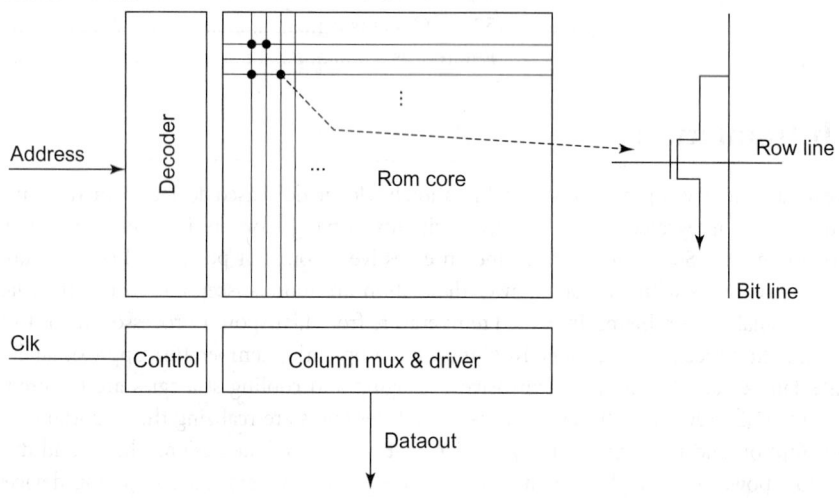

FIGURE 57.1   Basic ROM architecture.

TABLE 57.1   Power Dissipation ROM 2 K × 18

Block **	Power (mW)	Percentage (%)
Decoder	0.06	2.1
ROM core	2.24	89
Control	0.18	7.2
Drivers	0.05	1.7

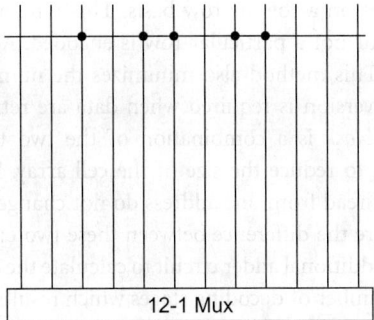

FIGURE 57.2 ROM BLs.

12-1 multiplexer and a BL with five transistors connected to it. This topology consumes excessive amounts of power, because four more BLs will switch instead of just one. The power dissipated in the decoder, control logic, and in drivers is due to the switching activity during the read and precharge cycles and generating control signals for the entire memory.

57.2.2 Low-Power ROMs

To significantly reduce the power consumption in ROMs, every part of the architecture has to be targeted and multiple techniques have to be applied. Several architectural improvements in the cell array that minimize energy waste and improve efficiency have been identified [8,13–16]. These techniques are:

- smaller cell arrays
- hierarchical wordline
- selective precharging
- minimization of nonzero terms
- inverted ROM core(s)
- row(s) inversion
- sign magnitude encoding
- sign magnitude and inverted block
- difference encoding
- three-dimensional decoding
- cascode sensing
- charge recycling and charge sharing

All of these methods result in a reduction of the capacitance and/or switching activity of bit- and rowlines. In applications where different bit sizes of data are needed, *smaller memory arrays* are useful to implement. If stored in a single memory array, its bit size is determined by the largest number. However, most of the bit positions in smaller numbers are occupied by nonzero values that would increase the bit and rowline capacitance. Therefore, by grouping the data to smaller memory arrays according to their size, significant savings in power can be achieved. A *hierarchical wordline* approach divides memory in separate blocks and run the block wordline in one layer and a global wordline in another layer. As a result, only the bit cells of the desired block are accessed. A *selective precharging* method addresses the problem of activating multiple BLs eventhough only a single memory location is being accessed. Using this method only those BLs are precharged which are being accessed. The hardware overhead for implementing this function is minimum. A *minimization of nonzero terms* reduces the total capacitance of bit- and rowlines, because zero terms do not switch BLs. This reduces also the number of transistors in the memory core. An *inverted ROM* applies to a memory with a large number of ones. In this case, the entire ROM array could be inverted and the final data will be inverted back in the output driver circuitry. Consequently, the number of transistors and the capacitance of bit and row lines is reduced. An *inverted row* method also minimizes nonzero terms but on a row by row basis. This type of encoding requires an extra bit (MSB), which indicates whether or not a particular row is encoded. A *sign and magnitude* encoding is used to store negative numbers. This method also minimizes the number of the ones in the memory. However, a two-complement conversion is required when data are retrieved from the memory. A *sign and magnitude and an inverted block* is a combination of the two techniques described previously. A *difference encoding* can be used to reduce the size of the cell array. In applications where a ROM is accessed sequentially and the data read from one address do not change significantly from the following address, the memory core can store the difference between these two entries instead of the entire value. The disadvantage is a need for an additional adder circuit to calculate the original value. *Three-dimensional decoding* significantly reduces a number of decoding stages which results in shorter delay as well as lower power consumption due to reduced area [13]. The block diagram of the circuit is shown in Figure 57.3.

The concept is as follows. The address lines are divided into three parts: lines A6, A5, A4 select the row number of the data core; lines A3, A2, and A1 determine which shared column to activate and lead to the upper decoder; line A0 is fed to the lower decoder. The upper and lower pass blocks are used to resolve nonnatural encoding problems. As a result, not only are the encoded ROM data in a natural order, but the design also reduces the number of transistors necessary for implementation. Compared with a traditional 2-D decoder structure, the 3-D concept improves the performance by almost 70% while maintaining power consumption level.

To support low-voltage operation, a *cascode sensing scheme* can be implemented [14]. The speed of a read operation is improved by using a dummy sense amplifier to control the BL precharging period despite the high dependence of programmed data on BL capacitance. The diagram of the sense-amplifying circuit is shown in Figure 57.4. The circuit consists of a read cascode sense amplifier (RSA), a dummy sense amplifier (DSA), a bitline (BL) booster, and a pull-down circuitry. Since BL capacitance increases with the number of zero-programmed cells per BL, it is difficult to precharge the BL to the optimum voltage level for sensing. Therefore, the BL booster is used to charge BL rapidly. The DSA monitors the BL level and stops charging through the BL booster as soon as the BL reaches the optimum level. This technique allows operation down to $V_{dd} = 0.8$ V in 0.25 μm CMOS technology for a fraction of area penalty.

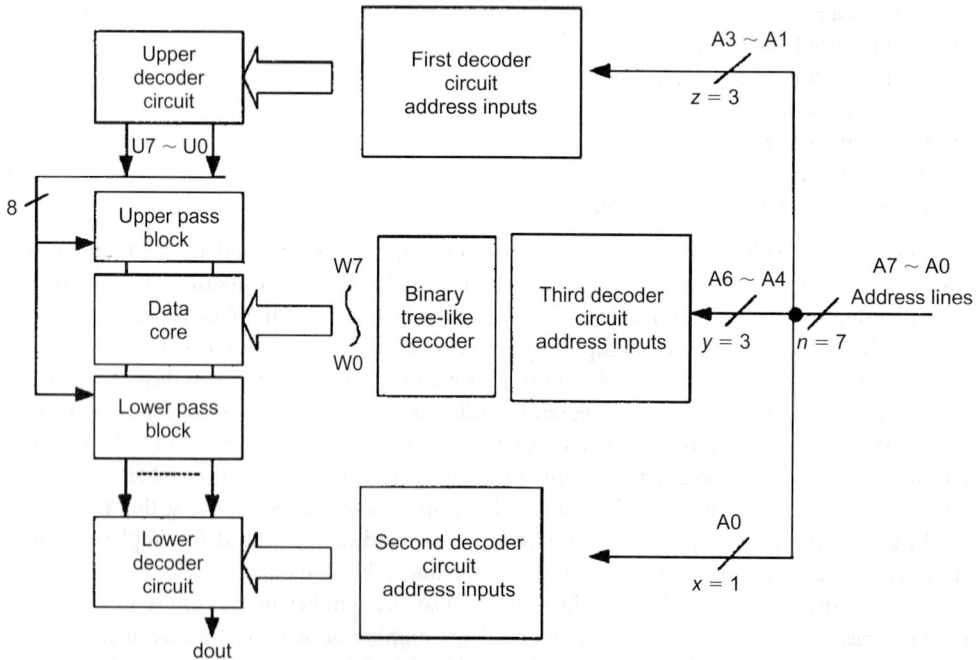

FIGURE 57.3 Block diagram of the 3-D decoding circuit for a 128 $\times$ 1 ROM.

In power reduction for ROMs, very efficient techniques are charge sharing and charge recycling [15,16]; specifically, charge recycling predecoder (CRPD), charge recycling wordline decoder (CRWD), charge recycling bitline (CRBL), and charge sharing bitline (CSBL). The concept of CRPD is shown in Figure 57.5. In the CRPD, the newly selected predecoder line is charged only to $V_{dd}/2$ (Figure 57.5[b]) as opposed to a full V_{dd} (Figure 57.5[a]) due to the charge sharing with the previously selected predecoder line. Figure 57.5(c)

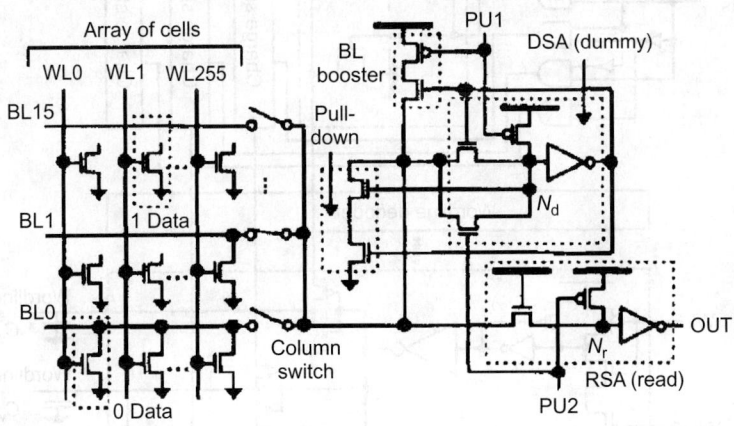

FIGURE 57.4 Circuit diagram of the cascode sensing circuit.

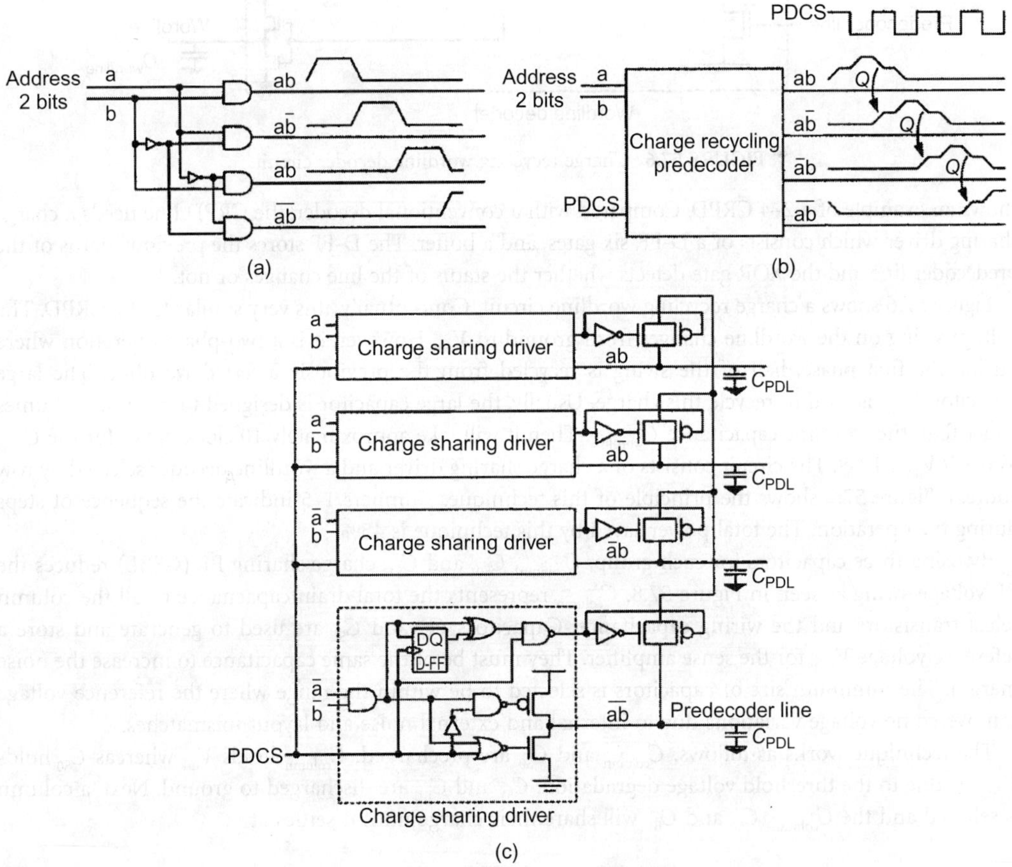

FIGURE 57.5 Charge recycling predecoder.

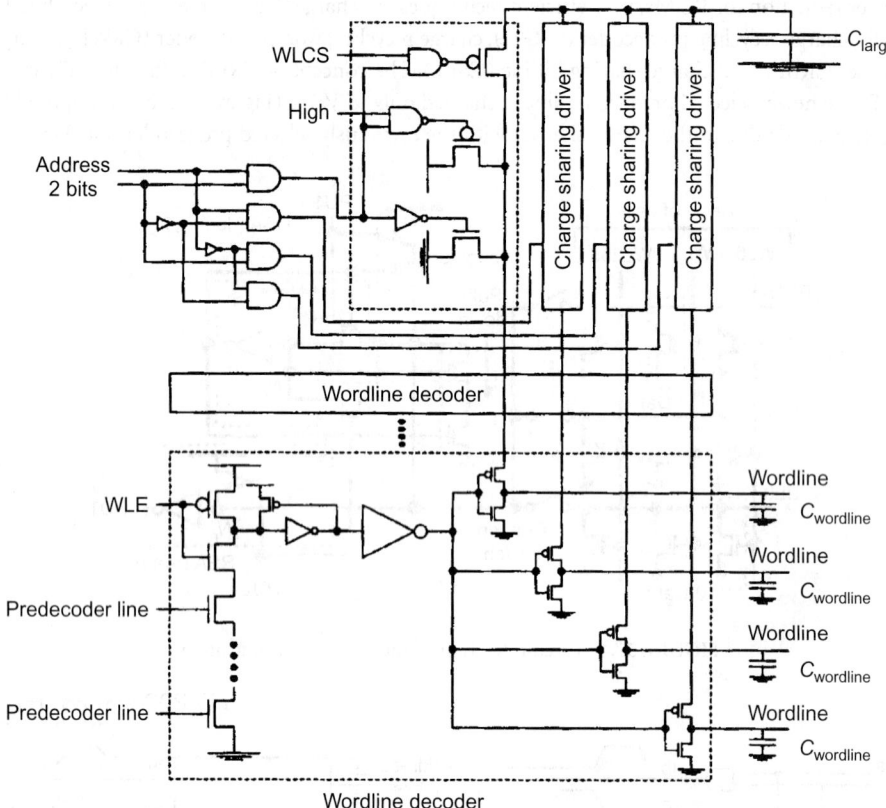

FIGURE 57.6 Charge recycling wordline decoder circuit.

shows an example of a 2–4 CRPD. Compared with a conventional decoder, the CRPD line needs a charge sharing driver which consists of a D-FF, six gates, and a buffer. The D-FF stores the previous status of the predecoder line and the XOR gate detects whether the status of the line changes or not.

Figure 57.6 shows a charge recycling wordline circuit. Conceptually, it is very similar to the CRPD. The voltage swing on the wordline changes from ground to V_{dd}. However, it is a two-phase operation where during the first phase, half of the swing is recycled from the previously asserted wordline. The large capacitor C_{large} is used to recycle this charge. Usually, the large capacitor is designed to be about 10 times larger than the wordline capacitance $C_{wordline}$. Then it will take approximately 10 clock cycles for the C_{large} to reach $V_{dd}/2$ level. The circuit consists of a charge sharing driver and a wordline decoder selected by row address. Figure 57.7 shows the principle of this technique. Numbers 1–5 indicate the sequence of steps during the operation. The total power saved by this technique is 45%.

By using three capacitors for each group, C_{column}, C_{S0}, and C_{S1}, charge sharing BL (CSBL) reduces the BL voltage swing as seen in Figure 57.8. C_{column} represents the total drain capacitance of all the column select transistors and the wiring capacitance. Capacitors C_{S0} and C_{S1} are used to generate and store a reference voltage V_{REF} for the sense amplifier. They must be of the same capacitance to increase the noise margin. The minimum size of capacitors is selected to be within the range where the reference voltage can overcome voltage variations due to internal and external noise and layout mismatches.

The technique works as follows. C_{column} and C_{S0} are precharged. C_{column} holds V_{dd} whereas C_{S0} holds $V_{dd} - V_t$ due to the threshold voltage degradation. C_{S1} and C_{BL} are discharged to ground. Next, a column is selected and the C_{column}, C_{S0} and C_{BL} will share their charge until it settles at

$$V_{CS} = [C_{column} V_{dd} + C_{S0} (V_{dd} - V_t)] / (C_{BL} + C_{column} + C_{S0})$$

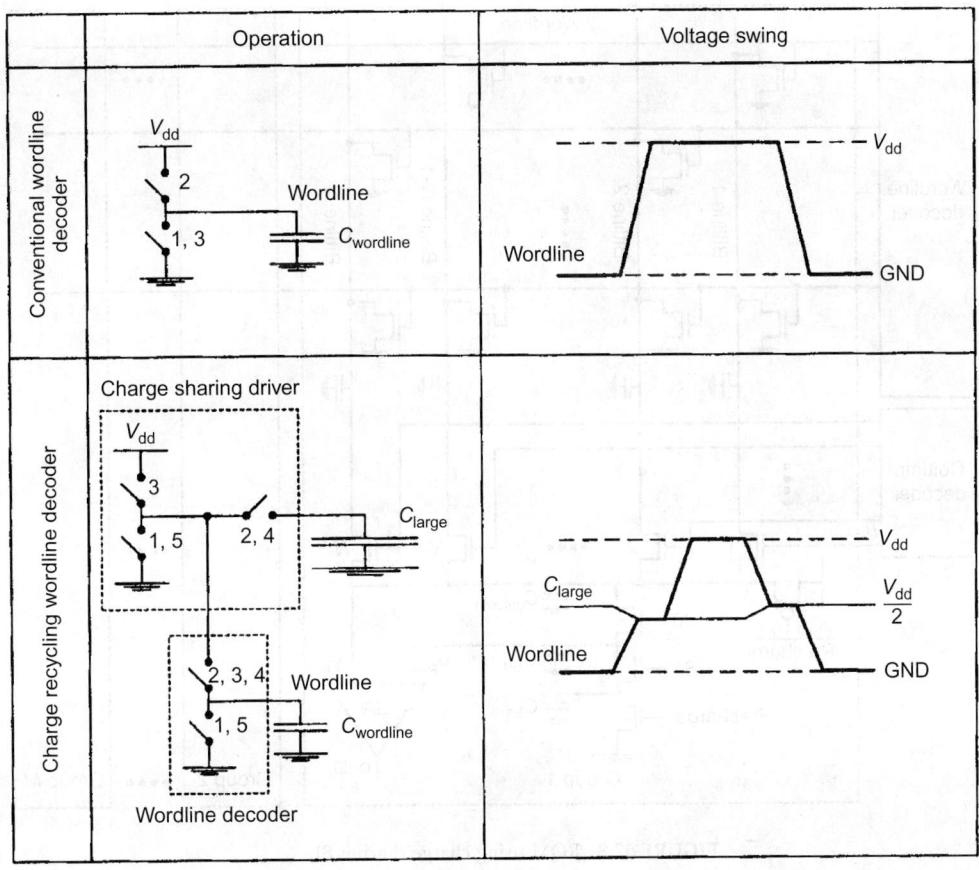

FIGURE 57.7 The concept of CRWD.

Afterwards, a wordline is selected. If the ROM data is "1," then the BL remains at V_{CS}; if it is "0", then the BL is discharged. V_{SS} (GND) to V_{CS} represents a small voltage swing. Therefore, a sense amplifier is needed to correctly read the data. Using CRPD, CRWD, and CSBL, the designer can expect to save on average 18, 28, and 36% of power, respectively.

Lastly, on the system level, CRBL technique can be implemented. BL swing voltage is reduced by charge recycling between BLs. The concept is illustrated in Figures 57.9 and 57.10. When N BLs recycle their charges, the voltage swing and the BL power decrease to $1/N$ and $1/N^2$, respectively.

BLs are grouped into pairs, a BL, and a complementary BL. The BLs within each pair are connected through a transistor switch controlled by *Equal* signal. During the *equalization* mode, the signal is asserted, all programmed connections between BL pairs are disconnected and the two BLs in each BL pair are connected. During this time, the connected BLs share their charges and the BL voltages equalize. When the ROM is in the *evaluation* phase, the BLs in each pair are disconnected and all BLs are connected to BLs in the neighboring BL pair by a transistor switch controlled by *Eval* signal. Figure 57.9 shows N BL pairs. The top BL of the top BL pair is charged to V_{dd}, whereas the complementary BL in the same pair is at $(N-1)/N\ V_{dd}$. Subsequently, the charge is recycled to the $N-1$ BL pair. The bottom BL of the bottom BL pair is at GND level. Figure 57.11 demonstrates the CRBL in ROM architecture. The power consumed by the BLs and the sense amplifiers is reduced by 72%.

On a circuit level, powerful techniques minimizing the power dissipation could be applied. The most common technique is reducing the power supply voltage to approximately $V_{dd} \sim 2V_t$ in a correlation with the architectural-based scaling. In this region of operation, the CMOS circuits achieve the maximum

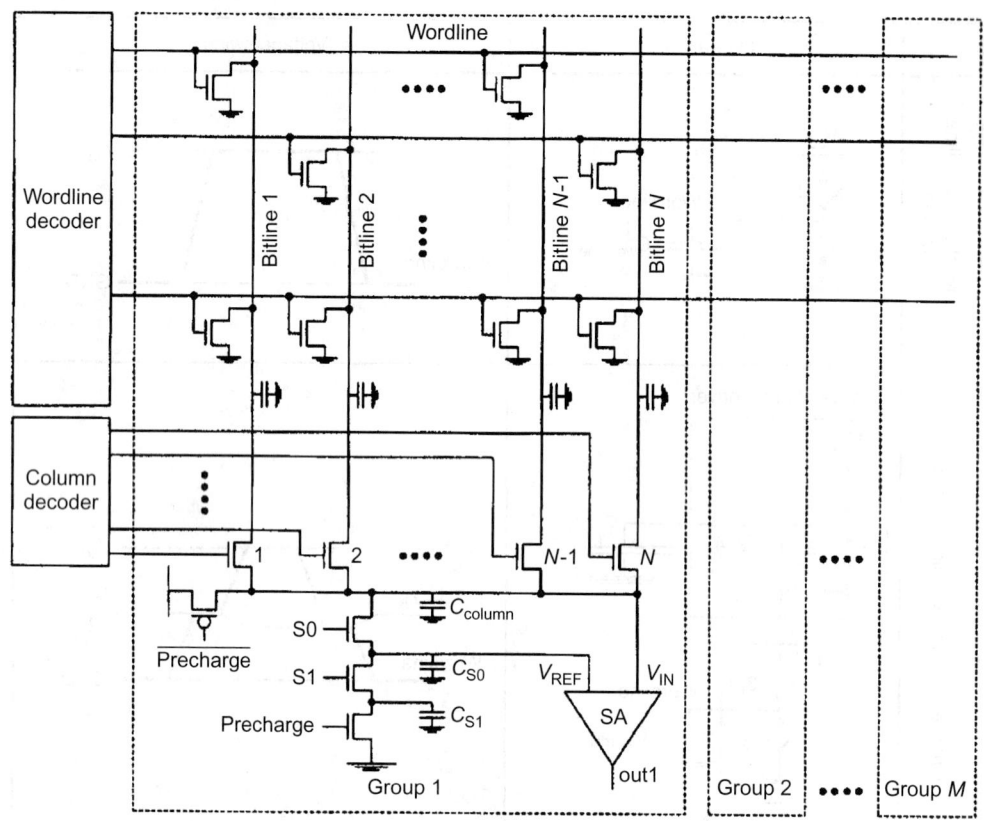

FIGURE 57.8 ROM using charge sharing BL.

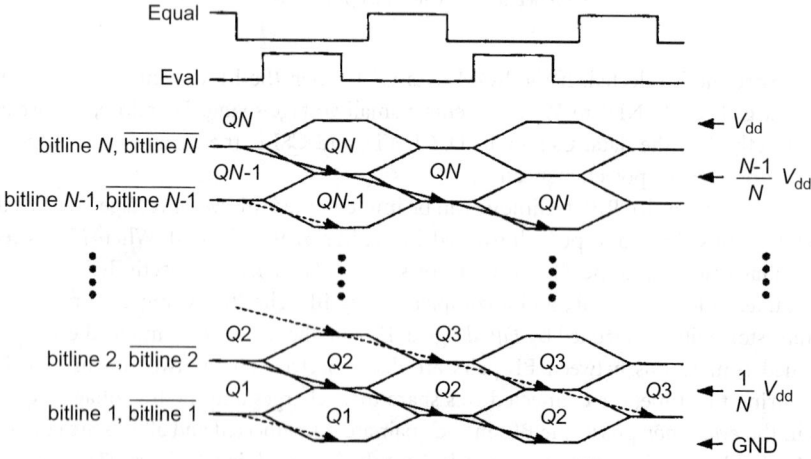

FIGURE 57.9 The concept of charge recycling BL.

power efficiency [9,10]. This results in large power savings because the power supply is a quadratic term in a well-known dynamic power equation. In addition, the static power and short-circuit power are also reduced. It is important that all the transistors in the decoder, control logic, and driver block be sized properly for low-power, low-voltage operation. Rabaey and Pedram [9] have shown that the ideal low-power sizing is when $C_d = C_L/2$, where C_d is a total parasitic capacitance from driving transistors and C_L

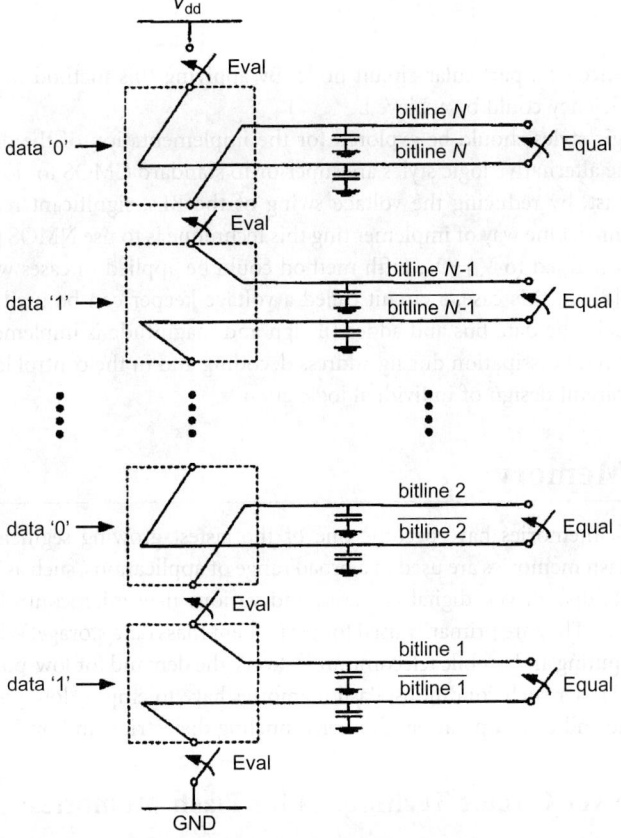

FIGURE 57.10 The structure of CRBL.

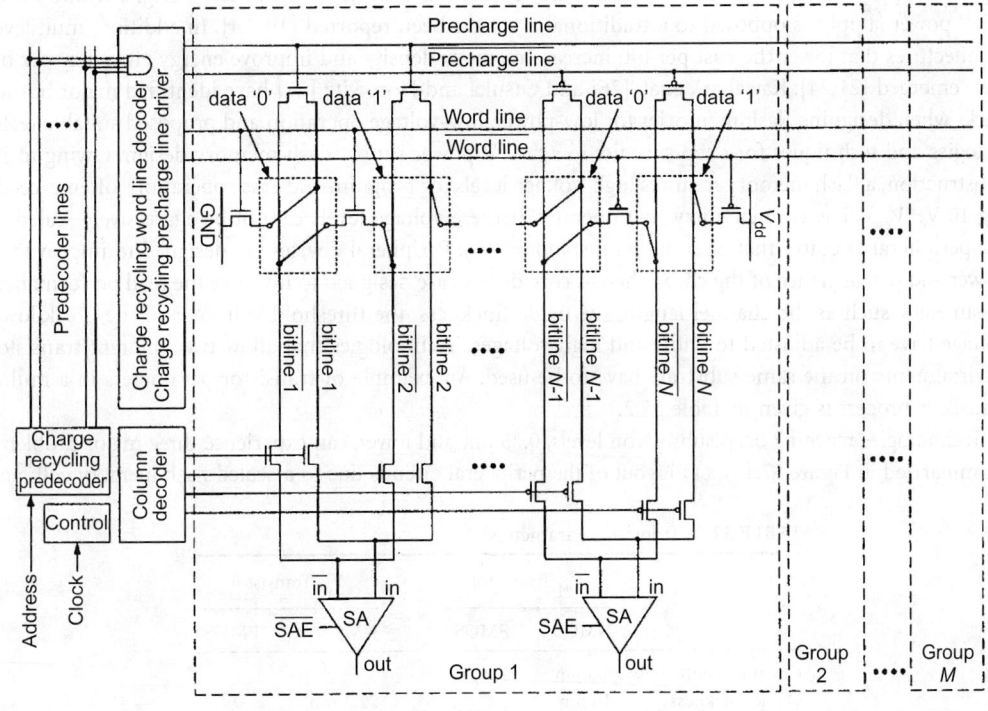

FIGURE 57.11 ROM architecture using CRBL.

is total load capacitance of a particular circuit node. By applying this method to every circuit node, a maximum power efficiency could be achieved.

Next, different logic styles should be explored for the implementation of the decoder, control logic, and the drivers. Some alternative logic styles are superior to standard CMOS for low-power, low-voltage operation [11,12]. Last, by reducing the voltage swing of the BLs, significant reduction in switching power could be obtained. One way of implementing this technique is to use NMOS precharge transistors. The BLs are then precharged to $V_{dd} - V_t$. Fifth method could be applied in cases when same location is accessed repeatedly [8]. In this case, a circuit called a voltage keeper can be used to store past history and avoid transitions in the data bus and adder (if sign and magnitude is implemented). Sixth method is in limiting short-circuit dissipation during address decoding and in the control logic and drivers. This can be achieved by careful design of individual logic circuits.

57.3 Flash Memory

In recent years, flash memories have become one of the fastest growing segments of semiconductor memories [13,14]. Flash memories are used in a broad range of applications, such as modems, networking equipment, PC BIOS, disk drives, digital cameras, and various new microcontrollers for leading edge embedded applications. They are primarily used for permanent mass data storage. With a rapidly emerging area of portable computing and mobile telecommunications, the demand for low-power, low-voltage flash memories increases. Under such conditions, flash memories have to employ low-power tunneling mechanisms for both write and erase operation, thinner tunneling dielectrics, and on-chip voltage pumps.

57.3.1 Low-Power Circuit Techniques for Flash Memories

To prolong the battery life in mobile devices, significant reductions of power consumption in all electronic components have to be achieved. One of the fundamental and most effective methods is a reduction of a power supply voltage. This method has been observed also in flash memories. Designs with a lower 3.3 V power supply as opposed to a traditional 5 V have been reported [19–24]. In addition, multilevel architectures that lower the cost per bit, increase memory density and improve energy efficiency per bit have emerged [21,24]. Kawahara et al. [26] and Otsuka and Horowitz [27] have identified major bottlenecks when designing flash memories for low-power, low-voltage operation and proposed suitable technologies and techniques for deep submicron sub-2 V power supply flash memory design. Owing to its construction, a flash memory requires high-voltage levels for program and erase operations often exceeding 10 V (V_{pp}). The core circuitry that operates at these voltage levels cannot be aggressively scaled as the peripheral circuitry that operates with standard V_{dd}. Peripheral devices are designed to improve the power and performance of the chip, whereas core devices are designed to improve the read performance. Parameters, such as the channel length, the oxide thickness, the threshold voltage, and the breakdown voltage have to be adjusted to withstand high voltages. Technologies that allow two different transistor environments on the same substrate have to be used. An example of transistor parameters in a multi-transistor process is given in Table 57.2.

Technologies reaching deep submicron levels, 0.25 μm and lower, can experience three major problems (summarized in Figure 57.12): (1) layout of the peripheral circuits due to a scaled flash memory cell; and

TABLE 57.2 Transistor Parameters

	V_{dd} Transistor		V_{pp} Transistor	
	NMOS	PMOS	NMOS	PMOS
Channel length	0.6 μm	1.2 μm		
Oxide thickness	10 nm		22.3 nm	
Threshold voltage	0.4 V		0.79 V	0.97 V

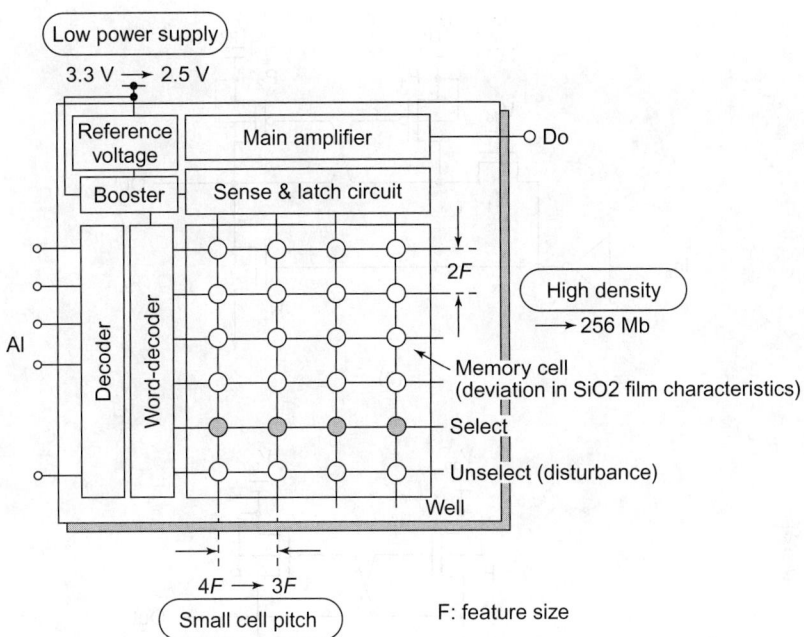

FIGURE 57.12 Quarter-micron flash memory.

(2) an accurate voltage generation for the memory cells to provide the required threshold voltage and narrow deviation; and (3) deviations in dielectric film characteristics caused by large number of memory cells. Kawahara et al. [26] have proposed several circuit enhancements that address these problems. They proposed a sensing circuit with a relaxed layout pitch, BL clamped sensing multiplex, and intermittent burst data transfer for a three times feature-size pitch. They also proposed a low-power dynamic bandgap generator with voltage boosted by using a triple-well bipolar transistors and a voltage-doubler charge pumping, for accurate generation of 10–20 V that operate at V_{dd} under 2.5 V. They demonstrated these improvements on a 128-Mb experimental chip fabricated using 0.25 μm technology.

On a circuit level, three problems have been identified by Otsuka and Horowitz [27]: (1) interface between peripheral and core circuitry; (2) sense circuitry and operation margin; and (3) internal high-voltage generation.

During program and erase modes, the core circuits are driven with higher voltage than the peripheral circuits. This voltage is higher than V_{dd} to achieve good read performance. Therefore, a level shifter circuit is necessary to interface between the peripheral and core circuitry. However, when a standard power supply (V_{dd}) is scaled to 1.5 V and lower, the threshold voltage of V_{pp} transistors will become comparable to one half of V_{dd} or less, which results in significant delay and poor operation margin of the level shifter and, consequently, degrades the read performance. A level shifter is necessary for the row decoder, column selection, and source selection circuit. Since the inputs to the level shifters switch while V_{pp} is at the read V_{pp} level, the performance of the level shifter needs to be optimized only for a read operation. In addition to a standard erase scheme, flash memories utilizing a negative-gate erase or program scheme have been reported [19,23]. These schemes utilize a single voltage supply which results in lower power consumption. The level shifters in these flash memories have to shift a signal from V_{dd} to V_{pp} and from GND to V_{bb}. Conventional level shifters suffer from delay degradation and increased power consumption when driven with low power supply voltage. There are several reasons attributed to these effects. First, at low V_{dd} (1.5 V) the threshold voltage of V_{pp} transistors is close to half of power supply voltage which results in an insufficient gate swing to drive the pull-down transistors (see Figure 57.13). This also reduces the operation margin of these shifters for the threshold voltage fluctuation of the V_{pp} transistor. Second, a rapid increase in power

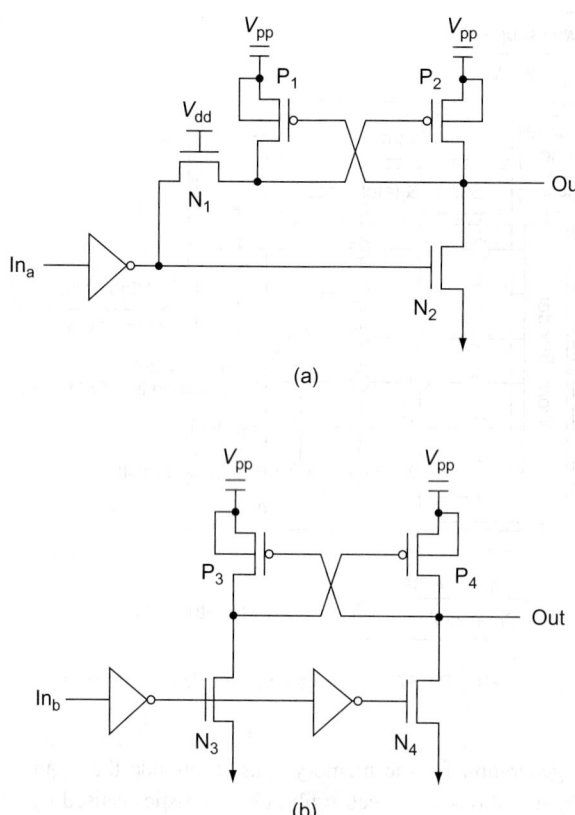

FIGURE 57.13 Conventional high-level shifter circuits with (a) feedback PMOS and (b) cross-coupled PMOS.

consumption at V_{dd} under 1.5 V is due to DC current leakage through V_{pp} to GND during the transient switching. At 1.5 V, 28% of the total power consumption of V_{pp} is due to DC current leakage. Two signal shifting schemes have been proposed, one for a standard flash memory and another for a negative-gate erase or program flash memories. The first proposed design is shown in Figure 57.14. This high-level shifter uses a bootstrapping switch to overcome the degradation due to a low input gate swing and improves the current drivability of both pull-down drivers. It also improves the switching delay and the power consumption at 1.5 V, because the bootstrapping reduces the DC current leakage during the transient switching. Consequently, the bootstrapping technique increases the operation margin. The layout overhead from the bootstrapping circuit, capacitors, and an isolated n-well is negligible compared with the total chip area because it is used only as the interface between the peripheral circuitry and the core circuitry. Figure 57.15 shows the operation of the proposed high-level shifter and Figure 57.16 illustrates the switching delay and the power consumption versus the power supply voltage of the conventional design and the proposed design. The second proposed design, shown in Figure 57.17, is a high-/low-level shifter that also utilizes a bootstrapping mechanism to improve the switching speed, reduce DC current leakage and improve operation margin. The operation of the proposed shifter is illustrated in Figure 57.18. At 1.5 V, the power consumption decreases by 40% compared with a conventional two-stage high-/low-level shifter (see Figure 57.19). The proposed level shifter does not require an isolated n-well and therefore the circuit is suitable for a tight-pitch design and a conventional well layout.

In addition to the more efficient level-shift scheme, Otsuka and Horowitz [27] also addressed the problem of sensing under very low power supply voltages ($\leq$1.5 V) and proposed a new self-bias BL-sensing method that reduces the delay's dependence on BL capacitance and achieves 19 ns reduction of the sense delay at low voltages. This enhances the power efficiency of the chip. On a system level,

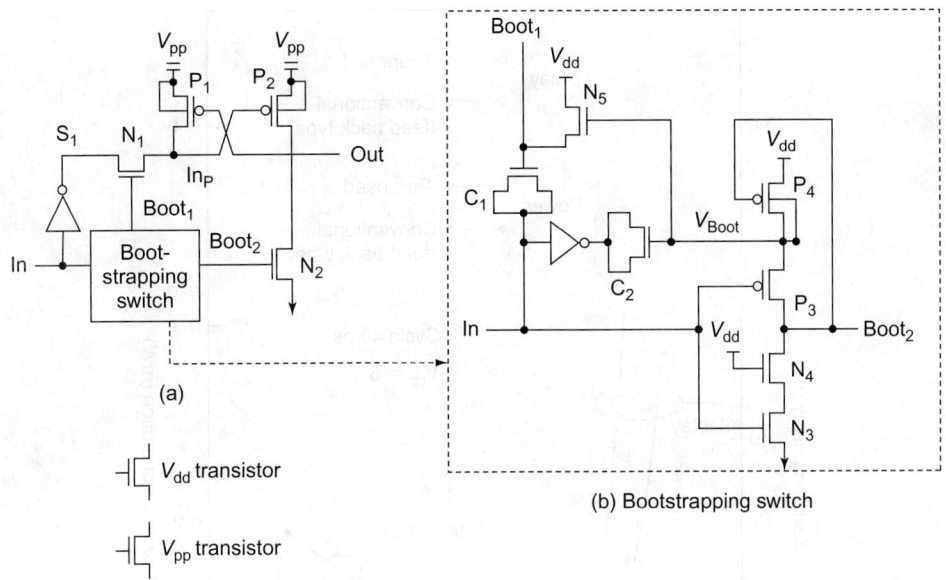

(a)

(b) Bootstrapping switch

FIGURE 57.14 A high-level shifter circuit with bootstrapping switch.

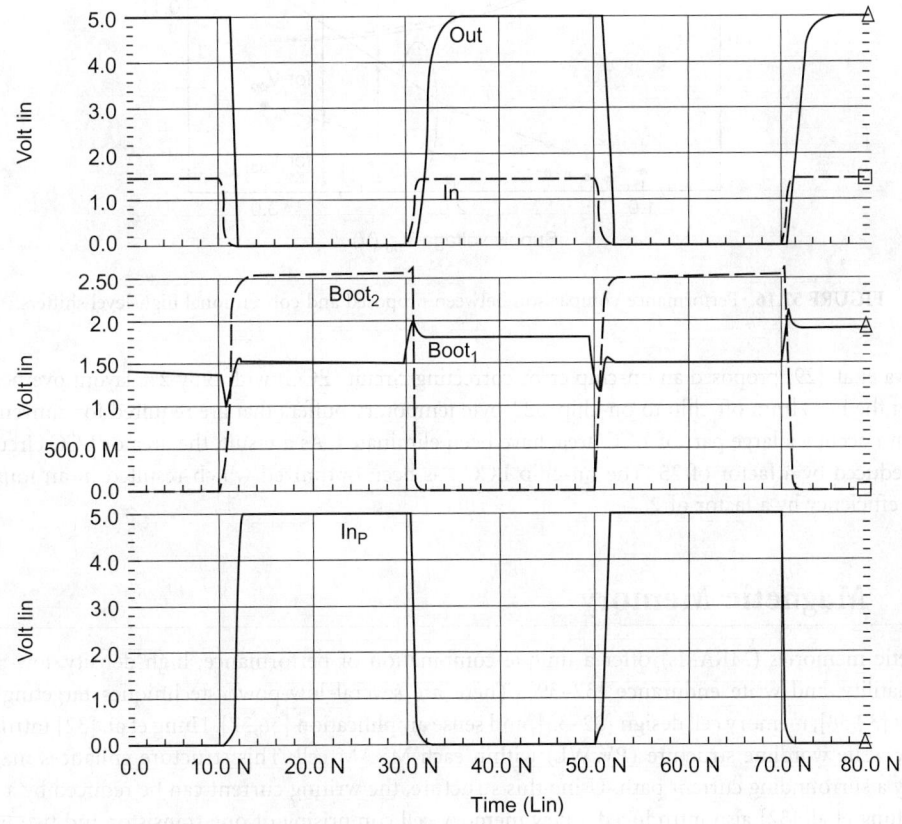

FIGURE 57.15 Operation of the proposed high-level shifter circuit.

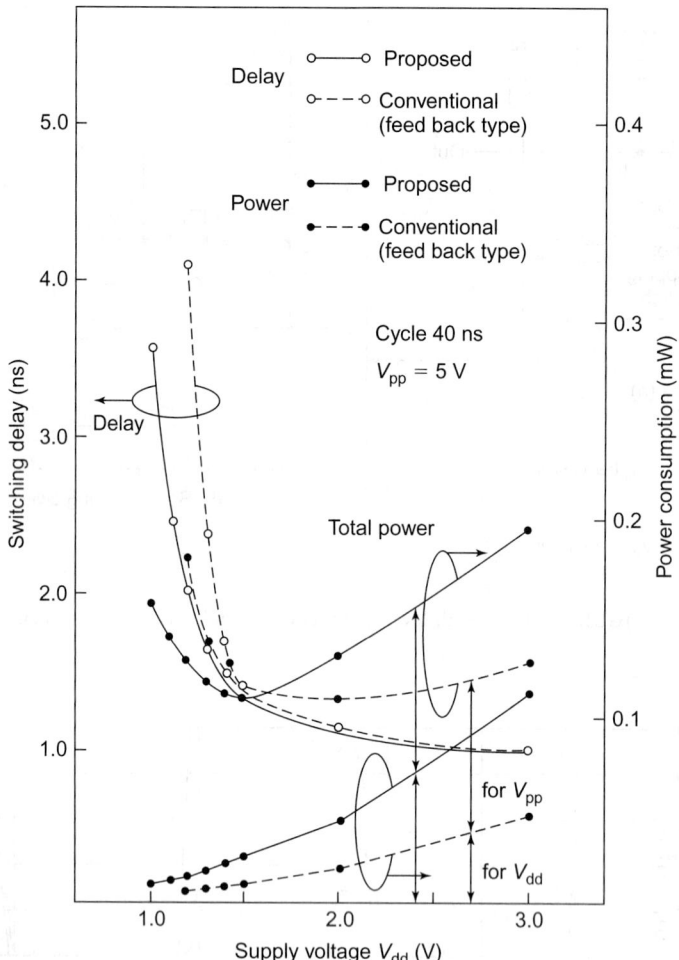

FIGURE 57.16 Performance comparison between proposed and conventional high-level shifters.

Tanzawa et al. [29] proposed an on-chip error correcting circuit (ECC) with only 2% layout overhead. By moving the ECC from off-chip to on-chip, 522-byte temporary buffers that are required for conventional ECC and occupy a large part of ECC area, have been eliminated. As a result, the area of ECC circuit has been reduced by a factor of 25. The on-chip ECC has been optimized which resulted in an improved power efficiency by a factor of 2.

57.4   Magnetic Memory

Magnetic memories (MRAMs) offer a unique combination of performance, high density, low-power, nonvolatility, and write endurance [32–39]. There are several low-power techniques targeting write current [32,38], memory cell design [32–35], and sense amplification [36,37]. Hung et al. [32] introduced a pillar write wordline structure (PWWL) within each MRAM cell. This structure enhances magnetic field by a surrounding current path. Using this structure, the writing current can be reduced by a factor of 2. Hung et al. [32] also introduced a new memory cell comprising of one transistor and two uneven magnetic tunnel junctions (1T2UMTJ). The implementation is shown in Figure 57.20. By electrically combining two memory bits in parallel that share one transistor, one transistor can be eliminated. Two

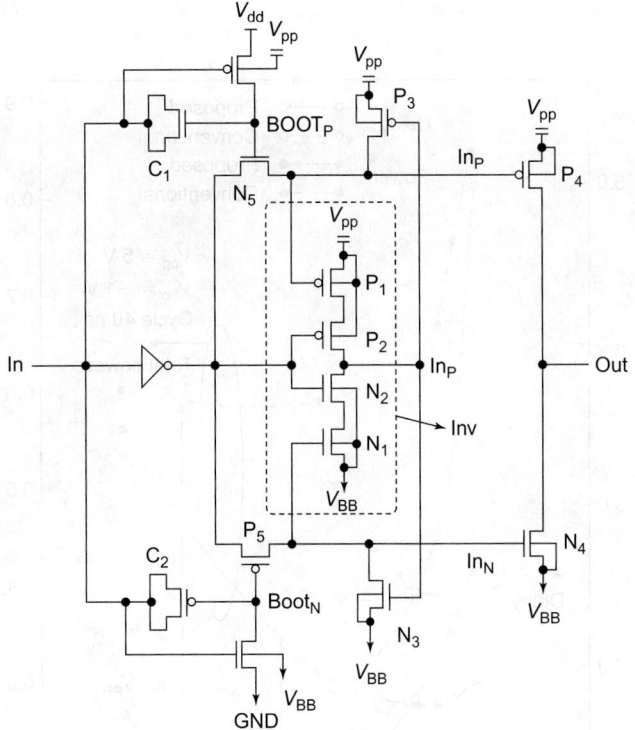

FIGURE 57.17 Proposed high-/low-level shifter circuit.

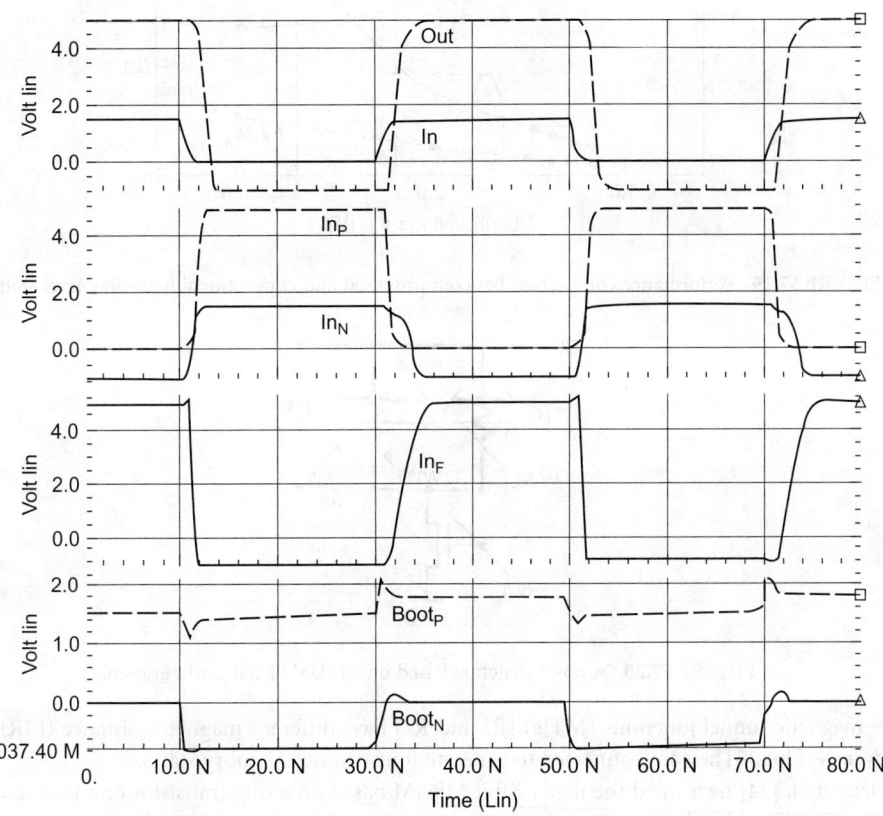

FIGURE 57.18 Operation of the proposed high-/low-level shifter circuit.

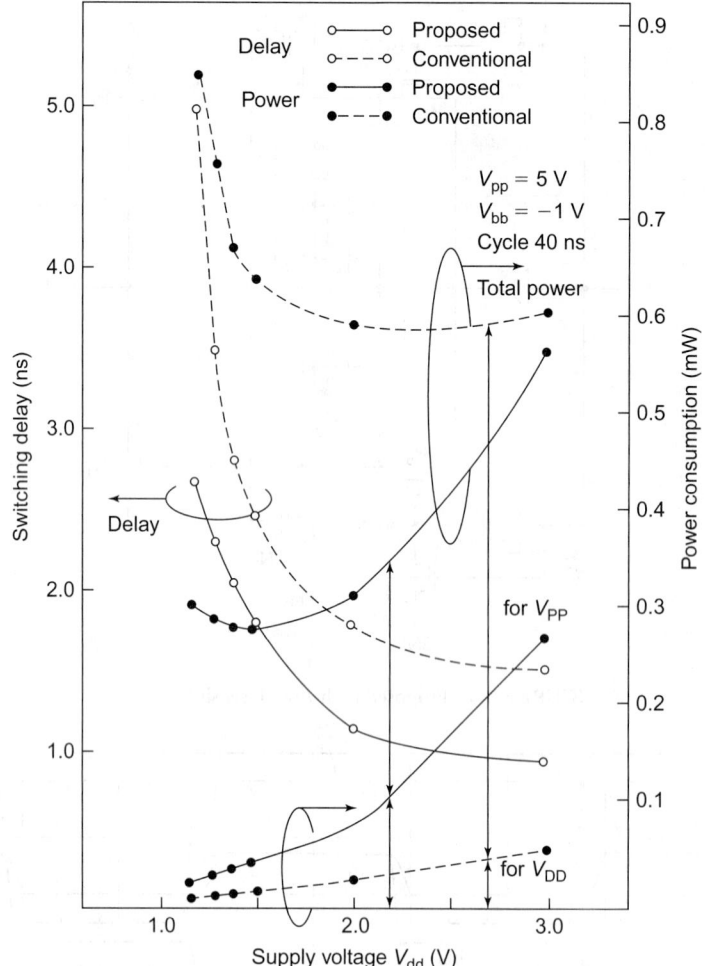

FIGURE 57.19 Performance comparison between proposed and conventional high-/low-level shifters.

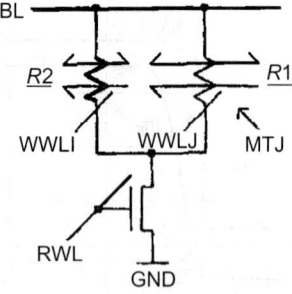

FIGURE 57.20 A novel structure based on 1T2UMTJ cell configuration.

uneven magnetic tunnel junctions (MTJs) ($R1$ and $R2$) have different magnetoresistance (MR) characteristics in R–I loop. They are connected to generate four distinct memory states.

Durlam et al. [34] developed the first 1 Mbit MRAM based on a one-transistor one-magnetic tunnel junction (1T1MTJ) bit cell. The cell architecture is based on the minimum-sized active transistor as the isolation device in conjunction with an MTJ. The MTJ has one electrode connected to the drain of a

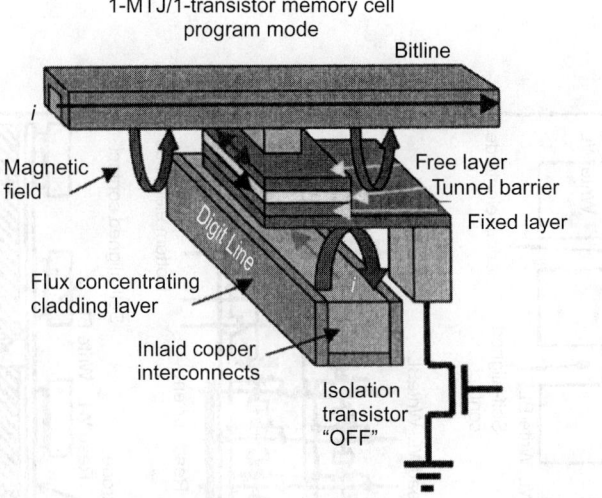

FIGURE 57.21 1T1MTJ memory cell during program mode.

pass transistor for isolation and the electrode connected to the BL (see Figure 57.21). The neighboring cells share the pass transistor source and isolation region to minimize the cell area. The total cell area is $9F^2$, where F is one half the metal pitch. The MRAM bit cell is programmed by the magnetic field, which is generated by a current flowing through two conductors. The transistor is turned off during this time. The direction of a BL current determines the direction of the polarization of the free layer. A special cladding layer is used to reduce the required current by a factor of 2. The group designed first 1-Mbyte MRAM with 50 ns access time consuming 24 mW in 0.6 μm CMOS process at 3.0 V power supply voltage.

Asao et al. [35] developed a new cross point cell with hierarchical BL architecture. This new design, shown in Figure 57.22, increases the cell density to $6F^2$ and reduces the sneak current that degrades the write signal. The required read current is also small. However, the performance is reduced. The group fabricated 1-Mbit MRAM in 0.13 μm CMOS running at 1.5 V.

Aoki et al. [36] proposed a new scalable MRAM cell built with one transistor and two MTJs. This new cell is capable of sensing voltage directly divided with resistance ratio of 2MTJs (Figure 57.23). The scheme uses a folded BL architecture with fixed reference voltage that creates a large sensing margin and thus leads into a stable operation.

Two MTJs are connected in series and the connection node is joined with a BL through a pass transistor. The MTJs are identical in size and are connected with WWL1 and WWL2. During write operation, a write current is applied to the DL and to both the WWL1 and the WWL2. They form a current loop. During the read, the voltage V_{read} is applied to MTJs. The voltage of the connection node is either more than $V_{read}/2$ when the resistance of MTJ1 is less than MTJ2, or less than $V_{read}/2$ when the resistance of MTJ1 is more than MTJ2. Two adjacent cells share a BL contact. The cell area is $8F^2$. Au et al. [37] introduced a new low power-sensing scheme for MTJ MRAM (Figure 57.24). Op-amps A1s are used to generate currents I_{mtj} and I_{ave}. The op-amps are two-stage Miller-compensated differential amplifiers (gain 70 dB and unity gain at 150 MHz). The output resistance of the opamps is comparable to MR. M_1s are current buffers necessary to drive the resistances. To improve the accuracy, the simple current mirrors can be replaced by cascode current mirrors. This sensor can operate at very low-power supply voltages; however, its sensitivity is limited to an MR ratio of 10%. By using the low-power current sense amplifier, the power consumption is reduced 1.46–3.33 times.

Last, Jeong et al. [33] introduced a novel reference cell-sensing scheme for 1T1MTJ cell structures that increases the sensing signal. This allows an increase in MR ratio which is necessary for a reliable MRAM operation. A diagram of the new sensing scheme is presented in Figure 57.25. As shown, the reference

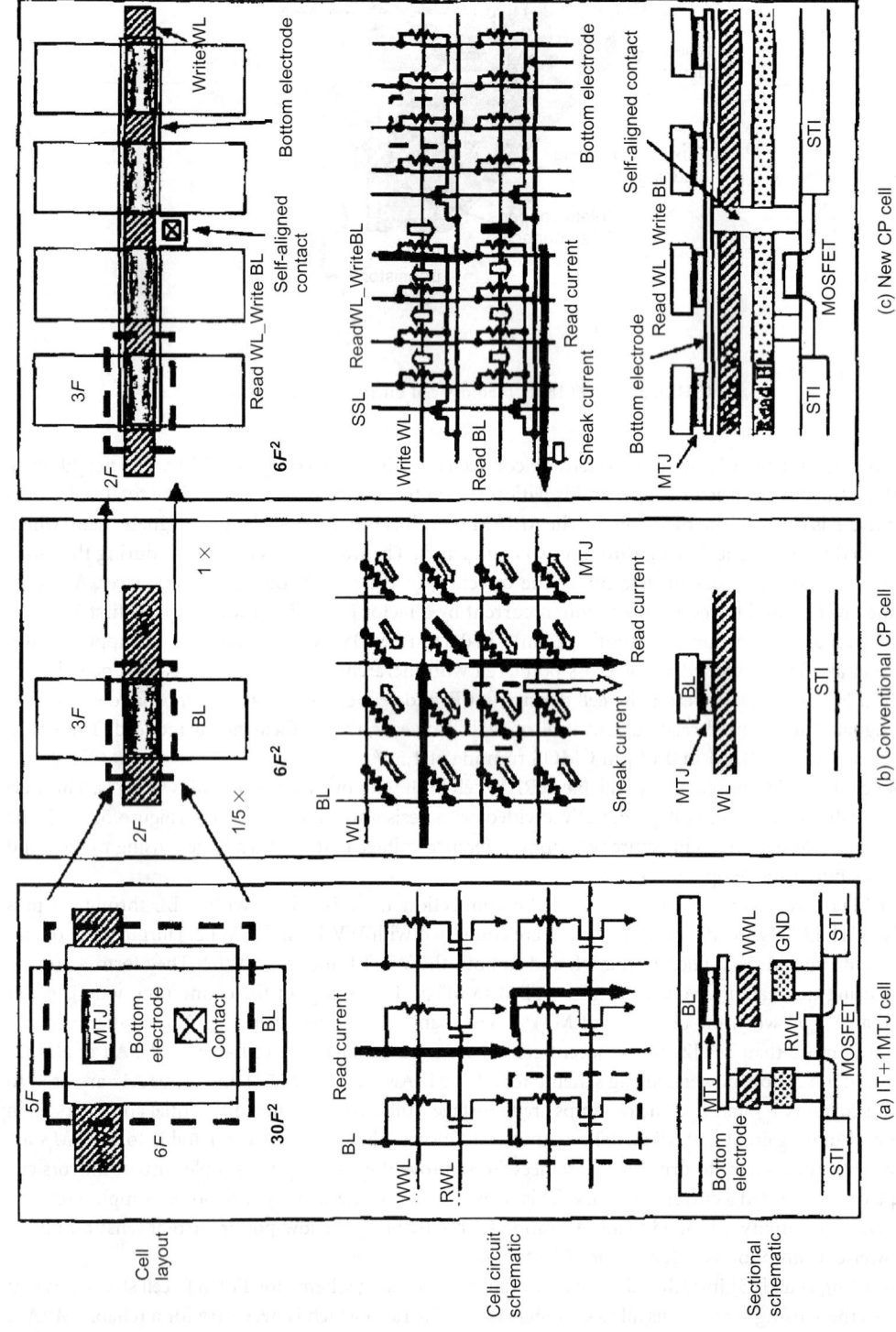

FIGURE 57.22 Cell layouts, the circuit schematic, and the sectional schematic of (a) 1T1MTJ, (b) conventional CP cell, and (c) new CP cell.

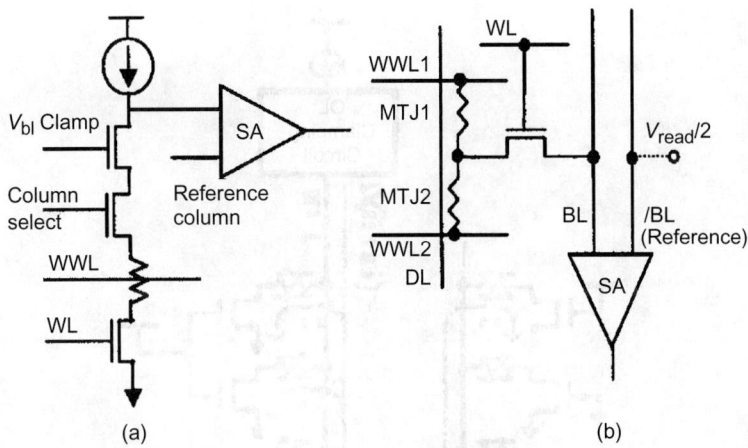

FIGURE 57.23 MRAM cell design: (a) conventional and (b) proposed by Aoki et al. (From M. Aoki et al., *Digest of Technical Papers of the IEEE Symposium on VLSI Circuits*, June 2005, pp. 170–171.)

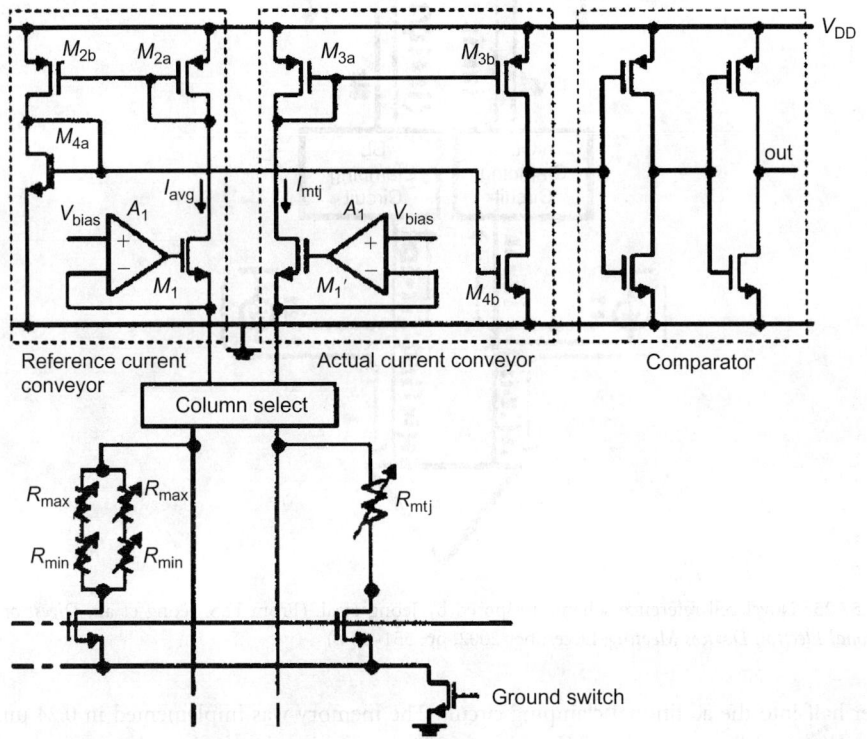

FIGURE 57.24 Low power-sensing circuit as proposed by Au et al. (From E.K.S. Au et al., *IEEE Transactions on Magnetics*, Vol. 40, No. 2, pp. 483–488, 2004.)

cell is built with a pair of memory cells connected in parallel and an additional BL clamping circuit. The clamping circuit draws half of the current flowing into the reference cell. The BL and the BL bar voltages are clamped at 0.4 V. This maintains high MR ratio during sensing operation. Memory is selected by the asserted address and by column decoder lines. The differential sense amplifier compares the currents flowing into the BLs. Half of the reference cell current flows into the sense amplifier and

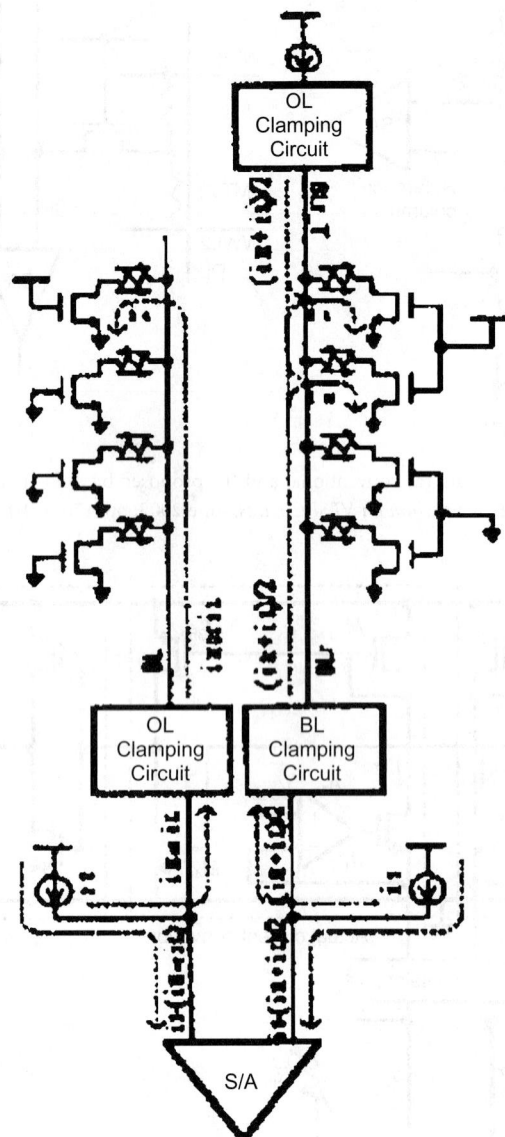

FIGURE 57.25 Novel cell reference scheme proposed by Jeong et al. (From H.S. Jeong et al., *Digest of the IEEE International Electron Devices Meeting*, December 2002, pp. 551–554.)

the other half into the additional clamping circuit. The memory was implemented in 0.24 μm CMOS technology, operated at 2 V, and an MR ratio of >30% was achieved.

57.5 Ferroelectric Memory

Ferroelectric memory (FeRAM) combines the advantages of a nonvolatile flash memory and the density and speed of a dynamic random-access memory (DRAM) memory. Advances in low-voltage, low-power design toward mobile computing applications have been seen in the literature [40,41]. Hirano et al. [40] reported a new 1-transistor 1-capacitor nonvolatile ferroelectric memory architecture that operates at 2 V with 100 ns access time. They achieved these results by using two new improvements, a BL-driven

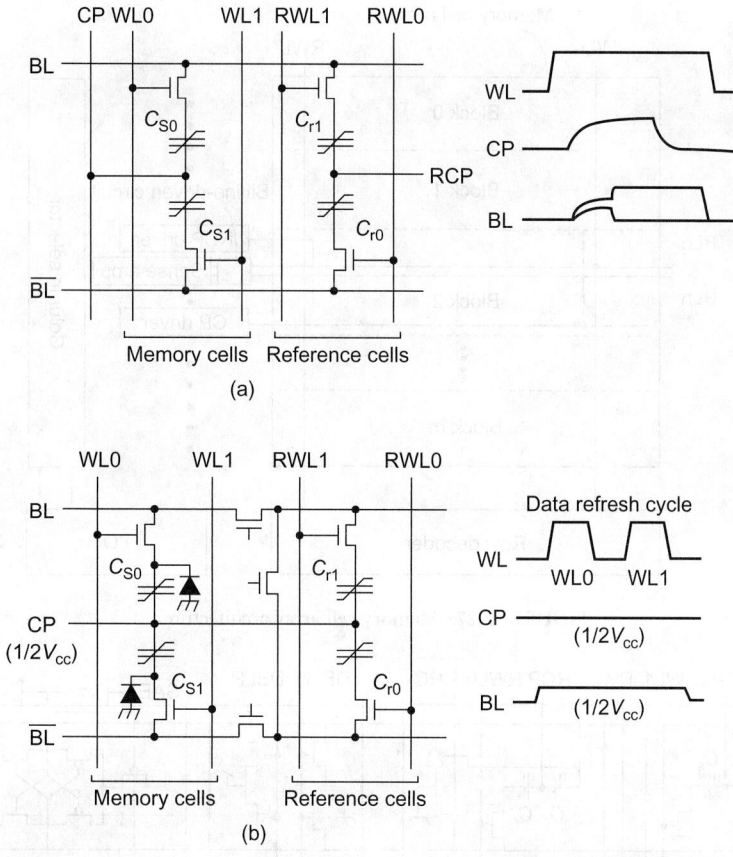

FIGURE 57.26 (a) Cell plate-driven read scheme and (b) noncell plate-driven read scheme.

read scheme and a nonrelaxation reference cell. In previous ferroelectric architectures, either a cell plate-driven or noncell plate-driven read scheme (Figure 57.26[a] and Figure 57.26[b]) was used [42,43]. Eventhough, the first architecture could operate at low supply voltages, the large capacitance of the cell plate, which connects to many ferroelectric capacitors and a large parasitic capacitor, would degrade the performance of the read operation due to large transient time necessary to drive the cell plate. The second architecture suffers from two problems. The first problem is a risk of losing the data stored in the memory due to the leakage current of a capacitor's storage node. The storage node of a memory cell is floating and the parasitic p–n junction between the storage node and the substrate leaks the current. Consequently, the storage node reaches the V_{ss} level and another node of the capacitor is kept at $1/2V_{dd}$ which causes the data destruction. Therefore, this scheme requires a refresh operation of a memory cell data.

The second problem arises from a low-voltage operation. Owing to a voltage across the memory cell capacitor being at $1/2V_{dd}$ under this scheme, the supply voltage must be twice as high as the coercive voltage of ferroelectric capacitors which prevents the low-voltage operation. To overcome these problems, Hirano et al. [40] have developed new BL-driven read scheme which is shown in Figure 57.27 and Figure 57.28. The BL-driven circuit precharges the BLs to supply V_{dd} voltage. The cell plateline is fixed at ground voltage in the read operation. An important characteristic of this configuration is that the BLs are driven while the cell plate is not driven. Also, the precharged voltage level of the BLs is higher than that of the cell plate. Figure 57.29 shows the limitations of previous schemes and the new scheme. During the read operation, the first previously presented scheme [42] requires a long delay time to drive the cell plateline (PL). However, the proposed scheme exhibits faster transient response because

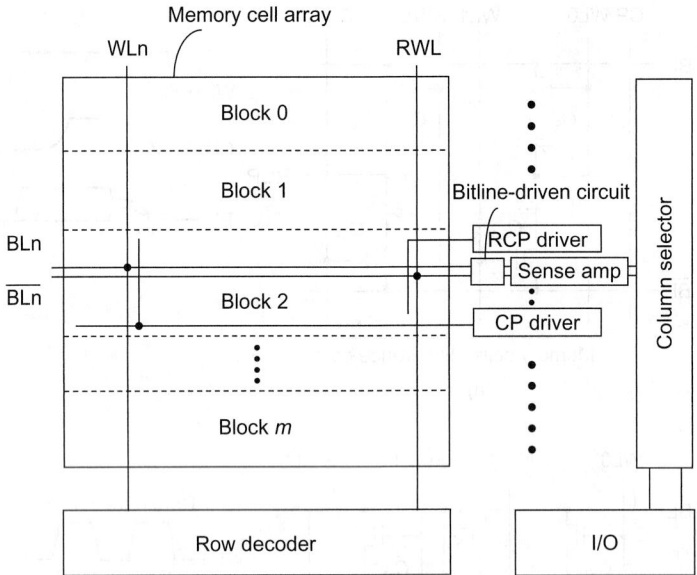

FIGURE 57.27 Memory cell array architecture.

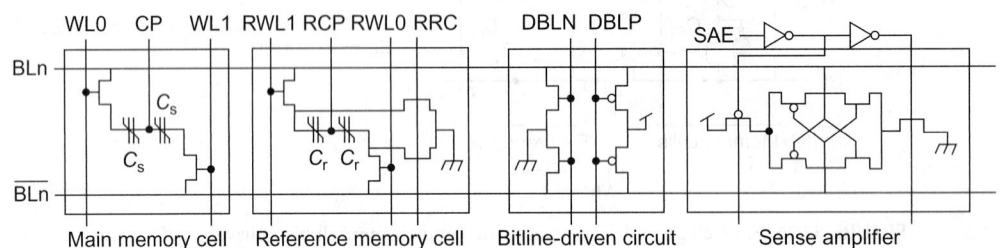

FIGURE 57.28 Memory cell and peripheral circuit with BL-driven read scheme.

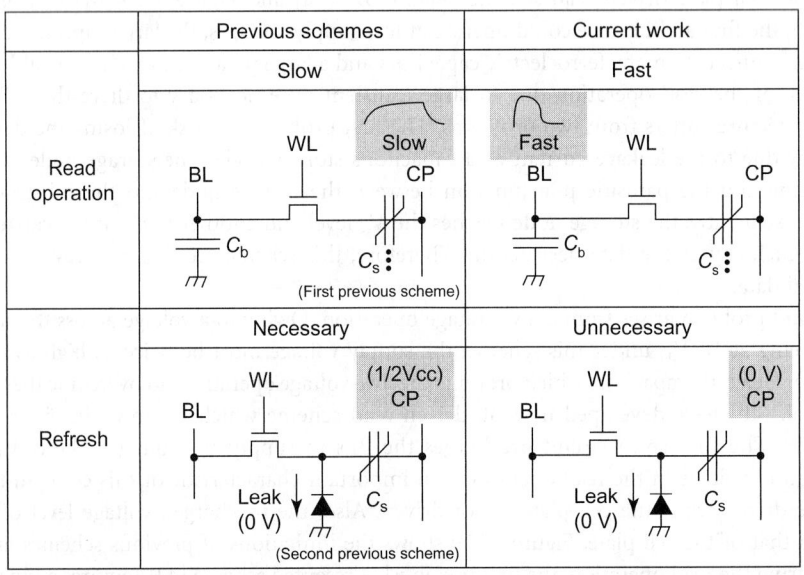

FIGURE 57.29 Limitations of previous schemes and proposed solutions.

the BL capacitance is less than 1/100 of the cell plateline capacitance. The second previously presented scheme [43] requires a data refresh operation to secure data retention.

The read scheme proposed by Hirano et al. [40] does not require any refresh operation since the cell plate voltage is at 0 V during the standby mode.

The reference voltage generated by a reference cell is a critical aspect of a low-voltage operation of ferroelectric memory. The reference cell is constructed with one transistor and one ferroelectric capacitor. While a voltage is applied to the memory cell to read the data, the BL voltage reading from the reference cell is set to about the midpoint of "H" and "L," which are read from the main-memory-cell data. The state of the reference cell is set to "Ref" (left side of Figure 57.30). However, a ferroelectric capacitor suffers from the relaxation effect which decreases the polarization (right side of Figure 57.30). As a result, each state of the main memory cells and the reference cell is shifted and the read operation of "H" data is marginal and prohibits the scaling of power supply voltage. Hirano et al. [40] have developed a reference cell that does not suffer from a relaxation effect, moves always along the curve from the "Ref" point and therefore enlarges the read operation margin for "H" data. This proposed scheme enables a low-voltage operation down to 1.4 V. Hirano et al. [48] successfully developed a low-power embedded 1 Mbit FeRAM operating down to 1.5 V with a ferroelectric capacitor operating down to 0.75 V. The group used two new techniques: a nondriven plate scheme with a nonrefresh operation and a selected driven BL scheme. The combined effect of these two schemes is significant. The memory size is reduced by 53% and the power consumption by 98%. The first scheme uses a reset circuit to reset storage nodes of the memory cells to the voltage of the PL, which isolates the storage nodes of the memory cells. As seen in Figure 57.31, the voltage of the PL and the reset node (VRST) is set to $V_{dd}/2$ (0.75 V). The second scheme is based on (a) a column select-driven BL and (b) divided BL concepts. The BL with a selectable memory cell is fully amplified until the power supply voltage and the other BLs reach $0.5V_{dd}$. The operation is shown in Figure 57.32.

The same group lead by Yamaoka et al. [49] achieved a very low-voltage operation using a novel reference voltage scheme and a multilayer shielded BL structure. The one-transistor one-capacitor (1T1C)-embedded memory operates at just 0.9 V. The new scheme overcomes three problems: first, reference voltage shift by imprint; second, wide distribution of the reference voltage; and third, read margin reduction due to the noise between the BLs in the high-density memory. The proposed reference scheme is shown in Figure 57.33. In the first phase, the stored Data H and L are transferred during read operation to C_{BH} and C_{BL}, respectively. In the second phase, the voltages BLB0 and BLB1 are equalized. As a result, the reference voltage shift can be reduced after imprint and the reference voltage can be adjusted by a reference voltage scheme with averaging multiple data. The imprint time is extended 20 times to 10 years compared with the conventional approach. Figure 57.34 shows the reference voltage-generating scheme. The voltage is generated by averaging m reference cells of the stored

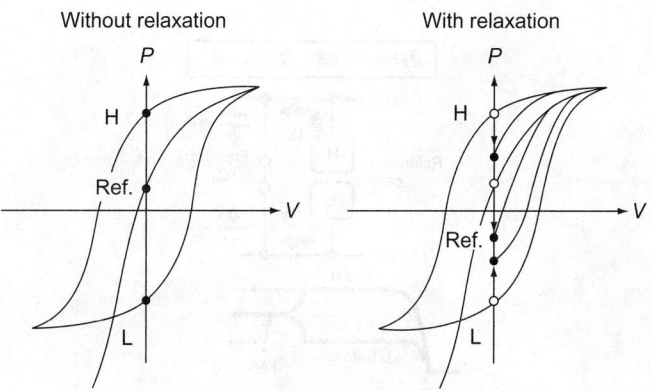

FIGURE 57.30 Reference cell proposed by Sumi et al. [42].

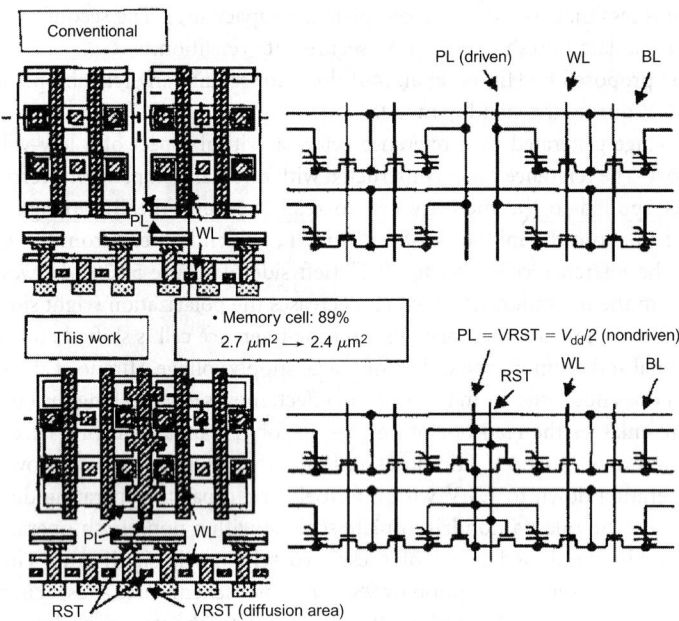

FIGURE 57.31 Organization of the memory and the circuit.

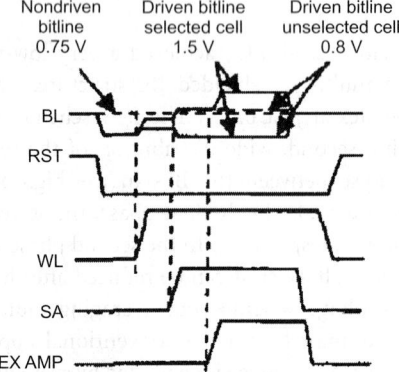

FIGURE 57.32 Timing diagram of the cell selected-driven BL scheme.

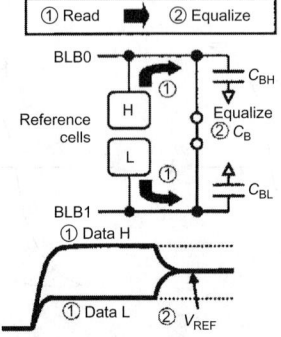

FIGURE 57.33 Reference voltage scheme proposed by Yamaoka et al. [50].

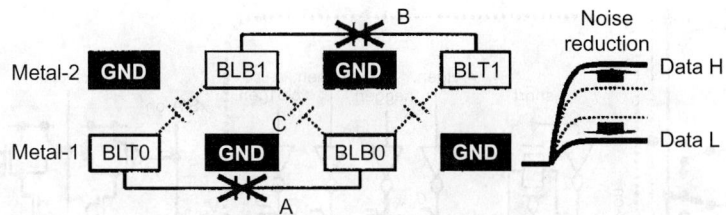

FIGURE 57.34   Reference voltage generating scheme.

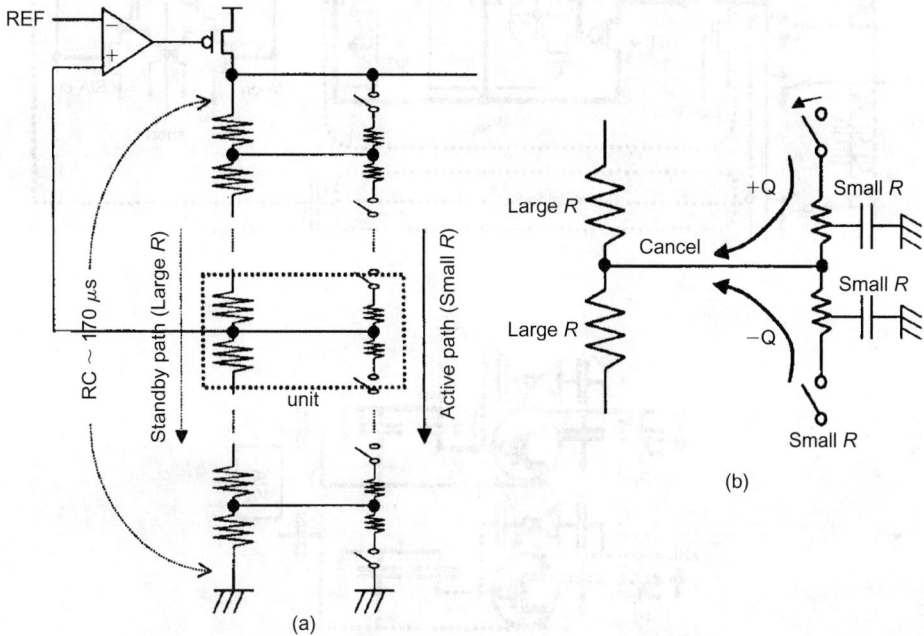

FIGURE 57.35   DC generator circuit for standby current suppression.

Data H and *n* reference cells of the stored Data L. The equalize transistors are located between each BL. The distribution of the reference voltage is reduced by averaging the BL voltages of *m* + *n* reference cells. Furthermore, the reference voltage can be adjusted to the midpoint between Data H and L by optimizing the ratio between *m* and *n* after imprint.

Shiratake et al. [45] demonstrated a 32-Mbyte chain ferroelectric memory using a compact memory cell block structure that eliminates the PL area and reduces the block selector area. In addition, the group used a segment/stitch array architecture which reduces the area row decoders and plate drivers. These two techniques address the power consumption indirectly. Finally, the group introduced a low standby current bias generator that suppresses the standby current to 3 μA. The DC generator circuit is shown in Figure 57.35. The feedback circuit enables switching between small resistance for fast response and large resistance for low standby current. The small resistance is used when the memory is in the standby mode. When the switches are turned off, the remaining charge in the resistor is distributed among the intermediate nodes causing voltage bounces. This can result in an unstable operation. Therefore, the active and standby paths are connected as seen in the figure. As a result of this symmetrical connection, the plus charge from the upper resistor in the active path compensates for the minus charge from the lower resistor when the switches are turned on and off. In this configuration, the redistribution takes less than the RC time of the small resistance.

Kawashima et al. [46] proposed a BL GND sensing (BGS) technique to increase the signal amplitude on the BLs in low-voltage FeRAMs. The sensing circuit is shown in Figure 57.36.

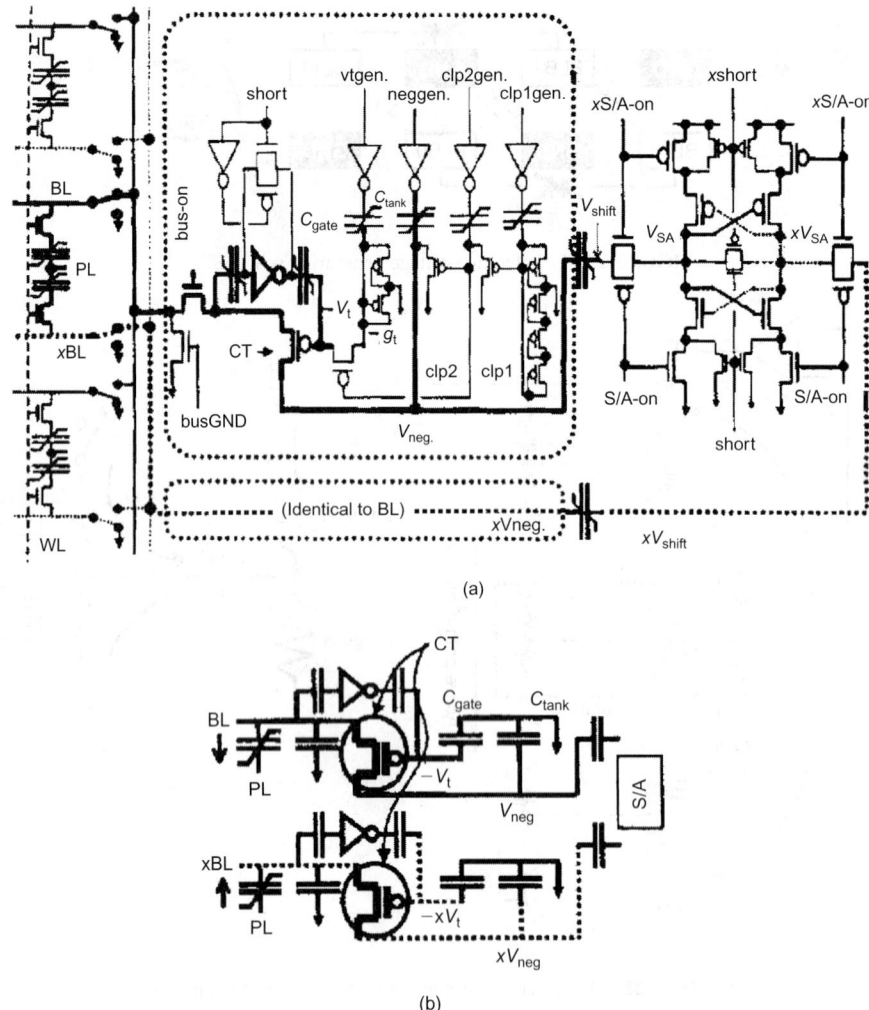

FIGURE 57.36 BL GND sensing circuit: (a) concept and (b) detail.

After the PL goes high, the negative charge sinks the polarization charge at the BL through a PMOS charge transfer (CT). V_{neg} will increase through charging by C_{tank}. The PMOS gate is biased at its threshold voltage for the source follower to keep the BL charge at the GND level. An optional inverter amplifier could be used to improve the feedback gain and the clipping of the BL at the GND voltage. The voltage difference between V_{neg} and xV_{neg} creates V_{SA}. Therefore, V_{SA} is a function of C_{tank}. Thus, a smaller C_{tank} and a larger V_{SA} signal can be obtained. If BL capacitance is ~1 pF (512 cells connected to BL), the V_{SA} swing can be 2.5 times larger using BGS technique compared with the traditional high impedance sensing.

Figure 57.37 and Figure 57.38 show a hierarchical BL sensing scheme and its timing diagram [47]. The cell array block is composed of multiple subcell array blocks with folded BL structure. To reduce capacitance of SBL, each cell array of SBL is limited to 64 WL rows. Reference BL is composed from neighbor cell array block like open BL architecture, with SW1 and SW2 signals controlling the switch devices. Therefore, the reference level is immune to coupling noise (CN). Each cell data-sensing voltage is protected from CN by the unused neighbor BL. During the active time period T0, SBL and MBL are coupled by activation to V_{dd} or V_{pp} level. The WL and PL signals are activated to V_{pp} level of near double the V_{dd} level. The sensing voltage of SBL is transferred to MBL. After the sense amplifier is activated, SBSW1 and PL signals are pulled down to GND level. The SBSW2 signal is switched to V_{pp} level.

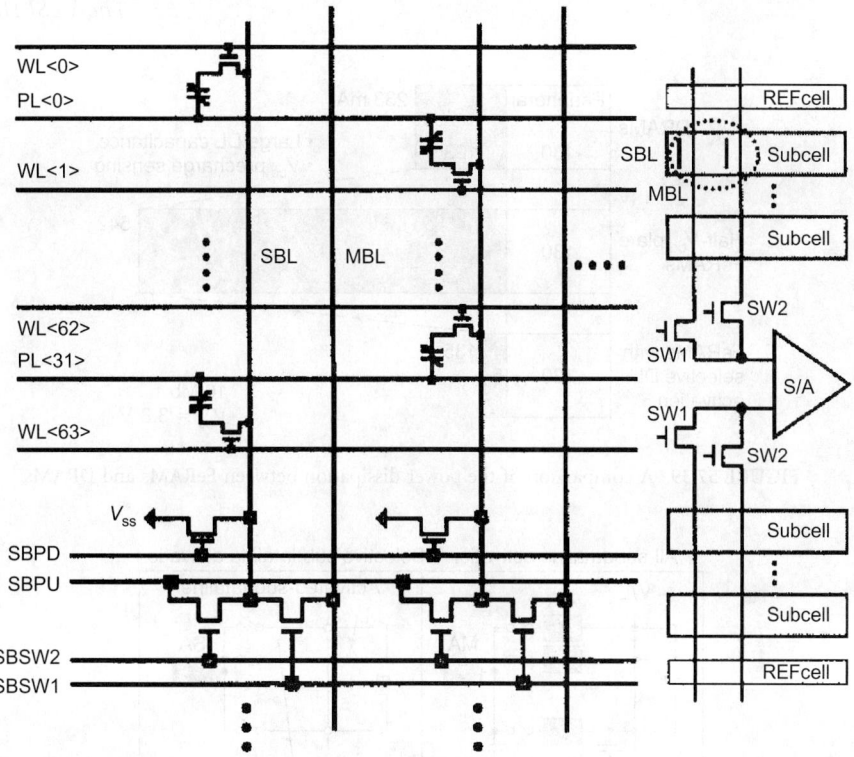

FIGURE 57.37 Hierarchy BL scheme and hybrid BL sensing. (From H.-B. Kang et al., *ISSCC Digest of Technical Papers*, Vol. 1, No. 3–7, pp. 158–159, 2002.)

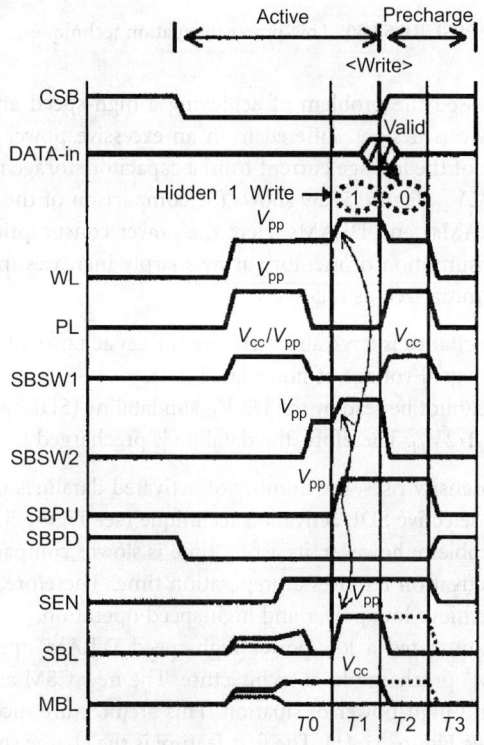

FIGURE 57.38 Timing diagram of the hierarchy BL scheme. (From H.-B. Kang et al., *ISSCC Digest of Technical Papers*, Vol. 1, No. 3–7, pp. 158–159, 2002.)

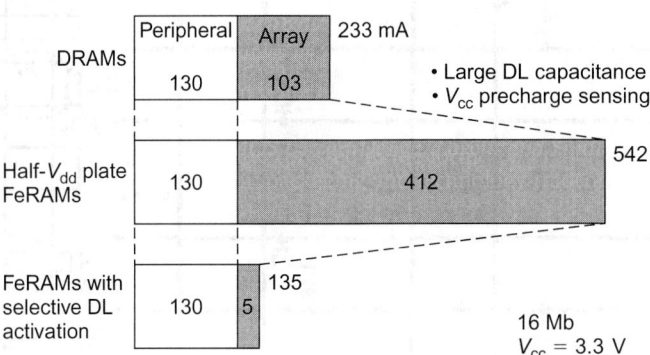

FIGURE 57.39 A comparison of the power dissipation between FeRAMs and DRAMs.

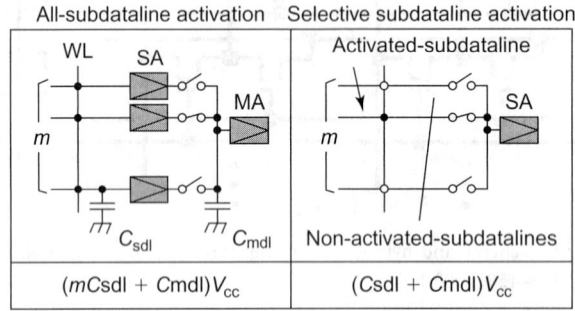

FIGURE 57.40 Low-power dissipation techniques.

Fujisawa et al. [41] addressed the problem of achieving a high-speed and low-power operation in ferroelectric memories. Previous designs suffered from an excessive power dissipation due to needed refresh cycle [42,43] because of the leakage current from a capacitor storage node to the substrate where the cell plates are fixed to $1/2V_{dd}$. Figure 57.39 shows the comparison of the power dissipation between ferroelectric memories (FeRAMs) and DRAMs. Here the power consumption of peripheral circuits is identical, but the power consumption of memory array sharply increases in the $1/2V_{dd}$ plate FeRAMs. These problems could be summarized as follows:

- The memory cell capacitance is large and therefore the capacitance of the dataline needs to be set larger to increase the signal voltage of nonvolatile data.
- The nonvolatile data cannot be read by the $1/2 V_{dd}$ subdataline (SDL) precharge technique because the cell plate is set to $1/2V_{dd}$. Therefore, the dataline is precharged to V_{dd} or GND.

When the memory cells' density rises, the number of activated datalines increases. This increases the array's power dissipation. A selective SDL activation technique (see Figure 57.40) proposed by Fujisawa et al. [41] overcomes this problem; however, its access time is slower compared with all-SDL activation because the selective SDL activation requires a preparation time. Therefore, neither of these two techniques can simultaneously achieve low-power and high-speed operation.

Fujisawa et al. [41] demonstrated a low-power high-speed FeRAM operation using an improved charge-share modified (CSM) precharge-level architecture. The new CSM architecture solves the problems of slow access speed and high-power dissipation. This architecture incorporates two features that reduce the sensing period (see Figure 57.41). The first feature is the charge sharing between the parasitic capacitance of the main dataline (MDL) and the SDL. During the standby mode, all SDLs and MDLs are precharged to $1/2 V_{dd}$ and V_{dd}, respectively. During the read operation, the precharge circuits are all

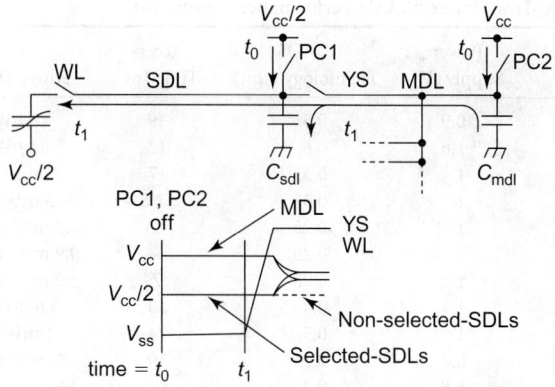

FIGURE 57.41 Principle of the CSM architecture.

cut off from the datalines (time t_0). After the y-selection signal (YS) is activated (time t_1), the charge in the parasitic capacitance of the MDL (C_{mdl}) is transferred to the selected parasitic capacitance of the SDL (C_{sdl}) and the selected SDL potential is raised by charge sharing. As a result, the voltage is applied only to a memory cell intersecting selected WL and YS. The second feature is a simultaneous activation of WL and YS without causing a loss of the readout voltage. During the write operation, only data of the selected memory cell are written whereas all the other memory cells keep their nonvolatile data.

Consequently, the power dissipation does not increase during this operation. The writing period is equal to the sensing period because WL and YS can be also activated simultaneously in the write cycle.

57.6 Static Random-Access Memory (SRAM)

SRAMs have experienced a very rapid development of low-power, low-voltage memory design during recent years due to an increased demand for notebooks, laptops, hand-held communication devices, and IC memory cards. Table 57.3 summarizes some of the latest experimental SRAMs for very low-voltage and low-power operation.

In this section, active and passive sources of power dissipation in SRAMs will be discussed and common low-power techniques will be analyzed.

57.6.1 Low-Power SRAMs

57.6.1.1 Sources of SRAM Power

There are different sources of active and standby (data retention) power present in SRAMs. The active power is the sum of the power consumed by the following components:

- decoders
- memory array
- sense amplifiers
- periphery (I/O circuitry, write circuitry, etc.) circuits

The total active power of an SRAM with $m \times n$ array of cells can be summarized by the expression [9,50,51]

$$P_{\text{active}} = (mi_{\text{active}} + m(n-1)i_{\text{leak}} + (n+m)fC_{\text{DE}}V_{\text{INT}} + mi_{\text{DC}}\,\Delta tf + C_{\text{PT}}V_{\text{INT}}f + I_{\text{DCP}}) * V_{\text{dd}} \qquad (57.1)$$

TABLE 57.3 Low-Power SRAMs Performance Comparison

Memory Size (Ref.)	Power Supply (V)	CMOS Technology (μm)	Access Time (ns)	Power Dissipation
4 Kb [40]	0.9	0.6	39	18 μW at 1 MHz
4 Kb [40]	1.6	0.6	12	64 μW at 1 MHz
32 Kb [44]	1	0.35	17	5 mW at 50 MHz
32 Kb [48]	1	0.35	11.8	3 mW at 10 MHz
32 Kb [49]	1	0.25	7.3	0.9 mW at 100 MHz
32 Kb [42]	1	0.25	—	0.9 mW at 100 MHz
32 Kb [55]	1	0.25	7	3.9 mW at 100 MHz
256 Kb [53]	1.4	0.4	60	3.6 mW at 5 MHz
1 Mb [50]	1	0.5	74	1 mW at 10 MHz
1 Mb [52]	0.8	0.35	10	5 mW at 100 MHz
4.5 Mb [51]	1.8	0.25	1.8	2.8 mW at 550 MHz
7.5 Mb [47]	3.3	0.6	6	8.42 mW at 50 MHz
7.5 Mb [58]	3.3	0.8	18	4.8 mW at 20 MHz

where i_{active} is the effective current of selected cells, i_{leak} the effective data retention current of the unselected memory cells, C_{DE} the output node capacitance of each decoder, V_{INT} the internal power supply voltage, i_{DC} the DC current consumed during the read operation, Δt the activation time of the DC current consuming parts (i.e., sense amplifiers), f the operating frequency, C_{PT} the total capacitance of the CMOS logic and the driving circuits in the periphery and the I_{DCP} the total static (DC) or quasi-static current of the periphery. Major sources of I_{DCP} are column circuitry and differential amplifiers on the I/O lines.

The standby power of an SRAM has a major source represented by $i_{leak} mn$, because the static current from other sources is negligibly small (sense amplifiers are disabled during this mode). Therefore, the total standby power can be expressed as

$$P_{stndby} = mni_{leak} * V_{dd} \tag{57.2}$$

57.6.1.2 Techniques for Low-Power Operation

To significantly reduce the power consumption in SRAMs all contributors to the total power must be targeted. The most efficient techniques used in recent memories are:

- Capacitance reduction of wordlines and the number of cells connected to them, datalines, I/O lines, and decoders
- DC current reduction by using new pulse operation techniques for wordlines, periphery circuits, and sense amplifiers
- AC current reduction by using new decoding techniques (i.e. multistage static CMOS decoding)
- Operating voltage reduction
- Leakage current reduction (in active and standby mode) by utilizing multiple threshold voltage (MT-CMOS) or variable threshold voltage technologies (VT-CMOS).

57.6.1.2.1 Capacitance Reduction

The largest capacitive elements in a memory are wordline, BLs, and datalines each with a number of cells connected to them. Therefore, reducing the size of these lines can have a significant impact on power consumption reduction. A common technique used often in large memories is called divided word line (DWL) which adopts a two-stage hierarchical row-decoder structure as shown in Figure 57.42 [51]. The number of subwordlines connected to one main wordline in the dataline direction is generally four, substituing the area of a main row decoder with the area of a local row decoder. DWL features two-step decoding for selecting one wordline, greatly reducing the capacitance of the address lines to a row decoder and the wordline RC delay.

A single BL cross-point cell activation (SCPA) architecture reduces the power further by improving the DWL technique [53]. The architecture enables the smallest column current possible without increasing the block division of the cell array thus reducing the decoder area and the memory core area. The cell architecture is shown in Figure 57.43. The Y-address controls the access transistors and the X-address. Since only one memory cell at the cross-point of X- and Y-address is activated, a column current is drawn only by the accessed cell. As a result, the column current is minimized. In addition, SCPA allows the number of blocks to be reduced because the column current is independent of the number of block divisions in the SCPA. The disadvantage of this configuration is that during the write "high" cycle, both X- and Y-lines have to be boosted using a wordline boost circuit.

Caravella [56,57] has proposed a similar subdivision technique as DWL which he demonstrated on 64×64 bit cell array. If C_j is a parasitic capacitance associated with a single bit cell load on a BL (junction and metal) and if C_{ch} a parasitic capacitance associated with a single bit cell on the wordline (gate, fringe, and metal), then the total BL capacitance is $64 * C_j$ and the total word capacitance is $64 * C_{ch}$. If the array is divided into four isolated subarrays of 32×32 bit cells, the total BL and wordline capacitances would be halved (see Figure 57.44). The total capacitance per read/write that would need to be discharged or charged is given $1024 * C_j + 32 * C_{ch}$ for the subarray architecture as opposed to $4096 * C_j + 64 * C_{ch}$ for the 64×64 array. This technique carries a penalty due to additional decode and control logic and routing.

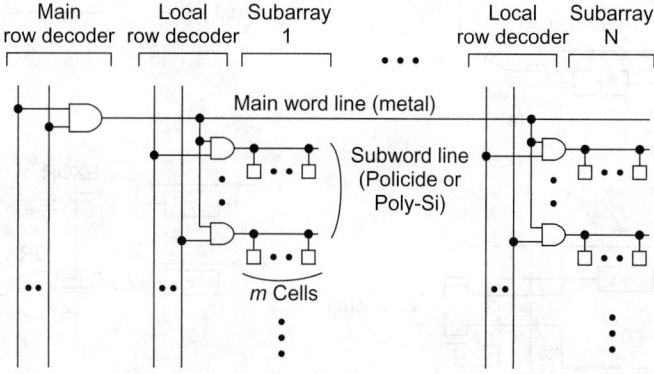

FIGURE 57.42 Divided wordline.

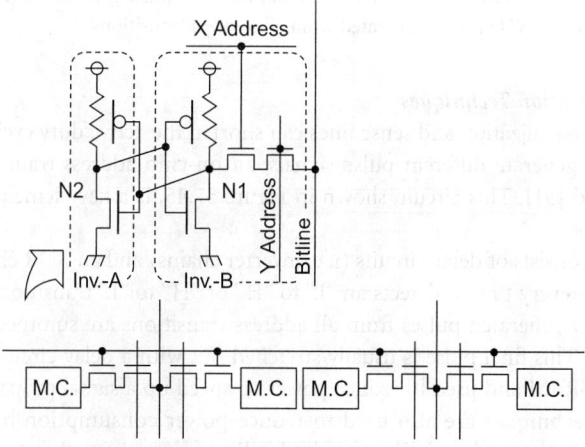

FIGURE 57.43 Memory cell used for SCPA architecture.

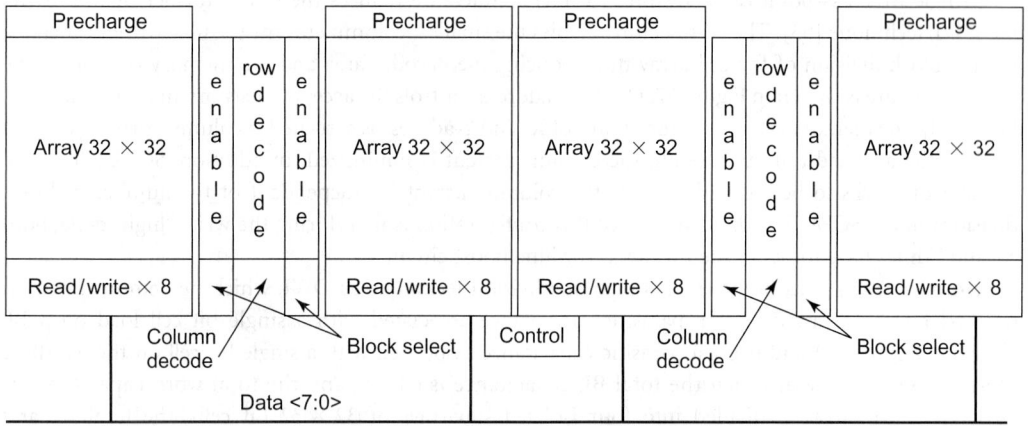

FIGURE 57.44 Memory architecture.

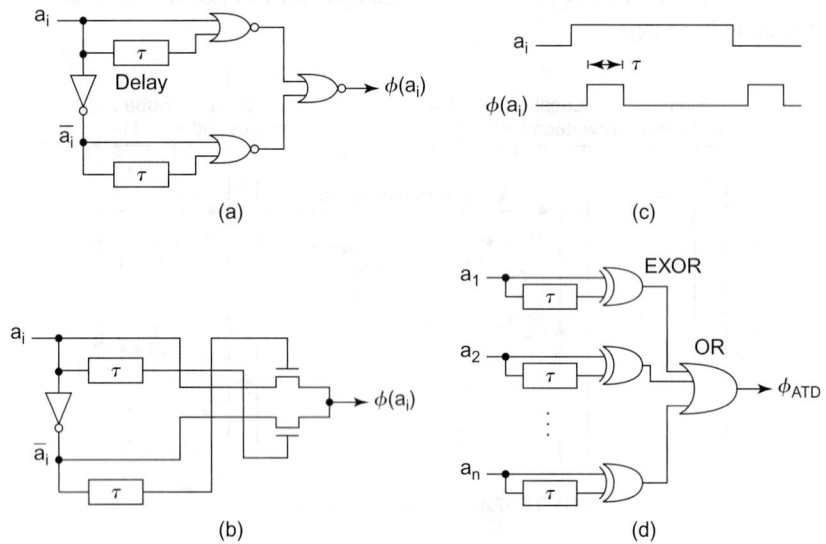

FIGURE 57.45 Address transition detection circuits: (a) and (b) ATD pulse generators; (c) ATD pulse waveforms; (d) a summation circuit of all ATD pulses generated from all address transitions.

57.6.1.2.2 Pulse Operation Techniques

Pulsing the wordlines, equalization and sense lines can shorten the active duty cycle and thus reduce the power dissipation. To generate different pulse signals, an on-chip address transition detection (ATD) pulse generator is used [51]. This circuit, shown in Figure 57.45, is a key element for the active power reduction in memories.

An ATD generator consists of delay circuits (i. e. inverter chains) and an XOR circuit. The ATD circuit generates a $\phi(a_i)$ pulse every time it detects an "L" to "H" or "H" to "L" transition on the input address signal a_i. Then all ATD-generated pulses from all address transitions are summed through an OR gate to a single pulse ϕ_{ATD}. This final pulse is usually stretched out with a delay circuit to generate different pulses needed in the SRAM and used to reduce power or speed up a signal propagation.

Pulsed operation techniques are also used to reduce power consumption by reducing the signal swing on high-capacitance predecode lines, write-buslines, and BLs without sacrificing the performance [54,59,67]. These techniques target the power that is consumed during write and decode

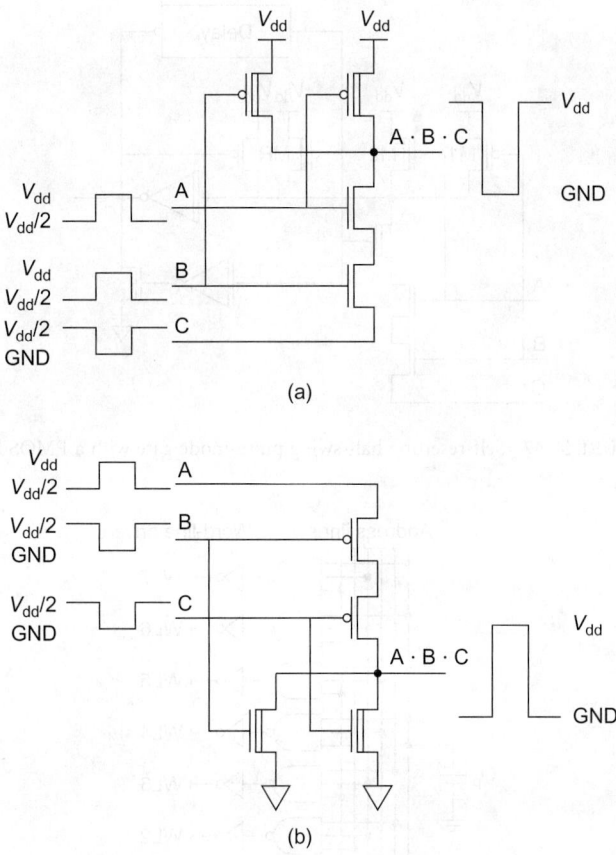

FIGURE 57.46   Half-swing pulse-mode AND gate: (a) NMOS style and (b) PMOS style.

operations. Most of the power savings come from operating the BLs from $V_{dd}/2$ rather than V_{dd}. This approach is based on new half-swing pulse-mode gate family. Figure 57.46 shows a half-swing pulse-mode AND gate. The principle of the operation is in a merger of a voltage-level converter with a logical AND. A positive half-swing (transitions from a rest state $V_{dd}/2$ to V_{dd} and back to $V_{dd}/2$) and a negative half-swing (transitions from a rest state $V_{dd}/2$ to GND and back to $V_{dd}/2$) combined with the receiver-gate logic style result in a full gate overdrive with negligible effects of the low-swing inputs on the performance of the receiver. This structure is combined with a self-resetting circuitry and a PMOS leaker to improve the noise margin and the speed of the output reset transition (see Figure 57.47).

Both negative and positive half-swing pulses can reduce the power consumption further by using a charge recycling. The charge used to produce the assert transition of a positive pulse can also be used to produce the reset transition of a negative pulse. If the capacitances of positive and negative pulses match, then no current would be drawn from the $V_{dd}/2$ power supply ($V_{dd}/2$ voltage is generated by an on-chip voltage converter).

Combining the half-swing pulse-mode logic with the charge recycling techniques, 75% of the power on high-capacitance lines can be saved [67].

57.6.1.2.3   AC Current Reduction

One of the circuit techniques that reduces AC current in memories is a multistage decoding. It is common that fast static CMOS decoders are based on OR/NOR and AND/NAND architectures. Figure 57.48 shows one example of a row decoder for a three-bit address. The input buffers drive the interconnect capacitance

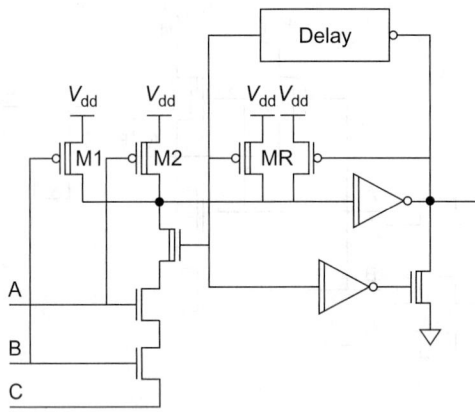

FIGURE 57.47   Self-resetting half-swing pulse-mode gate with a PMOS leaker.

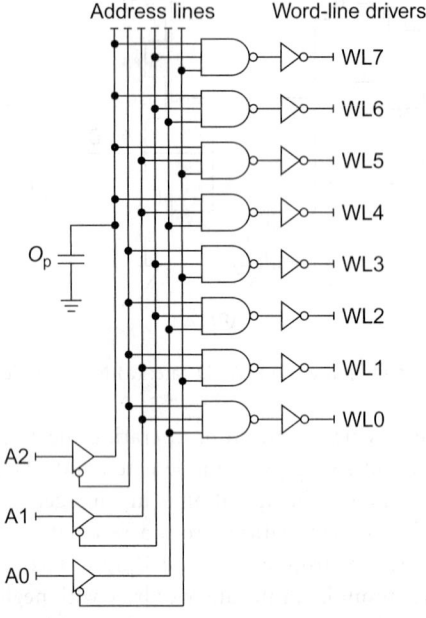

FIGURE 57.48   A row decoder for a three-bit address.

of the address line and also the input capacitance of the NAND gates. By using a two-stage decode architecture, the number of transistors, fanin, and the loading on the address input buffers is reduced (see Figure 57.49). As a result, both speed and power are optimized. The signal ϕ_x, generated by the ATD pulse generator, enables the decoder and secures pulse-activated wordline.

57.6.1.2.4 *Operating Voltage Reduction and Low-Power Sensing Techniques*

Operating voltage reduction is the most powerful method for power conservation. Power supply voltage reductions down to 1 V [52,59,61,63,66–68,73] and below [57,70,71] have been reported. This aggressively scaled environment requires news skills in new fast speed and low-power sensing schemes. A charge-transfer sense amplifying scheme combined with a dual-V_t CMOS circuit achieves a fast sensing speed and a very low power dissipation at 1 V power supply [61,73]. At this voltage level, the "roll-off" on threshold voltage versus gate length, the shortest gate length causes the V_{th} mismatch between the pairs of MOSFETs in the differential sense amplifier. Figure 57.50 shows the CT sense amplifier. The CT

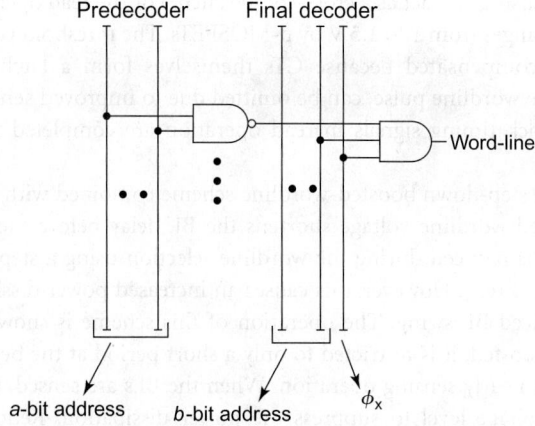

FIGURE 57.49 A two-stage decoder architecture.

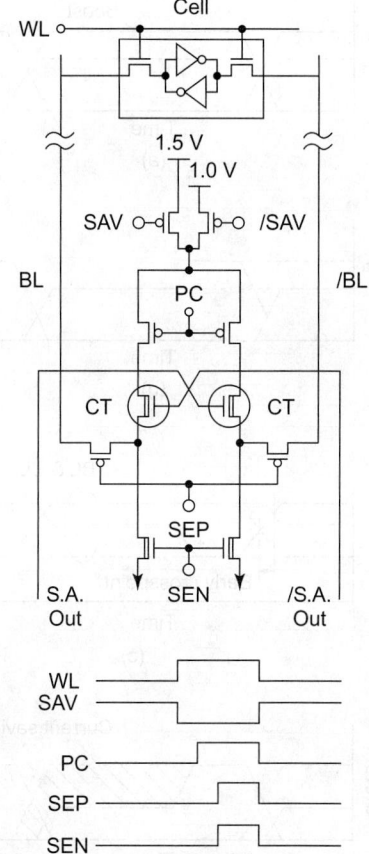

FIGURE 57.50 Charge-transfer sense amplifier.

transistors perform the sensing and act as a cross-couple latch. For the read operation, the supply voltage of the sense amplifiers changes from 1 to 1.5 V by p-MOSFETs. The threshold voltage mismatch between two CTs is completely compensated because CTs themselves form a latch. Consequently, the BL precharge time, before the wordline pulse, can be omitted due to improved sensitivity. The cycle time is shortened because all clock timing signals in read operation are completed within the width of the wordline pulse.

Another method is the step-down boosted-wordline scheme combined with current-sensing amplification. Boosting a selected wordline voltage shortens the BL delay before the stored data are sensed. The power consumption is reduced during the wordline selection using a stepping down technique of selected worldline potential [63]. However, this causes an increased power dissipation and a large transition time due to enhanced BL swing. The operation of this scheme is shown in Figure 57.51. After the selected wordline is boosted, it is restricted to only a short period at the beginning of the memory-cell access. This enables an early sensing operation. When the BLs are sensed, the wordline potential is reduced to the supply voltage level to suppress the power dissipation. Reduced signals on the BLs are sufficient to complete the read cycle with the current sensing. A fast read operation is obtained with little power penalty. The step-down boosting method is also used for write operation. The circuit diagram of this method is shown in Figure 57.52.

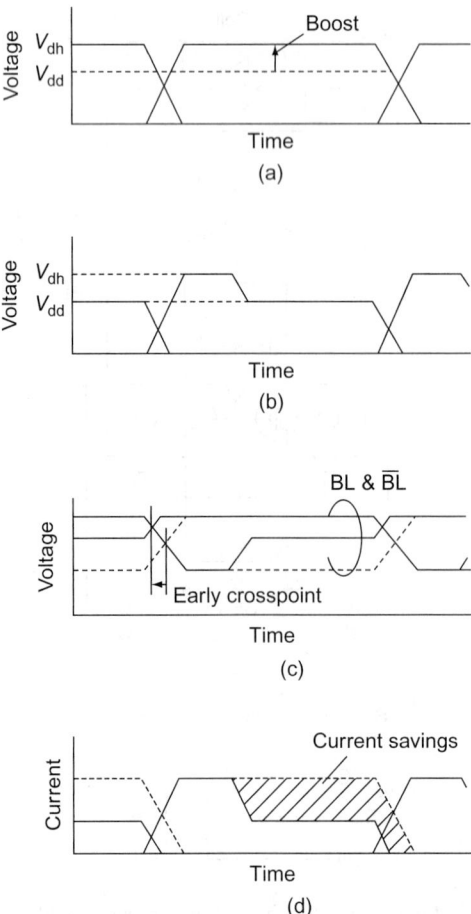

FIGURE 57.51 Step-down boosted wordline scheme: (a) conventional, (b) step-down boosted wordline, (c) BL transition, and (d) current consumption of a selected memory cell.

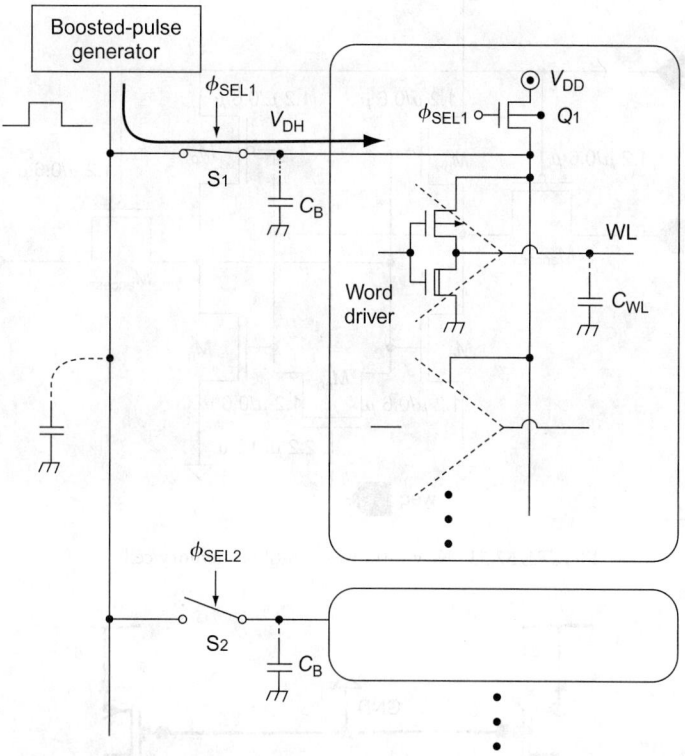

FIGURE 57.52 Circuit schematic of step-down boosted wordline method.

Word drivers are connected to the boosted-pulse generator via switches S1 and S2. These switches separate the parasitic capacitance C_B from the boosted line, thus reducing its capacitance. NMOS transistors are more suitable for implementing these switches because they do not require a level-shift circuit. Transistor Q1 is used for the stepping down function. During the boost, the gate electrode is set to V_{dd}. If the wordline charge exceeds $V_{dd} + |V_{tp}|$, then Q1 ($|V_{tp}|$ is a threshold voltage of Q1) turns on and the wordline is clamped. After the stepping-down process, ϕ_{SEL} switches low and Q1 guarantees V_{dd} voltage on the wordline.

An efficient method for reducing the AC power of BLs and datalines is to use the current-mode read and write operations based on new current-based circuit techniques [64,74,75]. Wang et al. proposed a new SRAM cell that supports current-mode operations with very small voltage swings on BLs and datalines. A fully current-mode technique consumes only 30% of the power consumed by a previous current-read only design. Very small voltage swings on BLs and datalines lead to a significant reduction of AC power. The new memory cell has seven transistors as shown in Figure 57.53. The additional transistor M_{eq} clears the content of the memory cell prior to the write operation. It performs the cell equalization. This transistor is turned off during the read operation so it does not disrupt the normal operation. An n-type current conveyor is inserted between the data input cell and the memory cell to perform a current-mode write operation which is a complementary way to read. The equalization transistor is sized to be as large as possible to improve fast equalization speed, but not to increase the cell size.

After a suitable sizing, the new 7-transistor (7T) cell is 4.3% smaller than its 6-transistor (6T) counterpart, as illustrated in Figure 57.55.

Wieckowski and Margala [65] introduced a 5-transistor (5T) SRAM cell that improves the noise margin by 60%, reduces the power consumption by 12% and the area by 6% compared with traditional 6T [65]. The new memory cell is shown in Figure 57.54. The main characteristics of the design are PMOS transistors that are directly connected to the BLs and an additional transistor M5 that couples the

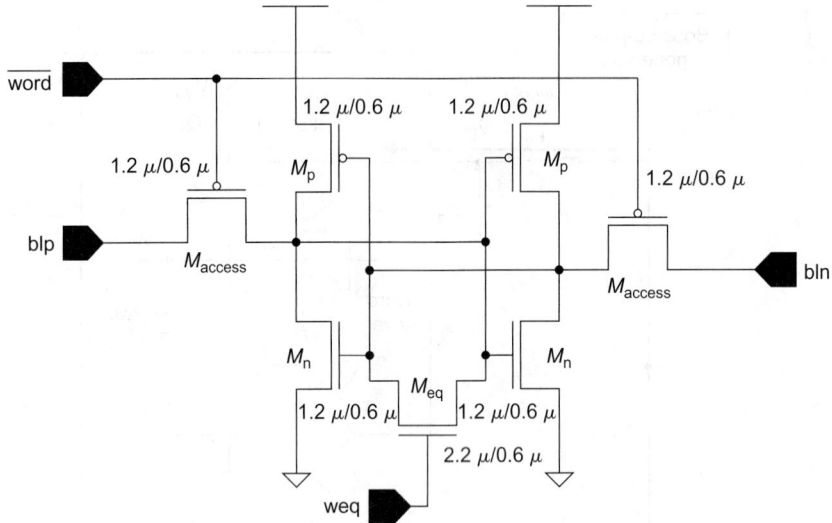

FIGURE 57.53 New 7-transistor SRAM memory cell.

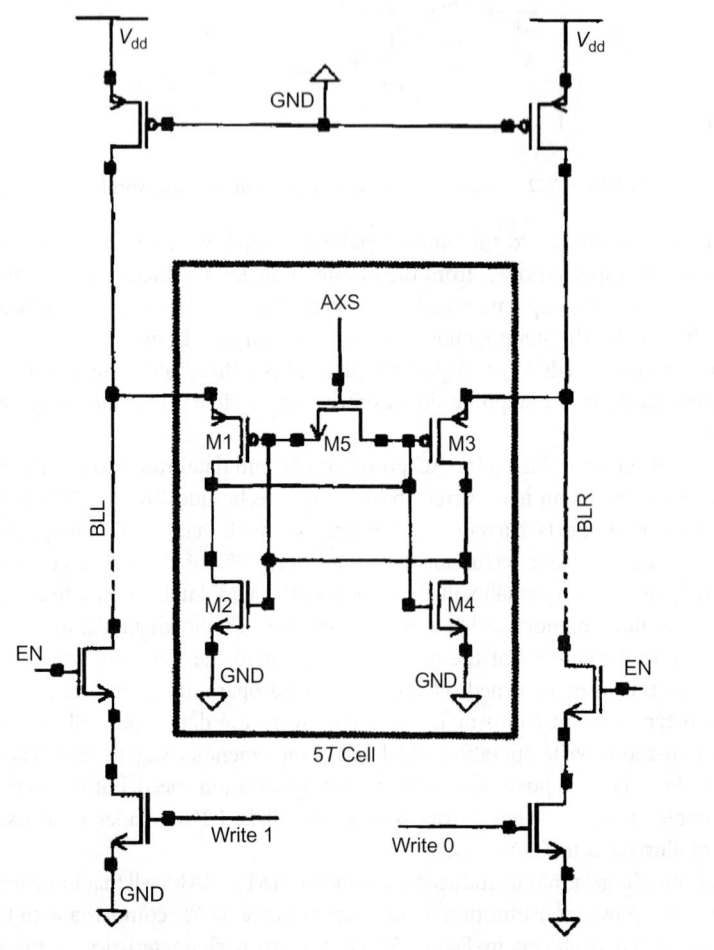

FIGURE 57.54 New 5-transistor SRAM memory cell.

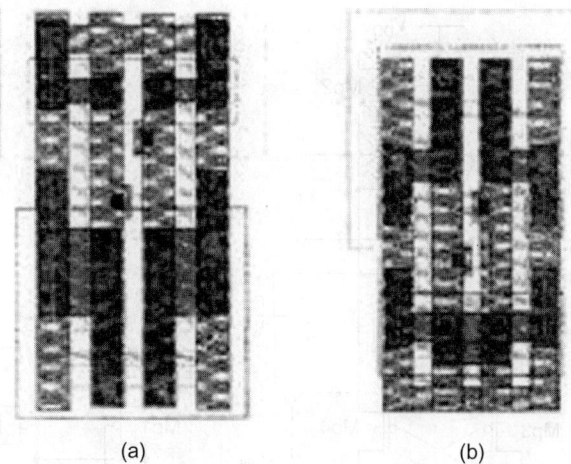

(a)   (b)

FIGURE 57.55 SRAM cell layout: (a) 6T cell and (b) new 7T cell.

inverters. In steady state, M5 is off. Data are preserved by the cross-coupled inverters, and no write signals are applied to the column NMOS transistors. The BLs provide power to the cells and are charged using precharge transistors at the top of each column. During the read operation, only the access (AXS) signal of the selected cell is asserted to turn on M5. M5 creates a current path from the BL to the GND through the cell. The added current drawn from PMOS transistors on the top of the cell column creates a differential voltage on the BL pair sufficient for standard sense amplifiers. Writing operation begins in a very similar way. The AXS signal is asserted and either Write 1 or Write 0 is activated to pull off the BLs to $2/3 V_{dd}$. This reduces the current flow through the cells and the voltage drop across M5.

Another new current-mode sense amplifier for 1.5 V power supply was proposed by Wang and Lee [75]. The new circuit overcomes the problems of a conventional sense amplifier with pattern dependency by implementing a modified current conveyor. Pattern-dependency problem limits the scaling of the operating voltage. Also, the circuit does not consume any DC power because it is constructed as a complementary device. As a result, the power consumption is reduced by 61–94% compared with a conventional design. The circuit structure of the modified current conveyor is similar to a conventional current conveyor design. However, an extra PMOS transistor Mp7, as seen in Figure 57.56, is used. The transistor is controlled by RX signal (a complement of CS). After every read cycle, transistor Mp7 is turned on and equalizes nodes RXP and RXN, which eliminates any residual differential voltage between these two nodes (limitation in conventional designs).

57.6.1.2.5 *Leakage Current Reduction*

To effectively reduce the dynamic power consumption, the threshold voltage is reduced along with the operating voltage. However, low threshold voltages increase the leakage current during both active and standby modes. The fundamental method for a leakage current reduction is a dual-V_{th} or a variable-V_{th} circuit technique. An example of one such technique is shown in Figure 57.57 [61,73]. Here, high V_{th} MOS transistors are utilized to reduce the leakage current during standby-mode. As the supply voltage for the word decoder (g) is lowered to 1 V, all transistors forming the decoder are low V_{th} to retain high performance. The leakage currents during the stand-by mode are substantially reduced by a cut-off switch (SWP, SWN). SWN consists of a high V_{th} transistor and SWP consists of a low V_{th} transistor. Both switches are controlled by a 1.5 V signal. Hence, the SWN gains a considerable conductivity. SWP can be quickly cut off because of the reverse biasing. The operating voltage of the local decoder (w) is boosted to 1.5 V. The high operating voltage gives sufficient drivability even to high V_{th} transistors.

This technique belongs to schemes that use dynamic boosting of the power supply voltage and the wordlines. However, in these schemes the gate voltage of MOSFETs is raised often to more than 1.4 V

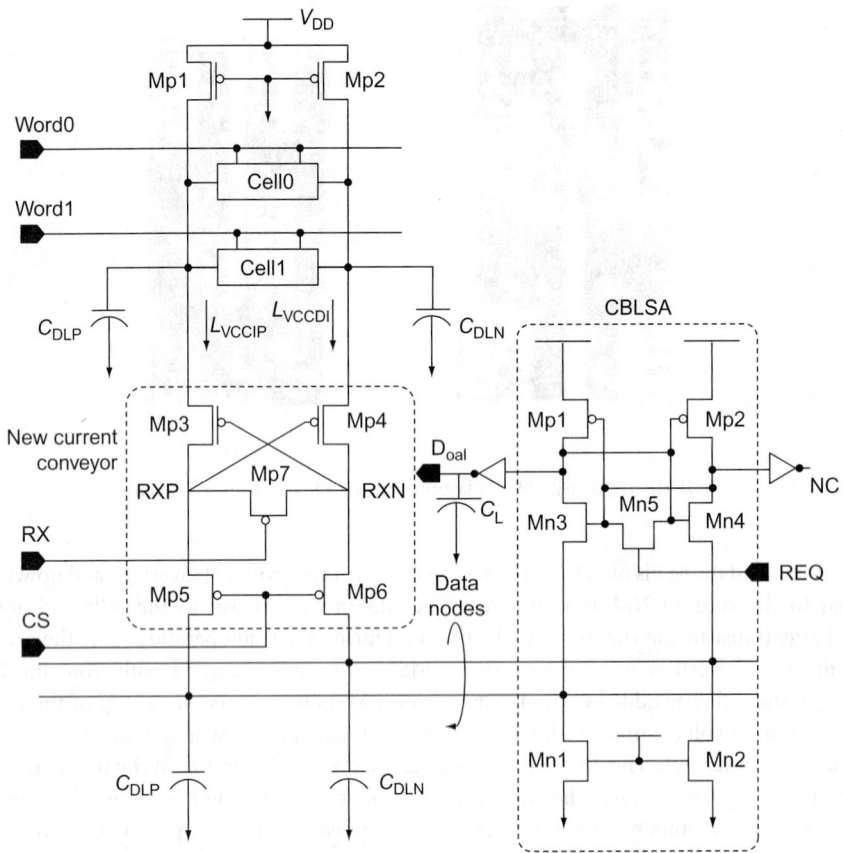

FIGURE 57.56 SRAM read circuitry with the new current-mode sense amplifier.

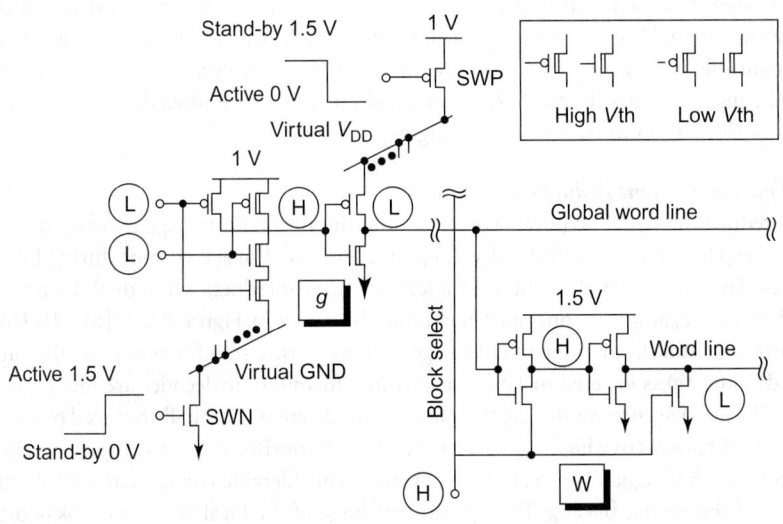

FIGURE 57.57 Dual V_{th} CMOS circuit scheme.

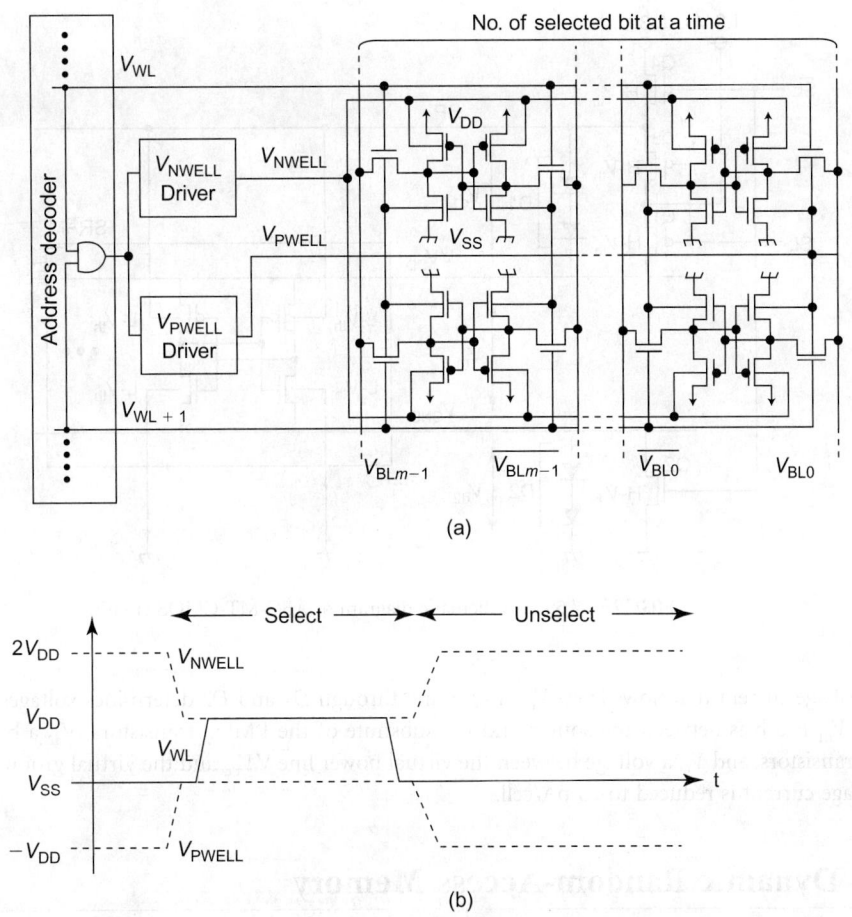

FIGURE 57.58 Circuit schematic of dynamic leakage cut-off and its operation.

eventhough the operating voltage is 0.8 V. This creates reliability problems. Kawaguchi et al. [72] introduced a new technique, a dynamic leakage cut-off (DLC) scheme. Operation waveforms are shown in Figure 57.58. A dynamic change of n-well and p-well bias voltages to V_{dd} and V_{ss}, respectively for selected memory cells is the key feature of this architecture. At the same time, the nonselected memory cells are biased with $\sim 2V_{dd}$ for V_{NWELL} and $\sim -V_{dd}$ for V_{PWELL}. After this, the V_{th} of the selected cells becomes low which aids in high drive, thus a fast operation is executed. In contrast, the V_{th} of the unselected memory cells is high enough to achieve low subthreshold current consumption. This technique is similar to the variable threshold CMOS (VT-CMOS) technique; however, the difference is in the synchronization signal of the well bias. While in VT-CMOS the well bias is synchronized with a standby signal, DLC technique is synchronized with the wordline signal.

Nii et al. [66] improved the MT–CMOS technique further and proposed the auto-backgate controlled (ABC) MT-CMOS method. The ABC MT-CMOS reduces significantly the leakage current during the "sleep" mode. The circuit diagrams of this method is shown in Figure 57.59. Transistors Q1–Q4 are high-threshold devices that act as switches to cut off the leakage current. The internal circuitry is designed with low-V_t devices. During the active mode, signal SL is pulled low and $\overline{SL}$ is pulled high. Q1, Q2, and Q3 turn on and Q4 turns off and virtual power supply VV_{dd} and the substrate bias BP become 1 V. During the sleep mode, signal SL is pulled high and $\overline{SL}$ is pulled low and Q1, Q2 and Q3 turn off, whereas Q4 turns on and BP becomes 3.3 V.

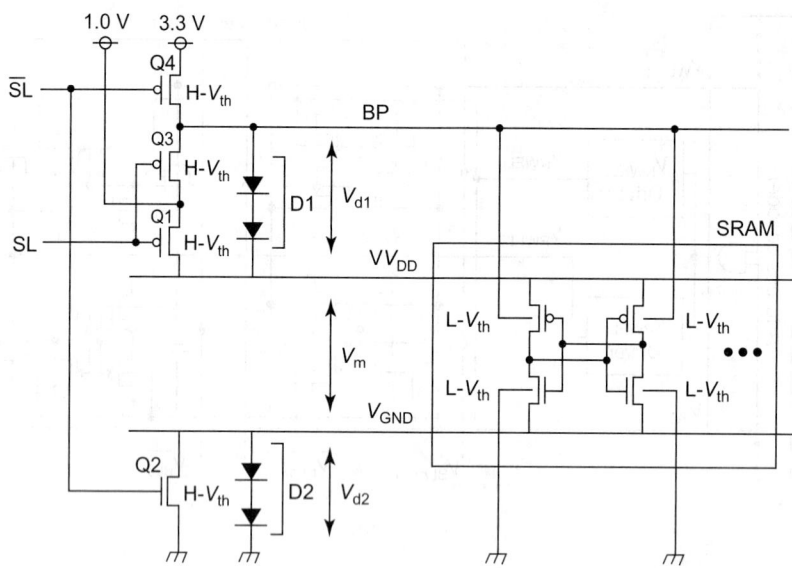

FIGURE 57.59 A schematic diagram of ABC-MT-CMOS circuit.

The leakage current that flows from V_{dd2} to ground through $D1$ and $D2$ determines voltages V_{d1}, V_{d2}, and V_m. V_{d1} is a bias between the source and the substrate of the PMOS transistors, V_{d2} a bias of the NMOS transistors, and V_m a voltage between the virtual power line VV_{dd} and the virtual ground VGND. The leakage current is reduced to 20 pA/cell.

57.7 Dynamic Random-Access Memory

Similar to all previous types of memories, DRAM has undergone a remarkable development toward higher access speed, higher density, and reduced power [51,79–82]. As for reducing power, a variety of techniques targeting various sources of power in DRAMs have been reported. In this section, sources of power consumption will be discussed and then several methods for the reduction of active and data retention power in DRAMs will be described.

57.7.1 Low-Power DRAM Circuits

57.7.1.1 Sources of DRAM Power

The total power dissipated in a DRAM has two components: *the active power* and *the data retention power*. Major contributors to the active power are decoders (row and column), memory array, sense amplifier, other circuits—DC current dissipation (a refresh circuitry, a substrate back-bias generator, a boosted level generator, a voltage reference circuit, a half-V_{dd} generator and a voltage down converter), remaining periphery circuits (main sense amplifier, I/O buffers, write circuitry, etc.) The total active power can be described as

$$P_{active} = [(mC_D\Delta V_D + C_{PT}V_{INT})f + I_{DCP}]V_{dd} \tag{57.3}$$

where C_D is the dataline capacitance, ΔV_D the dataline voltage swing ($0.5V_{dd}$), m the number of cells connected to the activated dataline, C_{PT} the capacitance of the periphery circuits, V_{INT} the internal supply voltage and I_{DCP} the static current.

The total data retention power is given as

$$P_{\text{retension}} = [(mC_D\Delta V_D + C_{PT}V_{INT})(n/t_{REF}) + I_{DCP}]V_{dd} \tag{57.4}$$

where n is the number of words that require refresh and $1/t_{REF}$ the frequency of the refresh operation (current).

57.7.1.2 Techniques for Low-Power Operation

To reduce power consumption during both modes of DRAM operation, many circuit techniques can be applied:

- Capacitance reduction, especially of datalines, wordlines, and shared I/O, using partial activation of multidivided datalines and partial activation of multidivided wordlines.
- Lowering of external and internal voltages.
- DC power reduction of peripheral circuits during the active mode by using static CMOS decoders, pulse techniques, and ATD circuit, similar to SRAMs.
- Refresh power reduction. In addition to capacitance reduction and operating voltages reduction which are applicable also to the refresh mode, decreasing the frequency of refresh cycle or decreasing the number of words n that require refresh affects the total refresh power.
- AC and DC power reduction of circuits such as a voltage down converter (VDC), a half-voltage generator (HVG), a boosted voltage generator (BVG) and a back-bias generator (BBG).

57.7.1.2.1 Capacitance Reduction

Charging and discharging large data and wordlines contribute to large amount of dissipated power in a DRAM [51,82]. Therefore, minimizing capacitance of these lines can accomplish significant gains in power savings. There are two fundamental methods used to reduce capacitance in DRAMs: *partial activation of multidivided dataline* and *partial activation of multidivided wordline*. The concept of both techniques is shown in Figure 57.60 and Figure 57.61.

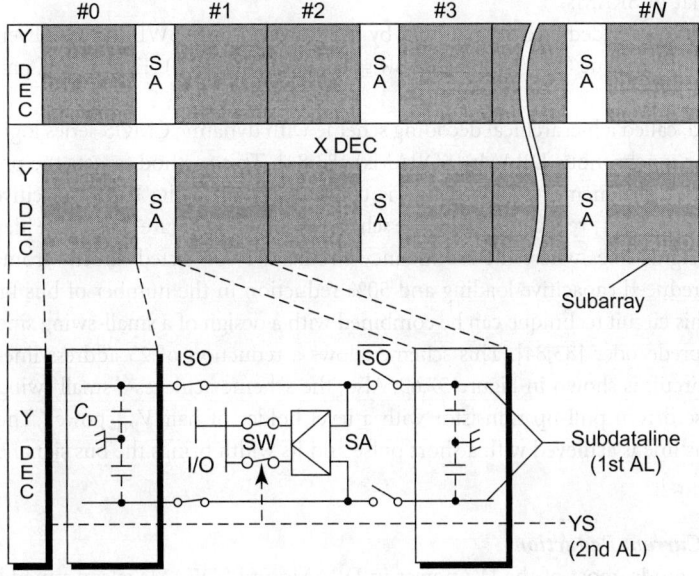

FIGURE 57.60 Multi-divided dataline architecture.

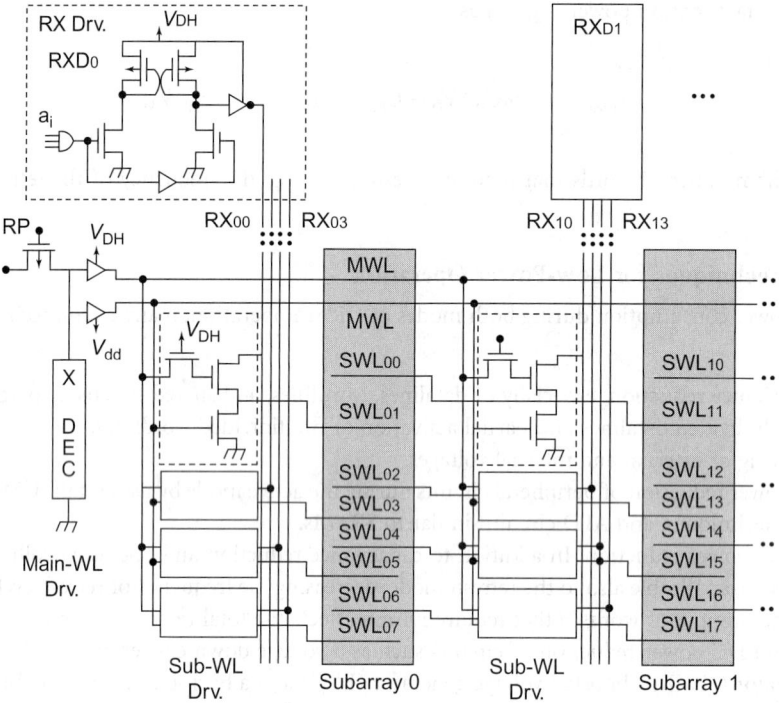

FIGURE 57.61 Hierarchical wordline architecture.

The foundation of partial activation of multidivided dataline (Figure 57.60) is in reducing the number of memory cells connected to an active dataline thus reducing its capacitance C_D. The datalines are divided into small sections with shared I/O circuitry and a sense amplifier. By sharing these resources, further reduction of C_D is achieved. The partial activation is performed by activating only one sense amplifier along the dataline. The principle of the partial activation of multidivided wordline (see Figure 57.61) is very similar to that of SRAMs.

A single wordline is divided into several ones by the subword-line (SWL) drivers. Every SWL has to be selected by the main wordline (MWL) and the row select line signal (RX). Thus, only a partial wordline will be activated.

Similar method, called a hierarchical decoding scheme with dynamic CMOS series logic predecoder, has been proposed for synchronous DRAMs (SDRAMs) [83,84]. This method targets the power losses in the peripheral region of the memory. This power is consumed due to the large capacitive loading of the datalines, address lines, and predecoder lines. The scheme is shown in Figure 57.62. The hierarchical decoder uses predecoded signal lines where the redundancy circuits are connected directly from the global lines. This results in a reduced capacitive loading and 50% reduction in the number of bus lines (column and row decoders). This circuit technique can be combined with a design of a small-swing single address driver with a dynamic predecoder [83,84]. This scheme allows a reduction of 23 address lines. The schematic diagram of this circuit is shown in Figure 57.63. Also, the scheme achieves a small swing in address lines with a short pulse driven pull-up transistor with a level holder of half-V_{INT} power. The pull-up for the reduced swing bus line is achieved with a short pulse and its width brings the bus signal close to the small swing voltage (V_{INTL}).

57.7.1.2.2   DC Current Reduction

During the active mode, most of the DC power in DRAMs and SDRAMs is consumed by the periphery circuits and I/O lines. The decoding and pulsed operation techniques based on an ATD circuit and similar

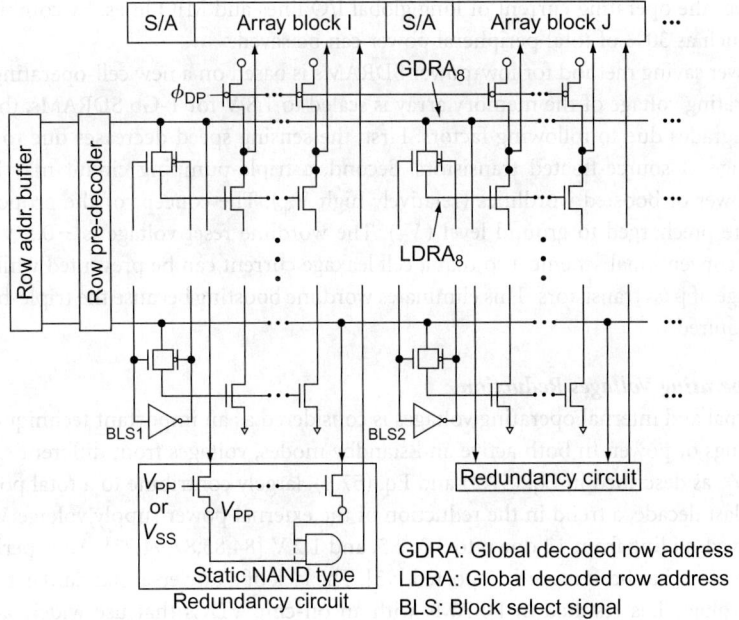

FIGURE 57.62 A decoding scheme with the hierarchical predecoded row signal and global signals shared with redundancy.

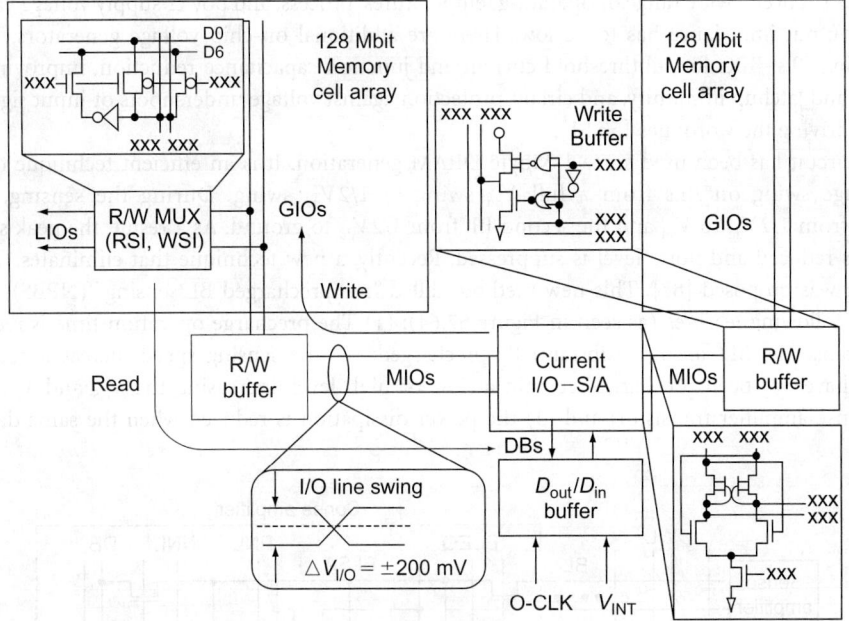

FIGURE 57.63 Block diagram of I/O datapath.

to those for SRAMs can be applied. To minimize power consumption of I/O lines in SDRAMs, two circuit techniques have been proposed [86]. As for the first technique, the extended small swing read operation ($\Delta V_{I/O} = \pm 200$ mV), the small-swing data paths (Local I/O and Global I/O) are extended up to the output buffer stages through Main I/O (MIO) lines (see Figure 57.63). Shared current sense amplifiers (I/O sense amplifiers) also reduce power consumption. In the second technique, the single I/O line driving write operation, halves the operating current of long global I/O lines and MIO lines. By combining these two methods, as much as 30% of total peripheral power can be saved.

Another power saving method for low-power SDRAMs is based on a new cell-operating concept [87]. When the operating voltage of the memory array is scaled to 1.8 V for 1-Gb SDRAMs, the performance significantly degrades due to following factors. First, the sensing speed decreases due to the noticeable threshold voltage of source-floated transistors. Second, a triple-pumping circuit may be required to increase the power of boosted wordlines (relatively high V_{pp}). The concept of the proposed method is that the BLs are precharged to ground level (V_{ss}). The wordline reset voltage is −0.5 V (as compared with $1/2 V_{dd}$ in conventional schemes) so that a cell leakage current can be prevented while lowering the threshold voltage of pass transistors. This eliminates wordline boosting because the triple-boosting circuit is no longer required.

57.7.1.2.3 *Operating Voltages Reduction*

Lowering external and internal operating voltages is considered as an important technique for achieving significant savings of power. In both active and standby modes, voltages from different sources, such as V_{dd}, V_{INT}, or ΔV_D, as described in Eq. (57.3) and Eq. (57.4), largely contribute to a total power consumption. Over the last decade, a trend in the reduction of the external power supply voltage V_{dd} for DRAMs has been observed, sliding from 12 down to 3.3, 2.5, and 1.2 V [84,85,87,94,97]. An experimental circuit with V_{dd} as low as 1 V has been recently reported [95]. The lack of a universal standard external operating power supply voltage has resulted in DRAMs with an on-chip VDCs that use widely accepted power supply voltages V_{dd}, such as 5 or lately 3.3 V, and lower the operating voltage for the memory core and thus gain power savings [50,51,91]. VDC is one of the most important DRAM circuits in achieving DRAM operation at battery voltage levels. In power-limited applications, VDC has to have a standby current <1 µA over a wide range of operating temperatures, process, and power supply voltage variations. Also its output impedance has to be low. There are additional on-chip voltage generators: HVG for precharging BLs; BBG for subthreshold current and junction capacitance reduction, improving device isolation and latchup immunity, and circuit protection against voltage undershoots of input signals; and BVG for driving the wordlines [50,51].

HVG circuit has been used since 1-Mbyte DRAM generation. It is an efficient technique to reduce the voltage swing on BLs from a full V_{dd} swing to $1/2 V_{dd}$ swing. During the sensing, one BL switches from $1/2 V_{dd}$ to V_{dd} and the second BL from $1/2 V_{dd}$ to ground. As a result, the peak switching current is reduced and noise level is suppressed. Recently, a new technique that eliminates $1/2 V_{dd}$ BL switching was proposed [88]. This new method, called "nonprecharged BL sensing" (NPBS), provides the three following features (as seen in Figure 57.64): (1) The precharge operation time is reduced by 78%, because the BLs are not substantially precharged; (2) the sensing speed increases because the BLs that have not been precharged remain at low or high level, increasing the V_{GS} and V_{DS} voltages for the sense amplifier transistor; and (3) the power dissipation is reduced when the same data occur

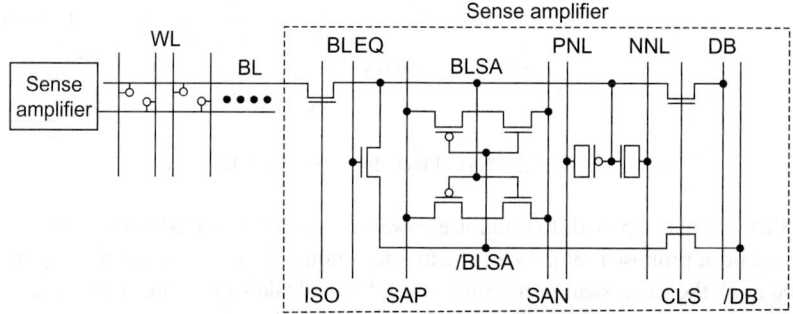

FIGURE 57.64 NPBS circuit and its operation.

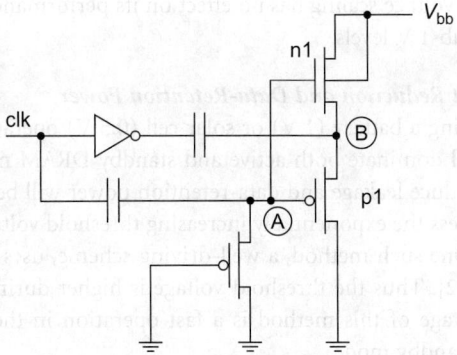

FIGURE 57.65 Low-voltage pumping circuit.

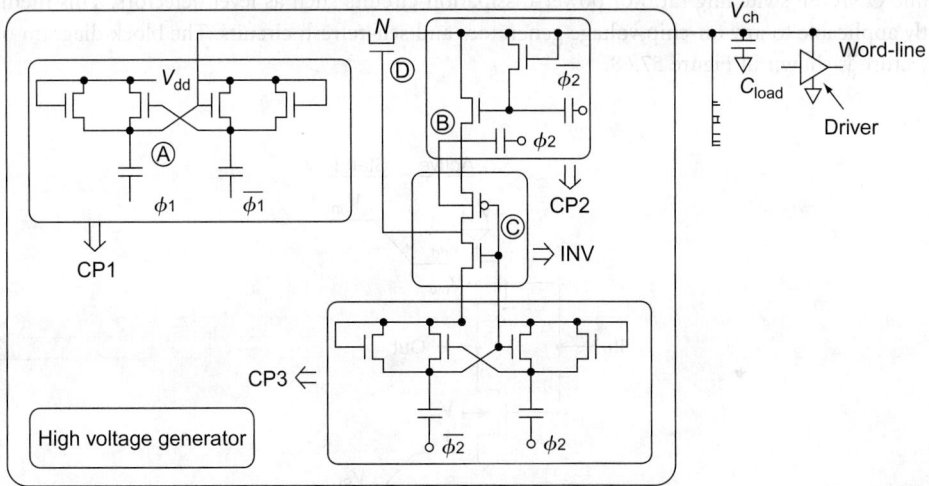

FIGURE 57.66 Boosted voltage generator.

on the BL. The power is reduced by ~43%. To maintain or improve the speed and reliability of DRAM operations, the threshold voltage V_t has to follow the same scaling pattern as the main power supply voltage. This scenario, however, results in a rapid increase of leakage currents in the entire memory during both active and standby modes. Therefore, an internal BBG circuit, also known as the charge pump, is needed to improve low-voltage, low-power operation by reducing the subthreshold currents. Figure 57.65 presents a pumping circuit that avoids the V_t losses [89]. When the clock (clk) is at logic low, the node voltage of the node A reaches "$|V_{tp}| - V_{dd}$." The PMOS transistor "p1" clamps the voltage of the node B to the ground level. The V_{bb} voltage settles at "$|V_{tp}| - V_{dd} - V_{tn}$." When clk changes to logic high, the node A changes to V_{tp} and the node B is capacitively coupled to $-V_{dd}$. As a result, V_{bb} voltage changes to $-V_{dd}$. This circuit requires triple-well technology to eliminate minority carrier injection of the "n1" transistor.

To limit the power consumption of this circuit during DRAM's standby mode, the frequency of the clk signal can be reduced. This is possible to implement with BBG's own ring oscillator controlled by BBG's enable signal.

A BVG is used in DRAMs to generate a power supply signal higher than V_{dd} for driving the word-lines. This wordline voltage is higher than V_{dd} by at least the threshold voltage. The boosted level cannot be directly applied to drive the load. An isolation transistor is necessary to separate the switching boosted voltage from the load. One such arrangement is shown in Figure 57.66 [90]. This particular circuit generates an output of $2V_{dd}$. Voltage scaling has no effect on its performance and therefore, it is suitable for V_{dd} reduction down to sub-1 V levels.

57.7.1.2.4 *Leakage Current Reduction and Data-Retention Power*

The key limitation in achieving a battery (1 V) or solar cell (0.5 V) operation will be the subthreshold power consumption that will dominate both active and standby DRAM modes. In this section, circuit techniques that drastically reduce leakage and data-retention power will be described.

Several methods that address the exponentially increasing threshold voltage in rapidly scaled technologies have been proposed. One such method, a well-driving scheme, uses a dynamic V_t by driving the well (see Figure 57.67) [82,92]. Thus the threshold voltage is higher during the standby mode than in the active mode. The advantage of this method is a fast operation in the active mode and a leakage current suppression in the standby mode.

To reduce the subthreshold currents in various DRAM voltage generators, a self-off-time detector circuit could be used [93]. It automatically evaluates the optimal off-time interval and controls the dynamic ON/OFF switching ratio of power-dissipation circuits such as level detectors. This method is directly applicable to any on-chip voltage generators and self-refresh circuits. The block diagram of this architecture is shown in Figure 57.68.

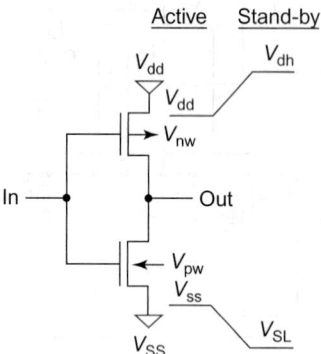

FIGURE 57.67 Low-voltage well-driving scheme.

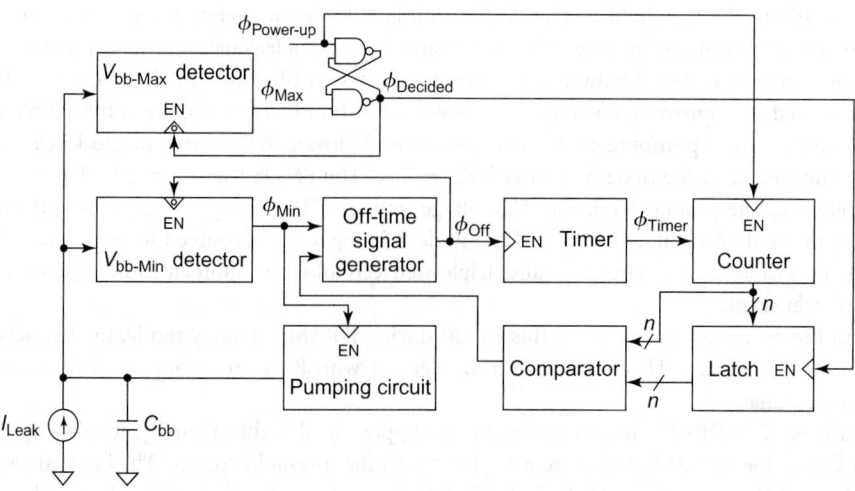

FIGURE 57.68 Block diagram of BBG circuit using the self-off-time detector.

A charge-transfer presensing scheme (CTPS) with $1/2V_{cc}$ BL precharge and a nonreset row block control (NRBC) scheme reduces the data-retention current by 75% [94]. The principle of CTPS technique is shown in Figure 57.69. The SA and the BL are separated by the transfer-gate (TG). The BL is precharged to $1/2V_{ccA}$ (power supply voltage for the array) and the sense amplifier node is precharged to a voltage higher than V_{ccA}. When TG is at a low level, the WL is activated and the data from the memory cell (MC) is transferred to the BL. A small voltage change appears on the BL pair. Then, TG voltage is set to the voltage for the CT condition and the charge of SA node is transferred to the BL. The transfer is complete when the BL voltage reaches "$V_{TG} - V_{tn}$." After that, a large variation of the readout voltage appears on the SA pair.

CTSP technique reduces the active array current and prolongs the data-retention time. The data-retention power can be reduced further by the NRBC scheme, which is used to reduce the charge/discharge number of row block control circuits to 1/128 of the conventional method. The NRBC architecture is shown in Figure 57.70. NRBC is a DWL structure where one SWL in the selected row block is activated if one MWL and one of four subdecode signals (SD0~3) are activated in this row block. Also the transfergates TG_L and TG_R are activated at both sides of this row block. After the data-retention mode is set,

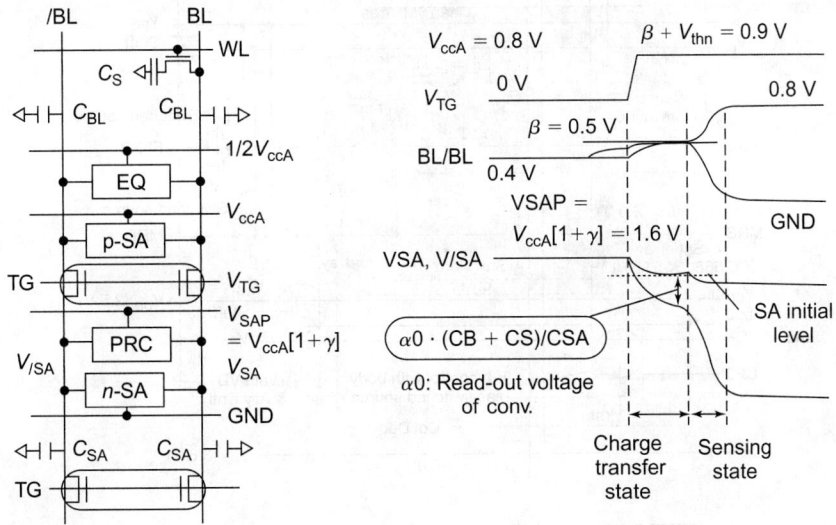

FIGURE 57.69 Concept of CTPS and its circuit organization; BL = $1/2V_{cc}$, V_{ccA} = 0.8 V.

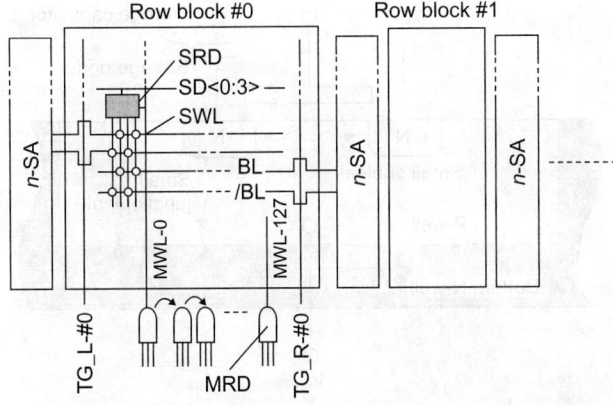

FIGURE 57.70 Basic circuits of the row block control in NRBC.

SD and TG signals do not swing fully at every cycle but only every 128 cycles for activating the same row block. As a result, the row control current is reduced by 70% compared with the conventional scheme.

Another effective method for leakage current reduction is "subthreshold leakage current suppression system" (SCSS) shown in Figure 57.71 [96]. The method features high drivability (I_{ds}) and low-V_t transistors. The principle of this method is to reduce the active mode leakage current with a body bias control and to reduce the standby mode current by body bias and switched-source impedance. PMOS transistors use the boosted wordline voltage as a body bias, whereas NMOS transistors use memory cell substrate voltage as a body bias.

In addition to leakage suppression techniques, extending the refresh time can also significantly reduce power consumption during the standby mode, as shown in Eq. (57.4) [85,98,99]. The refresh time is determined from the time needed for the stored charge in the memory cell to keep enough margin against leakage at high temperature. To achieve long refresh characteristics for a low-voltage operation, a negative wordline method can be applied [85]. Figure 57.72 shows the concept of this method. A negative gate-source voltage V_{gs} is applied which decreases MC transistor's subthreshold

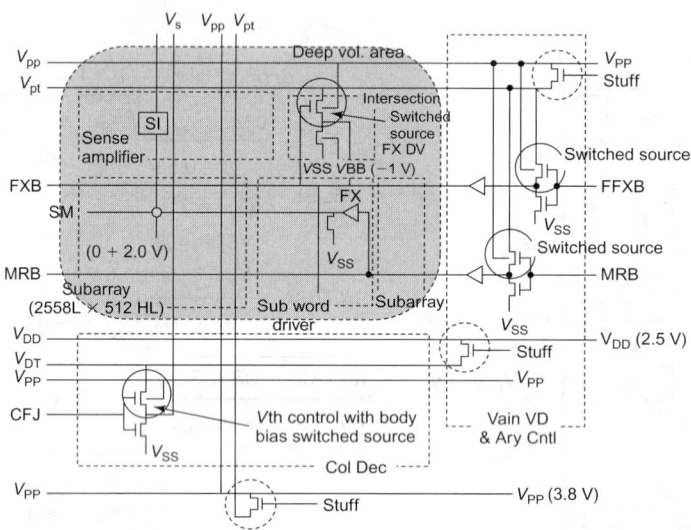

FIGURE 57.71 Subthreshold leakage current suppression system.

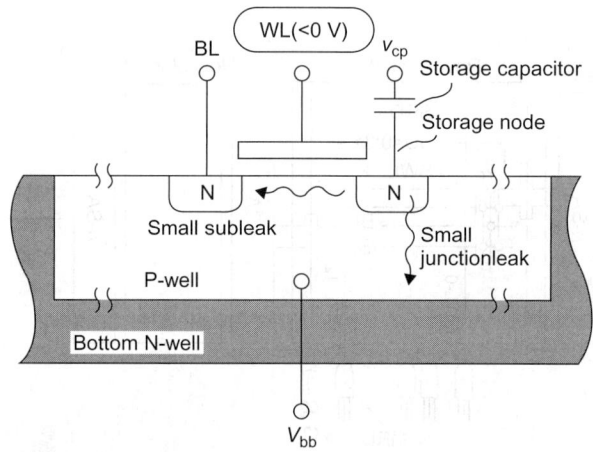

FIGURE 57.72 Principle of the negative voltage wordline technique.

current and provides a noise-free dynamic refresh. It also enables the shallow back-bias voltage V_{bb} that reduces the electrical field between the storage node and the p-well region under the memory cell and results in a small junction leakage current. This achieves longer static refresh time. Figure 57.73 shows an example of the negative voltage wordline driver.

Dual-period self-refresh (DPS-refresh) scheme is a method that can extend the refresh time by 4–6 times [98]. The principle of DPS-refresh scheme is shown in Figure 57.74 and the corresponding timing diagram in Figure 57.75. The key concept is to use two different internal self-refresh periods. All wordlines are separated into two groups according to retention test data that are stored in a PROM mode register implemented in the chip periphery. The short period t_1 corresponds to a conventional self-refresh period determined by the minimum retention time in a chip. The long period t_2 is set to the optimum refresh value. If all memory cells connected to a specific wordline have a retention time longer than t_2, they are called long period wordline cells (LPWL) and are refreshed in the long period of t_2. Otherwise, they are called short period wordline cells (SPWL) and the wordline is refreshed in the short period t_1. DPS-refresh operation is then achieved by periodically skipping refresh cycles for LPWLs. The operation is composed of T_1 periods repeated $(n - 1)$ times followed by a T_2. For a refresh cycle during T_1 period, the $\overline{inhibit\_k}$, where "k" is from 0 to 3, goes low if the wordline selected in the array block "k" is an LPWL and disables all AND-gated MSi signals. As a result, the refresh operation is not executed. However,

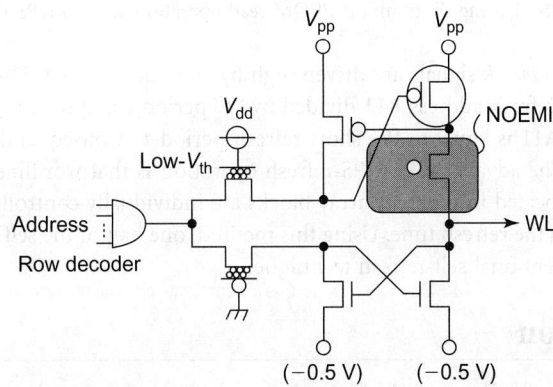

FIGURE 57.73 Negative voltage wordline driver.

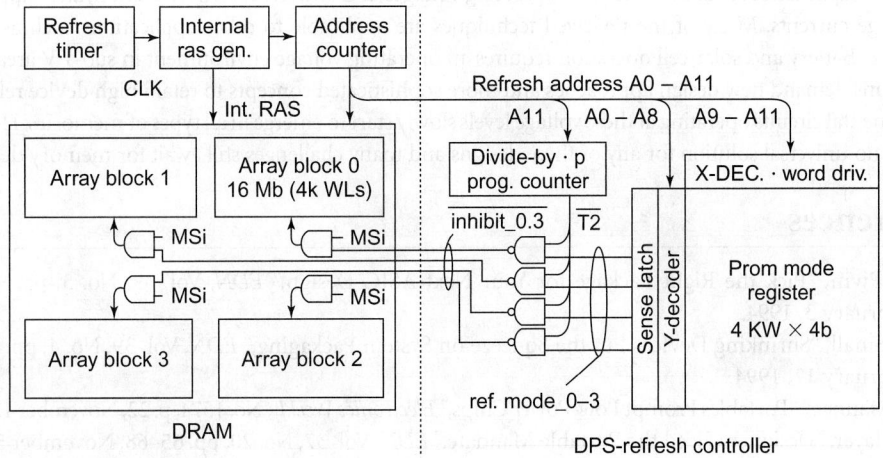

FIGURE 57.74 A schematic diagram of mode-register controlled DPS-refresh method.

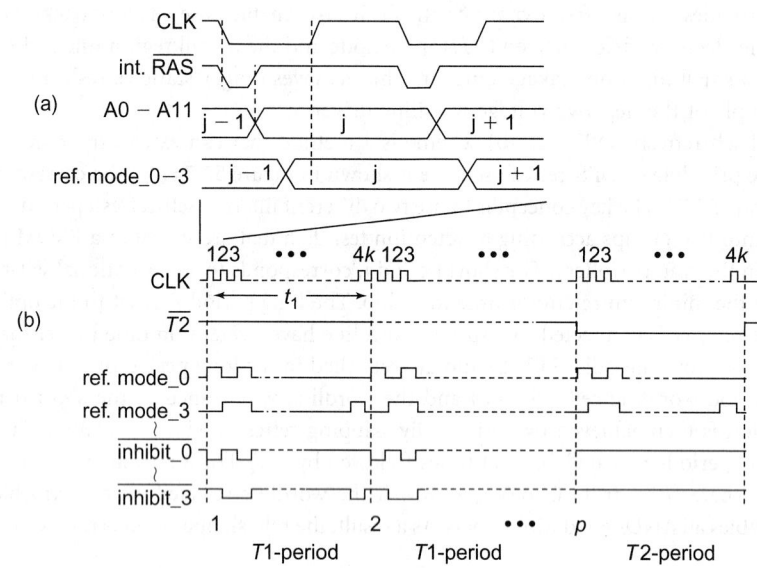

FIGURE 57.75 Timing diagram: (a) PROM read operation and (b) DPS-refresh operation.

during the T_2 period, $\overline{inhibit\_k}$ signals are driven high by $\overline{T2}$ clock signal. This signal is generated by the most significant bit refresh address A11 divided by "p" period using the programmable divide-by-p counter. The period of A11 is equal to the short refresh period t_1. Consequently, LPWLs are refreshed every "$p \times t_1$" periods. The advantage of DPS-refresh operation is that wordlines which have the same refresh address but are located in different array blocks are individually controlled by $\overline{inhibit\_k}$ signals, which aids in prolonging the refresh time. Using this method, one half of the self-refresh current is saved compared with the conventional self-refresh technique.

57.8 Conclusion

In this section, the latest developments in low-power circuit techniques and methods for ROMs, Flash memories, MRAMs, FeRAMs, SRAMs, and DRAMs were reviewed. All major sources of power dissipation in these memories were analyzed. Key techniques for drastic reduction of power consumption were identified. These are capacitance reduction, very low-operating voltages, DC and AC current reduction, and suppression of leakage currents. Many of the reviewed techniques are applicable to other applications such as ASICs, DSPs, etc. Battery and solar-cell operation requires an operating voltage environment in sub-1 V area. These conditions demand new design approaches and more sophisticated concepts to retain high device reliability. Experimental circuits operating at these voltage levels slowly start to emerge in all types of memories. However, there is no universal solution for any of these designs and many challenges still await for memory designers.

References

1. D. Pivin, "Pick the Right Package for Your Next ASIC Design," *EDN*, Vol. 39, No. 3, pp. 91–108, February 3, 1994.
2. C. Small, "Shrinking Devices Put the Squeeze on System Packaging," *EDN*, Vol. 39, No. 4, pp. 41–46, February 17, 1994.
3. D. Manners, "Portables Prompt Low-Power Chips," *Electronics Weekly*, No. 1574, p. 22, November 13, 1991.
4. J. Mayer, "Designers Heed the Portable Mandate," *EDN*, Vol. 37, No. 20, pp. 65–68, November 5, 1992.
5. R. Stephany et al., "A 200 MHz 32b 0.5 W CMOS RISC Microprocessor," *ISSCC Digest of Technical Papers*, pp. 15.5-1–15.5-2, San Francisco, CA, February, 1998.

6. H. Igura et al., "An 800MOPS 100 mW 1.5 V Parallel DSP for Mobile Multimedia Processing," *ISSCC Digest of Technical Papers*, pp. 18.3-1–18.3-2, San Francisco, CA, February, 1998.

7. A.K. Sharma, Semiconductor Memories—Technology, Testing and Reliability, *IEEE Press*, New York, 1997.

8. E. de Angel and E.E. Swartzlander Jr., "Survey of Low Power Techniques for ROMs," *Proceedings of ISLPED'97*, pp. 7–11, August, 1997.

9. J. Rabaey and M. Pedram, Eds., "Low-Power Methodologies," Kluwer Academic Publishers, Dordrecht, 1996.

10. M. Margala and N.G. Durdle, "Noncomplementary BiCMOS Logic and CMOS Logic Styles for Low-Voltage Low-Power Operation—A Comparative Study," *IEEE Journal of Solid-State Circuits*, Vol. 33, No. 10, pp. 1580–1585, October 1998.

11. M. Margala and N.G. Durdle, "1.2 V Full-Swing BiNMOS Logic Gate," *Microelectronics Journal*, Vol. 29, No. 7, pp. 421–429, July 1998.

12. M. Margala and N.G. Durdle, "Low-Power 4-2 Compressor Circuits," *International Journal of Electronics*, Vol. 85, No. 2, pp. 165–176, August 1998.

13. C.-C. Wang et al., "An Area-Saving Decoder Structure for ROMs," *IEEE Transactions on Very Large Scale Integration (VLSI) Systems*, Vol. 11, No. 4, pp. 581–589, August 2003.

14. R. Sasagawa et al., "High-Speed Cascode Sensing Scheme for 1.0 V Contact-programming Mask ROM," *Digest of Technical Papers of the IEEE Symposium on VLSI Circuits*, pp. 95–96, June 1999.

15. B.-D. Yang and L.-S. Kim, "A Low-Power Charge-Recycling ROM Architecture," *IEEE Transactions on Very Large Scale Integration (VLSI) Systems*, Vol. 11, No. 4, pp. 590–600, August 2003.

16. B.-D. Yang and L.-S. Kim, "A Low-Power ROM Charge Recycling and Charge Sharing Architecture," *IEEE Journal of Solid-State Circuits*, Vol. 38, No. 4, pp. 641–653, April 2003.

17. S. Grossman, "Future Trends in Flash Memories," *Proceedings of MTDT'96*, pp. 2–3, August 1996.

18. R. Verma, "Flash Memory Quality and Reliability Issues", *Proceedings of MTDT'96*, pp. 32–36, August 1996.

19. M. Ohkawa et al., "A 98 mm² Die Size 3.3-V 64-Mb Flash Memory with FN–NOR Type Four-Level Cell," *IEEE Journal of Solid-State Circuits*, Vol. 31, No. 11, pp. 1584–1589, November 1996.

20. J.-K. Kim et al., "A 120-mm² 64-Mb NAND Flash Memory Achieving 180 ns/Byte Effective Program Speed," *IEEE Journal of Solid-State Circuits*, Vol. 32, No. 5, pp. 670–679, May 1997.

21. T.-S. Jung et al., "A 117-mm² 3.3-V only 128-Mb Multilevel NAND Flash Memory for Mass Storage Applications," *IEEE Journal of Solid-State Circuits*, Vol. 31, No. 11, pp. 1575–1583, November 1996.

22. M. Hiraki et al., "A 3.3 V 90 MHz Flash Memory Module Embedded in a 32b RISC Microcontroller," *Advanced Program of ISSCC'99*, p. 17, San Francisco, CA, November 1998.

23. S. Atsumi et al., "A 3.3 V-only 16 Mb Flash Memory with row-decoding scheme," *ISSCC Digest of Technical Papers*, pp. 42–43, San Francisco, CA, February 1996.

24. K. Takeuchi et al., "A Multipage Cell Architecture for High-Speed Programming Multilevel NAND Flash Memories," *IEEE Journal Solid-State Circuits*, Vol. 33, No. 8, pp. 1228–1238, August 1998.

25. K. Takeuchi et al., "A Negative Vth Cell Architecture for Highly Scalable, Excellently Noise Immune and Highly Reliable NAND Flash Memories," *Digest of Technical Papers of Symposium on VLSI Circuits*, pp. 234–235, June 1998.

26. T. Kawahara et al., "Bit-Line Clamped Sensing Multiplex and Accurate High Voltage Generator for Quarter-Micron Flash Memories," *IEEE Journal of Solid-State Circuits*, Vol. 31, No. 11, pp. 1590–1600, November 1996.

27. N. Otsuka and M. Horowitz, "Circuit Techniques for 1.5-V Power Supply Flash Memory," *IEEE Journal of Solid-State Circuits*, Vol. 32, No. 8, pp. 1217–1230, August 1997.

28. M. Mihara et al., "A 29 mm² 1.8 V-only 16 Mb DINOR Flash Memory with Gate-Protected Poly-Diode Charge Pump," *Advanced Program of ISSCC'99*, p. 17, San Francisco, CA, November 1998.

29. T. Tanzawa, et al., "A Compact On-Chip ECC for Low Cost Flash Memories," *IEEE Journal of Solid-State Circuits*, Vol. 32, No. 5, pp. 662–669, May 1997.

30. A. Nozoe et al., "A 256Mb Multilevel Flash Memory with 2MB/s Program Rate for Mass Storage Application," *Advanced Program of ISSCC'99*, p. 17, San Francisco, CA, November 1998.

31. K. Imamiya et al., "A 130 mm2 256 Mb NAND Flash with Shallow Trench Isolation Technology," *Advanced Program of ISSCC'99*, p. 17, San Francisco, CA, November 1998.
32. C.C. Hung et al., "High Density and Low Power Design of MRAM," *Digest of the IEEE International Electron Devices Meeting*, pp. 575–578, December 2004.
33. H.S. Jeong et al., "Fully Integrated 64Kb MRAM with Novel Reference Cell Scheme," *Digest of the IEEE International Electron Devices Meeting*, pp. 551–554, December 2002.
34. M. Durlam et al., "A 1-Mbit MRAM based on 1T1MTJ Bit Cell Integrated with Copper Interconnects," *IEEE Journal of Solid-State Circuits*, Vol. 38, No. 5, pp. 769–773, May 2003.
35. Y. Asao et al., "Design and Process Integration for High-Density, High-Speed, and Low-Power 6F² Cross Point MRAM Cell," *Digest of the IEEE International Electron Devices Meeting*, pp. 571–574, December 2004.
36. M. Aoki et al., "A Novel Voltage Sensing 1Y/2MTJ Cell with Resistance Ratio for Highly Stable and Scalable MRAM," *Digest of Technical Papers of the IEEE Symposium on VLSI Circuits*, pp. 170–171, June 2005.
37. E.K.S. Au et al., "A Novel Current-Mode Sensing Scheme for Magnetic Tunnel Junction MRAM," *IEEE Transactions on Magnetics*, Vol. 40, No. 2, pp. 483–488, March 2004.
38. H. Miyatake et al., "Optimizing Write Current and Power Dissipation in MRAMs by Using an Astroid Curve," *IEEE Transactions on Magnetics*, Vol. 40, No. 3, pp. 1723–1731, May 2004.
39. W.J. Gallagher et al., "Recent Advances in MRAM Technology," *Proceedings of the IEEE Symposium on VLSI-TSA Technology*, pp. 72–73, April 2005.
40. H. Hirano et al., "2-V/100 ns 1T/1C Nonvolatile Ferroelectric Memory Architecture with Bitline-Driven Read Scheme and Nonrelaxation Reference Cell," *IEEE Journal of Solid-State Circuits*, Vol. 32, No. 5, pp. 649–654, May 1997.
41. H. Fujisawa et al., "The Charge-Share Modified (CSM) Precharge-Level Architecture for High-Speed and Low-Power Ferroelectric Memory," *IEEE Journal of Solid-State Circuits*, Vol. 32, No. 5, pp. 655–661, May 1997.
42. T. Sumi et al., "A 256 Kb nonvolatile ferroelectric memory at 3 V and 100 ns," *ISSCC Digest of Technical Papers*, pp. 268–269, San Francisco, CA, February 1994.
43. H. Koike et al., "A 60-ns 1-Mb Nonvolatile Ferroelectric Memory with a Nondriven Cell Plate Line Write/Read Scheme," *IEEE Journal of Solid-State Circuits*, Vol. 31, No. 11, pp. 1625–1634, November 1996.
44. R. Womack et al., "A 16-kb ferroelectric nonvolatile memory with a bit parallel architecture," *ISSCC Digest of Technical Papers*, pp. 242–243, San Francisco, CA, February 1989.
45. S. Shiratake et al., "A 32-Mb Chain FeRAM with segment/stitch array architecture," *IEEE Journal of Solid-State Circuits*, Vol. 38, No. 11, pp. 1911–1919, November 2003.
46. S. Kawashima et al., "Bitline GND sensing technique for low-voltage operation FeRAM," *IEEE Journal of Solid-State Circuits*, Vol. 37, pp. 592–598, May 2002.
47. H.-B. Kang et al., "A Hierarchy Bitline Boost Scheme for sub-1.5 V Operation and Short Precharge Time on High Density FeRAM," *ISSCC Digest of Technical Papers*, Vol. 1, No. 3–7, pp. 158–159, San Francisco, CA, February 2002.
48. H. Hirano et al., "High Density and Low Power Nonvolatile FeRAM with Non-Driven Plate and Selected Driven Bit-line Scheme," *Digest of Technical Papers of the IEEE Symposium on VLSI Circuits*, pp. 446–447, June 2004.
49. K. Yamaoka et al., "A 0.9-V 1T1C SBT-based Embedded Nonvolatile FeRAM With a Reference Voltage Scheme and Multilayer Shielded Bit-Line Structure," *IEEE Journal of Solid-State Circuits*, Vol. 40, No. 1, pp. 286–292, January 2005.
50. A. Bellaouar and M.I. Elmasry, *Low-Power Digital VLSI Design, Circuits and Systems*, Kluwer Academic Publishers, Dordrecht, 1996.
51. K. Itoh et al., "Trends in Low-Power RAM Circuit Technologies," *Proceedings of the IEEE*, pp. 524–543, April 1995.
52. H. Morimura and N. Shibata, "A 1-V 1-Mb SRAM for Portable Equipment," *Proceedings of ISLPED'96*, pp. 61–66, August 1996.

53. M. Ukita et al., "A Single Bitline Cross-Point Cell Activation (SCPA) Architecture for Ultra Low Power SRAMs," *ISSCC Digest of Technical Papers*, pp. 252–253, San Francisco, CA, February 1994.

54. B.S. Amrutur and M.A. Horowitz, "Techniques to Reduce Power in Fast Wide Memories," *Proceedings of SLPE'94*, pp. 92–93, 1994.

55. H. Toyoshima et al., "A 6-ns, 1.5-V, 4-Mb BiCMOS SRAM," *IEEE Journal of Solid-State Circuits*, Vol. 31, No. 11, pp. 1610–1617, November 1996.

56. J.S. Caravella, "A 0.9 V, 4 K SRAM for Embedded Applications," *Proceedings of CICC*, pp. 119–122, May 1996.

57. J.S. Caravella, "A Low Voltage SRAM For Embedded Applications," *IEEE Journal of Solid-State Circuits*, Vol. 32, No. 3, pp. 428–432, March 1997.

58. Y. Haraguchi et al., "A Hierarchical Sensing Scheme (HSS) of High-Density and Low-Voltage Operation SRAMs," *Digest of Technical Papers of Symposium on VLSI Circuits*, pp. 79–80, June 1997.

59. T. Mori et al., "A 1 V 0.9 mW at 100 MHz 2kx16b SRAM utilizing a Half-Swing Pulsed-Decoder and Write-Bus Architecture in 0.25 μm Dual-Vt CMOS," *ISSCC Digest of Technical Papers*, pp. 22.4-1–22.4-2, San Francisco, CA, February 1998.

60. J.B. Kuang et al., "SRAM Bitline Circuits on PD SOI: Advantages and Concerns," *IEEE Journal of Solid-State Circuits*, Vol. 32, No. 6, pp. 837–843, June 1997.

61. S. Kawashima et al., "A Charge-Transfer Amplifier and an Encoded-Bus Architecture for Low-Power SRAM's," *IEEE Journal of Solid-State Circuits*, Vol. 33, No. 5, pp. 793–799, May 1998.

62. B.S. Amrutur and M.A. Horowitz, "A Replica Technique for Wordline and Sense Control in Low-Power SRAM's," *IEEE Journal of Solid-State Circuits*, Vol. 33, No. 8, pp. 1208–1219, August 1998.

63. H. Morimura and N. Shibata, "A Step-Down Boosted-Wordline Scheme for 1-V Battery Operated Fast SRAM's, *IEEE Journal of Solid-State Circuits*, Vol. 33, No. 8, pp. 1220–1227, August 1998.

64. J.-S. Wang et al., "Low-Power Embedded SRAM Macros with Current-Mode Read/Write Operations," *Proceedings of ISLPED*, pp. 282–287, August 1998.

65. M. Wieckowski and M. Margala, "A Novel Five-Transistor (5T) SRAM Cell for High Performance Cache," *Proceedings of IEEE International SOC Conference*, pp. 101–102, September 2005.

66. K. Nii et al., "A Low Power SRAM using Auto-Backgate-Controlled MT-CMOS," *Proceedings of ISLPED*, pp. 293–298, August 1998.

67. K.W. Mai et al., "Low-Power SRAM Design Using Half-Swing Pulse-Mode Techniques," *IEEE Journal of Solid-State Circuits*, Vol. 33, No. 11, pp. 1659–1671, November 1998.

68. H. Sato et al., "A 5-MHz, 3.6 mW, 1.4-V SRAM with Nonboosted, Vertical Bipolar Bit-Line Contact Memory Cell," *IEEE Journal of Solid-State Circuits*, Vol. 33, No. 11, pp. 1672–1681, November 1998.

69. H. Nambu et al., "A 1.8-ns Access, 550-MHz, 4.5-Mb CMOS SRAM," *IEEE Journal of Solid-State Circuits*, Vol. 33, No. 11, pp. 1650–1658, November 1998.

70. H. Yamauchi et al., "A 0.8 V/100 MHz/sub-5 mW-Operated Mega-bit SRAM Cell Architecture with Charge-Recycle Offset-Source Driving (OSD) Scheme," *Digest of Technical Papers of Symposium on VLSI Circuits*, pp. 126–127, June 1996.

71. K. Itoh et al., "A Deep Sub-V, Single Power-Supply SRAM Cell with Multi-Vt Boosted Storage Node and Dynamic Load," *Digest of Technical Papers of Symposium on VLSI Circuits*, pp. 132–133, June 1996.

72. H. Kawaguchi et al., "Dynamic Leakage Cut-off Scheme for Low-Voltage SRAM's," *Digest of Technical Papers of Symposium on VLSI Circuits*, pp. 140–141, June 1998.

73. I. Fukushi et al., "A Low-Power SRAM Using Improved Charge Transfer Sense Amplifiers and a Dual-Vth CMOS Circuit Scheme," *Digest of Technical Papers of Symposium on VLSI Circuits*, pp. 142–143, June 1998.

74. M. Khellah and M.I. Elmasry, "Circuit Techniques for High-Speed and Low-Power Multi-Port SRAMS," *Proceedings of ASIC*, pp. 157–161, September 1998.

75. J.-S. Wang and H.Y. Lee, "A New Current-Mode Sense Amplifier for Low-Voltage Low-Power SRAM Design," *Proceedings of ASIC*, pp. 163–167, September 1998.

76. K.J. Shultz et al., "Low-Supply-Noise Low-Power Embedded Modular SRAM," *IEEE Proceedings—Circuits, Devices and Systems*, Vol. 143, No. 2, pp. 73–82, April 1996.

77. P. van der Wagt et al., "RTD/HFET Low Standby Power SRAM Gain Cell," *Texas Instruments Research Web-site*, 4pp., 1997.

78. J. Greason et al., "A 4.5 Megabit, 560 MHz, 4.5 GByte/s High Bandwidth SRAM," *Digest of Technical Papers of Symposium on VLSI Circuits*, pp. 15–16, June 1997.

79. M. Aoki and K. Itoh, "Low-Voltage and Low-Power ULSI Circuit Techniques," *IEICE Transactions on Electronics*, Vol. E77-C, No. 8, pp. 1351–1360, August 1994.

80. T. Suzuki et al., "High-Speed Circuit Techniques for Battery-Operated 16 MBit CMOS DRAM," *IEICE Transactions on Electronics*, Vol. E77-C, No. 8, pp. 1334–1342, August 1994.

81. K. Lee et al., "Low-Voltage, High-Speed Circuit Designs for Gigabit DRAM's," *IEEE Journal of Solid-State Circuits*, Vol. 32, No. 5, pp. 642–648, May 1997.

82. K. Itoh et al., "Limitations and Challenges of Multigigabit DRAM Chip Design," *IEEE Journal of Solid-State Circuits*, Vol. 32, No. 5, pp. 624–634, May 1997.

83. K.-C. Lee et al., "A 1GBit SDRAM with an Independent Sub-Array Controlled Scheme and a Hierarchical Decoding Scheme," *Digest of Technical Papers of Symposium on VLSI Circuits*, pp. 103–104, June 1997.

84. K. Lee et al., "A 1GBit SDRAM with an Independent Sub-Array Controlled Scheme and a Hierarchical Decoding Scheme," *IEEE Journal of Solid-State Circuits*, Vol. 33, No. 5, pp. 779–786, May 1998.

85. T. Tsuruda et al., "High-Speed/High-Bandwidth Design Methodologies for On-Chip DRAM Core Multimedia System LSI's," *IEEE Journal of Solid-State Circuits*, Vol. 32, No. 3, pp. 477–482, March 1997.

86. J.-H. Joo et al., "A 32-Bank 1 Gb Self-Strobing Synchronous DRAM with 1 GByte/s Bandwidth," *IEEE Journal of Solid-State Circuits*, Vol. 31, No. 11, pp. 1635–11644, November 1996.

87. S. Eto et al., "A 1-Gb SDRAM with Ground-Level Precharged Bit Line and Nonboosted 2.1-V Word Line," *IEEE Journal of Solid-State Circuits*, Vol. 33, No. 11, pp. 1697–1702, November 1998.

88. Y. Kato et al., "Non-Precharged Bit-Line Sensing Scheme for High-Speed Low-Power DRAMs," *Digest of Technical Papers of Symposium on VLSI Circuits*, pp. 16–17, June 1998.

89. Y. Tsikikawa et al., "An Efficient Back-Bias Generator with Hybrid Pumping Circuit for 1.5V DRAMs," *Digest of Technical Papers of Symposium on VLSI Circuits*, pp. 85–86, May 1993.

90. Y. Nakagome et al., "An Experimental 1.5-V 64-Mb DRAM," *IEEE Journal of Solid-State Circuits*, Vol. 26, No. 4, pp. 465–471, April 1991.

91. H. Tanaka et al., "A Precise On-Chip Voltage Generator for a Giga-Scale DRAM with a Negative Word-Line Scheme," *Digest of Technical Papers of Symposium on VLSI Circuits*, pp. 94–95, June 1998.

92. K. Seta et al., "50% active power savingwithout speed degradation using standby power reduction (SPA) circuit," *ISSCC Digest of Technical Papers*, pp. 318–319, San Francisco, CA, February 1995.

93. H.J. Song, "A Self-Off-Time Detector for Reducing Standby Current of DRAM," *IEEE Journal of Solid-State Circuits*, Vol. 32, No. 10, pp. 1535–1542, October 1997.

94. M. Tsukude et al., "A 1.2- to 3.3-V Wide Voltage-Range/Low-Power DRAM with a Charge-Transfer Presensing Scheme," *IEEE Journal of Solid-State Circuits*, Vol. 32, No. 11, pp. 1721–1727, November 1997.

95. K. Shimomura et al., "A 1-V 46-ns 16-Mb SOI-DRAM with Body Control Technique," *IEEE Journal of Solid-State Circuits*, Vol. 32, No. 11, pp. 1712–1720, November 1997.

96. M. Hasegawa et al., "A 256 Mb SDRAM with Subthreshold Leakage Current Suppression," *ISSCC Digest of Technical Papers*, pp. 5.5-1–5.5-2, San Francisco, CA, February 1998.

97. T. Okudi and T. Murotani, "A Four-Level Storage 4-Gb DRAM," *IEEE Journal of Solid-State Circuits*, Vol. 32, No. 11, pp. 1743–1747, November 1997.

98. Y. Idei et al., "Dual-Period Self-Refresh Scheme for Low-Power DRAM's with On-Chip PROM Mode Register," *IEEE Journal of Solid-State Circuits*, Vol. 33, No. 2, pp. 253–259, February 1998.

99. T. Tanizaki et al., "Practical Low Power Design Architecture for 256 Mb DRAM," *Proceedings of ESSCIRC'97*, pp. 188–191, September 1997.

Section VII

Analog Circuits

Bang-Sup Song
University of California, San Diego

58

Nyquist-Rate ADC and DAC

Bang-Sup Song
University of California at San Diego

CONTENTS

58.1   Introduction

The rapidly growing electronics field has witnessed the digital revolution that started with the digital telephone switching system in the early 1970s. The trend continued with digital audio in the 1980s and with digital video in the 1990s. The digital technique is expected to prevail in the coming multimedia era and to influence even future digital wireless systems. All electrical signals in the real world are analog in nature, and their waveforms are continuous in time. Since most signal processing is done numerically in discrete time, devices that convert an analog waveform into a stream of discrete digital numbers, or vice versa, have become technical necessities in implementing high-performance digital processing systems. The former is called an analog-to-digital converter (ADC or A/D converter), and the latter is called a digital-to-analog converter (DAC or D/A converter).

Typical systems in this digital era can be grouped and explained as in Figure 58.1. The processed data are stored and recovered later using magnetic or optical media such as tape, magnetic disc, or optical disc. The system can also transmit or receive data through communication channels such as telephone switch, cable, optical fiber, and wireless RF media. Through the Internet computer networks, even compressed digital video images are now made accessible from anywhere and at any time.

58.1.1   Resolution

Resolution is a term used to describe a minimum voltage or current that an ADC/DAC can resolve. The fundamental limit is a quantization noise due to the finite number of bits used in the ADC/DAC. In an N-bit ADC, the minimum incremental input voltage of $V_{\text{ref}}/2^N$ can be resolved with a full-scale input range of V_{ref}. That is, limited 2^N digital codes are available to represent the continuous analog input. Similarly, in an N-bit DAC, 2^N input digital codes can generate distinct output levels separated by $V_{\text{ref}}/2^N$ with a full-scale output range of V_{ref}. The *signal-to-noise ratio* (SNR) is defined as the power ratio of the maximum signal to the in-band uncorrelated noise. The spectrum of the quantization noise is evenly distributed within the *Nyquist bandwidth* (half the sampling frequency). This inband rms noise decreases

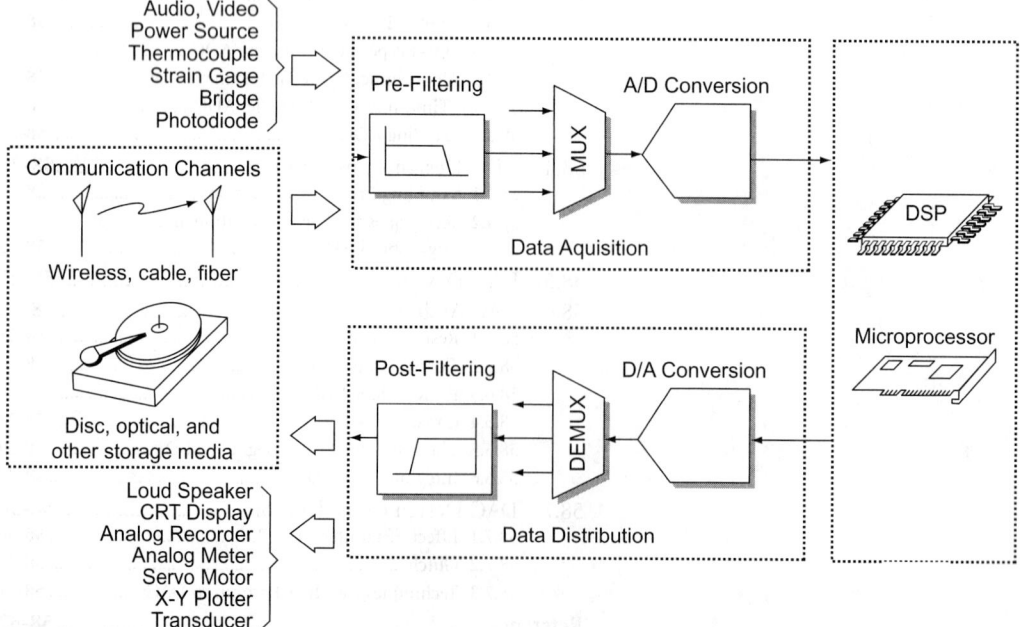

FIGURE 58.1   Information processing systems.

by 3 dB when the oversampling ratio is doubled. This implies that, when oversampled, the SNR within the signal band can be made higher. The SNR of an ideal *N*-bit ADC/DAC is approximated as

$$SNR = 1.5 \times 2^{2N} \approx 6.02N + 1.76 \text{ (dB)} \tag{58.1}$$

The resolution is usually characterized by the SNR, but the SNR accounts only for the uncorrelated noise. The real noise performance is better represented by the *signal-to-noise and distortion ratio* (SNDR, SINAD, or TSNR), which is the ratio of the signal power to the total inband noise including harmonic distortion. Also, a slightly different term is often used in place of the SNR. The useful signal range or *dynamic range* (DR) is defined as the power ratio of the maximum signal to the minimum signal. The minimum signal is defined as the smallest signal for which the SNDR is 0 dB, while the maximum signal is the full-scale signal. Therefore, the SNR of the non-ideal ADC/DAC can be lower than the ideal DR because the noise floor can be higher with a large signal present. In practice, performance is not only limited by the quantization noise but also by non-ideal factors such as noises from circuit components, power supply coupling, noisy substrate, timing jitter, settling, and nonlinearity, etc. An alternative definition of the resolution is the *effective number of bits* (ENOB), which is defined by

$$ENOB = \frac{SNDR - 1.76}{6.02} \text{ (bits)} \tag{58.2}$$

Usually, the ENOB is defined for the signal at half the sampling frequency.

58.1.2 Linearity

The input/output ranges of an ideal *N*-bit ADC/DAC are equally divided into 2^N small units, and one least significant bit (LSB) in the digital code corresponds to the analog incremental voltage of $V_{ref}/2^N$. Static ADC/DAC performance is characterized by *differential nonlinearity* (DNL) and *integral nonlinearity* (INL). The DNL is a measure of deviation of the actual ADC/DAC step from the ideal step for one LSB, and the INL is a measure of deviation of the ADC/DAC output from the ideal straight line drawn between two end points of the transfer characteristic. Both DNL and INL are measured in the unit of an LSB. In practice, the largest positive and negative numbers are usually quoted to specify the static performance. The examples of these DNL and INL definitions for ADC are explained in Figure 58.2.

However, several different definitions of INL may result, depending on how two end points are defined. In some architectures, the two end points are not exactly 0 and V_{ref}. The non-ideal reference point causes

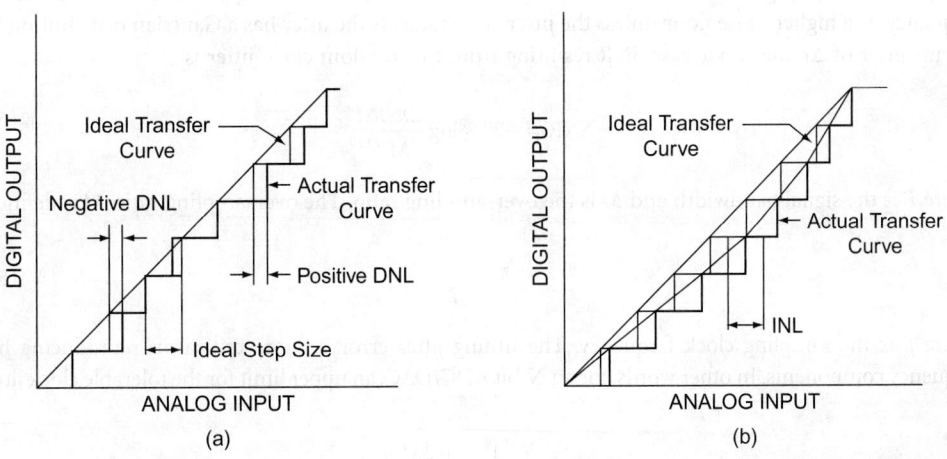

FIGURE 58.2 Definition of ADC nonlinearities: (a) DNL and (b) INL.

an offset error, while the non-ideal full-scale range gives rise to a gain error. In most applications, these offset and gain errors resulting from the non-ideal end points do not matter, and the integral linearity can be better defined in a relative measure using a straight-line linearity concept rather than the end-point linearity. The straight line can be defined as two end points of the actual transfer function, or as a theoretical straight line adjusted for best fit. The former definition is sometimes called end-point linearity, while the latter is called best-straight-line linearity.

Unlike ADC, the output of a DAC is a sampled-and-held step waveform held constant during a word clock period. Any deviation from the ideal step waveform causes an error in the DAC output. High-speed DACs which usually have a current output are either terminated with a 50 to 75-Ω low-impedance load or buffered by a wideband transresistance amplifier. The linearity of a DAC is often limited dynamically by the non-ideal settling of the output node. Anything other than ideal exponential settling results in linearity errors.

58.1.3 Monotonicity

In both the ADC and the DAC, the output should increase over its full range as the input increases. That is, the negative DNL should be smaller than one LSB for any ADC/DAC to be monotonic. Monotonicity is critical in most applications, in particular digital control or video applications. The source of non-monotonicity is an inaccuracy in binary weighting of a DAC. For example, the most significant bit (MSB) has a weight of half the full range. If the MSB weight is not accurate, the full range is divided into two non-ideal half ranges, and a major error occurs at the midpoint of the full scale. The similar non-monotonicity can take place at the quarter and one-eighth points. In DACs, monotonicity is inherently guaranteed if a DAC uses thermometer decoding. However, it is impractical to implement high-resolution DACs using thermometer codes since the number of elements grows exponentially as the number of bits increases. Therefore, to guarantee monotonicity in practical applications, DACs have been implemented using either a segmented DAC or an integrator-type DAC. Oversampling interpolative DACs also achieve monotonicity using a pulse-density modulated bitstream filtered by a lossy integrator or by a low-pass filter. Similarly, ADCs using slope-type, subranging, or oversampling architectures are monotonic.

58.1.4 Clock Jitter

Jitter is loosely defined as a timing error in analog-to-digital and digital-to-analog conversions. The clock jitter greatly affects the noise performance of both ADCs and DACs. For example, in ADCs, the right signal sampled at the wrong time is the same as the wrong signal sampled at the right time. Similarly, DACs need precise timing to correctly reproduce an analog output signal. If an analog waveform is not generated with the identical timing with which it is sampled, distortion will result because the output changes at the wrong time. This in turn introduces either spurious components related to the jitter frequency or a higher noise floor unless the jitter is periodic. If the jitter has a Gaussian distribution with an rms jitter of Δt, the worst-case SNR resulting from this random clock jitter is

$$\text{SNR} = -20 \times \log \frac{2\pi B \Delta t}{M^{1/2}} \tag{58.3}$$

where B is the signal bandwidth and M is the oversampling ratio. The oversampling ratio M is defined as

$$M = \frac{f_s}{2B} \tag{58.4}$$

where f_s is the sampling clock frequency. The timing jitter error is more critical in reproducing high-frequency components. In other words, for an N-bit ADC/DAC, an upper limit for the tolerable clock jitter is

$$\Delta t \leq \frac{1}{2\pi B 2^N} \left(\frac{2M}{3} \right)^{1/2} \tag{58.5}$$

This implies that the error power induced in the baseband by clock jitter should be no larger than the quantization noise. For example, a Nyquist-rate 16-b ADC/DAC with a 22-kHz bandwidth should have a clock jitter of less than 90 ps.

58.1.5 Nyquist-Rate vs. Oversampling

In recent years, high-resolution ADCs and DACs at the low end of the spectrum such as for digital audio, voice, and instrumentation are dominantly implemented using oversampling techniques. Although Nyquist-rate techniques can achieve comparable resolution, such techniques are in general sensitive to non-ideal factors such as process, component matching, and even environmental changes. The inherent advantage of oversampling provides a unique solution in the digital VLSI environment. The oversampling technique achieves high resolution by trading speed for accuracy. The oversampling lessens the effect of quantization noise and clock jitter. However, the quantization or regeneration of a signal above MHz using oversampling techniques is costly even if possible. Therefore, typical applications for high-sampling rates require sampling at a Nyquist rate.

58.2 ADC Design Arts

The conversion speed of the ADC is limited by the time needed to complete all comparator decisions. Flash ADCs make all the decisions at once, while successive-approximation ADCs make one-bit decisions at a time. Although it is fast, the complexity of the flash ADC grows exponentially. On the other hand, the successive-approximation ADC is simple but slow since the bit decisions are made in sequence. Between these two extremes, there exist many architectures resolving a finite number of bits at a time, such as pipeline and multi-step ADCs. They balance complexity and speed. Figure 58.3 shows recent

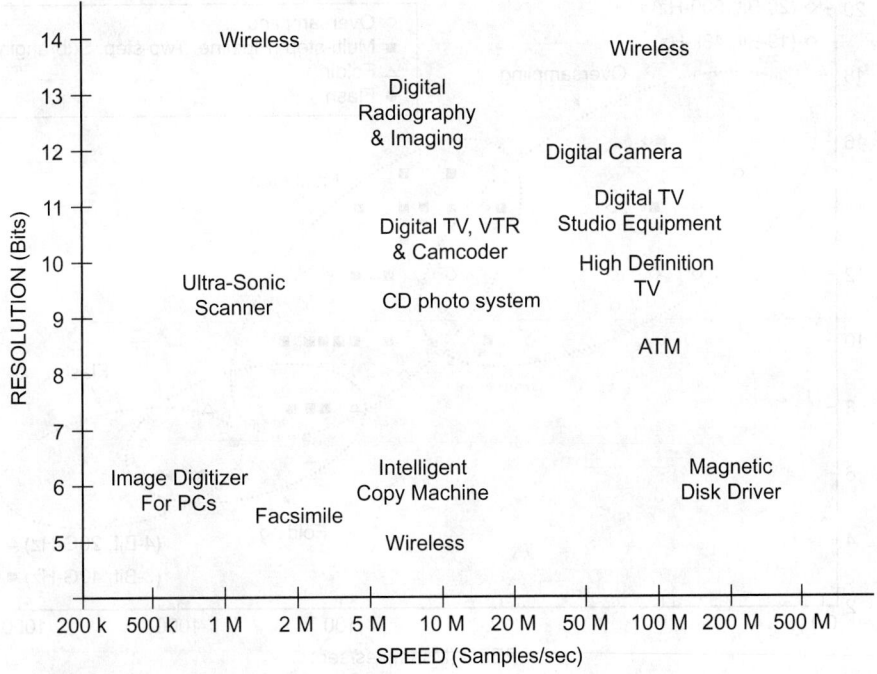

FIGURE 58.3 Recent high-speed ADC applications.

high-speed ADC applications in the resolution-versus-speed plot. ADC architecture depends on system requirements. For example, with IF (intermediate frequency) filters, wireless receivers need only 5 to 6 b ADC at a few MHz sampling rate. However, without IF filters, the dynamic range of 12 to 14 b is required for the IF sampling depending on IF as shown in Figure 58.3.

58.2.1 State of the Art

Some architectures are preferred to others for certain applications. Three architectures stand out for three important areas of applications. For example, the oversampling converter is exclusively used to achieve high resolution above the 12-b level at low frequencies. The difficulty in achieving better than 12-b matching in conventional techniques gives a fair advantage to the oversampling technique. For medium speed with high resolution, pipeline or multi-step ADCs are promising. At extremely high frequencies, only flash and folding ADCs survive, but with low resolution. Figure 58.4 is a resolution-versus-speed plot showing this trend. As both semiconductor process and design technologies advance, the performance envelope will be pushed further. The demand for higher resolution at higher sampling rates is a main driver of this trend.

58.2.2 Technical Challenge in Digital Wireless

In digital wireless systems, a need to quantize and to create a block of spectrum with low intermodulation has become the single most challenging problem. Implementing IF filters digitally has already become a necessity in wireless cell sites and base stations. Even in hand-held units, placing data conversion blocks closer to the RF (radio frequency) has many advantages. A substantial improvement in system cost and complexity of the RF circuitry can be realized by implementing high selectivity function digitally, and the digital IF can increase immunity to adjacent and alternate channel interferences. Furthermore, the RF transceiver architecture can be made independent of the system and can be adapted to different

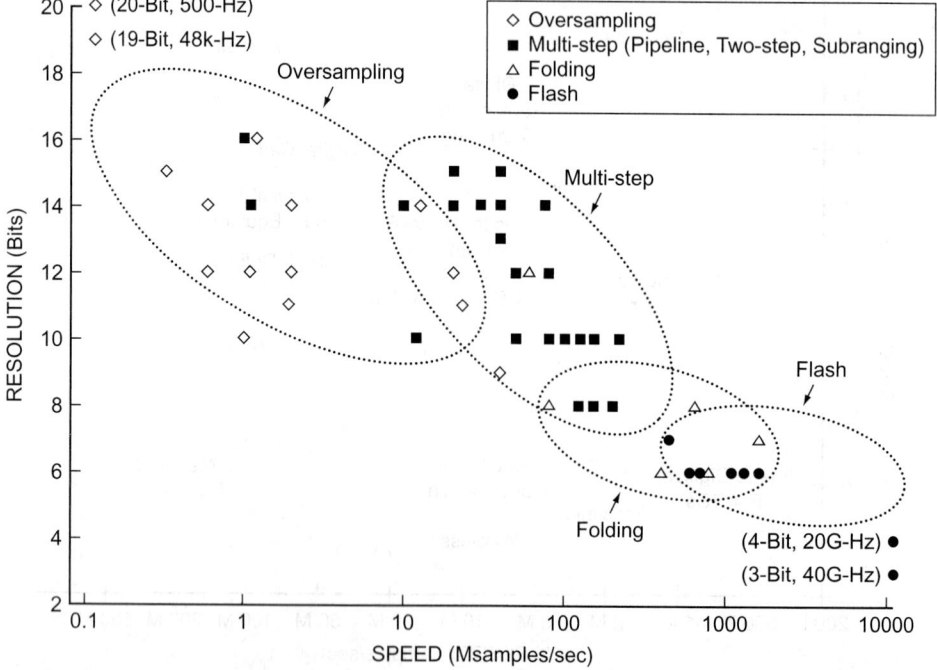

FIGURE 58.4 Performance of recently published ADCs: resolution versus speed.

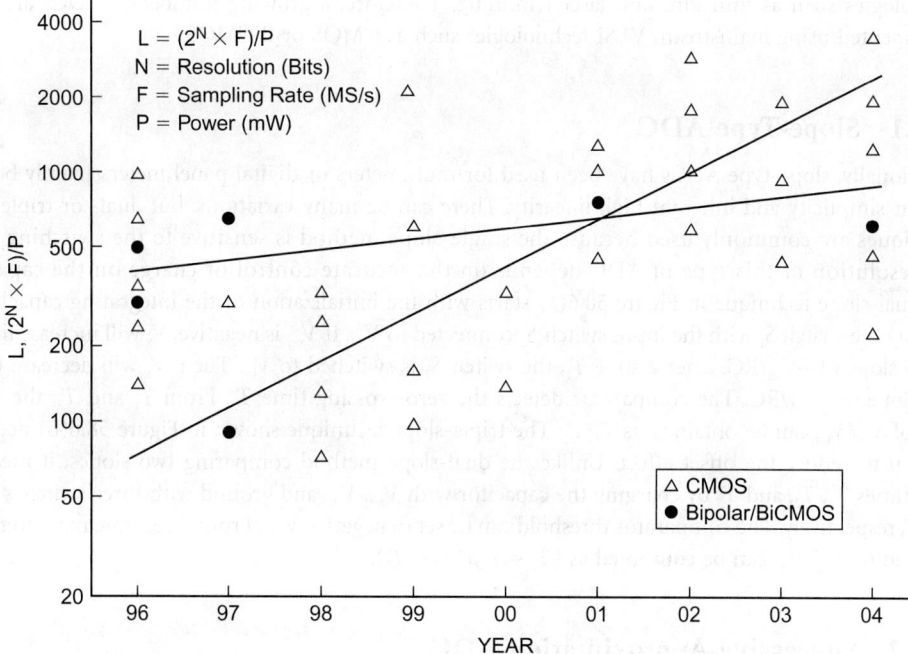

FIGURE 58.5 Figure of merit (L) versus year.

standards using software. Low-spurious, low-power data converters are key components in this new software radio environment.

The fundamental limit in quantizing IF spectrum is the crosstalk and overload, and the system performance heavily depends on the SFDR (spurious-free dynamic range) of the sampling ADC. To meet this growing demand, low-spurious data conversion blocks are being actively developed in ever-increasing numbers. For a 14b-level ideal dynamic range while sampling at 50 MHz, it is necessary to control the sampling jitter below a picosecond, which is within the current arts of the CMOS technology. However, unlike nonlinearity that causes interchannel mixing, the random jitter in IF sampling increases only the random noise floor. As a result, the random jitter is not considered fundamental in this application. This challenging new application for digital IF processing will lead to the implementation of data converters with very wide SFDR of more than 90 dB. Two high-speed candidate architectures, pipeline (or multi-step) and folding, are potential candidate architectures to challenge these limits with new system approaches.

58.2.3 ADC Figure of Merit

The ADC performance is often represented by a figure of merit L which is defined as $L = 2^N \times f_s/P$, where N is the number of bits, f_s is the sampling rate in Msamples/s, and P is the power consumption in mW. The higher the L is, the more bits are obtained at higher speed with lower power. The plot of L versus year shown in Figure 58.5 shows the low-power and high-speed trend both for leading integrated CMOS and bipolar/BiCMOS ADCs published in the last decade.

58.3 ADC Architectures

In general, the main criteria of choosing ADC architectures are resolution and speed, but auxiliary requirements such as power, chip area, supply voltage, latency, operating environment, or technology often limit the choices. The current trend is toward low-cost integration without using expensive discrete technologies such as thin film and laser trimming. Therefore, a growing number of ADCs are being implemented using mainstream VLSI technologies such as CMOS or BiCMOS.

58.3.1   Slope-Type ADC

Traditionally, slope-type ADCs have been used for multimeters or digital panel meters mainly because of their simplicity and inherent high linearity. There can be many variations, but dual- or triple-slope techniques are commonly used because the single-slope method is sensitive to the switching error. The resolution of this type of ADC depends on the accurate control of charge on the capacitor. The dual-slope technique in Figure 58.6(a) starts with the initialization of the integrating capacitor by opening the switch S_1 with the input switch S_2 connected to V_{ref}. If V_{ref} is negative, V_x will increase linearly with a slope of $-V_{ref}/RC$. After a time T_1, the switch S_2 is switched to V_{in}. Then, V_x will decrease with a new slope of $-V_{in}/RC$. The comparator detects the zero-crossing time T_2. From T_1 and T_2, the digital ratio of V_{in}/V_{ref} can be obtained as T_1/T_2. The triple-slope technique shown in Figure 58.6(b) needs no op-amp to reduce the offset effect. Unlike the dual-slope method comparing two slopes, it measures three times T_1, T_2, and T_3 by charging the capacitor with V_{ref}, V_{in}, and ground with three switches S_1, S_2, and S_3, respectively. The comparator threshold can be set to negative V_{TH}. From three time measurements, the ratio of V_{in}/V_{ref} can be computed as $(T_2 - T_3)/(T_1 - T_3)$.

58.3.2   Successive-Approximation ADC

The simplest concept of A/D conversion is comparing analog input voltage with an output of a DAC. The comparator output is fed back through the DAC as explained in Figure 58.7. The successive-approximation register (SAR) performs the most straightforward binary comparison. The sampled input is compared with

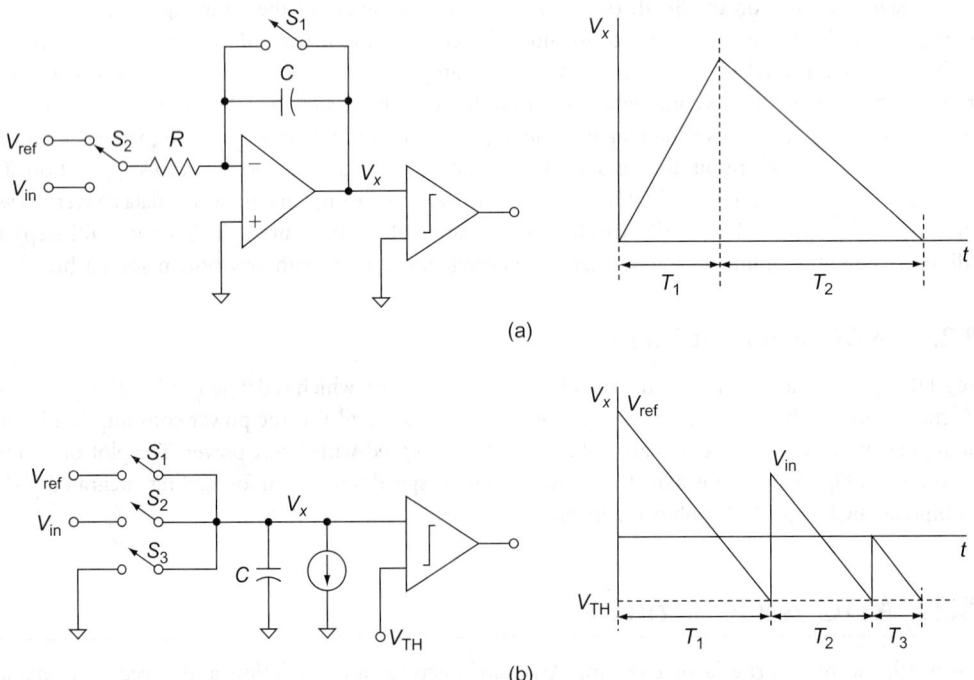

(a)

(b)

FIGURE 58.6   (a) Dual-slope and (b) triple-slope ADC techniques.

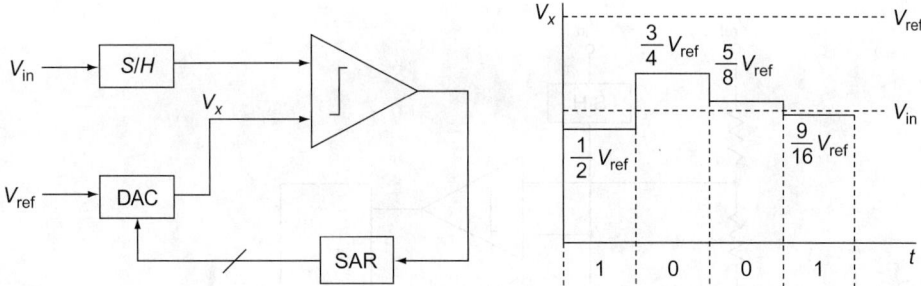

FIGURE 58.7 Successive-approximation ADC technique.

the DAC output by progressively dividing the range by two as explained in the 4-b example. The conversion starts by sampling input, and the first MSB decision is made by comparing the sample-and-hold (S/H) output with $V_{ref}/2$ by setting the MSB of the DAC to 1. If the input is higher, the MSB stays as 1. Otherwise, it is reset to 0. In the second bit decision, the input is compared with $3V_{ref}/4$ in this example by setting the second bit to 1. Note that the previous decision set the MSB to 1. If the input is lower, as in the example shown, the second bit is set to 0, and the third bit decision is done by comparing the input with $5V_{ref}/8$. This comparison continues until all the bits are decided. Therefore, the N-bit successive-approximation ADC requires $N+1$ clock cycles to complete one sample conversion.

The performance of the successive-approximation ADC is limited by the DAC resolution and the comparator accuracy. The commonly used DACs for this architecture are a resistor-string DAC and a capacitor-array DAC. Although binary-weighted capacitors have a 10b-level matching in MOS [1], diffused resistors have poor matching and high voltage coefficient. If differential resistor-string DACs are used, performance can be improved to the capacitor-array DAC level [2]. In general, the capacitor DAC exhibits poor DNL while the resistor-string DAC exhibits poor INL.

58.3.3 Flash ADC

The most straightforward way of making an ADC is to compare the input with all the divided levels of the reference simultaneously. Such a converter is called a flash ADC, and the conversion occurs in one step. The flash ADC is the fastest among all ADCs. The flash ADC concept is explained in Figure 58.8, where divided reference voltages are compared to the input. The binary encoder is needed because the output of the comparator bank is thermometer-coded. The resolution is limited both by the accuracy of the divided reference voltages and by the comparator resolution. The metastability of the comparator produces a sparkle noise when the comparator is indecisive. The reference division can be done using capacitor dividers [3,4] or transistor sizing [5] for small-scale flash ADCs. However, only resistor-string DACs can provide references as the number of bits grows.

In practical implementations, the limit is the exponential growth in the number of comparators and resistors. For example, an N-bit flash needs $2^N - 1$ comparators and 2^N resistors. Furthermore, for the Nyquist-rate sampling, the input needs a S/H to freeze the input for comparison. As the number of bits grows, the comparator bank presents a significant loading to the input S/H, diminishing the speed advantage of this architecture. Also, the control of the reference divider accuracy and the comparator resolution degrades, and the power consumption becomes prohibitively high. As a result, flash converters with more than 10-b resolution are rare. Flash ADCs are commonly used as coarse quantizers in the pipeline or multi-step ADCs. The folding/interpolation ADC, which is conceptually a derivative of the flash ADC, reduces the number of comparators by folding the input range [6].

For high resolution, the flash ADC needs a low-offset comparator with high gain, and the comparator is often implemented in a multi-stage configuration with offset cancelation. The front-end of the multistage comparator is called a preamplifier. A technique called *interpolation* saves the number of preamplifiers by interpolating the adjacent preamplifier outputs as shown in Figure 58.9(a), where two preamplifier

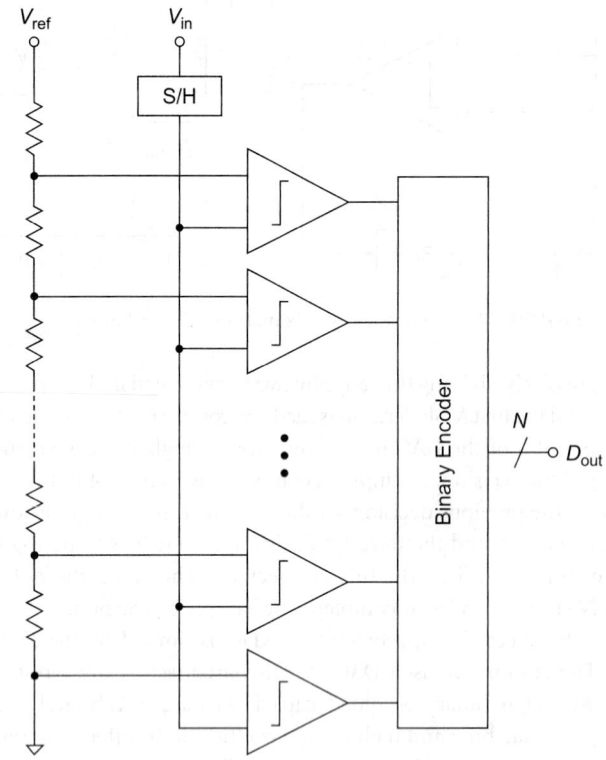

FIGURE 58.8 Flash ADC technique.

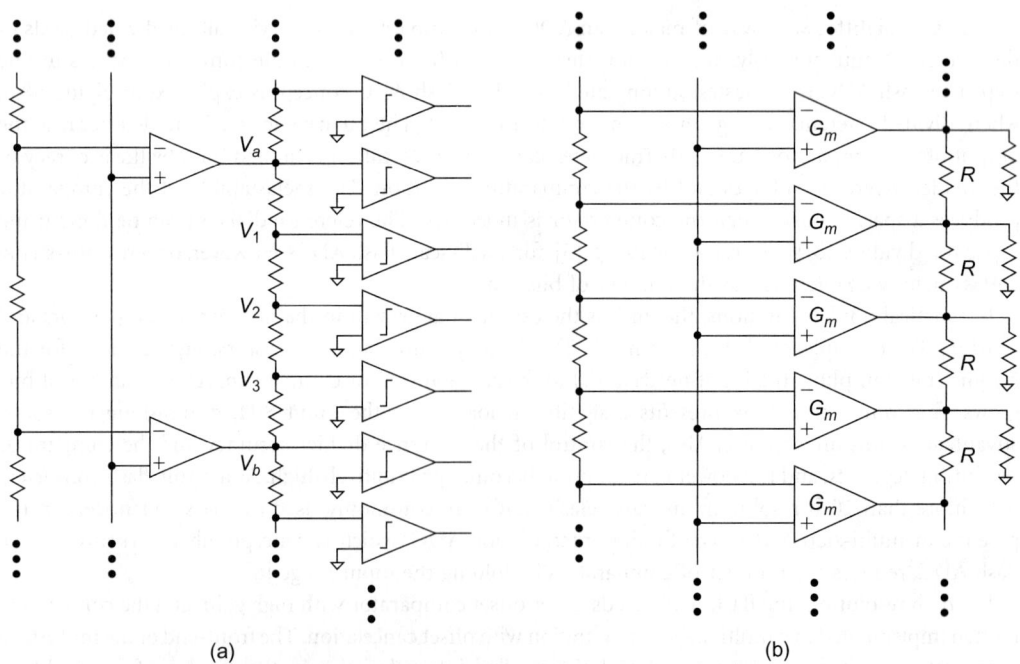

(a)   (b)

FIGURE 58.9 (a) Interpolation and (b) averaging techniques.

outputs V_a and V_b are used to generate three more outputs V_1, V_2, and V_3 using a resistor divider. The interpolation can improve the DNL within the interpolated range, but the overall DNL and INL are not improved. Interpolating any arbitrary number of levels is possible by making more resistor taps. The interpolation is usually done using resistors, but it is also possible to interpolate using capacitors and current sources. However, interpolating with independent current sources does not improve the DNL.

Another technique called *averaging*, as explained in Figure 58.9(b) is often used to average out the offsets of the neighboring preamplifiers as well as to enhance the accuracy of the reference divider [7]. The idea is to couple the outputs of the preamplifier transconductance (G_m) stage so that the offset errors can be spread over the adjacent preamplifier outputs as explained. For example, if the coupling resistor value is infinite, there exists no averaging. As the coupling resistor value decreases, one preamplifier output becomes the weighted sum of the outputs of its neighboring preamplifiers. Therefore, the overall DNL and INL can improve significantly [8]. However, for the case in which errors to average have the same polarity, the averaging is not that effective. In practice, both the interpolation and the averaging concepts are often combined.

58.3.4 Subranging ADC

Although the interpolation and averaging techniques simplify the flash ADC, the number of comparators stays the same. Instead of making all the bit decisions at once, resolving a few bits at a time makes the system simpler and more manageable. It also enables us to use a digital error correction concept. The simplest subranging ADC concept is explained in Figure 58.10 for the two-step conversion case. It is a straightforward subranging since one subrange out of 2^M subranges is chosen in the coarse M-bit decision. Once one subrange is selected, the N-bit fine decision can be made using a fine reference ladder interpolating the selected subrange.

Note that the subrange after the coarse decision is $V_{ref}/2^M$ and the fine comparators should have a resolution of $M + N$ bits. Unless the digital error correction with redundancy is used, the coarse comparators should also have a resolution of $M + N$ bits.

58.3.5 Multi-Step ADC

The tactic of making a few bit decisions at a time as shown in the subranging case can be generalized. A slight modification of the subranging architecture shown in Figure 58.11(a) to include a residue

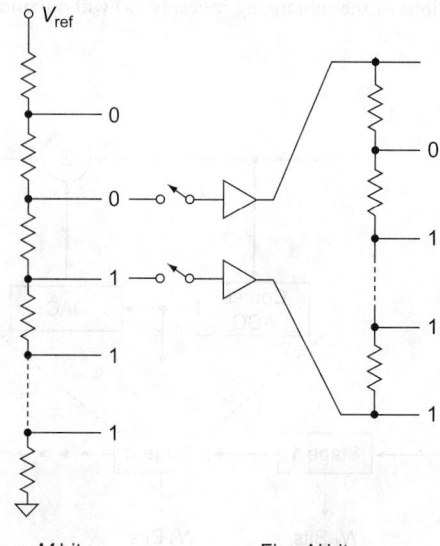

FIGURE 58.10 Coarse and fine reference ladders for two-step subranging ADC.

amplifier with a gain of 2^M results in Figure 58.11(b). The residue is defined as the difference between the input and the nearest DAC output lower than the input. The difference between the two concepts is subtle, but including one residue amplifier drastically changes the system requirements. The obvious advantage of using the residue amplifier is that the fine comparators do not need to be accurate because the residue from the coarse decision is amplified by 2^M. That is, the subrange after the coarse decision is no longer $V_{ref}/2^M$. The disadvantage is the accuracy and settling of the high-gain residue amplifier.

   Whether the residue is amplified or not, the subranging block consists of a coarse ADC, a DAC, a residue subtractor, and an amplifier. In theory, this block can be repeated as shown in Figure 58.12. How many times it is repeated determines the number of steps. So, in general terms, the *n*-step ADC

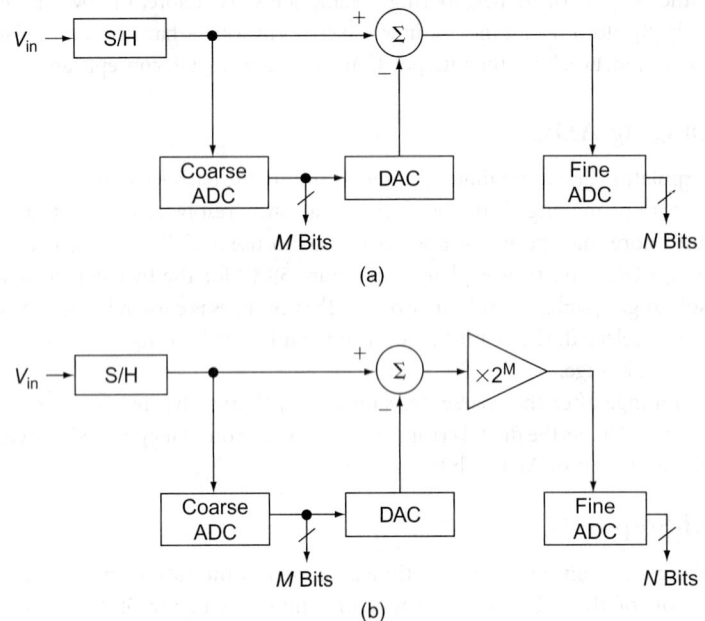

FIGURE 58.11 Variations of the subranging concepts: (a) without and (b) with residue amplifier.

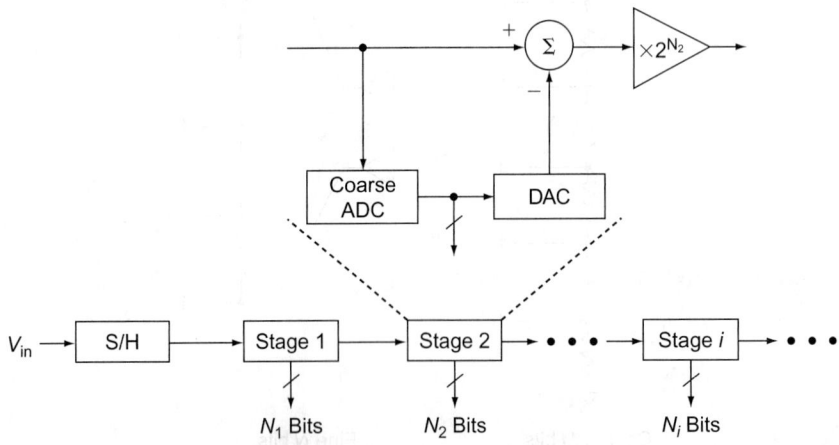

FIGURE 58.12 Multi-step ADC architecture.

has n–1 subranging blocks. To complete a conversion in one cycle, usually poly-phase subdivided clocks are needed. Due to the difficulty in clocking, the number of steps for the multi-step architecture is usually limited to two, which does not incur a speed penalty and needs the standard two-phase clocking.

There are many variations in the multi-step architecture. If no ploy-phase clocking is used, it is called a *ripple ADC*. Also in the two-step ADC, if one ADC is repeatedly used both for the coarse and fine decisions, it is called a *recycling ADC* [9]. In this ADC example, the capacitor-array multiplying DAC (MDAC) also performs the S/H function in addition to the residue amplification. This MDAC, with either a binary-ratioed or thermometer-coded capacitor array, is a general form of the residue amplifier. The same capacitor array has been used with a comparator to implement a charge-redistribution successive-approximation ADC [1]. This MDAC is suited for MOS technologies, but other forms of the residue amplification are possible using resistor-strings or current DACs.

58.3.6 Pipeline ADC

The complexity of the two-step ADC, although manageable and simpler than the flash ADC, still grows exponentially as the number of bits to resolve increases. Specifically, for high resolution above 12 b, the complexity reaches about the maximum, and a need to pipeline subranging blocks arises. The pipeline ADC architecture shown in Figure 58.13 is the same as the subranging or multi-step ADC architecture shown in Figure 58.12 except for the interstage S/H. Since the S/Hs are clocked by alternating clock phases, each stage needs to perform the decision and the residue amplification in each clock phase. Pipelining the residue greatly simplifies the ADC architecture. The complexity grows only linearly with the number of bits to resolve. Due to its simplicity, the pipeline ADCs have been gaining popularity in the digital VLSI environment.

In the pipeline ADC, each stage resolves a few bits quickly and transfers the residue to the following stage so that the residue can be resolved further in the subsequent stages. Therefore, the accuracy of the interstage residue amplifier limits the overall performance. The following four non-ideal error sources can affect the performance of the multi-step or pipeline ADCs: ADC resolution, DAC resolution, gain error of the residue amplifier, and inaccurate settling of the residue amplifier. The offset of the residue amplifier does not affect the linearity, but it appears as a system offset. Among these four error sources, the first three are static, but the residue amplifier settling is dynamic. If the residue amplifier is assumed to settle within one clock phase, three static error sources are limiting the linearity performance.

Figure 58.14 explains the residue from the 2-b stage in the systems shown in Figures 58.12 and 58.13. In the ideal case, as the input is swept from 0 to the full range V_{ref}, the residue change from 0 to V_{ref} repeats each time V_{ref} is subtracted at the ideal locations of the 2-b ADC thresholds, which are $V_{ref}/4$ apart. In this case, the 2-b stage does not introduce any nonlinearity error. However, in the other cases

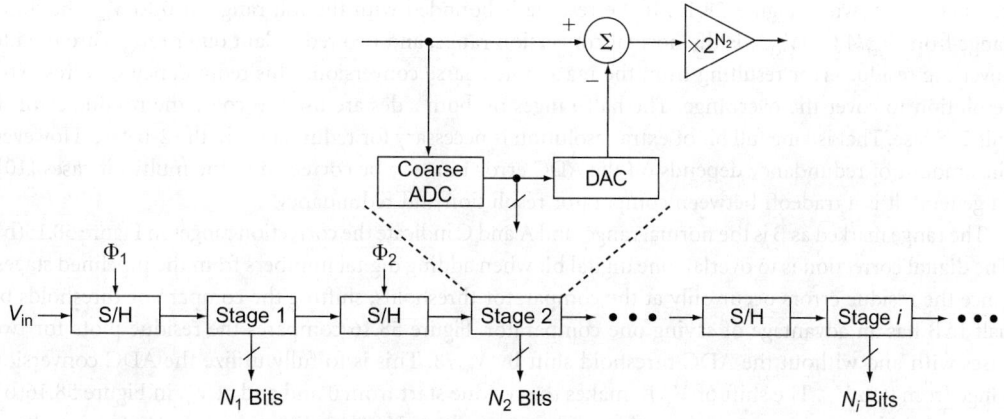

FIGURE 58.13 Pipeline ADC architecture.

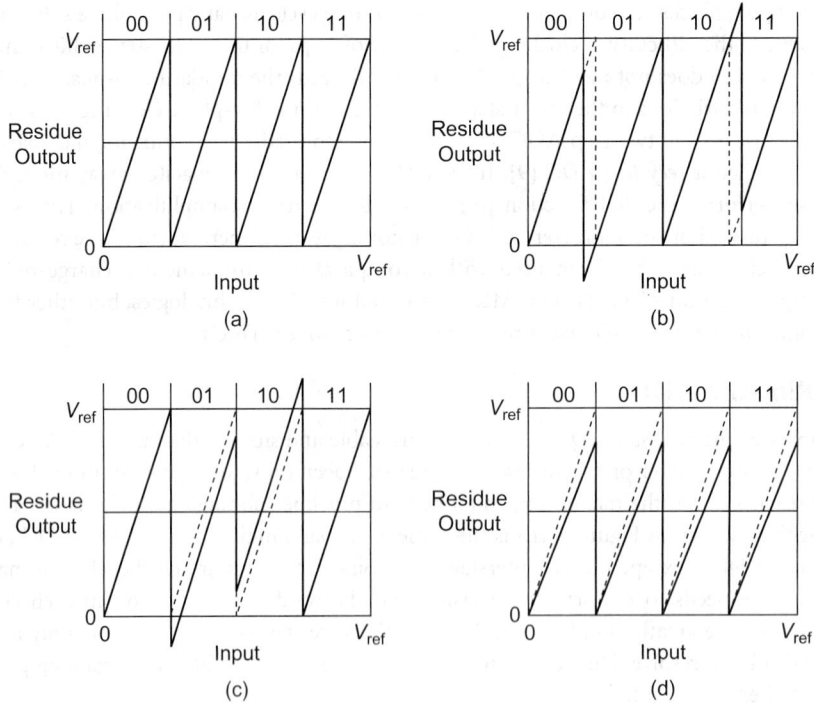

FIGURE 58.14 2b residue versus input: (a) ideal, and with (b) ADC, (c) DAC, and (d) gain errors.

with ADC, DAC, and gain errors, the residue ranges do not match with the ideal full-scale V_{ref}. If the residue range is smaller than the full range, missing codes are generated; and if the residue goes out of bounds, excessive codes are generated at the ADC thresholds. Unlike the DAC and gain errors, the ADC error appears as a shift of residue by V_{ref} as long as the DAC and the residue amplifier are ideal. This implies that the ADC error can be corrected digitally by restoring the ideal range of V_{ref}.

58.3.7 Digital Error Correction

Any multi-step or pipeline ADC system can be made insensitive to the ADC error if the ADC error is digitally corrected. The out-of-range residue can still be digitized by the following stage if the residue amplifier gain is reduced. That is, if the residue amplifier gain is set to 2^{N-1} instead of 2^N, the residue plots are as shown in Figure 58.15. If the residue is bounded with the full range of 0 to V_{ref}, the inner range from $V_{ref}/4$ to $3V_{ref}/4$ is the normal conversion range, and two redundant outer ranges are used to cover the residue error resulting from the inaccurate coarse conversion. This redundancy requires extra resolution to cover the overrange. The half ranges on both sides are used to cover the residue error in this 2-b case. That is, one full bit of extra resolution is necessary for redundancy in the 2-b case. However, the amount of redundancy depends on the ADC error range to be corrected in the multi-bit cases [10]. In general, it is a tradeoff between comparator resolution and redundancy.

The range marked as B is the normal range, and A and C indicate the correction ranges in Figure 58.15(b). The digital correction is to overlap one digital bit when adding digital numbers from the pipelined stages. Since the residue errors occur only at the comparator thresholds, shifting the comparator thresholds by half LSB has an advantage of saving one comparator. Figure 58.16 compares the residue plots for two cases with and without the ADC threshold shift by $V_{ref}/8$. This is to fully utilize the ADC conversion range from 0 to V_{ref}. The shift of $V_{ref}/8$ makes the residue start from 0 and end at V_{ref} in Figure 58.16(b), contrary to the previous case where the residue starts from $V_{ref}/4$ and ends at $3V_{ref}/4$. This results in saving one comparator. The former case needs 2^N-1 comparators, while the latter case needs 2^N-2.

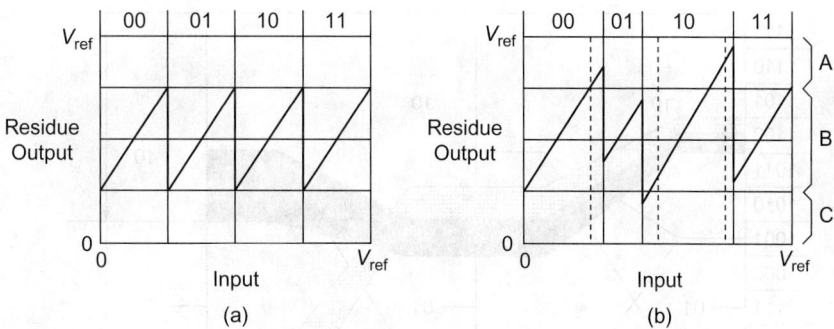

FIGURE 58.15 Over-ranged 2-b residue versus input: (a) ideal and with (b) ADC and DAC errors.

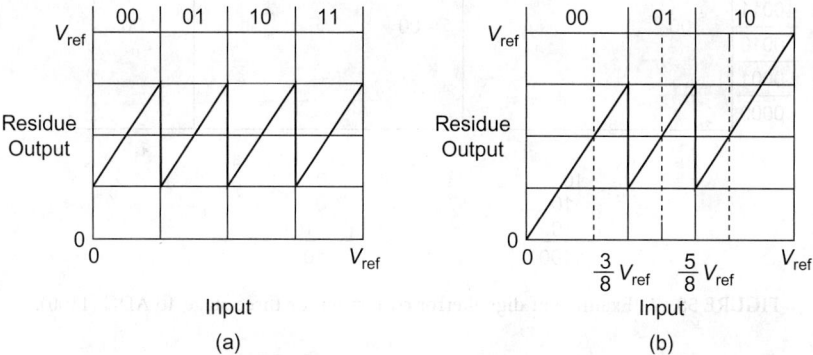

FIGURE 58.16 Half-bit shifted 2-b residue versus input: (a) with and (b) without ADC threshold-level shift.

Digital correction in the half-bit-shifted case is explained in the 4-b ADC example, made of three stages using one-bit correction per stage in Figure 58.17. The vertical axis marks the signal and residue levels as well as ADC decision levels. The dotted and shaded areas follow the residue paths when the ADC error occurs, but the end results are the same after digital correction. This half interval shift is valid for stages resolving any number of bits. Overall, the digital error correction enables fast data conversion using inaccurate comparators. However, the DAC resolution and the residue gain error still remain as the fundamental limits in multi-step and pipeline ADCs. The currently known ways to overcome these limits are either trimming or self-calibration.

58.3.8 One-Bit per Stage Pipeline ADC

The degenerate case of the pipeline ADC is when only one bit is resolved per stage as shown in Figure 58.18. Each stage multiplies its input V_{in} by two and subtracts the reference voltage V_{ref} to generate the residue voltage. If the sign of $2V_{in} - V_{ref}$ is positive, the bit is 1 and the residue goes to the next stage. Otherwise, the bit is 0 and V_{ref} is added back to the residue before it goes to the next stage. However, in reality, it is more desirable if the reference restoring time is saved. In the non-restoring algorithm, the previous bit decision affects the polarity of the reference voltage to be used in the current bit decision. If the previous bit is 1, the residue voltage is $2V_{in} - V_{ref}$, as in the restoring algorithm. But if the previous bit is 0, the residue voltage is $2V_{in} + V_{ref}$.

The switched-capacitor implementation of the basic functional block performing $2V_{in} \pm V_{ref}$ is explained using two identical capacitors and one op-amp in Figure 58.19. During the sampling phase, the bottom plates of two capacitors are switched to the input, and the top plate is connected either to

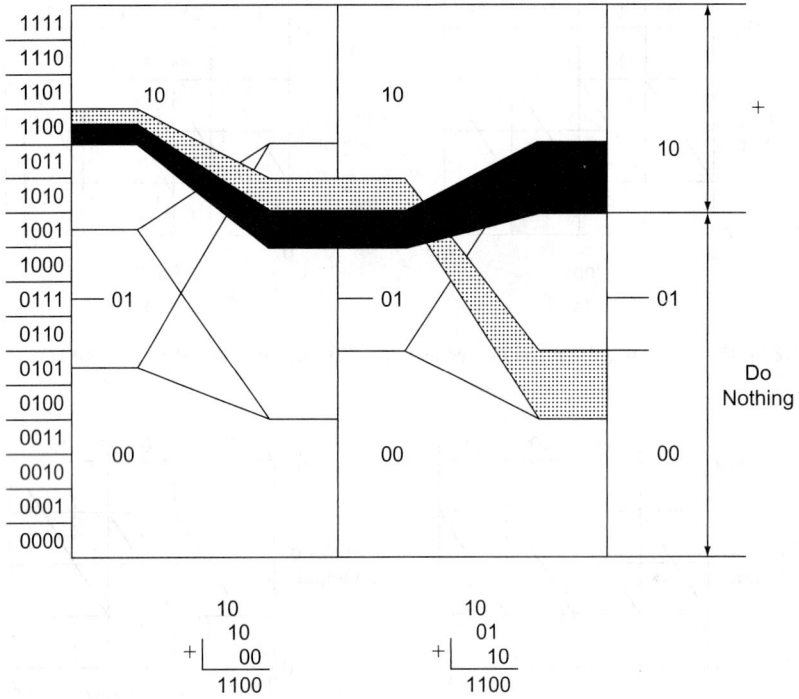

FIGURE 58.17   Example of digital error correction for three-stage 4b ADC (1100).

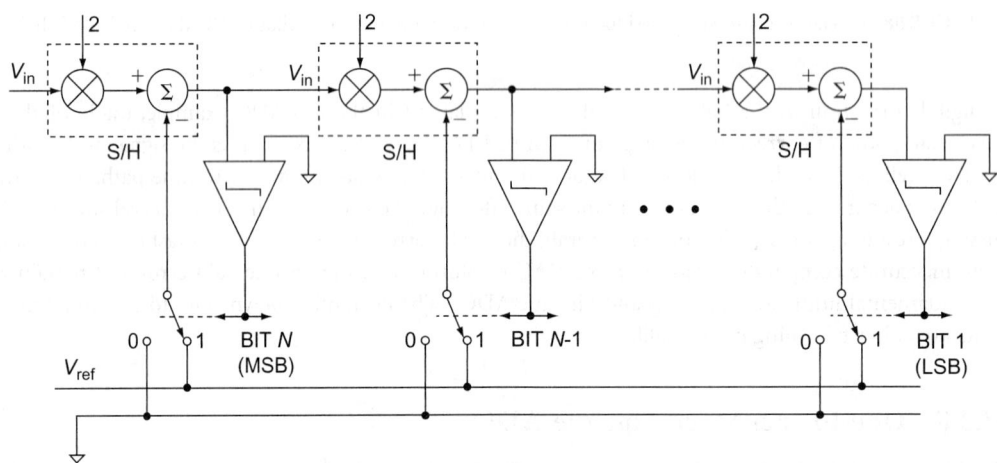

FIGURE 58.18   One-bit per stage pipeline ADC architecture.

the op-amp output or to the op-amp input common-mode voltage. During the amplification phase, one of the capacitors is connected to the output of the op-amp for feedback, but the other is connected to $\pm V_{\text{ref}}$. Then, the output of the op-amp will be $2V_{\text{in}} - V_{\text{ref}}$ and $2V_{\text{in}} + V_{\text{ref}}$, respectively, after the op-amp settles.

However, this simple one-bit pipeline ADC is of no use if the comparator resolution is limited. If any redundancy is used for digital correction, at least two bits should be resolved. A close look at Figure 58.16(b) gives a clue to using the simple functional block shown in Figure 58.19 for the 2-b residue amplification. The case explained in Figure 58.16(b) is sometimes called 1.5-b rather than 2-b because it needs only

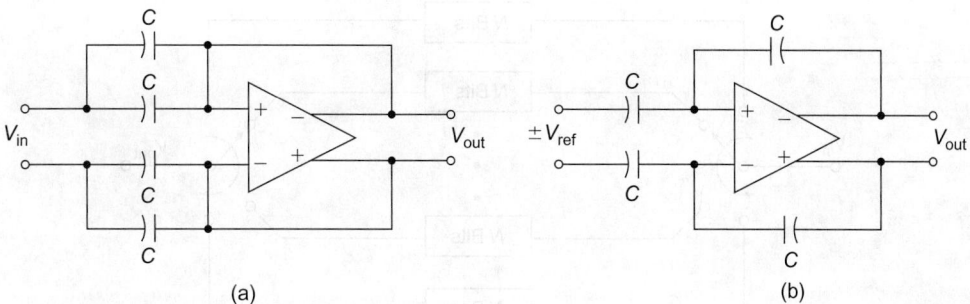

FIGURE 58.19   The simplest two-level MDAC: (a) sampling phase and (b) amplification phase.

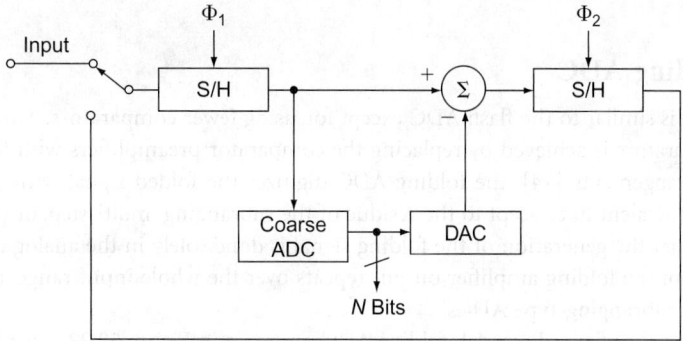

FIGURE 58.20   Algorithmic, cyclic, or recursive ADC architecture.

three DAC levels rather than four. The functional block in Figure 58.19 can have a two-level DAC subtracting $\pm V_{ref}$. However, in differential architecture, by shorting the input, one midpoint can be interpolated. Using the tri-level DAC, the 1.5-b per stage pipeline ADC can be implemented with the following algorithm [12]. If the input V_{in} is lower than $-V_{ref}/4$, the residue output is $2V_{in} + V_{ref}$. If the input is higher than $V_{ref}/4$, the residue output is $2V_{in} - V_{ref}$. If the input is in the middle, the output is $2V_{in}$.

58.3.9   Algorithmic, Cyclic, or Recursive ADC

The interstage S/H used in the multi-step architecture provides a flexibility of the pipeline architecture. In the pipeline structure, the same hardware repeats as shown in Figure 58.13. That is, the throughput rate of the pipeline is fast while the overall latency is limited by the number of stages. Instead of repeating the hardware, using the same stage repeatedly greatly saves hardware, as shown in Figure 58.20. That is, the throughput rate of the pipeline is directly traded for hardware simplicity. Such a converter is called an *algorithmic, cyclic,* or *recursive ADC*. The functional blocks used for the algorithmic ADC are identical to the ones used in the pipeline ADC.

58.3.10   Time-Interleaved Parallel ADC

The algorithmic ADC just described sacrifices the throughput rate for small hardware. However, the *time-interleaved parallel ADC* takes quite the opposite direction. It duplicates more hardware in parallel for higher throughput rates. The system is shown in Figure 58.21, where the throughput rate increases by the number of parallel paths multiplexed. Although it significantly improves the throughput rate and many refinements have been reported, it suffers from many problems [13]. Due to the multiplexing, even static nonlinearity mismatch between paths appears as a fixed pattern noise. Also, it is difficult to generate clocks with exact delays, and inaccurate clocking increases the noise floor.

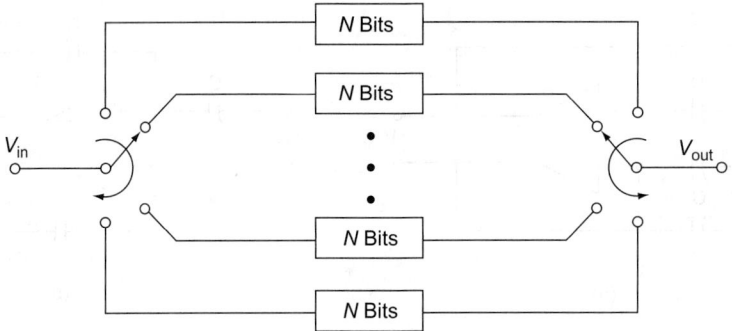

FIGURE 58.21 Time-interleaved parallel ADC architecture.

58.3.11   Folding ADC

The folding ADC is similar to the flash ADC except for using fewer comparators. This reduction in the number of comparators is achieved by replacing the comparator preamplifiers with folding amplifiers. In its original arrangements [14], the folding ADC digitizes the folded signal with a flash ADC. The folded signal is equivalent in concept to the residue of the subranging, multi-step, or pipeline ADC, but the difference is that the generation of the folding signal is done solely in the analog domain. Since the digitized code from the folding amplifier output repeats over the whole input range, a coarse coding is required, as in all subranging-type ADCs.

Consider a system configured as a 4-b folding ADC as shown in Figure 58.22. Four folding amplifiers can be placed in parallel to produce four folded signals. Comparators check the outputs of the folding amplifiers for zero crossing. If the input is swept, the outputs of the fine comparators show a repeating pattern, and eight different codes can be obtained by the four comparators. Because there are two identical fine code patterns, one comparator is needed to distinguish them. However, if this coarse comparator is misaligned with the fine quantizer, missing codes will result. A digital correction similar to that for the multi-step or pipeline ADC can be employed to correct the coarse quantizer error. For this system example, one-bit redundancy is used by adding two extra comparators in the coarse quantizer. The shaded region in the figure is where errors occur.

Having several folded signals instead of one has many advantages in designing high-speed ADCs. The folding amplifier requires neither linear output nor accurate settling. This is because in the folding ADC, the zero-crossings of the folded signals matter, but not their absolute values. Therefore, the offset of the folding amplifiers becomes the most important design issue. The resolution of the folding ADC can be further improved using the interpolation concept. When the adjacent folded signals are interpolated by I times, the number of zero-crossing points are also increased by I times. So, the resolution of the final ADC is improved by $\log_2 I$ bits. The higher bound for the degree of interpolation is set by the comparator resolution, the gain of the folding amplifiers, the linearity of folded signals, and the interpolation accuracy. Since the folding process increases the internal signal bandwidth by the number of foldings, the folding ADC performance is limited by the folding amplifier bandwidth. To increase the number of foldings while maintaining the reasonable comparator resolution, the folding amplifier's gain should be high. Since the higher gain limits the amplifier bandwidth, it is necessary to cascade the folding stages [6,8].

58.4   ADC Design Considerations

In general, multi-step ADCs are made of cascaded low-resolution ADCs. Each low-resolution ADC stage provides a residue voltage for the subsequent stage, and the accuracy of the residue voltage limits the resolution of the converter. One of the residue amplifiers commonly used in CMOS is a switched-capacitor

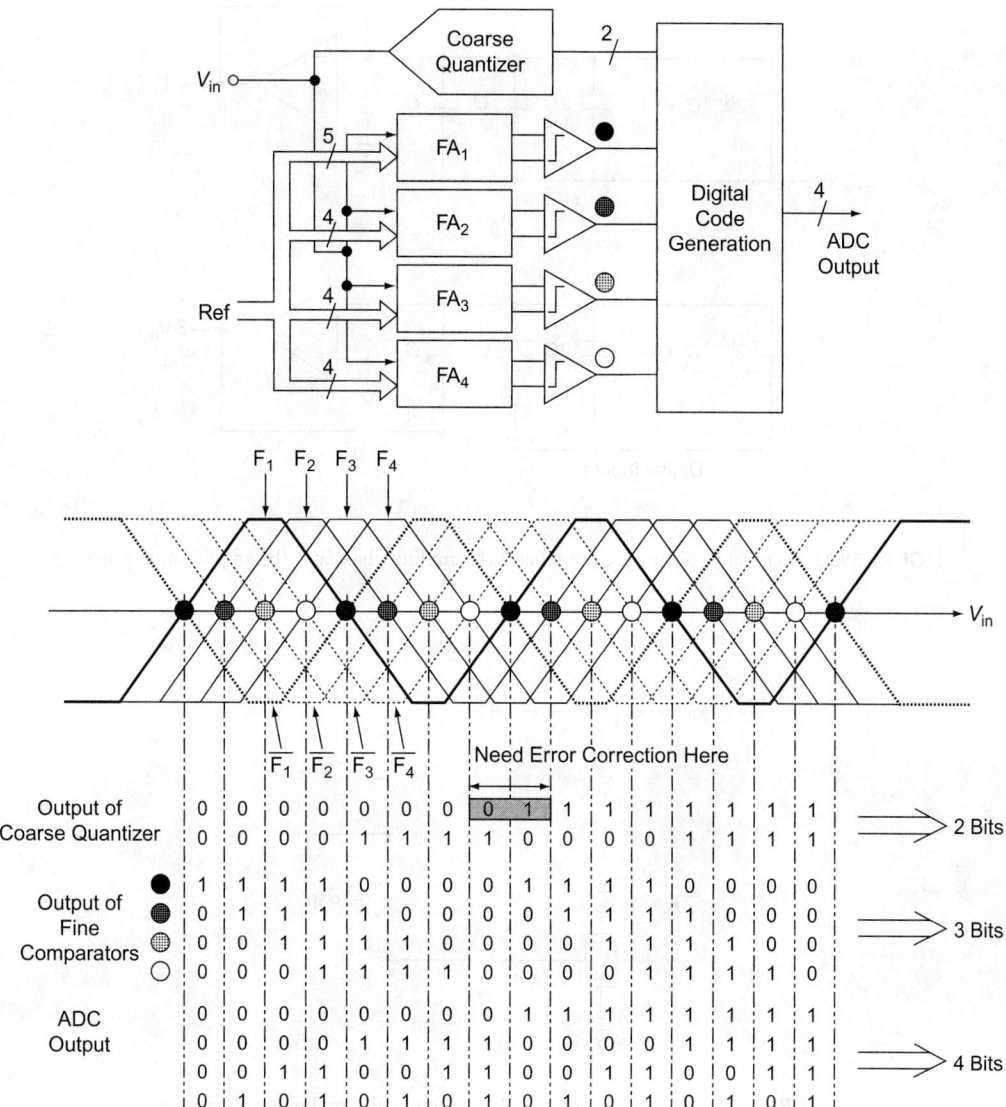

FIGURE 58.22 A 4-b folding ADC example with digital correction.

MDAC, whose connections during two clock phases are illustrated in Figure 58.23 for an *N*-bit case [9]. An extra capacitor *C* is usually added to double the feedback capacitor size so that the residue voltage may remain within the full-scale range for digital correction.

58.4.1 Sampling Error Considerations

Since the ADC works on a sampled signal, the accuracy in sampling fundamentally limits the system performance. It is well known that the noise power to be sampled on a capacitor along with the signal is KT/C. It is inversely proportional to the sampling capacitor size. The sampled rms voltage noise is 64 μV with 1 pF, but decreases to 20 μV with 10 pF. For accurate sampling, sampling capacitors should be large, but sampling on large capacitors takes time. The speed of the ADC is fundamentally limited by the sampling KT/C noise.

In sampling, there exists another important error source. Direct sampling on a capacitor suffers from switch feedthrough error due to the charge injection when switches are opened. A common way to reduce

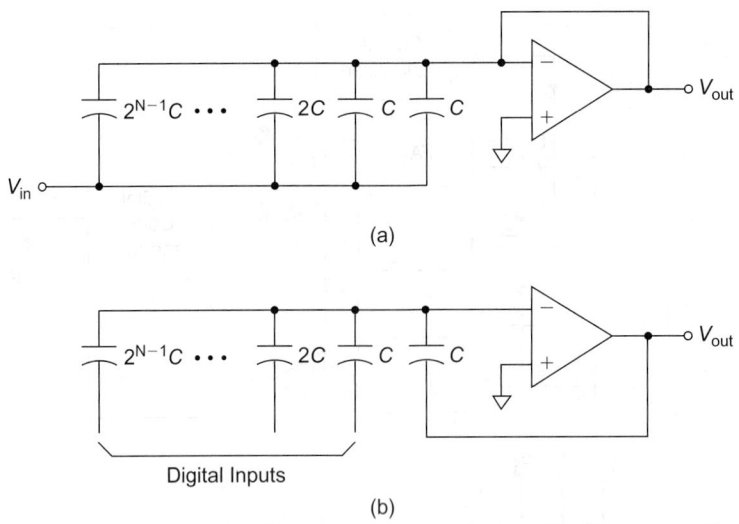

(a)

(b)

Digital Inputs

FIGURE 58.23 General N-bit residue amplifier: (a) sampling phase and (b) amplification phase.

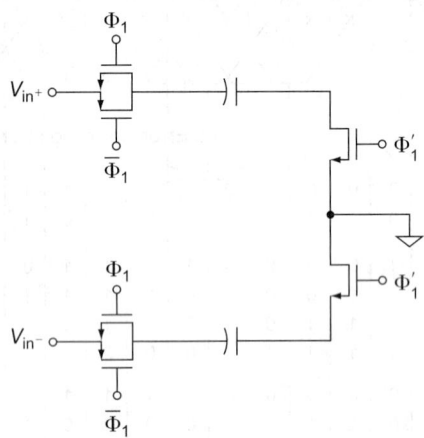

FIGURE 58.24 Open-loop bottom-plate differential sampling on capacitors.

this charge feedthrough error is to turn off the switches connected to the sampling capacitor top plate slightly earlier than the switches connected to the bottom plate. This is explained in Figure 58.24. Usually, the top plate is connected to the op-amp summing or comparator input node. The top plate is switched with one MOS transistor with a clock phase marked as Φ'_1 which makes a falling transition earlier than other clocks. The bottom plate is switched with a CMOS switch (both NMOS and PMOS) with clocks marked as Φ_1 and $\bar{\Phi}_1$. These clocks make falling transitions after the prime clock does. The net effect is that the feedthrough voltage stays constant because the top plate samples the same voltage repeatedly. The differential sampling using two capacitors symmetrically is known to provide the most accurate sampling known to date.

Unless limited by speed, the sampling error as well as the low-end spectrum of the sampled noise can be eliminated using a correlated double sampling (CDS) technique. The system has been used to remove the flicker noise or slowly-varying offset such as in charge-coupled device (CCD). The CDS needs two sampling clocks. The idea is to subtract the previously sampled sampling error from the new sample after one clock delay. The result is to null the sampling error spectrum at every multiple of the sampling frequency f_s. The CDS is effective only for the low-frequency spectrum.

58.4.2   Techniques for High-Resolution and High-Speed ADCs

Considering typical requirements, three representative ADC architectures are compared in Table 58.1. To date, all techniques known to improve ADC resolution are as follows: trimming, dynamic matching, ratio-independent technique, capacitor-error averaging, walking reference, and self-calibration. However, the trimming is irreversible and expensive. It is only possible at the factory or with precision equipments. The dynamic matching technique is effective, but it generates high-frequency noise. The ratio-independent techniques either require many clock cycles or are limited to improve differential linearity. The latter case is good for monotonicity, but it also requires accurate comparison. The capacitor-error averaging technique requires three clock cycles, and the walking reference is sensitive to clock sampling error. The self-calibration technique requires extra hardware for calibration and digital storage, but its compatibility with existing proven architectures may provide potential solutions both for high resolution and for high speed.

The ADC self-calibration concepts originated from the direct code-mapping concept using memory. The calibration is to predistort the digital input to the DAC so that the DAC output can match the ideal level from the calibration equipment. Due to the precision equipment needed, this method has limited use. The first self-calibration concept applied to the binary-ratioed successive-approximation ADC is to internally measure capacitor DAC ratio errors using a resistor-string calibration DAC as shown in Figure 58.25 [15]. Later, an improved concept of the digital-domain calibration was developed for the multistep or pipeline ADCs [16].

TABLE 58.1   Three Dominant ADC Architectures

	Interpolated Flash	Multi-step	Pipeline
Matching	Least	Medium	Most critical
Feedthrough	Most critical	Medium	Least
Bandwidth	Least	Most critical	Medium
Settling	Least	Medium	Most critical
Gain	Least	Medium	Most critical
Speed	Fast	Slow	Medium
Complexity	Complex	Medium	Simple
Problems	Clock jitter	Low loop gain	Matching
	Time skew		Wide bandwidth
	Sampling error		High gain

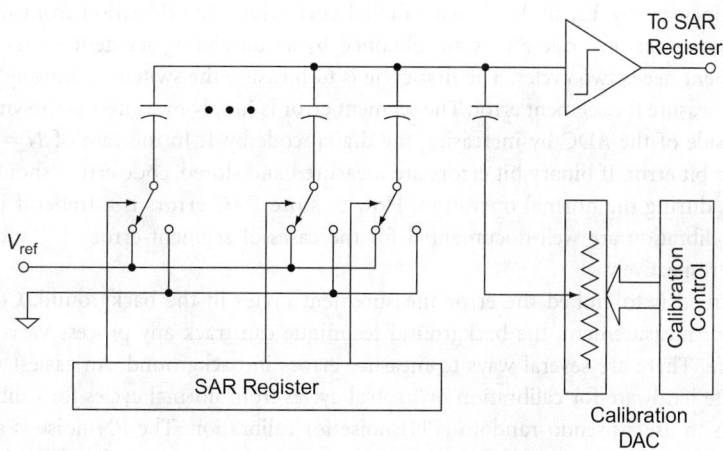

FIGURE 58.25   Self-calibrated successive-approximation ADC.

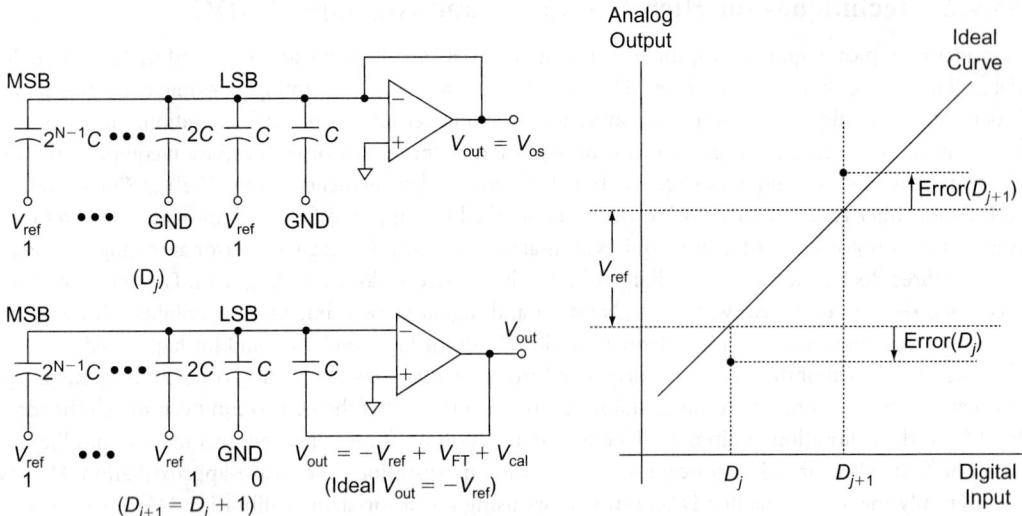

FIGURE 58.26 Segment-error measuring technique for digital self-calibration.

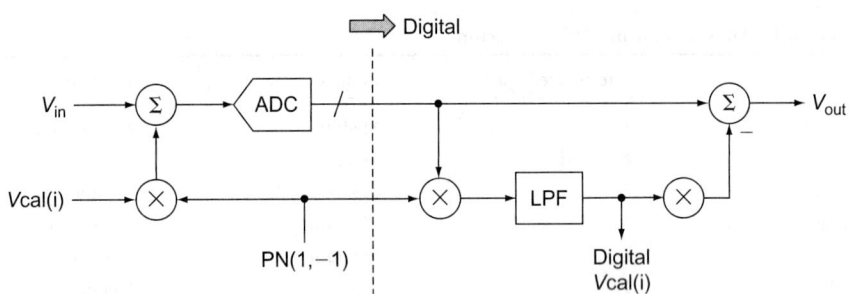

FIGURE 58.27 Background segment-error measurement scheme by dithering.

The general concept of the digital-domain calibration is to measure code or bit errors, to store them in the memory, and to subtract them during the normal operation. The concept is explained in Figure 58.26 using a generalized N-bit MDAC with a capacitor array. If the DAC code increases by 1, the MDAC output should increase by V_{ref} or $V_{ref}/2$ with digital correction. Any deviation from this ideal step is defined as a segment error. Code errors are obtained by accumulating segment errors. This segment-error measurement needs two cycles. The first cycle is to measure the switch feedthrough error, and the next cycle is to measure the segment error. The segment error is simply measured as shown in Figure 58.26 using the LSB-side of the ADC by increasing the digital code by 1. In the case of $N = 1$, the segment error becomes a bit error. If binary bit errors are measured and stored, code errors should be calculated for subtraction during the normal operation. How to store DAC errors is a tradeoff issue. Examples of the digital calibration are well documented for the cases of segment-error [17] and bit-error [18] measurements, respectively.

The recent trend is to embed the error measurement cycles in the background. Compared to the foreground error measurement, the background technique can track any process variations over time and temperature. There are several ways to measure errors in background. An easiest way is either to retire a duplicate hardware for calibration or to steal cycles from normal cycles for calibration. A more elaborate way is to use a pseudo-random (PN) noise for calibration. The PN noise is a random pulse sequence of 1 and −1 with a zero mean. When calibrating the i-th stage, a known calibration voltage $V_{cal}(i)$ is PN-modulated and injected into the i-th stage as shown in Figure 58.27. The injected $V_{cal}(i)$ is quantized by the following ADC stages, and demodulated by the same PN sequence to generate the digital $V_{cal}(i)$. The segment error can be obtained from this digitally measured $V_{cal}(i)$ by comparing it to the ideal digital V_{cal}. The injected dither for calibration is a random noise spread in the Nyquist band, but it can be digitally subtracted. The only penalty is that the signal swing is reduced to accommodate a large dither added to the signal. For this reason, a fraction of V_{ref} is used as V_{cal}.

58.5   DAC Design Arts

There are many different circuit techniques used to implement DACs, but the popular ones widely used today are of the parallel type in which all bits change simultaneously upon the application of an input code word. Serial DACs, on the other hand, produce an analog output only after receiving all digital input data in a sequential form. When DACs are used as stand-alone devices, their output transient behaviors limited by glitch, slew rate, word clock jitter, settling, etc. are of paramount importance, but used as subblocks of ADCs, DACs need only to settle within a given time interval. An output S/H, usually called a *deglitcher*, is often used for better transient performance. Three of the most popular architectures of DACs are resistor string, ratioed current sources, and a capacitor array. The current-ratioed DAC finds most applications as a stand-alone DAC, while the resistor-string and capacitor-array DACs are mainly used as ADC subblocks.

For speeds over 100 MHz, most state-of-the-art DACs employ current sources switched directly to output resistors [19–23]. Furthermore, owing to the high bit counts (12 to 16 b), segmented architectures are employed, with the current sources broken into two or three segments. The CMOS design has the advantages of lower power, smaller area, and lower manufacturing costs. In all cases, it is of interest to note that the dynamic performance of the DACs degrades rapidly as input frequencies increase, and true dynamic performance is not attained except at low frequencies. Since a major application of wideband-width, high-resolution DACs is in communications, poor dynamic performance is undesirable, owing to the noise leakage from frequency multiplexed channels into other channels. The goal of better dynamic performance continues to be a target of ongoing research and development.

58.6   DAC Architectures

An N-bit DAC provides a discrete analog output level, either voltage or current, for every level of 2^N digital words that is applied to the input. Therefore, an ideal voltage DAC generates 2^N discrete analog output voltages for digital inputs varying from 000…00 to 111…11. In the unipolar case, the reference point is 0 when the digital input is 000…00; but in bipolar or differential DACs, the reference point is the midpoint of the full scale when the digital input is 100…00. Although purely current-output DACs are possible, voltage-output DACs are common in most applications.

58.6.1   Resistor-String DAC

The simplest voltage divider is a resistor string. Reference levels can be generated by connecting 2^N identical resistors in series between V_{ref} and ground. Switches to connect the divided reference voltages to the output can be either 1-out-of-2^N decoder or binary tree decoder as shown in Figure 58.28 for the 3-b example. Since it requires a good switch, the stand-alone resistor-string DAC is easier to implement using CMOS. However, the lack of switches does not limit the application of the resistor string as a voltage reference divider subblock for ADCs in other process technologies.

Resistor strings are widely used as an integral part of the flash ADC as a reference divider. All resistor-string DACs are inherently monotonic and exhibit good differential linearity. However, they suffer from poor integral linearity and also have the drawback that the output resistance depends on the digital input code. This causes a code-dependent settling time when charging the capacitive load. This non-uniform settling time problem can be alleviated by adding low-resistance parallel resistors or by compensating the MOS switch overdrive voltages.

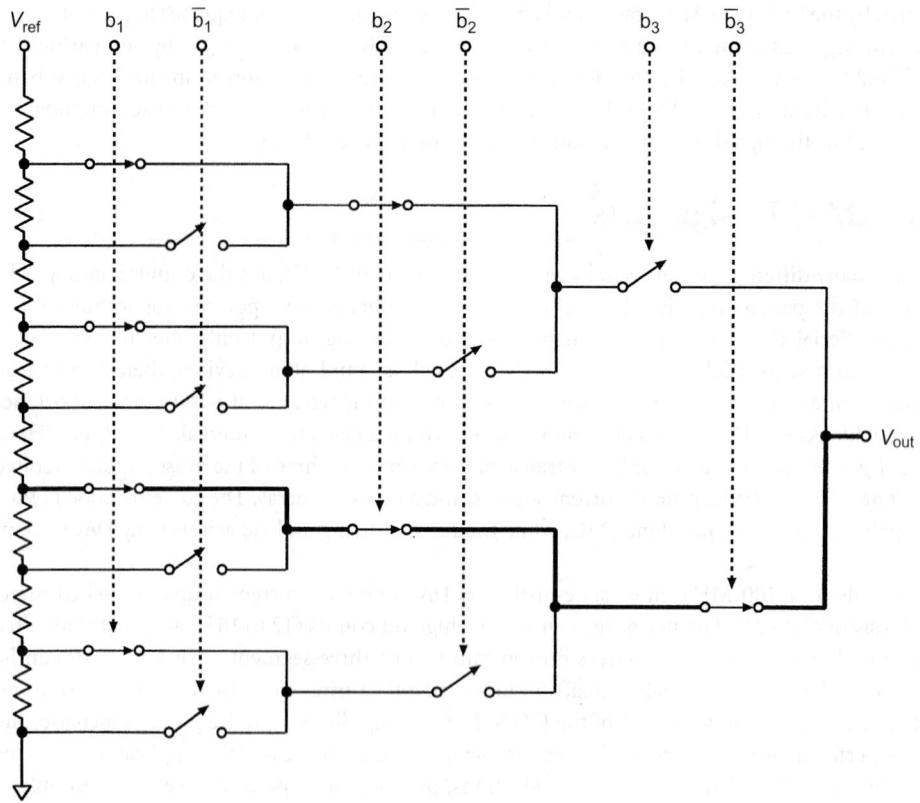

FIGURE 58.28   Resistor-string DAC.

58.6.2   Current-Ratioed DAC

The most popular stand-alone DACs in use today are *current-ratioed DACs*. There are two types: one is a weighted-current DAC and the other is an *R-2R* DAC. The weighted-current DAC shown in Figure 58.29 is made of an array of switched binary-weighted current sources and the current summing network. In bipolar technology, the binary weighting is achieved by ratioed transistors and emitter resistors with binary related values of R, $R/2$, $R/4$, etc., while in MOS technology, only ratioed transistors are used. DACs relying on active device matching can achieve an 8b-level performance with a 0.2 to 0.5% matching accuracy using a 10- to 20-μm device feature size, while degeneration with thin-film resistors gives a 10b-level performance. The current sources are switched on or off by means of switching diodes or emitter-coupled differential pairs (source-coupled pairs in CMOS). The output current summing is done by a wideband transresistance amplifier; but in high-speed DACs, the output current directly drives a resistor load for maximum speed. The weighted-current design has the advantage of simplicity and high speed, but it is difficult to implement a high-resolution DAC because a wide range of emitter resistors and transistor sizes are used, and very large resistors cause problems with both temperature stability and speed.

58.6.3   R-2R Ladder DAC

This large resistor ratio problem is alleviated by using a resistor divider known as an *R-2R ladder*, as shown in Figure 58.30. The *R-2R* network consists of series resistors of value R and shunt resistors of value $2R$. The top of each shunt resistor of value $2R$ has a single-pole double-throw electronic switch

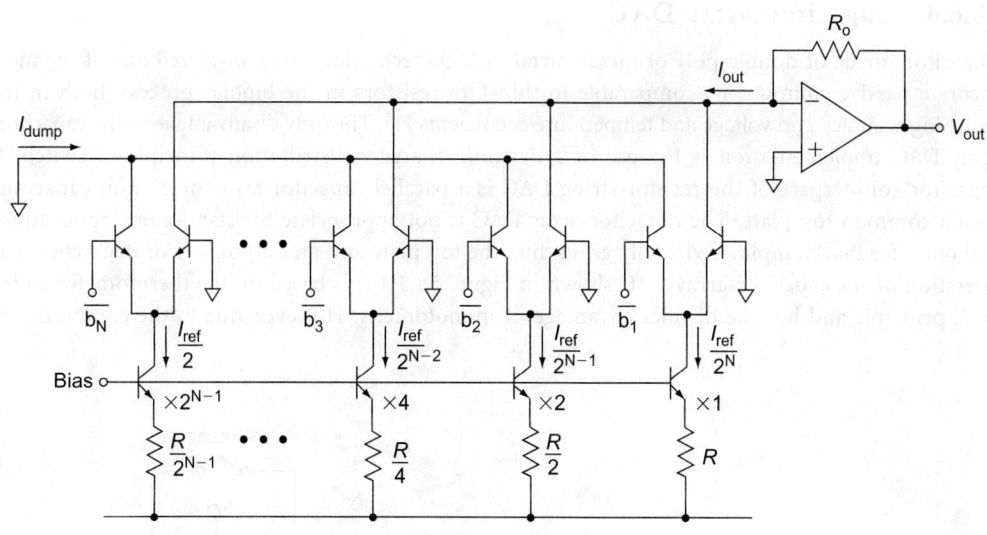

FIGURE 58.29 Current-ratioed DAC.

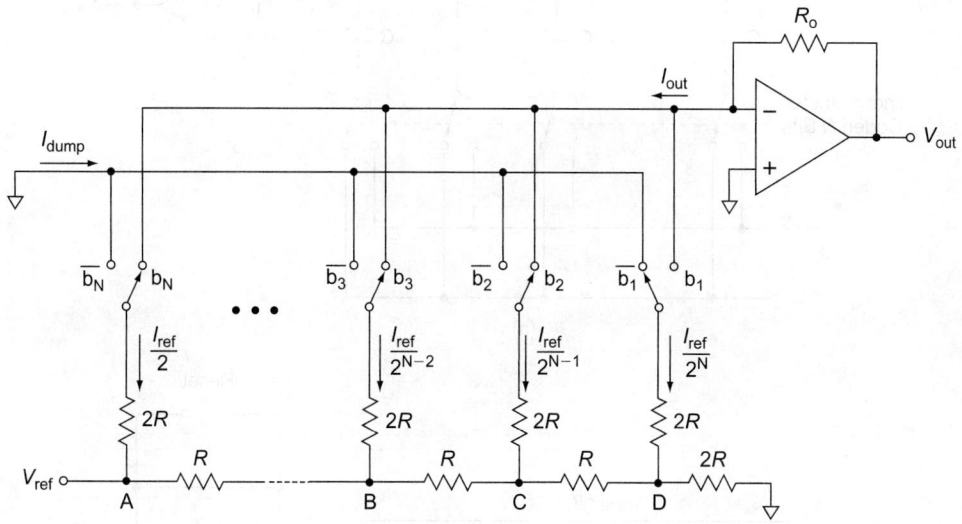

FIGURE 58.30 R-2R DAC.

that connects the resistor either to ground or to the current summing node. The operation of the R-$2R$ ladder network is based on the binary division of current as it flows down the ladder. At any junction of series resistor of value R, the resistance looking to the right side is $2R$. Therefore, the input resistance at any junction is R, and the current splits into two equal parts at the junction since it sees equal resistances in both directions. As a result, binary-weighted currents flow into shunt resistors in the ladder. The digitally controlled switches direct the currents to either ground or to the summing node. The advantage of the R-$2R$ ladder method is that only two values of resistors are used, greatly simplifying the task of matching or trimming and temperature tracking. In addition, for high-speed applications, relatively low resistor values can be used. Excellent results can be obtained using laser-trimmed thin-film resistor networks. Since the output of the R-$2R$ DAC is the product of the reference voltage and the digital input word, the R-$2R$ ladder DAC is often called an MDAC.

58.6.4 Capacitor-Array DAC

Capacitors made of double-poly or metal–metal in MOS technology are considered one of the most accurate passive components comparable to thin-film resistors in the bipolar process, both in the matching accuracy and voltage and temperature coefficients [1]. The only disadvantage in the capacitor-array DAC implementation is the use of a dynamic charge redistribution principle. A switched-capacitor counterpart of the resistor-string DAC is a parallel capacitor array of 2^N unit capacitors with a common top plate. The capacitor-array DAC is not appropriate for stand-alone applications without a feedback amplifier virtually grounding the top plate and an output S/H or deglitcher. The operation of the capacitor-array DAC shown in Figure 58.31(a) is based on the thermometer-coded DAC principle and has the distinct advantage of monotonicity. However, due to the complexity of

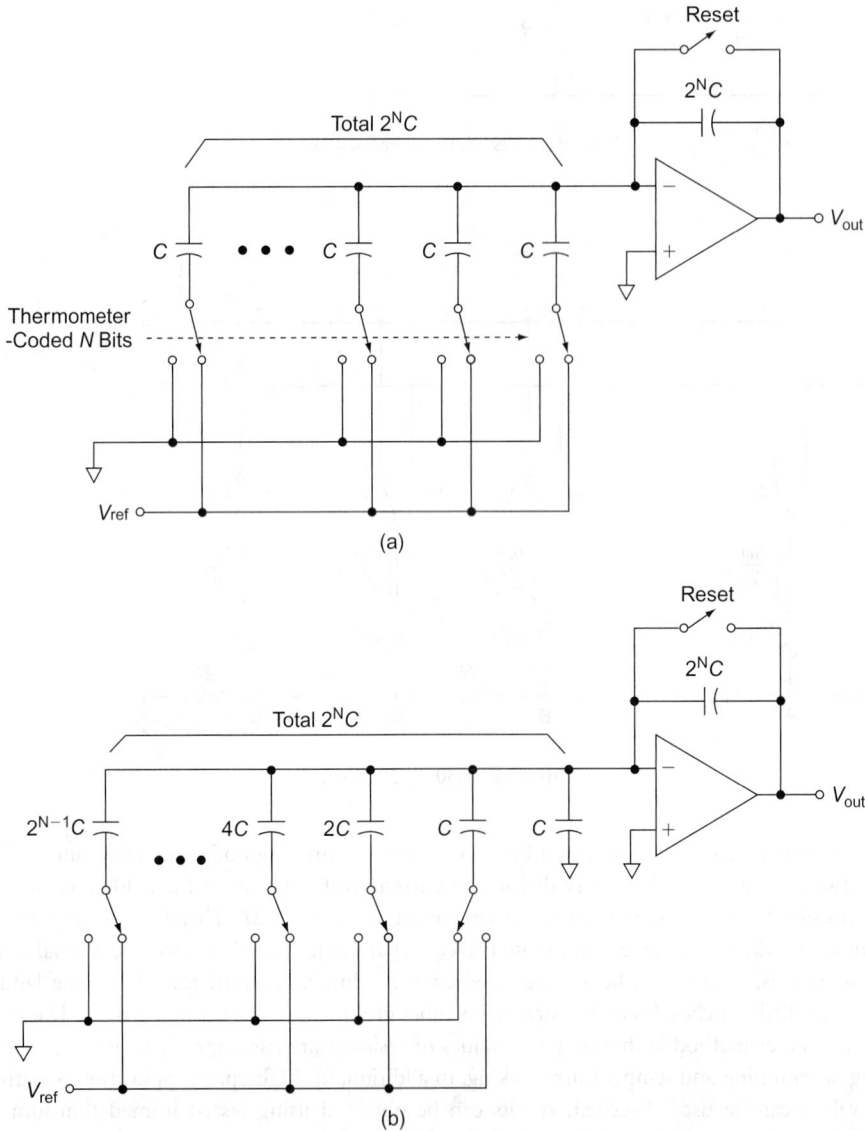

FIGURE 58.31 Capacitor-array DACs: (a) thermometer-coded and (b) binary-weighted.

handling the thermometercoded capacitor array, a binary-weighted capacitor array is often used, as shown in Figure 58.31(b) by grouping unit capacitors in binary ratio values. One important application of the capacitor-array DAC is as a reference DAC for ADCs. As in the case of the *R*-2*R* MDAC, the capacitor-array DAC can be used as an MDAC to amplify residue voltages for multi-step or pipeline ADCs.

58.6.5   Thermometer-Coded Segmented DAC

Applying a two-step conversion concept, a DAC can be made in two levels using coarse and fine DACs. The fine DAC divides one coarse MSB segment into fine LSBs. If one fixed MSB segment is subdivided to generate LSBs, matching among MSB segments creates a non-monotonicity problem. However, if the next MSB segment is subdivided instead of the fixed segment, the segmented DAC can maintain monotonicity regardless of the MSB matching. This is called the next-segment approach. The most widely used segmented DAC is a current-ratioed DAC, whose MSB DAC is made of identical elements for the next-segment approach, except that the LSB DAC is a current divider as shown in Figure 58.32. To implement a segmented DAC using two resistor-string DACs, voltage buffers are needed to drive the LSB DAC without loading the MSB DAC. Although the resistor-string MSB DAC is monotonic, overall monotonicity is not guaranteed due to the offsets of the voltage buffers. The use of a capacitor-array LSB DAC eliminates the need for voltage buffers.

58.6.6   Integrator-Type DAC

As mentioned, monotonicity is guaranteed only in a thermometer-coded DAC. The thermometer coding of a DAC output can be implemented either by repeating identical DAC elements many times or by using the same element over and over. The former requires more hardware, but the latter requires more time. In the continuous-time integrator-type DAC, the integrator output is a linear ramp and the time to stop integration can be controlled digitally. Therefore, monotonicity can be maintained. Similarly, the discrete-time integrator can integrate a constant amount of charge repeatedly and the number of integrations can be controlled digitally. The integration approach can give high accuracy, but its disadvantage is that its slow speed limits its applications.

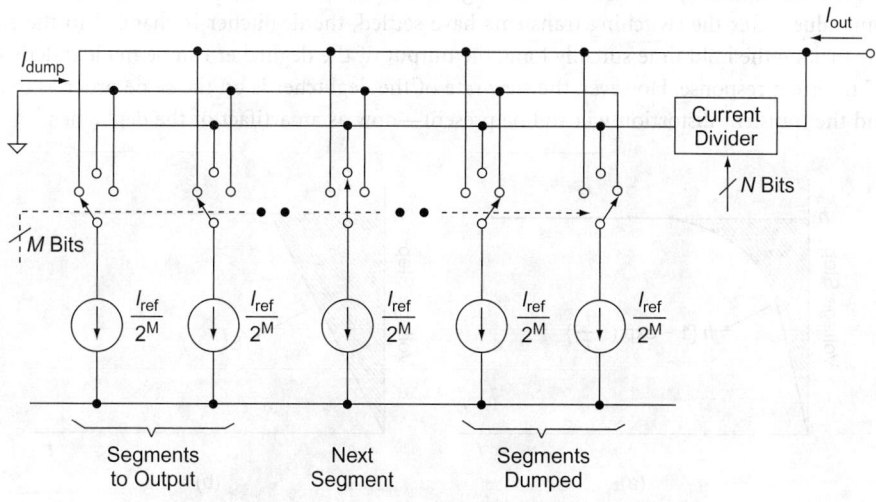

FIGURE 58.32 Thermometer-coded segmented DAC.

58.7   DAC Design Considerations

Figure 58.33 illustrates two step responses of a DAC when it settles with a time constant τ and when it slews with a slew rate S. The transient errors given by the shaded areas are $h\tau$ and $h^2/2S$, respectively. This implies that a single time-constant settling of the former case only generates a linear error in the output, which does not affect the DAC linearity, but the slew-limited settling generates a nonlinear error. Even in the single-time constant case, the code-dependent settling time constant can introduce a nonlinearity error because the settling error is a function of the time constant τ. This is true for a resistor-string DAC, which exhibits a code-dependent settling time because the output resistance of the DAC depends on the digital input.

58.7.1   Effect of Limited Slew Rate

The slew-rate limit is a significant source of nonlinearity since the error is proportional to the square of the signal, as shown in Figure 58.33(b). The height and width of the error term change with the input. The worst-case harmonic distortion (HD) when generating a sinusoidal signal with a magnitude V_o with a limited slew rate of S is [24]:

$$\mathrm{HD}_k = 8\,\frac{\sin^2 \dfrac{\omega T_c}{2}}{\pi k\,(k^2-4)} \times \frac{V_o}{ST_c}, \quad k = 1,3,5,7\ldots \tag{58.6}$$

where T_c is the clock period. For a given distortion level, the minimum slew rate is given. Any exponential system with a bandwidth of ω_o gives rise to signals with the maximum slew rate of $\omega_o V_o$. Therefore, by making $S > \omega_o V_o$, the DAC system will exhibit no distortion due to the limited slew rate.

58.7.2   Glitch

Glitches are caused by small turn-on and turn-off time difference when switching DAC elements. Take, for example, the major code transition at half-scale from 011...11 to 100...00. Here, the MSB current source turns on while all other current sources turn off. The small difference in switching times results in a narrow half-scale glitch, as shown in Figure 58.34. Such a glitch, for example, can produce distorted characters in CRT display applications. To alleviate both glitch and slew-rate problems related to transients, a DAC is followed by a deglitcher. The deglitcher stays in the hold mode while the DAC changes its output value. After the switching transients have settled, the deglitcher is changed to the sampling mode. By making the hold time suitably long, the output of the deglitcher can be made independent of the DAC transient response. However, the slew rate of the deglitcher is on the same order as that of the DAC, and the transient distortion will still be present—now as an artifact of the deglitcher.

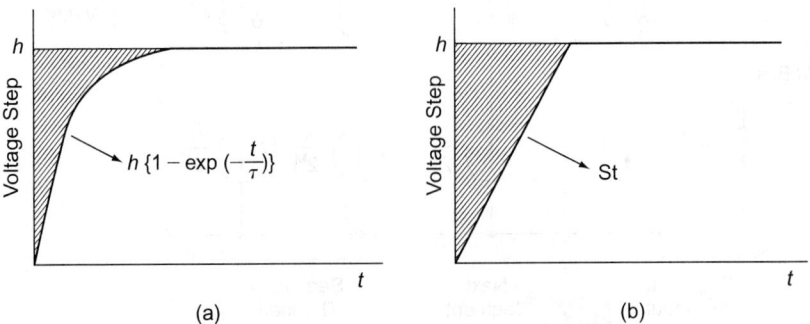

FIGURE 58.33 DAC settling cases: (a) exponential and (b) slew-limited case.

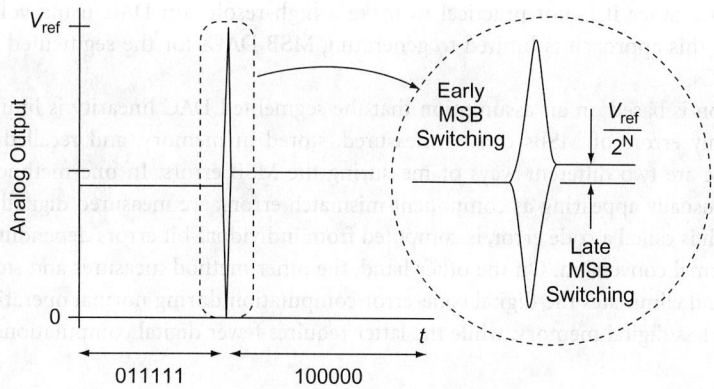

FIGURE 58.34 DAC output glitch.

58.7.3 Techniques for High-Resolution DACs

The following methods are often used to improve the linearity of DACs: Laser trimming, off-chip adjustment, common-centroid layout technique, dynamic element matching technique, voltage or current sampling, and electronic calibration techniques. The trend is toward more sophisticated and intelligent electronic solutions that overcome and compensate for some of the limitations of conventional trimming techniques. *Electronic calibration* is a general term to describe various circuit techniques, which usually predistort the DAC transfer characteristic so that the DAC linearity can be improved. The self-calibration is to incorporate all the calibration mechanisms and hardware on the DAC as a built-in function so that users can recalibrate whenever necessary.

The application of *dynamic element matching* to the binary-weighted current DAC is a straightforward switching of two complementary currents [25]. Its application to the binary voltage divider using two identical resistors or capacitors requires exchanging resistors or capacitors. This can be easily achieved by reversing the polarity of the reference voltage for the divide-by-two case. However, in the general case of N-element matching, the current division is inherently simpler than the voltage division. In general, to match the N independent elements, a switching network with N inputs and N outputs is required. The function of the switching network is to connect any input out of N inputs to one output with an average duty cycle of $1/N$. The simplest one is a barrel shifter rotating the input-output connections in a predetermined manner. This barrel shifter generates a low-frequency modulated error when N gets larger because the same pattern repeats every N clocks. A more sophisticated randomizer with the same average duty cycle can distribute the mismatch error over the wider frequency range.

The *voltage or current sampling concept* is an electronic alternative to direct mechanical trimming. The voltage sampler is usually called a S/H, while the current sampler is called a current copier. The voltage is usually sampled on the input capacitor of a buffer amplifier, and the current is usually sampled on the input capacitor of a transconductance amplifier such as MOS transistor gate. Therefore, both voltage and current sampling techniques are ultimately limited by their sampling accuracy.

The idea behind the voltage or current sampling DAC is to use one voltage or current element repeatedly. One example of the voltage sampling DAC is a discrete-time integrating DAC. The integrator integrates a constant charge repeatedly, and its output is sampled. This is equivalent to generating equally spaced reference voltages by stacking identical unit voltages [26]. The fundamental problem associated with this sampling voltage DAC approach is the accumulation of the sampling error and noise in generating larger voltages. Similarly, the current sampling DAC can sample a constant current on current sources made of MOS transistors [27]. Since one reference current is copied on other identical current samplers, the matching accuracy can be maintained as long as the sampling errors are kept constant. Since it is not practical to make a high-resolution DAC using voltage or current sampling alone, this approach is limited to generating MSB DACs for the segmented DAC or for the subranging ADCs.

Self-calibration is based on an assumption that the segmented DAC linearity is limited by the MSB DAC so that only errors of MSBs can be measured, stored in memory, and recalled during normal operation. There are two different ways of measuring the MSB errors. In one method, individual-bit non-linearities, usually appearing as component mismatch errors, are measured digitally [15,18], and a total error, which is called a code error, is computed from individual-bit errors depending on the output code during normal conversion. On the other hand, the other method measures and stores digital code errors directly and eliminates the digital code-error computation during normal operation [16,17]. The former requires less digital memory, while the latter requires fewer digital computations.

References

1. J. McCreary and P. Gray, All-MOS charge redistribution analog-to-digital conversion techniques-part I, *IEEE J. Solid-State Circuits*, vol. SC-10, pp. 371–379, Dec. 1975.
2. S. Ramet, A 13-bit 160kHz differential analog to digital converter, *ISSCC Dig. Tech. Papers*, pp. 20–21, Feb. 1989.
3. C. Mangelsdorf, H. Malik, S. Lee, S. Hisano, and M. Martin, A two-residue architecture for multistage ADCs, *ISSCC Dig. Tech. Papers*, pp. 64–65, Feb. 1993.
4. W. Song, H. Choi, S. Kwak, and B. Song, A 10-b 20-Msamples/s low power CMOS ADC, *IEEE J. Solid-State Circuits*, vol. 30, pp. 514–521, May 1995.
5. T. Cho and P. Gray, A 10-bit, 20-Msamples/s, 35-mW pipeline A/D converter, *IEEE J. Solid-State Circuits*, vol. 30, pp. 166–172, Mar. 1995.
6. P. Vorenkamp and R. Roovers, A 12-bits, 60MSPS cascaded folding & interpolating ADC, *IEEE J. Solid-State Circuits*, vol. 32, pp. 1876–1886, Dec. 1997.
7. K. Kattmann and J. Barrow, A technique for reducing differential nonlinearity errors in flash A/D converters, *ISSCC Dig. Tech. Papers*, pp. 170–171, Feb. 1991.
8. K. Bult and A. Buchwald, An embedded 240-mW 10-bit 50MS/s CMOS ADC in 1-mm^2, *IEEE J. Solid-State Circuits*, vol. 32, pp. 1887–1895, Dec. 1997.
9. B. Song, S. Lee, and M. Tompsett, A 10b 15-MHz CMOS recycling two-step A/D converter, *IEEE J. Solid-State Circuits*, vol. SC-25, pp. 1328–1338, Dec. 1990.
10. S. Lewis and P. Gray, A pipelined 5-Msamples/s 9-bit analog-to-digital converter, *IEEE J. Solid-State Circuits*, vol. SC-22, pp. 954–961, Dec. 1987.
11. B. Song, M. Tompsett, and K. Lakshmikumar, A 12-bit 1-Msample/s capacitor error-averaging pipelined A/D converter, *IEEE J. Solid-State Circuits*, vol. SC-23, pp. 1324–1333, Dec. 1988.
12. S. Lewis, S. Fetterman, G. Gross Jr., R. Ramachandran, and T. Viswanathan, A 10-b 20-Msample/s analog-to-digital converter, *IEEE J. Solid-State Circuits*, vol. SC-27, pp. 351–358, Mar. 1992.
13. C. Conroy, D. Cline, and P. Gray, An 8-b 85-MS/s parallel pipeline A/D converter in 1-μm CMOS, *IEEE J. Solid-State Circuits*, vol. SC-28, pp. 447–454, Apr. 1993.
14. R. Plassche and R. Grift, A high speed 7 bit A/D converter, *IEEE J. Solid-State Circuits*, vol. SC-14, pp. 938–943, Dec. 1979.
15. H. Lee, D. Hodges, and P. Gray, A self-calibrating 15-bit CMOS A/D converter, *IEEE J. Solid-State Circuits*, vol. SC-19, pp. 813–819, Dec. 1984.
16. S. Lee and B. Song, Digital-domain calibration of multistep analog-to-digital converters, *IEEE J. Solid-State Circuits*, vol. SC-27, pp. 1679–1688, Dec.1992.
17. S. Kwak, B. Song, and K. Bacrania, A 15-b, 5-Msamples/s low spurious CMOS ADC, *IEEE J. Solid-State Circuits*, vol. 32, pp. 1866–1875, Dec. 1997.
18. A. Karanicolas, H. Lee, and K. Bacrania, A 15-b 1-Msample/s digitally self calibrated pipeline ADC, *IEEE J. Solid-State Circuits*, vol. 28, pp. 1207–1215, Dec. 1993.

19. D. Mercer, A 16-b D/A converter with increased spurious free dynamic range, *IEEE J. Solid-State Circuits*, vol. 29, pp. 1180–1185, Oct. 1994.

20. B. Tesch and J. Garcia, A low glitch 14-b 100-MHz D/A converter, *IEEE J. Solid-State Circuits*, vol. 32, pp. 1465–1469, Sept. 1997.

21. D. Mercer and L. Singer, 12-b 125 MSPS CMOS D/A designed for special performance, *Proc. IEEE Int. Symp. Low Power Electronics and Design*, pp. 243–246, Aug. 1996.

22. C. Lin and K. Bult, A 10b 250MSample/s CMOS DAC in 1mm^2, *ISSCC Dig. Tech. Papers*, pp. 214–215, Feb. 1998.

23. A. Marques, J. Bastos, A. Bosch, J. Vandenbusche, M. Steyaert, and W. Sansen, A 12b accuracy 300M sample/s update rate CMOS DAC, *ISSCC Dig. Tech. Papers*, pp. 216–217, Feb. 1998.

24. D. Frumar, clearing difusion in digitial analog conversion, vol. 25, pp. 178–183, Apr. 1977.

25. R. Plassche, Dynamic element matching for high accuracy monolithic D/A converters, *IEEE J. Solid-State Circuits*, vol. SC-11, pp. 795–800, Dec. 1976.

26. D. Kerth, N. Sooch, and E. Swanson, A 12-bit 1-MHz two-step flash ADC, *IEEE J. Solid-State Circuits*, vol. SC-24, pp. 250–255, Apr. 1989.

27. D. Groeneveld, H. Schouwenaars, H. Termeer, and C. Bastiaansen, A self-calibration technique for monolithic high-resolution D/A converters, *IEEE J. Solid-State Circuits*, vol. SC-24, pp. 1517–1522, Dec. 1989.

59

Oversampled Analog-to-Digital and Digital-to-Analog Converters

John W. Fattaruso

Texas Instruments, Inc.

Louis A. Williams III

Texas Instruments, Inc.

CONTENTS

59.1 Introduction

In the absence of some form of calibration or trimming, the precision of the Nyquist rate converters described in the previous chapter is strictly dependent on the precision of the VLSI components that comprise the converter circuits. Oversampled data converters are a means of exchanging the speed and data-processing capability of modern submicron integrated circuits for precision that would otherwise not be readily attainable [1,2]. The precision of an oversampled data converter can exceed the precision of its circuit components by several orders of magnitude.

In this chapter, the basic operation and design techniques of the most widely used class of over-sampled data converters, sigma-delta* modulators, are described. In the next section, the basic theory of sigma-delta modulators is presented, using both time- and frequency-domain approaches. The issue of nonharmonic tones is also discussed. In Section 59.3, more complex sigma-delta architectures are described, including higher-order cascaded, and bandpass architectures. Filtering techniques unique to sigma-delta modulators are presented in Section 59.4. In Section 59.5, the basic circuit building blocks for sigma-delta modulators are described, and in Section 59.6, circuit design issues specific to sigma-delta-based data converters are discussed.

59.2 Basic Theory of Operation

Oversampled data conversion techniques have their roots in the design of signal coders for communication systems [3–6]. Oversampling techniques differ from Nyquist techniques in that their comprehension and design procedures draw equally from time and frequency domain representations of signals, whereas Nyquist techniques are readily understood in just the time domain.

In general, the function of data conversion by oversampling is typically performed by a serial connection of a modulator and various filter blocks. In analog-to-digital (A/D) conversion, shown in Figure 59.1, the analog input signal $x_i(t)$ is first bandlimited by an anti-alias filter, then sampled at a rate f_S. This sampling rate is M times faster than a comparable Nyquist rate converter with the same signal bandwidth; the value of M is called the oversampling ratio. The sampled signal $x[n]$ is coded by a modulator block that quantizes the data into a finite number of discrete levels. The resulting coded signal $y[n]$ is downsampled, or decimated, by a factor of M to produce an output that is comparable to a Nyquist rate converter. Digital-to-analog oversampled data conversion is basically the reverse of analog-to-digital conversion. As shown in Figure 59.2, the Nyquist rate digital samples are oversampled by an interpolation filter, coded by a modulator, and then reconstructed in the analog domain by an analog filter.

In both analog-to-digital and digital-to-analog data conversion, the block with the most unique signal-processing properties is the modulator. The remainder of this section and the subsequent section focus on the properties and architectures for oversampled data modulators.

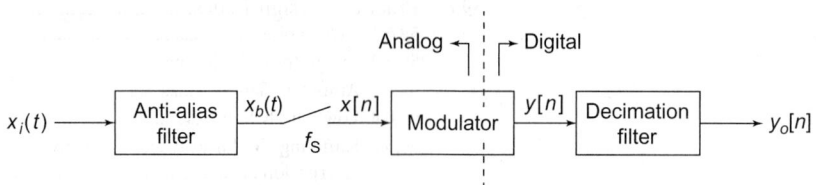

FIGURE 59.1 Oversampled A/D conversion.

<mark>*</mark>The reader will find functionally identical modulator blocks in the literature named either "sigma-delta" modulators or "delta-sigma" modulators, with the choice of terminology largely up to personal preference. We use the former term here.

59.2.1 Time-Domain Representation

The simplest modulator that would perform the requisite conversion to discrete output levels is the quantization function $Q(x)$ shown in Figure 59.3. This quantization can be thought of as merely the sum of the original signal $x[n]$ with a sampled error signal $e[n]$, as illustrated in Figure 59.4. In Nyquist rate converters, the error is reduced by using a large number of small steps in the quantizer characteristic. In oversampled data converters, specifically sigma-delta modulators, the error is corrected by a feedback network. This correction is made by estimating the error in advance and subtracting it from the input as shown in Figure 59.5, where $\hat{e}[n]$ is the error estimate. If this estimate were perfect, $\hat{e}[n]$ would equal $e[n]$ and the output $y[n]$ would equal the input $x[n]$. However, since the error is not known until it is made, $e[n]$ is not known when $\hat{e}[n]$ is needed. Therefore, some means must be found to estimate the error. In the case of sigma-delta converters, the error can be estimated by exploiting some knowledge of the frequency-domain behavior of the input signal. Specifically, it is assumed that the signal is changing very slowly from sample to sample, or equivalently, its bandwidth is much less than the sampling rate.

For exceedingly slow signals, a first-order estimate of the error to be committed in quantization can be formed. The first-order estimate of the current error $e[n]$ is simply the previous error $e[n-1]$. This error may be found simply by a subtraction across the quantization block as shown in Figure 59.6, and the output $y[n]$ is

$$y[n] = x[n] + e[n] - e[n-1] \qquad (59.1)$$

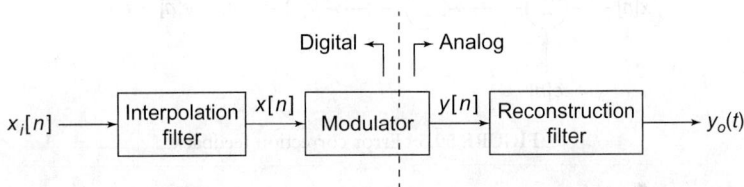

FIGURE 59.2 Oversampled D/A conversion.

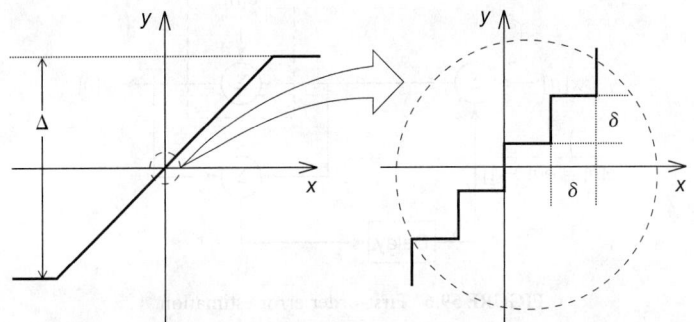

FIGURE 59.3 Quantizer transfer function.

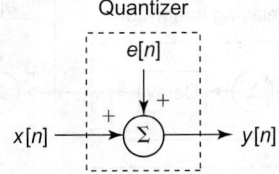

FIGURE 59.4 Quantization error.

The essential property of this structure is that if an error is committed that is not large enough to be corrected by a displacement to another quantizer level on the next sample, then the history of successive errors accumulate in the feedback loop until they eventually push the quantizer into another level. In this manner the output of the quantizer will, over time, correct the errors committed in previous samples, increasing the precision of the information being generated as a time sequence of samples.

As will be shown in Section 59.5, the most convenient and accurate sampled-data circuit building block in practice is an integrator. With a few straightforward steps, the system of Figure 59.6 can be transformed into that of Figure 59.7, where the delay element is now immersed in an integrator feedback loop. The output of this transformed modulator is

$$y[n] = x[n-1] + e[n] - e[n-1] \tag{59.2}$$

Comparing Eq. (59.1) with Eq. (59.2), it is evident that this transformation does require the addition of a single clock delay block in the input path $x[n]$, but this extra clock cycle of latency has no effect on the

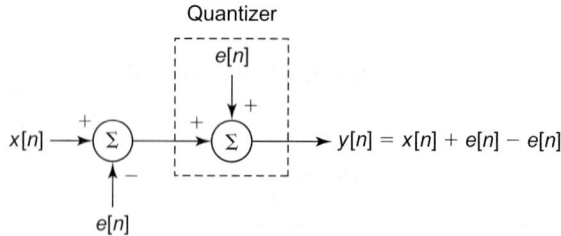

FIGURE 59.5 Error correction feedback.

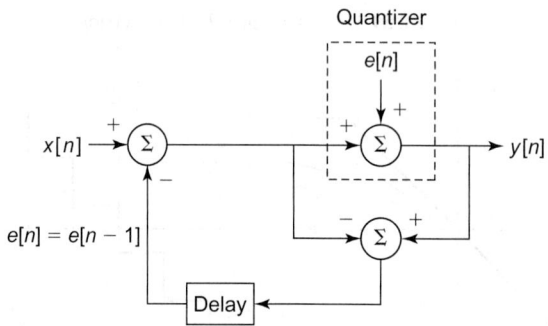

FIGURE 59.6 First-order error estimation.

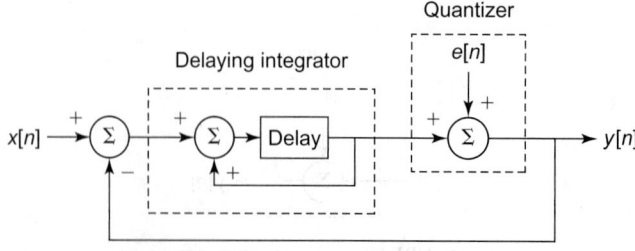

FIGURE 59.7 First-order equivalent modulator.

precision or frequency response of the modulator. The structure in Figure 59.7 is generally known as a first-order sigma-delta modulator [7–9].

An increase in precision may be obtained by using more accurate estimates of the expected quantizer error [6]. A second-order estimate of $e[n]$ may be formed by assuming that the error $e[n]$ varies linearly with time. In this case, an estimate of the current error $e[n]$ may be computed by changing the previous error $e[n-1]$ by an amount equal to the change between $e[n-2]$ and $e[n-1]$. The second-order error estimate is thus

$$e[n] = e[n-1] + (e[n-1] - e[n-2]) = 2e[n-1] - e[n-2] \tag{59.3}$$

and is illustrated in Figure 59.8. The output of the second-order estimation modulator is

$$y[n] = x[n] + e[n] - 2e[n-1] + e[n-2] \tag{59.4}$$

It can be shown, after a number of steps, that the modulator in Figure 59.8 can be transformed into a modulator in which the feedback loop delays are again immersed in practical integrator blocks. This second-order sigma-delta modulator [10–12] is shown in Figure 59.9; the output of this transformed modulator is

$$y[n] = x[n-2] + e[n] - 2e[n-1] + e[n-2] \tag{59.5}$$

which is entirely equivalent to that given by Eq. (59.4) except for the addition of two inconsequential delays of the input signal $x[n]$.

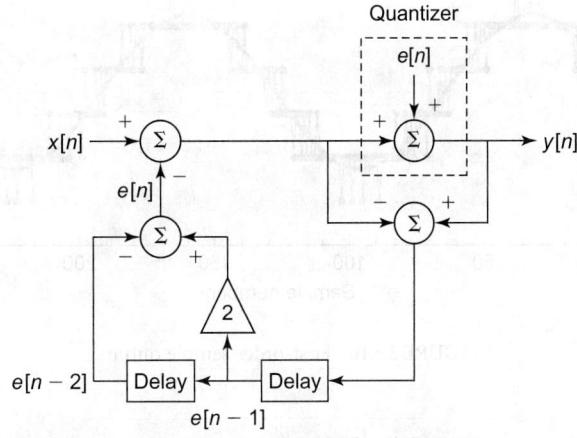

FIGURE 59.8 Second-order error estimation.

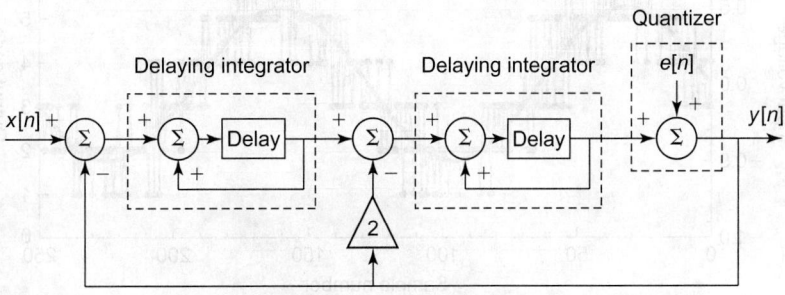

FIGURE 59.9 Second-order equivalent modulator.

A further increase in precision can be obtained using even higher-order estimates of the quantizer error such as quadratic or cubic. These high-order error-estimate modulators can also be transformed into a series of delaying integrators in a feedback loop. Unfortunately, as discussed in Section 59.3.1, practical difficulties emerge for orders greater than two, and alternative architectures are generally needed.

Computer simulation of modulator systems is straightforward, and Figure 59.10 shows the simulated output of the first-order modulator of Figure 59.7 when fed with a simple sinusoidal input. The resolution of the quantizer in the modulator loop was assumed to be eight levels. (The modulator output is drawn with continuous lines to emphasize the oscillatory nature of the modulator output, but the quantities plotted have meaning only at each sample time.) The coarsely quantized output code generally follows the input, but with occasional transitions that track intermediate values over local sample regions.

A second-order modulator with an eight-level quantizer exhibits the simulated behavior shown in Figure 59.11. Note that the oscillations by which the loop attempts to minimize quantization error appear "busier" than in the first-order case of Figure 59.10. It will be shown in the frequency domain that, for a given signal bandwidth, the more vibrant output code oscillations in Figure 59.11 actually represent the input signal with higher precision than the first-order case in Figure 59.10.

A special case that is of practical significance is the second-order modulator with a two-level quantizer, that is, simply a comparator closing the feedback loop. A simulation of such a modulator is shown in Figure 59.12. Although the quantized representation at the output appears crude, the continuous nature of the input level is expressed in the density of output codes. When the input level is high, around sample

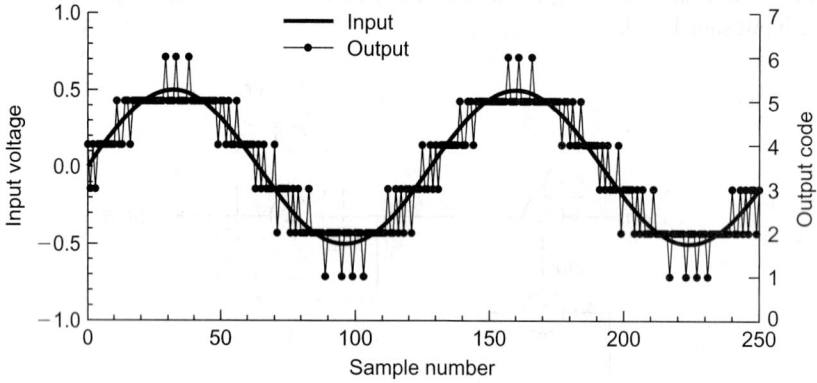

FIGURE 59.10   First-order sample output.

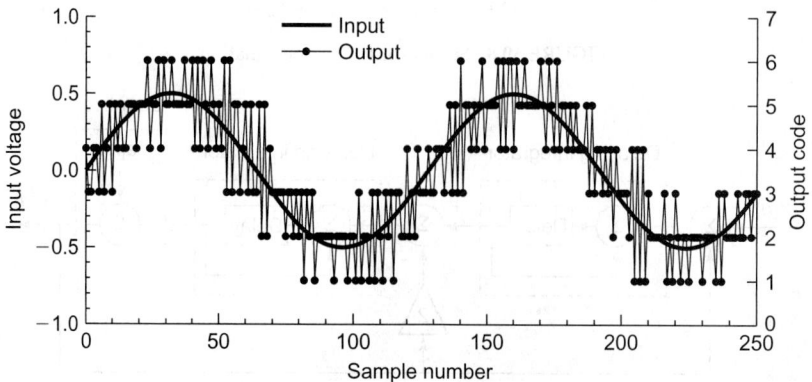

FIGURE 59.11   Second-order sample output.

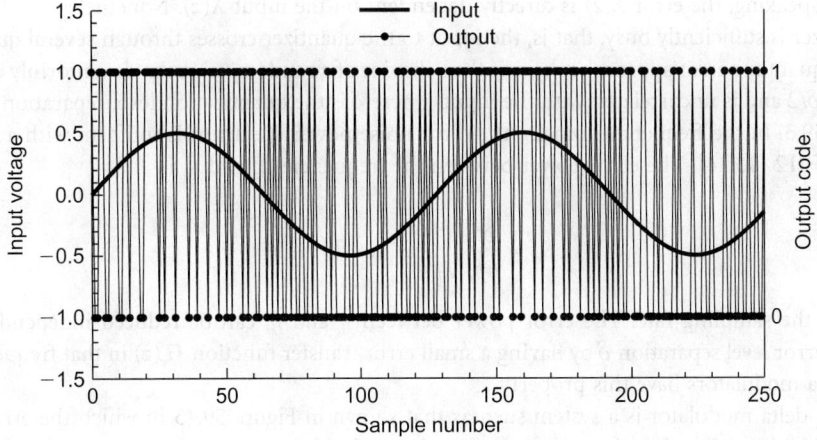

FIGURE 59.12 Second-order one-bit modulator sample output.

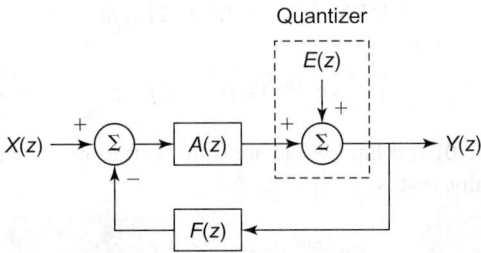

FIGURE 59.13 Generalized sigma-delta modulator.

numbers 32 and 160, there is a greater preponderance of "1" output codes, and at low swings of the input, around sample numbers 96 and 224, the "0" output code dominates.

The examples in Figures 59.10–59.12 demonstrate that information generated from the modulator expresses, in a high-speed coded form, the coarsely quantized input signal and the deviation between the signal and the quantization levels. Although the time-domain-coded modulator output looks somewhat unintelligible, the output characteristics are clearer in the frequency domain.

59.2.2 Frequency-Domain Representation

The modulators in Figure 59.7 and Figure 59.9 can be generalized as the sampled-data system shown in Figure 59.13, where the time-domain signals $x[n]$, $y[n]$, and $e[n]$ are written as their frequency-domain equivalents $X(z)$, $Y(z)$, and $E(z)$. The modulator output in Figure 59.13 can be written in terms of the input $X(z)$ and the quantizer error $E(z)$ as

$$Y(z) = H_x(z)X(z) + H_e(z)E(z) \tag{59.6}$$

where the input transfer function $H_x(z)$ is

$$H_x(z) = \frac{A(z)}{1 + A(z)F(z)} \tag{59.7}$$

and the error transfer function $H_e(z)$ is

$$H_e(z) = \frac{1}{1 + A(z)F(z)} \tag{59.8}$$

Strictly speaking, the error $E(z)$ is directly dependent on the input $X(z)$. Nonetheless, if the input to the quantizer is sufficiently busy, that is, the input to the quantizer crosses through several quantization levels, the quantizer error approaches having the behavior of a random value that is uniformly distributed between $\pm\delta/2$ and is uncorrelated with the input, where δ is the quantization level separation illustrated in Figure 59.3. In the frequency domain, the error noise power spectrum is uniform with a total error power of $\delta^2/12$ [13,14]. The error power between the frequencies f_L and f_H is

$$S_{ee} \approx \frac{\delta^2}{6f_S} \int_{f_L}^{f_H} |H_e(e^{j2\pi f/f_s})|^2 df \qquad (59.9)$$

where f_S is the sampling rate. The error power between f_L and f_H can be reduced independent of the quantizer error level separation δ by having a small error transfer function $H_e(z)$ in that frequency band. Sigma-delta modulators have this property.

A sigma-delta modulator is a system such as that shown in Figure 59.13 in which the error transfer function $H_e(z)$ is small and the input transfer function $H_x(z)$ is about unity for some band of frequencies. That is,

$$|H_e(e^{j2\pi f/f_s})| \ll 1, \quad f_L \le f \le f_H \qquad (59.10)$$

$$|H_x(e^{j2\pi f/f_s})| \approx 1, \quad f_L \le f \le f_H \qquad (59.11)$$

The requirements in Eq. (59.10) and Eq. (59.11) are equivalent to requiring that the loop gain be large and the feedback gain be unity, that is

$$|A(e^{j2\pi f/f_s})| \gg 1, \quad f_L \le f \le f_H \qquad (59.12)$$

$$|F(e^{j2\pi f/f_s})| \approx 1, \quad f_L \le f \le f_H \qquad (59.13)$$

There are many system designs that have the sigma-delta properties of high loop gain and unity feedback gain. The previous examples in Figure 59.7 and Figure 59.9 are part of an important class of modulator architectures called noise-differencing modulators that are particularly well suited to VLSI implementation. The forward path in a noise-differencing sigma-delta modulator consists of a series of delaying integrators. The *order* of the modulator is defined as the number of integrators. The forward gain of an Lth-order modulator is

$$A(z) = \left(\frac{z^{-1}}{1-z^{-1}}\right)^L \qquad (59.14)$$

The feedback gain in a noise-differencing modulator is designed such that the modulator open-loop gain is

$$A(z)F(z) = \frac{1}{(1-z^{-1})^L} - 1 \qquad (59.15)$$

From Eq. (59.14) and Eq. (59.15) it follows that the output for an Lth-order noise-differencing sigma-delta modulator is

$$Y(z) = z^{-L}X(z) + (1-z^{-1})^L E(z) \qquad (59.16)$$

The simulated frequency response for a second-order noise-differencing sigma-delta modulator with a sinusoidal input is shown in Figure 59.14. The large spike in the center is the original input signal. It

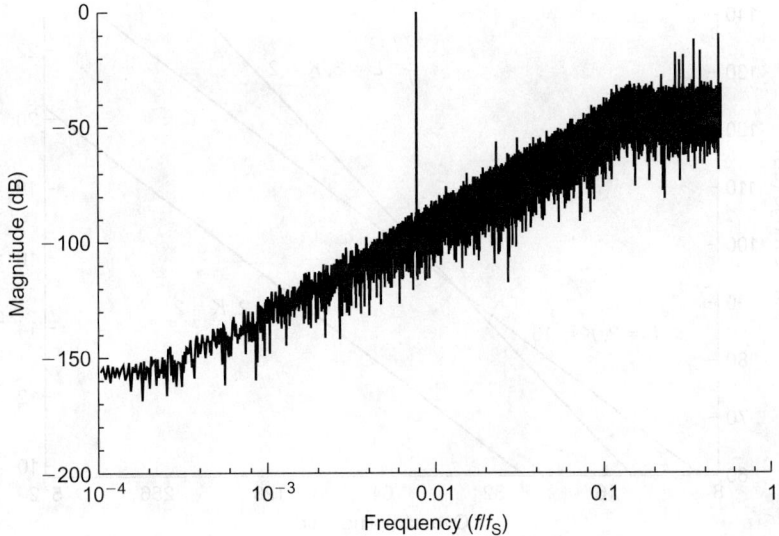

FIGURE 59.14 Second-order simulated frequency response.

is clear from the plot that the noise energy is lowest at low frequencies. Noise-differencing modulators are designed to reduce the quantization noise in the baseband, that is, $f_L = 0$ and $f_H \ll f_S$. The oversampling ratio, M, is

$$M = \frac{f_S}{2f_H} \tag{59.17}$$

(In a Nyquist rate converter, $M = 1$.) The baseband noise power for a noise-differencing modulator is

$$S_{ee} \approx \frac{\delta^2}{6f_S} \int_0^{f_H} \left| \left(1 - e^{-j2\pi f/f_S}\right)^L \right|^2 df \approx \frac{\delta^2}{12} \frac{\pi^{2L}}{2L+1} \frac{1}{M^{2L+1}} \tag{59.18}$$

One important measure of a sigma-delta modulator is its dynamic range, defined here as the ratio of the maximum sinusoidal input power to the noise power. With the quantizer output limited, as shown in Figure 59.3, to a range of Δ, the maximum sinusoidal signal power is $\Delta^2/8$. The quantizer range Δ is related to the quantizer level separation δ by the number of quantization levels K, where

$$\delta = \frac{\Delta}{K-1} \tag{59.19}$$

The dynamic range of a noise-differencing sigma-delta modulator is then

$$DR = \frac{3}{2} \frac{2L+1}{\pi^{2L}} M^{2L+1}(K-1)^2 \tag{59.20}$$

Because the dynamic range is such a strong function of the oversampling ratio, the number of bits required to achieve a given dynamic range is substantially less in a sigma-delta modulator than in a Nyquist rate converter. To illustrate this, the dynamic range, as given by Eq. (59.20), is shown in Figure 59.15 as a function of the oversampling ratio, M, for three combinations of modulator order, L, and number of quantization levels, K. The equivalent resolution in bits that would be required of a Nyquist rate converter to achieve the same dynamic range is shown in the right-hand axis of this figure. It can be inferred from Eq. (59.20) that a large dynamic range can be obtained even with only two quantization levels.

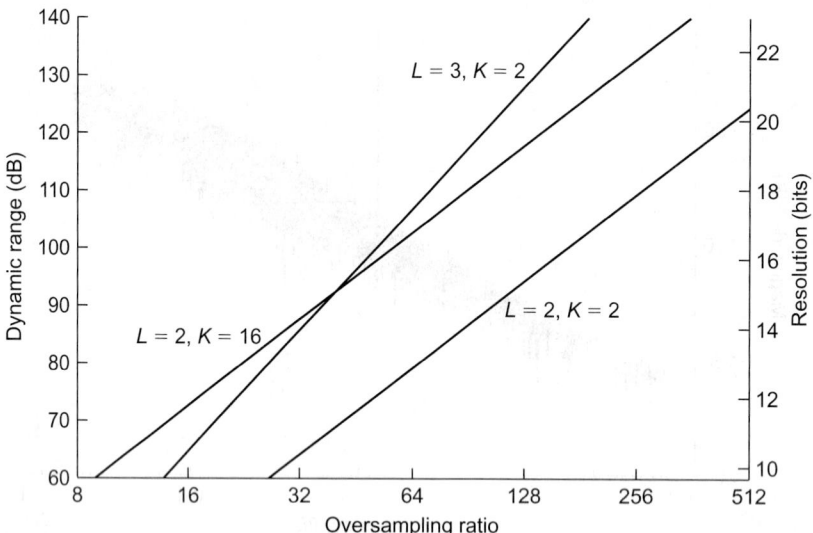

FIGURE 59.15 Calculated dynamic range versus oversampling ratio.

This is important when circuit imperfections in actual sigma-delta data converter implementations are considered.

59.2.3   Sigma-Delta Modulators in Data Converters

The generalized modulator shown in Figure 59.13 must be subtly modified when applied to the A/D and D/A converters in Figure 59.1 and Figure 59.2. In an A/D converter, the quantizer is actually a coarse K-level A/D converter (ADC), having an analog input and a digital output. (Since K is generally small, the K-level ADC is usually just a small flash converter as described in the previous chapter.) The quantized digital code is fed back into the analog $F(z)$ through a K-level D/A converter (DAC) as shown in Figure 59.16. Imperfections in this K-level DAC will introduce an additional error term $D_{AD}(z)$ as shown in Figure 59.17. With the addition of this DAC error term, the modulator output is

$$Y_{AD}(z) = H_x(z)[X(z) - F(z)D_{AD}(z)] + H_e(z)E(z) \tag{59.21}$$

Since the feedback transfer function, $F(z)$, is unity in the band of interest (see Eq. [59.13]), the DAC error is indistinguishable from the input. If there are more than two quantization levels, any mismatch between the level separations in the DAC will manifest itself as distortion because the DAC input is signal-dependent. In contrast, if there are only two quantization levels, there is only one level separation, and errors in the DAC levels will not cause distortion. (At worst, DAC errors in a two-level modulator will introduce a DC offset and a gain error.) Thus, with two-level sigma-delta modulators, it is possible to achieve low distortion and low-noise performance without precise component matching.

Unfortunately, most, if not all, of the statistical conditions that led to Eq. (59.9) and the subsequent equations are violated when a two-level single-threshold quantizer is used in a sigma-delta modulator [15]. Furthermore, the effective gain of the quantizer, which in Figure 59.3 is implied to be unity, is undefined for a single-threshold quantizer. Nonetheless, empirical evidence has indicated that Eq. (59.20) is still a reasonable approximation for two-level noise-differencing sigma-delta modulators, and is useful as a design guideline for the amount of oversampling needed to achieve a specific dynamic range for a given modulator order [16].

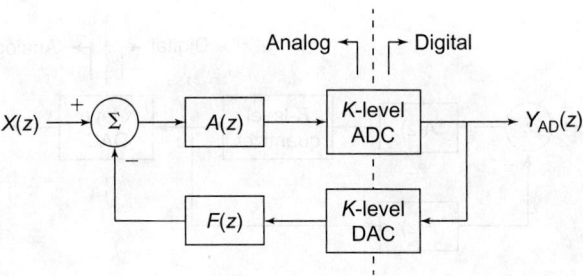

FIGURE 59.16 Sigma-delta modulator for A/D conversion.

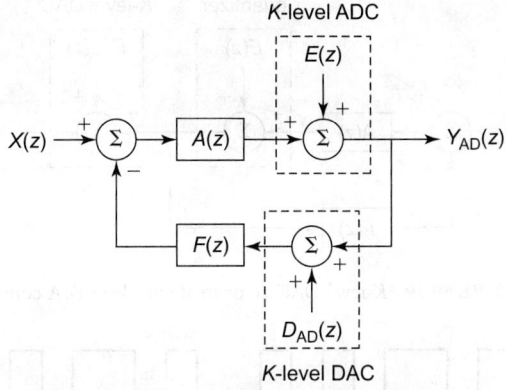

FIGURE 59.17 *K*-level DAC error in sigma-delta A/D converter.

As in sigma-delta A/D converters, there is also a DAC error term in sigma-delta-based D/A converters. In a sigma-delta D/A, the modulator loop is implemented digitally, and the output of that loop is applied to a coarse *K*-level DAC that provides the analog input for the reconstruction filter, as shown in Figure 59.18. Imperfections in the *K*-level DAC will introduce an error term $D_{DA}(z)$ as shown in Figure 59.19. With the addition of this error term, the modulator output is

$$Y_{DA}(z) = H_x(z)X(z) + D_{DA}(z) + H_e(z)E(z) \tag{59.22}$$

Since the input transfer function $H_x(z)$ is unity in the band of interest (see Eq. [59.11]), the DAC error is indistinguishable from the input, just as in the A/D case. Once again, two-level quantization can be used to avoid DAC-introduced distortion.

59.2.4 Tones

One problem with the simplified noise model of sigma-delta modulators that led to Eq. (59.20) is the failure to predict nonharmonic tones. This is especially true for two-level modulators. Repetitive patterns in the coded modulator output that cause discrete spectral peaks at frequencies not harmonically related to any input frequency can occur in sigma-delta modulators [10,16–18]. These tones can manifest themselves as odd "chirps" or "pops," and they exist even in ideal sigma-delta modulators; they are not caused by circuit imperfections [2].

The origin of sigma-delta tones is illustrated in the following example. Consider a first-order sigma-delta modulator, such as that shown in Figure 59.7, with a DC input of 0.0005. Let the quantizer have two output levels, +0.5 and −0.5. The output of such a modulator will be a sequence of +0.5s and −0.5s such that the time average of the outputs is 0.0005. To achieve this average, the output of the first-order

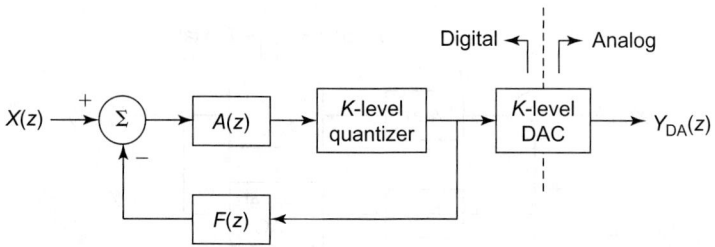

FIGURE 59.18 Sigma-delta modulator for D/A conversion.

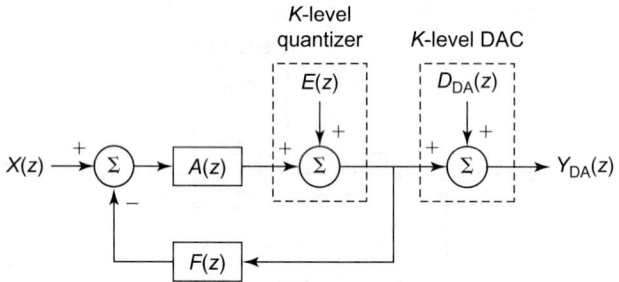

FIGURE 59.19 K-level DAC error in sigma-delta D/A converter.

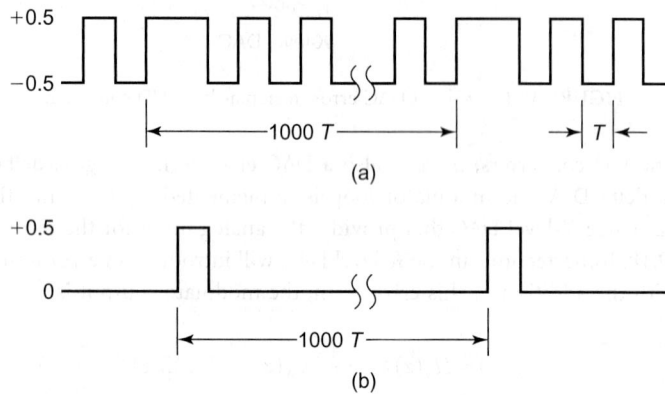

FIGURE 59.20 (a) Output sequence with average of 0.0005. (b) Running average of (a).

modulator will be a stream of alternating +0.5s and –0.5s, with an extra +0.5 every 1000 clock cycles. This is illustrated in Figure 59.20a, where T is the clock period ($T = 1/f_s$). The two-cycle running average of this output is shown in Figure 59.20b. For the most part, this running average is zero, except that at every 1000 clock cycles there is a one-clock cycle pulse. This repetitive pulse produces a tone in the output spectrum at a frequency of

$$f_P = \frac{f_s}{1000} = \frac{M}{500} f_H \tag{59.23}$$

If the oversampling ratio M is less than 500, this tone will appear in the baseband spectrum.

In sigma-delta modulators with more active input signals, the output sequence is typically more complex than that illustrated in Figure 59.20. Nonetheless, the concept underlying tone behavior is that

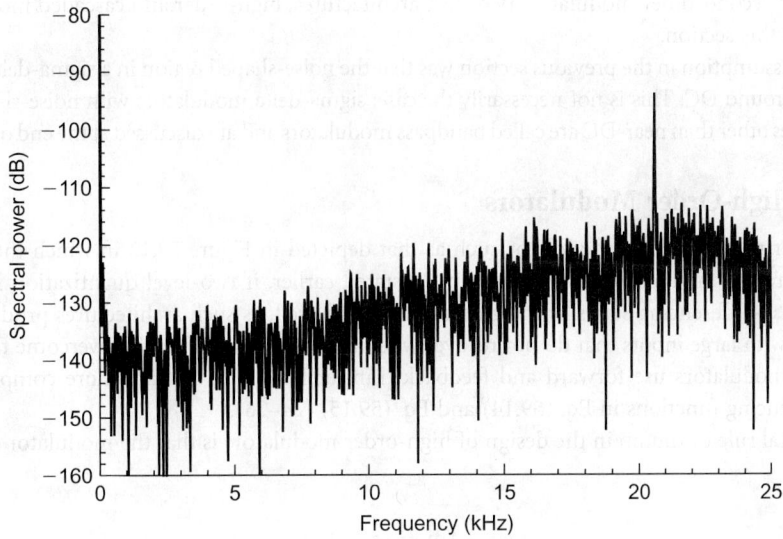

FIGURE 59.21   Measured tone in second-order modulator with DC input.

repeating patterns in the quantizer output cause nonuniformity in the quantizer error spectrum, which in the worst case is a discrete spectral peak. A measured tone for a second-order modulator with a DC input is shown in Figure 59.21 [19].

Several means of mitigating sigma-delta tones have been used. The first rule is to avoid using first-order sigma-delta modulators. Aside from having inferior noise-shaping properties compared to other modulator architectures, first-order modulators have terrible tone properties [15,16]. The situation improves dramatically with second- and higher-order modulators. In fact, the presence of tones may only be a perceptual or marketing concern, as the total tone power is usually less than the broadband quantization noise power [20].

When tone magnitudes must be reduced, several techniques have proven effective. These include dither, cascaded architectures, and multilevel quantization. Of these three, dither is the only technique whose sole benefit is the reduction of tones. The simplest type of dither is to add a moderate amplitude out-of-band signal, such as a square wave, to the input [9,21,22]. This dither signal is attenuated by the same filter that attenuates the quantization noise. The purpose of this dither is to keep the sigma-delta modulator out of modes that produce patterns, and for some types of tones this technique is effective. A more rigorously effective technique is to add a large amplitude pseudo-random noise signal at the quantizer input [23]. This noise is spectrally shaped just like the quantization noise, and is the most effective dither scheme for eliminating tones. Its drawbacks are the expense in silicon area of the random noise generator and the 2–3 dB reduction in dynamic range caused by the dither noise.

The other two tone mitigation techniques, cascaded architectures and multilevel quantization, are simply more complex sigma-delta architectures that happen to have improved tone properties over simple noise-differencing two-level sigma-delta modulators. These techniques are covered in Sections 59.3.2. and 59.6.7, respectively.

59.3   Alternative Sigma-Delta Architectures

Eq. (59.20) appears to indicate that the order of the modulator, L, can be any value, and that increasing L would be beneficial. However, one further problem with two-level sigma-delta modulators is that two-level noise-differencing modulators of order greater than 2 can exhibit unstable behavior [10]. For this reason, only first- and second-order modulators were discussed in Section 59.2. Nonetheless, there have been acceptably stable practical alternative architectures that achieve quantization noise shaping that is superior to a second-order modulator. Two such architectures, high-order and cascaded modulators, are discussed in this section.

Another assumption in the previous section was that the noise-shaped region in a sigma-delta modulator is centered around DC. This is not necessarily the case; sigma-delta modulators with noise-shaped regions at frequencies other than near-DC are called bandpass modulators and are discussed at the end of this section.

59.3.1 High-Order Modulators

A high-order modulator is a modulator such as that depicted in Figure 59.13 in which there are more than two zeros in the noise transfer function. As stated earlier, if two-level quantization is employed, a simple noise-differencing series of integrators cannot be used, as such architectures produce unstable oscillations with large inputs that do not recover when the input is removed. To overcome this problem, high-order modulators use forward and feedback transfer functions that are more complex than the noise-differencing functions in Eq. (59.14) and Eq. (59.15) [24–26].

The general rule of thumb in the design of high-order modulators is that the modulator can be made stable if

$$\lim_{z \to \infty} H_e(z) = 1 \qquad (59.24)$$

$$\left| H_e(z) \right| \le A \quad \text{for } |z| = 1 \qquad (59.25)$$

and the integrator outputs are clipped and scaled to prevent self-sustaining instability [26,27]. The maximum error gain A is about 1.5, but the value used represents a trade-off between noise attenuation and modulator stability. These rules cover a broad class of filter types and modulator architectures, and the type of filter used generally follows the traditions of the previous designers in an organization.

As an example, consider a fourth-order modulator with a highpass Butterworth error transfer function having a maximum gain, A, of 1.5, and a cutoff frequency set such that Eq. (59.24) is satisfied. The error spectrum of the Butterworth filter is shown in Figure 59.22, along with the error transfer function of an ideal fourth-order difference. While the Butterworth filter holds the maximum gain to 1.5 (3.5 dB), and while both filters have a fourth-order noise-shaping slope in the baseband (27 dB/octave), the error power

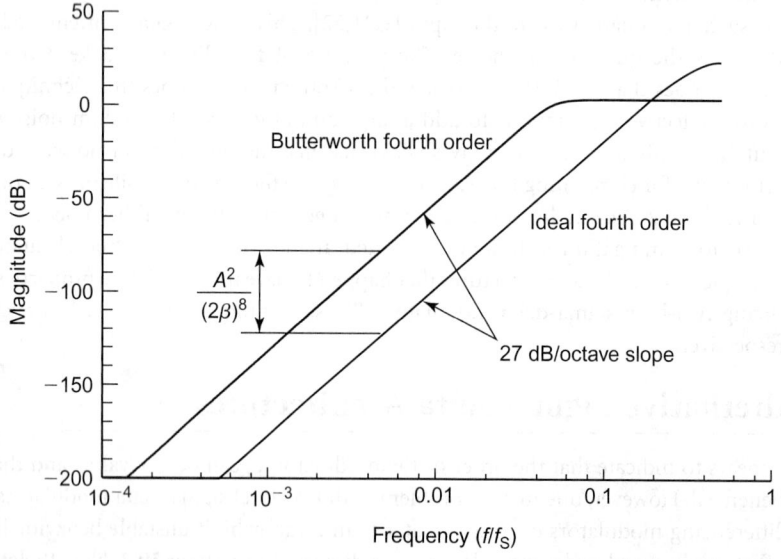

FIGURE 59.22 Fourth-order error spectrum.

in the baseband is 44 dB higher with the Butterworth filter than with the ideal noise-differencing filter. This error penalty is typical of high-order designs; there is usually a direct trade-off between stability and noise reduction.

Consider the more general case of an Lth-order highpass Butterworth error transfer function. The error transfer function of such a filter around the unit circle is

$$\left| H_e(e^{j\omega}) \right|^2 = \frac{A^2 \left(\dfrac{1}{\beta} \tan \dfrac{\omega}{2} \right)^{2L}}{1 + \left(\dfrac{1}{\beta} \tan \dfrac{\omega}{2} \right)^{2L}} \tag{59.26}$$

The filter coefficients for $H_e(z)$ needed to satisfy Eq. (59.26) can be computed using standard digital filter design techniques [28]. For a given filter order, L, and gain, A, the parameter β must be chosen to satisfy Eq. (59.24). (The condition in Eq. [59.25] is always satisfied when Eq. [59.26] is true.) These solutions can be computed numerically, and it is found empirically that

$$\beta = A^e \beta_N \tag{59.27}$$

where the values for β_N are tabulated in Table 59.1. The loss in dynamic range relative to an ideal noise-differencing modulator, given by $A^2/(2\beta)^{2L}$, is also tabulated. In spite of this loss, high-order modulators can still achieve better noise performance than second-order modulators. However, because of the compromise in dynamic range required to stabilize high-order modulators, third-order modulators are generally not worth the effort. More common are fourth- and fifth-order modulators.

The noise penalty required to stabilize high-order modulators can be mitigated to some extent by alternate zero placement [25]. Classic noise-differencing modulators place all of the zeros of the error transfer function at DC ($z = 1$). This causes most of the noise power to be concentrated at the highest baseband frequencies. If, instead, the zeros are distributed throughout the baseband, the total noise in the baseband can be reduced, as illustrated in Figure 59.23. The amount by which zero placement can improve the noise transfer function is summarized in Table 59.1. Also tabulated is the net loss in dynamic range of a high-order Butterworth modulator that uses zero placement relative to an ideal noise-differencing modulator that has zeros at DC.

59.3.2 Cascaded Modulators

Cascaded, or multistage, architectures are an alternative means of achieving higher-order noise shaping without the stability problems of the high-order modulators described in the previous section [29,30]. In a cascaded modulator, two or more stable first- or second-order modulators are connected in series, with the input of each stage being the error from the previous stage, as illustrated in Figure 59.24.

TABLE 59.1 High-Order Butterworth Gain Factors and Dynamic Range (DR) Loss

L	β_N	β	Loss in DR (dB)	Zero Placement DR Improvement (dB)	Net Loss in DR (dB)
3	0.052134	0.1570	33.7	8.0	25.8
4	0.051709	0.1557	44.1	12.8	31.2
5	0.033866	0.1020	72.6	17.9	54.7
6	0.034903	0.1051	84.8	23.2	61.6
7	0.025390	0.0764	117.7	28.6	89.1

Note: Except for β_N, all values are calculated for $A = 1.5$.

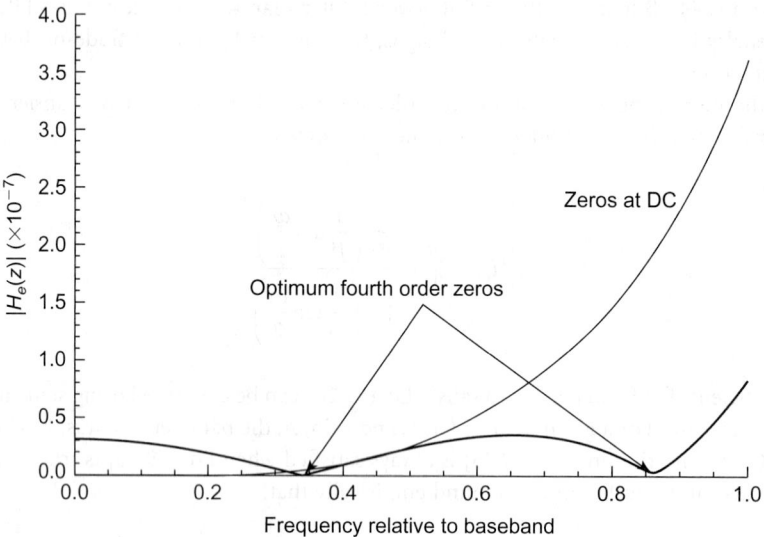

FIGURE 59.23 Fourth-order distributed zeros.

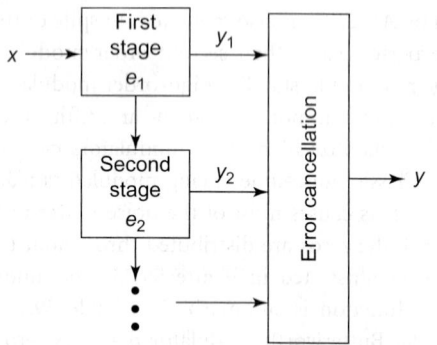

FIGURE 59.24 Cascaded sigma-delta modulators.

Referring to this illustration, the first stage of the cascade has two outputs, y_1 and e_1. The output y_1 is an estimate of the input x. The error in this estimate is e_1. The second stage has as its input the error from the first stage, e_1, and its outputs are y_2 and e_2. The second-stage output y_2 is an estimate of the first-stage error e_1. By subtracting this estimate of the first-stage error from the output of the first stage, y_1, only the second-stage error remains. Thus, the error cancellation network uses the output of one stage to cancel the error in the previous stage.

For example, in a cascaded architecture comprising a second-order noise-differencing modulator followed by a first-order noise-differencing modulator, the transforms of the output of the two stages, as given by Eq. (59.16), are

$$Y_1(z) = z^{-2}X(z) + (1-z^{-1})^2 E_1(z) \tag{59.28}$$

$$Y_2(z) = z^{-1}E_1(z) + (1-z^{-1})E_2(z) \tag{59.29}$$

If the error cancellation network combines the two outputs such that

$$Y(z) = z^{-1}Y_1(z) - (1-z^{-1})^2 Y_2(z) \tag{59.30}$$

then the error in the first stage will be cancelled, and the output will be

$$Y(z) = z^{-3}X(z) - (1 - z^{-1})^3 E_2(z) \qquad (59.31)$$

The final output of this cascaded modulator is third-order noise shaped. As a general rule, the noise shaping of a cascaded architecture is comparable to a single-stage modulator whose order is the sum of all the orders in the cascade.

The extent to which the errors in a cascaded modulator can be cancelled depends on the matching between the stages. The earliest multistage modulators were cascades of three first-order stages, often called the MASH architecture [29]. The disadvantage of this structure is that to achieve third-order performance, the error in the first stage, which is only first-order-shaped, must be cancelled. Cancelling this relatively large error places a stringent requirement on inter-stage matching. An alternative architecture that has much more relaxed matching requirements is the cascade of a second-order modulator followed by a first-order modulator. This architecture, like the MASH, ideally achieves third-order noise shaping. Its advantage is that the matching can be 100 times worse than a MASH and still achieve better noise-shaping performance [31].

An additional benefit of cascaded modulators is improved tone performance. It has been shown both analytically and experimentally that the error spectra of the second and subsequent stages in a cascade are not plagued by the spectral tones that can exist in single-stage modulators [19,32]. To the extent that the first-stage error is cancelled, any tone in the first-stage error spectrum is attenuated, and the final output of the cascaded modulator is nearly tone-free.

59.3.3 Bandpass Modulators

The aforementioned sigma-delta architectures, called herein baseband modulators, all have zeros at or near DC, that is, at frequencies much less than the modulator sampling rate. It is also possible to group these zeros at some other point in the sampling spectrum; such architectures are called bandpass modulators. Bandpass architectures are useful in systems that need to quantize a narrow band signal that is centered at some frequency other than DC. A common example of such a signal is the intermediate frequency (IF) signal in a communications receiver.

The simplest method for designing a bandpass modulator is by applying a transformation to an existing baseband modulator architecture. The most common transformation is to replace occurrences of z with $-z^2$ [2]. Such an architecture has zeros at $f_S/4$ and is stable if the baseband modulator is stable [33]. A comparison of the error transfer function of a baseband and bandpass modulator is shown in Figure 59.25. Note that a bandpass modulator generated through this transformation has twice the order of its equivalent baseband counterpart. For example, a fourth-order bandpass modulator is comparable to a second-order baseband modulator.

The noise-shaping properties of a bandpass modulator generated through the $-z^2$ transformation are equivalent to the baseband modulator that was transformed. Thus, the approximation in Eq. (59.20) can be used where L is the order of the baseband modulator that was transformed and M is the effective oversampling ratio, which in a bandpass modulator is the sampling rate divided by the signal bandwidth.

There are advantages and disadvantages to bandpass modulators when compared with traditional down-conversion and baseband modulation. One advantage of the bandpass modulator is its insensitivity to $1/f$ noise. Since the signal of interest is far from DC, at high frequencies where the factor $1/f$ is small, $1/f$ noise if often insignificant. Another advantage of bandpass modulation applies specifically to bandpass modulators having zeros at $f_S/4$ that are used in quadrature I and Q demodulation systems. If the narrowband IF signal is to be demodulated by a cosine and sine waveform, as shown in Figure 59.26, the demodulation operation becomes a simple multiplication by 1, −1, or 0 when the demodulation frequency is $f_S/4$ [34]. Furthermore, because a single modulator is used, the bandpass modulator is free of the I/Q path mismatch problems that can exist in baseband demodulation approaches.

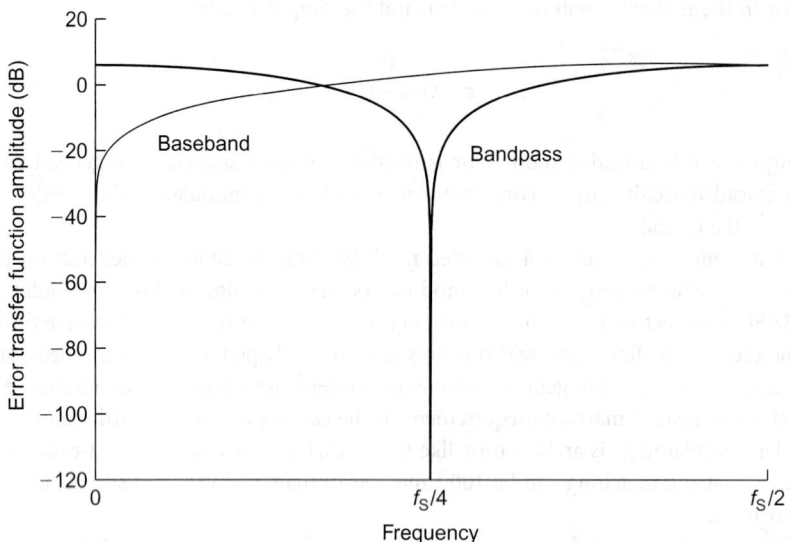

FIGURE 59.25 Bandpass noise transfer function.

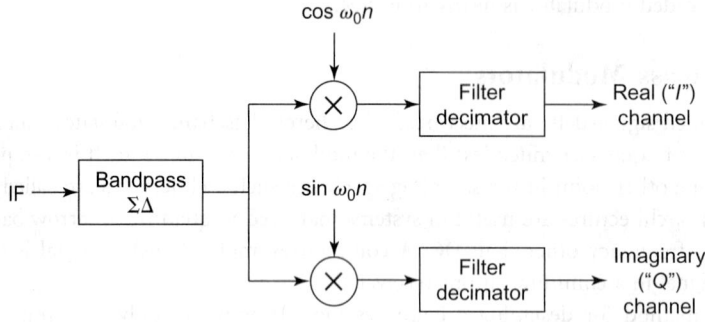

FIGURE 59.26 IQ demodulation with a bandpass modulator.

Two disadvantages of bandpass modulators involve the sampling operation. Sampling in a bandpass modulator has linearity requirements that are comparable to a Nyquist rate converter sampling at the same IF frequency; this is much more severe than the linearity requirements of the sampling operation in a baseband converter with the same signal bandwidth. Also, because of the higher signal frequencies, the sampling in bandpass modulators is much more sensitive to clock jitter. To date, the state of the art in bandpass modulators has about 20 dB less in dynamic range than comparable baseband modulators [2]. While the remainder of this chapter focuses once again on baseband modulators, many of the techniques are applicable to bandpass modulators as well.

59.4 Filtering for Sigma-Delta Modulators

In Sections 59.2 and 59.3, the discussion focused on the operation of the sigma-delta modulator core. While this core is the most unique aspect of sigma-delta data conversion, there are also filtering blocks that constitute an important part of sigma-delta A/D and D/A converters. In this section, the non-modulator components in baseband sigma-delta converters, namely the analog and digital filters, are described. First, the requirements of the analog anti-alias and reconstruction filters are described.

Second, typical architectures for the decimation and interpolation filters are discussed. While much of the design of these filters use standard techniques covered elsewhere in this volume, there are aspects of these filters that are specific to sigma-delta modulator applications.

59.4.1 Anti-Alias and Reconstruction Filters

The purpose of the anti-alias filter, shown in Figure 59.1 at the input of the sigma-delta A/D converter, is, as the name would indicate, to prevent aliasing. The sampling operation maps, or aliases, all frequencies into the range bounded by $\pm f_S/2$ [28]. Specifically, all signals within a baseband bandwidth of multiples of the sampling rate are mapped into the baseband. This is generally undesirable, so the anti-alias filter is designed to attenuate this aliasing to some tolerable level. One advantage of sigma-delta converters over Nyquist rate converters is that this anti-aliasing filter has a relatively wide transition region. As illustrated in Figure 59.27, the passband region for this filter is the signal bandwidth f_B, while the stopband region for this filter is only within f_B of the sampling rate. Thus, the transition region is $2(M-1)f_B$, and since $M \gg 1$, the transition region is relatively wide. A wide transition region generally means a simple filter design. The precise nature of the anti-alias filter is application-dependent, and can be designed using any number of standard analog filter techniques [35].

The reconstruction filter, shown in Figure 59.2 at the output of the sigma-delta D/A converter, is also an analog filter. Its primary purpose is to remove unwanted out-of-band quantization noise. The extent to which this noise must be removed varies widely from system to system. If the analog output is to be applied to an element that is naturally bandlimited, such as a speaker, then very little attenuation may be necessary. In contrast, if the output is applied to additional analog circuitry, care must be taken lest the high-frequency noise distort and map itself into the baseband. Circuit techniques for this filter are addressed further in Section 59.5.5.

59.4.2 Decimation and Interpolation Filters

In general, the filter characteristics of the decimation filter, shown in Figure 59.1 at the output of the sigma-delta A/D converter, are much sharper than those of the anti-alias filter, that is, the transition region is narrower. The saving grace is that the filter is implemented digitally, and modern submicron processes have made complex digital filters economically feasible. Nonetheless, care must be taken or the filter architecture will become more computationally complex than is necessary.

The basic purpose of the decimation filter is to attenuate quantization noise and unwanted signals outside the baseband so that the output of the decimation filter can be down-sampled, or decimated, without significant aliasing. Normally, the most efficient means of accomplishing this is to apply a multirate filter architecture, such as that illustrated in Figure 59.28 [36,37]. The comb filter is a relatively

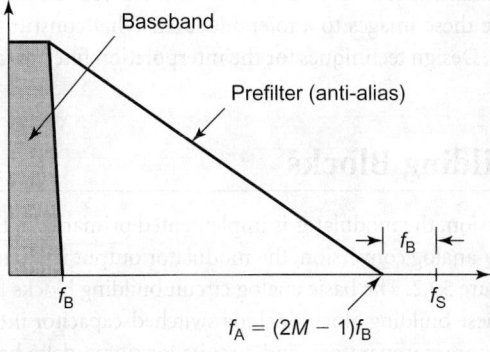

FIGURE 59.27 Anti-alias filter for sigma-delta A/D converters.

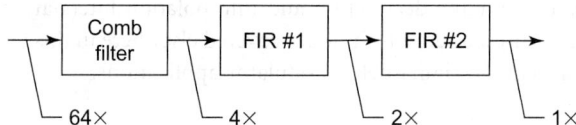

FIGURE 59.28 Typical decimation filter architecture.

crude, but easy to implement, filter that has zeros equally spaced throughout the sampled spectrum. The frequency response of an Nth-order comb filter, $H_C(z)$, is

$$H_C(z) = \left(\frac{1}{R} \frac{1 - z^{-R}}{1 - z^{-1}} \right)^N \tag{59.32}$$

where R is the impulse response length of the comb filter. If R is set equal to the decimation ratio of the comb filter (the comb filter input rate divided by its output rate), then the filter zeros will occur at every point that would alias to DC [38,39]. If the filter order N is one more than the modulator order, then the comb filter will be adequate to attenuate the out-of-band quantization noise to the point where it does not adversely increase the baseband noise after decimation [40].

Following the comb filter is typically a series of one or more FIR filters. Since the sample rates of these FIR filters are much slower than the oversampled clock rate, each filter output can be computed over many clock cycles. Also, since the output of each filter is decimated, only the samples that will be output need to be calculated. These properties can be exploited to devise computationally efficient structures for decimation filtering [41].

In the example in Figure 59.28, the first FIR filter is decimating from 4× to 2× oversampling. Since the output of this filter is still oversampled, the transition region is relatively wide and the attenuation at midband need not be very high. Thus, an economical half-band filter (a filter in which every other coefficient is zero) can be used [37].

The final FIR filter is by far the most complex. It usually has to have a very sharp transition region, and for strict anti-alias performance it cannot be a halfband filter. In high-performance sigma-delta modulators, this filter is often in the range of 50–200 taps in length. Standard digital filter design techniques can be used to select that tap weights for this filter [28]. Since it is going to be a complex filter anyway, it can also be used to compensate for any frequency droop in the previous filter stages.

The interpolation filter, shown in Figure 59.2 at the input of the sigma-delta D/A converter, upsamples the input digital words to the oversampling rate. In many ways, this filter is the inverse of a decimation filter, typically comprising a complex upsampling FIR filter, optionally followed by one or more simple FIR filters, followed by an upsampling comb filter. The upsampling operation, without this filter, would produce images of the baseband spectrum at multiples of the baseband frequency. The purpose of the interpolation is to attenuate these images to a tolerable level. What constitutes tolerable is very much a system-dependent criterion. Design techniques for the interpolation filter parallel those of the decimation filter discussed above.

59.5   Circuit Building Blocks

For analog-to-digital conversion, the modulator is implemented primarily in the analog domain as shown in Figure 59.16. In digital-to-analog conversion, the modulator output if filtered by an analog reconstruction filter as depicted in Figure 59.2. The basic analog circuit building blocks for these data converters are described in this section. These building blocks include switched-capacitor integrators, the amplifiers that are imbedded in the integrators, comparators, and circuits for sigma-delta based D/A conversion. At the end of this section, the techniques for continuous-time sigma-delta modulation are briefly discussed.

59.5.1   Switched-Capacitor Integrators

Switched-capacitor integration stages are commonly used to perform the signal-processing functions of integration and summation required for realization of the discrete-time transfer functions $A(z)$ and $F(z)$ in Figure 59.16. The circuit techniques outlined herein are drawn from a rich literature of switched-capacitor filters [42–45] that is detailed elsewhere in this volume.

Figure 59.29 is a typical integrator stage for the case of single bit feedback [11,20], and is designed to perform the discrete-time computation

$$V_{OUT}(z) = K \frac{z^{-1}}{1 - z^{-1}} (V_{IN}(z) - V_{DAC}(z)) \tag{59.33}$$

independent of the parasitic capacitances associated with the capacitive devices shown. The curved line in the capacitor symbol is the device terminal with which the preponderance of the parasitic capacitance is associated. For example, this will be the bottom plate of a stacked planar capacitance structure, where the parasitic capacitance is that between the bottom plate and the IC substrate. The circuit's precision stems from the conservation of charge at the two input nodes of the operational amplifier, and the cyclic return of the potential at those nodes to constant voltages. More details may be found in the Chapter 62, "Switched-Capacitor Filters."

Fully differential circuits will be shown here, as these are almost universally preferred over single-ended circuits in monolithic implementations owing to their greatly improved power supply rejection, MOS switch feedthrough rejection, and suppression of even-order nonlinearities. The switches shown in Figure 59.29 are generally full CMOS switches, as detailed in Figure 59.30. However, integrators with very low-power supply voltages may necessitate the use of only one polarity of switch device, possibly with a switch gate voltage boosting arrangement [46]. Sampling capacitors CSP and CSM are designed with the same capacitance C_S, and the effect of slight fabrication mismatches between the two will be mitigated by the common-mode rejection of the amplifier. Similarly, integration capacitors CIP and CIM are designed to be identical with capacitance C_I.

The discrete-time signal to be integrated is applied between the input terminals V_{IN+} and V_{IN-}, and the output is taken between V_{OUT+} and V_{OUT-}. V_{INCM} is the common-mode input voltage required by the amplifier. The single-bit DAC feedback voltage is applied between V_{DAC+} and V_{DAC-}. The stage must be clocked by two nonoverlapping signals, ϕ_1 and ϕ_2. During the ϕ_1 phase, the differential input voltage is

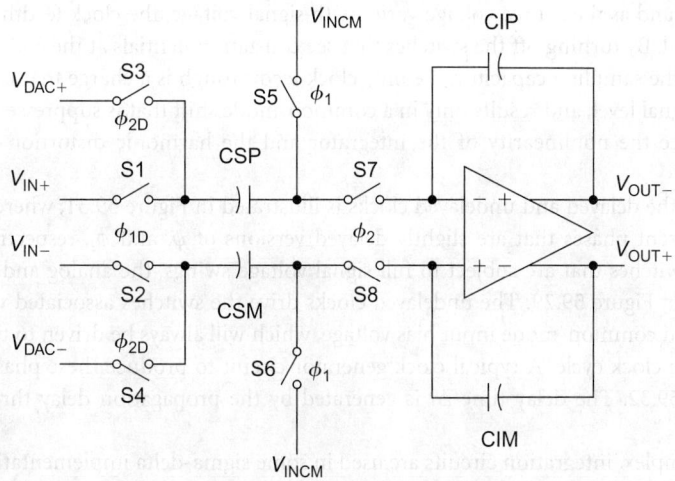

FIGURE 59.29 Typical integrator stage.

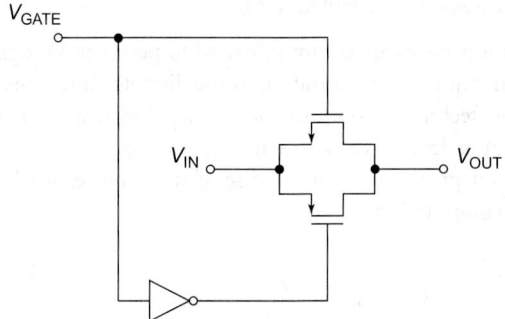

FIGURE 59.30 Full CMOS switch.

sampled on the bottom plates of CSP and CSM, while their top plates are held at the amplifier common-mode input level. During this phase, the amplifier summing nodes are isolated from the capacitor network, and the amplifier output will remain constant at its previously integrated value. During the ϕ_2 phase the bottom plates of the sampling capacitors CSP and CSM experience a differential potential shift of $(V_{DAC} - V_{IN})$, while the top plates are routed into the amplifier summing nodes. By forcing its differential input voltage to a small level, the amplifier will effect a transfer of a charge of $C_S(V_{IN} - V_{DAC})$ to the integration capacitors, and therefore the differential output voltage will shift to a new value by an increment of $(C_S/C_I)(V_{IN} - V_{DAC})$. Since this output voltage shift will accumulate from cycle to cycle, the discrete-time transfer function will be that of Eq. (59.33), with

$$K = \frac{C_S}{C_I} \qquad (59.34)$$

Over several cycles of initial operation, the amplifier input terminals will be driven to the common-mode level that is precharged onto the top plates of the sampling capacitors.

To suppress any signal-dependent clock feedthrough from the switches, it is helpful to slightly delay the clock phases that switch variable signal voltages with respect to the phases that switch current into constant potentials. The channel charge in each turned-on switch device can potentially dissipate onto the sampling capacitors when the switches are turned off, producing an error in the sampled charge. This channel charge is dependent on the difference between the switch gate to source voltage and its threshold voltage, and as the source voltage varies with signal voltage, the clock feedthrough charge will vary with the signal. By turning off the switches that see constant potentials at the end of each cycle first, and thus floating the sampling capacitor, the only clock feedthrough is a charge that is to the first-order independent of signal level, and results only in a common-mode shift that is suppressed by the amplifier. This acts to reduce the nonlinearity of the integrator and the harmonic distortion generated by the modulator.

The timing for the delayed and undelayed clocks is illustrated in Figure 59.31, where the clock phases ϕ_{1D} and ϕ_{2D} represent phases that are slightly delayed versions of ϕ_1 and ϕ_2, respectively. The delayed clocks drive the switches that are subject to full signal voltage swings, the analog and reference voltage inputs, as shown in Figure 59.29. The undelayed clocks drive the switches associated with the amplifier summing node and common-mode input bias voltage, which will always be driven to the same potential by the end of each clock cycle. A typical clock generator circuit to produce these phase relationships is shown in Figure 59.32. The delay time Δt is generated by the propagation delay through two CMOS inverters.

Other more complex, integration circuits are used in some sigma-delta implementations, for example, to suppress errors due to limited amplifier gain [47,48] or to effectively double the sampling rate of the integrators [49,50]. For the modulator structures discussed in Section 59.3 that are more elaborate than

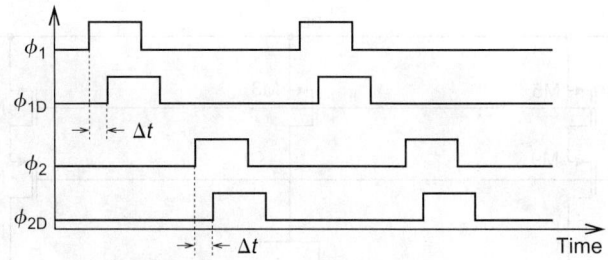

FIGURE 59.31 Delayed clock timing.

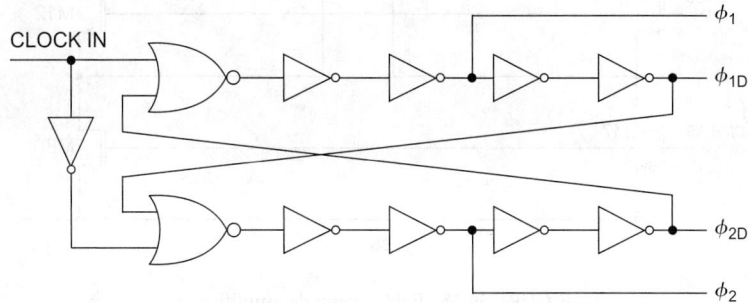

FIGURE 59.32 Nonoverlapping clock generator with delayed clocks.

a second-order loop, more complex switched-capacitor filtering is required. These may still, however, be designed with the same basic integrator architecture as in Figure 59.29, but with extra sampling capacitors feeding the amplifier summing node to implement additional signal paths [26,33,51]. Consult the Chapter 62 in this volume on switched-capacitor filtering for more information.

59.5.2 Operational Amplifiers

Embedded in the switched-capacitor integrator shown in Figure 59.29 is an operational amplifier. There are three major types of operational amplifiers typically used in switched-capacitor integrators [52]: the folded cascode amplifier [42], shown in Figure 59.33, the two-stage amplifier [43], shown in Figure 59.34, and the class AB amplifier [45], shown in Figure 59.35.

When the available supply voltage is high enough to permit stacking of cascode devices to develop high gain, a folded cascode amplifier is commonly used. A typical topology is shown in Figure 59.33. The input devices are PMOS, since most IC processes feature PMOS devices that exhibit lower $1/f$ noise than their NMOS counterparts [53]. The input differential pair M1 and M2 is biased with the drain current of M3. FETs M5, M6, M11, and M12 function as current sources, and M7, M8, M9, and M10 form cascode devices that boost the output impedance. The amplifier is compensated for stability in the integrator feedback loop by the dominant pole that is formed at its output node with the high output impedance and the load capacitance. In an integrator stage, the amplifier will be loaded with the load capacitance of the following stage sampling capacitance as well as its own integration capacitance. The nondominant pole at the drains of M1 and M2 limit the unity-gain frequency, which can be quite high.

When the power supply voltage is limited, and cascode devices cannot be stacked and still preserve adequate signal swing, a two-stage amplifier is a common alternative to the folded-cascode amplifier. As shown in Figure 59.34, the input differential pair of M1 and M2 now feed the active load current sources of M9 and M10 to form the first stage. The second stage comprises common-source amplifiers M7 and M8, loaded with current sources M5 and M6. Owing to the presence of two poles from the two stages of roughly comparable frequencies, compensation is generally achieved with a pole-splitting RC local feedback network

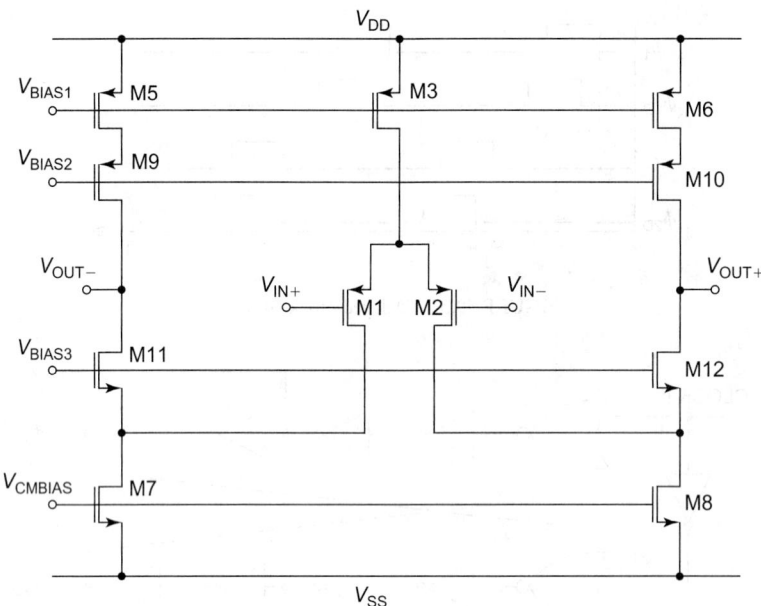

FIGURE 59.33 Folded cascode amplifier.

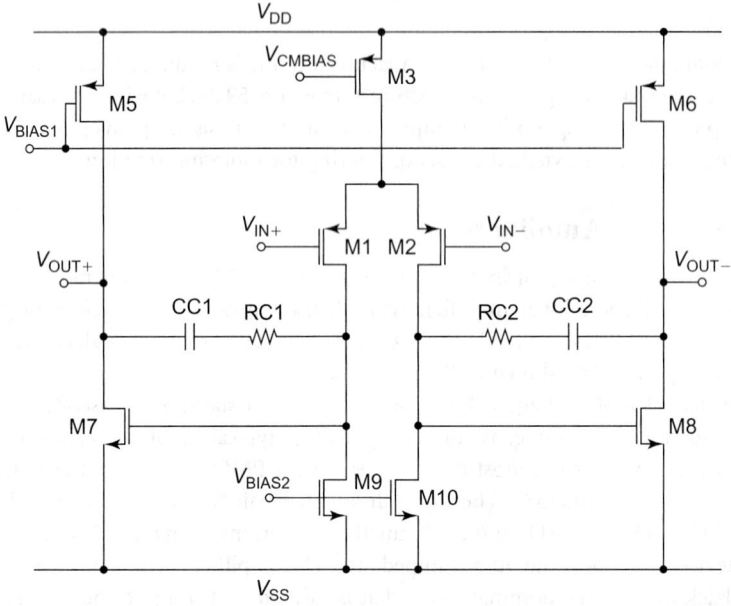

FIGURE 59.34 Two-stage amplifier.

as shown [52]. Often the resistors RC1 and RC2 are actually implemented as NMOS devices biased into their ohmic region by tying their gates to V_{DD}. In this arrangement the effective resistance of RC1 and RC2 will approximately track any drift in mobility of M7 and M8 over temperature and processing variations, preserving the compensated phase margin. For a given process, the bandwidth of a two-stage amplifier is less than what can be achieved than by a folded cascode design, but because the two-stage amplifier has no stacked cascode devices, the signal swing is higher.

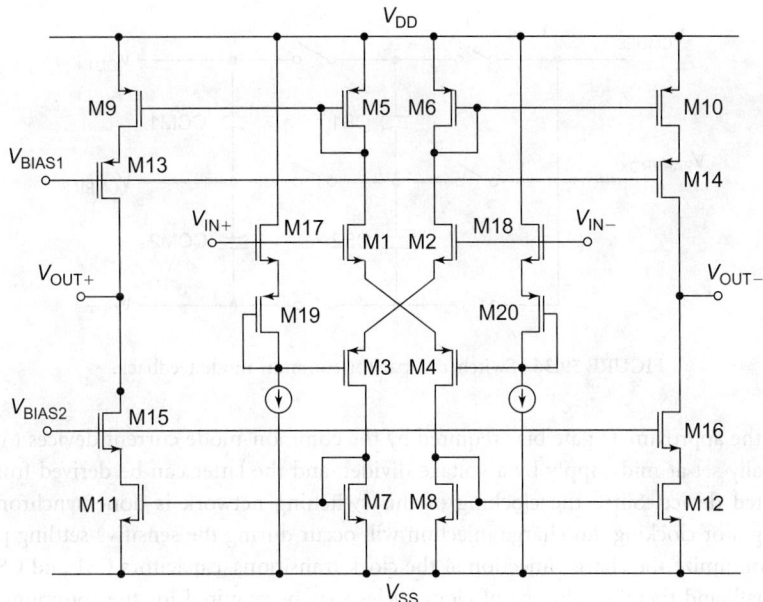

FIGURE 59.35   Class AB amplifier.

In the case of modulators with higher clock speeds, both folded-cascode and two-stage amplifiers may have unacceptably long settling times; in these amplifiers, the maximum slewing current that can be applied to charge or discharge the load capacitance is limited by fixed current sources. This slewing limitation can be overcome by a class AB amplifier topology that can supply a variable amount of output current and is capable of providing a large pulse of current early in the settling cycle when the differential input error voltage is high. A typical class AB amplifier topology is shown in Figure 59.35. The input differential pair from the folded-cascode and two-stage designs is replaced by M1 through M4, and their drain currents are mirrored to the output current sources M9–M12 by diode connected devices M5–M8. Cascode devices M13–M16 enhance the output impedance and gain. As with the folded cascode design, frequency compensation is accomplished by a dominant pole at the output node. The input voltage is fed directly to the NMOS input devices and to the PMOS input devices through the level shifting source follower and diode combination M17–M20. This establishes the quiescent bias current through the input network M1–M4, and therefore through the output devices as well.

In each of the three amplifier topologies discussed above there is either one or a set of two matched current sources driving both differential outputs. These current sources are controlled by a gate bias line labeled V_{CMBIAS}. The current output of these devices will determine the common-mode output voltage of the amplifier independent, to the first order, of the amplified differential signal. The appropriate potential for V_{CMBIAS} is determined by a feedback loop that is only operable in the common mode and is separate from the differential feedback instrumental in the charge integration process.

Since a discrete-time modulator is, by its nature, clocked periodically, a natural choice for the implementation of this common-mode feedback loop is the switched capacitor network of Figure 59.36 [44,45]. Capacitors CCM1 and CCM2 act as a voltage divider for transient voltages that derives the average, or common mode, voltage of the amplifier output terminals. This applies corrective negative feedback transients to the V_{CMBIAS} node to stabilize the feedback loop during each clock period while the amplifier is differentially settling.

A DC bias is then maintained on CCM1 and CCM2 by the switched capacitor network on the left side of the figure. This will slowly transfer the charge necessary to establish and maintain a DC level shift that makes up the difference between the common-mode level desired at the amplifier output terminals

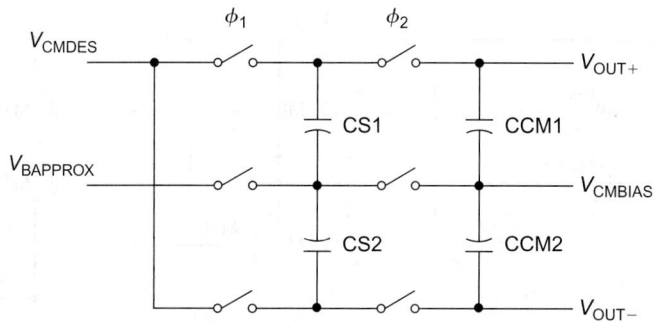

FIGURE 59.36 Switched-capacitor common-mode feedback.

(V_{CMDES}) and the approximate gate bias required by the common-mode current devices (V_{BAPPROX}). The former is usually set at mid supply by a voltage divider, and the latter can be derived from a matched diode-connected device. Since the clocking of this switching network is done synchronously to the amplifier integrator clocking, no charge injection will occur during the sensitive settling process of the amplifier. To minimize the charge injection at the clock transitions, capacitors CS1 and CS2 are usually made very small, and therefore dozens of clock cycles may be required for the common-mode bias to settle and the modulator to become operable.

59.5.3 Comparators

The noise-shaping mechanism of the modulator feedback loop allows the loop behavior to be tolerant of large errors in circuit behavior at locations closer to the output end of the network. Modulators are generously tolerant of large offset errors in the comparators used in the A/D converter forming the feedback path. For this reason, almost all modulators use simple regenerative latches as comparators. No preamp is generally needed, as the small error from clock kickback can easily be tolerated. Simulations show that offset errors that are even as large as 10% of the reference level will not degrade modulator performance significantly.

The circuit of Figure 59.37 is typical [54]. This is essentially a latch composed of two cross-connected CMOS inverters, M1–M4. Switch devices M5–M8 will disconnect this network when the clock input is low, and throw the network into a regenerative mode with the rising edge of the clock. The state in which the regeneration will settle may be steered by the relative strengths of the bias current output by devices M9 and M10, which in turn depend on the differential input voltage.

59.5.4 Complete Modulator

Figure 59.38 illustrates a complete second-order, single bit feedback modulator assembled from the components discussed above [11]. The discrete-time integrator gain factors that are derived in Sections 59.2 and 59.3 are realized by appropriate ratios between the integration and sampling capacitors in each stage. Since the single bit feedback DAC is only responsible for generating two output levels, it may be implemented by simply switching an applied differential reference voltage $V_{\text{REF+}}$ to $V_{\text{REF-}}$ in a direct or reversed sense to the sampling capacitor bottom plates during the amplifier integration phase, ϕ_2.

59.5.5 D/A Circuits

For the D/A converter system shown in Figure 59.2, the oversampled bit stream is generated by straight-forward digital implementations of the modulator signal flow graphs discussed in Section 59.2. The remaining analog components are the low-resolution DAC block and the reconstruction filter. Integrated sigma-delta D/A implementations are often employ two-level quantization, and the DAC block may

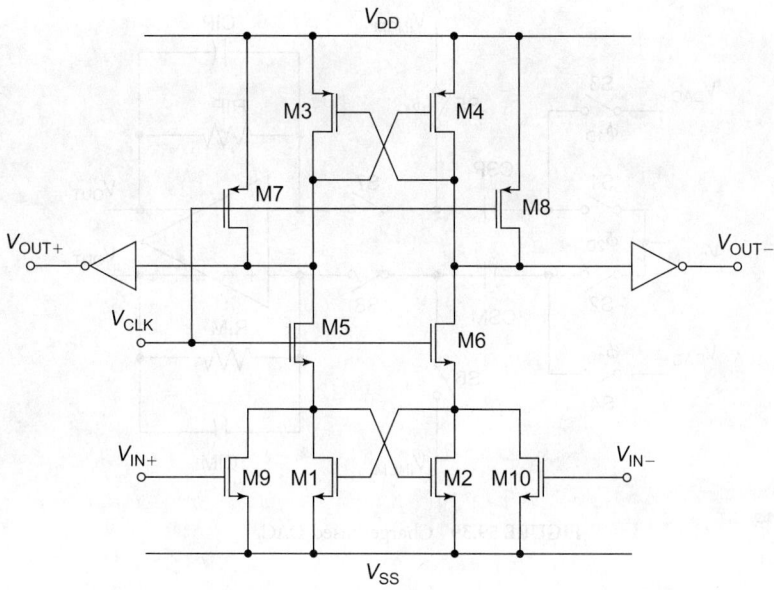

FIGURE 59.37 Typical modulator comparator.

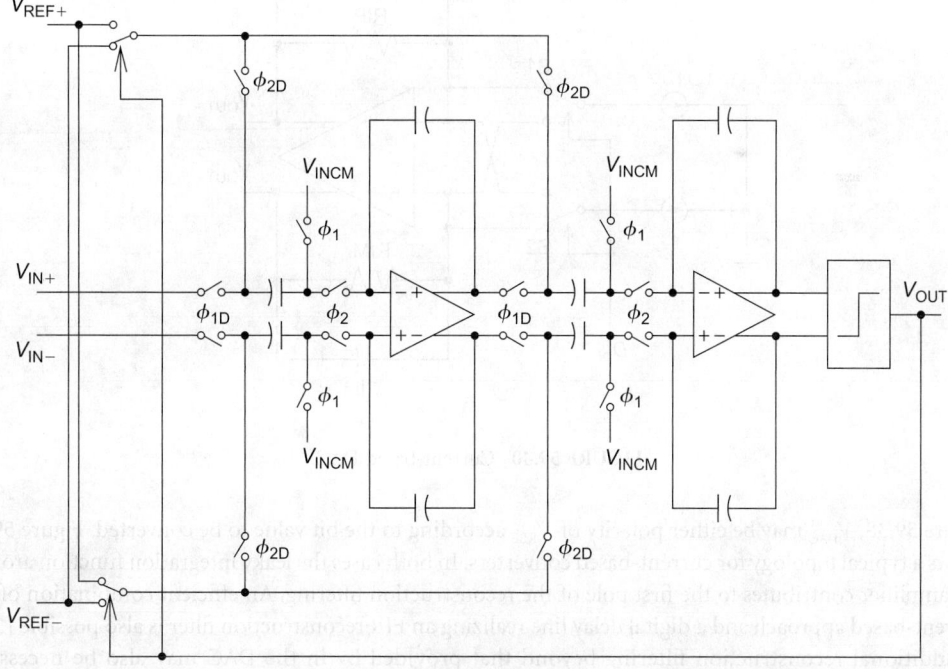

FIGURE 59.38 Complete second-order sigma-delta modulator.

either be designed as charge-based [55] or current-based [56]. Multilevel DAC approaches are also used, but for harmonic content less than about 60 dB below the reference some form of dynamic element matching must be added, as discussed in Section 59.6.7.

The charge-based approach for sigma-delta D/A conversion is illustrated in Figure 59.39, which is similar to the switched capacitor integrator of Figure 59.29, but without an analog signal input. As in

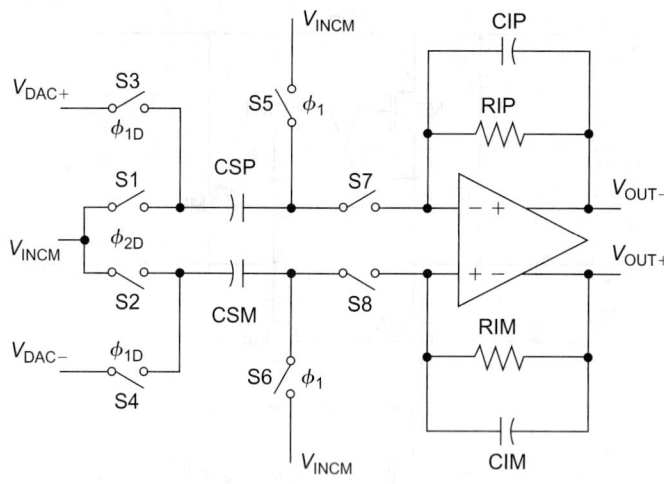

FIGURE 59.39 Charge-based DAC.

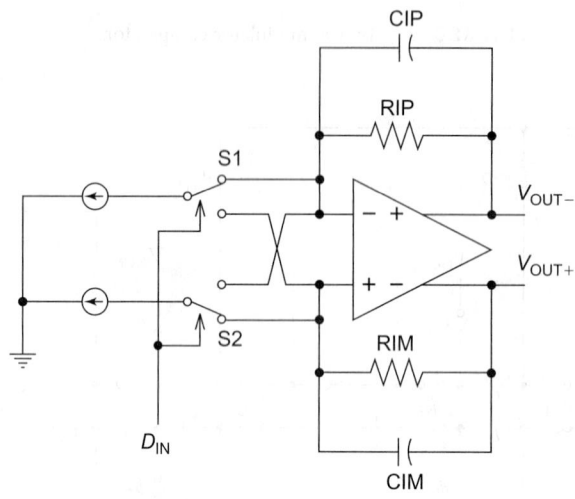

FIGURE 59.40 Current-based DAC.

Figure 59.38, V_{DAC} may be either polarity of V_{REF} according to the bit value to be converted. Figure 59.40 shows a typical topology for current-based converters. In both cases the leaky integration function around the amplifier contributes to the first pole of the reconstruction filtering. An efficient combination of the current-based approach and a digital delay line realizing an FIR reconstruction filter is also possible [57].

Additional reconstruction filtering beyond that provided by in the DAC may also be necessary. This is accomplished using the appropriate analog sampled-data filtering techniques described in Chapter 62.

59.5.6 Continuous-Time Modulators

In general, the amplifiers contained in the switched-capacitor integrators in a sampled-data sigma-delta data converter dissipate the majority of the analog circuit power. Since the integrator sections must settle accurately within each clock period at the oversampled rate, the amplifiers must often be designed with

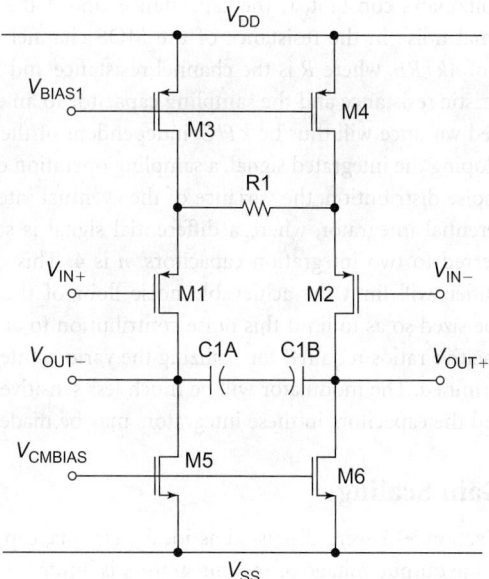

FIGURE 59.41 Gm-C integrator.

a unity-gain frequency much higher than the oversampled rate; typical unity-gain frequencies are in hundreds of MHz.

In applications in which dissipating the lowest possible power is important, sigma-delta modulators may also be implemented using continuous-time integrators. In these continuous-time modulators, the analog signal is not sampled until the quantizer at the back of the modulator loop [58]. Owing to the typical means employed for the DAC feedback, continuous-time modulators tend to be more sensitive to sampling clock jitter, but the influences of any aliasing distortion and nonlinearity at the sampler take place late in the loop where noise shaping is steepest, and as a consequence the anti-aliasing filter of Figure 59.1 may often be omitted [59]. The power advantage comes from the relaxed speed requirement of the integrator stages, which now need only have unity gain frequencies on the order of the oversampled clock frequency.

Instead of switched-capacitor discrete-time integrators, the continuous-time modulators generally use active Gm-C integrators. Circuits like the one shown in Figure 59.41 are typical [59]. The input differential pair M1 and M2 is degenerated by source resistance R1 to improve linearity. The output analog voltage is developed across capacitor C1, which may be split as shown to place the bottom plate parasitic capacitance at a common-mode node. As the integrator is now unclocked, continuous-time common-mode feedback must be used, as discussed in the literature for continuous-time filtering [60].

59.6 Practical Design Issues

As with any design involving analog components, there are a number of circuit limitations and trade-offs in sigma-delta data converter design. The design considerations discussed in this section include kT/C noise, integrator scaling, amplifier gain, and sampling nonlinearity. Also discussed in this section are the techniques of integrator reset and multilevel feedback.

59.6.1 kT/C Noise

In switched capacitor-based modulators, one fundamental nonideality associated with using a MOS device to sample a voltage on a capacitor is the presence of a random variation of the sampled voltage after the MOS switch opens [61–63]. This random component has a gaussian distribution with a variance of kT/C, where k is the Boltzman's constant, C the capacitance, and T the absolute temperature. The variation stems from thermal noise in the resistance of the MOS channel as it is opening. The noise voltage has a mean power of $4kTRB$, where R is the channel resistance and B the bandwidth. It is low-pass filtered by its characteristic resistance and the sampling capacitor to an equivalent noise bandwidth of $1/RC$. The total integrated variance will thus be kT/C, independent of the resistance of the switch.

If, in the process of developing the integrated signal, a sampling operation on n capacitors is used, then since we assume gaussian noise distribution, the variance of the eventual integrated value will be nkT/C. In the case of a fully differential integrator, where a differential signal is sampled onto two sampling capacitors and then transferred to two integration capacitors, n is 4. This effect, along with the input referred noise of the amplifier, will limit the achievable noise floor of the modulator. The first-stage sampling capacitors must be sized so as to limit this noise contribution to an acceptable level. From this starting point, and the capacitive ratios required for realizing the various integrator gains, the remaining capacitor sizes may be determined. The modulator will be much less sensitive to kT/C noise generated in integrators past the first, and the capacitors in these integrators may be made considerably smaller.

59.6.2 Integrator Gain Scaling

The integration stages in Section 59.2 were discussed as ideal elements, capable of developing any real output voltage. In practice, the output voltage of real integrators is limited to at most the supply voltage of the embedded amplifier. To ensure that this limitation does not adversely affect the modulator performance, a survey of the likely limit of integrator output voltages must be made for a given value of the DAC reference voltage. The modulator may be simulated over a large number of samples with a representative sinusoidal input, and a histogram of all encountered output voltages tabulated. These histograms may be expected to scale linearly with the reference voltage level. In general, this statistical survey will show that a modulator designed to realize the integrator gain constants in the ideal topologies of Sections 59.2 and 59.3 will have different ranges of expected output voltages from each of its integrators. For example, Figure 59.42 and Figure 59.43 show the simulated output voltages at the two integrators in a second-order modulator with eight- and two-level feedback, respectively. Since the largest value possible of reference level will generally mean the best-available signal-to-noise ratio for a given circuit power consumption, the integrator gain constants may be adjusted from their straightforward values so that the overall modulator transfer function remains the same, but the output voltages are scaled so that no

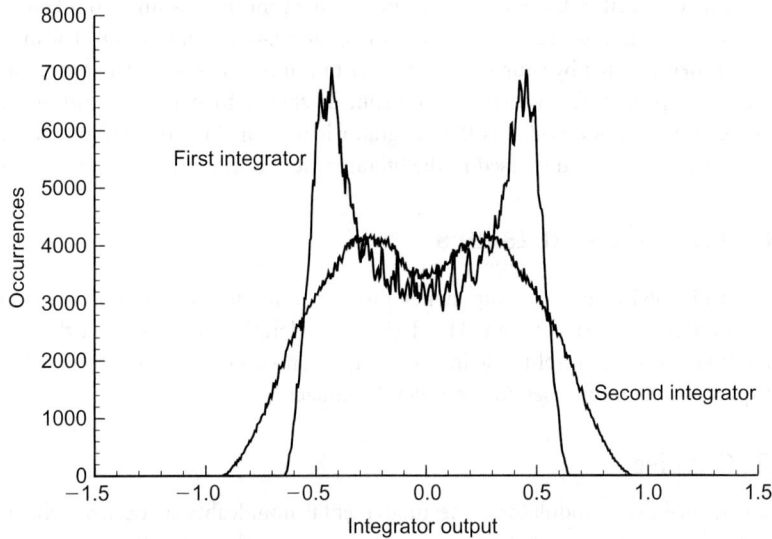

FIGURE 59.42 Integrator output distribution for an eight-level modulator.

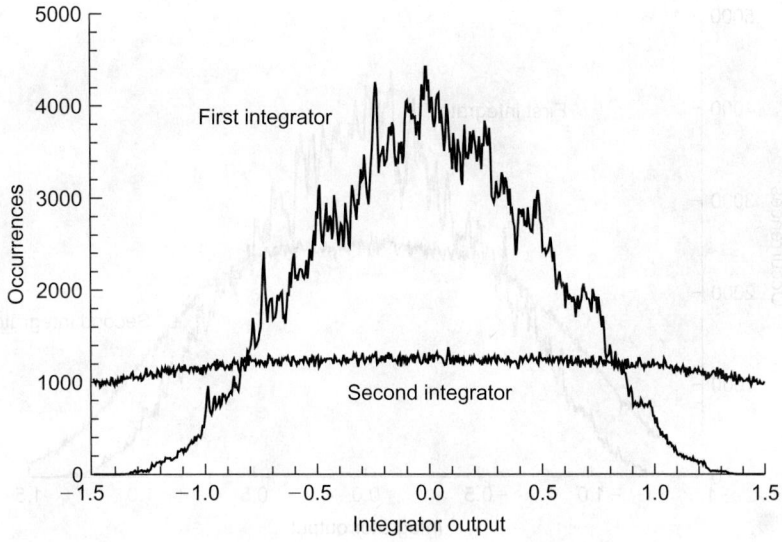

FIGURE 59.43 Integrator output distribution for a two-level modulator.

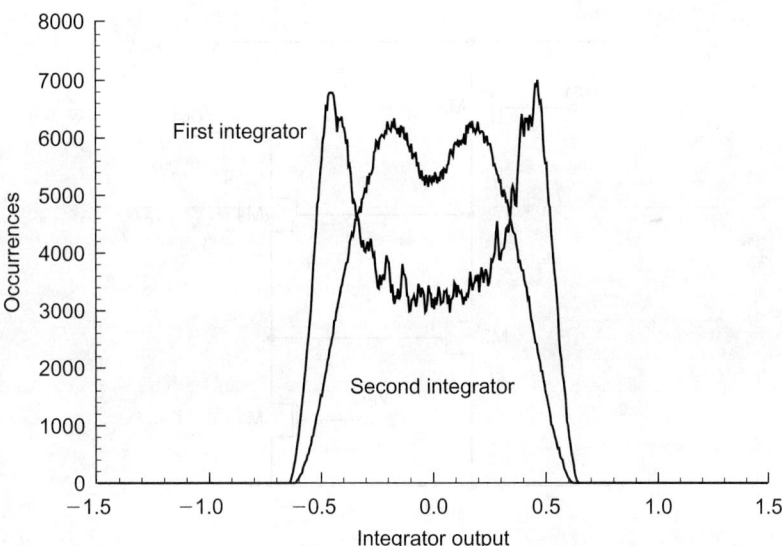

FIGURE 59.44 Integrator output distribution for a scaled eight-level modulator.

integrator limits the signal swing markedly before the other [11]. Figure 59.44 and Figure 59.45 illustrate the properly scaled second-order modulator examples.

59.6.3 Amplifier Gain

Another mechanism by which the actual characteristic of the integrator circuits fall short of the ideal is the limitation of finite amplifier gain. A study of many simulations of modulators with various amplifier gains [11] has shown that a modulator needs amplifiers with gains about numerically equal to the decimation ratio of the filter that follows it to avoid significant noise-shaping errors. At least this is the result with perfectly linear amplifiers, and in practice, amplifier gains often need to be at least 10 times this high to avoid distortion in the integrator characteristic due to the nonlinearity of the amplifier gain characteristic.

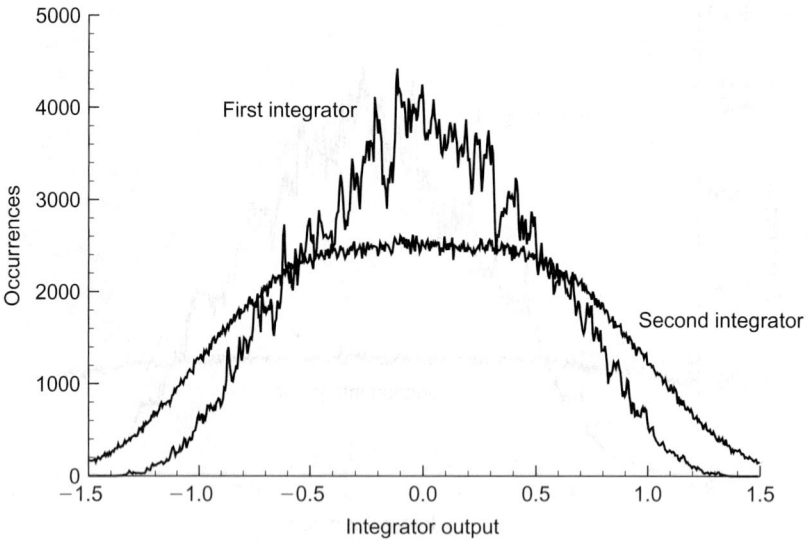

FIGURE 59.45 Integrator output distribution for a scaled two-level modulator.

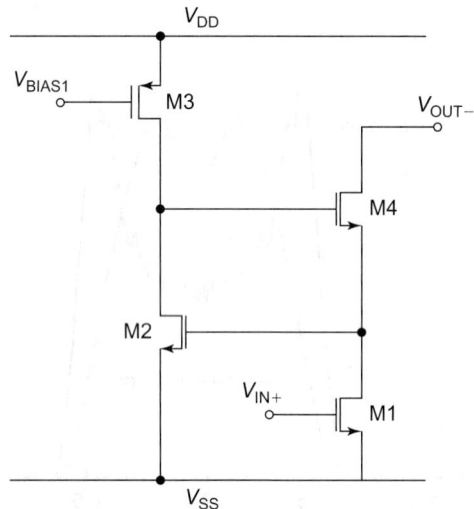

FIGURE 59.46 Regulated cascode topology.

One approach used when the simple circuits of Section 59.5.2 do not develop enough gain in a given process is the regulated cascode gain enhancement [64,65]. Figure 59.46 illustrates a typical circuit topology. This subcircuit may be substituted for the output common source amplifier stages in the amplifiers of Figures 59.33–59.35 if the power supply voltage can accommodate its somewhat increased requirement for headroom.

59.6.4 Low Supply Voltages

Oversampled modulators are typically integrated on a larger chip that includes at least the digital decimation filter, and often more complex digital signal processing systems. Given the trend to lower supply voltages for systems with significant digital content as a power reduction tactic, it is natural to extend the design of the analog modulator to allow operation at quite low supply voltages.

The circuit block that presents a large design challenge at low supply voltages is the complementary CMOS switch of Figure 59.30. As the supply voltage approaches the sum of the n- and p-channel threshold voltages, there will be a range of signal voltages around the center of the supply range that will encounter large enough resistance through both switch devices. This effect tends to increase sampled signal distortion to objectionable levels.

In the complete modulator schematic of Figure 59.38, judicious choices of the various reference and common-mode voltage levels can be made to place them close to one or the other supply rail. This will allow the use of single polarity switch devices with low resistance. The exception to this, however, are the switches at the outputs of opamps. As can be seen in the simulation results in Section 59.2, the opamp output switches must handle signals which may cover almost the entire supply range in normal modulator operation.

To address this limitation, the switched opamp design technique has been developed [66,77]. Figure 59.47 shows a revision of the modulator of Figure 59.38 where the first integrator opamp has been replaced by a switchable opamp. The clock line controlling the opamp determines when the amplifier will function normally and when it will be placed in an inactive state where the signal current generated at the output terminals will be shut off. This essentially performs the function of the switches at the output of the first opamp in Figure 59.38, but without requiring any analog switch block that is required to pass signals throughout the entire supply range.

An example switchable opamp design is shown in Figure 59.48 [68]. It is similar to the two-stage opamp of Figure 59.34, but includes clocked switch devices in the sources of M7 and M8. In addition, clocked switches are also inserted in series with the pole-splitting compensation capacitors. This freezes the state of charge on these capacitors while the opamp is in its inactive state so that the operating point may be quickly restored when returning to the active state.

An alternative approach to eliminating the need for full-range switches at the output of opamps is closing reset switches around the opamp to form a negative feedback loop in the inactive state [69].

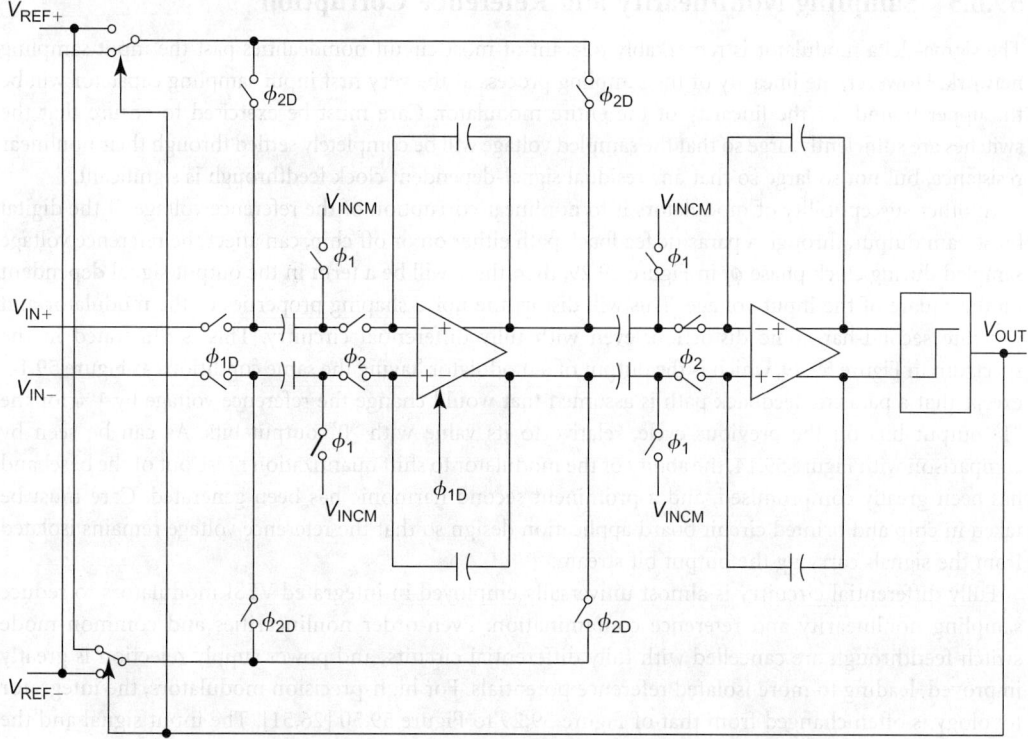

FIGURE 59.47 Switched opamp sigma-delta modulator.

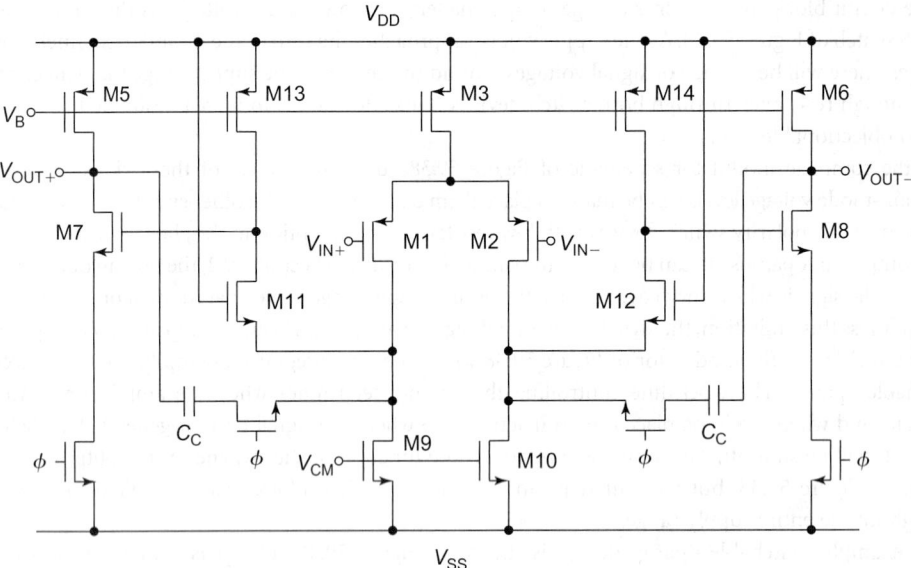

FIGURE 59.48 Switchable opamp.

Instead of switching off the opamp output current, this approach uses the gain properties of the opamp to clamp its output voltage in an inactive state. In some cases this approach can allow faster clock rates than the switched opamp approach.

59.6.5 Sampling Nonlinearity and Reference Corruption

The sigma-delta modulator is remarkably tolerant of most circuit nonidealities past the input sampling network. However, the linearity of the sampling process at the very first input sampling capacitor will be the upper bound for the linearity of the entire modulator. Care must be exercised to ensure that the switches are sufficiently large so that the sampled voltage will be completely settled through their nonlinear resistance, but not so large so that any residual signal-dependent clock feedthrough is significant.

Another susceptibility of modulators is to nonlinear corruption of the reference voltage. If the digital bit stream output, through a parasitic feedback path either on or off chip, can affect the reference voltage sampled during clock phase ϕ_2 in Figure 59.29, then there will be a term in the output signal dependent on the square of the input voltage. This will distort the noise-shaping properties of the modulator and generate second-harmonic distortion, even with fully differential circuitry. This is illustrated in the spectrum in Figure 59.49, which is the output of a modulator having the same conditions as Figure 59.14, except that a parasitic feedback path is assumed that would change the reference voltage by 1% for the "1" output bits on the previous cycle, relative to its value with "0" output bits. As can be seen by comparison with Figure 59.14, the ability of the modulator to shift quantization noise out of the baseband has been greatly compromised, and a prominent second harmonic has been generated. Care must be taken in chip and printed circuit board application design so that the reference voltage remains isolated from the signals carrying the output bit stream.

Fully differential circuitry is almost universally employed in integrated VLSI modulators to reduce sampling nonlinearity and reference contamination. Even-order nonlinearities and common-mode switch feedthrough are cancelled with fully differential circuits, and power supply rejection is greatly improved, leading to more isolated reference potentials. For high-precision modulators, the integrator topology is often changed from that of Figure 59.29 to Figure 59.50 [26,51]. The input signal and the DAC output voltage are sampled independently during phase ϕ_1, and then both discharged together into the summing node during ϕ_2. At the expense of additional area for capacitors and higher kT/C

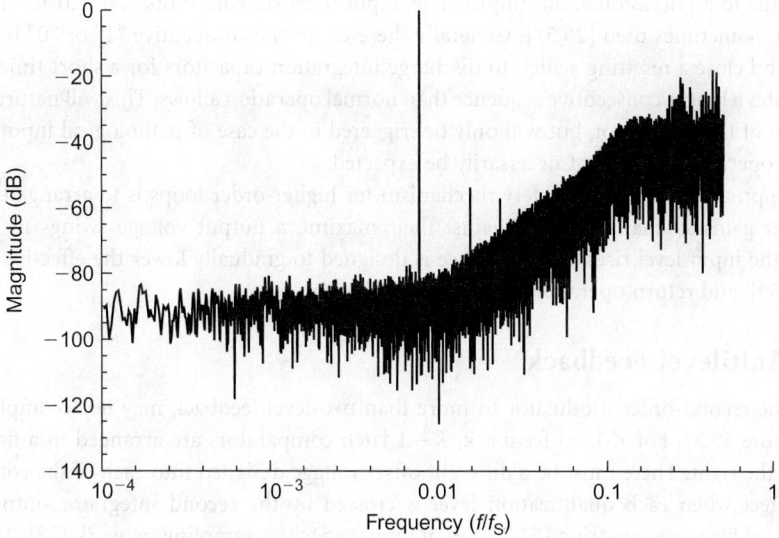

FIGURE 59.49 Output spectrum with reference corruption.

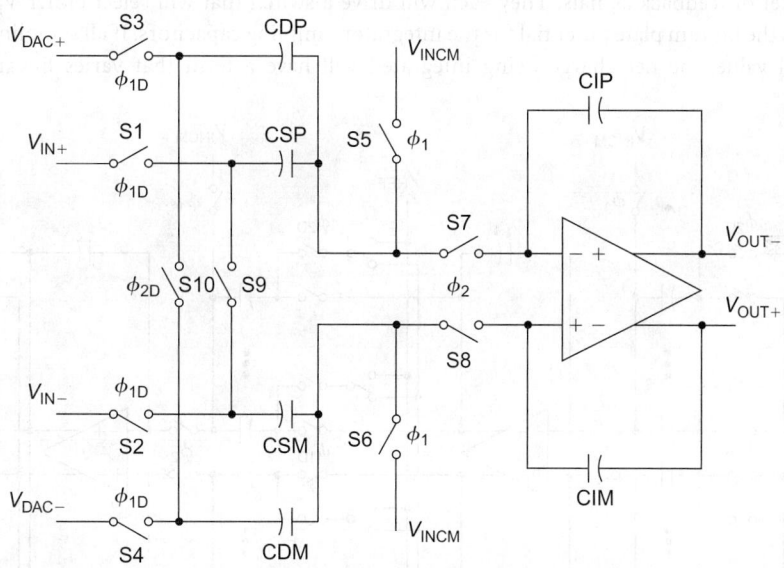

FIGURE 59.50 Integrator with separate DAC feedback capacitor.

noise, this arrangement insures that the same charge is drawn from the reference supply onto the DAC sampling capacitors CDP and CDM and then discharged into the summing node each cycle. Thus, a potential undesirable mechanism for reference supply loading that is dependent on the output bit history is eliminated [70].

59.6.6 High-Order Integrator Reset

Although careful design of the loop filter for higher-order modulators, as discussed in Section 59.3.1, will yield a generally stable design, their stability cannot be mathematically guaranteed as in the case of second-order loops. To protect against the highly undesirable state of low-frequency limit-cycle oscillations due to an occasional, but improbable, input overload condition, some form of forced integrator reset is sometimes used [26,51]. Generally these count the consecutive "1" or "0" bits out of the modulator, and close a resetting switch to discharge integration capacitors for a short time if the modulator generates a longer consecutive sequence than normal operation allows. This will naturally interrupt the operation of the modulator, but will only be triggered in the case of pathological input patterns for which linear operation would not necessarily be expected.

Another approach to a stability safety mechanism for higher-order loops is to arrange the scaling of the integrator gains so that they clip against their maximum output voltage swings in a prescribed sequence as the input level rises. The sequence is designed to gradually lower the effective order of the modulator [59], and return operation to a stable mechanism.

59.6.7 Multilevel Feedback

Expanding the second-order modulator to more than two-level feedback may be accomplished by the circuit in Figure 59.51. For K-level feedback, $K-1$ latch comparators are arranged in a flash structure as shown on the right. There must be a different offset voltage designed into each of the comparators so that they detect when each quantization level is crossed by the second integrator output. This can be implemented by a resistor string [54,71] or an input capacitive sampling network [72]. The output of the $K-1$ comparators is then a thermometer code representation of the integrator output. This may be translated into binary for the modulator output, but the raw thermometer code is the most convenient to use as a set of feedback signals. They each will drive a switch that will select either V_{REF+} or V_{REF-} to be used as the bottom plate potential for the integrator sampling capacitors. If all sampling capacitors are of equal value, the net charge being integrated will have a term that varies linearly with the

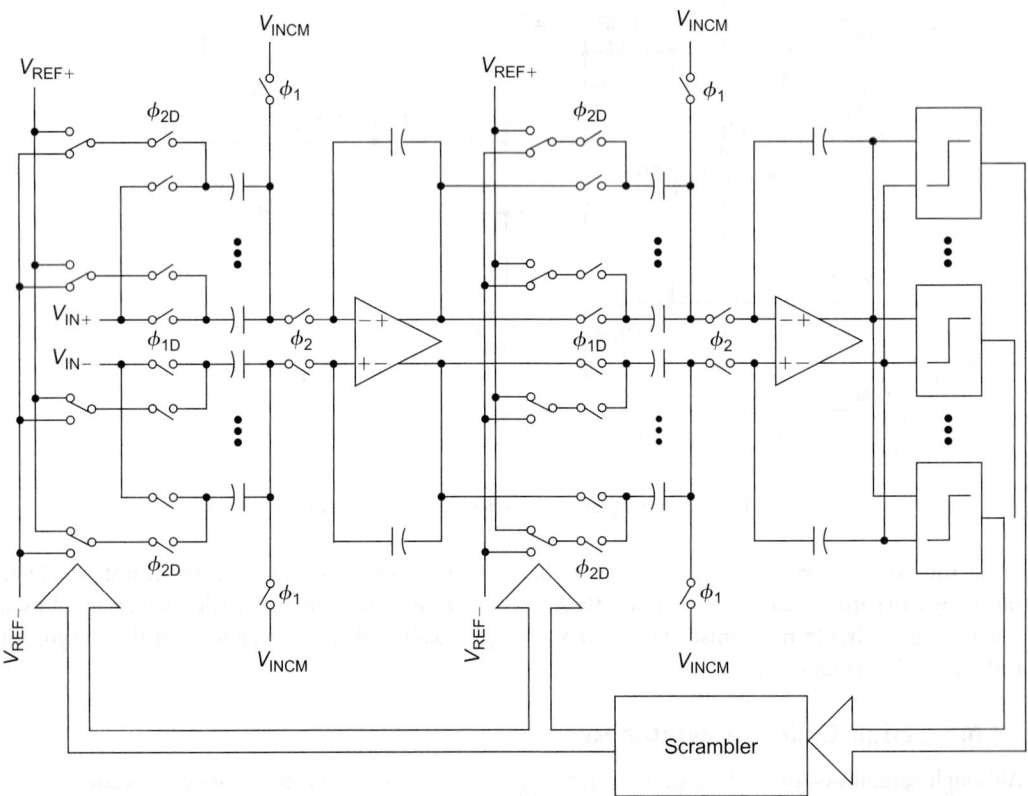

FIGURE 59.51 Multibit second-order modulator.

quantization level. Each comparator output drives two switches, so there are $2(K-1)$ switches and capacitors in the sampling array.

In any practical integrated structure, even if careful common-centroid layout techniques are used, the precision with which the various capacitors forming the sampling array will be matched is typically limited to 0.1 or 0.2%. As discussed in Section 59.2.3, this will limit the harmonic distortion that is inherent in the modulator to about −60 dB or higher. However, by varying the assignment of which sampling switch is driven by which comparator output dynamically as the modulator is clocked, much of this distortion may be traded off for white or frequency-shaped noise at the modulator output. This technique is referred to as dynamic element matching.

One simple way of implementing dynamic element matching is indicated in Figure 59.51 with the block labeled "scrambler." This block typically comprises an array of switches that provide a large number of permutations in the way the comparator output lines can be mapped onto sampling switch lines. A multitude of scrambler algorithms have been proposed in the literature. Some approaches attempt to randomize the mismatch errors, converting them into white noise [72,73]. Other approaches attempt to shape the mismatch error spectrum to achieve advantages similar to the way that a sigma-delta modulator shapes its quantization error. Some of these approaches are described below.

One conceptually simple approach for noise-shaping dynamic element matching is called individual level averaging [74,75]. As illustrated in Figure 59.52, separate counters (R_k) are maintained for each quantization level. When the kth quantization level is to be used, the counter R_k points to a cell in the capacitor array, and the next k cells are used to generate the quantizer output. The counter is then incremented by k using wrap-around arithmetic before the next sampling time. The individual level averaging algorithm insures that all of the capacitor cells are used equally when averaged over time. Furthermore, simulations and experimental data show that mismatch errors in the array are first-order noise shaped such that much of the error energy appears outside the baseband and is suppressed by the decimation filtering. The only real disadvantage of individual level averaging is that by requiring a separate counter for each quantization level, the implementation can become unwieldy as the number of quantization levels increases.

A simplification of individual level averaging, called data-weighted averaging [76], is much less complex than individual level averaging for large or even modest numbers of quantization levels. As illustrated in Figure 59.53, a single pointer is used for all quantization values. Like individual level averaging, the pointer is updated after every sample such that all of the quantizer array cells are used equally, and the

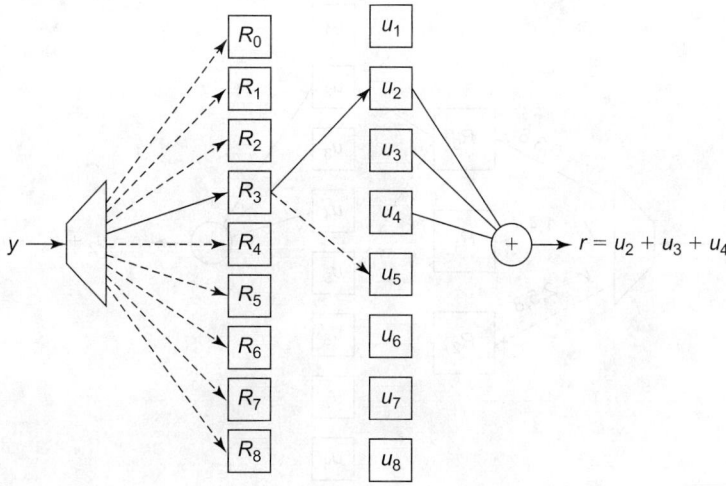

FIGURE 59.52 Individual level averaging example for $K = 9$, $k = 3$.

mismatch errors are first-order noise shaped. The disadvantage of data-weighted averaging is that, with only one pointer, data patterns can be created, which, in combination with array mismatch errors, can produce spectral tones.

A compromise in complexity between individual level averaging and data-weighted averaging is called grouped level averaging [77]. In grouped level averaging, the number of unit-cell pointers is more than one, but less than the number of quantization levels, as shown in Figure 59.54. The pointers are indexed such that no two adjacent quantization levels use the same pointer. If at least three pointers are used, simulations and experimental data show that the tone problems of data-weighted averaging are greatly diminished without the complexity penalties of individual level averaging.

Many other variations of dynamic element matching, too numerous to be catalogued here, have been published, ranging from complex algorithms that achieve better than first-order noise shaping of the unit-cell mismatch errors to simple algorithms that simply whiten the mismatch noise spectrum. Of particular note is a multiple stage butterfly network used to scramble the choice of array cells [78].

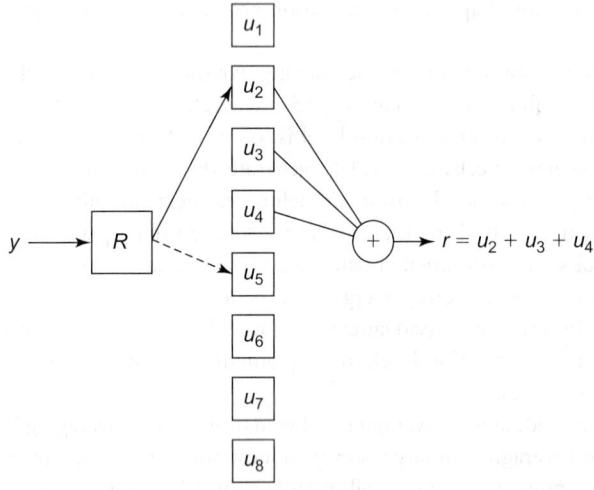

FIGURE 59.53 Data-weighted averaging example for $K = 9$, $k = 3$.

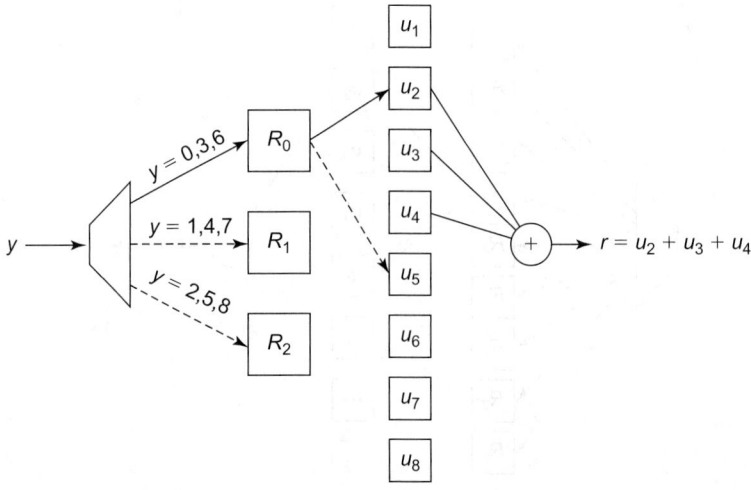

FIGURE 59.54 Grouped level averaging example for $K = 9$, $k = 3$.

59.7 Summary

In this chapter, a brief overview of sigma-delta data converters has been presented. Sigma-delta data conversion is a technique that effectively trades speed for resolution. High-linearity data conversion can be accomplished in modern IC processes without expensive device trimming or calibration. For a far more detailed treatment of this topic, refer to Norsworthy et al. [2]. For a compilation of some of the seminal papers that helped establish sigma-delta modulation as a mainstream technique, refer to Candy and Temes [1].

References

1. J. Candy and G. Temes, *Oversampling Delta-Sigma Data Converters*, IEEE Press, 1992.
2. S. Norsworthy, R. Schreier, and G. Temes, *Delta-Sigma Data Converters: Theory, Design, and Simulation*, IEEE Press, 1996.
3. C. Cutler, "Transmission systems employing quantization," U.S. Patent 2,927,962, March 8, 1960.
4. H. Spang III and P. Schultheiss, "Reduction of quantizing noise by use of feedback," *IRE Transactions on Communication Systems*, pp. 373–380, 1962.
5. H. Inose and Y. Yasuda, "A unity bit coding method by negative feedback," *Proc. IEEE*, vol. 51, pp. 1524–1535, 1963.
6. S. Tewksbury and R. Hallock, "Oversampled, linear predictive and noise-shaping coders of order N>1," *IEEE Trans. Circuits Systems*, vol. CAS-25, pp. 436–447, 1978.
7. J. Candy, "A use of limit cycle oscillations to obtain robust analog-to-digital converters," *IEEE Trans. on Commun.*, vol. COM-22, pp. 298–305, 1974.
8. H. Fiedler and B. Hoefflinger, "A CMOS pulse density modulator for high-resolution A/D converters," *IEEE J. Solid-State Circuits*, vol. SC-19, pp. 995–996, 1984.
9. B. Leung, R. Neff, P. Gray, and R. Brodersen, "Area-efficient multichannel oversampled PCM voice-band coder," *IEEE J. Solid-State Circuits*, vol. SC-23, pp. 1351–1357, 1988.
10. J. Candy, "A use of double integration in sigma delta modulation," *IEEE Trans. Commun.*, vol. COM-33, pp. 249–258, 1985.
11. B. Boser and B. Wooley, "The design of sigma-delta modulation analog-to-digital converters," *IEEE J. Solid-State Circuits*, vol. 23, pp. 1298–1308, 1988.
12. V. Friedman, D. Brinthaupt, D. Chen, T. Deppa, J. Elward, E. Fields, J. Scott, and T. Viswanathan, "A dual-channel voice-band PCM codec using $\Sigma\Delta$ modulation technique," *IEEE J. Solid-State Circuits*, vol. 24, pp. 274–280, 1989.
13. W. Bennett, "Spectra of quantized signals," *Bell Sys. Tech. J.*, vol. 27, pp. 446–472, 1948.
14. B. Widrow, "Statistical analysis of amplitude quantized sampled-data systems," *Trans. AIEE*, vol. 79, pp. 555–568, 1961.
15. R. Gray, "Oversampled sigma-delta modulation," *IEEE Trans. Commun.*, vol. COM-35, pp. 481–489, 1987.
16. J. Candy and O. Benjamin, "The structure of quantization noise from sigma-delta modulation," *IEEE Trans. Communications*, vol. COM-29, pp. 1316–1323, 1981.
17. B. Boser and B. Wooley, "Quantization error spectrum of sigma-delta modulators," *IEEE International Symposium on Circuits and Systems*, pp. 2331–2334, 1988.
18. R. Gray, "Quantization noise spectra," *IEEE Trans. Info. Theory*, vol. 36, pp. 1220–1244, 1990.
19. L. Williams and B. Wooley, "A third-order sigma-delta modulator with extended dynamic range," *IEEE J. Solid-State Circuits*, vol. 29, pp. 193–202, March 1994.
20. B. Brandt, D. Wingard, and B. Wooley, "Second-order sigma-delta modulation for digital-audio signal acquisition," *IEEE J. Solid-State Circuits*, vol. 26, pp. 618–627, 1991.
21. J. Everard, "A single-channel PCM codec," *IEEE J. Solid-State Circuits*, vol. SC-14, pp. 25–37, 1979.
22. M. Hauser, P. Hurst, and R. Brodersen, "MOS ADC-filter combination that does not require precision analog components," *ISSCC Digest Technical Papers*, pp. 80–82, February 1985.

23. S. Norsworthy, "Effective dithering of sigma-delta modulators," *Proceedings of the 1992 IEEE International Symposium on Circuits and Systems*, pp. 1304–1307, May 1992.

24. D. Welland, B. Del Signore, E. Swanson, T. Tanaka, K. Hamashita, S. Hara, and K. Takasuka, "A stereo 16-bit delta-sigma A/D converter for digital audio," *J. Audio Eng. Soc.*, vol. 37, pp. 476–486, 1989.

25. K. Chao, S. Nadeem, W. Lee, and C. Sodini, "A higher order topology for interpolative modulators for oversampling A/D converters," *IEEE Trans. Circuits Syst.*, vol. 37, pp. 309–318, 1990.

26. R. Adams, P. Ferguson, A. Ganesan, S. Vincelette, A. Volpe, and R. Libert, "Theory and practical implementation of a fifth-order sigma-delta A/D converter," *J. Audio Eng. Soc.*, vol. 39, pp. 515–528, 1991.

27. R. Schreier, "An empirical study of high-order single-bit delta-sigma modulators," *IEEE Trans. Circuits Syst.—II: Analog and Digital Signal Processing*, vol. 40, pp. 461–466, August 1993.

28. A. Oppenheim and R. Schafer, *Discrete-Time Signal Processing*, Prentice-Hall, 1989.

29. Y. Matsuya, K. Uchimura, A. Iwata, T. Kobayashi, M. Ishikawa, and T. Yoshitome, "A 16-bit oversampling A-to-D conversion technology using triple-integration noise shaping," *IEEE J. Solid-State Circuits*, vol. SC-22, pp. 921–929, 1987.

30. L. Longo and M. Copeland, "A 13 bit ISDN-band oversampling ADC using two-stage third order noise shaping," *IEEE 1988 Custom Integrated Circuits Conference*, pp. 21.2.1–21.2.4, 1988.

31. L. Williams and B. Wooley, "Third-order cascaded sigma-delta modulators," *IEEE Trans. Circuits Syst.*, vol. 38, pp. 489–498, 1991.

32. P.-W. Wong and R. Gray, "Two stage sigma-delta modulation," *IEEE Trans. Acoustics, Speech, Signal Process.*, vol. 38, pp. 1937–1952, 1990.

33. L. Longo and B.-R. Horng, "A 15b 30kHz bandpass sigma-delta modulator," *1993 IEEE International Solid-State Circuits Conference*, pp. 226–227, February 1993.

34. R. Schreier and W. M. Snelgrove, "Decimation for bandpass sigma-delta analog-to-digital conversion," *1990 IEEE International Symposium on Circuits and Systems*, vol. 3, pp. 1801–1804, May 1990.

35. R. Gregorian and G. Temes, *Analog MOS Integrated Circuits for Signal Processing*, Wiley, 1986.

36. D. Goodman and M. Carey, "Nine digital filters for decimation and interpolation," *IEEE Trans. Acoustics Speech Signal Process.*, vol. ASSP-25, pp. 121–126, 1977.

37. R. Crochiere and L. Rabiner, *Multirate Digital Signal Processing*, Prentice-Hall, 1983.

38. E. Hogenauer, "An economical class of digital filters for decimation and interpolation," *IEEE Trans. Acoustics Speech Signal Process.*, vol. ASSP-29, pp. 155–162, 1981.

39. S. Chu and C. Burrus, "Multirate filter designs using comb filters," *IEEE Trans. Circuits Syst.*, vol. CAS-31, pp. 913–924, 1984.

40. J. Candy, "Decimation for sigma delta modulation," *IEEE Trans. Commun.*, vol. COM-34, pp. 72–76, 1986.

41. B. Brandt and B. Wooley, "A low-power, area-efficient digital filter for decimation and interpolation," *IEEE J. Solid-State Circuits*, vol. 29, pp. 679–687, 1994.

42. T.C. Choi, R. Kaneshiro, R.W. Brodersen, P.R. Gray, W. Jett and M. Wilcox, "High-frequency CMOS switched-capacitor filters for communications applications", *IEEE J. Solid-State Circuits*, vol. 18, pp. 652–664, 1983.

43. W.C. Black, Jr., D.J. Allstot and R.A. Reed, "A high-performance low power CMOS channel filter", *IEEE J. Solid-State Circuits*, vol. 15, pp. 929–938, 1980.

44. D. Senderowicz, S.F. Dreyer, J.H. Huggins, C. Rahim and C.A. Laber, "A family of differential NMOS analog circuits for a PCM codec filter chip", *IEEE J. Solid-State Circuits*, vol. 17, pp. 1014–1023, 1982.

45. R. Castello and P.R. Gray, "A high-performance micropower switched-capacitor filter", *IEEE J. Solid-State Circuits*, vol. 20, pp. 1122–1132, 1985.

46. T.B. Cho and P.R. Gray, "A 10b, 20 Msample/s, 35 mW pipeline A/D converter," *IEEE J. Solid-State Circuits*, vol. 30, pp. 166–172, 1995.

47. K. Nagaraj, J. Vlach, T.R. Viswanathan and K. Singhal, "Switched-capacitor integrator with reduced sensitivity to amplifier gain", *Electron. Lett.*, vol. 22, p. 1103, 1986.

48. K. Huag, G.C. Temes and L. Martin, "Improved offset-compensation scheme for SC circuits," *IEEE International Symposium on Circuits and Systems*, pp. 1054–1057, 1984.

49. P.J. Hurst and W.J. McIntyre, "Double sampling in switched-capacitor delta-sigma A/D converters", *IEEE International Symposium on Circuits and Systems*, pp. 902–905, May 1990.

50. D. Senderowicz, G. Nicollini, S. Pernici, A. Nagari, P. Confalonieri and C. Dallavalle, "Low voltage double-sampled sigma-delta converters", *IEEE J. Solid-State Circuits*, vol. 32, pp. 1907–1919, 1997.

51. P. Furguson, Jr., A. Ganesan, R. Adams, S. Vincelette, R. Libert, A. Volpe, D. Andreas, A. Carpentier and J. Dattorro, "An 18b, 20kHz dual sigma-delta A/D converter", *IEEE International Solid-State Circuits Conference*, pp. 68–69, February 1991.

52. P.R. Gray and R.G. Meyer, "MOS operational amplifier design—A tutorial overview", *IEEE J. Solid-State Circuits*, vol. 17, pp. 969–981, 1982.

53. A. Abidi, C. Viswanathan, J. Wu, and J. Wikstrom, "Flicker noise in CMOS: A unified model for VLSI processes", 1987 Symposiun on VLSI Technology, pp. 85–86, May 1987.

54. A. Yukawa, "A CMOS 8-bit high speed A/D converter IC", *IEEE J. Solid-State Circuits*, vol. 20, pp. 775–779, 1985.

55. B. Kup, E. Dijkmans, P. Naus, and J. Sneep, "A bit-stream digital-to-analog converter with 18-b resolution," *IEEE J. Solid-State Circuits*, vol. 26, pp. 1757–1763, 1991.

56. R. Adams, K.Q. Nguyen, and K. Sweetland, "A 113-dB SNR oversampled DAC with segmented noise-shaped scrambling", *IEEE J. Solid-State Circuits*, vol. 33, pp. 1871–1878, 1998.

57. D. Su and B. Wooley, "A CMOS oversampling D/A converter with a current-mode semidigital reconstruction filter," *IEEE J. Solid-State Circuits*, vol. 28, pp. 1224–1233, 1993.

58. R. Schreier and B. Zhang, "Delta-sigma modulators employing continuous-time circuitry," *IEEE Trans. Circuits Systems—I: Fundamental Theory and Applications*, vol. 44, pp. 324–332, 1996.

59. E.J. van der Zwan and E.C. Dijkmans, "A 0.2-mW CMOS sigma-delta modulator for speech coding with 80dB dynamic range," *IEEE J. Solid-State Circuits*, vol. 31, pp. 1873–1880, 1996.

60. Y.P. Tsividis, "Integrated continuous-time filter design—an overview," *IEEE J. Solid-State Circuits*, vol. 29, pp. 166–176, 1994.

61. K.C. Hsieh, "Noise limitations in switched-capacitor filters", Ph.D. dissertation, University of California, Berkeley, 1981.

62. C. Gobet and A. Knob, "Noise analysis of switched-capacitor networks", *1981 International Symposium on Circuits and Systems*, April 1981.

63. C. Gobet and A. Knob, "Noise generated in switched-capacitor networks," *Electron. Lett.*, vol. 19, 1980.

64. E. Säckinger and W. Guggenbühl, "A high-swing, high-impedance MOS cascode circuit," *IEEE J. Solid-State Circuits*, vol. SC-25, pp. 289–298, 1990.

65. K. Bult and G.J.G.M. Geelen, "A fast-settling CMOS opamp for SC circuits with 90-dB DC gain," *IEEE J. Solid-State Circuits*, vol. SC-25, pp. 1379–1384, 1990.

66. J. Crols and M. Steyaert, "Switched-opamp: An approach to realize full CMOS switched-capacitor circuits at very low power supply voltages," *IEEE J. Solid State Circuits*, vol. 29, pp. 936–942, 1994.

67. V. Peluso, P. Vancorenland, A. Marques, M. Steyaert, and W. Sansen, "A 900-mV low-power DS A/D converter with 77-dB dynamic range," *IEEE J. Solid State Circuits*, vol. 33, pp. 1887–1897, 1998.

68. A. Baschirotto and R. Castello, "A 1-V 1.8MHz CMOS switched-opamp SC filter with rail-to-rail output swing," *IEEE J. Solid State Circuits*, vol. SC-32, pp. 1979–1986, 1997.

69. E. Bidari, M. Keskin, F. Maloberti, U. Moon, J. Steensgaard and G.C. Temes, "Low-voltage switched capacitor circuits", *Proceedings IEEE of the International Symposium on Circuits and Systems*, May 1999.

70. D. Ribner, R. Baertsch, S. Garverick, D. McGrath, J. Krisciunas, and T. Fuji, "A third-order multistage sigma-delta modulator with reduced sensitivity to nonidealities," *IEEE J. Solid-State Circuits*, vol. 26, pp. 1764–1774, 1991.

71. B. Brandt and B. Wooley, "A 50-MHz multibit sigma-delta modulator for 12-b 2-MHz A/D conversion," *IEEE J. Solid-State Circuits*, vol. 26, pp. 1746–1756, 1991.

72. J.W. Fattaruso, S. Kiriaki, M. de Wit and G. Warwar, "Self-calibration techniques for a second-order multibit sigma-delta modulator," *IEEE J. Solid-State Circuits*, vol. 28, pp. 1216–1223, 1993.

73. L. Carley, "A noise-shaping coder topology for 15+ bit converters," *IEEE J. Solid-State Circuits*, vol. 24, pp. 267–273, 1989.

74. B. Leung and S. Sutarja, "Multibit Σ-Δ A/D converter incorporating a novel class of dynamic element matching techniques," *IEEE Trans. Circuits Syst. II: Analog and Digital Signal Processing*, vol. 39, pp. 35–51, 1992.

75. F. Chen and B. Leung, "A high resolution multibit sigma-delta modulator with individual level averaging," *IEEE J. Solid-State Circuits*, vol. 30, pp. 453–460, 1995.

76. R. Baird and T. Fiez, "Linearity enhancement of multibit ΔΣ A/D and D/A converters using data weighted averaging," *IEEE Trans. Circuits Syst. II: Analog and Digital Signal Processing*, vol. 42, pp. 753–762, 1995.

77. L. Williams III, "An audio DAC with 90dB linearity using MOS to metal-metal charge transfer," *IEEE International Solid-State Circuits Conference*, pp. 58–59, February 1998.

78. T. Kwan, R. Adams, and R. Libert, "A stereo multi-bit ΣΔ D/A with asynchronous master-clock interface," *IEEE J. Solid-State Circuits*, vol. 31, pp. 1881–1887, 1996.

60

RF Communication Circuits

Michiel Steyaert

Wouter De Cock and

Patrick Reynaert

Katholieke Universiteit Leuven

CONTENTS

60.1 Introduction

During the last decade of the last century, the world of wireless communications started to grow rapidly. Today, cellular handsets are the largest consumer market in the world. The main trigger was the introduction of digital coding and digital signal processing in wireless communications. The aggressive scaling of CMOS process technology driven by the memory and microprocessor market made CMOS a logical choice for integration of digital signal processing in wireless applications. The development of these high-performance, low-cost CMOS technologies allowed integration of enormous amount of digital functionality on one chip. This enabled the use of sophisticated modulation schemes, complex demodulation algorithms, and high-quality error detection and correction to produce high data rate communication channels bringing the Shannon limit in sight [1].

The RF front-ends are the interface between the antenna and the digital modem of the wireless transceiver. They have to detect very weak signals (microvolts) that come in at a very high frequency (10 GHz), and at the same time transmit high power levels (up to several Watts) at the same high frequencies. This requires high-performance analog circuits, like filters, amplifiers, and mixers that translate the incoming modulated data between the antenna and the A/D conversion and digital signal processing. Consumer electronic markets are mainly driven by low cost and low power consumption. This makes the RF front-ends the bottleneck for future wireless applications. Low cost and low power are both linked to high integration level. A high level of integration renders a significant space, cost, weight, and power reduction. A higher degree of integration requires less discrete components reducing the bill of cost of materials. Keeping signals on chip greatly reduces power consumption since less I/O drivers are needed. Many different techniques to obtain a higher degree of integration have been presented over the years [2–5]. This chapter introduces and analyzes some advantages and disadvantages and their fundamental limitations.

Parallel to the trend for further integration, there is the trend to integrate RF circuitry in CMOS technologies. While digital baseband processing has already been implemented in CMOS technology in several product generations, CMOS RF has only recently made progress. For a long time many design houses believed that complicated mixed-signal RF CMOS chips were impossible to realize. The main objective against CMOS RF was the lack of high-Q passive components and its poor noise performance. It took the persistence of some academic institutions and some pioneering firms to prove them wrong. It is clear that the full potential of RF CMOS would not have been unfolded, if only stand-alone radios were developed. CMOS RF systems-on-chip (SoC) today implement all radio building blocks including phase-locked loop (PLL), low-noise amplifier (LNA), power amplifier (PA), up- and downconversion mixers, filters, and antenna switch. Furthermore, they include all digital baseband processing circuitry and ROM memory [6,7]. This reveals the real strength of CMOS RF over other "better suited" technologies like Si bipolar, BiCMOS, and silicon germanium (SiGe). Putting together RF and baseband in one chip permits compensation of lower radio performance with less-expensive digital signal-processing circuits, making its performance competitive with SiGe radios. Together with a possible 75% reduction of discrete components, RF CMOS offers the cheapest solution if one pursues the ultimate goal: a single chip including the physical layer (PHY) as well as the media access control (MAC) together with a MAC processor, memory, and I/O such as USB ports or PCI interfaces.

RF CMOS is not a matter of just replacing bipolar transistors with their CMOS counterpart. It requires a whole new range of architectures, techniques, and a high integration level. When compared to CMOS, SiGe requires less power for a certain gain and achieves a lower noise figure. The biggest drawback of CMOS is its inferior $1/f$ noise performance. This will only increase with the introduction of high-K dielectric materials in the gate of future CMOS technology nodes. CMOS design engineers therefore went looking for new topologies to reduce the impact of $1/f$ noise on the radio performance. Another problem that had to be overcome was the lack of high-Q passive components in CMOS technology. Extra processing steps as well as innovative layout and design techniques solved this problem. First, this chapter will analyze some concepts, trends, limitations, and problems posed by technology for high-frequency design. Next, we will discuss a variety of architectures used in modern RF CMOS transceivers. In the rest of the chapter, we will take a closer look at the different building blocks that appear in a typical RF transceiver. We will split this into downconversion, upconversion, and frequency synthesis.

In a final section, we will take a look at RF CMOS's last barrier: RF power transmission. As CMOS gate lengths shrink, lower voltages are tolerated at the transistor terminals. High-quality impedance converters must therefore be placed between the antenna and the transistor's drain for high power transmissions. These are not available yet in integrated form. One of the major bottlenecks in CMOS PAs is combining high efficiency with high linearity. For high-power transmission, designers are obliged to bias the PA high in its saturation region where linearity is low. Therefore, today's integrated PAs are limited to constant envelope modulation schemes like GSM. High-efficiency PAs still remain out of reach for modulation schemes with large peak-to-average power ratios like OFDM. This chapter will discuss some circuit techniques to circumvent this bottleneck bringing the ultimate goal of a single-chip CMOS solution that is compatible with all standards and is capable of adapting itself a step closer to reality.

60.2 System Level RF Design

60.2.1 General Overview

One of the main challenges facing the RF design engineer originates from the transmission medium used by RF systems. RF systems communicate through air by means of electromagnetic waves. Using air as transmission medium has one great advantage: it gives the transceiver the ability to be mobile. However, there are some disadvantages to this high degree of freedom. There exists only one medium i.e., air which is consequently used by numerous applications. An overview of these applications and the part of the spectrum they use can be found on the website of the National Telecommunications and Information Administration (NTIA) [8]. As a result, RF systems operate in a filled spectrum. Receivers will not only detect the desired signals indigenous to the application, but will also pick up other signals that will consequently be amplified and detected. These unwanted detected signals are called interferers. If the interferer is sufficiently large, it can corrupt the desired signals preventing them from being properly demodulated and understood. On the transmit side of the application, unwanted signals are generated and transmitted. They are picked up by other applications and can distort their performance. These unwanted transmitted signals are called spurious signals. It is the designer's responsibility to keep these inteferers and spurious signals as low as possible. Based on the former discussion, it is clear that one needs a regulator to manage the use of this spectrum. In the United States this is done using a dual organizational structure; NTIA manages the Federal Government's use of the spectrum while the Federal Communications Commission (FCC) [9] manages all other uses.

Signals travelling through air also suffer from attenuation. There are several mechanisms causing attenuation such as free-space dispersion, fading, and multipath. These mechanisms depend heavily on the distance between the transmitter and receiver, the frequency of transmission, and the environment. Discussion of these mechanisms, however, is beyond the scope of this text. More information concerning these topics can be found in [10,11]. As a result of these mechanisms, one can expect the received signal power to have a large variation since the distance between the transmitter and receiver can change considerably owing to the mobility. Performance of RF communication systems is also degraded by thermal noise. Noise, like in other communication systems, is the limiting factor when dealing with weak signals. The noise energy consists of two contributors. First, there is thermal noise which is determined by temperature and bandwidth and is beyond the control of the designer. Second, there is system noise. This kind of noise can, within limits, be controlled by the designer to allow a certain minimum level of signal power to be detected by the system.

In the following sections we will take a closer look at the challenges described in the former discussion. First, we will take a brief look at the tools and metrics RF designers use to describe and control the performance of their system in the presence of interferers and noise. We will end this section with a discussion of some commonly used transceiver architectures.

60.2.2 RF System Performance Metrics

As described in Section 60.2.1, the lowest signal power level that can be detected correctly by a receiver is limited by noise. The lowest power level that can be detected is usually called the receiver sensitivity. The receiver sensitivity is related to the signal-to-noise ratio (SNR) at the end of the receiver chain (baseband). The SNR at the baseband is determined by the bit error rate (BER) required by the application. It is usually expressed in terms of E_b/N_o. E_b is the energy per received bit and N_o is the noise power density received together with the bit. The relation between E_b/N_o and BER depends on the modulation scheme used in the application (e.g., GMSK in GSM) and is beyond the scope of this text. More information can be found in Ref. [12]. The SNR can be expressed as a function of E_b/N_o as follows:

$$\text{SNR} = \frac{S}{N} = \frac{E_b}{N_o} \times \frac{f_b}{B} \tag{60.1}$$

where f_b is the bit rate and B the receiver noise bandwidth. Note that the overall system noise at baseband N is the sum of the thermal and circuit noise. This leads to a figure of merit that describes the circuit's performance. It is called noise figure when expressed in decibel and noise factor otherwise. Noise factor or figure is a measure of the excess noise that is contributed by the circuit to the overall noise and is defined as the ratio between the SNR at the input of the receiver (SNR$_i$) and the SNR at the output of the receiver (SNR$_o$):

$$NF = \frac{SNR_i}{SNR_o} = \frac{(S/N)_i}{(S/N)_o} \tag{60.2}$$

If the receiver consists of different building blocks, one may want to know the noise figures of the different blocks and not only the overall noise figure. One can prove that in case of a series connection [12]

$$NF_{total} = NF_1 + \frac{NF_2 - 1}{G_1} + \frac{NF_3 - 1}{G_1 G_2} + \ldots \tag{60.3}$$

in which NF$_i$ are the noise factors of successive building blocks and G_i is their respective power gain. One can easily conclude from Eq. (60.3) that building blocks earlier in the receiver chain have a larger contribution to the overall noise figure than blocks at the end of the chain. This is the reason behind the use of an LNA at the input of an RF receiver. The large power gain combined with a low noise figure will relax the noise specifications for the following blocks. This principle is explained in Figure 60.1. If an LNA is omitted and the mixer is put directly behind the antenna, the signal is drowned in the mixer noise and the sensitivity will be low. The power gain of the LNA however pushes the antenna signal above the noise floor of the mixer. As long as the output noise of the LNA is greater than the input noise of the mixer, the sensitivity is fully determined by the NF of the LNA.

RF systems often operate in an interference-limited environment. Interference can also reduce receiver sensitivity. It is therefore more correct to describe the receiver sensitivity by its signal to (noise plus interference) ratio S/(N + I) also known as the signal to noise and distortion ratio (SNDR). One of the mechanisms by which interference limits the performance is nonlinearity. It can reduce the signal power as well as increase interference. Large signals can saturate the receiver resulting in a gain compression,

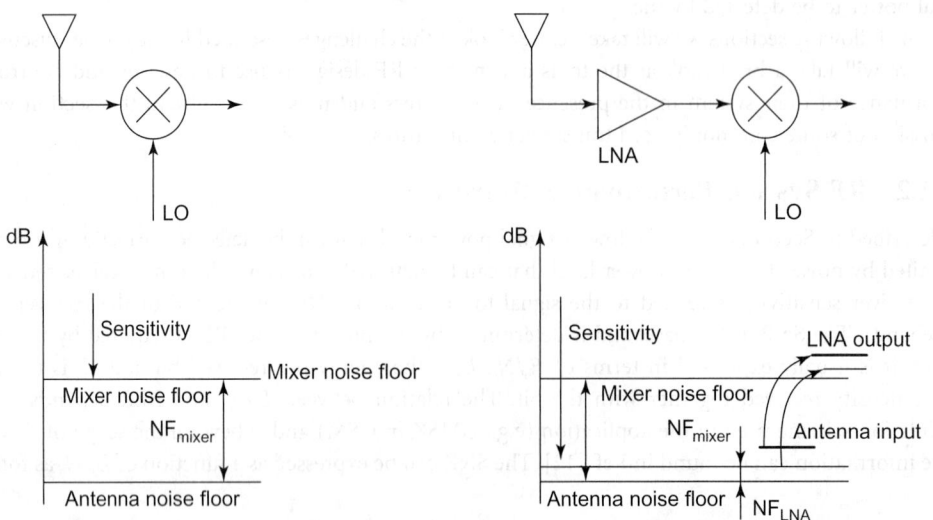

FIGURE 60.1 Benefit of using an LNA.

which reduces the signal power S. In contrast, two large interfering signals can, owing to nonlinearity produce cross-product terms that overlap with the desired signal, increasing the interference I. This cross-product generation is called intermodulation distortion (IMD). Nonlinearity performance is typically characterized by small-signal linearity described by second- and third-order intercept points (IP2 and IP3) and large-signal linearity described by the 1 dB compression point. Usually, balanced topologies are used in attenuating the second-order harmonics. Consequently, third-order nonlinearity will become the limiting factor. These concepts will be explained with help of Figure 60.2. Gain compression is characterized by the 1 dB compression point (P_{-1dB}) and is used to evaluate the ability of the system to cope with strong input or interference signals often referred to as blockers. It is defined as the input power for which the gain drops by 1 dB. By identifying the strongest signals at each stage of the design, one can calculate the required 1 dB compression point for each block in a receiver chain. As mentioned earlier, nonlinearity not only causes gain compression, but also generates IMD. This is produced by any pair of blockers that lie near the desired signal. If two tones at f_1 and f_2 are applied to a nonlinear block, frequencies are produced not only at f_1 and f_2, but also at $2f_1 - f_2$, $2f_2 - f_1$, $3f_1$, $3f_2$, and so on. f_1, f_2, $3f_1$, and $3f_2$ are not important since they lie far outside the frequency band of interest and can therefore be filtered out. $2f_1 - f_2$ and $2f_2 - f_1$ however are potential problems as they can overlap with the desired signal band and remain unaffected by filtering. The ratio of any of the two cross products is called third-order IMD (IMD3). The output power of the intermodulation products grows at a faster rate than that of the desired signal itself. Therefore, it follows that at a certain input power, the output power of the intermodulation signals will surpass the desired signal. The input power level, where this takes place is called the input-referred third-order intercept point (IIP3). The output power at this point is called the output-referred third-order intercept point (OIP3). Note that this is an imaginary point since gain compression occurs before this point is reached. If the receiver consists of different building blocks, one may want to know the contribution of the different building blocks to the overall linearity performance. One can prove that in case of a cascaded system

$$\frac{1}{IIP3_{total}} = \frac{1}{IIP3_1} + \frac{G_1}{IIP3_2} + \frac{G_1 G_2}{IIP3_3} + \ldots \tag{60.4}$$

where $IIP3_i$ are the input-referred third-order intercept points of the successive building blocks and G_i their respective power gain. One can conclude that contrary to noise (see Eq. (60.3)), the last blocks in the receiver chain has the largest influence on the overall linearity of the receiver. Eqs. (60.3) and (60.4) reveal a first trade-off. High gain at the input reduces noise constraints in the rest of the chain, but increases the linearity requirements.

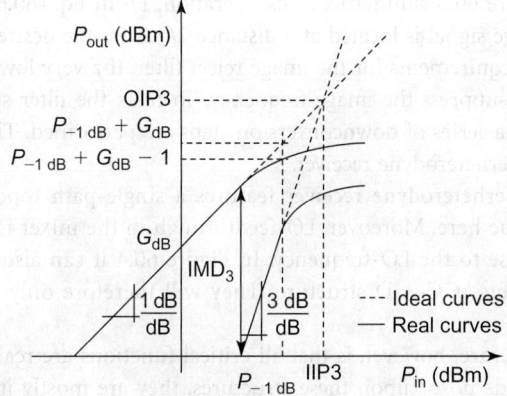

FIGURE 60.2 First- and third-order intermodulation as a function of the input power.

A last origin of distortion is due a nonideal local oscillator (LO) signal driving the mixers. In practice, the spectrum of an oscillator is never pure. There is always a certain amount of energy present close to the ideal LO frequency at $\omega_0 + \Delta\omega$. This can translate nearby frequency signals overlapping with the desired signal also deteriorating the SNDR of the system. A figure of merit to describe this nonideal LO behavior is called the LO phase noise and is defined as the ratio of the power present in a 1 Hz band at a certain offset frequency $\Delta\omega$ from the carrier frequency ω_0 to the carrier power:

$$\mathcal{L}\{\Delta\omega\} = 10 \log \left(\frac{\text{Noise power in a 1 Hz band at } \omega_0 + \Delta\omega}{\text{Carrier power}} \right) \tag{60.5}$$

60.2.3   RF Transceiver Architectures

In this section, a brief overview of some common transceiver structures will be discussed and contrasted with one another. The discussion will be restricted to the heterodyne transceiver, the zero-intermediate frequency (IF) or direct-conversion transceiver, and the low-IF transceiver. There exist numerous other types of transceivers, but their properties can be understood by looking at these three structures as they are all variations or combinations of the same. First, the different receiver architectures will be discussed followed by their transmitter equivalents.

The heterodyne receiver has been the dominant choice among RF systems for many decades. The reason behind this is its high performance and adaptability to different standards. Figure 60.4 shows the operation of a heterodyne receiver. The broadband antenna signal is first fed to a highly selective RF filter (band select filter), that suppresses all interferers outside the desired application band. An LNA boosts the desired signal above the mixer noise floor and an LO generates a signal located at an offset frequency f_{IF} from the desired signal. The result is that the following signals are downconverted by the mixer to f_{IF}:

$$f_{\text{desired}} = f_{LO} - f_{IF} \tag{60.6}$$

$$f_{\text{image}} = f_{LO} + f_{IF} \tag{60.7}$$

Not only is the desired signal mapped onto IF but also another signal called the image or mirror signal. This signal can corrupt the information content is such a way that the information is irreparable. To avoid this, an image reject filter is inserted before the mixer. This way, a highly attenuated version of the image signal overlaps with the desired signal, preventing the irreparable corruption of the information content of the signal. Figure 60.3 summarizes this operation. From Eq. (60.6) and (60.7), one can see that the center of the image signal is located at a distance $2f_{IF}$ from the desired signal. The choice of f_{IF} therefore determines the requirements for the image reject filter. If a very low f_{IF} is chosen, a very high-quality filter is needed to suppress the image frequency. To relax the filter specifications, f_{IF} is usually chosen relatively high and a series of downconversion steps are performed. The heterodyne structure is then referred to as the superheterodyne receiver.

The heterodyne or superheterodyne receiver features a single-path topology. Mismatch between different parts is not a issue here. Moreover, LO-feedthrough in the mixer is not a problem, since the desired signal is never close to the LO-frequency. In Figure 60.4 it can also be seen that the channel selection is done before the AGC-A/D structure. They will therefore only need to handle a limited dynamic range.

A drawback of the structure, however, is that all critical functions are realized with passive devices. Owing to the high demands posed upon these structures, they are mostly implemented off-chip. The integratability of the heterodyne transceiver is therefore rather low. This incurs an additional material cost. Moreover, the insertion loss of the passive filters needs to be compensated by a higher gain on-chip

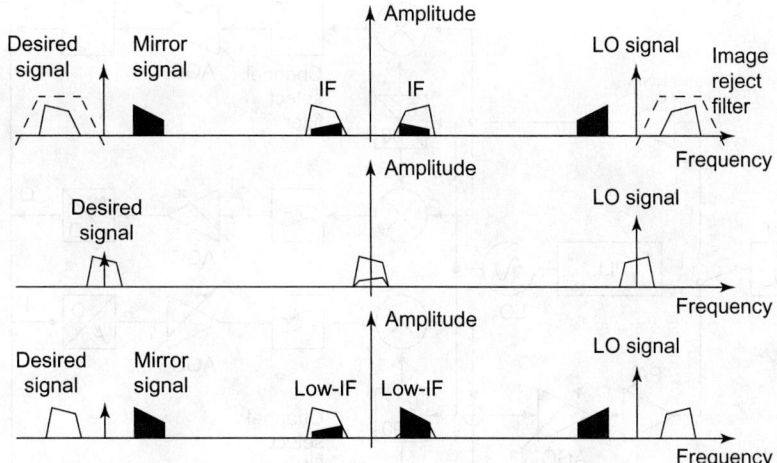

FIGURE 60.3 Downconversion process in an IF, zero-IF, and a low-IF receiver.

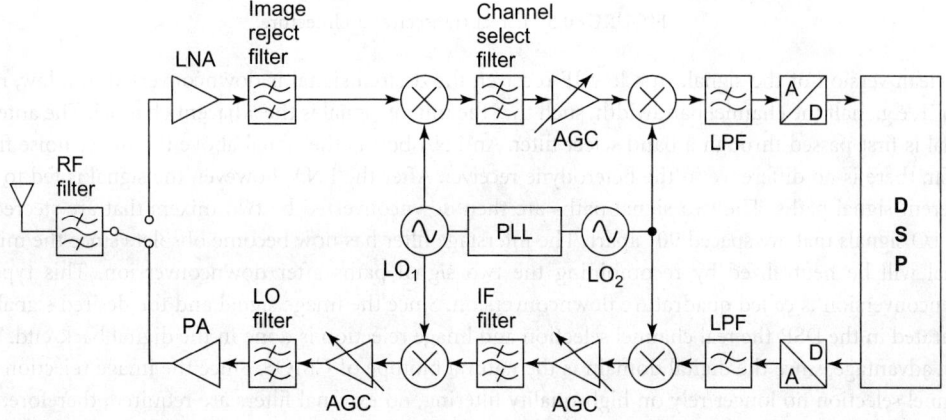

FIGURE 60.4 Heterodyne transceiver architecture.

to retain the required SNR. Since the filters need to be driven at low impedance (e.g., 50 Ω), one has the choice between using complex impedance transformation structures or using low-output impedance buffers. Using low-output impedance drivers, however, comes at the cost of an extra amount of power consumption.

The integratability, however, can dramatically be improved if one could find a way of getting rid of the external high-quality filters. This means looking for a way of suppressing the image frequency without filters. A first solution to this problem is obvious. Make the image signal the desired signal or chose $f_{IF} = 0$. This solution is called the zero-IF receiver or direct conversion receiver [13,14]. Another solution is related to the first one and is called the low-IF topology [3]. This topology takes advantage of the fact that the channels in the direct neighborhood of the desired channel—the adjacent channels—are usually much weaker than the desired signal and the signals lying further away. Furthermore, these frequency bands are usually regulated in the application specifications or by the FCC. So, if an IF-frequency is chosen such that the image frequency falls into this lower power bands, less image rejection is needed to retain the required SNR. Figure 60.5 shows the architecture of both a direct or zero-IF receiver and a low-IF receiver. The only difference between the two can be found in the choice of IF-frequency. In a zero-IF receiver, the desired channel is converted to DC and a mirrored version of the channel itself is superimposed onto

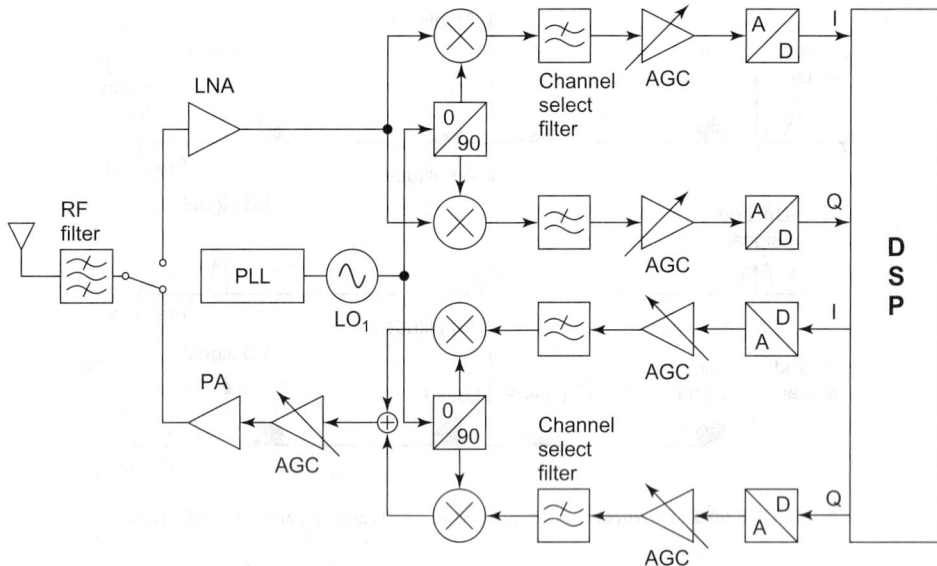

FIGURE 60.5 Direct transceiver architecture.

the clean version of the signal. In a low-IF receiver, the desired signal is downconverted to a low, non-zero IF, e.g., half the channel bandwidth, such that the mirror signal is the adjacent channel. The antenna signal is first passed through a band select filter. An LNA boosts the signal above the mixer noise floor. So far, there is no difference in the heterodyne receiver. After the LNA, however, the signal is fed to two different signal paths. The two signal paths are then downconverted by two mixers that are steered by two LO signals that are spaced 90° apart. The interstage filter has now become obsolete since the mirror signal will be neutralized by recombining the two signal paths after downconversion. This type of downconversion is called quadrature downconversion. Since the image signal and the desired signal are separated in the DSP, the real channel selection and image rejection is done in the digital back-end. This is an advantage, since the digital domain is the natural biotope of CMOS. Since the image rejection and channel selection no longer rely on high-quality filtering, no external filters are required, therefore, one does not have to cope with their inevitable loss and one does not need low-impedance drivers. This allows low-power operation. However, the spreading of the signal over two independent signal paths has some drawbacks. The topology relies heavily on the symmetry between the two paths, every mismatch between the two paths will lead to a deterioration of the image suppression and an increased corruption of the desired information content. Although one could think that image rejection requirements are more relaxed for a zero-IF receiver since the image signal is a mirrored version of the desired signal, this is not exactly true. For low-IF receivers, the image signal can be considered as noise for the desired signal, since there is no correlation at all between the two bands. For zero-IF receivers, there is a strong correlation between the image and the desired signal leading to a distortion of the desired signal. The required image suppression is therefore dependent on the type of modulation that is used in the system. When a QAM-type modulation is used, one can calculate that the required image rejection for zero-IF is 20–25 dB while 32 dB rejection is required for low-IF systems [15]. As the desired signals in both receivers are located at low frequencies (DC in case of zero-IF), the signal is susceptible to $1/f$ noise and DC-offset. Complicated feedback structures can get rid of the DC-offset, however, owing to the finite time constants in those loops, part of the signal is also canceled by the feedback. This can corrupt the signal in an unacceptable way. Low-IF topologies are less vulnerable. As long as the DC-offset does not saturate the A/D converters, there is no signal degradation. Owing to the absence of filtering in the RF part, the A/D converters, however, have to deal with larger dynamic ranges. Fortunately, as the signals are at low frequencies, oversampled converters that allow higher accuracies can be used.

The same topologies exist for the transmitter side of the transceiver. The heterodyne as well as the direct upconversion transmitter will be discussed. They are depicted in Figure 60.4 and Figure 60.5. The early upconversion architectures were in fact multistage architectures. They employed a number of mixing stages and intermediate frequencies. The main advantage of this type of systems is that only one D/A converter is needed. Quadrature modulated signals are therefore generated in the digital domain. This topology puts high demands on the D/A converter since it must deliver signals at a higher IF frequency. In contrast, the DSP must be able to deliver perfectly matched I/Q signals. This approach requires the use of high-quality passives, multiple LOs. The same conclusions can be drawn as in the receiver. Owing to the large number of external components, integratability is limited and power consumption will be high. Another implementation of this multistage architecture includes the use of two D/A converters. Quadrature modulated signals are then generated in the analog domain. Since they are generated at low frequencies, quadrature matching is superior. However, multiple RF filters are still needed, giving rise to a higher cost and power consumption. The topology, however, is not vulnerable to one of the main problems in the direct conversion architecture, oscillator pulling caused by the PA owing to the fact that the PA output spectrum is far away from the voltage controlled oscillator (VCO) frequency. Thus, the main problem in direct upconversion circuits is addressed. In direct conversion transmitters, the transmitted carrier frequency is equal to the LO frequency. As can be seen in Figure 60.5, modulation and upconversion occur in the same circuit. The I/Q quadrature modulator takes the baseband (or low-IF) input signal and upconverts it directly to the desired RF frequency. This eliminates the need for RF passives and limits the number of amplifiers, mixers, and LOs. The simplicity of the architecture makes it an obvious choice when high integration levels are demanded. However, as mentioned before, the circuit suffers from one major drawback, the disturbance of the LO by the PA. This phenomenon is explained in more detail in Refs. [16,17]. As the LO frequency lies in the transmit band, high demands are put on the LO/RF isolation. The system is also susceptible to I/Q mismatch errors, even the least phase mismatch or amplitude difference between I and Q path will result in distortion in the spectrum. However, the elimination of the IF stage in the transmitter leads to large saving in material cost and increases the robustness of the system as the number of discrete components that could fail is reduced. There is not only a cost saving in material cost, the direct upconverter architecture also allows a reduction in equipment size. This makes the circuit the first choice for applications with stringent space constraints [18] .

60.3   Technology

60.3.1   Active Devices

Since all high- or system-level designs in the end need to be implemented in terms of actual active and passive components, it is not surprising that the performance of the transistor is of major importance for the overall system performance. It is therefore imperative to know the performance limitations of the technology that one is working with and to be aware of the shortcomings of the model that one is using. It is clear that conformity between measurements and simulation results will strongly depend on the accuracy of the models used with respect to the actual behavior of the devices. Although several compact models exist to describe MOSFET transistors, the BSIM [19] is considered as the de-facto standard because it is the model that is generally provided by silicon foundries. Most models are quite accurate for low frequencies; however, most models fail when higher frequencies are to be modeled. "High frequency" means operating frequencies around one-tenth of the transistor's cutoff frequency f_t. Figure 60.6 gives an overview of f_t for different technology nodes. For example, for a standard 0.18 μm CMOS technology with an f_t of ~50 GHz, 5 GHz is considered to be a high frequency. Another parameter is plotted in Figure 60.6, f_{3dB} reflects the speed limitation of a transistor in a practical configuration. It is defined as the 3 dB point of a diode-connected transistor [20] and takes into account the parasitic speed limitation owing to overlap capacitances, drain-bulk junction, and gate-source capacitance while f_t only models the parasitic effect of the gate-source capacitance. In Ref. [21] an extended transistor model that can be used for circuit simulation

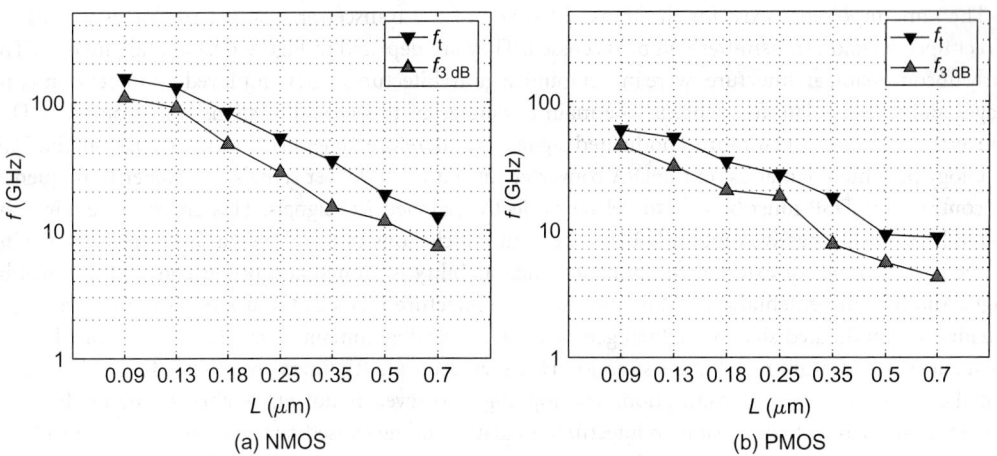

FIGURE 60.6 Maximum operating frequencies for different technology nodes.

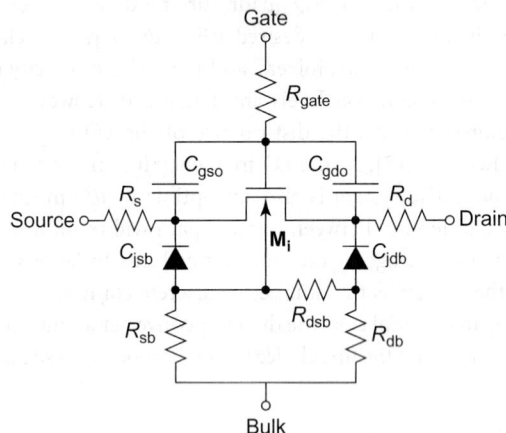

FIGURE 60.7 Extended RF transistor model.

at RF frequencies is presented. It is shown in Figure 60.7. All the extrinsic components are pulled out of the MOS transistor model, so that the MOS transistor symbol only represents the intrinsic part of the device. This allows to have access to internal nodes and model extrinsic components such as series resistances and overlap capacitances in a way different from what is available in the complete model. The source and drain series resistors are added outside the MOS model since the series resistances internal to the compact model is only used in the calculation of the I–V characteristic to account for the DC voltage drop across the source and drain. They do not add any poles and are therefore invisible for AC simulation. The gate resistance is usually not part of a MOSFET model, but plays a fundamental role in RF circuits and is therefore of utmost importance. The substrate resistors R_{dsb}, R_{sb}, and R_{db} have been added to account for the signal coupling through the substrate.

Apart from the extra components added in the extended transistor model presented in Ref. [21], another point deserves some attention. The classical transistor model is based on the so-called quasi-static assumption. This means that any positive (negative) change in charge at the gate is immediately compensated by a negative (positive) change of charge in the channel. In reality, however, there will always be a delay in the charge buildup in the channel. Individual electrons (holes), will need a finite time to travel from the bulk to the channel. This effect is called the nonquasi-static effect and has been described in Refs. [22–24]. This effect can be modelled by adding a resistance in series with the gate-source capacitance, introducing an extra time constant in the model:

$$\tau_{gs} = \frac{C_{gs}}{5g_m} = \frac{1}{5\omega_t} \quad (60.8)$$

This model is valid in strong inversion and within the long-channel approximation. Although one could think that this effect is negligible at realistic operating frequencies much lower than f_t, in bandpass applications, the gate-source capacitance can be tuned away by an inductor making the input impedance of the transistor purely resistive.

60.3.2 Passive Devices

For a long time, CMOS RF integration was believed to be impossible owing to the poor quality of passive devices. Smaller CMOS geometries and innovative design and layout [25–27], however, have enabled high-quality passive components at high frequency to be integrated on chip. Four passive devices (resistors, inductors, capacitors, and varactors) will be discussed. First, one needs figure of merit to qualify these passive devices. In general, the Q-factor is used for this purpose. Although there exist several definitions for the Q-factor, the most fundamental definition is based on the ratio between the maximum energy storage and the average power dissipation during one cycle in the device:

$$Q = \frac{\omega W_{max}}{P_{diss}} \quad (60.9)$$

For an overview of other definitions of the Q-factor, the reader is referred to Ref. [28]. For a purely reactive element (capacitor or inductor), current through the element and voltage over the element are 90° out of phase. Hence no power is dissipated in it. In reality, however, a certain amount of power will always be dissipated. Power dissipation assumes the presence of a resistance and a resistance always generates thermal noise. The Q-factor consequently is also a way of describing the pureness of a reactive device. Figure 60.8 shows some very common structures used in the modeling of reactive components used in RF circuits together with their Q-factor according to definition (60.9). Low-ohmic resistors are commonly available now in all CMOS technologies and their parasitic capacitance is such that they allow for more than high enough bandwidth. A more important passive device is the capacitor. In RF circuits, capacitors can be used for AC coupling. This enables DC-level shifting between different stages resulting

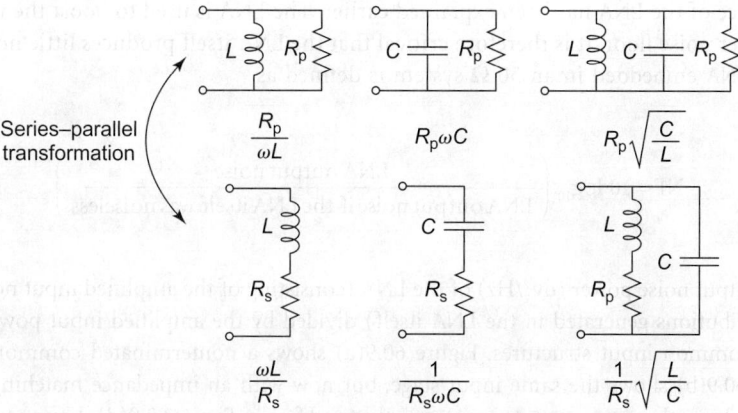

FIGURE 60.8 Quality factors of some common circuits.

in an extra degree of freedom enabling an optimal design of each stage. It also offers the possibility of lowering the power supply voltages. Another field, although not completely RF, where capacitors are commonly used is to implement switched capacitor circuits or arrays. This is more favorable than using common resistors since capacitors in general offer better matching properties than resistors do. The quality of an integrated capacitor is mainly determined by the ratio between the capacitance value and the value of the parasitic capacitance to the substrate. Too high a parasitic capacitor loads the transistor stages, thus reducing their bandwidth, and it causes an inherent signal loss owing to a capacitive division.

The passive device however that got the most attention in the past is the inductor. It was long believed that high-quality integrated inductors were simply impossible in standard CMOS processes [29] and were rather avoided if possible. However, owing to the use of hollow spiral inductors and slightly altered process technology (thick top metal layer), one is now able to produce high-Q inductors in CMOS. The use of inductors on chip allows a further reduction of the power supply and offers compensation for parasitic capacitors by tuning them away resulting in higher operating frequencies.

To be able to use integrated inductors in actual designs, an accurate model is needed. Ref. [30] introduces such a model. One of the problems faced when modeling an inductor is how to model the substrate. One of the major drawbacks of inductors are the losses introduced by the substrate underneath the coil by capacitive coupling and eddy currents. This reduces the quality factor of the inductor.

A last passive component that is often encountered in RF CMOS designs is the varactor. It is mostly used for implementing tunable RF filters and VCOs. Varactors can be divided into two classes: junctions and MOS capacitors. The latter can be used in accumulation and in inversion mode. For all cases, the devices have to be placed in a separate well to be able to use the well potential as the tuning voltage. For a standard n-well process, the available configurations are therefore limited to p^+/n^- junction diodes and PMOS capacitors. When comparing the different varactor types, one should look at the following specifications: the varactor should offer a high Q-factor, the tuning range over which the capacitance can be varied should be compatible with the supply voltages used in the design, the physical structure should be as compact as possible to limit die area, and its capacitance variation should be uniform over the complete tuning range as this makes feedback design more easy. For an extended discussion on the different types of varactor and their performance, the reader is referred to Ref. [27].

60.4 Receiver

60.4.1 LNA

The importance of the LNA has been explained earlier. The LNA is used to boost the received signal above the mixer noise floor. It is therefore critical that the LNA itself produces little noise. The noise figure of an LNA embedded in an 50 Ω system is defined as

$$\text{NF} = 10 \log_{10} \left(\frac{\text{LNA output noise}}{\text{LNA output noise if the LNA itself was noiseless}} \right) \qquad (60.10)$$

i.e., the real output noise power (dv^2/Hz) of the LNA (consisting of the amplified input noise power and all noise contributions generated in the LNA itself) divided by the amplified input power. Figure 60.9 shows some common input structures. Figure 60.9(a) shows a nonterminated common-source input stage. Figure 60.9(b) shows the same input stage, but now with an impedance matching at the input. Figure 60.9(c) shows the common-gate input structure and finally, Figure 60.9(d) shows a transimpedance amplifier structure that is commonly used for wideband applications. Their respective noise figures can

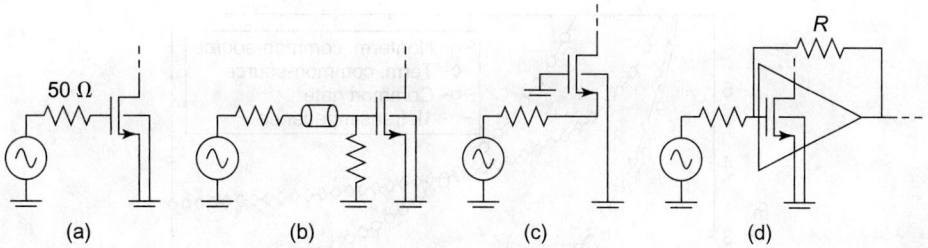

FIGURE 60.9 Some common LNA topologies.

be approximated with the following equations:

Common-source nonterminated (Figure 60.9[a]): $\quad \mathrm{NF} = 1 + \dfrac{1}{50 g_\mathrm{m}}$ (60.11)

Common-source terminated (Figure 60.9[b]): $\quad \mathrm{NF} = 2 + \dfrac{1}{50 g_\mathrm{m}}$ (60.12)

Common-gate (non)terminated (Figure 60.9[c]): $\quad \mathrm{NF} = \left[\dfrac{1 + 50 g_\mathrm{m}}{50 g_\mathrm{m}}\right]^2 + \dfrac{1}{50 g_\mathrm{m}}$ (60.13)

Common-source transimpedance (Figure 60.9[d]): $\quad \mathrm{NF} = 1 + \dfrac{1}{50 g_\mathrm{m}}\left[\dfrac{R + 50}{R}\right]^2 + \dfrac{50}{R}$ (60.14)

Figure 60.10 compares the noise figures of the different topologies. It is clear that the transimpedance structure and the nonterminated common-source circuit are far more superior than the other structures as far as noise is concerned. For those circuits the NF can be approximated as

$$\mathrm{NF} - 1 \approx \frac{1}{50 g_\mathrm{m}} = \frac{(V_\mathrm{gs} - V_\mathrm{t})}{(2)(50)I}$$ (60.15)

indicating that a low-noise figure requires a large transconductance in the first stage. To generate this transconductance with high power efficiency, we need to bias the transistor in the region with a large transconductance efficiency i.e., low $V_\mathrm{gs} - V_\mathrm{t}$. This, however, will result in a large gate-source capacitance limiting the bandwidth of the circuit. Together with the 50 Ω source resistance, the achievable bandwidth is limited by

$$f_\mathrm{3dB} = \frac{1}{2\pi 50 C_\mathrm{gs}}$$ (60.16)

When using the well-known approximative expression for the cutoff frequency of a transistor f_T,

$$f_\mathrm{T} = \frac{g_\mathrm{m}}{2\pi C_\mathrm{gs}}$$

one can conclude that

$$\mathrm{NF} - 1 = \frac{f_\mathrm{3dB}}{f_\mathrm{T}}$$ (60.17)

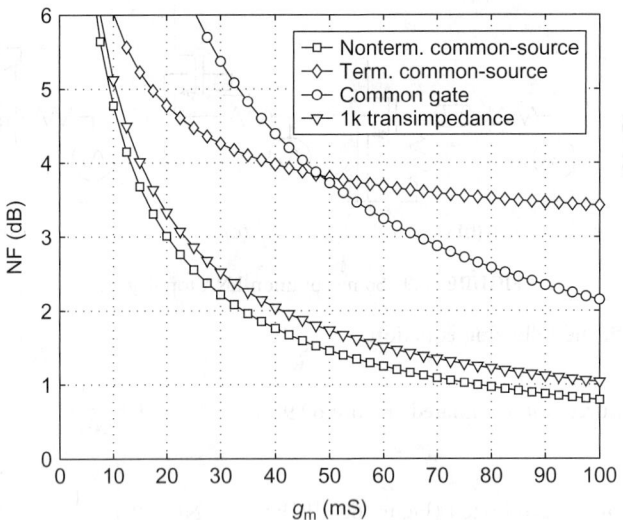

FIGURE 60.10 LNA input structure performance comparison.

This means that a low-noise figure can only be achieved by making a large ratio between the frequency performance of a transistor, represented by f_T and the theoretical bandwidth f_{3dB} of the circuit. Note that the f_{3dB} used here is not the same as the one used is Section 60.3. Since f_T is proportional to $V_{gs} - V_t$, a low-noise figure requires a large $V_{gs} - V_t$ and associated with it a large power drain. Only by going to deep submicron technologies will f_T become large enough to achieve low-noise figures for gigahertz operation with low power consumption. In practice, the noise figure is further optimized by using noise and source impedance matching. These matching techniques often rely on inductors to cancel out parasitics by creating resonant structures. This boosts the maximum operation frequency to higher frequencies. More information concerning the design and optimization of common source LNAs can be found in Refs. [31,15].

At high antenna input powers, the signal quality mainly degrades owing to in-band distortion components that are generated by third-order intermodulation in the active elements. Long-channel transistors are generally described by a quadratic model. Consequently, a one-transistor device ideally only suffers from second-order distortion and produces no third-order intermodulation products. As a result, high IIP3 values should easily be achieved. When transistor lengths shrink however, third-order intermodulation becomes more important.

To start the analysis of the main mechanisms behind third-order intermodulation one needs an approximate transistor model. A drain current equation that is strongly related to the SPICE level 2 and 3 models is

$$I_{ds} = \frac{\mu_0 C_{ox}}{2n} \frac{W}{L} \frac{(V_{gs} - V_t)^2}{1 + \Theta(V_{gs} - V_t)} \tag{60.18}$$

with

$$\Theta = \theta + \frac{\mu_0}{L_{eff} v_{max} n} \tag{60.19}$$

where θ stands for the mobility degradation owing to transversal electrical fields (surface scattering at the oxide–silicon interface) and the $\mu_0/(L_{eff} v_{max} n)$ models, the degradation owing to longitudinal fields (electrons reaching the thermal saturation speed). As the θ-term is small in today's technologies, it can often be neglected relative to the longitudinal term. It can be seen from Eq. (60.18) that for large values of $V_{gs} - V_T$, the current becomes a linear function of $V_{gs} - V_T$. The transistor is then conducting in the velocity saturation region. For smaller values of $V_{gs} - V_T$, the effect of Θ consists apparently in linearizing the quadratic relationship, but in reality, the effect results in an intermodulation behavior that is worse than in the case of quadratic transistors. The second-order modulation will be lower, but is realized at the cost of a higher third-order intermodulation. The following equations can be found by calculating the Taylor expansions of the drain current around a certain $V_{gs} - V_T$ value [32]:

$$\text{IIP2} \cong 10 + 20 \log_{10}((V_{gs} - V_T)(1 + r)(2 + r)) \tag{60.20}$$

$$\text{IIP3} \cong 11.25 + 10 \log_{10}((V_{gs} - V_T)V_{sv}(1 + r)^2(2 + r)) \tag{60.21}$$

where

$$V_{sv} = \frac{1}{\Theta} \tag{60.22}$$

represents the transit voltage between strong inversion and velocity saturation and

$$r = \frac{V_{gs} - V_T}{V_{sv}} \equiv \Theta(V_{gs} - V_T) \tag{60.23}$$

denotes the relative amount of velocity saturation. The transit voltage V_{sv} depends only on technology parameters. For deep submicron processes, this voltage becomes even smaller than 300 mV, which is very close to the $V_{gs} - V_T$ at the boundary of strong inversion. The expressions for IIP2 and IIP3 are normalized to 0 V db m, the voltage that corresponds to a power of 0 dB in a 50 Ω resistor. For a given L_{eff}, the IIP3-value of a transistor is only a function of the gate overdrive voltage. Figure 60.11 plots the IIP2 and IIP3 as a function of the gate overdrive voltage for different values of Θ. It can be seen that for a certain value of $V_{gs} - V_T$, the IIP2 increases for increasing Θ (decreasing gate lengths), which proves

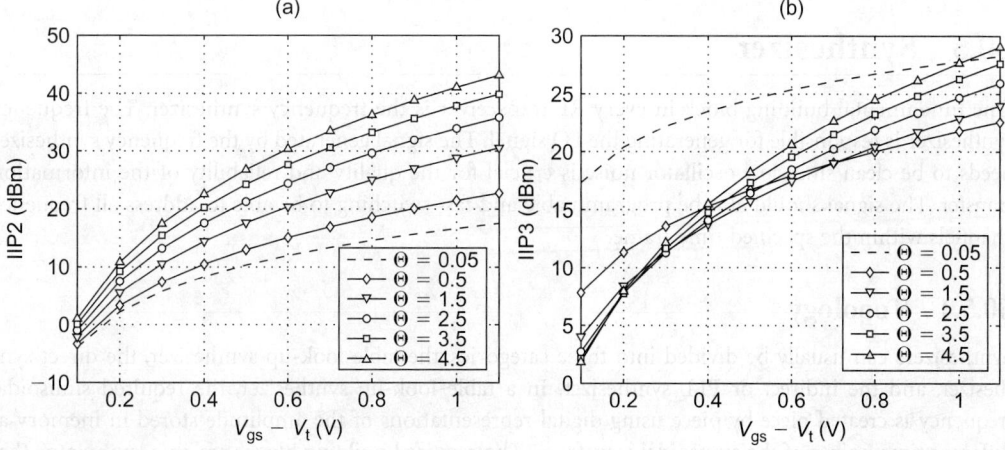

FIGURE 60.11 Linearity as a function of the gate overdrive voltage: (a) second-order intermodulation point; (b) third-order intermodulation point.

the former statements. The picture becomes a bit more complicated when looking at the IIP3 plot. For practical values of Θ, one can distinguish two regions in the $V_{gs} - V_T$ domain. For high gate overdrive voltages, deep submicron transistors clearly exhibit better linearity because the saturation voltage becomes lower and the transistor will reach velocity saturation earlier. Short-channel transistors therefore offer a maximum amount of linearity at a given power supply and require minimum $V_{gs} - V_T$ for a given IIP3. In contrast, for low overdrive voltages, short-channel transistors perform worse. Thus, to ensure a certain amount of linearity, one has to bias the transistors at a high enough overdrive voltage or apply some linearizing feedback technique (e.g., source degeneration). It can be shown that for the same equivalent g_m and the same distortion level, the required DC current is lower when local feedback is provided at the source. It is realized, however, at the cost of a larger transistor and this can compromise the amplifier bandwidth.

60.4.2 Downconverter

The most frequently used topology for a multiplier is the multiplier with cross-coupled variable transconductance differential stages. The use of this topology or related topologies (e.g., based on the square law) in CMOS is limited for high-frequency applications. Two techniques are used in CMOS: the use of the MOS transistor as a switch and the use of the MOS transistor in the linear region.

The technique often used in CMOS downconversion for its ease of implementation is subsampling on a switched-capacitor amplifier [33,34]. Here, the MOS transistor is used as a switch with a high input bandwidth. The desired signal is commutated via these switches. Subsampling is used to be able to implement these structures with a low-frequency op-amp. The switches and the switched-capacitor circuit run at a much lower frequency (comparable to an IF frequency or even lower). The clock jitter must however be low such that the high-frequency signals can be sampled with a high enough accuracy. The disadvantage of subsampling is that all signals and noise on multiples of the sampling frequency are folded upon the desired signal. The use of a high-quality HF filter in combination with the switched-capacitor subsampling topology is therefore absolutely necessary.

In Ref. [3] a fully integrated quadrature downconverter is presented. The circuit requires no external components, nor does it require tuning or trimming. It uses a double-quadrature structure, which renders a very high performance in quadrature accuracy. The downconverter topology is based on the use of MOS transistors in the linear region. By creating a virtual ground, a low-frequency op-amp can be used for downconversion. The MOS transistor in the linear region results in a very high linearity for both the RF and the LO signal.

60.5 Synthesizer

One fundamental building block in every RF transceiver is the frequency synthesizer. The frequency synthesizer is responsible for generating the LO signal. The signal generated by the frequency synthesizer needs to be clean since low oscillator noise is crucial for the quality and reliability of the information transfer. The signal should also be programmable and fast switching to be able to address all frequency channels within the specified time frame.

60.5.1 Topology

Synthesizers can usually be divided into three categories: the table look-up synthesizer, the direct synthesizer, and the indirect or PLL synthesizer. In a table look-up synthesizer, the required sinusoidal frequency is created piece by piece using digital representations of the amplitude stored in memory at different time points of the sinusoidal waveform. The required building blocks are an accumulator that keeps track of the time, a memory containing a sine, a digital-to-analog converter (DAC), and a low-pass filter to perform interpolation of the waveform to remove high-frequency spurs. This type of synthesis is limited is frequency owing to the access time of the memory and owing to the maximum operation frequency of the high-accuracy DAC. Moreover, high-frequency spurs, generated owing to the sampling behavior of the system tend to corrupt the spectral purity of the signal. The direct frequency synthesizer employs multiplication, division, and mixing to generate the desired frequency from a single reference. By repeatedly mixing and dividing, any level of accuracy is possible. The output spectrum is as clean as the reference frequency spectrum. Very fast-frequency hopping is possible. The main disadvantages of this type of system is the difficult layout of the system, the high power consumption owing to the numerous components present, and the corrupting of the spectral purity by cross coupling between stages. For generating high frequencies, the indirect or PLL type of frequency synthesizer is often the best choice. In a PLL, the synthesized frequency is generated by locking a locally generated frequency to an external frequency. The external frequency originates from a low-frequency high-quality crystal oscillator. To generate a local signal in the PLL, a VCO is used. A simple PLL topology is shown in Figure 60.12. A PLL includes the following building blocks: a VCO, a phase/frequency detector (PD/PFD), a loop filter, and frequency divider or prescaler. The last building block is needed to derive a low frequency signal from the LO. This allows the signal to be locked to the external frequency by means of the phase detector. The phase detector is a circuit that compares the external frequency phase with the locally generated frequency phase and outputs an error voltage proportional to the phase difference. After filtering, this error signal is fed back to the VCO. This constitutes a control system. Under lock conditions, the external frequency and the locally generated frequency have a constant phase relationship.

$$F_{\text{out}} = NF_{\text{ref}} \qquad\qquad (60.24)$$

The two signals are locked to each other, hence the name PLL. Even when a low-quality LO signal is generated, a high-quality signal can be synthesized. Owing to the phase relationship between the input and the output frequency, the output signal will have the same spectral purity as the high-quality input signal. This is due to the fact that the loop remains locked to the input phase and therefore follows the phase deviations of that signal thus taking over its phase noise. This however is only true as long as the loop dynamics can follow the input signal. The loop dynamics are mainly determined by the bandwidth of the loop. For offset frequencies below the loop bandwidth, the phase noise is determined by the phase noise of the reference signal, for frequency offsets above the loop bandwidth, the output phase noise will be determined by the phase noise of the locally generated signal.

When a programmable frequency divider is used in the loop, one can see that a set of frequencies can be synthesized. Assuming that the frequency by which the output signal is divided can be varied between N_1 and N_2, the output becomes

$$F_{\text{out}} = N1F_{\text{ref}}, (N1+1)F_{\text{ref}}, \ldots, N2F_{\text{ref}} \qquad\qquad (60.25)$$

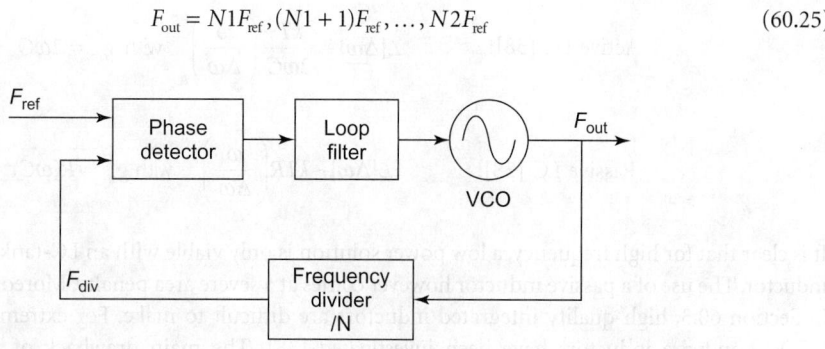

FIGURE 60.12 PLL-based frequency synthesizer.

The PLL synthesizer is inherently slower than the other two types of synthesizers. The switching speed between two frequencies in Eq. (60.25) is mainly determined by the loop bandwidth. Fast switching is only possible if a high loop bandwidth is implemented. Note that the loop bandwidth will also determine the phase noise performance. One however cannot indefinitely enlarge the loop bandwidth for stability reasons. A rule of thumb is that the loop bandwidth may not exceed 10% of the reference frequency to maintain stability. The loop bandwidth will also be limited by phase noise constraints. Spurious suppression and in-band phase noise levels will ultimately determine the loop bandwidth. When a low bandwidth has to be implemented, large capacitors will be needed. The total capacitance value is mainly determined by the need for implementing a stabilizing low-frequency zero in the loop filter. This makes integration difficult as it will blow up the silicon area and therefore increases the cost. One must therefore find ways to implement small bandwidth without having to use large capacitors. One obvious way of doing this is creating a low-frequency pole through the use of a large resitance. This however will increase phase noise. Other techniques however exist. In Ref. [35] a dual-path loop filter is used. The filter consist of one active path and one passive path. Combining both will create a low-frequency zero without the need for an extra resistor and capacitor. In Ref. [36] another technique is used to create the low-frequency zero. It is created in the digital domain. The signal in the loop filter is combined with a sampled and delayed version of itself. If the required switching speed is not achieved with a PLL configuration, one can make a combination of the direct synthesizer with the indirect synthesizer. In this topology, a number of PLLs is implemented and the outputs of all are combined with mixers. Thus, it is possible to synthesize a wide frequency range with a fast switching speed. This technique has recently been adopted for use in ultrawide band systems [37]. The major drawback of this technique, however, is that single sideband mixers have to be used. This requires accurate quadrature phases in all PLLs, low harmonic distortion, and well-matched mixers.

60.5.2 Oscillator

As was mentioned above, the VCO is the main source of the phase noise outside the loop bandwidth. Therefore, its design is one of the critical parts of a PLL design. For the design of subgigahertz VCO, two oscillator types are often used: ring oscillators and oscillators based on a resonant tank composed of an inductor and a capacitor. The latter is referred to as an LC-tank VCO. The inductor in an LC-tank VCO can be implemented in two ways: an active implementation and a passive implementation. It can be shown [38,39] that the phase noise is inversely proportional to the power consumption. In LC-tank VCOs, the power consumption is proportional to the quality factor of the tank. Eqs. (60.26)–(60.28) show this relationship:

$$\text{Ring oscillator [39]:} \quad L\{\Delta\omega\} \sim kTR\left(\frac{\omega}{\Delta\omega}\right)^2 \quad \text{with } g_m = \frac{1}{R} \tag{60.26}$$

$$\text{Active LC [38]:} \quad L\{\Delta\omega\} \sim \frac{kT}{2\omega C}\left(\frac{\omega}{\Delta\omega}\right)^2 \quad \text{with } g_m = 2\omega C \tag{60.27}$$

$$\text{Passive LC [38]:} \quad L\{\Delta\omega\} \sim kTR\left(\frac{\omega}{\Delta\omega}\right)^2 \quad \text{with } g_m = R(\omega C)^2 \tag{60.28}$$

It is clear that for high frequency, a low power solution is only viable with an LC-tank VCO with a passive inductor. The use of a passive inductor however comes at a severe area penalty. Moreover, as was discussed in Section 60.3, high-quality integrated inductors are difficult to make. For extremely low phase noise VCOs, bondwire inductors have been investigated [38]. The main drawback of using bondwires as inductors lies in reliability and yield. It is very difficult to make two bondwires exactly the same and reproduce this several times.

60.5.3 Prescaler

Several structures can be used as programmable divider. Programmable counters are the easiest solutions and are available in standard cell libraries. They are however limited in operation frequency. When high frequencies need to be synthesized, one can use a so-called prescaler. A prescaler divides by a fixed ratio and can therefore operate at high frequencies because it does not have to allow for delays involved with counting and presetting. A few high-speed prescaler stages lower the speed used in the subsequent counter stages. The disadvantage is that for a certain frequency resolution, the reference frequency has to be lowered. This slows the loop down as a lower bandwidth has to be implemented to maintain stability in the loop. A solution to this resolution problem is the use of dual or multimodulus prescalers. This circuit extends the prescaler with some extra logic to allow the prescaler to divide by N and $N + 1$ in case of a dual-modulus prescaler and by N to $N + x$ in case of a multimodulus prescaler. The reduction in speed of this extra circuitry can usually be limited. Figure 60.13 shows two possible implementations of a dual-modulus prescaler. Implementation (a) is a straight-forward implementation based on d-flip-flops. The critical path consists of a NAND gate and a d-flip-flop. Implementation (b) is a more complex implementation. It is based on the 90° phase relationship between the outputs of a master/slave toggle flip-flop. It contains no additional logic in the high-frequency path. The dual-modulus prescaler is as fast as an asynchronous fixed divider.

60.5.4 Fractional-N Synthesis

As can be concluded from Eq. (60.25), the minimal frequency resolution that can be achieved when using the topologies described previously is equal to F_{ref}. In GSM, for example, the channels are 200 kHz spaced apart, this means that we need a frequency resolution of 200 kHz and therefore a low reference frequency.

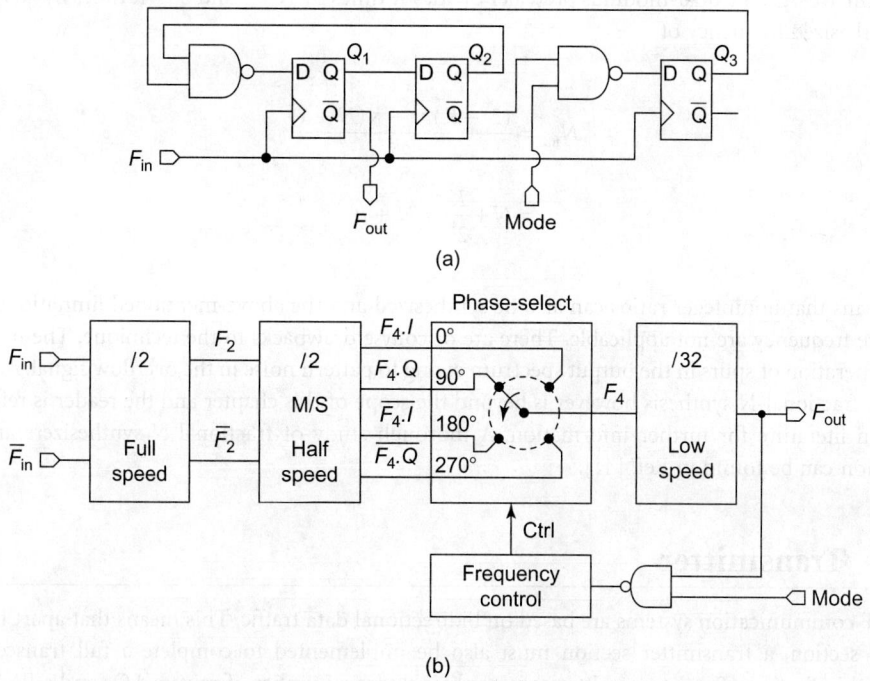

FIGURE 60.13 Dual-modulus prescaler architecture: (a) D-flip-flop based, (b) phase select topology (from Craninckx, J. and Steyaert, M., *IEEE J. Solid-State Circuits*, vol. 30, no. 12, pp. 1474–1482, 1995).

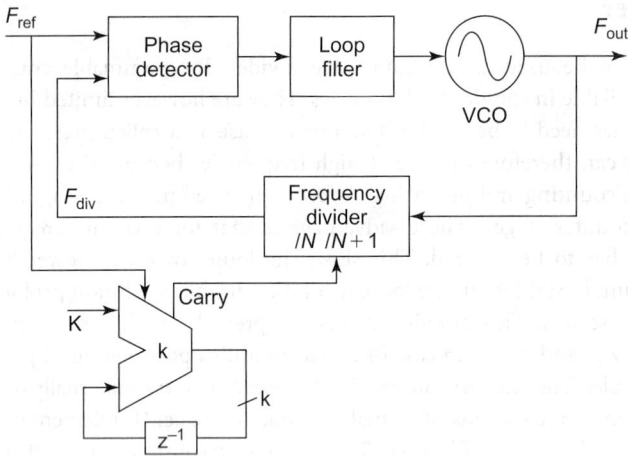

FIGURE 60.14 Fractional-N principle.

This results in high division ratios. The in-band phase noise of a PLL is however proportional to the division ratio, large ratios mean high in-band noise. As is already explained, a low reference frequency will also result in a low PLL bandwidth and therefore a slow loop. Therefore we need a technique that enables us to use a high reference frequency and still achieve the required frequency resolution. Fractional-N synthesizers solve this problem. Figure 60.14 elucidates this. A basic fractional-N synthesizer consists, besides the standard PLL building blocks, of an accumulator and a dual modulus prescaler. By switching fast between the two division ratios, fractional divisions can be synthesized. The accumulator increases its value every reference clock cycle with a certain amount $K = n2^k$. The dual-modulus prescaler is controlled by the accumulator overflow bit. If the accumulator overflows, the division ratio is $N + 1$, else it is N. On average, the dual-modulus prescaler divides K times by $N + 1$ and $2^k - K$ times by N, resulting in a synthesized frequency of

$$N_{frac} = \frac{(2^k - K)N + K(N + 1)}{2^k}$$

$$= N + \frac{K}{2^k} = N + n$$

(16.29)

This means that noninteger ratios can also be synthesized and the above-mentioned limitations on the reference frequency are not applicable. There are of course drawbacks to the technique. The major one is the generation of spurs in the output spectrum owing to pattern noise in the overflow signal. A detailed study of fractional-N synthesis however is beyond the scope of this chapter and the reader is referred to the open literature for further information. A thorough study of fractional-N synthesizers and their simulation can be found in Ref. [41].

60.6 Transmitter

Most RF communication systems are based on bidirectional data traffic. This means that apart from the receiver section, a transmitter section must also be implemented to complete a full transceiver. As explained in Section 60.2, a transmitter commonly includes a number of mixers, LOs and a PA. The LO has been dealt with in a previous section; this section will therefore only describe the mixer and the PA used in upconversion or transmitter systems.

60.6.1 Up versus Downconversion

Although a large amount of literature exists on the downconversion process, upconversion has been neglected for a long time. This is rather surprising. When looking at Figure 60.5, one immediately notices the parallelism between the receiver and the transmitter. The same functionality occurs. Both paths contain an amplifier (LNA, PA), both contain an interface to the digital domain (A/D, D/A), both contain a mixer, and both are steered by the same LO system. The nature of the signals in both paths (input and output) have a significant influence on the circuit implementation. This seems logical for the LNA/PA analogy or the A/D and D/A converter. Both have completely different topologies. Although the mixers in the upconversion path and the downconversion path face the same signals, there is typically not a great difference between the up- and downconversion mixer topology. Most implementations are variations on the four-quadrant mixer topology, better known as the Gilbert cell [42]. There are, however, fundamental differences between up- and downconversion. The first fundamental difference is located in the input signals of the mixer. In case of a downconversion mixer, the input usually is a high-frequency, low-power signal surrounded by large blocking signals. In case of an upconversion, the input signal is a locally generated large baseband signal with a clean spectrum. At the output side, the situation is the opposite. A downconverted signal is a low-frequency signal. It is therefore relatively easy to filter or apply feedback to cope with unwanted signals. At the transmitter side, however, a large and linear signal has to be processed within the technology-dependent limited frequency range. Every extra building block placed between the mixer and the PA has to deal with high-frequency signals. Filtering is therefore impossible behind the upconversion mixer as it will require a large amount of power. Therefore LO-leakage and other unwanted signals like intermodulation products have to be limited. A last, but not least difference lies within one of the design specifications of a mixer, the conversion gain G_c. It is defined as the ratio between the input power of the mixer and the output power. At the receiver side, the mixer input power is a design constraint as it is determined by the application. At the transmitter side, both input and output power are design variables. They can both be chosen freely. As it is easier and more power friendly to amplify a low-frequency signal, a large baseband signal is preferred.

60.6.2 CMOS Mixer Topologies

60.6.2.1 Switching Modulators

Many mixer implementations are based on the traditional variable transconductance multiplier with cross-coupled differential modulator stages [42]. It is depicted in Figure 60.15. The circuit was originally implemented in a bipolar technology and therefore based on its inherent translinear behavior. The MOS counterpart however can only be effectively used in the switching mode. This involves the use of large LO-driving signals and results in large LO-feedthrough and power consumption. Moreover, when using a square-wave-type modulation signal, a lot of energy is located at the third harmonic. This unwanted

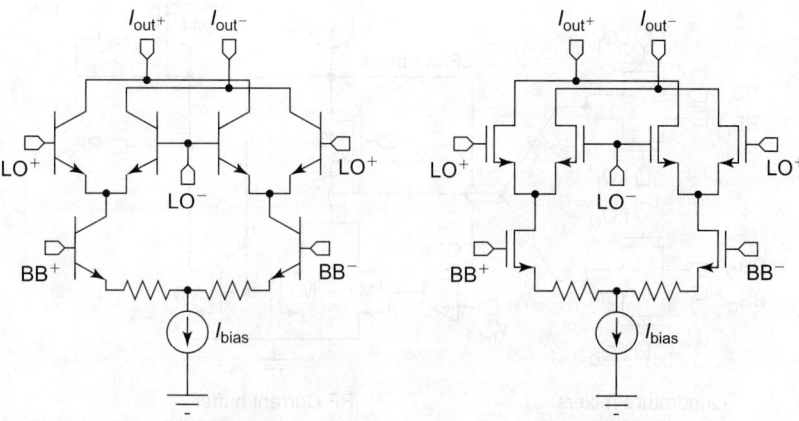

FIGURE 60.15 Bipolar and CMOS version of the Gilbert cell mixer.

signal can only be filtered out by an extra blocking filter at the output. In CMOS the variable transconductance is typically implemented using a differential pair biased in the saturation region. To avoid distortion problems, large $V_{gs} - V_T$ values or a large source degeneration resistor is needed. This results in a large power consumption and noise problems. For upconversion, one also has to be aware that the high-frequency current has to run through the modulating transistors. The source degeneration is therefore limited by bandwidth constraints. These problems can be circumvented by replacing the bottom differential pair with a pseudodifferential topology biased in the linear region [43].

60.6.2.2 Linear MOS Mixers

Figure 60.16 presents a linear CMOS mixer topology together with an output driver [44,45]. The circuit implements a real single-ended output topology avoiding the use of external combining. Some basic design ideas and some guidelines to optimize the circuit will be presented. The circuit is based on an intrinsically linear mixer topology. The circuit features four mixer transistors biased in the linear region. Each mixer converts a quadrature LO voltage and a baseband signal to a linearly modulated current. The expression for the source-drain current for a MOS transistor in the linear region is given by

$$I_{DS} = \beta \left[(V_{GS} - V_T)V_{DS} - \frac{V_{DS}^2}{2} \right]$$

(60.30)

This equation can be rewritten in terms of a DC and an AC term as follows:

$$I_{DS} = \beta(V_{DS} + v_{ds})\left(V_{GS} - V_T - \frac{V_D - V_S}{2} + v_g - \frac{v_d + v_s}{2} \right)$$

$$= \underbrace{\beta V_{DS}\left(V_{GS} - V_T - \frac{V_D - V_S}{2} \right)}_{\text{DC component}}$$

(60.31)

$$\underbrace{+ \beta v_{ds}\left(V_{GS} - V_T - \frac{V_D - V_S}{2} \right) + \beta V_{DS}\left(v_g - \frac{v_d + v_s}{2} \right) + \beta v_{ds}\left(v_g - \frac{v_d + v_s}{2} \right)}_{\text{AC component}}$$

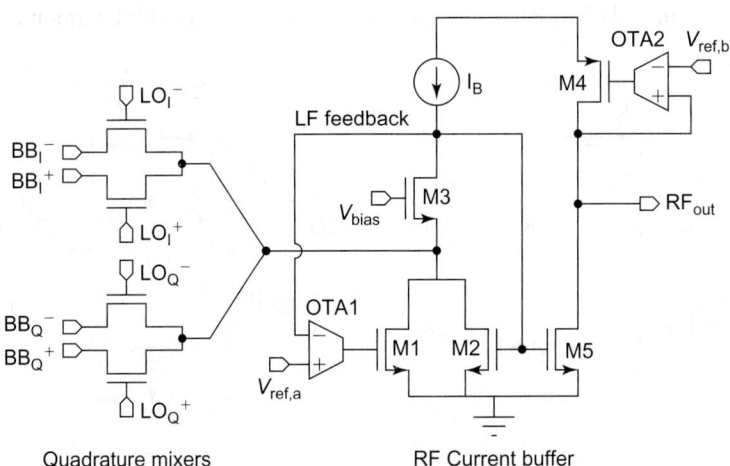

FIGURE 60.16 Schematic of a linear up-conversion mixer with output driver.

Two signals have to be applied to a mixer transistor, the low-frequency baseband signal and the high-frequency LO signal. Applying these signals may only result in the desired high-frequency currents. Based on Eq. (60.31), some conclusions can be drawn.

If the LO signal is applied to the drain/source of the mixer transistor, a product term

$$\beta v_{ds}\left(V_{GS} - V_T - \frac{V_D - V_S}{2}\right)$$

is formed. As this contains the product of a DC voltage with the oscillator signal, this component is located at LO frequency. It is preferable to avoid the formation of this frequency component. Therefore, the LO signal should not be applied to this node. Applying the LO signal to the gate of the mixer transistors results in the desired behavior. According to Eq. (60.31), only the high-frequency components are formed by

$$\beta v_g(V_{DS} + v_{ds})$$

By applying a zero-DC voltage between the source and drain, only the high-frequency mixer product is generated. The voltage to current conversion is perfectly balanced. The current of the four mixer paths are immediately added at the output of the mixers at a common node. This requires a virtual ground at that point that is achieved owing to the low-impedance input of the buffer stage (Figure 60.16). The total current flowing into the output buffer is given by

$$I_{MIX} = \beta\left(v_{bb_I}^2 + v_{bb_Q}^2 + 2v_{LO_I}v_{bb_I} + 2v_{LO_Q}v_{bb_Q}\right) \tag{60.32}$$

Eq. (60.32) shows two frequency components in the modulated waveform. $\beta v_{LO}v_{bb}$ is the desired signal. To prevent intermodulation products of the low-frequency baseband signal $\beta v_{bb_I}^2$ with the desired RF signal, the LF signal has to be suppressed at the current summing node. This is achieved by a low-frequency feedback loop in the output buffer.

The low-frequency feedback loop consists of OTA1 and transistors M1 and M3. It suppresses the low-frequency signals resulting in a higher dynamic range of the output stage and decreases unwanted intermodulation products. It also lowers the input impedance of the output stage at low-frequencies. The structure in fact separates the high-and low-frequency component of the input current and prevents the low-frequency component from being mirrored to the output stage. The RF current buffer also ensures a low-impedance at high-frequencies at the mixer current summing node and therefore provides the necessary virtual ground.

60.6.2.3 Non-Linearity and LO-Feedthrough Analysis

The difficulty in integrating IF filters is one of the reasons why direct conversion transmitters are implemented. This implies that the LO is at the same frequency as the RF signal and cannot be filtered out. To minimize the spurious-signal components at the LO frequency, one has to isolate the origins of the unwanted frequency components. They can be categorized into three topics: capacitive feedthrough owing to gate-source and gate-drain parasitic overlap capacitances, intrinsic nonlinearity of the mixers, and mixer products owing to a nonideal virtual ground.

When an ideal virtual ground is provided at the output of the mixer, capacitive LO-feedthrough is canceled. This cancellation is never perfect however, owing to technology mismatch. The capacitive LO-feedthrough for a single-mixer transistor, biased in the linear region is therefore given by

$$I_{LO} = 2\pi f v_{LO} WL\left(\frac{C_{ox}}{2} + \frac{C_{ov}}{L}\right) \tag{60.33}$$

where c_{ox} is the oxide capacitance, C_{ov} the gate-drain/source overlap capacitance, v_{LO} the amplitude of the LO signal and f its frequency. Based on Eq. (60.32) and Eq. (60.33), the ratio between the LO-feedthrough current and the modulated current is given by

$$\frac{i_{signal}}{\Delta(i_{LO})} = \frac{2\mu C_{ox}\dfrac{W}{L}v_{bb}v_{LO}}{\delta(i_{LO})2\pi f v_{LO}WL\left(\dfrac{C_{ox}}{2}+\dfrac{C_{ov}}{L}\right)}$$

$$= \frac{\mu C_{ox}v_{bb}}{\delta(i_{LO})\pi f L^2\left(\dfrac{C_{ox}}{2}+\dfrac{C_{ov}}{L}\right)} \tag{60.34}$$

where $\delta(i_{LO})$ accounts for the relative difference in LO-feedthrough for the different mixer transistors owing to mismatch. Eq. (60.34) shows that the ratio between the modulated current and the LO-feedthrough current is independent of the LO amplitude and the transistor width. Feedthrough will be less if shorter transistor lengths are used. The relative matching between the different mixer transistors will become worse however when shorter lengths are used [46]. One must therefore use Eq. (60.34) with care. The δ will increase for a smaller transistor length. With proper design and optimization, one should however be able to achieve a 30 dBc signal to LO-feedthrough ratio even if two LO-feedthrough currents are added instead of being canceled by the virtual ground ($\delta(i_{LO}=1$). When more realistic numbers of mismatch are considered (e.g., 10%), 50 dBc is easily achieved. The presented equations can therefore be used by the experienced designer to estimate the matching requirements and check if these requirements are realistic.

Another problem one faces is a possible DC-offset between the source and drain terminal of the mixer transistor. Eq. (60.31) explains the problem. Ideally, no DC is present. The mixer then shows the required behavior. When a DC is present, however, one can see that components are generated at DC, the LO frequency owing to multiplication with v_g and a component at baseband. While the low-frequency components can be filtered out by the low-frequency feedback in the output buffer, the component at the LO frequency remains. This component will therefore set the requirements for the tolerated DC-offset. A possible solution for this problem is measuring the DC-offset between the source and drain. The offset is then controllable. The offset requirement is translated into an offset specification on the op-amps used in the feedback loops in the output buffer.

If the common mixer node is not a ideal virtual ground, the modulated current will be converted to a voltage dependent on the impedance seen on that node. The spectrum of the modulated signal will therefore be a combination of the modulated current spectrum and the frequency dependence of the impedance. When a impedance Z_c is considered at the common mixer node, the modulated current is given as the result of a second-order equation

$$\left(\beta Z_c^2\right)I^2 - (1 + 2\beta Z_c(V_{GS}-V_T))I + 2\beta v_1 v_g + \beta v_1^2 = 0 \tag{60.35}$$

It can be noticed that when $Z_c = 0$, Eq. (60.35) is reduced to Eq. (60.32). As Eq. (60.35) is a second-order equation, it is a possible origin of distortion and therefore has to be taken into account. Note that only currents that are not canceled out by the differential character of the mixer are converted in a voltage. This advantage of a balanced structure however is not valid for a nonideal voltage source at the input of the mixer transistors. If a nonideal voltage source is used at this node, each frequency component of the modulated current will be converted into a voltage according to the specific frequency-dependent impedance. These voltages are then, similar to the baseband signal upconverted to the LO frequency. It is therefore essential to keep this node a low-impedance one as far as possible.

Eq. (60.32) is only valid if a very low impedance is seen at the source and drain terminals of the mixer transistors. If this condition is fulfilled, no unwanted high-frequency mixing components are present in the modulated signal. However, both in measurements and in simulations, a significant unwanted signal is noticed at $f_{LO} \pm 3f_{bb}$. One expects this component to originate from a $v_{LO}v_{bb}^3$ product term. However, Eq. (60.35) only shows a second-order relationship. The observed product term must therefore find its origin in another effect. It is proved to be a result of short-channel effects in a MOS transistor. Both the effective mobility and the threshold voltage are affected by the gate-source and drain-source voltage. The calculated impact of the threshold voltage modulation cannot explain the observed effect, it is therefore assumed that it is a result of the mobility modulation. After some calculations, one can prove that the effective mobility is

$$\mu_{eff} = \frac{\mu_0}{1 + \theta(V_{GS} - V_T)_{DC} + \theta\left(v_{LO} - \dfrac{v_{bb}}{2}\right) + \dfrac{\mu_0}{V_{max}L} + \dfrac{\theta}{2}|v_{bb}|} \tag{60.36}$$

Substituting v_{bb} with $A \sin(\omega_{bb}t)$ and making a Fourier series expansion of $|v_{bb}|$ results in

$$\mu_{eff} = \frac{\mu_0}{B\left(1 + \dfrac{\theta}{B}v_{LO} - \dfrac{\theta A}{2B}\sin(\omega_{bb}t) + C\cos(2\omega_{bb}t) + D\cos(4\omega_{bb}t) + \cdots\right)} \tag{60.37}$$

with

$$A \sin(\omega_{bb}t) = \text{the baseband signal}$$

$$B \approx 1 + (V_{GS} - V_T)$$

$$C = \frac{A}{B}\frac{4}{3\pi}\left(\frac{\mu_0}{V_{max}L} + \frac{\theta}{2}\right)$$

$$D = \frac{A}{B}\frac{4}{(5)(3\pi)}\left(\frac{\mu_0}{V_{max}L} + \frac{\theta}{2}\right)$$

Eq. (60.37) shows that a second-order baseband frequency component $\cos(2\omega_{bb}t)$ appears. In the DC reduction factor B, the third term is an order of magnitude smaller than 1. Hence it appears that the magnitude C of the second-order component has a first-order relationship to the baseband signal amplitude A. In the voltage to current relationship, μ_{eff} is multiplied with $v_{LO}v_{bb}$. As a result a mixing component at $f_{LO} \pm 3f_{bb}$ occurs. In the amplitude C of this distortion component, $\mu_0/(V_{max}L)$ is dominant to $\theta/2$ for most submicron technologies. It is also important to notice that the distortion is inversely proportional to the gate length. This indicates that this effect will become even more apparent as gate lengths continue to scale down.

60.6.3 PA

60.6.3.1 CMOS Power Amplification

The integration of PAs in a CMOS technology is impeded by the low supply voltage of the current deep-submicron and nanometer technologies. Apart from this, the relative high parasitic capacitances of the MOS transistor, at least compared to the GaAs or SiGe transistors, and the relative low-quality factor of on-chip inductors, further hinders the integration. On the other hand, the digital MOS transistor is optimized for switching and as such a lot of switching amplifiers have been integrated in CMOS with great success recently [47–53]. Furthermore, CMOS RF amplifiers are capable of breaking the 1-W barrier of output power performance [54]. In this section, the topic of switching RF amplifier is discussed first. In the second part, some linearization techniques will be discussed.

60.6.3.2 Switching Class E Amplifier

The Class E amplifier was invented in 1975 [55], but the first implementation of this amplifier in CMOS was reported in 1997 [47]. In contrast to the Class A, B, C, and F amplifiers, the Class E amplifier is designed in the time domain. Theoretically, the Class E amplifier is capable of achieving an efficiency of 100%. To achieve this, the transistor and output network are designed in such a way that the drain through the transistor is separated in time from the voltage across the transistor. This avoids power dissipation in the transistor, a necessary requirement to achieve a high efficiency. If all other elements are assumed to be lossless, the amplifier is then indeed capable of achieving an efficiency of 100%. Figure 60.17 depicts the basic circuit of a CMOS Class E amplifier. The nMOS transistor should act as a switch and therefore it is driven by a square wave between zero and the maximum permissible gate voltage, which is normally equal to V_{DD}, the supply voltage of the technology. As such the nMOS transistor can be modeled by an ideal switch with a series resistance R_{on}. Inductor L_1 can be seen as the DC feed inductance, and in the original Class E theory, this inductor is assumed to be very large, and can be replaced by an ideal current source. Finally, inductor L_x and capacitor C_1 are the two crucial elements that create the Class E waveform at the drain of the nMOS transistor.

In a fully integrated CMOS implementation, the DC feed inductor L_1 cannot be made very large. First, this would require a huge silicon area, but more importantly the relative high power loss of CMOS integrated inductors does not allow for such a large value. As such, the value of L_1 has to be reduced and the current through the latter will not be a constant. The amplifier can still be designed to meet the Class E conditions, even with a small value of L_1. In fact, reducing the value of L_1 will result in a larger value for C_1 and a smaller one for L_x.

The capacitor C_1 and inductor L_x are constrained by the two Class E requirements, given as follows:

$$\text{Class E} \Longleftrightarrow \begin{cases} v_{DS}(t = t_1) = 0 \\ \dfrac{dv_{DS}(t)}{dt}\bigg|_{t=t_1} = 0 \end{cases}$$

In Figure 60.18, the drain voltage and current for Class E operation are shown. Solving the two Class E equations will give a value for C_1 and L_x. Finally, the value of the load resistance is constrained by the required output power. To achieve sufficient output power in a low-voltage CMOS technology, an

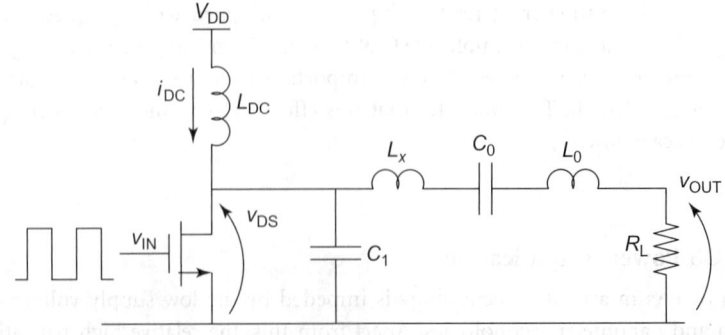

FIGURE 60.17 Basic Class E PA.

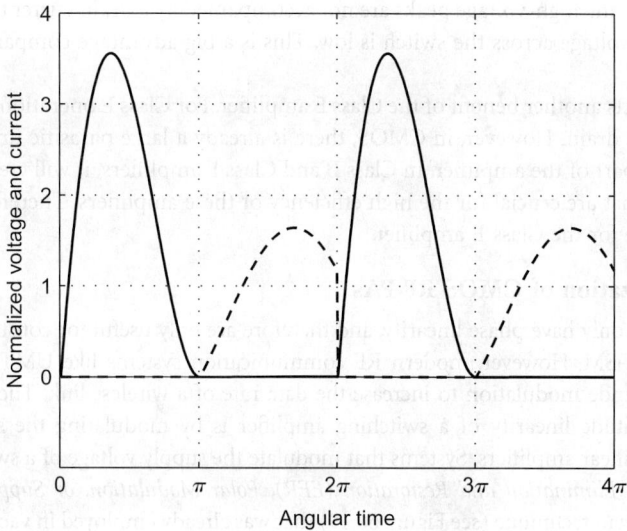

FIGURE 60.18 Voltage (solid line) and current (dashed line) of the Class E PA.

impedance matching network is required between the 50 Ω load or antenna impedance and the Class E amplifier. The on-resistance of the nMOS transistor can be written as

$$R_{on} = \frac{L}{\mu_n C_{ox} W (V_{GS} - V_T)}$$ (60.38)

The lower the on-resistance, the higher the efficiency of the amplifier, and thus it is beneficial to increase the width of the nMOS transistor. However, that large transistor cannot be directly connected to the upconversion mixer, and several amplifying stages are needed between them. If the nMOS transistor has a large gate width, more power will be consumed by the driver stages and thus the overall efficiency of the amplifier, defined as

$$\eta_{oa} = \frac{P_{out}}{P_{DC,PA} + P_{DC,DRV}}$$ (60.39)

will have a maximum value for a specific transistor width. The overall efficiency is not always a good figure to compare PAs, since it can never reach 100%, even if each of the stages have a conversion efficiency of 100%. After all, the power consumed by the driver stages will never flow to the output load, but is only needed to switch on and switch off the next stage in line.

The power added efficiency (PAE) defined as

$$PAE = \frac{P_{out} - P_{in}}{P_{DC}}$$ (60.40)

is a useful definition for stand-alone PAs that have an input matched to 50 Ω. However, one should know whether the DC power consumption of the driver stages is included in P_{DC}.

Another important aspect of switching amplifiers is the reliability. A drawback of the Class E amplifier, at least compared to the Class B and F ones, is that the drain voltage goes up to several times the supply voltage of the amplifier. This might cause reliability problems. On the other hand, the switching nature of the amplifier alleviates this. After all, owing to the switching, the voltage and current are separated in time. In other words, the high-voltage peaks are not accompanied by a drain current, and when the drain current is high, the voltage across the switch is low. This is a big advantage compared to other types of amplifiers.

Figure 60.17 depicts another benefit of the Class E amplifier. For Class E operation, a shunt capacitance C_1 is required at the drain. However, in CMOS, there is already a large parasitic drain capacitance, and it can now become part of the amplifier. In Class B and Class F amplifiers, it will create a low impedance for the harmonics that are crucial for the high efficiency of these amplifiers. Therefore, CMOS seems to be the logical choice for the Class E amplifier.

60.6.3.3   Linearization of CMOS RF PAs

Switching amplifiers only have phase linearity, and therefore are only useful for constant envelope systems like Bluetooth and GSM. However, modern RF communication systems like UMTS, CDMA-2000, and WLAN, allow amplitude modulation to increase the data rate of a wireless link. The only way to recover or restore the amplitude linearity of a switching amplifier is by modulating the supply voltage or by combining two nonlinear amplifiers. Systems that modulate the supply voltage of a switching amplifier are denoted as *Envelope Elimination and Restoration (EER), Polar Modulation,* or *Supply Modulation.* They originate from the Khan technique (see Figure 60.19) that was already employed in vacuum tube amplifiers. In CMOS, one can make use of the availability of digital signal processing to directly create the amplitude and phase signal, and as such, the limiter and envelope detector of Figure 60.19 can be avoided. Furthermore, AM-AM and AM-PM predistortion is relatively easy to implement. The general picture of polar modulation is shown in Figure 60.20. Another advantage of polar modulated amplifiers is that the entire phase path carries a constant envelope signal and thus one can use nonlinear or saturated blocks in the upconversion path. Furthermore, amplitude and phase feedback are relatively easy to implement. Another group of techniques combine the output of two constant envelope amplifiers that have a difference in phase. The two

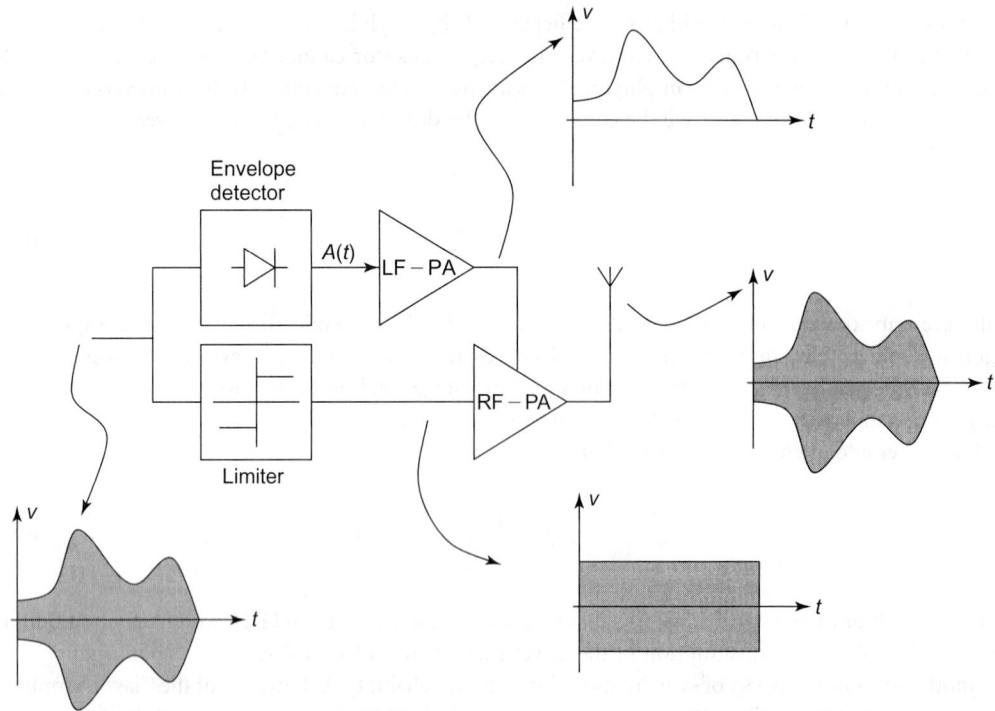

FIGURE 60.19   Kahn Technique linearized power amplifier.

amplifiers are combined through a transformer, a power combiner, or through transmission lines, and the output is, in general, the sum of the two amplifiers. These systems are called *Outphasing or LINC* (Linear amplification with nonlinear components), depending on the used combiner. Depending on their phase difference, the resulting output envelope can be higher or lower, and thus has amplitude modulation, as shown in Figure 60.21. The major drawback of these techniques is the difficulty to implement the power combiner in CMOS. Also, feedback is not as easy to implement in these systems. On the other hand, these systems allow to efficiently amplify signals that have a very high modulation bandwidth.

Apart from the two groups discussed in this section, several other techniques exist to amplify an amplitude-modulated signal. There is no *ideal* solution for CMOS integration. An alternative solution or approach is to use a linear amplifier with an efficiency improvement technique such as the Doherty amplifier or the bias adaption technique. However, the linearization of nonlinear amplifiers has the advantage that switching or nonlinear amplifiers that can be used are easier to implement in CMOS. Furthermore, the RF driver stages and all the blocks preceding the RF amplifier can be nonlinear as well. Needless to say, this is a significant advantage in low-voltage technologies.

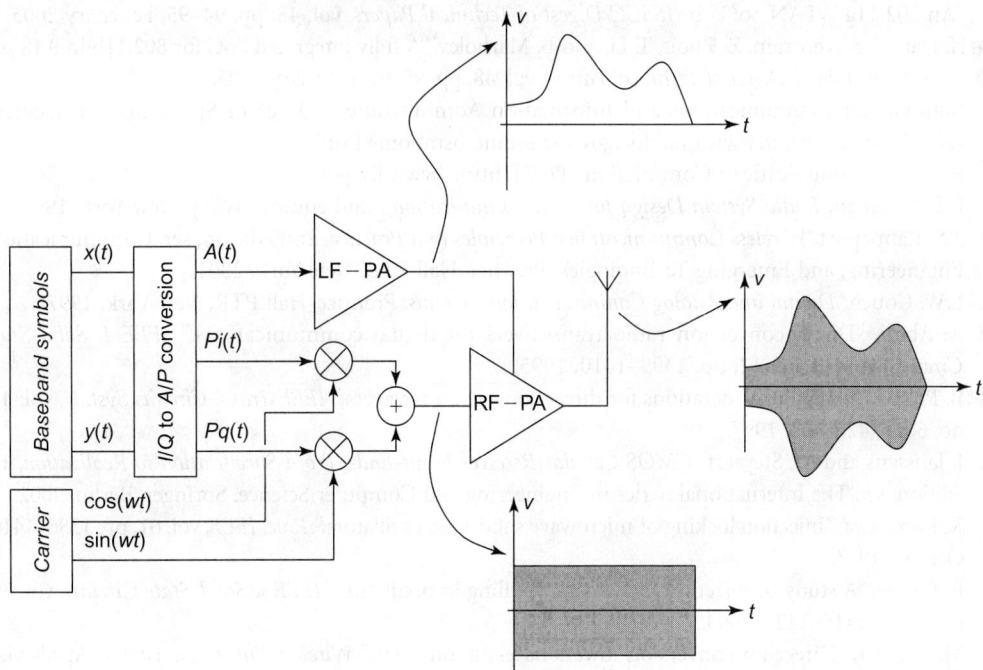

FIGURE 60.20 DSP-based polar modulation architecture.

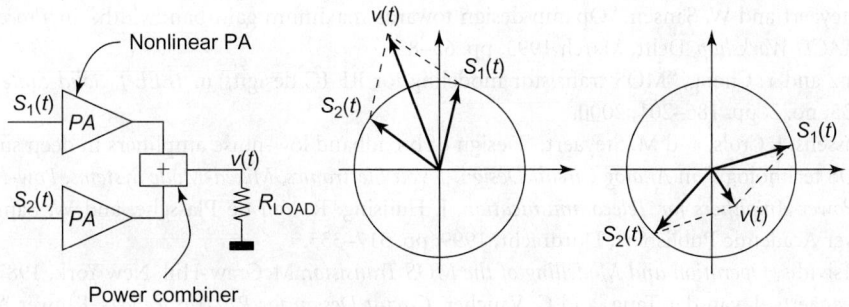

FIGURE 60.21 LINC or outphasing architecture.

References

1. C.E. Shannon, "Communication in the presence of noise," in *Proc. IRE*, vol. 37, pp. 10–21, January 1949.

2. J. Sevenhans, A. Vanwelsenaers, J. Wenin, and J. Baro, "An integrated Si bipolar transceiver for a zero-IF 90 MHZ GSM digital mobile radio front-end of a hand portable phone," in *Proceedings of the Custom Integrated Circuits Conference, CICC*, Tokyo, Japan, May 1991, pp. 7.7.1–7.7.4.

3. J. Crols and M. Steyaert, "A single chip 900 MHz CMOS receiver front-end with a high-performance low-IF topology," *IEEE J. Solid-State Circuits*, vol. 30, no. 12, pp. 1483–1492, 1995.

4. P.R. Gray and R.G. Meyer, "Future directions in silicon ICs for RF personal communications," in *Proceedings of the Custom Integrated Circuits Conference, CICC*, Santa Clara, CA, USA, May 1995, pp. 91–95.

5. A.A. Abidi, "Low-power radio-frequency ICs for portable communications," *Proc. IEEE*, vol. 83, no. 4, pp. 544–569, 1995.

6. S. Mehta, D. Weber, M. Terrovitis, K. Onodera, M. Mack, B. Kaczynski, H. Samavati, S. Jen, W. Si, M. Lee, K. Singh, S. Mendis, P. Husted, N. Zhang, B. McFarland, D. Su, T. Meng, and B. Wooley, "An 802.11g WLAN SoC," in *ISSCC Digest of Technical Papers*, vol. 48, pp. 94–95, February 2005.

7. H. Darabi, S. Khorram, Z. Zhou, T. Li, and B. Marholev, "A fully integrated SoC for 802.11b in 0.18 μm CMOS," in *ISSCC Digest of Technical Papers*, vol. 48, pp. 96–97, February 2005.

8. National Telecommunications and Information Administration, Office of Spectrum Management (NTIA-OSM), http://www.ntia.doc.gov/osmhome/osmhome.html.

9. Federal Communications Commission (FCC), http://www.fcc.gov/.

10. R.L. Freeman, *Radio System Design for Telecommunications*, 2nd edition. Wiley, New York, 1997.

11. T.S. Rappaport, *Wireless Communications: Principles and Practice*, 2nd edition, ser. Communication, Engineering, and Emerging Technologies. Prentice-Hall PTR, New York, 2001.

12. L.W. Couch, *Digital and Analog Communication Systems*. Prentice-Hall PTR, New York, 1997.

13. A. Abidi, "Direct conversion radio transceivers for digital communications," *IEEE J. Solid-State Circuits*, vol. 30, no. 12, pp. 1399–1410, 1995.

14. B. Razavi, "Design considerations for direct conversion receivers," *IEEE Trans. Circuits Syst. II*, vol. 44, no. 6, pp. 428–453, 1997.

15. J. Janssens and M. Steyaert, *CMOS Cellular Receiver Front-Ends: From Specification to Realization*, 1st edition, ser. The International Series in Engineering and Computer Science. Springer, Berlin, 2002.

16. K. Kurokawa, "Injection locking of microwave solid-state oscillators," *Proc. IEEE*, vol. 61, pp. 1386–1410, October 1973.

17. B. Razavi, "A study of injection locking and pulling in oscillators," *IEEE J. Solid-State Circuits*, vol. 39, no. 9, pp. 1415–1424, 2004.

18. M. Feulner, "Direct up-conversion lowers base-station costs," *Wireless Europe Magazine*, April–May 2005, wireless.iop.org.

19. W. Liu, X. Jin, X. Xi, J. Chen, M.-C. Jeng, Z. Liu, Y. Cheng, K. Chen, M. Chan, K. Hui, J. Huang, R. Tu, P. K. Ko, and C. Hu, *BSIM3v3.3 MOSFET Model, Users' Manual*. University of California, Berkeley, 1999.

20. M. Steyaert and W. Sansen, "Opamp design towards maximum gain-bandwidth," in *Proceedings of the AACD Workshop*, Delft, March 1993, pp. 63–85.

21. C. Enz and Y. Cheng, "MOS transistor modeling for RF IC design," in *IEEE J. Solid-State Circuits*, vol. 35, no. 2, pp. 186–201, 2000.

22. J. Janssens, J. Crols, and M. Steyaert, "Design of broadband low-noise amplifiers in deep submicron CMOS technology," in *Analog Circuit Design. 1 Volt Electronics, Mixed-Mode Systems, Low-Noise and RF Power Amplifiers for Telecommunication*, J. Huijsing, R. Van de Plassche, and W. Sansen, Eds. Kluwer Academic Publishers, Dordrecht, 1999, pp. 317–335.

23. Y.P. Tsividis, *Operation and Modelling of the MOS Transistor*. McGraw-Hill, New York, 1987.

24. D. Leenaerts, J. van der Tang, and C. Vaucher, *Circuit Design for RF Transceivers*. Kluwer Academic Publishers, Dordrecht, 2001.

25. J. Craninckx and M. Steyaert, "A 1.8 GHz low-phase noise CMOS VCO using optimized hollow spiral inductors," *IEEE J. Solid-State Circuits*, vol. 32, no. 5, pp. 736–744, May 1997.

26. N. Itoh, B.D. Muer, and M. Steyaert, "Low supply voltage integrated CMOS VCO with three terminals spiral inductor," in *Proceedings of the European Solid State Circuits Conference, ESSCIRC*, Duisburg, Germany, September 1999, pp. 194–197.

27. A.-S. Porret, T. Melly, C.C. Enz, and E.A. Vittoz, "Design of high-Q varactors for low-power wireless applications using a standard CMOS process," *IEEE J. Solid-State Circuits*, vol. 35, no. 3, pp. 337–345, 2000.

28. K.O, "Estimation methods for quality factors of inductors fabricated in silicon integrated circuit process technologies," *IEEE J. Solid-State Circuits*, vol. 33, no. 8, pp. 1249–1252, 1998.

29. C.S. Meyer, D.K. Lynn, and D.J. Hamilton, *Analysis and Design of Integrated Circuits*. McGraw-Hill, New York, 1968.

30. J. Crols, P. Kinget, J. Craninckx, and M. Steyaert, "An analytical model of planar inductors on low doped silicon substrates for high-frequency analog design uo to 3 GHz," in *Digest of Technical Papers, Symposium on VLSI Circuits*, Honolulu, Hawaii, USA, 1996.

31. P. Leroux and M. Steyaert, *LNA-ESD Co-Design for Fully Integrated CMOS Wireless Receivers*, ser. The International Series in Engineering and Computer Science. Springer, Berlin, 2005, vol. 843 .

32. J. Rabaey and J. Sevenhans, "The challenges for analog circuit design in mobile radio VLSI chips," in *Proceedings of the AACD Workshop*, Leuven, Belgium, 1993, vol. 2, pp. 225–236.

33. D. Shen, H. Chien-Meen, B. Lusignan, and B. Wooley, "A 900 MHz integrated discrete-time filtering RF front-end," in *ISSCC Digest of Technical Papers*, San Francisco, USA, February 1996, pp. 54–55, 417.

34. S. Sheng, L. Lynn, J. Peroulas, K. Stone, I. O'Donnell, and R. Brodersen, "A low-power CMOS chipset for spread spectrum communications," in *ISSCC Digest of Technical Papers*, San Francisco, USA, February 1996, pp. 346–347, 471.

35. J. Craninckx and M. Steyaert, "A fully integrated CMOS DCS-1800 frequency synthesizer," in *ISSCC Digest of Technical Papers*, San Francisco, USA, February 1998, pp. 372–373.

36. B. Zhang, P. Allen, and J. Huard, "A fast switching PLL frequency synthesizer with an on-chip passive discrete-time loop filter in 0.25 μm CMOS," *IEEE J. Solid-State Circuits*, vol. 38, no. 6, pp. 855–865, 2003.

37. D. Leenaerts1, R. van de Beek1, G. van der Weide, H. Waite, J. Bergervoet, K. Harish, Y. Zhang, C. Razzell, and R. Roovers, "SiGe BiCMOS 1 ns fast hopping frequency synthesizer for UWB radio," in *ISSCC Digest of Technical Papers*, San Francisco, USA, February 2005.

38. J. Craninckx and M. Steyaert, "Low-noise voltage controlled oscillators using enhanced LC-tanks," *IEEE Trans. Circuits Syst. II*, vol. 42, no. 12, pp. 794–804, 1995.

39. B. Razavi, "Analysis, modeling and simulation of phase noise in monolithic voltage controlled oscillators," in *Proceedings of the Custom Integrated Circuits Conference, CICC*, May 1995, pp. 323–326.

40. J. Craninckx and M. Steyaert, "A 1.8 GHz low-phase noise voltage controlled oscillator with prescaler," *IEEE J. Solid-State Circuits*, vol. 30, no. 12, pp. 1474–1482, 1995.

41. B. de Muer and M. Steyaert, *CMOS Fractional-N Synthesizers: Design for High Spectral Purity and Monolithic Integration*. Springer, Berlin, 2003.

42. B. Gilbert, "A precise four-quadrant multiplier with sub-nanosecond response," *IEEE J. Solid-State Circuits*, vol. 3, no. 4, pp. 365–373, 1968.

43. A. Rofougaran, J.Y.-C. Chang, M. Rofougaran, S. Khorram, and A.A. Abidi, "A 1 GHz CMOS RF front-end IC with wide dynamic range," in *Proceedings of the European Solid State Circuits Conference, ESSCIRC*, September 1995, pp. 250–253.

44. M. Borremans and M. Steyaert, "A 2 V low distortion 1 GHz CMOS up-converter mixer," *IEEE J. Solid-State Circuits*, vol. 33, no. 3, pp. 359–366, 1998.

45. M. Borremans, M. Steyaert, and T. Yoshitomi, "A 1.5 V wide band 3 GHz CMOS quadrature direct up-converter for multi-mode wireless communications," in *Proceedings of the Custom Integrated Circuits Conference, CICC*, May 1998, pp. 79–82.

46. M. Pelgrom, A. Duinmaijer, and A. Welbers, "Matching properties of mos transistor," *IEEE J. Solid-State Circuits*, vol. 24, no. 5, pp. 1433–1439, 1989.

47. D. Su and W. McFarland, "A 2.5 V, 1-W monolithic CMOS RF power amplifier," in *Proceedings of the Custom Integrated Circuits Conference, CICC*, May 1997, pp. 189–192.

48. K.C. Tsai and P.R. Gray, "1.9-GHz 1-W CMOS RF power amplifier for wireless communication," *IEEE J. Solid-State Circuits*, vol. 34, no. 7, pp. 962–970, 1999.

49. C. Yoo and Q. Huang, "A common-gate switched 0.9-W class-E power amplifier with 41% PAE in 0.25-μm CMOS," *IEEE J. Solid-State Circuits*, vol. 36, no. 5, pp. 823–830, 2001.

50. T. Sowlati and D. Leenaerts, "A 2.4 GHz 0.18 μm CMOS self-biased cascode power amplifier with 23 dBm output power," in *ISSCC Digest of Technical Papers*, San Francisco, USA, February 2002, pp. 294–295.

51. C. Fallesen and P. Asbeck, "A 1 W 0.35 μm CMOS power amplifier for GSM-1800 with 45% PAE," in *ISSCC Digest of Technical Papers*, San Francisco, USA, February 2001, pp. 158–159.

52. A. Shirvani, D.K. Su, and B.A. Wooley, "A CMOS RF power amplifier with parallel amplification for efficient power control," in *ISSCC Digest of Technical Papers*, San Francisco, USA, February 2001.

53. V.R. Vathulya, T. Sowlati, and D. Leenaerts, "Class 1 bluetooth power amplifier with 24 dBm output power and 48% PAE at 2.4 GHz in 0.25 μm CMOS," in *Proceedings of the European Solid State Circuits Conference, ESSCIRC*, Villach, Austria, 2001.

54. I. Aoki, S.D. Kee, D.B. Rutledge, and A. Hajimiri, "Fully integrated CMOS power amplifier design using the distributed active transformer architecture," *IEEE J. Solid-State Circuits*, vol. 37, no. 3, pp. 1–13, 2002.

55. N.O. Sokal and A.D. Sokal, "Class E—A new class of high-efficiency tuned single-ended switching power amplifiers," *IEEE Journal of Solid-State Circuits*, vol. 10, no. 3, June 1975, pp. 168–176.

61

PLL Circuits

Muh-Tian Shiue
National Central University

Chorng-kuang Wang
National Taiwan University

CONTENTS

61.1 Introduction

61.1.1 What's and Why Phase-Locked?

Phase-locked loop (PLL) is a circuit architecture that causes a particular system to track with another one. More precisely, PLL synchronizes a signal (usually a local oscillator output) with a reference or an input signal in frequency as well as in phase.

Phase locking is a useful technique that can provide effective synchronization solutions in many data transmission systems such as optical communications, telecommunications, disk drive systems, and local networks, in which data are transmitted in baseband or passband. In general, only data signals are transmitted in most of these applications, namely, clock signals are not transmitted to save hardware cost. Therefore, the receiver should have some mechanisms to extract the clock information from the received data stream to recover the transmitted data. The scheme is called a timing recovery or clock recovery.

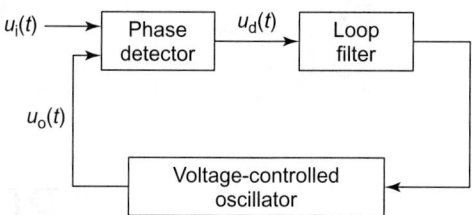

FIGURE 61.1   Basic block diagram of the PLL.

The cost of electronic interfaces in communication systems increases as the data rate gets higher. Hence, high-speed circuits are the critical issue of the high data rate systems implementation, and the advanced VLSI technology plays an important role in cost reduction for the high-speed communication systems.

61.1.2   Basic Operation Concepts of PLLs

Typically, as shown in Figure 61.1, a PLL consists of three basic functional blocks: a phase detector (PD), a loop filter (LF), and a voltage-controlled oscillator (VCO). PD detects the phase difference between the VCO output and the input signal, and generates a signal proportional to the phase error. The PD output contains a DC component and an AC component, the former is accumulated and the latter is filtered out by the loop filter. The loop filter output that is near a DC signal is applied to the VCO. This almost DC control voltage changes the VCO frequency toward a direction to reduce the phase error between the input signal and the VCO. Depending on the type of loop filter used, the steady-state phase error will be reduced to zero or to a finite value.

PLL has an important feature, which is the ability to suppress both the noises superimposed on the input signal and generated by the VCO. In general, the more narrow bandwidth the PLL has, the more effectively the filtering of the superimposed noises can be achieved. Although a narrow bandwidth is better for rejecting large amounts of the input noise, it also prolongs the settling time in the acquisition process. Then, the error of the VCO frequency cannot be reduced rapidly. So there is a trade-off between jitter filtering and fast acquisition.

61.1.3   Classification of PLL Types

Different PLL types have been built from different classes of building blocks. The first PLL ICs appeared around 1965 and were purely analog devices. In the so-called linear PLLs (LPLLs), an analog multiplier (four-quadrant) is used as the PD, the loop filter is built of a passive or an active resistor-capacitor (RC) filter, and the VCO is used to generate the output signal of the PLL. In most cases, the input signal to this linear PLL is a sine wave, whereas the VCO output signal is a symmetrical square wave.

The classical digital PLL (DPLL) uses a digital PD such as an XOR gate, a JK-flipflop, or a phase-frequency detector (PFD). The remaining blocks are still the same as LPLL. In many aspects, the DPLL performance is similar to the LPLL.

The function blocks of the all digital PLL (ADPLL) is implemented by purely digital circuits, and the signals within the loop are digital too. Digital versions of the PD are the same as DPLL. The digital loop filter is built of an ordinary up-downcounter, N-before-M counter or K-counter [1]. The digital counterpart of the VCO is the digital-controlled oscillator (DCO) [2,3].

In analogy to filter designs, PLLs can be implemented by software such as a microcontroller, microcomputer or digital signal processing (DSP), this type of PLL is called software PLL (SPLL).

61.2 PLL Techniques

61.2.1 Basic Topology

A PLL is a feedback system that operates and minimizes the phase difference between two signals. The PD works as a phase-error detector and an amplifier. It compares the phase of the VCO output signal $u_o(t)$ with the phase of the reference signal $u_i(t)$ and develops an output signal $u_d(t)$ that is proportional to the phase error θ_e. Within a limited range, the output signal can be expressed as

$$u_d(t) = k_d \theta_e \tag{61.1}$$

where k_d (in V/rad) represents the gain of the PD.

The output signal $u_d(t)$ of the PD consists of a DC component and a superimposed AC component. The latter is undesired and removed by the loop filter (LPF). Thus, the LPF generates an almost DC control voltage for the VCO to oscillate at the frequency equal to the input frequency.

How the building blocks of a basic PLL work together will be explained in the following. At first, assume both the waveforms of input signal and VCO output are rectangular. Furthermore, it is assumed that the angular frequency ω_i of the input signal $u_i(t)$ is equivalent to the central frequency ω_o of the VCO signal $u_o(t)$. Now a small positive frequency step is applied to $u_i(t)$ at $t = t_0$ (shown in Figure 61.2). $u_i(t)$ accumulates the phase increments faster than $u_o(t)$ of VCO does. If the PD can response wider pulses increasingly, a higher DC voltage is accordingly generated at the LPF output to increase the VCO frequency. Depending on the type of the loop filter that will be discussed later, the final phase error will be reduced to zero or a finite value.

It is important to note from the descriptions above that the loop locks only after the two conditions are satisfied: (1) ω_i and ω_o are equal and (2) the phase difference between the input $u_i(t)$ and the VCO output $u_o(t)$ settles to a steady-state value. If the phase error varies with time so fast that the loop is unlocked, the loop must keep on the transient process, which involves both "frequency acquisition" and "phase acquisition."

To design a practical PLL system, it is required to know the status of the responses of the loop if (1) the input frequency is varied slowly (tracking process), (2) the input frequency is varied abruptly (lock-in process), and (3) the input and the output frequencies are not equal initially (acquisition process). Using linear PLL as an example, these responses will be shown in Sections 61.2.3–61.2.5.

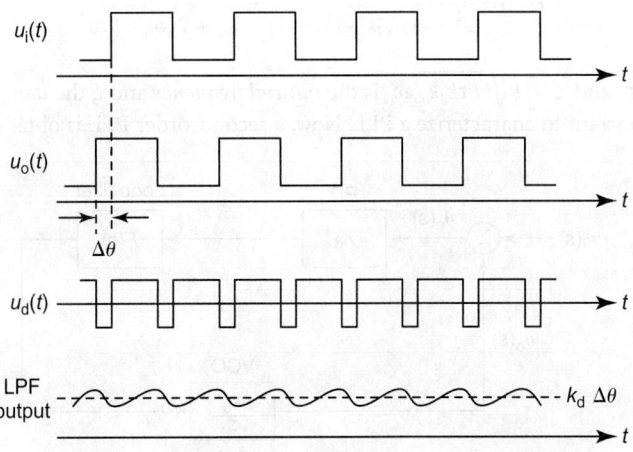

FIGURE 61.2 Waveforms in a PLL.

61.2.2 Loop Orders of the PLL

Figure 61.3 shows the linear model of a PLL. According to the control theory, the close-loop transfer function of PLL can be derived as

$$H(s) \triangleq \frac{\theta_o(s)}{\theta_i(s)} = \frac{k_d k_o F(s)}{s + k_d k_o F(s)} \qquad (61.2)$$

where k_d (in V/rad) is called the phase-detector gain, k_o (in rad/s V) the VCO gain factor. In addition to the phase transfer function, a phase-error transfer function $H_e(s)$ is derived as follows:

$$H_e(s) \triangleq \frac{\theta_e(s)}{\theta_i(s)} = \frac{s}{s + k_d k_o F(s)} \qquad (61.3)$$

The loop order of the PLL depends on the characteristics of the loop filter. Therefore, the loop filter is a key component that affects the PLL dynamic behavior. A PLL with a loop filter consisted of simple amplifier or attenuator is called a first-order PLL. As shown in Figure 61.3, set $F(s) = 1$ and the close-loop transfer function can be derived as

$$H(s) = \frac{k}{s + k} \qquad (61.4)$$

where $k = k_d k_o$. If it is necessary to design a high DC loop gain for fast tracking, then the bandwidth of the PLL must be wide enough because the DC loop gain k is the only parameter available, which is not suitable for noise suppression. Therefore, fast tracking and narrow bandwidth are incompatible in a first-order loop.

Another commonly used loop filter is the passive lag filter. The transfer function is

$$F(s) = \frac{1}{1 + s\tau} \qquad (61.5)$$

The close-loop transfer function can be derived as

$$H(s) = \frac{k_d k_o (1/\tau)}{s^2 + (1/\tau)s + (k_d k_o/\tau)} = \frac{\omega_n^2}{s^2 + 2\zeta\omega_n s + \omega_n^2} \qquad (61.6)$$

where $\omega_n = \sqrt{k_d k_o/\tau}$ and $\zeta = \frac{1}{2}\sqrt{1/\tau k_d k_o}$, ω_n is the *natural frequency* and ζ the *damping factor*. These two parameters are important to characterize a PLL. Now, a second-order PLL is obtained and there are two

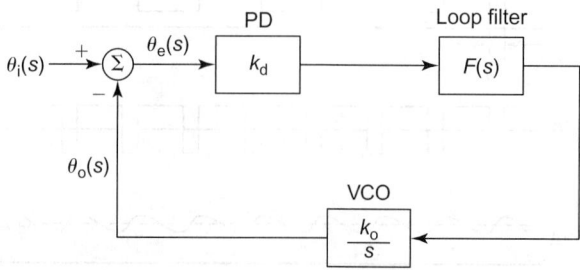

FIGURE 61.3   Linear model of PLL.

parameters (τ and $k = k_o k_d$) available to achieve fast tracking as well as the noise suppression. Then three loop parameters (ω_n, ζ, k) must be determined. If it is necessary to have a large DC loop gain and a very narrow bandwidth, the loop will be severely underdamped and the transient response will be poor.

In practice, there are three basic types of loop filter: passive lead-lag filter, active lead-lag filter, and active PI (proportional and integral) filter. The characteristics of the three types of loop filter and their effects on the PLL will be described in Section 3.3. Besides, a high-order filter is used for critical applications, because it provides better noise filtering, initial acquisition, and fast tracking. However, it is diffcult to design a high-order loop due to some problems such as loop stability.

61.2.3 Tracking Process

The linear model of a PLL shown in Figure 61.3 is suitable for analyzing the tracking performance of a PLL that is almost in lock, only with a small phase error. If the phase error changes too abruptly, the PLL fails to lock, and a large phase error is induced even though the change happens only momentarily. The unlock condition is a nonlinear process that cannot be analyzed via the linear model. The acquisition process will be described in Section 61.2.5.

At first, consider that a step phase error expressed as $\theta_i(t) = \Delta\theta_u(t)$ is applied to the input. The Laplace transform of the input is $\theta_i(s) = \Delta\theta/s$ that is substituted into Eq. (61.3) to get

$$\theta_e(s) = \frac{\Delta\theta}{s} \frac{s^2}{s^2 + 2\zeta\omega_n s + \omega_n^2} \tag{61.7}$$

According to the final value theorem of the Laplace transform,

$$\lim_{t\to\infty} \theta_e(t) = \lim_{s\to 0} s\theta_e(s) = 0$$

In another word, the loop will eventually track on the step phase change without steady-state phase error.

When a step change of frequency $\Delta\omega$ is applied to the input, the input phase change is a ramp, i.e., $\theta_i(t) = \Delta\omega t$, therefore $\theta_i(s) = \Delta\omega/s^2$. Substituting $\theta_i(s)$ in Eq. (61.3) and applying the final value theorem, then

$$\theta_v = \lim_{t\to\infty} \theta_e(t) = \lim_{s\to 0} s\theta_e(s)$$

$$= \lim_{s\to 0} \frac{\Delta\omega}{s + k_d k_o F(s)}$$

$$= \frac{\Delta\omega}{k_d k_o F(0)}$$

$$= \frac{\Delta\omega}{k_v} \tag{61.8}$$

where θ_v is called the *velocity error* or *static phase error* [4]. In practice, the input frequency almost never agrees exactly with the VCO free-running frequency, that is, usually there is a frequency difference $\Delta\omega$ between the two. From Eq. (61.8), if the PLL has a high DC loop gain, that is, $k_d k_o F(0) \gg \Delta\omega$, the steady-state phase error corresponding to a step frequency error input approaches to zero. This is the reason that a high gain loop has a good tracking performance. Now the advantage of a second-order loop using an active loop filter with high DC gain is evident. The active lead-lag loop filter with a high DC gain will make the steady-state phase error approach to zero and the noise bandwidth narrow simultaneously, which is impossible in a first-order loop.

If the input frequency is changed linearly with time at a rate of $\Delta\omega$, that is, $\theta_i(t) = \frac{1}{2}\Delta\omega t^2$, $\theta_i(s) = \Delta\omega/s^3$. According to a high gain loop and applying the final value theorem of Laplace transform, it is derived that

$$\theta_a = \lim_{t\to\infty}\theta_e(t) = \lim_{s\to 0} s\theta_e(s)$$

$$= \frac{\Delta\omega}{\omega_n^2} \tag{61.9}$$

where θ_a is called an *acceleration error* (sometimes calls *dynamic tracking error* or *dynamic lag*) [4].

In some applications, PLL needs to track an accelerating phase error without static tracking error. When frequency ramp is applied, the static phase error will be

$$\theta_e(s) = \lim_{s\to 0}\frac{\Delta\omega}{s(s + k_d k_o F(s))} \tag{61.10}$$

For θ_e to be zero, it is necessary to make $F(s)$ be a form of $G(s)/s^2$, where $G(0) \neq 0$. $G(s)/s^2$ implies that the loop filter has two cascade integrators. This results in a third-order loop. To eliminate the static acceleration error, a third-order loop is very useful for some special applications such as satellite and missile systems.

On the basis of Eq. (61.9), a large natural frequency ω_n is used to reduce the static tracking phase error in a second-order loop, however, a wide natural frequency has an undesired noise filtering performance. In the contrast, the zero tracking phase error for a frequency ramp error is concordant with a small loop bandwidth in a third-order loop.

All the preceding analysis on the tracking process is under the assumption that the phase error is relatively small and the loop is linear. If the phase error is large enough to make the loop drop out of lock, the linear assumption is invalid. For a sinusoidal-characteristic PD, the exact phase expression of Eq. (61.8) should be

$$\sin\theta_v = \frac{\Delta\omega}{k_v} \tag{61.11}$$

The sine function has solutions only when $\Delta\omega \leq k_v$. However, there is no solution if $\Delta\omega > k_v$. This is the case the loop loses lock and the output of the PD will be beat notes signal rather than a DC control voltage. Therefore, k_v can be used to define the *hold range* of the PLL, that is

$$\Delta\omega_H = \pm k_v = k_o k_d F(0) \tag{61.12}$$

The hold range is the frequency range in which a PLL is able to maintain lock *statically*. Namely, if input frequency offset exceeds the hold range statically, the steady-state phase error would drop out of the linear range of the PD and the loop loses lock. k_v is the function of k_o, k_d and $F(0)$. The DC gain $F(0)$ of the loop filter depends on the filter type. Therefore, it is important to make a loop filter have a high DC gain for extending the hold range. Referring to the characteristics of the three basic types of loop filter described in Section 61.3.3, the hold range $\Delta\omega_H$ can be $k_o k_d$, $k_o k_d k_a$, and ∞ for passive lead-lag filter, active lead-lag filter, and active PI filter, respectively. The hold range expressed in Eq. (61.12) is not correct when some other components in PLL are saturated earlier than the PD. When the PI filter is used, the real hold range is actually determined by the control range of the VCO.

Considering the dynamic phase error θ_a in a second-order loop, the exact expression for a sinusoidal characteristic PD is

$$\sin\theta_a = \frac{\Delta\omega}{\omega_n^2} \tag{61.13}$$

which implies that the maximum change rate of the input frequency is ω_n^2. If the rate exceeds ω_n^2, the loop will fall out of lock.

61.2.4 Lock-in Process

The *lock-in* process is defined as PLL locks within one single beat note between the input and the output (VCO output) frequency. The maximum frequency difference between the input and the output that PLL can lock within one single beat note is called the *lock-in range* of the PLL.

Figure 61.4 shows a case of PLL lock-in process that a frequency offset $\Delta\omega$ is less than the lock-in range, and the lock-in process happens. Then PLL will lock within one single beat note between ω_i and ω_o. In Figure 61.4, the frequency offset $\Delta\omega$ between input (ω_i) and output (ω_o) is larger than the lock-in range, hence the lock-in process will not take place, at least not instantaneously.

Suppose the PLL is unlocked initially. The input frequency ω_i is $\omega_o + \Delta\omega$. If the input signal $v_i(t)$ is a sine wave and given by

$$v_i(t) = A_i \sin(\omega_o t + \Delta\omega t) \tag{61.14}$$

And the VCO output signal $v_o(t)$ is usually a square wave written as a Walsh function [5]

$$v_o(t) = A_o W(\omega_o t) \tag{61.15}$$

$v_o(t)$ can be replaced by the Fourier series,

$$v_o(t) = A_o \left[\frac{4}{\pi} \cos(\omega_o t) + \frac{4}{3\pi} \cos(3\omega_o t) + \cdots \right] \tag{61.16}$$

So the PD output v_d is

$$v_d(t) = v_i(t) v_o(t) = A_i A_o \left[\frac{2}{\pi} \sin(\Delta\omega t) + \cdots \right] \tag{61.17}$$

$$= k_d \sin(\Delta\omega t) + \text{high} - \text{frequency terms}$$

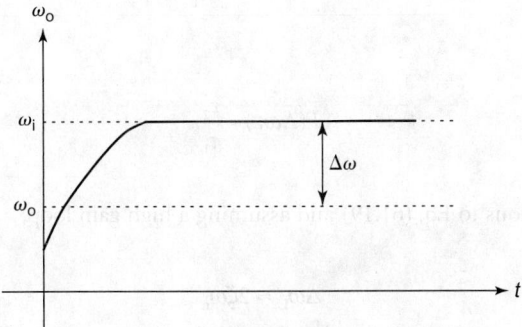

FIGURE 61.4 Lock-in process of the PLL.

The high-frequency components can be filtered out by the loop filter. The output of the loop filter is given by

$$v_f(t) \approx k_d \, | F(\Delta\omega)| \sin(\Delta\omega t) \qquad (61.18)$$

The peak frequency deviation based on Eq. (61.18) is equal to $k_d k_o |F(\Delta\omega)|$. If the peak deviation is larger than the frequency error between ω_i and ω_o, the lock-in process will take place. Hence the lock-in range is consequently given by

$$\Delta\omega_L = k_d k_o \, | F(\Delta\omega_L)| \qquad (61.19)$$

The lock-in range is always larger than the corner frequency $1/\tau_1$ and $1/\tau_2$ of the loop filter in practical cases. An approximation of the loop filter gain $F(\Delta\omega_L)$ is shown as follows:

for the passive lead-lag filter,

$$F(\Delta\omega_L) \approx \frac{\tau_2}{\tau_1 + \tau_2}$$

for the active lead-lag filter,

$$F(\Delta\omega_L) \approx k_a \frac{\tau_2}{\tau_1}$$

for the active PI filter,

$$F(\Delta\omega_L) \approx \frac{\tau_2}{\tau_1}$$

τ_2 is usually much smaller than τ_1, the $F(\Delta\omega_L)$ can be further approximated as follows:

for the passive lead-lag filter,

$$F(\Delta\omega_L) \approx \frac{\tau_2}{\tau_1}$$

for the active lead-lag filter,

$$F(\Delta\omega_L) \approx k_a \frac{\tau_2}{\tau_1}$$

for the active PI filter,

$$F(\Delta\omega_L) \approx \frac{\tau_2}{\tau_1}$$

Substituting above equations to Eq. (61.19) and assuming a high gain loop,

$$\Delta\omega_L \approx 2\zeta\omega_n \qquad (61.20)$$

can be obtained for all three types of loop filter shown in Figure 61.7.

61.2.5   Acquisition Process

Suppose that the PLL does not lock initially, the input frequency is $\omega_i = \omega_o + \Delta\omega$, where ω_o is the initial frequency of VCO. If the frequency error $\Delta\omega$ is larger than the lock-in range, the lock-in process will not occur. Consequently, the output signal $u_d(t)$ of the PD shown in Figure 61.5(a) is a sine wave that has the frequency of $\Delta\omega$. The AC PD output signal $u_d(t)$ passes through the loop filter. Then the output $u_f(t)$ of the loop filter modulates the VCO frequency. As shown in Figure 61.5(b), when ω_o increases, the frequency difference between ω_i and ω_o becomes smaller and vice versa. Therefore, the phase detector output $u_d(t)$ becomes asymmetric when the duration of positive half-periods of the PD output is larger than the negative ones. The average value $\overline{u_d(t)}$ of the PD output therefore becomes positive slightly. Then the frequency of VCO will be pulled up until it reaches the input frequency. This phenomenon is called a *pull-in process*.

Because the pull-in process is a nonlinear behavior, the mathematical analysis is quite complicated. According to the results of Ref. [1], the pull-in range and the pull-in time depend on the type of loop filter. For an active lead-lag filter with a high gain loop, the pull-in range is

$$\Delta\omega_P \approx \frac{4\sqrt{2}}{\pi}\sqrt{\zeta\omega_n k_o k_d} \tag{61.21}$$

and the pull-in time is

$$T_P \approx \frac{\pi^2}{16}\frac{\Delta\omega_0^2 k_a}{\zeta\omega_n^3} \tag{61.22}$$

where $\Delta\omega_0$ is the initial frequency error. Eq. (61.21) and Eq. (61.22) should be modified for different types of PDs [1].

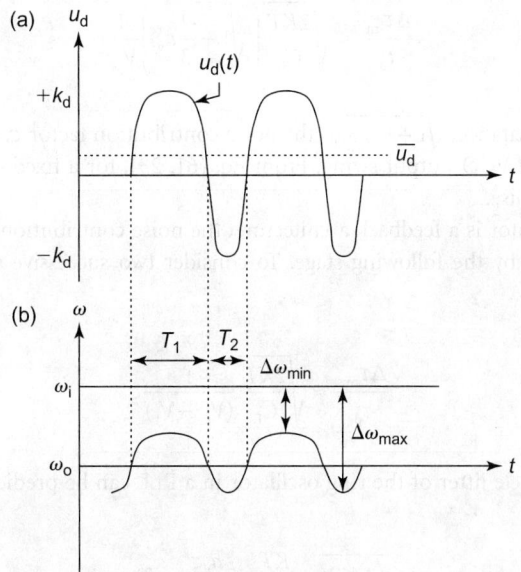

FIGURE 61.5 Pull-in process of the PLL.

61.2.6 Aided Acquisition

The PLL bandwidth is always too narrow to lock a signal with large frequency error. Furthermore, the frequency acquisition is slow and impractical. Therefore, there are aided frequency-acquisition techniques to solve this problem such as the frequency-locked loop (FLL) and the bandwidth-widening methods.

The FLL, which is very much similar to a PLL, is composed of a frequency discriminator, a loop filter, and a VCO. PLL is a coherent mechanism to recover a signal buried in noise. An FLL, in contrast, is a noncoherent scheme that cannot distinguish the phase error between input signal and VCO signal. Therefore, an FLL can only be useful to provide the signal frequency which exactly synchronizes with the reference frequency (RF) source.

The major difference between PLL and FLL is the PD and the frequency discriminator. The frequency discriminator is the frequency detector in the FLL. It generates a voltage proportional to the frequency difference between the input and the VCO. The frequency difference will be driven to zero in a negative feedback fashion. If a linear frequency detector is employed, it can be shown that the frequency-acquisition time is proportional to the logarithm of the frequency error [6]. In the literature, some frequency detectors like quadricorrelator [7], balance quadricorrelator [8], rotational frequency detector [9], and frequency delimiter [10] are disclosed.

61.2.7 PLL Noise Performance

In high-speed data recovery applications, a better performance of the VCO and the overall PLL itself is desired. Consequently, the random variations of the sampling clock, so-called jitter, is the critical performance parameter.

Jitter sources of PLL in the case of using a ring VCO mainly come from the input and the VCO itself. The ring oscillator jitter is associated with the power supply noise, the substrate noise, $1/f$ noise, and the thermal noise. The former two noise sources can be reduced by fully differential circuit structure. $1/f$ noise, in contrast, can be rejected by the tracking capability of the PLL. Therefore, the thermal noise is the worst noise source. From the analysis of Ref. [11], the one-stage RMS timing jitter error of the ring oscillator normalized to the time delay per stage can be shown as

$$\frac{\Delta\tau_{rms}}{t_d} \approx \sqrt{\frac{2KT}{C_L}\left(\sqrt{1+\frac{2}{3}a_v}\right)\frac{1}{V_{pp}}} \tag{61.23}$$

where C_L is the load capacitance, $\sqrt{1+(2/3)a_v}$ the noise contribution factor ς, a_v the small-signal gain of the delay cell, and V_{pp} the VCO output swing. From Eq. (61. 23), for a fixed output bandwidth, higher gain contributes larger noise.

Because the ring oscillator is a feedback architecture, the noise contribution of a single delay cell may be amplified and filtered by the following stage. To consider two successive stages, Eq. (61.23) can be rearranged as [11]

$$\frac{\Delta\tau_{rms}}{t_d} \approx \sqrt{\frac{2KT}{C_L}}\frac{1}{(V_{gs}-V_t)}\varsigma \tag{61.24}$$

Therefore, the cycle-to-cycle jitter of the ring oscillator in a PLL can be predicted by [11]

$$\overline{(\Delta\tau_N)^2} = \frac{KT}{I_{ss}}\frac{a_v\varsigma^2}{(V_{gs}-V_t)}T_o \tag{61.25}$$

where I_{ss} is the rail current of the delay cell, T_o the output period of the VCO. On the basis of Eq. (61.25), designing a low jitter VCO, ($V_{gs} - V_t$) should be as large as possible. For fixed delay and fixed current, a lower gain of each stage is better for jitter performance, but the loop gain must satisfy the Barkhausen criterion. From the viewpoint of VCO jitter, a wide bandwidth of PLL can correct the timing error of the VCO rapidly [12]. If the bandwidth is too wide, the input noise jitter may be so large that dominates the jitter performance of the PLL. Actually this is a trade-off.

For a PLL design, the natural frequency and the damping factor are the key parameters to be determined by designers. If the input signal-to-noise ratio $(SNR)_i$ is defined, then the output signal-to-noise ratio $(SNR)_o$ can be obtained [4]:

$$(SNR)_o = (SNR)_i \frac{B_i}{2B_L} \tag{61.26}$$

where B_i is the bandwidth of the prefilter and B_L the noise bandwidth. Hence the B_L can be derived using Eq. (61.26), and the relationship of B_L with ω_n and ζ is

$$B_L = \frac{\omega_n}{2}\left(\zeta + \frac{1}{4\zeta}\right) \tag{61.27}$$

Therefore ω_n and ζ can be designed to satisfy the $(SNR)_o$ requirement.

Beside the system and the circuit designs, jitter can be reduced in the board level design. Board jitter can be alleviated by better layout and noise-decoupling schemes, such as appending proper decouple and bypass capacitances.

61.3 Building Blocks of PLL Circuit

61.3.1 Voltage-Controlled Oscillators

The function of a VCO is to generate a stable and periodic waveform whose frequency can be varied by an applied control voltage. The relationship between the control voltage and the oscillation frequency depends upon the circuit architecture. A linear characteristic is generally preferred because of its wider applications. As a general classification, VCO can be categorized roughly into two types by the output waveforms: (i) harmonic oscillators that generate nearly sinusoidal outputs and (ii) relaxation oscillators that provide square or triangle outputs.

In general, a harmonic oscillator is composed of an amplifier that provides an adequate gain and a frequency-selective network that feeds a certain output frequency range back to the input. LC tank oscillators and crystal oscillators belong to this type. Generally, the harmonic oscillators have the following advantages: (1) superior frequency stability, which includes the frequency stability with temperature, power supply, and noise; (2) good frequency accuracy control, because the oscillation frequency is determined by a tank circuit or a crystal.

Essentially, harmonic oscillators are not compatible with monolithic IC technology and their frequency-tuning range is limited. On the contrary, relaxation oscillators are easy to be implemented in monolithic ICs. Since frequency is normally proportional to a controlled-current or -voltage and inversely proportional to timing capacitors, the frequency of oscillation can be varied linearly over a very wide range. In contrast, the ease of frequency tuning brings in drawbacks, such as poor frequency stability and frequency inaccuracy.

Relaxation oscillators are the most commonly used oscillator configuration in monolithic IC design, because they can operate in a wide frequency range with a minimum number of external components. According to the mechanism of the oscillator topology employed, relaxation oscillators can be further categorized into three types: (1) grounded capacitor VCO [13], (2) emitter-coupled VCO, and (3) delay-based

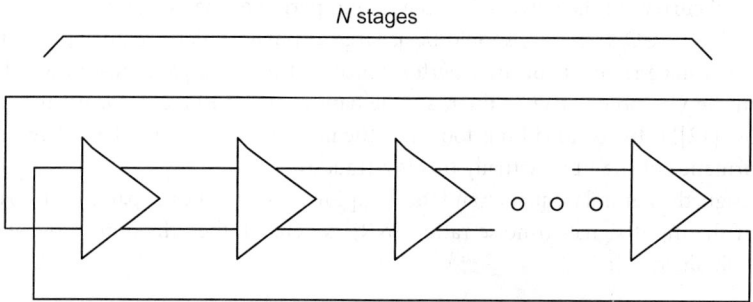

FIGURE 61.6 Ring oscillator.

ring VCO [14]. The operation of the first two oscillators are similar in the sense that time duration spent in each state is determined by the timing components and charge/discharge currents. The delay-based ring VCO operates quite differently since the timing relies on the delay in each gain stages that are connected in a ring configuration.

Recently, ring oscillator has received considerable attentions in high-frequency PLL applications for clock synchronization and timing recovery. Since they can provide high-frequency oscillation with simple digital-like circuits that are compatible with digital technology, they are suitable for VLSI implementations.

To achieve high rejection of power supply and substrate noises, both the signal path and the control path of a VCO must be fully differential. A common ring oscillator topology in monolithic PLLs is shown in Figure 61.6. The loop oscillates with a period equal to $2NT_d$, where T_d is the delay of each stage. The oscillation can be obtained when the total phase shift is zero and the loop gain is greater or equal to unity at a certain frequency. To vary the frequency of oscillation, the effective number of stages or the delay of each stage must be changed. The first approach is called "delay interpolating" VCO [14], where a shorter and a longer delay path are used in parallel. The total delay is tuned by increasing the gain of one path and decreasing the other, and the total delay is a weighted sum of the two delay paths. The second approach is to vary the delay time of each stage to adjust the oscillation frequency. The delay of each stage is tuned by varying the capacitance or the resistance seen at the output node of each stage. Because the tuning range of the capacitor is small and the maximum oscillation frequency is limited by the minimum value of the load capacitor, the "resistive tuning" is a better alternative technique. Resistive-tuning method provides a large, uniform frequency tuning range and leads itself easily to a differential control. In Figure 61.7(a), the on-resistance of the triode PMOS loads are adjusted by V_{cont}. The more V_{cont} decreases, the more the delay of the stage drops, because the time constant at the output node is decreased. However, the small-signal gain decreases as well when V_{cont} decreases. The circuit eventually fails to oscillate when the loop gain at the oscillation frequency is less than unity. In Figure 61.7(b), the delay of gain stage is tuned by adjusting the tail current, but the small-signal gain remains constant. So the circuit is better than Figure 61.7(a). As shown in Figure 61.7(c) [15], the PMOS current source with a pair of cross-coupled diode loads provide a differential load impedance that is independent of common-mode voltage. This makes the cell delay insensitive to common-mode noise. Figure 61.7(d) is a poor delay cell for a ring oscillator, because the tuning range is very small.

The minimum number of stages that can be used while maintaining a reliable operation is an important issue in a ring oscillator design. When the number of stages decreases, the required phase shift and DC gain per stage increases. Two-stage bipolar ring oscillator can be designed reliably [16], but CMOS implementations are not. Thus, CMOS ring oscillators utilize three or more stages typically.

61.3.2 Phase and Frequency Detectors

The PD type has the influence on the dynamic range of PLLs. Hold range, lock-in range, and pull-in range are analyzed in Sections 61.2.3, 61.2.4, and 61.2.5, respectively based on the multiplier phase detector. Most of the other types of PDs have a greater linear output span and a larger maximum output swing than a sinusoidal characteristic PD. A larger tracking range and a larger lock limit are available if the linear output range of PD increases. The three widely used PDs are XOR PD, edge-triggered JK-flipflop, and PFD. The characteristics of these PDs are plotted in Figure 61.8.

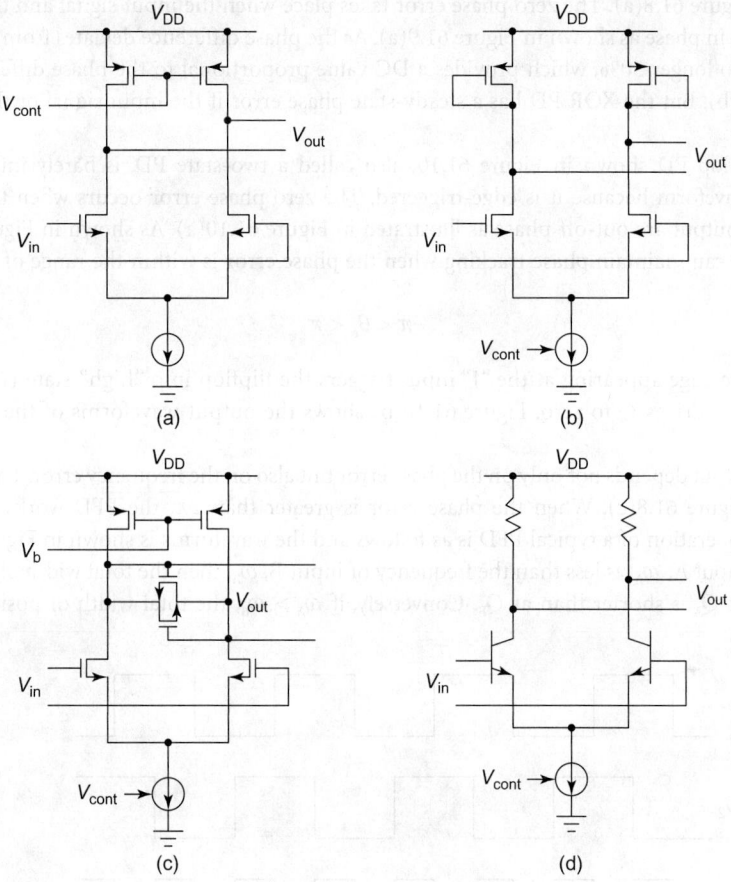

FIGURE 61.7 The gain stages using resistive tuning.

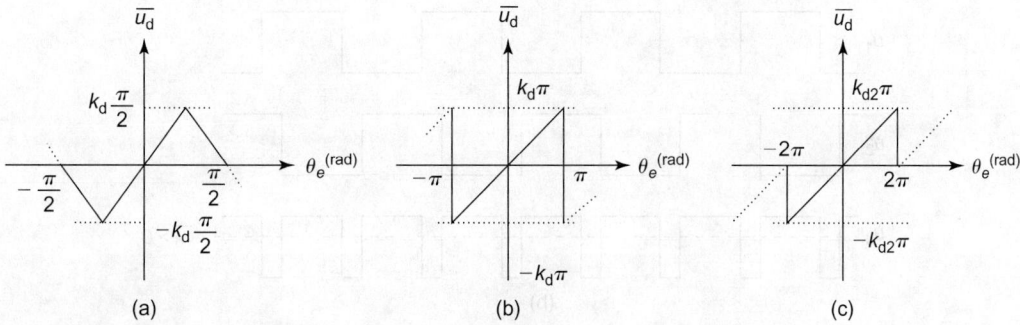

FIGURE 61.8 Phase detector characteristics of (a) XOR, (b) JK-flipflop, and (c) PFD.

The XOR PD can maintain phase tracking when the phase error θ_e is confined to the range

$$\frac{-\pi}{2} < \theta_e < \frac{\pi}{2}$$

as shown in Figure 61.8(a). The zero phase error takes place when the input signal and the VCO output are quadrature in phase as shown in Figure 61.9(a). As the phase difference deviates from $\pi/2$, the output duty cycle is no longer 50%, which provides a DC value proportional to the phase difference as shown in Figure 61.9(b). But the XOR PD has a steady-state phase error if the input signal or the VCO output are asymmetric.

The JK-flipflop PD shown in Figure 61.10, also called a two-state PD, is barely influenced by the asymmetric waveform because it is edge-triggered. The zero phase error occurs when the input signal and the VCO output are out-off phase as illustrated in Figure 61.10(a). As shown in Figure 61.8(b), the JK-flipflop PD can maintain phase tracking when the phase error is within the range of

$$-\pi < \theta_e < \pi$$

Here, a positive edge appearing at the "J" input triggers the flipflop into "high" state ($Q = 1$), and the rising edge of u_2 drives Q to zero. Figure 61.10(b) shows the output waveforms of the JK-flipflop PD for $\theta_e > 0$.

The PFD output depends not only on the phase error but also on the frequency error. The characteristic is shown in Figure 61.8(c). When the phase error is greater than 2π, the PFD works as a frequency detector. The operation of a typical PFD is as follows and the waveforms is shown in Figure 61.11. If the frequency of input A, ω_A, is less than the frequency of input B, ω_B, then the total width of positive pulses that appears at Q_A is shorter than at Q_B. Conversely, if $\omega_A > \omega_B$, the total width of positive pulses that

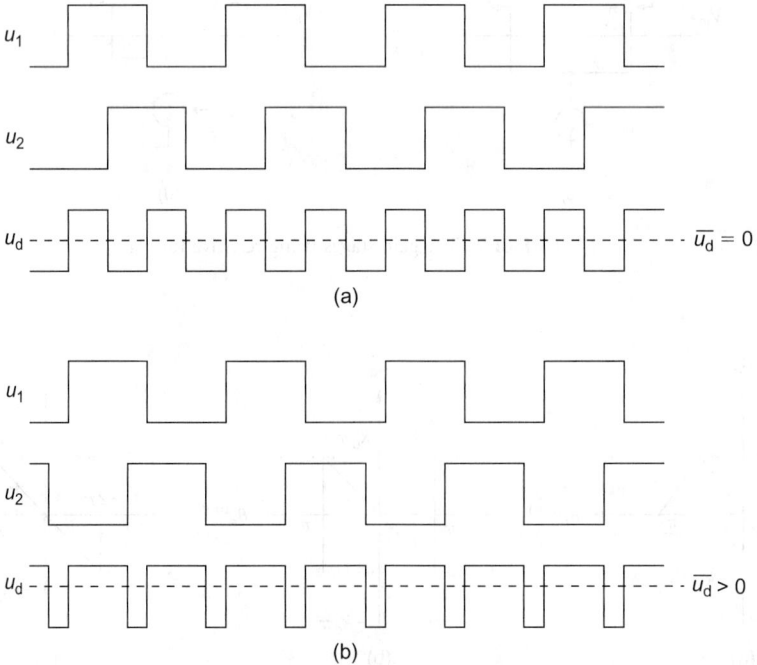

FIGURE 61.9 Waveforms of the signals for the XOR phase detector: (a) waveforms at zero phase error; (b) waveforms at positive phase error.

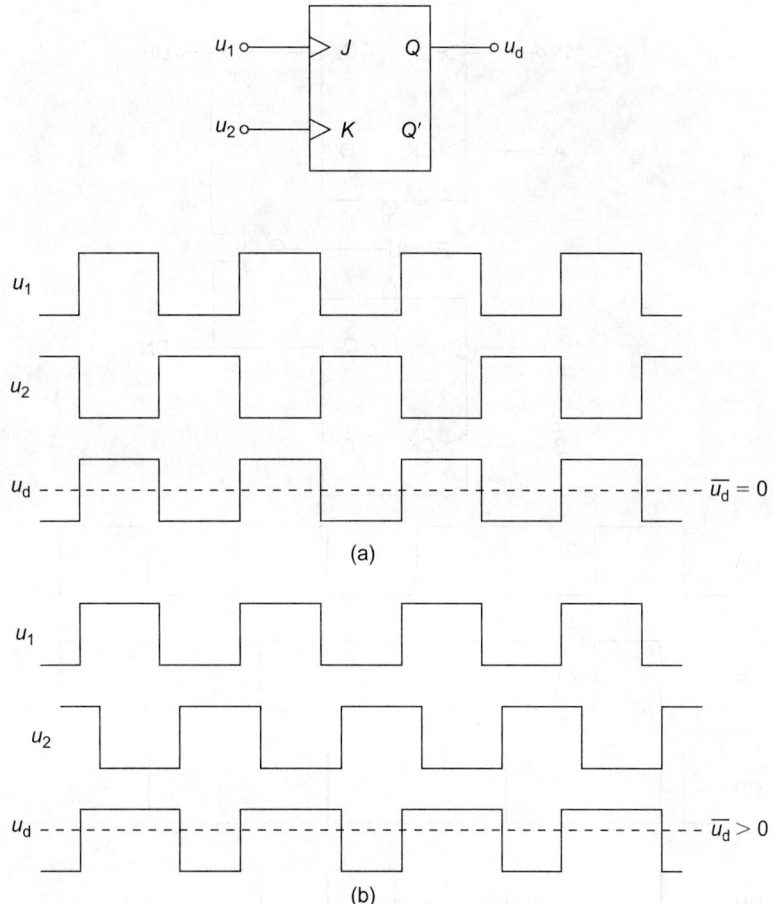

FIGURE 61.10 Waveforms of the signals for the JK-flipflop phase detector: (a) waveforms at zero phase error; (b) waveforms at positive phase error.

appears at Q_A is longer than at Q_B. If $\omega_A = \omega_B$, then the PFD generates pulses at either Q_A or Q_B with a width equal to the phase difference between the two inputs. The outputs Q_A and Q_B are usually called the "up" and "down" signals, respectively. If the input signal fails, which usually occurs at the nonreturn-to-zero (NRZ) data recovery applications during missing or extra transmissions, the output of the PFD would stick on the high state (or low state). This condition may cause VCO to oscillate fast or slow abruptly, which results in noise jitter or even losing lock. This problem can be remedied by additional control logic circuits to make the PFD output to toggle back and forth between the two logic level with 50% duty cycle [17], the loop is interpreted as zero phase error. The "rotational FD" described by Messerschmitt [9] can also solve this issue. The output of a PFD can be converted into a DC control voltage by driving a three-state charge-pump which will be described in Section 61.3.4.

61.3.3 Loop Filters

For a PLL with given VCO and PD, loop filter is the critical component to determine the PLL characteristics, such as the damping factor that determines the relative stability, open-loop/closed-loop bandwidths that relate to the convergence speed in the initial state and the tracking capability in the steady state, and so on. Various types of loop filter will be introduced in this section.

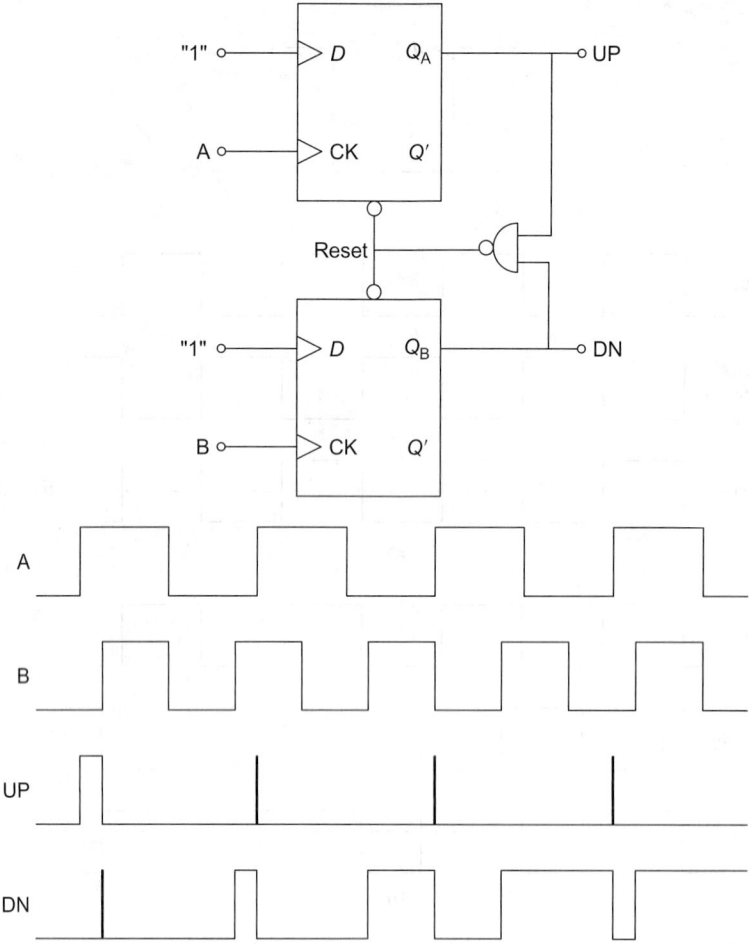

FIGURE 61.11 (a) PFD diagram. (b) Input and output waveforms of PFD.

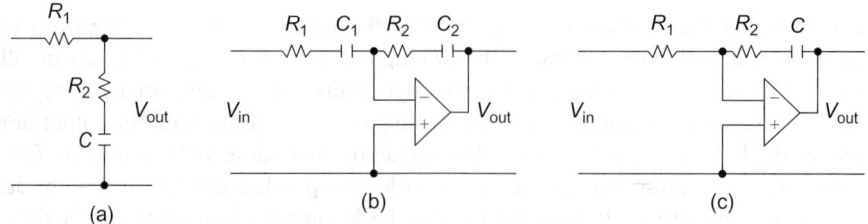

FIGURE 61.12 (a) Passive lead-lag filter; (b) active lead-lag filter; and (c) active PI filter.

61.3.3.1 Continuous-Time Loop Filters

Figure 61.12 shows three types of loop filters that are widely used. Figure 61.12(a) is a passive lead-lag filter with transfer function $F(s)$ given by

$$F(s) = \frac{1 + s\tau_2}{1 + s(\tau_1 + \tau_2)} \tag{61.28}$$

where $\tau_1 = R_1C$ and $\tau_2 = R_2C$. Figure 61.12(b) shows an active lead-lag filter, whose transfer function is

$$F(s) = k_a \frac{1 + s\tau_2}{1 + s\tau_1} \tag{61.29}$$

where $\tau_1 = R_1C_1$, $\tau_2 = R_2C_2$, and $k_a = -C_1/C_2$. A "*PI*" filter is shown in Figure 61.12(c), where "PI" stands for the "*proportional and integral*" action. The transfer function is given by

$$F(s) = \frac{1 + s\tau_2}{s\tau_1} \tag{61.30}$$

where $\tau_1 = R_1C$ and $\tau_2 = R_2C$. Their Bode diagrams are shown in Figures 61.13(a)–(c). High-order filters could be used in some applications, but additional filter poles introduce a phase shift. In general, it is not trivial to maintain the stability of high-order systems.

The transfer functions of the loop filters shown in Figure 61.12 are substituted for $F(s)$ in Eq. (61.2) to analyze the phase transfer function. We obtain the phase transfer functions as follows: for the passive lead-lag filter,

$$H(s) = \frac{k_d k_o \frac{1 + s\tau_2}{\tau_1 + \tau_2}}{s^2 + s\frac{1 + k_d k_o \tau_2}{\tau_1 + \tau_2} + \frac{k_d k_o}{\tau_1 + \tau_2}} = \frac{\omega_n\left(2\zeta - \frac{\omega_n}{k_d k_o}\right)s + \omega_n^2}{s^2 + 2s\zeta\omega_n + \omega_n^2} \tag{61.31}$$

for the active lead-lag filter,

$$H(s) = \frac{k_d k_a k_o \frac{1 + s\tau_2}{\tau_1}}{s^2 + s\frac{1 + k_d k_a k_o \tau_2}{\tau_1} + \frac{k_d k_a k_o}{\tau_1}} = \frac{\omega_n\left(2\zeta - \frac{\omega_n}{k_d k_a k_o}\right)s + \omega_n^2}{s^2 + 2s\zeta\omega_n + \omega_n^2} \tag{61.32}$$

and for the active PI filter,

$$H(s) = \frac{k_d k_o \frac{1 + s\tau_2}{\tau_1}}{s^2 + s\frac{k_d k_o \tau_2}{\tau_1} + \frac{k_d k_o}{\tau_1 + \tau_2}} = \frac{2\zeta\omega_n s + \omega_n^2}{s^2 + 2s\zeta\omega_n + \omega_n^2} \tag{61.33}$$

If the condition $k_d k_o \gg \omega_n$ or $k_d k_o k_a \gg \omega_n$ is true, this PLL system is called a *high-gain loop*. If the reverse is true, the system is a *low-gain loop*. Most practical PLLs are high-gain loops for good tracking performance. For a high-gain loop, Eqs. (61.31)–(61.33) become approximately

$$H(s) \approx \frac{2\zeta\omega_n s + \omega_n^2}{s^2 + 2s\zeta\omega_n + \omega_n^2} \tag{61.34}$$

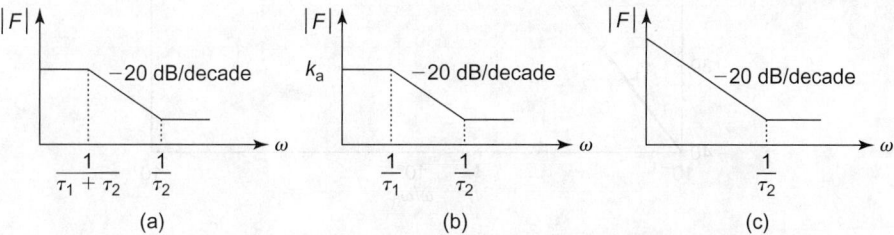

FIGURE 61.13 Bode diagrams of (a) passive lead-lag filter; (b) active lead-lag filter; (c) active PI filter.

Similarly, assuming a high-gain loop, the approximate expression of the phase-error transfer function $H_e(s)$ for all three loop filter types becomes

$$H_e(s) \approx \frac{s^2}{s^2 + 2s\zeta\omega_n + \omega_n^2} \tag{61.35}$$

The magnitude frequency responses of $H(s)$ for a high-gain loop with several values of damping factor are plotted in Figure 61.14. It exhibits that the loop performs a low-pass filtering on the input phase signal. That is, the second-order PLL is able to track both phase and frequency modulations of the input signal as long as the modulation frequency remains within the frequency band roughly between zero and ω_n.

The magnitude frequency responses of $H_e(s)$ are plotted in Figure 61.15. A high-pass characteristic is observed. It indicates that the second-order PLL tracks the low-frequency phase error but cannot track high-frequency phase error.

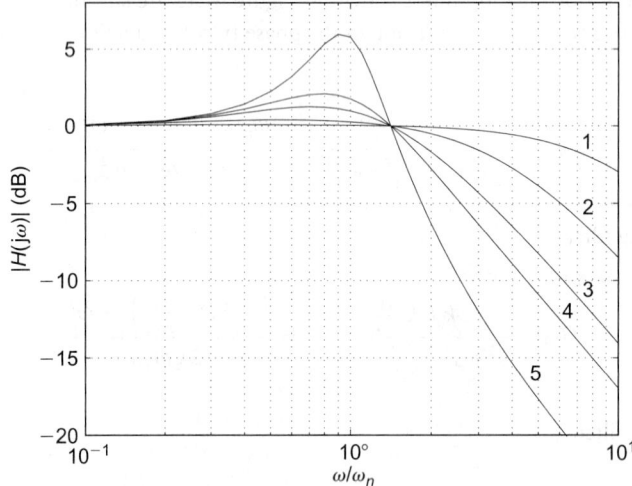

FIGURE 61.14 Frequency responses of the phase transfer function $H(j\omega)$ for different damping factors. Trace1, $\zeta = 5$; Trace2, $\zeta = 2$; Trace3, $\zeta = 1$; Trace4, $\zeta = 0.707$; Trace5, $\zeta = 0.3$.

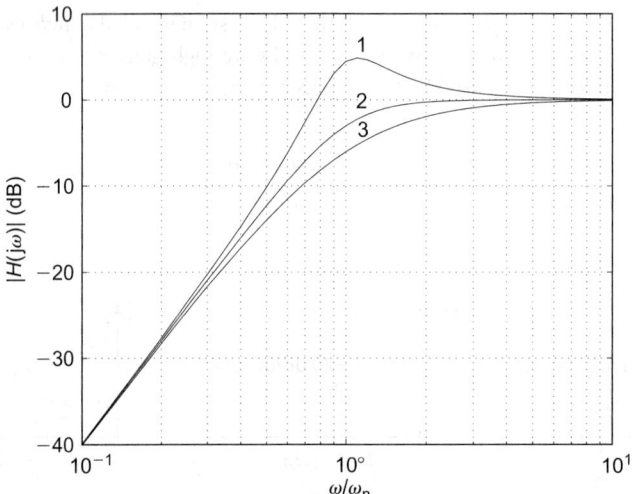

FIGURE 61.15 Frequency responses of the phase-error transfer function $H_e(j\omega)$ for different damping factors. Trace1, $\zeta = 0.3$; Trace2, $\zeta = 0.707$; Trace3, $\zeta = 1$.

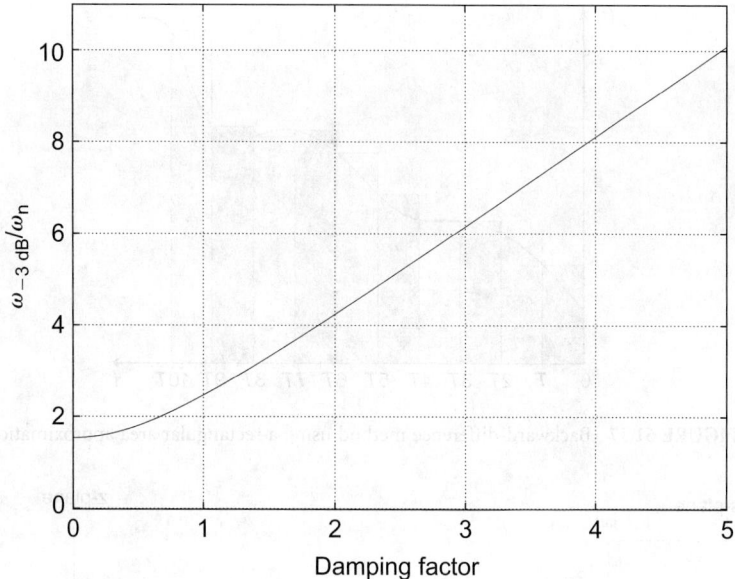

FIGURE 61.16 $\omega_{-3\text{dB}}$ Bandwidth of a second-order loop versus different damping factors.

The transfer function $H(s)$ has a -3 dB frequency, $\omega_{-3\text{dB}}$, which stands for the close-loop bandwidth of the PLL. The relationship between $\omega_{-3\text{dB}}$ and ω_n is presented here to provide a comparison with a familiar concept of bandwidth. In a high-gain loop case, by setting $|H(j\omega)| = 1/2$ and solving for ω, we can find

$$\omega_{-3\text{dB}} = \omega_n \left[2\zeta^2 + 1 + \sqrt{(2\zeta^2 + 1)^2} \right]^{1/2} \tag{61.36}$$

The relationship between $\omega_{-3\text{dB}}$ and ω_n for different damping factors is plotted in Figure 61.16 [4].

61.3.3.2 Transformations from Continuous Domain to Discrete Domain

As mentioned in the Section 61.1.3, the function blocks of the ADPLL is implemented by purely digital circuits, and the signals within the loop are digital too. In addition, SPLL implemented by a microcontroller, microcomputer, or DSP is another type of PLL in discrete domain. Therefore, the analysis and design of a PLL had to be better in discrete domain. The basic types of loop filter and their features have been described in the previous section. Here, the corresponding discrete-time version of the three basic types of loop filter will be described after the introduction of transformations from continuous domain to discrete domain. There are two popular methods to transform a filter from continuous domain to discrete domain: backward-difference method and bilinear transformation method. Figure 61.17 shows the principle of the backward-difference method that we approximate the area under each segment of continuous curve by a rectangular area. Referring to Figure 61.17, the backward-difference method means to approximate the integration areas of $\int_{(k-1)T}^{kT} y(t) \, dt$ by $y(kT)T$. On the basis of the backward-difference method, the z-domain equivalent transfer function $H(z)$ of an s-domain transfer function $H(s)$ is simple and obtained by the substitution

$$H(z) = H(s)\Big|_{s=(1-z^{-1})/T_s} \tag{61.37}$$

One of the advantages of the backward-difference method is that it will produce a stable discrete-time filter for a stable continuous-time filter. Figure 61.18 shows the mapping of the left half of the s-plane

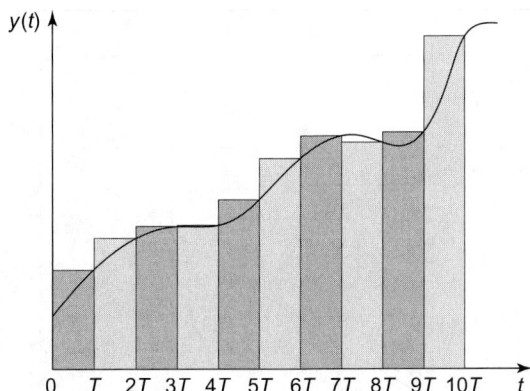

FIGURE 61.17 Backward-difference method using a rectangular area approximation.

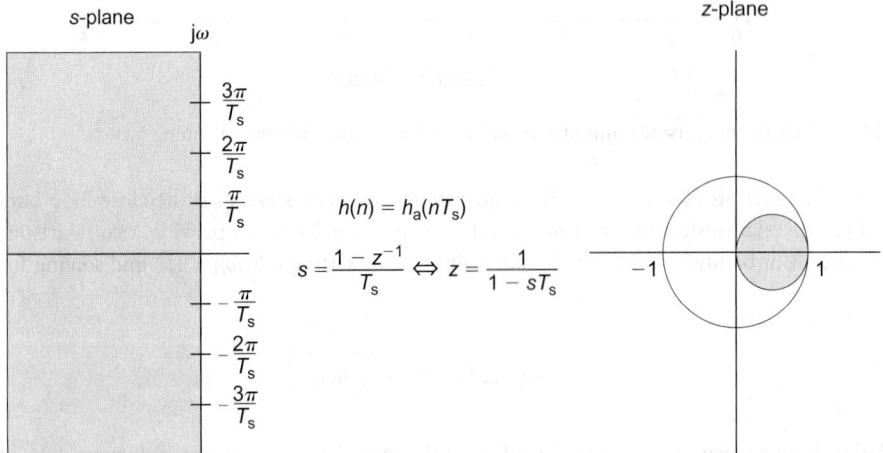

FIGURE 61.18 Mapping of the left half of the *s*-plane into the *z*-plane by the backward-difference method.

into the *z*-plane by the backward-difference method. However, there is considerable distortion in the transient and frequency response characteristics of the discrete-time filter obtained in this method since the stable region is mapped into only a circle within the unit circle of *z*-plane.

From calculus and Figure 61.17, a good approximation is obtained only if the continuous-time signal changes very slowly over the sampling interval T_s. In other words, the signal bandwidth has to be much smaller than the sampling rate since the mapping from *s*-domain to *z*-domain should become distorted while the sampling period is too long. To reduce the distortion, it is desired to use a faster sampling frequency, that is, a smaller sampling period.

Figure 61.19 shows the principle of the bilinear transformation method that we approximate the area under each segment of continuous curve by a trapezoidal area. Therefore, the bilinear transformation method is also called the trapezoidal integration method to approximate the integration areas $\int_{(k-1)T}^{kT} y(t)\,dt$ by $\frac{1}{2}[y(kT) + y((k-1)T)]T$. Thus, the *z*-domain equivalent transfer function $H(z)$ of a continuous-time filter $H(s)$ is obtained by

$$H(z) = H(s)\Big|_{s=2/T_s\ (1-z^{-1})/(1+z^{-1})} \tag{61.38}$$

By means of bilinear transformation method, the entire left half of the *s*-plane is mapped into the unit circle with center at the origin of the *z*-plane as shown in Figure 61.20. Hence, the bilinear transformation

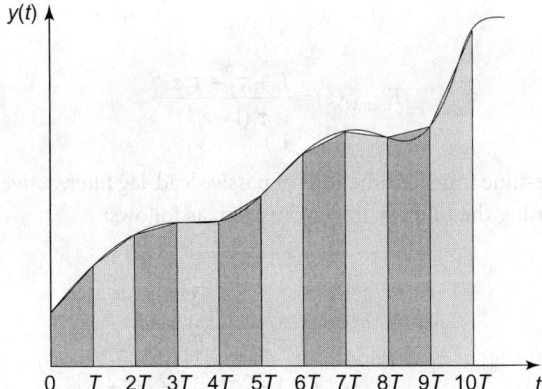

FIGURE 61.19 Bilinear transformation method using trapezoidal area approximation.

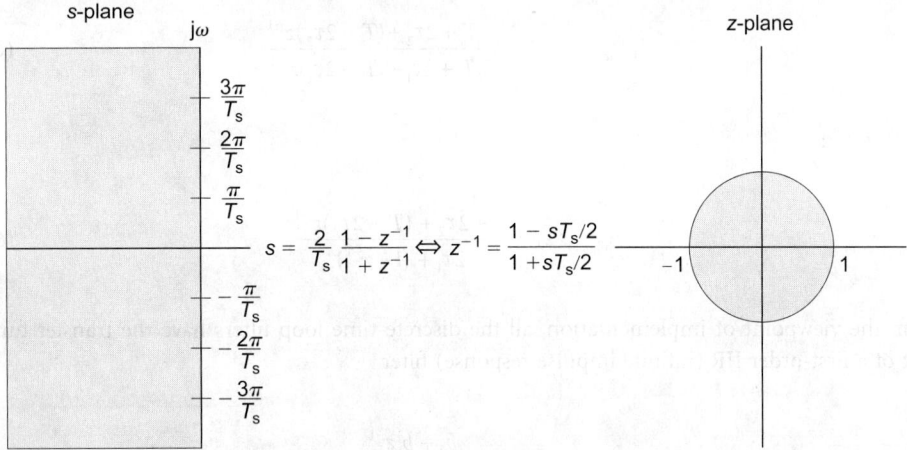

FIGURE 61.20 Mapping of the left half of the *s*-plane into the *z*-plane by the bilinear transformation method.

method produces a stable discrete-time filter for a stable continuous-time filter. Furthermore, there is no frequency folding by means of the bilinear transformation method since it maps the entire *jw* axis of the *s*-plane into one complete revolution of the unit circle in the *z*-plane.

61.3.3.3 Discrete-Time Loop Filters

As mentioned in Section 61.3.3.1, there are three typical loop filters for a phase-locked filter. Using backward-difference transformation, the discrete-time transfer functions of passive lead-lag filter, active lead-lag filter, and active PI filter can be obtained as follows:

Passive lead-lag filter:

$$F_{\text{back,PLL}}(z) = \frac{T_s + \tau_2 - \tau_2 z^{-1}}{T_s + \tau_1 + \tau_2 - (\tau_1 + \tau_2)z^{-1}} \tag{61.39}$$

Active lead-lag filter:

$$F_{\text{back,ALL}}(z) = k_a \frac{T_s + \tau_2 - \tau_2 z^{-1}}{T_s + \tau_1 - \tau_1 z^{-1}} \tag{61.40}$$

Active PI filter:

$$F_{\text{back,PI}}(z) = \frac{T_s + \tau_2 - \tau_2 z^{-1}}{\tau_1(1 - z^{-1})} \tag{61.41}$$

In contrast, the discrete-time transfer functions of passive lead-lag filter, active lead-lag filter, and active PI filter can be written, using the bilinear transformation, as follows:

Passive lead-lag filter:

$$F_{\text{bilinear,PLL}}(z) = \frac{T_s + 2\tau_2 + (T_s - 2\tau_2)z^{-1}}{T_s + 2\tau_1 + 2\tau_2 + (T_s - 2\tau_1 - 2\tau_2)z^{-1}} \tag{61.42}$$

Active lead-lag filter:

$$F_{\text{bilinear,ALL}}(z) = k_a \frac{T_s + 2\tau_2 + (T_s - 2\tau_2)z^{-1}}{T_s + 2\tau_1 + (T_s - 2\tau_1)z^{-1}} \tag{61.43}$$

Active PI filter:

$$F_{\text{bilinear}}(z) = \frac{T_s + 2\tau_2 + (T_s - 2\tau_2)z^{-1}}{2\tau_1 + (1 - z^{-1})} \tag{61.44}$$

From the viewpoint of implementation, all the discrete-time loop filters have the transfer function format of a first-order IIR (infinite impulse response) filter

$$F_{\text{LF}}(z) = \frac{b_0 + b_1 z^{-1}}{1 - a_1 z^{-1}} \tag{61.45}$$

The differences in hardware requirements are small, but the system characteristics and performance are dramatic. Using an approach similar to the Weiner filter theory, Jaffe and Rechtin [18] investigated the optimal loop filters for PLLs with different inputs. For a frequency step input, the form of active PI filter is shown to be optimal.

61.3.4　Charge-Pump PLL

A charge-pump PLL usually consists of four major blocks as shown in Figure 61.21. The PD is a purely PFD. The charge-pump circuit converts the digital signals UP, DN and null (neither up nor down) generated by the PD into a corresponding charge-pump current I_p, $-I_p$, and 0. The loop filter is usually a passive RC circuit converting the charge-pump current into an analog voltage to control VCO. The purpose of the "charge-pump" is to convert the logic state of the phase-frequency detector output into an analog signal suitable for controlling the VCO. The schematic of the charge-pump circuit and the loop filter is shown in Figure 61.22. The linear model shown in Figure 61.3 can be employed to describe a charge-pump PLL. k_d is the equivalent gain of a charge-pump circuit. If the loop bandwidth is much smaller than the input frequency, the detailed behavior within a single cycle can be ignored. Then the state of a PLL can be assumed to be only changed by a small amount during each input cycle. Actually

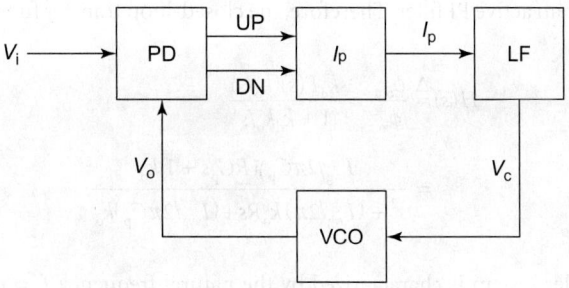

FIGURE 61.21 Charge-pump PLL diagram.

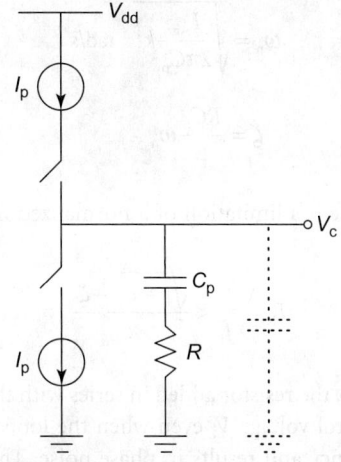

FIGURE 61.22 The schematic of loop filter.

the "average" behavior over many cycles is what we are interested in. The average current charging the capacitor is given by

$$I_{avg} = \frac{Q}{T} = \frac{I\Delta t}{T}$$

$$= \frac{I(\phi_e/2\pi)T}{T}$$

$$= \frac{I\phi_e}{2\pi} \qquad (61.46)$$

and the average k_d is

$$k_d \triangleq \frac{I_{avg}}{\phi_e} = \frac{I_{avg}}{2\pi} \qquad (61.47)$$

The charge-pump current is transferred to the control voltage of the following VCO by the loop filter, and consisted of a resistor and a capacitor as shown in Figure 61.22. The impedance (transfer function) of the RC loop filter is given by

$$F(s) = R + \frac{1}{C_p s} = \frac{1 + RC_p s}{C_p s} \qquad (61.48)$$

which has the format of an active PI filter. Therefore, the closed-loop transfer function can be obtained as

$$H(s) \triangleq \frac{\phi_{\text{out}}}{\phi_{\text{in}}} = \frac{k_{\text{d}}F(s)(k_{\text{o}}/s)}{1 + k_{\text{d}}k_{\text{o}}/s}$$

$$= \frac{(I_{\text{avg}}/2\pi C_{\text{p}})(RC_{\text{p}}s + 1)k_{\text{o}}}{s^2 + (I_{\text{avg}}/2\pi)k_{\text{o}}Rs + (I_{\text{avg}}/2\pi C_{\text{p}})k_{\text{o}}} \tag{61.49}$$

Generally, a second-order system is characterized by the natural frequency $f_{\text{n}} = \omega_{\text{n}}/2\pi$ and the damping factor ζ, and can be expressed as follows:

$$\omega_{\text{n}} = \sqrt{\frac{I_{\text{avg}}}{2\pi C_{\text{p}}}k_{\text{o}}} \quad \text{rad/s}$$

$$\zeta = \frac{RC_{\text{p}}}{2}\omega_{\text{n}} \tag{61.50}$$

For the stability consideration, there is a limitation of a normalized natural frequency F_{N} [19],

$$F_{\text{N}} \triangleq \frac{f_{\text{n}}}{f_{\text{i}}} < \frac{\sqrt{1 + \zeta^2} - \zeta}{\pi} \tag{61.51}$$

In the single-ended charge pump, the resistor added in series with the capacitor shown in Figure 61.22 may introduce "ripple" in the control voltage V_{c} even when the loop is locked [20]. The ripple control voltage modulates the VCO frequency and results in phase noise. This effect is especially undesired in frequency synthesizers. To suppress the ripple, a second-order loop filter, as shown in Figure 61.22 with a shunt capacitor in dotted line, is used. This configuration introduces a third pole in the PLL. Furthermore stability issues must be taken care of. Gardner [20] provides criteria for the stability of the third-order PLL.

An important property of any PLLs is the static phase error that arises from a frequency offset $\Delta\omega$ between the input signal and the free-running frequency of the VCO. According to the analysis in Ref. [20], the static phase error is

$$\theta_{\text{v}} = \frac{2\pi\Delta\omega}{k_{\text{o}}I_{\text{p}}F(0)} \quad \text{rad} \tag{61.52}$$

To eliminate the static phase error in conventional PLLs, an active loop filter with a high DC gain ($F(0)$ is large) is preferred. Nevertheless, the charge-pump PLL allows zero static phase error without the need of a large DC gain of the loop filter. This effect arises from the input open circuit during the "null" state (charge-pump current is zero). Real circuits will impose some resistive loading R_{s} in parallel to the loop filter. Therefore, the static phase error, from Eq. (61.52), will be

$$\theta_{\text{v}} = \frac{2\pi\Delta\omega}{k_{\text{o}}I_{\text{p}}R_{\text{s}}} \quad \text{rad} \tag{61.53}$$

The shunt resistive loading most likely comes from the input of a VCO control terminal. Compared with the static phase error of a conventional PLL as expressed in Eq. (61.8), the same performance can be obtained from a charge-pump PLL without a high DC-gain loop filter [18].

61.3.5 PLL Design Considerations

A PLL design usually starts by specifying the key parameters such as natural frequency ω_n, lock-in range $\Delta\omega_L$, damping factor ζ, and the frequency control range which mostly depend on applications. Design procedures based on a practical example are described as follows:

Step 1. Specify the damping factor ζ. The damping factor determines the relative stability of a PLL. ζ should be considered as a critical parameter to achieve fast response, small overshoot, and minimum noise bandwidth B_L. If ζ is very small, large overshoot occurs and the overshoot causes phase jitter [17]. If ζ is too large, the response becomes sluggish.

Step 2. Specify the lock-in range $\Delta\omega_L$ or the noise bandwidth B_L. As shown in Eq. (61.20) and Eq. (61.27), the natural frequency ω_n depends on $\Delta\omega_L$ and ζ (or B_L and ζ). If the noise is not the key issue of the PLL, we may ignore the noise bandwidth and specify the lock-in range. If the noise is concerned, we should specify B_L first, and keep the lock-in range of PLL.

Step 3. Calculate the ω_n according to step 2. If the lock-in range has been specified, Eq. (61.20) indicates

$$\omega_n = \frac{\Delta\omega_L}{2\zeta} \tag{61.54}$$

If the noise bandwidth has been specified, Eq. (61.27) indicates the natural frequency as

$$\omega_n = \frac{2B_L}{\zeta + 1/4\zeta} \tag{61.55}$$

Step 4. Determine the VCO gain factor k_o and the PD gain k_d. k_o and k_d are both characterized by circuit architectures. They must achieve the requirement of the lock-in range specified in step 2. For example, if k_o or k_d is too small, the PLL will fail to achieve the desired lock-in range.

Step 5. Choose the loop filter. Different types of the loop filter are available as shown in Figure 61.11. According to Eqs. (61.31)–(61.33), ω_n and ζ specified above are used to derive the time constants of the loop filter.

61.4 PLL Applications

61.4.1 Clock and Data Recovery

In data transmission systems such as optical communications, telecommunications, disk drive systems, and local networks, data are transmitted on baseband or passband. In most of these applications, only data signals are transmitted by transmitter, but clock signals are not transmitted to save hardware cost. Therefore, the receiver should have some schemes to extract the clock information from the received data stream and to regenerate transmitted data using the recovered clock. This scheme is called timing recovery or clock recovery.

To recover the data correctly, the receiver must generate a synchronous clock from the input data stream, the recovered clock must synchronize with the bit rate (the baud of data). The PLL can be used to recover the clock from the data stream, but there are some special design considerations. For example, because of the random nature of data, the choice of PFDs is restricted. In particular, three-state PD is not proper, because when there are no transitions in the data stream, the PD interprets that the VCO frequency is higher than the data frequency and remains its output in the "down" state, which makes the PLL to lose lock as shown in Figure 61.23. Thus, the choice of PFD for random binary data requires a careful examination over whether data transitions are absent. One useful method is the rotational frequency detector described in Ref. [9]. The random data also cause the PLL to introduce undesired phase variation in the recovered clock; it is called timing jitter and is an important issue in the clock recovery.

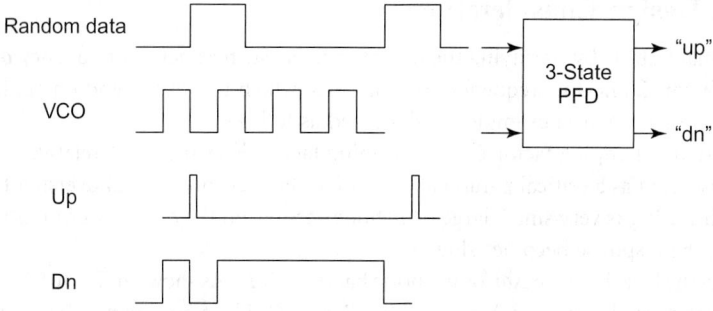

FIGURE 61.23   Response of a three-state PD to random data.

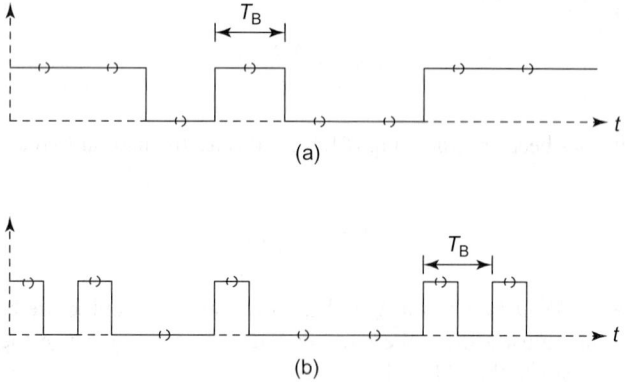

FIGURE 61.24   (a) NRZ data; (b) RZ data.

61.4.1.1   Data Format

Binary data are usually transmitted in an NRZ format as shown in Figure 61.24(a) because of the consideration of bandwidth efficiency. In NRZ format, each bit has a duration of T_B (bit period). The signal does not tend to zero between adjacent pulses representing 1's. It is shown in Ref. [21] that the corresponding spectrum has no line component at $f_B = 1/T_B$, and most of the spectrum of this signal lines below $f_B/2$. The term "non-return-to-zero" distinguishes itself from another data type called "return-to-zero"(RZ) as shown in Figure 61.24(b), where the signal tends to zero between consecutive bits. Therefore, the spectrum of RZ data has a frequency component at f_B. For a given bit rate, RZ data need wider transmitting bandwidth, and therefore NRZ data are preferable when channel or circuit bandwidth is a concern.

Owing to the lack of a spectral component at the bit rate of NRZ format, a clock recovery circuit may lock to spurious signals or fail to lock at all. Thus, a nonlinear process for the NRZ data is essential to create a frequency component at the baud rate.

61.4.1.2   Data Conversion

One way to recover the clock signal from the NRZ data is to convert it into an RZ-like data that has a frequency component at the bit rate, and then recover clock from data using a PLL. Transition detection is one of the methods to convert NRZ data into RZ-like data. As illustrated in Figure 61.25(a), the edge detection requires a mechanism to sense both positive and negative data transitions. In Figure 61.25(b), NRZ data is delayed and compared with itself by an exclusive-OR gate, therefore the transition edges are detected. In Figure 61.26, the NRZ data V_i is first differentiated to generate pulses corresponding to each transition. These pulses are made to be all positive by squaring the differentiated signal V_i. The result is that the signal V_i' looks just like RZ data where pulses are spaced at an interval of T_B.

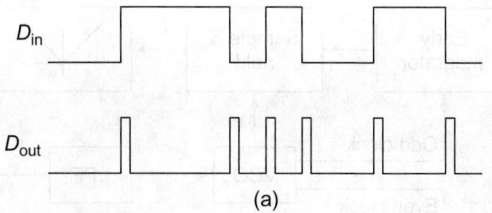

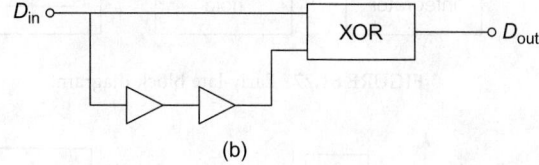

FIGURE 61.25 Edge detection of NRZ data.

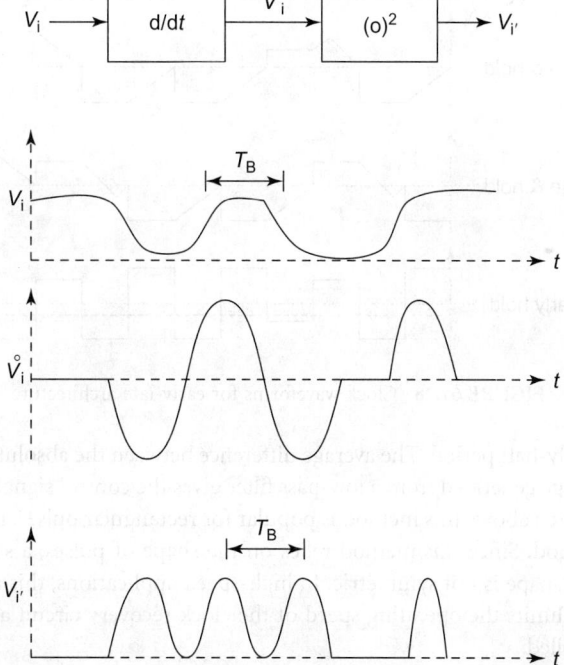

FIGURE 61.26 Converting NRZ into RZ-like signal.

61.4.1.3 Clock Recovery Architecture

On the basis of different PLL topologies, there are several clock recovery approaches. Here, the early-late and the edge-detector-based methods will be described.

Figure 61.27 shows the block diagram of the early-late method. The waveforms for the case in which the input lags the VCO output are shown in Figure 61.28, where the early integrator integrates the input signal for the early-half period of the clock signal, and holds it for the remainder of the clock signal. In contrast, the late integrator integrates the input signal for the late-half period of the clock signal and

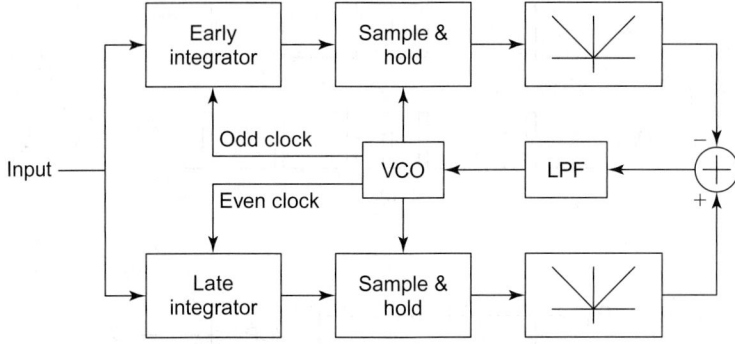

FIGURE 61.27 Early-late block diagram.

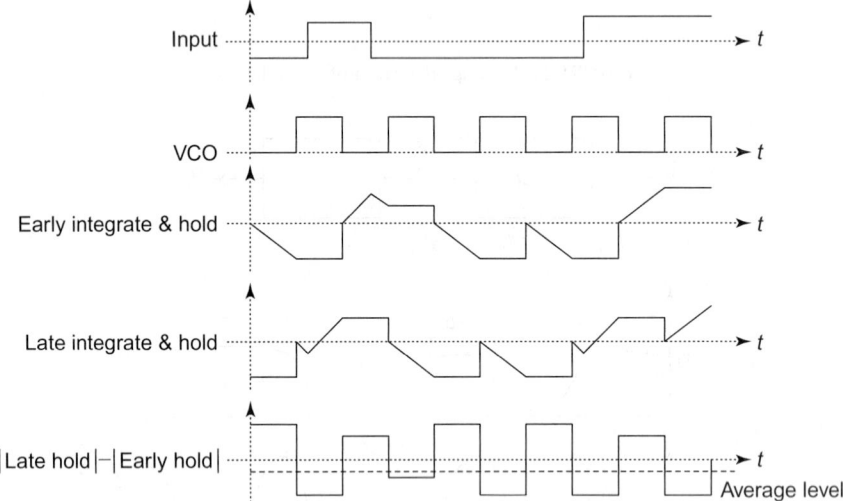

FIGURE 61.28 Clock waveforms for early-late architecture.

holds it for the next early-half period. The average difference between the absolute values of the late hold and the early hold voltage generated from a low-pass filter gives the control signal to adjust the frequency of the VCO. As mentioned above, this method is popular for rectangular pulses. However, there are some drawbacks in this method. Since this method relies on the shape of pulses, a static phase error can be introduced if the pulse shape is not symmetric. In high-speed applications, this approach requires a fast settling integrator that limits the operating speed of the clock recovery circuit and the acquisition time cannot be easily controlled.

The most widely used technique for clock recovery in high-performance, wideband data transmission applications is the edge-detection-based method. The edge-detection method is used to convert data format such that the PLL can lock the correct baud frequency. More details have been described in Section 61.4.1.2. There are many variations of this method depending on the exact implementation of each PLL loop component. The "quadricorrelator" introduced by Richman [7] and modified by Bellisio [22] is a frequency-difference discriminator and has been implemented in a clock recovery architecture. Figure 61.29 [23] is a phase-FLL using edge-detection method and quadricorrelator to recover timing information from NRZ data. As shown in Figure 61.29, the quadricorrelator follows the edge-detector with a combination of three loops sharing the same VCO. Loops I and II form a FLL that contains the quadricorrelator for frequency detection. Loop III is a typical PLL for phase alignment. Since the phase-locked loops and FLLs share the same VCO, the interaction between two loops is a very important issue.

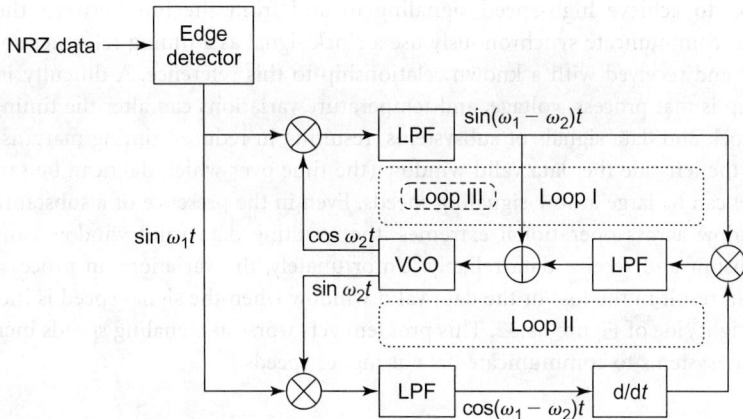

FIGURE 61.29 Quadricorrelator.

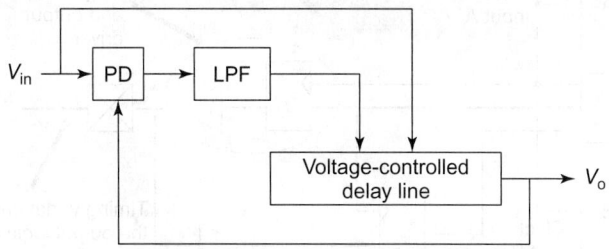

FIGURE 61.30 DLL block diagram.

As described in Ref. [23], when $\omega_1 \approx \omega_2$, the DC feedback signal produced by loops I and II approaches zero and loop III dominates the loop performance. A composite FLL and PLL is a good method to achieve fast acquisition and a narrow PLL loop bandwidth to minimize the VCO drift. Nevertheless, because the wideband FLL can response to noise and spurious components, it is essential to disable FLL when the frequency error gets into the lock-in range of the PLL to minimize the interaction. More clock recovery architectures are described in Refs. [14,16,17,24–26].

61.4.2 Delay-Locked Loop

Two major elements for adjusting the timing are VCO and voltage-controlled delay line (VCDL). Figure 61.30 shows a typical delay-locked loop (DLL) [27,28] that replaces the VCO of a PLL with a VCDL. The input signal is delayed by an integer multiple of the signal period because the phase error is zero when the phase difference between V_{in} and V_o approaches multiple of the signal periods. The VCDL usually consists a number of cascaded gain stages with variable delay. Delay lines, unlike ring oscillators, cannot generate a signal, therefore it is diffcult to make frequency multiplication in a DLL.

In a VCO, the output "frequency" is proportional to the input control voltage. The phase transfer function contains a pole, which is $H(s) = k_o/s$ (k_o is the VCO gain). In a VCDL, the output "phase" is proportional to the control voltage, and the phase transfer function is $H(s) = k_{VCDL}$. So the DLL can be easily stabilized with a simple first-order loop filter. Consequently, DLLs have much more relaxed trade-offs among gain, bandwidth, and stability. This is one of the two important advantages over PLLs. Another advantage is that delay lines typically introduce much less jitter than VCO [12]. Because a delay chain is not configured as a ring oscillator, there is no jitter accumulation since the noise does not contribute to the starting point of the next clock cycle.

A typical application of DLL is to synchronize the clock edges of subsystems within a digital system to access the bus between subsystems. Figure 61.31 shows modern digital systems that use synchronous communication to achieve high-speed signaling to and from the bus between the subsystems. Subsystems that communicate synchronously use a clock signal as a timing reference so that data can be transmitted and received with a known relationship to this reference. A diffculty in maintaining this relationship is that process, voltage, and temperature variations can alter the timing relationship between the clock and data signals of subsystems, resulting in reduced timing margins. Figure 61.32 shows that on the left side the data valid window (the time over which data can be sampled reliably by the receiver) can be large at low-signaling speeds. Even in the presence of a substantial shift in the data valid window across operational extremes, the resulting data valid window can still be large enough to transmit and receive data reliably. Unfortunately, the variations in process, voltage, and temperature can result in the loss of the data valid window when the signal speed is increased as also shown on the right side of Figure 61.32. This problem gets worse as signaling speeds increase, limiting the ability of subsystems to communicate data at higher speeds.

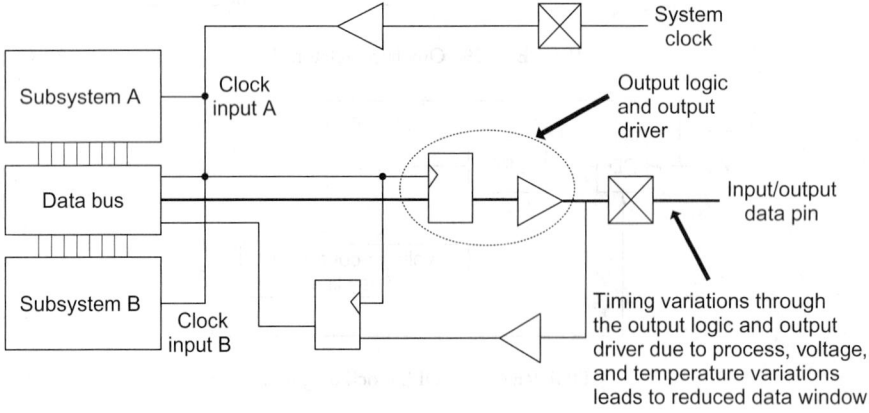

FIGURE 61.31 Modern digital systems use synchronous communication to achieve high-speed signaling to and from the bus between the subsystems.

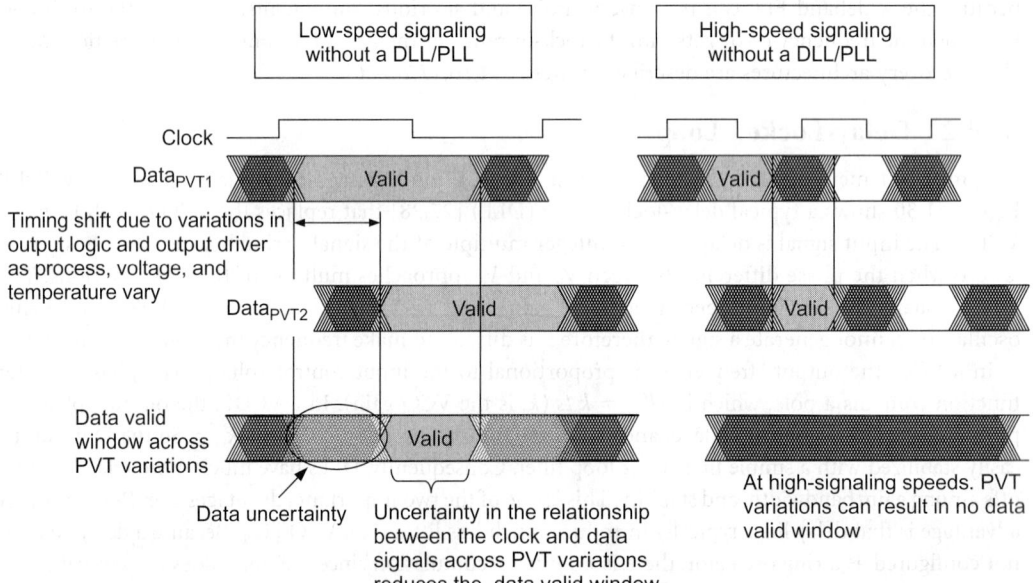

FIGURE 61.32 The timing relationships between the clock and data signals of subsystems in a conventional digital system.

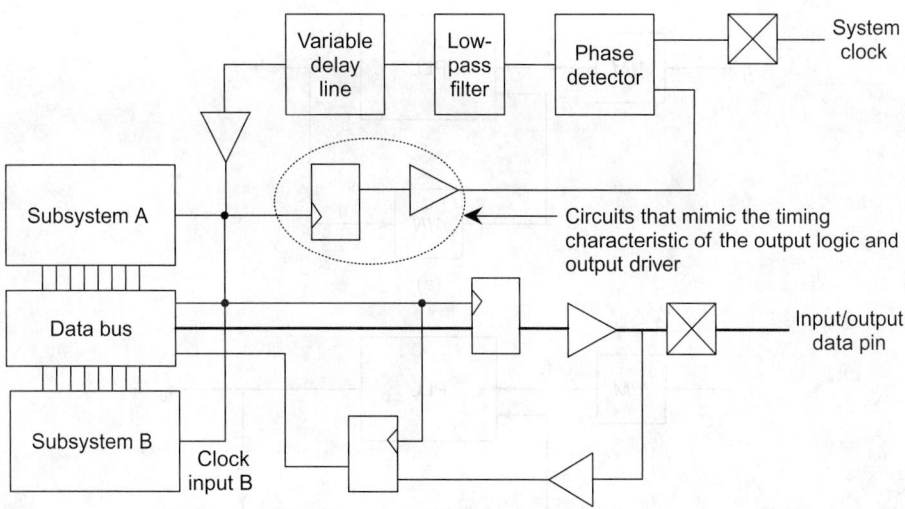

FIGURE 61.33 DLL-on-chip to maintain the timing relationship between a clock signal and an output data signal.

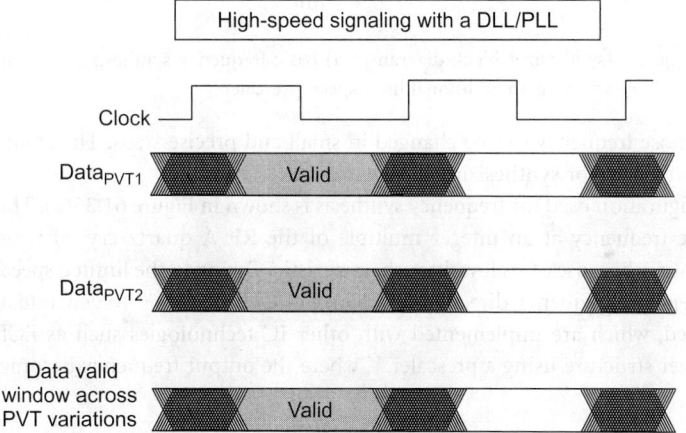

FIGURE 61.34 Timing relationships between subsystems when a DLL is employed for synchronizing the bus access.

The function of DLLs and PLLs to synchronize a signal with a reference or an input signal in frequency as well as in phase can be used to maintain such a fixed timing relationship between signals of subsystems. Figure 61.33 shows how a DLL is used to maintain the timing relationship between a clock signal and an output data signal. The PD detects phase differences between the clock and output data and sends control information through a low-pass filter to a variable delay line that adjusts the timing of the internal clock to maintain the desired timing relationship. The PD must account for the timing characteristics of the output logic and output driver. This is important since it estimates the phase differences between the clock and the data driven by the output driver, where the timing relationships of subsystems are changed over time due to the process, voltage, and temperature variations. Maintaining the timing relationships between the clock and output data with DLLs and PLLs results in improved timing margins as shown in Figure 61.34. Then, the important limitation to increasing signaling speeds is addressed.

61.4.3 Frequency Synthesizer

A frequency synthesizer generates any of the number of frequencies by locking a VCO to an accurate frequency source such as a crystal oscillator. For example, RF systems usually require a high-frequency

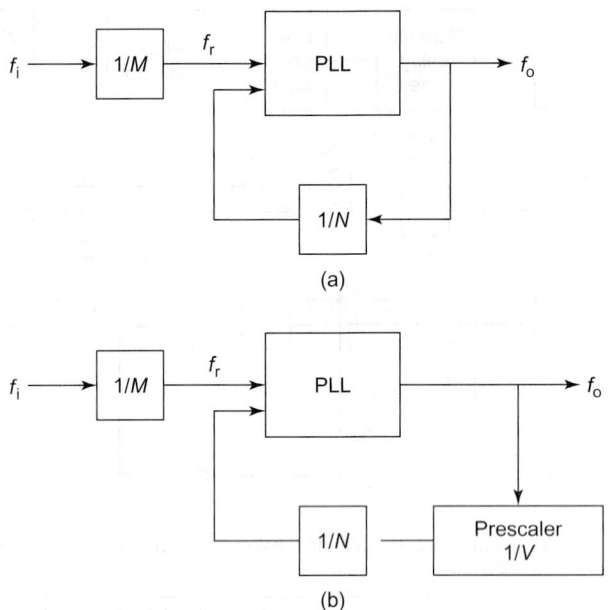

FIGURE 61.35 Frequency-synthesizer block diagrams: (a) basic frequency-synthesizer system; (b) system extends the upper frequency range by using an additional high-speed prescaler.

local oscillator whose frequency can be changed in small and precise steps. The ability of multiplying a RF makes PLLs attractive for synthesizing frequencies.

The basic configuration used for frequency synthesis is shown in Figure 61.35(a). The system is capable of generating the frequency at an integer multiple of the RF. A quartz crystal is usually used as the reference clock source because of its low jitter characteristic. Owing to the limited speed of CMOS device, it is difficult to generate frequency directly in the range of GHz or more. To generate higher frequencies, prescalers are used, which are implemented with other IC technologies such as ECL. Figure 61.35(b) shows a synthesizer structure using a prescaler V, where the output frequency becomes

$$f_{out} = \frac{NV f_i}{M} \tag{61.56}$$

Because the scaling factor V is obviously $\gg 1$, it is no longer possible to generate any desired integer multiple of the reference frequency. This drawback can be circumvented by using a so-called dual-modulus prescaler as shown in Figure 61.36. A dual-modulus prescaler is a divider whose division can be switched from one value to the other by a control signal. The following shows that the dual-modulus prescaler makes it possible to generate a number of output frequencies that are spaced only by one RF. The VCO output is divided by $V/V+1$ dual-modulus prescaler. The output of the prescaler is fed into a "program counter" $1/N$ and a "swallow counter" $1/A$. The dual-modulus prescaler is set to divide by $V+1$ initially. After "A" pulses out of the prescaler, the swallow counter is full and changes the prescaler modulus to V. After additional "$N-A$" pulses out of the prescaler, the program counter changes the prescaler modulus back to $V+1$ and restarts the swallow counter. Then the cycle is repeated. In this way, the VCO frequency is equal to $(V+1)A + V(N-A) = VN + A$ times the RF. Note that N must be higher than A. If this is not the case, the program counter would be full earlier than the $1/A$ and both counters would be reset. Therefore, the dual-modulus prescaler would never be switched from $V+1$ to V. For example, if $V = 64$, then A must be in the range 0–63 such that $N_{min} = 64$. The smallest realizable division ratio is

$$(N_{tot})_{min} = N_{min}V = 4096 \tag{61.57}$$

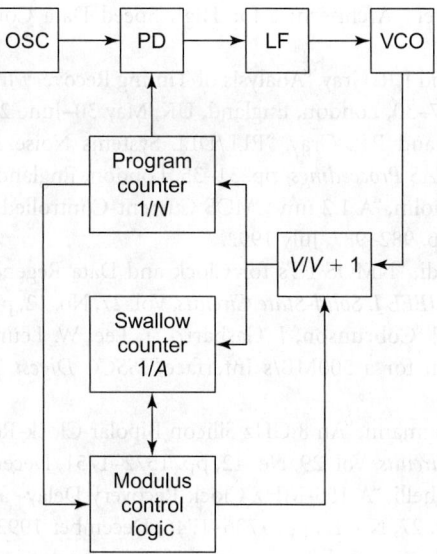

FIGURE 61.36 The block diagram of dual-modulus frequency synthesizer.

The synthesizer of Figure 61.36 is able to generate all integer multiples of the RF starting from $N_{tot} = 4096$. For extending the upper frequency range of frequency synthesizers but still allowing the synthesis of lower frequency, the four-modulus prescaler is a solution [1].

On the basis of the above discussions, the synthesized frequency is an integer multiple of a RF. In RF applications, the RF is usually larger than the channel spacing for loop dynamic performance considerations, in which the wider loop bandwidth for a given channel spacing allows faster settling time and reduces the phase jitter requirements to be imposed on the VCO. Therefore a "fractional" scaling factor is needed. Fractional division ratios of any complexity can be realized. For example, a ratio 3.7 is obtained if a counter is forced to divide by 4 in seven cycles of each group of ten cycles and by 3 in the remaining three cycles. On average, this counter divides the input frequency by 3.7 effectively.

References

1. R.E. Best, *Phase-Locked Loops Theory, Design, Applications*, McGraw-Hill, New York, 1984.
2. T.G. Donald and J.D. Gallia, *Digital Phase-Locked Loop Design Using SN54/74LS297*, Application Note AN 3216, Texas Instruments Inc., Dallas, TX.
3. W.B. Rosink, "All-Digital Phase-Locked Loops Using the 74HC/HCT297 IC," Designer's Guide, Philips Components, 1989.
4. F.M. Gardner, *Phaselock Techniques*, 2nd ed., John Wiley & Sons, New Jersey, 1979.
5. S.G. Tzafestas and G. Spyros, *Walsh Functions in Signal and Systems Analysis and Design*, Van Nostrand, New York 1985.
6. F.M. Gardner, "Acquisition of Phaselock," *Conference Record of the International Conference on Communications*, Vol. I, pp. 10-1–10-5, 1976 International Conference on Communications. York, NY, USA, June 1976.
7. D. Richman, "Color Carrier Reference Phase Synchronization Accuracy in NTSC Color Television," *Proc. IRE*, Vol. 42, pp. 106–133, January 1954.
8. F.M. Gardner, "Properties of Frequency Difference Detector," *IEEE Trans. on Commun.*, Vol. COM-33, No. 2, pp. 131–138, February 1985.
9. D.G. Messerschmitt, "Frequency Detectors for PLL Acquisition in Timing and Carrier Recovery," *IEEE Trans. on Commun.*, Vol. COM-27, No. 9, pp. 1288–1295, September 1979.

10. R.B. Lee, "Timing Recovery Architecture for High Speed Data Communication System," Master thesis, 1993.
11. T.C. Weigandt, B. Kim, and P.R. Gray, "Analysis of Timing Recovery Jitter in CMOS Ring Oscillator," *ISCAS Proceedings*, pp. 27–30, London, England, UK, May 30–June 2, 1994.
12. B. Kim, T.C. Weigandt, and P.R. Gray, "PLL/DLL Systems Noise Analysis for Low Jitter Clock Synthesizer Design," *ISCAS Proceedings*, pp. 31–35, London, England, UK, May 30–June 2, 1994.
13. M.P. Flyun and S.U. Lidholm, "A 1.2 um CMOS Current-Controlled Oscillator," *IEEE J. Solid-State Circuits*, Vol. 27, No. 7, pp. 982–987, July 1992.
14. S.K. Enam and A.A. Abidi, "NMOS IC's for Clock and Data Regeneration in Gigabit-per-Second Optical-Fiber Receivers," *IEEE J. Solid-State Circuits*, Vol. 27, No. 12, pp. 1763–1774, December 1992.
15. M. Horowitz, A. Chan, J. Cobrunson, J. Gasbarro, T. Lee, W. Leung, W. Richardson, T. Thrush and Y. Fujii, "PLL Design for a 500MB/s Interface," *ISSCC Digest Technical Paper*, pp. 160–161, February 1993.
16. A. Pottbacker and U. Langmann, "An 8 GHz Silicon Bipolar Clock-Recovery and Data-Regenerator IC," *IEEE J. Solid-State Circuits*, Vol. 29, No. 12, pp. 1572–1751, December 1994.
17. T.H. Lee and J.F. Bulzacchelli, "A 155-MHz Clock Recovery Delay- and Phase-Locked Loop," *IEEE J. Solid-State Circuits*, Vol. 27, No. 12, pp. 1736–1746, December 1992.
18. R. Jaffe and E. Rechtin, "Design and Performance of Phase-Lock Circuits Capable of Near-Optimal Performance Over a Wide Range of Input Signal and Noise Levels," *IRE Trans. Inf. Theory*, Vol. IT-1, pp. 66–76, March 1955.
19. M.V. Paemel, "Analysis of a Charge-Pump PLL: A New Model," *IEEE Trans. on Commun.*, Vol. 42, No. 7, pp. 131–138, February 1994.
20. F.M. Gardner, "Charge-pump Phase-Locked Loops," *IEEE Trans. On Commun.*, Vol. COM-28, pp. 1849–1858, November 1980.
21. B.P. Lathi, *Modern Digital and Analog Communication System*, HRW, Philadelphia, 1989.
22. J.A. Bellisio, "A New Phase-Locked Loop Timing Recovery Method for Digital Regenerators," *IEEE Int. Comm. Conf. Rec.*, Vol. 1, pp. 10-17–10-20, 1976 International Conference on Communications. York, NY, USA, June 1976.
23. B. Razavi, "A 2.5-Gb/s 15-mW Clock Recovery Circuit," *IEEE J. Solid-State Circuits*, Vol. 31, No. 4, pp. 472–480, April 1996.
24. R.J. Baumert, P.C. Metz, M.E. Pedersen, R.L. Pritchett, and J.A. Young, "A Monolithic 50–200 MHz CMOS Clock Recovery and Retiming Circuit," *IEEE Custom Integrated Circuits Conference*, pp. 14.5.5–14.5.4, San Diego, USA, 15–18 May, 1989.
25. B. Lai and R.C. Walker, "A monolithic 622Mb/s clock extraction data retiming circuit," *IEEE International Solid-State Circuits Conference*, pp. 144–145, San Francisco, USA, February 1991.
26. B. Kim, D.M. Helman, and P.R. Gray, "A 30 MHz Hybrid Analog/Digital Clock Recovery Circuit in 2-μm CMOS," *IEEE J. Solid-State Circuits*, Vol. 25, No. 6, pp. 1385–1394, December 1990.
27. M. Bazes, "A Novel Precision MOS Synchronous Delay Lines," *IEEE J. Solid-State Circuits*, Vol. 20, pp. 1265–1271, December 1985.
28. M.G. Johnson and E.L. Hudson, "A Variable Delay Line PLL for CPU-coprocessor Synchronization," *IEEE J. Solid-State Circuits*, Vol. 23, pp. 1218–1223, October 1988.

62

Switched-Capacitor Filters

Andrea Baschirotto
University of Lecce

CONTENTS

62.1 Introduction 62-2
62.2 Sampled-Data Analog Filters 62-2
62.3 The Principle of SC Technique 62-4
62.4 First-Order SC Stages 62-6
    62.4.1 The Active SC Integrators....................................62-6
    62.4.2 The Summing Integrator 62-7
    62.4.3 The Active Damped SC Integrator.........................62-8
    62.4.4 A Design Example 62-8
62.5 Second-Order SC Circuit 62-10
    62.5.1 The Fleischer&Laker Biquad 62-10
    62.5.2 Design Methodology...62-10
    62.5.3 Design Example 62-12
    62.5.4 A Biquadratic Cell for High
        Sampling Frequency 62-13
    62.5.5 High-Order Filters...62-15
62.6 Implementation Aspects 62-15
    62.6.1 Integrated Capacitors 62-16
    62.6.2 MOS Switches...62-17
    62.6.3 Transconductance Amplifier................................62-18
62.7 Performance Limitations 62-18
    62.7.1 Limitation Due to Switches 62-19
    62.7.2 Limitation Due to the Opamp 62-20
    62.7.3 Noise in SC Systems...62-22
62.8 Compensation Technique (Performance
    Improvements)..62-23
    62.8.1 CDS Offset-Compensated SC Integrator..............62-23
    62.8.2 Chopper Technique 62-24
    62.8.3 Finite-Gain-Compensated SC Integrator 62-25
    62.8.4 The Very-Long Time-Constant Integrator 62-26
    62.8.5 Double-Sampling Technique 62-28
62.9 Low-Voltage Switched-Capacitor Circuits.........62-29
    62.9.1 Processing a Reduced Signal Amplitude
        with Standard SC Solutions 62-29
    62.9.2 Processing a Rail-to-Rail Signal Amplitude
        with Novel Solutions 62-32
References 62-38

© 2006 by CRC Press LLC

62-1

62.1   Introduction

The accuracy of the absolute value of integrated passive devices (R and C) is very poor. As a consequence, the frequency response accuracy of integrated active-RC filters is poor and they are not feasible when high-accuracy performance is needed. A possible solution for the implementation of analog filters with accurate frequency response was given by the switched-capacitor (SC) technique since the late 1970s [1,2]. Their popularity has further increased since they can be realized with the same standard CMOS technology used for digital circuits. In this way, fully integrated, low-cost, high-flexibility, mixed-mode systems have become possible. The main reasons of the large popularity of SC networks can be summarized as follows:

1. The basic requirements of SC filters fit the popular MOS technology features. In fact, the infinite input impedance of the operational amplifier (opamp) is obtained using a MOS input device. MOS transconductance amplifiers can be used since only-capacitive load is present, precise switches are realized with MOS transistor, and capacitors are available in the MOS process.
2. SC filter performance accuracy is based on the matching of integrated capacitors (and not on their absolute values). In a standard CMOS process, the capacitor matching error can be less than 0.2%. As a consequence, SC systems guarantee very accurate frequency response without component trimming. For the same reason, temperature and aging coefficients track, reducing performance sensitivity to temperature and aging variations.
3. It is possible to realize SC filters with long time constants without using large capacitors and resistors. This means a chip area saving with respect to active-RC filter implementations.
4. SC systems operate with closed-loop structures; this allows to process large swing signals and to achieve large dynamic range.

In contrast the major drawbacks of SC technique are as follows:

1. To process a fully analog signal, an SC filter has to be preceded by an anti-aliasing (AA) filter and followed by a smoothing filter, which complicate the overall system and increase power and die size.
2. The opamps embedded in an SC filter have to perform a large DC gain and a large unity-gain bandwidth, much larger than the bandwidth of the signal to be processed. This limits the maximum signal bandwidth.
3. The power of noise of all the sources in the SC filter is folded in the band $[0 - F_s/2]$. Thus their noise power density is increased by factor $(F_s/2)/F_b$, where F_s is the sampling frequency and F_b the noise bandwidth at the source.

From its first proposals the SC technique has been highly developed. Many different circuit solutions have been realized with SC technique not only in analog filtering, but also in analog equalizers, analog-to-digital and digital-to-analog conversion (including in particular the oversampled $\Sigma\Delta$ converters), Sample&Hold, Track&Hold, etc.

In this chapter an overview of the main aspects of the SC technique is given; the reader can refer to more details in the literature. Few advanced solutions feasible for future SC systems are given in the last section.

62.2   Sampled-Data Analog Filters

A SC filter is a continuous-amplitude, sampled-data system. This means that the amplitude of the signals can assume any value within the possible range in a continuous manner. In contrast, these values are assumed at certain time instants and then they are held for all the sampling period. Thus the resulting waveforms are not continuous in time but looks like a staircase.

In an SC filter, the continuous-time input signal is first sampled at sampling frequency F_s and then processed through the SC network. This sampling operation results in a particular feature of the frequency response of the SC system. In the following the aspects relative to the sampling action are illustrated from an intuitive point of view, while a more rigorous description can be found in Ref. [3].

The sampling operation extracts from the continuous-time waveform the values of the input signal at the instant kT_s ($k = 1, 2, 3, \ldots$), where $T_s = 1/F_s$ is the sampling period. This is shown in Figure 62.1 for a single-input sine wave at $f_o = 0.16F_s$ (i.e., with $f_o < F_s/2$).

If the input sine wave is at $F_s + f_o$, the input sequence of analog samples is exactly equal to that previously obtained with f_o as input frequency (see Figure 62.2). Both sequences should then be processed exactly in the same way by the SC network, and the overall filter output sequence should then result to be again identical. The two input sine waves result to be indistinguishable after the sampling action. This effect is called *aliasing*. It can be demonstrated that a sine wave at frequency f_o in the range $[0-F_s/2]$ is aliased by the components at frequency f_{al} given by

$$f_{al} = kF_s \pm f_o \quad (k = 1, 2, 3, \ldots) \tag{62.1}$$

As a consequence, to avoid frequency aliasing (which means signal corruption), the input signal band of a sample data system must be limited in the $[0-F_s/2]$ range. The range $[0-F_s/2]$ is called baseband and the above limitation is an expression of the Nyquist theorem.

After the sampling, the SC network processes the sequence of samples, independently of how they have been produced. Since all the frequencies given in Eq. (62.1) produce the same sequence of samples, the gain for all of them results to be the same. This concept results in the fact that the transfer function of a sampled-data system is periodical with period equal to F_s, and it is symmetrical in its period.

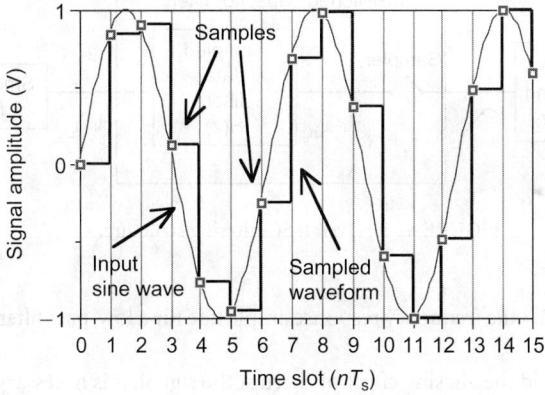

FIGURE 62.1 Sampling of the input signal.

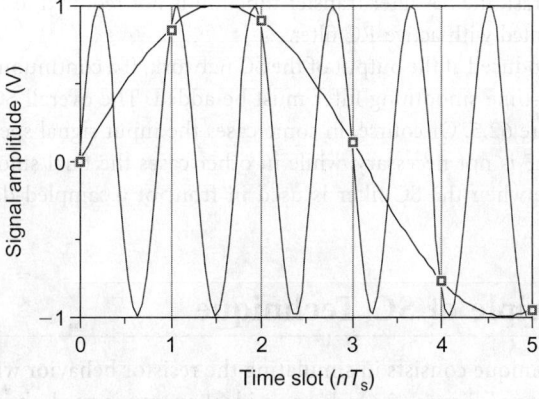

FIGURE 62.2 Aliasing between f_o and $F_s + f_o$.

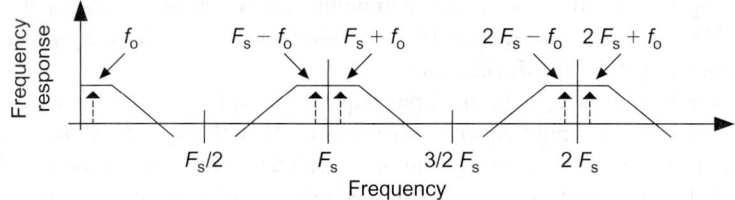

FIGURE 62.3 Periodicity of the sampled-data system transfer function.

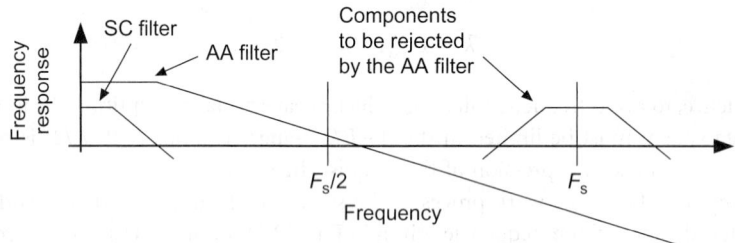

FIGURE 62.4 Transfer functions of switched-capacitor and anti-aliasing filters.

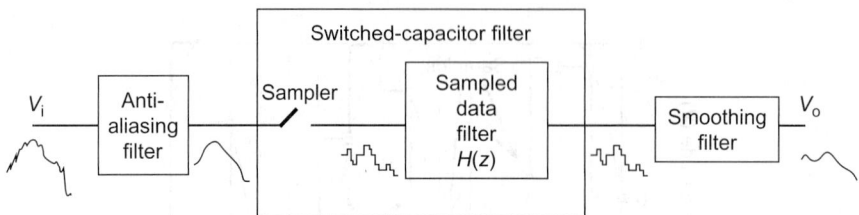

FIGURE 62.5 Overall SC filtering structure.

For instance, in Figure 62.3 the frequency response amplitude for a low-pass filter is shown for frequencies higher than F_s.

As stated above, to avoid the aliasing effect to corrupt the signal, it is necessary to limit the input signal bandwidth. This function is performed by the AA filter, which is placed in front of the SC filter and operates in the continuous-time domain (Figure 62.4). From a practical point of view, the poles of the SC filter are typically much smaller than $F_s/2$ and the frequency response is required to be accurate only in the passband. In contrast, the AA filter transfer function is not required to be accurate. Thus the AA filter is usually implemented with active-RC filter.

A staircase signal is produced at the output of the SC network. If a continuous-time output waveform is needed, a continuous-time smoothing filter must be added. The overall SC filter-processing chain results are given in Figure 62.5. Of course, in some cases the input signal spectrum is already limited to $F_s/2$ and then the AAF is not necessary, while in other cases the final smoothing filter is no more necessary, as in the case when the SC filter is used in front of a sampled-data system (for instance an ADC).

62.3   The Principle of SC Technique

The principle of SC technique consists in simulating the resistor behavior with a switched-capacitor structure. In the structure of Figure 62.6, where an ideal opamp is used, the resistor R_{eq} is connected between V_i and a zero-impedance, zero-voltage node (as a virtual ground is). This means that

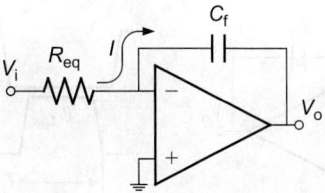

FIGURE 62.6 Basic RC integrator.

a continuous-time current I flows from V_i through R_{eq} into the virtual ground. This current is equal to:

$$I = V_i / R_{eq} \tag{62.2}$$

The alternative SC structure is shown in Figure 62.7(a). It is composed of an input sampling capacitor C_s (connected through four switches to the input signal V_i, to the opamp input node and to two ground nodes), an opamp, and a feedback (integrating) capacitor C_f. The clock phases driving the switches are shown in Figure 62.7(b). A switch is close (conductive) when its driving phase is high. It is necessary that the two clock phases are nonoverlapping, to connect each capacitor plate to only one low-impedance node for each time slot.

During phase ϕ_1, capacitor C_s is discharged. During phase ϕ_2, C_s is connected between V_i and the virtual ground. So a charge $Q = -C_s V_i$ is collected on its right-hand plate. Owing to the charge conservation law applied at the virtual ground node, this charge collection corresponds to an injection in virtual ground of the same amount of charge but with the opposite sign, given by

$$Q_{inj} = C_s V_i \tag{62.3}$$

Notice that this charge injection is independent of the component in the opamp feedback path. This charge injection occurs every clock period. Observing these effects for a long time slot T, the total charge injection Q_{tot} is given by

$$Q_{tot} = C_s V_i \frac{T}{T_s} \tag{62.4}$$

This corresponds to a mean current I_{mean} equal to

$$I_{mean} = \frac{Q_{tot}}{T} = C_s \frac{V_i}{T_s} \tag{62.5}$$

Equating Eq. (62.2) with Eq. (62.5), the following relationship holds:

$$R_{eq} = \frac{T_s}{C_s} \tag{62.6}$$

This means that the structure composed of C_s and the four switches operated at F_s is equivalent to a resistor R_{eq}. This approximation is valid for V_i equal to a DC value, as it is the case of the proposed example, and it is still valid for V_i *slowly variable with respect to the clock period*, otherwise the quantitative average operation of Eq. (62.5) is no more valid. The limits of the approximation between a resistor and an SC structure as expressed in Eq. (62.6) implies fundamental differences in the exact design of SC filters when derived from active-RC filters in a one-to-one correspondence.

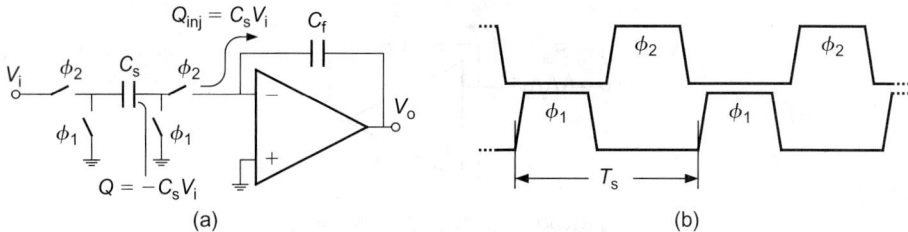

FIGURE 62.7 Basic SC integrator.

The synthesis of active filters is based on the use of some elementary blocks interconnected in different ways depending on the type of the adopted design philosophy. The different strategies and approaches for designing analog filters are well known from the continuous-time domain and they are also used in the case of SC filters, although the sampled-data nature of the SC filters can be profitably used either for simplifying or improving the design itself. In any case, basic building blocks are used to compose high-order filters.

In the following the main basic blocks are described. They implement first- (active integrators, undamped and damped, summers) and second-order (biquads) transfer function in the z-domain (z is the state variable in the sample data domain).

62.4 First-Order SC Stages

62.4.1 The Active SC Integrators

In Figure 62.8(a)–(c) the standard integrators normally used in SC designs are shown. For each integrator, the transfer function in the z-domain is reported, assuming that the input signal is sampled during phase ϕ_1 and is held to this value until the end of phase ϕ_2, while the output is read during phase ϕ_2.

$$H_a(z) = \frac{V_o(z)}{V_i(z)} = \frac{C_s}{C_f} \frac{z^{-1}}{1 - z^{-1}} \tag{62.7}$$

$$H_b(z) = \frac{V_o(z)}{V_i(z)} = -\frac{C_s}{C_f} \frac{1}{1 - z^{-1}} \tag{62.8}$$

$$H_c(z) = \frac{V_o(z)}{V_i(z)} = 2\frac{C_s}{C_f} \frac{1 + z^{-1}}{1 - z^{-1}} \tag{62.9}$$

The third integrator is called bilinear since it implements the bilinear mapping of the s-to-z transformation (s is the state variable in the continuous-time domain).

In all the above transfer functions, only capacitor ratios appear. For this reason the SC filter transfer functions are sensitive only to the capacitor ratios (i.e., to the capacitor matching) and independent of absolute capacitor value. This is a remarkable advantage of all the SC networks.

An important feature of all of these integrators is their insensitivity to parasitic capacitance. This can be verified observing that any parasitic capacitance connected to the capacitor left-hand plate is not connected to the virtual ground and therefore does not contribute to the amount of injected charge. In contrast, the stray capacitance connected to the capacitor right-hand armature could contribute to the charge transfer, but this capacitance is switched between two nodes (ground and virtual ground) at the same potential and thus no charge injection results.

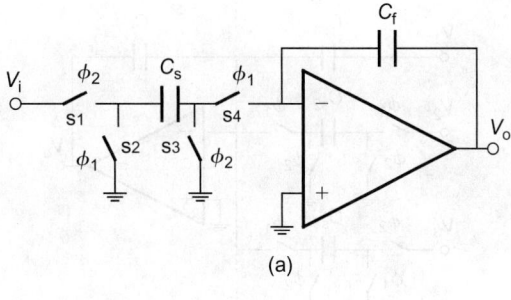

(a)

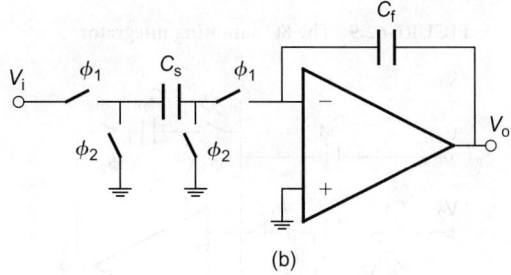

(b)

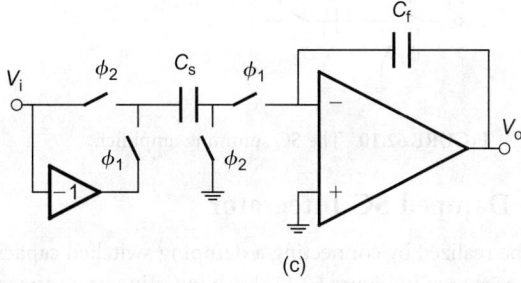

(c)

FIGURE 62.8 SC Integrators: (a) noninverting, (b) inverting, (c) bilinear.

62.4.2 The Summing Integrator

The SC operation is based on charge transfer. It is therefore easy to make weighted sum of multiple inputs by connecting different input branches to the same virtual ground. This concept is shown in the summing integrator of Figure 62.9. The transfer function from the three input signals to the output is given in Eq. (62.10).

$$V_o = \frac{C_{s1}(1 - z^{-1})V_1 - C_{s2}V_2 + C_{s3}z^{-1}V_3}{C_f(1 - z^{-1})} \tag{62.10}$$

If the integrating feedback capacitor (C_f) is replaced by a feedback switched capacitor (C_{sw}), the structure does not maintain memory of its past evolution and a simple summing amplifier is obtained. The resulting structure is shown in Figure 62.10, with the corresponding transfer function given in Eq. (62.11). This is the basic building block for the construction of SC filters implementing FIR frequency response [4].

$$V_o = \frac{1}{C_{sw}}(C_{s1}(1 - z^{-1})V_1 - C_{s2}V_2 + C_{s3}z^{-1}V_3) \tag{62.11}$$

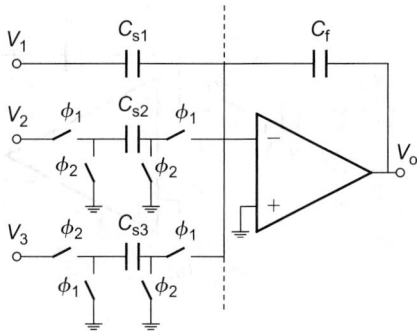

FIGURE 62.9 The SC summing integrator.

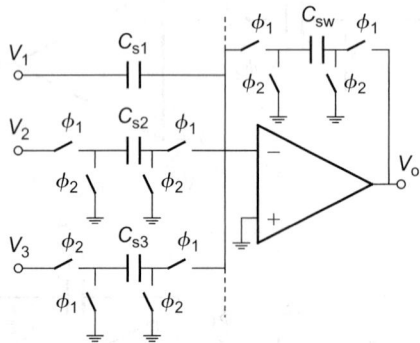

FIGURE 62.10 The SC summing amplifier.

62.4.3 The Active Damped SC Integrator

A damped integrator can be realized by connecting a damping switched capacitor (C_d) in parallel to the integrating capacitor (C_f), as shown in Figure 62.11. Both inverting or noninverting circuits are possible, depending on the type of the input sampling structure. Eq. (62.12) is valid for the clock phases not in parenthesis, while Eq. (62.13) is valid for the clock phases within parenthesis.

$$H_1(z) = \frac{C_d z^{-1}}{(C_d + C_f) - C_f z^{-1}} \tag{62.12}$$

$$H_2(z) = -\frac{C_d}{(C_d + C_f) - C_f z^{-1}} \tag{62.13}$$

62.4.4 A Design Example

As an example, the design of a damped SC integrator (Figure 62.11) is given. A possible design approach is to derive the capacitor values of the SC structure from the R and C values in the equivalent continuous-time prototype, which is shown in Figure 62.12, by Eq. (62.6). It results that $C_s = T_s/R_s$, and $C_d = T_s/R_d$. For instance, to have the pole frequency at 10 kHz with unitary DC gain, a possible solution is $R_s = R_d = 159.15$ kΩ, and $C_f = 10$ pF. Using $F_s = 1$ MHz, $C_s = C_d = 0.628$ pF.

The frequency response of the continuous-time and the SC damped integrator are shown in Figure 62.13(a). Line I refers to the damped integrator of Figure 62.11, line II to a damped integrator with bilinear input branch (see Figure 62.8[c]), and line III to the active-RC integrator of Figure 62.11. In the passband the

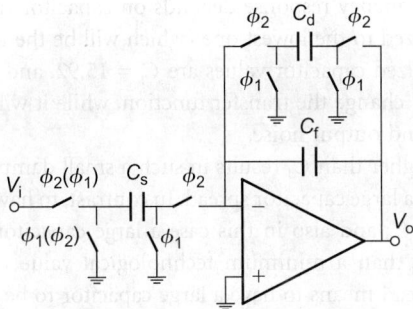

FIGURE 62.11 The dumped SC integrator.

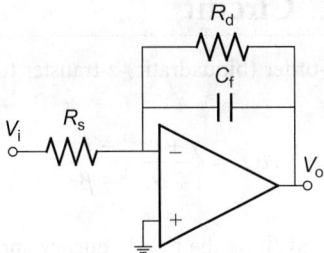

FIGURE 62.12 Continuous-time damped RC integrator.

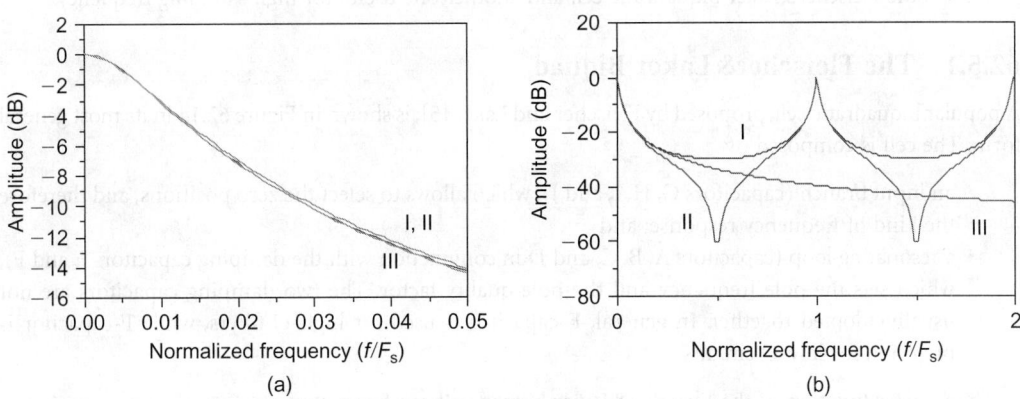

(a)   (b)

FIGURE 62.13 Frequency response comparison for different integrators.

frequency responses track very well. Increasing the input frequency, a difference becomes evident (as stated by the fact that Eq. (62.6) is valid for slowly variant signals). This is more pronounced if the frequency response is plotted up to $2F_s$ (see Figure 62.13[b]), where the periodic behavior of the sampled-data system frequency response is evident. Moreover, the key point of a sampled-data filter is the fact that the frequency response fixes the ratio between sampling frequency and pole frequency, i.e., with the above capacitor values the pole frequency is 10 kHz for $F_s = 1$ MHz, while it decreases to 1 kHz if $F_s = 100$ kHz is used (i.e., the ratio f_p/F_s remains constant). For this reason the frequency response is plotted as a function of the normalized frequency f/F_s.

A limited stopband attenuation results for line I. This attenuation is improved using the bilinear input branch which implements a zero at $F_s/2$. This does not affect the frequency response in the passband, while a larger attenuation is obtained in the stopband (line II).

For the SC networks, the frequency response depends on capacitor ratios. Thus the above extracted capacitor value can be normalized to the lowest one (which will be the unit capacitance). For this first-order cell example, the normalized capacitor values are $C_f = 15.92$, and $C_s = C_d = 1$. The chosen value of the unit capacitance will not change the transfer function, while it will affect other filter features like die size, power consumption, and output noise.

In general, using C_f much higher than C_d results in such a small damping impedance, as it is the case of high-Q filters. This results in a large capacitor spread. In contrast, to have a small time constant requires to have C_s much smaller than C_f, and also in this case a large capacitor spread occurs. Since the unit capacitance cannot be smaller than a minimum technological value (to achieve a certain matching accuracy), a large capacitor spread means to have a large capacitor to be driven and so a large power to the opamp to operate. In addition a large chip area is also needed. Thus to avoid large capacitor spread, possible solutions are proposed as follows.

62.5 Second-Order SC Circuit

The general expression of a second-order (biquadratic) z-transfer function can be written in the form

$$H(z) = \frac{\gamma + \varepsilon z^{-1} + \delta z^{-2}}{1 + \alpha z^{-1} + \beta z^{-2}} \tag{62.14}$$

The denominator coefficients (α and β) fix the pole frequency and quality factor, while the numerator coefficients (γ, ε, and δ) define the types of filter frequency response. Several SC biquadratic cells have been proposed in the literature to implement the above t.f. (transfer function). In the following two of them are presented: the Fleischer&Laker biquadratic cell and another one useful for high sampling frequency.

62.5.1 The Fleischer&Laker Biquad

A popular biquadratic cell, proposed by Fleischer and Laker [5], is shown in Figure 62.14 in its most general form. The cell is composed of:

- an input branch (capacitors G, H, I, and J), which allows to select the zero positions, and therefore the kind of frequency response; and
- a resonating loop (capacitors A, B, C, and D in conjunction with the damping capacitors E and F), which sets the pole frequency and the pole quality factor. The two damping capacitors are not usually adopted together. In general, E-capacitor is used for high-Q filters, while F-capacitor is preferred for low-Q circuits.

The transfer function of the Fleischer&Laker biquad cell can be written as

$$H(z) = \frac{V_o}{V_i} = -\frac{DI + (AG - D(I+J))z^{-1} + (JD - AH)z^{-2}}{D(B+F) - (D(2B+F) - A(C+E))z^{-1} + (DB - AE)z^{-2}} \tag{62.15}$$

This cell allows to synthesize any kind of transfer function by using parasitic-insensitive structures, which ensures performance accuracy. The key observation related to this biquad cell is that the first opamp operates in cascade to the second one; therefore its settling is longer than that of the second one.

62.5.2 Design Methodology

At the first order, the SC circuits can be derived from continuous-time circuits implementing the desired frequency response, by a proper substitution of each resistor with the equivalent SC structures, as shown for the first-order cell. An alternative approach is to optimize the transfer function in the z-domain, and

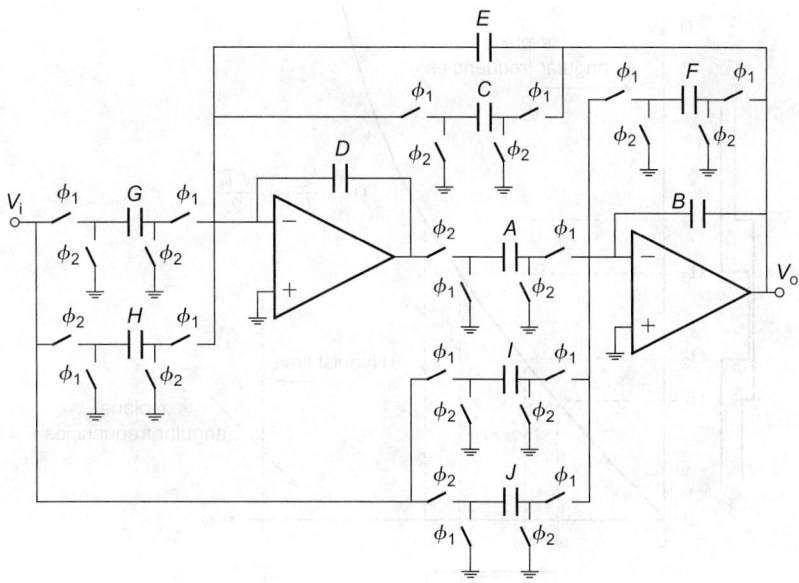

FIGURE 62.14 Fleischer&Laker biquadratic cell.

to numerically fit the desired transfer function with the transfer function implemented by the second-order cell. A third possibility is to adapt the *s*-domain transfer function to the *z*-domain signal processing of the SC structures. This procedure will be used in the following.

Let us consider the case of a given transfer function in *s*-domain, for instance when an approximation table (Butterworth, Chebichev, Bessel, etc.) is used. The *s*-domain transfer function to be implemented is written as:

$$H(s) = \frac{a_2 s^2 + a_1 s + a_0}{s^2 + b_1 s + b_0} \tag{62.16}$$

This *s*-domain transfer function is transformed in a *z*-domain transfer function through the use of the bilinear *s*-to-*z* transformation:

$$s = \frac{2}{T_s} \frac{1 - z^1}{1 + z^1} \tag{62.17}$$

This transformation also operates a warping of the frequency of interest. To avoid this error a characteristic frequency for the given design (i.e., the frequencies ω_i of interest for the final filter mask) should be "prewarped" according to the relationship between the angular frequencies in the *s*-domain (Ω_i) and in the *z*-domain (ω_i):

$$\Omega_i = \frac{2}{T_s} \tan\left(\frac{\omega_i T_s}{2}\right) \tag{62.18}$$

as it is indicated in Figure 62.15 for a band-pass-type response.

The characteristic frequency can be the −3 dB frequency for a low-pass filter. The $H'(s)$ that will satisfy the "prewarped" filter mask, will be automatically transformed by Eq. (62.17) into a *z*-domain transfer function whose frequency response satisfies the desired filter mask in the ω domain. Obviously, if $\omega_i T_s << 1$, no predistortion is needed, being $\Omega_i \approx \omega_i$. Assuming that $\omega_i T_s << 1$, $H'(s) \approx H(s)$ and Eq. (62.17)

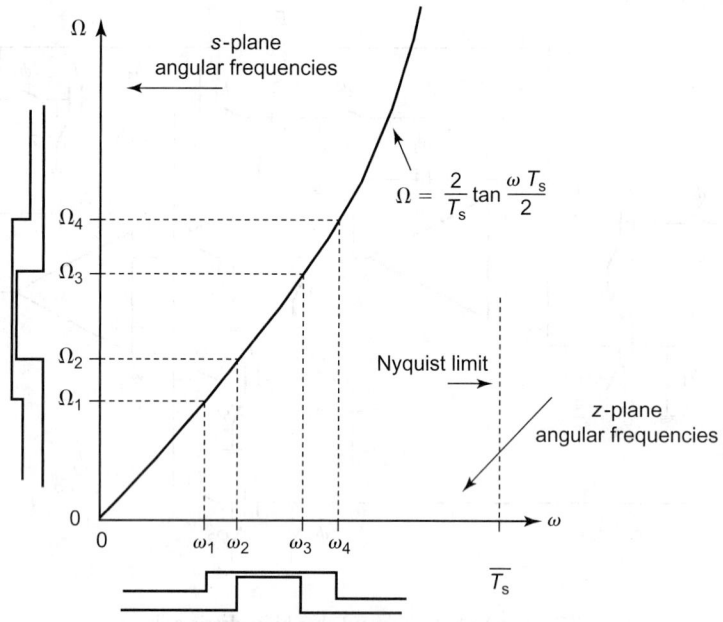

FIGURE 62.15 Bilinear mapping between continuous-time (Ω) and sampled-data (ω) frequency.

is substituted directly into Eq. (62.16); the resulting coefficients of terms z^{-i} in the denominator are then equated to the corresponding ones in Eq. (62.15). Assuming $A = B = D = 1$, the capacitor values for the E- and F-types can be extracted as follows:

$$F = 0: \quad E = \frac{b_1 T_s}{1 + \dfrac{b_1 T_s}{2} + \dfrac{b_0 T_{s^2}}{4}}, \; C = \frac{b_0 T_{s^2}}{1 + \dfrac{b_1 T_s}{2} + \dfrac{b_0 T_{s^2}}{4}} \qquad (62.19)$$

$$E = 0: \quad F = \frac{b_1 T_s}{1 - \dfrac{b_1 T_s}{2} + \dfrac{b_0 T_{s^2}}{4}}, \; C = \frac{b_0 T_{s^2}}{1 - \dfrac{b_1 T_s}{2} + \dfrac{b_0 T_{s^2}}{4}} \qquad (62.20)$$

When the bilinear transform is used, a second-order numerator is obtained, which can be written as in the following equation, where simple solutions for the input capacitors are also given:

Low-Pass: $K(1 + z^{-1})^2$, $I = J = IKI, G = 4IKI, H = 0$

Band-Pass: $K(1 + z^{-1})(1 - z^{-1})$, $I = G = H = IKI, J = 0$ (62.21)

High-Pass: $K(1 - z^{-1})^2$, $I = J = IKI, G = H = 0$

62.5.3 Design Example

As design example for the second-order cell, a bandpass response with $Q = 5$, $f_o = 20$ kHz, and $F_s = 1$ MHz is considered (i.e., $f_o/F_s = 0.02$). The transfer functions in the s-domain and in the z-domain (using bilinear transformation) are the following:

$$H(s) = \frac{s}{s^2 + \dfrac{2\pi f_o}{Q} s + (2\pi f_o)^2} = \frac{2.5133 \times 10^4 s}{s^2 + 2.5133 \times 10^4 s + 1.5791 \times 10^{10}} \qquad (62.22)$$

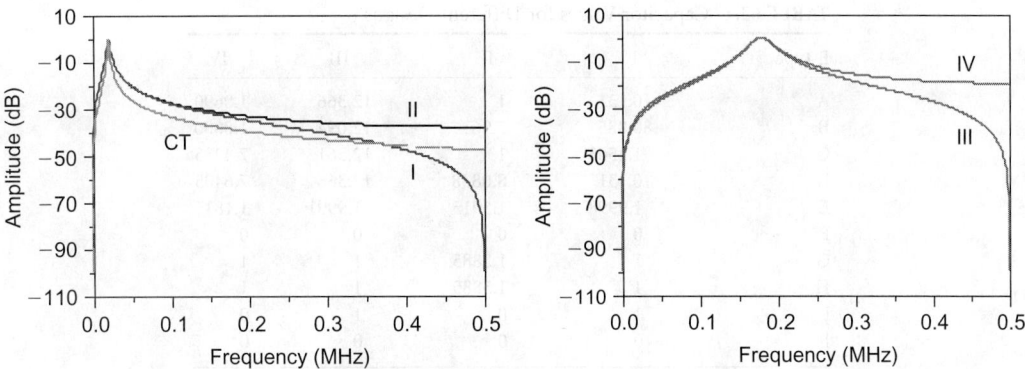

FIGURE 62.16 Frequency responses for different designs.

$$H(z) = \frac{z^{-2} - 1}{z^{-2} - 1.9719z^{-1} + 0.98755} \qquad (62.23)$$

No frequency has been prewarped, since, applying Eq. (62.18) to f_o, the prewarped pole frequency should result 19.765 kHz, with a negligible deviation of about 0.1%.

For the bandpass response, using the bilinear s-to-z mapping, the zero positions are at $\{z = 1, z = -1\}$. The frequency response is shown in Figure 62.16 with line I. The normalized capacitance value, obtained equating the transfer function of Eq. (62.23) with the transfer function of the biquadratic cell of Eq. (62.15), are given in Table 62.1, for the E- and F-type structures, in column I. A very large capacitor spread (>78) is needed. This results in large die area, and large power consumption. The capacitor spread could be reduced with a slight modification of the transfer function to the one given in the following:

$$H(z) = \frac{z^{-1} - 1}{z^{-2} - 1.9719z^{-1} + 0.98755} \qquad (62.24)$$

With respect to the bilinear transformation of Eq. (62.23), in this case, the zero at DC is maintained, while the zero at Nyquist frequency (at $F_s/2$, i.e., at $z = -1$) is eliminated. The normalized capacitor values are indicated again in Table 62.1, in Column II. It can be seen that a large reduction of the capacitor spread is obtained (from 80 to 8, for the E-type). The obtained frequency response is reported in Figure 62.16 with line II. In the passband no significant changes occurs; in contrast in the stopband, the maximum signal attenuation is about −35 dB. In some applications, this solution is acceptable, also in consideration of the considerable capacitor spread reduction. For this reason, if not strictly necessary, the zero at $F_s/2$ can be eliminated. However reducing the factor f_o/F_s results in reducing the stopband attenuation. For instance, for $f_o = 200$ kHz (i.e., $f_o/F_s = 0.2$), the frequency responses with and without the Nyquist zero are reported in Figure 62.16 with lines III and IV, respectively. In this case the stopband attenuation is reduced to −22 dB and therefore the Nyquist zero could be strongly needed. The relative normalized capacitor values are indicated in Table 62.1 in Column III (with zeros at $\{z = 1, z = -1\}$) and in Column IV (with zeros at $z = 1$).

62.5.4   A Biquadratic Cell for High Sampling Frequency

In the previous biquadratic cell, the two opamps operate in cascade during the same clock phase. This requires that the second opamp in cascade waits for the complete settling of the first opamp to complete its settling. This, of course, reduces the maximum achievable sampling frequency, or, alternatively, for a given sampling frequency increases the required power consumption since opamp with larger bandwidth

TABLE 62.1 Capacitor Values for Different Designs

E-type	I	II	III	IV
A	10.131	1	12.366	1.0690
B	**80.885**	7.963	12.094	1.0000
C	1.2565	1	12.561	7.4235
D	10.131	**8.0838**	12.366	7.6405
E	1.9998	1.5915	1.9991	1.1815
F	0	0	0	0
G	1	1.5885	1	1
H	1	1.5885	1	1
I	1	0	1	0
J	0	0	0	0

F-type	I	II	III	IV
A	10.006	5.1148	11.305	5.9919
B	**78.885**	**39.446**	10.095	5.0497
C	1.2565	1	12.561	7.4235
D	10.006	8.1404	11.305	7.0793
E	0	0	0	0
F	1.9998	1	1.9991	1
G	1	1.5885	1	1
H	1	1.5885	1	1
I	1	0	1	0
J	0	0	0	0

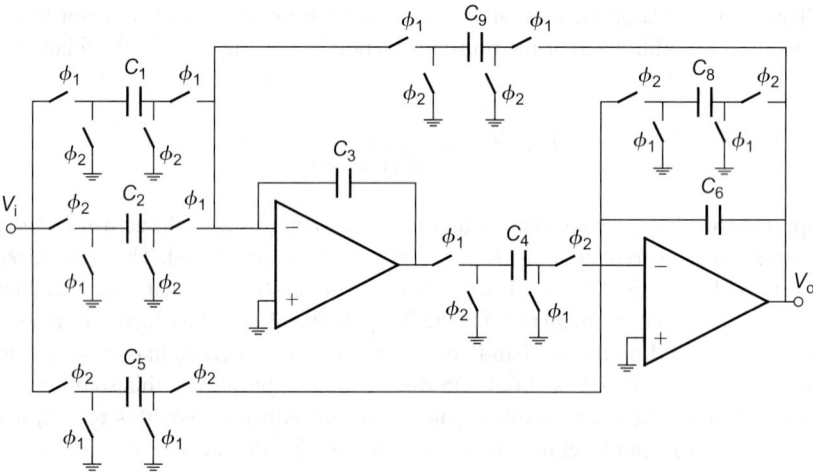

FIGURE 62.17 High-frequency biquadratic cell.

are needed. To avoid this aspect, the biquad shown in Figure 62.17 can be used. In this scheme the two opamps settle in different clock phase and thus they have the full clock phase time slot to settle.

The transfer function of the biquadratic cell is given in Eq. (62.25). As seen a limitation occurs in the possible transfer function, since the term in z^{-2} is not present.

$$H(z) = -\frac{C_3C_5 + C_1C_4 - (C_3C_5 - C_2C_4)z^{-1}}{C_3(C_8 + C_6) + (C_4C_9 - C_3C_8 - 2C_3C_6)z^{-1} + C_3C_6z^{-2}} \qquad (62.25)$$

62.5.5 High-Order Filters

The previous first- and second-order cells can be used to build up high-order filters. The main architectures are taken from the theory for the active-RC filters. Some of the most significant ones are: ladder [6] (with good amplitude response robustness with respect to component spread), cascade of first- and second-order cells (with good phase response robustness with respect to component spread), follow-the-leader feedback (for low-noise systems, such as reconstruction filters in oversampled DAC).

62.6 Implementation Aspects

The arguments presented so far have to be implemented in actual integrated circuits. Such implementations have to minimize the effects of the nonidealities of the actual blocks, which are capacitors, switches, and opamps. The capacitor behavior is quite stable, apart capacitance nonlinearities which affect the circuit performance only as a second-order effect.

In contrast, switches and opamp must be properly designed to operate in the SC system. The switches must guarantee a minimum conductance to ensure a complete charge transfer within the available time slot. For the same reason, the opamps must ensure large-DC gain, large unity-gain bandwidth, and large slew rate. For instance, in Figure 62.18 the ideal output waveform of an SC network is shown by a solid line, while the more realistic actual waveform is illustrated by a dotted line. The output sample is updated during phase ϕ_1, while it is held (at the value achieved at the end of phase ϕ_1) during phase ϕ_2. In phase ϕ_1, the output value moves from its initial to its final value. The slowness of this movement is affected by switches conductance, opamp slew rate, and opamp bandwidth.

The transient response of the system could be studied using the linear model of Figure 62.19, where the conductive switches are replaced by their on-resistance R_{on}, and the impulsive charge injection is replaced by a voltage step. The assumption of a complete linear system should allow to study exactly the system evolution. In this case the circuit time-constants depend on input branch ($\tau_{\text{in}} = 2R_{\text{on}}C_s$), opamp frequency response and feedback factor.

Nonlinear analysis is however necessary when opamp slew rate occurs. This analysis is difficult to be carried out and optimum performance can be achieved using computer simulations. Usually, for typical device models, 10% of the available time slot (i.e., $T_s/2$) is used for slew rate, while 40% is used for linear settling.

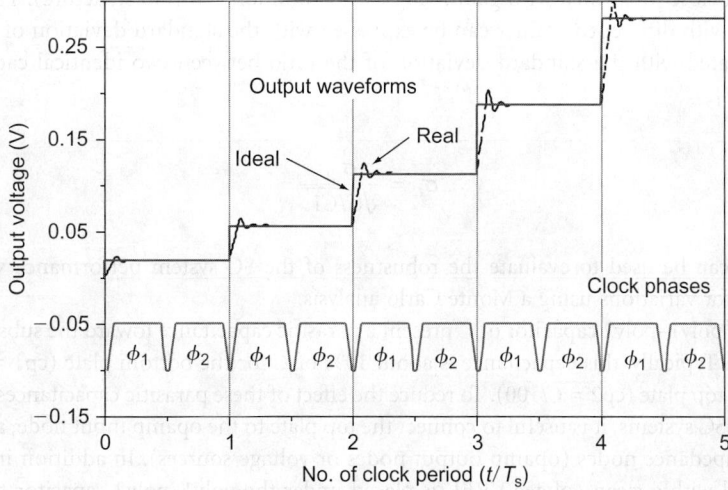

FIGURE 62.18 Output waveform evolution.

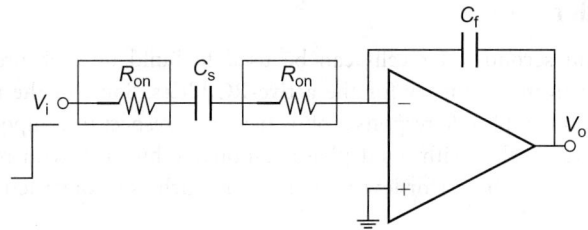

FIGURE 62.19 Linear model for transient analysis.

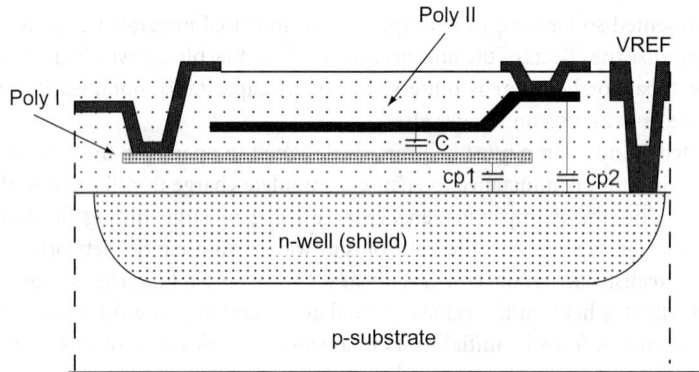

FIGURE 62.20 Poly1–poly2 capacitor cross section.

62.6.1 Integrated Capacitors

Integrated capacitors in CMOS technology for SC circuits are mainly realized using poly1–poly2 structure, whose cross section is shown in Figure 62.20. This capacitor implementation guarantees linear behavior over a large signal swing. The main drawbacks of integrated capacitors are related to their absolute and relative inaccuracy, and to their associated parasitic capacitance.

The absolute value of integrated capacitors can change ±30% from its nominal value. However the matching between equal capacitors can be the order of 0.2%, provided that proper layout solutions are adopted (in close proximity, with guard rings, with common centroid structure). The matching of two capacitors with different C values can be expressed with the standard deviation of their ratio σ_C, which is correlated with the standard deviation of the ratio between two identical capacitors σ_{C1} by the equation [7]

$$\sigma_C = \frac{\sigma_{C1}}{\sqrt{C/C1}} \tag{62.26}$$

This model can be used to evaluate the robustness of the SC system performance with respect to random capacitor variations using a Monte Carlo analysis.

The plates of poly1–poly2 capacitor of C present a parasitic capacitance toward the substrate, as shown in Figure 62.20. Typically, this capacitance is about 10% of C for the bottom plate (cp1 = $C/10$), and is 1% of C for the top plate (cp2 = $C/100$). To reduce the effect of these parasitic capacitances in the transfer function of the SC systems, it is useful to connect the top plate to the opamp input node, and the bottom plate to low impedance nodes (opamp output nodes or voltage sources). In addition in Figure 62.20, a n-well, biased with a clean voltage VREF, is placed under the poly1–poly2 capacitor to reduce noise coupling from the substrate, through parasitic capacitance.

62.6.2   MOS Switches

The typical situation during sampling operation is shown in Figure 62.21(a) (this is the input branch of the integrator of Figure 62.8[a]). The input signal V_i is sampled on the sampling capacitor C_s to have $V_c = V_i$.

In Figure 62.21(b) the switches are replaced by a single-NMOS device which operates in triode region with an approximately zero voltage drop between drain and source. The switch on-resistance R_{on} can be expressed as

$$R_{on} = \frac{1}{\mu n \, C_{ox} \dfrac{W}{L}(V_{GS} - V_{TH})} = \frac{1}{\mu n \, C_{ox} \dfrac{W}{L}(V_G - V_i - V_{TH})} \qquad (62.27)$$

where V_G is the amplitude of the clock driving phase, μn the electron mobility, C_{ox} the oxide capacitance, and W and L are the width and length of MOS device. Using $V_{DD} = 5$ V (i.e., $V_G = 5$ V), the dependence of R_{on} on the input voltage is plotted in Figure 62.22(a). This means that if R_{on} is required by the capacitor value be lower than a given value (to implement a low $R_{on}C_s$ time constant), a limitation in the possible input swing is given. For instance, if the maximum possible R_{on} is 2.5 kΩ, the maximum input signal swing is 0–0.5 V.

To avoid this limitation a complementary switch can be used. It consists of a NMOS and a PMOS device in parallel, as shown in Figure 62.23. The PMOS switch presents a R_{on} behavior complementary to that of the NMOS, as plotted in Figure 62.22(b). The complete switch R_{on} is then given by the parallel of the two contributions which is sufficiently low for all the signal swing.

To use this solution, it requires to distribute double clock lines controlling the NMOS and the PMOS. This could be critical for SC filters operating at high-sampling frequency, also considering

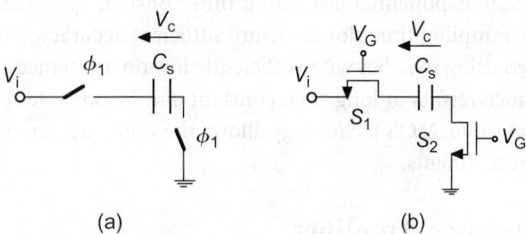

(a)   (b)

FIGURE 62.21   (a) Ideal sampling structure. (b) Sampling structure with NMOS switches.

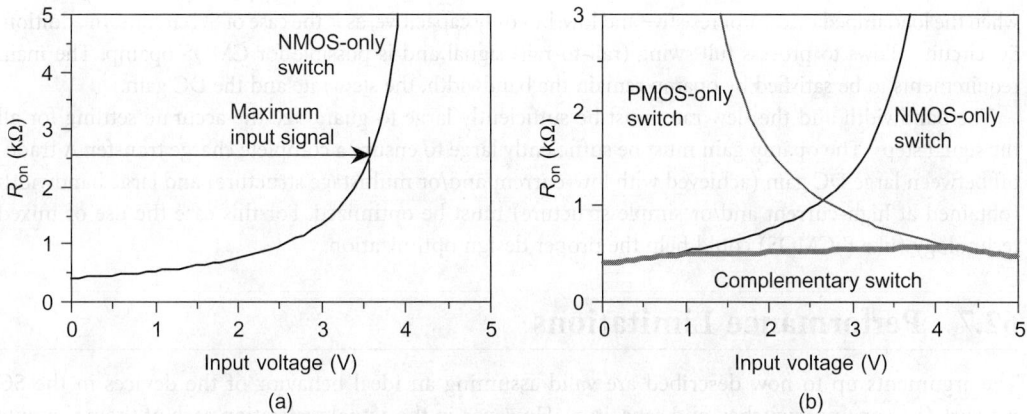

FIGURE 62.22   Switches on-resistance.

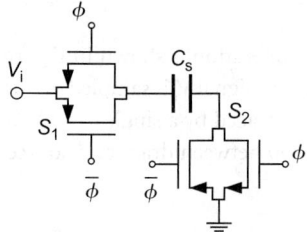

FIGURE 62.23 Sampling structure with complementary switches.

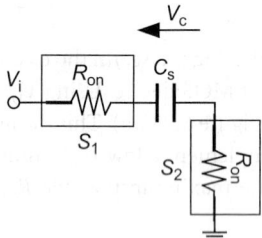

FIGURE 62.24 Sampling operation linear model.

the synchronization of the two phases and of the digital noise from the distribute clocks which could reduce the dynamic range.

Once a minimum conductance is guaranteed, the structure can be studied using the linear model for the MOS devices S_1 and S_2 that operate in triode region, resulting in the circuit of Figure 62.24. In this case V_c follows V_i, through an exponential law with a time constant $\tau_{in} = C_s 2 R_{on}$. Typically at least $6\tau_{in}$ must be guaranteed in the sampling time slot to ensure sufficient accuracy. For a given sampling capacitance value, this is achieved using switches with sufficiently low on-resistance and no voltage drop across its nodes. Large on-resistance results in long time constant and incomplete settling, while voltage drop results in an incorrect final value. MOS technology allows the implementation of analog switches satisfying both the previous requirements.

62.6.3 Transconductance Amplifier

The SC technique appears to be the natural application of available CMOS technology design features. This is true also for the case of the opamp design. In fact, SC circuits require an infinite input opamp impedance, as it is the case of opamp using a MOS input device. In contrast, CMOS opamps are particularly efficient when the load impedance is not resistive and low, but only capacitive, as is the case of SC circuits. In addition, SC circuits allows to process full swing (rail-to-rail) signal and is possible for CMOS opamp. The main requirements to be satisfied by opamp remain the bandwidth, the slew rate and the DC gain.

The bandwidth and the slew rate must be sufficiently large to guarantee the accurate settling for all the signal steps. The opamp gain must be sufficiently large to ensure a complete charge transfer. A trade-off between large DC gain (achieved with low-current and/or multistage structure) and large bandwidth (obtained at high-current and/or simple structure) must be optimized. For this case the use of mixed technology (like BiCMOS) could help the proper design optimization.

62.7 Performance Limitations

The arguments up to now described are valid assuming an ideal behavior of the devices in the SC network (i.e., opamp, switches, and capacitor). However, in the actual realization each of them presents nonidealities which reduce the performance accuracy of the complete SC circuit. The main limitations and their effects are described in the following. Finally, considerations about noise in SC systems conclude this section.

62.7.1 Limitation Due to Switches

As described before, CMOS switches satisfy both low on-resistance and zero voltage-drop requirements. However they introduce some performance limitations due to their intrinsic CMOS realization. The cross section of a NMOS switch in its on-state is shown in Figure 62.25. The connection between its nodes N1 and N2 is guaranteed by the presence of the channel, made up of the charge Q_{ch}. The amount of charge Q_{ch} can be written as

$$Q_{ch} = (WL)\, C_{ox}(V_G - V_i - V_{TH}) \tag{62.28}$$

where V_i is the channel (input) voltage. Both nodes N1 and N2 are at voltage V_i (no voltage drop between the switch nodes). In addition, the gate oxide that guarantees infinite MOS input impedance constitutes a capacitive connection between gate and both source and drain. This situation results in two nonideal effects: the charge injection and the clock feedthrough.

62.7.1.1 Charge Injection

At the switch turn-off, the charge Q_{ch} given in Eq. (62.28) is removed from the channel and is shared between the two nodes connected to the switch, with a partition depending on the node impedance level (Figure 62.26).

The charge kQ_{ch} is injected into $N2$ and collected on a capacitor C_c. A voltage variation ΔV_c across the capacitor arises, which is given by

$$\Delta V_c = k\frac{Q_{ch}}{C_c} \tag{62.29}$$

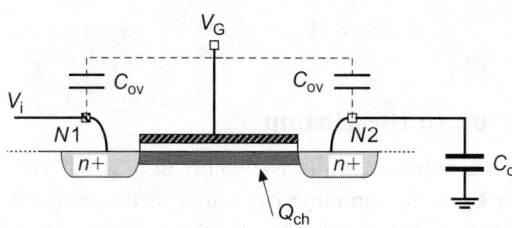

FIGURE 62.25 Switch charge profile of an NMOS switch in the on-state.

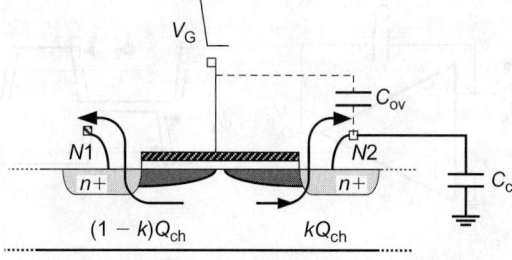

FIGURE 62.26 Charge displacement during turn-off.

For all the switches of a typical SC integrator this effect is important. For instance, for the switch S4 connected to the opamp virtual ground, the charge injection into the virtual ground is collected in the feedback capacitor and is processed as an input signal. The amount of this charge injection depends on different parameters (see Eq. [62.28] and Eq. [62.29]). Charge Q_{ch} depends on switch size W, which however cannot be reduced beyond a certain level; otherwise the switch on-resistance should increase. Thus a trade-off between charge injection and on-resistance is present. In addition, charge Q_{ch} depends on the voltage V_i to which the switch is connected to. For the switches S2, S3, and S4, the injected charge is proportional to $(V_G - V_{gnd})$ and is always fixed; as a consequence it can be considered like an offset. In contrast, for the switch S1 connected to the signal swing the channel charge Q_{ch} is dependent on $(V_G - V_i)$, i.e., from the signal amplitude and thus also the charge injection is signal-dependent. This creates an additional signal distortion.

Possible solutions for the reduction of the charge injection are use of dummy switches, use of slowly variable clock phase, use of differential structures, use of delayed clock phases [8], and use of signal-dependent charge pump [9].

Dummy switches operate with complementary phases to sink the charge rejected by the original switches. The use of differential structures reduces the offset charge injection to the mismatch of the two differential paths. For the signal-dependent charge injection, the delayed phases of Figure 62.27(b) are applied to the integrator of Figure 62.27(a). This clock phasing is based on the concept that at the turn-off, S3 is open before S1. In this way, when S1 opens, the impedance toward C_s is infinite and no signal-dependent charge injection occurs into C_s.

62.7.1.2 Clock Feedthrough

The clock feedthrough is the amount of signal that is injected into the sampling capacitor C_c from the clock phase through the MOS overlap capacitor (C_{ov}) path, shown in Figure 62.26, which is then proportional to the area of the switches. Using large switches, to reduce on-resistance, results in large charge injection and large clock feedthrough. This error is typically constant (it depends on capacitance partition) and therefore it can be greatly reduced by using differential structures. The voltage error ΔV_c across a capacitance C_c due to the feedthrough of the clock amplitude ($V_{DD} - V_{SS}$) can be written as

$$\Delta V_c = (V_{DD} - V_{SS}) \frac{C_{ov}}{C_{ov} + C_c} \qquad (62.30)$$

62.7.2 Limitation Due to the Opamp

The operation of SC networks is based on the availability of a "good" virtual ground, which ensures a complete charge transfer from the sampling capacitors to the feedback capacitor. Whenever this charge transfer is incompleted, the SC network performance derives from its nominal behavior. The main nonideality causes from the opamp are finite DC gain, finite bandwidth, finite slew rate, and gain nonlinearity.

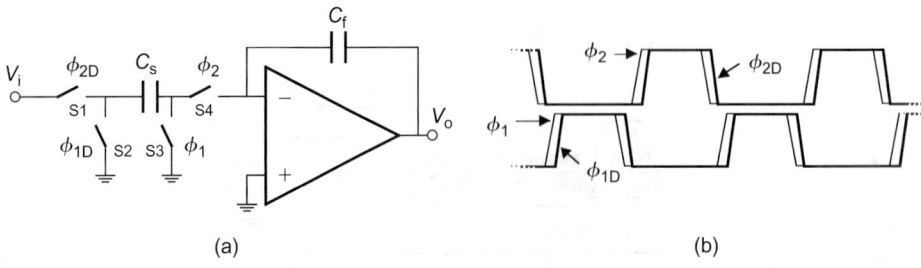

(a)   (b)

FIGURE 62.27   Clocking scheme for signal-dependent charge injection reduction.

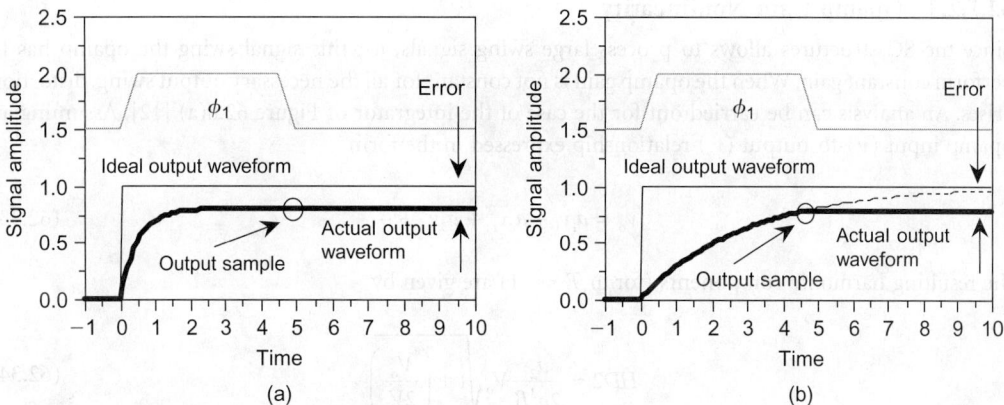

FIGURE 62.28 Opamp induced errors. (a) Finite gain error. (b) Finite bandwidth error.

62.7.2.1 Finite Opamp DC-Gain Effects

The opamp finite gain results in a deviation of output voltage at the end of the sampling period from the ideal one, as shown in Figure 62.28(a) [10,11]. This output sample deviation can be translated into SC system performance deviation. For the finite gain effect, an analysis, which correlates the opamp gain A_o with SC network performance deviation, can be carried out under the hypothesis that the opamp bandwidth is sufficiently large to settle within the available time slot.

For the case of the summing amplifier of Figure 62.10, it can be demonstrated that the effect of the finite opamp DC-gain (A_o) is only in a overall gain error. For this reason SC FIR filters (based on this scheme) exhibit a low sensitivity to opamp finite DC gain. In contrast, for the case of SC integrators, the finite gain effect results in pole and gain deviation. For instance, the transfer function of the integrator of Figure 62.8(a) becomes

$$H_a(z) = \frac{V_o(z)}{V_i(z)} = \frac{C_s}{C_f} \frac{z^{-1}}{1 + \frac{1}{A_o}\left(1 + \frac{C_s}{C_f}\right) - \left(1 + \frac{1}{A_o}\right)z^{-1}} \tag{62.31}$$

For a biquadratic cell, the opamp finite gain results in pole frequency and pole quality factor deviations. The actual frequency and quality factor of the pole (f_{oA} and Q_A) are correlated to their nominal value (f_o and Q) by the relationship

$$f_{oA} = \frac{A_o}{1 + A_o} f_o, \qquad Q_A = \frac{1}{\frac{1}{Q} + \frac{2}{A_o}} \approx \left(1 - \frac{2Q}{A_o}\right) Q \tag{62.32}$$

62.7.2.2 Finite Bandwidth and Slew Rate

Also opamp finite bandwidth and slew rate results in incomplete charge transfer, which still corresponds in deviation of the output sample with respect to its nominal value [10,11]. For the case of only finite bandwidth the effect is shown in Figure 62.28(b). An analysis similar to that of the finite gain for the finite bandwidth and slew-rate effect is not easy to be extracted. This is also due to the fact that incomplete settling is caused by the correlation of a linear effect (e.g., the finite bandwidth) and nonlinear effect (e.g., the slew rate). In addition, this case is worsened by the fact that in some structures several opamps are connected in cascade and then each opamp (a part the first one) has to wait for the operation conclusion of the preceding one.

62.7.2.3 Opamp Gain Nonlinearity

Since the SC structures allows to process large swing signals, for this signal swing the opamp has to perform constant gain. When the opamp gain is not constant for all the necessary output swing, distortion arises. An analysis can be carried out for the case of the integrator of Figure 62.8(a) [12]. Assuming an opamp input (v_i)-to-output (v_o) relationship expressed in the form

$$v_o = a_1 v_i + a_2 v_i^2 + a_3 v_i^3 + \cdots \tag{62.33}$$

the resulting harmonic components (for $\omega_o T_s \ll 1$) are given by

$$HD2 \approx \frac{a_2}{2a_1^3 \beta} V_o \sqrt{1 + \left(\frac{V_o}{2V_i}\right)^2} \tag{62.34}$$

$$HD3 \approx \frac{a_3}{2a_1^4 \beta} V_o^2 \left(1 + \frac{V_o}{3V_i}\right) \tag{62.35}$$

The distortion can then be reduced by making constant low gain (i.e., reducing a_2 and a_3) or using a very large opamp gain (i.e., increasing a_1). This second case is usually the adopted strategy.

62.7.3 Noise in SC Systems

In SC circuits the main noise sources are in the switches (thermal noise) and in the opamp (thermal noise and $1/f$ noise) [13,14]. These noise sources are processed by the SC structure as an input signal, i.e., they are sampled (with consequent folding) and transferred to the output with a given transfer function. As explained for the signal, the output frequency range of an SC filter is limited to the band $[0 - F_s/2]$. This means that for any noise source, its sampled noise band, independently of how large it is at the source before sampling, is limited to the $[0 - F_s/2]$ range. In contrast, the total power of noise remains constant after sampling; this means that the power density of sampled noise is increased by the factor $F_b/(F_s/2)$, where F_b is the noise band at its source. This can be seen for the switch noise in the following simple example. Let us consider the structure in Figure 62.29, where the resistance represents the switch on-resistance. Its associated noise spectral density is given by $v_n^2 = 4kTR_{on}$, where k is the Boltzmann's constant and T the absolute temperature. The transfer function to the output node (the voltage over the capacitor) is $H(s) = 1/(1 + sR_{on}C_s)$. The total ouput noise can be calculated from the expression

$$n_o^2 = \int_0^\infty v_n^2 \left| H(s)^2 \right| df = \frac{kT}{C_s} \tag{62.36}$$

The total sampled noise results then to be given by kT/C_s, and presents a bandwidth of $F_s/2$. This means that the output noise power density is $kT/C_s \, 2/F_s$.

The same folding concept can be applied to the opamp noise (Figure 62.30).

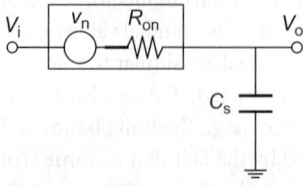

FIGURE 62.29 Sampling the noise on a capacitor.

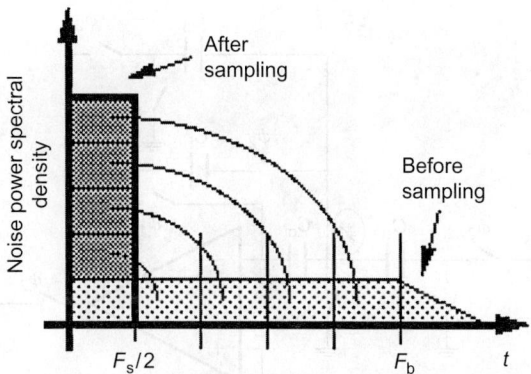

FIGURE 62.30 Folding of the noise power spectral density.

For the opamp $1/f$ noise, the corner frequency is usually lower than $F_s/2$. Therefore the $1/f$ noise is not modified by the sampling. In contrast, the white noise presents definitely a bandwidth $F_b > F_s/2$. This means that this noise component is modified in noise sources of bandwidth $F_s/2$ and noise power density multiplied by the factor $F_b/(F_s/2)$. When the noise sources are evaluated in this way, the output noise of an SC cell can be evaluated summing the different components properly weighted by their transfer functions from their source position to the output node.

Few considerations follow for the noise performance of a SC cell.

For the switch noise this is independent of R_{on}, since its dependence is cancelled in the bandwidth dependence. Thus this noise source is dependent only on C_s. Noise reduction is achieved by increasing C_s. This however trades with power increase necessary to drive the enlarged capacitance. Of course, even if R_{on} does not appear in the noise expression, as the capacitor in enlarged, the R_{on} must be adequately decreased to guarantee a proper sampling accuracy.

For the opamp noise, the noise band is usually correlated to the signal bandwidth. Therefore a good opamp settling (achieved with a large signal bandwidth) is in contrast with low-noise performance (achieved with reduced noise bandwidth). Therefore in low-noise systems, the bandwidth of the opamp is designed to be the minimum that guarantees proper settling.

62.8 Compensation Technique (Performance Improvements)

SC systems usually operate with a two-phase clock in which the opamp is "really" active only during one phase, while during the other phase is "sleeping". Provided that the opamp output node is not read during the second phase, this nonactive phase could be used to improve the performance of the SC systems, as shown in the following [15]. $1/f$ noise and offset can be reduced with correlated double sampling (CDS) or chopper technique. Similar structures are also able to compensate the error due to a finite gain of the operational amplifier. In contrast, proper structures are able to reduce the capacitor spread occurring in particular situations (high-Q or large time constant filters). Finally, double-sampled technique can be used to increase the sampling frequency of the SC system by a factor of 2.

62.8.1 CDS Offset-Compensated SC Integrator

The extra phase available in a two-phase SC system can be used to reduce opamp offset and $1/f$ noise effect at the output. A possible scheme is shown in Figure 62.31 [16], and operates as follows. Capacitor C_{of} is used to sample during ϕ_1 the offset V_{off} as it appears in the inverting node of the opamp close with unitary feedback. During ϕ_2 the inverting node is still at a voltage very close to V_{off}, since the bandwidth of V_{off} is assumed to be very small with respect to the sampling frequency. Capacitor C_{of} maintains the charge on its armatures and acts like a battery. Thus node X is a good virtual ground, independent of the opamp offset. In the same way also the output signal, read only during ϕ_2, is offset-independent. The

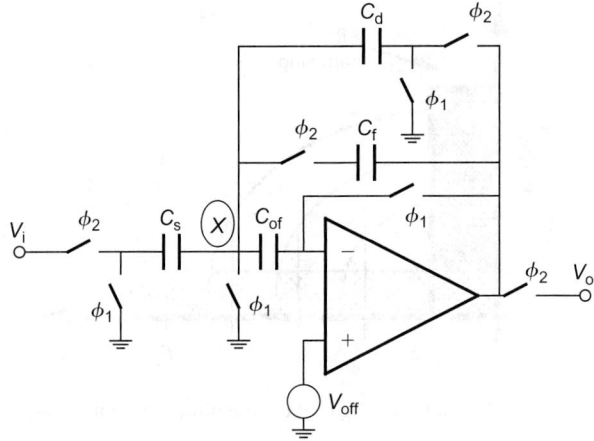

FIGURE 62.31 Offset-compensated SC integrator.

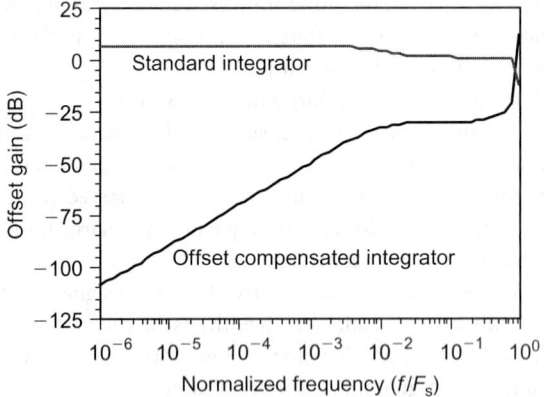

FIGURE 62.32 Offset-compensated SC integrator performance.

effect of this technique can be simulated using the value of the first-order cell of the previous example (i.e., $C_f = 15.92$, and $C_s = C_d = 1$), and $C_{of} = 1$. The transfer function V_o/V_{off} is shown in Figure 62.32. At low frequency, the V_{off} is highly rejected, while this is not the case of the standard (uncompensated) integrator. The main problem of this solution is due to the unity feedback operation of the structure during phase ϕ_1. This requires the stability of the opamp, which could consume high power.

62.8.2 Chopper Technique

An alternative solution to reduce offset $1/f$ noise at the output is given by the chopper technique. It consists of placing one SC mixer for frequency $F_s/2$ at the opamp input and one similar at the opamp output. This action does not affect white noise. In contrast, offset and $1/f$ noise are shifted to around $F_s/2$, not affecting anymore the frequencies around DC, where the signal to be processed is supposed to be.

This concept is shown for a fully differential opamp in Figure 62.33. In Figure 62.34 the input-referred noise power spectral density (PSD) without and with chopper modulation are shown. The white noise level (wnl) is not affected by the chopper operation and remains constant. It will be modified by the folding of the high-frequency noise, as described previously.

This technique is particularly advantageous for SC systems since the mixer can be efficiently implemented with SC technique as shown in Figure 62.35.

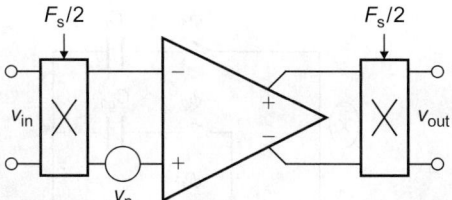

FIGURE 62.33 Opamp with chopper.

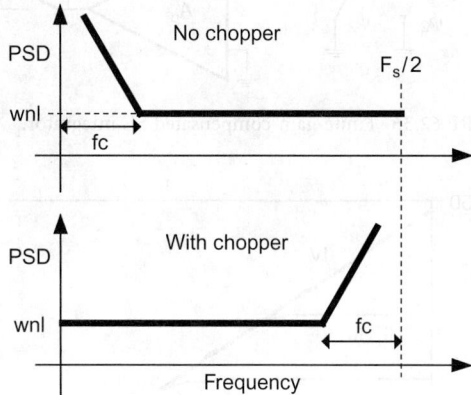

FIGURE 62.34 Input-referred noise power spectral density (PSD) with and without chopper technique.

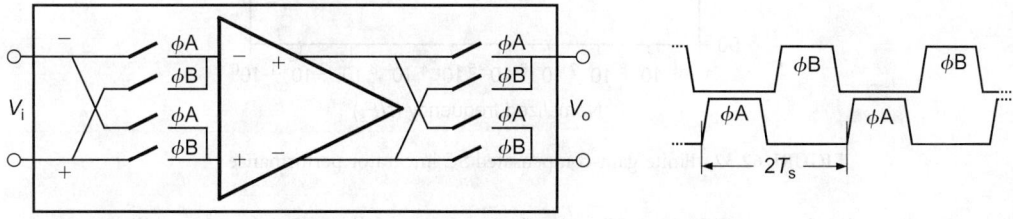

FIGURE 62.35 Opamp with SC chopper configuration.

62.8.3   Finite-Gain-Compensated SC Integrator

In the opamp design, a trade-off between opamp DC gain and bandwidth exists. Therefore, when a large bandwidth is needed, a finite DC gain necessarily occurs, reducing SC filter performance accuracy. To avoid this the available extra phase can be used to self-calibrate the structure with respect to the error due to the opamp finite gain. In the literature several techniques have been proposed. The major part of them are based on the concept of using a preview of the future output samples to precharge a capacitor placed in series to the opamp inverting input node to create a "good" virtual ground (as for offset cancellation). The various approaches differ on how they get the preview and how they calibrate the new virtual ground. For the different cases, they can be effective on a large bandwidth [17,18], on a small bandwidth [19,20], or on a passband bandwidth [21]. As an example for this kind of compensation, one of the earliest proposed scheme is reported in Figure 62.36.

The opamp finite gain makes the opamp inverting input node to be different from the virtual ground ideal behavior and to assume the value $-V_o/A_o$, where V_o is the output value and A_o the opamp DC gain. In the scheme of Figure 62.36, the future output sample is assumed to be close to the previous sample, sample of C_{g1}. This limits the effectiveness of this scheme to signal frequencies f for which this

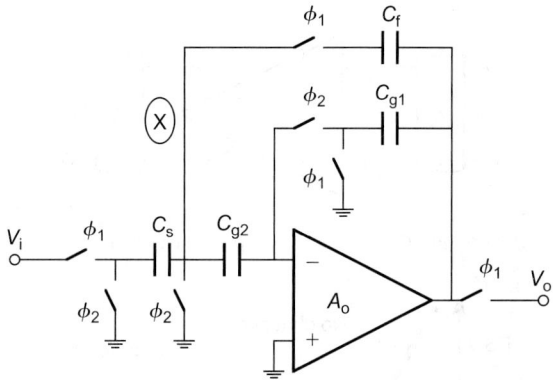

FIGURE 62.36 Finite-gain-compensated SC integrator.

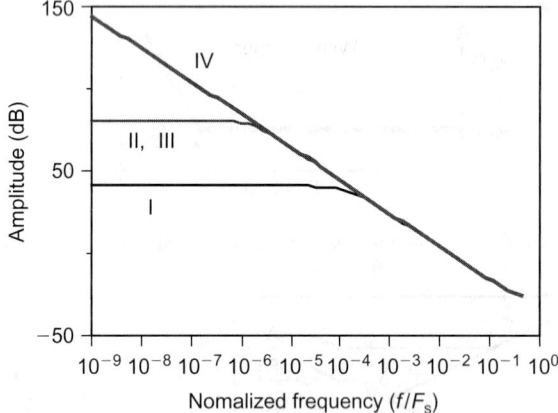

FIGURE 62.37 Finite-gain-compensated SC integrator performance.

assumption is valid, i.e., for $f/F_s \ll 1$. The circuit operates as follows. During ϕ_1, auxiliary capacitor C_{g1} samples the output while during ϕ_2, C_{g1} is used to precharge C_{g2} to $-V_o/A_o$, generating a good virtual ground at node X.

In Figure 62.37 the frequency response of different integrators are compared. Line I refers to an uncompensated integrator with $A_o = 100$; line II refers to the uncompensated integrator with $A_o = 10,000$. This line matches with line III, which corresponds to the compensated integrator with $A_o = 100$. Finally, line IV shows the frequency response of the ideal integrator. From this comparison, the compensation effect is to achieve A_o performance with an opamp gain similar to that achieved with an opamp gain A_{o2}.

Alternative solutions to the opamp gain compensation are based on the use of a replica amplifier matched with the main one. Also in this way the effectiveness of the solution is to achieve performance accuracy relative to an opamp DC gain of A_{o2}.

62.8.4 The Very-Long Time-Constant Integrator

In the design of very-long time-constant integrators using the scheme of Figure 62.8(a), typical key points to be considered are the following:

- *The capacitor spread*: if the pole frequency f_p is very low with respect to the sampling frequency F_s, then the capacitor spread $S = C_f/C_s$ of a standard integrator (Figure 62.8[a]) will be very large. This results in large die area and reduces performance accuracy for poor matching.

- *The sensitivity to the parasitic capacitances*: proper structure can reduce capacitor spread. They however suffer from the presence of parasitic capacitance. Parasitic-insensitive or at least parasitic-compensated designs should then be considered.
- *The offset of the operational amplifier*: offset compensations are needed when the opamp offset contribution cannot be tolerated.

In the literature, several SC solutions have been proposed, mainly oriented in reducing the capacitor spread rather than in compensating either the parasitics or the opamp offset.

A first solution is based on the use of a capacitive T-network in a standard SC integrator, as shown in Figure 62.38 [22]. The operation of the sampling T-structure is to realize a passive charge partition with the capacitors C_{s1}, and $C_{s2} + C_{s3}$. The final result is that only the charge on C_{s3} is injected into the virtual ground. Therefore the effect of this scheme is that C_s is replaced with the C_{s_equiv} given by the expression

$$C_{s\_equiv} = C_{s3} \frac{C_{s1}}{C_{s1} + C_{s2} + C_{s3}} \tag{62.37}$$

The net gain of this approach is that, using $C_{s2} = \sqrt{S}\, C_{s1} = \sqrt{S}\, C_{s3}$, the capacitor spread is reduced to $\sqrt{S}$. For example, an integrator with $C_s = 1$ and $C_f = 40$ can be realized with $C_{s1} = 1$, $C_{s2} = 6$, $C_{s3} = 1$, and $C_f = 5$, i.e., with the capacitor spread reduced to 6.

$$\frac{V_o}{V_i} = -\frac{C_{s1}}{C_{s1} + C_{s2} + C_{s3}} \frac{C_{s3}}{C_f} \frac{1}{1 - z^{-1}} \tag{62.38}$$

The major problem of the circuit in Figure 62.38 is due to the fact that the T-network is sensitive to the parasitic capacitance C_p (due to C_{s1}, C_{s2}, and C_{s3}) in the middle node of the T-network, which is added to C_{s2}, reducing frequency response accuracy.

A parasitic-insensitive circuit is the one proposed by Nagaraj [23], shown in Figure 62.39. In this case, the transfer function is given by Eq. (62.39). Also in this case $C_s = C_x = \sqrt{S}C_f$ are usually adopted to reduce

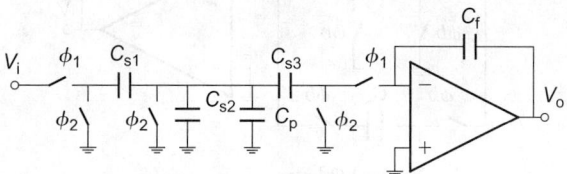

FIGURE 62.38 A T-network long-time-constant SC integrator.

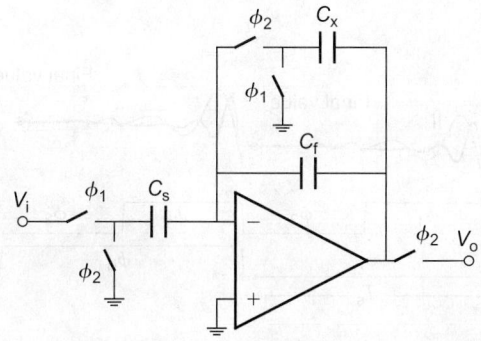

FIGURE 62.39 The Nagaraj's long-time-constant SC integrator.

the standard spread from S to $\sqrt{S}$. However, for the Nagaraj's integrator the opamp is used on both phases, disabling the possibility of using double-sampled structure.

$$\frac{V_o}{V_i} = \frac{C_f}{C_s} \frac{C_x}{C_f + C_x} \frac{1}{1 - z^{-1}} \tag{62.39}$$

It is, finally, possible to combine a long-time-constant scheme with an offset-compensated scheme to obtain a long-time-constant offset-compensated SC integrator [15].

62.8.5 Double-Sampling Technique

If the output value of the SC integrator of Figure 62.8(b) is read only at the end of ϕ_2, the requirement for the opamp to settle can be relaxed. For the integrator of Figure 62.8(b), the time available for the opamp to settle is $T_s/2$. The equivalent double-sampled structure is shown in Figure 62.40. The capacitor values for the two structures are the same, and thus they implement the same transfer function. The time evolution for the two structures are compared in Figure 62.41. For the double-sampled SC integrator, the time available for the opamp to settle is doubled.

This advantage can be used in two ways. First, when a high sampling frequency is required, if the opamp cannot settle in $T_s/2$, an extra time allows to reach the speed requirement (i.e., the double-sampling technique is used to increase the sampling frequency). Second, also at low sampling frequency when the power consumption must be strongly reduced, a smaller bandwidth to be guaranteed by the opamp reduces its power consumption.

The cost of the double-sampled structure is the doubling of all the switched capacitors. In addition, in the case of a small mismatch between the two parallel paths, mismatch energy could be present around $F_s/4$ [24].

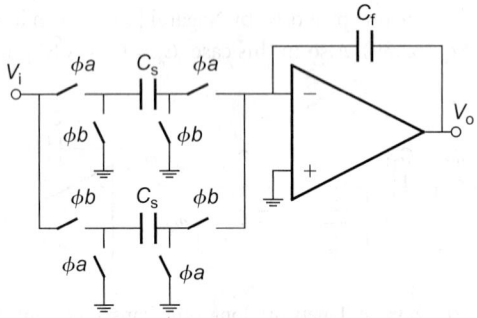

FIGURE 62.40 Double-sampled SC integrator.

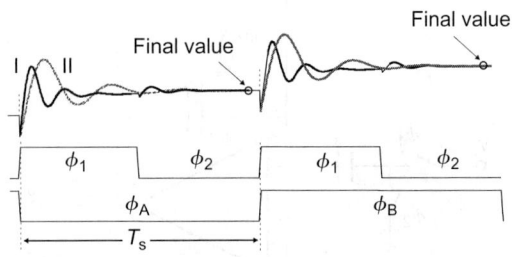

FIGURE 62.41 Double-sampled operation.

62.9   Low-Voltage Switched-Capacitor Circuits

The portable device market requires low-voltage and low-power integrated circuits. This requirement comes also from the technological evolution of the device shrink (which lowers the breakdown voltage of the devices), which forces to reduce the power supply voltage, as in the forecast of the ITRS roadmap of Figure 62.42.

In the typical SC integrator scheme, shown in Figure 62.43, it is indicated that different supply voltages can be used to bias the opamps and the switches (and their relative driving circuits).

62.9.1   Processing a Reduced Signal Amplitude with Standard SC Solutions

The fundamental limitation to the operation of SC circuits at low voltage is due to switche's operations. Figure 62.44 shows that at low supply voltage there is no conductance at $V_{DD}/2$; using standard complementary switches. This disables any rail-to-rail signal swing. A signal swing is possible only in two regions: one is closed to ground and the other is closed to V_{DD}. In these region use of complementary switches is no more advantageous, and so single MOS switches (NMOS-only or PMOS-only) could be used. An NMOS-only switch may then be adopted for signal swing closed to ground, while a PMOS-only switch would be used for signal swing closed to V_{DD}. This however makes the possible signal swing to depend

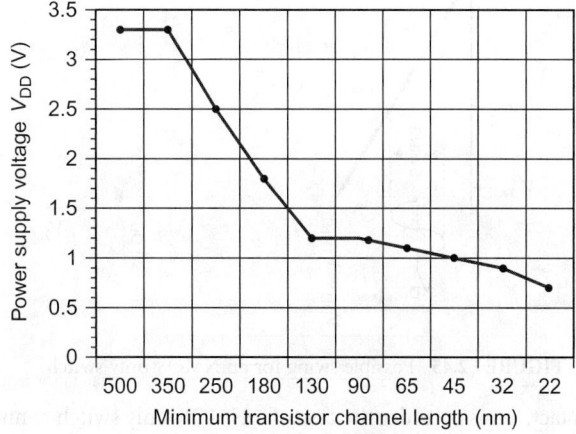

FIGURE 62.42   Evolution of the power supply voltage as a function of the technology node according to the ITRS roadmap.

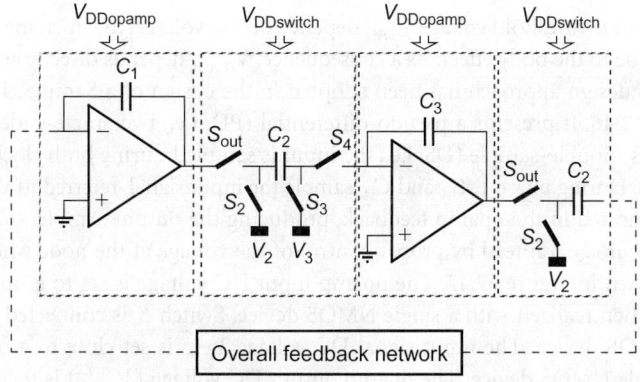

FIGURE 62.43   Typical SC integrator.

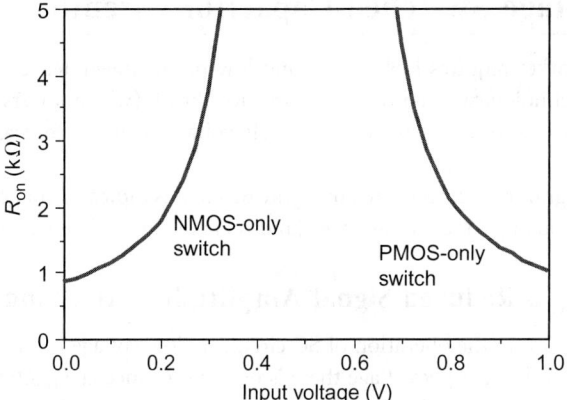

FIGURE 62.44 Switch R_{on} with $V_{DD} = 1$ V.

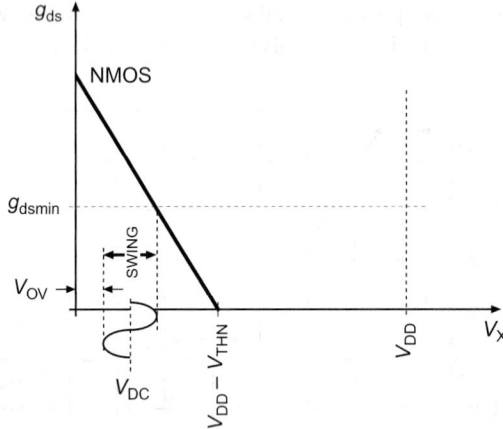

FIGURE 62.45 Possible swing for an NMOS-only switch.

on the power supply. In fact, let us consider the case of a NMOS-only switch connected to a signal biased at V_{DC} and with a signal swing V_{SW}. The resulting switch conductance is shown in Figure 62.45.

In this case the minimum supply voltage V_{DDmin} is given by

$$V_{DDmin} = V_{DC} + V_{SW} + V_{THN}(V_{DC} + V_{SW}) + V_{ov} \qquad (62.40)$$

Notice that the NMOS threshold voltage V_{THN} depends on the voltage, to which the switch is connected to, i.e., $(V_{DC} + V_{ov})$ due to the body effect. As a consequence, V_{DDmin} depends directly and indirectly on V_{SW}. This low-voltage SC design approach has been adopted in the design of a Sample&Hold whose scheme is shown in Figure 62.46. It presents a pseudo-differential (PD; i.e., two single-ended structures driven with opposite signals) double-sample (DS; i.e., the input is sampled during both clock phases) structure. It operates as follows. During phase 1, C_{1P} and C_{1M} sample the input signal, referred to V_{DD}. During phase 2, C_{1P} and C_{1M} are connected in the opamp feedback, producing the output sample.

Switch operations are guaranteed by proper control of the voltage at the node where the switches are connected to, as shown in Figure 62.47. The opamp input DC voltage is set to ground by the feedback action. Switch S_2 is then realized with a single NMOS device. Switch S_4 is connected to V_{DD}, and then is realized using a PMOS device. The input signal DC voltage V_{in_dc} is set close to ground; this allows to realize S_1 with a single NMOS device. The opamp output DC voltage (V_{out_dc}) is then fixed at the value:

$$V_{out\_dc} = V_{DD} - V_{in\_dc} \qquad (62.41)$$

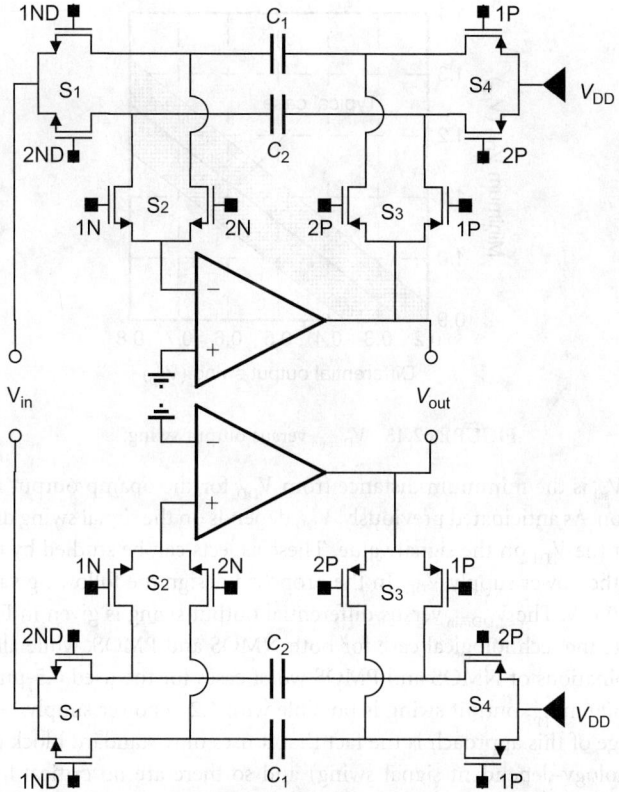

FIGURE 62.46 PD-DS S&H circuit.

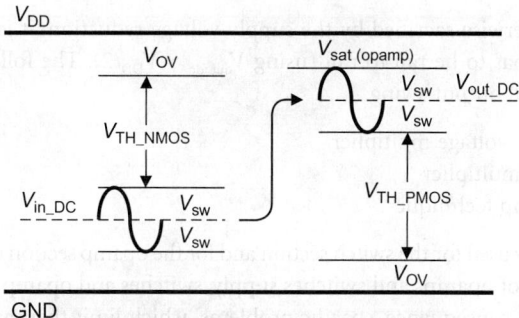

FIGURE 62.47 Voltage drop across the switches.

the results close to $V_{DD}S_3$ are then realized with a PMOS device. The minimum supply required by the structure is then fixed by proper operation of switches S_1 and S_3, and is given by

$$V_{DDn} > V_{in\_dc} + V_{sw} + V_{THN}(V_{in\_dc} + V_{sw}) + V_{ov} \tag{62.42}$$

$$V_{DDp} > V_{sat} + 2V_{sw} + V_{ov} + V_{THP}(V_{out\_oc} - V_{sw}) \tag{62.43}$$

where V_{sw} is the peak of the single-ended signal amplitude, V_{THN} and V_{THP} are the maximum values of the NMOS and PMOS threshold voltages obtained for the body effect evaluated at the maximum level

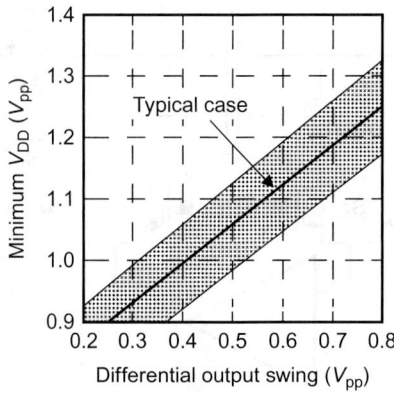

FIGURE 62.48 V_{DDmin} versus output swing.

of the signal swing. V_{sat} is the minimum distance from V_{DD} for the opamp output node before it enters in the saturation region. As anticipated previously, V_{DD} depends on the signal swing directly and indirectly, by the dependence of the V_{TH} on the signal value. These aspects can be studied by plotting the available output swing versus the power supply V_{DD}. In the proposed design the following values have been used: $V_{ov} = 50$ mV, $V_{sat} = 80$ mV. The V_{DDmin} versus differential output swing is given in Figure 62.48. The line with the stars indicate the technological case for both NMOS and PMOS, while all other lines indicate all the possible combinations of NMOS and PMOS worst cases for the used 0.5-μm CMOS technology. For the typical case 600 mV$_{pp}$, output swing is possible with 1.2 V power supply.

The main advantage of this approach is the fact that it uses only standard block design (at the cost of a reduced and technology-dependent signal swing) and so there are no critical limitations to operate with a high sampling frequency.

62.9.2 Processing a Rail-to-Rail Signal Amplitude with Novel Solutions

To maximize the DR, otherwise sacrified by the supply voltage reduction, it is mandatory to maximize the output swing, which has to be rail to rail (using $V_{out\_dc} = V_{DD}/2$). The following design approaches allow to achieve rail-to-rail output swing:

- The on-chip supply voltage multiplier
- The on-chip clock multiplier
- The switched-opamp technique

They differ on the supply used for the switch section and for the opamp section (as shown in Figure 62.43). Depending on this choice of opamps and switches supply, switches and opamp operation are guaranteed in different ways and, as a consequence, specific problems, which limit their application, arise.

62.9.2.1 On-Chip Supply Voltage Multiplier

The reuse of the SC circuit know-how is possible if an auxiliary supply V_{DDmult} to be used to power the complete SC filter is generated on-chip. In this way the SC filter is designed using the available analog cells for opamps and switches, operating from the multiplied supplied voltage (i.e., with $V_{DDswitch} = V_{DDopamp} = V_{DDmult}$).

The on-chip supply voltage multiplier suffers from the following limitations:

- The technology robustness: the scaled-down technology presents the maximum acceptable electric field between gate and channel (for gate oxide breakdown) and between drain and source (for hot electrons damage) must be reduced and this results in an absolute limit to the value of the multiplied supply voltage.
- The need to supply a DC current from the multiplied supply forces to use an external capacitor: an additional cost, not feasible for other system considerations.

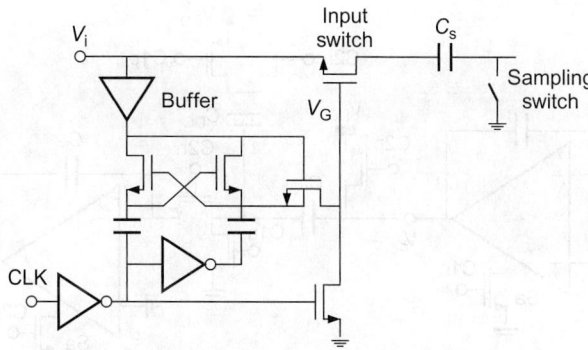

FIGURE 62.49 A possible clock multiplier.

- The conversion efficiency of the charge pump cannot be 100% and this could limit the application of this approach in battery-operated portable systems.

For these arguments this approach appears the least feasible for future applications and will not be discussed any further.

62.9.2.2 On-Chip Clock Voltage Multiplier

A second and more feasible alternative to operate low-voltage SC filters is the use of on-chip clock multiplier to drive only the switches, while the opamps operate from the low-supply voltage (Figure 62.49). Thus the voltage multiplier has only to drive the capacitive load due to the switch gates, while it is not required to supply any DC current to the opamp. No external capacitor is then required and the SC filter is fully integrated. Using this design approach the switches can operate as in standard SC circuit working with higher power supply.

In contrast, the opamp has to be properly designed to operate from the reduced supply voltage. In particular, the opamp input DC voltage is necessary to be set to $V_{in\_DC} = 0$ (this will be explained later), while the opamp output DC voltage is set to $V_{out\_DC} = V_{DD}/2$ to achieve rail-to-rail output swing. These DC levels are not equal and so a voltage level shift must be implemented. Such a level shift can be efficiently implemented using SC technique. In this way the operation is possible due to the full functionality of the switches at any input voltage using the multiplied clock supply. In the scheme of Figure 62.43, assuming $V_2 = V_{out\_DC}$ and $V_3 = V_{in\_DC}$ gives the proper DC voltage at the opamp input and at the output nodes. This design approach, as in the previous one, suffers from the technology limitation associated to the gate oxide breakdown. Even with these problems, this approach is very popular since it allows the filter to operate at high sampling frequency. Using this approach, tens of Ms/s sampling frequencies in pipeline A/D converters have been reported. This design solution can be improved by driving all the switches with a fixed V_{ov}. In this case, a constant switch conductance is ensured and this also reduces signal-dependent distortion. It however requires a specific charge pump for each switch, increasing area, power consumption, and noise injection.

62.9.2.3 Switched-OpAmp Technique

The "switched-opamp" (SOA) technique does not need any voltage multiplier. The basic considerations leading to the SOA technique are the following:

- The optimum condition for the switches driven with a low supply voltage is to be connected either to ground or to V_{DD}. Switch S_4 in Figure 62.43 is connected to virtual ground. As a consequence, the opamp input DC voltage has to be either ground or V_{DD}. This also allows to minimize the required opamp supply voltage.
- Biasing the opamp DC output voltage at $V_{DD}/2$ allows to achieve rail-to-rail output swing.

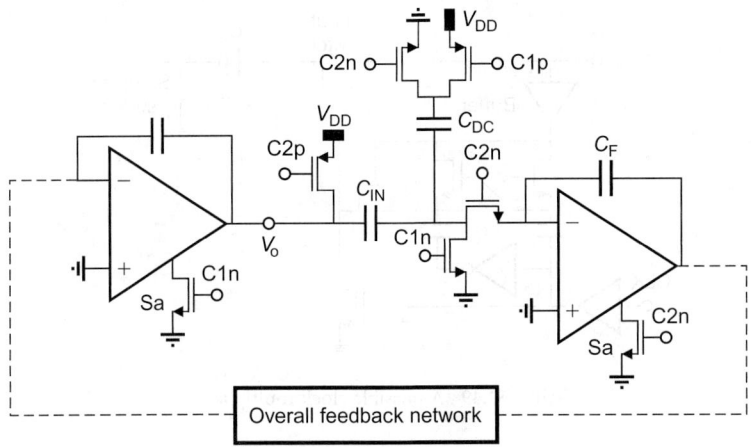

FIGURE 62.50 Switched-opamp SC integrator.

- Proper operations of switch S_{out} connected to the opamp output node are not guaranteed if the supply voltage is reduced. Its functionality has then to be guaranteed in some other way.

The use of the SOA solution fits all the previous points. Figure 62.50 shows the SOA SC integrator [14–16]. In this scheme the critical switch S_{out} is removed and its function is replaced by turning on and off the opamp driving force through S_a. The switch critical problem is then moved to the opamp design. The opamp input DC voltage is biased to ground; this minimizes the opamp supply requirements and guarantees proper operation for $S4$ in Figure 62.50.

In this scheme all the switches are connected to ground (and realized with a single NMOS device) or to V_{DD} (and realized with a single PMOS device). In this way all the switches are driven with the maximum overdrive, i.e., $V_{DD} - V_{TH}$. The minimum supply voltage required for proper operation of the switches is then given by

$$V_{DDmin} = V_{TH} + V_{ov} \tag{62.44}$$

where V_{TH} is the larger of the two threshold voltages (N- and P-type). V_{DDmin} is of the same order as the minimum supply voltage for the digital CMOS circuits operation.

As described previously, for the SOA technique it is necessary to implement a level shift due to the difference between the opamp input and output DC voltages. This is efficiently implemented in the scheme in Figure 62.50 with the switched-capacitor C_{DC}, which gives a fixed charge injection into virtual ground. The charge balance at the opamp input node allows to evaluate the amount of the level shift as

$$-C_{IN}V_{out\_DC} - C_{DC}V_{DD} = C_{IN}(V_{in\_DC} - V_{DD}) + C_{DC}V_{in\_DC} = 0 \tag{62.45}$$

Since V_{in_DC} is set to ground, the opamp output DC voltage V_{out_DC} is fixed at

$$V_{out\_DC} = V_{DD}\frac{C_{IN}}{C_{DC}} \tag{62.46}$$

To set $V_{out\_DC} = V_{DD}/2$ it is necessary to design $C_{DC} = C_{IN}/2$. This satisfies all the points previously stated with the scheme of Figure 62.50. The key advantage of the SC technique is that the proper phasing of C_{DC} realizes a negative impedance, and this allows to fix $V_{in\_DC} = 0$ (Figure 62.43). The main problems presented by the SOA can be summarized as follows:

- Only the noninverting and delayed SC integrator has been up to now proposed in the literature. Thus, a sign change must be properly implemented to close the basic two-integrator loop and to

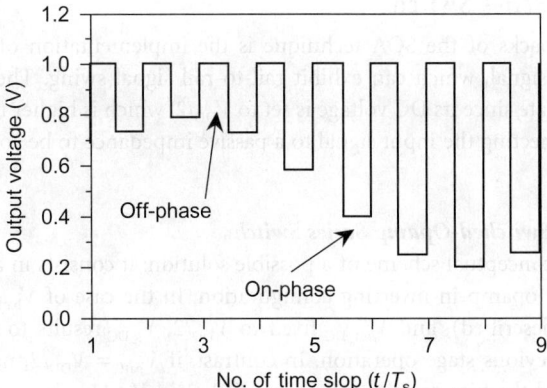

FIGURE 62.51 Switched-opamp SC integrator.

build high-order filters. This problem is still open for the single-ended structure and the only proposed solution is using an extra inverting stage.

- Any inaccuracy in the C_{DC} size gives an extra offset at the output node, which limits the output swing.
- Any noise and disturbance present on V_{DD} is injected into the signal path.

A fully differential structure can alleviate all of the above problems. In fact:

- A fully differential architecture provides both signal polarities at each node, which allows to build high-order structures without any extra elements (e.g., inverting stage).
- Any disturbance (offset or noise) injected by C_{DC} results in a common-mode signal which is largely rejected by the fully differential operation.

In addition to the above advantages, fully differential structures have a drawback, in that they require a common mode feedback (CMFB) circuit, which becomes critical at low supply voltage as discussed in the previous sections. In addition to this, the SOA design approach still suffers for the following open problems:

- A SOA structure uses an opamp which is turned on and off. The opamp turn on time results to be the main limitation in the increasing the sampling frequency.
- The output signal of a SOA structure is available only during one clock phase, because during the other clock phase the output is set to zero. If the output signal is read as a continuous-time waveform the zero-output phase has two effects: a gain loss of a factor of 2, and an increased distortion. This second drawback is due to the large output steps resulting in slew-rate-limited signal transient and glitches. However, when the SOA integrator precedes a sampled-data system (like an ADC) the SOA output signal is sampled only when it is valid and both the above problems are cancelled.
- The input coupling switch sees the entire voltage swing and so is still critical: only AC-coupling through a capacitor appears a good solution, up to now.

62.9.2.4 Turn-on Time Reduction: The Unity-Gain Feedback Technique

The unity-gain feedback technique allows to reduce the turn-on time of the opamp. This technique does not completely turn-off the opamp during the off phase, but it biases it in a stand-by condition.

The relative scheme for the basic SC integrator is shown in Figure 62.52. In the off phase, the output nodes are driven by a battery in the feedback to $V_{DD} - V_{sat}$, without turning-off the opamp. This dramatically reduces the turn-on time, allowing the use of higher sampling frequency. A possible drawback of this approach is the residual signal at the output during the off-phase.

62.9.2.5 The Input Series Switch

One of the main drawbacks of the SOA technique is the implementation of the series switch to be connected at the input signal, which can exhibit rail-to-rail signal swing. The input signal cannot be directly connected to a gate since its DC voltage is set to $V_{DD}/2$, which is higher than V_{TH}. Thus a possible solution consists in connecting the input signal to a passive impedance to be connected to some kind of virtual ground.

62.9.2.5.1 The Active Switched-Opamp Series Switch

Figure 62.53 shows the conceptual scheme of a possible solution: it consists in a switched buffer, implemented with a switched opamp in inverting configuration. In the case of $V_{batt} = 0$ V and V_{in_DC} set to ground (as previously described), and V_{out_DC} fixed to $V_{DD}/2$, V_{s_DC} results to be set to $-V_{DD}/2$, not a feasible value for the previous stage operation. In contrast, if $V_{batt} = V_{DD}/2$, node X acts like a virtual ground set to $V_{DD}/2$, and the bias condition becomes $V_{out\_DC} = V_{DD}/2 - V_{s\_DC}$. Using $V_{s\_DC} = V_{DD}/2$ for rail-to-rail input swing of the preceding stage, V_{o_DC} is set to $V_{DD}/2$. Notice that in R_1 and R_2 only signal current flows and, with $R_1 = R_2$, V_o follows V_s with negative unitary gain.

Figure 62.54 shows the complete circuit. In this circuit the battery V_{batt} is implemented with SC technique and operates as follows. During phase 1, capacitor C_1 is charged to V_{DD}, while capacitor C_2 has no charge on its nodes since they are both connected to V_{DD}. During phase 2, capacitors C_1 and C_2 are

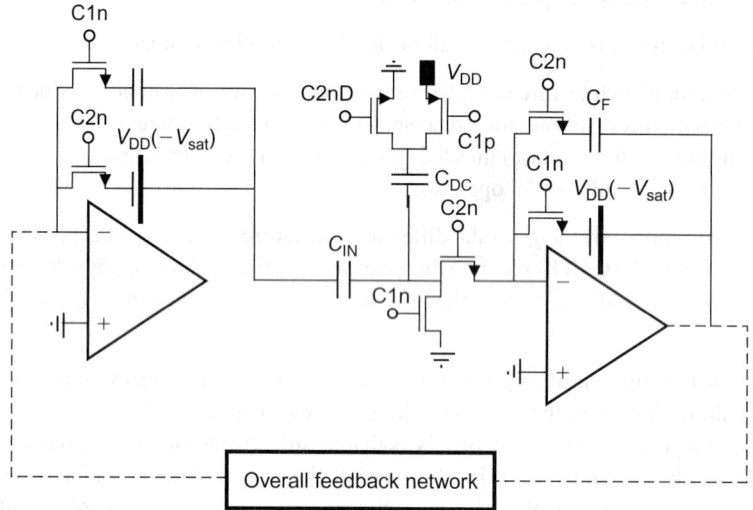

FIGURE 62.52 The unity-gain-feedback technique SC integrator.

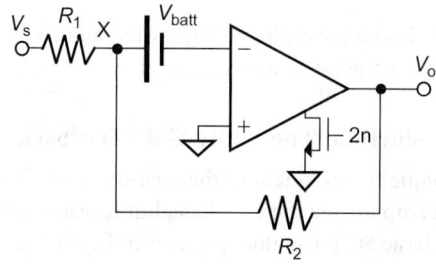

FIGURE 62.53 The series switch scheme.

connected in parallel. Using $C_1 = C_2$, across both capacitors a voltage equal to $V_{DD}/2$ results. In this phase no charge is added to C_1 and C_2, which then act like a battery from opamp input node (set to ground) to node X which results to be set to $V_{DD}/2$, as required. During phase 2 the feedback network (R_1–R_2) is active since the opamp inverting input node is set to ground. This forces V_X to be set to $V_{DD}/2$, as required, and $V_o(2)$ follows $V_s(2)$. The value of $V_o(2)$ is sampled on C_s, which is the input capacitor for the following stage in which it injects its charge during the following phase 1.

62.9.2.5.2 The Switched-RC Technique

An alternative solution for the input series switch is given by the switched-RC technique, as shown in Figure 62.55.

In this case, the opamp driving force is never turned-off. However, canceling the series switch would result in a large output current when the output node is connected to the reference voltage (ground or V_{DD}). Thus, at the driving opamp output node a series resistance is connected to limit the output current.

The main drawback of this technique is that when the series switch would be turned-off, there is a residual signal (given by the resistive partition of R_1 with the on-resistance of M_{SP}), which is continuously injected in the integrator. And this signal partition is also nonlinear due to the switch on-resistance nonlinearity.

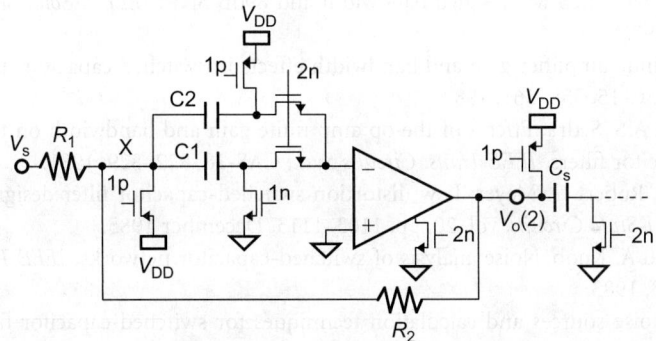

FIGURE 62.54 The switched-opamp buffer.

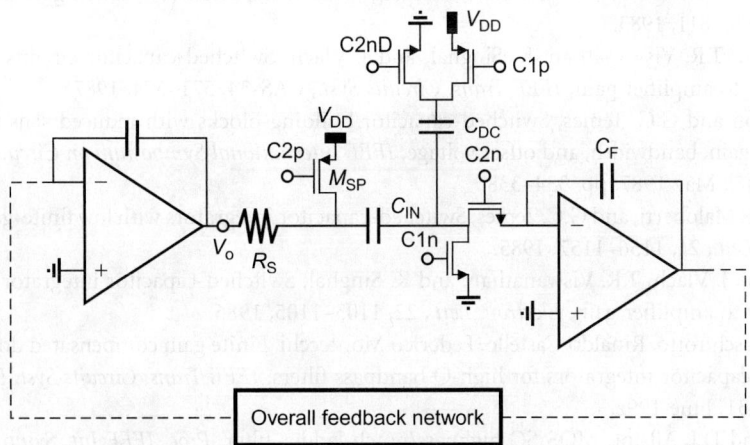

FIGURE 62.55 The switched-RC integrator technique.

References

1. B.J. Hosticka, R.W. Brodersen, and P.R. Gray, MOS sampled-data recursive filters using switched-capacitor integrators, *IEEE J. Solid-State Circuits*, SC-12, 600–608, 1977.

2. J.T. Caves, M.A. Copeland, C.F. Rahim, and S.D. Rosenbaum, Sampled analog filtering using switched capacitors as resistor equivalents, *IEEE J. Solid-State Circuits*, SC-12, 592–599, 1977.

3. R. Gregorian and G.C. Temes, *Analog MOS Integrated Circuits for Signal Processing*, Wiley, New York, 1986.

4. Gregory T. Uehara, Paul R. Gray; A 100 MHz A/D interface for PRML magnetic disk read channels, *IEEE Journal of Solid-State Circuits,* vol. 29, pp. 1606–1613, December 1994.

5. P.E. Fleischer and K.R. Laker, A family of active switched-capacitor biquad building blocks, *Bell Syst. Tech. J.*, 58, 2235–2269, 1979.

6. G.M. Jacobs, D.J. Allstot, R.W. Brodersen, and P.R. Gray, Design techniques for MOS switched-capacitor ladder filters, *IEEE Trans. Circuits Syst.*, CAS-25, 1014–1021, 1978.

7. J.B. Shyu, G.C. Temes, and F. Krummenacher, Random error effects in matched MOS capacitors and current sources, *IEEE J. Solid-State Circuits*, SC-19, 948–955, 1984.

8. D.G. Haigh and B. Singh, "A switching scheme for switched-capacitor filters which reduces the effects of parasitic capacitances associated with switch control terminals," *Proc. IEEE Int. Symp. on Circuits and Syst.*, Apr. 1983, pp. 586–589.

9. T. Brooks, D.H. Robertson, D.F. Kelly, A. DelMuro, and S.W. Harston, A cascaded sigma-delta pipeline A/D converter with 1.25 MHz signal bandwidth and 89dB SNR, *IEEE J. Solid-State Circuits*, SC-32, 1896–1906, 1997.

10. G.C. Temes, Finite amplifier gain and bandwidth effects in switched capacitor filters, *IEEE J. Solid-State Circuits*, SC-15, 358–361, 1980.

11. K. Martin and A.S. Sedra, Effects of the op amp finite gain and bandwidth on the performance of switched-capacitor filters, *IEEE Trans. Circuits Syst.*, CAS-28, 822–829, 1981.

12. Kuang-Lu Lee, Robert G. Meyer; Low-distortion switched-capacitor filter design techniques, *IEEE Journal of Solid-State Circuits*, vol. 20, pp. 1103–1113, December 1985.

13. C.A. Gobet and A. Knob, Noise analysis of switched-capacitor networks, *IEEE Trans. Circuits Syst.*, CAS-30, 37–43, 1983.

14. J.H. Fischer, Noise sources and calculation techniques for switched-capacitor filters, *IEEE J. Solid State Circuits*, SC-17, 742–752, 1982.

15. C. Enz and G.C. Temes, Circuit techniques for reducing the effects of opamp imperfections: Autozeroing, correlated double sampling, and chopper stabilization, *Proc. IEEE*, 84, 1584–1614, 1996.

16. K.K.K. Lam and M.A. Copeland, Noise-cancelling switched-capacitor (SC) filtering technique, *Electron. Lett.*, 19, 810–811, 1983.

17. K. Nagaraj, T.R. Viswanathan, K. Singhal, and J. Vlach, Switched-capacitor circuits with reduced sensitivity to amplifier gain, *IEEE Trans. Circuits Syst.*, CAS-34, 571–574, 1987.

18. L.E. Larson and G.C. Temes, Switched-capacitor building-blocks with reduced sensitivity to finite amplifier gain, bandwidth, and offset voltage, *IEEE International Symposium on Circuits and Systems (ISCAS '87)*, May 1987, pp. 334–338.

19. K. Haug, F. Maloberti, and G.C. Temes, Switched-capacitor integrators with low finite-gain sensitivity, *Electron. Lett.*, 21, 1156–1157, 1985.

20. K. Nagaraj, J. Vlach, T.R. Viswanathan, and K. Singhal, Switched-capacitor integrator with reduced sensitivity to amplifier gain, *Electron. Lett.*, 22, 1103–1105, 1986.

21. Andrea Baschirotto, Rinaldo Castello, Federico Montecchi, Finite gain compensated double-sampled switched-capacitor integrators for high-Q bandpass filters, *IEEE Trans. Circuits Syst. I*, vol. CAS-39, pp. 425–431, June 1992.

22. T. Huo and D.J. Allstot, MOS SC highpass/notch ladder filter, *Proc. IEEE Int. Symp. Circuits Syst.* (ISCAS1980), May 1980, pp. 309–312.

23. K. Nagaraj, A parasitic insensitive area efficient approach to realizing very large time constant in switched-capacitor circuits, *IEEE Trans. Circuits Syst.*, CAS-36, 1210–1216, 1989.
24. J.J.F. Rijns and H. Wallinga, Spectral analysis of double-sampling switched-capacitor filters, *IEEE Trans. Circuits Syst.*, 38, 1269–1279, 1991.
25. Andrea Baschirotto, Federico Montecchi, Rinaldo Castello; A 15 MHz 20mW BiCMOS switched-capacitor biquad operating with 150 Ms/s sampling frequency, *IEEE Journal of Solid-State Circuits*, vol. 30, pp. 1357–1366, December 1995.
26. A. Baschirotto, Considerations for the design of switched-capacitor circuits using precise-gain operational amplifiers, *IEEE Trans. Circuits Syst.-II*, 43, 827–832, 1996.
27. R. Castello, F. Montecchi, F. Rezzi, and A. Baschirotto, Low-voltage analog filter, *IEEE Trans. Circuits Syst.-II*, 827–840, 1995.
28. A. Baschirotto and R. Castello, 1V switched-capacitor filters, Workshop on Advances in Analog Circuit Design, Copenaghen, 28–30 April 1998.
29. J.F. Dickson, On-chip high-voltage generation in MNOS integrated circuits using an improved voltage multiplier technique, *IEEE J. Solid-State Circuits*, SC-11, 374–378, 1976.
30. J. Crols and M. Steyaert, Switched-Opamp: An approach to realize full CMOS switched-capacitor circuits at very low power supply voltages, *IEEE J. Solid-State Circuits*, SC-29, 936–942, 1994.
31. Andrea Baschirotto, Rinaldo Castello; A 1-V 1.8-MHz CMOS switched-opamp SC filter with rail-to-rail output swing, *IEEE Journal of Solid-State Circuits*, vol. 32, pp. 1979–1986, December 1997.

Section VIII

Microprocessor and ASIC

Steve M. Kang
University of California at Santa Cruz

63

Timing and Signal Integrity Analysis

Abhijit Dharchoudhury

Motorola, Inc.
Austin, Texas

David Blaauw

Department of Electrical
Engineering and Computer Science
University of Michigan
Ann Arbor, Michigan

Shantanu Ganguly

Intel Corporation
Austin, Texas

CONTENTS

63.1 Introduction

Microprocessors are rapidly moving into deep submicron dimensions, gigahertz clock frequencies, and transistor counts in excess of 10 million transistors. This trend is being fueled by the ever-increasing demand for more powerful computers on one side and by rapid advances in process technology, architecture, and circuit design on the other side. At these small dimensions and high speeds, timing and signal integrity analyses play a critical role in ensuring that designs meet their performance and reliability goals.

Timing analysis is one of the most important verification steps in the design of a microprocessor because it ensures that the chip is meeting speed requirements. Timing analysis of multi-million transistor microprocessors is a very challenging task. This task is made even more challenging because in the deep submicron regime, transistor-level and interconnect-centric analyses become vital. Therefore, timing analysis must satisfy the two conflicting requirements of accurate low-level analysis (so that deep sub-micron designs can be handled) and efficient high-level abstraction (so that large designs can be handled).

The term *signal integrity* typically refers to analyses that check that signals to not assume unintended values due to circuit noise. *Circuit noise* is a broad term that applies to phenomena caused by unintended circuit behavior such as unintentional coupling between signals, degradation of voltage levels due to leakage currents and power supply voltage drops, etc. Circuit noise does not encompass physical noise effects (e.g., thermal noise) or manufacturing faults (e.g., stuck-at faults). Signal integrity is also becoming a very critical verification task. Among the various signal integrity-related issues, noise induced by coupling between adjacent wires is perhaps the most important one. With the scaling of process tech-nologies, coupling capacitances between wires are become a larger fraction of the total wire capacitances. Coupling capacitances are also larger because a larger number of metal layers are now available for routing, and more and more wires are running longer distances across the chip. As operating frequencies increase, noise induced on signal nets due to coupling is much greater. Noise-related functional failures are increasing as dynamic circuits become more prevalent, with circuit designers looking for increased performance at the cost of noise immunity.

Another important problem in submicron high-performance designs is the integrity of the power grid that distributes power from off-chip pads to the various gates and devices in the chip. Increased operating frequencies result in higher current demands from the power and ground lines, which in turn increases the voltage drops seen at the devices. Excessive voltage drops reduce circuit performance and inject noise into the circuit, which may lead to functional failures. Moreover, with reductions in supply voltages, problems caused by excessive voltage drops become more severe. The analysis of the power and ground distribution network to measure the voltage drops at the points where the gates and devices of the chip connect to the power grid is called *IR-drop* or *power grid analysis*.

In this chapter, we will briefly discuss the important issues in static timing analysis, noise analysis with particular emphasis on coupling noise, and IR-drop analysis methods. Additional information on these topics is available in the literature and the reader is encouraged to look through the list of references.

63.2   Static Timing Analysis

Static timing analysis (TA) [1–4] is a very powerful technique for verifying the timing correctness of a design. The power of this technique comes from the fact that it is pattern independent, implicitly verifies all signal propagation paths in the design, and is applicable to very large designs. Further, it lends itself easily to higher levels of abstraction, which makes it even more computationally feasible to perform full-chip timing analysis. The fundamental idea in static timing analysis is to find the *critical paths* in the design. Critical paths are those signal propagation paths that determine the maximum operating frequency of the design. It is easiest to think of critical paths as being those paths from the inputs to the outputs of the circuit that have the longest delay. Since the smallest clock period must be larger than the longest path delay, these paths dictate the operating frequency of the chip. In very simple terms, static TA determines these long paths using breadth-first search as follows. Starting at the inputs, the latest time at which signals arrive at a node in the circuit is determined from the arrival times at its fan-in nodes. This latest arrival time is then propagated toward the primary outputs. At each primary output, we obtain the latest possible arrival time of signals and the corresponding longest path. If the longest path does not meet the timing constraints imposed by the designer, then a violation is detected. Alter-natively, if the longest path meets the timing constraints, then all other paths in the circuit will also satisfy the timing constraints. By propagating only the latest arrival time at a node, static TA does not have to explicitly enumerate all the paths in the design.

Historically, simulation-based or *dynamic timing analysis* techniques had been the most common timing analysis technique. However, with increasing complexity and size of recent microprocessor designs, static timing analysis has become an indispensable part of design verification and much more popular than dynamic approaches. Compared to dynamic approaches, static TA offers a number of advantages for verifying the timing correctness of a design. Dynamic approaches are pattern-dependent. Since the possible paths and their delays are dependent on the state of the circuit, the number of input patterns that are required to verify all the paths in a circuit is exponential with the number of inputs. Hence, only a subset of paths can be verified with a fixed number of input patterns. Only moderately large circuits can be verified because of the computational cost and size limitations of transient simulators. Static TA, on the other hand, implicitly verifies all the longest paths in the design without requiring input patterns. Dynamic timing analysis is still heavily used to verify complex and critical circuitry such as PLLs, clock generators, and the like. Dynamic simulation is also used to generate timing models for block-level static timing analysis. Dynamic timing analysis technique rely on a circuit simulator (e.g., SPICE [5]) or on a fast timing simulator (e.g., ILLIADS [6], ACES [7], TimeMill [8]) for performing the simulations. Because of the importance of static techniques in verifying the timing behavior of microprocessors, we will restrict the discussion below to the salient points of static TA.

63.2.1 DCC Partitioning

The first step in transistor-level static TA is to partition the circuit into *dc connected components* (DCCs), also called *channel-connected components*. A DCC is a set of nodes which are connected to each other through the source and drain terminals of transistors. The transistor-level representation and the DCC partitioning of a simple circuit is shown in Figure 63.1. As seen in the diagram, a DCC is the same as the gate for typical cells such as inverters, NAND and NOR gates. For more complex structures such as latches, a single cell corresponds to multiple DCCs. The inputs of a DCC are the primary inputs of the circuit or the gate nodes of the devices that are part of the DCC. The outputs of a DCC are either primary outputs of the circuit or nodes that are connected to the gate nodes of devices in other DCCs. Since the gate current is zero and currents flow between source and drain terminals of MOS devices, a MOS circuit can be partitioned at the gates of transistors into components which can then be analyzed independently. This makes the analysis computationally feasible since instead of analyzing the entire circuit, we can analyze the DCCs one at a time. By partitioning a circuit into DCCs, we are ignoring the current conducted by the MOS parasitic capacitances that

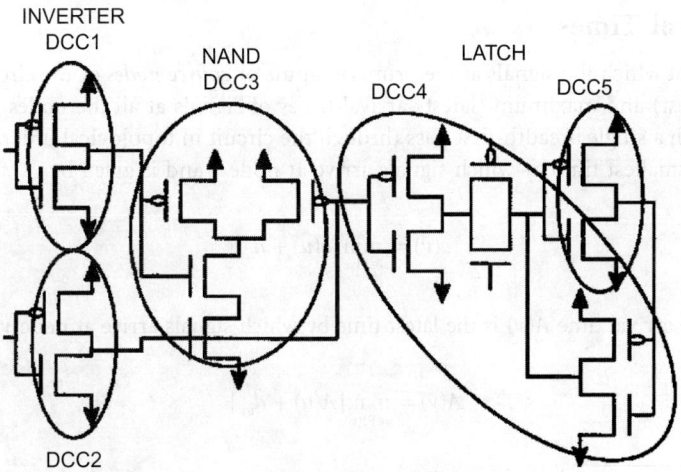

FIGURE 63.1 Transistor-level circuit partitioned into DCCs.

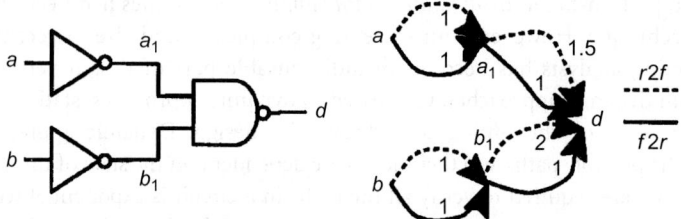

FIGURE 63.2 A simple digital circuit: (a) gate-level representation, and (b) timing graph.

couple the source/drain and gate terminals. Since this current is typically small, the error is small. As mentioned above, DCC partitioning is required for transistor-level static TA. For higher levels of abstraction, such as gate-level static TA, the circuit has already been partitioned into gates, and their inputs are known. In such cases, one starts by constructing the timing graph as described in the next section.

63.2.2   Timing Graph

The fundamental data structure in static TA is the *timing graph*. The timing graph is a graphical representation of the circuit, where each vertex in the graph corresponds to an input or an output node of the DCCs or gates of the circuit. Each edge or timing arc in the graph corresponds to a signal propagation from the input to the output of the DCC or gate. Each timing arc has a polarity defined by the type of transition at the input and output nodes. For example, there are two timing arcs from the input to the output of an inverter: one corresponds to the input rising and the output falling, and the other to the input falling and the output rising. Each timing arc in the graph is annotated with the propagation delay of the signal from the input to the output. The gate-level representation of a simple circuit is shown in Figure 63.2(a) and the corresponding timing graph is shown in Figure 63.2(b). The solid line timing arcs correspond to falling input transitions and rising output transitions, whereas the dotted line arcs represent rising input transitions and falling output transitions.

   Note that the timing graph may have cycles which correspond to feedback loops in the circuit. Combinational feedback loops are broken and there are several strategies to handle sequential loops (or cycles of latches) [5]. In any event, the timing graph becomes acyclic and the vertices of the graph can be arranged in topological order.

63.2.3   Arrival Times

Given the times at which the signals at the primary inputs or *source nodes* of the circuit are stable, the minimum (earliest) and maximum (latest) arrival times of signals at all the nodes in the circuit can be calculated with a single breadth-first pass through the circuit in topological order. The early arrival time $a(v)$ is the smallest time by which signals arrive at node v and is given by

$$a(v) = \min_{u \in \text{FI}(v)} [a(u) + d_{uv}] \tag{63.1}$$

Similarly, the late arrival time $A(v)$ is the latest time by which signals arrive at node v and is given by

$$A(v) = \max_{u \in \text{FI}(v)} [A(u) + d_{uv}] \tag{63.2}$$

In the above equations, FI(v) is the set of all fan-in nodes of v, i.e., all nodes that have an edge to v and d_{uv} is the delay of an edge from u to v. Equations 63.1 and 63.2 will compute the arrival times at a node v from the arrival times of its fan-in nodes and the delays of the timing arcs from the fan-in nodes to v. Since the timing graph is acyclic (or has been made acyclic), the vertices in the graph can be arranged in topological order (i.e., the DCCs and gates in the circuit can be *levelized*). A breadth-first pass through the timing graph using Eqs. 63.1 and 63.2 will yield the arrival times at all nodes in the circuit.

Considering the example of Figure 63.2, let us assume that the arrival times at the primary inputs a and b are 0. From Eq. 63.2, the maximum arrival time for a rising signal at node a_1 is 1, and the maximum arrival time for a falling signal is also 1. In other words, $A_{a1,r} = A_{a1,f} = 1$, where the subscripts r and f denote the polarity of the signal. Similarly, we can compute the maximum arrival times at node b_1 as $A_{b1,r} = A_{b1,f} = 1$, and at node d as $A_{d,r} = 2$ and $A_{d,f} = 3$.

In addition to the arrival times, we also need to compute the *signal transition times* (or slopes) at the output nodes of the gates or DCCs. These transition times are required so that we can compute the delay across the fan-out gates. Note that there are many timing arcs that are incident at the output node and each gives rise to a different transition time. The transition time of the node is picked to be the transition time corresponding to the arc that causes the latest (earliest) arrival time at the node.

63.2.4   Required Times and Slacks

Constraints are placed on the arrival times of signals at the primary output nodes of a circuit based on performance or speed requirements. In addition to primary output nodes, timing constraints are automatically placed on the clocked elements inside the circuit (e.g., latches, gated clocks, domino logic gates, etc.). These timing constraints check that the circuit functions correctly and at-speed. Nodes in the circuit where timing checks are imposed are called *sink nodes*.

Timing checks at the sink nodes inject required times on the earliest and latest signal arrival times at these nodes. Given the required times at these nodes, the required times at all other nodes in the circuit can be calculated by processing the circuit in reverse topological order considering each node only once. The late required time $R(v)$ at a node v is the required time on the late arriving signal. In other words, it is the time by which signals are required to arrive at that node and is given by

$$R(v) = \max_{u \in \mathrm{FO}(v)} [R(u) - d_{uv}] \tag{63.3}$$

Similarly, the early required time $r(v)$ is the required time on the early arriving signal. In other words, it is the time after which signals are required to arrive at node v and is given by

$$r(v) = \min_{u \in \mathrm{FO}(v)} [r(u) - d_{uv}] \tag{63.4}$$

In these equations, FO(v) is the set of fan-out nodes of v (i.e., the nodes to which there is a timing arc from node v) and d_{uv} is the delay of the timing arc from node u to node v. Note that $R(v)$ is the time *before* which a signal must arrive at a node, whereas $r(v)$ is the time *after* which the signal must arrive.

The difference between the late arrival time and the late required time at a node v is defined as the *late slack* at that node and is given by

$$S_1(v) = R(v) - A(v) \tag{63.5}$$

Similarly, the *early slack* at node v is defined by

$$S_e(v) = a(v) - r(v) \tag{63.6}$$

Note that the late and early slacks have been defined in such a way that a negative value denotes a constraint violation. The overall slack at a node is the smaller of the early and late slacks; that is,

$$S(v) = \min S_1(v), S_e(v) \tag{63.7}$$

Slacks can be calculated in the backward traversal along with the required times. If the slacks at all nodes in the circuit are positive, then the circuit does not violate any timing constraint. The nodes with the smallest slack value are called *critical nodes*. The most *critical path* is the sequence of critical nodes that connect the source and sink nodes.

Continuing with the example of Figure 63.2, let the maximum required time at the output node d be 1. Then, the late required time for a rising signal at node a_1 is $R_{a1,r} = -0.5$ since the delay of the rising-to-falling timing arc from a_1 to d is 1.5. Similarly, the late required time for a falling signal at node a_1 is $R_{a1,f} = R_{d,r} - 1 = 0$. The required times at the other nodes in the circuit can be calculated to be: $R_{b1,r} = -1$, $R_{b1,f} = 0$, $R_{a,r} = -1$, $R_{a,f} = -1.5$, $R_{b,r} = -1$, and $R_{b,f} = -2$. The slack at each node is the difference between the required time and the arrival time and are as follows: $S_{d,r} = -1.5$, $S_{d,f} = -2$, $S_{a1,r} = -1.5$, $S_{a1,f} = -1$, $S_{b1,r} = -2$, $S_{b1,f} = -1$, $S_{a,r} = -1$, $S_{a,f} = -1.5$, $S_{b,r} = -1$, and $S_{b,f} = -2$. Thus, the critical path in this circuit is b falling—$b1$ rising—d falling, and the circuit slack is -2.

63.2.5 Clocked Circuits

As mentioned earlier, combinational circuits have timing checks imposed only at the circuit primary outputs. However, for circuits containing clocked elements such as latches, flip-flops, gated clocks, domino/precharge logic, etc., timing checks must also be enforced at various internal nodes in the circuit to ensure that the circuit operates correctly and at-speed. In circuits containing clocked elements, a separate recognition step is required to detect the clocked elements and to insert constraints. There are two main techniques for detecting clocked elements: pattern recognition and clock propagation.

In *pattern recognition*-based approaches, commonly used sequential elements are recognized using simple topological rules. For example, back-to-back inverters in the netlist are often an indication of a latch. For more complex topologies, the detection is accomplished using templates supplied by the user. Portions of a circuit are typically recognized in the graph of the original circuit by employing subgraph isomorphism algorithms [9]. Once a subcircuit has been recognized, timing constraints are automatically inserted. Another application of pattern-based subcircuit recognition is to determine logical relationships between signals. For example, in pass-gate multiplexors, the data select lines are typically one-hot. This relationship cannot be obtained from the transistor-level circuit representation without recognizing the subcircuit and imposing the logical relationships for that subcircuit. The logical relationship can then be used by timing analysis tools. However, purely pattern recognition-based approaches can be restrictive and may necessitate a large number of templates from the user for proper functioning.

In *clock propagation*-based approaches, the recognition is performed automatically by propagating clock signals along the timing graph and determining how these clock signals interact with data signals at various nodes in the circuit. The primary input clocks are identified by the user and are marked as (simple) clock nodes. Starting from the primary clock inputs and traversing the timing arcs in the timing graph, the type of the nodes are determined based on simple rules. These rules are illustrated in Figure 63.3, where we show the transistor-level subcircuits and the corresponding timing subgraphs for some common sequential elements.

- A node that has only one clock signal incident on it and no feedback is classified as a *simple clock node* (Figure 63.3(a)).
- A node that has one clock and one or more data signals incident on it, but no feedback, is classified as a *gated clock node* (Figure 63.3(b)).
- A node that has multiple clock signals (and zero or more data signals) incident on it and no feedback is classified as a *merged clock node* (Figure 63.3(c)).

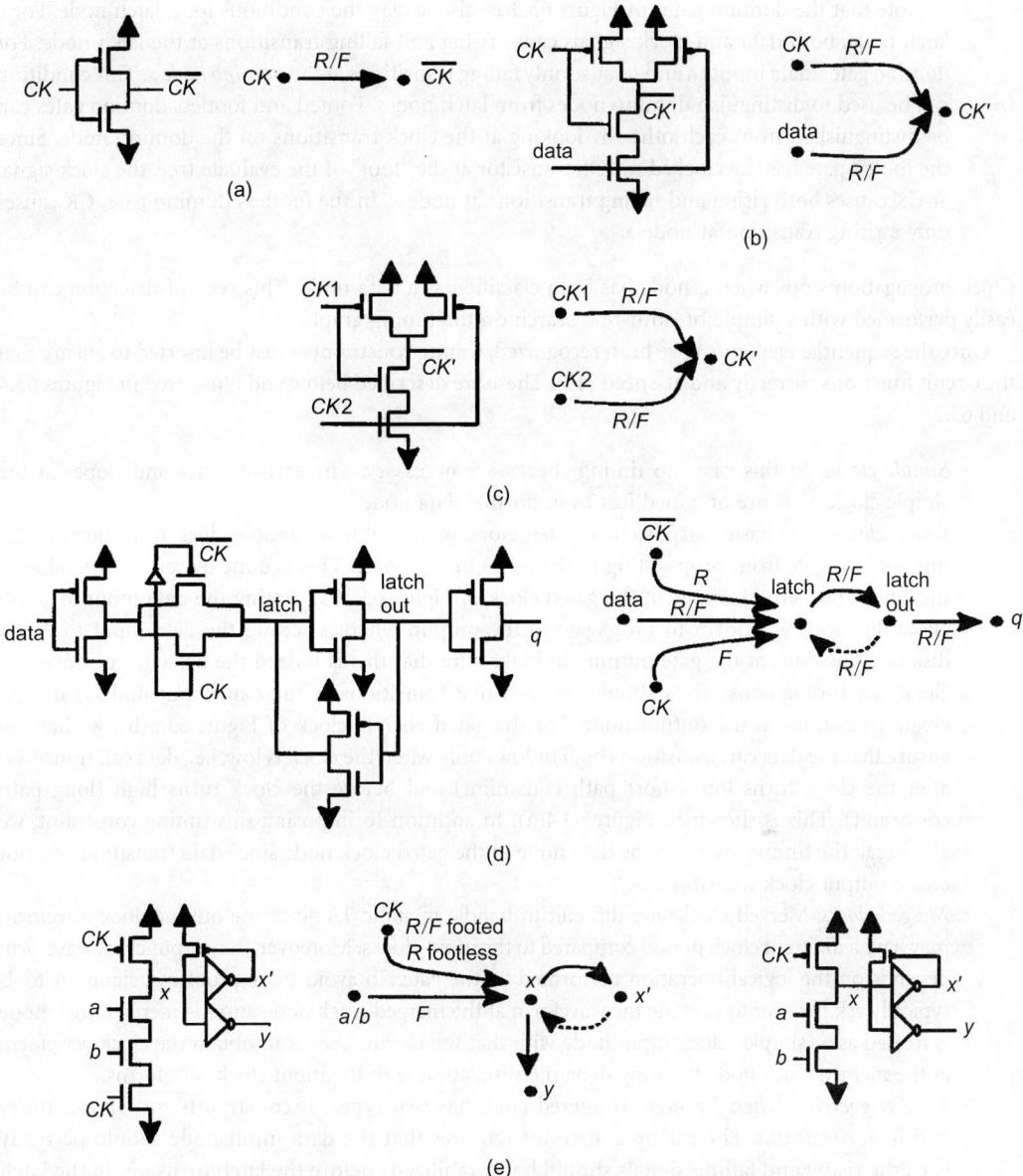

FIGURE 63.3 Sequential element detection: (a) simple clock, (b) gated clock, (c) merged clock, (d) latch node, and (e) footed and footless domino gates. Broken arcs are shown as dotted lines. Each arc is marked with the type of output transition(s) it can cause (e.g., R/F: rise and fall, R: rise only, and F: fall only).

- A node that has at least one clock and zero or more data signals incident on it and has a feedback of length two (i.e., back-to-back timing arcs) is classified as a *latch node* (Figure 63.3(d)). The other node in the two node feedback is called the *latch output node*. A latch node is of type data. The timing arc(s) from the latch output node to the latch is (are) broken.

  Latches can be of two types: *level-sensitive* and *edge-triggered*. To distinguish between edge-triggered and level-sensitive latches, various rules may be applied. These rules are usually design-specific and will not be discussed here. It is assumed that all latches are level-sensitive unless the user has marked certain latches to be edge-triggered.

Note that the domino gates of Figure 63.3(e) also satisfy the conditions for a latch node. For a latch node, both data and clock signals cause rising and falling transitions at the latch node. For domino gates, data inputs *a* and *b* cause only falling transitions at the *domino node x*. This condition can be used to distinguish domino nodes from latch nodes. Footed and footless domino gates can be distinguished from each other by looking at the clock transitions on the domino node. Since the footed gate has the clocked nMOS transistor at the "foot" of the evaluate tree, the clock signal at *CK* causes both rising and falling transitions at node *x*. In the footless domino gate, *CK* causes only a rising transition at node *x*.

Clock propagation stops when a node has been classified as a data node. This type of detection can be easily performed with a simple breadth-first search on the timing graph.

Once the sequential elements have been recognized, timing constraints must be inserted to ensure that the circuit functions correctly and at-speed [10]. These are described below and illustrated in Figures 63.4 and 63.5.

- *Simple clocks*: In this case, no timing checks are necessary. The arrival times and slopes at the simple clock node are obtained just as in normal data node.
- *Gated clocks*: The basic purpose of a gated clock is to enable or disable clock transitions at the input of the gate from propagating to the output of the gate. This is done by setting the value of the data input. For example, in the gated clock of Figure 63.3(b), setting the data input to 1 will allow the clock waveform to propagate to the output, whereas setting the data input to 0 will disable transitions at the gate output. To make sure that this is indeed the behavior of the gated clock, the timing constraints should be such that transitions at the data input node(s) do not create transitions at the output node. For the gated NAND clock of Figure 63.3(b), we have to ensure that the data can transition (high or low) only when the clock is low, i.e., data can transition after the clock turns low (short path constraint) and before the clock turns high (long path constraint). This is shown in Figure 63.4(a). In addition to imposing this timing constraint, we also break the timing arc from the data node to the gated clock node since data transitions cannot create output clock transitions.
- *Merged clocks*: Merged clocks are difficult to handle in static TA since the output clock waveform may have a different clock period compared to the input clocks. Moreover, the output clock waveform depends on the logical operation performed by the gate. To avoid these problems, static TA tools typically ask the user to provide the waveform at the merged clock node and the merged clock node is treated as a (simple) clock input node with that waveform. Users can obtain the clock waveform at the merged clock node by using dynamic simulation with the input clock waveforms.
- *Edge-triggered latches*: An edge-triggered latch has two types of constraints: *set-up constraint* and *hold constraint*. The set-up constraint requires that the data input node should be ready (i.e., the rising and falling signals should have stabilized) before the latch turns on. In the latch shown in Figure 63.3(d), the latch is turned on by the rising edge of the clock. Hence, the data should arrive some time before the rising edge of the clock (this time margin is typically referred to as the *set-up time* of the latch). This constraint imposes a required time on the latest (or maximum) arrival time at the data input of the latch and is therefore a long path constraint. This is shown in Figure 63.4(b). The hold constraint ensures that data meant for the current clock cycle does not accidentally appear during the on-phase of the previous clock cycle. Looking at Figure 63.4(b), this implies that the data should appear some time after the falling edge of the clock (this time margin is called the *hold time* of the latch). The hold time imposes a required time on the early (or minimum) arrival time at the data input node and is therefore a short path constraint. As the name implies, in edge-triggered latches, the on-edge of the clock causes data to be stored in the latch (i.e., causes transitions at the latch node). Since the data input is ready before the clock turns on, the latest arrival time at the latch node will be determined only by the clock signal. To make sure that this is indeed the behavior of

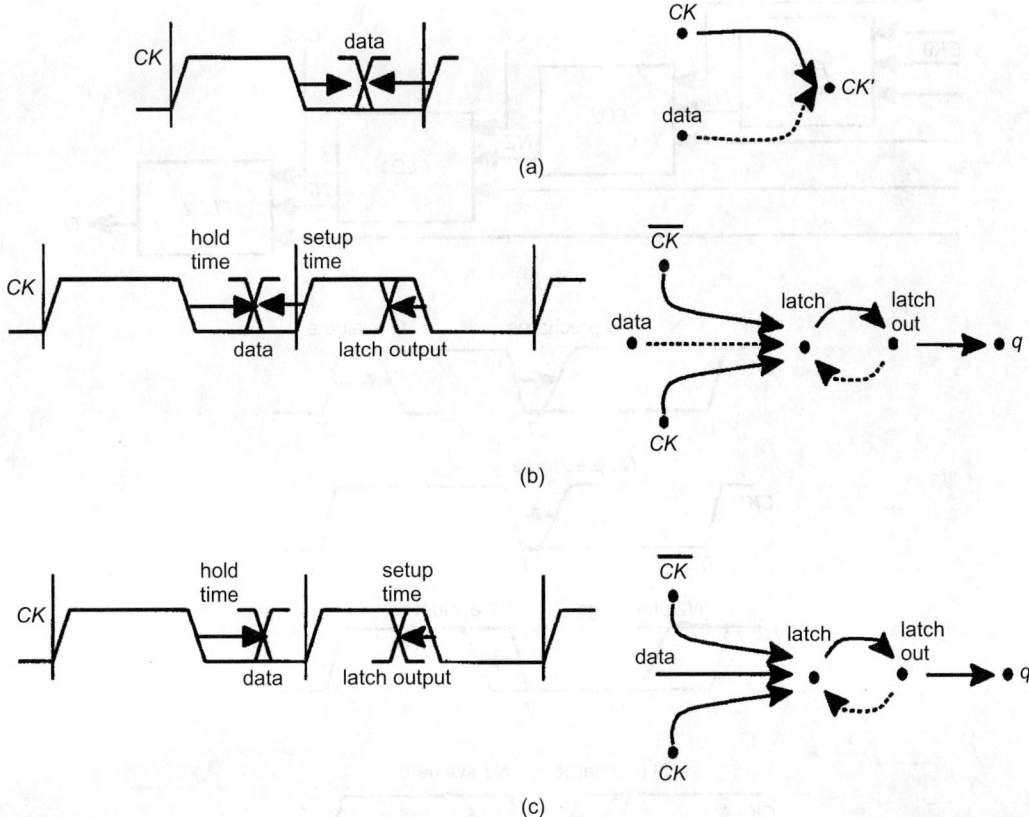

FIGURE 63.4 Timing constraints and timing graph modifications for sequential elements: (a) gated clock, (b) edge-triggered latch, and (c) level-sensitive latch. Broken arcs are shown as dotted lines.

the latch, the timing arc from the data input node to the latch node is broken, as shown in Figure 63.4(b). One additional set of timing constraints is imposed for an edge-triggered latch. Since data is stored at the latch (or latch output) node, we must ensure that the data gets stored before the latch turns off. In other words, signals should arrive at the latch output node before the off-edge of the clock.

- *Level-sensitive latches*: In the case of level-sensitive latches, the data need not be ready before the latch turns on, as is the case for edge-triggered latches. In fact, the data can arrive after the on-edge of the clock — this is called *cycle stealing* or *time borrowing*. The only constraint in this case is that the data gets latched before the clock turns off. Hence, the set-up constraint for a level-sensitive latch is that signals should arrive at the latch output node (not the latch node itself) before the falling edge of the clock, as shown in Figure 63.4(c). The hold constraint is the same as before; it ensures that data meant for the current clock cycle arrives only after the latch was turned off in the previous clock cycle. This is also shown in Figure 63.4(c). Since the latest arriving signal at the latch node may come from either the data or the clock node, timing arcs are not broken for a level-sensitive latch. Since data can flow through the latch, level-sensitive latches are also referred to as *transparent latches*.
- *Domino gates*: Domino circuits have two distinct phases of operation: *precharge* and *evaluate* [11]. Looking at the domino gate of Figure 63.3(e), we see that in the precharge phase, the clock signal is low and the domino node x is precharged to a high value and the output node y is pre-discharged to a low value. During the evaluate phase, the clock is high and if the values of the gate inputs establish a path to ground, domino node x is discharged and output node y turns

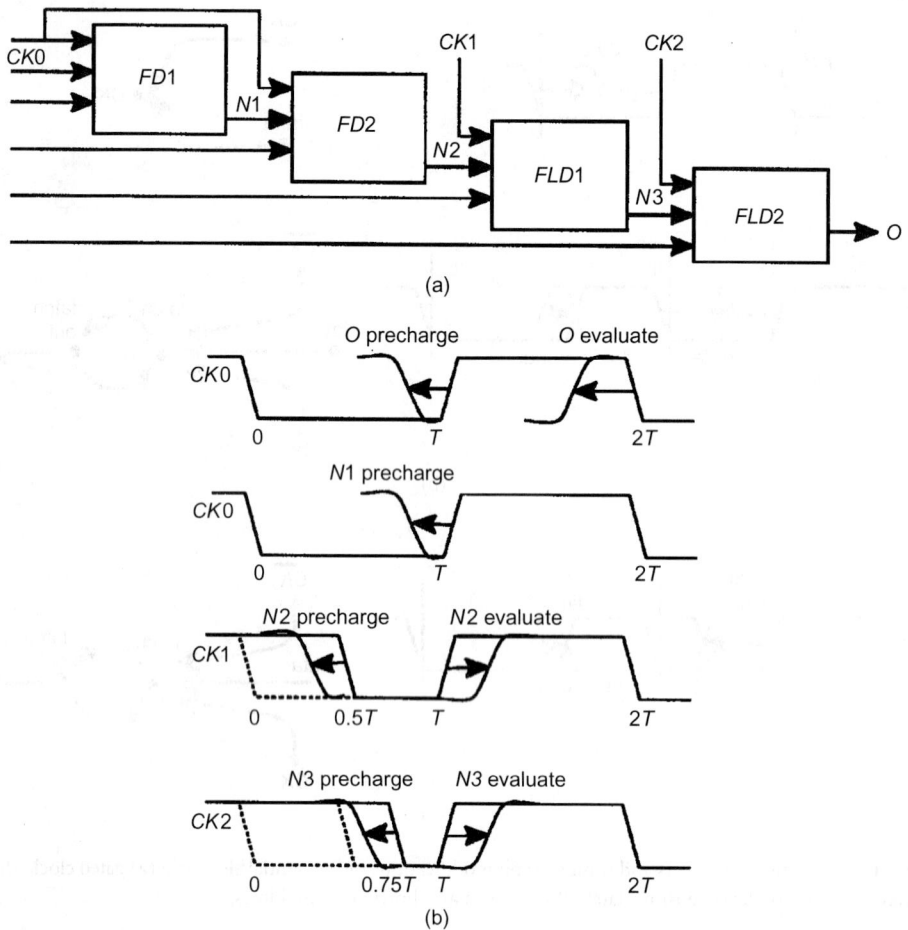

FIGURE 63.5 Domino circuit: (a) block diagram, and (b) clock waveforms and precharge and evaluate constraints. Note precharge implies the phase of operation (clock); the signals are falling.

high. The difference between *footed* and *footless domino gates* is the clocked nMOS transistor at the "foot" of the nMOS evaluate tree. To demonstrate the timing constraints imposed on domino circuits, consider the domino circuit block diagram and the clock waveforms shown in Figure 63.5. The footed domino blocks are labeled *FD1* and *FD2*, and the footless blocks are labeled *FLD1* and *FLD2*. From Figure 63.5(b), note that all three clocks have the same period $2T$, but the falling edge of *CK2* is $0.25T$ after the falling edge of *CK1* which in turn is $0.5T$ after the falling edge of *CK0*. Therefore, the precharge phase for *FD1* and *FD2* is T, for *FLD1* is $0.5T$, and for *FLD2* is $0.25T$. The various timing constraints for domino circuits are illustrated in Figure 63.5 and discussed below.

1. We want the output O to evaluate (rise) before the clock starts falling and to precharge (fall) before the clock starts rising.
2. Consider node $N1$, which is an output of *FD1* and an input of *FD2*. $N1$ starts precharging (falling) when *CK0* falls, and the constraint on it is that it should finish precharging before *CK0* starts rising.
3. Next, consider node $N2$, which is an input to *FLD1* clocked by *CK1*. Since this block is footless, $N2$ should be low during the precharge phase to avoid short-circuit current. $N2$ starts precharging (falling) when *CK0* starts falling and should finish falling before *CK1* starts falling.

Note that the falling edges of *CK0* and *CK1* are 0.5*T* apart, and the precharge constraint is on the late or maximum arrival time of *N2* (long path constraint). Also, *N2* should start rising only after *CK1* has finished rising. This is a constraint on the early or minimum arrival time of *N2* (short path constraint). In this example, *N2* starts rising with the rising edge of *CK0* and, since all the clock waveforms rise at the same time, the short path constraint will be satisfied trivially.

4. Finally, consider node *N3*. Since *N3* is an input of *FLD2*, it must satisfy the short-circuit current constraints. *N3* starts precharging (falling) when *CK1* starts falling and it should fall completely before *CK2* starts falling. Since the two clock edges are 0.25*T* apart, the precharge constraint on *N3* is tighter than the one on *N2*. As before, the short path constraint on *N3* is satisfied trivially.

The above discussion highlights the various types of timing constraints that must be automatically inserted by the static TA tool.

Note that each relative timing constraint between two signals is actually composed of two constraints. For example, if signal *d* must rise before clock *CK* rises, then (1) there is a required time on the late or maximum rising arrival time at node *d* (i.e., $A_{d,r} < A_{CK,r}$), and (2) there is a required time on the early or minimum rising arrival time at the clock node *CK* (i.e., $a_{CK,r} < a_{d,r}$). There is one other point to be noted. Set-up and hold constraints are fundamentally different in nature. If a hold constraint is violated, then the circuit will not function at any frequency. In other words, hold constraints are *functional constraints*. Set-up constraints, on the other hand, are *performance constraints*. If a set-up constraint is violated, the circuit will not function at the specified frequency, but it will function at a lower frequency (lower speed of operation). For domino circuits, precharge constraints are functional constraints, whereas evaluate constraints are performance constraints.

63.2.6 Transistor-Level Delay Modeling

In transistor-level static TA, delays of timing arcs have to be computed on the fly using transistor-level delay estimation techniques. There are many different transistor-level delay models which provide different tradeoffs between speed and accuracy. Before reviewing some of the more popular delay models, we define some notations. We will refer to the delay of a timing arc as being its *propagation delay* (i.e., the time difference between the output and the input completing half their transitions). For a falling output, the fall times is defined as the time to transition from 90% to 10% of the swing; similarly, for a rising output, the rise time is defined as the time to transition from 10% to 90% of the swing. The transition time at the output of the timing arc is defined to be either the rise time or the fall time. In many of the delay models discussed below, the transition time at the input of a timing arc is required to find the delay across the timing arc. At any node in the circuit, there is a transition time corresponding to each timing arc that is incident on that node. Since for long path static TA, we find the latest arriving signal at a node and propagate that arrival time forward, the transition time at a node is defined to be the output transition time of the timing arc which produced the latest arrival time at the node. Similarly, for short path analysis, we find the transition time as the output transition time of the timing arc that produced the earliest arrival time at the node.

Analytical closed-form formulae for the delay and output transition times are useful for static TA because of their efficiency. One such model was proposed in Hedenstierna et al., [12] where the propagation delay across an inverter is expressed as a function of the input transition time s_{in}, the output load C_L, and the size and threshold voltages of the NMOS and PMOS transistors. For example, the inverter delay for a rising input and falling output is given by

$$t_d = k_0 \frac{C_L}{\beta_n} + s_{in}(k_1 + k_2 V_{tn}) \tag{63.8}$$

where β_n is the NMOS transconductance (proportional to the width of the device), V_{tn} is the NMOS threshold voltage, and k_0, k_1, and k_2 are constants. The formula for the rising delay is the same, with PMOS device parameters being used. The output transition time is considered to be a multiple of the propagation delay and can be calibrated to a particular technology. More accurate analytical formulae for the propagation delay and output transition time for an inverter gate have been reported in the literature [13,14]. These methods consider more complex circuit behavior such as short-circuit current (both NMOS and PMOS transistors in the inverter are conducting) and the effect of MOS parasitic capacitances that directly couple the inputs and outputs of the inverter. More accurate models of the drain current and parasitic capacitances of the transistor are also used. The main shortcoming of all these delay models is that they are based on an inverter primitive; therefore, arbitrary CMOS gates seen in the circuit must be mapped to an equivalent inverter [15]. This process often introduces large errors.

A simpler delay model is based on replacing transistors by linear resistances and using closed-form expressions to compute propagation delays [16,17]. The first step in this type of delay modeling is to determine the charging/discharging path from the power supply rail to the output node that contains the switching transistor. Next, each transistor along this path is modeled as an effective resistance and the MOS diffusion capacitances are modeled as lumped capacitances at the transistor source and drain terminals. Finally, the Elmore time constant [18] of the path is obtained by starting at the power supply rail and adding the product of each transistor resistance and the sum of all downstream capacitances between the transistor and the output node. The accuracy of this method is largely dependent on the accuracy of the effective resistance and capacitance models. The effective resistance of a MOS transistor is a function of its width, the input transition time, and the output capacitance load. It is also a function of the position of the transistor in the charging/discharging path. The position variable can have three values: *trigger* (when the input at the gate of the transistor is switching), *blocking* (when the transistor is not switching and it lies between the trigger and the output node), and *support* (when the transistor is not switching and lies between the trigger and the power supply rail). The simplest way to incorporate these effects into the resistance model is to create a table of the resistance values (using circuit simulation) for various values of the transistor width, the input transition, and the output load. During delay modeling, the resistance value of a transistor is obtained by interpolation from the calibration table. Since the position is a discrete variable, a different table must be stored for each position variable. The effective MOS parasitic capacitances are functions of the transistor width and can also be modeled using a table look-up approach. The main drawbacks of this approach are the lack of accuracy in modeling a transistor as a linear resistance and capacitance, as well as not considering the effect of parallel charging/discharging paths and complementary paths. In our experience, this approach typically gives 10–20% accuracy with respect to SPICE for standard gates (inverters, NANDs, NORs, etc.); for complex gates, the error can be greater. These methods do not compute the transition time or slope at the output of the DCC. The transition time at the output node is considered to be a multiple of the propagation delay. Note that the propagation delay across a gate can be negative; this is the case, for example, if there is a slow transition at the input of a strong but lightly loaded gate. As a result, the transition time would become negative, giving a large error compared to the correct value.

Yet another method of modeling the delay from an input to an output of a DCC (or gate) is based on running a circuit simulator such as SPICE [5], or a fast timing simulator such as ILLIADS [6] or ACES [7]. Since the waveform at the switching input is known, the main challenge in this method is to determine the assertions (whether an input should be set to a high or low value) for the side inputs which gives rise to a transition at the output of the DCC [19]. For example, let us consider a rising transition at the input causing a falling transition at the output. In this case, a valid assertion is one that satisfies the following two conditions: (1) before the transition, there should be no conducting path between the output node and *Gnd*, and (2) after the transition, there should be at least one conducting path between the output node and *Gnd* and no conducting path between the output node and V_{dd}. The sensitization condition for a rising output transition is exactly symmetrical. The valid assertions are usually determined using a binary decision diagram [20]. For a particular input-output transition, there may be many valid assertions; these valid assertions may have different delay values since the primary charging/discharging path may be different or different node capacitances in the side paths may be charged/discharged. To find the assertion that causes the worst-case (or best-case) delay, one may resort to explicit simulations of all the valid assertions or employ other heuristics to prune out certain assertions. The main advantage of this type of delay modeling is that very accurate delay and transition time estimates can be obtained since the underlying simulator is accurate. The added accuracy is obtained at the cost of additional runtime.

Since static timing analyzers typically use simple delay models for efficiency reasons, the top few critical paths of the circuit should be verified using circuit simulation [21,22].

63.2.7 Interconnects and Static TA

As is well known, interconnects are playing a major role in determining the performance of current microprocessors, and this trend is expected to continue in the next generation of processors [23]. The effect of interconnects on circuit and system performance should be considered in an accurate and efficient manner during static timing analysis. To illustrate interconnect modeling techniques, we will use the example shown in Figure 63.6(a) of a wire connecting a driving inverter to three receiving inverters.

The simplest interconnect model is to lump all the interconnect and receiver gate capacitances at the output of the driver gate. This approximation may greatly overestimate the delay across the driver gate since, in reality, all of the downstream capacitances are not "seen" by the driver gate because of resistive shielding due to line resistances. A more accurate model of the wire as a distributed RC line is shown in Figure 63.6(b). This is the wire model output by most commercial RC extraction tools. In Figure 63.6(b),

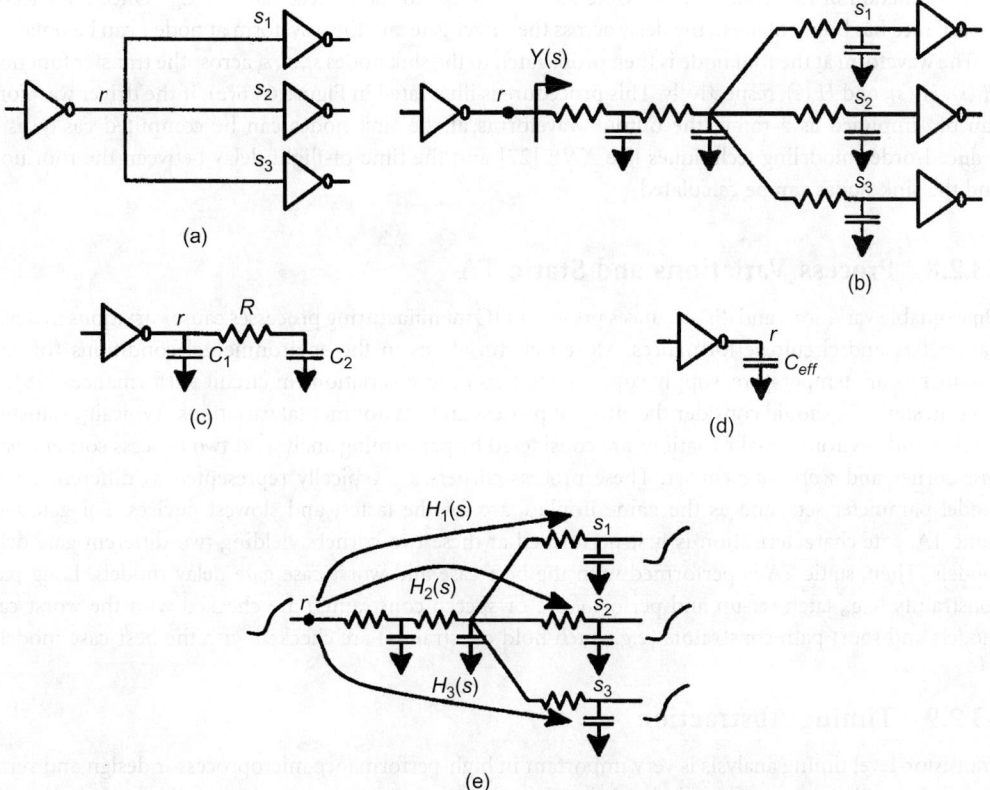

FIGURE 63.6 Handling interconnects in static TA: (a) a typical interconnect, (b) distributed RC model of interconnect, (c) reduced π-model to represent the loading of the interconnect, (d) effective capacitance loading, and (e) propagation of waveform from root to sinks.

node *r* is called the *root* of the interconnect and is driven by the driver gate, and the other end points of the wire at the inputs of the receiver gate are called *sinks* of the interconnect and are labeled s_1, s_2, and s_3. Interconnects have two main effects: (1) the interconnect resistance and capacitance determines the effective load seen by the driving gate and therefore its delay, and (2) due to non-zero wire resistances, there is a non-zero delay from the root to the sinks of the interconnect—this is called the *time-of-flight delay*.

To model the effect of the interconnect on the driver delay, we first replace the metal wire with a π-model load as shown in Figure 63.6(c) [24]. This is done by finding the first three moments of the admittance $Y(s)$ of the interconnect at node *r*. It can be shown that the admittance is given by $Y(s) = m_1 s + m_2 s^2 + m_3 s^3 + \cdots$. Next, we obtain the admittance of the π-load as $\hat{Y}(s) = s(C_1 + C_2) - s^2 R C_2^2 + s^3 R^2 C_2^3 + \cdots$, where R, C_1, and C_2 are the parameters of the π-load model. To obtain the parameters of the π-load, we equate the first three moments of $Y(s)$ and $\hat{Y}(s)$. This gives us the following equations for the parameters of the π-load model:

$$C_2 = \frac{m_2^2}{m_3}, \quad C_1 = m_1 - \frac{m_2^2}{m_3}, \quad \text{and} \quad R = -\frac{m_3^2}{m_2^3} \tag{63.9}$$

Now, if we are using a transistor-level delay model or a pre-characterized gate-level delay model that can only handle purely capacitive loading and not π-model loads, we have to determine an effective capacitance C_{eff} that will accurately model the π-load. The basic idea of this method [25,26] is to equate the average current drawn by the π-model load to the average current drawn by the C_{eff} load. Since the average current drawn by any load is dependent on the transition time at the output of the gate and the transition time is itself a function of the load, we have to iterate to converge to the correct value of C_{eff}. Once the effective capacitance has been obtained, the delay across the driver gate and the waveform at node *r* can be obtained.

The waveform at the root node is then propagated to the sink nodes s_1, s_2, s_3 across the transfer functions $H_1(s)$, $H_2(s)$, and $H_3(s)$, respectively. This procedure is illustrated in Figure 63.6(e). If the driver waveform can be simplified as a ramp, the output waveforms at the sink nodes can be computed easily using reduced-order modeling techniques like AWE [27] and the time-of-flight delay between the root node and the sink nodes can be calculated.

63.2.8 Process Variations and Static TA

Unavoidable variations and disturbances present in IC manufacturing processes cause variations in device parameters and circuit performances. Moreover, variations in the environmental conditions (of such parameters are temperature, supply voltages, etc.) also cause variations in circuit performances [28]. As a result, static TA should consider the effect of process and environmental variations. Typically, statistical process and environmental variations are considered by performing analysis at two process corners: *best-case* corner and *worst-case* corner. These process corners are typically represented as different device model parameter sets, and as the name implies, are for the fastest and slowest devices. For gate-level static TA, gate characterization is first performed at these two corners yielding two different gate delay models. Then, static TA is performed with the best-case and worst-case gate delay models. Long path constraints (e.g., latch set-up and performance or speech constraints) are checked with the worst-case models and short path constraints (e.g., latch hold constraints) are checked with the best-case models.

63.2.9 Timing Abstraction

Transistor-level timing analysis is very important in high-performance microprocessor design and verification since a large part of the design is hand-crafted and cannot be pre-characterized. Analysis at the transistor level is also important to accurately consider interconnect effects such as gate loading, charge-sharing, and clock skew. However, full-chip transistor-level analysis of large microprocessor designs is computationally infeasible, making timing abstraction a necessity.

63.2.9.1 Gate-Level Static TA

A straightforward extension of transistor-level static TA is to the gate level. At this level of abstraction, the circuit has been partitioned into gates, and the inputs and outputs of each gate have been identified. Moreover, the timing arcs from the inputs to the outputs of a gate are typically pre-characterized. The gates are characterized by applying a ramp voltage source at the input of the gate and an explicit load capacitance at the output of the gate. Then, the transition time of the ramp and the value of the load capacitance is varied, and circuit simulation (e.g., SPICE) is used to compute the propagation delays and output transition times for the various settings. These data points can be stored in a table or abstracted in the form of a curve-fitted equation. A popular curve-fitting approach is the k-factor equations [26], where the delay t_d and output transition time t_{out} are expressed as non-linear functions of the input transition time s_{in} and the capacitive output load C_L:

$$t_d = (k_1 + k_2 C_L) s_{in} + k_3 C_L^2 + k_4 C_L + k_5 \qquad (63.10)$$

$$t_{out} = (k_1' + k_2' C_L) s_{in} + k_3' C_L^2 + k_4' C_L + k_5'. \qquad (63.11)$$

The various coefficients in the k-factor equations are obtained by curve fitting the data. Several modifications, including more complex equations and dividing the plane into a number of regions and having equations for each region, have been proposed.

The main advantage of gate-level static TA is that costly on-the-fly delay and output transition time calculations can be replaced by efficient equation evaluations or table look-ups. This is also a disadvantage since it requires that all the timing arcs in the design are pre-characterized. This may be a problem when parts of the design are not complete and the delays for some timing arcs are not available. This problem can be avoided if the design flow ensures that at early stages of a design, estimated delays are specified for all timing arcs which are then replaced by characterized numbers when the design gets completed. To apply gate-level TA to designs that contain a large amount of custom circuits, timing rules must be developed for the custom circuits also. Gate-level static TA is still at a fairly low level of abstraction and the effects of interconnects and clock skew can be considered. Moreover, at the gate level, the latches and flip-flops of the design are visible and so timing constraints can be inserted directly at those nodes.

63.2.9.2 Black-Box Modeling

At the next higher level of abstraction, gates are grouped together into blocks and the entire design (or chip) now consists of these blocks or "boxes." Each box contains combinational gates as well as sequential elements such as latches as shown in Figure 63.7(a). Timing checks inside the block can be verified using static TA at the transistor or gate level. At the chip level, the internal nodes of the box are no longer visible and its timing behavior must be abstracted at the input, output, and clock pins of the box. In black-box modeling, we assume that the first and last latch along any path from input to output of the box are edge-triggered latches; in other words, cycle stealing is not allowed across these latches (cycle stealing may be allowed across other transparent latches inside the box). The first latch along a path from input to output is called an *input latch* and the last latch is called an *output latch*. With this assumption, there can be two types of paths to the outputs of the box. First, paths that originate at box inputs and end at box outputs without traversing through any latches. These paths are represented as input-output arcs in the block-box with the path delays annotated on the arcs. Second, there are paths that originate at the clock pins of the output edge-triggered latches and end at the box outputs. These paths are represented as clock-to-input arcs in the black-box and the paths delays are annotated on the arcs. Finally, the set-up and hold time constraints of the input latches are translated to constraints between the box inputs and clock pins. These constraints will be checked at the chip-level static TA. The constraints

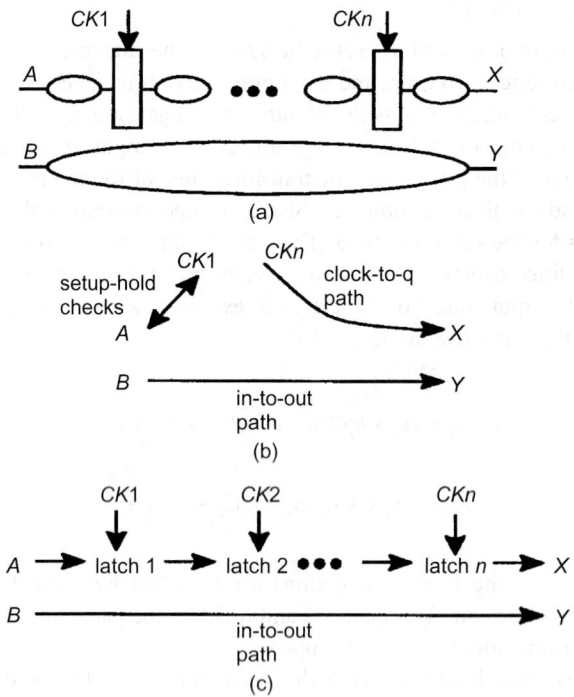

FIGURE 63.7 High-level timing abstraction: (a) a block containing combinational and sequential elements, (b) black-box model, and (c) gray-box model.

and the arcs are shown in Figure 63.7(b). Note that the timing checkpoints inside a block have been verified for a particular set of clocks when the black-box model is generated. Since these timing checkpoints are no longer available at the chip level, a black-box model is valid only for a particular frequency. If a different clock frequency (or different clock waveforms) is used, then the black-box model must be regenerated.

63.2.9.3 Gray-Box Modeling

Gray-box modeling removes the edge-triggered latch restrictions of black-box modeling. All latches inside the box are allowed to be level-sensitive and therefore have to be visible at the top level so that the constraints can be checked and cycle-stealing is allowed through these latches. As shown in Figure 63.7(c), the gray-box model consists of timing arcs from the box inputs to the input latches, from latches to latches, and from the output latches to the box outputs. The clock pins of each of the latches are also visible at the chip level, and so the set-up and hold timing constraints for each latch in the box is checked at the chip level. In addition to these timing arcs, there can also be direct input-output timing arcs. Note that since the timing checkpoints internal to the box are available at the chip level, the gray-box model is frequently independent— unlike the black-box model.

63.2.10 False Paths

To find the critical paths in the circuit, static TA propagates the arrival times from the timing inputs to the timing outputs. Then, it propagates the required times from the outputs back to the inputs and computes the slacks along the way. During propagation, static TA does not consider the logical functionality of the circuit. As a result, some of the paths that it reports to the user may be such that they cannot be activated by any input vector. Such paths are called false paths [29–31]. An example of a false path is shown in Figure 63.8(a). For x to propagate to a, we must set $y = 1$, which is the non-controlling value

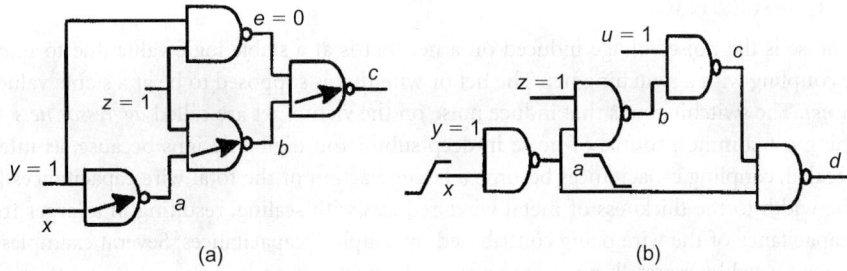

FIGURE 63.8 False path examples: (a) static false path, and (b) dynamic false path.

of the NAND gate. Similarly, for a to propagate to b, we set $z = 1$. Now, since $y = z = 1$, $e = 0$ (the controlling value for a NAND gate), and there can be no signal propagation from b to c. Therefore, there can be no propagation from x to c (i.e., $x - a - b - c$ is a false path). False paths that arise due to logical correlations are called *static false paths* to distinguish them from dynamic false paths, which are caused by temporal correlations.

A simple example of a *dynamic false path* is shown in Figure 63.8(b). Suppose we want to find the critical path from node x to the output d. It is clear that there are two such paths, $x - a - d$ and $x - a - b - c - d$, of which the latter has a larger delay. In order to sensitize the longer path $x - a - b - c - d$, we would set the other inputs of the circuit to the non-controlling values of the gates (i.e., $y = z = u = 1$). If there is a rising transition on node x, there will be a falling transition on nodes a and c. However, because of the propagation delay from a to c, node a will fall well before node c. As soon as node a falls, it will set the primary output d to be 1 (since the controlling value of a NAND gate is 0). Because node a always reaches the controlling value before node c, it is not possible for a transition at node c to reach the output. In other words, the path x rising – a falling – b rising – c falling – d rising is a dynamic false path. Note that if we add some combinational logic between the output of the first NAND gate and the input of the last NAND gate to slow the signal a down, then the transition on c could propagate to the output. The example shown above is for purposes of illustration only and may appear contrived. However, dynamic false paths are very common in carry-lookahead adders [32].

Finding false paths in a combinational circuit is an NP-complete problem. There are a number of heuristic approaches that find the longest paths in a circuit while determining and ignoring the false paths [29–31]. Timing analysis techniques that can avoid false paths specified by the user have also been reported [33,34].

63.3 Noise Analysis

In digital circuits, nodes that are not switching are at the nominal values of the supply (logic 1) and ground (logic 0) rails. In a digital system, noise is defined as a deviation of these node voltages from their stable high or low values. Digital noise should be distinguished from physical noise sources that are common in analog circuits (e.g., shot noise, thermal noise, flicker noise, and burst noise) [35]. Since noise causes a deviation in the stable logic voltages of a node, it can be classified into four categories: (1) *high undershoot* noise reduces the voltage of a node that is supposed to be at logic 1; (2) *high overshoot* noise which increases the voltage of a logic 1 node above the supply level (V_{dd}); (3) *low overshoot* noise increases the voltage of a node that is supposed to be at logic 0; and (4) *low undershoot* noise which reduces the voltage of a logic 0 node below the ground level (*Gnd*).

63.3.1 Sources of Digital Noise

The most common sources of noise in digital circuits are crosstalk noise, power supply noise, leakage noise and charge-sharing noise [36].

63.3.1.1 Crosstalk Noise

Crosstalk noise is the noise voltage induced on a net that is at a stable logic value due to interconnect capacitive coupling with a switching net. The net or wire that is supposed to be at a stable value is called the *victim net*. The switching nets that induce noise on the victim net are called *aggressor nets*. Crosstalk noise is the most common source of noise in deep submicron digital designs because, as interconnect wires get scaled, coupling capacitances become a larger fraction of the total wire capacitances [23]. The ratio of the width to the thickness of metal wires reduces with scaling, resulting in a larger fraction of the total capacitance of the wire being contributed by coupling capacitances. Several examples of functional failures caused by crosstalk noise are given in the next section.

63.3.1.2 Power Supply Noise

This refers to noise on the power supply and ground nets of a design that is passed onto the signal nets by conducting transistors. Typically, the power supply noise has two components. The first is produced by IR-drop on the power and ground nets due to the current demands of the various gates in the chip (discussed in the next section). The second component of the power supply noise comes from the RLC response of the chip and package to current demands that peak at the beginning of a clock cycle. The first component of power supply noise can be reduced by making the wires that comprise the power and ground network wider and denser. The second component of the noise can be reduced by placing on-chip decoupling capacitors [37].

63.3.1.3 Charge-Sharing Noise

Charge-sharing noise is the noise induced at a dynamic node due to charge redistribution between that node and the internal nodes of the gate [32]. To illustrate charge-sharing noise, let us again consider the two-input domino NAND gate of Figure 63.9(a). Let us assume that during the first evaluate phase shown in Figure 63.9(b), both nodes x and x_1 are discharged. Then, during the next precharge phase, let us assume that the input a is low. Node x will be precharged by the PMOS transistor MP, but x_1 will not and will remain at its low value. Now, suppose CK turns high, signaling the beginning of another evaluate phase. If during this evaluate phase, a is high but b is low, nodes x and x_1 will share charge resulting in the waveforms shown in Figure 63.9(b): x will be pulled low and x_1 will be pulled high. If the voltage on x is reduced by a large amount, the output inverter may switch and cause the output node y to be wrongly set to a logic high value. Charge-sharing in a domino gate is avoided by precharging the internal nodes in the NMOS evaluate tree during the precharge phase of the clock. This is done by adding an anti-charge sharing device such as MNc in Figure 63.9(c) which is gated by the clock signal.

63.3.1.4 Leakage Noise

Leakage noise is due to two main sources: *subthreshold conduction* and *substrate noise*. Subthreshold leakage current [32] is the current that flows in MOS transistors even when they are not conducting (off). This

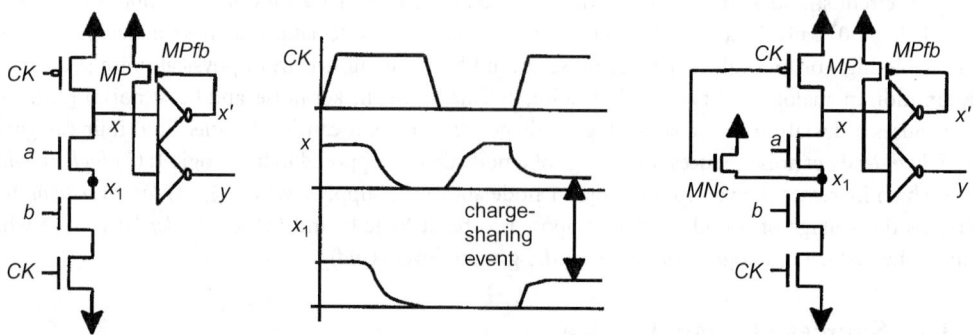

FIGURE 63.9 Example of charge-sharing noise: (a) a two-input domino NAND gate, (b) waveforms for charge-sharing event, and (c) anti charge-sharing device.

current is a strong function of the threshold voltage of the device and the operating temperature. Subthreshold leakage is an important design parameter in portable devices since battery life is directly dependent on the average leakage current of the chip. Subthreshold conduction is also an important noise mechanism in dynamic circuits where, for a part of the clock cycle, a node does not have a strong conducting path to power or ground and the logic value is stored as a charge on that node. For example, suppose that the inputs a and b in the two-input domino NAND gate of Figure 63.9(a) are low during the evaluate phase of the clock. Due to subthreshold leakage current in the NMOS evaluate transistors, the charge on node x may be drained away, leading to a degradation in its voltage and a wrong value at the output node y. The purpose of the half latch device *MPfb* is to replenish the charge that may be lost due to the leakage current.

Another source of leakage noise is minority carrier back injection into the substrate due to bootstrapping. In the context of mixed analog-digital designs, this is often referred to as *substrate noise* [38]. Substrate noise is often reduced by having guard bands, which are diffusion regions around the active region of a transistor tied to supply voltages so that the minority carriers can be collected.

63.3.2 Crosstalk Noise Failures

In this section, we provide some examples of functional failures caused by crosstalk noise. Functional failures result when induced noise voltages cause an erroneous state to be stored at a memory element (e.g., at a latch node or a dynamic node). Consider the simple latch circuit of Figure 63.10(a) and let us

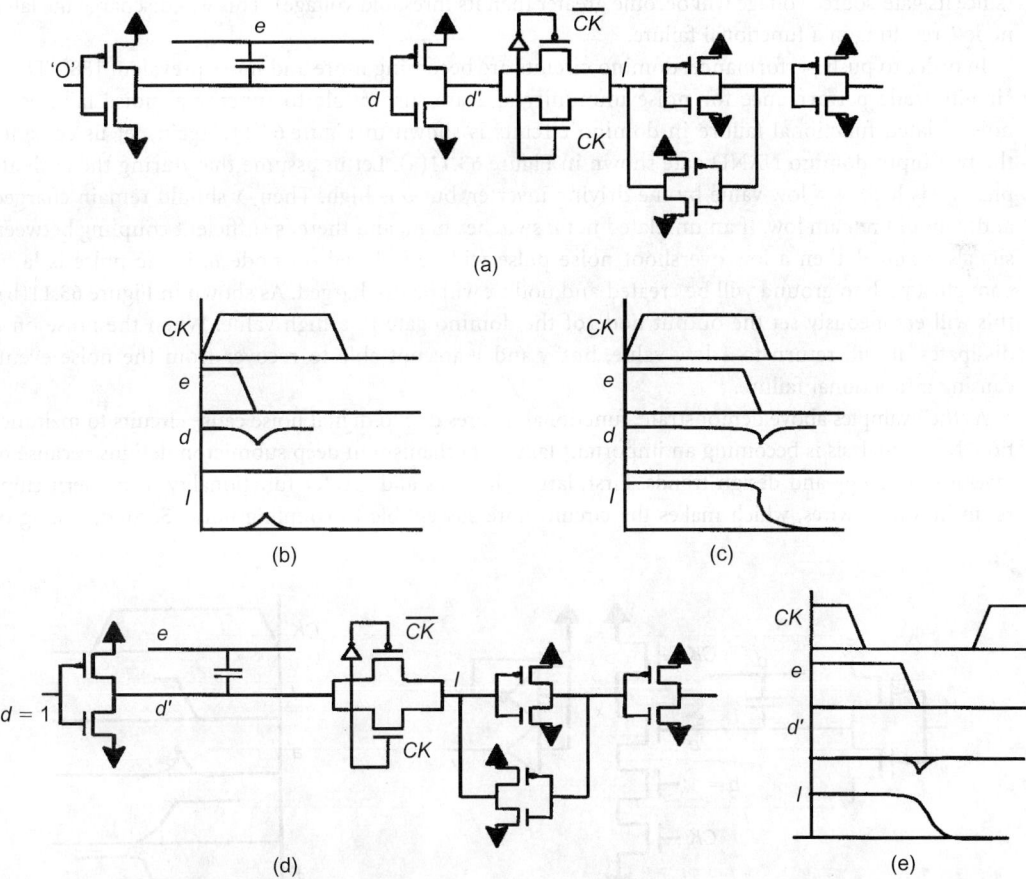

FIGURE 63.10 Crosstalk noise-induced functional failures: (a) latch circuit; (b) high undershoot noise on d does not cause functional failure in (b) but does cause failure in (c); (d) same latch circuit with noise induced on an internal node; and (e) low undershoot noise causing a failure.

assume that the data input *d* is a stable high value and the latch *l* has a stable low value. If the net corresponding to node *d* is coupled to another net *e* and there is a high to low transition on net *e*, net *d* will be pulled low. When *e* has finished switching, *d* will be pulled back to a high value by the PMOS transistor driving net *d* and the noise on *d* will dissipate. Thus, the transition on net *e* will cause a noise pulse on *d*. If the amplitude of this noise pulse is large enough, the latch node *l* will be pulled high. Depending on the conditions under which the noise is injected, it may or may not cause a wrong value to be stored at the latch node. For example, let us consider the situation depicted in Figure 63.10(b), where *CK* is high and the latch is open. If the noise pulse on *d* appears near the middle of the clock phase, then the latch node will be pulled high; but as the noise on *d* dissipates, latch node *l* will return to its correct value because the latch is open. However, if the noise pulse on *d* appears near the end of the clock phase as shown in Figure 63.10(c), the latch may turn off before the noise on *d* dissipates, the latch node may not recover, and a wrong value will be stored. A similar unrecoverable error may occur if noise appears on the clock net turning the latch on when it was meant to be off. This might cause a wrong value to be latched.

Now, let us consider the latch circuit of Figure 63.10(d) where the wire between the input inverter and the pass gate of the latch is long and subject to coupling capacitances. Suppose the latch is turned off (*CK* is low), the data input is high so that the node *d'* is low, and a high value is stored at the latch node. If net *e* transitions from a high to a low value, a low undershoot noise will be introduced on *d'*. If this noise is sufficiently large, the NMOS pass transistor will turn on even through its gate voltage is zero (since its gate-source voltage will become greater than its threshold voltage). This will discharge the latch node *l*, resulting in a functional failure.

In order to push performance, domino circuits are becoming more and more prevalent [88]. These circuits trade performance for noise immunity and are susceptible to functional noise failures. A noise-related functional failure in domino circuits is shown in Figure 63.11. Again, let us consider the two-input domino NAND gate shown in Figure 63.11(a). Let us assume that during the evaluate phase, *a* is held to a low value by the driving inverter, but *b* is high. Then, *x* should remain charged and *y* should remain low. If an unrelated net *d* switches high, and there is sufficient coupling between signals *a* and *d*, then a low overshoot noise pulse will be induced on node *a*. If the pulse is large enough, a path to ground will be created and node *x* will be discharged. As shown in Figure 63.11(b), this will erroneously set the output node of the domino gate to a high value. When the noise on *a* dissipates, it will return to a low value, but *x* and *y* are not able to recover from the noise event, causing a functional failure.

As the examples above demonstrate, functional failures due to digital noise cause circuits to malfunction. Noise analysis is becoming an important failure mechanism in deep submicron designs because of several technology and design trends. First, larger die sizes and greater functionality in modern chips result in longer wires, which makes the circuit more susceptible to coupling noise. Second, scaling of

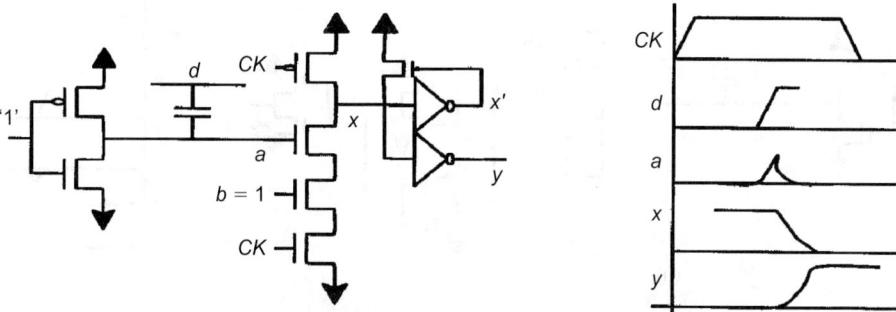

FIGURE 63.11 Functional failure in domino gates: (a) two-input NAND gate, and (b) voltage waveforms when input noise causes a functional failure.

interconnect geometries has resulted in increased coupling between adjacent wire [23]. Third, the drive for faster performance has increased the use of faster non-restoring logic families such as domino logic. These circuit families have faster switching speeds at the expense of reduced noise immunity. False switching events at the inputs of these gates are catastrophic since precharged nodes may be discharged and these nodes cannot recover their original state when the noise dissipates. Fourth, lower supply voltage levels reduce the magnitudes of the noise margins of circuits. Finally, in state-of-the-art microprocessors, many functional units located in different parts of the chip are operating in parallel and this causes a lot of switching activity in long wires that run across different parts of the chip. All of these factors make noise analysis a very important task to verify the proper functioning of digital designs.

63.3.3   Modeling of Interconnect and Gates for Noise Analysis

Let us consider the example of Figure 63.12(a) where three wires are running in parallel and are capacitively coupled to each other. Suppose that we are interested in finding the noise that is induced on the middle net by the adjacent nets switching. The middle net is called the *victim net* and the two neighboring nets are called *aggressors*. Consider the situation when the victim net is held to a stable logic zero value by the victim driver and both the aggressor nets are switching high. Due to the coupling between the nets, a low overshoot noise will be induced on the victim net as shown in Figure 63.12(a). If the noise pulse is large and wide enough, the victim receiver may switch and cause a wrong value at the output of the inverter.

The circuit-level models for this system are explained below and shown in Figure 63.12(b).

1. The (net) complex consisting of the victim and aggressor nets is modeled as a coupled distributed RC network. The coupled RC lines are typically output by a parasitic extraction tool.
2. The non-linear victim driver is holding the victim net to a stable value. We model the non-linear driver as a linear holding resistance. For example, if the victim driver holds the output to logic 0 (logic 1), we determine an effective NMOS (PMOS) resistance. The value of the holding resistance for a gate can be obtained by pre-characterization using SPICE.

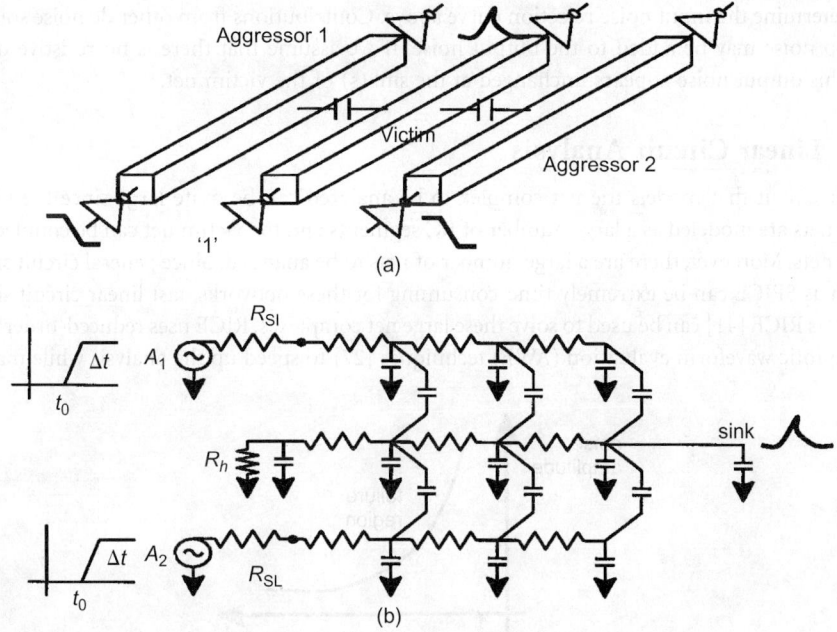

FIGURE 63.12   (a) A noise pulse induced on the victim net by capacitive coupling to adjacent aggressor nets, and (b) linearized model for analysis.

3. The aggressor driver is modeled as a Thevenin voltage source in series with a switching resistance. The Thevenin voltage source is modeled as a shifted ramp, where the ramp starts switching at time t_0 and the transition time is Δt. The switching resistance is denoted by R_s.
4. The victim receiver is modeled as a capacitor of value equal to the input capacitance of the gate.

These models convert the non-linear circuit into a linear circuit. The multiple sources in this linear circuit can now be analyzed using *linear superposition*. For each aggressor, we get a noise pulse at the sink(s) of the victim net, while shorting the other aggressors. These noise pulses have different amplitudes and widths; the amplitude and width of the *composite noise waveform* is obtained by aligning these noise pulses so that their peaks line up. This is a conservative assumption to simulate the worst-case noise situation.

63.3.4 Input and Output Noise Models

As mentioned earlier, noise creates circuit failures when it propagates to a charge-storage node and causes a wrong value to be stored at the node. Propagating noise across non-linear gates [39] makes the noise analysis problem complex. In this discussion, a more conservative simple model will be discussed. With each input terminal of a victim receiver gate, we associate a noise rejection curve [40]. This is a curve of the noise amplitude versus the noise width that produces a predefined amount of noise at the output. If we assume a triangular noise pulse at the input of the victim receiver, the noise rejection curve defines the amplitude-width combination that produces a fixed amount of noise at the output of the receiver. A sample noise rejection curve is shown in Figure 63.13. As the width becomes very large, the noise amplitude tends toward the dc noise margin of the gate. Due to the lowpass nature of a digital gate, very sharp noise pulses are filtered out and do not cause any appreciable noise at the output. When the noise pulse at the sink(s) of the victim net have been obtained, the pulse amplitude and width are compared against the noise rejection curve to determine if a noise failure occurs.

Since we do not propagate noise across gates, noise injected into the victim net at the output of the victim driver must model the maximum amount of noise that may be produced at the output of a gate. The output noise model is a dc noise that is equal to the predefined amount of output noise that was used to determine the input noise rejection curve above. Contributions from other dc noise sources such as IR-drop noise may be added to the output noise. If we assume that there is no resistive dc path to ground, this output noise appears unchanged at the sink(s) of the victim net.

63.3.5 Linear Circuit Analysis

The linear circuit that models the net complex to be analyzed can be quite large since the victim and aggressor nets are modeled as a large number of RC segments and the victim net can be coupled to many aggressor nets. Moreover, there are a large number of nets to be analyzed. Since general circuit simulation tools such as SPICE can be extremely time-consuming for these networks, fast linear circuit simulation tools such as RICE [41] can be used to solve these large net complexes. RICE uses reduced-order modeling and asymptotic waveform evaluation (AWE) techniques [27] to speed up the analysis while maintaining

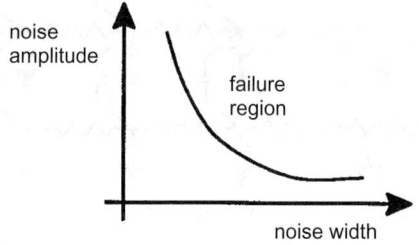

FIGURE 63.13 A typical noise rejection curve.

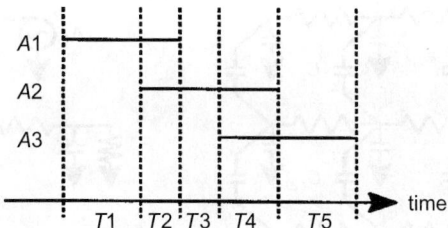

FIGURE 63.14 Effect of timing windows on aggressor selection for noise analysis.

sufficient accuracy. Techniques that overcome the stability problems in AWE, such as Pade via Lancszos (PVL) [42], Arnoldi-based techniques [43], congruence transform-based techniques (PACT) [44], or combinations (PRIMA) [45], have been proposed recently.

63.3.6   Interaction with Timing Analysis

Calculation of crosstalk noise interacts tightly with timing analysis since timing analysis lets us determine which of the aggressor nets can switch at the same time. This reduces the pessimism of assuming that for a victim net, all the nets it is coupled to can switch simultaneously and induce noise on it. Timing analysis defines timing windows by the earliest and latest arrival times for all signals. This is shown in Figure 63.14 for three aggressors $A1$, $A2$, and $A3$ of a particular victim net of interest. Based upon these timing windows, we can define five different scenarios for noise analysis where different aggressors can switch simultaneously. For example, in interval $T1$, only $A1$ can switch; in $T2$, $A1$, and $A2$ can switch; in $T3$, only $A2$ can switch; and so on. Note that in this case, all three aggressors can never switch at the same time. Without considering the timing windows provided by timing analysis, we would have overestimated the noise by assuming that all three aggressors could switch at the same time.

63.3.7   Fast Noise Calculation Techniques

Any state-of-the-art microprocessors will have many nets to be analyzed, but typically only a small fraction of the nets will be susceptible to noise problems. This motivates the use of extremely fast techniques that provably overestimate the noise at the sinks of a net. If a net passes the noise test under this quick analysis, then it does not need to be analyzed any further; if a net fails the noise test, then it can be analyzed using more accurate techniques. In this sense, these fast techniques can be considered to be *noise filters*. If these noise filters produce sufficiently accurate noise estimates, then the expectation is that a large number of nets would be screened out quickly. This combination of fast and detailed analysis techniques would therefore speed up the overall analysis process significantly. Note that noise filters must be provably pessimistic and that multiple noise filters with less and less pessimism can be used one after the other to successively screen out nets.

Let us consider the net complex shown in Figure 63.15(a), where we have modeled the net as distributed RC lines, the victim driver as a linear holding resistance, and the aggressors as voltage ramps and linear resistances. The grounded capacitances of the victim net is denoted as C_{gv}, and the coupling capacitances to the two aggressors are denoted as C_{c1} and C_{c2}. In Figures 63.15(b–d), we show the steps through which we can obtain a circuit which will provide a provably pessimistic estimate of the noise waveform. In Figure 63.15(b), we have removed the resistances of the aggressor nets. This is pessimistic because, in reality, the aggressor waveform slows down as it proceeds along the net. By replacing it with a faster waveform, more noise will be induced on the victim net. In Figure 63.15(c), the aggressor waveforms are capacitively coupled directly into the sink net; for each aggressor, the coupling capacitance is equal to the sum of all the coupling capacitances between itself and the victim net. Since the aggressor is directly coupled to the sink net, this transformation will result in more induced noise. In Figure 63.15(d),

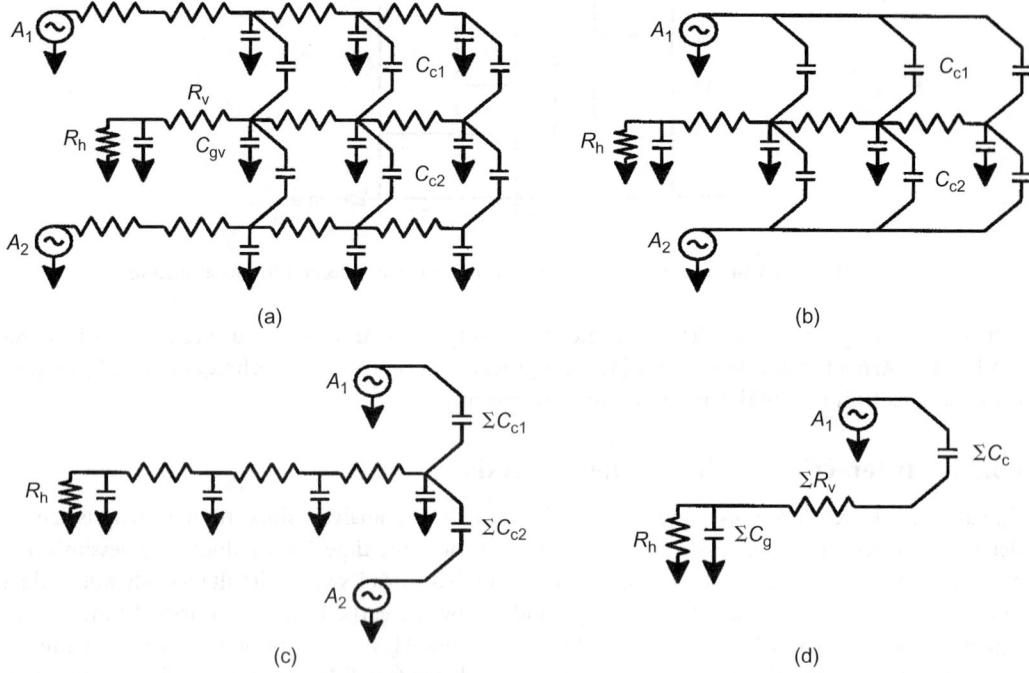

FIGURE 63.15 Noise filters: (a) original net complex with distributed RC models for aggressors and victims, (b) aggressor lines have only coupling capacitances to victim, (c) aggressors are directly coupled to sink of victim, and (d) single (strongest) aggressor and all grounded capacitors of victim moved away from sink.

we have made two modifications; first, we replaced the different aggressors by one capacitively coupled aggressor and, second, we moved all the grounded capacitors on the victim net away from the sink node. The composite aggressor is just the fastest aggressor (i.e., the aggressor that has the smallest transition time) and it is coupled to the victim net by a capacitor whose value is equal to the sum of all the coupling capacitances in the victim net. To simplify the victim net, we sum all the grounded capacitors and insert it at the root of the victim net and sum all the net resistances. By moving the grounded (good) capacitors away from the sink net, we increase the amount of coupled noise. This simple network can now be analyzed very quickly to compute the (pessimistic) noise pulse at the sink.

An efficient method to compute the peak noise amplitude at the sink of the victim net is described by Devgan [46]. Under infinite ramp aggressor inputs, the maximum noise amplitude is the final value of the coupled noise. For typical interconnect topologies, these analytical computations are simple and quick.

63.3.8 Noise, Circuit Delays, and Timing Analysis

Circuit noise, especially crosstalk noise, significantly affects switching delays. Let us consider the example of Figure 63.16(a), where we are concerned about the propagation delay from *A* to *C*. In the absence of any coupling capacitances, the rising waveform at *C* is shown by the dotted line of Figure 63.16(b). However, if net 2 is switching in the opposite direction (node *E* is rising as in Figure 63.16(b)), then additional charge is pumped into net 1 due to the coupling capacitors causing the signals at nodes B_1 and B_2 to slow down. This in turn causes the inverter to switch later and causes the propagation delay from *A* to *C* to be much larger, as shown in the diagram. Note that if net 2 switched in the same direction as net 1, then the delay from *A* to *C* would be reduced. This implies that delays across gates and wires depend on the switching activity on adjacent coupled nets. Since coupling capacitances are a large fraction of the total capacitance of wires, this dependence will be significant and timing analysis should account

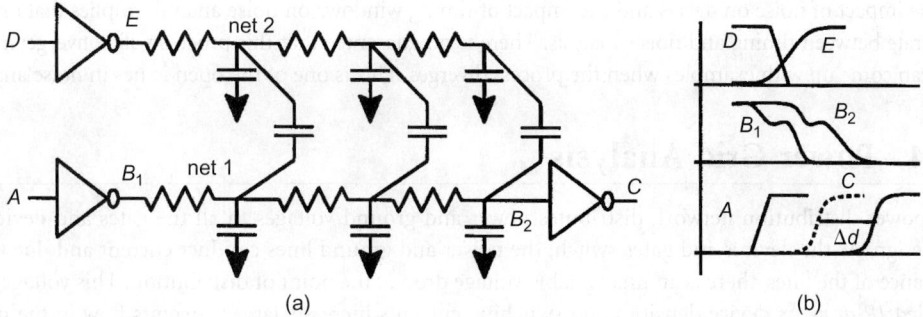

FIGURE 63.16   Effect of noise on circuit delays: (a) victim and aggressor nets, and (b) typical waveforms.

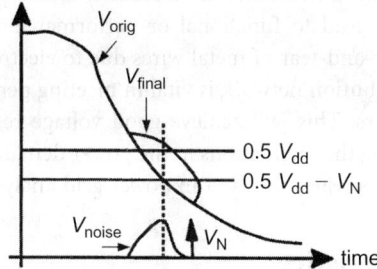

FIGURE 63.17   Aligning the composite noise waveform with the original waveform to produce worst-case delay.

for this behavior. Using the same terminology as crosstalk noise analysis, we call the net whose delay is of primary interest (net 1 in the above example) the *victim net* and all the nets that are coupled to it are called *aggressor nets*.

A model that is commonly used to approximate the effect of coupling capacitors on circuit delays is to replace each coupling capacitor by a grounded capacitor of twice the value. This model is accurate only when the victim and aggressor nets are identical and the waveforms on the two nets are identical, but switching in opposite directions. For some cases, doubling the coupling capacitance may be pessimistic, but in many cases it is not—the effective capacitance is much more than twice the coupling capacitance. Note that the effect on the propagation delay due to coupling will be strongly dependent on how the aggressor waveforms are aligned with respect to each other and to the victim waveform. Hence, one of the main issues in finding the effect of noise on delay is to determine the aggressor alignments that cause the worst propagation delay.

A more accurate model for considering the effect of noise on delay is described by Dartu et al. [47]. In this approach, the gates are replaced by linearized models (e.g., the Thevenin model of the gate consists of a shifted ramp voltage source in series with a resistance). Once the circuit has been linearized, the principle of linear superposition is applied. The voltage waveform at the sink of the victim net is first obtained by assuming that all aggressors are "quiet." Then the victim net is assumed to be quiet and each aggressor is switched one at a time and the resultant noise waveforms at the victim sink node is recorded. These noise waveforms are offset with respect to each other because of the difference in the delays between the aggressors to the victim sink node. Next, the aggressor noise waveforms are shifted such that the peaks get lined up and a composite noise waveform is obtained by adding the individual noise waveforms. The remaining issue is to align the composite noise waveform with the noise-free victim waveform to obtain the worst delay. This process is described in Figure 63.17, where we show the original noise-free waveform V_{orig} and the (composite) noise waveform V_{noise} at the victim sink node. Then, the worst case is to align the noise such that its peak is at the time when $V_{orig} = 0.5V_{dd} - V_N$, where V_N is the peak noise [47,48]. The final waveform at C is marked V_{final}.

The impact of noise on delays and the impact of timing windows on noise analysis implies that one has to iterate between timing and noise analysis. There is no guarantee that this process will converge; in fact, one can come up with examples when the process diverges. This is one of the open issues in noise analysis.

63.4 Power Grid Analysis

The power distribution network distributes power and ground voltages to all the gates and devices in the design. As the devices and gates switch, the power and ground lines conduct current and due to the resistance of the lines, there is an unavoidable voltage drop at the point of distribution. This voltage drop is called *IR-drop*. As device densities and switching currents increase, larger currents flow in the power distribution network causing larger IR-drops. Excessive voltage drops in the power grid reduce switching speeds of devices (since it directly affects the current drive of devices) and noise margins (since the effective rail-to-rail voltage is lower). Moreover, as explained in the previous section, IR-drops inject dc noise into circuits which may lead to functional or performance failures. Higher average current densities lead to undesirable wear-and-tear of metal wires due to electromigration [49]. Considering all these issues, a robust power distribution network is vital in meeting performance and reliability goals in high-performance microprocessors. This will achieve good voltage regulation at all the consumption points in the chip, notwithstanding the fluctuations in the power demand across the chip. In this section, we give a brief overview of various issues involved in power grid analysis.

63.4.1 Problem Characteristics

The most important characteristic of the power grid analysis problem is that it is a global problem. In other words, the voltage drop in a certain part of the chip is related to the currents being drawn from that as well as other parts of the chip. For example, if the same power line is distributing power to several functional units in a certain part of the chip, the voltage drop in one functional unit depends on the currents being drawn by the other functional units. In fact, as more and more of the functional units switch together, the IR-drop in all the functional units will increase because the current supply demand on the power line is more.

Since IR-drop analysis is a global problem and since power distribution networks are typically very large, a critical issue is the large size of the network. For a state-of-the-art microprocessor, a number of nodes in the power grid is on the order of millions. An accurate IR-drop analysis would simulate the non-linear devices in the chip, together with the non-ideal power grid, making the size of the network even more unmanageable. In order to keep IR-drop analysis computationally feasible, the simulation is done in two steps. First, the non-linear devices are simulated assuming perfect supply voltages, and the power and ground currents drawn by the devices are recorded (these are called *current signatures*). Next, these devices are modeled as independent time-varying current sources for simulating the power grid and the voltage drops at the consumption points (where transistors are connected to power and ground rails) are measured. Since voltage drops are typically less than 10% of the power supply voltage, the error incurred by ignoring the interaction between the device currents and the actual supply voltage is usually small. The linear power and ground network is still very large and hierarchy has to be exploited to reduce the size of the analyzed network. Hierarchy will be discussed in more detail later.

Yet another characteristic of the IR-drop analysis problem is that it is dependent on the activity in the chip, which in turn is dependent on the vectors that are supplied. An important problem in IR-drop analysis is to determine what this input pattern should be. For IR-drop analysis, patterns that produce maximum instantaneous currents are required. This topic has been addressed by a few papers [50–52], but will not be discussed here. However, the fact that vectors are important means that transient analysis of the power grid is required. Since each solution of the network is expensive and since many simulations are necessary, dynamic IR-drop analysis is very expensive. The speed and memory issues related to linear system solution techniques becomes important in the context of transient analysis. An important issue in transient analysis is related to the capacitances (both parasitic and intentional decoupling) in the power grid. Since capacitors prevent instantaneous changes in node voltages, IR-drop analysis without considering capacitors will be more pessimistic. A pessimistic analysis can be done by ignoring all power grid capacitances, but a more accurate analysis with capacitances may require additional computation time for solving the network.

Yet another issue is raised by the vector dependence. As mentioned earlier, the non-linear simulation to determine the currents drawn from the power grid is done separately (from the linear network) using the supplied vectors. Since the number of transistors in the whole chip is huge, simultaneous simulation of the whole chip may be infeasible because of limitations in non-linear transient simulation tools (e.g., SPICE or fast timing simulators). This necessitates partitioning the chip into blocks (typically corresponds to functional units, like floating point unit, integer unit, etc.) and performing the simulation one block at a time. In order to preserve the correlation among the different blocks, the blocks must be simulated with the same underlying set of chip-wide vectors. To determine the vectors for a block, a logic simulation of the chip is done, and the signals at the inputs of the block are monitored and used as inputs for the block simulation.

Since dynamic IR-drop analysis is typically expensive (especially since many vectors are required), techniques to reduce the number of simulations are often used. A commonly used technique is to compress the current signatures from the different clock cycles into a single cycle. The easiest way to accomplish this is to find the maximum envelope of the multi-cycle current signature. To find the maximum envelope over N cycles, the single-cycle current signature is computed using

$$i_{sc}(t) = \max i_{orig}(t + kT), \ \ 1 \le k \le N, \ 0 \le t \le T \tag{63.12}$$

where $i_{sc}(t)$ is the single-cycle, $i_{orig}(t)$ is the original current signature, and T is the clock period. Since this method does not preserve the correlation among different current sources (sinks), it may be overly pessimistic.

A final characteristic of IR-drop analysis is related to the way in which the analysis is typically done. Typically, the analysis is done at the very last stages of the design when the layout of the power network is available. However, IR-drop problems that could be revealed at this stage are very expensive or even impossible to fix. IR-drop analysis that is applicable to all stages of a microprocessor design has been addressed by Dharchoudhury et al. [53].

63.4.2 Power Grid Modeling

The power and ground grids can be extracted by a parasitic extractor to obtain an R-only or an RC network. Extraction implies that the layout of the power grid is available. To insert the transistor current sources at the proper nodes in the power grid, the extractor should preserve the names and locations of transistors. Power grid capacitances come from metal wire capacitances (coupling and grounded), device capacitances, and decoupling capacitors inserted in the power grid to reduce voltage fluctuations. Several interesting issues are raised in the modeling of power grid capacitances. The power or ground net is coupled to other signal nets and since these nets are switching, the effective grounded capacitance is difficult to compute. The same is true for capacitances of MOS devices connected to the power grid. Making the problem worse, the MOS capacitances are voltage dependent. These issues have not been completely addressed as yet. Typically, one resorts to worst-case analysis by ignoring coupling capacitances to signal nets and MOS device capacitances, but considering only the grounded capacitances of the power grid and the decoupling capacitors.

There are three other issues related to power grid modeling. First, for electromigration purposes, via arrays should be extracted as resistance arrays so that current crowding can be modeled. Electromigration problems are primarily seen in the vias and if the via array is modeled as a single resistance, such problems could be masked. Second, the inductance of the package pins also creates a voltage drop in the power grid. This drop is created by the time-varying current in the pins ($v = L\,di/dt$). This effect is typically handled by adding a fixed amount of drop on top of the on-chip IR-drop estimate. Third, a word of caution about network reduction or crunching. Most commercial extraction tools have options to reduce the size of an extracted network. This reduction is typically performed using reduced-order modeling techniques with interconnect delay being the target. This reduction is intended for signal nets and is done so that errors in the interconnect delay is kept below a certain threshold. For IR-drop analysis, such crunching should not be done since we are not interested in the delay. Moreover, during the reduction the nodes at which transistors hook up to the power grid could be removed.

63.4.3 Block Current Signatures

As mentioned above, accurate modeling of the current signatures of the devices that are connected to the power grid is important. At a certain point in the design cycle of a microprocessor, different blocks may be at different stages of completion. This implies that multiple current signature models should be available so that all the blocks in the design can be modeled at various stages in the design [53].

The most accurate model is to provide transient current signatures for all the devices that are connected to the supply or ground grid. This assumes that the transistor-level representation of the entire block is available. The transient current signatures are obtained by transistor-level simulation (typically with a fast transient simulator) with user-specified input vectors. As mentioned earlier, in order to maintain correlation with other blocks, the input vectors for each block must be derived from a common chip-wide input vector set. At the chip-level, the vectors are usually hot loops (i.e., the vectors try to turn on as many blocks as possible). The block-level inputs for the transistor-level simulation are obtained by monitoring the signal values at the block inputs during a logic simulation of the entire chip with the hot loop vectors.

At the other end of the spectrum, the least accurate current model for a block is an area-based dc current signature. This is employed at early stages of analysis when the block design is not complete. The average current consumption per unit area of the block can be computed from the average power consumption specification for the chip and the normal supply voltage value. Since the peak current can be larger than the average current, some multiple of the average per-unit-area current is multiplied by the block area to compute the current consumption for the block.

An intermediate current model can be derived from a full-chip gate-level power estimation tool. Given a set of input vectors, this tool computes the average power consumed by each block over a cycle. From the average power consumption, an average current can be computed for each cycle. Again, to account for the difference between the peak and average currents, the average current can be multiplied by a constant factor. Hence, one obtains a multi-cycle dc current signature for the block in this model.

63.4.4 Matrix Solution Techniques

The large size of power grids places very stringent demands on the linear system solver, making it the most important part of an IR-drop analysis tool. The power grids in typical state-of-the-art microprocessors usually contain multiple layers of metal (processes with up to six layers of metal are currently available) and the grid is usually designed as a mesh. Therefore, the network cannot usually be reduced significantly using a tree-link type of transformation. In older-generation microprocessors, the power network was often "routed" and therefore more amenable to tree-link type reductions. In networks of this type, significant reduction in the size can typically be obtained [54].

In general, matrix solution techniques can be categorized into two major types: direct and iterative [55]. The size and structure of the conductance matrix of the power grid is important in determining the type of linear solution technique that should be used. Typically, the power grid contains millions of nodes, but the conductance matrix is very sparse (typically, less than five entries per row or column of the matrix). Since it is a conductance matrix, the matrix will also be symmetric positive definite — for a purely resistive grid, the conductance matrix may be ill-conditioned.

Iterative solution techniques apply well to sparse systems, but their convergence can be slowed down by ill-conditioning. Convergence can usually be improved by applying pre-conditioners. Another important advantage of iterative methods is that they do not suffer from size limitations as much as direct techniques. Iterative techniques usually need to store the sparse matrix and a few iteration vectors during the solution. The disadvantage of iterative techniques is in transient solution. If constant time steps are used during transient simulation, the conductance matrix remains the same from one time point to another and only the right-hand side vector changes. Iterative techniques depend on the right-hand side and so a fresh solution is required for each time point during transient simulation. The solution from previous time points cannot be reused. The most widely used iterative solution technique for IR-drop analysis is the conjugate gradient solution technique. Typically, a pre-conditioner such as incomplete Cholesky pre-conditioning is also used in conjunction with the conjugate gradient scheme.

Direct techniques rely on first factoring the matrix and then using these factors with the right-hand side vector to find the solution. Since the matrix is symmetric positive definite, one can apply specialized direct techniques such as Cholesky factorization. The main advantage of direct techniques in the context of IR-drop analysis is in transient analysis. As explained earlier, transient simulation with constant time steps will result in the linear solution of a fixed matrix. Direct techniques can factor this matrix once and the factors can be reused with different right-hand side vectors to give some efficiency. The main disadvantage of direct techniques is memory usage to store the factors of the conductance matrix. Although the conductance matrix is sparse, its factors are not and this means that the memory usage will be $O(n^2)$, were n is the size of the matrix.

63.4.5 Exploiting Hierarchy

From the discussions above, it is clear that IR-drop analysis of large microprocessor designs can be limited by size restrictions. The most effective way to reduce the size is to exploit the hierarchy in the design. In this discussion, we will assume a two-level hierarchy consisting of the chip and its constituent blocks. This hierarchy in the blocks also partitions the entire power distribution grid into two parts: the global grid and the intra-block grid. The global grid distributes power from the chip pads to tap points in the various blocks (these are called block ports) and the intra-block grid distributes power from these tap points to the transistors in the block. This partitioning allows us to apply hierarchical analysis. First, the intra-block power grid can be analyzed to find the voltages at the transistor tap points. This analysis assumes that the voltages at the block ports are equal to ideal supply (V_{dd}) or ground (0). The intra-block analysis must also determine a macromodel for the block which is then used for analyzing the global grid. A block admittance macromodel will consist of a current source at each port and an admittance matrix relating the currents and voltages among the ports. The size of the admittance matrix will be equal to the number of ports and each entry will model the effect of the voltage at one port to the current at some other port. In other words, the off-diagonal entries in the admittance matrix will model current redistribution between the ports of the block. Note that, in general, the admittance matrix will be dense and have p^2 entries if p is the number of ports. If n is the number of nodes in the intra-block grid, this block would have contributed a sparse submatrix of size n to the global grid during flat analysis. For hierarchical analysis, this block contributes a dense submatrix of size p. If $p \ll n$, hierarchical analysis will be more efficient than a flat analysis, both in terms of computational time and memory usage.

For exact equivalence with flat analysis, the admittance between every pair of ports must be modeled, resulting in a dense admittance matrix for the block. This will reduce the sparsity of the global conductance matrix and adversely affect solution speed. However, if a block is large, the effective resistance between two ports that are far away will be very large and so the corresponding entry in the admittance matrix can be zeroed with very little loss in accuracy. In fact, the simplest block model will consist of current sources at the ports and a diagonal admittance matrix. For chip-level analysis, the error from this assumption can be kept small if the blocks themselves are small. There is one other source of error in hierarchical analysis and that is the dependence of the block currents on the port voltages. Again, if the voltage drops to the blocks are small (as it will be in a well-designed grid), the error due to this assumption will be small.

References

1. R.B. Hitchcock, G.L. Smith, and D.D. Cheng, Timing analysis of computer hardware, *IBM J. Res. Develop.*, 26(1), 100–105, Jan. 1982.
2. N.P. Jouppi, Timing analysis and performance improvement of MOS VLSI designs, *IEEE Trans. Computer-Aided Design*, 6(4), 650–665, July 1987.
3. K.A. Sakallah, T.N. Mudge, and O.A. Olukotun, check T_c and min T_c: Timing verification and optimal clocking of synchronous digital circuits, *Proc. IEEE Intl. Conf. Computer-Aided Design*, pp. 552–555, Nov. 1990.
4. T. Burks, K.A. Sakallah, and T.N. Mudge, Critical paths in circuits with level-sensitive latches, *IEEE Trans. Very Large Scale Integration Systems*, 3(2), 273–291, June 1995.
5. L.W. Nagel, SPICE 2: A computer program to simulate semiconductor circuits, Technical Report ERL-M520, Univ. of California, Berkeley, May 1975.
6. Y.H. Shih, Y. Leblebici, and S.M. Kang, ILLIADS: A fast timing and reliability simulator for digital MOS circuits, *IEEE Trans. Computer-Aided Design*, pp. 1387–1402, Sept. 1993.
7. A. Devgan and R.A. Rohrer, Adaptively controlled explicit simulation, *IEEE Trans. Computer-Aided Design*, pp. 746–762, June 1994.
8. TimeMill Reference Manual, Epic Design Technology, 1996.
9. Generalized recognition of gates, Bull Worldwide Information Systems, Sept. 1994.
10. N. Weste and K. Eshragian, *Principles of CMOS VLSI Design*, Addison-Wesley, 1990.
11. A. Dharchoudhury, D. Blaauw, J. Norton, S. Pullela, and J. Dunning, Transistor-level sizing and timing verification of domino circuits in the PowerPC™ microprocessor, *Proc. Intl. Conf. Computer Design*, pp. 143–148, 1997.
12. N. Hedenstierna and K.O. Jeppson, CMOS circuit speed and buffer optimization, *IEEE Trans. Computer-Aided Design*, 6(2), 270–281, Mar. 1987.
13. T. Sakurai and A.R. Newton, Alpha-power law MOSFET model and its applications to CMOS inverter delay and other formulas, *IEEE J. Solid-State Circuits*, 25(2), 584–594, April 1990.
14. A.I. Kayssi, K.A. Sakallah, and T.M. Burks, Analytical transient response of CMOS inverters, *IEEE Trans. Circuits. Syst.*, 39(1), 42–45, Jan. 1992.
15. A. Nabavi-Lishi and N.C. Rumin, Inverter models of CMOS gates for supply current and delay evaluation, *IEEE Trans. Computer-Aided Design*, 13(10), 1271–1279, Oct. 1994.
16. J. Rubinstein, P. Penfield, and M.A. Horowitz, Signal delay in RC tree networks, *IEEE Trans. Computer-Aided Design*, 2(3), 202–211, July 1983.
17. J. Cherry, Pearl: A CMOS timing analyzer, *Proc. ACM/IEEE Design Automation Conf.*, pp. 148–153, 1988.
18. W.C. Elmore, The transient response of damped linear networks with particular regard to broadband amplifiers, *J. Applied Physics*, 19(1), 55–63, Jan. 1948.
19. T. Burkes and R.E. Mains, Incorporating signal dependencies into static transistor-level delay calculation, *Proc. TAU 97*, pp. 110–119, Dec. 1997.
20. R. Bryant, Graph-based algorithms for boolean function manipulation, *IEEE Trans. Computers*, 35(8), 677–691, Aug. 1986.
21. M. Desai and Y.T. Yen, A systematic technique for verifying critical path delays in a 300 MHz Alpha CPU design using circuit simulation, *Proc. Design Automation Conf.*, pp. 125–130, 1996.
22. S. Savithri, D. Blaauw, and A. Dharchoudhury, A three tier assertion technique for SPICE verification of transistor-level timing analysis, *Proc. Intl. VLSI'99*, Jan. 1999.
23. H. Bakoglu, *Circuits, Interconnection and Packaging for VLSI*, Addison-Wesley, Reading, MA, 1990.
24. P.R. O'Brien and T.L. Savarino, Modeling the driving point characteristics of resistive interconnect for accurate delay estimation, *Proc. IEEE Intl. Conf. Computer-Aided Design*, pp. 512–515, Nov. 1989.
25. J. Qian, S. Pullela, and L.T. Pillage, Modeling the effective capacitance for the RC interconnect of CMOS gates, *IEEE Trans. Computer-Aided Design*, pp. 1526–1555, Dec. 1994.

26. F. Dartu, N. Menezes, J. Qian, and L.T. Pileggi, A gate-delay model for high-speed CMOS circuits, *Proc. ACM/IEEE Design Automation Conf.*, 1994.

27. L.T. Pillage and R.A. Rohrer, Asymptotic waveform evaluation for timing analysis, *IEEE Trans. Computer-Aided Design*, 9, 352–366, April 1990.

28. J.C. Zhang and M.A. Styblinski, *Yield and Variability Optimization of Integrated Circuits*, Kluwer Academic: Boston, 1995.

29. D.H.C. Du, S.H.C. Yen, and S. Ghanta, On the general false path problem in timing analysis, *Proc. Design Automation Conf.*, pp. 555–560, 1989.

30. P.C. McGeer and R.K. Brayton, Efficient algorithms for computing the longest viable path in a combinational network, *Proc. Design Automation Conf.*, pp. 561–567, 1989.

31. Y. Kukimoto, W. Gost, A. Saldanha, and R. Brayton, Approximate timing analysis of combinational circuits under the XBD0 model, *Proc. ACM/IEEE Conf. Computer-Aided Design*, pp. 176–181, 1997.

32. M. Shoji, *CMOS Digital Circuit Technology*, Prentice-Hall: Englewood Cliffs, NJ, 1988.

33. K.P. Belkhale and A.J. Seuss, Timing analysis with known false sub graphs, *Proc. ACM/IEEE Intl. Conf. Computer-Aided Design*, pp. 736–740, Nov. 1995.

34. D. Blaauw and T. Edwards, Generating false path free timing graphs for circuit optimization, *Proc. TAU99*, March 1999.

35. D.A. Hodges and H.G. Jackson, *Analysis and Design of Digital Integrated Circuits*, McGraw-Hill: New York, NY, 1988.

36. K.L. Sheppard and V. Narayanan, Noise in deep submicron digital design, *Proc. ACM/IEEE Design Automation Conf.*, pp. 524–531, 1996.

37. H.C. Chen, Minimizing chip-level simultaneous switching noise for high-performance microprocessor design, *Proc. IEEE Intl. Symp. Circuits Syst.*, 4, 544–547, 1996.

38. P.K. Su, M.J. Loinaz, S. Masui, and B.A. Wooley, Experimental results and modeling techniques for substrate noise in mixed-signal integrated circuits, *IEEE J. Solid-State Circuits*, 28(4), 420–430, 1993.

39. K.L. Sheppard, V. Narayana, P.C. Elmendorf, and G. Zheng, Global harmony: Coupled noise analysis for full-chip RC interconnect networks, *Proc. Intl. Conf. Computer-Aided Design*, pp. 139–146, 1997.

40. J. Lohstroh, Static and dynamic noise margins of logic circuits, *IEEE J. Solid-State Circuits*, SC-14, 591–598, June 1979.

41. C.L. Ratzlaff, N. Gopal, and L.T. Pillage, RICE: Rapid interconnect circuit evaluator, *IEEE Trans. Computer-Aided Design*, 13(6), 763–776, 1994.

42. P. Feldman and R.W. Freund, Efficient linear circuit analysis by Pade approximation via the Lanczos process, *IEEE Trans. Computer-Aided Design*, 14(5), 639–649, May 1995.

43. L.M. Elfadel and D.D. Ling, Block rational Arnoldi algorithm for multipoint passive model-order reduction of multiport RLC networks, *Proc. IEEE/ACM Intl. Conf. Computer-Aided Design*, pp. 66–71, Nov. 1997.

44. K.J. Kerns, I.L. Wemple, and A.T. Yang, Stable and efficient reduction of substrate model networks using congruence transforms, *Proc. IEEE/ACM Intl. Conf. Computer-Aided Design*, pp. 207–214, 1995.

45. A. Odabasioglu, M. Celik, and L.T. Pileggi, PRIMA: Passive reduced-order interconnect macromodeling algorithm, *Proc. Intl. Conf. Computer-Aided Design*, pp. 58–65, 1997.

46. A. Devgan, Efficient coupled noise estimation for on-chip interconnects, *Proc. IEEE Intl. Conf. Computer-Aided Design*, pp. 147–151, Nov. 1997.

47. F. Dartu and L.T. Pileggi, Calculating worst-case gate delays due to dominant capacitance coupling, *Proc. ACM/IEEE Design Automation Conf.*, pp. 46–51, June 1997.

48. P. Gross, R. Arunachalam, K. Rajgopal, and L.T. Pileggi, Determination of worst-case aggressor alignment for delay calculation, *Proc. Intl. Conf. Computer-Aided Design*, pp. 212–219, Nov. 1998.

49. J.R. Black, Electromigration failure modes in aluminum metalization for semiconductor devices, *Proc. IEEE*, pp. 1587–1594, Sept. 1969.

50. S. Chowdhury and J.S. Barkatullah, Estimation of maximum currents in MOS IC logic circuits, *IEEE Trans. Computer-Aided Design*, 9(6), 642–654, June 1990.

51. H. Kriplani, F. Najm, and I. Hajj, Pattern independent maximum current estimation in power and ground buses of CMOS VLSI circuits, *IEEE Trans. Computer-Aided Design*, 14(8), 998–1012, Aug. 1995.

52. A. Krstic and K.T. Cheng, Vector generation for maximum instantaneous current through supply lines for CMOS circuits, *Proc. ACM/IEEE Design Automation Conf.*, pp. 383–388, 1997.

53. A. Dharchoudhury, R. Panda, D. Blaauw, R. Vaidyanathan, B. Tutuianu, and D. Bearden, Design and Analysis of power distribution networks in Power PC™ microprocessors, *Proc. ACM/IEEE Design Automation Conf.*, pp. 738–743, 1998.

54. D. Stark, Analysis of power supply networks in VLSI circuits, Research Report 91/3, Western Research Lab, Digital Equipment Corp., Apr. 1991.

55. G. Golub and C. Van Loan, *Matrix Computations*, Johns Hopkins Univ. Press: Baltimore, MD, 1989.

64

Microprocessor Design Verification

Vikram Iyengar

IBM Microelectronics

CONTENTS

64.1   Introduction

The task of verifying that a microprocessor implementation conforms to its specification across various levels of design hierarchy is a major part of the microprocessor design process. Design verification is a complex process which involves a number of levels of abstraction (e.g., architectural, RTL, and gate), several aspects of design (e.g., timing, speed, functionality, and power), as well as different design styles [1]. With the high complexity of present-day microprocessors, the percentage of the design cycle time required for verification is often >50%.

The increasing complexity of designs has led to a number of approaches being used for verification. Simulation and formal verification are widely recognized as being at opposite ends of the design verification spectrum, as shown in Figure 64.1 [2]. Simulation is the process of simulating a software model of the design in an environment that models the actual hardware system. The values of internal and output signals are obtained for a given set of inputs and are compared with expected results to determine whether the design is behaving as specified. Formal verification, on the other hand, uses mathematical formulae on an abstracted version of the design to *prove* that the design is correct or that particular aspects of the design are correct.

Formal verification includes equivalence checking, model checking, and theorem proving. Equivalence checking verifies whether one description of a design is functionally equivalent to another. Model checking verifies that specified properties of a design are true, i.e., that certain aspects of the design always work as intended. In theorem proving, the entire design is expressed as a set of mathematical assumptions. Theorems are expressed using these assumptions and are then proven. Formal verification is particularly useful at lower levels of abstraction, e.g., to verify that a gate-level model matches its RTL specification. Formal verification is becoming popular as a means of achieving 100% coverage, at least for specific areas of the design, and is described more fully elsewhere in this book.

There are several problems inherent in applying formal verification to large microprocessor designs. While equivalence checking ensures that no functional errors are inserted from one design iteration to the next, it does not guarantee that the design meets the designer's specifications. Model checking is useful to check consistency with specifications; however, the assertions to be verified must be manually written in most cases. The size of the circuit or state machine that can be formally verified is severely limited owing to the problem of state-space explosion. Finally, formal techniques cannot be used for performance validation because timing-dependent circuits such as oscillators rely on analog behavior that is not handled by mathematical representations.

Simulation is therefore the primary commercial design verification methodology in use, especially for large microprocessor designs. Simulation is performed at various levels in the design hierarchy, including at the register transfer, gate, transistor, and electrical levels, and is used for both functional verification and performance analysis. Timing simulation is becoming critical for ultradeep submicron designs because the problems of power grid IR drops, interconnect delays, clock skews, and signal electromigration intensify with shrinking process geometries and affect circuit performance adversely [3]. Timing verification involves performing 2D or 3D parasitic RC extraction on the layout followed by back-annotating the capacitance values obtained onto the netlist. A combination of static and dynamic timing analysis is performed to find critical paths in the circuit. Static analysis involves analyzing delays using a structural model of the circuit, while dynamic analysis uses vectors to simulate the design to locate

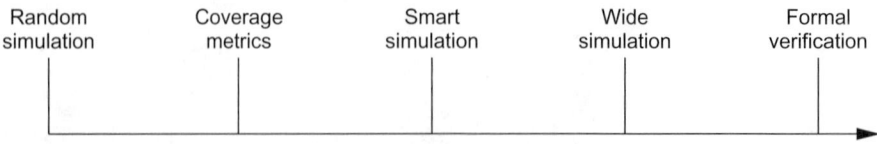

FIGURE 64.1   The spectrum of design verification techniques, which range from simulation to formal verification (from Dill, D.A., *Proc. Design Automation Conference*, p. 839, 1996).

critical paths [3]. Accurate measurements of the critical path delays can then be obtained using SPICE. Techniques for timing verification are described elsewhere in this book.

Pseudorandom vector generation is the most popular form of generating instruction sequences for functional simulation. Random test generators provide the ability to generate test programs that lead to multiple simultaneous events, which would be extremely time consuming to write by hand [4]. Furthermore, the amount of computation required to generate random instruction sequences is low. However, random simulation often requires a very long time to achieve a suitable level of confidence in the design. This has given rise to the use of a number of semiformal metrics to estimate and improve simulation coverage. These methods combine the advantages of simulation and formal verification to achieve a higher coverage, while avoiding the scaling and methodology problems inherent in formal verification. In this chapter, we focus on the tools and techniques used to generate instruction sequences for a simulation-based verification environment.

The chapter is organized as follows. We begin with a description of the design verification environment in Section 64.2. Random and biased-random instruction generation, which lie at the simulation end of the spectrum, are discussed in Section 64.3. Section 64.4 describes three popular correctness checking methods that are used to determine the success or failure of a simulation run. Coverage metrics, which are used to estimate simulation coverage, are presented in Section 64.5. In Section 64.6, we move to the middle of the design verification spectrum and discuss *smart* simulation which is used to generate vectors satisfying semiformal metrics. *Wide* simulation, which refers to the use of formal assertions to derive vectors for simulation is described in Section 64.7. Having covered the spectrum of semiformal verification methods, we conclude with a description of hardware emulation in Section 64.8. Emulation uses dynamically configured hardware to implement a design, which can be simulated at high speeds.

64.2 Design Verification Environment

In this section, we present a design verification environment that is representative of many commercial verification methodologies. This environment is illustrated in Figure 64.2, and the typical levels of design abstraction are shown in Figure 64.3. We describe the different parts of the environment and the role each part plays in the verification process.

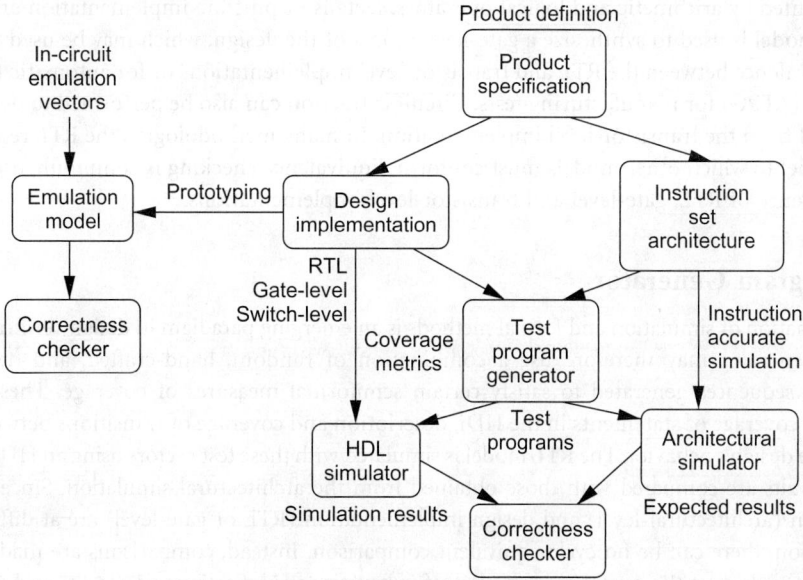

FIGURE 64.2 A representative design verification environment and verification process flow.

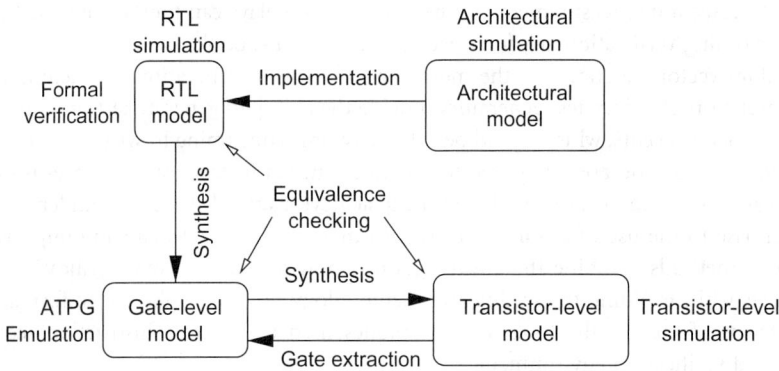

FIGURE 64.3 Different levels of design abstraction.

Architectural Model

A high-level specification of the microprocessor is first derived from the product requirements and from the requirement of compatibility with previous generations. An architectural simulation model and an RTL model are then implemented based on the product specification. The architectural model, often written in C or C++, includes the programmer-visible registers and the capability to simulate the execution of an instruction sequence. This model emphasizes simulation speed and correctness over implementation detail and therefore does not represent pipeline stages, parallel functional units, or caches. This model is instruction accurate, but not clock cycle accurate [1]. A typical architectural model executes over 100 times faster than a detailed RTL model [4].

RTL Model

The RTL model, implemented in a hardware description language (HDL) such as VHDL or Verilog, is more detailed than the architectural model. Data is stored in register variables and transformations are represented by arithmetic and logical operators. Details of pipeline implementation are included. The RTL model is used to synthesize a gate-level model of the design, which may be used to formally verify equivalence between the RTL and transistor-level implementations or for automatic test pattern generation (ATPG) for manufacturing tests. Circuit extraction can also be performed to derive a gate-level model from the transistor-level implementation. In many methodologies, the RTL represents the *golden* model to which other models must conform. Equivalence checking is commonly used to verify the equivalence of RTL, gate-level and transistor-level implementations.

Test Program Generator

The combination of simulation and formal methods is an emerging paradigm in design verification. A test program generator may therefore use a combination of random, hand-crafted, and deterministic instruction sequences generated to satisfy certain semiformal measures of coverage. These measures include the coverage of statements in the HDL description and coverage of transitions between control states in the design's behavior. The RTL model is simulated with these test vectors using an HDL simulator and the results are compared with those obtained from the architectural simulation. Since the design specification (architectural-level) and design implementation (RTL or gate-level) are at different levels of abstraction, there can be no cycle-equivalent comparison. Instead, comparisons are made at special checkpoints such as at the completion of a set of instructions [5]. Sections 64.3, 64.6, and 64.7 discuss the most popular techniques used for test generation.

HDL Simulator

HDL simulation consists of two stages. In the *compile* stage, the design is checked for errors in syntax or semantics and is converted into an intermediate representation. The design representation is then reduced to a collection of signals and processes. In the *execute* stage, the model is simulated by initializing values on signals and executing the sequential statements belonging to the various processes. This can be achieved in two ways: *event-driven* simulation and *cycle-based* simulation. Event-driven simulation is based on determining changes (events) in the value of each signal in a clock cycle and may incorporate various timing models. A process is first simulated by assigning a change in value to one or more of its inputs. The process is then executed, and new values for other signals are calculated. If an event occurs on another signal, other processes which are sensitive to that signal are executed. Events are processed in the order of the time at which they are expected to occur according to the timing model used. In this manner, all events occurring in a clock cycle are calculated. Cycle-based simulators, in contrast, limit calculations by determining simulation results only at clock edges and ignoring interphase timing. Cycle-based simulators focus only on the design functionality by performing zero-delay, two-valued simulation (memory elements are assumed to be initialized to known values) and they offer an improvement in speed of up to $10\times$ while utilizing one-fifth of the memory required for event-driven simulation. However, cycle-based simulators are inefficient in verifying asynchronous designs, and event-driven simulators must be used to derive initializing sequences and for timing calculations. Simulation techniques used at various levels of design abstraction are discussed more fully in this book.

Emulation Model

Hardware emulation is a means of embedding a dynamically configured prototype of the design in its final environment. This hardware prototype, known as the emulation model, is derived from the gate-level implementation of the design. The prototype can execute both random vectors and software application programs faster than conventional software logic simulators. It is also connected to a hardware environment, known as an *in-circuit* facility, to provide it with a high throughput of test vectors at appropriate speeds. Hardware emulation executes from three to six orders of magnitude faster than simulation and subsequently requires considerably less verification time. However, hardware emulators have limitations on the sizes of the circuits they can handle.

Table 64.1 presents the results of a survey conducted by 0-In Design Automation on verification techniques currently used in industry [6]. Columns 1 and 3 in the table list the different techniques, while columns 2 and 4 show the percentage of surveyed engineers currently using a particular approach. While formal methods are becoming popular as a means to more exhaustively cover the design, pseudorandom simulation is still a vital part of the verification engineer's repertoire. In Section 64.3 we review some conventional verification techniques that use pseudorandom and biased-random test programs for simulation.

TABLE 64.1 0-In Bug Survey Results

Stimulus Techniques	Percentage Used (%)	Advanced Verification Techniques	Percentage Used (%)
System stimulation	94	Cycle-based simulation	25
Directed tests	89	Equivalence checking	19
Regression tests	88	Hardware/software	15
Pseudorandom	82	co-design	
Prototype silicon	58	Model checking	13
Emulation	49		

Note: Percentages of various validation techniques used by design verification engineers (May 1997–May 1998).

Source: 0-In Design Automation, Bug survey results, http://www.0-In.com.

64.3   Random and Biased-Random Instruction Generation

Random vector simulation is the primary verification methodology used for microprocessors today. New designs as well as changes made to existing designs are subjected to a battery of simulation and regression tests involving billions of pseudorandom vectors before focused testing is performed. Random test generation, also known as black-box testing, produces more complex combinations of instructions than can be manually written by the design verification engineer. A large number of test programs are generated randomly. Each test program consists of a set of register and memory initializations and a sequence of instructions. It may also contain the expected contents of the registers and memory after execution of the instructions, depending on the implementation. The expected contents of the registers and memory are obtained using an architectural model of the design. The test programs are translated to assembler or machine-language code that is supported by the HDL simulator and are simulated on the RTL model. However, purely random test programs are not ideal because the instruction sequences developed may not exercise a sufficient number of corner cases; thus, millions of vectors and days of simulation are required before reasonable levels of coverage can be achieved. In addition, random vectors may violate constraints on memory addressing, thus causing invalid instruction execution.

64.3.1   Biased-Random Testing

Biasing is the manipulation of the probability of selecting instructions and operands during instruction generation. Biased-random instruction generation is used to create test programs that have a higher probability of leading to execution hazards for the processor. For example, the biasing scheme in Ref. [7] utilizes knowledge of the Alpha 21264 architecture to favor the generation of instructions that test architecture-specific corner cases, specifically those affecting control-flow, out-of-order processing, superscalar structures, cache transactions, and illegal instructions.

Constraint solving, another biasing technique, identifies output conditions or intermediate values that are important to verify Ref. [8]. The instruction generator identifies input values that would lead to these conditions and generates instructions that utilize these "biased" input values. Constraint solving is useful because it improves the probability of exercising certain corner cases. Both of these schemes have biases hard-coded into the test generation algorithm based on the instruction type.

64.3.2   Static and Dynamic Biasing

Biasing can be classified as being either static or dynamic. Static biasing of test vectors involves randomly initializing the registers and memory, generating the biased-random test program and applying it to the architectural and RTL models, e.g., the RIS tool from Motorola [9]. A major complication of this method is that the test generator must construct a test that does not violate the acceptable ranges for data and memory addresses. The solution to this problem is to constrain biasing within a restricted set of choices that define a constrained model of the environment, e.g., to reserve certain registers for indexed addressing [1].

Dynamically biased test generators use knowledge of the current processor states, memory state, and user bias preferences to generate more effective test programs. In dynamic instruction generation, the states of the programmer model in the test generator are updated to reflect the execution of the instruction after each instance of instruction generation [8,10]. The test generator interacts with a tightly coupled functional model of the design to update current state information.

Drawbacks of random and biased-random testing include the vast amount of simulation time required to achieve acceptable levels of coverage and the lack of effective biasing methodologies. Determining when an acceptable level of coverage has been achieved is a major concern. Semiformal verification techniques have therefore become popular as a means to monitor simulation coverage as well as improve coverage by generating vectors to cover test cases that have not been exercised by random simulation.

In Section 64.4, we discuss several correctness checking techniques that are used to determine whether the simulation test was successful. Later, in Section 64.5, we review some of the common metrics used to evaluate the coverage of test programs.

64.4 Correctness Checking

Correctness checking is the process of isolating a design error by determining whether the simulation test was successful. In this section, we discuss three techniques for correctness checking: self-checking, reference model comparison, and assertion checking. The three methods are complementary and are often used in conjunction to achieve the highest coverage. Figure 64.4 illustrates the three correctness checkers in the verification flow of the Alpha 21164 microprocessor [4].

64.4.1 Self-checking

Self-checking is the simplest way to determine success for focused, hand-coded tests. The test program sets up a combination of conditions and then checks to see if the RTL model reacted correctly to the simulated situation [11]. However, this approach is time consuming, prone to error, and intrusive at the register transfer level. The test generator may be required to maintain an extensive amount of state. Furthermore, the technique is often not useful beyond a single focused test.

64.4.2 Reference Model Comparison

An alternative to self-checking is to compare the traces generated by the RTL model with the simulation traces of an architectural reference model, as illustrated in Figure 64.2. This technique, known as reference model comparison, obviates the need for constantly checking the state of the processor being simulated. The reference model is an abstraction of the design architecture written in a high-level language such as C++. It represents all features visible to software, including the instruction set and support for memory and I/O space [4].

Several correctness checks may be performed using the reference model, of which the simplest is *end-of-state* comparison. When simulation completes, the contents of memory locations are accessed, and the final state of the register files are compared. However, end-of-state comparison is not very useful for lengthy simulations because it may be difficult to identify incorrect intermediate results, which are overwritten during simulation. Comparing intermediate results during simulation is a solution; however, this requires the reference model to match the timing of the RTL model, and is not easily implemented. Additional comparisons that can be made include checking the PC flow and checking writes to integer and floating-point registers. Incorrect values here will signal problems with control-flow and data manipulation instructions.

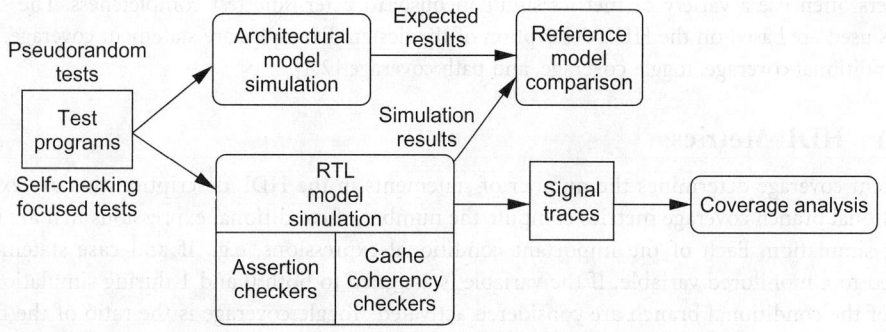

FIGURE 64.4 Verification flow and correctness checking for the Alpha 21164 microprocessor (from Kantrowitz, M. and Noack, L.M., *Process Design Automation Conference*, p. 325, 1996).

TABLE 64.2 Effectiveness of Correctness Checks Used in the Verification of the Alpha 21264 Microprocessor

Origin of Bug	Bugs Introduced[a] (%)	Correctness Checker	Bugs Detected[a] (%)
Implementation error	78	Assertion checker	25
Programming mistake	9	Register miscompare	22
Matching model to schematics	5	Simulation "no progress"	15
		PC miscompare	14
Architectural conception	3	Memory state miscompare	8
Other	5	Manual inspection	6
		Self-checking test	5
		Cache coherency check	3
		SAVES check	2

[a]Percentage of total design errors.
Source: Taylor, S. et al., *Proc. Design Automation Conference*, p. 638, 1998.

64.4.3 Assertion Checking

Assertion checking, another popular means to check correctness, is the process of adding segments of code to the RTL model to verify that certain properties of design behavior always hold true under simulation. Examples of simple assertion checking include monitoring illegal states and invalid transitions. More complex checking involves monitoring queue overruns and violation of the bus protocol [7]. An example of a specialized assertion checker is the cache coherency checker used in the verification of the Alpha 21164 microprocessor [4]. The system supports three levels of caching, with the second- and third-level caches being write-back. Cache coherency checking was activated at regular intervals during simulation to ensure that coherency rules were not violated. Table 64.2 presents the origins of bugs introduced into the design and the percentages of bugs detected by the various correctness checking mechanisms for the Alpha 21264 microprocessor [7]. Assertion checkers were the most successful; however, when viewed collectively, 44% of errors were found by reference model comparison.

With the correctness checking problem examined, the next major issue in simulation-based verification is determining whether acceptable levels of coverage have been achieved by the simulation vectors. In the next section, we look at several techniques for coverage analysis.

64.5 Coverage Metrics

Coverage analysis provides information on how thoroughly a design has been exercised during simulation. Coverage metrics are used to evaluate the effectiveness of random simulation vectors as well as guide the generation of deterministic tests. A number of coverage metrics have been proposed and verification engineers often use a variety of metrics simultaneously to determine test completeness. The simplest metrics used are based on the HDL description of the design. Examples are statement coverage, branch and conditional coverage, toggle coverage, and path coverage [2,4].

64.5.1 HDL Metrics

Statement coverage determines the number of statements in the HDL description that are executed. Conditional branch coverage metrics compute the number of conditional expressions that are toggled during simulation. Each of the important conditional expressions, e.g., **if** and **case** statements, is assigned to a monitored variable. If the variable is assigned to both 0 and 1 during simulation, both paths of the conditional branch are considered activated. Toggle coverage is the ratio of the number of signals that experienced 1-to-0 and 0-to-1 transitions during simulation to the total number of effective signals. The number of effective signals is adjusted to include only those that can possibly be toggled in the fault-free model. Another recently proposed HDL metric is based on error observability at the primary outputs of the design [12]. Observability is computed by tagging variables during simulation and checking whether the tags are propagated to the outputs. A tag calculus, similar to that of the D-algorithm used for ATPG, is introduced. Coverage is measured as the percentage of tags visible at the design outputs. The method provides a stricter measure of coverage than does HDL-line coverage. However, while HDL-based metrics are useful, they are generally not effective measures of whether logic is being functionally exercised.

64.5.2 Manufacturing Fault Models

A second class of coverage metrics is based on manufacturing fault models [13,14]. These metrics characterize a class of design errors analogous to faults in hardware testing and measure coverage through fault simulation. Logic design errors, such as gate substitution, missing gates, and extra inverters are injected randomly into the design. The design is then simulated using a stuck-at fault simulator. This approach has been used for measuring the coverage of ATPG tests for embedded arrays in microprocessors. ATPG is the process of automatically generating test patterns for manufacturing tests and is typically performed at the gate level. The problems with using manufacturing fault models to estimate coverage are that fault simulation is often computationally intensive, and faults introduced during the manufacturing process do not always model design errors.

64.5.3 Sequence and Case Analysis

Other metrics widely used in industry include sequence analysis, occurrence analysis, and case analysis [4,7]. Sequence analysis monitors sequences of events during simulation, e.g., request-acknowledge sequences, interrupt assertions, and traps. Occurrence analysis determines the presence of one-time events such as a carry-out being generated for every bit in an adder. The absence of such an event could signal a failure. Finally, case analysis consists of collecting and studying simulation statistics such as exerciser type, cycles simulated, issue rate, and instruction types [7].

64.5.4 State Machine Coverage

A more formal way to evaluate coverage is to look at an abstraction of the design in the form of a finite state machine (FSM) [5,15–17]. The control logic of the design is extracted as an FSM, which has a smaller state space than the original design, but which exhibits the design's control behavior. Coverage is typically estimated by the fraction of different states reached by the FSM or the fraction of state transitions exercised during simulation. FSM coverage metrics are also used to generate test programs with high coverage, as described in Section 64.6. Binary decision diagrams (BDDs), borrowed from formal verification, are used to describe and traverse the implementation state space. A BDD is a way of efficiently representing a set of binary-valued decisions (scalar boolean functions) in the form of a tree or directed acyclic graph.

A method of transforming the high-level description of the circuit into a reduced FSM that has far fewer states than the original design is proposed in Ref. [12]. Simulation coverage is estimated by relating the fraction of transitions in the state graph traversed by this reduced FSM to the number of HDL statements exercised in the high-level description.

More recently, Moundanos et al. [17] describe the extraction of the control logic of the design from its HDL description. The control logic is extracted in the form of an FSM, which represents the control space of the entire circuit. The vectors whose coverage is to be evaluated are simulated on the FSM. Simulation coverage is estimated by the following two ratios:

$$\text{State coverage metric (SCM)} = \frac{\text{Number of states visited}}{\text{Total number of reachable states}}$$

$$\text{Transition coverage metric (TCM)} = \frac{\text{Number of transitions traversed}}{\text{Total number of reachable transitions}}$$

A similar approach to evaluating coverage is used in Ref. [16]. Since only a subset of state variables directly controls the datapath, the noncontrolling independent state variables are removed from the state graph of the FSM. This reduced state graph is called a *control event* graph, and each reachable state is a control event. Coverage is evaluated in terms of the number of control events visited and the number of transitions exercised in the control event graph.

Further along the spectrum toward formal verification lie techniques dubbed smart simulation [2], which not only evaluate coverage of the given vectors, but also *generate* new functional tests using coverage metrics. In Section 64.6, we discuss several such techniques that use semiformal coverage metrics to derive simulation vectors. We begin with techniques based on identifying hazard patterns and later discuss more formal methods that use state machine coverage.

64.6   Smart Simulation

Deterministic or smart simulation uses vectors that cover a certain aspect of the design's behavior using details of its implementation. We first describe ad hoc techniques such as hazard-pattern enumeration, which target specific blocks in the processor, and then describe more general techniques aimed at verifying the entire chip.

64.6.1   Hazard Pattern Enumeration

Ad hoc techniques typically target a specific block in the design such as a pipeline [18,19] or cache controller [20]. Errors in the pipeline mechanism represent only a small fraction of the total errors. In a study undertaken in Ref. [19], it was shown that only 2.79% of the total errors in a commercial 32-bit CPU design was related to the pipeline interlock controller. However, these errors are widely acknowledged as being the hardest to detect and are therefore targeted by ad hoc methods.

Pipeline hazards are situations that prevent the next instruction from executing in its designated clock cycle. These are classified as structural hazards, data hazards, and control hazards. Structural hazards occur when two instructions in different pipeline stages attempt to access the same physical resource simultaneously. Data hazards are of three types: read-after-write (RAW), write-after-write (WAW), and write-after-read (WAR) hazards. The most common are RAW hazards, in which the second instruction attempts to read the result of the first instruction before it is written. Control hazards are treated as RAW hazards in the program counter (PC). An algorithm that enumerates all the structural, data, and control hazard patterns for each common resource in the pipeline is presented in Ref. [18]. Test programs that include all the patterns that can cause the pipeline to stall are then generated.

Lee and Siewiorek [19] define the set of state variables read by an instruction as its *read state* and the set written by the instruction as its *write state*. A conflict exists between two instructions if at least one of them is a write and the intersection of their read/write or write/write states is not empty. A dependency graph is constructed with nodes representing all the possible read/write instructions and edges (or dependency arcs) representing conflicts between instructions. Test programs are generated to cover all the dependency arcs in the graph and the dependency arc coverage is calculated.

In Ref. [20], a cache controller is verified using a model of the memory hierarchy, a set of cache coherence protocols, and enumeration capabilities to generate test programs for the design.

The problem inherent in ad hoc techniques is that pipeline behavior after the detection of a hazard is usually not considered [21]. Test cases reachable only after a hazard has occurred are therefore not covered. We next discuss more general test generation techniques, which are applicable to a larger part of the design.

64.6.2   ATPG

An important class of verification techniques is based on the use of test programs generated by ATPG tools. Coverage is measured as the fraction of design errors detected. These methods have been used in industry to verify the equivalence between the gate- and transistor-level models, e.g., in the verification of PowerPC™ arrays [13,14]. In this approach, a gate-level model is created from the transistor-level implementation, and tests generated at the gate-level view are simulated at the transistor level to verify equivalence. However, these techniques, while effective at lower levels of abstraction, do not provide a good measure of the extent to which the design has been exercised.

64.6.3   State and Transition Traversal

Tests generated by traversing the design's state space work on the principle that verification will be close to complete if the processor either visits all the states or exercises all the transitions of its state graph during simulation [15,17,22]. Since memory limitations make it impossible to examine the state graph of the entire design, the design behavior is usually abstracted in the form of a reduced state graph. Test sequences are then generated which cause this reduced FSM to exercise all the transitions. Figure 64.5 illustrates the verification flow for this technique. The first step is to extract the control logic of the design in the form of an FSM. The datapath is usually not considered because most designs have datapaths of substantial sizes, which can lead to an unmanageable state space. Furthermore, errors in the datapath usually result from incorrect implementation, not incorrect design, and can be easily tested by conventional simulation [17].

   A method to extract the control logic of the design in the form of an FSM can be found in Ref. [17]. This is illustrated in Figure 64.6. The data variables in the design are made nondeterministic by including them in the set of primary inputs to the FSM. Since the datapath is to be excluded from consideration, the inputs to the data variables are excluded. This is represented by the dotted lines in Figure 64.6. The support set of the primary outputs and control state variables is now determined in terms of the primary inputs, control state variables, and data variables. This support set forms the new set of primary inputs to the FSM. Data variables that are not a part of the support set are excluded from the FSM. In this manner, the effect of the data variables on the control flow is taken into account, even though the data registers are abstracted.

   After the FSM has been extracted, state enumeration is performed to determine the reachable states and a state graph which details the behavior of the FSM is generated. Since coverage is typically evaluated by the number of states visited or the number of transitions exercised, a *state* or *transition tour* of the state graph is found. A state (transition) tour of a directed state graph is a sequence of transitions that traverses every state (transition) at least once. Several polynomial-time algorithms have been developed for finding transition tours in nonsymmetric, strongly connected graphs, since this problem (the Chinese Postman problem) is frequently encountered in protocol conformance testing [22]. The transition tour is translated into an instruction sequence that will cause the FSM to exercise all transitions.

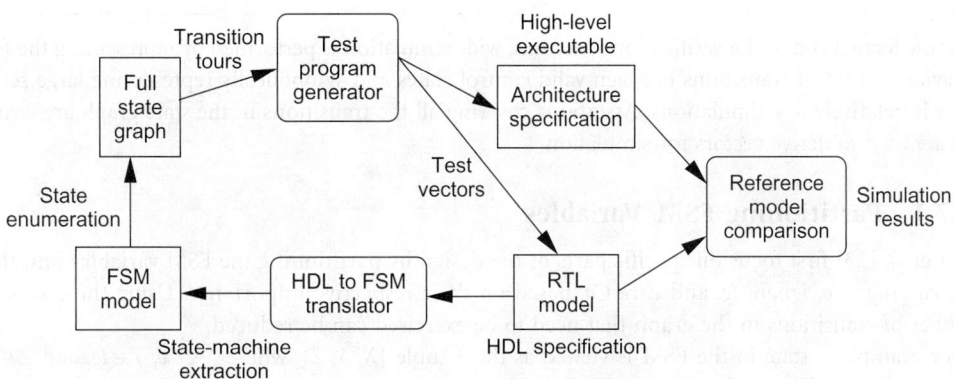

FIGURE 64.5   Verification flow for a representative state machine traversal technique (from Ho, R.C. et al., *International Symposium on Computer Architecture*, p. 404, 1995).

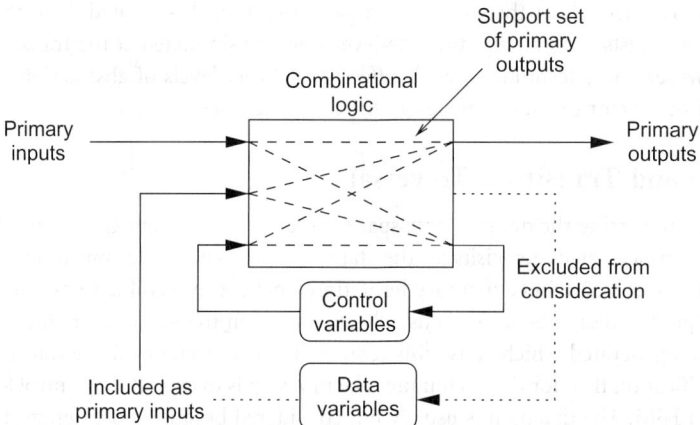

FIGURE 64.6 Extraction of the control flow machine (from Moundanos, D. et al., *IEEE Transactions on Computers*, vol. 47, no. 1, p. 2, 1998).

Cheng and Krishnakumar [15] use exhaustive coverage of the reduced FSM to generate test programs guaranteeing that all statements in the original HDL representation are exercised. A test generation technique based on visiting all states in the state graph is presented in Ref. [21]. Test cases are developed based on enumerating hazard patterns in the pipeline and are translated into sequences of states in the state graph. Simulation vectors that satisfy all test cases are generated. A more general transition-traversal technique is given in Ref. [22]. A translator is used to convert the HDL representation into a set of interacting FSMs. A full-state enumeration of the FSMs is performed to find all reachable states from reset. This produces a complete state graph, which is used to generate vectors that cause the processor to take a transition tour.

Finally, several classes of processors for which transition coverage is effective are identified in Ref. [5]. The authors demonstrate that under a given set of conditions, transition tours of the state graph can be used to completely validate a large class of processors.

State-space explosion is currently a key problem in computing state machine coverage. As designs get larger and considerably more complex, the maximum size of the state machine that can be handled is the major limiting factor in the use of formal methods. However, research is currently being undertaken to deal with state explosion, and we foresee an increasing use of formal coverage metrics in the future.

64.7 Wide Simulation

Near the formal end of the verification spectrum, wide simulation is performed by representing the FSM behavior as a set of transitions between valid control states and symbolically representing large sets of states in relatively few simulations. Assertions covering all the transitions in the state graph are written and are used to derive vectors for simulation.

64.7.1 Partitioning FSM Variables

Geist et al. [23] first focus on specific parts of the design by partitioning the FSM variables into three sets, *coverage Co*, *ignore Ig*, and *care Ca*, based on their respective importance. Using these sets, the number of transitions in the graph that need to be exercised can be reduced.

For example, a state in the FSM is viewed as the 3-tuple $\{X, Y, Z\}$, where $X \in Co$, $Y \in Ig$, and $Z \in Ca$. Two transitions, $T_1((X_1, Y_1, Z_1), (X_2, Y_2, Z_2))$ and $T_2((X_3, Y_3, Z_3), (X_4, Y_4, Z_4))$, which differ in the value of a coverage variable, are distinct and require separate tests; e.g., if $X_1 \neq X_3$ or $X_2 \neq X_4$, then T_1 and T_2 require different tests. However, two transitions that differ only in the value of an ignored variable are equivalent. Therefore, if $X_1 = X_3$, $X_2 = X_4$, $Z_1 = Z_3$ and $Z_2 = Z_4$, then T_1 and T_2 are equivalent and a vector that tests T_1 will also test T_2. Finally, two transitions that differ in the value of a care variable do not necessarily require different tests [23].

In this manner, the state graph is represented as the set of all valid transitions T, of which only a few must be exercised, based on the equivalence relations. Next, formal assertions are written for each transition. An assertion is a temporal logic expression of the form *antecedent* $\rightarrow$ *consequent*, where both antecedent and consequent can consist of complex logical expressions [13]. The first step in the test generation process is to choose a valid transition $T_1(v, v') \in T$ and write an assertion of the form $state(v) \rightarrow next(-state(v'))$, which means that if the FSM is in state v, then the next state cannot be v'.

64.7.2 Deriving Simulation Tests from Assertions

A model checker may be used to generate sequences of input vectors which satisfies the assertion [23]. A model checker is a formal verification tool that is used either to prove that a certain property is satisfied by the system or to generate a counterexample to show that the property does not always hold true. The model checker reports that the assertion $state(v) \rightarrow next(-state(v'))$ is false and that the transition is indeed valid. The model checker also outputs a symbolic sequence of states and input patterns which lead to state v. This symbolic (high-level) sequence of patterns is then translated into a simulation vector sequence and is used to verify the design. The transition T_1 and all transitions equivalent to T_1 are removed from T, and the process is repeated [23].

Wang and Abadir [13,14] use tools to automatically generate formal assertions for PowerPC™ arrays from the RTL model. Symbolic trajectory evaluation, a descendant of symbolic simulation, is used to formally prove that all assertions are true. After the design has been formally verified, simulation vectors are derived from the assertions and are used for simulating the design. The methods used to derive these vectors are as follows. The symbolic values used in the antecedent of each assertion are replaced with a set of vectors based on each condition specified in the consequent. First, symbolic address comparison expressions are replaced with address marching sequences, e.g., to test large decoders. Next, symbolic data comparison expressions are replaced with data marching sequences, e.g., in testing comparators. Stand-alone symbolic values representing sets of states or input patterns are replaced with random vectors. Assertion decision trees are constructed and tests are generated to cover all branches, e.g., in testing control logic. Finally, control signal decision trees are constructed in order to generate tests that cover abnormal functional space [13].

Finally, we have reached the *formal* end of our discussion on verification techniques, which range from random simulation to semiformal verification. Formal verification, which uses mathematical formulae to prove correctness, is described by Levitan in Chapter 11. In Section 64.8, we describe emulation, which is a means to implement a design using programmable hardware, with performance several orders of magnitude faster than conventional software simulators. Emulation has become popular as a means to test a processor against real-world application programs, which are impossibly slow to run using simulation.

64.8 Emulation

The fundamental difference between simulation and emulation is that simulation models the design in software on a general-purpose host computer, while emulation actually implements the design using dynamically configured hardware. Emulation, performed in addition to simulation has several advantages. It provides up to six orders of magnitude improvement in execution performance and enables several tests that are too complex to simulate to be performed prior to tapeout. These include power-on-self-tests, operating system boots, and running software applications, e.g., OpenWindows [24]. Finally, emulation reduces the number of silicon iterations that are needed to arrive at the final design, because errors caught by emulation can be corrected before committing the design to silicon.

64.8.1 Preconfiguration

The emulation process consists of four major phases: preconfiguration, configuration, testbed preparation, and in-circuit emulation (ICE) [24]. In the preconfiguration phase, the different components of the design are assembled and converted into a representation that is supported by the emulation vendor. For example, in the K5 emulation, each custom library cell was expressed in terms of primitives that could be mapped to a field-programmable gate array (FPGA) [25]. An FPGA is a simple programmable logic device that allows users to implement multilevel logic. Several thousand FPGAs must be connected together to prototype a complex microprocessor. Once the cell libraries have been translated, the various gate-level netlists are converted to a format acceptable to the configuration software. This can be complicated because the netlists obtained from standard cell and datapath designers are often in a variety of formats [24].

There is often no FPGA equivalent for complex transistor-level megacells, which are commonly used in full custom processors. Gate-level emulation models for megacells must therefore be created. These gate-level blocks are implemented in the programmable hardware and are verified against simulation vectors to ensure that each module performs correctly according to the simulation model.

64.8.2 Full Chip Configuration

In this phase, the design netlists and libraries are combined with control and specification files and downloaded to program the emulation hardware. In the first stage of configuration, the netlists are parsed for semantic analysis and logic optimization [24]. The design is then partitioned into a number of logic board modules (LBMs) to satisfy the logic and pin constraints of each LBM. The logic assigned to each LBM is flattened, checked for timing and connectivity, and further partitioned into clusters to allow the mapping of each cluster to an individual FPGA [25]. Finally, the interconnections between the LBMs are established, and the design is downloaded to the emulator.

64.8.3 Testbed and In-Circuit Emulation

The testbed is the hardware environment in which the design to be emulated will finally operate. This consists of the target ICE board, logic analyzer, and supporting laboratory equipment [24]. The target ICE board contains PROM sockets, I/O ports, and headers for the logic analyzer probes.

Verification takes place in two modes, the simulation mode and ICE. In the simulation mode, the emulator is operated as a fast simulator. Software is used to simulate the bus master and other hardware devices and the entire simulation test suite is run to validate the emulation model [25]. An external monitor and logic analyzer are used to study results at internal nodes and determine success. In the ICE mode, the emulator pins are connected to the actual hardware (application) environment. Initially, diagnostic tests are run to verify the hardware interface. Finally, application software provides the emulation model with billions of vectors for high-speed functional verification.

In Section 64.9, we conclude our discussion on design verification and review some of the areas of current research.

64.9 Conclusions

Microprocessor design teams use a combination of simulation and formal verification to verify presilicon designs. Simulation is the primary verification methodology in use, since formal methods are applicable mainly to well-defined parts of the RTL or gate-level implementation. The key problem in using formal verification for large designs is the unmanageable state space.

Simulation typically involves the application of a large number of pseudorandom or biased-random vectors in the expectation of exercising a large portion of the design's functionality. However, random instruction generation does not always lead to certain highly improbable (corner case) sequences, which are the most likely to cause hazards during execution. This has led to the use of a number of semiformal methods, which use knowledge derived from formal verification techniques to cover more fully the design behavior. For example, techniques based on HDL statement coverage ensure that all statements in the HDL representation of the design are executed at least once. At a more formal level, a state graph of the design's functionality is extracted from the HDL description, and formal techniques are used to derive test sequences that exercise all transitions between control states. Finally, formal methods based on the use of temporal logic assertions and symbolic simulation can be used to automatically generate simulation vectors. We next describe some current directions of research in verification.

64.9.1 Performance Validation

With an increasing sophistication in the art of functional validation, ensuring the lack of performance bugs in microprocessors has become the next focus of verification. The fundamental hurdle to automating performance validation for microprocessors is the lack of formalism in the specification of error-free pipeline execution semantics [26]. Current validation techniques rely on focused, handwritten test cases with expert inspection of the output. In Ref. [26], analytical models are used to generate a controlled class of test sequences with *golden* signatures. These are used to test for defects in latency, bandwidth, and resource size implementations coded into the processor model. However, increasing the coverage to include complex, context-sensitive parameter faults and generating more elaborate tests to cover the cache hierarchy and pipeline paths remain open problems.

64.9.2 Design for Verification

Design for verification (DFV) is the new buzzword in microprocessor verification today. With the costs of verification becoming prohibitive, verification engineers are increasingly looking to designers for *easy-to-verify* designs. One way to accomplish DFV is to borrow ideas from design for testability (DFT), which is commonly used to make manufacturing testing easier. Partitioning the design into a number of modules and verifying each module separately is one such popular DFT technique. DFV can also be accomplished by adding extra modes to the design behavior to suppress features such as out-of-order execution during simulation. Finally, a formal level of abstraction, which expresses the micro-architecture in a formal language that is amenable to assertion checking, would be an invaluable aid to formal verification.

References

1. Pixley, C., Strader, N., Bruce, W., Park, J., Kaufmann, M., Shultz, K., Burns, M., Kumar, J., Yuan, J. and Nguyen, J., Commercial design verification: Methodology and tools, in *Proceedings of the International Test Conference*, p. 839, 1996.
2. Dill, D.A., What's between simulation and formal verification?, in *Proceedings of the Design Automation Conference*, p. 328, 1998.
3. Saleh, R., Overhauser, D. and Taylor, S., Full-chip verification of UDSM designs, in *Proceedings of the International Conference on Computer-Aided Design*, p. 453, 1998.
4. Kantrowitz, M. and Noack, L.M., I'm done simulating; now what? Verification coverage analysis and correctness checking of the chip 21164 Alpha microprocessor, in *Proceedings of the Design Automation Conference*, p. 325, 1996.
5. Gupta, A., Malik, S. and Ashar, P., Toward formalizing a validation methodology using simulation coverage, in *Proceedings of the Design Automation Conference*, p. 740, 1997.
6. 0-In Design Automation, Bug survey results, http://www.0-In.com.
7. Taylor, S., Quinn, M., Brown, D., Dohm, N., Hildebrandt, S., Huggins, J. and Ramey, C., Functional verification of a multiple-issue, out-of-order, superscalar alpha processor—The Alpha 21264 microprocessor, in *Proceedings of the Design Automation Conference*, p. 638, 1998.

8. Chandra, A., Iyengar, V., Jameson, D., Jawalekar, R., Nair, I., Rosen, B., Mullen, M., Yoon, J., Armoni, R., Geist, D. and Wolfsthal, Y., AVPGEN—A test generator for architecture verification, *IEEE Transactions on Very Large Scale Integration Systems*, vol. 3, no. 2, p. 188, 1995.

9. Freeman, J., Duerden, R., Taylor, C. and Miller, M., The 68060 microprocessor functional design and verification methodology, in *Proceedings of the On-Chip Systems Design Conference*, p. 10-1, 1995.

10. Aharon, A., Bar-David, A., Dorfman, B., Gofman, E., Leibowitz, M. and Schwartzburd, V., Verification of the IBM RISC system/6000 by a dynamic biased pseudo-random test program generator, *IBM Systems Journal*, vol. 30, no. 4, p. 527, 1991.

11. Hosseini, A., Mavroidis, D. and Konas, P., Code generation and analysis for the functional verification of microprocessors, in *Proceedings of the Design Automation Conference*, p. 305, 1996.

12. Fallah, F., Devadas, S. and Keutzer, K., OCCOM: Efficient computation of observability-based code coverage metrics for functional verification, *IEEE Transactions on Computer-Aided Design of Integrated Circuits and Systems*, vol. 20, p. 1003, 2001.

13. Wang, L-C. and Abadir, M.S., Experience in validation of PowerPC™ microprocessor embedded arrays, *Journal of Electronic Testing: Theory and Applications*, vol. 15, no. 1, p. 191, 1999.

14. Wang, L-C. and Abadir, M.S., On measuring the effectiveness of various design validation approaches for PowerPC™ microprocessor embedded arrays, *ACM Transactions on Design Automation of Electronic Systems*, vol. 3, no. 4, p. 524, 1998.

15. Cheng, K-T. and Krishnakumar, A.S., Automatic generation of functional vectors using the extended finite state machine model, *ACM Transactions on Design Automation of Electronic Systems*, vol. 1, p. 57, 1996.

16. Ho, R.C. and Horowitz, M.A., Validation coverage analysis for complex digital designs, in *Proceedings of the Internation Conference on Computer-Aided Design*, p. 146, 1996.

17. Moundanos, D., Abraham, J.A. and Hoskote, Y.V., Abstraction techniques for validation coverage analysis and test generation, *IEEE Transactions on Computers*, vol. 47, no. 1, p. 2, 1998.

18. Iwashita, H., Nakata, T. and Hirose, F., Integrated design and test assistance for pipeline controllers, *IEICE Transactions on Information and Systems*, vol. E76-D, no. 7, p. 747, 1993.

19. Lee, D.C. and Siewiorek, D.P., Functional test generation for pipelined computer implementations, in *Proceedings of the International Symposium on Fault-Tolerant Computing*, p. 60, 1991.

20. O'Krafka, B., Mandyam, S., Kreulen, J., Raghavan, R., Saha, A. and Malik, N., MTPG: A portable test generator for cache-coherent multiprocessors, in *Proceedings of the Phoenix Conference on Computers and Communications*, p. 38, 1995.

21. Iwashita, H., Kowatari, S., Nakata, T. and Hirose, F., Automatic test program generation for pipelined processors, in *Proceedings of the International Conference on Computer-Aided Design*, p. 580, 1994.

22. Ho, R.C., Yang, C.H., Horowitz, M.A. and Dill, D.A., Architecture validation for processors, in *Proceedings of the International Symposium on Computer Architecture*, p. 404, 1995.

23. Geist, D., Farkas, M., Landver, A., Lichtenstein, Y., Ur, S. and Wolfsthal, Y., Coverage-directed test generation using symbolic techniques, in *Proceedings of the International Test Conference*, p. 143, 1996.

24. Gateley, J., Blatt, M., Chen, D., Cooke, S., Desai, P., Doreswamy, M., Elgood, M., Feierbach, G., Goldsbury, T., Greenley, D., Joshi, R., Khosraviani, M., Kwong, R., Motwani, M., Narasimhaiah, C., Nicolino, S.J., Jr., Ozeki, T., Peterson, G., Salzmann, C., Shayesteh, N., Whitman, J. and Wong, P., UltraSPARC™-I emulation, in *Proc. Design Automation Conference*, p. 13, 1995.

25. Ganapathy, G., Narayan, R., Jorden, G., Fernandez, D., Wang, M. and Nishimura, J., Hardware emulation for functional verification of K5, in *Proceedings of the Design Automation Conference*, p. 315, 1996.

26. Bose, P., Performance test case generation for microprocessors, in *Proceedings of the VLSI Test Symposium*, p. 54, 1998.

65

Microprocessor Layout Method

Tanay Karnik
Intel Corporation

65.1 Introduction

This chapter presents various concepts and strategies employed to generate a layout of a high-performance, general-purpose microprocessor. The layout process involves generating a physical view of the microprocessor that is ready for manufacturing in a fabrication facility (fab) subject to a given target frequency. The layout of a microprocessor differs from ASIC layout because of the size of the problem, complexity of today's superscalar architectures, convergence of various design styles, the planning of large team activities, and the complex nature of various, sometimes conflicting, constraints.

In June 1979, Intel introduced the first 8-bit microprocessor with 29,000 transistors on the chip with 8-MHz operating frequency [1]. Since then, the complexity of microprocessors has been closely following Moore's law, which states that the number of transistors in a microprocessor will double every 18 months [2]. The number of execution units in the microprocessor is also increasing with generations. The increasing die size poses a layout challenge with every generation. The challenge is further augmented by the ever-increasing frequency targets for microprocessors. Today's microprocessors are marching toward the GHz frequency regime with more than 10 million transistors on a die. Table 65.1 includes some statistics of today's leading microprocessors[1]:

---

[1]The reader may refer to Refs. 3 through 10 for further details about these processors.

TABLE 65.1 Microprocessor Statistics

Manufacturer	Part Name	# Transistors (millions)	Frequency (MHz)	Die Size (mm²)	Technology (μm)
Compaq	Alpha 21264	15.2	600	314	0.35
IBM	PowerPC	6.35	250	66.5	0.3
HP	PA-8000	3.8	250	338	0.5
Sun	Ultrasparc-I	5.2	167	315	0.5
Intel	Pentium II	7.5	450	118	0.25

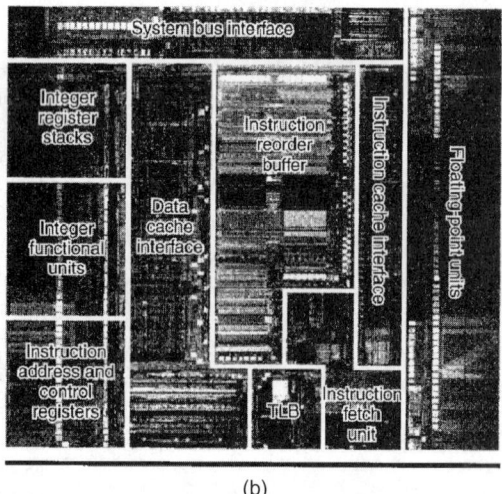

(a)   (b)

FIGURE 65.1 Chip micrographs. (a) Compaq Alpha 21264; (b) HP PA-8000. ((a) Courtesy of IEEE, Ref. 4; (b) Courtesy of IEEE, Ref. 6. With permission.)

In order to understand the magnitude of the problem of laying out a high-performance microprocessor, refer to the sample chip micrographs in Figure 65.1. Various architectural modules, such as functional blocks, datapath blocks, memories, memory management units, etc. are physically separated on the die. There are many layout challenges apparent in this figure. The floorplanning of various blocks on the chip to minimize chip-level global routing is done before the layout of the individual blocks is available. The floorplanning has to fit the blocks together to minimize chip area and satisfy the global timing constraints. The floorplanning problem is explained in Section 65.4.1 (Floorplanning). As there are millions of devices on the die, routing power and ground signals to each gate involves careful planning. The power routing problem is described in Section 65.4.2 (Clock Planning). The microprocessor is designed for a particular frequency target. There are three key steps to high performance. The first step involves designing a high-performance circuit family, the second one involves design of fast storage elements, and the third is to construct a clock distribution scheme with minimum skew. Many elements need to be clocked to achieve synchronization at the target frequency. Routing the global clock signal exactly from an initial generator point to all of these elements within the given delay and skew budgets is a hard task. Section 65.4.3 (Power Planning) includes the description of clock planning and routing problems. There are various signal buses routed inside the chip running among chip I/Os and blocks. A 64-bit datapath bus is a common need in today's high-performance architectures, but routing that wide a bus in the presence of various other critical signals is very demanding, as explained in Section 65.4.4 (Bus Routing).

The problems identified by looking at the chip micrographs are just a glimpse of a laborious layout process. Before any task related to layout begins, the manufacturing techniques need to be stabilized and the requirements have to be modeled as simple design rules to be strictly obeyed during the entire design process. The manufacturing constraints are caused by the underlying process technology (Section 65.3.2, Technology Process) or packaging (Section 65.3.1, Packaging).

Another set of decisions to be taken before the layout process involve the circuit style(s) to be used during the microprocessor design. Examples of such styles include full custom, semi-custom, and automatic layout. They are described in Section 65.2. The circuit styles represent circuit layout styles, but there is an orthogonal issue to them, namely, circuit family style. The examples of circuit families include static CMOS, domino, differential, cascode, etc. The circuit family styles are carefully studied for the underlying manufacturing process technology and ready-to-use cell libraries are developed to be used during the block layout. The library generation is illustrated in Section 65.5.

Major layout effort is required for the layout of functional blocks. The layout of individual blocks is usually done by parallel teams. The complex problem size prompts partitioning inside the block and reusability across blocks. Cell libraries as well as shared mega-cells help expedite the process. Well-established methodologies exist in various microprocessor design companies. Block-level layout is usually done hierarchically. The steps for block-level layout involve partitioning, placement, routing, and compaction. They are detailed in Section 65.6.

65.1.1 CAD Perspective

The complexity of microprocessor design is growing, but there is no proportional growth in design team sizes. Historically, many tasks during the microprocessor layout were carefully hand-crafted. The reasons were twofold. The size of the problem was much smaller than what we face today. The second reason was that computer-aided design (CAD) was not mature. Many CAD vendors today are offering fast and accurate tools to automatically perform various tasks such as floorplanning, noise analysis, timing analysis, placement, and routing. This computerization has enabled large circuit design and fast turn-around times. References to various CAD tools with their capabilities have been added throughout this chapter.

CAD tools do not solve all of the problems during the microprocessor layout process. The regular blocks, like datapath, still need to be laid out manually with careful management of timing budgets. Designers cannot just throw the netlist over the wall to CAD to somehow generate a physical design. Manual effort and tools have to work interactively. Budgeting, constraints, connectivity, and interconnect parasitics should be shared across all levels and styles. Tools from different vendors are not easily interoperable due to a lack of standardization. The layout process may have proprietary methodology or technology parameters that are not available to the vendors. Many microprocessor manufacturers have their own internal CAD teams to integrate the outside tools into the flow or develop specific point tools internally. This chapter attempts to explain the advantages as well as shortcomings of CAD for physical layout.

Invaluable information about Physical Design Automation and related algorithms is provided in Refs. 11 and 12. These two textbooks cover a wide range of problems and solutions from the CAD perspective. They also include detailed analyses of various CAD algorithms. The reader is encouraged to refer to Refs. 13 to 15 for a deep understanding of digital design and layout.

65.1.2 Internet Resources

The Internet is bringing the world together with information exchange. Physical design of microprocessors is a widely discussed topic on the internet. The following web sites are a good resource for advanced learning of this field.

The key conference for physical design is the International Symposium on Physical Design (ISPD), held annually in April. The most prominent conference in Electronic Design Automation (EDA) community is the ACM/IEEE Design Automation Conference (DAC), (www.dac.com). The conference features an exhibit program consisting of the latest design tools from leading companies in design automation. Other related conferences are International Conference on Computer Aided Design (ICCAD) (www.iccad.com), IEEE International Symposium on Circuits and Systems (ISCAS) (www.iscas.nps.navy.mil), International Conference on Computer Design (ICCD), IEEE Midwest Symposium on Circuits and Systems (MSCAS), IEEE Great Lakes Symposium on VLSI (GLSVLS) (www.eecs.umich.edu/glsvlsi), European Design Automation Conference (EDAC), International Conference on VLSI Design (vcapp.csee.usf.edu/vlsi99/), and Microprocessor Forum. There are several journals which are dedicated to the field of VLSI Design Automation and include broad coverage of all topics in physical design. They are *IEEE Transactions on CAD of Circuits and Systems* (akebono.stanford.edu/users/nanni/tcad), *Integration, IEEE Transactions on Circuits and Systems, IEEE Transactions on VLSI Systems,* and the *Journal of Circuits, Systems and Computers.* Many other journals occasionally publish articles of interest to physical design. These journals include *Algorithmica, Networks, SIAM Journal of Discrete and Applied Mathematics,* and *IEEE Transactions on Computers.*

An important role of the Internet is through the forum of newsgroups. comp.lsi.cad is a newsgroup dedicated to CAD issues, while specialized groups such as comp.lsi.testing and comp.cad.synthesis discuss testing and synthesis topics. The reader is encouraged to search the Internet for the latest topics. *EE Times* (www.eet.com) and *Integrated System Design* (www.isdmag.com) magazines provide latest information about physical design and they are both online publications. Finally, the latest challenges in physical design are maintained at (www.cs.virginia.edu/pd_top10/). The current benchmark problems for comparison of PD algorithms are available at www.cbl.ncsu.edu/www/.

We describe various problems involved throughout the microprocessor layout process in the Section 65.2.

65.2 Layout Problem Description

The design flow of a microprocessor is shown in Figure 65.2. The architectural designers produce a high-level specification of the design, which is translated into a behavioral specification using function design, structural specification using logic design, and a netlist representation using circuit design. In this chapter, we discuss the microprocessor layout method called *physical design.* It converts a netlist into a mask layout

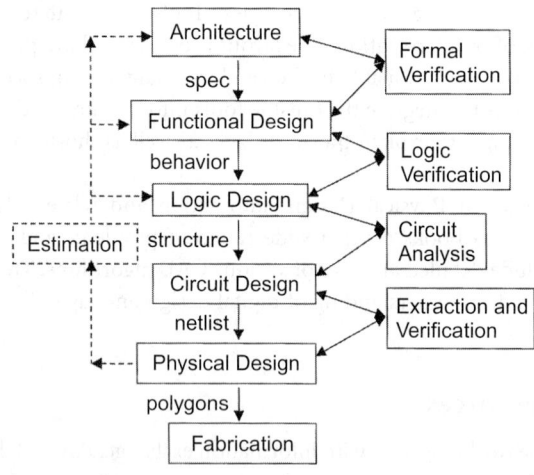

FIGURE 65.2 Microprocessor design flow.

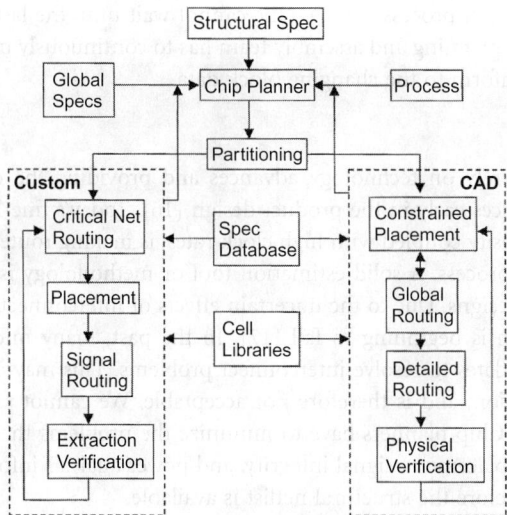

FIGURE 65.3 Microprocessor physical design flow.

consisting of physical polygons, which is later fabricated on silicon. The boxes on the right side of Figure 65.2 depict the need for verification during all stages of the design. Due to high frequencies and shrinking die sizes, estimation of eventual physical data is required at all stages before physical design during the microprocessor design process. The estimation may not be absolutely necessary for other types of designs.

Let us consider the physical design process. Given a netlist specification of a circuit to be designed, a layout system generates the physical design either manually or automatically and verifies that the design conforms to the original specification. Figure 65.3 illustrates the microprocessor physical design flow.

Various specifications and constraints have to be handled during microprocessor layout. Global specs involve the target frequency, density, die size, power, etc. Process specs will be discussed in Section 65.3. The chip planner is the main component of this process. It partitions the chip into blocks, assigns blocks for either full custom (manual) layout or CAD (automatic) layout and assembles the chip after block-level layout is finished. It may also iterate this process for better results. Full custom and CAD layout differ in the approach to handle critical nets. In the custom layout, critical nets are routed as a first step of block layout. In the CAD approach, the critical net requirements are translated into a set of constraints to be satisfied by placement and routing tools. The placement and global routing have to work in an iterative fashion to produce a dense layout. The double-sided arrow in CAD box represents this iteration. In both layout styles, iterations are required for block layout to completely satisfy all the specs. Some microprocessor teams employ a semi-custom approach which takes advantage of careful hand-crafting and power savings on the full custom side, and the efficiency and scalability of the CAD side.

65.2.1 Global Issues

The problems specific to individual stages of physical design are discussed in the following sections. This section attempts to explain the problems that affect the whole design process. Some of them may be applicable to the pre-layout design stages and post-layout verification.

65.2.1.1 Planning

There has to be a global flow to the layout process. The flow requires consistency across all levels and support for incremental re-design. A decision at one level affects almost all the other levels. The chip planning and assembly are the most crucial tasks in the microprocessor layout process. The chip is partitioned into blocks. Each block is allotted some area for layout. The allotment is based on estimation based on past experience. When the blocks are actually laid out, they may not fit in the allotted area.

The full microprocessor layout process is long. One cannot wait until the last moment to assemble the blocks inside the chip. The planning and assembly team has to continuously update the flow, chip plans, and block interfaces to conform to the changing block data.

65.2.1.2 Estimation

New product generations rely on technology advances and providing the designer with a means of evaluating technology choices early in the product design [16]. Today's fine-line geometries jeopardize timing. Massive circuit density coupled with high clock rates, is making routed interconnects hardest to gauge early in the design process. A solid estimation tool or methodology is needed to handle today's complex microprocessor designs. Due to the uncertain effects of interconnect routing, the wall between logical and physical design is beginning to fall [17]. In the past, many microprocessor layout teams resorted to post-layout updates to resolve interconnect problems. This may cause major re-design and another round of verification, and is therefore not acceptable. We cannot separate logical design and physical design engineers. Chip planners have to minimize the problems that interconnect effects may cause. Early estimation of placement, signal integrity, and power analysis information is required at the floorplanning stage even before the structural netlist is available.

65.2.1.3 Changing Specifications

Microprocessor design is a long process. It is driven by market conditions, which may change during the course of the design. So, architectural specs may be updated during the design. During physical design, the decisions taken during the early stages of the design may prove to be wrong. Some blocks may have added functionalities or new circuit families, which may need more area. The global abstract available to block-level designers may continuously change, depending on sibling blocks and global specs. Hence, the layout process has to be very flexible. Flexibility may be realized at the expense of performance, density, or area—but it is well worth it.

65.2.1.4 Die Shrinks and Compactions

The easiest way to achieve better performance is process shrinks. Optical shrinks are used to convert a die from one process to a finer process. Some more engineering is required to make the microprocessor work for the new process. A reduction in feature size from 0.50 μm to 0.35 μm results in an increase of approximately 60% more devices on a similarly sized die [3]. Layouts designed for a manufacturing process should be scalable to finer geometries. The decisions taken during layout should not prohibit further feature shrinks.

65.2.1.5 Scalability

CAD algorithms implemented in automatic layout tools must be applicable to large sizes. The same tools must be useful across generations of microprocessor. Training the designers on an entirely new set of CAD tools for every generation is impractical. The data representation inside the tools should be symbolic so that the process numbers can be updated without a major change in tools.

65.2.2 Explanation of Terms

There are many terms related to microprocessor layout used in the following sections. The definitions and explanation of those terms is provided in this section.

> **Capacitance**: A time-varying voltage across two parallel metal segments exhibits capacitance. The voltage (v) and current (i) relation across a capacitor (C) is

$$i = C \frac{dv}{dt}$$

Closely spaced unconnected metal wires in layout can have significant cross-capacitance. Capacitance is very significant at 0.5-μm process and beyond [18].

Inductance: A time-varying current in a wire loop exhibits inductance. If the current through a power grid or large signal buses changes rapidly, this can have inductive effects on adjacent metal wires. The voltage (v) and current (i) relation across an inductor (L) is

$$v = L\frac{di}{dt}$$

Inductance is not a local phenomenon like capacitance.

Parasitics: The shrinking technology and increasing frequencies are causing analog physical behavior in digital microprocessors [19]. The electrical parameters associated with final physical routes are called interconnect parasitics. The parasitic effects in the metal routes on the final silicon need to be estimated in the early phases of the design.

Design rules: The process specification is captured in an easy-to-use set of rules called design rules.

Spacing: If there is enough spacing between metal wires, they do not exhibit cross-capacitance. Minimum metal spacing is a part of the design rules.

Shielding: The power signal is routed on a wide metal line and does not have time-varying properties. In order to reduce external effects like cross-capacitance, on a critical metal wire, it is routed between or next to a power wire. This technique is called shielding.

Electromigration: Also known as metal migration, it results from a conductor carrying too much current. The result is a change in conductor dimensions, causing high resistive spots and eventual failure. Aluminum is the most commonly used metal in microprocessors. Its current density (current per width) threshold for electromigration is

$$2\,\frac{mA}{\mu m}$$

65.3 Manufacturing

Manufacturing involves taking the drawn physical layout and fabricating it on silicon. A detailed description of fabrication processes is beyond the scope of this book. Elaborate descriptions of the fabrication process can be found in Refs. 11 and 13. The reader may be curious as to why manufacturing has to be discussed before the layout process. The reality is that all of the stages in the layout flow need a clear specification of the manufacturing technology. So, the packaging specs and design rules must be ready before the physical design starts.

In this section, we present a brief overview of chip packaging and technology process. The reader is advised to understand the assessment of manufacturing decisions (see Ref. 16). There is a delicate balancing of the system requirements and the implementation technology. New product generation relies on technology advances and providing the designer with a means of evaluating technology choices early in the product design.

65.3.1 Packaging

ICs are packages into ceramic or plastic carriers usually in the form of a pin grid array (PGA) in which pins are organized in several concentric rectangular rows. These days, PGAs have been replaced by surface-mount assemblies such as ball grid arrays (BGAs) in which an array of solder balls connects the package to the board. There is definitely a performance loss due to the delays inside the package. In many microprocessors, naked dies are directly attached to the boards. There are two major methods

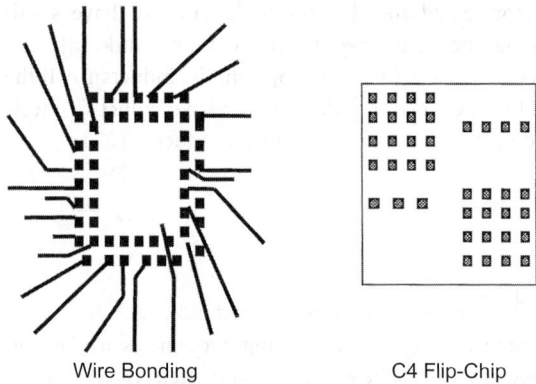

Wire Bonding   C4 Flip-Chip

FIGURE 65.4   Die attachment styles.

of attaching naked dies. In wire bonding, I/O pads on the edge of the die are routed to the board. The active side of the die faces away from the board and the I/Os of the die lie on the periphery (peripheral I/Os). The other die attachment, control collapsed chip connection (C4) is a direct connection of die I/Os and the board. The I/O pins are distributed over the die and a solder ball is placed over each I/O pad (areal I/Os). The die is flipped and attached to the board. The technology is called C4 flip-chip. Figure 65.4 provides an abstract view of the two styles.

There is a discussion about practical issues related to packaging available in Ref. 20. According to the Semiconductor Industry Association's (SIA) roadmap, there should be 600 I/Os per package in 2507 rows, 7 μm package lines/space, 37.5 μm via size, and 37.5 μm landing pad size by the year 1999. The SIA roadmap lists the following parameters that affect routing density for the design of packaging parameters:

- **Number of I/Os**: This is a function of die size and planned die shrinks. The off-chip connectivity requires more pins.
- **Number of rows**: The number of rows of terminals inside the package.
- **Array shape**: Pitch of the array, style of the array (i.e., full array, open at center, only peripheral).
- **Power delivery**: If the power and ground pins are located in the middle, the distribution can be made with fewer routing resources and more open area is available for signals. But then, the power cannot be used for shielding the critical signals.
- **Cost of package**: This includes the material, processing cost, and yield considerations. The current trend in packaging indicates a package with 1500 I/O on the horizon and there are plans for 2000 I/Os.

There is a gradual trend toward the increased use of areal I/Os. In the peripheral method, the I/Os on the perimeter are fanned out until the routing metal pitch is large enough for the chip package and board to handle it. There may be high inductance in the wire bonding. Inductance causes current time delay at switching, slow rise time, and ground bounce in which the ground plane moves away from 0 V, noise, and timing problems. These effects have to be handled during a careful layout of various critical signals. Silicon array attachments and plastic array packages are required for high I/O densities and power distribution. In microprocessors, the packaging technology has to be improvised because of the growth in bus widths, additional metal layers, less current capacity per wire, more power to be distributed over the die, and the growing number of data and control lines due to bus widths. The number of I/Os has exceeded the wire bonding capacity. Additionally, there is a limit to how much a die can be shrunk in the wire bonding method. High operating frequencies, low supply voltage, and high current requirements manifest themselves into a difficult power distribution across the whole die. There are assembly issues with fine pitches for wire bonds. Hence, the microprocessor manufacturers are employing C4 flip-chip technologies. Areal packages reduce the routing inside the die but need more routing on the board.

The effect of area packaging is evident in today's CAD tools [21]. The floorplanner has to plan for areal pads and placement of I/O buffers. Area interconnect facilitates high I/O counts, shorter interconnect rates, smaller power rails, and better thermal conductivity. There is a need for an automatic area pad planner to optimize thousands of tightly spaced pads. A separate area pad router is also desired. The possible locations for I/O buffers should be communicated top-down to the placement tool and the placement info should be fed back to the I/O pad router. After the block level layout is complete and the chip is assembled, the area pad router should connect the power pads to inner block-level power rails.

Let us discuss some industry microprocessor packaging specs. The packaging of DEC/Compaq's Alpha 21264 has 587 pins [4]. This microprocessor contains distributed on-chip decoupling capacitors (decap) as well as a 1-μm package decap. There are 144-bit (128-bit data, 16-bit ECC) secondary cache data interface and 72-bit system data interface. Cache and system data pins are interleaved for efficient multiplexing. The vias have to arrayed orthogonal to the current flow. HP's PA-8000 has a flip-chip package, which enables low resistance, less inductance, and larger off-chip cache support. There are 704 I/O signals and 1200 power and ground bumps in the 1085-pin package. Each package pin fans out to multiple bumps [6]. PowerPC™ has a 255-pin CBGA with C4 technology [7]. 431 C4's are distributed around the periphery. There are 104 VDD and GND internal C4's. The C4 placement is done for optimal L2 cache interface.

There is a debate about moving from high-cost ceramic to low-cost plastic packaging. Ceramic ball grid arrays suffer from 50% propagation speed degradation due to high dielectric constant (10). There is a trend to move toward plastic. However, ceramic is advantageous in thermal conductivity and it supports high I/O flip-chip packaging.

65.3.2 Technology Process

The whole microprocessor layout is driven by the underlying technology process. The process engineers decide the materials for dielectric, doping, isolation, metal, via, etc. and design the physical properties of various lithographic layers. There has to be close cooperation between layout designers and process engineers. Early process information and timely updates of technology parameters are provided to the design teams, and a feedback about the effect of parameters on layout is provided to the process teams. Major process features are managed throughout the design process. This way, a design can be better optimized for process, and future scaling issues can be uncovered.

The main process features that affect a layout engineer are metal width, pitch and spacing specs, via specs, and I/O locations. Figure 65.5(a) shows a sample multi-layer routing inside a chip. Whenever two metal rails on adjacent layers have to be connected, a via needs to be dropped between them. Figure 65.5(b) illustrates how a via is placed. The via specs include the type of a via (stacked, staggered), coverage of via (landed, unlanded, point, bar, arrayed), bottom layer enclosure, top layer enclosure, and the via width. In today's microprocessors, there is a need for metal planarization. Some manufacturers are actually adding planarization metal layers between the usual metal layers forr fabrication as well as shielding. Aluminum

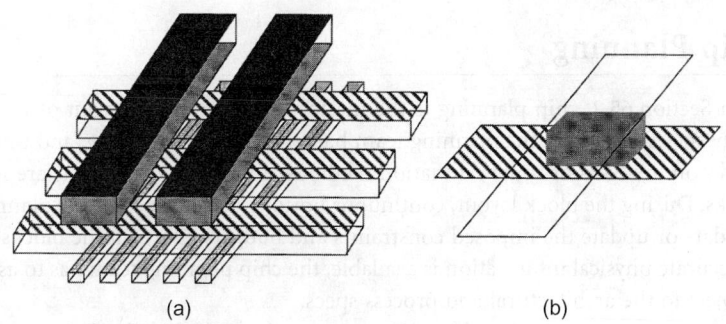

(a)   (b)

FIGURE 65.5 A view of multi-layer routing and a simple via.

was the most common metal for fabrication. IBM has been successful in getting copper to work instead of aluminum. The results show a 30% decrease in interconnect delay.

The process designers perform what-if analyses and design sensitivity studies of all of the process parameters on the basis of early description of the chip with major datapath and bus modeling, net constraints, topology, routing, and coupled noise inside the package. The circuit speed is inversely proportional to the physical scale factor. Aggressive process scaling makes manufacturing difficult. On the other hand, slack in the parameters may cause the die size to increase. We have listed some of the process numbers in today's leading microprocessors in this section. The feature sizes are getting very small and many unknown physical effects have started showing up [22]. The processes are so complicated to correctly obey during the design, an abstraction called design rules is generated for the layout engineers. Design rules are constraints imposed on the geometry or topology of layouts and are derived from basic physics of circuit operation such as electromigration, current carrying capacity, junction breakdown, or punch-through, and limits on fabrication such as minimum widths, spacing requirements, misalignments during processing, and planarization. The rules reflect a compromise between fully exploiting the fabrication process and producing a robust design on target [5].

As feature sizes are decreasing, optical lithography will need to be replaced with deep-UV, X-ray, or electron beam techniques for features sizes below 0.15 μm [20]. It was feared that quantum effects will start showing up below 0.1 μm. However, IBM has successfully fabricated a 0.08-μm chip in the laboratory without seeing quantum effects. Another physical limit may be the thickness of the gate oxide. The thickness has dropped to a few atoms. It is soon going to hit a fundamental quantum limit.

Alpha 21264 has 0.35-μm feature size, 0.25-μm effective channel length, and 6-nm gate oxide. It has four metal layers with two reference planes. All metal layers are AlCu. Their width/pitches are 0.62/1.225, 0.62/1.225, 1.53/2.8, and 1.53/2.8 μm, respectively [4]. Two thick aluminum planes are added to the process in order to avoid cycle-to-cycle current variations. There is a ground reference plane between metal2 and metal3, and a VDD reference plane above metal4. Nearly the entire die is available for power distribution due to the reference planes. The planes also avoid inductive and capacitive coupling [8].

PowerPC™ has 0.3-μm feature size, 0.18-μm effective channel length, 5-nm gate oxide thickness, and a five-layer process with tungsten local interconnect and tungsten vias [7]. The metal widths/pitches are 0.56/0.98, 0.63/1.26, 0.63/1.26, 0.63/1.26, and 1.89/3.78 μm, respectively.

HP-8000 has 0.5-μm feature size and 0.29-μm effective channel length [6]. There is a heavy investment in the process design for future scaling of interconnect and devices. There are five metal layers, the bottom two for local fine routing, metal3 and metal4 for global low resistive routing, and metal5 reserved for power and clock. The author could not find published detailed metal specs for this microprocessor.

Intel Pentium II is fabricated with a 0.25-μm CMOS four-layer process [23]. The metal width/pitches are 0.40/1.44, .64/1.36, .64/1.44, and 1.04/2.28 μm, respectively. The two lower metal layers are usually used in block-level layout, metal3 is primarily used for global routing, and metal4 is used for top-level chip power routing.

65.4   Chip Planning

As explained in Section 65.2, chip planning is the master step during the layout of a microprocessor. During the early stages of design, the planning team has to assign area, routing, and timing budgets to individual blocks on the basis of some estimation methods. Top-down constraints are imposed on the individual blocks. During the block layout, continuous bottom-up feedback to the planner is necessary in order to validate or update the imposed constraints and budgets. Once all the blocks have been laid out and their accurate physical information is available, the chip planning team has to assemble the full chip layout subject to the architectural and process specs.

Chip planning involves partitioning the microprocessor into blocks. The finite state machines are considered random control logic and partitioned into automatically synthesizable blocks. Regular structures like arrays, memories, and datapath require careful signal routing and pitch matching. They have to be partitioned into modular and regular blocks that can be laid out using full-custom or semi-custom techniques.

IBM adopted a two-level hierarchical approach for the G4 processor [24]. They identified groups of 10,000 to 20,000 non-array transistors as macros. Macros were individually laid out by parallel teams. The macro layouts were simplified and abstracted for floorplanning, place and route, and global extraction. The shapes of individual blocks varied during the design process. The chip planner performed the layouts for global interconnects and physical design of the entire chip. The global environment was abstracted down to the block level. A representation of global wires was added overlaying a block. That included global timing at block interfaces, arrival times with phase tags at primary inputs (PI), required times with phase tags at primary outputs (PO), PI resistances, and PO capacitances. Capacitive loading at the outputs was based on preliminary floorplan analysis. Each block was allowed sufficient wiring and cell area. The control logic was synthesized with high-performance standard cell library; datapaths were designed with semi-custom macros. Caches, Memrory Management Unit (MMU) arrays, branch unit arrays, Phase-Locked Loop (PLL), and Delay-Locked Loop (DLL) were all full-custom layouts [7]. There were three distinct physical design styles optimizing for different goals, namely, full custom for high performance and density, structured custom for datapath, and fully automated for control logic. The floorplan was flexible throughout the methodology. There are 44% memory arrays, 21% datapath, 15% control, 11% I/O, 9% miscellaneous blocks on the die. Final layout was completely hierarchical with no limits on the levels of hierarchy involved inside a block. The block layouts had to conform to a top abstracted global shadow of interconnects and blockages. The layout engineers performed post-placement re-tuning and post-placement optimization for clock and scan chains.

For the 1-GHz integer PowerPC™ microprocessor, the planning team at IBM enforced strict partitioning on latch boundaries for global timing closure [5]. The planning team constructed a layout description view of the mega-cells containing physical shape data of the pads, power buses, clock spine, and global interconnects. At the block level, pin locations, capacitance, and blockages were available. The layouts were created by hand due to the very high-performance requirements of the chip.

We describe the major steps during the planing stages, namely, floorplanning, power planning, clock planning, and bus routing. These steps are absolutely essential during microprocessor design. Due to the complicated constraints, continuous intelligent updates, and top-down/bottom-up communication, manual intervention is required.

65.4.1 Floorplanning

Floorplanning is the task of placing different blocks in the chip so as to fit them in the minimum possible area with minimum empty space. It must fill the chip as close to the brim as possible. Figure 65.6 shows an example of floorplanning. The blocks on the left hand side are fitted inside the chip on the right. The reader can see that there is very little empty space on the chip. The blocks may be flexible and their

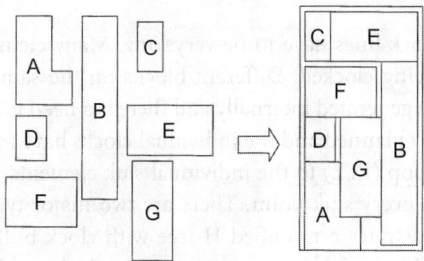

FIGURE 65.6 An example of floorplanning.

TABLE 65.2 CAD Tools Available for Floorplanning

Company	Internet	Product	Description
Avant!	www.avanticorp.com	Planet	Timing-driven hierarchical floorplanner
Cadence	www.cadence.com	Preview	Mixed-level floorplanning and analysis environment
Compass	www.compass-da.com	ChipPlanner-RTL	Timing constraint satisfaction before logic synthesis
HLD	www.hlds.com	Physical DP	Constraint-driven floorplanning
HLD	www.hlds.com	Top-down DP	RTL-level timing analysis for pre-synthesis; internal estimation tool
SVR	www.svri.com	FloorPlacer	Timing and routability analysis with floorplanning

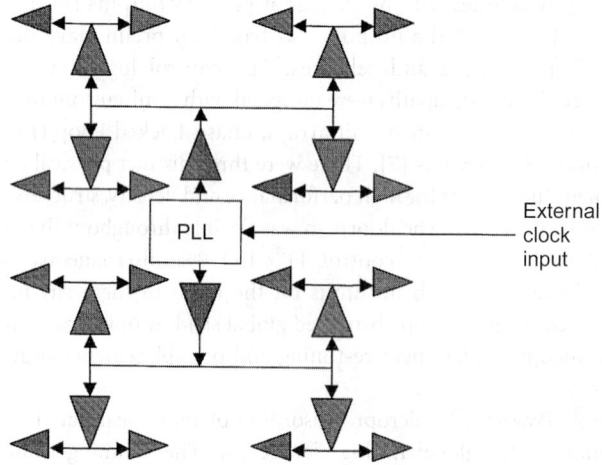

FIGURE 65.7 A sample global clock buffered H-tree.

orientation not fixed. Due to the dominance of interconnect in the overall delay on the chip, today's floorplanning techniques also try to minimize global connectivity and critical net lengths.

There are many CAD tools available for floorplanning from the EDA vendors. The survey of all such tools is available [25]. The tools are attempting to bridge the gap between synthesis and layout. All of the automatic tools are independent of IC design style. There are two types of floorplanners. Functional floorplanners operate at the RTL level for timing management and constraints generation. The goal of physical floorplanners is to minimize die size, maximize routability, and optimize pin locations. Some physical floorplanners perform placement inside floorplanning. As explained in the routing section, when channel routing is used, the die size is unpredictable. The floorplanners cannot estimate routing accurately. Hence, channel allocation on the die is very difficult. Table 65.2 summarizes the CAD tools available for floorplannning.

65.4.2 Clock Planning

Clock is a global signal and clock lines have to be very long. Many elements in high-frequency microprocessors are continuously being clocked. Different blocks on the same die may operate at different frequencies. Multiple clocks are generated internally and there is a need for global synchronization. Clock methodology has to be carefully planned and the individual clocks have to be generated and routed from the chip's main phase-locked loop (PLL) to the individual sink elements. The delays and skews (defined later) have to exactly match at every sink point. There are two major types of clock networks, namely, trees and grids. Figure 65.7 illustrates a modified H-tree with clock buffers. Figure 65.8 shows a clock grid used in Alpha processors. Most of the power consumption inside today's high-frequency processors is in their clock networks. In order to reduce the chip power, there are architectural modifications to

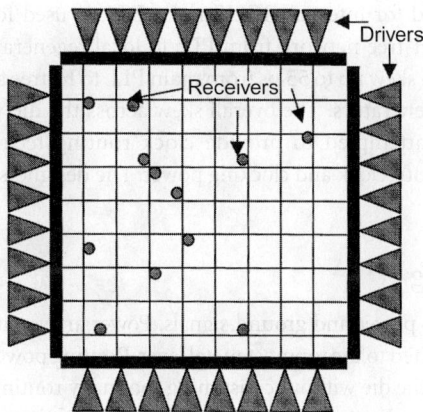

FIGURE 65.8 A sample clock grid.

shut off some part of the chip. This is achieved by clock gating. The clock gator routing has become an integral part of clock routing.

Let us explain the some terms used in clock design. Clock skew is the temporal variation of the same clock edge arriving at various locations on the die. Clock jitter is the temporal variation of consecutive clock edges arriving at the same location. Clock delay is the delay from the source PLL to the sink element. Both skew and jitter have a direct relation to clock delay. Globally synchronous behavior dictates minimum skew, minimum jitter, and equal delay.

Clock grids, being perfectly symmetric, achieve very low skews, but they need high routing resources and stacked vias, and cause signal reflections. The wire loading on driving buffers feeding to the grid is also high. This requires large buffer arrays that occupy significant device area. Electrical analysis of grids is more difficult than trees. Buffered trees are preferred in high-performance microprocessors because they achieve acceptable skews and delays with low routing resource usage.

Ideally, the skew should be 0. However, there are many unknowns due to processing and randomness in manufacturing. Instead of matching the clock receivers exactly, a skew budget is assigned. In high-performance microprocessor designs, there is usually a global clock routing scheme (GCLK) that spawns into multiple matched clock points in various regions on the chip. Inside the region, careful clock routing is performed to match the clock delay within assigned skew budgets.

Alpha 21264 has a modified H-tree. On-chip PLL dissipates power continuously, 40% of the chip power dissipation was measured to be in the clocking network. Reduction of clock power was a primary concern to reduce overall chip power [26]. There is a GCLK network that distributes clock to local clock buffers. GCLK is shielded with VCC or VSS throughout the die [4]. GCLK skew is 70 ps, with 50% duty cycle and uniform edge rate [8]. The clock routing is done on metal3 and metal4. In earlier Alpha designs, a clock grid was used for effective skew minimization. The grid consumed most of the metal3 and metal4 routing resources. In 21264, there is a savings of 10 W power over previous grid techniques. Also, significantly less metal3 and metal4 is used for clock routing. This proved that a less aggressive skew target can be achieved with a sparser grid and smaller drivers. The new technique also helped power and ground networks by spreading out the large clock drivers across the die.

HP-8000 also has a modified H-tree for clock routing [6,18]. External clock is delivered to the chip PLL through a C4 bump. The microprocessor has a three-level clock network. There is a modified H-tree that routes GCLK from PLL to 12 secondary buffers strategically placed at various critical locations in various regions on the chip. The output of the receiver is routed to matched wire lengths to a second level of clock buffers. The third level involves 7000 clock gators that gate the clock routing from the buffers to local clock receivers. There are many flavors of gated clocks on the chip. There is a 170-ps skew across the die. Due to a large die, PA8000 buffers were designed to minimize process variations.

In PowerPC™, a PLL is used for internal GCLK and a DLL is used for external SRAM L2 interface [7]. There is a semi-balanced H-tree network from PLL to local regenerators. Semi-balanced means the design was adjusted for variable skew up to 55 ps from main PLL to H-tree sinks. There are three variations of masking 486 local clock regenerators. The overall skew across the die was 300 ps.

Many CAD vendors have attempted to provide clock routing technologies. The microprocessor community is very paranoid about clock and clocking power. The designers prefer hand-crafting the whole clock network.

65.4.3   Power Planning

Every gate on the die needs the power and ground signals. Power arrives at many chip-level input pins or C4 bumps and is directly connected to the topmost metal layer. Routing power and ground from the topmost layer to each and every gate on the die without consuming too many routing resources, not causing voltage drops in the power network, and using effective shielding techniques constitutes the power planning problem. A high-performance power distribution scheme must allow for all circuits on the die to receive a constant power reference. Variation in the reference will cause noise problems, subthreshold conduction, latch-up, and variable voltage swings.

The switching speed of CMOS circuits in the first order is inversely proportional to the drain-to-source current of the transistor (I_{ds}), in the linear region:

$$t = C \int \frac{dV}{I_{ds}}$$

where, C is the loading capacitance, V is the output voltage, and t is the switching delay. I_{ds}, in turn, depends on the IR drop (V_{drop}) as:

$$I_{ds} \propto \left(V_{gs} - V_t - V_{drop} \right)$$

where V_{gs} is the gate to source voltage and V_t is the threshold voltage of the MOS transistor. Therefore, achieving the highest switching speed requires distributing the power network from the pads at the periphery of the die or C4 bumps to the sources of the transistors with minimal IR drop due to routing. The problem of reducing V_{drop} is modeled in terms of minimum allowable voltage at the source and the difference between V_{dd} and V_{ss} acceptable at the sinks. All physical stages from pads to pins have to be considered. Some losses, like tolerance of the power supply, the tester guardband, and power drop in the package, are out of the designer's control. The remaining IR drop budget is divided among global and local power meshes.

The designers at Motorola have provided a nice overview of power routing in Ref. 27. Their design of PowerPC™ power grid continued across all design stages. A robust grid design was required to handle the possible switching and large current flow into the power and ground networks. Voltage drops in power grid cause noise, degrading performance, high average current densities, and undesirable wearing of metal. The problem was to design a grid achieving perfect voltage regulation at all demand points on the chip, irrespective of switching activities and using minimum metal layers. The PowerPC™ processor family has a hierarchy of five or six metal layers for power distribution. Structure, size, and layout of the power grid had to be done early in the design phase in the presence of many unknowns and insufficient data. The variability continued until the end of design cycle. All commercial tools depend on post-layout power grid analysis after the physical data is available. One cannot change the power plan at that stage because too much is at stake toward the end. Hence, Motorola designers used power analysis tools at every stage. They generated applicable constant models for every stage. There are millions of demand points in a typical microprocessor. One cannot simulate all non-linear devices with a non-ideal power grid. Therefore, the approach was as follows. They simulated non-linear devices with fixed power, converted all devices to current sources, and then analyzed the power grid. There was still a large linear system to handle. So, a hierarchical approach was used. Before the floorplaning stage, the locations of clean VCC/GND pads and power grid widths/pitches were decided on the basis of design rules and via styles (point or bar vias). After the floorplan was fixed, all blocks were given block power service terminals. Wires that connect global power to block power were also modeled in the service terminals. Power was routed inside the blocks and PowerMill simulations were used for validation.

Alpha 21264 operates at a high frequency and has a large die as listed in Table 65.1. The large die and high frequency lead to high power supply currents. This has a serious effect on power, clock, and ground networks [3,4]. Power dissipation was the sole factor limiting chip complexity and size; 198 out of 587 chip-level pins are VDD and VSS pins. Supply current has doubled during every generation of Alpha microprocessor. Hence, a very complex power distribution was required. In order to meet very large cycle-to-cycle current variations, two thick low-resistance aluminum planes were added to the process [8]. One plane was placed between metal2 and metal3 connected to VSS, and the other above the topmost metal4 connected to VDD. Nearly the entire die area was available for power distribution. This helped in inductive and capacitive decoupling, reduced on-chip crosstalk, and presented excellent current returns paths for analysis and minimized inductive noise.

UltraSparce-I™ has 288 power and ground pins out of 520 [9]. The methodology involved an early identification of excessive voltage drop points and seamless integration of power distribution and CAD tools. Correct-by-construction power grid design was done throughout the design cycle. The power networks were designed for cell libraries and functional blocks. They were reliability-driven designs before mask generation. This enabled efficient distribution of the V_{dd} and V_{ss} networks on a large die. Minimization of area overhead, as well as IR drop for power distribution, was considered throughout the design cycle. Parts of power distribution network are incorporated into the standard cell library layouts. CAD tools were used for the composition of standard cell and datapath with correct-by-construction power interconnections. The methodology was designed to be scalable to future generations. Estimation and budgeting of IR drops was done across the chip. Metal4 was the only over-the-block routing layer. It was used for routing power from peripheral I/O pads to individual functional units. It was the primary means of distributing power. The power distribution should not constrain the floorplan. Hence, two meshes were laid out: a top-down global mesh and an in-cell local mesh. This enabled block movement during placement because they have only local mesh. As long as the local power mesh crosses the global mesh, the power can be distributed inside the block. Metal3 local power routes have to be orthogonal to global metal4 power. The direction of metal1 and metal2 do not matter. The global chip is divided into two parts. In part 1, metal3 was vertical and metal4 was horizontal. The opposite directions were selected for the second part. A block could be moved half the die distance because of two types of regions for power on the chip. The power grid on three metal layers with interconnections, number of vias, and via types was simulated using HSPICE to determine the widths, spacings, and number of vias of the power grid. Vias had to be arrayed orthogonal to the current flow. There was a 90-mV IR drop from M3-M4 via to the source of a cell. Additional problems existed because the metal2 width is fixed in UltraSparc™. Up to a certain drive strength, the metal2 power rail was 2.5 μm. Beyond that, additional rail of 1 μm was added. The locations of clock receivers changed throughout the design process. They had to be shifted to align power.

65.4.4 Bus Routing

The author considers bus routing a critical problem and it needs the same attention as power or clock routing. The problem arises due to today's superscalar, large bit-width microprocessor architectures. The chip planners design the clock and power plans and floorplan the chip very efficiently to minimize empty space on the die, but leave limited routing resources on the top layers to route busses. There is a simple analogy to understand this problem. Whenever a city is being planned, the roads are constructed before the individual buildings. In microprocessor layout, busses must be planned before the blocks are laid out.

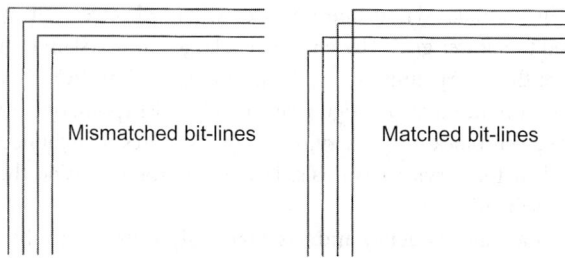

FIGURE 65.9   Bus interleaving.

A bus, by nature, is bi-directional and must have matching characteristics at all data bits. There should be a matching RC delay viewed from both ends. It connects a wide datapath to another. If it is routed straight from one datapath block to another, then the characteristics match; but it is not always feasible on the die to achieve straight routes. Whenever there is a directional change, via delay comes into picture. The delays due to via and uneven lengths for all the bit-lines in the bus cause a mismatch across the bits of the bus. Figure 65.9 depicts a simple technique called bus interleaving, employed in today's microprocessors, to achieve matching lengths.

The problems do not end there. Bus interleaving may match the lengths across the bit-widths, but it does not guarantee matching environment for all the bit-lines. Crosstalk due to adjacent layers or busses may cause mismatch among the bit-lines. In differential circuits, very low voltage busses are routed with long routing lengths. Alpha designers had to carefully route low swing busses in 21264 to minimize all differential noise effects [3]. These types of busses need shielding to protect the low-voltage signals. If all bits in a bus switch simultaneously, large current variations inject inductive noise into the neighboring signal lines. Hence, other signals also need to be shielded from active busses.

65.4.5   Cell Libraries

A major step toward high performance is the availability of a fast ready-to-use circuit library. Due to large and complex circuit sizes, transistor-level layout is formidable. All microprocessor teams design a family of logic gates to perform certain logic operations. These gates become the bottom level units in the netlist hierarchy. They serve as a level of abstraction higher than a basic transistor. Predefined logic functions help in automatic synthesis. The gates may differ in their circuit family, logic functions, drive strength, power consumption, internal layout, placement of cell interface ports, power rails, etc. The number of different cells available in the design libraries can be as high as 2000. The libraries offer the most common predefined building blocks of logic and low-level analog and I/O functions. Complex designs require multiple libraries. The libraries enable fast time to market, aid synthesis in logic minimization, and provide an efficient representation of logic in hardware description languages.

Block-level layout tools support cell-based layout. They need the cells to be of a certain height and perform fast row-based layout. The block-level layout tools are very mature and fast. Many microprocessor design teams design their libraries to be directly usable by block-level layout tools. There are many CAD tools available for cell designs and cell-based block designs. The most common approach is to develop a different library for each process and migrate the design to match the library. Process-specific libraries lead to small die size with high performance. There are tools available on the market for automatic process porting, but the portability across processes causes performance and area degradation.

Microprocessor manufacturers have their in-house libraries designed and optimized for proprietary processes. The cell libraries have to be designed concurrently with the process design and they must be ready before the block-level design begins. The libraries for datapath and control can differ in styles, size, and routing resource utilization. As datapath is considered crucial to a microprocessor, datapath libraries may not support porosity, but the control logic library has to provide porosity for neighboring datapath cells to use some of its routing resources. Thus, datapath libraries are designed for higher performance

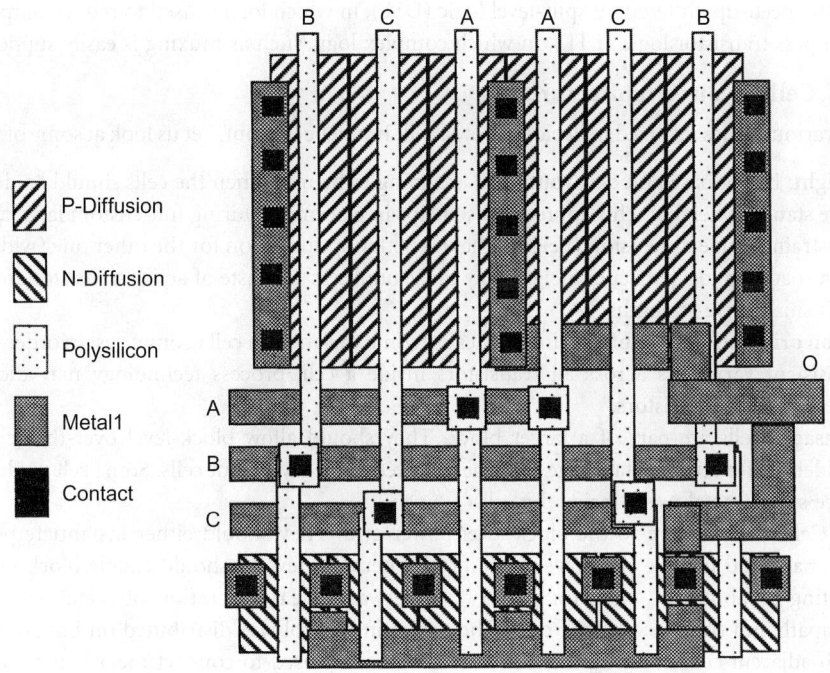

FIGURE 65.10 A three-input CMOS NOR layout.

than control. In UltraSparc-I™ processor, the design team at Sun Microsystems used separate standard cells for datapath and control [9].

In this section, we present various layout aspects of cell library design. The reader is requested to refer to Refs. 13–15 for circuit aspects of libraries.

65.4.5.1 Circuit Family

The most common circuit family is CMOS. They are very popular because of the static nature. It is a fully restored logic in which output either sets at V_{dd} or V_{ss}. The rise and fall times are of the same order. This family has almost zero static power dissipation. The main advantage in layout is its symmetric nature, nice separation of n and p transistors, and ability to produce regular layouts. Figure 65.10 shows a three-input CMOS NOR library cell.

The other popular circuit family in high-performance microprocessors is that of dynamic circuits. The inputs feed into the n-stack and not the p-stack. There is a precharge p-transistor and a smaller keeper p-transistor in the p-stack. So, the number of transistors in p-stack is exactly 2. The dynamic circuits need careful analysis and verification, but allow wide OR structures, less fan-in and fan-out capacitance. The switching point is determined by the nMos threshold and there is no crossover current during output transition. As there is less loading on the inputs, this circuit family is very fast. As one can see in Figure 65.7, the area occupied by the p-stack is very large compared to the n-stack in static CMOS. Domino logic families have a significant area advantage over static if the same static netlist can be synthesized in monotonic domino gates. However, layout of domino gates is not trivial. Every gate needs a clock routed to it. As the family does not support fully restoring logic, the domino gate output needs to be shielded from external noise sources. Additional circuitry may be required to avoid charge-sharing and noise problems.

Other circuit families include bipolar complementary metal oxide semiconductor (BiCMOS), in which bipolar transistors are used for high speed and CMOS transistors are used for low power, high-density gates; differential cascode voltage switch logic (DVSL), in which differential output logic uses positive feedback for speed-up; differential split-level logic (DSL), in which load is used to reduce output voltage swing; and pass transistor logic (PTL), in which complex logic such as muxing is easily supported.

65.4.5.2 Cell Layout Architecture

There are various issues involved in deciding how a cell should be laid out. Let us look at some of the issues.

Cell height: If row-based block layout tools are going to be used, then the cells should be designed to have standard heights. This approach also helps in placement during full-custom layout. Basically, constraining one dimension (height) enables better optimization for the other one (width). However, snapping to a particular height may cause unnecessary waste of active transistor area for cells with small drive strengths.

Diffusion orientation: Manufacturing may cause some variation in cell geometries. In order to achieve consistent variations across all transistors inside a cell, process technology may dictate fixed orientation of transistors.

Metal usage: Cells are part of a larger block. They should allow block-level over-the-cell routing. Guidelines for strict metal usage must be followed while laying out cells. Some cell guidelines may force single-metal usage inside the cell.

Power: Cells must adhere to the block-level power grid. They should either instantiate power pins internally and include the power pins in the interface view, or should enable block-level power routing by abutment. In UltraSparc-I™, there was a clear separation of metal usage between datapath and control standard cells. The power in control was distributed on horizontal metal1 with adjacent cells abutting the rails. Metal2 was only used to connect metal1 to metal3 power. Metal2 power hook-up could have been longer for better power delivery, but it would consume routing resources. The datapath library had vertical metal2 abutting for power and it was directly connected to metal3 power grid [9].

Cell abstraction: Internal layout details of a cell are not required at the block level. Cells should be abstracted to provide a simplified view of interface pins (ports), power pins, and metal obstructions. Design guidelines may have requirements for coherent cell abstract views. Multiple cell families may differ in their internal layout, but there may be a need for generating consistent abstract views for easy placement and routing.

Port placement: If channel routers are used, then interface ports must lie at the cell boundaries. For area routers, the ports can be either at the boundary or at internal locations where there is enough space to drop a via from a higher metal layer passing over the cell.

Gridding: All geometries inside the cell must lie on the manufacturing grid. Some automatic tools may enforce gridding for cell abstracts. In that case, the interface ports must be on a layout routing grid dictated by the tools.

Special requirements: These can include family-specific constraints. A domino cell may need specific clock placement; a different logic cell may need strict layout matching for differential signals, etc.

Stretchability: Consider two versions of the CMOS NOR3 gate as shown in Figure 65.11. As we can see, the widths of the transistors changed, but the overall layout looks very similar. This is the idea behind stretchability and soft libraries. Generate new cells from a basic cell, depending on the drive strength required. In the G4 processor, IBM design team used a continuously tunable, parameterized standard cell library with logic functions chosen for performance [24]. The cells were available in discrete levels or sizes. The rules were continuously tunable. Parameterization was done for delay, not size. They also had a parameterized domino library. Beta and gain tuning enabled delay optimization during placement, even after initial placement. Changes due to actual routing were handled as engineering change orders (ECOs). The cell layouts were generated from soft libraries. The automatic generator concentrated on simple static cells. The most complex cell was a 2×2 AO/OA. The soft library also allowed customization of cell images. The cell generator generated a standard set of sizes, which were selected and used over the entire chip. This approach loses the cell library notion. So, the layout was completely flattened. Some cells

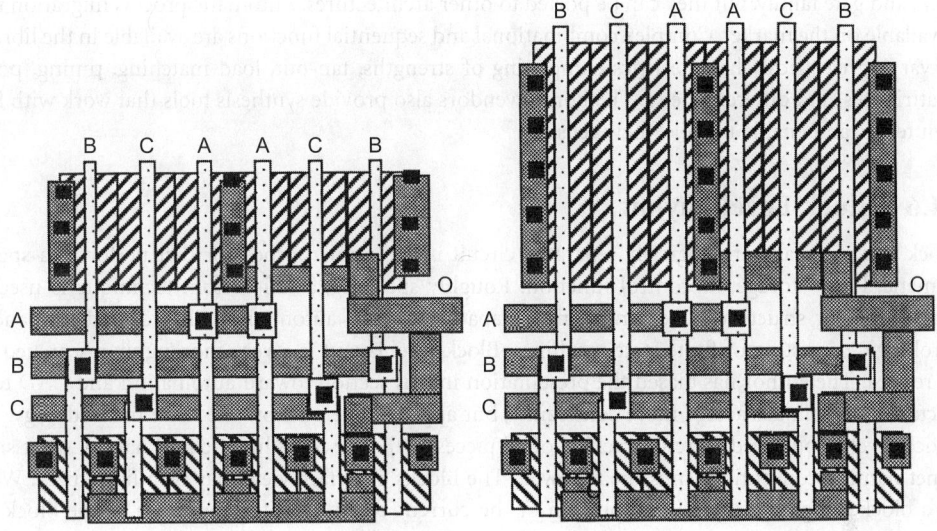

FIGURE 65.11 Cell stretching.

were also non-parameterized. Schematics were generated on the basis of tuned library and flattened layout. This basically led to a block-level mega-cell just like a standard cell.

Characterization: As we mentioned before, circuit aspects of cell design are out of the scope of this section. However, we briefly explain characterization of the cell because it impacts layout. The detailed electrical parasitics of cell layout are extracted and the behavior of each library cell is individually characterized over a range of output loads and input rise/fall times. The parameters tracked during this process are propagation delay, output rise/fall times, and peak/average current. The characterization can be represented as a closed-form equation of input rise/fall times, output loading, and device characteristics inside the cell. Another popular method involves generating look-up table models for the equations. The tables need interpolation methods. Using the process data and electromigration limits, the width of signal/supply rails and minimum number of contacts were determined in UltraSparc-I™. These values are formulated as a set of layout verification rules for post-layout checks [9]. In the PowerPC microprocessor, all custom circuits and library elements were simulated over various process corners and operating conditions to guarantee reliable operation, sufficient design margin, and sufficient scalability [7].

Mega-cells: Today's superscalar microprocessors have regular and modular architectures. Not only standard cells, but large layout blocks such as clock drivers, ROMs, and ALUs can also be repeated at several locations on the die. Mega-cells is a concept that generalizes standard cells to a larger size. This automatically converts logic function to a datapath function. Automatic layout is not recommended for mega-cells because of the internal irregularity. Layout optimization of a mega-cell is done by full-custom technique, which is time-consuming; but if it is used multiple times on the die, the effort pays off.

65.4.5.3 Cell Synthesis

As mentioned earlier in this section, there are CAD vendors supporting library generation tools. Cadabra (www.cadabratech.com) is a leading vendor in this area with its CLASSIC tool suite. Another notable vendor tool is Tempest-Cell from Sycon Design Inc. (www.sycon-design.com). A very good overview of such tools and external library vendors is available in Ref. 28. The idea of external libraries originated from IC databooks. In the past, ready-to-use ICs were available from various vendors with fully detailed electrical characteristics. Now, the same concept is applied to cell libraries, which are not ICs, but ready-to-use layouts that can be included in bigger circuits. The libraries are designed specific to a particular process and gate family, but they can be ported to other architectures. Automatic process migration tools are available on the market. Complex combinational and sequential functions are available in the libraries with varying electrical characteristics comprising of strengths, fan-out, load matching, timing, power, area attributes, and different views. The library vendors also provide synthesis tools that work with logic design teams and enable usage of new cells.

65.4.6 Block-Level Layout

A block is a physically and logically separated circuit inside a microprocessor that performs a specific arithmetic, logic, storage, or control function. Roughly speaking, a full-custom technique is used for layout of regular structures, like arrays and datapath; whereas, automatic tools are used for random control logic consisting of finite state machines. Block-level layout is a very thoroughly researched and mature area. The author has biased the presentation in this section toward automation and CAD tools. Full-custom techniques accept more constraints, but approximately follow the same methodology.

Block-level layout needs careful tracking of all pieces [29]. Due to its hierarchical nature, strict signal and net naming conventions must be followed. The blocks' interface view may be a little fuzzy. Where does a block design end? At the output pin of the current block or at the input pin of the block it is feeding to? There may be some logic that cannot be classified into any of the types and it is not large enough to be considered a separate block of its own. Such logic is called glue logic. Glue logic at the chip level may actually be tightly coupled to lower-level gates. It needs physical proximity to the lower level. Every block may be required to include some part of such glue logic during layout.

In IBM's G4 microprocessor, custom layout was used for dataflow stacks and arrays. A semi-custom cell-based technique was used for control logic [24]. Capacitive loading at the block outputs was based on preliminary floorplan analysis. During the early phase of the design, layout-dependent device models were used for block-level optimization. For UltraSparc™, layout of mega-cells and memory cells was done in parallel with RTL design [30]. Initial layout iterations were performed with estimated area and boundaries. There were concurrent chip and block-level designs as well as concurrent datapath and standard cell designs. The concurrency yielded faster turn-around time for logical-physical design iterations. Critical net routing and detailed routing was done after the block-level layout iterations converged.

A survey of CAD tools available on the market for block-level layout is included in Table 65.3. The author presents various steps in the block-level layout process in the following sections. Constraints associated with different block types are also included in the individual sections, wherever applicable.

65.4.6.1 Placement

The chip planner partitions the circuit into different blocks. Each block consists of a netlist of standard cells or subblocks, whose physical and electrical characteristics are known. For the sake of simplicity, let us only consider a netlist of cells inside the block. The area occupied by each block can be estimated and the number of block-level I/Os (pins) required by each block is known. During the placement step, all of the movable pins of the block and internal cells are positioned on the layout surface, in such fashion that no two cells are overlapping and enough space is left for interconnection among the cells.

Figure 65.12 illustrates an example placement of a netlist. The numbers next to the pins of the cells on the left side specify the nets they are connected to. The placement problem is stated as follows: given an electrical circuit consisting of cells, and a netlist interconnecting terminals on these cells and on the periphery of the block itself, construct a layout indicating positions of these blocks such that all the nets can be routed and the total layout area of the block is minimized. For high-performance microprocessors, an alternative objective is chosen where the placement is optimized to minimize the total delay of the circuit by minimizing lengths of all critical paths subject to a fixed block area constraint. In full-custom style, the placement problem is a packing problem where cells of different sizes and shapes are packed inside the block area.

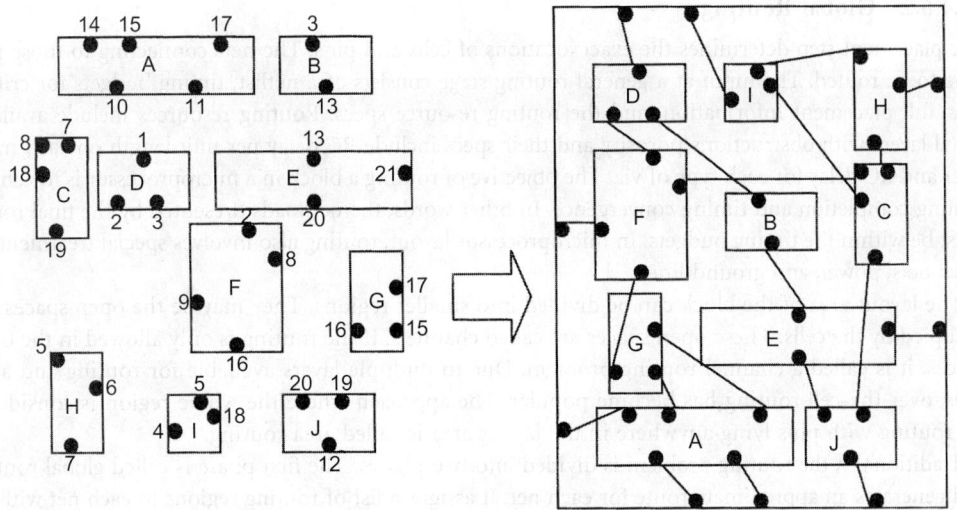

FIGURE 65.12   Example of placement.

Various factors affect the decisions taken during placement. We discuss some of the factors. All microprocessor designers may face many additional constraints due to the circuit families, types of libraries, layout methodology, and schedule.

Shape of the cells: In automatic placement tools, the cell are assumed to be rectangular. If the real cell is not rectangular, it may be snapped to an overlapping rectangle. The snapping tends to increase block area. Cells may be flexible and different aspect ratios may be available for each cell. Row-based placement approaches also need standardized height for all the cells.

Routing considerations: All of the tools and algorithms for placement are routing driven. Their objective is to estimate routing lengths and congestions at the placement stage and avoid unroutability. The cells have to be spaced to allow routing completion. If over-the-cell (OTC) routes are used, then the spacing may be avoided.

Performance: For high-performance circuits, critical nets must be routed within their timing budgets. The placement tool has to operate with a fast and accurate timing analyzer to evaluate various decisions taken during placement. This approach is called performance-driven placement. It forces cells connected to critical nets to be placed very close to each other, which may leave less space for routing that critical net.

Packaging: When the circuit is operational, all cells generate heat. The heat dissipated should be uniform over the entire layout surface of the block. The high power-consuming cells will have to be spaced apart. This approach may directly conflict with performance-driven placement. C4 bumps and power grids may cause some restrictions on allowable locations for some of the cells.

Pre-placed cells: In some cases, the locations of some cells may be fixed or a region may be specified for their placement. For instance, a block-level clock buffer must be at the exact location specified by the clock planner to achieve minimum skew. The placement approach must follow these restrictions.

Special considerations: In microprocessor designs, the placement methodology may be expected to place and sometimes reorder the scan chain. Parts of blocks may be allowed to overlap. Block-level pins may be ordered but not fixed. If the routing plan separates chip and block-level routing layers, there may be areal block-level I/Os in the middle of the layout area.

The CAD algorithms for placement have been thoroughly studied over many decades. The algorithms are classified into simulated annealing-based, partitioning-based, genetic algorithm-based, and mathematical programming-based approaches. All of these algorithms have been extended to performance-driven techniques for microprocessor layouts. For an in-depth analysis of these algorithms, please refer to Refs. 11 and 12.

65.4.6.2 Global Routing

The placement step determines the exact locations of cells and pins. The nets connecting to those pins have to be routed. The input at a general routing stage consists of a netlist, timing budgets for critical nets, full placement information, and the routing resource specs. Routing resources include available metal layers with obstructions/porosity and their specs include RC delay per unit length on each metal layer and RC delay for each type of via. The objective of routing a block in a microprocessor is to achieve routing completion and timing convergence. In other words, the net loads presented by the final routes must be within the timing budgets. In microprocessor layout, routing also involves special treatment for clock nets, power, and ground lines.

The layout area of the block can be divided into smaller regions. They may be the open spaces not occupied by the cells. These open spaces are called channels. If the routing is only allowed in the open spaces, it is called a channel routing problem. Due to multiple layers available for routing and areal I/Os, over-the-cell routing has become popular. The approach where the whole region is considered for routing with pins lying anywhere in the layout area is called area routing.

Traditionally, the routing problem is divided into two phases. The first phase is called global routing and generates an approximate route for each net. It assigns a list of routing regions to each net without specifying the actual geometric layout of wires. The second phase, called detailed routing, will be discussed in the next subsection.

Global routing consists of three phases: region definition, region assignment, and pin assignment. During definition, the regions are decided by partitioning the routing space into different regions. Each region has a capacity, which means the maximum number of nets that can pass through that region on a layer in a direction. The routing capacity of a region is a function of design rules and wire geometries. During the second phase, nets or parts if the nets, are assigned to various regions, depending on the current occupancy and the net criticality. This phase identifies a sequence of regions through which a net will be routed. Once the region assignment is done, pins are assigned at the boundary of the regions so that the detailed routing can proceed on each region independently. As long as the pins are fixed at the region boundaries, the whole layout area will be fully connected by abutment.

There is a slight difference between full-custom and automatic layout styles for global routing. In full custom, since regions can be expanded, some violations of region capacities is allowed. However, too many violations may enforce a re-placement.

Some of the factors affecting the decisions taken at global routing are:

Block I/O: Location of block I/Os and their distribution along the periphery may affect region definitions. Areal I/Os need special considerations because they may not lie at a region boundary.
Nets: Multi-terminal nets need special consideration during global routing. There is a different class of algorithms to handle such nets.
Pre-routes: There may be pre-routed nets, like clock, already occupying region capacities. A completely unconnected bus may be passing through the block. Such pre-routes have to be correctly modeled in the region definition.
Performance: Critical nets may have a length and via bound. The number of vias must be minimized for such nets. Critical nets may also need shielding, so they have to be routed next to a power route. Some nets may have spacing requirements with respect to other nets. Some nets may be wider than others, and the region occupancy must include the extra resources required for wide routes.
Detailed router: The type and style of detailed routing affects the decisions taken during the global routing. The detailed router may be a channel router, for which pins must be placed on the opposite sides of the region. In some cases, the detailed router may need information about via bounds from the global router.

Global routing is typically studied as a graph problem. There are three types of graph models to represent regions and their capacities, namely, the grid graph model, the checker board model, and the channel intersection graph model. For two terminal nets, there are three types of global routing algorithms: maze routing, line-probe, and shortest path based. For multi-terminal routing, Steiner tree-based approaches are very popular. There are some mathematical formulations for global routing; however, they provide solutions on small blocks only.

65.4.6.3 Detailed Routing

Global routing uses the original net information and separates the routing problem into a set of restricted region routing problems. A routing region can be a channel (pins on opposite sides), a 2-D switchbox (pins on all sides in 2-D), or a 3-D switchbox (pins on all faces in 3-D). The detailed router places the actual wire segments within the regions, thus completing the required connection between the cells. There is a limited scope for the regions to expand into other regions. A detailed router has to intelligently order the regions to be routed, depending on the occupancy and criticality. Factors affecting detailed routing are:

Metal layers: Traditionally, two or three routing layers were available at the block-level detailed routing. There are numerous techniques published for two- or three-layer detailed routing. Today's microprocessors consist of four or five metal layers. The number of layers is likely to increase to ten in the near future. A detailed router should fully utilize the available layers. Their widths, spacing, pitch, and electrical requirements must be obeyed. Obstructions must be handled on all metal layers.

Via: The via count is of major concern in detailed routing and must be minimized to improve performance and area. Vias impact manufacturability, cause RC delays, signal reflections, and transmission line effects. They also make post-layout compaction difficult.

Nets: Traditionally, a multi-terminal net is decomposed into a set of two terminal nets for ease of routing. Current approaches handle multi-terminal nets directly. Variable-width nets need special attention during detailed routing. In high-performance designs, nets may also be tapered, that is, the same routing segment of a net may have variable widths. The detailed router should support tapering. Due to the criticality, some nets may be required to be routed across all the regions before the rest of the nets. This breaks the paradigm for sequential region routing, unless such nets are modeled as pre-routes.

Region specs: Depending on the type of the region, pins may be located at various boundaries or faces. Regions may be flexible to some extent. However, the detailed router must try not to exceed the region bounds.

Gridding: A detailed router may assume wire gridding, implying that the pitch of wires on any metal layer is considered fixed. All pins in the regions and on the cell are on the routing grid specified by the detailed router. The layout area can be modeled as an array of grid points. Hence, the routing is very fast. Gridding hinders routing with variable-width variable spacing of metal layers. It can be accomplished at the cost of area. Hence, non-gridded routers are used in microprocessors for critical net routing.

Until the process technology advanced to the point when over-the-cell (OTC) routing became feasible, channel routing was the most popular area of research for CAD. The channel routing approaches are classified into algorithms for a single layer, a single row, two layers, and three layers. Multi-layer channel routing algorithms have also been published. Channel routing approaches can also be extended to switchboxes. The switchbox routing is not guaranteed to complete. A rip-up and re-route utility is added to the detailed routers for switchboxes.

Let us understand some of the routing tools and methodologies followed internal to various microprocessor companies. IBM developed a grid-based router to connect blocks together [5]. For the G4 processor, they employed two strategies. In the first method, chip-level routing was performed without any blockages from the block level [24]. Then, the block level routes tap the chip-level shadows appropriately. This approach was used only where wiring resources were limited. In the alternative method, the wiring tracks were divided between chip and block level. The negative image of each level was available at the other level. Pre-routes were also supported. The second method enables parallel routing effort while the first enables efficient use of wiring resources. Long routes were split at appropriate places and buffers (repeaters) were placed to minimize delays.

In HP's PA-8000, the block router is really pushing the limits of technology. It achieves high routing completion, supports multi-width wires, optimizes the ratio of wire area/block area, has a fast turnaround time, and strictly follows a rigid placement model [31]. The router was originally a channel router with blocks and channels, but it was modified for multiple layers. The placement of C4 I/O bumps is fixed. Changes in locations of bumps may cause alpha-particle emission. Hence, metal5 was not included with other layers during automatic routing. Routing channels were not expandable, but they could be moved. An electrical model of the block I/Os was supplied to the router. The area routing problem was converted to channels with blockages so that an in-house channel router could be used. L-shaped blocks were cut into two rectangular blocks, but intelligent port placement and constraints bound them together so that the same block router was used. In earlier HP processors, the ports were at the block boundary. In PA-8000, over-the-block (OTB) routing was supported. Blocks were considered black-boxes at the chip level and no internals were supplied to the router; however, an abstract virtual grid model of each block was available. The grid model enabled the lowest cost path of a global net to traverse through any region over a block. The router minimized jogging and distributed unavoidable jogs to reduce congestion. A sophisticated net flow optimizer was developed for obstacles, ports inside the block, jog allocation, and optimal exit points to avoid jogging. A density estimator was used for close estimation of detailed routing. It had port models and net characteristics for multi-terminal net routing. The topology of ports and obstacles was negotiated between the chip and block layouts. The OTB router supported variable widths and spacing. A graph theoretic approach was used to allocate trunks in channels with obstacles. The routers did not support crosstalk or delay modeling. When these violations occurred, jog insertion and wrong-side segmenting was employed. The router always finished routing under constrained placement and reported spacing problems.

65.4.6.4 Compaction

The original idea behind compaction was to improve layout productivity. The designers were free to explore alternative layout strategies and generate a topological design without geometrical details. The compaction tool was expected to produce a correct geometrical design from the topological design that completely satisfied all of the design rules of the manufacturing process [32]. The approaches employing hierarchical compaction helped in chip planning and assembly because the compactors had flexibility to choose interconnections, abutment, routing area, etc.

Today, compactors are used to minimize layout area after detailed routing. They are used as automatic tools or layout aids. Due to excessive area allotment by the chip planner, sub-optimal layout algorithms, or local optimization of internal layout, some vacant space is present in the block layout area. The goal of compaction is to minimize layout without violating design rules, without significant changes to the existing layout topology, and without violating the designer specified constraints [11]. The main idea is to reduce the space between features as much as possible without violating spacing design rules. Compaction can also be used when scaling down a design to a new set of process rules. The features can be regenerated to the new process spec and the empty area around the features can be recovered using compaction [12].

A compactor needs three things: the initial layout representation, technology information, and a compaction strategy. The same approach can be applied to full-custom and automatic layout styles because there is no apparent difference between the three inputs generated by both styles.

The initial layout is represented as a constraint graph or a virtual grid. The former represents connection and separation rules as linear inequalities, which can be modeled as a weighted directed graph. A separation constraint leads to one inequality, while a connection constraint leads to two. Shadow propagation and scanlines are two examples of techniques to generate constraint graphs. The latter representation requires that each component be attached to a grid line on the layout grid. The minimum distance between grid lines is the maximum separation required between any two features occupying the grid lines. This representation leads to very fast and simple algorithms, but does not produce as good results as the constraint graph representation. All compactors allow the designers to specify additional constraints specific to a circuit.

The most popular strategy is 1-D compaction. The layout is compacted along the x-direction, followed by a compaction in the y-direction. Longest path or network flow methods are commonly used for 1-D compaction. As the full 2-D view is not available, the results may be inferior to 2-D strategy. The reader should note that the 2-D compaction problem is proven to be NP-complete. The 2-D problem is solved by an integer linear programming technique, whose complexity is exponential. So the 2-D approach is impractical even for moderate-sized circuits. There are 1H-D approaches employing zone refinement techniques, but they change the original topology of the layout.

Hierarchical compaction strategies are used to compact a full chip or large blocks. In this approach, hierarchical input representation is generated at each level of the hierarchy from the bottom up. Initially, leaf-level individual blocks or subblocks are compacted and then layout of group of blocks is compacted. Finally, a flat level compactor can also be used for generating a compact cell library.

65.4.6.5 CAD Tools

Surveys of the latest CAD tools for block-level layout are available in Refs. 25 and 33. The routers are classified into three stages. Stage 1 routing means point-to-point single-width routing without any electrical info; stage 2 means routing with geometric data and design rules, and stage 3 means interconnect RC aware routing. All tools interact with the floorplan. They consider length, timing, routability, and use automatic cell padding to minimize congestion. Some tools also perform scan chain reordering. Placement with estimated global routing is a very common feature. The tools are very mature and widely used. However, some physical design problems stem less from the technical challenge than from the lack of industry standards. Except for GDSII, there are no standard data formats. One cannot easily represent block boundaries, dimensions, ports, channel locations, connection points, open spaces for OTC across all the tools. Microprocessor layout teams go through strenuous processes to integrate point tools from various vendors to work as a common tool suite.

There are three types of constraint driven routing tools: channel routing, area routing, and hybrid routing. In channel routing, the die size is unknown. Hence, they force an additional floorplanning iteration. Area routers try to finish routing even if they violate design rules.

The major vendor for block-level placement and routing tools is Cadence (www.cadence.com). It is supplying fundamentally new engines. There is a new timing-driven flow with no need to re-synthesize. Buffer optimization is done during placement. It will soon include an extraction capability and analysis of crosstalk, electromigration, and hot electron effects. The new Warp router eliminates clock skew. Cadence also supplies a detailed router, IC craftsman, capable of shape-based routing. It is a stage 3 router. The warp router will have the same capability soon. Currently available block-level layout tools are presented in Table 65.3. The reader should note that all of the automatic tools also support manual editing. So they can be used as layout editors for full custom techniques.

65.4.7 Physical Verification

Let us re-visit the physical design flow described earlier. The chip planner partitions the chip into blocks, the blocks are floorplanned, critical signals are routed, the blocks are laid out, and finally the chip is assembled. A large database of polygons representing the physical features inside the chip is generated. The chip layout represented in the database must be verified against the high-level architectural goals of the microprocessor, such as frequency, power, manufactuarability, etc. Post-silicon debug is an expensive process. In some cases, editing the manufactured die may be impossible. Physical verification is the last, but very important step during microprocessor layout method. If a serious design rule or timing violation is observed, the entire layout process may have to be re-visited, followed by re-verification.

TABLE 65.3 Currently Available Block-Level Tools

Company	Internet	Tool	Block Type	Description
Arcadia Design Systems	www.arcadiadesign.com	Mustang	Datapath	Regularity extraction and placement
Avanti Corp.	www.avanticorp.com	Apollo	Control, mega-blocks	All path timing-driven place and route
Cadence	www.cadence.com	Silicon Ensemble	Control, mega-blocks	Timing-driven place and route
Cadence	www.cadence.com	IC Craftsman	All	Detailed routing
Duet Technologies	www.duettech.com	Epoch	Control	Placement and timing-driven routing
Everest Design Automation	www.everest-da.com	(Under development)	Control	Interconnect design, physical floorplannig, gridless routing
Gambit Automated Design	www.gambit.com	Grandmaster	Control	Parallel processing-based place and route
Mentor Graphics Corp.	www.mentorg.com	IC Station	Control, mega-blocks	Cell-based place and route
Snaketech, Inc.	www.snaketech.com	Cellsnake	Control	For cell-based ICs
Stanza Systems, Inc.	www.stanzas.com	PolarSLE	All	Custom layout editor with router
Sycon Design, Inc.	www.sycon-design.com	Tempest-Cell	All	Layout synthesis, structured custom style or block-level place and route
Tanner EDA	www.tanner.com	Tanner Tools Pro	Control	Editing, placement, routing, simulation
Timberwolf Systems, Inc.	www.twolf.com	TimberWolf	Control	Placement, global routing, detailed routing

The reader may be aware of commonly used terms during physical verification: post-layout performance verification (PLPV), design rule checking (DRC), electrical rule checking (ERC), and layout verification system (LVS). ERC and PLPV involve extracting the layout in the form of electrical elements and analyzing the electrical representation of the circuit by simulation methods. Some CAD vendors and microprocessor design teams are investing in new tools to reveal the full effects of a circuit's parasitic coupling, delays, degradation, signal integrity, crosstalk, IR drops, hot spots from thermal build-up, charge accumulation, electromigration, etc. Simulation and electrical analysis is beyond the scope of this chapter.

There are two types of design rules checked during DRC. The first type are composition rules, which describe how to construct components and wires from the layers that can be fabricated. The other type are spacing rules, which describe how far apart objects in the layout must be for them to be reliably built [32]. Adherence to both types is required during DRC. The rules are checked by expanding the components and wires into rectangles as specified by their design rule views.

Due to the confidential nature of manufacturing process, the exact details of the verification methods are proprietary to the microprocessor manufacturers. There is a significant gap between silicon capabilities and CAD tools on the market [29]. The high-performance requirements need verification to be done at greater levels of detail and accuracy. Due to the large number of transistors in a microprocessor, there is an explosion of layout data. To solve this problem, verification should provide a close interaction between front-end design and back-end layout. It should be able to operate on approximate data available at various stages of the layout to identify potential problems related to power, signal integrity, electromigration, electromagnetic interference, reliability, and thermal effects.

The challenges involved in physical verification and available vendor tools for automatic verification are presented in Ref. 33. These tools are modified inside the microprocessor design teams to conform to the confidential manufacturing and architectural specification. The basic problem suffered by all tools is too much data from accurate physical analysis. In a typical microprocessor, there may be

500,000 nets, which lead to 21 million coupling capacitors and 2.5 million resistances. Hence, fast and accurate verification is a problem. The number of parasitic effects and circuit data is growing with every microprocessor generation. Unless efficient physical verification tools are available, over-engineering will continue to compensate for the uncertainty in final parasitics. Process shrinks are causing more layers, more interconnect, 3-D capacitive effects, and even inductive effects. The lack of efficient verification tools prohibits further feature shrinks. Verification has to be a complex set of algorithms handling large data. There is a need for incremental and hierarchical systems that have new parasitic extractors, circuits analyzers, and optimizers. Some microprocessor layout designers have employed automatic updates of routed edges, non-uniform etching, and remedies for the antenna effect.

Let us discuss some verification approaches followed by leading microprocessor manufacturers. Alpha 21264 includes very high-speed circuits and the layout was full-custom [8]. It needed careful and detailed post-layout electrical verification. No CAD tools capable of handling this were available. Therefore, an internally developed simulator was used. It is non-logic; that is, it checks timing behavior, electrical hazards, reliability, charge sharing, IR noise, interconnect capacitance, noise-induced minority carrier injection, circuit topology violations, dynamic nodes, latches, stack height minimization, leaker usage, fan-in-fan-out restrictions, wireability, beta ratios, races, edge rates, and delays.

The verification for the G4 microprocessor at IBM was divided between chip level and block level [24]. The modeling had three levels of accuracy: namely, statistical, Steiner, and detailed RC. Pathmill[2] was used for timing analysis. The verification tool extracted and analyzed the layout and inserted decoupling capacitors, wide wires, and repeaters automatically. If a full-chip long net was found not to meet its timing, a repeater had to be inserted on the net. IBM observed a problem with the repeater insertion methodology. What if the die does not have a space at the location of the repeater to be inserted? Some space had to be deliberately created for this problem.

In UltraSparc-I™, the power network was extensively verified using an internal tool called PGRID [9]. The block-level layout was translated into a schematic model for the chip-level verification. The voltages at four corners of a block were extracted from HSPICE runs. Finally, a graphical error map for electromigration and IR drop violations was generated at all levels of the layout.

References

1. T. Jamil, Fifth-generation microprocessors, *IEEE Potentials*, 15(5), 33, Dec. 1996–Jan. 1997.
2. R.N. Noyce, Microelectronics, *Scientific American*, 237(3), 65, Sept. 1977.
3. M.K. Gowan, L.L. Biro, and D.B. Jackson, Power considerations in the design of the alpha 21264 microprocessor, *Proceedings of Design Automation Conference*, pp. 726–731, 1998.
4. M. Matson et al., Circuit Implementation of a 600 MHz Superscalar RISC Microprocessor, *ICCD 98*, pp. 104–110, 1998.
5. S. Posluszny et al., Design Methodology for a 1.0 GHz microprocessor, *ICCD*, pp. 17–23, 1998.
6. A. Kumar, The HP PA-8000 RISC CPU, *IEEE Micro.*, 17, 27, 1997.
7. G. Gerosa, A 250 MHz 5-W PowerPC microprocessor with on-chip L2 cache controller, *IEEE Journal of Solid State Circuits*, 32, 11, 1997.
8. Gronowski et al., High-performance microprocessor design, *IEEE Journal of Solid-State Circuits*, 33(5), 676, 1998.
9. A. Dala, L. Lev, and S. Mitra, Design of an efficient power distribution network for the UltraSPARC-I™ Microprocessor, *Proceedings of ICCD*, pp. 118–123, 1995.
10. K. Diefendorff, K7 Challenges *Intel. Microprocessor Report*, 12, Oct. 26, 1998.
11. N. Sherwani, *Algorithms for VLSI Physical Design Automation*, 2nd ed., Kluwer Academic Publishers, 1995.

---

[2]A tool from Synopsys.

12. S.M. Sait and H. Youssef, *VLSI Physical Design Automation Theory and Practice*, McGraw-Hill, 1995.

13. N.H.E. Weste and K. Eshraghian, *Principles of CMOS VLSI Design—A Systems Perspective*, 2nd ed., Addison Wesley, 1993.

14. S.M. Kang and Y. Leblebici, *CMOS Digital Integrated Circuits Analysis and Design*, McGraw-Hill, 1996.

15. R.J. Baker, H.W. Li, and D.E. Boyce, *CMOS Circuit Design, Layout and Simulation*, IEEE Press, 1998.

16. D.P. LaPotin, Early assessment of design, packaging and technology tradeoffs, *International Journal of High Speed Electronics*, 2(4), 209, 1991.

17. G. Bassak, Focus Report: IC physical design tools, *Integrated System Design Magazine*, Nov. 1998.

18. P.J. Dorweiler, F.E. Moore, D.D. Josephson, and G.T. Colon-Bonet, Design methodologies and circuit design tradeoffs for the HP PA 8000 processor, *Hewlett-Packard Journal*, 48, 16, Aug. 1997.

19. E. Malavasi, E. Charbon, E. Feit, and A. Sangiovanni-Vincentelli, Automation of IC layout with analog constraints, *IEEE Transactions on CAD*, 15, 923, Aug. 1996.

20. D. Trobough, IC design drives array packages, *Integrated System Design Magazine*, Aug. 1998.

21. Farbarik et al., CAD tools for area-distributed I/O pad packaging, *Proceedings of 1997 IEEE Multi-Chip Module Conference*, pp. 125–129, 1997.

22. B.T. Preas and M.J. Lorenzetti, Physical design automation of VLSI Systems, *Introduction to Physical Design Automation*, Benjamin Cummings, Menlo Park, CA, 1988.

23. N. Sherwani, Panel Discussion, *International Symposium on Physical Design*, Monterey, CA, Apr. 1998.

24. K.L. Sheperd et al., Design methodology for the high performance G4 S/390 microprocessor, *ICCAD*, pp. 232–240, 1997.

25. [Schultz 97].

26. H. Fair and D. Bailey, Clocking design and analysis for a 600 MHz alpha microprocessor, *ISSCC Digest of Technical Papers*, pp. 398–399, Feb. 1998.

27. A. Dharchoudhury, R. Panda, D. Blauuw, and R. Vaidyanathan, Design and analysis of power distribution networks in PowerPC microprocessors, *Proceedings of Design Automation Conference*, pp. 738–743, 1998.

28. R.T. Maniwa, Focus report: design libraries, *Integrated System Design Magazine*, Aug. 1997.

29. T. Maniwa, Physical verification: challenges and problems for new designs, *Integrated System Design Magazine*, Nov. 1998.

30. A. Cao et al., CAD Methodology for the design of UltraSPARC™-I Microprocessor at Sun Microsystems, Inc., *Proceedings of 32nd Design Automation Conference*, pp. 19–22, 1995.

31. J.C. Fong, H.K. Chan, and M.D. Kruckenberg, Solving IC interconnect routing for an advanced PA-RISC processor. *Hewlett-Packard Journal*, 48(4), 40, Aug. 1997.

32. W.J. Wolf and A.E. Dunlop, Symbolic layout and compaction, Chapter 6 in *Physical Design Automation of VLSI Systems*, Benjamin Cummings, Menlo Park, CA, 1988.

33. G. Bassak, Focus report: physical verification tools, *Integrated System Design Magazine*, Feb. 1998.

66
Architecture

Daniel A. Connors
University of Colorado

Wen-mei W. Hwu
*University of Illinois
at Urbana-Champaign*

CONTENTS

The microprocessor industry is divided into the computer and embedded sectors. Both computer and embedded microprocessors share aspects of computer system design, instruction set architecture, organization, and hardware. The term *architecture* is used to describe these fundamental aspects, and more directly refers to the hardware components in a computer system and the flow of data and control information among them. In this section, various types of microprocessors will be described, fundamental architecture mechanisms relevant in the operation of all microprocessors will be presented, and microprocessor industry trends discussed.

66.1 Types of Microprocessors

Computer microprocessors are designed for use as the central processing units (CPUs) of computer systems such as personal computers, workstations, servers, and supercomputers. Although microprocessors started as humble programmable controllers in the early 1970s, virtually all computer systems built after 1990 use microprocessors as their CPUs. The dominating architecture in the computer microprocessor domain today is the Intel 32-bit Architecture, also known as IA-32 or X86. Other high-profile architectures in the computer microprocessor domain include Intel Itanium Processor Family (IPF), HP PA-RISC, SUN Microsystems SPARC, and IBM/Motorola PowerPC.

Embedded microprocessors are increasingly used in consumer and telecommunication products to satisfy the demands for quality and functionality. Major product areas that require embedded microprocessors include digital TV, DVD players, digital camera, network switches, high-speed modems, digital cellular phones, video games, laser printers, and automobiles. Future improvements in energy consumption, fabrication cost, and performance will further enable new applications such as intelligent hearing aid. Many experts expect that embedded microprocessors will form the fastest growing sector of the semiconductor business in the next decade [1].

Embedded microprocessors have been categorized into DSP processors and embedded CPUs due to historic reasons. DSP processors have been designed and marketed as special-purpose devices that are mostly programmed by hand to perform digital signal processing computations. A recent trend in the DSP market is to use compilers to alleviate the need for tedious hand coding in DSP development. Leading DSP microprocessor vendors include Texas Instruments, Lucent Technologies, Motorola, and ST.

Embedded CPUs are often used to execute operating system, networking, and user-interface code in a consumer and telecommunication products. They have been traditionally derived from out-of-date computer microprocessors. Embedded CPUs often reuse the compiler and related software support developed for their computer cousins. Recycling the microprocessor design and compiler software minimizes engineering cost. Major vendors of embedded CPUs include IBM, Motorola, ARM, and MIPS.

An important recent trend in the embedded microprocessor market is toward integrating an embedded CPU, a DSP processor, and application-specific logic to form a single-chip solution. This approach is enabled by the ever-increasing transistor density achieved by the semiconductor fabrication technology. The major benefit is reduced system cost and energy consumption. In these designs, embedded CPU and DSP processor vendors no longer market microprocessor chips. Rather, they make their designs available for licensing by solution developers such as a cell phone vendor. The embedded CPU and the DSP processors thus incorporated into these single-chip solutions are called embedded CPU cores and DSP processor cores. For example, MIPS customized its embedded CPU core for use in Nitendo64. IBM, ARM, ST, NEC, and Hitachi offer similar products and services. Owing to an increasing need to perform DSP computation in consumer and telecommunication products, an increasing number of embedded CPUs have extensions to enable more effective DSP computation.

There are several ways in which the needs of embedded computing differ from those of more traditional general-purpose computing systems. Constraints on the code size, weight, and power consumption place stringent requirements on embedded microprocessors and the software they execute. Also constraints rooted in real-time requirements are often a significant consideration in many embedded systems. Furthermore, cost is a severe constraint on designing and manufacturing embedded processors.

In spite of the different constraints and product markets, both computer and embedded microprocessors share many main elements in their design. These main elements will be described. Additionally, over the past decade, substantial research has gone into the design of microprocessors embodying parallelism at the instruction level as well as aggressive compiler optimization and analysis techniques for harnessing this opportunity. Exploitation of parallelism is a very effective approach to reducing the power consumption under given performance requirements. Much of this effort has since been validated through the proliferation of high-performance general-purpose microprocessors, such as the Intel Itanium II processor, based on these technologies. Nevertheless, growing demand for high performance in embedded computing systems is creating new opportunities to leverage these techniques in application-specific domains. The research of instruction-level parallelism (ILP) has developed distinct architecture methodologies referred to as very long instruction word (VLIW) and explicitly parallel instruction computing (EPIC) technology. Overall, these techniques represent fundamental, substantial changes in computer architecture.

66.2   Major Components of a Microprocessor

The main hardware of a microprocessor system can be divided into sections according to their functionalities. A popular approach is to divide a system into four subsystems: the central processor, the memory subsystem, the input/output (I/O) subsystem, and the system interconnect. Figure 66.1 shows the connection between these subsystems. The main components and characteristics of these subsystems will be described.

66.2.1   Central Processor

A modern microprocessor system's central processor is typically further divided into control unit and data path.

66.2.1.1   Control Unit

The control unit of a microprocessor generates the control signals to orchestrate the activities in the data path. There are two major types of communication lines between the control unit and the data path: the control lines and the condition lines. The control lines deliver the control signals from the control unit to the data path. Different signal values on these lines trigger different actions in the data path. The condition lines carry the status of the execution from data path to the control unit. These lines are needed to test conditions involving the registers in the data path to make future control decisions. Note that the decision is made in the control unit but the registers are in the data path. Therefore, the conditions regarding the register contents are formed in the data path and then communicated to the control unit for decision making. A control unit can be implemented with hardwiring, microprogramming, or a combination of both.

A hardwired control unit is designed as a finite-state machine that is realized with registers, combinational logic, and wires. Once constructed, the design can be changed only through physically rewiring the unit. Therefore, the resulting circuits are called hardwired control units. Design optimizations are typically performed on the combinational logic to minimize component count and maximize operation speed, which makes the resulting circuits exhibit little structure. The lack of structure makes it very difficult to design and debug complicated control units with this technique. Therefore, hardwiring is normally used when the control unit is relatively simple.

Most of the design difficulties in the hardwired control units are due to the effort of optimizing the combinational logic. One popular alternative to hardwired control unit design is to use either read only memory (ROM) or random access memory (RAM) to implement the combinational logic. A control unit whose combinational logic is realized by the use of ROM or RAM is called a *microprogrammed control unit*. The memory used is called *control memory (CM)*. The practice of realizing the combinational circuit in a control unit with ROM/RAM is called *microprogramming*. The concept of microprogramming was first introduced by Maurice V. Wilkes.

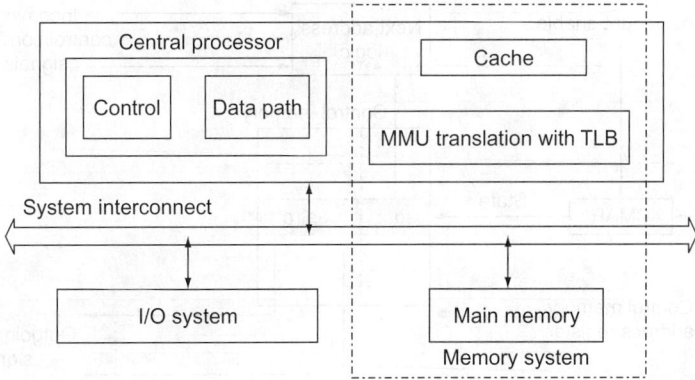

FIGURE 66.1   Subsystems of a computer system.

The idea of using a memory to implement a combinational circuit can be illustrated with a simple example. Assume that we are to implement a logic function with three input variables, as described in the truth table illustrated in Figure 66.2(a). A common way to realize this function is to use Karnaugh maps to derive highly optimized logic and wiring. The result is shown in Figure 66.2(b). The same function can also be realized using a memory with eight 1-bit locations to retain the eight possible combinations of the three input variables. Location *i* contains an *F* value corresponding to the *i*th input combination. For example, location 3 contains the *F* value (0) for the input combination 011. The three input variables are then connected to the address input of the memory to complete the design (Figure 66.2[c]). In essence, the memory contains the entire truth table. Considering the decoding logic and storage cells involved in a 8 × 1 memory, it is obvious that the memory approach uses a lot more hardware components than the Karnaugh map approach. However, the design is much simpler in the memory approach.

Figure 66.3 illustrates the general model of a microprogrammed control unit. Each control memory location consists of an address field and some control fields. The address field plus the next address logic implements the combinational logic for generating the next state value. The control fields implement the combinational logic for generating the control signals. The state register/counter is referred to as *control memory address register (CMAR)* for an obvious reason: the contents of the register are used as the address input to the control memory. An important insight is that the CMAR stores the state of the control unit.

In modern microprocessor designs, hardwiring and microprogramming are often used in conjunction with each other. In such a design, hardwiring is used to handle common, simpler cases of instruction execution whereas microprogramming is used to handle complex cases such as string move instructions in the IA-32 microprocessors.

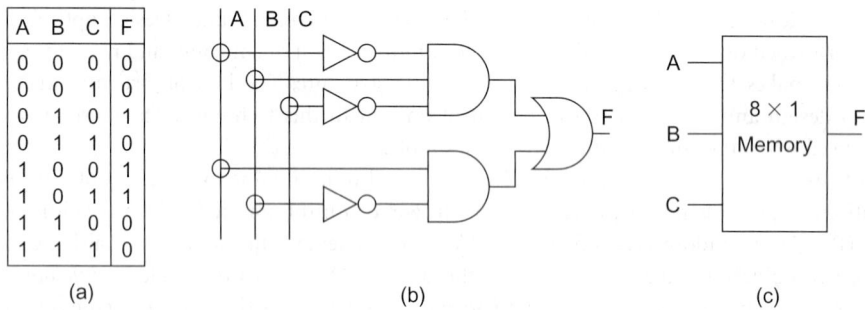

FIGURE 66.2   Using memory to simplify combinational logic design: (a) truth table, (b) Karnaugh map approach, and (c) memory approach.

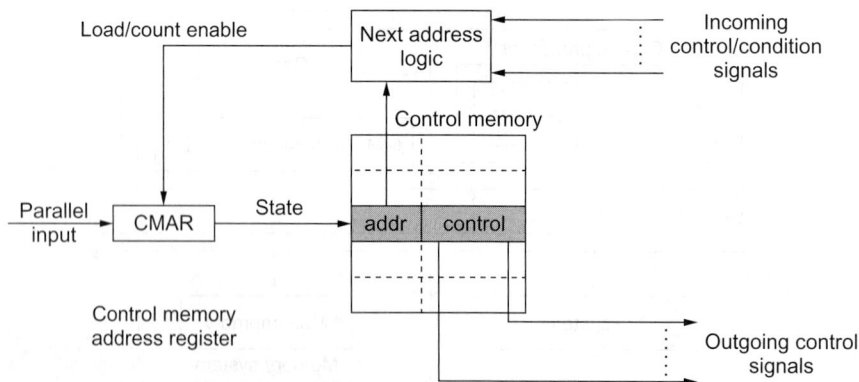

FIGURE 66.3   Basic model of microprogrammed control units.

66.2.1.2 Data Path

The data path of a microprocessor contains the main arithmetic and logic execution units required to execute instructions. Designing the data path involves analyzing the function(s) to be performed, then specifying a set of hardware registers to hold the computation state, and designing computation steps to transform the contents of these registers into the final result. In general, the functions to be performed are divided into steps each of which can be done with a reasonable amount of logic in one clock cycle. Each step brings the contents of the registers closer to the final result. The data path must be equipped with sufficient amount of hardware to allow these computation steps in one clock cycle. The data path of a typical microprocessor contains integer and floating-point register files, 10 or more functional units for computation and memory access, and pipeline registers. The concept of pipelining and pipeline registers will be introduced later in this chapter.

66.2.2 Memory Subsystem

The memory system serves as a repository of information in a microprocessor system. The processing unit retrieves information stored in memory, operates on the information, and returns new information back to memory. The memory system is constructed from basic semiconductor DRAM units called modules or banks.

There are several properties of memory, including speed, capacity, and cost that play an important role in the overall system performance. The speed of a memory system is the key performance parameter in the design of the microprocessor system. The latency (L) of the memory is defined as the time delay from when the processor first requests data from memory until the processor receives the data. Bandwidth is defined as the rate at which information can be transferred to and from the memory system. Memory bandwidth and latency are related to the number of outstanding requests (R) that the memory system can service:

$$\text{BW} = \frac{R}{L} \tag{66.1}$$

Bandwidth plays an important role in keeping the processor busy with work. However, technology trade-offs to optimize latency and improve bandwidth often conflict with the need to increase the capacity and reduce the cost of the memory system.

66.2.2.1 Cache Memory

Cache memory, or simply cache, is a fast memory constructed using semiconductor SRAM. In modern computer systems, there is usually a hierarchy of cache memories. The top level (Level-1 or L1) cache is closest to the processor and the lowest level is closest to the main memory. Each lower-level cache is about 3 to 5 times slower than its predecessor level. The purpose of a cache hierarchy is to satisfy most of the processor memory accesses in one or a small number of clock cycles. The L1 cache is often split into an instruction cache and a data cache to allow the processor to perform simultaneous accesses for instructions and data. Cache memories were first used in the IBM mainframe computers in the 1960s. Since 1985, cache memories have become a standard feature for virtually all microprocessors.

Cache memories exploit the principle of locality of reference. This principle dictates that some memory locations are referenced more frequently than others based on two program properties. *Spatial locality* is the property that an access to a memory location increases the probability that the nearby memory location will also be accessed. Spatial locality is predominantly based on sequential access to program code and structured data. To exploit spatial locality, memory data are placed into cache in multiple-word units called *cache lines*. *Temporal locality* is the property that access to a memory location greatly increases the probability that the same location will be accessed in the near future. Together, the two properties assure that most memory references will be satisfied by the cache memory.

Set associativity refers to the flexibility in placing a memory data into a cache memory. If a cache design allows each memory data to be placed into any of N cache locations, it is referred to as an N-way set-associative cache. A one-way set-associative cache is also called a direct-mapped cache, as is shown in Figure 66.4(a). A two-way set-associative cache, as shown in Figure 66.4(b), allows a memory data to

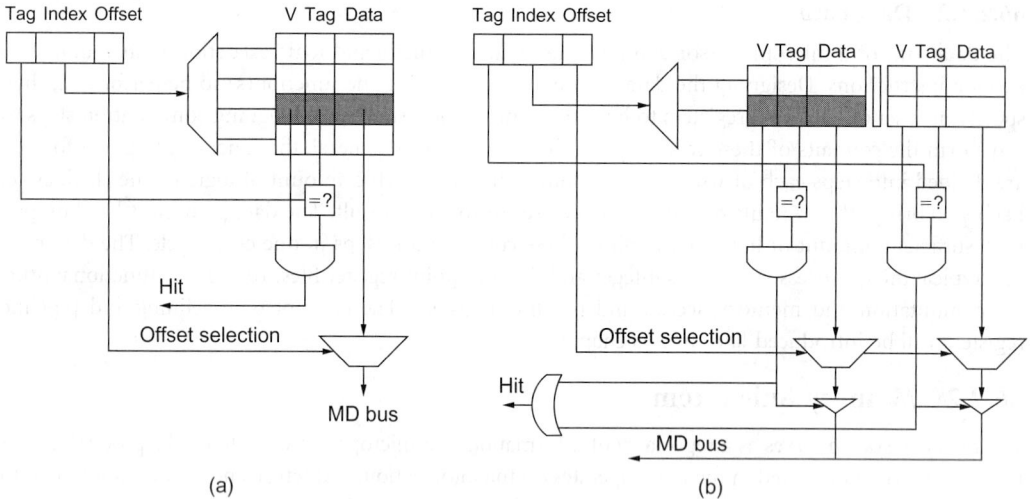

FIGURE 66.4   Cache memory: (a) direct mapped design and (b) two-way set associative design.

reside in one of two locations in the cache. Finally, an extreme design called fully associative cache allows a memory data to be placed anywhere in the cache.

Cache misses occur when the data requested does not reside in any of the possible cache locations. Misses in caches can be classified into three categories: conflict, compulsory, and capacity. Conflict misses are misses that would not occur for fully associative caches with least recently used (LRU) replacement. Compulsory misses are misses required in cache memories for initially referencing a memory location. Capacity misses occur when the cache size is not sufficient to keep data in the cache between references. Complete cache miss definitions are provided in Ref. [4].

The latency in cache memories is not fixed and depends on the delay and frequency of cache misses. A performance metric that accounts for the penalty of cache misses is *effective latency*. Effective latency depends on the two possible latencies, hit latency (L_{HIT}), the latency experienced for accessing data residing in the cache and miss latency (L_{MISS}), the latency experienced when accessing data not residing in the cache. Effective latency also depends on the *hit rate* (H), the percentage of memory accesses that are hits in the cache, and *miss rate* (M or $1 - H$), percentage of memory accesses that miss in the cache. Effective latency in a cache system is calculated as

$$L_{\text{effective}} = L_{HIT} * H + L_{MISS} * (1 - H) \qquad (66.2)$$

In addition to the base cache design and size issues, there are several other dimensions of cache design that affect the overall cache performance and miss rate in a system. The main memory update method dictates when the main memory will be updated by store operations. In a *write-through* cache, each write is immediately reflected to the main memory. In a *write-back* cache, the writes are reflected to main memory only when the respective cached data is purged from cache to make room for other memory data. Cache allocation designates whether cache locations are allocated on writes and/or reads. Lastly, cache replacement algorithms for associative structures can be designed in various ways to extract additional cache performance. These include LRU, least frequently used (LFU), random, and FIFO (first in, first out). These cache management strategies attempt to exploit the properties of locality. Traditionally, when caches service misses they would *block* all new requests. However, *nonblocking* cache can be designed to service multiple miss requests simultaneously, thus alleviating delay in accessing memory data.

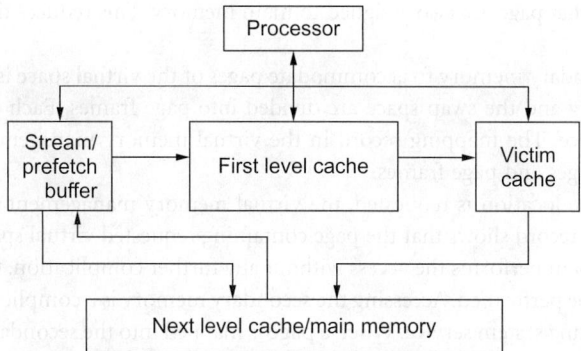

FIGURE 66.5   Advanced cache memory system.

In addition to the multiple levels of cache hierarchy, additional memory buffers can be used to improve cache performance. Two such buffers are a streaming/prefetch buffer and a victim cache [2]. Figure 66.5 illustrates the relation of the streaming buffer and victim cache to the primary cache of a memory system. A streaming buffer is used as a prefetching mechanism for cache misses. When a cache miss occurs, the streaming buffer begins prefetching successive lines starting at the miss target. A victim cache is typically a small fully associative cache loaded only with cache lines that are removed from the primary cache. In the case of a miss in the primary cache, the victim cache may hold the missed data. The use of a victim cache can improve performance by reducing the number of conflict misses. Figure 66.5 illustrates how cache accesses are processed through the streaming buffer into the primary cache on cache requests and from the primary cache through the victim cache to the secondary level of memory on cache misses.

Overall, cache memory is constructed to hold the most important portions of memory. Techniques using either hardware or software can be used to select which portions of main memory to store in cache. However, cache performance is strongly influenced by program behavior and numerous hardware design alternatives.

66.2.2.2   Virtual Memory

Cache memory illustrated the principle that the memory address of data can be separate from a particular storage location. Similar address abstractions exist in the two-level memory hierarchy of main memory and disk storage. An address generated by a program is a called a *virtual address* which needs to be translated into a *physical address* or location in main memory. Virtual memory management is a mechanism which provides the programmers with a simple, uniform method to access both main and secondary memories. With virtual memory management, the programmers are given a virtual space to hold all the instructions and data. The virtual space is organized as a linear array of locations. Each location has an address for convenient access. Instructions and data have to be stored somewhere in the real system; these virtual space locations must correspond to some physical locations in the main and secondary memory. Virtual memory management assigns (or maps) the virtual space locations into the main and secondary memory locations. The mapping of virtual space locations to the main and secondary memory is managed by the virtual memory management mechanism. The programmers are not concerned with the mapping.

The most popular memory management scheme today is demand paging virtual memory management where each virtual space is divided into pages indexed by the page number (PN). Each page consists of several consecutive locations in the virtual space indexed by the page index (PI). The number of locations in each page is an important system design parameter called page size. Page size is usually defined as a power of two so that the virtual space can be divided into an integer number of pages. Pages are the basic unit of virtual memory management. If any location in a page is assigned to the main memory, the other locations in that page are also assigned to main memory. This reduces the size of the mapping information.

The part of the secondary memory to accommodate pages of the virtual space is called the swap space. Both the main memory and the swap space are divided into page frames. Each page frame can host a page of the virtual space. The mapping record in the virtual memory management keeps track of the association between pages and page frames.

When a virtual space location is requested, the virtual memory management looks up the mapping record. If the mapping record shows that the page containing requested virtual space location is in main memory, the management performs the access without any further complication. Otherwise a secondary memory access has to be performed. Accessing the secondary memory is a complicated task and is usually performed as an operating system service. When a page is mapped into the secondary memory, the virtual memory management has to request a service in the operating system to transfer the requested page into the main memory, update its mapping record, and then perform the access. The operating system service thus performed is called the page fault handler.

The core process of virtual memory management is a memory access algorithm. A one-level memory access algorithm is illustrated in Figure 66.6. At the start of the memory access, the algorithm receives a virtual address in a memory address register (MAR), looks up the mapping record, requests an operating system service to transfer the required page if necessary, and performs the main memory access. The mapping is recorded in a data structure called the page table located in main memory at a designated location marked by the page table base register (PTBR).

Each page is mapped by a page table entry, which occupies a fix number of bytes in the page table. Thus one can easily multiply the page number by the size of each PTE to form a byte index into the page table. The byte index of the PTE is then added to the PTBR to form the physical address (PAPTE) of the required PTE. Each PTE includes two fields: a hit/miss bit and a page frame number. If the hit/miss (H/M) bit is set (hit), the corresponding page is in main memory. In this case, the page frame hosting the requested page is pointed to by the page frame number (PFN). The final physical address (PAD) of the requested data is then formed by concatenating the PFN and PI. The data is returned and

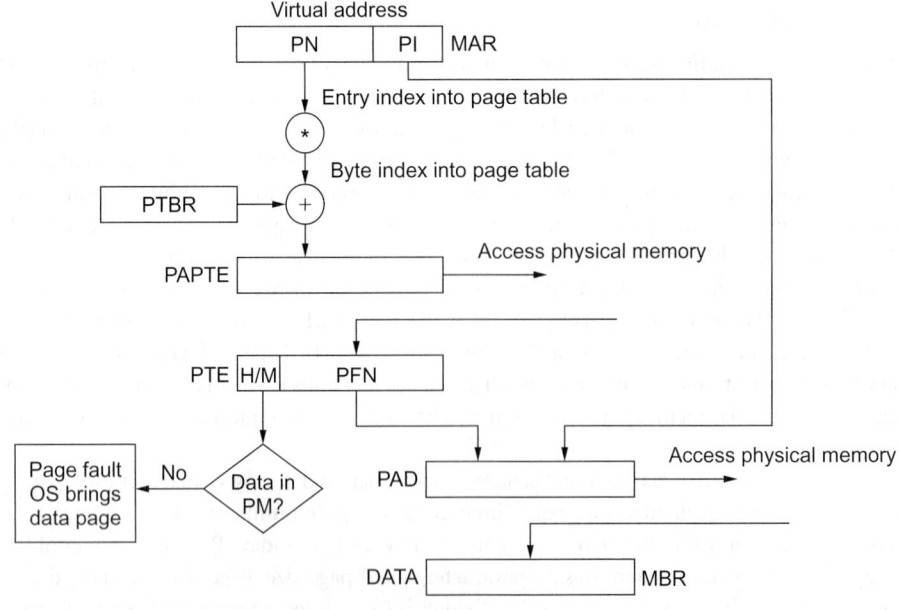

FIGURE 66.6   A simple virtual memory access algorithm.

placed in the memory buffer register (MBR) and the processor is informed of the completed memory access. Otherwise (miss), a secondary memory access has to be performed. In this case, the page frame number should be ignored. The fault handler has to be invoked to access the secondary memory. The hardware component that performs the address translation part of the memory access algorithm is called the memory management unit (MMU).

The complexity of the algorithm depends on the mapping structure. A very simple mapping structure is used in this section to focus on the basic principles of the memory access algorithms. However, more complex two-level schemes are often used when the page table becomes a large portion of the main memory. There are also requirements for such designs in a multiprogramming system, where there are multiple processes active at the same time. Each processor has its own virtual space and therefore its own page table. As a result, these system need to keep multiple page tables at the same time. It usually takes too much main memory to accommodate all the active page tables. Again, the natural solution to this problem is to provide additional levels of mapping where a second-level page table is used to map the main page table. In such designs, only the second-level page table is stored in a reserved region of main memory, while the first page table is mapped just like the code and data between the secondary storage and the main memory.

66.2.2.3 Translation Lookaside Buffer

Hardware support for virtual memory management generally includes a translation lookaside buffer (TLB) to accelerate the translation of virtual addresses into physical addresses. A TLB is a cache structure, which contains the frequently used page table entries for address translation. With a TLB, address translation can be performed in a single clock cycle when TLB contains the required page table entries (TLB hit). The full address translation algorithm is performed only when the required page table entries are missing from the TLB (TLB miss).

Complexities arise when a system includes both virtual memory management and cache memory. The major issue is whether address translation is done before accessing the cache memory. In *virtual* cache systems, the virtual address directly accesses cache. In a *physical* cache system, the virtual address is translated into a physical address before cache access. Figure 66.7 illustrates both the *virtual* and *physical* cache approaches.

A physical cache system typically overlaps the cache memory access and the access to the TLB. The overlap is possible when the virtual memory page size is larger than the cache capacity divided by the degree of cache associativity. Essentially, since the virtual page index is the same as the physical address index, no translation for the lower indexes of the virtual address is necessary. These lower index bits can be used to access the cache storage while the page number (PN) bits go through the TLB. The PFN bits that come out of the TLB are compared with the tag bits of the cache storage output, as shown in Figure 66.6, to determine the hit/miss status of cache memory access. Thus the cache storage can be accessed in parallel with the TLB.

Virtual caches have their pros and cons. Typically, with no TLB logic between the processor and the cache, access to cache can be achieved at lower cost in virtual cache systems. This is particularly true in multiaccess-per-cycle cache systems, where a multiported TLB is needed for a physical cache design. However, the virtual cache alternative introduces virtual memory consistency problems. The same virtual address from two different processes mean different physical memory locations. Solutions to this form of aliasing are to attach process identifier to the virtual address or to flush cache contents on context switches. Another potential alias problem is that different virtual addresses of the same process may be mapped into the same physical address. In general, there is no easy solution to this second form of aliasing; typical solutions involve reverse translation of physical addresses to virtual addresses.

Physical cache designs are not always limited by the delay of the TLB and cache access. In general there are two solutions to allow large physical cache design. The first solution, employed by companies with past commitments to page size, is to increase the set associativity of cache. This allows the lower index portion of the address to be used immediately by the cache in parallel with virtual address translation.

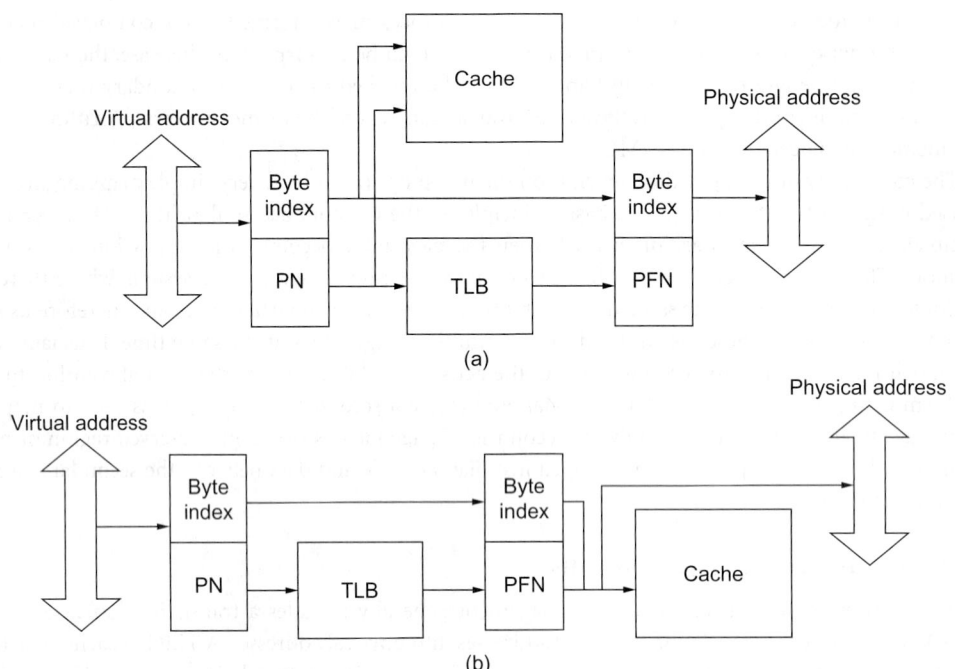

FIGURE 66.7   Translation lookaside buffer (TLB) architectures: (a) virtual cache and (b) physical cache.

However, large set associativity is very difficult to implement in a cost-effective manner. The second solution, employed by companies without past commitment is to use a larger page size. The cache can be accessed in parallel with the TLB access similar to the other solution. In this solution there are fewer address bits that are translated through the TLB, potentially reducing the overall delay. With larger page sizes, virtual caches do not have advantage over physical caches in terms of access time. With the advance of semiconductor fabrication processes, in particular the increasing levels of metals, it has become increasingly inexpensive to have highly set-associative caches and multiported TLB. As a result, physical caches have become much more favored solution today.

66.2.3   Input/Output Subsystem

The input/output (I/O) subsystem transfers data between the internal components (CPU and main memory) and the external devices (disks, networks, printers, keyboards, pointing devices, and scanners).

66.2.3.1   Peripheral Controllers

The CPU usually controls the I/O subsystem by reading from and writing into the I/O (control) registers. There are two popular approaches for allowing the CPU to access these I/O register, I/O instructions, and memory-mapped I/O. In an I/O instruction approach, special instructions are added to the instruction set to access I/O status flags, control registers, and data buffer registers. In a memory-mapped I/O approach, the control registers, the status flags, and the data buffer registers are mapped as physical memory locations.

Owing to the increasing availability of chip area and pins, microprocessors are increasingly including peripheral controllers on chip. This trend is especially clear for embedded microprocessors.

66.2.3.2   Direct Memory Access Controller

A direct memory access (DMA) controller is a peripheral controller which can directly drive the address lines of the system bus. The data is directly moved from the I/O data buffer registers to the main memory, rather than from the data buffer registers to a CPU register, then from CPU register to main memory. DMA controllers greatly increase the achieved transfer bandwidth of I/O operations and have become a standard feature of microprocessor systems today.

66.2.4 System Interconnect

System interconnect is the facilities that allow the components within a computer system to communicate with each other. There are numerous logical organizations of these system interconnect facilities.

Dedicated links or point-to-point connections enable dedicated communication between components. There are different system interconnect configurations based on the connectivity of the system components. A complete connection configuration, requiring $N(N - 1)/2$ links, is created when there is one link between every possible pair of components. A *hypercube* configuration assigns a unique n-tuple $\{1,0\}$ as the coordinate of each component and constructs a link between components whose coordinates differ only in one dimension, requiring $N \log N$ links. A *mesh* connection arranges the system components into a N-dimensional array and has connections between immediate neighbors, requiring $2N$ links.

Switching networks are a group of switches that determine the existence of communication links among components. A cross-bar network is considered the most general form of switching network and uses a $N \times M$ two-dimensional array of switches to provide an arbitrary connection between N components on one side to M components on another side using NM switches and $N + M$ links. Another switching network is the multistage network which employs multiple stages of shuffle networks to provide a permutation connection pattern between N components on each side by using $N \log N$ switches and $N \log N$ links.

Shared buses are single links that connect all components to all other components and are the most popular connection structure. The sharing of buses among the components of a system requires several aspects of bus control. First there is a distinction between bus masters, the units controlling bus transfers (CPU, DMA) and bus slaves, the other units (memory, programmed I/O interface).

Bus interfacing and bus addressing are the means to connect and disconnect units on the bus. Bus arbitration is the process of granting the bus resource to one of the masters. Arbitration typically uses a selection scheme based on some type of priority assignment. Fixed-priority arbitration gives every master a fixed priority, and dynamic-priority arbitration such as round-robin ensures that every master becomes the most favorable at one point in time.

Bus timing refers to the method of communication among the system units and can be classified as either synchronous or asynchronous. Synchronous bus timing uses a shared clock that defines the time other bus signals change and stabilize. Clock sharing by all units allows the bus to be monitored at agreed time intervals and action taken accordingly. However, the synchronous system bus must operate at the speed of the slowest component. Asynchronous bus timing allows units to use different clocks, but the lack of a shared clock makes it necessary to use extra signals to determine the validity of bus signals. The use of these signals to determine the validity of bus signals is called handshaking protocols, which typically reduce the achievable transfer bandwidth via the bus.

66.3 Performance-Enhancing Hardware Techniques

66.3.1 Pipelining

In the 1970s, only supercomputers and mainframe computers were pipelined. Today, virtually all microprocessors are pipelined. In fact, pipelining has been a major reason why microprocessors today outperform supercomputer CPUs built less than 10 years ago. Pipelining is a technique to coordinate parallel processing of operations [2]. This technique has been used in assembly lines of major industries for more than a century. The idea is to have a line of workers specializing in different pieces of work required to finish a product. A conveying belt carries products through the line of workers. Each worker performs a small piece of work on each product. Each product is finished after it is processed by all the workers in the assembly line.

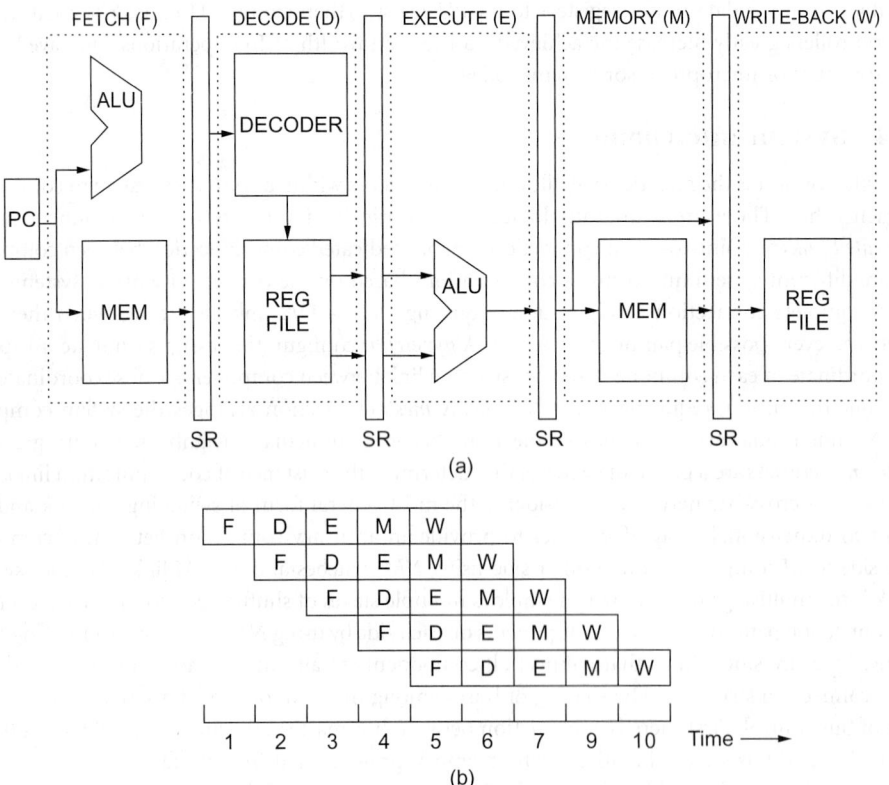

FIGURE 66.8   Pipeline architecture: (a) machine and (b) overlapping instructions.

The obvious advantage of pipelining is to allow one worker to immediately start working on a new product after finishing the work on a current product. The same methodology is applied to instruction processing in microprocessors. Figure 66.8(a) shows an example five-stage pipeline dividing instruction execution into fetch (F), decode (D), execute (E), memory (M), and write-back (W) operations, each requiring various stage-specific logic. Between each stage is a stage register (SR), or pipeline register, used to hold the instruction information necessary to control the instruction. A very basic principle of pipelining is that the work performed by each stage must take about the same amount of time. Otherwise, the efficiency will be significantly reduced because one stage becomes a bottleneck of the entire pipeline. That is, the time duration of the slowest pipeline stage determines the overall clock frequency of the processor. Owing to this constraint and the often-time slower memory speeds, some of the principle five stages are often divided into smaller stages. For instance, the memory stage may be divided into three stages, allowing memory accesses to be pipelined and the overall processor clock speed to be a fraction of the memory access latency.

The time required to finish N instructions in a pipeline with K stages can be calculated. Assume a cycle time of T for the overall instruction completion, and an equal T/K processing delay at each stage. With a pipeline scheme, the first instruction completes the pipeline after T, and there will be a new instruction out of the pipeline per stage delay T/K. Therefore the delays of executing N instructions without (1) and with (2) pipelining, respectively, are

$$T*(N) \tag{66.3}$$

$$T + (T/k)*(N-1) \tag{66.4}$$

There is an initial delay in the pipeline execution model before each stage has operations to execute. The initial delay is usually called *pipeline startup delay*, and is equal to total execution time of one instruction (T). The speedup of a pipelined machine relative to a nonpipelined machine is calculated as

$$\frac{T * N}{T + (T/k) * (N-1)} \tag{66.5}$$

When N is much larger than the number of pipe stages k, the ideal speedup approaches k. This is an intuitive result since there are k parts of the machine working in parallel, allowing the execution to go about k times faster in ideal conditions.

The overlap of sequential instructions in a processor pipeline is shown in Figure 66.8(b). The processing of each instruction is shown as a five-stage process that goes from left to right; each stage is marked by the initial of its name: F(etch), D(ecode), E(xecute), M(emory), W(riteback). A new instruction is initiated in every clock cycle. The instruction pipeline becomes full after the pipeline delay of $k = 5$ cycles. Although the pipeline configuration executes operations in each stage of the processor, two important mechanisms are constructed to insure correct functional operation between dependent instructions in the presence of data hazards. Data hazards occur when instructions in the pipeline generate results that are necessary for later instructions that are already started in the pipeline. In the pipeline configuration of Figure 66.8(a), register operands are initially retrieved during the decode stage. However, the execute and memory stage can define register operands and contain the correct current value but are not able to update the register file until the later write-back execution stage. Forwarding (or register bypassing) is the action of retrieving the correct operand value for an executing instruction between the initial register file access and any pending instruction's register file updates. Interlocking is the action of stalling an operation in the pipeline when conditions cause necessary register operand results to be delayed. It is necessary to stall early stages of the machine so that the correct results are used, and the machine does not proceed with incorrect values for source operands. The primary causes of delay in pipeline execution are initiated due to instruction fetch delay and memory latency.

66.3.2 Branch Handling

Branch instructions pose serious problems for pipelined processors by causing hardware to fetch and execute incorrect instructions until the branch instructions are completed. Executing incorrect instructions can result in severe performance degradation through the introduction of wasted cycles into the execution process.

There are several methods for dealing with pipeline stalls caused by branch instructions. The simplest performance scheme handles branches by marking branch instructions as either *taken* or *not-taken*. The compiler marks each branch by using a special branch opcode for each case. The designation allows the pipeline to fetch subsequent instructions according to the compiler's choice of opcode. However, the fetched instruction may need to be discarded and the instruction fetch restarted when the branch outcome is inconsistent with the compiler's designation.

Delayed branching is a scheme which treats the set of sequential instructions following a branch as delay slots. The delay-slot instructions are executed whether or not the branch instruction is taken. Limitations on delayed-branches are caused by the compiler and program characteristics being unable to identify numerous instructions that execute independent of the branch direction. Improvements have been introduced to provide *nullifying* branches, which include a predicted direction for the branch. When the prediction is incorrect, the delay-slot instructions are nullified.

A more modern approach to reducing branch penalties uses hardware to dynamically predict the outcome of a branch. Branch prediction strategies reduce overall branch penalties by allowing the hardware to continue processing instructions along the predicted control path, thus eliminating wasted

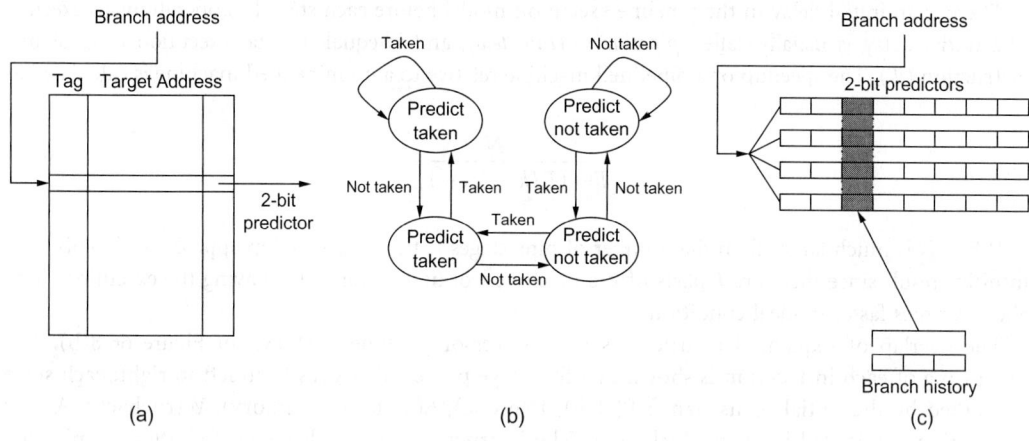

FIGURE 66.9   Branch prediction.

cycles. Efficient execution can be maintained while branch targets are correctly predicted. However, performance penalty is incurred when a branch is mispredicted. Branch target buffer is a cache structure that is accessed in parallel with the instruction fetch. It records the past history of branch instructions so that a prediction can be made while the branch is fetched again. The prediction method adapts the branch prediction to the run-time program behavior, generating a high prediction accuracy. The target address of each branch is also saved in the buffer so that the target instruction can be fetched immediately if a branch is predicted taken.

Several methodologies of branch target prediction have been constructed [3]. Figure 66.9 illustrates the relation of several general branch prediction schemes to the processor pipeline. The table in Figure 66.9(a) retains history information for each branch. The history includes the previous branch directions for making predictions on future branch directions. The simplest history is last-taken which uses 1 bit to recall whether the branch condition was taken or not-taken during its most recent execution. A more effective branch predictor, as shown in Figure 66.9(b), uses a 2-bit saturating state history counter to determine the future branch outcome. Two bits rather than one bit allows each branch to be tagged as strongly or weakly taken or not taken. Every correct prediction reinforces the prediction, while an incorrect prediction weakens it. It takes two consecutive mispredictions to reverse the direction (whether taken or not taken) of the prediction.

Recently, more complex two-level adaptive branch prediction schemes have been built which use two levels of branch history to make predictions [6]. An example of such a predictor is shown in Figure 66.9(c). The first level is the branch outcome history of the last branches encountered. The second level is the branch behavior for the last occurrences of a specific pattern of branch histories, commonly maintained as a 2-bit saturating counters. There are alternative ways of constructing both levels of adaptive branch prediction schemes, the mechanisms can contain information that is either based on individual branches, groups (set-based), and all (global). Individual information contains the branch history for each branch instruction. Set-based information groups branches according to their instruction address, thereby forming sets of branch history. Global information uses a global history containing all branch outcomes. The second level containing branch behaviors can also be constructed using any of the three types. In general, the first-level branch history pattern is used as an index into the second-level branch history.

66.3.3   Dynamic Instruction Execution

A major limitation of pipelining techniques is the use of in-order instruction execution. When an instruction in the pipeline stalls, no further instructions are allowed to proceed to insure proper execution of in-flight instruction. This problem is especially serious for multiple issue machines, where each stall cycle potentially costs work of multiple instructions. However, in many cases, an instruction could execute properly if no data dependence exists between the stalled instruction and the instruction waiting to execute. Dynamic instruction execution, also known as out-of-order execution, is an approach that uses hardware to rearrange the instruction execution to reduce the effect of stalls. The concept of dynamic execution uses hardware to detect dependences in the in-order instruction stream sequence and rearrange the instruction sequence in the presence of detected dependences and stalls.

Today, most modern superscalar microprocessors use dynamic execution to increase the number of instructions executed per cycle. Such microprocessors use basically the same concept: all instructions pass through an issue stage in order, are executed out of order, and are retired in order. There are several functional elements of this common sequence which have developed into computer architecture concepts. The first functional concept is *scoreboarding*. Scoreboarding is a technique for allowing instructions to execute out of order when there are available resources and no data dependences. Scoreboarding originates from the CDC 6600 machine's issue logic, named the scoreboard. The overall goal of scoreboarding is to execute every instruction as early as possible.

A more aggressive approach to dynamic execution is *Tomasulo's algorithm*. This scheme was employed in the IBM 360/91 processor. Although there are many variation in this scheme, the key concept of avoiding write-after-read (WAR) and write-after-write (WAW) dependences during dynamic execution is attributed to Tomasulo. In Tomasulo's scheme, the functionality of the scoreboarding is provided by *reservation stations*. Reservations stations buffer the operands of instructions waiting to issue as soon as they become available. The concept is to issue new instructions immediately when all source operands become available instead of accessing such operands through the register file. As such, waiting instructions designate the reservation station entry that will provide their input operands. This action removes WAW dependences caused by successive writes to the same register by forcing instructions to be related by dependences instead of by register specifiers. In general, renaming of register specifiers for pending operands to the reservation station entries is called *register renaming*. Overall, Tomasulo's scheme combines scoreboarding and register renaming. Tomasulo [7] provides complete details of his scheme in the context of a floating-point unit design. Patt et al. [8] first described the extensions required to implement complete modern CPUs with Tomasulo's algorithm.

66.4 Instruction Set Architecture

There are several elements that characterize an instruction set architecture, word size, instruction encoding, and architecture style.

66.4.1 Word Size

Programs often differ in the size of data they prefer to manipulate. Word processing programs operate on 8- or 16-bit data that correspond to characters in text documents. Many applications require 32-bit integer data to avoid frequent overflow in arithmetic calculation. Scientific computation often require 64-bit floating-point data to achieve desired accuracy. Operating systems and data bases may require 64-bit integer data to represent a very large name space with integers. As a result, the processors are usually designed to access data of a variety of sizes from memory systems. This is a well-known source of complexity in microprocessor design.

The endian convention specifies the numbering of bytes within a memory word. In the little endian convention, the least significant byte in a word is numbered byte 0. The number increases as the positions increase in significance. The DEC VAX and X86 architectures follow the little endian convention. In the big endian convention, the most significant byte in a word is numbered 0. The number decreases as the positions decrease in significance. The IBM 360/370, HP PA-RISC, SUN SPARC, and Motorola 680X0 architectures follow the big endian convention. The endian convention determines how the word is stored in the address space or in a binary file. The difference usually manifest itself when users try to transfer binary files between machines using different endian conventions.

66.4.2   Instruction Encoding

Instruction encoding plays an important role in the code density and performance of microprocessors. Traditionally, the cost of memory capacity was the determining factor in designing either a fixed-length or variable-length instruction set. Fixed-length instruction encoding assigns the same encoding size to all instructions. Fixed-length encoding is generally a product of the increasing advancements in memory capacity.

Variable length instruction set is the term used to describe the style of instruction encoding that uses different instructions lengths according to addressing modes of operands. Common addressing modes included either register or methods of indexing memory. Figure 66.10 illustrates two potential designs found in modern use of decoding variable length instructions. The first alternative in Figure 66.10(a) involves an additional instruction decode stage in the original pipeline design. In this model, the first stage is used to determine instruction lengths and steer the instructions to the second stage where the actual instruction decoding is performed. The second alternative in Figure 66.10(b) involves predecoding and marking instruction lengths in the instruction cache. The primary advantage of this scheme is the simplification of the number of decode stages in the pipeline design. However, the method requires a larger instruction cache structure for holding the resolved instruction information. Both design methodologies have been effectively used in decoding X86 variable instructions [5].

66.4.3   Architecture Style

Several instruction set architecture styles have existed over the past three decades of computing. First, complex instruction set computers (CISC) characterized designs with variable instruction formats, numerous memory addressing modes, and large numbers of instruction types. The original CISC philosophy was to create instructions sets that resembled high-level programming languages in an effort to simplify compiler technology. In addition, the design constraint of small memory capacity also led to the development of CISC. Two examples of the CISC model are the Digital Equipment Corporation VAX and Intel X86 architecture families.

Reduced instruction set computers (RISC) gained favor with the philosophy of uniform instruction lengths, load-store instruction sets, limited addressing modes, and reduced number of operation types. RISC concepts allow the micro-architecture design of machines to be more easily pipelined, reducing

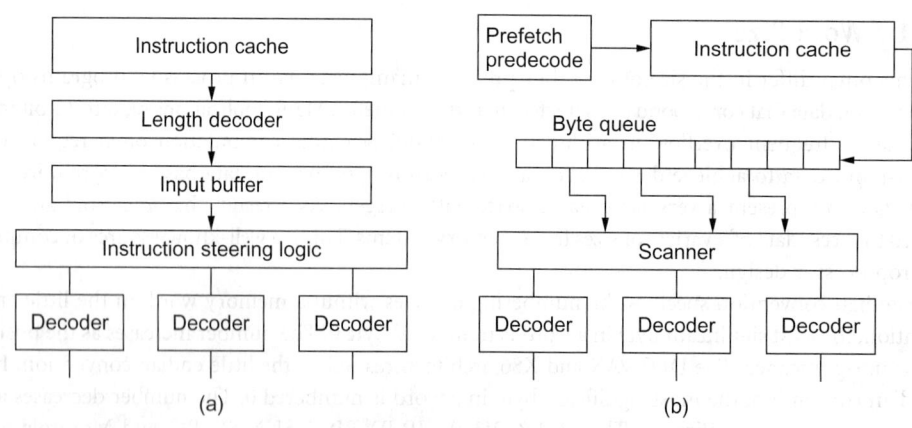

FIGURE 66.10   Variable-sized instruction decoding: (a) staging and (b) predecoding.

the processor clock cycle frequency and the overall speed of a machine. The RISC concept resulted from improvements in compiler technology and memory size. The HP PA-RISC, Sun SPARC, IBM Power PC, MIPS, and DEC Alpha machines are examples of RISC architectures.

Architecture models that specify multiple operations to issue in a clock cycle are VLIW (very long instruction word). VLIWs issue a fixed number of operations conveyed as a single long instruction and place the responsibility of creating the parallel instruction packet on the compiler. Early VLIW processor suffered from code expansion due to unfilled operation slots in the long instructions. Examples of VLIW technology are the Multiflow Trace, Cydrome Cydra machines, and TI-C6X. Explicitly parallel instruction computing (EPIC) is similar in concept to VLIW in that both use the compiler to explicitly group instructions for parallel execution. In fact, many of the ideas for EPIC architectures come from previous RISC and VLIW machines. In general, the EPIC concept solves the excessive code expansion and scalability problems associated with VLIW models by providing encoding mechanisms to reduce the need to represent unfilled operation slots in long instructions. Also, the trend of compiler-controlled architecture mechanisms such as predicated execution and speculative execution to be described later in this section are generally considered part of the EPIC style architecture domain. The Intel IA-64, Philips Trimedia, and Lucent/Motorola StarCore are examples of EPIC machines.

66.5 Instruction Level Parallelism

EPIC processors are equipped with instruction set architecture mechanisms designed to facilitate compiler's effort to arrange for the parallel execution of many operations in each clock cycle. These mechanisms, referred to as instruction level parallelism (ILP) features, are new instruction set architecture concepts for improving microprocessor performance.

66.5.1 Predicated Execution

Branch instructions are recognized as a major impediment to exploiting ILP. Branches force the compiler and hardware to make frequent predictions of branch directions in an attempt to find sufficient parallelism. Branch prediction strategies reduce this problem by allowing the compiler and hardware to continue processing instructions along the predicted control path, thus eliminating these wasted cycles. However, misprediction of these branches can result in severe performance degradation through the introduction of wasted cycles into the instruction stream.

Predicated execution provides an effective means to eliminate branches from an instruction stream. Predicated execution refers to the conditional execution of an instruction based on the value of a boolean source operand, referred to as the predicate of the instruction. This architectural support allows the compiler to use an *if-conversion* algorithm to convert conditional branches into predicate defining instructions, and instructions along alternative paths of each branch into predicated instructions [9]. Predicated instructions are fetched regardless of their predicate operand value. Instructions whose predicate operands are true are executed normally. Conversely, instructions whose predicate operands are false are nullified, and thus are prevented from modifying the processor state. Predicated execution allows the compiler to trade instruction fetch efficiency for the capability to expose ILP to the hardware along multiple execution paths.

Predicated execution offers the opportunity to improve branch handling in microprocessors. Eliminating frequently mispredicted branches may lead to a substantial reduction in branch prediction misses. As a result, the performance penalties associated with mispredicting these branches are removed. Eliminating branches also reduces the need to handle multiple branches per cycle for wide-issue processors. Finally, predicated execution provides an efficient interface for the compiler to expose multiple execution paths to the hardware. Without compiler support, the cost of maintaining multiple execution paths in hardware can be prohibitive.

The essence of predicated execution is the ability to suppress the modification of the processor state based upon some execution condition. There must be a way to express this condition and a way to express when the condition should affect execution. Full predication cleanly supports this through a combination of instruction set and micro-architecture extensions. These extensions can be classified as support for suppression of execution and expression of condition. The result of the condition which determines if an instruction should modify state is stored in a set of 1-bit registers. These registers are collectively referred to as the predicate register file. The values in the predicate register file are associated with each instruction in the extended instruction set through the use of an additional source operand. This operand specifies which predicate register will determine whether the operation should modify processor state. If the value in the specified predicate register is 1, or true, the instruction is executed normally; if the value is 0, or false, the instruction is suppressed.

Predicate register values may be set using predicate define instructions, as described in the HPL Playdoh architecture [10]. There is a predicate define instruction for each comparison opcode in the original instruction set. The major difference with conventional comparison instructions is that these predicate defines have up to two destination registers and that their destination registers are predicate registers. The instruction format of a predicate define is shown below.

$$\text{Pred\_}<cmp>\ \text{Pout1}_{<type>},\ \text{Pout2}_{<type>},\ \text{src1},\ \text{src2}\ (\text{P}_{in})$$

This instruction assigns values to *Pout1* and *Pout2* according to a comparison of *src1* and *src2* specified by $<cmp>$. The comparison $<cmp>$ can be: equal (eq), not equal (ne), greater than (gt), etc. A predicate $<type>$ is specified for each destination predicate. Predicate defining instructions are also predicated, as specified by P_{in}.

The predicate $<type>$ determines the value written to the destination predicate register given the result of the comparison and of the input predicate, P_{in}. For each combination of comparison result and P_{in}, one of three actions may be performed on the destination predicate. It can write 1, write 0, or leave it unchanged. There are six predicate types which are particularly useful, the unconditional (*U*), *OR*, and *AND* type predicates and their complements. Table 66.1 contains the truth table for these predicate definition types.

Unconditional destination predicate registers are always defined, regardless of the value of P_{in} and the result of the comparison. If the value of P_{in} is 1, the result of the comparison is placed in the predicate register (or its compliment for $\overline{U}$). Otherwise, a 0 is written to the predicate register. Unconditional predicates are utilized for blocks which are executed along only one path in the if-converted code region.

The *OR* type predicates are useful when execution of a block can be enabled along multiple paths, such as logical AND (&&) and OR (||) constructs in C. *OR* type destination predicate registers are set if P_{in} is 1 and the result of the comparison is 1 (0 for $\overline{OR}$), otherwise the destination predicate register is unchanged. Note that *OR* type predicates must be explicitly initialized to 0 before they are defined and used. However, after they are initialized multiple *OR* type predicate defines may be issued simultaneously and in any order on the same predicate register. This is true since the *OR* type predicate either writes a 1 or leaves the register unchanged which allows implementation as a wired logical *OR* condition. *AND* type predicates, are analogous to the *OR* type predicate. *AND* type destination predicate registers are cleared if P_{in} is 1 and the result of the comparison is 0 (1 for $\overline{AND}$), otherwise the destination predicate register is unchanged.

Figure 66.11 contains a simple example illustrating the concept of predicated execution. Figure 66.11(a) shows a common programming if-then-else construction. The related control flow representation of that

TABLE 66.1 Predicate Definition Truth Table

P_{in}	Comparison	P_{out}					
		U	$\overline{U}$	OR	$\overline{OR}$	AND	$\overline{AND}$
0	0	0	0	—	—	—	—
0	1	0	0	—	—	—	—
1	0	0	1	—	1	0	—
1	1	1	0	1	—	—	0

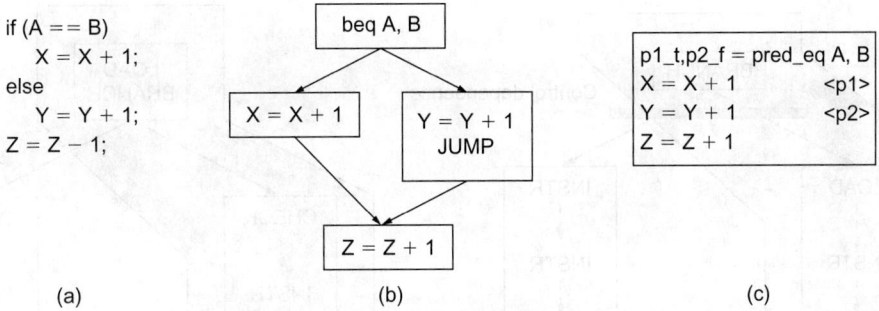

FIGURE 66.11   Instruction sequence: (a) program code, (b) traditional execution, and (c) predicated execution.

programming code is illustrated in Figure 66.11(b). Using if-conversion, the code in Figure 66.11(b) is then transformed into the code shown in Figure 66.11(c). The original conditional branch is translated into a *pred_eq* instructions. Predicate register *p1* is set to indicate if the condition $(A = B)$ is true, and *p2* is set if the condition is false. The "then" part of the if statement is predicated on *p1* and the "else" part is predicated on *p2*. The *pred_eq* simply decides whether the addition or subtraction instruction is performed and ensure that one of the two parts is not executed.

66.5.2   Speculative Execution

The amount of ILP available within basic blocks, defined as consecutive instruction sequences without branching in or out, is extremely limited in most programs. As such, processors must optimize and schedule instructions across basic block boundaries to achieve higher performance. In addition, future processors must contend with both long latency load operations and long latency cache misses. When load data is need by subsequent dependent instructions, the processor execution must wait until the cache access is complete.

In these situations, our-of-order processors dynamically perform branch prediction and reorder the instruction stream to execute nondependent instructions. As a result, they have ability of exploiting parallelism between instructions before and after a correctly predicted branch instruction. However, this approach requires complex circuitry at the cost of chip die space as well as additional power consumption. Similar performance gains can be achieved using static compile-time speculation methods without complex, power hungry out-of-order logic. Speculative execution, a technique for executing an instruction before knowing its execution is required is an important technique for exploiting ILP in programs. Speculative execution is best known for its use in hiding memory latency.

A compiler can utilize speculative code motion to achieve higher performance in several ways. First, in regions of code where insufficient ILP exists to fully utilize the processor resources, useful instructions from other regions may be executed. Second, instructions at the beginning of long dependence chains may be executed early to reduce the computation's critical path. Finally, long latency instructions may be initiated early to overlap their execution with other useful operations. Figure 66.12 illustrates a simple example of code before and after a speculative compile-time transformation is performed to execute a load instruction above a conditional branch.

Figure 66.12(a) shows how the branch instruction and its implied control flow define a control dependence that restricts the load operation from being scheduled earlier in the code. Cache miss latencies would halt the processor unless out-of-order execution mechanisms were used. However, with speculation support, Figure 66.12(b) can be used to hide the latency of the load operation.

The solution requires the load to be speculative or nonexcepting. A speculative load will not signal exceptions such as address alignment errors or address space access violations. Essentially, the load remains silent for these exceptions. The additional check instruction in Figure 66.12(b) enables these signals to be detected when the execution does reach the original location of the load. When the other

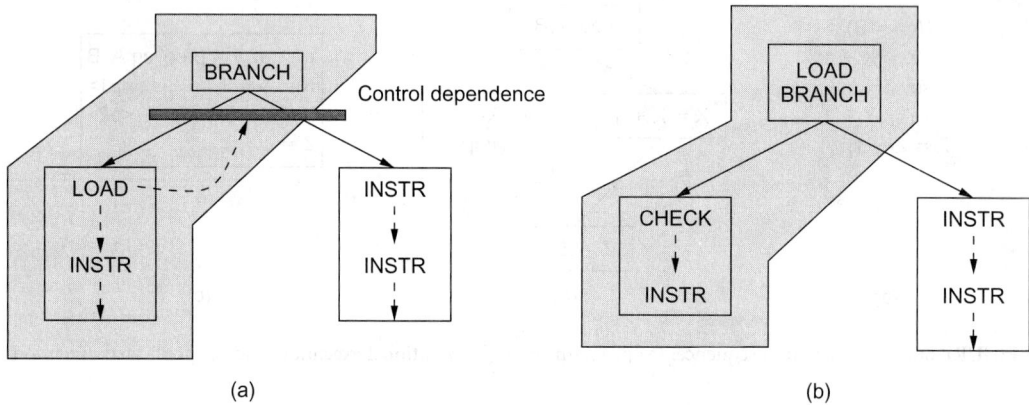

FIGURE 66.12   Instruction sequence: (a) traditional execution and (b) speculative execution.

path of branch's execution is taken, such silent signals are meaningless and can be ignored. Using this mechanism, the load can be placed above all existing control dependences, providing the compiler with the ability to hide load latency. Details of compiler speculation can be found in Ref. [11].

66.6   Industry Trends

The microprocessor industry is one of the fastest-moving industry today. Healthy demands from the market place have stimulated strong competition, which in turn has resulted in great technical innovations.

66.6.1   Computer Microprocessor Trends

Recent trends of computer microprocessors include deep pipelining, high clock frequency, wide instruction issue, speculative and out-of-order execution, predicated execution, multimedia data types, large on-chip caches, floating-point capabilities, and multiprocessor support. In the area of pipelining, the Intel Pentium 4 processor is pipelined approximated twice as deeply as its predecessor Pentium 3. The deep pipeline has allowed the clock of Pentium 4 processor to run at a much higher clock frequency than Pentium 3. This trend has, however, been reversed in 2005 due to power budget limitations. The pipeline depth of the Intel IA32 microprocessors that succeed Pentium 4 have been reduced toward that of Pentium 3.

In the area of wide instruction issue, the Pentium 4 processor can decode and issue up to three X86 instructions per clock cycle, compared to the two-instruction issue bandwidth of Pentium. More recently, the Intel Itanium and Itanium 2 processors can issue up to six instructions per clock cycle. Wide instruction issue, however, requires multiported register and multiple cache access ports that can significantly increase power consumption. As a result, future microprocessors will likely maintain or reduce the issue width compared to their recent predecessors.

Pentium 4 has dedicated a very significant amount of chip area to branch history table, branch target buffer, reservation stations, load-store queue, and reorder buffer to support speculative and out-of-order execution. These structures together allow the Pentium 4 processor to maintain a large instruction window within which it performs aggressive speculative and out-of-order execution. All these structures, however, consume power in an intensive manner. As a result, the trend of larger instruction windows has also slowed due to power budget limitations.

One important trend of the computer microprocessors in general is the slowdown of the increase in complexity, size, and clock frequency of processor cores. Rather, the industry is moving into incorporating multiple processor cores on the same chip. If all the cores can be productively used, such model can achieve much higher performance than a single core given the same power and chip area budget. This however, places much more burden on the programmer, compiler, and operating system than traditional single core models.

In the area of predicated execution, Pentium 4 supports a conditional move instruction that was not available in Pentium. This trend is furthered by the next-generation IA-64 architecture where all instructions can be conditionally executed under the control of predicate registers. This ability will allow future microprocessors to execute control-intensive programs much more efficiently than their predecessors.

In the area of data types, the multimedia instructions from Intel and AMD have become a standard feature of all X86 microprocessors. These instructions take advantage of the fact that multimedia data items are typically represented with a smaller number of bits (8–16 bits) than the width of an integer data path today (32–64 bits). Based on an observation the same operation is often repeated on all data items in multimedia applications, the architects of multimedia instructions specify that each such instruction performs the same operation on several multimedia data items packed into one register word. Intel first proposed MMX instructions that process several integer data items simultaneously to achieve significant speedup in targeted applications. In 1998, AMD proposed the 3DNow! instructions to address the performance needs of 3-D graphics applications. The 3DNow! instructions are designed on the basis of the concept that 3-D graphics data items are often represented in single precision floating-point format and they do not require the sophisticated rounding and exception handling capabilities specified in the IEEE standard format. Thus, one can pack two graphics floating-point data into one double-precision floating-point register for more efficient floating-point processing of graphics applications. Note that MMX and 3DNow! are similar in concepts applied to integer and floating-point domains. More recently, Intel proposed the SSE instructions to compete with AMD 3DNow!.

In the area of large on-chip caches, the popular strategies used in computer microprocessors are either to enlarge the first-level caches or to incorporate second-level and sometimes third-level caches on chip. For example, the AMD K7 microprocessor has a 64-kB first-level instruction cache and a 64-kB first-level data cache. These first-level caches are significantly larger than those found in the previous generations. For another example, the Intel Celeron microprocessor has a 128-kB second-level combined instruction and data cache. These large caches are enabled by the increased chip density that allows much more transistors on the chip. The Compaq Alpha 21364 microprocessor has both: a 64-kB first-level instruction cache, a 64-kB first-level data cache, and a 1.5-MB second-level combined cache. The recent Intel Itanium processors have up to 9-MB third-level combined cache on chip.

In the area of floating-point capabilities, the computer microprocessors in general have much stronger floating-point performance than their predecessors. For example, the Intel Pentium 4 processor achieves several times of floating-point performance improvements of the Pentium processor. For another example, most RISC and EPIC microprocessors now have floating-point performance that rival supercomputer CPUs built just a few years ago.

Owing to the increasing demand of multiprocessor enterprise computing servers, many computer microprocessors now seamlessly support cache coherence protocols. For example, the AMD K7 microprocessor provides direct support for seamless multiprocessor operation when multiple K7 microprocessors are connected to a system bus. This capability was not available in its predecessor AMD K6. The more recent AMD Opteron processors further support HyperTransport protocol to allow each processor in a multiprocessor system to have much higher communication bandwidth than what the traditional memory controllers can support.

66.6.2 Embedded Microprocessor Trends

There are three clear trends in embedded microprocessors. The first trend is to integrate a DSP core with an embedded CPU/controller core. Embedded applications increasingly require DSP functionalities such as data encoding in disk drives and signal equalization for wireless communications. These functionalities enhance the quality of services of their end consumer products. At the 1999 Embedded Microprocessor Forum, ARM, Hitachi, and Siemens all announced products with both DSP and embedded microprocessors [12].

Three approaches exist in the integration of DSP and embedded CPUs. One approach is to simply have two separate units placed on a single chip. The advantage of this approach is that it simplifies the development of the microprocessor. The two units are usually taken from existing designs. The software development tools can be directly taken from each unit's respective software support environments. The disadvantage is that the application developer needs to deal with two independent hardware units and two software development environments. This usually complicates software development and verification.

An alternative approach to integrating DSP and embedded CPUs is to add the DSP as a coprocessor of the CPU. This CPU fetches all instructions and forwards the DSP instructions to the coprocessor. The hardware design is more complicated than the first approach due to the need to more closely interface the two units, especially in the area of memory accesses. The software development environment also needs to be modified to support the co-processor interaction model. The advantage is that the software developers now deal with a much more coherent environment.

The third approach to integrating DSP and embedded CPUs is to add DSP instructions to a CPU instruction set architecture. This usually require brand new designs to implement the fully integrated instruction set architecture. The benefit is that software developers need to deal with just one development environment.

The second trend in embedded microprocessors is to support the development of single chip solutions for large volume markets. Many embedded microprocessor vendors offer designs that can be licensed and incorporated into a larger chip design that includes the desired input/output peripheral devices and application-specific integrated circuit (ASIC), and field programmable gate array (FPGA) design. This paradigm is referred to as system-on-a-chip design. A microprocessor that is designed to function in a such a system is often referred to as a licensable core.

The third major trend in embedded microprocessors is aggressive adoption of high-performance techniques. Traditionally, embedded microprocessors are slow to adopt high-performance architecture and implementation techniques. They also tend to reuse software development tools such as compilers from the computer microprocessor domain. However, owing to the rapid increase of required performance in embedded markets, the embedded microprocessor vendors are now making fast moves in adopting high-performance techniques. This trend is especially clear in the DSP microprocessors. Texas Instruments, Motorola/Lucent, and Analog Devices have all been shiping aggressive EPIC/VLIW DSP microprocessors and associated compilers.

66.6.3 Microprocessor Market Trends

Readers who are interested in market trends for microprocessors are referred to *Microprocessor Report*, a periodic publication by MicroDesign Resources (www.MDRonline.com). In every issue, there is a summary of microarchitecture features, physical characteristics, availability, and pricing of microprocessors.

References

1. J. Turley, RISC volume gains but 68K stll reigns, *Microprocessor Rep.*, vol. 12, pp. 14–18, 1998.
2. J.L. Hennessy and D.A. Patterson, *Computer Architecture A Quantitative Approach*. Morgan Kaufman, San Francisco, CA, 1990.
3. J.E. Smith, A study of branch prediction strategies, in *Proceedings of the 8th International Symposium on Computer Architecture*, pp. 135–148, May 1981.
4. W.W. Hwu and T.M. Conte, The susceptibility of programs to context switching, *IEEE Trans. Comput.*, vol. C-43, pp. 993–1003, 1994.
5. L. Gwennap, Klamath extends P6 family, *Microprocessor Rep.*, vol. 1, pp. 1–9, 1997.
6. T.Y. Yeh and Y.N. Patt, A comprehensive instruction fetch mechanism for a processor supporting speculative execution, in *Proceedings of the 25th International Symposium on Microarchitecture*, pp. 129–139, December 1992.

7. R.M. Tomasulo, An efficient algorithm for exploiting multiple arithmetic units, *IBM J. Res. Dev.*, vol. 11, pp. 25–33, 1967.

8. Y.N. Patt, W.-M. Hwu, and M. Shebanow, HPS, a new microarchitecture: Rationale and introduction, in *Proceedings of the 18th Annual Workshop on on Microprogramming*, pp. 103–106, December 1985.

9. J.R. Allen, K. Kennedy, C. Porterfield, and J. Warren, Conversion of control dependence to data dependence, in *Proceedings of the 10th ACM Symposium on Principles of Programming Languages*, pp. 177–189, January 1983.

10. V. Kathail, M.S. Schlansker, and B.R. Rau, HPL PlayDoh architecture specification: Version 1.0, Technical Report HPL-93–80, Hewlett-Packard Laboratories, Palo Alto, CA, February 1994.

11. S.A. Mahlke, W.Y. Chen, R. Bringmann, R. Hank, W.W. Hwu, M. Schlansker, and B. Rau, Sentinel scheduling: A model for compiler-controlled speculative execution, *ACM Trans. Comput. Syst.*, vol. 11, November 1993.

12. *Embedded Microprocessor Forum*, San Jose, CA, October 1998.

67

Logic Synthesis for Field Programmable Gate Array (FPGA) Technology

John Lockwood

Washington University in St. Louis

CONTENTS

67.1 Introduction

Field programmable gate arrays (FPGAs) enable rapid implementation of complex digital circuits. FPGA devices have the added advantage that they can be reprogrammed and reused, allowing the same hardware to implement entirely new designs or to allow existing hardware systems to implement a circuit with revised logic. While many general techniques used for traditional IC logic synthesis methods are used in the computer-aided design tools for FPGA hardware, FPGA circuits have unique characteristics that affect the synthesis process.

The FPGA device consists of a number of configurable logic blocks (CLBs) interconnected by a routing matrix. Pass transistors are used in the routing matrix to connect segments of metal lines. There are three major types of CLBs: those based on Programmable Logic Arrays (PLAs), those based on multiplexers, and those based on table look-up (TLU) functions. Automated logic synthesis tools optimize the mapping of the Boolean network to the physical circuits within the FPGA device. FPGA synthesis extends the methods used to solve the general problem of multilevel logic synthesis. FPGA logic synthesis is usually solved in two phases. A technology-independent phase uses a general multilevel logic optimization tool (such as Berkeley's MIS) to reduce the complexity of the Boolean network. Next, a technology-dependent optimization phase optimizes the logic for the particular type of device. In the case of the TLU-based FPGA, each CLB can implement an arbitrary logic function of a limited number of variables. Different FPGA optimization algorithms aim to optimize different objectives that include the number of CLBs used, the logic depth, and the routing density.

The Chortle algorithm, for example, is a direct method that performs dynamic programming to map the logic into TLU-based CLBs. It converts the Boolean network that describes the function of the circuit into a forest of directed acyclic graphs (DAGs); then it evaluates and records the optimal subsolutions to the logic mapping problem as it traverses the DAG. Two-step algorithms operate by first decomposing the nodes, and then performing a node elimination. Later sections of this chapter discuss and detail the Xmap, Hydra, and MIS-pga algorithms.

FPGA devices are fabricated using the same submicron geometries as other silicon devices. As such, the devices benefit from the rapid advances in device technology. The overhead of the programming bits, general function generators, and general routing structures, however, reduce the total amount of logic available to the end user.

67.2   FPGA Structures

An FPGA consists of reconfigurable logic elements, flip-flops, and a reprogrammable interconnect structure. The logic elements are typically arranged in a matrix. The interconnect is arranged as a mesh of variable-length metal wires and pass transistors to interconnect the logic elements. The logic elements are programmed by downloading binary control information from an external ROM, a built-in EPROM, or a host processor. After the download, the control information is stored in the device and used to determine the function of the logic elements and the state of the pass transistors. Unlike a PLA, the FPGA can be configured to implement multilevel logic functions.

The granularity of an FPGA refers to the complexity of the individual logic elements. A fine-grain logic block appears to the user to be much like a standard mask-programmable gate array. Fine-grain logic blocks implement simple functions of a few variables. A course-grain logic block (such as those in devices from Xilinx, Altera, Actel, and Quicklogic) provides more general functions of a larger number of variables. A Xilinx look up table (LUT), for example, can implement any Boolean function of five variables, or two Boolean functions of four variables.

It has been found that the course-grain logic blocks generally provide better performance than the fine-grain logic blocks. Course-grained devices require less space for interconnect and routing by combining multiple logic functions into one logic block. In particular, it has been shown that a four-input logic block uses the minimal chip area for a large variety of benchmark circuits [1]. The expense of a few extra underutilized logic blocks outweighs the area required for the larger number of fine-grained logic blocks and their associated larger interconnect matrix and pass transistors. This chapter focuses on the logic synthesis for course-grained logic elements.

A course-grained CLB can be implemented using a PLA-based AND/OR elements, multiplexers, or SRAM-based LUT elements. These configurations are described below in detail.

67.2.1   Look-Up Table-Based CLB

The basic unit of LUT-based FPGAs is the CLB, implemented as an SRAM of size $2^n \times 1$. Each CLB can implement any arbitrary logic function of n variables, for a total of 2^n functions.

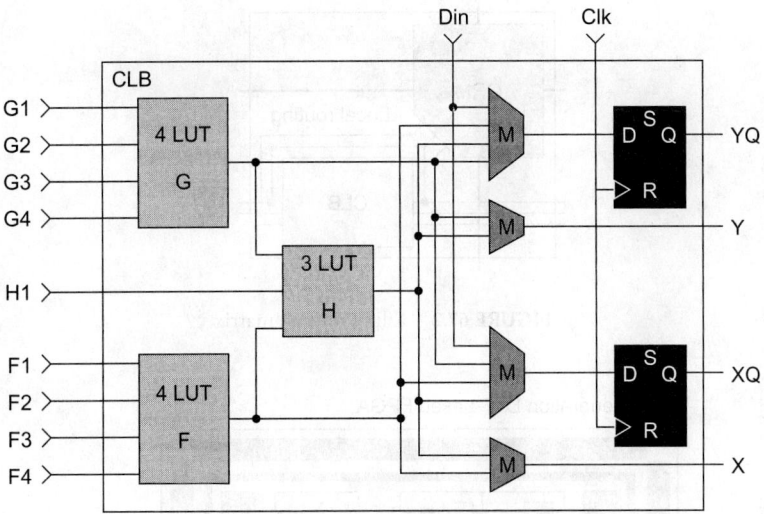

FIGURE 67.1   Xilinx CLB.

An example of an LUT-based FPGA is the Xilinx FPGA, as illustrated in Figure 67.1. Each CLB has three LUT generators, and two flip-flops [2]. The first two LUTs implement any function of four variables, while the third LUT implements any function of three variables. Separately, each CLB can implement two functions of four variables. Combined, each CLB can implement any one function of five variables, or some restricted functions of nine variables (such as AND, OR, XOR).

67.2.2   PLA-Based CLB

PLA-based FPGA devices evolved from the traditional PLDs. In a PLD, each basic logic block is an AND–OR block consisting of wide fan-in AND gates feeding a few-input OR gate. The advantage of this structure is that many logic functions can be implemented using only a few levels of logic, due of the large number of literals that can be used at each block. It is, however, difficult to make efficient use of all inputs to all gates. Even so, the amount of wasted area is minimized by the high packing density of the wired-AND gates.

To further improve the density in a PLD, another type of logic block, called the logic expander, has been introduced. It is a wide-input NAND gate whose output could be connected to the input of the AND–OR block. While its delay is similar, the NAND block uses less area than the AND–OR block, and thus increases the effective number of product terms available to a logic block.

67.2.3   Multiplexer-Based CLB

Multiplexer-based FPGAs utilize a multiplexer to implement different logic function by connecting each input to a constant or a signal [3]. The ACT-1 logic block, for example, has three multiplexers and one logic gate. Each block has eight inputs and one output, implementing:

$$f = (\overline{s_3 + s_4})(\overline{s_1}w + s_1 x) + (s_3 + s_4)(\overline{s_2}y + s_2 x) \tag{67.1}$$

Multiplexer-based FPGAs can provide a large degree of functionality for a relatively small number of transistors. Multiplexer-based CLBs, however, place high demands on routing resources due to the large number of inputs.

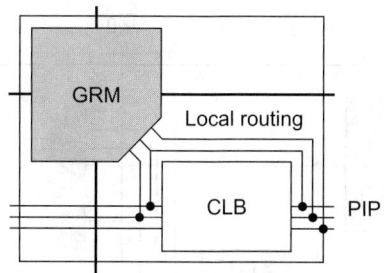

FIGURE 67.2   Xilinx routing matrix.

Third generation LUT-based FPGA

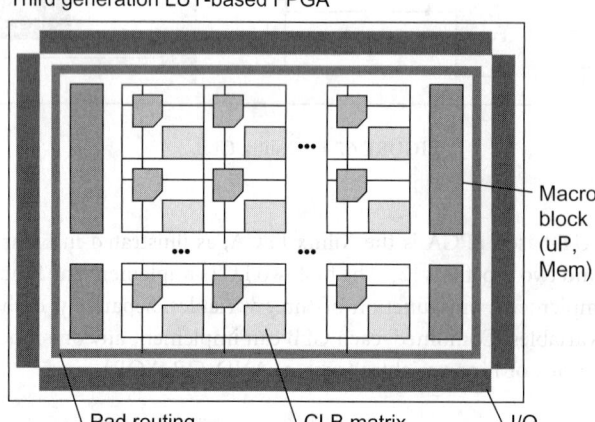

FIGURE 67.3   FPGA chip layout.

67.2.4 Interconnect

In all structures, a reprogrammable routing matrix interconnects the configurable logic blocks. A portion of the routing matrix in a Xilinx 4000-series FPGA, for example, is illustrated in Figure 67.2. Local interconnects are used to join adjacent CLBs. Global routing modules are used to route signals across the chip.

The routing and placement issues for the FPGAs are somewhat different from those of custom logic. For a large fan-out node, for example, an optimal placement for the elements for the fan-out would be along a single row or column, where the routing could be implemented using a long line. For custom logic, the optimal placement would be as a cluster, where the optimization attempted to minimize the distance between nodes. For the FPGA, the routing delay is influenced more by the number of pass transistors for which the signal must cross rather than by the length of the signal line.

The power of the FPGA comes from the flexibility of the interconnect. A block diagram of a typical third-generation FPGA device is shown in Figure 67.3. The CLB matrix and the mesh of the interconnect occupy most of the chip real area. Macro blocks, when present, implement functions such as high-density memory block, multiplier, digital signal processor, microprocessor, or Gigabit-rate SERializer/DESerializer (SERDES) cores. The I/O blocks surround the chip and provide connectivity to external devices.

67.3 Logic Synthesis

Logic synthesis is typically implemented as a two-phase process: a technology-independent phase, followed by a technology-mapping phase [4]. The first phase attempts to generate an optimized abstract representation

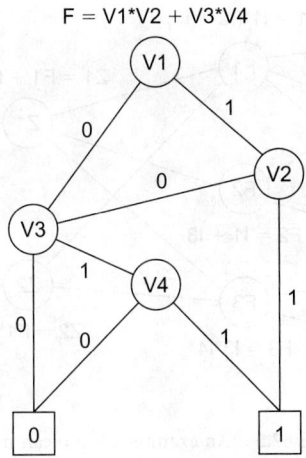

F = V1*V2 + V3*V4

FIGURE 67.4 Binary decision diagram.

of the target circuit, and the second-phase determines the optimal mapping of the optimized abstract representation onto a particular type of device, such as an FPGA. The second-phase optimization may drastically alter the circuit to optimize the logic for a particular technology. The technology-dependent FPGA optimization typically minimizes the amount of logic as measured by the number of LUTs.

The abstract representation of a combination logic function f is not unique. For example, f may be expressed by a truth table, a sum of products (SOP) (such as $f = ab + cd + e'$), a factored form (such as $f = (a + b)(c + (e'(f + g'))))$, a binary decision diagram (BDD) DAG, an if-then-else DAG, or any combination of the above forms.

The BDD is a DAG where the logic function is associated with each node, as shown in Figure 67.4. It is canonical because, for a given function and a given order of the variables along all the paths, the BDD DAG is unique. A BDD may contain a great deal of redundant information, however, as the subfunctions may be replicated in the lower portions of the tree.

The if-then-else DAG consists of a set of nodes, each with three children. Each node is a two-to-one selector, where the first child is connected to the control input of the selector and the other two are connected to the signal inputs of the node.

67.3.1 Technology-Independent Optimization

In the technology-independent synthesis phase, the combinational logic function is represented by the Boolean network, as illustrated in Figure 67.5. The nodes of the network are initially general nodes, which can represent any arbitrary logic function. During optimization, these nodes are usually mapped from the general form to a generic form, which only consists of AND, OR, and NOT logic nodes [4]. At the end of the first synthesis phase, the complexity and number of nodes of the Boolean network has been reduced. Two classes of operations — network restructuring and node minimization — are used to optimize the network. Network restructuring operations modify the structure of the Boolean network by introducing new nodes, eliminating others, and adding and removing arcs. Node minimization simplifies the logic equations associated with nodes [5].

67.3.1.1 Restructuring Operations

Decomposition reduces the support of the function, F (denoted as $\sup(F)$). The support of the function refers to the set of variables that F explicitly depends on. The cardinality of a function (denoted by $|\sup(F)|$), represents the number of variables that F explicitly depends on.

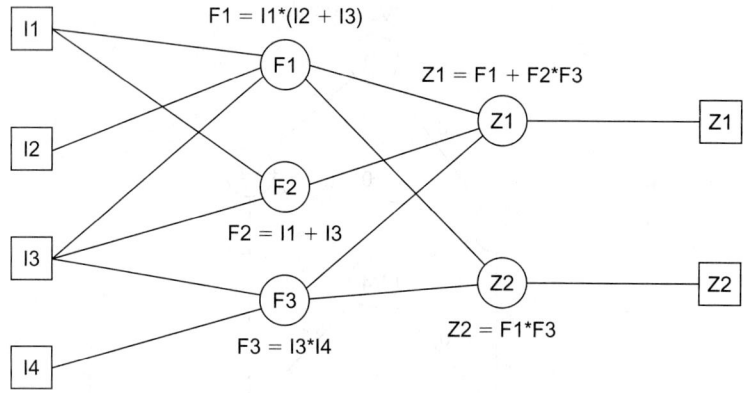

FIGURE 67.5   An example of Boolean network.

Factoring is used to transform the SOP form of a logic function into a factored form. Substitution expresses one given logic function in terms of another. Elimination merges a subfunction, G, into the function, F, so that F is expressed only in terms of its fan-in nodes of F and G (not in terms of G itself).

The efficiency of the restructuring operations depends on finding a suitable divisor, P, to factor the function, that is, given functions F, choose a divisor P, and find the functions Q and R such that $F = PQ + R$. The number of possible divisors is hopelessly large; thus, an effective procedure is needed in practice to restrict the search subspace for a good divisor. The Brayton and McMullen kernel matching technique has proven effective at this task.

The kernels of a function F are the set of expressions: $K(F) = \{g \mid g \subset D(F)\}$, where g is cube-free and $D(F)$ are the primary divisors.

A cube is a logic function given by the product of literals. A cube of a function F is a cube whose on-set does not have vertices in the off-set of F (e.g., if $F = ab(c + d)$, ab is a cube of F). An expression F is cube-free if no cube divides the expression evenly [6]. For example, $F = ab + c$ is cube-free, while $F = ab + ac$ is not cube-free. Finally, the primary divisors of F are the set of expression: $D(F) = F/C \mid C$ is a cube [7].

Kernel functions can be computed effectively by several fast algorithms. Based on the kernel functions extracted, the restructuring operations can generate acceptable results usually within a reasonable amount of time [4]. Speed/quality tradeoffs are still needed, however, as is the case with MIS, which is a multilevel logic synthesis system [8].

67.3.1.2   Node Minimization

Node minimization attempts to reduce the complexity of a given network by using Boolean minimization techniques on its nodes.

A two-level logic minimization with consideration of the don't-care inputs and outputs can be used to minimize the nodes in the circuit. Two types of don't-care sets — satisfiability don't care (SDC) and observability don't care (ODC) — are used in the two-level minimizer. The SCD set represents combinations of input variables that can never occur because of the structure of the network itself, while the ODC set represents combinations of variables that will never be observed at outputs. If the SDCs and ODCs are too large, a practical running time for the algorithm can only be achieved by using a limited subset of SDCs and ODCs [8].

Another technique to reduce the complexity of a Boolean network uses a tautology checker to determine if two Boolean networks are equivalent. The result is determined by computing the XNOR of the circuit's primary outputs [9]. A node is first tentatively simplified by deleting either variables or cubes. If the result of tautology check is 1 (equivalent), then this deletion is performed. As with the first method, an exhaustive search of all possible networks is usually not feasible because of the high computational cost of the tautology check.

67.3.2 Technology Mapping

The technology-mapping phase for an FPGA attempts to realize the Boolean network using a minimal number of CLBs. Synthesis algorithms fall into two main categories: algorithmic approaches and rule-based techniques.

By expressing the optimized AND/OR/NOT network as a subject graph (a network of two-input NAND gates), and a library of potential mappings as a pattern graphs, the algorithmic approach converts the mapping problem into a covering problem with the goal of finding the minimum-cost cover of the subject graph by the pattern graphs. The problem is NP-hard; thus, heuristics must be used. If the Boolean network is not a tree, a step of decomposition into forest of trees is performed; then the mapping problem is solved as a tree-covering-by-tree problem.

The rule-based technique traverses the Boolean network and replaces subnetworks with patterns in the library when a match is found. It is slow compared to the algorithmic method, but can generate better results. Hybrid approaches which perform a tree-covering step followed by a rule-based clean-up step are used in industry.

67.4 Look-Up Table Synthesis

Approaches to synthesize FPGAs with LUTs are summarized in Figure 67.6. Beginning with an optimized AND/OR/NOT Boolean network generated by a general-purpose multilevel logic minimizer, such as MIS-II, these algorithms attempt to minimize the number of LUTs needed to realize the logic network.

67.4.1 Library-Based Mapping

Library-based algorithms were originally developed for use in the synthesis of standard cell designs. It was assumed that there was a small number of predesigned logic elements. The goal of the mapping function was to optimize the use of these blocks. MIS is one such library-based approach that performs multilevel logic minimization. It existed long before the conception of FPGAs and has been used for TLU logic synthesis. Nonequivalent functions in MIS are explicitly described in terms of two-input NAND gates. Therefore, an optimal library needs to cover all functions that can be implemented by the TLU. Library-based algorithms are generally not appropriate for TLU-based FPGAs due to their large number of functions which each CLB can implement.

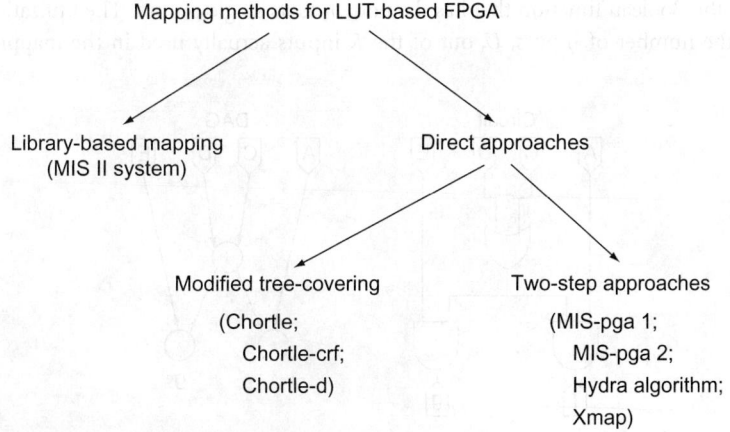

FIGURE 67.6 Approaches to synthesize FPGAs based on LUTs.

67.4.2 Direct Approaches

Direct approaches generate the optimized Boolean network directly, without the explicit construction of library components. Two classes of method are used: modified tree-covering algorithms (i.e., Chortle and its improved versions) and two-step methods.

67.4.2.1 Modified Tree-Covering Approaches

The modified tree-covering approach begins with an AND/OR representation of the optimized Boolean network. Chortle, and its extensions (Chortle-crf and Chortle-d), first decompose the network into a forest of trees by clipping the multiple-fan-out nodes. An optimal mapping of each tree into LUTs is then performed using dynamic programming, and the results are assembled together according to the interconnection patterns of the forest. The details of the Chortle algorithms are given in Section 67.5.

67.4.2.2 Two-Step Approaches

Instead of processing the mapping in one direct step, the two-step methods handle the mapping by node decomposition followed by node elimination. The decomposition operation yields a network that is feasible. The node-elimination step reduces the number of nodes by combining nodes based on the particular structure of a CLB.

A Boolean network is feasible if every intermediate node is realized by a feasible function. A feasible function is a function that satisfies $|\text{sup}(f)| \leq K$, or informally, can be realized by one CLB.

Different two-step approaches have been proposed and implemented, including MIS-pga1 and MIS-pga2 from University of California, Berkeley, Xmap from University of California, Santa Cruz, and Hydra from Stanford University. Each algorithm has its own advantages and drawbacks. Details of these methods are given in Section 67.6. Comparisons among the direct and two-step methods are given in Section 67.7.

67.5 Chortle

The Chortle algorithm is specifically designed for TLU-based FPGAs. The input to the Chortle algorithm is an optimized AND/OR/NOT Boolean network. Internally, the circuit is represented as a forest of DAGs, with the leaves representing the inputs and the root representing the output, as shown in Figure 67.7. The internal nodes represent the logic functions AND/OR. Edges represent inverting or noninverting signal paths.

The goal of the algorithm is to implement the circuit using the fewest number of K-input CLBs in minimal running time. Efficient running time is a key advantage of Chortle, as FPGA mapping is a computationally intensive operation in the FPGA synthesis procedure.

The terminology of the Chortle algorithm defines the mapping of a node, n, in a tree as the circuit of look-up tables rooted at that node that extends to the leaf nodes. The root look-up table of node n is the mapping of the Boolean function that has the node n as its single output. The utilization of a look-up table refers to the number of inputs, U, out of the K inputs actually used in the mapping. Finally, the

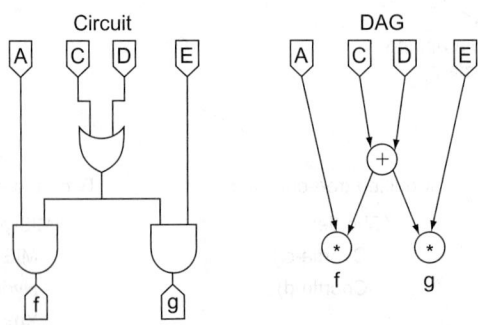

FIGURE 67.7 Boolean network and DAG representation.

utilization division, μ, is a vector that denotes the distribution of the inputs to the root look-up table among subtrees. For example, a utilization vector of $\mu = \{2,1\}$ would refer to a table look-up function that has two of the K inputs from the left logic subtree, and one input from the right subtree.

67.5.1 Tree-Mapping Algorithm

The first step of the Chortle algorithm is to convert the input graph into forest of fan-out-free trees, where each logic function has exactly one output. As illustrated in Figure 67.8, node n has a fan-out degree of two; thus, two new nodes, n_1 and n_2, are created that implement the same Boolean equation of node n. Each subtree is then evaluated independently.

Chortle uses a postorder traversal of each DAG to determine the mapping of each node. The logic functions connecting the inputs (leaves) are processed first; the logic functions connecting those functions are processed next, and so on until reaching the output node (root).

Chortle's tree-mapping algorithm is based on dynamic programming. Chortle computes and records the solution to all subproblems, proceeding from the smallest to the largest subproblem, avoiding recomputation of the smaller subproblems. The subproblem refers to computation of the minimum-cost mapping function of the node n in the tree. For each node n_i, the subproblem, $minMap(n_i, U)$, is solved for each value of U, ranging from 2, ..., K ($U = K$ refers to a look-up function that is fully utilized, while $U = 2$ refers to a TLU with only two inputs).

In general, for the same value of U, multiple utilization vectors, $\mu(u_1, u_2, ..., u_f)$, are possible, such that $\sum_{i=1}^{f} \mu_i = U$. The utilization vector determines how many inputs are to be used from each of the previous optimal subsolutions. Chortle examines each possible mapping function to determine this node's minimum-cost mapping function, $cost(minMap(n, U))$. For each value of $U \in \{2, ..., K\}$, the utilization division of the minimum-cost mapping function is recorded [10].

67.5.2 Example

The Chortle mapping function is best illustrated by an example, as illustrated in Figure 67.9. For this example, we will assume that each CLB may have as many as four inputs (i.e., $K = 4$). The inputs, $\{A,B,C,D,E,F\}$, perform the logic function: $A * B + (C * D) E + F$.

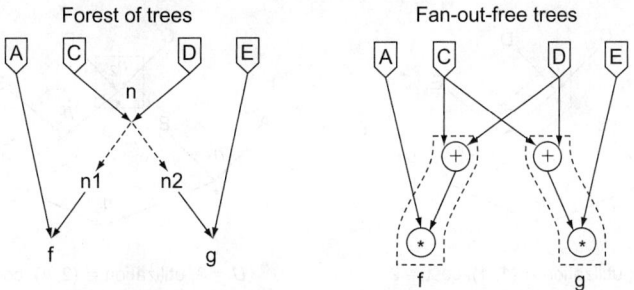

FIGURE 67.8 Forest of fan-out-free trees.

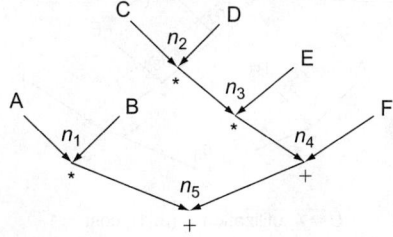

FIGURE 67.9 Chortle mapping example.

In the postorder traversal n_1 is visited first, followed by $n_2 \ldots n_5$. For n_1, there is only one possible mapping function, namely, $U = 2$, $\mu = \{1,1\}$. The same is true for n_2.

When n_3 is evaluated, there are two possibilities, as illustrated in Figure 67.10. First, the function could be implemented as a new CLB with two inputs ($U = 2$), driven from the outputs of n_2 and E. This subgraph would use two TLBs; thus, it would have a cost function of 2. For $U = 3$, only one utilization vector is possible, namely, $\mu = \{2,1\}$. All three primary inputs C, D, and E are grouped into one CLB, thus producing a cost function of 1. We store only the utilization vectors and cost functions for minMax(n_3,2) and minMax(n_3,3).

When n_4 is evaluated, there are many possibilities, as illustrated in Figure 67.11. With $U = 2$ ($\mu = \{1,1\}$), a two-input CLB would combine the optimal result for n_3 with the primary input F, producing a function with a cost of 2. For $U = 3$ ($\mu = \{2,1\}$), a three-input CLB would combine the optimal result for n_3: $U = 2$ with both inputs E and F, also at a cost of two CLBs. Finally, for $U = 4$, a single CLB would implement the function $(C * D) * (E + F)$, at a cost of 1. We store the utilization vectors and cost functions for minMax(n_4,2), minMax(n_4,3), and minMax(n_4,4).

Finally, we evaluate the output node, n_5, as illustrated in Figure 67.12. We see that there are four possible mappings and, of those, two minimal mappings are possible. Chortle may return either of the mappings where two CLBs implement: $n_5 = (A * B) + n_3 + F$ and $n_3 = (C * D) * E$.

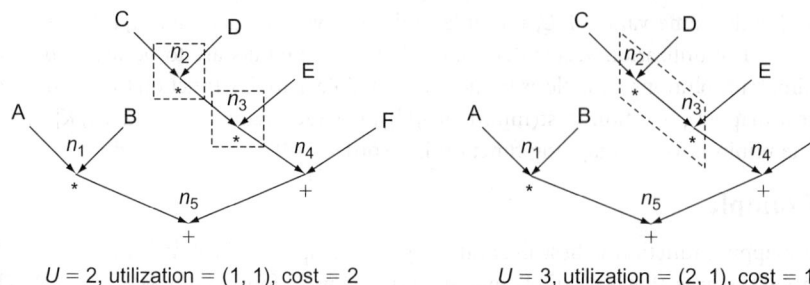

$U = 2$, utilization = (1, 1), cost = 2 $U = 3$, utilization = (2, 1), cost = 1

FIGURE 67.10 Mapping of node 3.

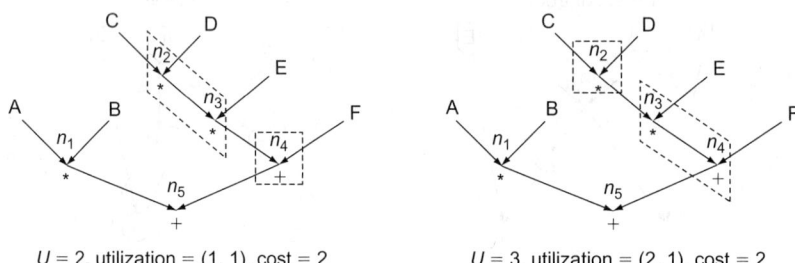

$U = 2$, utilization = (1, 1), cost = 2 $U = 3$, utilization = (2, 1), cost = 2

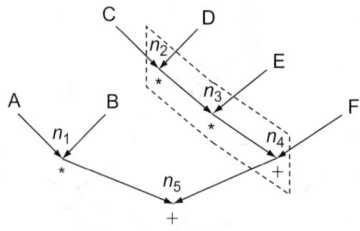

$U = 4$, utilization = (3, 1), cost = 1

FIGURE 67.11 Mapping of node 4.

67.5.3   Chortle-crf

The Chortle-crf algorithm is an improvement of the original Chortle algorithm. The major innovation with Chortle-crf involves the method for choosing gate-level node decomposition. The other improvements involve the algorithm's response to reconvergent and replicated logic. The name, Chortle-crf, is based on the new command line options (−crf) that may be given when running the program (−c for constructive bin-packing for decomposition, −r for reconvergent optimization, and −f for replication optimization) [11]. Each of the optimizations are detailed below.

67.5.3.1   Decomposition

Decomposition involves splitting a node and introducing intermediate nodes. Decomposition is required if the original circuit has a fan-in greater than K. In this case, no one CLB could implement the entire function. In general, the decomposition of a node may yield a circuit that uses fewer CLBs. Consider, for example, implementations with four-input CLBs ($K = 4$) of the circuit shown in Figure 67.13. Without decomposition, the output node forces the suboptimal use of the first two function generators (i.e., A * B and C * D are implemented as individual CLBs). With decomposition, however, the output node OR gate is decomposed to form a new node, which implements the function (A * B) + (C * D), which can be implemented in one CLB.

The original Chortle algorithm used an exhaustive search of all possible decompositions to find the optimal decomposition for the subcircuit, causing the running time at a node to increase exponentially as

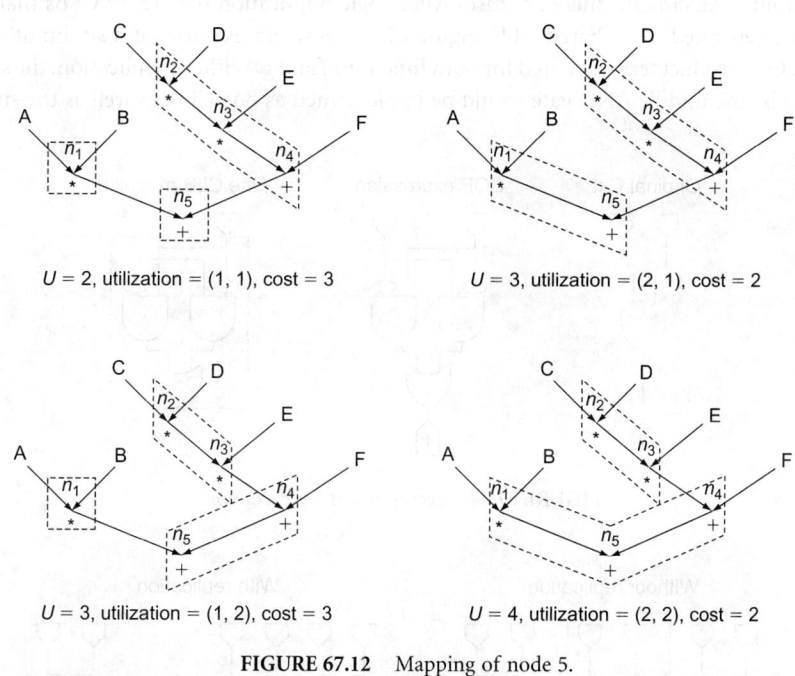

$U = 2$, utilization $= (1, 1)$, cost $= 3$ $U = 3$, utilization $= (2, 1)$, cost $= 2$

$U = 3$, utilization $= (1, 2)$, cost $= 3$ $U = 4$, utilization $= (2, 2)$, cost $= 2$

FIGURE 67.12   Mapping of node 5.

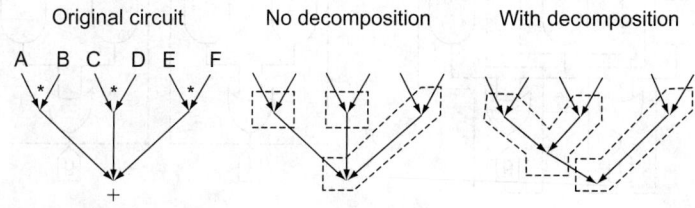

FIGURE 67.13   Decomposition example.

the fan-in increased. As a heuristic within the original Chortle algorithm, nodes would be arbitrarily split if the fan-in to a node exceeded 10, allowing each subfunction to be computed in a reasonable amount of time. If a node was split, however, the solution was no longer guaranteed to be optimal.

The improved Chortle-crf algorithm uses first-fit-decreasing bin-packing algorithm to solve the decomposition problem. Large fan-in nodes are decomposed into smaller subnodes with smaller fan-in. Next, the look-up tables for the input functions are bin-packed into CLBs. A look-up table with k inputs is merged into the first CLB that has at least $K - k$ unused inputs remaining. A new CLB is generated, if needed, to accommodate the k inputs.

67.5.3.2 Reconvergent Logic

Reconvergent logic occurs when a signal is split into multiple function generators, and then those output signals merge at another generator. An example of reconvergent logic is shown in Figure 67.14. When the XOR gate was converted into a SOP format by the technology-independent minimization phase, two AND gates and an OR gate were generated. Both AND gates share the same inputs. If the total number of distinct inputs is less than the size of the CLB, it is possible to map these functions into one CLB. The Chortle-crf algorithm finds all local reconvergent paths, and then examines the effect of merging those signals into one CLB.

67.5.3.3 Replicated Logic

For multioutput logic circuits, there are cases when logic duplication uses fewer CLBs than logic that uses subterms generated by a shared CLB. Figure 67.15 shows an example of a six-input circuit with two outputs. One product term is shared for both functions f and g. Without replication, the subfunction implemented by the middle AND gate would be implemented as one CLB, as well as the subfunctions

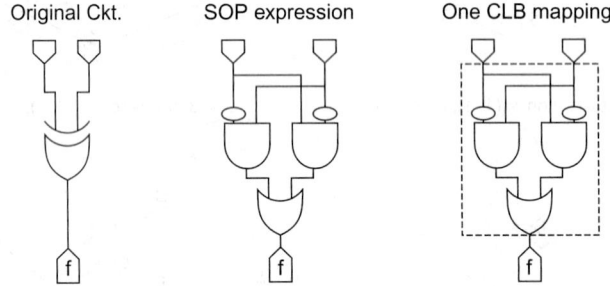

FIGURE 67.14 Reconvergent logic example.

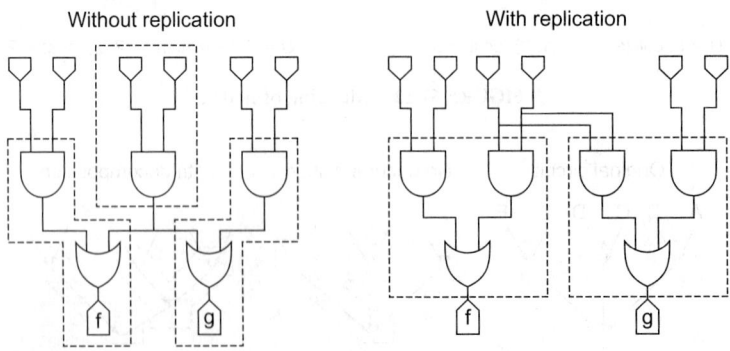

FIGURE 67.15 Replicated logic example.

for *f* and *g*. In this case, however, the middle AND gate can be replicated, and mapped into both function generators, thus allowing the entire circuit to be implemented using two CLBs, rather than three.

When a circuit has a fan-out greater than one, Chortle may implement the node explicitly or implicitly. For an explicit node, the subfunction is generated by a dedicated CLB, and this output signal is treated as an input to the rest of the logic. For an implicit node, the logic is replicated for each fan-out subcircuit. The algorithm computes the cost of the circuit, both with replication and without. Logic replication is chosen if this reduces the number of CLBs used to implement the circuit.

67.5.4 Chortle-d

The primary goal of Chortle-d is to reduce the depth of the logic (i.e., the largest number of CLBs for any signal path through combinational logic) [12]. By minimizing the longest paths, it is possible to increase the frequency at which the circuit can operate. Chortle-d is an enhancement of the Chortle-crf algorithm. Chortle-d, however, may use more look-up tables than Chortle-crf to implement a circuit with a shorter depth.

The Chortle-d algorithm separates logic into strata. Each stratum contains logic at the same depth. When nodes are decomposed, the outputs of the tables with the deepest stratum are connected to those at the next level. Chortle-d also employs logic replication, where possible. Replication often reduces the depth of the logic, as illustrated in Figure 67.15.

The depth optimization is only applied to the critical paths in the circuit. The algorithm first minimizes depth for the entire circuit to determine the maximum target depth. Next, the Chortle-crf algorithm is employed to find a circuit that has minimum area. For paths in the area-optimized circuit that exceed the target depth, depth-minimization decomposition is performed. This has the effect of equalizing the delay through the circuit.

It was found that for the 20 circuits in the MCNC logic synthesis benchmark, the chortle-d algorithm constructed circuits with 35% fewer logic levels, but at the expense of 59% more look-up tables.

67.6 Two-Step Approaches

As with Chortle, the two-step methods start with an optimized network in which the number of literals is minimized. The network is decomposed to be feasible in the first step; then the number of nodes is reduced in the second step. If the given network is already feasible, the first step is skipped.

67.6.1 First Step: Decomposition

For a given FPGA device, with a *k*-input TLU, all nodes of the network with more than *k* inputs must be decomposed. Different methods decompose the network in different ways.

67.6.1.1 MIS-pga 1

MIS-pga 1 was developed at Berkeley for FPGA synthesis, as an extension of MIS-II. It uses two algorithms, kernel decomposition and Roth-Karp decomposition, to decompose the infeasible nodes separately; then it selects the better result.

Kernel decomposition decomposes an infeasible node n_i by extracting a kernel function, k_i, and splitting n_i based on k_i and its residue, r_i. The residue r_i, of a kernel k_i, of a function F, is the expression for F with a new variable substituted for all occurrences of k_i in F; for example, if $F = x_1x_2 + x_1x_3$, then $k_i = x_2 + x_3$, and $r_i = x_1k_i$. As there may be more than one kernel function that exists for a node, a cost function is associated with each kernel: $cost(k_i) = |sup(k_i) \cap sup(r_i)|$. The kernel with minimum cost is chosen. A kernel decomposition is illustrated in Figure 67.16.

Splitting infeasible nodes by kernel functions minimizes the number of new edges generated. Therefore, the considerations of wiring resources and logic area are integrated together. This procedure is applied recursively until all nodes are feasible. If no kernels can be extracted for a node, an AND–OR decomposition is applied.

Roth-Karp decomposition is based on the classical decomposition of Ashenhurst and Curtis [13]. Instead of building a decomposition chart whose size grows exponentially, as it does with the original method, a compact cover representation of the on-set and the off-set of the function is used. The Roth-Karp algorithm avoids the expensive computation of the best solution by accepting the first bound set. As with kernel decomposition, the AND/OR decomposition is used as a last resort.

67.6.1.2 Hydra Decomposition

The Hydra algorithm, developed at Stanford University, is designed specifically for two-output TLU FPGAs [14]. Decomposition in Hydra is performed in three stages. The first and third stages are AND–OR decompositions, while the second stage is a simple-disjoint decomposition, which is defined as the following:

Given a function, F, and its support, S, with $F = G(H(S^a), S^b)$, where S^a, $S^b \subseteq S$ and $S^a \cup S^b = S$; If $S^a \cap S^b = 0$, then G is a disjoint decomposition of F.

The first stage is executed only if the number of inputs to the nodes in the given network is larger than a given threshold. Without performing the first stage, the efficiency of the second stage would be reduced. The last stage is applied only if the resulting network is still infeasible.

In the second stage, the algorithm searches for all the function pairs that have common variables. It then applies the simple-disjoint decomposition on those function pairs. As a result, two CLBs with the same fan-ins can be merged into one two-output CLB. The rationale is illustrated in Figure 67.17.

A weighted graph $G(V, E, W)$ that represents the shared-variable relationship is constructed based on the given Boolean network. In $G(V, E, W)$, V is the node set corresponding to that of the Boolean network; edge, $e_{ij} \subset E$, exists for any pair of nodes, $\{v_i, v_j\} \subset V$, if they share variables; and weight, $w_{ij} \subset W$, is the number of variables shared correspondingly. Edges are first sorted by weight and then traversed in

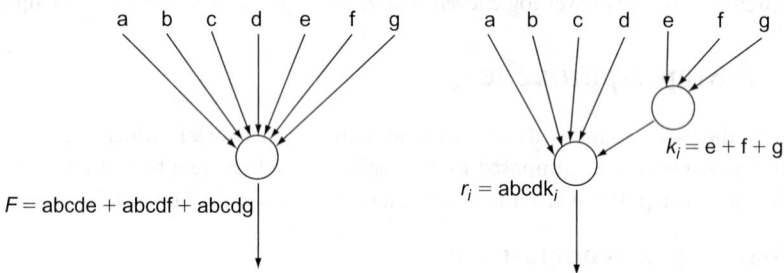

FIGURE 67.16 Example of kernel decomposition.

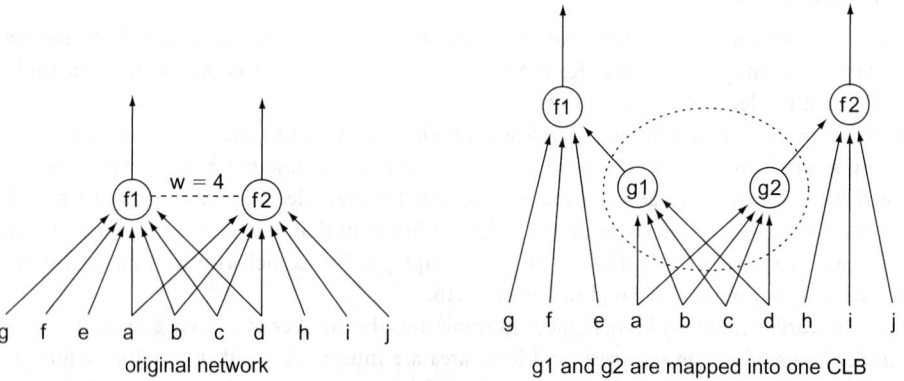

FIGURE 67.17 CLB mapping example.

decreasing order to check for simple-disjoint decomposition. A cost function, which is the linear combination of the number of the shared inputs and the total number of variables in the extracted functions, is computed to decide whether or not to accept a certain simple decomposition.

67.6.1.3 Xmap Decomposition

The Xmap decomposes the infeasible network by converting the SOP form from MIS-II into an if-then-else DAG representation [15]. The terms of the SOP network are collected in a set, T; then, variables are sorted in decreasing order of the frequency of their appearance in T; finally, the if-then-else DAG is formed by the following recursive function:

- Let V be the most frequently used variable in the current set, T.
- Sort the terms in T into subsets $T(V_d)$, $T(V_1)$, according to $VT(V_d)$ is the subset in which V does not appear, $T(V_1)$ is the onset of V, and $T(V_0)$ is the offset of V.
  - Delete V from all terms in T; then apply the same procedure recursively to the three subsets until all variables are tested.

The resulting if-then-else DAG after the first iteration is given in Figure 67.18. A circuit that has been mapped to an if-then-else DAG is immediately suited for use with multiplexer-based CLBs [16]. Additional steps are used to optimize the DAG for use with TLU functions.

67.6.2 Second Step: Node Elimination

Three approaches have been proposed for node elimination: local elimination, covering, and merging.

67.6.2.1 Local Elimination

The operation used for local elimination is collapsing, which merges node n_i into node n_j whenever n_i is a fan-in node to n_j and the new node obtained is feasible. The Hydra algorithm accepts local eliminations as soon as they are found. MIS-pga 1, however, first orders all possible local eliminations as a function of the increase in the number of interconnections resulting from each elimination, and then greedily selects the best local eliminations.

The number of nodes can be reduced by local elimination, but its myopic view of the network causes local elimination to miss better solutions. Additionally, the new node created by merging multifan-out nodes may substantially increase the number of connections among TLUs and hence make the wiring problem more difficult. This problem is more severe in Hydra than in MIS-pga 1.

67.6.2.2 Covering

The covering operation takes a global view of the network by identifying clusters of nodes that could be combined into a single TLU. The operation is a procedure of finding and selecting supernodes.

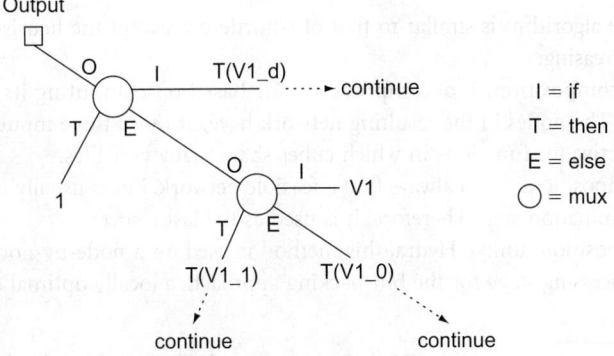

FIGURE 67.18 Result of first iteration.

A supernode, S_i, of a node n_i, is a cluster of nodes consisting of n_i and some other nodes in the transitive fan-in of n_i such that the maximum number of inputs to S_i is k. Obviously, more than one supernode may exist for a node.

In MIS-pga 1, the covering operation is performed in two stages. In the first stage, the supernodes are found by repeatedly applying the maxflow algorithm at each node. In the second stage, an optimal subset of the supernodes that can cover the whole network using a minimum number of supernodes is selected by solving a binate covering problem whose constrains are: first, all intermediate nodes should be included in at least one supernode; second, if a supernode S_i is selected, some supernodes that supply the inputs of S_i must be selected (the ordinary [unate], covering problem just has the first constraint).

Hydra examines the nodes of the network in order of decreasing number of inputs. An unassigned node with the maximal number of inputs is chosen first. A second node is then chosen such that the two nodes can be merged into the same TLU and the cost function (same cost function as was used in decomposition step) is maximized. This greedy process stops when all unexamined nodes have been considered.

For Xmap, the logic blocks to be found are sub-DAGs of the if-then-else DAG for the entire circuit. The algorithm traverses the if-then-else DAG from inputs to outputs and keeps a log of inputs in the paths (called signals set) that can be used to compute the function of the node under consideration. Nodes in the signals set could be a marked node or a clean node. A marked node isolates its inputs to the current node, while a clean node exposes all its fan-ins. An overflow node, a node whose signals set is larger than k (the number of inputs of the TLU), a marking procedure is executed to reduce the fan-in. Xmap first marks the high-fan-out descendants of the node, and then marks the children of the node in decreasing order of the size of their signals set. The more inputs Xmap can isolate from the node under consideration, the better. The marking process cuts the if-then-else into pieces, each of which can be mapped into one CLB.

67.6.2.3 Merging

The purpose of the merging step is to combine nodes that share some inputs to exploit some of the particular features of FPGA architecture. For example, each CLB in the Xilinx XC4000 device has two four-input TLUs and a third TLU combining them with the ninth input (Section 67.3). In the three approaches discussed above, a postprocessing step is performed to merge pairs of nodes after the covering operation. The problem is formulated as a maximum cardinality matching problem.

67.6.3 MIS-pga 2: A Framework for TLU-Logic Optimization

MIS-pga 2 is an improved version of MIS-pga 1. It combines the advantageous features of Chortle-crf, MIS-pga 1, Xmap, and Hydra. In each step, Mis-pga 2 tries different algorithms and chooses the best [17]. Four decomposition algorithms are executed in the decomposition step:

1. Bin-packing: the algorithm is similar to that of Chortle-crf, except the heuristic of MIS-pga 2 is the Best-Fit Decreasing.
2. Cofactoring decomposition: it decomposes a node based on computing its Shannon cofactor ($f = f_1 f_2 + f_1' f_3$). The nodes in the resulting network have, at most, three inputs. This approach is particularly effective for functions in which cubes share many variables.
3. AND/OR decomposition: it can always find a feasible network, but is usually not a good network for the node-elimination step. Therefore, it is used as the last resort.
4. Disjoint decomposition: unlike Hydra, this method is used on a node-by-node basis. When it is used as a preprocessing stage for the bin-packing approach, a locally optimal decomposition can be found.

MIS-pga 2 interweaves some operations of the two-step methods. For example, the local elimination operation is applied to the original infeasible network as well as to the decomposed, feasible network. This same operation is referred to as partial collapse when applied before decomposition. Unlike MIS-pga 1, which separates the covering and the merging operations, these two operations are combined together to solve a single, binate covering problem.

Because MIS-pga 2 does a more exhaustive decomposition phase, and because the combined covering/merging phase has a more global view of the circuit, MIS-pga 2 results are almost always superior to those of Chortle-crf, MIS-pga 2's results are almost always superior to those of Chortle-crf, MIS-pga 1, Hydra, and Xmap. For the same reason, MIS-pga 2 is relatively slow, as compared to the other algorithms.

67.7 Conclusion

By understanding how FPGA logic is synthesized, hardware designers can make the best use of their development tools to implement complex, high-performance circuits. Synthesis of FPGA logic devices makes use of the algorithms of Chortle and its extensions: Xmap, Hydra, MIS-pga 1, and MIS-pga 2. Each of these methods starts with an optimized Boolean network and then maps the logic into the configurable logic blocks of a field-programmable gate array circuit. Because the optimal covering problem is NP-hard, heuristic approaches are used to find a near-optimal solution in reasonable running time. Understanding these tradeoffs is key to rapidly prototyping logic with FPGA technology.

References

1. J. Rose, A.E. Gamal, and A. Sangiovanni-Vincentelli, Architecture of field-programmable gate arrays, *Proc. IEEE*, vol. 81, pp. 1013–1029, 1993.
2. Xilinx, Inc., *The Programmable Logic Data Book*, 2006.
3. ACTEL, *FPGA Data Book and Design Guide*, 1996.
4. A. Sangiovanni-Vincentelli, A.E. Gamal, and J. Rose, Synthesis methods for field programmable gate arrays, *Proc. IEEE*, vol. 81, pp. 1057–1083, 1993.
5. R.K. Brayton, G.D. Hachtel, and A. Sangiovanni-Vincentelli, Multilevel logic synthesis, *Proc. IEEE*, vol. 78, pp. 264–300, 1990.
6. R. Brayton, R. Rudell, A. Sangiovanni-Vincentelli, and A. Wang, Multi-level logic optimization and the rectangular covering problem, *IEEE International Conference on Computer-Aided Design*, Santa Clara, CA, pp. 62–65, 1987.
7. R. Murgai, Y. Nishizaki, N. Shenoy, R.K. Brayton, and A. Sangiovanni-Vincentelli, Logic synthesis for programmable gate arrays, *ACM/IEEE Design Automation Conference*, Orlando, FL, pp. 620–625, 1990.
8. R.K. Brayton, R. Rudell, A. Sangiovanni-Vincentelli, and A.R. Wang, MIS: A multiple-level logic optimization system, *IEEE Trans. Computer-Aided Design*, vol. CAD-6, pp. 1062–1081, 1987.
9. D. Bostick, G.D. Hachtel, R. Jacoby, M.R. Lightner, P. Moceyunas, C.R. Morrison, and D. Ravenscroft, The boulder optimal logic design system, *IEEE International Conference on Computer-Aided Design*, Santa Clara, CA, pp. 62–69, 1987.
10. R.J. Francis, J. Rose, and K. Chung, Chortle: A technology mapping program for look-up table-based field programmable gate arrays, *ACM/IEEE Design Automation Conference*, Orlando, FL, pp. 613–619, 1990.
11. R.J. Francis, J. Rose, and Z. Vranesic, Chortle-crf: Fast technology mapping for look-up table-based FPGAs, *ACM/IEEE Design Automation Conference*, San Francisco, CA, pp. 227–233, 1991.
12. R.J. Francis, J. Rose, and Z. Vranesic, Technology mapping of look-up table-based FPGAs for performance, *IEEE International Conference on Computer-Aided Design*, Santa Clara, CA, pp. 568–575, 1991.
13. T. Luba, M. Markowski, and B. Zbierzchowski, Logic decomposition for programmable gate arrays, *Euro ASIC '92*, Paris, France, pp. 19–24, 1992.
14. D. Filo, J.C.-Y. Yang, F. Mailhot, and G.D. Micheli, Technology mapping for a two-output RAM-based field programmable gate array, *European Design Automation Conference*, Amsterdam, Netherlands, pp. 534–538, 1991.

15. K. Karplus, Xmap: A technology mapper for table-lookup field programmable gate arrays, *ACM/IEEE Design Automation Conference*, San Francisco, CA, pp. 240–243, 1991.

16. R. Murgai, R.K. Brayton, and A. Sangiovanni-Vincentelli, An improved systhesis algorithm for multiplexor-based pga's, *ACM/IEEE Design Automation Conference*, Anaheim, CA, pp. 380–386, 1992.

17. R. Murgai, N. Shenoy, R.K. Brayton, and A. Sangiovanni-Vincentelli, Improved logic synthesis algorithms for table look up architectures, *IEEE International Conference on Computer-Aided Design*, Santa Clara, CA, pp. 564–567, 1991.

Section IX

Testing of Digital Systems

Nick Kanopoulos
Atmel Corp.

68

CAD DFT and Test Architectures

Dimitri Kagaris
Southern Illinois University

Nick Kanopoulos
Atmel Corp.

Spyros Tragoudas
Southern Illinois University

CONTENTS

Design for testability (DFT) constitutes a set of design rules that must be followed in the design phase to improve the testability of faults that model physical defects. They are further distinguished among techniques that are applied at the circuit level or at the system level and are discussed in Sections 68.1 and 68.2 below.

68.1 Circuit-Level DFT: Scan Designs

Circuit-level DFT is either *empirical* or *automated* [1,2]. The former consists of a list of practices that have been applied successfully in testing. Following such empirical rules, circuits are designed so that each embedded flip-flop must be initializable, a gate should not have large number of fan-ins because the inputs are difficult to control and the output is difficult to observe, direct control during testing must be provided for embedded signals that are difficult to control. Unless such methods are supplemented by automated methods, such empirical methods do not perform well on test that are provided by automatic test pattern generation (ATPG) tools. In most cases, they require manual test generation which is impractical in VLSI.

In automated DFT methods circuitry is added to allow testing under a predetermined manner. They are further distinguishable as *built-in self-test* (BIST) methods and *scan-based* methods. BIST is examined in a later chapter because it involves elaborate methods to automate the design of circuitry that generates the test patterns and analyzes the circuit responses. This section studies the principle of scan where an extra mode, referred to as the *test mode*, allows the flip-flops to form one or more registers, referred to as the *scan registers*. This is the *conventional scan* method. It has been suggested that this operation can be virtually performed as a Random Access Memory in which case the approach is referred to as *random access scan*. Both the conventional and random access scan are distinguished as *full scan* where all flip-flops participate in the test function or *partial scan* where only a subset of flip-flops participate.

A popular method to select a subset of flip-flops for scan is the *structural partial scan method*. In this approach, the *structure graph* of the sequential circuit is constructed [2]. In this graph the very node corresponds to a flip-flop, and a directed edge from node *a* to node *b* indicates that there is a directed path from flip-flop *a* to flip-flop *b* that consists of combinational components. Then nodes are removed from this graph, and the respective flip-flops are scanned. A minimum number of nodes must be removed so that the structure graph becomes acyclic (self-loops need not be eliminated), and then the longest path in the acyclic graph is less than a predefined bound. The latter quantity is also referred to as the sequential depth, and determines an upper bound on the sequence of patterns that must be applied to detect faults that model physical defects. The smaller the sequential depth, the smaller is the length of the sequence.

In scan-based designs only D-type master–slave flip-flops should be used, as in Figure 68.1. At least one primary input pin must be available for test. If more than one pins is available, then multiple scan registers are formed. It is also important that all clock signals are controllable from primary inputs and that they should not feed data inputs of flip-flops. If the latter is not followed, then races may occur in the operational mode of the circuit.

Figure 68.2 shows a scan flip-flop design with only clock signal (CK). When the test control signal (T) is 1, the input data (D) are stored in the flip-flop; when T is 0, the shift data (S) are stored.

Figure 68.3 shows the level-sensitive scan design (LSSD) flip-flop where there are two nonoverlapping clock signals, the master clock (M_CK) and the slave clock (S_CK). When M_CK is 1, D is stored in the master latch and when S_CK is high, D is stored in the slave latch. They are never both set to 1. To operate in scan mode, M_CK is set to 0 and then data S are stored using T and S_CK as master and slave clocks, respectively [1,2]. This design reduces the performance degradation owing to the added multiplexer required to implement the scan mode.

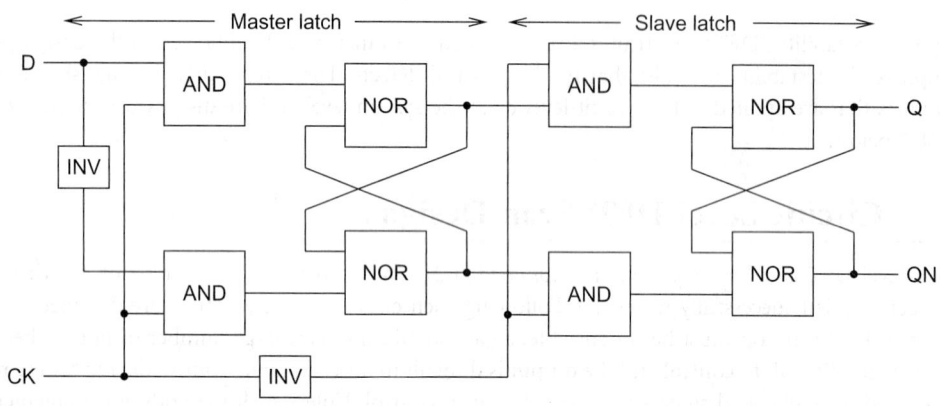

FIGURE 68.1   Schematic of a D flip-flop.

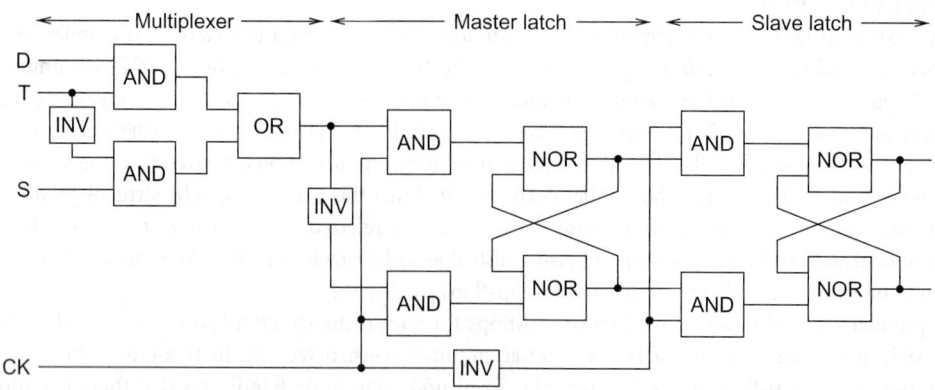

FIGURE 68.2   Schematic of a single clock scan flip-flop.

Testing of circuits with scan is done in two phases. First, a toggle sequence is shifted into the scan register to test for defects in the shift register that may impact its correctness as far as the shift operation is concerned. The sequence must produce the two transitions and 2-bit stable value subsequences in each flip-flop. Its length is the number of scanned flip-flops plus four.

In the second phase, faults in the combinational logic are targeted. Assume that the test set consists of a collection of test patterns that must be applied to the combinational logic. Each pattern consists of the input portion (bit values that must be applied to the primary inputs) and the scan portion (bit values that must be stored into the scan flip-flops). Each pattern is applied in the operational mode, but prior to that the test mode is used to shift-in the scan portion. This requires, in the worst case, as many clocks as the number of flip-flops in the scan register. Once the pattern is applied, using a single operational clock, the next pattern is shifted-in, while the latched values of the previously applied pattern are scanned-out.

When testing for delay defects, the test set collection consists of test patterns that must be applied in pairs. Figure 68.4 shows an enhancement of the scan flip-flop, called enhanced scan flip-flop, where an additional latch allows for any pair of input vectors to be applied to the combinational logic. A pair of patterns is applied with the following sequence of operations:

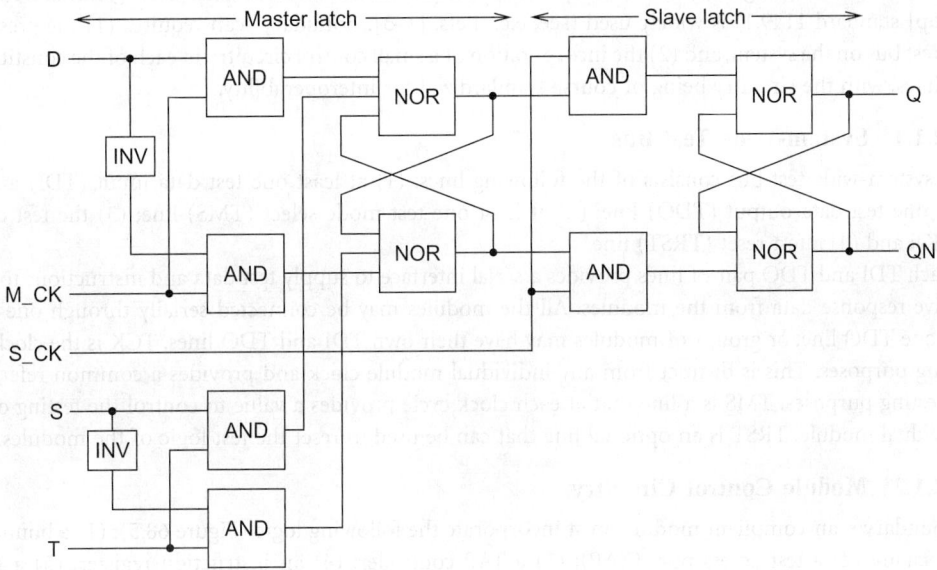

FIGURE 68.3 Schematic of an LSSD scan flip-flop.

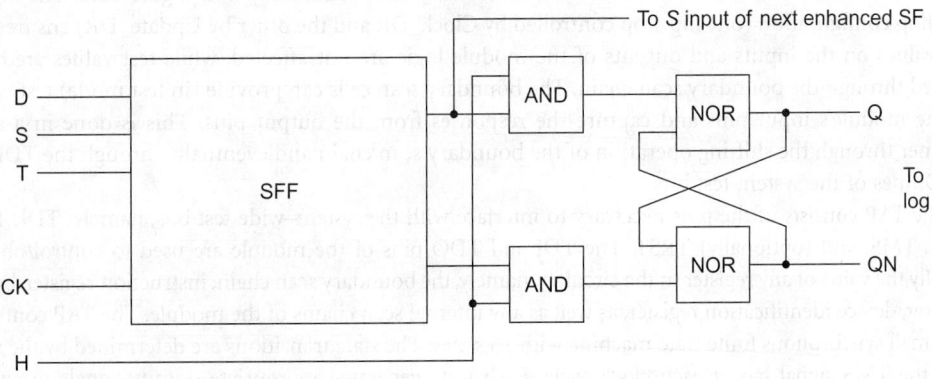

FIGURE 68.4 Schematic of an enhanced scan flip-flop.

First H = T = 0, and, using CK, the scan portion of the first pattern is scanned-in. Then H = T = 1, and the input portion of the first pattern is applied. Then the above procedure is repeated for the second pattern. Then H = 0 and CK is applied to capture the output of the combinational logic that drives the scan flip-flops. Then T = 0 and the contents of the flip-flops are scanned-out.

68.2 Test Architectures for System-Level DFT

68.2.1 Boundary Scan Standard

Digital logic modules are typically made parts of bigger systems by being mounted (packaged as chips) on printed circuit board (PCB) or by being integrated (as "bare die") in multichip modules (MCMs). The resulting systems present their own difficulties for testing even if all participating modules have been tested prior to assembly. The main problems are faults on module interconnects and faults in the system-level behavior of the individual modules (such as insufficient driving capability). In addition, it is desirable to easily test each module under a lifetime testing policy, and also to easily isolate any faulty module(s).

To accomplish these goals, the boundary scan technique (standardized as IEEE JTAG [Joint Test Action Group] standard 1149.1) is widely used (see, e.g., Refs. [1–3]. Boundary scan requires (1) the presence of a test bus on the system, and (2) the incorporation of a small control circuitry in each of the constituent modules, with the circuitry being of course standardized for interoperability.

68.2.1.1 System-Wide Test Bus

The system-wide test bus consists of the following lines: (1) at least one test data input (TDI) and at least one test data output (TDO) line; (2) at least one test mode select (TMS) line; (3) the test clock (TCK); and (4) a test reset (TRST) line.

Each TDI and TDO pair of lines provides a serial interface to supply test data and instructions to and receive response data from the modules. All the modules may be connected serially through one TDI and one TDO line, or groups of modules may have their own TDI and TDO lines. TCK is the clock for testing purposes. This is distinct from any individual module clock and provides a common reference for testing purposes. TMS is a line that at each clock cycle provides a value to control the testing of an individual module. TRST is an optional line that can be used to reset the test logic of the modules.

68.2.1.2 Module Control Circuitry

A boundary scan-compliant module must incorporate the following logic (Figure 68.5): (1) a boundary scan chain; (2) a test access port (TAP); (3) a TAP controller; (4) an instruction register; (5) a 1-bit bypass register; and (6) a device identification register plus other optional registers.

The boundary scan chain consists of individual boundary scan cells, one cell for each input and each output pin of the module. The structure of a boundary scan cell is shown in Figure 68.6. The double flip-flop configuration (one flip-flop controlled by Clock_DR and the other by Update_DR) ensures that the values on the inputs and outputs of the module logic are not affected, while test values are being shifted through the boundary scan chain. The boundary scan cells can provide (in test mode) test values to the module's input pins and capture the responses from the output pins. This is done in a serial manner through the shifting operation of the boundary scan chain and eventually through the TDI and TDO lines of the system test bus.

The TAP consists of the pins necessary to interface with the system–wide test bus, namely: TDI, TDO, TCK, TMS, and (optionally) TRST. The TDI and TDO pins of the module are used to control/observe serially the value of any register in the circuitry, namely, the boundary scan chain, instruction register, bypass register, device identification register, as well as any internal scan chains of the module. The TAP controller is a small synchronous finite state machine with 16 states. The state transitions are determined by the value that the TMS signal has at each clock cycle. Each state generates appropriate output signals to control the function of the instruction register, the boundary scan chain, and any required test data register.

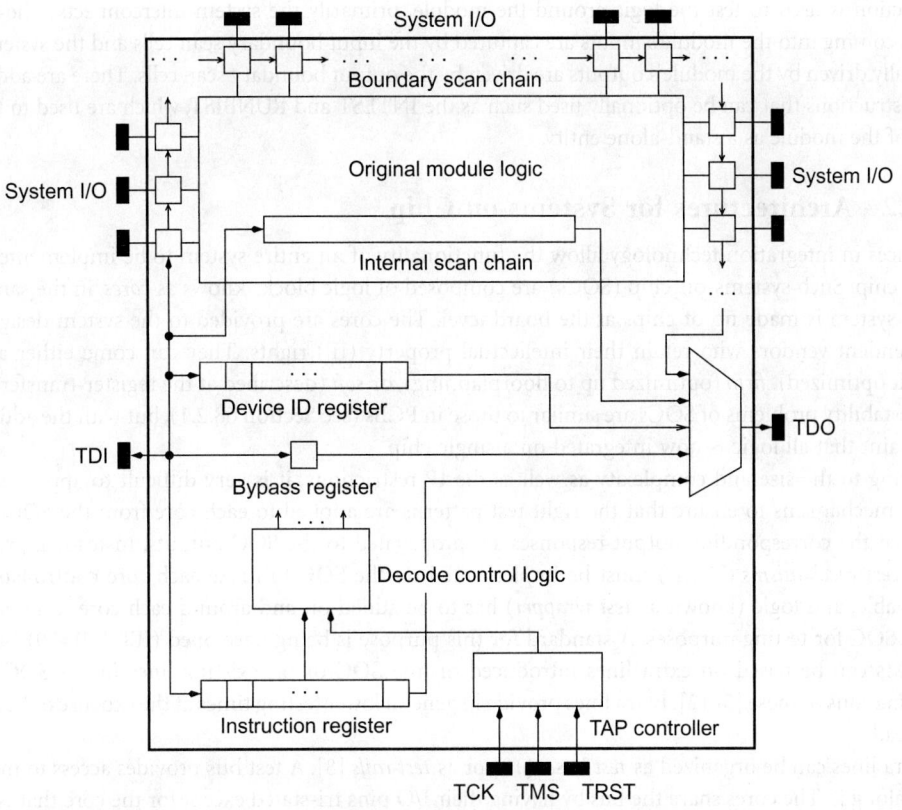

FIGURE 68.5 Boundary scan module control logic.

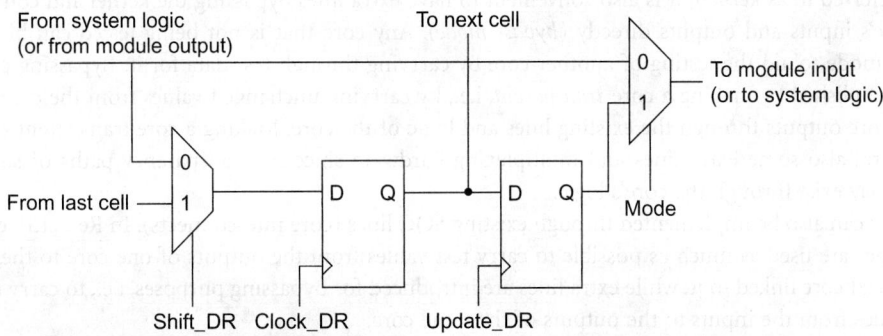

FIGURE 68.6 Boundary scan cell structure.

The instruction register holds the test instruction (serially supplied through the TDI pin) which each time controls the operation of the module's test logic. The mandatory test instructions are BYPASS, SAMPLE/ PRELOAD, and EXTEST. The BYPASS instruction connects the bypass register alone between the module's TDI and TDO pins without the intervention of the boundary scan chain. This is very useful in reducing the length of the system–wide boundary scan chain and thus delivers faster test data to and from the module under test. The SAMPLE/PRELOAD instruction allows the transfer of the module I/O pin values to the boundary scan cells without disrupting normal operation. It also allows preloading of the boundary scan chain by shifting in values to the flip-flops controlled by Clock_DR (Figure 68.6). The EXTEST instruction is used to test the logic around the module, primarily the system interconnects. The system values coming into the module's inputs are captured by the input boundary scan cells and the system lines normally driven by the module's outputs are driven by the output boundary scan cells. There are additional test instructions that can be optionally used such as the INTEST and RUNBIST, which are used to test the logic of the module as a stand-alone entity.

68.2.2 Architectures for Systems-on-Chip

Advances in integration technology allow the functionality of an entire system to be implemented on a single chip. Such systems-on-chip (SOCs) are composed of logic blocks knows as *cores*, in the same way that a system is made up of chips at the board level. The cores are provided to the system designer by independent vendors who retain their intellectual property (IP) rights. They can come either as *hard* (layout optimized), *firm* (optimized up to floorplanning), or *soft* (described at the register-transfer level). The testability problems of SOCs are similar to those in PCBs (see Section 68.2.1), but with the additional constraint that all logic is now integrated on a single chip.

Owing to the size and complexity as well as the IP restrictions, it is very difficult to apply standard ATPG mechanisms to ensure that the right test patterns are applied to each core from the SOC inputs and that the corresponding output responses are propagated to the SOC outputs. Instead, appropriate *test access mechanisms* (TAMs) must be implemented on the SOC to make each core controllable and observable, and logic (known as *test wrapper*) has to be added on and around each core to interface it to the SOC for testing purposes. A standard for this purpose is being developed (IEEE P1500) [4].

TAMs can be based on extra lines introduced on the SOC or on existing lines in the SOC or on combinations of these [5–12]. Extra lines provide in general lower testing times at the expense of hardware overhead.

Extra lines can be organized as *test buses* [10] or as *test rails* [8]. A test bus provides access to multiple cores along it. The cores share the bus by having their I/O pins tri-stated except for the core that is being tested at each time. A test control bus must also be provided on the SOC along with the logic (test controller) that controls it. In contrast, test rails are extra interconnects from one core to the next. Since the test rails on core inputs and the test rails on its outputs are separated by the application logic of the core (referred to as *kernel*), it is also convenient to have extra lines bypassing the kernel and connecting the core's inputs and outputs directly (*bypass mode*). Any core that is not being tested can be in this bypass mode to aid the testing of another core by carrying through test data for it. Bypassing can also be accomplished by making a core *transparent*, i.e., by carrying unchanged values from the core inputs to the core outputs through the existing lines and logic of the core. Making a core transparent requires in general also some extra lines and multiplexing hardware since no transparency paths of sufficient width may exist through the core's logic.

TAMs can also be implemented through existing SOC lines (core interconnects). In Ref. [13], existing SOC lines are used as much as possible to carry test values from the outputs of one core to the inputs of the next core linked to it, while extra lines are introduced for bypassing purposes, i.e., to carry though test values from the inputs to the outputs of the same core.

In conjunction with the TAM, the terminals of each core must be appropriately interfaced to it. This is achieved by logic around the core, known as *core test wrapper*. Examples of core test wrappers include the test collar [10], the test shell [8,40], the InTeRail wrapper [13], and the isolation ring (a boundary scan-like structure).

The basic functions of a core test wrapper include: (1) normal mode, where the wrapper is configured to allow the original core interconnections for normal operation; (2) kernel test mode, where the wrapper is configured to allow the testing of the core kernel itself; (3) interconnect test mode, where the wrapper is configured to make the core output pins be controllable and the core input pins be observable; and (4) bypass mode, where the wrapper is configured to bypass the core kernel.

The test wrapper should also provide for *width adaptation*. That is, the TAM width may actually be smaller than the number of input or output terminals of a core. In that case, some of the core inputs and outputs must be configured as a small shift register that is driven by a single TAM line, while the wrapper does serial-to-parallel conversion for inputs and parallel-to-serial conversion for outputs. In addition, the wrapper may allow for *dynamic reconfigurability* [14]. That is, the wrapper may provide the SOC designer with the capability to customize the number and length of the above shift registers as well as the number and length of any internal scan chains that the core kernel is equipped with to achieve particular trade-offs between test time and TAM hardware overhead.

Design automation aspects in SOC design typically focus primarily on minimizing the SOC test time. Issues include the determination of optimal number of TAMs, partitioning the TAM width among cores, assignment of the cores to TAMs, determination of optimal wrapper designs for the cores, use of test compression, and concurrent testing of the cores. Algorithms for these purposes can be found in Refs. [13,15–17,20].

The problem of test scheduling for concurrent core testing is actually intertwined with the problem of test resource allocation (TAM lines, wrapper width). It can be formulated in its general form as follows:

Let W be the total number of SOC pins and let N be the total number of cores in the SOC. Each core $C_i, 1 \leq i \leq N$, has w_i pins to interface it (through the wrapper) to the rest of the SOC for testing purposes. Notice that the number of functional pins of the core may be larger than w_i, but as mentioned above the wrapper provides for width adaptation when in testing mode. Each core C_i also has a testing time T_i associated with it, which depends on the available width w_i (T_i is a nonincreasing function of w_i). Additional values may be associated with a core C_i such as power consumption P_i for the application of the test set.

The objective in the test scheduling/resource allocation problem is to determine a quadruple (w_i, S_i, T_i, t_i) for each core $C_i, 1 \leq i \leq N$, where w_i is the number of test access pins of core C_i, S_i the set of the specific SOC pins assigned to access C_i through the TAM ($|S_i| = w_i$), T_i the testing time corresponding to w_i, and t_i the scheduled starting time for the testing of C_i such that (1) $\Sigma w_i \leq W$, (2) $\max(T_i + t_i)$ is minimized, (3) all cores being tested concurrently at any time instant have disjoint S_i sets, and (4) all cores being tested concurrently at any time instant also satisfy a number of other (optional) constraints such as that the sum of their power consumptions P_i be less than a prescribed bound P, etc.

An algorithmic framework for solving the above problem is provided by the two-dimensional packing problem (also known as rectangle packing). Each core C_i is associated with a rectangle of length T_i (horizontal dimension) and width w_i (vertical dimension). All cores have to be packed in a "bin" of width W and of minimum length (testing time), so that no rectangles overlap and all rectangles intersected by a vertical line satisfy the other (optional) constraints in (4) above. Algorithms for this purpose can be found in Refs. [15,16]. An alternative framework based on network flows can be found in Ref. [14]. Other approaches with restrictions of various degrees appear in Refs. [17–19], among others.

68.2.3 Testing Systems with Networks on Chip

As mentioned in the previous section, an SoC integrates several cores in a single chip for use in the communications field, multimedia, and general electronics. According to the International Technology Roadmap for Semiconductors, by the end of the decade, SoCs using 50 nm transistors and operating below 1 V, will grow to 4 billion transistors running at 10 GHz. A major challenge in such systems is to provide reliable means for component interaction, guarantee global synchronization of the chip, and tackle the growing wire delays that become the dominant factor of the overall delay.

The interconnects of traditional SoC architectures that were reviewed in the previous section are either dedicated channels or shared buses. Dedicated channels offer the best communication performance, but have poor reusability and are thus undesirable as the number of cores in the SoC grows. A shared bus is reusable, but supports one communication at a time (time division multiplexing). Its bandwidth is shared among all the cores in the system and its operating frequency decreases with the system's growth. As SoCs grow in size, cross talk effects between wires, electromagnetic interferences, and even radiation-induced charge injection [21] render the internal communication between the cores unreliable.

A network on chip (NoC) solves such interconnect issues on SoC architectures. NoCs decouple the communication tasks from the computation tasks and offer well–defined protocols to eliminate contention in the channels. Ideas from computer networks have been borrowed to provide interconnections and communication among on-chip cores. A study presented in Ref. [22] demonstrates that NoC interconnects have better communication performance than the traditional bus interconnects for chips having as low as eight cores when the communication load is heavy. In case of a lighter communication load, the central bus architecture performs better than an NoC for chips of up to 16 cores. Clearly, as the number of cores per chip increases and the communication workload increases, NoCs become an appealing solution. In nanometer-scale technology, the performance of the communication in an NoC is more predictable than in bus–based SoCs because the geometry is regular, submicron effects are better contained, and, finally, there is separation between communication and computation tasks. NoCs can act in a globally asynchronous locally synchronous manner.

One important feature of an NoC is its reconfigurability and adaptability. Reconfigurability is ensured by the reuse of the communication network, and the reuse of the design, simulation, and prototype environment. The communication network can easily be reused and its channels can provide significantly better quality of service guarantees than an SoC bus. Reuse of design, simulation, and prototype environment makes it possible for many products to be based on the same NoC platform and reduces the expenses associated with the design and simulation of an NoC. Adding a new resource in a shared-bus system has a deep effect on the performance of the rest of the system. In contrast, the NoC is adaptable and neither its performance nor the system's performance is degraded when adding new resources to the SoC.

Another important characteristic of systems with an NoC is that little or no additional hardware is needed to test the embedded cores, the on–chip micronetwork, and the network interface. Bus–based SoC architectures use TAMs such as in Refs. [13,23] to ensure accessibility and controllability of each core during testing.

In a system with an NoC the test patterns are applied via the existing networking infrastructure. To that respect, NoCs are an effective mechanism to reduce the overall DFT overhead of the SoC. All existing methods for testing the embedded cores of an NoC apply the test patterns in a bus-like manner [24–28] which requires a data transmission mode that is only recommended for jitter-free transmission. The test data for a core are submitted along the same route, and the test application is guided by scheduling formulations that resemble the case of bus–based SoCs [24,25].

This is a simple and effective approach. However, such a direction does not utilize significant features of the NoC infrastructure. The remaining of the section overviews existing on-chip micronetwork architectures and transmission–related characteristics that may be used to successfully eliminate contention in the channels to reduce the test application time and increase the quality of testing.

NoCs were introduced in Refs. [21,29]. The early work was on new interconnect architecture and data exchanges between the cores in the form of structured packets. In Ref. [30], a hybrid routing algorithm for on-chip communication that combines the advantages of both the deterministic and adaptive routing schemes is given. Results demonstrate that it significantly reduces the communication delay under heavy traffic.

Several NoC architectures have been proposed. AEthereal is Philips' NoC architecture [31,32,41]. Routers may be connected with a flexible topology and intermodule communication is done with real packet switching, called the best effort (BE) mode, where data between different source–destination pairs share channels. A guaranteed throughput (GT) mode is also provided. It is used only when jitter–free transmission is required since it is known that BE transmission is more efficient [33].

The NoC is composed of the routers, the interconnects, and the network interfaces (NIs). Each NI acts as a bridge between an embedded core and an adjacent router. Data are transmitted using packets. Packets are composed of flits, the minimum transmission unit. Each flit can have up to three words of data. The header contains the amount of credit (used for flow control), the queue id (corresponds to the next router on the path), and the path to the destination (sequence of routers). Each flit in the packet

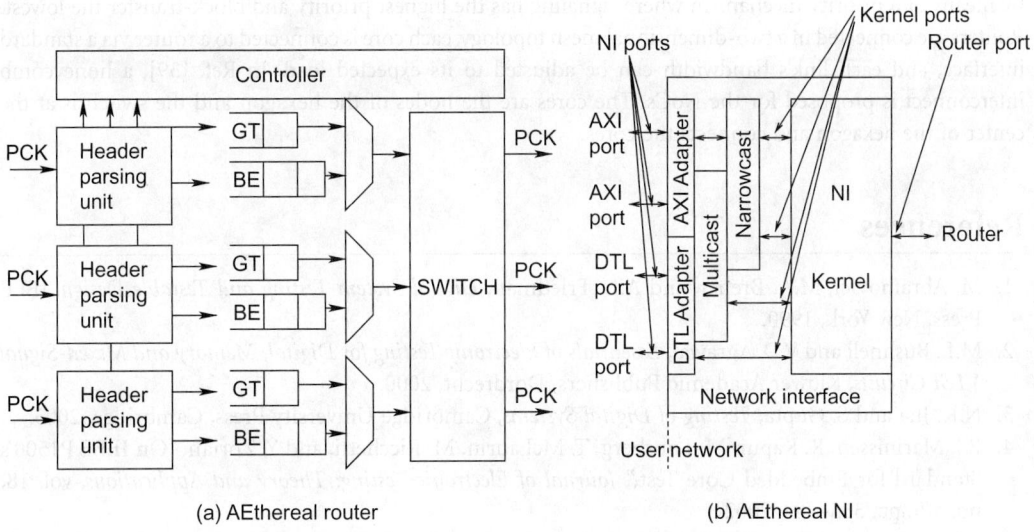

(a) AEthereal router   (b) AEthereal NI

FIGURE 68.7   AEthereal router and network interface module.

has an id field used to specify if the data is BE or GT, a size field which contains the number of data words in the flit and an end of packet flag.

Figure 68.7(a) shows that the building blocks of the AEthereal router are the controller (has separate control units for BE and GT routing), the header parsing units, the queues, and the switch. The header parsing unit receives input flits from the NI attached to the router or from another router in the NoC. It parses the flits, sends them to GT or BE queues, and notifies the controller of the arrivals. The controller is responsible for scheduling the flits. BE flits are scheduled in a round-robin fashion after scheduling any GT flits. The switch receives a signal from the controller and connects the appropriate queue to the appropriate port. For GT data, a circuit-switched path is reserved from the NI module of the source router. In the case of BE data, end-to-end credit-based flow control is implemented to avoid network overflow and congestion. Each outgoing port in the AEthereal architecture has separate queues for the GT and BE data.

The NI consists of the kernel and the shells (also see Figure 68.7[b]). The kernel receives the message from the core, packetizes it, schedules the packets to the routers, implements the end-to-end flow control and performs clock domain conversions between the core and the router. It has one message queue for outgoing messages and another one for incoming messages. The clock domain conversions are carried out at the queues. The routing path is configured at the NI and is added to the packet header. The shells implement narrowcast or multicast connections that involve the active NI port (starts the connection) and one or more (passive) NI ports. In narrowcast connection, the communication involves only one passive port at a time, while in multicast, the request is duplicated and sent to every passive NI port simultaneously. Shells also provide conversions to other protocols.

Another NoC architecture is the SoCBUS [34,35]. The interconnect is a two-dimensional mesh topology of switches. Each switch has five ports, four to connect to adjacent switches and the fifth one to connect to the local core through a wrapper. The wrapper has a functionality similar to the NI of AEthereal. In Ref. [36], a fat-tree interconnect architecture is proposed with the routers as the internal nodes and the cores as the leaves. This topology is cost effective for VLSI realization [37]. In Ref. [38], the intercore communication traffic is classified into four kinds of service: Signaling has the highest priority and is used for control signals and urgent messages. Real-time service is used for delay constrained data and could be achieved by enforcement circuits in the network. Read/Write (RD/WR) service is used for short memory and register data access. Block-transfer service is used for large data bursts. The service levels are implemented by means of a priority mechanism where signaling has the highest priority and block-transfer the lowest. Routers are connected in a two-dimensional mesh topology, each core is connected to a router via a standard interface, and each link's bandwidth can be adjusted to its expected load. In Ref. [39], a honeycomb interconnect is proposed for the NoCs. The cores are the nodes of the hexagon and the switch is at the center of the hexagon and connects the cores.

References

1. M. Abramovici, M.A. Breuer, and A.D. Friedman, *Digital System Testing and Testable Design*, IEEE Press, New York, 1990.
2. M.L. Bushnell and V.D. Agrawal, *Essentials of Electronic Testing for Digital, Memory and Mixed-Signal VLSI Circuits*, Kluwer Academic Publishers, Dordrecht, 2000.
3. N.K. Jha and S. Gupta, *Testing of Digital Systems*, Cambridge University Press, Cambridge, 2003.
4. E.J. Marinissen, R. Kapur, M. Lousberg, T. McLaurin, M. Ricchetti, and Y. Zorian, "On IEEE P1500's Standard for Embedded Core Test," *Journal of Electronic Testing: Theory and Applications*, vol. 18, no. 4/5, pp. 365–383, 2002.
5. E.J. Marinissen, "An Industrial Approach to Core–Based System Chip Testing," *SOC Design Methodologies*, M. Robert, B. Rouzeyre, C. Piguet, M.-L. Flottes (Eds), pp. 389–400, Kluwer Academic Publishers, Dordrecht, 2002.
6. S. Bhatia, T. Gheerwala, and P. Varma, "A Unifying Methodology for Intellectual Property and Custom Logic Testing," *Proceedings of the International Test Conference*, pp. 639–648, 1996.
7. N. Touba and B. Pouya, "Testing Embedded Cores using Partial Isolation Rings," *Proceedings of the IEEE VLSI Test Symposium*, pp. 10–16, 1997.
8. E.J. Marinissen, R. Andersen, G. Bos, H. Digemanse, M. Lousberg, and C. Wouters, "A Structured and Scalable Mechanism for Test Access to Embedded Reusable Cores," *Proceedings of the International Test Conference*, pp. 284–293, 1998.
9. V. Immaneni and S. Raman, "Direct Access Test Scheme–Design of Block and Core Cells for Embedded ASICs," *Proceedings of the International Test Conference*, pp. 448–492, 1990.
10. P. Varma and S. Bhatia, "A Structured Test Reuse Methodology for Core–Based System Chips," *Proceedings of the International Test Conference*, pp. 294–302, 1998.
11. I. Ghosh, S. Dey, and N.K. Jha, "A Fast and Low–Cost Testing Technique for Core–Based System–Chips," *IEEE Transaction on CAD*, vol. 19, no. 8, pp. 863–877, 2000.
12. Y. Zorian, E. Marinissen, and S. Dey, "Testing Embedded Core–Based System Chips," *Proceedings of the International Test Conference*, pp. 130–143, 1998.
13. D. Kagaris, S. Tragoudas, and S. Kuriakose, "InTeRail: A Test Architecture for Core–Based SOCs," *IEEE Transactions on Computers*, vol. 55, no. 2, pp. 137–149, 2006.
14. S. Koranne, "Design of Reconfigurable Access Wrappers for Embedded Core Based SoC Test," *IEEE Transactions on VLSI Systems*, vol. 11, no. 5, pp. 955–960, 2003.
15. V. Iyengar, K. Chakrabarty, and E. Marinissen, "Test Access Mechanism Optimization, Test Scheduling, and Tester Data Volume Reduction for System-on-Chip," *IEEE Transactions on Computer-Aided Design of Integrated Circuits and Systems*, vol. 52, no. 12, pp. 1619–1632, 2003.
16. Y. Huang, W.-T. Cheng, C.-C. Tsai, N. Mukherjee, and S.M. Reddy, "Static Pin Mapping and SOC Test Scheduling for Cores with Multiple Test Sets," *Proceedings of the International Symposium on Quality of Electronic Design*, 2003, pp. 99–104.
17. K. Chakrabarty, "Optimal Test Access Architectures for System-On-Chip," *ACM Transactions of Design Automation of Electronic Systems*, vol. 6, no. 1, pp. 26–49, 2001.
18. C.P. Ravikumar, G. Chandra, and A. Verma, "Simultaneous Module Selection and Scheduling for Power-Constrained Testing of Core-Based Systems," *Proceedings of the International Conference on VLSI Design*, pp. 462–467, 2000.

19. M. Nourani and C. Papachristou, "Parallelism in Structural Fault Testing of Embedded Cores," *Proceedings of the IEEE VLSI Test Symposium*, pp. 15–20, 1997.

20. V. Iyengar, K. Chakrabarty, and E. Marinissen, "Test Wrapper and Test Access Mechanism Co–optimization for System–On–Chip," *Proceedings of the International Test Conference*, vol. 35, pp. 1023–1032, 2001.

21. L. Benini, and G. De Micheli, "Networks on Chips: a New SOC Paradigm," *IEEE Computer*, vol. 35, pp. 70–78, January 2002.

22. C. Zeferino, M. Kreutz, L. Carro, and A. Susin, "A Study on Communication Issues For Systems-on-Chip," *Symposium on Integrated Circuits and Systems Design*, pp. 121–126. IEEE Computer Society, Washington, DC, USA, September 2002.

23. E.J. Marinissen, R. Andersen, G. Bos, H. Digemanse, M. Lousberg, and C. Wouters, "A Structured and Scalable Mechanism for Test Access to Embedded Reusable Cores," *Proceedings of the International Test Conference*, pp. 284–293, 1998.

24. E. Cota, M. Kreutz, C.A. Zeferino, L. Carro, M. Lubaszewski, and A. Susin, "The Impact of NoC Reuse on the Testing on the Testing of core-based Systems," *Proceedings of the 21st IEEE VLSI Test Symposium (VTS'03)*, pp. 128–133, April 2003.

25. E. Cota, L. Carro, F. Wagner, and M. Lubasewski, "Power-aware NoC Reuse on the Testing of Core-based Systems," *Proceedings of the International Test Conference (ITC03)*, vol. 1, pp. 612–621, September 2003.

26. A. Jantsch and H. Tenhunen, *Networks on Chip*, Kluwer Academic Publishers, Dordrecht, 2003, pp. 131–152.

27. M. Nahvi and A. Ivanov, "Indirect Test Architecture for SoC Testing," *Transactions on Computer-Aided Design of Integrated Circuits and Systems*, vol. 23, no. 7, pp. 1128–1142, 2004.

28. B. Vermeulen, J. Dielissen, K. Goossens, and C. Ciordas, "Bringing Communication Networks On Chips: Test and Verification Implications," *IEEE Communications Magazine*, vol. 4, no. 9, pp. 74–81, 2003.

29. W.J. Daly, and B. Towles, "Route Packets, Not Wires: On-Chip Interconnection Networks," *Design Automation Conference*, pp. 684–689, Las Vegas, NV, June 2001.

30. J. Hu and R. Marculescu, "DyAD – Smart Routing for Networks-on-Chip," *Proceedings of the Design Automation Conference (DAC)*, pp. 260–263, 2004.

31. J. Dielissen, A. Radulescu, K. Goossens, and E. Rijpkema, "Concepts and Implementation of the Philips Network-on-Chip," http://www.us.design-reuse.com/articles/article6958.html.

32. E. Rijpkema, K. Goosens, A. Radulescu, J. Dielissen, J. van Meerbergen, P. Wielage, and E. Waterlander, "Trade Offs in the Design of a Router with Both Guaranteed and Best-Effort Services for Networks on Chip," *Proceedings of the Design Automation and Test in Europe (DATE)*, pp. 294–302, 2003.

33. T. Sheldon (Ed.), "Packets and Packet Switched Networks," *Encyclopedia of Networking and Communications*, McGraw Hill, New York, 2001.

34. D. Wiklund and D. Liu, "Design of a System-on-Chip Switched Network and its Design Support," *Proceedings of the International Conference on Communications, Circuits and Systems (ICCCAS)*, vol. 2, pp. 1279–1283, June 2002.

35. D. Wiklund and D. Liu, "Socbus: Switched Network on Chip for Hard Real Time Systems," *Proceedings of the International Parallel and Distributed Processing Symposium (IPDPS)*, April 2003.

36. P. Guerrier and A. Greiner, "A Generic Architecture for On-Chip Packet-Switched Interconnections," *Design Automation and Test in Europe*, pp. 250–256, March 2000.

37. C. Leiserson, "Fat-Trees: Universal Networks for Hardware-Efficient Supercomputing," *IEEE Transactions on Computers*, vol. C-34, no. 10, pp. 892–901, 1985.

38. E. Bolotin, I. Cidon, R. Ginosar, and A. Kolodny, "QNOC: QoS Architecture and Design Process for Network on Chip," *Journal of Systems Architecture*, special issue on Networks on Chip, vol. 50, pp. 105–128, February 2004.

39. A. Hemani, A. Jantsch, S. Kumar, A. Postula, J. Oberg, M. Millberg, and D. Lindqvist, "Network on a Chip: An Architecture for the Billion Transistor Era," *Proceedings of the NorChip 2000*, November 6–7, 2000, Turku, Finland.

40. E. Marinissen, S. Goel, and M. Lousberg, "Wrapper Design for Enbedded Core Test," *Proceedings of the International Test Conference*, pp. 911–920, 2000.

41. A. Radulescu, J. Dielissen, S.G. Pestana, and O.P. Gangwal, E. Rijpkema, P. Wielage, and K. Goossens. "An Efficient On-Chip NI Offering Guaranteed Services, Shared-Memory Abstraction, and Flexible Network Configuration," *IEEE Transactions on Computer-Aided Design of Integrated Circuits and Systems*, vol. 24, no. 1, pp. 4–17, 2005.

69

Automatic Test Pattern Generation

Spyros Tragoudas
Southern Illinois University

CONTENTS

Automatic test pattern generation (ATPG) is the process of obtaining a set of input stimuli, called test patterns that detect possible faulty behavior of a circuit after its fabrication. This may be due to fabrication defects, fabrication errors, or physical failures. Such physical defects in a circuit are detected using fault models that are related to the modeling of the circuit so that we detect the impact of these faults in the operation of the circuit. Only permanent defects that are always present after their occurrence are examined. Sections 69.1 and 69.2 present ATPG techniques for defects in random logic at the gate level or register transfer level. Defects during fabrication can affect the state of logic signals in the circuit independent of its operating frequency. These defects are modeled by *logical faults* and are discussed in Section 69.1. Some manufacturing defects may impact the functionality of the circuit only under high operating frequency and are modeled by *delay faults*. Delay faults are examined in Section 69.2. The challenge in random logic ATPG is due to the difficulty in controlling and observing logic values at the fault site. The ATPG problem is intractable [1]. Section 69.3 considers memory testing and reviews fault models and ATPG methods. In memory testing the logic values are directly controllable and observable at the fault site. The ATPG process is complicated due to the complex fault models used and the very large number of memory cells.

69.1 Logical Faults

Defects in a circuit that are modeled by logical faults include breaks on wires, open transistors, shorted transistors, and technology-specific faults. The most common fault model used in practice is the *stuck-at mode* where lines in a gate- or register-transfer-level description of a circuit are assumed to be set permanently to a "1" or "0" value in the presence of a fault. An additional restriction is that the modeled faults cause only one line in the circuit to have a stuck-at value. The fault model is called the *single stuck-at fault (SSF) model*. Patterns generated under the SSF model have been shown in practice to cover many faults under more complex fault models, such as the multiple-stuck-at fault model where more than one line may be permanently set to value 0 or 1 [2,3].

Given a list of faults, the primary goal of ATPG is to generate a test pattern for each of these faults, and additionally to keep the overall number of test patterns generated as small as possible. The latter is required for reducing the time/cost of applying the test patterns to the circuit. This section presents some algorithms for finding a test pattern for a single fault. Aspects for accelerating the ATPG process for all the faults and reducing the number of generated test patterns are also described. The methods are described assuming the SSF model.

69.1.1 ATPG Algorithms

Let l s-a-v denote the target fault of line l being stuck at value v in a combinational circuit. The fault function of l s-a-v is the function that contains all possible test patterns for this fault. One method for generating a pattern is to form the test function and then identify a cube (product term) in the function. Recent advances in function representations and manipulations have sparked interest in this direction. Compact graph-based data structures, such as reduced order binary decision diagrams (ROBDDs), store functions in a canonical form, in which case a cube can be selected very fast [3]. Alternatively, efficient methods have been proposed for identifying a cube of a function that is expressed as a product of sums [4].

The fault function can be generated in many ways, and it has been also shown how to convert it into a compact product-of-sums form if desirable. One approach is via a virtual circuit construction. First, the faulty circuit is formed, a simplification of the circuit under test (CUT) by fixing value v on line l. Then the CUT and the faulty circuit are merged so that they are driven by a common set of inputs. In addition, the outputs of the two circuits are pair-wise-connected to two-input exclusive-OR (XOR) gates. The fault function is then formed by applying a Boolean OR operation on the functions at the XOR gates.

The process can be accelerated by generating a test pattern without generating the test function since ATPG typically requires that only one cube of the test function be returned. Such ATPG algorithms manipulate the structure of the CUT and may revisit any circuit line several times in an attempt to generate a pattern such that (1) the pattern brings l to have a value v (*fault activation*) and (2) the same pattern carries over the effect of the fault to a primary output (*fault propagation*). A path from line l to a primary output along each line of which the effect of the fault is carried over is called a sensitized path.

The case of a line l having a value of 1 in the correct circuit and a value of 0 in the circuit under the fault l s-a-v is denoted by the symbol D and, similarly, the opposite case is denoted by D′. Given the symbols D and D′, the basic Boolean operations AND, OR, and NOT can be extended in a straightforward manner. For example, AND(1, D) = D, AND(l, D) = D, AND(O, D) = 0, AND(O, D) = 0, AND(x, D) = x, AND(x, D) = x (where x denotes the don't-care case), etc.

A basic TPG algorithm for combinational circuits is the D-algorithm [5]. This algorithm works as follows: all values are initially assigned a value of x, except line l which is assigned a value of D if the fault is l s-a-0, and a value of D′ if the fault is l s-a-1. Let G be the gate whose output line is l. The algorithm first selects an assignment for the inputs of G out of all possible assignments that produce the appropriate D-value (i.e., a D or D′) at the output of G. This step is known as *fault activation*. All possible assignments are fixed for each gate type and are referred to as the *primitive d-cubes for the fault* (*pdcfs*) of the gate. For example, the *pdcfs* of a two-input AND gate are 0xD′, x0D′, and 11D, and the *pdcfs of* a two-input OR gate are lxD, x1D, and 00D′ (using the notation abc for a gate with input values a and b and output value c).

Subsequently, the algorithm repeatedly selects a gate from the set of gates whose output is currently x but has at least one input with a D-value. This set of gates is known as the D-frontier. Then it selects an assignment for the inputs of that gate out of all possible assignments that set the output to a D-value. All possible assignments are fixed for each gate type and are referred to as the *propagation d-'cubes (pdcs)* of the gate. For example, the *pdcs* of a two-input AND gate are 1DD, D1D, 1D′D′, D′1D′, DDD, and D′D′D′. By repeated application of this step, a D-value is eventually propagated to a primary output. This step is known as *fault propagation*.

In a third step, the algorithm finds an assignment of values for the primary inputs that establishes the candidate values required in steps (1) and (2). This step is known as *line justification*. For each value that is not currently accounted for, the line justification process tries to establish ("justify") the value by (a) assigning binary values (and no D-values) on the inputs of the corresponding gate, working its way back to the primary inputs (this process is referred to as *backtracing*; and (b) determining all values that are imposed by all candidate assignments thus far (*implication*) and checking for any inconsistencies (*consistency check*).

If an inconsistency is found during the line justification step, then the computation is restored to its state at the last decision point. This process is known as *backtracking*. A decision point can be (a) the decision in step (1) of which *pdcf* to select; (b) the decisions in step (2) of which gate to select from the D-frontier and which *pdc* to select for that gate; or (c) the decision in step (3) of which binary combination to select for each value that has to be justified.

If line justification is eventually successful after zero or more backtrackings, then the existing values on the primary inputs (some of which may well be *x*) constitute a test pattern for the fault. Otherwise, no pattern can be found to test the given fault and that fault is thus shown to be redundant.

The order of the fault propagation and line justification steps may be interchanged, or even the two steps may be interspersed, in an attempt to reduce the running time, but the discovery or not of a pattern is not affected by such changes.

Consider the circuit in Figure 69.1 and G s-a-1. To establish G = D, the *pdcf* CD = 00 is chosen and the D-frontier becomes {J} (gates are named by their output line). Then, gate J is considered and the *pdc* setting I = 1 is selected with result J = D and new D-frontier {M, N}. Assume gate M is selected. Then, the *pdc* setting H = 0 is selected with result M = D'. The justification of current values H = 0 and I = 1 results in conflict, so the algorithm backtracks and tries the next *pdc* for gate M which sets H = D. This cannot be justified, the algorithm backtracks once more and selects gate N from the D-frontier. The assignment E = 1 is made, which results in N = D. The values E = 1 and I = 1 can be justified without conflict. The algorithm terminates successfully and generates test pattern ABCDE = 11001.

Sometimes the algorithm must sensitize two paths simultaneously from the fault site to a PO to detect the fault. This is referred to as *multipath sensitization*. Consider the circuit in Figure 69.2 and the fault B s-a-l. The assignment B = 0 is made and the D-frontier becomes {F, G}. Assume that gate F is selected. The *pdc* A = 1 and G = 0 are selected to propagate fault D to line H. This results in conflict, as B (and E) are required to be 0. The algorithm backtracks and tries the next available *pdc* of gate H which sets G = D. This value is justified by setting C = 1. The resulting test pattern is ABC = 101.

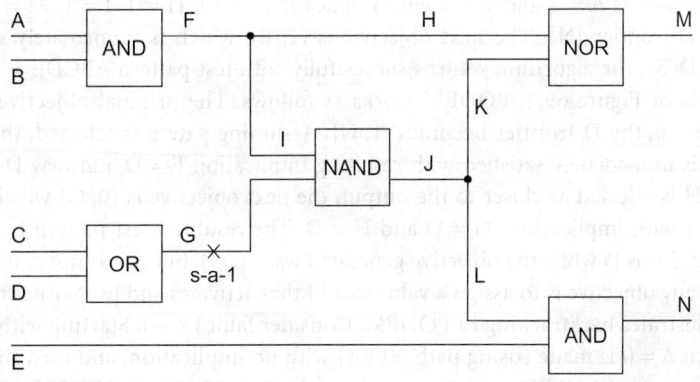

FIGURE 69.1 Illustration of the ATPG algorithm.

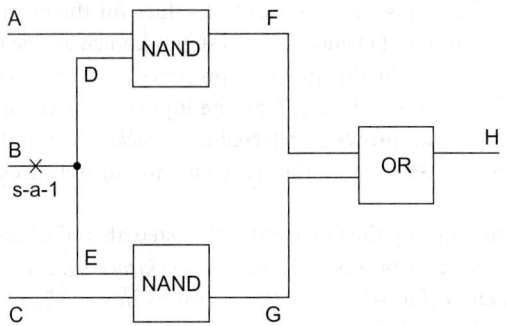

FIGURE 69.2 Illustration of the multipath sensitization.

Another basic TPG algorithm is called PODEM, and also uses the five-valued logic (0, 1, *x*, D, D̄) [6]. In PODEM, all lines are initialized to value *x* except line *l*, which is assigned a value of D if the fault is *l* s-a-O, and a value of D′ if the fault is *l* s-a-l. The algorithm at each step tries to satisfy an *objective* (*v*, *l*), defined as a desired value *v* at a line *l* by making assignments only to primary inputs (PIs), one PI at a time. The mapping of an objective to a single PI value is done heuristically. The initial objective is (*v*′, *l*). All implications of the current pattern of values assigned to PIs are computed, and the algorithm terminates when the effect of the fault is propagated to a primary output (PO). If a conflict occurs and the fault cannot be activated or cannot be propagated to a PO, then the algorithm backtracks to the previous decision point, which is the last assignment to a PI. If no conflict occurs but the fault has not been activated or not been propagated to a PO because the currently implied values on the lines involved are *x*, then the algorithm continues with the same objective (*v*, *l*) if the fault is still not activated, or with an objective (*c*′, *k*) if the fault has been activated but not propagated, where *k* is an input line of a gate from the D-frontier that has currently assigned a value of *x* on it, and *c* the controlling value of that gate.

The determination of which single PI to be selected and which value to assign to it given an objective (*v*, *l*) is done heuristically. A simple heuristic is to select a path from line *l* to a PI, and assign to that PI the value *v* (*v*′) if the total number of inverting gates (such as NOT, NAND, and NOR) along that path is even (odd). Concerning the selection of a gate from the D-frontier, a simple heuristic is to select the gate that is closest to a PO. As an example of the application of PODEM, consider the circuit of Figure 69.1 and the fault G s-a-l. The initial objective is (0, G). The chosen PI assignment is C = 1 for which there are no implications. The objective remains the same, the PI assignment becomes D = 0, and this implies G = D. The D-frontier becomes {J}, and the next objective is (1, I). This eventually results in PI assignments A = 1 and B = 1 with implications F = l, H = 1, I = 1, M = 0, J = D, K = D, L = D, and new D-frontier {N}. The next objective is (1, E), which is immediately satisfied and has implication N = D. So, the algorithm returns successfully with test pattern ABCDE = 11001.

In the example of Figure 69.2, PODEM works as follows. The original objective is (0, B). With PI assignment B = 0, the D-frontier becomes {F, G}. Assuming gate F is selected, the next objective is (1, A), which is immediately satisfied with resulting implication F = D and new D-frontier {G, H}. Given that gate H is selected as closer to the output, the next objective is (0, G) which leads to the PI assignment C = 1 with implications G = D and H = D. The resulting test pattern is ABC = 101. The implied value for G was D while the objective generated was (1, G), but this is not considered a conflict since the goal of any objective is to assign a value to a PI that activates and propagates the fault to a PO.

Figure 69.3 illustrates backtracking in PODEM. Consider fault J s-a-l. Starting with objective (0, J), the PI assignment A = 0 is made (using path HFEA) with no implication, and then the PI assignment B = 0 is made (using path HFEB) with implications E = 0, F = 0, G = 0, H = 0, I = 1, J = 1. The last assignment results to a conflict, and the algorithm backtracks trying PI assignment A = 1. The implications of this assignment are E = 1, F = 1, G = 1. The fault at J is still not activated and objective

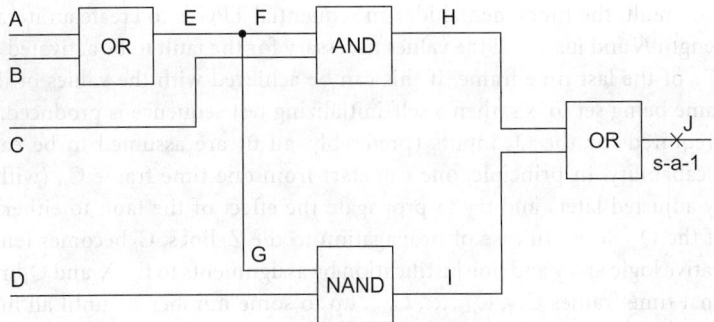

FIGURE 69.3 Illustration of backtracking in PODEM.

(1, B) is generated (using path HFEB), which is satisfied immediately but has no new implications. Then the objective (0, C) is generated (using path HCJ), which is satisfied immediately and has implication H = 0. Finally, the objective (1, D) is generated (using path ID), which is satisfied immediately and has implications I = 0 and J = 0. Since the fault is now activated and (trivially) propagated, the algorithm terminates successfully with test pattern ABCD = 1101.

Several extensions to PODEM have been proposed, such as working with more than one objective each time and stopping backtracking before reaching PIs. The FAN algorithm maintains a list of multiple objectives and stops backtracking at headlines rather than just PI lines [7]. A *headline* is a line that is driven by a subcircuit containing no line that is reachable from some fan-out stem, and, therefore, can be justified at the end with no conflicts. The fault in Figure 69.3 is activated when H and I are 0. The objectives (H, 0) and (I, 0) are now both taken into consideration. To achieve objective (H, 0), the assignment E = 0 can be selected because line E is a headline. But to achieve objective (I, 0), the assignment E = 1 is required. PODEM assigns C = 0 on PI C for objective (0, H), and then selects the assignment E = 1 on headline E and D = 1 on PI D for objective (0, I), which results in success. The justification of results to either ABCD = 1x00 or ABCD = x100.

There are many ATPG algorithms based on various strategies (see, e.g., Ref. [5] for more information). It is noted that in the worst case, all ATPG algorithms may take exponential time. In fact, the test pattern generation problem has been shown to be NP-complete.

So far it was assumed that the CUT is combinational. Generating a pattern to detect a fault in sequential circuits is much more difficult than for combinational circuits. Testing of synchronous ASICs is typically reduced to combinational testing using the previously described methods, and the combinational core is tested using *full-scan design* [3], a *Design for Testability (DFT)* technique that was described in the previous chapter. If full scan is not available, a sequence of patterns is generally required for each fault. ATPG techniques for combinational circuits can still be applied to sequential circuits by considering the iterative logic array model of the sequential circuits. This model applies to both synchronous and asynchronous sequential circuits, although it is more complex for the latter.

Given a current state vector Q and a current input vector X, the function of a sequential circuit is specified as a mapping from (X, Q) to (Q$^+$, Z), where Q$^+$ is the next state vector and Z the resulting output. In the iterative logic array representation, the sequential circuit is modeled as a series of combinational circuits C_0, C_1, ..., C_N, where N is the length of the current input pattern sequence applied to the sequential circuit [2,3]. Each circuit C_i, referred to as a *time frame*, is an identical copy of the sequential circuit but with all feedback removed, and has inputs X_i and Q_i, and outputs Q$^+$ and Z_i. Inputs X_i are driven by the ith pattern applied to the sequential circuit and inputs Q_i are driven by the outputs Q_{i-1}^+, of the previous time frame for $i > 0$, with Q_0 being set to the original initial state of the sequential circuit. All outputs Z_i are ignored except for the outputs Z_N of the last time frame, which constitute the output of the sequential circuit resulting from the specific input sequence and initial state.

Given a stuck-at fault, the fundamental idea in sequential TPG is to create an iterative logic array of appropriate length N and justify all the values necessary for the fault to be activated and propagated to the outputs Z_N of the last time frame. If this can be achieved with the values of the Q_0 inputs of the first time frame being set to 'x's, then a self-initializing test sequence is produced. Otherwise, the specific values required for the Q_0 inputs (preferably, all 0) are assumed to be easily established through a reset capability. In principle, one can start from one time frame C_i, (with the index i to be appropriately adjusted later) and try to propagate the effect of the fault to either some of the Z_i lines or some of the Q_i^+ lines. In case of propagation to the Z_i lines, C_i becomes tentatively the last frame in the iterative logic array and line justification by assignments to the X_i and Q_i lines is repeatedly done in additional time frames $C_{i-1}, C_{i-2}, \ldots, C_{i-m}$, up to some number m, until all lines are justified with either Q_{i-m}, being set to all 'x's or to an initial state that can be reset. In case of propagation to the Q_i lines, additional time frames $C_{i+1}, C_{i+2}, \ldots, C_{i+k}$, for some number k, are considered until the effect of the fault, is propagated to the Z_k lines. Notice that because each time frame contains the same fault, the propagation can be done from any of the $C_{i-1}, C_{i-2}, \ldots, C_{i-m}$ time frames to the Z_k lines. Then, line justification is again attempted as above. In case of conflict during the justification process, backtracking is attempted to the last decision point, and this backtracking can reach as far as the C_{i-m} frame.

To reduce the storage required for the computation status as well as the time requirements of this process, algorithms that consider only backward justification and no forward fault propagation have been proposed. For example, the *extended backtrace* (EBT) algorithm selects a path from the fault site to a primary output which may involve several time frames and then tries to justify all values for the sensitization of this path (along with the requirements for the initial state) by working with time frames $C_{i-1}, C_{i-2}, \ldots, C_{i-m}$ [8]. It has also been shown by Muth [9] that four additional logic values are required besides the five values 0, 1, D, D', and x, especially where state initialization may be affected by the fault. A differing signal may be 0 at the fault-free circuit and x in the faulty circuit. In the nine-valued algebra this is denoted by a new value $0/x$ whereas the five-valued algebra used in combinational ATPG cannot denote it as a differing signal and simply denotes it by x. Similarly, $1/x$, $x/1$ and $x/0$ are defined. Using this nine-valued algebra test patterns may be detected for faults that cannot be detected using the five-valued algebra.

A fault for which no test pattern exists is referred to as *redundant* if the circuit is combinational and *untestable* if the circuit is sequential. If a sequential circuit is tested using full scan, then the set of redundant faults is only a subset of the untestable faults in the CUT.

69.1.2 A Complete Test Set

Any one of the presented algorithms can be used repeatedly to generate a set of patterns for all the faults. This section focuses on methods to accelerate the process and result to a smaller set of test patterns. It describes three commonly used techniques: *fault collapsing, elimination of randomly testable faults,* and *fault simulation.* More sophisticated methods to reduce the total number of required test patterns (*test set compaction*) can be found in the literature but such methods do not necessarily accelerate the ATPG process and are typically applied after a set of test patterns that detect all the faults has been generated. The first two techniques, fault collapsing and elimination of randomly testable faults, are typically applied prior to generating any test pattern. The latter is applied whenever a test pattern is generated. The objective in all three described techniques is to quickly reduce the number of faults on which the algorithms of the previous subsection apply. It is assumed that the CUT is combinational.

For a circuit with n lines in total, there are $2n$ possible stuck-at faults to consider. Fault collapsing reduces this initial number by taking advantage of equivalence and dominance relations among faults. Two faults are said to be *functionally equivalent* if all patterns that detect the one detect also the other. Given a set of functionally equivalent faults, only one fault from that set has to be considered for test generation. A fault j is said to *dominate* a fault h if all patterns that detect h detect also j, and there is at least one pattern that detects j but not h. Then only h needs to be considered for test generation. It can be shown that the fault s-a-(c XOR *i*) on the output of a gate is functionally equivalent with the fault s-a-c on any of the gate inputs and that the fault s-a-(c XOR *i*) on the output of a gate dominates the fault *s-a-c* on any of the gate inputs, where c is the controlling value of the gate and *i* is 1 (0) if the gate is inverting (noninverting). As an example, using these relations on the circuit of Figure 69.1, we obtain that (F-s-0, A-s-0, B-s-0), (G-s-1, C-s-1, D-s-1), (J-s-1, G-s-0, I-s-0), (M-s-0, H-s-1, K-s-1), (N-s-0, E-s-0, 1-s-0) are functionally equivalent sets of faults, and that F-s-1 dominates A-s-l and B-s-1, G-s-0 dominates C-s-0 and D-s-0, J-s-0 dominates G-s-l and I-s-1, M-s-1 dominates H-s-0 and K-s-0, and N-s-1 dominates E-s-1 and L-s-1. Given these relations, only the set of faults {A-s-1, B-s-1, C-s-0, D-s-0, G-s-1, I-s-1 H-s-0, K-s-0, E-s-1, 1-s-1, F-s-0, M-s-0, N-s-0} need be considered and the number of target stuck-at faults is reduced from 28 to 13. In general, more faults can be eliminated if fault equivalence and fault dominance are examined using the test function for the stuck-at faults. However, such an approach is time consuming and may be useful only for test set compaction.

A very simple way of eliminating faults from a target fault list is to generate test patterns at random and verify, by *fault simulation*, which target faults (if any) each generated pattern detects. Techniques for fault simulation are reviewed later. First we focus on the random pattern generation process. Such patterns are generated by a *pseudorandom* method, that is, an algorithmic method whose behavior under specific statistical criteria seems close to random. Eliminating all faults by pseudorandom test pattern generation generally requires a very large number of patterns but the ATPG is accelerated because fault simulation process is very fast compared to ATPG for one SSF. Under the assumption of uniform input distribution and independent test pattern generation, the smallest number of patterns to detect with probability P, a fault whose detection probability is d is $N = \ln(P)/\ln(1 - d)$. In general, faults with small detection probability are referred to as *randomly untestable* or *hard-to-detect faults*, whereas faults with high detection probability are referred to as *randomly testable* or *easy-to-detect* faults. For example, in a circuit consisting of a single *k-input* AND gate with output line *I*, the fault *I* s-a-0 is a hard-to-detect fault as only one out of 2^k patterns can detect it, whereas the fault *I* s-a-1 is an easy-to-detect fault as 2^{k-1} out of 2^k patterns can detect it. In practice, an acceptable number of pseudorandom test patterns are generated and simulated to drop many easy-to-detect faults from the target fault list, with all remaining faults given over to a *deterministic* (as opposed to pseudorandom) TPG tool [2,3].

The remaining of the section reviews algorithms for fault simulation which is applied whenever a test pattern has been generated. The goal is to determine quickly all the faults that the test pattern detects and remove them for the list of not-yet-targeted faults. The fault simulation process is composed of one or more topological circuit traversals, and requires polynomial time in contrast to the ATPG process which requires (in the worst-case) exponential time per fault.

The simplest form of simulation is called *single-fault propagation*. After a test pattern is simulated, the stuck-at faults are inserted one after the other. The values of every faulty circuitry are compared with the error-free values. A faulty value needs to be propagated from the line where the fault occurs. The propagation process continues line-by-line, in a topological search manner, until there is no faulty value that differs from the respective good one. If the latter condition is not satisfied, the fault is detected.

In an alternative approach, called *parallel-fault propagation*, the goal is to simulate *n* test patterns in parallel using *n*-bit memory. Gates are evaluated using Boolean instructions operating on *n*-bit operands. The problem with this type of simulation is that events may occur only in a subset of the *n* patterns while at a gate. If on average a fraction *B* of gates have events on their inputs in one test pattern, the parallel simulator will simulate $1/B$ more gates than an event-driven simulator. Since *n* patterns are simulated in parallel, the approach is more efficient when *n* is not less than $1/B$, and the speed-up is *nB*. Single and parallel fault propagation can also be combined efficiently.

In deductive fault simulation only one topological circuit traversal is required. The idea is to maintain for each line *l* a list L_l of SSFs that have been propagated by the test patterns up to that line. While examining a gate *g* with inputs *a*, *b* and output *c*, list Lc is constructed by manipulating the contents of L_a and L_b. Assume that g is an AND gate and that the test pattern brings values $a = 0$, $b = 1$. L_c contains c-s-1 since $c = 0$. Additional faults in L_c include faults in L_a because b has a noncontrolling value. However, only those faults in L_a that are not part of L_b can propagate to line c; faults in L_a that also propagate to line b essentially change its value to controlling and cannot propagate. Such logical arguments can be used to generate the list L_c at the output of gate g, under any assignment of binary values to its inputs. This simulation method is faster than the single fault simulation method when the contents of the fault lists are small on many lines in the CUT.

An approach to further accelerate the fault simulation process is to report many but not all the SSFs that the test pattern detects. The most popular representative of *approximate fault simulation* is the *critical path tracing approach*. For every test pattern, the approach first simulates the fault-free circuit and then determines the detected faults by determining which lines have *critical values*. Such lines are called *critical*. A line has critical value 0 (1) in pattern t if and only if test pattern t detects the fault stuck-at 0 (1) at the line. Therefore, finding the critical lines in pattern t amounts to finding the stuck-at faults that are detected by t. Critical lines are found by backtracking from the primary outputs. Such a backtracking process determines paths of critical lines that are called *critical paths*. The process of generating critical paths uses the concept of *sensitive inputs* of a gate with two or more inputs (for a test pattern t). This is determined as follows: If only input l has the controlling value of a gate, then it is sensitive. On the other hand, if all the inputs of a gate have noncontrolling value, then they are all sensitive. There is no other condition for labeling some input line of a gate as sensitive. Thus, the sensitive inputs of a gate can be identified during the fault-free simulation of the circuit.

The operation of the critical path tracing algorithm is based on the observation that when a gate output is critical, then all its sensitive inputs are critical. On fan-out free circuits, critical path tracing is a simple traversal that applies recursively to the above observation. The situation is more complicated when there exist re-convergent fan-outs. This is illustrated in Figure 69.4. In Figure 69.4(a), starting from g, we determine critical lines g, e, b, and $c1$ as critical, in that order. To determine whether c is critical, we need additional analysis. The effects of the fault stuck-at 0 on line c propagate on re-convergent

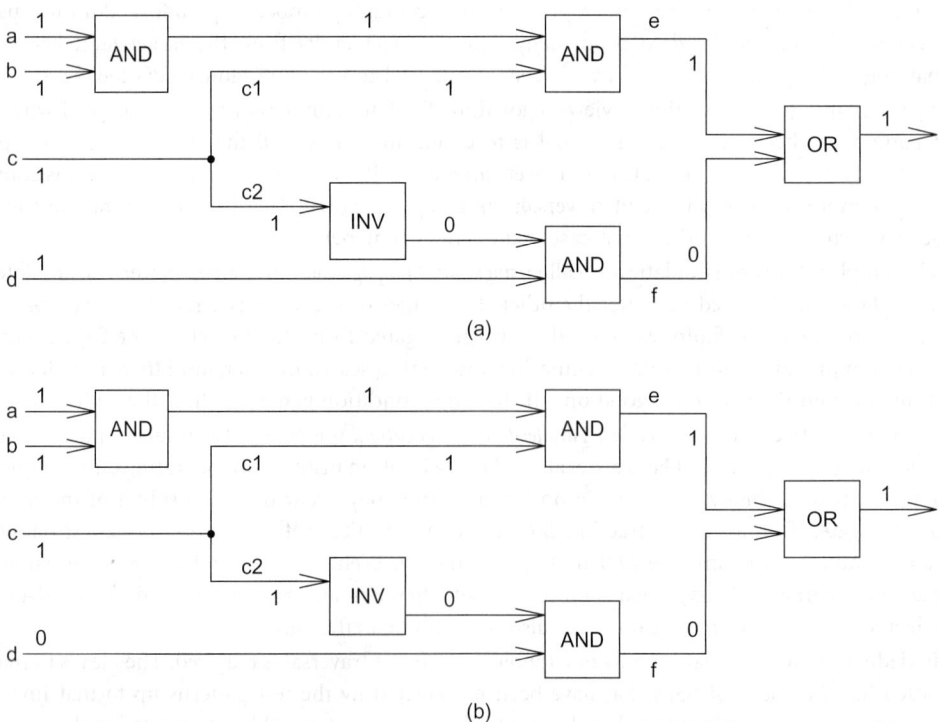

(a)

(b)

FIGURE 69.4 Critical path tracing.

paths with different parities which cancel each other when they re-converge at gate *g*. This is called *self-masking*. Self-masking does not occur in Figure 69.4(b) because the fault propagation from c2 does not reach the re-convergent point. In Figure 69.4(b), c is critical.

Therefore, the problem is to determine whether self-masking occurs or not at the stem of the circuit. Let 0 (1) be the value of a stem *l* under test pattern *t*. A solution is to explicitly simulate the SSF c-s-1 (c-s-0), and if *t* detects this fault, then *l* is marked as critical. Instead, the approach uses bottlenecks in the propagation of faults that are called *capture lines*. Let *a* be a line sensitized to SSF fault *f* with a pattern *t*. If every path sensitized to *f* either goes through *a* or does not reach any other line with a greater topological level, then *a* is a capture line of *f* under pattern *t*. Such a line is common to all paths on which the effects of *f* can propagate to the primary output under pattern *t*. The capture lines of a fault form a transitive chain. Therefore, a test *t* detects fault *f* if and only if all the capture lines of *f* under test pattern *t* are critical. Thus, to determine whether a stem is critical, the algorithm does not propagate the effects of the fault step up to the primary output; it only propagates the fault effects up to the capture line that is closest to the stem.

69.2 Delay Faults

Defects may cause the propagation time along paths in the CUT to exceed the dock period. Such types of faults are called delay faults. Several fault models have been proposed to detect delay faults. Independent of the model used, a test for a fault requires that transition must be propagated from the fault site to an observable point. Therefore more than one test pattern must be applied to detect a delay fault. A pair of patterns is required if the CUT is combinational. The pattern pair must also generate a transition in at least one input of the CUT.

In sequential logic, an enhanced scan design is required so that any pair of test patterns can be applied to the CUT. This DFT approach is discussed in the previous chapter. If traditional scan is used, then the ATPG process must be modified so that the second pattern in the pair is obtained from the first one. Two methods have been proposed, the *broad-side delay test* and the *scan-shift delay test* [3].

In the broad-side test, the flip-flop portion of the second pattern must be obtained when applying the first pattern. The first pattern is scanned-in and then the application of the primary input bits produces the flip-flop portion of the second pattern. Then an application of the clock in the normal mode and the application of the primary input portion of the second pattern brings a transition generated at some inputs of the embedded combinational core. The CUT is kept in normal mode, and the outputs are observed or scanned-out after one rated-clock period.

In scan-shift test, the second pattern is generated by scanning the first pattern by one bit. The scan-in of the first pattern is followed by a slow-clock cycle while the CUT remains in scan mode. When the second pattern is applied the mode changes from scan to normal, and the outputs are latched with one application of a rated clock. They are directly observed or scanned-out as in the previous approach.

The first pattern can be applied with a slow clock so that the logic values on all lines settle to their final value before the second pattern is applied. This method, also called slow-fast clock application requires that the Automatic Test Equipment (ATE) has two test clocks. Alternatively, both patterns can be applied with the rated-clock. This is called at-speed testing. Although at-speed tests require less expensive ATE, they do not cover as many faults as the variable clock-tests.

If the sequential logic does not have full-scan support, then more than two patterns must be applied. All of them can be applied at the rated speed (at-speed nonscan sequential test) or some of them can be applied at a slow clock and the other at the rated-speed (variable clock nonscan sequential test).

The remaining of this section assumes that the CUT is combinational or that enhanced scan is provided. In addition, it is assumed that the first pattern is applied using a slow clock and the second using a rated clock.

A suitable fault model to detect gross delay defects is the *transition fault (TF) model*. There exist two possible faults per line, a slow-rising and a slow-falling transition. Thus, the total number of TF faults is twice as many as the number of lines in the CUT. This number is the same as the number of SSFs. A test for a single stuck-at 0 on that line can be used effectively when generating a pair of test patterns to detect a slow-rising TF. This test sets the line to 1 in the fault-free circuit, and propagates an error to an output or, equivalently, ensures that any change at the fault site is observable to an output. This test will be used as the second test pattern in the pair. The first pattern must set the line to 0. Similarly, one can derive a pair of test patterns for detecting a slow-falling TF so that the first pattern sets the line to 0 and second pattern is a test for a single stuck-at 1 on that line.

Similar to the ATPG process, fault grading techniques for SSFs can be modified to apply for TFs. The fundamental assumption for TF tests is that hazards do not interfere with the observation of the propagation of the delayed transition from the fault site to the output. The delay defect must also be large enough so that it is detected even if it propagates along short paths. This fault model is not appropriate for detecting distributed delay defects, where many small delay defects in several gates and interconnects may cause failures along a long enough path. For this reason, a more refined fault model has been proposed for delay fault testing. In this model, faults are associated with paths instead of lines, and for this reason is it called the *path delay fault (PDF)* model.

The following assume a fully scanned sequential circuit with enhanced scan where path delays are examined in the embedded combinational logic. A PDF is a path where a rising or a falling transition propagates along every line on it. In practice, only long enough paths, referred to as *critical PDFs*, are examined. However, in many circuits the majority of PDFs are long enough to be categorized as critical, and for simplicity the following assume that all the PDFs are critical. Therefore, for every physical path in the circuit, there exist two PDFs. The first PDF is associated with a rising transition on the first line on the path. The second PDF is associated with a falling transition on the first line on the path.

The fundamental difficulty of the PDF model is that the number of faults may be, and often is, exponential to the size of the CUT. For this reason both fault coverage and ATPG must be done implicitly, i.e., in a fault nonenumerative manner. In ATPG the goal is to generate a set of pairs of test patterns so that the *number* of tested PDFs is high. Note that the PDF model assumes that all patterns that test a PDF bring the same propagation delay along the PDF. In fault coverage, the goal is to determine *how many* PDFs are tested (simulated) by a given set of pairs of patterns. An ATPG or a fault coverage algorithm cannot determine the number the tested PDFs unless the tested PDFs are kept implicitly using appropriate data structures. In this case, it can be determined in polynomial time whether a particular PDF is tested or not. Unfortunately, implicit (nonenumerative) fault coverage (and therefore ATPG) is not easy. The implicit PDF coverage problem has been shown to be intractable.

There exist different ways under which a PDF can be *sensitized*, i.e., the transition propagate throughout the path, by a pair of test patterns. The most appropriate PDF sensitization condition is the *robust* condition. Robust tests guarantee detection of the PDF independent of any delays in the rest of the circuit. This is the most desirable way of testing a PDF. However, a robust test must satisfy strict constraints *for each gate on the PDF* to test it robustly. The input line of a gate in the PDF is referred to as its *on-path input*, and the remaining inputs as its *off-path inputs*. If the on-path input is a transition to a noncontrolling value (referred to as −ncv), then the off-inputs can be either stable at the noncontrolling value (referred to as sncv) or may have a transition to the noncontrolling value (−ncv). If the on-path input is a transition to the controlling value (referred to as −cv), then all off-inputs must be stable to sncv.

A PDF fault is tested nonrobustly if there is a test pattern that detects it if it is the only PDF in the CUT. This is a more relaxed fault sensitization condition because the only constraint is that the off-inputs have a final ncv value, i.e., can be either −ncv or −ncv. A nonrobust test for a PDF may not guarantee that the transition propagates along the path. Assume that the on-input transition at a gate is −cv and the off-input is −ncv. The on-path transition actually propagates as a static hazard through the gate only if it arrives at the gate long enough after the off-input transition, for the width of the hazard to be detectable. Thus, a path must be tested nonrobustly only if there is no robust test for it. However, in some cases robust tests may exist to guarantee that no off-path inputs have delayed transitions. A nonrobust test for the PDF that is applied together with the latter tests is called *validatable nonrobust* because it sensitizes the PDF since the fault model assumes that all pattern for the off-input PDFs bring the same propagation delay [3].

Functionally sensitizable PDFs can be tested in the presence of multiple path delays. Such tests (referred to as *functional*) detect PDFs that cannot be tested by the previous methods. They detect a set of faults that is a superset of those detected by nonrobust test vectors. In particular, this sensitization condition allows for one or more gates to have a –cv transition on the on-input and the off-inputs also to have –ncv transitions. The PDF is actually sensitized by the test only when all such off-input transitions are also delayed. It has been shown that FT faults have better probability to be detected when the maximum off-input slack (or, simply, slack) is a small integer. (The slack of an off-input is defined as the difference between the stable time of the on-input signal and the stable time of the off-input signal.) PDFs that cannot be detected by any of the above conditions are referred to as *functionally unsensitizable* [10].

Other classifications of PDFs have been recently proposed in the literature, but they are not presented here [11,12]. Systematic PDF classification is very important when considering test pattern generation. For example, nonrobust or functional test pattern generation should be considered only when the PDF is not robustly testable. It is important to observe that ATPG does not consider actual delays on the gates under any of the existing classifications. The PDF model (as well as the TF model) generate test patterns independent of actual gate delay (or interconnect delay) values. This is appropriate in the presence of delay defects and the monotone speed-up effect [13], but also because in deep submicron it is difficult to estimate gate and interconnect delays accurately. However, ATPG tools can be enhanced so that nonrobust and functional tests are generated by considering a range on the gate and interconnect delay values. Such enhancements are not considered here.

69.2.1 ATPG for Path Delay Faults

The conventional approach for generating a pair of test patterns for a PDF P is a modification of an SSF ATPG tool. Initially, transitions are assigned on the lines of path P. This is called the *path sensitization phase*. Then, a modified ATPG for stuck-at fault is executed to ensure that constraints on the on-path inputs as well as on the off-path inputs (according to the followed PDF sensitization condition) are justified. This is the *line justification phase*. It should be noted that the sncv and scv constraints on some lines are also handled recursively. For example, for the output of a gate to have a stable *noncontrolled* value all inputs must have an sncv, and to have a stable *controlled* value it suffices that one of its inputs has an scv.

The problem with this conventional approach is that ATPG will be executed as many times as the number of PDFs, which is an exponential quantity to the size of the circuit. Any practical ATPG tool must be able to generate a polynomial number of pairs of test patterns. Thus, ATPG process must be modified as follows: For each pair of patterns to be generated, the path sensitization phase must be able to sensitize multiple paths. Then the line justification phase must be able to justify the assigned line constraints so that many PDFs remain sensitized.

One way to construct a *non-PDF enumerative* (referred to as nonenumerative) ATPG tool is to generate a pair of patterns that sensitizes and justifies the transitions on *all* the lines of a subcircuit that contains many paths. A necessary condition for the paths of a subcircuit to be simultaneously sensitized is to be *structurally compatible* with respect to the parity (on the number of inverters) between any two re-convergent nodes in the subcircuit. This concept is illustrated in Figure 69.5.

Consider the circuit on the top portion of Figure 69.5. The subgraph induced by the thick edges consists of two structurally compatible paths. These two paths share two OR gates. The two subpaths that share the same OR gate endpoints have even parity. Any graph that constraints structurally compatible graphs is called a *structurally compatible* (SC) graph. The methods in Refs. [14,15] consider a special case of SC graphs with a single primary input and a single primary output. Such an SG graph is called a *primary compatible* (PC) graph.

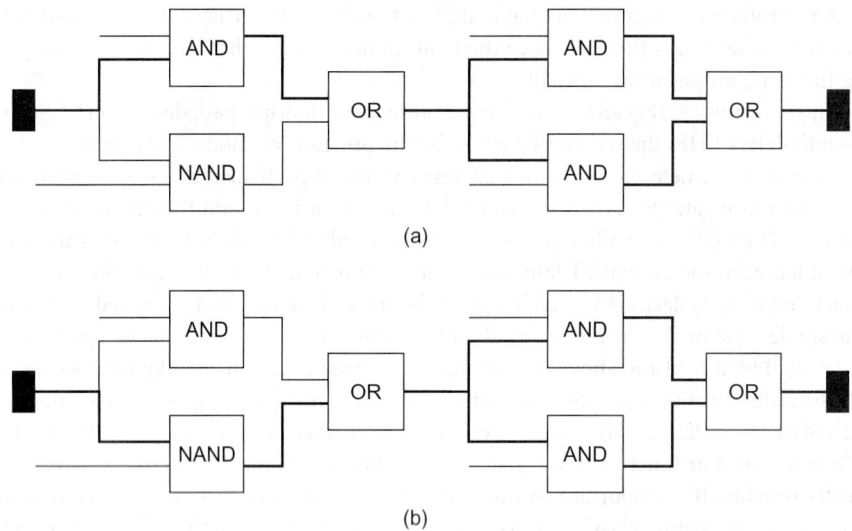

FIGURE 69.5 Test generation for more than one PDF. (a) A primary compatible graph. (b) A sibling primary compatible graph.

For the same pair of primary input and output nodes in the circuit, there may be many different PC graphs, which are called sibling PC graphs. Sibling PC graphs contain mutually incompatible paths. The subgraph induced by the thick edges on the bottom portion of Figure 69.5 shows a PC graph that is sibling to the one on the top portion. This graph also contains two paths (the ones induced by the thick edges).

The ATPG tool in Ref. [15] generates large sibling PC graphs for every pair of primary input and output nodes in the circuit. Robust PDF sensitization is considered. The size of each returned PC graph is measured in terms of the number of structurally compatible paths that satisfy the requirements for robust propagation described earlier. Experimentation in Ref. [15] shows that the line justification phase satisfies the constraints along paths in a manner proportional to the size of the graph returned by the path sensitization phase.

Given a pair of primary input and primary output nodes, a large sibling PC graphs as follows. Initially, a small number of lines in the circuit are removed so that the subcircuit between the selected primary inputs and outputs is a series–parallel graph. A polynomial time algorithm is applied on the series–parallel graph which finds the maximum number of structurally compatible paths that satisfy the conditions for robust propagation. An intermediate tree structure is maintained, which helps extract many such large sibling PC graphs for the same pair of primary input and output nodes. Finally, many previously deleted edges are inserted so that the size of the sibling PC graphs is increased further by considering paths that do not necessarily belong on the previously constructed series–parallel graph.

Once a pair of patterns is generated, fault simulation must be done so that the number of robust paths detected by the generated pair of patterns can be determined. The fault simulation problem for the PDF model is not as easy as for the SSF model and its methods are reviewed later. The difficulty relies on the fact that the number of PDFs is not necessarily a polynomial quantity.

Each generated pair of patterns targets robust PDFs in a particular sibling PC graph. It is shown that these PDFs can be detected easily. It may, however, detect robust PDFs in the portion of the circuit outside the targeted PCG. This complicates the fault simulation process. Thus, faults are simulated only within the current PCG in which case a simple topological graph traversal suffices to detect them.

Another ATPG approach for PDFs was proposed recently in Ref. [16]. In an initial step, static implication procedures are used to remove unsensitizable PDFs. This is illustrated in Figure 69.6. If line d receives value 1 then line g receives value 1 and h receives value 0. However, 0 is the controlling

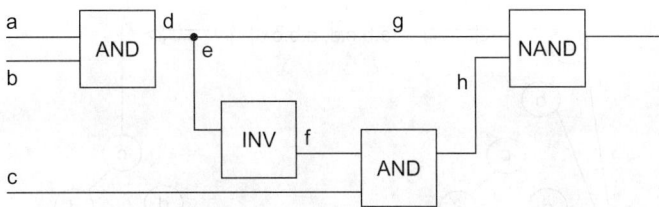

FIGURE 69.6 Static implication.

value of the NAND gate and, therefore, no PDF with a rising transition on lines d and g can be sensitized robustly in the CUT. All such PDFs are then removed implicitly form further consideration. Such static implication rules were proposed in earlier works, [17–19] but Ref. [16] implements static implication rules implicitly using functions. Although function-based SSF ATPG is slower than structural-based SSF ATPG methods, fault-implicit PDF ATPG function-based methods have shown to outperform structural implicit PDF ATPG methods.

PDFs can be stored implicitly using canonical data structures. The most efficient approach appears to be a directed acyclic graph for storing and manipulating collections of sparse combinational sets, called the zero-suppressed binary decision diagram (ZBDD). Another data structure is presented in Ref. [20] but the technique ZBDD-based method is reviewed here. Since ZBDD is a decision diagram, each variable is represented by one or more node with two outgoing edges corresponding to a true and a false assignment on the variable, each function is a pointer to some variable node in the diagram, and its minterms are paths that originate from that node a terminate to a designated node labeled 1. When using ROBDDs, a function is further suppressed by removing variable nodes whose both outgoing edges point to the same node, thus suppressing don't-cares. See also the ROBDD in Figure 69.7 for function $a'b'cd + 'b'c'd$. In contrast, the ZBBD represents the function by suppressing the absence of a variable in each minterm. See also the ZBDD in Figure 69.7 for the same function. The collection of paths or PFS in a CUT can be represented efficiently in a ZBDD since each path of PDF is a sparse set. Since ZBDD are intended for sparse set manipulations, built-in operations such as set union, intersection, difference have been implemented.

All the PDFs in the CUT (sensitizable or not) are initially stored in a ZBDD with a topological traversal, and by using built-in ZBDD operations. Subsequently, each static implication is implemented by removing implicitly all unsensitized PDFs using built-in ZBDD operation during topological traversals on the CUT.

Next, the test functions for PDFs are derived via test functions for PDF portions called *segments*. A segment starts and ends at any two consecutive (topological-wise) stems in the CUT. There is a polynomial number of segments in the CUT. The test functions of all sensitized segments are stored in a ROBDD. All PDFs that contain an unsensitizable segment are removed implicitly from the ZBDD. Then the algorithm recursively manipulates the test functions of sensitized segments. If the concatenation of two or more test functions for segments is empty, then all PDFs that contain all respective segments are removed implicitly. For space efficiency, only the test functions of segments are kept in the ROBDD.

As an example, consider the circuit of Figure 69.6. Let F_1 denote the test function for segment ad with a rising transition on line a, and F_2 denote the test function for segment dgi with rising transition on line d. The test function for PDF adgi with rising transition on a is the concatenation of F_1 and F_2. However, if either F_1 is empty both PDFs adgi and bdgi with rising transition on their input lines are untestable.

This step of the algorithm is not truly implicit since only unsensitizable PDFs are removed implicitly. However, in all existing benchmarks circuits, the majority of the PDFs are unsensitizable and the approach is able to identify all sensitizable PDFs for very large benchmarks [16]. The method can be accelerated by considering only critical PDFs, i.e., PDFs that contain a large number of gates and long interconnects. That way, all sensitizable critical PDFs can be identified even for very large benchmarks.

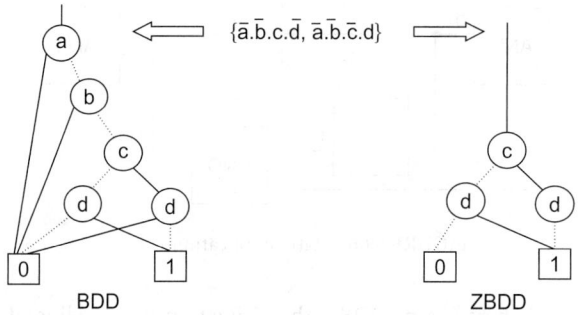

FIGURE 69.7 BDD and ZBBB.

69.2.2 Path Delay Fault Simulation

The exact *number* of PDFs detected by a given set of pair of test patterns is an intractable problem. This is independent of the sensitization criterion used (robust, nonrobust, and functional). The only way to ensure exact implicit PDF coverage is to use canonical data structures that store compactly PDFs and allow for their quick identification. An exact PDF simulation tool stores and manipulates PDFs on a data structure called the path status graph [20]. The drawback of the approach is that it may require exponential space for many benchmarks. As was described earlier, ZBDD is a promising data structure for storing a very large number of PDFs. An implicit and exact fault coverage method has been introduced and it has been shown that exact fault coverage can be obtained on any existing benchmark and any test set [21]. The method is fast and the memory requirement is not prohibitive.

To further accelerate the fault coverage process or if the memory requirements do not allow for the use of exact methods the number of PDFs that are covered (tested) by a set of pair of test patterns must be approximated. The remaining of this section describes methods CAD for obtaining lower bounds on the number of detected PDFs by a given set of n pairs of patterns. These approaches apply to any type of PDF sensitization and are referred to as fault estimation schemes.

In Ref. [22], every time a pair of patterns is applied, the CAD tool examines whether there exists at least one line where either a rising or falling transition has not been encountered by the previously applied pairs of test patterns. Let E_i denote the set of lines for which either a rising or a falling transition occurs for the first time when the pair of patterns P_i is applied.

When $|E_i|$ is greater than zero, a new set of PDFs is detected by pattern P_i. These are the paths that contain lines in E_i. A simple topological search of the combinational circuit suffices to detect their number. If for some P_i we have $|E_i| = 0$, the approach does not detect any PDF. The approach in Ref. [22] is nonenumerative but returns a conservative lower bound to the number of detected paths.

Figure 69.8 illustrates a case where a PDF may not be counted. Assume that the PDFs in all three patterns start with a rising transition. Furthermore, assume that the first pair of patterns detects PDFs along all the paths of the subcircuit which is covered by thick edges. Let the second pair of patterns detect PDFs on all the paths of the subcircuit covered by dotted edges, and let the dashed path indicate a PDF detected by the third pair of patterns. Clearly, the latter PDF cannot be detected by the approach in Ref. [22].

For this reason, Ref. [22] suggests that fault simulation is done by virtually partitioning the circuit into subcircuits. The subcircuits should contain disjoint paths. One implementation for such a partitioning scheme is to consider lines that are independent in the sense that there is no physical path in the circuit that contains any two selected lines. Once a line is selected, we form a subcircuit that consists of all lines that depend on the selected line. In addition, the selected lines must form a cut separating the inputs from the outputs so that every physical path contains one selected line. This way, every PDF belongs to exactly one subcircuit. Figure 69.9 shows three selected lines (the thick lines) of the circuit in Figure 69.8 that are independent and also separate the inputs from the outputs.

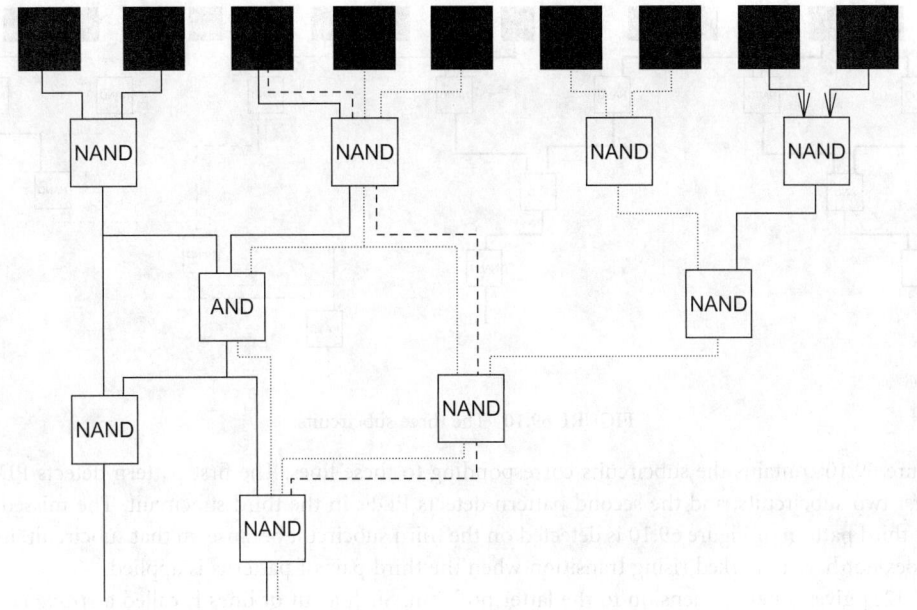

FIGURE 69.8 A nonsimulated PDF.

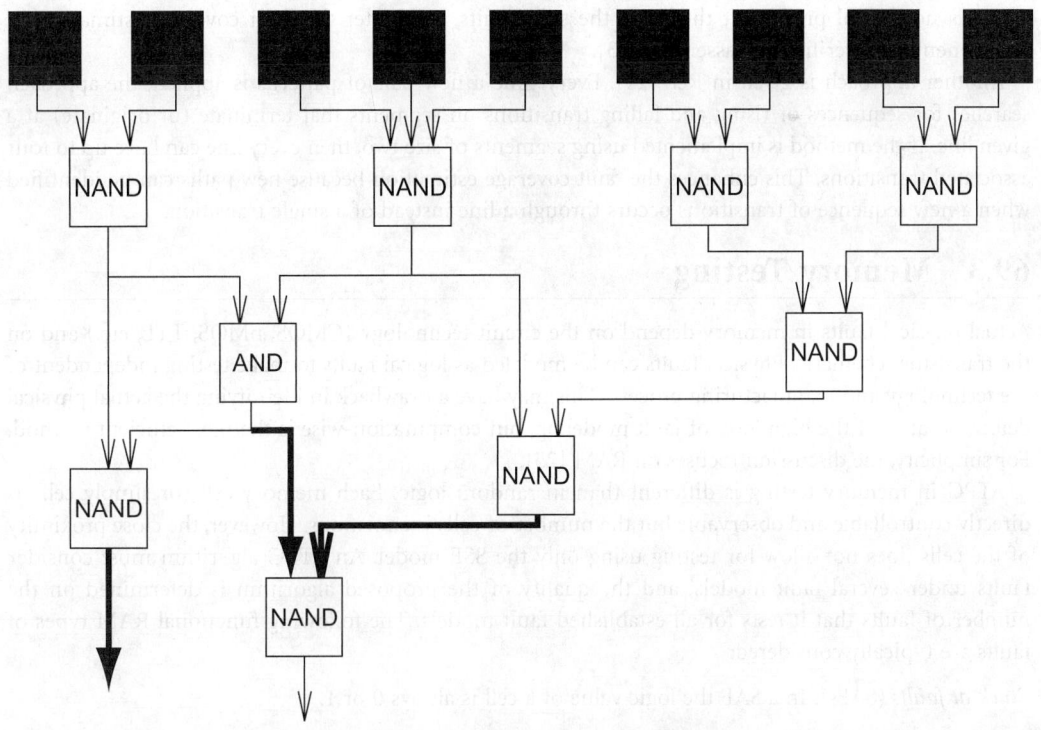

FIGURE 69.9 The independent lines that form a cut.

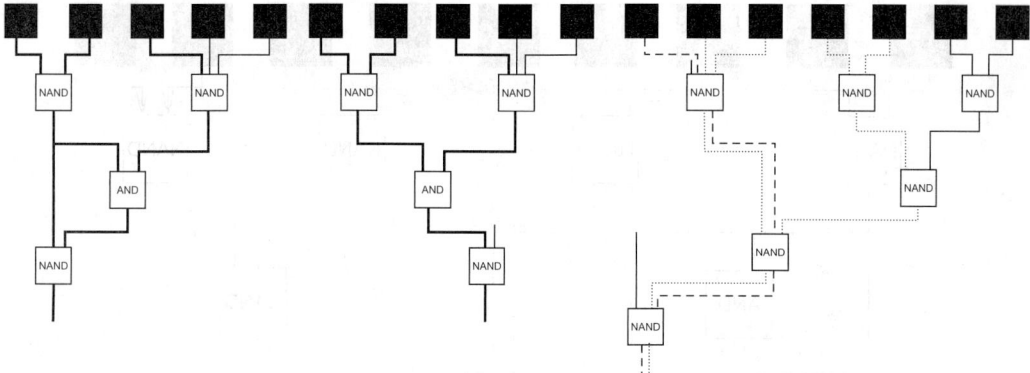

FIGURE 69.10 The three subcircuits.

Figure 69.10 contains the subcircuits corresponding to these lines. The first pattern detects PDFs in the first two subcircuits, and the second pattern detects PDFs in the third subcircuit. The missed PDF by the third pattern of Figure 69.10 is detected on the third subcircuit because, in that subcircuit, its first line does not have a marked rising transition when the third pair of patterns is applied.

Ref. [23] gives a new dimension to the latter problem. Such a cut of lines is called a *strong cut*. The idea is to find a maximum strong cut that allows for a maximum collection of subcircuits where fault coverage estimation can take place. A method is presented in Ref. [29] that returns such a maximum cardinality strong cut. The problem reduces to that of finding a maximum weighted independent set in a comparability graph, which is solvable in polynomial time using a minimum flow technique. There is no formal proof that the more the subcircuits, the better the fault coverage estimation is. Experimentation verifies this assertion [23].

Another approach is given in Ref. [24]. Every time a new pair of patterns is applied, the approach searches for sequences of rising and falling transitions on segments that terminate (or originate) at a given line. If the method is implemented using segments of size two, then every line can have up to four associated transitions. This enhances the fault coverage estimation because new paths can be identified when a new sequence of transitions occurs through a line instead of a single transition.

69.3 Memory Testing

Actual physical faults in memory depend on the circuit technology (CMOS, nMOS, TTL, etc.) and on the transistor schematic. Physical faults can be modeled as logical faults to make testing independent of the technology and manufacturing process. This may have a drawback in identifying the actual physical defect, because of the high level of fault modeling, but computation-wise is the most efficient method. For simplicity, the discussion focuses on RAM [25].

ATPG in memory testing is different than in random logic. Each memory cell (or simply cell) is directly controllable and observable but the number of cells is enormous. However, the close proximity of the cells does not allow for testing using only the SSF model. An ATPG algorithm must consider faults under several fault models, and the quality of the proposed algorithm is determined on the number of faults that it tests for all established fault models. The following functional RAM types of faults are typically considered:

Stuck-at faults (SAFs): In a SAF the logic value of a cell is always 0 or 1.

Transition faults (TFs): A special case of a SAF where a cell fails to make a rising or falling transition when it is written.

Coupling faults (CFs): A transition in cell a causes a nonintended change in another cell b. They are further distinguished as *inversion coupling faults* (ICFs) where a rising or falling transition in cell a inverts the contents of cell *b*, dynamic coupling faults (DCFs) where a read or write operation on cell *a* forces the contents of cell *b* to either 0 or 1, and idempotent coupling faults (IdCFs) where a rising or falling transition in cell *a* sets cell *b* to value 0 or 1.

Bridging faults (BFs): Represent a short circuit between two (or more) cells, and either cell can affect the other. In an *AND bridging fault* (ABF), the bridge value is the AND of the shorted cells. Likewise the *OR bridging fault* (OBF) is defined.

State coupling faults (SCFs): The coupling cell when in some state (0 or 1) forces the coupled cell into a state (0 or 1).

Neighborhood pattern sensitive faults (NPSFs): The content of cell *a* (the base cell) is impacted by the contents of other neighboring cells (the neighborhood) when they are set to a specific state (the pattern). An NPSF is *active* (ANPSFs) where the base cell is either set to a specific state or its state changes due to a transition in only one cell in the neighborhood. It is *passive* (PNPSF) when a certain state in the neighborhood prevents the base cell from changing. It is *static* (SPNSF) when the basic cell is stuck at a state under a specific state in the neighborhood.

Address decoder faults (AFs): Each such fault represents an address decoding error. There are fault types of AFs: (a) No cell is accessed for a specific address. (b) No address accesses a specific cell. (c) More than one cell is accessed for a specific address. (d) A specific cell is accessed by more than one address.

Any number of different faults can also occur simultaneously. Faults also may be *linked*, i.e., a fault may influence the behavior of other faults. The SAF involves a cell and only one SAF can happen at a time. Therefore SAF cannot be linked with other SAFs. Likewise, linked TFs cannot occur. Linked faults are not necessarily restricted within the same fault model. For example, AFs can be linked with all other types of faults.

The most common ATPG algorithms for logical faults is the *march tests*. The march test is applied to each cell before proceeding to the next cell. Thus, if a specific test pattern is applied to one cell then this pattern must be applied to all other $n-1$ cells. This is done either in increasing order (from cell 0 to cell $n-1$) or in decreasing order. In a march test the address order may be irrelevant. By default, the address order in applying two consecutive test patterns must be reversed in any march test.

Not all types of faults can be detected by march tests. For example, not all linked ICFs can be detected by march tests [25]. However march tests are very powerful. Consider the following notation:

R: a read operation, R0: read a 0 from a cell, R1: read 1 from a cell, W: a write operation, W0: write 0 to a cell, W1: write 1 to a cell, I: increasing memory addressing order, D: decreasing memory addressing order, ID: addressing order can be either increasing or decreasing.

It has been shown that all ADFs can be detected by a march test that contains two test patterns. One pattern is I(R*x*, ..., W*x'*) and the other D(R*x'*, ..., W*x*). In the above, the symbol ... denotes any number of read or write operations, *x* can be value 0 or value 1, and the symbols (,) are used to define a set of read or write operations that comprises the test pattern.

The march algorithm applies test pattern ID(W0) followed by test pattern ID(R0,W1) followed by test pattern ID(R1). This algorithm is also known as the MATS algorithm and detects SAFs and some AF [25].

Algorithm MATS++ is a little more time consuming. The first test pattern is the same as the first of MATS. The second is the same as the second of MATS but the application order is I. The last pattern is D(R1,W0,R0). Despite the little amount of modification, MATS++ detects all SAFs, all AFs and all TFs [25].

References

1. M.R. Garey and D.S. Johnson, *Computers and Intractability—A Guide to the Theory of NP-Completeness*, W.H. Freeman, New York, 1979.
2. M. Abramovici, M.A. Breuer, and A.D. Friedman, *Digital Systems Testing and Testable Design*, Computer Science Press, 1990.

3. M.L. Bushnell and V.D. Agrawal, *Essentials of Electronic Testing for Digital, Memory and Mixed-Signal VLSI Circuits*, Springer, Berlin, 2000.

4. T. Larrabee, Test pattern generation using Boolean satisfiability, *IEEE Trans. Computer Computer-Aided Design*, 11, 4, 1992.

5. J.P. Roth, W.B. Bouricious, and P.R. Schneider, Programmed algorithms to compute tests to detect and distinguish between failures in logic circuits, *IEEE Trans. Electron. Comput.*, 16, 157, 1967.

6. P. Goel, An implicit enumeration algorithm to generate tests for combinational logic circuits, *IEEE Trans. Electron. Comput.*, 30, 215, 1981.

7. H. Fujiwara and T. Shimono, On the acceleration of test generation algorithms, *IEEE Trans. Comput.*, 32, 1137, 1983.

8. R.A. Marlett, EBT: A comprehensive test generation technique for highly sequential circuits, *Proceedings of the 15th Design Automation Conference*, Las Vegas, NV, 335, 1978.

9. P. Muth, A nine-valued circuit model for test generation, *IEEE Trans. Comput.*, 6, 630, 1976.

10. K.T. Cheng and H.-C. Chen, Delay testing for robust untestable faults, *Proceedings of the International Test Conference*, Las Vegas, NV, 954, 1993.

11. W.K. Lam, A. Saldhana, R.K. Brayton, and A.L. Sangiovanni-Vincentelli, Delay fault coverage and performance tradeoffs, *Proceedings of the Design Automation Conference*, Las Vegas, NV, 446, 1993.

12. M.A. Gharaybeh, M.L. Bushnell, and V.D. Agrawal, Classification and test generation for path delay faults using stuck-fault tests, *Proceedings of the International Test Conference*, Las Vegas, NV, 139, 1995.

13. G. De Michelli, *Synthesis and Optimization of Digital Circuits*, McGraw Hill, Hightown, 1994.

14. L. Pomeranz, S.M. Reddy, and P. Uppaluri, NEST: An nonenumerative test generation method for path delay faults in combinational circuits, *IEEE Trans. Computer-Aided Design*, 14, 1505, 1995.

15. D. Karayiannis and S. Tragoudas, ATPD: An automatic test pattern generator for path delay faults, *Proceedings of the International Test Conference*, Las Vegas, NV, 443, 1996.

16. S. Padmanaban and S. Tragoudas, Efficient identification of (critical) testable path delay faults using decision diagrams, *IEEE Trans. Computer-Aided Design*, 24, 77, 2005.

17. K. Heragu, J.H. Patel, and V.D. Agrawal, Fast identification of untestable delay faults using implications, *Proceedings of the International Conference on CAD*, Las Vegas, NV, 642, 1997.

18. Z.C. Li, R.K. Brayton, and Y. Min, Efficient identification of non-robustly untestable path delay faults, *Proceedings of the International Test Conference*, Las Vegas, NV, 992, 1997.

19. Y. Shao, S.M. Reddy, S. Kajihara, and I. Pomeranz, An efficient method to identify untestable path delay faults, *Proceedings of the Asian Test Symposium*, 2001.

20. M.A. Gharaybeh, M.L. Bushnell, and V.D. Agrawal, An exact non-enumerative fault simulator f~r path-delay faults, *Proceedings of the International Test Conference*, Las Vegas, NV, 276, 1996.

21. S. Padmanaban, M. Michael and S. Tragoudas, Exact path delay fault coverage with fundamental ZBDD operations, *IEEE Trans. Computer Aided Design*, 22, 305, 2003.

22. I. Pomeranz and S.M. Reddy, An efficient nonenumerative method to estimate the path delay fault coverage in combinational circuits, *IEEE Trans. Computer-Aided Design*, 13, 240, 1994.

23. D. Kagaris, S. Tragoudas, and D. Karayiannis, Improved nonenumerative path delay fault coverage estimation based on optimal polynomial time algorithms, *IEEE Trans. Computer-Aided Design*, 3, 309, 1997.

24. K. Heragu, V.D. Agrawal, M.L. Bushnell, and J.H. Patel, Improving a nonenumerative method to estimate path delay pault coverage, *IEEE Trans. Computer-Aided Design*, 7, 759, 1997.

25. A.J. van de Goor, *Testing Semiconductor Memories: Theory and Practice*, Wiley, Chichester, UK, 1991.

70

Built-In Self-Test

Dimitri Kagaris
Southern Illinois University

CONTENTS

To make the testing of a very large-scale integration (VLSI) circuit easier, several design-for-testability (DFT) criteria must be taken into account together with the other traditional design criteria of cost, delay, area, and power. For example, transforming a sequential circuit into combinational parts by linking in a "test mode" all its flip-flops into a shift register so that patterns to initialize the flip-flops can be easily loaded and the responses observed is a common DFT technique known as *full-scan*. Built-in Self-Test (BIST) is an ultimate DFT technique [8,17,21] in which extra circuitry is introduced on-chip to provide test patterns to the original circuit under test (CUT) and verify its output responses. The aim is to provide a faster and more economic alternative to external testing. The difficulty in the BIST approach is the discovery of schemes which have very low hardware overhead and provide the required test quality to justify their inclusion on-chip. The two principal parts in a BIST design are the test pattern generation (TPG) logic and the response verification logic.

70.1 Built-In TPG Mechanisms

Test pattern generators (TPGs) for BIST are usually some hardware-efficient kinds of autonomous linear finite-state machines (see, e.g., Ref. [1]). The most popular of these are linear feedback shift registers (LFSRs) and cellular automata (CA). Figure 70.1 shows examples of an LFSR of the external-XOR type (Figure 70.1[a]), an LFSR of the internal-XOR type (Figure 70.1[b]) and a CA (Figure 70.1[c]). The behavior of each TPG mechanism is fully described by its *transition matrix M*. If s_t is the current state of the TPG (represented as a $d \times 1$ binary vector) then the next state is given by $s_{t+1} = Ms_t$. The *characteristic polynomial* of the TPG is defined as $P(x) = |M + Ix|$. A TPG mechanism with d cells has the potential to produce a maximum number of $2^d - 1$ nonzero states. For this to happen, the characteristic polynomial must be *primitive* (see, e.g., [1,2]). The construction of an LFSR can be directly derived from its characteristic polynomial. If the polynomial is $P(x) = x^d + p_{d-1}x^{d-1} + \cdots + p_2x^2 + p_1x + 1$, then for an external-XOR LFSR, cells c_{d-1-i} for each $p_i \neq 0$, $0 \leq i \leq d-1$, should be XORed (with the output of the XOR gate driving cell c_0),

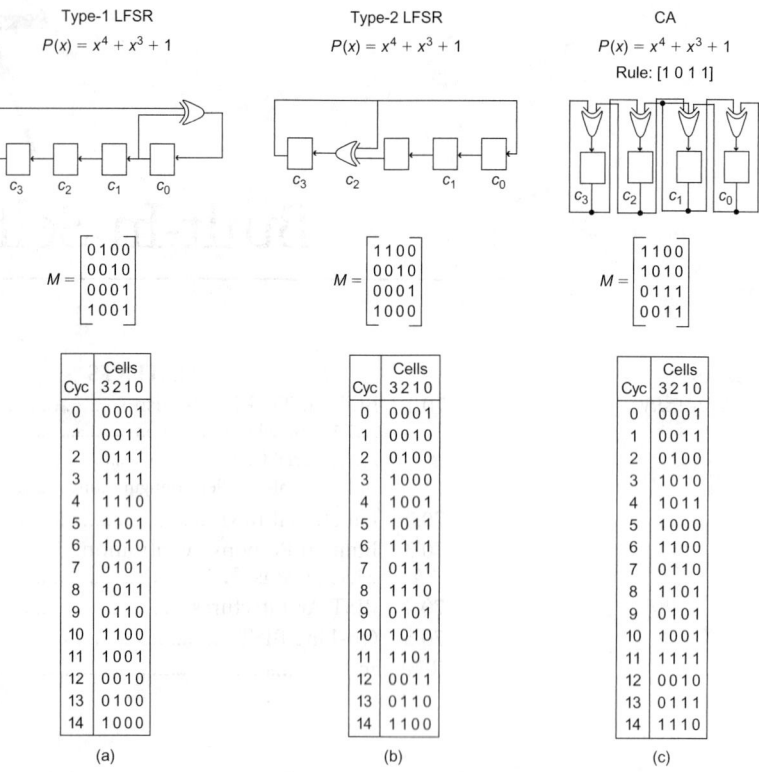

FIGURE 70.1 (a) An internal-XOR LFSR; (b) an External-XOR LFSR; (c) a CA.

while for an internal-XOR LFSR, each cell c_{i-1} for $p_i \neq 0$, $1 \le i \le d-1$, should be XORed with the feedback value of cell c_{d-1} (with the output of each XOR gate driving the next cell c_i). The construction of a CA bears no direct relationship to its characteristic polynomial. Rather, it is directly determined by its rule vector r, where bit r_i, $1 \le i \le d$, means that the next state of cell c_i of the CA is given as $c_i(t+1) = c_{i-1}(t) \oplus (r_i c_i(t)) \oplus c_{i+1}(t)$ (c_0 and c_{d+1} are considered to be constant 0). The rule vector coincides with the diagonal of the transition matrix of the CA (see, e.g., Figure 70.1[c]).

For any TPG mechanism with primitive characteristic polynomial, the bit sequence produced in each of the d stages also has maximum period $2^d - 1$. These bit sequences have the property that any of them is a shifted version of a principal sequence, referred to as the *characteristic sequence*. The prime attribute of this sequence is that it appears to be random (*pseudorandom*). Any TPG with the same characteristic polynomial has the same characteristic sequence, but the bit sequences in the individual stages of each TPG are shifted versions of the characteristic sequence by different amounts depending on the particular TPG structure. Several other kinds of LFSM-based TPGs such as Galois LFSRs [3], ring generators [4], and others exist.

BIST TPG mechanisms can be used as a source of test patterns in two basic configurations: (a) parallel and (b) serial. In parallel, the states (d-bit vectors) of the TPG are used (directly or indirectly) as the test patterns. In serial, the bit sequence from a specified stage (usually the leftmost or rightmost) is used to form the test patterns.

70.1.1 Parallel Configuration

In a parallel configuration, the number d of TPG stages equals the number of the CUT inputs. At each clock cycle, a different pattern is applied to the CUT, constituting what is known as a *test-per-clock* scheme. The fundamental issues in applying test patterns in parallel are discussed below:

70.1.1.1 Pseudorandom Scenario

A straightforward way to test a CUT with d inputs is to drive it by a TPG with d cells and let the TPG run for some desired number of cycles B starting from some initial state (seed). If $B = 2^d - 1$, this is known as *exhaustive testing*, but is, of course, practical only for small values of d. Usually, $2^d - 1$ is a large number and B is a number lesser than that. The B consecutive states constitute a *pseudorandom* test set that hopefully covers many of the modeled faults. Estimation formulas of the value of B for attaining a given level of fault coverage have been given in the literature. In general, pseudorandom testing is effective for the so-called easy-to-detect faults. It is usually used as a preprocessing step before applying some deterministic scheme (see below) for the remaining "hard-to-detect" faults.

Two ways to improve the effectiveness of the basic pseudorandom scheme are:

Weighted random testing. The probability of a "1" or "0" bit in each stage of a pseudorandom TPG is, in general, 0.5. A weighting logic can be appended to the TPG so that these initial probabilities are changed to other desired values depending on a precomputed test set for the CUT. For example, by ORing two TPG stages one obtains a bit sequence in which the probability of 1 is close to 0.75.

Phase shifting logic. The bit sequences produced by successive stages of a TPG mechanism exhibit correlations owing to the fact that each one is a shifted version of the other. These correlations may impair the pseudorandom test capability of the TPG [5,6]. The situation is more aggravated for external-XOR LFSRs since each bit sequence is a shift-by-1 version of the other. To reduce such correlations, *phase shifters* can be appended to the TPG mechanism. A phase shifter is a multiinput XOR gate driven by specific TPG stages. According to the shift-and-add property, the output of the XOR gate produces a bit sequence that is a shifted version of the characteristic sequence. Given a requested phase shift k for a particular stage (cell position) with respect to a reference sequence, an effective technique to find which TPG stages should drive the phase shifter so that the resulting bit sequence is a shifted version by k positions of the reference sequence, is given in Ref. [7].

70.1.1.2 Embedding Scenario

Given a CUT and a set of modeled faults, a test set can be obtained by an Automatic Test Pattern Generation (ATPG) algorithm to cover all faults in consideration or the random pattern resistant faults (also known as "hard-to-detect" faults). Given such a test set (also known as a *deterministic* test set) the question is how to reproduce it within reasonable hardware and time limits. Note that the reproduction of the test set would be trivial if an expensive module such as a ROM were allowed.

Test pattern mapping. A straightforward solution to the test set embedding problem is to append to the TPG mechanism a mapping logic that maps a subset of TPG states to the patterns in the given test matrix T. The approach requires the determination of an initial state (seed) and a length L so that a subset of states in the subsequence of length L maps economically to the patterns in T. In general, the mapping logic tends to be large (see, e.g., [23]).

Reseeding. Given a test set T, another approach is to start the TPG mechanism from different seeds $s_1, s_2, \ldots, s_k$ and let it run for $B_1, B_2, \ldots, B_k$ cycles, respectively to cover the patterns in T. The number of seeds k must be small since the seeds must then be stored (see, e.g., [24]) or otherwise reproduced on-chip (in a fully BIST scheme). The numbers B_i, $1 \leq i \leq k$, must also be small to keep the total number of cycles low.

70.1.2 Serial Configuration

In the serial configuration all test-phase inputs of the CUT are assumed to be organized in a chain (shift register) which is driven by a single stage of the TPG mechanism. The *test-phase* inputs comprise the primary inputs of the CUT as well as any other inputs (such as scan flip-flops and bypass storage cells) inserted by a DFT technique to facilitate testing. The chain is commonly referred to as *scan chain*. The length L of the scan chain (equivalently, the number of bits in each applied test pattern) is typically much larger than the number d of the TPG stages. Under a test-per-clock scheme, a new test pattern can be obtained at every clock cycle (in which case the next test pattern is a shift-by-1 version of the previous), while under a *test-per-scan scheme,* a new test pattern is generated every L clock cycle. Two fundamental issues in serial configuration have been discussed below.

Linear dependencies. As the test bits are shifted through the scan chain, some subcircuits of the CUT may receive patterns whose bits are not random but dependent. Owing to the way the characteristic sequence of the TPG mechanism is formed, *linear dependencies* occur among several bits in a maximal-length bit sequence. This is detrimental not only for pseudorandom application, but also for *pseudo-exhaustive* application. In the latter, it is assumed that all cones (subcircuits of the CUT with a single test-phase output and multiple test-phase inputs) have a number of inputs less than or equal to the number of stages of the TPG mechanism in use. If this is the case and no linear dependencies are present in any subcircuit, then all subcircuits (i.e., the whole CUT) can be tested in parallel in time $2^d - 1 + L$, where L is the length of the scan chain. For example, consider a scan chain that is driven by an LFSR with characteristic polynomial $P(x) = x^4 + x^3 + 1$ (Figure 70.2). Cell C_6 in the figure is a scanned flip-flop, while cell C_5 is a bypass storage cell (bsc) [8] that was inserted to make the number of inputs of any cone be no more than the degree of $P(x)$. As it can be observed, the bit sequence received by cell C_6 is always the XOR (starting at cycle 4) of the bit sequences of cells C_2 and C_3, that is, the cone of the last NAND gate of the circuit driven by C_2, C_3, C_5, and C_6 is forced to receive only half of all possible test patterns.

In either pseudorandom or pseudoexhaustive contexts, the linear dependencies must be minimized. A mathematical formula relating the linear dependencies and the characteristic polynomial $P(x)$ of the TPG mechanism in use has been established in Refs. [9,10] for the case of an LFSR of the external-XOR type. The relation states: Given an external-XOR LFSR with characteristic polynomial $P(x)$ of degree d driving a single scan chain with a total number of L cells, a cone A with k test-phase inputs $\{a_1, a_2, \ldots, a_k\}, 0 \leq a_1, a_2, \ldots, a_k \leq L - 1$, has no linear dependencies if and only if the k polynomials $P_i = x^{a_i} \bmod P'(x), 1 \leq i \leq k$, are linearly independent, where $P'(x)$ depends on cell numbering and is equal to either $P(x)$, if the cells are numbered so that the cell with index $L - 1$ coincides with the 0th LFSR cell [9], or $\tilde{P}(x)$, if the cells are numbered so that the cell with index 0 coincides with the 0th LFSR cell [10] ($\tilde{P}(x)$ is the reciprocal polynomial of $P(x)$ defined as $\tilde{P}(x) = x^d P(1/x)$).

The relation has been extended in Ref. [11] to apply to any kind of linear TPG including in particular internal-XOR LFSRs and CA.

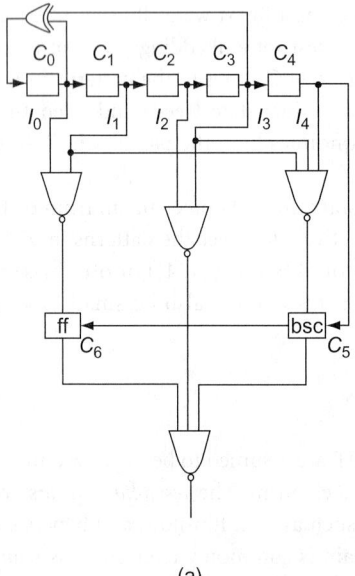

Cyc	Cells						
	0	1	2	3	4	5	6
0	1	0	0	0	0	0	0
1	1	1	0	0	0	0	0
2	1	1	1	0	0	0	0
3	1	1	1	1	0	0	0
4	0	1	1	1	1	0	0
5	1	0	1	1	1	1	0
6	0	1	0	1	1	1	1
7	1	0	1	0	1	1	1
8	1	1	0	1	0	1	1
9	0	1	1	0	1	0	1
10	0	0	1	1	0	1	0
11	1	0	0	1	1	0	1
12	0	1	0	0	1	1	0
13	0	0	1	0	0	1	1
14	0	0	0	1	0	0	1
15	1	0	0	0	1	0	0
16	1	1	0	0	0	1	0
17	1	1	1	0	0	0	1
18	1	1	1	1	0	0	0

(a)   (b)

FIGURE 70.2   (a) Example circuit. (b) Table of states.

Seed encoding. In the deterministic context in a serial configuration, a set T of L-bit patterns are given to be applied to the CUT, where L is the total length of the scan chain. To generate these patterns on-chip by a TPG mechanism whose number of stages d is much smaller than L (typically d is around 30, whereas L is of the order of hundreds or thousands), a d-bit vector has to be identified for each L-bit test pattern in T so that if the TPG is initialized by that vector (seed), the resulting bit sequence after L cycles will match the pattern in T. It has been estimated that if s is the maximum number of specified (no don't-care) bits in any pattern in T, then an LFSR with $d = s + 20$ stages suffices to reduce the probability of not finding a seed for a pattern to $<10^{-6}$. The computation of each required seed can be done by solving a system of linear equations. But even if all seeds have been computed, they have to be stored on-chip or be provided externally. So the problem can in fact be viewed as a problem of test set compression (see, e.g., [19,20,22]): the original $k \times L$ matrix T of k L-bit patterns is compressed to another set with total bit count less than kL (e.g., k seeds of $s + 20$ bits each, or k seeds of variable bit length, or $k' < k$ seeds). For any such compression scheme, there is also extra hardware logic that must be used to help with the on-chip decompression.

70.1.3 Serial–Parallel Configuration

In a combination of serial and parallel configuration, multiple scan chains instead of just a single one can be used to speed up the loading of the test patterns. The STUMPS architecture [12] is a prototype of this configuration. Each stage of the TPG now drives a scan chain of length $l = L/d$. Each test pattern (which is still of length L) is loaded in l cycles (test-per-scan scheme). A test pattern can also be applied at every clock cycle in a test-per-clock scheme. The issues in this serial–parallel configuration are the same as those in linear configuration (namely, linear dependencies and seed encoding).

70.2 Functional BIST

To economize on the extra hardware needed for BIST for TPG and response compaction, functional modules such as arithmetic and logic units (ALUs) that may already be present in the CUT can be used to perform both these tasks. An additional advantage is the minimization of extra delays along critical paths. Typically, an accumulator is used in additive mode with an appropriately selected constant so as to generate a maximum length sequence of states (see, e.g., [18,25]). Notice that the operation is no more linear. In particular, the bit sequence produced in some of the stages of such a mechanism may not have the maximum period even if the sequence of states is of maximum period (as an example, consider the least significant bit positions of a standard binary counter).

70.3 Built-In Response Verification Mechanisms

Verification of the output responses of a circuit under a set of test patterns consists in principle of comparing each resulting output value against the correct one, which has been precomputed and prestored for each test pattern. However, for built-in output response verification, such an approach cannot be used (at least for large test sets) because of the associated storage overhead. Rather, practical built-in output response verification mechanisms rely on some form of *compression* of the output responses so that only the final compressed form needs to be compared against the (precomputed and prestored) compressed form of the correct output response. Some representative built-in output response verification mechanisms based on compression are given below.

(1) *Count of "ones."* In this scheme, the number of times that each output of the circuit is set to "1" by the applied test patterns is counted by a binary counter and the final count is compared against the corresponding count in the fault-free circuit.

(2) *Transition count.* In this scheme, the number of transitions (i.e., changes from both $0 \rightarrow 1$ and $1 \rightarrow 0$) that each output of the circuit goes through when the test set is applied is counted by a binary counter and the final count is compared against the corresponding count in the fault-free circuit. (These counts must be computed under the same ordering of the test patterns.)

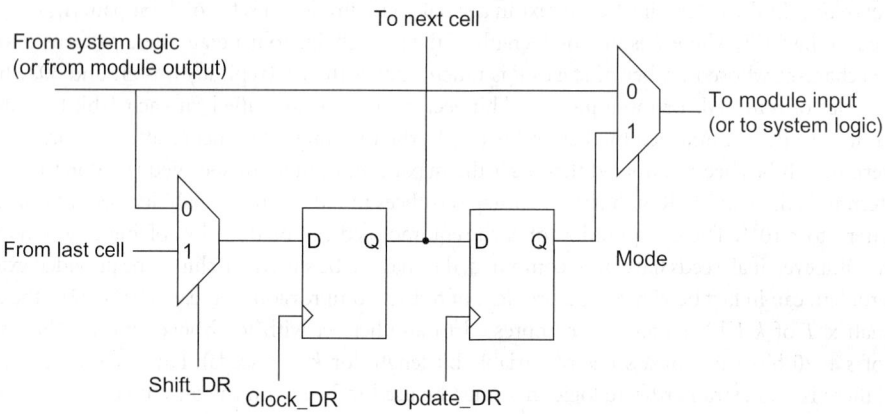

FIGURE 70.3   General structure of an MISR.

(3) *Signature analysis.* In this scheme, the specific bit sequence of responses of each output is represented as a polynomial $R(x) = r_0 + r_1x + r_2x^2 + \cdots + r_{s-1}x^{s-1}$, where r_i is the value that the output takes under pattern t_i, $0 \leq i \leq s$, and s is the total number of patterns. Then this polynomial is divided by a selected polynomial $G(x) = g_0 + g_1x + g_2x^2 + \cdots + g_mx^m$ of degree m for some desired value m, and the remainder of this division (referred to as *signature*) is compared against the remainder of the division by $G(x)$ of the corresponding fault-free response $C(x) = c_0 + c_1x + c_2x^2 + \cdots + c_{s-1}x^{s-1}$. Such a division is done efficiently in hardware by an LFSR structure. In practice, the responses of all outputs are handled together by an extension of the division circuit, known as *multiple-input signature register* (*MISR*). The general form of an MISR is shown in Figure 70.3.

In all compression techniques it is possible for the compressed forms of a faulty response and the correct one to be the same. This is known as *aliasing* or *fault masking*. For example, the effect of aliasing in 1s count output response verification is that faults that cause the overall number of 1s in each output to be the same as in the fault-free circuit are not going to be detected after compression, although the appropriate test patterns for their detection have been applied. In general, signature analysis offers a very small probability of aliasing. This is due to the fact that if the correct response is $C(x)$, and the current response is $R(x) = C(x) + E(x)$, where $E(x)$ represents the error pattern, an LFSR with characteristic polynomial $G(x)$ will produce the same signature as the correct response $C(x)$, if and only if $E(x)$ is a multiple of $G(x)$. Mathematical expressions for the aliasing probability in LFSRs and MISRs have been given in Refs. [13,14].

70.4   BIST Architectures

BIST strategies for systems composed of combinational logic blocks and registers generally rely on partial modifications of the register structure of the system to economize on the cost of the required mechanisms for TPG and output response verification. For example, in the Built-In Logic Block Observer (BILBO) scheme [15], each register that provides input to a combinational block and receives the output of another combinational block is transformed into a multipurpose structure that can act as an LFSR (for TPG), an MISR (for output response verification), a shift register (for scan chain configurations), and also a normal register. An implementation of the BILBO structure for a 4-bit register is shown in Figure 70.4. In this example, the characteristic polynomial for the LFSR and MISR is $P(x) = x^4 + x + 1$.

By setting $B_1B_2B_3 = 001$, the structure acts like an LFSR. By setting $B_1B_2B_3 = 101$, the structure acts like an MISR. By setting $B_1B_2B_3 = 000$, the structure acts like a shift register (with serial input SI and serial output SO). By setting $B_1B_2B_3 = 11x$, the structure acts like a normal register, and by setting $B_1B_2B_3 = 01x$, the register can be cleared.

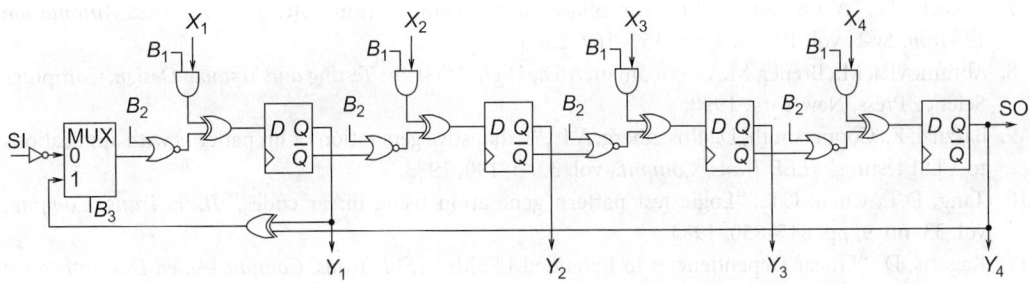

FIGURE 70.4   BILBO structure for a 4-bit register.

As two more representatives of system BIST architectures, we mention here the STUMPS scheme [12], where multiple scan paths are used in which each one is driven by a cell of the same LFSR and drives a cell of the same MISR and the LOCST scheme [16], where there is a single boundary scan chain for inputs and a single boundary scan chain for outputs, with an initial portion of the input chain configured as an LFSR and a final portion of the output chain configured as an MISR.

70.5   On-Line BIST

A special form of BIST is the design of *self-checking* circuits in which no explicit test patterns are provided, but the operation of the circuit is tested *on-line* by identifying any *invalid* output responses, i.e., responses that can never occur under fault-free operation. If, however, there is a fault that can cause one valid response to be changed into another one, then that fault cannot be detected. The identification of faulty behavior is done by a special built-in circuit called *checker*. For example, in a $k : 2^k$ decoder a checker can check if exactly one of the 2^k output lines has a value of 1 each time. If the number of 1s in the output pattern is zero or more than one, then an error is detected. If however, a fault in the decoder causes an input pattern to assert only one output line but not the correct one, then the fault cannot be detected by such a checker. In general, the design of self-checking circuits is based on coding theory. The checker has to encode all output responses of the circuit under fault-free operation to distinguish between valid and invalid responses. For example, using the single-bit parity code, a checker can compute the parity of the actual response of the circuit for the current input and also the parity of the (known) correct output response corresponding to that input, and compare the two parities.

Faults in the checker can beat the purpose of fault detection in the original circuit. However, the assumption is that the logic of the checker is much simpler than the circuit it checks, and, therefore, can be tested far more easily. Research on the design of *self-checking checkers* seeks to minimize the logic that is not self-testable.

References

1. Stone, H.S., *Discrete Mathematical Structures and Their Applications*, Science Research Associates, Chicago, IL, 1973.
2. Peterson, W.W., Weldon, E.J., Jr., *Error-Correcting Codes, MIT Press*, Cambridge, MA, 1972.
3. Pradhan, D.K., Chatterjee, M., "GLFSR—a new test pattern generator for built-in-self-test," *IEEE Trans. Comput.-Aided Des. Integrated Circuits Syst.*, vol. 18, no. 2, pp. 238–247, 1999.
4. Mrugalski, G., Rajski, J., Tyszer, J., "Ring generators—new devices for embedded test applications," *IEEE Trans. Comput.-Aided Des. Integrated Circuits Syst.*, vol. 23, no. 9, pp. 1306–1320, 2004.
5. Rajski, J., Tamarapalli, N., Tyszer, J., "Automated synthesis of phase shifters for built-in self-test applications," *IEEE Trans. Comput.-Aided Des. Integrated Circuits Syst.*, vol. 19, no. 10, pp. 1175–1188, 2000.
6. Mrugalski, G., Rajski, J., Tyszer, J., "Cellular automata-based test pattern generators with phase shifters," *IEEE Trans. Comput.-Aided Des. Integrated Circuits Syst.*, vol. 19, no. 8, pp. 878–893, 2000.

7. Kagaris, D., "A unified method for phase shifter computation," *ACM Trans. Des. Automation Electron. Syst.*, vol. 10, no. 1, pp. 157–167, 2005.

8. Abramovici, M., Breuer, M.A., Friedman, A.D., *Digital Systems Testing and Testable Design*, Computer Science Press, New York, 1990.

9. Barzilai, Z., Coppersmith, D., Rosenberg, A.L., "Exhaustive generation of bit patterns with applications to VLSI testing," *IEEE Trans. Comput.*, vol. 32, p. 190, 1983.

10. Tang, D.T., Chen, C.L., "Logic test pattern generation using linear codes," *IEEE Trans. Comput.*, vol. 33, no. 9, pp. 845–850, 1984.

11. Kagaris, D., "Linear Dependencies in Extended LFSMs," *IEEE Trans. Comput.-Aided Des. Integrated Circuits Syst.*, vol. 21, no. 7, pp. 852–859, 2002.

12. Bardell, P.H., McAnney, W.H., Savir, J., *Built-in Test for VLSI*, Wiley, New York, 1987.

13. Damiani, M., Olivo, P., Ercolani, S., Ricco, B., "An analytical model for the aliasing probability in signature analysis testing," *IEEE Trans. Comput.-Aided Des. Integrated Circuits Syst.*, vol. 11, no. 8, pp. 1133–1144, 1989.

14. Pradhan, D.K., Gupta, S.K., "A new framework for analyzing BIST techniques and zero aliasing compression," *IEEE Trans. Comput.*, vol. 40, no. 6, pp. 743–763, 1991.

15. Koenemann, B., Mucha, J., Zwiehoff, G., Built-in test for complex digital integrated circuits, *IEEE J. Solid-State Circuits*, vol. 15, p. 315, 1980.

16. LeBlanc, J., LOCST: a built-in self-test technique, *IEEE Des. and Test of Comput.*, vol. 1, p. 42, 1984.

17. Bushnell, M.L., Agrawal, V.D., *Essentials of Electronic Testing for Digital, Memory, and Mixed-Signal VLSI Circuits*, Springer, Boston, 2000.

18. Dorsch, R., Wunderlich, H.-J., "Accumulator-based deterministic BIST," *Proceedings of the International Test Conference*, 1998, pp. 412–421.

19. Hellebrand, S., Rajski, J., Tarnick, S., Venkataraman, S., Courtois, B., "Built-in test for circuits with scan based on reseeding of multiple-polynomial linear feedback shift registers," *IEEE Trans. Comput.*, vol. 44, no. 2, pp. 223–233, 1995.

20. Hua-Guo, L., Hellebrand, S., Wunderlich, H.-J., "Two-dimensional test data compression for scan-based deterministic BIST," *Proceedings of the International Test Conference*, 2001, pp. 894–902.

21. Jha, N.K., Gupta, S., *Testing of Digital Systems*, Cambridge University Press, London, 2003.

22. Krishna, C.V., Jas, A., Touba, N.A., "Test vector encoding using partial LFSR reseeding," *Proceedings of the International Test Conference*, 2001, pp. 885–893.

23. Touba, N.A., McCluskey, E.J., "Synthesis of mapping logic for generating transformed pseudo-random patterns for BIST," *Proceedings of the International Test Conference*, 1995, pp. 674–682.

24. S. Chiusano, S. Di Carlo, P. Prinetto, H.-J. Wunderlich, "On Applying the Set Covering Problem to Reseeding," *Proc. Design Automation and Test in Europe*, 2001, pp. 156–160.

25. S. Gupta, J. Rajski, J. Tyszer, "Test Pattern Generation Based on Arithmetic Operations," *IEEE/ACM International Conference on Computer-Aided Design*, Nov. 1994, pp. 117–124.

Section X

Compound Semiconductor Integrated Circuit Technology

Stephen I. Long
University of California at Santa Barbara

0-8493-XXXX-X/04/$0.00+$1.50
© 2006 by CRC Press LLC

71

Compound Semiconductor Materials

Stephen I. Long
University of California at Santa Barbara

CONTENTS

71.1 Introduction

Very high-speed or high-frequency integrated circuit (IC) design is a multidisciplinary challenge. First, there are several IC technologies available for these applications. Each of these claims to offer unique benefits to the user. To choose the most appropriate or cost-effective technology for a particular application or system, the designer must understand the materials, the devices, the limitations imposed by the process on yields, and the thermal limitations due to power dissipation.

Second, very high-speed ICs present design challenges if the inherent performance of the devices is to be retained. At the upper limits of speed, there are no digital circuits, only analog circuits. Circuit design techniques formerly thought to be exclusively in the domain of analog IC design are effective in optimizing digital IC designs for the highest performance. The performance of high-frequency analog ICs, often referred to as Radio Frequency IC (RFICs) or Microwave Monolithic IC (MMICs), is often limited by the on-chip passive components. Good models for these components must be derived as well as for the active devices.

Finally, when using the highest speed technologies, system integration presents an additional challenge. Interconnections and clock and power distribution both on-chip and off-chip require much care, need accurate modeling, and often restrict the achievable performance of an IC in a system.

The entire scope of very high-speed IC design is much too vast to be covered in a single chapter. Therefore, this chapter and the subsequent three chapters will provide some useful tools for the designer. The focus will be primarily on compound semiconductor technologies to restrict the scope. SiGe bipolar ICs will be included in this chapter, but not the standard Si CMOS or bipolar ones. However, the device operation and circuit design principles presented are independent of technology.

In this chapter on materials, a brief introduction to compound semiconductor materials is presented. The transport properties of typical III–V materials are compared with silicon and SiGe alloys. The use of nonsilicon materials is primarily needed for the highest speed, highest frequency, and lowest noise applications. Chapter 72 describes a technology-independent description of device operation for high-speed or high-frequency applications. The charge control methodology provides insight and connects the basic material properties and device geometry with performance. Chapter 73 gives an overview of the typical circuit design techniques used for high-frequency circuits for RF and microwave applications. Finally, Chapter 74 presents a technology-independent analog circuit design methodology that can be used to predict the performance of digital ICs or analog RFICs.

71.2 Silicon and Silicon–Germanium

There is no denying that silicon is the workhorse of the semiconductor industry. Large, high-quality substrates are relatively inexpensive, a highly stable oxide can be grown with low interface state density, and a highly advanced processing technology has enabled extremely large circuit density and extremely fine lines to be achieved with low parasitic capacitances. Its greatest disadvantage in electronic device applications is the relatively low electron velocity and mobility. These intrinsic properties lead to higher transit times and access resistances, respectively, a limitation on high-frequency device performance. The deeply scaled submicron technology has compensated for these deficiencies to some degree by aggressive reduction in gatelength or basewidth. In addition, replacing Si with a SiGe alloy for the base of bipolar transistors improves high-frequency performance by grading the composition across the base as discussed in Section 71.4 and Chapter 72. Also, p-SiGe has higher hole mobility than p-Si, so access resistance can be improved. And, using the strain induced by local depositions of SiGe in MOSFETs increases electron and hole mobilities. However good Si IC technology may be, there exist compound semiconductor materials whose intrinsic electron velocity and mobility are considerably superior to Si and so can potentially offer higher frequency, higher speed, or higher power performance.

The III–V FET and bipolar device technology can provide the highest frequency and lowest noise circuit applications. Its main limitation is density. Device footprints are often significantly larger than those of similar Si devices. Thus, the high intrinsic performance of these devices is achieved in circuits of relatively low complexity.

71.3 Defining III–V Compound Semiconductors

The compound semiconductor family, as traditionally defined, is composed of Group III and Group V elements shown in Table 71.1. Each semiconductor is formed from at least one Group III and one Group V element. Group IV elemental semiconductors such as C, Si, and Ge are used as dopants whose activity can be either n or p type depending on the host material and conditions of growth. There are also several Group II and Group VI dopant elements such as Be or Mg for p type and S, Te, and Se for n type.

Binary semiconductors such as GaAs and InP can be grown in large single-crystal ingot form using the liquid-encapsulated Czochralski method [1] and are the materials of choice for substrates. Table 71.2 summarizes the properties of typical III–V substrate materials. At the present time, GaAs wafers with diameters of 100 and 150 mm are the most widely used. InP is still limited to 75–100 mm diameter. SiC substrates are even more difficult to grow, but 50 and 75 mm diameters are presently available [2].

TABLE 71.1 Compound Semiconductor Family

II	III	IV	V	VI
Be	B	C	N	O
Mg	Al	Si	P	S
Zn	Ga	Ge	As	Se
Cd	In	Sn	Sb	Te

Note: Column III, IV, and V elements are normally associated with compound semiconductors.

TABLE 71.2 Comparison of Substrate Properties

Material	Thermal Conductivity (W/cm K)	Dielectric Constant	Wafer Size (mm)	Electrical Conductivity	Cost
Silicon	1.45	11.7	200–300	n or p	Low
GaAs	0.45	13.1	100–150	n, p, or semi-insulating	Medium
InP	0.68	12.4	75–100	n, p, or semi-insulating	High
Sapphire	0.42	9.4	200	Insulating	Low
SiC	3.0–3.8	9.8	50–75	n, p, or semi-insulating	High

The binary substrate materials can be doped n or p type, although p-type doping efficiency is very low in SiC. Unique to III–V materials, semi-insulating crystals with very high resistivity up to 10^8 Ω cm are possible. The high resistivity reduces losses in passive components such as inductors and transmission lines. Of particular note is the high thermal conductivity of SiC. Temperature rise in transistors due to power dissipation limits their performance and reliability. GaAs and InP are fairly poor thermal conductors, thus power devices often use very thin substrates to minimize thermal resistance. The high conductivity of SiC make it an excellent substrate material for SiC MESFET and GaN HEMT power amplifier devices. The lower dielectric constant of SiC is also beneficial for reducing capacitive loading on the transistors and reducing parasitic capacitances of passive components.

Three or four elements are often mixed together when grown as thin *epitaxial* films on top of the binary substrates. The alloys thus formed allow electronic and structural properties such as bandgap and lattice constant to be varied as needed for device purposes. Junctions between different semiconductors can be used to further control charge transport as discussed in Section 71.4.

71.3.1 Why III–V Semiconductors?

The main reason for using the III–V compound semiconductors for device applications is their superior electronic properties when compared with those of the dominant semiconductor material, silicon. Figure 71.1 is a plot of steady-state *electron velocity* of several n-type semiconductors versus electric field. From this graph, we see that at low electric fields the slope of the III–V semiconductor curves (*mobility*) is higher than that of silicon. High mobility means that the semiconductor resistivity will be less for III–V n-type materials, and it will therefore be easier to achieve lower access resistance. *Access resistance* is the series resistance between the device contacts and the internal active region. An example would be the base resistance of a bipolar transistor. Lower resistance will reduce some of the fundamental device time constants to be described in Chapter 72 which often dominate the high-frequency performance of the device. Figure 71.1 also shows that the peak electron velocity is higher for the Group III–V elements, and the peak velocity can be achieved at much lower electric fields. High velocity reduces *transit time*, the time required for a charge carrier to travel from its source to its destination, and improves the high-frequency performance of the device, also discussed in Chapter 72. Achieving this high velocity at lower electric fields means that the devices will reach their peak performance at lower voltages, useful for low-power, high-speed applications. Higher velocity of electrons also increases the current density of a device since current is the product of charge and velocity. Mobility and peak velocities of several semiconductors are compared in Table 71.3.

The higher velocities are a consequence of the band structure of III–V materials. Figure 71.2(a) shows the valence and conduction bands of silicon. Since Si is an indirect bandgap material, conduction electrons reside in a high effective mass conduction band (CB). Mobility is dominated by the high effective mass.

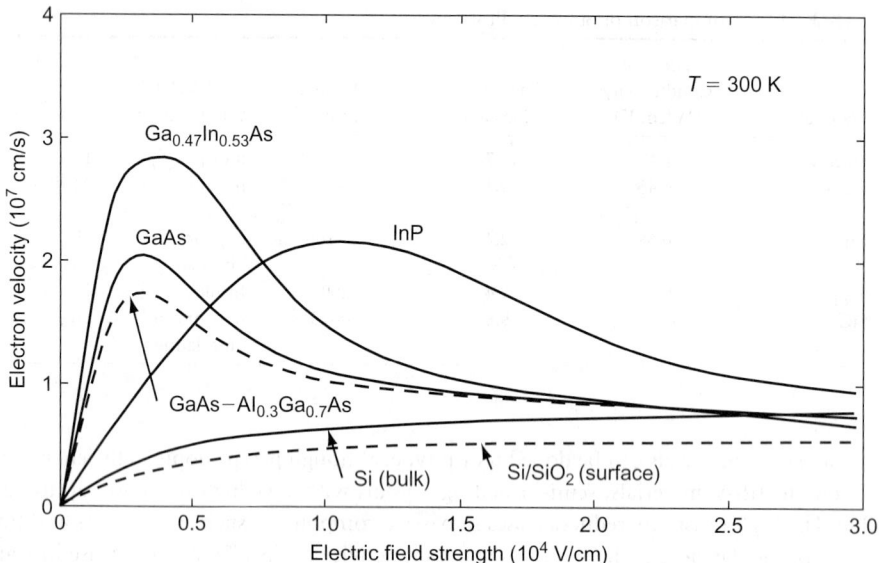

FIGURE 71.1 Electron velocity versus electric field for several n-type semiconductors.

TABLE 71.3 Electronic Properties of Compound Semiconductors Compared with Si and Ge

Semiconductor	E_G (eV)	ε_r	Electron Mobility (cm^2/V s)	Hole Mobility (cm^2/V s)	Peak Electron Velocity (cm/s)
Si (bulk)	1.12	11.7	1450	450	N.A.
Ge	0.66	15.8	3900	1,900	N.A.
InP	1.35 D	12.4	4600	150	2.1×10^7
GaAs	1.42 D	13.1	8500	400	2×10^7
$Ga_{0.47}In_{0.53}As$	0.78 D	13.9	11,000	200	2.7×10^7
InAs	0.35 D	14.6	22,600	460	4×10^7
$Al_{0.3}Ga_{0.7}As$	1.80 D	12.2	1000	100	—
AlAs	2.17	10.1	280	—	—
$Al_{0.48}In_{0.52}As$	1.92 D	12.3	800	100	—
GaN	3.39 D	9.0	1500	30	$2.5–2.7 \times 10^7$
SiC (4H)	3.26	9.8	500		2.2×10^7

Note: In bandgap energy column the symbol "D" indicates direct bandgap, otherwise, indirect bandgap.

At high electric fields, the optical phonon generation process limits the maximum achievable electron drift velocity.

GaAs, on the other hand, is a direct bandgap material, and Figure 71.2(b) illustrates that the electron mobility is high because of the lower energy, low effective mass CB where conduction electrons are confined at low fields. Figure 71.2(c) demonstrates that the average electron velocity will be reduced at higher electric fields owing to scattering into the higher mass CB. This produces a saturated drift velocity less than the peak drift velocity, typical of the direct-gap Group III–V semiconductors.

To obtain significant transit velocity improvement over silicon, one must use a ternary III–III–V semiconductor such as InGaAs. Figure 71.3 illustrates this point. The high effective mass CB is separated by 50% of the bandgap for InGaAs, whereas for GaAs it was only 20%. Thus, the peak velocity in InGaAs can be much higher than GaAs because more energy can be transferred to the conduction electrons

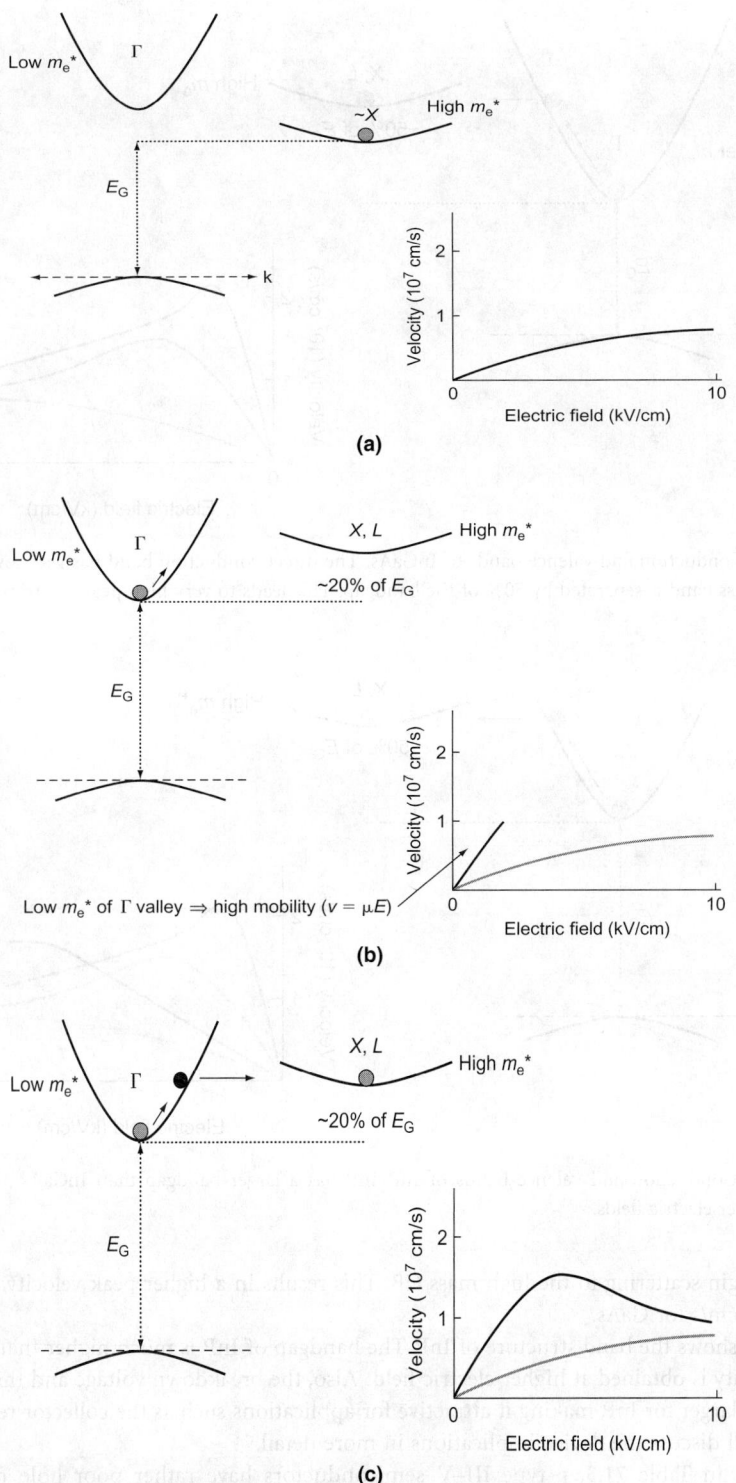

FIGURE 71.2 (a) Conduction and valence bands of silicon; (b) Gallium arsenide with a direct bandgap. At low electric field, electrons are confined to the low effective mass conduction band and mobility is large; (c) Gallium arsenide at high electric fields. Carriers transfer to high effective mass satellite conduction band valley causing net reduction in electron velocity.

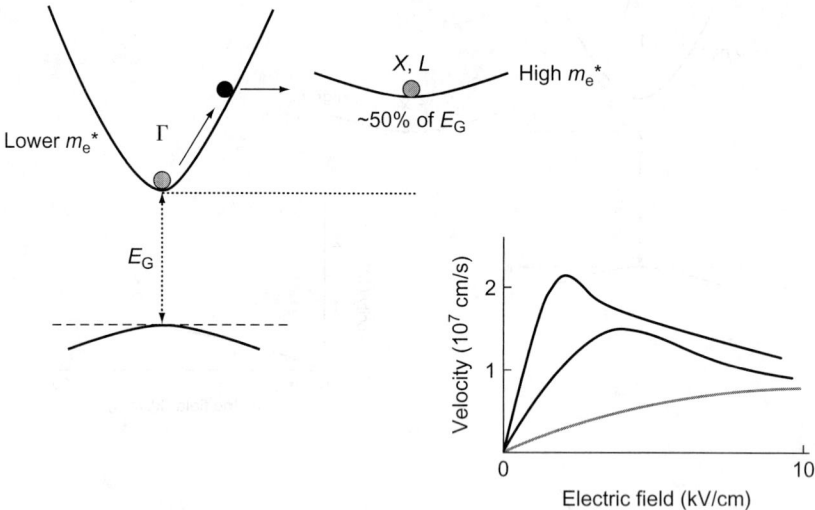

FIGURE 71.3 Conduction and valence bands of InGaAs. The direct conduction band has very low effective mass, and the high mass band is separated by 50% of the bandgap. This leads to very high peak electron velocity.

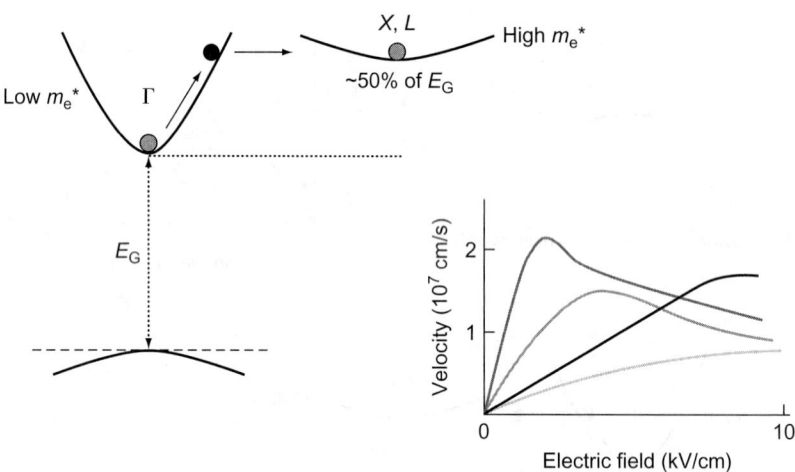

FIGURE 71.4 Conduction and valence bands of InP. InP has a larger bandgap than InGaAs, thus the velocity saturates at higher electric fields.

before they begin scattering to the high mass CB. This results in a higher peak velocity, 2.7×10^7 cm/s versus 2×10^7 cm/s for GaAs.

Figure 71.4 shows the band structure of InP. The bandgap of InP is much higher than InGaAs, thus the peak velocity is obtained at higher electric field. Also, the breakdown voltage and thermal conductivity is much larger for InP, making it attractive for applications such as the collector region of HBTs. Chapter 72 will discuss the device implications in more detail.

Also shown in Table 71.3, p-type III–V semiconductors have rather poor hole mobility when compared with elemental semiconductor materials such as silicon or germanium. The holes also reach their peak velocities at much higher electric fields than the electrons. p-Type III–V materials are used in the base of a bipolar transistor, for example. The base thickness must be extremely small to reduce transit time. Under this condition, the small hole mobility can lead to excessive access resistance unless the lateral distance across the base is very small and doping very high. In spite of this, the maximum operating frequencies of InP-based HBT circuits exceed that of SiGe circuits owing to the superior transit time and lower access resistance on aggressively scaled devices [3,4]. CMOS-like complementary FET technologies have also been developed, but their performance has been marginal, limited by the poor speed of the p-channel devices, and this approach seems to have been abandoned owing to the superior performance of deep submicron CMOS.

71.3.2 Wide-Bandgap Compound Semiconductors

In recent years there has been increasing interest in the wide-bandgap compound semiconductors, SiC and GaN (and associated alloys of Al/In/Ga with N). The electron device applications have been primarily for microwave power applications because the wider bandgap increases breakdown voltage while the band structure allows for high electron peak velocities in both materials [5,6]. Table 71.4 compares the fundamental physical properties of the wide-gap compound semiconductors with GaAs and Si. It should be noted that there is no uniform agreement on the wide-gap parameter values from one reference to the next, but the numbers presented are representative of the current literature. As seen in this table, thermal conductivity is very high for both SiC and GaN, allowing for effective removal of heat from power devices. In fact, at room temperature, SiC has a higher thermal conductivity than any metal.

Figure 71.5 compares the electron velocity of GaN and SiC with GaAs and silicon [7]. The peak velocity of GaN is reached at electric fields above 150 kV/cm. Both SiC and GaN retain their good

TABLE 71.4 Comparison of Electrical and Thermal Properties of Wide-Bandgap Compound Semiconductors GaN and SiC with GaAs and Si

Material	Bandgap (eV)	Mobility (cm²/V s)	E_c (V/cm)	Saturated Drift Velocity (cm/s)	Thermal Conductivity (W/cm K)
n-SiC (4H)	3.26	500	2.2×10^6	2×10^7	3.0–3.8
n-GaN	3.39	1500	3×10^6	1.5×10^7	2.2
n-GaAs	1.4	5000	3×10^5	0.6×10^7	0.45
n-Si	1.1	1300	2.5×10^5	1×10^7	1.45

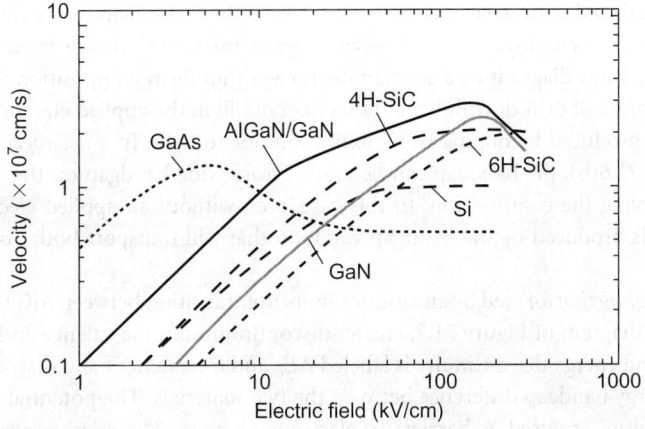

FIGURE 71.5 Electron velocity versus electric field of GaN and SiC compared with silicon and GaAs. (From Trew, R.J., *Proc. IEEE*, Special Issue on Wide-Bandgap Semiconductors, vol. 90, pp. 1032–1047, 2002, with permission.)

TABLE 71.5   Lattice Constant and Bandgap of the Nitrides

Material	Lattice Constant (Å)	Bandgap (eV)
InN	3.55	0.8
GaN	3.2	3.39
AlN	3.12	6.1

transport properties for high-power applications. Table 71.4 shows that GaN has high electron mobility as well, which helps to reduce parasitic source resistance. Hole mobility for wide-bandgap compound semiconductors is quite low, however, generally <50 cm²/V s.

Lattice constant versus bandgap for the nitride family is shown in Table 71.5. There is significant lattice mismatch between a GaN channel in a heterojunction FET and the AlGaN barrier layer. However, the strain caused by this mismatch produces polarization and piezoelectric effects that induce large-sheet charge densities, above 10^{13} cm⁻² in the channel, beneficial for high current density operation of these devices [7]. This level of charge is about five times higher than what can be induced in GaAs channels in the AlGaAs/GaAs heterostructure (to be discussed in Chapter 72).

Figures of merit are often employed when comparing materials for microwave power amplifier applications. Johnson's FOM [8] has units of power–frequency.

$$\text{JFOM} = \frac{E_c v_{sat}}{2\pi}$$

E_c is the maximum or critical electric field for breakdown and v_{sat} the saturated drift velocity at high electric fields. This expresses the electronic merits of the material, but neglects to consider thermal conductivity, also of importance in power electronics. Nevertheless, based on the electronic properties alone, GaN and SiC have a JFOM approximately 18 times greater than Si or GaAs. If the superior thermal conductivity is also considered, it becomes clear that these materials are extremely well suited for microwave power.

71.4   Heterojunctions

In the past, most semiconductor devices were composed of a single semiconductor element such as silicon or gallium arsenide, and employed n- and p-type doping to control charge transport. Figure 71.6(a) illustrates an energy band diagram of a semiconductor with uniform composition that is in an applied electric field. Electrons will drift downhill and holes drift uphill in the applied electric field. The electrons and holes could be produced by doping or by ionization due to light. In a *heterogeneous* semiconductor as shown in Figure 71.6(b), the bandgap can be graded from wide bandgap on the left to a narrow one on the right by varying the composition. In this case, even without an applied electric field, a built-in quasi-electric field is produced by the bandgap variation that will transport both holes and electrons in the *same* direction.

The abrupt *heterojunction* formed by an atomically abrupt transition between AlGaAs and GaAs, shown in the energy band diagram of Figure 71.7, creates discontinuities in the valence and conduction bands. The conduction band energy discontinuity is labeled ΔE_C and the valence band discontinuity, ΔE_V. Their sum equals the energy bandgap difference between the two materials. The potential energy steps caused by these discontinuities are used as barriers to electrons or holes. The relative sizes of these potential barriers depend on the composition of the semiconductor materials on each side of the heterojunction. In this example, an electron barrier in the conduction band is used to confine carriers into a narrow

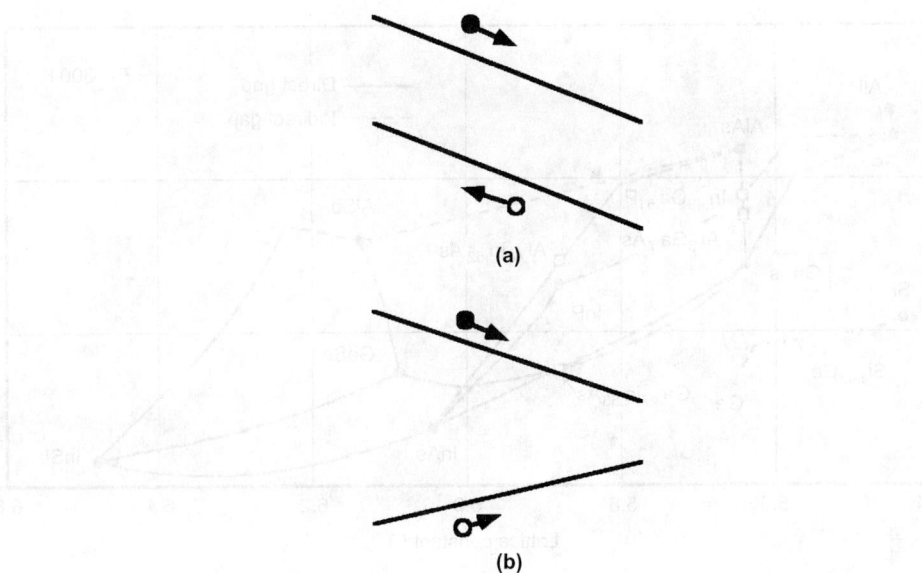

FIGURE 71.6 (a) Homogeneous semiconductor in uniform electric field; (b) heterogeneous semiconductor with graded energy gap. No applied electric field.

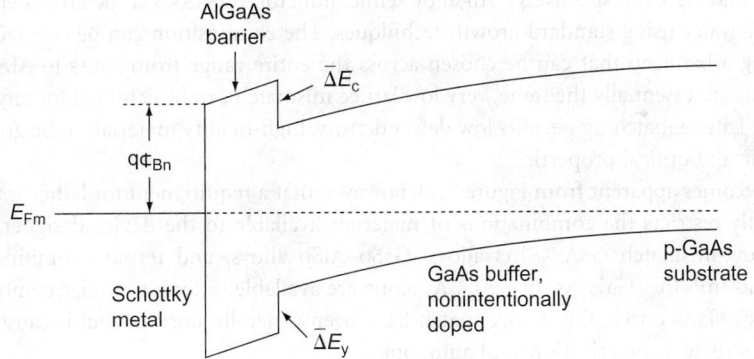

FIGURE 71.7 Energy band diagram of an abrupt heterojunction.

potential energy well with triangular shape. Quantum well structures such as these are used to improve device performance through two-dimensional charge transport channels, similar to the role played by the inversion layer in MOS devices. The structure and operation of heterojunctions in FETs and BJTs will be described in Chapter 72.

The overall principle of the use of heterojunctions is summarized in a *central design principle*:

Heterostructures use energy gap variations in addition to electric fields as forces acting on holes and electrons to control their distribution and flow [9,10].

The energy barriers can control motion of charge both across the heterojunction and in the plane of the heterojunction. In addition, heterojunctions are most widely used in light-emitting devices since the compositional differences also lead to either a stepped or graded refractive index, which can be used to confine, refract, and reflect light. The barriers also control the transport of holes and electrons in the light-generating regions.

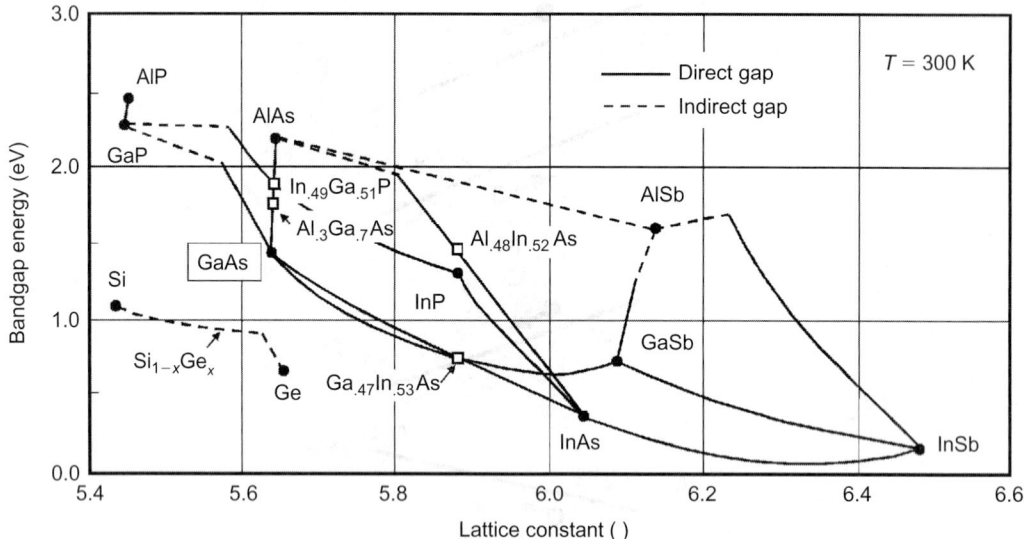

FIGURE 71.8 Energy bandgap versus lattice constant for compound semiconductor materials.

Figure 71.8 shows a plot of bandgap versus lattice constant for many of the III–V semiconductors. Consider GaAs for example. GaAs and AlAs have the same lattice constant (~0.56 nm), but different bandgaps (1.4 and 2.2 eV, respectively). An alloy semiconductor, AlGaAs, can be grown epitaxially on a GaAs substrate wafer using standard growth techniques. The composition can be selected by the Al to Ga ratio giving a bandgap that can be chosen across the entire range from GaAs to AlAs. Since both lattice constants are essentially the same, very low lattice mismatch can be achieved for any composition of $Al_xGa_{1-x}As$. Lattice matching permits low defect density, high-quality materials to be grown that have good electronic and optical properties.

It quickly becomes apparent from Figure 71.8, however, that a requirement for lattice matching to the substrate greatly restricts the combinations of materials available to the device designer. For electron devices, the low-mismatch GaAs–AlAs alloys, GaSb–AlSb alloys, and ternary combinations GaAs–$Ga_{0.49}In_{0.51}P$ and InP–$In_{0.53}Ga_{0.47}As$–$In_{0.52}Al_{0.48}As$ alone are available. Efforts to utilize combinations such as GaP on Si or GaAs on Ge that lattice match have been generally unsuccessful because of problems with interface structure, polarization, and autodoping.

For several years, lattice matching was considered to be a necessary condition if mobility-damaging defects were to be avoided. This barrier was later broken when it was discovered that high-quality semiconductor materials could still be obtained although lattice-mismatched if the thickness of the mismatched layer is sufficiently small [11,12]. This technique, called *pseudomorphic* growth, opened another dimension in III–V device technology, and allowed device structures to be optimized over a wider range of bandgap for better electron or hole dynamics and optical properties.

Two of the pseudomorphic systems that have been very successful in high-performance millimeter-wave FETs are the InAlAs/InGaAs/GaAs and InAlAs/InGaAs/InP systems. The $In_xGa_{1-x}As$ layer is responsible for the high electron mobility and velocity both of which improve as the In concentration x is increased. Up to $x = 0.25$ for GaAs substrates and $x = 0.80$ for InP substrates have been demonstrated and result in great performance enhancements when compared to lattice-matched combinations [13].

InP substrates, however, are more expensive, smaller, and more easily broken than GaAs. Moreover, the 3.8% lattice mismatch would seem to be too great for direct epitaxy of $In_{0.53}Ga_{0.47}As$ on GaAs substrates. It has been demonstrated, however, that good-quality devices can be obtained using the *metamorphic* growth technique. A thick InP transition layer or a graded InGaP layer is grown directly upon a GaAs substrate. The defects caused by the lattice mismatch are largely contained in this layer, and low defect layers can be obtained when grown upon this transitional buffer layer. In addition, the higher thermal conductivity of InP is beneficial in reducing the thermal resistance of devices [3,4,14,15].

71.5 Conclusion

While the mainstream semiconductor device and circuit technology is defined by silicon and its related materials, the superior electron transport properties of compound semiconductor materials offer unique advantages in applications requiring the highest frequency, speed, and power. The range of possibilities with heterojunctions is far greater, and this has led to new device types with high-performance electronic and photonic applications.

This chapter has given a brief overview of the reasons why compound semiconductors perform differently than silicon. Chapter 72 will describe the physics and structure of these devices and present current performance examples. Chapter 73 will then illustrate specific circuit examples using these devices. Chapter 74 presents a technology-independent methodology for designing high-speed, high-frequency digital and analog circuits.

References

1. Ware, R., Higgins, W., O'Hearn, K., and Tiernan, M., "Growth and Properties of Very large Crystals of Semi-Insulating Gallium Arsenide," presented at the *18th IEEE GaAs IC Symposium*, Orlando, FL, 1996, p. 54.
2. www.cree.com.
3. Griffith, Z. et al., "InGaAs-InP metamorphic DHBTs grown on GaAs with lattice-matched device performance and fT, $f$max > 268 GHz," *IEEE Electron Dev. Lett.*, vol. 25, no. 10, pp. 675–677, 2004.
4. Griffith, Z. et al., "InGaAs/InP DHBTs with 120 nm collector having simultaneously high fT, $f$max > 450 GHz," *IEEE Electron Dev. Lett.*, vol. 26, no. 8, pp. 530–532, 2005.
5. Gelmont, B., Kim, K., and Shur, M., "Monte-Carlo simulation of electron transport in gallium nitride," *J. Appl. Phys.*, vol. 74, pp. 1818–1821, 1993.
6. F. Schwierz, "An electron mobility model for wurtzite GaN," *Solid State Elect.*, vol. 49, pp. 889–895, 2005.
7. R.J. Trew, "SiC and GaN Transistors: Is There One Winner for Microwave Power Applications?" *Proc. IEEE*, Special Issue on Wide-Bandgap Semiconductors, vol. 90, pp. 1032–1047, 2002.
8. Johnson, E.O., "Physical limitations on frequency and power parameters of transistors," *RCA Rev.*, vol. 26, no. 2, pp. 163–177, 1965.
9. Kroemer, H., "Heterostructures for everything: Device principles of the 1980's?," *Japanese J. Appl. Phys.*, vol. 20, no. 9, pp. 9–13, 1981.
10. Kroemer, H., "Heterostructure bipolar transistors and integrated circuits," *Proc. IEEE*, vol. 70, no. 13, pp. 13–25, 1982.
11. Matthews, J.W. and Blakeslee, A.E., "Defects in epitaxial multilayers, III. Preparation of almost perfect layers," *J. Crystal Growth*, vol. 32, p. 265, 1976.
12. Matthews, J.W. and Blakeslee, A.E., "Coherent strain in epitaxially grown films," *J. Crystal Growth*, vol. 27, p. 118, 1974.
13. Kok, Y.L. et al., "160–190 GHz monolithic low-noise amplifiers," *IEEE Microwave Guided Wave Lett.*, vol. 9, no. 8, pp. 311–313, 1999.
14. Hoke, W.E. et al., "Properties of metamorphic materials and devices," *Molecular Beam Epitaxy 2002 International Conference*, pp. 69–70, September 2002.
15. Schlechtweg, M. et al., "Millimeter-wave and mixed-signal integrated circuits based on advanced metamorphic HEMT technology," *16th International Conference on Indium Phosphide and Related Materials*, pp. 609–614, 2004.

72

Compound Semiconductor Devices for Analog and Digital Circuits

Donald B. Estreich
Agilent Technologies

CONTENTS

In this chapter we discuss the most important active devices used in RF and microwave circuits and applications. This includes devices based upon gallium arsenide (GaAs) and indium phosphide (InP) and the newly emerging gallium nitride (GaN) active devices. After a brief historical background we present the charge control viewpoint of active devices followed by a description of the operation of field-effect transistors (FETs) and bipolar junction (and heterojunction) transistors.

72.1 Historical Background of Compound Semiconductor RF and Microwave Devices

Germanium and silicon were the first semiconductors to be used in the making of transistors. Germanium had practical fabrication* and operational temperature limitations (from its smaller bandgap) and by the mid-1950s it was clear that silicon was the material of choice for the emerging field of semiconductor electronics. Today, silicon is by far the dominant semiconductor material fueling the electronics industry, in fact, since the late 1950s it has dominated. However, in the early 1950s speculation began about the possibility of using compound semiconductors for making electronic devices. The major obstacle in developing compound semiconductor devices has always been finding a source for material of sufficient quality, wafer size, and purity to build good devices. New materials are often intriguing because of different material parameters and the possibility of being able to build new or higher performance active devices.

Process technology and materials technology have played a critical role in the progress of both silicon and compound semiconductor devices. In the early 1950s the first suggestions [1] appeared that compound semiconductors were possible candidates for high-performance devices. At that time growing high-quality elemental semiconductor crystals such as germanium and silicon was far more advanced than growing compound semiconductors such as GaAs and InP. In fact, it took a long time to bring crystal growing technology for GaAs and InP to the point where reliable, high-performance devices could be repeatedly fabricated. Compound semiconductor device development benefited immensely from the equipment developed for the silicon integrated circuit (IC) industry. This is very significant because without the advanced processes and equipment for silicon, the emergence of compound semiconductor devices would have most certainly been significantly delayed. Progress in silicon device performance has been driven by process development; however, compound semiconductor device progress has been more of an exploitation of their material properties. Epitaxial growth by molecular beam epitaxy (MBE) and metal organic chemical vapor deposition (MOCVD) has been of great importance for the formation of heterojunctions and has led to bandgap engineering of many different and novel structures. This will be evident from the advanced FET structures described below.

It is interesting to view the historical context of compound semiconductor devices relative to that of the mainstream silicon devices. Table 72.1 lists some of the important milestones in the development of transistor concepts and silicon semiconductor devices and ICs. The FET concept actually dates back to 1926 (the filing date of Lilienfeld's patent [2]). It does not appear that Lilienfeld succeeded in building a working device however. Lilienfeld's concept was essentially today's GaAs metal-semiconductor FET (MESFET) which was not physically realized until 1966 by Mead [3] at the California Institute of Technology. The transistor was invented and first demonstrated in point contact form in 1947 at the Bell Telephone Laboratories by John Bardeen and Walter Brattain. William Shockley [4] followed this with the theory of the bipolar junction transistor (BJT) in 1948 with the related patent [5] being issued in 1951. It was in 1953 that the first silicon junction field-effect transistor (JFET) [6] was demonstrated at Bell Telephone Laboratories. Although Shockley was the first to suggest (one short paragraph) the possibility of using a wide bandgap emitter to form the heterojunction bipolar transistor (HBT) in his 1951 patent, Herbert Kroemer [7] in 1957 first presented a detailed and specific theory of the device that has become technologically important today.

From 1958 through 1961 considerable progress in silicon device was made with the development of the *planar process* [8,9] and invention of the IC [10]. The complementary metal-oxide-semiconductor (CMOS) circuit concept was put forth in 1963 at Fairchild Semiconductor [11]—CMOS presently

---

*Silicon had the advantage of being able to grow a high quality oxide that could be used for masking and passivation. Aluminum metal easily formed a very good ohmic contact with silicon which gave silicon a tremendous advantage in terms of fabrication technology. When compared to silicon, germanium had no chance of becoming the material of choice.

TABLE 72.1 Some Important Material and Device Milestones for Silicon Device Technology

Device Milestone	Year	Where
J. Lilienfeld's 1930 patent (filed 1926—proposed MESFET)	1930	Univ. Leipzig
J. Lilienfeld's 1933 patent (filed 1928—proposed TF MOSFET)	1933	Amrad
PN junction demonstrated (Russel Ohl)	1940	Bell Labs
Point contact transistor (J. Bardeen and W. Brattain)	1947	Bell Labs
Shockley bipolar transistor theory	1948	Bell Labs
p-n junction patent and wide bandgap emitter (W. Shockley)	1951	Bell Labs
First silicon crystal growth (Gordon Teal)	1952	TI
"Electronics in solid block" concept—IC (Geoffrey W.A. Drummer)	1952	unknown
Silicon Junction FET (G.C. Dacey and I.M. Ross)	1953	Bell Labs
Silicon oxide masking demonstrated	1954	Bell Labs
First commercial silicon transistor	1954	TI
Kroemer's HBT theory paper (*Proceedings of the IRE*—November)	1957	RCA
Integrated circuit invented (Noyce and Kilby)	1958	Fairchild/TI
Stable silicon dioxide (Atalla)	1959	Bell Labs
Planar process (J. Hoerni, R. Noyce, and K. Lehovec)	1959	FSC/Sprague
Al/SiO$_2$/silicon MOSFET (Kahng and Atalla)	1960	Bell Labs
First commercial planar process transistor—2N1613	1960	Fairchild
First commercial silicon integrated circuits	1961	TI & Fairchild
CMOS invented (F. Wanlass and C.T. Sah)	1963	Fairchild
First commercial silicon MOSFET	1964	FSC/RCA
"Moore's Law" suggested (Gordon Moore)	1965	Fairchild
Silicon-gate self-aligned MOSFET (Kerwin, Klein, and Sarace)	1967	Bell Labs
First silicon-germanium films grown with MBE (E. Kasper et al.)	1975	AEG
Silicon-germanium HBT demonstrated	1987	IBM

dominates IC applications[†] by a large margin and in the last several years has become a major contender for many high-volume RF applications. In the mid-1970s the first silicon–germanium (SiGe) films were grown using MBE [12]. This eventually led to IBM first demonstrating the SiGe HBT in 1987 [13]. SiGe HBT ICs have advanced so significantly since 1987 such that numerous high-performance RF and microwave ICs are now produced with this technology. It also means that silicon has entered the realm of compound semiconductors with its SiGe composition base region. The compound semiconductor SiGe HBT is compatible with silicon processes, hence, it has the advantage of being combined with CMOS circuitry and it leverages the huge infrastructure built for silicon processing.

Table 72.2 lists some key milestones in the development of compound semiconductor for high-frequency devices. The earliest work on GaAs devices was on light-emitting diodes and solid-state lasers. More important for this discussion is the first demonstrated GaAs active device, the Schottky-barrier MESFET [3]. Shortly thereafter microwave performance for a GaAs MESFET was demonstrated at Fairchild Semiconductor by Hooper and Lehrer [14]. This was followed by a number of companies starting work on GaAs MESFET development. The first GaAs monolithic microwave integrated circuits (MMICs) demonstrated in the 1970s were built with MESFETs. Today the GaAs MESFET is considered a mature technology with moderate performance and mostly built on 6 in. wafers making it the most cost-effective compound semiconductor technology.

In the 1970s very high electron mobilities were demonstrated in quantum well structures. Especially noteworthy is the pioneering work at Bell Telephone Laboratories by Dingle [15]. This led to the realization of the high electron mobility transistor[‡] (HEMT) by Takashi Mimura at Fujitsu [16,17]. Mimura used a gallium arsenide/aluminum gallium arsenide heterojunction to form his device. Investigations on pseudomorphic HEMT (p-HEMT) followed this work in 1983 at Sandia National Laboratories [18]. The

---

[†]Today CMOS is used for approximately 80–90% of all semiconductor shipments. Of course, this is largely because digital ICs dominate the volume when compared with analog ICs.

[‡]The HEMT has also been known as the modulation doped FET (MODFET), two-dimensional electron gas FET (TEGFET), or the selectively-doped FET (SDFET). HEMT now seems to be the preferred name.

TABLE 72.2 Some Important Material and Device Milestones for GaAs, InP, and GaN Compound Semiconductors

Device Milestone	Year	Where
V.M. Goldschmidt first creates GaAs material	1929	
Heinrich Welker points out potential of III–V Semiconductors	1952	Siemens
First GaAs infrared LED (M.I. Nathan et al.)	1962	IBM
First GaAs semiconductor laser (Hall and Redhiker)	1962	GE & MIT LL
GaAs Cr-doped semi-insulating substrate (Cronin and Haisty)	1964	TI
GaAs Gunn Diode (J.B. Gunn)	1964	IBM
GaAs MESFET demonstrated (Carver Mead)	1966	CalTech
GaAs MESFET with microwave performance (Hooper and Lehrer)	1967	Fairchild
AlGaAs/GaAs VPE Heterojunction transistor (Dumke et al.)	1972	IBM
Dingle et al. show high mobility with Modulation Doping	1978	Bell Labs
AlGaAs/GaAs HEMT (also known as MODFET or SDHT or TEGFET)	1980	Fujitsu
GaAs heterojunction transistor (Asbeck et al.)	1981	Rockwell
GaAs heterojunction transistor (Yuan et al.)	1982	TI
GaAs pseudomorphic HEMT (Zipperian et al.)	1983	Sandia
Vertical InP HBT (Tabatabaie-Alavi et al.)	1983	Bell Labs
AlGaN/GaN HEMT (Khan et al.)	1994	APA Optics

early 1980s saw a great amount of work at universities and industry on HEMT-like structures such as the use of indium phosphide/indium gallium arsenide heterojunctions to build InP-based HEMTs [19]. The InP HEMT has demonstrated the highest frequency performance and lowest noise figures. The next major step in HEMT evolution has been the application of GaN and aluminum gallium nitride/gallium nitride heterojunctions to produce HEMT structures capable of much higher output powers. The first GaN-based HEMT was demonstrated by Khan in 1994 [20].

We now return to the development of bipolar transistor. The first reported GaAs-based HBT was grown by Dumke, Woodall, and Rideout (IBM) [21] in 1972 using liquid phase epitaxy. The first practical HBTs used in IC demonstrations were in the early 1980s by Asbeck and his team at Rockwell [22,23] and Yuan at Texas Instruments [24]. The Rockwell work focused upon emitter-coupled logic ICs, whereas the Texas Instruments work produced a gate array with integrated-injection logic (also known as HI^{2}L) cells, both with GaAs-based HBT devices. The HBT today has reached very high performance with all the advantages of the bipolar device over the unipolar device. Silicon–germanium HBTs are competing with GaAs and InP HBTs for capturing application sockets. Before discussing specific performance details, we next present the charge control viewpoint on active devices.

72.2 Unifying Principle for Active Devices: Charge Control Principle

An active device is an electron device such as a transistor capable of delivering power amplification by converting DC bias power into time-varying signal power. It delivers a greater energy to its load than if the device were absent. The charge control framework [25–27] discussed below presents a unified understanding of the operation of all electron devices and simplifies the comparison of the several active devices used in compound semiconductor analog and digital ICs.

Consider a generic electron device as represented in Figure 72.1. It consists of three electrodes encompassing a charge transport region. The transport region is capable of supporting charge flow (electrons as shown in the figure) between an emitting electrode and a collecting electrode. A third electrode, called the control electrode, is used to establish the electron concentration within the transport region. Placing a control charge, Q_C, on the control electrode establishes a controlled charge, denoted as $-Q$, in the transport region. The operation of active devices depends upon the charge control principle [25]:

Each charge placed upon the control electrode can at most introduce an equal and opposite charge in the transport region between the emitting and collecting electrodes.

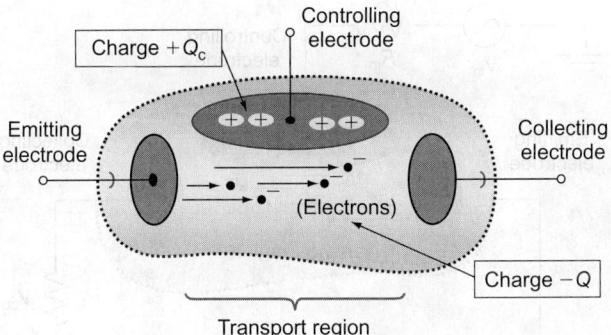

FIGURE 72.1 Generic charge control device consisting of three electrodes embedded around a charge transport region.

At most we have the relationship, $|-Q| = |Q_C|$. Any parasitic coupling of the control charge to charge on the other electrodes or remote parts of the device will decrease the controlled charge in the transport region, that is, $|-Q| < |Q_C|$ more generally. For example, charge coupling between the control electrode and the collecting electrode forms a feedback or output capacitance, say C_o. Time variation of Q_C leads to the modulation of the current flow between the emitting and collecting electrodes.

The generic structure in Figure 72.1 could represent any one of a number of active devices (e.g., vacuum tubes, unipolar transistors, bipolar transistors, and photoconductors). Hence, charge control analysis is very broad in scope and it applies to all electronic transistors.

Starting with the charge control principle, we associate two characteristic time constants with an active device, thereby, leading to a first-order description of its behavior. Application of a potential difference between the emitting and collecting electrodes, say V_{CE}, establishes an electric field in the transport region, although this applied field is not always needed when diffusion or internal fields from doping profiles are effective. Electrons in the transport region respond to the electric field and move across this region with a transit time τ_r. The transit time[§] is the first of the two important characteristic times used in charge control modeling. With charge $-Q$ in the transit region, the static (DC) current I_o between the emitting and collecting electrodes is

$$I_o = -Q/\tau_r = Q_c/\tau_r \tag{72.1}$$

A simple interpretation of τ_r is as follows: τ_r is equal to the length l of the transport region divided by the average velocity of transit (i.e., $\tau_r = l/\langle v \rangle$). From this perspective a charge of $-Q$ (coulombs) is swept out of the collecting electrode every τ_r seconds.

Consider Figure 72.2 showing the common-emitting electrode connection of the active device of Figure 72.1 connected to input and output (i.e., load) resistances, say R_{in} and R_L, respectively. The second characteristic time of importance can now be defined—it is the "lifetime" time constant and we denote it by symbol τ. It is a measure of how long a charge placed on the control electrode will remain on the control terminal. The lifetime time constant is established in one of several ways depending upon the physics of the active device and its connection environment. The controlling charge may "leak away" by (1) discharging through the external resistor R_{in} as it typically happens with FET devices, (2) recombining with intermixed oppositely charged carriers within the device (e.g., base recombination in a bipolar transistor), or (3) discharging through an internal shunt leakage path within the device. The DC current flowing to replenish the lost control charge is

$$I_{in} = -Q/\tau = Q_c/\tau \tag{72.2}$$

[§]The transit time τ_r is best interpreted as an average transit time per carrier (in our case, the electron). We note that $1/\tau_r$ is common to all devices—it is related to a device's ultimate capability of processing information.

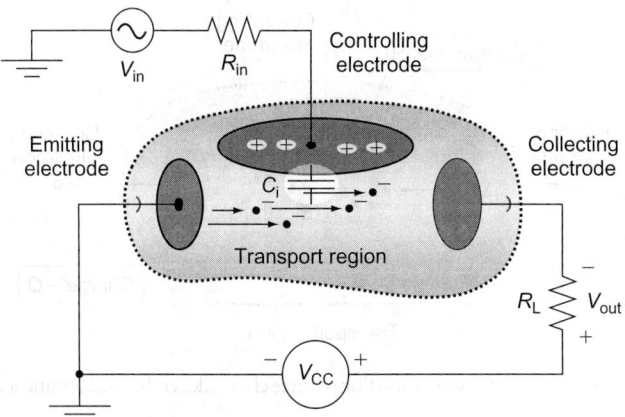

FIGURE 72.2 Generic charge control device of Figure 72.1 connected to input and output resistors, R_{in} and R_L, respectively, with bias voltage and input signal applied.

The static (DC) current gain G_I of a device is defined as the current delivered to the output divided by the current replenishing the control charge during the same time period, where in τ seconds charge $-Q$ is both lost and replenished, and charge Q_c times the ratio τ/τ_r is supplied to the output resistor R_L. In symbols, the static current gain is

$$G_I = I_o/I_{in} = \tau/\tau_r \tag{72.3}$$

provided $|-Q| = |Q_C|$ holds.

In the dynamic case the process of small-signal amplification consists of an incremental variation of the control charge Q_c directly resulting in an incremental change in the controlled charge, $-Q$. The resulting variation in output current flowing in the load resistor translates into a time-varying voltage v_o. The charge control formalism holds just as well for large-signal situations. In the large-signal case the changes in control charge are no longer small incremental changes. Charge control analysis under large charge variations is less accurate owing to the simplicity of the model, but still very useful for approximate switching calculations in digital circuits.

An important dynamic parameter is the input capacitance C_i of the active device. Capacitance C_i is a measure of the work required to introduce a charge carrier in the transport region. Capacitance C_i is given by the change in charge Q for a corresponding change in input voltage v_{in}. It is desirable to maximize C_i in an active device. The transconductance g_m is calculated from

$$g_m = \left(\frac{\partial I_o}{\partial v_{in}} \right)_{v_o} = \left(\frac{\partial I_o}{\partial Q} \right) \left(\frac{\partial Q}{\partial v_{in}} \right) \tag{72.4}$$

The first partial derivative on the right-hand-side of Eq. (72.4) is simply $(1/\tau_r)$ and the second partial derivative is C_i. Hence, the transconductance g_m is the ratio

$$g_m = \frac{C_i}{\tau_r} \tag{72.5}$$

A physical interpretation of g_m is the ratio of the work required to introduce a charge carrier to the average transit time of a charge carrier in the transport region. The transconductance is one of the most commonly used device parameters in circuit design and analysis.

In addition to C_i another capacitance, say C_o, is introduced and associated with the collecting electrode. Capacitance C_o accounts for charge on the collecting electrode coupled to either static charge in the

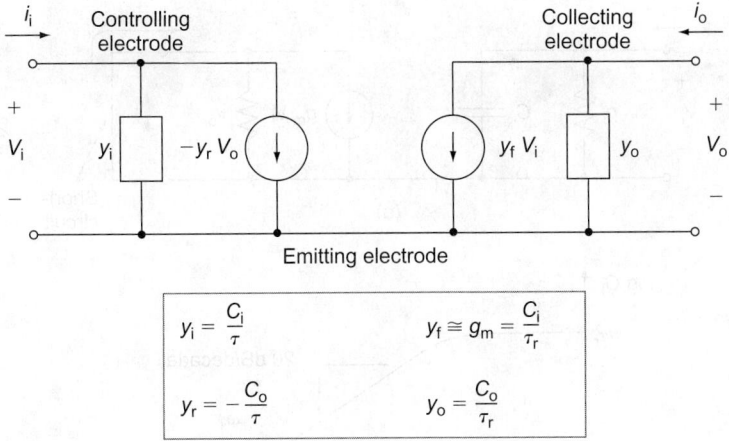

FIGURE 72.3 Two-port, small-signal, admittance charge control model with the emitting electrode selected as the common terminal to both input and output.

transport region or charge on the control electrode. A nonzero C_o indicates that the coupling between the controlling electrode and the charge in transit is less than unity (i.e., $|-Q| < |Q_C|$).

For small-signal analysis the capacitance parameters usually take fixed numbers evaluated about the device's bias state. When using charge control in the large-signal case, the capacitance parameters must include the voltage dependencies. For example, the input capacitance C_i can be strongly dependent upon the control electrode to emitting electrode and collecting electrode potentials. Hence, during the change in bias state within a device the magnitude of the capacitance C_i is time varying. This variation can dramatically affect the switching speed of the active device. Parametric dependencies upon the instantaneous bias state of the device are at the heart of accurate modeling of large-signal or switching behavior of active devices.

We introduce the small-signal admittance charge control model shown in Figure 72.3. This model uses the emitting electrode as the common terminal in a two-port connection. The transconductance g_m is the magnitude of the real part of the forward admittance y_f and is represented as a voltage-controlled current source positioned from the collecting to emitting electrode. The input admittance, denoted by y_i, is equivalent to (C_i/τ), where τ is the control charge 'lifetime' time constant. Parameter y_i can be expressed in the form $(g_i + sC_i)$, where $s = j\omega$. An output admittance, similarly denoted by y_o, is given by (C_o/τ_r) where τ_r is the transit time and, in general $y_o = (g_o + sC_o)$. Finally, the output-to-input feedback admittance y_r is included using a voltage-controlled current source at the input. Often y_r is small enough to approximate as zero (the model is then said to be unilateral).

Consider the frequency dependence of the dynamic (AC) current gain G_i. The low-frequency current gain is interpreted as follows: an incremental charge q_c is introduced on the control electrode with lifetime τ. This produces a corresponding incremental charge $-q$ in the transport region. Charge $-q$ is swept across the transport region every "transit time" τ_r seconds. In time τ, charge $-q$ crosses the transit region τ/τ_r times, which is equal to the low-frequency current gain.

The lifetime τ associated with the control electrode arises from charge "leaking off" the controlling electrode. This is modeled as an RC time constant at the input of the equivalent circuit shown in Figure 72.4(a) with τ equal to $R_{in}C_i$. The break frequency ω_B associated with the control electrode is

$$\omega_B = \frac{1}{\tau} = \frac{1}{R_{in}C_i} \tag{72.6}$$

When the charge on the control electrode varies at a rate ω less than ω_B, G_i is given by τ/τ_r because charge leaks off the controlling electrode faster than $1/\omega$. Alternatively, when ω is greater than ω_B, G_i decreases

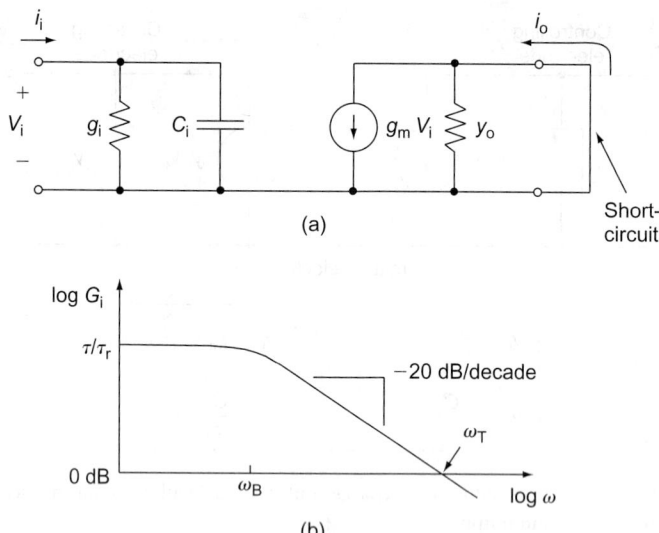

(a)

(b)

FIGURE 72.4 (a) Small-signal admittance model with output short-circuited, and (b) magnitude of the small-signal current gain G_i plotted as a function of frequency. The unity current gain crossover (i.e., $G_i = 1$) defines the parameter f_T (or ω_T).

with increasing ω because the applied signal charge varies upon the control electrode more rapidly than $1/\tau$. In this case G_i is inversely proportional to ω,

$$G_i = \frac{1}{\omega \tau_r} = \frac{\omega_T}{\omega} \tag{72.7}$$

where ω_T is the common-emitter unity current gain frequency. At $\omega = \omega_T \ (=2\pi f_T)$ the AC current gain equals unity as illustrated in Figure 72.4(b).

Consider the current gain-bandwidth product $G_i \Delta f$. Purely capacitive input impedance cannot define a bandwidth. However, a finite real impedance always appears at the input terminal in any practical application. Let R_i be the effective input resistance if the device (i.e., R_i will be equal to $[1/g_i]$) is parallel to the external resistance R_{in}. Since the input current is equal to q_c/τ and the output current is equal to q/τ_r, the current gain-bandwidth product becomes

$$G_i \Delta f = \frac{q/\tau_r \, \omega}{q_c /\tau \, 2\pi} \tag{72.8}$$

For $\omega \gg \omega_B$, at $\tau = 1/\omega$, and assuming $|q_c| = |-q|$,

$$G_i \Delta f = \frac{1}{2\pi\tau_r} = \frac{\omega_T}{2\pi} = f_T \tag{72.9}$$

f_T (or ω_T) is a widely quoted parameter used to compare or "benchmark" active devices. Sometimes f_T (or ω_T) is interpreted as a measure of the maximum speed a device can drive a replica of itself. It is easy to compute and historically has been easy to measure with bridges and later using S parameters. However, f_T does have interpretative limitations because it is defined as current into a short-circuit output. Hence, it ignores input resistance and output capacitance effects upon actual circuit performance.

TABLE 72.3 Charge Control Relations for All Active Devices [25]

Parameter	Symbol	Expression
Transconductance	g_m	$\dfrac{C_i}{\tau_r} \Leftrightarrow \omega_T C_i$
Current amplification	G_i	$\dfrac{1}{\omega \tau_r} \Leftrightarrow \dfrac{\omega_T}{\omega}$
Voltage amplification	G_v	$\dfrac{1}{\omega \tau_r} \dfrac{C_i}{C_o} \Leftrightarrow \dfrac{\omega_T}{\omega} \dfrac{C_i}{C_o}$
Power amplification	$G_p = G_i G_v$	$\dfrac{1}{\omega^2 \tau_r^2} \dfrac{C_i}{C_o} \Leftrightarrow \dfrac{\omega_T^2}{\omega^2} \dfrac{C_i}{C_o}$
Current gain-bandwidth product	$G_i \cdot \Delta f$	$\dfrac{1}{\tau_r} \Leftrightarrow \omega_T$
Voltage gain-bandwidth product	$G_v \cdot \Delta f$	$\dfrac{1}{\tau_r} \dfrac{C_i}{C_o} \Leftrightarrow \omega_T \dfrac{C_i}{C_o}$
Power gain-bandwidth product	$G_p \cdot \Delta f^2$	$\dfrac{1}{\tau_r^2} \dfrac{C_i}{C_o} \Leftrightarrow \omega_T^2 \dfrac{C_i}{C_o}$

Note: $R_i C_i = R_o C_o$.

Likewise, voltage and power gain expressions can be derived. It is necessary to define the output impedance before either can be quantified. Let R_o be the effective output resistance at the output terminal of the active device. Assuming both the input and output RC time constants to be identical, that is, $R_i C_i = R_o C_o$. The voltage gain G_v can be expressed in terms of G_i,

$$G_v = G_i \frac{R_o}{R_i} = G_i \frac{C_i}{C_o} \tag{72.10}$$

where R_o is the parallel equivalent output resistance from all resistances at the output node.

The power gain G_p is computed from the product of $G_i G_v$ along with the power gain-bandwidth product. These results are listed in Table 72.3 as summarized from Johnson and Rose [25]. These simple expressions are valid for all devices as interpreted from the charge control perspective. They provide for a first-order comparison, in terms of a few simple parameters, among the active devices commonly available. From an examination of Table 72.3, it is evident that maximizing C_i and minimizing τ_r leads to higher transconductance, higher parametric gains, and greater frequency response. This is an important observation in understanding how to improve upon the performance of any active device.

Whereas f_T has limitations, the frequency at which the maximum power gain extrapolates to unity, denoted by f_{max}, is often a more useful indicator of device performance. The primary limitation of f_{max} is that it is very difficult to calculate accurately and is usually extrapolated from S-parameter measurements in which the extrapolation is approximate at best.

72.3 Comparing Unipolar Transistors and Bipolar Transistors

Unipolar transistors are active devices that operate using only a single charge carrier type, usually electrons, in their transport region. FETs fall into the unipolar classification. FETs utilizing channel electrons are referred to as n channel, whereas those using holes are referred to as p channel. In contrast, bipolar transistors depend upon positively and negatively charged carriers, that is, both majority and minority carriers, within the transport region. A fundamental difference arises from the relative locations

TABLE 72.4   Family Tree Showing Two Major Classes of Transistors: (a) Unipolar Transistors (commonly known as FETs) and (b) Bipolar Transistors (BJT and HBT)

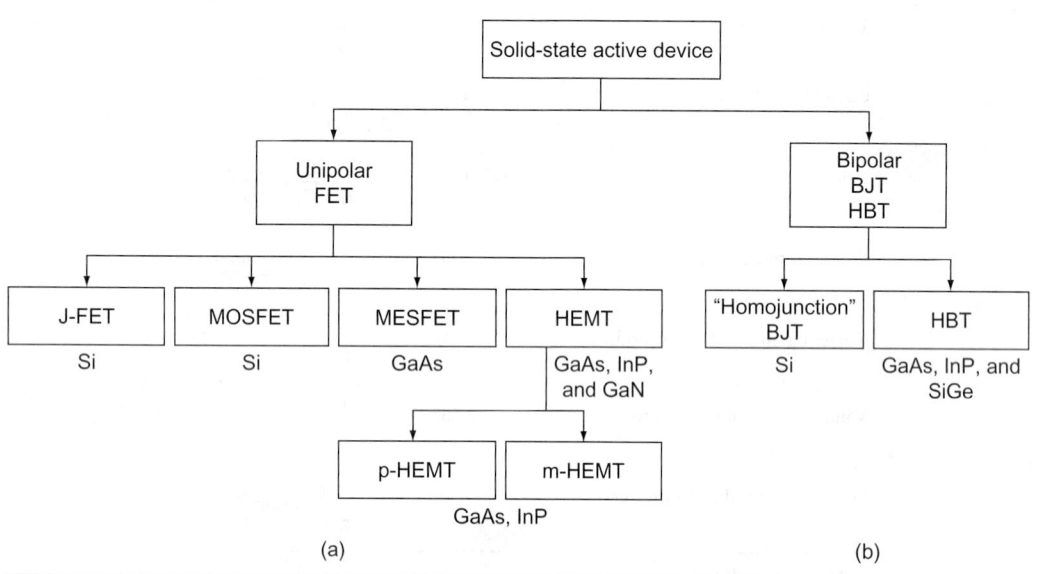

of the control electrode and transport region—in unipolar devices they are physically separated (i.e., gate and channel), whereas in bipolar devices, they merge into the same physical region (i.e., the base region). Bipolar transistors depending upon electron transport within the base region are called npn transistors, but those depending upon hole transport are called pnp transistors. The notation of "npn" and "pnp" refers to the doping type of the emitter–base–collector regions, respectively. There are two types of bipolar transistors—BJT and HBT. The homojunction transistor depends upon doping level differences to minimize minority carrier injection into the emitter (and provide high current gain) while the hetero-junction utilizes bandgap differences to achieve this. Table 72.4 presents an inverted-tree classification scheme for transistor active devices, showing the unipolar and bipolar branches along with a further breakdown by transistor structure. Before reviewing the physical operation of each, transport in semiconductors is briefly reviewed.

72.3.1   Charge Transport in Semiconductors

Bulk semiconducting materials are useful because their conductivity can be controlled over many orders of magnitude by changing the doping level. Both electrons and holes [28–30] can conduct current in semiconductors and this is very important for bipolar transistors. In the formation of ICs, various metal, semiconductor, and insulator layers are used together in precisely positioned shapes, sizes, and thicknesses to form useful device and circuit functions. It is the ability to form unique combinations of layers with wide ranges of conductivities that has been so incredibly useful in advanced electronics.

Figure 72.5 illustrates the behavior of electron velocity as a function of local electric field strength for several important semiconducting materials. Two characteristic regions of behavior can be identified: a linear or ohmic region at low electric fields, and a velocity saturated region at high fields. At low fields, current transport is proportional to the carrier's mobility. Mobility is a measure of how easily carriers move through a material [28]. At high fields carriers saturate in velocity, hence, current levels will correspondingly saturate in active devices. The data in Figure 72.5 assumes low doping levels (i.e., $N_x < 10^{15}$ cm^{-3}). The dashed curve represents transport in a GaAs quantum well formed adjacent to an $Al_{0.3}Ga_{0.7}As$ layer—in

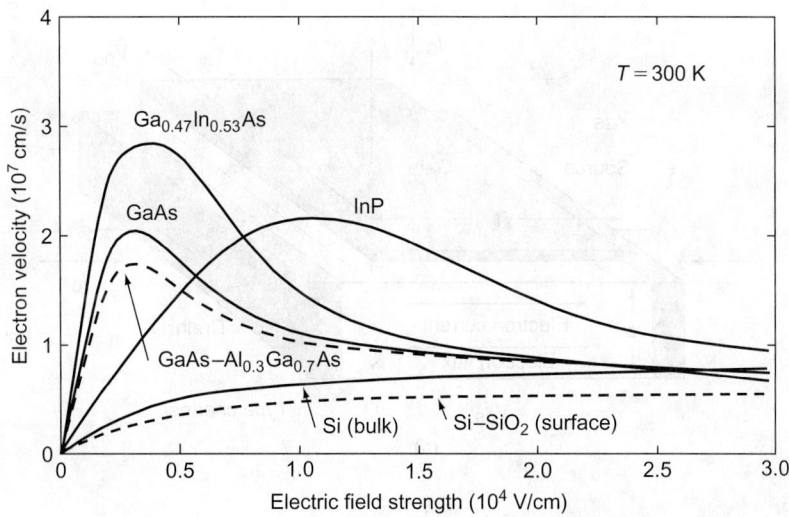

FIGURE 72.5   Electron velocity versus electric field plot for several semiconductor materials.

TABLE 72.5   Semiconductor Material Parameters at a Temperature of 300 K and in the "Weak Doping" Limit

Semiconductor	E_G (eV)	ε_r	Electron Mobility (cm²/V s)	Hole Mobility (cm²/V s)	Peak Electron Velocity (cm/s)
Si (bulk)	1.12	11.7	1,450	450	N.A.
Ge	0.66	15.8	3,900	1,900	N.A.
InP	1.35 D	12.4	4,600	150	2.1×10^7
GaAs	1.42 D	13.1	8,500	400	2×10^7
$Ga_{0.47}In_{0.53}As$	0.78 D	13.9	11,000	200	2.7×10^7
InAs	0.35 D	14.6	22,600	460	4×10^7
$Al_{0.3}Ga_{0.7}As$	1.80 D	12.2	1,000	100	—
AlAs	2.17	10.1	280	—	—
$Al_{0.48}In_{0.52}As$	1.92 D	12.3	800	100	—
GaN	3.49 D	9.0	1,500	30	2×10^7

Note: Symbol D indicates direct bandgap; all others are indirect bandgap.

this case interface scattering lowers the mobility. A similar situation is found for transport in silicon at a semiconductor–oxide interface such as found in metal-oxide–semiconductor (MOS) devices.

Several general conclusions can be extracted from this data:

1. Compound semiconductors generally have higher electron mobilities than silicon or silicon–germanium compounds.
2. At high fields (for example, $E > 20,000$ V/cm) saturated electron velocities tend to converge to about 1×10^7 cm/s.
3. Many compound semiconductors show a transition region between low and high electric field strengths with a negative differential mobility owing to electron transfer from the Γ ($k = 0$) valley to conduction band valleys with higher effective masses (by the way, this gives rise to Gunn Effect [31]).

Hole mobilities are much lower than electron mobilities in all semiconductors. Saturated velocities of holes are also lower at higher electric fields. This is why n-channel FETs show better performance than the p-channel ones, and why npn bipolar transistors show better performance than pnp bipolar transistors. Table 72.5 compares electron and hole mobilities for several semiconducting materials.

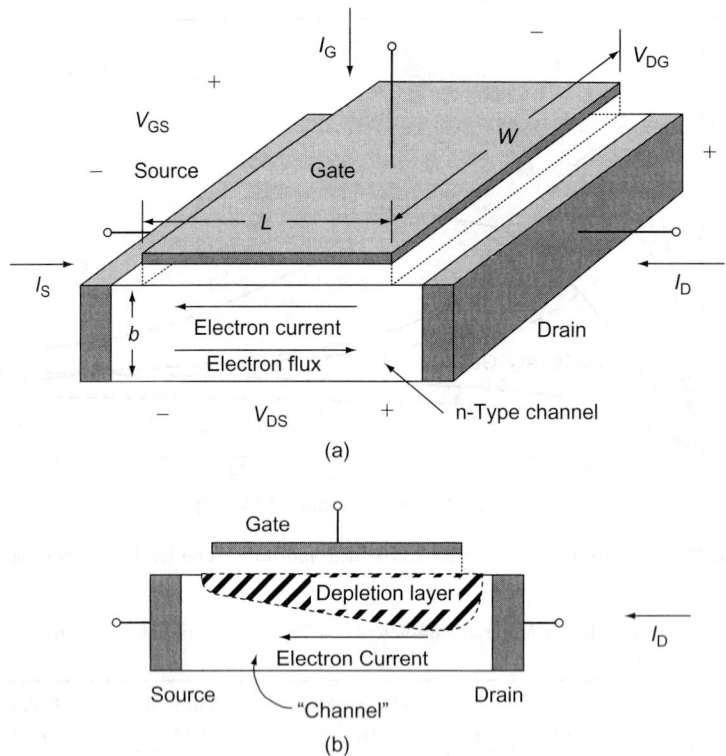

FIGURE 72.6   (a) Conceptual view of a FET with the channel sandwiched between source and drain ohmic contacts and a gate control electrode in close proximity; (b) cross-sectional view of the FET with a depletion layer shown such as would be present in a compound semiconductor MESFET.

72.3.2   Field-Effect (Unipolar) Transistors

Figure 72.6(a) shows a conceptual view of an n-channel FET [32–34]. As shown the n-type channel is a homogeneous semiconducting material of thickness b, with electrons supporting the drain-to-source current. A p-type channel would rely on mobile holes for current transport and all voltage polarities would be exactly reversed from those shown in Figure 72.6(a). The control charge in the gate region (of length L and width W) establishes the number of conduction electrons per unit area in the channel by electrostatic attraction or exclusion. The cross section on the FET channel in Figure 72.6(b) shows a depletion layer, a region devoid of electrons, as an intermediary agent between the control charge and the controlled charge. This depletion region is present in the JFET and Schottky barrier junction (this is a metal–semiconductor junction) MESFET structures.

In all FET structures the gate is physically separated from the channel. By physically separating the control charge from the controlled charge, the gate-to-channel impedance can be very large at low frequencies. The gate impedance is predominantly capacitive and, typically, very low gate leakage currents are observed in high-quality FETs. This is a distinguishing feature of the FET—the high input impedance is desirable for many circuit applications.

The FET's channel, positioned between the source and drain ohmic contacts, forms a resistor whose resistance is modulated by the applied gate-to-channel voltage. We know that the gate potential controls the channel charge by the charge control relation. Time variation of the gate potential translates into a corresponding time variation of the drain current (as well as the source current). Therefore, transconductance g_m is the natural parameter to describe the FET from this viewpoint.

Figure 72.7(a) shows the I_D–V_{DS} characteristic of the n-channel FET in the common-source connection with constant electron mobility and a long channel assumed. Two distinct operating regions appear in

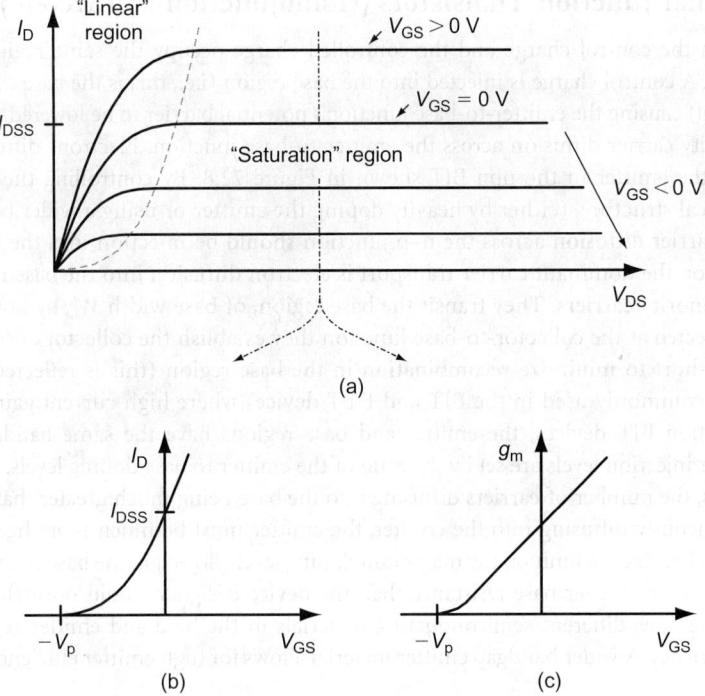

FIGURE 72.7   (a) FET drain current (I_D) versus drain-to-source voltage (V_{DS}) characteristic with the gate-to-source voltage (V_{GS}) as a stepped parameter; (b) I_D versus V_{GS} "transfer curve" for a constant V_{DS} in the saturated region of operation, revealing its "square-law" behavior; (c) transconductance g_m versus V_{GS} for a constant V_{DS} in saturated region of operation corresponding to the transfer curve in part (b). These curves assume constant mobility, no velocity saturation, and the "long-channel FET approximation."

Figure 72.7(a)—the linear (i.e., nonsaturated) region and the saturated region, separated by the dashed parabola. The origin of current saturation corresponds to the onset of channel pinch-off owing to carrier exclusion at the drain end of the channel. Pinch-off occurs when the drain voltage is positive enough to deplete the channel completely of electrons at the drain end—this corresponds to a gate-to-source voltage equal to the pinch-off voltage, denoted as $-V_p$ in Figure 72.7(b) and Figure 72.7(c). For constant V_{DS} in the saturated region, the I_D versus V_{GS} transfer curve approximates the "square law" behavior; mathematically this can be approximated by

$$I_D = I_{DSS}\left(1 - \frac{V_{GS}}{(-V_p)}\right)^2 \quad \text{for } -V_p \le V_{GS} \le \varphi \tag{72.11}$$

where I_{DSS} is the drain current when $V_{GS} = 0$, and φ is a built-in potential associated with the gate-to-channel junction or interface (e.g., a metal–semiconductor Schottky barrier as in the MESFET). The symbol $I_{D,sat}$ denotes the drain current in the saturated region of operation. Transconductance g_m is linear with V_{GS} for the saturation transfer characteristic of Eq. (72.11) and is approximated by

$$g_m = \frac{\partial I_D}{\partial V_{GS}} \cong 2\frac{I_{DSS}}{V_p}\left(1 - \frac{V_{GS}}{(-V_p)}\right) \quad \text{for } -V_p \le V_{GS} \le \varphi \tag{72.12}$$

Eq. (72.11) and Eq. (72.12) are plotted in Figure 72.7(b) and Figure 72.7(c), respectively.

72.3.3   Bipolar Junction Transistors (Homojunction and Heterojunction)

In the BJT both the control charge and the controlled charge occupy the same region (i.e., the base region) [35–37]. A control charge is injected into the base region (i.e., this is the base current flowing in the base terminal) causing the emitter-to-base junction's potential barrier to be lowered. Barrier lowering results in majority carrier diffusion across the emitter-to-base junction. Electrons diffuse into the base and holes into the emitter in the npn BJT shown in Figure 72.8. By controlling the emitter-to-base junction's physical structure, (either by heavily doping the emitter or using a wider bandgap emitter) the dominant carrier diffusion across the n–p junction should be injection into the base region. For the npn transistor, the dominant carrier transport is electron diffusion into the base region where the electrons are minority carriers. They transit the base region, of base width W_b, by both diffusion and drift. When collected at the collector-to-base junction they establish the collector current I_C. The base width must be short to minimize recombination in the base region (this is reflected in the current gain parameter commonly used in the BJT and HBT devices where high current gain is desirable).

In homojunction BJT devices, the emitter and base regions have the same bandgap energy. The respective carrier injection levels are set by the ratio of the emitter to base doping levels. For high emitter efficiency, that is, the number of carriers diffusing into the base being much greater than the number of carriers simultaneously diffusing into the emitter, the emitter must be much more heavily doped than the base region. This places a limit on the maximum doping level allowed in the base of the homojunction BJT, thereby leading to higher base resistance than the device designer would normally desire [35]. In contrast, the HBT uses different semiconducting materials in the base and emitter regions to achieve high emitter efficiency. A wider bandgap emitter material allows for high emitter efficiency while allowing

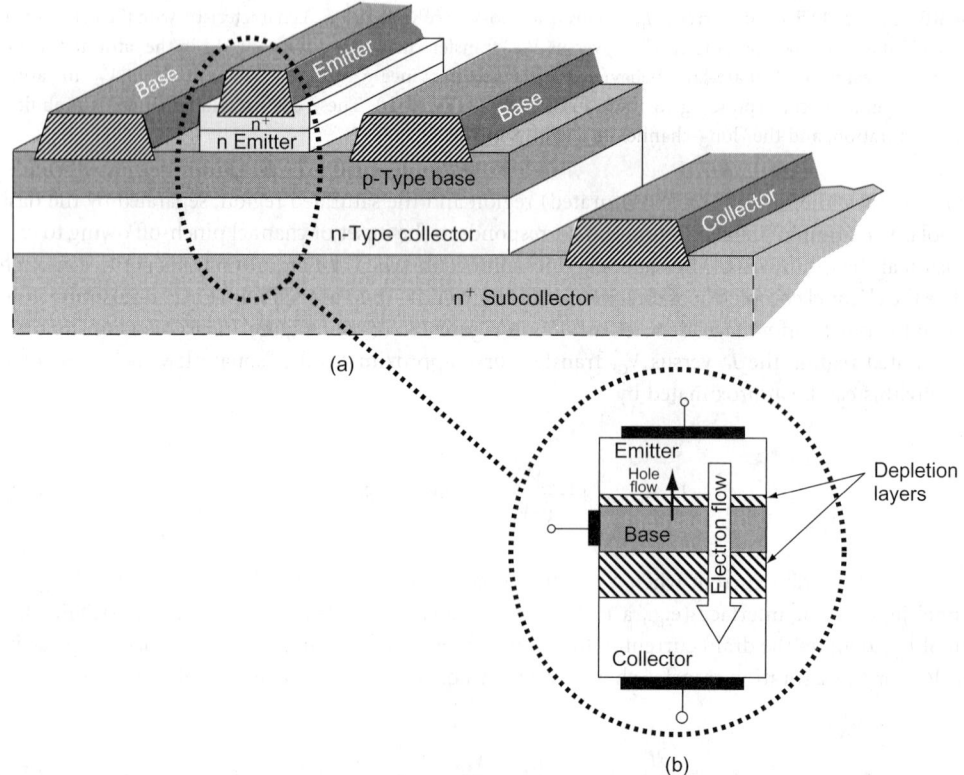

FIGURE 72.8   (a) Conceptual view of a bipolar junction transistor with the base region sandwiched between emitter and collector regions. Structure is representative of a compound semiconductor heterojunction bipolar transistor; (b) simplified cross-sectional view of a vertically structured BJT device with primary electron flow represented by large arrow.

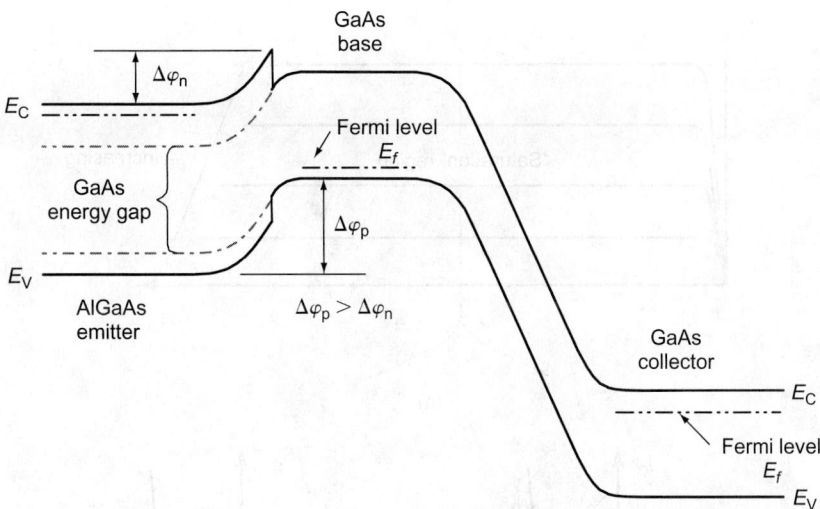

FIGURE 72.9   The bandgap diagram for an HBT AlGaAs/GaAs device with the wider bandgap for the AlGaAs emitter (solid line) compared to a homojunction GaAs BJT emitter (dot–dash line). The double dot–dashed line represents the Fermi level in each region as it was positioned in equilibrium.

for higher base doping levels which in turn lower the parasitic base resistance. An example of a wider bandgap emitter transistor is shown in Figure 72.9. In this example, the emitter is AlGaAs whereas the base and collector are formed with GaAs. Figure 72.9 shows the band diagram under normal operation with the emitter–base junction forward biased and the collector–base junction reverse biased. The discontinuity in the valence band edge at the emitter–base heterojunction is the origin of the reduced diffusion into the emitter region. The injection ratio determining the emitter efficiency depends exponentially upon this discontinuity. If ΔE_g is the valence band discontinuity, the injection ratio is proportional to the exponential of ΔE_g normalized to the thermal energy kT,

$$\frac{J_n}{J_p} \propto \exp(-\Delta E_g / kT) \tag{72.13}$$

For example, when ΔE_g is equal to $8kT$, an exponential factor of approximately 8000 is obtained, thereby leading to an emitter efficiency of nearly unity as desired. The use of the emitter-base band discontinuity is a very efficient way to hold high emitter efficiencies. The exponential factor in Eq. (72.13) is the enhancement factor for the minority carrier density in the base region, the collector current density, the current gain of the device, and its current gain × early voltage "figure of merit".¶

In bipolar devices the collector current I_C is given by the exponential of the base-emitter forward voltage V_{BE} normalized to the thermal voltage, kT/q:

$$I_C = I_S \exp(qV_{BE}/kT) \tag{72.14}$$

The saturation current I_S is given by a quantity that depends upon the structure of the device; it is inversely proportional to the base doping charge Q_{BASE} and proportional to the device's area A, namely

$$I_S = \frac{qADn_i^2}{Q_{BASE}} \tag{72.15}$$

¶The early voltage is a measure of the flatness of the collector current versus collector-emitter voltage, or how constant the output conductance is over output voltage excursions. The higher the value of the early voltage, the lower the output conductance (or higher output resistance).

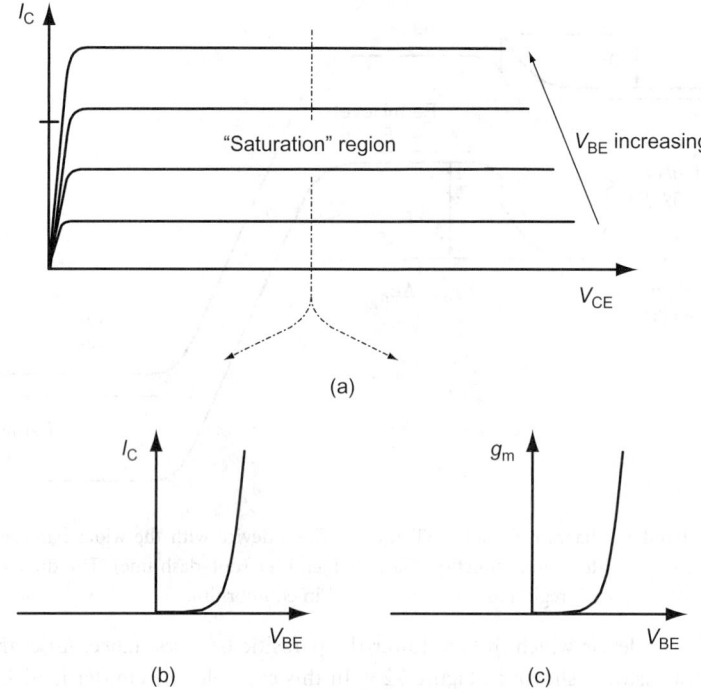

FIGURE 72.10 (a) Collector current (I_C) versus collector-to-emitter voltage (V_{CE}) characteristic curves with the base-to-emitter voltage (V_{BE}) as stepped parameter; (b) I_C versus V_{BE} "transfer curve" for a constant V_{CE} in saturated region of operation shows exponential behavior; (c) transconductance g_m versus V_{BE} for a constant V_{CE} in the saturated region of operation corresponding to the transfer curve in part (b).

where the other symbols have their usual meanings (D is the minority carrier diffusion constant in the base, n_i is the intrinsic carrier concentration of the semiconductor, and q is the electron's charge). In actual devices often junction edge defects result in leakage currents that deviate from Eq. (72.15) in the reverse-biased state.

A typical collector current versus collector-emitter voltage characteristic, for several increasing values of (forward-biased) emitter-base voltages, is shown in Figure 72.10(a). Note the similarity to Figure 72.7(a), with the BJT having a quicker turn-on for low V_{CE} values compared with the softer knee for the FET at comparable V_{DS} values. The transconductance of the BJT and HBT is found by taking the derivative of Eq. (72.14), thus

$$g_m = \frac{\partial I_C}{\partial V_{BE}} = \frac{qI_S}{kT} \exp(qV_{BE}/kT) \tag{72.16}$$

Both I_C and g_m are of exponential form as observed in Figure 72.10—Eq. (72.14) and Eq. (72.16) are plotted in Figure 72.10(b) and Figure 72.10(c), respectively. The transconductance of the BJT/HBT is generally much larger than that of the best FET devices (this can be verified by comparing Eq. (72.12) with Eq. (72.16) with typical parameter values inserted). This has significant circuit design advantages for the BJT/HBT devices over the FET devices because high transconductance is needed for high current drive to charge load capacitance in digital circuits. In general, higher g_m values allow a designer to use feedback to a greater extent in design and this provides for greater tolerance to process variations.

72.3.4 Comparing FET and BJT/HBT Parameters

Table 72.6 compares some of the more important features and parameters of the BJT/HBT device with these of the FET device. For reference, the common-source FET configuration is compared with the

TABLE 72.6 Comparison of Electrical Parameters of the BJT/HBT and the FET

Parameter	BJT/HBT	FET
Input impedance Z	Low Z owing to forward biased junction—large diffusion capacitance C_{be}	High Z owing to reverse biased junction or insulator—small depletion layer capacitance C_{gs}
Turn-on voltage	Forward voltage V_{BE} highly repeatable—Set by thermodynamics	Pinch-off voltage V_p not very repeatable—Set by process variation
Transconductance	High g_m $(= I_C/(kT/q))$	Low g_m $(\cong v_{sat}C_{gs})$
Current gain	β (or h_{FE}) = 50–150; β is important owing to low input impedance	Not meaningful at low frequencies and falls as $1/\omega$ at high frequencies
Unity current gain cutoff frequency f_T	$f_T = g_m/2\pi C_{BE}$ is usually lower than for FETs	$f_T = g_m/2\pi C_{gs}$ $(= v_{sat}/2\pi L_g)$ higher for FETs
Maximum frequency of oscillation f_{max}	$f_{max} = (f_T/(8\pi r_b C_{bc}))^{1/2}$	$f_{max} = f_T (r_{ds}/R_{in})^{1/2}$
Feedback capacitance	C_{bc} large because of large collector junction	Usually C_{gd} is much smaller than C_{bc}
$1/f$ Noise	Low in BJT/HBT	Very high $1/f$ noise corner frequency
Thermal behavior	Thermal runaway and second breakdown	No thermal runaway
Other		Backgating is problem in semi-insulating substrates

common-emitter BJT/HBT configuration. One of the more striking differences is the input impedance parameter. A FET has high input impedance at low- to mid-range frequencies because it is essentially a capacitor. As frequency increases, the magnitude of the input impedance decreases by $1/\omega$ because the capacitive reactance varies as $|C_{gs}/\omega|$. The BJT/HBT emitter-base junction is a forward-biased pn junction, which is inherently a low-impedance structure because of its rapid variation in carriers crossing the junction with changing applied voltage. The BJT/HBT input is also capacitive (from the large diffusion capacitance owing to stored charge primarily in the base), but a large conductance (i.e., small resistance) appears in parallel assuring a low input impedance even at low frequencies.

BJT/HBT devices are known for their higher transconductance g_m which is proportional to collector current. It is limited by parasitic resistance in the device's structure and from natural limits in how high the operating current can be. A FET's g_m is proportional to the saturated velocity v_{sat} and its input capacitance C_{gs}. Thus, device structure and material parameters set the performance of the FET whereas thermodynamics plays the key role in establishing the magnitude of g_m in a BJT/HBT.

Thermodynamics establishes the magnitude of the turn-on voltage (this follows simply from Eq. (72.14)) in the BJT/HBT device. For digital circuits turn-on voltage, or threshold voltage, is important in terms of repeatability and consistency for circuit robustness. The BJT/HBT is clearly superior to the FET in this regard because doping concentration and physical structure establish a FET's turn-on voltage. In general, these variables are less controllable. However, the forward turn-on voltage in the AlGaAs/GaAs HBT is higher (~1.4 V) because of the band discontinuity at the emitter-base heterojunction. For InP-based HBTs the forward turn-on voltage is lower (~0.8 V) than that of the AlGaAs/GaAs HBT and comparable to the approximate 0.7 V found in silicon BJTs. This is important in digital circuits because reducing the signal swing allows for faster circuit speed and lowers power dissipation from reduced power supply voltages.

For BJT/HBT devices, current gain (usually referred to by the symbol β or h_{FE}) is a meaningful and important parameter. "Good" BJT devices inject little current into the emitter and, hence, operate with low base current levels. The current gain is defined as the collector current divided by the base current and is therefore a measure of the quality of the device (i.e., traps and defects, both surface and bulk, degrade the current gain owing to higher recombination currents). At low to mid-range frequencies, current gain is not especially meaningful for the high input impedance FET device because of the capacitive input.

The intrinsic gain of an HBT is higher because of its higher early voltage V_A. The early voltage is a measure of the intrinsic output conductance of a device (this is the slope of the saturated $I-V$ characteristic in a plot of I_C versus V_{CE}). In the HBT the change in the collector voltage has very little effect on the modulation of the collector current. This is true because the band discontinuity dominates the establishment of the current collected at the collector-base junction. A figure of merit is the intrinsic voltage gain of an active device, given by the product $g_m V_A$, and the HBT has higher values compared to silicon BJTs and compound semiconductor FETs.

It is important to have a dynamic figure of merit or parameter to assess the usefulness of an active device for high-speed operation. Both the unity current gain cutoff frequency f_T and maximum frequency of oscillation f_{max} have been discussed in the charge control section above. The higher the value of both parameters, the better the high-speed circuit performance will be. This is not the whole story because in digital circuits other factors such as output node-to-substrate capacitance, external load capacitances, and interconnect resistance also play an important role in determining the speed of a circuit. Both these figures of merit are used because they are simple and can generally be correlated to circuit speed.

Generally $1/f$ noise is much higher in the FET devices than in the BJT/HBT devices. This is usually of more importance in analog applications and for oscillators in particular. Thermal behavior in high-speed devices is important as designers push circuit performance. Bipolar devices are more susceptible to thermal runaway than FETs because of the positive feedback associated with the forward-biased pn junction (i.e., a smaller forward voltage is required to maintain the same current at higher temperatures). This is not true in the FET; in fact, FETs generally have negative feedback under common biases used in practical applications. Both GaAs and InP have poorer thermal conductivity than silicon, with GaAs being about one-third of silicon and InP being about one-half of silicon.

Finally, circuits built on GaAs or InP semi-insulating substrates are susceptible to backgating. Backgating is similar to the backgate-bias effects in MOS transistors, only it is not as predictable or repeatable as the well-known backgate-bias effect is in silicon MOSFETs on silicon lightly doped substrates. Interconnect traces with negatively applied voltages, and located adjacent to active devices, can change their threshold voltage (or turn-on voltage) of unipolar devices. It turns out that BJT/HBT devices do not suffer from backgating and this is one of their advantages. Of course, semi-insulating substrates[ll] are nearly ideal for microstrip or coplanar waveguide transmission lines positioned on top of the substrate because of their very low loss. Silicon substrates are much more lossy in comparison and this is a decided advantage when using GaAs and InP substrates for ICs.

72.4 Device Structures and Performance

In this section, a few representative device structures are described. We begin with FET device structures and then follow with BJT/HBT structures. Table 72.4 lists the most important variants, but by no means all of them.

72.4.1 FET Structures

In the silicon VLSI world, the metal-oxide semiconductor field-effect transistor (MOSFET) dominates. In the MOSFET the channel is formed by an inversion layer at the oxide-semiconductor interface upon applying a voltage to the gate that attracts carriers to this interface [38,39]. The thin layer of mobile carriers forms a two-dimensional sheet of carriers. One of the limitations with the MOSFET is that the oxide-semiconductor interface scatters the carriers in the channel and degrades the performance of the MOSFET. This is evident in Figure 72.5 where the lower electron velocity at the Si–SiO$_2$ interface is compared with electron velocities in bulk silicon and several compound semiconductors. For many years device physicists have looked for device structures and materials which increase electron velocity. Recently, it has been found that strained silicon layers can lead to enhanced electron velocity in MOSFETs [40,41].

[ll]The closest equivalent in silicon structures is "silicon-on-insulator".

One way to enhance mobility is to introduce tensile strain in the silicon located in the channel layer by growing silicon on a "relaxed" silicon–germanium layer. The strain is introduced by the lattice mismatch because the silicon layer is stretched in the plane of the interface of the epitaxially grown silicon upon the silicon–germanium layer. The strain in the silicon reduces the carrier effective mass thereby enhancing the carrier mobility. Another way to accomplish this is to grow strained silicon directly on insulator (SSDOI) which eliminates the need for the added silicon–germanium layer [42,43]. Still another approach to enhancing MOSFET performance is to utilize the fact that mobility varies with silicon orientation [44]. Band-engineered structures can enhance carrier mobility by perhaps as much as a factor of 2 beyond bulk silicon for electrons and by up to 8 or 10 for holes, but production implementation of strain enhancement of mobility falls short of these numbers so far.

FET structures using compound semiconductors have led to much faster devices such as the MESFET and the HEMT. Historically, the MESFET was demonstrated several years before the HEMT and today the MESFET is still more widely used. The MESFET uses a thin doped channel (almost always n type because electrons are much more mobile in semiconductors) with a reverse-biased Schottky barrier for the gate control. The cross section of a typical MESFET is shown in Figure 72.11(a). A recessed gate is used along with a highly doped n^+ layer at the surface to reduce the series resistance at the source as well as the drain connections. The gate length and electron velocity in the channel dominate in determining the high-speed performance of a MESFET. Much work has gone into developing processes that form shorter gate structures. For digital devices lower breakdown voltages are permissible and therefore shorter gate lengths and higher channel doping are more compatible with such devices. For a given semiconductor material, a device's breakdown voltage BV_{GD} times its unity current gain cutoff frequency f_T is approximately a constant. Therefore it is possible to trade off BV_{GD} for f_T in device design. A high f_T is required in high-speed digital circuits because devices with a high f_T over their logic swing will have a high g_m/C ratio for large-signal operation. A high g_m/C ratio translates into a device's ability to drive load capacitances. However, in the analog domain higher breakdown voltages are often required. This is especially true for power amplifiers where 50 Ω systems set the voltage swings that must be accommodated for the

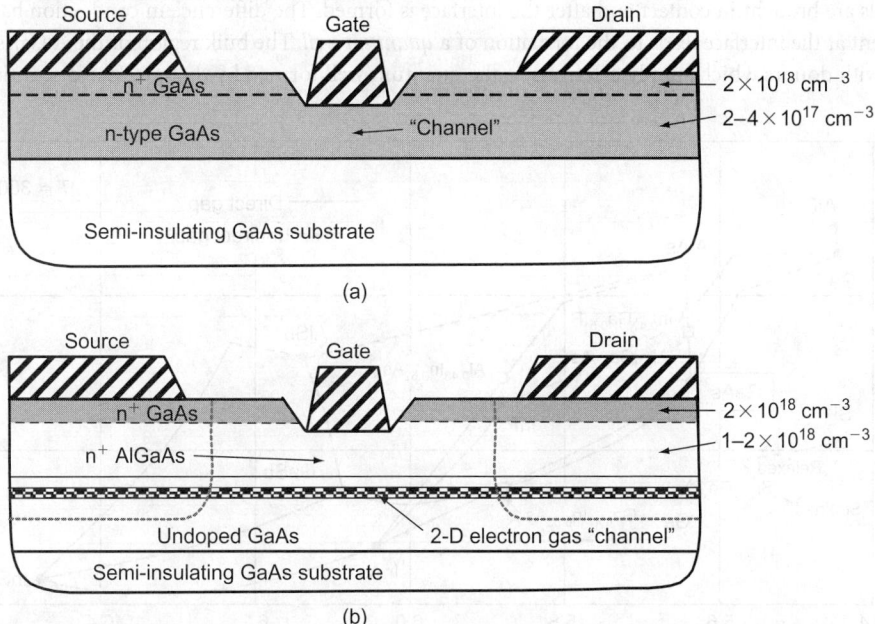

FIGURE 72.11 Representative FET cross sections for (a) GaAs MESFET device with doped channel; (b) AlGaAs/GaAs HEMT device with single quantum well containing two-dimensional electron gas. Donor doping resides in AlGaAs layer that contributes the electrons which spill over into the quantum well.

power output required. So for power amplifiers both high current drive and high voltage breakdowns are simultaneously required.

Compound semiconductor FET devices are clearly superior to silicon MOSFETs with respect to the breakdown voltage requirement. Lightly doped or lateral double diffused MOSFET devices [45], known initially as DMOS and now as LDMOS, have been developed to address this need, but are presently restricted to frequencies of <4 or 5 GHz. In LDMOS the higher voltage is achieved by forming the channel region using two diffusions through the same oxide opening with the channel length being the lateral difference in the diffusions. A low-doped drift region, into which the diffusions penetrate, allows the drain and gate to be widely separated for high breakdown voltage. This drift region is depleted in operation so that the lateral electric field is greatly reduced. However, the low doping levels in LDMOS leads to high "on" resistance R_{on} and lower saturation current $I_{D,sat}$. LDMOS devices have found application in RF power amplifiers such as in bay station transmitters for cellular communications. Competing against LDMOS for high-power applications is the newer gallium nitride HEMT device which is discussed below.

To achieve higher current performance we must maximize the charge in the channel per unit gate area. This allows for higher currents per unit gate width and greater ability to drive large capacitive loads. The higher current per unit gate width also favors greater IC layout density. In the MESFET the doping level of the channel sets this limit. MESFET channels are usually ion-implanted and the added lattice damage further reduces the electron mobility.

To achieve still higher currents per gate width and even higher figures of merit (such as f_T and f_{max}), the HEMT structure has evolved [46,47]. The HEMT is similar to the MESFET except that the doped channel is replaced with a two-dimensional quantum well containing electrons (sometimes referred to as a 2-D electron gas or 2DEG). The quantum well is formed by a discontinuity in conduction band edges between two different semiconductors (such as AlGaAs and GaAs). From Figure 72.12, which is reproduced from Chapter 71 for convenience, we see that GaAs and $Al_{0.3}Ga_{0.7}As$ have nearly identical lattice constants, but with somewhat different bandgaps. One compound semiconductor can be easily grown (i.e., using MBE or MOCVD techniques) upon a different compound semiconductor if the lattice constants are matched. Figure 72.13 shows a band diagram of the GaAs/AlGaAs heterojunction before the materials are brought in contact and after the interface is formed. The difference in conduction band edge alignment at the interface leads to the formation of a *quantum well*. The bulk region of the AlGaAs layer is doped with donors which supply electrons to the quantum well formed by the interface discontinuity in

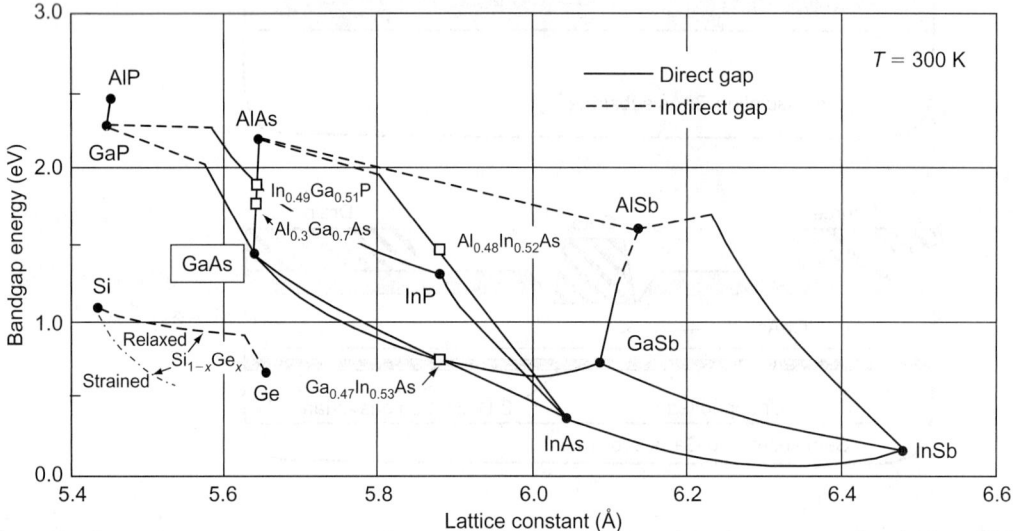

FIGURE 72.12 Bandgap energy versus lattice constant for various compound semiconductors. Reproduced from Chapter 71.

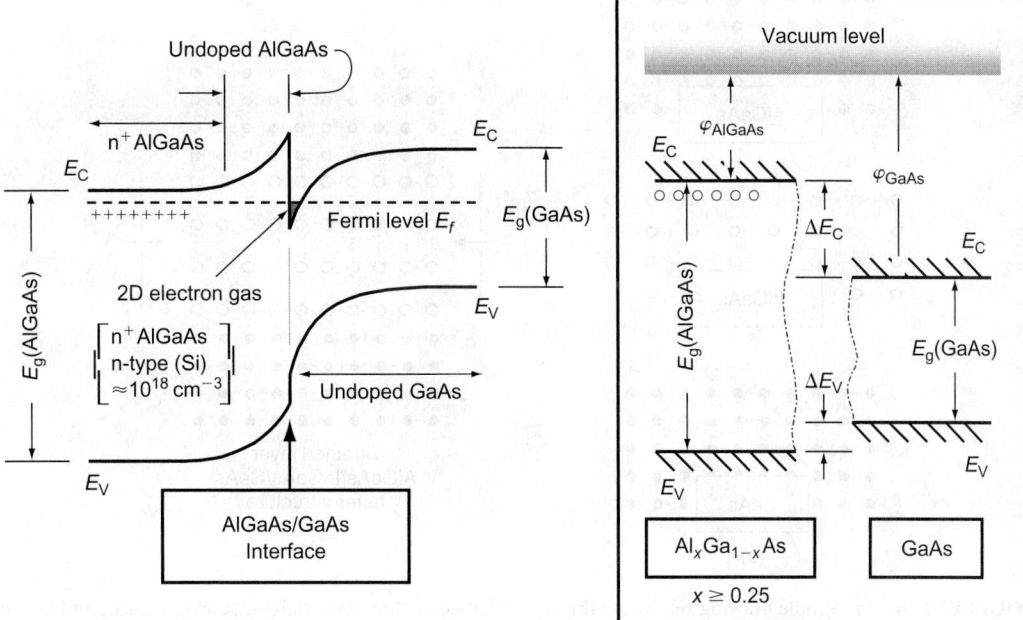

FIGURE 72.13   Band diagram showing the alignment of GaAs to AlGaAs forming a heterojunction. Band bending at this interface forms the quantum well which is a minimum in potential for electrons.

conductance band edges and the band bending required to align the Fermi levels at equilibrium. The greater the edge misalignment or offset, the deeper the quantum well can be, and the greater the number of carriers per unit area the quantum well can hold. From the prior charge control theory the charge per unit area that a quantum well can hold directly translates into greater current per unit gate width. Thus, the information in Figure 72.12 can be used to *bandgap engineer* different material interfaces that can be combined in lattice matched layers. Examples of currently used lattice matched systems are $Al_xGa_{1-x}As/GaAs$, $In_{0.51}Al_{0.49}As/GaAs$, and $In_{0.52}Al_{0.48}As/In_{0.53}Ga_{0.47}As/InP$.

Quantum wells can be formed using semiconductor materials having higher electron velocity and mobility than the substrate material (e.g., GaAs or InP). For example, $In_{0.53}Ga_{0.47}As$ has an electron mobility of approximately 11,000 cm²/V s compared to 4,600 cm²/V s for InP. The HEMT cross section in Figure 72.11(b) shows the dopant atoms positioned in the wider bandgap AlGaAs layer. When the donors ionize the electrons they contribute spill into the quantum well because of its lower energy. Higher electron mobility is possible because the ionized donors are not located in the quantum well layer thereby eliminating ionized impurity scattering. A recessed gate is positioned over the quantum well, usually upon a semiconductor layer such as the AlGaAs layer in Figure 72.11(b), allowing modulation of the charge in the quantum well channel. In summary, we want the quantum well carrier density per unit area n_s to be large and the mobility μ to be large—a figure of merit might be to maximize the $n_s\mu$ product for comparing quantum well structures.

There are only a few lattice-matched structures possible. However, semiconductor layers for which the lattice constants are not matched are possible if the layers are thin enough (critical thickness t_{crit} is of the order of a few nanometers). MBE and MOCVD make it possible to grow layers of a few atomic layers in thickness. Good quality heterostructures must be grown with care. Thin layers formed from mismatched materials lead to lattice deformation—these layers are called strained layers or pseudomorphic layers. The concept of the strained layer is illustrated in Figure 72.14. The amount of strain is proportional to the magnitude of lattice mismatch. If a strained layer is grown such that it exceeds its critical thickness t_{crit}, then the layer will relax and dislocations are generated at the mismatched interface (e.g., for the $In_{0.3}Ga_{0.7}As/GaAs$ interface $t_{crit} \cong 13$ nm [46]). Using such structures to form FET devices produces what

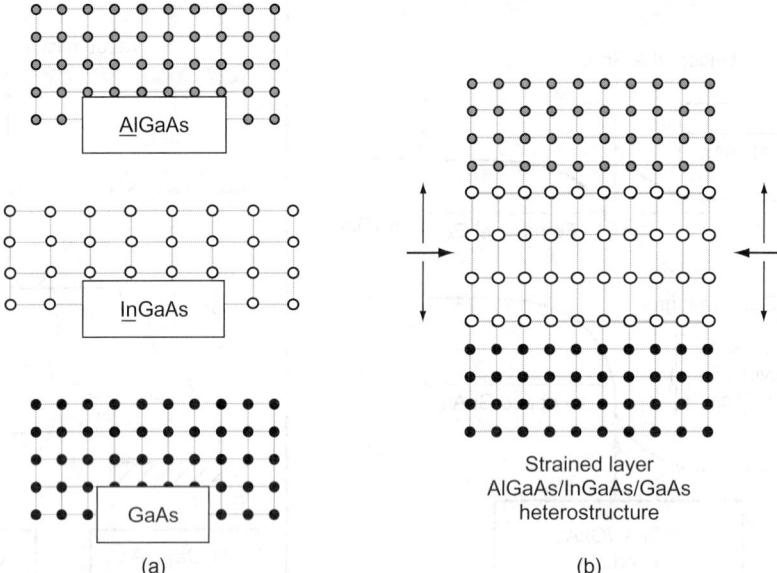

FIGURE 72.14 (a) Simple drawing represents the crystal lattice of three materials—AlGaAs, InGaAs, and GaAs; (b) formation of a thin strained layer (not drawn to scale) of InGaAs when the materials are brought together (elastic strain has t_{crit}).

are called pseudomorphic HEMT (p-HEMT) devices [48]. The use of strained layers gives more flexibility in engineering quantum well systems which hold greater charge per unit area and have higher electron velocities and mobilities. For this reason, the pseudomorphic HEMT produces higher performance than the MESFET or the lattice-matched HEMT. Examples of strained layer systems currently used include $Al_xGa_{1-x}As/In_zGa_{1-z}AsAs$, $In_{0.52}Al_{0.48}As/In_xGa_{1-x}As/InP$ with $x > 0.53$, $Al_xGa_{1-x}N/GaN$, and Si_xGe_{1-x}/Si. Table 72.7 lists a few heterojunctions with the conduction band discontinuity ΔE_c, carrier density per unit area n_s, mobility μ, and peak saturation velocity v_{pk}.

Bandgap engineering can easily grade these layers to reduce the strain in the layers. By grading the composition of the layer, rather than abruptly changing the composition, the strain is relaxed and much thicker layers can be grown to accommodate even greater lattice mismatch between materials. Such layers are called metamorphic layers and can be used to grow favorable quantum well systems on inexpensive substrates** such as GaAs. The graded buffer layer allows the quantum well to be formed with a high indium content InGaAs channel layer [49]. An example of this might be an $Al_{0.48}In_{0.52}As/In_yGa_{1-y}As$ quantum well interface with a thick graded buffer of $Al_xGa_{1-x}As$ (x varies from 0 to a value matching the lattice of the channel layer) on a semi-insulating GaAs wafer. Other buffer layers reported include $In_yAl_xGa_{1-x-y}As$ and $Al_xGa_{1-x}As_ySb_{1-y}$ (and there are many other possibilities).

The first AlGaN/GaN HEMT devices were explored in the early 1990s. Gallium nitride is attractive because it has a large bandgap, moderate electron mobility, high electron saturation velocity, and very high critical breakdown field. Most of this work was carried out on sapphire and silicon carbide*† substrates (GaN wafers were not available until recently). Recently, some work has used silicon wafers to grow these structures for economic reasons. The AlGaN layer is used as a spacer and reservoir for donor atoms to populate the AlGaN/GaN quantum well once the donors are ionized. This quantum well can hold a high carrier sheet density n_s with values typically reported in the range $1–2 \times 10^{13}$ cm^{-2}. This

---

**InP substrates are more expensive and more brittle than the GaAs ones. Moreover, GaAs wafers are larger than the InP ones.

*†Silicon carbide (SiC) has a very high thermal conductivity but is very expensive. Today 3-in SiC substrates are readily available (but compare this to 6-in GaAs wafers giving four times the useable area).

TABLE 72.7　Comparison of Several Heterojunction Structures

Heterojunction Units	Type	Substrate	ΔE_c (eV)	n_S (cm^{-2})	μ (cm²/V s)	V_{pk} (cm/s)
Al$_{0.3}$Ga$_{0.7}$As/GaAs	LM	GaAs	0.22	1×10^{12}	8,500	2×10^7
In$_{0.52}$Al$_{0.48}$As/ In$_{0.53}$Ga$_{0.47}$As	LM	InP	0.51	3×10^{12}	10,000	2.7×10^7
Al$_{0.3}$Ga$_{0.7}$As/ In$_{0.15}$Ga$_{0.85}$As	PM	GaAs	0.42	2.5×10^{12}	9,000	2.3×10^7
Al$_{0.25}$Ga$_{0.75}$As/ In$_{0.22}$Ga$_{0.78}$As	PM	GaAs	—	3.3×10^{12}	6,000	—
In$_{0.40}$Al$_{0.60}$As/ In$_{0.65}$Ga$_{0.35}$As	PM	InP	—	2.7×10^{12}	10,100	—
In$_{0.35}$Al$_{0.65}$As/ In$_{0.35}$Ga$_{0.65}$As	MM	GaAs	—	3.5×10^{12}	6,200	—
In$_{0.32}$(AlGa)$_{0.68}$As/ In$_{0.47}$Ga$_{0.53}$As	MM	GaAs	—	3×10^{12}	8,800	—

Note: LM, lattice matched; PM, pseudomorphic; MM, metamorphic.

allows AlGaN/GaN HEMT devices to support large currents while achieving high breakdown voltages [50–52]. This is what is required for high-power amplifiers and why AlGaN/GaN HEMTs are rapidly becoming competitive with silicon LDMOS devices. Today, GaN devices are quite expensive and much work is focused on demonstrating long-term reliability and bringing the cost down.

72.4.2　FET Performance

All currently used FET structures are n-channel because hole velocities are very low compared to electron velocities. Typical gate lengths range from 0.5 μm down to sub-0.1 μm for the fastest devices. Consider Figure 72.15 comparing five classes of FET-active devices where a "figure of merit" frequency, either f_T or f_{max}, is plotted against the gate-to-drain breakdown voltage. This plot primarily addresses analog applications, especially power amplifier applications rather than digital applications. In Figure 72.15 two different device classes stand out—InP HEMT devices produce the highest frequency performance and AlGaN/GaN HEMT devices are capable of the highest operating voltages. Both classes are relatively expensive to manufacture however. The circles in the figure are only approximate and a few specific cases for each group have demonstrated performance beyond the drawn boundaries.

The GaAs MESFET (ca. 1968) was the first compound semiconductor FET structure and is still used today because of its simplicity, maturity, and low cost of manufacture. GaAs MESFET devices have f_T values in the 20–100 GHz range corresponding to gate lengths of 0.5 μm down to 0.1 μm, and f_{max} values from ~60 to 160 GHz for the same gate length range. Transconductance values are commonly in the range of 180–400 mS/mm. These devices typically have I_{DSS} currents in the range of 200–400 mA/mm, where the parameter I_{DSS} is the common-source drain current with a zero gate voltage applied in the saturated state of operation. At a drain voltage of 10 V output powers of <12 GHz are of the order of 0.5–0.85 W/mm (gate width). Noise figures of <1 dB up to 26 GHz have been demonstrated with very short gate lengths of 0.1 μm or less and at a drain current of approximately 0.2 I_{DSS}. Noise figure at 12 GHz is about 0.5 dB for the same device.

In comparison, the first HEMT used an AlGaAs/GaAs material system. These devices are higher performance devices compared with the GaAs MESFET (e.g., given an identical gate length, the AlGaAs/GaAs HEMT has a f_T about 50–100% higher depending upon the details of the device structure and quality of material). Correspondingly higher currents are achieved in the AlGaAs/GaAs HEMT devices. The extrinsic transconductance ranges typically range over 250–550 mS/mm. The best power density reported is 1.7 W/mm (this device used a field plate to increase the breakdown voltage) [53]. With the GaAs pseudomorphic HEMT the performance is higher. For example, the transconductance range is now 300–800 mS/mm and the maximum channel current range is 500–800 mA/mm. The cutoff frequency f_T

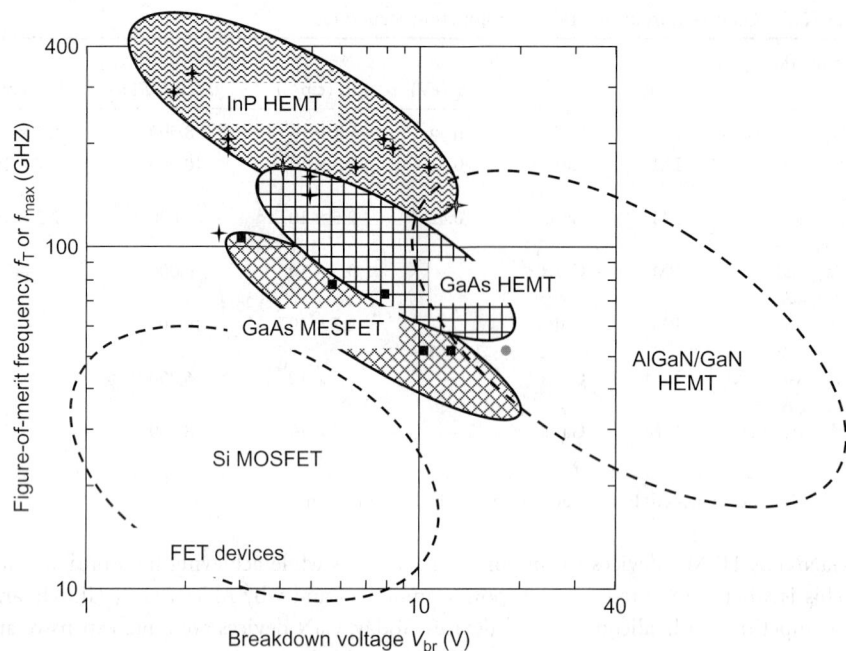

FIGURE 72.15 Figure-of-merit frequency (f_T and f_{max}) versus gate-to-drain breakdown voltage for five different classes of FET devices.

can be of the order of 150 GHz and f_{max} can be ~300 GHz. But reported power densities up to 1.6 W/mm are comparable to GaAs lattice-matched devices. GaAs metamorphic HEMT devices have demonstrated transconductance values of 600–1000 mS/mm and maximum channel currents from 600–900 mA/mm. For comparison, f_T frequencies are of the order of 200 GHz and f_{max} values have been reported up to 400 GHz.

But the highest performance is achieved using InP-based HEMTs. For example, InP HEMT devices have yielded f_T values of the order of 400 GHz with gate lengths of 0.1 μm. Furthermore, such devices have maximum current values approaching 1200 mA/mm and high transconductances up to 1600 mS/mm. InP HEMT devices have demonstrated extremely low noise figures. For example, noise figures of ~0.7 dB at 62 GHz and 1.2 dB at 94 GHz—no other compound semiconductor active device achieves noise figure values lower than these. Silicon MOSFETs have the highest noise figures of all the FET classes discussed in this section.

GaN HEMT devices excel in high power. However, GaN HEMTs have lower transconductances, typically in the range of 200–300 mS/mm, but with high maximum currents of the order of 1000 mA/mm. Gate-to-drain breakdown voltages BV_{GD} typically fall into the range of 60–200 V (sometimes greater). Cutoff frequencies are typically lower with the better f_T values of ~100 GHz and f_{max} near 150 GHz. Figure 72.16 compares the output power per unit gate width versus frequency for the FET classes we have been discussing. Clearly, the GaN HEMT is an order of magnitude higher in output power capability relative to the GaAs and InP HEMTs. The highest output power curve corresponds to the use of a field plate [54,55] over the gate-to-drain region to preserve high gate-to-drain breakdown voltages. Table 72.8 summarizes some recent power amplifier results with AlGaAs/GaN HEMT devices. Work at present is aimed at demonstrating improved reliability and lowering its cost—the maximum frequency of operation is also being increased.

72.4.3   Heterojunction Bipolar Structures

Practical HBT devices [56,57] built with silicon now join the ranks of compound semiconductor devices with the introduction of the silicon–germanium HBT. MBE is used to grow the doped layers making up the vertical semiconductor structure in GaAs and InP HBTs. In fact, HBT structures were not really practical until the advent of MBE (and MOCVD) even though the idea behind the HBT dates back to

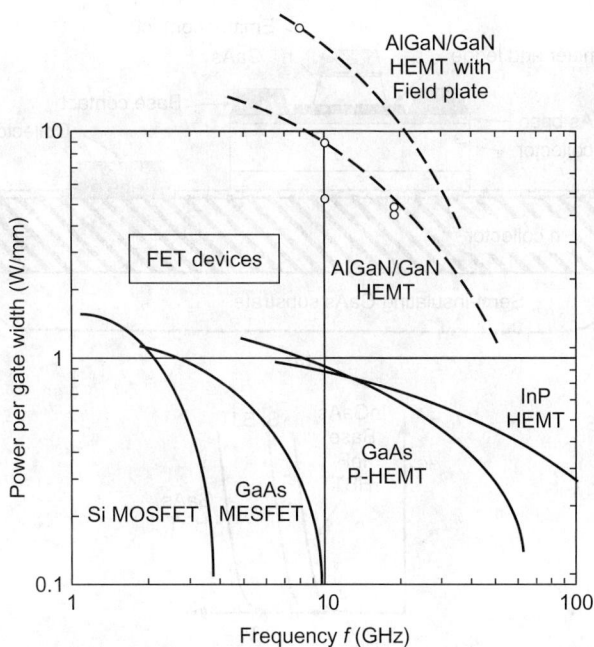

FIGURE 72.16   Power output per unit gate width versus frequency for several different FET devices.

TABLE 72.8   Some Recent AlGaAs/GaN HEMT Power Amplifier Results

Reference Units	Gate Length (μm)	Substrate	Saturated Power (W/mm)	At Frequency (GHz)	Efficiency η (%)	g_m (mS/mm)	f_{max} (GHz)
Kumar et al. (2003), University of Illinois	0.25	Sapphire	4.65	20	30	354	89
Kumar et al. (2002), University of Illinois	0.12	SiC	4.2	20	26.5	314	162
Jessen et al. (2003), Wright-Patterson	0.14	SiC	4.6	10	46	320	122
Johnson et al. (2004), Nitronix	0.70	Si (111)	12	2.14	52.7	~300	31
Y.-F. Yu et al. (2004), Cree	0.5–0.6	SiC[a]	30.6	8	49.6	b	68
K.K. Chu et al. (2004), BEA and Cree	b	GaN	9.4	10	40	220	b
L. Chen et al. (2001), University of California Santa Barbara	0.25	GaN	8.4	8	28	200	56

[a]Device had *field plate* giving greater current–voltage swing resulting in greater power output.
[b]Values not available.

around 1950 [5]. The vastly superior compositional control and layer thickness control with MBE is what made HEMT and HBT devices possible. The first HBT devices used an AlGaAs/GaAs heterojunction with the wider bandgap AlGaAs layer forming the emitter region. An example of such a mesa GaAs HBT structure is shown in Figure 72.17(a). Compound semiconductor HBT devices are often mesa structures, as opposed to the more nearly planar structures used in silicon bipolar technology, because top surface contacts must be made to the collector, base, and emitter regions. MBE grows the stack of layers over the entire wafer, whereas, in silicon VLSI CMOS processes selective implantations and oxide masking localize the doped regions. Hence, etching down to the respective layers allows for contact to the base and collector regions. More recently, HBTs have been formed with InGaP emitters primarily for improved

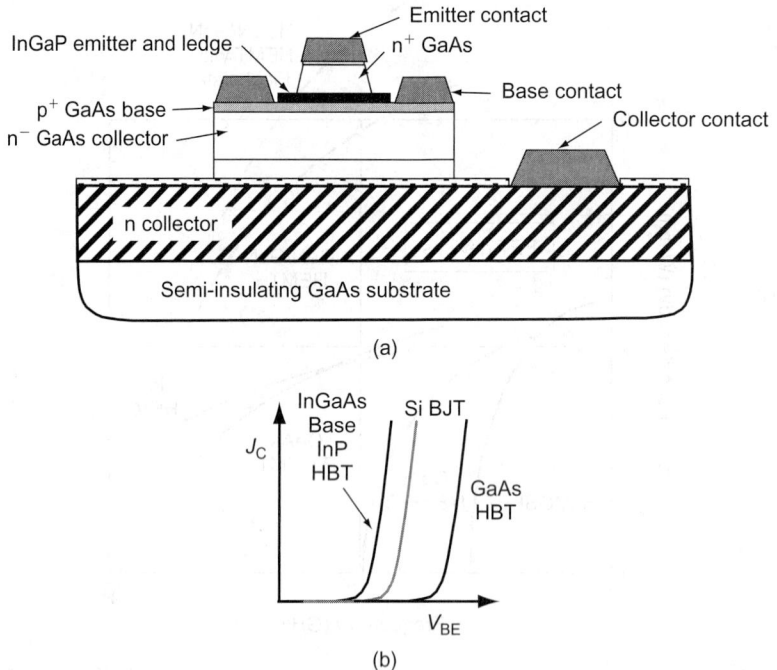

FIGURE 72.17 (a) Cross section of an HBT device with carbon-doped p+ base and an InGaP emitter [57]. Note the commonly used mesa structure where selective layer etching is required to form contacts to the base and collector regions; (b) emitter-base forward voltage versus collector current for InP, GaAs, and Si HBT transistors. Note that InP HBT has a lower turn-on voltage whereas GaAs has the highest turn-on voltage.

reliability [58] over the AlGaAs emitter. The advantages of the GaAs HBT over silicon are derived from the wide bandgap emitter and the higher electron velocity in GaAs. With a wide bandgap emitter a higher base doping can be used and still maintain high emitter injection efficiency. This leads to lower base resistance which is very desirable in bipolar transistors—base resistance lowers f_{max}, increases noise figure, and reduces RF/microwave gain. The early voltage is higher in the HBT and high-level injection effects are delayed until higher current densities are reached. With the reduced emitter doping level, a lower emitter-base capacitance can be achieved. Of course, the use of GaAs gives the added benefit of having a semi-insulating substrate with its reduced parasitic capacitance.

Recently, InP-based HBTs [59] have emerged as candidates for use in high-speed circuits. The two dominant heterojunctions are InP/InGaAs and AlInAs/InGaAs in InP devices. The small but significant bandgap difference between AlInAs directly upon InP greatly limits its usefulness. InP-based HBT device structures are similar to those of GaAs-based devices and the reader may refer to Jalali and Pearson [59] for specific fabrication information on InP HBT devices. Generally, InP has advantages of lower surface recombination (higher current gain results), better electron transport, lower forward turn-on voltage, and higher substrate thermal conductivity.

In the last several years silicon–germanium HBT devices [60] have received much attention and RFIC products are now available from several companies*‡. The SiGe HBT uses a strained layer for the base region consisting of a Si_xGe_{1-x} compound semiconductor with x often in the range of $0.6< x <0.8$. These SiGe layers are not grown with MBE, as used with GaAs and InP heterostructures, but rather using chemical vapor deposition (CVD) because it produces more uniform layers with a lower particulate density. Sometimes the SiGe base layer is uniform in fractional composition (a true HBT structure); however, a higher performance can be achieved with a graded composition base as shown in Figure 72.18. As the germanium fraction is

---

*‡As of 2004 there were at least 11 companies offering SiGe HBT processes.

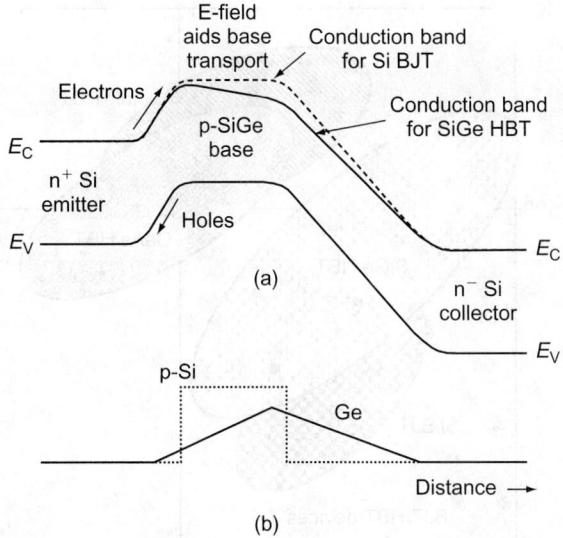

FIGURE 72.18   (a) Bandgap diagram of a SiGe HBT transistor with (b) the doping profiles for the p-type dopant and germanium doping. Note that bandgap narrowing occurs in the base and collector-base depletion regions as a result of the addition of the germanium.

increased, the bandgap is reduced relative to the bandgap of silicon (see Figure 72.12 for the SiGe system). The graded base has the advantage that it reduces the base transit time (τ_r in charge control analysis) and reduces the output conductance (i.e., increases early voltage V_A) [61,62]. IBM uses the graded base HBT and has published extensively on its capabilities. Major advantages of the SiGe HBT are (1) it can be combined with silicon CMOS processes for BiCMOS circuits, and (2) it leverages high-yielding silicon processes using either 8 or 12 in diameter wafers. Many classes of RF circuits are more cost effective using SiGe HBTs. A major disadvantage is that very high mask costs and access to large silicon foundries can be very expensive.

72.4.4   HBT Performance

The "figure-of-merit" frequency, either f_T or f_{max}, versus breakdown voltage for GaAs, InP, and SiGe HBTs is shown in Figure 72.19. The silicon homojunction BJT has also been included for comparison. Again these ranges are only approximate and indicate relative positioning. Clearly, the GaAs HBT with its higher breakdown voltages stands out for power amplifier applications.

Where the HBT really excels is in being able to generate much higher values of transconductance compared to the FET devices. This is a clear advantage in driving larger load capacitances and delivering higher output power. The ultimate limitation as to how high the transconductance g_m can be is thermal. The power dissipation in the collector–base junction is highly localized leading to high local temperatures. BJT/HBT devices degrade in performance as their temperature rises. In fact, it is possible to have thermal runaway with bipolar devices and silicon devices sometimes experience second breakdown. It is very important to have good thermal management solutions for BJT/HBT applications so that the heat generated is adequately removed. A very rough rule of thumb in comparing HBT to FET compound semiconductor devices is that HBT transconductance is about 3–4 times that of FET in typical application situations.

Typical f_T values for GaAs HBT processes in manufacturing are in the 50–150 GHz range and the f_{max} range is 100–300 GHz. For example, the HBT example [58] in Figure 72.17(a), with a 2 μm $\times$ 2 μm emitter, the f_T is approximately 65 GHz at a current density of 0.6 mA/μm^2 and its DC current gain is around 50. In GaAs HBT devices the parameter f_{max} is generally greater than its f_T value (e.g., for the device shown in Figure 72.17(a), f_{max} is about 75 GHz). Base resistance (refer to Table 72.5 for equation) is the dominant limiting factor in establishing f_{max}. It is common for HBT devices to have f_{max} values only

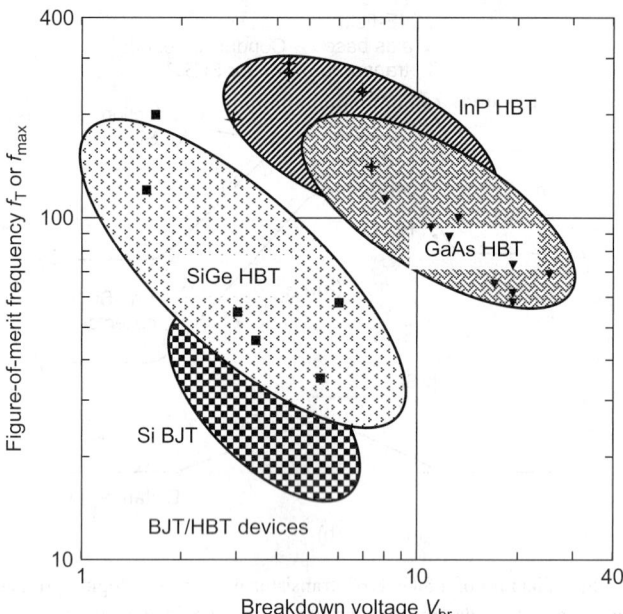

FIGURE 72.19   Figure-of-merit frequency (f_T and f_{max}) versus collector-to-emitter breakdown voltage for four different classes of BJT/HBT devices.

slightly higher than their f_T values. In comparison, MESFET and HEMT devices typically have a higher f_{max}/f_T ratios.

For GaAs HBT devices the best output power performance reported is approximately 10 W at 1 GHz and about 1 W at 35 GHz. For BJT/HBT devices it is convenient to quote power performance per unit emitter area—below 10 GHz. GaAs HBT power densities are as high as 10 mW/μm^2 and more typically in the range of 2–4 mW/μm^2 from 10–30 GHz. These power densities are limited by the considerable self-heating inherent in BJT/HBT physical structures. GaAs HBT devices do not have the best low-noise performance with InP HBTs being superior to GaAs HBTs below 5 GHz in general.

For InP HBT devices the best f_T and f_{max} values extend up to about 300 GHz. Noise figures around 0.5 dB have been reported around 1–2 GHz. Generally, the output power is not as good as reported for GaAs HBTs; however, less work has been done on InP HBT power applications and recent work looks promising. Clearly, the InP HEMT has a much superior noise figure performance than the InP HBT.

There is one very impressive performance exception to the above performance numbers—this is the work of Rodwell's group at the University of California, Santa Barbara [63]. Using a substrate transfer process to greatly reduce the extrinsic (parasitic) elements, an InGaAs-base HBT has been demonstrated with an f_T approaching 300 GHz and f_{max} close to 1.1 THz (terahertz is equivalent to 1000 GHz). This work demonstrates how limiting the extrinsic elements are to BJT/HBT performance in general. This level of performance was achieved by reducing base thickness and reducing the RC time constants as a result of decreasing the base and collector junction widths. The most severe parasitic element was identified as the collector–base junction under the base ohmic contacts. Using their substrate transfer process the Santa Barbara group was able to make dramatic reductions in junction widths. Unfortunately, the process used is not close to being a production-ready process.

The use of SiGe HBT technology is rapidly growing in RF and microwave applications [64]. For example, Sirenza recently disclosed that they have shipped over 100 million RFIC parts in an SiGe HBT process [65]. Probably, the most watched and discussed SiGe HBT processes are those available from IBM [66]. They are currently shipping with IBM's fourth-generation BiCMOS process—the 8HP process using

130 nm CMOS [65]. The 8HP process gives SiGe HBT performance characterized by an f_T of 210 GHz and an f_{max} of 185 GHz. The collector-emitter breakdown voltage BV_{CEO} is listed at 1.9 V. IBM has hinted at their future-generation 9HP BiCMOS process as having an $f_T > 250$ GHz.

72.4.5 Summary

In this chapter, we used the charge control viewpoint of active devices to describe the operation of compound semiconductor devices. Also included is a brief history of the compound semiconductor active devices targeted for RF and microwave applications. The technologically most important unipolar and bipolar active devices are described in the later part of the chapter. Compound semiconductor technology has produced very high-performance devices extending well into the millimeter wave frequency range. The battle between GaAs, InP, and GaN devices versus Si devices and SiGe HBT devices will vigorously continue into the foreseeable future.

References

1. N.H. Welker, "Uber neue halberstende Verbindungen," *Zeitschrift Naturforsch*, A**7a**, 1952, 744; A**8a**, 1953, 248. (Verfahren zur Herstellung eines Halbleiterkristalls aus einer A III-B V-verbindung mit Zonen verschiedenen Leitungstype, German patent DBP 976 791, 12c, 2.)
2. J.E. Lilienfeld, "Method and Apparatus for Controlling Electric Currents," U.S. Patent # 1,745,175, filed October 8, 1926, and granted January 28, 1930.
3. C.A. Mead, "Schottky Barrier Gate Field-Effect Transistor," *Proceedings of the IEEE*, **54**, February 1966, 307–308.
4. W. Shockley, "The Theory of pn Junctions in Semiconductors and pn Junction Transistors," *Bell System Technical Journal*, **28**, 1949, 345–389.
5. W. Shockley, "Current Element Utilizing Semiconductor Material," U.S. Patent # 2,569,347, filed June 26, 1948, and granted September 25, 1951.
6. G.C. Dacey and I.M. Ross, "Unipolar Field-Effect Transistors," *Proceedings of the IRE*, **41**, August 1953, 970–979.
7. H. Kroemer, "Theory of a Wide-Gap Emitter for a Transistor," *Proceedings of the IRE*, **45**, November 1957, 1535–1537.
8. J.A. Hoerni, "Planar Silicon Diodes and Transistors," *IRE Transactions on Electron Devices*, **8**, March 1961, 178.
9. J.A. Hoerni, "Method of Manufacturing Semiconductor Devices," U.S. Patent # 3.025,589, filed on May 1, 1959, and granted on March 20, 1962.
10. C.M. Melliar-Smith et al., "Key Steps to the Integrated Circuit," *Bell Labs Technical Journal*, **2**, Autumn 1997, 15–28.
11. F.M. Wanlass and C.T. Sah, "Nanowatt Logic Using Field-Effect Metal-Oxide-Semiconductor Triodes," *International Solid-State Circuits Conference Digest of Technical Papers*, February 1963, pp. 62–63.
12. E. Kasper et al., "A One-Dimensional SiGe Superlattice Grown by UHV Epitaxy," *Applied Physics A: Materials Science and Processing*, **8**, November 1975, 199–205.
13. S.S. Iyer et al., "Silicon-Germanium Base Heterojunction Bipolar Transistor by Molecular Beam Epitaxy," *International Electron Device Meeting Technical Digest*, December 1987, pp. 874–876.
14. W. Hooper and W. Lehrer, "An Epitaxial GaAs Field-Effect Transistor," *Proceedings of the IEEE*, **55**, 1967, 1237–1238.
15. R. Dingle et al., "Electron Mobilities in Modulation-Doped Semiconductor Heterojunction Superlattices," *Applied Physics Letters*, **33**, October, 1978, 665–667.
16. T. Mimura et al., "A New Field-Effect Transistor with Selectively Doped GaAs/n-Al$_x$Ga$_{1-x}$ As Heterojunctions," *Japanese Journal of Applied Physics*, **19**, 1980, L225–L227. See also, T. Mimura, "Development of High Electron Mobility Transistor," Invited Review Paper, *Japanese Journal of Applied Physics*, **44**, December 2005, 8263–8268.

17. T. Mimura, "High Electron Mobility Single Heterojunction Semiconductor Devices," U.S. Patent # 4,424,525, filed December 29, 1980, and granted January 3, 1984.

18. T.E. Zipperian et al., "An $In_{0.2}Ga_{0.8}As$/GaAs Modulation Doped Strained Layer Superlattice Field-Effect Transistor," *International Electron Device Meeting Technical Digest*, 1983, pp. 696–699.

19. D.L. Lile, "The History and Future of InP-Based Electronics and Optoelectronics," *10th International Conference on Indium Phosphide and Related Compounds*, May 11–15, 1998, Tusukuba, Japan, pp. 6–9.

20. M.A. Khan et al., "Microwave performance of a 0.25 um AlGaN/GaN heterostructure filed effect transistor," *Applied Physics Letters*, **65**, August, 1994, 1121–1123.

21. W.P. Dumke et al., GaAs-GaAlAs Heterojunction Transistor for High Frequency Operation," *Solid-State Electronics*, **15**, December 1972, 1329–1334.

22. P.M. Asbeck et al., "Emitter Coupled Logic Circuits Implemented with Heterojunction Bipolar Transistors," *IEEE GaAs IC Symposium Technical Digest*, Phoenix, AZ, October 25–27, 1983, pp. 170–173.

23. D.L. Miller et al., "(GaAl)As/GaAs heterojunction bipolar transistors with graded composition in the base," *Electronics Letters*, **19**, May, 1983, 367–368.

24. H.T. Yuan, "GaAs Bipolar Gate Array Technology," *IEEE GaAs IC Symposium Technical Digest*, New Orleans, LA, November 9–11, 1982, pp. 100–103.

25. E.O. Johnson and A. Rose, "Simple general analysis of amplifier devices with emitter, control, and collector functions," *Proceedings of the IRE*, **47**, 1959, 407.

26. E.M. Cherry and D.E. Hooper, *Amplifying Devices and Low-Pass Amplifier Design*, Wiley, New York, 1968, Chapters 2 and 5.

27. R. Beaufoy and J.J. Sparkes, "The junction transistor as a charge-controlled device," *ATE Journal*, **13**, 1957, 310.

28. W. Shockley, *Electrons and Holes in Semiconductors*, Van Nostrand, New York, 1950.

29. D.K. Ferry, *Semiconductors*, Macmillan, New York, 1991.

30. M. Lundstrom, *Fundamentals of Carrier Transport*, Addison-Wesley, Reading, 1990.

31. S.M. Sze, *Physics of Semiconductor Devices*, 2nd ed., Wiley, New York, 1981, Chapter 11.

32. E.S. Yang, *Fundamentals of Semiconductor Devices*, McGraw-Hill, New York, 1978, Chapter 7.

33. M.A. Hollis and R.A. Murphy, "Homogeneous field-effect transistors," in Sze, S.M., (editor), *High-Speed Semiconductor Devices*, Wiley-Interscience, New York, 1990, Chapter 4.

34. S.J. Pearton and N.J. Shah, "Heterostructure field-effect transistors," in Sze, S.M., (editor), *High-Speed Semiconductor Devices*, Wiley-Interscience, New York, 1990, Chapter 5.

35. R.S. Muller and T.I. Kamins, *Device Electronics for Integrated Circuits*, 2nd ed., Wiley, New York, 1986, Chapters 6 and 7.

36. P.E. Gray, D. Dewitt, A.R. Boothroyd, and J.F. Gibbons, *Physical Electronics and Circuit Models of Transistors*, Wiley, New York, 1964, Chapter 7.

37. S.M. Sze, *Physics of Semiconductor Devices*, 2nd ed., Wiley, New York, 1981, Chapter 3.

38. J.R. Brews, "The submicron MOSFET," in Sze, S.M., (editor), *High-Speed Semiconductor Devices*, Wiley-Interscience, New York, 1990, Chapter 3.

39. Y. Tsividis, *Operation and Modeling of the MOS Transistor*, 2nd ed., McGraw-Hill, New York, NY, 1999.

40. M.L. Lee et al., "Strained Si, SiGe, and Ge Channels for High-Mobility Metal-Oxide-Semiconductor Field-Effect Transistors," *Journal of Applied Physics*, **97**, January, 2005, 011101-1–011101-27.

41. K. Rim et al., "Mobility Enhancement in Strained Si NMOSFETs with HfO_2 Gate Dielectrics," *2002 Symposium on VLSI Technology Digest of Technical Papers*, Honolulu, HI, June 11–13, 2002, pp. 12–13.

42. C.W. Liu et al., "Mobility-Enhancement Technologies," *IEEE Circuits & Devices Magazine*, **21**, May–June 2005, 21–36.

43. K. Rim et al., "Fabrication and Mobility Characteristics of Ultra-Thin Strained Silicon Directly on Insulator (SSDOI) MOSFETs," *International Electron Device Meeting Technical Digest*, December 2003, pp. 49–52.

44. M. Ieong et al., "Silicon Device Scaling to the Sub-10-nm Regime," *Science*, **306**, December, 2004, 2057–2060.

45. H.J. Sigg et al., "D-MOS Transistor for Microwave Applications," *IEEE Transactions on Electron Devices*, **ED-19**, January 1972, 45–53.

46. L.D. Nguyen, L.E. Larson, and U.K. Mishra, "Ultra-high-speed modulation-doped field-effect transistors: A tutorial review," *Proceedings of the IEEE*, 80;494, 1992, 494–518.

47. F. Schwierz and J.J. Liou, *Modern Microwave Transistors Theory, Design, and Performance*, Wiley, New York, NY, 2003. HEMTs are covered primarily in Chapter 5.

48. T. Henderson et al., "Microwave performance of a quarter-micrometer gate low-noise pseudomorphic InGaAs/AlGaAs modulation-doped field-effect transistor," *IEEE Electron Devices Letters*, **EDL-7**, December 1986, 649–651.

49. A. Cappy et al., "Status of Metamorphic $In_xAl_{1-x}As/In_xGa_{1-x}As$ HEMTs," *1999 GaAs Integrated Circuit Symposium Digest of Technical Papers*, November 1999, pp. 217–220.

50. U.K. Mishra et al., "AlGaN/GaN HEMTs—An Overview of Device Operation and Applications," *Proceedings of the IEEE*, **90**, June 2002, 1022–1031.

51. C.H. Oxley, "Gallium Nitride: the promise of high RF power and low microwave noise performance in S and I band," *Solid-State Electronics*, **48**, 2004, 1197–1203.

52. R.J. Trew et al., "Microwave AlGaN/GaN HFETs," *IEEE Microwave Magazine*, March 2005, 56–66.

53. K. Asano et al., "Novel High Power AlGaAs/GaAs HFET with a Field-Modulating Plate Operated at 35 V Drain Voltage," *IEEE International Electron Devices Meeting Digest of Technical Papers*, 1998, pp. 59–63.

54. Y.-F. Wu et al., "30-W/mm GaN HEMTs by Field Plate Optimization," *IEEE Electron Devices Letters*, **25**, March 2004, 117–119.

55. S. Karmalkar et al., "Field-Plate Engineering for HFETs," *IEEE Transactions on Electron Devices*, **52**, December 2005, 2534–2540.

56. P.M. Asbeck, "Bipolar transistors," in Sze, S.M., (editor), *High-Speed Semiconductor Devices*, Wiley-Interscience, Wiley, New York, 1990, Chapter 6.

57. H. Kroemer, "Heterostructure Bipolar Transistors and Integrated Circuits," *Proceedings of the IEEE*, **70**, January 1982, 13–25.

58. T.S. Low et al., "Migration from an AlGaAs to an InGaP emitter HBT IC process for improved reliability," *IEEE GaAs IC Symposium Technical Digest*, Atlanta, GA, November 1–4, 1998, pp. 153–156.

59. B. Jalali and S.J. Pearson (editors), *InP HBTs Growth, Processing and Applications*, Artech House, Boston, MA, 1995.

60. D.J. Paul, "Si/SiGe heterostructures: from material and physics to devices and circuits," *Semiconductor Science and Technology*, **19**, 2004, R75–R108.

61. J.S. Dunn et al., "Foundation of rf CMOS and SiGe BiCMOS technologies," *IBM Journal of Research & Development*, **47**, March/May 2003, 101–137.

62. S.S. Iyer et al., "Silicon-Germanium base heterojunction bipolar transistors by molecular beam epitaxy," *IEEE International Electron Device Meeting Digest of Technical Papers*, 1987, pp. 874–876.

63. M. Rodwell et al., "Submicron Scaling of HBTs," *IEEE Transactions on Electron Devices*, **48**, November 2001, 2606–2624.

64. J.D. Cressler, "SiGe HBT Technology: A New Contender for Si-Based RF and Microwave Circuit Applications," *IEEE Transactions on Microwave Theory & Techniques*, **46**, May 1998, 572–589.

65. R. Szweda, "SiGe set for expansion," *III-Vs Review*, **18**, November 2005, 40–41.

66. IBM Press Release, "IBM Announces Next Generation Silicon Germanium Technology," East Fishkill, NY, August 5, 2005.

73

Compound Semiconductor RF Circuits

Donald B. Estreich

Agilent Technologies

CONTENTS

In this chapter we describe some of the circuits that benefit from the high performance of compound semiconductor devices. Silicon has been dominant in transistor manufacturing since the 1950s and integrated circuits (ICs) continued this dominant position. However, compound semiconductors have long held out the promise of superior performance for some applications. The next section discusses the potential of GaAs, InP, and GaN in compound semiconductor devices.

73.1   Why Compound Semiconductors Instead of Silicon

In Chapters 71 and 72 the most widely used compound semiconductor properties were summarized and the most important compound semiconductor active devices were presented. The early work on compound semiconductors was largely, but not exclusively, focused upon gallium arsenide (GaAs). Its attraction was predominantly its (1) high electron mobility and (2) high electron drift or saturation velocity. Also important were (3) the ability to form reliable Schottky barriers on GaAs and (4) the very high resistivity semi-insulating substrate in GaAs. The semi-insulating substrate was very significant because low parasitic capacitances could be achieved with GaAs substrates and low-loss transmission lines, usually in the form of microstrip and coplanar waveguide transmission lines that could be built directly on the top surface of the GaAs substrate. It is hard to form low-loss transmission lines at RF and microwave frequencies directly upon the silicon substrate because even residual doping levels render the substrate too conductive for low-loss transmission lines. Today, silicon IC structures form low-loss transmission lines only with the added complexity of including insulating layers and multiple metallization layers. Another important aspect of GaAs is its ability to form various heterojunction structures to form very high-performance devices. This is accomplished with epitaxial layer growth using either *molecular beam epitaxy* (MBE) or *metalo-organic chemical vapor deposition* (MOCVD). The same is true for indium phosphide (InP), although the InP work followed that of GaAs. More recently, gallium nitride (GaN) has attracted a great deal of attention because of its superior high voltage and high power possibilities. Our interest here is in RF, microwave and millimeter-wave active devices, and ICs and we will not discuss the use of GaAs to produce light-emitting diodes (LEDs) or solid-state lasers [1]. LEDs and lasers have played the most important role in the history of material growth technology for compound semiconductors and, of course, form a very large and important technological market for compound semiconductors.

73.2   Compound Semiconductor Applications

The early 1970s saw the emergence of a number of companies commercializing GaAs FETs followed in the 1980s by work on GaAs ICs. Most of the early work focused upon military markets for military communications, radar, and weapons systems. The Department of Defense investigation and development funding was crucial in getting the compound semiconductor industry off the ground. But manufacturing volumes were (and still are) somewhat limited for military applications. Of course, there were numerous other niche applications where compound semiconductors were very important. For example, Hewlett-Packard built an internal compound semiconductor facility to provide FETs and ICs for RF and microwave instrumentation. This effort started in the mid-1970s and continues today to be of great value at Agilent Technologies. But the volumes of instrumentation have never been sufficient to support the large volume of business as required by the compound semiconductor industry to grow and become cost effective.

The wireless explosion in the late 1990s, and accelerating to the present, has been vital to the compound semiconductor industry. It provided substantial growth opportunities for this industry. As an example, consider the simplified W-CDMA transceiver block diagram shown in Figure 73.1. Two components in which compound semiconductors have been widely used are for switches and power amplifiers. Switches are discussed in more detail in Section 73.3.5 below. GaAs power amplifiers are essentially dependent upon the properties of the active devices used. Three parameters are especially important for power amplifiers: (1) low distortion at the power output required, (2) operating efficiency must be high because many wireless applications are battery powered, and (3) lowest cost among any of the acceptable solutions. For some time the GaAs MESFET has filled this role for cell phone handsets; however, silicon–germanium heterojunction bipolar transistors are emerging

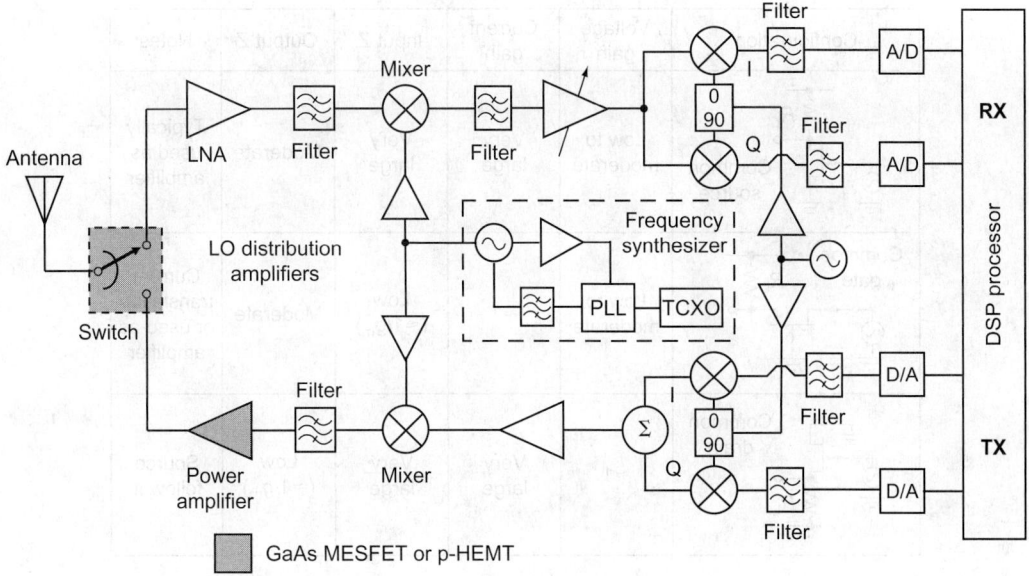

FIGURE 73.1 Typical RF cellphone transceiver showing blocks, where compound semiconductor devices dominate in this application.

contenders. Wireless application component sockets are being challenged by silicon, especially by silicon–germanium HBT ICs [2–4] and by RF CMOS [5–7]. Of course, compound semiconductor solutions for many functions, such as mixers, low-noise amplifiers, and wideband amplifiers, to name only some, exist and are important. The above discussion has centered upon RF wireless applications—this area is more competitive because silicon has high performance. As the frequency increases into the microwave and millimeter-wave bands, compound semiconductor solutions do excel in performance and are less sensitive to cost.

73.3 RF and Microwave Amplifiers

All electronic and communication systems require amplification to boost signal levels. Undoubtedly, amplifiers are the most common functional block in electronic systems. There are several common amplifier topologies widely used for RF and microwave amplifiers that we discuss in this section. The foundation of all amplifiers is the active device around which they are designed. Generally, amplifiers require fast devices with high f_T and f_{max} values that must be properly biased and impedance matched to provide for efficient power transfer. Power amplifiers also require specific devices that are capable of delivering sufficient current to their load (often this means driving a 50 Ω transmission line) and swinging sufficient voltage without experiencing channel or junction breakdown. We begin the discussion with the circuit properties of some of the commonly used gain blocks.

73.3.1 Single-Transistor Connections

Consider first the unipolar or field-effect transistor (FET) which can be connected in three unique two-port connections. These are the *common source*, *common gate*, and *common drain* connections appearing in Figure 73.2. This table summarizes selected key parameters such as the voltage gain, current gain, and port impedances. Typically, the common-source and common-gate configurations are used for gain blocks, whereas the common drain stage or *source follower* is an impedance transforming stage used to drive capacitive or low-impedance loads. For the bipolar junction transistor (BJT/HBT) Figure 73.3 presents analogous information for the *common emitter*, *common base*, and *common collector* (or *emitter follower*) configurations. Together, these two tables make it easy to compare the parameters of the FET with those of the BJT/HBT.

Configuration	Voltage gain	Current gain	Input Z	Output Z	Notes
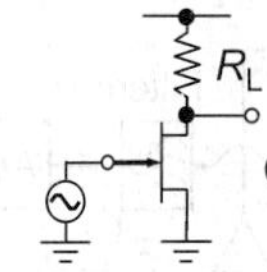 Common source	Low to moderate	Very large	Very large	Moderate	Typically used as amplifier
Common gate 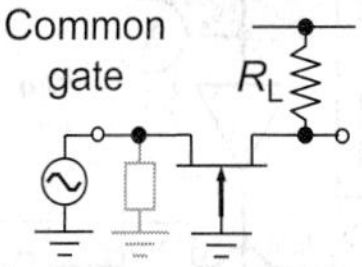	Low to moderate	≈ 1	Low $(\approx 1/g_m)$	Moderate	Current translation or used as amplifier
Common drain 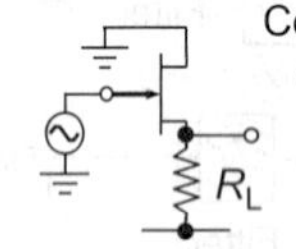	≈ 1	Very large	Very large	Low $(\approx 1/g_m)$	Source follower

FIGURE 73.2 Table of the three commonly used FET configurations comparing voltage gain, current gain, input impedance, and output impedance among all three configurations. The table presents small-signal parameters at low-to-intermediate frequencies.

Configuration	Voltage gain	Current gain	Input Z	Output Z	Notes
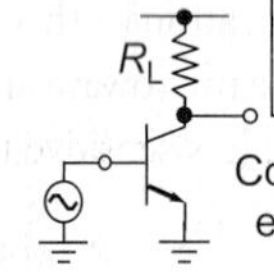 Common emitter, $g_m = \dfrac{qI_c}{kT}$	Large	Moderate $= \beta$ (or h_{FE})	Low to moderate $\dfrac{\beta}{g_m} + \beta r_o + r_b$	Large	Typically used as amplifier
Common base 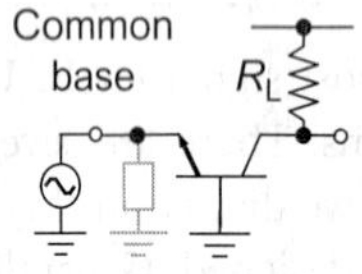	Large	≈ 1	Low $\dfrac{1}{g_m} + \dfrac{r_b}{\beta} + r_e$ $(\approx 1/g_m)$	Large	Current translation or used as amplifier
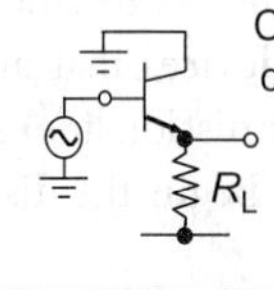 Common collector	≈ 1	Moderate $= \beta$ (or h_{FE})	Very large	Low $\dfrac{1}{g_m} + \dfrac{r_b}{\beta} + r_e$ $(\approx 1/g_m)$	Emitter follower

FIGURE 73.3 Table of the three commonly used BJT/HBT configurations comparing voltage gain, current gain, input impedance, and output impedance among all three configurations. The table presents small-signal parameters at low-to-intermediate frequencies.

73.3.2 Two-Transistor Gain Blocks

There are combinations of transistors that form attractive building blocks for RF circuits [8]. Some of these are shown in Figure 73.4. The *cascode stage* is shown in Figure 73.4(a) and Figure 73.4(b) for both FET and BJT/HBT devices. The cascode cell has two primary advantages over using a single common source or common emitter device. First, the overall output-to-input internal feedback capacitance is reduced by using two transistors instead of one. Second, its output impedance is higher resulting in

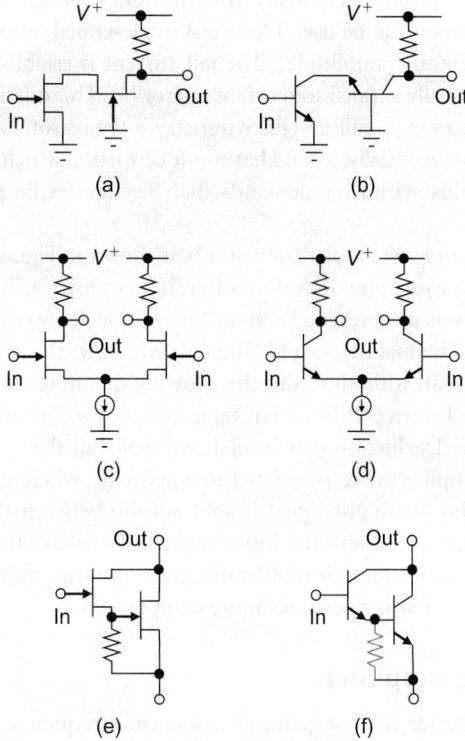

FIGURE 73.4 Various two-transistor gain cells. (a) FET cascode configuration, (b) BJT/HBT cascode configuration, (c) FET differential configuration, (d) BJT/HBT differential configuration, (e) FET f_T-doubler, and (f) BJT/HBT Darlington connection.

higher voltage gain. The transconductance g_m of the cascode cell is primarily determined by the input device because the output device has unity current gain. For a single transistor with voltage gain equal to $g_m R_{load}$, the internal feedback capacitance C_{FB} (specifically, it is C_{gd} for a FET and C_{bc} for the BJT/HBT) is multiplied by the *Miller effect* factor $(1 + g_m R_{load})$, which is typically much greater than unity. The Miller capacitance is positioned at the input node of the device leading to reduced bandwidth. However, in the cascode cell the voltage gain of the input device is small* because the transistor's load is the input impedance of a common gate or common base transistor which is usually lower than R_{load}. The feedback capacitance for the cascode connection is further reduced owing to the feedback capacitors of both active devices connected in series. This is why the overall feedback capacitance is much reduced in the cascode configuration. It has important stability advantages in RF and microwave circuits, where the feedback capacitance can produce positive feedback under certain conditions. The output resistance of the cascode cell is higher because the output resistance of the input device is multiplied by the intrinsic voltage gain of the output device. This can be quite large in some cases. The disadvantages of the cascode cell are the requirement of a larger voltage and the bias and layout complexity of two devices rather than one device.

Another very useful cell is the *differential pair* or *emitter coupled* (and source coupled) pair. Differential cells are used to amplify the difference of two signals, that is double-ended signals, and reject signals common to both input terminals. The differential cell is illustrated in Figure 73.4(c) for FET devices and in Figure 73.4(d) for BJT/HBT devices. The bias circuitry for setting the terminal voltages and currents is not

---

*For a FET cascode (both FETs are the same size) the resistance presented to the input FET is approximately $1/g_m$, therefore, the voltage gain of the input FET is approximately equal to $g_m(1/g_m) \cong 1$. The same is true for the BJT/HBT device.

shown. The output is generally taken differentially from the nodes connecting the device outputs to their load resistors, although at times it can be useful to take a single-ended output (of course, with a penalty of it being one-half the differential amplitude). The tail current is established with the current source shown in Figure 73.4(d) is typically a transistor current mirror [9]. This differential gain cell is an excellent example of the use of symmetry in circuit design. Symmetry is the reason for the differential cell's ability to reject even-order distortion (especially, second harmonic distortion which is often the largest distortion signal present). Of course, this symmetry demands that the devices be parameter and performance matched.

Another gain cell is the *Darlington* bipolar transistor pair shown in Figure 73.4(f). Although a resistor is shown across the emitter-base junction of the lower BJT/HBT transistor, in the original implementation by S. Darlington this resistor was not present. Without the resistor the overall current gain is the product of the current gains of the individual transistors. The resistor allows the input transistor to be operated at a higher collector current than without it, thereby allowing the input transistor to be biased where it has a higher f_T value. With FET devices it is not advantageous to work in terms of current gain; however, a configuration related to the Darlington pair is what we shall call the f_T-*multiplier*. This is shown in Figure 73.4(e). In the f_T-multiplier the resistor is set to equal $1/g_m$, where g_m is the transconductance of the input FET. With this value the input signal is split equally between the two FET devices (we are assuming that they are of equal size) and the input capacitance is effectively halved owing to voltage division at the input. The current multiplication comes from the two drains being connected together. Moreover, with two devices the biasing becomes more complicated.

73.3.3 Broadbanding Amplifiers

The response of any active device is a flat gain out to a corner frequency f_B and thereafter declines at −20 dB/decade as the frequency increases. This is illustrated in Figure 73.5 with the small-signal current gain versus frequency plotted for the equivalent circuit in Figure 73.5(a). The current gain $h_f(f)$ response into a short-circuit load is plotted on logarithmic axes in Figure 73.5(b). Recall from the charge control discussion in a prior chapter that the low-frequency current gain h_{fo} is simply the ratio of the two charge control time constants, namely (τ/τ_r). An important and often quoted figure of merit

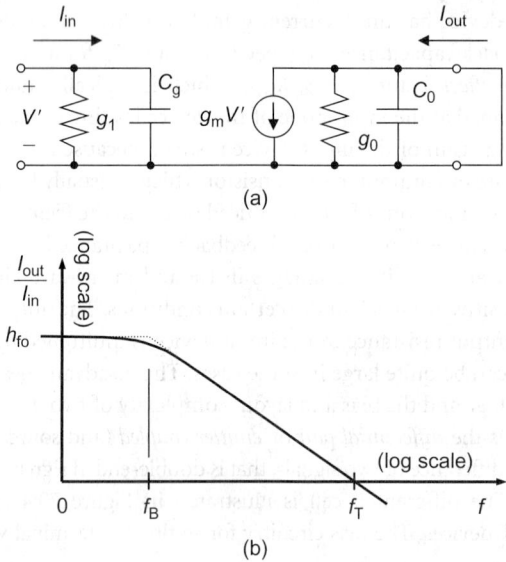

FIGURE 73.5 (a) Small-signal equivalent circuit for FET or BJT with short-circuited output and (b) log–log plot of current gain versus frequency showing definition of f_T as the current gain extrapolating to unity.

is the *current gain-bandwidth product*, denoted commonly by f_T, being the frequency where h_f(at $f = f_T$) = unity. The input capacitance previously described in the charge control presentation is the origin of this behavior—this is the capacitor labeled C_g in the equivalent circuit in Figure 73.5(a). For the FET device, $f_T = [g_m/(2\pi C_g)]$ and for a BJT/HBT device $f_T = [g_m/(2\pi C_\pi)]$, where the internal feedback capacitance has been ignored in these approximations.

We also come to the same result for a cascade of identical stages each loaded with resistor R_L. In this case, we consider voltage gain rather than the above analysis using current gain. If we use the very simple equivalent circuit, where the input is capacitor C_{in} and the output is the transconductance g_m driven by the voltage across the capacitor, then the voltage gain per stage is just $-g_m R_L$. This implies a bandwidth BW = $(2\pi R_L C_{in})^{-1}$ because this is just the effective RC time constant at the node. Multiplying the voltage gain times the bandwidth gives the *gain-bandwidth product* equal to $[g_m/(2\pi C_{in})]$. But this is just the cutoff frequency f_T because C_{in} is essentially C_g or C_π.

A transistor can be presented with a variety of loads as suggested in Figure 73.6. Both passive and transistor or active loads can be connected. In addition to the common passive loads, an active FET load is shown where the gate and source are tied together. An active load has the advantage of allowing for a higher resistance (hence, higher voltage gain) than would be possible if a simple passive resistor were used. Furthermore, active loads can be quite small in area and have layout advantages owing to this.

Consider an amplifier with a single resistor load, denoted by R_L, and parasitic load capacitance[†] C as shown in Figure 73.7(a). In this case, the bandwidth is generally limited by the RC time constant at the output node. The addition of an inductor can be used in conjunction with the load resistor to extend the amplifier's bandwidth by producing peaking in the frequency response or transient response. An example of *resonant peaking* or *series peaking* [10] is shown in Figure 73.7(b). For this discussion define a dimensionless factor $m = (L/R_L^2 C)$—note that m is proportional to L by this

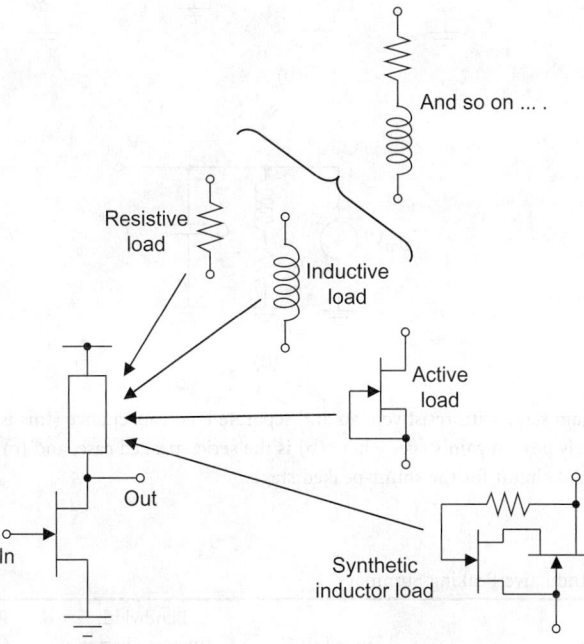

FIGURE 73.6 Five different representative loads with the resistive, active, and resistor-inductor loads being the most common. A variety of loads may be presented to an active device.

[†]Sometimes capacitance C is the input capacitance of the following stage.

definition. In series peaking selecting $m = 0.5$ extends the bandwidth by a factor of the square root of two (i.e., by 1.414) compared to the case of $m = 0$ with no inductor present. However, it is possible to do even better with resonant peaking. *Shunt peaking* [10] is illustrated in Figure 73.7(c), where now the inductor is placed in series with the resistive load. The maximum bandwidth extension possible with shunt peaking is 1.85 corresponding to $m = 0.71$. However, the penalty is a 19% peaking in response above the low-frequency gain of the amplifier. For some applications such peaking may not be acceptable; hence to reduce peaking, lower values of m can be chosen. For example, choosing $m = 0.5$ reduces the bandwidth extension to 1.80, but now the peaking response is only 3%. The $m = 0.5$ case corresponds to the magnitude of the inductive reactance equaling the resistance R_L. Reducing the value of m to 0.41 essentially eliminates the peaking, while still giving a bandwidth boost of 1.72. This case is often called the *maximally flat amplitude* setting. Finally, for *maximally flat delay* or *optimum linear phase* m is set to 0.32 resulting in a bandwidth boost of 1.60. For analytical purposes the equivalent circuit shown in Figure 73.7(d) is adequate to quantify shunt peaking. A simple way to look at shunt peaking is that the inductor delays the current flow in the resistor so that momentarily more current is available to charge and discharge the capacitor. All of the above information is summarized in Table 73.1.

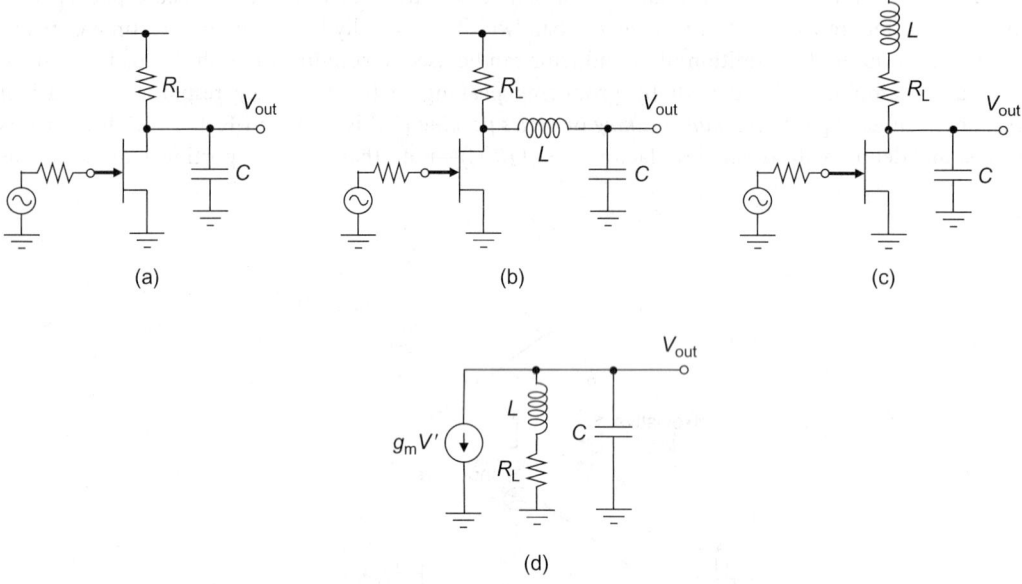

FIGURE 73.7 (a) FET gain stage with resistive load and separate load capacitance (this is taken as reference). Two commonly used inductively peaked gain stages where (b) is the series-peaked case, and (c) is the shunt-peaked gain stage. (d) Simple equivalent circuit for the shunt-peaked stage.

TABLE 73.1 Inductive Peaking Summary

Load Condition	$m = L/R_L^2 C$	Bandwidth Increase Factor	Percent (%) Peaking
Resistor only load (Reference)	0.00	1.00	None
Maximum bandwidth	0.71	1.85	19%
$|Z_L| = R_L @ f_{RC} = (2\pi R_L C)^{-1}$	0.50	1.80	3%
Maximum flat magnitude	0.41	1.72	0%
Maximum flat delay	0.32	1.60	0%

There are more sophisticated combinations of inductances leading to still greater bandwidth boosts than the simple series or shunt peaking described above. Using inductance to separate the capacitance of the active device from the load capacitance is the principle used here for extending bandwidth. Of course, the trade-off is always that separation of the capacitance to be charged and discharged results in additional delay. Partitioning the capacitance leads to serial charging of these separate capacitors. When greater bandwidth is more important than delay of the signal transmission, then inductive partitioning is a good trade-off. The final example we present is the *bridged-T coil* network [10,11] shown in Figure 73.8. It is a transformer-coupled network presenting a constant resistance R_L at the input node (i.e., the node connected to the drain of the FET) when properly designed. Although we show the use of the bridged-T coil at the output of the gain stage, this network is also useful at the input to drive the input capacitance of the active device. The bridged-T coil can boost the bandwidth by a factor of approximately 2.8 in a maximally flat amplitude condition (i.e., *Butterworth*) and approximately 2.7 for the maximally flat delay condition. A requirement for the bridged-T coil is that $C_b \ll C$. Design equations are given by both Lee [10] and Feucht [11].

73.3.4   Common RF and Microwave Amplifiers

Next we briefly discuss four commonly used RF and microwave amplifier topologies (but not all such topologies of course). Figure 73.9 shows (a) the *reactive matched* amplifier, (b) the *lossy match* amplifier, (c) the *feedback* amplifier, and (d) the *distributed* (sometimes also called the *additive* or *traveling-wave*) amplifier.

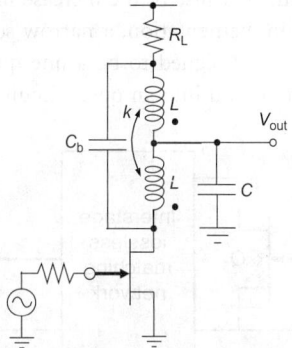

FIGURE 73.8   Bridged-T coil peaking network for bandwidth extension. This is also known as an all-pass network or a constant-R network.

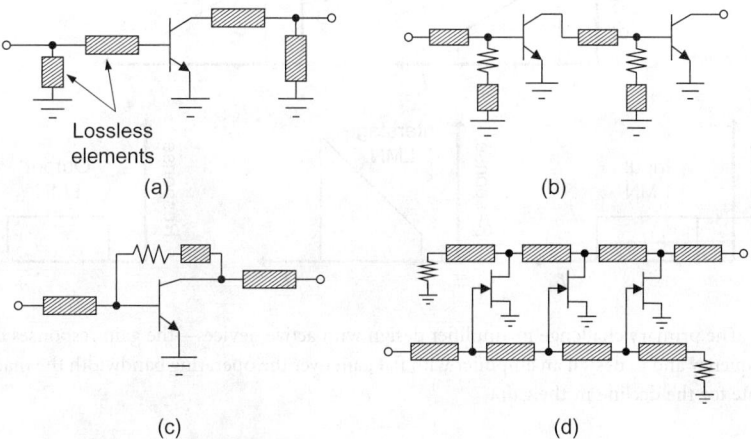

FIGURE 73.9   Four commonly used microwave amplifiers. (a) The reactive match amplifier, (b) lossy match amplifier, (c) feedback amplifier, and (d) distributed (also known as the additive or traveling wave) amplifier.

Reactively matched amplifiers have traditionally been widely used[‡] because of their high gain and low-noise figure when designed accordingly [12–14]. Inductors and capacitors are the reactive elements used in various combinations, either lumped at lower RF frequencies or distributed at higher frequencies, to impedance match at the desired frequency band of operation. Generally, this type of amplifier is used for narrow-band applications (less than or at most an octave) because of the matching restrictions set by the Bode and Fano criteria [15]. Excluding resistors in the matching networks reduces signal transmission loss and eliminates thermal noise that would be added by resistive components. The general behavior of active devices in the RF and microwave region is a falling response at the rate of −20 dB/decade (as shown in Figure 73.5) because the charge control capacitance dominates as described in the previous chapter. Linear matching networks are designed to compensate for this falling response as schematically shown in Figure 73.10. A two-stage amplifier is illustrated in Figure 73.10 with input, interstage, and output matching networks. Typically, these matching networks are designed to cancel the response of the active device by having a +20 dB/decade response per device. In the figure the entire compensation is accomplished in the interstage matching network, but in general this can be parceled out among the individual linear matching networks [16].

Lossy matched amplifiers are capable of being designed to have flat responses over much wider frequency bands than the reactively matched amplifiers. The shunt network used in lossy matched amplifiers is a series inductor and resistor. This is shown in Figure 73.11 for both the lumped element and distributed element implementations. The essential idea is for the resistor to set the amplifier's gain at low to mid-range frequencies, while the reactive element removes the effect of the resistor at the high end of the band. This occurs because of the increase in reactive impedance $|\omega L|$ at the high end of the band. For the distributed implementation, a narrow section of transmission line is chosen because it is inductive and the length is designed to be a one-quarter wavelength at the upper band frequency—this rotates the short termination into an open circuit at this upper band frequency.

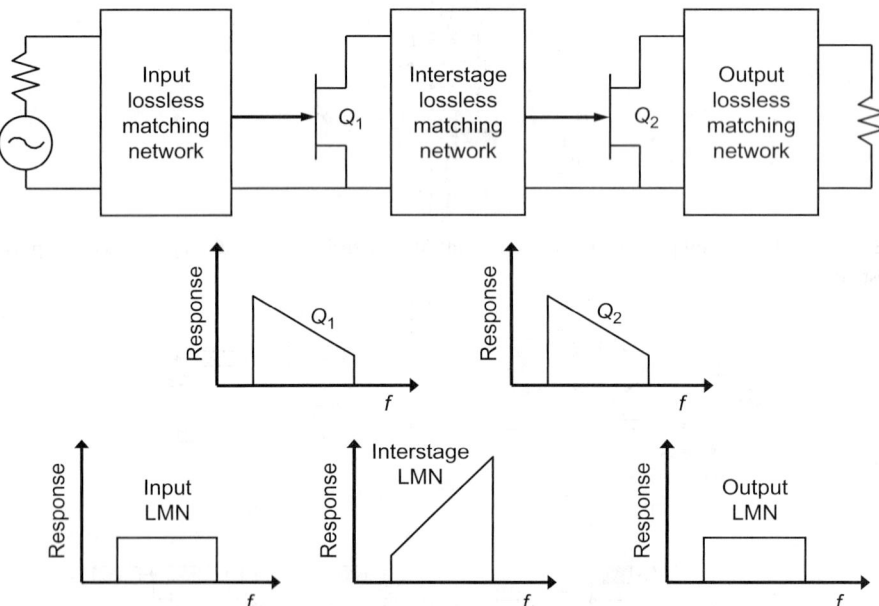

FIGURE 73.10   The primary challenge in amplifier design with active devices—the gain responses of active devices decline with frequency and to design an amplifier with flat gain over the operating bandwidth the matching networks must compensate for the decline in the gain.

[‡]Many books and a multitude of papers have been published on the reactively matched amplifier. For this reason we do not discuss the design details.

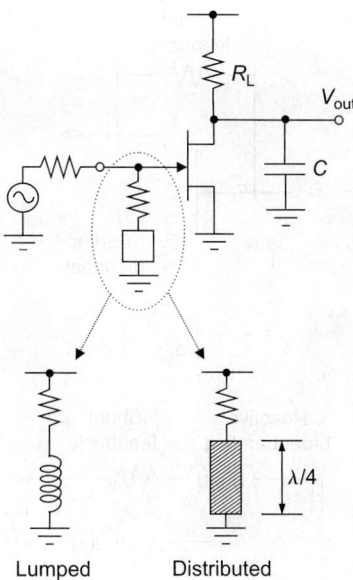

FIGURE 73.11   Lossy match gain stage showing two implementations—lumped *R–L* branch and distributed inductance branch.

Feedback [17] is commonly used in amplifiers for several reasons. These include (1) to flatten out the gain response for wider bandwidth, (2) to reduce sensitivity to variation in active device parameters, (3) to reduce distortion (and sometimes noise), (4) to control impedances at the input and output, and (5) to practically realize driving-point immittance[§] and transfer functions. There are some disadvantages of feedback including (1) more complexity in the circuitry, (2) gain reduction at lower frequencies when extending the bandwidth (but not really a problem if the open-loop gain is high), and (3) instability. If the phase of the feedback signal is not carefully controlled, it is possible to have *positive feedback* leading to unwanted oscillations. For amplifiers there is a very close relationship among gain, bandwidth, amount of feedback, and stability. An example of feedback is shown in Figure 73.12(a) where both *series feedback* are connected to the source terminal and *shunt feedback* connected from drain to gate. Either one can be applied individually and are often used together. Putting all of this together Figure 73.12(b) shows a single-stage FET amplifier with feedback, reactive broadbanding with inductors, and input and output matching networks, although sometimes it is hard to distinguish between impedance matching and inductive broadbanding.

73.3.5   Distributed Amplifiers

The last amplifier type considered in this section is the distributed amplifier. We begin with a brief discussion of *artificial transmission lines* [18]. This is really the issue of the distributed element versus lumped element construction of a transmission line. A transmission line such as a coaxial cable, microstrip, or coplanar waveguide formed upon an insulating substrate can be modeled by a ladder network of inductances and capacitances as shown in Figure 73.13(a). The equivalent circuit for the transmission line is partitioned into short segments of length Δx, with series inductance L_l and shunt capacitance C_l values assigned to each segment proportional to length Δx. For the purely distributed line segment length Δx is allowed to become infinitesimally small. This gives a characteristic impedance, denoted by the symbol Z_o, for the transmission line that is real, assuming losses are ignored, and equal to $[L_l/C_l]^{1/2}$. An ideal distributed transmission line has an infinite bandwidth. The delay per section is denoted by τ_D and is equal to $[L_lC_l]^{1/2}$.

[§]Impedance or admittance, whichever is appropriate for the situation.

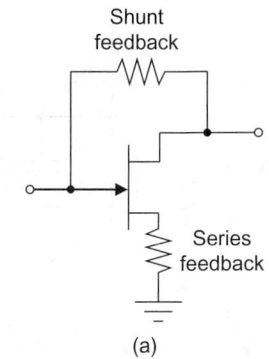

(a)

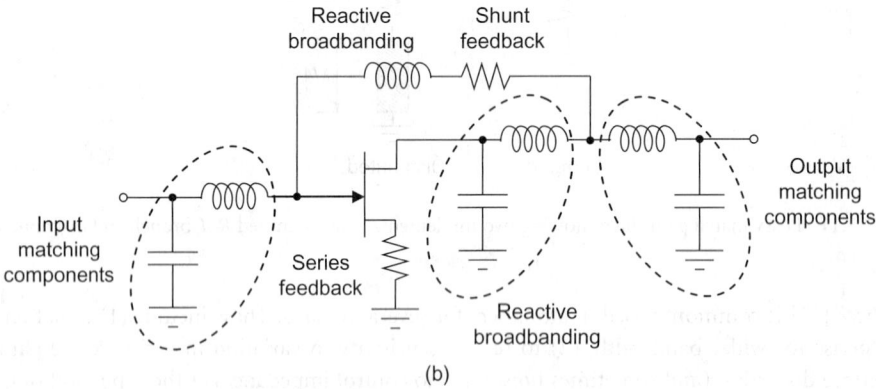

(b)

FIGURE 73.12 (a) Feedback amplifier with two feedback paths. (b) Common-source gain stage with shunt and series resistive feedback, reactive broadbanding and lumped input and output matching *L*-networks.

When Δx is not infinitesimal, we have a lumped transmission line. Such transmission lines are used for filters and delay lines in practice. Lumped element transmission lines have a cutoff frequency as opposed to an infinite bandwidth of the purely distributed line. The *cutoff frequency* is

$$f_{\text{cutoff}} = 1/\pi (L_l C_l)^{1/2} \tag{73.1}$$

At the cutoff frequency, no energy can propagate down the transmission line because of the abrupt discontinuity in the phase velocity. It is analogous to Bragg scattering in crystals because of the periodic planes of atoms. Only standing waves exist at the cutoff frequency. A signal will propagate down the lumped or artificial transmission line below the cutoff frequency, but not above it. The characteristic impedance of the lumped transmission line also varies with frequency and is discontinuous at the cutoff frequency. The original work relating to artificial transmission lines, specifically on electric wave filters, dates back to 1910 by George Campbell[§] at AT&T. These transmission lines are often synthesized using the theory of constant-*k* and *m*-derived filters [17,18]. For example, an *m*-derived section with $m \cong 0.6$ allows for the transmission line to be used to ~0.8–0.9 times the cutoff frequency [19]. The bridged-T coil section mentioned above can also be used—this is sometimes referred to as an *all-pass filter* or *constant-R* network section.

Next consider lumped-element periodic loading of the transmission line with the capacitance from the connecting to the active devices. Suppose an active device has a total node capacitance of C_D and we

---

[§]Campbell eventually field for a patent in 1915 for electric wave filters which was granted in 1917 (U. S. Patent 1,227,113).

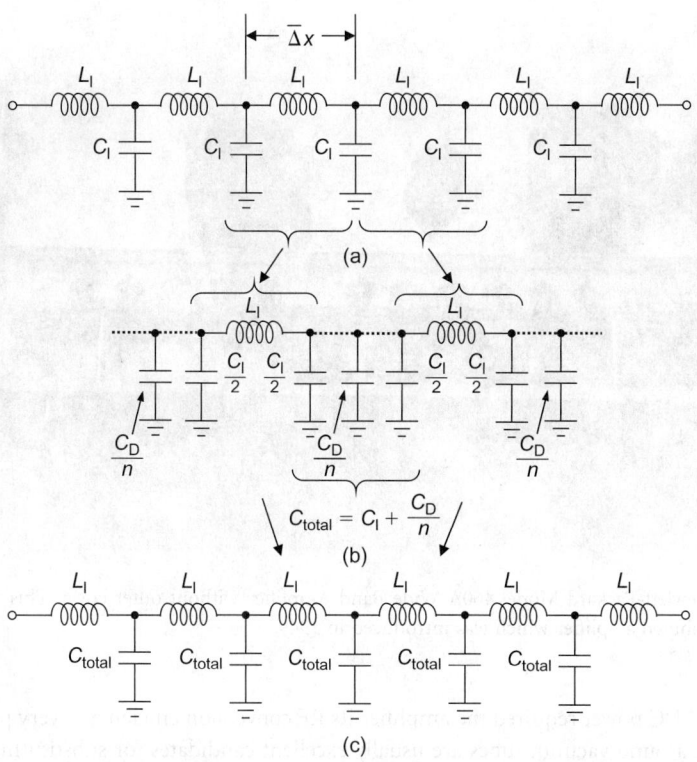

FIGURE 73.13 (a) Lumped-element transmission line-equivalent circuit with n sections of length Δx, (b) adding lumped capacitors (value C_D/n at nodes), and (c) new artificial transmission line after lumped capacitors have been added.

want to split this device into n equal segments so as to distribute the device along the transmission line. The capacitive loading at each connection point along a transmission line will be C_D/n because it is $1/n$th of the active device's node capacitance. Figure 73.13(b) shows the addition of capacitance C_D/n at each node of the lumped transmission line of Figure 73.13(a). This gives a new lumped transmission line with parameters C_{total} and L_l as illustrated in Figure 73.13(c). In a monolithic form the loaded transmission line can be formed in microstrip with segments of narrow (i.e., high Z_o) microstrip being short enough to be inductive. The capacitance comes from the microstrip conductor-to-ground capacitance for each section plus the loading capacitance from the active devices being connected at their respective nodes.

A distributed amplifier is a circuit approach that pushes the bandwidth versus delay trade-off to its limit. In fact, it allows an active device to be used beyond its cutoff frequency f_T with the highest bandwidth being limited only by f_{max} of the device. The distributed amplifier circuit idea was first patented in 1937 by William Percival [20] in the United Kingdom. However, distributed amplifiers generated strong interest following an important paper[||] by Ginzton et al. in 1948 [21]. The first commercial product using the distributed amplifier concept was a vacuum tube distributed amplifier featured in the *HP Journal* (1:1) in September 1949 [22]. The Hewlett-Packard Model 460A Wide Band Amplifier appears in Figure 73.14 without its case. The vacuum tubes (twelve 6AK5 tubes) and the wound coils and disk capacitors comprising the lumped-element artificial transmission line are clearly visible. It was primarily marketed as a pulse amplifier and had a Gaussian bandwidth of about 140 MHz with a gain of 20 dB into 200 Ω and the rise time was 2.9 ns with no overshoot. Its maximum output voltage into 330 Ω was about 4.75 V

[||]This paper was quite comprehensive covering principles, equalization of the response with m-derived networks and bridged-T coil networks, proper line terminations, the use of tapered output lines, the effect of line losses, and noise behavior.

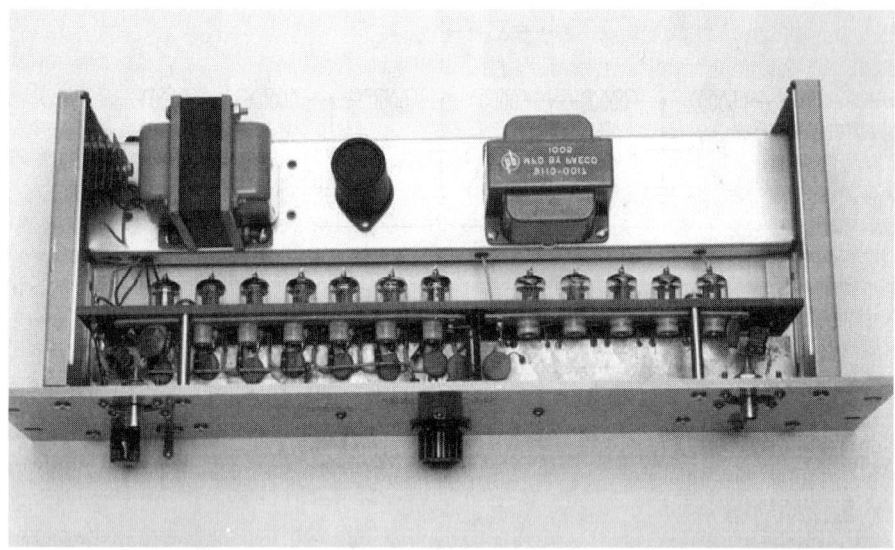

FIGURE 73.14 Hewlett-Packard Model 460A Wide Band Amplifier without outer cover. This is an example of a vacuum tube distributed amplifier which was introduced in 1949.

peak—with 35 W DC power required the amplifier its RF conversion efficiency is very poor. Incidentally, circuits designed around vacuum tubes are usually excellent candidates for substituting semiconductor FETs, benefiting especially from their much higher frequency response.

The earliest reported work (in 1981 and 1982) on distributed amplifiers using GaAs monolithic platforms were by Strid and Gleason [23] and by Ayasli et al. [24,25]. The work of Strid and Gleason at Tektronix involved spiral inductors interconnected to six gain cells, each consisting of 300 μm × 0.7 μm MESFET devices, thereby forming the distributed amplifier all in an area of <1 mm². It operated from DC −12 GHz with 7–9 dB of gain and had a pulse risetime of about 40 ns. The work at Raytheon by Yalcin Ayasli and team produced a 1–13 GHz, four-cell distributed amplifier. The four gain cells were MESFET devices, 300 μm × 0.8 μm each, embedded in a microstrip layout producing approximately 9 dB of gain. This performance was quickly extended to 2–20 GHz and reported the following year by Ayasli [26]. More recently, even bipolar transistor and CMOS devices have been used to construct distributed amplifiers.

A distributed amplifier suitable for monolithic form is shown schematically in Figure 73.15. Seven cascoded gain cells are equally spaced along the transmission line. Figure 73.16 shows a photomicrograph of the MMIC realization of this 2–26.5 GHz distributed amplifier [27] in Hewlett-Packard's MMIC GaAs 0.4 μm gate length MESFET process [28]. This process had an MBE doped layer at a doping level of 2.5×10^{17} donors per cm³, with recessed gate and silicon nitride passivation.

In the design of distributed amplifiers the L/C ratio of the artificial transmission line sets the characteristic impedance, usually 50 Ω so it easily interfaces with other RF components. The L/C ratio is constrained because the LC product must allow for a high enough cutoff frequency to meet the bandwidth requirement of the amplifier. Lower LC product values result in higher cutoff frequencies, but require smaller devices to keep C small. But smaller devices limit the output power that can be delivered to a load. For a distributed amplifier with n sections (i.e., gain cells) and a transconductance of g_{m} per section, the total voltage gain A_{v} is approximated by

$$A_{\mathrm{v}} = \frac{n g_{\mathrm{m}} Z_{\mathrm{o}}}{2} \qquad (73.2)$$

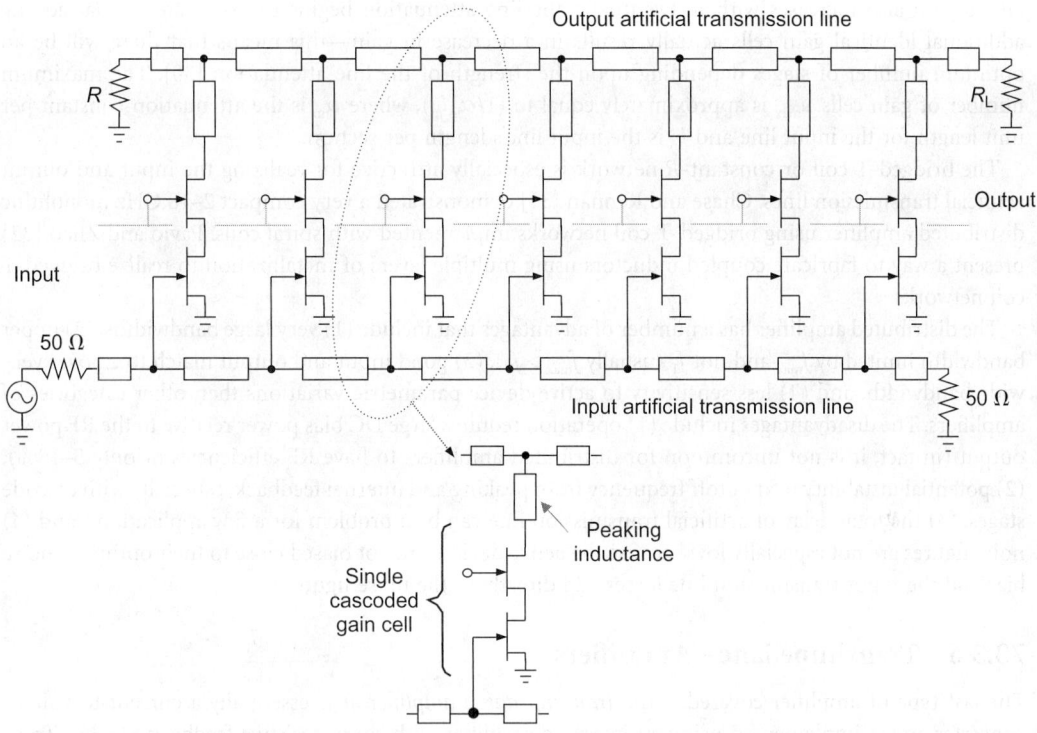

FIGURE 73.15 Equivalent circuit of seven-cell distributed amplifier using cascoded gain cells. The short sections of transmission lines are distributed inductive elements.

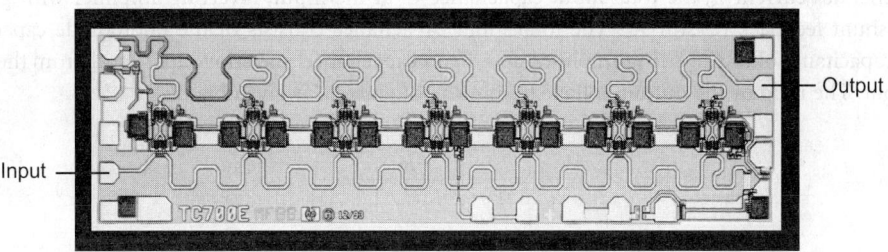

FIGURE 73.16 Hewlett-Package monolithic, 2–26.5 GHz distributed amplifier built with a 0.4 μm gate MMIC process.

where each section adds or contributes $\frac{1}{2}g_m Z_o$ with n gain cells adding together. The impedance presented to the output node of each device is equal to $Z_o/2$. The key point here is that this is an additive architecture as compared to the multiplicative architecture of most other classes of amplifiers. It is essential that the delay between gain cells be identical for both the input and output transmission lines. As we add additional gain cells the total gain increases by factor n and the total delay also increases by factor n. Most active devices have much larger input capacitances than output capacitances. Sometimes, a short series-connected transmission line at the device's output node is used to compensate for this difference. This is shown in Figure 73.15 and is labeled as a "peaking inductance." Eq. (73.2) suggests that continuing to add gain cells allows the voltage gain to be as high as we would like. However, we have neglected losses in both input and output transmission lines [29]. While adding gain cells initially increases the voltage gain by factor n, the total transmission line length also increases by factor n, and transmission line attenuation also increases with *n*. Eventually, the line attenuation begins to dominate so that adding additional identical gain cells actually results in a decrease in gain—this means that there will be an optimum number of stages depending upon the strength of the line attenuation [30]. The maximum number of gain cells n_{max} is approximately equal to $(1/\alpha_{in}l_{in})$, where α_{in} is the attenuation constant per unit length for the input line and l_{in} is the input line's length per section.

The bridged-T coil or constant-*R* network is especially attractive for realizing the input and output artificial transmission lines. Chase and Kennan [31] demonstrated a very compact 2–18 GHz monolithic distributed amplifier using bridged-T coil networks implemented with spiral coils. Pavio and Zhao [32] present a way to fabricate coupled inductors using multiple layers of metallization to realize bridged-T coil networks.

The distributed amplifier has a number of advantages that include (1) very large bandwidths, (2) upper bandwidth limited by f_{max} and not f_T (usually $f_{max} > f_T$), (3) good input and output match to Z_o over very wide bandwidth, and (4) less sensitivity to active device parametric variations than other categories of amplifiers. The disadvantages include (1) operation requires large DC bias power relative to the RF power output (in fact, it is not uncommon for distributed amplifiers to have RF efficiencies of only 5–10%), (2) potential instability near cutoff frequency from peaking and internal feedback, especially with cascode stages, (3) the total delay of artificial transmission line can be a problem for a few applications, and (4) noise figures are not especially low because the active devices are not biased close to their optimum noise bias and the input transmission line losses add directly to the noise figure.

73.3.6 Transimpedance Amplifiers

The last type of amplifier covered is the *transimpedance amplifier*. It is essentially a current-to-voltage converter and is implemented using an inverting amplifier with shunt resistive feedback [33,34]. Transimpedance amplifiers are used for optical receivers where an avalanche phototdiode or PIN diode is used to convert the optical signal to a current at the input of the transimpedance amplifier. The primary elements in the transimpedance amplifier are illustrated in Figure 73.17(a). These include the photodiode that generates current I_S, the total input capacitance C_t at the input, inverting amplifier with gain $-A$, and a shunt feedback resistor R_F. The total input capacitance consists of the photodiode capacitance, input capacitance of the inverting amplifier, a feedback capacitance, and stray capacitance from the wiring or layout. The ratio of the output voltage to input photocurrent is given by [35]

$$\frac{V_{out}}{I_S} = \frac{-R_F}{1 + j\omega R_F C_t/(1 + A)} \tag{73.3}$$

with a corner frequency f_c of $(1 + A)/2\pi R_F C_t$.

A practical realization of the transimpedance amplifier is shown in Figure 73.17(b). The input active device is a FET with capacitive input impedance. The output of this common-source device drives two source followers, one of which is connected to the feedback resistor R_F ($=200\ \Omega$ in this example) and the other is a larger sized source follower device which drives the external load.

73.4 RF and Microwave Mixers

RF and microwave electronic mixers [36,37] are primarily used for frequency translation and fundamentally employ signal multiplication to perform this function. Mixers are three-port devices with (1) an RF signal input, (2) an oscillator (LO) input, and (3) an output labeled as the intermediate frequency (IF). Frequency translation via a mixer is typically presented using the simple example of two sinusoidal signals, say *X* and *Y*, of differing frequencies driving a multiplier device with output *W*. Suppose we have two signals, $X = A_{RF}\sin(2\pi f_{RF}t)$ and $Y = B_{LO}\sin(2\pi f_{LO}t)$, where A_{RF} and B_{LO} are constants quantifying the amplitudes of the two signals and f_{RF} and f_{LO} the respective frequencies. The subscripts "RF" and "LO"

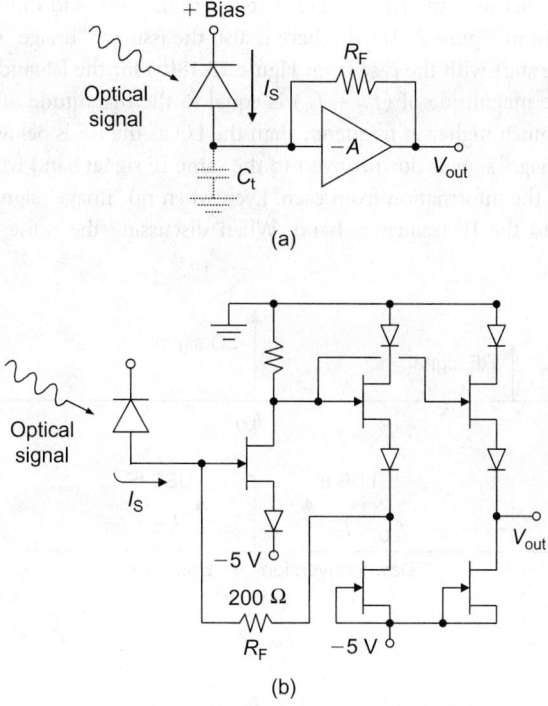

(a)

(b)

FIGURE 73.17 (a) Block diagram of a transimpedance amplifier with photodiode at its input providing a current drive at its input. (b) Circuit schematic of one possible implementation of a transimpedance amplifier with FET active devices.

stand for "radio frequency" and "local oscillator," where the RF is the signal carrying information and the LO signal is chosen to translate the multiplier's output to a specific frequency or frequency band, labeled IF for "intermediate frequency." The IF frequency is chosen to make it more convenient to process or recover the information carried by the incoming RF signal. Performing the multiplication of X and Y, and ignoring any scale factor from the multiplication operation, gives

$$W = XY = A_{RF} \sin(2\pi f_{RF} t) B_{LO} \sin(2\pi f_{LO} t)$$

$$= \tfrac{1}{2} A_{RF} B_{LO} [\sin(2\pi (f_{RF} + f_{LO}))t + \sin(2\pi (f_{RF} - f_{LO}))t] \qquad (73.4)$$

Note that the multiplication operation generates two output signals, one at the sum (or addition) frequency and the other at the difference frequency of the RF and LO frequencies. This is a consequence of the quadratic (or square-law) form assumed of the multiplication nonlinearity. This is a simplification because most device nonlinearities are not so ideal. A square-law device is ideal because a square-law mixer generates the fewest number of undesired frequencies. We show below that actual devices generate many additional frequencies beyond the ideal square-law mixer.

From this relationship we see that we have the flexibility to select the LO frequency so as to place the IF frequency (e.g., say we want the difference frequency signal) at a desired frequency where filters and/or amplifiers of lower cost, or more readily available, can be used for further signal processing. Figure 73.18 shows the spectral lines (frequency domain) of the sum and difference IF signals for two cases. First, when the RF and LO are widely separated, as represented in Figure 73.18(a), we see the sum and difference IF signals closely positioned about the LO frequency. The difference frequency IF signal is sometimes referred to as the lower sideband (LSB) and the sum frequency IF signal is then referred to as the upper sideband

(USB). As the separation between the RF and LO is reduced, the sum and difference frequencies spread further apart as illustrated in Figure 73.18(b). There is also the issue of "image" signals. This is illustrated in Figure 73.19 where we start with the case from Figure 73.18(b) for the RF and LO signals. Now add an "image" signal where the magnitude of $(f_{LO} - f_{RF})$ is equal to the magnitude of $(f_{image} - f_{LO})$, that is, the "image" frequency is as much higher in frequency than the LO as the RF is below the LO. The problem is that both the RF and "image" signals downconvert to the same IF signal band which leads to the problem of being able to separate the information from each. Even when no "image" signal is present, noise about f_{image} is downconverted to the IF frequency band. When discussing the noise behavior of mixers it is

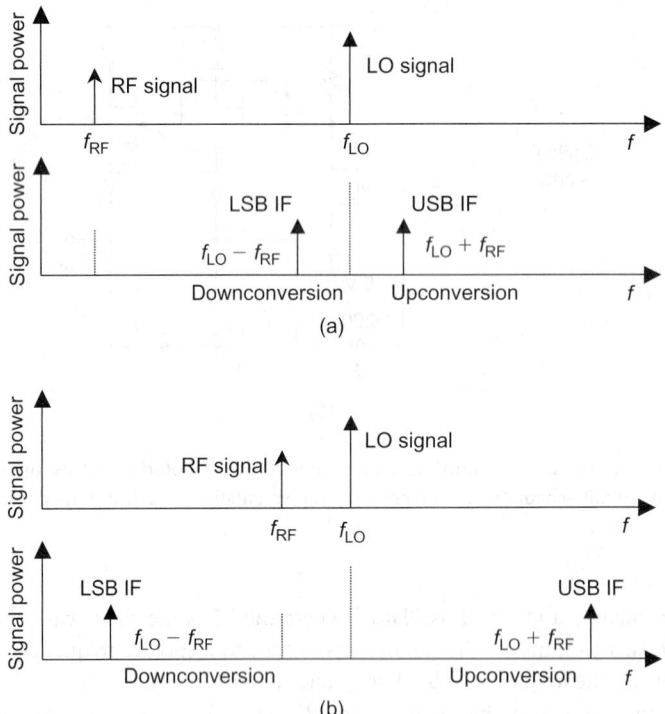

FIGURE 73.18 (a) Frequency spectrum of "widely separated" RF and LO signals mixing to generate USB and LSB products. (b) Frequency spectrum of "narrowly separated" RF and LO signals mixing to generate USB and LSB products. For case (b) the IF products are now widely separated. The mixer is a perfect square law for this example.

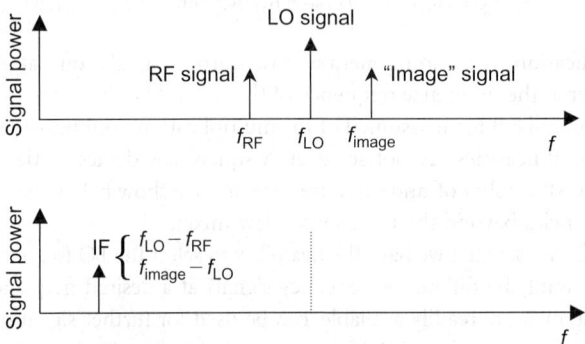

FIGURE 73.19 Frequency spectrum of LO, RF, and RF "image" signals mixing to produce overlapping IF signals. This occurs because $(f_{LO} - f_{RF})$ and $(f_{image} - f_{LO})$ are of equal magnitude. The problem is the resolution of overlapping signal products.

important to specify single-sideband or double-sideband noise figure. This degrades the sensitivity of the mixer owing to the added noise and is an important consideration for the noise figure definition of a mixer. This is why much effort has gone into designing image rejection mixers for applications that are sensitive to image responses.

The above example assumed an ideal quadratic or square-law mixer nonlinearity for the multiplication function. Nonlinearities in electronic devices generally deviate from square-law behavior over their wide operating ranges. The one active device that does approximate the square-law behavior is a long-channel FET such as the silicon junction FET (JFET). The JFET drain current versus its gate-to-source voltage transfer characteristic, with constant V_{DS} in the saturated region, is

$$I_D = I_{DSS}\left(1 - \frac{V_{GS}}{(-V_P)}\right)^2 \quad \text{for} -V_P \leq V_{GS} \leq \varphi \tag{73.5}$$

where I_{DSS} is the drain current at $V_{GS} = 0$ and φ a built-in potential associated with the gate-to-channel junction. However, even the best JFET devices only approximate the square-law behavior, so Eq. (73.5) is not exact.

73.4.1 General Device Nonlinearity

A Taylor series expansion is often a good way to model the nonlinearity of an electronic device. Let the device be represented as a two-port with v_i being the input voltage and v_o the output voltage of the two-port. Then the relationship between v_o and v_i is the series

$$v_o = V_{dc} + Gv_i + Av_i^2 + Bv_i^3 + \cdots \tag{73.6}$$

where V_{dc} is a constant (or offset) voltage, G the linear "gain" (but often is less than unity), A the coefficient of the quadratic or square-law term, and B the coefficient of the third-order or cubic term, and so on with higher order terms. Hopefully, the higher order terms become small rapidly so that not too many terms are required to faithfully model the nonlinearity. Consider the frequency generation behavior of each of the terms. The linear term, namely Gv_i, cannot generate new frequencies. Thus, for input v_i with frequencies of f_{RF} and f_{LO}, the output v_o can only contain frequencies f_{RF} and f_{LO}. The quadratic term, namely Av_i^2, generates four new frequencies, $(f_{RF} + f_{LO})$, $(f_{RF} - f_{LO})$, $2f_{RF}$, and $2f_{LO}$. Next, the third-order term, namely Bv_i^3, generates six new frequencies—$(2f_{RF} + f_{LO})$, $(2f_{RF} - f_{LO})$, $(2f_{LO} + f_{RF})$, $(2f_{LO} - f_{RF})$, $3f_{RF}$, and $3f_{LO}$. The spectrum $S(f)$ for the linear, quadratic, and cubic terms is shown in Figure 73.20 assuming f_{RF} to be 0.8 and f_{LO} to be 1.0. Higher order terms generate new frequencies rapidly and obviously it becomes very complicated. The use of trigonometric identities when expanding terms in Eq. (73.6), setting the input $v_i = A_{RF} \sin(2\pi f_{RF}t) + B_{LO} \sin(2\pi f_{LO}t)$, is the most direct way to find these new frequencies although it is tedious to do so.

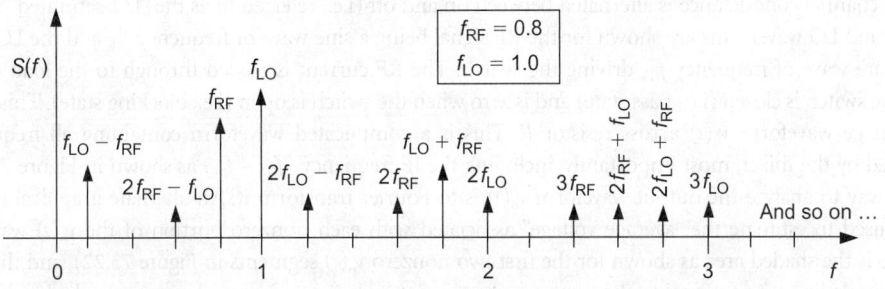

FIGURE 73.20 Frequency spectrum of a nonsquare-law mixer showing the higher order mixing products along with the RF and LO harmonics. This is more representative of actual mixers and the spectrum they typically generate.

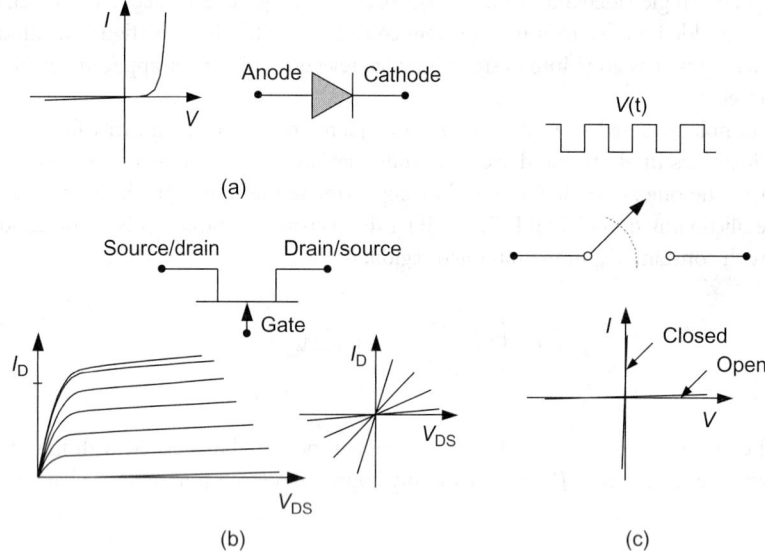

FIGURE 73.21 Examples of nonlinear components or devices. (a) Diode with nonsymmetric $I–V$ characteristic (forward bias gives familiar exponential $I–V$ curve). (b) FET with typical $I–V$ characteristic in common-source configuration and linear $I–V$ "voltage-controlled" resistor characteristic with source and drain are operated symmetrically. (c) Ideal switch and its $I–V$ two-state characteristic.

The multiplication function required to make a mixer can be achieved using components that either exhibit a nonlinear transfer characteristic or a time-varying behavior. This is illustrated in Figure 73.21 showing both a diode and FET as examples of components with nonlinear $I–V$ characteristics (Figure 73.21[a] and Figure 73.21[b]) and a periodically driven switch as an example of a time-varying component (Figure 73.21[c]). Diodes of course, have a strongly nonlinear exponential $I–V$ characteristic. Active devices such as a FET can be operated either over the nonlinear saturation region, or used in the resistive region to perform a switching function with the gate being driven at the commutating frequency and the source and drain connected in a symmetrical configuration. The time-varying switch in Figure 73.21(c) is often realized using FETs operated in their resistive region. An example of this is presented below with the widely used four-FET ring resistive mixer.

73.4.2 Switch Mixer

The simplest mixer is a switch passing the RF signal with the timing set by the LO signal. The switch mixer is shown in Figure 73.22 with a load resistor R used to establish the output voltage $V(t)$. Although the schematic shows a generic switch symbol, in practice this might be a FET positioned with the LO diving its gate such that the channel conductance is alternated between on and off (i.e., referred to as the "LO-saturated" mode). The RF and LO waveforms are shown for the RF signal being a sine wave of frequency f_{RF} and the LO signal is a square wave of frequency f_{LO} driving the switch. The RF current is passed through to the load resistor when the switch is closed (i.e., pass state) and is zero when the switch is open (i.e., blocking state). This results in a voltage waveform $v_o(t)$ across resistor R. This is a complicated waveform containing all frequencies generated by the mixer, most importantly, including the IF frequency ($f_{LO} - f_{RF}$) as shown in Figure 73.22.

One way to analyze the output waveform $v_o(t)$ is to Fourier transform it. An alternate graphical method can be used to estimate the "average voltage" associated with each nonzero portion of the $v_o(t)$ waveform (i.e., this is the shaded area as shown for the first two nonzero $v_o(t)$ segments in Figure 73.22), and then plot an amplitude line proportional to the area of each segment and centered on each segment calculated. This is the plot labeled "IF" placed immediately below the $v_o(t)$ waveform in Figure 73.22. It appears as a time sequence of lines proportional to the amplitude of each nonzero segment of $v_o(t)$. This plot of time-sequenced amplitudes

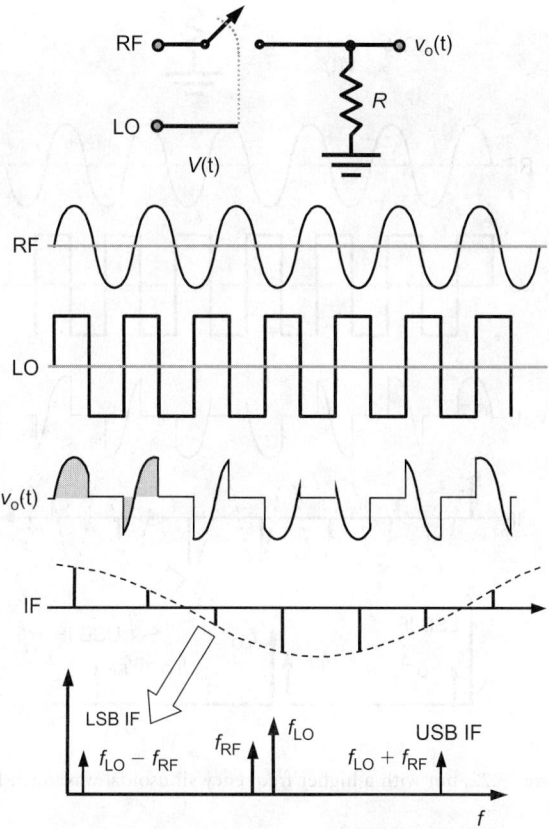

FIGURE 73.22 Operation of ideal switch as a mixer. Both the square-wave LO waveform and a sine wave RF signal are shown. The voltage waveform across resistor R is shown as chopped up waveform. Immediately below is the IF signal with discrete lines proportional to the area under each portion of the nonzero IF waveform (this is analogous to sampling). Finally, a sinusoidal waveform is fitted to the discrete lines of the IF signal corresponding to the $(f_{LO} - f_{RF})$ LSB IF product.

looks like the process of sampling. A sampler is a circuit that records a value (voltage, or charge, or some other variable) at specific intervals (in the case above, the interval is just the period associated with the LO signal frequency of period $T = 1/f_{LO}$). Generally, in the sampling process the "sampling interval" or "aperture time" is short compared to the sampling period, but for the example in Figure 73.22 the "aperture time" is one-half the sampling period. The next step is to try to fit sinusoidal waveforms to the sampled sequence of amplitudes. The dashed-line curve in the figure shows the lowest frequency that can be fitted to this sequence of amplitudes. It corresponds to the difference frequency, namely $f_{LO} - f_{RF}$, which is the IF signal of the square-law mixing process. We can fit another sine wave to this sequence of amplitudes as shown in Figure 73.23. This corresponds to the sum frequency $f_{LO} + f_{RF}$. The graphical analysis in Figure 73.22 and Figure 73.23 is of limited accuracy because it requires very precise graphing and does not easily yield amplitude information.

The multiplication operation is equivalent to "sampling and filtering" [38]. Consider the spectrum of a periodic pulse train $V(t)$ shown in Figure 73.24(a) where the pulse train represents the LO drive in the above example. Let T be the "interpulse interval" (equivalent to the "sampling period" above) and τ the "aperture time" of the waveform $V(t)$. The frequency domain spectrum of $V(t)$ can be found using the Fourier transform.

$$C_n(f) = \frac{1}{T} \int_{-T/2}^{T/2} V(t) \exp(-2\pi n f_S \tau)\, dt \qquad (73.7)$$

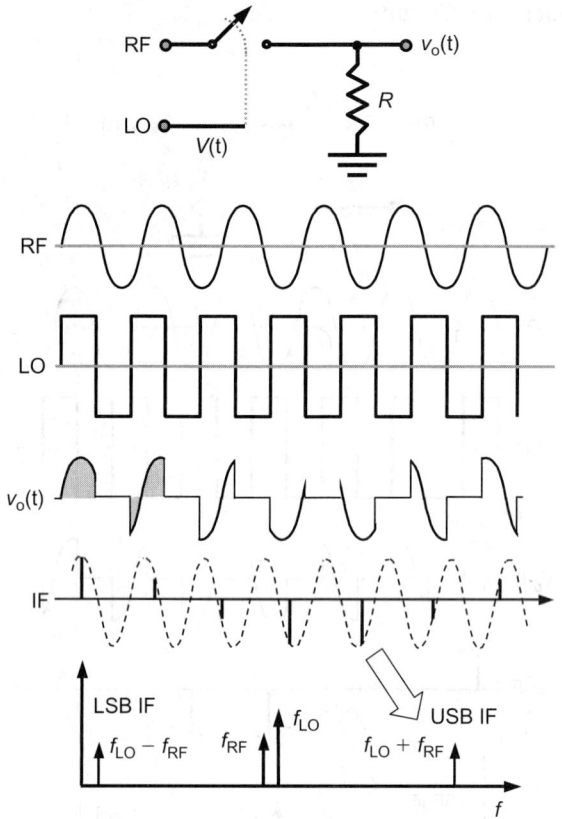

FIGURE 73.23 As in Figure 73.22, but with a higher frequency sinusoidal waveform fitted corresponding to the $(f_{LO} + f_{RF})$ USB IF product.

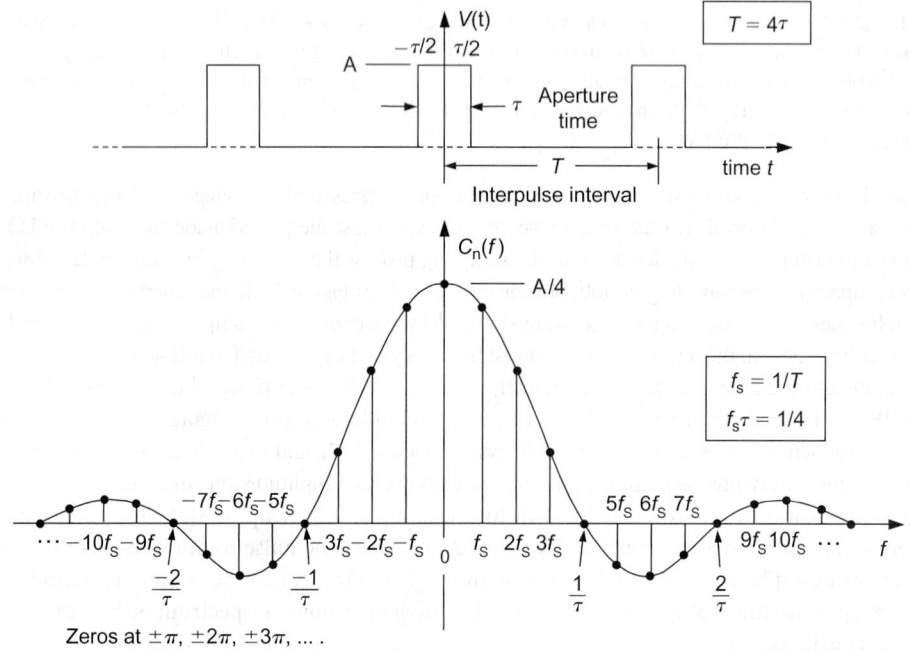

Zeros at $\pm\pi, \pm2\pi, \pm3\pi, \dots$.

FIGURE 73.24 A pulse train of pulses with of width (aperture time) τ, strength A, with pulse interval T, gives rise to Fourier series with amplitudes in the lower curve. The frequencies of the Fourier expansion terms are multiplies of frequency f_s which are equal to $1/T$.

The spectrum of $V(t)$ is shown Figure 73.24(b) for the case of $T = 4\tau$. It is a comb of harmonically related frequencies with amplitudes $C_n(f)$ weighted by the "sampling function" $Sa(n\pi f_S \tau)$,

$$Sa(n\pi f_S \tau) = \left(\frac{\sin(n\pi f_S \tau)}{n\pi f_S \tau} \right) \tag{73.8}$$

where f_S is the "sampling frequency" and is equal to the reciprocal of period T. This function is sometimes called the "sinc" function.** The spectrum amplitudes $C_n(f)$ are

$$C_n(f) = A f_S \tau \times Sa(n\pi f_S \tau) \tag{73.9}$$

where A is a constant proportional to the amplitude of the waveform.

Suppose we have an information-carrying, band-limited waveform $g(t)$ with Fourier transform $G(f)$ calculated from

$$G(f) = \frac{1}{T} \int_{-\infty}^{\infty} g(t) \exp(-2\pi n f_S \tau) \, dt \tag{73.10}$$

Multiplying $g(t)$ by $V(t)$ performs the frequency translation or mixing function. This can be examined by applying the Convolution Theorem [39]

$$g(t)V(t) \Leftrightarrow \frac{1}{2\pi} G(f) * C_n(f) \tag{73.11}$$

where the symbol $*$ denotes the convolution operation. This convolution operation is illustrated in Figure 73.25, with $g(t)$ and $V(t)$ having Fourier transforms $G(f)$ and $C_n(f)$ in Figure 73.25(a), and the convolution operation appearing in Figure 73.25(b). We assume that $G(f)$ is band-limited to a narrow frequency range as shown. Graphically, convolution is performed by taking $G(f)$, reversing it in frequency to form $G(-f)$, then sliding it over $C_n(f)$ from left to right, and multiplying them together during this translation. The convolution operation generates the spectrum appearing at the bottom graph of Figure 73.25(b). This can be interpreted as frequency translating the spectrum $G(f)$ to each nonzero spectral line associated with $C_n(f)$. The spectrum $G(f)$ is replicated about f_S and all of its harmonics for both positive and negative frequencies. Of course, filtering is required to extract the band-limited spectrum $G(f)$ centered at frequency $n f_S$, where f_S is the sampling frequency and n is the integer chosen to select the desired center frequency.

73.4.3 Mixer Performance Parameters

Before discussing various mixer configurations and presenting examples of mixers, we briefly summarize key electrical parameters used to describe mixer operation and performance.

1. *Conversion gain* (or *conversion loss*) is defined as the ratio of the desired IF signal power to the RF input power. By desired IF output we mean that portion of the output from the RF input and not from other signals such as the image input discussed above. Passive mixers have loss and active mixers using transistors can have conversion gain if appropriately biased. Conversion gain results from amplification within the mixer itself.

**The sinc(x) function is also referred to as the filtering or interpolating function. It has wide application in engineering and its properties can be found in Bracewell [39] and other books covering the Fourier transform.

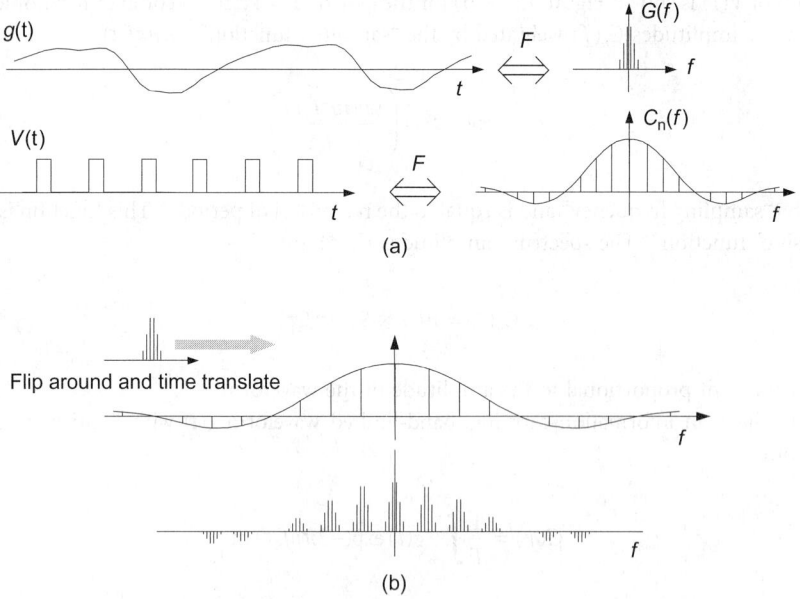

FIGURE 73.25 Convolution example with signal $g(t)$ and pulse train $V(t)$. (a) Fourier transform of $g(t)$ is $G(f)$ and Fourier coefficients of Fourier series are denoted by $C_n(f)$. (b) Convolution carried out by changing from $G(f)$ to $G(-f)$ and multiplying by $C_n(f)$. Final frequency spectrum is shown at the bottom of the figure.

2. *Noise figure* is the signal-to-noise power ratio at the RF input divided by the signal-to-noise power ratio at the IF output port.[*†] It is important to designate if the RF input includes the image band (in which case it is a double-sideband [DSB] noise figure) or not (then it is a single-sideband [SSB] noise figure). From the measurement viewpoint the SSB noise figure will be greater than the DSB noise figure because although both have the same IF noise power, the SSB noise figure includes signal power from the RF signal, but not the image signal. It is common practice to report DSB noise figures when discussing mixers because it is lower in value. When in doubt, assume that a DSB noise figure is intended.

3. *Port-to-port isolation* quantifies the feedthrough between the various ports of the mixer. We typically do not want the RF and LO signals to leak through to the IF port, especially the LO signal because it is usually the largest signal present. For example, the *RF-to-IF isolation* is defined as the signal power at the IF port at frequency f_{RF} divided by the input RF signal power. Isolation between the other combinations of ports and signals is defined analogously.

4. *Output power compression point* is an often cited metric. In a well-behaved mixer we would like the IF signal to be proportional to the desired RF input signal. For most mixers this generally holds at low signal levels, however, as the input power increases a power level is reached where the desired IF output power begins to saturate or flatten. Saturation is the consequence of odd-order distortion with third-order distortion being especially significant. The parameter *conversion compression* allows us to quantify the ratio of RF input power level to the IF output power level where distortion begins to cause signal power to be shifted to the distortion products. The saturation effect is shown in Figure 73.26(a) on the plot of IF output power versus RF input power. At low signal levels the IF output linearly tracks the RF input level as expected.

[*†] Note the noise figure is defined as the input over output and not output of input as is done for power gain and other parameters.

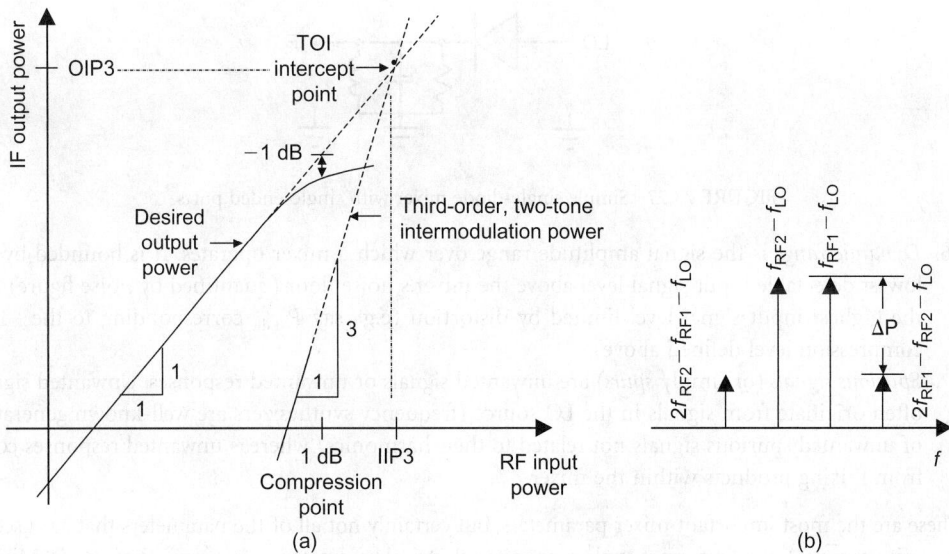

FIGURE 73.26 (a) Mixer IF output power versus RF input power for fundamental output power and third-order output product power. Where the two power lines cross is by definition the third-order intercept (TOI). (b) Two-tone RF signals (denoted by f_{RF1} and f_{RF2}) are used to quantify third-order distortion. All signal powers are typically stated in units of dBm.

As RF input power is increased a level is reached where the output no longer increases as rapidly as the input power is increased. It is common practice to designate the RF input power level at which the IF output is 1 dB below what it would have been had the linear increase continued—this is designated the 1 dB *compression point.*

5. *Third-order intermodulation intercept* (TOI) is a commonly cited mixer parameter for characterizing mixer distortion. Intermodulation distortion (IMD) involves mixing between RF signals and the harmonics of those signals. Given two RF input signals, say f_{RF1} and f_{RF2}, the third-order intermodulation products take the general form $[(\pm m_1 f_{RF1} \pm m_2 f_{RF2}) \pm m_3 f_{LO}]$, where m_1, m_2, and m_3 are all nonzero positive integers. The third-order IMD products are the combinations of $[(\pm 2f_{RF1} \pm f_{RF2}) \pm f_{LO}]$ and $[(\pm f_{RF1} \pm 2f_{RF2}) \pm f_{LO}]$. They are said to be "third-order" because $m_1 + m_2 = 3$. For a change in RF input power of 1 dB, an nth-order IMD product will increase n dB, so with third-order IMD products each of them increases by 3 dB. The problem is that these IMD products are often downconverted into or close to the IF band. A two-tone measurement scheme using two closely spaced RF input signals (again f_{RF1} and f_{RF2}) is often used to characterize and determine the third-order intercept point in a mixer (and other networks such as amplifiers). Figure 73.26 illustrates the behavior that motivates the measurement. The point labeled TOI *intercept point* in Figure 73.26 is never reached in practice because of mixer saturation, so it must be extrapolated or computed from measured data at lower powers. The third-order intercept can be defined in terms of either the input power, denoted by IIP3, or the output power, denoted by OIP3. For mixers it is more common to use IIP3, but for other components it is more common to use OIP3. They are related by the relationship OIP3 = IIP3 + conversion gain (loss) of the mixer. The significance of TOI is that it quantifies the mixer's usable range bounded by the condition that the third-order IMD products are less than the IF signal's power level (the difference is denoted by ΔP as indicated in Figure 73.26[b]). Note that TOI = $P_{IF} + (\Delta P/2)$, where P_{IF} is the IF signal power and ΔP is the power difference between the IMD product and P_{IF}. Other IMD intercept points can also be defined, but third-order IMD is generally the most important parameter for mixers.

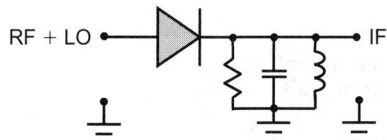

FIGURE 73.27 Simple single-diode mixer with single-ended ports.

6. *Dynamic range* is the signal amplitude range over which a mixer operates. It is bounded by the lowest detectable input signal level above the mixer's noise floor (quantified by noise figure) and the highest input signal level limited by distortion (e.g., say $P_{-1\,\text{dB}}$ corresponding to the −1 dB compression level defined above).

7. *Spurious signals* (or simply *spurs*) are unwanted signals or unwanted responses. Unwanted signals often originate from signals in the LO source (frequency synthesizers are well-known generators of unwanted spurious signals not related to their harmonics) whereas unwanted responses come from mixing products within the mixer.

These are the most important mixer parameters, but certainly not all of the parameters that are used to characterize mixers. A good reference explaining mixer electrical parameters is Synergy Microwave's "Mixers" application note [40].

73.4.4   Single-Diode Mixer

The simplest and least expensive mixer we can imagine is that of a single diode (think of substituting a diode for the switch in Figure 73.22). A single-ended (i.e., unbalanced) connection of the single-diode[*‡] with an RLC filter network at its output is shown in Figure 73.27. With single-ended input and output there is no input-to-output isolation and because the diode is a passive device there can be no conversion gain. The single-diode mixer has been used in many applications. It is the detector used in the crystal radios[*§] built by millions. This mixer was effectively used for radar detectors during World War II and in very early UHF television tuners. At the output all mixing products appear in addition to the feedthrough of both RF and LO signals. This places special importance on precise filtering to select out the desired IF signal, which is further complicated by the desired IF signal sometimes being smaller in amplitude than the feedthrough and possibly some of the IMD products. Modern communication systems now have such stringent bandwidth and adjacent signal rejection requirements that the unbalanced single-diode mixer is not acceptable because the supporting hardware to extract the desired signal would be too expensive or cumbersome.

In preparation for discussing balanced mixers we next briefly discuss networks for converting single-ended or unbalanced signals into differential or balanced signals.

73.4.5   Balanced versus Unbalanced Signals

The transmission of signals on balanced transmission lines, or balanced networks, has important advantages, especially for mixers. Figure 73.28 illustrates what we mean by balanced transmission line or balanced network by evolving a low-pass filter network from its single-ended or unbalanced configuration to a pure balanced configuration. The single-ended or unbalanced network in Figure 73.28(a) shows one terminal as "ground." Usually, we think of "ground" in the ideal sense that it is the same potential everywhere in the circuit. But in the physical layout of RF and microwave circuits maintaining this

---

[*‡] A "hopelessly unsophisticated" mixer as described by Proffessor Tom Lee of Stansford University [41].

[*§] Real crystal radios use a point contact to a crystal such as galena (lead sulfide) or fools gold (iron pyrite). It is also possible to use a diode, but for better sensitivity a germanium diode such as the 1N34A is recommended over a silicon diode.

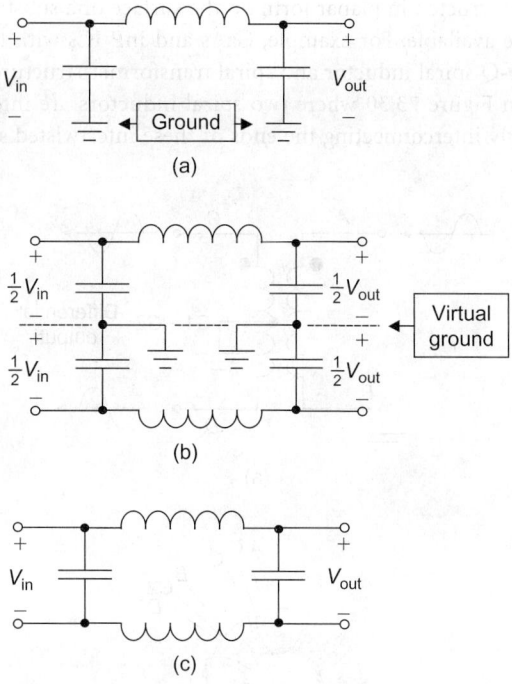

FIGURE 73.28 (a) Single-ended *L-C* pi-section of filter; (b) use of symmetry to duplicate filter section turning it into a double-ended realization; and (c) full double-ended (i.e., differential) realization of *L-C* filter.

"ground" potential across the circuit is very difficult, and becomes increasingly difficult as the frequency continues to increase. Parasitic resistance and inductance of vias and other interconnects causes problems in maintaining the "ground" potential at all common nodes. Connecting the filter from Figure 73.28(a) into a "push-pull" configuration (as drawn in Figure 73.28[b]) is a key step in progressing to the balanced configuration. Here the terminal voltages swing symmetrically about the "ground" potential. Removing the "ground" reference altogether in Figure 73.28[c] is the final step in achieving a true balanced filter network. In a balanced configuration the signal is the difference in voltage between the two conductors and neither voltage need be referenced to "ground." Common-mode signals, arising from ground loops and the presence of parasitic elements in the ground path, do not degrade the integrity of a differential signal. This is key to signal rejection in a mixer where we desire to exclude RF and LO signals from the IF port, or exclude the LO signal from the RF port, and so on. Transformers and baluns are commonly used to do this conversion which is our next topic.

73.4.6 Transformers and Baluns

Transformers are easily configured to convert single-ended (unbalanced) signals to differential (balanced) signals. At lower frequencies, say below a few hundred megahertz, transformers are commonly used to perform this transformation. Figure 73.29(a) shows a transformer schematic with unbalanced input, but balanced output. Using reciprocity, it is conceptually easy to reverse the input and output to convert from balanced back to unbalanced signal. Figure 73.29(b) shows a trifilar transformer wound on a ferrite toroid forming a commonly used RF transformer [42,43]. Wire-wound transformers are generally large and bulky and limited in their upper frequency limit. An alternative to lumped element transformers is to use distributed transformers which make direct use of transmission line structures. Distributed elements can be used at much higher frequencies. Of course, distributed transformers are physically large at the lower RF frequencies, but at microwave frequencies they become more compact and become more attractive as the frequency of operation increases.

Transformers can be constructed in planar form on the surface of a substrate when several layers of interconnect or wiring are available. For example, GaAs and InP ICs with their semi-insulating substrates allow for moderate-Q spiral inductor and spiral transformer structures. One way to build such a transformer is shown in Figure 73.30 where two spiral inductors are interleaved to form a transformer section. By correctly interconnecting the ends of these intertwisted structures a center-tapped

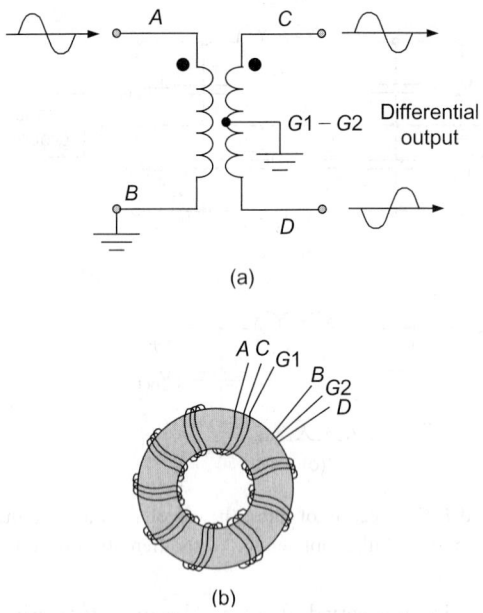

(a)

(b)

FIGURE 73.29 Use of transformer to convert from a single-ended signal into double-ended or differential signal. (b) Trifilar wound toroid realization of transformer for lower RF frequencies.

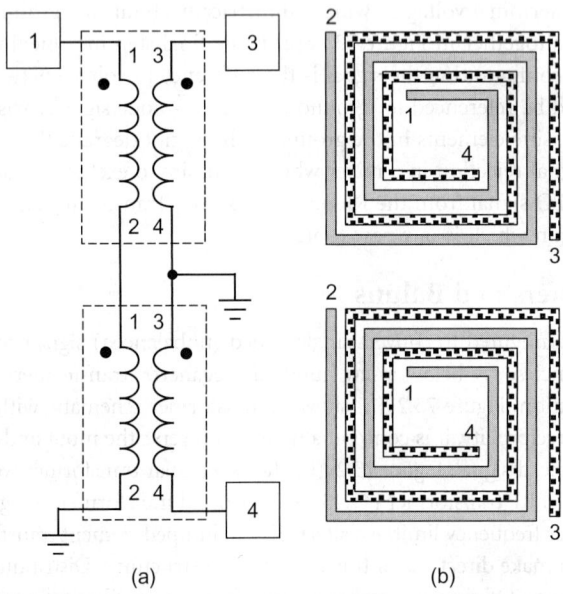

(a)   (b)

FIGURE 73.30 (a) Transformer performing single- to double-ended conversion schematic, and (b) monolithic IC realization using a pair of interwound spiral inductors when properly connected (but not shown).

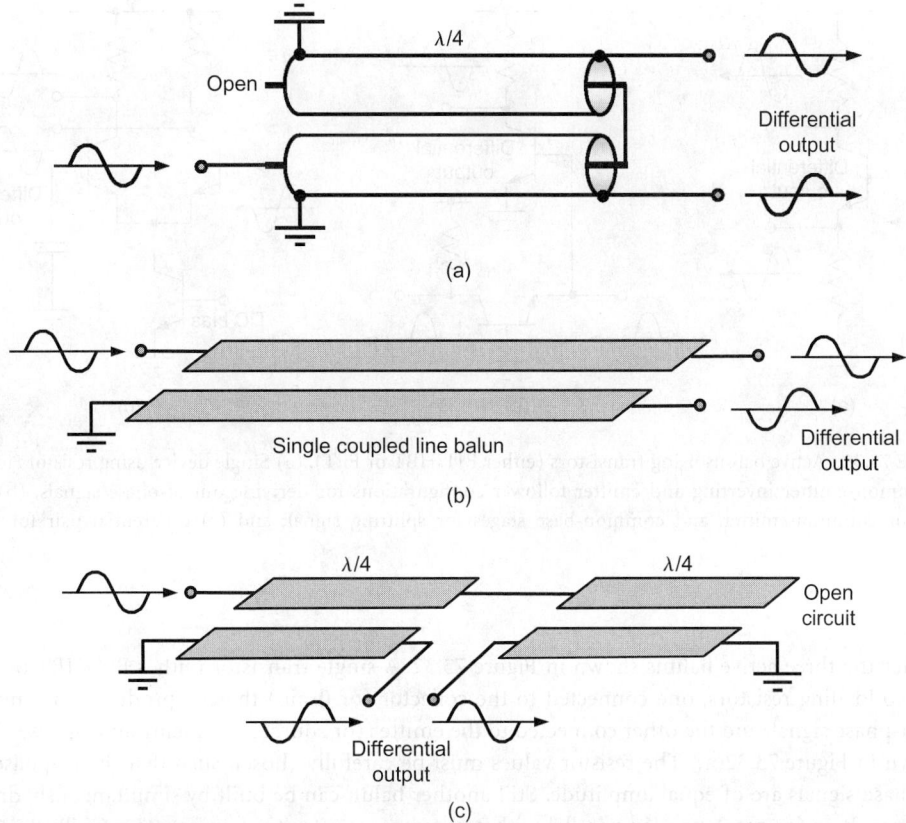

FIGURE 73.31 Some examples of baluns. (a) Coaxial line realization, (b) two coupled microstrip lines, and (c) Marchand balun again using coupled transmission lines.

transformer structure can be realized. Such transformers have been built and characterized to ~30 GHz and beyond [44].

Distributed transmission line transformers and *baluns* (the word is derived from "balanced to unbalanced") can be easily constructed. Here we focus on distributed balun structures using quarter-wavelength ($\lambda/4$) sections of transmission lines. Figure 73.31(a) shows a coaxial transmission line balun. The use of coupled transmission lines built from quarter-wavelength transmission line sections will perform the same function. Using microstrip transmission lines with edge-coupling, a distributed coupled-line balun is illustrated in Figure 73.31(b). A coupled-line balun can also be constructed with stripline (which requires that multiple layers of interconnect be available) or other forms of transmission lines that can be physically coupled. Although straight transmission line sections appear in the figure, meandering the layout of the transmission line is useful for reducing the overall size. The major limitation of distributed baluns is that they are narrowband (perhaps ±30% around the center frequency and more if performance specifications are relaxed). The final example of a distributed balun is the *Marchand balun* [45] in Figure 73.31(c). The Marchand balun uses an open-circuited, half-wavelength line coupled to two quarter-wavelength lines, each shorted at their distant ends. This produces a near-ideal balanced signal when driven by an unbalanced signal at one end of the half-wavelength transmission line. The Marchand balun is widely used and can be broadbanded (up to a decade) by using multisection pieces of transmission lines in place of each quarter-wavelength line.

In RF ICs very compact baluns can be realized with active devices such as transistors. These *active baluns* can function from DC to very high RF frequencies—functionally they are broadband baluns.

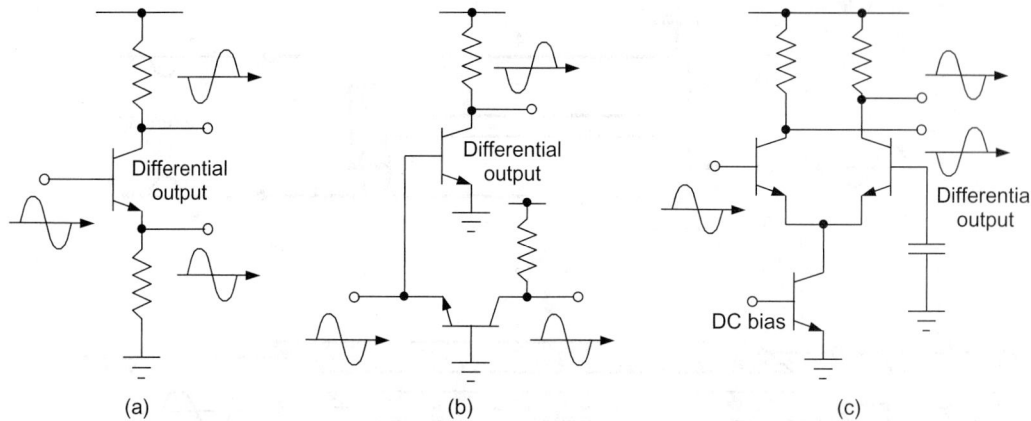

FIGURE 73.32 Active baluns using transistors (either BJT/HBT or FET). (a) Single device using resistors in both the common-emitter inverting and emitter follower configurations for deriving out-of-phase signals; (b) two-transistor common-emitter, and common-base stages for splitting signal; and (c) differential pair for signal splitting.

Consider the three active baluns shown in Figure 73.32. A single transistor (either BJT/HBT or FET) with two loading resistors, one connected to the collector (or drain) thereby producing an inverted or para-phase signal, and the other connected to the emitter (or source) producing an in-phase signal, is shown in Figure 73.32(a). The resistor values must be carefully chosen such that the in-phase and para-phase signals are of equal amplitude. Still another balun can be built by simultaneously driving a common base (or gate) stage in parallel with a common-emitter (or source) stage as illustrated in Figure 73.32(b). Again, careful selection of the resistor values is required so that the in-phase and para-phase output signals are of equal amplitude. However, the most versatile balun uses two transistors in a symmetrical differential pair configuration (now using identical resistor values) as shown in Figure 73.32(c). The resistor values can be chosen to set the gain of the differential stage. The differential stage can also be driven with a balanced input (both base terminals are then driven differentially), but a single resistor is used to extract an unbalanced output.

We can use the configuration of Figure 73.32(c) in a distributed amplifier format to achieve much greater bandwidths. In a prior section we discussed the very broadband advantages of the artificial transmission line of the distributed amplifier. The input transmission line can be shared to drive two parallel-connected distributed amplifiers to form a *power splitter* (note that both outputs are in-phase) as shown in Figure 73.33(a). If one of the two distributed amplifiers is now changed from common-source (CS) gain cells (inverting) to common-gate (CG) gain cells (noninverting), a balun is formed. The transmission line parameters must be modified to accommodate the CG gain cells because the impedance presented to the artificial transmission line is different from the usual CS configuration. Figure 73.33(b) shows the CS/CG distributed balun schematic [46]. The main advantage of this approach is the extremely wide bandwidths achievable, but the disadvantages include larger layout area consumption and a greater DC power requirement than the much simpler stages shown in Figure 73.32.

73.4.7 Single Balanced versus Double Balanced

Balanced mixers dominate in the world of mixer applications. Double-balanced mixers have superior isolation, but sometimes a single-balanced mixer is used. The advantage of a single-balanced mixer is

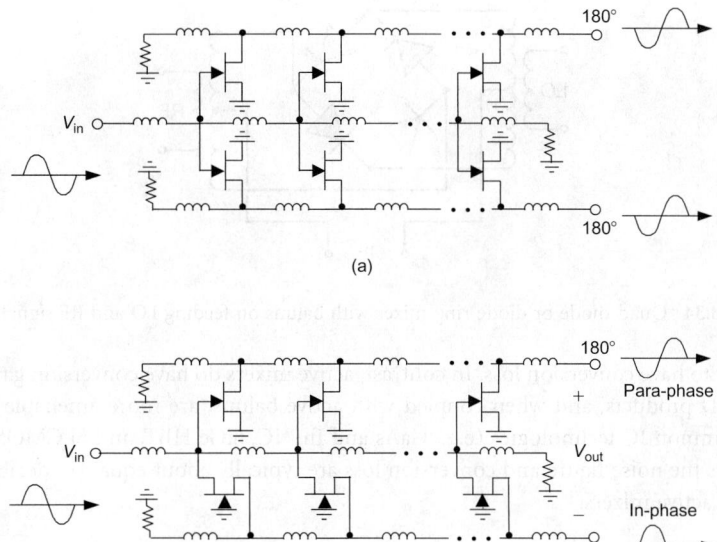

FIGURE 73.33 (a) Parallel distributed amplifier (input line in common with parallel amplifiers) for signal splitter, and (b) single-ended to differential signal transformation using an inverting distributed amplifier in parallel with a noninverting distributed amplifier.

TABLE 73.2 Comparing Unbalanced, Single- and Double-Balanced Mixer Performance

	Mixer classification			
	Unbalanced	LO Balanced	RF Balanced	Double Balanced
LO-to-RF Isolation	Poor	Yes	Poor	Yes
LO-to-IF Isolation	Poor	Poor	Yes	Yes
RF-to-IF Isolation	Poor	Yes	Poor	Yes
LO Harmonic Rejection	None	Even only	Even & Odd	Even & Odd
RF Harmonic Rejection	None	Even & Odd	Even only	Even & Odd
2-Tone, 2nd Order Product Rejection	None	None	Yes	Yes

that it uses one less balun and sometimes fewer diodes or transistors. Table 73.2 compares the isolation and rejection properties of unbalanced (without baluns), single balanced (two versions because either the RF port or LO port can be connected to the balun), and double balanced (two or three baluns). The examples presented below are all double-balanced mixers.

73.4.8 Passive Mixer versus Active Mixer

Mixers can be categorized into passive and active ones. Passive mixers typically use switches, diodes, and transistors operated in their linear region of operation. Active mixers use combinations of active devices such as FET and BJT/HBT transistors biased such that they are capable of either current or voltage gain. Passive mixers typically are simple, capable of wide bandwidths,[*] typically have lower IMD products,

---

[*] Generally, it is the balun that limits the bandwidth of a mixer. Passive mixers can have extremely wide operational bandwidth.

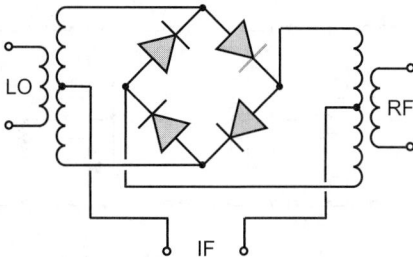

FIGURE 73.34 Quad-diode or diode ring mixer with baluns on feeding LO and RF signals to mixer.

but are confined to have conversion loss. In contrast, active mixers do have conversion gain, but typically have poorer IMD products, and when coupled with active baluns, are more amenable for integration using several common IC technologies (e.g., GaAs and InP IC, SiGe HBT, and Si CMOS technologies). In passive mixers the noise figure and conversion loss are typically about equal (in decibel), but usually lower than with active mixers.

73.4.9 Double-Balanced Passive Diode-Ring Mixer

We first look at the widely-used *diode-ring mixer*. A double-balanced mixer using the diode ring appears in Figure 73.34. A pair of transformers or baluns feeds the RF and LO signals to the mixer. By using the symmetry of balanced operation the ports are effectively isolated. Although not shown, sometimes a balun is connected to the two terminals of the IF port. The diode-ring mixer that is passive at best has a theoretical conversion loss of $2/\pi$ or 3.9 dB (this assumes no loss in the circuit itself and the diode's turn-on voltage equals zero as if it were a perfect switch which commonly adds 2–4 dB to the conversion loss). The diodes act as on/off switches that are commutated at the LO frequency f_{LO}. The LO power must be large enough to turn the diodes on and off and is the largest signal in the mixer because the LO drive level must be greater than the RF level. Without sufficient LO drive both the conversion loss and noise figure of this mixer rapidly degrade. Its operation in terms of current circulation in the circuit is illustrated in Figure 73.35. The arrows in Figure 73.35 show the IF current path for one polarity of RF, but both polarities of the LO signal. The time averages of the LO as well as the RF signals do not appear at the IF port resulting in good isolation. Typical conversion losses for the diode-ring mixer range from 6–8 dB and isolations are typically around 25–30 dB. It is a very simple and small structure allowing for low cost realization, and with its good performance, it is widely used in the RF and microwave bands.

When diodes are periodically switched on and off, their impedances are not well defined because they are time varying. Therefore, it is not possible to actually impedance match the ports. That is not to say that just any impedance will be acceptable, but rather it is necessary to empirically determine the best impedances to present to the mixer ports. That includes reasonable matching at frequencies where non-IF signals are at high levels because their reflections back into the mixer often produces unwanted results.

Discrete versions of the diode-ring are commercially available. High-quality diodes are required for good mixer performance. Integration of this mixer into an IC process is possible only if high-quality diodes can be fabricated within the process. Typically, GaAs or InP FET gate-to-channel (Schottky barrier) diodes are not especially good for mixer applications.

73.4.10 Double-Balanced Passive FET-Ring Mixer

A mixer that is similar to the diode-ring mixer is the passive *FET-ring mixer* shown in Figure 73.36. Here FETs (either MESFET or HEMT or MOS) devices are substituted for the diodes in the diode-ring mixer. Since the mixer does not require high-quality diodes it is more amenable to integration. The FET devices are operated as voltage-controlled resistors that are turned on or off by the LO signal being directly applied to the gates of the FETs. The FET-ring mixer is usually more linear than the diode-ring mixer in

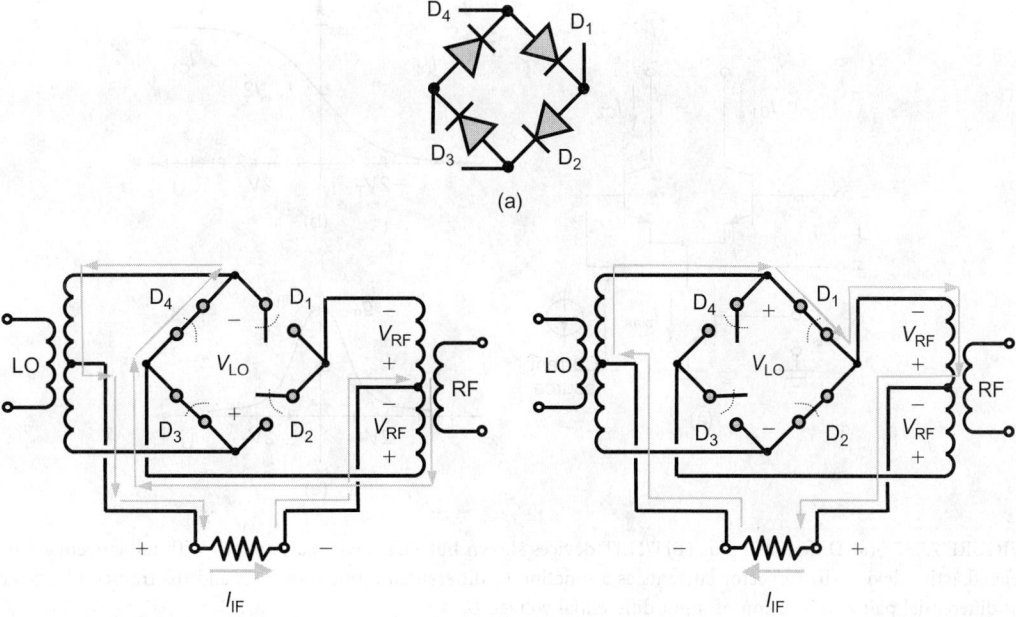

FIGURE 73.35 Operation of the (a) diode ring mixer, showing (b) current flow through IF resistor during one phase of the LO signal, and (c) the current flow during the opposite phase of the LO signal.

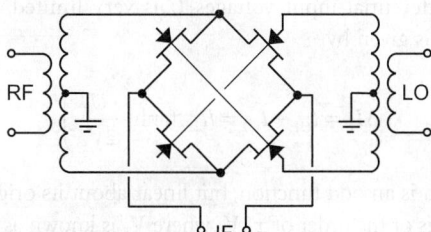

FIGURE 73.36 Analogous to the quad-diode mixer is the quad-FET mixer. The FETs are operated in their linear region as voltage-controlled resistors.

terms of the RF to IF translation. It has a conversion loss of about the same magnitude as the diode-ring mixer. Again, a relatively large LO drive is required for operation as with the diode-ring mixer. Gilbert realized in the 1960s that the four diodes could be replaced by transistors to form an active mixer. We discuss this next under the topic of Gilbert mixers.

73.4.11 Gilbert Active Mixer

Differential transistor pairs can be used to make mixers (and modulators). Here we discuss what is commonly called the *Gilbert mixer*, named after Barrie Gilbert [47], although it was first patented (filed in 1963) by H.E. Jones [48]. Both unipolar and bipolar active devices can be used in forming the Gilbert mixer. The Gilbert mixer is attractive for a number of reasons: (1) it is easily fabricated in monolithic form with active baluns, (2) it can give conversion gain, (3) it requires only a moderate LO drive to operate, (4) gives good isolation between the three ports, and (5) its performance is not highly sensitive to the impedances presented to its ports. The distortion of the Gilbert mixer is primarily odd-order by symmetry in the design.

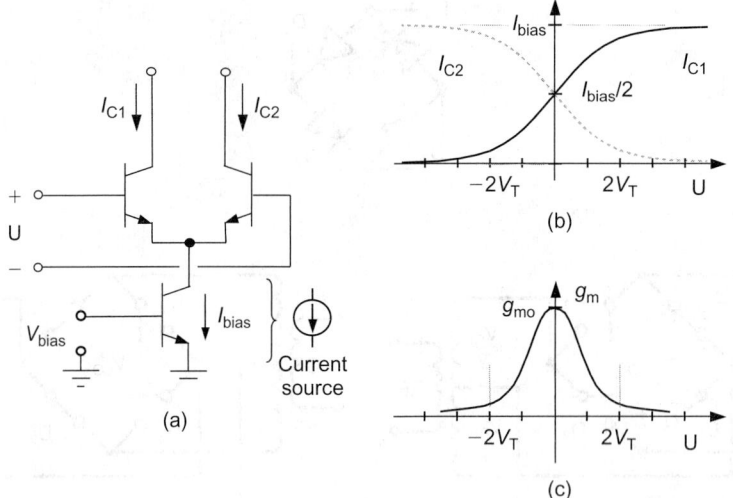

(a)

(b)

(c)

FIGURE 73.37 (a) Differential pair (BJT/HBT devices shown but FET devices also work) with tail current set by biased active device. (b) Collector currents as a function of differential input voltage U, and (b) transconductance of differential pair as a function of input differential voltage U.

Figure 73.37(a) shows a bipolar differential pair or emitter-coupled pair. This is often said to be a *transconductance stage* because the output is a current signal for an input voltage as shown in the figure. Owing to the exponential change in collector current from a change in the base-emitter voltage transfer characteristic, the range of differential input voltages U is very limited. The transfer characteristic is shown in Figure 73.37(b) and is given by

$$\Delta I_{\mathrm{C}} = I_{\mathrm{C1}} - I_{\mathrm{C2}} = I_{\mathrm{bias}} \tanh\left(\frac{U}{2V_{\mathrm{T}}}\right) \tag{73.12}$$

The hyperbolic tangent function is an odd function, but linear about its origin (i.e., limited to small values of U). The input voltage range is of the order of $\pm 2V_{\mathrm{T}}$ where V_{T} is known as the thermal voltage[ll] ($= kT/q$). Near room temperature V_{T} is about 25 mV, therefore, the overall range of U is about 100 mV as the differential stage's collector current begins to saturate. The currents of the two collectors of a differential pair are out of phase, that is, when I_{C1} increases, I_{C2} decreases. The differential pair's transconductance g_{m} is given by

$$g_{\mathrm{m}} = I_{\mathrm{bias}} \operatorname{sech}^2\left(\frac{U}{2V_{\mathrm{T}}}\right) \tag{73.13}$$

This is shown in Figure 73.37(c) with the stage's transconductance g_{m} as a function of differential input voltage U illustrating the narrow input range. The 1-dB compression points occur near $U = \pm 18$ mV.

The Gilbert mixer is formed using three differential pairs as shown in Figure 73.38. Two identical pairs form the *mixer core* (i.e., transistors Q1 through Q4 in Figure 73.38) which is driven by the LO signal. The tail currents to these two differential pairs are set by another differential pair, transistors Q5 and Q6, forming the transconductance stage. The transconductance stage is driven by the RF signal. Connecting load resistors R_{L} to the two current terminals allows a voltage output to be derived from the collector currents. With proper choice of resistor values conversion gain is possible.

[ll] The thermal voltage is Boltzmann's constant k times the temperature T divided by the electron's charge q.

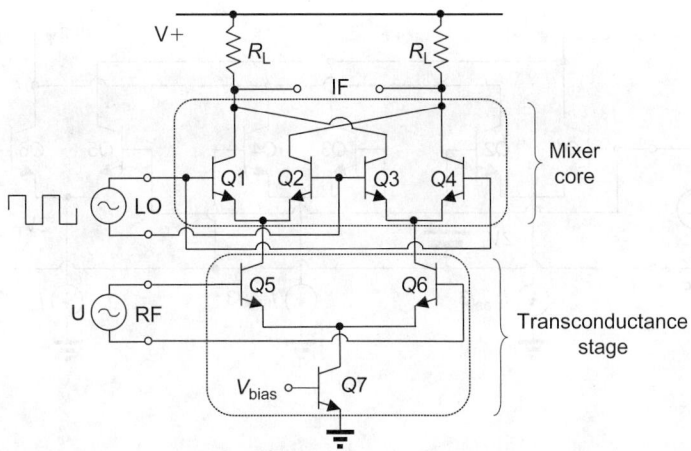

FIGURE 73.38 Gilbert cell mixer consisting of three differential pair and a current source (transistor $Q7$). The LO drive is strong enough to drive pairs $Q1$–$Q2$ and $Q3$–$Q4$ into saturation so that core mixer is operates as commuting switches. Transconductance stage is differential pair $Q5$–$Q6$ with input being the RF signal to mixer.

The dynamic range of the mixer can be increased by extending the voltage capacity of the transconductance cell thus allowing for larger input voltage swings. This simultaneously results in the transconductance becoming more linear over the extended range. There are several methods to do this. The simplest method is to use *emitter degeneration* by placing a resistor in series with each emitter. The principle is to use "local negative feedback" because the voltage across the emitter resistor generated by changes in the emitter current is anti-phase to that of the applied input signal. To quantify this effect, the 1-dB compression threshold increases by a factor of $(1 + 1.7\gamma)$ as R_E is increased, where γ is defined to be $(I_{bias}R_E/2V_T)$ and R_E is the series emitter resistance [49]. This also reduces the effective transconductance of the cell, but has the positive benefit of making the input transfer characteristic more linear.

An elegant approach as developed by Gilbert [50] is the *multi-tanh method* as illustrated in Figure 73.39. Here the transconductance stage is split into several parallel pieces, each offset in voltage from the others. In the figure a triplet of pairs is shown with offsets in multiples of $\pm 2V_T$, but other offsets can also be used to tailor the response. These voltage offsets are shown as "batteries" in Figure 73.39(a); more practical solutions would use resistors or resistor ratios to generate the internal offsets, or better yet, in monolithic form scale the size of the emitters to generate the offsets. The widening of the transconductance g_m is shown in the plot in Figure 73.39(b). Still another approach by Gilbert [51] is the so-called "micromixer" where the Gilbert cell is still the core mixer, but the transconductance stage (or RF stage) is modified to use a common base stage (or common gate stage if FETs are used) and a current mirror cell. In this way, much larger currents are accommodated at the RF stage. Gilbert's original paper not only explains the micromixer's operation in detail, but gives a good overview of basic Gilbert mixer operation and multi-tanh techniques in comparison.

The Gilbert mixer has been used in many RF and microwave applications (such as in the field of terrestrial communications and satellite TV tuners to mention only two). Compound semiconductor and Si/SiGe BJT/HBT devices are used to fabricate these mixers. Table 73.3 lists a sampling of microwave "broadband" Gilbert mixers recently reported in the literature. The table lists the operational bandwidth and gain of the mixer along with the year and a reference for further information. Although not included in the table, Gilbert mixers have been developed for narrowband mixer applications well beyond 40 GHz.

73.4.12 Image Rejection Mixers

In heterodyne receivers filters are used to select out the desired signals and to reject unwanted responses. As the RF spectrum becomes more crowded, and precise bands are squeezed in bandwidth and spacings

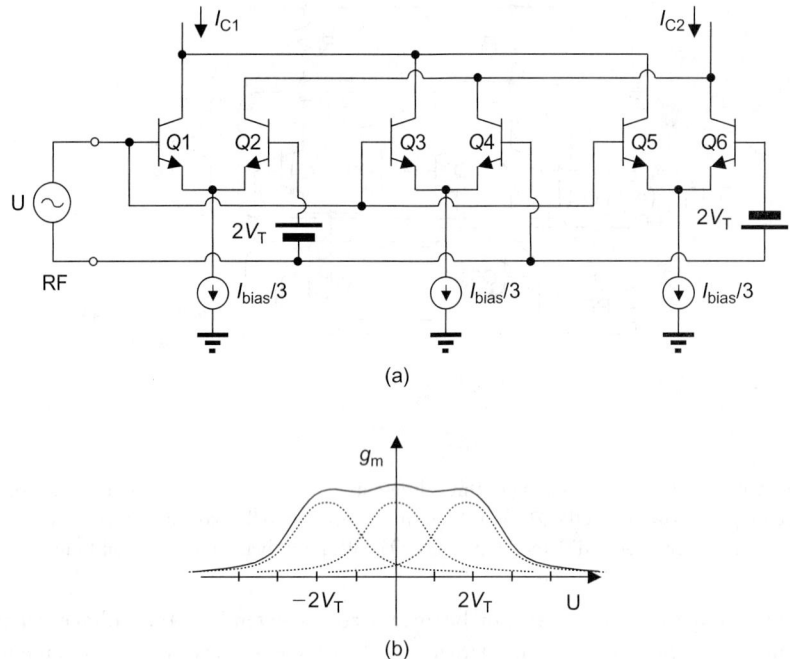

(a)

(b)

FIGURE 73.39 Wider operating range of Gilbert cell mixer can be achieved by paralleling differential pairs of the transconductance stage and offsetting the DC input biasing of the multiple input pairs.

TABLE 73.3 Comparison of Recent Results with Wide Bandwidth Gilbert Cell Mixers

Operating Band	Mixer Gain	Technology	Reference	Year
1–17 GHz	9.3 dB	GaAs HBT $f_T = 40$ GHz	[64]	2002
DC–20 GHz	15 dB	InP HBT $f_T = 70$ GHz	[65]	2000
DC–20 GHz	5 dB	GaAs HBT $f_T = 90$ GHz	[66]	1991
DC–12 GHz	5.5 dB	SiGe HBT $f_T = 47$ GHz	[67]	1998
DC–8 GHz	15 dB	Silicon BJT $f_T = 20$ GHz	[68]	1998
DC–40 GHz	12 dB	InP HBT $f_T = 70$ GHz	[69]	2003

relative to adjacent bands, the realization of low-cost filtering becomes ever more difficult. This has prompted engineers to seek techniques to suppress image signals in mixers rather than to rely on preselection filters.

The mixers discussed above can be combined into architectures used for image rejection mixers. Sometimes such mixers are categorized as *complex mixers* because several mixers are combined. Figure 73.40 illustrates such an image rejection mixer. This type of circuit requires the generation of a quadrature phase shift (and generally we require equal amplitudes for both the in-phase and quadrature-phase signals). Figure 73.41 shows both a lumped circuit and two distributed circuits for quadrature generation. The RC–CR network of Figure 73.41(a) generates a 90° shift between the two output terminals when $|\tan^{-1}(2\pi fRC)| = \pi/4 = 45°$. The amplitudes of signals V_{out1} and V_{out2} are equal, and in

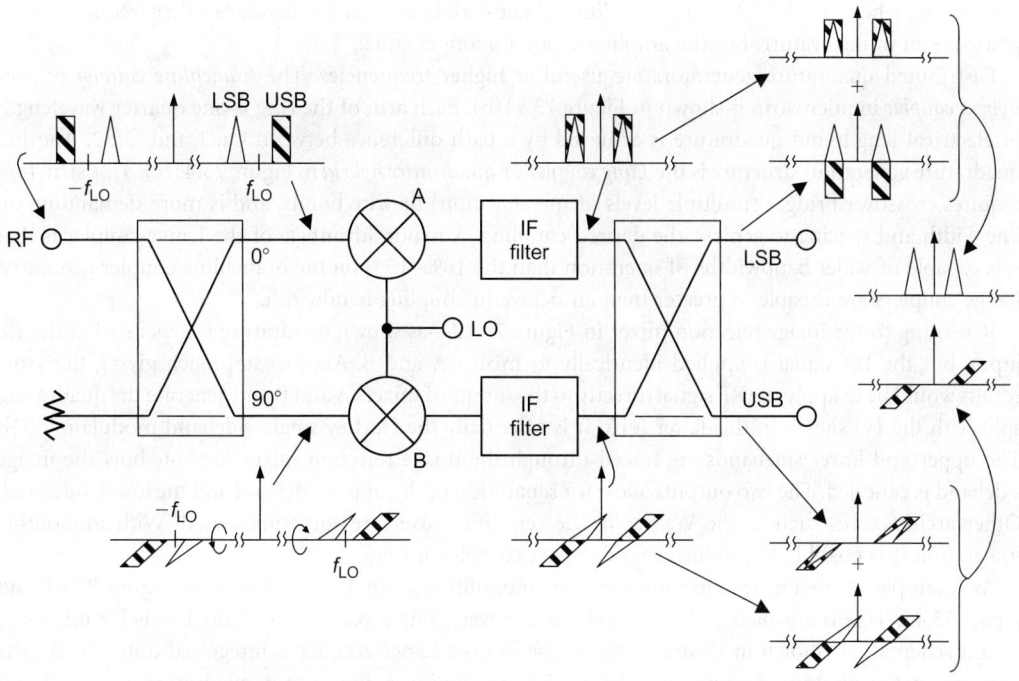

FIGURE 73.40 Image-rejection mixer can be constructed from two mixers and driving them in quadrature. The USB and LSB IF signals are traced through the mixer to illustrate how the image-rejection mixer operates. Quadrature hybrid couplers are used to generate the quadrature phase shifts between signals as required for mixer operation.

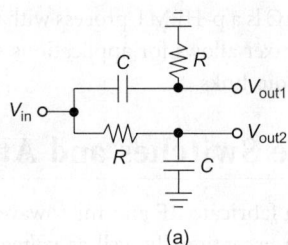

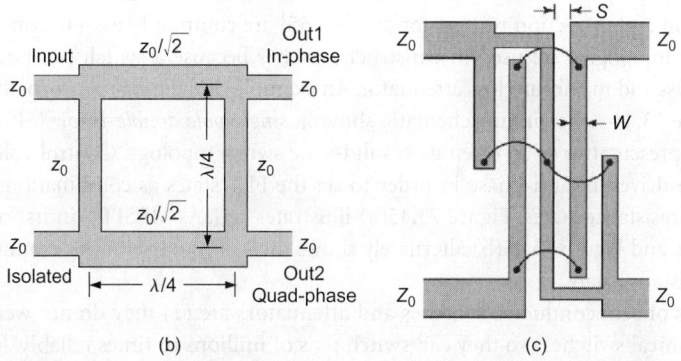

FIGURE 73.41 Three methods to generate quadrature signals. (a) Lumped element network is in quadrature when amplitudes are equal; (b) quadrature ring hybrid circuit in microstrip; and (c) Lange coupler in microstrip with wire bonds for cross-over connections.

quadrature, when $f = 1/(2\pi RC)$, thus the *RC–CR* network is very narrowband. At other frequencies the phase is still in quadrature, but the amplitudes are no longer equal.

Distributed quadrature generators are useful at higher frequencies. The *branchline coupler* or *ring hybrid coupler* in microstrip is shown in Figure 73.41(b). Each arm of the ring is one quarter wavelength of electrical length and quadrature is achieved by a path difference between Out1 and Out2. Another quadrature generation structure is the *Lang coupler* or *quadrature hybrid* in Figure 73.41(c). This structure requires crossover bridges (multiple levels of metallization) or wire bonds, and is more demanding on line width and spacing to achieve the desired coupling. A major advantage of the Lange coupler is that it is capable of wider bandwidths of operation than the 10%–15% on the branchline coupler geometry. Lange couplers are capable of greater than an octave in coupling bandwidth.

Returning to the image rejection mixer in Figure 73.40—as shown quadrature is generated at the RF input, but the LO signal is applied identically to mixers A and B. Another approach giving the same results would be to apply the RF signal directly to the inputs of mixers A and B and generate the quadrature split with the LO signal. In this latter form it is essentially the Hartley single-sideband modulator [52]. The upper and lower sidebands are traced through the image rejection mixer to show how the image sideband is canceled. The two outputs allow for separation of the upper sideband and the lower sideband. Other architectures such as the Weaver image rejection mixer are sometimes used. With monolithic integration it is possible to produce very compact complex mixers.

An example of an image rejection mixer in monolithic form [53] is shown in Figure 73.42 and Figure 73.43. This is a broadband image rejection mixer using a pair of distributed resistive mixers. A circuit schematic is shown in Figure 73.42. At the input a Lange coupler is integrated onto the chip as revealed in the photomicrograph (see Figure 73.43). However, the output quadrature hybrid is not integrated onto the IC chip and must be placed on adjacent circuitry. The bandwidth of operation is very wide spanning 6–30 GHz and its USB conversion loss is −10 dB and a 15 dB image rejection ratio with a +10 dBm LO drive and −20 dBm RF input signal. The LO-to-RF isolation is >20 dB, the LO-to-IF isolation >22 dB, and the RF-to-IF isolation >10 dB because of RF signal filtering limitations at the IF port. The process used to fabricate the IC is a p-HEMT process with 0.15 µm gate length and $f_T = 85$ GHz. The wide range of operation of this mixer allows for applications such as local multipoint distribution service (LMDS) and point-to-point radio links.

73.5   RF and Microwave Switches and Attenuators

GaAs and InP FETs are widely used to fabricate RF and microwave switches and step attenuators. Both MESFET and HEMT devices perform exceptionally well as voltage variable resistors (i.e., passive elements) in their linear operating region [54]. Semi-insulating substrates are used to fabricate these devices giving exceptionally low parasitic capacitance which is important for operation at very high frequencies. T-section, L-section and Pi-section resistor topologies [55] are commonly used to construct attenuators. The same resistor topologies are used to construct switches because a switch can be viewed as a two-state, minimum loss and maximum loss attenuator. An example of the use of a two-path L-section switch is shown in Figure 73.44—the circuit schematic shows a *single-pole double-throw* (SPDT[+]) passive FET switch which is representative of an often used solid-state switch topology. Control voltage lines, labeled Sel1 and Sel2, are driven in anti-phase in order to set the FET states as combinations of "on" or "off," or low- and high-resistance states. Figure 73.45(a) illustrates a GaAs MESFET in its "on" state with low channel resistance, and Figure 73.45(b) alternately shows the "off" state with the channel region pinched off in its high resistance state.

The advantages of semiconductor switches and attenuators are (1) they do not wear out (as do RF/microwave mechanical switches) so they can switch tens of millions of times reliably, (2) they have very fast switching speeds without contact bounce, (3) they are small, compact, and light weight, and

---

[+]Pole refers to the number of switch contact sets and throw refers to the number of conducting positions.

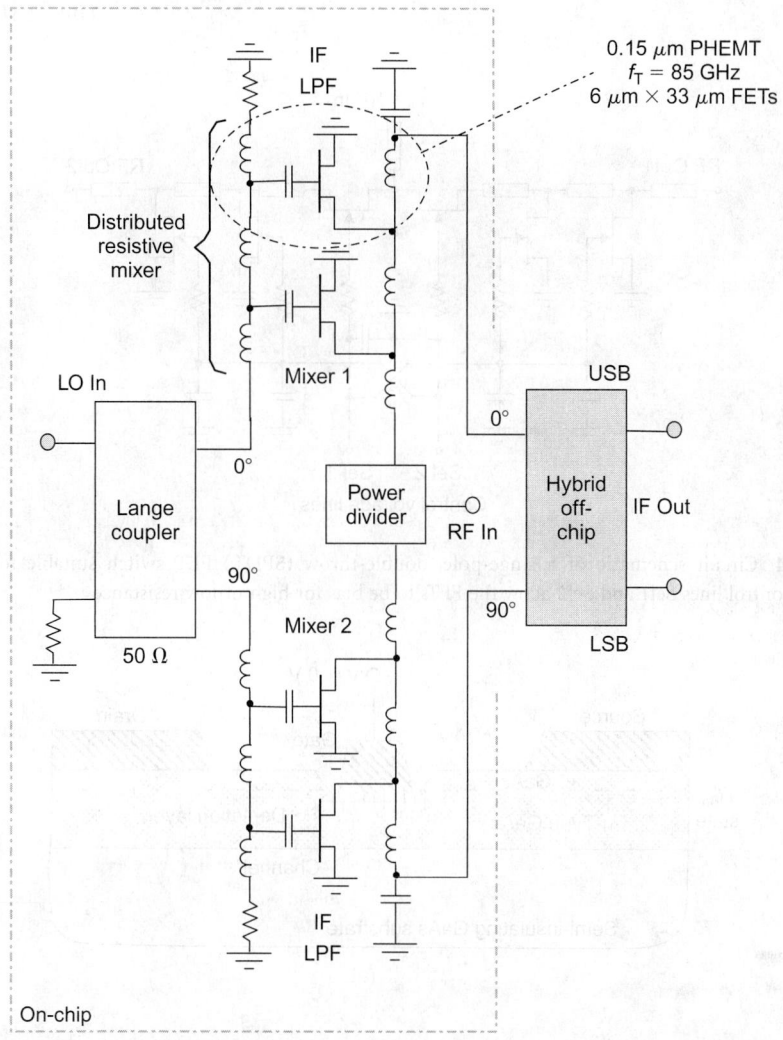

FIGURE 73.42 MMIC image-rejection resistive mixer schematic (Agilent Technologies).

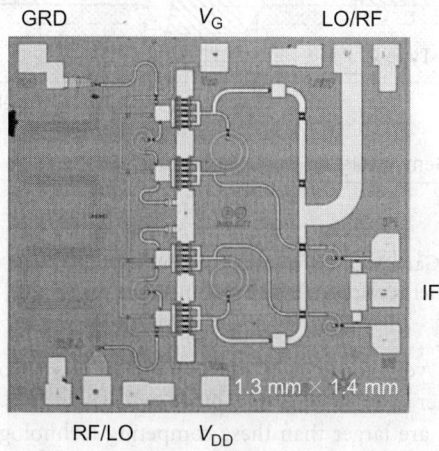

FIGURE 73.43 Photomicrograph of the MMIC image-rejection resistive mixer shown in Figure 73.42. Pad connections are indicated.

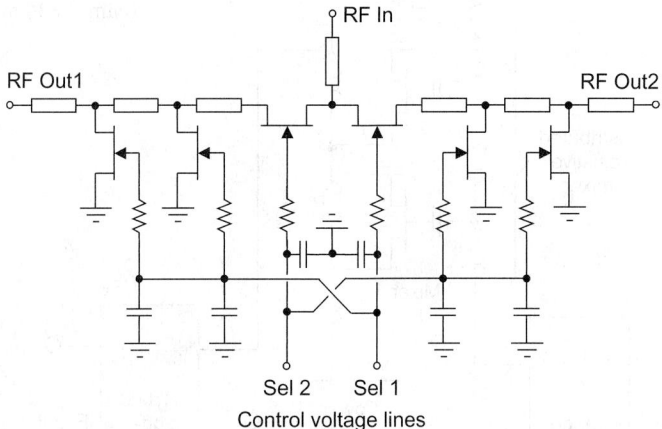

FIGURE 73.44 Circuit schematic of a singe-pole, double-throw (SPDT) FET switch suitable for monolithic integration. Control lines Sel1 and Sel2 allow the FETs to be bias for high or low resistance.

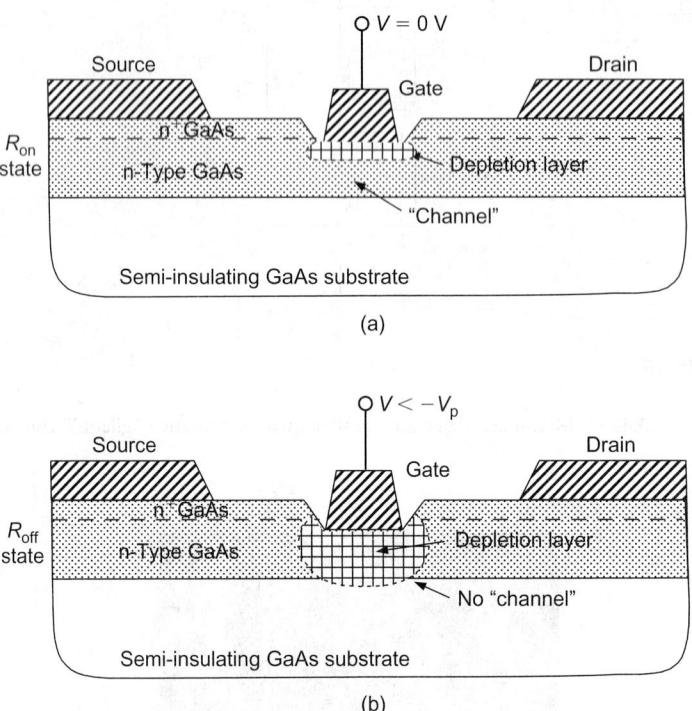

FIGURE 73.45 Cross section of GaAs MESFET showing (a) "on" state for lowest resistance and (b) "off" state for highest resistance (MESFET channel is completely depleted under the gate).

(4) switch activation requires very low power because the FET is a voltage-controlled device. The disadvantages include (1) higher insertion loss than other switches such as mechanical switches, because typically FET "on" resistances are larger than these competing technologies, (2) compound semiconductor devices generally have surface and interface traps (i.e., meaning charge trapping) resulting in slow transients or long settling times in their switching waveforms, and (3) greater distortion at high power levels because of the nonlinear behavior present in FET devices. MESFET and HEMT devices

have rather limited ranges over which their linear region of operation are valid, thereafter saturation sets in as voltages are increased.

Switches and attenuators generically fall into the category of RF signal or power control circuits. Switches are further categorized as being either *reflective* or *nonreflective*, depending upon whether they are matched to 50 Ω or simply reflect RF power in their "off" or blocking state. Attenuators are categorized as either *step attenuators* or *variable attenuators*. Variable attenuators are more challenging because the gate voltages must be carefully coordinated to give precise attenuation levels—the relationship is nonlinear and requires feedback electronics and careful calibration. An attenuator operated at only its minimum and maximum attenuation states essentially functions as a switch.

73.5.1 Switch and Attenuator Parameters

The primary parameters used to characterize the performance of switches and attenuators are as follows:

1. *Insertion loss* is the ratio of the output power to the incident input power. It is typically expressed in decibels and computed from IL(dB) = $10 \log_{10}$ (output power/input power). For switches the insertion loss should be low in the "on" state and very high for the "off" state. Generally the "on" resistance R_{on} of the FET is the primary limitation on the insertion loss in its transmit or "on" state.
2. *Isolation* for a switch is the insertion loss in the "off" state, or the ratio of signal powers when comparing the signal output power levels among the various output ports. Parasitic capacitances largely determine the isolation limits in switches and step attenuators.
3. *Return loss* is a measure of the match of a port relative to the characteristic impedance of transmission line connecting to the port. It is defined as the reflected power at the port to the incident input power and is calculated using RL(dB) = $10 \log_{10}$ (reflected power/input power).
4. *Output power compression* (or *1 dB compression power*) is essentially identical to the definition given for mixer parameters. It is a measure of the power saturation characteristic of the power transmitted through the switch or attenuator and is defined as the output power at which the output power is lower by 1 dB than what it would be if a linear relationship between output power versus input power continued to hold true. A plot of the output power versus input power gives a straight line in the linear region of operation and the 1 dB compression point is the deviation from the linear line by an amount of 1 dB.
5. *Intermodulation distortion* such as third-order intermodulation is again identical to the definition given for mixer parameters. Both second- and third-order intermodulation are often cited for switches and attenuators. Extrapolation of the fundamental signal power, second-order intermodulation power, third-order intermodulation power, and so on, to their intercept points defines the IM power levels as shown in Figure 73.26.
6. *Switching speed* measures how fast a switch or attenuator responds to a request to fully change its state. It is usually measured by monitoring the time change in RF power at the output relative to a very fast pulse or step voltage change to the control lines. The control voltage change can be very fast and is often assumed to be almost instantaneous for analytical purposes.

73.5.2 Simple FET Model for Switches

When operated in its linear region a FET can be modeled very simply. This is illustrated in Figure 73.46 using the cross section of a MESFET for illustration. In the "on" state a FET in its linear region is a low-valued resistor between source and drain. The channel "on" resistance is denoted by R_{on} and is connected between source and drain in Figure 73.46(a). The impedance of R_{on} is sufficiently low so that the FET capacitances can safely be ignored resulting in considerable simplification to the model used for switch and attenuator applications. For variable attenuator applications the gate-to-channel voltage would be varied to change the

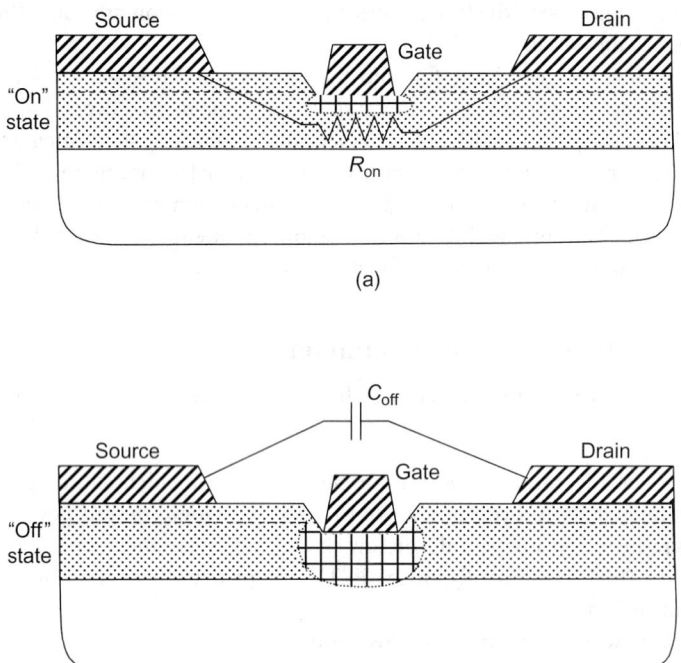

FIGURE 73.46   Cross section of GaAs MESFET showing (a) "on" state channel resistor R_{on} and (b) "off" state drain-to-source capacitance C_{off}.

source-to-drain resistance (we denote R_{on} as the source-to-drain resistance with maximum channel opening giving minimum channel resistance). In the "off" state the channel is fully pinched off such that the source-to-drain resistance is very high and now the FET capacitances will become the most important. Figure 73.46(b) shows the capacitance C_{off} connected from source-to-drain, where C_{off} is a combination of the source to-drain electrode capacitance and with the series combination of the gate-to-source capacitance and gate-to-drain capacitance included. Capacitance C_{off} is quite small for FETs fabricated on GaAs and InP substrates because of their semi-insulating properties. These two parameters allow a figure of merit to be defined which is very useful for comparing FET technologies and processes in signal control applications. This figure of merit is the *cutoff frequency* and is defined as $f_c = (2\pi R_{on} C_{off})^{-1}$. For GaAs HEMT processes f_c is typically 300–450 GHz and for InP processes it is higher ranging from 450–600 GHz. GaN HEMT devices have lower f_c values in the range of 125–280 GHz, although GaN HEMT devices have much higher inter-modulation distortion powers making them very attractive for higher power switches and attenuators.

To demonstrate the use of the simple FET two-component model, consider Figure 73.47 which is the equivalent circuit of the SPDT switch drawn in Figure 73.44. Branch 1 is in the "on" or transmit state while branch 2 is in the "off" or blocking state. FETs with a gate bias such that they are "on" have R_{on} inserted in their position in the circuit and FETs in the "off" state have C_{off} inserted. Reversing the Sel1 and Sel2 voltages changes the transmit branch from branch 1 to branch 2.

73.5.3   Slow-Switching Transients

A problem with compound semiconductors is charge trapping at the surface and at internal interfaces such as semiconductor interfaces or in the semi-insulating substrate. These traps hold electrons for long times, ranging from hundreds of microseconds to tens of seconds or longer. Charge trapping modulates the source-to-drain resistance thereby leading to a time variation in the insertion loss or low-value attenuation settings. Figure 73.48(a) shows the MESFET cross section with charge trapped at the surfaces in the regions between

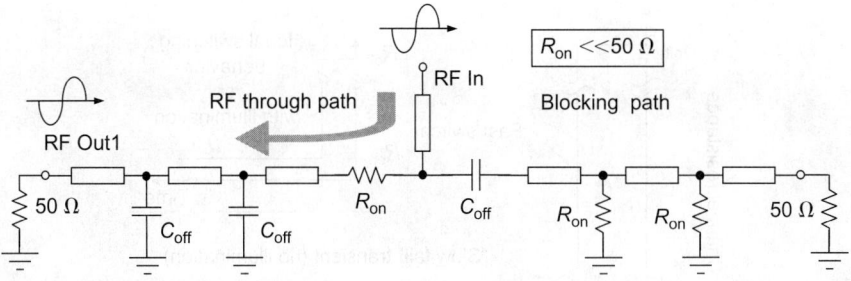

FIGURE 73.47 Equivalent circuit of the SPDT FET switch using the "on" state and "off" state element values from Figure 73.46 with branch one passing RF signal and branch two in its "blocking" state.

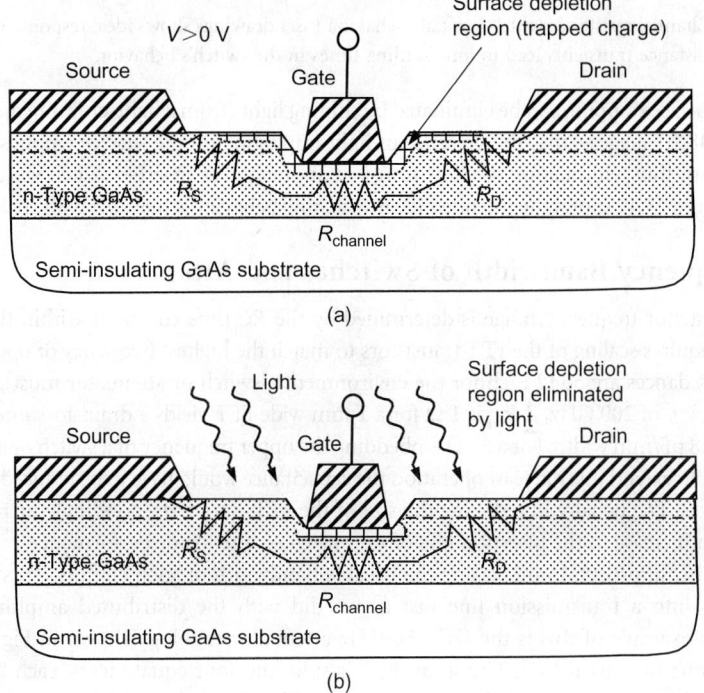

FIGURE 73.48 (a) "On" state resistance R_{on} ($=R_S + R_{channel} + R_D$) with surface trapping. R_S and R_D are larger owing to reduced cross section. (b) Lower R_{on} when light shines on the extrinsic channel's surface thereby removing trapped surface charge leading to a reduction of the parasitic resistances R_S and R_D. R_S and R_D are smaller owing to increased cross section.

the gate and drain and the gate and source. This trapped charge sets up a depletion layer that extends into the conductive regions between the gate and drain and the gate and source contacts. The total source-to-drain resistance is the sum of the channel resistance $R_{channel}$, parasitic source resistance R_S, and parasitic drain resistance R_D. The extent of the depletion layer's penetration determines the resistance values of R_S and R_D owing to the reduction in resistive cross section. As the trap occupation changes with time, both parasitic resistances also change with time. Figure 73.49 shows the time-varying behavior of the FET channel resistance. The "slow tail" behavior is the result of the surface traps not being fully depleted of carriers at the instant the control line voltages are switched on (say at time t_{SW}). Switches and attenuators depend critically on the precise value of R_{on} so this is easily observed with a fast responding power meter or oscilloscope driven by a detector. This is a major problem for some applications. However, the slow transient

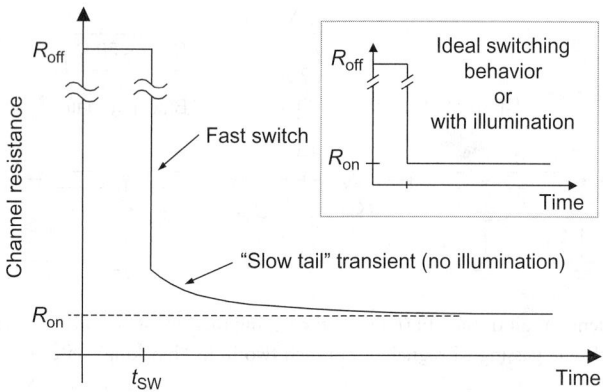

FIGURE 73.49   Channel resistance with "slow tail" behavior. Inset drawings shows ideal response without "slow tail" behavior. Such resistance transients lead to long settling times in the switch's behavior.

behavior from surface trapping can be eliminated by shining light (from a light-emitting diode for example) [56]. Figure 73.48(b) suggests that the presence of light impinging on the surface empties these traps, and keeps them empty, thus, eliminating the slow transient anomaly. The inserted switch waveform in the box appearing in Figure 73.49 shows the intended ideal switching behavior achievable with such illumination.

73.5.4   Frequency Bandwidth of Switches and Attenuators

Switch and attenuator frequency range is determined by the RC time constants within the circuit layout. Circuit design requires scaling of the FET transistors to match the highest frequency of operation. The port terminating impedances are 50 Ω each for the environment a switch or attenuator must work within. For a cutoff frequency f_c of 200 GHz, if R_{on} is 1 Ω for a 1 mm wide FET yields a drain-to-source capacitance of approximately 0.8 pF/mm width. For a 50 Ω embedding the upper frequency of a switch or attenuator would be about 4 GHz. To extend this range of operation the capacitance would have to be reduced by using smaller width FET devices. The penalty in reducing the size of the FET is to correspondingly increase the value of R_{on} leading to an increase in the insertion loss of the switch or attenuator performance.

   There is another approach to increasing the useable bandwidth of such circuits. This is to absorb the capacitance C_{off} into a transmission line just as we did with the distributed amplifier presented in Section 73.3. An example of this is the DC −50 GHz attenuator topology shown in Figure 73.50 where a T-section attenuator is used [57]. The shunt FET is split into four equal pieces, each being 200 μm in width, and embedded into a section of transmission line. The absorption of the C_{off} into the LR structure increases the attenuation of the attenuator network at higher frequencies. Of course, C_{off} also affects the minimum insertion loss of the attenuator. The two series FETs are both 750 μm in width and have a 50 Ω resistor placed across the source/drain path. These resistors minimize the effect of the parasitic capacitance of the series FETs at the maximum attenuation setting and also serve to reduce the harmonic distortion at high attenuation settings resulting in better linearity. The process used for this MMIC variable attenuator was a 0.4 μm gate MESFET process. The measured insertion loss performance [57] was 1.8 dB at 26.5 GHz and 2.6 dB at 40.0 GHz. Recent work [58] has demonstrated SPST HEMT switches using the traveling-wave concept to at least 80 GHz with <3 dB insertion loss using a 0.15 μm HEMT process. In addition, Mizutani and Takayama [59] described a traveling-wave SPST switch covering DC to 110 GHz using a 0.15 μm HFET process.

73.5.5   Gallium Nitride Switches and Attenuators

Previous work with AlGaN/GaN HEMT devices [60–62] suggests that these devices are excellent in the making of switches because of their ability to swing large voltages. The higher voltage breakdowns, higher

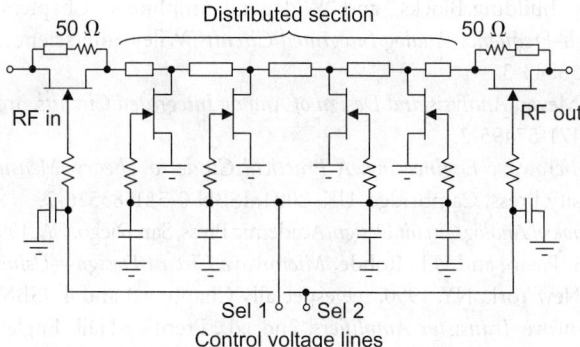

FIGURE 73.50 Distributed attenuator (or switch) using distributed section for the shunt resistance. By distributing the capacitance from the shunt FETs the bandwidth of operation can be greatly extended.

saturation currents because of the greater quantum well sheet charge, and generally higher pinch-off voltage of GaN HEMT devices accounts for the higher power handling capability of switches and attenuators built with these devices. Whereas InP and GaAs switches can handle power densities of the order of 1 W/mm, GaN switches can support power densities ten to twenty times that of InP- or GaAs-based switches and attenuators. GaN circuits built on either SiC or Al_2O_3 substrates benefit from a lower capacitance resulting from the use of an insulating substrate.

Tanaka [63] recently reported results on an SPDT AlGaN/GaN switch using a single-series FET and a single-shunt FET with a 1-dB compressed output power of 43 W at 1 GHz. The gate width of the GaN devices was 2 mm with extracted device parameters of R_{on} = 1.86 Ω mm and C_{off} = 0.35 pF/mm. Measured insertion loss was a low 0.26 dB and the isolation was 27 dB. The overall chip size was 0.48 mm².

73.5.6 Summary

This chapter has discussed several important classes of RF and microwave circuits which perform especially well using compound semiconductor devices. The major categories of circuits included broadband amplifiers, mixers, and switches and attenuators. Major trends in the future will be the use of GaN HEMT devices for RF and microwave power amplifiers. Compound semiconductor MESFET and HEMT devices perform especially well as power control devices for switching and attenuation. Finally, SiGe HBT devices and BiCMOS will play a greater role in such applications in the future.

References

1. E. Fred Schubert, *Light-Emitting Diodes*, Cambridge University Press, Cambridge, UK, 2003. ISBN 0-521-33351-1.
2. P.A. Houston, "High-Frequency Heterojunction Bipolar Transistor Device Design and Technology," *IEE Electron. and Commun. Engng. J.*, **12**, October 2000, 220–228.
3. B.S. Meyerson, "Silicon:Germanium Base Mixed-Signal Technology for Optimization of Wired and Wireless Telecommunications," *J. IBM Res. Dev.*, **44**, May 2000, 391–407.
4. J.D. Cressler, "SiGe Technology: A New Contender for Si-Based RF and Microwave Circuit Applications," *IEEE Trans. Microwave Theory & Techniques*, **46**, May 1998, 572–589.
5. T.H. Lee, *The Design of CMOS Radio-Frequency Integrated Circuits*," 2nd ed., Cambridge University Press, Cambridge, UK, 2003, ISBN 0-521-83539-9.
6. A.A. Abidi, "RF CMOS Comes of Age," *IEEE Microwave Magazine*, **4**, December 2003, 47–60.
7. C.-Y. Chang et al., "RF CMOS Technology for MMIC," *Microelectron. Reliab.*, **42**, April–May 2002, pp. 721–733.

8. D.B. Estreich, "Basic Building Blocks," and "Wideband Amplifiers," Chapters 5 and 6 appearing in R. Goyal (Ed.), *High-Frequency Analog Integrated Circuits*, Wiley-Interscience, Wiley, New York, NY, 1995, ISBN 0-471-53043-3.

9. P.R. Gray and R.G. Meyer, *Analysis and Design of Analog Integrated Circuits*, 3rd ed., Wiley, New York, NY, 1993, ISBN 0-471-57495-3.

10. T.H. Lee, *Planar Microwave Engineering: A Practical Guide to Theory, Measurements and Circuits,"* Cambridge University Press, Cambridge, UK, 2004, ISBN 0-521-83526-7.

11. D.L. Feucht, *Handbook of Analog Circuit Design*, Academic Press, San Diego, CA, 1990, ISBN 0-12 254240-1.

12. G.D. Vendelin, A.M. Pavio, and U.L. Rohde, *Microwave Circuit Design—Using Linear and Nonlinear Techniques*, Wiley, New York, NY, 1990, see especially Chapters 3 and 4, ISBN 0-471-60276-0.

13. G. Gonzalez, *Microwave Transistor Amplifiers*, 2nd ed., Prentice-Hall, Englewood Cliffs, NJ, 1997, ISBN 0-132-54335-4.

14. T.T. Ha, *Solid-State Microwave Amplifier Design*, Wiley-Interscience, Wiley, New York, NY, 1981, ISBN 0-471-08971-0.

15. R.M. Fano, "Theoretical Limitations on the Broadband Matching of Arbitrary Impedances," *J. Franklin Institute*, **249**, January/February 1950, 57–83, 139–154.

16. D.J. Mellor and J.G. Linvill, "Synthesis of Interstage Networks of Prescribed Gain versus Frequency Slopes," *IEEE Trans. Microwave Theory & Techniques*, **MTT-23**, December 1975, 1013–1020.

17. M.S. Ghausi, *Principles and Design of Linear Active Circuits*, McGraw-Hill, New York, NY, 1965.

18. J.M. Petit and M.M. McWhorter, *Electronic Amplifier Circuits Theory and Design*, McGraw-Hill, New York, NY, 1961.

19. J.B. Beyer et al., "MESFET Distributed Amplifier Design Guidelines," *IEEE Trans. Microwave Theory & Techniques*, **MTT-32**, March 1984, pp. 268–275.

20. W.C. Percival, "Thermionic Value Circuits," U.K. patent no. 460,562, filed July 24, 1935 and granted on January 25, 1937.

21. E.L. Ginzton, W.R. Hewlett, J.H. Jasberg and J.D. Noe, "Distributed Amplification," *Proc. IRE*, **36**, August 1948, 956–969.

22. N.B. Schrock, "Distributed Amplifier," *HP Journal*, 1:1, September 1949, 1–4.

23. E.W. Strid and K.R. Gleason, "A DC-12 GHz Monolithic GaAs FET Distributed Amplifier," *IEEE Trans. Microwave Theory & Techniques*, **MTT-30**, July 1982, 969–975.

24. Y. Ayasli et al., "A Monolithic GaAs 1–13-GHz Traveling-Wave Amplifier," *IEEE Trans. Microwave Theory & Techniques*, **MTT-30**, July 1982, 976–981.

25. Y. Ayasli, "Distributed Amplifier," U.S. patent no. 4,486,719, filed July 1, 1982 and granted on December 4, 1984.

26. Y. Ayasli et al., "2–20 GHz GaAs Traveling-Wave Power Amplifier," *IEEE Microwave and Millimeter-Wave Monolithic Circuits Symposium Technical Digest*, 1983, 67–70.

27. J. Orr, "A Stable 2–26.5 GHz Two-Stage Dual-Gate Distributed MMIC Amplifier," *1986 IEEE MTT-S International Microwave Symposium Digest*, Baltimore, MD, June 2–4, 1986, pp. 817–820.

28. D.C. D'Avanzo et al., "A Manufacturable, 26 GHz GaAs MMIC Technology," *Proceedings of the GaAs IC Symposium*, Nashville, TN, November 6–9, 1988, pp. 317–320.

29. Y. Ayasli et al., "Capacitively Coupled Traveling-Wave Power Amplifier," *IEEE Trans. Microwave Theory & Techniques*, **MTT-32**, December 1984, pp. 1704–1709.

30. Y. Ayasli et al., "2–20 GHz GaAs Traveling-Wave Power Amplifier," *IEEE Trans. Microwave Theory & Techniques*, **MTT-32**, March 1984, 290–294.

31. E.M. Chase and W. Kennan, "A Power Distributed Amplifier Using Constant-R Networks," *IEEE Microwave and Millimeter-Wave Monolithic Circuits Symposium Technical Digest*, 1986, pp. 13–17.

32. A.M. Pavio and L. Zhao, "Tapered Constant "R" Network for Use in Distributed Amplifiers," U.S. patent no. 6,714,095, filed June 18, 2002 and granted March 30, 2004.

33. B.L. Kasper, "Receiver Design," in *Optical Fiber Communications II*, S.E. Miller and I.P. Kaminow, Eds., Academic Press, Boston, MA, 1988, pp. 689–722.

34. R.G. Smith and S.D. Personick, "Receiver Design for Optical Fiber Communication Systems," in *Semiconductor Devices for Optical Communication*, H. Kressel, ed., Springer, Berlin, Germany, 1982, pp. 89–160.

35. D.B. Estreich, "Wideband Amplifiers," in *High-Frequency Analog Integrated Circuit Design*, R. Goyal, Ed., Wiley-Interscience, Wiley, New York, NY, 1995, pp. 230–237.

36. S.A. Maas, *Microwave Mixers*, 2nd ed., Artech House, Norwood, MA, 1993.

37. G.D. Vendelin, A.M. Pavio, and U.L. Rohde, *Microwave Circuit Design—Using Linear and Nonlinear Techniques*, Wiley, New York, NY, 1990, see especially Chapter 7 titled, "Microwave Mixer Design." ISBN 0-471-60276-0.

38. P.J. Nahin, *The Science of Radio*, 2nd ed., Springer, New York, NY, 2001. ISBN 0-387-95150-4.

39. R.N. Bracewell, *The Fourier Transform and Its Applications*, 3rd ed., McGraw-Hill, New York, NY, 1999. ISBN 0-07-303938-1.

40. Synergy Microwave Corporation, "Mixers," Paterson, NJ, available at http://www.synergymwave.com/Articles/PDF/Mixers.pdf.

41. T.H. Lee, *Planar Microwave Engineering A Practical Guide to Theory, Measurement and Circuits*, Cambridge University Press, Cambridge, UK, 2004. ISBN 0-521-83526-7.

42. C.L. Ruthroff, "Some Broad-band Transformers," *Proc. IRE*, **47**, August 1959, 1337–1342.

43. J. Sevick, *Transmission Line Transformers*, 4th ed., Noble Publishing Corp., Thomasville, GA, 2001. ISBN 1-884932-18-5.

44. P.-S. Wu et al., "Compact and Broad-Band Millimeter-Wave Monolithic Transformer Balanced Mixers," *IEEE Trans. Microwave Theory & Techniques*, 53, October 2005, pp. 3106–3114.

45. N. Marchand, "Transmission Line Conversion Transformers," *Electronics*, **17**, December 1944, 142–145.

46. R.M. Waugh and M. Kumar, "Monolithic Double Balanced Mixer With High Third-Order Intercept Point Employing an Active Distributed Balun," U.S. patent no. 5,060,298, filed December 9, 1988 and granted October 22, 1991.

47. B. Gilbert, "A Precise Four-Quadrant Multiplier with Subnanosecond Response," *IEEE J. Solid-State Circuits*, **SC-3**, December 1968, 365–373.

48. H.E. Jones, "Dual Output Synchronous Detector Utilizing Transistorized Differential Pairs," U.S. patent no. 3,241,078, filed June 18, 1963 and granted March 15, 1966.

49. B. Gilbert and R. Baines, "Fundamentals of Active Mixers," *Applied Microwave & Wireless*, winter 1995, 10–27.

50. B. Gilbert, "The Multi-tanh Principle: A Tutorial Overview," *IEEE J. Solid-State Circuits*, **33**, January 1998, 2–17.

51. B. Gilbert, "The MICROMIXER: A Highly Linear Variant of the Gilbert Mixer Using a Bisymmetric Class-AB Input Stage," *IEEE J. Solid-State Circuits*, **32**, September 1997, 1412–1423.

52. Ralph V.L. Hartley, "Modulation System," U.S. patent no. 1,666,206, filed January 15, 1925 and granted April 17, 1928.

53. K. Fujii and H. Morkner, "A 6–30 GHz Image-Rejection Distributed Resistive MMIC Mixer in a Low-Cost Surface Mount Package," *IEEE MTT-Symp. International Microwave Symposium Digest*, Long Beach, CA, June 2005.

54. R. Goyal (Ed.), *Monolithic Microwave Integrated Circuits: Technology and Design*, Artech House, Norwood, MA, 1989, pp. 527–548. ISBN 0-89006-309-5.

55. H.P. Westman et al. (Ed.), *Reference Data for Radio Engineers*, 5th ed., Howard W. Sams and Company, Inc., Indianapolis, IN, 1968, see especially Chapter 10, "Attenuators," pp. 10-1 to 10-9.

56. D.B. Nicholson & E.R. Ehlers, "Faster Switching GaAs FET Switches by Illuminating with High Intensity Lamp," U.S. patent no. 5,808,322, filed April 1, 1997 and granted September 15, 1998.

57. H. Kondoh, "DC-50 GHz MMIC Variable Attenuator with a 30 dB Dynamic Range," *1998 MTT-Symposium Technical Digest*, June 1998, pp. 499–502.

58. K.-Y. Lin et al., "Millimeter-Wave MMIC Passive HEMT Switches Using Traveling-Wave Concept," *IEEE Trans. Microwave Theory & Techniques*, **52**, August 2004, 1798–1808.

59. H. Mizutani and Y. Takayama, "DC-110-GHz MMIC Traveling-Wave Switch," *IEEE Trans. Microwave Theory & Techniques*, **48**, May 2000, 840–845.

60. R.H. Caverly, N.V. Drozdovski, and M.J. Quinn, "Gallium Nitride-based Microwave and RF Control Devices," *Microwave Journal*, **44**, February 2001, 112–124.

61. E. Alekseev et al., "Broadband AlGaN/GaN HEMT MMIC Attenuators with High Dynamic Range," *30th European Microwave Conference*, September 2000.

62. M. Kameche and N.V. Drozdovski, "GaAs-, InP- and GaN HEMT-Based Microwave Control Devices: What is Best and Why," *Microwave Journal*, **48**, May 2005, 164–180.

63. T. Tanaka, "Current Status of AlGaN/GaN HFETs and MMICs for RF Applications," presented at the *Short Course on GaN Technology and Its Wireless-Microwave-Millimeter-Wave IC Applications*, 2005 IEEE Compound Semiconductor Integrated Circuit Symposium, Palm Springs, CA, October 30, 2005.

64. B. Tzeng et al., "A 1–17 GHz InGaP-GaAs HBT MMIC Analog Multiplier and Mixer with Broad-Band Input-Matching Networks," *IEEE Trans. Microwave Theory & Techniques*, **50**, November 2002, 2564–2568.

65. K.W. Kobayashi et al., "A DC-20 GHz InP HBT Balanced Analog Multiplier for High-Data-Rate Direct-Digital Modulation and Fiber Optic Applications," *IEEE Trans. Microwave Theory & Techniques*, **48**, February 2000, 194–202.

66. K. Osafune and Y. Yamauchi, "20-GHz 5-dB-Gain Analog Multipliers with AlGaAs/GaAs HBTs," *IEEE MTT-Symp International Microwave Symposium Digest*, Boston, MA, 1991, pp. 1258–1288.

67. J. Glenn et al., "12-GHz Gilbert Mixers Using a Manufacturable Si/Si-Ge Epitaxial-Base Bipolar Technology," *Proc. IEEE Bipolar/BiCMOS Circuits Technology Meeting Digest*, Minneapolis, MN, 1998, pp. 186–189.

68. P. Weger, "Gilbert Multiplier as an Active Mixer with Conversion Gain Bandwidth of Up to 17 GHz," *Electron. Lett.*, **27**, March 1998, 570–571.

69. K.W. Kobayashi, "A DC-40 GHz InP HBT Gilbert Multiplier," *GaAs IC Symposium Technical Digest*, 2003, pp. 251–254.

74

High-Speed Circuit Design Principles

Stephen I. Long

*University of California at
Santa Barbara*

CONTENTS

The digital and analog circuits used with high-speed compound semiconductor devices must satisfy the same essential conditions for design robustness and performance as integrated circuits (ICs) fabricated in other technologies. For example, the static or DC design of a logic cell must guarantee adequate voltage and current gain to restore the signal levels in a chain of similar cells. A minimum noise margin must be provided for tolerance against process variation, temperature, and induced noise from ground bounce, cross talk, and EMI so that functional circuits and systems are produced with good electrical yield. Techniques for static logic design have been clearly described in other references [1] and are basically the same for compound semiconductor IC technologies. The major difference, and this includes CMOS and SiGe BJT designs as well, occurs when speed and bandwidth are of primary concern. In that situation, propagation delays, rise and fall times, maximum clock frequencies of digital circuits, and bandwidth and transient response of analog circuits must be determined as a function of extrinsic loading and power dissipation. Compound semiconductor designs emphasize speed, so logic voltage swings are generally low—low τ_r so that transconductances and f_T are high, and device access resistances are made as low as possible to minimize the lifetime time constant τ. This combination makes circuit performance very sensitive to parasitic R, L, and C, especially when the highest operation frequency is desired. The following sections will describe some of the techniques that can be used for estimating the high-frequency performance of digital and wideband analog ICs.

The most effective methods for guiding the design are those providing insight that helps to identify the dominant time constants, which determine circuit performance. These are not necessarily the most accurate methods, but are highly useful because they allow the designer to determine what part of the circuit or device is limiting the speed. Circuit simulators are far more accurate (at least to the extent that the device models are valid), but do not provide much insight into performance limitations. Without simple analytical techniques to guide the design, performance optimization becomes a trial-and-error exercise with no guarantee of finding an optimal solution. The following sections will describe some of these techniques with particular emphasis on the devices that provide insight and can be used for estimating the high-frequency performance of logic and wideband analog ICs. It is assumed that layout parasitics will be estimated as well as part of the design process and included in the analysis. Because these are highly process-and application-dependent, they will not be considered as part of the design examples presented in the later sections.

74.1 Technology-Independent Design Methodologies

74.1.1 Zeroth-Order Delay Estimate

The first technique, useful for some digital circuit applications, uses the simple relationship between voltage and current in a capacitor

$$I = C_L \frac{dV}{dt} \tag{74.1}$$

This equation is relevant when circuit performance is dominated by wiring or fanout capacitance. This will be the case if the delay predicted by this equation caused by the total loading capacitance, C_L, significantly exceeds the intrinsic delay of a basic inverter or logic gate. To apply this approach, determine the average current available from the driving logic circuit for charging (I_{LH}) and discharging (I_{HL}) the load capacitance. The logic swing ΔV is known, so low-to-high (t_{PLH}) and high-to-low (t_{PHL}) propagation delays can be determined from Eq. (74.1). These delays represent the time required to charge or discharge the circuit output to 50% of its final value. Thus, t_{PLH} is given by

$$t_{PLH} = \frac{C_L \Delta V}{2 I_{LH}} \tag{74.2}$$

where I_{LH} is the average charging current during the output transition from logic low voltage V_{OL} to $V_{OL} + \Delta V/2$. The net propagation delay is given by

$$t_P = \frac{t_{PLH} + t_{PHL}}{2} \tag{74.3}$$

At this limit, where speed is dominated by the ability to drive load capacitance, we see that increasing the currents will reduce t_P. In fact, the product of power (proportional to current) and delay (inversely proportional to current) is nearly constant under this situation. Increases in power lead to reduction of delay until the interconnect distributed RC delays or electromagnetic propagation delays become comparable to t_p. The equation also shows that the small voltage swing ΔV is good for speed if the noise margin and drive current are not compromised. This means that the devices with high transconductance are faster in such applications.

This method can be effective in estimating delays in large digital ICs where wiring capacitance is dominant. In the case of deeply scaled submicron circuits, distributed RC delay on interconnect lines must also be considered. Because the emphasis in this chapter is on small, very high-speed circuits, other methods that are more dependent on device parasitics and intrinsic delays will be developed next. There are many references that give more information on the simple delay estimation methods [1].

74.1.2   Time Constant Delay Methods: Elmore Delay and Risetime

Time constant delay estimation methods are very useful when the wiring capacitance is quite small or the charging current is quite high. In this situation, typical of very high-speed SSI and MSI circuits that push the limits of the device and process technology, the circuit delays are dominated by the devices themselves. The method to be described relies upon a large-signal equivalent circuit model of the transistors, an approximation that is dubious at best. The construction of the large-signal equivalent circuit requires averaging of nonlinear model elements such as transconductance and certain device capacitances over the appropriate signal voltage range. But the objective of the technique is not absolute accuracy. That is much less important than being able to identify the dominant contributors to the delay and risetime, since more accurate but less intuitive solutions are easily available through circuit simulation.

The propagation delay definition described above, the delay required to reach 50% of the logic swing, must be relaxed slightly to apply methods based on linear system analysis. It was first shown by Elmore [2] in 1948, and apparently rediscovered by Ashar [3] in 1964, that the delay time T_D between an impulse function $\delta(0)$ applied at $t = 0$ to the input of a network and the centroid or "center of mass" of the impulse response $f(t)$ at the output is in many cases quite close to the 50% delay t_p. The Elmore delay is called T_D to avoid confusion with the propagation delay t_p defined above. The Elmore delay is illustrated in Figure 74.1 and can be calculated by the normalized value of the first moment of the impulse response.

$$T_D = \frac{\displaystyle\int_0^\infty t\, f(t)\, dt}{\displaystyle\int_0^\infty f(t)\, dt} \tag{74.4}$$

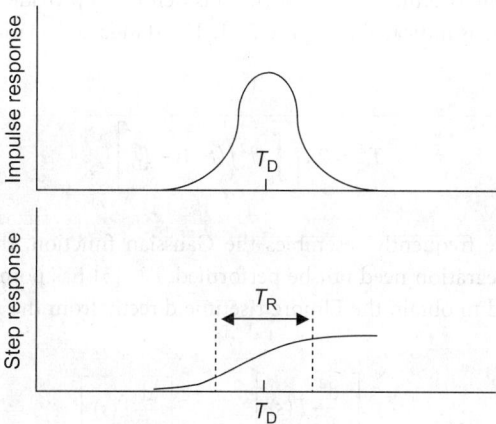

FIGURE 74.1   Elmore delay T_D is defined by the center of mass or centroid of the impulse response [2]. The risetime T_R of the step response is approximated by the standard deviation of the impulse response.

Two conditions must be satisfied to use this approach. First, the step response of the network is monotonic. This implies that the impulse response is purely a positive function. Monotonic step response is valid only when the circuit poles are all negative and real or the circuit is heavily damped. Owing to feedback through device and circuit elements and the underdamping effects caused by a source or emitter follower driving a common source or emitter stage, this condition is seldom completely correct. Complex poles often exist. But, strongly underdamped circuits are seldom useful for reliable logic circuits because their transient response will exhibit ringing, so efforts to compensate or damp such oscillations are needed in these cases anyway. Then, the circuit becomes heavily damped or at least dominated by a single pole and fits the above requirement more precisely. In any event, the method will yield a conservative estimate of delay and risetime.

Second, the correspondence between T_D and t_p is improved if the impulse response is symmetric in shape as in Figure 74.1. It is shown in Ref. [2] that cascaded stages with similar time constants have a tendency to approach a Gaussian-shaped distribution as the number of stages becomes large. Most logic systems require several cascaded stages, so this condition is often true as well.

Assuming that these conditions are approximately satisfied, the Laplace transform of the impulse response of a circuit is the same as the network function $F(s)$ in the complex frequency $s = \sigma + j\omega$. Then, the Elmore delay, T_D, can be determined by

$$T_D = \frac{\int_0^\infty t f(t)\, dt}{\int_0^\infty f(t)\, dt} = \lim_{s \to 0} \frac{\int_0^\infty t f(t) e^{-st}\, dt}{\int_0^\infty f(t) e^{-st}\, dt} = \left[\frac{-\dfrac{dF(s)}{ds}}{F(s)} \right]_{s=0} \tag{74.5}$$

T_D can be obtained directly from the network function $F(s)$ as shown. But, the network function must be calculated from the large-signal equivalent circuit of the device including all important parasitics, driving impedances, and load impedances. This is notoriously difficult if the circuit includes a large number of capacitances or inductances. Tien [4] has described an approach for calculating the Elmore delay strictly from $F(s)$ by breaking up the circuit into subnetworks. However, the algebraic labor is still considerable.

74.1.2.1 Risetime

The standard definition of risetime is the 10–90% time delay of the step response of a network. While convenient for measurement, this definition is analytically unpleasant to derive for anything except simple, first-order circuits. Elmore [2] demonstrated that the standard deviation of the impulse response could be used to estimate the risetime of a network. This definition provides estimates that are close to the usual risetime definition as indicated in Figure 74.1. The standard deviation of the impulse response can be calculated using

$$T_R^2 = 2\pi \left[\int_0^\infty t^2 f(t)\, dt - T_D^2 \right] \tag{74.6}$$

Since the impulse response frequently resembles the Gaussian function, the integral could be easily evaluated; however, the integration need not be performed. Lee [5] has pointed out that the transform techniques can also be used to obtain the Elmore risetime directly from the network function, $F(s)$.

$$T_R^2 = 2\pi \left[\frac{\dfrac{d^2}{ds^2} F(s)}{F(s)} \right]_{s=0} - 2\pi \left[\frac{\dfrac{d}{ds} F(s)}{F(s)} \right]_{s=0}^2 \tag{74.7}$$

This result can also be used to show that the risetimes of cascaded networks add as the square of the individual risetimes. If two networks are characterized by risetimes T_{R1} and T_{R2}, the total risetime, $T_{R,total}$, is given by the root mean square (RMS) sum of the individual risetimes

$$T_{R,total}^2 = T_{R1}^2 + T_{R2}^2 \tag{74.8}$$

74.1.3 Time Constant Methods: Open-Circuit Time Constants

The frequency domain/transform methods for finding delay and risetime are particularly valuable for design optimization because they identify dominant time constants. Once the time constants are found, the designer can make efforts to change biases, component values, or optimize the design of the transistors themselves to improve the performance through addressing the relevant bottleneck in performance. The drawback in the above technique is that a network function must be derived. This becomes tedious and time consuming if the network is of even modest complexity. An alternate technique was developed [5,6] that also can provide reasonable estimates for delay, but with much less computational difficulty. The open-circuit time constant (OCTC) method is widely used for the analysis of the bandwidth of analog electronic circuits only for this reason. It is just as applicable for estimating the delay of very high-speed digital circuits.

The basis for this technique again comes from the transfer or network function $F(s) = V_o(s)/V_i(s)$. Considering low-pass transfer functions containing only poles, the function can be written as

$$F(s) = \frac{a_0}{b_n s^n + b_{n-1} s^{n-1} + \cdots + b_1 s + 1} \tag{74.9}$$

The denominator comes from the product of n factors of the form $(\tau_j s + 1)$ where τ_j is the time constant associated with the jth pole in the transfer function. The b_1 coefficient can be shown to be equal to the sum

$$b_1 = \sum_{j=1}^{n} \tau_j \tag{74.10}$$

of the time constants and b_2 the product of all the time constants. Often, the first-order term dominates the frequency response. In this case, the 3 dB bandwidth is then estimated by $\omega_{3\,dB} = 1/b_1$. The higher order terms are neglected. The accuracy of this approach is good when the circuit has a dominant pole. If all poles have the same frequency, the error in delay predicted by this method is about 25%. Much worse errors can occur however if the poles are complex or if there are zeros in the transfer function as well. This case will be discussed later.

Elmore [2] has once again provided the connection needed to obtain delay and risetime estimates from the network function. In the more general case where there could be a first-order zero in the numerator, a_1, the Elmore delay is given by

$$T_D = b_1 - a_1 \tag{74.11}$$

Elmore also proves that you can estimate the risetime by

$$T_R^2 = b_1^2 - a_1^2 + 2(a_2 - b_2) \tag{74.12}$$

In this equation, a_2 and b_2 correspond to the coefficients of the second-order zero and pole, respectively.

At this point, it would appear that nothing has been gained since finding the time constants associated with the poles and zeros is well known to be difficult. Fortunately, it is possible to obtain the b_1 and b_2

coefficients directly by a much simpler method, OCTC or method of time constants (MOTC). It has been shown that [5,6]

$$b_1 = \sum_{j=1}^{n} R_{jo}C_j = \sum_{j=1}^{n} \tau_{jo} \tag{74.13}$$

the sum of the time constants τ_{jo}, defined as the product of the effective open-circuit resistance R_{jo} across each capacitor C_j when all other capacitors are open-circuited, equals b_1. These time constants are very easy to calculate since open-circuiting all other capacitors greatly simplifies the network by decoupling many other components. Consider Figure 74.2(a). A network is represented by the box and all internal capacitors are brought to the outside of the box. Figure 74.2(b) then illustrates how the driving point resistance, $R_{1o} = V_1/I_1$, is determined for each capacitor with all others removed from the network.

Dependent sources must be considered in the calculation of the R_{jo} open-circuit resistances. Note that these open-circuit time constants are not equal to the pole time constants, but their sum gives b_1. It should also be noted that the individual OCTCs give the time constant of the network if the jth capacitor were the only capacitor. Thus, each time constant provides information about the relative contribution of that part of the circuit to the bandwidth or the delay [5]. If one of these is much larger than the rest, this is the place to begin working on the circuit to improve its speed.

The b_2 coefficient can also be found by a similar process [7], taking the sum of the product of time constants of all possible pairs of capacitors. A short-circuit time constant is also required. Figure 74.3 illustrates how the resistance is calculated while short-circuiting one of the other ports. For example, in a three-capacitor circuit, b_2 is given by

$$b_2 = R_{1o}C_1 R_{2s}^1 C_2 + R_{1o}C_1 R_{3s}^1 C_3 + R_{2o}C_2 R_{3s}^2 C_3 \tag{74.14}$$

where the R_{js}^i resistance is the resistance across capacitor C_j calculated when capacitor C_i is *short*-circuited and all other capacitors *open*-circuited. The superscript indicates which capacitor is to be shorted. So,

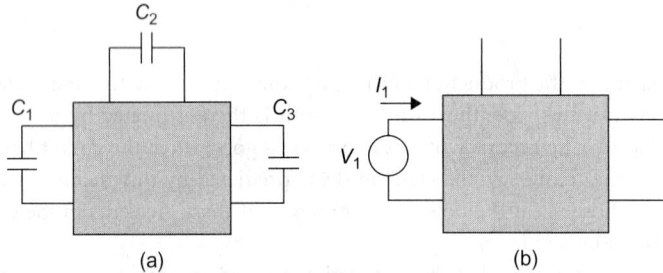

(a)   (b)

FIGURE 74.2   (a) A network represented by the box and all internal capacitors brought to the outside of the box. (b) The driving point resistance determined for each capacitor by removing all others from the network.

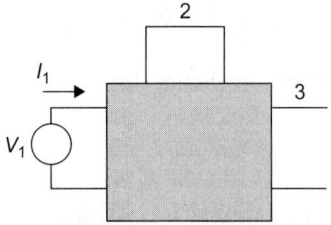

FIGURE 74.3   The "short-circuit" resistance at port 1, R_{1s}^2, calculated by short-circuiting port 2 and open-circuiting all other ports.

R_{3s}^2 is the resistance across C_3 when C_2 is short-circuited and C_1 is open-circuited. Note that the first time constant in each product is an open-circuit time constant that has already been calculated. In addition, for any pair of capacitors in the network, we can find an OCTC for one and an SCTC for the other. The order of choice does not matter because

$$R_{io}C_i R_{js}^i C_j = R_{jo}C_j R_{is}^j C_i \qquad (74.15)$$

so one can choose whichever combination minimizes the computational effort [7].

74.1.4 Time Constant Methods: Complications

As attractive as the time constant delay and risetime estimates are computationally, the user must beware of complications that will degrade the accuracy by a large margin. First, consider that both methods depend on a restrictive assumption regarding monotonic risetime. In many cases, however, it is not unusual to experience complex poles. This can occur owing to feedback which leads to inductive input or output impedances and emitter or source followers which also have inductive output impedance. When combined with a predominantly capacitive input impedance, complex poles will generally result unless the circuit is well damped. The time constant methods ignore the complex pole effects which can be quite significant if the poles are split and $\sigma \ll j\omega$. In this case, the circuit transient response will exhibit ringing, and time constant estimates of bandwidth, delay, and risetime will be in serious error. Of course, the ringing will show up in the circuit simulation, and if present, must be dealt with by adding damping resistances at appropriate locations.

An additional caution must be given for circuits that include zeros. Although Elmore's equations can modify the estimates for T_D and T_R when there are zeros, the OCTC method provides no help in finding the time constants of these zeros. Zeros often occur in wideband amplifier circuits that have been modified through the addition of inductance for shunt peaking, for example. The addition of inductance, either intentionally or accidentally, can also produce complex pole pairs. Zeros are intentionally added for the optimization of speed in very high-speed digital ICs; however, the large area required for the spiral inductors when compared with the area consumed by active devices tends to discourage the use of this method in all but the simplest (and fastest) designs where transmission lines can be used as inductors [5].

74.2 Time Constant Methods in High-Speed Digital Design

The time constant methods described above are applicable to digital and analog designs. They are particularly useful in finding the dominant circuit elements that control bandwidth, delay, or risetime. This can give guidance in choosing bias currents, device areas, and passive circuit element values. The time constants also provide a window on device design, showing which internal device parameters need to be improved through design or process innovations. In this section, the time constant method will be used to estimate the delay of a static ECL frequency divider. This analysis will be used to compare two technologies used for high-speed static frequency divider implementation.

74.2.1 Evaluating Device Equivalent Circuit Model Elements
for Time Constant Analysis

The time constant analysis is only possible when equivalent circuit device model element values can be determined for computation. Digital circuit analysis requires determination of large-signal equivalents of these elements in most cases because the voltages and currents generally are varying over a wide range during the complete logic voltage or current swing. An ECL inverter, whose schematic is shown

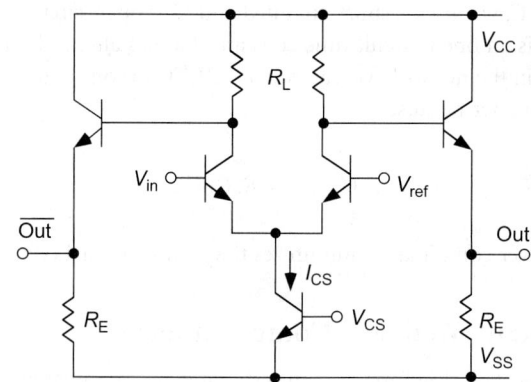

FIGURE 74.4   Schematic of the ECL inverter.

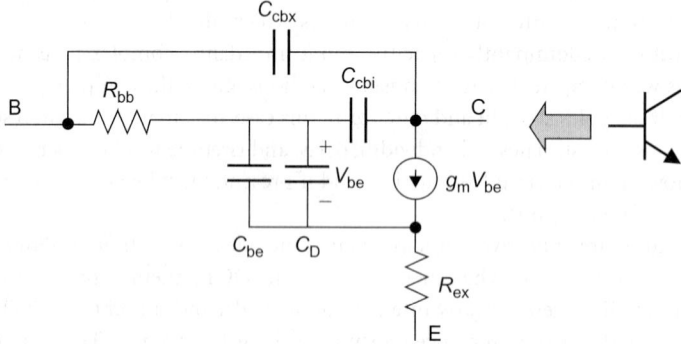

FIGURE 74.5   Modified hybrid pi equivalent circuit model.

in Figure 74.4, is selected for illustration because it is representative of many digital circuit problems. The analysis is based on work described in more detail in Ref. [8].

The first step is to construct the large-signal equivalent circuit model. The hybrid-pi bipolar junction transistor (BJT) model shown in Figure 74.5 has been used with several simplifications. The dynamic input resistance, r_π, has been neglected because other circuit resistances are typically much smaller. The output resistance, r_o, has also been neglected for the same reason. The collector-to-substrate capacitance, C_{CS} has been neglected because in III–V technologies, semi-insulating substrates are typically used. The capacitance to substrate is quite small compared to other device capacitances. It may be necessary to include this in some silicon device implementations. Retained in the model are resistances R_{bb}, the extrinsic and intrinsic base resistance, and R_{EX}, the parasitic emitter resistance. Both are very critical for optimizing high-speed performance. Base–emitter junction capacitance, C_{be}, and diffusion capacitance, C_D, are separated in this figure because they must be calculated independently.

Often C_{CB} is split into two portions for more accurate modeling as seen in Figure 74.5. This split separates the intrinsic collector–base capacitance, C_{CBi}, under the emitter from the extrinsic capacitance, C_{CBx}, under the base contact region. The base resistance R_{bb} connects these two elements. Rather than calculating these through physical areas of the intrinsic and extrinsic base, the division of capacitance is better determined by best fitting the measured and modeled f_{max}.

Figure 74.6(a) is a schematic of the ECL inverter in Figure 74.4 where the half-circuit approximation has been used owing to the inherent symmetry of differential circuits [9]. In Figure 74.6(b), the equivalent circuit model has been inserted into the circuit in Figure 74.6(a). R_{IN} is the sum of the driving

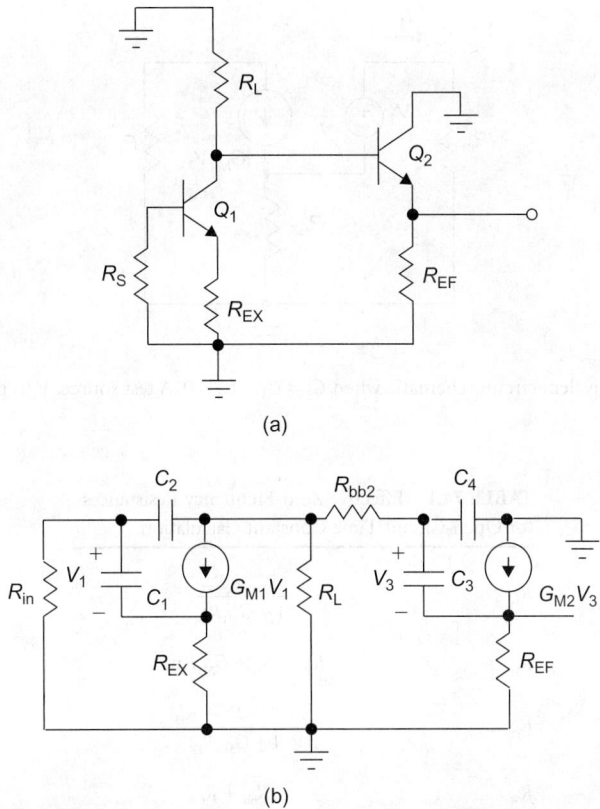

(a)

(b)

FIGURE 74.6 (a) Half-circuit model of ECL inverter; (b) simplified equivalent circuit model of the BJT inserted into the circuit in Figure 74.6(a).

point resistance from the previous stage, probably an emitter follower output and R_{bb1} of Q_1. R_L is the collector load resistor, whose value is determined by the single-ended output voltage swing, ΔV, and the DC emitter current, I_{CS}. $R_L = \Delta V/I_{CS}$. The R_{EX} of the emitter follower is included in R_{EF}. To simplify the example, the collector–base capacitances have been merged into single capacitors, C_2 and C_4 in Figure 74.6(b).

Now, calculate open-circuit resistances seen by each of the four capacitors in the circuit. C_1 is the combined base–emitter diffusion and depletion capacitance of Q_1. C_2 is the collector–base depletion capacitance of Q_1. C_3 and C_4 are the corresponding base–emitter and base–collector capacitances of Q_2. Figure 74.7 represents the equivalent circuit schematic when $C_2 = C_3 = C_4 = 0$. A test source, V_1, is placed at the C_1 location. $R_{1o} = V_1/I_1$ is determined by circuit analysis to be

$$R_{1o} = \frac{R_{IN} + R_{EX}}{1 + G_{M1}R_{EX}} \tag{74.16}$$

Table 74.1 shows the result of similar calculations for R_{2o}, R_{3o}, and R_{4o}.

Considering the results in Table 74.1, it can be seen that there are many contributors to the time constants, and that it will be possible to determine the dominant terms after evaluating the model and circuit parameters. Next, estimates must be made of the device transconductances and capacitances to evaluate the time constant terms.

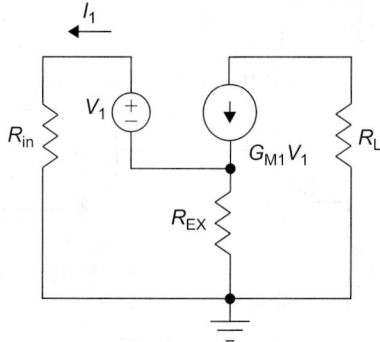

FIGURE 74.7   The equivalent circuit schematic when $C_2 = C_3 = C_4 = 0$. A test source, V_1 is placed at the C_1 location. $R_{1o} = V_1/I_1$.

TABLE 74.1   Effective Zero-Frequency Resistances for Open-Circuit Time Constant Calculation

R_{1o}	$\dfrac{R_{IN} + R_{EX}}{1 + G_{M1}R_{EX}}$
R_{2o}	$R_{IN} + R_L + G_{M1}R_{IN}R_L$
R_{3o}	$\dfrac{R_{bb} + R_L + R_{EF}}{1 + G_{M2}R_{EF}}$
R_{4o}	$R_{bb} + R_L$

Note: Calculation is for the circuit of Figure 74.6(b).

74.2.2   Large-Signal Transconductance

Device transconductance, the change in collector (or drain current) with change in the input voltage, must be determined by the circuit context for the particular device. That is, if the input voltage varies little, as in the case of an emitter follower, then a small signal g_m is most relevant. For a bipolar transistor

$$g_m = \frac{\partial I_C}{\partial V_{BE}} = \frac{qI_C}{kT} \tag{74.17}$$

For a FET, one must derive this from the DC device characteristics. The differential amplifier device, however, experiences a large change in V_{BE} (relative to kT/q) and so a large-signal transconductance must be defined.

$$G_M = \frac{\Delta I_C}{\Delta V_{BE}} \tag{74.18}$$

Referring again to the circuit model of the ECL differential pair, Figure 74.4, the single-ended logic swing is given by

$$\Delta V_{Logic} = I_{CS}R_L \tag{74.19}$$

and the current in each device, ΔI_C, varies from zero to I_{CS} since the device switches between cutoff and I_{CS}. Because G_M is governed by the internal device V_{BE} and there is always some parasitic emitter resistance, R_{EX}, then large signal G_M is given by

$$G_M = \frac{I_{CS}}{\Delta V_{Logic} - I_{CS}R_{EX}} = \frac{1}{R_L - R_{EX}} \tag{74.20}$$

Note that G_M will be considerably smaller than the small signal g_m because the voltage swing ΔV_{Logic} is much greater than kT/q. This is beneficial for delay as shown in the next section.

74.2.3 Large-Signal Capacitances

C_1 and C_3 of Figure 74.6(b) consist of the parallel combination of the *depletion (space charge) layer capacitance*, C_{be}, and the *diffusion capacitance*, C_D. The large-signal diffusion capacitance C_D can be found from

$$C_D = G_M \tau_f \tag{74.21}$$

where τ_f is the forward transit delay, $\tau_B + \tau_C$, as defined in Section 74.2. Note that this large-signal diffusion capacitance is reduced by the large-signal transconductance. The base requires charge to be delivered only while the collector current varies from 0 to I_{CS}. This occurs within a V_{BE} range of approximately $2kT/q$ or 50 mV. The overall voltage swing ΔV_{Logic} is much greater than $2kT/q$, typically by factors of 2 or 3. While depletion capacitance must be charged during the entire logic swing, the diffusion capacitance charging occurs over a smaller subset of voltage. Thus, for devices with small transit times, the effect of C_D on delay is generally much smaller than that of C_{be}, the base–emitter depletion capacitance.

The diffusion capacitance can be larger in cases such as the emitter follower, Q_2 of Figure 74.6(a), C_3 of Figure 74.6(b), where the V_{BE} varies over a relatively smaller range. In this instance, the small signal g_m is appropriate for calculating C_D.

The base–emitter depletion capacitances must also be added to C_1 and C_3. Also, C_2 and C_4 are the base–collector depletion capacitances. Depletion capacitances are voltage varying according to

$$C(V) = C_{j0}\left(1 - \frac{V}{\phi}\right)^{-m} \tag{74.22}$$

where C_{j0} is the capacitance at zero bias, ϕ the built-in voltage, and m grading coefficient. An equivalent large-signal depletion capacitance can be calculated by

$$C_{EQ} = \frac{\Delta Q}{\Delta V} = \frac{Q_2(V_2) - Q_1(V_1)}{V_2 - V_1} = K_{EQ}C_{j0} \tag{74.23}$$

Q_i is the charge at the initial (1) or final (2) state corresponding to the voltages V_i. $Q_2 - Q_1 = \Delta Q$, and

$$\Delta Q = \int_{V_1}^{V_2} C(V)\,dV \tag{74.24}$$

K_{EQ} is an effective large-signal coefficient that is determined for the particular voltage swing that a capacitor sees [1]. Using Eq. (74.22),

$$K_{EQ} = \frac{-\phi^m}{(V_2 - V_1)(1 - m)}[(\phi - V_2)^{1-m} - (\phi - V_1)^{1-m}] \tag{74.25}$$

74.2.4 Delay Estimates

The b_1 coefficient (first-order estimate of T_D) can now be found from the sum of the OCTCs.

$$b_1 = R_{1o}C_1 + R_{2o}C_2 + R_{3o}C_3 + R_{4o}C_4 \tag{74.26}$$

First, however, we need to inspect the circuit to see whether there are any forward-path zeros in the network. These capacitors are identified by tracing the signal path from V_{in} to V_{out}. Two such capacitors can be seen in Figure 74.6(b). The C_{BC} of Q_1 passes the signal directly across what should be an inverting device, thus increasing delay. The C_{BE} of Q_2 reduces delay, passing the signal more quickly through the emitter follower stage. The a_1 term can then be shown as in Ref. [8] to be

$$a_1 = \frac{C_{BE2}}{g_m} - \frac{C_{BC1}}{G_M} \tag{74.27}$$

Finally, the Elmore propagation delay is given by Eq. (74.11)

$$T_D = b_1 - a_1$$

The Elmore risetime estimate Eq. (74.12) requires the calculation of b_2. Since there are four capacitors in the large-signal equivalent circuit, six terms will be necessary.

$$b_2 = R_{1o}C_1 R_{2s}^1 C_2 + R_{1o}C_1 R_{3s}^1 C_3 + R_{1o}C_1 R_{4s}^1 C_4 + R_{2o}C_2 R_{3s}^2 C_3 + R_{2o}C_2 R_{4s}^2 C_4 + R_{3o}C_3 R_{4s}^3 C_4 \tag{74.28}$$

R_{2s}^1 will be calculated to illustrate the procedure. The remaining short-circuit equivalent resistances are shown in Table 74.2. Referring to Figure 74.8, the equivalent circuit for calculation of R_{2s}^1 is shown. This is the resistance seen across C_2 when C_1 is shorted. If C_1 is shorted, $V_1 = 0$ and the dependent current source is dead. It can be seen from inspection that

$$R_{2s}^1 = R_{IN} \| R_{EX} + R_L \tag{74.29}$$

Once b_2 is calculated, Eq. (74.12) can be used to estimate T_R, the Elmore risetime.

TABLE 74.2 Effective Resistances Derived for Short-Circuit Time Constant Calculation

R_{2s}^1	$R_{IN} \| R_{EX} + R_L$
R_{3s}^1	R_{3o}
R_{4s}^1	R_{4o}
R_{3s}^2	$\dfrac{\left(\dfrac{1}{G_{M1}} \| R_{IN} \| R_L\right) + R_{bb} + R_{EF}}{1 + G_{M2}R_{EF}}$
R_{4s}^2	$\left(\dfrac{1}{G_{M1}} \| R_{IN} \| R_L\right) + R_{bb}$
R_{4s}^3	$(R_L + R_{bb}) \| R_{EF}$

Note: R_{2s}^1 is derived from the circuit of Figure 74.8.

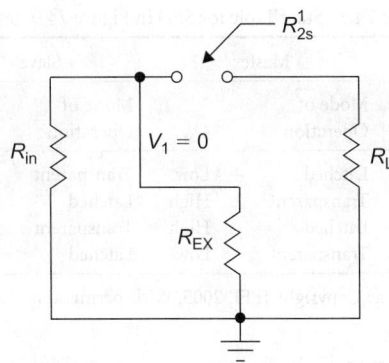

FIGURE 74.8 The equivalent circuit for calculation of R_{2s}^1.

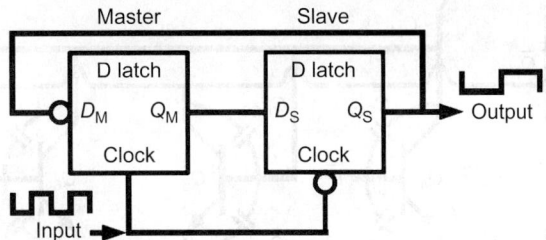

FIGURE 74.9 An SFD composed of two identical latches configured as an MSFF Ref. [8] (Copyright IEEE 2005. With permission.)

74.2.5 Static Frequency Divider

A static frequency divider (SFD) is composed of two identical latches configured as a master–slave flip-flop (MSFF), with the output, Q_S, inverted and fed back to the input, as in Figure 74.9 [8]. The divider is driven by an applied clock signal, and Q_S changes on each falling edge of the clock. There are four distinct states that the divider may occupy, arbitrarily labeled states (i) through (iv). The states are characterized by two features: the mode of operation of the latches (transparent or latched) and the outputs of the latches (high or low). These are shown in Table 74.3. On every edge of the clock the divider changes state. The slave states are identical to the master states, delayed by half of a clock cycle. To complete a cycle, the divider must undergo transitions between all four states. The maximum speed of operation of the circuit can be determined by the sum of the delays of each transition.

The latches have two basic operations. Referring to Figure 74.10, the first is a current steering operation in the Q_1, Q_2 differential pair, moving between latched and transparent settings. The second is a voltage operation that can only occur after the current steering, changing the output voltage at Q_M and Q_S. Both operations introduce delay into the circuit and limit the maximum operating speed of the divider. Table 74.3 [8] shows that the master performs the first operation during the transition from state (ii) to state (iii) (going from transparent to latched), and performs both operations during the transition from state (iii) to state (iv). When both the master and slave are examined during the (iii)–(iv) transition, it is seen that the master will complete the transition more slowly than the slave, because it must perform both operations, whereas the slave is only performing the first operation (going from transparent to latched). Since both latches are transitioning simultaneously, and the master takes longer, only the delay contribution of the master's transition need be considered.

Once the delay of the master in the (iii)–(iv) transition is found, it can be doubled to find the delay associated with the entire divider over the two transitions. Since the analysis is based on equal rise and fall times, the delay of the (i)–(ii), and (ii)–(iii) transitions can be assumed to be equal to the delay of (iii)–(iv) and (iv)–(i) transitions. The four transitions correspond to two input clock cycles, therefore

TABLE 74.3 State Table for SFD in Figure 74.9 Ref. [8]

	Master		Slave	
State	Mode of Operation	Q_M	Mode of Operation	Q_S
(i)	Latched	Low	Transparent	High
(ii)	Transparent	High	Latched	High
(iii)	Latched	High	Transparent	Low
(iv)	Transparent	Low	Latched	Low

Source: Copyright IEEE 2005. With permission.

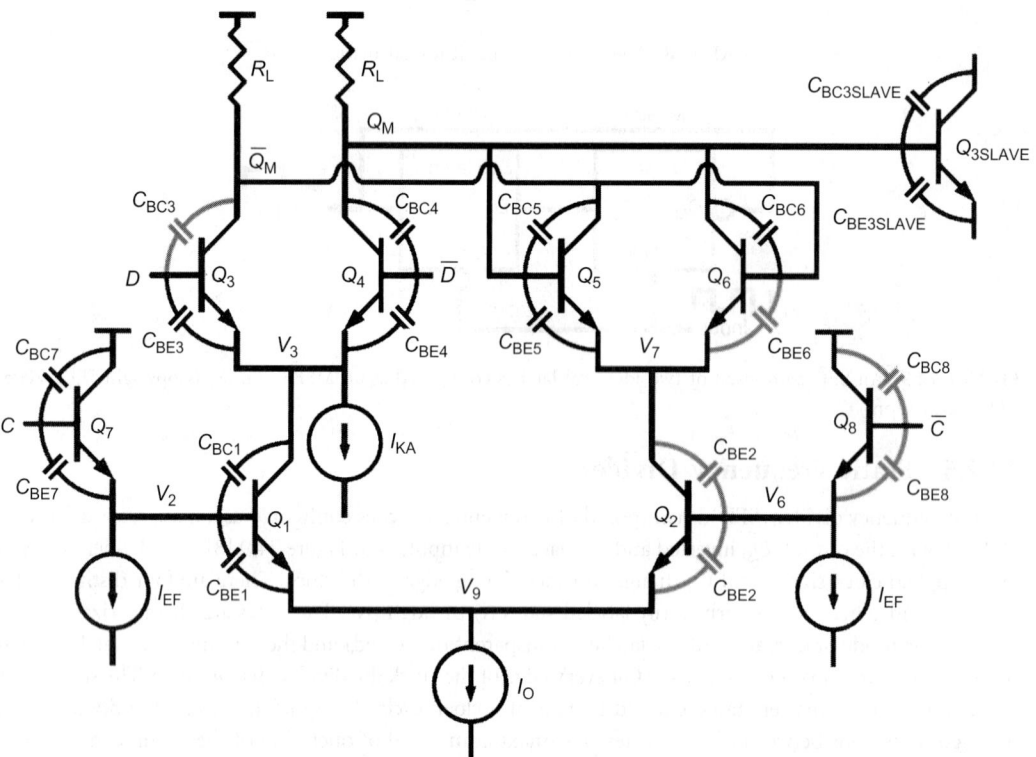

FIGURE 74.10   Schematic diagram of the MSFF Ref. [8] (Copyright IEEE 2005. With permission.)

the maximum clock speed of operation is half their sum, which is just double the (iii)–(iv) transition delay. This corresponds to the clock-to-Q delay of the master stage.

Therefore, the signal path has been identified, as shown in Figure 74.10 and the OCTC analysis can be used to identify the time constants for each capacitor loading the device nodes. Voltage swings at each node must be determined, and large-signal device parameters can be calculated as described above. The reader should refer to Ref. [8] for the detailed analysis.

As an example of the result of such an analysis, an InP HBT-based SFD is compared with a SiGe-based SFD in Ref. [8], and the results are shown in Table 74.4. The contribution of each intrinsic circuit element can be identified in the table. It can be seen that the largest contributors to the delay for InP are R_L and C_{CB} and for SiGe, R_L and C_{BE}.

$$R_L = \Delta V_{\text{Logic}}/I_{CS} \tag{74.30}$$

TABLE 74.4 Relative Contribution of Circuit Elements and Device Equivalent Circuit Components to the Total Elmore Delay

Parameter	T_D Contribution	
	InP(%)	SiGe(%)
C_{CB}	45.1	31.6
C_{JE}	37.9	38
C_{DIFF}	17	30.4
R_{EX}	3.4	4.8
R_{BB}	20.1	25.2
R_C	6.3	6.1
$r_E(EF)$	1.3	7.7
R_L	68.9	56.2

Note: T_D is the total Elmore delay of the SFD of Figure 74.10. An SFD fabricated in an InP HBT process is compared with the same circuit in an SiGe BiCMOS process.

Thus, the $R_L C_{CB}$ and $R_L C_{BE}$ time constants must be the dominant ones. This indicates that the performance can be improved if the logic swing could be reduced or the tail current increased. The logic swing is limited by the noise margin and cannot be reduced below roughly

$$6kT/q + R_{EX}I_{CS} \tag{74.31}$$

Here, the indirect importance of R_{EX} can be seen; R_{EX} must be minimized to reduce the logic voltage swing without compromising noise margin.

The current is limited by the maximum device current density, J_{MAX}, generally set by the Kirk effect, which increases collector transit delays at high current density. This is determined by the saturated drift velocity, v_{sat}, the collector–base voltage, V_{CB}, the junction potential, ϕ, and the collector thickness, T_C.

$$J_{MAX} = 2\varepsilon v_{sat}(V_{CB} + V_{CB,min} + 2\phi)/T_C^2 \tag{74.32}$$

V_{CB} is limited by breakdown and cannot be arbitrarily increased. But the collector thickness can be reduced. When C_{CB} and J_{MAX} are considered, we find that

$$C_{CB}\Delta V_{Logic}/I_C = (\varepsilon A_C/T_C)(\Delta V_{Logic}/I_C)$$

$$= \left(\frac{\Delta V_{Logic}}{V_{CE} + V_{CE,min}}\right)\left(\frac{A_C}{A_E}\right)\left(\frac{T_C}{2v_{sat}}\right) \tag{74.33}$$

where A_C and A_E are collector and emitter areas, respectively. So, it is seen that the collector capacitance charging time is reduced by decreasing the collector thickness and increasing the current [10]. Unfortunately, decreasing C_{BE} is not simple as the depletion thickness and doping are critical to the conduction band grading at the heterointerface and thus injection efficiency of the junction. But, larger current will reduce the time constant as well.

The time constant technique is also useful for estimating the optimum areas of the emitter follower and differential pair transistors since their respective contributions to total delay can be separated [8].

74.2.6 Additional Techniques for Performance Improvement

Very high-speed static dividers often can be made to clock at higher frequencies by small modifications in the circuit. Two of these, shunt peaking and "keep-alive" current, will be discussed in this section.

The method of shunt peaking is an old one, dating from the vacuum tube era, where high-frequency performance was lacking [5,11]. A zero is added to the amplifier transfer function by including a series inductance with the collector or drain resistance. This can in principle, compensate for a lower frequency pole and extend bandwidth and improve risetime. Figure 74.11 presents an example of an amplifier with shunt peaking. This approach can be extended to the load resistances of an MSFF frequency divider. Walker and Wallmann [11] show that the risetime can be significantly increased, but at the expense of some overshoot. Some overshoot can generally be tolerated in high-speed digital circuits, so the approach has been used successfully to improve speed. Defining a parameter M in Eq. (74.34),

$$M = \frac{R^2 C_{\mathrm{L}}}{L} \tag{74.34}$$

Table 74.5 shows that risetime can be reduced by a factor of 2.1 if about 11% overshoot is acceptable.

Static frequency dividers were designed and fabricated in a 60 GHz f_{T} SiGe and 170 GHz f_{T}InP HBT technologies as described in Ref. [8]. Versions with and without shunt peaking inductance were evaluated. In both cases, a moderate increase in maximum clock frequency was observed as shown in Table 74.6.

A second technique to improve maximum clock frequency is the use of keep-alive current. The keep-alive current, a small current source, I_{KA}, is connected to the emitters of the Q_3, Q_4 differential pair in Figure 74.10. Recall that the (iii)–(iv) transition determines the circuit delay. The data inputs, D and $DBAR$, steer current in the Q_3, Q_4 pair. Since the D and $DBAR$ inputs have settled to a stable value before the clock switches Q_1, it is the clock transition that controls the delay of the track stage. By keeping a small but constant current on the Q_3/Q_4 emitter node, the V_{BE} change produced by the clock transition is reduced. This has the effect of increasing G_M and consequently C_D, but the charging resistance looking into the emitter node is decreased by the same amount. The net benefit comes from reducing the charging time of C_{BE} since that is not strongly dependent on the voltage swing. A detailed explanation of this effect can be found in Ref. [8].

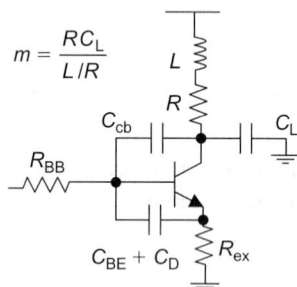

FIGURE 74.11   Example of an amplifier with shunt peaking.

TABLE 74.5 Shunt Peaking

M	Risetime Reduction Factor	Overshoot (%)
1.66	2.1	11.4
2.0	1.9	6.7
2.4	1.7	3.1
4.0	1.4	0
∞	1.0	0

Note: Shown in Figure 74.11. It can be used to decrease risetime if some overshoot can be tolerated.

TABLE 74.6 Increase in SFD Maximum Measured Clock Frequency by Addition of Shunt Peaking and Keep-Alive Current

	Maximum Clock Frequency (GHz)		
Process	Without Shunt Peaking	With Shunt Peaking	With Keep-Alive Current and Shunt Peaking
SiGe	21	24	26.5
InP	62	65	67

Static frequency dividers with and without keep-alive current were also fabricated and tested [8]. The results shown in Table 74.6 also indicate that a small improvement in maximum clock frequency can be gained by the combined use of the shunt peaking and keep-alive current methods.

74.3 Time Constant Methods in High-Speed Analog Design

Analog circuits, with the exception of power amplifiers, operate in the small-signal regime. Thus, the device small-signal transconductance is used in analysis, and the device capacitances are evaluated at a single bias point.

The OCTC methodology described above was originally used for estimating high-frequency performance of analog circuits. In these applications, both frequency and time domain responses are important. Properties of bandwidth, gain peaking and phase margin in frequency domain and risetime, ringing, and overshoot in time domain can be predicted by the tools described above. And, most importantly, the method provides guidance on what parameters must be changed to avoid signal distortion.

First, assume that the circuit transfer function, $F(s)$, is a low-pass with only poles in the denominator. If there are zeros present, the result will be modified somewhat. Then, rewrite the denominator in the standard form typically used in control theory textbooks

$$1 + b_1 s + b_2 s^2 = 1 + \frac{2\zeta s}{\omega_n} + \frac{s^2}{\omega_n^2} \tag{74.35}$$

The natural frequency, ω_n, is related to the coefficients by

$$\omega_n = b_2^{-1/2} \tag{74.36}$$

and the damping factor ζ by

$$\zeta = \frac{b_1 \omega_n}{2} \tag{74.37}$$

For a well-behaved frequency or transient response, the system should be overdamped ($\zeta > 1$) to minimize ringing or gain peaking. This will help to provide a flatter group delay and hence less pulse distortion. This condition applies to the case where the poles are both on the negative real axis. Table 74.7 presents equations that can be used to predict the frequency and time domain behavior of a circuit subject to the restrictions indicated in the table.

The second pole frequency can be estimated by the separated pole approximation

$$F(s) = \frac{V_{\text{out}}(s)}{V_{\text{in}}(s)} = \frac{k}{(1 + b_1 s)(1 + b_2 s / b_1)} \tag{74.38}$$

TABLE 74.7 Second-Order Low-Pass System Frequency and Step Response [16]

Frequency response

3 dB bandwidth
$$\omega_{3dB} = \omega_n \left[1 - 2\zeta^2 + \sqrt{2 - 4\zeta^2 + 4\zeta^4} \right]^{1/2}$$

Gain peaking
$$M_p = \frac{1}{2\zeta\sqrt{1-\zeta^2}} \quad \zeta < 0.707$$

Step response

Risetime (10–90%)
$$t_r = 2.2 / \omega_{3dB}$$

Overshoot (%)
($\zeta < 0.707$)
$$100 \exp\left(\frac{-\pi\zeta}{\sqrt{1-\zeta^2}} \right)$$

Ringing frequency
($\zeta < 0.707$)
$$\omega_r = \omega_n \sqrt{1-\zeta^2}$$

Here, it is assumed that the third-order coefficient, b_3, and higher terms are small, and that $b_1 \gg b_2/b_1$. If the latter condition is not met, then the quadratic formula should be used instead. Note that the second pole must be much higher in frequency than the dominant pole to maintain a reasonably well-damped circuit.

When feedback is used, the method is still applicable in many cases. Feedback zeros will not affect the response. But the predictions of the response in Table 74.7 will not be valid if there are zeros in the forward path, sometimes used for compensation purposes. In this case, use of a tool such as MATLAB [12] can be used to gain intuition. The final analysis will in any case require the use of a circuit simulator such as SPICE or ADS [13].

74.3.1 Example of a Transimpedance Amplifier

Transimpedance amplifiers are often used as photodiode preamplifiers in optical communication links. They are also used as a component in certain wideband, fast transient-response amplifiers such as the Cherry–Hooper design [14]. One of the simplest ways to implement a transimpedance amplifier is with the dual feedback (series–series and shunt–shunt) approach shown in Figure 74.12. This design has the capability of using feedback to not only exchange gain for bandwidth but also to provide a wideband low-impedance match at the input and output. For this example, let us suppose that the amplifier is designed to have equal input and output impedance and a transimpedance gain of 6 dB. An InP HBT device is used to produce realistic simulations and time constant numbers for this example. Z_{IN} and Z_{OUT} were 50 Ω, and a bias current of 10 mA was selected.

The design equations for the voltage gain, A_v and input and output impedances of this amplifier are shown below.

$$A_v = \frac{V_O}{I_{in}R_S} = -\frac{R_L}{R_E}\left[\frac{R_F - R_E}{R_F + R_L} \right] \tag{74.39}$$

$$Z_O = Z_{in} = \frac{R_F}{1 - A_v} \tag{74.40}$$

The same procedure as described in Sections 74.1 and 74.2 was used to find the time constants. In this case, the design is small signal, so the device transconductance and capacitances did not require linearization. Once the bias voltages and currents are determined, these elements can be calculated and their respective time constants found. The device model was simplified to include only C_{BE}, C_D,

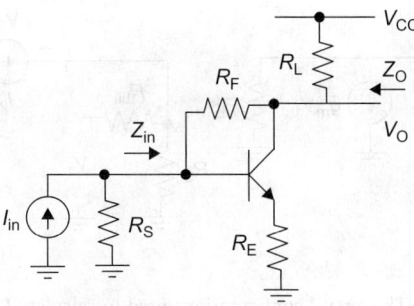

FIGURE 74.12 Schematic of the dual feedback (series–series and shunt–shunt) transimpedance amplifier.

and C_{CB} capacitances. R_{bb} was lumped as one resistor and $R'_E = R_E + R_{EX}$. The contribution of the depletion and diffusion base–emitter capacitance were combined as $C_\pi = C_{BE} + C_D$. To simplify the calculations further, the transconductance, g_m, and C_π were degenerated as shown in Eq. (74.41) and Eq. (74.42) so that the emitter feedback resistor is absorbed, and a common-emitter equivalent circuit as shown in Figure 74.13 is produced. This eliminates the difficulties in analysis that the series feedback otherwise produces.

$$\tilde{g}_m = \frac{g_m}{1 + g_m R'_E} \tag{74.41}$$

$$\tilde{C}_\pi = \frac{C_\pi}{1 + g_m R'_E} \tag{74.42}$$

Figure 74.14(a) and Figure 74.14(b) show the circuit configurations used to calculate R_{1o} and R_{2o}, and Eqs. (74.13)–(74.15) are used to derive b_1 and b_2. Then

$$R_{1o} = \frac{V_1}{I_1} \| r_\pi \tag{74.43}$$

$$R_{2o} = \frac{V_2}{I_2} \| R_F \tag{74.44}$$

and

$$b_1 = R_{1o} \tilde{C}_\pi + R_{2o} C_{CB} \tag{74.45}$$

$$b_2 = R_{1o} \tilde{C}_\pi R^1_{2s} C_{cb} \tag{74.46}$$

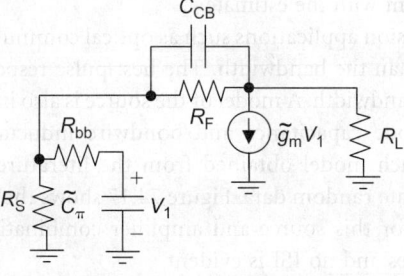

FIGURE 74.13 The transconductance, g_m, and capacitance C_π are degenerated and the emitter feedback resistor is absorbed into a common-emitter equivalent circuit of the dual feedback amplifier.

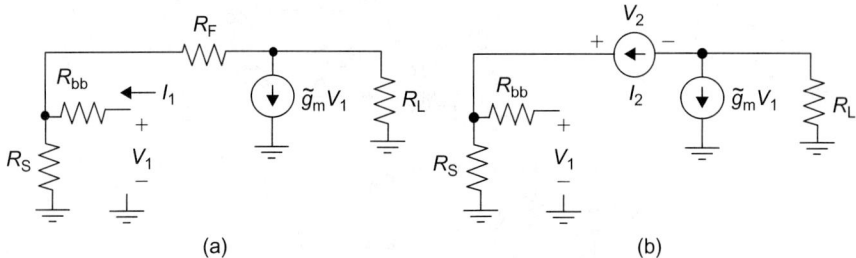

FIGURE 74.14 (a), (b) The circuit configurations used to calculate R_{1o} and R_{2o}, respectively.

TABLE 74.8 Open- and Short-Circuit Time Constants and Coefficients b_1 and b_2 Calculated for the Transimpedance Amplifier

$R_{1o}\tilde{C}_{\pi}$	$R_{2o}C_{cb}$	$R_{1o}\tilde{C}_{\pi}R_{2s}^{1}C_{cb}$	b_1
2.6 ps	2.8 ps	1.3×10^{-24}	5.4 ps

Note: Calculation based on the examples of Figure 74.12, Figure 74.13, and Figure 74.14 (a) and Figure 74.14(b). An InP HBT device model was used to derive these numbers.

TABLE 74.9 Prediction of Estimated Pole Frequencies, Natural Frequency, Damping Factor, and 3 dB Bandwidth by Analysis of the Transimpedance Amplifier

f_{P1} (GHz)	f_{P2} (GHz)	ω_n (GHz)	ζ	$f_{3\,dB}$ (GHz)	S_{21} (dB)
29	401	141	2.43	27.8	6.2

Note: Transducer gain S_{21} is also obtained by small-signal analysis.

Table 74.8 shows the two main OC time constants and the SC time constant derived for this design and evaluated using small-signal model parameters for the InP HBT. The b_1 and b_2 coefficients that were calculated from these time constants are also shown in this table. The pole frequencies estimated by Eq. (74.38), Eq. (74.45), and Eq. (74.46) are found to be 29 and 401 GHz. Thus, the widely separated pole approximation is valid, and since a dominant pole is present, one would expect a good, well-damped transient response for this amplifier with little or no gain peaking in the frequency response. Table 74.9 shows the ω_n, ζ, and 3 dB bandwidth predicted by the equations in Table 74.8. Since $\zeta > 1$, we do not have complex-conjugate poles, so the method should provide reasonably good estimates. The simulated frequency response of the same amplifier is shown in Figure 74.15. The simulated bandwidth is about 30 GHz, in reasonable agreement with the estimate.

In addition, in data transmission applications such as optical communication, the transient response is also very critical, more so than the bandwidth. The best pulse response requires a constant group delay over the required signal bandwidth. A model of the source is also important to include in transient simulations, because photodiode capacitance and bondwire inductances will affect the response. Figure 74.16 illustrates one such model obtained from the literature [15]. A linear feedback shift register source is used to simulate random data. Figure 74.17 shows the output voltage of the amplifier presented as an eye diagram for this source and amplifier combination. There is a little overshoot caused mainly by the bondwires and no ISI is evident.

The performance seems adequate for a 40 Gb/s NRZ application. But can the performance be improved further to possibly use this technology for a RZ application which would require higher

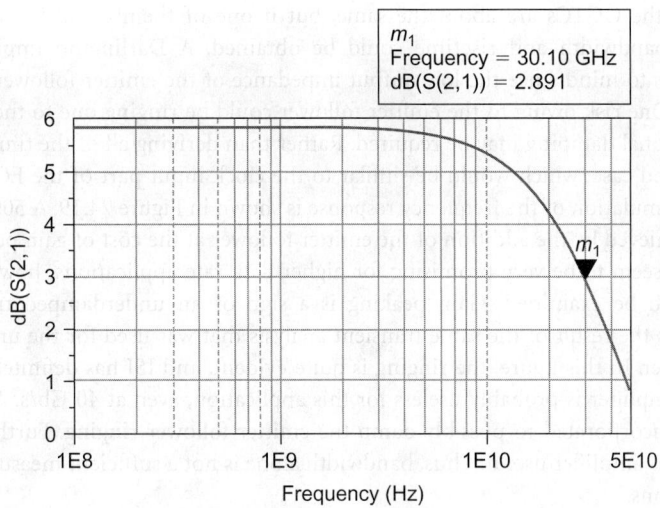

FIGURE 74.15 Simulated frequency response of the dual feedback amplifier.

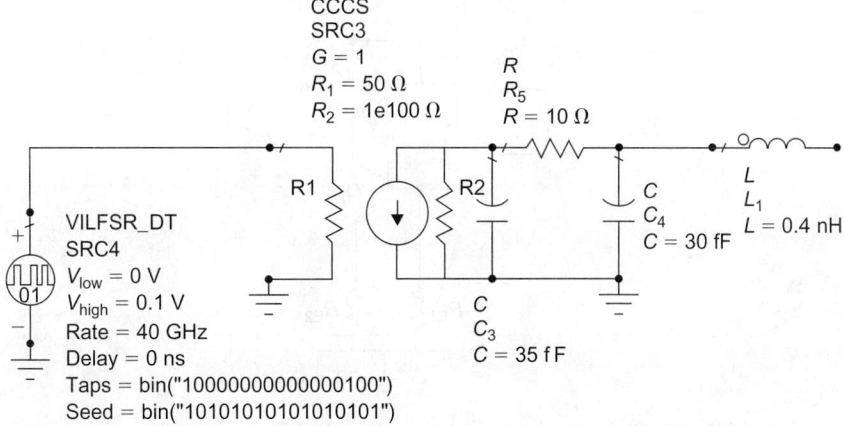

FIGURE 74.16 A model of the source included in transient simulations, because photodiode capacitance and bondwire inductances will affect the response. The output resistance (R2) of the current-controlled current source is assumed to be negligible.

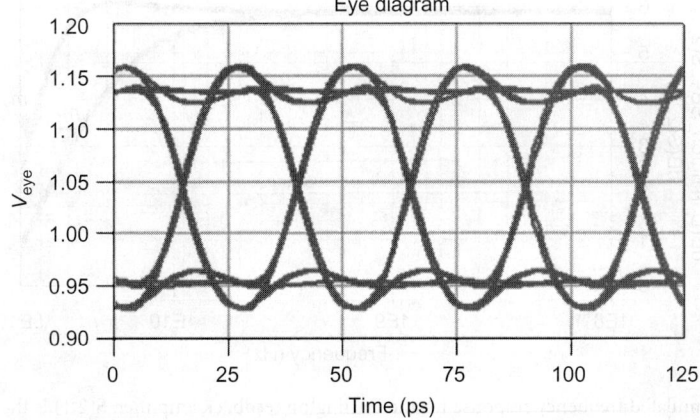

FIGURE 74.17 Simulated output voltage eye diagram for a linear feedback shift register source and amplifier combination.

bandwidth? Both the OCTCs are about the same, but if one of them could be reduced, a possible improvement in bandwidth and risetime could be obtained. A Darlington implementation as in Figure 74.18 comes to mind since the low output impedance of the emitter follower could reduce the $R_{11}^o \tilde{C}_\pi$ contributor. One risk owing to the emitter follower could be ringing due to the inductive output impedance. Additional damping may be required. Rather than deriving all of the time constant predictions for this second case, which would be similar to the clock input part of the ECL SFD analysis in Section 74.2, the simulation of the frequency response is shown in Figure 74.19. A 50% increase in 3 dB bandwidth was achieved by the addition of the emitter follower at the cost of gain peaking. The higher bandwidth would seem to be very promising for higher data rate applications; however, the transient response must also be examined. Gain peaking is a sign of an underdamped transient response. Figure 74.20 shows the result of the same transient analysis that was used for the unmodified 30 GHz bandwidth amplifier. In this figure, the ringing is quite evident, and ISI has definitely increased to the point where the amplifier is probably useless for this application, even at 40 Gb/s. Additional modifications could be incorporated to possibly damp the emitter follower ringing. Further design work is needed to make this amplifier useful. Thus, bandwidth alone is not a sufficient measure of performance for pulse applications.

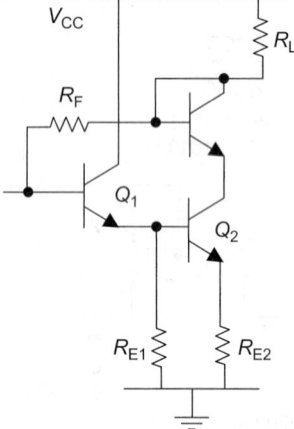

FIGURE 74.18   Darlington implementation of the dual-feedback amplifier.

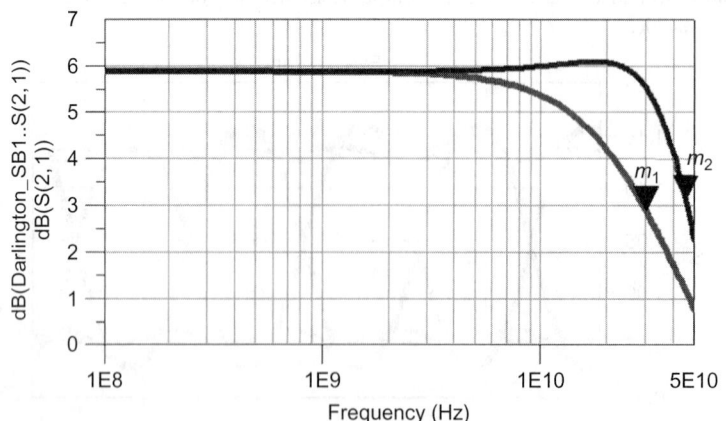

FIGURE 74.19   Simulated frequency response for the Darlington feedback amplifier. S(2,1) is the transducer power gain, expressed in dB. Note the improvement in bandwidth as expected.

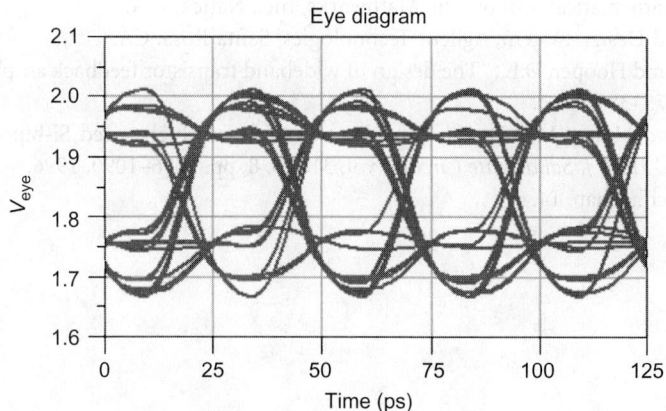

FIGURE 74.20 Simulated eye diagram of the Darlington feedback amplifier output voltage. Excessive ringing makes this unusable for high-speed data communication applications.

74.4   Conclusion

This chapter has described general methods of analysis of high-speed circuits. Time constant methods are shown to provide design guidance for both digital and analog applications. The analysis is not used as a substitute for circuit simulation methods, which give more accurate performance prediction, but are used in addition to simulation. Knowing which circuit elements and device parameters most strongly limit speed and bandwidth provides design insight that most simulation tools lack.

References

1. Rabaey, J.M., *Digital Integrated Circuits: A Design Perspective*, second ed., Prentice-Hall, New York, 2004.
2. Elmore, W.C., "The transient response of damped linear networks with particular regard to wideband amplifiers," *J. Appl. Phys.*, vol. 19, p. 55, 1948.
3. Ashar, K.G., "The method of estimating delay in switching circuits and the figure of merit of a switching transistor," *IEEE Trans. Elect. Dev.*, vol. ED-11, p. 497, 1964.
4. Tien, P.K., "Propagation delay in high-speed silicon bipolar and GaAs HBT digital circuits," *Int. J. High Speed Elect.*, vol. 1, p. 101, 1990.
5. Lee, T.H., *The Design of CMOS Radio-Frequency Integrated Circuits*, second ed., Cambridge University Press, Cambridge, U.K., 2004, Chap. 8.
6. Gray, P.E. and Searle, C.L., *Electronic Principles: Physics, Models, and Circuits*, Wiley, New York, 1969, p. 531.
7. Millman, J. and Grabel, A., *Microelectronics*, second ed., McGraw-Hill, New York, 1987, p. 482.
8. Collins, T.E., Manan, V., and Long, S., "Design analysis and circuit enhancements for high-speed bipolar flip-flops," *IEEE J. Solid State Circuits*, vol. 40, no. 5, pp. 1166–1174, 2005.
9. Gray, P. and Meyer, R., *Analysis and Design of Analog Integrated Circuits*, third ed., Wiley, New York, 1993, Chap. 3.
10. Rodwell, M., private communication.
11. Walker, R.M., and Wallman, H., "High Fidelity Pulse Amplifiers", in *Vacuum Tube Amplifiers*, vol. 18, MIT Radiation Lab Series, McGraw-Hill, New York, 1948, Chap. 3.

12. MATLAB, a mathematical tool by The Mathworks, Inc., Natick, MA.
13. ADS, Advanced Design System, Agilent Technologies, Santa Rosa, CA.
14. Cherry, E.M., and Hooper, D.E., "The design of wideband transistor feedback amplifiers," *Proc. IEEE*, vol. 110, pp. 375–389, 1963.
15. Rein, H.M., and Moller, M., "Design considerations for very-high-speed Si-bipolar IC's operating up to 50 Gb/s," *IEEE J. Solid-State Circuits*, vol. 31, no. 8, pp. 1076–1090, 1996.
16. Lee, T.H., op. cit., Chap. 14.

Section XI

XI

Design
Automation

Wai-Kai Chen
University of Illinois at Chicago

75

Internet-Based Micro-Electronic Design Automation (IMEDA) Framework

Moon Jung Chung
Michigan State University

Heechul Kim
Hankuk University of Foreign Studies

CONTENTS

75.1 Introduction

As the complexity of VLSI systems continues to increase, the micro-electronic industry must possess an ability to reconfigure design and manufacturing resources and integrate design activities so that it can quickly adapt to the market changes and new technology. Gaining this ability imposes a two-fold challenge: (1) to coordinate design activities that are geographically separated and (2) to represent an immense amount of knowledge from various disciplines in a unified format. The Internet can provide the catalyst by abridging many design activities with the resources around the world not only to exchange information but also to communicate ideas and methodologies.

In this chapter, we present a collaborative engineering framework that coordinates distributed design activities through the Internet. Engineers can represent, exchange, and access the design knowledge and carry out design activities. The crux of the framework is the formal representation of process flow using the process grammar, which provides the theoretical foundation for representation, abstraction, manipulation, and execution of design processes. The abstraction of process representation provides mechanisms to represent hierarchical decomposition and alternative methods, which enable designers to manipulate the process flow diagram and select the best method. In the framework, the process information is layered into separate specification and execution levels so that designers can capture processes and execute them dynamically. As the framework is being executed, a designer can be informed of the current status of design such as updating and tracing design changes and be able to handling exception. The framework can improve design productivity by accessing, reusing, and revising the previous process for a similar. The cockpit of our framework interfaces with engineers to perform design tasks and to negotiate design tradeoff. The framework has the capability to launch whiteboards that enable the engineers in a distributed environment to view the common process flows and data and to concurrently execute dynamic activities such as process refinement, selection of alternative process, and design reviews. The proposed framework has a provision for various browsers where the tasks and data used in one activity can be organized and retrieved later for other activities.

One of the predominant challenges for micro-electronic design is to handle the increased complexity of VLSI systems. At the turn of the century, it is expected that there will be 100 million transistors in a single chip with 0.1 micron features, which will require an even shorter design time (Spiller, 1997). This increase of chip complexity has given impetus to trends such as system on a chip, embedded system, and hardware/software co-design. To cope with this challenge, industry uses custom-off-the-shelf (COTS) components, relies on design reuse, and practices outsourcing design. In addition, design is highly modularized and carried out by many specialized teams in a geographically distributed environment. Multi-facets of design and manufacturing, such as manufacturability and low power, should be considered at the early stage of design. It is a major challenge to coordinate these design activities (Fair, 94). The difficulties are caused by due to the interdependencies among the activities, the delay in obtaining distant information, the inability to respond to errors and changes quickly, and general lack of communications. At the same time, the industry must contend with decreased expenditures on manufacturing facilities while maintaining rapid responses to market and technology changes.

To meet this challenge, the U.S. government has launched several programs. The Rapid Prototyping of Application Specific Signal Processor (RASSP) Program was initiated by the Department of Defense to bring about the timely design and manufacturing of signal processors. One of the main goals of the RASSP program was to provide an effective design environment to achieve a four-time improvement in the development cycle of digital systems (Chung, 1996). DARPA also initiated a program to develop and demonstrate key software elements for Integrated Product and Process Development (IPPD) and agile manufacturing applications. One of the foci of the earlier program was the development of infrastructure for distributed design and manufacturing. Recently, the program is continued to Rapid Design Exploration & Optimization (RaDEO) to support research, development, and demonstration of enabling technologies, tools, and infrastructure for the next generation of design environments for complex electro-mechanical systems. The design environment of RaDEO is planned to provide cognitive support to engineers by vastly improving their ability to explore, generate, track, store, and analyze design alternatives (Lyons, 1997).

The new information technologies, such as the Internet and mobile computing, are changing the way we communicate and conduct business. More and more design centers use PCs, and link them on the Internet/intranet. The web-based communication allows people to collaborate across space and time, between humans, humans and computers, and computers in a shared virtual world (Berners-Lee, 1994). This emerging technology holds the key to enhance design and manufacturing activities. The Internet can be used as the medium of a virtual environment where concepts and methodologies can be discussed, accessed, and improved by the participating engineers. Through the medium, resources and activities can be reorganized, reconfigured, and integrated by the participating organizations. This new paradigm certainly impacts the traditional means for designing and manufacturing a complex product. Using Java, programs can be implemented in a platform-independent way so that they can be executed in any machine with a Web browser. Common Object Request Broker Architecture (CORBA) (Yang, 1996) provides distributed services for tools to communicate through the Internet (Vogel). Designers may be able to execute remote tools through the Internet and see the visualization of design data (Erkes, 1996; Chan, 1998; Chung, 1998).

Even though the potential impact of this technology will be great on computer aided design, Electronic Design Automation (EDA) industry has been slow in adapting this new technology (Spiller, 1997). Until recently, EDA frameworks used to be a collection of point tools. These complete suites of tools are integrated tightly by the framework using their proprietary technology. These frameworks have been suitable enough to carry out a routine task where the process of design is fixed. However, new tools appear constantly. To mix and match various tools outside of a particular framework is very difficult. Moreover, tools, expertise, and materials for design and manufacturing of a single system are dispersed geographically. Now we have reached the stage where a single tool or framework is not sufficient enough to handle the increasing complexity of a chip and emerging new technology. A new framework is necessary which is open and scalable. It must support collaborative design activities so that designers can add new tools to the framework, and interface them with other CAD systems. There are two key functions of the framework: (1) managing the process and (2) maintaining the relationship among many design representations. For design data management, refer to (Katz, 1987). In this chapter, we will focus on the process management aspect.

To cope with the complex process of VLSI system design, we need a higher level of viewing of a complete process, i.e., the abstraction of process by hiding all details that need not to be considered for the purpose at hand. As pointed out in National Institute of Standards and Technology reports (Schlenoff, 1996; Knutilla, 1998), a "unified process specification language" should have the following major requirements: abstraction, alternative task, complex groups of tasks, and complex sequences.

In this chapter, we first review the functional requirements of the process management in VLSI system design. We then present the Internet-based Micro-Electronic Design Automation (IMEDA) System. IMEDA is a web-based collaborative engineering framework where engineers can represent, exchange, and access design knowledge and perform the design activities through the Internet. The crux of the framework is a formal representation of process flow using process grammar. Similar to the language grammar, production rules of the process grammar map tasks into admissible process flows (Baldwin, 1995a). The production rules allow a complex activity to be represented more concisely with a small number of high-level tasks. The process grammar provides the theoretical foundation for representation, abstraction, manipulation, and execution of design and manufacturing processes. It facilitates the communication at an appropriate level of complexity. The abstraction mechanism provides a natural way of browsing the process repository and facilitates process reuse and improvement. The strong theoretical foundation of our approach allows users to analyze and predict the behavior of a particular process. The cockpit of our framework interfaces with engineers to perform design tasks and to negotiate design tradeoff. The framework guides the designer in selecting tools and design methodologies, and it generates process configurations that provide optimal solutions with a given set of constraints. The just-in-time binding and the location transparency of tools maximize the utilization of company resources. The framework is equipped with whiteboards so that engineers in a distributed environment can view the common process flows and data and concurrently execute dynamic activities such as process refinement, selection of alternative processes, and design reviews. With the grammar, the framework gracefully handles exceptions and alternative productions. A layered approach is used to separate the specification of design process and execution parameters. One of the main advantages of this separation is freeing designers from the over-specification and graceful exception handling. The framework, implemented using Java, is open and extensible. New process, tools, and user-defined process knowledge and constraints can be added easily.

75.2 Functional Requirements of Framework

Design methodology is defined as a collection of principles and procedures employed in the design of engineering systems. Baldwin and Chung (Baldwin, 1995a) define design methodology management as selecting and executing methodologies so that the input specifications are transformed into desired output specifications. Kleinfeldt (1994), states that "design methodology management provides for the definition, presentation, execution, and control of design methodology in a flexible, configured way." Given a methodology, we can select a process or processes for that particular methodology.

Each design activity, whether big or small, can be treated as a task. A complex design task is hierarchically decomposed into simpler subtasks, and each subtask in turn may be further decomposed. Each task can be considered as a transformation from input specification to output specification. The term *workflow* is used to represent the details of a process including its *structure in terms of all the required tasks and their interdependencies.* Some process may be ill-structured, and capturing it as a workflow may not be easy. Exceptions, conditional executions, and human involvement during the process make it difficult to model the process as a workflow.

There can be many different tools or alternative processes to accomplish a task. Thus, a design process requires design decisions such as selecting tools and processes as well as selecting appropriate design parameters. At a very high level of design, the input specifications and constraints are very general and may even be ill-structured. As we continue to decompose and perform the tasks based on design decisions, the output specifications are refined and the constraints on each task become more restrictive. When the output of a task does not meet certain requirements or constraints, a new process, tools, or parameters must be selected. Therefore, the design process is typically iterative and based on previous design experience. Design process is also a collaborative process, involving many different engineering activities and requiring the coordination among engineers, their activities, and the design results.

Until recently, it was the designer's responsibility to determine which tools to use and in what order to use them. However, managing the design process itself has become difficult, since each tool has its own capabilities and limitations. Moreover, new tools are developed and new processes are introduced continually. The situation is further aggravated because of incompatible assumptions and data formats between tools. To manage the process, we need a framework to monitor the process, carry out design tasks, support cooperative teamwork, and maintain the relationship among many design representations (Chiueh, 1990; Katz, 1987). The framework must support concurrent engineering activities by integrating various CAD tools and process and component libraries into a seamless environment. Figure 75.1 shows the RASSP enterprise system architecture (Welsh, 1995). It integrates tools, tool frameworks, and data management functions into an enterprise environment. The key functionality of the RASSP system is managing the RASSP design methodology by "process automation", that is, controlling CAD program execution through workflow.

75.2.1 The Building Blocks of Process

The lowest level of a building block of a design process is a tool. A *tool* is an unbreakable unit of a CAD program. It usually performs a specific task by transforming given input specifications into output specifications. A *task* is defined as design activities that include information about what tools to use and how to use them. It can be decomposed into smaller subtasks. The simplest form of the task, called an *atomic task*, is the one that cannot be decomposed into subtasks. In essence, an atomic task is defined as an encapsulated tool. A task is called *logical* if it is not atomic. A workflow of a logical task describes the details of how the task is decomposed into subtasks, and the data and control dependencies such as the relationship between design data used in the subtasks. For a given task, there can be several workflows, each of which denotes a possible way of accomplishing the task. A *methodology* is a collection of workflow supported together with information on which workflow should be selected in a particular instance.

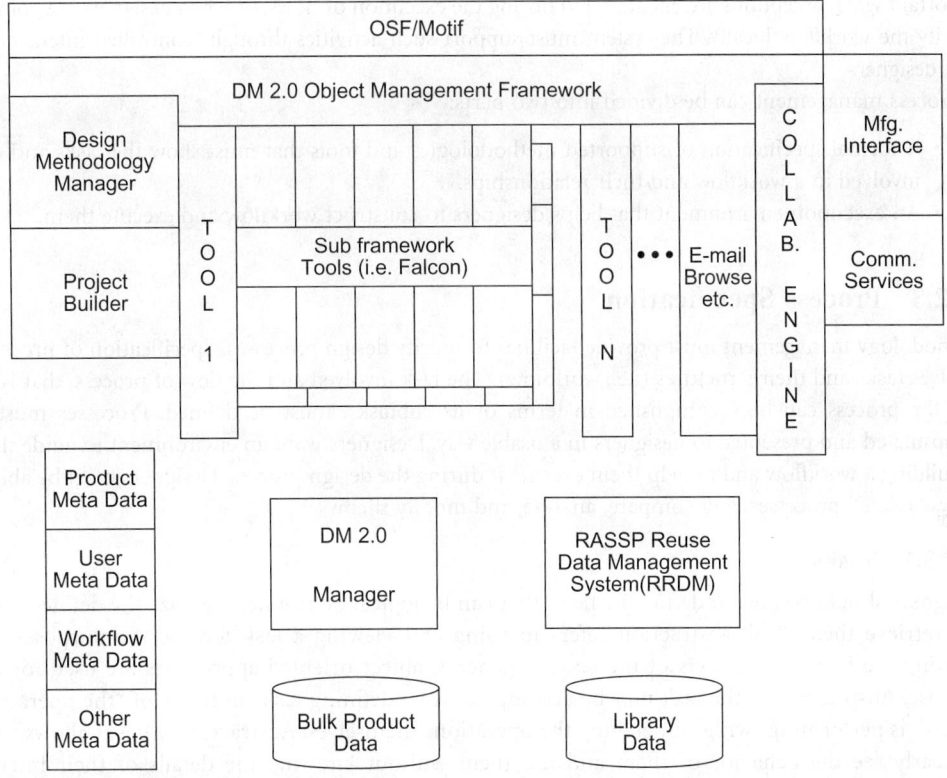

FIGURE 75.1 RASSP enterprise system architecture.

75.2.2 Functional Requirements of Workflow Management

To be effective, a framework must integrate many design automation tools and allow the designer to specify acceptable methodologies and tools together with information such as when and how they may be used. Such a framework must not only guide the designer in selecting tools and design methodologies, but also aid the designer in constructing a workflow that is suitable to complete the design under given constraints. The constructed workflow should guarantee that required steps are not skipped; built-in design checks are incorporated into the workflow. The framework must also keep the relationships between various design representations, maintain the consistency between designs and support cooperative teamwork, and allow the designer to interact with the system to adjust design parameters or to modify the previous design process. The framework must be extendible to accommodate rapidly changing technologies and emerging new tools. Such a framework can facilitate developing new hardware systems as well as redesigning a system from a previous design.

During a design process, a particular methodology or workflow selected by a designer must be based on available tools, resources (computing and human), and design data. For example, a company may impose a rule that if input is a VHDL behavioral description, then designers should use Model Technology's VHDL simulator, but if the input is Verlig, they must use ViewLogic simulator. Or, if a component uses Xilinx, then all other components must also use Xilinx. Methodology must be driven by local expertise and individual preference, which in turn, are based on the designer's experience.

The process management should not constrain the designer. Instead, it must free designers from routine tasks, and guide the execution of workflow. User interaction and a designer's freedom are especially important when exceptions are encountered during the execution of flows, or when designers are going to modify the workflow locally. The system must support such activities through "controlled interactions" with designers.

Process management can be divided into two parts:

- A formal specification of supported methodologies and tools that must show the tasks and data involved in a workflow and their relationships.
- An execution environment that helps designers to construct workflow and execute them.

75.2.3   Process Specification

Methodology management must provide facilities to specify design processes. Specification of processes involves tasks and their structures (i.e., workflow). The task involved and the flow of process, that is the way the process can be accomplished in terms of its subtasks, must be defined. Processes must be encapsulated and presented to designers in a usable way. Designers want an environment to guide them in building a workflow and to help them execute it during the design process. Designers must be able to browse related processes, and compare, analyze, and modify them.

75.2.3.1   Tasks

Designers should be able to define the tasks that can be logical or atomic, organize the defined tasks, and retrieve them. Task abstraction refers to using and viewing a task for specific purposes and ignoring the irrelevant aspects of the task. In general, object-oriented approaches are used for this purpose. Abstraction of the task may be accomplished by defining tasks in terms of "the operations the task is performing" without detailing the operations themselves. Abstraction of tasks allows users to clearly see the behavior of them and use them without knowing the details of their internal implementations. Using the generalization–specialization hierarchy (Chung, 1990), similar tasks can be grouped together. In the hierarchy, a node in the lower level inherits its attributes from its predecessors. By inheriting the behavior of a task, the program can be shared, and by inheriting the representation of a task (in terms of its flow), the structure (workflow) can be shared. The Process Handbook (Malone, in press) embodies concepts of specialization and decomposition to represent processes.

There are various approaches associated with binding a specific tool to an atomic task. A tool can be bound to a task statically at the compile time, or dynamically at the run time based on available resources and constraints. When a new tool is installed, designers should be able to modify the existing bindings. The simplest approach is to modify the source code or write a script file and recompile the system. The ideal case is plug and play, meaning that CAD vendors address the need of tool interoperability, e.g., the Tool Encapsulation Specification (TES) proposed by CFI (CFI, 1995).

75.2.3.2   Workflow

To define a workflow, we must specify the tasks involved in the workflow, data, and their relationship. A set of workflows defined by methodology developers enforces the user to follow the flows imposed by the company or group. Flows may also serve to guide users in developing their own flows. Designers would retrieve the cataloged flows, modify them, and use them for their own purposes based on the guidelines imposed by the developer. It is necessary to generate legal flows. A blackboard approach was used in (Lander, 1995) to generate a particular flow suitable for a given task. In Nelsis (Bosch, 1991), branches of a flow are explicitly represented using "or" nodes and "merge" nodes. A task can be accomplished in various ways. It is necessary to represent alternative methodologies for the task succinctly so that designers can access alternative methodologies and select the best one based on what-if analysis. IDEF3.X (IDEF) is used to graphically model workflow in RASSP environment. Figure 75.2 shows an example of workflow using IDEF3.X. A node denotes a task. It has inputs, outputs, mechanism, and

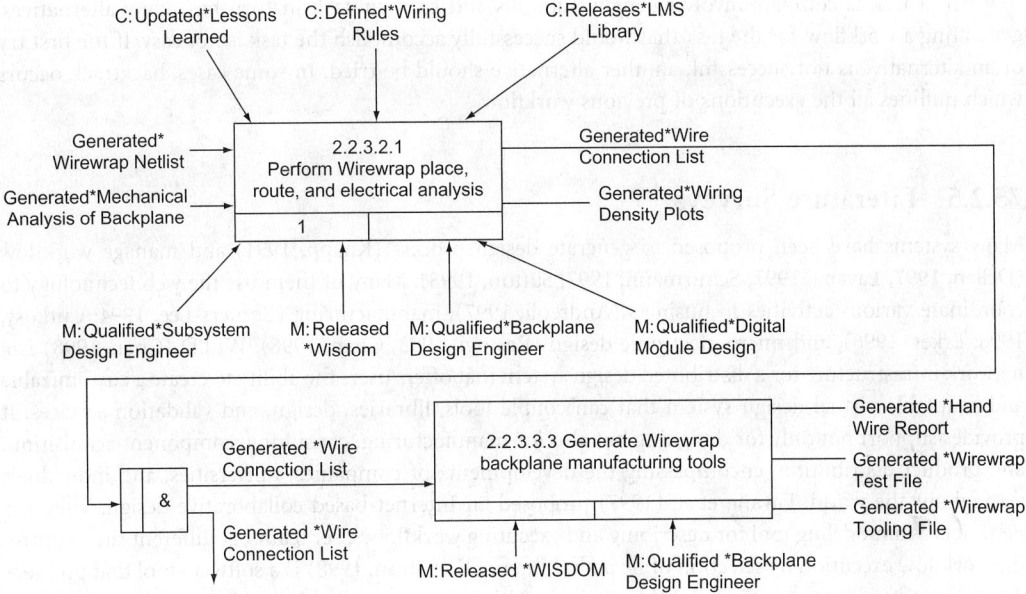

FIGURE 75.2 Workflow example using IDEF definition.

conditions. IDEF definition that has been around for 20 years mainly to capture flat modeling such as a shop floor process. IDEF specification, however, requires complete information such as control mechanisms and scheduling at the specification time, making the captured process difficult to understand. In IDEF, "or" nodes are used to represent the alternative paths. It does not have an explicit mechanism to represent alternative workflow. IDEF is ideal only for documenting the current practice and not suitable for executing iterative process which is determined during the execution of the process. Perhaps, the most important aspect missing from most process management systems is the abstraction mechanism (Schlenoff, 1996).

75.2.4 Execution Environment

The execution environment provides dynamic execution of tasks and tools and binds data to tools, either manually or automatically. Few frameworks separate the execution environment from the specification of design process. There are several modes in which a task can be executed (Kleinfeldth, 1994): manual mode, manual execution of flow, automatic flow execution, and automatic flow generation. In manual flow execution, the environment executes a task in the context of a flow. In an automatic flow execution environment, tasks are executed based on the order specified on the flow graph. In automatic flow generation, the framework generates workflow dynamically and executes them without the guidance of designers. Many frameworks use blackboard- or knowledge-based approaches to generate workflow. However, it is important for designers to be able to analyze the workflow created and share it with others. That is, repeatability and predictability are important factors if frameworks support dynamic creation of workflow.

Each task may be associated with pre- and post-conditions. Before a task is executed, the pre-condition of the task is evaluated. If the condition is not satisfied, the framework either waits until the condition is met, or aborts the task and selects another alternative. After the task is executed, its post-condition is evaluated to determine if the result meets the exit criteria. If the evaluation is unsatisfactory, another alternative should be tried.

When a task is complex involving many subtasks and each subtask in turn has many alternatives, generating a workflow for the task that would successfully accomplish the task is not easy. If the first try of an alternative is not successful, another alternative should be tried. In some cases, backtrack occurs which nullifies all the executions of previous workflow.

75.2.5 Literature Surveys

Many systems have been proposed to generate design process (Knapp, 1991) and manage workflow (Dellen, 1997; Lavana, 1997; Schurmann, 1997; Sutton, 1998). Many of them use the web technology to coordinate various activities in business (Andreoli, 1997), manufacturing (Berners-Lee, 1994; Cutkosy, 1996; Erkes, 1996), and micro-electronic design (Rastogi, 1993, Chan, 1998). WELD (Chan, 1998) is a network infrastructure for a distributed design system that offers users the ability to create a customizable and adaptable virtual design system that can couple tools, libraries, design, and validation services. It provides support not only for designing but also for manufacturing, consulting, component acquisition, and product distribution, encompassing the developments of companies, universities, and individuals throughout the world. Lavana et al. (1997) proposed an Internet-based collaborative design. They use Petri nets as a modeling tool for describing and executing workflow. User teams, at different sites, control the workflow execution by selection of its path. Minerva II (Sutton, 1998) is a software tool that provides design process management capabilities serving multiple designers working with multiple CAD frameworks. The proposed system generates design plan and realizes unified design process management across multiple CAD frameworks and potentially across multiple design disciplines. ExPro (Rastogi, 1993) is an expert-system-based process management system for the semiconductor design process.

There are several systems that automatically determine what tools to execute. OASIS (OASIS, 1992) uses Unix make file style to describe a set of rules for controlling individual design steps. The Design Planning Engine of the ADAM system (Knapp, 1986; Knapp, 1991) produces a plan graph using a forward chaining approach. Acceptable methodologies are specified by listing pre-conditions and post-conditions for each tool in a lisp-like language. Estimation programs are used to guide the chaining. Ulysses (Bushnell, 1986) and Cadweld (Daniel, 1991) are blackboard systems used to control design processes. A knowledge source, which encapsulates each tool, views the information on the blackboard and determines when the tool would be appropriate. The task management is integrated into the CAD framework and Task Model is interpreted by a blackboard architecture instead of a fixed inference mechanism. Minerva (Jacome, 1992) and the OCT task manager (Chiu, 1990) use hierarchical strategies for planning the design process. Hierarchical planning strategies take advantage of knowledge about how to perform abstract tasks which involve several subtasks.

To represent design process and workflow, many languages and schema have been proposed. NELSIS (Bosch, 1991) framework is based on a central, object-oriented database and on a flow management. It uses a dataflow graph as Flow Model and provides the hierarchical definition and execution of design flow. PLAYOUT (Schurmann, 1997) framework is based on separate Task and Flow Models which are highly interrelated among themselves and the Product Model. In (Barthelmann, 1996), graph grammar is proposed in defining the task of software process management. Westfechtel (1996) proposed "process-net" to generate the process flow dynamically. However, in many of these systems, the relationship between task and data is not explicitly represented. Therefore, representing the case in which a task generates more than one datum and each of them goes to a different task is not easy. In (Schurmann, 1997), Task Model (describing the I/O behavior of design tools) is used as a link between the Product Model and the Flow Model. The proposed system integrates data and process management to provide traceability. Many systems use IDEF to represent a process (Chung, 1996; IDEF; Stavas). IDEF specification, however, requires complete information such as control mechanisms and scheduling at the specification time, making the captured process difficult to understand.

Although there are many other systems that address the problem of managing process, most proposed system use either a rule-based approach or a hard-coded process flow. They frequently require source code modification for any change in process. Moreover, they do not have mathematical formalism. Without the formalism, it is difficult to handle the iterative nature of the engineering process and to simulate the causal effects of any changes in parameters and resources. Consequently, coordinating the dynamic nature of processes is not well supported in most systems. It is difficult to analyze the rationale how an output is generated and where a failure has occurred. They also lack a systematic way of generating all permissible process flows at any level of abstraction while providing means to hide the details of the flow when they are not needed. Most systems have the tendency to over-specify the flow information, requiring complete details of a process flow before executing the process. In most real situations, the complete flow information may not be known after the process has been executed: they are limited in their ability to address the underlying problem of process flexibility. They are rather rigid and not centered on users, and do not handle exceptions gracefully. Thus, the major functions for the collaborative framework such as adding new tools and sharing and improving the process flow cannot be realized. Most of them are weak in at least one of the following criteria suggested by NIST (Schlenoff et al., 1996): process abstraction, alternative tasks, complex groups of tasks, and complex sequences.

75.3   IMEDA System

The Internet-based Micro-Electronic Design Automation (IMEDA) System is a general management framework for performing various tasks in design and manufacturing of complex micro-electronic systems. It provides a means to integrate many specialized tools such as CAD and analysis packages, and allows the designer to specify acceptable methodologies and tools together with information such as when and how they may be used. IMEDA is a collaborative engineering framework that coordinates design activities distributed geographically. The framework facilitates the flow of multimedia data sets representing design process, production, and management information among the organizational units of a virtual enterprise. IMEDA uses process grammar (Baldwin, 1995) to represent the dynamic behavior of the design and manufacturing process. In a sense, IMEDA is similar to agent-based approach such as Redux (Petrie, 1996). Redux, however, does not provide process abstraction mechanism or facility to display the process flow explicitly.

The major functionality of the framework includes

- Formal representation of the design process using the process grammar that captures a complex sequence of activities of micro-electronic design.
- Execution environment that selects a process, elaborates the process, invokes tools, pre- and post-evaluates the productions if the results meet the criterion, and notifies designers.
- User interface that allows designers to interact with the framework, guides the design process, and edits the process and productions.
- Tool integration and communication mechanism using Internet Socket and HTTP.
- Access control that provides a mechanism to secure the activity and notification and approval that provide the mechanisms to disperse design changes to, and responses from, subscribers.

IMEDA is a distributed framework design knowledge, including process information, manager programs, etc., are maintained in a distributed fashion by local servers. The following Figure 75.3 illustrates how IMEDA links tools and sites for distributed design activities. The main components of IMEDA are

- **System Cockpit:** It controls all interactions between the user and the system and between the system components. The cockpit will be implemented as a Java applet and may be executable on any platform for which a Java enabled browser is available. It keeps track of the current design status and informs the user of possible actions. It allows users to collaboratively create and edit process flows, production libraries, and design data.
- **Manager Programs:** These encapsulate design knowledge. Using pre-evaluation functions, managers estimate the possibility of success for each alternative. They invoke tools and call post-evaluation

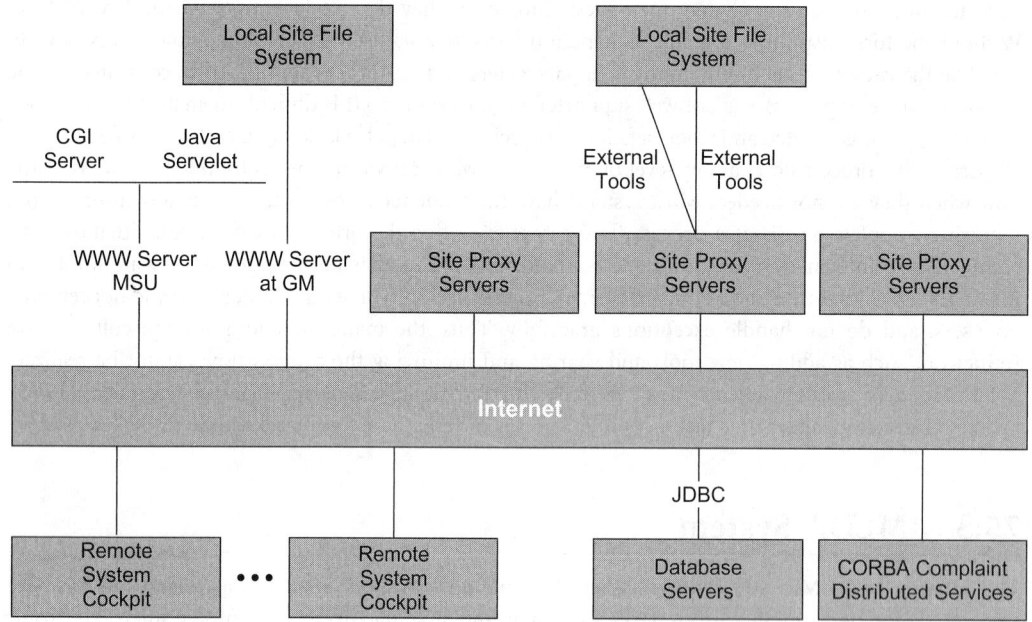

FIGURE 75.3 The architecture of IMEDA.

functions to determine if a tool's output meets the specified requirements. The interface servers allow cockpits and other Java-coded programs to view and manipulate production, task and design data libraries. Manager programs must be maintained by tool integrators to reflect site-specific information such as company design practices and different ways of installing tools.

- **Browsers:** The task browser organizes the tasks in a generalization-specialization (GS) hierarchy and contains all the productions available for each task. The data-specification browser organizes the data-specifications in a GS hierarchy and contains all the children.
- **External Tools:** These programs are the objects invoked by the framework during DM activities. Each atomic task in a process flow is bound to an external tool. External tools are written typically by the domain experts.
- **Site Proxy Server:** Any physical site that will host external tools must have a site proxy server running. These servers provide an interface between the cockpit and the external tools. The site server receives requests from system cockpits, and invokes the appropriate tool. Following the tool completion, the site server notifies the requesting cockpit, returning results, etc.
- **CGI Servers and Java Servlets:** The system cockpit may also access modules and services provided by CGI servers or the more recently introduced Java servlets. Currently, the system integrates modules of this type as direct components of the system (as opposed to external tools that may vary with the flow).
- **Database Servers:** Access to component data is a very important function. Using an API called JBDC, the framework can directly access virtually any commercially available database server remotely.
- **Whiteboard:** The shared cockpit, or "whiteboard" is a communication medium to share information among users in a distributed environment. It allows designers to interact with the system and guides the design process collaboratively. Designers will be able to examine design results and current process flows, post messages, and carry out design activities both concurrently and collaboratively. Three types of whiteboards are the process board, the chat board, and the freeform drawing board. Their functionality includes: (i) process board to the common process flow graph indicating the current task being executed and the intermediate results arrived at before the current task; (ii) drawing board

to load visual design data, and to design and simulate process; and (iii) chat board to allow participants to communicate with each other via text-based dialog box.

IMEDA uses a methodology specification based on a process flow graphs and process grammars (Baldwin, 1995). Process grammars are the means for transforming high-level process flow graphs into progressively more detailed graphs by applying a set of substitution rules, called productions, to nodes that represent logical tasks. It provides not only the process aspect of design activities but also a mechanism to coordinate them. The formalism in process grammar facilitates abstraction mechanisms to represent hierarchical decomposition and alternative methods, which enable designers to manipulate the process flow diagram and select the best method. The formalism provides the theoretical foundations for the development of IMEDA.

IMEDA contains the database of admissible flows, called process specifications. With the initial task, constraints, and execution environment parameters, including personal profile, IMEDA guides designers in constructing process flow graphs in a top-down manner by applying productions. It also provides designers with the ability to discover process configurations that provide optimal solutions. It maintains consistency among designs and allows the designer to interact with the system and adjust design parameters, or modify the previous design process. As the framework is being executed, a designer can be informed of the current status of design such as updating and tracing design changes and be able to handling exception.

Real-world processes are typically very complex by their very nature; IMEDA provides designers the ability to analyze, organize, and optimize processes in a way never before possible. More importantly, the framework can improve design productivity by accessing, reusing, and revising the previous process for a similar design.

The unique features of our framework include

Process Abstraction/Modeling—Process grammars provide abstraction mechanism for modeling admissible process flows. The abstraction mechanism allows a complex activity to be represented more concisely with a small number of higher-level tasks, providing a natural way of browsing the process repository. The strong theoretical foundation of our approach allows users to analyze and predict the behavior of a particular process. With the grammar, the process flow gracefully handles exceptions and alternative productions. When a task has alternative productions, backtracking occurs to select other productions.

Separation of Process Specification and Execution Environment—Execution environment information such as complex control parameters and constraints is hidden from the process specification. The information of these two layers is merely linked together to show the current task being executed on a process flow. The represented process flow can be executed in both automatic and manual modes. In the automatic mode, the framework executes all possible combinations to find a solution. In the manual mode, users can explore design space.

Communication and Collaboration—To promote real-time collaboration among participants, the framework is equipped with the whiteboard, a communication medium to share information. Users can browse related processes, compare them with other processes, analyze, and simulate them. Locally managed process flows and productions can be integrated by the framework in the central server. The framework manages the production rules governing the higher level tasks, while lower level tasks and their productions are managed by local servers. This permits the framework to be effective in orchestrating a large-scale activity.

Efficient Search of Design Process and Solution—IMEDA is able to select the best process and generate a process plan, or select a production dynamically and create a process flow. The process grammar easily captures design alternatives. The execution environment selects and executes the best one. If the selected process does not meet the requirement, then the framework backtracks and selects another alternative. This backtrack occurs recursively until a solution is found. If you allow a designer to select the best solution among many feasible ones, the framework may generate many multiple versions of the solution.

Process Simulation—The quality of a product depends on the tools (maturity, speed, and special strength of the tool), process (or workflow selected), and design data (selected from the reuse library). Our framework predicts the quality of results (product) and assesses the risk and reliability. This information can be used to select the best process/work flow suitable for a project.

Parallel Execution of Several Processes and Multiple Versions—To reduce the design time and risk, it is necessary to execute independent tasks in parallel whenever they are available. Sometimes, it is necessary to investigate several alternatives simultaneously to reduce the design time and risk. Or the designer may want to execute multiple versions with different design parameters. The key issue in this case is scheduling the tasks to optimize the resource requirements.

Life Cycle Support of Process Management—The process can be regarded as a product. A process (such as airplane designing or shipbuilding) may last many years. During this time, it may be necessary for the process itself to be modified because of new tools and technologies. Life cycle support includes updating the process dynamically, and testing/validating the design process, version history and configuration management of the design process. Tests and validations of the design processes, the simulation of processes, and impact analysis are necessary tools.

75.4 Formal Representation of Design Process[1]

IMEDA uses a methodology specification based on a process flow graphs and process grammars (Baldwin, 1995). The grammar is an extension of graph grammar originally proposed by Ehrig (1979) and has been applied to interconnection network (Derk, 1998) and software engineering (Heiman, 1997).

75.4.1 Process Flow Graph

A process flow graph depicts tasks, data, and the relationships among them, describing the sequence of tasks for an activity. Three basic symbols are used to represent a process flow graph. Oval nodes represent Logical Tasks, two-concentric oval nodes represent Atomic Tasks, rectangular nodes represent Data Specifications and diamond nodes represent Selectors. A task that can be decomposed into subtasks is called *logical*. Logical task nodes represent abstract tasks that could be done with several different tools or tool combinations. A task that cannot be decomposed is *atomic*. An atomic task node, commonly called a tool invocation, represents a run of an application program.

A *selector* is a task node that selects data or parameter. Data specifications are design data, where the output specification produced by a task can be consumed by another task as an input specification. Each data specification node, identified by a rectangle, is labeled with a data specification type. Using the graphical elements of the flow graph, engineers can create a process flow in a top-down fashion. These elements can be combined into a process flow graph using directed arcs. The result is a bipartite acyclic directed graph that identifies clearly the task and data flow relationships among the tasks in a design activity. The set of edges indicates those data specifications used and produced by each task. Each specification must have at most one incoming edge. Data specifications with no incoming edges are inputs of the design exercise. $T(G)$, $S(G)$, and $E(G)$ are the sets of task nodes, specification nodes, and edges of graph G, respectively. Figure 75.4 shows a process flow graph that describes a possible rapid prototyping design process, in which a state diagram is transformed into a field-programmable gate array (FPGA) configuration file.

---

[1]Materials in this section are excerpted from R. Baldwin and M.J. Chung, *IEEE Computer*, Feb. 1995. With permission.

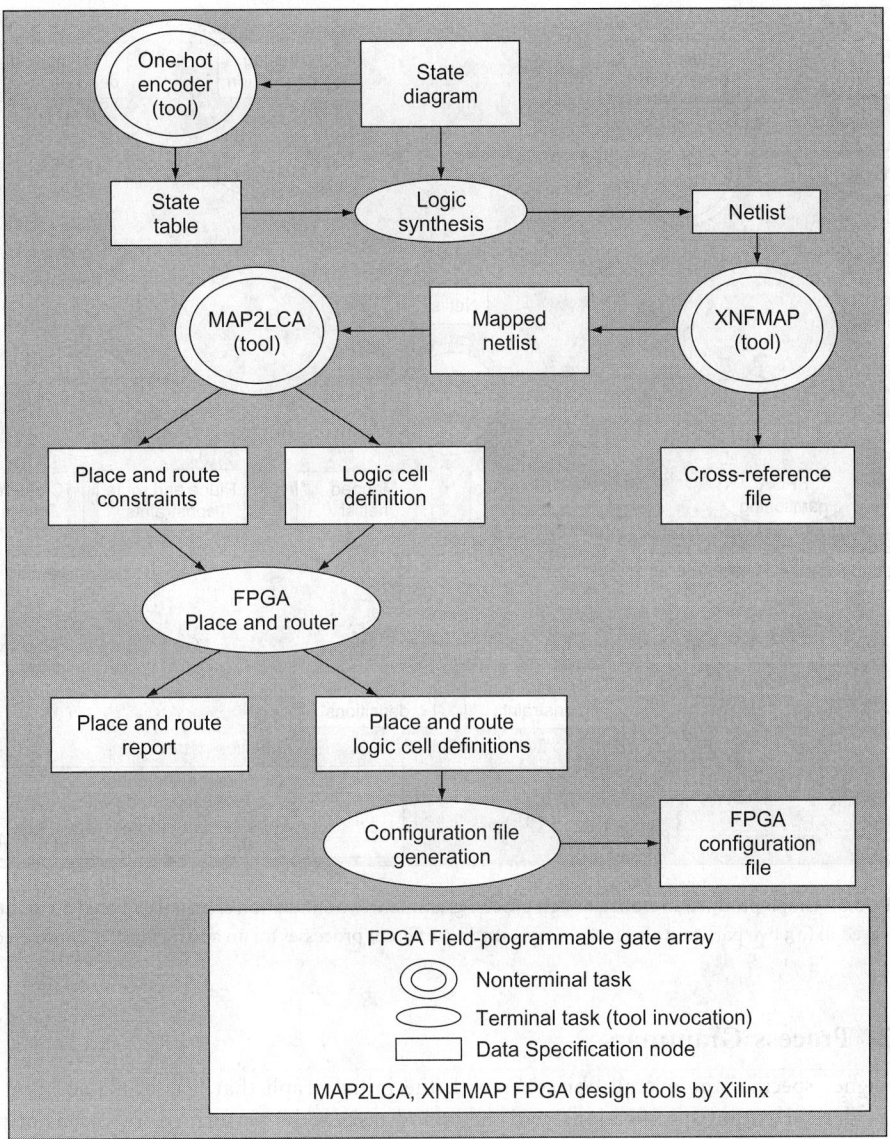

FIGURE 75.4 A sample process flow graph in which a state diagram is transformed into a field-programmable gate array configuration file.

The various specification types form a class hierarchy where each child is a specialization of the parent. There may be several incompatible children. For example, VHDL and Verilog descriptions are both children of simulation models. We utilize these specification types to avoid data format incompatibilities between tools (see Figure 75.5a). Process flow graphs can describe design processes to varying levels of detail. A graph containing many logical nodes abstractly describes what should be done without describing how it should be done (i.e., specifying which tools to use). Conversely, a graph in which all task nodes are atomic completely describes a methodology.

In our prototype, we use the following definitions: $\text{In}(N)$ is the set of input nodes of node N: $\text{In}(N) = \{M \mid (M,N) \in E\}$. $\text{Out}(N)$ is the set of output nodes of node N: $\text{Out}(N) = \{M \mid (N,M) \in E\}$. $I(G)$ is the set of input specifications of graph G: $\{N \in S(G) \mid \text{In}(N) = \varnothing\}$.

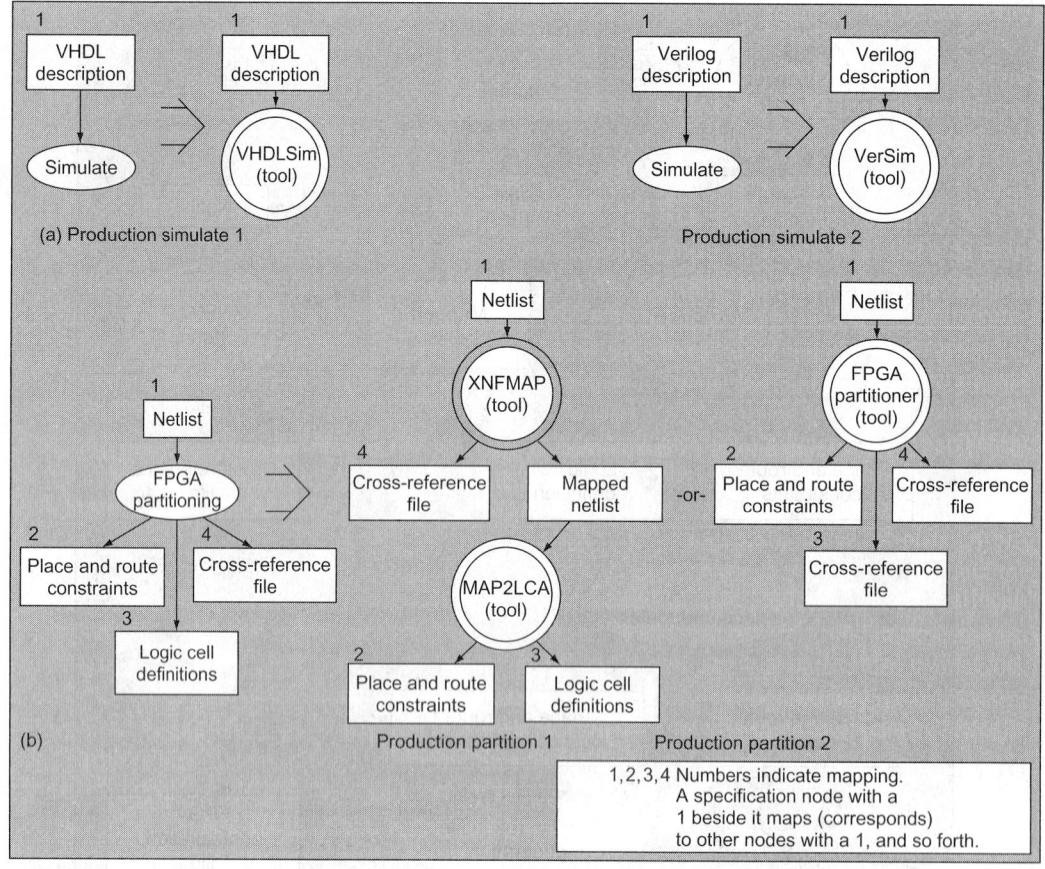

FIGURE 75.5 Graph production from a design process grammar. Two simulation alternatives based on input format are portrayed in (a); two partition alternatives representing different processes for an abstract task are protrayed in (b).

75.4.2 Process Grammars

The designer specifies the overall objectives with the initial graph that lists available input specifications, desired output specifications, and the logical tasks to be performed. By means of process grammars, logical task nodes are replaced by the flows of detailed subtasks and intermediate specifications. The output specification nodes are also replaced by nodes that may have a child specification type.

The productions in a graph grammar permit the replacement of one subgraph by another. A production in a design process grammar can be expressed formally as a tuple $P = (G_{LHS}, G_{RHS}, \sigma_{in}, \sigma_{out})$, where G_{LHS} and G_{RHS} are process flow graphs for the left side and the right side of the production, respectively, such that (i) G_{LHS} has one logical task node representing the task to be replaced, (ii) σ_{in} is a mapping from the input specifications $I(G_{LHS})$ to $I(G_{RHS})$, indicating the relationship between two input specifications (each input specification of $I(G_{RHS})$ is a subtype of $I(G_{LHS})$), and (iii) σ_{out} is a mapping from the output specifications of G_{LHS} to output specifications of G_{RHS} indicating the correspondence between them. (each output specification must be mapped to a specification with the same type or a subtype). Figure 75.5 illustrates productions for two tasks, `simulate` and `FPGA partitioning`. The mappings are indicated by the numbers beside the specification nodes. Alternative productions may be necessary to handle different input specification types (as in Figure 75.5a), or because they represent different processes- separated by the word "or"—for performing the abstract task (as in Figure 75.5b).

Let A be the logical task node in G_{LHS} and A' be a logical task node in the original process flow graph G such that A has the same task label as A'. The production rule P can be applied to A', which means that A' can be replaced with G_{RHS} only if each input and output specifications of A' matches to input and output specifications of G_{LHS}, respectively. If there are several production rules with the same left side flow graph, it implies that there are alternative production rules for the logical task. Formally, the production matches A' if

(i) A' has the same task label as A.
(ii) There is a mapping ρ_{in}, from In(A) to In(A'), indicating how the inputs should be mapped. For all nodes $N \in$ In(A), $\rho_{in}(N)$ should have the same type as N or a subtype.
(iii) There is a mapping, ρ_{out}, from Out(A') to Out(A), indicating how the outputs should be mapped. For all nodes $N \in$ Out(A'), $\rho_{out}(N)$ should have the same type as N or a subtype.

The mappings are used to determine how edges that connected the replaced subgraph to the remainder should be redirected to nodes in the new subgraph. Once a match is found in graph G, the production is applied as follows:

(i) Insert $G_{RHS} - I(G_{RHS})$ into G. The inputs of the replaced tasks are not replaced.
(ii) For every N in $I(G_{RHS})$ and edge (N,M) in G_{RHS}, add edge $(\rho_{in}(\sigma_{in}(N)),M)$ to G. That is to connect the inputs of A' to the new task nodes that will use them.
(iii) For every N in Out (A') and edge (N,M) in G, replace edge (N,M) with edge $(\sigma_{out}(\rho_{out}(N)),M)$. That is to connect the new output nodes to the tasks that will use them.
(iv) Remove A' and Out (A') from G, along with all edges incident on them.

Figure 75.6 illustrates a derivation in which the FPGA partitioning task is planned, using a production from Figure 75.5b.

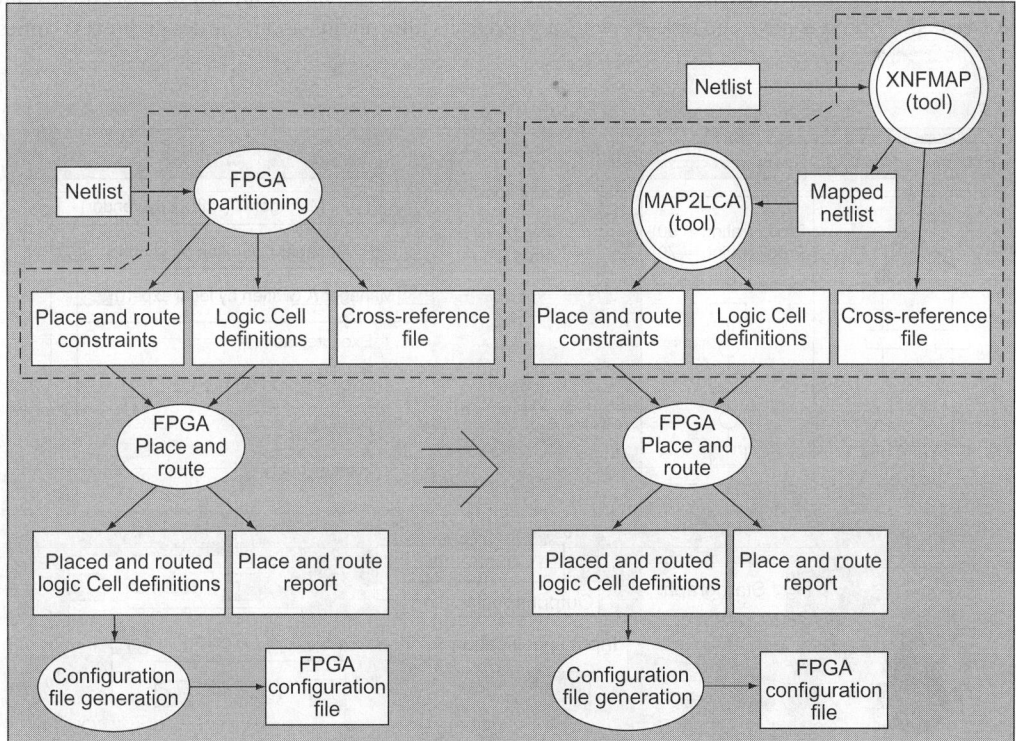

FIGURE 75.6 A sample graph dertivation. Nodes in the outlined region, left, are replaced with nodes in the outlined region, right, according to production partition 1 in Figure 75.5.

The process grammar provides mechanism of specifying alternative methods for a logical task. A high-level flow graph can then be decomposed into detailed flow graphs by applying *production rules* to a logical task. A *production rule* is a substitution that permits the replacement of a logical task node with a flow graph that represents a possible way of performing the task. The concept of applying productions to logical tasks is somewhat analogous to the idea of productions in traditional (i.e., non-graph) grammars. In this sense, logical tasks correspond to *logical symbols* in grammar, and atomic tasks correspond to *terminal symbols.*

75.5 Execution Environment of the Framework

Figure 75.7 illustrates the architecture of our proposed system, which applies the theory developed in the previous section. Decisions to select or invoke tools are split between the designers and a set of manager programs, where manager programs are making the routine decisions and the designers make decisions that requires higher-level thinking. A program called Cockpit coordinates the interaction among manager programs and the designers. Tool sets and methodology preferences will differ among sites and over time. Therefore, our assumption is that each unit designates a person (or group) to act as system integrator, who writes and maintains the tool-dependent code in the system. We provide the tool-independent code and template to simplify the task of writing tool-dependent code.

75.5.1 The Cockpit Program

The designer interacts with Cockpit, a program which keeps track of the current process flow graph and informs the designer of possible actions such as productions that could be applied or tasks that could be executed. Cockpit contains no task-specific knowledge; its information about the design process comes

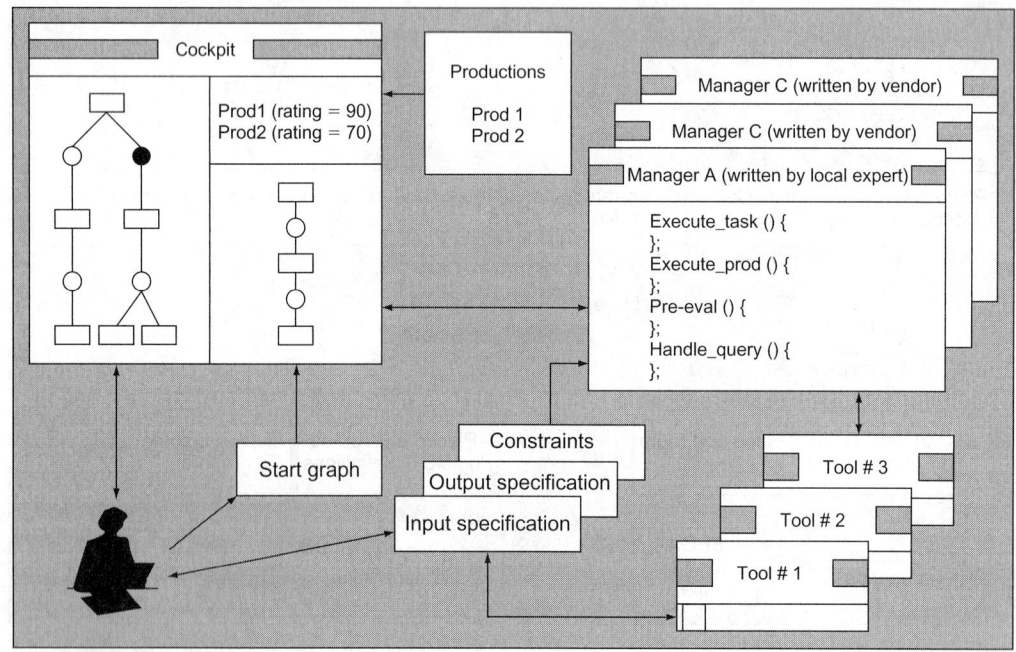

FIGURE 75.7 The proposed system based on Cockpit.

entirely from a file of graph productions. When new tools are acquired or new design processes are developed, the system integrator modifies this file by adding, deleting, and editing productions.

To assist the designer in choosing an appropriate action, Cockpit interacts with several manager programs which encapsulate design knowledge. There are two types of manager programs: task managers and production managers. Task managers invoke tools and determine which productions to execute for logical task nodes. Production managers provide ratings for the productions and schedule the execution of tasks on the right-had side of the production. Managers communicate with each other using messages issued by Cockpit.

Our prototype system operates as follows. Cockpit reads the initial process flow graph from an input file generated by using a text editor. Cockpit then iteratively identifies when productions can be applied to logical task nodes and requests that the production managers assign the ratings to indicate how appropriate the productions are for those tasks. The process flow graph and the ratings of possible production applications are displayed for the designer, who directs Cockpit through a graphical user interface to apply a production or execute a task at any time. When asked to execute a task, Cockpit sends a message to a task manager. For an atomic task node, the task manager simply invokes the corresponding tool. For a logical task, the task manager must choose one or more productions, as identified by a Cockpit. The Cockpit applies the production and requests that the production manager executes it.

75.5.2 Manager Programs

Manager programs must be maintained by system integrators to reflect site-specific information, such as company design practices and tool installation methods. Typically, a manager program has its own thread. A Cockpit may have several manager programs, and therefore multi-threads. We define a communication protocol between Cockpit and manager programs and provide templates for manager programs. The manager programs provide five operations: pre-evaluation, tool invocation, logical task execution, production execution, and query handling. Each operation described below corresponds to a C++ or Java function in the templates, which system integrators can customize as needed.

Pre-evaluation: Production managers assign ratings to help designers and task managers select the most appropriate productions. The rating indicates the likelihood of success from applying this production. The strategies used by the system integrator provide most of the code to handle the rating. In some cases, it may be sufficient to assign ratings statically, based on the success of past productions. These static ratings can be adjusted downward when the production has already been tried unsuccessfully on this task node (which could be determined using the query mechanism). Alternatively, the ratings may be an arbitrarily complex function of parameters obtained through the query mechanism or by examining the input files. Sophisticated manager programs may continuously gather and analyze process metrics that indicate those conditions leading to success, adjust adjusting ratings accordingly.

Tool Invocation: Atomic task mangers must invoke the corresponding software tool when requested by Cockpit, then determine whether the tool completed successfully. In many cases, information may be predetermined and entered in a standard template, which uses the tool's result status to determine success. In other cases, the manager must determine tool parameters using task-specific knowledge or determine success by checking task-specific constraints. Either situation would require further customization of the manager program.

Logical Task Execution: Logical task managers for logical tasks must select productions to execute the logical task. Cockpit informs the task manager of available productions and their ratings. The task manager can either direct Cockpit to apply and execute one or more productions, or it can decide that none of the productions is worthwhile and report failure. The task manager can also request that the productions be reevaluated when new information has been generated that might influence the ratings, such as a production's failure. If a production succeeds, the task manager checks any constraints; if they are satisfied, it reports success.

Production Execution: Production managers execute each task on the right-hand side of the production at the appropriate time and possibly check constraints. If one of the tasks fails or a constraint is violated, backtracking can occur. The production manager can use task-specific knowledge to determine which tasks to repeat. If the production manager cannot handle the failure itself, it reports the failure to Cockpit, and the managers of higher level tasks and productions attempt to handle it.

Query Handling: Both production and task managers participate in the query mechanism. A production manager can send queries to its parent (the task manager for the logical task being performed) or to one of its children (a task manager of a subtask). Similarly, a task manager can send a query to its parent production manager or to one of its children (a production manager of the production it executed). The manager templates define C functions, which take string arguments, for sending these queries. System integrators call these functions but do not need to modify them. The manager templates also contain functions which are modified by system integrators for responding to queries. Common queries can be handled by template code; for example, a production manager can frequently ask its parent whether the production has already been attempted for that task and whether it succeeded. The manager template handles any unrecognized query from a child manager by forwarding it to the parent manager. Code must be added to handle queries for task-specific information such as the estimated circuit area or latency.

75.5.3 Execution Example

Now we describe a synthesis scenario that illustrates our prototype architecture in use. In this scenario, the objective is to design a controller from a state diagram, which will ultimately be done following the process flow graph in Figure 75.4. There are performance and cost constraints on the design, and the requirement to produce a prototype quickly. The productions used are intended to be representative but not unique. For simplicity, we assume that a single designer is performing the design with, therefore, only one Cockpit.

The start graph for this scenario contains only the primary task, chip synthesis, and specification nodes for its inputs and outputs (like the graph in the left in Figure 75.8). Cockpit tells us that the production of Figure 75.8 can be applied. We ask Cockpit to apply it. The chip synthesis node is then replaced by

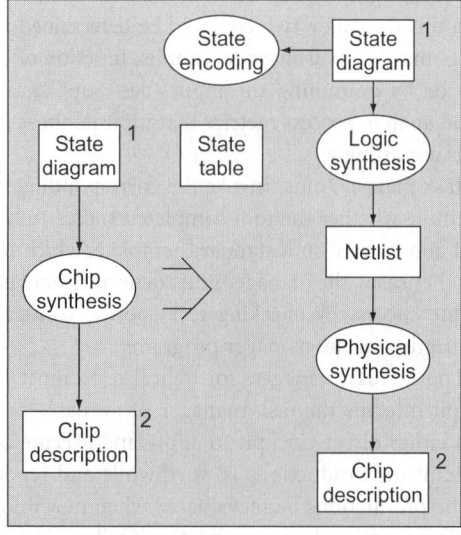

FIGURE 75.8 Productions for chip synthesis.

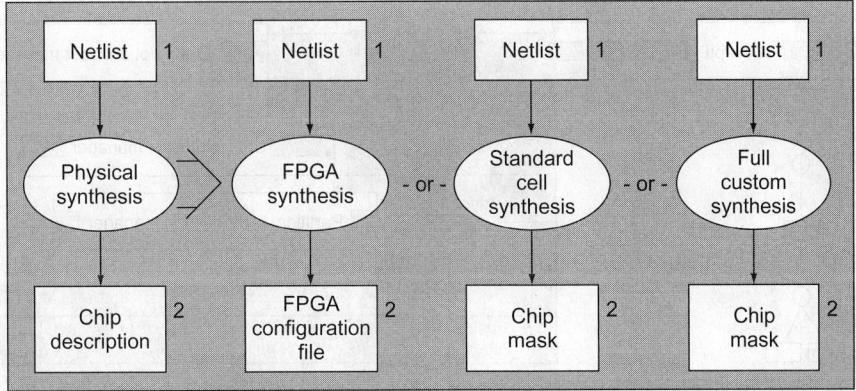

FIGURE 75.9 Productions for physical synthesis.

nodes for state encoding, logic synthesis, and physical synthesis, along with intermediate specification nodes. Next, we want to plan the physical synthesis task. Tasks can be planned in an order other than they are to be performed. Cockpit determines that any of the productions shown in Figure 75.9 may be applied, then queries each production's task manager program asking it to rate the production's appropriateness in the current situation. Based on the need to implement the design quickly, the productions for standard cell synthesis and full custom synthesis are rated low while the production for FPGA synthesis is rated high. Ratings are displayed to help us decide.

When we plan the state encoding task, Cockpit finds two productions: one to use the tool Min-bits Encoder and the other to use the tool One-hot Encoder. One-hot Encoder works well for FPGAs, while Minbits Encoder works better for other technologies. To assign proper ratings to these productions, their production managers must find out which implementation technology will be used. First, they send a query to their parent manager, the state encoding task manager. This manager forwards the message to its parent, the chip synthesis production manager. In turn, this manager forwards the query to the physical synthesis task manager for an answer. All messages are routed by Cockpit, which is aware of the entire task hierarchy. This sequence of actions is illustrated in Figure 75.10.

After further planning and tool invocations, a netlist is produced for our controller. The next step is the FPGA synthesis task. We apply the production in Figure 75.11 and proceed to the FPGA partitioning task. The knowledge to automate this task has already been encoded into the requisite manager programs, so we direct Cockpit to execute the FPGA partitioning task. It finds the two productions illustrated in Figure 75.5b and requests their ratings, Next, Cockpit sends an execute message, along with the ratings, to the FPGA partitioning task manager. This manager's strategy is to always execute the highest-rated production, which in this case is production Partition 1. (Other task managers might have asked that both productions be executed or, if neither were promising, immediately reported failure.) This sequence of actions is shown in Figure 75.12.

Because the Partition 1 manager used an as-soon-as-possible task scheduling strategy, it asks Cockpit to execute XNFMAP immediately. The other subtask, MAP2LCA, is executed when XNFMAP complete successfully. After both tasks complete successfully, Cockpit reports success to the FPGA partitioning task manager. This action sequence is illustrated in Figure 75.13.

75.5.4 Scheduling

In this subsection, we describe a detailed description and discussion of auto-mode scheduling, including the implementation of the *linear scheduler*. The ability to search through the configuration space of a design process for a design configuration that meets user-specified constraints is important. For example, assume that a user has defined a process for designing a digital filter with several different alternative ways

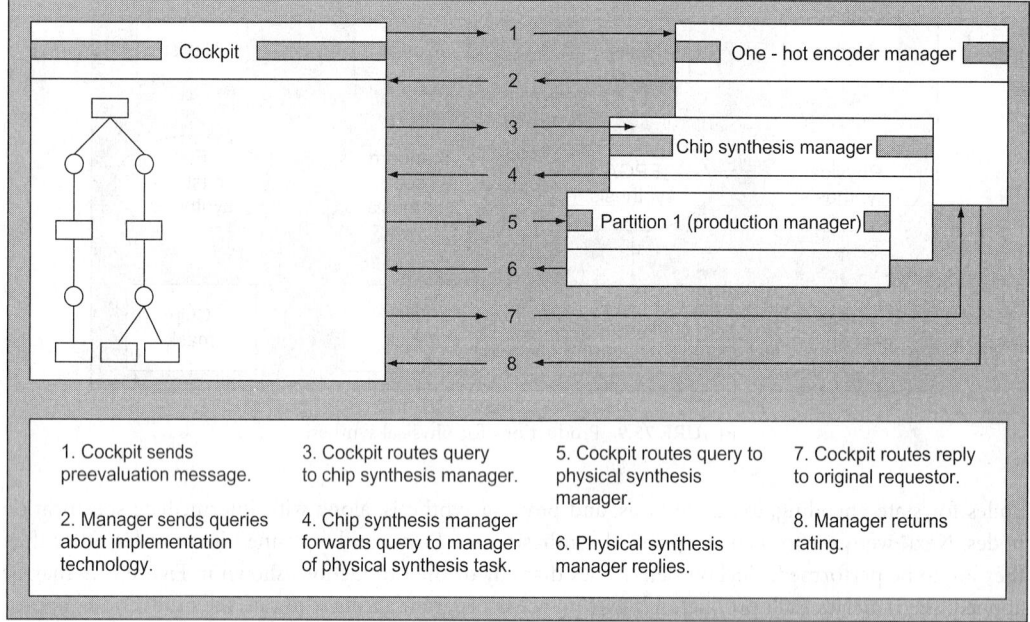

FIGURE 75.10 Sequence of actions for query handling.

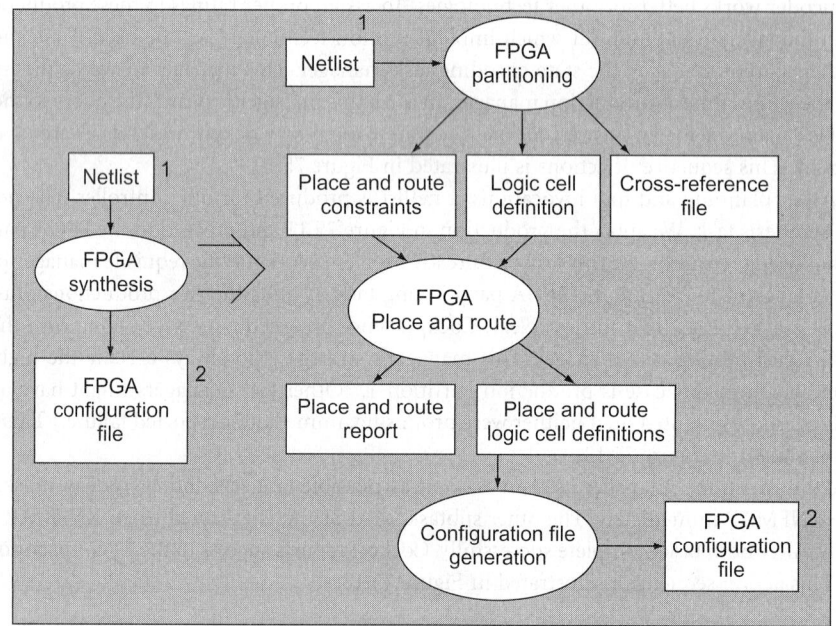

FIGURE 75.11 Production for field-programmable gate array synthesis.

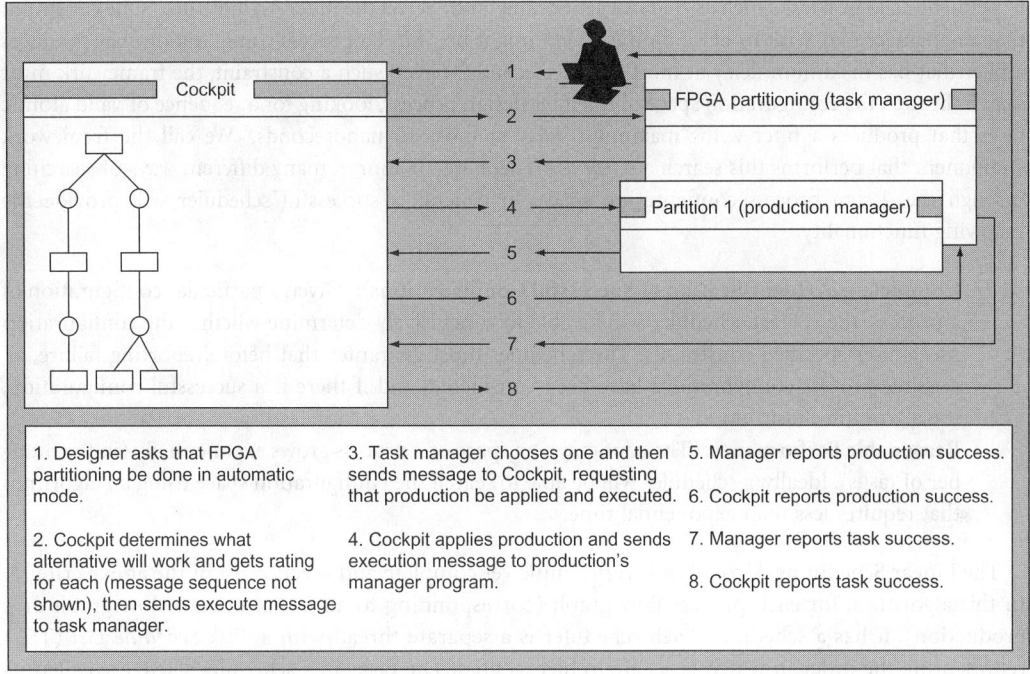

1. Designer asks that FPGA partitioning be done in automatic mode.

2. Cockpit determines what alternative will work and gets rating for each (message sequence not shown), then sends execute message to task manager.

3. Task manager chooses one and then sends message to Cockpit, requesting that production be applied and executed.

4. Cockpit applies production and sends execution message to production's manager program.

5. Manager reports production success.

6. Cockpit reports production success.

7. Manager reports task success.

8. Cockpit reports task success.

FIGURE 75.12 Sequence of action during automatic task execution.

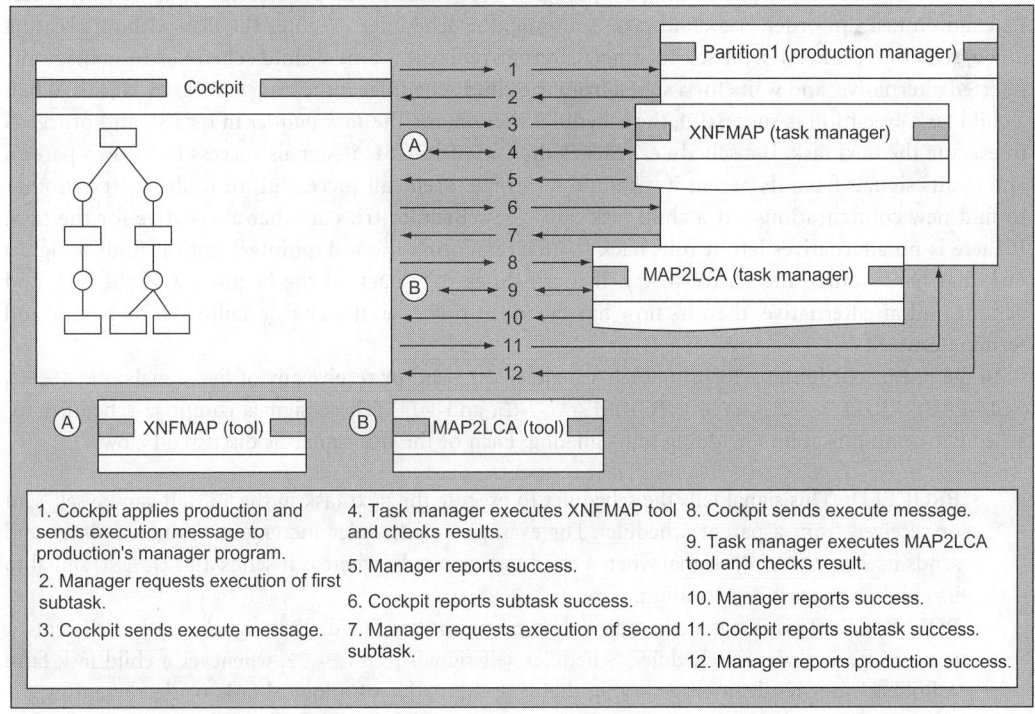

1. Cockpit applies production and sends execution message to production's manager program.

2. Manager requests execution of first subtask.

3. Cockpit sends execute message.

4. Task manager executes XNFMAP tool and checks results.

5. Manager reports success.

6. Cockpit reports subtask success.

7. Manager requests execution of second subtask.

8. Cockpit sends execute message.

9. Task manager executes MAP2LCA tool and checks result.

10. Manager reports success.

11. Cockpit reports subtask success.

12. Manager reports production success.

FIGURE 75.13 Sequence of actions during automatic production execution.

of performing logical tasks such as "FPGA Partitioning" and "Select the Filter Architecture." One constraint that an engineer may wish to place on the design might be: "Find a process configuration that produces a filter that has maximum delay at most 10 nanoseconds." Given such a constraint, the framework must search through the configuration space of the filter design process, looking for a sequence of valid atomic tasks that produces a filter with "maximum delay at most 10 nanoseconds." We call the framework component that performs this search a *scheduler*. There are, of course, many different ways of searching through the design process configuration space. In general, a successful scheduler will provide the following functionality:

- **Completeness (Identification of Successful Configurations):** Given a particular configuration of a process, the correct scheduler will be able to conclusively determine whether the configuration meets user-specified constraints. The scheduler must guarantee that before reporting failure, all possible process configurations have been considered, and if there is a successful configuration, the algorithm must find it.
- **Reasonable Performance:** The configuration space of a process grows exponentially (in the number of tasks). Ideally, a scheduler will be able to search the configuration space using an algorithm that requires less than exponential time.

The Linear Scheduling Algorithm is very simple yet complete and meets most of the above criteria. In this algorithm, for each process flow graph (corresponding to an initial process flow graph or a production), it has a scheduler. Each scheduler is a separate thread with a *Task Schedule List* (TSL) representing the order in which tasks are to be executed. The tasks in a scheduler's TSL are called its children tasks. A scheduler also has a *task pointer* to indicate the child task being executed in the TSL. The algorithm is recursive such that with each new instantiation of a production of a given task, a new scheduler is created to manage the flow graph representing the production selected. A liner scheduler creates a TSL by performing a topological sort of the initial process flow graph and executes its children tasks in order. If a child task is atomic, the scheduler executes the task without creating a new scheduler; otherwise, it selects a new alternative, creates a new child scheduler to manage the selected alternative, and waits for a signal from the child scheduler indicating success or failure. When a child task execution is successful, the scheduler increments the *task pointer* in its TSL and proceeds to execute the next task. If a scheduler reaches the end of its TSL, it signals success to its own parent, and awaits signals from its parent if it should terminate itself (all successful) or rollback (try another to find new configurations). If a child task fails, the scheduler tries another alternative for the task. If there is no alternatives left, it rolls back (by decrementing the task pointer) until it finds a logical task that has another alternative to try. If a scheduler rolls back to the beginning of the TSL and cannot find an alternative, then its flow has failed. In this case, it signals a failure to its parent and terminates itself.

In the linear scheduling algorithm, each scheduler can send or receive any of five signals: PROCEED, ROLLBACK, CHILD-SUCCESS, CHILD-FAILURE, and DIE. These signals comprise scheduler-to-scheduler communication, including self-signaling. Each of the five signals is discussed below.

- **PROCEED:** This signal tells the scheduler to execute the next task in the TSL. It can be self-sent or received from a parent scheduler. For example, a scheduler increments its task pointer and sends itself a PROCEED signal when a child task succeeds, whereas it sends a PROCEED signal to its children to start its execution.
- **ROLLBACK:** This is signaled when a task execution has failed. This signal may be self-sent or received from a parent scheduler. Scheduler self-signals ROLLBACK whenever a child task fails. A Rollback can result in either trying the next alternative of a logical task, or decrementing the task pointer and trying the previous task in the TSL. If rollback results in decrementing the task pointer to point to a child task node which has received a success-signal, the parent scheduler will send a rollback signal to that child task scheduler.

- **CHILD-SUCCESS:** A child scheduler sends a `CHILD-SUCCESS` to its parent scheduler if it has successfully completed the execution of all of the tasks in its `TSL`. After sending the child-success signal, the scheduler remains active, listening for possible rollback signals from the parent. After receiving a child-success signal, parent schedulers self-send a proceed signal.
- **CHILD-FAILURE:** A child-failure signal is sent from a child scheduler to its parent in the event that the child's managed flow fails. After sending a child-failure signal, children schedulers terminate. Upon receiving child-failure signals, parent scheduler self-send a rollback signal.
- **DIE:** This signal may be either self-sent, or sent from parent schedulers to their children schedulers.

75.6 Implementation

In this section, a high level description of the major components of IMEDA and their organization and functionality will be presented. Detailed explanations of the key concepts involved in the architecture of the Process Management Framework will also be discussed, including external tool integration, the tool invocation process, the Java File System, and state properties.

75.6.1 The System Cockpit

The System Cockpit, as its name suggests, is shown in Figure 75.14 where nearly all user interaction with the framework takes place. It is here that users create, modify, save, load, and simulate process flow graphs

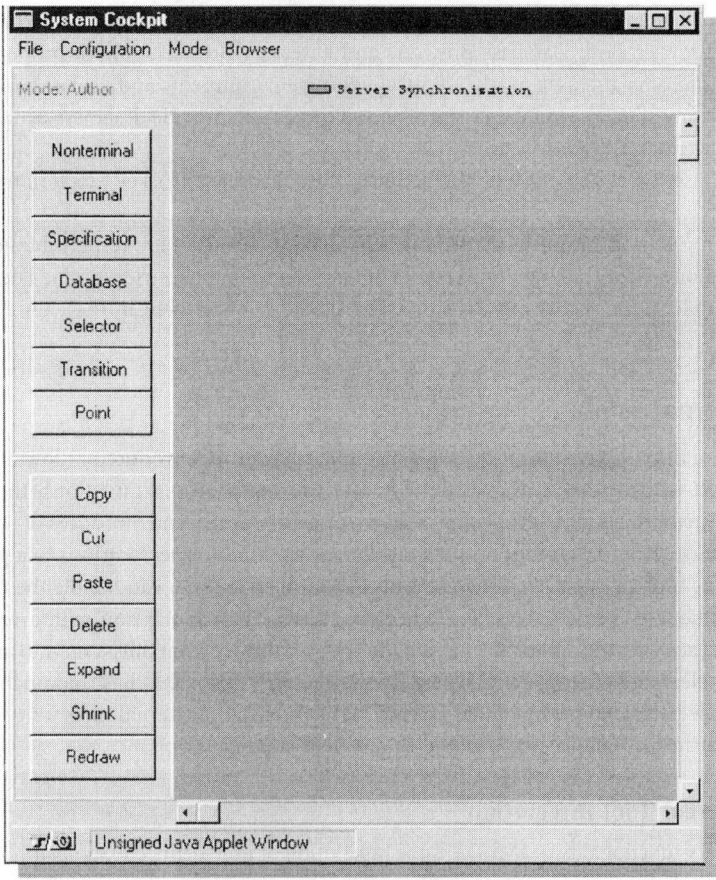

FIGURE 75.14 A system cockpit window.

representing design processes. This system component has been implemented as a Java applet. As such, it is possible to run the cockpit in any Java-enabled Web browser such as Netscape's Navigator or Microsoft's Internet Explorer. It is also possible to run the cockpit in some Java-enabled operating systems such as IBM's OS/2.

Each cockpit component also has the following components:

- **Root Flow.** Every cockpit has a Root Flow. The Root Flow is the flow currently being edited in the Cockpit's Flow Edit Panel. Notice that the Root Flow may change as a result of applying a production to a flow graph, in which case the Root Flow becomes a derivation of itself.
- **Flow Edit Panel.** The Flow Edit Panel is the interactive Graphical User Interface for creating and editing process flow graphs. This component also acts as a display for animating process simulations performed by various schedulers such as the manual or automode linear scheduler.
- **Class Directory.** The Cockpit has two Class Directories: Task Directory and the Specification Directory. These directories provide the "browser" capabilities of the framework, allowing users to create reusable general-to-specific hierarchies of task classes. Class Directories are implemented using a *tree* structure.
- **Production Database.** The Production Database acts as a warehouse for logical task productions. These productions document the alternative methods available for completing a logical task. Each Production Database has a list of Productions. The Production Database is implemented as a tree-like structure, with Productions being on the root trunk, and Alternatives being leaves.
- **Browser.** Browsers provide the tree-like graphical user interface for users to edit both Class Directories and Databases. There are three Browsers: Database Browser for accessing the Production Database, Directory Browser for accessing the Task Directory, and Directory Browser for accessing the Spec Browser. Both Database Browsers and Directory Browsers inherit properties from object Browser, and offer the user nearly identical editing environments and visual representations. This deliberate consolidation of Browser interfaces allowed us to provide designers with an interface that was consistent and easier to learn.
- **Menu.** A user typically performs and accesses most of the system's key function from the cockpit's Menu.
- **Scheduler.** The cockpit has one or more schedulers. Schedulers are responsible for searching the configuration space of a design process for configurations that meet user specified design constraints. The Scheduler animates its process simulations by displaying them in the Flow Edit Panel of the Cockpit.

75.6.2   External Tools

External Tools are the concrete entities to which atomic tasks from a production flow are bound. When a flow task object is expanded in the Cockpit Applet (during process simulation), the corresponding external tool is invoked. The external tool uses a series of inputs and produces a series of outputs (contained in files). These inputs and outputs are similarly bound to specifications in a production flow. Outputs from one tool are typically used as inputs for another. IMEDA can handle the transfer of input and output files between remote sites. The site proxy servers, in conjunction with a remote file server (also running at each site) automatically handle the transfer of files from one system to another. External tools may be implemented using any language, and on any platform that has the capability of running a site server. While performing benchmark tests of IMEDA, we used external tools written in C, Fortran, Perl, csh (a Unix shell script), Java applications, and Mathematica scripts.

75.6.2.1   External Tool Integration

One of the primary functionality of IMEDA is the integration of user-defined external tools into an abstract process flow. IMEDA then uses these tools both in simulating the process flow to find a flow configuration that meets specific constraints, and in managing selected flow configurations during actual design execution.

There are two steps to integrating tools with a process flow defined in IMEDA: *association* and *execution*. Association involves "linking" or "binding" an abstract flow item (e.g., an atomic task) to an external tool. Execution describes the various steps that IMEDA takes to actually invoke the external tool and process the results.

75.6.2.2 Binding Tools

External tools may be bound to three types of flow objects: Atomic Tasks, Selectors, and Multiple Version Selectors. Binding an external tool to a flow object is a simple and straightforward job, involving simply defining certain properties in the flow object.

The following properties must be defined in an object that is to be bound to an external tool:

- **SITE.** Due to the fact that IMEDA can execute tools on remote systems, it is necessary to specify the site where the tool is located on. Typically, a default SITE will be specified in the system defaults, and making it unnecessary to define the site property unless the default is to be overriden. Note that the actual site ID specified by the SITE property must refer to a site that is running a Site Proxy Server listening on that ID. See the "Executing External Tools" section below for more details.

- **CMDLINE.** The CMDLINE property specifies the command to be executed at the specified remote site. The CMDLINE property should include any switches or arguments that will always be sent to the external tool. Basically, the CMDLINE argument should be in the same format that would be used if the command were executed from a shell/DOS prompt.

- **WORKDIR.** The working directory of the tool is specified by the WORKDIR property. This is the directory in which IMEDA will actually execute the external tool, create temporary files, etc. This property is also quite often defined in the global system defaults, and thus may not necessarily have to be defined for every tool.

- **WRAPPERPATH.** The JDK 1.0.2 does not allow Java Applications to execute a tool in an arbitrary directory. To handle remote tool execution, a wrapper is provided. It is a "go-between" program that would simply change directories and then execute the external tool. This program can be as simple as a DOS/NT batch file, a shell script, or a perl program. The external tool is wrapped in this simple script, and executed. Since IMEDA can execute tools at remote and heterogeneous sites, it was very difficult to create a single wrapper that would work on all platforms (WIN32, Unix, etc.). Therefor, the wrapper program may be specified for each tool, defined as global default, or a combination of the two.

Once the properties above have been defined for a flow object, the object is said to be "bound" to an external tool. If no site, directory, or filename is specified for the outputs of the flow object, IMEDA automatically creates unique file names, and stores the files in the working directory of the tool on the site that the tool was run. If a tool uses as inputs data items that are not specified by any other task, then the data items must be bound to static files on some site.

75.6.2.3 Executing External Tools

Once flow objects have been bound to the appropriate external tools, IMEDA can be used to perform process simulation or process management. IMEDA actually has several "layers" that lie between the Cockpit (a Java applet) and the external tool that is bound to a flow being viewed by a user in the Cockpit. A description of each of IMEDA components for tool invocations is listed below.

- **Tool Proxy.** The tool proxy component acts as a liaison between flow objects defined in Cockpits and the Site Proxy Server. All communication is done transparently through the communication server utilizing TCP/IP sockets. The tool proxy "packages" information from Cockpit objects (atomic tasks, selectors, etc.) into string messages that the Proxy Server will recognize. It also listens for and processes messages from the Proxy Server (through the communications server) and relays the information back to the Cockpit object that instantiated the tool proxy originally.

- **Communications Server.** Due to various security restrictions in the 1.0.2 version of Sun Micro-system's Java Development Kit (JDK), it is impossible to create TCP/IP socket connections between a Java applet and any IP address other than the address from which the applet was loaded. Therefore, it was necessary to create a "relay server" in order to allow cockpit applets to communicate with remote site proxy servers. The sole purpose of the communications server is to receive messages from one source and then to rebroadcast them to all parties that are connected and listening on the same channel.
- **Site Proxy Server.** Site Proxy Servers are responsible for receiving and processing invocation requests from tool proxies. When an invocation request is received, the site proxy server checks to see that the request is formatted correctly, starts a tool monitor to manage the external tool invocation, and returns the exit status of the external tool after it has completed.
- **Tool Monitors.** When the site proxy server receives an invocation request and invokes an external tool, it may take a significant amount of time for the tool to complete. If the proxy server had to delay the handling of other requests while waiting for each external tool to complete, IMEDA would become very inefficient. For this reason, the proxy server spawns a tool monitor for each external tool that is to be executed. The tool monitor runs as a separate thread, waiting on the tool, storing its stdout and stderr, and moving any input or output files that need moving to their appropriate site locations, and notifying the calling site proxy server when the tool has completed. This allows the site proxy server to continue receiving and processing invocation requests in a timely manner.
- **Tool Wrapper.** Tool wrapper changes directories into the specified WORKDIR, and then executes the CMDLINE.
- **External Tool.** External tools are the actual executable programs that run during a tool invocation. There is very little restriction on the nature of the external tools.

75.6.3 Communications Model

The Communications Model of IMEDA is perhaps the most complex portion of the system in some respects. This is where truly *distributed* communications come into play. One system component is communicating with another via network messages rather than function calls.

The heart of the communications model is the Communications Server. This server is implemented as a broadcast server. All incoming messages to the server are simply broadcast to all other connected parties. FlowObjects communicate with the Communications Server via ToolProxys. A ToolProxy allows a FlowObject to abstract all network communications and focus on the functionality of invoking tasks. A ToolProxy takes care of constructing a network message to invoke an external tool. That message is then sent to the Communications Server via a Communications Client. The Communication Client takes care of the low-level socket based communication complexities. Finally, the Communications Client sends the message to the Communications Server, which broadcasts the message to all connected clients. The client for which the message was intended (typically a Site Proxy Server) decodes the message and, depending on its type, creates either a ToolMonitor (for an Invocation Message) or an External Redraw Monitor (for a Redraw Request).

The Site Proxy Server creates these monitors to track the execution of external programs, rather than monitoring them itself. In this way, the Proxy Server can focus on its primary job – receiving and decoding network messages. When the Monitors invoke an external tool, they must do so within a Wrapper. Once the Monitors have observed the termination of an external program, they gather any output on *stdout* or *stderr* and return these along with the exit code of the program to the Site Proxy Server. The Proxy Server returns the results to the Communications Server, then the Communications Client, then the ToolProxy, and finally to the original calling FlowObject.

75.6.4 User Interface

The Cockpit provides both the user interface and core functionality of IMEDA. While multiple users may use different instances of the Cockpit simultaneously, there is currently no provision for direct collaboration between multiple users. Developing efficient means of real-time interaction between IMEDA users is one of the major thrusts of the next development cycle.

Currently the GUI of the cockpit provides the following functionality:

- **Flow editing.** Users may create and edit process flows using the flow editor module of the Cockpit. The flow editor provides the user with a simple graphical interface that allows the use of a template of tools for "drawing" a flow. Flows can be optimally organized via services provided by a remote Layout Server written in Perl.
- **Production Library Maintenance.** The Cockpit provides functionality for user maintenance of collections of logical task productions, called libraries. Users may organize productions, modify input/output sets, or create/edit individual productions using flow editors.
- **Class Library Maintenance.** Users are provided with libraries of task and specification classes that are organized into a generalization-specialization hierarchy. Users can instantiate a class into an actual task, specification, selector, or database when creating a flow by simply dragging the appropriate class from a class browser and dropping it onto a flow editor's canvas. The Cockpit provides the user with a simple tree structure interface to facilitate the creation and maintenance of class libraries.
- **Process Simulation.** Processes may be simulated using the Cockpit. The Cockpit provides the user with several scheduler modules that determine how the process configuration space will be explored. The schedulers control the execution of external tools (through the appropriate site proxy servers) and simulation display (flow animation for user monitoring of simulation progress). There are multiple schedulers for the user to choose from when simulating a process, including the manual scheduler, comprehensive linear scheduler, etc.
- **Process Archival.** The Cockpit allows processes to be archived on a remote server using the Java File System (JFS). The Cockpit is enabled by a JFS client interface to connect to a remote JFS server where process files are saved and loaded. While the JFS system has its clear advantages, it is also awkward to not allow users to save process files, libraries, etc. on their local systems. Until version 1.1 of the Java Development Kit, local storage by a Java applet was simply not an option—the browser JVM definition did not allow access to most local resources. With version 1.1 of the JDK, however, comes the ability to electronically sign an applet. Once this has been done, users can grant privileged resource access to specific applets after a signature has been verified.

75.6.4.1 Design Flow Graph Properties

Initially, a flow graph created by a user using GUI is not associated with any system-specific information. For example, when a designer creates an atomic task node in a flow graph, there is initially no association with any external tool. The framework must provide a mechanism for users to bind flow graph entities to the external tools or activities that they represent.

We have used the concept of *properties* to allow users to bind flow graph objects to external entities. In an attempt to maintain flexibility, properties have been implemented in a very generic fashion. Users can define any number of properties for flow object. There are a number of key properties that the framework recognizes for each type of flow object. The user defines these properties to communicate needed configuration data to the framework.

A property consists of a *property label* and *property contents*. The label identifies the property, and consists of an alpha-numeric string with no white space. The contents of a property is any string. Currently users define properties using a freeform text input dialog, with each line defining a property. The first word on a line represents the property label, and the remainder of the line constitutes the property contents.

75.6.4.2 Property Inheritance

To further extend the flexibility of flow object properties, the framework requires that each flow object be associated with a *flow object class*. Classes allow designers to define properties that are common to all flow objects that inherit from that flow object class. Furthermore, classes are organized into a general-to-specific hierarchy, with children classes inheriting properties from parent classes.

Therefore, the properties of a particular class consist of any properties defined locally for that object, in addition to properties defined in the object's inherited class hierarchy. If a property is defined in both the flow object and one of its parent classes, the property definition in the flow object takes precedence. If a property is defined in more than one class in a class hierarchy, the "youngest" class (e.g., the child in a parent-child relationship) takes precedence.

Classes are defined in the Class Browsers of IMEDA. Designers that have identified a clear general-to-specific hierarchy of flow object classes can quickly create design flow graphs by dragging and dropping from class browsers onto flow design canvases. The user would then need only to overload those properties in the flow objects that are different from their respective parent classes. (See Figure 75.15.)

For example, consider a class hierarchy of classes that all invoke the same external sort tool, but pass different flags to the tool, based on the context. It is likely that all of these tools will have properties in common, such as a common working directory and tool site. By defining these common properties in a common ancestor of all of the classes, such as *Search*, it is unnecessary to redefine the properties in the children classes.

Of course, children classes can define new properties that are not contained in the parent classes, and may also overload property definitions provided by ancestors. Following these rules, class *Insertion* would have the following properties defined: WORKDIR, SITE, WRAPPERPATH, and CMDLINE. (See Figure 75.16.)

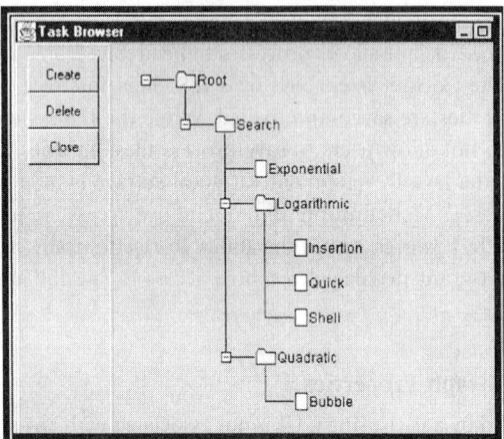

FIGURE 75.15 A task browser.

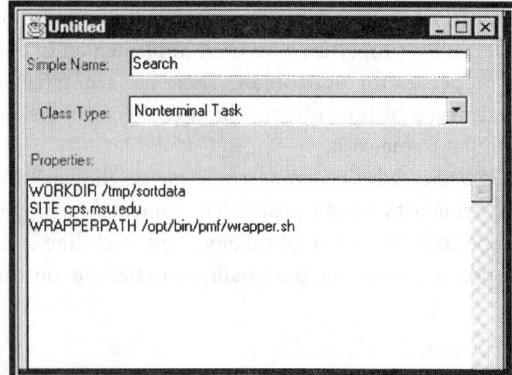

FIGURE 75.16 A property window and property inheritance.

75.6.4.3 Macro Substitution

While performing benchmarks on IMEDA, one cumbersome aspect of the framework that users often pointed out was the need to re-enter properties for tasks or specifications if, for example, a tool name or working directory changed. Finding every property that needed to be changed was a tedious job, and prone to errors.

In an attempt to deal with this problem, we came up with the idea of *property macros*. That is, a property macro is any macro that is not a key system macro. A macro is a textual substitution rule that can be created by users. By using macros in the property databases of flow objects, design flows can be made more flexible and more amiable to future changes.

As an example, consider a design flow that contains many atomic tasks bound to an external tool. Our previous example using searches is one possible scenario. On one system, the path to the external tool may be "/opt/bin/sort," while on another system the path is "`/user/keyesdav/public/bin/sort`." Making the flow object properties flexible is easy if a property macro named SORTPATH is defined in an ancestor of all affected flow objects. (See Figure 75.17.) Children flow objects can then use that macro in place of a static path when specifying the flow object properties. As a further example, consider a modification to the previous "Search task hierarchy" where we define a macro SORTPATH in the Search class, and then use that macro in subsequent children classes, such as the Insertion class.

In the highlighted portion of the Property Database text area, a macro called "SORTPATH" is defined. In subsequent class' Property Databases, this macro can be used in place of a static path. This makes it easy to change the path for all tools that use the SORTPATH property macro—just the property database dialog where SORTPATH is originally defined needs to be modified. (See Figure 75.18.)

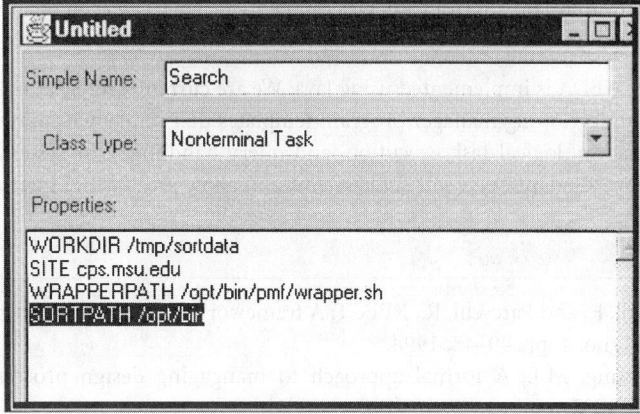

FIGURE 75.17 Macro definition.

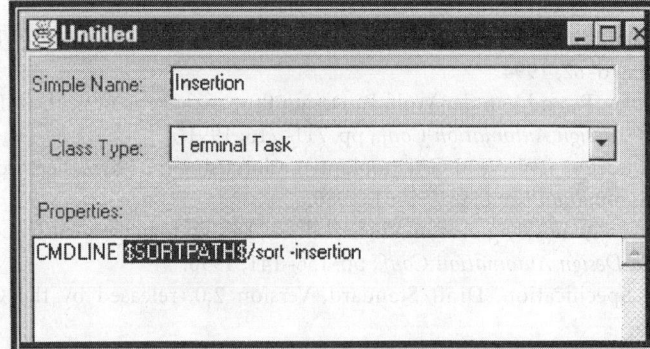

FIGURE 75.18 Macro substitution.

75.6.4.4 Key Framework Properties

In our current implementation of IMEDA, there are a number of key properties defined. These properties allow users to communicate needed information to the framework in a flexible fashion. Most importantly, it allows system architects to define or modify system properties quickly. This is an important benefit when working with evolving software such as IMEDA.

75.7 Conclusion

Managing the design process is the key factor to improve the productivity in the micro-electronic industry. We have presented an *Internet-based Micro-Electronic Design Automation* (IMEDA) framework to manage the design process. IMEDA uses a powerful formalism, called design process grammars, for representing design processes. We have also proposed an execution environment that utilizes this formalism to assist designers in selecting and executing appropriate design processes. The proposed approach is applicable not only in rapid prototyping but also in any environment where a design is carried out hierarchically and many alternative processes are possible.

The primary advantages of our system are

- *Formalism*: A strong theoretical foundation enables us to analyze how our system will operate with different methodologies.
- *Parallelism*: In addition to performing independent tasks within a methodology in parallel, our system also allows multiple methodologies to be executed in parallel.
- *Extensibility*: New tools can be integrated easily by adding productions and manager programs.
- *Flexibility*: Many different control strategies can be used. They can even be mixed within the same design exercise.

The prototype of IMEDA is implemented using Java. We are currently integrating more tools into our prototype system and developing manager program templates that implement more sophisticated algorithms for pre-evaluation, logical task execution, and query handling. Our system will become more useful as CAD vendors to adapt open software systems and allow greater tool interoperability.

References

Andreoli, J.-M., Pacull, F., and Pareschi, R., XPECT: A framework for electronic commerce, *IEEE Internet Comput.*, vol. 1, no. 4, pp. 40–48, 1998.

Baldwin, R. and Chung, M.J., A formal approach to mangaging design processes, *IEEE Comput.*, pp. 54–63, Feb. 1995a.

Baldwin, R. and Chung, M.J., Managing engineering data for complex products, *Res. Eng. Design*, 7, pp. 215–231, 1995b.

Barthelmann, K., Process specification and verification, *Lect. Notes Comput. Sci.*, 1073 pp. 225–239, 1996.

Berners-Lee, T., Cailliau, R., Luotonen, A., Nielsen, H. F., and Secret, A., The World-Wide Web, *Commun. ACM*, 37, 8, pp. 76–82, 1994.

ten Bosch, K.O., Bingley, P., and Van der Wolf, P., Design flow management in the NELSIS CAD framework, *Proc. 28th Design Automation Conf.*, pp. 711–716, 1991.

Bushnell, M.L. and Director, S.W., VLSI CAD tool integration using the Ulysses environment, *23rd ACM/IEEE Design Automation Conf.*, pp. 55–61, 1986.

Casotto, A., Newton, A.R., and Snagiovanni-Vincentelli, A., Design management based on design traces, *27th ACM/IEEE Design Automation Conf.*, pp. 136–141, 1990.

Tool Encapsulation Specification, Draft Standard, Version 2.0, released by the CFI TES Working Group, 1995.

Chan, F.L., Spiller, M.D., and Newton, A.R., WELD—An environment for web-based electronic design, *35th ACM/IEEE Design Automation Conf.*, June 1998.

Chiueh, T.F. and Katz, R.H., A history model for managing the VLSI design process, *Int. Conf. Comput. Aided Design*, pp. 358–361, 1990.

Chung, M.J., Charmichael, L., and Dukes, M., Managing a RASSP design process, *Comp. Ind.*, 30, pp. 49–61, 1996.

Chung, M.J. and Kim, S., An object-oriented VHDL environment, *27th Design Automation Conf.*, pp. 431–436, 1990.

Chung, M.J. and Kim, S., Configuration management and version control in an object-oriented VHDL environment, *ICCAD 91*, pp. 258–261, 1991.

Chung, M.J. and Kwon, P., A web-based framework for design and manufacturing a mechanical system, *1998 DETC*, Atlanta, GA, Sept. 1998.

Cutkosy, M.R., Tenenbaum, J.M., and Glicksman, J., Madefast: collaborative engineering over the Internet, *Commun. ACM*, vol. 39, no. 9, pp. 78–87, 1996.

Daniell, J. and Director, S.W., An object oriented approach to CAD tool control, *IEEE Trans. Comput.-Aided Design*, pp. 698–713, June 1991.

Dellen, B., Maurer, F., and Pews, G., Knowledge-based techniques to increase the flexibility of workflow management, in *Data and Knowledge Engineering*, North-Holland, 1997.

Derk, M.D. and DeBrunner, L.S., Reconfiguartion for fault tolerance using graph grammar, *ACM Trans. Comput. Syst.*, vol. 16, no. 1, pp. 41–54, Feb. 1998.

Ehrig, H., Introduction to the algebraic theory of graph grammars, *1st Workshop on Graph Grammars and Their Applications to Computer Science and Biology*, pp. 1–69, Springer, LNCS, 1979.

Erkes, J.W., Kenny, K.B., Lewis, J.W., Sarachan, B.D., Sobololewski, M.W., and Sum, R.N., Implementing shared manufacturing services on the World-Wide Web, *Commun. ACM*, vol. 39, no.2, pp. 34–45, 1996.

Fairbairn, D.G., 1994 Keynote Address, *31st Design Automation Conference*, pp. xvi–xvii, 1994.

Hardwick, M., Spooner, D.L., Rando, T., and Morris, K.C., Sharing manufacturing information in virtual enterprises, *Commun. ACM*, vol. 39, no. 2, pp. 46–54, 1996.

Hawker, S., SEMATECH Computer Integrated Manufacturing(CIM) framework Architecture Concepts, Principles, and Guidelines, version 0.7.

Heiman, P. et al., Graph-based software process management, *Int. J. Software Eng. Knowledge Eng.*, vol. 7, no. 4, pp. 1–24, Dec. 1997.

Hines, K. and Borriello, G., A geographically distributed framework for embedded system design and validation, *35th Annual Design Automation Conf.*, 1998.

Hsu, M. and Kleissner, C., Objectflow: towards a process management infrastructure, *Distributed and Parallel Databases*, 4, pp. 169–194, 1996.

IDEF *http://www.idef.com*.

Jacome, M.F. and Director, S.W., A formal basis for design process planning and management, *IEEE Trans. Comput.-Aided Design of Integr. Circuits Syst.*, vol. 15, no. 10, pp. 1197–1211, October 1996.

Jacome, M.F. and Director, S.W., Design process management for CAD frameworks, *29th Design Automation Conf.*, pp. 500–505, 1992.

Di Janni, A., A monitor for complex CAD systems, *23rd Design Automation Conference*, pp. 145–151, 1986.

Katz, R.H., Bhateja, R., E-Li Chang, E., Gedye, D., and Trijanto, V., Design version management, *IEEE Design and Test*, 4(1) pp. 12–22, Feb. 1987.

Kleinfeldth, S., Guiney, M., Miller, J.K., and Barnes, M., Design methodology management, *Proc. IEEE*, vol. 82, no. 2, pp. 231–250, Feb. 1994.

Knapp, D. and Parker, A., The ADAM design planning engine, *IEEE Trans. Comput. Aided Design Integr. Circuits Syst.*, vol. 10, no. 7, July 1991.

Knapp, D.W. and Parker, A.C., A design utility manager: the ADAM planning engine, *23rd ACM/IEEE Design Automation Conf.*, pp. 48–54, 1986.

Kocourek, C., An architecture for process modeling and execution support, *Comput. Aided Syst. Theor.—* EUROCAST, 1995.

Kocourek, C., Planning and execution support for design process, *IEEE Interantional symposium and workshop on systems engineering of computer based system proceedings*, 1995.

Knutilla, A., Schlenoff, C., Ray, S., Polyak, S.T., Tate, A., Chiun Cheah, S., and Anderson, R.C., Process specification language: an analysis of existing representations, NISTIR 6160, National Institute of Standards and Technology, Gaithersburg, MD, 1998.

Lavana, H., Khetawat, A., Brglez, F., and Kozminski, K., Executable workflows: a paradigm for collaborative design on the Internet, *34th ACM/IEEE Design Automation Conf.*, June 1997.

Lander, S.E., Staley, S.M., and Corkill, D.D., Designing integrated engineering environments: blackboard-based integration of design and analysis tools, *Proc. IJCAI-95 Workshop Intelligent Manuf. Syst.*, AAAI, 1995.

Lyons, K., RaDEO Project Overview, http://www.cs.utah.edu/projects/alpha1/arpa/mind/index.html.

Malone, T.W., Crowston, K., Lee, J., Pentland, B.T., Dellarocas, C., Wyner, G., Quimby, J., Osborne, C., Bernstein, A., Herman, G., Klein, M., and O'Donnell, E., in press.

OASIS Users Guide and Reference Manual, MCNC, Research Triangle Park, North Carolina, 1992.

Petrie, C.J., Agent Based Engineering, the Web, and Intelligence, *IEEE Expert*, Dec. 1996.

Rastogi, P., Koziki, M., and Golshani, F., ExPro-an expert system based process management system, *IEEE Trans. Semiconductor Manuf.*, vol. 6, no. 3, pp. 207–218.

Schlenoff, C., Knutilla, A., and Ray, S., Unified process specification language: requirements for modeling process, *NISTIR 5910,* National Institute of Standards and Technology, Gaithersburg, Maryland, 1996.

Schurmann, B. and Altmeyer, J., Modeling design tasks and tools—the link between product and flow model, *Proc. 34th ACM/IEEE Design Automation Conf.*, June 1997.

Sutton, P.R. and Director, S.W., Framework encapsulations: a new approach to CAD tool interoperability, *35th ACM/IEEE Design Automation Conf.*, June 1998.

Sutton, P.R. and Director, S.W., A description language for design process management, *33rd ACM/IEEE Design Automation Conf.*, pp. 175–180, June 1996.

Spiller, M.D. and Newton, A.R., EDA and Network, *ICCAD*, pp. 470–475, 1997.

Stavas, J. et al., Workflow modeling for implementing complex, CAD-based, design methodologies.

Toye, G., Cutkosky, M.R., Leifer, L.J., Tenenbaum, J.M., and Glicksman, J., SHARE: a methodology and environment for collaborative product development, *Proc. Second Workshop Enabling Technol.: Infrastruct. Collaborative Enterprises*, Los Alamitos, California, IEEE Computer Society Press, pp.33–47, 1993.

Vogel, A. and Duddy, K., *Java Programming with CORBA*, Wiley Computer Publishing, New York.

Welsh, J., Kalathil, B., Chanda, B., Tuck, M.C., Selvidge, W., Finnie, E., and Bard, A., Integrated process control and data management in RASSP enterprise system, *Proc. of 1995 RASSP Conf.*, 1995.

Westfechtel, B., Integrated product and process management for engineering design applications, *Integr. Comput.-Aided Eng.*, vol. 3, no. 1, pp. 20–35, 1996.

Yang, Z. and Duddy, K., CORBA: a platform for distributed object computing *ACM Operating Syst. Rev.*, vol. 30, no. 2, pp. 4–31, 1996.

76

System-Level Design

Alice C. Parker
Yosef Tirat-Gefen and
Suhrid A. Wadekar
University of Southern California

CONTENTS

76.1 Introduction

A *system* is a collection of interdependent operational components that together accomplish a complex task. Examples of systems range from cellular phones to camcorders to satellites. Projections point to a continuous increase in the complexity of systems in the coming years.

The term system, when used in the digital design domain, connotes many different entities. A system can consist of a processor, memory, and input/output, all on a single integrated circuit (IC), or can consist of a network of processors, geographically distributed, all performing a specific application. There can be a single clock, with modules communicating synchronously and multiple clocks with asynchronous communication or entirely asynchronous operation. The design can be general purpose, or specific to a given application—i.e., application-specific. The above variations together constitute the *system style*. System style selection is determined to a great extent by the physical technologies used, the environment in which the system operates, designer experience, and corporate culture, and is not automated to any great extent.

System-level design covers a wide range of design activities and design situations. It includes the more specific activity *system engineering*, which involves the requirements, development, test planning, sub-system interfacing, and end-to-end analysis of systems. System-level design is sometimes called *system architecting*, a term used widely in the aerospace industry.

General-purpose system-level design involves the design of programmable digital systems including the basic modules containing storage, processors, input/output, and system controllers. At the system level, the design activities include determining the following:

- the *power budget* (the amount of power allocated to each module in the system),
- the cost and performance budget allocated to each module in the system,
- the interconnection strategy,
- the selection of commercial off-the-shelf (COTS) modules,
- the packaging of each module,
- the overall packaging strategy,
- the number of processors, storage units, and input/output interfaces required, and
- the overall characteristics of each processor, storage unit, and I/O interface.

For example, memory system design focuses on the number of memory modules required, how they are organized, and the capacity of each module. A specific system-level decision in this domain can be how to partition the memory between the processor chip and the off-chip memory. At a higher level, a similar decision might involve configuration of the complete storage hierarchy, including memory, disk drives, and archival storage.

For each general-purpose system designed, many more systems are designed to perform specific applications. *Application-specific system design* involves the same activities as described above, but can involve many more decisions, since there are usually more custom logic modules involved. Specifications for application-specific systems contain not only requirements on general capabilities, but also contain the functionality required in terms of specific tasks to be executed. Major application-specific system-level design activities include not only the above general-purpose system design activities, but also the following activities:

- partitioning an application into multiple functional modules,
- scheduling the application tasks on shared functional modules,
- allocating functional modules to perform the application tasks,
- allocating and scheduling storage modules to contain blocks of data as they are processed,
- determining the implementation styles of functional modules,
- determining the word lengths of data necessary to achieve a given accuracy of computation, and
- predicting resulting system characteristics once the system design is complete.

Each of the system design tasks given in the two lists above will be described in detail below. Since the majority of system design activities are application specific, this section will focus on system-level design of application-specific systems. Related activities, hardware–software codesign, verification, and simulation are covered in other sections.

76.1.1 Design Philosophies and System-Level Design

Many design tools have been constructed with a *top-down design* philosophy. Top-down design represents a design process whereby the design becomes increasingly detailed until final implementation is complete. Considerable prediction of resulting system characteristics is required to make the higher-level decisions with some degree of success.

Bottom-up design, on the other hand, relies on designing a set of primitive elements, and then forming more complex modules from those elements. Ultimately, the modules are assembled into a system. At each stage of the design process, there is complete knowledge of the parameters of the lower-level elements. However, the lower-level elements may be inappropriate for the tasks at hand.

System designers in industry describe the design process as being much less organized and considerably more complex than the top-down and bottom-up philosophies suggest. There is a mixture of top-down and bottom-up activities with major bottlenecks of the system receiving detailed design consideration while other parts of the system still exist only as abstract specifications. For this reason, the system-level design activities we present in detail here support such a complex design situation. Modules, elements, and components used to design at the system level might exist, or might only exist as abstract estimates along with requirements. The system can be designed after all modules have been designed and manufactured, prior to any detailed design, or with a mixture of existing and new modules.

76.1.2 The System Design Space

System design, like data path design, is quite straightforward as long as the constraints are not too severe. However, most designs must solve harder problems than problems solved by existing systems. Designers must race to produce working systems faster than competitors, systems that are also less expensive. More variations in design are possible than ever before and such variations require a large *design space* to be explored. The dimensions of the design space (its axes) are system properties such as cost, power, design time, and performance. The design space contains a population of designs, each of which possesses different values of these system properties. There are literally millions of system designs for a given specification, each of which exhibits different cost, performance, power consumption, and design time. Straightforward solutions that do not attempt to optimize system properties are easy to obtain, but may be inferior to designs that are produced by system-level design tools and have undergone many iterations of design. The complexity of system design is not because system design is an inherently difficult activity, but because so many variations in design are possible and time does not permit exploration of all of them.

76.2 System Specification

Complete system specifications contain a wide range of information, including

- constraints on the system power, performance, cost, weight, size, and delivery time,
- required functionality of the system components,
- any required information about the system structure,
- required communication between system components,
- the flow of data between components,
- the flow of control in the system, and
- the specification of input precision and desired output precision.

Most systems specifications that are reasonably complete exist first in a natural language. However, natural language interfaces are not currently available with commercial system-level design tools.

More conventional system specification methods used to drive system-level design tools include formal languages, graphs, and a mixture of the two. Each of the formal system specification methods described here contains some of the information found in a complete specification, i.e., most specification methods are incomplete. The designer can provide the remaining information necessary for full system design interactively, can be entered later in the design process, or can be provided in other forms at the same time the specification is processed. The required design activities determine the specification method used for a given system design task.

There are no widely adopted formal languages for system-level hardware design although System-Level Design Language (SLDL) was developed by an industry group. Hardware descriptive languages such as VHDL [1] and Verilog [2] are used to describe the functionality of modules in an application-specific system. High-level synthesis tools can then synthesize such descriptions to produce register-transfer designs. Extensions of VHDL and Verilog have been proposed to encompass more system-level design properties. Apart from system constraints, VHDL specifications can form complete system descriptions. However, the level of detail required in VHDL and to some extent in Verilog requires the designer to make some implementation decisions. In addition, some information that is explicit in more abstract specifications such as the flow of control between tasks is implicit in HDLs.

Graphical tools have been used for a number of years to describe system behavior and structure. *Block diagrams* are often used to describe system structure. Block diagrams assume that tasks have already been assigned to basic blocks and their configuration in the system has been specified. Block diagrams generally cannot represent the flow of data or control, or design constraints. The processor memory switch (PMS) notation invented by Bell and Newell was an early attempt to formalize the use of block diagrams for system specification [3].

Petri nets have been used for many years to describe system behavior using a *token-flow* model. A token-flow model represents the flow of control with tokens, which flow from one activity of the system to another. Many tokens can be active in a given model concurrently, representing asynchronous activity and parallelism, important in many system designs. Timed Petri nets have been used to model system performance, but Petri nets cannot easily be used to model other system constraints, system behavior, or any structural information.

State diagrams and graphical tools such as *State Charts* [4] provide alternative methods for describing systems. Such tools provide mechanisms to describe the flow of control, but do not describe system constraints, system structure, data flow, or functionality.

Task-flow graphs, an outgrowth from the control/data-flow graphs (CDFG) used in high-level synthesis are often used for system specification. These graphs describe the flow of control and data between tasks. When used in a hierarchical fashion, task nodes in the task-flow graph can contain detailed functional information about each task, often in the form of a CDFG. Task flow graphs contain no mechanisms for describing system constraints or system structure.

Spec charts [5] incorporate VHDL descriptions into state-chart-like notation, overcoming the lack of functional information found in state charts.

Figure 76.1 illustrates the use of block diagrams, Petri nets, task-flow graphs, and spec charts.

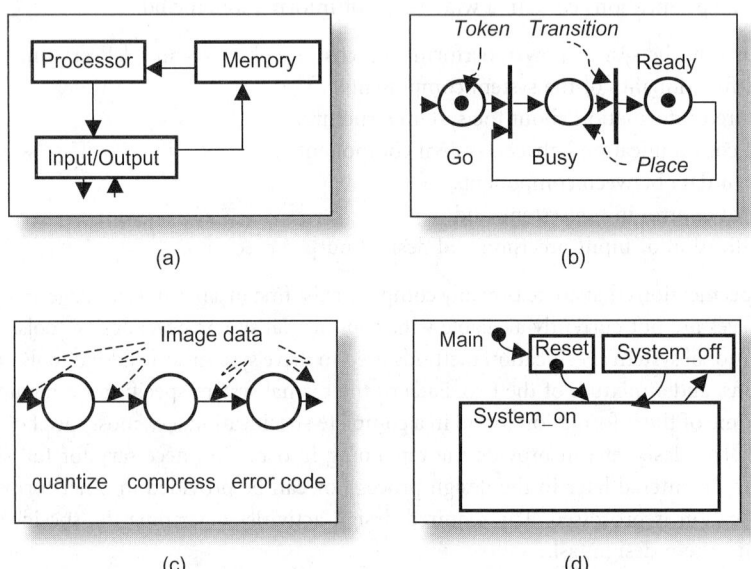

(a) (b)

(c) (d)

FIGURE 76.1 Use of block diagrams, Petri nets, task-flow graphs, and spec charts, shown in simplified form. (a) Block diagram; (b) Petri net; (c) Task-flow graph; and (d) Spec chart.

76.3 System Partitioning

Most systems are too large to fit on a single substrate. If the complexity of the system tasks and the capacities of the system modules are of the same order, then partitioning is not required. All other systems must be partitioned so that they fit into the allowed dies, packages, boards, multichip modules, and cases. *Partitioning determines the functions, tasks, or operations in each partition of a system.* Each partition can represent a substrate, package, multichip module (MCM), or larger component. Partitioning is performed with respect to a number of goals, including minimizing cost, design time or power, or maximizing performance. Any of these goals can be reformulated as specific constraints such as meeting given power requirements.

When systems are partitioned, resulting communication delays must be taken into account, affecting performance. Limitations on interconnection size must be taken into account, affecting performance as well. Pin and interconnection limitations force the multiplexing of inputs and outputs, reducing performance and sometimes affecting cost. Power consumption must also be taken into account. Power balancing between partitions and total power consumption might both be considerations. To meet market windows, system partitions can facilitate the use of COTS, programmable components, or easily fabricated components such as gate arrays. To meet cost constraints, functions that are found in the same partition might share partition resources. Such functions or tasks cannot execute concurrently, affecting performance.

Partitioning is widely used at the logic level as well as on physical designs. In these cases, much more information is known about the design properties and the interconnection structure has been determined. System partitioning is performed when information about properties of the specific components might be uncertain, and the interconnection structure undetermined. For these reasons, techniques used at lower levels must be modified to include predictions of design properties not yet known and prediction of the possible interconnection structure as a result of the partitioning.

The exact partitioning method used depends on the type of specification available. If detailed CDFG or HDL specifications are used, the partitioning method might be concerned with which register-transfer functions (e.g., add, multiply, and shift) are found in each partition. If the specification primitives are tasks, as in a task-flow graph specification, then the tasks must be assigned to partitions. Generally, the more detailed the specification, the larger the size of the partitioning problem. Powerful partitioning methods can be applied to problems of small size ($n < 100$). Weaker methods such as incremental improvement must be used when the problem size is larger.

Partitioning methods can be based on *constructive* partitioning or *iterative improvement*. Constructive partitioning involves taking an unpartitioned design and assigning operations or tasks to partitions. Basic constructive partitioning methods include *bin packing* using a first-fit decreasing heuristic, *clustering* operations into partitions by assigning nearest neighbors to the same partition until the partition is full, *random* placement into partitions, and *integer programming* approaches.

76.3.1 Constructive Partitioning Techniques

Bin packing involves creating a number of bins equal in number to the number of partitions desired and equal in size to the size of partitions desired. One common approach involves the following steps: The tasks or operations are sorted by size. The largest task in the list is placed in the first bin, and then the next largest is placed in the first bin, if it will fit, or else into the second bin. Each task is placed into the first bin in which it will fit, until all tasks have been placed in bins. More bins are added if necessary. This simple heuristic is useful to create an initial set of partitions to be improved iteratively later.

Clustering is a more powerful method to create partitions. Here is a simple clustering heuristic. Each task is ranked by the extent of "connections" to other tasks either owing to control flow, data flow, or physical position limitations. The most connected task is placed in the first partition and then the tasks connected to it are placed in the same partition, in the order of the strength of their connections to the first task. Once the partition is full, the task with the most total connections remaining outside a partition is placed in a new partition, and other tasks are placed there in the order of their connections to the first task. This heuristic continues until all tasks are placed.

Random partitioning places tasks into partitions in a greedy fashion until the partitions are full. Some randomization of the choice of tasks is useful in producing a family of systems, each member of which is partitioned randomly. This family of systems can be used successfully in iterative improvement techniques for partitioning, as described later in this section.

The most powerful technique for constructive partitioning is mathematical programming. Integer and mixed integer-linear programming (MILP) techniques have been used frequently in the past for partitioning. Such powerful techniques are computationally very expensive and are successful only when the number of objects to be partitioned is small. The basic idea behind integer programming used for partitioning is the following: an integer, $TP(i,j)$, is used to represent the assignment of tasks to partitions. When $TP = 1$, task i is assigned to partition j. For each task in this problem, there would be an equation

$$\sum_{j=1}^{\text{Partition total}} TP(i,j) = 1 \qquad (76.1)$$

This equation states that each task must be assigned to one and only one partition. There would be many constraints of this type in the integer program, some of which were inequalities. There would be one function representing cost, performance, or other design property to be optimized. The simultaneous solution of all constraints, given some minimization or maximization goal, would yield the optimal partitioning.

Apart from the computational complexity of this technique, the formulation of the mathematical programming constraints is tedious and error prone if performed manually. The most important advantage of mathematical programming formulations is the discipline it imposes on the CAD programmer in formulating an exact definition of the CAD problem to be solved. Such problem formulations can prove useful when applied in a more practical environment, as described below in the following section.

76.3.2 Iterative Partitioning Techniques

Of the many iterative partitioning techniques available, two have been applied most successfully at the system level. These are *min-cut partitioning*, first proposed by Kernigan and Lin, and *genetic algorithms*.

Min-cut partitioning involves exchanging tasks or operations between partitions to minimize the total amount of "interconnections" cut. The interconnections can be computed as the sum of data flowing between partitions, or as the sum of an estimate of the actual interconnections that will be required in the system. The advantage of summing the data flowing is that is provides a quick computation, since the numbers are contained in the task-flow graph. Better partitions can be obtained if the required physical interconnections are taken into account since they are related more directly to cost and performance than is the amount of data flowing. If a partial structure exists for the design, predicting the unknown interconnections allows partitioning to be performed on a mixed design, one that contains existing parts as well as parts under design.

Genetic algorithms, highly popular for many engineering optimization problems, are especially suited to the partitioning problem. The problem formulation is similar in some ways to mathematical programming formulations. A simple genetic algorithm for partitioning is described here. In this example, a chromosome represents each partitioned system design, and each chromosome contains genes, representing information about the system. A particular gene $TP(i,j)$ might represent the fact that task i is contained in partition j when it is equal to 1, and is set to 0 otherwise. A family of designs created by some constructive partitioning technique then undergoes mutation and crossover as new designs evolve. A *fitness function* is used to check the quality of the design and the evolution is halted when the design is considered fit or when no improvement has occurred after some time. In the case of partitioning, the fitness function might include the estimated volume of interconnections, the predicted cost or performance of the system, or other system properties.

The reader might note some similarity between the mathematical programming formulation of the partitioning problem presented here and the genetic algorithm formulation. This similarity allows the CAD developer to create a mathematical programming model of the problem to be solved, find optimal solutions to small problems, and then create a genetic algorithm version. The genetic algorithm version can be checked against the optimal solutions found by the mathematical program. However, genetic algorithms can take into account many more details than can mathematical program formulations, can handle nonlinear relationships better, and can even handle *stochastic parameters.**

Partitioning is most valuable when there is a mismatch between the sizes of system tasks and the capacities of system modules. When the system tasks and system modules are more closely matched, then the system design can proceed directly to *scheduling* and *allocating* tasks to processing modules.

76.4 Scheduling and Allocating Tasks to Processing Modules

Scheduling and allocating tasks to processing modules involves the determination of how many processing modules are required, which modules execute which tasks, and the order in which the tasks are processed by the system. In the special case where only a single task is processed by each module, the scheduling becomes trivial. Otherwise, if the tasks share modules, the order in which the tasks are processed by the modules can affect system performance or cost. If the tasks are ordered inappropriately, some tasks might wait too long for input data, and performance might be affected. Alternatively, to meet performance constraints, additional modules must be added to perform more tasks in parallel, increasing system cost.

A variety of modules might be available to carry out each task, with differing cost and performance parameters. As each task is allocated to a module, that module is selected from a set of modules available to execute the task. This is analogous to the task *module selection*, which occurs as part of high-level synthesis. For the system design problem considered here, the modules can be general-purpose processors, special-purpose processors (e.g., signal-processing processors), or special-purpose hardware. If all (or most) modules used are general purpose, the systems synthesized are known as heterogeneous application-specific multiprocessors.

A variety of techniques can be used for scheduling and allocation of system tasks to modules. Just as with partitioning, these techniques can be constructive or iterative. Constructive scheduling techniques for system tasks include greedy techniques such as ASAP (as soon as possible) and ALAP (as late as possible). In ASAP scheduling, the tasks are scheduled as early as possible on a free processing module. The tasks scheduled first are the ones with the longest paths from their outputs to final system outputs or system completion. Such techniques, with variations, can be used to provide starting populations of system designs to be further improved iteratively. The use of such greedy techniques for system synthesis differs from the conventional use in high-level synthesis, where the system is assumed to be synchronous, with tasks scheduled into time steps. System task scheduling assumes no central clock, and tasks take a wide range of time to complete. Some tasks could even complete stochastically, with completion time being a random variable. Other tasks could complete basic calculations in a set time, but could perform a finer grain (more accurate) of computations if more time were available. A simple task-flow graph is shown in Figure 76.2, along with a Gantt chart illustrating the ASAP scheduling of tasks onto two processors. Note that two lengthy tasks are performed in parallel with three shorter tasks and that no two tasks take the same amount of time.

---

*Stochastic parameters represent values that are uncertain. There is a finite probability of a parameter taking a specific value that varies with time, but that probability is less than one, in general.

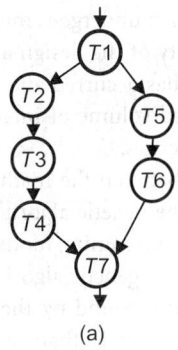

(a)

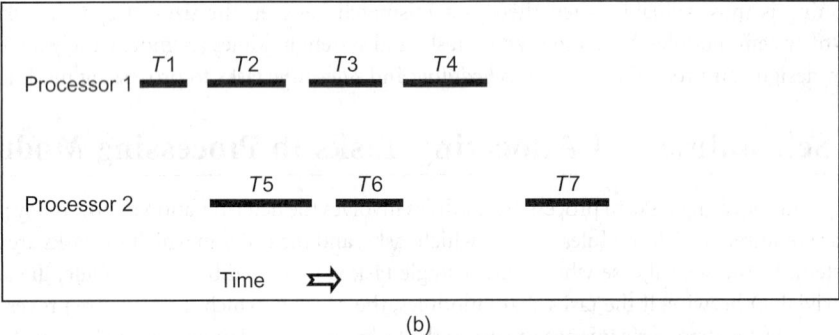

(b)

FIGURE 76.2 An example of task-flow graph and schedule. (a) Task-flow graph; (b) Gantt chart showing schedule.

Similar to partitioning, scheduling, allocation and module selection, can be performed using mathematical programming. In this case, since the scheduling is asynchronous, time becomes a linear rather than integer quantity. Therefore, MILP is employed to model system-level scheduling and allocation. A typical MILP timing constraint is the following:

$$T_{OA}(i) + C_{delay} \leq T_{IR}(j) \tag{76.2}$$

where $T_{OA}(i)$ is the time the output is available from task i, C_{delay} is the communication delay, and $T_{IR}(j)$ is the time the input is required by task j. Unfortunately, the actual constraints used in scheduling and allocation are mostly more complex than this, because the design choices have yet to be made. Here is another example:

$$T_{OA}(i) \geq T_{IR}(i) + \sum_{k} [P_{delay}(k) * M(i,k)] \tag{76.3}$$

This constraint states that the time an output from task i is available is greater than or equal to the time necessary inputs are received by task i, and a processing delay P_{delay} has occurred. $M(i, k)$ indicates that task i is allocated to module k. P_{delay} can take on a range of values, depending on which of the k modules is being used to implement task i. The summation is actually a linearized select function that picks the value of P_{delay} to use depending on which value of $M(i, k)$ is set to 1.

As with partitioning, mathematical programming for scheduling and allocation is computationally intensive, and impractical for all but the smallest designs, but it does provide a baseline model of design that can be incorporated into other tools.

The most frequent technique used for iterative improvement in scheduling and allocation at the system level is a genetic algorithm. The genes can be used to represent task allocation and scheduling. To represent asynchronous scheduling accurately, time is generally represented as a linear quantity in such genes rather than an integer quantity.

76.5   Allocating and Scheduling Storage Modules

In digital systems, all data require some form of temporary or permanent storage. If the storage is shared by several data sets, the use of the storage by each data set must be scheduled. The importance of this task in system design has been overlooked in the past, but has now become an important system-level task.

Modern digital systems usually contain some multimedia tasks and data. The storage requirements for multimedia tasks sometimes result in systems where processing costs are dwarfed by storage costs, particularly caching costs. For such systems, storage must be scheduled and allocated either during or after task scheduling and allocation. If storage is scheduled and allocated concurrently with task scheduling and allocation, the total system costs are easier to determine and functional module sharing can be increased if necessary to control total costs. Alternatively, if storage allocation and scheduling are performed after task scheduling and allocation, then both programs are simpler, but the result may not be as close to optimal.

Techniques similar to those used for task scheduling and allocation can be used for storage scheduling and allocation.

76.6   Selecting Implementation and Packaging Styles for System Modules

Packaging styles can range from single-chip dual-in-line packages (DIPs) to MCMs, boards, racks, and cases. Implementation styles include general-purpose processor, special-purpose programmable processor (e.g., signal processor), COTS modules, field programmable gate arrays (FPGAs), gate array, standard cell, and custom integrated circuits. System cost, performance, power, and design time constraints determine selection of implementation and packaging styles for many system designs. Tight performance constraints favor custom integrated circuits, packaged in MCMs. Tight cost constraints favor off-the-shelf processors and gate array implementations, with small substrates and inexpensive packaging. Tight power constraints favor custom circuits. Tight design time constraints favor COTS modules and FPGAs. If a single design property has high priority, the designer can select the appropriate implementation style and packaging technology. If, however, design time is crucial, but the system to be designed must process video signals in real time, then trade-offs in packaging and implementation style must be made. The optimality of system cost and power consumption might be sacrificed: the entire design might be built with FPGAs, with much parallel processing and at great cost and large size. Because time-to-market is so important, early market entry systems may sacrifice the optimality of many system parameters initially, and then improve them in the next version of the product.

Selection of implementation styles and packaging can be accomplished by adding some design parameters to the scheduling and allocation program, if that program is not already computationally intensive. The parameters added would include a variable indicating that a particular

- functional module was assigned a certain implementation style,
- storage module was assigned a certain implementation style,
- functional module was assigned a certain packaging style, and
- storage module was assigned a certain packaging style.

Some economy of processing could be obtained if certain implementation styles precluded certain packaging styles.

76.7 Interconnection Strategy

Modules in a digital system are usually interconnected in some carefully architected, consistent manner. If point-to-point interconnections are used, they are used throughout the system, or in a subsystem. In the same manner, buses are not broken arbitrarily to insert point-to-point connections or rings. For this reason, digital system design programs usually assume an interconnection style and determine the system performance relative to that style. The most common interconnection styles are bus, point-to-point, and ring.

76.8 Word Length Determination

Functional specifications for system tasks are frequently detailed enough to contain the algorithm to be implemented. To determine the implementation costs of each system task, knowledge of the word widths to be used is important as system cost varies almost quadratically with word width.

Tools to automatically select task word width are currently experimental, but the potential for future commercial tools exists.

In typical hardware implementations of an arithmetic-intensive algorithm, designers must determine the word lengths of resources such as adders, multipliers, and registers. Wadekar and Parker [6] in a recent publication, present algorithm-level optimization techniques to select distinct word lengths for each computation. These techniques meet the desired accuracy and minimize the design cost for the given performance constraints. The cost reduction is possible by avoiding unnecessary bit-level computations that do not contribute significantly to the accuracy of the final results. At the algorithm level, determining the necessary and sufficient precision of an individual computation is a difficult task since the precision of various predecessor/successor operations can be traded off to achieve the same desired precision in the final result. This is achieved using a mathematical model [7] and a genetic selection mechanism [6]. There is a distinct advantage to word-length optimization at the algorithmic level. The optimized operation word lengths can be used to guide high-level synthesis or designers to achieve an efficient utilization of resources of distinct word lengths and costs. Specifically, only a few resources of larger word lengths and high cost may be needed for operations requiring high precision to meet the final accuracy requirement. Other relatively low-precision operations may be executed by resources of smaller word lengths. If there is no timing conflict, a large word length resource can also execute a small word length operation, thus improving the overall resource utilization further. These high-level design decisions cannot be made without the knowledge of word lengths prior to synthesis.

76.9 Predicting System Characteristics

In system-level design, early prediction gives designers the freedom to make numerous high-level choices (such as die size, package type, and latency of the pipeline) with confidence that the final implementation will meet power and energy as well as cost and performance constraints. These predictions can guide power budgeting and subsequent synthesis of various system components, which is critical in synthesizing systems that have low power dissipation, or long battery life. The use by synthesis programs of performance and cost lower bounds allows smaller solution spaces to be searched, which leads to faster computation of the optimal solution.

System cost, performance, power consumption, and design time can be computed if the properties of each system module are known. System design using existing modules requires little prediction. If system design is performed prior to design of any of the contained system modules, however, their properties must be predicted or estimated. Owing to the complexities of prediction techniques, describing these techniques is a subject worthy of an entire chapter. A brief survey of related readings is found in the next section.

The register-transfer and subsequent lower level power prediction techniques such as gate- and transistor-level techniques are essential for validation before fabricating the circuit. However, these techniques are less efficient for system-level design as a design must be generated before prediction can be done.

76.10   A Survey of Research in System Design

Many researchers have investigated the problem of system design, dating back to the early 1970s. This section highlights work that is distinctive, along with tutorial articles covering relevant topics. Much good research is not referenced here, and the reader is reminded that the field is dynamic, with new techniques and tools appearing almost daily.

Issues in top-down versus bottom-up design approaches were highlighted in the design experiment reported by Gupta et al. [8].

76.10.1   System Specification

System specification has received little attention historically except in the specific area of software specifications. Several researchers have proposed natural language interfaces capable of processing system specifications and creating internal representations of the systems that are considerably more structured. Of note is the work by Granacki [9] and Cyre [10]. One noteworthy approach is the design specification language (DSL), found in the design analysis and synthesis environment [11]. One of the few books on the subject concerns the design of embedded systems—systems with hardware and software designed for a particular application set [12]. In one particular effort, Petri nets were used to specify the interface requirements in a system of communicating modules, which were then synthesized [13]. The SIERA system designed by Srivastava, Richards, and Broderson [14] supports specification, simulation, and interactive design of systems.

The Rugby model [15] represents hardware/software systems and the design process, using four dimensions to represent designs: time, computation, communication, and data. da Silva [16] describes the system data structure (SDS), used for internal representation of systems. da Silva also presents an external language for system descriptions called OSCAR. SDS is general and comprehensive, covering behavior and structure. OSCAR is a visual interface to SDS with a formal semantics and syntax based on a visual coordination language paradigm. It is used to capture the behavior and the coordination aspects of concurrent and communicating processes.

SystemC [17] provides hardware-oriented constructs as a class library implemented in standard C++. Although the use of SystemC is claimed to span design and verification from concept to implementation in hardware and software, the constructs provided are somewhat of lower level than most of the system design activities described in this chapter. Extensions to VHDL and Verilog also support some system design activities at the lower levels of system design.

76.10.2   Partitioning

Partitioning research covers a wide range of system design situations. Many early partitioning techniques dealt with assigning register-level operations to partitions. APARTY, a partitioner designed by Lagnese and Thomas, partitions CDFG designs for single-chip implementation to obtain efficient layouts [18]. Vahid [19] performed a detailed survey of techniques for assigning operations to partitions. CHOP assigns CDFG operations to partitions for multi-chip design of synchronous, common clocked systems [20]. Vahid and Gajski developed an early partitioner, SpecPart, which assigns processes to partitions [21]. Chen and Parker reported on a process-to-partition technique called ProPart [22].

76.10.3   Nonpipelined Design

Although research on *system design* spans more than two decades, most of the earlier works focus on single aspects of design like task assignment, and not on the entire design problem. We cite some representative works here. These include graph theoretical approaches to task assignment [23,24], analytical modeling approaches for task assignment [25], and probabilistic modeling approaches for task partitioning [26,27], scheduling [28], and synthesis [29]. Two publications of note cover application of heuristics to system design [30,31].

Other publications of note include mathematical programming formulations for task partitioning [32] and communication channel assignment [33]. Early efforts include those done by soviet researchers since the beginning of the 1970s such as Linsky and Kornev [34] and others, where each model only included a subset of the entire synthesis problem. Chu et al. [35] published one of the first MILP models for a subproblem of system-level design, scheduling. The program Synthesis of Systems (SOS) including a compiler for MILP models [36,37] was developed, based on a comprehensive MILP model for system synthesis. SOS takes a description of a system described using a task-flow graph, a processor library, and some cost and performance constraints, and generates an MILP model to be optimized by an MILP solver. The SOS tool generates MILP models for the design of nonperiodic (nonpipelined) heterogeneous multiprocessors. The models share a common structure, which is an extension of the previous work by Hafer and Parker for high-level synthesis of digital systems [38].

Performance bounds of solutions found by algorithms or heuristics for system-level design are proposed in many papers, including the landmark papers by Fernandez and Bussel [39], Garey and Graham [40], and more recent publications [41].

The work of P. Gupta et al. [8] reported the successful use of system-level design tools in the development of an application-specific heterogeneous multiprocessor for image processing. R. Gupta and Zorian [42] describe the design of systems using cores, silicon cells with at least 5000 gates. The same issue of *IEEE Design and Test of Computers* contains a number of useful articles on the design of embedded core-based systems. Li and Wolf [43] report on a model of hierarchical memory and a multiprocessor synthesis algorithm, which takes into account the hierarchical memory structure.

A major project, RASSP, is a rapid-prototyping approach whose development is funded by the US Department of Defense [44]. RASSP addresses the integrated design of hardware and software for signal-processing applications.

An early work on board-level design, MICON, is of particular interest [45]. Other research results solving similar problems with more degrees of design freedom include the research by C.-T. Chen [46] and D.-H. Heo [47]. GARDEN, written by Heo, finds the design with the shortest estimated time to market that meets cost and performance constraints.

All the MILP synthesis works cited up to this point address only the nonperiodic case.

Synthesis of application-specific heterogeneous multiprocessors is a major activity in the general area of system synthesis. One of the most significant system-level design efforts is Lee's Ptolemy project at the University of California, Berkeley. Representative publications include papers by Lee and Bier describing a simulation environment for signal processing [48] and the paper by Kalavede et al. in 1995 [49]. Another prominent effort is the SpecSyn project of Gajski et al. [50] which is a system-level design methodology and framework.

76.10.4   Macro-Pipelined Design

Macro-pipelined (periodic) multiprocessors execute tasks in a pipelined fashion, with tasks executing concurrently on different sets of data. Most research work on design of *macro-pipelined* multiprocessors has been restricted to homogeneous multiprocessors having negligible communication costs. This survey divides the previous contributions according to the execution mode: preemptive or *nonpreemptive*.

76.10.4.1   Nonpreemptive Mode

The nonpreemptive mode of execution assumes that each task is executed without interruption. It is used quite often in low-cost implementations. Much research has been performed on system scheduling for the nonpreemptive mode. A method to compute the minimum possible value for the initiation interval for a task-flow graph given an unlimited number of processors and no communication costs was found by Renfors and Neuvo [51].

Wang and Hu [52] use heuristics for the allocation and *full static scheduling* (meaning that each task is executed on the same processor for all iterations) of generalized perfect-rate task-flow graphs on homogeneous multiprocessors. Wang and Hu apply planning, an artificial intelligence method, to the task scheduling problem. The processor allocation problem is solved using a *conflict-graph* approach.

Gelabert and Barnwell [53] developed an optimal method to design macro-pipelined homogeneous multiprocessors using *cyclic-static scheduling*, where the task-to-processor mapping is not time-invariant as in the full static case, but is periodic, i.e., the tasks are successively executed by all processors. Gelabert and Barnwell assume that the delays for *intraprocessor* and *interprocessor* communications are the same, which is an idealistic scenario. Their approach is able to find an optimal implementation (minimal iteration interval) in exponential time in the worst case.

Tirat-Gefen [54], in his doctoral thesis, extended the SOS MILP model to solve for optimal macro-pipelined application-specific heterogeneous multiprocessors. He also proposed an ILP model allowing simultaneous optimal retiming and processor/module selection in high- and system-level synthesis [55].

Verhauger [56] addresses the problem of periodic multidimensional scheduling. His thesis uses an ILP model to handle the design of homogeneous multiprocessors without communication costs implementing data-flow programs with nested loops. His work evaluates the complexity of the scheduling and allocation problems for the multidimensional case, which were both found to be NP-complete. Verhauger proposes a set of heuristics to handle both problems.

Passos, Sha, and Bass [57] evaluate the use of multidimensional retiming for synchronous data-flow graphs. However, their formalism can only be applied to homogeneous multiprocessors without communication costs.

76.10.4.2 The Preemptive Mode of Execution

Feng and Shin [58] address the optimal static allocation of periodic tasks with precedence constraints and preemption on a homogeneous multiprocessor. Their approach has an exponential time complexity. Ramamritham [59] developed a heuristic method that has a more reasonable computational cost. **Rate-monotonic scheduling** (RMS) is a commonly used method for allocating periodic real-time tasks in distributed systems [60]. The same method can be used in homogeneous multiprocessors.

76.10.5 Genetic Algorithms

Genetic algorithms are becoming an important tool for solving the highly nonlinear problems related to system-level synthesis. The use of genetic algorithms in optimization is well discussed by Michalewicz [61] where formulations for problems such as bin packing, processor scheduling, traveling salesman, and system partitioning are outlined.

Research works involving the use of genetic algorithms to system-level synthesis problems are starting to be published, as for example the results of

- Hou et al. [62]—scheduling of tasks in a homogeneous multiprocessor without communication costs.
- Wang et al. [63]—scheduling of tasks in heterogeneous multiprocessors with communication costs, but not allowing cost versus performance trade-off, i.e., all processors have the same cost.
- Ravikumar and Gupta [64]—mapping of tasks into a reconfigurable homogeneous array processor without communication costs.
- Tirat-Gefen and Parker [65]—a genetic algorithm for design of application-specific heterogeneous multiprocessors (ASHM) with nonnegligible communications costs specified by a nonperiodic task-flow graph representing both control and data flow.
- Tirat-Gefen [54]—introduced a full-set of genetic algorithms for system-level design of ASHMs incorporating new design features such as imprecise computation and probabilistic design.

76.10.6 Imprecise Computation

The main results in imprecise *computation* theory are due to Liu et al. [66] who developed polynomial time algorithms for optimal scheduling of preemptive tasks on homogeneous multiprocessors without communications costs. Ho et al. [67] proposed an approach to minimize the total error, where the error of a task being imprecisely executed is proportional to the amount of time that its optional part was not allowed to execute, i.e., the time still needed for its full completion. Polynomial time-optimal algorithms were derived for some instances of the problem [66].

Tirat-Gefen et al. [68] presented in 1997 a new approach for ASHM design that allows trade-offs between cost, performance, and data-quality through incorporation of imprecise computation into the system-level design cycle.

76.10.7 Probabilistic Models and Stochastic Simulation

Many probabilistic models for solving different subproblems in digital design have been proposed recently. The problem of task and data-transfer scheduling on a multiprocessor when some tasks (data transfers) have nondeterministic execution times (communication times) can be modeled by PERT networks, which were introduced by Malcolm et al. [69] along with the critical path method (CPM) analysis methodology.

A survey on PERT networks and their generalization to conditional PERT networks is done by Elmaghraby [70]. In system-level design, the completion time of a PERT network corresponds to the *system latency*, whose cumulative distribution function (c.d.f.) is a nonlinear function of the probability density distributions of the computation times of the tasks and the communication times of the data transfers in the task-flow graph.

The exact computation of the cumulative probability distribution function (c.d.f.) of the completion time is computationally expensive for large PERT networks, therefore it is important to find approaches that approximate the value of the expected time of the completion time and its c.d.f. One of the first of these approaches was due to Fulkerson [71], who derived an algorithm in 1962 to find a tight estimate (lower bound) of the expected value of the completion time. Robillard and Trahan [72] proposed a different method using the characteristic function of the completion time in approximating the c.d.f. of the completion time.

Mehrotra et al. [73] proposed a heuristic for estimating the moments of the probabilistic distribution of the system latency t_c. Kulkarni and Adlakha [74] developed an approach based on Markov processes for the same problem. Hagstrom [75] introduced an exact solution for the problem when the random variables modeling the computation and communication times are finite discrete random variables. Kamburowski [76] developed a tight upper bound on the expected completion time of a PERT network.

An approach using random graphs to model distributed computations was introduced by Indurkhya et al. [26], whose theoretical results were improved by Nicol [27]. Purushotaman and Subrahmanyam [77] proposed formal methods applied to concurrent systems with a probabilistic behavior. An example of modeling using queueing networks instead of PERT networks is given by Thomasian and Bay [78]. Estimating errors owing to the use of PERT assumptions in scheduling problems is discussed by Lukaszewicz [79].

Tirat-Gefen developed a set of genetic algorithms using stratified stochastic sampling allowing simultaneous probabilistic optimization of the scheduling and allocation of tasks and communications on ASHM with nonnegligible communication costs [54].

76.10.8 Performance Bounds Theory and Prediction

Sastry [80] developed a stochastic approach for estimation of wireability (routability) for gate arrays. Kurdahi [81] created a discrete probabilistic model for area estimation of VLSI chips designed according to a standard cell methodology. Küçükçakar [82] introduced a method for partitioning of behavioral specifications onto multiple VLSI chips using probabilistic area/performance predictors integrated into a package called BEST (behavioral estimation). BEST provides a range of prediction techniques that can be applied at the algorithm level and includes references to prior research. These predictors provide information required by Tirat-Gefen's system-level probabilistic optimization methods [54].

Lower bounds on the performance and execution time of task-flow graphs mapped to a set of available processors and communication links were developed by Liu and Liu [83] for the case of heterogeneous processors, but no communication costs and by Hwang et al. [84] for homogeneous processors with communication costs. Tight lower bounds on the number of processors and execution time for the case of homogeneous processors in the presence of communication costs were developed by Al-Mouhamed [85]. Yen and Wolf [86] provide a technique for performance estimation for real-time distributed systems.

At the system and register-transfer level, estimating power consumption by the interconnect is important [87]. Wadekar et al. [88] reported "Freedom," a tool to estimate system energy and power that accounts for functional-resource, register, multiplexer, memory, input/output pads, and interconnect power. This tool employees a statistical estimation technique to associate low-level, technology-dependent, physical, and electrical parameters with expected circuit resources and interconnect. At the system level, Freedom generates predictions with high accuracy by deriving an accurate model of the load capacitance for the given target technology—a task reported as critical in high-level power prediction by Brand and Visweswariah [89]. Methods to estimate power consumption prior to high-level synthesis were also investigated by Mehra and Rabaey [90]. Liu and Svensson [91] reported a technique to estimate power consumption in CMOS VLSI chips. The reader is referred to an example of a publication that reports power prediction and optimization techniques at the register-transfer level [92].

76.10.9 Word-Length Selection

Many researchers studied word-length optimization techniques at the register-transfer level. A few example publications are cited here. These techniques can be classified as statistical techniques applied to digital filters [93], simulated annealing-based optimization of filters [94], and simulation-based optimization of filters, digital communication, and signal-processing systems [95]. Sung and Kum reported a simulation-based word-length optimization technique for fixed-point digital signal-processing systems [96]. The objective of these particular architecture-level techniques is to minimize the number of bits in the design that is related to, but not the same as the overall hardware cost.

76.10.10 Embedded Systems

Embedded systems are becoming ubiquitous. The main factor differentiating embedded systems from other electronic systems is the focus of attention on the application rather than on the system as a computing engine. Typically, the I/O in an embedded system is to end users, sensors, and actuators. Sensors provide information on environmental conditions, for example, and actuators control the mechanical portion of the system. In an autonomous vehicle, an embedded system inputs the vehicle's GPS location via a sensor, and outputs control information to actuators that control the acceleration, braking and steering. Embedded systems are typically real-time systems, falling into the classes hard real-time systems, where failure is catastrophic, and soft real-time systems, where failure is not catastrophic.

In an embedded system, there is typically at least one general-purpose processor or micro-controller, and one or more coprocessors that are commercial off-the-shelf chips, DSPs, FPGAs, or custom VLSI chips. The main tasks to be performed for embedded systems are partitioning into hardware/software tasks, assigning tasks to processors, scheduling the tasks, and simulation.

76.10.11 System on Chip (SoC) and Network on Chip (NoC)

The level of integration of microelectronics has allowed entire systems to be fabricated on a single integrated circuit. The physical design of on-chip processing elements called cores can be obtained from vendors. Thus, along with the traditional system design issues, there are issues of intellectual property (IP) for such cores. SoC tend to be used for embedded systems, where hardware/software codesign is a major activity.

Researchers and organizations are also examining the connection of large quantities of cores on a single chip, and proposing to interconnect the cores with various types of interconnection networks. Benini and DeMicheli [97] proposed the use of packet switched on-chip networks, and there has been much publication activity in this area since. Raghavan proposed a hierarchical, heterogeneous approach, and described an alternative network architecture with torus topology and token ring protocol [98].

References

1. *IEEE Standard VHDL Language Reference Manual*, IEEE Std. 1076, IEEE Press, New York, 1987.
2. Bhasker, J., *A Verilog HDL Primer*, Star Galaxy Press, Allentown, PA, 1997.
3. Bell, G. and Newell, A., *Computer Structures: Readings and Examples*, McGraw-Hill, New York, 1971.
4. Harel, D., Statecharts: A visual formalism for complex systems, *Science of Computer Programming*, 8, 231–274, June 1987.
5. Vahid, F., Narayan, S., and Gajski, D.D., SpecCharts: a VHDL front-end for embedded systems, *IEEE Transactions on CAD*, 14, 694, 1995.
6. Wadekar, S.A. and Parker, A.C., Accuracy sensitive word-length selection for algorithm optimization, in *Proceedings of the International Conference on Circuit Design [ICCD]*, 1998, 54.
7. Wadekar, S.A. and Parker, A.C., Algorithm-level verification of arithmetic-intensive application-specific hardware designs for computation accuracy, in *Digest Third International High-Level Design Validation and Test Workshop*, 1998.
8. Gupta, P., Chen, C.T., DeSouza-Batista, J.C., and Parker, A.C., Experience with image compression chip design using unified system construction tools, in *Proceedings of the 31st Design Automation Conference*, 1994.
9. Granacki, J. and Parker, A.C., PHRAN—Span: a natural language interface for system specifications, in *Proceedings of the 24th Design Automation Conference*, 1987, 416.
10. Cyre, W.R., Armstrong, J.R., and Honcharik, A.J., Generating simulation models from natural language specifications, *Simulation*, 65, 239, 1995.
11. Tanir, O. and Agarwal, V.K., A specification-driven architectural design environment, *Computer*, 6, 26–35, 1995.
12. Gajski, D.D., Vahid, F., Narayan, S., and Gong, J., *Specification and Design of Embedded Systems*, Prentice Hall, Englewood Cliffs, NJ, 1994.
13. de Jong, G. and Lin, B., A communicating Petri net model for the design of concurrent asynchronous modules, *ACM/IEEE Design Automation Conference*, June 1994.
14. Srivastava, M.B., Richards, B.C., and Broderson, R.W., System-level hardware module generation, *IEEE Transactions on Very Large-Scale Integration [VLSI] Systems*, 3, 20, 1995.
15. Jantsch, A., Kumar, S., and Hemani, A., The RUGBY model: a conceptual frame for the study of modeling, analysis and synthesis concepts of electronic systems, in *Proceedings of the Conference on Design, Automation and Test in Europe*, Munich, Germany, 72–78, 1999.
16. da Silva, D., Jr., A comprehensive framework for the specification of hardware/software systems, Ph.D. Dissertation, University of Southern California, Department of Electrical Engineering, December 2001.
17. Grötker, T., Liao, S., Martin, G., and Swan, S., *System Design with SystemC*, Kluwer Academic Publishers, Boston, 2002.
18. Lagnese, E. and Thomas, D., Architectural partitioning for system level synthesis of integrated circuits, *IEEE Transactions on Computer-Aided Design*, 1991.
19. Vahid, F., A survey of behavioral-level partitioning systems, Technical Report TR ICS 91-71, University of California, Irvine, 1991.
20. Kucukcakar, K. and Parker, A.C., Chop: a constraint-driven system-level partitioner, in *Proceedings of the 28th Design Automation Conference*, 1991, 514.
21. Vahid, F. and Gajski, D.D., Specification partitioning for system design, in *Proceedings of the 29th Design Automation Conference*, 1992.
22. Parker, A.C., Chen, C.-T., and Gupta, P., Unified system construction, in *Proceedings of the SASIMI Conference*, 1993.
23. Bokhari, S.H., Assignment problems in parallel and distributed computing, Kluwer Academic Publishers, Dordrecht, 1987.

24. Stone, H.S. and Bokhari, S.H., Control of distributed processes, *Computer*, 11, 97, 1978.

25. Haddad, E.K., Optimal load allocation for parallel and distributed processing, Technical Report TR 89-12, Department of Computer Science, Virginia Polytechnic Institute and State University, April 1989.

26. Indurkhya, B., Stone, H.S., and Cheng, L.X., Optimal partitioning of randomly generated distributed programs, *IEEE Transactions on Software Engineering*, SE-12, 483, 1986.

27. Nicol, D.M., Optimal partitioning of random programs across two processors, *IEEE Transactions on Software Engineering*, 15, 134, 1989.

28. Lee, C.Y., Hwang, J.J., Chow, Y.C., and Anger, F.D., Multiprocessor scheduling with interprocessor communication delays, *Operations Research Letters*, 7, 141, 1988.

29. Tirat-Gefen, Y.G., Silva, D.C., and Parker, A.C., Incorporating imprecise computation into system-level design of application-specific heterogeneous multiprocessors, in *Proceedings of the 34th Design Automation Conference*, 1997.

30. DeSouza-Batista, J.C., and Parker, A.C., Optimal synthesis of application-specific heterogeneous pipelined multiprocessors, in *Proceedings of the International Conference on Application-Specific Array Processors*, 1994.

31. Mehrotra, R. and Talukdar, S.N., Scheduling of tasks for distributed processors, Technical Report DRC-18-68-84, Design Research Center, Carnegie-Mellon University, December 1984.

32. Agrawal, R. and Jagadish, H.V., Partitioning techniques for large-grained parallelism, *IEEE Transactions on Computers*, 37, 1627, 1988.

33. Barthou, D., Gasperoni, F., and Schwiegelshon, U., Allocating communication channels to parallel tasks, *Environments and Tools for Parallel Scientific Computing*, Elsevier Science Publishers B.V., Amsterdam, 1993, 275.

34. Linsky, V.S. and Kornev, M.D., Construction of optimum schedules for parallel processors, *Engineering Cybernetics*, 10, 506, 1972.

35. Chu, W.W., Hollaway, L.J., and Efe, K., Task allocation in distributed data processing, *Computer*, 13, 57, 1980.

36. Prakash, S. and Parker, A.C., SOS: synthesis of application-specific heterogeneous multiprocessor systems, *Journal of Parallel and Distributed Computing*, 16, 338, 1992.

37. Prakash, S., Synthesis of application-specific multiprocessor systems, PhD Thesis, Department of Electrical Engineering and Systems, University of Southern California, January 1994.

38. Hafer, L. and Parker, A., Automated synthesis of digital hardware, *IEEE Transactions on Computers*, C-31, 93, 1981.

39. Fernandez, E.B. and Bussel, B., Bounds on the number of processors and time for multiprocessor optimal schedules, *IEEE Transactions on Computers*, C-22, 745, 1975.

40. Garey, M.R. and Graham, R.L., Bounds for multiprocessor scheduling with resource constraints, *SIAM Journal of Computing*, 4, 187, 1975.

41. Jaffe, J.M., Bounds on the scheduling of typed task systems, *SIAM Journal of Computing*, 9, 541, 1991.

42. Gupta, R. and Zorian, Y., Introducing core-based system design, *IEEE Design and Test of Computers*, October–December, 15, 1997.

43. Li, Y. and Wolf, W., A task-level hierarchical memory model for system synthesis of multiprocessors, in *Proceedings of the Design Automation Conference*, 1997, 153.

44. *IEEE Design and Test*, Vol. 13, no. 3, Fall, 1996.

45. W. Birmingham, and D. Siewiorek, MICON: a single board computer synthesis tool, in *Proceedings of the 21st Design Automation Conference*, 1984.

46. Chen, C-T., System-level design techniques and tools for synthesis of application-specific digital systems, Ph.D. thesis, Department of Electrical Engineering and Systems, University of Southern California, January 1994.

47. Heo, D.H., Ravikumar, C.P., and Parker, A., Rapid synthesis of multi-chip systems, *in Proceedings of the 10th International Conference on VLSI Design*, 1997, 62.

48. Lee, E.A., and Bier, J.C., Architectures for statically scheduled dataflow, *Journal of Parallel and Distributed Computing*, 10, 1990, 333–348.

49. Kalavede, A., Pino, J.L., and Lee, E.A., Managing complexity in heterogeneous system specification, simulation and synthesis, *Proceedings of the International Conference on Acoustics, Speech and Signal Processing (ICASSP)*, May, 1995.

50. D.D. Gajski, F. Vahid, and S. Narayan, A design methodology for system-specification refinement, in *Proceedings of the European Design Automation Conference*, 1994, 458.

51. Renfors, M. and Neuvo, Y., The maximum sampling rate of digital filters under hardware speed constraints, *IEEE Transactions on Circuits and Systems*, CAS-28, 196, 1981.

52. Wang, D.J. and Hu, Y.H., Multiprocessor implementation of real-time DSP algorithms, *IEEE Transactions on Very Large-Scale Integration (VLSI) Systems*, 3, 393, 1995.

53. Gelabert, P.R., Barnwell, T.P., Optimal automatic periodic multiprocessor scheduler for fully specified flow graphs, *IEEE Transactions on Signal Processing*, 41, 858, 1993.

54. Tirat-Gefen, Y.G., Theory and practice in system-level of application-specific heterogeneous multiprocessors, Ph.D. dissertation, Dept. of Electrical and Computer Engineering, University of Southern California, Los Angeles, 1997.

55. CasteloVide-e-Souza, Y.G., Potkonjak, M. and Parker, A.C., Optimal ILP-based approach for throughput optimization using algorithm/architecture matching and retiming, *Proceedings of the 32nd Design Automation Conference*, June 1995.

56. Verhauger, W.F., Multidimensional periodic scheduling, Ph.D. thesis, Eindhoven University of Technology, Holland, 1995.

57. Passos, N.L., Sha, E.H., and Bass S.C., Optimizing DSP flow-graphs via schedule-based multidimensional retiming, *IEEE Transaction on Signal Processing*, 44, 150, 1996.

58. Feng, D.T. and Shin, K.G., Static allocation of periodic tasks with precedence constraints in distributed real-time systems, in *Proceedings of the 9th International Conference of Distributed Computing*, 1989, 190.

59. Ramamritham, K., Allocation and scheduling of precedence-related periodic tasks, *IEEE Transactions on Parallel and Distributed Systems*, 6, 1995.

60. Ramamritham, K., Stankovic, J.A., and Shiah, P.F., Efficient scheduling algorithms for real-time multiprocessors systems, *IEEE Transaction on Parallel and Distributed Systems*, 1, 184, 1990.

61. Michalewicz, Z., *Genetic Algorithms + Data Structures = Evolution Programs*, Springer, Berlin, 1994.

62. Hou, E.S.H, Ansari, N., and Ren, H., A Genetic algorithm for multiprocessor scheduling, *IEEE Transactions on Parallel and Distributed Systems*, 5, 113, 1994.

63. Wang, L., Siegel, H.J., and Roychowdhury, V.P., A genetic-algorithm-based approach for task matching and scheduling in heterogeneous computing environments, in *Proceedings of the Heterogeneous Computing Workshop, International Parallel Processing Symposium*, 1996, 72.

64. Ravikumar, C.P. and Gupta, A., Genetic algorithm for mapping tasks onto a reconfigurable parallel processor, *IEEE Proceedings in Computing Digital Technology*, 142, 81, 1995.

65. Tirat-Gefen, Y.G. and Parker, A.C., MEGA: an approach to system-level design of application-specific heterogeneous multiprocessors, in *Proceedings of the Heterogeneous Computing Workshop, International Parallel Processing Symposium*, 1996, 105.

66. Liu, J.W.S., Lin, K.-J., Shih, W.-K., Yu, A.C.-S., Chung, J.-Y., and Zhao, W., Algorithms for scheduling imprecise computations, *IEEE Computer*, 24, 58, 1991.

67. Ho, K., Leung, J.Y-T., and Wei, W-D., Minimizing maximum weighted error for imprecise computation tasks, Technical Report UNL-CSE-92-017, Department of Computer Science and Engineering, University of Nebraska, Lincoln, 1992.

68. Tirat-Gefen, Y.G., Silva, D.C., and Parker, A.C., Incorporating imprecise computation into system-level design of application-specific heterogeneous multiprocessors, in *Proceedings of the 34th Design Automation Conference*, 1997.

69. Malcolm, D.G., Roseboom, J.H., Clark, C.E., and Fazar, W., Application of a technique for research and development program evaluation, *Operations Research*, 7, 646, 1959.

70. Elmaghraby, S.E., The theory of networks and management science: part II, *Management Science*, 17, B.54, 1970.

71. Fulkerson, D.R., Expected critical path lengths in pert networks, *Operations Research*, 10, 808, 1962.

72. Robillard, P. and Trahan, M., The completion time of PERT networks, *Operations Research*, 25, 15, 1977.

73. Mehrotra, K., Chai, J., and Pillutla, S., A study of approximating the moments of the job completion time in PERT networks, Technical Report, School of Computer and Information Science, Syracuse University, New York, 1991.

74. Kulkarni, V.G. and Adlakha, V.G., Markov and Markov-regenerative pert networks, *Operations Research*, 34, 769, 1986.

75. Hagstrom, J.N., Computing the probability distribution of project duration in a PERT network, *Networks*, 20, 1990, 231.

76. Kamburowski, J., An upper bound on the expected completion time of PERT networks, *European Journal of Operational Research*, 21, 206, 1985.

77. Purushothaman, S. and Subrahmanyam, P.A., Reasoning about probabilistic behavior in concurrent systems, *IEEE Transactions on Software Engineering*, SE-13, 740, 1987.

78. Thomasian, A., Analytic queueing network models for parallel processing of task systems, *IEEE Transactions on Computers*, C-35, 1045, 1986

79. Lukaszewicz, J., On the estimation of errors introduced by standard assumptions concerning the distribution of activity duration in PERT calculations, *Operations Research*, 13, 326, 1965.

80. Sastry, S. and Parker, A.C., Stochastic models for wireability analysis of gate arrays, *IEEE Transactions on Computer-Aided Design*, CAD-5, 1986.

81. Kurdahi, F.J., Techniques for area estimation of VLSI layouts, *IEEE Transaction on Computer-Aided Design*, 8, 81, 1989.

82. Küçükçakar, K. and Parker, A.C., A methodology and design tools to support system-level VLSI design, *IEEE Transactions on Very Large-Scale Integration [VLSI] Systems*, 3, 355, 1995.

83. Liu, J.W.S. and Liu, C.L., Performance analysis of multiprocessor systems containing functionally dedicated processors, *Acta Informatica* 10, 95, 1978.

84. Hwang, J.J., Chow, Y.C., Ahnger, F.D., and Lee, C.Y., Scheduling precedence graphs in systems with interprocessor communication times, *SIAM Journal of Computing*, 18, 244, 1989.

85. Mouhamed, M., Lower bound on the number of processors and time for scheduling precedence graphs with communication costs, *IEEE Transactions on Software Engineering*, 16, 1990.

86. Yen, T.-Y. and Wolf, W., Performance estimation for real-time embedded systems, in *Proceedings of the International Conference on Computer Design*, 1995, 64.

87. Landman, P.E. and Rabaey, J.M., Activity-sensitive architectural power analysis, *IEEE Trans. on CAD*, 15, 571, 1996.

88. Wadekar, S.A., Parker, A.C., and Ravikumar, C.P., FREEDOM: Statistical behavioral estimation of system energy and power, in *Proceedings of the Eleventh International Conf. on VLSI Design*, 1998, 30.

89. D. Brand, and C. Visweswariah, Inaccuracies in power estimation during logic synthesis, in *Proceedings of the European Design Automation Conference (EURO-DAC)*, 1996, 388.

90. Mehra, R. and Rabaey, J., Behavioral level power estimation and exploration, in *Proceedings of the First International Workshop on Low Power Design*, 1994, 197.

91. Liu, D. and Svensson, C., Power consumption estimation in CMOS VLSI chips, *IEEE Journal of Solid-State Circuits*, 29, 663, 1994.

92. Landman, P.E. and Rabaey, J.M., Activity-sensitive architectural power analysis, *IEEE Transactions on Computer-Aided Design*, 15, 571, 1996.

93. Zeng, B. and Neuvo, Y., Analysis of floating point roundoff errors using dummy multiplier coefficient sensitivities, *IEEE Transactions on Circuits and Systems*, 38, 590, 1991.

94. Catthoor, F., Vandewalle, J., and De Mann, H., Simulated annealing based optimization of coefficient and data word lengths in digital filters, *International Journal of Circuit Theory Applications*, 16, 371, 1988.

95. Grzeszczak, A., Mandal, M.K., Panchanathan, S., and Yeap, T., VLSI implementation of discrete wavelet transform, *IEEE Transactions on VLSI Systems*, 4, 421, 1996.

96. Sung, W. and Kum, Ki-II., Simulation-based word-length optimization method for fixed-point digital signal processing systems, *IEEE Transactions on Signal Processing*, 43, 3087, 1995.

97. Benini, L. and De Micheli, G., Networks on chip: a new paradigm for systems on chip design, *in Proceedings of the Conference on Design, Automation and Test in Europe*, 2002, 418–419.

98. Raghavan, D., *Extending the design space for networks on chip*, M.S. thesis, University of Southern California Electrical Engineering Department, May 2004.

77

Performance Modeling and Analysis Using VHDL and SystemC

Robert H. Klenke
Virginia Commonwealth University

Jonathan A. Andrews
Virginia Commonwealth University

James H. Aylor
University of Virginia

CONTENTS

77.1   Introduction

It has been noted by the digital design community that the greatest potential for additional cost and iteration cycle time savings is through improvements in tools and techniques that support the early stages of the design process [1]. As shown in Figure 77.1, decisions made during the initial phases of a product's development cycle determine up to 80% of its total cost. The result is that accurate, fast analysis tools must be available to the designer at the early stages of the design process to help make these decisions. Design alternatives must be effectively evaluated at this level with respect to multiple metrics, such as performance, dependability, and testability. This analysis capability will allow a larger portion of the design space to be explored yielding higher quality as well as lower cost designs.

In 1979, Hill and vanCleemput [2] of the SABLE environment identified three or four stages or phases of design, and noted that each traditionally uses their own simulator. While the languages used have changed and some of the phases can be simulated together these phases still exist in most industry design flows: first an initial high-level simulation, then an intermediate level roughly at the instruction set simulation level, and then a gate-level phase. The authors add a potential fourth stage to deal with custom integrated circuit (IC) simulation. While acknowledging the potential for more efficient simulation at a given level due to potential optimization, they point out five key disadvantages to having different simulators and languages for the various levels. They are:

1. Design effort is multiplied by the necessity of learning several simulator systems and recoding a design in each.
2. The possibility of error is increased as more human manipulation enters the system.
3. Each simulator operates at just one level of abstraction. Because it is impossible to simulate an entire computer at a low level of abstraction, only small fragments can be simulated at any one time.
4. Each fragment needs to be driven by a supply of realistic data and its output needs to be interpreted. Often, writing the software to serve these needs is more effortful than developing the design itself.
5. As the design becomes increasingly fragmented to simulate it, it becomes difficult for any designer to see how his own particular piece fits into the overall function of the system.

While SABLE was developed almost three decades ago, it is worth noting that today designers are still struggling to address the same basic issues expressed above. Much of the industry has at least two or three design phases requiring separate models and simulators, which still results in a multiplication of design effort. This multiplication of design effort still increases the possibility of error due to inherently fallible human manipulation. While it may not be impossible to simulate an entire computer at one low level of

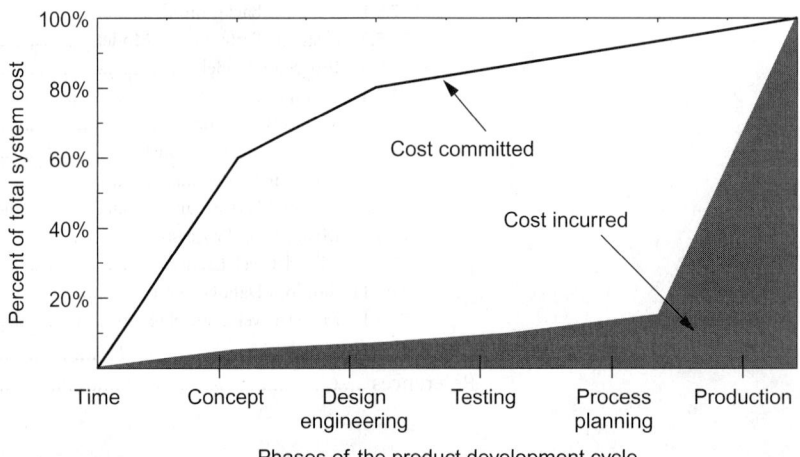

FIGURE 77.1   Product costs over the development cycle.

abstraction, it is still impractical and inefficient. Thus, designs are often still fragmented for development and simulation, these fragments still need realistic data, and generating it is still time-consuming. Designers working on a fragment often have difficulty seeing how his or her piece fits into overall function, and this often leads to differing assumptions between fragments and expensive design revision.

There are a number of current tools and techniques that support analysis of these metrics at the system level to varying degrees. A major problem with these tools is that they are not integrated into the engineering design environment in which the system will ultimately be implemented. This problem leads to a major disconnect in the design process. Once the system-level model is developed and analyzed, the resulting high-level design is specified on paper and thrown "over the wall" for implementation by the engineering design team, as illustrated in Figure 77.2. As a result, the engineering design team has to often interpret this specification to implement the system, which often leads to design errors. It also has to develop their own initial "high-level" model from which to begin the design process in a top down manner. Additionally, there is no automated mechanism by which feedback on design assumptions and estimations can be provided to the system design team by the engineering design team.

For systems that contain significant portions of both application-specific hardware and software executing on embedded processors, design alternatives for competing system architectures and hardware/ software (HW/SW) partitioning strategies must be effectively and efficiently evaluated using high-level performance models. Additionally, the selected hardware and software system architecture must be refined in an integrated manner from the high-level models to an actual implementation to avoid implementation mistakes and the associated high redesign costs. Unfortunately, most existing design environments lack the ability to model and design a system's hardware and software in the same environment. A similar wall to that between the system design environment and the engineering design environment exists between the hardware and the software design environments. This results in a design path as shown in Figure 77.3, where the hardware and software design process begins with a common system requirement and specification, but proceeds through a separate and isolated design process until final system integration. At this point, assumptions on both sides may prove to be drastically wrong resulting in incorrect system function and poor system performance.

A unified, cooperative approach in which the hardware and software options can be considered together is required to increase the quality and decrease the design time for complex HW/SW systems. This approach is called *hardware/software codesign*, or simply *codesign* [3–5]. Codesign leads to more efficient implementations and improves overall system performance, reliability, and cost effectiveness [5]. Also, because decisions regarding the implementation of functionality in software can

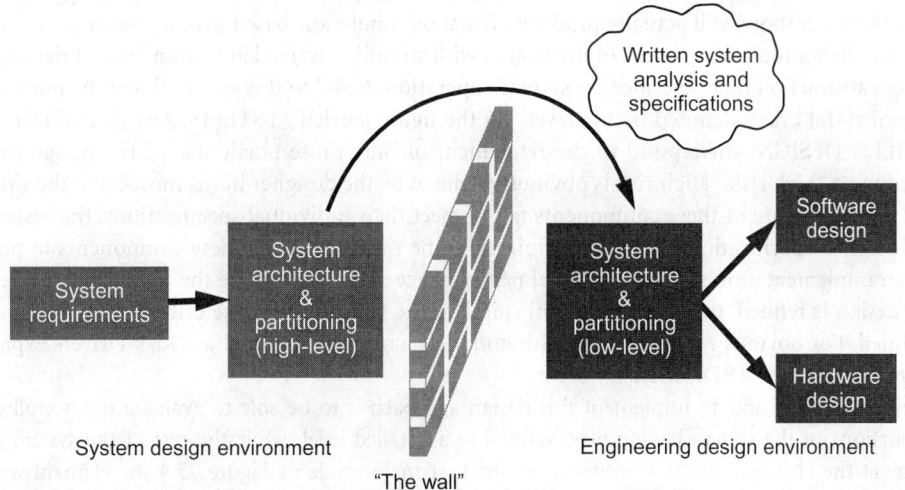

FIGURE 77.2 The disconnect between system-level design environments and engineering design environments.

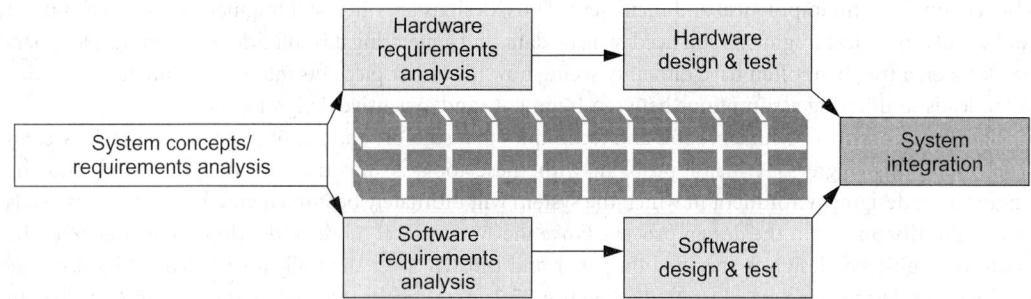

FIGURE 77.3 Current hardware/software system development methodology.

impact hardware design (and vice versa), problems can be detected and changes made earlier in the development process [6].

Codesign can especially benefit the design of *embedded systems* [7], which contain hardware and software tailored for a particular application. As the complexity of these systems increases, the issue of providing design approaches that scale up to more complicated systems becomes of greater concern. A detailed description of a system can approach the complexity of the system itself [8], and the amount of detail present can make analysis intractable. Therefore, decomposition techniques and abstractions are necessary to manage this complexity.

What is needed is a design environment in which the capability for performance modeling of HW/ SW systems at a high level of abstraction is fully integrated into the engineering design environment. To completely eliminate the "over the wall" problem and the resulting model discontinuity, this environment must support the incremental refinement of the abstract system-level performance model into an implementation-level model. Using this environment, a design methodology based on incremental refinement can be developed.

The design methodology illustrated in Figure 77.4 was proposed by Lockheed Martin Advanced Technology Laboratory as a new way to design systems [9]. This methodology uses the level of the risk of not meeting the design specifications as the metric for driving the design process. In this spiral-based design methodology, there are two iteration cycles. The major cycles (or spirals), denoted as CYCLE 1, CYCLE 2, ..., CYCLE N in the figure, correspond to the design iterations where major architectural changes are made in response to some specification metric(s) and the system as a whole is refined and more design detail is added to the model. Consistent with the new paradigm of system design, these iterations will actually produce virtual or simulation-based prototypes. A virtual proto-type is simply a simulatable model of the system with stimuli described at a given level of design detail or design abstraction that describes the system's operation. Novel to this approach are the mini spirals. The mini spiral cycles denoted by the levels on the figure labeled SYSTEMS, ARCHITECTURE, and DETAILED DESIGN, correspond to the refinement of only those portion(s) of the design that are deemed to be "high risk." High risk is obviously defined by the designer but is most often the situation where if one or more of these components fail to meet their individual specifications, the system will fail to meet its specifications. The way to minimize the risk is to refine these components to possibly make an implementation so that the actual performance is known. Unlike the major cycles where the entire design is refined, the key to the mini spirals is the fact that only the critical portions of design are refined. For obvious reasons, the resulting models have been denoted as "Risk Driven Expanding Information Models" (RDEIMs).

The key to being able to implement this design approach is to be able to evaluate the overall design with portions of the system having been refined to a detailed level, while the rest of the system model remains at the abstract level. For example, in the first major cycle of Figure 77.4 the element with the highest relative risk is fully implemented (detailed design level) while the other elements are described at more abstract levels (system level or architectural level). If the simulation of the model shown in the first

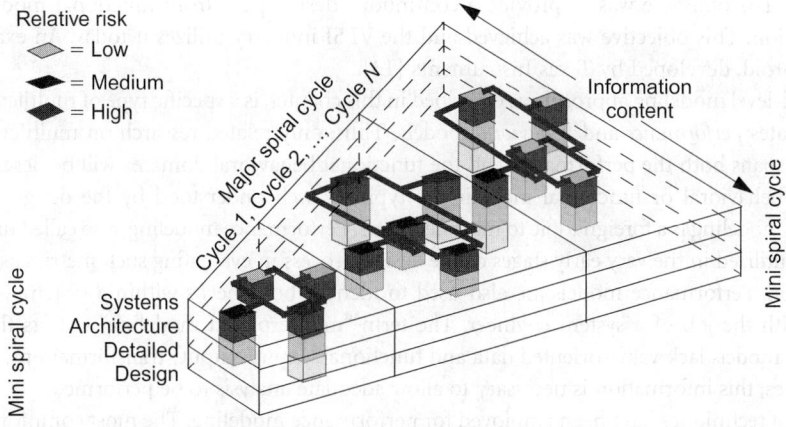

FIGURE 77.4 Risk-Driven Expanding Information Model (RDEIM).

major cycle detects that the overall system will not meet its performance requirements, then the "high risk" processing element could be replaced by two similar elements operating in parallel. This result is shown in the second major cycle and at this point, another element of the system may become the new "bottleneck," i.e., the highest relative risk, and it will be refined in a similar manner.

Implied in the RDEIM approach is a solution to the "over the wall" problem including hardware/software codesign. The proposed solution is to fully integrate performance modeling into the design process.

Obviously, one of the major capabilities necessary to implement a top-down design methodology such as the RDEIM is the ability to cosimulate system models which contain some components that are modeled at an abstract performance level (uninterpreted models) and some that are modeled at a detailed behavioral level (interpreted models). This capability to model and cosimulate uninterpreted models and interpreted models is called *mixed-level modeling* (sometimes referred to as hybrid modeling). Mixed-level modeling requires the development of interfaces that can resolve the differences between uninterpreted models that, by design, do not contain a representation of all of the data or timing information of the final implementation, and interpreted models which require most, or possibly all, data values and timing relationships to be specified. Techniques for systematic development of these mixed-level modeling interfaces and resolution of these differences in abstraction is the focus of some of the latest work in mixed-level modeling.

In addition to the problem of mixed-level modeling interfaces, another issue that may have to be solved is that of different modeling languages being used at different levels of design abstraction. While VHDL, and to some extend, various extensions to Verilog, can be used to model at the system level, many designers constructing these types of models prefer to use a language with a more programming language-like syntax. As a result of this and other factors, the SystemC language [12] was developed. SystemC is a library extension to C++ that builds key concepts from existing hardware description languages (HDLs)—primarily a timing model that allows simulation of concurrent events—into an intrinsically object oriented HDL. Its use of objects, pointers, and other standard C++ items makes it a versatile language. Particularly the ability to describe and use complex data types, and the concepts of interfaces and channels make it much more intuitive for describing abstract behaviors than existing HDLs such as VHDL or Verilog. Since SystemC is C++, existing C or C++ code that models behavior can be imported with little or no modification.

77.1.1 Multilevel Modeling

The need for multilevel modeling was recognized almost three decades ago. Multilevel modeling implies that representations at different levels of detail coexist within a model [8,10,11]. Until the early 1990s, the term multilevel modeling was used for integrating behavioral or functional models with lower level models. The objective was to provide a continuous design path from functional models down to implementation. This objective was achieved and the VLSI industry utilizes it today. An example is the tool called Droid, developed by Texas Instruments [13].

The mixed-level modeling approach, as described in this chapter, is a specific type of multilevel modeling which integrates *performance* and *behavioral* models. Thus, only related research on multilevel modeling systems that spans both the performance and the functional/behavioral domains will be described.

Although behavioral or functional modeling is typically well understood by the design community, performance modeling is a foreign topic to most designers. Performance modeling, also called uninterpreted modeling, is utilized in the very early stages of the design process in evaluating such metrics as throughput and utilization. Performance models are also used to identify bottlenecks within a system and are often associated with the job of a system engineer. The term "uninterpreted modeling" reflects the view that performance models lack value-oriented data and functional (input/output) transformations. However, in some instances, this information is necessary to allow adequate analysis to be performed.

A variety of techniques have been employed for performance modeling. The most common techniques are Petri nets [14–16] and queuing models [17,18]. A combination of these techniques, such as a mixture of Petri nets and queuing models [19], has been utilized to provide more powerful modeling capabilities. All of these models have mathematical foundations. However, models of complex systems constructed using these approaches can quickly become unwieldy and difficult to analyze.

Examples of a Petri net and a queuing model are shown in Figure 77.5. A queuing model consists of queues and servers. Jobs (or customers) arrive at a specific arrival rate and are placed in a queue for service. These jobs are removed from the queue to be processed by a server at a particular service rate. Typically, the arrival and service rates are expressed using probability distributions. There is a queuing discipline, such as first-come-first-serve, which determines the order in which jobs are to be serviced. Once they are serviced, the jobs depart and arrive at another queue or simply leave the system. The number of jobs in the queues represents the model's state. Queueing models have been used successfully for modeling many complex systems. However, one of the major disadvantages of queuing models is their inability to model synchronization between processes.

As a system modeling paradigm, Petri nets overcome this disadvantage of queuing models. Petri nets consist of places, transitions, arcs, and a marking. The places are equivalent to conditions and hold tokens, which represent information. Thus, the presence of a token in the place of a Petri net corresponds to a particular condition being true. Transitions are associated with events, and the "firing" of a transition indicates that some event has occurred. A marking consists of a particular placement of tokens within the places of a Petri net and represents the state of the net. When a transition fires, tokens are removed from the input places and are added to the output places, changing the marking (the state) of the net and allowing the dynamic behavior of a Petri net to be modeled.

Petri nets can be used for performance analysis by associating a time with the transitions. Timed and stochastic Petri nets contain deterministic and probabilistic delays, respectively. Normally, these Petri nets are uninterpreted, since no *interpretation* (values or value transformations) are associated with the

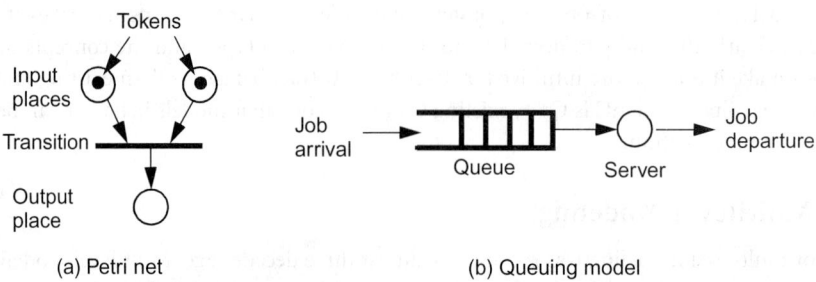

(a) Petri net   (b) Queuing model

FIGURE 77.5   Petri net and queuing model.

tokens or transitions. However, values or value transformations can be associated with the various elements of Petri net models as described below.

Petri nets that have values associated with tokens are known as colored Petri nets (CPNs). In the colored Petri nets, each token has an attached "color," indicating the identity of the token. The net is similar to the basic definition of the Petri net except that a functional dependency is specified between the color of the token and the transition firing action. In addition, the color of the token produced by a transition may be different from the color of the token(s) on the input place(s). Colored Petri nets have an increased ability to efficiently model real systems with small nets which are equivalent to much larger plain Petri nets due to their increased descriptive powers.

Numerous multilevel modeling systems exist based on these two performance modeling techniques. Architecture Design and Assessment System (ADAS) is a set of tools specifically targeted for high-level design [20]. ADAS models both hardware and software using directed graphs based on timed Petri nets. The flow of information is quantified by identifying discrete units of information called tokens. ADAS supports two levels of modeling. The more abstract level is a dataflow description and is used for performance estimation. The less abstract level is defined as a behavioral level but still uses tokens which carry data structures with them. The functionality is embedded into the models using C or Ada programs. The capability of generating high-level VHDL models from the C or Ada models is provided. These high-level VHDL models can be further refined and developed in a VHDL environment but the refined models cannot be integrated back into the ADAS performance model. The flow of information in these high-level VHDL models is still represented by tokens. Therefore, implementation-level components cannot be integrated into an ADAS performance model. Another limitation is that all input values to the behavioral node must be contained within the token data structure.

Scientific and engineering software (SES)/Workbench is a design specification modeling and simulation tool [21]. It is used to construct and evaluate proposed system designs and to analyze their performance. A graphical interface is used to create a structural model which is then converted into a specific simulatable description (SES/sim). SES/Workbench enables the transition across domains of interpretation by using a *user node*, in which C-language and SES/sim statements can be executed. Therefore, SES/Workbench has similar limitations to ADAS; the inability to simulate a multilevel model when input values of behavioral nodes are not fully specified and the inadequacy of simulating components described as implementation-level HDLs (the capability of integrating VHDL models has been introduced later in the next paragraph). In addition, multiple simulation languages (both SES/sim and C) are required for multilevel models.

The Reveal Interactor is a tool developed by Redwood Design Automation [22]. A model constructed in Reveal is aimed at the functional verification of RTL level VHDL and Verilog descriptions and, therefore, does not include a separate transaction-based performance modeling capability. However, Reveal can work in conjunction with SES/Workbench. By mixing models created in Reveal and SES/Workbench, a multilevel modeling capability exists. Again, these multilevel models are very limited due to the fact that all the required information at the lower level part of the model must be available within the higher level model.

Integrated Design Automation System (IDAS) is a multilevel design environment which allows for rapid prototyping of systems [23]. Although the behavioral specifications need to be expressed as Ada, C, or Fortran programs, IDAS provides the capability of automatically translating VHDL description to Ada. However, the user cannot create abstract models in which certain behavior is unspecified. Also, it does not support classical performance models (such as queuing models and Petri nets) and forces the user to specify a behavioral description very early in the design process.

Transcend claims to integrate multilevel descriptions into a single environment [24,25]. In the more abstract level, T-flow models are used, in which tokens are used to represent flow of data. The capability of integrating VHDL submodels into a T-flow model is provided. However, interfacing between the two models requires a "C++ like" language, which map variables to/from VHDL signals, resulting in a heterogeneous simulation environment. Although their approach is geared toward the same objective as mixed-level modeling, the T-flow model must also include all the necessary data to activate the VHDL submodels. Therefore, the upper-level model cannot be "too abstract" and must include lower level details.

Methodology for integrated design and simulation (MIDAS) supports the design of distributed systems via iterative refinement of partially implemented performance specification (PIPS) models [26]. A PIPS model is a partially implemented design where some components exist as simulation models and others as operational subsystems (i.e., implemented components). Although they use the term "hybrid model" in this context it refers to a different type of modeling. MIDAS is an "integrated approach to software design" [26]. It supports the performance evaluation of software being executed on a given machine. It does not allow the integration of components expressed in an HDL into the model.

The Ptolemy project is an academic research effort being conducted at the University of California at Berkeley [27,28]. Ptolemy, a comprehensive system prototyping tool, is actually constructed of multiple domains. Most domains are geared toward functional verification and have no notion of time. Each domain is used for modeling a different type of system. They also vary in the modeling level (level of abstraction). Ptolemy provides limited capability of mixing domains within one design. The execution of a transition across domains is accomplished with a "wormhole." A wormhole is the mechanism for supporting the simulation of heterogeneous models. Thus, a multilevel modeling and analysis capability is provided. There are two major limitations to this approach compared with the mixedlevel modeling approach being described. The first one is the heterogeneity—several description languages. Therefore, translation between simulators is required. The second one is that the interface between domains only translates data. Therefore, all the information required by the receiving domain must be generated by the transmitting domain.

Honeywell Technology Center (HTC) conducted a research effort that specifically addressed the mixed-level modeling problem [29]. This research had its basis in the UVa mixedlevel modeling effort. The investigators at HTC developed a performance modeling library (PML) [30,31] and added a partial mixedlevel modeling capability to this environment. The PML is used for performance models at a relatively low level of abstraction. Therefore, it assumes that all the information required by the interpreted element is provided by the performance model. In addition, their interface between uninterpreted and interpreted domains allows for bidirectional data flow.

Transaction-level modeling is a model paradigm that combines key concepts from interface-based design [32] with the system level modeling to develop a refineable model of the system as a whole. Essentially, the system as a whole is represented with the model of computation done in components separated as much as possible from the model of communication between components.

As explained by Cai and Gajski [33], "In a transaction-level model (TLM), the details of communication among computation components are separated from the details of computational components. Communication is modeled by channels, while transaction requests take place by calling interface functions of these channel models. Unnecessary details of communication and computation are hidden in a TLM and may be added later. TLMs speed up simulation and allow exploring and validating design alternatives at the higher level of abstraction."

There are a number of advantages to such an approach. Of course you get the advantages of a system-level model, but the separation of the modeling of the communication from the modeling of the computation also has a number of advantages. The primary advantage is that it allows the designer to focus on one aspect of the design at a time. The separation limits the complexity that the designer must contend with just as encapsulation and abstraction of component behaviors do. This also allows the designer to develop or refine the communication model of the system in the same way that they would do for the computation model.

To summarize, numerous multilevel modeling efforts exist. However, they are being developed, not addressing the issue of lack of information at the transition between levels of abstraction. The solution of this problem is essential for true stepwise refinement of the performance models to behavioral models. In addition, integration of performance modeling level and behavioral modeling level was mostly performed by mixing different simulation environments, which results in heterogeneous modeling approach.

77.2 The ADEPT Design Environment

A unified end-to-end design environment based solely on VHDL that allows designers to model systems from the abstract level to the gate level as described above has been developed. This environment supports the development of system-level models of digital systems that can be analyzed for multiple metrics like performance and dependability, and can then be used as a starting point for the actual implementation. A tool called Advanced design environment prototype tool (ADEPT) has been developed to implement this environment. ADEPT actually supports both system-level performance and dependability analysis in a common design environment using a collection of predefined library elements. ADEPT also includes the capability to simulate both system- and implementation-level (behavioral) models in a common simulation environment. This capability allows the stepwise refinement of system-level models into implementation-level models.

Two approaches to creating a unified design environment are possible. An evolutionary solution is to provide an environment that "translates" data from different models at various points in the design process and creates interfaces for the noncommunicating software tools used to develop these models. With this approach, users must be familiar with several modeling languages and tools. Also, analysis of design alternatives is difficult and is likely to be limited by design time constraints.

A revolutionary approach, the one being developed in ADEPT, is to use a single modeling language and mathematical foundation. This approach uses a common modeling language and simulation environment which decreases the need for translators and multiple models, reducing inconsistencies and the probability of errors in translation. Finally, the existence of a mathematical foundation provides an environment for complex system analysis using analytical approaches.

Simulators for hardware description languages accurately and conveniently represent the physical implementation of digital systems at the circuit, logic, register-transfer, and algorithmic levels. By adding a system-level modeling capability based on extended Petri nets and queuing models to the hardware description language, a single design environment can be used from concept to implementation. The environment would also allow for the mixed simulation of both *uninterpreted* (performance) models and *interpreted* (behavioral) models because of the use of a common modeling language. Although it would be possible to develop the high-level performance model and the detailed behavioral model in two different modeling languages and then develop some sort of "translator" or foreign language interface to hook them together, a better approach is to use a single modeling language for both models. A single modeling language that spans numerous design phases is much easier to use, encouraging more design analysis and consequently better designs.

ADEPT implements an end-to-end unified design environment based upon the use of the VHSIC Hardware Description Language (VHDL), IEEE Std. 1076 [34]. VHDL is a natural choice for this single modeling language in that it has high-level language constructs, but unlike other programming languages, it has a built-in timing and concurrency model. VHDL does have some disadvantages in terms of simulation execution time, but techniques have been developed to help address this problem [35].

ADEPT supports the integrated performance and dependability analysis of system-level models and includes the capability to simulate both uninterpreted and interpreted models through mixed-level modeling. ADEPT also has a mathematical basis in Petri nets thus providing the capability for analysis through simulation or analytical approaches [36].

In the ADEPT environment, a system model is constructed by interconnecting a collection of predefined elements called ADEPT modules. The modules model the information flow, both data and control, through a system. Each ADEPT module has a VHDL behavioral description and a corresponding mathematical description in the form of a CPN based on Jensen's CPN model [37]. The modules communicate by exchanging *tokens*, which represent the presence of information, using a fully interlocked, four-state handshaking protocol [38]. The basic ADEPT modules are intended to be building blocks from which useful modeling functionality can be constructed. In addition, custom modules can be developed by the user if required and incorporated into a system model as long as the handshaking protocol is adhered to. Finally, some libraries of application-specific, high-level modeling modules such as Multiprocessor Communications Network Modeling Library [39] have been developed and included in ADEPT.

The following sections discuss the VHDL implementation of the token data type and transfer mechanism used in ADEPT, and the modules provided in the standard ADEPT modeling library.

77.2.1   Token Implementation

The modules defined in this chapter have been implemented in VHDL, and use a modified version of the token passing mechanism defined by Hady [38]. Signals used to transport tokens must be of the type *token*, which has been defined as a record with two fields. The first field, labeled STATUS, is used to implement the token passing mechanism. The second field, labeled COLOR, is an array or integers that is used to hold user-defined color information in the model. The ADEPT tools allow the user to select from a predefined number of color field options. The tools then automatically link in the proper VHDL design library so that the VHDL descriptions of the primitive modules operate on the defined color field. The default structure of the *data-type* token used in the examples discussed in this document is shown in Figure 77.6.

There are two types of outputs and two types of inputs in the basic ADEPT modules. *Independent* outputs are connected to *control* inputs, and the resulting connection is referred to as a *control-type* signal. *Dependent* outputs are connected to *data* inputs, and the resulting connection is referred to as a *data-type* signal. To make the descriptions more intuitive, outputs are often referred to as the "source side" of a signal, and inputs are referred to as the "sink side" of a signal.

Tokens on *independent* outputs may be written over by the next token, so the "writing" process is *independent* of the previous value on the signal. In contrast, new tokens may not be placed on *dependent*

```
type handshake is (removed, acked, released, present);
type token_fields is (status,
                      tag1, tag2, tag3, tag4, tag5,
                      tag6, tag7, tag8, tag9, tag10,
                      tag11, tag12, tag13, tag14, tag15,
                      boole1, boole2, boole3,
                      fault, module_info,
                      index, act_time, color);

type color_type is array (token_fields range tag1 to act_time) of integer;

type token is
     record
         status   : handshake;
         color    : color_type;
     end record;

type op_fields is (add, sub, mul, div);
type cmp_fields is (lt, le, gt, ge, eq, ne);
type tkf_array is array (1 to 25) of token_fields;-- used in file_read


type token_vector is array (integer range <>) of token;

--   resolution function for token
function protocol (input : token_vector) return token;
subtype token_res is protocol token;
```

FIGURE 77.6   Token type definition.

outputs until the previous token has been removed, so the "writing" process is *dependent* on the previous value on the signal. Data-type signals use the four-step handshaking process to ensure that tokens do not get overwritten, but no handshaking occurs on *control-type* signals.

The STATUS field of a token signal can take on four values. Signals connecting *independent* outputs to *control* inputs (control-type signals) make use of only two of the four values. *Independent* outputs place a value PRESENT on the status field to indicate that a token is present. Since *control* inputs only copy the token but do not remove it, the *independent* output only needs to change the value of the status field to RELEASED to indicate that the token is no longer present on the signal, and the *control* input can no longer consider the signal to contain valid information. The signals connecting *dependent* outputs to *data* inputs (data-type signals) need all four values (representing a fully-interlocked hand-shaking scheme) to ensure that a dependent output does not overwrite a token before the data input to which it is connected has read and removed it. This distinction is important since a token on control-type signals represents the presence or absence of a condition in the network while tokens on data-type signals, in contrast, represent information or data that cannot be lost or overwritten. In addition, fanout is not permitted on *dependent* outputs and is permitted on *independent* outputs.

The signals used in the implementation are of type *token*. The token passing mechanism for data-type signals is implemented by defining a VHDL bus resolution function. This function is called each time a signal associated with a dependent output and a data input changes value. The function essentially looks at the value of the STATUS field at the ports associated with the signal and decides the final value of the signal. At the beginning of simulation the STATUS field is initialized to REMOVED. This value on a signal corresponds to an idle link in the nets defined by Dennis [40]. This state of a signal indicates that a signal is free and a request token may be placed on it. Upon seeing the REMOVED value, the requesting module may set the value of the signal to PRESENT. This operation corresponds to the ready signal in Dennis' definition. The requested module acknowledges a ready signal (PRESENT) by setting the signal value to ACKED, after the requested operation has been completed. Upon sensing the value of ACKED on the signal, the requesting module sets the value of the signal to RELEASED, which in turn causes the requested module to place the value of REMOVED on the signal. This action concludes one request/acknowledge cycle, and the next request/acknowledge cycle may begin.

In terms of Petri Nets, the models described here can be described as one-safe Petri nets since, at any given time, no place in the net can contain more than one token. The correspondence between the signal values and transitions in a Petri net may be defined, in general, in the following manner:

(1) *PRESENT.* A token arrives at the input place of a transition.
(2) *ACKED.* The output place of a transition is empty.
(3) *RELEASED.* The transition has fired.
(4) *REMOVED.* The token has been transferred from the input place to the output place.

The modules to be described in this chapter may be defined in terms of Petri nets. As an example, a Petri net description of the Wye module is shown in Figure 77.7.

The function of the Wye module is to copy the input token to both outputs. The input token is not acknowledged until both output tokens have been acknowledged. In the Petri net of Figure 77.7, the "r" and "a" labels correspond to "ready" and "acknowledge," respectively. When a token arrives at the place labeled "0r," the top transition is enabled and a token is placed in the "1r," "2r," and center places. The first two places correspond to a token being placed on the module outputs. Once the output tokens are acknowledged (corresponding to tokens arriving at the "1a" and "2a" places, the lower transition is enabled and a token is placed in "0a," corresponding to the input token being acknowledged. The module is then ready for the next input token. Note that since this module does not manipulate the color fields, no color notation appears on the net.

The specific CPN description used is based on the work of Jensen [37]. The complete CPN descriptions of each of the ADEPT building blocks can be found in Ref. [41].

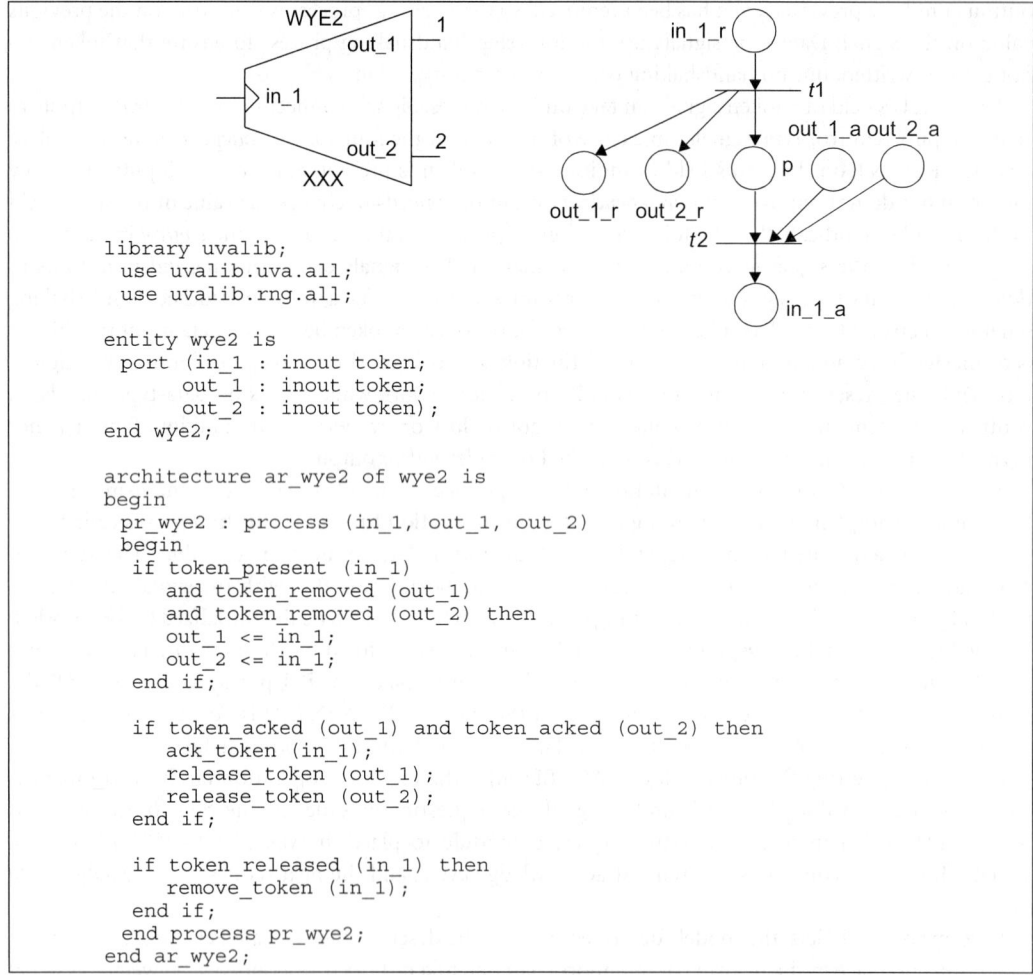

```
library uvalib;
 use uvalib.uva.all;
 use uvalib.rng.all;

entity wye2 is
 port (in_1 : inout token;
       out_1 : inout token;
       out_2 : inout token);
end wye2;

architecture ar_wye2 of wye2 is
begin
 pr_wye2 : process (in_1, out_1, out_2)
 begin
  if token_present (in_1)
     and token_removed (out_1)
     and token_removed (out_2) then
     out_1 <= in_1;
     out_2 <= in_1;
  end if;

  if token_acked (out_1) and token_acked (out_2) then
     ack_token (in_1);
     release_token (out_1);
     release_token (out_2);
  end if;

  if token_released (in_1) then
     remove_token (in_1);
  end if;
 end process pr_wye2;
end ar_wye2;
```

FIGURE 77.7 Wye module ADEPT symbol, its behavioral VHDL description, and its CPN representation.

77.2.2 ADEPT Handshaking and Token Passing Mechanism

Recall that the ADEPT standard token has a *status* field which can take on four values: *PRESENT, ACKED, RELEASED,* and *REMOVED.* These values reflect which stage of the token passing protocol is currently in progress. Several examples will now be presented to show how the handshaking and token passing occurs between ADEPT modules. Figure 77.8 illustrates the handshaking process between a Source module connected to a Sink module.

Figure 77.8(A) describes the "simplified" event sequence. Here, we can think of the four-step handshaking as consisting of simply a forward propagation of the token and a backward propagation of the token acknowledgment. Thinking of the handshaking in this simplified manner, Table A in Figure 77.8 shows the simplified event sequence and times. At time 0 ns, the Source module places a token on A, corresponding to Event 1. The token is immediately acknowledged by the Sink module, corresponding to Event 2. Since the Source module has a *step* generic of 5 ns, and there is no other delay along the path to the Sink, the next token will be output by the Source module at time 5 ns.

Figure 77.8(B) details the entire handshaking process. Table B in Figure 8 lists the detailed event sequence for this example. Since handshaking occurs in zero simulation time, the events are listed by *delta cycles* within each simulation time. The function of delta cycles in VHDL is to provide a means for

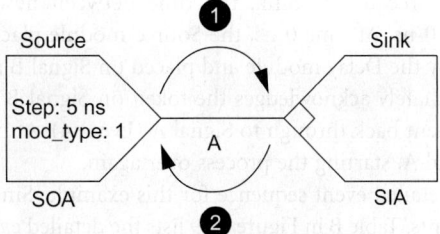

Time between new tokens on A = step + path delay = 5 ns

TABLE A Simplified Event Sequence

Event	Time (ns)	Description
1	0	Source module places token on signal A
2	0	Sink module acknowledges token on A
1	5	Source module places next token on signal A

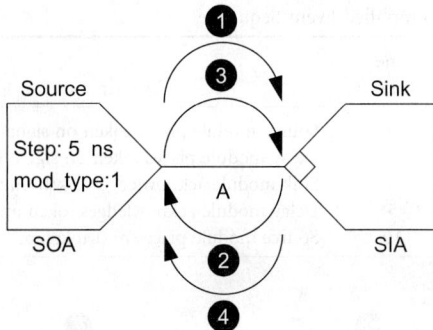

TABLE B Detailed Event Sequence

Event	Time (ns)	Delta	Description	Resolved Signal A
1	0	1	Source module places token on A	Present
2	0	2	Sink module acknowledges token on A	Acked
3	0	3	Source module releases token on A	Released
4	0	4	Sink module removes token on A	Removed
1	5	1	Source module places token on A	Present

FIGURE 77.8 Two module handshaking examples.

ordering and synchronizing events (such as handshaking) which occur in zero time. The actual event sequence for this example consists of four steps, as shown in Figure 77.8(B). Event 1 is repeated at time 5 ns, which starts the handshaking process over again.

77.2.2.1 A Three Module Example

To illustrate how tokens are passed through intermediate modules, a three module example will now be examined. Consider the case where a Fixed Delay module is placed between a Source and Sink. This situation is illustrated in Figure 77.9.

Figure 77.9(A) shows the simplified event sequence, where we can think of the four-step handshaking as consisting of simply a forward propagation of the token and a backward propagation of the token acknowledgment. Table A in Figure 77.9 lists this simplified event sequence. Notice that since now there is a path delay from the Source to the Sink, the time between new tokens from the Source is *step* + *path_delay* = 5 + 5 = 10 ns. At time 0 ns, the Source module places the first token on Signal A (Event 1). This token is read by the Delay module and placed on Signal B after a delay of 5 ns (Event 2). The Sink module then immediately acknowledges the token on Signal B (Event 3). The Delay module then passes this acknowledgment back through to Signal A (Event 4). At time 10 ns, the Source module places the next token on Signal A, starting the process over again.

Figure 77.9(B) shows the detailed event sequence for this example. Since there are two signals, there are a total of eight detailed events. Table B in Figure 77.9 lists the detailed event sequence for this example. Again since the handshaking occurs in zero simulation time, the events are listed by delta cycles within

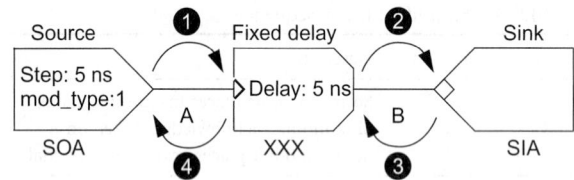

TABLE A   Simplified Event Sequence

Event	Time (ns)	Description
1	0	Source module places token on signal A
2	5	Delay module places token on signal B
3	5	Sink module acknowledges token on signal B
4	5	Delay module acknowledges token on signal A
1	10	Source module places next token on signal A

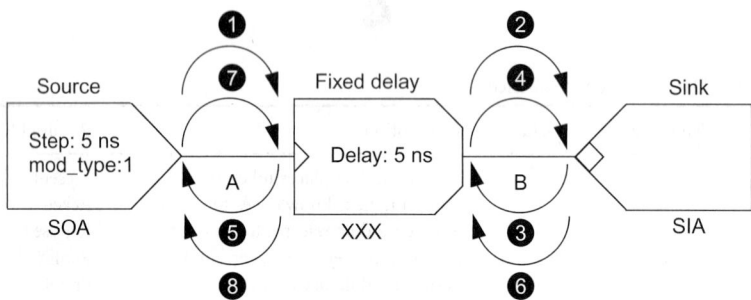

TABLE B   Detailed Event Sequence

Event	Time (ns)	Delta	Description	Resolved Signal A	Resolved Signal B
1	0	1	Source module places token on A	Present	Removed
2	5	1	Delay module places token on B	—	Present
3	5	2	Sink module acknowledges token on B	—	Acked
4	5	3	Delay module releases token on B	—	Released
5			Delay module acknowledges token on A	Acked	—
6	5	4	Sink module removes token on B	—	Removed
7			Source module releases token on A	Released	—
8	5	5	Delay module removes token on A	Removed	—
1	10	1	Source module places token on A	Present	—

FIGURE 77.9   Three module handshaking examples.

each simulation time. Table B lists the resolved values on both Signal A and Signal B, where a value of "----" indicates no change in value. Notice that the sequence of events on each signal proceeds in this order: place the token (present) by the "source side," acknowledge the token (acked) by the "sink side," release the token (released) by the "source side," and finally remove the token (removed) by the "sink side." The important concept to note in this example is the ordering of events. Notice that the Delay module releases the token on its output (Event 4) before it acknowledges the token on its input (Event 5), even though the two events occur in the same simulation time. These actions trigger the Sink module to then remove the token on Signal B (Event 6) and the Source module to release the token on Signal A (Event 7), both in the next delta cycle. Thus Signal B is ready for another token (removed) a full delta cycle before Signal A is ready.

77.2.2.2 Token Passing Example

To illustrate how tokens propagate in a larger system, consider the model shown in Figure 77.10. This figure shows the simplified event sequence, where again we can think of the four-step handshaking as consisting of simply a forward propagation of the token and a backward propagation of the token acknowledgment. In this system, the Source module provides tokens to the Sequence A (SA) module (Event 1), who first passes them to the Sequence B (SB) module (Event 2). The Sequence B (SB) module first passes the token to the Fixed Delay A (FDA) module (Event 3), who passes it to the Sink A (SIA) module after a delay of 5 ns (Event 4). The SIA module then immediately acknowledges the token (Event 5), which causes the FDA module to pass the acknowledgment back to the SB module (Event 6). At this point the Sequence B (SB) module places a copy of the token on its second output (Event 7), where it is passed through the Union A and onto its output (Event 8). The Read Color A module then places the token on its out_1 output (Event 9) and simultaneously places the token on its *independent* output. The Sink B module then acknowledges the token on Signal H (Event 10). Upon seeing this acknowledgment, the Read Color A module releases the token on its *independent* output since the Read Color module has no "memory" (the *release?* generic equals *true*). The acknowledgment is then propagated back to the Sequence B module (Events 11 and 12). At this point, the Sequence B module

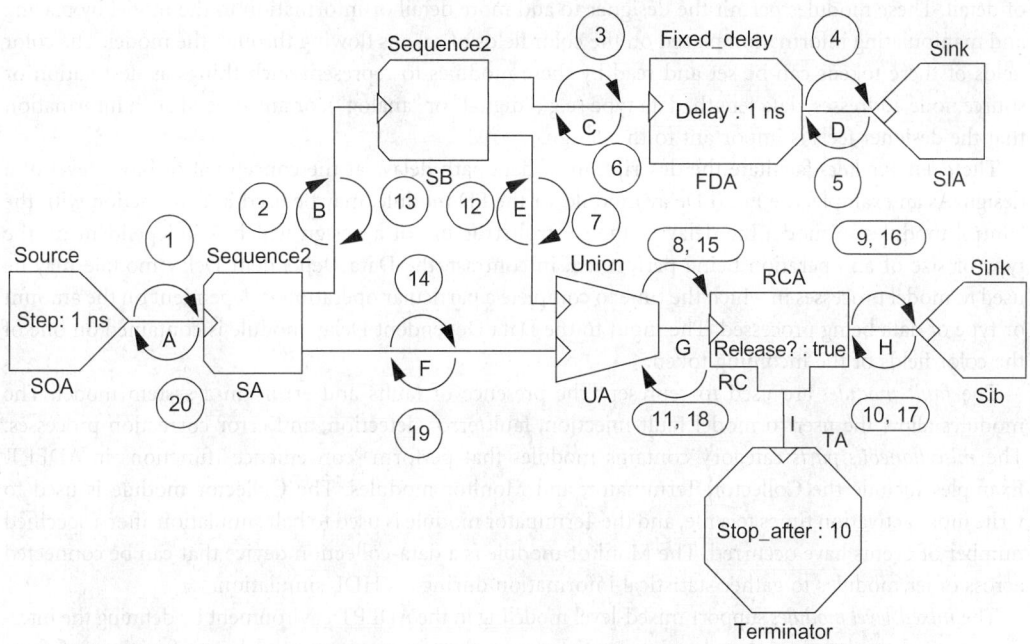

FIGURE 77.10 Token passing example showing simplified event sequence.

acknowledges its input token (Event 13), freeing the Sequence A module to place a token on its second output (Event 14). This token is then passed through the Union A, Read Color A, and Sink module (Events 15 and 16) and the acknowledgment is returned (Events 17–19). The Sequence A module can then acknowledge the token on its input (Event 20), allowing the Source module to generate the next token 5 ns (*step*) later.

If we examine the activity on Signal I (the *independent* output of the Read Color A module), we see that it goes present after Event 8, released after Event 10, present after Event 15, and released again after Event 17. Consider the operation of the Terminator module attached to this signal. Since the Terminator module halts simulation after the number of active events specified by the *stop_after* generic, if we were to simulate this example, the simulation would halt after Event 15 (the second active event).

77.2.3   Module Categories

The ADEPT primitive modules may be divided into six categories: *control modules, color modules, delay modules, fault modules, miscellaneous parts modules,* and *mixed-level modules.*

As the name implies, the *control modules* are used to control the flow of tokens through the model. As such, the control modules form the major portion of the performance model. The control modules operate only on the STATUS field of a signal and do not alter the color of a token on a signal. Further, no elapsed simulation time results from the activation of the control modules since they do not have any delay associated with them. There are several control modules whose names end in a numeral, such as "union2." These modules are part of a family of modules that have the same function, but differ in the number of inputs or outputs that they possess. For example, there are eight "union" modules, "union2," "union3," "union4," ...,"union8." With the exception of the Switch, Queue, and logical modules, the control modules have been adapted from those proposed by Dennis [40].

The control modules described above process colored tokens but do not alter the color fields of the tokens passing through them. Manipulation of the color field of a token is reserved to the *color modules.* These color modules permit various operations on the color fields such as allowing the user to read and write the color fields of the tokens. The color modules also permit the user to compare the color information carried by the tokens and to control the flow of the tokens based on the result of the comparisons. The use of these color modules enables high-level modeling of systems at different levels of detail. These modules permit the designer to add more detail or information to the model by placing and manipulating information placed on the color fields of tokens flowing through the model. The color fields of these tokens can be set and read by these modules to represent such things as destination or source node addresses, data length, data type (e.g., "digital" or "analog"), or any type of such information that the designer feels is important to the design.

The *delay modules* facilitate the description of data path delays at the conceptual or block level of a design. As an example, the Fixed Delay module, or the FD module, may be used in conjunction with the control modules to model the delay in the control structure of a design which is independent on the type or size of an operation being performed. In contrast, the Data Dependent Delay module may be used to model processes in which the time to complete a particular operation is dependent on the amount or type of data being processed. The input to the Data Dependent Delay module is contained on one of the color fields of the incoming token.

The *fault modules* are used to represent the presence of faults and errors in a system model. The modules allow the user to model fault injection, fault/error detection, and error correction processes. The *miscellaneous parts* category contains modules that perform "convenience" functions in ADEPT. Examples include the Collector, Terminator, and Monitor modules. The Collector module is used to write input activation times to a file, and the Terminator module is used to halt simulation after a specified number of events have occurred. The Monitor module is a data-collection device that can be connected across other modules to gather statistical information during a VHDL simulation.

The *mixed-level modules* support mixed-level modeling in the ADEPT environment by defining the interfaces around the interpreted and uninterpreted components in a system model. The functions of these modules and their use in creating mixed-level models is described in more detail in Sections 77.4.2 and 77.4.3.

The complete functionality of each primitive module is defined and the generics associated with each of the modules is described in the *ADEPT Library Reference Manual* [42]. The standard ADEPT module symbol convention is also explained in more detail in the *ADEPT Library Reference Manual*.

In addition to the primitive modules, there are libraries of more complex modules included in ADEPT. In general, these libraries contain modules for modeling systems in a specific applications area. These libraries are also discussed in more detail in the *ADEPT Library Reference Manual*.

77.2.4 ADEPT Tools

The ADEPT system is currently available on Sun platforms using Mentor Graphics' *Design Architect* as the front end schematic capture system, or on Windows PCs using OrCAD's *Capture* as the front end schematic capture system. The overall architecture of the ADEPT system is shown in Figure 77.11.

The schematic front end is used to graphically construct the system model from a library of ADEPT module symbols. Once the schematic of the model has been constructed, the schematic capture system's netlist generation capability is used to generate an electronic design interchange format (EDIF) 2.0.0 netlist of the model. Once the EDIF netlist of the model is generated, the ADEPT software is used to translate the model into a structural VHDL description consisting of interconnections of ADEPT modules. The user can then simulate the structural VHDL that is generated using the compiled VHDL behavioral descriptions of the ADEPT modules to obtain performance and dependability measures.

In addition to VHDL simulation, a path exists that allows the CPN description of the system model to be constructed from the CPN descriptions of the ADEPT modules. This CPN description can then be translated into a Markov model using well-known techniques and then solved using commercial tools to obtain reliability, availability, and safety information.

77.3 A Simple Example of an ADEPT Performance Model

This section presents a simple example of the usage of the primitive building blocks for performance modeling. The example is a three computer system wherein the three computers share a common bus. The example also presents simulation results and system performance evaluations.

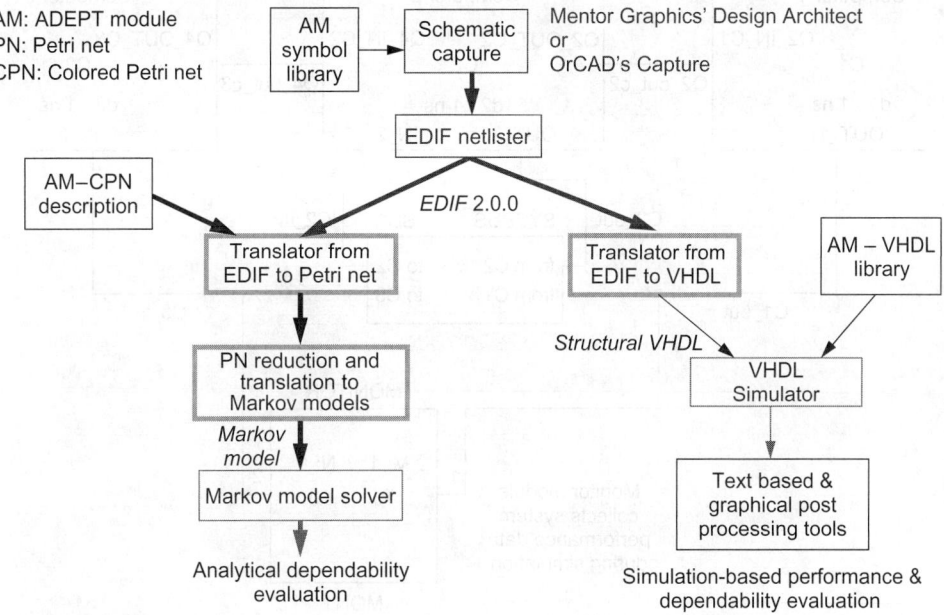

FIGURE 77.11 ADEPT design flow.

77.3.1 A Three Computer System

This section discusses a simple example to illustrate how the modules discussed previously may be interconnected to model and evaluate the performance of a complete system. The system to be modeled consists of three computers communicating over a common bus, as shown in Figure 77.12. Each block representing a computer can be thought to contain its own processor, memory and peripheral devices. The actual ADEPT schematic for the three computer system is shown in Figure 77.13.

Computer C1 contains some sensors and preprocessing capabilities. It collects data from the environment, converts it into a more compact form and then sends it to computer C2 via the bus. The computer C2 further processes the data and passes it to computer C3 where the data are appropriately utilized. It is assumed that data are transferred in packets and each packet of data are of varying length. In the example described here, computers C1 and C2 receive packets whose sizes are uniformly distributed between 0 and 100. The packet

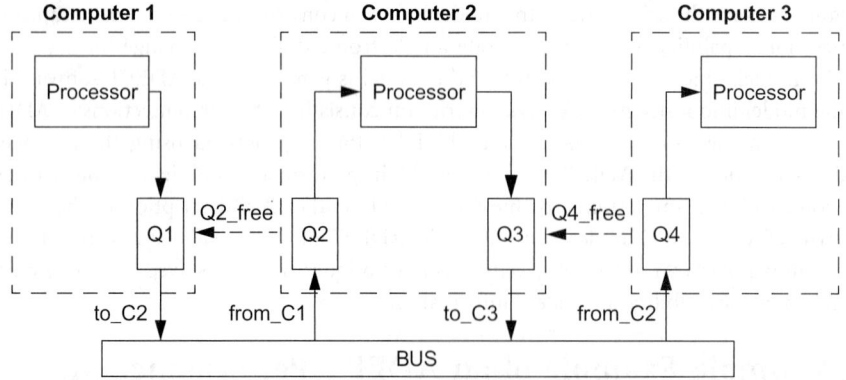

FIGURE 77.12   A three computer system.

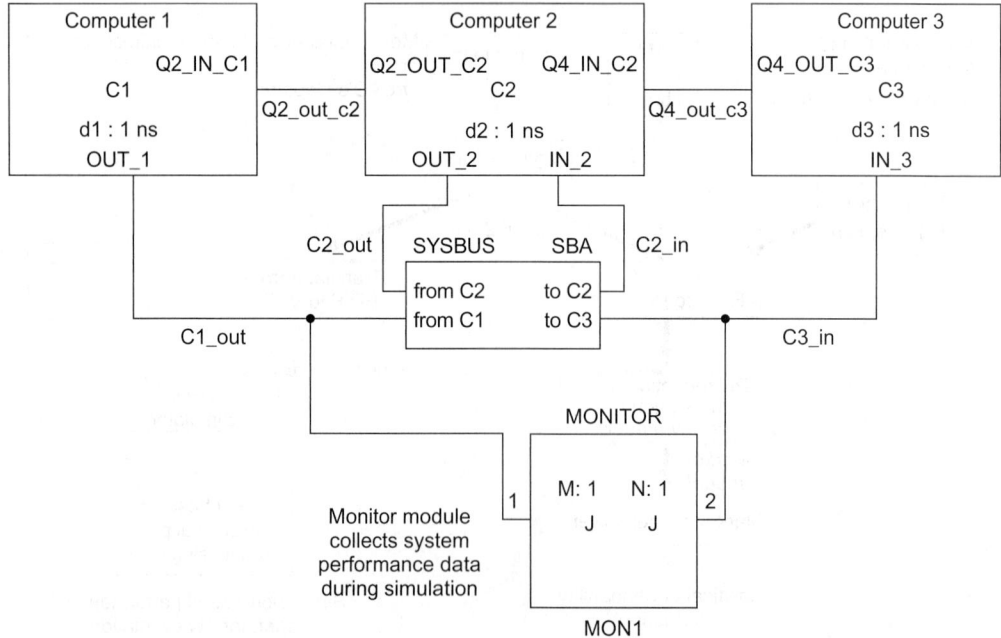

FIGURE 77.13   Three computer system ADEPT schematic.

size of computer C3 is uniformly distributed between 0 and 500. The external environment in this example is modeled by a Source module in C1 and a Sink module in C3.

C1 has an output queue, C2 has both an input queue and an output queue, while C3 has one input queue. All queues in this example are assumed to be of length 8. If the input queues of C2 or C3 are full the corresponding Q_free signal is released (value=RELEASED). This interconnection prevents the computer writing into the corresponding queue from placing data on the bus when the queue is full. This technique not only prevents the bus from being unnecessarily held up but also eliminates the possibility of a deadlock.

The ADEPT model of Computer 1 (C1) is shown in Figure 77.14. The Source A (SOA) module along with the Sequence A (SA), Set_Color A (SCA), and Random A (RA) modules generates tokens whose *tag1* field is set according to a distribution which is representative of the varying data sizes that C1 receives. The Data-Dependent Delay A (DDA) models the processing time of C1 which is directly proportional to the packet size. The *unit_step* delay for the DDA module is passed down as a generic *d1*1 ns. As soon

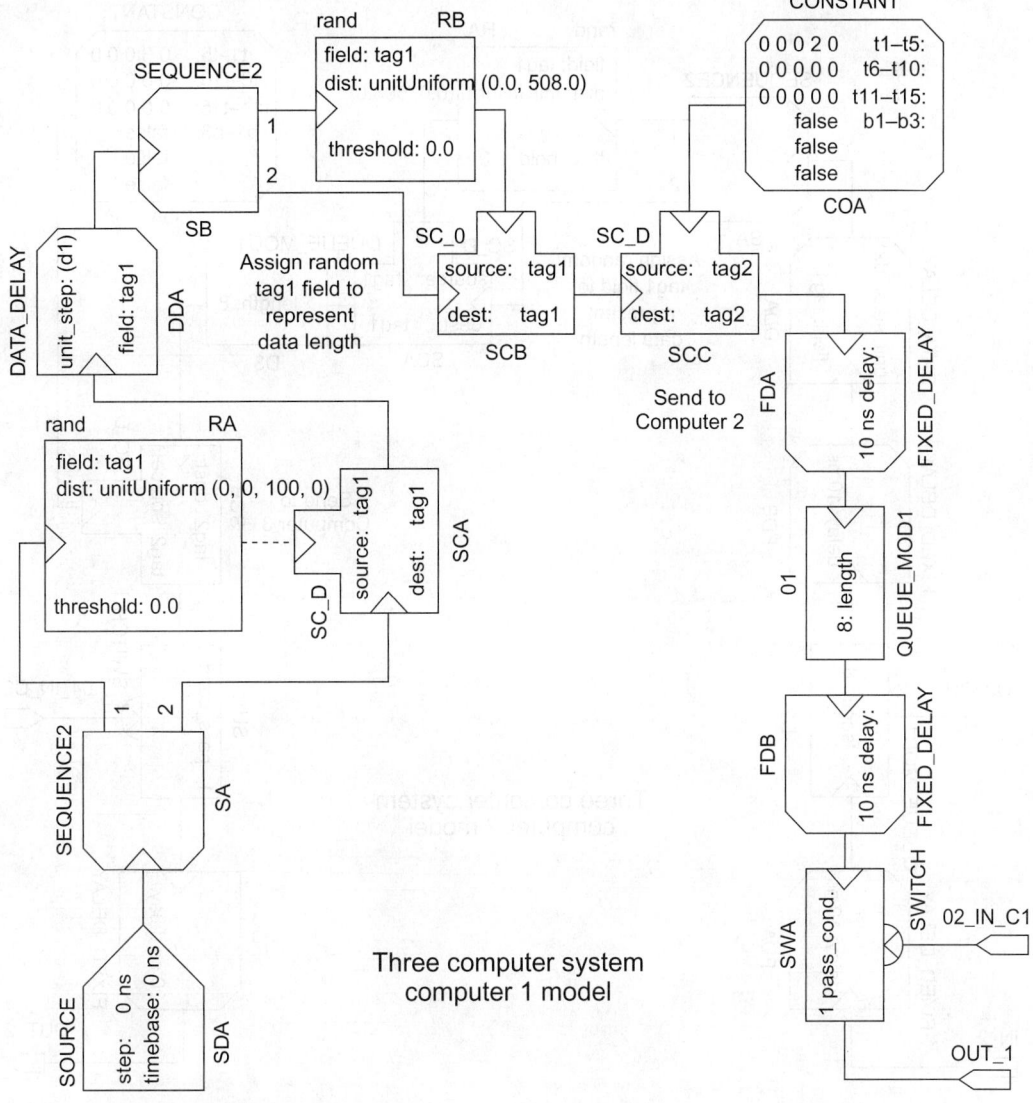

FIGURE 77.14 ADEPT model of Computer 1.

as a token, representing one packet of data, appears at the output of the DDA module, the Sequence B (SB), Set_Color B (SCB), and Random B (RB) modules together set the *tag1* field of the token to represent the packet size after processing. The Constant A (COA) and Set_Color C (SCC) modules set the *tag2* field of the token to 2. This coloring indicates to the bus arbitration unit that the token is to be transferred to C2. The token is then placed in the output Queue (Q1) of C1 and the token is acknowledged. This acknowledge signal is passed back to the Source A module which then produces the next token. The Fixed Delay (FDA & FDB) modules represent the read and write times associated with the Queue. The Switch A (SWA) element is controlled by the Q_free signal from C2 and prevents a token from the output queue of C1 from being placed on the bus if the incoming Q_free signal is inactive.

Figure 77.15 shows the ADEPT model of Computer 2 (C2). When a token arrives at the *data* input of C2 the token is placed at the input of the Queue (Q2). The *control* output of the Queue becomes the Q2_Free output of C2. The remaining modules perform the same function as in C1 except that the tag2 field of the tokens output by C2 is set to 3 which indicates to the bus arbitration unit that the token is

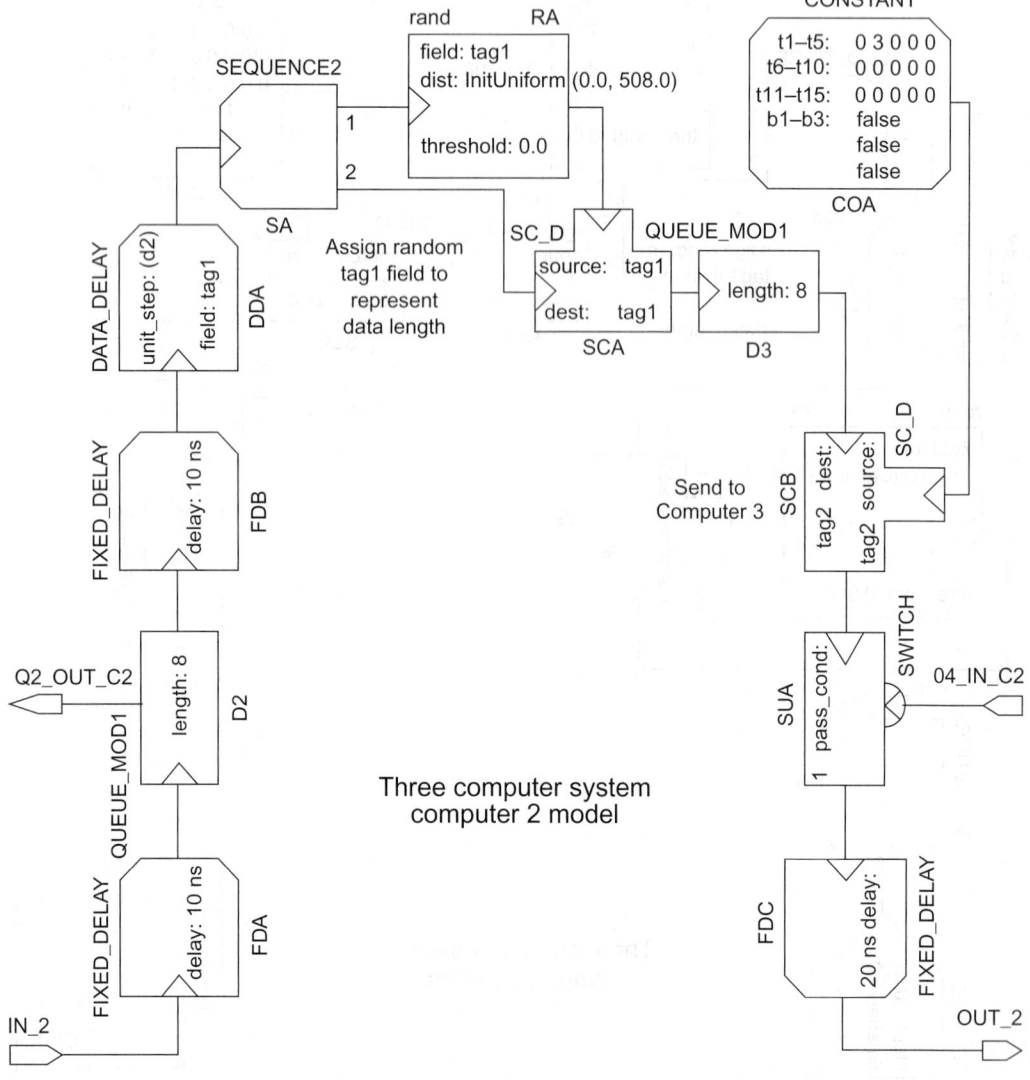

FIGURE 77.15   ADEPT model of Computer 2.

to be sent to C3. The DDA module represents the relative processing speed of Computer C2. The *unit_step* delay for the DDA module is passed down as a generic $d2*1$ ns.

The modules defining Computer 3 (C3) are shown in Figure 77.16. A token arriving at the input is placed in the queue. The DDA element reads one token at a time and provides the delay associated with the processing time of C3 before the Sink A (SIA) module removes the token. The *unit_step* delay for the DDA module is passed down as a generic $d3*1$ ns.

The bus is shown in Figure 77.17. Arbitration is provided to ensure that both C1 and C2 do not write onto the bus at the same time. The Arbiter A (AA) element provides the required arbitration. Since the output of C2 is connected to the IN_1(1) input of the AA element, it has a higher priority over C1. The Union A (UA) element passes a token present at either of its inputs to its output. The output of the UA element is connected to the Data-Dependent Delay A (DDA) element, which models the packet transfer delay associated with moving packets over the bus. The delay through the bus is dependent on the size of the packet of information being transferred (size stored on the *tag1* field). Note that in this analysis, the bus delay was set to 0. The *independent* output of the Read_Color A(RCA) element is connected to

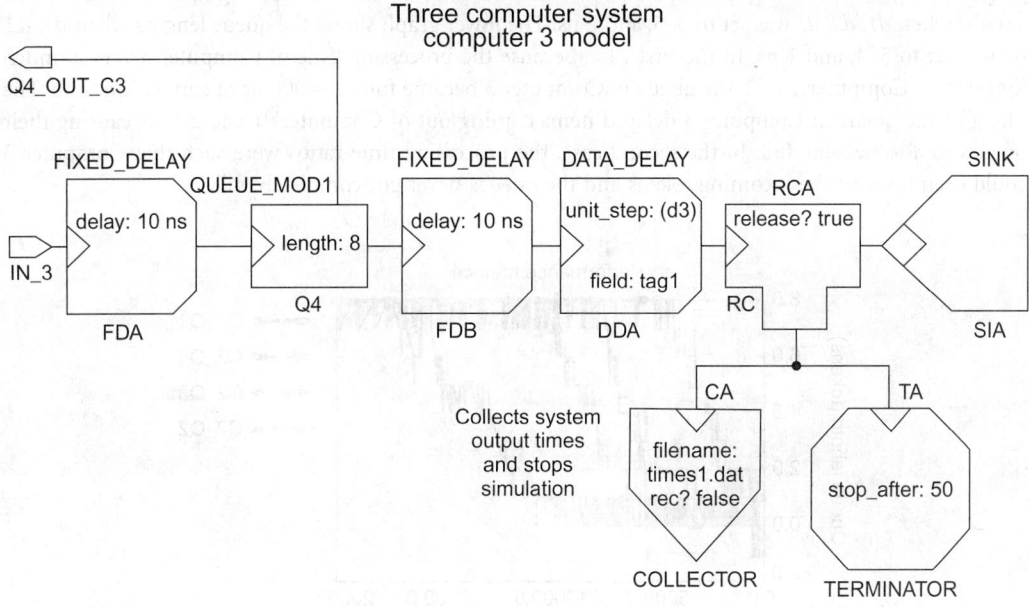

FIGURE 77.16 ADEPT model of Computer 3.

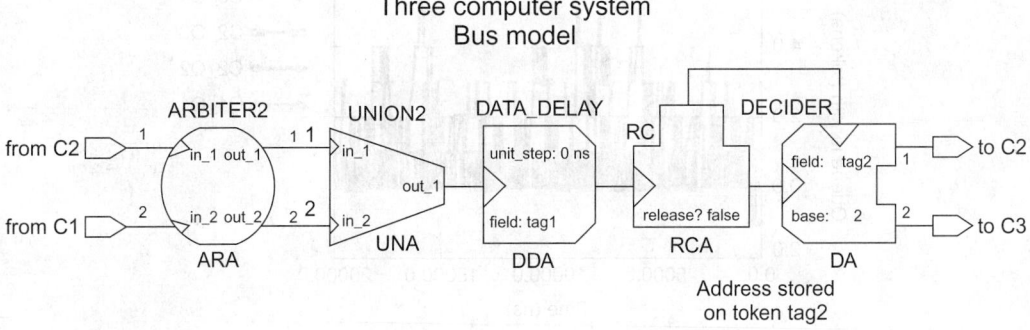

FIGURE 77.17 ADEPT model of system bus.

the *control* input of the Decider A (DA) module. The base of the DA element is set to 2. Since the *tag2* field of the token is set to 2 or 3 depending on whether it originated from C1 or C2, the first output of the Decider A module is connected to C2 and the second output is connected to C3. This technique ensures that the tokens are passed on to the correct destination, C2 or C3.

77.3.2   Simulation Results

This section presents simulation results that illustrate the queuing model capabilities of the ADEPT system. The model enables the study of the overall speed of the system in terms of the number of packets of information transferred in a given time. It also allows the study of the average number of tokens present in each queue during simulation, and the effect of varying system parameters on the number of items in the queues and overall throughput of the system. The generic unit_step delay of the DD elements (*d1*, *d2*, and *d3*) associated with the three computers is representative of the processing speed of the computers. The optimal relative speeds of the computers may also be deduced by simulation of this model. Figure 77.18 shows graphs of the number of items in each queue versus simulation time. These graphs were generated using the BAARS postsimulation analysis tool. The upper graph shows the queue lengths when *d1*, *d2*, *d3* was set to 5, 5, and 2 ns. The lower graph shows the queue lengths when *d1*, *d2*, *d3* was set to 5, 4, and 1 ns. In the first case, because the processing time of Computer 3 was so much longer than Computer 1 or 2, the queue in Computer 3 became full at ~4000 ns of simulation time. The filling of the queue in Computer 3 delayed items coming out of Computers 1 and 2 thus causing their queues to also become full. In the second case, the processing time ratios were such that Computer 3 could keep up with the incoming tokens and the queues never got completely full.

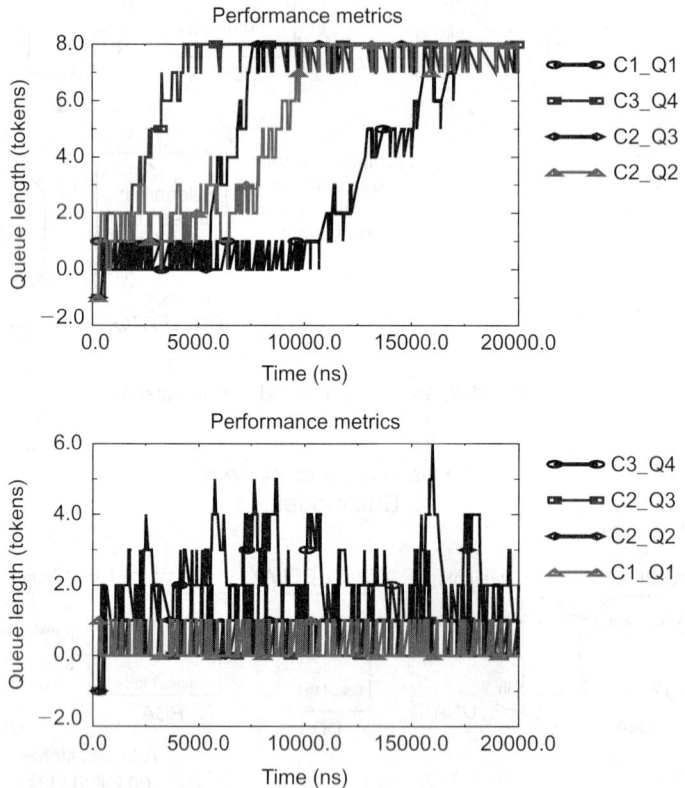

FIGURE 77.18   Queue lengths versus simulation time for various processing delay times.

TABLE 77.1 System Throughput versus Computer Delay Times

C1 Delay (ns)	C2 Delay (ns)	C3 Delay (ns)	Packets per 100,000 Time Units (ns)
8	8	3	131
9	10	2	194
5	5	2	197
3	5	2	197
2	2	2	197
5	6	1	321
5	4	1	322
4	3	1	384
3	2	1	385
2	2	1	386

Table 77.1 summarizes the effect of relative speeds of the computers on the number of packets transferred. Since the size of the packets received by C3 is uniformly distributed between 0 and 500 while the size of the packets received by C1 and C2 is uniformly distributed between 0 and 100, it is intuitively obvious that the overall throughput of the system is largely determined by the speed of Computer C3. The results do indicate this behavior. For example, when the relative instruction execution times for C1, C2, and C3 are 5, 5, and 2, respectively, a total of 197 packets are transferred. By increasing the instruction execution time of C2 by one time unit and decreasing the instruction execution time of C3 by one time unit, it is seen that a total of 321 packets are transferred, an increase of slightly over 60%.

This example has illustrated the use of the various ADEPT modules to model a complete system. Note how the interconnections between modules describing a component of the system are similar to a flow chart describing the behavior of the component. This example also demonstrated that complex systems at varying levels of abstraction and interpretation can easily be modeled using the VHDL-based ADEPT tools.

77.4 Mixed-Level Modeling

As described earlier, performance (uninterpreted) modeling has been previously used primarily by systems engineers when performing design analysis at the early stages of the design process. Although most of the design detail is not included in these models since this detail does not yet exist, techniques such as token coloring can be used to include that design detail that is necessary for an accurate model. However, most of the detail, especially very low-level information such as word widths and bit encoding, are not present. In fact, many additional design decisions must be made before an actual implementation could ever be constructed. In performance models constructed in ADEPT, abstract tokens are used to represent this information (data and control) and its flow in the system. An illustration of such a performance model with its analysis was given the three computer system described in Section 77.3.1. In contrast, behavioral (interpreted) models can be thought of as including much more design detail often to the point that an implementation could be constructed. Therefore, data values and system timing are available and variables typically take on integer, real, or bit values. For example, a synthesizable model of a carry-lookahead adder would be one extreme of a behavioral model.

It should be obvious that the two examples described above are extremes of the two types of models. Extensive design detail can be housed in the tokens of a performance model and information in a behavioral model can be encapsulated in a very abstract type. However, there is always a difference in the abstraction level of the two modeling types. Therefore, to develop a model that can include both performance and behavioral models, interfaces between the different levels of design abstraction, called *mixed-level interfaces,* must be included in the overall model to resolve differences between these two modeling domains.

The mixed-level interface is divided into two primary parts which function together; the part that handles the transition from the uninterpreted domain to the interpreted domain (U/I) and the part that handles the transition from the interpreted domain to the uninterpreted domain (I/U). The general structure of a mixed-level model is shown in Figure 77.19.

In addition to the tokens-to-values (U/I) and values-to-tokens (I/U) conversion processes, the mixed-level interface must resolve the differences in design detail that naturally exist at the interface between uninterpreted and interpreted elements. These differences in detail appear as differences in data and timing abstraction across the interface. The differences in timing abstraction across the interface arise because the components at different levels model timing events at different granularities. For example, the passing of a token that represents a packet of data being transferred across a network in a performance model may correspond to hundreds or thousands of bus cycles for a model at the behavioral level.

The differences in data abstraction across the interface are due to the fact that typically, performance models do not include full functional details whereas behavioral models require full functional data (in terms of values on their inputs) to be present before they will execute correctly. For example, a performance level modeling token arriving at the inputs to the mixed-level interface for a behavioral floating point coprocessor model may contain information about the operation the token represents, but may not contain the data upon which the operation is to take place. In this case, the mixed-level interface must generate the data required by the behavioral model and do it in a way that a meaningful performance metric, such as best- or worst-case delays, is obtained.

77.4.1 Mixed-Level Modeling Taxonomy

The functions that the mixed-level interface must perform and the most efficient structure of the interface is affected by several attributes of the system being modeled and the model itself. To partition the mixed-level modeling space and better define the specific solutions, a taxonomy of mixed-level model classes has been developed [43]. The classes of mixed-level modeling are defined by those model attributes which fundamentally alter the development and the implementation of the mixed-level interface. The mixed-level modeling space is partitioned according to three major characteristics:

(1) the evaluation *objective* of the mixed-level model,
(2) the *timing mechanism* of the uninterpreted model, and
(3) the *nature* of the interpreted element.

For a given mixed-level model, these three characteristics can be viewed as attributes of the mixed-level model and the analysis effort. Figure 77.20 summarizes the taxonomy of mixed-level models.

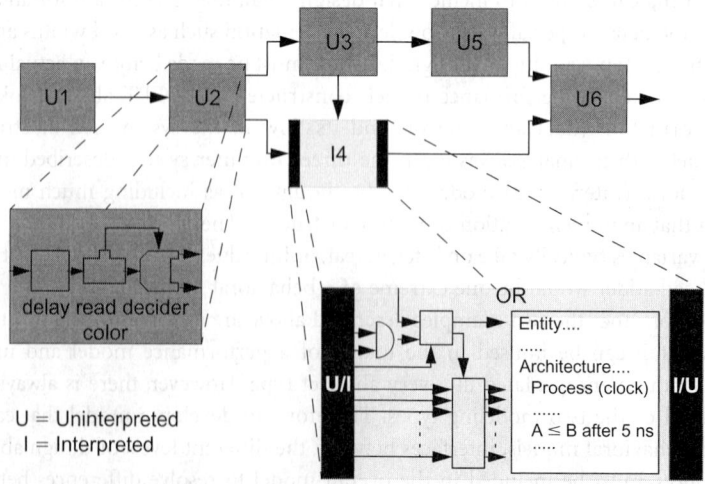

FIGURE 77.19   General structure of a mixed-level model.

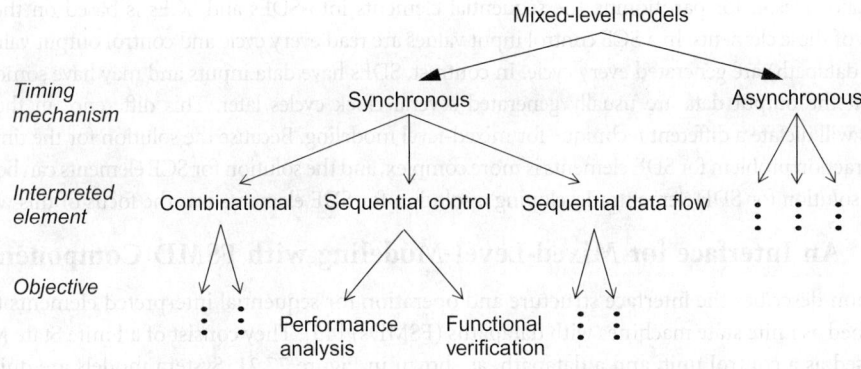

FIGURE 77.20   Mixed-level modeling categories.

Mixed-level modeling objectives. The structure and the functionality of the mixed-level interface are strongly influenced by the objective that the analysis of the mixed-level model will be targeted toward. For the purposes of this work, these objectives were broken down into two major categories:

1. *Performance analysis* and *timing verification.* To analyze the performance of the system (as defined previously) and verify that the specific component(s) under consideration meet system timing constraints. Note that other metrics, such as power consumption, could be analyzed using mixed-level models, but these were outside the scope of this classification.
2. *Functional verification.* To verify that the function (input-to-output value transformation) of the interpreted component is correct within the context of the system model.

Timing mechanisms. Typically, performance (uninterpreted) models are asynchronous in nature and the flow of tokens depends on the handshaking protocol. However, in modeling a system that is globally synchronous (all elements are synchronized to a global clock) a mechanism to synchronize the flow of tokens across the model can be introduced. This synchronization of the performance model will require different mixed-level modeling approaches. Thus the two types of system models that affect the mixed-level interface are:

1. *Asynchronous models.* Tokens on independent signal paths within the system model move asynchronously with respect to each other and arrive at the interface at different times.
2. *Synchronous models.* The flow of tokens in the system model is synchronized by some global mechanism and they arrive at the interface at the same time, according to that mechanism.

Interpreted component. The mixed-level modeling technique strongly depends upon the type of the interpreted component that is introduced into the performance model. It is natural to partition interpreted models into those that model combinational elements and sequential elements. Techniques for constructing mixed-level interfaces for models of combinational interpreted elements have been developed previously [44]. In the combinational element case, the techniques for resolving the timing across the interface were more straightforward because of the asynchronous nature of the combinational elements, and the data-abstraction problem was solved using a methodology similar to the one presented here. However, using sequential interpreted elements in mixed-level models requires more complex interfaces to solve the synchronization problem and different specific techniques for solving the data-abstraction problem. Further research into the problem of mixed-level modeling with sequential elements suggested that interpreted components be broken down into three classifications as described below:

1. *Combinational elements*: Unclocked (with no states) elements.
2. *Sequential control elements* (SCEs): Clocked elements (with states) that are used for controlling data flow, e.g., a control unit or a controller.
3. *Sequential data flow elements* (SDEs): Elements that include datapath elements *and* clocked elements that control the data flow, e.g., control unit and datapath.

The major reason for partitioning the sequential elements into SDEs and SCEs is based on the timing attributes of these elements. In a SCE control input values are read every cycle and control output values (that control a datapath) are generated every cycle. In contrast, SDEs have data inputs and may have some control inputs but the output data are usually generated several clock cycles later. This difference in the timing attributes will dictate a different technique for mixed-level modeling. Because the solution for the timing and data-abstraction problem for SDE elements is more complex, and the solution for SCE elements can be derived from the solution for SDE elements, developing a solution for SDE elements was the focus of this work.

77.4.2   An Interface for Mixed-Level Modeling with FSMD Components

This section describes the interface structure and operation for sequential interpreted elements that can be described as finite state machines with datapaths (FSMDs) [45]. They consist of a Finite State Machine (FSM) used as a control unit, and a datapath, as shown in Figure 77.21. System models are quite often naturally partitioned to blocks which adhere to the FSMD structure. Each of these FSMDs process the data and have some processing delays associated with them. These FSMDs indicate the completion of a processing task to the rest of the system either by asserting a set of control outputs or by the appearance of valid data on their outputs. This characteristic of being able to determine when the data-processing task of the FSMD is completed is a key property in the methodology of constructing mixed-level models with FSMD interpreted components.

The functions performed by the U/I operator include placing the proper values on the inputs to the FSMD interpreted model, and generating a clock signal for the FSMD interpreted model. The required values to be placed on the inputs to the FSMD interpreted model are either contained on the incoming token's information, or "color" fields, are supplied by the modeler via some outside source, or are derived using the techniques described in Section 77.4.2.1. The clock signal is either generated locally if the system-level model is globally asynchronous, or converted from the global clock into the proper format if the system-level model is globally synchronous. The functions performed by the I/U operator include releasing the tokens back into the performance model at the appropriate time and, if required, coloring them with new values according to the output signals of the interpreted element. The structure of these two parts of the mixed-level interface is shown in Figure 77.22. The U/I operator is composed of the following blocks: a Driver, an Activator, and a Clock_Generator. The I/U operator is composed of an Output_Condition_Detector, a Colorer, and a Sequential_Releaser.

In the U/I operator, the Activator is used to detect the arrival of a new token (packet of data) to the interpreted element, inform the Driver of a new token arrival, and to drive control input/s of the interpreted element. The Activator's output is also connected to the I/U operator for use in gathering information on the delay through the interpreted element. The Driver reads information from the token's

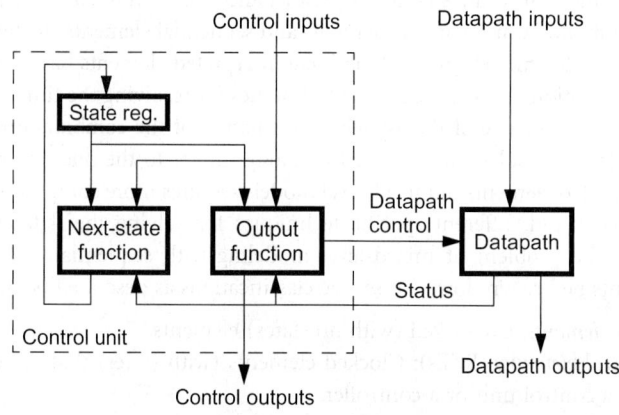

FIGURE 77.21   Generic FSMD block diagram.

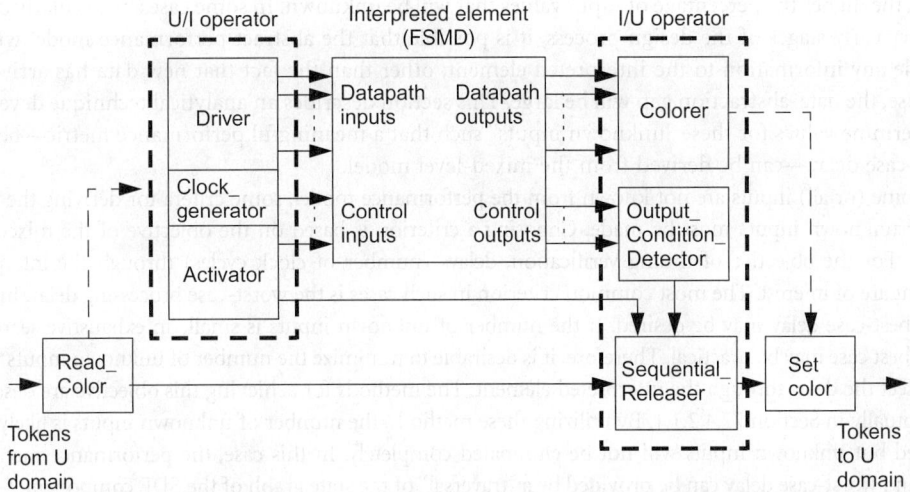

FIGURE 77.22   The mixed-level element structure.

color fields and drives the proper input signals to the interpreted element according to predefined assignment properties. The Clock_Generator generates the clock signal according to the overall type of system-level model, either synchronous or asynchronous.

In the I/U operator, the Output_Condition_Detector detects the completion of the interpreted element data-processing operation, as discussed earlier, by comparing the element's outputs to predefined properties. The Colorer samples the datapath outputs and maps them to color fields according to predefined binding properties. The Sequential_Releaser, which "holds" the original token, releases it back to the uninterpreted model upon receiving the signal from the Output_Condition_Detector. The information carried by the token is then updated by the Colorer and the token flows back to the uninterpreted part of the model.

Given this structure, the operation of the mixed-level model can be described using the general example in Figure 77.22. Upon arrival of a new token to the mixed-level interface (U/I operator), the Read_Color module triggers the Activator component and passes the token to the Sequential_Releaser, where it is stored until the interpreted component is finished with its operation. Once triggered by the Read_Color module, the Activator notifies the Driver to start the data-conversion operation on a new packet of data. In the case of a globally asynchronous system-level model, the Activator will notify the Clock_Generator to start generating a clock signal. Since the interpreted element is a sequential machine, the Driver may need to drive sequences of values onto the inputs of the interpreted element. This sequence of values is supplied to the interpreted element, while the original token is held by the Sequential_Releaser. This token is released back to the uninterpreted model only when the Output_Condition_Detector indicates the completion of the interpreted element operation. The Output_Condition_Detector is parameterized to recognize the completion of data processing by the particular FSMD interpreted component. Once the Output_Condition_Detector recognizes the completion of data processing, it signals the Sequential_ Releaser to release the token into the performance model. Once the token is released, it passes through the Colorer which maps the output data of the interpreted element onto color fields of the token. The new color information on the token may be used by the uninterpreted model for such things as delays through other parts of the model or for routing decisions.

77.4.2.1   Finding Values for the "Unknown Inputs" in an FSMD-Based Mixed-Level Model

As described previously, because of the abstract nature of a performance model, it may not be possible to derive values for all of the inputs to the interpreted component from the data present in the performance model. Typically, the more abstract the performance model (i.e., the earlier in the design cycle), the higher the percentage of input values that will be unknown. In some cases, particularly during the very early stages of the design process, it is possible that the abstract performance model will not provide any information to the interpreted element, other than the fact that new data has arrived. In this case, the data-abstraction gap will be large. This section describes an analytical technique developed to determine values for these "unknown inputs" such that a meaningful performance metric—best- or worst-case delay—can be derived from the mixed-level model.

If some (or all) inputs are not known from the performance model, some critera for deriving the values on the unknown inputs must be made. Choosing a criterion is based on the objective of the mixed-level model. For the objective of timing verification, delays (number of clock cycles) through the interpreted element are of interest. The most common criterion in such cases is the worst-case processing delay. In some cases, best-case delay may be desired. If the number of unknown inputs is small, an exhaustive search for worst/best case may be practical. Therefore, it is desirable to minimize the number of unknown inputs which can affect the delay through the interpreted element. The methods for achieving this objective are described conceptually in Section 77.4.2.1.1. By utilizing these methods, the number of unknown inputs is likely to be reduced but unknown inputs will not be eliminated completely. In this case, the performance metrics of best- and worst-case delay can be provided by a "traversal" of the state graph of the SDE component. Section 77.4.2.1.2 describes the traversal method developed for determining the delays through a sequential element.

Note that in the algorithms described below, the function of the SDE component is represented by the state transition graph (STG) or state table. These two representations are essentially equivalent and are easily generated from a behavioral VHDL description of the SDE, either by hand, or using some of the automated techniques that have been developed for formal verification.

77.4.2.1.1 Reducing the Number of Unknown Inputs

Although it is not essential, reducing the number of unknown inputs can simplify the simulation of a mixed-level model significantly. Since the FSMD elements being considered have output signals that can be monitored to determine the completion of data processing as discussed previously, other outputs may not be significant for performance analysis. Therefore, the "nonsignificant" (insignificant) outputs can be considered as "don't cares." By determining which inputs do not affect the values on the significant outputs, it is possible to minimize the number of unknown delay affecting inputs (DAIs).

The major steps in the DAI detection algorithm are:

Step 1. Select the "insignificant" outputs (in terms of temporal performance).
Step 2. In the state transition graph (STG) of the machine, replace all values for these outputs with a "don't-care" to generate the modified state machine.
Step 3. Minimize the modified state machine and generate the corresponding state table.
Step 4. Find the inputs which do NOT alter the flow in the modified state machine by detecting identical columns in the state table, and combining them by implicit input enumeration.

This method is best illustrated by an example. Consider the state machine which is represented by the STG shown in Figure 77.23. This simple example is a state machine with two inputs, X_1 and X_2, and two outputs, Y_1 and Y_2. This machine cannot be reduced, i.e., it is a minimal state machine. Assume that this machine is the control unit of an FSMD block and that the control output Y_1 is the output which indicates the completion of the "data processing" when its value is 1. Thus, output Y_2 is an "insignificant output" in terms of delay in accordance with Step 1. Therefore, a "don't-care" value is assigned to Y_2 as per Step 2. The modified STG is shown in Figure 77.24. As per Step 3, the modified STG is then reduced. In this example, states A, C, and E are equivalent (can be replaced by a single state, K) and the minimal machine consists of three states, K, B, and D. This minimal machine is described by the state table shown in Table 77.2.

All possible input combinations appear explicitly in Table 77.2. However, it can be seen that the first two columns of the table are identical (i.e., the same next state and output value for all possible present states). Similarly, the last two columns of the table are identical. Therefore, in accordance with Step 4, these columns can be combined yielding the reduced state table shown in Table 77.3.

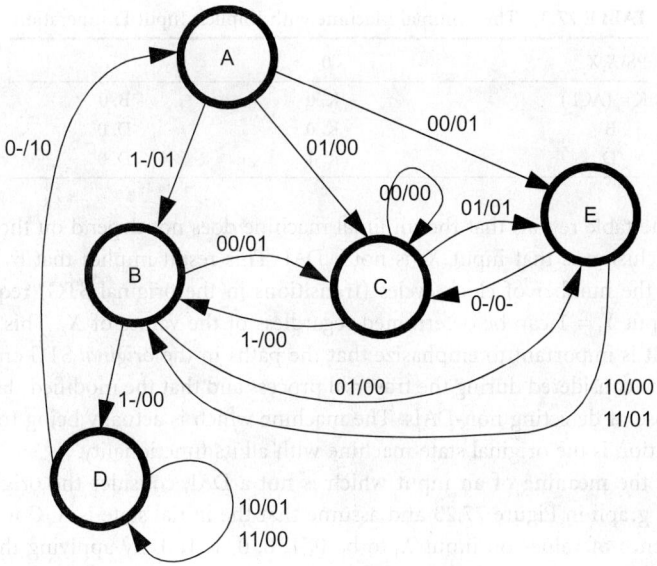

FIGURE 77.23   STG of a two-inputs two-outputs state machine.

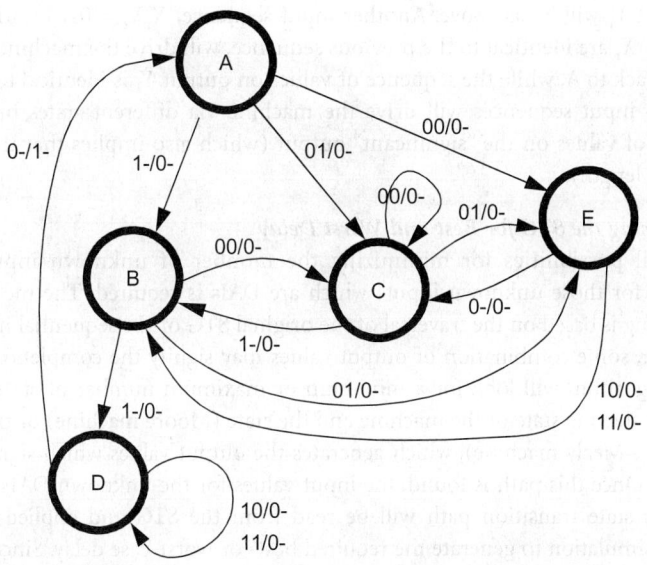

FIGURE 77.24   STG with "significant" output values only.

TABLE 77.2   Next State and Output Y_1 of the Minimal Machine

PS\$X_1 X_2$	00	01	10	11
K = (ACE)	K, 0	K, 0	B, 0	B, 0
B	K, 0	K, 0	D, 0	D, 0
D	K, 1	K, 1	D, 0	D, 0

TABLE 77.3   The Minimal Machine with Implicit Input Enumeration

$PS\backslash X_1X_2$	0-	1-
K = (ACE)	K, 0	B, 0
B	K, 0	D, 0
D	K, 1	D, 0

The reduced state table reveals that the minimal machine does not depend on the value of input X_2. Therefore, the conclusion is that input X_2 is not a DAI. This result implies that by knowing only the value of input X_1, the number of clock cycles (transitions in the original STG) required to reach the condition that output $Y_1 = 1$ can be determined regardless of the values of X_2. This is the case for any given initial state. It is important to emphasize that the paths in the *original* STG and their lengths are those which must be considered during the traversal process and that the modified state machine is used only for the purpose of detecting non-DAIs. The machine which is actually being traversed during the mixed-level simulation is the original state machine with all its functionality.

To demonstrate the meaning of an input which is not a DAI, consider the original state machine represented by the graph in Figure 77.23 and assume that the initial state is A. Consider, for example, one possible sequence of values on input X_1 to be 0, 1, 0, 0, 1, 1, 0. By applying this input sequence, the sequence of values on output Y_1 is 0, 0, 0, 0, 0, 0, 1, regardless of the values applied to input X_2. Therefore, two input sequences which differ only in the values of the non-DAI input X_2 will produce the same sequence of values on the "significant" output Y_1. For example, the sequence $X_1X_2 = 00, 10, 00, 01, 10, 10, 01$ will drive the machine from state A to E, B, C, E, B, D, and back to A, and the sequence of values on output Y_1 will be as above. Another input sequence, $X_1X_2 = 01, 11, 01, 00, 11, 11, 00$, in which the values of X_1 are identical to the previous sequence, will drive the machine from state A to C, B, E, C, B, D, and back to A, while the sequence of values on output Y_1 is identical to the previous case. Therefore, the two input sequences will drive the machine via different states but will produce an identical sequence of values on the "significant" output (which also implies that the two paths in the STG have an equal length).

77.4.2.1.2 Traversing the STG for Best and Worst Delay

After extracting all possibilities for minimizing the number of unknown inputs, a method for determining values for those unknown inputs which are DAIs is required. The method developed for mixed-level modeling is based on the traversal of the original STG of the sequential interpreted element. As explained earlier, some combination of output values may signify the completion of processing the data. The search algorithm will look for a minimum or maximum number of state transitions (clock cycles) between the starting state of the machine and the state (Moore machine) or transition (state and input combination—Mealy machine), which generates the output values which signify the completion of data processing. Once this path is found, the input values for the unknown DAIs that will cause the SDE to follow this state transition path will be read from the STG and applied to the interpreted component in the simulation to generate the required best- or worst-case delay. Since the state machine is represented by the STG, this search is equivalent to finding the longest or shortest path between two nodes in a directed graph (digraph).

The search for the shortest path utilizes a well-known algorithm. Search algorithms exist for both single-source shortest path and all-pairs shortest path. One of the first and most commonly used algorithm is Dijkstra's algorithm [46], which finds the shortest path from a specified node to any other node in the graph. The search for all-pairs shortest path is also a well-investigated problem. One such algorithm by Floyd [47] is based on work by Warshall [48]. Its computation complexity is $O(n^3)$ when n is the number of nodes in the graph, which makes it quite practical for moderate-sized graphs. The implementation of this algorithm is based on Boolean matrix multiplication and the actual realization of all-pairs shortest paths can be stored in an $n \times n$ matrix. Utilizing this algorithm required some enhancements to make it applicable to mixed-level modeling. For example, if some of the inputs to the interpreted element are known (from the performance model), then the path should include transitions that include these known input values.

In contrast, the search for the longest path is a more complex task. It is an NP-complete problem that has not attracted significant attention. Since most digraphs contain cycles, the cycles need to be handled during the search to prevent a path from containing an infinite number of cycles. One possible restriction that makes sense for many state machines that might be used in mixed-level modeling is to construct a path that will not include a node more than once. Given a digraph $G(V, E)$ that consists of a set of vertices (or nodes) $V = \{v_1, v_2, \ldots\}$ and a set of edges (or arcs) $E = \{e_1, e_2, \ldots\}$, a *simple path* between two vertices v_{init} and v_{fin} is a sequence of alternating vertices and edges $P = v_{init}, e_n, v_m, e_{n+1}, v_{m+1}, e_{n+2}, \ldots, v_{fin}$ in which each vertex does not appear more than once. Although an arbitrary choice was made to implement the search allowing each vertex to appear in the path only once, the same algorithm could be easily modified to allow the appearance of each vertex a maximum of N times.

Given an initial node and a final node, the search algorithm developed for this application starts from the initial node and adds nodes to the path in a depth-first-search (DFS) fashion, until the final node is reached. At this point, the algorithm backtracks and continues looking for a longer path. However, since the digraph may be cyclic, the algorithm must avoid the possibility of increasing the path due to a repeated cycle, which may produce an infinite path.

The underlying approach for avoiding repeated cycles in the algorithm *dynamically* eliminates the cycles while searching for the longest simple path. Let u be the node that the algorithm just added to the path. All the in-arcs to node u can be eliminated from the digraph at this stage of the path construction. The justification for this dynamic modification of the graph is that, while continuing in this path, the simple path cannot include u again. While searching forward, more nodes are being added to the path and more arcs can be removed temporarily from the graph. At this stage, two things may happen (1) either the last node being added to the path is the final node or (2) the last node has no out-arcs in the dynamically modified graph. These two cases are treated in the same way except that in the first case the new path is checked to see if it is longer than the longest one found so far. If it is, the longest path is updated. However, in both cases the algorithm needs to backtrack.

Backtracking is performed by removing the last node from the path, hence decreasing the path length by one. During the process of backtracking, the in-arcs to a node being removed from the path must be returned to the current set of arcs. This process will enable the algorithm to add this node when constructing a new path. At the same time, whenever a node is removed from the path, the arc that was used to reach that node is marked in the dynamic graph. This process will eliminate the possibility that the algorithm repeats a path that was already traversed. Therefore, by dynamically eliminating and returning arcs from/to the graph, a cyclic digraph can be treated as if it does not contain cycles. The process of reconnecting nodes, i.e. arcs being returned to the dynamic graph, requires that the original graph be maintained. A more detailed description of this search algorithm can be found in Ref. [49].

The restriction that a longest path not include any *node* in the graph more than once may not be appropriate for all cases. In some mixed-level modeling cases, a more realistic restriction might be that the longest path not include any *transition* (arc) more than once. A longest-path with no repeated arcs may include a node multiple times as long as it is reached via different arcs. In the case of more than one transition that meets the condition on the output combination, a search for the longest path should check all paths between the initial state and all of these transitions. However, such a path should include any of these transitions only once, and it should be the last one in the path.

Performing a search with the restriction that no arc is contained in the path more than once requires maintaining information on arcs being added or removed from the path. Maintaining this information during the search makes the algorithm and its implementation much more complicated relative to the search for the longest path with no repeated nodes. Therefore, a novel approach to this problem was developed. The approach used is to map the problem to the problem of searching for the longest path with no repeated nodes. This mapping can be accomplished by transforming the digraph to a new digraph, to be referred to as the transformed digraph (or Tdigraph). Given a digraph, $G(V, E)$, which consists of a set of nodes $V = \{v_1, v_2, \ldots, v_k\}$ and a set of arcs $E = \{e_1, e_2, \ldots, e_l\}$ the transformation τ maps $G(V, E)$ into a Tdigraph $TG(N, A)$, where N is its set of nodes and A its set of arcs. The transformation is defined as $\tau(G(V, E)) = TG(N, A)$, and contains the following steps:

Step 1. $\forall(e_i \in E)$ generate a node $n_i \in N$.

Step 2. $\forall(v \in V)$ and $\forall(e_p \in d_{in}^v, e_q \in d_{out}^v)$ generate an arc $a \in A$ such that $a : n_p \to n_q$.

The first step is used to create a node in the Tdigraph for each arc in the original digraph. This one-to-one mapping defines the set of nodes in the Tdigraph to be $N = \{n_1, n_2, \ldots, n_i\}$, which has the same number of elements found in the set E.

The second step creates the set of arcs, $A = \{a_1, a_2, \ldots, a_u\}$, in the Tdigraph. For each node in the original digraph and for each combination of in-arcs and out-arcs to/from this node, an arc in the Tdigraph is created. For example, given a node v with one in-arc e_i and one out-arc e_j, e_i is mapped to a node n_i, e_j is mapped to a node n_j, and an arc from n_i to n_j is created. In Step 2 of the transformation process, it is guaranteed that all possible connections in the original digraph are preserved as transitions between nodes in the Tdigraph. As a result of this transformation, the restriction on not visiting an arc more than once in the original digraph is equivalent to not visiting a node more than once in the Tdigraph. Therefore, by using this transformation, the problem of searching for the longest path with no repeated arcs in the original digraph is mapped to a search for the longest path with no repeated nodes in the Tdigraph. The algorithm described above can then be used to search the Tdigraph.

This transformation is best illustrated by a simple example. Consider the digraph shown in Figure 77.25(A). The arcs in this digraph are labeled by numbers 0–6. The first step of the transformation is to create a node for each arc in the original digraph. Therefore, there will be seven nodes, labeled 0–6, in the Tdigraph as shown in Figure 77.25(B). The next step is to create the arcs in the Tdigraph. As an illustration of this step, consider node C in the original digraph. Arc "2" is an in-arc to node C while arcs "3" and "4" are out-arcs from node C. Applying Step 2 results in an arc from node "2" to node "3" and a second arc from node "2" to node "4" in the Tdigraph. Considering node B in the original digraph, the Tdigraph will include an arc from node "1" to node "2" and an arc from node "5" to node "2." This process is repeated for all the nodes in the original digraph until the Tdigraph, as shown in Figure 77.25(B), is formed. A search algorithm can now be executed to find the longest path with no repeated nodes.

A mixed-level modeling methodology, which is composed of all the methods described above, has been integrated into the ADEPT environment. The steps for minimizing the unknown inputs can be

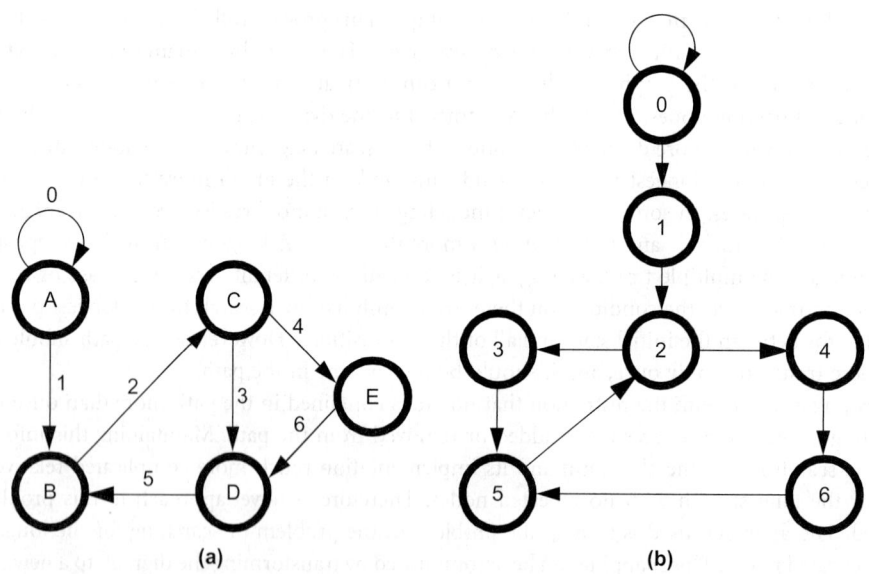

(a)   (b)

FIGURE 77.25 Transformation of a digraph.

performed prior to simulating the mixed-level model. On the other hand, the search for longest/shortest possible delay must be performed during the simulation itself. This requirement arises because each token may carry different information which may alter the known input values and, therefore, alter the search of the STG. The STG traversal process has been integrated into the ADEPT modeling environment using the following steps: (1) when a token arrives to the mixed-level interface, the simulation is halted and the search for minimum/maximum number of transitions is performed and (2) upon completion of the search, the simulation continues while applying the sequence of inputs found in the search operation. The transfer of information between the VHDL simulator and the search program, which is implemented in C is done by using the simulator's VHDL/C interface. A component called the Stream_Generator has been created that implements the STG search process via this interface.

Since mixed-level models are part of the design process and are constructed by refining a performance model, it is likely that many tokens will carry identical relevant information. This information may be used for selective application of the STG search algorithm, hence increasing the efficiency of the mixed-level model simulation. For example, if several tokens carry exactly the same information (and assuming the same initial state of the FSM), the search is performed only once, and the results can be used for the following identical tokens.

77.4.2.2 An Example of Mixed-Level Modeling with an FSMD Component

This section presents an example of the construction of a mixed-level model with an FSMD interpreted component. The example is based on the performance model of an execution unit of a particular processor. This execution unit is composed of an integer unit (IU), a floating-point unit (FPU), and a load-store unit (LSU). These units operate independently although they receive instructions from the same queue (buffer of instructions). If the FPU is busy processing one instruction and the following instruction requires the FPU, it is buffered, waiting for the FPU to be free again. Meanwhile, instructions which require the IU can be consumed and processed by IU at an independent rate. Both the FPU and the IU have the capability of buffering only one instruction. Therefore, if two or more consecutive instructions are waiting for the same unit, the other units cannot receive new instructions (since the second instruction is held in the main queue). One practical performance metric that can be obtained from this model is the time required for the execution unit to process a given sequence of instructions.

Because the FPU was identified as the most complex and time critical portion of the design, a behavioral description of a potential implementation of it was developed. At this point, a mixed-level model, in which the behavioral description of the FPU is introduced into the performance model, was constructed using the interface described above. The mixed-level model is shown in Figure 77.26. The mixed-level interface is constructed around the interpreted block which is the behavioral description of the FPU. This FPU is an FSMD type of element, and the interpreted model consists of a clock cycle-accurate VHDL behavioral description of this component. The inputs to the FPU include the operation to be performed (Add, Sub, Comp, Mul, MulAdd, and Div), the precision of the operation (single or double), and some additional control information. The number of clock cycles requires to complete any instruction depends on these inputs.

Figure 77.27 shows the execution unit performance derived from the mixed-level model for three different instruction traces. In this case, only 40% of the inputs have values that are supplied by the abstract performance model, the remainder of the values for the inputs are derived using the techniques described in Section 77.4.2.1. The performance value is normalized by defining unity to be the amount of time required to process a trace according to the initial uninterpreted performance model. In this example, the benefit of the simulation results of the mixed-level model is clear. It provides performance bounds in terms of best- and worst-case delays, for the given implementation of the FPU.

77.4.3 An Interface for Mixed-Level Modeling with Complex Sequential Components

A methodology and components for constructing a mixed-level interface involving general sequential interpreted components that can be described as FSMDs was detailed in the previous section. However,

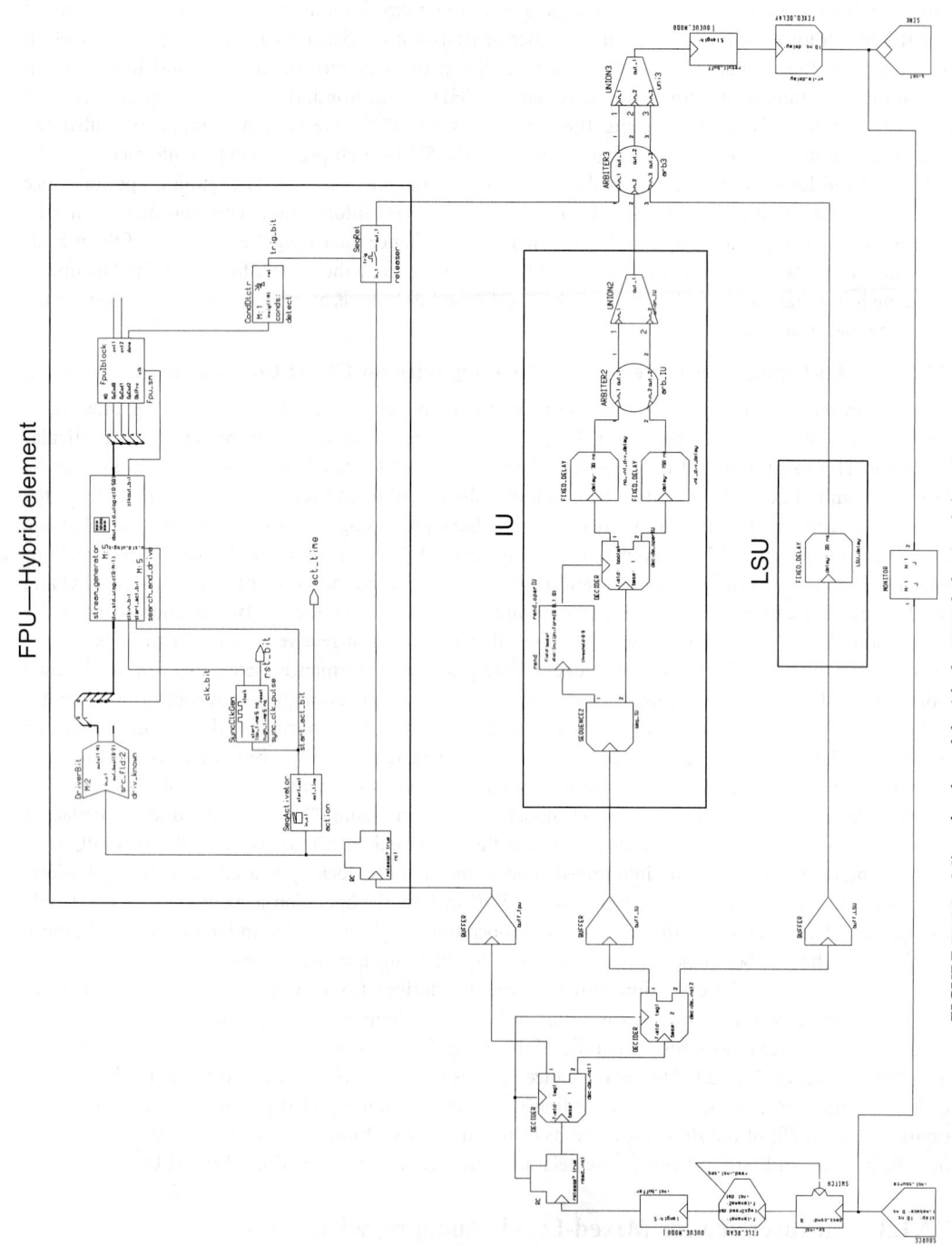

FIGURE 77.26 Mixex-level model with the FPU behavioral description.

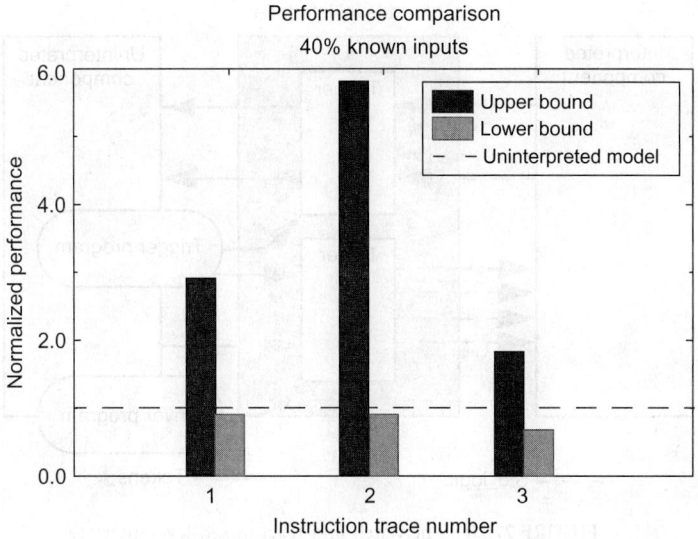

FIGURE 77.27   Performance comparison of the execution unit for three instruction traces.

many useful mixed-level models can be constructed that include sequential interpreted components that are too complex to be represented as FSMDs, such as microprocessors, complex coprocessors, network interfaces, etc. In these cases, a more "programmable" mixed-level interface that is able to deal with the additional complexity in the timing abstraction problem was needed. This section describes the "watch-and-react" interface that was created to be a generalized, flexible interface between these complex sequential interpreted components and performance models.

The two main elements in the watch-and-react interface are the *Trigger* and the *Driver*. Figure 77.28 illustrates how the Trigger and Driver are used in a mixed-level interface. Both elements have ports that can connect to signals in the interpreted components of a model. Collectively, these ports are referred to as the *probe*. The primary job of the Trigger is to detect events on the signals attached to its probe, while the primary job of the Driver is to force values onto the signals attached to its probe. The Driver decodes information carried in tokens to determine what values to force onto signals in the interpreted model (the U/I interface) and the Trigger encodes information about events in the interpreted model onto tokens in the performance model (the I/U interface).

The Trigger and Driver were designed to be as generic as possible. A command language was designed that specifies how the Trigger and Driver should behave to allow users to easily customize the behavior of the Trigger and Driver elements without having to modify their VHDL implementation. This command language is interpreted by the Trigger and Driver during simulation. Because the language is interpreted, changes can be made to the Trigger and Driver programs without having to recompile the model, thus minimizing the time required to make changes to the mixed-level model.

The schematic symbol for the Trigger is shown in Figure 77.29. The primary job of the Trigger is to detect events in the interpreted model. The probe on the Trigger is a bus of std_logic signals `probe_size` bits wide, where `probe_size` is a generic on the symbol. There is one token output called `out_event_token` and one token output called `out_color_token`. Tokens generated by the Trigger when events are detected are placed on the `out_event_token` port. The condition number (an integer) of the event that caused the token to be generated is placed on the `condition` tag field, which is specified as a generic on the symbol. Also, each time a signal changes on the probe, the color of the token on the `out_color_token` port changes appropriately regardless of whether an event was detected or not. The probe value is placed on the `probe_value` tag field of the `out_color_token` port. The `sync` port is used to synchronize the actions of the Trigger element with the Driver element as explained below.

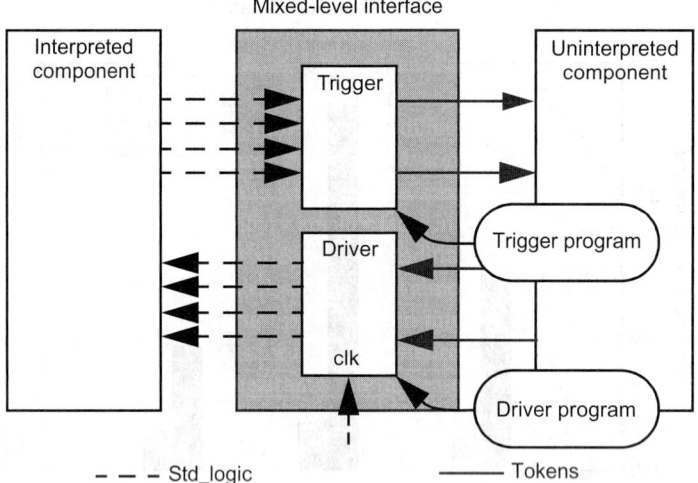

FIGURE 77.28   The watch-and-react mixed-level interface.

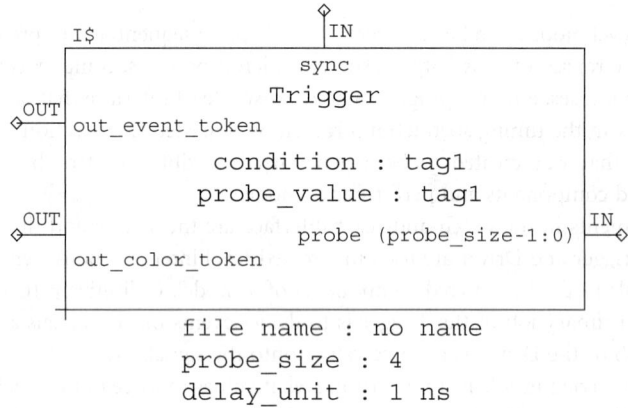

FIGURE 77.29   Schematic symbol for the Trigger.

The name of the file containing the Trigger's program is specified by the `filename` generic on the symbol. The `delay_unit` generic on the symbol is a multiple that is used to resolve the actual length of an arbitrary number of delay units specified by some of the interface language statements.

The schematic symbol for the Driver element is shown in Figure 77.30. The primary job of the Driver is to create events in the interpreted model by driving values on its probe. These values come from either the command program, or from the values of tag fields on the input tokens to the Driver. The probe on the Driver is a bus of std_logic signals `probe_size` bits wide, where the `probe_size` is a generic on the symbol. There is one token input called `in_event_token`, one token input called `in_color_token`, and a special input for a std_logic type clock signal called `clk`. The `clk` input allows Driver to synchronize its actions with an external interpreted clock source. The `sync` port is used to synchronize the actions of the Driver element with the Trigger element. The `filename` and `delay_unit` generic on the symbol function the same as for the Trigger.

As discussed previously, a command language was developed to allow the user to program the actions of the Trigger and Driver. This command language is read by the Trigger and Driver at the beginning of the simulation. Constructs are available within the command language to allow waiting on events of various signals and driving values or series of values on various signals, either asynchronously, or synchronous

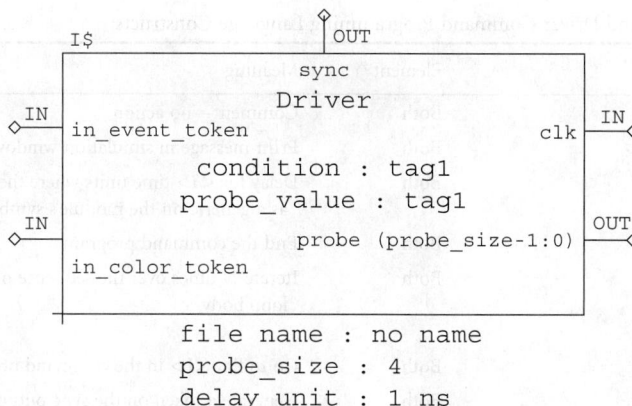

FIGURE 77.30 Schematic symbol for the Driver.

with a specified clock signal. In addition, several looping and go-to constructs are available to implement complex behaviors more easily. A summary of the syntax of the command language constructs is shown in Table 77.4.

77.4.3.1 Example of Mixed-Level Modeling with a Complex Sequential Element

This section presents an example which demonstrates how the Trigger and Driver elements can be used to interface an interpreted model of a complex sequential component with an uninterpreted model. In this example, the interpreted model is a microprocessor-based controller and the uninterpreted model is that of a motor control system including a motor controller and a motor. The motor controller periodically asserts the microcontroller's interrupt line. The microcontroller reacts by reading the motor's current speed from a sensor register on the motor controller, calculating the new control information, and writing the control information to the motor controller.

The microcontroller system consists of interpreted models of eight-bit, RISC-like microprocessor, RAM, memory controller, I/O controller, and clock. The memory controller handles read and write requests issued by the processor to the RAM, while the I/O controller handles read and write request issued by the processor to an I/O device. In the system model, the I/O device is the uninterpreted model of a motor controller. A schematic of the model is shown in Figure 77.31.

Three Triggers and two Drivers are used to construct the mixed-level interface for the control system model. One of the Triggers is used to detect when the I/O controller is doing a read or write. The other two Triggers are used to collect auxiliary information about the operation, such as the address on the address bus and data on the data bus. One of the Drivers is used to create a microcontroller interrupt and the other Driver is used to force data onto the data bus when the processor reads from the speed sensor register on the motor controller.

The interrupt Driver's program is listed in Figure 77.32. The program begins by forcing the interrupt line to 'Z' and then waiting for a token from the uninterpreted model of the motor controller to arrive. Once a token arrives, the program forces the interrupt line high for 10 clock cycles. This condition is accomplished by using a `for-next` statement with a `wait_on_rclk` as the loop body. After 10 clock cycles, the program jumps to line 10 where the cycle begins again.

The data Driver's program is listed in Figure 77.33. The program begins by also waiting for a token from the uninterpreted model of the motor controller to arrive. If the value on the condition tag field of the token is 1, then "ZZZZZZZZ" is forced onto the data bus. If the condition tag field value is 3, then the value on the `probe_value` tag field of the `in_color_token` input is forced on the data bus. This process is repeated for every token that arrives.

The I/O Trigger's program is listed in Figure 77.34. This Trigger waits until there is a change on the probe. Once there is a change, the program checks to see if the I/O device is being unselected, written

TABLE 77.4 Trigger and Driver Command Programming Language Constructs

Command	Element	Meaning
--<comment>	Both	Comment—no action
alert_user	Both	Print message in simulation window
delay_for <T>	Both	Delay for <T> time units where the time unit is specified as a generic on the module's symbol
end	Both	End the command program
for <N> <loop body> **next**	Both	Iterate N times over the sequence of statement in the loop body
goto <L>	Both	Go to line <L> in the command program
output_sync	Both	Generate a token on the sync output of the module
wait_on_sync	Both	Wait for an occurrence of a token on the sync port of the module

case_probe_is
  when <STD_LOGIC_VAL>
  <sequence of statements>
  when. . .
end_case
	Trigger	Conditionally execute some sequence of statements depending on the STD_LOGIC value of the probe signal

output <INTEGER_VAL> **after** <T>	Trigger	Generate a token on the Trigger's output with the value of <INTEGER_VAL> on the tag field specified by the generic on the symbol after <T> time units
trigger	Trigger	Must appear as the first statement in the Trigger's command program
wait_on <STD_LOGIC_VAL>	Trigger	Wait until the probe signal takes on the specified STD_LOGIC value
wait_on_probe	Trigger	Wait until there is ANY event on the probe signal

case_token_is
  when <INTEGER_VAL>
  <sequence of statements>
  when . . .
end_case
	Driver	Conditionally execute some sequence of statements depending on the integer value of the input token's tag field specified by the generic on the symbol

driver	Driver	Must appear as the first statement in the Driver's command program
dynamic_output_after <T>	Driver	Force the value from the specified tag field of the input token onto the probe after <T> time units
output <STD_LOGIC_VAL> **after** <T>	Driver	Force the specified STD_LOGIC value onto the probe signal after <T> time units
wait_on <INTEGER_VAL>	Driver	Wait until a token with the given integer value on the tag field specified by the generic on the symbol arrives on the in_event_token signal
wait_on_fclk	Driver	Wait until the falling edge of the clock occurs
wait_on_rclk	Driver	Wait until the rising edge of the clock occurs
wait_on_token	Driver	Wait until a token arrives on the in_event_token signal

to, or read from. If one of the when statements matches the probe value, then its corresponding output statement is executed. An output of 1 corresponds to the I/O device not being selected. An output of 2 corresponds to the processor writing control information to the motor controller. An output of 3 corresponds to the processor reading the motor's speed from the sensor register on the motor controller.

Figure 77.35 shows the results from the mixed-level model as a plot of the sensor output and the processor's control response. Some random error was introduced to the sensor's output to reflect variations in the motor's load as well as sensor noise. The target speed for the system was 63 ticks per sample time (a measure of the motor's RPM). The system oscillates slightly around the target value because of the randomness introduced into the system.

77.5 Performance and Mixed-Level Modeling Using SystemC

This section describes a performance-modeling environment capable of mixed-level modeling that is based on the SystemC language [50]. The environment is intended to model the system at the Processor Memory Switch level much like the Honeywell PML environment described earlier. The goal of this work was to show how SystemC could be used to construct a mixed-level modeling capability.

In the SystemC-based PBMT (Performance-Based Modeling Tool), the user begins by describing the functions executed by the system as a task graph. A task graph is a representation of the flow of execution through an application. The nodes in a task graph represent computational tasks, and the edges in a task graph represent the flow of control, or the actual transfer of data, between tasks. An example of a task graph for a simple application is shown in Figure 77.36. Note that the topology shown in the figure, the example application has the opportunity for some tasks, such as like Task 2, Task 3, and Task 4, to be executed in parallel if the system architecture upon which the application is to be executed, allows for it.

Once the task graph model is constructed, the user then selects a system architecture on which the application will execute. The system architecture is specified by the number of processors in the architecture, and an interconnect topology used to provide communications between them. The available interconnect topologies include a bus (a single, shared communications resource), a crossbar switch (a partially shared communications resource), or fully-connected (a completely nonshared communication resource). Note that in this high-level architecture model, what actually constitutes a processor in the system is not specified. That is, a processor is simply modeled as a computational resource and may in implementation be a general purpose processor (of any clock speed), a special purpose processor like a DSP, or custom hardware for a specific task.

Once the system architecture is specified, all that remains is for the user to specify upon which processor each of the tasks is to execute and what the total execution time for that task on the specific processor will be. This delay value may be either fixed, or dependent on the amount of data that is passed into the task by the previous task in the graph. Once this task-to-processor mapping and delay specification is done, the complete SystemC model is constructed and simulated using either the reference simulator, included as part of the SystemC distribution available in Ref. [12], or the commercial Mentor Graphics ModelSim simulator which includes the capability to co-simulate SystemC models along with Verilog or VHDL models.

Figure 77.37 shows the results of executing the task graph of Figure 77.36 on three different system architectures. All of the architectures utilize a single shared bus for communications. The first result is for an architecture with only a single processor. In this case, the obvious result is that all of the tasks execute in sequence on the single processor and the run time is simply the sum of the individual task execution times. The second result is for a three-processor architecture. In this case, after Task 1 completes, some latency can be seen before Tasks 3 and 4 begin execution. This accounts for the communication time required to send data from the processor that executed Task 1 to the processors that are executing Tasks 3 and 4. In addition, the graph shows that Task 4 begins execution after Task 3 because of the contention for the single shared bus communications resource. Likewise, the latency between the end of execution for Tasks 5–7, and the start of execution for Task 8 accounts for the time required to communicate Task 6 and 7's results back to the single processor that is schedule to execute Task 8. The overall run time for this configuration is much less than the first example because the inherent parallelism in the application is being exploited by the selected architecture. Finally, the third result is for a four-processor architecture. In this simulation, Tasks 1 and 8 are allocated to the fourth processor, separate from the other Tasks 2–7. This results in additional communications time being required using the single bus to transfer all of the data from, and back to, that extra processor. Thus, as can be seen from the graph this architecture actually takes longer than the three processor architecture to execute the application.

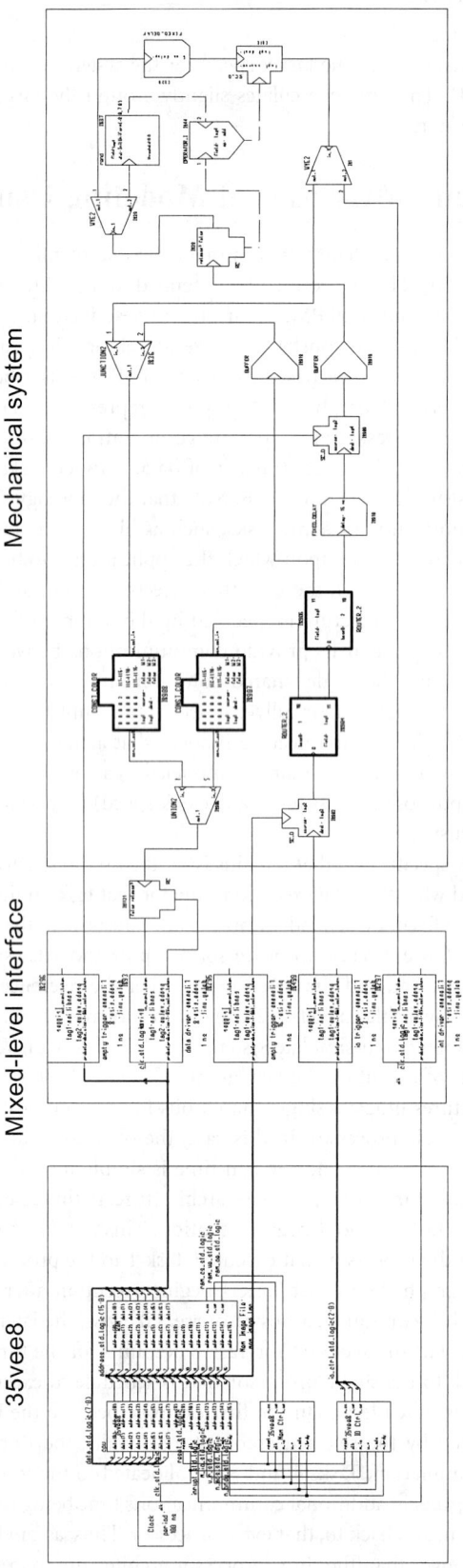

FIGURE 77.31 Schematic of control system model.

```
         driver
      10 output Z after 0
      20 wait_on_token
      30 output 1 after 0
      40 for 10
      50    wait_on_rclk
      60 next
      70 goto 10
      80 end
```

FIGURE 77.32   The interrupt Driver's program for the control system.

```
      driver
   10 wait_on_token
   20 case_token_is
   30     when 1
   40         -- sensor not selected
   50         output ZZZZZZZZ after 0
   60     when 3
   70         -- sensor selected for reading
   80         dynamic_output_after 0
   90 end_case
  100 goto 10
  110 end
```

FIGURE 77.33   The data Driver's program for the control system.

```
      trigger
   10 wait_on_probe
   20 case_probe_is
   30     when 111
   40         -- sensor not selected
   50         output 1 after 0
   60     when 001
   70         -- sensor selected for writing
   80         output 2 after 0
   90     when 010
  100         -- sensor selected for reading
  110         output 3 after 0
  120 end_case
  130 goto 10
  140 end
```

FIGURE 77.34   The I/O Trigger's program for the control system.

77.5.1   SystemC Background

This section describes some of the basic concepts of SystemC and how it is used to model systems at a high level. SystemC is a library extension to C++. Essentially, SystemC uses a set of classes and predefined functions to build a new "language" on top of C++. The basic unit of a SystemC model is the SC_MODULE class. The SC_MODULE is roughly equivalent to an entity in VHDL. It can have input, output, and bidirectional ports. However, it is a C++ class and as such it can, and does, have member functions. Any member function that is declared as public can be accessed just like the member function of any other class. Every SC_MODULE must have at least one public member function, its constructor. In the constructor the module is given a SystemC name, and declares if it has any processes, and does anything else necessary to get the model ready for simulation. There are two types of processes in SystemC, the SC_THREAD and the SC_METHOD. The behavior of a process must first be declared as a

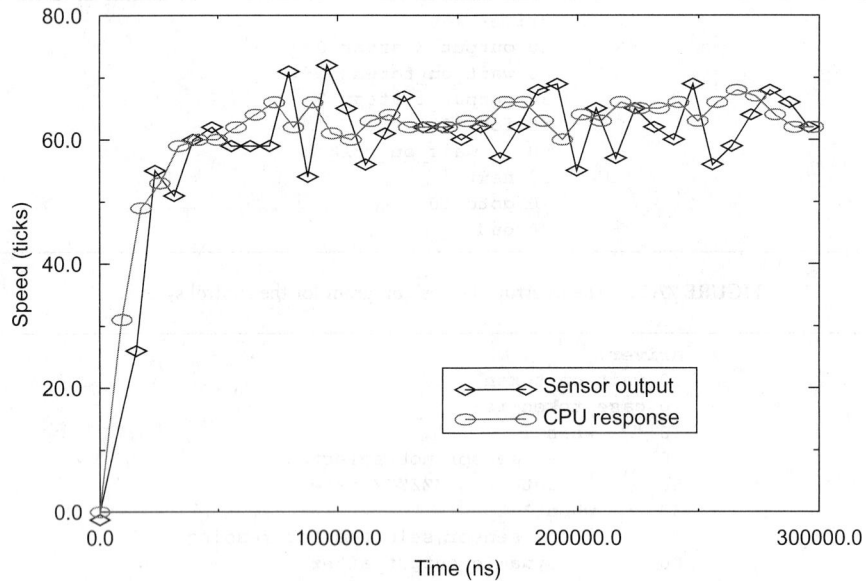

FIGURE 77.35   Sensor and processor outputs for the control system.

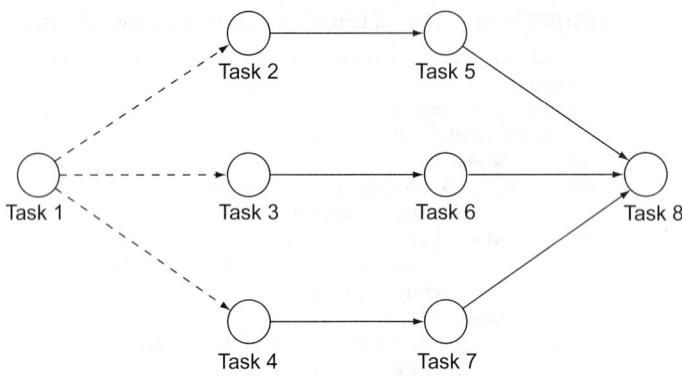

FIGURE 77.36   Example task graph.

member function, then that member function can be declared to be either an SC_THREAD or an SC_METHOD in the constructor. The primary differences are how they behave when they reach the end of their definition, and when the scheduler activates them.

An SC_THREAD will terminate, and never be activated again, if it reaches the end of its description. Typically SC_THREADs are infinite loops with one or more wait statements to break the execution, and wait on some signal change or other event. An SC_METHOD will be activated any time something it is sensitive to changes, and will run once through to the end of its description. If something an SC_METHOD is sensitive to changes again, it will run again. SC_THREADs do have the ability to use a form of the wait command that is not available to SC_METHODs. They may use a wait command with no parameters which will cause them to be reactivated and continue execution with the line after the wait statement when something in their sensitivity list has an event.

As indicated by the discussion of processes, SystemC uses an event-driven simulation methodology. The library provides basic signal classes that have a notify-update sequence much like VHDL. As mentioned in the Transaction Level Modeling discussion above, SystemC has a concept of channels. A channel is generally

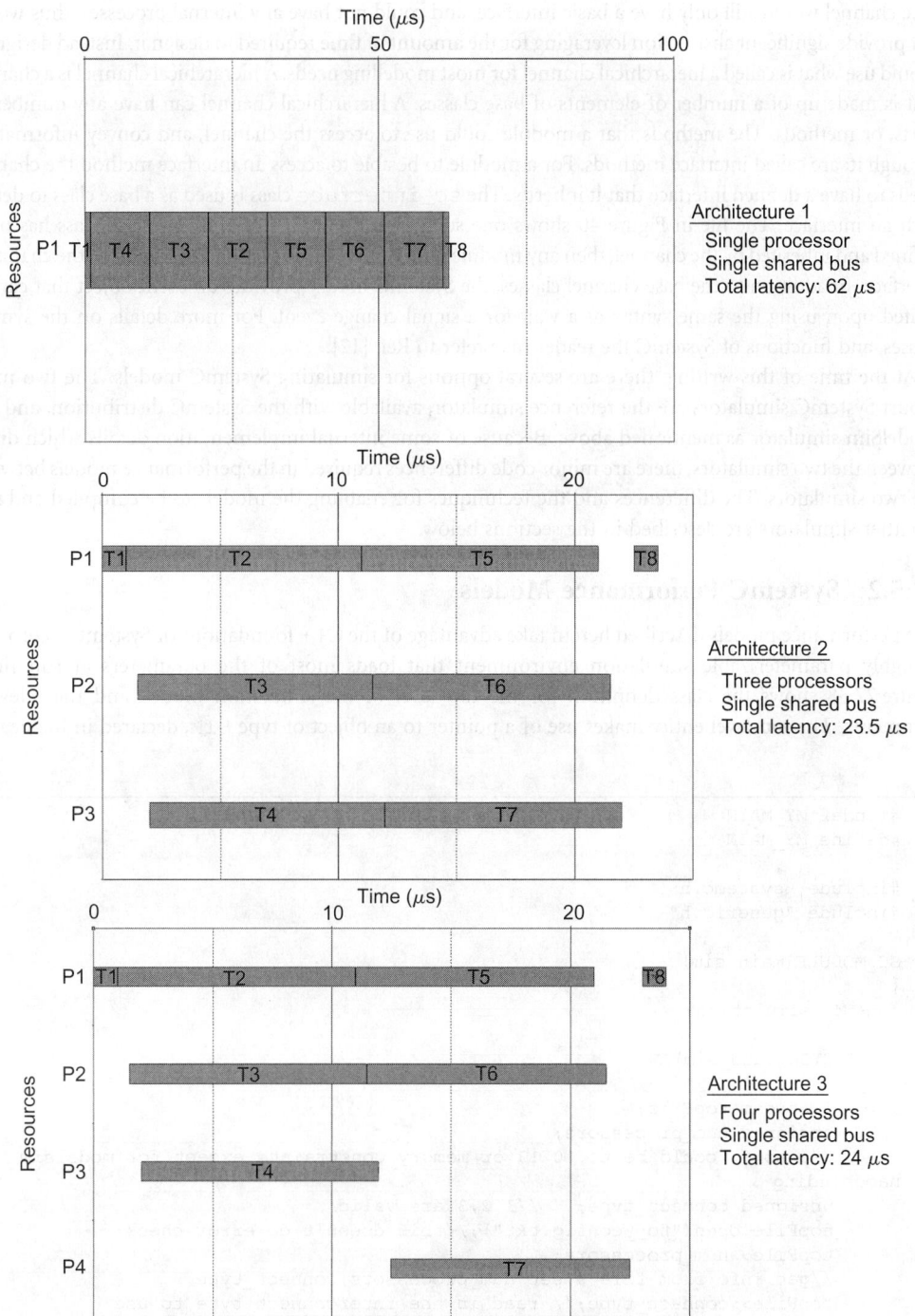

FIGURE 77.37 Parallel example shared bus timelines.

some means of moving information. The basic signal classes provided are base channels. A designer could potentially design some new base channel type that has the same interfaces as the existing base channels. However, building a new base channel is rather involved since its implementation must interact directly with the scheduler to implement the notify/update semantics of a signal. Additionally, such a user-designed base channel would still only have a basic interface, and could not have any internal processes, thus would not provide significant abstraction leveraging for the amount of time required to design it. Instead designers should use what is called a hierarchical channel for most modelling needs. A hierarchical channel is a channel that is made up of a number of elements of base classes. A hierarchical channel can have any number of ports, or methods. The methods that a module could use to access the channel, and convey information through it, are called interface methods. For a module to be able to access an interface method the channel needs to have a defined interface that it inherits. The `sc_interface` class is used as a base class to define such an interface. The file in Figure 40 shows one such interface class. Once an interface class has been defined and inherited by the channel, then any module with a port of that type can be bound to the channel's interface. In addition to the base channel classes, the SystemC library provides an event object that can be waited upon using the same syntax as a wait for a signal change event. For more details on the syntax, classes, and functions of SystemC the reader may refer to Ref. [12].

At the time of this writing, there are several options for simulating SystemC models. The two most robust SystemC simulators are the reference simulator, available with the SystemC distribution, and the ModelSim simulator as mentioned above. Because of some internal implementation details which differ between the two simulators, there are minor code differences required in the performance models between the two simulators. The differences and the techniques for enabling the models to be compiled and run in either simulators are described in the sections below.

77.5.2 SystemC Performance Models

The performance models described herein take advantage of the C++ foundations of SystemC to provide a highly parameterizable simulation environment that loads most of the parameters at run time. Figure 77.38 shows the class definition for the top level of the simulation model, and the relevant comments. The top-level entity makes use of a pointer to an object of type `SIM`, declared in the header

```
#ifndef MY_MAIN
#define MY_MAIN

#include <systemc.h>
#include "generic.h"

SC_MODULE(main_sim)
{
   SIM *simulation;

   SC_CTOR(main_sim)
   {
      ifstream topFile;
      unsigned num_processors;
      //1 to 9, could be to 10^13 or memory constraints except for modelsim
namebinding
      unsigned connect_type;   //1,2,3 are valid
      topFile.open("top_config.txt");//this doesn't do error check!
      topFile>>num_processors;
      //get info from file & set num_processors,connect_type
      topFile>>connect_type;// read in the interconnect type to use
      simulation = new SIM("simulation",num_processors,connect_type);
      // systemC name,number of processor ,connection class
   };

};
#endif
```

FIGURE 77.38 Top-level class definition.

file *generic.h*, to allow the constructor arguments to be read in from the *top_config.txt* file rather than statically specified in the source code. When the simulation is loaded the SystemC constructor (SC_CTOR in the code) will be called. The constructor will create a streambuf object, a number of processors variable, a connection type variable, as well as a pointer variable to point to an object of type SIM. It will then open the *top_config.txt* file and associate it with the streambuf variable. Then it will read an unsigned value into the number of processors variable, then read an unsigned value into the connection type variable. Once those values are read in then the constructor uses the 'new' command to instantiate a new object of type SIM. The "new" command allows passing variable constructor arguments to the objects constructor, so the new SIM object's constructor builds the object with the value of num_ processors processors, and the value of connect_type interconnect number. If the number of processors or interconnect type was specified via a template argument, or a fixed constructor value (e.g.: "simulation = new SIM("it,9,1);") in the source code then the simulation would have to be recompiled every time something was changed.

Notice that all the implementation details of the SystemC modules are declared in header files. This is the recommended way of describing models to compile the SystemC code for the ModelSim simulator. The included guards are also a must for any SC_MODULE definitions to ensure they are not included multiple times by the ModelSim SystemC compiler, SCCOM. The actual C++ file that is compiled to generate the simulation models is shown in Figure 77.39. The SC_MODULE_EXPORT(module_ class_name) is the function used to create the ModelSim model for the specified entity. The MTI_SYSTEMC compiler definition is defined specifically for compilation using SCCOM, and allows having a single set of code for both the ModelSim and reference simulator. Any ModelSim specific code can be placed inside a #ifdef MTI_SYSTEMC compiler directive statement so that the compiler only uses it when compiling for ModelSim. The #else statement prevents the ModelSim compiler from trying to compile the reference simulator-specific portions, and the #endif closes out the if else statement.

The *generic.h* header file defines the class SIM. It makes use of the shared base class my_basic_ rw_port to declare a pointer that will allow assignment of an instantiation of any of the three channel types developed for the environment. The my_basic_rw_port base class definition is shown in Figure 77.40. Notice the use of the keyword virtual. Since the method is declared as virtual, all classes that inherit from this class must either provide, or inherit, an implementation for the read, write, and non_blk_write methods. This common interface is also what allows the type compatibility for pointer assignment used in the *generic.h* file. This is described in more detail in Section 77.5.4.

77.5.3 Processor Model

The processor model has a fairly simple structure. There are three methods in this model, the constructor, and two member functions. The first member function describes the performance only behavior of the

```
#include <systemc.h>
#include "generic_main.h"
#ifdef MTI_SYSTEMC
SC_MODULE_EXPORT(main_sim);
#else
//for OSCI reference simulator
int sc_main(int ac, char *av[] )
{
    main_sim my_sim("my_sim");
    sc_start(500, SC_NS);
    sc_close_vcd_trace_file(main_trace);
    return 0;
};
#endif
```

FIGURE 77.39 Top-level C++ file.

```
#ifndef INTERFACE_TYPE
#define INTERFACE_TYPE
include <systemc.h>
/* this header describes the interface type...*/


class my_basic_rw_port
  : public virtual sc_interface
{
public:
  // basic read/write interface
  //virtual bool read(int address, int data) = 0; //not used anymore!
  virtual bool read(int address, int source_address, int data,
                    int dest_task) = 0;
  virtual bool write(const int source, int address, int data,
                     int dest_task) = 0;
  virtual bool non_blk_write(const int dest_address, int source,
                             int data, int dest_task)=0;
}; // end class
#endif
```

FIGURE 77.40   "interface_type.h."

processor, the second describes the mixed-level behavior. The mixed-level functionality will be described in detail in Section 77.5.9.

In addition to those methods, the processor model has a number of objects that are members of the class. It has three integer variables for passing command arguments to the interface methods, a pointer for a command_in object that opens the command file and parses the model execution commands for the processor model, a command_type object that is used to return the commands from the command_in object, a signal of enumerated type action_type to display the current action, an unsigned signal to display the current task number, and a pointer for a refined computation model. Figure 77.41 graphically shows the objects in the processor model. On the far right is the IO port. It is mapped to the interconnect interface. Next on the right are the three integer variables used to pass information to the interconnect calls. Below the variables are boxes representing the two signals that allow the state of the processor to be viewed in the ModelSim waveform window. Then to the left there are the performance and mixed-level descriptions. One of these member functions will be turned into an SC_THREAD and will control the model's behavior during simulation. On the top labeled as refined model, is an outlined box representing the pointer to a refined model object. Below is a storage location (labeled command) that the command_in object will return values in. In the bottom left is a dashed box representing a pointer to a command_in object. In the top left is a box representing the constructor. The io-port and constructor are both on the edge of the processor model because they are the only ones that interact with other objects in the simulation.

The constructor receives a processor number, and creates a command_in object with the appropriate processor number for the processor the current instance represents. The command_in object creates a string with the proper processor number in the middle, and uses it to open the processor's command file. It is called by the active behavior to read in the next command once the previous one has executed. Its primary function is to remove command parsing from the processor model. Having it as a separate object makes changing how the commands for the processor are read in or generated a simple matter of including a different implementation of the object. Once the command_in object is created and initialized, the processor model constructor then instantiates either the performance only or mixed-level implementation, and opens its log file. It does this by registering the proper member function with the simulation kernel as an SC_THREAD. During simulation the processor model essentially reads in a command from a file, then uses a case statement to perform whatever command was read in and repeats until it reaches a done command, an end of file, or some command it does not recognize. Figure 77.42 shows the framework of the main processor loop.

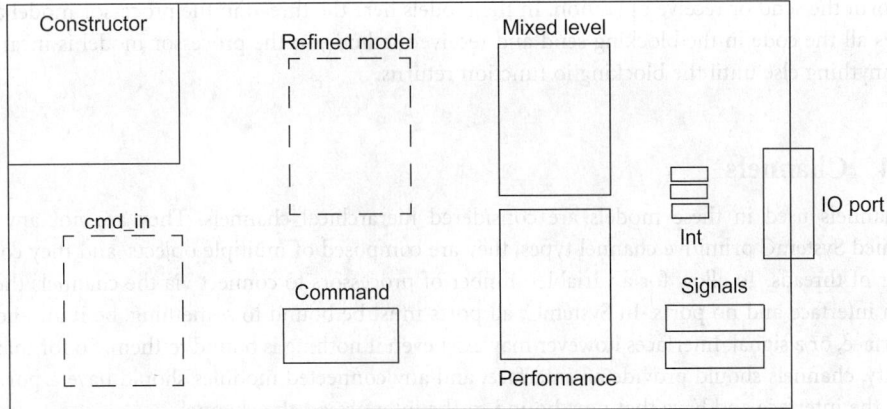

FIGURE 77.41 A graphical representation of processor model.

```
void tlm_behav()
{
    int temp2;
    double temp_time;
    while(1)
    {
        cmd_input->get_command(&this_command);
        switch (this_command.command)
        {
        case 1://ie send
            {...
                break;}
        case 2: //ie receive
            {...
                break;};
        case 0: //ie compute
            {...
                break;};
        case 3: //non_blocking_send
            {...
                break;};
        case 4: //io done
            {...
                break;};
        case 5: //loop/branch
            {...
                break;};
        default: //idle the processor
            {...
                break;};
        };//end of switch
        wait(SC_ZERO_TIME); //to break things up in the text output...
    };//end of infinite while loop
};
```

FIGURE 77.42 Main processor loop.

Notice in the source code that the processor has a port of the same type as the base type for the interconnect channels my_basic_rw_port. This port is bound to the interface of the channel object. If the command read in from the command file is a send or receive command, then the processor uses the port as a pointer to the interface to the channel object and accesses the appropriate interface method to perform the send or receive operation. In the models here the thread in the processor model actually executes all the code in the blocking send and receive methods, so the processor model is incapable of doing anything else until the blocking io function returns.

77.5.4   Channels

The channels used in these models are considered hierarchical channels. They are not any of the predefined SystemC primitive channel types, they are composed of multiple objects, and they contain a number of threads. To allow for a variable number of processors to connect via the channels they have only an interface and no ports. In SystemC, all ports must be bound to something, be it another port, an interface, or a signal. Interfaces however may exist even if nothing is bound to them. So for maximum flexibility, channels should provide an interface, and any connected modules should have a port of the type of the interface and have that port bound to the interface on the channel.

Since multiple ports can be bound to a single interface, the channel object can have any number of processors bound to it. However, the crossbar and fully connected channels' behavior is determined in part by the number of processors present, so all channels are passed a constructor argument that tells them how many processors are present in the simulation. The channel models are the most extensive models since their behavior is an abstract representation of all of the characteristics of an interconnect topology. They model the arbitration, data transfer, and blocking/nonblocking characteristics of the interconnect without restricting the designer to a particular implementation. Since nonblocking sends are allowed they also implicitly model a sending queue.

All the channel models are all based on the `comm_medium` class. The `comm_medium` class provides a logical connection between processors, with signals for the source processor number, the destination processor number, and transaction type, as well as blocking and nonblocking read and write methods, and an arbitration thread. In addition, the `comm_medium` class provides two threads to allow for nonblocking reads and writes. Figure 77.43 shows the members of the `comm_medium` class. The four signals shown in the top left are signals to show the current state of the logical connection in the waveform window. To the right are the two wait queues, one for write request, and one for read requests. New transaction requests received via the interface methods are placed into these queues. The boolean `no_match` variable maintains whether there is currently a match between read operations and received data, the integers below it are used to store the values of the current processor for the sender and receiver, and to store the location in the queue of the current send request being executed, and the current receive request being executed. Below the integers is a dummy variable whose sole purpose is to fix an existing bug in the implementation of the SystemC simulator. In the bottom left of the figure are the member functions of the object. The functions all the way to the left are the functions intended to be accessed by other objects, the remaining four are intended for internal use only, though they are declared as public and thus visible to other objects. The functions with a star after them are registered with the simulation scheduler as SC_THREADs. The top two events in the bottom right of the figure are used by the read and write methods to notify the arbitrator process that there may be new pairs of requests that could be activated to communicate. The event in the very bottom is used to coordinate the execution of two threads when they communicate. In the top right of the figure are two transaction pointers. These pointers are used by the arbitrator to keep track of which two transactions it is dealing with. The event pointers in the bottom center of the figure are also used by the arbitrator. When the arbitrator has selected two transaction requests to communicate it does not actually handle the communication. There is a write thread, and read thread that execute the `transact()` code in each of the two transaction objects. For any blocking transaction objects the calling processor's thread is suspended in the interface call waiting on the transaction to notify it that the transaction is complete. Any nonblocking transaction objects have already had their calling thread return to the processor model code. The arbitrator has no need to check for any potential request pairs until after the two threads have completed their transaction.

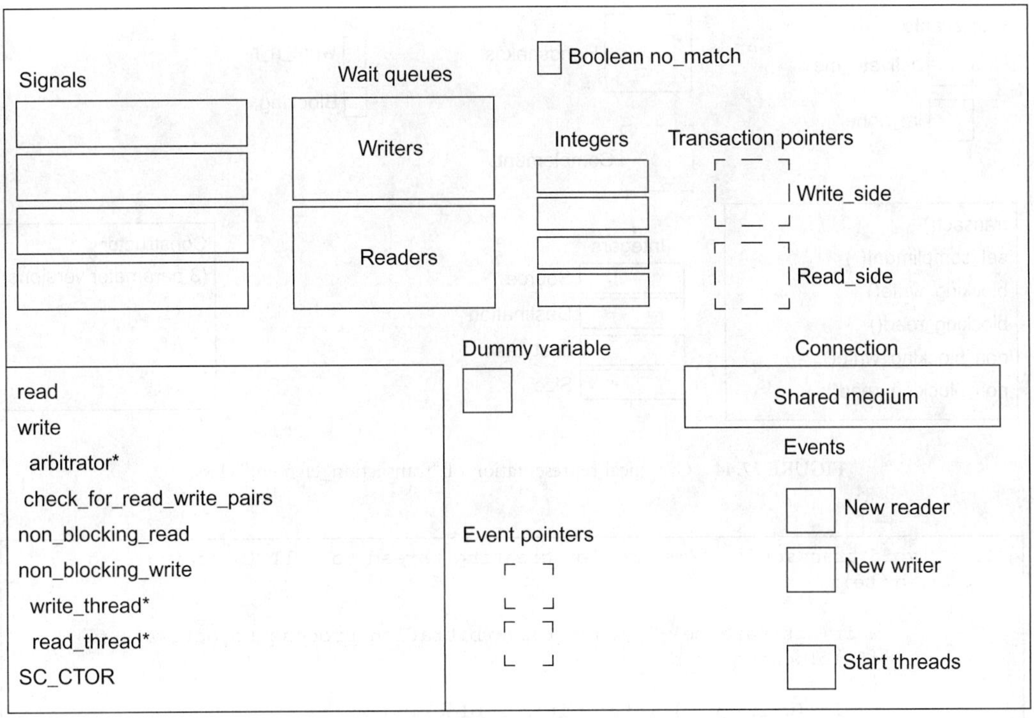

FIGURE 77.43 Graphical representation of "comm_medium" class.

These two pointers are set to allow the arbitrator to suspend until both threads have completed, rather than wasting simulation resources polling to see if they are done.

While support exists in the comm_medium object for nonblocking reads, the channels do not have methods to give access to that functionality to the processor models. This was done on purpose to avoid having to check data dependencies before beginning a computation. This also keeps the simulation simple, and more efficient in terms of simulation time. The functionality was built into the comm_medium object because it was easy to do and makes adding nonblocking reads at some future point much easier. The arbitration scheme provided is a longest waiting first scheme. As soon as at least one transaction pair, a matching send and receive, is present the pair with the largest sum of positions in the wait queues is selected to transact next. The crossbar channel uses a variant of the comm_medium class. In the crossbar variant there are pointers to the transaction wait queues which allow a single set of queues to be used by all of the logical channels.

The comm_medium class also makes use of the class transaction_wait_queue, which is a specialized linked list to allow for a large number of waiting transactions without allocating a fixed large amount of memory. The elements of the linked list are of the class transaction_element, which contains all the essential information about a transaction request. The only item of importance from the linked list is the class that actually holds the transaction information. This transaction_element class contains all of the information about the transaction request. Figure 77.44 graphically depicts the key elements of the transaction_element class.

The integers in the bottom center contain the source processor number, the destination processor number, the destination task's id number and the size, in nanoseconds, associated with the transaction. The boolean variables in the top right tell whether the transaction element is a write or read, and whether it is blocking or not. The handshakes object in the upper middle is a set of four events that are used by the transact method to logically "perform" the transaction. The complement pointer below the handshakes is a pointer to a transaction_element object. To communicate two transaction_element objects must be paired up the channel's arbitrator process. It does this pairing by setting a read and a

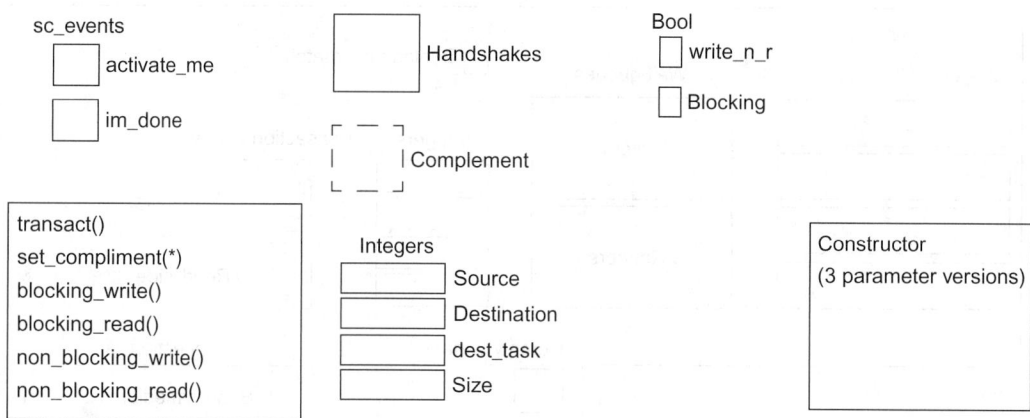

FIGURE 77.44   Graphical representation of "transaction_element" class.

```
    void transact()  //method for creating thread to call (blocking read
or write)
    {
        wait(activate_me); //wait for arbitration process to activate me
        if (blocking)
        {
            if (write_n_r) //then it's a blocking write
                blocking_write(); //call the blocking write routine
            else //it's a blocking read
                blocking_read();  //call the blocking read routine
        }
        else
        {
            if (write_n_r) //then it's a non-blocking write
                non_blocking_write(); //call the non-blocking write routine
            else //it's a non-blocking read
                non_blocking_read();  //call the non-blocking read routine
        };
        im_done.notify(SC_ZERO_TIME);
        return;
    };
```

FIGURE 77.45   Transaction_element transact() method.

write's compliment pointer to each other. The `activate_me` event in the top left is used by the arbitrator to activate the thread executing the element's side of a transaction. While the `im_done` event is used to notify the arbitrator, and in the case of a blocking transaction the requesting processor's thread, that the transaction is complete. The constructor for this object depicted in the bottom left has three different implementations with different parameter lists, the first constructor implementation that sets all of the transaction values is the one used in the current version of the models. The others were left to maintain backward compatibility with previous versions, and may be useful for future versions. The purpose of the methods displayed in the bottom left of the figure are self-explanatory. The `transact` method is what controls the actual behavior of a transaction pair once it has been scheduled. The nondebugging parts are repeated below in Figure 77.45. The blocking and nonblocking versions of the read and write routines are the same in this version of the models.

The nondebugging version of the blocking read and write methods are shown in Figure 77.46. These methods show how the four-way handshaking is implemented. The use of a full four-way handshaking

```
    void blocking_write()
    {
        compliment->handshakes.start_write.notify(SC_ZERO_TIME);
        wait(handshakes.start_ack);
        compliment->handshakes.write_done.notify(SC_ZERO_TIME);
        wait(handshakes.done_ack);
    };

    void blocking_read()
    {
        wait(handshakes.start_write);

compliment->handshakes.start_ack.notify(size,SC_NS);//SC_ZERO_TIME);
        wait(handshakes.write_done);
        compliment->handshakes.done_ack.notify(SC_ZERO_TIME);
    };
```

FIGURE 77.46 Transaction_element write and read methods.

```
    void blocking_write()
    {
        wait(handshakes.start_ack);
    };

    void blocking_read()
    {

compliment->handshakes.start_ack.notify(size,SC_NS);//SC_ZERO_TIME);
        wait (size,SC_NS)
    };
```

FIGURE 77.47 Alternative write and read methods.

in the channel model is somewhat arbitrary, but it makes incremental refinement of the channel easier. However, with the abstract behavior described here a single line for the write method and two for the read method would be sufficient. Figure 77.47 shows how the methods could be implemented in this way.

All of the channel models also read in parameter information from the *channel_param.txt* file located in the directory that the simulation is running in. This file contains two lines. The first line is the bus speed in megabytes per second. The second line is the fixed communication overhead per communication transaction. In the top-level channel models the data size parameter passed to read interface method is run through a `data_to_delay` function that return the delay in nanoseconds that the communication should take based on the specified bandwidth and communication overhead.

77.5.5 Shared Bus Model

The shared bus architecture consists of a single logical communication medium, which is an object of class `comm_medium`. The behavior is encapsulated in the `shared_bus_io` class. This class inherits the virtual interface functions from the `my_basic_rw_port` class, and must provide an implementation for them. The implementation for these functions is shown in Figure 77.48. The functions essentially map the interface functions to the methods of the `comm_medium` object. The `data_to_delay` function takes the `data_size` passed to the read method, and calculates the required transaction time in nanoseconds based on the bandwidth, and then adds the fixed communication overhead that the channel's constructor reads in from the channel parameter file.

```
    inline bool read(int dest_address, int source_address, int data_size, int
dest_task)
    //blocking read function with writer's address
       //read function takes data to be size of data transmission
    {
       int temp_size;
       temp_size = data_to_delay(data_size);
       shared_bus.read(dest_address, source_address, temp_size, dest_task);
       return 1;
    };
    inline bool write(const int dest_address, int source, int data,
                      int dest_task)
       //blocking write function here
    {
       shared_bus.write(dest_address, source, data, dest_task);
       return 1;
    };
    inline bool non_blk_write(const int dest_address, int source, int data,
                      int dest_task)
    //non-blocking write function here
    {
       shared_bus.non_blocking_write(dest_address,source,data,dest_task);
       return 1;
    };
```

FIGURE 77.48   Implementation of inherited virtual interface functions.

77.5.6   Fully Connected Model

The fully connected architecture is built around the same comm_medium object as the Shared Bus model. The fully connected architecture creates a comm_array object which contains all the logical connections, and copies the addresses of the comm_medium objects into a two-dimensional array that it uses to map a processor's request to the overall communication architecture to the appropriate logical connection. The fully connected architecture passes the number of processors to the comm_array object which then instantiates the number of comm_medium objects needed to provide a dedicated shared bus between each pair of processors in the simulation.

77.5.7   Crossbar Model

The crossbar architecture is similar to the fully connected architecture except that it only requires four cross_comm_medium objects since with nine processors only four concurrent connections are allowed. As mentioned earlier, it uses its own version of the basic comm_medium object. This is necessary because the logical connections in the crossbar are not associated with any particular processor, and the communication requests are not associated with any of the logical connections. Just to clarify in the fully connected architecture there is a logical connection between every processor modeled by a comm_medium object. The fully connected architecture model simple directed the requests it received to the correct logical connection. In the fully connected architecture there is only one connection that all requests are intended for, but in the crossbar the number of logical connections is equal to the number of processors divided by two, and every request could potentially communicate over any of them.

77.5.8   SystemC Performance Modeling Examples

This section contains a number of examples of performance models constructed using the SystemC modeling modules described above. The examples are presented as a demonstration that the models execute correctly and also that they demonstrate performance of the system they are intended to model. The first example is a trivial example with a set of four tasks all executing on one processor.

Since data communication inside a processor is assumed to take no time, the description should take simulation time equal to the sum of the computation time of all the tasks. The second example is the same four tasks allocated to two processors such that each task must send the data over the communication channel to the next task. This second example should take longer, with three 100 byte sends being sent over the communication channel. The third example is the same task graph description with varying bus parameters. The fourth and fifth examples have bus contention, to show that contention is handled properly. Each example lists the simulated latency, and a timeline showing the simulation results.

77.5.8.1 Single Processor

Figure 77.49 shows the simple sequential task graph for the first example. Here all the tasks are allocated to processor 0. Each task has a compute value of 10 (μs), and each edge has a data value of 100 (bytes). Since communication within a processor is assumed to take no time, the latency for this description should be the sum of the compute times, which is 40 μs.

In addition to the processing timeline shown previously, the models generate a text output stream that describes the actions each module is taking at a given simulation time. The text output for this simulation is shown below. Note that the final task is completed at 40 μs of simulation time which is exactly as expected.

> *shared bus architecture*
> *0 s proc#0 Task number 1 Computing for 10000 ns*
> *0 μs proc#0 Task number 2 Computing for 10000 ns*
> *20 μs proc#0 Task number 3 Computing for 10000 ns*
> *30 μs proc#0 Task number 4 Computing for 10000 ns*
> *40 μs proc#0 done!*
> *sim done!?*

77.5.8.2 Dual Processor

The task graph for the second example is shown in Figure 77.50. The graph also shows the allocation of the tasks to two processors. Notice that the sequential tasks are on different processors so the data must be transferred across the communication channels before the computations can begin. Here the communication channel's bandwidth determines how long a communication transaction should take to complete. The length of time is the data size in bytes divided by the bandwidth in megabytes per second. The communication channel can also take into account communication overhead, in nanoseconds, if it

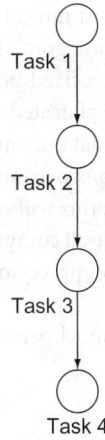

FIGURE 77.49 Single processor example task graph.

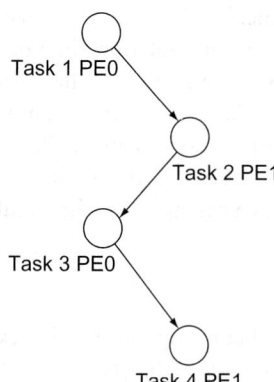

FIGURE 77.50 Dual processor example task graph.

is specified. The channel bandwidth, and communication overhead are read in from a file. If this file is not present or an item is missing it will take on its default value. The value specified for this example is 100 Mbyte/s for bandwidth and 5 ns for channel overhead.

Since all the communication is of the same size, the expected latency for this example can be determined using the following equation:

$$\text{Computation time} + \text{numtrans} * \left(\frac{\text{Data size (byte)}}{\text{Bandwidth (Mbyte/s)}} + \text{Communication overhead (ns)} \right)$$

Which, for this example, evaluates as follows:

$$40 \ (\mu s) + 3 * \left(\frac{100 \ \text{byte}}{10 \ (\text{Mbyte/s})} + 5 \ (\text{ns}) \right) = 40 \ (\mu s) + \frac{300}{100 * 10^6}(\text{s}) + 15 \ (\text{ns}) = 40 \ (\mu s) + \frac{300}{100} \ (\mu s) + 15 \ (\text{ns})$$

Thus the expected latency is 40 μs for computation plus 3 μs for the actual data transmission, plus 15 ns for communication overhead. That gives a total latency of 43015 ns. The timeline output for this simulation is shown in Figure 77.51. Note that the final task, Task 4, completes at 43 μs on the graph.

77.5.8.3 Parallel Communications Example

The next example shows the effect of various communications topologies on an application with requirements for simultaneous communications. The task graph for this example is shown in Figure 77.52. Each task (called nodes in this graph) computes for a fixed period of time and then sends data to a second task causing it to begin execution. The tasks are allocated to processors such that after completion of the first set of tasks, all processors attempt to send data to another processor. Because the first tasks all have the same execution time, all the communications become ready to begin at the same time. Thus, if an architecture has parallel communication paths, this will result in a decrease in the total application run time. The four start tasks (nodes) in this example all compute for 10 μs, then attempt to do a nonblocking send of size 100 byte to a task allocated to another processor. They then move on to start the read required to begin their next task.

For all of the results discussed below, the channel parameters are set to a bus bandwidth of 1 Mbyte/s and a communication overhead of 0 ns.

77.5.8.3.1 *Shared Bus Simulation Results*

The first set of results are for a system with a single shared bus. On this system Tasks 0–3 all execute in parallel on the four processors. At this point there will be four-way contention for the single system bus.

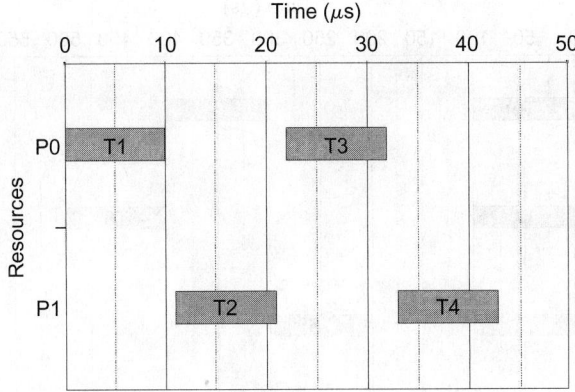

FIGURE 77.51 Dual processor example timeline.

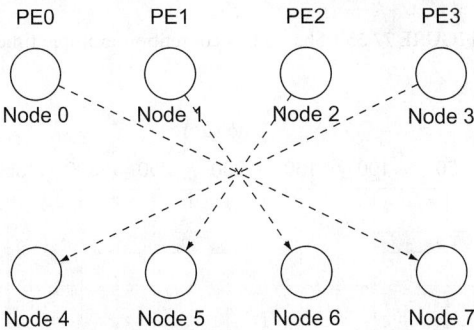

FIGURE 77.52 Bus contention example task graph.

The communications operations will be assigned priority on a first come first serve basis. In the current implementation of SystemC, the task that will get first priority to communicate its data cannot be determined ahead of time, however the tasks will all run in the same order every time the simulation is run.

With a shared bus architecture, the latency is 100 μs for all processors to compute in parallel plus $4 * (\text{Data size/Bandwidth}) * 1000$ (ns), or 400 μs, for each of the four sends to occur in series, plus 100 μs for the last receiver to compute after completing their receive. Thus the overall latency should be 600 μs.

The timeline for this example is shown in Figure 77.53. The timeline shows all the tasks beginning their blocking read then having to wait for the arbitrator to select them to communicate across the bus. Once the individual communications have taken place, the destination task, Tasks 4–7, execute. The timeline correctly shows the last task completing execution at 600 μs.

77.5.8.3.2 Fully Connected Simulation Results

In the fully connected architecture there is a dedicated communications channel between each pair of processors. However, in this architecture, it was decided to model a system where a processor cannot both send and receive a message from the same processor at the same time. Because of the connectivity of the task graph for this application, after the first set of tasks execute in parallel, each processor needs to send and receive a message before it can execute the next task. For example, processor zero cannot send to three and receive from three at the same time. Rather it must do one, then the other. Thus, for this example, each channel in the fully connected architecture is effectively a half duplex connection.

During execution the run time is 100 μs for all processors to compute the first four tasks in parallel plus 100 μs for the first set of sends, plus 100 μs for the second set of sends—during which the tasks started by the first set of sends also execute, then finally 100 μs for the last two tasks to compute in

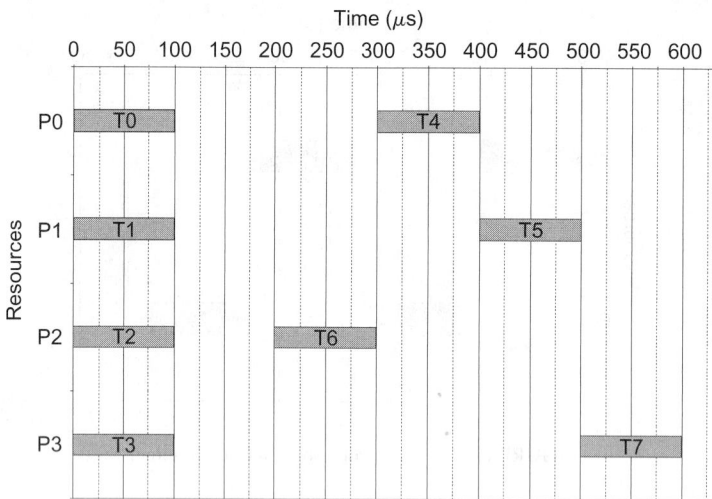

FIGURE 77.53   Shared bus contention example timeline.

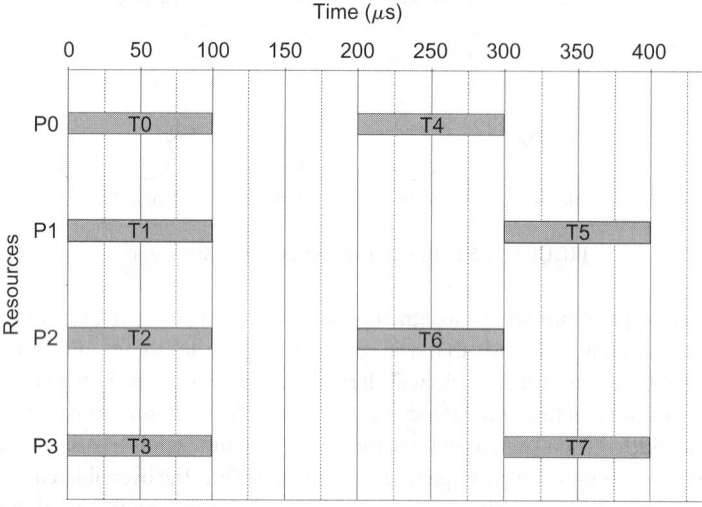

FIGURE 77.54   Fully connected contention example timeline.

parallel. Thus the overall latency for the fully connected architecture should be 400 μs. The timeline for this example is shown in Figure 77.54.

77.5.8.3.3   *Crossbar Simulation Results*

As mentioned above, the crossbar architecture behaves like a fully connected architecture where the maximum number of connections is limited to the number of processors divided by two. Thus for this four processor example, the crossbar architecture will only allow two communications at a time. This characteristic means that for this example, the crossbar architecture will have the same latency as the fully connected architecture for this example. This result is shown in Figure 77.55.

77.5.8.4   Second Contention Example

This second example expands on the previous example by showing a slightly different set of contention conditions. In the first example, the communication requirements specified by the task graph required

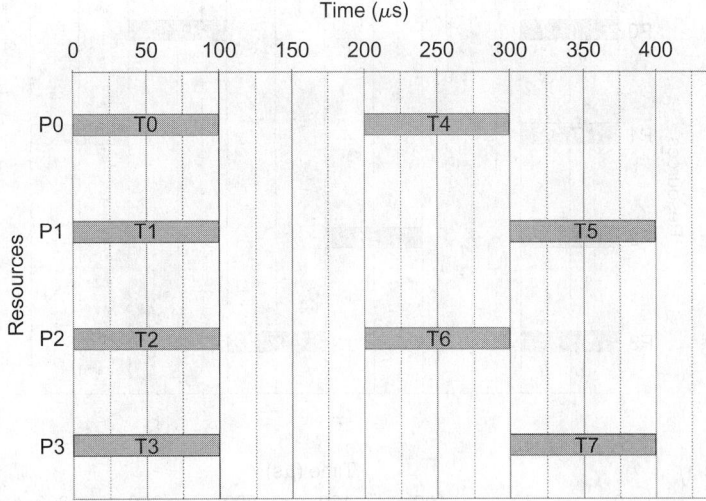

FIGURE 77.55   Crossbar contention example timeline.

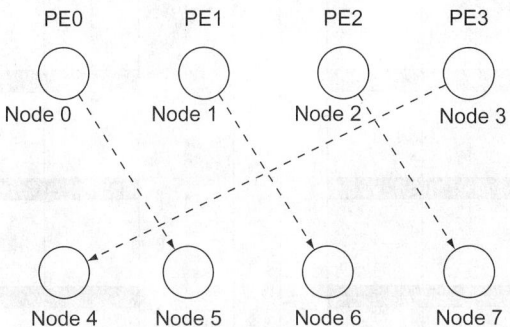

FIGURE 77.56   Second contention example task graph.

the processors to send and receive data from the same processor. This effectively allowed for only two active communication transactions on the fully connected architecture. In this example, as shown in Figure 77.56, the processors will be sending and receiving data from different processors during the communications portion of the application. This set of communication requirements will allow all of the available communication channels to be used concurrently with the fully connected architecture.

Figure 77.57 shows the results for this example for the shared bus, fully connected, and crossbar architectures. Note that in this example, in the fully connected architecture, all of the communication operations occur in parallel which allows the entire application to execute in 300 µs.

77.5.9   Mixed Level Processor Model

Although simplistic in nature, the above examples show that the SystemC-based performance modeling methodology can be used to model the execution of different applications on various system architectures. In addition to this capability, as mentioned earlier the, SystemC module for the processor has the ability to replace the performance only computation delay with a refined computation model that is described in VHDL or Verilog. This mixed-level modeling capability allows the model to be refined to a lower level in a step-wise fashion.

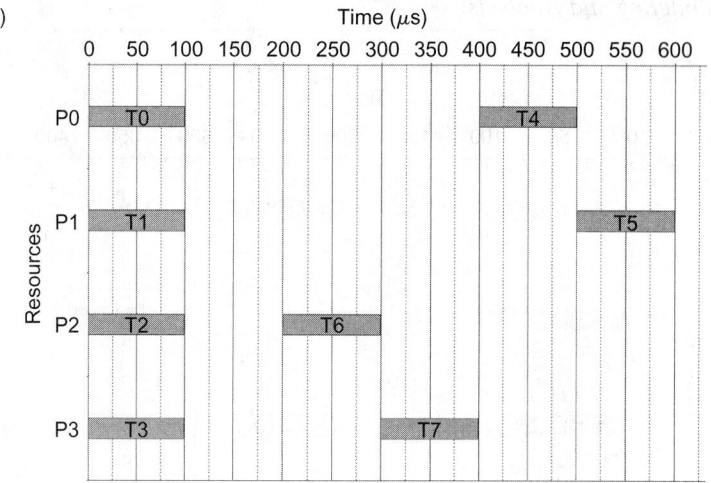

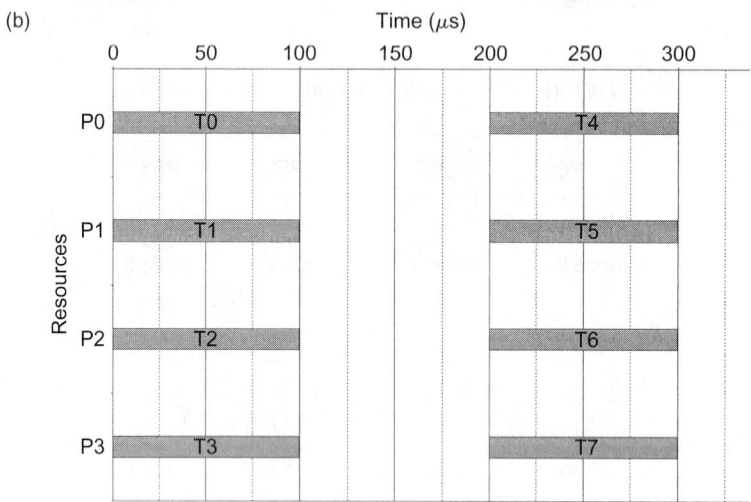

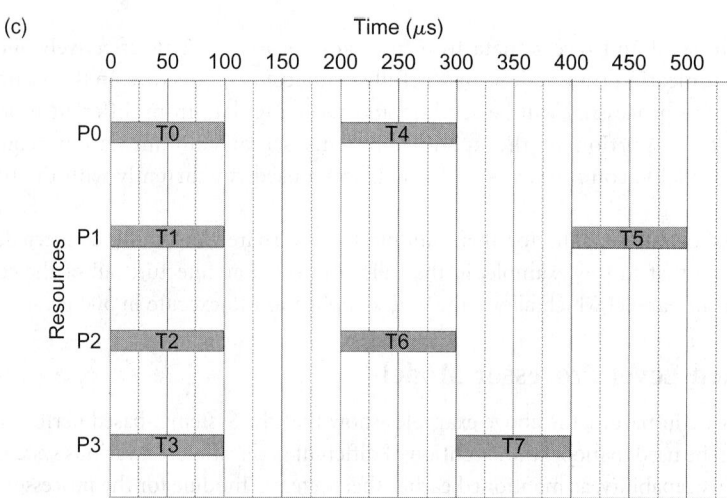

FIGURE 77.57 Second contention example timelines (a) shared bus, (b) fully connected, and (c) crossbar.

The replacement of an abstract processor model with a refined (RTL or gate-level model) is accomplished by using the ModelSim SC_FORIEGN_MODULE syntax. The SC_FORIEGN_MODULE provided by Model-Sim allows a non-SystemC model that has been compiled for ModelSim to be instantiated by a SystemC model. Incidentally, it also allows an already compiled SystemC module to be loaded in the same manner.

The processor model opens its processor command file and looks at the first line during the execution of its constructor. If the first line is "mixed" then the processor model knows that it should run as a mixed-level model with a refined computation model. If the model is to be a mixed-level one, the constructor of the processor model will then look for a *mixed_processor.txt* text file that specifies the ModelSim path for the refined computation model to use. The relevant part of the constructor that instantiates the refined model is shown in Figure 77.58.

The refined_computation object is the one that opens the *mixed_processorX.txt* file, where *X* is the processor number. Once this file is opened, the constructor uses the path contained within it to open the refined model object. In this example, it instantiates an object of class rng_comp_tb that is shown in Figure 77.59.

This class is essentially a wrapper for the actual precompiled model VHDL or Verilog model. This class/module definition would map to a VHDL entity definition like the one shown in Figure 77.60. Note that this entity corresponds to what is effectively a test bench for the refined model.

The intent for this interface is for it to be easy to integrate into an existing test bench for the refined model of computation. The start and done signals are active high. So when the start goes high to a logical '1', the refined model should start a "computation." This computation is effectively the test bench applying a set of predefined stimulus waveforms to the refined model. As described below, these stimulus waveforms

```
        cmd_input = new command_in(proc_num);
        string_temp.assign(cmd_input->get_proc_type());
        if(/*1st_line*/string_temp=="mixed")
        {
            SC_THREAD(mixed_behav);
            //only make the relevant behavior a thread!
            hardware_compute=new
    refined_computation("name",proc_num);
            //may need to add an argument this class...
        }
```

FIGURE 77.58 Refined computation part of processor model constructor.

```
        class rng_comp_tb : public sc_foreign_module
        {
        public:
            sc_in<sc_logic> start;
            sc_out<sc_logic> done;

            rng_comp_tb(sc_module_name nm, const char* hdl_name)
              : sc_foreign_module(nm, hdl_name),
                start("start"),
                done("done")
            {
                //cout<<"Building hardware model\n";
            };

            ~rng_comp_tb()
            {}

        };
```

FIGURE 77.59 Code for "rng_comp_tb" class.

```
ENTITY rng_comp_tb IS
   port(
      start : in std_logic;
      done : out std_logic
      );

END rng_comp_tb;
```

FIGURE 77.60   Sample VHDL refined computation entity declaration.

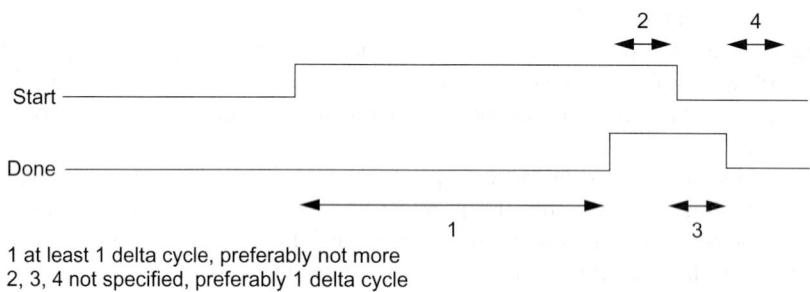

1 at least 1 delta cycle, preferably not more
2, 3, 4 not specified, preferably 1 delta cycle

FIGURE 77.61   Refined computation model interface-timing diagram.

can be derived for the specific refined model in a number of different ways depending on the objectives for the mixed-level model.

When the refined model has finished its "computation" it should raise the done signal. Both signals should be low at the start of the simulation. When the processor model puts the start signal high, the refined model begins its computation. The presumption is that the refined model is something like a test bench, with a model of the actual hardware, or some other more detailed model, instantiated inside of it. The refined model presents any required data to the detailed model, and watches for whatever condition indicates that it has completed the computation. Once the computation is completed, it raises the done signal telling the abstract processor model it is done. The processor model will then lower the start signal, and the refined model will respond by lowering the done signal. Figure 77.61 shows the basic timing diagram for the interface.

Figure 77.62 shows the portion of the processor model that raises and lowers the start signal and waits for the done signal. This code is located in the `refined_computation` object. The `refined_computation` object is instantiated by the processor model's constructor when it reads in from the command file and determines that it should be a mixed-level model.

Figure 77.63 shows a portion of sample VHDL test bench that implements the refined computation model's side of the start/done interface. The start signal from the start/done interface activates the process. The process uses the go signal to cause the test bench to perform a computation, and the test bench raises the `tb_done` signal when it is finished with a single computation.

77.5.10   Mixed-Level Examples

The following examples are mixed-level examples where the computation part of the processor is modeled in more detail. Since the rest of the simulation is at a more abstract level and does not have all of the stimuli needed for the refined model, the stimuli need to be created in some way. The following examples focus mainly on different ways of generating the data that the various refined models need to function.

77.5.10.1   Fixed Cycle Length

The first example is one where an abstract processor is replaced with a random number generator. The model for the random number generator is an RTL level discrete digital model of a random number generator described in Ref. [51]. The model attempts to describe a number of elements that are extremely sensitive to initial conditions, and thus in reality exhibit more random behavior than can be modeled with

```
        bool start_compute()
        {/* basic algorithm is:
          send start signal
          wait for done signal
          return
          */
#ifdef MTI_SYSTEMC
          temp= true; //'1';
          start = temp;
          while (done != temp)  //ie while done != 1
             wait(done.value_changed_event());//wait until event and
loop..
          temp = false;
          start = temp;
#endif
          return 0;
        };
```

FIGURE 77.62 SystemC start/done interface code.

```
          -----------------------------------------------------------------
          -- Process to respond to start and create go and done. JA
          -----------------------------------------------------------------
          CompBench: PROCESS (start, tb_done)
          Begin
            if (start='1') then
              if ((tb_done'event) AND (tb_done='1')) then
                 go   <= '0';
                 done <= '1';
              ELSE
                 go   <= '1';
                 done <= '0';
              end if;
            end if;
          End process CompBench;
```

FIGURE 77.63 Sample VHDL start/done interface code.

a solely digital model. As it is, the model always generates the same nonrepeating sequence of values. For this example, the existing test bench was modified to put the generator through its reset cycle, then through a single random number generation. The only inputs to the refined model are a set of control signals and a clock. When the processor, that it is the refined computation model for, gets a compute command; it will send the start signal to the test bench, which will then go through the reset and generate phases, and signal back with the done signal once a number has been generated. In this particular example the internal signals continue to oscillate between compute commands. Since this model takes a fixed number of cycles using the refined model, and the values generated are not passed elsewhere, this example is less efficient and no more accurate than putting in the actual compute time for the abstract compute command.

77.5.10.2 Variable Cycle Length

The rest of the mixed-level examples presented use an RTL level booth multiplier model. The booth multiplier takes a variable number of cycles to complete the binary multiplication. The number of cycles required depends on the numbers being multiplied. The inputs to the model are the two numbers to be multiplied, and a clock, the outputs are the result and a control signal indicating that the multiplication is complete. There is a slight propagation delay for the result to appear on the output after the done signal appears. If the input clock continues to cycle after the multiplication is complete, then the result will become invalid after a clock cycle. Thus the test bench allows a half cycle to elapse before considering the computation done. Note that the effect of this refined model is that the computational delay is dependent on the data being applied to it for the computation. This delay mechanism is a more accurate representation of how a refined model would be used in, and add additional accuracy to, a system-level performance model. However, as discussed in the section above on VHDL-based mixed-level modeling, the important question is how to generate the data that is input to the refined model in such a way as to accurately represent the performance of the refined component in the real system. Typically, this can be done by either presenting the refined model with random data to develop a statistical representation of average system performance, or presenting the refined model with predefined data. This data can be generated by the designer to represent typical system performance, or to exercise the best-, or worst-case delay scenarios.

77.5.11   Random Data Generation

In this example, the booth multiplier is presented with two random numbers generated by two other entities instantiated in the test bench. Since random number generation is not present in standard VHDL library, ModelSim's mixed language ability is utilized to allow a SystemC random number generator module to be used. The SystemC module uses the C++ rand() function to generate a pseudo random number, which is passed back to the VHDL test bench. Since the VHDL test bench is instantiated by the SystemC performance simulation, this is in fact a SystemC–VHDL–SystemC hierarchy. ModelSim's mixed language interface allows the designer to use whatever language is best suited for the task at hand. Here the random number must be passed back to the VHDL test bench as an sc_logic vector, which is automatically translated into a VHDL std_logic_vector by the simulator. Since both numbers are pseudo random, the compute will take a variable amount of time to complete. Although it is possible that the use of the rand() function will result in repeating sequence of numbers, most simulations will not run long enough for this to be noticeable. If a more random distribution, or a particular type of distribution is desired, a specialized random number generator package can be used in the model. Since most of the computations at the system level, will consist of multiple operations, to make this example more typical, a means of generating a set multiple numbers to be multiplied was needed. While it is possible to generate a fixed size data set this example goes just a little farther and generates a pseudorandom size data set of pseudorandom numbers. The generation of the random size of the data set is done using the same SystemC module that generates the random data itself. A few changes to the test bench needed to be made so that it did not send the done signal back until all of the multiplications were completed. Figure 77.64 shows two waveforms from two different mixed-level simulations, each with two different size sets of random numbers multiplied together.

77.5.11.1   Data Set From File

As described above, when a sample set of data that exercises a specific scenario for the refined model is available, it can be advantageous to use it rather than generating a new set that may or may not be close to the actual data. Using this predefined sample data set ensures accurate performance for that data set, and removes any guesswork as to what a realistic data set might be. Since VHDL has standard file access capabilities that are easy to use, the test bench for this example reads the values directly from the input file, and does not send the done signal until it reaches the end of file. Unfortunately, the current implementation ModelSim used for this example did not allow passing generic information across the language boundary, thus specification of the file to use had to be done in the VHDL code. An extension to this approach would be to read the filename to use from a configuration file similar to what is done to specify the refined model to use in the SystemC performance code. Figure 77.65 shows the waveform of the mixed-level model simulation where the data for the refined component were read in from a file.

77.5.12   Mixed-Level Example Summary

Mixed-level modeling techniques in SystemC-based performance models have been demonstrated. These examples utilized simple pseudorandom data generation, variable size pseudorandom data set generation, reading a dataset from single file, and reading multiple data sets from multiple files. Multiple variations and combinations of the above approaches can be used to generate stimuli for a wide variety of refined

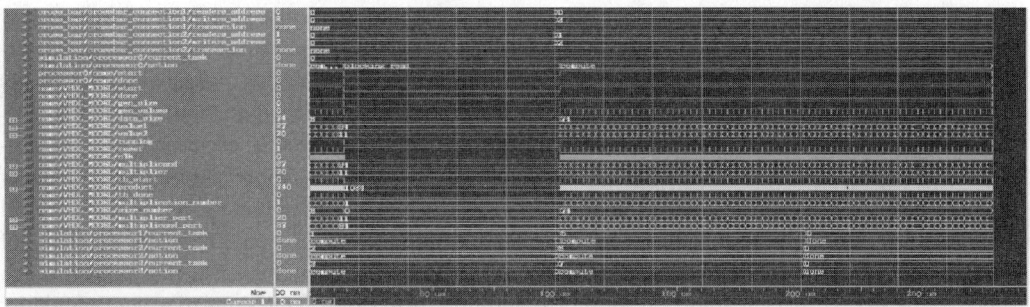

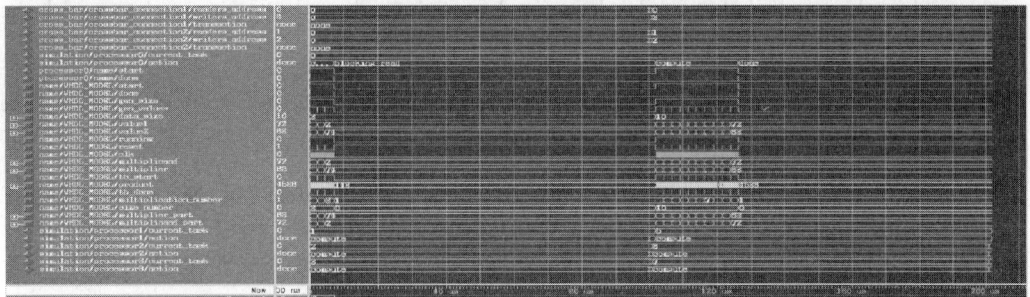

FIGURE 77.64 First- and second-random data set waveforms.

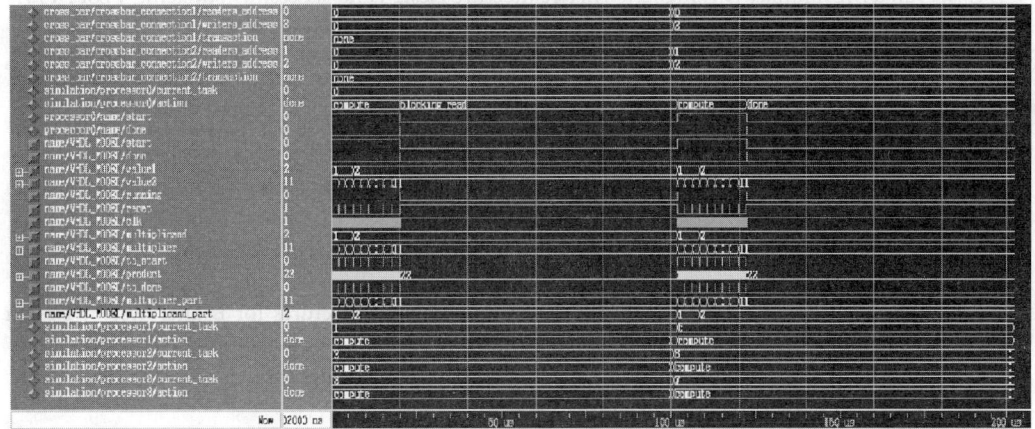

FIGURE 77.65 Data from file example waveform.

computation models. Anything from randomly selecting a file from which to read a data set, to using a file to parameterize random data generation is possible depending on whether realistic data are available and on the simulation objective. It is clear to see, in any case, that including refined behavioral RTL or gate-level components in a performance model in SystemC is fairly easily done and can increase the overall accuracy of the performance model, just as was possible in the VHDL-based performance model.

77.6 Conclusions

Integration of performance modeling into the design process such that it can actually drive the refinement of the design into an implementation has clear advantages in terms of design time and quality. The capability to cosimulate detailed behavioral models and abstract system-level models is vital to the development of a design environment that fully integrates performance modeling into the design process. One methodology and implementation for cosimulating behavioral models of individual components with an abstract performance model of the entire system was presented. This environment results in models that can provide estimates of the performance bounds of a system that converge as the refinement of the overall model increases.

This chapter has only scratched the surface on the possible improvements that performance- or system-level modeling can have on the rapid design of complex VLSI systems. As more and more functionality can be incorporated into the embedded VLSI systems and these systems find their way into safety-critical applications, measures of dependability such as reliability and safety at the system-level are becoming of great interest. Tools and techniques are being developed that can use the performance model from which to derive the desired dependability measures. In addition, behavioral fault simulation and testability analysis are finding their way into the early phases of the design process. In summary, the more attributes of the final implementation that can be determined from the early and often incomplete model, the better the resulting design and shorter the design cycle.

References

1. ASIC & EDA Magazine, January 1993.
2. D. Hill, W. vanCleemput, "SABLE: A Tool for Generating Structured, Mulit-Level Simulations," *Proceedings of the 16th ACM/IEEE Design Automation Conference*, 1979, pp. 272–279.
3. D.W. Franke, M.K. Purvis, "Hardware/Software Codesign: A Perspective," *Proceedings of the 13th International Conference on Software Engineering*, 1991, pp. 344–352.
4. D.W. Franke, M.K. Purvis, "Design Automation Technology for Codesign: Status and Directions," *International Symposium on Circuits & Systems*, 1992, pp. 2669–2671.
5. G.A. Frank, D.L. Franke, W.F. Ingogly, "An Architecture Design and Assessment System," *VLSI Design*, pp. 30–50, August 1985.
6. C. Terry, "Concurrent Hardware and Software Design Benefits Embedded Systems," *EDN*, pp. 148–154, July 1990.
7. S. Napper, "Embedded-System Design Plays Catch-Up," *IEEE Computer*, Vol. 31, No. 8, pp. 118–120, August 1998.
8. F.W. Zurcher, B. Randell, "Iterative Multi-level Modeling—A Methodology for Computer System Design," *Proceedings of IFIP Congress '68*, 1968, pp. 867–871.
9. Martin Marietta Laboratories, RASSP First Annual Interim Technical Report (CDRL A002), Moorestown, October 31, 1994.
10. P.A. Wilsey, S. Dasgupta, "A Formal Model of Computer Architectures for Digital System Design Environments," *IEEE Transaction on Computer-Aided Design*, Vol. 9, No. 5, pp. 473–486, May 1990.
11. C.A. Giumale, H.J. Kahn, "Information Models of VHDL," *Proceedings of the 32nd Design Automation Conference*, San Francisco, CA, 1995, pp. 678–683.
12. Open SystemC Initiative Website, www.systemC.org.
13. P. Kollaritsch, S. Lusky, D. Matzke, D. Smith, P. Stanford, "A Unified Design Representation Can Work," *Proceedings of the 26th Design Automation Conference*, 1989, pp. 811–813.
14. J. Peterson, "Petri Nets," *Computing Surveys*, Vol. 9, No. 3, pp. 223–252, September 1997.
15. M.K. Molloy, "Performance Analysis using Stochastic Petri Nets," *IEEE Transactions on Computers*, Vol. C-31, No. 9, pp. 913–917, September 1982.
16. M.A. Holliday, M.K. Vernon, "A Generalized Timed Petri Net for Performance Analysis," *IEEE Transactions on Software Engineering*, Vol. SE-13, No. 12, pp. 1297–1310, December 1987.
17. L. Kleinrock, *Queuing Systems, Vol. 1: Theory*, Wiley, New York, 1975.
18. G.S. Graham, "Queuing Network Models of Computer System Performance," *Computing Surveys*, Vol. 10, No. 3, pp. 219–224, September 1978.
19. G. Balbo, S.C. Bruell, S. Ghanta, "Combining Queuing Networks and Generalized Stochastic Petri Nets for Solutions of Complex Models of System behavior," *IEEE Transactions on Computers*, Vol. 37, No. 10, pp. 1251–1268, October 1988.

20. G. Frank, "Software/Hardware Codesign of Real-Time Systems with ADAS," *Electronic Engineering*, pp. 95–102, March 1990.

21. SES/workbench User's Manual, Release 2.0, Scientific and Engineering Software Inc., Austin, TX, January 1991.

22. L. Maliniak, "ESDA Boosts Productivity for High-Level Design," *Electronic Design*, pp. 125–128, May 1993.

23. IDAS Integrated Design Automation System, JRS Research Laboratories Inc., Orange, CA, 1988.

24. Application Note: TRANSCEND/VANTAGE Optium Cosimulation, TD Technologies, pp. 1–33.

25. Application Note for TRANSCEND Structural Ethernet Simulation, TD Technologies, August, 1993, pp. 1–13.

26. R. Bargodia, C. Shen, "MIDAS: Integrated Design and Simulation of Distributed Systems," *IEEE Transactions on Software Engineering*, Vol. 17, No. 10, pp. 1042–1058, October 1991.

27. E. Lee et al., "Mini Almagest," University of California at Berkeley, Department of Electrical Engineering and Computer Science, April, 1994.

28. E.A. Lee, D.G. Messerschmitt, "Synchronous Data Flow," *Proceedings of the IEEE*, Vol. 75, No. 9, pp. 1235–1245, September 1987.

29. VHDL Hybrid Models Requirements, Honeywell Technology Center, Version 1.0, December 27, 1994.

30. Honeywell Technology Center, "VHDL Performance Modeling Interoperability Guideline," Version Draft, August, 1994.

31. F. Rose, T. Steeves, T. Carpenter, "VHDL Performance Models," *Proceedings of the First Annual RASSP Conference*, pp. 60–70, Arlington, VA, August 1994.

32. J.A. Rowsen, A. Sangiovanni-Vincentelli, "Interface-Based Design," Proceedings of the 34th ACM/IEEE Design Automation Conference (DAC-97), 1997, pp.178–183.

33. L. Cai, D. Gajski, *Transaction Level Modeling: An Overview*, CODES+ISSS'03, 2003.

34. IEEE, "IEEE Standard VHDL Language Reference Manual," IEEE Std. 1076–1993, New York, NY, June 6, 1994.

35. A.P. Voss, R.H. Klenke, J.H. Aylor, "The Analysis of Modeling Styles for System Level VHDL Simulations," *Proceedings of the VHDL International Users Forum*, Fall 1995, pp. 1.7–1.13.

36. J.H. Aylor, R. Waxman, B.W. Johnson, R. D. Williams, "The Integration of Performance and Functional Modeling in VHDL," in *Performance and Fault Modeling with VHDL*, J.M. Schoen (Ed.), Prentice-Hall, Englewood Cliffs, NJ, 1992, pp. 22–145.

37. K. Jensen, "Colored Petri Nets: A high level language for system design and analysis," in *High-level Petri Nets: Theory and application*, K. Jensen and G. Rozenberg (Eds.), Springer, Berlin, 1991, pp. 44–119.

38. F.T. Hady, "A Methodology for the Uninterpreted Modeling of Digital Systems in VHDL," Master's thesis, Department of Electrical Engineering, University of Virginia, January 1989.

39. A.P. Voss, "Analysis and Enhancements of the ADEPT Environment," Master of Science (Electrical Engineering) Thesis, University of Virginia, May 1996.

40. J.B. Dennis, "Modular, Asynchronous Control Structures for a High Performance Processor," *ACM Conference Record, Project MAC*, MA, 1970, pp. 55–80.

41. G. Swaminathan, R. Rao, J. Aylor, B. Johnson, "Colored Petri Net Descriptions for the UVa Primitive Modules," CSIS Technical Report # 920922.0, University of Virginia, September 1992.

42. ADEPT Library Reference Manual, CSIS Technical Report No. 960625.0, University of Virginia, June 6, 1996.

43. M. Meyassed, R. McGraw, J. Aylor, R. Klenke, R. Williams, F. Rose, J. Shackleton, "A Framework for the Development of Hybrid Models," *Proceedings of the 2nd Annual RASSP Conference*, Arlington, VA, July 1995.

44. R.A. MacDonald, Williams. R., Aylor, J., "An Approach to Unified Performance and Functional modeling of Complex Systems," *IASTED Conference on Modeling and Simulation*, April 1995.

45. D. Gajski, N. Dutt, A. Wu, S. Lin, 'HIGH-LEVEL SYNTHESIS: Introduction to Chip and System Design,' Kluwer Academic Publishers, Dordrecht 1992.

46. E.W. Dijkstra, "A note on two problems in connection with graphs," *Numerische Mathematik,* Vol. 1, pp. 269–271, 1959.
47. R. Floyd, "Algorithm 97 (shortest path)," *Communications of the ACM*, Vol. 5, No. 6, p. 345, 1962.
48. S. Warshall, "A Theorem on Boolean Matrices," *Journal of the ACM*, Vol. 9 No. 1, pp. 11–12, 1962.
49. M. Meyassed, "System-Level Design: Hybrid Modeling with Sequential Interpreted Elements," Ph.D. dissertation, Department of Electrical Engineering, University of Virginia, January 1997.
50. Hein, J.J., J.H. Aylor, R.H. Klenke, "A Performance-Based Design Tool for Hardware/Software Systems," *Proceedings of the IEEE Workshop on Rapid System Prototyping*, June 2003.
51. Mitchum, S.T., R.H. Klenke, "Design and Fabrication of a Digitally Synthesized, Digitally Controlled, Ring Oscillator," *Proceedings of the IASTED Circuits Signals and Systems Conference*, 2005.

78

Embedded Computing Systems and Hardware/Software Co-Design

Wayne Wolf
Princeton University

CONTENTS

78.1   Introduction

This chapter describes embedded computing systems that make use of microprocessors to implement part of the system's function. It also describes hardware/software co-design, which is the process of designing embedded systems while simultaneously considering the design of its hardware and software elements.

78.2   Uses of Microprocessors

An embedded computing system (or more simply an embedded system) is any system which uses a programmable processor but itself is not a general purpose computer. Thus, a personal computer is not an embedded computing system (though PCs are often used as platforms for building embedded systems), but a telephone or automobile which includes a CPU is an embedded system. Embedded systems may offer some amount of user programmability—3Com's PalmPilot, for example, allows users to write and download programs even though it is not a general-purpose computer—but embedded systems generally run limited sets of programs. The fact that we know the software that we will run on the hardware allows us to optimize both the software and hardware in ways that are not possible in general-purpose computing systems.

Microprocessors are generally categorized by their word size, since word size is associated both with maximum program size and data resolution. Commercial microprocessors come in many sizes; the term microcontroller is used to denote a microprocessor which comes with some basic on-chip peripheral devices, such as serial input/output (I/O) ports. Four-bit microcontrollers are extremely simple but capable of some basic functions. Eight-bit microcontrollers are workhorse low-end microprocessors. Sixteen- and 32-bit microprocessors provide significantly more functionality. A 16/32-bit microprocessor may be in the same architectural family as the CPUs used in computer workstations, but microprocessors destined for embedded computing often do not provide memory management hardware. A digital signal processor (DSP) is a microprocessor tuned for signal processing applications. DSPs are often Harvard architectures, meaning that they provide separate data and program memories; Harvard architectures provide higher performance for DSP applications. DSPs may provide integer or floating-point arithmetic.

Microprocessors are used in an incredible variety of products. Furthermore, many products contain multiple microprocessors. Four- and eight-bit microprocessors are often used in appliances: for example, a thermostat may use a microcontroller to provide timed control of room temperature. Automatic cameras often use several eight-bit microprocessors, each responsible for a different aspect of the camera's functionality: exposure, shutter control, etc. High-end microprocessors are used in laser and ink-jet printers to control the rendering of the page. Many printers use two or three microprocessors to handle generation of pixels, control of the print engine, and so forth. Modern automobiles may use close to 100 microprocessors, and even inexpensive automobiles generally contain several. High-end microprocessors are used to control the engine's ignition system—automobiles use sophisticated control algorithms to simultaneously achieve low emissions, high fuel economy, and good performance. Low-end microcontrollers are used in a number of places in the automobile to increase functionality: for example, four-bit microcontrollers are often used to sense whether seat belts are fastened and turn on the seat belt light when necessary.

Microprocessors may replace analog components to provide similar functions, or they may add totally new functionality to a system. They are used in several different ways in embedded systems. One broad application category is signal conditioning, in which the microprocessor or DSP performs some filtering or control function on a digitized input. The conditioned signal may be sent to some other microprocessor for final use. Signal conditioning allows systems to use less-expensive sensors with the application of a relatively inexpensive microprocessor. Beyond signal conditioning, microprocessors may be used for more sophisticated control applications. For example, microprocessors are often used in telephone systems to control signaling functions, such as determining what action to take based on the reception of dial tones, etc. Microprocessors may implement user interfaces; this requires sensing when buttons, knobs, etc. are used, taking appropriate actions, and updating displays. Finally, microprocessors may perform data processing, such as managing the calendar in a personal digital assistant.

There are several reasons why microprocessors make good design components in such a wide variety of application areas. First, digital systems often provide more complex functionality than can be created using analog components. A good example is the user interface of a home audio/video system, which provides more information and is easier use than older, non-microprocessor-controlled systems. Microprocessors also allow related products much more cost-effectively. An entire product family, including models at various price and feature points, can be built around a single microprocessor-based platform. The platform includes both hardware components common to all the family members and software running on the microprocessor to provide functionality. Software elements can easily be turned on or off in various family members. Economies of scale often mean that it is cheaper to put the same hardware in both expensive and cheap models and to turn off features in the inexpensive models rather than to try to optimize the hardware and software configurations of each model separately. Microprocessors also allow design changes to be made much more quickly. Many changes may be possible simply by reprogramming; other features may be made possible by adding memory or other simple hardware changes along with some additional programming. Finally, microprocessors aid in concurrent engineering. After some initial design decisions have been made, hardware and software can be designed in parallel, reducing total design time.

While embedded computing systems traditionally have been fabricated at the board level out of multiple chips, embedded computing systems will play an increasing role in integrated circuit design as well. As VLSI technology moves toward the ability to fabricate chips with billions of transistors, integrated circuits will increasingly incorporate one or several microprocessors executing embedded software. Using microprocessors as components in integrated circuits increases design productivity, since CPUs can be used as large components which implement a significant part of the system functionality. Single-chip embedded systems can provide much higher performance than board-level equivalents, since chip-to-chip delays are eliminated.

78.3 Embedded System Architectures

Although embedded computing spans a wide range of application areas, from automotive to medical, there are some common principles of design for embedded systems. The application-specific embedded software runs on a hardware platform. An example hardware platform is shown in Figure 78.1. It contains a microprocessor, memory, and I/O devices. When designing on a general-purpose system such as a PC, the hardware platform would be predetermined, but in hardware/software co-design the software and hardware can be designed together to better meet cost and performance requirements.

Depending on the application, various combinations of criteria may be important goals for the system design. Two typical criteria are speed and manufacturing cost. The speed at which computations are made often contributes to the general usability of the system, just as in general-purpose computing. However, performance is also often associated with the satisfaction of deadlines—times at which computations must be completed to ensure the proper operation of the system. If failure to meet a deadline causes a major error, it is termed a hard deadline. And missed deadlines, which result in tolerable but unsatisfactory degradations are called soft deadlines. Hard deadlines are often (though not always) associated with safety-critical systems. Designing for deadlines is one of the most challenging tasks in embedded system design. Manufacturing cost is often an important criteria for embedded systems. Although the hardware components ultimately determine manufacturing cost, software plays an important role as well. First, the size of the program determines the amount of memory required, and memory is often a significant component of the total component cost. Furthermore, the improper design of software can cause one to require higher-performance, more-expensive hardware components than are really necessary. Efficient utilization of hardware resources requires careful software design. Power consumption is becoming an increasingly important design metric. Power is certainly important in battery-operated devices, but it can be important in wall socket-powered systems as well—lower power consumption means smaller, less-expensive power supplies and cooling and may result in environmental ratings that are advantageous in the marketplace. Once again, power consumption is ultimately determined by the hardware, but software plays a significant role in power characteristics. For example, more efficient use of on-chip caches can reduce the need for off-chip memory access, which consumes much more power than on-chip cache references.

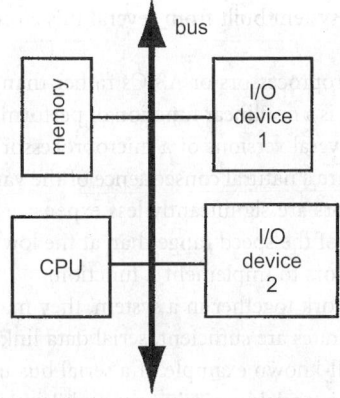

FIGURE 78.1 Hardware structure of a microprocessor system.

Figure 78.1 shows the hardware architecture of a basic microprocessor system. The system includes the CPU, memory, and some I/O devices, all connected by a bus. This system may consist of multiple chips for high-end microprocessors or a single-chip microcontroller. Typical I/O devices include analog/digital (ADC) and digital/analog (DAC) converters, serial and parallel communication devices, network and bus interfaces, buttons and switches, and various types of display devices. This configuration is a complete, basic, embedded computing hardware platform on which application software can execute.

The embedded application software includes components for managing I/O devices and for performing the core computational tasks. The basic software techniques for communicating with I/O devices are polling and interrupt-driven. In a polled system, the program checks each device's status register to determine if it is ready to perform I/O. Polling allows the CPU to determine the order in which I/O operations are completed, which may be important for ensuring that certain device requests are satisfied at the proper rate. However, polling also means that a device may not be serviced in time if the CPU's program does not check it frequently enough. Interrupt-driven I/O allows a device to change the flow of control on the CPU and call a device driver to handle the pending I/O operation. An interrupt system may provide both prioritized interrupts to allow some devices to take precedence over others and vectored interrupts to allow devices to specify which driver should handle their request.

Device drivers, whether polled or interrupt-driven, will typically perform basic device-specific functions and hand-off data to the core routines for processing. Those routines may perform relatively simple tasks, such as transducing data from one device to another, or may perform more sophisticated algorithms such as control. Those core routines often will initiate output operations based on their computations on the input operations.

Input and output may occur either periodically or aperiodically. Sampled data is a common example of periodic I/O, while user interfaces provide a common source of aperiodic I/O events. The nature of the I/O transactions affects both the device drivers and the core computational code. Code which operates on periodic data is generally driven by a timer which initiates the code at the start of the period. Periodic operations are often characterized by their periods and the deadline for each period. Aperiodic I/O may be detected either by an interrupt or by polling the devices. Aperiodic operations may have deadlines, which are generally measured from the initiating I/O event. Periodic operations can often be thought of as being executed within an infinite loop. Aperiodic operations tend to use more event-driven code, in which various sections of the program are exercised by different aperiodic events, since there is often more than one aperiodic event which can occur.

Embedded computing systems exhibit a great deal of parallelism which can be used to speed up computation. As a result, they often use multiple microprocessors which communicate with each other to perform the required function. In addition to microprocessors, application-specific ICs (ASICs) may be added to accelerate certain critical functions. CPUs and ASICs in general are called processing elements (PEs). An example multiprocessor system built from several PEs along with I/O devices and memory is shown in Figure 78.2.

The choice of several small microprocessors or ASICs rather than one large CPU is primarily determined by cost. Microprocessor cost is a nonlinear function of performance, even within a microprocessor family. Vendors generally supply several versions of a microprocessor which run at different clock rates; chips which run at varying speeds are a natural consequence of the variations in the VLSI manufacturing process. The slowest microprocessors are significantly less expensive than the fastest ones, and the cost increment is larger at the high end of the speed range than at the low end. As a result, it is often cheaper to use several smaller microprocessors to implement a function.

When several microprocessors work together in a system, they may communicate with each other in several different ways. If slow data rates are sufficient, serial data links are commonly used for their low hardware cost. The I²C bus is a well-known example of a serial bus used to build multi-microprocessor embedded systems; the CAN bus is widely used in automobiles. High-speed serial links can achieve moderately high performance and are often used to link multiple DSPs in high-speed signal processing

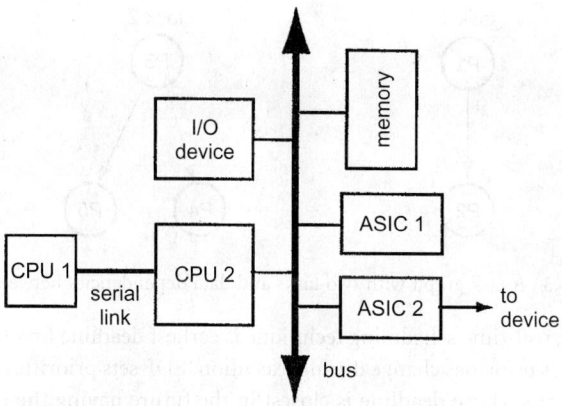

FIGURE 78.2   A heterogeneous embedded multiprocessor.

systems. Parallel data links provide the highest performance thanks to their sheer data width. High-speed busses such as PCI can be used to link several processors.

The software for an embedded multiprocessing system is often built around processes. A process, as in a general-purpose computing system, is an instantiation of a program with its own state. Since problems complex enough to require multiprocessors often run sophisticated algorithms and I/O systems, dividing the system into processes helps manage design complexity. A real-time operating system (RTOS) is an operating system specifically designed for embedded, and specifically real-time applications. The RTOS manages the processes and device drivers in the system, determining when each executes on the CPU. This function is termed scheduling. The partitioning of the software between application code which executes core algorithms and an RTOS which schedules the times to which those core algorithms are executed is a fundamental design principle in computing systems in general and is especially important for real-time operation.

There are a number of techniques which can be used to schedule processes in an embedded system— that is, to determine which process runs next on a particular CPU. Most RTOSs use process priorities in some form to determine the schedule. A process may be in any one of three states: currently executing (there can obviously be only one executing process on each CPU); ready to execute; or waiting. A process may not be able to execute until, for example, its data has arrived. Once its data arrives, it moves from waiting to ready. The scheduler chooses among the ready processes to determine which process runs next. In general, the RTOS's scheduler chooses the highest-priority ready process to run next; variations between scheduling methods depend in large part on the ways in which priorities are determined. Unlike general-purpose operating systems, RTOSs generally allow a process to run until it is preempted by a higher-priority process. General-purpose operating systems often perform time-slicing operations to maintain fair access of all the users on the system, but time-slicing does not allow the control required for meeting deadlines.

A fundamental result in real-time scheduling is known as rate-monotonic scheduling. This technique schedules a set of processes which run independently on a single CPU. Each process has its own period, with the deadline happening at the end of each period. There can be arbitrary relationships between the periods of the processes. It is assumed that data does not in general arrive at the beginning of the period, so there are no assumptions about when a process goes from waiting to ready within a period. This scheduling policy uses static priorities—the priorities for the processes are assigned before execution begins and do not change. It can be shown that the optimal priority assignment is based on period— the shorter the period, the higher the priority. This priority assignment ensures that all processes will meet their deadlines on every period. It can also be shown that at most, 69% of the CPU is used by this scheduling policy. The remaining cycles are spent waiting for activities to happen—since data arrival times are not known, it is not possible to utilize 100% of the CPU cycles.

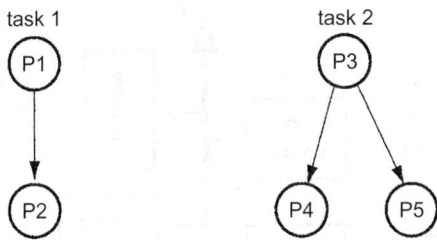

FIGURE 78.3   A task graph with two tasks and data dependencies between processes.

Another well-known, real-time scheduling technique is earliest deadline first (EDF). This is a dynamic priority scheme—process priorities change during execution. EDF sets priorities based on the impending deadlines, with the process whose deadline is closest in the future having the highest priority. Clearly, the rate of change of process priorities depends on the periods and deadlines. EDF can be shown to be able to utilize 100% of the CPU, but it does not guarantee that all deadlines will be met. Since priorities are dynamic, it is not possible in general to analyze whether the system will be overloaded at some point.

Processes may be specified with data dependencies, as shown in Figure 78.3, to create a task graph. An arc in the data dependency graph specifies that one process feeds data to another. The sink process cannot become ready until all the source processes have delivered their data. Processes which have no data dependency path between them are in separate tasks. Each task can run at its own rate. Data dependencies allow schedulers to make more efficient use of CPU resources. Since the source and sink processes of a data dependency cannot execute simultaneously, we can use that information to eliminate some combinations of processes which may want to run at the same time. Narrowing the scope of process conflicts allows us to more accurately predict how the CPU will be used.

A real-time operating system is often designed to have a small memory footprint, since embedded systems are more cost-sensitive than general-purpose computers. RTOSs are also designed to be more responsive in two different ways. First, they allow greater control over the order of execution of processes, which is critical for ensuring that deadlines are met. Second, they are designed to have lower context-switching overhead, since that overhead eats into the time available for meeting deadlines. The kernel of an RTOS is the basic set of functions that is always resident in memory. A basic RTOS may have an extremely small kernel of only a few hundred instructions. Such microkernels often provide only basic context-switching and scheduling facilities. More complex RTOSs may provide high-end operating system functions such as file systems and network support; many high-end RTOSs are POSIX (a Unix standard) compliant. While running such a high-end operating system requires more hardware resources, the extra features are useful in a number of situations. For example, a controller for a machine on a manufacturing line may use a network interface to talk to other machines on the factory floor or the factory coordination unit; it may also use the file system to access a database for the manufacturing process.

78.4   Hardware/Software Co-Design

Hardware/software co-design refers to any methodology which takes into account both hardware and software during the design of an embedded computing system. When the hardware and software are designed together, the designer has more opportunities to optimize the system by making tradeoffs between the hardware and software components. Good system designers intuitively perform co-design, but co-design methods are increasingly being embodied in computer-aided design (CAD) tools. We will discuss several aspects of co-design and co-design tools, including models of the design, co-simulation, performance analysis, and various methods for architectural co-synthesis. We will conclude with a look at design methodologies that make use of these phases of co-design.

78.4.1 Models

In designing embedded computing systems, we make use of several different types of models at different points in the design process. We need to model basic functionality. We must also capture nonfunctional requirements: speed, weight, power consumption, manufacturing cost, etc.

In the earliest stages of design, the task graph is an important modeling tool. The task graph does not capture all aspects of functionality, but it does describe the various rates at which computations must be performed and the expected degrees of parallelism available. This level of detail is often enough to make some important architectural decisions. A useful adjunct to the task graph are the technology description tables, which describe how processes can be implemented on the available components. One of the technology description tables describes basic properties of the processing elements, such as cost and basic power dissipation. A separate table describes how the processes may be implemented on the components, giving execution time (and perhaps other function-specific parameters like precise power consumption) on a processing element of that type. The technology description is more complex when ASICs can be used as processing elements, since many different ASICs at differing price/performance points can be designed for a given functionality, but the basic data still applies.

A more detailed description is given by either high-level language code (C, etc.) for software or hardware description language code (VHDL, Verilog, etc.) for software components. These should not be viewed as specifications—they are, in fact, quite detailed implementations. However, they do provide a level of abstraction above assembly language and gates and so can be valuable for analyzing performance, size, etc. The control-data flow graph (CDFG) is a typical representation of a high-level language: a flowchart-like structure describes the program's control, while data flow graphs describe the behavior within expressions and basic blocks.

78.4.2 Co-Simulation

Simulation is an important tool for design verification. The simulation of a complete embedded system entails modeling both the underlying hardware platform and the software executing on the CPUs. Some of the hardware must be simulated at a very fine level of detail—for example, buses and I/O devices may require gate-level simulation. On the other hand, the software can and should be executed at a higher level of abstraction. While it would be possible to simulate software execution by running a gate-level simulation of the CPU and modeling the program as residing in the memory of the simulated CPU, this would be unacceptably slow.

We can gain significant performance advantages by running different parts of the simulation at different levels of detail: elements of the hardware can be simulated in great detail, while software execution can be modeled much more directly. Basic functionality aspects of a high-level language program can be simulated by compiling the software on the computer on which the simulation executes, allowing those parts of the program to run at the native computer speed. Aspects of the program which deal with the hardware platform must interface to the section of the simulator which deals with the hardware. Those sections of the program are replaced by stubs which interface to the simulator. This style of simulation is a multi-rate simulation system, since the hardware and software simulation sections run at different rates: a single instruction in the software simulation will correspond to several clock cycles in the hardware simulation. The main jobs of the simulator are to keep the various sections of the simulation synchronized and to manage communication between the hardware and software components of the simulation.

78.4.3 Performance Analysis

Since performance is an important design goal in most embedded systems, both for overall throughput and for meeting deadlines, the analysis of the system to determine its speed of operation is an important element of any co-design methodology. System performance—the time it takes to execute a particular aspect of the system's functionality—clearly depends both on the software being executed and the underlying hardware platform. While simulation is an important tool for performance analysis, it is not sufficient, since simulation does not determine the worst-case delays. Since the execution times of most programs are data-dependent, it is necessary to give the simulation of the program the proper set of inputs to observe worst-case delay. The number of possible input combinations makes it unlikely that one will find those worst-case inputs without the sort of analysis that is at the heart of performance analysis.

In general, performance analysis must be done at several different levels of abstraction. Given a single program, one can place an upper bound on the worst-case execution time of the program. However, since many embedded systems consist of multiple processes and device drivers, it is necessary to analyze how these programs interact with each other, a phase which makes use of the results of single-program performance analysis.

Determining the worst-case execution time of a single program can be broken into two subproblems: determining the longest execution path through the program and determining the execution time of that program. Since there is at least a rough correlation between the number of operations and the actual execution time, we can determine the longest execution path without detailed knowledge of the instructions being executed—the longest path depends primarily on the structure of conditionals and loops. One way to find the longest path through the program is to model the program as a control-flow graph and use network flow algorithms to solve the resulting system.

Once the longest path has been found, we need to look at the instructions executed along that path to determine the actual execution time. A simple model of the processor would assume that each instruction has a fixed execution time, independent of other factors such as the data values being operated on, surrounding instructions, or the path of execution. In fact, such simple models do not give adequate results for modern high-speed microprocessors. One problem is that in pipelined processors, the execution time of an instruction may depend on the sequence of instructions executed before it. An even greater cause of performance variations is caching, since the same instruction sequence can have variable execution times, depending on whether the code is in the cache. Since cache miss penalties are often 5X or 10X, the cost of mischaracterizing cache performance is significant. Assuming that the cache is never present gives a conservative estimate of worst-case execution time, but one that is so over-conservative that it distorts the entire design. Since the performance penalty for ignoring the cache is so large, it results in using a much faster, more expensive processor than is really necessary. The effects of caching can be taken into account during the path analysis of the program—path analysis can determine bound how often an instruction present in the cache.

There are two major effects which must be taken into account when analyzing multiple-process systems. The first is the effect of scheduling multiple processes and device drivers on a single CPU. This analysis is performed by a scheduling algorithm, which determines bounds on when programs can execute. Ratemonotonic analysis is the simplest form of scheduling analysis—the utilization factor given by ratemonotonic analysis tells one an upper limit on the amount of active CPU time. However, if data dependencies between processes are known, or some knowledge of the arrival times of data is known, then a more accurate performance estimate can be computed. If the system includes multiple processing elements, more sophisticated scheduling algorithms must be used, since the data arrival time for a process on one processing element may be determined by the time at which that datum is computed on another processing element.

The second effect which must be taken into account is interactions between processes in the cache. When several programs on a CPU share a cache, or when several processing elements share a second-level cache, the cache state depends on the behavior of all the programs. For example, when one process is suspended by the operating system and another process starts running, that process may knock the first program out of the cache. When the first process resumes execution, it will initially run more slowly, an effect which cannot be taken into account by analyzing the programs independently. This analysis clearly depends in part on the system schedule, since the interactions between processes depends on the order in which the processes execute. But the system scheduling analysis must also keep track of the cache state—which parts of which programs are in the cache at the start of execution of each process. Good accuracy can be obtained with a simple model which assumes that a program is either in the cache or out of it, without considering individual instructions; higher accuracy comes from breaking a process into several sub-processes for analysis, each of which can have its own cache state.

78.4.4 Hardware/Software Co-Synthesis

Hardware/software co-synthesis tries to simultaneously design the hardware and software for an embedded computing system, given design requirements such as performance as well as a description of the functionality. Co-synthesis generally concentrates on architectural design rather than detailed component design—it concentrates on determining such major factors as the number and types of processing elements required and the ways in which software processes interact.

The most basic style of co-synthesis is known as hardware/software partitioning. As shown in Figure 78.4, this algorithm maps the given functionality onto a template architecture consisting of a CPU and one or more ASICs communicating via the microprocessor bus. The functionality is usually specified as a single program. The partitioning algorithm breaks that program into pieces and allocates pieces either to the CPU or ASICs for execution. Hardware/software partitioning assumes that total system performance is dominated by a relatively small part of the application, so that implementing a small fraction of the application in the ASIC leads to large performance gains. Less performance-critical sections of the application are relegated to the CPU.

The first problem to be solved is how to break the application program into pieces; common techniques include determining where I/O operations occur and concentrating on the basic blocks of inner loops. Once the application code is partitioned, various allocations of those components must be evaluated. Given an allocation of program components to the CPU or ASICs, performance analysis techniques can be used to determine the total system performance; performance analysis should take into account the time required to transfer necessary data into the ASIC and to extract the results of the computation from the ASIC. Since the total number of allocations is large, heuristics must be used to search the design space. In addition, the cost of the implementation must be determined. Since the CPU's cost is known in advance, that cost is determined by the ASIC cost, which varies as to the amount of hardware required to implement the desired function. High-level synthesis can be used to estimate both the performance and hardware cost of an ASIC which will be synthesized from a portion of the application program.

Basic, co-synthesis heuristics start from extreme initial solutions: We can either put all program components into the CPU, creating an implementation which is minimal cost but probably does not meet

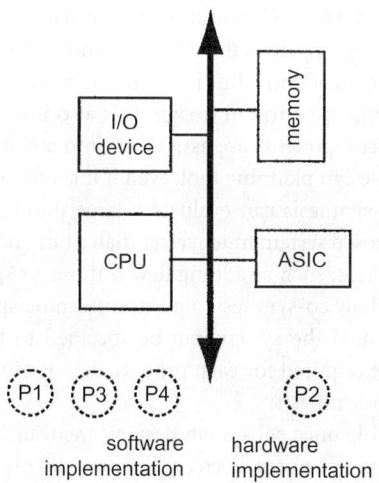

FIGURE 78.4 Hardware/software partitioning.

performance requirements, or put all program elements in the ASIC, which gives a maximal-performance, maximal-expense implementation. Given this initial solution, heuristics select which program component to move to the other side of the partition to either reduce hardware cost or increase performance, as desired. More sophisticated heuristics try to construct a solution by estimating how critical a component will be to overall system performance and choosing a CPU or ASIC implementation accordingly. Iterative improvement strategies may move components across the partition boundary to improve the design.

However, many embedded systems do not strictly follow the one CPU, one bus, n ASIC architectural template. These more general architectures are known as distributed embedded systems. Techniques for designing distributed embedded systems rest on the foundations of hardware/software partitioning, but they are generally more complicated, since there are more free variables. For example, since the number and types of CPUs is not known in advance, the co-synthesis algorithm must select them. If the number of busses or other communication links is not known in advance, those must be selected as well. Unfortunately, these decisions are all closely related. For example, the number of CPUs and ASICs required depends on the system schedule. The system schedule, in turn, depends on the execution time of each of the components on the available hardware elements. But those execution times depend on the processing elements available, which is what we are trying to determine in the first place. Co-synthesis algorithms generally try to fix several designs and vary only one or a few, then check the results of a design decision on the other parameters. For example, the algorithm may fix the hardware architecture and try to move processes to other processing elements to make more efficient use of the available hardware. Given that new configuration of processes, it may then try to reduce the cost of the hardware by eliminating unused processing elements or replacing a faster, more expensive processing element with a slower, cheaper one.

Since the memory hierarchy is a significant contributor to overall system performance, the design of the caching system is an important aspect of distributed system co-synthesis. In a board-level system with existing microprocessors, the sizes of second-level caches is under designer control, even if the first-level cache is incorporated on the microprocessor and therefore fixed in size. In a single-chip embedded system, the designer has control over the sizes of all the caches. Co-synthesis can determine hardware elements such as the placement of caches in the hardware architecture and the size of each cache. It can also determine software attributes such as the placement of each program in the cache. The placement of a program in the cache is determined by the addresses used by the program—by relocating the program, the cache behavior of the program can be changed. Memory system design requires calculating the cache state when constructing the system schedule and using the cache state as one of the factors to determine how to modify the design.

78.4.5 Design Methodologies

A co-design methodology tries to take into account aspects of hardware and software during all phases of design. At some point in the design process, the hardware and software components are well-specified and can be designed relatively independently. But it is important to consider the characteristics of both the hardware and software components early in design. It is also important to properly test the system once the hardware and software components are assembled into a complete system.

Co-synthesis can be used as a design planning tool, even if it is not used to generate a complete system architectural design. Because co-synthesis can evaluate a large number of designs very quickly, it can determine the feasibility of a proposed system much faster than a human designer. This allows the designer to experiment with what-if scenarios, such as adding new features or speculating on the effects of lower component costs in the future. Many co-synthesis algorithms can be applied without having a complete program to use as a specification. If the system can be specified to the level of processes with some estimate of the computation time required for each process, then useful information about architectural feasibility can be generated by co-synthesis.

Co-simulation plays a major role once subsystem designs are available. It does not have to wait until all components are complete, since stubs may be created to provide minimal functionality for incomplete components. The ability to simulate the software before completing the hardware is a major boon to software development and can substantially reduce development time.

79

Design Automation Technology Roadmap

Donald R. Cottrell
Silicon Integration Initiative, Inc.

CONTENTS

79.1 Introduction

No invention in the modern age has been as pervasive as the semiconductor and nothing has been more important to its technological advancement than has electronic design automation (EDA). EDA began in the 1960's both for the design of electronic computers and because of them. It was the advent of the computer that made possible the development of specialized programs that perform the complex management, design, and analysis operations associated with electronics and electronic systems. At the same time, it was the design, management, and manufacture of the thousands (now tens of millions) of devices that make up a single electronic assembly that made EDA an absolute requirement to fuel the semiconductor progression. Today, EDA programs are used in electronic packages for all business markets from computers to games, telephones to aerospace guidance systems, and toasters to automobiles. Across these markets, EDA supports many different package types such as integrated circuit (IC) chips, multichip modules (MCMs), printed circuit boards (PCBs), and entire electronic system assemblies.

No electronic circuit package is as challenging to EDA as the IC. The growth in complexity of ICs has placed tremendous demands on EDA. Mainstream EDA applications such as simulation, layout, and test generation have had to improve their speed and capacity characteristics with this ever-increasing growth in the number of circuits to be processed. New types of design and analysis applications, new methodologies, and new design rules have been necessary to keep pace. Yet, even with the technological breakthroughs that have been made in EDA across the past four decades, it is still having difficulty keeping up with the breakthroughs being made in the semiconductor technology progression that it fuels. Decrease in size and spacing of features on the chip is causing the number of design elements per chip to increase at a tremendous rate. The decrease in feature size and spacing coupled with the increase in operating frequency is causing additional levels of complexity to be approximated in the models used by design and analysis programs.

In the period from 1970 to the present semiconductor advances such as the following have had a great impact on EDA technology (Figure 79.1):

- IC integration has grown from tens of transistors on a chip, to beyond tens of millions.
- The feature size on production ICs has shrunk from 10 μm to 90 nm and smaller.
- On-chip clock frequency has increased from a few megahertz to many gigahertz.

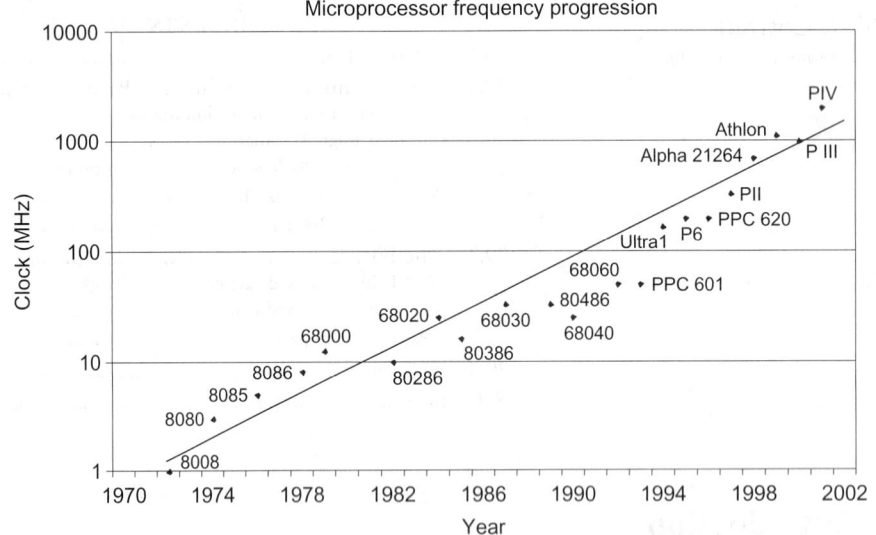

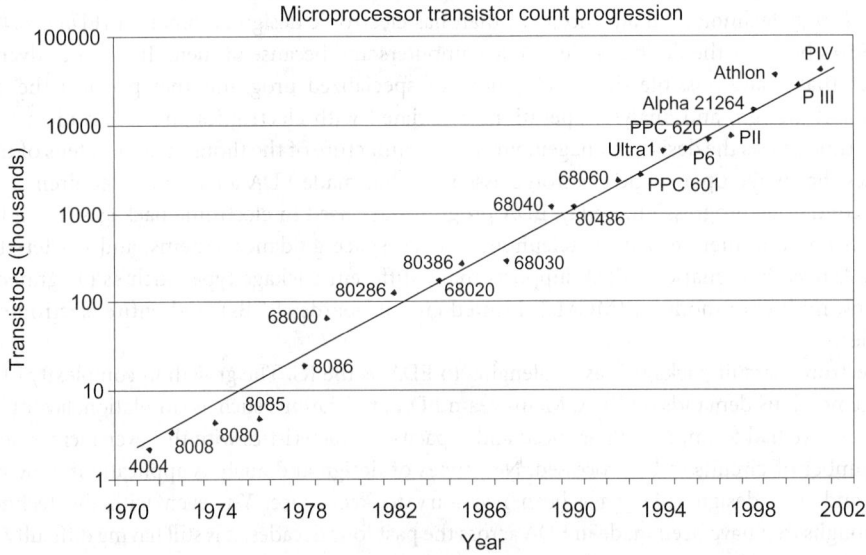

FIGURE 79.1 Microprocessor development.

Playing an essential part in the advancement of EDA have been advances in computer architectures that run the EDA applications. These advances have included the following:

- Computer CPU speed: from less than a million instructions per second (MIPS) of shared mainframe to hundreds of MIPS on a dedicated workstation
- Computer memory: from <32 KBytes to >500 GBytes
- Data archive: from voluminous reels of (rather) slow-speed tape to virtually limitless amounts of high-speed electronic storage.

However, these major improvements in computing power alone would not have been sufficient to meet the EDA needs of semiconductor advancement. Major advances have also been made to fundamental EDA algorithms, and entirely new design techniques and design paradigms have been invented and developed to support semiconductor advancement. This chapter will trace the more notable advancements made in EDA across its history for electronics design and discuss important semiconductor technology trends predicted across the next decade along and the impact they will have on EDA for the future. It is important to understand these trends and projections, because if the EDA systems cannot keep pace with the semiconductor projections, then these projections cannot be realized. Although it may be possible to build foundries that can manufacture ultradeep submicron wafers and even to acquire the billions of dollars of capital required for each, without the necessary EDA support these factories will never be fully utilized. SEMATECH reports that chip design productivity has increased at a compounded rate somewhere between 21 and 30%, while Moore's Law predicts the number of transistors on a chip to increase at a compound rate of 56%. This means that at some time, bringing on-line new foundries that can produce smaller, denser chips may reach a point of diminishing returns, because the ability to design and yield chips with that many transistors may not be possible.

79.2 Design Automation: Historical Perspective

79.2.1 The 1960s: The Beginnings of Design Automation

Early entries into design automation were made in the areas of design (description) records, PCB wiring, and manufacturing test generation. A commercial EDA industry did not exist and developments were made within companies with the need, such as IBM [1] and Bell Labs, on mainframe computers such as the IBM 7090. The 7090 had addressable 36-bit words and a limit of 32,000 words of main storage (magnetic cores). This was far less than the typical 256+ Mbytes of RAM on the average notebook PC, and certainly no match for a high-function workstation with gigabytes of main store. Though computer limitations continue to be an on-going challenge for EDA development, the limitation of the computers of the 1960s was particularly acute.

In retrospect, the limitation of computers in the 1960s was a blessing for the development of design automation. Because of these limitations, design automation developers were forced to invent highly creative algorithms that operated on very compact data structures. Many of the fundamental concepts developed during this period are still in use within commercial EDA systems today. Some notable advances during this period were:

- The fundamental "stuck-at" model for manufacturing test and a formal algebra for the generation of tests and diagnosis of faults.
- Parallel fault simulation, which provided simulation of many fault conditions in parallel with the good-machine (nonfaulty circuit) to reduce fault-simulation run times.
- A three-valued algebra for simulation which yields accurate results using simple delay models, even in the presence of race conditions within the design.
- Development of fundamental algorithms for the placement and wiring of components.
- Checking of designs against prescribed electrical design requirements (rules).

Also, there was development of fundamental heuristics for placement and wire routing, and for divide and conquer concepts supporting both. One such concept was the hierarchical division of a wiring image into cells, globally routing between cells, and then performing detailed routing within cells, possibly subdividing them further. Many of these fundamental concepts are still applied today, although the complexities of physical design (PD) of today's large-scale integration (LSI) is vastly more complex.

The 1960s represented the awakening of design automation and provided the proof of its value and need for electronics design. It would not be until the end of this decade when the explosion of the number of circuits designed on a chip would occur and the term LSI would by coined. EDA development in the 1960s was primarily focused on printed circuit assemblies, but the fundamental concepts developed for design entry, test generation, and PD provided the basics for EDA in the LSI era.

79.2.1.1 Design Entry

Before the use of computers in electronics design, the design schematic was a hard-copy drawing. This drawing was a draftsman's rendering of the notes and sketches provided by the circuit designer. The drawings provided the basis for manufacturing and repair operations in the field. As automation developed, it became desirable to store these drawings on storage media usable by computers so that the creation of input to the automated processes could, itself, be automated. So, the need to record the design of electronic products and assemblies in computers was recognized in the late 1950s and early 1960s. In the early days of design automation, the electronics designer would develop the design using paper and pencil and then transcribe it to a form suitable for keyed entry to a computer. Once keyed into the computer, the design could be rendered in a number of different formats to support the manufacturing and field operations. It was soon recognized that these computerized representations of the circuit design drawing could also drive design processes such as the routing of printed circuit traces or the generation of manufacturing test patterns. Finally, from there, it was but a short step to the use of computers to generate data in the form required to drive automated manufacturing and test equipment.

Early design entry methods involved the keying of the design description onto punch cards that were read into the computer and saved on a persistent storage device. This became known as the design's database, and is the start of the design automation system. From the database, schematic diagrams and logic diagrams were rendered for use in engineering, manufacturing, and field support. This was typically a two-step process, whereby the designer drew the schematic by hand and then submitted it to another for conversion to the transcription records, keypunch, and entry to the computer. Once in the computer, the formal automated drawings were generated, printed, and returned to the designer. Although this process seems archaic by today's standards, it did result in a permanent record of the design in computer readable format. This could be used for many forms of records management, engineering change history, and as input to design, analysis, and manufacturing automation that would soon follow.

With the introduction of the alphanumeric terminal, the keypunch was replaced as the window into the computer. With this, new design description languages were developed, and the role of the transcription operator began to move back to the designer. Although these description languages still represented the design at the device or gate level, they were free format and keyword oriented and design engineers were willing to use them. The design engineer now had the tools to enter design descriptions directly into the computer thus eliminating the inherent inefficiencies of the "middleman." Thus, a paradigm shift began to evolve in the method by which design was entered to the EDA system. Introduction of the direct access storage devices (disks) in the late 1960s also improved the entry process as well as the entire design system by providing on-line, high-speed direct access to the entire design or any portion of it. This was also important to the acceptance of design entry by the designer as the task was still viewed as a necessary overhead rather than a natural part of the design task. It was necessary to get access to the other evolving design automation tools, but typically, the real design thought process took place with pencil and paper techniques. Therefore, any change that made the entry process faster and easier was eagerly accepted.

The next shift occurred in the later part of the 1970s with the introduction of graphics terminals. With these, the designer could enter design into the database in schematic form. This form of design entry was a novelty at first, but not a clear performance improvement. In fact, until the introduction of the workstation with dedicated graphics support, graphic entry was detrimental to design productivity in many cases. Negative effects such as less than effective transaction speed, time lost in making the schematic esthetically pleasing and the low level of detail all added to less than obvious advances. In contrast, use of the graphics display to view the design and make design changes proved extremely effective and was a great improvement over the red-lined hard-copy prints. For this reason, use of computer graphics represented a major advance and this style of design entry took off with the introduction of the workstation in the 1980s. In fact, the graphic editor's glitz and capability was often a major selling point in the decision to use one EDA system over another. Further, to be considered a commercially viable system, graphics entry was required. Nevertheless, as the density of ICs grew, graphics entry of schematics would begin to yield to the productivity advantages of (alphanumeric) description languages. As EDA design and analysis tool technology advanced, entry of design at the register-transfer level (RTL) level would become commonplace and today the design engineer is able to represent his design ideas at many levels of abstraction and throughout the design process. System-level and RTL design languages have been introduced and the designer is able to verify design intent much earlier in the design cycle.

By the 1990s and after the introduction of synthesis automation, design entry using RTL descriptions was the generally accepted approach for entry of design, although schematic entry remained the accepted method for PCB design and many elements of custom ICs. There is no doubt that the introduction of graphics into the design automation system represents a major advance and a major paradigm shift. The use of graphics to visualize design details, wiring congestion, and timing diagrams is of major importance. The use of graphics to perform edit functions is standard operating procedure.

Large system design, often entails control circuitry, dataflow, and functional modules. Classically, these systems span across several chips and boards and employ several styles of entry for the different physical packages. These may include:

- Schematics—graphic
- RTL and behavorial level languages—alphanumeric
- Timing diagrams—graphic
- State diagrams—alphanumeric
- Flowcharts—graphic.

Today, these entry techniques can be found in different EDA tools and each is particularly effective for different types of design problems. Schematics are effective for the design of "glue" logic that interconnects functional design elements such as modules on a PCB and for custom IC circuitry. Behavorial languages are useful for system-level design, and particularly effective for dataflow behavior. Timing diagrams lend themselves well to describe the functional operations of "black-box" components at their I/Os without needing to describe their internal circuitry. State diagrams are a convenient way to express the logical operation of combinational circuits. Flowcharts are effective for describing the operations of control logic, much like use of flowcharts for specification of software program flow. With technology advances, the IC is engulfing more and more of the entire system and all of these forms of design description may be prevalent on a single chip. It is even expected that the design of "black-box" functions will be available from multiple sources to be embedded onto the chip similar to the use of modules on a PCB. Thus, it is likely that future EDA systems will support a mixture of design description forms to allow the designer to represent sections of the design in a manner most effective to each. After all, design is described in many forms by the designer outside the design system.

79.2.1.2 Test Generation

Testing of manufactured electronic subassemblies entails the use of special test hardware that can provide stimulus (test signals) to selected (input) pins of the part under test and measure for specified responses on selected (output) pins. If the measured response matches the specified response, then the part under test has passed that test successfully. If some other response is measured, then the part has failed that test and the presence of a defect is indicated. Manufacturing test is the successive application test patterns that cause some measurable point on the part under test to be sensitized to the presence of a manufacturing defect. That is, some measurable point on the part under test will result in a certain value if the fault is present and a different one if no fault were present. The collection of test patterns causes all (or almost all) possible manufacturing failures to render a different output response than would the nondefective part. For static DC testers, each stimulus is applied and after the part under test settles to a steady state, the specified outputs are measured and compared with the expected results for a nondefective part.

To bound the test generation problem, a model was developed to represent possible defects at the abstract gate level. This model characterizes the effects of defects as stuck-at values. This model is fundamental to most of the development in test generation and is still in use today. It characterizes defects as causing either a stuck-at-one or a stuck-at-zero condition at pins on a logic gate. It assumes hard faults (that is, if present, a fault remains throughout the test) and that only one fault occurs at a time. Thus, this model became known as the single stuck-at fault model. Stuck-at fault testing assumes that the symptom of any manufacturing defect can be characterized by the presence of a stuck-at fault some place within the circuit and that it can be observed at some point on the unit under test. By testing for the presence of all possible stuck-at faults that can occur, all possible manufacturing hard-defects in the logic devices can thus be tested.

The stuck-at fault models for NAND and NOR gates are shown in Figure 79.2. For the NAND gate, the presence of an input stuck-at-one defect can be sensitized (made detectable) at the gate's output pin by setting the good-machine state for that input to zero, and setting the other input values to 1. If a zero is observed at the gate's output node, then a fault (stuck-at-one) on the input pin set to zero is detected. A stuck-at-zero condition on any specific input to the NAND gate is not distinguishable from a stuck-at-zero on any other input to the gate, thus is not part of the model. However, a stuck-at-one is modeled for the gate's output to account for such a fault or a stuck-at-one fault on the gate's output circuitry. Similarly, a stuck-at-zero is modeled on the gate's output.

The fault model for the NOR gate is similar except that here input faults are modeled as stuck-at-zero, as the stuck-at-one defect cannot be isolated to a particular input.

Later in time, additional development would attack defects not detectable with this stuck-at fault model; for example, bridging faults where nodes are shorted together, and delay faults where the output response does not occur within the required time. The stuck-at fault model cannot detect these fault types and they became important as the development of CMOS progressed. Significant work was

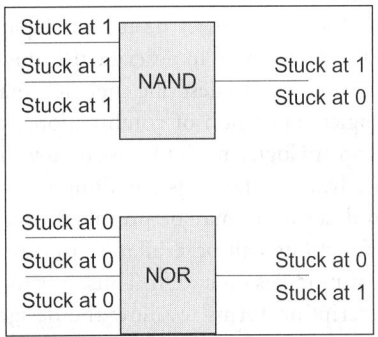

FIGURE 79.2   Stuck-at fault models.

performed in both these areas beginning in the 1970s, but it did not have the impact on test generation development that the stuck-at fault model did.

The creation of the fault model was very important to test generation as it established a realistic set of objectives to be met by automated test set generation that could be achieved in a realistic amount of computational time. A formal algebra was developed by Roth [2] called the D-ALG that formalized an approach to test generation and fault diagnosis.

The test generation program could choose a fault based on the instances of gates within the design and the fault models for the gates. It could then trace back from that fault to the input pins of the design and, using the D-ALG calculus, it could find a set of input states that would sensitize the fault. Then, it could trace forward from that fault to the design's output pins, sensitizing the path (causing the good-machine value to be the opposite of the stuck-at-fault value along that path) to at least one observable pin (Figure 79.3).

The use of functional patterns as the test patterns in lieu of heuristic stuck-at test generation was another approach for manufacturing test. However, this required an extreme number of tests to be applied and provided no measurable objective to be met (i.e., 100% of all possible stuck-at faults) thus, depended on the experience and skill of the designer to create quality tests. The use of fault models and automatic test generation produced a minimum set of tests and greatly reduced intensive manual labor.

The exhaustive method to assure coverage of all the possible detectable defects would be to apply all possible input states to the design under test and compare the measured output states with the simulated good-machine states. For a design with n input pins, however, this may theoretically require the simulation of 2^n input patterns for combinatorial logic and at least 2^{n+m} for sequential logic (where m is the number of independent storage elements). For even a relatively small number of input pins, this amount of simulation would not be possible even on today's computers, and the time to apply this number of patterns at the tester would be grossly prohibitive.

The amount of time that a part resides on the tester is critical in the semiconductor business, as it impacts a number of parts that can be produced in a given amount of time and the capital cost for testers. Thus, the number of test patterns that need to be applied at the tester should be kept to a minimum. Early work in test generation attacked this problem in two ways. First, when a test pattern was generated for a specific fault, it was simulated against all possible faults one at a time. In many cases, the application of a test pattern that is targeted for one specific fault will also detect several other faults at the same time. The presence of a specific fault may be detectable on one output pin, for example, but at the same time

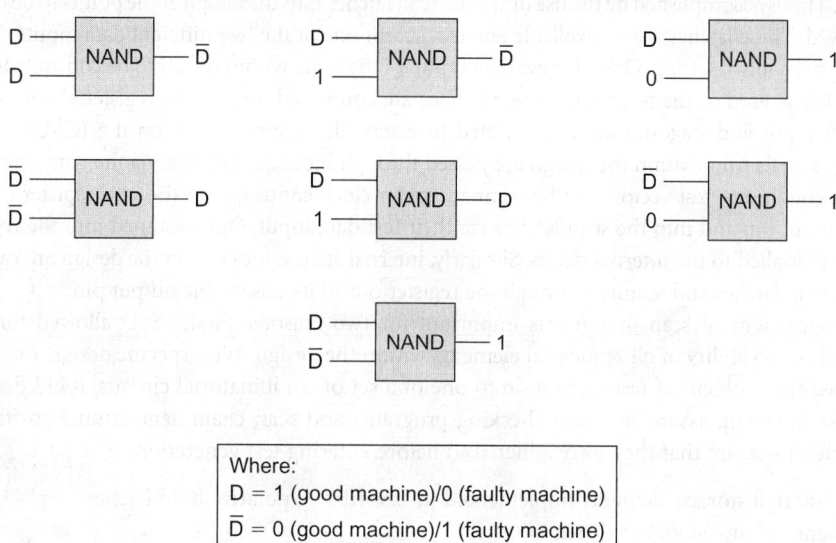

Where:
D = 1 (good machine)/0 (faulty machine)
$\overline{D}$ = 0 (good machine)/1 (faulty machine)

FIGURE 79.3 D-ALG calculus.

additional faults may be observable at the other output pins. The use of fault simulation (discussed in Section 79.2.1.3) detected these cases and provided a mechanism to mark as tested those faults that were "accidentally" covered. This meant that the test generation algorithm did not have to generate a specific test for those faults. Second, schemes were developed to merge test patterns together into a smaller test pattern set to minimize the number of patterns required for detection of all faults. This is possible when two adjacent test patterns require the application of specific values on different input pins, each allowing all the other input pins to be at a do not-care state. In these cases, the multiple test patterns can be merged into one. With successive analysis in this way, all pairs of test patterns (pattern n with $n + 1$, or the merger of m and $m + 1$ with pattern $m + 2$) are analyzed and merged into a reduced set of patterns.

Sequential design elements severely complicate test generation, as they require the analysis of previous states and the application of sequences of patterns. Early work in test generation broke feedback nets, inserted a lumped delay on them, and analyzed the design as a combinatorial problem using a Huffman model. Later work attempted to identify the sequential elements within the design using sophisticated topological analysis and then used a Huffman model to analyze each unique element. State tables for each unique sequential element were generated and saved for later use as lookup tables in the test generation process. The use of three-value simulation within the analysis reduced the analysis time as well as guaranteed that the results were always accurate. Huffman analysis required 2^x simulations (where x is the number of feedback nets) to determine if critical hazards existed in the sequential elements. Using three-valued simulation [3] (all value transitions go through an X state), this was reduced to a maximum of $2x$ simulations.

This lumped delay model did not account for the distribution of delays in the actual design, thus it often caused pessimistic results. Often simulated results yielded do not-know (X-state) conditions when a more accurate model could yield a known state. This made it difficult to generate patterns that would detect all faults in the model. As the level of integration increased, the problems associated with automatic test generation for arbitrary sequential circuits became unwieldy. This necessitated that the designer be called back into the problem of test generation most often to develop tests that would detect those that were missed by the test generation program. New approaches that ranged from random pattern generation to advanced algorithms and heuristics, which use a combination of different approaches were developed. However, by the mid-1970s the need to design-for-test was becoming evident to many companies.

During the mid-1970s the concept of scan design such as IBM's Level Sensitive Scan Design (LSSD), was developed [4]. Scan design provides the ability to externally control and observe internal state variables of the design. This is accomplished by the use of special scan latches into the design at the points to be controlled and observed. These latches are controllable and can accept one of the two different data inputs depending on an external control setting. One of these data inputs is the node within the IC to be controlled/observed and the other is used as the test input. These latches are connected into a shift register chain that has its stage-0 test input and stage-n output connected to externally accessible pins on the IC. Under normal conditions, signals from within the design are passed through individual latches via the data input/output. Under test conditions, test vectors can be scanned, under clock control, onto the scan register's externally accessible input pin and into the scan latches via their test data input. Once scanned into the register, the test vector is applied to the internal nodes. Similarly, internal state values within the design are captured in individual scan latches and scanned through the register out to its observable output pin.

The development of scan design was important for two reasons. First, LSSD allowed for external control and observability of all sequential elements within the design. With specific design-for-test rules, this reduced the problem of test generation to one of a set of combinatorial circuits. Rigid design rules such as the following assure this, and checking programs and scan chain generation algorithms were implemented to assure that they were adhered to before entering test generation:

- All internal storage elements implemented in hazard-free polarity-hold latches
- Absence of any global feedback loops
- Latches may not be controlled by the same clock that controls the latches feeding them
- Externally accessible clocks control all shift register latches.

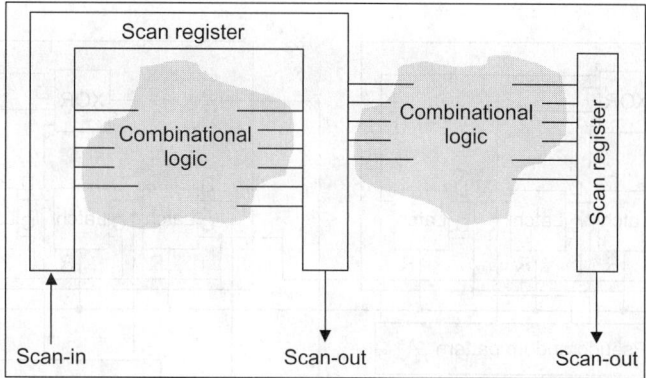

FIGURE 79.4 Scan design.

Second, scan design allowed for external control and observability at otherwise nonprobable points between modules on an MCM or PCB. During the 1980s an industry standard was developed, called Boundary Scan (IEEE 1149.1 Joint Test Action Group) [5], which uses scan design techniques to provide control and observability for all chip or module I/O's from pins of their next level package (MCM or PCD, respectively).

Scan design requires additional real estate in the IC design and has a level of performance overhead. However, with the achievable transistor density levels on today's ICs and the test and diagnosis benefits accrued, these penalties are easily justified in all but the most performance critical designs (Figure 79.4).

During the 1980s, IC density allowed for the design of built-in self-test (BIST) circuitry on the chip itself. Operating at hardware speed, it became feasible to generate complete exhaustive tests and large numbers of random tests never possible with software and stored program testers. BISTs are generated on the tester by the device under test, reducing the management of data transfer from design to manufacturing. A binary counter is used or linear feedback shift registers (LFSR) are used to generate the patterns for exhaustive or random tests, respectively. In the latter case, a pseudorandom bit sequence is formed by the exclusive-OR of the bits on the LFSR and this result is then fed back into the LFSR input. Thus, a pseudorandom bit sequence whose sequence length is based on the number of LFSR stages and the initial LFSR state can be generated. The design can be simulated to determine the good-machine state conditions for the generated test patterns, and these simulated results are compared with the actual device-under-test results observed at the tester.

BIST techniques have become common for the test of on-chip RAM and ROS. BIST is also used for logic sections of chips either with fault-simulated weighted random test patterns or good machine-simulated exhaustive patterns. In the latter case, logic is partitioned into electrically isolated regions with a smaller number of inputs to reduce the number of test patterns. Partitioning of the design reduces the number of exhaustive tests from 2^n (where n is the total number of inputs) to

$$\sum_{i=1}^{m} 2^{n^i}$$

where m is the number of partitions and n^i ($n^i < n$) is the number of inputs on each logic partition.

Since the use of BIST implies an extremely large number of tests, the simulated data transfer and test measurement time is reduced greatly by the use of a compressed signature to represent the expected and actual test results. Thus, only a comparison of the simulated signatures for each BIST region needs to be made with the signatures derived by the on-chip hardware, rather than the results of each individual test pattern. This is accomplished by feeding the output bit sequence to a single input serial input LFSR after

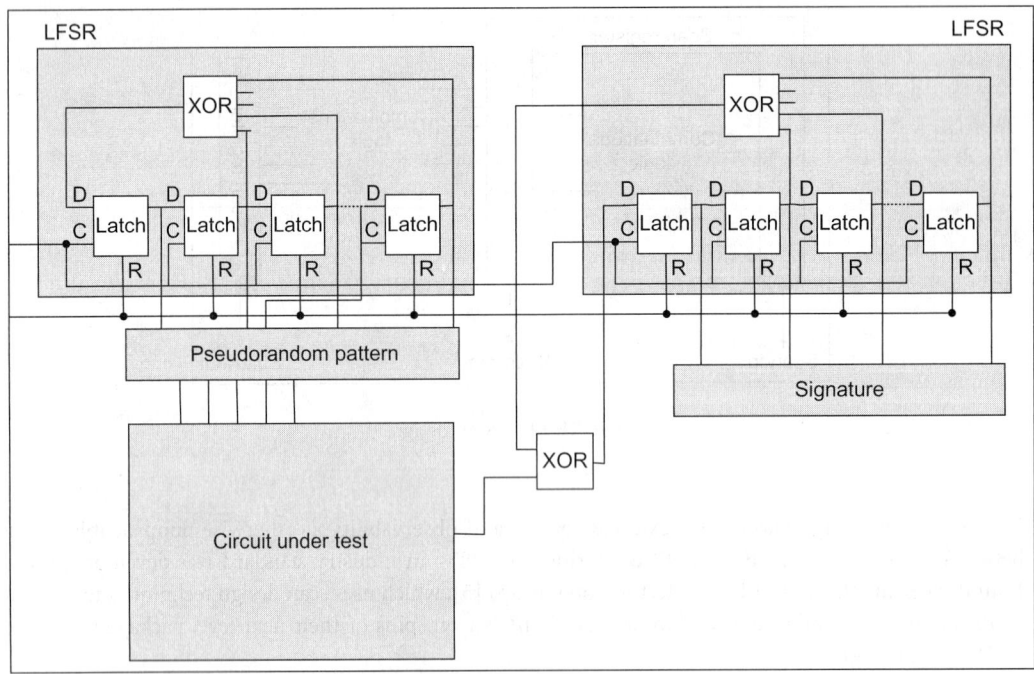

FIGURE 79.5 BIST logic.

exclusive-OR with the pseudorandom pattern generated by that LFSR. In this way, a unique pattern can be observed for a sequence of test results, which is a function of the good-machine response and a pseudorandom number (Figure 79.5).

Today, testing of ICs typically consists of combinations of different test strategies. These may include stored program stuck-at fault tests, BIST, delay (or at-speed) test (testing for signal arrival times in addition to state), and I_{DDQ} tests (direct drain quiescent current testing checks for CMOS defects that cause excessive current leakage). The latter two test techniques are used to detect defect conditions not identified by stuck-at fault tests such as shorts (bridging faults) between signal nets or gate oxide defects that cause incorrect device operation and parametric variability resulting from process variations in the wafer foundry. However, the stuck-at fault model and scan techniques have been and will continue to be fundamental ingredients for the manufacturing test recipe.

79.2.1.3 Fault Simulation

Fault simulation is used to predict the state of a design at observable points when in the presence of a defect. This is used for manufacturing test and for field diagnostic pattern generation. Early work in fault simulation relied on the stuck-at fault model and performed successive simulations of the design with each single fault independent of any other fault; thus, a single stuck-at-fault model was assumed. Because even these early designs consisted of thousands of faults, it was too time consuming to simulate each fault serially and it was necessary to create high-speed fault simulation algorithms.

For manufacturing test, fault simulation took advantage of three-valued zero-delay simulation. The simulation model of the design was levelized and compiled into an executable program. Levelization assured that driver gates were simulated before the gates receiving the signals; thus, allowing a state resolution in a single simulation pass. Feedback loops were cut and the X-transition of three-valued simulation resolved race conditions. The inherent instruction set of the host computer (e.g., AND, OR, and XOR.) allowed the use of a minimum set of instructions to simulate a gate's function.

The parallel-fault simulation algorithm that allowed many faults to be simulated in parallel was developed during the 1960s. Using the 32-bit word length of the IBM 7090 computer architecture, for example, simulation of 31 faults in a single pass (using the last bit for the good-machine) was possible. For each gate in the design, two host machine words were assigned to represent its good-machine state and the state for 31 single stuck-at faults. The first bit position of the first word was set to one or zero representing the good-machine state for the node, and each of the successive bit position was set to the 1/0 state that would occur if the simulated faults were present (each bit-position representing a single fault). The corresponding bit position in the second word was set to zero if it was a known state and one if it was an *X*-state. The entire fault list was divided into *n* partitions of 31 faults and each partition was then simulated against all input patterns. In this way, the run time for simulation is a function of

$$\sum_{i=1}^{n=F/31} Patterns$$

where *F* is the total number of single stuck-at faults in the design.

Specific faults are injected within the word representing their location within the design by the insertion of a mask that is AND'd or OR'd at the appropriate word. For stuck-at-one conditions the mask contains a 1-bit in the position representing the fault (and 0-bits at the others) and it is OR'd to the gate's memory location. For stuck-at-zero faults the mask contains a 0-bit at the positions representing the fault (and 1-bits at the others) and it is AND'd with the gate's memory location.

As the level of integration increased, so did the number of faults. The simulation speed improvement realized from parallel-fault simulation was limited to the number of faults simulated in parallel and because of the algorithm overhead, it did not scale with the increase in faults.

Deductive-fault simulation was developed early in the 1970s and required only one simulation pass per test pattern. This is accomplished by simulating only the good-machine behavior and using deductive techniques to determine each fault that is detectable along the simulated paths. Note here that fault detection became the principal goal and fault isolation (for repair purposes) was ignored, as by now the challenge was to isolate from a wafer bad chips that were not repairable. Because lists of faults detectable at every point along the simulated path need to be kept, this algorithm requires extensive memory, far more than the parallel-fault simulation one. However, with increasing memory on host computers and the inherent increase in fault simulation speed, this technique won the favor of many fault simulators.

Concurrent-fault simulation refined the deductive algorithm by recognizing the fact that paths in the design quickly become insensitive to the presence of most faults, particularly after some initial set of test patterns is simulated (an observation made in the 1970s was that a high percentage of faults is detected in a low percentage of the initial test patterns, even if these patterns are randomly generated). The concurrent-fault simulation algorithm simulates the good machine and concurrently simulates a number of faulty machines. Once it is determined that a particular faulty machine state is the same as the good-machine one, simulation for that fault ceases. Since on logic paths most faults will become insensitive rather close to the point where the fault is located, the amount of simulation for these faulty machines was kept small. This algorithm required even more memory, particularly for the early test patterns; however, host machine architectures of the late 1970s were supporting, what then appeared as, massive amounts of addressable memory.

With the introduction of scan design in which all sequential elements are controllable from the tester, the simulated problem is reduced to that of a combinatorial circuit whose state is deterministic based on any single test pattern, and is not dependent on previous patterns or states. Parallel-pattern fault simulation was developed in the late 1970s to take advantage of this, by simulating multiple test patterns in parallel against a single fault. A performance advantage is achieved because compiled simulation could again be utilized as opposed to the more costly event-based approach. In addition, because faults not detected by the initial test patterns are typically only detectable by a few patterns and for these a sensitized path often disappears within a close proximity to the fault's location, simulation of many patterns does not require a complete pass across the design.

Because of the increasing number of devices on chips, the test generation and fault simulation problem continued to face severe challenges. With the evolution of BIST however, the use of fault simulation was relaxed. With BIST, only the good-machine behavior needs to be simulated and compared with the actual results at the tester.

79.2.1.4   Physical Design

As with test generation, PD has evolved from early work on PCBs. PD automation programs place devices and generate the physical interconnect routing for the nets that connect them into logic paths assuring that electrical and physical constraints are met. The challenge for PD has become ever greater since its early development for PCBs where the goal was simply to place components and route nets typically looking for the shortest path. Any nets that could not be auto-routed were routed (embedded) manually, or as a last resort with nonprinted (yellow) wires. As the problem moved on to the ICs, the ability to use nonprinted wires to finish routing was no more. Now, all interconnects had to be printed circuits and anything less than a 100% solution is unacceptable. Further, as the IC densities increased, so did the number of nets which necessitated the invention of smarter wiring programs and heuristics. Even a small number of incomplete (overflow) routes became too complex of a task for manual solutions.

As IC device sizes shrank and the gate delays decreased, the delay caused by interconnect wiring also became an important factor for a valid solution. No longer was any wiring solution a correct solution. Complexity increased by the need to find wiring solutions that fall within acceptable timing limits. Thus, the wiring lengths and thickness needed to be considered. As IC features become packed closer together cross-coupled capacitance (cross talk) effects between them is also an important consideration and for the future, wiring considerations will expand into a three-dimensional space. PD solutions must consider these complex factors and still achieve a 100% solution that meets the designer-specified timing for IC designs that contain hundreds of millions of nets.

Because of these increasing demands on the PD, major paradigm changes have taken place in the design methodology. In the early days, there was a clear separation of logic design and PD. The logic designer was responsible for creating a netlist that correctly represented the logic behavior desired. Timing was a function of the drive capability of the driving circuit and the number of receivers. Different power levels for drivers could be chosen by the logic designer to match the timing requirements based on the driven circuits. The delay imposed by the time of flight along interconnects and owing to the parasitics on the interconnect was insignificant. Therefore, the logic designer could hand off the PD to another, more adept at using the PD programs and manually embedding overflow wires. As the semiconductor technology progressed, however, there needed to be more interactions between the logic designer and the physical designer, as the interconnect delays became a more dominant factor across signal paths. The logic designer had to give certain timing constraints to the physical designer and if these could not be met, the design was often passed back to the logic designer. The logic designer, in turn, then had to choose different driver gates or a different logical architecture to meet his design specification. In many cases, the pair had to become a team or there was a merger of the two previously distinct operations into one "IC designer".

This same progression of merging logic design and PD into one operational responsibility has also begun at the EDA system architecture level. In the 1960s and 1970s, front-end (design) programs were separate from back-end (physical) programs. Most often they were developed by different EDA development teams and designs were transferred between them by means of data files. Beginning in the 1980s, the data transferred between the front-end programs and the back-end ones included specific design constraints that must be met by the PD programs—the most common being a specific amount of allowed delay across an interconnect or signal path. Moreover, as the number of constraints that must be met by the PD programs increases so does the challenge to achieve a 100% solution. Nonetheless, many of the fundamental wiring heuristics and algorithms used by PD today spawned from work done in the 1960s for PCBs.

Early placement algorithms were developed to minimize the total length of the interconnect wiring using Steiner Trees and Manhattan wiring graphs. In addition, during these early years, algorithms were developed to analyze wiring congestion that would occur as a result of placement choices and minimize it to give routing a chance to succeed. Later work in the 1960s led to algorithms that performed a hierarchical division of the wiring image and performed global wiring between these subdivisions (then called cells) before routing within the cells [6]. This divide-and-conquer approach simplified the problem and led to quicker and more complete results. Min-cut placement algorithms often used today are a derivative of this divide-and-conquer approach. The image is divided into partitions and the placements of these partitions are swapped to minimize interconnect length between them and possible wiring congestion. Once a global solution for the cells is found, placement is performed within them using the same objectives. Many current placement algorithms are based on these early techniques, although they now need to consider more physical and electrical constraints.

Development of new and more efficient routing algorithms progressed. The Lee algorithm [7] finds a solution by emitting a "wave" from both the source and target points to be wired. This wave is actually an ordered identification of available channel positions—where the available positions adjacent to the source or destination are numbered 1, and the available positions adjacent to them are numbered 2, etc. Successive moves and sequential identification is made (out in all directions as would a wave) until the source and destination moves meet (the waves collide). Then a backtrace is performed from the intersecting position in reverse sequential order along the numbered track positions back to the source and destination. At points where a choice is available (that is, there are two adjacent points with the same order number), the one which does not require a change in direction is chosen.

The Hightower Line-Probe technique [8], also developed during this period and speeded up routing, by use of emanating lines rather than waves. This algorithm emanated a line from both the source and destination points, toward each other. When either line encounters an obstacle, then another line is emanated from a point just missing the edge of the obstacle on the original line at a right angle to the original line, and toward the target or source. Thus, the process is much like walking blindly in an orthogonal line toward the target and changing direction only after bumping into a wall. This process continues until the lines intersect at which time the path is complete.

In today's ICs, the challenge of completed wiring that meets all constraints is of crucial importance. Unlike test generation, which can be considered successful when a very high percentage of the faults are detected by the test patterns, 100% complete is the only acceptable answer for PD. Further, all of the interconnects must fall within the required electrical and physical constraints. Nothing <100% is acceptable! Today these constraints include timing, power consumption, noise, and yield and this list will become more complex as IC feature sizes and spacing are reduced further.

79.2.2 The 1970s: The Awaking of Verification

Before the introduction of large-scale integrated (LSI) circuits in the 1970s, it was common practice to build prototype hardware to verify the design correctness. PCB packages containing discrete components, single gates, and small-scale integrated modules facilitated engineering rework of real hardware within the verification cycle. Prototype PCBs were built and engineering used test stimulus drivers and oscilloscopes to determine whether the correct output conditions resulted from input stimuli. As design errors were detected, they were repaired on the PCB prototype, validated, and recorded for later engineering into the production version of the design. Use of the wrong logic function within the design could easily be repaired by replacing component(s) in error with the correct one(s). Incorrect connections could easily be repaired by cutting a printed circuit and replacing it with a discrete wire. Thus, design verification (DV) was a sort of trial-and-error process using real hardware.

The introduction of LSI circuits drastically changed the DV paradigm. Although the use of software simulation to verify system design correctness began during the 1960s, it was not until the advent of the LSI circuits that this concept became widely accepted. With large-scale integration, it became impossible to use prototype hardware or to repair a faulty design after it was manufactured. The 1970s are best represented by a quantum leap into verification before manufacture through the use of software modeling (simulation). This represented a major paradigm shift in electronics design and was a difficult change for some to accept. DV on hardware prototypes resulted in a tangible result that could be touched and held. It was a convenient tangible, which could be shown by management to represent real progress. Completed DV against a software model did not produce the same level of touch and feel. Further, since the use of computer models was a relatively new concept, it met with the distrust of many. However, the introduction of LSI circuits demanded this change and today, software DV is a commonplace practice used in all levels of electronic components, subassemblies, and systems.

Early simulators simulated gate-level models of the design with two-valued (1 and 0) simulation. Since gates had nearly equal delays and the interconnect delay was insignificant by comparison, the use of a unit of delay for each gate with no delay assigned to interconnects was common. Later simulators exploited the use of three-valued simulation (1, 0, and unknown) to resolve race conditions and identify oscillations within the design more quickly. With the emergence of LSI circuits, however, these simple models had to become more complex and, additionally, simulators had to become more flexible and faster—much faster. In the first half of the 1970s, important advances were made to DV simulators that included

- the use of abstract (with respect to the gate-level) models to improve simulation performance and enable verification throughout the design cycle (not just at the end)
- more accurate representations of gate and interconnect delays to enhance the simulated accuracy.

In the latter half of the decade, significant contributions were made to facilitate separation of function verification from timing verification and formal approaches for verification. However, simulation remains a fundamental DV tool and the challenge to make simulators faster and more flexible continues even today.

79.2.2.1 Simulation

Though widely used for DV, simulation has a couple of inherent problems. First, unlike test generation or PD, there is no precise measure of completeness. Test generation has the stuck-at fault model and PD has a finite list of nets that must be routed. However, there is no equivalent metric to determine when verification of the design is complete, or when enough simulation is done. During the 1970s research began to develop a metric for verification completeness, but to this day none has been generally accepted. Use is made of minimum criteria such as all nets must switch in both directions, and statistical models using random patterns. Recent work in formal verification is applying algorithmic approaches to validate coverage of paths and branches within the model. However, the generally accepted goal is to "do more".

Second, it is a hard and time-consuming task to create effective simulation vectors to verify a complex design. DV simulators typically support a rich high-level language for the stimulus generation, but still require the thought, experience, and ingenuity of the DV engineer to develop and debug these programs. Therefore, it is desirable to simulate the portion of the design being verified in as much of the total system environment as possible and have the simulation create functional stimulus for the portion to be verified.

Finally, it is tedious to validate the simulated results for correctness, making it desirable to simulate the full system environment where it is easier to validate results. Ideally, the owner of an ASIC chip being designed for a computer could simulate that chip within a model of all of the computer's hardware, the microcode, the operating system, and use example software programs as the ultimate simulation experiment. To even approach this goal, however, simulator technology must continuously strive for rapid and more rapid techniques.

Early DV simulators often used compiled models (where the design is represented directly as a computer program), but this technique gave way to interpretative event-driven simulation. Compiled simulators have the advantage of higher speeds because host machine instructions are compiled in-line to represent the design to be verified and are directly executed with minimum simulator overhead. Event-based simulators require additional overhead to manage the simulation operations, but provide a level of flexibility and generality not possible with the compiled model. This flexibility was necessary to provide for simulation of timing characteristics as well as function, and to handle general sequential designs. Therefore, this approach was generally adopted for DV simulators.

79.2.2.1.1 *Event-Based Simulation*

There are four main concepts to an event-based simulator.

- The netlist, which provides the list of blocks (gates at first, but any complex function later), connections between blocks, and delay characteristics of the blocks.
- Event time queues, which are lists of events that need to be executed (blocks that need to be simulated) at specific points in (simulation) time. Event queues contain two types of events— update and calculate. Update-events change the specified node to the specified value, then schedule calculate-events for the blocks driven from that node. Calculate-events call the simulation behavior for the specified block and, on return from the behavior routine, schedule update-events to change the states on the output nodes to the new values at the appropriate (simulation) time.
- Block Simulation Behavior (the instructions that will compute that block's output(s) states when there is a change to its input(s) state—possibly also scheduling some portion of the block's behavior to be simulated at a later time).
- Value list—the current state of each node in the design.

Simulation begins by scheduling update-events in appropriate time queues for the pattern sequence to be simulated. After the update-events are stored, the first-time event queue is traversed and, one-by-one, each update-event in the queue is executed. Update events update the node in the value list and, if the value new value is different from the current value (originally set to unknown), schedule calculate-events for blocks driven from the updated node. These calculate-events are saved back in time queues based on delay specifications for later execution—which will be discussed later. After all update-events are executed and removed from the queue, the simulator selects calculate-events in the queue sequentially, interprets its function, and passes control to the appropriate block simulation behavior for calculation.

The execution of a calculate-event causes simulation of a block to take place. This is accomplished by passing control to the simulation behavior of the block with pointers to the current state-values on its inputs (in the value list). When complete, the simulation routine of the block will pass control back to the simulator with the new state condition for its output(s). The simulator then schedules the block output(s) value update by placing an update-event for it in the appropriate time queue. The decision of which time queue the update-events are scheduled within is based on what the delay value is for the block.

Once all calculate-events are executed and removed from the queue, the cycle begins again at the next time queue and the process of executing update-events followed by calculate-events for the current time queue repeats. This cycle repeats until there are no more events or until some specified maximum simulation time. To keep update-events separated from calculate-events within the linked-list queues, it is common to add update-events at the top of the linked-list and calculate-events at the bottom, updating the chain-links accordingly.

Because it is not possible to predetermine how many events will reside in a queue at any time, it is common to create these as linked lists of dynamically allocated memory. Additionally, time queues are linked since the required number of time queues cannot be determined in advance and similar to events, new time queues can be inserted into that chained list as required (Figure 79.6).

A number of techniques have been developed to make queue management fast and efficient as they are the heart of the simulator.

Because of its generality, the event-based algorithm was the clear choice for DV. One advantage of event-based simulation over the compiled approach is that it easily supports the simulation of delay. Delays can be simulated for blocks and nets and even early simulators typically supported a very complete delay model, even before sophisticated delay calculators were available to take advantage of it.

Treatment of delay has had to evolve and expand since the beginnings of software DV. Unit delay was the first model used—where each gate in the design is assigned one unit of delay and interconnect delays are zero. This was a crude approximation, but allowed for high-speed simulations because of the simplifying assumptions and it worked reasonably well. As the IC evolved, however, the unit delay model

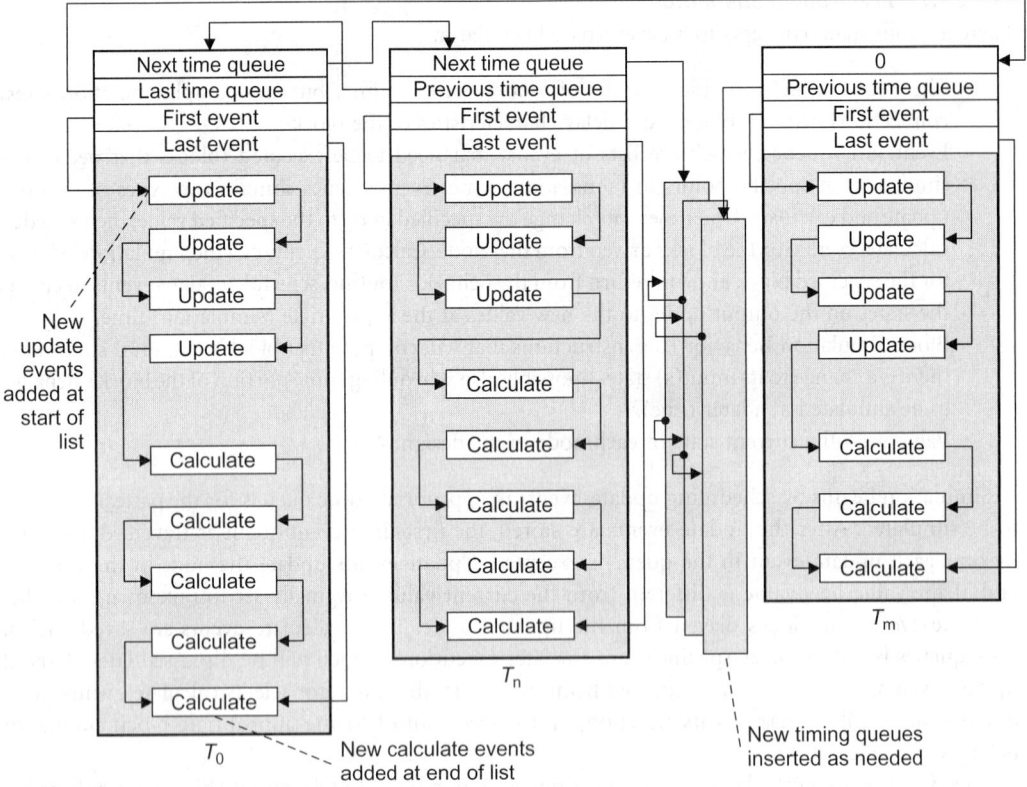

FIGURE 79.6 Event queues.

was replaced by a lumped delay model in which, each gate could be assigned a unique value for delay—actually, a rise-delay and a fall-delay. This was assigned by the technologist based on some average load assumption. At this time, also, the beginnings of development of delay calculators began. These early calculators used simple equations, adding the number of gates driven by the gate being calculated, and then adding the additional delay to the intrinsic delay values of that gate. As interconnect wiring became a factor in the timing of the circuit the pin-to-pin delay came into use. Though delay values used for simulation in the 1970s was crude by today's norm, the delay model Figure 79.7 was rich and supported specification and simulation for

- intrinsic block delay (T_{block})—the time required for the block output to change state relative to the time that a controlling input to that block changed state.
- Interconnect delay (T_{int})—the time required for a specific receiver pin in a net to change state relative to the time that the driver pin changed state.
- Input-output delay (T_{io})—the time required for a specific block output to change state relative to the time the state changed on a specific input to that block.

Today's ICs, however, require delay calculation based on very accurate distributed RC models for interconnects, as these delays have become more significant with respect to gate delays. Future ICs will require even more precise modeling for delays, as will be discussed later in this chapter, and consider inductance (L) in addition to the RC parasitics. Additionally, transmission line models will be necessary for the analysis of certain critical global interconnects. However, the delay model defined in the early years of DV and the capabilities of the event-based simulation algorithm stand ready to meet this challenge.

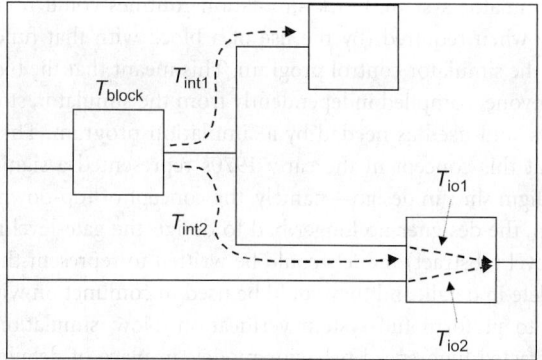

FIGURE 79.7 Simulation delay types.

Another significant advantage of the event-based simulation is that it easily supports simulation of blocks in the netlist at different levels of abstraction. Recall that one of the fundamental components of event simulation is the block simulation behavior. A calculation-event for a block passes control to the subroutine that simulates the behavior using the block's input states. For gate-level simulation, these behavior subroutines are quite simple—AND, OR, NAND, and NOR. However, since the behavior is actually a software program, it can be arbitrarily complex as well. Realizing this, early work took place to develop description languages that could be coded by designers and compiled into block simulation behaviors to represent arbitrary sections of a design as a single (netlist) block. For example, a block simulation behavior might look like the following:

```
FUNCTION (CNOR);
/* Complimentary output NOR function with 3 inputs. Input delay = 2,
   intrinsic delay for true output = 10, intrinsic delay for complement
   output = 12*/

INPUT (a,b,c); /*Declare for inputs a, b, c*/
OUTPUT (x,y);
DECLARE input1, input2, input3; /*Declare storage for inputs*/
DECLARE out, cout;              /*Declare storage for outputs*/

GET a,input1, b,input2, c,input3;

DELAY 2, Entry1;
    Entry1: out = input1|input2|input3;
    cout = ¬out;

    DELAY 10, (x=out); /*Schedule this update-event for 10 time units
                        later */
    DELAY 12 (y=cout); /*Schedule this update-event for 12 time units
                        later */
    END Entry1;

ENDCNOR;
```

The sophisticated use of these behavorial commands supported the direct representation not only for simple gates, but also for complex functions such as registers, MUXs, RAM, and ROS. In doing so, simulation performance was improved because what was treated before as a complex interconnection of gates could now be simulated as a single block.

By generalizing the simulator system, block simulation routines could be loaded from a library at simulation runtime only when required (by the use of a block with that function in the netlist), and dynamically linked with the simulator control program. This meant that the block simulation behaviors could be developed by anyone, compiled independently from the simulator, stored in a library of simulation behavioral models, and used as needed by a simulation program. This is common practice in DV simulators today, but this concept in the early 1970s represented a significant breakthrough and supported a major paradigm shift in design—namely, the concept of top-down-design and verification.

With top-down design, the designer no longer had to design the gate-level netlist before verification could take place. High-level (abstract) models could be written to represent the behavior of sections of the design not yet complete in detail, and they could be used in conjunction with gate-level descriptions of the completed parts to perform full system verification. Now, simulation performance could be improved by use of abstracted high-speed behavior models in place of detailed gate-level descriptions for portions of the overall design. Now, concurrent design and verification could take place across design teams. Now, there was a formal method to support design reuse of system elements without the need to expose internal design details for reusable elements. In the extreme case, the system designer could write a single functional model for the entire system and verify it with simulation. The design could then be partitioned into subsystem elements and each could be described with an abstract behavorial model before being handed off for detailed design. During the detailed design phase, individual designers could verify their subsystem in the full system context even before the other subsystems were completed. Also, the behavorial model concept was particularly valuable for generation of the functional patterns to be simulated as they could now be generated by the simulation of the other system components. Thus, verification of the design no longer had to wait until the end, it could now be a continuous process throughout the design.

Throughout the period, improvements were made to DV simulators to improve performance and in the formulation and capability of behavorial description languages. In addition, designers found increasingly novel ways to use behavorial models to describe full systems containing nondigital system elements and peripherals such as I/O devices. Late in the 1970s and during the 1980s two important formal languages for describing system behavior were developed:

- VHDL [9] (very high-speed integrated circuit [VHSIC] high-level description language), sponsored by DARPA, and
- Verilog [10], a commercially developed RTL-level description language.

These two languages are now accepted industry-standard design description languages.

79.2.2.1.2 Compiled Simulation

Synchronous design such as that used for scan-based design which emerged in the 1970s provides for a clean separation of timing verification and functional verification of combinatorial circuits. Because of this, the use of compiled simulation returned for high-speed verification for designs meeting constraints. A simulation technique called cycle simulation that was developed yielded a major performance advantage over event-based simulation. Cycle-simulation treats the combinational sections of a scan-based design as compiled zero-delay models moving data between them at clock-cycle boundaries. The simulator executes each combinatorial section at each simulated cycle by passing control to the compiled routine for it along with its input states. The resulting state values at the outputs of these sections are assumed to be correctly captured in their respective latch positions. That is, the clock circuitry and path delays are assumed correct, and are not simulated during this phase of verification. The (latched) output values are used as the input states to the combinatorial sections they drive at the next cycle, and the process repeats for each simulated cycle. Each simulation pass across the compiled models represents one cycle of the design's system clock, starting with an input state and resulting in an output state. To assure that only one pass is required, the gates or RTL-level statements for the combinational sections are levelized before compilation into the host-machine instructions. This assures that value updates occur before calculations and only one pass across a section of the model is required to achieve the correct state response at the outputs.

Simulation performance was greatly improved with cycle simulation because of the compiled model and because the clock circuitry did not have to be simulated repeatably with each simulated-machine cycle. Cycle simulation did not replace event simulation even for constrained synchronous designs as the clock circuitry needs to be verified; however, it proved to be effective for many large systems and these early developments provided the foundations for modern compiled simulators used today.

79.2.2.1.3 Hardware Simulators

During the 1980s, research and development took place on custom hardware simulators and accelerators. Special-purpose hardware-simulators use massively parallel instruction processors with much customized instruction sets to simulate gates. These provide simulation speeds that are orders of magnitude faster than the software simulation on general-purpose computers. However, they are expensive to build, lack the flexibility of software simulators, and the hardware technology they are built in soon becomes outdated (although general parallel architectures may allow the incremental addition of additional processors). Hardware accelerators use custom hardware to simulate portions of design in conjunction with the software simulator. These are more flexible, but still have the inherent problems that their big brothers have. Nonetheless, use of custom hardware to tackle the simulation performance demands has gained acceptance in many companies and they are commercially available today using both gate-level and hardware description language (HDL) design descriptions.

79.2.2.2 Timing Analysis

The practice of divide and conquer in DV started in the 1970s with the introduction of the behavorial model. Another divide-and-conquer style born in the 1970s, which achieved wide popularity in the 1990s, is to separate verification of the design's function from its timing. With the invention of scan design, it became possible to verify logic as a combinational circuit using high-speed compiled cycle simulators. Development of path tracing algorithms to verify timing resulted in a technique to verify timing without simulation; thus providing a complete solution which is not dependant on completeness of any input stimulus as required by simulation. For this reason alone, this technique coined static timing analysis (STA) was a major contribution to EDA—one that became key to the notion of "signoff" to the wafer foundry.

STA is used to analyze projected versus required timing along signal paths from primary inputs to latches, latches to latches, latches to primary outputs, and primary inputs to primary outputs. This is done without the use of simulation, but by summing the min–max delays along each path. At each gate, the STA program computes the min–max time in which that gate will change in state based on the min–max arrival times of its input signals. STA tools do not simulate the gate function, they only add its contribution to the path delay, although the choice of using rise or fall times for the gate is based on whether it has a complementary output, or not. Because the circuitry between the latches is combinational, only one pass needs to be made across the design. The addition can be based on the minimum rise or fall delay for gates or both, providing a min–max analysis. The designer specifies the required arrival times for paths at the latches or primary outputs, and the STA program compares these with the actual arrival times. The difference between the required arrival and the actual arrival is defined as slack. The STA tool computes the slack at the termination of each path, sorts them numerically, and provides a report. The designer then verifies the design correctness by analysis of all negative slacks.

Engineering judgement is applied during the analysis, including the elimination of false-path conditions. A false-path is a signal transition that will never occur in the real operation of the design. Since the STA tool does not simulate the behavior of the circuit, it cannot automatically eliminate all false paths. Through knowledge of the signal polarity, STA can eliminate false paths caused by the fan-out and reconvergence of certain signals. However, other forms of false-paths are only identifiable by the designer.

79.2.2.3 Formal Verification

Development of formal methods to verify the equivalence of two different representations of a design began during the 1970s. Boolean verification analyzes a circuit against a known good reference and provides a mathematical proof of equivalence. An RTL model, for example, of the reference circuit is verified using standard simulation techniques. The Boolean verification program then compiles the known-good reference design into a canonical NAND–NOR equivalent circuit. This equivalent circuit is compared with the gate-level hardware design using sophisticated theorem provers to determine equivalence. To reduce processing times, formal verification tools may preprocess the two circuits to create an overlapping set of smaller logic *cones* to be analyzed. These *cones* are simply the set of logic traversed by backtracing across the circuit from an output node (latch positions or primary outputs) to the controlling input nodes (latch positions or primary inputs). User controls specify the nodes that are supposed to be logically equivalent between the two circuits.

Early work in this field explored the use of test generation algorithms to prove equivalence. Two *cones* to be compared can be represented as $F_{\text{cone1}} = f(a, b, c, X, Y, Z, \ldots)$ and $F_{\text{cone2}} = f(d, e, f, \ldots, X', Y', Z', \ldots)$. F_{cone1} and F_{cone2} are user defined output nodes to be compared for equivalence. The terms a, b, c and d, e, f represent the set of input nodes for the function and X, Y, Z are the computational subfunctions. User inputs define the equivalence between a, b, c and d, e, f. If F_{cone1} and F_{cone2} are functionally equivalent, then the value of G must be 0 for all possible input states. If the two cones are equivalent then, use of D-ALG test generation techniques for $G = F_{\text{cone1}}$ XOR F_{cone2} will be unable to derive a test for the stuck-at-zero fault on the output of G. Similarly, the use of random pattern generation and simulation can be applied to prove equivalence between *cones* by observing the value of G for all input states.

Research during the 1980s provided improvements in Boolean equivalence checking techniques and models (such as binary decision diagrams), and modern Boolean equivalence checking programs may employ a number of mathematical and simulation algorithms to optimize the overall processing. However, Boolean equivalence checking methods require the existence of a reference design against which equivalence is proved. This implies there must be a complete validation of the reference design against the design specification. Validating the reference design has typically been a job for simulation and, thus, is vulnerable to the problems of assuring coverage and completeness of the simulation experiment. Consequently, formal methods to validate the correctness of functional-level models have become an important topic in EDA resesarch. Modern design validation tools use a combination of techniques to validate the correctness of a design model. These typically include techniques used in software development to measure completeness of simulation test cases such as

- Checking for coverage of all instructions (in the model)
- Checking to assure that all possible branch conditions (in the model) were exercised.

They may also provide more formal approaches to validation such as

- Checking (the model) against designer-asserted conditions (or constraints) that must be met
- Techniques that construct a proof that the intended functions are realized by the design.

These concepts and techniques continue to be the subject of research and will gain more importance as the size of IC designs stretch the limits of simulation-based techniques.

79.2.2.4 Verification Methodologies

With the use of structured design techniques, the design to be verified (Figure 79.8) can be treated as a set of combinational designs. With the arsenal of verification concepts that began to emerge during this period, the user had many verification advantages not previously available. Without structured design, delay simulation of the entire design was required. Use of functional models intermixed with gate-level descriptions of subsections of the design provided major improvements, but was still very costly and time consuming. Further, to be safe, the practical designer would always attempt to delay simulate the entire design at the gate level.

With structured design techniques, the designer could do the massive simulations at the RTL focusing of the logic function only, regardless of the timing. Cycle simulation improved the simulation performance by 1 to 2 orders of magnitude by eliminating repetitive simulations of the clock circuitry and use of compiled (or table lookup) simulation. For some, the use of massively parallel hardware simulators offered even greater simulation speeds—the equivalent of hundreds of millions events per second. Verification of the logic function of the gate-level design could be accomplished by formally proving its equivalence with the simulated RTL model. STA provided the basis to verify the timing of all data paths

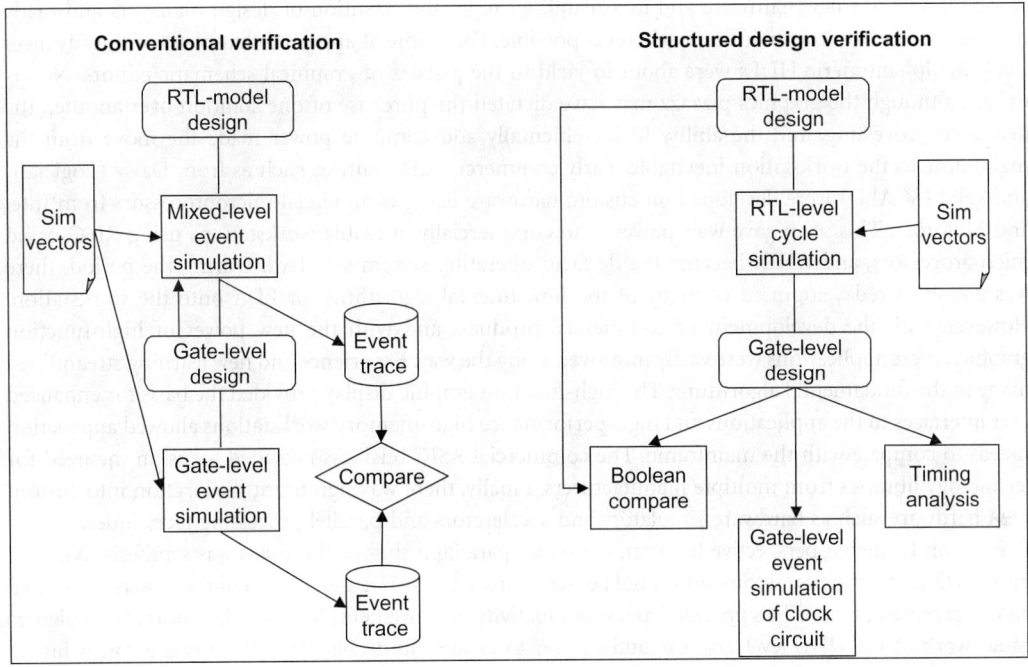

FIGURE 79.8 Verification methodology.

in a rigorous and methodical manner. Functional and timing verification of the clock circuitry could be accomplished with the standard delay-simulation techniques using event simulation. Collectively, these tools and techniques provided major design productivity improvements. However, the IC area and performance overhead required for these structured design approaches limited the number of designs taking advantage of them. During the 1980s as the commercial EDA business developed, these new verification tools and techniques remained as in-house tools in a few companies. Commercial availability did not emerge until the 1990s when the densities and complexities of ICs began to demand the change.

79.2.3 The 1980s: Birth of the Industry

Up to the 1980s design automation was, for the most part, developed in-house by a small number of large companies for their own proprietary use. Almost all EDA tools operated against large mainframe computers using company-specific interfaces. High-level description languages were unique and most often proprietary, technology rule formats were proprietary, and user interfaces were unique. As semiconductor foundries made their manufacturing lines available for customer-specific chip designs; however, the need for access to EDA tools grew. With the expansion of the application-specific integrated circuit (ASIC), the need for commercially available EDA tools exploded. Suddenly, a number of commercial EDA companies began to emerge and electronics design companies had the choice of developing tools in-house or purchasing them from a variety of vendors. The EDA challenge now often became that of integrating tools from multiple suppliers into a homogeneous system. Therefore, one major paradigm shift of the 1980s was the beginnings of EDA standards to provide the means to transfer designs from one EDA design system to another or from a design system to manufacturing. VHDL and Verilog matured to become IEEE industry-standard HDLs. Electronic data interchange format (EDIF) [11] was developed as an industry standard for the exchange of netlist and GDSII (developed in the 1970s at Calma) became a standard interface for transferring mask pattern data. (In 2003, SEMI introduced a new mask pattern exchange standard, OASIS, that compacts mask pattern data to one-tenth or less of the bytes used by GDSII.)

A second paradigm shift of the 1980s was the introduction of the interactive workstation as the platform for EDA tools and systems. Although some may view it as a step backward, the "arcade" graphics capabilities of this new hardware and its scalability caught the attention of design managers and made it a clear choice over the mainframe wherever possible. For a time, it appeared that many of the advances made in alphanumeric HDLs were about to yield to the pizzazz of graphical schematic editors. Nevertheless, although the graphics pizzazz may have dictated the purchase of one solution over another, the dedicated processing, and the ability to incrementally add compute power made the move from the mainframe to the workstation inevitable. Early commercial EDA entries such as from Daisy (Logician) and Valid (SCALD) were developed on custom hardware using commercial microprocessors from Intel and Motorola. This soon gave way, however, to commercially available workstations using RISC-based microprocessors, and UNIX became the de facto operating system standard. During the period, there was a rush of redevelopment of many of the fundamental algorithms for EDA onto the workstation. However, with the development of commercial products and with the new power of high-function graphics, these applications were vastly improved along the way. Experience and new learning streamlined many of the fundamental algorithms. The high-function graphic display provided the basis for enhanced user interfaces to the applications and high-performance high-memory workstations allowed application speeds to compete with the mainframe. The commercial ASIC business provided focus on the need for technology libraries from multiple manufacturers. Finally, there was significant exploration into custom EDA hardware such as hardware simulators and accelerators and parallel processing techniques.

From an IC design perspective however, the major paradigm shift of the 1980s was synthesis. With the introduction of synthesis, automation could be used to reduce an HDL description of the design to the final hardware representation. This provided major productivity improvements for ASIC design, as chip designers could work at the HDL level and use automation to create the details. Also, there was a much higher probability that the synthesis-generated design would be correct than for manually created schematics. The transgression from the early days of IC design to the 1980s is similar to what occurred earlier in software. Early computer programming was done at the machine language level. This could provide optimum program performance and efficiency, but at the maximum labor cost. Programming in machine instructions proved too inefficient for the vast majority of software programs, thus the advent of assembly languages. Assembly language programming offers a productivity advantage over machine instructions because the assembler abstracts up several of the complexities of machine language. Thus, the software designer works with less complexity, using the assembler to add the necessary details and build the final machine instructions. The introduction of functional-level program languages (such as FORTRAN then, and C++ now) provided even more productivity improvements by providing a set of programming statements for functions that would otherwise require many machine or assembler instructions to implement. Thus, the level at which the programmer could now work was even higher, allowing him/her to construct programs with far fewer statements. The analog in IC design is the progression of transistor-level design (machine level), to gate-level design (assembler level), to HDL-based design. Synthesis provided the basis for HDL-based design and its inherent productivity improvements, and major changes to IC design methodologies.

79.2.3.1 Synthesis

Fundamentally, synthesis is a three-step process:

- Compile an HDL description of a design into an equivalent NAND–NOR description
- Optimize the NAND–NOR description based on design targets
- Map the resulting NAND–NOR description to the technology building blocks (cells) supported for the wafer foundry (process) to be used.

Although work on synthesis techniques has occurred on and off since the beginnings of design automation and back to Transistor-Transistor Logic (TTL)-based designs, it was not until gate array style ICs reached a significant density threshold that it found production use. In the late 1970s and 1980s, considerable research and development in industry and universities took place on the high-speed algorithms and heuristics for synthesis. In the early 1980s, IBM exploited the use of synthesis on the ICs used in the 3090 and AS400 computers. These computers used chips that were designed from qualified sets

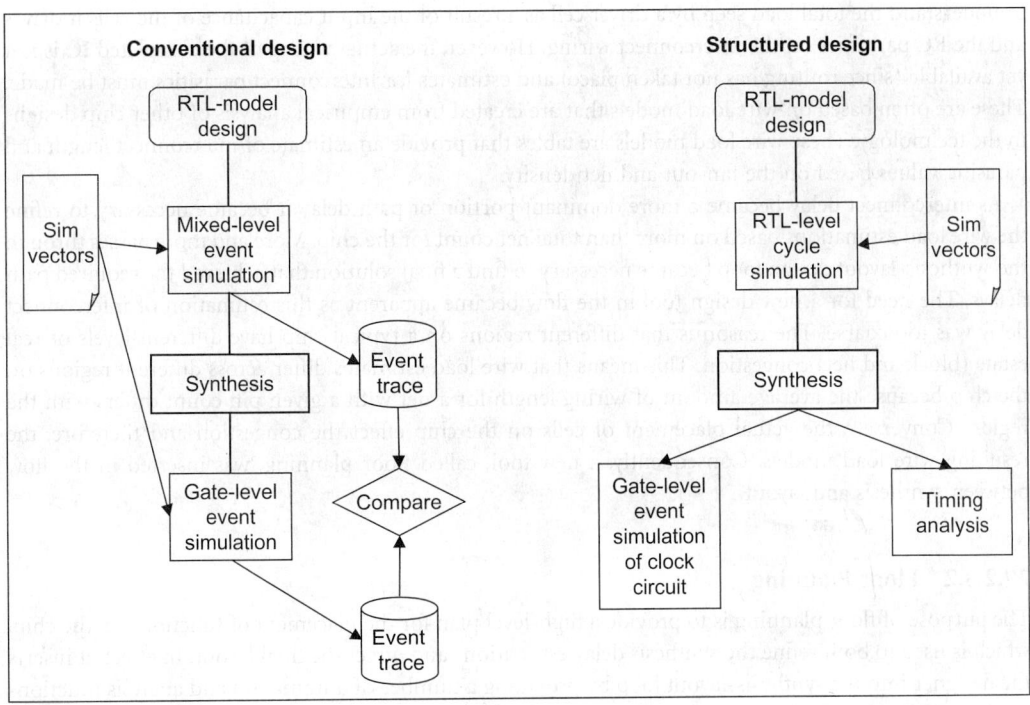

FIGURE 79.9 Synthesis methodology.

of predesigned functions interconnected by personalized wiring. These functions (now called cells) represented the basic building blocks for each specific IC family. The computers used a high number of uniquely personalized chip designs so it was advantageous to design at the HDL level and use automation to compile to the equivalent cell-level detail. The result was significant improvements to the overall design productivity.

Today, synthesis is a fundamental cornerstone of the design methodology (Figure 79.9) for ASICs supporting both VHDL and Verilog as the standard input.

As the complexities of IC design have increased, so have the challenges for synthesis. Early synthesis had relatively few constraints to be observed in its optimization phase. Based on user controls, the design was optimized for minimum area, minimum fanout (minimum power), or maximum fanout (maximum performance). This was accomplished by applying a series of logic reduction algorithms (transforms) that provide different types of reduction and refinement [12]. Cell behavior could also be represented in primitive-logic equivalent form and a topological analysis could find matches between the design gate patterns and equivalent cells. The mapping phase would then select the appropriate cells based on function and simple electrical and physical parameters for each cell in the library such as drive strength and area. Previously, synthesis was not overly concerned with signal delays in the generated design nor for other electrical constraints that the design may be required to meet. As IC feature sizes decreased, however, these constraints became critical in synthesis and automation became required to consider them in its solution.

The design input to modern synthesis tools is not the functional HDL design alone, but now includes constraints such as the maximum allowed delay along a path between two points in the design. This complicates the synthesis decision process as it must now generate a solution that meets the required function with a set of cells and interconnections that will fall within the required timing constraints. Therefore, additional trade-offs must be made between the optimization and mapping phases, and the effects of interconnection penalty need to be considered. Additionally, synthesis must now have technology characteristics available to it such as delay for cells and wiring. To determine this delay it is necessary to understand the total load seen by a driver cell as a result of the input capacitance of the cells it drives and the RC parasitics on the interconnect wiring. However, the actual wiring of the completed IC is not yet available (since routing has not taken place) and estimates for interconnect parasitics must be made. These are often based on wire load models that are created from empirical analysis of other chip designs in the technology. These wire load models are tables that provide an estimate of interconnect length and parasitic values based on the fan-out and net density.

As interconnect delay became a more dominant portion of path delay it became necessary to refine the wire load estimations based on more than total net count for the chip. More and more passes through the synthesis-layout design loop became necessary to find a final solution that achieved the required path delays. The need for a new design tool in the flow became apparent as this estimation of interconnect delay was too coarse. The reason is that different regions on a typical chip have different levels of real estate (block and net) congestion. This means that wire load estimates differ across different regions on the chip because the average amount of wiring length for a net with a given pin count differs with the region. Conversely, the actual placement of cells on the chip effect the congestion and therefore, the resulting wire load models. Consequently, a new tool, called floor planning, was inserted in the flow between synthesis and layout.

79.2.3.2 Floor Planning

The purpose of floor planning is to provide a high-level plan for the placement of functions on the chip, which is used to both refine the synthesis delay estimations and direct the final layout. In effect, it inserts the designer into the synthesis-layout loop by providing a number of automation and analysis functions used to effectively partition and place functional elements on the chip. As time progressed, the number of analysis and checking functions integrated into the floorplanner grew to a point where today, it is more generally thought of as design planning [13]. The initial purpose of the floor planner was to provide the ability to partition the chip design functions and develop a placement of these partitions that optimized the wiring between them. Optimization of the partitioning and placement may be based on factors such as minimum wiring or minimum timing across a signal path. A typical scenario for the use of a design planner is as follows:

- Run synthesis on the RTL description and map to the cell level
- Run the design planner against the cell-level design with constraints such as path delays and net priorities
  - Partition and floor plan the design, either automatically or manually, with the design planner
  - Create wire load models based on the congestion within the physical partitions and empirical data
- Rerun synthesis to optimize the design at the cell level using the partitioning, global route, and wire load model data.

User-directed graphics and autoplacement-wiring software algorithms are used to place the partitions and route the interpartition (global) nets. One key to the planner is the tight coupling of partitioning and routing capability with electrical analysis tools such as power analysis, delay calculation, and timing analysis. Checking of the validity of a design change (such as placement) can be made immediately on that portion of the design that changed. Another key is that it needs to be reasonably tightly connected to synthesis, as it is typical to pass through the synthesis-planner-synthesis loop a number of times before reaching a successful plan for the final layout tools.

After a pass in the floor planner, more knowledge is available to synthesis. With this new knowledge, the optimization and mapping phases can be rerun against the previous results to produce a refined solution. The important piece of knowledge now available is the gate density within different functions in the design (or regions on the chip). With this, the synthesis tool can be selective about which wire load table to use and develop a more accurate estimate of the delay resulting from a specific solution choice.

Floor planning has been a significant enhancement to the design process. In addition to floor planning, modern design planners include support for clock tree, design power, bus design, I/O assignment, and a wealth of electrical analysis tools. As will be discussed later, semiconductor technology trends will place even more importance on the design planner and dictate an ever-tightening integration between it with both synthesis and PD. Today, synthesis, design planning, and layout are three discrete steps in the typical design flow. Communication between these steps is accomplished by file-based interchange of the design data and constraints. As designs become more dense and with the growing complexities of electrical analysis required, it will become necessary to integrate these three steps into one tight process as are the functions within the modern design planners.

79.2.4 The 1990s: The Age of Integration

Before EDA tools were commercially available, software architecture standards could be established to assure smooth integration of the tools into complex EDA systems supporting the entire design flow. Choice of languages, database and file exchange formats, host hardware and operating systems, and user interface conventions could be made solely by the developing company. Moreover, the need for comprehensive EDA systems supporting the entire flow through design and manufacturing release in a managed process is of paramount importance. Design of ICs with hundreds of thousands of gates (growing to hundreds of millions) across a design team requires gigabytes of data contained within thousands of files. All these data must be managed and controlled to assure its correctness and consistency with other design components. Design steps must be carried out and completed in a methodical manner to assure correctness of the design before release to manufacturing. The EDA system that encompasses all of this must be efficient, effective, and manageable. This was always a challenge for EDA system integrators, but the use of EDA tools from multiple external vendors compounded the problem. EDA vendors may develop their tools to different user interfaces, database and file formats, computer hardware, and operating systems. In turn, it is the EDA system integrator who must find a way to connect them in a homogeneous flow that is manageable, yet provides the necessary design efficiency across the tools. By the late 1980s it was said that the cost to integrate an EDA tool into the in-house environment was often over twice that of the tool itself. Consequently, the CAD Framework Initiative (CFI) was formed whose charter was to establish software standards for critical interfaces necessary for EDA tool integration. Thus began the era of the EDA Framework.

The EDA Framework [14] is a layer of software function between the EDA tool and the operating system, database, or user. For this software layer CFI defined a set of functional elements, each of which had a standard interface and provided a unique function in support of communication between different tools. The goal was to provide EDA system integrators with the ability to choose EDA tools from different vendors and be able to plug them into a homogeneous system to provide an effective system operation across the entire design flow. The framework components included

- Design information access—a formal description of design elements and a standard set of software functions to access each, called an application program interface (API)
- Design data management—a model and API to provide applications the means to manage concurrent access and configuration of design data
- Intertool communication (ITC)—an API to communicate control information between applications and a standard set of controls (called messages)
- Tool encapsulation (TES)—a standard for specifying the necessary information elements required to integrate a tool into a system such as input files and output files
- Extension language (EL)—specification of a high-level language that can be used external to the EDA tools to access information about the design and perform operations on it.
- Session management—a standard interface used by EDA tools to initiate and terminate their operation and report errata

- Methodology management—APIs used by tools so that they could be integrated into software systems to aid the design methodology management
- User interface—a set of APIs to provide consistency in the graphical user interface (GUI) across tools.

Each of these functional areas needs to be considered for integration of EDA tools and to provide interoperability between them. In its first release, CFI published specifications for design representation (DR) and access, ITC, TES, and EL. Additionally, the use of MOTIF and X-Windows as specified as the GUI toolkit. However, for many practical, business, and human nature reasons, these standards did not achieve widespread adoption across the commercial EDA industry and the "plug-and-play" motto of CFI was not realized. However, the early work at CFI did serve to formalize the science of integration and communication, and a number of the information models developed for the necessary components are used today, but often, slightly modified and within a proprietary interface.

As learned, it was not only expensive to change EDA tools to use a different internal interface or data structures, but it was also not considered good business to allow tools to be integrated into commercial system offerings from competitors. For a period, several framework offerings became available, but these quickly became the product end in themselves rather than a means to an end. These products were marketed as a means to bring outside EDA tools into the environment, but few vendors modified their tools to support this by adopting the CFI specifications. Instead, to meet the need for tool integration, the commercial industry focused on data flow only and chose an approach that provides communication between tools by standard data file formats and sequential data translation. Consequently, EDA flows of the 1990s were typically a disparate set of heterogeneous tools stitched together by the industry-standard data files. These files are created from one tool and translated by another tool (within the design flow) to its internal data structures. Many of these formats evolved from proprietary formats developed within commercial companies that were later transferred or licensed for use across the industry. Common EDA file formats that evolved were

- Electronic data interchange format (EDIF)—netlist exchange (EIA → IEC)
- Physical data exchange format (PDEF)—floor plan exchange (Synopsys → IEEE)
- Standard delay file (SDF)—delay data exchange (Cadence Design Systems → IEEE)
- Standard parasitic exchange file (SPEF)—parasitic data exchange (Cadence Design Systems → IEEE)
- Library exchange file (LEF)—physical characteristics information for a circuit family (Cadence → OpenEDA)
- Data exchange format (DEF)—physical data exchange (Cadence → OpenEDA)
- Library data format (.lib)—ASIC cell characterization (Synopsys)
- ASIC library format (ALF)—ASIC cell characterization (Open Verilog International)
- VISIC high-level design language (VHDL)—behavorial design description language (IEEE)
- Verilog—RTL-level design description language (Cadence Design Systems → IEEE)

These "standard" formats have provided the desired engineering and business autonomy necessary to get adoption and their use across the industry has improved the ability to integrate tools into systems drastically compared with the 1980s. It has proved to be an effective method for the design of today's commodity ICs above 0.25 μm. However, the inherent ambiguities, translation requirements, rapidly increasing file sizes (owing to increasing IC density), and the fundamental nonincremental sequential nature of design flows based on them will soon give way to IC technology advances.

79.3 The Future

It may be said that the semiconductor industry is the most predictable industry on earth. Whether it was an astute prediction or the cause, the prediction of Gordon Moore (now known as Moore's Law) has held steadily over the past two decades. Every 18 months, the number of transistors or bits on an

IC doubles. To accomplish this, feature sizes are shrinking, spacing between features is shrinking, and chip die sizes are increasing. In addition, the fundamental physics that govern the electrical properties of these devices and between them has been documented in electrical engineering textbooks for some time. However, there are points within this progression where paradigm shifts occur in the elemental assumptions and models required to characterize these circuits, and where fundamental EDA design and analysis must change. Earlier, we discussed paradigm shifts in DV because of the inability to repair. We also discussed shifts resulting from the number of elements in a design versus the necessary tool performance and designer productivity. Through the past three decades, we have seen shifts in test, verification, design abstraction, and design methodologies. When the 0.25 μm node was passed, another paradigm shift was required. Feature packing, decreased rise times, increased clock frequencies, increased die sizes, and the explosion of the number of switching transistors are all interacting to place new demands of models, tools, and EDA systems. After that and including the present, new challenges came about in design tools, rules, systems, and schools for problematic areas of delay, signal integrity, power, test, and manufacturability.

79.3.1   International Technology Roadmap for Semiconductors

SEMATECH periodically publishes a report called the "International Technology Roadmap for Semiconductors" [15] (ITRS). This report is jointly sponsored by the European Semiconductor Industry Association, Japan Electronics and Information Technology Industries Association, Korea Semiconductor Industry Association, Taiwan Semiconductor Industry Association, and the Semiconductor Industry Association. The objective of the ITRS "is to ensure advancements in the performance of integrated circuits." Thus ITRS characterizes planned semiconductor directions and advances across the next decade and the necessary technology advancements in areas including

- Design and test
- Process integration devices and structures
- Front-end (manufacturing) processes
- Lithography
- Interconnect
- Factory integration
- Assembly and packaging
- Environment, safety, and health
- Defect reduction
- Metrology
- Modeling and simulation.

ITRS provides a wealth of information on the semiconductor future as well as the problematic areas that will arise and areas where inventive new technology will be required. Some relevant ITRS-projected semiconductor advancements for MPU/ASIC ICs are shown in Table 79.1 and compared with previous ITRS numbers for year 2001. The reader is encouraged to study the technology characteristics in the ITRS, as the following is just a summary of certain points for the purpose of illustration.

To understand some of the implications of the ITRS data, it is convenient to review the effects of scaling on a number of electrical parameters [16].

Gate delay is a function of the gate capacitance and transistor resistance. The gate capacitance ($C = k\varepsilon_o A/T_{gox}$) is a function of the gate area (A), and the gate-oxide thickness (T_{gox}). Assuming ideal scaling (all features scale down in direct proportion to the change in transistor size), gate capacitance scales in direct proportion to transistor-size scaling (ΔSize) since

$$\Delta C_{gate} \propto (\Delta W_{gate}\Delta L_{gate})/\Delta T_{gox}$$

TABLE 79.1 2004 ITRS Microprocessor Roadmap

Characteristic	2001	2004	2007	2010	2013	2016
Transistor gate length (nm)	90	53	35	25	18	13
Chip size (mm²)	310	310	310	310	310	310
Million transistors/mm²	0.89	1.78	3.57	7.14	14.27	28.54
Million transistors/chip	276	553	1106	2212	4424	8848
Wire pitch (nm), metal	350	210	152	104	76	54
Clock frequency (GHz) (on-chip)	1.684	4.171	9.285	15.079	22.980	39.683
Supply voltage (V) (low power)	1.1	.9	0.8	0.7	0.6	0.5
Maximum power (W) (high-performance)	130	158	189	198	198	198

therefore

$$\Delta C_{gate} \propto \Delta \text{Size}$$

The transistor resistance (R_{tr}) is proportional to supply voltage and inversely proportional to current both of which scale down proportionally, so the R_{tr} is constant. Thus, the gate delay (D_{gate}) decreases in direct proportion to ΔSize since

$$D_{gate} = R_{tr}C_{gate},$$

therefore

$$\Delta D_{gate} \propto 1 * \Delta \text{Size} = \Delta \text{Size}$$

Thus, gate delays scale down in proportion to feature size. However, the effects of interconnect delay must be considered to understand the impact of scaling on timing. For $R_{int} < R_{tr}$ interconnect effects have little affect on signal path timing. As R_{int} increases, however, delay analysis must also take into account the interconnect parasitics.

Interconnect delay is a function of the *RC*-time constant, which is a result of resistance and distributed capacitance of the interconnect. Assuming ideal scaling, the interconnects' cross-sectional area scales in direct proportion to ΔSize.

Interconnect resistance is proportional to the length and inversely proportional to the cross-sectional area of the interconnect. Since

$$R_{int} \propto L_{int}/(W_{int}H_{int}),$$

therefore

$$\Delta R_{int} \propto L_{int}/ \Delta \text{Size}^2$$

where, L_{int}, H_{int}, and W_{int} are the length, height, and width of the interconnect, respectively.

The self-capacitance ($C_{self} = k\varepsilon_o A/t$) of the interconnect is proportional to its plate area ($A = W_{int}L_{int}$) and inversely proportional to dielectric layer thickness (t). Assuming ideal scaling for t,

$$\Delta C_{self} \propto \Delta L_{int}\Delta \text{Size}/\Delta \text{Size}$$

therefore

$$\Delta C_{\text{self}} \propto \Delta L_{\text{int}}$$

The resulting interconnect *RC* constant therefore, scales as

$$\Delta R_{\text{int}} \Delta C_{\text{self}} \propto (\Delta L_{\text{int}}/\Delta \text{Size}^2)(\Delta L_{\text{int}}) = \Delta L_{\text{int}}^2/\Delta \text{Size}^2$$

For *local nets* the interconnect length (L_{int}) scales with ΔSize thus, the *RC* constant remains relatively constant

$$\Delta (R_{\text{int}} C_{\text{self}})_{\text{local}} \propto \Delta L_{\text{int}}^2/\Delta \text{Size}^2 = \Delta \text{Size}^2/\Delta \text{Size}^2 = 1$$

Local interconnect delay does not scale down in proportion to gate delay on successive technology generations. Therefore, its relative impact on the overall timing across a signal path is increasing in importance.

For *global nets*, the interconnect length does not scale down with the feature size. Here, the die size (S_{die}) is the more predominant determinant. Therefore, the *RC* constant for global nets scales as

$$\Delta (R_{\text{int}} C_{\text{self}})_{\text{global}} \propto \Delta L_{\text{int}}^2/\Delta \text{Size}^2 = 1/\Delta \text{Size}^2$$

Global interconnect delay increases in indirect proportion to gate delay for successive technology generations. Thus, global interconnect routing and delay analysis has become a critically important area for EDA.

For complete analysis on interconnect delay, the cross-coupling effect of closely spaced features on the chip must also be considered. Capacitance results whenever two conducting bodies are charged to different electric potentials and may be viewed as the reluctance of voltage to build or decay quickly in response to injected power. The self-capacitance between the interconnect surfaces and ground is only a part of the overall capacitance to be considered. Another important consideration is the mutual capacitance between interconnects in close proximity. To get an accurate measure of the total capacitance along any interconnect it is necessary to consider all conducting bodies within a sufficiently close proximity. This includes not only the ground plane, but also adjacent interconnects as there is a mutual capacitance between these. This mutual capacitance is a function of interconnect sidewall area, the distance between it and adjacent interconnects, and the dielectric constant of the dielectric separating them.

Figure 79.10 depicts three equally spaced adjacent interconnect wires on a signal plane. An approximate formula for the capacitance [17] of the center wire is

$$C = \text{self-capacitance} + \text{mutual capacitance} = C_{\text{int-ground}} + 2C_{\text{m}}$$

The mutual capacitance is a function of the plate area along the sides of the interconnects and the spacing between them and may be simply approximated as

$$C_{\text{m}} = k\varepsilon_{\text{o}} L_{\text{int}} H_{\text{int}}/S_{\text{int}}$$

where, S_{int} is the distance between the interconnects.

Therefore (for a constant *k*-value), the change in mutual capacitance as a function of ideal scaling is given by

$$\Delta C_{\text{m(Local)}} \propto \Delta \text{Size}^2/\Delta \text{Size} = \Delta \text{Size} \quad \text{and} \quad \Delta C_{\text{m(Global)}} \propto \Delta \text{Size}/\Delta \text{Size} = 1$$

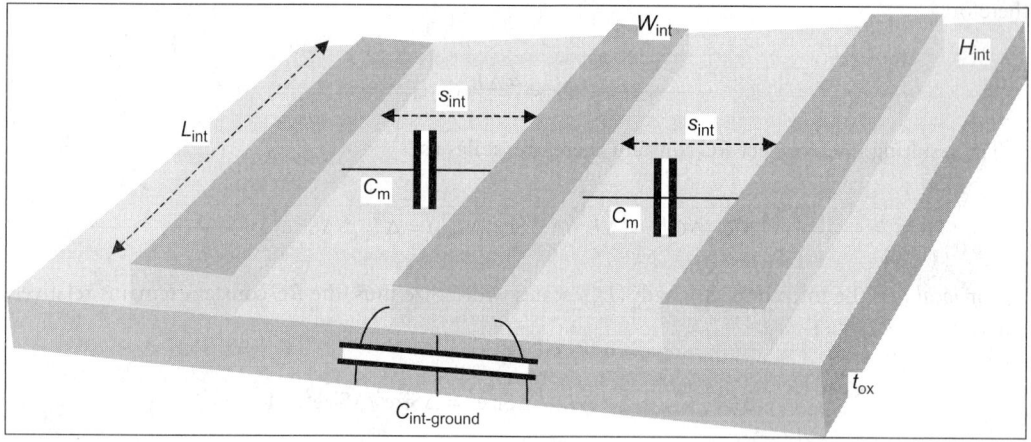

FIGURE 79.10 Mutual capacitance.

Thus, mutual capacitance for local interconnects scales with feature size, while for global interconnects, it does not. Because of increased frequencies, electrical cross talk effects between interconnects must be considered. Further, mutual capacitance between only adjacent interconnects may not be complete. Full analysis may require consideration of all interconnects within a sphere around the interconnect being analyzed or even on wiring layers above or below.

Noise is the introduction of unwanted voltage onto a signal line, and is proportional to mutual capacitance between them and inversely proportional to the rate at which the voltage potential between them changes. Therefore, because the full effect of mutual capacitance is also a function of the voltage potential between interconnects (Miller effect) analysis of the simultaneous signal transitions on each becomes an important consideration for delay. Mutual capacitance injects current into adjacent (victim) signal lines proportional to the rate of change in voltage on the exciting (aggressor) signal line. The amount of induced current can be approximated by the following equation [18]:

$$I_{\text{mC}} = C_{\text{m}}(dV_{\text{aggressor}}/dt)$$

The amount of noise induced onto the victim net is proportional to the resistance on the victim net and the mutual capacitance, and inversely proportional to the rise time (τ_{r}) of the aggressor waveform

$$V_{\text{mC}} = I_{\text{m}}Z_{\text{victim}} = C_{\text{m}}(\Delta V_{\text{aggressor}}/\tau_{\text{r}})R_{\text{victim}}$$

Noise can increase or even decrease [19] the signal delay on the victim net. Voltage glitches induced can also give rise to faulty operation in the design (e.g., if the glitch is of sufficient amplitude and duration to cause a latch to go to an unwanted logic state).

Inductance can also cause noise when large current changes occur in very short times according to the following relationship:

$$V_{\text{L}} = L(dI/dt)$$

Sudden changes in large amounts of current can be seen on the power grid resulting from a large number of circuits switching at the same time. The inductance in the power grid and this large current draw over short periods of time (di/dt) results in a self-induced electromagnetic force whose voltage is equal to Ldi/dt. This induced voltage causes the power supply level to drop, resulting in a voltage glitch that is proportional to the switching speed, the number of switching circuits, and the effective inductance in the power grid.

79.3.2 EDA Impact

Scaling effects have impact on EDA beyond delay and timing, and the list of DF*x* (design-for-*x*) topics is increasing beyond design-for-test and design-for-timing (closure). Scaling down V_{dd} decreases power for a transistor, but the massive scaling up of total transistors on the IC dictates extensive analysis of high currents and leakage (design for power). Lithographic limitations (discussed later) now require consideration of diffraction physics (design for manufacturability [DFM]). Further, while the design complexities owing to electrical effects such as these are increasing, the number of circuits on the IC continues to increase. The fundamental challenge for EDA is more analysis on many more circuit elements with no negative effect on design cycle times.

79.3.2.1 EDA System Integration

It is now necessary to use EDA applications with increased accuracy and scope for electrical analysis to assure that designs will meet intended specifications and to integrate EDA design flows in ways that contain design cycle times. EDA systems must now take a data-centric view of the design process as opposed to the tool-centric view of the past. Data must be organized and presented in ways that allow EDA to exploit: abstraction and hierarchy, shared access by design tools, incremental processes (whereby only portions of a design that have changes need be reanalyzed), and concurrent execution of design and analysis tools within the design flow.

EDA systems have evolved or are evolving toward integration and database technology that meets the above requirements to one degree or another. EDA systems have evolved from vendor-specific sets of design and analysis applications supporting data exchange between custom, often proprietary, file formats to systems that support interchange between vendors through industry-standard files. Over the past decade they have moved toward systems of applications communicating through vendor-specific integrated database technology supporting intersystem exchange via the same industry-standard files. Because of the increasing need for new and additional EDA applications companies performing IC design continue to require the ability to use EDA applications from many vendors and develop flows that provide efficient and complete intervendor integration. With increasing feature counts and complexities for successive technology generations, this dictates the need to reduce the overhead and possible data loss caused by file-based interchange (Figure 79.11). In 2000, a multicompany effort under Si2 was initiated to develop an industry-open data model that could be used by commercial EDA companies and for university research and (EDA customer) proprietary EDA development. This effort resulted in the publication of an EDA data model specification, which includes an application program software interface, and development of a production-quality *reference* database that is compliant with that specification. Collectively, this specification and database is called OpenAccess (see www.si2.org) and was made available to the industry in 2003 on a royalty-free basis. OpenAccess has a excellent chance of being the foundation on which the next point of EDA systems' evolution will be founded. At that point, highly efficient EDA flows will be built using EDA tools from multiple sources and integrated around a single database meeting the fundamental requirements mentioned above.

79.3.2.2 Delay

Simplifying assumptions and models for delay have provided the foundation for high-speed event-driven delay simulation since its initial use in the 1970s. More accurate waveform simulation such as that provided by SPICE implementations has played a crucial role in the characterization of IC devices during over 3 decades and continues to do so. SPICE-level simulations are important for characterization and qualification of ASIC cells and custom macros of all levels of complexity. However, the runtimes required for this level of device modeling are too large to support its use across the entire ICs. SPICE is often used to analyze the most critical signal paths, but SPICE-level simulation on circuits with hundreds of millions of transistors is not practical today. Consequently, simulation and STA at the abstracted gate-level, or higher, is essential for IC timing analysis. However, simplifying models used to characterize delay for discrete-event simulation and STA have become more complex as feature sizes on ICs have shrunk, and

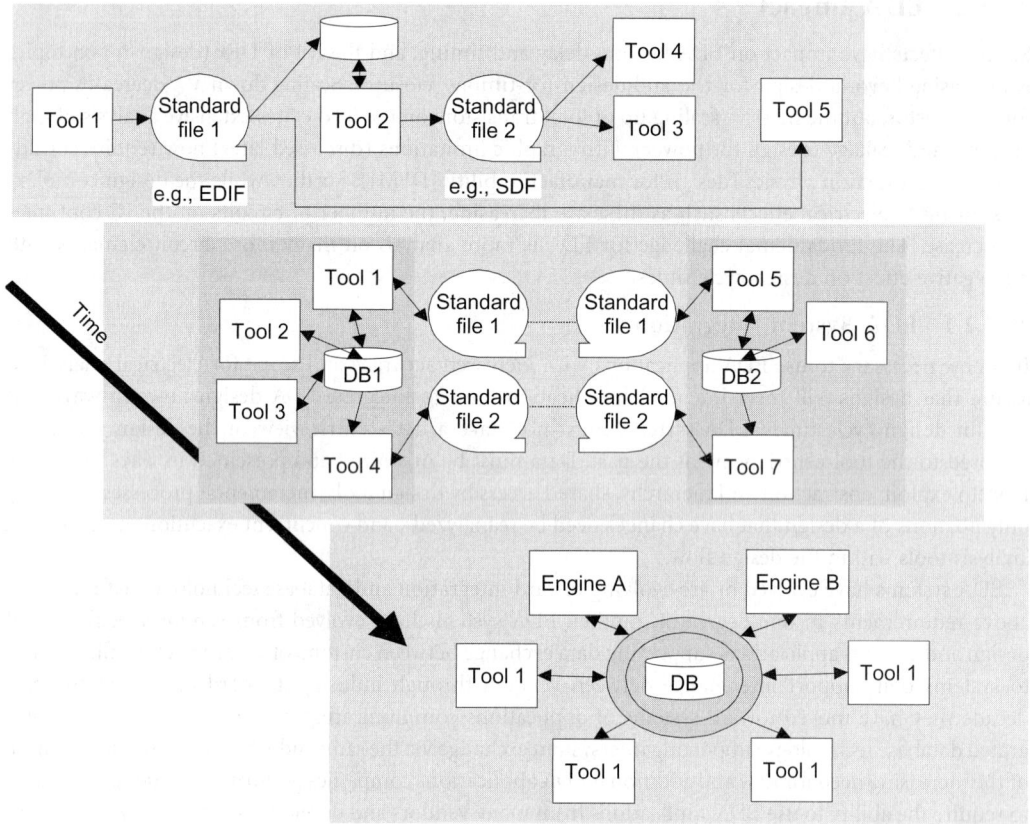

FIGURE 79.11   EDA system architecture progression.

the importance of interconnect resistance and cross talk increased. In the future, these models will need to improve even more (Figure 79.12).

When ICs were considerably >1 μm, gate delay was the predominant factor in determination of timing. Very early simulation of TTL logic used a simple model which assigned a fixed unit of delay to each gate and assumed no delay across the interconnects. Timing was based only on the number of gates through which a signal traversed along its path. As LSI advanced delay was based on a foundry-specified gate delay value (actually a rise delay and a fall delay) that was adjusted based on the capacitive load of the gates it fanned out to (receiver gates). At integration levels above 1 μm, the load seen by a gate was dominated by the capacitance of its receivers, so this delay model was sufficiently accurate. As feature sizes crossed below 1 μm, however, the parasitic resistance and capacitance along the interconnects became a significant factor. More precise modeling of the total load effect on gate delay and interconnect delay had to be taken into account. By 0.5 μm the delay attributed to global nets almost equaled that of gates, and by 0.35 μm the delay attributed to short nets equaled the gate delay.

Today, a number of models are used to represent the distributed parasitics along interconnects using lumped-form equations. The well-known π-model, for example, is a popular method to model these distributed RCs. Different degrees of accuracy can be obtained by adding more sections to the equivalent-lumped RC model for interconnect.

As the importance of timing-driven design tools grows, integration of multiply sourced design tools into an EDA flow becomes problematic. As more EDA vendors use custom built-in models for computation of gate and interconnect delays, the calculated values may differ across different design tools.

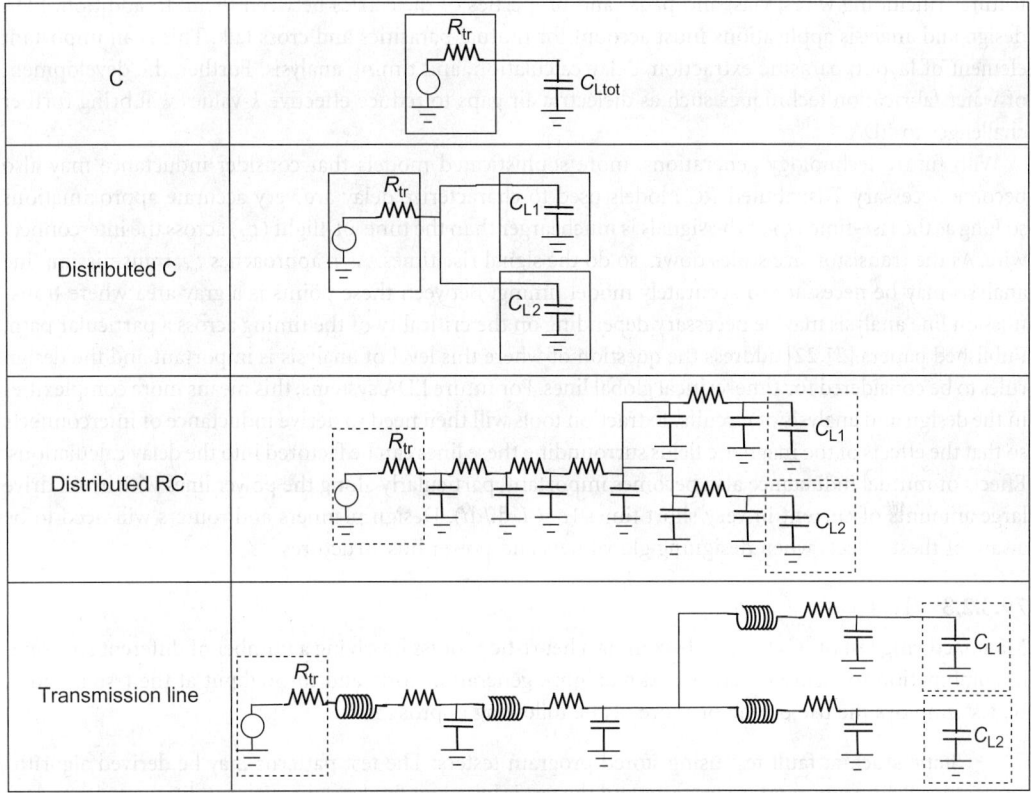

FIGURE 79.12 Delay models.

This results in difficult and time-consuming analysis on the designers' part to correlate the different results. In mid-1994, an industry effort began to develop an open architecture supporting common delay-calculation engines [20]. The goal of this effort was to provide a standard language usable by ASIC suppliers to specify the delay calculation models and equations (or table lookups) for their technology families. In this way, the complexities of deciding appropriate models to be used to characterize circuit delay and the resulting calculation expressions could be placed in the hands of the semiconductor supplier rather than across the EDA vendors. These models could then be compiled into a form that could be directly used by any EDA application requiring delay calculation and in a way that would protect the intellectual property contained within them. By allowing the semiconductor supplier to provide a single software engine for the calculation of delay, all applications in the design flow could provide consistency in the computed results. The Delay Calculation Language (DCL) technology, originally developed by IBM Corporation, was contributed to industry as the basis for this. Today, DCL has been extended to cover power calculation in addition to delay and it has been ratified by the IEEE as an open industry standard (Delay and Power Calculation System [DPCS], IEEE 1481–1999). For a number of technical and business reasons, adoption of this standard has failed to take hold. Although release and adoption of the Synopsys .lib format, used to provide technology parameters used by delay calculation in a standard form, greatly improved this situation, problems with accuracy and consistency of delay calculation remain. To that end, a renewed interest in developing an industry-wide solution came about in 2005 and the Open Modelling Coalition (OMC) has been formed under Si2 to readdress the challenge with industry partners.

At 0.25 μm, cross talk noise resulting from mutual capacitance between signal lines begins to have a significant effect on delay. At this technology generation, it became necessary to consider effects of mutual capacitance between interconnects. Today, sophisticated extraction tools must analyze the proximity of features (including wires, vias, and pads) and properties of dielectrics between them. In addition, EDA design and analysis applications must account for mutual parasitics and cross talk. This is an important element of layout, parasitic extraction, delay calculation, and timing analysis. Further, the development of wafer fabrication techniques such as dielectric air gaps to reduce effective k-values will bring further challenges to EDA.

With future technology generations, more sophisticated models that consider inductance may also become necessary. Distributed RC models used to characterize delay are very accurate approximations so long as the rise-time (t_r) of the signals is much larger than the time-of-flight (t_{of}) across the interconnect wire. As the transistor size scales down, so do the signal rise times. As t_r approaches t_{of}, transmission line analysis may be necessary to accurately model timing. Between these points is a gray area where transmission line analysis may be necessary depending on the criticality of the timing across a particular path. Published papers [21,22] address the question of where this level of analysis is important and the design rules to be considered for time-critical global lines. For future EDA systems, this means more complexities in the design and analysis of circuits. Extraction tools will then need to derive inductance of interconnects so that the effects of the magnetic fields surrounding these lines can be factored into the delay calculations. Effects of mutual inductance also becomes important, particularly along the power lines which will drive large amounts of current in very short times ($e = L \, di/dt$). Design planners and routers will need to be aware of these effects when designing global nets and power bus structures.

79.3.2.3 Test

Manufacturing test of ICs today is becoming a heuristic process involving a number of different strategies in combination to assure coverage and maximize generation costs and throughput at the tester. Today, IC test employs the use of one or more of the following approaches:

- Static stuck-at fault test using stored program testers: The test patterns may be derived algorithmically or from random patterns and the test is based on final steady-state conditions independent of arrival times. The goal of stuck-at fault test is to achieve 100% detection of all stuck-at faults (except for redundancies and certain untestable design situations that can result from the design). Algorithmic test generation for all faults in general sequential circuits is often not possible, but test generation for combinational circuits is. Therefore the use of design-for-test strategies (such as scan design) is becoming more accepted.
- Delay (At-Speed) Test: Patterns for delay testing may be algorithmically generated or functional patterns derived from DV. The tester measures output pins for the logic state at the specified time after the test pattern is applied.
- BIST: BIST tests are generated at the tester by the device under test and output values are captured in scan latches for comparison with simulated good-machine responses. BIST tests may be exhaustive or random patterns and the expected responses are determined by simulation. BIST requires special circuitry on the device-under-test (LFSR and scan latches) for pattern generation and result capture. To improve tester times, output responses are typically compressed into a signature and comparison of captured results and simulated results is made only at the end of sequences of tests.
- Quiescent current test (I_{DDQ}): I_{DDQ} measures quiescent current draw (current required to keep transistors at their present state) on the power bus. This form of testing can detect faults not observable by stuck-at fault tests (such as bridging faults) that may cause incorrect system operation.

Complicated semiconductor trends that will affect manufacturing test in the future are the speed at which these future chips operate and the quiescent current required for the large density of transistors on the IC. To perform delay test, for example, the tester must operate at speeds of the device-under-test. That is, if a signal is to be measured at a time n after the application of the test pattern, then the tester must be able to cycle in n units of time. It is possible that the hardware costs required to build testers that operate at frequencies above future IC trends will be prohibitive. I_{DDQ} tests are a very effective means to quickly identify faulty chips as only a small number of patterns are required. Further, these tests find faults not otherwise detected by stuck-at fault test. Electric current measurement techniques are far more precise than voltage measurements. However, the amount of quiescent current draw required for the millions of transistors in future ICs may make it impossible to detect the small amount of excess current resulting from small numbers of faulty transistors. New inventions and techniques may be introduced into the test menu. However, at present it appears that in the future manufacturing test must rely on static stuck-at fault tests and BIST. Therefore, it is expected that more and more application of scan design will be prevalent in future ICs.

79.3.2.4 Design Productivity

The ITRS points out that future ICs will contain hundreds of millions of transistors. Even with the application of the aforementioned architectural features and new and enhanced EDA tools, it is unclear that design productivity (number of good circuits designed per person year) will be able to keep pace with semiconductor technology capability. Designing and managing anything containing over 100,000,000 subelements is almost unthinkable. Doing it right and within the typical 18-month product cycles of today seems impossible. Yet, if this is not accomplished, semiconductor foundries may run at less than full production, and the possibility of obtaining returns against the exorbitant capital expenditures required for new foundries will be low. Ultimately, this could negate the predictions in the ITRS.

Over the history of IC design, a number of paradigm shifts enhanced designer productivity, the most notable being high-level design languages and synthesis. Use of design rules such as LSSD to constrain the problem has also represented productivity advances. Algorithm advances and faster processing computers will necessarily play a crucial role, as the design cycle time is a function of computer resources required. Many of the EDA application algorithms, however, are not linear with respect to design size and processing times can increase as an exponential function of transistors. Therefore exploitation of hierarchy, abstraction, shared, incremental, and concurrent EDA system architectures will play an important role to overall productivity. Even with all of this, there is a major concern that industry will not be able to design a sufficient number of good circuits fast enough. Consequently, there is a major push in the semiconductor industry toward design reuse. There is a major activity in the EDA industry (Virtual Sockets Interface Alliance [VSIA] www.vsia.org) to define the necessary standards, formats, and test rules to make design reuse a reality with ICs.

Design reuse is not new to electronic design or EDA. Early TTL modules were an example of reuse, as are standard cell ASICs and PCB-level MSI modules. With each, the design of the component is done once, then qualified, then reused repeatedly in application-specific designs. Reuse is also common where portions of a previous design are carried forward to a new system or technology node, and where a common logical function used multiple times across the system. However, the ability to embed semiconductor designs qualified for one process into chips manufactured on another, or for which the logical design was developed by one company to be used by another one results in new and unique challenges. This is the challenge that the VSIA is addressing.

The VSIA has defined the following three different types of reusable property for ICs:

1. Hard macros—these functions have been designed and verified, and have a completed layout. They are characterized by being a technology-fixed design and a mapping of manufacturing processes is required to retarget them to another fabrication line. The most likely business model for hard macros is that they will be available from the semiconductor vendor for use in application-specific designs being committed to that supplier's fabrication line. In other words, hard macros will most likely not be generally portable across foundries except in cases where special business partnerships are established. The complexities of plugging a mask-level design for one process into another process line are a gating factor for further exploitation at present.
2. Firm macros—can be characterized as reusable parts that have designed down to the cell level through partitioning and floorplanning. These are more flexible than hard macros since they are not process dependent and can be retargeted to other technology families for manufacture.
3. Soft macros—these are truly portable design descriptions, but are only completed down through the logical design level. No technology mapping or PD is available.

To achieve reuse at the IC level, it will be necessary to define or otherwise establish a number of standard interfaces for design data. The reason for this is that it will be necessary to integrate design data from other sources into the design and EDA system being used to develop the IC. VSIA was established to determine where standard interfaces or data formats are necessary and choose the right standard. For soft macros these interfaces will need to include behavorial descriptions, simulation models, and timing models. For firm macros the scope will additionally include cell libraries, floorplan information, and global wiring information. For hard macros GDSII may be supplied.

To reuse designs that were developed elsewhere (hereafter called intellectual property (IP) blocks), the first essential requirement is that the functional and electrical characteristics of the available IP blocks be available. Since internal construction of firm and hard IP may be highly sensitive and proprietary, it will become more important to describe the functional and timing characteristics at the I/Os. This will drive the need for high-level description methods such as use of VHDL behavorial models, DCL, I/O Buffer Information Specification (IBIS) models, and dynamic timing diagrams. Further, the need to test these embedded IP blocks will mandate more use of scan-based test techniques such as Joint Test Action Group (JTAG) boundary scan. Standard methods such as these for encapsulating IP blocks will be of paramount importance for broad use across many designs and design systems.

There are risks, however, and reuse of IP will not cover all design needs. Grand schemes for reusable software yielded less than desired results. Extra effort to generalize the IP block design points for broad use, and describe its characteristics in standard formats is compounded by the ability to identify an IP block that fits a particular design need. The design and characterization of reusable IP will need to be robust in timing, power, reliability, and noise in addition to function and cost. Further, the broad distribution of IP information may conflict with business objectives. Moreover, even if broadly successful, use of reusable IP will not negate the fundamental need for major EDA advances in the areas described. First, tools are needed to design the IP. Second, tools are needed to design the millions of circuits that will interconnect IP. Finally, tools and systems will require major advances to accurately design and analyze the electrical interactions between, over, and under IP and application-specific design elements on the IC. Standards are essential, but they are not by themselves sufficient for success. Tools, rules, and systems will be the foundation for future IC design as they have been over the past. Nevertheless, the potential rewards are substantial and there is an absence of any other clear EDA paradigm shift. Consequently, new standards will emerge for design and design systems, and new methods for characterization, distribution, and lookup of IP will become necessary.

79.3.2.5 DFM

79.3.2.5.1 *Lithography*

Because of the physical properties of diffraction, as printed features shrink, the ability to print them with the required degree of fidelity becomes a challenge for mask making. The scaling effects for lithographic resolution can be generally viewed using the Rayleigh equation for a point light source.

$$\text{Smallest resolvable feature} = K(\lambda/\text{NA})$$

Here, the value of K is a function of the photoresist process used, λ is the wavelength of the coherent light source used for the exposure and NA the numerical aperture of the lens system. As feature shrink continues, the ITRS projects that design wafer foundry lithographic systems will keep pace by moving to light sources of smaller wavelength, photoresist systems with lower values for K, and immersion lithography. (Immersion lithography increases the maximum achievable numerical aperture (NA = $I \sin a$) beyond that in air (in air, the refractive index $(I) = 1$, thus $\text{NA}_{max} = 1$), because of higher refractive indices for water and other fluids). However, the physics of light diffraction is now affecting the design of photomasks, and this will dramatically increase in importance with future technology generations.

Without attempting to discuss the physics behind them, two complications need to be accounted for in the design of a mask. First, the intensity of the light source passing through the mask and onto the photoresist is nonlinear. That is, the intensity at the edges of small features is less than that in the center.

Thus, edges of small features do not print with fidelity. To compensate for this, for example on the corners of lines (which will expose as rounded edges) and line ends (which will be rounded and short), a series of reticle (mask) enhancement techniques is used. These techniques add subresolution features to the pattern data. Some examples are the addition of serifs on corners and hammerheads on line-ends to extend as well as square them.

The second complication is the interference of diffracted light between adjacent features. That is, the light used to expose densely packed features may interfere with the exposure of other features in close proximity. This may be corrected for by phase-shift mask techniques or be accounted for by adjusting feature placement (spreading densely packed areas and using scatter bars to fill sparsely packed areas) and biasing feature widths to adjust for the interference.

Though these corrections are necessary for two distinct physical properties, they are generally collectively performed by optical proximity correction (OPC) programs. OPC in today's design systems is generally performed a step after the design is complete and ready for tape-out. The desired design pattern data is topographically analyzed by OPC against a set of rules or models that determine if and what corrections need to be made in the mask pattern to assure wafer fidelity. These corrections result in the incorporation of additional features to be cut into the mask such as serifs, hammerheads, and scatter bars.

With successive technology generations, the number of OPC corrections for a mask set will increase and cause mask pattern data to increase beyond that of Moore's Law for features. This, in turn, causes increased time (and resulting costs) in capital-intensive mask manufacturing. Therefore, future design systems will need to consider lithography limitations early in the design process (synthesis and PD) rather than at the end, and new OPC methods to optimize the mask pattern volume will be required.

79.3.2.5.2 *Yield*

Traditionally, wafer yield has been viewed as the responsibility of the wafer foundry alone. Hard defects caused by random particles or incorrect process settings are under control of the foundry manager and defects in feature parametrics were sufficiently bounded by design rules to be insignificant. As features shrink, small variations in print fidelity or process parameters have an increasingly important impact on feature parasitics and this can be a significant factor in the yield of properly operating chips. Traditionally, parasitic extraction performed on the PD-generated geometry was sufficient to analyze a parasitic effect on, for example, timing. In future, because of lithography limitations, the analysis of parasitics may require extraction based on the simulated geometries that are projected to print on the wafer. Traditionally, EDA design applications can operate within the specific bounds of the design rules to yield good design. Future design applications may require that statistical analysis of their decisions be performed to account for intrachip parametric variations and to provide a design known to be good within a certain degree of probability. The topic of yield will become an important aspect of the next EDA paradigm shift.

Future design system flows will require a closer linkage between the classical EDA flow and Technology Computer Aided Design (TCAD) systems and mask making. In addition, the need for rich models that can be used to accurately predict the impacts of design choices on the results of foundry and mask manufacturing processes will become important. New design tools such as Statistical Static Timing Analysis (SSTA) have become available and are a topic of much discussion. How to incorporate the effects of variability on yield in, for example PD and synthesis will be a topic for future research. High-speed lithographic simulation, capable of full chip analysis, will be an important field for EDA R&D. Finally, database and integration technology to support this increased level of collaboration between IC designers, mask manufacturers, and the fabricating engineers will be critical.

79.4 Summary

Over the past four decades, EDA has become a critical science for electronics design. Where it may have begun as a productivity enhancement, it is now a fundamental requirement for the design and manufacture of electronic components and products. The criticality of EDA for semiconductors is the focus of much attention because of the expanding semiconductor advancements. Fundamental approaches in test, simulation, and PD are being constantly emphasized by these semiconductor advancements. New EDA disciplines are opening up. Fundamental electrical models are becoming more important and at the same time, more complex. New problems in design and IC failure modes will surely surface.

The next decade of EDA will prove to be as challenging and exciting as its past!

References

1. Case, P.W., Correia, M., Gianopulos, W., Heller, W.R., Ofek, H., Raymond, T.C., Simel, R.L., Stieglitz, C.B., Design automation in IBM, *IBM Journal of Research and Development,* vol. 25, no. 5, p. 631, 1981.
2. Roth, J.P., Diagnosis of automata failures: a calculus and a method, *IBM Journal of Research and Development,* vol. 10, p. 278, 1966.
3. Eichelberger, E.B., Hazard detection in combinational and sequential switching circuits, *IBM Journal of Research and Development,* vol. 9, p. 90, 1965.
4. Eichelberger, E.B., Williams, T.W., A logic design structure for LSI testability, *Proceedings of the 14th Design Automation Conference,* p. 462, 1977.
5. IEEE Std 1149.1-1990, *IEEE Standard Test Access Port and Boundary-Scan Architecture,* IEEE, Washington, 1990.
6. Hitchcock, R.B., Cellular wiring and the cellular modeling technique, *Proceedings of the 6th Annual Design Automation Conference,* p. 25, 1969.
7. Lee, C.Y., An Algorithm for path connection and its applications, *IRE Transactions on Electronic Computers,* p. 346, 1961.
8. Hightower, D.W., The intelrconnection problem, a tutorial, *Proceedings of the 10th Annual Design Automation Workshop,* p. 1, 1973.
9. *1076-1993 IEEE Standard VHDL Language Reference Manual,* IEEE, Washington, 1993.
10. *1364-2005 IEEE Standard Hardware Description Language Based on the Verilog® Hardware Description Language,* IEEE, Washington, 2006.
11. Electronic Industry Associates, *Electronic Design Interchange Format (EDIF),* IHS (www.globalihs.com).
12. Stok, L., Kung, D.S., Brand, D., Drumm, A.D., Sullivan, A.J., Reddy, L.N., Heiter, N., Geiger, D.J., Chao, H.H., Osler, P.J., BooleDozer: Logic synthesis for ASICs, *IBM Journal of Research and Development,* vol. 40, no. 4, p. 407, 1996.
13. Sayah, J.Y., Gupta, R., Sherlekar, D.D., Honsinger, P.S., Apte, J.M., Bollinger, S.W., Chen, H.H., DasGupta, S., Hseih, E.P., Huber, A.D., Hughes, E.J., Kurzum, Z.K.M., Rao, V.B., Tabteing, T., Valijan, V., Yang, D.Y., Design planning for high-performance ASICs, *IBM Journal of Research and Development,* vol. 40 no. 4, p. 431, 1996.
14. Barnes, T.J., Harrison, D., Newton, R.A., Spickelmier, R.L., *Electronic CAD Frameworks,* Kluwer Academic Publishers, Dordrecht, Chapter 10, 1992.
15. Semiconductor Industry Association, *International Technology Roadmap for Semiconductors, Technology Needs, 2001 Edition,* www.sematech.org.
16. Bakoglu, H.B., *Circuits, Interconnections, and Packaging for VLSI,* Addison-Wesley, Reading, MA, Chapters 1–7, 1990, ISBN 0-201-06008-6.
17. Goel, A.K., *High-Speed VLSI Interconnections, Modeling, Analysis & Simulation,* Wiley, New York, Chapter 2, 1994, ISBN 0-471-57122-9.
18. Johnson, H.W., Graham, M., *High-Speed Digital Design, A Handbook of Black Magic,* Prentice-Hall PTR, New York, Chapter 1, 1993, ISBN 0-13-395724-1.
19. Pandini, D., Scandolaro, P., Guardiani, C., Network reduction for crosstalk analysis in deep submicron technologies, *ACM/IEEE International Workshop on Timing Issues in the Specification and Synthesis of Digital Systems,* p. 280, December 1997.

20. Cottrell, D.R., Delay calculation standard and deep submicron design, *Integrated Systems Design*, July 1996.
21. Fisher, P.D., Clock cycle estimates for future microprocessor generations, *IEEE 1997 ISIS: International Conference on Innovative Systems in Silicon*, October 1997.
22. Deutsh, A., Kopcsay, G.V., Restle, P., Katopis, G., Becker, W.D., Smith, H., Coteus, P.W., Surovic, C.W., Rubin, B.J., Dunne, R.P., Gallo, T., Jenkins, K.A., Terman, L.M., Dennard, R.H., Sai-Halasz, G.A., Knebel, D.R., When are transmission-line effects important for on-chip interconnections, *Electronic Components and Technology Conference*, p. 704, 1997.

Section
XII

VLSI Signal Processing

W. Kenneth Jenkins
The Pennsylvania State University

Introduction

When digital signal processing (DSP) began to emerge as a discipline in the 1960's the hardware that was available at that time for implementation was based on digital computer technology of the 1950's. In general DSP functions were implemented on large general-purpose digital computers that were programmed using punch cards in standard programming languages (e.g. assembly languages, Basic, Fortran, etc.), to achieve off-line processing of digital data that was stored on magnetic or paper tapes. In special cases where dedicated hardware was needed to implement well-defined DSP functions in real time, it was typically constructed from small scale integrated (SSI) or medium scale integrated (MSI) circuits. While SSI circuits were very limited integrated circuits that interconnected clusters of transistor switches to form integrated logic gates (AND, OR, NAND, NOR, etc.), MSI chips were more highly integrated circuits that contained rudimentary arithmetic units, such as latches, shift registers, full adders, and bit slices of digital array multipliers. The most common logic families at the time were based on RTL logic, TTL logic, or ECL logic, families that spanned the range of low-power-low-speed to high-power high-speed logic. In those days computer memories were implemented with ferrite core technology, and the concept of large-scale semiconductor memories was being developed in corporate research laboratories around the world.

During the 1960's the theory of digital signal processing began to emerge as a discipline in its own right, appearing largely in research papers and special reports from institutions such as Bell Laboratories, Lincoln Labs, and the IBM Thomas Watson Research Center. In 1959 when Hoff and Widrow published the Least Mean Squares (LMS) adaptive filtering algorithm the state-of-the-art in digital hardware was not sufficiently advanced for engineers to consider practical implementations of an adaptive filter in purely digital form. Their first implementation of the adaptive line enhancer (Adeline) was implemented with a combination of analog circuits and complicated arrangements of analog relays that performed the switching necessary to adjust the filter tap weights. It was not until over ten years later, in 1971, that Hoff, Mazor, and Faggin invented the first "computer-on-a-chip" at Intel, which was the Intel 4004 4-bit microprocessor originally produced as a computational core for the calculator industry. During the following two decades the microcomputer underwent rapid development, and by the 1980's the need for special purpose DSP chips began to be recognized. In response to this need the semiconductor industry (Motorola, Texas Instruments, American Micro Devices, etc.) began to produce new generations of computationally powerful "DSP chips" (e.g. the TI 320-XXX family) that enabled the real time implementation of many advanced DSP concepts that previously existed only in theory.

Not surprisingly these rapid developments in integrated circuit technology spurred increased theoretical research, resulting in the prolific publication of the theory of digital signal processing. While the theory of DSP appeared primarily in research publications (archival journals and conference proceedings) during the 1960's, the 1970's witnessed the first wave of reference books and textbooks that treated DSP as a stand-alone discipline. For example two IEEE Press edited volumes were at the forefront of these new publications, "Digital Signal Processing I, 1972 and, "Digital Signal Processing II, 1976. The classic reference text *Theory and Applications of Digital Signal Processing* published by Rabiner and Gold (Prentice Hall, 1975), and the classic graduate level text book *Digital Signal Processing* by Oppenheim and Schafer (Prentice Hall, 1975) led this new wave of DSP textbooks that continued to appear during the next two decades. The theory of DSP was developing rapidly and the discipline of DSP began to be recognized by universities as academic core material that needed to be integrated into both undergraduate and graduate curricula.

The history of integrated circuit technology and digital signal processing theory over the past fifty years demonstrates how science and technology can stimulate each other over extended periods of time, each driving the other toward new advancements, and each providing new capabilities that allow the opposite forces to prosper. In the case of the VLSI–DSP technology race, it is clear that advancements in DSP theory continually begged for more speed, higher densities, and greater computational power in order to bring new DSP theories into practice. Similarly, the steady advances in VLSI circuit technology and the increased computational power that resulted from these advances allowed DSP theoreticians to develop more advanced signal processing techniques with the expectation that their most sophisticated concepts would one day find a way into real products and systems.

This Section on VLSI Signal Processing consists of five chapters that deal with important engineering applications where VLSI signal processing has exerted a strong influence on advancing the state-of-the-art. Chapter 80 presents some basic concepts on "Computer Arithmetic for VLSI Signal Processing." As described above, until the 1970's nearly all signal processing was done with analog circuits, whereas now it is done mostly with digital circuits. The change was a result of two technology developments: mixed signal (analog and digital) circuits (mostly A/D and D/A converters) and VLSI arithmetic elements (mostly adders and multipliers). It examines arithmetic elements at both algorithm and logic design levels. Different algorithms may be used to vary the speed of an arithmetic element by an order of magnitude or more while the complexity varies by less than 50%. It also examines number systems for signal processing. In particular the implementation of both fixed point arithmetic elements and floating point arithmetic elements are both treated in detail.

Chapter 81 focuses on "VLSI Architectures for JPEG 2000 EBCOT: Design Techniques and Challenges." JPEG 2000, as new image compression standard, provides a rich set of features that are not available in existing JPEG standards. These new benefits are mainly contributed by application of the Discrete Wavelet Transform (DWT) and Embedded Block Coding with Optimized Truncation (EBCOT) tier-1. In this section, the EBCOT tier-1 algorithm including both serial and parallel coding modes is introduced. Two categories of VLSI architectures for implementing these algorithms are discussed: Parallel Bit-plane Coding Scheme (ParaBCS) and Serial Bit-plane Coding Scheme (SeriBCS). In either ParaBCS or SeriBCS the architecture is based on BC (Bit-plane Coding), AE (Arithmetic Encoding), and a FIFO that connects BC with AE and balances the different throughput between them. In this section several design techniques for BC, FIFO, and AE are introduced and two case studies (one is based on SeriBCS and the other is based on ParaBCS) are presented.

Chapter 82 deals with "VLSI Architectures for Forward Error-Control Decoders." The publication of Shannon's ground breaking paper in 1948 identifying the fundamental limits of information transfer capacity of noisy communication channels resulted in the birth of information theory and coding theory as two distinct areas of research. Coding theory involves the investigation of efficient error control codes, the study of performance bounds, and the determination of computationally efficient decoding algorithms. In many regards the present moment is historic as the tremendous advances in theory and practice of coding theory converge with the semiconductor industry progress embodied by Moore's Law. The latter has enabled the implementation of decoders that that were considered too complex to implement just a few years ago. This chapter discusses architectural issues in the design of high-speed decoders for forward error-control techniques in broadband communication systems. The design of high-speed and low-power Reed-Solomon decoders, Viterbi decoders, turbo decoders, and low-density parity check decoders is presented.

Chapter 83 presents "An Exploration of Hardware Architectures for Face Detection." Face detection is a very important application in the field of machine vision. Face detection algorithms have been proposed by multiple researchers, but so far most algorithms have been implemented in software. While software algorithms are effective in terms of detection rate and ease of use, they still cannot achieve accurate detection on a real-time video signal. With the increased demands in security and control applications, the need for hardware implementation of face detection is evident.

In this chapter the problem of face detection is developed as it evolves from the software domain into hardware platforms, and major challenges and issues are identified in implementing face detection in hardware. The authors demonstrate the porting of two state-of-the-art algorithms to hardware, and examine how the targeted design platform affects the performance of the detection. An FPGA implementation and an ASIC implementation of one of the algorithms are presented.

Finally, Chapter 84 discusses the fundamental concepts of "Multidimensional Logarithmic Number System." In this chapter the multidimensional logarithmic number system (MDLNS) is introduced as a generalization of the classical 1-D logarithmic number system (LNS) and analyzed for use in DSP applications. A major drawback of the LNS is the "difficult" operations of addition, subtraction and conversion. These operations either require the use of very large ROM arrays or other techniques such as bipartite function evaluation in order to reduce ROM table sizes. The MDLNS, however, allows exponential reduction of the size of the ROMs used without affecting the speed of computing the "difficult" operations. Moreover, the calculations over different bases and digits are completely independent, which makes this particular representation perfectly suitable for massively parallel DSP architectures. Several examples are presented showing the computational advantages gained by the proposed approach.

80

Computer Arithmetic for VLSI Signal Processing

Earl E. Swartzlander, Jr.
University of Texas

CONTENTS

80.1 Introduction

Until the 1970s nearly all signal processing was done with analog circuits. Now it is mostly digital. The change is the result of two technology developments: mixed-signal (analog[A] and digital[D]) circuits (mostly, A/D and D/A converters) and VLSI arithmetic elements (mostly, adders and multipliers).

This chapter examines the arithmetic elements at the algorithm and logic design levels. Different algorithms may be used to vary the speed of an arithmetic element by an order of magnitude or more while the complexity varies by <50%.

This chapter examines number systems for signal processing in Section 80.2. The implementation of fixed-point arithmetic elements are examined in Section 80.3. Finally, Section 80.4 briefly considers the implementation of floating-point arithmetic elements.

Regarding notation, capital letters represent digital numbers (i.e., words), whereas subscripted lowercase letters represent bits of the corresponding word. The subscripts range from $n - 1$ to 0 to indicate the bit position within the word (x_{n-1} is the most significant bit of X, x_0 is the least significant bit of X, etc.). The logic designs in this chapter are based on positive logic with AND, OR, and invert operations. Depending on the technology used for implementation, different operations (such as NAND and NOR) may be used, but the basic concepts do not change.

80.2 Fixed-Point Number Systems

In digital signal processing, most arithmetic is performed with fixed-point binary numbers that have constant scaling (i.e., the position of the binary point is fixed). The numbers can be interpreted as fractions, integers, or mixed numbers, but fractions are the most commonly used.

Pairs of fixed-point numbers are used to create floating-point numbers, as discussed in Section 80.4.

Fixed-point binary numbers are generally represented using the two's complement number system. This choice has prevailed over the sign magnitude and one's complement number systems, because the frequently performed operations of addition and subtraction are the easiest to perform on two's complement numbers. Sign magnitude numbers are more efficient for multiplication and division, but the lower frequency of multiplication and the development of Booth's efficient two's complement multiplication algorithm have resulted in the nearly universal selection of the two's complement number system for most applications. The algorithms presented in this chapter assume the use of two's complement numbers.

Fixed-point number systems represent numbers, for example, A, by n bits: a sign bit and $n-1$ data bits. By convention, the most significant bit a_{n-1} is the sign bit, which is 1 for negative numbers and 0 for positive numbers. The $n-1$ data bits are $a_{n-2}, a_{n-3}, \ldots, a_1, a_0$. In the following material, fixed-point fractions will be described for both the two's complement and the sign magnitude number systems.

Two's Complement. In the two's complement fractional number system, the value of a number is the sum of $n-1$ positive binary fractional bits and a sign bit, which has a weight of -1

$$A = -a_{n-1} + \sum_{i=0}^{n-2} a_i 2^{i-n+1} \tag{80.1}$$

Two's complement numbers are negated by complementing all bits and adding a 1 to the least significant bit (lsb) position. For example, to form $-3/8$,

	+3/8	=	0011		
	invert all bits	=	1100		
	add 1 lsb		0001		
			1101	=	−3/8
Check:	invert all bits	=	0010		
	add 1 lsb		0001		
			0011	=	+3/8

Sign/Magnitude. Sign magnitude fractional numbers consist of a sign bit and $n-1$ bits that express the magnitude of the number,

$$A = (1 - 2a_{n-1}) \sum_{i=0}^{n-2} a_i 2^{i-n+1} \tag{80.2}$$

Sign magnitude numbers are negated by complementing the sign bit. For example, to form $-3/8$,

	+3/8	=	0011		
	invert sign bit	=	1011	=	−3/8
Check:	invert sign bit	=	0011	=	+3/8

Table 80.1 compares 4-bit fractional fixed-point numbers in the two number systems. Note that the sign magnitude number system has two zeros and that the two's complement number system is capable of representing -1. For positive numbers both systems have identical representations.

TABLE 80.1   Example of 4-bit Fractional Fixed-Point Numbers

Number	Two's Complement	Sign Magnitude
+7/8	0111	0111
+3/4	0110	0110
+5/8	0101	0101
+1/2	0100	0100
+3/8	0011	0011
+1/4	0010	0010
+1/8	0001	0001
+0	0000	0000
−0	N/A	1000
−1/8	1111	1001
−1/4	1110	1010
−3/8	1101	1011
−1/2	1100	1100
−5/8	1011	1101
−3/4	1010	1110
−7/8	1001	1111
−1	1000	N/A

A significant difference between the two's complement and sign magnitude number systems is their behavior under truncation. Figure 80.1 shows the effect of truncating high-precision fixed-point fractions X, to form 4-bit fractions $T(X)$. Truncation of two's complement numbers never increases the value of the number (i.e., the truncated numbers have values that are unchanged or shift toward negative infinity), as can be seen from Eq. (80.1) where any truncated bits have positive weight. This bias can cause an accumulation of errors for computations that involve summing many truncated numbers (which may occur in signal-processing applications). In the sign magnitude number system, truncated numbers are unchanged or shifted toward zero, so that if approximately half of the numbers to be added are positive and half are negative, the errors will tend to cancel.

80.3   Fixed-Point Arithmetic Elements

80.3.1   Adders

Addition is performed by summing the corresponding bits of two numbers, including the sign bits. Subtraction is performed by summing the corresponding bits of the minuend and the two's complement of the subtrahend. Overflow is detected in a two's complement adder by comparing the carry signals into and out of the most significant adder stage (i.e., the stage which computes the sign bit). If the carries differ, an overflow has occurred and the result is invalid. Alternatively, if adding two numbers of like sign, if the sign of the result is different, then overflow has occurred. If adding numbers of unlike sign (or subtracting numbers of like sign) overflow cannot occur.

Full adder. The full adder is the fundamental building block of most arithmetic circuits. Its operation is defined by the truth table shown in Table 80.2.

The sum and the carry outputs are described by the following equations:

$$s_k = a_k \oplus b_k \oplus c_k \tag{80.3}$$

$$c_{k+1} = a_k b_k + a_k c_k + b_k c_k$$

where $\oplus$ denotes the exclusive-OR logic operation, a_k, b_k, and c_k are the inputs to the kth fu[ll adder] and s_k and c_{k+1} the sum and carry outputs, respectively.

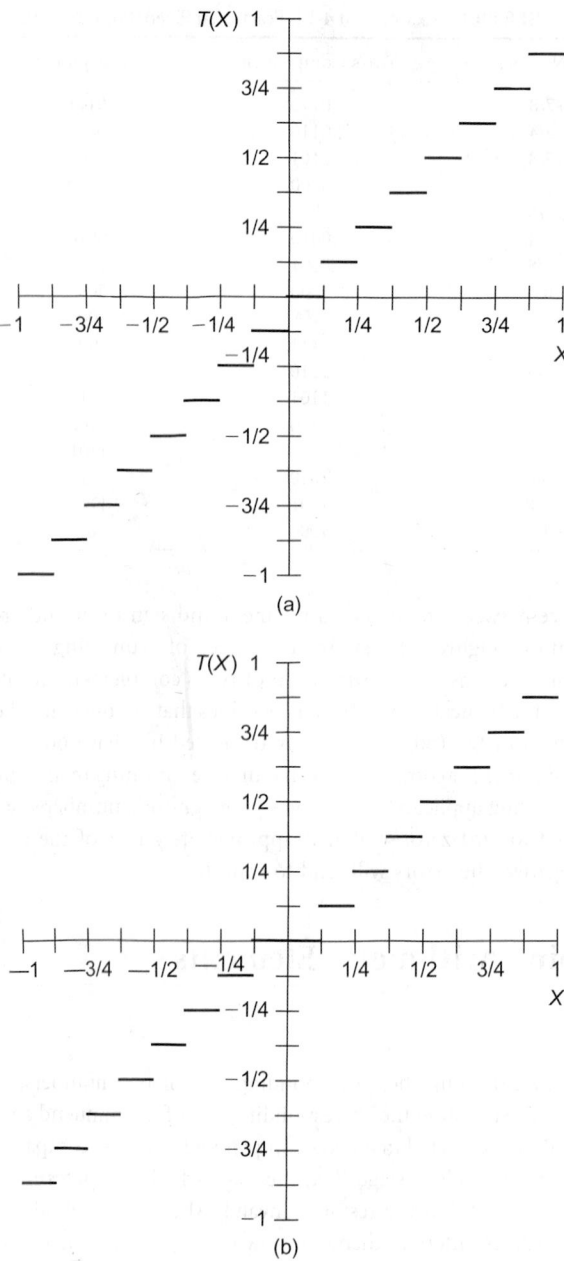

FIGURE 80.1   Behavior of fixed-point fractions under truncation: (a) two's complement and (b) sign magnitude.

In evaluating the relative complexity of implementations, it is useful to assume a nine gate realization of the full adder, as shown in Figure 80.2. In this implementation, the exclusive-OR logic operation is performed with two AND gates, an OR gate, and an inverter. For this full adder, the delay from either a_k or b_k to s_k is six gate delays, the delay from either a_k or b_k to c_{k+1} is five gate delays, the delay from c_k to c_{k+1} is two gate delays and the delay from c_k to s_k is three gate delays. In some technologies, such as CMOS, inverting gates (e.g., NAND and NOR gates) are more efficient than the noninverting gates that are used here. Circuits with equivalent speed and complexity can be constructed with inverting gates.

TABLE 80.2 Truth Table for a Full Adder

Inputs			Outputs	
a_k	b_k	c_k	c_{k+1}	s_k
0	0	0	0	0
0	0	1	0	1
0	1	0	0	1
0	1	1	1	0
1	0	0	0	1
1	0	1	1	0
1	1	0	1	0
1	1	1	1	1

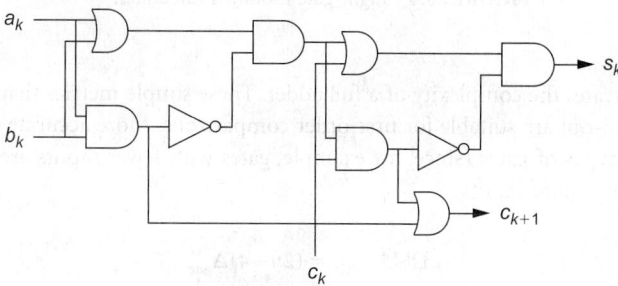

FIGURE 80.2 Nine-gate full adder.

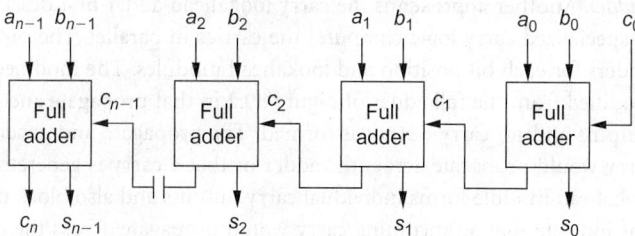

FIGURE 80.3 Ripple carry adder.

Ripple carry adder. A ripple carry adder for n-bit numbers is constructed by concatenating n full adders as shown in Figure 80.3. At the kth-bit position, the kth bits of operands A and B and a carry signal from the $(k-1)$th stage are used to form the kth bit of the sum, s_k, and the carry, c_{k+1}, to the next adder stage. This is called a ripple carry adder, since the carry signals "ripple" from the lsb position to the most significant bit position. If the ripple carry adder is implemented by concatenating n of the nine gate full adders, as shown in Figure 80.2 and Figure 80.3, an n-bit ripple carry adder requires $2n + 4$ gate delays to produce the most significant sum bit (5 gate delays to form c_1 in the least significant full adder, 2 gate delays to form c_{i+1} from c_i in each of the $n-2$ intermediate full adders and 3 gate delays to form s_{n-1} from c_{n-1} in the most significant full adder) and $2n + 3$ gate delays to produce the most significant carry output. A total of $9n$ logic gates are required to implement the n-bit ripple carry adder.

In comparing adders, the delay from data input to the most significant sum output and the complexity (i.e., the gate count) will be used. These will be denoted by DELAY and GATES (subscripted by RCA to indicate ripple carry adder), respectively. In Eq. (80.5) Δ_{gate} indicates the delay of a logic gate. Similarly,

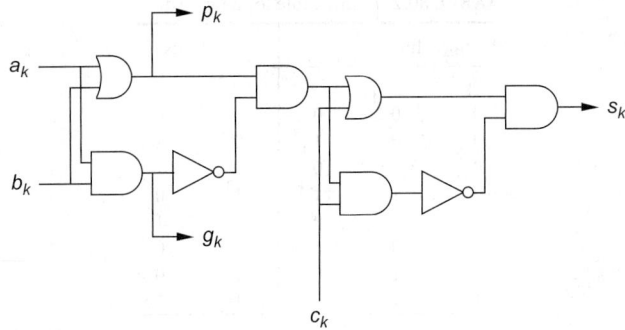

FIGURE 80.4   Eight-gate modified full adder.

in Eq. (80.6), β_{FA} indicates the complexity of a full adder. These simple metrics that ignore the effects of varying fan-in and fan-out are suitable for first-order comparisons. More accurate comparisons require consideration of the types of gates (since, for example, gates with fewer inputs are generally faster and smaller than those with more inputs).

$$\text{DELAY}_{RCA} = (2n+4)\Delta_{\text{gate}} \tag{80.5}$$

$$\text{GATES}_{RCA} = n\,\beta_{FA} \tag{80.6}$$

Carry lookahead adder. Another approach is the carry lookahead adder first described by Weinberger and Smith [1]. Here specialized carry logic computes the carries in parallel. The carry lookahead adder uses modified full adders for each bit position and lookahead modules. The modified full adders shown in Figure 80.4 are modified from the full adder of Figure 80.2 in that propagate and generate signals are made available as outputs and no carry output is formed. The propagate and generate signals indicate that an incoming carry would propagate across the adder or that a carry is generated within the adder, respectively. Each lookahead module forms individual carry outputs and also block propagate and block generate outputs that indicate that an incoming carry would propagate across the data block or that a carry is generated within the data block, respectively.

Rewriting Eq. (80.5) using propagate and generate variables, $p_k = a_k + b_k$ and $g_k = a_k b_k$, respectively:

$$c_{k+1} = g_k + p_k c_k \tag{80.7}$$

This equation explains the concept of carry generation and carry propagation. At a given stage, a carry is generated irrespective of the carry into the stage if g_k is true (i.e., both a_k and b_k are 1), and a stage propagates a carry from its input to its output if p_k is true (i.e., either a_k or b_k is a 1). Actually, $p_k = a_k \oplus b_k$, but as noted in Ref. [2] the alternative $p_k = a_k + b_k$ gives correct results in Eq. (80.7), is easier to realize, and provides redundancy in the calculation of c_{k+1}.

Extending Eq. (80.7) to a second stage,

$$c_{k+2} = g_{k+1} + p_{k+1} c_{k+1}$$

$$= g_{k+1} + p_{k+1}(g_k + p_k c_k)$$

$$= g_{k+1} + p_{k+1} g_k + p_{k+1} p_k c_k \tag{80.8}$$

Eq. (80.8) results from evaluating Eq. (80.7) for the $(k + 1)$th stage and substituting the value for c_{k+1} from Eq. (80.7). Carry c_{k+2} exits from stage $k + 1$ if: a carry is generated there; or if a carry is generated in stage k and propagates across stage $k + 1$; or if a carry enters stage k and propagates across both stages k and $k + 1$.

Extending Eq. (80.7) to a third stage,

$$c_{k+3} = g_{k+2} + p_{k+2}c_{k+2}$$

$$= g_{k+2} + p_{k+2}(g_{k+1} + p_{k+1}g_k + p_{k+1}p_kc_k)$$

$$= g_{k+2} + p_{k+2}g_{k+1} + p_{k+2}p_{k+1}g_k + p_{k+2}p_{k+1}p_k c_k \tag{80.9}$$

Although it would be possible to continue this process indefinitely, each additional stage increases the fan-in (i.e., the number of inputs) of the logic gates. Four inputs (as required to implement Eq. (80.9)) is frequently the maximum number of inputs per gate for current technologies. To continue the process, generate and propagate signals are defined over 4-bit blocks (stages k to $k+3$), $g_{k+3:k}$ and $p_{k+3:k}$, respectively.

$$g_{k+3:k} = g_{k+3} + p_{k+3}g_{k+2} + p_{k+3}p_{k+2}g_{k+1} + p_{k+3}p_{k+2}p_{k+1}g_k \tag{80.10}$$

and

$$p_{k+3:k} = p_{k+3}p_{k+2}p_{k+1}p_k \tag{80.11}$$

Eq. (80.7) can be expressed in terms of the 4-bit block generate and propagate signals,

$$c_{k+4} = g_{k+3:k} + p_{k+3:k} c_k \tag{80.12}$$

Thus, the carry out from a 4-bit-wide block can be computed in only four gate delays (the first to compute p_i and g_i for $i = k$, $k + 1$, $k + 2$, and $k + 3$, the second to evaluate $p_{k+3:k}$ using Eq. (80.11), the second and third to evaluate $g_{k+3:k}$ using Eq. (80.10), and the third and fourth to evaluate c_{k+4} using Eq. (80.12)).

An n-bit carry lookahead adder requires $\lceil (n - 1)/(r - 1) \rceil$ lookahead blocks, where r is the "width" of the block and where $\lceil X \rceil$ indicates the smallest integer greater than or equal to X. A 4-bit lookahead block is a direct implementation of Eqs. (7)–(11) requiring 14 logic gates. In general, an r-bit lookahead block requires $\frac{1}{2}(3r + r^2)$ logic gates.

Figure 80.5 shows the interconnection of 16 modified full adders and 5 lookahead logic blocks to realize a 16-bit carry lookahead adder. The sequence of events which occurs during an add operation is as follows: (1) apply A, B, and the least significant carry in, c_0; (2) the 16 modified full adders computes p_i and g_i; (3) the first-level lookahead logic blocks compute the 4-bit propagate and generate signals; (4) the second-level lookahead logic block computes c_4, c_8, and c_{12}; (5) the first-level lookahead logic blocks compute the individual carries, c_1–c_3, c_5–c_7, c_9–c_{11} and c_{13}–c_{15}; and (6) each modified full adder computes the sum outputs. This process may be extended to larger adders by subdividing the large adder into 16-bit blocks and using additional levels of carry lookahead (e.g., a 64-bit adder requires three levels).

The delay of carry lookahead adders is evaluated by recognizing that an adder with a single level of carry lookahead (for r-bit words) realized with 8 gate modified full adders as shown in Figure 80.4 has six gate delays and that each additional level of lookahead increases the maximum word size by a factor of r and adds four gate delays. More generally [3, pp. 83–88], the number of lookahead levels for an n-bit adder is $\lceil \log_r (n) \rceil$, where r is the number of bits per lookahead module. Since an r-bit carry lookahead adder has six gate delays and there are four additional gate delays per carry lookahead level after the first,

$$\text{DELAY}_{\text{CLA}} = \left(2 + 4\lceil \log_r (n) \rceil\right)\Delta_{\text{gate}} \tag{80.13}$$

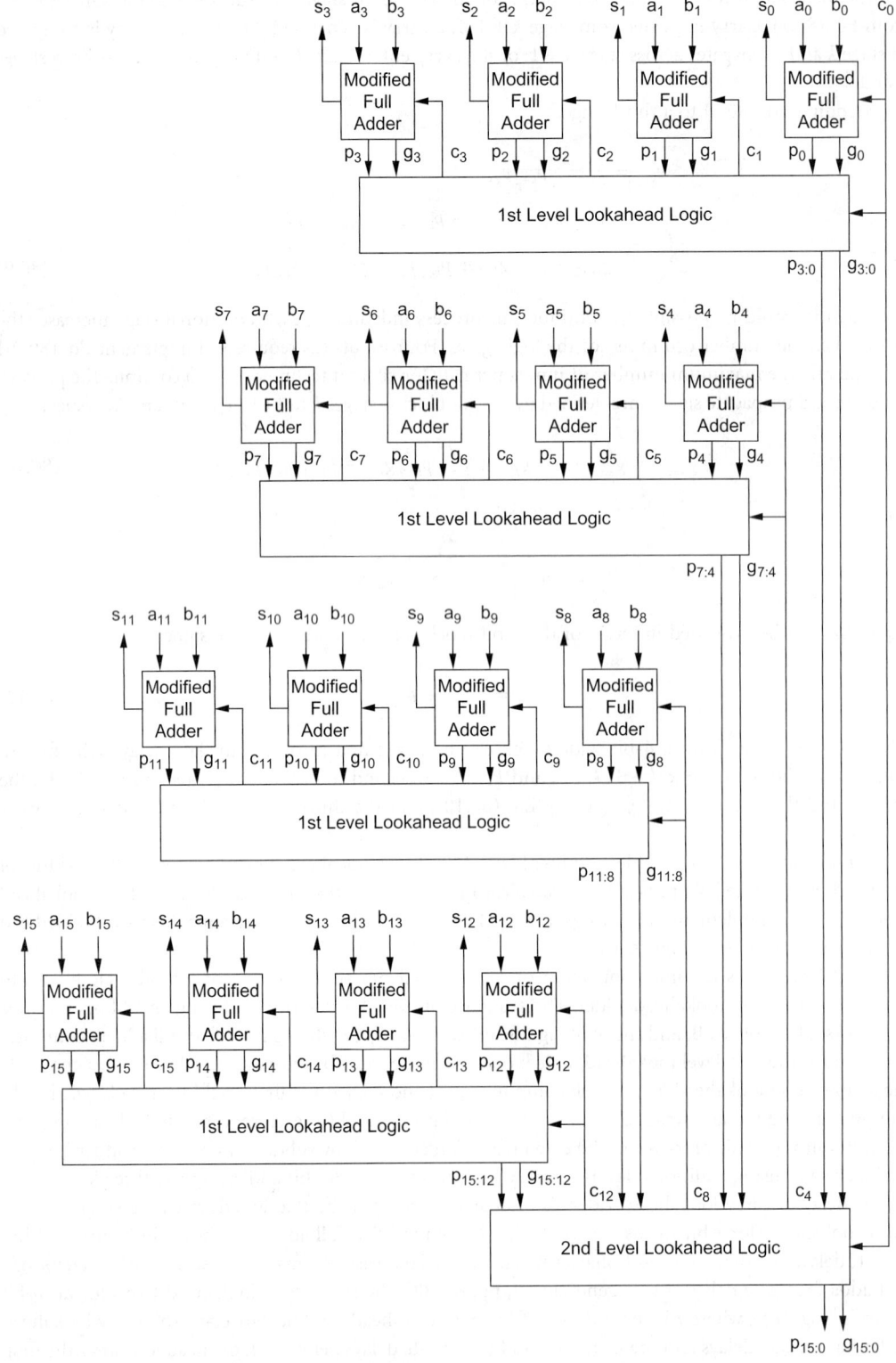

FIGURE 80.5 Sixteen-bit carry lookahead adder.

The complexity of an n-bit carry lookahead adder implemented with r-bit lookahead blocks is n modified full adders (each of which has a complexity of β_{MFA} and $\lceil (n-1)/(r-1) \rceil$ lookahead logic blocks (each of which requires $\frac{1}{2}(3r + r^2)$ logic gates with fan-in ranging from 2 to r:

$$\text{GATES}_{CLA} = n\beta_{MFA} + \left(\frac{1}{2}(3r + r^2) \left\lceil \frac{n-1}{r-1} \right\rceil \right)\beta_{gate} \tag{80.14}$$

For the currently common case of $r = 4$ and $\beta_{MFA} = 8$,

$$\text{GATES}_{CLA} = (12\tfrac{2}{3}n - 4\tfrac{2}{3})\beta_{gate} \tag{80.15}$$

The carry lookahead approach reduces the delay of adders from increasing linearly with the word size (as for ripple carry adders) to increasing as the logarithm of the word size. As with ripple carry adders, the carry lookahead adder complexity grows linearly with the word size. For $r = 4$, $\beta_{MFA} = 8$, and $\beta_{FA} = 9$, the gate count of a carry lookahead adder is 40% greater than that of a ripple carry adder.

Carry select adder. Another approach is the carry select adder first described by Bedrij [4]. The carry select adder divides the words to be added into blocks and forms two sums for each block in parallel (one with a carry in of ZERO and the other with a carry in of ONE). As shown for a 16-bit carry select adder on Figure 80.6, the carry out from the previous block controls a multiplexer that selects the appropriate sum. The carry out is computed using Eq. (80.7), since the block propagate signal is the carry out of an adder with a carry input of ONE and the block generate signal is the carry out of an adder with a carry input of ZERO.

If a constant block width of k is used and if ripple carry adders constructed from the nine gate full adder of Figure 80.2 are used, there will be n/k blocks and the delay to generate the sum is $2k + 3$ gate delays to form the carry out of the first block, 2 gate delays for each of the $\frac{n}{k} - 2$ intermediate blocks, and 3 gate delays (for the multiplexer) in the final block. The total delay, DELAY_{SEL}, is thus

$$\text{DELAY}_{SEL} = \left(2k + 2\left\lceil \frac{n}{k} \right\rceil + 2\right)\Delta_{gate} \tag{80.16}$$

The optimum block size is found by taking the derivative of DELAY_{SEL} with respect to k, setting it to zero, and solving for k. The result is

$$k = \sqrt{n} \tag{80.17}$$

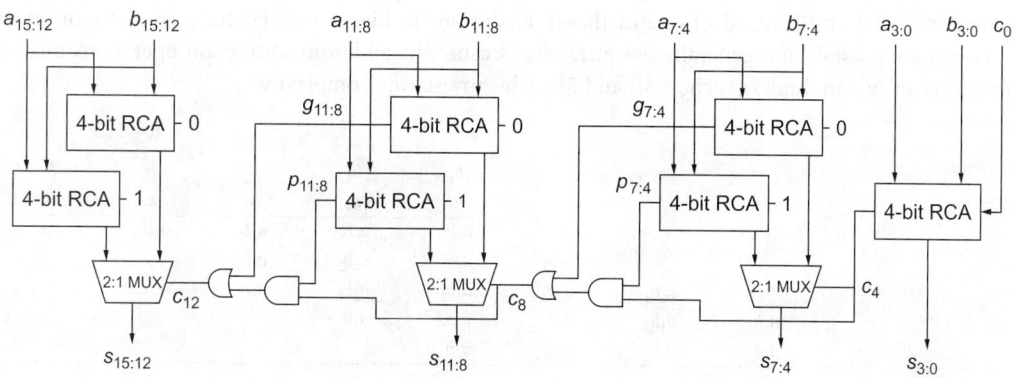

FIGURE 80.6 Sixteen-bit carry select adder.

and

$$\text{DELAY}_{\text{SEL}} = (2 + 4\sqrt{n})\Delta_{\text{gate}} \tag{80.18}$$

Slightly better results can be obtained by varying the width of the blocks. In this case the optimum is to make the two least significant blocks of the same size and make each succeeding block one bit larger. For this configuration, the delay for each block's most significant sum bit will equal the delay to the multiplexer control signal for that block [5, p. A-38].

The complexity of the carry select adder is $2n - k$ ripple carry adder stages, the intermediate carry logic, and $(\lceil \frac{n}{k} \rceil - 1)k$ bit wide 2:1 multiplexers.

$$\text{GATE}_{\text{SEL}} = \left(9(2n - k) + 2\left(\left\lceil \frac{n}{k} \right\rceil - 2\right) + 3(n - k) + \left\lceil \frac{n}{k} \right\rceil - 1 \right) \beta_{\text{gate}}$$

$$= \left(21n - 12k + 3\left\lceil \frac{n}{k} \right\rceil - 5 \right) \beta_{\text{gate}} \tag{80.19}$$

This is somewhat more than twice the complexity of a ripple carry adder.

80.3.2 Multipliers

The bit product matrix of an $n \times n$ (in this case 5×5) multiplier for unsigned operands is shown in Figure 80.7. The two operands, A and B, are shown at the top, then there are n rows (each consisting of n bit products) that comprise the bit product matrix. Finally, the product ($2n$ bits wide) is at the bottom.

There are several ways to implement a multiplier. One of the oldest methods is to use an n bit wide adder to sum the rows of the bit product matrix in a row by row fashion. This can be quite slow as $n - 1$ cycles (each long enough to complete an n bit addition) are required. If a ripple carry adder is used, the time to multiply two n bit numbers is proportional to n^2. If a fast adder such as a carry lookahead adder is used, the time is proportional to $n \log_2(n)$.

Booth multiplier. The Booth multiplier [6] and the modified Booth multiplier are attractive for two's complement multiplication, since they accept two's complement operands, produce a two's complement product and are easy to implement. The sequential Booth multiplier requires n cycles to form the product of a pair of n bit numbers, where each cycle consists of an n-bit addition and a shift, an n-bit subtraction and a shift, or a shift without any other arithmetic operation. The radix-4 modified Booth multiplier [2] takes half the number of cycles as the "standard" Booth multiplier although the operations performed during a cycle are slightly more complex (since it is necessary to select one of five possible addends, namely, $\pm 2B$, $\pm B$, or 0 instead of one of three). Extensions to higher radices that examine more than three bits are possible, but generally not attractive because the addition/subtraction operations involve nonpower of two multiples (such as 3B and 5B) which raises the complexity.

					b_4	b_3	b_2	b_1	b_0
					a_4	a_3	a_2	a_1	a_0
					a_0b_4	a_0b_3	a_0b_2	a_0b_1	a_0b_0
				a_1b_4	a_1b_3	a_1b_2	a_1b_1	a_1b_0	
			a_2b_4	a_2b_3	a_2b_2	a_2b_1	a_2b_0		
		a_3b_4	a_3b_3	a_3b_2	a_3b_1	a_3b_0			
	a_4b_4	a_4b_3	a_4b_2	a_4b_1	a_4b_0				
p_9	p_8	p_7	p_6	p_5	p_4	p_3	p_2	p_1	p_0

FIGURE 80.7 Five-bit by five-bit multiplier for unsigned operands.

						b_4	.	b_3	b_2	b_1	b_0
						a_4	.	a_3	a_2	a_1	a_0
					1	$\overline{a_0b_4}$		a_0b_3	a_0b_2	a_0b_1	a_0b_0
					$\overline{a_1b_4}$	a_1b_3		a_1b_2	a_1b_1	a_1b_0	
				$\overline{a_2b_4}$	a_2b_3	a_2b_2		a_2b_1	a_2b_0		
			$\overline{a_3b_4}$	a_3b_3	a_3b_2	a_3b_1		a_3b_0			
1	a_4b_4		$\overline{a_4b_3}$	$\overline{a_4b_2}$	$\overline{a_4b_1}$	$\overline{a_4b_0}$					
p_9	p_8	.	p_7	p_6	p_5	p_4		p_3	p_2	p_1	p_0

FIGURE 80.8   Five-bit by five-bit multiplier for two's complement operands.

Array multiplier. An alternative approach to multiplication involves the combinational generation of all bit products and their summation with an array of adders. To accommodate two's complement operands, the bit product matrix of Figure 80.7 is modified as shown by Figure 80.8. The modifications include inverting the bit products that comprise the most significant column and the most significant row of the matrix (but not the single bit product at the most significant column and row of the matrix) and adding two ONES to the matrix one in column $(n + 1)$ and one in column $2n$ [7, p. 179].

As shown in Figure 80.8, a product that is $2n$ bits wide is produced with two integer bits. This growth in word size is necessary to accommodate the case of -1^2 (an artifact of the nonsymmetric nature of the two's complement number system) that requires an extra integer column for its representation since $+1$ cannot be represented in the "normal" two's complement fractional number system. In most digital signal-processing applications, an n bit two's complement result is desired. This can be accomplished by removing the $(n-1)$th product bits and the $2n$th product bit. The $n-1$ bits can be removed by either truncation (where the $n-1$ least significant columns are not formed) or by rounding (where the $n-1$ least significant columns are formed, a ONE is added in column $(n-1)$ and after the product is formed, the $n-1$ least significant bits are removed). In either case, column $2n$ is not formed, so the case of attempting to compute -1^2 must be prevented at the system level.

Figure 80.9 shows an implementation of the 5 bit $\times$ 5 bit multiplier for two's complement operands of Figure 80.8. It uses a 5×5 array of cells to form the bit products and four adders (at the bottom of the array) to complete the evaluation of the product. Only a few types of cells are used in this implementation: gate cells (marked AND or NAND) which use a single gate to form the logic AND or NAND of the a and b inputs to the cell, half adder cells (marked HA) which sum the second input to the cell with the logic AND of the a and b inputs to the cell and full adder cells (marked FA) which sum the second and third inputs to the cell with the logic AND (or NAND for the a_4 row next to the bottom of the array) of the a and b inputs to the cell. Standard full adders are used in the bottom row of the multiplier. There is a special modified half adder, marked MHA in the bottom row. It forms the carry and sum corresponding to one plus the carry and sum of the two operands that are input to the MHA. Its position at the right end of the bottom row adds the one shown at the top of the p_5 column of Figure 80.8. The one shown in the p_9 column of Figure 80.8 is not implemented (it is assumed that either a five-bit single-precision product or a nine-bit double-precision product is desired). One of the adders in the next to the top row is a full adder, if the extra input to this adder is a one, a rounded single precision product is available at outputs $p_8 \cdot p_7$, p_6, p_5, and p_4.

This implementation of an $n \times n$ array multiplier requires $2n-1$ gate cells, $n-2$ half adders, 1 modified half adder, and $(n-1)^2$ full adders. Of the half and full adder cells $(n-1)^2$ have an extra AND or NAND gate. Thus the total complexity is

$$\text{GATES}_{\text{ARRAY MPY}} = n^2\beta_{\text{gate}} + (n-1)^2\beta_{\text{FA}} + (n-1)\beta_{\text{HA}} \tag{80.20}$$

If a full adder is comprised of 9 gates and a half adder (either regular or modified) is comprised of 4 gates, the total complexity of an n-bit $\times$ n-bit array multiplier is

$$\text{GATES}_{\text{ARRAY MPY}} = (10n^2 - 14n + 1)\beta_{\text{gate}} \tag{80.21}$$

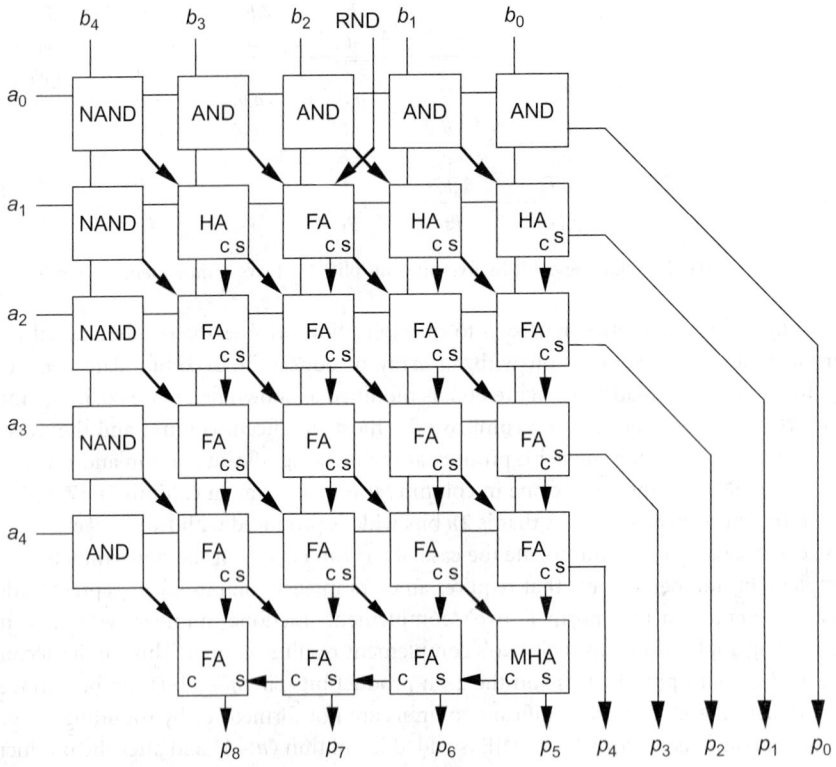

FIGURE 80.9 Two's complement 5-bit × 5-bit array multiplier.

The delay of the array multiplier is evaluated by following the pathways from the inputs to the outputs. The longest path starts at the upper left corner, progresses to the lower right corner, and then across the bottom to the lower left corner. If it is assumed that the delay from any adder input (for either half or full adders) to any adder output is Δ_{adder}, then the total delay of an n-bit × n-bit array multiplier is

$$\text{DELAY}_{\text{ARRAY MPY}} = \Delta_{\text{gate}} + (2n - 2)\Delta_{\text{adder}} \tag{80.22}$$

If the adders in the bottom row of the array multiplier are replaced with an $n - 1$ bit carry lookahead adder, the resulting modified array multiplier is as shown in Figure 80.10. Assuming the use of 4-bit lookahead blocks:

$$\text{GATES}_{\text{MODIFIED ARRAY MPY}} = \left(10n^2 - 9\tfrac{1}{3}n - 4\tfrac{1}{3}\right)\beta_{\text{gate}} \tag{80.23}$$

and from Eq. (80.13):

$$\text{DELAY}_{\text{MODIFIED ARRAY MPY}} = \left(3 + 4\left\lceil \log_4(n - 1)\right\rceil\right)\Delta_{\text{gate}} + (n - 1)\Delta_{\text{adder}} \tag{80.24}$$

Thus, the modified array multiplier is a bit more complex and also faster than the original array multiplier. For a 16-bit × 16-bit multiplier the complexity increases by a few percent while the delay drops by $\tfrac{1}{3}$ or more (depending on the delay of an adder).

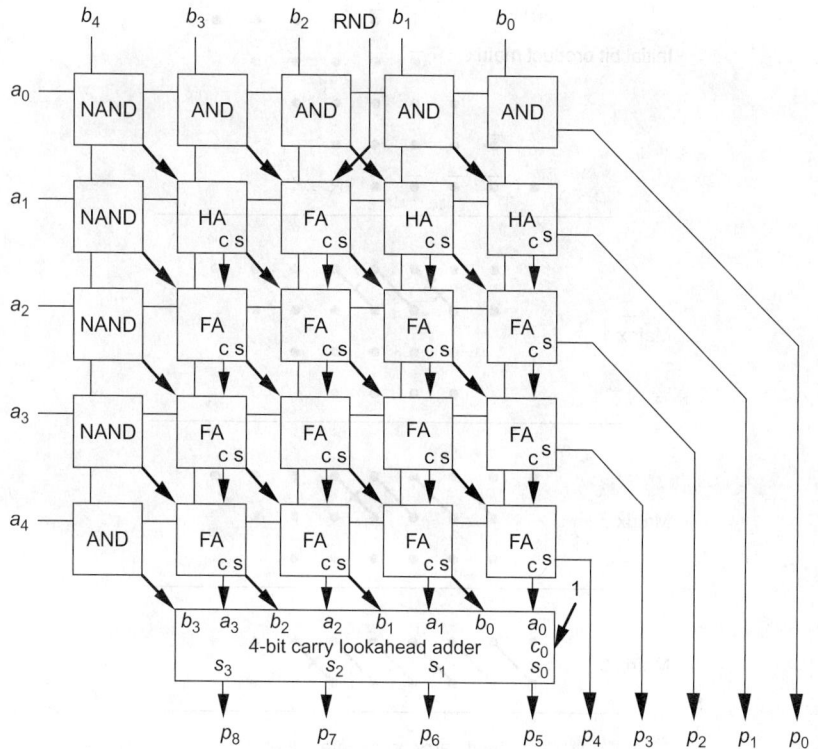

FIGURE 80.10 Two's complement 5-bit × 5-bit modified array multiplier.

Array multipliers are easily laid out in a cellular fashion, making them quite attractive for VLSI implementation, where minimizing the design effort may be more important than maximizing the speed.

Dadda fast multiplier. A method for fast multiplication was developed by Wallace [8] and refined by Dadda [9]. With this method, a three-step process is used to multiply two numbers: (1) the bit products are formed, (2) the bit product matrix is "reduced" to an equivalent two row matrix, where the sum of the two rows equals the sum of the rows in the initial bit product matrix, and (3) the two numbers are summed with a fast adder to produce the product. Although this may seem to be a complex process, it yields multipliers with delay proportional to the logarithm of the operand word size which is "faster" than the Booth multiplier, the modified Booth multiplier, or array multipliers all of which have delays proportional to the word size.

The fast multiplication process is shown for a 5-bit × 5-bit two's complement Dadda multiplier in Figure 80.11. An input matrix of dots (each dot represents a bit product, each dot with an overbar represents the logical complement of the corresponding bit product, the 1 indicates a fixed logic ONE and R indicates a bit that is set to logic ONE when a rounded 5-bit product is desired and is set to logic ZERO when a double-precision 9-bit product is desired) is shown as the initial bit product matrix.

The height of the matrices in the reduction process is determined by working back from the final (two row) matrix and limiting the height of each matrix to the largest integer that is no more than 1.5 times the height of its successor. The first several terms of the sequence are 2, 3, 4, 6, 9, 13, 19, 28, 42, 63, 94,....

Since the height of the initial bit product matrix in Figure 80.11 is 5, columns having more than four dots (or that will grow to more than four dots owing to carries) are reduced by the use of half adders (each half adder takes in two dots from a column of the bit product matrix and outputs one

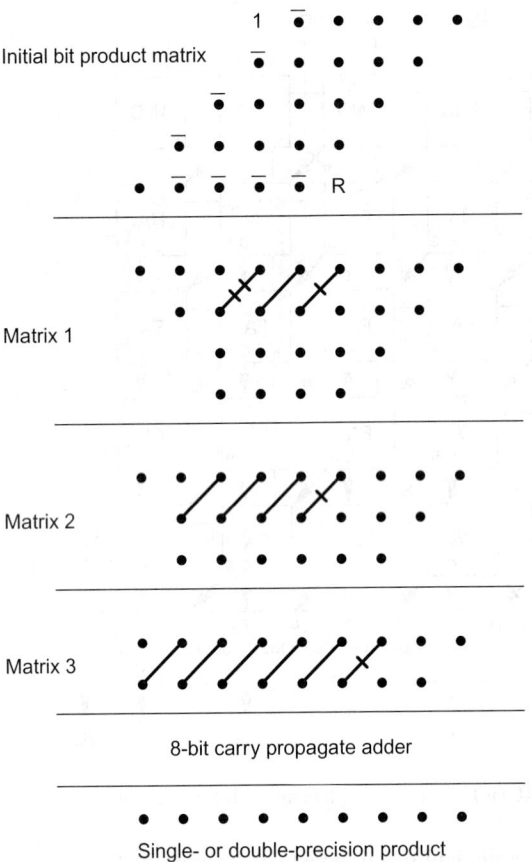

Initial bit product matrix

Matrix 1

Matrix 2

Matrix 3

8-bit carry propagate adder

Single- or double-precision product

FIGURE 80.11   Two's complement 5-bit × 5-bit Dadda multiplier.

dot in the same column of the next matrix and one dot in the next more significant column of the next matrix), a modified half adder and full adders (each full adder takes in three dots from a column of the bit product matrix and outputs one dot in the same column of the next matrix and one dot in the next more significant column of the next matrix) so that no column in the next matrix will have more than four dots.

The outputs of half adders are shown by a "crossed" line in the succeeding matrix, the outputs of the modified half adder are shown by a double crossed line in the succeeding matrix and the outputs of full adders are shown by an uncrossed line in the succeeding matrix. In each case the rightmost dot of the pair that is connected by a line is in the column from which the inputs were taken in the preceding matrix for the adder. In the succeeding steps reduction to Matrix 2 with no more than three dots per column, and finally Matrix 3 with no more than two dots per column is performed.

Each matrix is produced from its predecessor in one adder delay. Since the number of matrices is logarithmically related to the number of rows in the initial bit product matrix which is equal to the number of bits in the words to be multiplied, the delay of the matrix reduction process is proportional to log *n*. Since the carry propagate adder that reduces the final two row matrix can be implemented as a carry lookahead adder (which also has delay that is proportional to the logarithm of the word size), the total delay for this Dadda multiplier is proportional to the logarithm of the word size.

The exact delay of the Dadda multiplier is evaluated by following the pathways from the inputs to the outputs. The longest path starts at the center column, progresses through the successive reduction matrices (there are approximately $\log_{1.44}(n) - 2$ stages), and finally through the $2n - 2$ bit adder. If a carry lookahead adder realized with 4-bit lookahead blocks is used for the carry propagate adder (with delay given by Eq. (80.13)), the total delay of an n-bit by n-bit Dadda multiplier is

$$\text{DELAY}_{\text{Dadda MPY}} = (\log_{1.44}(n) - 2)\Delta_{\text{adder}} + \left(3 + 4\left\lceil \log_r(2n-2)\right\rceil\right)\Delta_{\text{gate}} \tag{80.25}$$

The Dadda multiplier complexity is determined by evaluating the complexity of its parts, n^2 gates ($2n - 2$ are NAND gates, the rest are AND gates) to form the bit product matrix, $(n-2)^2$ full adders, a few (slightly less than n) half adders and one modified half adder for the matrix reduction, and a $2n - 2$ bit carry propagate adder for the addition of the final two-row matrix.

$$\text{GATES}_{\text{Dadda MPY}} = n^2\beta_{\text{gate}} + (n-2)^2\beta_{\text{FA}} + n\,\beta_{\text{HA}} + (2n-2)\beta_{\text{CPA}} \tag{80.26}$$

If a carry lookahead adder (implemented with 4-bit lookahead blocks) is used for the carry propagate adder and if a full adder is comprised of 9 gates and a half adder (either regular or modified) is comprised of 4 gates, then Eq. (80.27) reduces to

$$\text{GATES}_{\text{Dadda MPY}} = (10n^2 - 6\tfrac{2}{3}n - 26)\beta_{\text{gate}} \tag{80.27}$$

Figure 80.12 shows a 16-bit × 16-bit two's complement Dadda multiplier. The reduction becomes concentrated in the center columns in the first stage, but it broadens and becomes more uniform as it progresses through the subsequent stages.

Wallace fast multiplier. The Wallace multiplier [8] is similar to the Dadda multiplier described above, except that a different approach is used to reduce the bit product matrix. In Dadda's approach the minimum reduction is done to each matrix to cause the succeeding matrix height to conform to the sequence of matrix heights. In the Wallace approach, the maximum reduction is done at each stage. This is illustrated by a 16-bit by 16-bit two's complement Wallace multiplier shown in Figure 80.13.

It is clear from Figure 80.13 that the Wallace approach does most of its reduction in the first few stages. For example, 113 of the 201 full adders are used in the first two stages. For a given word size the width of the carry propagate adder of the Wallace multiplier is smaller than that of the Dadda multiplier (if implemented with a carry lookahead adder, for some word sizes the final adder for the Dadda multiplier will require an extra level of lookahead logic). The number of reduction stages is the same for Wallace and Dadda multipliers, although the height of the matrices is not the same at all stages. For most sizes the delay is the same for Wallace and Dadda multipliers (for some word sizes the Dadda multiplier will require an extra level of lookahead logic for its carry propagate adder adding an extra four gate delays). The Wallace multiplier uses more full adders than the Dadda multiplier does, but its carry propagate adder is smaller by the same amount, so that the combined adder complexity is the same. Although the Wallace multiplier does use quite a few more half adders, its complexity is only slightly greater than that of Dadda multipliers.

80.3.3 Dividers

There are two major types of dividers: digit recurrence dividers that compute a fixed number of bits of the quotient on each iteration and functional approximation dividers that compute an increasingly accurate approximation to the quotient on each iteration.

Digit recurrence dividers use a sequence of shift, subtract, and compare operations to compute the quotient. The comparison operation is significant: it results in an inherently serial process which is not amenable to parallel implementation.

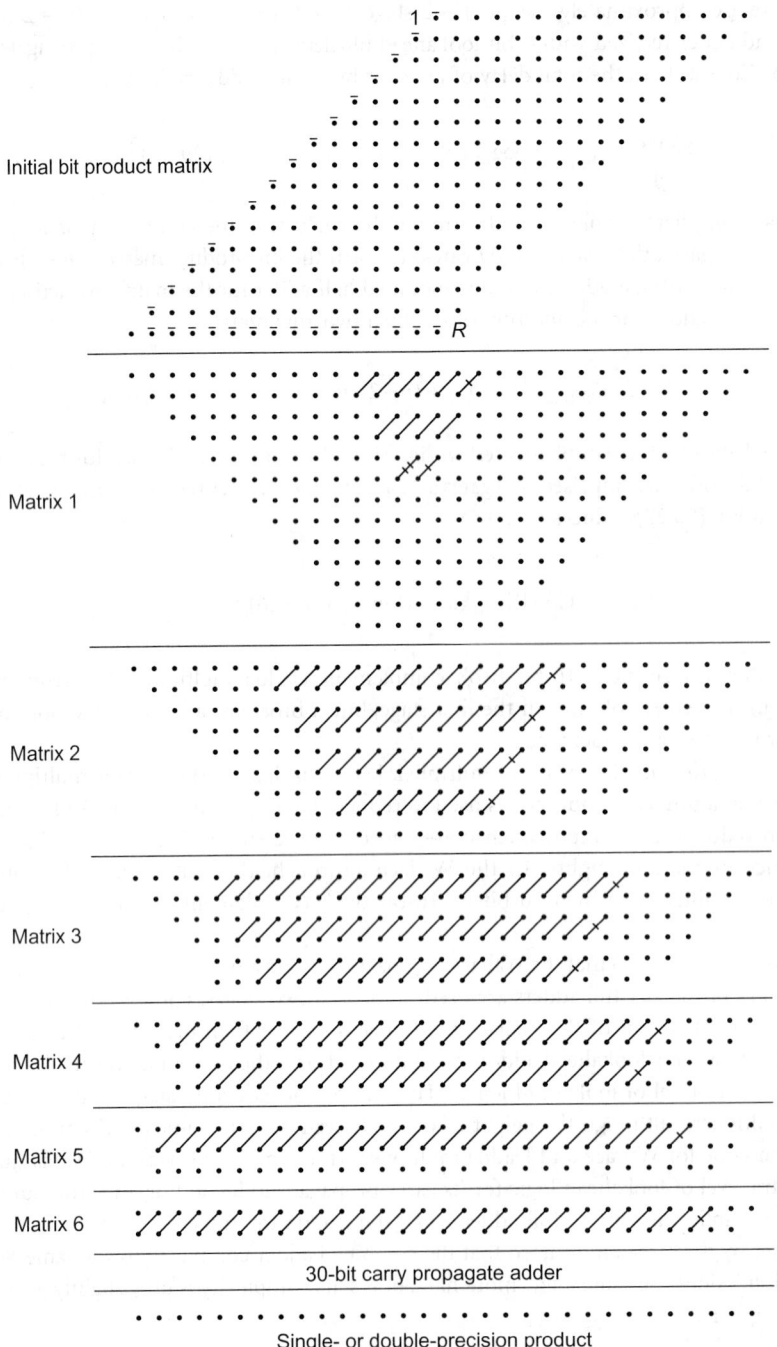

FIGURE 80.12 Two's complement 16-bit × 16-bit Dadda multiplier.

The digit recurrent algorithms [10] are based on selecting digits of the quotient Q (where $Q = (N/D)$) to satisfy the following equation:

$$P_{k+1} = rP_k - q_{n-k-1}D \quad \text{for } k = 1, 2, \ldots, n-1 \tag{80.28}$$

where P_k is the partial remainder after the selection of the kth quotient digit, $P_0 = N$ (subject to the constraint $|P_0| < |D|$), r is the radix, q_{n-k-1} is the kth quotient digit to the right of the binary point, and D

$$P_0 = N = \tfrac{5}{8}$$

$$D = \tfrac{7}{8}$$

$$n = 4$$

$k=0, TP_1 = 2P_0 - D = \tfrac{3}{8}$ since $TP_1 > 0$, then $q_3 = 1$ and $P_1 = TP_1 = \tfrac{3}{8}$

$k=1, TP_2 = 2P_1 - D = \tfrac{-1}{8}$ since $TP_2 < 0$, then $q_2 = 0$ and $P_2 = TP_2 + D = \tfrac{3}{4}$

$k=2, TP_3 = 2P_2 - D = \tfrac{5}{8}$ since $TP_3 > 0$, then $q_1 = 1$ and $P_3 = TP_3 = \tfrac{5}{8}$

$k=3, TP_4 = 2P_3 - D = \tfrac{3}{8}$ since $TP_4 > 0$, then $q_0 = 1$ and $P_4 = TP_4 = \tfrac{3}{8}$

$$Q = 0.1011$$

FIGURE 80.15 Example of restoring division.

$$\text{If } -0.5 < P_k < 0.5, q_{n-k-1} = 0 \quad \text{and} \quad P_{k+1} = 2P_k \tag{80.30}$$

$$\text{If } P_k \geq 0.5, q_{n-k-1} = -1 \quad \text{and} \quad P_{k+1} = 2P_k + D \tag{80.31}$$

The basic scheme is shown in Figure 80.16. Block 1 initializes the algorithm. In steps 3 and 5 P_k, is compared to ± 0.5. If $P_k \geq 0.5$, in step 4 the quotient digit is set to 1 and $P_{k+1} = 2P_k - D$. If $P_k \leq -0.5$, in step 6 the quotient digit is set to -1 and $P_{k+1} = 2P_k + D$. If the value of P_k is between -0.5 and 0.5, step 7 sets $P_{k+1} = 2P_k$. Finally, step 8 tests whether all bits of the quotient have been formed and goes to step 2 if more need to be computed. Each pass through steps 2–8 forms one digit of the quotient. As shown in Figure 80.16, each pass through steps 2–8 requires one or two comparisons in steps 3 and 5, and may require an addition (in step 4 or step 6). Thus computing an n-bit quotient will involve up to n additions and from n to $2n$ comparisons. In practice, most implementations use logic to inspect the top few bits of P_k so the comparisons that are shown in series in Figure 80.16 are performed in parallel in a single step.

The signed digit number (comprised of ± 1 digits) can be converted into a conventional binary number by subtracting, NEG, the number formed by the $\bar{1}$s (with ZEROS where there are +ONES or ZEROS and ONES where there are $\bar{1}$s in Q) from, POS, the number formed by the +ONES (with ONES where there are +ONES and ZEROS where there are $\bar{1}$s or ZEROS in Q). For example:

$$Q = 0.11\bar{1}\bar{1}01 = \tfrac{37}{64}$$

$$POS = 0.110001$$

$$NEG = 0.001100$$

$$Q = 0.110001 - 0.001100$$

$$Q = 0.110001 + 1.110100$$

$$Q = 0.100101 = \tfrac{37}{64}$$

Higher radix SRT divider. The higher radix SRT division process is similar to the binary SRT algorithms. Radix 4 is the most common higher radix SRT division algorithm with either the minimally redundant digit set of $\{\pm 2, \pm 1, 0\}$ or the maximally redundant digit set of $\{\pm 3, \pm 2, \pm 1, 0\}$. The operation of the algorithm is similar to the binary SRT algorithm shown in Figure 80.16 except that P_k and Q are applied

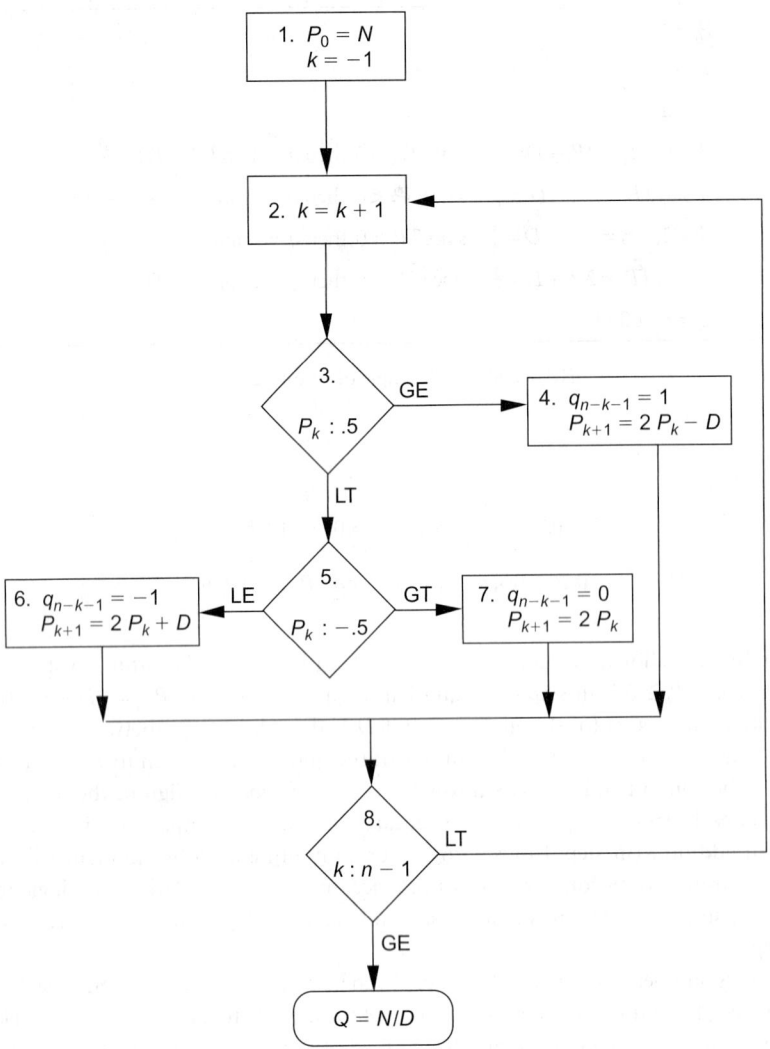

FIGURE 80.16 Binary SRT division.

to a lookup table or a programmable logic array to determine the quotient digit. A research monograph provides a detailed treatment of SRT division [10].

Newton–Raphson divider. A second division technique uses a form of Newton–Raphson iteration to derive a quadratically convergent approximation to the reciprocal of the divisor that is then multiplied by the dividend to produce the quotient. In systems which include a fast multiplier, this process is may be faster than digit recurrence division [10].

The Newton–Raphson division algorithm to compute $Q = \dfrac{A}{B}$ consists of three basic steps:

(1) Calculate a starting estimate of the reciprocal of the divisor, X_0. If the divisor, B, is normalized (i.e., $\frac{1}{2} \le B < 1$), then $X_0 = 3 - 2B$ is exactly correct at $B = 0.5$ and $B = 1$ and exhibits maximum error (of approximately 0.17) at $B = \sqrt{\frac{1}{2}}$. Adjusting X_0 downward by half the maximum error gives

$$X_0 = 2.915 - 2B \qquad (80.32)$$

This produces an initial estimate that is within about 0.087 of the correct value for all points in the interval $\frac{1}{2} \leq B < 1$.

(2) Compute successively more accurate estimates of the reciprocal by the following iterative procedure:

$$X_{i+1} = X_i(2 - BX_i) \quad \text{for } i = 0,1, \ldots, k \tag{80.33}$$

(3) Form the quotient by multiplying the dividend times the reciprocal of the divisor.

$$Q = A X_{k+1} \tag{80.34}$$

Figure 80.17 illustrates the operation of the Newton–Raphson algorithm. With this algorithm, the error decreases quadratically, so that the number of correct bits in each approximation is roughly twice the number of correct bits on the previous iteration. Thus, from a roughly four-bit initial approximation from Eq. (80.32), two iterations of Eq. (80.33) produce a reciprocal estimate accurate to about 16 bits, three iterations produce a reciprocal estimate accurate to about 32 bits, and four iterations produce a reciprocal estimate accurate to about 64 bits.

The efficiency of this process is dependent on the availability of a fast multiplier, since each iteration of Eq. (80.33) requires two multiplications and a subtraction. The complete process for the initial estimate, two iterations, and the final quotient determination requires three subtracts and five multiples to produce a 16-bit quotient. This is faster than a conventional nonrestoring divider if multiplication is roughly as fast as addition, a condition which may be satisfied for systems which include hardware multipliers.

```
A = .625
B = .75

X₀ = 2.915 − 2 • B                                      1 Subtract
   = 2.915 − 2 • .75
X₀ = 1.415

X₁ = X₀(2 − B X₀)                                       2 Multiplies, 1 Subtract
   = 1.415(2 − .75 • 1.415)
   = 1.415 • .95875
X₁ = 1.32833125

X₂ = X₁(2 − B X₁)                                       2 Multiplies, 1 Subtract
   = 1.32833125(2 − .75 • 1.32833125)
   = 1.32833125 • 1.00375156
X₂ = 1.3333145677

X₃ = X₂(2 − B • X₂)                                     2 Multiplies, 1 Subtract
   = 1.3333145677(2 − .75 • 1.3333145677)
   = 1.3333145677 • 1.00001407
X₃ = 1.3333333331

Q = A • X₃                                              1 Multiply
Q = .83333333319
```

FIGURE 80.17 Example of Newton–Raphson division.

80.4 Floating-Point Arithmetic

Recent advances in VLSI have increased the feasibility of hardware implementations of floating-point arithmetic units. The main advantage of floating-point arithmetic is that its wide dynamic range virtually eliminates overflow for most applications.

Floating-point number systems. A floating-point number, A, consists of a significand (also called a fraction or a mantissa), S_a, and an exponent, E_a. The value of a number, A, is given by the equation

$$A = S_a r^{E_a} \tag{80.35}$$

where r is the radix (or base) of the number system. Use of the binary radix (i.e., $r = 2$) gives maximum accuracy, but may require more frequent normalization than higher radices.

The IEEE standard 754 single-precision (32-bit) floating-point format which is widely implemented, has an 8-bit biased integer exponent which ranges between 0 and 255 [11]. The exponent is expressed in excess 127 code so that its effective value is determined by subtracting 127 from the stored value. Thus, the range of effective values of the exponent is −127 to 128, corresponding to stored values of 0 to 255, respectively. A stored exponent value of ZERO (E_{min}) serves as a flag for ZERO (if the significand is ZERO) and for denormalized numbers (if the significand is non-ZERO). A stored exponent value of 255 (E_{max}) serves as a flag for Infinity (if the significand is ZERO) and for "Not a Number" (if the significand is non-ZERO). The significand is a 25-bit sign magnitude mixed number (the binary point is to the right of the most significant bit). The most significant bit is always a ONE except for denormalized numbers. More detail on floating-point formats and on the various considerations that arise in the implementation of floating-point arithmetic units are given in [7,12].

Floating-point addition. A flow chart for floating-point addition is shown in Figure 80.18. For this flowchart, the operands are assumed to be "unpacked" and normalized with magnitudes in the range [1, 2]. On the flow chart, the operands are (E_a, S_a) and (E_b, S_b), the result is (E_s, S_s), and the radix is 2. In step 1 the operand exponents are compared; if they are unequal, the significand of the number with the smaller exponent is shifted right in step 3 or 4 by the difference in the exponents to properly align the significands. For example, to add the decimal operands 0.867×10^5 and 0.512×10^4, the latter would be shifted right by 1 digit and 0.867 added to 0.0512 to give a sum of 0.9182×10^5. The addition of the significands is performed in step 5. Steps 6–8 test for overflow and correct if necessary by shifting the significand one position to the right and incrementing the exponent. Step 9 tests for a zero significand. The loop of steps 10–11 scales unnormalized (but non-ZERO) significands upward to normalize the result. Step 12 tests for underflow.

Floating-point subtraction is implemented with a similar algorithm. Many refinements are possible to improve the speed of the addition and subtraction algorithms, but floating-point addition will, in general, be much slower than fixed-point addition as a result of the need for preaddition alignment and postaddition normalization.

Floating-point multiplication. The algorithm for floating-point multiplication forms the product of the operand significands and the sum of the operand exponents. For radix 2 floating-point numbers, the significand values are ≥ 1 and < 2. The product of two such numbers will be ≥ 1 and < 4. At most a single right shift is required to normalize the product.

Floating-point division. The algorithm for floating-point division forms the quotient of the operand significands and the difference of the operand exponents. The quotient of two normalized significands will be ≥ 0.5 and < 2. At most a single left shift is required to normalize the quotient.

Floating-point rounding. All floating-point algorithms may require rounding to produce a result in the correct format. A variety of alternative rounding schemes have been developed for specific applications. Round to the nearest, round toward plus infinity, round toward negative infinity, and round toward ZERO are required for implementations of the IEEE floating-point standard.

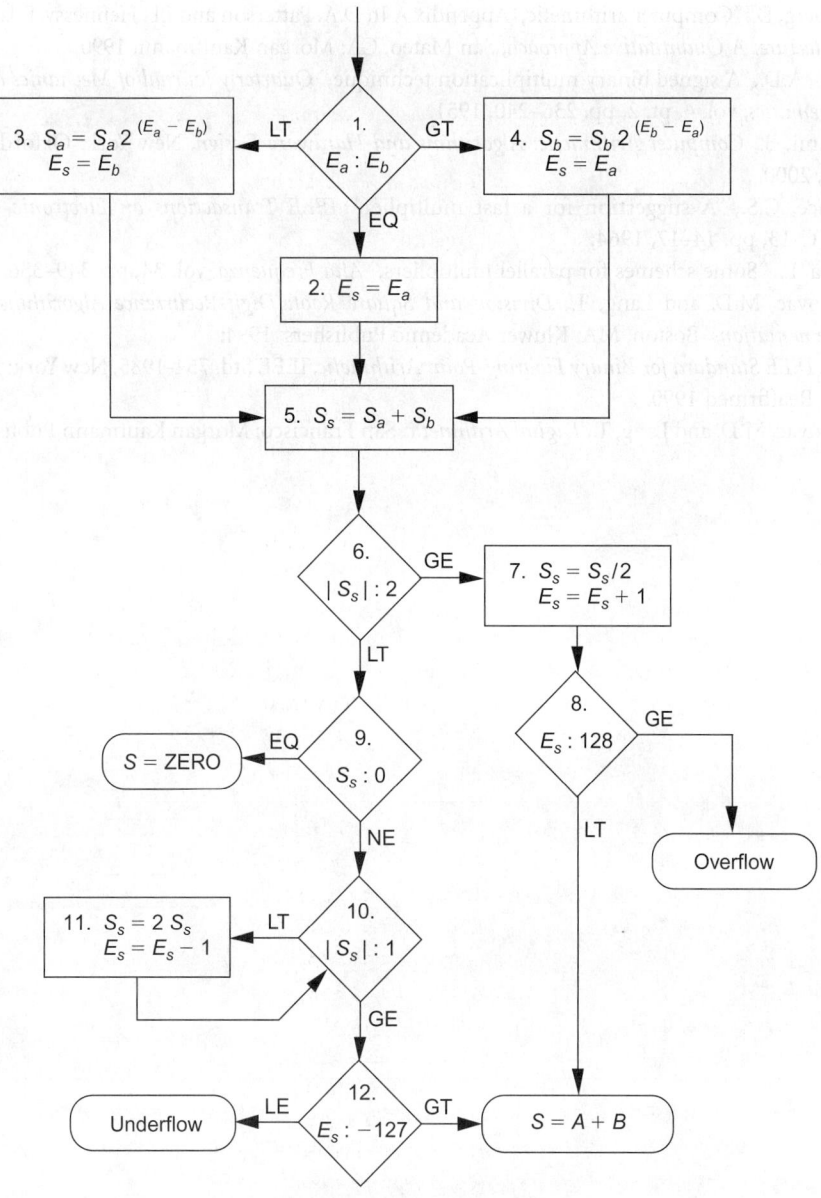

FIGURE 80.18 Floating-point addition.

References

1. Weinberger, A. and Smith, J.L., "A logic for high-speed addition," *National Bureau of Standards Circular 591*, Washington, DC: National Bureau of Standards, pp. 3–12, 1958.
2. MacSorley, O.L., "High-speed arithmetic in binary computers," *Proceedings of the IRE*, vol. 49, pp. 67–91, 1961.
3. Waser, S. and Flynn, M.J., *Introduction to Arithmetic for Digital Systems Designers*, New York: Holt, Rinehart and Winston, 1982.
4. Bedrij, O.J., "Carry-select adder," *IRE Transactions on Electronic Computers*, vol. EC-11, pp. 340–346, 1962.

5. Goldberg, D., "Computer arithmetic," Appendix A In D.A. Patterson and J.L. Hennessy, Eds., *Computer Architecture: A Quantitative Approach*, San Mateo, CA: Morgan Kauffmann, 1990.

6. Booth, A.D., "A signed binary multiplication technique," *Quarterly Journal of Mechanics and Applied Mathematics*, vol. 4, pt. 2, pp. 236–240, 1951.

7. Parhami, B., *Computer Arithmetic: Algorithms and Hardware Design*, New York: Oxford University Press, 2000.

8. Wallace, C.S., "A suggestion for a fast multiplier," *IEEE Transactions on Electronic Computers*, vol. EC-13, pp. 14–17, 1964.

9. Dadda, L., "Some schemes for parallel multipliers," *Alta Frequenza*, vol. 34, pp. 349–356, 1965.

10. Ercegovac, M.D. and Lang, T., *Division and Square Root: Digit-Recurrence Algorithms and Their Implementations*, Boston, MA: Kluwer Academic Publishers, 1994.

11. IEEE, *IEEE Standard for Binary Floating-Point Arithmetic*, IEEE Std. 754-1985, New York: IEEE Press, 1985, Reaffirmed 1990.

12. Ercegovac, M.D. and Lang, T., *Digital Arithmetic*, San Francisco: Morgan Kaufmann Publishers, 2004.

81

VLSI Architectures for JPEG 2000 EBCOT: Design Techniques and Challenges

Yijun Li
University of Louisiana at Lafayette

Magdy Bayoumi
University of Louisiana at Lafayette

CONTENTS

81.1 Introduction

In January 2001, JPEG 2000 was introduced by ISO/IEC JTC1/SC20/WG1 as a new image compression standard [1]. This new standard supports a rich set of features that are not available in other JPEG standards, such as excellent low bit-rate performance, both lossy and lossless encoding in one algorithm, random code-stream access, precise single-pass rate control, region-of-interest coding, and improved error resiliency. All the above features intend to meet the continual expansion of multimedia and Internet applications, from consumer applications (e.g., digital cameras and PDA) to client/server communication, including both wired and wireless. More specially, to achieve the efficient transmission of JPEG 2000 imagery over an error-prone wireless network, JPEG 2000 wireless (JPWL), i.e., Part 11 of JPEG 2000 standard, extends the elements in the core coding system described in Part 1 of JPEG 2000 standard with mechanisms for error protection and correction. Besides the support of still images, the support of motion sequences is described in Motion JPEG 2000 (Part 3 of the JPEG 2000 standard) [2]. Motion JPEG 2000 preserves the best image quality after compression for use in high-quality images systems like medical imaging systems, HDTV cameras, or digital cinema cameras.

The block diagram of JPEG 2000 is shown in Figure 81.1. The original image data are divided into nonoverlapping rectangular tiles. Then either (5,3) discrete wavelet transform (DWT) supporting lossless compression or (9,7) DWT supporting lossy compression is performed on the tiles by filtering each row and column of the image tiles with a high-pass and low-pass filter. Filtering the image in the DWT phase creates a set of DWT subbands (LL, HL, LH, and HH). If lossy compression is chosen, the wavelet coefficients in DWT subbands are scalar-quantized. Each wavelet subband is divided into code blocks. Then the wavelet coefficients in code blocks are entropy coded by using EBCOT tier-1 algorithm. Finally, data ordering and rate control organize the compressed data into a feature-rich codestream, i.e., the compressed image.

EBCOT tier-1, the entropy encoder of JPEG 2000 standard, is complicated, full of bit operations, and cannot be implemented efficiently in software. To evaluate performance of EBCOT tier-1, profiling techniques are used to evaluate software implementation of JPEG 2000 standard. The run-time profile for JPEG 2000 encoder VM 7.2 is shown in Table 81.1 [3]. The results show that the EBCOT tier-1 algorithm is a huge time-consuming part (typically more than 50%) [3,4].

As a result, EBCOT tier-1 becomes a bottleneck for tremendous data throughput that are often required by multimedia systems, especially, real-time applications. While system performance is essential for the

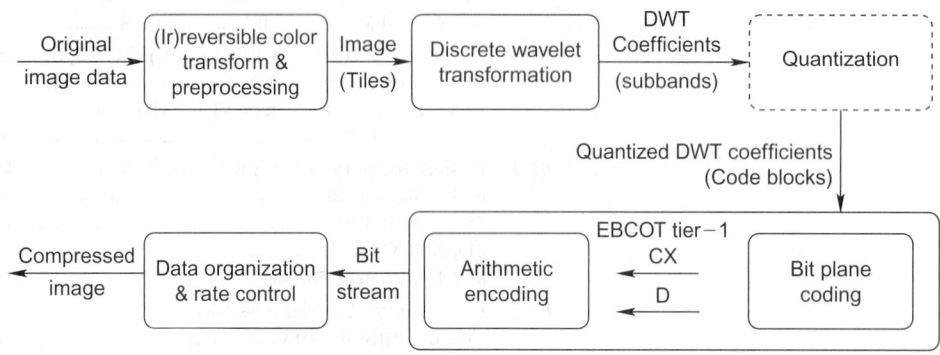

FIGURE 81.1   Block Diagram of JPEG2000.

TABLE 81.1   Run-Time Profile for JPEG 2000 VM 7.2 [5]

Image	Gray Scale		Color Scale	
Operation	Lossless	Lossy	Lossless	Lossy
Color Transform	NA	NA	0.91%	14.12%
DWT (%)	10.81	26.38	11.90	23.97
Quantization	NA	6.42%	NA	5.04%
EBCOT tier-1 (%)	71.63	52.26	69.29	43.85
Pass 1 (%)	14.89	14.82	13.90	12.39
Pass 2 (%)	10.85	7.00	10.94	5.63
Pass 3 (%)	26.14	16.09	25.12	13.77
A.E. (%)	19.75	14.35	19.33	12.06
EBCOT tier-2 (%)	17.56	14.95	17.90	13.01

Image size 1792 × 1200, five-level wavelet decomposition, one layer, profiling platform is a PIII-733 PC with 128 MRAM, Microsoft Visual C++ 6.0 and Windows ME.

real-time application, reducing power consumption has become more and more important in system design. With technology scaling, power density dramatically increases, leading to higher temperature that reduces noise immunity and system reliability. For example, hot-spots may malfunction, even destroy devices. Also, battery capacity is far behind the speed of digital circuit integration. With limited battery capacity, higher power consumption shortens battery life when battery life is expected to be as long as possible for portable system. Considering that specific hardware design is usually power-efficient, bit-based operations are more suitable for hardware implementation, EBCOT tier-1 hardware implementation could be a possible solution for all the bottleneck mentioned above.

In contrast, optimizing the individual components only may lead the overall encoding system to suffer from performance degradation [6,7], because different components have different I/O bandwidths and buffers. Memory issues in JPEG 2000 system illustrate this. In general, the larger the tile size parameter to perform JPEG 2000 compression, the higher is the compression ratio. But more memory is required. The tile memory occupies more than 50% of area in conventional JPEG 2000 architecture [8]. These bottlenecks are mainly caused by the different coding flow between the DWT and EBCOT tier-1 processes since the DWT process requires an entire time memory to carry out the subband transformation [9] and EBCOT tier-1 divides each subband into several code blocks and performs entropy coding. So EBCOT tier-1 hardware design should be considered under the context of the entire JPEG 2000 system.

EBCOT tier-1 hardware architectures can be divided into two categories: parallel bit-plane coding scheme (ParaBCS), where all bit-planes in a code block are coded in parallel, and serial bit-plane coding scheme (SeriBCS), where all bit-planes in a code-block are coded bit-plane by bit-plane. ParaBCS has more parallelism and does not need state memories required by SeriBCS. In either ParaBCS or SeriBCS, the architecture comprises of Bit-plane Coding (BC), Arithmetic Encoding (AE), and first-in, first-out (FIFO) buffer that connects BC with AE and balances the different throughputs between them. To clearly understand these VLSI architectures, several design techniques for BC, FIFO, and AE will be discussed. Then, two case studies are presented: one case study is based on SeriBCS and the other on ParaBCS.

This chapter is organized as follows. Section 81.2 describes EBCOT tier-1 algorithm in detail. Section 81.3 compares SeriBCS and ParaBCS and presents various VLSI architectures for EBCOT in JPEG 2000. Design techniques used by these architectures are discussed in Section 81.4. Two case studies are introduced in Section 81.5 and Section 81.6. Finally, Section 81.7 concludes this chapter.

81.2 EBCOT Tier-1 Algorithm in JPEG 2000

As shown in Figure 81.1, EBCOT tier-1 algorithm [1] is divided into two phases: BC phase and adaptive AE phase. BC phase uses the neighbor information of the current bit to construct context information, i.e., context (CX) and decision (D). Following BC, AE uses CX to estimate the probability of occurrence of D. With this probability, AE adaptively encodes the decision (D) bit by bit to achieve high coding efficiency.

81.2.1 Bit-plane Coding

The DWT coefficients in a code-block memory are stored as binary numbers. Each binary number can be seen as a sequence of bits with position index from n to 1, where n is the number of bits and the binary number has n bits. All the bits at the same position compose a bit-plane. For example, all the least significant bits (LSBs) of the DWT coefficients in a code-block compose a bit-plane because they are on the same position 1. Each bit plane is divided into stripes that are continuous four rows of a bit-plane, i.e., a bit-plane with size $N \times N$ should have $N/4$ stripes. So one column of a stripe has four bits. Bit-plane, stripe, column, and bit are associated with coding order (pass). There are three passes in bit-plane coding phase: significant pass (pass 1), magnitude refinement pass (pass 2) and clean-up pass (pass 3). Each pass should scan a bit-plane in order: bit by bit, column by column, and stripe by stripe. Then a code block is scanned bit-plane by bit-plane. Figure 81.2 provides a topview (Figure 81.2[a]) and a sideview (Figure 81.2[b]) for a code block and scanning order of passes.

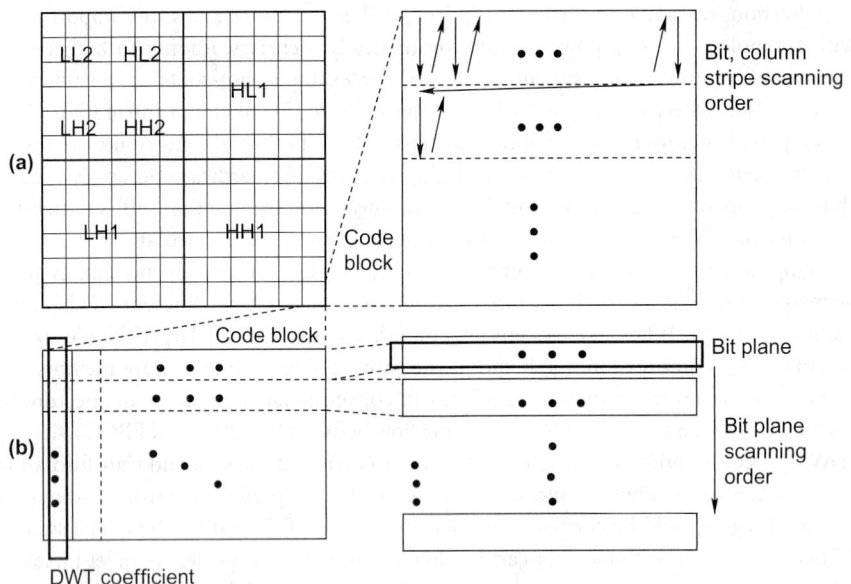

FIGURE 81.2   EBCOT coding order of a code block.

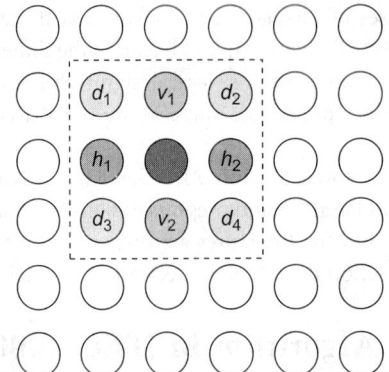

FIGURE 81.3   Calculation of NC (*D*, *H*, and *V*).

A DWT coefficient is associated with a significance state (SigState). Before starting the three passes, all the SigStates of a code-block are initialized as insignificant. During pass scanning, the SigState of a wavelet coefficient is changed from the insignificant state to the significant one whenever the most significant bit (MSB) of the wavelet coefficient is coded. So all the bits of a DWT coefficient before MSB coding are considered insignificant for all the three passes and all the bits of a DWT coefficient after MSB coding are considered significant for all the three passes. As shown in Figure 81.3, each bit has 8 neighbors, i.e., $d_1 - d_4$, v_1, v_2, h_1, h_2, where *d*, *h*, *v* stand for diagonal bits, horizontal bits, and vertical bits, respectively. The SigState of one bit and its 8 neighbors determine which pass a bit is coded in. If one bit's SigState is insignificant, the bit is insignificant. Otherwise, the bit is significant. If a bit is insignificant and any neighbor is significant, the bit is coded in pass 1. If a bit is significant, the bit is coded in pass 2. The rest of all the bits are coded in pass 3. During the scanning of the three passes, any bit can only be coded in one of three passes (pass 1, pass 2, or pass 3). Besides, the SigState of one bit and its 8 neighbors are used to calculate both CX and D through primitive encoders. Primitive encoders use neighbor contributions (NC) (*D, H, V*) that are calculated by using the formula, $D = d_1 + d_2 + d_3 + d_4$; $H = h_1 + h_2$; $V = v_1 + v_2$.

81.2.2   AE Algorithm

The BC phase provides AE phase with the context information (CX and D). CX represents the significance information for one bit and its neighbors. Depending on the bit's location in a bit-plane, the bit and its neighbors may have various SigState. So 19 types of CX are used to indicate the different combinations of these SigStates. D is 1-bit symbol (0 or 1) that could be the current bit value or the binary results from run-length coding. CX is used to adaptively tune the probability at which a certain D (0 or 1) occurs under the CX-indexed neighbor significance conditions. By using this probability, D is encoded into codeword. The coding process goes as follows. Initially, the current probability interval is initialized to 1. Then, the current probability interval is partitioned into two subintervals: more probable subinterval (MPS) and less probable subinterval (LPS). One of the subintervals will be selected to initialize the current probability interval for the next iteration and the codeword is modified so that it points to the base of the selected probability subinterval; in other words, the codeword is represented by the base of the selected probability subinterval. Since the coding process involves addition of probability subinterval (binary fraction) rather than concatenation of binary decision, the more probable binary decisions can often be coded at a cost of much less than one bit per decision. The accuracy to estimate the probability of occurrence of D directly affects AE coding efficiency. To adaptively estimate the probability at which D takes place, a finite-state machine (FSM) is defined in JPEG 2000 standard. The FSM defines an estimated probability for each state, the transition from one state to the next state, and how to alternate the current MPS status. Even though only one FSM is defined, each type of CX has its own FSM instance. Only CX in the same type shares one FSM instance and can affect states of the FSM instance. So totally, 19 FSM instances are needed for all CXs.

The AE algorithm described above is shown in more detail in Figure 81.4, where A is the current interval size, C the codeword, Q_e the estimated probability, MPS the current MPS status and $\wedge$ means XOR operation. Probability (Q_e) and MPS status are searched in FSM by CX indexing. Q_e and MPS status are used to update A and C. After A and C are updated, renormalization procedure is executed, i.e., left-shifting both A and C until the MSB of A is located in the leftmost position. During renormalization procedure, the bits shifted out are stored in a buffer and the number of bit shifting operations is counted by a counter. Whenever the counter is counted down to 0, the counter restores its initial value, the byte-out procedure is called, and a byte is outputted with bit-stuffing techniques. Then the next CX is fetched and processed using the same procedure above.

81.2.3   AE Termination

Since A and C accumulate a history of all the coding bits, AE is a strictly serial process. Hence, AE termination is used to remove the dependency of two bit-streams to favor error-resilient bit-stream. By using various AE terminations, EBCOT tier-1 coding can be processed in either serial pass mode (SeriPM) or parallel pass mode (ParaPM). If coding in SeriPM, three passes scan a bit-plane in serial. After all the bit planes of a code-block are scanned, AE terminates the bit stream. If coding in ParaPM, three passes scan a code-block in a manner similar to the serial mode. But AE terminates the bit stream at the end of each pass to reduce the dependency among passes. To remove the dependency of the current stripe coding on the next stripe, stripe-causal mode is adopted, i.e., the NC from the next stripe are always considered as 0. This termination pattern and stripe-causal mode remove the dependency between passes completely. As a result, ParaPM may allow three passes to scan a bit-plane simultaneously, since the AE probability intervals of three passes do not depend on each other. The parallel mode is more error-resilient, although its performance is slightly poorer than the serial mode (average degradation of PSNR is only 0.19 dB) [10,11].

81.2.4   ParaPM versus SeriPM

In ParaPM, three passes scan a bit-plane in parallel. In SeriPM, three passes scan a bit-plane in serial. Figure 81.5 shows pseudocodes for the coding order and illustrates the difference between ParaPM and SeriPM.

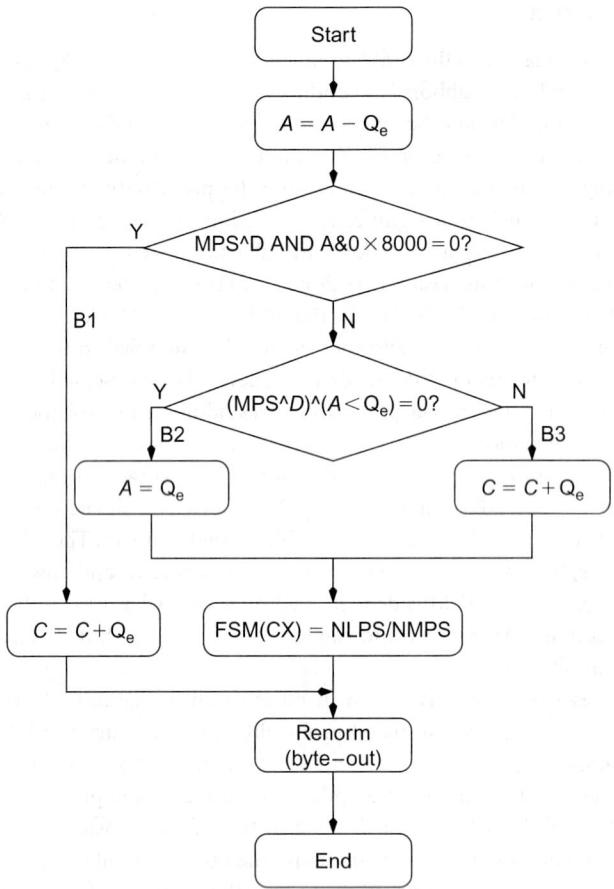

FIGURE 81.4 AE algorithm flow chart.

```
for code-blocks{
    for bit-planes {
        for passes    {
            for columns{
                for bits    {
                    bit coding
}}}}}
```
(a)

```
for code-blocks{
    for bit-planes {
        for columns {
            for bits    {
                for passes {
                    bit coding
}}}}}
```
(b)

FIGURE 81.5 Coding order for ParaPM and SeriPM. (a) Serial pass mode. (b) Parallel pass mode.

ParaPM provides with more parallelism, since three passes could be coded simultaneously instead of a serial process of three passes. ParaPM is also more error-resilient. That is an important characteristic in wireless application. In ParaPM, AE terminates each pass on each bit-plane, while in SeriPM AE terminates when a code-block has finished coding. Therefore, in ParaPM, errors can only spread in smaller space (one pass) instead of one code-block in SeriPM once errors occur in a coding stream. However, all these advantages accrue in ParaPM at the cost of degradation of PSNR performance.

To evaluate this degradation of PSNR performance, the differences in PSNR versus bit-rate between ParaPM and SeriPM are shown in Figure 81.6, when integer mode and various block sizes are applied. The results show that the bigger the block size, the better is the performance for ParaPM and SeriPM.

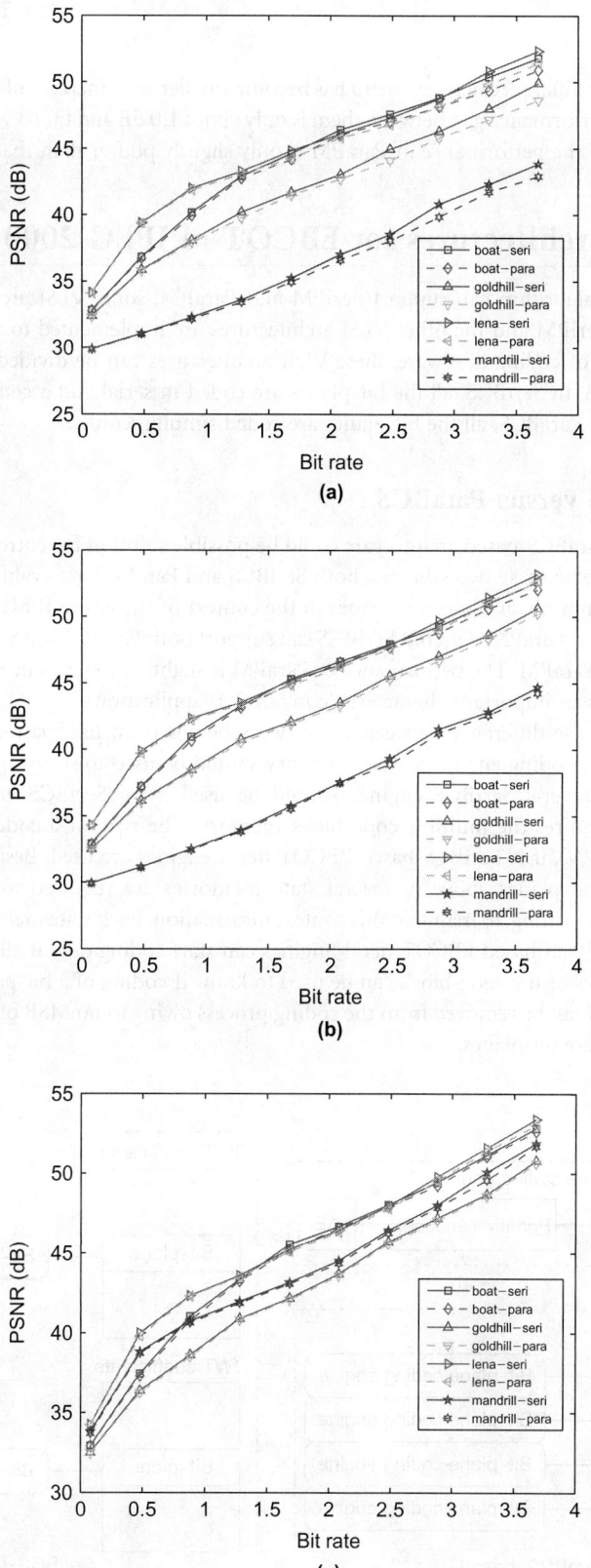

FIGURE 81.6 Comparison between SeriPM and ParaPM. (a) 16 × 16 code-block. (b) 32 × 32 code-block. (c) 64 × 64 code-block.

Also, the performance difference between them has become smaller with increase of block size. In 16×16 code block size, the performance gap between them is only about 1.0 dB and for 64×64, the performance gap is only 0.2 dB. So the performance of ParaPM is only slightly poorer than that of SeriPM.

81.3 VLSI Architectures for EBCOT of JPEG 2000

Since EBCOT tier-1 algorithm can support SeriPM and ParaPM, some VLSI architectures are implemented to support SeriPM and the other VLSI architectures are implemented to support ParaPM. But from structural view of coding hardware, these VLSI architectures can be divided into two categories: SeriBCS and ParaBCS. In SeriBCS, all the bit-planes are coded in serial and a code-block is coded bit-plane by bit-plane. In ParaBCS, all the bit-planes are coded simultaneously.

81.3.1 SeriBCS versus ParaBCS

Both SeriBCS- and ParaBCS-based architecture could be possible solution for entropy encoding in JPEG 2000 system. To compare these two schemes, both SeriBCS and ParaBCS are evaluated in PSNR performance, power consumption, and memory issues in the context of the entire JPEG 2000 system [12].

ParaBCS can support ParaPM only, but SeriBCS can support both ParaPM and SeriPM. This is because ParaBCS is based on ParaPM. The performance of ParaPM is slightly poorer than SeriPM, but it is more error-resilient (that is an important characteristic in wireless application).

Figure 81.7 shows the difference between these two schemes from hardware design view. SeriBCS contains one bit-plane coding engine, so state memory should be used to store SigState. To have a high system throughput, several SeriBCS engines should be used. (two SeriBCS engines are shown in Figure 81.7). This requires the multiple code-block memory. The size for a code-block is commonly $0.5N$ kB ($64 \times 64 \times N$), if N SeriBCS-based EBCOT tier-1 engines are used. Besides, since a bit-plane depends on all the bit-planes above it, several state memories are required to store states such as SigState, first MR, and visiting, to maintain this context information. Each state memory size is commonly 64×64 bits, since SeriBCS-based EBCOT tier-1 engines can start coding only if all data in a code-block are ready. The statistics of the code-block can be used to know if coding of a bit-plane is needed or not. Then some bit-planes can be removed from the coding process owing to no MSB of any DWT coefficient being above or on these bit-planes.

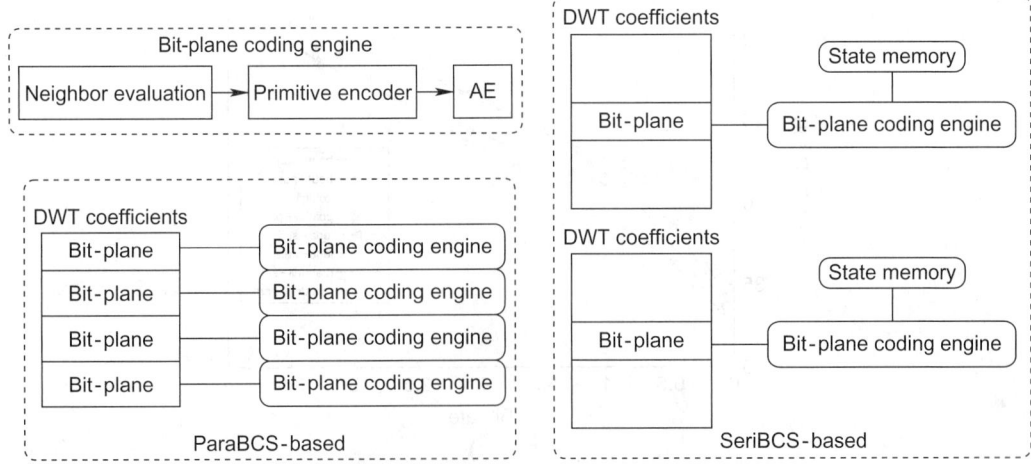

FIGURE 81.7   Hardware comparison between ParaBCS and SeriBCS.

ParaBCS contains as many bit-plane coding engines as the number of bit-planes in a code-block, so state memory can be removed by predicting SigState. With this prediction, ParaBCS can start coding whenever all data in a column are available. After coding, the memory for the column can be released for DWT coding engine to store its outputs. By taking advantages of this method, Refs. [8,13] proposed a stripe pipeline scheme, where DWT and EBCOT tier-1 coding engines work based on stripes. The buffers between DWT and EBCOT tier-1 coding engines are of sizes similar to that of the stripe. DWT and EBCOT tier-1 coding engines can switch among these buffers to process coding. So it optimized the buffers between DWT coding and EBCOT tier-1 coding, dramatically reduced memory size, and achieved memory-efficient JPEG 2000 system, where the memory requirements are reduced to only 8.5% compared with conventional architectures. But, since all data in a code-block are unknown while the coding process starts, the statistics of the whole code-block cannot be used to decide which bit-plane is not coded in the current code-block. To do a correct coding, the worst case must be considered, i.e., all the bit-planes in a code-block are coded although some bit-planes are probably not coded. This results in a big redundancy in computation that just wastes power consumption.

Note that ParaBCS-based architecture is comprised of N bit-plane coding engines to achieve a system throughput N times as fast as the system throughput of SeriBCS-based architecture. To fairly evaluate hardware implementation, we compared these two architectures under the condition of the same system throughput, i.e., the hardware required to code one bit-plane. Obviously, ParaBCS require 0.6 kbits memory while SeriBCS requires 16 kbits.

Finally, SeriBCS and ParaBCS are compared with respect to power consumption. Since the distinct difference between them is that SeriBCS contains memories that are not in ParaBCS, a SeriBCS-based architecture is implemented to evaluate the effect of memories. Since Xilinx FPGA prototyping supports memory IP cores and can provide rough power analysis by XPower tools, SeriBCS-based architecture is prototyped in Xilinx Vertex II pro device. The prototyped architecture works at 50 MHz and the device utilization is summarized in Table 81.2. Its power analysis is shown in Figure 81.8. The experimental results show that memory access takes much power (~47%) in the entire EBCOT tier-1 coding engine. While ParaBCS-based architecture does not have power consumption in memory

TABLE 81.2 VIRTEX II PRO Utilization after Place and Route

Components	RAM16	SLICE	External IOB	BUFGMUX
Utilization (%)	75	24	64	6

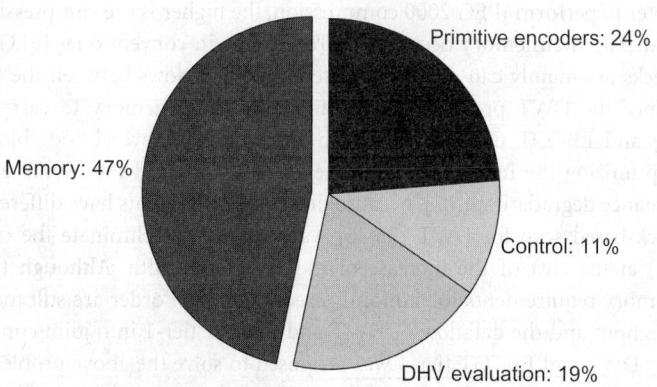

FIGURE 81.8 Power distribution for SeriBCS.

access like SeriBCS-based architecture, ParaBCS introduces redundancy in computation that consumes more power.

81.3.2 Further Discussion on VLSI Architectures for EBCOT

VLSI architectures for EBCOT support either SeriPM or ParaPM of EBCOT tier-1 algorithm. An architecture implementing SeriPM was introduced in Ref. [3], where three passes scan a bit-plane in serial. In this architecture, fetching operation is column based, i.e., in a clock cycle, one column (4 bits) is fetched from memory instead of one bit to reduce the number of memory access. Pixel-skipping techniques were adopted to skip the unnecessary bit evaluation (the bit evaluation for bits that are not coded in the current pass). But pixel-skipping techniques cannot remove the unnecessary bit evaluation completely and still waste some clock cycles. So two architectures implementing ParaPM were proposed in Refs. [10,14], where three passes scan a bit-plane in parallel and column-based fetching operation was adopted similar to that in Ref. [3]. Architectures implementing ParaPM evaluate one exact bit in a clock cycle and no clock cycle is wasted as in architectures implementing SeriPM. In Ref. [14], the parallelism between four coding bits was introduced, where two bits from different passes may be coded in the same clock cycle, leading to improvement of system throughput. All the architectures mentioned above are based on SeriBCS, i.e., all the bit-planes are coded in serial and a code-block is coded bit-plane by bit-plane. So a state memory that associates SigState with the current bit-plane is required to maintain the historical coding information. To remove state memory, a ParaBCS was proposed in Ref. [11]. This architecture does not need the memory for SigState and all the bit-planes of a code-block are coded together in parallel. But four bits of a column are coded bit by bit and only one bit is coded in one clock cycle. By uncovering that parallelism among four bits of a column one can significantly increase system throughput; the parallelism was embedded into ParaBCS and the proposed three-level parallel architecture to achieve high system throughput for multimedia real-time applications [15].

AE, as one part of EBCOT tier-1, is a strictly serial process and consumes the context information constructed by bit-plane coding phase. So low performance of AE can significantly limit the performance of high-throughput bit-plane coding, leading to low performance of EBCOT. Multicycle architecture for AE obviously has low system throughput. To achieve high system throughput, some pipelined AE architectures were proposed in Refs. [3,16–20].

While all the above architectures focus on how to overcome the computation complexity of EBCOT tier-1, many architectures for the entire JPEG 2000 system [6,7,21–25] were proposed. They can be classified into two categories: (1) to process multiple code-blocks in parallel by using multiple SeriBCS-based EBCOT tier-1 engines [6,7,21,22]; (2) to process one single code-block by using a single ParaBCS-based EBCOT tier-1 engine [23–25]. Since ParaBCS-based EBCOT tier-1 engine has higher system throughput than the SeriBCS-based one, the above two categories could meet the system throughput requirement. For the entire JPEG 2000 system, memory issues are also key factors. In general, the larger the tile size parameter to perform JPEG 2000 compression, the higher is the compression ratio. But more memory is required. The tile memory occupies >50% of area in conventional JPEG 2000 architecture [8]. These bottlenecks are mainly caused by the different coding flows between the DWT and EBCOT tier-1 processes, since the DWT process requires an entire time memory to carry out the subband transformation [9] and EBCOT tier-1 divides each subband into several code-blocks and performs entropy coding. Optimizing the individual components only may lead the overall encoding system to suffer from performance degradation [6,7], because different components have different I/O bandwidths and buffers. A block-based scan for DWT [23,26] was proposed to eliminate the use of tile memory (commonly 96 kB) at the cost of the increase of memory bandwidth. Although the tile memory is eliminated, the memory requirements for nonoptimized block scan order are still too high. In Ref. [8], by taking the throughput and the dataflow of DWT and EBCOT tier-1 into joint consideration, a stripe pipeline scheme for DWT and EBCOT tier-1 was proposed to solve the above problems. The main idea was to match the throughput and the dataflow of the two modules so that the size of local buffers between the two modules was minimized.

81.4 Design Techniques for EBCOT of JPEG 2000

In this section, several design techniques used by the architectures mentioned above are discussed in more detail. Since EBCOT architectures are comprised of BC, FIFO, and AE, design techniques are presented separately.

81.4.1 BC in SeriBCS

(1) *Memory access pattern* [3]. In SeriBCS, state memory is used to store significance state. Memory access and arrangement is a nontrivial issue for BC. In bit-based design, nine significant state variables are required, including variables of the sample under coding and the eight neighboring samples. In a column-based design, three columns, including current and two neighboring columns, are needed at the same time. With six bits for each column, totally 18 significant and sign variables are required. The column-based operation is shown in Figure 81.9. The rectangles stand for a shift register array for significant state variables. Black rectangles represent current coding samples and the white ones represent neighbor samples. In addition, adder trees are used to provide the sum of significant neighbors of each sample in current coding column simultaneously and to share hardware cost. After a column is coded, variables in the register array are shifted to the nearby column on the left, the original variables in the left column are abandoned, and new variables are loaded from memory and stored in the right column.

Since significance states from neighboring stripes are needed, it is quite a challenge to arrange and access memory. Three feasible memory arrangements (A, B, and C) are shown in Figure 81.10 and indicate how memory arrangement can affect system performance. For memory arrangement A, four significance states of a column are grouped into a word. These words are interleaved-placed in three memories, i.e., three stripes are interleaved-placed in three memories. Every time a new column is processed, three words (12 bits), including the current coding column, the upper one, and the lower one, have to be loaded from three memories, one word per memory. During read access, although only six variables are needed for coding a new column, 12 variables are loaded from memory. During write access, if there are any changes in the significant state, only the current column has to be written back. Therefore, there are 6 redundant bits in read cycle and no redundancy in write cycle. For memory arrangement B, variables are interleaved-placed in two memories. Every time a new column is processed, two words (8 bits) are loaded from two different memories. During read access, although only six variables are needed for coding a new column, eight variables are loaded from memory. During write access, when data have to be written back to memory, there will be four redundant bits. Therefore, there are two redundant variables in read cycle and four redundant variables in write cycle. In summary, arrangement A reads six redundant bits in every 12 bits, but writes no redundant bits, while configuration B reads two redundant bits and writes four redundant bits.

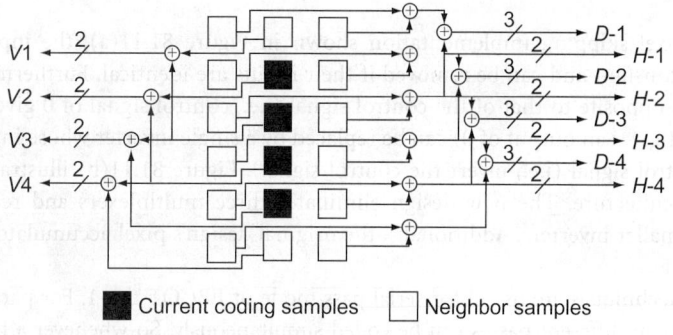

■ Current coding samples   □ Neighbor samples

FIGURE 81.9 State variable PE.

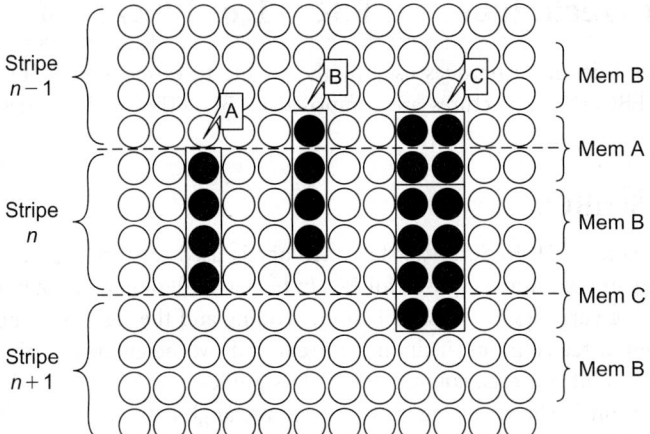

FIGURE 81.10 Feasible memory arrangements.

In memory arrangement C, every four variables in two nearby columns are grouped to be a word, and words are placed in three memories in an interleaving format (A, B, C, B, A, B, C, B, C, B, A). During read access, the variables of two nearby columns are loaded into the register together in one cycle and the memory access frequency is reduced to half. Although the memory bus width is still 12 bits, no redundant variables are loaded. During write access, there will be 4 redundant bits for the write back of data of two columns. Compared with arrangement A and B, arrangement C has two redundant bits access for one read and write of one column, while arrangement A as well as B have six redundant bits.

(2) *Pixel skipping* [3,4]. Four bits in a column could be coded in different passes. So if four bits are evaluated in every pass, many clock cycles are wasted. The pixel skipping technique is used to identify only those coefficients in a column that *need to be coded* (NBC) for the current pass. An "NBC pixel" variable is used as input to the pixel skipping circuitry. This input has 4 bits that indicate which bits in the column need to be encoded during the current pass. For example, if the NBC pixel input is equal to $(1010)_2$, then the *zeroth* and second bits (reading from left to right) are to be encoded, and the NBC indexes produced should be $(00)_2$ and $(10)_2$, respectively. A diagram of two-pixel skipping circuitry is presented in Figure 81.11.

The uppermost adder in the figure tallies the bits in the NBC pixel input to determine the total number of bits to be coded for the column (e.g., 1010_2 implies two bits are to be encoded). A pixel accumulator is incremented by one in each cycle to keep track of how many bits in the column have been coded thus far. These values are used as inputs to a comparator to determine when coding of the entire column is complete. At this time, a ChangeColumn signal is set, and non-NBC bits are skipped.

In the basic pixel skipping implementation shown in Figure 81.11(a), the inputs to the initial multiplexers are constant and can be removed if their inputs are identical. Furthermore, multiplexers with input values opposite to that of the control signal (i.e., control signal of 0 gives an output of 1; control signal of 1 gives an output of 0) can be replaced by a single inverter whose input is the original multiplexer's control signal (i.e., invert the control signal). Figure 81.11(b) illustrates the simplified pixel skipping architecture. The new design eliminates three multiplexers and reduces two of the multiplexers to smaller inverters. Additionally, the original design's pixel accumulator can be replaced with a simple DFF.

Pixel skipping techniques are useful for serial pass mode of EBCOT tier-1. For parallel pass mode of EBCOT tier-1, bits in different passes can be coded simultaneously. So whenever a bit is accessed and evaluated, it is assigned to one pass and coded immediately.

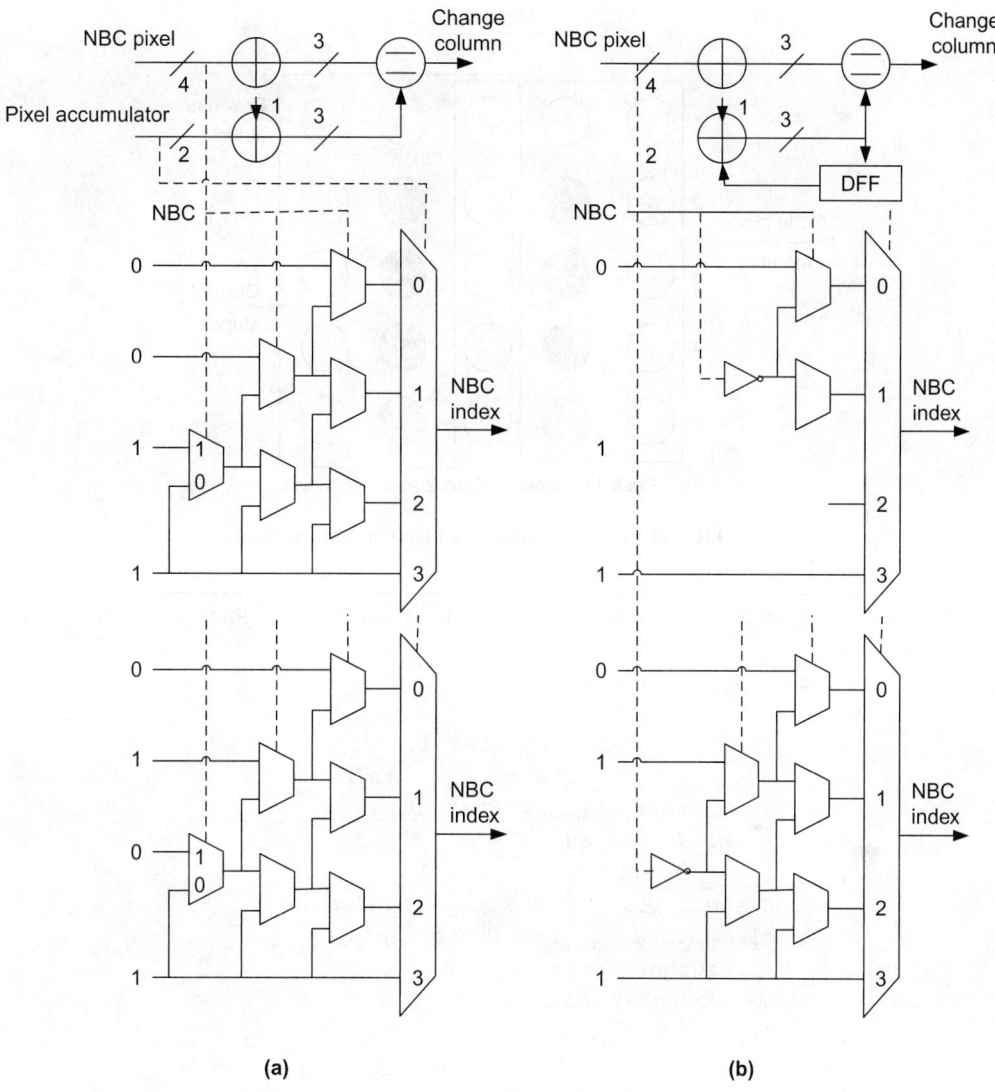

FIGURE 81.11 Pixel skipping architecture.

81.4.2 BC in ParaBCS

Supporting three-level parallelism [15]. Since ParaBCS is based on ParaPM, ParaBCS can adopt more parallelism. To support the multilevel parallelism, removing data dependency is essential. The data dependency among three passes can be removed if the parallel mode of EBCOT is adopted. The data dependency among bit-planes and among coding bits can be removed if SigState can be predicted before three pass scannings. The change of SigState can occur either in pass 1 or pass 3. As a result, pass 1 is self-dependent, pass 2 depends on pass 1, and pass 3 depends on pass 1 and is also self-dependent.

To remove these dependencies, two context windows shown in Figure 81.12 were adopted, where one column is mapped to the fourth bit from the previous stripe and four bits from the current stripe. Columns A, B, and C form the context window 1. Columns C, D, and E form the context window 2. These two context windows are implemented as 5×5 register array. Context window 1 is used to indicate pass 1 coding state, i.e., whether the bits are coded in pass 1 or not. It can be predicted by using initial

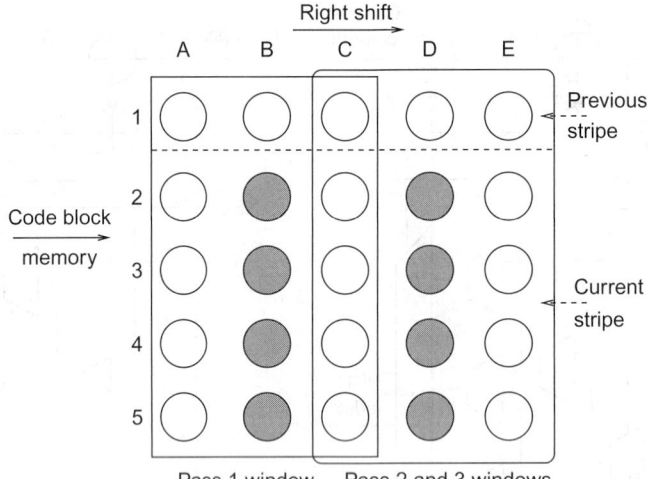

FIGURE 81.12   Column-based DHV generator.

```
(1) Calculate the SigState after Pass 1 scanning or Pass 3 scanning.
    (a)   For C₁, D₁, E₁,
          State_afterPass1 = State_initial + Value
          For other bits,
          (i)    if it is coded in Pass 1,
                 State_afterPass1 = State_initial + Value
          (ii)   else State_afterPass1 = State_initial
    (b)   For D₂, D₃, D₄, D₅ and E₂, E₃, E₄, E₅
          (i)    if it is coded in Pass 3,
                 State_afterPass3 = State_initial + Value
          (ii)   else State_afterPass3 = State_afterPass1
(2) Calculate the neighbor contributions
    (a)   If Dᵢ is coded in Pass 1
          (i)    (Cᵢ₋₁, Cᵢ, Cᵢ₊₁, Dᵢ₊₁), State_initial
          (ii)   (Dᵢ₋₁, Eᵢ₋₁, Eᵢ, Eᵢ₊₁), State_afterPass1
    (b)   If Dᵢ is coded in Pass 2, all the neighbors
          should be State_afterPass1
    (c)   If Dᵢ is coded in Pass 3
          (i)    (Cᵢ₋₁, Cᵢ, Cᵢ₊₁, Dᵢ₊₁), State_afterPass1
          (ii)   (Dᵢ₋₁, Eᵢ₋₁, Eᵢ, Eᵢ₊₁), State_afterPass3
```

FIGURE 81.13   NC calculation algorithm.

SigState and its bit-value. If a bit is insignificant and any neighbor is significant, it is coded in pass 1. After one column finishes coding, all values are right shifted by one column.

The pass 1 coding states are shifted to context window 2. In context window 2, all the NC and pass coding states (i.e., which pass the bit belongs to) can be calculated by using initial SigState from load logic model, the bit values and pass 1 coding states. Pass 1 coding states indicate the bits coded in pass 1. The bit is coded in pass 2, if the bit is initialized as significant. The rest of all the bits are coded in pass 3. The neighbor contribution calculation algorithm is shown in Figure 81.13, where A, B, C, D, E, 1, 2, 3, 4, 5 are used to identify the location of bits as shown in Figure 81.12.

Parallelism among four bits in one column can benefit reduction of power consumption in two ways. First, this parallelism permits four bits to be evaluated simultaneously and optimization can be made for all four bits together. So, adding this parallelism can reduce memory access, share circuits to evaluate neighbor information, and reduce the movements of values in registers. Second, the increased throughput from this parallelism can be used for trade-off with power consumption. The computation may finish quickly and stay in sleep mode longer.

81.4.3 FIFO

Depth of FIFO depends on burst rate of inputs and reading rate of outputs. If reading rate of output is fixed, and the higher the burst rate of inputs is, the deeper FIFO is. For a specific application, an optimized depth exists. If FIFO is too long, memory space is wasted. If FIFO is too short, data overflow will happen.

(1) *Combining bit-planes to reduce depth of FIFO* [11,15]. Assume that the number of bit-planes for a code-block is N, the MSB has an index N, and the LSB has an index 1. Bit-plane scanning order is from MSB to LSB. In the bit-plane indexing with bigger number, pass 3 dominates the coding process. In the bit-plane indexing with smaller number, pass 1 or pass 2 dominates the coding process. This behavior results from the BC process. At the beginning, the entire significance state is initialized as *insignificant* and bits either have higher probability to be coded in pass 3 or do not need coding. With scanning going from top to bottom, there is more significance state change from *insignificant* to *significant* and bits have higher probability to be coded in pass 1 or pass 2. Figure 81.14 shows the number of bits coded in three passes.

Note that pass 3 can use run-length coder to generate CX–D pair. In the best case, four bits could be coded together into one CX–D pair only. But in pass 1 or pass 2, one bit is coded into at least one CX–D pair. So the higher the index of bit-plane, the lower the burst rate of the bit-plane. To save FIFO space, outputs from two bit-planes (one is the lower-level bit-plane with index i and the other is the higher-level bit-plane with index $N - i$, where N is the total number of bit-planes) are combined together and stored into one FIFO. Thus, the data rates from two bit-planes are averaged and the total burst rate will be reduced, leading to reduced depth of FIFO.

81.4.4 Arithmetic Encoder

AE is a strictly serial process. To increase system throughput, multicycle and pipelined architectures were proposed in literature. Multicycle architecture has an obvious disadvantage in system throughput. So pipeline architecture is usually adopted to enhance system throughput. In this section, pass switching technique and forwarding technique are discussed. Pass switching arithmetic encoder can support ParaPM efficiently. Forwarding technique is combined to pipelined AE to reduce power consumption.

(1) *Pass Switching Arithmetic Encoder* [10]. In ParaPM, three passes are coded simultaneously. To consume these simultaneous CX–D pairs in three passes, it seems that three AE coders are required. Since AE coder

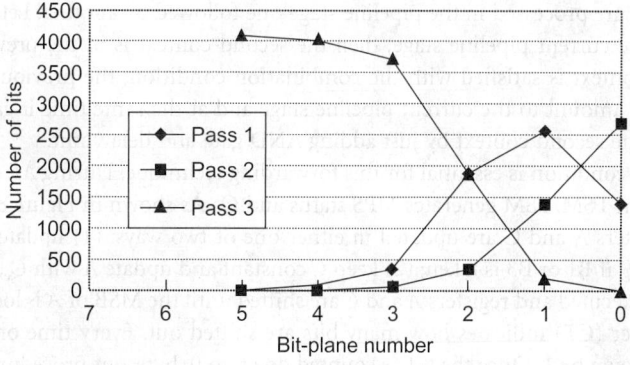

FIGURE 81.14   Number of bits coded in three passes.

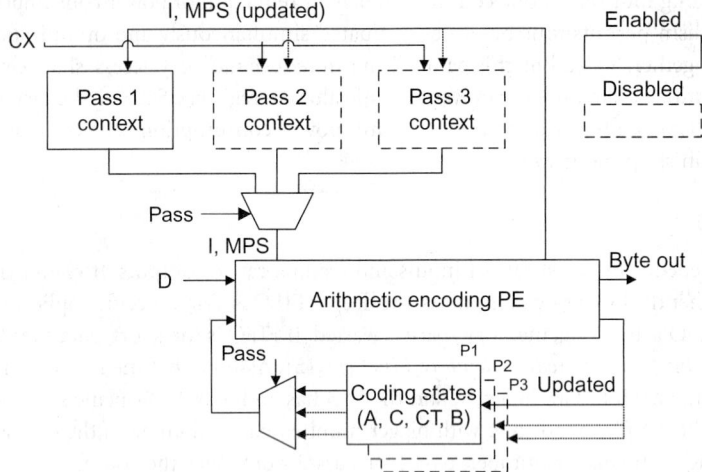

FIGURE 81.15   Pass switching arithmetic encoder.

is usually costly, three AE codes make the pass-parallel architecture impractical. Fortunately, instead of using three AE coders, a low-cost pass switching arithmetic encoders (PSAE) as shown in Figure 81.15 can be used. Like normal AE coders, PSAE includes the functionalities of coding state transition, interval and code register calculating, renormalization, bit-stuffing, and encoder termination. PSAE differs from the normal AE coders in that it has the capability of coding CX–D pairs in different coding passes without any conflicts. This is achieved by using three sets of context registers as well as coding state registers, which are allocated for three passes respectively. The pass signal delivered from BC selects one of the three sets of registers. Therefore, the much expensive part (i.e., the arithmetic coding PE) is shared for three coding passes. Although more registers are required for PSAE, the number of contexts registers required for PSAE is not the triple of 18 contexts. There are only 13 possible contexts generated in pass 1, 3 in pass 2, and 15 in pass 3.

(2) *Combining forwarding technique to pipelined AE* [12]. To reduce the power consumption of AE, forwarding technique combined with clock gating technique can be applied. One distinct characteristic of this scheme is that two symbols can be coded as one symbol in the last two pipeline stages. As a result, only two pipeline stages are required instead of four pipeline stages for two symbols. So the switching activities in two pipeline stages are removed without decreasing the system throughput. AE encoding procedure is inherently a serial process with high dependency, but we still observe the possibility of combining two continuous contexts. If the second context is encoded in B2 in Figure 81.4, the codeword is not changed and only shift operation is needed. So if this shift operation can be forwarded, these two continuous contexts can be processed as one context with only attached shift amount. In AE, two continuous contexts are processed in the pipeline stage one followed by another. Let us suppose that the first context is in the current pipeline stage, then the second context is in the previous pipeline stage. When the second context is satisfied with the combination condition, the previous pipeline stage can forward the shifting amount to the current pipeline stage and at the same time it gates out the last two pipeline stages for the second context by just adding AND gate and delay unit.

The combination condition is essential for this forwarding technique. During AE coding, a pair of CX and D is read into the FSM. FSM generates MPS status and Q_e. As shown in Figure 81.4, no matter what the inputs are, registers A and C are updated in either one of two ways: (1) update C with $C + Q_e$ and update A with $A - Q_e$, if B1 or B3 is taken; (2) keep C constant and update A with Q_e, if B2 is taken. Then renormalization is executed and registers A and C are shifted until the MSB of A is located in the leftmost position. The counter (CT) indicates how many bits are shifted out. Every time one bit is shifted out, the CT is counted down by 1. Once the CT is counted down to 0, byte-out procedure is called to output one byte into the bitstream buffer. Then the CT is reinitialized as 7 or 8 according to bit-stuffing technique.

```
If (MSB of (A−Qₑ) is 1) AND (MPS ⊗ D) XOR (A < 2Qₑ << 1)
    A = Qₑ, C = C
else
    A = A - Qₑ, C = C + Qₑ
if (MPS ⊗ D) AND (MSB of (A - Qₑ) is 1)
    FSM skip to the next state
else
    FSM stay at the current state
```

FIGURE 81.16   Simplifying AE algorithm.

```
(1)    Fetch CX, D
(2)    For CX and D,
       if B2 is taken, this CX can be combined with the previous one
       together, forward the shift amount to the next pipeline and gate
       out the clock cycle for the current pipeline
       else one CX is coded
(3)    Calculate C as the original algorithm
       Calculate A as the original algorithm
(4)    Renorm procedure is executed as the original algorithm
```

FIGURE 81.17   The AE algorithm.

Note that if (2) happens, only shift operation is needed for codeword and the number of shifting is determined by Q_e only. So two continuous contexts can be processed as one context except for the increased number of shifting.

To decrease the complexity of circuits, an efficient evaluation is needed. Some simplified operations are adopted for this evaluation by modifying the algorithm in Figure 81.4 such as $A = A - Q_e$ is removed, $A\&0x8000$ is replaced with $A[14:0] < Q_e[14:0]$ and $A < Q_e$ is replaced with $A < Q_e \ll 1$. So A and C are updated as in Figure 81.16.

The algorithm is shown in Figure 81.17. The four-stage pipelined architecture is as shown in Figure 81.18. In stage 1, context information (CX and D) are fetched in one clock cycle. CX is fed into FSM to generate the corresponding Q_e and the possible CX updates, i.e., the next state that FSM will skip to. In stage 2, the updated CX is fed back to stage 1 to update the FSM. If the CX is processed in branch B2 shown in Figure 81.4, the shifting amount is forwarded to stage 3 and the clock cycles for stages 3 and 4 are gated out, since the context is combined with the previous one already. Otherwise, the CX is coded as the original algorithm. C register accumulation is done by using 28-bit adder. To avoid a long time delay, the 28-bit adder is divided into 16 bit adder in stage 3 and 12-bit adder in stage 4. The lower 16 bits of C register are updated by using 16-bit adder and shifting operation. In stage 4, after the higher 12 bits of C register are updated by using 12-bit adder and shift operation, the renormalization and byte-out procedure are executed.

81.5 Case Study: A Serial Pass Mode Architecture for BPC

A simplified block diagram of the serial pass mode EBCOT tier-1 architecture [3,4] is presented in Figure 81.19. To begin, the "Memories" block actually contains three individual memories, each the same dimensions as the code-block and 1-bit per entry. The *significant status memory* (σ) maintains

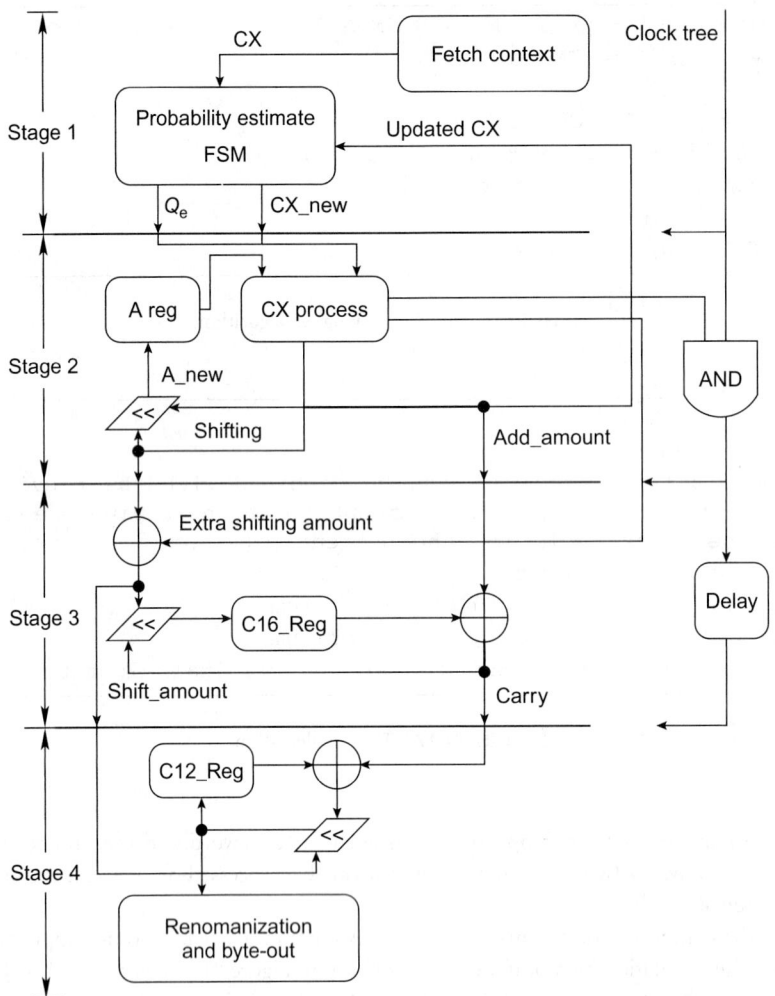

FIGURE 81.18   The pipelined AE architecture.

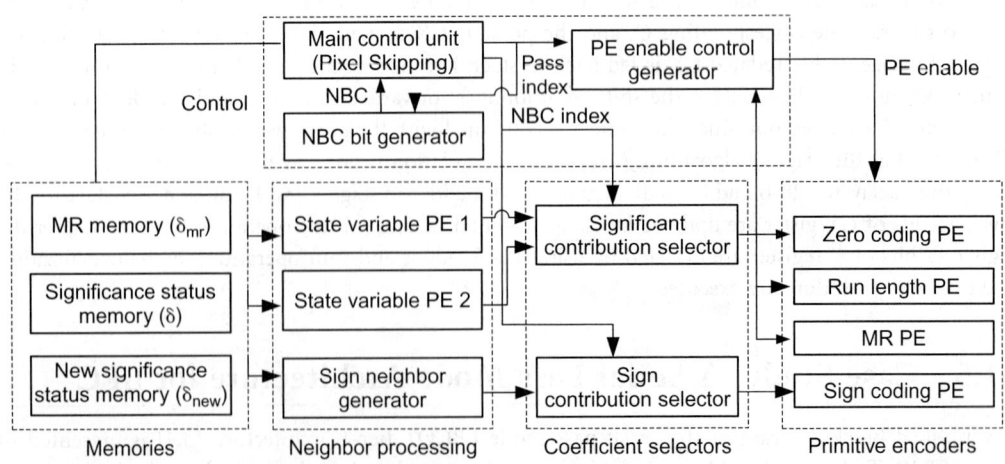

FIGURE 81.19   Serial Pass Mode Architecture

TABLE 81.3 Synopsis Results for Serial EBCOT Architecture

Module	Area	Delay (ns)	Power (mW)
State variable PE	322.99	6.35	118.93
Control unit	491.89	1.57	84.33
Selector	259	1.08	336.46
Sign coding PE	40	1.41	22.65
Zero coding PE	140.33	1.41	51.98
Magnitude refinement PE	28.33	1.41	26.70

information on the significance status of each bit in the current bit-plane of the code-block. The second memory, *first refinement status memory* (σ_{mr}), uses a single bit to indicate whether or not the coefficient has been processed during an earlier magnitude refinement pass. The final memory subblock, *new significant status memory* (σ_{new}), contains the most recent updates to the coefficient's significance status and is similar to σ. For example, once a coefficient is found significant, the corresponding bit in σ_{new} is set to indicate its new significance status. At the start of a bit-plane, the contents of σ_{new} are copied into σ to reflect all status changes that occurred in the previous bit-plane. Using the column-processing concept, the "neighbor processing" block reads the data from the memories and calculates *H*, *V*, and *D* for each bit in a column simultaneously. These values along with the significance information of the coefficient are used to determine which coefficients are coded during the current pass (NBC) and thus enable pixel skipping. The "context formation" block contains the circuitry for generating the context labels for those NBC bits. The control unit included several submodules that manage the operation of the entire architecture.

One obstacle encountered with column processing is that the state variable processing element (PE) produces the NC for each of the four coefficients in parallel. However, the PEs for context formation operate in serial. To resolve this discrepancy, the selector block uses the NBC information to serially choose which of the coefficients will be sent to the context formation PEs. It is possible to implement parallel processing in the context formation blocks; however, this requires several duplications of the hardware, which may be underutilized when only a few coefficients from a column are coded during a particular pass.

The components of this architecture were implemented using Synopsis and 0.5 μm technology. The results are shown in Table 81.3.

81.6 Case Study: A Three-Level Parallel High-Speed Low-Power Architecture for EBCOT-Tier-1

Since EBCOT tier-1 is a bottleneck component of the entire JPEG 2000 system, hardware design should be considered in the interactions of EBCOT tier-1 with other components as well as the improvement of EBCOT tier-1 performance itself, i.e., considering not only memory design issue in the context of the entire JPEG 2000 system but also the system throughput and power consumption of the single EBCOT component. Besides, there are trade-offs between system throughput and power consumption. Reducing the supply voltage is a popular power saving technique. But it reduces system performance by increasing delay. In contrast, in submicro field, drain leakage will increase with *V*t dropping to maintain noise margins and meet frequency demands. It leads to excessive battery draining standby power consumption. By increasing system throughput and finishing subtask quickly, the system can have more time to sleep and save standby power.

With consideration of all the issues mentioned above, ParaBCS-based three-level parallel high-speed low-power architecture [12] is discussed here. To reduce memory size, this architecture can fit into stripe-based pipeline scheme (mentioned in Section 81.3.2).* To achieve high system throughput, three

---

*The proposed EBCOT tier-1 architecture can be used for any level pipeline scheme although we only show stripe-level pipeline scheme here.

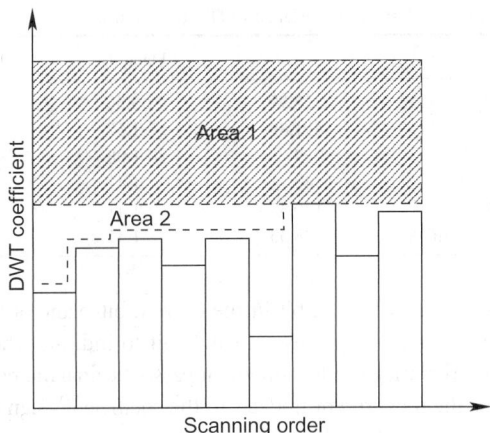

FIGURE 81.20   Reduce computation in ParaBCS.

levels of parallelism in Section 81.4.2 are adopted: (1) the parallelism among bit-planes: all the bit-planes can be processed simultaneously and SigState memory can be removed by predicting the SigState of each bit; (2) the parallelism among pass scanning: three passes scan one bit-plane in parallel, bits are coded immediately after they are evaluated and no processing time is wasted; and (3) the parallelism among coding bits: four bits in a column can be coded in the same clock cycle simultaneously by adding extra primitive encoders.

Since this architecture is based on ParaBCS, the power consumption in memory access in SeriBCS is reduced. To further reduce the power consumption, computation in coding process is reduced without decreasing the system throughput. In ParaBCS, all the bit-planes in a code-block are coded in parallel together. In fact, for some bit-planes, no MSB of any DWT coefficient is located on or above them, so their coding is not needed. For stripe-based pipeline scheme, it is not possible to find out these redundant bit-planes because all data in a code-block are not ready when coding starts and the worst case has to be considered. So a detecting technique is adopted to detect these redundancies and disable the coding components for them. To simply put this idea, scanning of 1-D DWT coefficients is shown in Figure 81.20, where the horizontal axis is the coding scanning order and the vertical axis is the magnitude of DWT coefficients. For the original ParaBCS, the worst case should be considered and the entire coding area, including area 1, area 2, and real DWT coefficients area, are coded. For the original SeriBCS, it is known that area 1 is not needed, so coding activities in areas 1 are removed in comparison with the original ParaBCS. By noticing that all coding activity for one DWT coefficient in areas 1 and 2 are the same and the coding results are identical fixed values. So bit-plane coding in areas 1 and 2 is not necessary. Here, the detecting techniques can detect area 1 as well as area 2 and remove computation in these areas.

Since EBCOT tier-1 includes bit-plane coding and AE coding which are connected by FIFO, computation can be reduced in bit-plane coding phase, FIFO, and AE coding phase. In the bit-plane coding phase and FIFO phase, for areas 1 and 2, CX are from run-length coder in Pass 3 and D is 0. So coding could be removed and only a counter is applied to indicate how many CX and D and a disable flag are applied to indicate if the counter is valid for the current bit-plane. AE phase is a strictly serial process and contains the historical information. So only the coding in the same scanning step of areas 1 and 2 have the same code-word. Keeping just the first AE coding, all other AE codings in areas 1 and 2 are disabled. Before scanning for one bit-plane that leaves from area 2 to real DWT coefficient area, the current AE is initialized from the first AE coding and works like normal coding. Besides, a forwarding technique is used to further reduce power consumption in AE. Some two continuous context labels can be combined in the late two pipeline stages and they work like one context label. So one clock cycle is gated

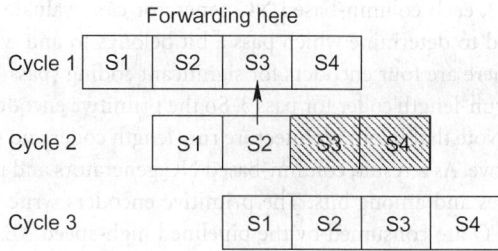

FIGURE 81.21   Forwarding technique in AE.

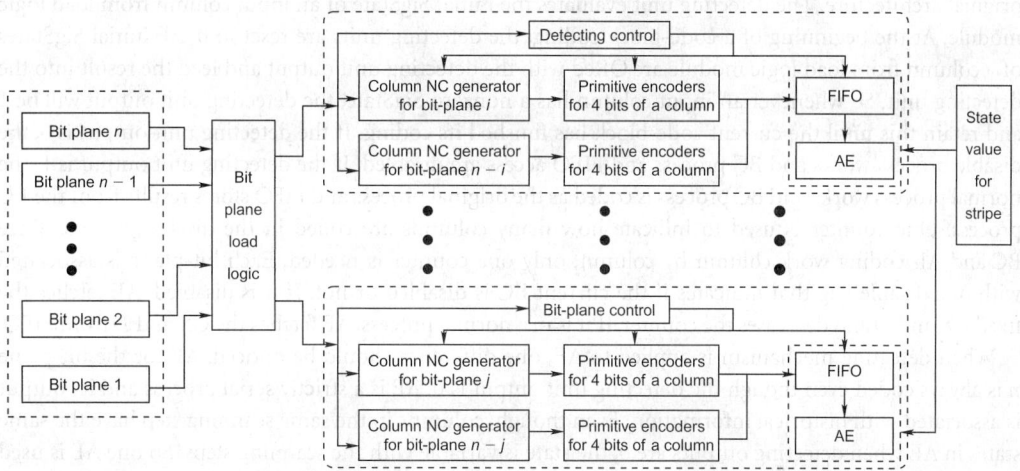

FIGURE 81.22   Three-level parallel EBCOT architecture.

out instead of two clock cycles being required, leading to reduction in power consumption. Figure 81.21 shows how forwarding technique reduces switching activities of the last two pipeline stages. So for cycle 2, the last two pipeline stages are removed after forwarding happens.

81.6.1   EBCOT Tier-1 Architecture

The VLSI architecture is shown in Figure 81.22. The bit-plane buffer banks contain all the DWT coefficients. The load logic model fetches four wavelet coefficients in one clock cycle from the code block memory. MSB of these DWT coefficients are used by load-logic model to initialize the SigState of bits. For each DWT coefficient, the bits above its MSB, including MSB, are initialized as insignificant and the others as significant. The initialized SigStates are fed into column-based NC generator models for evaluation of NC. Note that these initial SigStates are not the final values that can be used to calculate the NC. More steps are needed to predict the final values. (Section 81.4.2 will discuss this in detail.)

The column-based NC generator for bit-plane i is associated with a column of the bit-plane i. So n column-based NC generators may code all the bit-planes simultaneously. Here, the bit-plane $n-i$ coding and the bit-plane i coding are combined together in FIFO model. (for simplicity, only two combinations are shown in Figure 81.22). The number of context labels from the bit plane $n-i$ is usually less than that from the bit plane i since the bit plane $n-i$ may have more bits coded by run-length coding. Run-length coding may encode more than 1 bit, but it generates only one context label. This combination will benefit FIFO (No big variance for data in FIFO).

As shown in Figure 81.12, each column-based NC generator can evaluate NC for four bits simulta-neously. All the NC are used to determine which pass a bit belongs to and what the context label is. In primitive encoder model, there are four encoders for significant coding (pass 1 or pass 3), four encoders for MR (pass 2 only) and a run-length coder for pass 3. So the primitive encoder model may concurrently code four bits in a column. Note that in our architecture run-length coder and sign-coder work in parallel with the primitive coder above. As a result, column-based NC generators and primitive encoders provide the parallelism among passes and among bits. The primitive encoders write their outputs (CX and D) into FIFO. CX and D in FIFO are consumed by the pipelined high-speed AE.

The detecting mechanism is implemented by using detecting control that is shown in Figure 81.23. Owing to detecting mechanism, two processes may happen: (1) disable process where BC, FIFO access, and AE coding are disabled; (2) normal process where all coding units work in the same way as the original architecture. The detecting unit evaluates the initial SigState of an input column from load logic module. At the beginning of a code-block coding, the detecting units are reset to 0. All initial SigStates of a column from load logic module are ORed with the detecting unit output and feed the result into the detecting unit. So whenever an input column has a nonzero SigState, the detecting unit output will be 1 and retain this until the current code-block has finished its coding. If the detecting unit output is 0, the disable process works and BC process and FIFO access are disabled. If the detecting unit output is 1, the normal process works and BC process is coded as the original process and FIFO stores results from the BC process. One counter is used to indicate how many columns are coded in the disable process. Since BC and AE coding work column by column, only one counter is needed. Each bit-plane is associated with one disable flag that indicates if the current BC is disabled or not. If it is disabled, AE fetches the fixed CX and D and decreases the counter. If it is in a normal process, AE fetches the CX and D from FIFO.

When detecting mechanism is applied to AE, one difference should be noticed. AE for the bit-plane n is always coded even though the detecting unit output is 0. AE is a strictly serial process and its output is associated with historical information. Even though, columns in the same scanning step have the same states in AE when detecting outputs are 0, the state is variable with the scanning steps. So one AE is used to trace the change of states. For other AEs except the one for the bit-plane n, they are disabled when their detecting outputs are 0. Whenever detecting unit output changes from 0 to 1, the current register A, register C, and byte-out are loaded from the AE for the bit-plane n.

This architecture can achieve higher system throughput and reduce power consumption. As shown in Table 81.4, this architecture can encode a bit-plane with one code block of size $N \times N$ in only $0.35 \sim 0.46 \times N \times N$ clock cycles and is four times as fast as the other architectures. The detecting technique is adopted to reduce power consumption. Figure 81.24 shows how the detecting techniques can efficiently reduce computation of bit coding in BC. Similar trends are found in FIFO and AE. Experimental results with standard test image benchmarks show that the forwarding techniques in AE and detection techniques retain

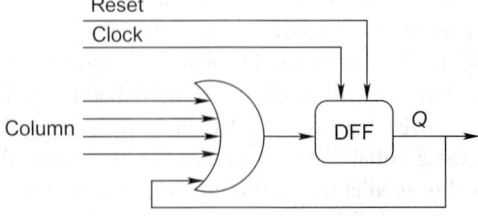

FIGURE 81.23   Detecting control unit.

TABLE 81.4   Processing Time for an $N \times N$ Code-Block

Architecture	[3]	[10]	[11]	Case study
Time ($\times N^2$ cycles)	$1.3 \times n$	n	1.2	$0.35 \sim 0.46$

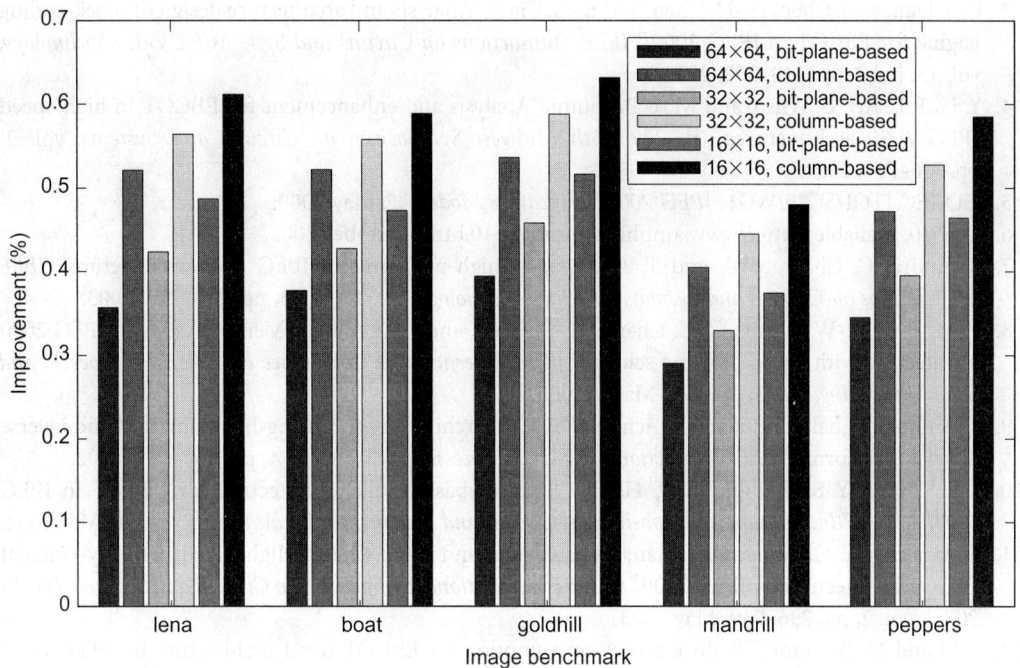

FIGURE 81.24 Enhancement with code-block size.

the same system throughput and achieve about 48, 16, and 20% improvement for BC, FIFO, and AE, respectively, in power consumption by comparison with the architecture based on the original ParaBCS.

81.7 Conclusion

There are two typical VLSI architectures in literature: ParaBCS, where all bit-planes in a code block are coded in parallel, and SeriBCS, where all bit-planes in a code-block are coded bit-plane by bit-plane. ParaBCS has more parallelism and does not need state memories required by SeriBCS. These two schemes are compared in system throughput, PSNP performance, memory size, and power consumption. In either ParaBCS or SeriBCS, the architecture comprises of BC, AE, and FIFO that connects BC with AE and balances the different throughputs between them. Subsequently, VLSI architectures in literature are evaluated and design techniques used by these architectures are discussed in more detail. Finally, two case studies (one based on SeriBCS and the other on ParaBCS) are presented.

Acknowledgments

The authors acknowledge the support of the Governor's Information Technology Initiative and the NSF, INF 6-001-006.

References

1. M. Boliek, C. Christopoulos, and E. Majani (Editors), *JPEG 2000 Part I Final Publication Draft*, ISO/IEC JTC1/SC29/WG1 N2678, July 2002.
2. S. Fossel, G. Fottinger, and J. Mohr, "Motion JPEG 2000 for high-quality video systems," *IEEE Transactions on Consumer Electronics*, vol. 49, no. 4, pp. 787–791, 2003.

3. C.-J. Lian, K.-F. Chen, H.-H. Chen, and L.-G. Chen, "Analysis and architecture design of block-coding engine for EBCOT in JPEG 2000," *IEEE Transactions on Circuits and Systems for Video Technology,* vol. 13, no. 3, pp. 219–230, 2003.

4. Y. Li, R.E. Aly, B. Wilso, and M.A. Bayoumi, "Analysis and enhancement for EBCOT in high-speed JPEG 2000 architectures," in *The 45th Midwest Symposium on Circuits and Systems,* vol. 2, pp. 736–739, 2002.

5. ISO/IEC JTC1/SC29/WG1, *JPEG 2000 Verification Model 7.2,* May 2000.

6. *CS6510,* available: http://www.amphion.com/cs6510.html, October 2002.

7. K. Andra, C. Chakrabarti, and T. Acharya, "A high-performance JPEG 2000 architecture," *IEEE Transactions on Circuits and Systems for Video Technology,* vol. 13, no. 3, pp. 209–218, 2003.

8. H.-C. Fang, Y.-W. Chang, C.-C. Cheng, C.-C. Chen, and L.-G. Chen, "Memory efficient JPEG 2000 architecture with stripe pipeline scheme," *IEEE International Conference on Accoustics, Speech, and Signal Processing,* vol. 5, pp. 1–4, March 2005.

9. K. Andra, C. Chakrabarti, and T. Acharya, "A VLSI architecture for lifting-based forward and inverse wavelet transform," *IEEE Transactions on Signal Processing,* vol. 50, no. 4, pp. 966–977, 2002.

10. J.-S. Chiang, Y.-S. Lin, and C.-Y. Hsieh, "Efficient pass-parallel architecture for EBCOT in JPEG 2000," in *IEEE International Symposium on Circuits and Systems, 2002,* vol. 1, pp. 773–776, May 2002.

11. H.-C. Fang, T.-C. Wang, C.-J. Lian, T.-H. Chang, and L.-G. Chen, "High-speed memory-efficient EBCOT architecture for JPEG 2000," in *IEEE International Symposium on Circuits and Systems (ISCAS 2003),* vol. 2, pp. 736–739, May 2003.

12. Y. Li and M. Bayoumi, "Reduce power consumption for EBCOT tier-1 architecture in JPEG 2000," in *IEEE International Symposium on Circuits and Systems,* May, 2006.

13. H.-C. Fang, T.-C. Wang, C.-J. Lian, T.-H. Chang, and L.-G. Chen, "High-speed memory-efficient EBCOT architecture for jpeg2000," *Proceedings of the IEEE International Symposium on Circuits and Systems,* vol. 2, pp. 736–739, May 2003.

14. Y. Li, R.E. Aly, M.A. Bayoumi, and S.A. Mashali, "Parallel high-speed architecture for EBCOT in JPEG 2000," in *2003 IEEE International Conference on Acoustics, Speech, and Signal Processing,* vol. 2, pp. 481–484, April 2003.

15. Y. Li and M. Bayoumi, "Three-level parallel high-speed architecture for EBCOT in JPEG 2000," in *IEEE International Conference on Accoustics, Speech, and Signal Processing,* vol. 5, pp. 5–8, March 2005.

16. K.-K. Ong, W.-H. Chang, Y.-C. Tseng, Y.-S. Lee, and C.-Y. Lee, "A high-throughput low-cost context-based adaptive arithmetic codec for multiple standards," in *IEEE International Conference on Image Processing,* vol. 1, pp. 872–875, September 2002.

17. J.-S. Chiang, C.-H. Chang, Y.-S. Lin, C.-Y. Hsieh, and C.-H. Hsia, "High-speed EBCOT with dual context-modeling coding architecture for JPEG 2000," in *IEEE International Symposium on Circuits and Systems,* vol. 3, pp. 865–868, May 2004.

18. H.H. Chen, C.J. Lian, K.F. Chen, and L.G. Chen, "Context-based adaptive arithmetic encoder design for JPEG 2000," in *Proceedings of Taiwan VLSI Design/CAD Symposium 2001, Section C1-10,* August 2001.

19. M. Tarui, M. Oshita, T. Onoye, and I. Shirakawa, "High-speed implementation of JBIG arithmetic coder," in *TENCON'99, Region-10,* vol. 2, pp. 1291–1294, 1999.

20. Y. Li, M. Elgamel, and M. Bayoumi, "A partial parallel algorithm and architecture for arithmetic encoder in JPEG 2000," in *IEEE International Symposium on Circuits and Systems,* pp. 5198–5201, May 2005.

21. *JPEG2K E,* available: http://www.alma-tech.com/, October 2002.

22. *ADV202,* available: http://www.analog.com/, October 2002.

23. H. Yamauchi, S. Okada, K. Taketa, T. Ohyama, T. Matsuda, T. Mori, S. Okada, T. Watanabe, Y. Matsuo, Y. Yamada, T. Ichikawa, and Y. Matsushita, "Image processor capable of block-noise-free JPEG 2000 compression with 30 frames/s for digital camera applications," in *ISSCC Digest of Technical Papers,* pp. 46–47, February 2003.

24. H. Yamauchi, K. Mochizuki, K. Taketa, T. Watanabe, T. Mori, Y. Matsuda, Y. Matsushita, A. Kabayashi, and S. Okada, "A 1440 × 1080 pixels 30 frames/s motion-JPEG 2000 codec for hd movie transmission," in *ISSCC Digest of Technical Papers*, pp. 326–327, February 2004.
25. H.-C. Fang, C.-T. Huang, Y.-W. Chang, T.-C. Wang, P.-C. Tseng, C.-J. Lian, and L.-G. Chen, "81 ms/s JPEG 2000 single chip encoder with rate-distoration optimization," in *ISSCC Digest of Technical Papers*, pp. 328–329, February 2004.
26. B.-F. Wu and C.-F. Lin, "Analysis and architecture design for high performance JPEG 2000 coprocessor," in *Proceedings of the IEEE International Symposium on Circuits and Systems*, vol. 2, pp. 225–228, May 2004.

82

VLSI Architectures for Forward Error-Control Decoders

Arshad Ahmed
Texas Instruments, Inc.

Seok-Jun Lee
Texas Instruments, Inc.

Mohammad Mansour
American University of Beirut

Naresh R. Shanbhag
University of Illinois at Urbana-Champaign

CONTENTS

82.1   Introduction

The publication of Shannon's ground breaking paper [1,2] in 1948, identifying the fundamental limits of information transfer capacity of noisy communication channels, resulted in the birth of information theory and coding theory as two distinct areas of research. Coding theory involves the investigation of efficient error control codes (ECC), the study of their performance bounds, and the determination of computationally efficient decoding algorithms. In fact, the present moment is historic in another sense as the tremendous advances in the theory and practice of coding theory converges with the semiconductor industry trend embodied by Moore's Law [3]. The latter has enabled the implementation of decoders that were considered too complex to implement just a few years ago. As a result, it is not uncommon to see codes such as low-density parity check (LDPC) codes [4] being "rediscovered" and incorporated in next generation communications standards.

A well-designed code can enable orders-of-magnitude reduction in the bit error-rate (BER) with a nominal cost in terms of power and latency. ECC can be said to have singularly resulted in the tremendous growth in communication infrastructure that has transformed modern society. This is evident in the pervasiveness of the Internet, cellular phones, modems, and wireless systems today and the continuing emergence of newer generations of communications standards ranging from digital video broadcast (DVB-H,T-DMB [5,6]), wireless metropolitan area network (MAN) (IEEE 802.16 [7]), wireless local area network (LAN) (IEEE 802.11g [8]), wireless personal area network (Bluetooth, [9]), 4G [10], fiber communications (OC-192 [11]), asymmetric digital subscriber lines (ADSL [12]), backplane/storage area networks (ANSI X3.230 [13]), and many others. The availability of a reliable, low-cost implementation substrate such as silicon has made integrated circuits (ICs) for communications an important area of research in industry as well as in academia.

A key component of any communications standards mentioned above is the ECC and a key requirement in being able to deliver low-cost silicon is a deep understanding of the fundamentals of ECC performance and their very large-scale integrated circuit (VLSI) architectures. This chapter presents the basics of the theory underlying some of the commonly employed ECCs as well as some of the more advanced ones such as LDPC along with discussion of issues involved in a VLSI implementation of ECC decoders. This chapter is organized as follows: in Section 82.2, we present Reed–Solomon (RS) decoders followed by convolutional codes, the Viterbi algorithm, and Viterbi decoder architectures in Section 82.3. In Section 82.4, we present turbo decoders that were discovered in the early 1990s and are considered to be a major breakthrough in coding theory, and since then have been incorporated in many wireless communication standards. Finally, in Section 82.5, we present LDPC decoders which are considered as a competitor to turbo codes and are being evaluated for inclusion in next-generation communication standards.

82.2 RS Decoder Architectures

RS codes [14–16] are employed in numerous communications systems. These include deep space, digital subscriber loops, data storage, wireless, and more recently, optical systems. The drive toward higher data rates makes it necessary to devise very high-speed implementations of decoders for RS codes.

An (n,k) RS code is a mapping from k *data symbols* to n *code-symbols*. The (255, 239) RS code is a good example. A data or code-symbol is an m-bit word where $m = 8$ (i.e., a byte) is typical. Such a symbol is said to lie in a finite field or a Galois field of size 2^m ($\mathrm{GF}(2^m)$), e.g., $\mathrm{GF}(256)$ is the set of all 8-bit numbers. Usually, $n = 2^m - 1$ and the construction of an (n,k) RS code guarantees that up to $t = \lfloor (n-k)/2 \rfloor$ symbol errors are correctable. For example, up to 8 symbol errors are correctable in a (255, 239) code.

Inclusion of an (n,k) RS code in a communication system results in code rate $R = k/n$ and an increase in line rate by a factor of $1/R$. For example, earlier in this decade, optical links at OC-192 data rates (9.953 Gb/s) needed to increase their data rates to 10.6 Gb/s to include the overhead owing to the inclusion of (255, 239) RS code. This coding overhead is considered quite acceptable because the code provides a few orders-of-magnitude reduction in the BER for the same signal-to-noise ratio (SNR).

82.2.1 Finite-Field Arithmetic

An RS encoder and decoder is composed of finite-field arithmetic units such as the finite-field adder, multiplier, and divider. Finite-field arithmetic units map operands from $\mathrm{GF}(2^m)$ back into $\mathrm{GF}(2^m)$ and thus differ from conventional integer arithmetic such as two's complement. Finite-field computations are elegantly described if we view elements of $\mathrm{GF}(2^m)$ as polynomials with degree at most $m - 1$, e.g., the element 10000101 in GF(256) can be represented as the polynomial $1 + x^5 + x^7$. Note that the coefficients in the polynomial representation belong to the binary set 0, 1. More importantly, these coefficients belong to GF(2) in which addition is simply an exclusive-OR (EXOR) operation and multiplication is an AND operation.

The addition of two GF(256) elements $A(x)$ and $B(x)$ is denoted by $A(x) \oplus B(x)$ and is simply the addition of the two polynomials. Thus, finite-field addition of two operands is a bit-wise EXOR of the two operands and hence a GF(256) adder is an array of 8 EXOR gates.

Finite-field multiplication of $A(x)$ and $B(x)$ is given as

$$Y(x) = [A(x)B(x)] \bmod P(x) \tag{82.1}$$

where $P(x)$ is a *primitive polynomial* with degree m. The root of $P(x)$ is denoted as α and holds special significance as all nonzero elements of GF(2^m) can be expressed as powers of α. A divider is simply multiplication by the inverse of the divisor.

In this section, we will consider finite-field arithmetic units as being combinational logic circuits. Numerous techniques and architectures exist for efficiently implementing finite-field arithmetic units (see Ref. [17] for details).

82.2.2 RS Encoder

The encoding and decoding processes are elegantly described if we view the k data symbols and n code symbols as polynomials over GF(2^m), i.e., polynomials whose coefficients are in GF(2^m).

The encoding process is best described in terms of the *data polynomial* $D(z) = d_{k-1}z^{k-1} + d_{k-2}z^{k-2} + \cdots + d_1 z + d_0$, where $(d_{k-1}, d_{k-2}, \ldots, d_1, d_0)$ are k m-bit data symbols, being transformed into a *codeword polynomial* $C(z) = c_{n-1}z^{n-1} + c_{n-2}z^{n-2} + \cdots + c_1 z + c_0$, where $(c_{n-1}, c_{n-2}, \ldots, c_1, c_0)$ are n m-bit code symbols. These code symbols are transmitted over the communication channel (or stored in memory).

All codeword polynomials $C(z)$ are polynomial multiples of $G(z)$, the *generator polynomial* of the code, which is defined as

$$G(z) = \prod_{i=0}^{2t-1} (z - \alpha^{m_0+i}) \tag{82.2}$$

where m_0 is typically 0 or 1 and α is the primitive element. However, other choices sometimes simplify the decoding process slightly. All communication standards specify $G(z)$, e.g., $G(z) = \prod_{i=0}^{2t-1}(z - \alpha^i)$ for wireless MAN systems [18].

Since $2t$ consecutive powers $\alpha^{m_0}, \alpha^{m_0+1}, \ldots, \alpha^{m_0+2t-1}$ of α are roots of $G(z)$, and $C(z)$ is a multiple of $G(z)$, it follows that

$$C(\alpha^{m_0+i}) = 0, \quad 0 \leq i \leq 2t - 1 \tag{82.3}$$

for all codeword polynomials $C(z)$. In fact, an arbitrary polynomial of degree less than n is a codeword polynomial if and only if it satisfies Eq. (82.3).

A *systematic* encoder produces codewords that are comprised of data symbols followed by *parity-check symbols*, and is obtained as follows: Let $Q(z)$ and $P(z)$ denote the quotient and remainder, respectively when the polynomial $z^{n-k}D(z)$ of degree $n - 1$ is divided by $G(z)$ of degree $2t = n - k$. Then, the codeword $C(z)$ is given by

$$(c_{n-1}, c_{n-2}, \ldots, c_1, c_0) = (d_{k-1}, d_{k-2}, \ldots, d_1, d_0, -p_{n-k-1}, -p_{n-k-2}, \ldots, -p_1, -p_0)$$

and consists of data symbols followed by parity-check symbols. The architecture of a systematic RS encoder is shown in Figure 82.1.

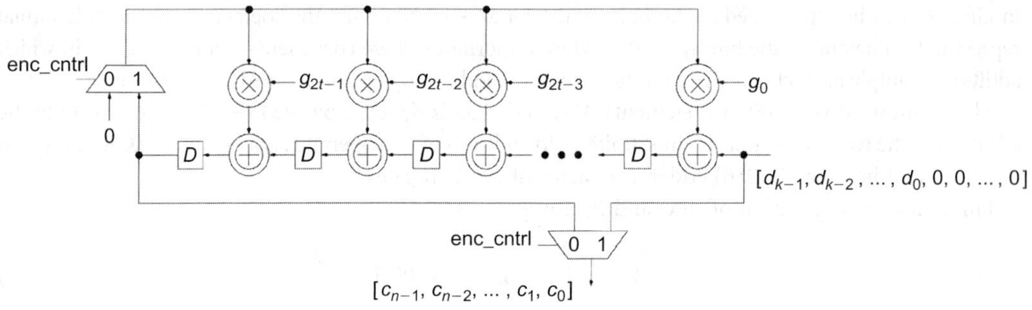

FIGURE 82.1 Systematic RS encoder architecture.

In broadband communication systems, the RS encoder is typically followed by a convolutional encoder. In the absence of a convolutional encoder, the code symbols are input to a *channel modulator*, which maps the code symbols on to a signal constellation [19]. The choice of the signal constellation depends on the transmit power, the channel frequency response and SNR, the data rate, and also the required BER at the receiver. Typical signal constellations include BPSK, QPSK, and QAM [19].

82.2.3 RS Decoder

RS decoders are classified into two categories: hard-decision RS decoders and soft-decision RS decoders. A hard-decision RS decoder receives one symbol per transmitted symbol and no other information. In contrast, a soft-decision decoder may receive one or more symbols for each transmitted code symbol in addition to information regarding the reliability of these symbols. We focus on hard-decision decoders in this section as they constitute the majority of decoders in current implementations. A brief introduction to soft-decision decoders is provided at the end of this section.

The input to the hard-decision decoder for each transmitted codeword is the set of n received symbols: $(r_{n-1}, r_{n-2}, \ldots, r_0)$ termed the *received word*. Analogous to the transmitted codeword polynomial $C(z)$, the received word can also be viewed as a polynomial $R(z)$. Any of the received symbols in $R(z)$ may be in error. Hence, the received word $R(z)$ can be written as

$$R(z) = C(z) + E(z) \tag{82.4}$$

where the number of nonzero coefficients in the *error-polynomial* $E(z)$ depends on the number of erroneous symbols v within the received word $R(z)$. Since the maximum number of decoder correctable errors t in a received word is usually much smaller than the code polynomial degree $n-1$, for typical high rate codes, $E(z)$ can be compactly expressed as

$$E(z) = Y_1 z^{i_1} + Y_2 z^{i_2} + \cdots + Y_v z^{i_v} \tag{82.5}$$

where $Y_1, Y_2, \ldots, Y_v$ are the *error values* and $X_1 = \alpha^{i_1}, X_2 = \alpha^{i_2}, \ldots, X_v = \alpha^{i_v}$ are the *error locations*. A hard-decision decoder can detect and correct all the erroneous symbols as long as $v \leq t$. An efficient RS decoder determines the error values and error locations in four stages, which are listed below:

1. Syndrome polynomial computation
2. Key equation solution
3. Chien formula-based *error location search*
4. Forney formula-based *error value evaluation*

The architectures for each of these steps will be discussed in the remainder of this section.

(1) *Architectures for syndrome polynomial computation.* Syndrome computation is the evaluation of the received word at the roots of the transmitted codeword or the generator polynomial $G(z)$. The syndromes s_i for $0 \le i < 2t$ are given by

$$s_i = R(\alpha^{m_0+i})$$

From Eq. (82.3) and Eq. (82.4), it follows that $R(\alpha^{m_0+i}) = E(\alpha^{m_0+i})$. Hence, syndrome computation is nothing but the evaluation of the error polynomial at $2t$ finite-field locations. Syndrome computation is typically performed using Horner's rule which computes $R(\alpha^{m_0+i})$ as

$$R(\alpha^{m_0+i}) = ((r_{n-1}\alpha^{m_0+i} + r_{n-2})\alpha^{m_0+i} + \cdots + r_1)\alpha^{m_0+i} + r_0 \tag{82.6}$$

The $n-1$ multiplications and n additions in Eq. (82.6) are recursively computed using a multiplier–accumulator (MAC). The architecture for computing $2t$ syndromes in parallel using $2t$ multipliers and adders with Horner's rule is shown in Figure 82.2(a).

This architecture requires n clock cycles to complete the computation. It should be noted that one of the operands to the multiplier is a constant. The complexity of constant input multipliers is roughly half that of a regular multiplier. One can further reduce area by pipelining the multipliers and folding more than one syndrome value computation onto the same MAC.

High-throughput syndrome computation architectures are obtained by using a two-stage implementation as shown in Figure 82.2(b). In the first stage, l input received symbols are passed through each MAC to compute a partial product of $R(\alpha^{m_0+i})$. In the second stage, another multiplier–adder pair multiplies and adds the partial products together. Both stages are based on Horner's rule.

Given the syndrome values $s_0, s_1, \ldots, s_{2t-1}$, the syndrome polynomial is defined as

$$S(z) = s_0 + s_1 z + \cdots + s_{2t-1} z^{2t-1}$$

The syndrome polynomial is input to the key equation solver which is discussed next.

(2) *Architectures for key equation solver.* The key equation solver block contains recursion and hence determines the throughput of the entire RS decoder. The outputs of the key equation solver are the *error*

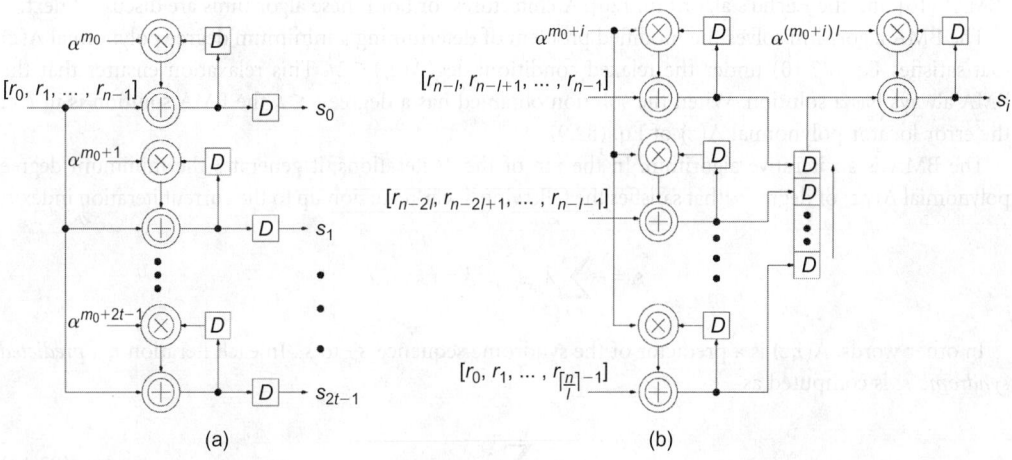

FIGURE 82.2 Syndrome computation architectures: (a) low-area architecture, and (b) high-throughput architecture.

locator polynomial $\Lambda(z)$, and the *error evaluator* polynomial $\Omega(z)$ defined as

$$\Lambda(z)=\prod_{j=1}^{v}(1-X_j z)=1+\lambda_1 z+\lambda_2 z^2+\cdots+\lambda_v z^v \tag{82.7}$$

$$\Omega(z)=\sum_{i=1}^{v}Y_i X_i^{m_0}\prod_{j=1,j\neq i}^{v}(1-X_j z)=\omega_0+\omega_1 z+\cdots+\omega_{v-1}z^{v-1} \tag{82.8}$$

where $v \leq t$ are the number of symbol errors and the v roots of the error locator polynomial are the error locations obtained via Chien search (described in Section 82.1.4). The error evaluator polynomial contains information regarding the error values that are obtained using Forney's formula as shown in Section 82.1.4.

The key equation relates the syndrome polynomial $S(z)$, the error locator polynomial $\Lambda(z)$, and the error evaluator polynomial $\Omega(z)$ as follows:

$$\Lambda(z)S(z) \equiv \Omega(z)\bmod z^{2t} \tag{82.9}$$

The key equation solver determines the polynomials $\Lambda(z)$ and $\Omega(z)$ using $S(z)$. From Eq. (82.9), it is easily seen that once $\Lambda(z)$ is determined, $\Omega(z)$ is obtained as the coefficients of the first $2t$ terms in the product $\Lambda(z)S(z)$. In addition, we have from the linear complexity property [16] of the finite-field Fourier transform that the error locator satisfies the following linear recursion:

$$s_k=-\sum_{j=1}^{v}\lambda_j s_{k-j}, \quad k=v,\ldots,2t-1 \tag{82.10}$$

In other words, the error locator polynomial $\Lambda(z)$ is a minimal degree predictor of the syndrome sequence $\{s_v, \ldots, s_{2t-1}\}$. Hence, $\Lambda(z)S(z)$ has zero coefficients for terms with degree v to $2t-1$. Consequently, $\Omega(z)$ is also guaranteed to have zero coefficients for degrees v to $2t-1$.

Hence using Eq. (82.10), the key equation can be rewritten as:

$$\Lambda(z)S(z) = \Omega(z) + z^{2t}\Omega^{(h)}(z) \quad \text{with } deg(\Omega(z)) \leq v \tag{82.11}$$

Solving the key equation comprises two separate problems: determining the least degree polynomial $\Lambda(z)$ (with degree v) such that the product $\Lambda(z)S(z)$ has zero coefficients from degree v to degree $2t-1$. Note that, since the maximum correctable errors is limited to t, all valid error locators have degree $v \leq t$. This is followed by determining the product $\Lambda(z)S(z)$ up to degree $v-1$. Typically, both the problems of determining the error locator and the error evaluator are solved together in an iterative process.

Two main iterative algorithms available for solving the key equation are: the Berlekamp–Massey algorithm (BMA) [16] and the Euclid's algorithm [20]. Architectures for both these algorithms are discussed next.

The BMA algorithm solves the modified problem of determining a minimum degree polynomial $\Lambda(z)$ that satisfies Eq. (82.10) under the relaxed condition: $deg(\Lambda(z)) \leq 2t$. This relaxation ensures that the BMA always has a solution. When the solution obtained has a degree $v \leq t$, the BMA solution is in fact the error locator polynomial $\Lambda(z)$ of Eq. (82.9).

The BMA is an iterative algorithm. In the rth of the $2t$ iterations, it generates the minimum degree polynomial $\Lambda(r,z)$ of degree v_r that satisfies the following linear recursion up to the current iteration index r:

$$s_k=-\sum_{j=1}^{v_r}\lambda_{j,r}s_{k-j}, \quad k=v_r,\ldots,r$$

In other words, $\Lambda(r,z)$ is a predictor of the syndrome sequence s_{v_r} to s_r. In each iteration r, a *predicted syndrome* $\hat{s}_r$ is computed as

$$\hat{s}_r=-\sum_{j=1}^{v_{r-1}}\lambda_{j,r-1}s_{r-j} \tag{82.12}$$

This is subtracted from the actual syndrome value s_r to obtain the *discrepancy coefficient* $\Delta_r = s_r - \hat{s}_r$. If $\Delta_r \neq 0$, then the error locator is updated by addition of a scaled version of the error locator from a previous iteration called the *scratch polynomial*. The scaling constant is chosen to reduce the discrepancy coefficient to zero. The scratch polynomial is updated every iteration to ensure that the minimum degree property of the error locator is maintained.

From an architectural perspective, the throughput bottleneck occurs in computing the discrepancy coefficient. This is because the critical path in computing Eq. (82.12) passes through at least one multiplier and a logarithmic number of adders implementing Eq. (82.12). Updating the error locator and error evaluator polynomials has a fixed complexity and involves multiplication of a polynomial with a scalar constant followed by component-wise addition. These operations have a critical path passing through a multiplier and an adder.

A systolic architecture to reduce the critical path has been described in [21]. The main idea behind this architecture is to iterate the discrepancy coefficients in a manner similar to the computation of the error locator polynomial. With this modification, the architecture becomes extremely regular. The critical path within the iteration is then reduced to a multiplier followed by an adder. This is turn increases the achievable throughput of the RS decoder.

The pseudocode of the reformulated inversionless Berlekamp–Massey (RiBM) algorithm from Ref. [20] is given in Figure 82.3. The systolic architecture implementing the pseudocode is shown in Figure 82.4. In the pseudocode, $\delta_i(0)$ (for $0 \leq i < 2t$) denotes the initial value of the discrepancy polynomial and $\theta_i(0)$ (for $0 \leq i < 2t$) is the initial value of the corresponding scratch polynomial. Note that the $\delta_{3t}(0)$ corresponds to the initial error locator polynomial (set to unity), and $\theta_{3t}(0)$ corresponds to its scratch polynomial (also initialized to unity). The architecture shown in Figure 82.4(a) consists of an array of $3t$ identical processing elements. The details of the processing elements are shown in Figure 82.4(b). As seen in this figure, the critical path of the architecture passes through one multiplier and an adder. With each iteration, all the terms of the error locator and the discrepancy polynomial move one position to the left. Hence, the discrepancy coefficient required in the current iteration (denoted by $\delta_0(r)$ in Figure 82.4) is always available at the leftmost processing element. At the end of $2t$ iterations, the error locator is available at processing elements from t to $2t$ and $\Omega^{(h)}(z)$ from Eq. (82.11) at processing elements numbered 0 to $t - 1$. More details on the reformulation leading to this algorithm and further architectural details are available in Ref. [21].

In addition to BMA, the key equation can also be solved by using the well-known Euclid's algorithm [20] for determining the greatest common divisor (GCD) of two polynomials. RS decoder implementation

```
Initialization: δ₃ₜ(0) =1, θ₃ₜ(0) =1;  δᵢ(0) =0  for  i=2t, ..., 3t-1  k(0)=0  γ(0) =1
Input: sᵢ, i=0,1, ...,2t-1
       δᵢ(0)  =  θᵢ(0)  =  sᵢ, (i  =  0,1, ...,2t-1)
    for  r=0  to  2t-1  do
       Step  RiBM.1  δᵢ(r+1)  =  γ(r)δᵢ₊₁(r)  -  δ₀(r)·θᵢ(r),(i  =  0,1, ...,3t)
       Step  RiBM.2  if  δ₀(r)  ≠  0  and  k(r)  ≥  0
          θᵢ(r+1)  =  δᵢ₊₁(r),  (i  =  0,1, ...,3t)
          γ(r+1)  =  δ₀(r)
          k(r+1)  =  -k(r)-1
       else
          θᵢ(r+1)  =  θᵢ₊₁(r),  (i  =  0,1, ...,3t)
          γ(r+1)  =  γ(r)
          k(r+1)  =  k(r)+1
       end if
    end for
Output:  λᵢ  =  δₜ₊ᵢ,(i  =  0,1, ...,t);  ωᵢ  =  δᵢ(2t),  (i  =  0,1, ...,t  -  1).
```

FIGURE 82.3 Pseudocode of the RiBM algorithm.

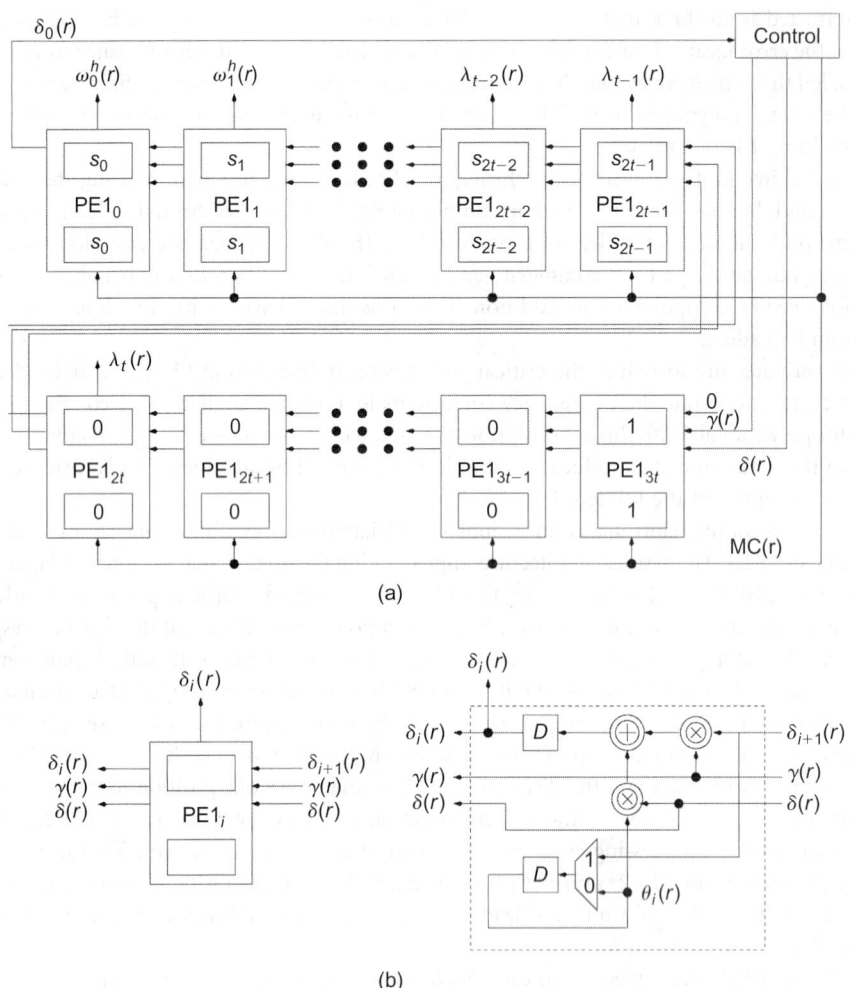

FIGURE 82.4 RiBM architecture: (a) systolic array, and (b) processing element.

based on the Euclid's algorithm is a well-researched problem [22–24]. In general, if $A(z)$ and $B(z)$ are two polynomials with coefficients over a field, and if $G(z)$ is the GCD of the two polynomials, then Euclid's algorithm expresses $G(z)$ as a function of $A(z)$ and $B(z)$, i.e.,

$$G(z) = R(z)A(z) + T(z)B(z) \tag{82.13}$$

Comparing Eq. (82.11) and Eq. (82.13), we see that the key equation is equivalent to finding the GCD $G(z)$ (error evaluator $\Omega(z)$) and $R(z)$ (error locator $\Lambda(z)$) given the syndrome polynomial $S(z) = A(z)$ and $B(z) = x^{2t}$. Euclid's algorithm finds the GCD through a process of iterative division of two polynomials and their remainders. This is followed by a traceback of the iterations to find $R(z)$. However, a modified form of the Euclid's algorithm called the *extended Euclid's algorithm* determines both the GCD (error evaluator) and the coefficient polynomial (error locator) in the same set of iterations. The pseudocode of the extended Euclid's' algorithm, based on Ref. [22] is shown in Figure 82.5. In this figure, a_i and b_i are the leading coefficients of $R_i(z)$ and $Q_i(z)$, respectively. This pseudocode can be implemented on a systolic array of $2t$ processing elements similar to that of the systolic BMA. The

```
Initialization: R₀(z) = S(z); Q₀(z) = z²ᵗ; Θ(z) = 0  Φ(z) = 1
    for  i = 0 to 2t-1 and deg(Rᵢ) > t do
        lᵢ = deg(Rᵢ(z)) - deg(Qᵢ(z))
        if (lᵢ ≥ 0)
            Rᵢ₋₁(z) = bᵢRᵢ(z) - zˡⁱaᵢQᵢ(z)
            Qᵢ₊₁(z) = Qᵢ(z)
            Θᵢ₊₁(z) = bᵢΘᵢ(z) - zˡⁱaᵢΦᵢ(z)
            Φᵢ₊₁(z) = Φᵢ(z)
        else
            Rᵢ₋₁(z) = aᵢQᵢ(z) - zˡⁱbᵢRᵢ(z)
            Θᵢ₊₁(z) = Rᵢ(z)
            Θᵢ₊₁(z) = aᵢΦᵢ(z) - z⁻ˡⁱbᵢΘᵢ(z)
            Φᵢ₊₁(z) = Θᵢ(z)
        end if
    end for
Output: Λ(z) = Θ(z);  Ω(z) = Rᵢ₊₁(z) .
```

FIGURE 82.5 Pseudocode of the extended Euclid's algorithm.

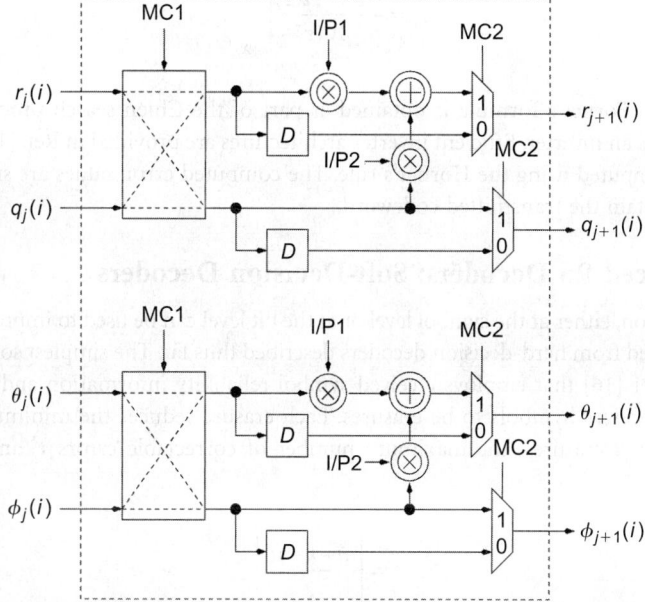

FIGURE 82.6 Processing element of the extended Euclid's architecture.

structure of one processing element [22] is shown in Figure 82.6. Further details on this implementation can be found in Ref. [22].

Both the extended Euclid's architecture and the **RiBM** architecture have the same critical path and require the same number of iterations for processing a syndrome polynomial. If the number of multipliers is taken as a representative of the hardware complexity, then the systolic **RiBM** architecture requires $6t$ multipliers whereas the systolic extended Euclid's architecture requires $8t$ multipliers. In addition, the control circuitry of the Euclid's architecture is more complex compared to the **RiBM** architecture. Pipelining the BM architecture and the extended Euclid's architecture is discussed in Ref. [21] and Ref. [23], respectively.

82.2.4 Chien Search and Forney's Formula

The roots of the error locator polynomial $\Lambda(z)$ can be obtained from the *Chien search* procedure [25]. The error values at the error locations can be determined from the *Forney formula* [26]. Chien search involves the evaluation of the error locator polynomial at all the elements of the finite field and the flagging of those elements at which the polynomial evaluated to zero. Note that, from Eq. (82.7), the roots of the error locator are X_i^{-1} for each error location $X_i = \alpha^i$. Since the error locator degree is constrained to a maximum of t, we require t multiplications and $t + 1$ additions to evaluate the error locator at one finite-field symbol value. Hence the complexity of Chien search is nt multiplications and $n(t + 1)$ additions. Note that Chien search can be implemented via the Horner's rule as discussed in connection with syndrome computation. The highest throughput is obtained when all the evaluations are performed in parallel using n MACs while minimum area implementations perform all the evaluations serially using one MAC. The Chien search architecture is chosen on the basis of the throughput of the other blocks such as the key equation solver architecture.

If $deg(\Lambda(z)) \leq t$ and all the roots of $\Lambda(z)$ lie in the Galois field of operation, then the error values at these error locations are found using the Forney algorithm. Given the error location X_i, the error locator $\Lambda(z)$, and the error evaluator $\Omega(z)$, the error value Y_i is given by Forney's formula [26] as

$$
Y_i = -\frac{z^{m_0-1}\Omega(z)}{\Lambda'(z)}\bigg|_{z=X_i^{-1}}
\tag{82.14}
$$

The denominator in Forney's formula is obtained as part of the Chien search procedure. The Forney formula still requires an inverter. Efficient inverter architectures are provided in Ref. [17]. The numerator in Eq. (82.14) is computed using the Horner's rule. The computed error values are subtracted from the received word to obtain the transmitted codeword.

82.2.5 Advanced RS Decoders: Soft-Decision Decoders

Reliability information, either at the symbol level or at the bit level can be used to improve on the bit-error performance obtained from hard-decision decoders described thus far. The simplest soft-decision decoder is an *erasure* decoder [16] that employs received symbol reliability information and a preset threshold to declare certain received symbols to be erasures. Each erasure reduces the minimum distance of the code by unity. With μ erasures, the maximum number of correctable errors t' among the unerased symbols is given by

$$
t' = \left\lfloor \frac{n-k-\mu}{2} \right\rfloor
$$

A hard-decision decoder can perform erasure decoding with modification to the initialization of the error locator and error evaluator. Hence, the complexity of a erasure decoder is similar to that of a hard-decision decoder. If the erased locations contain some of the erroneous received symbols, then the total number of errors that can be corrected is greater than that of the hard-decision decoder. The maximum number of received symbols that can be erased is however limited by the minimum distance of the code.

An extension of the erasure decoding idea is the *generalized minimum distance* (GMD) algorithm [27], that iteratively increases the number of erasures and checks for decodability at each iteration. This algorithm performs t erasure decoding trials with the number of erasures increased by two for each trial. The complexity of GMD decoding can be reduced by using the results of the previous trial as an initialization of the current trial. This version of the GMD decoding algorithm is discussed in Ref. [28].

Another soft-decision decoding algorithm primarily employed for RS codes over small fields is the *ordered statistics decoding* (OSD) algorithm [29,30]. OSD uses bit-level reliabilities to reduce the parity check matrix of the code to a form having unit weight columns for low-reliability bits. This is followed by flipping low-reliability bits that do not satisfy the binary parity check equations thereby providing an initial estimate of the transmitted codeword. A search is conducted over sets of high-reliability bits to obtain a codeword closer to the received word. The size of the search set of high-reliability bits is typically limited to two or three as the search complexity is exponential in the number of these bits.

Algebraic soft-decision decoding [31] is a recently developed algorithm for soft-decoding RS codes. This technique assumes more than one received symbol and corresponding reliability information for each transmitted symbol. Coordinate pairs are then formed using the code-evaluation point and the received symbol. A bivariate (two-dimensional) finite–field polynomial is generated that passes through each of these coordinate points. The number of times the polynomial passes through a particular coordinate point is determined by the reliability of the received symbol. The curve is then factorized to obtain the transmitted data polynomial. Architectures for algebraic soft-decoding have been proposed in Refs. [32,33] and are presently an active area of research.

82.3 Viterbi Decoder Architectures

Convolutional codes find use in a wide range of applications such as magnetic disk-drive channels, satellite, and in wireless communications. These codes are typically low-rate codes, i.e., $R = 1/2$ or $R = 1/3$ for most commonly employed codes. The Viterbi algorithm [34] is an efficient algorithm for decoding convolutional codes. The complexity of a Viterbi decoder is inversely proportional to the code rate unless the code rate is increased via *puncturing*.

Convolutional codes can be viewed as linear block codes with memory, i.e., the output of the linear encoder depends not only on the present information symbols but also on previous information symbols. Thus, a convolutional encoder is a finite-state machine (FSM) and the Viterbi algorithm is a particularly efficient method of estimating the state sequence of an FSM from a sequence of noisy observations of its output.

82.3.1 Convolutional Encoder

Consider an (n_o, k_o, K) convolutional code where the encoder is an FSM with $K - 1$ memory elements that encodes a k_o-bit data symbol into an n_o-bit code symbol as shown in Figure 82.7(a). The parameter K is called the constraint length of the code. At time index k, the $N = 2^{K-1}$ possible encoder states are denoted as $s_i(k)$ where $i = 0, \dots, N - 1$ and the branch connecting states $s_i(k)$ with $s_j(k + 1)$ is denoted as $b_{i,j}(k)$. Note that $s_i(k)$ and $b_{i,j}(k)$ are labels and not variables.

Given the current state $s_i(k)$ and an input symbol $u(k)$, the encoder transitions to a unique next state $s_j(k + 1)$ and generates a unique output symbol $c(k)$. Hence, for an input sequence of L data symbols $\mathbf{u} \triangleq (u(0), u(1), \dots, u(L - 1))$, the encoder starts from an initial state $s_0(0)$ and performs L state transitions

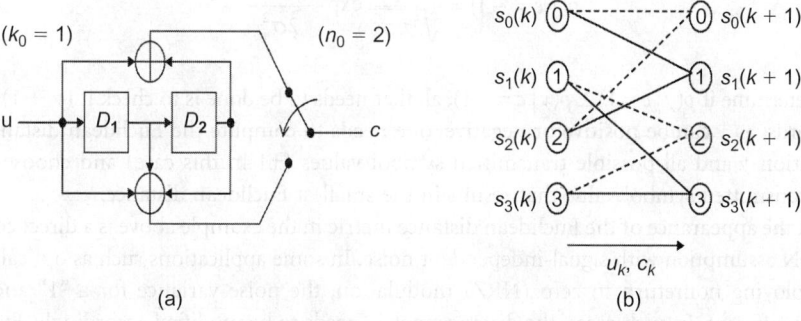

(a)   (b)

FIGURE 82.7 Example of a convolutional encoder: (a) a (2,1,3) convolutional encoder with 2 memory elements and modulo 2 adders, and (b) a trellis section resulting from state transitions of the encoders in (a).

while generating L, n_o-bit code symbols $\mathbf{c} \triangleq (c(0), c(1), \ldots, c(L-1))$. An efficient method of describing all possible state transitions in an encoder is by using a *trellis diagram*. A section of the trellis of the encoder of Figure 82.7(a) is depicted in Figure 82.7(b), where the solid edges correspond to $u(k) = 1$ and dashed edges correspond to $u(k) = 0$. Note that the number of branches emerging from or ending in any state is 2^{k_o}.

The received sequence $\mathbf{y} = (y(0), y(1), \ldots, y(L-1))$ at the receiver is simply a noisy version of the transmitted code sequence $\mathbf{c}$. Thus, the convolutional decoding problem can be defined as estimating the input sequence $\mathbf{u}$ given the noisy sequence $\mathbf{y}$. Two major approaches for estimating the transmitted bits $u(k)$ from $\mathbf{y}$ are: (1) maximum likelihood (ML) decoding, and (2) maximum a posteriori probability (MAP) decoding. MAP is known to be much more complex than ML decoding. Furthermore, the results of MAP and ML decoding are identical when the probability distribution of the input $u(k)$ is uniform. In 1966, Andrew Viterbi [34] proposed a computationally efficient procedure to implement ML decoding of convolutional codes. As a result, the Viterbi algorithm is almost universally applied whenever there is a convolutional code in a communication link. Hence, we will focus on the Viterbi algorithm and the ML decoding approach in this section.

Next, we provide a brief description of the ML decoding principle and then describe the Viterbi algorithm.

82.3.2   Maximum Likelihood Decoding and the Viterbi Algorithm

Maximum likelihood decoding involves the estimation of the transmitted symbol sequence $\mathbf{u}$ by determining the path in the trellis which maximizes the probability $p(\mathbf{y} \mid \mathbf{u})$. Hence, this approach is also called ML sequence estimation (MLSE). The ML principle is quite general though and can be applied equally well to the problem of estimating individual symbols, i.e., for symbol-by-symbol detection. For example, consider the case of ML symbol-by-symbol detection when the transmitted symbols c are ± 1, and the received symbols are $y = c + \eta$, where η is a zero mean additive white Gaussian noise (AWGN) with a probability density function (PDF) given by

$$p(\eta) = \frac{1}{\sqrt{2\pi\sigma_n^2}} \exp\frac{-\eta^2}{2\sigma_n^2} \tag{82.15}$$

where σ_n^2 is the noise variance. Clearly, $p(y \mid c)$ is given by

$$p(y|c = 1) = \frac{1}{\sqrt{2\pi\sigma_n^2}} \exp\frac{-(y-1)^2}{2\sigma_n^2} \tag{82.16}$$

$$p(y|c = -1) = \frac{1}{\sqrt{2\pi\sigma_n^2}} \exp\frac{-(y+1)^2}{2\sigma_n^2} \tag{82.17}$$

Hence, to determine if $p(y \mid c = 1) \geq p(y \mid c = -1)$, all that needs to be done is to check if $(y-1)^2 \leq (y+1)^2$. In other words, as y can be positive or negative, one needs to compute the Euclidean distance between the observation y and all possible transmitted symbol values (± 1 in this case) and choose as the ML result the transmitted symbol value that results in the smallest Euclidean distance.

Note that the appearance of the Euclidean distance metric in the example above is a direct consequence of the AWGN assumption with signal-independent noise. In some applications such as optical communications employing nonreturn to zero (NRZ) modulation, the noise variance for a "1" and a "0" are significantly different. In such cases, the distance metric needs to be modified accordingly. Furthermore, in situations where the noise is not Gaussian, there may not exist a distance metric and instead the PDFs themselves need to be estimated and employed in the decoding process.

The above description can be extended to the observed sequence **y** and the code sequence **c**, where the latter is known to be the output of the encoder FSM that takes **u** as the input. The decoder knows the number of states in the FSM and the state-transition matrix, but it does not know what state sequence has been traversed at any given time index. The job of the ML decoder is to determine the most likely state sequence traversed based on the observation **y** from which it can then determine the most likely code sequence **c** and more importantly, the most likely input sequence **u**. The ML decoder is said to execute an MLSE procedure.

As mentioned earlier, the Viterbi algorithm [34] was first proposed as an efficient procedure to compute the ML solution to the problem of decoding convolutional codes.

82.3.3   Viterbi Decoder Architectures

The Viterbi decoding algorithm determines the ML transmitted information sequence **u**. Recall that the receiver knows the encoder trellis and the noisy observations **y**. The Viterbi decoder consists of three types of computations: (1) *branch metric* computation implemented in a branch metric unit (BMU), (2) *state/path metric* computation implemented in a path metric unit (PMU), and (3) *survivor path* computation implemented in the survivor memory unit (SMU). Thus, a generic Viterbi decoder architecture will take the form shown in Figure 82.8.

We now describe each of the computations in more detail.

(1) *Branch Metric Computation.* We compute the *branch metric* $B_{i,j}(k)$ for the branch connecting states $s_i(k)$ and $s_j(k+1)$ and do so for every branch in the trellis. The branch metric is a measure of the probability of traversing the branch $b_{i,j}(k)$ given that the encoder is in state $s_i(k)$ and the received value is $y(k)$.

If the demodulator provides hard decisions such that $y(k)$ is binary, then the branch metric is the Hamming distance between $y(k)$ and the code symbol $c(k)$ associated with the branch. If $y(k)$ is a real number and the noise in the channel is AWGN, then the ML principle described in Section 82.3.2 results in the branch metric computation to be identical to the computation of the Euclidean distance given by

$$\sum_{l=1}^{n_o}(y_l(k)-c_l(k))^2 \tag{82.18}$$

where l indexes the individual components of $y(k)$ and $c(k)$, i.e., the samples and bits of $y(k)$ and $c(k)$, respectively. The BMU has a feed-forward architecture and hence can be pipelined easily and therefore is usually not in the critical path.

(2) *State/Path Metric Computation.* We compute the *state/path metric* $S_i(k)$ for each state in the trellis. The state metric $S_i(k)$ is a measure of the probability of the encoder arriving at state $s_i(k)$ given the observations $\{y(0), y(1), \dots, y(k-1)\}$. In computing $S_i(k)$, the Viterbi algorithm employs the branch metrics. Hence, it has knowledge of the most likely path that leads to the state $s_i(k)$. This path is called the *survivor path* and hence the state metric is also called the survivor path metric or simply the path metric. The survivor path needs to be stored in memory.

To understand the state metric computation, consider the two-state trellis shown in Figure 82.9 where the state metrics $S_0(k)$ and $S_1(k)$ corresponding to states $s_0(k)$ and $s_1(k)$, respectively, are employed as labels for the trellis nodes. The branches corresponding to $u(k)=0$ are dotted and the solid branches correspond to $u(k)=1$. The trellis branches are labeled with the corresponding branch

FIGURE 82.8   Block diagram of a generic Viterbi decoder.

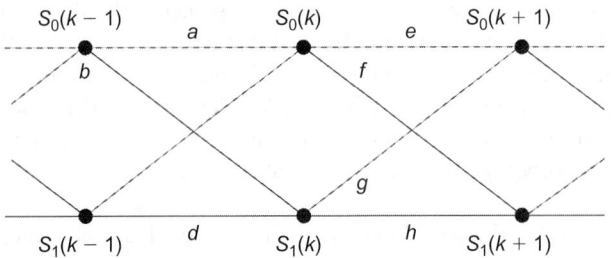

FIGURE 82.9 Two sections of a two-state trellis.

metrics. From Figure 82.9 we see that

$$S_0(k+1) = \min[S_0(k-1) + a + e, \ S_0(k-1) + b + g, \ S_1(k-1) + c + e, \ S_1(k-1) + d + g]$$

$$= \min[\min\{S_0(k-1) + a, \ S_1(k-1) + c\} + e, \ \min\{S_0(k-1) + b, \ S_1(k-1) + d\} + g]$$

$$= \min[S_0(k) + e, S_1(k) + g] \tag{82.19}$$

where $S_0(k) = \min[S_0(k-1) + a, \ S_1(k-1) + c]$ and $S_1(k) = \min[S_0(k-1) + b, \ S_1(k-1) + d]$. From Eq. (82.19), it is clear that the basic operation in determining the state metrics involves *addition* of the branch metric to the previous state metric followed by a *comparison* with other such products, and finally followed by a *selection* of the minimum of all possible path metrics. Therefore, the state metrics are computed as

$$S_j(k+1) = \min_i \ \{S_i(k) + B_{i,j}(k)\}$$
$$d_{s_j}(k) = u(k - K + 1) \tag{82.20}$$

where the min is over all 2^{k_o} states at time index k that are connected to state $s_j(k+1)$ and $d_{s_j}(k)$ referred to as the *decision*, is the value held by the earliest bit in the encoder shift-register. Note that the availability of the decision (earliest) bit makes it possible to determine the state at time index $k-1$ from which the survivor path was extended to state $s_j(k)$. This information is useful in tracing back the survivor path to generate the decoded bits. For each state at time k, indexing the 2^{k_o} incoming branches with the earliest encoder bits at time index $k-1$ enables the decision bit to also indicate the branch of the survivor path. Eq. (82.20) is referred to as the add-compare-select (ACS) operation and is implemented in an ACS unit (ACSU) that forms a key processing kernel in the PMU. The PMU is usually the throughput determining block of a Viterbi decoder and hence we will discuss the design of this block in somewhat greater detail.

The PMU can be designed in a state-parallel implementation where all the N states metrics are computed in parallel using N ACSUs, or serially or a hybrid. A direct implementation of an ACSU is shown in Figure 82.10(a) which includes a 2^{k_o}-input CS block. In this figure, $\hat{S}_{i,j}(k)$ for $0 \le i < 2^{k-o}$ are the state metrics of the 2^{k_o} states connected to state j. The 2^{k_o}-input CS block can be implemented as a tree of two-input CS blocks as shown in Figure 82.10(b) (for $k_0 = 2$). The two input CS block compares the MSB of the two inputs and proceeds down to the LSB. If any two input bits differ, then the CS block flags a decision and selects the smaller of its inputs as the output. The critical path of the ACSU consists of a carry ripple in the adder (LSB to MSB) followed by a compare ripple (MSB to LSB). The critical path of the ACSU can be reduced by using redundant carry-save MSB first computation [35].

Two algorithmic techniques for increasing throughput are *parallel processing* and *higher radix processing* [36]. In parallel processing, the input sequence is divided into subblocks and more than one subblock is processed at a time. The other option for increasing throughput is to process more than one trellis section at a time. This process is termed higher radix processing. Figure 82.11, shows the concept of merging two trellis sections of the encoder shown in Figure 82.7(a). This leads to radix-4 processing.

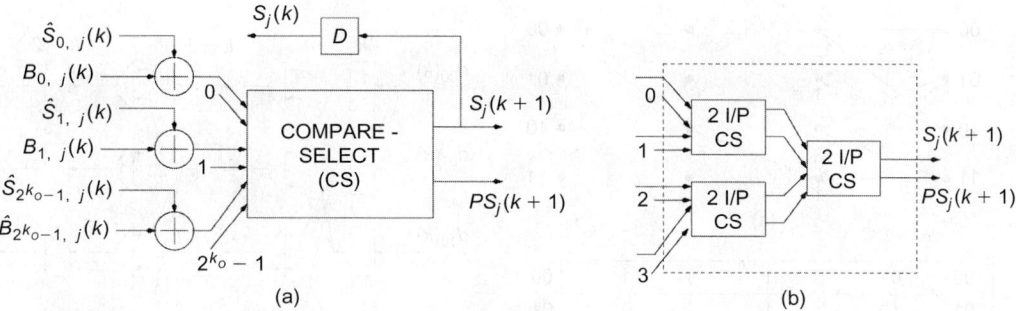

FIGURE 82.10 The ACSU: (a) block diagram; (b) tree-based implementation of the CS processing element.

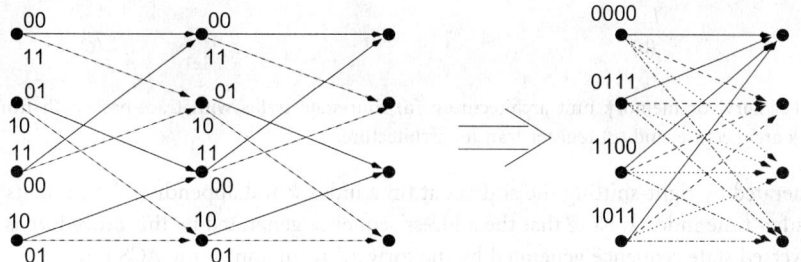

FIGURE 82.11 Modification of trellis diagram from radix-2 to radix-4.

The output code symbols on the branches going to the zeroth state are shown in the radix-4 trellis in Figure 82.11. It is clear that the complexity of BMU doubles in going from conventional radix-2 to radix-4 processing; twice as many branches now exist and each branch metric computation requires the processing of twice the number of inputs. Further, the number of comparisons in the PMU also doubles. However, the additions are performed outside the ACS loop leading to a net speed up of the PMU which in turn leads to a throughput increase compared to radix-2 processing. The achievable throughput increase has been observed to lie between 1 and 2 for radix-4 processing.

(3) *Survivor path computation.* Each path metric update in the PMU results in the extension of the survivor paths to each of the N states $s_j(k)$ from one of the N states $s_i(k-1)$. These survivor paths need to be stored and processed to determine the decoder output. Typically, as a rule of thumb, the survivor path length or the survivor memory depth L is chosen to be four to five times the constraint length K. Doing so results in the survivor paths of all the states converging to a single path for time index smaller than $k - L$ with a probability approaching one. This unique path is the maximum likelihood path and the input labels or the decisions $d_{s_i}(k - L - l)(l \geq 1)$ (see Eq. (82.20)) are the decoded outputs of the Viterbi decoder.

There are two main approaches to implementing the SMU: (1) trace-back [37] and (2) register-transfer [38]. Figure 82.12(a) shows a 4-state trellis with the branch labels indicating the decisions and the states being marked as shown. In this figure, the MSB of the state-index represents the earliest encoder bit and the LSB represents the most recent encoder bit. It is seen that the branch labels correspond to the MSB of the state from which the branch originates.

The survivor path memory in Figure 82.12(b) is organized in an N row-by L column format where each row corresponds to a state and each column corresponds to a time index in the range k to $k - L$. Thus, in each time index, the vector of decisions from Eq. (82.20) are stored in the corresponding column. In trace-back, the survivor path of any state at time k is chosen to be traced back. The trace-back procedure involves recursively reading the contents of the survivor path memory shown in Figure 82.12(b) from time index k to $k - L$. The address, which is essentially the row index of the memory, for time index

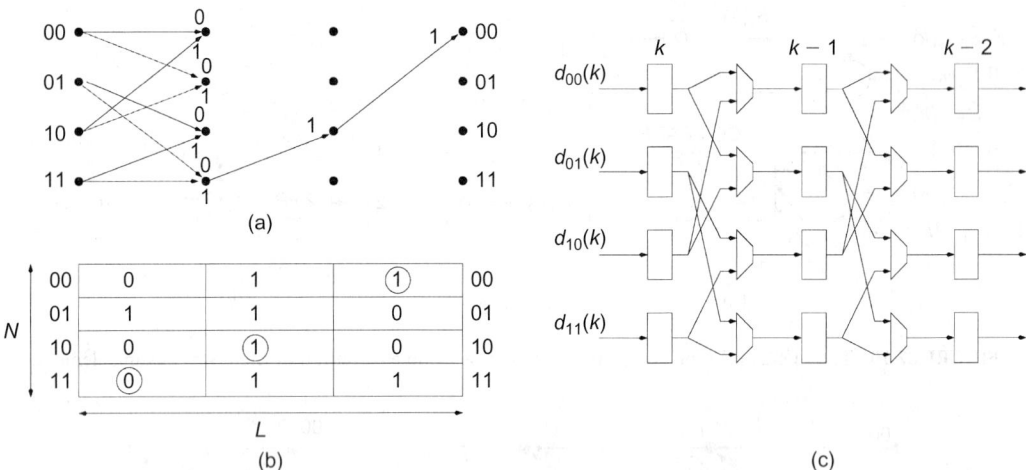

FIGURE 82.12 Survivor memory unit architectures: (a) four-state trellis with trace-back path from zero state, (b) trace-back architecture, and (c) register-transfer architecture.

$k - 1$ is generated by right-shifting the address at time index k and appending the contents of memory location read at time index k. Note that the address sequence generated by this procedure is identical to the time-reversed state sequence generated by the forward recursion of the ACS unit.

Figure 82.12(c) shows the architecture of a register-transfer architecture of the SMU. In this approach, the survivor paths and not the vector of decisions, are stored in an L bit register (*path register*) for every state. At each time index k, the survivor path gets updated based on the decision made by the ACSU. The most recent bit in the register for a state $s_i(k)$ is always equal to $d_{s_j}(k)$. However, the other bits in the path register may need to be replaced entirely with the contents of the path register of another state. This occurs if the survivor path for state $s_i(k)$ is obtained by extending the survivor path from another state $s_j(k - 1)$. Register transfer is fast, but power hungry owing to the extensive data movement needed to update the survivor path register.

Hybrid approaches for designing the SMU [39] exist where trace-back and register-exchange approaches are combined. In general, the trace-back approach is suitable for decoders with a large number of states ($N \geq 8$) and low data rates (<2 Mb/s) while the register-transfer works well when the number of states is small and data rates are high (>10 Mb/s). Though most applications in the past fell neatly into these two categories, many modern-day applications such as ultra-wideband [40] have data rates in the 100's of Mb/s along with a large number of states such as $N = 64$. Thus, the design of high-speed, low-power, large-state Viterbi decoders continues to be an important area of research [41,42].

82.3.4 The Soft-Output Viterbi Algorithm

The Viterbi decoder described so far generates the most likely transmitted bit sequence based on channel observations and the knowledge of the encoder state transition table. The soft-output Viterbi algorithm (SOVA) [43] provides the reliability of each bit in the decoded sequence in addition to the decoded sequence. Reliability of a bit $u(k)$ is quantified by the log-likelihood ratio (LLR) which is given by

$$L(u(k)) = \ln\left[\frac{p_e(u(k))}{1 - p_e(u(k))}\right] \tag{82.21}$$

where $p_e(u(k))$ is the probability of error of bit $u(k)$. The LLRs are referred to as soft outputs. These soft outputs improve decoding performance in both serial and parallel concatenated decoders such as turbo decoders, which will be described in Section 82.4.

It is easy to show that the probability of the encoder state sequence being the same as the survivor path of state $s_i(k)$ is given by

$$p(s_i(k)) \approx e^{-S_i(k)}, \quad 0 \le i \le N$$

For $k_0 = 1$, the ACS unit decides between two paths, *path* 1 and *path* 2. Assume that path 1 is chosen. The probability of selecting the wrong survivor path, i.e., the *path error* probability is then given by

$$p_e(s_i(k)) = \frac{e^{-S_i'(k)}}{e^{-S_i(k)} + e^{-S_i'(k)}} = \frac{1}{1 + e^{S_i'(k) - S_i(k)}} = \frac{1}{1 + e^{\delta}}$$

where $\delta = S_i'(k) - S_i(k)$, $S_i'(k)$ is the state metric corresponding to path 2 in Eq. (81.20). Thus, $S_i(k) \le S_i'(k)$. Hence, with a probability $p_e(s_i(k))$, the Viterbi algorithm can make errors in all the positions where the information bits $u(k)$ of path 2 differ from those of path 1. Using the path-error probability $p_e(s_i(k))$, the formula for bit-error probability $p_e(u(l),k)$ for bit $u(l)$ at time index k is given by the recursion

$$p_e(u(l), k) \leftarrow p_e(u(l), k-1)(1 - p_e(s_i(k))) + (1 - p_e(u(l), k-1))p_e(s_i(k)) \qquad (82.22)$$

From Eq. (82.21) and Eq. (82.22), the LLR of $u(l)$ can be refined until $l \le k - L$, i.e., the bit is decoded. The above recursion can also be applied directly in the log-likelihood domain in which case the recursion is then given by

$$L(u(l), k) \leftarrow f(L(u(l), k-1), \delta)$$

where the function $f(L(u(l), k - 1), \delta) \approx \min(L(u(l), k - 1), \delta)$. It has been shown [44] that the hardware complexity of SOVA is roughly twice that of the Viterbi decoder.

82.4 Turbo Decoder Architectures

Turbo decoders are composed of two or more constituent soft-input soft-output (SISO) decoders, which correspond to the component codes employed in the transmitter, and an interconnection of these constituent decoders through an interleaver/deinterleaver [45,46]. The decoding algorithm employed in the constituent decoders is the MAP algorithm or SOVA, but it is well known that MAP-based turbo decoders outperform SOVA-based turbo decoders. The MAP algorithm provides the LLR of the transmitted code symbols $u(k)$. The LLRs are iteratively updated by the constituent decoders. However, the use of iterative processing results in a large computational and storage complexity, and hence high-power dissipation in the receiver. Therefore, low-power and high-throughput architectures for turbo decoders have recently been investigated [36, 47–64] for wireless and broadband applications. In this section, we first describe the algorithm and VLSI architecture of SISO MAP decoding and then present high-throughput and low-power turbo decoder architectures.

82.4.1 MAP Decoder

In this section, the algorithm for the SISO MAP decoder is described, followed by a description of a baseline VLSI architecture for a SISO MAP decoder.

(1) *The MAP Algorithm.* Unlike the Viterbi algorithm, the MAP algorithm determines each of the transmitted information symbols $u(k)$ independently by maximizing the *a posteriori* probability $P(u(k) | \mathbf{y})$, where $\mathbf{y}$ denotes the received encoded bits used to determine u_k. The BCJR algorithm [48]

solves the MAP decoding problem efficiently and its log-domain version, i.e., the log-MAP algorithm
has the advantage that it can be formulated in terms of additions instead of multiplications.

A log-MAP algorithm estimates the LLR of data symbol $u(k)$ denoted as $\Lambda(u(k))$ defined below

$$\Lambda(u(k)) = \ln \frac{p(u(k)=1|\mathbf{y})}{p(u(k)=0|\mathbf{y})} = \ln \frac{\sum_{\mathbf{u}:u(k)=1} p(\mathbf{u},\mathbf{y})}{\sum_{\mathbf{u}:u(k)=0} p(\mathbf{u},\mathbf{y})} \tag{82.23}$$

where the summations in the numerator and the denominator are over all trellis branches where $u(k) = 1$
and $u(k) = 0$, respectively, and the observation is made over L trellis sections, i.e., the sequence $\mathbf{y} = \{y(0),
y(1), \ldots, y(L-1)\}$. We can rewrite $\Lambda(u(k))$ as

$$\Lambda(u_k) = \ln \sum_{\mathbf{u}:u(k)=1} e^{\ln p(\mathbf{u},\mathbf{y})} - \ln \sum_{\mathbf{u}:u(k)=0} e^{\ln p(\mathbf{u},\mathbf{y})} \tag{82.24}$$

Typically, one term will dominate in each summation leading to the following approximation

$$\Lambda(u(k)) \approx \max_{\mathbf{u}:u(k)=1} \ln p(\mathbf{u},\mathbf{y}) - \max_{\mathbf{u}:u(k)=0} \ln p(\mathbf{u},\mathbf{y}) \tag{82.25}$$

The log-domain probability $\ln p(\mathbf{u},\mathbf{y})$ can be further decomposed [49] as follows:

$$\begin{aligned}
\ln p(\mathbf{u},\mathbf{y}) &= \ln\{p(s_i(k-1), y(0:k-1))p(y(k), s_j(k)|s_i(k-1))p(y(k+1:L-1)|s_j(k))\} \\
&= \ln p(s_i(k-1), y(0:k-1)) + \ln p(y(k), s_j(k)|s_i(k-1)) \\
&\quad + \ln p(y(k+1:L-1|s_j(k)) \\
&= \alpha(si(k-1)) + \lambda(s(k-1), s(k)) + \beta(s_j(k))
\end{aligned} \tag{82.26}$$

where $y(0{:}k-1)$ and $y(k+1:L-1)$ are the observed sequence before and after the kth trellis section.
The first term $\alpha(s_i(k-1))$ is referred to as the *forward metric* and is equal to the probability that the
trellis reaches state $s_i(k-1)$ given the past observation $y(0: k-1)$. The forward metric $\alpha(s_i(k))$ plays
the same role as the state metric $S_i(k)$ in a Viterbi decoder and is computed recursively as

$$\alpha(s_j(k)) = \max_{s_i(k-1)}\{\alpha(s_i(k-1)) + \lambda(s_i(k-1), s_j(k))\} \tag{82.27}$$

where the max operation is performed over all states $s_i(k-1)$ that are connected to state $s_j(k)$. The $\alpha(s_j(k))$
update is performed recursively starting from $s_i(0)$ and is referred to as forward iteration. The second term
in Eq. (82.26) $\lambda(s_i(k-1), s_j(k))$, is referred to as the *branch metric* and is related to the probability that
a transition from $s_i(k-1)$ to $s_j(k)$ occurs. The branch metric $\lambda(s_i(k-1), s_j(k))$ is computed from the
channel output, noise statistics, and the error-free output of the branch connecting $s_i(k-1)$ and $s_j(k)$ at
time k. Note that each trellis section has 2^K possible transitions. The third term $\beta(s_j(k))$, is referred to as
the *backward metric* and is equal to the probability that the trellis reaches state $s_j(k)$ given the future
observations $y(k+1:L-1)$. The backward metric $\beta(s_j(k))$ is computed recursively as

$$\beta(s_i(k)) = \max_{s_j(k+1)}\{\beta(s_j(k+1)) + \lambda(s_i(k), s_j(k+1))\} \tag{82.28}$$

where the max operation is performed over all states $s_j(k+1)$ that are connected to state $s_i(k)$. The $\beta(s_i(k))$
update is performed recursively starting from $s_j(L-1)$ and is referred to as backward iteration.

Hence, the decoding process consists of three steps. First, the branch metrics in each trellis section are computed. Second, the forward and backward metrics $\alpha(s_j(k))$ and $\beta(s_i(k))$ are computed recursively via Eq. (82.27) and Eq. (82.28). Third, the LLR $\Lambda(u(k))$ is computed as

$$\Lambda(u(k)) = \max_{u:u_k=1} \{\alpha(s_i(k-1)) + \lambda(s_i(k-1), s_j(k)) + \beta(s_j(k))\}$$
$$- \max_{u:u_k=0} \{\alpha(s_i(k-1)) + \lambda(s_i(k-1), s_j(k)) + \beta(s_j(k))\} \tag{82.29}$$

It is known that the function $\overset{*}{\max}$ defined as

$$\overset{*}{\max}\{x, y\} = \max\{x, y\} + \ln(1 + e^{|-x-y|}), \tag{82.30}$$

can be employed instead of the max, in which case the performance of the log-domain MAP algorithm approaches that of the BCJR algorithm [49] to within 0.05 dB. The second term in Eq. (82.30) is referred to as the *correction factor.*

(2) *MAP decoder architectures.* Numerous architectures can be employed to implement the log-MAP algorithm [50]. However, the trellis sweep over all L observed symbols requires large memory to hold the forward and backward metrics until they are used in the LLR computation in Eq. (82.29). Hence, the sliding window log-MAP algorithm has become popular as it minimizes the metric storage requirements [49].

The sliding window log-MAP decoding algorithm is derived via the property that the forward and backward metrics α and β converge after a few constraint lengths have been traversed in the trellis, independent of the initial conditions [49]. We refer to this property as the *warm-up* property and the warm-up period is assumed to have a duration of L symbols. Owing to the warm-up property, the state metrics (α and β) computation can be partitioned into windows of size L. Further, the computations in each window can be done in parallel. Figure 82.13 shows an example of a decoding flow where the warm-up property is employed only for computing backward metrics. The warm-up or initialization period is depicted by dashed lines and the computation period by solid lines. This warm-up property will be exploited later in deriving parallel and block-interleaved pipelined (BIP) architectures.

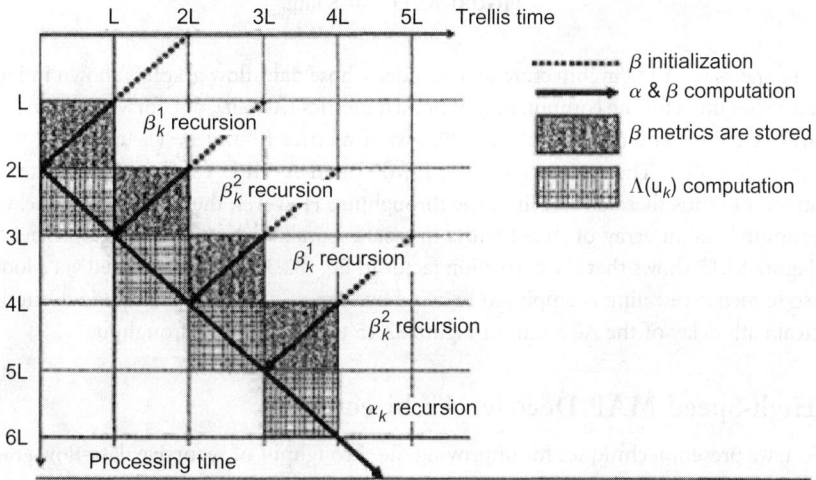

FIGURE 82.13 Scheduling of state metric recursions for α and β in the sliding window log-MAP. For simplicity, it is assumed that the computation and warm-up period equals L. Shadowed region indicates the computation and storage of β. The gridded region indicates the computation of α followed by the computation of $\Lambda(u(k))$. Here, β^1 and β^2 are the first and second β-recursion outputs, respectively.

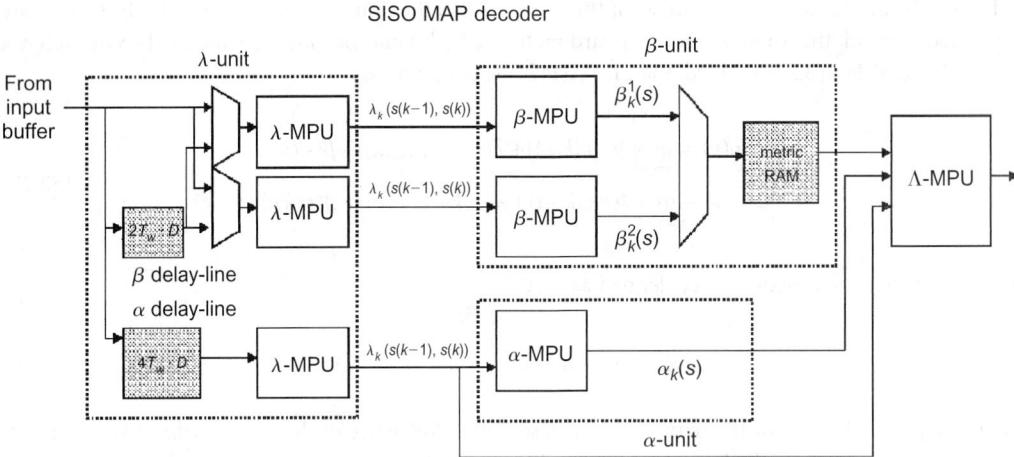

FIGURE 82.14   Baseline VLSI architecture of a sliding window log-MAP decoder.

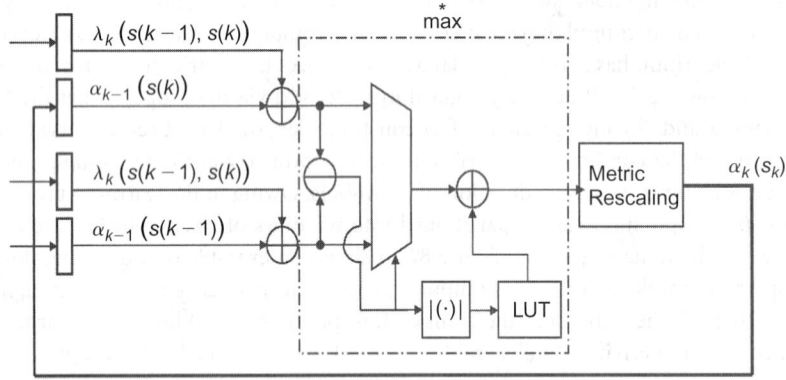

FIGURE 82.15   ACS unit.

Figure 82.14 shows the VLSI architecture of a decoder whose data-flow graph is shown in Figure 82.13. The architecture has units for the computation of branch metrics (λ-unit), one forward recursion (α-unit), two backward recursions and a buffer to store backward metrics β^1 and β^2 (β-unit), and the Λ metric processing unit (Λ-MPU). The computations in λ-MPU and Λ-MPU can be implemented in a feed-forward manner and thus these do not limit the throughput. However, the forward and backward recursions are computed via an array of ACS kernels in a state-parallel manner. The ACS kernel for a MAP decoder in Figure 82.15 shows that the correction factor in Eq. (82.30) is implemented via a look-up-table (LUT) and state metric rescaling is employed to avoid overflows [47]. As is the case in a Viterbi decoder, it is the critical path delay of the ACS unit in Figure 82.15 that limits the throughput.

82.4.2 High-Speed MAP Decoder Architectures

In this section, we present techniques for improving the throughput of recursive data-flow graph present in the MAP decoder, in particular in the ACS kernel. First, we review existing techniques of parallel processing [51,52] and look-ahead transform [36,53]. Then, we present the block-interleaved pipelining (BIP) technique [54–56].

In general, pipelining or parallel processing becomes difficult for a recursive data-path. However, if the data is being processed in blocks and the processing satisfies the following two properties: (1) computation

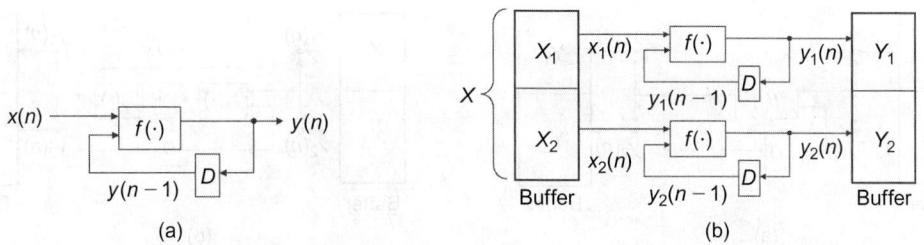

FIGURE 82.16 Parallel processing: (a) original, and (b) subblock parallel architecture.

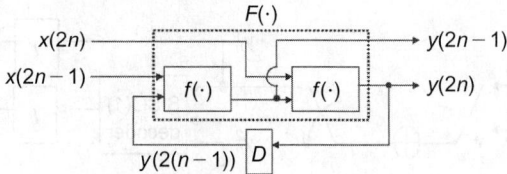

FIGURE 82.17 Look-ahead transform.

between blocks are independent and (2) computation within a block is recursive, then, a block parallel processing architecture can be achieved. Further, if a block can be segmented into computationally independent subblocks, parallel processing can be applied at the subblock level. This leads to the high-throughput MAP decoder architectures presented in Refs. [51,52,54–56].

(1) *Parallel Processing.* Consider the recursive architecture in Figure 82.16(a). Note that the architecture in this figure cannot be easily pipelined or parallelized owing to the presence of the feedback loop. However, if a data block of length B is processed independently of other blocks and the computations within a block can be segmented into computationally independent subblocks, then one can parallelize the architecture as shown in Figure 82.16(b), where the parallelization factor $M = 2$ and a block $\mathbf{X}$ is divided into $M = 2$ subblocks, $\mathbf{X}_1$ and $\mathbf{X}_2$. It is obvious that the critical path is not affected and the throughput is increased by a factor of M at the expense of a factor of M increase in hardware complexity.

(2) *Look-ahead transform.* Another transform to achieve high-throughput for recursive data-flow graphs is look-ahead computation [53]. Look-ahead leads to an increase in the number of symbols processed at each time step as shown in Figure 82.17, where two symbols are processed in one clock cycle. If $\mathbf{x}(n) = [x(2n - 1), x(2n)]$ and $\mathbf{y}(n) = [y(2n - 1), y(2n)]$, then look-ahead results in the output being expressed as $\mathbf{y}(n) = F(\mathbf{x}(n), y(2(n - 1)))$. Note that $F(\cdot)$ will have a longer critical path delay than the original computation of Figure 82.16(a). However, it has been shown that the function $F(\cdot)$ can be optimized via logic minimization so that an overall increase in throughput can be achieved. For example, as mentioned earlier, in the context of Viterbi decoding, it has been shown that a 1.7$\times$ increase in throughput is feasible via radix-4 computation [36].

(3) *BIP.* The parallel architecture in Figure 82.16(b), where the level of parallelism is equal to 2, has two identical computation units processing two independent input symbols. Therefore, the hardware complexity increases linearly with the level of parallelism M. A comparatively area-efficient architecture is obtained by using the BIP technique proposed in Ref. [58]. First, the data-flow of Figure 82.16(b) is folded [53] on to a single computation unit as shown in Figure 82.18(a), where two independent computations are carried out in a single computational unit. Note that the resulting BIP architecture in this figure is inherently pipelined. Therefore, an application of retiming [53] (see Figure 82.18(b)) results in reduction of the critical path delay by a factor of two over that of the original architecture in Figure 82.16(a). It is clear that the retimed BIP architecture in Figure 82.18(b) leads to high throughput at the cost of a marginal increase in memory owing to pipelining latches when compared to the architecture in Figure 82.16(a).

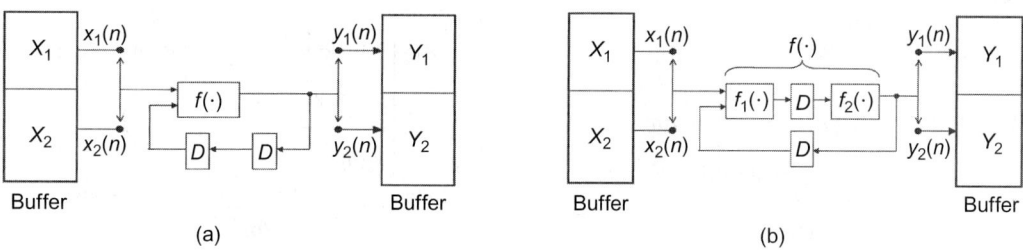

FIGURE 82.18   BIP: (a) BIP architecture, and (b) retimed BIP architecture.

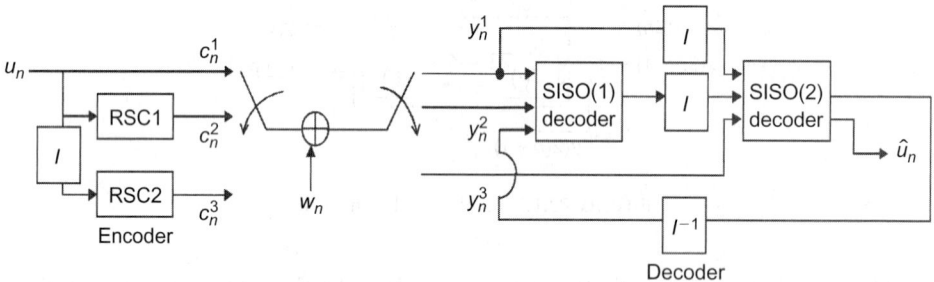

FIGURE 82.19   Block diagram of a PCCC encoder and a turbo decoder. Here I and I^{-1} denote block interleaving and deinterleaving, respectively.

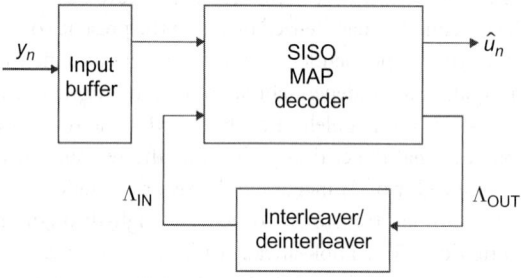

FIGURE 82.20   Serial VLSI turbo decoder architecture.

82.4.3   High-Speed Turbo Decoder Architectures

The turbo code considered in this section is made up of two recursive systematic convolutional (RSC) encoders concatenated in parallel as shown in Figure 82.19. The bit sequences transferred from one encoder to the other are permuted by an interleaver. The decoder contains two SISO MAP decoders which are associated with the two RSC encoders as depicted in this figure. The decoding of the observed sequences is performed iteratively via the exchange of soft output information ($\Lambda(u(k))$) between the constituent decoders. The decoding process is repeated iteratively until an appropriate stopping criterion is satisfied.

The turbo decoder can be implemented via a serial architecture as shown in Figure 82.20, where one SISO MAP decoder is time-shared. Hence, increasing the throughput of the SISO MAP decoder directly leads to an overall improvement in throughput of the turbo decoder.

High-throughput turbo decoder architectures are designed using the MAP decoder architectures presented in Section 82.4.2. Table 82.1 summarizes the key parameters for the parallel, BIP, and look-ahead BIP architectures for two codes with encoder polynomials $[5,7]_8$ ($L = 16$) and $[13,15]_8$ ($L = 32$), with constraint length (encoder memory) $K = 3$ and 4, respectively. These parameters are obtained from

TABLE 82.1 Turbo Decoder Application Results

Parameter	K	Parallel	BIP	Look-ahead BIP
Critical path delay (ns)	3	13.994	7.126	13.011
	4	17.767	9.793	15.517
Speed-up (η)	3	2	1.96	2.15
	4	2	1.81	2.29
Normalized area	3	2	1.63	2.03
	4	2	1.76	1.83
Measured area (mm²)	3	5.3(2.87/2.43)	4.32(2.87/1.45)	5.4(2.36/3.04)
(Memory/logic)	4	10.9(7.46/3.44)	9.62(7.46/2.16)	9.99(5.58/4.41)

an implementation using 2.5 V, 0.25 μm CMOS technology standard cell library using *Synopsys Design Compiler*-based synthesis, *Cadence Silicon Ensemble*-based place and route, and *Pathmill*-based postlayout simulations. The parallelization factor $M = 2$ for all the architectures. From Table 82.1, we see that the BIP-based turbo decoder architecture delivers the same speedup as compared to parallel processing with a comparatively smaller area. In particular, the BIP technique reduces the logic complexity nearly by a factor of M keeping the memory complexity the same as that of parallel processing. Further details can be found in Ref. [55].

82.4.4 Low-Power Turbo Decoder Architectures

The application of BIP, folding, and retiming reduces the critical path delay in the ACS kernel of MAP decoders with marginal area overhead. Subsequent application of voltage scaling can result in savings in power keeping the same throughput. A low-power decoder architecture can be obtained by processing M subblocks of size B/M bits via subblock interleaved computations, where B denotes the block length of information bits. In this section, we present results related to achievable power savings in turbo decoders by combining voltage scaling and BIP.

To reduce power, we scale the supply voltage of the block-interleaved architecture such that the block processing time is made equal to that of the conventional architecture. The conventional architecture requires $B + 2L$ cycles to process a block, where L is the warm-up depth. The proposed architecture requires

$$M \times \left(\frac{B}{M} + 2L \right) = B + 2ML \tag{82.31}$$

cycles for processing one block. Equating the block processing times of the two architectures we get

$$\tau_{\text{cri,p}} = \tau_{\text{cri,s}} \frac{(B + 2L)}{(B + 2ML)} \tag{82.32}$$

where $\tau_{\text{cri,p}}$ and $\tau_{\text{cri,s}}$ are the critical path delays of the block-interleaved and conventional architectures, respectively. Thus, we can reduce the supply voltage such that $\tau_{\text{cri,p}}$ is equal to $\tau_{\text{cri,s}} \times B + 2L/B + 2ML$.

Figure 82.21 depicts the critical path delay ratio, area overhead ratio, and power savings for $K = 3$, $B = 1024$, and $L = 16$ for velocity saturation index values of $a = 1$ and 2. It is observed that at a certain point there is no further power savings owing to the area overhead. Further, as the number of states ($= 2^{K-1}$) is increased, the power savings decrease because the area overhead increases rapidly for large values of K (see Figure 82.22).

82.5 Low-Density Parity Check (LDPC) Decoder Architectures

LDPC codes were introduced by Gallager in his seminal work in 1963 [4], and have largely been ignored since then until the successful introduction of turbo codes by Berrou et al., in 1993 [65]. With this renewed interest, several researchers rediscovered LDPC codes and began to investigate codes on graphs in conjunction with

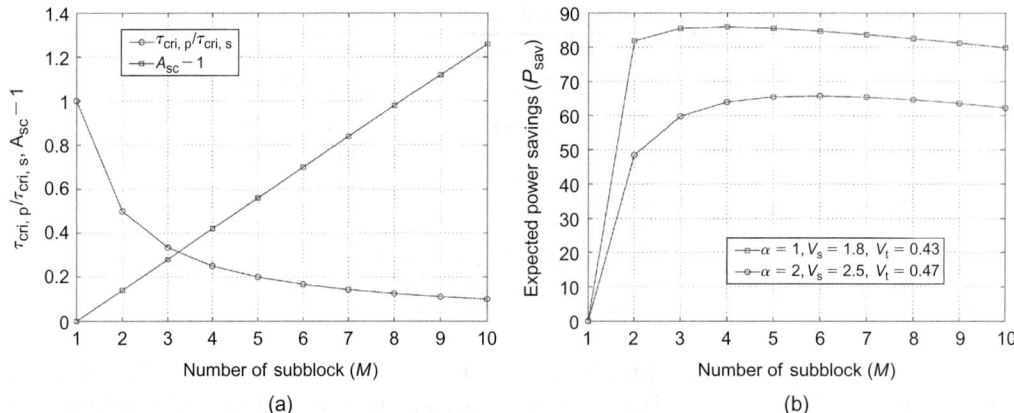

FIGURE 82.21   (a) Delay ratio ($\tau_{cri,p}/\tau_{cri,s}$) and area increase factor ($A_{sc} - 1$) versus the number of subblocks, and (b) power savings (P_{sav}) versus the number of subblocks.

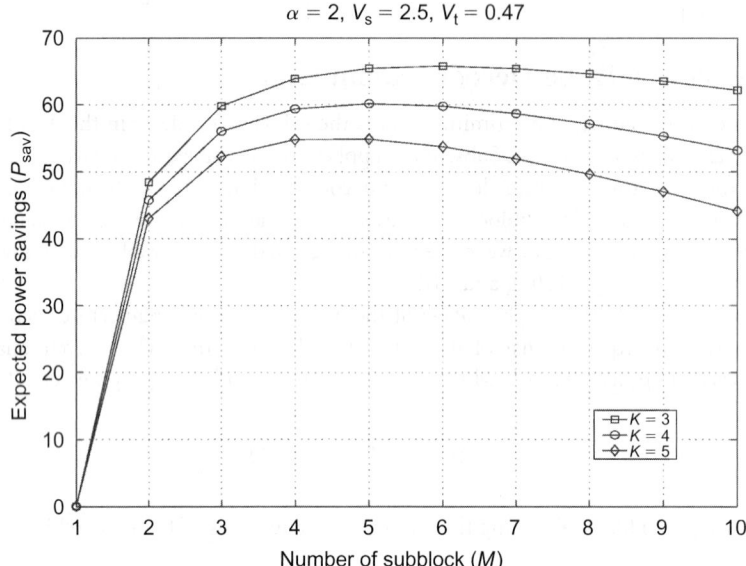

FIGURE 82.22   Power savings (P_{sav}) versus different K values.

iterative decoding. Long LDPC codes with iterative decoding have been shown to almost achieve capacity within a fraction of decibels [66], [67]. These discoveries have promoted LDPC codes as strong competitors to turbo codes in many communication and storage systems where high reliability is required.

This section focuses on the VLSI design aspects of LDPC decoders. After introducing LDPC codes in Section 82.5.1, two iterative decoding message-passing algorithms for LDPC codes are discussed in Section 82.5.2. In Section 82.5.3, several architectures for the message computation kernels employed by the decoding algorithm are presented and used in the decoder architectures presented in Section 82.5.4. Finally, Section 82.5.5 extends the discussion to repeat-accumulate codes.

82.5.1   LDPC Codes

An LDPC code is a linear block code defined by a *sparse* parity-check matrix $\mathbf{H}_{m \times n}$ with m parity-check equations on n codeword bits with $k = n - m$ information bits. An LDPC code is typically described by a bipartite Tanner graph [68] whose adjacency matrix is **H** having n *bit nodes* and m *check nodes* corresponding to the n columns and m rows of **H**, respectively (see Figure 82.23). A bit node u_j is connected to r_j check nodes and a check node v_i is connected to c_i bit nodes, where r_j and c_i are called the node degrees. Equivalently, the jth column of **H** has r_j ones, and the ith row has c_i ones. If all bit-node degrees and all check-node degrees are uniform, the code is said to be *regular*, otherwise it is called an *irregular* LDPC code. In general, the graph edges are defined randomly by a $p \times p$ "edge" permuter, where $p = \sum_{j=1}^{n} r_j$. Two graph parameters relevant to performance are the *girth* or length of the shortest cycle in the graph and *diameter* or maximum length of the shortest path in the graph. The code length n and the randomness of the edge permuter play an important role in the decoder design of an LDPC code. The encoding complexity of LDPC codes is quadratic in the code length n.

Much of the research on LDPC codes has focused on explicit construction methods of bipartite Tanner graphs with large girth. Architecture-aware LDPC (AA-LDPC) codes are a class of structured LDPC codes with large girth oriented toward an efficient decoder implementation [69]. The parity-check matrix of an AA-LDPC code is divided into $S \times S$ submatrices, where each submatrix is either the zero matrix or a binary permutation matrix as shown in Figure 82.24(a). Equivalently, the bit nodes and check nodes in an AA-LDPC Tanner graph are grouped into clusters of size S such that if one node from a cluster connects to a node from another cluster, then there exist distinct connections between all nodes from

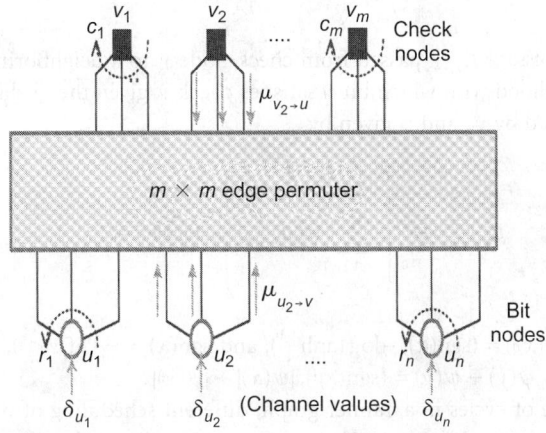

FIGURE 82.23   Tanner graph of an LDPC code.

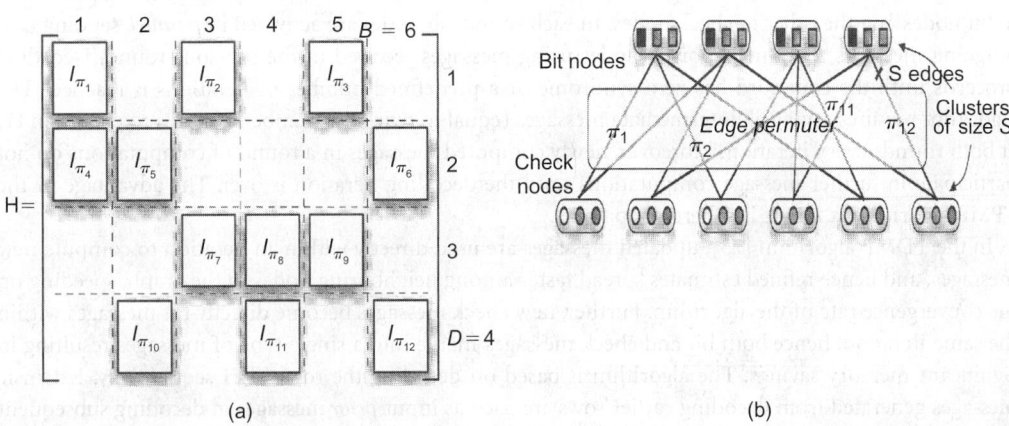

FIGURE 82.24   Parity-check matrix and Tanner graph of an AA-LDPC code with $B = 6$ bit clusters and $D = 4$ check clusters.

both clusters (see Figure 82.24(b)). The connections between two clusters are done according to the permutation of the corresponding submatrix in **H**.

82.5.2   LDPC Decoding Algorithms

LDPC codes are decoded iteratively using a suboptimal message-passing algorithm, called the sum–product algorithm [70–72], which closely approximates maximum likelihood decoding. The algorithm operates on the Tanner graph of an LDPC code by *computing* messages for every node and *communicating* these messages along the edges in the graph. These messages represent estimates of the transmitted bits together with a measure of the reliability of these estimates. The message associated with a transmitted bit u_j with noisy received bit $\tilde{u}_j$ is given by the log-likelihood ratio $\delta_{u_j} = \log P(u_j = 0|\tilde{u}_j)/P(u_j = 1|\tilde{u}_j)$, and is called the *channel observation* or *intrinsic* message. The bit message $\mu_{u_j \to v}$ passed from bit node u_j to a neighboring check node $v \in \mathcal{N}(u_j)$, where $\mathcal{N}(u_j)$ is the set of all check nodes connected to u_j, is proportional to the likelihood of bit u_j in favor of zero given the information obtained from check nodes other than v and is given by

$$\mu_{u_j \to v} = \delta_{u_j} + \sum_{\substack{w \in \mathcal{N}(u_j) \\ w \neq v}} \mu_{w \to u_j}, \quad \text{for all } v \in \mathcal{N}(u_j) \tag{82.33}$$

Similarly, the check message $\mu_{v_i \to u}$ passed from check node v_i to a neighboring bit node $u \in \mathcal{N}(v_i)$ is proportional to the likelihood with which bit u satisfies check v_i given the likelihood in favor of zero of the remaining bits checked by v_i, and is given by

$$\mu_{v_i \to u} = \psi^{-1} \left(\sum_{\substack{s \in \mathcal{N}(v_i) \\ |s \neq u}} \psi(\mu_{s \to v_i}) \right), \quad \text{for all } u \in \mathcal{N}(v_i) \tag{82.34}$$

where $\psi(x) = (\text{sgn}(x), |\psi(x)|) = (\text{sgn}(x), -\log \tanh \frac{|x|}{2})$, and $\text{sgn}(x) = -1$ if $z < 0$, $+1$ otherwise. The sum $\psi(x) + \psi(y)$ is defined as $\psi(y) + \psi(y) \triangleq (\text{sgn}(xy), |\psi(x)| + |\psi(y)|)$.

Owing to the presence of cycles in a Tanner graph, different scheduling of node computations yield nonequivalent message-passing algorithms. The two main scheduling techniques for message-passing are the *two-phase message-passing* (TPMP) and the *turbo-decoding message-passing* (TDMP) algorithms. In the TPMP algorithm [4], a decoding iteration is divided into two rounds of computations, one pertaining to bit nodes and the other to check nodes. In each round, all nodes are activated in *parallel*, sending new outgoing messages to all neighboring nodes using messages received in the previous round. Decoding proceeds until the codeword has zero syndrome or a predefined number of iterations is reached. The algorithm requires saving all intermediate messages (equal to twice the number of nonzero entries in **H**) at both rounds every iteration. Moreover, newly computed messages in a round of computations do not participate in further message computations until the decoding iteration is over. The advantage of the TPMP algorithm is that it is inherently parallel.

In the TDMP algorithm [73], updated messages are used directly within an iteration to compute new messages, and hence refined estimates spread faster among neighboring nodes in the graph, speeding up the convergence rate of the algorithm. Further, new check messages become directly bit messages within the same iteration, hence both bit and check messages merge into a single type of messages resulting in significant memory savings. The algorithm is based on decoding the rows of **H** sequentially. Extrinsic messages generated from decoding earlier rows are used as input *prior* messages in decoding subsequent rows. Each row i in **H** is associated with a vector of *extrinsic* messages λ^i corresponding to the nonzero entries in that row. For each bit the sum of all messages generated by the rows in which that bit participates are stored in the vector γ of n *posterior* messages. Decoding the ith parity-check row involves first reading the extrinsic messages λ^i and the posterior messages $\gamma(I_i)$, where I_i denotes the set of indices of the ones in row i. Next, prior messages $\rho = [\rho_1, \ldots, \rho_{w_i}] = \gamma(I_i) - \lambda$ are computed by subtracting λ^i from $\gamma(I_i)$ to reduce correlation between messages generated from decoding earlier rows. Output messages $\Lambda^i = [\Lambda^i_1, \ldots, \Lambda^i_{w_i}]$ are then computed using Eq. (82.34) with ρ as input

$$\Lambda^i_j = \psi^{-1}\left(\sum_{l:l \neq j} \psi(\rho_l)\right), \quad i = 1, \ldots, n, \quad j = 1, \ldots, w_i \tag{82.35}$$

where w_i is the number of ones in row i. Equivalently, Λ^i_j can be computed more efficiently as

$$\Lambda^i_j = Q(\rho_1, \rho_2, \ldots, \rho_{j-1}, \rho_{j+1}, \ldots, \rho_{w_i}), \quad i = 1, \ldots, n, \quad j = 1, \ldots, w_i \tag{82.36}$$

The Q-function is defined recursively as

$$Q(x_1, x_2, \ldots, x_w) \triangleq Q(\ldots(Q(Q(x_1, x_2), x_3), \ldots), x_w) \tag{82.37}$$

and $Q(x_1, x_2)$ approximates the difference between two logsum operations [74]

$$Q(x_1, x_2) \approx \log(e^{x_1} + e^{x_2}) - \log(e^{x_1 + x_2} + 1)$$

$$= \max(x_1 + x_2, 0) - \max(x_1, x_2) + \max\left(\frac{5}{8} - \frac{|x_1 + x_2|}{4}, 0\right) - \max\left(\frac{5}{8} - \frac{|x_1 - x_2|}{4}, 0\right)$$

Finally, the vector Λ^i replaces the old extrinsic messages λ^i, and the posterior messages for the bits located at positions indicated by I_i are updated by adding Λ^i to ρ: $\gamma(I_i) = \rho + \lambda^i$. These steps constitute a decoding *subiteration*, while a round of subiterations over all rows of **H** constitutes a decoding *iteration*.

The TPMP algorithm requires memory storage for $2\sum_{j=1}^{n} r_j + n$ messages, while the TDMP algorithm requires storage for only $\sum_{j=1}^{n} r_j + n$ messages, resulting in a memory savings of $\sum_{j=1}^{n} r_j / 2\sum_{j=1}^{n} r_j + n \times 100\%$. For uniform column degrees r, this amounts to $r/(2r + 1) \times 100\%$. Moreover, the TDMP algorithm requires on average 50% less iterations to converge for moderate to high SNR compared to the TPMP algorithm as determined by simulations [73].

82.5.3 Architectures for Message Computation Kernels

The message computation kernels of Eq. (82.33) and Eq. (82.34) compute distinct messages to all the neighbors of each node in a Tanner graph. These kernel computations from a source node to a neighboring destination node can be implemented by first computing the total sums in Eq. (82.33) and Eq. (82.34) and then subtracting out the component pertaining to the destination node to reduce correlation between messages. Figure 82.25(a) and 82.25(b) show serial dataflow graphs implementing Eq. (82.33) and Eq. (82.34) assuming a node having six neighbors. The latency of these serial architectures for a general node degree d is equal to $d + 1$. These message computation kernels are sometimes referred to as constituent SISO decoders. Figure 82.25(c) shows a two-way stack-based architecture implementing kernel equations (82.36) and (82.37) of the TDMP algorithm using the Q-function whose logic schematic is shown in Figure 82.25. A pair of Q-function blocks compute the recursions in Eq. (82.37) both from the left (ρ_1 up to ρ_5) and the right (ρ_6 down to ρ_2). Intermediate results from the left and the right are pushed onto a pair of stacks until both recursions have reached the half way point. After that, the stored results from the left (up to $\rho_i - 1$) are popped from the stack and fed together with the results of the running recursion from the right into a new Q-function block that computes the output messages Λ_3, Λ_2,

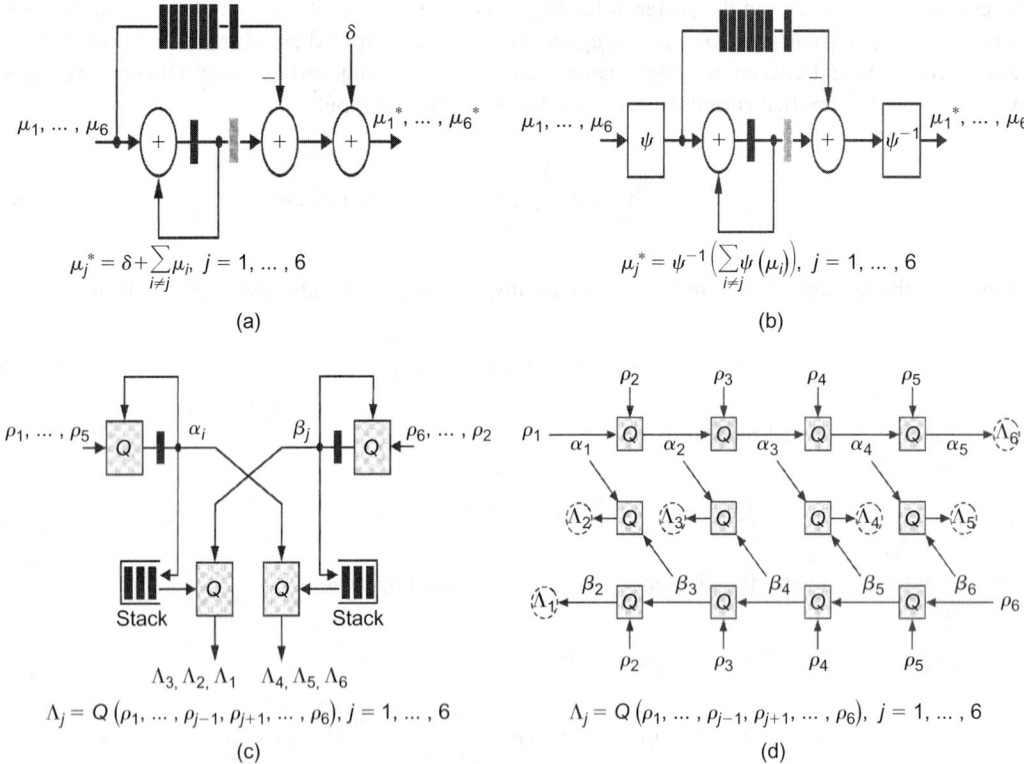

FIGURE 82.25   Architectures for message processing units: (a) serial dataflow graphs for computing the bit message, and (b) check messages, (c) serial and (d) parallel message computation kernels using the Q-function.

and Λ_1. Similarly, the stored results in the right stack are combined with the results of the running recursion from the left to generate output messages Λ_4, Λ_5, and Λ_6. The latency of the architecture in Fig. 82.25(c) is 6 clock cycles. In general, the latency is d clock cycles.

Figure 82.25(d) shows a parallel dataflow graph using 12 Q-function blocks. The upper blocks process recursion (82.37) from the left, while the lower blocks process the recursion from the right. The middle row of blocks combines the results from both recursions to generate output messages. Note that the inputs must be skewed for prosper operation. In general, this parallel architecture requires $3(d-1)$ Q-function blocks.

82.5.4   LDPC Decoder Architectures

LDPC decoder architectures are typically classified as parallel, serial, or partly-parallel (scalable) architectures with respect to allocating resources for message computation and message communication. A *parallel* LDPC TPMP-decoder simply implements the Tanner graph of the code, with bit function units (BFUs) allocated to compute bit messages, check function units (CFUs) to compute check messages, and a physical interconnect of wires that connect these units according to the edge permuter as shown in Figure 82.26(a). To sustain the high bandwidth of message communication between function units, the edge permuter must be implemented as a physical interconnect of wires. While such decoder fully exploits the inherent parallelism of the TPMP algorithm achieving high-throughput with low-power consumption, it is constrained by chip area and interconnect complexity, and hence does not scale gracefully with code length. Long on-chip interconnect wires present implementation challenges in terms of placement, routing, and buffer-insertion to achieve timing closure. For example, the average interconnect wire length of the rate-1/2, 1024-bit LDPC decoder of Ref. [75] is 3 mm using 0.16 μm CMOS technology, and has a chip area of 7.5 mm × 7 mm of which only 50% is utilized owing to routing congestion.

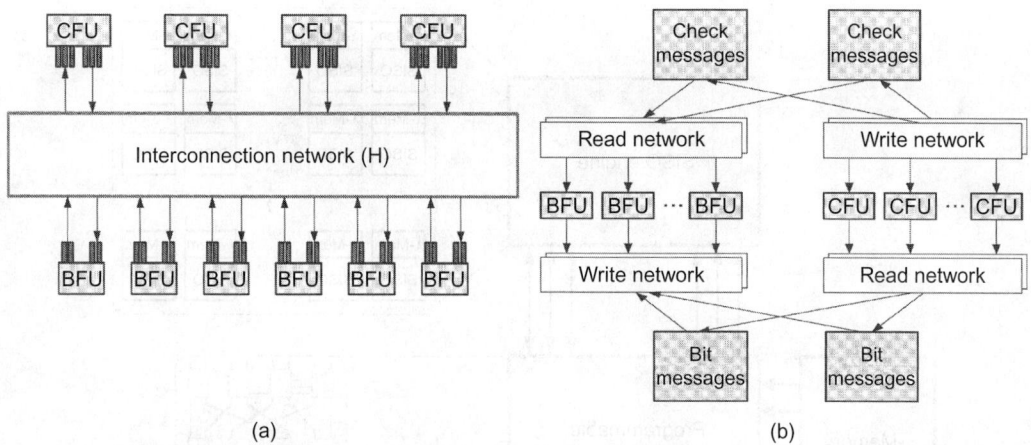

(a)   (b)

FIGURE 82.26 LDPC decoder architectures: (a) parallel and (b) serial LDPC decoder architectures.

In contrast, in *serial* decoders [76] computations are folded onto a small set of bit and CFUs that communicate through memory instead of a complex interconnect as shown in Figure 82.26(b). Each function unit requires separate read/write networks with complex control to access messages in memory. The BFUs read check messages and generate updated bit messages, while CFUs read bit messages and generate updated check messages. Computations proceed until all bit node and check node operations complete. To avoid overlapping messages in memory, two memory modules are used for storing bit messages where one module is read while the other is being written, and similarly for check messages. At the end of every iteration, the read and write order from/to the memory modules is swapped. Serial decoders require a substantial memory overhead that amounts to four times the number of nonzero elements in **H**, and their throughput is constrained by the number of function units and read/write networks to access memory. The advantage of serial decoders is their small area.

Partly-parallel or *scalable* decoders combine the high-throughput characteristics of parallel decoders and the area efficiency of serial decoders. In the following, partly-parallel TDMP decoders for AA-LDPC codes will be described assuming a regular LDPC code of length n having check and bit node degrees r and c, respectively, and whose **H** matrix is decomposed into $S \times S$ submatrices with D block rows and B block columns. Referring to Figure 82.24(a), the S rows of ones in each row of submatrices in an AA-LDPC matrix **H** are nonoverlapping, and hence can be processed in parallel using S SISO decoders. Decoder s processes row s in each row of submatrices, for a total of D rows, and maintains its extrinsic messages in a local λ-memory as shown in Figure 82.27. Posterior messages are stored in a global γ-memory and communicated in parallel to and from the decoders using a network that implements the factored edge permuter. The advantage of the AA structure is evident in that it enables an efficient implementation of the permuter using a programmable multistage interconnection network of switches (e.g., Benes network [77]). The switches are programmed to route the various permutations when processing the rows of submatrices in **H** using π-memory. The network in general has $2 \log(S) - 1$ stages, and each stage contains $S/2$ 2×2 switches. The parameter S is a parallelism factor that determines the throughput Θ of the decoder. The total number of λ-messages that need to be stored is $cSD = c(n - k)$, while the total size of γ-memory is $n = BS$ messages. The size of π-memory is $\frac{S}{2}(2 \log S - 1)$ bits. The SISO decoders complete one pass over all rows of **H** (one iteration) in cD clock cycles assuming serial SISO decoders similar to Figure 82.25(c). For I iterations and clock frequency f, the throughput attained by the decoder is

$$\Theta = \frac{nf}{cDI} = \frac{f}{rI}S \quad \text{bits/s}$$

Note that the architecture shown in Figure 82.27 can decode any AA-LDPC code with the same parameters n, r, c, and S, but having different structure and location of permutation submatrices in **H**.

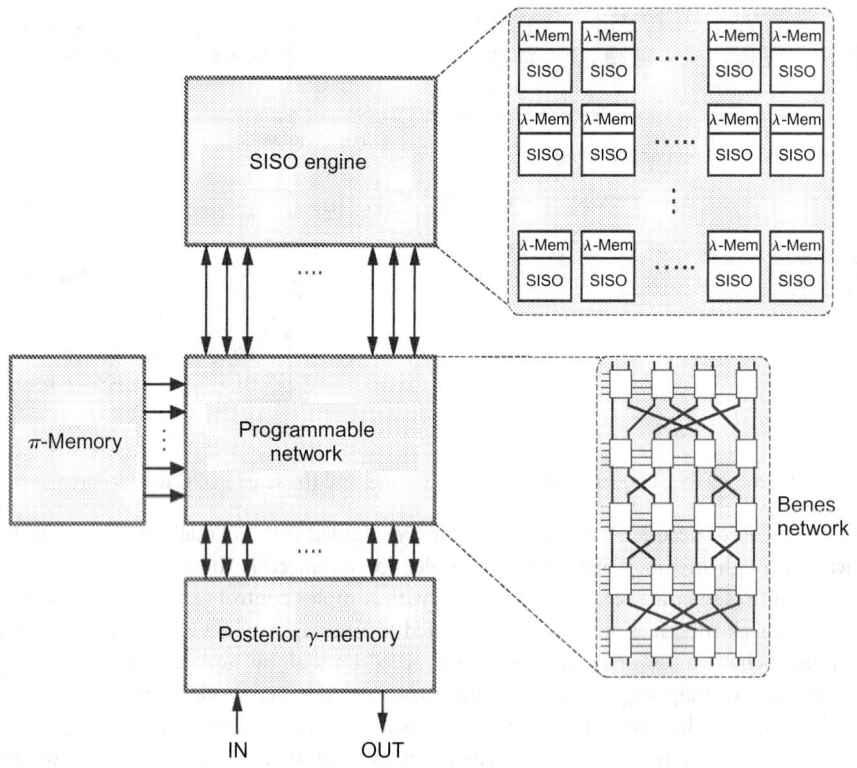

FIGURE 82.27   TDMP decoder architecture.

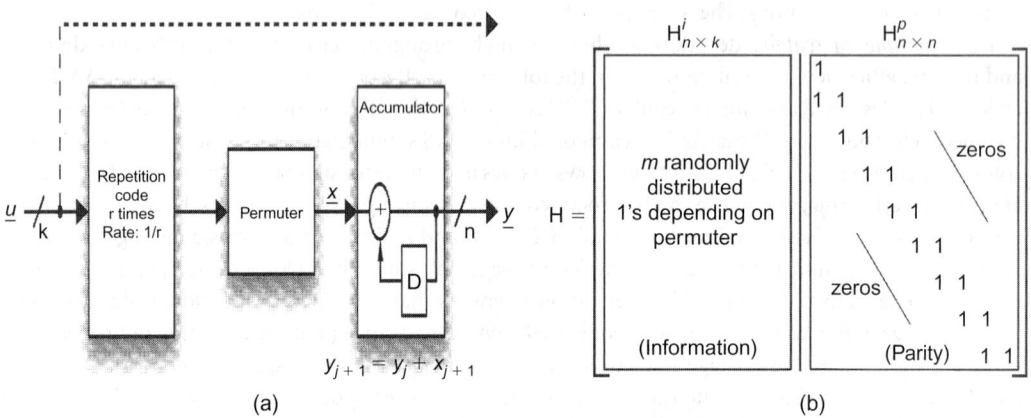

FIGURE 82.28   RA codes: (a) encoder, and (b) parity-check matrix.

82.5.5 Extensions to Repeat-Accumulate (RA) Codes

RA codes [78] are a subclass of LDPC codes that can be encoded in linear time. Formally, an RA code
is a linear block code composed of a repetition code concatenated through a permuter with a rate-1
memory-1 convolutional code. An information message is encoded by repeating each bit r times, and
then accumulating the repeated bits as shown in Figure 82.28(a). Typically, the information bits are
transmitted together with the parity bits in an RA code. Equivalently, an RA code can also be defined
algebraically as the null-space of a sparse parity-check matrix $\mathbf{H}_{n \times (n+k)} = [\mathbf{H}^i_{n \times k} | \mathbf{H}^p_{n \times n}]$ as shown in
Figure 82.28(b). The left portion of $\mathbf{H}$ corresponds to the information bits with ones distributed according

to the permuter structure. The right portion corresponds to the parity bits (accumulator) having ones on the main diagonal and the first lower off-diagonal. Given the sparse-matrix representation of RA codes, the previous discussion on LDPC codes and their decoder architectures applies to RA codes as well. Similarly, AA RA codes can be defined by decomposing the parity-check matrix into permutation submatrices and the TDMP algorithm naturally applies. However, the parity portion needs some modification. Instead of accumulating parities using adjacent check nodes in the Tanner graph, parities are accumulated using checks from adjacent clusters in parallel by connecting the rightmost check node cluster to the leftmost parity node cluster. The reader is referred to Ref. [79] for more details.

82.6 Conclusions

The subject of coding theory and ECC architectures is an active area of research today and is expected to remain so for the foreseeable future. New applications of communication systems combined with the unrelenting scaling of feature sizes in modern semiconductor processes makes for a rich interplay between the areas of coding theory and VLSI architectures.

While Moore's Law has enabled the widespread deployment of communications systems, there are questions regarding the ability of the semiconductor industry to scale feature sizes well into the nanometer regime in a cost-effective manner. These questions arise from the emergence of nonidealities such as noise, process variations, soft errors, and many others, in modern silicon process technology. Hence, system-on-a-chip subsystems such as datapath and control (computation), buses (communication) and memory (storage) are beginning to resemble noisy communication networks. This fact points to a unique opportunity for extending and applying ECC and ECC-type techniques to enable the design of reliable and efficient ICs in general and communication ICs in particular. Examples of such research can be found in [80]–[82] where communication-centric concepts such as equalization, coding, and joint equalization and coding have been applied to on-chip computation and communication. This is an area replete with possibilities and can lead to an era where coding theory extends if not rescues Moore's Law from obsolescence.

References

1. C.E. Shannon, "A mathematical theory of communication (Part 1)," *Bell System Technical Journal*, vol. 27, pp. 379–423, 1948.
2. C.E. Shannon, "A mathematical theory of communication (Part 2)," *Bell System Technical Journal*, vol. 27, pp. 623–656, 1948.
3. G.E. Moore, "Progress in digital integrated electronics", Technical Digest of the Integrated Electronic Devices Meeting, pp. 11–13, 1975.
4. R.G. Gallager, *Low-Density Parity-Check Codes*, MIT Press, Cambridge, MA, 1963.
5. ETSI EN 300 744, "Digital video broadcasting (DVB); Framing structure, channel coding and modulation for digital terrestrial television," 1997.
6. Telecommunications Group Standard, "Radio broadcasting systems; VHF digital multimedia broadcasting (DMB) to mobile, portable, and fixed receivers," October 2003.
7. IEEE Std. 802.16-2004, "IEEE Standard for Local and Metropolitan Area Networks, Part 16: Air Interface for Fixed Broadband Wireless Access Systems," 2004.
8. IEEE Std. 802.11g-2003, "IEEE Standard for IT-Telecom, and information exchange between systems-local and metropolitan area networks," 2003.
9. J. Haartsen, "The bluetooth radio system," *IEEE Personal Communications*, vol. 7, pp. 28–36, February 2000.
10. A. Bria et. al., " Fourth-generation wireless infrastructures: scenarios and research challenges," *IEEE Personal Communications*, vol. 8, pp. 25–31, December 2001.
11. SONET OC-192, "Transport system generic criteria," Bellcore, GR-1377-CORE, no. 4, March 1998.
12. ANSI T1.413-1995, "Network and customer installation-asymmetric digital subscriber line (ADSL) metallic interface," 1995.

13. ANSI X3.230, "Fiber channel physical and signalling interface (FC-PH)," 1994.

14. I.S. Reed and G. Solomon, "Polynomial codes over certain finite fields," *Journal of the Society of Industrial and Applied Mathematics*, vol. 8, pp. 300–304, 1960.

15. E.R. Berlekamp, *Algebraic Coding Theory*, McGraw-Hill, New York, 1968.

16. R.E. Blahut, *Algebraic Codes for Data Transmission*, Cambridge University Press, Cambridge, UK, 2002.

17. C. Paar, *Efficient VLSI architectures for bit-parallel arithmetic in Galois fields*, Ph.D. thesis, University of Essen, Germany, 1994.

18. IEEE Std 802.16-2004, "IEEE standard for local and metropolitan area networks, Part 16: Air interface for fixed broadband wireless access systems," October 2004.

19. J.G. Proakis, *Digital Communications*, McGraw-Hill, New York, fourth edition, 2000.

20. J.J. Rotman, *Advanced Modern Algebra*, Prentice-Hall, New Jersey, 2002.

21. D.V. Sarwate and N.R. Shanbhag, "High-speed architectures for Reed-Solomon decoders," *IEEE Transactions on VLSI Systems*, vol. 9, pp. 641–655, October 2001.

22. H.M. Shao, T.K. Truong, L.J. Deutsch, J.H. Yuen, and I.S. Reed, "A VLSI design of a pipeline Reed-Solomon decoder," *IEEE Transactions on Computers*, vol. C-34, pp. 393–403, May 1985.

23. W. Wilhelm, "A new scalable VLSI architecture for Reed-Solomon decoders," *IEEE Journal on Solid State Circuits*, vol. 34, pp. 388–396, March 1999.

24. E.R. Berlekamp, G. Seroussi, and P. Tong, "A hypersystolic Reed-Solomon decoderdecoding," in *Reed-Solomon codes and their applications*, S.B. Wicker and V.K. Bhargava, Eds. 1994, pp. 205–241, IEEE Press.

25. R.T. Chien, "Cyclic decoding procedures for Bose-Chaudhuri-Hocquenghem codes," *IEEE Transactions on Information Theory*, vol. IT-10, pp. 357–363, 1964.

26. G.D. Forney, "On decoding BCH codes," *IEEE Transactions on Information Theory*, vol. IT-11, pp. 549–557, 1965.

27. G.D. Forney, "Generalized minimum distance decoding," *IEEE Transactions on Information Theory*, vol. 12, no. 2, pp. 125–131, 1966.

28. R. Koetter, "Fast generalized minimum-distance decoding of Algebraic-Geometric and Reed-Solomon codes," *IEEE Transactions on Information Theory*, vol. 42, no. 3, pp. 721–737, 1996.

29. M.P.C. Fossorier and S. Lin, "Soft-decision decoding of linear block codes based on ordered statistics," *IEEE Transactions on Information Theory*, vol. 41, no. 5, pp. 1379–1396, 1995.

30. M.P.C. Fossorier and S. Lin, "Computationally efficient soft-decision decoding of linear block codes based on ordered-statistics," *IEEE Transactions on Information Theory*, vol. 42, no. 3, pp. 738–750, 1996.

31. R. Koetter and A. Vardy, "Algebraic soft-decision decoding of Reed-Solomon codes," *IEEE Transactions on Information Theory*, vol. 49, pp. 2809–2825, 2003.

32. A. Ahmed, R. Koetter, and N.R. Shanbhag, "VLSI architectures for soft-decoding Reed-Solomon codes," in *Proceedings of the IEEE International Conference on Communications*, Paris, France, June 2004, pp. 81–87.

33. A. Ahmed, N.R. Shanbhag, and R. Koetter, "Systolic interpolation architectures for soft-decoding Reed-Solomon codes," in *Proceedings of the IEEE Workshop on Signal Processing Systems*, Seoul, S. Korea, August 2003, pp. 81–87.

34. G.D. Forney, "The Viterbi algorithm," *Proceedings of the IEEE*, vol. 61, no. 3, pp. 268–278, 73.

35. Behrooz Parhami, *Computer Arithmetic: Algorithms and Hardware Designs*, Oxford University Press, New York, 2000.

36. P.J. Black and T.H. Meng, "A 140-Mb/s, 32-state, radix-4 Viterbi decoder," *IEEE Journal of Solid-State Circuits*, vol. 27, pp. 1877–1885, December 1992.

37. C. Rader, "Memory management in a Viterbi decoder," *IEEE Transactions on Communications*, vol. 29, no. 9, pp. 1399–1401, 1981.

38. G.C. Clark and J.B. Cain, *Error Correction Coding for Digital Communication*, Plenum, New York, 1981, p. 262.

39. P.J. Black and T.H. Meng, "Hybrid survivor path architectures for Viterbi decoders," in *Proceedings of the IEEE International Conference on Acoustics, Speech, and Signal Processing*, Minneapolis, MN, April 1993, pp. 433–436.

40. IEEE Std. 802.15.3, "IEEE Standard for telecom and information exchange between systems-LAN/MAN specific requirements, Part 15.3", 2005.

41. Y.N. Chang, H. Suzuki, and K.K. Parhi, "A 2-Mb/s 256-state 10-mW rate-1/3 Viterbi decoder," *IEEE Journal of Solid State Circuits*, vol. 35, pp. 826–834, June 2000.

42. T. Gernmeke, M. Gansen, and T.G. Noll, "Implementation of scalable, power and area efficient high-throughput Viterbi decoders," *IEEE Journal of Solid State Circuits*, vol. 37, pp. 941–948, July 2002.

43. J. Hagenauer and P. Hoeher, "A Viterbi algorithm with soft-decision outputs and its applications," in *Proceedings of the Global Telecommunications Conference*, Dallas, TX, vol. 3, November 1989, pp. 1680–1686.

44. C. Berrou, P. Adde, E. Angui, and S. Faudeil, "A low complexity soft-output Viterbi decoder architecture," in *Proceedings of the IEEE International Conference on Communications*, vol. 2, May 1993, pp. 737–740.

45. C. Berrou, A. Glavieux, and P. Thitimajshima, "Near Shannon limit error-correcting coding and decoding: Turbo codes," in *Proceedings of the IEEE International Conference on Communications*, Geneva, May 1993, pp. 1064–1070.

46. S. Benedetto et al., "A soft-input soft-output maximum a posteriori (MAP) module to decode parallel and serial concatenated codes," TDA Progress Report 42-127, JPL, November 1996, vol. 6, pp. 507–511, September–October 1995.

47. Z. Wang, H. Suzuki, and K.K. Parhi, "VLSI implementation issues of turbo decoder design for wireless applications," in *Proceedings of the IEEE Signal Processing Systems* (SiPS): *Design and Implementation*, October 1999, pp. 503–512.

48. L. Bahl et al., "Optimal decoding of linear codes for minimizing symbol error rate," *IEEE Transactions on Information Theory*, vol. 20, pp. 284–287, March 1974.

49. A.J. Viterbi, "An intuitive justification and a simplified implementation of the MAP decoder for convolutional codes," *IEEE Journal on Selected Areas in Communication*, vol. 16, no. 2, pp. 260–264, 1998.

50. G. Bauch and V. Franz, "A comparison of soft-in/soft-out algorithms for turbo detection," in *Proceedings of the IEEE International Conference on Telecommunications* (*ICT '98*), June 1998, pp. 259–263.

51. Z. Wang, Z. Chi, and K. Parhi, "Area-efficient high-speed decoding schemes for turbo decoders," *IEEE Transactions on VLSI systems*, vol. 10, no. 6, pp. 902–912, 2002.

52. J. Hsu and C. Wang, "A parallel decoding scheme for turbo codes," in *Proceedings of the 1998 IEEE International Conference on Circuits and Systems*, 1998, vol. 4, pp. 445–448.

53. K.K. Parhi, *VLSI Digital Signal Processing Systems: Design and Implementation*, Wiley, New York, 1999.

54. S. Lee, N. Shanbhag, and A. Singer, "A low-power VLSI architecture for turbo decoding," in *Proceedings of the 2003 International Symposium on Low Power Electronics and Design (ISLPED '03)*, Seoul, S. Korea, August 2003, pp. 366–371.

55. S. Lee, N. Shanbhag, and A. Singer, "Area-Efficient High-Throughput MAP Decoder Architectures," *IEEE Transactions on VLSI Systems*, vol. 13, pp. 921–933, no. 8, 2005.

56. S. Lee, N. Shanbhag, and A. Singer, "A 285-MHz pipelined MAP decoder in 0.18-mum CMOS," *IEEE Journal of Solid-State Circuits*, vol. 40, no. 8, pp. 1718–1725, 2005.

57. S. Lee, N. Shanbhag, and A. Singer, "Low-power turbo equalizer architecture," in *Proceedings of the IEEE Signal Processing Systems (SiPS): Design and Implementation*, October 2002, pp. 33–38.

58. S. Lee, N. Shanbhag, and A. Singer, "Area-efficient high-throughput VLSI architecture for MAP-based turbo equalizer," in *Proceedings of IEEE Signal Processing Systems (SiPS): Design and Implementation*, August 2003, pp. 87–92.

59. M.M. Mansour and N.R. Shanbhag, "Design methodology for high-speed iterative decoder architectures," in *Proceedings of the 2002 IEEE International Conference on Acoustics, Speech, and Signal Processing*, 2002, vol. 3, pp. 3085–3088.

60. M.M. Mansour and N.R. Shanbhag, "VLSI architectures for SISO-APP decoders," *IEEE Transactions on VLSI Systems*, vol. 11, no. 4, pp. 627–650, 2003.

61. A. Giulietti et al., "Parallel turbo code interleavers: avoiding collisions in accesses to storage elements," *Electronics Letters*, vol. 38, no. 5, pp. 232–234, 2002.

62. M. Bickerstaff, L. Davis, C. Thomas, D. Garret, and C. Nicol, "A 24 Mb/s radix-4 LogMAP turbo decoder for 3GPP-HSDPA mobile wireless," *ISSCC Digest Technical Papers*, pp. 150–151, 2003.

63. E. Yeo, S. Augsburger, W.R. Davis, and B. Nikolic, "Implementation of high throughput soft output Viterbi decoders," in *Proceedings of the IEEE Signal Processing Systems (SiPS): Design and Implementation*, October 2002, pp. 146–151.

64. F. Cathoor, S. Wuytack, E. de Greef, F. Balasa, L. Nachtergaele, and A. Vandecapelle, *Custom Memory Management Methodology, Exploration of Memory Organization for Embedded Multimedia System Design*, Kluwer Academic Publishers, Dordrecht 1998.

65. C. Berrou, A. Glavieux, and P. Thitimajshima, "Near Shannon limit error-correcting coding and decoding: Turbo codes," in *IEEE International Conference on Communications*, 1993, pp. 1064–1070.

66. T. Richardson, M. Shokrollahi, and R. Urbanke, "Design of capacity-approaching irregular low-density parity-check codes," *IEEE Transactions on Information Theory*, vol. 47, no. 2, pp. 619–637, 2001.

67. S.-Y. Chung et al., "On the design of low-density parity-check codes within 0.0045 dB of the Shannon limits," *IEEE Communication Letters*, vol. 5, no. 2, pp. 58–60, 2001.

68. R.M. Tanner, "A recursive approach to low complexity codes," *IEEE Transactions on Information Theory*, vol. IT-27, pp. 533–547, September 1981.

69. M.M. Mansour and N.R. Shanbhag, "Architecture-aware low-density parity-check codes," in *Proceedings of the IEEE International Symposium on Circuits and Systems 2003 (ISCAS '03)*, Bangkok, Thailand, May 2003, vol. 2, pp. 57–60.

70. J. Pearl, *Probabilistic Reasoning in Intelligent Systems: Networks of Plausible Inference*, Morgan Kaufmann, San Mateo, CA, 1988.

71. R.J. Mcliece, D.J.C. Mackay, and J-F. Cheng, "Turbo decoding as an instance of Pearl's "belief propagation" algorithm," *IEEE Journal on Selected Areas of Communications*, vol. 16, pp. 140–152, February 1998.

72. F.R. Kschischang and B.J. Frey, "Iterative decoding of compound codes by probability propagation in graphical models," *IEEE Journal on Selected Areas of Communications*, vol. 16, pp. 219–230, February 1998.

73. M.M. Mansour and N.R. Shanbhag, "Turbo decoder architectures for low-density parity-check codes," in *Proceedings of the IEEE Global Telecommunications Conference 2002 (GLOBECOM '02)*, Taipei, Taiwan, November 2002, pp. 1383–1388.

74. M.M. Mansour and N.R. Shanbhag, "High-throughput LDPC decoders," *IEEE Transactions on VLSI Systems*, vol. 11, no. 6, pp. 976–996, 2003.

75. A. Blanksby and C. Howland, "A 690-mW 1-Gb/s 1024-b, rate-1/2 low-density parity-check decoder," *IEEE Journal of Solid-State Circuits*, vol. 37, no. 3, pp. 404–412, 2002.

76. E. Yeo et al., "VLSI architectures for iterative decoders in magnetic recording channels," *IEEE Transactions on Magnetics*, vol. 37, no. 2, pp. 748–755, 2001.

77. V.E. Benes, "Optimal rearrangeable multistage connecting networks," *Bell System Technical Journal*, vol. 43, pp. 1641–1656, 1964.

78. D. Divsalar, H. Jin, and R.J. McEliece, "Coding theorems for turbo-like codes," in *Proceedings of the 36th Allerton Conference on Communications, Control, and Computing*, September 1998, pp. 201–210.

79. Mohammad M. Mansour, "High-performance decoders for regular and irregular repeat-accumulate codes," in *Proceedings of the IEEE Global Telecommunications Conference, Dallas, Texas*, November–December 2004, pp. 201–210.

80. R. Hegde and N.R. Shanbhag, "Toward achieving energy efficiency in presence of deep submicron noise," *IEEE Transactions on VLSI Systems*, vol. 8, pp. 379–391, August 2000.

81. L. Wang and N.R. Shanbhag, "Energy-efficiency bounds for deep submicron VLSI systems in the presence of noise," *IEEE Transactions on VLSI Systems*, vol. 11, pp. 254–269, April 2003.

82. S.R. Sridhara and N.R. Shanbhag, "Coding for system-on-chip networks: a unified framework," *IEEE Transactions on VLSI Systems*, vol. 13, pp. 655–667, June 2005.

83

An Exploration of Hardware Architectures for Face Detection

T. Theocharides
University of Cyprus, Cyprus

C. Nicopoulos
K. Irick
N. Vijaykrishnan
M.J. Irwin
Pennsylvania State University

CONTENTS

83.1 Introduction

Face detection is defined as the process of identifying all image regions that contain a face regardless of the position, the orientation, and the environment conditions in the image. A common mistake people make is to confuse it with face recognition. However, when recognizing faces, the recognition process already knows that an image contains a face. The problem now shifts into identifying the person to whom the particular face belongs [1–5]. As a result face detection has been a major research topic both in academia and industry, and the popularity of the topic appears in a wide range of applications and fields. From security to identification systems, face detection plays a primary role. It is the primary step toward face recognition [6] and serves as a forestep toward multiple applications such as identification, monitoring, and tracking. Face detection algorithms have been developed through the years, and have improved drastically both in terms of performance and speed. However, with today's design technology, we are given the chance to perform face detection at a higher level, in the real-time domain and independent of image and environment variations. Face detection so far has been extensively done using software. With the transistor technology entering the nanometer era however, and the improvement of the detection algorithms, we are able to shift the detection stage in the hardware domain, which offers several advantages [5]. Hence, a fast hardware implementation that can be integrated either on a generic processor or as part of a larger system, directly attached to the video source, such as a security camera or a robot's camera becomes desirable.

Software face detection methods have reached a very high level of both effectiveness and detection rate, as well as a condition-invariant level, where detection can be performed under harsh environments. However, the state-of-the-art software face detection can achieve up to 15 image frames per second (fps) [7] under favorable circumstances, and as such is not quite suitable for real-time deployment. A problem with almost all of these detection algorithms for real-time support is the complexity of the preprocessing and filtering stages that the image goes through before the detection stage [5,7–9]. With the improvement of digital cameras and digital image-processing algorithms, however, the task of hardware face detection is attainable.

This chapter outlines the advances in hardware face detection and introduces two widely accepted face detection algorithms and their respective hardware implementations. We present an FPGA- and an application specific integrated circuit (ASIC)-based implementation of a neural network-based face detection algorithm, and an architecture for the extremely fast AdaBoost algorithm proposed by Viola and Jones [7]. Through the description of both algorithm implementations, this chapter illustrates the issues and challenges in designing both hardware platforms. Section 83.2 introduces the transformation from software implementations to the hardware implementations, highlighting the design decisions that need to be made. Section 83.3 illustrates the differences in the design of a face detection system targeting different hardware platforms, by presenting two alternate implementations of a neural network-based face detection, an FPGA-based implementation, and an ASIC-based implementation. Section 83.4 shows how the algorithm choice impacts the design decision-making by illustrating the design of an architecture to implement the AdaBoost algorithm. The AdaBoost algorithm differs largely from the neural network algorithm, as it is a feature-based algorithm rather than an image-based algorithm. Section 83.5 summarizes and concludes the chapter by emphasizing the important design issues and comparing the two algorithmic approaches in hardware.

83.2   Face Detection—From Software to Hardware

83.2.1   Face Detection: A General Background

This section gives a short background of the face detection process and the operations associated with it. First, let us take a look at the general problem of face detection. We assume that a unit receives an input image frame and detects faces present on that frame. Source images can be of various resolutions and formats, however, the detection technique typically accepts a predefined size and format; training of the detection method is essentially based on a fixed image size and format. Typically, the size of the frame that the detector operates is much smaller than the size of the input image frame. Hence, most of the face detection algorithms operate in three stages, two of which are common to most algorithms [5]. The first stage is image pyramid generation (IPG). The role of the IPG stage is to receive an input image frame and produce subimages (search windows) with the prefixed size that the detection algorithm operates on. On a large source image frame however, a face might be larger than the size of a search window, hence the source image frame needs to be scaled down and more search windows generated. If the downscaled image is still larger than the detection size, it can still contain faces larger than the detection window size. Hence, further downscaling is necessary, until the input image frame size matches the size of the detection window. The IPG operation requires extensive data movement and is typically the bottleneck of the computation [5]. Several methods have been proposed to speed the operation up, such as reducing the number of scales that a source image is searched at the expense of accuracy. Additionally, increasing the search window size reduces the number of searched regions, at the expense of the detection algorithm, which has to operate on a larger amount of data (as the detection frame size increases) [5]. Another approach, based on assumption that within an image frame human faces typically have some small separation between them, is to perform searches by reducing the search overlap by two consecutive search windows [7,10,11] thus reducing the number of searched regions.

The second stage is the preprocessing stage. This stage is also common to all algorithms; however it varies among algorithms in the way it performs its task. The task of this stage is to eliminate as much environment and lighting variations from an input image as possible, by performing various filtering functions over the image, reducing the image variance. The filtering functions used during this stage depend on the robustness of the detection stage, which is the third and final stage. However, recent advances in camera technology and image processing algorithms embedded on digital video cameras provide great image quality for pattern recognition and detection algorithms, and reduce the importance of this stage. It must be noted however, that a clear image with emphasized features and edges always enhances the probabilities of accurate detection [12].

The generalized task of this stage is to take an image segment that has been filtered for adjustment to maintain environment and pose invariance, and to output whether or not the segment contains a face. This stage is initialized with a training set of data, that is a database of features that are used to match the presence of a face or not. The training set should contain both positive and negative examples that are both images with faces, and without faces [7,10,11]. Figure 83.1 illustrates a simple face detection framework, where three classifiers are used in conjunction to detect a face using as input preprocessed $M \times M$ search windows generated from the original source image.

Face detection algorithms can be classified into two major categories: Feature Based/Template Matching Approach, and Image Based/Pattern Classification Approach. The first approach deals with face detection by searching for features that are unique to faces, or uses image features such as color and geometric models to search for a face. The second approach employs classification—it treats the image as data to be classified into containing a face or not. This second approach is more commonly employed, and includes neural networks, Support Vector Machines and linear subspace methods such as PCA and ICA [10,11]. As mentioned earlier, generic face detection unit can be partitioned into the three major stages: image search, image processing, and detection. The first two stages require most of the data manipulation, as well as large-scale mathematical operation such as downscaling and filtering. The third stage operates in a different

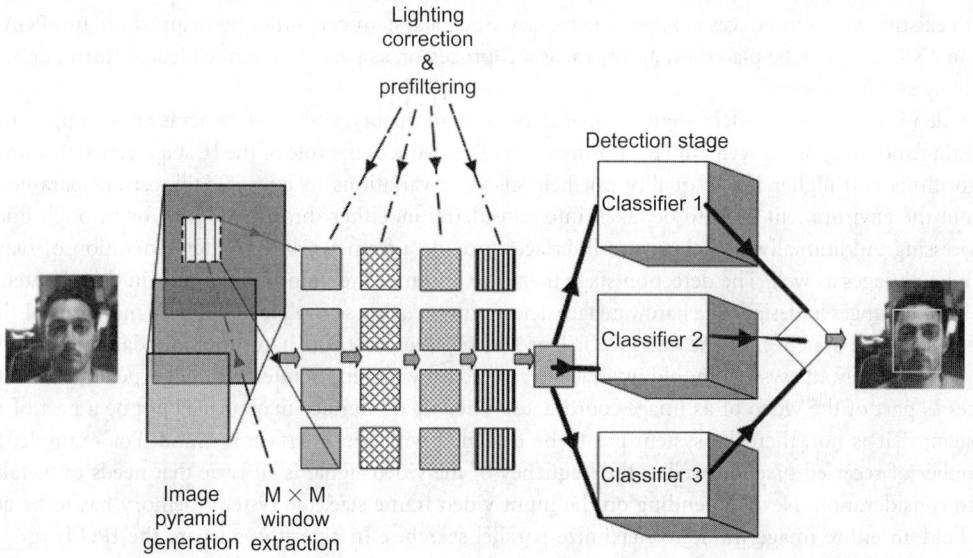

FIGURE 83.1 Face detection outline.

manner however, and that makes it special—given that the detection unit classifies an image as a potential face or not, it has to have two different modes of operation—training and recognition. During the training mode, the detection unit is given data that contain faces and those that do not contain faces. During the detection stage, the unit uses the already data from the training phase to classify the image as a face or not.

83.2.2   Hardware Implementations: Challenges and Issues

There are several design challenges in designing a hardware face-detection system. There has been extensive research in the field, ranging mostly in the software domain [1–4,7–9,11–18]. In recent years, however, a few attempts at hardware implementations that implement face detection on multiple FPGA boards and multiprocessor platforms using programmable hardware [2,3,19,20] have been proposed. Many of the proposed hardware solutions though, are not compact and do not meet mobile environments. Additionally, these implementations require specific interfacing criteria, and do not fill the plug-and-play approach of modern hardware platforms. For example, the FPGA implementations utilize up to nine boards or utilized a general-purpose processor assisted with a coprocessor which performs certain algorithmic computations [2,19]. Some of the attempted hardware implementations generally feature algorithms that are not as effective as traditional software approaches, such as competitive feature approach [1], which does not perform as well as a neural network or other state-of-the-art algorithms. An exception is the implementation on an embedded platform using neural networks in Ref. [16]. That implementation though is not purely on hardware; rather than on a reconfigurable multiprocessor platform integrated with embedded software, which achieves 10 fps.

Other implementations utilizing embedded hardware have surfaced recently. In Ref. [20], the authors explain an architectural methodology for mapping algorithms in hardware, and using an embedded development platform (Xilinx ML-310 Board) they show that they can achieve a frame rate of 12 fps, which is acceptable for certain applications. The authors use a shape-based approach where edge detection is used to detect elliptic shapes, and further template matching is used in conjunction to detect the presence of a face in the input image. Such embedded platforms can be built as add-on expansion cards to a general-purpose computer, or as individual components to perform detection. An embedded face detection algorithm, which achieves up to 4 fps is presented in Ref. [21]. The algorithm uses a neural network approach and is built in a cell-phone embedded processor. The application targets low-performance and ultra-low power cameras, hence the small frame rate is acceptable. As evidenced from the existing work, the need for a stand-alone system, which meets real-time frame rate of 30 fps and can interface with existing video interfaces is beyond necessary. Such a system can either be mapped on an FPGA or as an ASIC, as it can be placed on a camera, as a coprocessor, as part of an embedded platform, or as an entirely stand-alone system.

Video frames from modern digital cameras are of high quality, and most cameras are equipped with lighting and image enhancement (IE) features [22]. This reduces the role of the IE stage as better training algorithms and higher image quality can help alleviate variations in images. Still, certain parameters about the environment need to be taken into consideration, either through training or through image processing. Additionally, digital camera interfaces provide a framework for better generation of search window images as well. The detection stage therefore becomes the point of emphasis in this chapter.

The challenges in designing a hardware face detection system are several. In addition to meeting real-time frame rate, the system must be energy efficient and reliable. First, the hardware boundaries need to be explored. Ideally, the system should interface to a standard video input interface, and export the detected faces as part of the video or as image coordinates. The video interface may or may not be a part of the system; if it is not, then the system has to be designed with the interface in mind. For example, the number of received pixels as well as the frequency of the video signal is an issue that needs to be taken into consideration. Next, depending on the input video frame size, the system memory has to be able to hold an entire image frame to maximize parallel searches. In algorithms where the IPG is used to generate search window images, the system memory must be designed in such a way so as to maximize the flow of data and consequently increase the parallelism of the system. Last but not least, the algorithm computation requirements vary with the chosen algorithm, and algorithms with high accuracy in software do not necessarily perform as accurately in hardware at real-time frame rate, due to the complexity of the computations. For example, an algorithm that consists of several floating-point computations and operations such as divisions and square roots might detect faces at a high rate, however when mapped in hardware, it will require extensive resources to detect the faces with a high accuracy and fast frame rate. Similarly, in floating- and fixed-point computations overflow results in loss of accuracy as well. As such, the choice of an appropriate algorithm is crucial. As in other image-processing applications, data movement and memory accesses dominate the computation, and as such the hardware design needs to be optimized for parallel data movement, and intelligent memory partitioning. The choice of the algorithm impacts the design typically in the detection stage; image manipulation is still necessary regardless of the chosen algorithm, and memory partitioning depends on the algorithmic data flow.

Another issue is handling of training data; whether the system can be trained in hardware or whether it uses predetermined training data, the data need to be stored and accessed in a way that it does not delay the computation. The choice of the algorithm obviously has a lot to do with the organization of the system components, as the access pattern of the image data changes with the algorithm. In image-based computations, image data can be accessed in any order and hence an order, which exploits parallelism, can be designed. However, data access in feature-based algorithms depends on the features used, and the order of operations is determined by the training data.

In this chapter, we explore both types of algorithms. First, we will look into the implementation of a neural network-based face detection in hardware, which is an image-based approach. Then we will approach the problem of face detection using a feature-based method, AdaBoost. Both methods provide certain advantages and disadvantages in designing hardware platforms, which we explore in detail in this chapter. First we start by presenting a neural network-based face detection, where the targeted implementation is both an FPGA and an ASIC.

83.3 Neural Network Hardware Face Detection

In this section, we present the implementation of a face detection algorithm based on the classification-based neural network algorithm proposed originally by Rowley et al. [8]. The algorithm provides high parallelism capabilities for hardware implementation, and achieves high detection rates when trained extensively. To illustrate however the challenges in a hardware implementation, we show how the targeted hardware platform can affect the implementation. As such, we present two different design approaches; the first targets an ASIC implementation, whereas the second targets an FPGA implementation.

The neural network algorithm consists of the three stages outlined in Section 83.2.1. The first stage is the IPG stage, and scans the image at different scales generating 20×20 search windows as shown in Figure 83.2 (right). A search window is defined as an image subregion searched for a human face. Each of the windows is treated as an individual image. The windows are then passed through an IE stage where histogram equalization and lighting correction is performed. Finally, the 20×20 window is evaluated for a face or not using three separate neural networks, each of which looks at the image in a different manner as shown in Figure 83.2. The first network looks at 4 10×10 regions, the second looks at sixteen 4×4 regions, and the third looks at six horizontal 5×20 overlapping strips. Note that the neural network structure can change to adapt to different image sizes [14], at the cost of a larger network size. General purpose neural network processors such as the ones proposed in Refs. [23,24] can be used to map neural networks which change in connectivity and size.

The hardware implementation can be partitioned into three stages, each stage designed individually and interfaced to each other. The implementation of the detection stage, however, is the one that we place more emphasis on, and in this section we illustrate its implementation when targeting different hardware platforms. We first address the design of an ASIC, which performs face detection and consists of all the three stages.

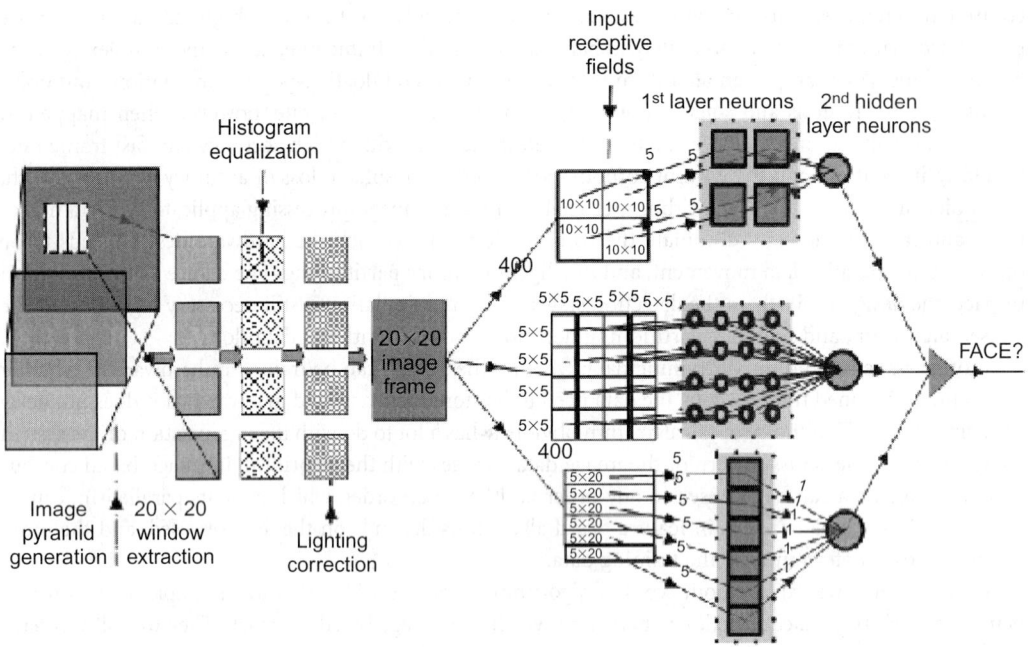

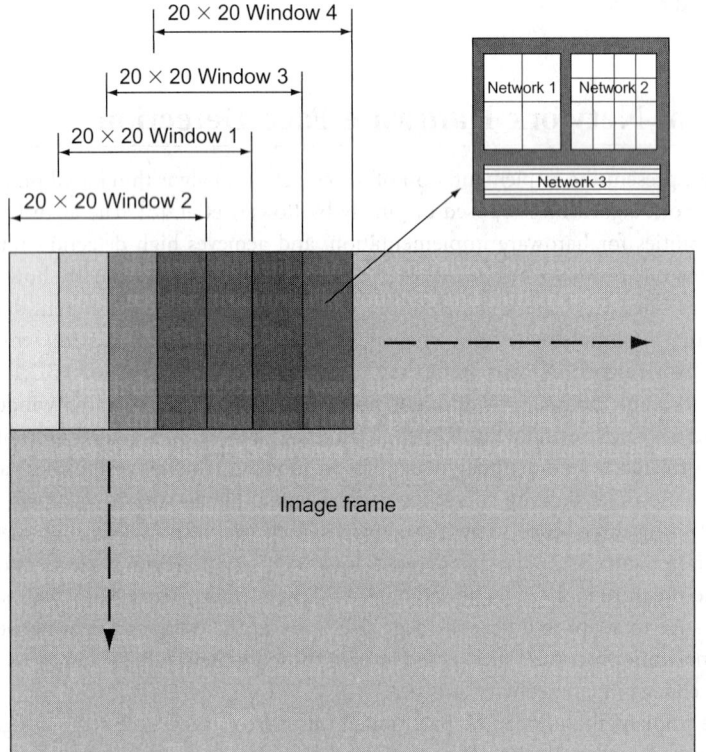

FIGURE 83.2 Neural network-based face detection block diagram and IPG process.

83.3.1 ASIC Implementation of a Neural Network-Based Face Detection

An ASIC implementation of the neural network-based face detection was first presented in our previous work in Ref. [25]. The implementation presented here is significantly of more area and is power-efficient. The entire face detection process with all three stages was mapped on a single chip. However, this section will emphasize the detection architecture, i.e., the neural network stage. The IPG and IE stages aim to enhance the detection process, and are, in fact, decoupled from the neural network block. A brief description of these two stages will be given below; for a more detailed analysis, the reader is referred to Refs. [25,26].

The IPG module is the interface of the system with the outside world. The image data source feeds data to the module via a 64-bit bus running at the system clock frequency. The incoming image is a 320×240-pixel grayscale frame. The image pyramid generator creates a large number of 20×20 subwindows. The original frame is also scaled down at various levels of magnification and each level generates further 20×20 subwindows. In this fashion, both large and small faces in the original image are guaranteed to be analyzed by the system. However, this method generates a total of 8050 subwindows per 320×240 frame, causing the IPG module to be memory-dominated. It requires 80 kB of memory, spread over 10 equally sized banks, two of which are dual ported. Windows are generated in raster-scan style and each window is individually and completely handed off to the IE module through a 32-bit interface and several hand-shaking control signals. The IPG module is throttled by the IE and neural network blocks to maintain a steady flow of data. The scaling of the original picture is performed using a subsampling technique, rather than a more complex matrix-multiplication-based affine transformation. This method reduces system complexity significantly, with only a modest reduction in accuracy. Because of its memory-dominated nature, IPG is by far the largest module of the system. Furthermore, the RAM modules used by the IPG also dictate the system clock frequency, which is 125 MHz.

The IE module is responsible for improving the overall image quality of the incoming 20×20 windows, thereby improving the chances of detection by the neural network block. There are several enhancement techniques that can be employed, such as lighting correction, edge sharpening, and histogram equalization. The aim is to minimize the impact of environmental variations within the picture on the detection process. The processed pictures should, ideally, have a uniform image quality. This implementation assumes use of modern-day cameras, which eliminate the need for dedicated lighting correction and sharpening units. Therefore, the only technique implemented was histogram equalization. Histogram equalization improves the contrast and intensity distribution of each window. The module works in two distinct phases: a cumulative distribution function (CDF) array construction (the main component of the histogram equalization technique) and output streaming to the neural network. The IE module performs histogram equalization on the complete 20×20 window in a single pass, which provides better image results than using smaller subwindows. The module requires 503 clock cycles to fully process a window and send it to the neural network block.

The neural network component of the face detection implementation bears the responsibility of detecting the presence of a face within the 20×20 search windows generated by the IPG unit and subsequently enhanced in the preprocessing stage (see Figure 83.1). The neural network (NN) module receives input from the preprocessing unit (i.e., the IE module) at a rate of 1 pixel (8 bits) per clock cycle. The unit produces a single bit output, which is asserted when a face is detected within a window. No handshaking is necessary between IE and NN; the control is left entirely to the IE unit which initiates a new window transfer by asserting a "Frame Start" signal. A constant stream of pixels then follows for 400 consecutive cycles (recall that 1 pixel is transmitted per clock cycle), which concludes the 20×20 (i.e., 400 pixels) window transfer. The neural network unit requires 513 clock cycles to complete the processing of a single window. By utilizing computational overlapping between the three major units of the system, this time is completely masked by the IPG and IE modules.

The NN module was architected to employ parallelism, but remains as small as possible by fully utilizing existing resources in all stages of its operation. This approach ensures minimal area and resource budgets. However, despite the benefits that such minimalist philosophy affords, it also spawns a major design challenge: timing and scheduling complexity. For this reason, the neural network unit is more control-intensive, rather than data or computation-intensive. The functionality of the module is shown in Figure 83.2. Each incoming pixel is multiplied by a predetermined weight value (obtained off-line through training), accumulated over an entire subwindow, and the resulting sum passed through an activation function, which determines the output value forwarded to the next layer of neurons. In this implementation, the neurons are distributed in three stages: the first stage operates directly on the incoming pixel values, the second stage operates on the outputs of the first stage, and the third stage on the outputs of the second stage. The third stage provides the final single-bit output indicating the presence or not of a face within the 20×20 window.

The accuracy of a hardware-implemented neural network depends heavily on the accuracy of the activation function implementation. The activation function used in this system is the hyperbolic tangent (tanh), implemented as a look-up table (LUT) in an SRAM. After investigating both fixed-point and floating-point arithmetic choices and several bit widths of precision, it was found that a 16-bit fixed-point implementation would achieve accuracy to within 0.1% of double-precision floating-point arithmetic (used in software like MATLAB). This way, the extra hardware complexity and delay overhead of a floating-point architecture were avoided, with negligible impact on accuracy. Figure 83.3 illustrates the hyperbolic tangent function and the significant increase in accuracy by moving from an 8-bit to a 16-bit implementation. Note that the stored LUT covers only domain values from 0 to 3. This is because tanh is odd and saturating. Hence, negative domain values are the same as the corresponding positive ones multiplied by -1, while values beyond 3 (-3) saturate to 1 (-1). These properties allow us to keep the LUT small, saving SRAM area and improving performance.

The overall architecture of the neural network detection block is shown in Figure 83.4. Stage 1 of the architecture (shown on the left-hand side of Figure 83.4) is the most complex. It consists of a 4 kB single-ported SRAM (SPRAM 1), which holds weight values and four multiply-accumulate (MAC) blocks. The stage is divided into three "virtual" neuron groups: the first group divides the 20×20 window into four 10×10 subwindow neurons (see Figure 83.2) and employs a single MAC block. The second group divides the 20×20 window into sixteen 5×5 subwindow neurons and also employs a single MAC block. The third group divides the main window into six 5×20 overlapping subwindow neurons. The overlapping nature of the neurons necessitates the use of two MAC blocks in this group. The incoming pixels come in a raster-scan form, which implies that the MAC operation constantly skips from neuron to neuron before completion. Therefore, the partial MAC results of each neuron group are stored in registers and swapped back and forth, allowing the use of a single MAC block by all neurons within the group. Careful selection of word length and ordering of read requests by the three neuron groups eliminates contention for SPRAM 1.

Stages 2 and 3 are shown on the right-hand side of Figure 83.4. They share a single MAC block and a 4 kB single-ported SRAM (SPRAM 2). This RAM holds the weight values for stages 2 and 3, the results of stages 1 and 2, and the activation function LUT. There are contention issues with SPRAM 2, because of simultaneous requests for the LUT from all three stages, compounded by requests for stage 2 and 3 weights

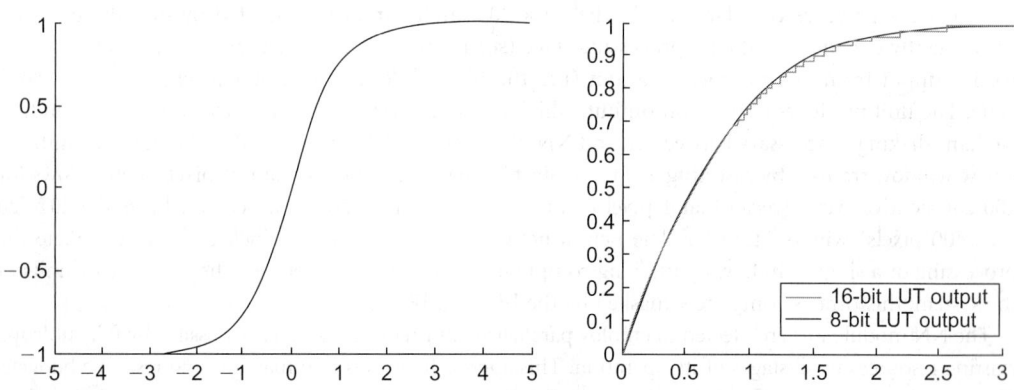

FIGURE 83.3   Double-precision tanh function (left); 8/16-bit stored tanh functions (right).

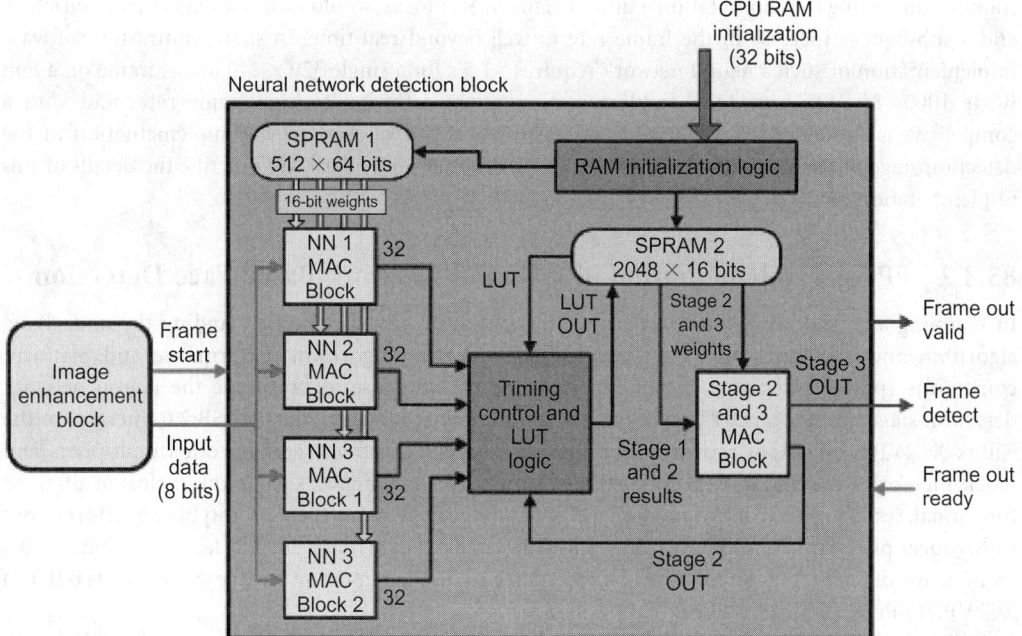

FIGURE 83.4 Neural network architectural overview.

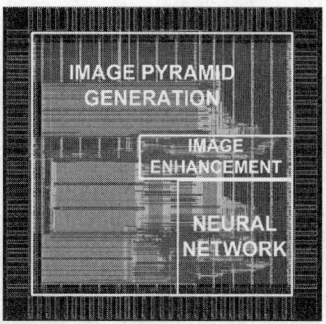

FIGURE 83.5 Face detection chip layout.

and stage 1 and 2 results. These issues are resolved by the Timing Control and LUT Logic unit, which coordinates the whole process with minimal delay overhead. Stage 1 is run in parallel with stages 2 and 3 to maximize throughput and resource utilization, consistent with our lightweight design methodology. To preserve accuracy, the bit widths of the internals of the MAC blocks were chosen so as to avoid overflow even in a worst-case scenario. This is imperative, since tricks like saturating addition reduce the detection accuracy, which is of utmost importance in a neural network.

The design also includes some RAM initialization logic, which provides an interface with an off-chip CPU, which stores the weight values (obtained during the off-line training phase) and the activation function in RAM upon power-up.

The entire face detection system was implemented in Verilog HDL and simulated in ModelSim to ensure correctness. Following functional verification of the architecture, the system was synthesized in Synopsys Design Compiler and underwent postsynthesis verification in ModelSim. Finally, Cadence Encounter was used for placement and routing. The final layout is shown in Figure 83.5.

The final chip is 7.3 mm² and consumes a total of 165 mW. The original goal of a lightweight design was achieved. The chip operates at 125 MHz and performs face detection at 24 fps (320 × 240-pixel frames). Increasing the computational units, i.e. the MAC blocks, would facilitate deeper parallelization and a subsequent increase in the frame rate to well beyond real time. In sharp contrast, a software implementation of such a neural network requires ~1.5 s for a single 320×240 image frame on a Sun Blade 1000 [8]. While an ASIC implementation offers detection at high frame rates and with a competitive accuracy, it is complicated and costly. As such, we examine the implementation of the detection stage on an alternative platform, an FPGA prototyping board. We describe the details of this implementation next.

83.3.2 FPGA Implementation of a Neural Network-Based Face Detection

In mapping any traditionally software-based algorithm to hardware, understanding the underlying algorithm and the target platform is critical in leveraging algorithm performance and platform constraints (power and area). In our implementation, we chose to prototype the neural network detection stage on the Xilinx XUP2V-Pro development board, which offers a USB 2.0 interface to the Xilinx XC2VP30 FPGA, as shown in Figure 83.6. This FPGA offers 136 embedded multipliers and block memories making it ideal for the MAC intense neural network. It is the inclusion of these functional resources that make modern FPGA architectures more than prototyping platforms but rather ideal platforms for implementing real-time imaging algorithms such as face detection. In this section, we describe the neural network face detector implementation on the Xilinx Virtex-II Pro XC2VP30 FPGA.

As Figure 83.7 shows, the structure of networks in the first, second, and third layer only differ by the number of internal neurons. The following discussion, therefore, will be limited to Network 1 of the first layer.

FIGURE 83.6 FPGA implementation board.

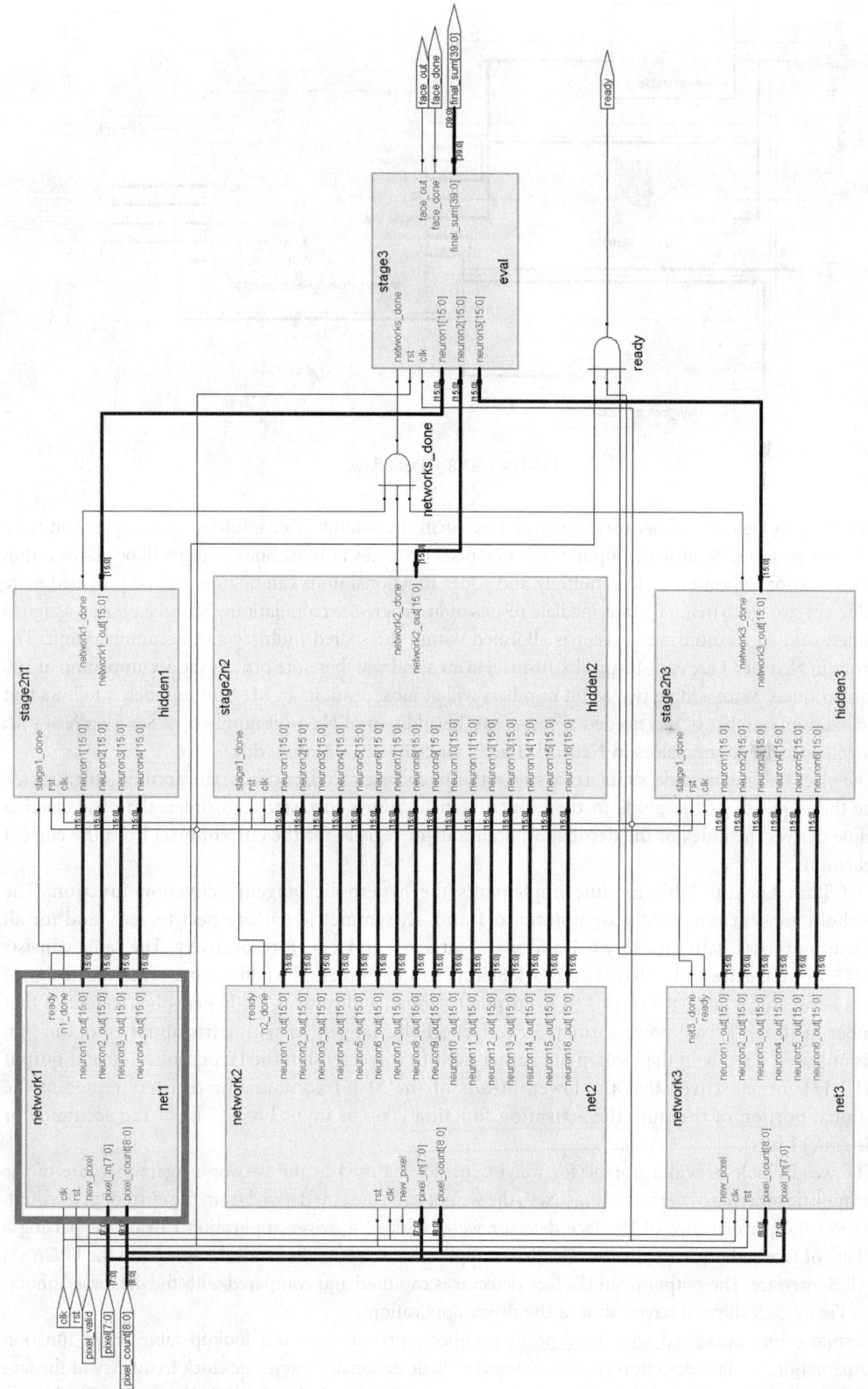

FIGURE 83.7 Proposed architecture for implementation of detection network on the Virtex-II FPGA.

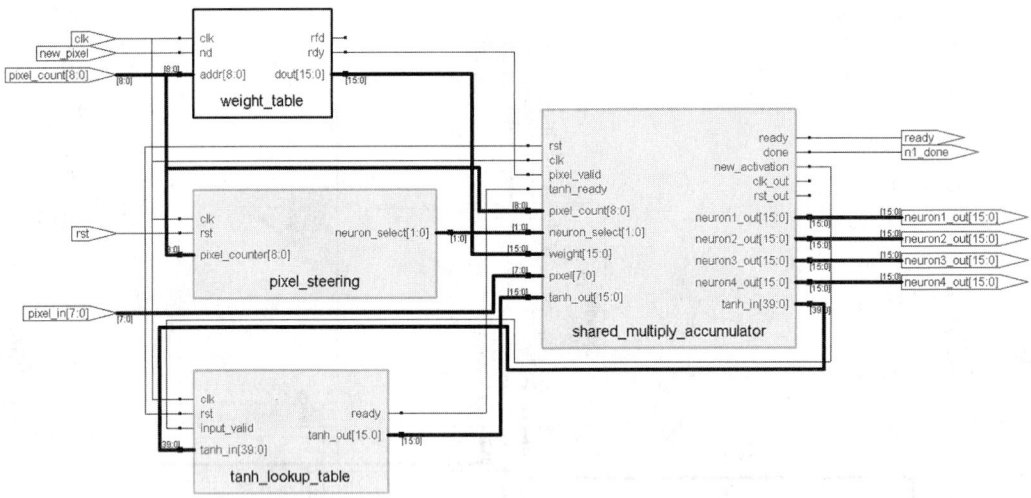

FIGURE 83.8 Data flow.

Referring to Figure 83.8, networks in layer 1 perform the Multiply Accumulate, MAC, operation for a number of neurons. Because the input rate is one pixel per cycle, at most one neuron will be active within the network for a given cycle. The multiply and adder functional units can be time-shared provided there is sufficient storage to maintain intermediate results of in-progress accumulations. For each neuron assigned to a network, an accumulator register is allocated within the shared multiply and accumulate unit. The neurons in Network 1 are each 10-pixel × 10-pixel in area and will therefore process the accumulation of 100 24-bit products. Since adding two M bit numbers will at most result in an M+1 bit number, it follows that the maximum number of bits needed to represent the addition of N M-bit numbers is $S = \lceil \log(N) \rceil + M$. Consequently the accumulators in Network 1 are $\lceil \log(100) \rceil + 24 = 31$ bits wide.

The pixel steering module, as its name suggests, directs pixels to the appropriate accumulator register. Since the sequence of the pixels in the input stream is always in raster scan order, the pixel steering module derives the index of the destination accumulator register for the current pixel from the current pixel count.

The Tanh Lookup Table module implements the hyperbolic tangent activation function. The hyperbolic tangent function is asymptotic to 1 and −1, symmetric with respect to zero, and for all practical purposes, saturates to y = 1 and y = −1 at x = 8 and x = −8 respectively. The table consists of 2048 16-bit entries and utilizes two Block SelectRAM resources. The address port receives unsigned values between 0.0 and +8.0 in {0.3.9} fixed-point format, and returns the hyperbolic tangent of that number in {0.1.14} fixed point format. Since Hyperbolic Tangent is symmetric about zero, the sign bit is not used directly in the lookup but rather used to correctly sign the Hyperbolic Tangent output in {1.1.14} format. Given that the lower 14-bits of the 31-bit accumulator registers represent the fractional portion of the sum, the activation function takes as input bits [17.6] of the accumulator registers (11 bits).

The weight table provides storage for weight coefficients used in the network. Again, because of the deterministic ordering of the incoming pixels, the weight table index is derived from the current pixel count.

To test the performance of the face detector we developed a driver application capable of parsing a database of face and non-face images. The driver application sequentially sends the images to the FPGA via the USB interface. The output from the face detector is captured and compared with the database annotations. Figure 83.9 shows a screen shot of the driver application.

Despite errors associated with fixed-point number representation and lookup table based function approximation, the face detection system achieved 94% detection accuracy. The clock frequency of the face detector is 100 MHz and it requires 813 cycles to process a single 20-pixel × 20-pixel sub window. The latency

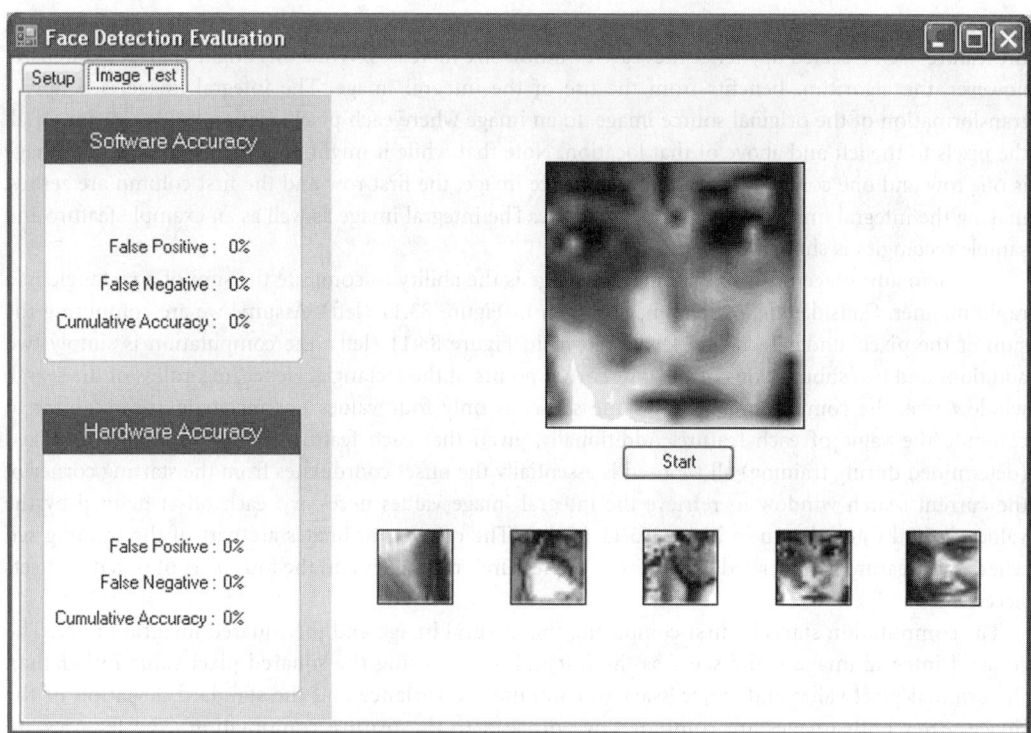

FIGURE 83.9   Ingress and accumulator node architecture.

of the system is therefore, 8.13 microseconds. A 320×240 image frame scaled four times—at reasonable scale factors—with five pixel overlaps will generate approximately 3000 sub windows. The face detector will process this frame in 24 milliseconds and will process 41 such frames in 1 second.

83.4   Algorithm Impact—Hardware Face Detection using AdaBoost

In section 83.3, we have seen how an image-based approach is implemented in hardware. Although the neural network itself is ideal for hardware implementation, the algorithm is not the most efficient one primarily because the detection stage has to look through the entire image for a face, which might occupy just a small area of the image. Consequently, several time and computation power is wasted in looking over areas that have no faces. Additionally, the IPG is a computationally intensive process. Viola and Jones [7] have proposed a feature-based algorithm, which uses a chain of classifiers to determine whether a location within an image contains a face. The approach uses the concept of weak and strong classifiers; a strong classifier consists of a series of weak classifiers, and the outcome of all the weak classifiers determines the outcome of the strong classifier. The approach essentially attempts to scan the entire image for locating face candidates. Locations that fail to produce evidence that there might be a face in the image are discarded and eliminated from further search. As the locations which have candidate faces decrease through the search methodology, the computation focuses only on locations which present strong evidence that a face exists. Eventually, if a location passes through a series of cascaded classifiers, it is said to have a face. The classifiers used in the algorithm are features; a feature is represented as a set of black and white rectangles located at specific offsets within a search window. The outcome of the feature equals the sum of the pixels which lie under the black rectangle, minus the sum of the pixels under the white rectangle. The feature size determines the size of the image window being searched.

As such, instead of scanning the image at various scales to find images which do not fit the search window, the feature itself is scaled up; hence the search window size increases. While this might seem very intensive however, the algorithm benefits from the use of the integral image. The integral image is simply a transformation of the original source image, to an image where each pixel location holds the sum of all the pixels to the left and above of that location. Note that while it might appear that the integral image is one row and one column larger than the source image, the first row and the first column are zeroes, making the integral image size same as the source. The integral image as well as an example feature and sample rectangles is shown in Figure 83.10.

The main advantage of using the integral image is the ability to compute the sum of a rectangle in a rapid manner. Consider the source image shown in Figure 83.11 (left). Assume we are computing the sum of the pixels under rectangle D. As shown in Figure 83.11 (left), the computation is simply two additions and two subtractions of the four corner points of the rectangle. Hence, regardless of the search window size, the computation remains the same, as only four values per rectangle are necessary to compute the value of each feature. Additionally, given that each feature and its rectangles are fixed (determined during training) all we need is essentially the offset coordinates from the starting corner of the current search window to retrieve the integral image values necessary, each offset defined by the values dx and dy as shown in Figure 83.11 (right). The offset coordinates are part of the training set, where each feature is associated with a list of the feature's rectangles and the four pairs of (dx, dy) offsets necessary.

The computation starts by first computing the integral image and the squared integral image. The squared integral image is the same as the integral image, using the squared pixel value rather than the original pixel value and is necessary to compute the variance and the standard deviation of the image. After both images are computed, we proceed to the feature computation. The features are computed in stages. Each feature is evaluated over a search window of equal size as the feature. In Viola and Jones' implementation, the starting feature size is 24×24 pixels; we will use this size as

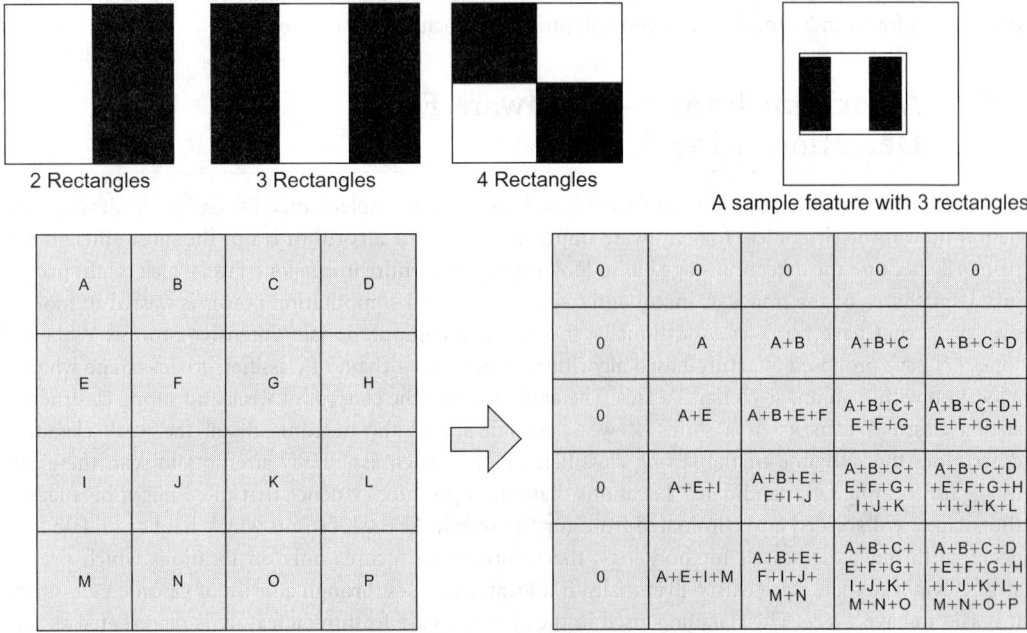

FIGURE 83.10   Feature rectangles, features and integral image.

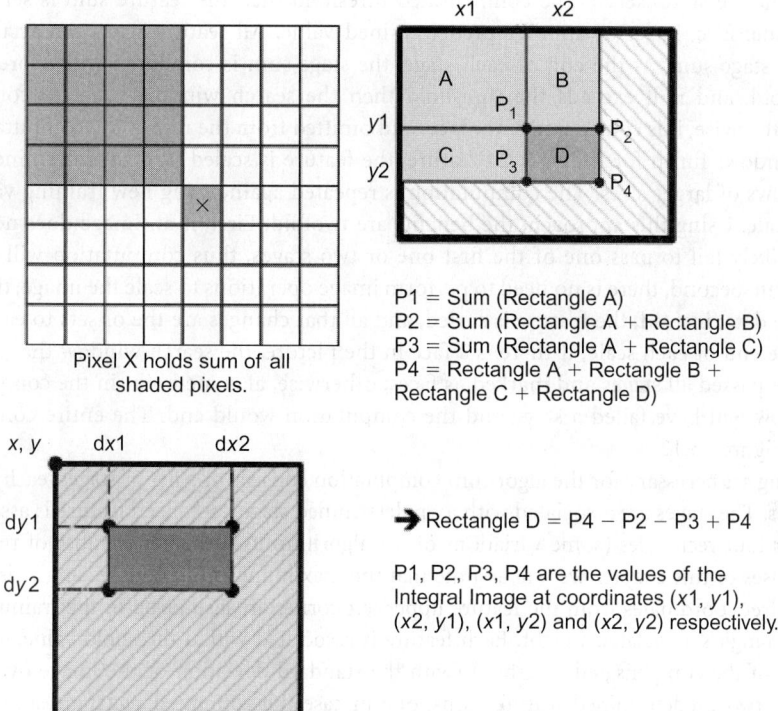

P1 = Sum (Rectangle A)
P2 = Sum (Rectangle A + Rectangle B)
P3 = Sum (Rectangle A + Rectangle C)
P4 = Rectangle A + Rectangle B +
Rectangle C + Rectangle D)

➔ Rectangle D = P4 − P2 − P3 + P4

P1, P2, P3, P4 are the values of the
Integral Image at coordinates (x1, y1),
(x2, y1), (x1, y2) and (x2, y2) respectively.

FIGURE 83.11 Rectangle computation in integral image (upper left and right) and offset values (dx, dy) (left). Each rectangle needs four offset (dx, dy) pairs.

the reference starting feature size. The location of the rectangles within each feature is predetermined from training, hence to evaluate each rectangle we need the offset dx and dy values from the starting coordinate of each feature. The first search window is set over the origin of the image, and computation starts. Features are grouped into stages, with each stage acting as the strong classifier and each feature acting as the weak classifier. Each stage has a number of features. The sum of a feature is computed by using the rectangle sums, and multiplying each rectangle sum with predetermined weights. Prior to evaluating each feature, the threshold is adjusted to take into consideration the standard deviation of the image, to compensate for brightness and contrast variations. The standard deviation can be computed from the image variance, as shown in Eq. (83.1). This is computed by evaluating the integral and integral squared images over the search window size, in the same manner that a feature rectangle is evaluated, using the four corners of the search window. It is only needed to be done once however, and all subsequent features evaluated over that search window can use the computed standard deviation value (σ), which is computed by computing the square root of the variance as follows:

$$VAR = E[X^2] - \{E[X]\}^2 \Rightarrow VAR = \sum_{i=0}^{\text{no. of pixels}} X_i^2 - \left[\frac{\sum_{i=0}^{\text{no. of pixels}} X_i}{\text{AREA}} \right]^2 \quad \text{and} \quad \sigma = \sqrt{VAR} \qquad (83.1)$$

Integral Squared Image   Integral Image

The standard deviation is multiplied with the predetermined threshold to obtain the *compensated threshold*. The sum of all weighted feature rectangles is used to determine the feature sum and if the weighted rectangle sum exceeds the compensated threshold then the feature sum is set to a prede-termined value. Else, is set to another predetermined value. All feature sums are accumulated to compute the stage sum. At the end of each stage, the stage sum is compared with a predetermined stage threshold, and if it exceeds the threshold then the search window is set to contain a face candidate. Otherwise, it is considered a nonface, and omitted from the rest of the computation. When all search windows finish for the 24 × 24 feature, the feature is scaled by a predetermined factor, to search windows of larger sizes. The computation is repeated again, using new training values set for the higher scale. Using this approach, the benefits are twofold. First, if an image does not contain a face, it will likely fail to pass one of the first one or two stages, thus computation will stop with a nonface output. Second, there is no need to perform image operations to scale the image, thus limiting loss of image data; instead, the feature is scaled, and all that changes are the offsets to each rectangle corner. At the end of each scale, if there is a face in the picture, the search window that contains the face will have passed all stages and marked as faces; otherwise, at some point in the computation the search window will have failed a stage and the computation would end. The entire computation is outlined in Figure 83.12.

The training set necessary for the algorithm computation consists of a list of stages, each stage with a list of features. The stages are associated with a predetermined stage sum. Each feature is associated with two, three, or four rectangles (some variations of the algorithm use a higher number of rectangles but for the purposes of this chapter we will assume that the maximum number of rectangles is four). Each rectangle's offset coordinates from the feature upper left corner are also given in the training set, along with each rectangle's associated weight. Each feature is associated with a threshold value, to be used in computation of the compensated threshold (with the standard deviation of the image over the search window) and two predetermined feature sums; one in case the computed rectangle sum exceeds the compensated threshold and one in case the sum is lower.

In the proposed architecture, the computation is performed over an array grid, where each node in the array holds the integral image and integral squared image data, and serves as computation and data transfer unit as well. The array provides the ability to perform all rectangle operations for the entire image array, each rectangle computed in parallel. The proposed architecture is explained in detail in the next section.

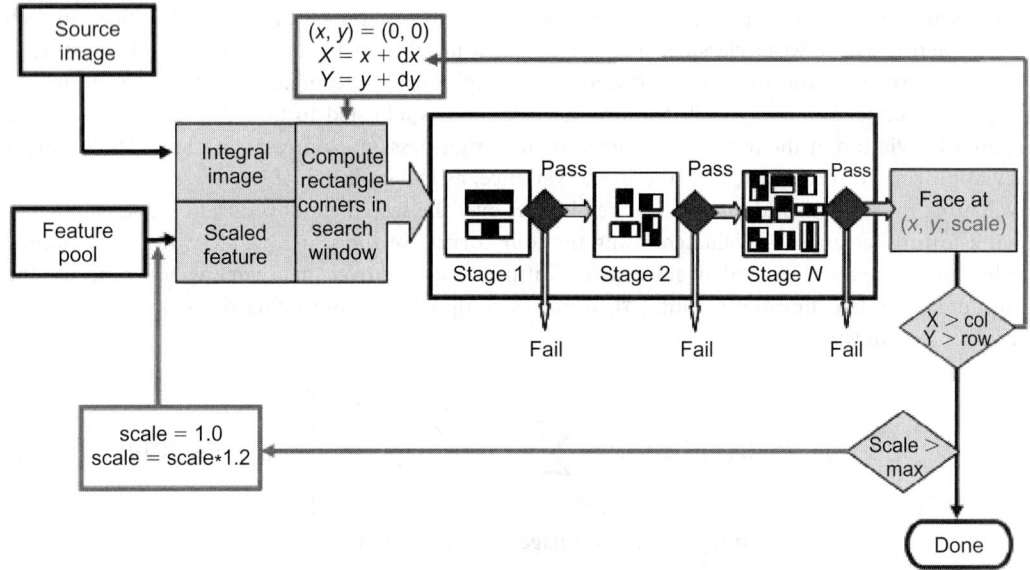

FIGURE 83.12   Algorithm outline.

83.4.1 Proposed Architecture

83.4.1.1 Data Storage Requirements

First we need to determine the hardware requirements for the algorithm in terms of storage and data flow. There are two places of interest: the training set storage and the computation stage. To evaluate the necessary storage requirements for the training parameters, we used the Intel Open Computer Vision Library (CV) [27] for the training data. The CV library provides a state-of-the-art software implementation of the AdaBoost detector, utilizing a very accurate pool of features. The training set uses a starting feature size of 24×24 pixels, and scales each feature by a factor of 1.2, resulting in 13 scaled feature sizes. The largest feature size is 214×214 pixels. The training set associated with the library provides us with necessary information about the accuracy and precision required for the hardware computation. From the training set, we can derive that we need 8 bits per rectangle weight, for each threshold value and for each predetermined feature sum. All these values are not integers; they are signed fixed-point numbers. Hence we need a sign bit and fixed-point representation of 2 integer bits and 5 decimal bits. The dynamic range supported is ± 3.96875, which reflects the values given in the CV training set. We also need up to 8 bits to store each rectangle offset from the feature corner, as the largest feature size used in the CV set is 224×224. As such, each rectangle needs 4×16 bits to store the dx and dy values, and 8 bits for its associated weight. Each feature has either two, three or four rectangles, and each stage has a number of features, ranging from 9 to 211 features per stage. The total number of features used in the reference training set is 2913, spread over 25 stages. The total number of rectangles is 6383. An important factor, however, is the frequency of each feature computation; due to the nature of the algorithm, ~80% of the computation occurs during features from stages 1 and 2, which are only 25 features for a total of 50 rectangles. Hence, our emphasis falls on providing rapid access to the data necessary to compute the first two stages as thereafter; only locations that have a very high probability of containing a face will be evaluated.

Next, we determine the necessary storage for the integral image and integral squared image, as well as the data flow parameters. The input image is considered to be an 8-bit per pixel grayscale image. We use a 320×240 input image frame size, with the maximum pixel value set at 255. As such we need to provide storage for the case where all pixels will be set at 255, an unlikely scenario, but necessary for correct operation. As such, the maximum integer value that can be stored in an integral image is $255 \times 320 \times 240$ and the maximum integer value that can be stored in an integral squared image is $(255)^2 \times 320 \times 240$. This requires 25 bits per integral image entry, and 33 bits per squared integral image entry. The rectangle sums need an accumulator, which can accommodate nonsaturating arithmetic, and the computed result needs at least 25 bits of storage. During the variance computation, the accumulator needs 33 bits to facilitate the squared integral image values. We describe next the proposed architecture.

83.4.1.2 Architecture Description

The algorithm operations are computed over an array grid processor as said earlier. The array consists of units which hold the integral and integral squared image values, and minimal hardware to propagate data in all directions in the array. Each unit is also equipped with hardware to perform additions and subtractions, so that computation of the integral and integral squared image can happen in a systolic manner, and so that the rectangle sum can also be computed within the array. Essentially the system consists of three major components: the collection and data transfer units (CDTUs), the multiplication and evaluation units (MEUs) and the control units (CUs). The units are organized in a grid-like manner, with the MEUs located at the left side at each pair of rows of the grid, and the CUs distributed evenly across the rows. The number of CUs and the distribution among rows depends on the size of the entire array and the delay/performance requirements to reduce the size of the control region. The number of CUs and the amount of CDTUs in each control region can vary according to the design budget and performance requirements. Each CDTU can communicate with each of its four neighbors via a data bus of 36 bits. A floorplan of the particular system is shown in Figure 83.13, illustrating the location of each unit and the data movement across the system. Next we discuss in detail the architecture of each unit.

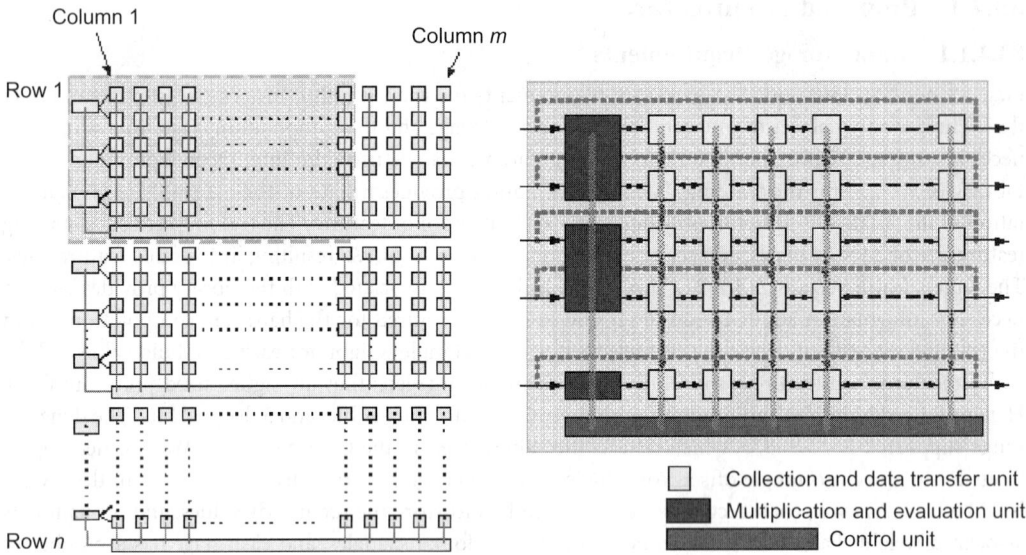

FIGURE 83.13 Architectural block diagram. The shaded rectangle on the left diagram is shown in detail in the right diagram.

83.4.1.3 Collection and Data Transfer Units

Each CDTU represents the starting upper left corner of a search window in the image, and holds certain data for that window (such as image standard deviation and whether or not the window contains a face). The CDTUs are responsible for data movement throughout the system, and collecting and accumulating image data to be used in the computation. Each unit is composed of an adder/subtractor, a local bus controller, and a register file. The register file is small and holds the image data necessary for the computation as well as data on travel to the collection points. Each CDTU acts as a collection point, collecting and accumulating data for each feature rectangle. The register file provides data storage for the integral image value, the squared integral image value, the collected rectangle sums (supports up to four rectangles per feature), the accumulated stage sum, the standard deviation of the image for the search window represented by the CDTU, and temporary registers used to store data in movement and during computation. Additionally, the CDTU holds a flag bit (FB), which is reset only when the search window, represented by the CDTU, does not contain a face. The bit is set at the beginning of every computation (either a new image or a new feature size) and is reset by the MEUs at the end of a stage computation. For data movement purposes, FB is moved with the accumulated stage sum. A detailed block diagram of the CDTU is shown in Figure 83.14.

The CDTUs are controlled by the controller units, which determine the action of CDTU. The actions performed at each CDTU are shifts to all four directions, addition and accumulation of incoming pixel values and squared pixel values, additions and accumulation of incoming rectangle points, and being idle when they are waiting on the MEUs. Each CDTU action is determined by a state machine in each of the controller units, and a global opcode of 4 bits is sent to all CDTUs by their respective controller unit for the CDTU to perform the appropriate action.

83.4.1.4 Multiplication and Evaluation Unit

The MEUs are located to the far left of the array, one for each two rows, and are equipped with a multiplier. The multiplier is the slowest component in the design; it can therefore be pipelined accordingly to increase the overall clock frequency. The MEUs receive data from the CDTUs, starting from the rectangle values, the standard deviation of the image and lastly the accumulated stage sum. The rectangle sums are multiplied with the rectangle weights, and the standard deviation is multiplied with the feature threshold to determine the feature sum to be added to the accumulated stage sum.

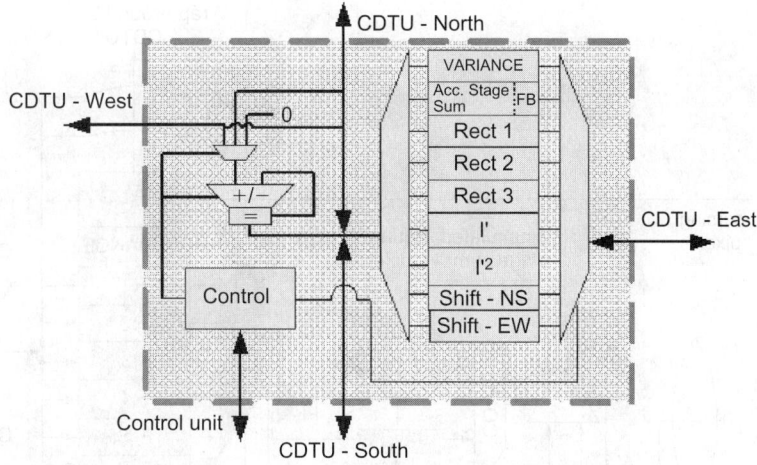

FIGURE 83.14 CDTU block diagram.

If the computed feature is the last of a stage, the accumulated stage sum is compared with the stage threshold and the FB is reset if the stage fails. Else, both the accumulated stage sum and flag bit are shifted out to the toroidal link into the far right CDTU. The MEU starts the computation when signaled from the CU; when it ends the computation it signals to the CU to proceed with the next feature. The CU in the meantime stalls shifting in the CDTUs while waiting on the MEU to complete. When a stage is evaluated, the MEU sends a signal to the CUs to reduce the face counter if a stage fails, or keep it if it passes. The face counter is located in each of the CUs, and it will be described in detail in the next section.

Each of the MEUs interfaces via an 8-bit bus to an external memory, which holds the training data necessary for the feature computation and in a FIFO manner reads feature data from the external memory to be used for computation. The MEU block diagram is shown in Figure 83.15. Each MEU is connected via a multiplexed bus to two CDTU rows. As such, the image is essentially searched on alternate rows rather than every successive row. The CUs can select which row they can propagate to the MEUs, and the row that is not evaluated is simply propagated around the toroidal link to maintain order of data flow. The search windows alternate every row, following the pattern used in the algorithm [7].

83.4.1.5 Control Units

The system is synchronized via 12 identical CUs, each of which generate control signals to drive the CDTUs and the MEUs. Although identical, the CUs are spread in the entire system to reduce the size of each CU's control region. Each unit consists of a finite state machine controlled by the training data, feedback signals from the MEU, and two counters; a global counter and a face counter. The global counter controls the flow starting from the computation of the integral image and rectangle collection.

The face counter is also used to control the operation and is used for early termination in case no search window in the image passes beyond a stage. The counter is updated on every MEU evaluation at the end of each stage computation. Since MEUs operate in parallel, and at most 240 MEUs can produce the outcome of a stage in a single cycle, a thermometer decoder with a priority encoder at each CU (receiving the 240 FB from the MEUs) determines the number of search windows that pass a certain stage. The thermometer decoder value is then used to update the face counter value. When an entire stage is computed the face counter is checked whether it has reached zero or not, and if it has, the CU simply outputs a no-face signal. Otherwise, it resumes with the computation of the next stage.

Essentially, the CUs are responsible for ensuring that during different computation stages, the CDTUs and the MEUs are in the right computation stage and synchronized to each other. The computation

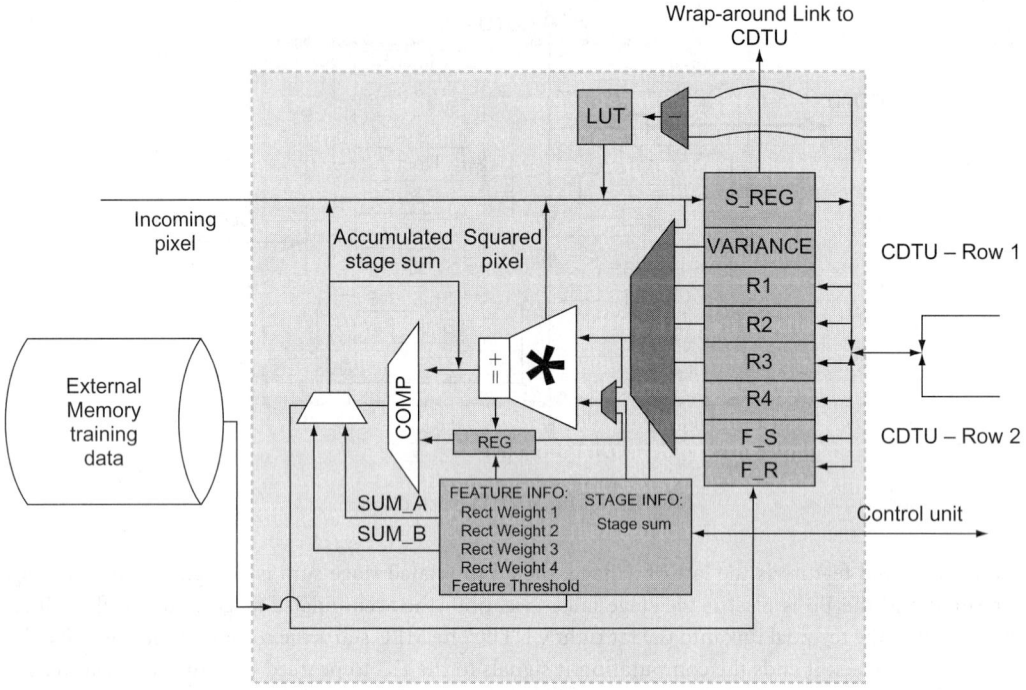

FIGURE 83.15   MEU block diagram.

states for each operation are listed in Table 83.1. For each computation state, each of the units performs certain actions. The action is determined in the CUs, and the global 4 bit opcode is transmitted every cycle to all CDTUs and MEUs. Each of the units decodes the opcode and performs the operation associated with each action. The computation is overviewed in detail in the following section.

83.4.2 Computation Overview

The operation essentially is partitioned into the following stages: computation of integral and integral squared images, computation of image standard deviation, computation of rectangles per feature, feature computation, stage evaluation, and configuration (see Table 83.1). When a feature size increases (i.e., when a search window size increases) the computation is essentially treated as a new one. When the image has been searched at all search window sizes, the system is ready for the next image frame.

In each case, all three units collaborate to perform the computation. Incoming pixels stream in the processor in parallel along all rows of the processor and are shifted in row-wise every cycle. First, the integral image is computed. The computation consists of horizontal and vertical shifts and additions. Incoming pixels are shifted inside the array on each row. Depending on the current pixel column, each of the computation units performs one of three operations; it either adds the incoming pixel value into the stored sum, or propagates the incoming value to the next-in-raw processing element while, either shifting and adding in the vertical dimension (downwards) the accumulated sum or simply doing nothing in the vertical dimension. The computation and data movement is outlined in Figure 83.16. To compute the squared integral image, the same procedure is followed. The incoming pixel passes through the multiplier in the MEU, which computes the square of the pixel value, and then that value alternates with the original pixel value as inputs to the array. It must be noted that since the multiplier present in the MEU allows for the result of an 8×8 multiplication to be available at a single clock cycle, by using a duplex multiplier [28]. As such, the integral and squared integral image are computed in alternate cycles with the entire computation taking $2 * [(m + (m - 1) + (n - 1)]$ cycles, for an input image of n rows by m columns.

TABLE 83.1 Computation Stages and the States Associated with Each Stage

	Computation Stages					
UNITS	Integral Image/ Integral Squared Image	Variance computation	Rectangle computation	Feature computation	Stage evaluation	Configuration/ Scale Change
CDTU	Shift right/add	Shift left	Shift left	Shift left (No. of rects, variance, flag bit + stage_sum)	Shift right (Stage_Sum & Flag Bit)	IDLE
	Shift right/shift down/add	Shift up	Shift up			
	Shift right/shift down	Shift left/add	Shift left/add	IDLE	IDLE	
	Shift right	Shift left/subtract	Shift left/subtract			
	Shift down	Shift up/add	Shift up/add			
	Shift down/add	Shift up/subtract	Shift up/subtract			
	IDLE	IDLE	IDLE			
MEU	Shift right	Receive Data from CDTU	IDLE	Receive Data from CDTU	Compute stage (compute feature + stage sum)	LOAD NEXT FEATURE INFO
	Shift right/ multiply					
	IDLE	Compute variance (compute difference and access LUT)		Compute feature (no data movement, computation for no. of rects, feature_sum)	Shift flag bit around	IDLE
		Shift acc. stage sum around		Shift acc. stage sum around	Increase/decrease face counter	
CU					Increase/decrease face counter	LOAD dx and dy values for stage 1/2 features
						LOAD no. of features left in current stage
						LOAD no. of rectangles/ feature no.
						LOAD dx/dy for each rectangle

The next step is to compute the image variance for each search location. The variance is computed by taking the sum of the mean square pixels over the search window and subtracting from it the squared sum of the mean pixels. Given that the search window over the integral and integral squared images can be viewed as a rectangle itself, it makes sense to follow the same approach as with the feature rectangles, using as corners the four search window size corners. Also, given that the search window size is known, there is no need to divide the sum of the pixels and over the size of the window; instead, that is incorporated into a constant when the variance is computed. The sum of the pixels is then squared, and subtracted from the value of the squared integral image, to give us the variance. To compute the standard deviation, we need the square root of the variance. The square root however is a tedious operation when it comes

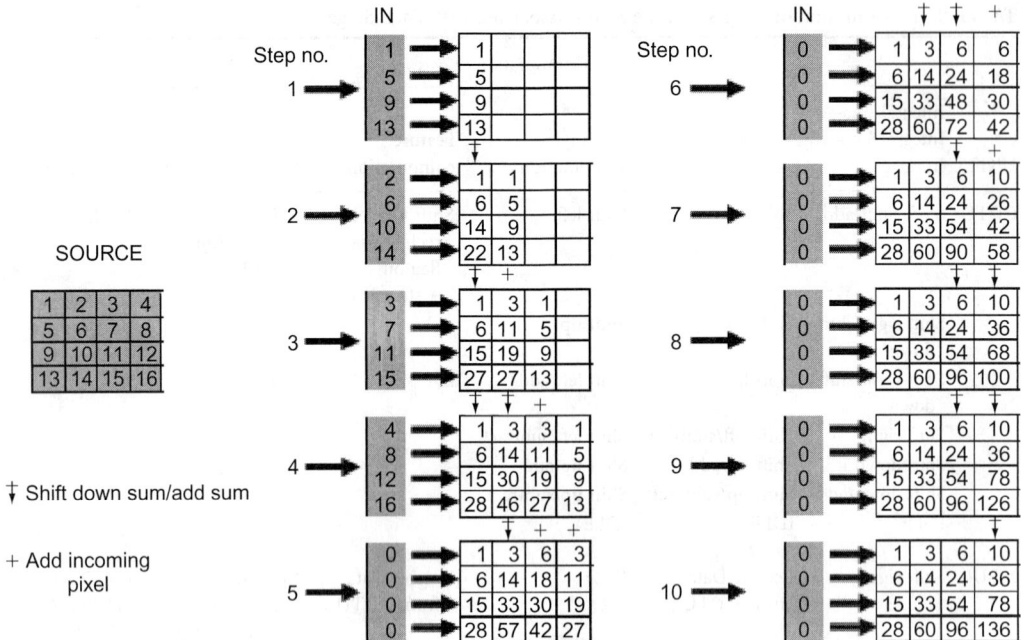

FIGURE 83.16   Computation of the integral image: Data movement.

to hardware, so a better solution must be found. Recall that to compute the compensated threshold (t) we need the original feature threshold (t_o), as shown in the paragraph following Eq. (83.1). The original feature threshold is part of the training set. The compensated threshold used in the feature evaluation is the product of the original threshold and the image standard deviation (σ_I), as shown in Eq. (83.2). If we therefore square both sides of the equation, the computation essentially becomes a multiplication of the variance, with the squared value of the original threshold (which can be precomputed during training and stored in the training set) to yield the squared value of the compensated threshold. In the original computation, the feature sum is compared with the compensated threshold, and depending on the outcome, a predetermined value is added to the accumulated stage sum. In the modified version, we need to compare the squared value of the feature sum instead; hence, instead of having to take a square root, we use the available multiplier in the MEU to compute the squared feature sum and compare that with the compensated threshold obtained from the multiplication of the variance with the squared value of the original threshold. It must be noted that the image variance is computed only once for each search window and it is stored into each CDTU for the entire algorithm at each scale. The variance value changes only when the feature size changes or a new frame arrives:

$$t = t_o * \sigma_I$$

$$\sigma_I = \sqrt{E[X^2] - \{E[X]\}^2} = \sqrt{VAR} \qquad (83.2)$$

$$\Rightarrow t^2 = (t_o)^2 * VAR$$

Next, the rectangle computation happens. Each feature has either two, three, or four rectangles that need to be computed. Recall from the algorithm that the sum of each rectangle is computed via two additions and two subtractions using the rectangle corner data. The computation happens in the following manner. Each CDTU acts as a starting corner for each search window (i.e. the left most top value). For each rectangle in each feature, each corner point is shifted toward the collection point. The points move one at a time, but in parallel for all rectangles/features in the array. At each collection point, the point is either added or subtracted to an accumulated sum, with the rectangle value computed when all points of each rectangle arrive at the collection point. As such, each point requires $dx + dy$ cycles to reach the collection point, where dx and dy are the offset coordinates of the point with respect to the upper left corner of the search window. The CDTUs support computation of up to four rectangles per feature. When finished collecting the rectangle sums for a single feature, the collected sums are stored in the CDTU that represents the starting corner for each feature. Next, all the collected sums are then shifted leftwards toward the MEUs, one sum at a time per MEU. From left to right, eventually all sums arrive in each MEU, where the rectangle weights are multiplied with the incoming sums, to evaluate the feature. It must be noted that each feature contains up to four rectangles, and to compute the stage sum we also need the accumulated feature sum from the previous feature computation and the variance. Hence each CDTU takes six shifts to transfer each rectangle and the accumulated stage sum and variance value to its neighboring CDTU. When each rectangle value enters the MEU, the rectangle weight is multiplied with each rectangle sum and accumulated together to compute the feature sum. The compensated threshold is then computed using the original threshold and the variance as described earlier. The feature sum is then squared using the multiplier, and compared with the compensated threshold to set the feature result. The partial stage sum is accumulated with the feature result and shifted with the FB in a toroidal fashion to the CDTU on the far right of the grid, to continue the computation. Eventually, when all feature results are computed, they are stored back into the CDTUs in the grid and the computation resumes with the next feature.

At the end of a stage, the stage sum is compared against the stage threshold, and if the threshold is not met, the location is flagged as a nonface by resetting the FB that is shifted with the stage sum. If the FB is reset, a counter at each controller unit is decremented to keep information about the number of face candidates in the image. If the counter is set to zero after the completion of a stage, the controller unit simply changes scale and proceeds to compute the next scale as no search window has passed the stage. Note that since data movement needs to happen however as computation happens in parallel for all locations, a location, although not containing a face, still has to move along with the rest to maintain the order of computation. However, when a location without a face arrives for computation at the MEUs, the MEU does not compute the feature sum, rather sends a request to the controller unit to simply shift the next location to the MEU, and the location around to the far right CDTU to resume computation. Additionally, there are certain CDTUs that do not take part in the collection of data points. For example, to scan the bottom of the integral image of a 240×320 frame, the search window starts at location $(0, 216)$; locations below row 216 will not be starting corners for a 24×24 search window, as the window will exceed the image size. Similarly, CDTUs for locations 296 and further right, there is also no starting search window corner. As such, the respective CDTUs are also treated as dummies and the MEU does not perform any computation on them; it instead resets their FB so that they are treated as nonface locations. Note that the face counter is initialized with the dummy CDTUs in mind and the dummy CDTUs are identical to the rest of the system.

When all stages complete for a single scale, the flagged locations contain a face. If the scale computed is the last one, the computation ends, and each search window with a face still has its FB set inside the representing CDTU. Each location that contains a face is shifted to the right and outside of the grid array, to the output of the processor for the host application to proceed.

83.4.3 Performance Evaluation

One of the advantages of the AdaBoost algorithm is the ability to distinguish between regions that do not contain a face and the ones that might contain a face, and stop and proceed to the next frame in the case that there is no strong evidence of a face present in the current frame. This results in two interesting observations when we transform the algorithm in hardware. The first observation has to do with the number of faces present in an image. In images where a single face is present, one of the search windows (at least) will have to be evaluated along all stages. Given that the image is searched in parallel however, the time taken to evaluate a single search window is the same time needed to evaluate all search windows, thus, unlike software, there is no increase in latency. However, when there are multiple faces present in the image, unlike software where the latency increases rapidly [7], the parallel implementation processes all faces in parallel, thus the delay remains constant. Another interesting case is the number of images with different sizes; when two or more faces of different size are present in the source image, detection will occur at each scale. Therefore, the worst-case scenario for detection faces would be at least one face in every detection scale. Practically however, this is extremely hard to happen within an image frame; a large face will cover most secondary smaller faces in an image, and will result in a large face with a number of smaller faces, or, similar sized faces spread throughout the source image [7,12]. Hence, it is reasonable to assume that the worst-case scenario will almost never happen. The second observation has to do with the cases where faces are not present in an image frame; the search windows will likely fail somewhere through the first few stages for all search windows at all scales, thus enabling a new image frame to be processed. In such cases, the frame rate increases. As such, sample frames containing a number of faces of different sizes were chosen, and by taking these observations into consideration the architecture was evaluated.

We use an 8-bit per pixel 320×240 grayscale image as our input frame. As such we need to provide storage for the case where all pixels will be set at 255, an unlikely scenario, but necessary for correct operation. Recall that the maximum integer value that can be stored in an integral image is $255 \times 320 \times 240$ and the maximum integer value that can be stored in an integral squared image is $(255)2 \times 320 \times 240$. This requires 25 and 33 bits, respectively. We design the architecture using 320×240 CDTUs, 120 MEUs, and 4 CUs. Each CDTU connects to its neighbors via a 33 bit bus.

To evaluate the performance of the proposed implementation, we designed and verified the architecture using Verilog HDL and the ModelSim® simulator. We then synthesized the architecture using a commercial 90 nm library and targeting a 500 MHz clock cycle. Our synthesized design indicates that the experimental architecture consumes an area of $\sim$115 mm^2. The area can be reduced depending on the desired image size however. We used several 320×240 images [27] containing a number of faces. Given the large size of the array, simulation of an entire 320×240 frame on ModelSim would be an extremely time-consuming operation and would require extensive amount of resources. As such, a prototype 24×24 array of CDTUs was designed, along with the corresponding 12 MEUs and a CU. The 24×24 size was chosen because it is the base feature size as proposed in [7]. Each 320×240 image was partitioned into 24×24 subimages. Each subimage was fed as input to the array, and using the ModelSim simulator, the computation proceeded for a 24×24 portion of the image. The total number of clock cycles until each 24×24 frame was completely processed was measured. However to detect faces of larger size (thus computing the time features of larger size compute), the corresponding software implementation was used. Each frame was run through the software implementation, and for every search window in the source image, the computation progressed was marked (indicating how far along a search window computation progressed in terms of the features and stages computed). The resulting endpoint for each search window was then used to compute the number of cycles for that search window if the operation would happen on the hardware architecture, and was used along with the computation time obtained through simulations to project the detection frame rate. The computed number of cycles was projected for an entire 320×240 frame, and the average number of cycles per test frame was then estimated, giving a rough estimate of 52 fps. Obviously, the frame rate depends on the number of faces of different sizes rather than the number of faces in the picture, a major advantage over the software implementation where both parameters affect the latency. Some experimental frames are shown in Figure 83.17. As the frames show, the detection accuracy is affected by the face orientation, as profile faces are much harder to detect and better training is desired. It must be noted, however, that the hardware implementation does not impact the accuracy of the computation compared with software implementation, as the same experimental frames had equal detection accuracy when modeled in software. A particular issue however concerns the amount of search windows marked as faces; the hardware platform returns whether a search window contains a face or not. Faces, which are detected by several windows (due to the face being

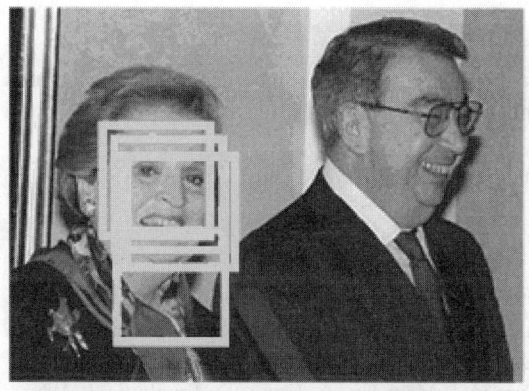

FIGURE 83.17 Experimental frames.

sufficiently small and the search windows overlapping), need to be identified as such (and not as two or more faces). This is however left upon the host application. Figure 83.17 shows such scenarios as well.

83.5 Conclusion

In this chapter we examined various architectures for hardware face detection. We have seen how each implementation varies, in terms of the frame rate, hardware resources, design methodology and most importantly, effectiveness. The presented architectures and designs show that hardware face detection can achieve very high detection frame rates. The gain in speed results also in a very interesting observation; face detection usually is part of a higher-level problem, such as face recognition, demographics, biomedical imaging applications, etc. By speeding up the face detection process, we can design complete systems either completely in hardware or as an embedded platform. Results from this chapter show that hardware face detection can achieve real-time frame rate detection, ranging from 24 to 52 fps depending on the chosen architecture.

The porting of algorithms such as face detection in hardware is a significant step toward the design of artificial intelligence. The algorithms used to detect faces can also be used to detect other objects as well. Pattern recognition in general has been a fundamental in artificial intelligence, and as technology progresses, more artificial intelligence algorithms can be ported in hardware, with the inherent benefits applied to several applications. Medical, control, security, space and aeronautics, and several other high end applications can benefit from hardware implementation of pattern recognition algorithms. Face detection, while a small part of pattern recognition, is one of the first and most fundamental algorithms, and as such provides opportunities for expanding the hardware architectures to implement such algorithms.

Acknowledgments

The authors would like to acknowledge the significant contributions of T. Richardson, J. Kim, J. Conner, S. Srinivasan, A. Mupid, B. Vaidyanathan, and P. Sundararajan in the ASIC implementation of the Neural Network based face detection.

This work was supported in part by grants from NSF CAREER 0093085 and GSRC—PAS.

References

1. S. Ben-Yacoub, B. Fasel, "Fast Multi-Scale Face Detection," *IDIAP*, Eurecom, Sofia-Antipolis, France, 1998.
2. R. McCready, "Real-Time Face Detection on a Configurable Hardware System," *International Symposium on Field Programmable Gate Arrays*, Monterey, CA, USA, 2000.
3. R. Herpers, G. Verghese, K. Derpanis, R. McCready, J. MacLean, A. Levin, D. Topalovic, L. Wood, A. Jepson, J.K. Tsotsos, "Detection and Tracking of Faces in Real Environments," Technical Report, University of Applied Sciences, Sankt Augustin, Germany.
4. Fraunhofer Institute of Integrated Circuits, http://www.iis.fraunhofer.de/, August 2005.
5. Erik Hjelmås, Boon Kee Low, "Face Detection: A Survey," *Computer Vision and Image Understanding*, Vol. 83, No. 3, pp. 236–274, September 2001.
6. E. Trucco, A. Veri, *Introductory Techniques for 3-D Computer Vision*, Prentice-Hall, New York, 1998.
7. P. Viola, M.J. Jones, "Robust Real-Time Face Detection," *International Journal Computer Vision*, Volume 57, No. 2, pp. 137–154, December 2003–January 2004.
8. H.A. Rowley, S. Baluja, T. Kanade, "Neural Network-Based Face Detection," *IEEE Trans. On Pattern Analysis and Machine Intelligence*, Vol. 20, No. 1, pp. 39–51, 1998.
9. H. Rowley, S. Baluja, and T. Kanade. Rotation invariant neural network-based face detector. *Proceedings of IEEE Conference on Computer Vision and Pattern Recognition*, pp. 38–44, Santa Barbara, CA, June 23–25, 1998.
10. Yang Ming-Hsuan, D.J. Kriegman, N. Ahuja, "Detecting faces in images: a survey," *IEEE Transactions on Pattern Analysis and Machine Intelligence*, Vol. 24, No. 1, pp. 34–58, January 2002.
11. K. Sung, T. Poggio, "Example-based learning for view-based human face detection," *IEEE Transactions on Pattern Analysis and Machine Intelligence*, Vol. 20, No. 1, pp. 39–51, January 1998.
12. K. Sung, "Learning and Example Selection for Object and Pattern Recognition," Ph.D. thesis, MIT, MIT Press, 1995.
13. B. Nagendra, "Pixel Statistics in Neural Networks for Domain Independent Object Detection," MSc thesis, RMIT University, Melbourne, Australia, 2001.
14. R. Feraund, O.J. Bernier, J. Viallet, M. Collobert, "A fast and accurate face detector based on neural networks," *IEEE Transactions on Pattern Analysis and Machine Intelligence*, Vol. 23, No. 1, pp. 42–53, January 2001.
15. Zhang Zhen Qiu, Zhu Long, S.Z. Li, Zhang HongJiang, "Real-time multi-view face detection" *Proceedings of the Fifth IEEE International Conference on Automatic Face and Gesture Recognition*, pp. 142–147, 20–21 May 2002.
16. B. Srijanto, "Implementing a Neural Network based face detection onto a reconfigurable computing system using Champion," M.S. thesis, University of Tennessee, Knoxville, August 2002.
17. Henry Schneiderman, Takeo Kanade, Jay Pujara, "*CMU Face Detection Algorithm Demonstration*", http://www.vasc.ri.cmu.edu/cgi-bin/demos/findface.cgi, 2004.
18. H. Schneiderman, T. Kanade, "Object Detection Using the Statistics of Parts," *International Journal of Computer Vision*, Vol. 56, No. 3, pp. 151–177, February–March 2004.
19. Martijn Reuvers, "Face Detection on the INCA+ System," Masters thesis, The University of Amsterdam, 2004.
20. V. Kianzad, S. Saha, J. Schlessman, G. Aggrarwal, S.S. Bhattacharyya, W. Wolf, "An Architectural Level Design Methodology for Embedded Face Detection," *Proceedings of the International Conference on Hardware/Software Codesign and System Synthesis*, New York, September 2005.

21. H. Broers, R. Kleinhorst, M. Reuvers, B. Kröse, "Face Detection and Recognition on a Smart Camera," *Proceedings of the Advanced Concepts for Intelligent Vision Systems Conference,* Belgium, August 2004.

22. Panasonic Corporation, http://www.panasonic.com.

23. J.L. Ayala, A.G. Lomena, M. Lopez-Vallejo, A. Fernandez, "Design of a Pipeline Hardware Architecture for Real-Time neural network computations," Technical report, Departmento de Ingeneria Electronica, Universidad Politechinca de Madrid, Spain, 2002.

24. R. Frischholz, "The Face Detection Homepage," http://home.t-online.de/home/Robert.Frischholz/face.htm, July 2005.

25. T. Theocharides, G. Link, N. Vijaykrishnan, M.J. Irwin, W. Wolf, "Embedded Hardware Face Detection," *Proceedings of the 17th International Conference on VLSI Design,* Mumbai, India, January 2004.

26. Intel Open Source Computer Vision Library, http://www.intel.com/technology/computing/opencv/index.htm, September 2005.

27. FERET Face Image Database, http://www.itl.nist.gov/iad/humanid/feret/feret_master.html, June 2003.

28. Synopsys Designware® IP Library, http://www.synopsys.com/products/designware/dwlibrary.html.

Multidimensional Logarithmic Number System

Roberto Muscedere
University of Windsor

Vassil S. Dimitrov
University of Calgary

Graham A. Jullien
University of Calgary

CONTENTS

84.1 Introduction

The logarithmic number system (LNS) [1–3] is an alternative to the binary representation and it has been a subject of some investigation [4,5], particularly in the field of digital signal processing (DSP) [6,7], where the computation of inner (dot) products is a major computational step. In the LNS, multiplication and division are easy operations, whereas addition and subtraction are difficult operations, traditionally implemented by making use of large ROM arrays [8,9] or other techniques [10,11]. It has been recognized that LNS architectures are perfectly suited for low-power, low-precision DSP problems.

Inner products computed in DSP algorithms are often between a predetermined set of coefficients (e.g., Finite Impulse Response (FIR) filters or discrete transform basis functions) and integer data. For fixed-point binary implementations, the uniform quantization properties are perfectly matched to the mapping of most input data (the mapping of input data for nonlinear hearing instrument processing is a counterexample), but often the predetermined coefficients are better suited to a nonuniform quantization mapping. A study of a large number of filter designs reveals a histogram that benefits from the quantization associated with a logarithmic mapping [12].

A logarithmic-like representation, referred to as the index calculus double-base number system (IDBNS), that promises implementation improvements over the LNS while maintaining a logarithmic quantization distribution was previously introduced [13] and there have been several papers published on special results from this number representation [14–20].

In this chapter we will generalize the LNS number system and present several results that demonstrate the efficiencies in using this representation over the classical LNS for typical DSP computations. We will also detail the hardware used to perform mapping from binary to the multidimensional logarithmic number system (MDLNS) along with implementation examples that demonstrate the efficiencies of using the MDLNS.

84.1.1 Multidimensional Representations

The IDBNS is based on a single-digit representation of the form

$$y = s2^a3^b \tag{84.1}$$

where $s \in \{-1, 0, 1\}$ and a and b are signed integers. In this case we have the following theorem [14]:

Theorem 84.1

For every $\varepsilon > 0$ and every nonnegative real number x, there exists a pair of integers a and b, such that the following inequality holds:

$$\left| x - 2^a3^b \right| < \varepsilon$$

We may therefore approximate, to arbitrary precision, every real number with the triple $\{s, a, b\}$. We may look at this representation as a *two-dimensional generalization* of the binary logarithmic number representation. The important advantage of this generalization in multiplication is that the binary and ternary indices are operated independently from each other, with an attendant reduction in complexity of the implementation hardware. As an example, a very large-scale integration (VLSI) architecture for inner product computation with the IDBNS, proposed in Refs. [14,15,20], has an area complexity dependent entirely on the dynamic range of the ternary exponents. Provided the range of the ternary exponent is smaller than the LNS dynamic range for equivalent precision, we have the potential for a large reduction in the IDBNS hardware compared to that required by the LNS. We can capitalize on this potential improvement by placing design constraints on the ternary exponent size. For example, if we want to represent digital filter coefficients in the IDBNS, then we can design the coefficients in such a way that the ternary exponent is minimized, an integer programming task [16,20]. Although this approach is sound, and can produce modest improvements, we can do better. In fact, the complexity of the architecture does not depend on the particular choice of the second base. Therefore, one may attempt to find a base, x, (or set of bases) such that the filter coefficients can be very well approximated of the form $s2^a x^b$, while keeping b (the nonbinary exponent) as small as possible.

We begin with a review of the necessary mathematical preliminaries, followed by a discussion of the nature of input data mapping by explaining both the error and nonerror-free representations using the MDLNS. We continue with information about the approximations to unity which play an important role in constraining the dynamic range of the exponents for calculations. Then, we discuss hardware complexity which includes an explanation of the single-digit computation unit (CU) followed by a generalization to an n-digit CU. We then demonstrate a filter design using MDLNS and provide an example where we can see which different bases (selected and optimal) can impact the physical implementation complexity. Finally, we show some fabricated MDLNS DSP inner-product computational unit designs and summarize the main results.

84.2 Mathematical Preliminaries

There are some well-established results on s-integers that we can build upon. We start with two basic definitions [21].

Definition 84.1
An s-integer is a number whose largest prime factor does not exceed the sth prime number.

For example, nonnegative powers of 2 are 1-integers; numbers of the form $2^a 3^b$ where a,b are nonnegative integers, are 2-integers, and so on.

Definition 84.2
Modified 2-integers are numbers of the form $2^a p^b$ where p is an odd integer.

Note that we do not impose restrictions on the signs of a and b in Definition 84.2.

The next definition offers the most general representation scheme we will consider in this chapter.

Definition 84.3
A representation of the real number x in the form

$$x = \sum_{i=1}^{n} s_i \prod_{j=1}^{b} p_j^{e_j^{(i)}} \tag{84.2}$$

where $s \in \{-1, 0, 1\}$ and $p_j, e_j^{(i)}$ are integers, is called a *multidimensional n-digit logarithmic (MDLNS) representation of x*. b is the number of bases used (at least two, the first one, that is, p_1, will always be assumed to be 2).

The next two definitions are special cases of Definition 84.3; the representation schemes defined by them will be used extensively in the chapter.

Definition 84.4
An approximation of a real number x as a signed modified 2-integer $s 2^a p^b$ is called a two-dimensional logarithmic representation of x.

Definition 84.5
An approximation of a real number x as a sum of signed modified 2-integers $\sum_{i=1}^{n} s_i 2^{a_i} p^{b_i}$ is called an *n-digit two-dimensional logarithmic representation of x*. $n = 2$ will be a special case.

It is important to note that an extension of the classical LNS to a multidigit (or multicomponent) representation does not provide any inherent advantages in terms of complexity reduction. Arnold et al. [4] were the first to consider a similar representation scheme in the case of classical LNS (we shall call it a 2-component LNS). Although it leads to some reduction of the dynamic range of the exponents (correspondingly, a reduction of the ROM sizes), the number of bits required by the larger exponent to store the integer number, x, is approximately $\log(x)$. The storage reduction in the 2-component LNS (as opposed to the 1-component LNS) comes from the observation that in the 1-component LNS one needs approximately $\log_2(x) + \log_2(\log(x))$ bits to encode x.

We will demonstrate that hardware complexity for the MDLNS is exponentially dependent on the size of the nonbinary-base(s) exponents; we clearly have a potential for quite a considerable hardware reduction provided that the dynamic range of the non-binary exponents is reduced as much as possible.

84.2.1 Mathematical Operations

To summarize, a two-dimensional logarithmic number system (2DLNS) representation provides a triple, $\{s_i, a_i, b_i\}$, for each digit, where s_i is the sign bit and a_i and b_i the exponents of the binary and nonbinary bases, and a number, x, is approximated by Definition 84.5.

84.2.1.1 Multiplication and Division

MDLNS multiplication and division are the simplest of the arithmetic operations. The equations for multiplication and division, given a single-digit 2DLNS representation of $x = \{s_x, a_x, b_x\}$ and $y = \{s_y, a_y, b_y\}$, are [14]

$$x \cdot y = \{s_x s_y, a_x + a_y, b_x + b_y\} \tag{84.3}$$

$$x \div y = \{s_x s_y, a_x - a_y, b_x - b_y\} \tag{84.4}$$

Eq. (84.3) and Eq. (84.4) show that single-digit 2DLNS multiplication can be implemented in hardware using two independent binary adders and simple logic for the sign correction. As we start to add digits to the representation we will face the equivalent of implementing multiplication with the addition of partial products. A 2-digit representation will produce four independent partial products that will have to be added, and since addition is an expensive operation we try to optimize this process as much as possible (we will show an optimized structure later).

84.2.1.2 Addition and Subtraction

Unfortunately, as with the classical LNS, addition and subtraction operations are not as simple as multiplication and division operations. Traditionally, addition and subtraction must be handled through a set of identities and lookup tables (LUTs). The identities are [14]

$$(2^{a_1} \cdot p^{b_1}) + (2^{a_2} \cdot p^{b_2}) = (2^{a_1} \cdot p^{b_1}) \cdot (1 + 2^{a_2 - a_1} \cdot p^{b_2 - b_1})$$

$$\approx (2^{a_1} \cdot p^{b_1}) \cdot \Phi(a_2 - a_1, b_2 - b_1)$$

$$(2^{a_1} \cdot p^{b_1}) - (2^{a_2} \cdot p^{b_2}) = (2^{a_1} \cdot p^{b_1}) \cdot (1 - 2^{a_2 - a_1} \cdot p^{b_2 - b_1})$$

$$\approx (2^{a_1} \cdot p^{b_1}) \cdot \Psi(a_2 - a_1, b_2 - b_1)$$

The operators Φ and Ψ are LUTs that store the precomputed 2DLNS values of

$$\Phi(x, y) = 1 + (2^x \cdot p^y) \tag{84.5}$$

$$\Psi(x, y) = 1 - (2^x \cdot p^y) \tag{84.6}$$

The use of large LUTs, implemented through the use of ROMs, for the evaluation of addition and subtraction operations is the most straightforward approach in systems such as the LNS [4] (other techniques for evaluating the functions of Eq. (84.5) and Eq. (84.6) can be found in Refs. [7,9]). The large ROM table technique is only feasible for very small ranges of 2DLNS numbers. It is more practical, in most cases, to convert the 2DLNS numbers into binary and perform the addition and subtraction using a binary representation.

  The conversions from 2DLNS to binary form will still require an LUT, but one that is much smaller than required for handling 2DLNS addition and subtraction. The LUT is used to convert the second base portion of the 2DLNS number into a binary representation. Therefore, the size of the LUT is dependent on the number of bits used to represent the second base exponent.

It should be noted that a new architecture is proposed in Ref. [22] which significantly reduces the size of the tables requires for 2DLNS addition and subtraction.

84.2.1.3 Multidigit MDLNS Arithmetic

Multidigit MDLNS arithmetic is simply an extension of the single-digit MDLNS arithmetic, and is necessary when numbers are represented by more than one MDLNS digit. When performing a computation using multidigit MDLNS each digit can be treated as an independent MDLNS number and the operations handled separately. For example, if X and Y are 2-digit MDLNS numbers such that $X = x_1 + x_2$ and $Y = y_1 + y_2$ then

$$X \cdot Y = (x_1 + x_2)(y_1 + y_2) = (x_1 \cdot y_1) + (x_1 \cdot y_2) + (x_2 \cdot y_1) + (x_2 \cdot y_2) \tag{84.27}$$

where x_i and y_i are single-digit MDLNS numbers. The independence of the arithmetic operations is very important as it naturally allows for parallel architectures.

84.2.2 Error-Free Integer Representations

As stated in the introduction, more often in DSP applications the input data has to be converted from analog to a fixed-point binary value with a uniform quantization error bound. Mapping to integers has a quantization error bounded by ±0.5 for all converted values. For a classical LNS representation (and also a one-digit MDLNS representation) we do not have this uniform quantization accuracy so we have to choose a sufficient number of bits so that we will be able to maintain this conversion accuracy for the larger data values. In the multidigit MDLNS we can mitigate this quantization problem; in fact, we can find certain MDLNS representations that are completely error-free.

Consider the case of the two odd prime bases, $(3, 5)$.

A representation of a real number into forms given in Definitions 84.3–84.5 is called *error-free* if the approximation error is zero. The next three theorems and one conjecture provide new and interesting results about the error-free two-dimensional logarithmic representation of numbers.

Theorem 84.2

Every real number x may have at most 14 different error-free two-digit two-dimensional logarithmic representations.

Proof

Let us assume that x is represented in the form of Definition 84.5:

$$x = \pm 2^a p^b \pm 2^c p^d \tag{84.7}$$

Clearly, x must be a rational number. Now we multiply the two sides of Eq. (84.7) by $z = 2^{-\min(a,c,0)} \, p^{-\min(b,d,0)}$. The left- and right-hand sides of the new equation will be integers. Divide by the greatest common divisor of $(2^{a-\min(a,c,0)} \, p^{b-\min(b,d,0)}, 2^{c-\min(a,c,0)} \, p^{d-\min(b,d,0)})$. Let us denote the left-hand side of the equation obtained as M. We may obtain only two types of equations:

$$M = \pm 1 \pm 2^e p^f \tag{84.8}$$

or

$$M = \pm 2^e p^f \tag{84.9}$$

Eq. (84.8) may have at the most one solution, owing to the fundamental theorem of arithmetic. Eq. (84.9) can be treated as follows. If the signs in Eq. (84.9) are different, we are in a position to apply the following result (recently proved by Bennett [23]): If a, b, and c are integers, a, $b \geq 2$, then the equation $a^x - b^y = c$ may have at most two different solutions in integers (x, y). Therefore, Eq. (84.9) may have at most four different solutions if the signs are different. If the signs are identical (say, positive, which corresponds to positive M), then we can do the following. Represent the exponents e and f with respect to modulo 3: $e = 3e_1 + e_2$ and $f = 3f_1 + f_2$, $e_2, f_2 \in \{0, 1, 2\}$. For the nine possible combinations of residues (e_2, f_2) we have nine Diophantine equations of the form

$$M = A2^{3e_1} + Bp^{3f_1} \tag{84.10}$$

where $A \in \{1, 2, 4\}$ and $B \in \{1, p, p^2\}$. We substitute $X = 2^{e_1}$ and $Y = p^{f_1}$. The final equation we have is

$$M = c_1 X^3 + c_2 Y^3 \tag{84.11}$$

where c_1 and c_2 are constants. Delone–Fadeev's theorem [24] about the cubic Diophantine equations states that Eq. (84.11) may have at most one solution in integers. Since we have nine Diophantine equations of this type, we may have at most nine different solutions of Eq. (84.10).

Therefore, the total number of different error-free representations is bounded from above by 4 (the maximal number of solutions of Eq. (84.9)) +9 (the maximal number of solutions of Eq. (84.10) plus one (the maximal number of solutions of Eq. (84.8)), that is, 14.

In some important special cases, this bound can be considerably improved.

Theorem 84.3

Every real number x may have at most seven different error-free two-digit two-dimensional logarithmic representations if the nonbinary base is 3.

Proof

Assume that x is represented in the form

$$x = \pm 2^a 3^b \pm 2^c 3^d \tag{84.12}$$

Proceeding in the same way as in the proof of Theorem 84.4, we obtain the following four equations:

$$M = \pm 1 \pm 2^e 3^f \tag{84.13}$$

$$M = 2^e - 3^f \tag{84.14}$$

$$M = -2^e + 3^f \tag{84.15}$$

$$M = 2^e + 3^f \tag{84.16}$$

Again, Eq. (84.13) may have at most one solution in integers and Eq. (84.14) and Eq. (84.15) may have at most four (totally) different solutions in integers, according to Bennett's theorem [23]. Eq. (84.16) can be treated by making use of the following theorem, due to Stroeker and Tijdeman [25]: All solutions of $2^x - 2^y = 3^z - 3^w$, $x > y > 0$, $z > w > 0$ in integers x, y, z, w are given by $(3, 1, 2, 1)$, $(5, 3, 3, 1)$, and $(8, 4, 5, 1)$. Therefore, only three numbers $(11, 35,$ and $259)$ may have two different representations as a sum of a power of two and a power of three. That is, the total number of error-free representation in this case (nonbinary base 3) is bounded from above by 7.

The upper bound proved in Theorem 84.2 can certainly be improved. We have not found any real number with more than five error-free two-digit LNS representations; here we report one case having exactly five error-free representations.

Let $x = 3.25$; then x can be represented with no error in a two-digit 2DLNS with odd base 3 as follows:

$$3.25 = (1,-2,2,1,0,0)$$

$$3.25 = (1,0,1,1,-2,0)$$

$$3.25 = (1,2,0,-1,-2,1)$$

$$3.25 = (1,1,1,-1,-2,2)$$

$$3.25 = (1,6,0,-1,-2,5)$$

The point of the theorem is to establish an effectively computable upper bound that could be a starting point for improvements. The example given shows that the lower bound for the maximal number of error-free representation is five.

Theorem 84.4

The smallest positive integer with no error-free two-digit 2DLNS representation in the case of odd base three is 103.

Proof

The proof is based on the following result proved by Ellison [26]: let $x > 11$, $x \neq 13,14,16,19,27$; then for all $x, y \in N$ the following inequality holds: $|2^x - 3^y| > e^{x(\ln 2 - 0.1)}$.

Up to 102 we can give proper examples found by computational experiments. A simple check shows that 103 is not a sum of integers of the form $2^a 3^b$. 103 is not divisible by 2 and 3, therefore, it must be a difference of a power of 2 and a power of 3 (or vice versa). Applying Ellison's theorem we have

$$103 = |2^x - 3^y| > e^{x(\ln 2 - 0.1)} \tag{84.17}$$

Therefore, x should be smaller than 8, and there are only 13 possible values for x, namely, $x = 1,2,3,4,5,6,7,11,13,14,16,19,27$. We now calculate, and in none of the cases do we find that the corresponding y is integer, therefore, 103 cannot be represented in an error-free manner in two-digit 2DLNS with bases 2 and 3.

Theorem 84.5

The smallest positive integer with no error-free two-digit two-dimensional logarithmic representation, in the case of nonbinary base 5, is 43.

Proof

The proof is based on a theorem, proved by Tijdeman [27], which has a much more general result than Ellison's result used in Theorem 84.4. Tijdeman states that if x and y are two consecutive s-integers, $s > 1$, then $|x - y| > y / (\log y)^c$, where c is an effectively computable constant. In the case of 2-integers c is estimated to be less than 64, based on recent results in transcendental number theory [28]. We apply this theorem to numbers of the form $2^a 5^b$.

Again, up to 42 inclusively, we can find appropriate error-free representations; 43 is not a sum of two numbers of this form, therefore it could only be a difference, so we may apply Tijdeman's theorem and obtain an upper bound for the possible solutions. Namely, if $43 = |x - y|$, x and y being of the form $2^a 5^b$, then the theorem implies that $43(\log y)^{64} > y$, therefore, $y < 2^{575}$. There are 296,371 numbers of the form $2^a 5^b$ less than 2^{575}, that is 296,371 potential values of x to be checked out; in none of these cases is the corresponding value for $x = y + 43$ a number of the form $2^a 5^b$. This shows that 43 is indeed the smallest positive integer without an error-free two-dimensional two-digit LNS representation for the case of nonbinary base 5.

We note that Tijdeman's theorem can be used for the proof of Theorem 84.3. We have, however, provided this new proof based on a result concerning the set of bases two and three to point out the difficulties that may arise if one applies general theorems from transcendental number theory.

The following conjecture is based on extensive numerical calculations.

Conjecture 84.1

The smallest positive integer with no error-free 3-digit two-dimensional logarithmic representation in the case of odd base 3 is 4985. Or, in the language of the exponential Diophantine equations, it can be posed as follows—the equations $\pm 2^a 3^b \pm 2^c 3^d \pm 2^e 3^f = 4985$ do not have solutions in integers.

It is important to note that such results will be available (and different) for every particular set of bases that we choose. In this case (i.e., a 3-digit two-dimensional logarithmic representation with odd base 3) we see that a 12-bit error-free mapping that is a useful dynamic range for many DSP applications is available. It should also be noted that the size of all of the exponents used (a, b, c, d, e, f) only require 3-bit unsigned integers for their representation.

84.2.3 NonError-Free Integer Representations

Clearly, error-free representations are special cases of the MDLNS, but the extra degree of freedom provided by the use of multiple digits can mitigate the nonuniform quantization properties of the classical LNS.

To illustrate this, we present numerical results for mapping 10-bit signed binary input data to the two-digit 2DLNS where we treat the nonbinary base as a parameter. To demonstrate the ability to closely match input data with very small exponents, we have restricted the odd base exponent to *3-bits only*. We are allowing the binary exponent to be unrestricted; however, owing to the 10-bit input range, the system automatically limits itself to 6-bits. We will see in the next section that this has very little bearing on the overall complexity of the inner product implementation (i.e., the hardware complexity is mainly driven by the dynamic range of the nonbinary exponents). As stated above, we require quantization errors to be <0.5 to match the quantization error of a binary representation. Table 84.1 shows the number of nonerror-free representations along with the worst quantization error for nonbinary bases in the set $\{3, 5, 7, 11, 13, 17, 47\}$.

The goal of applying this approximation scheme is to reduce as much as possible the size of the nonbinary exponent(s). For example, with a nonbinary base of 47, $x = 0101001110_2 = 334_{10}$ is represented as $334_{10} = 0101001110_2 \rightarrow 2^9 47^{-1} + 2^{25} 47^{-3} = 334.082429$. In this case we have used only three bits for the nonbinary exponents, that is they are restricted to the set $\{-4, -3, -2, -1, 0, 1, 2, 3\}$. Although a base of 47 only has 2 nonerror-free representations for a 10-bit signed range, it is possible to select a noninteger base that will provide completely error-free representations. We will see an example of this in Section 84.5.2.

To compare these results with an implementation using a classical LNS representation, we need to determine the number of bits of the logarithm to produce an absolute error of <0.5. A previous study [4] has found that we require $n + \log_2(n)$ bits for the logarithm to achieve this accuracy for an n-bit positive number [26]. We have, in fact, checked this for the case of $n = 9$ (used in our two-digit 2DLNS study) and 12-bits are required for the logarithm to satisfy the same accuracy. If we assume that the hardware complexity of the classical LNS representation is driven by the number of bits in the logarithm, then we can see a potential for an enormous reduction in the implementation complexity of the two-digit 2DLNS versus the classical LNS.

84.2.4 Approximations to Unity

A very fundamental difference between the classical LNS and MDLNS is the possibility of finding nontrivial approximations of one in the MDLNS. They can be used to constrain the dynamic range of exponents during general computations.

TABLE 84.1 Number of NonError-Free Representations and Worst Quantization Error for Different Bases

Base	3	5	7	11	13	17	47
# ≥ 0.5	10	58	18	20	4	6	2
Worst error	0.777778	1.232	0.724698	0.840909	0.960401	0.714431	0.5

TABLE 84.2 Near Unity Approximants

a	b	$2^a 3^b$	$a_n - a_{n-1}$	$b_n - b_{n-1}$
122	−77	0.971232	233	−147
38	−24	0.973262	−84	53
−46	29	0.975296	−84	53
103	−65	0.984482	149	−94
19	−12	0.98654	−84	53
−65	41	0.988603	−84	53
84	−53	0.997914	149	−94
0	0	1	−84	53
−84	53	1.00209	−84	53
65	−41	1.011529	149	−94
−19	12	1.013643	−84	53
−103	65	1.015762	−84	53
46	−29	1.025329	149	−94
−38	24	1.027473	−84	53
−122	77	1.02962	−84	53

From Theorem 84.1, we know that unity can be approximated with arbitrary precision as a 2-integer. In fact, *both* bases can be changed and the theorem will still remain valid. Here we expand the discussion of these approximants within the MDLNS, and introduce new results.

As an example with 8-bit exponents, consider the generation of a sequence of successive values of possible one-digit MDLNS (2-integer) values. Table 84.2 shows a small subset of values (around unity) obtained from such a sequence. The fourth and fifth columns show the difference between successive binary and ternary exponents within the sequence. In this small subset of the complete sequence we observe that the differences are limited to only 3 sets of 2-integers with indices $(233, -147)$, $(149, -94)$, and $(-84, 53)$. Each of the 2-integers represents a close approximation to unity, multiplication by which generates the next value in the sequence.

The usefulness of the existence of good approximations of unity, for general computations within dynamic constraints on the exponents, can be seen from the following example:

Example 84.1

Calculate x^2 by using 9-bit fixed-point arithmetic, where $x = (180, -115)$ in 2DLNS with odd base 3. The actual value of x is 0.207231. Clearly, $x^2 = (360, -230)$, which would cause overflow in 9-bit arithmetic. However, if we multiply in advance by a (properly selected) good approximation of unity, then the result obtained will have much smaller binary and ternary exponents; consequently, there will not be any risk of overflow. In our case, if we multiply x by $(-84, 53)$ we obtain $(96, -62)$ and now the squaring can be achieved in 9-bit arithmetic without overflow.

More to the point, if, at any stage of the computational process, one obtains a pair of large exponents, they can be reduced to within the required exponent dynamic range by multiplying the number obtained by a suitably good approximation of unity.

84.3 Hardware Complexity

To provide complexity results for our MDLNS inner product CU, we expand on the inner product processor architecture initially developed for the one-digit 2DLNS [14]. The processor can be used in a systolic array for 1D convolution.

84.3.1 Single-Digit Computational Unit

Figure 84.1 shows the structure of the proposed single-digit CU. Since we do not need to retain the MDLNS representation of the accumulated output, and since the CU is used only in feed-forward

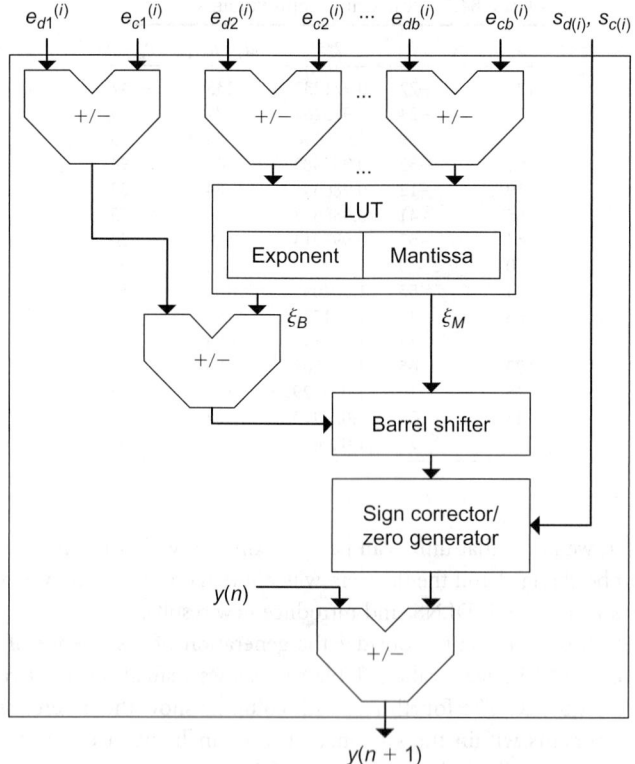

FIGURE 84.1 One-digit MDLNS inner product CU.

architectures, we can use the MDLNS domain for the coefficient multiplication and a binary representation for the accumulated output.

The multiplication is performed by small, parallel adders for each of the data and coefficient base exponents. The addition output for each of the $b - 1$ odd bases is concatenated into an address for a lookup table (ROM). This table produces an equivalent floating-point (FP) value for the product of the odd bases raised to the exponent sum, as shown below:

$$\prod_{j=2}^{b} p_j^{\left(e_{d_j}^{(i)} + e_{c_j}^{(i)}\right)} \approx 2^{\xi_B} \cdot \xi_M \tag{84.18}$$

We note that since the size of the exponents of each odd base in an MDLNS representation (where there are at least two digits and two bases) can be very small (<4 bits), the maximum address input to the ROM is given by $4 \cdot (b - 1)$ bits. This is an 8-bit address table for a 3DLNS. For large-dimensional LNS, we can also consider the use of unity approximants to reduce the output of each odd-base adder to the number of bits of the input exponents (or even less if we are willing to accept the increased mapping error). This reduction process stores a small number of unity approximants that can be added in parallel to the output of the odd-base adders. The reduced input to the ROM is selected from these parallel results. The ROM input address size is now reduced by $(b - 1)$ bits.

84.3.2 *n*-digit Computational Unit

The *n*-digit computational unit is a simple parallel extension of the one-digit unit. Each of the units computes the binary output for one of the digit combinations. As an example, consider multiplying an accumulating sequence, y, with a coefficient x, $z = xy$, where

$$y = \sum_{i=1}^{2} S_i^{[y]} \prod_{j=2}^{b} p_j^{e_j^{[y](i)}}, \quad x = \sum_{i=1}^{2} S_i^{[x]} \prod_{j=2}^{b} p_j^{e_j^{[x](i)}}$$

We can perform this with 4 parallel one-digit units, where the (u, v) unit computes

$$z_{u,v} = S_u^{[y]} S_v^{[x]} \prod_{j=2}^{b} p_j^{\left(e_j^{[y](u)} + e_j^{[x](v)} \right)} \tag{84.19}$$

Clearly there are n^2 such units in an n-digit MDLNS.

The parallel outputs are summed at the end of the systolic arrays using an adder tree.

The biggest advantage of the use of more than one digit for the input data and the filter coefficients is that one can obtain extremely accurate representations with very small nonbinary exponents. But the price that has to be paid is that the number of computational channels required is increased to at least 4.

84.4 Binary to MDLNS Conversion

A logarithmic representation, a, of the number, x, is given by the relationship in the following equation, where s is the sign of the number and the base is r (usually 2):

$$x = sr^a \tag{84.20}$$

The MDLNS provides more degrees of freedom than the LNS by use of the orthogonal bases and the ability to use multiple digits. However, these extra features introduce new complexities in the binary conversion process. The binary-to-LNS conversion process is simplified owing to the monotonic relationship between x and a. Unfortunately, this solution is not applicable to the MDLNS since there is no monotonic relationship between x and the multiple digits/bases.

The technique initially proposed for binary-to-MDLNS conversion used simple LUTs [14]; however, the input data range for the target application area of video processing was only 8 bits. Although an LUT offers a simple and fast binary-to-MDLNS conversion scheme, the size of the LUTs required is exponentially dependent on the input binary dynamic range. The LUT sizes further depend on the number of digits and bases in the MDLNS representation. For example, an error-free (where the absolute representation error is <0.5) 12-bit unsigned range with 3-digits and 2-bases [2] would require a direct mapping LUT of the size 4096 × 33 or 136 kbit, a reasonably sized VLSI component. However, if an error-free 23-bit unsigned range were needed, the mapping LUT would be 8388608 × 48 or 403 Mbit, which is not reasonable at all for VLSI fabrication.

In this section we will describe several hardware implementation techniques for converting a binary representation into a 2DLNS representation. The techniques are based on the reversal of a previously published MDLNS-to-binary converter [14] with the aid of a new memory device [29].

To simplify the presentation of the binary-to-MDLNS process we will restrict ourselves to a subset of the MDLNS with only two bases (an n-digit 2DLNS representation) and we will assume that the exponent of the second (or nonbinary) base has a predefined finite precision (equivalent to limiting the number of bits of precision in a classical LNS). The simplified representation of an input, x, as an n-digit 2DLNS is shown in the following equation:

$$x = \sum_{i=1}^{n} s_i 2^{a_i} D^{b_i} \tag{84.21}$$

The second base, D, is a suitably chosen number (relatively prime to 2), $s_i \in \{-1, 0, +1\}$, and the exponents are integers. R is the number of bits of the second base exponent (i.e., $b_i \in \{-2^{R-1}, \ldots, 2^{R-1} - 1\}$) and it directly affects the complexity of the MDLNS system. The precision of the binary exponent is B bits (i.e., $a_i \in \{-2^{B-1},\ldots, 2^{B-1}-1\}$). Unlike R, B does not directly affect the complexity of the system.

84.4.1 Binary to One-Digit 2DLNS

Since the table method for converting single-digit 2DLNS-to-binary (shown in Figure 84.2, simplified from section 3) is quite fast and can be implemented efficiently in hardware, it seems only logical to reverse this process: i.e., to convert a binary 2's complement representation into a single-digit 2DLNS. This is a three-step process where the first step is not easily reversible, while the last two steps of shifting and sign correction can easily be performed on either 2's complement or FP formats.

Table 84.3 shows the contents of the LUT for $D = 3$ and $R = 3$. The number of table rows is 2^R (8, for this example). The precision of the mantissa is $C = 10$, where C is the number of bits of the fractional part.

84.4.1.1 Sign Determination

For 2's complement conversion, the sign of the binary input, x, is generally the high-order bit. If x is in an FP format (e.g., IEEE–754), the sign is determined from the FP sign bit (unless $x = 0$, a special FP

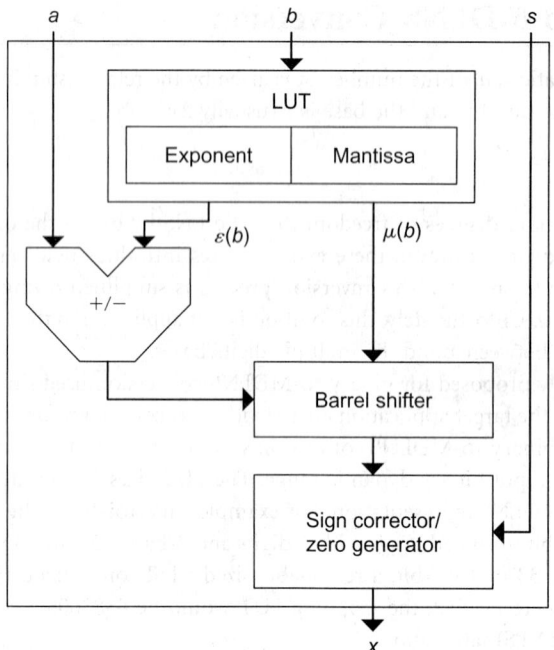

FIGURE 84.2 One-digit 2DLNS to binary converter.

TABLE 84.3 Single-Digit 2DLNS-to-binary
Conversion LUT for $D = 3$, $R = 3$, and $C = 10$

Input	Output	
b	$\mu(b)$(base 2)	$\varepsilon(b)$
0	1.0000000000	0
1	1.1000000000	1
2	1.0010000000	3
3	1.1011000000	4
−4	1.1001010010	−7
−3	1.0010111101	−5
−2	1.1100011100	−4
−1	1.0101010101	−2

case). In the event where $s = 0$, the 2DLNS exponents will also be set to 0 to minimize the chance of arithmetic overflow when performing simple 2DLNS operations (multiplication and division). Once the sign is extracted, $|x|$ is found.

84.4.1.2 Normalization

From the MDLNS to binary conversion, a shifter is used to set the output properly based on the sum of the input binary exponent and the LUT pseudo-FP exponent. For the reverse case we are trying to find the value of the binary exponent. Originally $shift = a + \varepsilon(b)$, but can be rearranged to $a = shift - \varepsilon(b)$ where $shift$ is the number of bits $|x|$ has to be shifted to normalize it (i.e., set it between 1 and 2). $\varepsilon(b)$ will be provided from the reverse of the LUT operation. For 2's complement this basically requires a priority encoder, whereas, for the case of FP, $|x|$ is already represented in a normalized form. The number of shifts can be determined from the exponent portion of the FP notation. However, this method is only valid for normalized FP notations. Denormalized FP notations (i.e., values $<2^{-126}$ for 32-bit FP and 2^{-1022} for 64-bit FP, including zero) will require additional hardware as in the 2's complement case.

84.4.1.3 Reverse LUT

The input to the single-digit 2DLNS-to-binary LUT is the second base exponent, b, and the outputs are the mantissa and the exponent. To reverse the process, the input to the LUT is now the mantissa, $\mu(b)$. Since the mantissa is not influenced by the exponent $\varepsilon(b)$, it can remain an output. Table 84.4 shows a preliminary binary-to-single-digit 2DLNS LUT for $D = 3$ and $R = 3$. The complexity of the unused portions of the LUT (the nonshaded area) is $O(2^C)$; however, since the table contains undefined entries for all possible input values except for the 2^R values that are exactly representable (shown as $\downarrow$ and referred to as ranges) the complexity of the required storage area is only $O(2^R)$. The shaded table entries, downward, are formed by mantissas doubling and exponents decrementing by 1. Shaded table entries, upward, have mantissas halving and exponents incrementing by 1. These complexity expressions exclude any calculations based on the output bit widths since they are considered a constant.

To reduce the LUT complexity to $O(2^R)$ (i.e., 2^R ranges) the undefined entries must be removed and the input range restricted to that of the normalized mantissa (the nonshaded area in the table). The latter is achieved by the normalizer, whereas the removal of the undefined entries can be achieved by rounding any input in an undefined range to the nearest representable value (i.e., using a mid-point function between the defined addresses). The complete LUT for $D = 3$, $R = 3$, and $C = 10$, is shown in Table 84.5. Note that an extra range is required owing to the possibility of rounding up numbers near 2.0. Thus, the number of ranges is $2^R + 1$, but the complexity is maintained at $O(2^R)$.

The following example demonstrates the use of Table 84.5 in performing a 2's complement binary-to-single-digit 2DLNS.conversion.

Example 1
Given $x = 1392$, find s, a, and b.
Solution
<ul><li>Since $x > 0$, $s = +1$</li><li>Normalize $|x| = 1392$ to $|x| = 1.359375 \cdot 2^{10}$ or $|x| = 1.010111_2 \cdot 2^{10}$ (10 shifts)</li><li>Find the mantissa 1.010111_2 in the LUT (Table 84.5)</li><li>The LUT returns $b = -1$ with an exponent of -2</li><li>$a = 10 - (-2) = 12$</li><li>Therefore, $+1 \cdot 2^{12} \cdot 3^{-1} = +1365$ is the nearest approximation with $R = 3$</li></ul>

TABLE 84.4 Preliminary Binary-to-Single-Digit 2DLNS Conversion LUT for $D = 3$, and $R = 3$

Input	Output	
$\mu(b)$	$\varepsilon(b)$	b
$\cdots$	$\cdots$	$\cdots$
$\downarrow$	?	?
0.8888888888	−3	−2
$\downarrow$	?	?
1.0000000000	0	0
$\downarrow$	?	?
1.1250000000	3	2
$\downarrow$	?	?
1.1851851851	−5	−3
$\downarrow$	?	?
1.3333333332	−2	−1
$\downarrow$	?	?
1.5000000000	1	1
$\downarrow$	?	?
1.5802469145	−7	−4
$\downarrow$	?	?
1.6875000000	4	3
$\downarrow$	?	?
1.7777777777	−4	−2
$\downarrow$	?	?
2.0000000000	−1	0
$\downarrow$	?	?
$\cdots$	$\cdots$	$\cdots$

Unfortunately, to build a conventional LUT (similar to Table 84.5), all possible values of the mantissa must be accommodated in the address decoder, which increases the complexity to $O(2^C)$. Clearly, the conventional LUT address decoder is not appropriate for this memory architecture.

84.4.1.4 Range-Addressable LUT (RALUT)

A standard LUT architecture is shown in Figure 84.3, where an address decoder is used to match the address to a unique stored value. The RALUT architecture of Figure 84.4 shows the new address decoder system that matches a stored value to a range of addresses. The decoder compares the input address, I, to a range of two neighboring monotone addresses (e.g., Addr(1) and Addr(2)). Only one of these comparisons will match the input and activate a word-enable line which drives the data patterns, Data, to the output, O, of the RALUT.

We can remove half of the comparators in the range decoder by noting that

$$(I < \text{Addr}(n)) = \overline{(\text{Addr}(n) \le I)}$$

$$(I \ge \text{Addr}(n)) \oplus (I \ge \text{Addr}(n+1)) = \text{Addr}(n) \le I < \text{Addr}(n+1)$$

Since the addresses are monotonic, if $(I \ge \text{Addr}(n+1))$ is true, then $(I \ge \text{Addr}(n))$ must also be true. Therefore we can reduce the XOR operator and rewrite the equation as

$$(I \ge \text{Addr}(n)) \cdot \overline{(I \ge \text{Addr}(n+1))} = \text{Addr}(n) \le I < \text{Addr}(n+1)$$

TABLE 84.5 Complete Binary-to-Single-Digit 2DLNS Conversion LUT for $D = 3$, $R = 3$, and $C = 10$

Input	Output	
$\mu(b)$ (base 2)	$\varepsilon(b)$	b
1.0000000000 ↓ 1.0000111111	0	0
1.0001000000 ↓ 1.0010011101	3	2
1.0010011110 ↓ 1.0100001000	−5	−3
1.0100001001 ↓ 1.0110101001	−2	−1
1.0110101010 ↓ 1.1000101000	1	1
1.1010001001 ↓ 1.1010001000	−7	−4
1.1010001001 ↓ 1.1011101101	4	3
1.1011101110 ↓ 1.1110001101	−4	−2
1.1110001110 ↓ 1.1111111111	−1	0

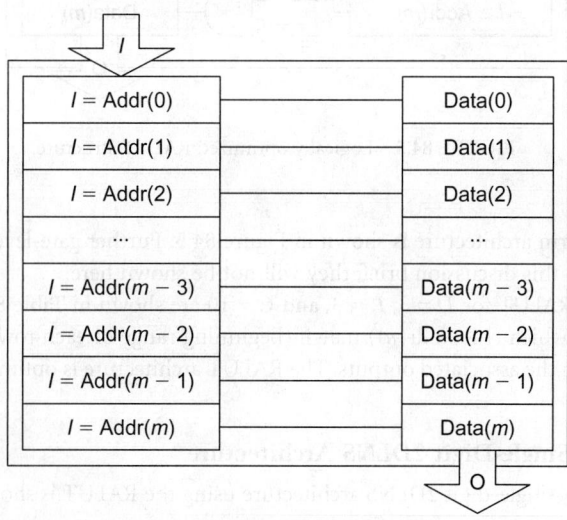

FIGURE 84.3 Standard LUT architectures.

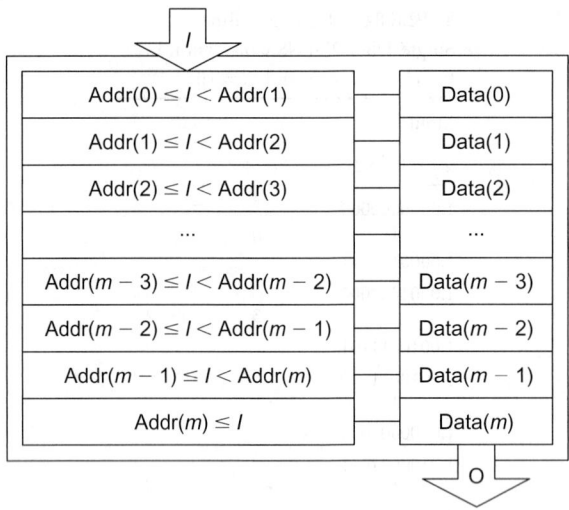

FIGURE 84.4 Range-addressable LUT architectures.

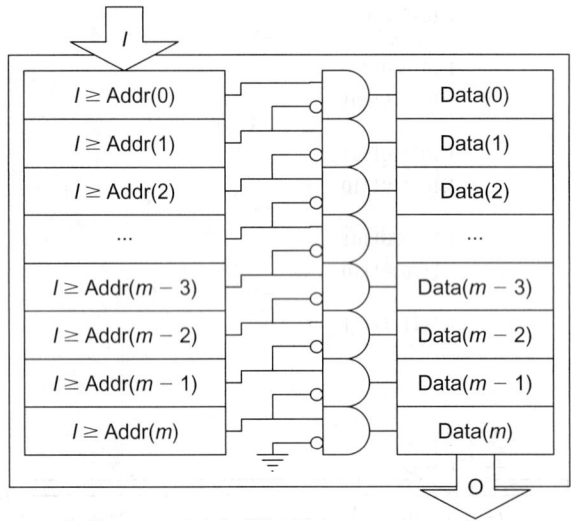

FIGURE 84.5 Logically optimized RALUT structure.

The optimized uniform architecture is shown in Figure 84.5. Further gate-level optimizations can be made; however, to keep this discussion brief, they will not be shown here.

The contents of the RALUT for $D = 3$, $R = 3$, and $C = 10$ are shown in Table 84.6. The input column contains the values for which $(I \geq \text{Addr}(n))$ match (beginning range of each row in Table 84.5) and the output columns contain the associated outputs. The RALUT architecture is optimal since it requires only $2^R + 1$ rows.

84.4.1.5   Binary-to-Single-Digit 2DLNS Architecture

The complete binary-to-single-digit 2DLNS architecture using the RALUT is shown in Figure 84.6. This structure can be implemented in a single (low latency) or multiprocess (pipelined, higher latency, lower power) circuit depending on the system constraints.

TABLE 84.6 Complete Binary-to-
Single-Digit 2DLNS Conversion
RALUT for $D = 3$, $R = 3$, and $C = 10$

Input	Output	
$\mu(b)$ (base 2)	$\varepsilon(b)$	b
1.0000000000	0	0
1.0001000000	3	2
1.0010011110	−5	−3
1.0100001001	−2	−1
1.0110101010	1	1
1.1000101001	−7	−4
1.1010001001	4	3
1.1011101110	−4	−2
1.1110001110	−1	0

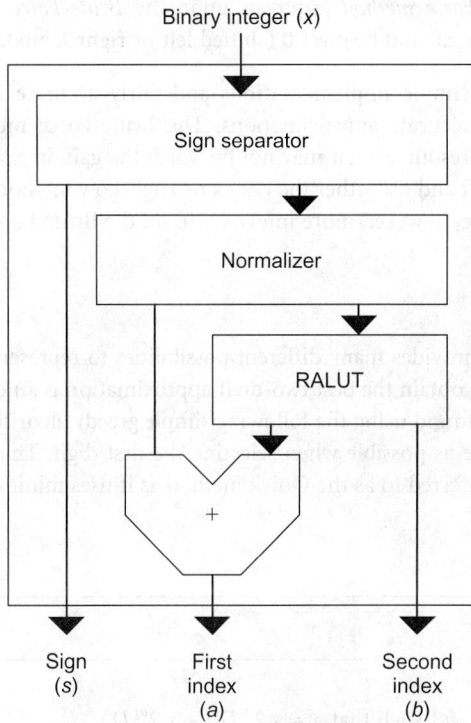

FIGURE 84.6 Binary-to-single-digit 2DLNS converter structure.

84.4.2 Binary to Two-Digit 2DLNS

From Eq. (84.21), we obtain the two-digit 2DLNS representation shown in the following equation:

$$x \approx s_1 2^{a_1} D^{b_1} + s_2 2^{a_2} D^{b_2} \qquad (84.22)$$

Conversion from a two-digit 2DLNS-to-binary is a fairly simple process. Both 2DLNS digits are converted separately using the single-digit method and their results accumulated to produce the final binary representation. There is only one binary representation of the result as binary is a positional number system.

The binary to two-digit 2DLNS process is more difficult as there can be many MDLNS representations for a single number. In Ref. [14], simple LUT tables were used to convert a binary 2's complement input into a MDLNS representation. The tables were generated by computing all possible MDLNS representations given the full range of B and R and selecting the nearest integer values. This method is clearly not feasible for a real-time hardware solution. There are four known methods for binary to two-digit 2DLNS conversion [29]:

1. The *Quick method* chooses the nearest first-digit to the target and generates the second-digit to reduce the error, a simple greedy algorithm.
2. The *High/Low method* chooses the two nearest approximations to the target as the first-digits, generates two associated second-digits for the error and selects the combination with the smaller error.
3. The *Brute-Force method* operates by selecting the combination with the smallest error, but it uses all possible mantissas of D^b as the first-digits instead of just one (*Quick*) or two (*High/Low*).
4. The *Extended-Brute-Force method* improves upon the *Brute-Force method* by using first-digit approximations above 2.0 and below 1.0 (shifted left or right L bits).

Each method ranges from simple implementations and fairly accurate approximations to difficult implementations and very accurate approximations. The Brute-Force methods, although realizable, require significant hardware resources that may not be worth the gain in accuracy. In some cases it may be more feasible to increase R and use either the Quick or High/Low methods. Therefore, only these two methods will be detailed here; however, more information on the Brute Force methods can be found in Ref. [29].

84.4.2.1 Quick Method

A two-digit approximation provides many different possibilities to represent a real number in 2DLNS. The only technique found to obtain the best two-digit approximation is an exhaustive search, but a very good approximation can be found using the following simple greedy algorithm, greedy in the sense that it takes as much of the value as possible when selecting the first digit. This method of binary-to-two-digit 2DLNS conversion is referred to as the Quick method as it uses minimal resources and is fast. The algorithm is shown below:

Quick Method Algorithm
Input: Real number, x.
Output: $\{s_1, a_1, b_1\}, \{s_2, a_2, b_2\}$ such that $x \approx s_1 2^{a_1} D^{b_1} + s_2 2^{a_2} D^{b_2}$.
Step 1: Find $\{s_1, a_1, b_1\}$ based on the binary-to-single-digit 2DLNS conversion of x.
Step 2: Generate $\tilde{x} = s_1 \mu(b_1) 2^{a_1 + \varepsilon(b_1)}$ using an additional output of the RALUT from Step 1.
Step 3: Determine the error $x - \tilde{x}$ using a matched mantissa output (matched by exponent) from the RALUT (see Table 84.7).
Step 4: Find $\{s_2, a_2, b_2\}$ based on a binary-to-single-digit 2DLNS conversion of $x - \tilde{x}$.

To determine the error for the second-digit approximation, we see that the RALUT output for the first-digit approximation must also include the matched mantissa (see Table 84.7). The size of the RALUT is therefore increased owing to this extra output, but it is a linear, not exponential growth.

TABLE 84.7 Complete Binary-to-Single-Digit
2DLNS Conversion RALUT with Mantissa Output
for $D = 3$, $R = 3$, and $C = 10$

Input			Output
$\mu(b)$ (base 2)	$\varepsilon(b)$	b	$\mu(b)$ (base 2)
1.0000000000	0	0	01.0000000000
1.0001000000	3	2	01.0010000000
1.0010011110	−5	−3	01.0010111101
1.0100001001	−2	−1	01.0101010101
1.0110101010	1	1	01.1000000000
1.1000101001	−7	−4	01.1001010010
1.1010001001	4	3	01.1011000000
1.1011101110	−4	−2	01.1100011100
1.1110001110	−1	0	10.0000000000

The following example illustrates the quick method technique.

Example 2
Given $x = 3840$, find s_1, a_1, b_1, s_2, a_2, and b_2
Solution
• Since $x = 3840 > 0$, $s_1 = +1$ • $|x| = 3840$ • Normalize $|x|$ • $|x| = 1.875 \cdot 2^{11}$ or $|x| = 1.111_2 \cdot 2^{11}$ • Find 1.111_2 in the RALUT (Table 84.7) • $b_1 = -2$ with $\varepsilon = -4$ and $\mu = 01.11000111_2$ • $a_1 = 11-(-4) = 15$ • Matching mantissa error: $(1.111_2 - 01.11000111_2) \cdot 2^{11} = 0.00011001_2 \cdot 2^{11}$
Second digit
• $x = 0.00011001_2 \cdot 2^{11}$, since $x > 0$ $s_2 = +1$, and $|x| = 0.00011001_2 \cdot 2^{11}$ • Normalized, $|x|$ is $|x| = 1.1001_2 \cdot 2^7$ (7 shifts) • From the RALUT (Table 84.7) $b_2 = -4$ with $\varepsilon = -7$ and $\mu = 1.100101_2$ • $a_2 = 7-(-7) = 14$
Final approximation
• $+1 \cdot 2^{15} \cdot 3^{-2} + 1 \cdot 2^{14} \cdot 3^{-4} = 3843$

84.4.2.2 Quick Method Architecture

For the second-digit approximation, the mantissa output of the RALUT is not required. In a serial
hardware implementation of this converter, only a single RALUT (with mantissa output) is needed. The
first-digit approximation uses the mantissa output while the second-digit approximation ignores it, as
in Figure 84.7.

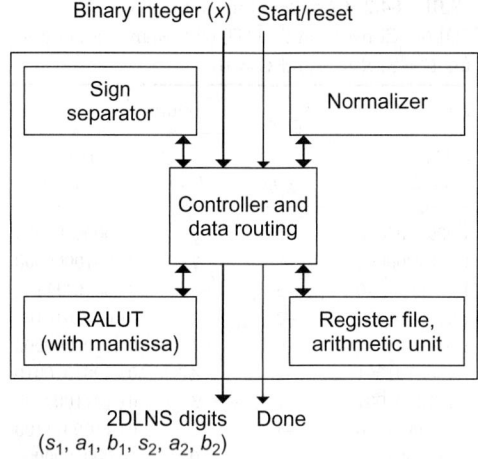

FIGURE 84.7   Quick binary-to-two-digit 2DLNS serial converter architecture.

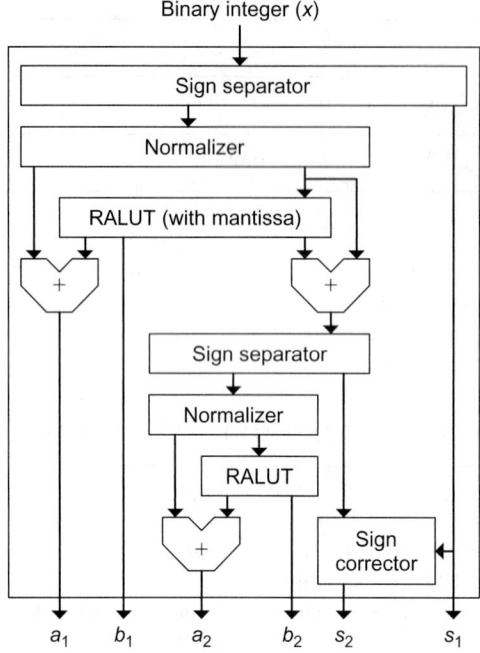

FIGURE 84.8   Quick binary-to-two-digit 2DLNS serial parallel architecture.

A pipelined hardware implementation of this converter can take advantage of this architecture by using two separate RALUTs (one with the mantissa output and one without) to minimize the area, as in Figure 84.8.

84.4.2.3 High/Low Method

Although the quick binary-to-two-digit 2DLNS conversion method operates efficiently, it does not always provide the most accurate result. Depending on the choice of R, there can be many possible representations for a given number. The Quick method selects the first digit to be above or below the target value, based on the nearest approximation, but independently of the selection of the second digit.

An alternative method, the High/Low method, generates two representations, one with the first digit below the target and the other with the first digit above the target. In either case, the choice of the two first digits are the two nearest approximations to the target; a modified greedy algorithm. The second digit is generated based on the error calculated from the first digits and the mantissas in the RALUT. The final representation is selected based on the minimum error. To do this, the second RALUT also requires the mantissa output (unlike the Quick method) to generate the final error. Note that one of these final representations will be the same as that produced by the Quick method. This technique is shown in the algorithm below:

High/Low Method Algorithm
Input: Real number, x.
Output: $\{s_1, a_1, b_1\}, \{s_2, a_2, b_2\}$ such that $x \approx s_1 \cdot 2^{a_1} \cdot D^{b_1} + s_2 \cdot 2^{a_2} \cdot D^{b_2}$.
Step 1: Find $\{s_1^-, a_1^-, b_1^-\}$ and $\{s_1^+, a_1^+, b_1^+\}$ based on the binary-to-single-digit 2DLNS conversion of $x^- \le x$ and $x^+ > x$, respectively.
Step 2: Generate $\tilde{x}^- = s_1^- \cdot \mu(b_1^-) 2^{a_1^- + \varepsilon(b_1^-)}$ and $\tilde{x}^+ = s_1^+ \mu(b_1^+) \cdot 2^{a_1^+ + \varepsilon(b_1^+)}$ using additional outputs of the RALUT from Step 1.
Step 3: Determine the errors $x - \tilde{x}^-$ and $x - \tilde{x}^+$ using the matched mantissa output (matched by exponent) from the RALUT (see Table 84.8).
Step 4: Find $\{s_2, a_2, b_2\}$ based on a binary-to-single-digit 2DLNS conversion of $x - \tilde{x}^-$ or $x - \tilde{x}^+$ which minimizes the error.

84.4.2.4 Modifying the RALUT for the High/Low Approximation

To implement the High/Low method, the RALUT has to store both the high and low approximations for the first digit. The RALUT addresses are modified to include the range from the low to the high value of each row. The RALUT contents for $D = 3$, $R = 3$, and $C = 10$ are shown in Table 84.8. Note that the high contents of each row are equal to the low contents of the next row. The number of table rows is 2^R, one row less than the previous tables, since cyclical connectivity is maintained by using the high element of the last row instead of requiring a new row.

TABLE 84.8 Binary-to-Single-Digit 2DLNS Conversion RALUT for the High/Low Method for $D = 3$, $R = 3$, and $C = 10$

Input	Low			High		
$\mu(b)$ (base 2)	$\varepsilon(b)$	b	$\mu(b)$ (base 2)	$\varepsilon(b)$	b	$\mu(b)$ (base 2)
1.0000000000	0	0	01.0000000000	3	2	01.0010000000
1.0010000000	3	2	01.0010000000	-5	-3	01.0010111101
1.0010111101	-5	-3	01.0010111101	-2	-1	01.0101010101
1.0101010101	-2	-1	01.0101010101	1	1	01.1000000000
1.1000000000	1	1	01.1000000000	-7	-4	01.1001010010
1.1001010010	-7	-4	01.1001010010	4	3	01.1011000000
1.1011000000	4	3	01.1011000000	-4	-2	01.1100011100
1.1100011100	-4	-2	01.1100011100	-1	0	10.0000000000

The following example illustrates the high/low binary-to-two-digit 2DLNS conversion method.

Example 3
Given $x = 3840$, find s_1, a_1, b_1, s_2, a_2, and b_2 for the most accurate representation using the high/low method
<div align="center">Solution</div>
• Since $x = 3840 > 0$, $s_1 = +1$ and $\lvert x\rvert = 3840$ • Normalize $\lvert x\rvert$: $\lvert x\rvert = 1.875 \cdot 2^{11}$ or $\lvert x\rvert = 1.111_2 \cdot 2^{11}$ • Find 1.111_2 in the RALUT (Table 84.8)

Low approximation, first digit	High approximation, first digit
• $b_1 = -2$ with $\varepsilon = -4$ and $\mu = 01.11000111_2$ • $a_1 = 11 - (-4) = 15$ • Matching mantissa error: $\quad (1.111_2 - 01.11000111_2) \cdot 2^{11} = 0.00011001_2 \cdot 2^{11}$	• $b_1 = 0$ with $\varepsilon = -1$ and $\mu = 10.0000000000_2$ • $a_1 = 11 - (-1) = 12$ • Matching mantissa error: $\quad (1.111_2 - 10.0_2) \cdot 2^{11} = -0.001_2 \cdot 2^{11}$

Low approximation, second digit	High approximation, second digit
• $x = 0.00011001_2 \cdot 2^{11}$, therefore $s_2 = +1$ and $\lvert x\rvert = 0.00011001_2 \cdot 2^{11}$ • Normalized, $\lvert x\rvert$ is $\lvert x\rvert = 1.1001_2 \cdot 2^7$ (7 shifts) • From the RALUT (Table 84.7) $b_2 = -4$ with $\varepsilon = -7$ and $\mu = 1.100101_2$ • $a_2 = 7 - (-7) = 14$ • Find the difference (or approximation error) between the two mantissas: $\quad (1.1001_2 - 1.100101_2) \cdot 2^7 = 0.000001_2 \cdot 2^7$ $\quad$ or $1.0_2 \cdot 2^1$ • $+1 \cdot 2^{15} \cdot 3^{-2} + 1 \cdot 2^{14} \cdot 3^{-4} = 3843$ with a computed error of 2 • Same solution as quick method	• $x = -0.001_2 \cdot 2^{11}$, therefore $s_2 = -1$ and $\lvert x\rvert = 0.001_2 \cdot 2^{11}$ • Normalized, $\lvert x\rvert$ is $\lvert x\rvert = 1.0_2 \cdot 2^8$ (8 shifts) • From the RALUT (Table 84.7) $b_2 = 0$ with $\varepsilon = 0$ and $\mu = 01.0_2$ • $a_2 = 8 - (0) = 8$ • Find the difference (or approximation error) between the two mantissas: $\quad (1.0_2 - 1.0_2) \cdot 2^8 = 0$ • $+1 \cdot 2^{12} \cdot 3^0 - 1 \cdot 2^8 \cdot 3^0 = 3840$ with a computed error of 0 • First-digit not greedy

Final approximation
• $+1.2^{12} \cdot 3^0 - 1 \cdot 2^8 \cdot 3^0 = 3840$

84.4.2.5   High/Low Method Architecture

For a pipelined hardware implementation, a single RALUT with dual outputs (as in Table 84.8) and two RALUTs with mantissa outputs are needed (see Figure 84.9). For a serial hardware implementation, it is possible to use only one RALUT (dual outputs). The input mantissa can be compared with the two output mantissas to determine the nearest approximation for the second digit. The implementation is similar to that of the quick method (see Figure 84.7) with some additional resources (storage and arithmetic).

84.4.3   Extending to More Bases

The conversion methods shown here can be easily extended into MDLNS representations with more than 2 bases (assuming one of the bases is still 2). Since all the methods shown rely only on information about the mantissa and exponent, multiple bases can be merged, generating a single mantissa and exponent. For example, Table 84.9 shows an MDLNS-to-binary conversion LUT using bases 2, 3, and 5, and a word length of 2 bits for the indices using bases 3 and 5. The nonbase 2 indices can be combined through

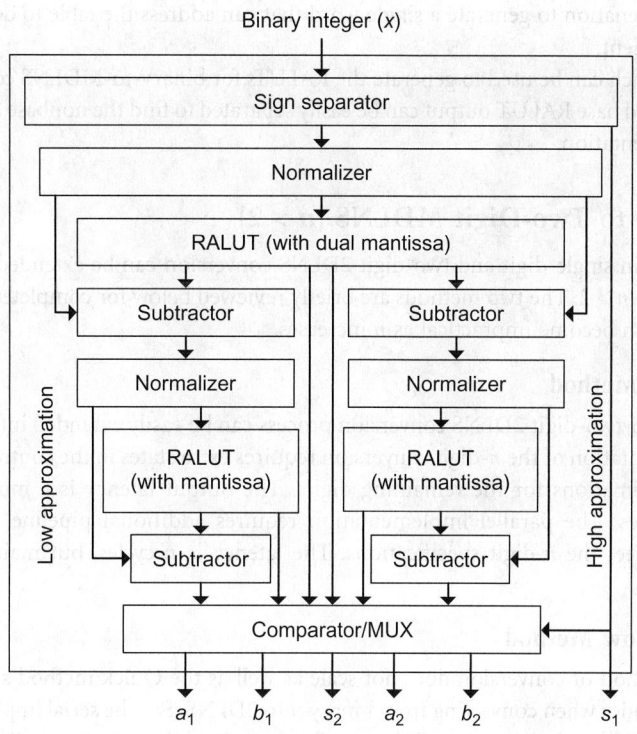

FIGURE 84.9 High/low binary-to-two-digit 2DLNS parallel converter architecture.

TABLE 84.9 MDLNS-to-binary Conversion LUT for $D_1 = 3$, $D_2 = 5$, and $R = 2$

Input		Output	
b_1	b_2	$\mu(b_1, b_2)$ (Combined Mantissas)	$\varepsilon(b_1, b_2)$ (Combined Exponent)
0	0	1.0000000000	0
−1	−1	1.0666666667	−4
−2	1	1.1111111111	−1
−2	−2	1.1377777778	−8
1	−1	1.2000000000	−1
0	1	1.2500000000	2
0	−2	1.2800000000	−5
−1	0	1.3333333333	−2
−2	−1	1.4222222222	−6
1	0	1.5000000000	1
0	−1	1.6000000000	−3
−1	1	1.6666666667	0
−1	−2	1.7066666667	−7
−2	0	1.7777777778	−4
1	1	1.8750000000	3
1	−2	1.9200000000	−4

simple word concatenation to generate a single word that can address the table to determine the proper mantissa and exponent.

This same approach can be used to generate the RALUTs for binary-to-MDLNS conversion. Similarly, the combined second base RALUT output can be easily separated to find the nonbase 2 indices to generate the MDLNS representation.

84.4.4 Binary to Two-Digit MDLNS ($n > 2$)

The methods used in single-digit and two-digit 2DLNS conversion can be extended to operate with an n-digit 2DLNS with $n > 2$. The two methods are briefly reviewed below for completeness, although some implementations can become impractical as n increases.

84.4.4.1 Quick Method

The quick binary-to-two-digit 2DLNS conversion process can be easily extended into an n-digit system. The serial implementation of the n-digit conversion requires extra states in the controlling state machine to produce approximations for the remaining digits. The output latency is a multiple of n, but the bandwidth decreases. The parallel implementation requires additional pipeline stages (normalizers and RALUTs) to meet the n-digit specifications. The latency is n cycles, but maintains its operating bandwidth.

84.4.4.2 High/Low Method

The High/Low method of conversion does not scale as well as the Quick method since there are many possibilities to consider when converting from binary into 2DLNS. For the serial implementation a stack-based state machine is used to traverse all the possible representations for the n-digit 2DLNS representation to find the best representation [22]. The parallel implementation requires double the number of conversion components for each pipeline stage from the previous stage (i.e., 1, 2, 4, 8, 16, …). In total, $2^n - 1$ conversion components will be needed. Just as in the Quick method, the latency will increase, but the bandwidth will remain the same. The final stage will require $2^{n-1} - 1$ comparators to determine the best n-digit 2DLNS approximation (see Figure 84.10).

84.5 FIR Digital Filter: A Case Study

84.5.1 Choice of the Nonbinary Base

A closer look at the architecture of Figure 84.1 shows the following feature. Assume that we are working with 2DLNS. In this case, we simply store the input data into the form $\pm 2^{a_1} D^{b_1} \pm 2^{a_2} D^{b_2}$, or, every input sample is stored as a combination of six integers $(s_1, a_1, b_1, s_2, a_2, b_2)$. The multiplication between the input data and the filter coefficients is replaced by addition of the corresponding exponents with the result transformed to a binary representation by making use of a ROM table which stores an FP-like representation of the powers of the second base D. The area complexity depends almost entirely on the size of the nonbinary exponents, b_1 and b_2; therefore, our main goal in approximating the digital filter coefficients is to minimize as much as possible the size of the largest nonbinary exponents used while maintaining the design constraints. The actual value of D can be selected to optimize the implementation without changing the overall complexity of the architecture; in fact, as we shall see, such an optimization offers a great potential for further reductions of the hardware complexity. It is also important to point out that D does not have to be an integer! While the expansion of the idea to noninteger bases leads to large theoretical hurdles (in terms of rigorous justification of the results obtained), it is equally clear that with this extension in hand one can vastly increase the chance to obtain an extremely good representation of the filter coefficients with very small exponents and only single-digit representations.

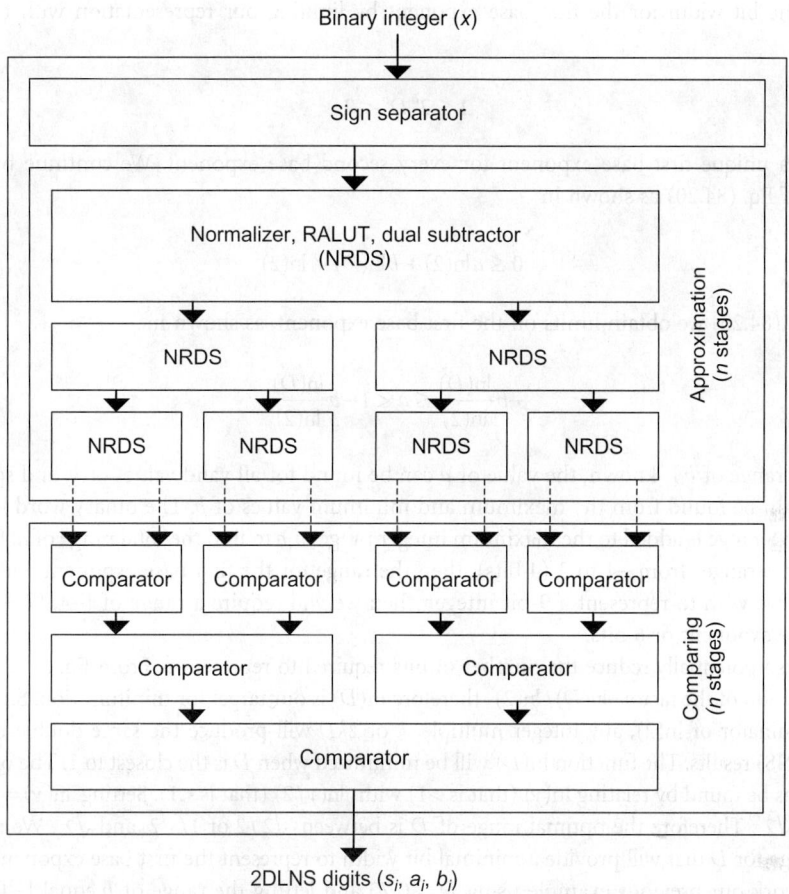

FIGURE 84.10   High/low binary-to-*n*-digit 2DLNS parallel converter structure.

84.5.1.1 Limiting the Nonbinary Base

We can limit the potential range of what could be considered to be an optimal value by analyzing the unsigned single-digit representation as shown in the following equation:

$$2^a D^b = 2^{a-b}(2D)^b = 2^{a+b}\left(\frac{D}{2}\right)^b \tag{84.23}$$

This shows that we can multiply or divide the unknown base by the first base and not change the computational result; however, the exponent of the first base will change. This simple relationship implies a restriction on the range of values of an optimal base. For example, if our search were to begin at $D = 3$, then it would be pointless to go outside of the range 3–6 as the results of the representation would simply repeat.

The relationship in Eq. (84.23) also shows that as the value of D is reduced by a multiple of 2, the exponent of the first base will increase when b is positive, but decrease when b is negative. A similar conclusion can be made for the case when D is multiplied by 2. Therefore, some representations may have large values for the first base exponent, and some may have small values. For a VLSI implementation, the bit width of the first base exponent should be minimized while maintaining the selected representation space. We can determine the bit width for the first base exponent by limiting our representation with the following equation:

$$1 \leq 2^a D^b < 2 \tag{84.24}$$

There is a unique first base exponent for every second base exponent. We continue by taking the logarithm of Eq. (84.20) as shown in

$$0 \leq a\ln(2) + b\ln(D) < \ln(2) \tag{84.25}$$

From Eq. (84.25) we obtain limits on the first base exponent, as shown in

$$-b\frac{\ln(D)}{\ln(2)} \leq a < 1 - b\frac{\ln(D)}{\ln(2)} \tag{84.26}$$

Since the range of b is known, the value of a can be found for all valid values of b, and so the integer range of a can be found from the maximum and minimum values of b. The binary word length of the usable 2DLNS range is added to the maximum integer range of a to find the total range of a. For example, if $D = 3$ and b ranges from −4 to 3 (4 bits), then the range for the first base exponent will be between −4 and 7. If we wish to represent a 9-bit integer, then we will require a range of $[-4, (7 + 9 = 16)]$ for the first base exponent or 6 bits.

We can also potentially reduce the number of bits required to represent a. From Eq. (84.26), the range of a is a function of the factor $\ln(D)/\ln(2)$, therefore $\ln(D)$ is our target for minimization. Since the factor has a denominator of $\ln(2)$, any integer multiple, k of $2kD$ will produce the same double-base number system (DBNS) results. The function $\ln(D)$ will be minimized when D is the closest to 1. The optimal range of D can thus be found by relating $\ln(y)$ (that is >1) with $\ln(y/2)$ (that is <1). Setting $\ln(y) = -\ln(y/2)$ we obtain $y = \sqrt{2}$. Therefore the optimal range of D is between $\sqrt{2}/2$ or $1/\sqrt{2}$ and $\sqrt{2}$. We now have an optimal range for D that will provide a minimal bit width to represent the first base exponent, a.

If we rework our previous example using $D = 0.75$ and letting the range of b equal $[-4, 3]$ (4 bits), then the range for the first base exponent will be between −1 and 2. To represent a maximum of a 9-bit integer, we will require a range of $[-1, (2 + 9 = 11)]$ for the first base exponent or 5 bits. This is a saving of 1 bit from the previous example, using $D = 3$, but with no change in the representation.

Finding the optimal value of the second base for a given digital filter specification is a challenging optimization problem even for a single-digit 2DLNS representation. It can be posed as follows:

Input: (1) Digital filter coefficients $(h_0, h_1, \ldots, h_{k-1})$ designed to satisfy specific digital filter constraints and (2) the number of bits, R, of the nonbinary exponents.

Output: A real number D and triples of integers (s_i, a_i, b_i), $i = 0, 1, \ldots, k-1$, such that: (1) $h_i \approx s_i 2^{a_i} D^{b_i} = h_i^{(D)}$; (2) $b_i = \{-2^{R-1}, \ldots, 2^{R-1} - 1\}$; (3) $s_i = \pm 1$; and (4) the approximated coefficients $h_i^{(D)}$ satisfy the constraints provided.

There are many methods for designing digital filters, each of which prioritize different output characteristics. To further reduce the complexity of this problem we will first generate the filter coefficients by using classical design techniques. In this case, we will minimize the pass band ripple, maximize the stopband attenuation, and maintain linear phase. We then optimize the mapping of the real coefficients into a 2DLNS representation. This approach can be used with many more filter design techniques or even for applications other than the filter design.

Note that we do not impose restrictions on the size of the binary exponents, since they have very little contribution to the overall complexity of the architecture proposed. We require, however, to know what their range will be.

To demonstrate how important it is to choose the optimum base D, we provide the following example of a 53-tap FIR filter described by the coefficients (coefficients 28–53 are a mirror of 1–26 to guarantee linear phase) shown in Table 84.10:

TABLE 84.10 Filter Coefficients
For a Sample 53-Tap Filter

#	Coefficient
1	0.000088386028
2	−0.000035439162
3	−0.000305517680
4	0.000011540218
5	0.000727598360
6	0.000107077480
7	−0.001490251000
8	−0.000496150920
9	0.002676588800
10	0.001374708700
11	−0.004370857000
12	−0.003084987300
13	0.006583184400
14	0.006077961200
15	−0.009253069000
16	−0.010984374000
17	0.012225248000
18	0.018760160000
19	−0.015265215000
20	−0.031184863000
21	0.018081726000
22	0.052638228000
23	−0.020371451000
24	−0.099135649000
25	0.021867630000
26	0.315925630000
27	0.477612000000

For the signal samples we will need to use two or more digits. In Ref. [12] the typical distribution of the coefficients of many different filters was found to be a Gaussian-like function centered on zero. Such a coefficient distribution is better represented by a logarithmic-like number system such as the MDLNS rather than a linear number representation (such as binary). Therefore we should be able to obtain very good single-digit approximations in the 2DLNS by making use of a carefully calculated second base. This allows us to reduce the number of computational channels by a factor of two. We shall also consider a two-digit 2DLNS representation.

The frequency response of the above filter is shown below in Figure 84.11 along with an implementation of a 1-digit and 2-digit DBNS filter responses, each using almost the same number of bits and 3 as the second base.

The size of the second base exponent plays an important role in the size of the hardware owing to the required LUTs; a 1-bit increase to the nonbinary exponent doubles the LUT size. An increase in the binary exponent adds minimal hardware. Any change to the second base, including to real numbers has no impact on the structure of the hardware. Therefore, hardware designed for a ternary base is easily converted into the optimal base.

84.5.1.2 Finding the Optimal Base

To determine the optimal base a program was created that performs a thorough search of real bases. We have already seen that the most efficient bases for hardware implementation lie in the range $[1/\sqrt{2}, \sqrt{2}]$. This limitation offers a practical start and endpoint for a range-type search. The program uses a dynamic step size which is applied by analyzing the change in the optimization score and controlling the test points for the base, from the start to the endpoint, to reduce the overall search time. The step size increases so long as the resulting scores are monotonically improving. The program retraces and decreases the step

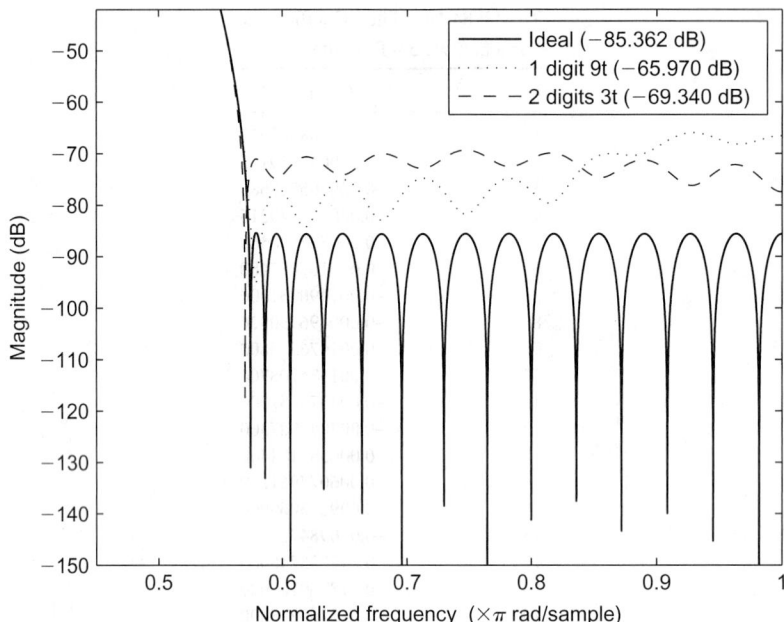

FIGURE 84.11   Ideal and mapped 2DLNS filter magnitude responses (second base=3).

size when the scores change in a nonmonotonic fashion. Each time a better result or score is found, it is added to a running list of "best scores." Once the entire range has been processed, each element in this list is finely adjusted by progressively smaller values. This optimization algorithm drastically improves search times by initially performing a coarse search followed by a finer search near the optimum point.

As the dynamic range of the nonbinary exponent is increased, we increase the efficiency of the coefficient mapping therefore increasing the performance of the implemented filter.

84.5.1.3   Results

By using a single-digit 2DLNS representation with an optimal base of 0.7278946656213228 the filter was able to achieve a stop-band attenuation of over 80 dB, and a two-digit 2DLNS representation with an optimal base of 0.7352545180621994 was able to achieve a stop-band attenuation of over 81.5 dB (see Figure 84.12). Using these optimum second bases we see a considerable performance improvement compared to filters designed using a second base of 3 (see Figure 84.11) that have a stop-band attenuation of 66 dB for single-digit and 69 dB for two-digit representation.

Table 84.11 shows the filter coefficients and the approximations for the optimal base. A closer look at the table shows the advantages of the proposed number system in terms of the dynamic range of the computations. For a one-digit 2DLNS we have used only 9 bits for the nonbinary exponents and, correspondingly, we have to use only a 512-word ROM (two for a parallel implementation). For a two-digit 2DLNS we need only 3 bits to represent the nonbinary exponents requiring only an 8-word ROM (four for a parallel implementation). The same accuracy can be achieved in the case of classical binary arithmetic if one uses a 16-bit dynamic range for the filter coefficients. It is also very important to note that the entire architecture is multiplier-free; it consists of only small adders and very small ROM arrays, several orders of magnitude smaller than the equivalent classical LNS design.

84.5.2   Applying an Optimal Base to Input Data

We have seen how applying an optimal base to the coefficients of a digital filter can significantly increase the accuracy of the mapping to 2DLNS. This same improvement can be seen when applied to the input

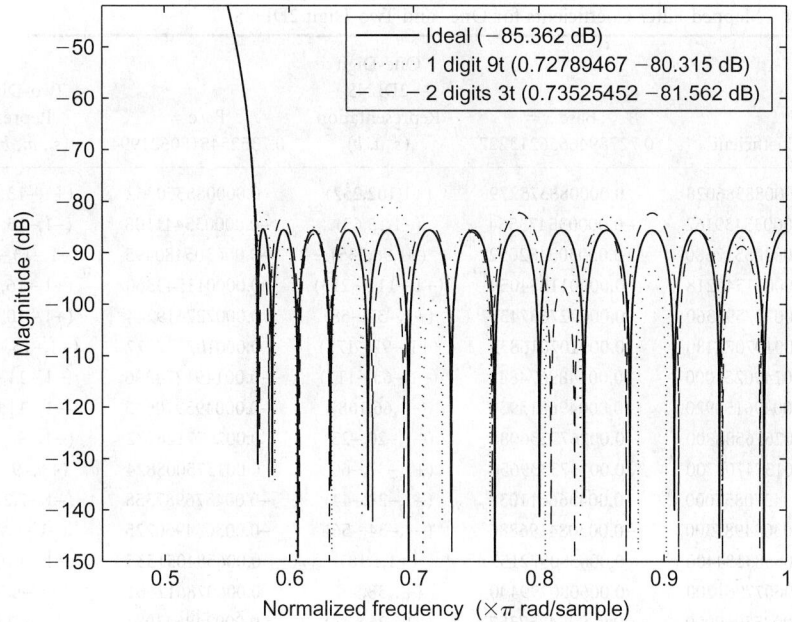

FIGURE 84.12 Ideal and mapped 2DLNS filters' magnitude response (optimal second base).

data of the filter; however, the nonbinary bases must be the same for both data and coefficients for the 2DLNS arithmetic to operate properly. For implementation efficiency the sum of the corresponding coefficient and input data indices should never overflow while performing a 2DLNS multiplication. We can guarantee that this situation will never happen by limiting the range on both the coefficient and data exponents. For example, with the earlier one-digit optimal base case, the nonbinary exponent range is 9 bits or $[-256, 255]$. To avoid overflow we could limit the range on the two-digit input data mapping nonbinary exponent from $[-32, 31]$ (or 6 bits), which would require a change to the range of the coefficients to $[-224, 224]$ to prevent overflow on the sum (9 bits only). Changing these limits requires that we recalculate the optimal base, which results in slightly lowering the stop-band attenuation of 77.741 dB from the original value of 80.315 dB. Our range of the input data nonbinary exponent influences the input data mapping. In this case we are unable to achieve a complete error-free mapping. Table 84.12 summarizes the results from three different input ranges for an optimal base selected from the coefficients. Another approach to limit the number of nonerror-free input representations is to find the optimal base for the inputs themselves. Doing this will significantly limit the stop-band attenuation as the coefficients are single digits and mapped using the optimal base from the input data. Table 84.12 also includes results from this approach. Clearly, for the one-digit coefficient/two-digit input data case one has to select what should be prioritized.

We will explore the same scenario with the two-digit case where the coefficient nonbinary exponent range is $[-4, 3]$. Since this range is very small, we need only limit the input data nonbinary exponent ranges, initially, to $[-28, 28]$ so that the range of the sum is $[-32, 31]$ (or 6 bits). Our stop-band attenuation does not change from 81.562 dB; however, the data mapping is far from being error-free. Table 84.13 summarizes the results from the same three input ranges as in the previous case. To try to improve the input data mapping, the nonbinary exponent ranges can be increased to $[-60, 60]$ which results in better mapping (see Table 84.13), but larger LUTs in the inner product processor. The alternative, again, is to optimize the base for the input data. In the single-digit case earlier, the stop-band attenuation suffered as a result of the coefficients not being mapped as well as before. To avoid this, we can increase the range on the coefficients to $[-8, 7]$; therefore, providing more possible representations and reducing

TABLE 84.11 Mapped Filter Coefficients for One- and Two-Digit 2DLNS

#	Coefficient	Base $x =$ 0.72789466562132277	One-Digit 2DLNS Representation (s, a, b)	Base = 0.7352545180621994	Two-Digit 2DLNS Representation $(s_1, a_1, b_1, s_2, a_2, b_2)$
1	0.000088386028	0.000088375279	(+1,102,252)	0.000088350362	(+1,−13,1,−1,−19,1)
2	−0.000035439162	−0.000035473564	(−1,15,65)	−0.000035441108	(−1,−15,−1,1,−16,3)
3	−0.000305517680	−0.000305482052	(−1,96,235)	−0.000305480495	(−1,−13,−3,1,−21,−4)
4	0.000011540218	0.000011534059	(+1,−114,−213)	0.000011543360	(+1,−16,1,1,−22,−1)
5	0.000727598360	0.000727637435	(+1,−37,−58)	0.000727619244	(+1,−10,1,1,−18,−3)
6	0.000107077480	0.000107016833	(+1,−92,−172)	0.000107017877	(+1,−15,−4,1,−19,−1)
7	−0.001490251000	−0.001489674843	(−1,−63,−117)	−0.001491274396	(−1,−11,−4,1,−12,1)
8	−0.000496150920	−0.000496013930	(−1,66,168)	−0.000495910645	(−1,−11,0,−1,−17,0)
9	0.002676588800	0.002677236086	(+1,−20,−25)	0.002677146132	(+1,−9,−1,1,−16,−1)
10	0.001374708700	0.001373339656	(+1,−37,−60)	0.001375008824	(+1,−9,1,−1,−14,0)
11	−0.004370857000	−0.004366611038	(−1,−28,−44)	−0.004376987358	(−1,−7,2,−1,−14,−3)
12	−0.003084987300	−0.003084196886	(−1,−34,−56)	−0.003084960225	(−1,−9,−2,1,−10,2)
13	0.006583184400	0.006580572157	(+1,1,18)	0.006584051535	(+1,−7,0,−1,−11,−3)
14	0.006077961200	0.006080299400	(+1,38,99)	0.006078611661	(+1,−6,3,−1,−12,2)
15	−0.009253069000	−0.009254257357	(−1,−26,−42)	−0.009248543981	(−1,−7,0,−1,−9,1)
16	−0.010984374000	−0.010986717246	(−1,−34,−60)	−0.010984583240	(−1,−7,−1,−1,−11,1)
17	0.012225248000	0.012226517020	(+1,−76,−152)	0.012227106720	(+1,−5,3,−1,−11,3)
18	0.018760160000	0.018759469892	(+1,41,102)	0.018753272390	(+1,−5,1,−1,−7,2)
19	−0.015265215000	−0.015264183280	(−1,21,59)	−0.015265989005	(−1,−6,0,1,−11,1)
20	−0.031184863000	−0.031179933849	(−1,−16,−24)	−0.031184008886	(−1,−5,0,1,−13,2)
21	0.018081726000	0.018081111094	(+1,2,17)	0.018081896930	(+1,−6,0,1,−10,−3)
22	0.052638228000	0.052644577257	(+1,4,18)	0.052636526952	(+1,−4,1,1,−9,−4)
23	−0.020371451000	−0.020367241670	(−1,−122,−254)	−0.020373197433	(−1,−7,−3,−1,−10,1)
24	−0.099135649000	−0.099138361541	(−1,−7,−8)	−0.099132593032	(−1,−3,1,−1,−8,−2)
25	0.021867630000	0.021875010939	(+1,−55,−108)	0.021865368722	(+1,−6,−1,1,−12,−3)
26	0.315925630000	0.315924361803	(+1,13,32)	0.315918851070	(+1,−3,−3,1,−9,1)
27	0.477612000000	0.477611998152	(+1,−84,−181)	0.477611998152	(+1,4,3,−1,−3,1)

TABLE 84.12 Data or Coefficient Optimal Base Results for One-Coefficient/Two-Data Digit System

Optimal Base Derived From	Conversion Method	Data Input Range	Data NonError-Free Representations	Data Maximum Error	Stop-Band Attenuation (dB)
Coefficient	Quick	−8192–8191	463	1.27	77.741
Coefficient	High/low	−8192–8191	225	1.11	77.741
Coefficient	High/low	−16384–16384	1900	2.11	77.741
Coefficient	High/low	−32768–32767	7759	4.92	77.741
Data	Quick	−8192–8191	0	0	58.647
Data	High/low	16384–16384	2	0.52	56.658
Data	High/low	−32768–32767	411	1.1	59.484

Non-binary exponent range (data: −32 to 31, coefficients: −224 to 224).

the input data nonbinary exponent range to [−24, 24] to maintain the same dynamic range at the output of the adder as before. This approach allows a completely error-free mapping for an input range of [−8192, 8191] with a simpler converter, but limits the size of the inner product LUTs. Table 84.13 summarizes the results for other ranges. For two-digit input data and coefficient architectures, this appears to be the best approach for minimizing hardware and computational error.

TABLE 84.13 Data or Coefficient Optimal Base Results for Two-Coefficient/Two-Data Digit System

Optimal Base Derived From	Conversion Method	Data Input Range	Data NonBinary Exponent Range	Data NonError-Free Representations	Data Maximum Error	Coefficient NonBinary Exponent Range	Stop Band Attenuation (dB)
Coefficient	Quick	−8192–8191	−28–28	1591	4.79	−4–3	81.562
Coefficient	High/low	−8192–8191	−28–28	1001	4.37	−4–3	81.562
Coefficient	High/low	−16384–16384	−28–28	3475	8.74	−4–3	81.562
Coefficient	High/low	−32768–32767	−28–28	10689	18	−4–3	81.562
Coefficient	Quick	−8192–8191	−60–60	4	0.69	−4–3	81.562
Coefficient	High/low	−8192–8191	−60–60	1	0.52	−4–3	81.562
Coefficient	High/low	−16384–16384	−60–60	23	1.06	−4–3	81.562
Coefficient	High/low	−32768–32767	−60–60	242	2.27	−4–3	81.562
Data	Quick	−8192–8191	−24–24	0	0	−8–7	79.465
Data	High/low	−16384–16384	−24–24	55	0.7	−8–7	81.085
Data	High/low	−32768–32767	−24–24	2268	1.42	−8–7	75.338

84.5.3 Single Sign-Bit Architecture

The data path of the MDLNS processor (shown in Figure 84.1) is affected significantly by the signs of the operands. The required sign correction operation comes at a cost of additional logic and power. Thus far, our particular filter architecture requires additional processing to be performed after the MDLNS processor such as summing all the channels. It is possible to use the common single sign-bit binary representation for the intermediate results. We have therefore developed a new MDLNS sign system to reduce the processing path of the MDLNS inner product CU while producing a single sign-bit binary representation.

Our original MDLNS notation uses two bits to represent the sign for each digit (−1, 0, and 1). There are only three of four states used, one of which (zero) represents only a single value. Using two sign bits results in having nearly 50% of the representation space unused. To improve this ratio, only a single sign bit is needed to represent the most used cases (−1 and 1). We now represent zero by setting the nonbinary exponents to their most negative values (i.e., if the range is $[-4, 3]$, then −4 is used to represent zero). This allows us to reduce the circuitry of the system while maintaining the independent processing of the exponents; this modification is easily integrated into the existing two-bit sign architecture. This special case for zero still leaves us with unused representation space, but not nearly as much as with the two sign-bit system.

By using the single sign-bit architecture for a four-channel filter, the word length for the 2DLNS representation of the coefficients and data are reduced by 2 bits. The 2DLNS processor is improved since it no longer needs to handle the negative or special zero case; only the absolute output is required. The coefficient and data signs are simply XORed to produce the output sign that is used along with the absolute output, to determine the final sum (see Figure 84.13).

In the case of a two-digit data and coefficient system, the four channels and output accumulation process is simplified with a single sign-bit by using only four adder/subtractor components and simple logic to coordinate the proper series of operations (see Figure 84.14). The processing delay from the LUTs is only three arithmetic operations and the overall logic is also reduced since the four adder/subtractor components are smaller than the four separate adder and 2's complement generator components.

84.5.4 Fabricated Designs

As an example of the applicability of the MDLNS representation to DSP applications, a variety of digital filters using the architecture proposed in Section 84.3.2 have been created. These designs have used both one- and two-digit 2DLNS representations with different bases.

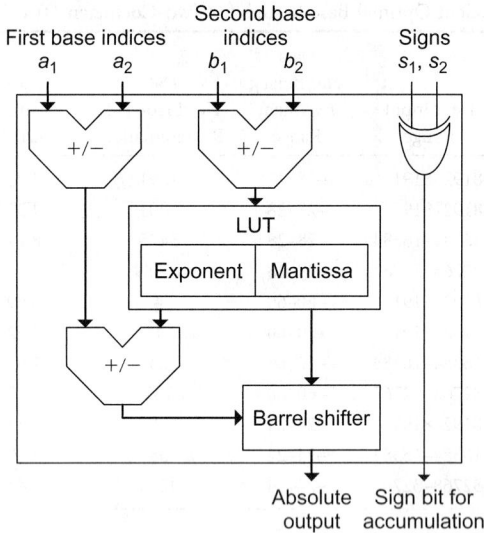

FIGURE 84.13 One-digit 2DLNS inner product CU with single sign-bit architecture.

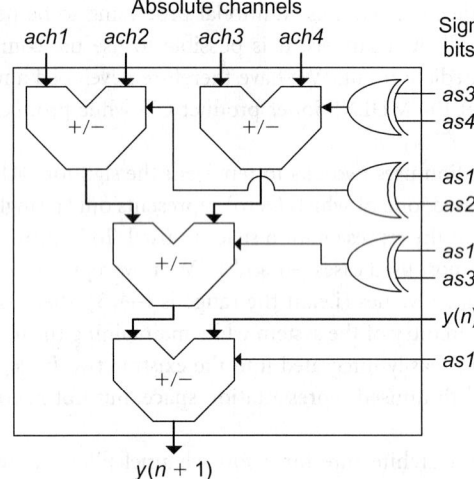

FIGURE 84.14 Four-channel accumulator for single sign-bit architecture.

84.5.4.1 Two-Digit Data, One-Digit Coefficient Parallel Design

A two-digit 2DLNS architecture for a 15-tap filter has been fabricated (TSMC 0.35 μm three-metal CMOS process) and successfully tested [12,20]. The input data (10-bit 2's complement converted via an LUT) uses the full two-digit representation, but the filter coefficients are designed using a one-digit MDLNS (a hybrid representation). This requires two inner product computational units per coefficient. The chip size is very large at 9 mm × 16 mm, owing to the 1024-word LUT in each of the inner product computational units which are implemented with logic gates and only three layers of metal. Although impractical in terms of the large area of silicon used, it was the first MDLNS circuit fabrication and also allowed design comparisons based solely on representation choices and not on design skill. The layout is shown in Figure 84.15 and the two parallel arrays are clearly visible.

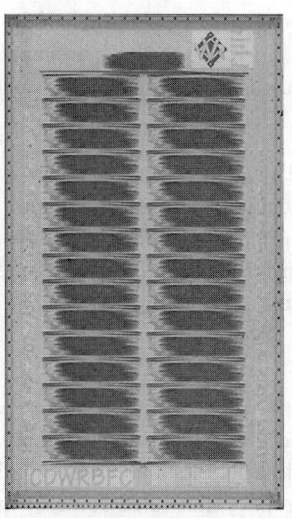

FIGURE 84.15 Hybrid 2DLNS 15-tap filter.

FIGURE 84.16 Eight-band 75-tap 2DLNS filterbank.

84.5.4.2 Two-Digit Data and Coefficient Serial Design

An eight-band 75-tap filterbank [30] designed for use in a low-power digital hearing instrument (fabricated in a TSMC 0.18 μm six-metal CMOS process) uses a two-digit data and coefficient 2DLNS system. The input binary data (16-bit 2's complement) is converted to 2DLNS using a two-digit 2DLNS high/low serial converter. The system uses a four-channel architecture with each channel utilizing a 2DLNS to binary LUT containing 32 words. The filter coefficient's optimal second base range is limited to only 2 bits. Since the filterbank is intended for processing speech and low-power operation, a serial implementation was selected to minimize both power and area. The design consumes 708 μW and the core size is 1 mm × 1 mm, half of which is occupied by the converter as the RALUT is built using standard cells (see Figure 84.16).

A second-generation unfabricated design [31] promises half the power (316 μW), almost half the logic cells, and one-third of the area (555 μm $\times$ 555 μm). These improvements are the result of utilizing an optimal base on the input data versus the filter coefficients, single-phase operation, single-port RAM, and a single sign-bit architecture.

84.5.4.3 Complexity Reduction

Although the two designs are fabricated in different technologies, one can see that the thirty 1024-word LUTs dominate the area of the 15-tap filter. Assuming a technology scaling factor of 2, over 200 of the second-generation 75-tap filter designs can be placed in the same area as that of the 15-tap filter. This clearly shows the benefit of using a full multiple-digit MDLNS.

84.6 Conclusions

In this chapter we have demonstrated that a multi-digit MDLNS has potential for considerable implementation advantages over the classical LNS for many DSP applications. This chapter has generalized previous studies of a DBNS, and we have illustrated some of the advantages of the extra degrees of freedom in both numbers of digits and the dimensionality of the representation. We have demonstrated, using FIR filter VLSI layouts, the huge complexity reduction obtainable by using a multiple-digit MDLNS representation.

References

 1. Koren, I., *Computer arithmetic algorithms*, Prentice-Hall, Upper Saddle River, NJ, 1993.
 2. Swartzlander, E.E. and Alexopoulos, A.G., The sign/logarithm number system, *IEEE Trans. Comput.*, 42, 1238, 1975.
 3. Taylor, F.J. et al., A 20-bit logarithmic number system processor, *IEEE Trans. Comput.*, 37, 190, 1988.
 4. Arnold, M.G. et al., Redundant logarithmic arithmetic, *IEEE Trans. Comput.*, 39(8), 1077, 1990.
 5. Muller, J.M., Scherbyna, A., and Tisserand, A., Semi-logarithmic number systems, *IEEE Trans. Comput.*, 47(2), 145, 1998.
 6. Kingsbury, N.G. and Rayner, P.J.W., Digital filtering using logarithmic arithmetic, *Electron. Lett.*, 7, 56, 1971.
 7. Lewis, D.M., 114 MFLOPS logarithmic number system arithmetic unit for DSP applications, *IEEE J. Solid-State Circuits*, 30, 1547, 1995.
 8. Lewis, D.M., Interleaved memory function interpolators with application to an accurate LNS arithmetic unit, *IEEE Trans. Comput.*, 43(8), 974, 1994.
 9. Coleman, J.N. et al., Arithmetic on the European logarithmic microprocessor, *IEEE Trans. Comput.*, 49(7), 702, 2000.
10. Lewis, D.M., An architecture for addition and subtraction of long word length numbers in the logarithmic number system, *IEEE Trans. Comput.*, 39(11), 1325, 1990.
11. Lewis, D.M., Complex logarithmic number system arithmetic using high-radix redundant CORDIC algorithms, *Proceedings of the 14th IEEE Symposium of Computer Arithmetic*, 194, 1999.
12. Eskritt, J. et al., A 2-digit DBNS filter architecture, *IEEE Workshop on Signal Processing*, 447, 2000.
13. Dimitrov, V.S. et al., A near canonic double-base number system with applications in DSP, SPIE Conference on Signal-Processing Algorithms, 2846, 14, 1996.
14. Dimitrov, V.S., Jullien, G.A., and Miller, W.C., Theory and applications of the double-base number system, *IEEE Trans. Comput.*, 48(10), 1098, 1999.
15. Sadeghi-Emamchaie, S. et al., Digital arithmetic using cellular neural networks, *IEEE. J. Circuits, Systems and Comput.*, 6(8), 515, 1998.
16. Jullien, G.A. et al., A hybrid DBNS processor for DSP computation, *Proceedings of the 1999 IEEE International Symposium on Circuits and Systems*, 1, 5, 1999.

17. Dimitrov, V.S., Jullien, G.A., and Miller, W.C., An algorithm for modular exponentiation, *Info. Process. Lett.*, 36(5), 155, 1998.

18. Muscedere, R., et al., Non-linear signal processing using index calculus DBNS arithmetic, *Advanced Signal Processing Algorithms, Architectures, and Implementations X*, F.T. Luk, Ed., The International Society for Optical Engineering, 2000, 247.

19. Muscedere, R., et al., On efficient techniques for difficult operations in one- and two-digit DBNS index calculus, *34th Asilomar Conference on Sig., Syst. and Comp.*, M.B. Mathews, Ed., 2000, 870.

20. Eskritt, J., Inner product computational architectures using the double base number system, M.A.Sc. Thesis, University of Windsor, 2001.

21. deWeger, B.M.M., *Algorithms for Diophantine Equations, CWI Tracts* 65, Amsterdam, 1989.

22. Muscedere, R., Difficult operations in the multi-dimensional logarithmic number system, Ph.D. Thesis, University of Windsor, 2003.

23. Bennett, M.A. On some exponential equation of S.S.Pillai, *Can. J. Math.*, 53(5), 897, 2001.

24. Delone, B.N. and Fadeev, D.K., *The theory of irrationalities of the third degree, Translation of Mathematical Monographs*, no. 10, American Mathematical Society, 1964.

25. Stroeker, R.J. and Tijdeman, R., Diophantine equations, in computational methods in number theory (*part II*), *CWI Tracts* 155, 321, 1987.

26. Ellison, W.J., On a theorem of Sivasankaranayana, Seminars on number theory, Technical Report No. 12, CNRS, Talene, 1971.

27. Tijdeman, R., On the maximal distance between integers composed of small primes, *Copositio Mathmeticae*, 28, 159, 1974.

28. Waldshmidt, M., Linear independence measures for logarithms of algebraic numbers, *International Mathematical Year*, Cetraro, 24, 2000.

29. Muscedere, R. et al., Efficient techniques for binary to multi-digit multi-dimensional logarithmic number system conversion using range addressable look-up tables, *IEEE Trans. Comput., Special Issue on Computer Arithmetic*, 54(3), 257–271, 2005.

30. Li, H. et al., The application of 2-D logarithms to low-power hearing-aid processor, *45th IEEE International Midwest Symposium on Circuits and Systems*, August 4–7, Tulsa, Oklahoma, vol. 3, 13–16, 2002.

31. Muscedere, R. et al., A low-power 2-digit multi-dimensional logarithmic number system filterbank architecture for a digital hearing aid, *EURASIP J. Appl. Signal Process., Special Issue on DSP in Hearing Aids and Cochlear Implants*, 18, 3015, 2005.

Section XIII

Design Languages

Zainalabedin Navabi
University of Tehran

0-8493-XXXX-X/04/$0.00+$1.50
© 2006 by CRC Press LLC

85

Languages for Design and Implementation of Hardware

Zainalabedin Navabi

Nanoelectronics Center of Excellence
School of Electrical and Computer Engineering
University of Tehran

CONTENTS

As the size and complexity of digital systems increase, more computer-aided design (CAD) tools are introduced into the hardware design process. Early simulation and primitive hardware generation tools have given way to sophisticated design entry, validation, high-level synthesis, formal verification, automatic hardware and layout generation, automatic test pattern generation, and various design and test utilities. Growth of design automation tools is largely due to languages for description and modeling hardware and design methodologies that are based on these languages. Languages in hardware design are used for description of hardware at various levels of abstraction as well as description of hardware for interfacing various design tools and equipment.

The category of hardware-related languages that are used for describing hardware are called hardware description languages (HDLs). On the basis of HDLs, new digital system CAD tools have been developed and are now widely utilized by hardware designers. At the same time new languages and formats for porting a hardware description from one tool to another and from tools to test and manufacturing equipment are designed. Work on languages for higher level design and easier and better verification and test of designs continues. In today's fast changing VLSI technology, the knowledge of hardware languages for VLSI designers is a necessity.

This chapter presents an overview of a modern hardware design process and based on this, discusses the languages needed for each stage of the design process. We discuss steps involved in taking a high-level hardware–software description from a system description to its implementation in hardware. Processes and terminologies are illustrated here.

85.1 System-Level Design Flow

For the design of a digital system, the design flow begins with specification of the system at a high level of abstraction and ends with files representing test data, layouts, and the final hardware in the form of a chip or a board. In this process, languages are used for data representation, design entry, or tool control. Figure 85.1 shows a high-level design flow and languages that are used in each step of the design.

85.1.1 System-Level Languages

At the highest level of abstraction, a design is entered as an interaction of processes and tasks. A designer at this level does not have to worry about what processes are done in hardware and what is done in software, rather, a designer studies his or her system from the point of view of tasks that have to be performed by the complete design and how these tasks interact with each other. Languages used for this purpose are UML, C, and C++. Chapter 86 of this handbook covers these languages. Since C and C++ are software languages that are too extensive to be covered as part of a chapter, our presentation in Chapter 86 will merely focus of terminologies needed for understanding other hardware languages that are derived from C and C++.

After the complete high-level analysis of a design, a designer's attention goes to partitioning the system description into hardware and software parts and defining communications between them. Communications describe how data are exchanged between various system components regardless of them being hardware or software components. Languages used for description of software parts of a system include C and C++. Although many other languages can effectively be used for this purpose, the strong link between C++ and its hardware-oriented derivates make this language a very popular language in hardware/software codesign environments.

After system partitioning and when lines are drawn between parts that are to be implemented in hardware and those that are to be implemented in software, hardware parts are described in transaction level modeling (TLM), in C++ for the behavioral components, or in standard register transfer level (RTL) hardware languages. SystemC is a library of functions based on C++. Because this language has strong ties with C++ it is often used for description of hardware in C++. TLM is a library that is written on top of SystemC and is used for description of systems that involve complex mechanisms for data communications. Chapter 86 discusses SystemC and TLM.

85.1.2 RTL Languages

Realization of a hardware described at the high level translates to extracting its RTL description from its high-level description. This process can be done manually, or with new electronic system level (ESL) tools it can be done semiautomatically. After completing this phase of design, translated hardware parts become available in VHDL, Verilog, RTL SystemC, and System Verilog. These languages are discussed in Chapters 87–90 of this handbook.

Like System Verilog, SystemC is considered both as a system-level design language and a powerful RTL language. SystemC covered in Chapter 86 looks at this language from a system point of view, while the presentation of Chapter 89 of SystemC merely focuses on its RTL features.

85.1.3 Description of Analog Parts

Often, a complete system involves several analog parts, and these components must participate in the simulation of the complete system if the simulation is to give a realistic prediction of what happens between analog and digital parts. The same way that SystemC is popular because it has links with its higher-level C++ language, a good analog description language must have links with hardware description languages that are used for the digital parts of a hardware description.

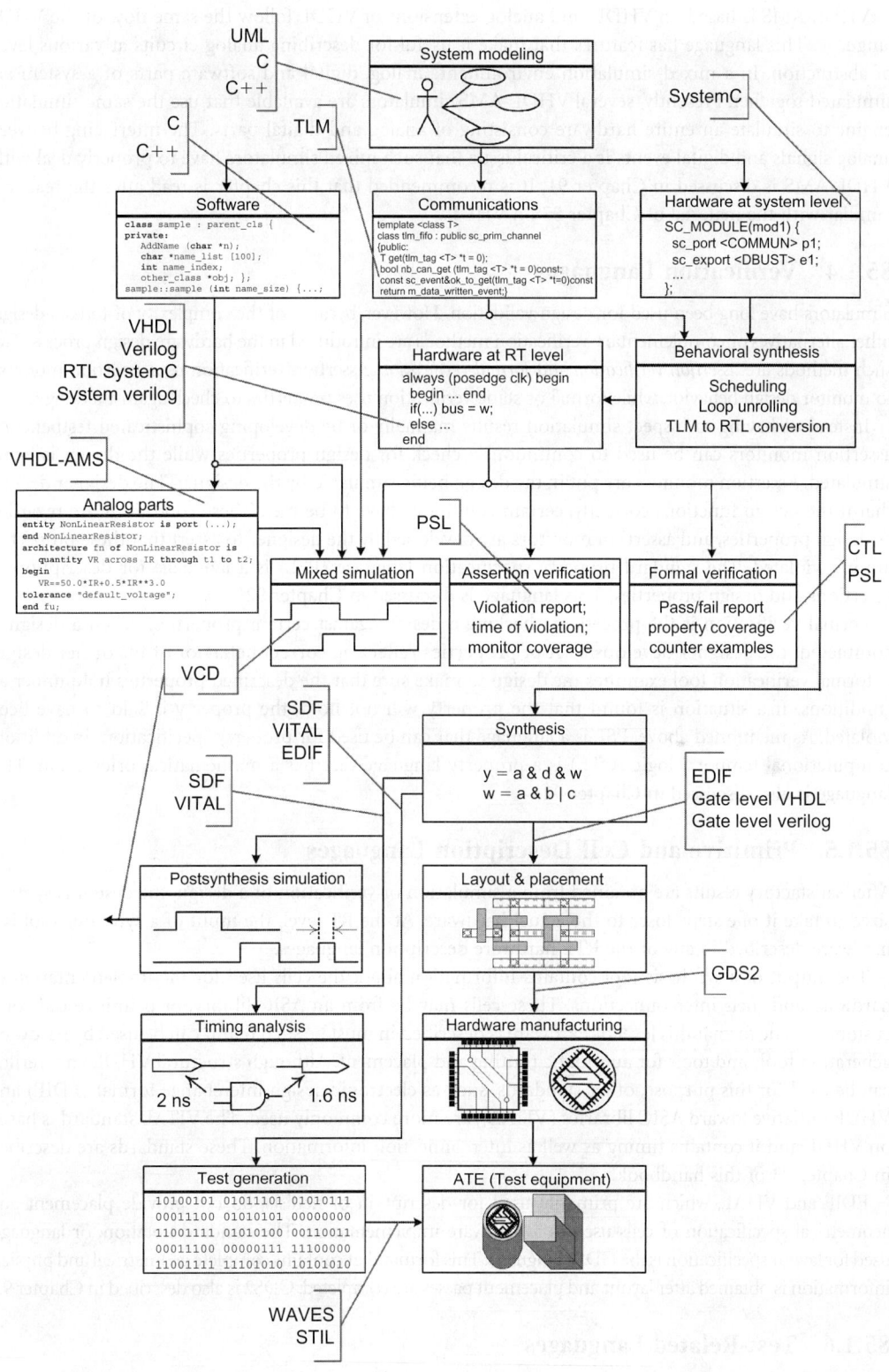

FIGURE 85.1   The design flow using HDLs.

VHDL-AMS is based on VHDL, and analog extensions of VHDL follow the same flow of the VHDL language. This language has features that make it useful for describing analog circuits at various levels of abstraction. In a mixed-simulation environment, analog, digital and software parts of a system are simulated together. Presently, several VHDL-AMS simulators are available that use the same simulation engine to simulate an entire hardware consisting of analog and digital parts. The interfacing between analog signals and digital events is a critical issue that such mixed simulators have to properly deal with. VHDL-AMS is discussed in Chapter 91. It is recommended that this chapter is read after the reader is familiar with the material of Chapter 87 on VHDL.

85.1.4   Verification Languages

Simulators have long been used for design validation. However, because of the complexity of today's design, other alternatives or complementary verification methods are introduced in the hardware design process. Two such methods are *assertion verification* and *formal verification*. Assertion verification uses assertion monitors to monitor design behavior, while formal or static verification uses properties to check against a design.

Instead of having to inspect simulation results manually or by developing sophisticated testbenches, assertion monitors can be used to continuously check for design properties while the design is being simulated. Assertion monitors are put in the design being simulated by the designer. The designer decides that if the design functions correctly, certain conditions have to be met. These conditions are regarded as design properties, and assertion monitors are developed by the designer to assert that these properties are not violated. The standard property specification language (PSL) is a language for description of assertions and design properties. This language is discussed in Chapter 92.

Formal verification is the process of checking a design against certain properties. When a design is completed, the designer develops a set of properties reflecting correct behavior of his or her design. A formal verification tool examines the design to make sure that the described properties hold under all conditions. If a situation is found that the property will not hold, the property is said to have been violated. As mentioned above, PSL is a language that can be used for property specification. In addition, computational temporal logic (CTL) is a property language that has a mathematical orientation. This language is also discussed in Chapter 92.

85.1.5   Primitive and Cell Description Languages

After satisfactory results are obtained from a simulation or verification of a design, the design is synthesized to take it one step closer to the actual hardware. At the RT level, the input of a synthesis tool is a hardware described in any of the RTL hardware description languages.

The output of a synthesis tool contains information about the cells used for the implementation of hardware and their interconnections. These cells may be from an ASIC library, or primitive cells of a custom IC. The format this level of hardware is described in must be such that it can be used by hardware generation tools and tools for automatic routing and placement. Although structural VHDL and Verilog can be used for this purpose, other standards, such as electronic design interchange format (EDIF) and VHDL initiative toward ASIC libraries (VITAL), are more commonly used. The VITAL standard is based on VHDL and it contains timing as well as interconnection information. These standards are described in Chapter 93 of this handbook.

EDIF and VITAL, which are primarily used for description of netlists, do not provide placement and geometrical specification of cells used in a hardware implementation. The standard notation, or language, used for layout specification is the GDS2 language. This format that contains complete geometrical and physical information is obtained after layout and placement passes are completed. GDS2 is also described in Chapter 93.

85.1.6   Test-Related Languages

Netlists generated by synthesis tools represent gate level details of hardware being designed. This level of abstraction contains proper information for test generation and fault simulation programs. An automatic test pattern generation (ATPG) program uses gate level information to generate data for testing the hardware after it is built. A fault simulation program uses test patterns generated by an ATPG to simulate the hardware for possible physical faults. In contrast, data generated by a postsynthesis simulation run can also be used for modeling the correct behavior of a manufactured circuit.

Waveform and vector exchange specification (WAVES) and standard test interface language (STIL) are used for test data representation. These languages that are described in Chapter 94 provide mechanisms for specification of stimuli and timing of data as well as data and timing of expected outputs. Automatic test equipment (ATE) machines use these formats to apply data to a circuit that is being tested and to analyze its responses. In addition, the value change dump (VCD) format (part of Chapter 95) that will be mentioned next is also used as an input format for ATE.

85.1.7 Timing Specification Languages

In addition to netlists, synthesis tools also generate formats that can be used for postsynthesis simulators and timing analyzers. Such formats describe timing of cells and interconnections used in netlists generated by a synthesis tool. One such format is the standard delay format (SDF) that is primarily used for representation of timing of cells used in a netlist. An HDL simulator uses an SDF file for reading timing of the cells used in the hardware structure being simulated. SDF is used by both VHDL and Verilog simulation engines. This format is described in Chapter 95.

Another form of timing representation is used in VCD. VCD that is discussed in Chapter 95 is generated by HDL simulators and is the textual form of waveforms. This makes it a standard way of representing input and output waveforms that are used for graphical waveform displays or for ATE, as discussed in the previous section.

85.2   Design Tools

Electronic design automation (EDA) tools use the various languages mentioned above for design validation, synthesis, test, and manufacturing of hardware. In the last chapter of Section 13 of this handbook, we briefly describe some of the existing tools that are used in design automation of hardware. This subject, however, needs a more thorough treatment that is not the subject of Section 13 of this handbook.

85.3   Summary

We have briefly described languages that are used in design automation of electronic digital systems. We have shown the role of each language and some environments and applications that they are used in. The chapters that follow discuss each language at a getting-started level. Many of the chapters are on design languages. The discussions in these chapters are detailed enough that after completing them, readers can start writing codes in the languages that are discussed. Other chapters discuss languages that are used between tools or equipment for exchange of information. Chapters on these languages provide basic knowledge of the language they are discussing enough for being able to read and understand existing codes.

86

System Level Design Languages

Shahrzad Mirkhani and
Zainalabedin Navabi
*Nanoelectronics Center of Excellence
School of Electrical and Computer Engineering
University of Tehran*

CONTENTS

Today, the complexity of designs has increased. This implies that the computational parts of the designs have become larger, and more important, communication between various design modules have become more complex. This complexity makes these designs harder to test and verify.

In contrast, most of the designs have software parts and hardware modules. In the traditional design methodologies, a software team has to postpone software tests to the time when the hardware models have been mapped to an actual hardware. This sequence delays the whole design and development process.

The need to have shorter time-to-market leads to the fact that designers should parallelize hardware and software development processes. This concept directs them to hardware/software codesign, coverification, and cosimulation.

Partitioning is another issue which becomes more critical when designs grow more complex. As the design complexity increases, the effect of module partitioning or hardware/software partitioning on intra-module communications becomes more important. If partitioning is not done properly, communication between modules becomes a bottleneck. Therefore, partitioning in a design plays a very important role in performance, speed, power dissipation, and area requirement of the final hardware.

The problems stated above (hardware test and verification, hardware/software codesign, and partitioning) can be eased by electronic system level (ESL) design. In recent years, designers have moved their design methodology to system level. The heart of the ESL design methodology is a high-level language that is used for specification of a complete system. This language can be C/C++, Java, SystemC, and perhaps other high-level programming languages. There are other languages known as system level design languages (SLDL), like Rosetta, for developing a design at the system level, which are now in their development stages. At the present time, most designers use C/C++ and its application-specific libraries (like SystemC, Handel-C, and ArchC) as their system level design language.

ESL Methodology—The point that makes system level design a more powerful methodology than register transfer level (RTL) is that in ESL, designers are not concerned with design details. Since most modules are described at a high level of abstraction, a designer has an overall view of the complete system. Therefore, the designer can decide on problems like partitioning, system functionality, and developing testbenches in a more organized fashion. Other problems like the exact delay of each part, signal glitches, and other problems, which need the details of a design can be solved during the conversion process from system level to RT level or in the RT (or even gate) level design.

Figure 86.1 shows the necessary steps in a system level design. It is important to note that the ESL design methodology in Figure 86.1 is not the only proposed methodology for ESL. A design team can have its own customized design methodology and the methodology in Figure 86.1 is only shown to clarify the ESL concept.

In Figure 86.1, most of the steps like *system partitioning* and *converting to RTL* could be performed automatically. However, the lack of ESL tools forces designers to perform such steps manually or semi-manually. Although many ESL tools have become available to the System on Chip (SoC) market, they do not completely support the ESL concepts. The reason is that an ESL design is a new concept and many of its aspects are still in the study phase (e.g., analyzing the design for partitioning and complete behavioral synthesis).

The use of several high-level languages, like UML 2.0, C/C++, and SystemC have become more common than other programming and hardware languages for description of an ESL system. Although SystemC is a library developed in C++ and it is not a separate language, it is generally regarded as a language by the hardware design community.

In what follows, we will describe UML 2.0, C/C++, and SystemC as ESL design languages in separate sections. In the final section of this chapter, an important concept in ESL known as Transaction Level Modeling (TLM) will be discussed. As stated above, the complexity of today's designs results in the complexity of the communication between various parts of a system. Using TLM, these communications can be modeled in higher levels of abstraction which will result in a faster simulation process and an easier verification.

86.1 Unified Modeling Language

As stated above, the necessity of having high-level languages for modeling large systems has become very important in the recent years. The complexity of software systems grows faster than hardware designs. Therefore, the need for a standard methodology for system design in software engineering became evident sooner than that of hardware designs.

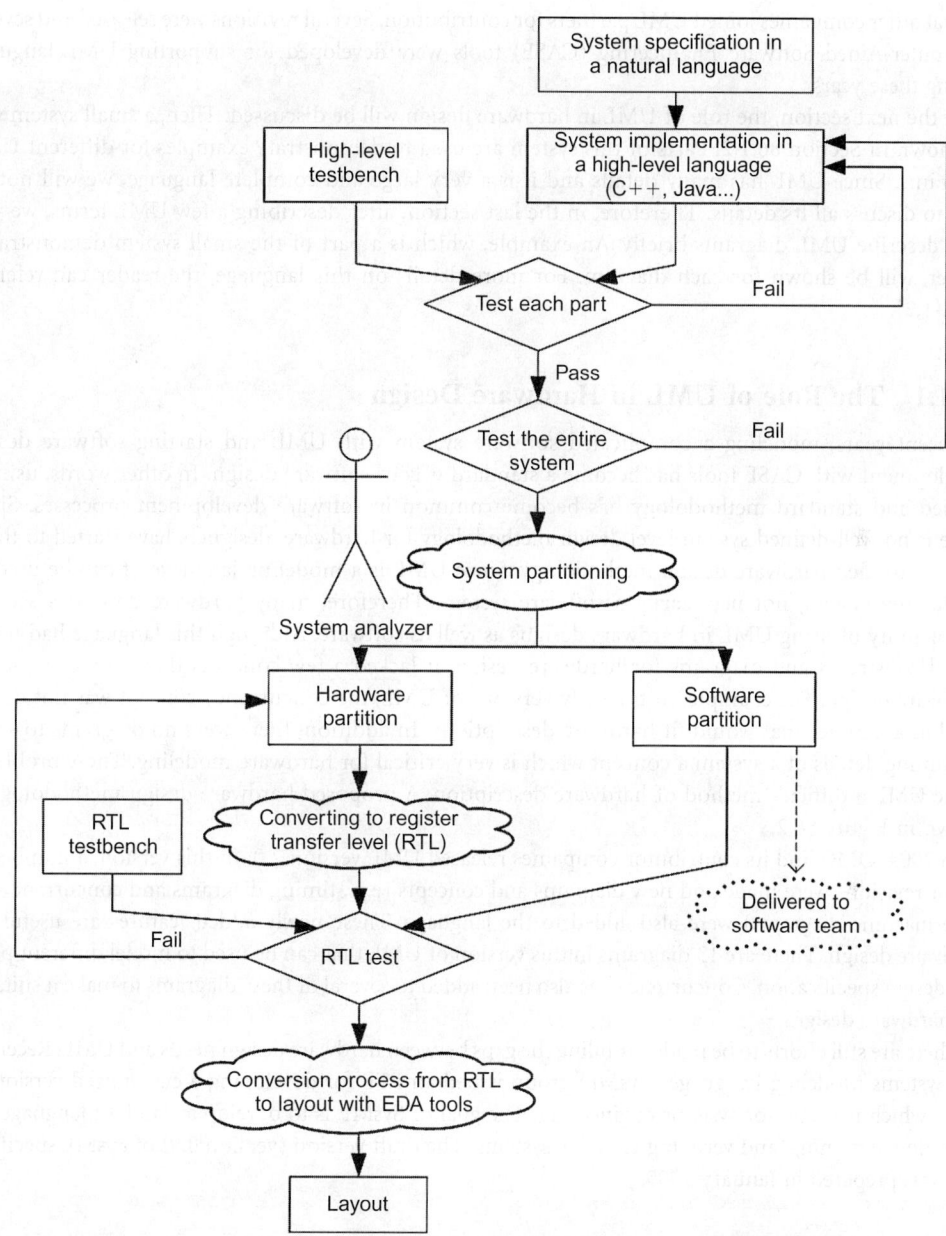

FIGURE 86.1 System level design flow.

In late 1994, Booch and Rumbach began the study of unifying Booch and object modeling technique (OMT) methods. This led to the generation of a graphical modeling language called Unified Modeling Language (UML). UML 0.9 and 0.91 were released in mid-1996. After receiving several feedbacks, a Request For Proposal (RFP) was issued by Object Management Group (OMG). As a result, organizations and companies produced a joint RFP response which led to the definition of UML 1.0. The companies which contributed most to the definition of UML 1.0 include Digital Equipment Corp., HP, i-Logix, IntelliCorp, IBM, ICON Computing, MCI SystemHouse, Microsoft, Oracle, Rational Software, Texas Instruments, and Unisys. UML 1.0, which was a well-defined and powerful modeling language, was submitted to OMG in 1997 as an initial RFP response. During the following years, these companies and several other companies joined UML partners for contribution. Several revisions were released and several Computer-Aided Software Engineering (CASE) tools were developed for supporting UML language during these years.

In the next section, the role of UML in hardware design will be discussed. Then, a small system will be shown in Section 86.1.2. Parts of this system are used to demonstrate examples for different UML diagrams. Since UML has many details and it is a very large and complete language, we will not be able to discuss all its details. Therefore, in the last section, after describing a few UML terms, we will only describe UML diagrams briefly. An example, which is a part of the small system demonstrated earlier, will be shown for each diagram. For more details on this language, the reader can refer to Refs.[1,2].

86.1.1   The Role of UML in Hardware Design

In recent years, modeling a complicated software system with UML and starting software design development with CASE tools has become a standard way of software design. In other words, using a unified and standard methodology has become common in software development processes. Since there is no well-defined system level design methodology for hardware, designers have started to think about a unified hardware design methodology. Since UML is a modeling language, it can be used to model any system, not necessarily a software system. Therefore, many hardware designers see the opportunity of using UML in hardware designs as well as software. Although this language had many useful constructs and diagrams for hardware design, it lacked a few concepts that were essential in hardware design. For example, in the early versions of UML, the concurrency concept was not developed in a manner that would fit hardware descriptions. In addition, there were no diagrams to show the timing details of a system a concept which is very critical for hardware modeling. These problems made UML a difficult method of hardware description. A proposed hardware design methodology is shown in Figure 86.2.

In 2004, OMG and its contributor companies released UML version 2.0. In this version, a number of major revisions were made and new diagrams and concepts (e.g., timing diagrams and concurrency for state machine diagrams) were also added to the language. These newly added features are useful for hardware design. There are 13 diagrams in this version of UML that can be used to model different parts of a design specification. Concurrency has also been added to several of these diagrams to make it suitable for hardware designs.

There are still efforts to be made for filling the gaps between hardware design needs and UML. Recently, the Systems Modeling Language (*SysML*) group issued an RFP for developing a customized version of UML which is useful for systems engineering. The goal of SysML is to develop a standard language for analyzing, designing, and verifying complex systems. The draft version (version 0.9) of SysML specification was prepared in January 2005.

86.1.2   An Example: SSimBoard

In this chapter, we will describe a simple system, used as an example to describe UML diagrams. In the hardware design environment, the increase in the size of designs and the number of design inputs cause the simulation process to become very time-consuming. Our example system is an accelerator board for improving speed of circuit model simulation (SPICE simulation).

Figure 86.3 shows a view of this system. As shown in this figure, this system consists of two main parts: a master and a slave. The master part in this system can be a general PC. The user, who needs to simulate a SPICE design, interacts with the master via a Graphic User Interface (GUI). The slave, in contrast, is a board including a CPU and a memory module. SPICE simulation task is performed by the slave via a SPICE simulation program that is installed on its CPU. The interconnection between master and slave is performed via a parallel port. Therefore, there are drivers for parallel ports in both master and slave sides.

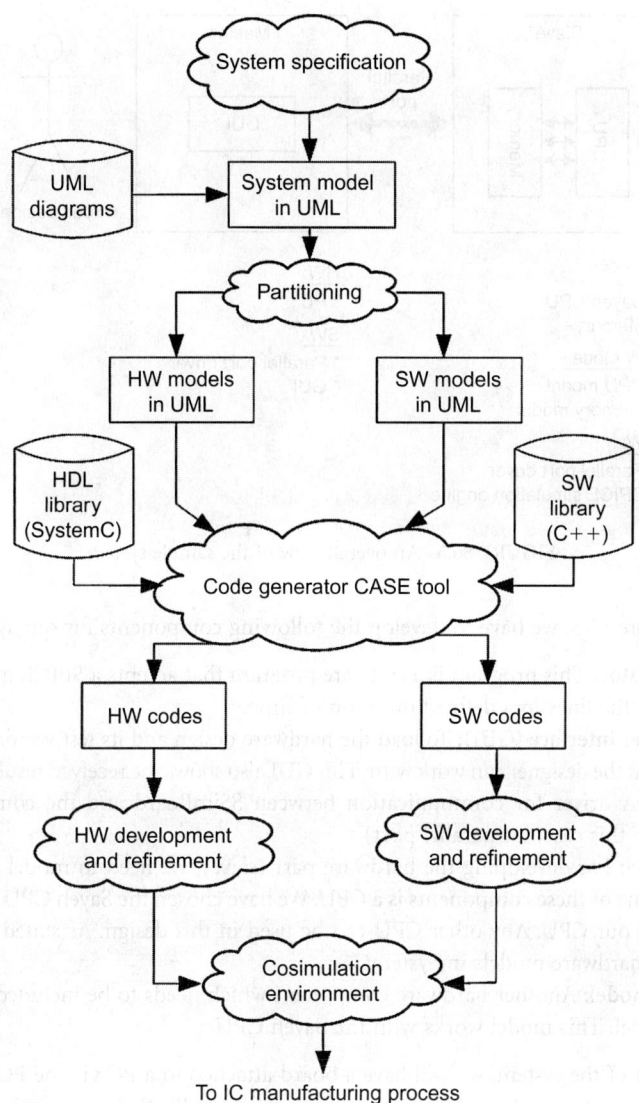

FIGURE 86.2 Hardware design methodology in UML.

The slave board, which we will refer to as *SSimBoard*, attaches to the computer system handling the simulation via a parallel port. The simulator program and necessary drivers are placed in the board's memory.

The designer sends his or her design file and the design input vectors to the slave board via the parallel port from the GUI in the master side. Then the simulation process starts by running the simulator program on the SSimBoard's processor. Since this processor has no other task to perform, it helps the designer to perform faster simulation runs. If the circuit to be simulated is very large, or if there are a large number of input stimuli, this board can either only simulate a part of the circuit or simulate the circuit for only a portion of input stimuli. The partitioning strategy must be defined by the master side via GUI.

We use C++ for the design of software parts (GUI and driver in the master side, and SPICE simulator and driver in the slave side), and SystemC for its hardware parts (CPU and memory modules in the slave side).

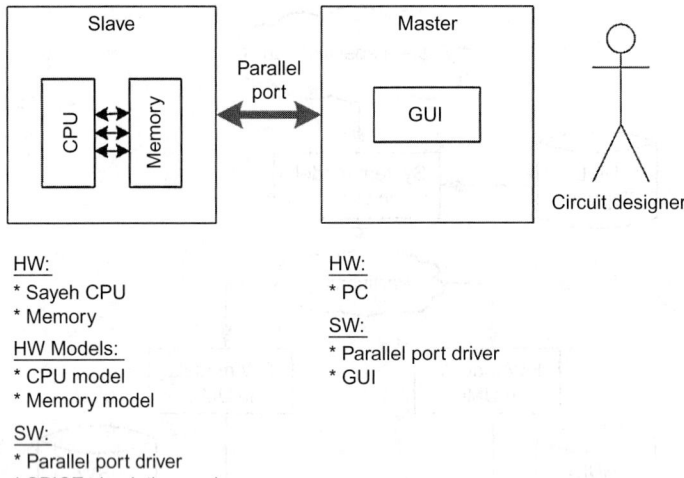

FIGURE 86.3   An overall view of the sample system.

As shown in Figure 86.3, we have to develop the following components for our system:

- **SPICE simulator:** This program is a software program that accepts a SPICE model and simulates the values on the lines in a defined duration of time.
- **Graphical user interface (GUI):** To load the hardware design and its test vectors, there needs to be a program that the designer can work with. This GUI also shows the received results of the simulation.
- **Port driver:** A driver for communication between SSimBoard and the computer needs to be developed (in this case, for parallel port).
- **A CPU model:** For developing the hardware part (slave), we need to model the components of this board. One of these components is a CPU. We have chosen the Sayeh CPU (a simple academic processor) as our CPU. Any other CPU can be used in this design. As stated earlier, we want to develop our hardware models in SystemC.
- **A memory model:** Another hardware component which needs to be included in SSimBoard is a memory model. This model works with the Sayeh CPU.

As the final result of the system, we will have a board attached to a PC via the PC's parallel port and a SPICE simulator is loaded on it, and a software program on the PC that communicates with this board in a user-friendly manner. The scenario a user has to follow to work with this system is listed in the following steps and is shown in Figure 86.4.

1. The hardware designer (who is the user of this system) describes his or her circuit in SPICE.
2. The designer develops test data for his or her SPICE model.
3. The designer connects SSimBoard to the computer and switches on the board.
4. The designer runs the GUI and sends a specified design and input vector files through this GUI to SSimBoard.
5. The designer starts the simulation process by issuing the start SPICE simulate command through the GUI.
6. The SPICE simulation process runs on the SSimBoard CPU. During this process, the user can perform other programs on the PC.
7. Completion of the SPICE simulation is announced by SSimBoard to the PC via the parallel port connecting the two systems. The user becomes aware of this through the GUI.
8. The user can receive the simulation result through the GUI.
9. The GUI will show the simulation result in a graphical mode.

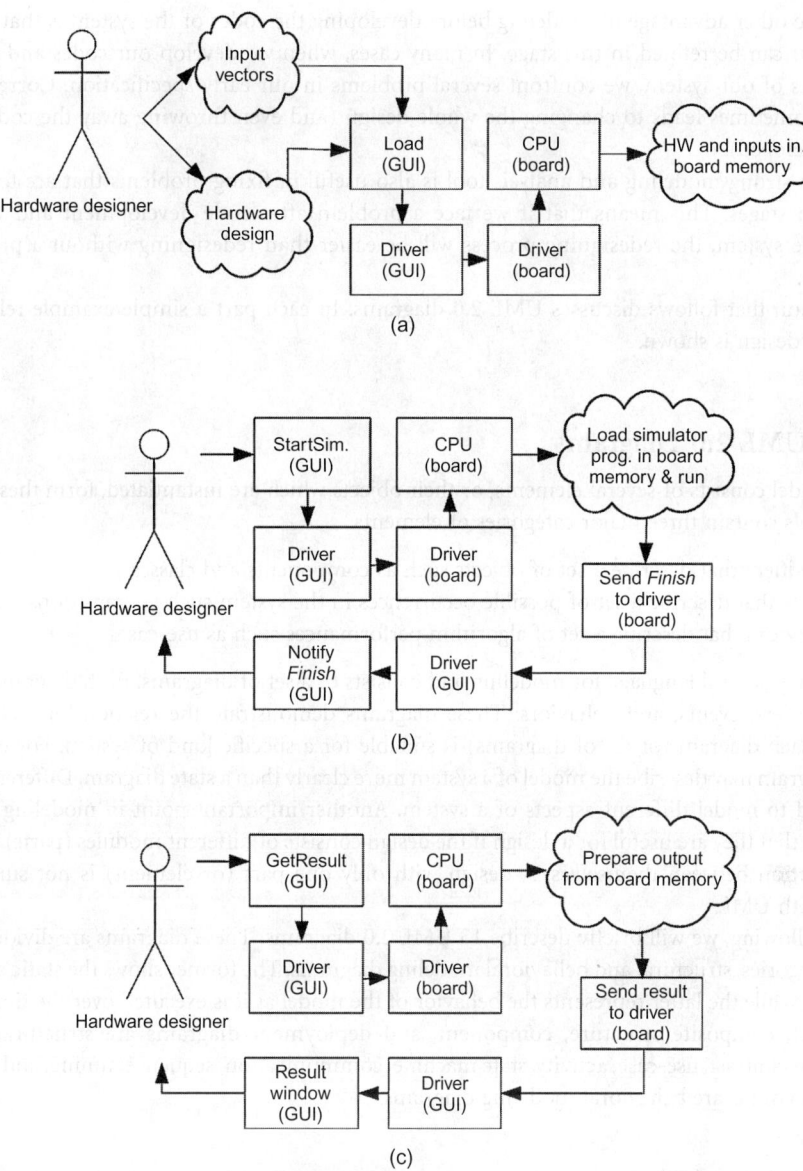

FIGURE 86.4 SSimBoard's scenario of use: (a) loading a design; (b) starting simulation; and (c) getting the results.

To design this system, we implement the necessary software components of SSimBoard and PC in C++, design the hardware of SSimBoard CPU and memory in SystemC, and cosimulate these two parts in a C++ environment. On the SSimBoard side the software components are its parallel port driver and a SPICE simulator; and on the PC side, a parallel port driver and GUI constitute its software components. After coverification of hardware and software parts, we can follow the hardware manufacturing steps (synthesis, postsynthesis simulation, layout generation, manufacturing the ICs, PCB generation, etc.) to create our SSimBoard.

The design process we are pursuing is shown in Figure 86.2. In this process, we model our system in UML and analyze different parts of the system with UML diagrams. With a suitable CASE tool, a code generator tool in this case, we automatically generate the structure of the necessary codes from our models. The other advantage of modeling before developing the codes of the system is that the system specification can be refined in this stage. In many cases, when we develop our codes and implement several parts of our system, we confront several problems in our early specification. Correcting these problems sometimes leads to changing the whole design (and even throwing away the codes we have already developed).

Having a strong modeling and analysis tool is also useful in fixing problems that are found in the later design stages. This means that if we face a problem after code development and we need to redesign the system, the redesigning process will be easier than redesigning without a proper set of CASE tools.

The section that follows discusses UML 2.0 diagrams. In each part a simple example related to the SSimBoard design is shown.

86.1.3   UML 2.0 Diagrams

A UML model consists of several elements, or their objects which are instantiated, form these elements. UML models contain three major categories of elements.

- **Classifiers** that describe a set of objects such as components and classes
- **Events** that describe a set of possible occurrences in the system such as transitions
- **Behaviors** that describe a set of algorithm performances such as use-cases

UML is a graphical language for modeling and consists of a set of diagrams. Each diagram includes a set of classifiers, events, and behaviors. These diagrams demonstrate the relationship between these elements. Each diagram (or set of diagrams) is suitable for a specific kind of system. For example, an activity diagram may describe the model of a system more clearly than a state diagram. Different diagrams can be used to model different aspects of a system. Another important point in modeling with UML diagrams is that they are useful for a design if the design consists of different modules (parts) which have communication between themselves. A design with only one part (or element) is not suitable to be modeled with UML.

In the following, we will briefly describe 13 UML 2.0 diagrams. These diagrams are divided into two general categories, structural and behavioral modeling diagrams. The former shows the static architecture of a model, while the latter represents the behavior of the model as it is executed over the time. Package, class, object, composite structure, component, and deployment diagrams are structural modeling diagrams. In contrast, use-case, activity, state machine, communication, sequence, timing, and interaction overview diagrams are behavioral modeling diagrams.

86.1.3.1   Use-Case Diagram

The use-case diagram shows the system requirements from a user's point of view. It gives an overall view of the system's functionality. This diagram consists of two elements: actors and use-cases.

An **actor** is an external entity for the system. It can either be a person (the user of the system, the operator, or the system administrator) or another system. In a use-case diagram, an actor can either be represented as a stick-man figure with its name above or under it, or as a rectangle with the ⟨⟨**actor**⟩⟩ stereotype as shown in Figure 86.5. Actors can have generalizations; this implies that an actor can be a special case of a more general actor.

A **use-case** is a simple unit of a meaningful task in a system. In our SSimBoard system, loading a circuit model into the board can be considered as a use-case. A use-case that has a name and a description can also have other use-cases inside it.

A use-case has a functional specification and usually the flow of events in textual mode. It can also have constraints which include preconditions, postconditions, and invariant conditions. Preconditions, like that of *StartSimulation* in Figure 86.6, are the conditions that must be true before entering

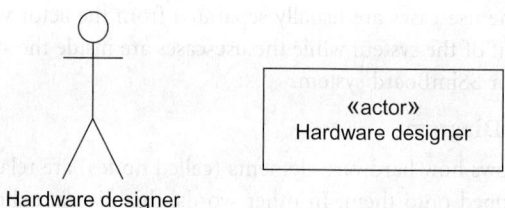

FIGURE 86.5 An actor.

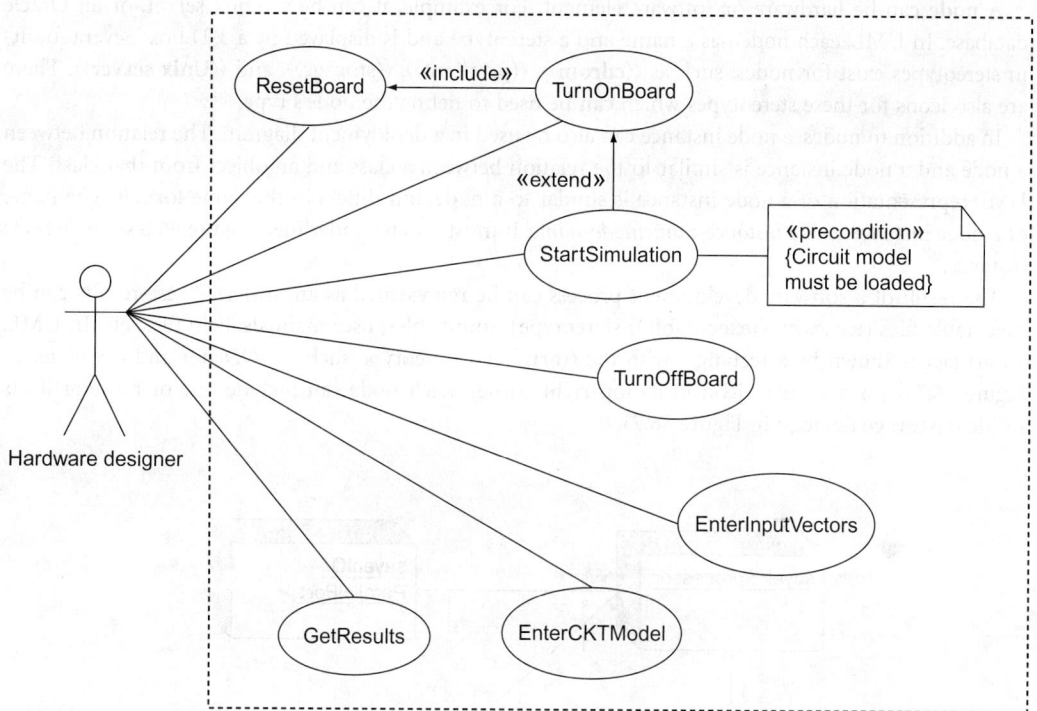

FIGURE 86.6 A use-case diagram.

a use-case, while postconditions must be true after a use-case is executed. Invariant conditions are the conditions that must be true during all the time of a use-case execution. In our first element categorization, use-cases can be considered as behaviors. In UML, each use-case is shown by an ellipse.

The relation between an actor and a use-case or two use-cases is shown by an arrow. There are two different kinds of relationships between two use-cases:

- **extend:** the extend relationship occurs between two use-cases (from the extending use-case to the extended use-case). It means that the behavior of the extending use-case can be extended by the behavior of the extended use-case under specific conditions and points. These specific points are named extension points and are defined in the extended use-case. The conditions can be added to the arrows between these use-cases. An extending relationship is defined by the ⟨⟨**extend**⟩⟩ stereotype in the diagram.
- **include:** the include relationship occurs between two use-cases (from the including use-case to the included use-case). It means that the including use-case contains the behavior of the included use-case. The including relationship is defined by the ⟨⟨**include**⟩⟩ stereotype in the diagram.

In use-case diagrams, the use-cases are usually separated from the actor with a boundary. This is to show that the actors are out of the system while the use-cases are inside the system. Figure 86.6 shows a simple use-case diagram for SSimBoard system.

86.1.3.2 Deployment Diagram

A deployment diagram shows how hardware elements (called **nodes**) are related to each other and how software elements are mapped onto them. In other words, this diagram shows the general hardware/ software configuration used for running a system. See Figure 86.7 for an example of a deployment diagram. A deployment diagram includes nodes, node instances, artifacts, and connectors between these elements.

A **node** can be hardware or software element. For example, it can be a Linux server, or an Oracle database. In UML, each node has a name and a stereotype and is displayed by a 3-D box. Several built-in stereotypes exist for nodes, such as ⟨⟨**cdrom**⟩⟩, ⟨⟨**pc client**⟩⟩, ⟨⟨**storage**⟩⟩, and ⟨⟨**Unix server**⟩⟩. There are also icons for these stereotypes which can be used to define the node's type.

In addition to nodes, a node instance can also be used in a deployment diagram. The relation between a node and a node instance is similar to the relation between a class and an object from that class. The UML representation of a node instance is similar to a node, but differs in the name format. The name of a node instance is like *instance-name:node-name*. It must also be underlined. Figure 86.8 shows a node instance.

The result of a software development process can be represented as an *artifact*. These results can be executable files (e.g., with ⟨⟨**executable**⟩⟩ stereotype), source files, user manuals, help files, etc. In UML, an artifact is shown by a rectangle with the ⟨⟨**artifact**⟩⟩ stereotype such as *IODriver* and *Simulator* in Figure 86.7, or a document icon in its top-right corner. Each node can include one or more artifacts inside (⟨⟨**storage device**⟩⟩ in Figure 86.7).

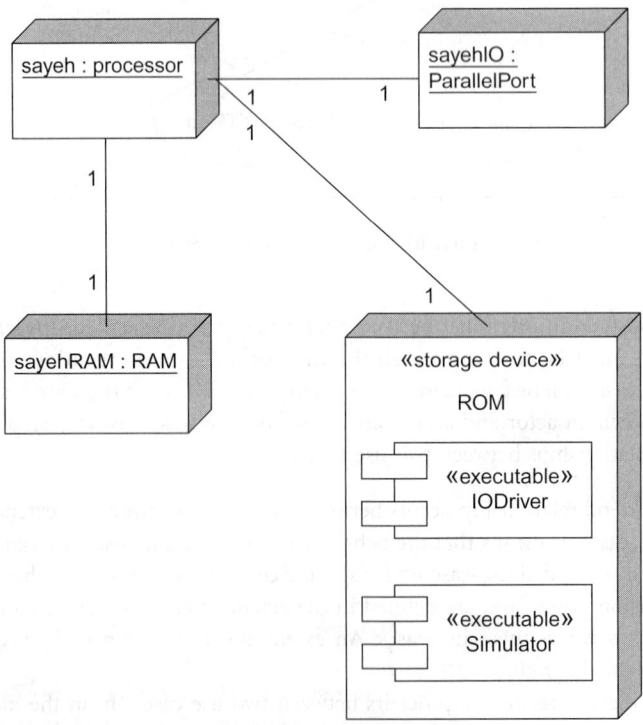

FIGURE 86.7 A deployment diagram.

System Level Design Languages

86-11

FIGURE 86.8 A node instance.

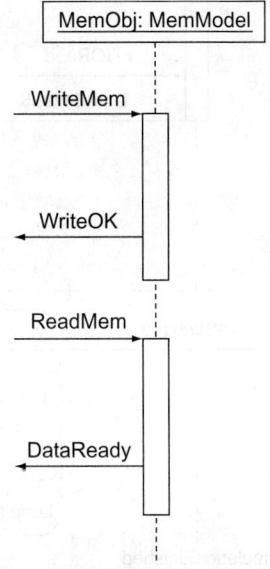

FIGURE 86.9 A sample lifeline.

A deployment diagram consists of nodes connected to each other by connectors. Each connector in this diagram can have a stereotype and a multiplicity at each end. Figure 86.7 demonstrates a deployment diagram for the SSimBoard system.

86.1.3.3 Sequence Diagram

In a system, various objects communicate with each other by sending and receiving messages. This gives a **lifeline** to an object, which begins from the time it is created to the time it receives no more messages. The lifeline of the *MemModel* object is shown in the example diagram of Figure 86.9. A sequence diagram shows the interaction between different objects in a system and depicts their lifelines. An example of a sequence diagram is shown in Figure 86.10; interaction between *BoardPanel, ParallelPort, Sayeh,* and *Memory* objects are shown in this figure. This diagram is in the category of the interaction diagrams. Other interaction diagrams include *communication, timing,* and *interaction overview* diagrams. Lifelines and messages construct a sequence diagram. The following paragraphs present more details of lifelines and messages.

As stated above, the **lifeline** of an object shows the duration in which the object sends/receives messages to/from other objects. In UML, an object lifeline is displayed with a vertical line. A thin rectangle is used in several sections of this vertical line, which is called an **execution occurrence**. This rectangle shows the times in which that object is active (working due to a message). The name of the object is displayed in a rectangle on top of the lifeline and message names are shown on horizontal lines leading to the lifeline. Time increases from top to bottom in this diagram. Figure 86.9 shows an example of a lifeline.

© 2006 by CRC Press LLC

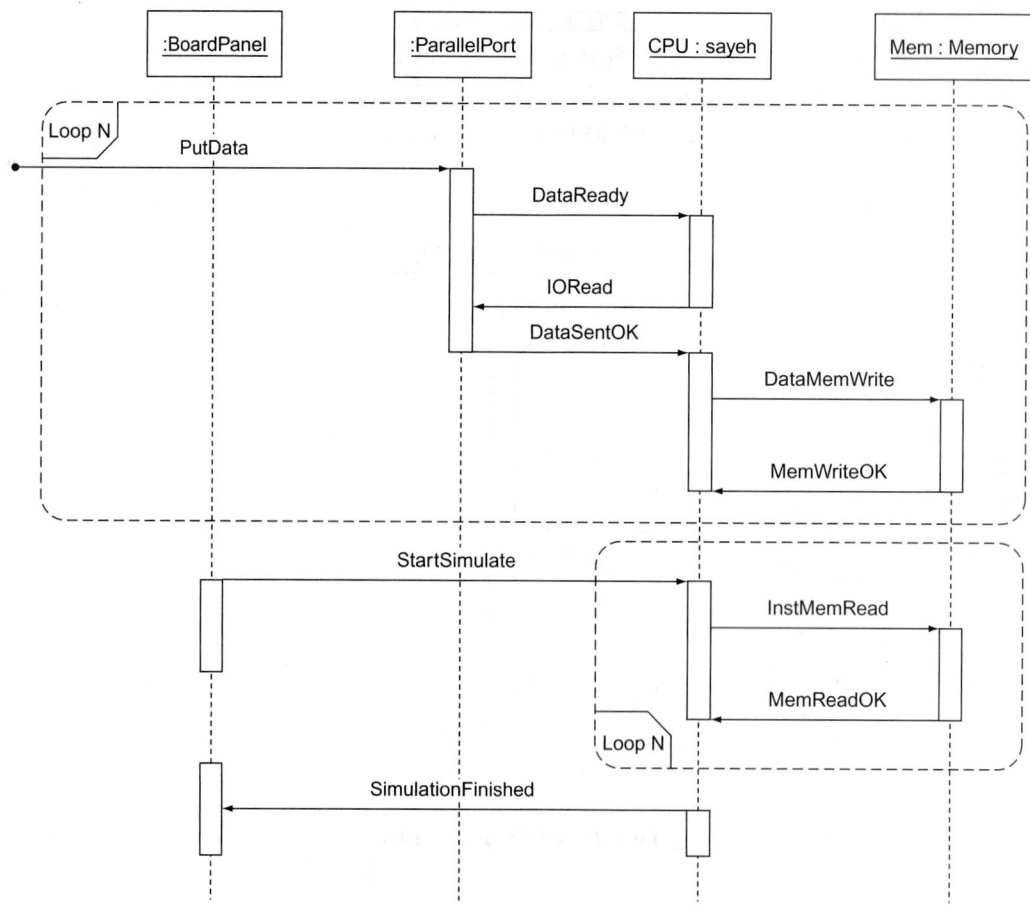

FIGURE 86.10 A sequence diagram.

A sequence diagram object can be used for elements such as an **actor**, a **boundary**, a **control**, or an **entity**. A lifeline is created and destroyed dynamically in sequence diagrams. It can be created (e.g., in C++ it can be *new*ed) or deleted by another object. Lifelines can have constraints. This means that they can have a condition that must be true at run-time for activating that object. An object can have more than one lifeline. For this purpose, an object must be partitioned into different parts, with each part having its own lifeline. Partitioning an object makes use of the intra-object message passing. The partitions are displayed by vertical lines in the object rectangle.

Messages in sequence diagrams show communications between two objects. These messages are displayed by arrows with the name (or description) of a message. Messages are of different kinds: *complete*, *lost*, or *found*. They can be synchronous or asynchronous. They can also be call or signal. A lost message is a message whose destination is either not specified in the sequence diagram or does not exist at all. Found messages are messages whose source object does not exist in the diagram. The source and destination of a message can be the same lifeline. These messages are called **self-messages**. They show the communication between different parts of an object or a recursion in an object.

In several cases, e.g., in real-time systems, it is critical for a message to have timeout, or to be processed in a certain time limit. Therefore, messages in sequence diagrams can have timing constraints. These messages are shown with sloping lines.

For more advanced and complex sequence diagrams, combined fragments can be used. A **combined fragment** groups portions of several lifelines into a fragment. A fragment uses a keyword such as **alt**,

opt, **par**, **seq**, **strict**, **neg**, **loop**, **assert**, etc. For example, a parallel fragment (denoted by **par**) models a portion of a sequence diagram, which has concurrent processing. In Figure 86.10, the areas with *loop N* on top of them are combined fragments. We can reference a sequence diagram in another sequence diagram by the **ref** keyword above the referenced diagram. The example of Figure 86.10 is a sequence diagram for a portion of the SSimBoard system.

86.1.3.4 Communication Diagram

Another category of interaction diagrams is a communication diagram. This diagram is very similar to the sequence diagrams and in the older versions it was referred to as a collaboration diagram. The difference between a communication and an interaction diagram is that in communication diagrams the main focus is on the relationship between objects and not on the sequence of the messages. The other difference between these two kinds of diagrams is that none of the structuring mechanisms in the sequence diagrams (such as combined fragments) can be found in communication diagrams. In UML, a communication diagram is represented by objects and messages.

In a communication diagram, an example of which is shown in Figure 86.11, **objects** are shown with rectangles. Between two related objects, a line is drawn. **Messages** are shown with small arrows on each line, and are identified by a number, e.g., *1a.1* in Figure 86.11 is the number for *EnterCKT* message. A sequence of messages has a numbering sequence that identifies the originating objects, and the complete path of objects receiving and replying to the message. The name of a message follows a colon that comes after its number sequence. An example message is *1a.2.1: CheckInput*. This message originates by the actor; it is message 2 for the actor, and message 1 that *GUI* has issued. The example of Figure 86.11 shows a communication diagram of the SSimBoard system.

86.1.3.5 Activity Diagram

Up to this point, we have discussed UML diagrams that deal with higher levels of system modeling. In this and the following sections, we will describe diagrams that express more details about designs. An activity diagram is one of the UML diagrams that model a system in more detail than use-case, sequence, deployment, and communication diagrams. It shows the flow of activities and the different

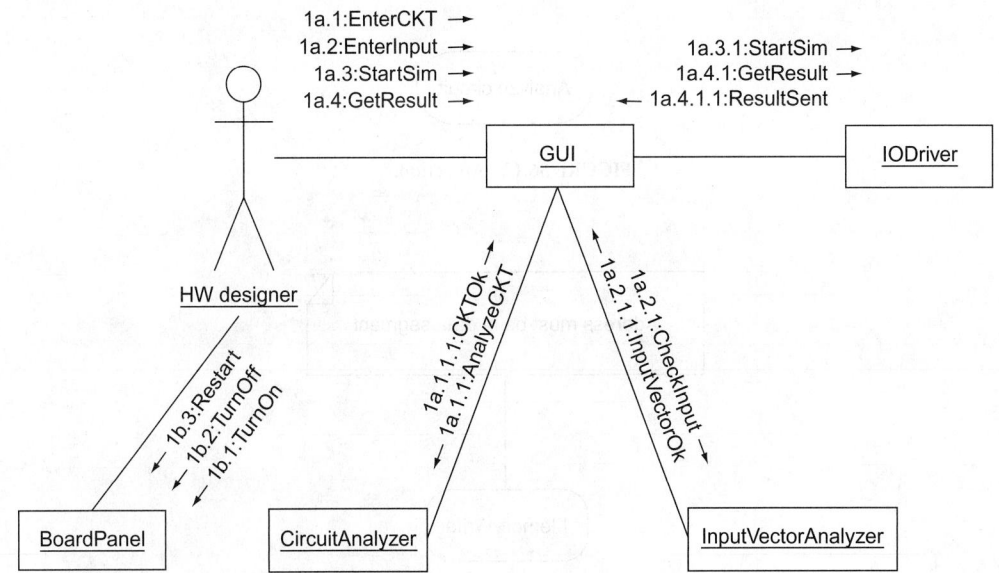

FIGURE 86.11 A communication diagram.

paths that may be passed at run-time in a system. It has a starting point and an end point. This diagram also has the capability to show that two different paths execute concurrently. The elements constructing activity diagrams include activities, actions, initial and end nodes, control flows, object and data flows, decision nodes, concurrency nodes, expansion regions, exception handlers, interruptible activity regions, and partitions. As evident from the elements mentioned, they are suitable for modeling hardware designs.

Figures 86.12–86.20 highlight various concepts used in this diagram. A complete activity diagram is shown in Figure 86.21. An activity is the specification of a sequence of behaviors. In UML, each **activity** is shown by a rounded rectangle containing the name of the activity. An activity includes several **actions**, each of which is considered a step in the activity. Figure 86.12 shows an action displayed in UML.

Constraints can be put on an action (Figure 86.13). These constraints can be preconditions describing the conditions that must be true before the flow reaches an action, or postconditions describing conditions that must be true after the flow reaches an action. For example, to write a data in a memory module, the address applied to that memory must be in a valid data segment of that memory. This condition depicted in Figure 86.13 is used as a precondition for the *MemoryWrite* action.

Activity diagrams start from a special node called the **initial node** displayed by a filled circle. An **end node** for an activity diagram (called **final node**) is shown with a circuit containing a filled circle inside. In addition to the activity final node, each flow can have its own final node. The final node of each flow is shown with a circle with a cross inside. The activity final node shows the end of all flows in an activity diagram.

An activity diagram consists of various **control** and **object flows**. The control flow is simply shown with an arrowed line between two actions. An object flow is used in an activity diagram to pass data and objects through actions. Two ways to demonstrate an object flow are demonstrated in Figure 86.14.

Decisions made in an activity diagram cause branching out to different flows. A decision, which is represented by a **decision node**, generates a different path in an activity diagram. All decisions must be merged in **merge nodes**. Decision and merge nodes are shown with diamond shaped nodes, illustrated in Figure 86.15.

Although it is possible to have several paths with decision and merge nodes, only one of these paths can have an active flow. For concurrent paths, a **fork-join node** must be used. A fork node receives a

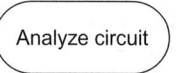

FIGURE 86.12   An action.

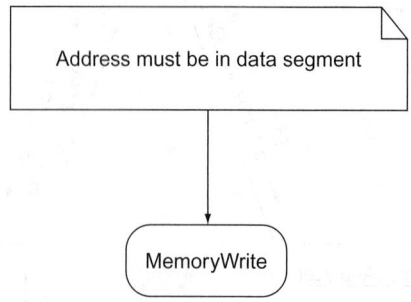

FIGURE 86.13   An example of preconditions in actions.

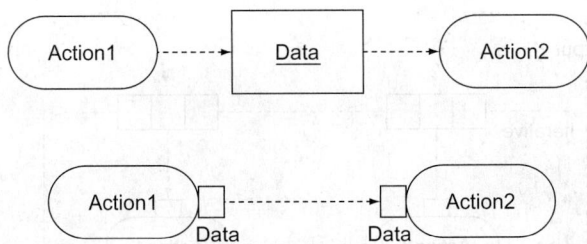

FIGURE 86.14 Two ways of object flow demonstration.

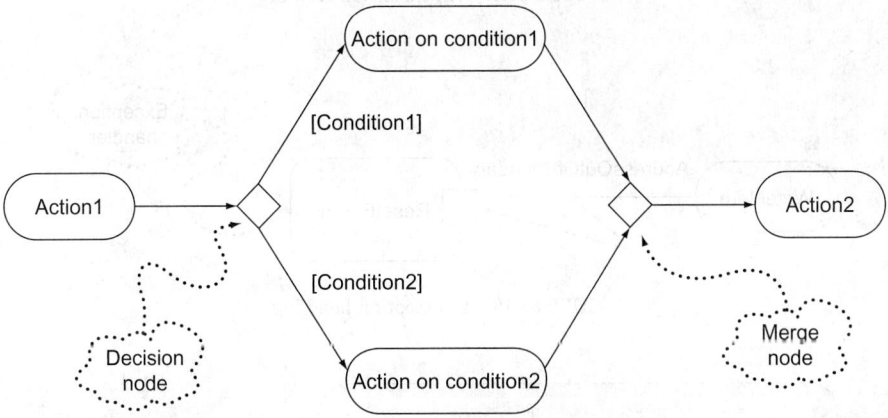

FIGURE 86.15 Decision and merge nodes.

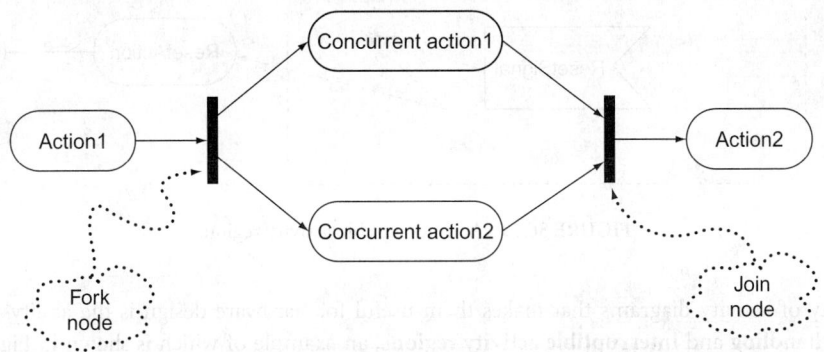

FIGURE 86.16 Fork and join nodes.

single flow and produces several concurrent flows. In contrast, a join node waits on several input flows and when it receives all these flows it generates a single output flow. Symbols for fork-join nodes are depicted in Figure 86.16.

An **expansion region** in an activity diagram is a region in which the activity must be executed more than one time. *Multiple times,* here, has several meanings. An activity can be executed in **iteration** mode, **parallel** mode, or in **stream** mode. The input and output in an expansion region are shown as three-box groups, representing multiple data sets. Figure 86.17 shows an expansion region in iterative mode.

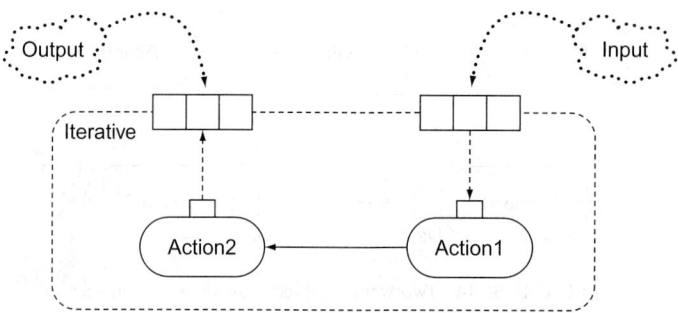

FIGURE 86.17   An expansion region.

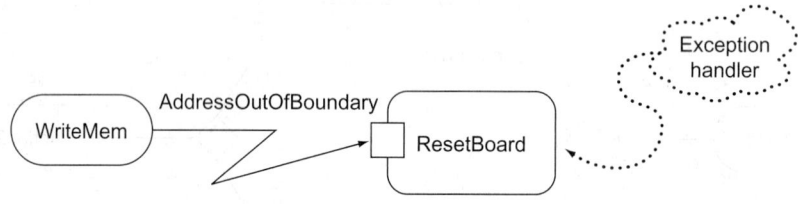

FIGURE 86.18   An exception handling.

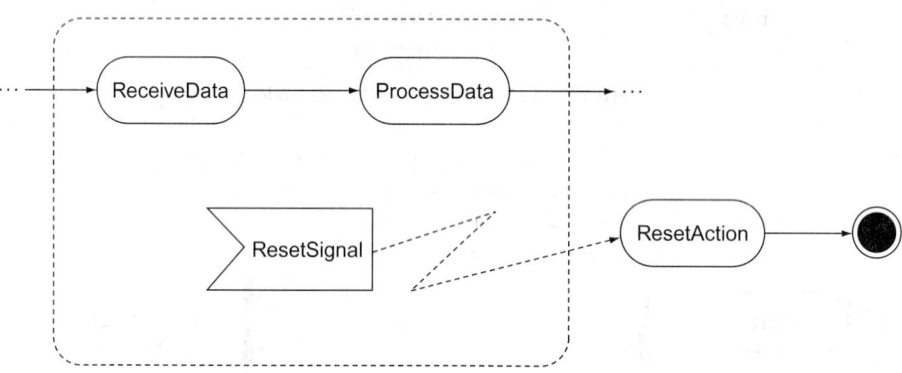

FIGURE 86.19   An interruptible activity region.

A facility of activity diagrams that makes them useful for hardware design is the ability to express **exception handling** and **interruptible activity regions**, an example of which is shown in Figure 86.18. Special actions are defined as exception handlers in an activity diagram (e.g., *ResetBoard*). If at any time a specified exception occurs (e.g., *AddressOutOfBoundary*) in a specified action (e.g., *WriteMem*), a specified handler will be executed.

Having an interruptible activity region is another facility of activity diagrams. If at any point of a flow in this region, a condition becomes true (an interrupt occurs), then the current flow will be broken and another predefined flow begins. The flow that has become active is usually outside the interruptible region. Figure 86.19 shows that while in the *ReceiveData–ProcessData* flow, if *ResetSignal* condition becomes true, the *ResetAction* is taken.

In some cases, different activities are done in different places or by different people in a system. In this situation, if an activity diagram is partitioned, its flow becomes clearer. An example of SSimBoard partition diagram is depicted in Figure 86.20.

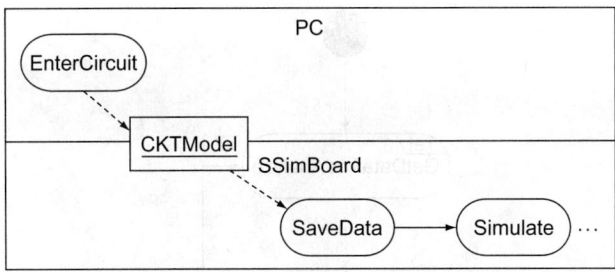

FIGURE 86.20 Partitioning in an activity diagram.

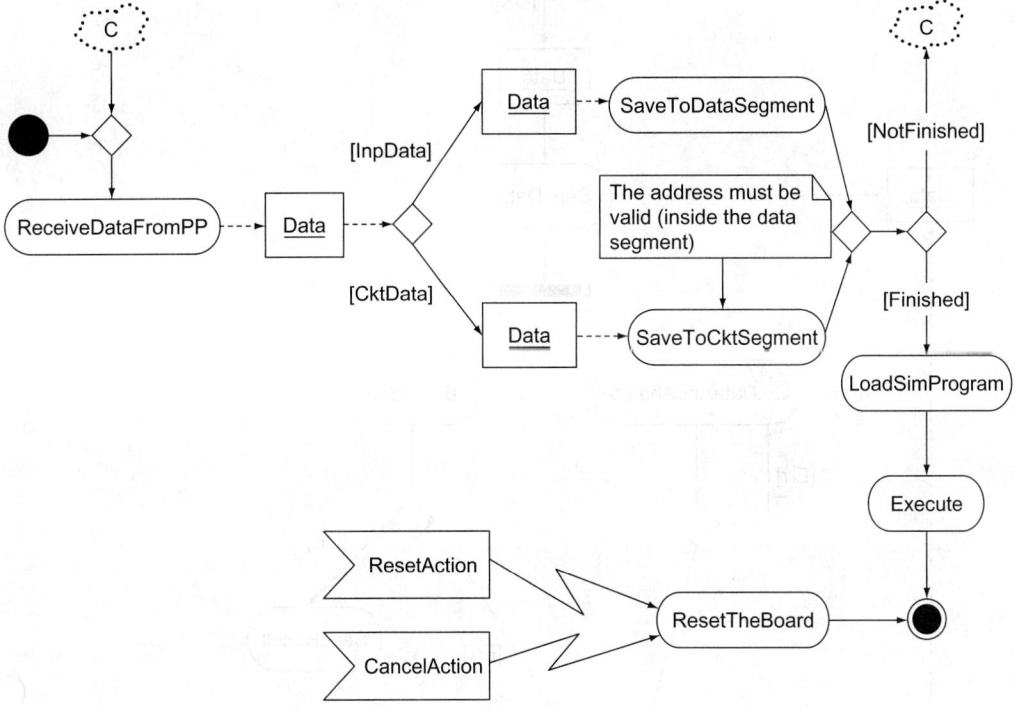

FIGURE 86.21 An activity diagram.

An activity diagram for our SSimBoard example system is shown in Figure 86.21. This diagram illustrates various structures of activity diagrams discussed in this section.

86.1.3.6 Interaction Overview Diagram

Another type of interaction diagram is the interaction overview diagram. This diagram is similar to the activity diagram, with the difference that the interaction overview diagrams accept interaction diagrams as their nodes. As discussed earlier in the section on activity diagrams, an interaction diagram includes four of UML 2.0 diagrams. An interaction overview diagram gives the designer a general view of a system and interaction between its various parts. The interaction overview diagram nodes can be represented as interaction occurrences or interaction elements.

An **interaction occurrence node** is a diagram node depicted as a reference to another existing diagram and is marked with the **ref** keyword.

An **interaction element node** is a diagram node the details of elements of which are inside it. Interaction element nodes are useful only for small diagrams. An example of an interaction overview diagram is shown in Figure 86.22 for the SSimBoard system. Each node in this figure is an activity diagram

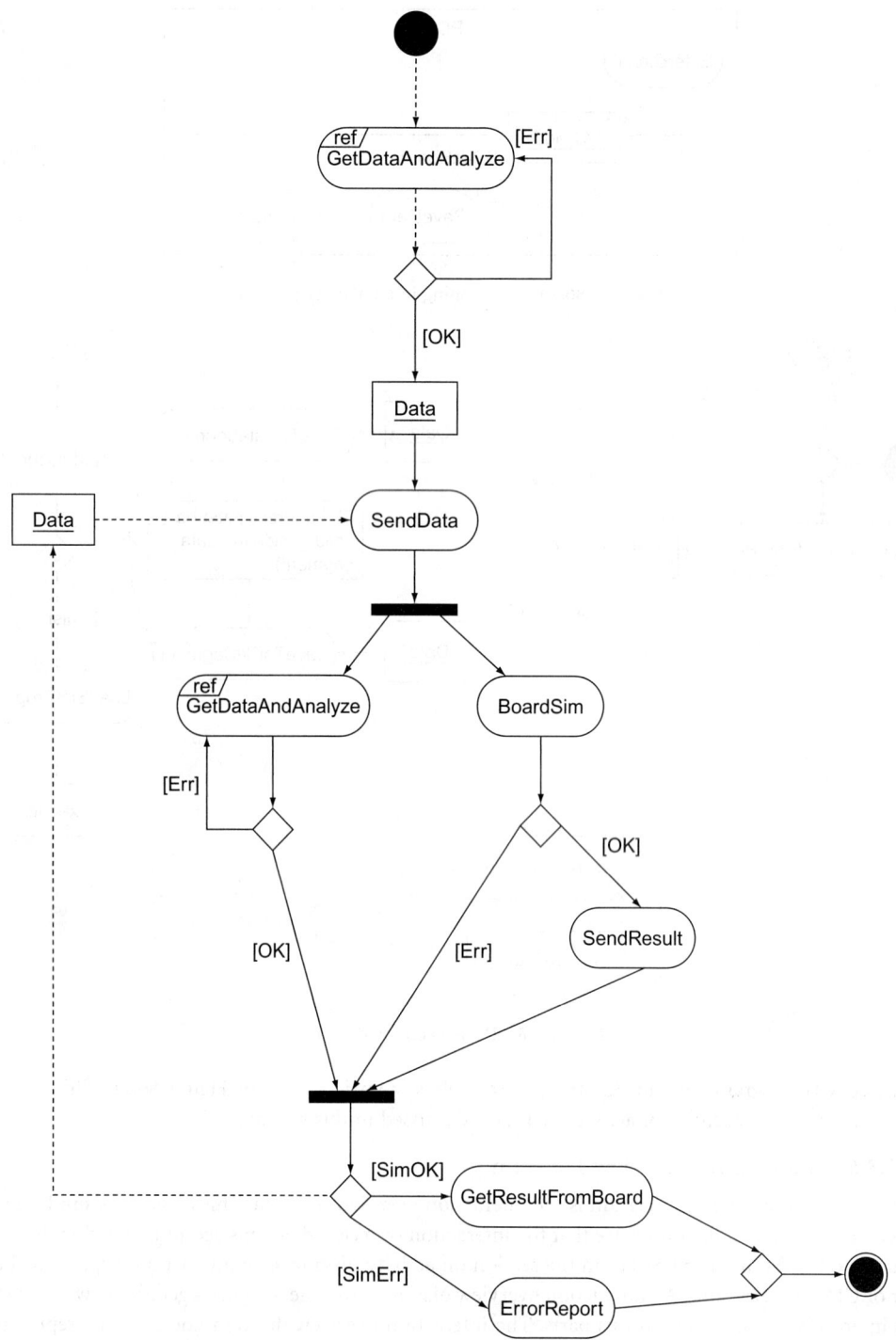

FIGURE 86.22   An interaction overview diagram.

from different parts of the system. For example the *GetDataAndAnalyze* activity diagram refers to the diagram of Figure 86.21.

86.1.3.7 Composite Structure Diagram

Elements of a composite structure diagram are shown in Figures 86.23–86.26. An example of a composite structure diagram is shown in Figure 86.27. A composite structure diagram is new in UML 2.0, and it emphasizes on interactions between classifiers in a system, how they collaborate with each other, and their interfaces. A structured classifier is one of the elements used in this diagram (Figure 86.23 and Figure 86.24). It has one or more ports, parts, and interfaces. Other elements used in this diagram are

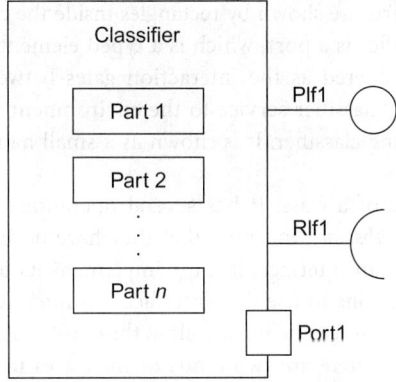

FIGURE 86.23 A sample composite classifier.

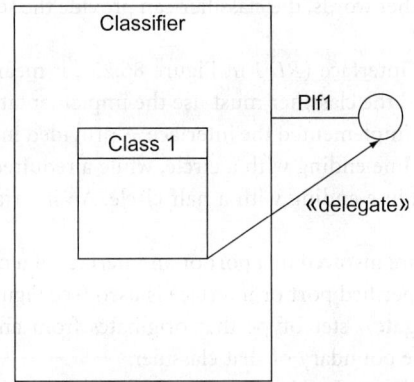

FIGURE 86.24 The delegate connector.

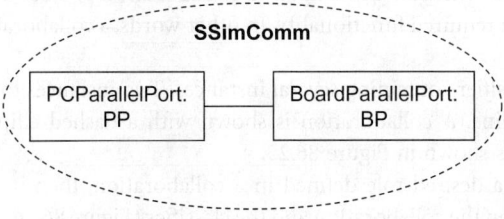

FIGURE 86.25 A sample collaboration.

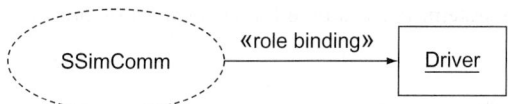

FIGURE 86.26   A role binding connection.

collaborations (Figure 86.25), and connectors between a collaboration (dotted eclipse) and a structured classifier (Figure 86.26).

In composite structure diagrams, a **classifier**, like that of Figure 86.23, contains **parts**, **ports**, and **interfaces**. Each part shows a set of instances. A part can be deleted before its parent is destroyed (e.g., the classifier in Figure 86.23). Parts are shown by rectangles inside the classifier.

Another element in the classifier is a port, which is a typed element that shows the visible parts of the classifier. Ports can be considered as the interaction gates between a classifier and its outside environment. A port can offer a classifier service to the environment it is in, and it can also provide the requirements of its containing classifier. It is shown as a small named rectangle on the boundary of its containing classifier.

An interface is a special case of a class. It has several operations which are all public for other classifiers. These operations are *abstract*, meaning that they have no implementation in the interface itself. If a classifier wants to use an interface, it must implement its operations itself, or it must use the other classifier's implementations to use that interface. An interface is similar to a set of rules in a design. A classifier that uses an interface must follow these rules. An interface operation can have more than one implementation. There are two kinds of interfaces in UML: **provided** and **required** (*PIF1* and *RIF1* in Figure 86.23).

If a classifier has a provided interface (the upper interface in Figure 86.23), the classifier has implemented the operation of that interface. This implies that a **realization** link exists between that classifier and the interface. In other words, the classifier can provide the services of that interface to other elements.

If a classifier has a required interface (*RIF1* in Figure 86.23), it means that the classifier needs the operations of that interface, and the classifier must use the implementation of that interface operation from another classifier that has implemented the interface. A provided interface link between a classifier and an interface is shown by a line ending with a circle, while a required interface between a classifier and an interface is shown by a line ending with a half circle. An interface is displayed as a class with the $\langle\langle$**interface**$\rangle\rangle$ stereotype.

To specify the elements that are involved in a port or an interface of a classifier, the delegate connector between that element and the specified port or interface is used (see Figure 86.24). A delegate connector is a connector with the $\langle\langle$**delegate**$\rangle\rangle$ stereotype that originates from an element in a classifier to the specified port of interface in the boundary of that classifier.

The preceding paragraphs discussed various elements of a classifier that can be used in a composite structure diagram. In what follows, we will discuss collaborations that are also used as elements in composite structure diagrams. Before defining a collaboration, we need to define roles. A **role**, in UML, specifies a set of features an instance must have. A **collaboration** is a set of roles, collaborating with each other, to accomplish a required functionality. In other words, a collaboration explains how a system (or part of it) works.

A collaboration is a classifier containing several instances. Two instances that have a collaboration are connected with a simple line. A collaboration is shown with a dashed ellipse and it has a name. An example of collaboration is shown in Figure 86.25.

If a classifier performs a desired role defined in a collaboration, then it is said that there is a **role binding** connection between the collaboration and the classifier (Figure 86.26). A role binding connection is shown by an arrow with the $\langle\langle$**role binding**$\rangle\rangle$ stereotype.

Other connector types between UML collaborations and classifiers are **represents** and **occurrence**.

A represents connection shows that the classifier uses the collaboration. It is shown by the ⟨⟨**represents**⟩⟩ stereotype. An occurrence connection shows that the collaboration represents the classifier. It is shown by the ⟨⟨**occurrence**⟩⟩ stereotype.

With what was discussed above, we are now ready to display a complete composite structure diagram for SSimBoard using the structured classifiers and the collaborations. This is shown in Figure 86.27.

86.1.3.8 Component Diagram

Any system can be considered as a set of components. These components can be software or hardware parts. A component diagram (Figure 86.28) shows the relationship between components of a system, and consists of components and interfaces between them.

A **component** is a classifier that can include several class instances in run-time. Usually, components use interfaces to communicate with their outside environment. **Interfaces** are the same as those defined in composite structure diagrams. In UML, a component is displayed with a rectangle with the ⟨⟨**component**⟩⟩ stereotype, or a component icon on the top-right part of the component rectangle. An important advantage of a component is its reusability. Components already implemented in a component diagram can be reused in new designs. As an example, consider the SSimBoard system. If the driver for parallel ports is already implemented in a component diagram in a system, it can be reused elsewhere in the system.

The connection between every two components in a component diagram is performed by **assembly connectors**. Assembly connectors that connect provided and required interfaces (discussed in Section 86.1.3.7). Similar to structured classifiers, components can also have **ports** to share their services and requirements with other components in the system.

86.1.3.9 Class Diagram

A class diagram is a form of a structured diagram. This diagram illustrates classes and the static relationship between them.

A **class**, an example of which is shown in Figure 86.29, is a named element with several **attributes** and **methods** (or operations). A class in UML resembles the **class** concept in object-oriented languages. Attributes and operations of a class can be of type **public** (shown by +), **private** (shown by −), or **protected** (shown by #). There are special types of classes in UML such as an interface, discussed in Section 86.1.3.7. Another special class type is a **table.** A table may consist of several columns (with the ⟨⟨**column**⟩⟩ stereotype), primary keys (with the ⟨⟨**PK**⟩⟩ stereotype), and foreign keys (with the ⟨⟨**FK**⟩⟩ stereotype). A table is defined by the ⟨⟨**table**⟩⟩ stereotype.

Another important element in a class diagram is the relationship between two classes. There are different kinds of relationships between classes in a class diagram. A class can be derived from another class (parent class). In this situation, a **generalization** relationship exists from the child class to its parent class.

One class may have an instance of another class among its attributes. In this situation, an **association** relationship exists from one class to another. Associations have multiplicity at both ends. If there are attributes or operations suit to neither of the associated classes, then they can be assigned to the association itself. This kind of association is called an association class. A class diagram is shown in Figure 86.30.

A class can be composed of smaller components (for example, in C++, it can contain pointers to other classes). In this situation, an **aggregation** relationship is drawn between this class and its smaller components (see *CKTAnalyzer* and *GUI* in Figure 86.30). If two classes have an aggregation relationship, then the composite object (the *GUI* in Figure 86.30) is responsible for allocating and deallocating the composed object (*CKTAnalyzer*). This implies that if a composite object is deleted, then the composed object will also be destroyed.

If operations of a class are implemented by another class, then there is a realization relationship from the implementer class to the implemented one. An interface is a good example for realization. The example depicted in Figure 86.30 is part of the SSimBoard class diagram.

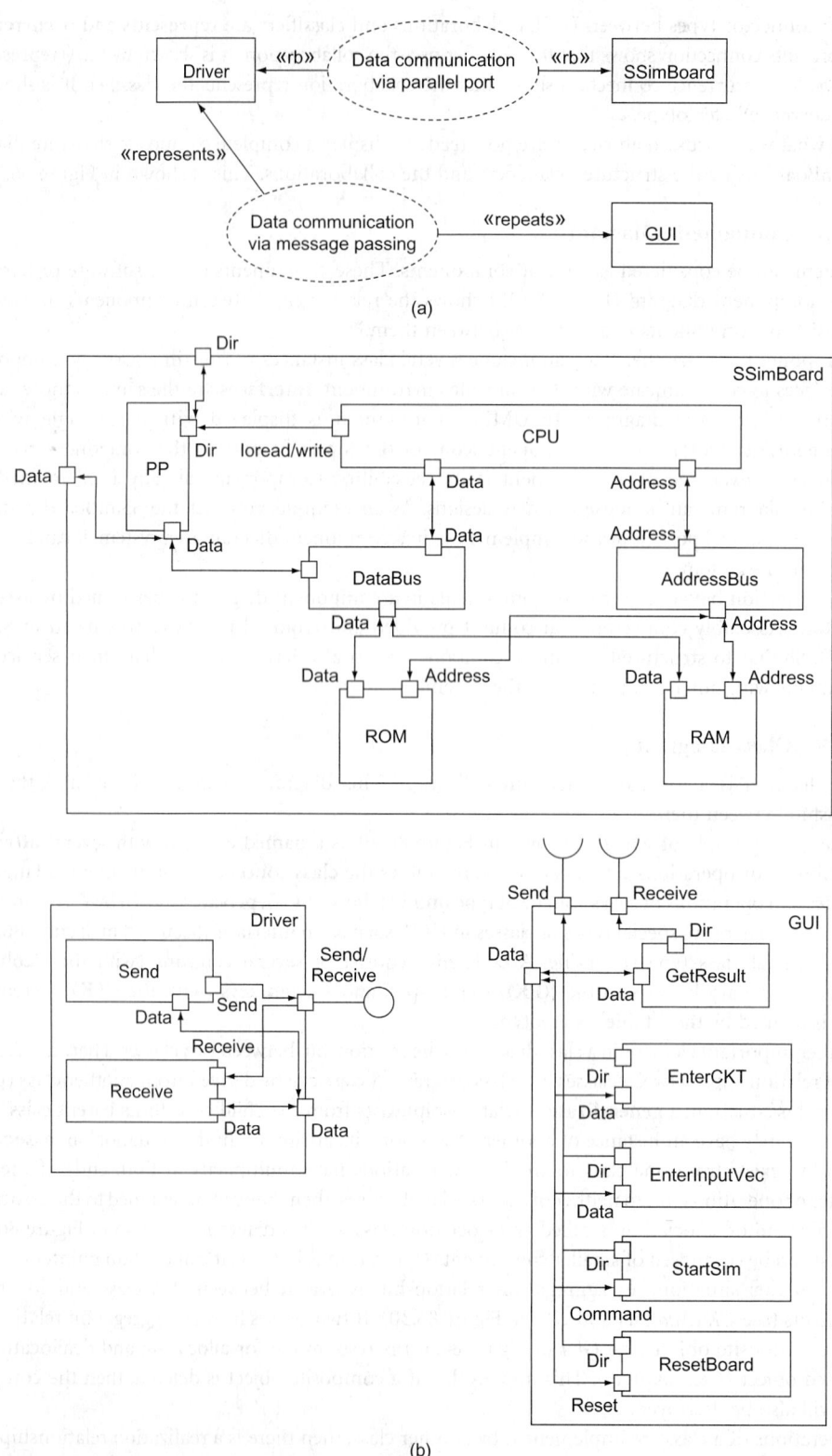

FIGURE 86.27 A composite structure diagram. (a) Collaborations and (b) composite classifiers.

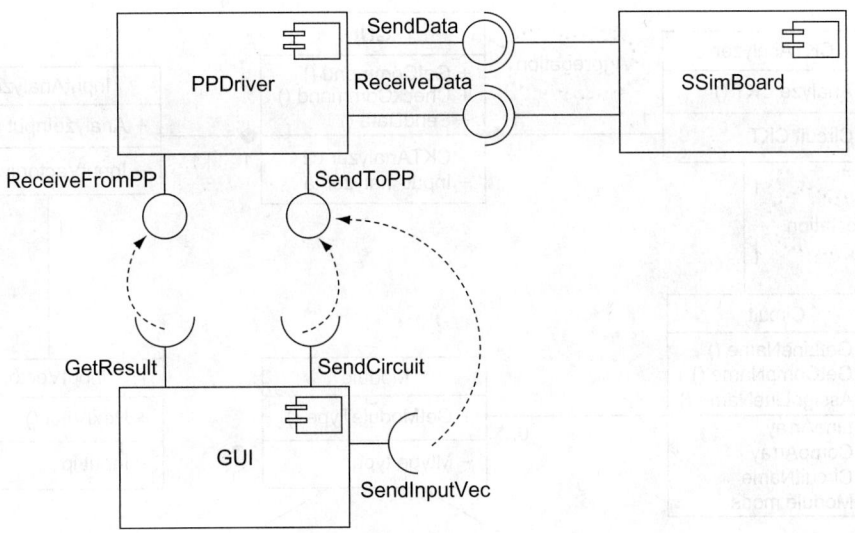

FIGURE 86.28 A component diagram.

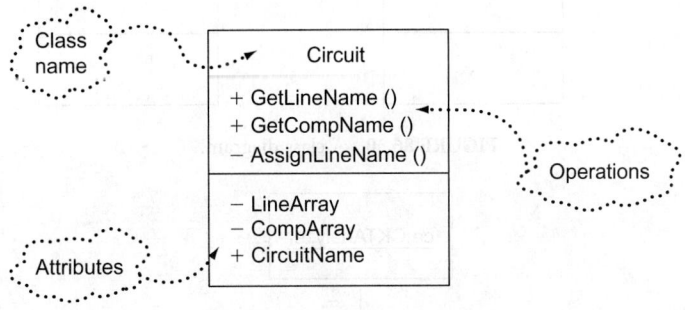

FIGURE 86.29 A sample class.

86.1.3.10 Object Diagram

An **object** is an instance of a class. An object diagram (Figure 86.31) can be considered as a special case of a class diagram. This diagram emphasizes on the relationship and the multiplicity of class instances in a specified time. Objects, in UML, are displayed with a rectangle with the name of the object followed by a colon followed by the name of its corresponding class. The name of an object must be underlined. In general, object elements cannot have attributes or operations, but in an object diagram they can show the contents of attributes or operation arguments in a specified time. A portion of an object diagram of the class diagram in Figure 86.30 is demonstrated in Figure 86.31.

86.1.3.11 Package Diagram

Packages are used to organize UML elements. For example, several related classes can be placed into a package. In this case, all of these classes are assumed to have one **namespace** (which implies that their names must be unique). Package diagrams show packages of a model and their relationships. They are generally used for organizing use-case and class diagrams.

A **package** usually shows a physical or logical relationship between a set of elements. For example, a set of classes with the same parent can be placed in a package. A package diagram related to our SSimBoard example is shown in Figure 86.32.

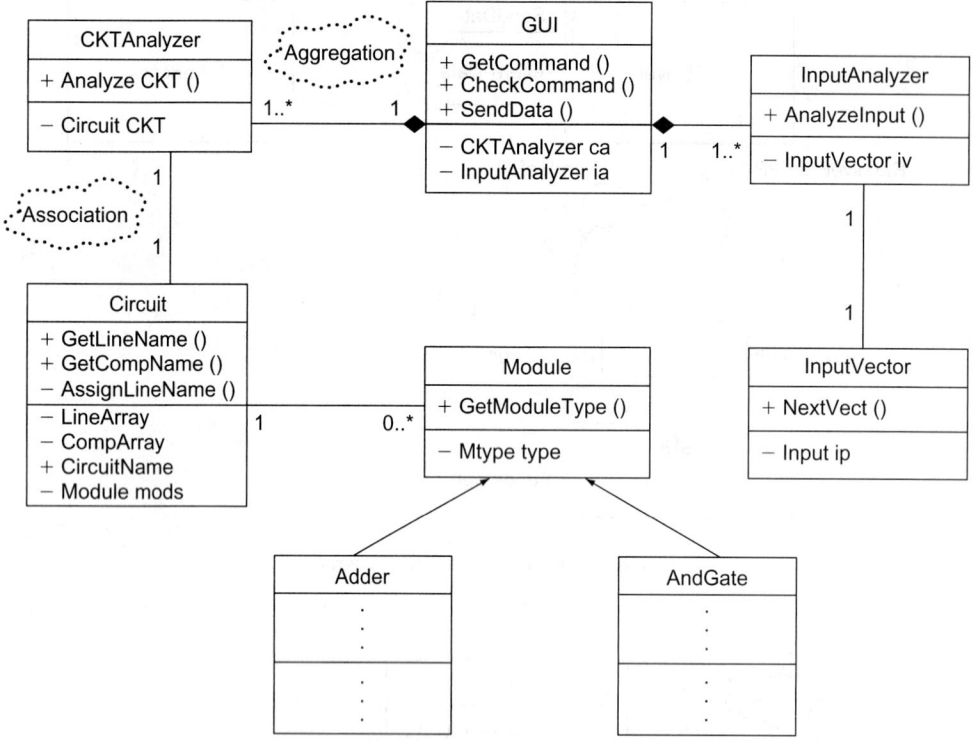

FIGURE 86.30   A class diagram.

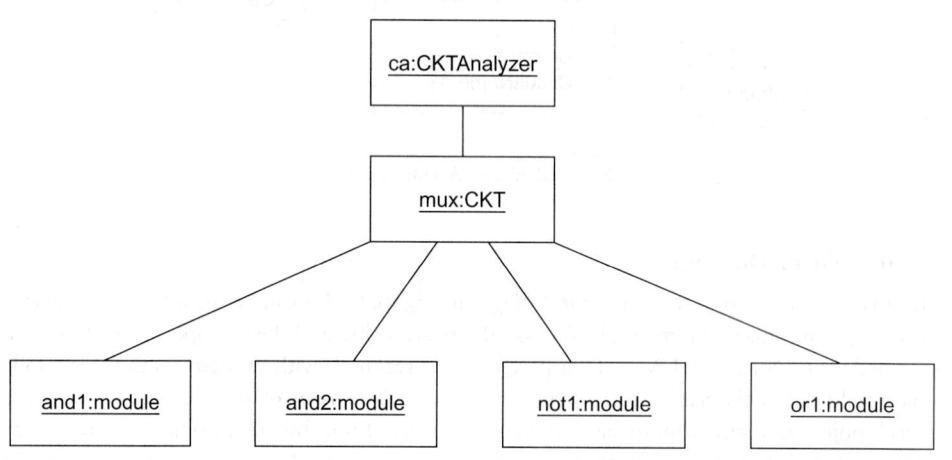

FIGURE 86.31   An object diagram.

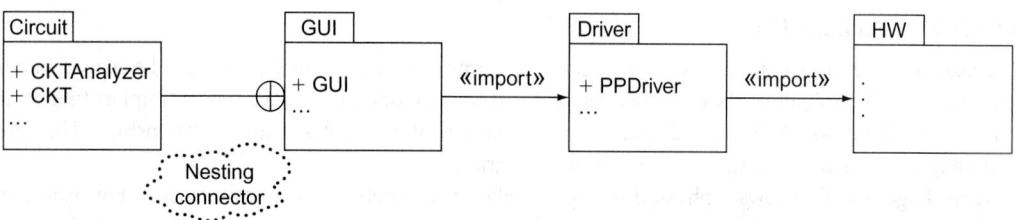

FIGURE 86.32   A package diagram.

The connector between packages, in a package diagram, has three stereotypes. A connector with the ⟨⟨**merge**⟩⟩ type shows that all common attributes and operations between the source and the destination package are expanded in the source package.

A connector with the ⟨⟨**import**⟩⟩ type imports the elements of the target package into the source package. For example *PPDriver* is imported to the *GUI* in Figure 86.32. The namespace of the source package can access the target's elements.

It is possible that a package contains another package. In this case **nesting** relationship exists between two packages. An example is the *GUI* and *Circuit* packages in Figure 86.32.

86.1.3.12   State Machine Diagram

A state machine diagram models the behavior of an object when different events happen at run-time. State machine diagrams include states and transitions.

States, in a state machine, illustrate the condition and status of objects. States are shown with a round-cornered rectangle. A state element in state machine diagrams can be another state machine diagram. In addition, there are a few **pseudo-states** which have special behaviors. Pseudo-states are shown in Figure 86.33 and Figure 86.34, and Figure 86.35 shows **transition** from *state1* to *state2*. Figure 86.36(a) and Figure 86.36(b) are complete examples of state machine diagrams.

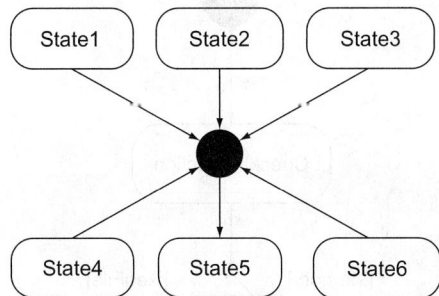

FIGURE 86.33   A junction pseudo-state.

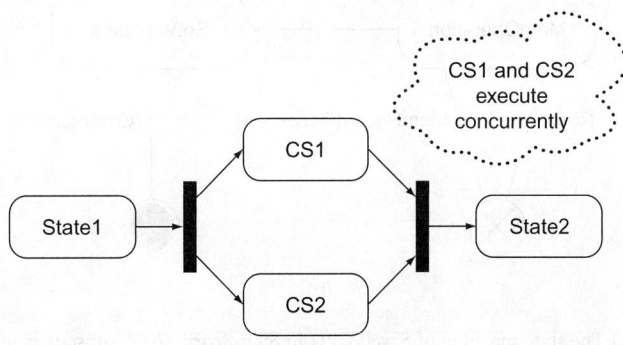

FIGURE 86.34   Concurrent regions in a state machine diagram.

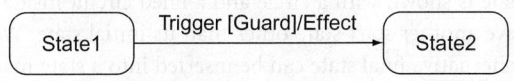

FIGURE 86.35   A transition.

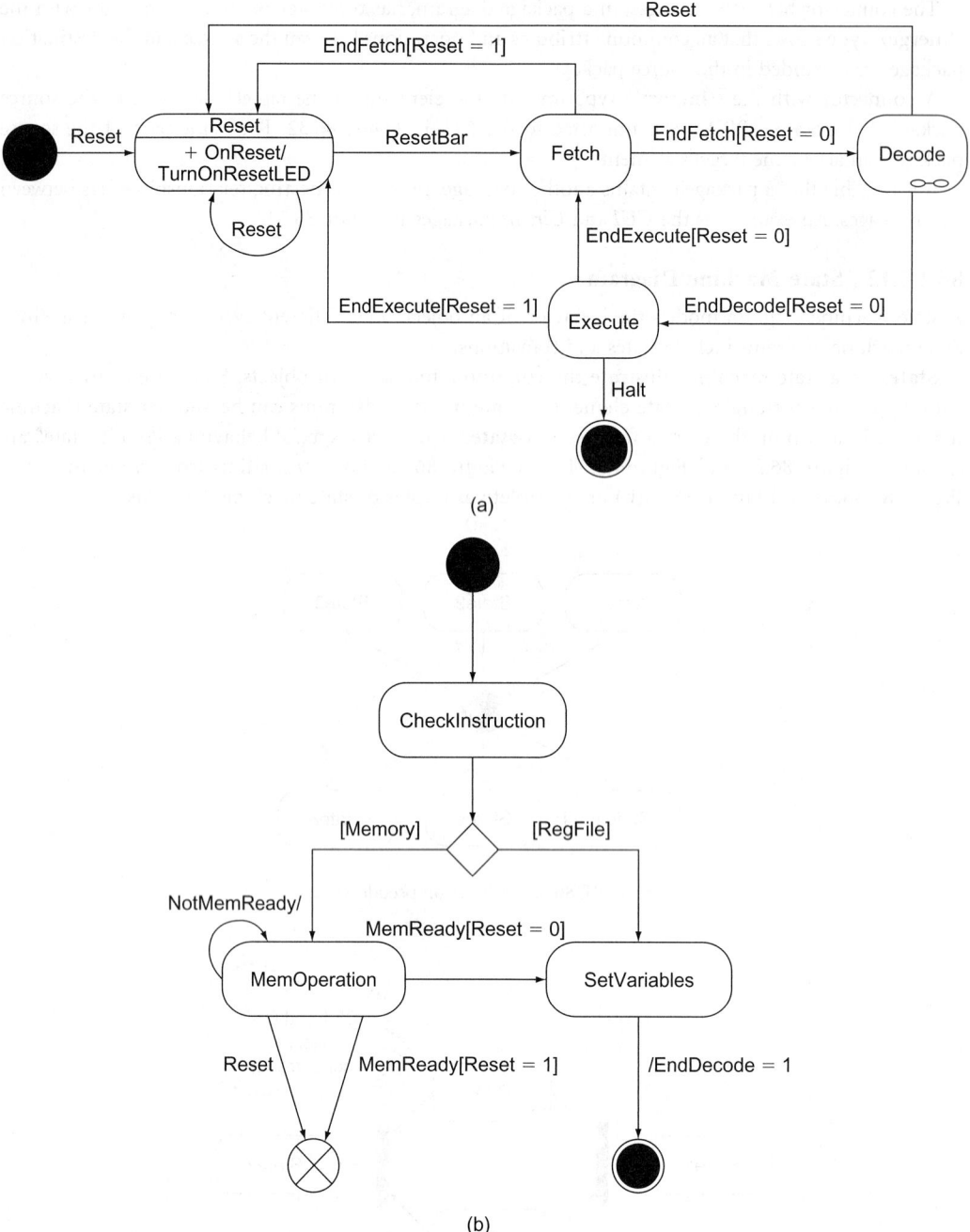

FIGURE 86.36 (a) The state machine of Sayeh CPU in SSimBoard. (b) A substate machine for *Decode*.

Initial state of a state machine diagram is a pseudo-state from where the state machine begins its transitions. It is depicted with a black circle. Similarly, the **final state** is a pseudo-state where the state machine halts. The final state is shown with a circle and a filled circuit inside it.

A state machine may have another start state other than its initial state. We call this pseudo-state an **entry point**. Similarly, an alternative final state can be inserted into a state machine that is referred to as an **exit point**. An entry point is shown with a hollow circle while an exit point is shown with a circle and a cross inside it.

In some cases, there must be different paths for different conditions on an event in a state machine. A **choice** pseudo-state makes several decision paths in a state machine. This is shown by a diamond like that of Figure 86.36(b).

A **terminate** pseudo-state shows that the lifeline of the source state is finished. Therefore, it is no longer sensitive to any events. This is illustrated by a simple cross.

Another pseudo-state used in state machine diagrams is the **junction** pseudo-state. Junction is used to make sequences of transitions. It has multiple inputs and multiple outputs. An example of a junction pseudo-state is depicted in Figure 86.33.

A **history** pseudo-state in a state machine diagram stores the most recent state of a system. It is useful when an interrupt occurs in the system.

A new feature added in UML 2.0 is *concurrency* in state machines. In **concurrent regions**, the states can be executed concurrently. Their symbols are similar to fork and join nodes in activity diagrams (see Section 86.1.3.5). Figure 86.34 shows two concurrent states, CS1 and CS2.

Another element used in a state machine diagram is a **transition** shown in Figure 86.35. A transition is depicted with an arrow between two states. It causes the current state of the state machine to change from the source state to the target state. A transition in UML is constructed from three parts that include **trigger**, **guard**, and **effect**. A trigger can be an event, a signal, a value change, or a timeout. A trigger is the main reason why a transition happens. A guard is the condition that must be true when a trigger happens in a transition. An effect is an action that must happen when a trigger with a true guard occurs. The effect part of a transition is sometimes placed in the target state (in this case the effect is called a state action).

Source and destination of a transition can be identical. This transition is called a **self transition**. The example state transition diagram of Figure 86.36(a) and Figure 86.36(b) is part of our SSimBoard example.

86.1.3.13 Timing Diagram

Another diagram, new in UML 2.0, is the timing diagram. In this diagram, changes of a state (in a state machine) or values of an element (for example, an attribute of a class instance) are depicted over time. This diagram is very useful for hardware design. The horizontal axis of timing diagram represents the time, while the list of possible states or the names of the variables are shown on the vertical axis.

State lifeline displays the changes of a state, while **value lifeline** shows the value changes of an element in the diagram. Figure 86.37(a) depicts a timing diagram (state lifeline) of the state machine shown in

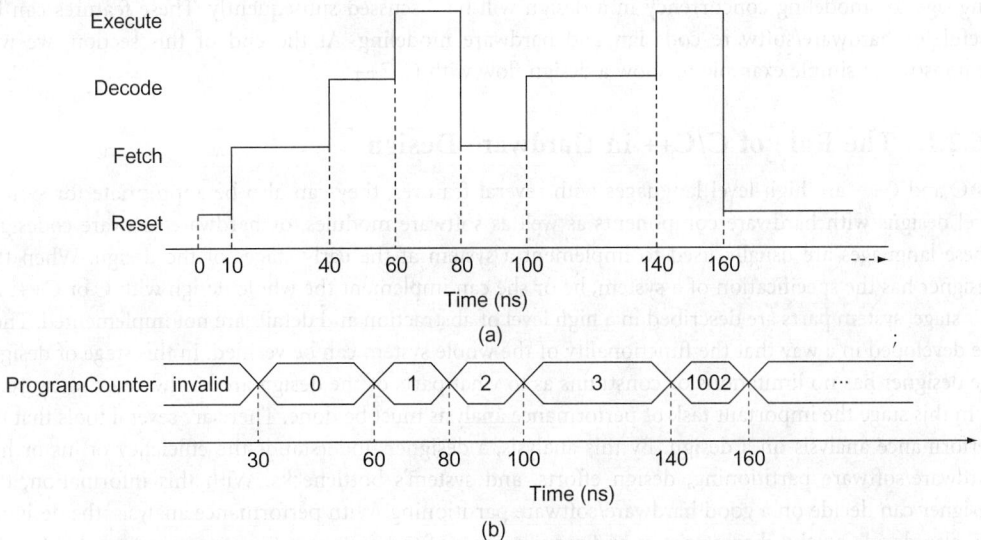

FIGURE 86.37 Timing diagrams. (a) A state lifeline timing diagram. (b) A value lifeline timing diagram.

Figure 86.36(a). A timing diagram (value lifeline) of the *program counter* attribute used in Sayeh processor is shown in Figure 86.37(b).

86.1.4   Our Coverage of UML

This section gave an overview of UML 2.0 diagrams. Using these diagrams, a hardware designer can develop a model from a design specification. He or she can use CASE tools to convert this model into codes in a high-level programming language. In the next section, we will discuss C/C++ high-level programming languages which can specify an early prototype of a system. This early prototype helps designers analyze their design for performance and partitioning.

86.2   C/C++

The C language was developed in early 1970s (1969–1973) by Dennis Ritchie at Bell Labs. The base of this language was the B language. It also has several features from Algol68. This language was developed when the UNIX operating system was designed. After developing C, almost the whole UNIX operating system was written by this language. Until late 1980s (1989), when C became a standard by ANSI (this version of C is called ANSI C), several versions of this language were developed and used by programmers. This language is a procedural language, which means that a system is broken into smaller and simpler subsystems (hierarchically) and each subsystem is implemented by a function in C. Therefore it is used for top-down design methodology.

In early 1980s (1983–1985) Bjarne Stroustrup proposed object-oriented version of the C language, which was called C++, at Bell Labs. In this language, several concepts like classes and objects, inheritance, operator overloading, etc. were added to the original C language. A number of these features were derived from Simula67 and Algol68 languages.

C is a subset of C++ but the philosophy of programming with C++ is different from that of C. As stated earlier, C is a procedural language, while C++ is a modular language. In other words, as the building blocks of a system are designed with classes in C++, the language can also be used for bottom-up design methodology.

In the next section, the role of high-level languages like C and C++ in hardware design will be discussed. Then, the basic concepts and features of C and C++ will be listed. Various advanced features of the C language for modeling concurrency in a design will be discussed subsequently. These features can be useful for hardware/software codesign and hardware modeling. At the end of this section, we will demonstrate a simple example to show a design flow with C/C++.

86.2.1   The Role of C/C++ in Hardware Design

As C and C++ are high-level languages with several features, they can also be appropriate for system level designs with hardware components as well as software modules, or hardware/software codesign. These languages are usually used to implement a system at the early stages of the design. When the designer has the specification of a system, he or she can implement the whole design with C or C++. At this stage, system parts are described in a high level of abstraction and details are not implemented. They are developed in a way that the functionality of the whole system can be verified. In this stage of design, the designer has no limitations or constrains as to what parts of the design are hardware or software.

In this stage the important task of performance analysis must be done. There are several tools that do performance analysis on a design. By this analysis, a designer understands the efficiency of his or her hardware/software partitioning, design efforts, and system's bottlenecks. With this information, the designer can decide on a good hardware/software partitioning. With performance analysis, the designer can also decide on the characteristics and requirements of the processor (or processors) needed for the implementation of software part(s).

When partitioning is known, the designer can perform C-synthesis (also known as behavioral synthesis) to convert high-level hardware parts into lower-level RTL modules. Figure 86.38 shows the design flow using C/C++. It is important to note that the diagram in Figure 86.38 is not the only possible C/C++ design methodology.

System behavior level in C/C++ is usually higher than RT or gate level. While these lower levels generally involve timings and events, higher-level descriptions see design at a more abstract level. In general, for RT or lower-level descriptions, VHDL, Verilog, or other such HDLs are more appropriate. In system level designs, the number of modules and events are fewer than those in RTL, and modules can also communicate by more abstract message passing mechanisms. In system level design with C/C++, we can partition the design into several tasks, develop each task by a function or develop classes and use their methods and attributes to perform each task and then continue the design step as illustrated in Figure 86.38.

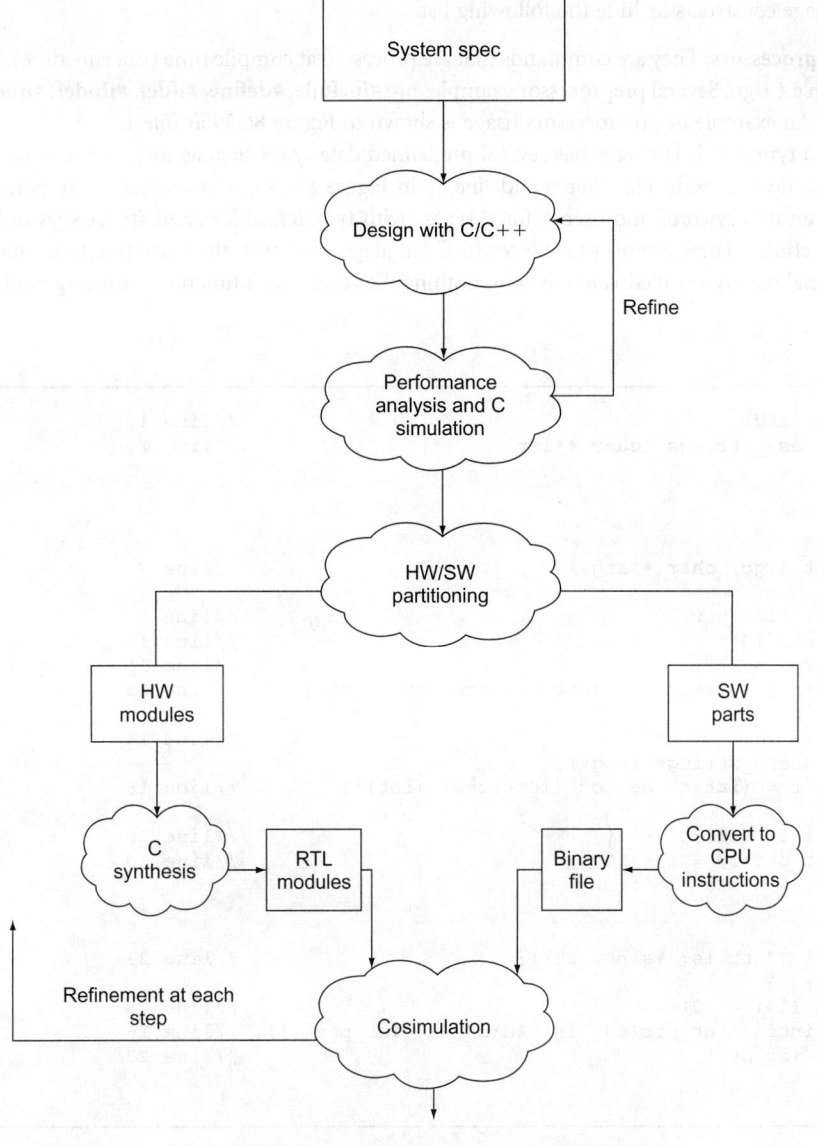

FIGURE 86.38 C/C++ design methodology.

Since C++ is an object-oriented and a more complete language than C for the design of hardware, it is recommended to use C++ and its corresponding libraries like SystemC or Handel-C. These libraries are geared toward hardware and provide a good interface between hardware thinking in a software environment.

86.2.2   Basic C Language Constructs

C and C++ are extensive, and too complex to be fully described here. The intention here is to discuss these languages from a hardware point of view and not get into many of their details. We assume here that the reader is familiar with the basics of these two languages, and the basic constructs of C are listed only as a reminder. In a few cases, the similar C++ syntax is also described. For further study on C language, refer to Ref. [3]. A sample C code is shown in Figure 86.39. However, complete C and C++ code examples can be found in Sections 86.3 and 86.4, and Chapter 89. Description of the C language constructs given below, becomes easier to comprehend by referring to this example code.

C language constructs include the following list:

- **Preprocessors:** They are commands that are processed at compile time (not run-time). They begin with a # sign. Several preprocessor examples are **#include, #define, #ifdef, #ifndef, #undef, #endif,** etc. An example of preprocessors usage is shown in Figure 86.39 at *line 1*.
- **Data types:** This language has several predefined data types such as **int, __int64, char, unsigned char, float, double,** etc. (*line 9* and *line 10* in Figure 86.39). It also accepts enumerations (with the **enum** keyword) and user-defined types (with **typedef, union,** and **struct** keywords).
- **Functions:** There are no procedures in C language. However, there are functions and there is a special data type called *void* meaning nothing. For example, a function returning **void** implies no

```
#include <stdio.h>                                      //line 1
void process_strings (char **inp)                       //line 2
{
   ...
}

main (int argc, char **argv)                            //line 7
{
   int i, *int_ptr;                                     //line 9
   char str[10];                                        //line 10
   if (argc > 2) {                                      //line 11
      printf ("Error in number of arguments!\n");       //line 12
      exit (-1);
   } else                                               //line 14
      process_strings (argv);
   int_ptr = (int *) malloc (10*sizeof (int));          //line 16
   ...
   for (i=0; i<10; i++) {                               //line 18
      int_ptr[i] = i * 1000;                            //line 19
      str[i] = 'a' + i;
   }
   ...
   printf ("str is: %s\n", str);                        //line 23
   i = 0;
   while (i++ < 10)                                     //line 25
      printf ("int_ptr(%d) is: %d\n", i, int_ptr[i]);   //line 26
   free (int_ptr);                                      //line 27
}
```

FIGURE 86.39   A sample C code.

return value for the function (*line 2* in Figure 86.39), and a function with **void** as its parameter is considered as a function with no parameters. A C program consists of several functions and a main function called **main()** (*line 7* in Figure 86.39). **main()** is the starting point of a program and thus the use of this function is mandatory in all C programs. It has two arguments called *argc* and *argv* that are used for the program's parameters. The former shows the number of program parameters (including the program itself), while the latter contains the parameters themselves. For example, if we have a program named *mycopy.exe* with two filenames as its parameters, and if we use the string "*mycopy fn1.c fn2.txt*" in the command line, then *argc* becomes 3 and *argv* is an array containing *mycopy*, *fn1.c*, and *fn2.txt*, respectively.

Another important point about functions is function prototyping. If function *f1* is to be used inside function *f2*, *f1* must appear before *f2*. This is because the compiler needs to know about *f1* when it is compiling *f2*. In such cases, compilation errors can be avoided by prototyping *f1* before *f2*. Prototyping is simply defining a function at the beginning of a C file. In prototyping, the function's name and optionally its arguments are listed followed by a semicolon. By prototyping, the compiler recognizes a function before reaching its implementation. It is recommended that all functions have a prototype.

- **Conditions and conditional statements:** A conditional statement partitions a portion of code to several parts. On some conditions one (or more) of these parts are executed. The conditional statements include **if-else** (*line 11* and *line 14* in Figure 86.39), **?-:**, and **switch-case** constructs. The conditions are Boolean statements (0 for false and nonzero value for true) and can be combined by other Boolean statements, and with **&&** (logical AND), **||** (logical OR), and the unary operator **!** (logical not).
- **Loops:** Loop and iteration constructs in C include **for**-loops (*line 18* in Figure 86.39), **while**-loops (*line 25* in Figure 86.39), and **do-while** loops. Examples for these constructs are listed below.

```
for (i=1; i<10; i++) {loop body}
while (counter < 10) {loop body}
do {loop body} while (counter < 10);
```

C also provides two statements for controlling program flow inside a loop: **break** and **continue**. When using **break**, the flow exits from its current block and continues from the next statement after the block. In contrast, **continue** causes the flow to exit from the current loop (or block) iteration and start the next iteration.

- **Variables:** Variables in C/C++ are of several kinds, including simple variables (*i* in Figure 86.39), static arrays (*str* in Figure 86.39), dynamic arrays and pointers (*int_ptr* in Figure 86.39). The syntax of a simple variable declaration is

```
type var_name [=initial_value];
```

With this syntax, an integer can be defined as

```
int int_var;
```

A static array is an array of a defined type. A precise amount of memory will be allocated when these variables are defined. For example, an array of 100 integers is defined by

```
int int_array [100];
```

If an integer uses two bytes of the memory, when the compiler reaches the above definition statement, it will allocate 200 bytes of memory for *int_array* variable.

Another type of variables, which is an important strength point of the C language, is a pointer. A pointer is an address that corresponds to a variable's address. Pointers are represented by a ∗ preceding variable names. For example, in **int**∗ *int_ptr*, *int_ptr*, is a variable that points to an integer.

Another operator that is related to pointers is the **&** (address-of) operator. It is used before a variable (for example &*int_var*) to show the address of that variable. For example if *int_var* is 1000 and if we assign &*int_var* to an integer pointer like *int_ptr*, since this integer pointer points to the address of *int_var*, *int_ptr is equal to 1000.

   C also allows two or more dimensional pointers. For example, a pointer to another pointer is defined by ** (*argv* in Figure 86.39). Pointers are a complicated concept in C/C++ and care must be taken in using them. It is strongly recommended to study pointers in more detail before starting to use them.

   Dynamic arrays are pointers, to which memory can be assigned at run-time. Allocating memory can be done by several memory allocation functions (such as *malloc* in C, shown in *line 16* in Figure 86.39, and *new* in C++). It is necessary to free an allocated memory of a variable after using it, because it will not be de-allocated automatically (like static arrays). Memory leaks are resulted from improper de-allocation. Memory de-allocation functions are *free* in C, shown in *line 27* in Figure 86.39, and *delete* in C++.

   Another interesting point about pointers and dynamic arrays is **void*** or a pointer to the **void** type. As stated above, **void** in C/C++ means *nothing*, but when it is used as a pointer, it means *anything*. If a variable of type **void*** is used as a function parameter, any type of pointer can be passed to that function for that parameter (for example, **char***, **int***, etc.). In addition if a **void*** variable is defined inside a function, memory can be allocated with any type to that variable. For example, if *v_ptr* is a **void***, what is shown below generates an array of 100 characters:

```
(char *)  malloc (100);
```

And an array of 50 integers is generated by

```
(int *)  malloc (100);
```

   An important point about variables in C/C++ is type casting (similar to *line 16* in Figure 86.39). Unlike languages like PASCAL, in C/C++ it is permitted to define a variable of type *t1* in a function and use it as a variable of type *t2* in the body of that function. Type casting is also used in parameter passing in function calls. As type casting (like pointers) is a very complicated concept in C/C++, it is recommended that the reader studies type casting in detail before starting to use it.

- **Bit-level operators:** Another feature of the C language, which can also be useful for hardware designs, is the bit-level operators. These operators consider their operands as a set of bits. They perform their operations on the corresponding bits of each operand. These operations include
  - **&** for bitwise AND of two variables
  - **|** for OR of two variables
  - **∧** for XOR of two variables
  - **~** for one's complement
  - **⟨⟨** for left shift
  - **⟩⟩** for right shift

   As an example, consider the case where there are two character variables *c1* (=7) and *c2* (=1). The '*c3* = *c1* **&** *c2*' statement in this program results in *c3* becoming 1 (00000111 & 00000001). This is because the **&** operator performs bit-by-bit ANDing of *c1* and *c2* and puts the result in bits of *c3*.

   The variable bit-length can also be defined in C/C++. For example, if we want to have a 4-bit variable, we can define it as

```
int fbit_var:4;
```

Defining bit-length for variables is only allowed in a **struct**.

- **Communication with the outside world:** Several library functions (usually defined in *stdio.h* and *stdlib.h* header files) are used for inputting data from the user or file and outputting the results of a program to the output display or a file. A number of useful functions for I/O tasks include *printf* (*line 12*, *line 23*, and *line 26* in Figure 86.39), *fprintf, scanf, fscanf, open, fopen, read, fread, write, fwrite, close*, and *fclose* functions are also available for working with different types of files. Functionalities of these functions are evident from their names.

In C++, two standard streams *cin* and *cout* are defined for input and output streams. Two operations ≫ and ≪ are also overloaded for inputting data to a specified variable, and sending the value of the specified variable to the output. These operations are defined in the *iostream* header file.

86.2.3 Basic Concepts Used in C++

As stated above, C++ can be considered as a superset of the C language. Since it is even more complicated than C, we cannot discuss this language in detail. Therefore, we will simply describe a few object-oriented concepts used in C++. References to several C++ texts appear at the end of this chapter. Referring to code of Figure 86.40 facilitates understanding of the discussion that follows. More complete C and C++ examples can be found at the end of Sections 86.3 and 86.4, and Chapter 89.

- **Class:** A **class** is a user-defined type (like *sample* in *line 1* of Figure 86.40). It can be realized as an item in the real world. It has characteristics and it can perform different tasks. For example, a **class** can describe a human being, a chair, a vehicle, a CPU, a cat, etc. Everything with a specified functionality and characteristics can be modeled with a **class**.

```
class sample : parent_cls {              //line 1
public:
    sample (int name_size);              //line 3
    ~sample ();                          //line 4
    RemoveName (char *n);                //line 5
    ProcessName (char *n);               //line 6
private:
    AddName (char *n);                   //line 8
    char *name;
    char *name_list[100];
    int name_index;
    other_class *obj;                    //line 12
};
sample::sample (int name_size)           //line 14
{
    name_index = 0;
    name_list[0] = new char(name_size);  //line 17
}
sample::~sample ()                       //line 19
{
    delete name_list[0];                 //line 21
}
sample::ProcessName (char *n)            //line 23
{
    AddName (n);
    // processing the member variable name
    ...
    // processing the member variable name_list
    ...
}
...
```

FIGURE 86.40 A sample C++ code.

Classes have *attributes* (like *name_index* in Figure 86.40) and *methods* (like *AddName* in Figure 86.40). Their characteristics are described with *attributes*, while their operations and tasks are described by *methods*. Attributes are implemented with variables, while methods are implemented by functions inside the **class** definition.

An attribute and a method can be of three types: **private**, **public**, or **protected**. All other classes can access **public** methods or variables of a class. In contrast, no other class can access a **private** method and variables of a class. Private members can only be accessed in other methods of that class. Accessibility of the **protected** attributes and methods (members) are discussed later in this section.

Unlike a normal variable in C/C++, class attributes cannot be initialized in the class body. They can be initialized in a special method called *constructor* (*line 3* in Figure 86.40). Attributes can be initialized in other methods, but it is recommended to initialize them in the constructor. Every class has a constructor and a *destructor* (*line 4* in Figure 86.40). In the constructor, which must have the same name as its class, all necessary initialization and memory allocations are performed (*line 14* to *line 17* in Figure 86.40). A constructor is executed automatically when a class is created (if defined by reference) or allocated (if defined as a pointer). On the contrary, a destructor, the name of which begins with ~ followed by its class name, is executed automatically just before an instance of a class is destroyed. The necessary memory de-allocations are usually done in the destructor.

Methods in a class are usually developed outside the class definition. In this case, each method definition in a class plays the role of a *prototype* for that function (*line 5, line 6,* and *line 8* in Figure 86.40). In the method implementation (*line 14, line 19,* and *line 23* in Figure 86.40), the name of the class followed by two colons must come before the function (method) name. However, the function implementation can be enclosed in the class definition. This is not recommended unless the method body is very small. A class and one of its methods' implementation are shown in Figure 86.41.

- **Generalization:** In the C language, classes can be derived from other classes. Consider a case where class *A* is derived from class *B*. In this case, *B* is called the parent of *A*, and *A* is considered the child of *B*. For example, in Figure 86.40, *parent_cls* is the parent of *sample*. The **public** and **protected** attributes and methods (members) of the parent class are inherited automatically to the child class. Here, we can define a **protected** member of a class. A **protected** member acts like a **public** member for its children and acts like a **private** member otherwise. *Multiple inheritance* is when a class has more than one parent. A class can also have several children classes.

```
class cls {
private:
    int  att1;
    char att2;
    void mtd1 ();
    void mtd2 (char);
public:
    cls ();
    ~cls ();
};
...
void cls::mtd2 (char ch)
{
    att2 = ch;
    return;
}
```

FIGURE 86.41 A sample class in C++.

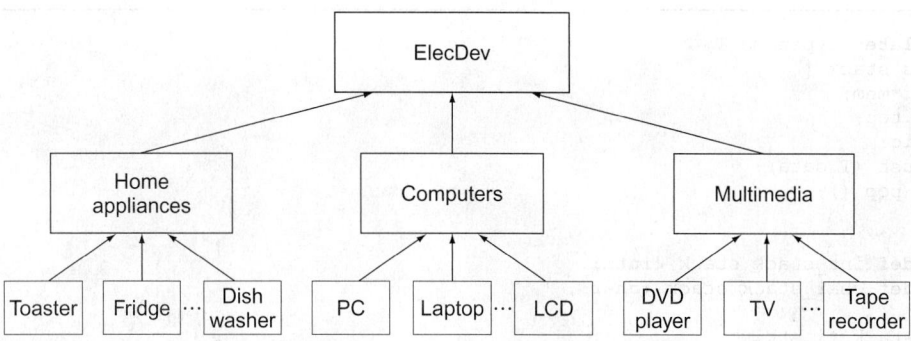

FIGURE 86.42 *ElecDev* class hierarchy.

As an example of classes, consider a class for electrical devices (named *ElecDev*). It is obvious that there are several electrical devices (like televisions, refrigerators, computers, and lamps). All of these devices have a number of common characteristics like the voltage they work with, their power dissipation, the electrical current they use, etc. These general characteristics can be defined in the *ElecDev* class, while each of them has its own special characteristics. These special characteristics can be defined as a separate class derived from the *ElecDev* class. Figure 86.42 shows the electrical devices class hierarchy.

- **Friend members:** Function *f* can access and use class *A*'s private members if *f* is defined as a friend for *A*. It is generally recommended not to use friend members except for the case when a global variable or public data members need to be used. A friend method is defined by the **friend** keyword before the method definition.
- **Object:** An *object* is a class instance (like *obj* in *line 12* of Figure 86.40). Class members can be used in a program through objects. The relationship between a class and an object is equivalent to the relationship between a data type and a variable. Therefore, similar to variables, an object can be defined statically (as a reference) or dynamically (as a pointer). When an object is defined statically, its constructor is executed after its definition and its destructor is executed before the program exits the current function. If an object is defined by a pointer to a class, the programmer is responsible for its memory allocation and deallocation. Memory for a pointer object is allocated by the *new* function as pointed out earlier. The constructor is executed after calling the *new* function. A pointer memory is destroyed by calling the *delete* function. The destructor of that class is similarly called before deleting this object.
- **Namespace:** Sets of elements in C++ (such as classes, variables, functions, etc.) that have a logical relationship with each other can be put in a *namespace*. All elements inside a namespace must have unique names. To reference an element inside a namespace, we need to use that namespace first (by writing '**using** *namespace_name*') and then use the name of that namespace followed by two colons together with the element that is intended to be accessed. Namespaces become useful for large programs.
- **Template:** Templates support *generic programming*. Using templates gives programmers the opportunity to use different data types as arguments. Suppose we want to develop a stack class. This stack must store many types of data. One way is to define a stack class and overload every method in this class for every needed data type. An alternative, more efficient method is to use *templates*. We can define a template class for *stack* and define the data type it stores, and the data of that type as arguments of the template. Then this template class can be instantiated with any type and its methods and attributes can be used in codes. Figure 86.43 presents a template class for a general *stack*. For more details on templates you can refer to Ref. [4].

```
template <typename T>
class stack {
    T *mem;
    T top;
public:
    push (T data);
    T pop ();
};
...
typedef int_stack stack <int>;
typedef char_stack stack <char>;
...
int_stack *i_stk;
//we can also use: stack <int> *i_stk;
```

FIGURE 86.43   A template class for *stack*.

86.2.4   Concurrency in C Language

Since concurrency is an important concept in hardware design, and C/C++ is a sequential program, we need mechanisms for modeling concurrency in this language.

In multiprocess operating systems (like UNIX), two or more processes can be executed independently. These processes (or threads which are *light-weighted* processes) can communicate with each other by several predefined data structures which can be shared between multiple processes. There are several functions for handling processes and their communications in the C language.

Functions for handling processes include *fork, exec, spawn, wait,* and *exit.* In addition, the data structures developed for interprocess communication (IPC) include *pipes, message queues, semaphores, shared memory,* and *sockets.* The following presents brief descriptions of the above functions and data structures.

86.2.4.1   Functions for Process Handling

A process can be created from another running process. A process has a unique identification number called *PID* and can be terminated by calling the *exit* function inside it. There are several functions that can create other processes. A number of these functions are listed below:

- **execl:** This function means *execute* and *leave.* It uses a program name as its arguments, creates a new process, runs the program specified in its arguments, and returns when the specified program terminates.
- **fork and spawn:** The functionalities of these functions are similar. However, *fork* is more flexible than *spawn* in changing the environment of the newly created process. These functions create new processes and return integers indicating every newly created process identification (PID). The returned integer is the PID of the newly created process (known as the child process). As an example, consider a system with *n* hardware modules for which we need to model concurrency to check the exact behavior of the system. For this purpose, we can use *fork* or *spawn* functions for each system module, and execute the functionality of the modules in the newly created processes.

  An example of the *fork* function is shown in Figure 86.44. This code has a *CPU,* a *memory* module, and an *IO device* which are working concurrently. The **main** function uses three *fork* functions with *pid1, pid2,* and *pid3* identifiers. This causes **main** to copy itself three times. In each instance of **main**, a PID is checked and appropriate functions related to that PID is executed.
- **wait:** This function causes the parent process to wait for the termination of its children. Its return value is the terminated child PID if it is greater than zero, or −1 if an error occurs. As an example,

```
main ()
{
    ...
    int pid1, pid2, pid3;
    Mem mem_module;
    CPU CPU_module;
    IO  IO_module;
    ...
    pid1 = fork (); // creates a process for memory module
    if (pid1 < 0)    // error
        exit (-1);
    else if (pid1 == 0) { // child process
        mem_module.run ();
        exit (0);
    }
    else { // positive pid means we are in the parent process. this is the child ID
        pid2 = fork (); // creates a process for CPU module
        if (pid2 < 0)
            exit (-2);
        else if (pid2 == 0) { // child process
            cpu_module.run ();
            exit (0);
        }
        else { //again we are in the parent process
            pid3 = fork (); // creates a process for IO device
            if (pid3 < 0)
                exit (-3);
            else if (pid3 == 0) { // child process
                IO_module.run ();
                exit (0);
            }
            else { //we are in parent process
                printf ("Three processes have been created and are working now! \n");
                exit (0);
            }
        }
    }
}//main
```

FIGURE 86.44 An example of *fork* function.

consider the parent process in Figure 86.44, which creates the three processes for *CPU, IO device*, and the *memory* module. This process which is shown in Figure 86.45 waits for the termination of each of these processes if necessary. Figure 86.45 shows a sample code that waits for the three processes, discussed above, to terminate.

There are also several other functions for process handling like *vfork, clone, waitpid, kill, killpg, system, execv, mutex* lock for threads, etc. For the description of these functions, refer to Ref. [5].

86.2.4.2 Interprocess Communication (IPC)

The discussion above illustrated the creating and handling of processes. These processes may need to exchange data between each other. Since a process has its own environment and memory segment, it cannot access variables belonging to other processes unless they follow specific protocols. In these protocols, there is a shared portion of memory (which the programmer can access implicitly or explicitly). Two or more processes can access this shared portion for reading and writing. In the following we have listed a number of these protocols with their implemented functions in C.

```
… // same as Fig. 86.44
else {
   printf ("Three processes have been created and are working now! \n");
   int tmp_pid, status;
   for (int i=0; i<3; i++)
      tmp_pid = wait (&status);
      if (tmp_pid == pid1)
         printf ("Memory terminated with status %d", status);
      else if (tmp_pid == pid2)
         printf ("CPU terminated with status %d", status);
      else if (tmp_pid == pid3)
         printf ("IO terminated with status %d", status);
}
```

FIGURE 86.45   An example of *wait* function.

- **Message queues:** A message queue is a shared memory between two or more processes. One process can send a message to a message queue and the other can receive the message from it. All processes that use the same message queue should share a common key. A message queue must first be initialized (by the *msgget* function). Then the processes can send data to the message queue (by the *msgsnd* function) or receive data from it (by the *msgrcv* function). Sending and receiving messages can be performed in nonblocking or blocking modes. In nonblocking mode, the *msgsnd* and *msgrcv* functions only return a value that shows the message is sent or received. In blocking mode, the process that calls one of these functions will be suspended until the send or receive function is finished.

- **Shared memory:** As is obvious from its name, shared memory is a portion of memory that one process creates and another can access. Its creation is performed by the *shmget* function. *shmctl* gives each process proper access to a created shared memory. The *shmctl* function can be performed only by the shared memory creator. A shared portion can be attached or detached to the address space of a process. These are done by *shmat* and *shmdt* functions, respectively. If a shared portion is attached to the memory space of a process (under proper permissions), that process can read from or write to that memory. Therefore, two processes can communicate via this portion of memory.

- **Pipe:** A pipe is a structure, similar to a file pointer or a descriptor, and makes the output of a process the input of another. There are two ways of creating a pipe. One way is to use the *popen* function which returns a *FILE* *. The two processes can use *fprintf* and *fscanf* functions to write or to read from a pipe. The *pclose* function is used for closing a pipe. Another way of creating a pipe is to use the *pipe* function that accepts an integer array with two elements. This function returns two file descriptors in this array. The first element is used for reading from a pipe, and the second element is used for writing to a pipe. This function returns 0 on success and −1 in the case of an error. The two processes can use this pipe by *read* and *write* functions. This pipe is closed by closing the two file descriptors (by the *close* function).

- **Semaphore:** Semaphore is a shared variable (simply an integer) that several processes can have access to. Each process waits for this variable to become 0. When it becomes 0, only one process can access this variable and increment it by 1. The process owning the semaphore is actually having access to a protected part of resources. When the process is finished with the resource, it releases the semaphore by decrementing it by 1. Then another process that was waiting on this semaphore will perform the same operations.

  Semaphores are useful when several processes need to access a protected resource. They are even useful for several (m) processes that need to access a limited number of protected resources (n);

here, $m > n$. In this situation, each process must wait on the semaphore to become equal or less than $n - 1$. Semaphores are initialized with the *semget* function. Their characteristics and permissions can be controlled by the *semctl* function, while their operations are performed by the *semop* function. Another set of functions for semaphores include *sem_open, sem_init, sem_close, sem_unlink, sem_destroy, set_getvalue, sem_wait, sem_trywait,* and *sem_post.* These functions are defined in the *semaphore.h* header file. Similar to message queues, the operations on semaphores can be done in blocking or nonblocking modes.

- **Socket:** Sockets are the most useful IPC elements. A socket generates a point-to-point and bidirectional communication between two processes. Sockets can also be used to connect two processes on different systems. In this situation, the socket address space is called the internet domain. This domain uses the TCP/IP protocol. Socket types include *stream* sockets, *datagram* sockets, *sequential packet* sockets, and *raw* sockets. Note that only sockets of the same type can be connected to each other. Sockets are created by the *socket* function. An internet address is assigned to them by the *bind* function. *Listen* and *connect* functions are used for their connection. One is used for the server side and the other for the client side. Data transfer can be performed by *read, write, send,* and *recv* functions. Functions *sendto, sendmsg, recvfrom,* and *recvmsg* are used for data transfer in a datagram socket type. *datagram* sockets do not require a connection to be established.

All of the communication types and protocols described above can be used in hardware design. For example, a CPU and a memory model can communicate via a message queue, or a bus arbiter can be modeled with semaphores.

86.2.5   An Example: A Data Acquisition System

Suppose we have a device that outputs data in periodic intervals. This output data is needed to be processed and analyzed by a Data Acquisition System (*DAQS*). The requirement is to design and implement this data acquisition system. There are questions that need to be answered before we are able to complete this design.

- What is the period of data preparation by the device? This specifies the worst-case delay of DAQS.
- What is the maximum budget for developing this system? This leads to the type of hardware we want to use in implementing DAQS.
- What are the main functions (operations) in this system? This is the most important implementation-related question.
- What is the system's memory limitation? This gives us an estimate of the size of our system's buffer and the number of registers and memory blocks.

Suppose that the data-sending rate is F, which means that the worst-case delay of DAQS is T (=$1/F$). Also assume that the maximum budget for the DAQS project is B. The functions that DAQS must perform in a reasonable time include:

- *FindMin:* finds the minimum value of data it has received at any point in time.
- *FindMax:* finds the maximum value of data it has received at any point in time.
- *FindVariance:* finds the variance of the data it has received at any point in time.
- *FindDifference:* calculates the difference between two consecutive data.
- *CheckRange:* compares the received data with the low and high thresholds. It will display an error message if the data is out of range.
- *FindAverage:* calculates the average of the received data at any point in time.
- *FindMean:* calculates the mean of a number of most recent data. The accuracy of the result of this task depends on the number of recently stored data.
- *Display:* displays the received data, the result of each operation, and the probable error messages.

Figure 86.46 shows a schematic view of DAQS.

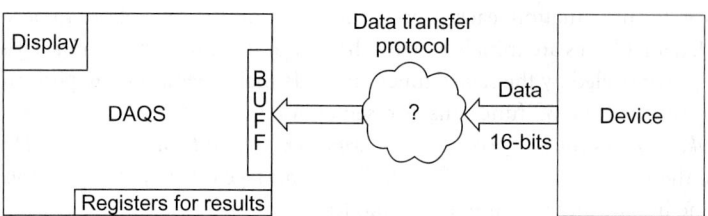

FIGURE 86.46 DAQS diagram.

```
int FindMin ()
{
   if (min > last_data)
      min = last_data;
   return min;
}
```

FIGURE 86.47 A sample function (*FindMin*) of DAQS implemented in C.

We can implement each operation by a separate function. For example, the *FindMin* function is implemented and shown in Figure 86.47.

After implementing all functions, we are ready to put all these functions together in the **main**() function. These functions can be placed one after another, or be executed in different processes.

When this model is completed we need to evaluate it. This can be done by applying large amounts of data to the DAQS C++ model and studying timing and memory usage of each of its functions. Alternatively, we can perform this analysis by performance analysis tools.

After analyzing the time and memory consumed for each part, and comparing them with T (time limit) and B (budget limit), we can decide and answer the following questions:

- What type of communication we should put between the device and DAQS?
- Can we implement DAQS on a general-purpose PC (fully software-based) or should we manufacture a special board for it (implement a few or all parts in hardware)?
- If we should construct a special-purpose board for DAQS (due to time or budget limitations), which parts can still be implemented in software?
- If we have a special-purpose board and we also have software parts, what processor should we use for our software parts? What existing CPU meets our requirements and limitations?
- Which parts should be executed in parallel and which parts can be executed sequentially?
- How much memory should we use for a required exactness in our calculations? How much memory can we use in DAQS?

After answering all of the above questions, if our specification still does not meet our requirements, we have to refine our system or change our limitations. By performance analysis, a set of questions and a set of critical decisions in the system design can be answered which could not be answered without a large amount of implementations being done. In the performance analysis process, high-level languages like C and C++ play a very important role to decrease the design time and increase the speed of defining a design specification and functional verification.

86.2.6 Our Coverage of C/C++

In this section, we discussed an overview of two high-level programming languages (C and C++). These two high-level languages are useful for developing an early prototype of a design from its model. This prototype is useful for performance analysis and partitioning of the design. The next step in a system level design can be adding more details to hardware parts defined in this step, and developing them in a more hardware-related language. The language used for this purpose can be SystemC that is discussed in the next section.

86.3 SystemC

This section gives a brief overview of SystemC, which has now become the IEEE Std-1666 standard [6]. Although many designers consider SystemC as a language, it is a library written in C++ language. About 10 years ago, in the mid-1990s, designers and engineers from Synopsys, University of California at Irvine, Frontier Design, and IMEC introduced the SystemC idea. Then in 1999 the beta version of this library (version 0.9) was released by the OSCI organization. OCSI (Open SystemC Initiative) is an independent, not-for-profit organization that contributes to development of SystemC and other related libraries. Many EDA companies and universities are working with OSCI for developing SystemC and other system-level-related libraries.

The first version of SystemC (version 1.0) was released in 2000. This version includes implementation for data types, processes, modules, ports, etc. The classes implemented for this library have a simulation kernel, which handles events. The kernel in this version was cycle-based. In the following year, in 2001, the second version of this library was released by OSCI. This version was more stable than the previous one, and its kernel, which was event-based, was the major difference between this version and the previous version. This version became version 2.0 and was much faster than version 1.0.

After releasing version 2.0, minor revisions, changes, and bug fixings were made in version 2.0.1. Then, in 2005, OSCI released SystemC version 2.1. In this version, a few problems of the previous versions were fixed, and a few concepts (like **sc_export**) are added to this library for system level design. After being finalized, OSCI submitted this version of SystemC to IEEE standardization. These days, OSCI is working on the standardization of this library. In December 2005, the IEEE institute standardized the SystemC languages as the IEEE Std-1666 [6].

SystemC version 3.0 that is being planned as the next version of SystemC will focus on operating system (OS) modeling and embedded software models. Both of these features help designers in hardware/software codesign.

The next section discusses the role of SystemC in hardware design. Then the syntax of SystemC (like its data types, module definitions, port definitions, and channel definitions) will be described followed by an example of a simple processor. The reader can refer to Refs. [6,7] for more details of SystemC syntax and constructs.

86.3.1 The Role of SystemC in Hardware Design

As stated above in Section 86.2, C++ can be used for system level modeling. SystemC is a C++ class library that has certain characteristics that lean more toward description of hardware. Because of these characteristics, SystemC has become a popular language for description of hardware when system level design is an issue. The main hardware-oriented parameters that cannot be seen explicitly in C++ and are implemented in SystemC are as follows:

- **Time:** The notion of simulation time (e.g., nanoseconds and picoseconds) has been added to this library.
- **Concurrency:** C++ is a sequential language, which means that every statement is executed after its previous one. However, hardware components in a circuit are concurrent. Therefore, to model hardware components, we need to have a means of expressing concurrency. This implies that

different portions of the hardware model must be simulated in one simulation time. This ability has been implemented in the SystemC library using processes such as **SC_THREAD** and **SC_METHOD**.

· **Hardware data types:** A software programming language has its own types such as integer, fixed point, and floating point real numbers. In contrast, hardware models have their own types for their interconnection wires. For example, **Z** (high impedance) is not defined in C++. These hardware-specific types and their operations (like AND and OR) have been implemented in the SystemC library.

The above facilities implemented in SystemC ease the use of this library as a hardware modeling language. But there can still be questions as to why we should use SystemC as opposed to other hardware-oriented HDLs like Verilog and VHDL. The original goal of developing a library in C++ was that designers needed a unified environment for developing hardware as well as software. This common language helps designers test and debug their designs that include hardware and software simultaneously.

Another goal of developing SystemC is system level design. Designers can develop their models using SystemC in any level of abstraction. These levels include system, register transfer, and even gate level. SystemC has the necessary constructs for each of these levels. But, as the name implies, most designers prefer to use SystemC at the system level. Many designers prefer to develop their RTL or gate-level designs in the traditional HDLs with which they are most familiar. Such codes are then automatically converted into SystemC enabling cosimulation and coverification of hardware and software parts.

As stated above, SystemC is widely used in system level designs, and simulates faster when compared to traditional HDLs (like VHDL and Verilog). This is because in RTL designs the number of events is increased dramatically (compared to system level designs). Simulation speed is very important at the early stages of the design, since it can be used in analyzing the whole system for partitioning. As HDLs do not have this capability, SystemC has a great advantage for being used at the system level.

The use of SystemC by designers for high-level system design is becoming more popular. After releasing this library, other facility libraries have been released by OSCI. These libraries include

· **Master-Slave library:** This library contains functions and classes for abstract communications. Different levels of communication can be modeled with this library (like bus-functional and cycle-accurate). This library is useful for designs that have one or more processors with bus-based communication between these processors and other parts of the design. Since this library models different levels of abstraction for bus communication, it can also be used for interface synthesis.

· **Verification library (SCV):** This library contains several methods and classes for helping designers verify their SystemC designs more rapidly and easily. It facilitates developing better testbenches. This library contains methods for assertions, random number generation, etc.

Figure 86.48 illustrates the flow of using SystemC for designs.

86.3.2 SystemC Language Description

In this section, we will describe a number of necessary SystemC constructs for hardware design. The constructs described are the basic constructs of SystemC version 2.1. More details on these constructs and more complex constructs can be found in Refs. [6,7].

The examples shown for each construct are parts of Sayeh CPU [8], which we will demonstrate its SystemC implementation at the end of this section.

86.3.2.1 Design Parts: Modules

A hardware design consists of several parts that work independently and concurrent with other modules. In SystemC, a hardware component description of which is referred to as a module, begins

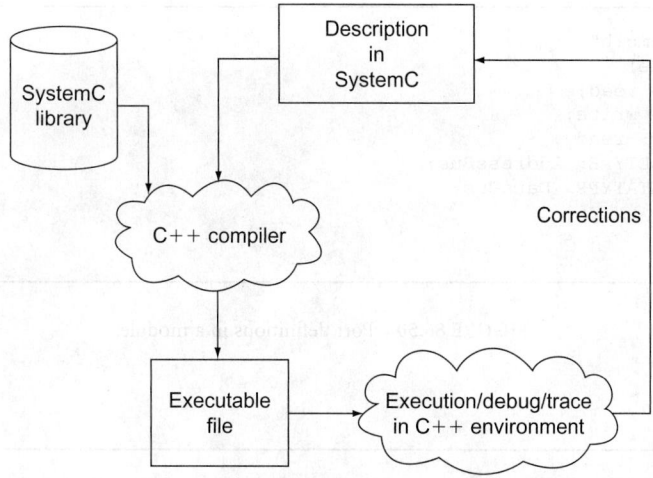

FIGURE 86.48 A process for using SystemC.

```
#include "systemc.h"
SC_MODULE (module_name)
{
    module-body
};
```

```
#include "systemc.h"
SC_MODULE (sayeh) {
    Port definitions
    Member data
    Other modules' instances
    Constructor
    Destructor
    Member functions
};
```

FIGURE 86.49 SC_MODULE syntax.

with the **SC_MODULE** keyword. The syntax and an example of the use of **SC_MODULE** are shown in Figure 86.49. A module defined by **SC_MODULE** is the smallest functional part of a design that can be instantiated hierarchically in other modules or in a testbench.

As illustrated in Figure 86.49, each module has a body. This body contains port definitions, member data instances, instantiations of other modules, constructor (which is mandatory), destructor, and member functions (including processes and helper functions). We will describe the syntax of each section below.

86.3.2.1.1 *Port Definitions*

In this part, communication between modules is defined. Figure 86.50 shows ports for *cache* module. As shown in this figure, the input and output ports in *cache* include the SystemC basic port types **sc_in** for input ports (e.g., *read*), **sc_out** for output ports (e.g., *ready*), and **sc_inout** for bidirectional ports (e.g., *AddressBus*). In addition to the above port types, there can be more general and more complex port types

```
#include "systemc.h"
SC_MODULE (cache) {
   sc_in <bool> read;
   sc_in <bool> write;
   sc_out <bool> ready;
   sc_inout <ADDTYPE> AddressBus;
   sc_inout <DATATYPE> DataBus;
   ...
}
```

FIGURE 86.50   Port definitions in a module.

```
     data-type member-instance;
(a)  data-type member-array[n];
     data-type *member-pointer;
```

```
     #include "systemc.h"
     SC_MODULE (sayeh) {
        ...
     public:
        bool m_halt_executed;
        ...
(b)  private:
        REG16 m_IR, m_PC; //registers
        bool  m_Carry, m_Zero; //flags
        REG16 m_RegFile[8]; //register file
        ...
     };
```

FIGURE 86.51   (a) Member data definition syntax. (b) An example.

defined in SystemC. These ports use **sc_port** in their definition. More details on ports in SystemC can be found in Section 86.3.2.4.1.

86.3.2.1.2   *Member Data Instances*
As **SC_MODULE** is simply a **struct** in C++, member data instances are the same as member variables in **class** and **struct** definitions. The syntax of member data instances is shown in Figure 86.51. The type of these data instances can be C/C++ or SystemC types. Those of SystemC types are discussed in Section 86.3.2.3.

As shown in the example of Figure 86.51(b), registers and CPU status flags are modeled as member data instances. In this example, *m_RegFile* is a block of eight 16-bit registers.

86.3.2.1.3   *Instantiations of Other Modules*
To support hierarchy in SystemC designs, the designer must have the ability to instantiate a module inside another. In SystemC, a pointer or reference to an instantiated module is defined in the definition of instantiating module.

```
           #include "systemc.h"
           SC_MODULE (sayeh) {
              ...
(a)           cache *internal_cache;
              ...
           }
```

```
           #include "systemc.h"
           SC_MODULE (sayeh) {
              ...
           public:
              sc_signal <bool> c_read, c_write;
              ...
(b)           SC_CTOR (sayeh) {
                 internal_cache = new cache;
                 internal_cache.read (c_read);
                 ...
              }
              ...
           }
```

FIGURE 86.52   (a) Module instantiation in another module. (b) An example.

Suppose in our Sayeh CPU example there is a need for an internal cache (a cache module inside the CPU). For this purpose, we need to develop a cache model in SystemC and implement it in an **SC_MODULE**. Then we have to instantiate this module in the *sayeh* module. The example is shown in Figure 86.52. As shown, *internal_cache* is an instance of the cache module.

Note that if we define the *internal_cache* in Figure 86.52 as a pointer, we have to allocate memory for it using the C++ *new* function (In this text, we will use new as a verb for such allocations). This has to be done in the *sayeh* module constructor before binding its ports. Figure 86.52(b) shows an example of such a binding.

86.3.2.1.4   *Constructor*
Similar to C++ classes, each module needs a special part where the necessary initializations are done. This part is called the constructor of a module, and is an essential part of every module. A constructor is called when the pointer to a module is *new*ed (in case of a pointer instance of a module) or when the module is defined (in case of a nonpointer instance of a module).

Constructors, in both C++ and SystemC, perform initializations. In SystemC the concurrent parts of a module are also defined in the constructor. These concurrent parts are named processes, which will be described in Section 86.3.2.2.

Like C++, constructors in SystemC modules may have arguments. If a constructor has arguments, it must be defined within its corresponding module with **SC_HAS_PROCESS** (Figure 86.53(a)).

In the case of no arguments, a constructor can be defined using **SC_CTOR** keyword (Figure 86.53(b)). In this case, its implementation must be included in the module definition. Otherwise, it can be implemented outside the module definition. As an example, in the Sayeh CPU, we can define the clock period and clock start time as the constructor arguments or as static definitions. The implementation of *sayeh* constructor of Figure 86.53(a) must be included elsewhere, such as a separate *.cpp* file. In a situation where a constructor is too complex to be included in a module definition, **SC_HAS_PROCESS** can be used without arguments.

86.3.2.1.5   *Destructor*
Destructors are responsible for necessary memory deallocations and pointer settings. Like C++, every module in SystemC may have a destructor. Having destructors is not mandatory. The syntax of destructors in SystemC is the same as destructors in C++ [4].

(a)
```
#include "systemc.h"
SC_MODULE (sayeh) {
   ...
   sayeh (sc_time clk, sc_time start);
   ...
   SC_HAS_PROCESS (sayeh);
   ...
};
```

(b)
```
#include "systemc.h"
#define CLK_PERIOD 4, SC_NS
#define CLK_START 5, SC_NS
SC_MODULE (sayeh) {
   ...
   SC_CTOR (sayeh) {
      sc_time p (CLK_PERIOD);
      m_clk_period = p;
      m_halt_executed = false;
      ...
   }
   ...
};
```

FIGURE 86.53   Constructors in a module. (a) SC_HAS_PROCESS example. (b) SC_CTOR example.

```
#include "systemc.h"
SC_MODULE (sayeh) {
   ...
private:
   void sayeh_func ();
   void reset ();
   void fetch ();
   void decoder_execute (REG16 instr);
   ...
};
```

FIGURE 86.54   An example of member function definitions in a module.

86.3.2.1.6   *Member Functions*

Functionality of a module is defined in terms of its member functions (Figure 86.54). These functions are defined in a module and can be implemented in a separate file. In SystemC, member functions are divided into two main categories. They can represent concurrent parts of a module or they can be utility functions, which are developed for making the implementation of the concurrent parts easier. We will describe concurrency in Section 86.3.2.2. These two categories are not syntactically different and it is the module constructor that differentiates the concurrent functions from the utility functions. The syntax of module member functions is similar to that of class member functions in C++, which is shown in Figure 86.54.

Up to this point, we have defined a module (the example is *sayeh* module). The characteristics of modules are in their definition. To define the functionality of a module, we need to implement each of its member functions. The type of each member function (whether it is a concurrent process or a helper function) specifies which SystemC constructs should or should not be used in the function body. The details of member function bodies will be discussed in Section 86.3.2.2.

86.3.2.2 Concurrent Parts: Processes

As mentioned before, to model hardware components we need to model concurrency. In VHDL, concurrent statements execute simultaneously with other statements. A SystemC process is defined by a member function in a module, and is registered within the module's constructor. When a member function is registered, we have the option of defining the signals it is sensitive to, and its activation mechanism. A concurrent process in SystemC can be defined as a method (**SC_METHOD**), a thread (**SC_THREAD**), or a clocked thread (**SC_CTHREAD**). Member functions defined as such in a constructor are executed concurrently with other methods, threads, or clocked threads.

86.3.2.2.1 *Methods*

A method, like reset in Figure 86.55, is a process in SystemC. A method is used for describing a concurrent process like a **PROCESS** statement with sensitivity list in VHDL. A method is defined with the **SC_METHOD** keyword in SystemC. A sensitivity list is a list of events on specified signals. If any of these events occur, the statements in the method will be executed. For example, to model a memory that can work with the Sayeh CPU, we can define its resetting operation as a simple method sensitive to the positive edge of the *Reset* signal. In Figure 86.55, every time the *Reset* signal is changed from false to true the message in the *reset* function, the implementation of which is shown in Figure 86.55(b), will be displayed on the output screen.

If a value is assigned to a local variable in a method, it will not remain in that local variable on its next invocation. Suppose that the *Reset* signal in Figure 86.55(a) is changed from false to true at times 9 and 17 ns. If the *reset* method of Figure 86.56 is used instead of that of Figure 86.55(b) at 9 ns, variable

```
        #include "systemc.h"
        SC_MODULE (memory) {
            ...
            sc_in <bool> Reset;
            ...
            SC_CTOR (memory) {
                ...
(a)             SC_METHOD (reset);
                sensitive << Reset.pos ();
                ...
            }
            ...
        private:
            void reset ();
            ...
        };
```

```
        void memory::reset ()
        {
(b)         MemDataReady = false;
            cout << "memory reset finished!"
        }
```

FIGURE 86.55 (a) SC_METHOD registration. (b) A method function.

```
void memory::reset ()
{
   int tmp = 1000;
   MemDataReady = false;
   tmp++;
}
```

FIGURE 86.56 An example of local variables in methods.

```
void memory::reset ()
{
   if (Reset.read () && m_count < 1000) {
      do_reset ();
      m_count++;
   }
   if (m_count == 1000) {
      cout << "Number of resets exceeds 1000. Use master reset!";
      next_trigger (MasterReset.posedge_event ());
   }
   if (MasterReset.read ()) {
      do_master_reset ();
      m_count = 0;
      next_trigger (Reset.posedge_event ());
      cout << "master reset done!";
   }
}
```

FIGURE 86.57 The next_trigger example.

tmp will become 1001. But at 17 ns, *tmp* will not become 1002, and will again change to 1001. If variables that retain their values are needed in methods, they have to be defined in the definition of the method's corresponding module.

Another important point about methods is that they cannot include **wait**, and can only have sensitivity lists. If a **wait** function is used in the body of a method, a run-time error will occur.

In spite of this, there is still a way to add dynamic sensitivity to a method. This can be done by the **next_trigger** function which will be discussed in the following section.

- **next_trigger:** If the **next_trigger** function is used in a method, the default sensitivity list will be overwritten. Consider the *reset* method in the memory module of Figure 86.55. The static sensitivity list of this method is the positive edge of the *Reset* signal. We rewrite this method as shown in Figure 86.57. If the *Reset* signal makes 1000 low-to-high transitions (the initial value of *m_count* is 0), the default sensitivity list will be replaced by the positive edge of *MasterReset* signal. After one master resetting, the sensitivity will again be overwritten to *reset.posedge_event* which is its original sensitivity (by executing the second **next_trigger** function).

Note that using **next_trigger** does not cause context switching in SystemC. It means that the statements after **next_trigger** will still be executed.

For example, in Figure 86.57, after running *next_trigger* (*Reset.posedge_event()*) the statement *cout* << *"Master reset done!"* will also be executed.

The format of the **next_trigger** arguments is similar to those of the **wait** function (see Section 86.3.2.2.2).

Another important point about methods is their initialization process. Each method is executed once at the beginning of the simulation. In a case where the initial execution should not occur, **dont_initialize** function is written in the constructor, after the method registration part.

86.3.2.2.2 Threads

A thread is a SystemC process that is used for describing concurrent processes with **wait** functions. Unlike methods, which are executed each time an event happens on their sensitivity list, threads are executed only once during the simulation. This means that if a thread process exits, it will never be executed again during the simulation, even if an event happens on its sensitivity list. That is why most of the functions defined as a thread use an explicit infinite loop.

A SystemC thread is defined by the **SC_THREAD** keyword and, similar to **SC_METHOD**, it is registered in the constructor of a module. Threads can use both static and dynamic sensitivity methods. Static sensitivity refers to the sensitivity list, while dynamic sensitivity is when the **wait** statement is used in the body of a thread. When a thread executes a **wait** statement, the control of the simulation returns to the simulation kernel. This means that **wait** suspends the execution of that thread. When the conditions of the arguments used in a **wait** statement are met, the execution of the thread will be continued from the statement after that **wait** statement. The reset method which was rewritten in Figure 86.57 is implemented with a thread, instead of a method, and is shown in Figure 86.58(a).

(a)
```
#include "systemc.h"
SC_MODULE (memory) {
    ...
    SC_THREAD (reset);
    sensitive << Reset.pos ();
    ...
};
```

(b)
```
void memory::reset()
{
    int local_count = 0;
    while (1) {
        wait (); //waits for the static sensitivity list events
        if (Reset.read () && local_count < 1000) {
            do_reset ();
            local_count++;
        }
        if (local_count == 1000) {
            cout << "Number of resets exceeds 1000. Use master reset!";
            wait (MasterReset.posedge_event ());
        }
        if (MasterReset.read ()) {
            do_master_reset ();
            local_count = 0;
            cout << "Master reset done!";
        }
    } //while(1)
}
```

FIGURE 86.58 (a) SC_THREAD registration. (b) An example.

As discussed above, threads can only work with **wait** statements to switch simulation to other modules. This kind of context switching makes threads more easy to use for designers. Although they are easier to use, methods are considered as light-weight processes compared to the threads. That is, they are simulated faster than threads. Since **wait** statements play an important role in developing threads, the following section describes this function in more details.

wait: As discussed earlier, when a thread executes a **wait** statement it will be suspended and the simulation control continues with other processes. The execution of a thread will be resumed when the *events* or *timeouts* specified as the **wait** arguments expire.

An *event* (defined by the **sc_event** keyword in SystemC) happens at a specified simulation time. Events have no values. There are several implemented functions for SystemC types which return **sc_event**. For example, **posedge_event** is used to indicate if a low to high transition occurs on a signal.

wait accepts several kinds of arguments. The different argument configurations of **wait** are listed below.

- **wait():** This function, with no arguments, suspends its enclosing thread, and that thread resumes when an event happens on its static sensitivity list.
- **wait(event):** The argument of this function is of type **sc_event**. For example, *wait(clock.negedge_event())* suspends its enclosing thread, and that thread resumes at the negative edge of line *clock*.
- **wait(timeout):** The argument of this function is of type **sc_time**. For example, *wait(12, SC_NS)* suspends its enclosing thread, and that thread resumes 12 ns after the current simulation time.
- **wait(event1 | event2 | ...):** This function suspends its enclosing thread, and that thread resumes if any of event1, event2, etc. happen.
- **wait(event1 & event2 & ...):** This function suspends its enclosing thread, and that thread resumes if all events event1, event2, etc. occur.
- **wait(timeout, event):** The first argument is of type **sc_time**, while the second argument is of type **sc_event**. For example, *wait(12, SC_NS, clock.negedge_event())* will either return because of a negative edge of *clock* or as a result of 12 ns timeout.
- **wait(timeout, combination_of_events):** The behavior of this function is similar to *wait(timout, event)* except that its second argument is not a single event. It is the OR'ed list or the AND'ed list of the events depending on the operator used.

An important point about the **wait** statements in a thread is that the values of the local variables are saved before context switching and they are restored when the thread is resumed. For example, in Figure 86.58(b) *local_count* is a local variable which is initialized to zero. Before the *wait(MasterReset.posedge_event())* statement the value of *local_count* is 1000 and it remains 1000 when *MasterReset* positive edge event happens. This shows another difference between **SC_METHOD** and **SC_THREAD**.

86.3.2.2.3 Clocked Threads

A clocked thread is another type of a process, which is a specialized version of **SC_THREAD**. This revised version is defined by the **SC_CTHREAD** keyword in SystemC. There are discussions for deprecating **SC_CTHREAD** in the latest version of SystemC library, but for deleting this concept, **SC_METHOD** and **SC_THREAD** must be revised. This is due to the fact there are facilities implemented in **SC_CTHREAD**, which cannot be found in the other two processes.

Clocked threads are threads, statically sensitive to a specified edge of a specified signal. An example of a clocked thread registration is shown in Figure 86.59.

As shown in Figure 86.59, the *controller* function of the RTL version of Sayeh is sensitive to the positive edge of *clock*.

Similar to threads, in clocked threads **wait** statements can be used for context switching. The acceptable arguments for **wait** in clocked threads are different from those acceptable for threads. The legal **wait** statements are

- **wait():** If we use **wait** without any arguments, the clocked thread will be suspended until the next specified edge of the specified clock signal.

```
#include "systemc.h"
SC_MODULE (RTLsayeh) {
   sc_in <bool> clock;
   ...
   SC_CTOR (RTLsayeh) {
      SC_CTHREAD (controller, clock.pos ());
      ...
   }
   ...
private:
   void controller ();
   ...
};
```

FIGURE 86.59 SC_CTHREAD registration.

```
SC_MODULE (RTLsayeh) {
// refer to Fig.86.59 for SC_CTHREAD registration
...
   SC_CTHREAD (controller, clock.pos ());
   reset_signal_is (Reset, true);
...
};

void RTLsayeh::controller ()
{
   cout << "This is the beginning of the control process …";
   while (1) {
      ...
   }
}
```

FIGURE 86.60 An example of reset_signal_is.

- **wait(N):** This version of **wait** uses an integer as its argument. This function suspends the clocked process for the N clock cycles.
- **wait_until(condition):** This version of **wait** statement suspends the clocked process until the condition on the specified edge of the clock is met. Note that each condition is checked only at the specified edge of the clock. Every signal used in the condition statement must be a delayed signal. In other words, signals used must be of format *signal.delayed()*. This function returns the value of *signal* at the end of a delta cycle. A delta cycle in SystemC is similar to delta time in VHDL.

If we use other formats of **wait** in a clocked thread, it will cause a run-time error. There are other utilities developed for clocked threads, including

- **reset_signal_is(signal, active_value):** This function determines the signal for resetting a clocked thread. The value of this signal is checked at the specified edge of the clock. If the value is equal to the specified active value (defined as the second argument), the clocked thread function will be executed from the beginning. For example, in Figure 86.60, each time the *Reset* signal is true at the positive edge of the *clock*, the statement *"This is the beginning of the control process …"* is displayed in the output window.

```
void RTLsayeh::controller ()
{
   while (1) {
      W_BEGIN
         watching (Reset.delayed () == true);
      W_DO
         //controller state machine
      W_ESCAPE
         cout << "Reset is high at this clock cycle!";
         do_reset ();
         wait ();
      W_END
   }
}
```

FIGURE 86.61 An example of local watching.

- **watching(signal.delayed() == active_value):** There are two types of the watching method that is used in clocked threads. They are global and local watchings, and an example of a local watching is shown in Figure 86.61. Global *watching* methods are used in the body of the module constructor after the clocked thread registration. This kind of watching method works similar to the **reset_signal_is** method. A local *watching* method defined in a clocked thread causes the thread to jump to a specified block, if its condition becomes true on the edge of the clock. A local *watching* method is defined and used with three blocks of code bracketed by **W_BEGIN**, **W_DO**, **W_ESCAPE**, and **W_END** keywords. Referring to Figure 86.61, these blocks are definition block, normal operation block, and watching operation block, that are described below.

  - **Definition block:** This block comes between **W_BEGIN** and **W_DO**. Local *watching* methods are defined in this block.

  - **Normal operation block:** This block comes between **W_DO** and **W_ESCAPE**. If the condition of the specified local *watching* methods is not met, this block will be executed at the invocation of the clocked thread.

  - **watching() operation block:** This block comes between **W_ESCAPE** and **W_END**. If the condition of the specified local *watching* methods becomes true, this block is executed instead of the normal operation block.

Figure 86.61 shows an example of a local watching related to the *controller* clocked thread that was defined in Figure 86.59.

The **halt** function is used to stop a clocked thread. Using this function in **SC_CTHREAD** is similar to the use of **return** in an **SC_THREAD**.

In this section, the basics needed for developing designs in SystemC were explained. The next section explains data types in SystemC.

86.3.2.3 SystemC Data Types

SystemC library has several predefined data types that can be used for modules' member variables, internal signals, and communication ports. In addition to these types, designers are free to use any of the allowed data types in C++, e.g., **int**, **char**, and user-defined types. A brief description of SystemC predefined types follows. For more details, and other SystemC types not presented here, readers should refer to Ref. [6].

```
sc_string name ("0 base [sign] number [e[+|-] exp]");
```

FIGURE 86.62 sc_string syntax.

86.3.2.3.1 Numeric and Arithmetic Data Types

- **sc_string:** This is a unified C-style representation for numbers. Figure 86.62 shows the **sc_string** syntax. Using this format, -2 in **sc_string** decimal format is represented by **-0d2**. The same number in 4-bit **sc_string** binary format is represented by **0b1110**, and in 4-bit **sc_string** unsigned binary by **0b0010**.
- **sc_int<n>, sc_uint<n>:** These types represent integers and unsigned integers. As stated earlier, the C++ **int** and **unsigned int** can also be used. But these SystemC data types have several utility functions that make them easier to use for a hardware design. You can also define the number of bits for these types. For example, **sc_uint<5>** is used for declaring a 5-bit unsigned integer with a maximum value of 31.
- **sc_bigint<n>, sc_biguint<n>:** These types are similar to **sc_int<n>** and **sc_uint<n>**, but they are used for very large numbers that C++ integer types cannot support. For example, **sc_biguint<70>** is used for integers with a maximum value of 2^{70}-**1**. This type is usually used for integer representation of wide data or address busses in a hardware design.
- **Fixed point types:** This group of data types is used for fixed-point numbers. It includes **sc_fixed**, **sc_ufixed**, **sc_fixed_fast**, **sc_ufixed_fast**, **sc_fix_fast**, and **sc_ufix_fast**.

86.3.2.3.2 Multivalue Data Types

- **sc_logic, sc_lv<n>:** **sc_logic** is a 4-valued data type. It can be used as a signal type. **sc_logic** values can be **SC_LOGIC_1**, **SC_LOGIC_0**, **SC_LOGIC_X**, and **SC_LOGIC_Z**. For easier use, '**1**', '**0**', '**X**', and '**Z**' can be used respectively. **sc_lv<n>** is the vector form of **sc_logic** and is equivalent to a vector of size *n* of **sc_logic** type.

All of the above data types have predefined overloaded operations. For example, ==, >=, <=, >, <, = (assignment), +, +=, −, −=, **&**, and I have been implemented for the above data types (if applicable). In addition, several utility functions for these predefined data types have been implemented that makes them easier to use than the C++ data types. Several utility functions for **sc_in<n>** type are **range**, **to_int**, **to_ulong**, **length**, **value**, and **to_string**. These functions help designers develop their system level designs faster and develop more readable codes.

The focus of the above SystemC overview was on concurrent parts and data types. In the next section, we will describe methods used for interconnecting various modules.

86.3.2.4 Communication between Modules

Communication between SystemC modules or concurrent processes is discussed here. First, ports and signals are described, channels and primitive channels will be discussed then, and at the end a new concept developed in SystemC 2.1, which is called **sc_export**, will be described.

86.3.2.4.1 Ports and Signals

For communicating with the outside world, a hardware module has to receive inputs and provide outputs. These inputs and outputs constitute the ports of that module. In SystemC, ports are generally defined with the **sc_port** class. **sc_port** syntax is shown in Figure 86.63.

As an example for a port, consider the partial code of Figure 86.64. In this figure, the **sc_port** shown is used in place of *MemRead*, *MemWrite*, and *MemDataReady* signals.

```
#include "systemc.h"
SC_MODULE (module_name) {
    sc_port <port_type> port_name;
    ...
};
```

FIGURE 86.63 sc_port syntax.

```
#include "systemc.h"
#include "mem_communication.h"
SC_MODULE (sayeh) {
    ...
    sc_port <COMM_TYPE> MemComm;
    //the above port is instead of MemRead, MemWrite, and MemDataReady signals
    ...
};
```

FIGURE 86.64 sc_port example.

```
      #include "systemc.h"
      SC_MODULE (module_name) {
         sc_in <type> input_port_name;
(a)      sc_out <type> output_port_name;
         sc_inout <type> input_output_port_name;
         ...
      };
```

```
      #include "systemc.h"
      SC_MODULE (sayeh) {
         sc_inout <ADDRESSTYPE> AddressBus;
         sc_inout <DATATYPE> DataBus;
(b)      sc_in <bool> ExternalReset, MemDataReady;
         sc_out <bool> MemRead, MemWrite;
         sc_out <bool> IORead, IOWrite;
         ...
      };
```

FIGURE 86.65 (a) Input and output port syntax. (b) An example.

There are also other port definitions that are simpler than **sc_port**. Although **sc_port** is used for system level implementations, these definitions are closer to the RTL descriptions. These port definitions are **sc_in<>**, **sc_out<>**, and **sc_inout<>**. Examples of these port definitions, showing their syntax are shown in Figure 86.65.

sc_in, **sc_out**, and **sc_inout** have several predefined methods that are used by the processes in modules for event passing. A number of these methods are **posedge_event**, **value_changed_event**, and

default_event methods belonging to the **sc_in** class. Note that **posedge_event** and **negedge_event** are used for special types of ports like **bool**. These functions are not valid for types like **int** or **sc_int<n>**.

Other classes developed for event passing in SystemC are **sc_signal** and **sc_buffer**. **sc_signal** is similar to **SIGNAL** in VHDL. **sc_signal**s are used to connect module ports in a higher level of hierarchy in a system. They are also used when there is a need to pass events inside a module. Figure 86.66 shows an example of **sc_signal** declaration and usage.

There are predefined methods for **sc_signal** that deal with events. Examples of **sc_signal** methods are **posedge_event**, **negedge_event**, and **value_changed_event**.

sc_buffer's behavior is similar to **sc_signal**, but differs in the notification of the **value_changed_event** method. This method in **sc_signal** is notified only when a *write* occurs and the value of that signal differs from its previous value. In contrast, this method in **sc_buffer** is notified when a *write* occurs, even if the value of the signal remains unchanged.

Like signals in VHDL, if a value is written to a signal of type **sc_signal** or **sc_buffer**, the signal value will be changed in the next delta cycle (**SC_ZERO_TIME**).

Another signal type in SystemC is **sc_resolved_signal**. This type is used when more than one module wants to write to a signal. This type is similar to VHDL STD_LOGIC or other resolved types.

86.3.2.4.2 *Channels*

Modules in a design can communicate with each other through events. But with complex handshakings between modules, using these events becomes risky since a number of them may be missed. For this reason, designers use SystemC primitives or their custom channels for complex data communications.

(a)
```
#include "systemc.h"
SC_MODULE (sayeh) {
    ...
    sc_signal <bool> internal_halt;
    ...
    SC_CTOR (sayeh) {
        SC_METHOD (emergency_halt);
        sensitive << internal_halt;
        ...
};
```

(b)
```
#include "systemc.h"
#include "sayeh.h"
...
sc_main (argc, argv)
{
    ...
    sc_signal <bool> external_reset;
    memory mem;
    sayeh cpu;
    ...
    memory.Reset (external_reset);
    ...
    cpu.Reset (external_reset);
    ...
}
```

FIGURE 86.66 (a) sc_signal declaration. (b) sc_signal usage.

SystemC primitive channels have several methods for safe use of shared data. The simplest primitive channels include the following classes:

- **sc_mutex:** Mutex is a shared resource to which several modules can access without any collisions. Each module must lock a mutex to use it. If at this time any other module wants to use the mutex, it must wait until the consumer module unlocks it. Modules can use SystemC mutexes (defined by **sc_mutex**) in two ways: blocking and nonblocking. If a module calls the blocking methods of a mutex, it will be suspended until the mutex is unlocked. In contrast, a nonblocking method will only return false if the mutex is locked. The methods for a mutex are **lock** for blocking lock of a mutex, **try_lock** for nonblocking lock, and **unlock** for unlocking a mutex. Figure 86.67 shows an arbiter example that uses a **sc_mutex** in blocking mode.

- **sc_fifo:** FIFO is a shared resource which simply works as a queue, meaning that the first data read from it is the first written to it. A FIFO is implemented in the **sc_fifo** class in SystemC. The default length of **sc_fifo** is 16. Like **sc_mutex**, there are blocking and nonblocking methods for accessing **sc_fifo**. Similarly, when a module wants to read from a FIFO, it will have to wait if the FIFO is empty. In the blocking mode, the module wanting to read the FIFO is suspended until it is no longer empty. In nonblocking mode the read function only returns false if the FIFO is empty and true if it is not empty. The methods implemented for **sc_fifo** include **write** for blocking write to a FIFO, **nb_write** for nonblocking write, **read** for blocking read from a FIFO, **nb_read** for nonblocking read, **num_available** for the number of written but not yet read elements, **num_free** for free elements (not yet written) of a FIFO, **data_written_event** for an event that shows the occurrence of a write operation, and **data_read_event** for an event that shows the occurrence of a read operation.

```
SC_MODULE (arbiter)
{
    ...
    resource1 *r1;
    resource2 *r2;
    ...
    sc_mutex arb_m;
    SC_CTOR (arbiter) {
        SC_THREAD (rf1);
        SC_THREAD (rf2);
        ...
    }
    ...
void rf1 ();
void rf2 ();
...
}
void arbiter::rf1 ()
{
    while (1) {
        arb_m.lock ();
        process_r1 ();
        arb_m.unlock ();
    }
}
...
```

FIGURE 86.67 sc_mutex example.

- **sc_semaphore:** Semaphore is a limited number of shared resources to which several modules can access. Semaphore is similar to mutex, but mutex has only one resource. In SystemC, semaphores are implemented in the **sc_semaphore** class. Several modules can access a semaphore, use one of its available resources, and free that resource. Similar to **sc_fifo** and **sc_mutex**, **sc_semaphore** has blocking and nonblocking modes. In the blocking mode if there are no free resources in a semaphore for a module, that module will be suspended until a resource in the semaphore becomes free. In the nonblocking mode, the function only returns false if all resources are busy. The functions of **sc_semaphore** are **wait** for blocking mode resource allocation, **try_wait** for nonblocking mode resource allocation, **get_value** for the number of available resources in a semaphore, and **post** to unlock a previously locked resource.

86.3.2.4.3 sc_export

A new method of communication has been developed in SystemC 2.1, which is also named *portless channel access*. This kind of channel access is necessary when module *m* of Figure 86.68(a) wants to access a member function of a channel instance in module *c* of this figure. In this case, there must be a communication between the channel instance *ch* of module *c* with module *m*. This communication that is implemented with **sc_export** is shown in Figure 86.68(b).

In this example, the internal channel *ch*, in module *c* is connected to its communication means with other modules (*c_ch*) in the constructor of module *c*. In contrast, module *m* needs to use module *c*'s internal channel in its *f* function through its *m_ch* port. In **sc_main**, this communication is established by *m_inst.m_ch (c_inst.c_ch)*; statement.

Another example of using **sc_export** is the memory communications in Sayeh CPU as shown in Figure 86.69. As we discussed, we can encapsulate the memory operations in a channel. When using **sc_ports**, we have to instantiate a channel in **sc_main** and use this channel instance in both memory and CPU (bind this instance to CPU and memory ports). If we instantiate our channel in the memory module, we need to use **sc_export** in the memory. By using this concept, the CPU can access the channel member functions directly (see Figure 86.69).

Another advantage of using **sc_export** is that it can be bound directly to another **sc_export** of a module in the hierarchy. Using **sc_port**s in contrast needs an explicit connection, while **sc_exports** do not require such connections.

86.3.2.5 Testbenches in SystemC

In SystemC the **sc_main** function usually acts as a testbench for the whole system. We can define and instantiate modules and connect their ports within this function. **sc_main** in SystemC is a function similar to the *main* function in C/C++. Generally, the simulation and signal tracing functions are called in **sc_main**. Every design in SystemC must have one **sc_main** function. **sc_main** arguments (*argc* and *argv*) are similar to those in the *main* function in C/C++. Several methods useful in developing testbenches in SystemC are described here.

86.3.2.5.1 Module Instantiation

To instantiate a developed module in **sc_main** (or any other module), two approaches are available: instantiation by position and instantiation by name.

These alternatives are as follows:

- **Instantiation by position**
  - Syntax: *module_name (signal_name1, signal_name2, …)*;
  - Example:
        sayeh cpu;
        sc_signal<ADDRESSTYPE> abus, …;
        cpu (abus, dbus, …);

- Instantiaion of *cpu* connects *abus*, *dbus*, and other declared signals to ports of the *sayeh* module. The order of signals in this instantiation must be the same as those of the *sayeh* module.

```
SC_MODULE (m) {
    ...
private:
    void f ();
    ...
};
void m::f ()
{
    ch->ch_func ();
    ...
}


SC_MODULE (c) {
    ...
    c_channel ch;
    ...
};
```
(a)

```
SC_MODULE (m) {
    sc_port <c_channel> m_ch;
    ...
};
void m::f () {
    m_ch->ch_func ();
    ...
}


SC_MODULE (c) {
    sc_export <c_channel> c_ch;
    c_channel ch;
    ...
    SC_CTOR (c) {
        c_ch (ch); //export binding
        ...
    }
    ...
};


sc_main () {
    m m_inst;
    c c_inst;
    ...
    m_inst.m_ch (c_inst.c_ch);
    ...
}
```
(b)

FIGURE 86.68 (a) Module *m* needs to access *c_channel*. (b) Communication between *m* and *c* using sc_export.

- **Instantiation by name**
  - Syntax: *module_name.port_name (signal_name_in_sc_main)*;
  - Example:
    ```
    sayeh cpu;
    sc_signal <DATATYPE> dbus, …;
    cpu.DataBus(dbus);
    …
    ```

  - In this way of instantiation, individual ports of the *cpu* instance must be connected to their respective local signals. The above example connects the declared *dbus* signal to *DataBus* port of *sayeh* module.

```
         #include "systemc.h"
         SC_MODULE (memory) {
             ...
             sc_export <COMM_TYPE> MemComm;
             COMM_TYPE ex_MemComm;
(a)          ...
             SC_CTOR (memory) {
                 MemComm (exMemComm);
                 ...
             }
             ...
         };


         sc_main (argc, argv)
         {
             memory mem;
             sayeh cpu;
(b)          ...
             cpu.MemComm (mem.MemComm);
             ...
         }
```

FIGURE 86.69 (a) Memory example using sc_export. (b) Port binding.

86.3.2.5.2 sc_start

This function starts the simulation process and returns when the simulation time limit is reached or when there are no more events left. **sc_start** can be used in each of the following formats:

- **sc_start ():** Simulation terminates when no events are left in the scheduler.
- **sc_start (double, sc_time_unit):** Simulation terminates when the simulation time reaches the specified time in the argument. An example is *sc_start (10, SC_NS)*.
- **sc_start (const sc_time&):** This format is the same as *sc_start (double, sc_time_unit)*. An example is *sc_start (sim_time);* where *sim_time* is a variable of type **sc_time**.

86.3.2.5.3 Trace Files

A trace file is a file where the sequence of events, happening on the system signals, is stored. SystemC supports the VCD format for its trace files. There are several predefined functions that help users create their own VCD trace files. These functions include:

- **sc_create_vcd_file:** This function creates an empty VCD file.
- **sc_close_vcd_file:** This function closes a previously generated and opened VCD file.
- **sc_write_comments:** This function writes a comment to a VCD file.
- **sc_trace:** This function has many overloads for accepting various data types. It writes the value of a specified signal or variable to a VCD trace file (in VCD format).

86.3.2.6 Other Useful Classes and Methods

This section provides a brief overview of SystemC. There are many predefined classes and functions in SystemC that were not discussed. The following list shows functions not discussed here. The reader is referred to Refs. [6,7] for more details of SystemC and an explanation of functions listed below.

- sc_clock
- sc_time
- sc_event_queue

- sc_report
- sc_spawn
- FORK/JOIN
- sc_event_finder
- sc_attribute
- sc_exception

86.3.3 An Example: System Level Modeling of Sayeh CPU

In this section we will put together all the previously discussed concepts in a comprehensive CPU example. The example is the system level implementation of a simple CPU called *Sayeh*. Sayeh was first used as a benchmark example in several papers on CPU testing. It was then used in Ref. [8]. The example description of this CPU and its Verilog description appear in this reference. For the sake of completeness, a brief overview will be given in the following paragraph.

Sayeh has a register file that is used for data-processing instructions. It has a 16-bit data bus and a 16-bit address bus. This processor has 8- and 16-bit instructions (it has 29 instructions). Short instructions contain shadow instructions, which effectively pack two 8-bit instructions into a 16-bit word. Long instructions have the *Immediate* field that short instructions do not. Sayeh uses its register file for most of its data instructions. Addressing modes of this processor also take advantage of this structure. Because of this, the addressing hardware of Sayeh is a simple one and the register file output is used in address calculations. Sayeh components that are used by its instructions include the standard registers such as the Program Counter, Instruction Register, the Arithmetic Logic Unit, and Status Register. In addition, this processor has a register file forming registers R0, R1, R2, and R3 as well as a Window Pointer that defines R0, R1, R2, and R3 within the register file.

The SystemC codes of Sayeh are shown in Figure 86.70 (header file) and Figure 86.71 (implementation file). The testbench is shown in Figure 86.72. The SystemC codes for memory and other parts are not demonstrated.

86.3.4 Our Coverage of SystemC

In this section, a brief overview of SystemC for system level design was given using several examples. An early hardware-oriented specification of hardware parts in a system can be achieved by this language. These hardware prototypes can be cosimulated with software parts of a system. Another important point in developing hardware prototypes, which are also mentioned in this section, is to define an efficient communication between different hardware parts of a system. The next section discusses efficient ways of designing communication protocols using TLM.

86.4 Transaction Level Modeling

In the recent years, the ESL industry has defined a level of abstraction in modeling and design of systems. This level of abstraction is generally higher than RTL in that the timing details are not considered and the communications between modules are modeled by high-level channels instead of a wire or a group of wires. This level of abstraction, which is going to become the starting point in system level design, is called transaction level and the act of modeling a system in this level is referred to as TLM. System level designers and innovators utilize the transaction concept in this level of system modeling. This concept has been used in networks for many years and in these days it has been closely related to network-on-chip (NoC) designs (see Chapter 16 for description of NoC's).

As a matter of fact, there is no special and unique definition for TLM between system level designers. Each EDA vendor or system level designer defines TLM by one of its aspects. In a general definition of TLM, the system is divided into two parts: *communication* and *computation* parts. In this definition, TLM is considered as modeling the communication parts of a system at a high level of abstraction (e.g., by functions). With this definition, the computation parts (modules) of a design can be at various levels of

```
#ifndef __SAYEH_H__
#define __SAYEH_H__
#include "systemc.h"
#include "types.h"
#include "mem_communication.h"

SC_MODULE (sayeh)
{
    sc_inout <ADDRESS_TYPE> AddressBus;
    sc_inout <DATA_TYPE> DataBus;
    sc_in    <bool> ExternalReset;
    sc_out   <bool> IORead, IOWrite;
    sc_port  <MEM_TYPE> MemComm;

    REG16 m_IR, m_PC;
    REG3 m_WP;
    SC_CTOR (sayeh)
    {
        sc_time p(clk_period);
        sc_time s(clk_start);
        m_clk_period = p;
        m_clk_start = s;
        m_PC = 0; m_IR = 0; m_WP = 0;
        m_halt_executed = false;
        m_instr8_runned = false;
        SC_THREAD (sayeh_func);
    }
private:
    sc_time m_clk_period, m_clk_start;
    //main function
    void sayeh_func();
    //cpu states
    void reset();
    void fetch();
    void decode_execute(REG16 instr);

    REG16 m_RegFile[8];
    //flags
    char m_zero, m_carry;
    bool m_halt_executed, m_instr8_runned;;
};
#endif
```

FIGURE 86.70 *sayeh* module definition.

abstraction. It is obvious that the higher level the modules are designed, the faster their simulation process and the easier their connection with communication parts are.

In the next section we give a brief history of the development of TLM. Then, communications in TLM are discussed. TLM advantages are also discussed in Section 86.4.3. Different levels in TLM will be listed in Section 86.4.4. Then, the weaknesses of TLM are described in Section 86.4.5. Sections 86.4.6 to 86.4.8 are devoted to TLM constructs and concepts. In Section 86.4.9, an example using TLM is shown, and the last section presents a summary of this section and the entire chapter.

86.4.1 TLM History

Open core protocol-international partnership (OCP-IP), which has several industrial partnerships, has been working on TLM models and libraries since 2002. These models were developed in C++ and SystemC. Later, open SystemC initiative (OSCI) organization with collaboration of many companies and

```
#include "sayeh.h"
#define REG_DST m_RegFile[(instruction.range(11, 10).to_uint()+m_WP)%8]
#define REG_SRC m_RegFile[(instruction.range(9, 8).to_uint()+m_WP)%8]

void sayeh::sayeh_func ()
{
    REG16 instruction;  __int64 next_clk;

    while (1) {
        if (ExternalReset.read ()) {
            reset ();
            m_instr8_runned = false;
            wait (ExternalReset.negedge_event ());
        } else {
            if (!m_instr8_runned) {
                fetch ();
                m_instr8_runned = true;
                if (ExternalReset.read ())
                    continue;
                m_IR = DataBus.read ();
                instruction = m_IR;
            } else {
                instruction *= 256;
                m_instr8_runned = false;
            }
            decode_execute (instruction);
            if (ExternalReset.read ())
                continue;

            if (m_halt_executed) {
                m_halt_executed = false;
                wait (ExternalReset.posedge_event ());
            }
            next_clk = sc_time_stamp ().value () - m_clk_start.value ();
            next_clk = m_clk_period.value () - (next_clk % m_clk_period.value ());
            next_clk = next_clk/1000; //for conversion to nanoseconds

            wait (next_clk, SC_NS);
        }
    } //while (1)
}

void sayeh::reset ()
{
    m_PC = 0;   m_IR = 0;    m_WP = 0;
    m_zero = 0; m_carry = 0;
}

void sayeh::fetch ()
{
    AddressBus = m_PC;
    m_PC += 1;
    MemComm->read_action ();
}

void sayeh::decode_execute (REG16 instruction)
{
    switch (instruction.range (15, 12).to_uint ())
    {
        case 0: switch (instruction.range (11, 8).to_uint ())
            {
                case 0: //nop
                    return;
                case 1: //halt
                    m_halt_executed = true;
                    return;
                … //other instructions
```

FIGURE 86.71 *sayeh* module implementation

```
        }
        break;
    ...
    case 2: //load addressed
        AddressBus = REG_SRC;
        MemComm->read_action ();
        REG_DST = DataBus.read ();
        break;
    ... other instructions
        return;
    }
}
```

FIGURE 86.71 (*Continued*).

```
#include "types.h"
#include "sayeh.h"
#include "reset.h"
#include "memory.h"
#include "mem_communication.h"

int sc_main ( int argc , char **argv ) {
    //cpu signals
    sc_signal <ADDRESS_TYPE> AddressBus;
    sc_signal <DATA_TYPE> DataBus;
    sc_signal <bool> ExternalReset;
    sc_signal <bool> IORead, IOWrite;
    MEM_TYPE  MemComm;

    //reset signal
    sc_time rtime (7, SC_NS);
    reset reset ("rst", rtime);
    reset.rst (ExternalReset);

    memory mem ("memory"); //memory should be defined before cpu!
    //if memory is defined after cpu, the first memory read request
    //event will be missed. this is because of the sequence of the
    //initializations
    sayeh cpu ("sayeh_cpu");

    cpu.AddressBus (AddressBus);
    cpu.DataBus (DataBus);
    cpu.ExternalReset (ExternalReset);
    cpu.IORead (IORead);
    cpu.IOWrite (IOWrite);
    cpu.MemComm (MemComm);

    mem.AddressBus (AddressBus);
    mem.DataBus (DataBus);
    mem.MemComm (MemComm);

    sc_start (450, SC_NS);

    cout << "simulation finished!" << endl;
    return 0;
}
```

FIGURE 86.72 sc_main for testing *sayeh* module.

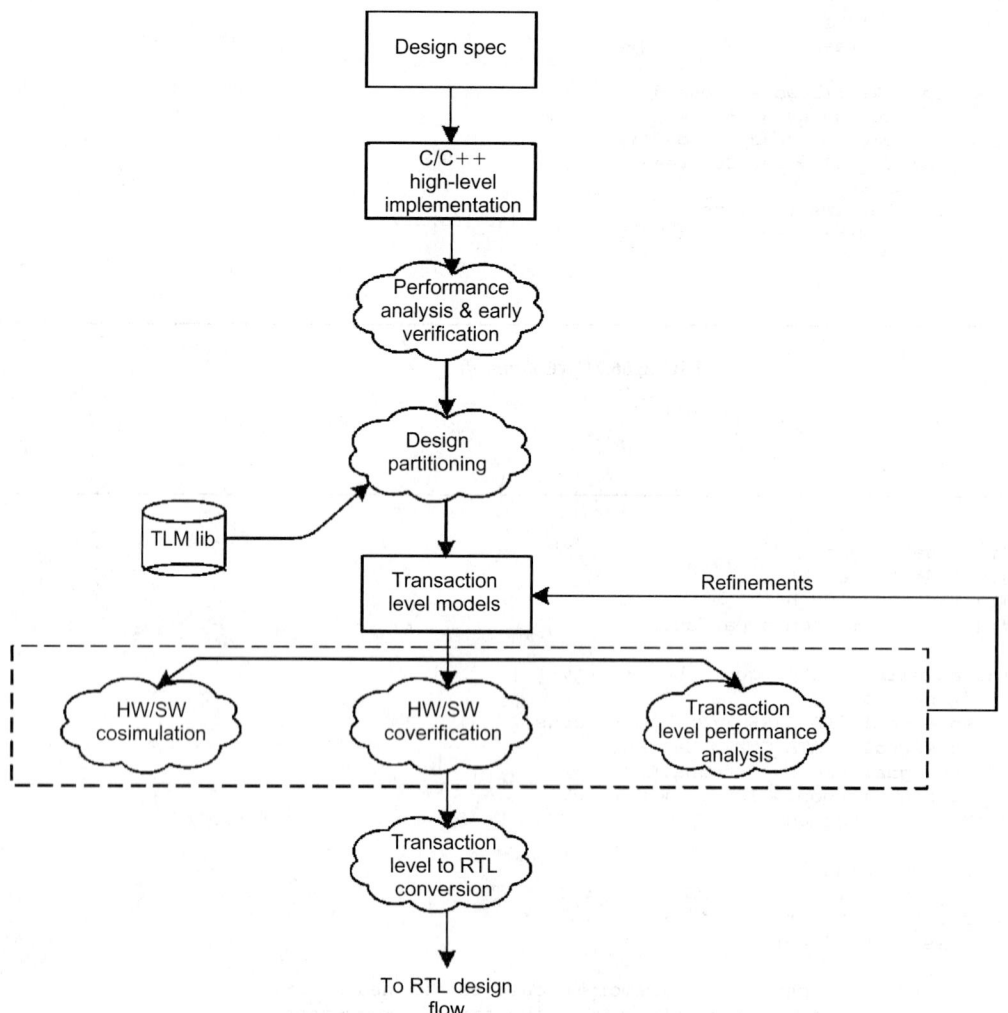

FIGURE 86.73 TLM design methodology.

EDA organizations (including OCP-IP), released TLM application program interfaces (APIs) and an open source library, in June 2005. As announced by the OSCI TLM working group chair, more than 50 members have been working on this library for more than 2 years. Although the expressions used in the OSCI library and the OCP-IP library differ, it has been tried to maintain the compatibility between these two libraries for more reliable use of TLM in ESL design.

The OSCI TLM 1.0 library has been developed in SystemC 2.1 in association with many companies including Cadence Design Systems, CoWare, Forte Design Systems, Mentor Graphics, Celoxica Ltd., ChipVision Design Automation AG, Synfora, Inc., and Summit Design Automation, Inc. Using this library helps designers develop an early prototype of their system and make the necessary decisions in early stages of design. Figure 86.73 shows the design flow using TLM library. In this figure, the conversion process from TLM to RTL can be done manually or automatically.

86.4.2 TLM Communication

As we can mix RTL/gate-level designs in RTL modeling, it is possible to develop designs that include modules in RT and transaction levels of abstraction. Since the kind of communication connections is

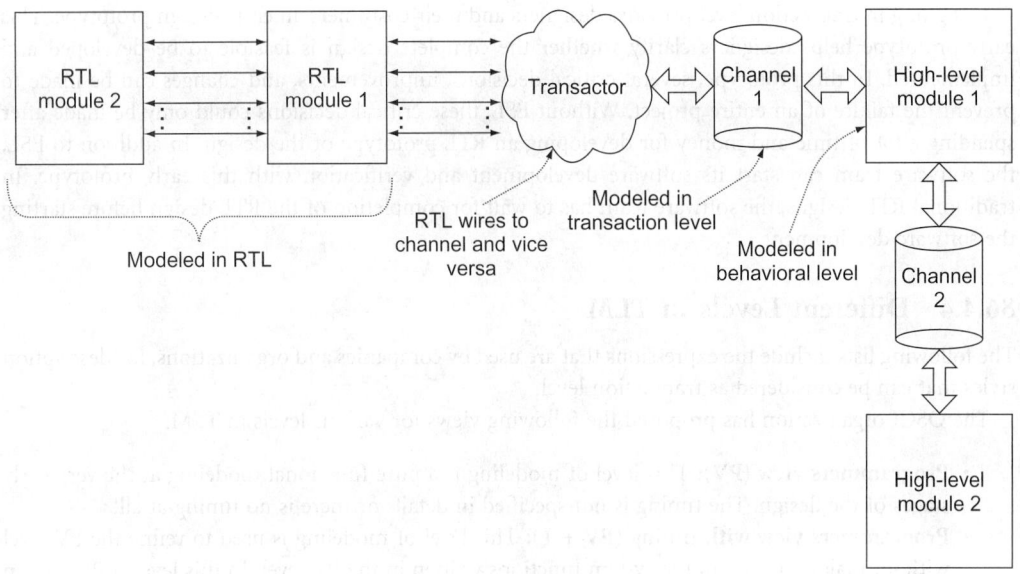

FIGURE 86.74 A mixed RTL/transaction level design.

different in these two levels, we have to put special modules, called *transactors*, around our RTL modules. Figure 86.74 shows a mixed RTL/transaction level design.

In this figure, there are two RTL modules and two higher-level modules. These high-level modules can be developed in any high-level language. As shown, these high-level modules are connected to each other by a transaction level channel. This means that communications between these modules are done by function calls instead of event passing or waiting on signals.

Also the communication between the *high-level module 1* and *RTL module 1* are modeled by transaction level channels. But since *RTL module 1* cannot communicate in this level, there must be a translator between this module and the connector channel. This translator (called a transactor) converts the events of the RTL module outputs into transactions and passes them to the channels and receives the transactions from the *high-level module 1* through the channel and converts them into input events for the RTL module. It is obvious that the communication speed in a channel between two high-level modules is higher than the communication speed in a channel between a transactor and a high-level module.

In Figure 86.74, the RTL modules communicate by event passing (the traditional way of communication in RTL). If we were to use a transaction level communication between these two modules, we had to put two transactors between them which may not be as efficient.

86.4.3 TLM Advantages

Features and advantages of TLM over RTL modeling have made designers to move their design starting point from RTL models to transaction level models. Although it can be used for faster modeling, the main purpose designers use TLM is its simulation speed. Since the communications between modules are modeled with efficient and high-level functions and the computational modules are also modeled in a higher level of abstraction, designers can gain over $1000\times$ simulation efficiency from TLM simulation over RTL simulation. The simulation efficiency leads to other TLM features including more efficient partitioning, faster and more precise functional verification, faster hardware/software coverification, and faster hardware/software cosimulation.

Some of the above features produce other benefits for the designers. For example, by modeling a system with TLM, the designer has a more comprehensible view for the design partitioning. A mature partitioning in a design can even affect the power consumption of the manufactured SoC.

Designing in transaction level provides designers and their costumers an early design prototype. This early prototype helps designers clarify whether the complete design is feasible to be developed and implemented. In this prototype, several critical decisions, improvements, and changes can be made to prevent the failure of an entire project. Without ESL, these critical decisions could only be made after spending a lot of time and money for developing an RTL prototype of the design. In addition to ESL, the software team can start its software development and verification with this early prototype. In traditional RTL designs, the software team has to wait for completion of the RTL design before starting the software development.

86.4.4   Different Levels in TLM

The following lists include the expressions that are used by companies and organizations, for description styles that can be considered as transaction level.

The OSCI organization has proposed the following views for various levels in TLM:

- **Programmers view (PV):** This level of modeling is a pure functional modeling at the very early stages of the design. The timing is not specified in details or there is no timing at all.
- **Programmers view with timing (PV + T):** This level of modeling is used to refine the PV level without major changes in the system functions written in the PV level. In this level, a PV and an interface (e.g., a FIFO interface) exist in the design.
- **Cycle callable (CC):** In this level of modeling, the design must be developed to work cycle-by-cycle. This level of modeling is cycle-accurate. The interfaces are developed to have a specified task in each cycle. Also the functional models are altered to work correctly with cycles. This level is still considered as a transaction level and it is still simulated faster than RTL. This is because the interfaces still use higher-level ports and interconnections than the ports in RTL. This implies that a cycle callable or cycle-accurate model is different from a pin-accurate model.

The OCP-IP organization has its own defined levels of modeling that are listed below.

- **Message Layer (L-3):** This layer is very similar to the PV level described above. In this level, the design is described at the highest level of abstraction. It has no timing and it is event-driven. Each event contains several data. This layer is very practical for being used as a proof-of-concept model and for the first versions of partitioning.
- **Transaction Layer (L-2):** This layer contains more details than layer L-3. It is usually used for hardware/software partitioning, hardware performance analysis, and generation of a testbench for the design developed in the L-1 layer. The designs in this layer have an approximate timing and are structurally more accurate than the models in layer L-3, which means that the structure of the designs is closer to a real design than those in layer L-3.
- **Transfer Layer (L-1):** This layer is very similar to CC layer defined by OSCI. In this level, where we have a clocked cycle-accurate model, the interfaces are changed to be mapped onto the chosen hardware interfaces and bus structures. Similar to the CC layer, this layer is still higher than the RT level. L-1 models are useful for developing testbenches for RTL models. They can also be used for performance comparisons with RTL models. The RTL layer in this hierarchy is called *layer L-0.*

The above two sets of definitions are very similar. There are also other definitions for transaction levels based on these two sets definitions and are listed below.

CoWare defines the following set of design layers:

- Programmers view (PV)
- Architects view (AV)
- Verification view (VV)

TABLE 86.1 Mapping of Different TLM Levels

General Definition	OSCI Definition	OCP-IP Definition	CoWare Definition	STMicro Definition
HLT	PV	L-3	PV	FV
MLT	PV + T	L-2	AV	AV
LLT	CC	L-1	VV	MV

There is also another set of design levels defined by ST-MicroElectronics [9]:

- SoC functional view (FV)
- SoC architecture view (AV)
- SoC micro-architecture view (MV)

Although the above groups and companies use different names and expressions for their definitions, they follow the same abstraction in each level. In a general definition, these abstractions include a functional level with poor or no timing, a lower-level architectural design with poor timing, and a cycle-accurate (but not necessarily pin-accurate) level with timing details. We can call these levels high-level transaction (HLT), medium level transaction (MLT), and low level transaction (LLT), respectively. All of these levels can be approximately mapped as shown in Table 86.1.

86.4.5 TLM Problems and Weaknesses

Although TLM speeds up the entire design process and verification of a system, it has weaknesses that limit its effective use in the design process. One of its main problems is that there is no defined standard for TLM. Design reuse is more feasible if design groups have the same point of view of the design levels and subsets.

The other problem is the lack of mature ESL tools to support TLM. Today, many ESL tool vendors are working on automation tools that can accept a design developed in transaction level. These tools include synthesis tools, verification tools, and debugging and simulation tools. A TLM synthesis tool can convert a transaction level design into its corresponding RTL design. The ESL tool vendors can develop their tools more efficiently if there is a specified and standard subset for TLM.

Although there are problems in utilizing TLM in the design process, there are companies that use TLM as their starting point in their design process. These companies use TLM to speed up their modeling phase, early verification process, and software development. With today's tools they may have to convert their high-level model into lower levels manually or semimanually, but their entire design process time is reduced by using TLM.

Another critical issue in TLM is the design and mapping of real-time operating systems (RTOS) on the design processors and DSPs, and this issue is in its early study phase.

86.4.6 TLM Basic Definitions

The basic vocabulary used in a transaction level design is discussed below. Refer to the diagram of Figure 86.75 for a visual representation of these terms.

- **Transaction:** The data exchange between any two modules is done by a transaction.
- **Module:** A meaningful part of a system which has a specified functionality. Different modules in a system can communicate via their I/O ports and channels.
- **Port:** Connection of a channel to a module is done via a port.
- **Channel:** The communication protocol between two modules is defined in a channel. Channels are categorized to primitive and hierarchical channels. A primitive channel is a channel which connects two modules with a simple functionality, and a hierarchical channel is a more complex channel which may include ports, modules, and even other channels.
- **Interface:** An interface is a set of methods needed for definition and implementation of a communication protocol.

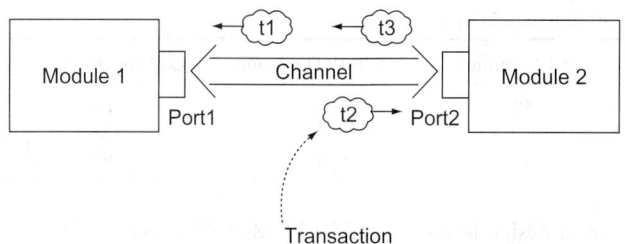

FIGURE 86.75   Basic components of TLM.

86.4.7   TLM Design Process

A design process begins with a high-level specification of the system being designed. The design step, which must be taken after this description becomes available, is to generate a transaction level model. For this purpose, we have to define our top-level modules and their means of communication. Module granularity and specification may be refined in the next stages of the design process. Accordingly, the functionality of communication channels and their interfaces will also be refined. Defining these interfaces and channels can take advantage of a standard TLM library. If any of the communication processes have complex functionalities, application-specific channels and interfaces must be implemented based on the TLM library primitive channels and interfaces. Figure 86.76 shows the process of developing a transaction level model.

The next section describes the structure of TLM library developed in SystemC by OSCI (TLM 1.0). A simple example in SystemC follows this discussion.

86.4.8   TLM Implemented in SystemC

As stated in the previous section, TLM is a modeling concept that can be implemented in any system level language (e.g., C++, System Verilog, and Java). The first implemented version of a TLM library for system level design was in SystemC.

At the present time, most universities and companies that work on TLM and system level design use this language. One reason for choosing SystemC is the wide use of C++ language by software developers. C++ is a powerful programming language with several constructs for both a system and a software design. For example, C++ class definition and inheritance capabilities make it useful for defining interfaces and channels. To make SystemC more useful for hardware description, basic channels and interfaces are described in this language. Other useful interfaces and channels can be built on these predefined interfaces and channels in SystemC.

The TLM library, discussed in this section, has been implemented in SystemC 2.1 by the OSCI organization.

In this version, the main focus is on implementing a generic set of TLM interfaces and channels. The object passing semantic is based on the **sc_fifo** message passing of SystemC. In this library raw C/C++ pointers are discouraged to avoid memory problems caused by pointers. In this library, most of the implemented classes are template classes. This feature provides generality for the classes.

Communication APIs in this library can be unidirectional or bidirectional. They can also be in blocking or nonblocking modes. When performing a blocking data transfer, the caller must wait until the data transfer is completed. In the nonblocking mode, the transfer function will return the status of the data transfer immediately (whether it is done or not). In SystemC, the blocking mode data transfer can only be done in **SC_THREAD** and **SC_CTHREAD** processes. Nonblocking data transfers can also be done in **SC_METHOD** processes (see Section 86.3).

The implementation of this library includes several subdirectories. Each directory contains one or more C++ header files for TLM functions and their necessary implementations. These directories are discussed below.

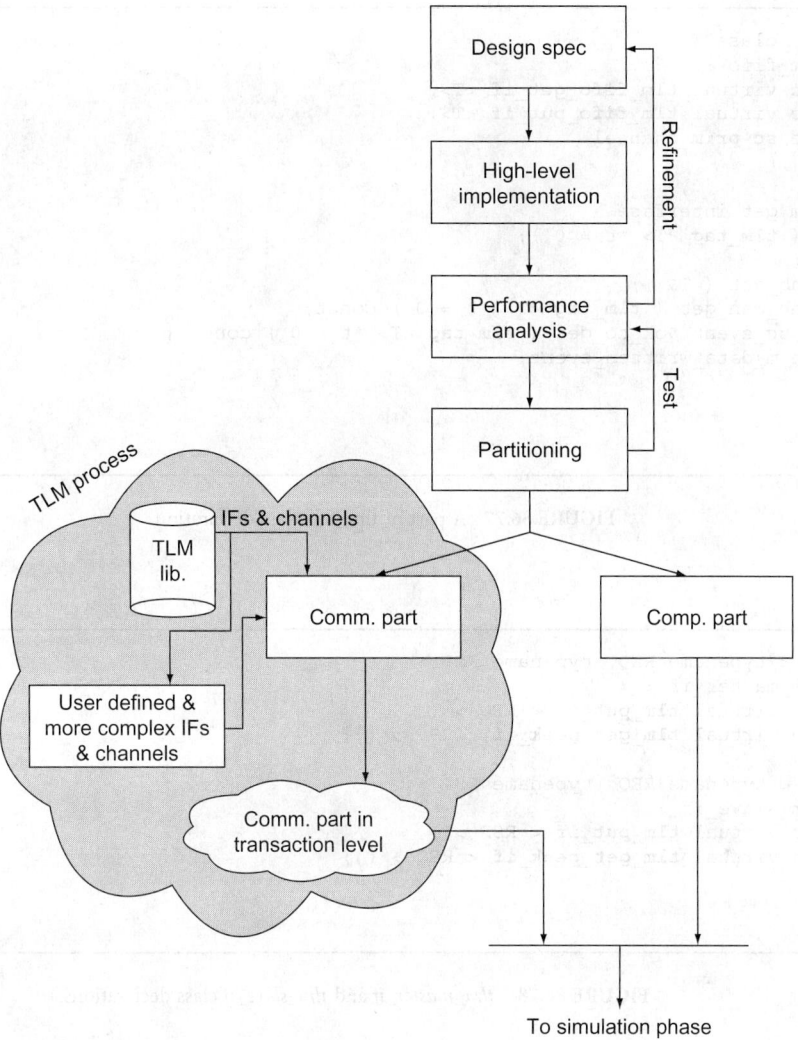

FIGURE 86.76 TLM design process.

- **tlm_adapters:** In this part, two classes named **tlm_transport_to_master** and **tlm_slave_to_transport** are defined. These classes are useful for transporting data to/from a master or a slave.
- **tlm_event_finder:** Classes implemented in this part are used in blocking and nonblocking ports. These classes provide static sensitivity for these ports. The defined classes in this part include **tlm_blocking_get_port**, **tlm_nonblokcing_peek_port**, **tlm_nonblocking_put_port**, and **tlm_event_finder_t** class for handling events.
- **tlm_fifo:** Operations and necessary classes for a FIFO channel in TLM are implemented in this part. One of the main classes defined in this part is **tlm_fifo**. This class, a part of which is shown in Figure 86.77, includes basic operations of a FIFO such as **get**, **put**, **peek**, **size**, and **kind**. These operations are implemented in both blocking and nonblocking modes.
- **tlm_interfaces:** Basic interfaces are implemented in this part. There are two main classes named **tlm_master_if** and **tlm_slave_if** for sending and receiving data. These classes are shown in Figure 86.78. Also there are these classes for FIFO interfaces: **tlm_fifo_debug_if**, **tlm_fifo_put_if**, **tlm_fifo_if**, and **tlm_fifo_config_size_if**.

```
template <class T>
class tlm_fifo :
   public virtual tlm_fifo_get_if <T>,
   public virtual tlm_fifo_put_if <T>,
   public sc_prim_channel
{
public:
   // tlm get interface
   T get( tlm_tag <T> *t = 0 );

   bool nb_get ( T& );
   bool nb_can_get ( tlm_tag <T> *t = 0 ) const;
   const sc_event &ok_to_get ( tlm_tag <T> *t = 0 ) const {
   return m_data_written_event;
   }
   ...
}
```

FIGURE 86.77   A part of tlm_fifo class declaration.

```
template < typename REQ, typename RSP >
class tlm_master_if :
   public virtual tlm_put_if < REQ > ,
   public virtual tlm_get_peek_if < RSP > {};

template < typename REQ, typename RSP >
class tlm_slave_if :
   public virtual tlm_put_if < RSP > ,
   public virtual tlm_get_peek_if < REQ > {};

};
```

FIGURE 86.78   *tlm_master_if* and *tlm_slave_if* class declarations.

The *tlm_core_ifs.h* file is also located in this directory. This file has the descriptions for unidirectional/bidirectional blocking/nonblocking interfaces (such as **tlm_nonblocking_get_if** and **tlm_transfer_if**).

- **tlm_req_rsp:** The basic channels of TLM are implemented in this directory. These channels are implemented as the **tlm_req_rsp_channel** and **tlm_transport_channel** classes. They have bidirectional and unidirectional master and slave channels. Several types of requests and responses can be passed to these classes via its template arguments *REQ* and *RSP.* A part of **tlm_req_rsp_channel** is shown in Figure 86.79.

86.4.9   An Example: A Cache-Memory System

The previous section gave a brief overview of classes implemented in OSCI TLM 1.0 library. In this library, several classes and methods are implemented for communication in an efficient and safe manner. In the following, we will use the above classes for a simple design. This design describes a system that contains an RTL cache model with a higher-level memory and a master (like a CPU) in behavioral level. The CPU sends *read* and *write* requests to the cache. Since the cache model is in RT level, we have to

```
template < typename REQ, typename RSP >
class tlm_req_rsp_channel : public sc_module
{
public:
   // uni-directional slave interface
   sc_export < tlm_fifo_get_if < REQ > > get_request_export;
   sc_export < tlm_fifo_put_if < RSP > > put_response_export;
   ...
protected:
   tlm_master_imp < REQ, RSP > master;
   tlm_slave_imp < REQ, RSP > slave;
   tlm_fifo <REQ> request_fifo;
   tlm_fifo <RSP> response_fifo;
}
```

FIGURE 86.79 A part of tlm_req_rsp_channel class definition.

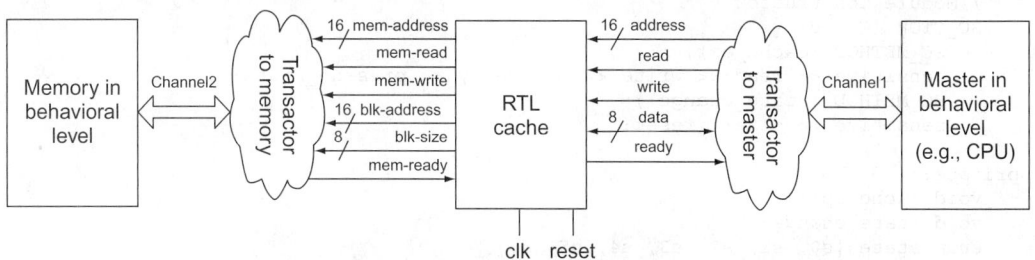

FIGURE 86.80 Cache-memory schematic view.

convert its signals to adopt them to system level channels for memory and CPU access. The general block diagram of this cache example is shown in Figure 86.80.

86.4.9.1 Cache Model

The header and the implementation of the cache model are shown in Figure 86.81 and Figure 86.82, respectively. As stated above, this system includes a master (e.g., a CPU) that accesses its memory through a cache.

As shown in Figure 86.81, the interface of the cache with the CPU includes *ready*, *write*, *read*, *data*, and *address* signals. Its interface with the memory includes *mem_ready*, *mem_write*, *mem_read*, *blk_address*, and *blk_size* signals. *blk_address* and *blk_size* signals are used when a *miss* happens in the cache and the cache needs to read a block from the memory. The cache header file also includes declarations for cache registers and states of the cache controller. Figure 86.82 implements the cache controller using a synthesizable RT level state machine style of coding.

The cache storage is modeled with an array of logic vector (**sc_lv** type).

The SystemC method *cache_op* in Figure 86.82 performs the main operation of the cache, and the *state_change* method performs state transitions of cache controller.

86.4.9.2 CPU Model

The header and the implementation of the CPU model are shown in Figure 86.83 and Figure 86.84, respectively. This interface-level model issues read and write requests to cache. The cache model accesses the memory on behalf of the CPU. The CPU model in this system plays the role of a testbench that tests the functionality of the cache, the memory, and their interconnections.

```
#include <systemc.h>
#include "sc_util.h"
#define  BLOCK_SIZE 16
SC_MODULE (Cache)
{
   sc_in <bool> clk;
   sc_in <bool> reset;
   // interface with cpu
   sc_in <sc_uint <16> > address;
   sc_in <bool> read;
   sc_in <bool> write;
   sc_inout <sc_lv <8> > data;
   sc_out <bool> ready;
   // interface with memory
   sc_out <sc_uint <16> > mem_address;
   sc_out <bool> mem_read;
   sc_out <bool> mem_write;
   sc_out <sc_uint <16> > blk_address;
   sc_out <sc_uint <8> > blk_size;
   sc_in <bool> mem_ready;
   //Module Constructor
   SC_CTOR (Cache) {
      SC_METHOD (cache_op);
      sensitive << read << write << mem_ready << present_state;
      SC_METHOD (state_change);
      sensitive << clk << reset;
   }
private:
   void cache_op();
   void state_change();
   enum state {s0, s1, s2, s3, s4, s5, s6};
   sc_signal <state> present_state;
   sc_signal <state> next_state;

   sc_lv <5> tag [128];
   sc_lv <7> block [128];
   bool dirty_bit [128];
};
```

FIGURE 86.81 Cache declaration.

In our implementation, the CPU uses the *master_port* bidirectional port of *basic_initiator_port* type. The *basic_initiator_port* is a class derived from **sc_port** and it is defined in the *examples* directory in the OSCI TLM 1.0 library. It has two ports named *basic_request* and *basic_response* for byte data transfer between the CPU and the cache transactor (see Figure 86.80 for the role of the transactors).

As shown in Figure 86.84, the *master_port* uses *read* and *write* functions of the *basic_initiator_port*. Data and address for reading and writing are used as arguments of these functions, e.g., *a* and *d*. The *run* thread defined in the master class performs *read* and *write* operations through the *master_port* channel.

As shown in Figure 86.84, in the CPU model, byte-level TLM function calls are:

```
master_port.write (a, d+a);
master_port.read (a, d);
```

86.4.9.3 Memory Model

The header and the implementation of the memory model are demonstrated in Figure 86.85 and Figure 86.86, respectively. The memory model is a high-level model of memory with two ports called

```
#include "cache.h"
#include "cache_mem.h"

void Cache::cache_op()
{
    int target_block;
    sc_lv <16> mem_address_temp;

    switch (present_state) {
        case s0:
            if (read.event () && read.read () == true) {
                next_state = s1;
                ready = false;
            }
            else …
            break;
        case s1:
            target_block = address.read().range(10, 4).to_uint() ;
            if (tag[target_block] == (sc_lv<5>) address.read().range(15, 11)) { // cache hit
                data = word[target_block].range(address.read().range(3, 0).\
                        to_uint()*8 + 7, address.read().range(3, 0).to_uint()*8);
                ready = true;
                next_state = s0;
            }
            else { // cache miss
                if (dirty_bit[target_block]) {
                    mem_write = true;
                    blk_address = target_block;
                    blk_size = BLOCK_SIZE;
                    mem_address_temp = (tag[target_block], block[target_block],\
                                    (sc_lv<4>)"0000");
                    mem_address = mem_address_temp.to_uint();
                    next_state = s3;
                }
                else {
                    mem_read = true;
                    mem_address = address.read ();
                    blk_address = target_block;
                    blk_size = BLOCK_SIZE;
                    next_state = s4;
                }
            }
            break;
        case s2:
            …
    }
}

void Cache::state_change ()
{
    if (reset == true) {
        present_state = s0;
        // clear all dirty bits
        for (int i = 0; i< 128; i++)
            dirty_bit[i] = false;
    }
    else if (clk.event () && clk == true) {
        cout << "next_state = " << next_state.read() << endl;
        present_state = next_state;
    }
}
```

FIGURE 86.82 Cache implementation.

request_port and *response_port.* These two ports are of type *block_basic_request* and *block_basic_response*, respectively. These classes are similar to the *basic_request* and *basic_response* classes defined in the *examples* part of the TLM library. They perform block-by-block instead of byte-by-byte data transfer.

```
#include "bus_types.h"
#include "basic_initiator_port.h"
using basic_protocol::basic_initiator_port;
namespace user
{
    class Master : public sc_module
    {
    public:
        Master (sc_module_name module_name);
        SC_HAS_PROCESS (Master);
        basic_initiator_port <ADDRESS_TYPE,DATA_TYPE> master_port;
    private:
        void run();
    };
};
```

FIGURE 86.83   CPU declaration.

```
#include "master.h"
using user::Master;

Master::Master (sc_module_name module_name) :
    sc_module( module_name ),
    master_port("iport")
{
    SC_THREAD (run);
}

void Master::run()
{
    DATA_TYPE d;

    d = 1000;
    for (ADDRESS_TYPE a=0; a<2; a++) {
        cout << "Writing Address " << a << " value " << d+a << endl;
        master_port.write (a, d+a);
    }
    for (ADDRESS_TYPE a=0; a<2; a++) {
        master_port.read(a, d);
        cout << "Read Address " << a << " got " << unsigned int(d) << endl;
    }
    wait();
}
```

FIGURE 86.84   CPU implementation.

Operations for putting and getting data from the memory ports are done in the blocking mode. Therefore, we have to define a thread (instead of a method) for the memory operations. This thread is named *run*, and it simply waits for a request on its *request_port*, processes that request by calling the private member function *process_request*, and puts the result to its *response_port*. Codes around segments described above are for partitioning of blocks of data to write and read.

```
#include "systemc.h"
#include "block_bus_types.h"
#include "block_basic_protocol.h"
#include "basic_slave_base.h"
using tlm::tlm_get_peek_if;
using tlm::tlm_blocking_put_if;
using block_basic_protocol::block_basic_request;
using block_basic_protocol::block_basic_response;

class MemoryModel :
    public sc_module
{
public:
    MemoryModel (sc_module_name module_name);
    SC_HAS_PROCESS (MemoryModel);

    sc_port <tlm_get_peek_if < block_basic_request \
            < POINTER_TYPE, SIZE_TYPE, ADDRESS_TYPE> > > request_port;
    sc_port <tlm_blocking_put_if <block_basic_response \
            < POINTER_TYPE, SIZE_TYPE > > > response_port;

    ~MemoryModel ();
private:
    void run();
    block_basic_response < POINTER_TYPE, SIZE_TYPE > process_request (const \
        block_basic_request < POINTER_TYPE, SIZE_TYPE, ADDRESS_TYPE > &req);

    ADDRESS_TYPE *memory;
};
```

FIGURE 86.85 Memory declaration.

In Figure 86.86, *get* and *put* functions of *req* and *rsp* ports in the *run* process perform the communication tasks. Block-level TLM functions for writing into the memory are implemented by the following part implemented in *process_request*:

```
for (SIZE_TYPE i=0; i< s; i++)
    memory[a+i] = word[p/16].range(i*8+7, i*8).to_uint();
response.p = request.p;
response.s = request.s;
response.status = block_basic_protocol::SUCCESS;
```

Likewise, block-level TLM functions for reading from the memory is implemented in *process_request*:

```
for (SIZE_TYPE i=0; i< s; i++)
    word[p/16].range(i*8+7, i*8) = memory[a+i];
response.p = request.p;
response.s = request.s;
response.status = block_basic_protocol::SUCCESS;
```

86.4.9.4 Transactors of the Cache Model

As stated earlier, our cache model is developed in RTL, while the other parts of the system have been modeled in a high level of abstraction. The cache model has RTL inputs and outputs, while the other parts of the system use transaction level channels and interfaces as their input/output ports. Therefore, there must be a translator (referred to as transactors) between the cache model and the high-level models on its two sides. One transactor in our design is for connecting the cache model to the CPU, and the other is for connecting the cache model to the memory.

```
#include "mem_fifo_slave.h"
using block_basic_protocol::block_basic_request;
using block_basic_protocol::block_basic_response;

MemoryModel::MemoryModel (sc_module_name module_name):
   sc_module ( module_name ) ,
   request_port ("in_port") , response_port ("out_port")
{
   SC_THREAD (run );
   memory = new ADDRESS_TYPE[16*1024];
}

void MemoryModel::run ()
{
   block_basic_request <POINTER_TYPE, SIZE_TYPE, ADDRESS_TYPE> req;
   block_basic_response <POINTER_TYPE, SIZE_TYPE> rsp;
   while (true) {
      request_port->get (req);
      rsp = process_request (req);
      response_port->put (rsp);
      wait (request_port->ok_to_get ());
   }
}

block_basic_response < POINTER_TYPE, SIZE_TYPE >
MemoryModel::process_request (const block_basic_request \
< POINTER_TYPE, SIZE_TYPE, ADDRESS_TYPE > &request)
{
   block_basic_response <POINTER_TYPE, SIZE_TYPE> response;

   POINTER_TYPE p = request.p;
   SIZE_TYPE    s = request.s;
   ADDRESS_TYPE a = request.a;
   response.type = request.type;

   switch( request.type ) {
   case block_basic_protocol::READ :

      for(SIZE_TYPE i=0; i< s; i++)
         word[p/16].range(i*8+7, i*8) = memory[a+i];
      response.p = request.p;
      response.s = request.s;
      response.status = block_basic_protocol::SUCCESS;
      break;
   case block_basic_protocol::WRITE:

      for(SIZE_TYPE i=0; i< s; i++)
         memory[a+i] = word[p/16].range(i*8+7, i*8).to_uint();
      response.p = request.p;
      response.s = request.s;
      response.status = block_basic_protocol::SUCCESS;
      break;
   default:
      response.status = block_basic_protocol::ERROR;
      break;
   }
   return response;
}
```

FIGURE 86.86 Memory implementation.

```
#include <systemc.h>
#include "tlm.h"
#include "bus_types.h"
#include "block_bus_types.h"
using tlm::tlm_nonblocking_get_if;
using tlm::tlm_nonblocking_put_if;

class CacheMemTrans : public sc_module
{
public:
    CacheMemTrans (sc_module_name module_name);

    sc_port < tlm_nonblocking_put_if < BLOCK_REQUEST_TYPE > > request_port;
    sc_port < tlm_nonblocking_get_if < BLOCK_RESPONSE_TYPE > > response_port;

    SC_HAS_PROCESS (CacheMemTrans);

    sc_in <bool> clk;
    sc_in <bool> reset;
    sc_in <sc_uint <16> > mem_address;
    sc_in <bool> mem_read;
    sc_in <bool> mem_write;
    sc_in <sc_uint<16> > b_address;
    sc_in <sc_uint<8> > b_size;
    sc_out <bool> mem_ready;
private:
    enum state {REQUEST, RESPONSE};
    void run();

    BLOCK_REQUEST_TYPE req;
    BLOCK_RESPONSE_TYPE rsp;

    bool got_response;
    sc_signal< state > m_state;
};
```

FIGURE 86.87 Cache-to-CPU transactor declaration.

The header and the implementation for the cache-to-master transactor are shown in Figure 86.87 and Figure 86.88, respectively. The header file has the RT level signals that are the same as those defined in the cache model for its CPU interface. These signals include *clk, reset, address, data, read, write,* and *ready* signals. In contrast, this transactor has two unidirectional ports named *request_port* and *response_port* (recall that a similar port was used in the CPU model for byte-by-byte level read and write).

These ports use nonblocking methods for transferring data. The transactor model uses **tlm_non-blocking_get_if** and **tlm_nonblocking_put_if** interfaces for its *request_port* and *response_port*, respectively.

The method of this transactor is *run*, which is sensitive to the positive edge of *clk*. It is a simple state machine which includes two states named *EXECUTE* and *WAIT*. This method decides on its *request_port* and the ready signal at every positive edge of the clock in the *EXECUTE* state. Then it sets the proper values to *read, write, address,* and *data* signals if the requested operation is *WRITE*, or reads the proper values from *read, write,* and *address* signals if the requested operation is *READ*. For the *READ* operation, data is read from the *data* signal. In the *WAIT* state, the transactor puts data obtained from the cache to *response_port*, and waits for it to be transferred.

```
#include "CacheMasterwrapper.h"
using block_basic_protocol::READ;
using block_basic_protocol::WRITE;
using block_basic_protocol::SUCCESS;
using block_basic_protocol::ERROR;

CacheMemTrans::CacheMemTrans (sc_module_name module_name):
    sc_module (module_name), clk ("clk"),
    reset ("reset"), mem_address ("mem_address"),
    mem_read ("mem_read"), mem_write ("mem_write"),
    b_size ("block_size"), mem_ready ("mem_ready")
{
    SC_METHOD (run);
    sensitive << clk.pos ();
    dont_initialize ();
}

void CacheMemTrans::run() {
    if (reset) {
        m_state = REQUEST;
        got_response = false;
        mem_ready = false;
    }
    else {
        switch (m_state) {
        case REQUEST:
            mem_ready = false;
            if (mem_write.read () == true) {
                req.type = WRITE;
                req.a = mem_address.read ();
                req.p = (POINTER_TYPE) b_address.read ();
                req.s = b_size.read ();
                request_port->nb_put (req);
                m_state = RESPONSE;
            }
            else if (mem_read.read() == true) {
                req.type = READ;
                req.a = mem_address.read ();
                req.p = (POINTER_TYPE) b_address.read ();
                req.s = b_size.read ();
                request_port->nb_put (req);
                m_state = RESPONSE;
            }
            break;
        case RESPONSE:
            got_response = response_port->nb_get (rsp);
            if (got_response) {
                mem_ready = true;
                m_state = REQUEST;
            }
            break;
        }
    }
}
```

FIGURE 86.88 Cache-to-CPU transactor implementation.

The header and the implementation for the cache-to-memory transactor are shown in Figure 86.89 and Figure 86.90, respectively.

This transactor is similar to the cache-to-master transactor in that it has several RTL ports that connect it to the cache and two unidirectional ports to convert RTL port data into the high-level memory

```
#include <systemc.h>
#include "tlm.h"
#include "bus_types.h"
using tlm::tlm_nonblocking_get_if;
using tlm::tlm_nonblocking_put_if;

class CacheMasTrans :    public sc_module
{
public:
   CacheMasTrans (sc_module_name module_name);

   sc_port < tlm_nonblocking_get_if < REQUEST_TYPE > > request_port;
   sc_port < tlm_nonblocking_put_if < RESPONSE_TYPE > > response_port;

   SC_HAS_PROCESS (CacheMasTrans);

   sc_in <bool> clk;
   sc_in <bool> reset;
   sc_out <sc_uint<16> > address;
   sc_out <bool> read;
   sc_out <bool> write;
   sc_inout <sc_lv<8> > data;
   sc_in <bool> ready;
private:
   enum state {EXECUTE, WAIT};

   void run();

   REQUEST_TYPE req;
   RESPONSE_TYPE rsp;
   bool got_request;
   bool put_status;
   sc_signal <state> m_state;
};
```

FIGURE 86.89 Cache-to-memory transactor declaration.

operations and vice versa. RTL signals are *clk, reset, mem_address, mem_read, mem_write, blk_address, blk_size,* and *mem_ready.*

On the memory side, TLM ports include a *request_port* which uses **tlm_nonblocking_put_if** methods and a *response_port* which uses and **tlm_nonblocking_get_if** methods.

Similar to the previous transactor, this transactor has a method named *run* that is sensitive to the positive edge of *clk.* This method implements a simple state machine with two states named *REQUEST* and *RESPONSE.* This method reads *mem_read* and *mem_write* signals from its cache side signals and makes the proper requests for sending the data obtained from the cache to the memory via its *request_port.* It then obtains memory response via its *response_port* and asserts the *mem_ready* signal.

86.4.9.5   Connecting the Models in main.cpp

The above sections described individual components of Figure 86.80. These models must be connected together to generate a complete system. This is done in the *main.cpp* file shown in Figure 86.91. There are two channels, *cache_mem_chan* (of type **tlm_req_rsp_channel**) and *cache_mas_chan* (of type **tlm_transport_channel**) in this code. These channels connect the *CacheMasterTrans* (an instance of cache-to-master transactor) and *CacheMemTrans* (an instance of cache-to-memory transactor) to the CPU and memory, respectively.

```
#include "wrapper.h"
using basic_protocol::READ;
using basic_protocol::WRITE;
using basic_protocol::SUCCESS;
using basic_protocol::ERROR;

CacheMasTrans::CacheMasTrans (sc_module_name module_name):
    sc_module (module_name), clk ("clk"),
    reset ("reset"), address ("address"),
    read ("read"), write ("write"),
    data ("data"), ready ("ready")
{
    SC_METHOD (run);
    sensitive << clk.pos ();
    dont_initialize ();
}

void CacheMasTrans::run () {
    if(reset) {
        m_state = EXECUTE;
        got_request = false;
        put_status = true;
        read = false;
        write = false;
    }
    else {
        switch (m_state) {
        case EXECUTE :
            if (ready == true) {
                m_state = WAIT;
                got_request = false;
                rsp.status = SUCCESS;
                if (read) {
                    rsp.type = READ;
                    rsp.d = (data.read ()).to_uint ();
                }
                else if (write) {
                    rsp.type = WRITE;
                }
                read = false;
                write = false;
                put_status = response_port->nb_put (rsp);
            }
            else {
                if (!got_request)
                    got_request = request_port->nb_get (req);
                if (got_request) {
                    address = req.a;
                    if( req.type == WRITE ) {
                        read = false;
                        write = true;
                        data = req.d;
                    }
                    else {
                        write = false;
                        read = true;
                    }
                }
            }
        break;
        case WAIT :
            read = false;
            write = false;
            if (!put_status)
                put_status = response_port->nb_put (rsp);
            if (put_status)
                m_state = EXECUTE;
            break;
        }
    }
}
```

FIGURE 86.90 Cache-to-memory transactor implementation.

There are also several RTL signals (e.g., *clk*, address, and *mem_ready*) that connect the *CacheMaster-Trans* and *CacheMemTrans* to the cache model. The necessary comments appear in the implementation in Figure 86.91.

```
#include "bus_types.h"
...
using tlm::tlm_transport_channel;
using tlm::tlm_req_rsp_channel;

int sc_main(int argc, char **argv) {
    //defining the clock period and reset signal
    sc_clock clk ("clock", sc_time (10, SC_NS));
    reset rst ("reset", 5);

    //this portion connects the cache to memory transactor to the memory model
    //the following definition defines channel1 in Fig. 89.80
    tlm_req_rsp_channel <block_basic_request <POINTER_TYPE, SIZE_TYPE, ADDRESS_TYPE>,
        block_basic_response <POINTER_TYPE, SIZE_TYPE > > cache_mem_chan;

    //the following definitions instantiate a transactor and a memory model
    MemoryModel mem ("Memory");
    CacheMemTrans cache_mem ("CacheMem");

    //the ports are interconnected in the following part
    cache_mem.request_port (cache_mem_chan.put_request_export);
    cache_mem.response_port (cache_mem_chan.get_response_export);
    mem.request_port (cache_mem_chan.get_request_export);
    mem.response_port (cache_mem_chan.put_response_export);

    //this portion connects the cache to master transactor to the master model
    //the following definition defines channel2 in Fig. 89.80
    tlm_transport_channel <REQUEST_TYPE, RESPONSE_TYPE > cache_mas_chan;
    //the following definitions instantiate a transactor and a master model
    Master mas ("Master");
    CacheMasTrans cache_mas ("CacheMas");

    //the ports are interconnected in the following part
    mas.master_port (cache_mas_chan.target_export);
    cache_mas.request_port (cache_mas_chan.get_request_export);
    cache_mas.response_port (cache_mas_chan.put_response_export);

    //in the following part the cache is instantiated and its
    //RTL signals are connected to its transactors master interface signals
    sc_signal <sc_uint<16> >   address;
    sc_signal <bool>           read;
    ...
    // memory interface signals
    sc_signal < sc_uint<16> > mem_address;
    sc_signal <bool> mem_read;
    ...
    rst.clk (clk);
    rst.rst (rst1);

    cache_mas.reset (rst1);
    cache_mas.address (address);
    ...
    Cache *cache;
    cache = new Cache ("cache");
    (*cache)(clk, rst1, address, ...);

    cache_mem.clk (clk);
    cache_mem.mem_address (mem_address);
    ...
    sc_start (1000, SC_NS);
    return 0;
}
```

FIGURE 86.91 sc_main for cache example.

86.4.10   Our Coverage of TLM

In this section, TLM techniques were discussed. These techniques are useful for efficient modeling of communication parts in a system. When size and complexity of a system grows, communications between various modules in a system can be a bottleneck for system performance. Therefore, having well-defined methods to model and simulate system communications helps designers develop more efficient designs.

Sections in this chapter discussed various steps in ESL. It began with high-level modeling of a design using UML 2.0. Then, developing a high-level early prototype of a system was discussed in the section on C/C++. After developing an early prototype and partitioning a system, the hardware partitions can be modeled in a hardware-based and high-level language. Therefore, SystemC was described in the section after C/C++. The last section of this chapter discussed TLM that has several predefined efficient ways to model communication in a system.

References

1. S.W. Ambler, *The Elements of UML™ 2.0 Style*, Cambridge University Press, New York, 2005.
2. M. Fowler, *UML Distilled: A Brief Guide to the Standard Object Modeling Language*, 3rd Edition, Addison-Wesley Professional, Massachusetts, 2003.
3. B.W. Kernighan, D. Ritchie, and D.M. Ritchie, *The C Programming Language*, 2nd Edition, Prentice Hall PTR, New Jersey, 1988.
4. B. Stroustrup, *The C++ Programming Language*, 3rd Edition, Addison-Wesley, Massachusetts, 1997.
5. W.R. Stevens, *UNIX Network Programming*, 2nd Edition, Prentice-Hall PTR, New Jersey, 1998.
6. *IEEE Standard SystemC Language Reference Manual*, IEEE Std 1666, Inst. of Elect. & Electronic, New York, 2005.
7. T. Grötker, S. Liao, G. Martin, and S. Swan, *System Design with SystemC*, Springer, Massachusetts, 2002.
8. Z. Navabi, *Verilog Digital System Design*, 2nd Edition, McGraw-Hill Professional, New York, 2005.
9. F. Ghenassia, *Transaction-Level Modeling with SystemC: TLM Concepts and Applications for Embedded Systems*, Springer, The Netherlands, 2005.

87

RT Level Hardware Description with VHDL

Mahsan Rofouei
Nanoelectronics Center of Excellence
School of Electrical and Computer Engineering
University of Tehran

Zainalabedin Navabi
Nanoelectronics Center of Excellence
School of Electrical and Computer Engineering
University of Tehran

CONTENTS

This chapter presents register transfer level VHDL for describing a design for performing behavioral simulation and synthesis. The purpose is to familiarize readers with the main topics of VHDL and present this language such that after reading this chapter readers can start writing VHDL code for actual designs. For this purpose, many of the complex VHDL constructs related to timing and fine modeling features of this language will not be covered here. Furthermore, we will only cover predefined standard types and not get into type definitions and complex language packages. The chapter first describes VHDL with emphasis on design using simple examples. We will cover the basics, just enough to describe our examples. In a later section, after a general familiarity with the language is gained, more complex features of this language with emphasis on testbench development will be described.

87.1 Design with VHDL

VHDL (VHSIC hardware description language) became the IEEE standard 1076 in 1987. Since then it has gone through several minor revisions, and its most common version is the IEEE 1076-1993. VHDL was an important factor in moving digital designs from the gate level to RTL.

VHDL is a rich language for having many features for description of designs from gate level to system level. However, as far as hardware design is concerned, only a small subset of this language is used. This subset includes synthesizable language constructs for description of combinational and sequential components.

For a beginner wanting to learn VHDL for design and synthesis, this chapter provides the necessary basic information. Although gate level descriptions are not often used for synthesis, to cover basic concepts of VHDL, this chapter discusses gate level design, but focuses on register transfer level. To provide designers with tools for testing circuits that they are designing, a section will be devoted to basics of testbenches in VHDL.

87.1.1   Entities and Architectures

Entities used in VHDL for description of hardware components are defined as a pair of entity and architecture declaration. The interface of the circuit is specified by its entity, while its operation is described by architecture bodies associated with that entity. The cause for allowing multiple architectures associated with an entity is having configurable designs for a given interface.

As shown in Figure 87.1 the interface specification of a circuit begins with the **entity** keyword and is followed by the name of the entity together with **is**. Entity declaration contains a list of the component's input–output ports and their types. In contrast, the architecture specification begins with the **architecture** keyword and describes the functionality of a defined entity. The functionality of the component is described using gate instantiations, signal assignments, and processes in the architecture body. The architecture body consists of two parts: the declarative part and the statement part. The declarative part is the section before the **begin** keyword, while the statement part is enclosed between **begin** and **end** of the architecture. The repeat of architecture or entity names after the **end** keyword is optional in both entity and architecture specifications.

A design may be described in a hierarchy of entities, which means that a module can be composed of several submodules. Component instantiation is the construct used for bringing a lower level entity into a higher level one. The connections between these submodules are defined within the architecture of the top-level module. Figure 87.2 shows a hierarchy of components.

The operation of an entity can be described in several ways. Architecture *simple1* in Figure 87.3 describes a circuit in the gate level. This is done by a component instantiation. Architecture *simple2* shows a signal assignment, while architecture *simple3* uses a *process statement* to describe the functionality of the design. A process is used for behavioral descriptions of the design. A process is recognized with the **process** keyword and includes sequential statements. The execution of a process is triggered by events. These events are either listed in a sensitivity list enclosed in a set of parenthesis or *wait statements* are used to control the process execution. However, we will be using the former throughout this chapter.

87.1.2   Entity Ports and Signals

Following the **port** keyword in an entity declaration is a set of parenthesis with a list of input–output ports. A port may be declared as **in**, **out**, **inout**, or **buffer**. **In** and **out** modes are used for declaring

```
entity entity_name is
    input and output ports
end entity_name;

architecture identifier of entity_name is
   declarative part
begin
   statement part
end identifier;
```

FIGURE 87.1   Interface and architecture specification.

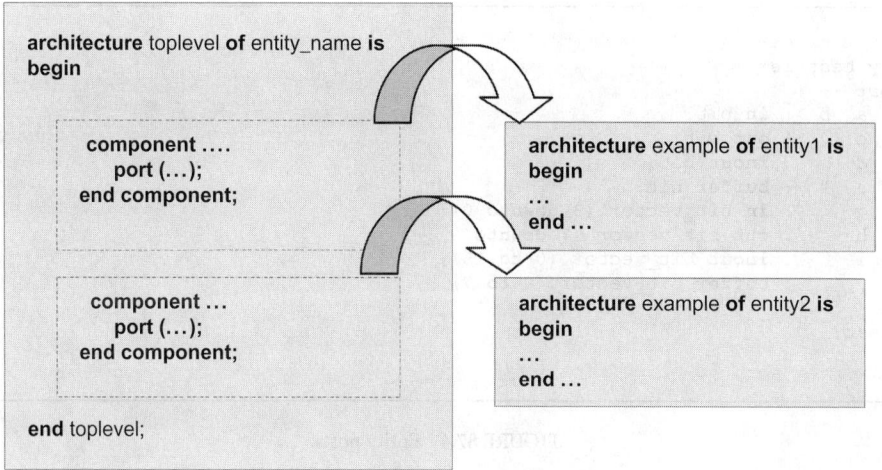

FIGURE 87.2 Module hierarchy.

```
architecture simple1 of entity1 is
   component AND2
      port ( in1, in2 : in bit;
                 w        : out bit);
   end component;

begin
   ANDa: AND2 port map (i1, i2, w);
end simple1;
```

```
architecture simple2 of entity1 is
begin
   w <= i1 and i2;
end simple2;
```

```
architecture simple3 of entity1 is
begin
   process ( i1, i2 )
   begin
      w <= i1 and i2;
   end process
end simple3;
```

FIGURE 87.3 Architecture definition alternatives.

input and output ports, respectively. The **inout** mode is mainly used for bidirectional lines, which implies both an input side and an output side. On the other hand, **buffer** is primarily used for output ports, which are read in the same body as they are driven.

As shown in Figure 87.4, each port also requires a type specification. Type **bit** is a predefined VHDL type which has been used in this example. We will be using this type for now but keep in mind that other standard and user-defined types are also allowed.

In addition to ports, signals can be declared as data carriers in an architecture. Signal declaration takes place in the declarative part of an architecture between the architecture header and the **begin** keyword. Similar to signals, variables are used as intermediate carriers with the difference of only being accessible in process bodies. Therefore in addition to ports, signals are the only data objects that can carry data between processes and other concurrent components.

```
...
entity test is
   port (
      a, b   : in bit;
      c      : out bit;
      d      : inout bit;
      e, f   : buffer bit;
      g      : in bit_vector (3 downto 0);
      h      : out bit_vector (7 downto 0);
      i      : inout bit_vector (0 to 15);
      y      : buffer bit_vector (0 to 7)
   );
end test;
...
...
```

FIGURE 87.4 Entity ports.

87.1.3 Logic Value System

The *std_logic* standard package defines a logic value system consisting of nine logic values. However, *std_logic* is not a part of the VHDL language and is only an IEEE standard utility package for the language. The nine values of *std_logic*, which are shown in Table 87.1, are used to represent high impedance, unknown, uninitialized, capacitive, and resistive **1** or **0**, and driven **1** or **0**. In most cases only four or five of these nine [2] values are sufficient to express the logic-level behavior of a circuit. U is considered as the default type.

Since the *std_logic* type includes all values of the **bit** type, we will be using it instead of **bit** type from now onward in the chapter. The *std_logic_unsigned* package, which will be used for unsigned types, contains a set of signed arithmetic, conversion, and comparison functions. Finally, it should be kept in mind that special care needs to be taken as VHDL has very strict type checking rules.

87.2 Combinational Circuits

A combinational circuit can be represented by its subcomponents at its gate level. VHDL does not provide primitive structures and thus any gate primitive (NAND, NOR, etc.) must be separately described using an entity–architecture pair. Figure 87.5 shows the VHDL code for several common gate level components.

87.2.1 Gate Level Combinational Circuits

This section describes the VHDL codes for individual gates of Figure 87.5. As for the 2-input AND gate component, the entity has been defined with *in1* and *in2* as its input pins of type *std_logic* and a single output named *out1*. For defining the component's functionality a signal assignment is used. A signal assignment statement assigns values to its left-hand-side signal or **output**, **buffer**, or **inout** ports. An event observed on *in1* or *in2*, right-hand-side signals of the signal assignment of the AND2 gate causes the evaluation of the right-hand-side expression. The VHDL code shows a 3 ns delay for the 2-input AND component. This results in the assignment of a new value to *out1* taking place 3 ns after the evaluation of the right-hand-side expression.

The BUFIF1 and BUFIF0 gates are implemented using conditional signal assignments which will be described later. Other primitive gates can be easily defined as shown above. These primitive gates provide a sufficient set of components for the description of larger structures. With the help of component instantiation, which will be shortly discussed, larger components can be described.

87.2.1.1 Majority Example

We will use the majority example of Figure 87.6 to illustrate how a defined component can be instantiated and used in a design. Figure 87.6 shows a structural description of a majority circuit consisting of four

TABLE 87.1 Logic Value System

Value	Representing
'U'	Uninitialized
'X'	Forcing unknown
'0'	Forcing 0
'1'	Forcing 1
'Z'	High impedance
'W'	Weak unknown
'L'	Weak 0
'H'	Weak 1
'-'	Don't care

```
entity AND2 is
    port (
        in1, in2 : in std_logic;
        out1     : out std_logic );
end AND2;

architecture example of AND2 is
begin
    out1 <= in1 and in2 after 3 ns;
end example;
```

```
entity OR3 is
    port (
        in1, in2, in3 : in std_logic;
        out1              : out std_logic );
end OR3;

architecture example of OR3 is
begin
    out1 <= in1 or in2 or in3 after 6 ns;
end example;
```

```
entity BUFIF1 is
    port (
        in1, en : in std_logic;
        out1    : out std_logic );
end BUFIF1;

architecture example of BUFIF1 is
begin
    out1 <= in1 when en='1' else 'Z' after 3 ns;
end example;
```

```
entity BUFIF0 is
    port (
        in1, en : in std_logic;
        out1    : out std_logic );
end BUFIF0;

architecture example of BUFIF0 is
begin
    out1 <= in1 when en='0' else 'Z' after 3 ns;
end example;
```

FIGURE 87.5 Basic primitives.

subcomponent instantiation. Components that are to be instantiated in this architecture need to be declared in its declarative part. A component declaration consists of the name of the component followed by its ports. In the statement part, these components are instantiated and interconnected. Each component instantiation starts with an arbitrary label followed by the name of the component and then by a mapping between the ports of the instantiated component and the actual component. The interconnection between these subcomponents is done with signals declared in the declarative part of the architecture. The components used in this example are described in Figure 87.5.

87.2.1.2 Multiplexer Example

The multiplexer example is a good example to show how a resolution function works for a signal. As shown in Figure 87.7 the 2-to-1 multiplexer is described using BUFIF1 and BUFIF0, which we defined in Figure 87.5.

When a line is derived with two drivers a resolution function needs to be specified for that signal. The *std_logic* type has a predefined resolution function in the IEEE Std 1164 package. Since wiring two signals, one with a 0 value and another with a 1 value results in an unknown value, driving 0 and 1 simultaneously into a resolved signal results in X. Although the *resolved* resolution function satisfies most design requirements, other resolution functions can be described in VHDL. Table 87.2 illustrates how the *resolved* function in the IEEE Std 1164 package works for five of the more commonly used values.

```
library ieee;
use ieee.std_logic_1164.all;

entity majority is
   port (
      a, b, c : in std_logic;
      y       : out std_logic);
end majority;

architecture sample of majority is
   component AND2
   port ( in1, in2 : in std_logic;
          out1     : out std_logic);
   end component;

   component OR3
   port( in1, in2, in3  : in std_logic;
         out1           : out std_logic);
   end component;

   signal im1, im2, im3 : std_logic;
begin
   ANDa: AND2 port map (a, b, im1);
   ANDb: AND2 port map (b, c, im2);
   ANDc: AND2 port map (a, c, im3);
   ORa : OR3  port map (im1, im2, im3, y);
end sample;
```

FIGURE 87.6   VHDL code for the majority circuit.

```
library ieee;
use ieee.std_logic_1164.all;

entity mux2to1 is
   port (
      a, b, s : in std_logic;
      y       : out std_logic);
end mux2to1;

architecture sample of mux2to1 is
   component BUFIF0
   port (
         in1, en : in std_logic;
         out1    : out std_logic);
   end component;

   component BUFIF1
   port (
         in1, en : in std_logic;
         out1    : out std_logic);
   end component;

begin
   BUFIF1a: BUFIF1 port map (b, s, y);
   BUFIF0a: BUFIF0 port map (a, s, y);
end sample;
```

FIGURE 87.7   VHDL code for the multiplexer circuit.

TABLE 87.2   A Subset of the *Resolved* Function

U	X	0	1	Z	U
U	U	U	U	U	U
X	U	X	X	X	X
0	U	X	0	X	0
1	U	X	X	1	1
Z	U	X	0	1	Z

TABLE 87.3   VHDL Operators

Boolean **Operators**	not and or nand nor xor
Comparison **Operators**	= /= < <= > >=
Arithmetic **Operators**	sign sign abs + − * / mod rem ** + −
Concatenation **Operators**	&

The resolved function is the only supported resolved function defined for *std_logic*, if needed, others such as wired-and and wired-or have to be defined by the user.

87.2.2   Description by Use of Equations

A combinational circuit may also be described by the use of Boolean, logical, and arithmetic expressions. The VHDL language has a set of standard operators that can be used for Boolean, logical, and arithmetic equations. Table 87.3 shows this set of operators. These operators can be used in assign statements to create the desired functionality for the circuit.

87.2.2.1   XOR Example

Consider the description of an XOR gate using Boolean equations in Figure 87.8. As discussed before, a signal assignment can be used to assign values to ports or other signals. Here the **xor** operator has been used in a simple signal assignment to assign the result of *in1* xored with *in2* to the output after a 3 ns delay.

So, it can be seen that instead of having to write our own gates in the way described in Section 87.2.1, we can simply use operators listed in Table 87.3 to produce any desired expression.

87.2.2.2   Full Adder Example

The code shown in Figure 87.9 corresponds to a single-bit full adder. A full adder can be simply described by two signal assign statements: one for sum and one for carry out. The *sum* output can be easily generated with a Boolean expression using two **xor** operators. The expression will be the same as the XOR example described above with the difference of having three operands this time. Therefore two **xor** operators will be used. As for the *carryOut* output, again another signal assignment with **and** and **or** operators is used. Note that the use of parenthesis is necessary in the carry signal assignment.

Another property of signal assignment statements is their concurrency. This means that all statements are concurrent and their order of appearance is not significant.

87.2.2.3   Comparator Example

Consider the code shown in Figure 87.10, which corresponds to a 4-bit comparator. To define 4-bit arrays, a predefined array type in the *std* package called *std_logic_vector* which represents a collection of

```
library ieee;
use ieee.std_logic_1164.all;

entity xor2 is
   port (
       in1, in2 : in  std_logic;
       out1     : out std_logic
   );
end xor2;

architecture behavioral of xor2 is
begin
   out1 <= in1 xor in2 after 3 ns;
end behavioral;
```

FIGURE 87.8   XOR VHDL code.

```
library ieee;
use ieee.std_logic_1164.all;

entity FA is
   Port (
       in1, in2, carryIn : in std_logic;
       sum, carryOut     : out std_logic
      );
end FA;

architecture behavioral of FA is
begin
   sum      <= in1 xor in2 xor carryIn after 3 ns ;
   carryOut <= (in1 and in2)or(in1 and carryIn)or(in2 and carryIn) after 3 ns;

end behavioral;
```

FIGURE 87.9   Full adder VHDL code.

```
library ieee;
use ieee.std_logic_1164.all;

entity comp_4bit is
   port (
       in1, in2 : in std_logic_vector (3 downto 0);
       eq       : out std_logic
   );
end comp_4bit;

architecture behavioral of comp_4bit is
   signal im : std_logic_vector (3 downto 0);
   function NOR_reduce (in1 : in std_logic_vector (3 downto 0)) return std_logic is
      variable result : std_logic ;
   begin
      result := not (in1(3) or in1(2) or in1(1) or in1(0)) ;
      return result ;
   end;
begin
   im <= in1 xor in2;
   eq <= NOR_reduce(im);
end behavioral;
```

FIGURE 87.10   4-Bit comparator VHDL code.

*std_logic*s, is used. Notice how the *std_logic_vector* is used in the definition of a port or a signal. (3 **downto** 0) is the range of indices for the vector (3 is the index of the leftmost element and 0 is for the rightmost element in the vector). Note that the range can also be declared as (0 **to** 3), but the former declaration is recommended since the indices correspond to bit position weights in an array.

First, the result of XORing bits of *in1* and *in2* are assigned to *im*, which is defined as a signal in the declaration part of the architecture. In the next line a function call is used to produce the *eq* output. A function is a subprogram consisting of sequential statements that can be called anywhere in a VHDL code. Functions can be declared in several places such as the declarative part of an architecture or a process. A function declaration starts with the **function** keyword and a function name followed by its parameters and finally a single return value type. In this case, our function has a 4-bit input of type *std_logic_vector* and a std_logic return type. This function performs bit-by-next-bit operation, giving a 1-bit result. The function is simply called in the architecture statement body with *im* as its input.

As shown in Figure 87.10, *result* is defined as a **variable**. Notice the use of := instead of <= in the assignment. There will be more on variables later in the chapter.

Another way to describe the comparator circuit is to use a *conditional signal assignment*. A conditional signal assignment is a signal assignment that takes place only when the condition stated after the **when** keyword is met. In the case of our comparator example, the below *conditional signal assignment* produces the correct *eq* value.

eq <= '1' **when** in1 = in2 **else** '0';

Note that it is possible to have another conditional part after the **else** keyword.

87.2.2.4 Multiplexer Example

The multiplexer example is another good example of the use of *conditional signal assignment*s. Figure 87.11 illustrates a 2-to-1 multiplexer.

87.2.2.5 Decoder Example

Figure 87.12 shows a 2-to-4 decoder, coded using a *selected signal assignment*. The behavior of a *selected signal assignment* is similar to the *case statement*. A *case statement* can only appear in VHDL sequential bodies where, *selected signal assignments* are for the concurrent bodies of VHDL, e.g., an architecture body. A selected signal assignment statement is very similar to a *conditional signal assignment* in the way that it chooses from a number of expressions based on a condition, but they differ in that only conditions relating to one signal are used in a selected signal assignment.

```
library ieee;
use ieee.std_logic_1164.all;

entity mux2to1 is
   port (
       in1, in2 : in  std_logic_vector (3 downto 0);
       sel      : in  std_logic;
       y        : out std_logic_vector (3 downto 0)
   );
end mux2to1;

architecture structural of mux2to1 is
begin
   y <= in1 when sel = '1' else in2;
end structural;
```

FIGURE 87.11   2-to-1 Multiplexer VHDL code.

```
library ieee;
use ieee.std_logic_1164.all;

entity dcd2to4 is
   port (
      sel : in std_logic_vector (1 downto 0);
      y   : out std_logic_vector (3 downto 0)
   );
end dcd2to4;

architecture structural of dcd2to4 is
begin
   with sel select
      y <= "0001" when "00",
           "0010" when "01",
           "0100" when "10",
           "1000" when "11",
           "0000" when others;
end structural;
```

FIGURE 87.12   2-to-4 Decoder VHDL code.

```
library ieee;
use ieee.std_logic_1164.all;
use ieee.std_logic_arith.all;

entity adder is
   port (
      a, b : in  std_logic;
      ci   : in  std_logic;
      s    : out std_logic;
      co   : out std_logic
   );
end adder;

architecture structural of adder is
   signal mid : unsigned (1 downto 0);
begin
   mid <= '0'&a + b + ci;
   (co, s) <= mid;
end structural;
```

FIGURE 87.13   Full adder VHDL code.

87.2.2.6   Adder Example

As another example, consider a full adder circuit with a carry-in and a carry-out output shown in
Figure 87.13. An assignment statement is used to set the two outputs to their values. The left-hand side
of this assignment is an aggregate. Aggregate is a collection of values. In this example an aggregate is
used to match the return 2-bit vector produced from adding a, b, and ci.

87.2.3   Description with Procedural Statements

As described earlier in the chapter, VHDL provides constructs such as different kinds of signal assignments,
component instantiation, and processes to model concurrency. In addition to these constructs, VHDL

also provides constructs for sequential description of hardware. This is referred to as the behavioral approach to describe hardware components. In this approach, the designer expresses the functionality or behavior of hardware instead of its structural details. Behavioral descriptions are supported with the *process statements* in VHDL.

A *process statement* runs concurrently with other concurrent components but its body consists of sequential statements. The *process statement* can appear in the architecture body just as signal assignments and, encloses in a **begin-end** pair. The sequential statements in a *process statement* run repeatedly during a simulation run. The **process** keyword starts the definition of a *process statement*. A list of signals enclosed in a pair of parenthesis can appear after the **process** keyword as its sensitivity list. Any event on any of these signals will cause the process to be executed again. If no sensitivity list is specified for a process statement, that process will run continuously until the end of the simulation time.

A *process statement* consists of a *declarative part* and a *statement part*. The *statement part* is where the sequential statements are listed while the *declarative part* consists of function, variable, and other declarations (signal declarations are not allowed in the *declarative part* of a process). Variables are used to store intermediate values within a process. Variable declaration is similar to signals and it takes place in the *declarative part* of a process. Assigning values to variables is done with :=. Initial values can also be assigned to variables in their declarations.

Although variables can be data carriers like signals but they differ from them in several ways. Signals do not change in a process execution and therefore cannot hold intermediate values within a process. This is where variables become handy. They can store intermediate values by a variable assignment. Declarations including variable declarations take place once at the beginning of a simulation run and a variable's value is retained between process iterations. Another difference between signals and variables is that only signal assignments are timed and the after clause cannot be used with variable statements.

Another issue regarding signal assignments in concurrent bodies is the delta delay. A simulation cycle is referred to as delta delay in VHDL. Consider two signal assignments in an architecture body, one of which its output causes the evaluation of the right-hand side of the other signal assignment. Since these are concurrent statements one expects both of their results to be ready at the same time. On the other hand, the first signal assignment causes an event on the right-hand side of the second signal assignments. So it is obvious that there is a time delay between these two statements. This delay is what is referred to as delay time or delta delay, and is shown by the δ symbol.

87.2.3.1 Majority Example

Figure 87.14 shows a majority circuit described by the use of a *process statement*. This *process statement* is sensitive to *a*, *b*, and *c* signals. Any event on these signals will cause the process block to run again and a new value will be assigned to the *y* output. No delay has been considered for this signal assignment. Therefore, the value of *y* will be updated δ time after any event on *a*, *b*, or *c*.

87.2.3.2 Majority Example with Delay

The VHDL code corresponding to a majority circuit with delay is shown in Figure 87.15. The code is very similar to the code described in Figure 87.14 with an after clause with a 5 ns delay value. In this case the result of the right-hand side will be placed on *y*'s driver with the 5 ns delay value.

87.2.3.3 Procedural Multiplexer Example

Figure 87.16 illustrates another example of procedural blocks using an *if statement*. VHDL allows the use of *if statements* in *process statements*. An *if statement* is similar to the *conditional signal assignment* construct, with the difference that it is only allowed in sequential bodies. Depending on the condition specified for the *if statement*, a corresponding branch is taken. Note that each *if statement* contains a corresponding **then** branch. An *if statement* ends with the **end if** keyword. There can be any number of **elsif** branches to an *if statement*. In contrast, a maximum of one else statement is allowed per each if statement. This optional else statement should be the last branch in an if statement.

```
library ieee;
use ieee.std_logic_1164.all;

entity maj3 is
    port (
        a,b,c : in std_logic;
        y     : out std_logic
    );
end maj3;

architecture behavioral of maj3 is
begin
    process(a, b, c)
    begin
        y <= (a and b) or (b and c) or (a and c);
    end process;
end behavioral;
```

FIGURE 87.14   Majority circuit VHDL code.

```
library ieee;
use ieee.std_logic_1164.all;

entity maj3 is
    port (
        a,b,c : in std_logic;
        y     : out std_logic
    );
end maj3;

architecture behavioral of maj3 is
begin
    process(a, b, c)
    begin
        y <= (a and b) or (b and c) or (a and c) after 5 ns;
    end process;
end behavioral;
```

FIGURE 87.15   Majority circuit with delay VHDL code.

In this example, if the selector is equal to zero the block after the **then** keyword is executed and *y* takes *i*0, otherwise the block of statements after the **else** are taken and therefore *y* takes *i*1.

87.2.3.4   Procedural ALU Example

A *case statement* is similar to *if statements*, described in the previous example, in branching on a condition. However a *case statement* chooses a branch based on the value of the *case expression*, which does not need to be a Boolean. *Case statements* are preferred over *if statements* when many choices exist. This is why we have used the *case statement* in Figure 87.17. A *case statement* is somehow similar to *selected signal assignment* described earlier. As discussed before, selected signal assignments and conditional signal assignments cannot be used in process statement bodies.

This ALU receives *a*, *b*, and *f* as its inputs. The *process statement*, shown in the VHDL code, is sensitive to all its inputs. The *case statement* used here selects the correct operation corresponding to *f* from the *case alternatives*. Consider the **when others** *case alternative* used in the last line of the *case statement*. This branch is taken when the condition does not match any of the other alternatives. In our example inputs containing **Z** or **X** will take this branch.

```
library ieee;
use ieee.std_logic_1164.all;

entity mux2to1 is
   port (
      i0,i1,s : in  std_logic;
      y       : out std_logic
   );
end mux2to1;

architecture behavioral of mux2to1 is
begin
   process (i0, i1, s)
   begin
      if s = '0'  then
         y <= i0;
      else
         y <= i1;
      end if;
   end process;
end behavioral;
```

FIGURE 87.16   Procedural multiplexer.

```
library ieee;
use ieee.numeric_bit.all;

entity ALU4bit is
   port (
      a, b : in  unsigned (3 downto 0);
      f    : in  unsigned (1 downto 0);
      y    : out unsigned (3 downto 0)
      );
end ALU4bit;

architecture behavioral of ALU4bit is
begin
   process (a, b, f)
   begin
      case f is
         when "00" => y <= a + b;
         when "01" => y <= a - b;
         when "10" => y <= a and b;
         when "11" => y <= a xor b;
         when others => y <= "0000" ;
      end case;
   end process;
end behavioral;
```

FIGURE 87.17   Procedural ALU.

87.2.4 Combinational Rules

Now that *if* and *case statement*s are covered, consider a case where there are conditions under which the output of a combinational circuit is not assigned a value. Obviously the output retains its previous value, which implies the latching behavior. This latching behavior is unwanted in describing combinational circuits. Therefore two rules can be considered in describing combinational circuits with *process statements*.

1. List all inputs of the combinational circuit in the sensitivity list of the *process statement* describing it.
2. Make sure all combinational circuit outputs receive some value regardless of how the program flows in the conditions of *if* or *case statements*. If there are too many conditions to check, set all outputs to their inactive values at the beginning of the *process statement*.

87.2.5 Bussing

Bus structures can be implemented by the use of multiplexers or three-state logic. Various methods of describing combinational logic can be used for the description of a bus, in VHDL.

Figure 87.18 shows the VHDL code for a three-state bus with three sources, *busin1, busin2,* and *busin3*. These sources are put on *busout* by active high enabling control signals: *en1, en2,* and *en3*.

Three *conditional signal assignments* are used for assigning values to *busout*. Each *conditional signal assignment* either selects a bus driver or a 4-bit **Z** value for *busout*. As mentioned before, multiple assignments to a signal produce multiple drivers for that circuit. Therefore, multiple *conditional signal assignments* are appropriate for representing busses in VHDL. Note that only one enable should be active at a time. Multiple driving values for *busout* are resolved using the *resolved* resolution function. Enabled inputs of *busout* produce logic values on their drivers, while inactive ones put all **Z** values on their drivers. Resolving **Z** with logic values results in a valid logic value, but resolving multiple logic values results in unknown. This is why we use high impedance for inactive enables.

87.3 Sequential Circuits

As with any digital circuit, a sequential circuit can be described in VHDL by the use of gates, Boolean expressions, or behavioral constructs (e.g., *process statements*). Although gate level description allows a more detailed description of timing and delays, because of the complexity of clocking, and register and flip-flop controls, sequential circuits are usually described by *process statements*. In the next section we will first discuss some basic elements at the gate level and then represent coding styles for more complex sequential circuits.

```
library ieee;
use ieee.std_logic_1164.all;

entity bussing is
    port (
        busin1, busin2, busin3  : in std_logic_vector( 3 downto 0);
        en1, en2, en3           : in  std_logic;
        busout                  : out std_logic_vector( 3 downto 0)
    );
end bussing;

architecture structural of bussing is
begin
    busout <= busin1 when en1 = '1' else (others => 'Z');
    busout <= busin2 when en2 = '1' else (others => 'Z');
    busout <= busin3 when en3 = '1' else (others => 'Z');
end structural;
```

FIGURE 87.18 Three-State bus implementation.

87.3.1 Basic Memory Elements at the Gate Level

A clocked D-latch latches its input data when clock is active. The latch structure retains its value until the next active clock cycle. This element is the basis of all static memory elements.

Figure 87.19 shows a simple implementation of the D-latch that uses cross-coupled NOR gates, while the corresponding VHDL code is shown in Figure 87.20. Notice the use of **buffer** for *q* and *q_b*. This is due to the use of these signals on the right-hand side of signal assignments. These signals could have been declared as **out** and two signals could have done the job, but in this case two additional signal assignments would have been added to the code.

This code can be coded in another way using a simple *conditional signal assignment* as below:

q <= d **when** c = '1' **else** q;

Using two such statements with complementary clock values describes a master–slave flip-flop. As shown in Figure 87.21, *qm* is the master output and *q* the flip-flop output.

87.3.2 Memory Elements Using Procedural Statements

Although latches and flip-flops can be described in gate level by component instantiation and signal assignments, describing more complex register structures cannot be done this way. This section presents coding styles for describing latches and flip-flops using *process statement*s. Later on, it will be shown that the same coding styles presented here can be used to describe memories with more complex control units as well as functional register structures such as counters and shift registers.

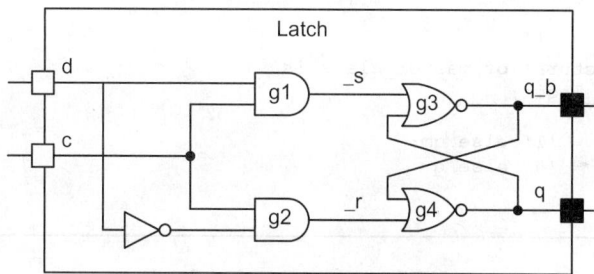

FIGURE 87.19   Clocked D-latch.

```
library ieee;
use ieee.numeric_bit.all;

entity latch is
   port (
      d, c  : in     std_logic;
      q, q_b : buffer std_logic
   );
end latch;

architecture structural of latch is
   signal s, r : std_logic;
begin
   s   <= c and d       after 6 ns;
   r   <= c and (not d) after 6 ns;
   q_b <= s nor q       after 4 ns;
   q   <= r nor q_b     after 4 ns;

end structural;
```

FIGURE 87.20   VHDL code for clocked D-latch.

87.3.2.1 Latches

Figure 87.22 illustrates a D-latch described with a *process statement*. This *process statement* is sensitive to the latch clock and data input represented by *c* and *d*, respectively. The *if statement* used in the *process statement* puts the data input into *q* when the latch clock is active. This implies that any change on *d* while *c* is 1 can be transparently seen on the output. This behavior is referred to as transparency and it is how latches work.

87.3.2.2 D Flip-Flop

Different to latches, the data on D flip-flop's input does not directly propagate to its output. This change will only have an affect on the specified edge of the clock. Consider Figure 87.23, which describes a positive-edge trigger D flip-flop. The *process statement* used is sensitive to changes on the clock input. Since the code describes a positive-edge flip-flop, it only transfers the data on the D-input to its output when *clk* has made a **0** to **1** transition. A *process statement* sensitive to the *clk* and an *if statement* that

```
library ieee;
use ieee.numeric_bit.all;

entity master_slave is
    port (
        d, c : in  std_logic;
        q      : buffer std_logic
    );
end master_slave;

architecture structural of master_slave is
  signal qm : std_logic;
begin
  qm <= d  when c = '1' else qm;
  q  <= qm when c = '0' else q;
end structural;
```

FIGURE 87.21   Master–slave flip-flop.

```
library ieee;
use ieee.std_logic_1164.all;

entity latch1 is
    port (
        d, c : in  std_logic;
        q    : out std_logic
    );
end latch1;

architecture behavioral of latch1 is
begin
    process (d, c)
    begin
        if c = '1' then
            q <= d;
        end if;
    end process;
end behavioral;
```

FIGURE 87.22   Procedural latch.

determines a **0** to **1** transition on *clk* specifies the positive edge of *clk*. Similarly, a negative-edge trigger D flip-flop is created by changing the *if statement* to "**if** clk = '0'."

87.3.2.3 Synchronous Control

The coding style presented for D flip-flops is general and can be easily expanded to describe many features found in flip-flops and memory structures. One of these features is having synchronous control for setting and resetting. The VHDL code of Figure 87.24, describes a D flip-flop with synchronous set and reset inputs.

```
library ieee;
use ieee.std_logic_1164.all;

entity DFF1 is
   port (
        d, clk : in std_logic;
        q      : out std_logic
   );
end DFF1;

architecture behavioral of DFF1 is
begin
   process(clk)
   begin
     if clk = '1' and clk'event then
        q <= d;
     end if;
   end process;
end behavioral;
```

FIGURE 87.23   A positive-edge D flip-flop.

```
library ieee;
use ieee.std_logic_1164.all;

entity DFF1 is
   port (
      d, clk, s, r : in std_logic;
      q            : out std_logic
   );
end DFF1;

architecture behavioral of DFF1 is
begin
   process (clk)
   begin
     if clk = '1' and clk'event then
        if s = '1' then
           q <= '1';
        elsif r = '1' then
           q <= '0';
        else
           q <= d;
        end if;
     end if;
   end process;
end behavioral;
```

FIGURE 87.24   VHDL code for D flip-flop with synchronous control.

The difference between this type of flip-flop and a simple D flip-flop is that on every positive edge of the clock, it first checks for the set and rest inputs and puts a **1** into the output if *s* is active and a **0** if *r* is active. Only when *s* and *r* are inactive the flip-flop places the D-input on its output. Note that the *s* input has been given a higher priority to *r* by first checking for *s* in the sequential *process statement*. The flip-flop structure corresponding to this description is shown in Figure 87.25.

Note that any other synchronous control can be added in the same way. For example, a clock enable input can be easily added by including another *if statement* in the **else** part of the last *if statement* before putting *d* into *q*.

87.3.2.4 Asynchronous Control

Consider the VHDL code of Figure 87.26 that describes a D flip-flop with asynchronous control. To have asynchronous control inputs we only need to add these signals to the process sensitivity list and change the ordering of *if statement* conditions. In the previous case when the *process statement* was only sensitive to *clk*, the flow of the *process statement* would only start if any change happened on the *clk* input. In other words, changes on control inputs did not have any effect on starting the flow of the *process statement*. In

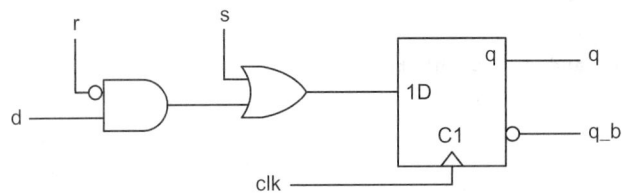

FIGURE 87.25   D flip-flop with synchronous control.

```
library ieee;
use ieee.std_logic_1164.all;

entity DFF1 is
    port (
        d, clk, s, r : in std_logic;
        q              : out std_logic
    );
end DFF1;

architecture behavioral of DFF1 is
begin
    process (clk, s, r)
    begin
        if s = '1' then
            q <= '1';
        elsif r = '1' then
            q <= '0';
        elsif clk = '1' and clk'event then
            q <= d ;
        end if;
    end process;
end behavioral;
```

FIGURE 87.26   D flip-flop with asynchronous control.

this case however, by adding control signals to the sensitivity list, the process block flow can start by any event seen on these signals or the *clk*.

87.3.3 Registers, Shifters, and Counters

As mentioned previously, the coding styles presented for flip-flops can be extended to be used in other sequential circuits such as registers, shift registers, and counters.

87.3.3.1 Registers

Figure 87.27 shows the VHDL code for an 8-bit register with synchronous set and reset control. An 8-bit register is defined similar to a D flip-flop but for a bit-vector of eight elements. Assigning **1**s (**0**s) to bits of q is done by an aggregate operation on the right-hand side of the assignment. This operation selects bits of the right-hand-side vector it is forming and associates values to these bits. The **others** keyword can be used in array aggregates to select all the indexes that have not been selected up to the appearance of **others**. A special case of this form, which is used in our example, is when no other indexes have been selected. In this case **others** selects all the indexes of the array.

87.3.3.2 Shift Registers

Consider Figure 87.28, which describes a 4-bit shift register with synchronous reset, right and left shift capability, and parallel loading.

Note that instead of having two variables q and q_t we could have defined q as buffer.

87.3.3.3 Counters

Figure 87.29 describes a 4-bit up-down counter with reset control input and parallel loading capability. The code for this counter is similar to the shift register described above. The only difference is that the

```
library ieee;
use ieee.std_logic_1164.all;

entity register8 is
   port (
      d             : in std_logic _vector (7 downto 0);
      clk, s, r  : in std_logic;
      q             : out std_logic _vector ( 7 downto 0)
   );
end register8;

architecture behavioral of register8 is
begin
   process (clk)
   begin
      if clk = '1' and clk'event then
         if s= '1' then
            q <= (others => '1');
         elsif r = '1' then
            q <= (others => '0');
         else
            q <= d;
         end if;
      end if;
   end process;
end behavioral;
```

FIGURE 87.27 An 8-bit register.

```
library ieee;
use ieee.std_logic_1164.all;

entity shift_reg4 is
   port (
        d                        : in std_logic_vector (3 downto 0);
        clk, ld, rst, l_r, s_in : in std_logic;
        q                        : out std_logic_vector ( 3 downto 0)
        );
end shift_reg4;

architecture behavioral of shift_reg4 is
begin
   process (clk)
      variable q_t: std_logic_vector (3 downto 0);
   begin
      if clk = '1' and clk'event then
         if rst= '1' then
            q_t := (others => '0');
         elsif ld = '1' then
            q_t := d;
         elsif l_r = '1' then
            q_t :=  q_t (2 downto 0) & s_in ;
         else
            q_t := s_in & q_t (3 downto 1);
         end if;
      end if;
      q <= q_t;
   end process;
end behavioral;
```

FIGURE 87.28   A 4-bit shift register.

counter performs arithmetic addition or subtraction instead of shifting. The *u_d* input determines whether the counter counts up or down. Upon activation of the *ld* input on the positive edge of the clock, the value on *d_in* will be loaded into the counter. Again, definition of *q* as a **buffer** would have eliminated the use of variable *t_q*.

87.3.4   State Machine Coding

Finite-state machines, which are commonly used in the controller part of a digital system that has been partitioned into a data path and a controller, can be modeled in VHDL. This section shows coding styles used for Moore and Mealy state machines. The examples presented here are simple sequence detectors; however they present coding styles for coding more complex controllers.

87.3.4.1   Moore Detector

A Moore machine is a finite-state machine, which indicates an output signal for each state independent of circuit's inputs.

Figure 87.30 shows the state diagram for a **101** Moore sequence detector. The machine looks for a sequence of **101** and when found, signals the output to become **1**.

Figure 87.31 shows the VHDL code for this state diagram. The *current* state is represented by an enumeration type. This type has a value for each state; therefore, for detecting **101** this type should have

```
library ieee;
use ieee.std_logic_1164.all;
use ieee.std_logic_unsigned.all;

entity counter is
   port(
      d_in                 : in unsigned ( 3 downto 0);
      clk, rst, ld, u_d : in std_logic;
      q                    : out unsigned ( 3 downto 0)
   );
end counter;

architecture behavioral of counter is
begin
   process (clk)
      variable t_q: unsigned (3 downto 0);
   begin
      if clk = '1' and clk'event then
         if rst= '1' then
            t_q := (others => '0');
         elsif ld = '1' then
            t_q := d_in;
         elsif u_d = '1' then
            t_q := t_q + 1;
         else
            t_q := t_q - 1;
         end if;
      end if;
      q <= t_q;
   end process;
end behavioral;
```

FIGURE 87.29   An up-down counter.

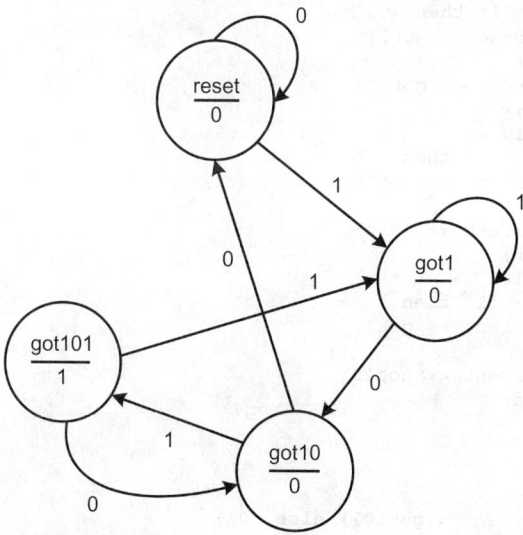

FIGURE 87.30   A Moore sequence detector.

four values including *init*. Figure 87.31, shows an enumeration-type declaration that declares *States* as a set of four states including *init*, *got1*, *got10*, and *got101*. Type declarations take place in the declarative part of the architecture body. As another example of enumeration types, consider the **bit** type of the standard package that is an enumeration of '0' and '1'.

The *process statement*, which is sensitive to *clk*, is in charge of state transitions and register clocking. Upon the positive edge of the clock, checked by "clk = '1' **and** clk'**event**" the *process statement* checks

```vhdl
library ieee;
use ieee.std_logic_1164.all;

entity Moore_detector_101 is
   port (
      clr, clk, x : in  std_logic;
      z           : out std_logic
   );
end Moore_detector_101;

architecture behavioral of Moore_detector_101 is
   type States is (init, got1, got10, got101);
   signal current: States;
begin
   process (clk)
   begin
      if clk = '1' and clk'event then
         if clr = '1' then
            current <= init;
         else
         case current is
            when init =>
               if x = '1' then
                  current <= got1;
               else
                  current <= init;
               end if;
            when got1 =>
               if x = '1' then
                  current <= got1;
               else
                  current <= got10;
               end if;
            when got10 =>
               if x = '1' then
                  current <= got101;
               else
                  current <= init;
               end if;
            when got101 =>
               if x = '1' then
                  current <= got1;
               else
                  current <= got10;
               end if;
         end case;
         end if;
      end if;
   end process ;
   z <= '1' when (current = got101) else '0';
end behavioral;
```

FIGURE 87.31   Moore machine VHDL code.

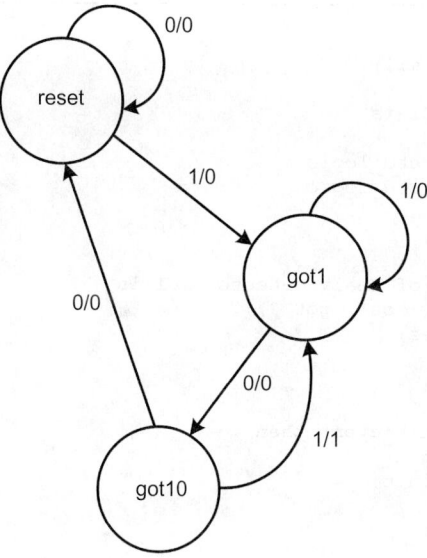

FIGURE 87.32 A Mealy sequence detector.

for the *clr* input. If *clr* is active then the next state becomes *init* by assigning *init* to *current*. Note that the value put into *current* in this pass gets checked in the next run of the *process statement*. On the other hand if the *clr* input is not active, the program flow reaches the *case statement* and takes one of the case alternatives according to *current*. A single conditional signal assignment is in charge of assigning values to *z*.

87.3.4.2 Mealy Detector

Finite-state machines can also be implemented as Mealy state machines whose output, unlike Moore state machines, is determined by the current state of the machine as well as the inputs of the circuit. Therefore the Mealy outputs are not fully synchronized with the clock input. Figure 87.32 shows the state diagram of a Mealy state machine that inspects a **101** sequence on its input. In this state diagram the outputs are determined by the edges coming out of the states.

Notice that we only have three states in this example, in contrast with the four states of the Moore implementation. Usually, there is an extra state in the Moore state machine since output '1' needs to be associated with a state. The VHDL code for this machine is shown in Figure 87.33.

As shown the *process statement* is only responsible for the state transitions. The output is determined with the conditional signal assignment in the architecture body. Notice that overlapping sequences are allowed in both examples.

87.3.4.3 Huffman Coding Style

The Huffman model for a digital system characterizes it as an interconnection of a combinational and a register block. In VHDL, a Huffman model can be described using two process blocks, one describing the register part and the other for the combinational part. This is shown in Figure 87.34.

Figure 87.35 shows the VHDL code for the Moore sequence detector described before according to the Huffman model. This time instead of *current*, present state (*p_state*) and next state (*n_state*) have been used. The sequential part of the circuit is coded using a *process statement* with *clr* and *clk* in its sensitivity list. On the positive edge of the *clk*, *p_state* is either set to *init* in case of *clr*, or loaded with *n_state*.

```
library ieee;
use ieee.std_logic_1164.all;

entity Mealy_detector_101 is
    port (
        clr, clk, x : in  std_logic;
        z           : out std_logic
    );
end Mealy_detector_101;

architecture behavioral of Mealy_detector_101 is
    type States is (init, got1, got10);
    signal current: States;
begin
    process (clk)
    begin
        if clk = '1' and clk'event then
            if clr = '1' then
                current <= init;
            else
            case current is
                when init =>
                    if x = '1' then
                        current <= got1;
                    else
                    current <= init;
                    end if;
                when got1 =>
                    if x = '1' then
                        current <= got1;
                    else
                        current <= got10;
                    end if;
                when got10 =>
                    if x = '1' then
                        current <= got1;
                    else
                        current <= init;
                    end if;
                end case;
                end if;
            end if;
    end process ;

    z <= '1' when (current = got10 and x = '1') else '0';
end behavioral;
```

FIGURE 87.33   Mealy machine VHDL code.

The combinational block however uses a *case statement* in the process block to decide the next state. This *case statement* has *case alternatives* for *init, got1, got10,* and *got101*. In each case, based on inputs, *ns* and *y* are assigned values.

87.3.5   Memories

Memories are represented with user-defined types in VHDL. Figure 87.36 shows a memory declaration and its corresponding block diagram together with several valid memory operations. As said before, type

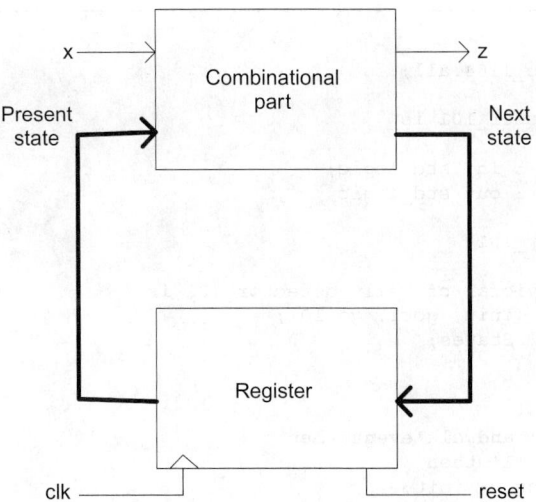

FIGURE 87.34 Huffman partitioning.

declarations take place in the declarative part of an architecture. Here, type *memory* is defined as an unconstrained array of 8-bit *std_logic* vectors. Then the *mem* signal is defined as a 1023-element array of *memory* elements.

A memory can be written into or read from with the use of signal assignments and indexing. A memory can be read by addressing it within its address range. A part of a word can be read directly from a memory using double indexing.

87.4   Writing Testbenches

Coding styles presented so far were for coding hardware structures, with the intention of synthesizability. On the other hand, testbenches do not need to have hardware correspondence. The VHDL code of Figure 87.31 which is a **101** Moore detector will be used as the design under test (DUT) in this section.

Figure 87.37 shows a testbench developed for the Moore detector of Figure 87.31. The testbench entity which instantiates design under test (DUT), *Moore_detector* in this example, has no ports. Initial values of DUT inputs are assigned in the signal declarations stated in the declarative part of this architecture. This testbench applies test data to DUT. This testbench uses three processes: one *conditional signal assignment* and two *process statements*. It should be kept in mind that all the processes run concurrently. The first process waits for 20 ns and then puts **0** into *clr_t*. It then executes a **wait** which suspends it forever.

Normally a process statement executes all its sequential statements and then repeats itself. However, a process may suspend itself by using a *wait statement*. *Wait statements* have several forms and can only be used in sequential blocks. Using **wait** without any parameter suspends the process forever, but when followed by **for** and a wait time, causes suspension of the process statement until this time elapses. There are other types of wait statements, such as **wait on** and **wait until** that wait on events in their sensitivity clause and until the wait condition in their condition clause becomes true, respectively.

Coming back to our test bench example, the second process is a *conditional signal assignment* and is responsible for producing the *clk* signal while the last process creates the *x* signal. The third process waits for a certain amount of time and then complements its variable, which generates periodic data.

```
library ieee;
use ieee.std_logic_1164.all;

entity Mealy_detector_101 is
    port (
        clr, clk, x : in  std_logic;
        z           : out std_logic
    );
end Mealy_detector_101;

architecture behavioral of Mealy_detector_101 is
    type States is (init, got1, got10);
    signal current: States;
begin
    process (clk)
    begin
        if clk = '1' and clk'event then
            if clr = '1' then
                current <= init;
            else
            case current is
                when init =>
                    if x = '1' then
                        current <= got1;
                    else
                    current <= init;
                    end if;
                when got1 =>
                    if x = '1' then
                        current <= got1;
                    else
                        current <= got10;
                    end if;
                when got10 =>
                    if x = '1' then
                        current <= got1;
                    else
                        current <= init;
                    end if;
                end case;
            end if;
        end if;
    end process ;

    z <= '1' when (current = got10 and x = '1') else '0';
end behavioral;
```

FIGURE 87.35   VHDL Huffman coding style.

87.5   Our Coverage of VHDL

This chapter presented the VHDL HDL language from a hardware design point of view. The chapter used complete design examples at various levels of abstraction for showing ways in which VHDL could be used in a design. We showed how timing details could be incorporated in cell descriptions. Aside from

```
............
............
architecture .........
............

    type memory is array (integer range <>) of std_logic_vector(7 downto 0);
    signal mem: memory(0 to 1023);
begin
process (mem)
    variable data: std_logic_vector(7 downto 0);
    variable short_data: std_logic_vector(3 downto 0);
begin
    ............
    ............
    data := mem(956);
    short_data := mem(931)(6 downto 3);
    ............
    ............
    mem(932) <= data ;
    mem(321)(5 downto 2) <= short_data;
    mem(940) <= "0000" & short_data ;
    ............
end process;
end .....
```

FIGURE 87.36   Memory representation.

```
library ieee;
use ieee.std_logic_1164.all;

entity test_Moore is
end test_Moore;

architecture test_bench of test_Moore is
    component Moore_detector_101
    port (
        clr, clk, x:  in  std_logic;
        z             : out std_logic
    );
    end component;
    signal clk_t, x_t, z_t : std_logic := '0';
    signal clr_t : std_logic := '0';

begin
    DUT: Moore_detector_101 port map (clr_t, clk_t, x_t, z_t);
    process
    begin
        wait for 20 ns;
        clr_t <= '0';
        wait;
    end process;

    clk_t <= not clk_t after 5 ns when now <= 150 ns else clk_t;

    process
    begin
        wait for 7 ns;
        x_t <= '1';
        wait for 7 ns;
        x_t <= '0';
    end process;
end test_bench;
```

FIGURE 87.37   Testbench example.

this discussion of timing, all examples that were presented had one-to-one hardware correspondence and were synthesizable. We have shown how combinational and sequential components can be described for synthesis and how a complete system can be put together using combinational and sequential blocks for it to be tested and synthesized. This chapter did not cover all of VHDL, but only the most often used parts of this language.

References

1. *IEEE Standard Multi-Value Logic System,* IEEE Std 1076-1993.5, IEEE, Inc., New York, 1993.
2. Z. Navabi, *VHDL: Analysis and Modeling of Digital Systems,* 2nd ed., McGraw-Hill Publishing, New York, 1998.

88

Register Transfer Level Hardware Description with Verilog

Zainalabedin Navabi

Nanoelectronics Center of Excellence
School of Electrical and Computer Engineering
University of Tehran

CONTENTS

This chapter presents Verilog for describing a design for performing behavioral simulation and synthesis. The purpose is to familiarize readers with the main topics of Verilog and present this language so that it can be used as a getting-started chapter, after completion of which readers can start writing Verilog codes for their basic designs. For this purpose, many of the complex Verilog constructs related to timing and fine modeling features of this language are not covered here. The chapter first describes Verilog with emphasis on design using simple examples. In a later section after a general familiarity with the language is gained, more complex features of the Verilog language with emphasis on testbench development will be described. Verilog covered in this chapter comply with Verilog 1999 and not Verilog 2001. More advanced features of Verilog 2001 are described in Chapter 90.

88.1 Verilog in Digital Design Flow

Verilog syntax and language constructs are designed to facilitate description of hardware components for simulation and synthesis. In addition, Verilog can be used to describe testbenches, specify test data, and monitor circuit responses. Figure 88.1 shows a simulation model that consists of a design and its testbench in Verilog. Simulation output is generated in the form of a waveform for visual inspection or data files for machine readability.

After a design passes basic functional validations, it must be synthesized into a netlist of components of a target library. Constructs used for verification of a design, or timing checks and timing specifications are not synthesizable. A Verilog design that is to be synthesized must use language constructs that have a clear hardware correspondence. Figure 88.2 shows a block diagram specifying the synthesis process.

The output of synthesis is a netlist of components of the target library. Often synthesis tools have an option to generate this netlist in Verilog. In this case, the same testbench prepared for presynthesis simulation can be used with the netlist generated by the synthesis tool.

88.1.1 Modules

The entity used in Verilog for description of hardware components is a **module**. A module can describe a hardware component as simple as a transistor or a network of complex digital systems. Modules begin with the **module** keyword and end with **endmodule**. A design may be described in a hierarchy of other modules. The top-level module is the complete design, and modules lower in the hierarchy are the design's components. Module instantiation is the construct used for bringing a lower level module into a higher level one. Figure 88.3 shows a hierarchy of several nested modules.

As shown in Figure 88.4, in addition to the **module** keyword, a module header also includes the module name and list of its ports. Following the module header, its ports and internal signals and variables are declared. Specification of the operation of a module follows module declarations.

Operation of a module can be described at the gate level, using Boolean expressions at the behavioral level or a mixture of various levels of abstraction. Figure 88.5 shows three ways of describing a module. Module *simple1a* in Figure 88.5 uses Verilog's gate primitives, *simple1b* uses concurrent statements, and *simple1c* uses a procedural statement.

The subsections that follow describe details of module ports and description styles. In the examples in this chapter, Verilog keywords and reserved words are shown in bold; Verilog is case-sensitive. It allows letters, numbers and special character "_" to be used for names. Names are used for modules, parameters, ports, variables, and instance of gates and modules.

88.1.2 Module Ports

Following the name of a module is a set of parenthesis with a list of module ports. This list includes inputs, outputs, and bidirectional input/output lines. Ports may be listed in any order. This ordering can

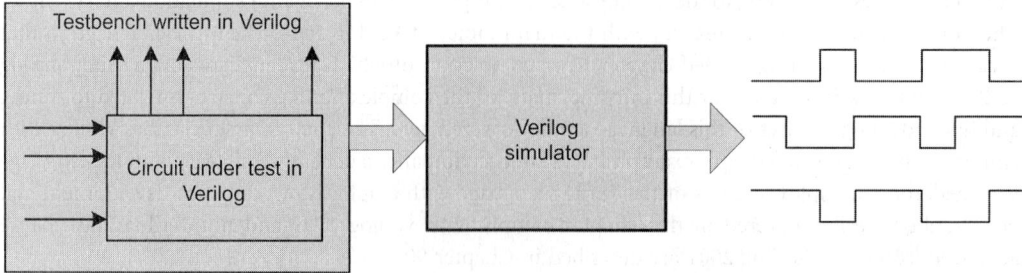

FIGURE 88.1 Simulation in Verilog.

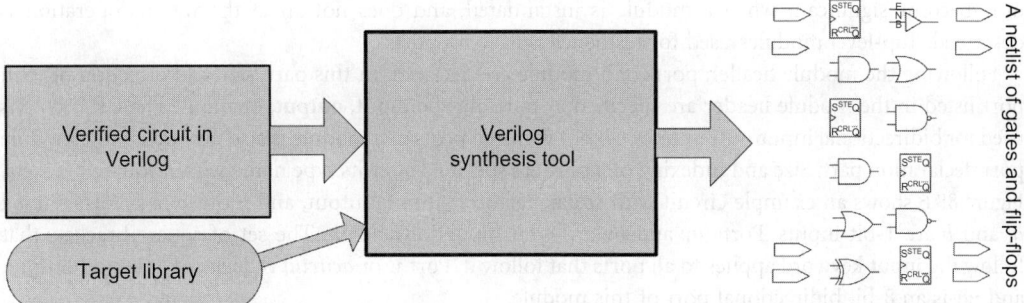

FIGURE 88.2 Synthesis.

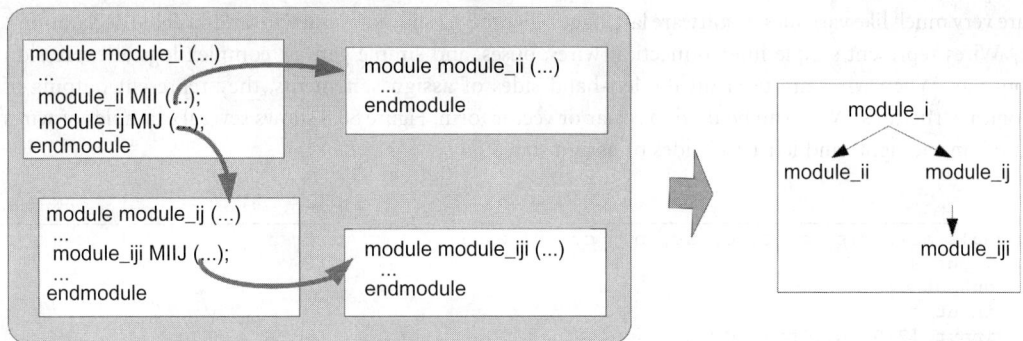

FIGURE 88.3 Module hierarchy.

```
module name (ports);
  port declarations;
  other declarations;
  . . .
  statements
  . . .
endmodule
```

FIGURE 88.4 Module outline.

```
module simple1a (i1, i2, i3, w1, w2);
  input i1, i2;
  output w1, w2;
  . . .
  and g1 (w1, i1, i
endmodule
```

```
module simple1b (i1, i2, i3, w1, w2);
  input i1, i2;
  output w1, w2;
  . . .
  assign w1 = i1 & i2;
  assign w2 = i1 & i2 | ~
endmodule
```

```
module simple1c (i1, i2, i3, w1, w2);
  input i1, i2;
  output w1, w2;
  reg w1;
  always @ (i1 or i2) begin
    if (i1 == 1) w1 = i2; else w1 = 0;
  end
endmodule
```

FIGURE 88.5 Module definition alternatives.

only become significant when a module is instantiated, and does not affect the way its operation is described. Top-level modules used for testbenches have no ports.

Following the module header, ports of a module are declared. In this part, size and direction of each port listed in the module header are specified. A port may be **input**, **output**, or **inout**. The latter type is used for bidirectional input/output lines. Size of vectored ports of a module is also declared in the module port declaration part. Size and indexing of a port are specified after its type name within square brackets. Figure 88.6 shows an example circuit with scalar, vectored, input, output, and inout ports. Ports named *a*, and *b* are 1-bit inputs. Ports *av* and *bv* are 8-bit inputs of *acircuit*. The set of square brackets that follow the **input** keyword applies to all ports that follow it. Port *w* of *acircuit* is declared as a 1-bit output, and *wv* is an 8-bit bidirectional port of this module.

In addition to port declarations, a module declarative part may also include wire and variable declarations that are to be used inside the module. Wires (that are called **net** in Verilog) are declared by their types, **wire**, **wand** or **wor**; variables are declared as **reg**. Wires are used for interconnections and have properties of actual signals in a hardware component. Variables are used for behavioral descriptions and are very much like variables in software languages. Figure 88.7 shows several wire and variable declarations.

Wires represent simple interconnection wires, buses, and simple gate or complex logical expression outputs. When wires are used on the left-hand sides of **assign** statements, they represent outputs of logical structures. Wires can be used in scalar or vector form. Figure 88.8 shows several examples of wires used on the right- and left-hand sides of **assign** statements.

```
module acircuit (a, b, c, av, bv, cv, w, wv);
  input a, b;
  output w;
  inout c;
  input [7:0] av, bv;
  output [7:0] wv;
  inout [7:0] cv;
  . . .
endmodule
```

FIGURE 88.6 Module ports.

```
module bcircuit (a, b, av, bv, w, wv);
  input a, b;
  output w;
  input [7:0] av, bv;
  output [7:0] wv;
  wire d;
  wire [7:0] dv;
  reg e;
  reg [7:0] ev;
  . . .
endmodule
```

FIGURE 88.7 Wire and variable declaration.

```
module vcircuit (av, bv, cv, wv);
  input [7:0] av, bv, cv;
  output [7:0] wv;
  wire [7 :0] iv, jv;
  assign iv = av & cv;
  assign jv = av | cv;
  assign wv = iv ^ jv;
endmodule
```

FIGURE 88.8 Using wires.

In the vector form, inputs, outputs, wires, and variables may be used as a complete vector, part of a vector, or a bit of the vector. The latter two are referred to as part-select and bit-select.

88.1.3 Logic Value System

Verilog uses a four-value logic value system. Values in this system are **0**, **1**, **Z**, and **X**. Value **0** is for logical **0**, which in most cases represents a path to ground (GND). Value **1** is logical **1** and it represents a path to supply (V_{dd}). Value **Z** is for float, and **X** is used for uninitialized, undefined, unknown, and value conflicts. Values **Z** and **X** are used for wired-logic, buses, initialization values, tri-state structures, and switch-level logic.

For more logic precision, Verilog uses strength values in addition to logic values. Our dealing with Verilog is for design and synthesis, and these issues will not be discussed here.

88.2 Combinational Circuits

A combinational circuit can be represented by its gates, its Boolean functionality, description of its behavior, or a mix of all these forms. At the gate level, interconnection of its gates are shown; at the functional level, Boolean expressions representing its outputs are written; and at the behavioral level, a software-like procedural description represents its functionality. This section shows these levels of abstraction for describing combinational circuits.

88.2.1 Gate-Level Combinational Circuits

Verilog provides primitive gates and transistors. Some of the more commonly used Verilog primitives and their logical representations are shown in Figure 88.9. In this figure, w is used for gate outputs, i for inputs, and c for control inputs.

Basic logic gates are **and**, **nand**, **or**, **nor**, **xor**, **xnor**. These gates can be used with one output and any number of inputs. The other two structures shown are **not** and **buf**. These gates can be used with one input and any number of outputs.

Another group of primitives shown in this figure are three-state (tri-state is also used to refer to these structures) gates. Gates shown have w for their outputs, i for data inputs, and c for their control inputs. These primitives are **bufif1**, **notif1**, **bufif0**, and **notif0**. When control c for such gates is active (**1** for first

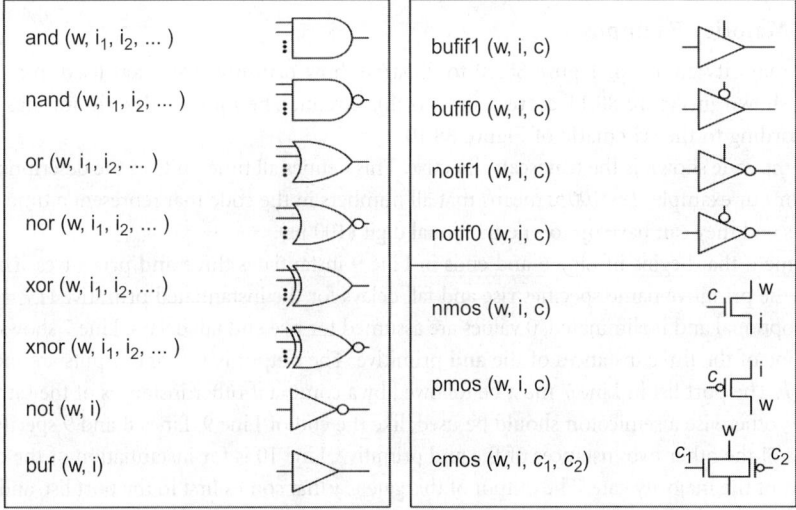

FIGURE 88.9 Basic primitives.

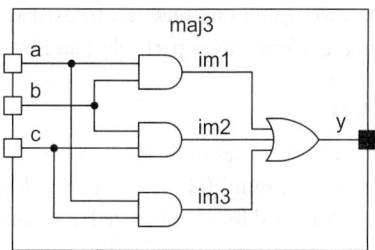

FIGURE 88.10 A majority circuit.

```
`timescale 1ns/100ps                          // Line 1
module maj3 ( a, b, c, y );
   input a, b, c;
   output y;

   and #(2,4)                                  // Line 6
      ( im1, a, b ),                           // Line 7
      ( im2, b, c ),                           // Line 8
      ( im3, c, a );                           // Line 9
   or  #(3,5) ( y, im1, im2, im3 );            // Line 10

endmodule
```

FIGURE 88.11 Verilog code for the majority circuit.

and third, and **0** for the others), the data input, *i*, or its complement appears on the output of the gate. When control input of a gate is not active, its output becomes high impedance, or **Z**.

Also shown in Figure 88.9 are NMOS, PMOS, and CMOS structures. These are switches that are used in switch-level description of gates, complex gates, and buses. The **nmos** (**pmos**) primitive is a simple switch with an active high (low) control input. The **cmos** switch is usually used with two complementary control inputs. These switches behave like the three-state gates. They are different in their output voltage levels and drive strengths. These parameters are modeled by wire strengths and are not discussed in this book.

88.2.1.1 Majority Example

We use the majority circuit of Figure 88.10 to illustrate how primitive gates are used in a design. The description shown in Figure 88.11 corresponds to this circuit. The module description has inputs and outputs according to the schematic of Figure 88.10.

Line 1 of the code shown is the **timescale** directive. This defines all time units in the description and their precision. For our example, *1ns/100ps* means that all numbers in the code that represent a time value are in nanoseconds and they can have up to one fractional digit (100 ps).

The statement that begins in Line 6 and ends in Line 9 instantiates three **and** primitives. The construct that follows the primitive name specifies rise and fall delays for the instantiated primitive ($t_{plh} = 2$, $t_{phl} = 4$). This part is optional and if eliminated, 0 values are assumed for rise and fall delays. Line 7 shows inputs and outputs of one of the three instances of the **and** primitive. The output is *im1* and inputs are module input ports *a* and *b*. The port list in Line 7 must be followed by a comma if other instances of the same primitive are to follow, otherwise a semicolon should be used, like the end of Line 9. Lines 8 and 9 specify input and output ports of the other two instances of the **and** primitive. Line 10 is for instantiation of the **or** primitive at the output of the majority gate. The output of this gate is *y* that comes first in the port list, and is followed by inputs of the gate. In this example, intermediate signals for interconnection of gates are *im1*, *im2*, and *im3*. Scalar interconnecting wires need not be explicitly declared in Verilog.

The three **and** instances could be written as three separate statements, such as instantiation of the **or** primitive. If we were to specify different delay values for the three instances of the **and** primitive, we had to have three separate primitive instantiation statements.

Three-state gates are instantiated in the same way as the regular logic gates. Outputs of three-state gates can be wired to form wired-and, wired-or, or wiring logic. For various wiring functions, Verilog uses **wire**, **wand**, **wor**, **tri**, **tri0**, and **tri1 net** types. When two wires (**net**s) are connected, the resulting value depends on the two **net** values as well as the type of the interconnecting **net**. Figure 88.12 shows **net** values for **net** types **wire**, **wand**, and **wor**.

The table shown in Figure 88.12 is called a **net** resolution table. Several examples of **net** resolutions are shown in Figure 88.13. The **tri net** type mentioned above is the same as the **wire** type. **tri0** and **tri1** types resolve to **0** and **1**, respectively, when driven by all **Z** values.

88.2.1.2 Multiplexer Example

Figure 88.14 shows a 2-to-1 multiplexer using three-state gates. The Verilog code of this multiplexer is shown in Figure 88.15.

Lines 6 and 7 in Figure 88.15 instantiate two three-state gates. Their output is y, and since it is driven by both gates a wired-net is formed. Since y is not declared, its **net** type defaults to **wire**. When s is **1**, *buf*if1 conducts and the value of b propagates to its output. At the same time, because s is **1**,

	Driving **net** values									
	Two driving **net** values									
net type	0,0	0, 1	0, Z	0, X	1, 1	1, Z	1, X	Z, Z	Z, X	X, X
wire	0	X	0	X	1	1	X	Z	X	X
wand	0	0	0	0	1	X	X	Z	X	X
wor	0	1	X	X	1	1	1	X	X	X

FIGURE 88.12 "net" Type resloutions.

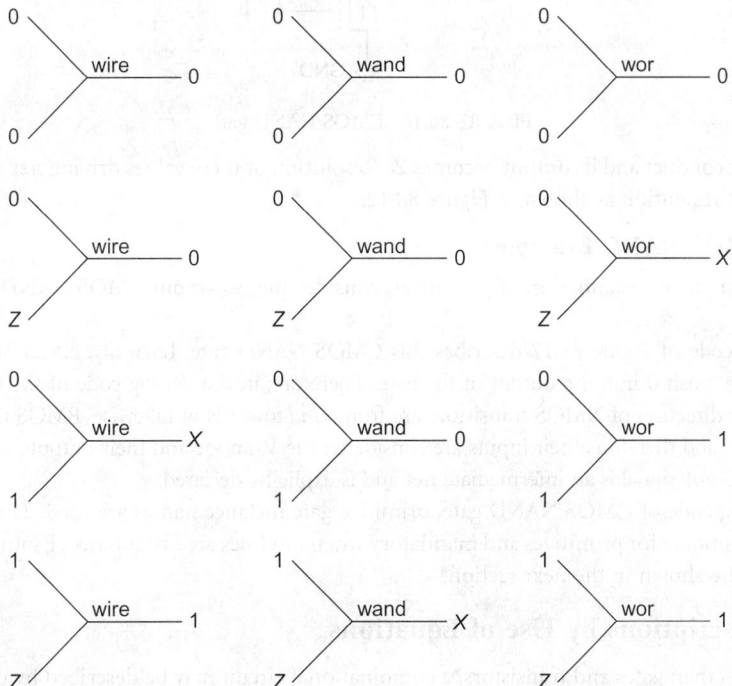

FIGURE 88.13 "net" Resolution examples.

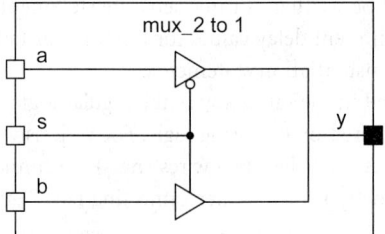

FIGURE 88.14 Multiplexer using three-state gates.

```
`timescale 1ns/100ps

module mux_2to1 ( a, b, s, y );
    input a, b, s;
    output y;
    bufif1 #(3) ( y, b, s );                    // Line 6
    bufif0 #(5) ( y, a, s );                    // Line 7
endmodule
```

FIGURE 88.15 Multiplexer Verilog code.

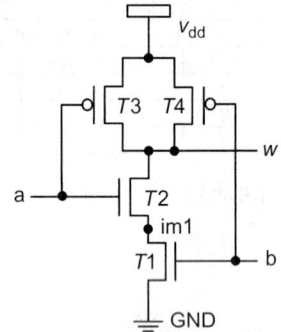

FIGURE 88.16 CMOS NAND gate.

*bufif*0 does not conduct and its output becomes **Z**. Resolution of these values driving **net** *y* is determined by the **wire net** resolution as shown in Figure 88.12.

88.2.1.3   CMOS NAND Example

As another example of instantiation of primitives, consider the two-input CMOS NAND gate shown in Figure 88.16.

The Verilog code of Figure 88.17 describes this CMOS NAND gate. Logically, NMOS transistors in a CMOS structure push **0** into the output of the gate. Therefore, in the Verilog code of the CMOS NAND, input to output direction of NMOS transistors are from *Gnd* towards *w*. Likewise, PMOS transistors push a **1** value into *w*, and therefore, their inputs are considered the V_{dd} node and their outputs are connected to the *w* node. The *im*1 signal is an intermediate **net** and is explicitly declared.

In the Verilog code of CMOS NAND gate, primitive gate instance names are used. This naming (*T*1, *T*2, *T*3, *T*4) is optional for primitives and mandatory when modules are instantiated. Examples of module instantiations are shown in the next section.

88.2.2   Descriptions by Use of Equations

At a higher level than gates and transistors, a combinational circuit may be described by use of Boolean, logical, and arithmetic expressions. For this purpose the Verilog concurrent **assign** statement is used. Figure 88.18 shows Verilog operators that can be used with **assign** statements.

```
module cmos_nand (a, b, w);
input a, b;
output w;
wire im1;
supply1 vdd;
supply0 gnd;

    nmos #(3, 4)
       T1 (im1, gnd, b),
       T2 (w, im1, a);

    pmos #(4, 5)
       T3 (w, vdd, a),
       T4 (w, vdd, b);

endmodule
```

FIGURE 88.17 CMOS NAND Verilog description.

Bitwise operators	&	|	^	~	~^	^~	
Reduction operators	&	~&	|	~|	^	~^	^~
Arithamatic operators	+	−	*	/	%		
Logical operators	&&	||	!				
Compare operatos	<	>	<=	>=	==		
Shift operators	>>	<<					
Concatenation operators	{}	{n{}}					

FIGURE 88.18 Verilog operators.

```
module xor3 ( a, b, c, y );
   input a, b, c;
   output y;

   assign y = a ^ b ^ c;

endmodule
```

FIGURE 88.19 XOR Verilog code.

88.2.2.1 XOR Example

As our first example for using an **assign** statement consider the description of an XOR gate as shown in Figure 88.19. The **assign** statement uses *y* on the left-hand side and equates it to Exclusive-OR of *a*, *b*, and c inputs.

Effectively, this **assign** statement is like driving *y* with the output of a three-input **xor** primitive gate. The difference is that, the use of an **assign** statement gives us more flexibility and allows the use of more complex functions than what is available as primitive gates. Instead of being limited to the gates shown in Figure 88.9, we can write our own expressions using operators of Figure 88.18.

88.2.2.2 Full-Adder Example

Figure 88.20 shows another example of using **assign** statements. This code corresponds to a full-adder circuit. The *s* output is the XOR result of *a*, *b*, and *ci* inputs, and the *co* output is an AND–OR expression involving these inputs.

A delay value of 10 ns is used for the *s* output and 8 ns for the *co* output. As with the gate outputs, rise and fall delay values can be specified for a **net** that is used on the left-hand side of an **assign** statement. This construct allows the use of two delay values. If only one value is specified, it applies to both rise and fall transitions.

Another property of **assign** statements that also corresponds to gate instantiations is their concurrency. The statements in the Verilog module of Figure 88.20 are concurrent. This means that the order in which they appear in this module is not important. These statements are sensitive to events on their right-hand

```
`timescale 1ns/100ps

module add_1bit ( a, b, ci, s, co );
    input a, b, ci;
    output s, co;
    assign #(10) s = a ^ b ^ ci;
    assign #(8) co = ( a & b ) | ( b & ci ) | ( a & ci );
endmodule
```

FIGURE 88.20   Full-adder Verilog code.

```
module mux2_1 ( i0, i1, s, y );
    input [3:0] i0, i1;
    input s;
    output [3:0] y;

    assign y = s ? i1 : i0;

endmodule
```

FIGURE 88.21   A 2-to-1 mux using conditional operator.

```
`timescale 1ns/100ps

module dcd2_4( a, b, d0, d1, d2, d3 );
    input a, b;
    output d0, d1, d2, d3;

    assign {d3, d2, d1, d0} =
    ( {a, b} == 2'b00 ) ? 4'b0001 :
    ( {a, b} == 2'b01 ) ? 4'b0010 :
    ( {a, b} == 2'b10 ) ? 4'b0100 :
    ( {a, b} == 2'b11 ) ? 4'b1000 :
        4'b0000;

endmodule
```

FIGURE 88.22   Decoder using?: and concatenation.

sides. When a change of value occurs on any of the right-hand side **net** or variables, the statement is evaluated and the resulting value is scheduled for the left-hand side **net**.

88.2.2.3   Multiplexer Example

Figure 88.21 shows a 2-to-1 multiplexer using a conditional operator. The expression shown reads as follows: **if** s is **1, then** y is $i1$ **else** it becomes $i0$. The right-hand side of the **assign** statement used in this example has a condition expression that uses the (**? :**) conditional operator. This operator is like if-then-else. In reading expressions that involve a conditional operator, **?** and **:** take places of **then** and **else**, respectively. The if-condition appears to the left of **?**.

88.2.2.4   Decoder Example

Figure 88.22 shows another example using the condition operator. In this example, a nesting of several **?:** operators are used to describe a decoder.

   The decoder description also uses the concatenation operator { } to form vectors from its scalar inputs and outputs. The decoder has four outputs, $d3$, $d2$, $d1$, and $d0$ and two inputs a and b. Input values **00, 01, 10,** and **11** produce **0001, 0010, 0100,** and **1000** outputs. To be able to compare a and b with their possible values, a 2-bit vector is formed by concatenating a and b. The {a, b} vector is then compared with the four possible values it can take using a nesting of **?:** operators.

   Similarly, to be able to place vector values on the outputs, the four outputs are concatenated using the { } operator and used on the left-hand side of the **assign** statement shown in Figure 88.22.

This example also shows the use of sized numbers. Constants for the inputs and outputs have the general format of **n`bm**. In this format, **n** is the number of bits, **b** the base specification, and **m** the number in base **b**. For calculation of the corresponding constant, number **m** in base **b** is translated to **n** bit binary. For example, $4`hA$ becomes 1010 in binary.

88.2.2.5 Adder Example

For another example using **assign** statements, consider an 8-bit adder circuit with a carry-in and a carry-out output. The Verilog code of this adder, shown in Figure 88.23, uses an **assign** statement to set concatenation of *co* on the left-hand side of *s* to the sum of *a*, *b* and *ci*. This sum results in nine bits with the left-most bit being the resulting carry. The sum is captured in the 9-bit left-hand side of the **assign** statement in {co, s}.

So far in this section we have shown the use of operators of Figure 88.18 in **assign** statements. A Verilog description may contain any number of **assign** statements and can use any mix of the operators discussed. The next example shows multiple **assign** statements.

88.2.2.6 ALU Example

As our final example of **assign** statements, consider an ALU that performs add and subtract operations and has two flag outputs *gt* and *zero*. The *gt* output becomes **1** when input *a* is greater than input *b*, and the *zero* output becomes **1** when the result of the operation performed by the ALU is **0**.

Figure 88.24 shows the Verilog code of this ALU. Used in this description are arithmetic, concatenation, condition, compare, and relational operations.

88.2.3 Descriptions with Procedural Statements

At a higher level of abstraction, Verilog provides constructs for procedural description of hardware instead of describing hardware with gates and expressions. Unlike gate instantiations and **assign** statements which correspond to concurrent substructures of a hardware component, procedural statements describe the hardware by its behavior. Also, unlike concurrent statements that appear directly in a module body, procedural statements must be enclosed in procedural blocks before they can be put inside a module.

```
module add_4bit ( a, b, ci, s, co );
   input [7:0] a, b;
   output [7:0] s;
   input ci;
   output co;

   assign {co, s} = a + b + ci;

endmodule
```

FIGURE 88.23 Adder with carry-in and carry-out.

```
module ALU ( a, b, ci, addsub, gt, zero, co, r );
   input [7:0] a, b;
   output [7:0] r;
   input ci;
   output gt, zero, co;

   assign {co, s} = addsub ? (a + b + ci) : (a - b - ci);
   assign gt = (a>b);
   assign zero = (r == 0);

endmodule
```

FIGURE 88.24 ALU Verilog code using a mix of operations.

The main procedural block in Verilog is the **always** block. This is considered a concurrent statement that runs concurrent with all other statements in a module. Within this statement, procedural statements like **if-else** and **case** statements are used and are executed sequentially. If there are more than one procedural statement inside a procedural block, they must be bracketed by **begin** and **end** keywords.

Unlike assignments in concurrent bodies that model driving logic for left-hand side wires, assignments in procedural blocks are assignments of values to variables that hold their assigned values until a different value is assigned to them. A variable used on the left-hand side of a procedural assignment must be declared as **reg**.

An event-control statement is considered a procedural statement, and is used inside an **always** block. This statement begins with an at-sign, and in its simplest form, includes a list of variables in the set of parenthesis that follow the at-sign, e.g., @ (*v1* **or** v2 …).

When the flow of the program execution within an **always** block reaches an event-control statement, the execution halts (suspends) until an event occurs on one of the variables in the enclosed list of variables. If an event-control statement appears at the beginning of an **always** block, the variable list it contains is referred to as the *sensitivity list* of the **always** block. For combinational circuit modeling all variables that are read inside a procedural block must appear on its sensitivity list.

Examples that follow show various ways combinational component may be modeled by procedural blocks.

88.2.3.1 Majority Example

Figure 88.25 shows a majority circuit described by the use of an **always** block. In the declarative part of the module shown, the *y* output is declared as **reg** since this variable is to be assigned a value inside a procedural block.

The **always** block describing the behavior of this circuit uses an event-control statement that encloses a list of variables that is considered as the sensitivity list of the **always** block. The **always** block is said to be sensitive to *a*, *b*, and *c* variables. When an event occurs on any of these variables, the flow into the **always** block begins and as a result, the result of the Boolean expression shown will be assigned to variable *y*. This variable holds its value until the next time an event occurs on *a*, *b*, or *c* inputs.

In this example, since the **begin** and **end** bracketing only includes one statement, its use is not necessary. Furthermore, the syntax of Verilog allows elimination of semicolon after an event-control statement. This effectively collapses the event control and the statement that follows it into one statement.

88.2.3.2 Majority Example with Delay

The Verilog code shown in Figure 88.26 is a majority circuit with a 5 ns delay. Following the **always** keyword, the statements in this procedural block are an event-control, a delay-control, and a procedural assignment. The delay-control statement begins with a sharp-sign and is followed by a delay value. This statement causes the flow into this procedural block to be suspended for 5 ns. This means that after an event on one of the circuit inputs, evaluation and assignment of the output value to *y* takes place after 5 ns.

Note in the description of Figure 88.26 that **begin** and **end** bracketing is not used. As with the event-control statement, a delay-control statement can collapse into its next statement by removing their separating semicolon. The event-control, delay-control, and assignment to *y* become a single procedural statement in the **always** block of *maj3* code.

```
module maj3 ( a, b, c, y );
   input a, b, c;
   output y;
   reg y;

   always @( a or b or c )
   begin
      y = (a & b) | (b &c) | (a & c);
   end

endmodule
```

FIGURE 88.25 Procedural block describing a majority circuit.

```
`timescale 1ns/100ps

module maj3 ( a, b, c, y );
   input a, b, c;
   output y;
   reg y;

   always @( a or b or c ) #5 y = (a & b) | (b &c) | (a & c);

endmodule
```

FIGURE 88.26 Majority gate with delay.

```
`timescale 1ns/100ps

module add_1bit ( a, b, ci, s, co );
   input a, b, ci;
   output s, co;
   reg s, co;

   always @( a or b or ci )
   begin
      s = #5 a ^ b ^ ci;
      co = #3 (a & b) | (b &ci) | (a & ci);
   end
endmodule
```

FIGURE 88.27 Full-Adder using procedural assignments.

88.2.3.3 Full-Adder Example

Another example of using procedural assignments in a procedural block is shown in Figure 88.27. This example describes a full-adder with sum and carry-out outputs.

The **always** block shown is sensitive to a, b, and ci inputs. This means that when an event occurs on any of these inputs, the **always** block wakes up and executes all its statements in the order that they appear. Since assignments to s and co outputs are procedural, both these outputs are declared as **reg**.

The delay mechanism used in the full-adder of Figure 88.27 is called an intrastatement delay that is different than that of the majority circuit of Figure 88.26.

In the majority circuit, the delay simply delays execution of its next statement. However, the intrastatement delay of Figure 88.27 only delays the assignment of the calculated value of the right-hand side to the left-hand side variable. This means that in Figure 88.27, as soon as an event occurs on an input, the expression $a \wedge b \wedge c$ is evaluated. But, the assignment of the evaluated value to s and proceeding to the next statement takes 5 ns.

Because assignment to co follows that to s, the timing of the former depends on that of the latter, and evaluation of the right-hand side of co begins 5 ns after an input change. Therefore, co receives its value 8 ns after an input change occurs. To remove this timing dependency and be able to define the timing of each statement independent of its previous one, a different kind of assignment must be used.

Assignments in Figure 88.27 are of the blocking type. Such statements block the flow of the program until they are completed. A different assignment is of the nonblocking type. A different version of the full-adder that uses this construct is shown in Figure 88.28. This assignment schedules its right-hand side value into its left-hand side to take place after the specified delay. Program flow continues into the next statement while propagation of values into the first left-hand side is still going on.

In the example of Figure 88.28, evaluation of the right-hand side of s is done immediately after an input changes. Evaluation of the right-hand side of co occurs 8 ns after that. To make s and co delays match those of Figure 88.27, an 8 ns delay is used for assignment to co.

```
`timescale 1ns/100ps

module add_1bit ( a, b, ci, s, co );
    input a, b, ci;
    output s, co;
    reg s, co;

    always @( a or b or ci )
    begin
        s <= #5 a ^ b ^ ci;
        co <= #8 (a & b) | (b &ci) | (a & ci);
    end
endmodule
```

FIGURE 88.28 Full-Adder using nonblocking assignments.

```
module mux2_1 ( i0, i1, s, y );
    input i0, i1, s;
    output y;
    reg y;

    always @( i0 or i1 or s )      begin
        if ( s==1'b0 )
            y = i0;
        else
            y = i1;
    end
endmodule
```

FIGURE 88.29 Procedural multiplexer.

Since our focus is on synthesizable coding and gate delay timing issues are not of importance, we will mostly use blocking assignments for combinational circuit descriptions.

88.2.3.4 Procedural Multiplexer Example

For another example of a procedural block, consider the 2-to-1 multiplexer of Figure 88.29. This example uses an **if-else** construct to set y to $i0$ or $i1$ depending on the value of s.

As in the previous examples, all circuit variables that participate in determining the value of y appear on the sensitivity list of the **always** block. Also since y appears on the left-hand side of a procedural assignment, it is declared as **reg**.

The **if-else** statement shown in Figure 88.29 has a condition part that uses an equality operator. If the condition is false (or equal to **0**), the block of statements that follow it will be taken, otherwise block of statements after the **else** are taken. In both cases, the block of statements must be bracketed by **begin** and **end** keywords if there is more than one statement in a block.

88.2.3.5 Procedural ALU Example

The **if-else** statement, used in the previous example, is easy to use, descriptive, and expandable. However, when many choices exist, a **case**-statement which is more structured may be a better choice. The ALU description of Figure 88.30 uses a **case** statement to describe an ALU with add, subtract, AND, and XOR functions.

The ALU has a and b data inputs and a 2-bit f input that selects its function. The Verilog code shown in Figure 88.30 uses a, b, and f on its sensitivity list. The **case**-statement shown in the **always** block uses f to select one of the **case** alternatives. The last alternative is the **default** alternative, which is taken when f does not match any of the alternatives that appear before it. This is necessary to make sure that unspecified input values (here, those that contain **X** and/or **Z**) cause the assignment of the default value to the output and not leave it unspecified.

```
module alu_4bit ( a, b, f, y );
    input [3:0] a, b;
    input [1:0] f;
    output [3:0] y;
    reg [3:0] y;

    always @ ( a or b or f )    begin
        case ( f )
            2'b00 : y = a + b;
            2'b01 : y = a - b;
            2'b10 : y = a & b;
            2'b11 : y = a ^ b;
            default: y = 4'b0000;
        endcase
    end
endmodule
```

FIGURE 88.30 Procedural ALU.

```
module bussing (busin1, busin2, busin3, en1, en2, en3, busout );
    input [3:0] busin1, busin2, busin3;
    input en1, en2, en3;
    output [3:0] busout;

    assign busout = en1 ? busin1 : 4'bzzzz;
    assign busout = en2 ? busin2 : 4'bzzzz;
    assign busout = en3 ? busin3 : 4'bzzzz;

endmodule
```

FIGURE 88.31 Implementing a three-state bus.

88.2.4 Combinational Rules

Completion of **case** alternatives or **if-else** conditions is an important issue in combinational circuit coding. In an **always** block, if there are conditions under which the output of a combinational circuit is not assigned a value, because of the property of **reg** variables the output retains its old value. The retaining of old value infers a latch on the output. Although, in some designs this latching is intentional, obviously it is unwanted when describing combinational circuits. With this, we have set two rules for coding combinational circuits with **always** blocks.

1. List all inputs of the combinational circuit in the sensitivity list of the **always** block describing it.
2. Make sure all combinational circuit outputs receive some value regardless of how the program flows in the conditions of **if-else** and/or **case** statements. If there are too many conditions to check, set all outputs to their inactive values at the beginning of the **always** block.

88.2.5 Busing

Bus structures can be implemented by use of multiplexers or three-state logic. In Verilog, various methods of describing combinational circuits can be used for the description of a bus.

Figure 88.31 shows Verilog coding of *busout* that is a three-state bus and has three sources, *busin1*, *busin2*, and *busin3*. Sources of *busout* are put on this bus by active high-enabling control signals, *en1*, *en2*, and *en3*. Using the value of an enabling signal, a condition statement either selects a bus driver or a **4**-bit **Z** value to drive the *busout* output.

Verilog allows multiple concurrent drivers for **net**s. However, a variable declared as a **reg** and used on a left-hand side in a procedural block (**always** block), can only be driven by one source. This makes the use of **net**s more appropriate for representing buses.

88.3   Sequential Circuits

As with any digital circuit, a sequential circuit can be described in Verilog by the use of gates, Boolean expressions, or behavioral constructs (e.g., the **always** statement). While gate-level descriptions enable a more detailed description of timing and delays, because of complexity of clocking and register and flip-flop controls, these circuits are usually described by the use of procedural **always** blocks. This section shows the various ways sequential circuits are described in Verilog. The following discusses primitive structures like latch and flip-flops, and then generalizes coding styles used for representing these structures to more complex sequential circuits including counters and state machines.

88.3.1   Basic Memory Elements at the Gate Level

A clocked D-latch latches its input data during an active clock cycle. The latch structure retains the latched value until the next active clock cycle. This element is the basis of all static memory elements.

A simple description for a D-latch uses two cross-coupled NOR gate instantiations. Such a description is at the gate level and not very efficient in simulation. Alternatively, the same latch can be described with an **assign** statement as shown below.

```
assign #(3) q = c ? d : q;
```

This statement simply describes what happens in a latch. The statement says that when c is **1**, the q output receives d, and when c is **0** it retains its old value. Using two such statements with complementary clock values a master–slave flip-flop is described. As shown in Figure 88.32, the qm **net** is the master output and q the flip-flop output.

This code uses two concurrent **assign** statements. As discussed before, these statements model logic structures with **net**-driven outputs (qm and q). The order in which the statements appear in the body of the *master_slave* **module** is not important.

88.3.2   Memory Elements Using Procedural Statements

Although latches and flip-flops can be described by primitive gates and **assign** statements, such descriptions are hard to generalize, and describing more complex register structures cannot be done this way. This section uses **always** statements to describe latches and flip-flops. We will show that the same coding styles used for these simple memory elements can be generalized to describe memories with complex control as well as functional register structures like counters and shift registers.

88.3.2.1   Latches

Figure 88.33 shows a D-latch described by an **always** statement. The outputs of the latch are declared as **reg** because they are being driven inside the **always** procedural block. Latch clock and data inputs (c and d) appear in the sensitivity list of the **always** block, making this procedural statement sensitive to

```
`timescale 1ns/100ps

module master_slave ( d, c, q );
    input d, c;
    output q;

    wire qm;

    assign #(3) qm = c ? d : qm;
    assign #(3) q = ~c ? qm : q;

endmodule
```

FIGURE 88.32   Master–slave flip-flop.

```
module latch ( d, c, q, q_b );
    input d, c;
    output q, q_b;
    reg q, q_b;

    always @ ( c or d )
        if ( c ) begin
            #4 q = d;
            #3 q_b = ~d;
        end
endmodule
```

FIGURE 88.33 Procedural latch.

```
`timescale 1ns/100ps

module d_ff ( d, clk, q, q_b );
    input d, clk;
    output q, q_b;
    reg q, q_b;

    always @ ( posedge clk )
    begin
    #4 q <= d;
    #3 q_b <= ~d;
    end
endmodule
```

FIGURE 88.34 A positive-edge D flip-flop.

c and d. This means that when an event occurs on c or d, the **always** block wakes up and it executes all its statements in the sequential order from beginning to end.

The **if**-statement enclosed in the **always** block puts d into q when c is active. This means that if c is **1** and d changes, the change on d propagates to the q output. This behavior is referred to as transparency, which is how latches work. While clock is active, a latch structure is transparent, and input changes affect its output.

Any time the **always** statement wakes up, if c is **1**, it waits 4 ns and then puts d into q. It then waits another 3 ns and then puts the complement of d into q_b. This makes the delay of the q_b output 7 ns.

88.3.2.2 D Flip-Flop

While a latch is transparent, a change on the D-input of a D flip-flops does not directly pass on to its output. The Verilog code of Figure 88.34 describes a positive-edge trigger D-type flip-flop.

The sensitivity list of the procedural statement shown includes **posedge** of *clk*. This **always** statement only wakes up when *clk* makes a **0** to **1** transition. When this statement does wake up, the value of d is put into q. Obviously, this behavior implements a rising-edge D flip-flop.

Instead of **posedge**, use of **negedge** would implement a falling-edge D flip-flop. After the specified edge, the flow into the **always** block begins. In our description, this flow is halted by 4 ns by the #4 delay-control statement. After this delay, the value of d is read and put into q. Following this transaction, the flow into the **always** block is again halted by 3 ns, after which $\sim d$ is put into qb. This makes the delay of q after the edge of the clock equal to 4 ns. The delay for q_b becomes the accumulation of the delay values shown, and it is 7 ns. Using nonblocking assignments (arrows in Figure 88.34) is recommended in clocked always statements.

88.3.2.3 Synchronous Control

The coding style presented for the above simple D flip-flop is a general one and can be expanded to cover many features found in flip-flops and even memory structures. The description shown in Figure 88.35 is a D-type flip-flop with synchronous set and reset (*s* and *r*) inputs.

```
module d_ff ( d, s, r, clk, q, q_b );
    input d, clk, s, r;
    output q, q_b;
    reg q, q_b;

    always @ ( posedge clk ) begin
        if ( s ) begin
            #4 q <= 1'b1;
            #3 q_b <= 1'b0;
        end else if ( r ) begin
            #4 q <= 1'b0;
            #3 q_b <= 1'b1;
        end else begin
            #4 q <= d;
            #3 q_b <= ~d;
        end
    end

endmodule
```

FIGURE 88.35 D flip-flop with synchronous control.

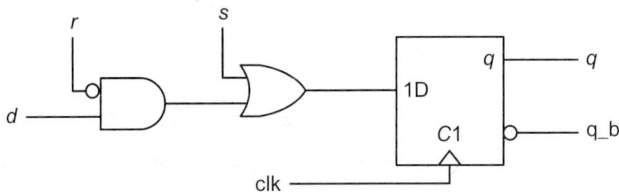

FIGURE 88.36 D flip-flop with synchronous control.

The description uses an **always** block that is sensitive to the positive-edge of *clk*. When *clk* makes a **0** to **1** transition, the flow into the **always** block begins. Immediately after the positive-edge, *s* is inspected and if it is active (**1**), after 4 ns *q* is set to **1** and 3 ns after that *q_b* is set to **0**. Following the positive-edge of *clk*, if *s* is not **1**, *r* is inspected and if it is active, *q* is set to **0**. If neither *s* nor *r* are **1**, the flow of the program reaches the last **else** part of the **if**-statement and assigns *d* to *q*.

The behavior discussed here only looks at *s* and *r* on the positive-edge of *clk*, which corresponds to a rising-edge trigger D-type flip-flop with synchronous active high set and reset inputs. Furthermore, the set input is given a higher priority over the reset input. The flip-flop structure that corresponds to this description is shown in Figure 88.36.

Other synchronous control inputs can be added to this flip-flop in a similar fashion. A clock enable (*en*) input would only require inclusion of an **if**-statement in the last **else** part of the **if**-statement in the code of Figure 88.35. Note that delays such as those of this flip-flop are ignored in synthesis.

88.3.2.4 Asynchronous Control

The control inputs of the flip-flop of Figure 88.35 are synchronous because the flow into the **always** statement is only allowed to start when the **posedge** of *clk* is observed. To change this to a flip-flop with asynchronous control, it is only required to include asynchronous control inputs in the sensitivity list of its procedural statement.

Figure 88.37 shows a D flip-flop with active high asynchronous set and reset control inputs. Note that the only difference between this description and the code of Figure 88.35 (synchronous control) is the inclusion of **posedge** *s* and **posedge** *r* in the sensitivity list of the **always** block. This inclusion allows the flow into the procedural block to begin when *clk* becomes **1** or *s* becomes **1** or *r* becomes **1**. The **if**-statement in this block checks for *s* and *r* being **1**, and if none are active (activity levels are high) then clocking *d* into *q* occurs.

```
module d_ff ( d, s, r, clk, q, q_b );
    input d, clk, s, r;
    output q, q_b;
    reg q, q_b;
    always @ (posedge clk or posedge s or posedge r )
    begin
        if ( s ) begin
            #4 q <= 1'b1;
            #3 q_b <= 1'b0;
        end else if ( r ) begin
            #4 q <= 1' b0;
            #3 q_b <= 1'b1;
        end  else  begin
            #4 q <= d;
            #3 q_b <= ~d;
        end
    end
endmodule
```

FIGURE 88.37 D flip-flop with asynchronous control.

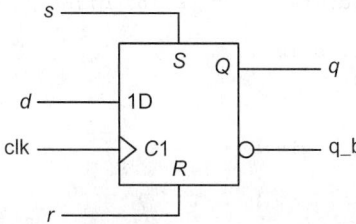

FIGURE 88.38 Flip-flop with asynchronous control inputs.

An active high (low) asynchronous input requires inclusion of **posedge** (**negedge**) of the input in the sensitivity list, and checking its **1** (**0**) value in the **if**-statement in the **always** statement. Furthermore, clocking activity in the flip-flop (assignment of *d* into *q*) must always be the last choice in the **if**-statement of the procedural block.

The graphic symbol corresponding to the flip-flop of Figure 88.37 is shown in Figure 88.38.

88.3.3   Registers, Shifters, and Counters

Registers, shifter-registers, counters and even sequential circuits with more complex functionalities can be described by simple extensions of the coding styles presented for the flip-flops. In most cases, the functionality of the circuit only affects the last **else** of the **if**-statement in the procedural statement of codes shown for the flip-flops.

88.3.3.1   Registers

Figure 88.39 shows an 8-bit register with synchronous set and reset inputs. The *set* input puts all **1**s in the register and the *reset* input resets it to all **0**s. The main difference between this and the flip-flop with synchronous control is the vector declaration of inputs and outputs.

88.3.3.2   Shift-Registers

A 4-bit shift-register with right- and left-shift capabilities, a serial-input, synchronous reset input, and parallel loading capability is shown in Figure 88.40. As shown, only the positive-edge of *clk* is included in the sensitivity list of the **always** block of this code, which makes all activities of the shift-register synchronous with the clock input. If *rst* is **1**, the register is reset, if *ld* is **1** parallel *d* inputs are loaded

```
module register (d, clk, set, reset, q);
    input [7:0] d;
    input clk, set, reset;
    output [7:0] q;
    reg [7:0] q;

    always @ ( posedge clk )
    begin
        if ( set )
            #5 q <= 8'b1;
        else if ( reset )
            #5 q <= 8'b0;
        else
            #5 q <= d;
    end

endmodule
```

FIGURE 88.39   An 8-bit Register.

```
module shift_reg (d, clk, ld, rst, l_r, s_in, q);
    input [3:0] d;
    input clk, ld, rst, l_r, s_in;
    output [3:0] q;
    reg [3:0]q;

    always @( posedge clk ) begin
        if ( rst )
            #5 q <= 4'b0000;
        else if ( ld )
            #5 q <= d;
        else if ( l_r )
            #5 q <= {q[2:0], s_in};
        else
            #5 q <= {s_in, q[3:1]};
    end

endmodule
```

FIGURE 88.40   A 4-bit shift-register.

into the register, and if none are **1** shifting left or right takes place depending on the value of the *l_r* input (**1** for left, **0** for right). Shifting in this code is done by use of the concatenation operator { }. For left-shift, *s_in* is concatenated to the right of *q[2:0]* to form a 4-bit vector that is put into *q*. For right-shift, *s_in* is concatenated to the left of *q[3:1]* to form a 4-bit vector that is clocked into *q[3:0]*. As before, delay values are only used for simulation and do not synthesize.

The style used for coding this register is the same as that used for flip-flops and registers presented earlier. In all these examples, a single procedural block handles function selection (e.g., zeroing, shifting, or parallel loading) as well as clocking data into the register output.

Another style of coding registers, shift-registers and counters is to use a combinational procedural block for function selection and another for clocking.

As an example, consider a shift-register that shifts *s_cnt* number of places to the right or left depending on its *sr* or *sl* control inputs (Figure 88.41). The shift-register also has an *ld* input that enables its clocked parallel loading. If no shifting is specified, i.e., *sr* and *sl* are both zero, then the shift register retains its old value.

The Verilog code of Figure 88.41 shows two procedural blocks that are identified by *combinational* and *register*. A block name appears after the **begin** keyword that begins a block and is separated from this keyword by use of a colon.

```
module shift_reg ( d_in, clk, s_cnt, sr, sl, ld, q );
    input [3:0] d_in;
    input clk, sr, sl, ld;
    input [1:0] s_cnt;
    output [3:0] q;
    reg [3:0] q, int_q;

    always @ ( d_in or s_cnt or sr or sl or ld ) begin: combinational
        if ( ld )   int_q = d_in;
        else if ( sr )   int_q = int_q >> s_cnt;
        else if ( sl )   int_q = int_q << s_cnt;
        else int_q = int_q;
    end

    always @ ( posedge clk )   begin: register
        q <= int_q;
    end

endmodule
```

FIGURE 88.41   Shift-register using two procedural blocks.

The *combinational* block is sensitive to all inputs that can affect the shift-register output. These include the parallel *d_in*, the *s_cnt* shift-count, *sr* and *sl* shift-control inputs, and the *ld* load-control input. In the body of this block an **if-else** statement decides on the value placed on the *int_q* internal variable. The value selection is based on values of *ld*, *sr*, and *sl*. If *ld* is **1**, *int_q* becomes *d_in* that is the parallel input of the shift register. If *sr* or *sl* is active, *int_q* receives the previous value of *int_q* shifted to right or left as many as *s_cnt* places. In this example, shifting is done by the use of >> and << operators. On the left, these operators take the vector to be shifted, and on the right they take the number of places to shift.

The *int_q* variable that is being assigned values in the *combinational* block is a 4-bit **reg** that connects the output of this block to the input of the register block.

The *register* block is a sequential block that handles clocking *int_q* into the shift-register output. This block (as shown in Figure 88.41) is sensitive to the positive edge of *clk* and its body consists of a single **reg** assignment. As before, we use nonblocking assignments for variables that represent registers.

Note in this code that both *q* and *int_q* are declared as **reg** because they are both receiving values in procedural blocks.

88.3.3.3   Counters

Any of the styles described for the shift-registers in the previous discussion can be used for describing counters. A counter counts up or down, while a shift-register shifts right or left. We use arithmetic operations in counting as opposed to shift or concatenation operators in shift-registers.

Figure 88.42 shows a 4-bit up-down counter with a synchronous *rst* reset input. The counter has an *ld* input for doing the parallel loading of *d_in* into the counter. The counter output is *q* and it is declared as **reg** since it is receiving values within a procedural statement.

Discussion about synchronous and asynchronous control of flip-flops and registers also apply to the counter circuits. For example, inclusion of ***posedge*** *rst* in the sensitivity list of the counter of Figure 88.42 would make its resetting asynchronous.

88.3.4   State Machine Coding

Coding styles presented so far can be further generalized to cover finite state machines of any type. This section shows coding for Moore and Mealy state machines. The examples we will use are simple sequence detectors. These circuits represent the controller part of a digital system that has been partitioned into a data path and a controller.

```
module counter (d_in, clk, rst, ld, u_d, q );
    input [3:0] d_in;
    input clk, rst, ld, u_d;
    output [3:0] q;
    reg [3:0] q;

    always @ ( posedge clk ) begin
        if ( rst )
            q <= 4'b0000;
        else if ( ld )
            q <= d_in;
        else if ( u_d )
            q <= q + 1;
        else
            q <= q - 1;
    end

endmodule
```

FIGURE 88.42 An up-down counter.

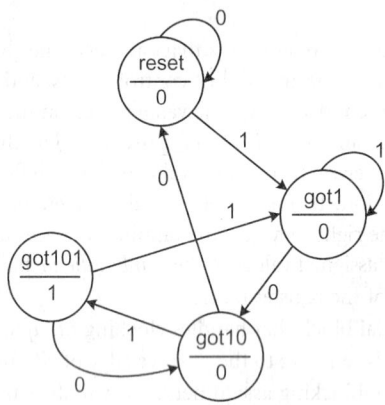

FIGURE 88.43 A Moore sequence detector.

88.3.4.1 Moore Detector

State diagram for a Moore sequence detector detecting **101** on its *x* input is shown in Figure 88.43. The machine has four states that are labeled: *reset*, *got1*, *got10*, and *got101*. Starting in *reset*, if the **101** sequence is detected, the machine goes into the *got101* state in which the output becomes **1**. In addition to the *x* input, the machine has a *rst* input that forces the machine into its *reset* state. The resetting of the machine is synchronized with the clock.

The Verilog code of the Moore machine of Figure 88.43 is shown in Figure 88.44. After the declaration of inputs and outputs of this module, **parameter** declaration declares four states of the machine as 2-bit parameters. The square-brackets following the **parameter** keyword specify the size of parameters being declared. Following parameter declarations in the code of Figure 88.44, the two-bit *current* **reg** type variable is declared. This variable holds the current state of the state machine.

The **always** block used in the module of Figure 88.44 describes state transitions of the state diagram of Figure 88.43. The main task of this procedural block is to inspect input conditions (values on *rst* and *x*) during the present state of the machine defined by *current* and set values into *current* for the next state of the machine.

The flow into the **always** block begins with the positive edge of *clk*. Since all activities in this machine are synchronized with the clock, only *clk* appears on the sensitivity list of the **always** block. Upon entry into this block, the *rst* input is checked and if it is active, *current* is set to *reset* (*reset* is a declared parameter and its

```
module moore_detector (x, rst, clk, z);
  input x, rst, clk;
  output z;
  parameter [1:0]  reset = 0, got1 = 1, got10 = 2, got101 = 3;
  reg [1:0] current;
  always @ ( posedge clk ) begin
    if ( rst ) begin
      current <= reset;
    end else case ( current )
      reset:  begin
              if ( x==1'b1 ) current <= got1;
              else current <= reset;
            end
      got1:  begin
              if ( x==1'b0 ) current <= got10;
              else current <= got1;
            end
      got10:  begin
              if ( x==1'b1 ) begin
                current <= got101;
              end else begin
                current <= reset;
              end
            end
      got101:  begin
              if ( x==1'b1 ) current <= got1;
              else current <= got10;
            end
      default:  begin
              current <= reset;
            end
    endcase
  end
  assign z = (current == got101) ? 1 : 0;
endmodule
```

FIGURE 88.44 Moore machine Verilog code.

value is **0**). The value put into *current* in this pass through the **always** block gets checked in the next pass with the next edge of the clock. Therefore this assignment is regarded as the next-state assignment. When this assignment is made, the **if-else** statements skip the rest of the code of the **always** block, and this **always** block will next be entered with the next positive edge of *clk*. An assignment statement outside of the **always** block is used for assigning values to the *z* output of the circuit. As shown here and as shown in the state diagram of this machine, the output becomes **1** when the *current* variable is *got101*.

Upon entry into the **always** block, if *rst* is not **1**, program flow reaches the **case** statement that checks the value of *current* against the four states of the machine. Figure 88.45 shows an outline of this **case**-statement. This statement has five **case**-alternatives. A **case**-alternative is followed by a block of statements bracketed by the **begin** and **end** keywords. In each such block, actions corresponding to the active state of the machine are taken. The **default** case alternative is taken when none of the other alternatives are true. This part is useful for avoiding unwanted latches in synthesis.

In this coding style, for every state of the machine there is a **case**-alternative that specifies the next state values. For larger machines, there will be more **case**-alternatives, and more conditions within an alternative. Otherwise, this style can be applied to state machines of any size and complexity.

This same machine can be described in Verilog in many other ways. We will show alternative styles of coding state machines by use of examples that follow.

88.3.4.2 A Mealy Machine Example

Unlike Moore machines the outputs of which are only determined by the current state of the machine, a Mealy machine, the outputs are determined by the state the machine is in as well as the inputs of the

```
case ( current )
    reset:   begin . . . end
    got1:    begin . . . end
    got10:   begin . . . end
    got101:  begin . . . end
    default: begin . . . end
endcase
```

FIGURE 88.45   Case-statement outline.

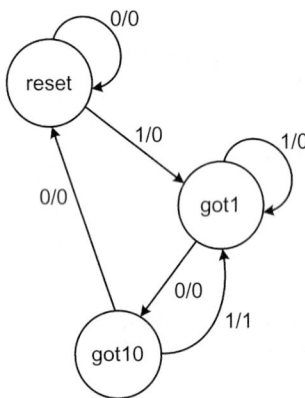

FIGURE 88.46   A 101 Mealy detector.

circuit. This makes Mealy outputs not fully synchronized with the circuit clock. In the state diagram of a Mealy machine the outputs are specified along the edges that branch out of the states of the machine.

Figure 88.46 shows a **101** Mealy detector. The machine has three states, *reset, got1,* and *got10.* While in *got10,* if the *x* input becomes **1** the machine prepares to go to its next state with the next clock. While waiting for the clock, its output becomes **1**. While on the edge that takes the machine out of *got10,* if the clock arrives the machine goes into the *got1* state. This machine allows overlapping sequences. The machine has no external resetting mechanism. A sequence of two zeros on input *x* puts the machine into the *reset* state in a maximum of two clocks.

The Verilog code of the **101** Mealy detector is shown in Figure 88.47. After input and output declarations, a **parameter** declaration defines bit patterns (state assignments) for the states of the machine. Note here that state value **3** or binary **11** is unused. As in the previous example, we use the *current* two-bit **reg** to hold the current state of the machine.

After the declarations, an **initial** block sets the initial state of the machine to *reset.* This procedure for initializing the machine is only good for simulation and is not synthesizable.

This example uses an **always** block for specifying state transitions and a separate statement for setting values to the *z* output. The **always** statement responsible for state transitions is sensitive to the circuit clock and has a **case** statement that has **case** alternatives for every state of the machine. Consider, for example, the *got10* state and its corresponding Verilog code segment, as shown in Figure 88.48.

The always statement used here specifies the next states of the machine based on conditions of the input and present state. Notice in this code segment that the **case** alternative shown does not have **begin** and **end** bracketing. Actually, **begin** and **end** keywords do not appear in blocks following **if** and **else** keywords either. Verilog only requires **begin** and **end** bracketing if there is more than one statement in a block. The use of this bracketing around one statement is optional. Since the **if** part and the **else** part each only contain one statement, **begin** and **end** keywords are not used. Furthermore, since the entire **if-else** statement reduces to only one statement, the **begin** and **end** keywords for the **case**-alternative are also eliminated.

```
module mealy_detector (x, clk, z);
  input x, clk;
  output z;
  parameter [1:0]
    reset  = 0,  // 0 = 0 0
    got1   = 1,  // 1 = 0 1
    got10  = 2;  // 2 = 1 0

  reg [1:0] current;

  initial current <= reset;
  always @ ( posedge clk )
  begin
    case ( current )
      reset:  if( x==1'b1 ) current <= got1;
              else current <= reset;
      got1:  if( x==1'b0 ) current <= got10;
              else current <= got1;
      got10:  if( x==1'b1 ) current <= got1;
                else current <= reset;
        default: current <= reset;
      endcase
  end
  assign z= ( current==got10 && x==1'b1 ) ? 1'b1 : 1'b0;

endmodule
```

FIGURE 88.47 Verilog code of 101 Mealy detector.

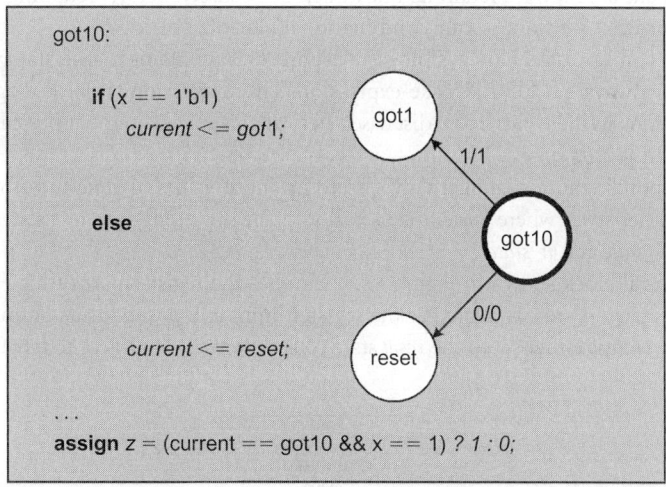

FIGURE 88.48 Coding a Mealy state.

The last **case**-alternative shown in Figure 88.47 is the **default** alternative. When checking *current* against all alternatives that appear before the **default** statement fail, this alternative is taken. There are several reasons that we use this default alternative. One is that, our machine only uses three of the possible four 2-bit assignments and **11** is unused. If the machine ever begins in this state, the default case makes *reset* the next state of the machine. The second reason why we use **default** is that Verilog assumes a four-value logic system that includes **Z** and **X**. If *current* ever contains a **Z** or **X**, it does not match any of the defined case alternatives, and the default case is taken. Another reason for use of **default** is that our machine does not have a hard reset and we are making provisions for it to go to the *reset* state. The last reason for **default** is that it is just a good idea to have it.

The last statement in the code fragment of Figure 88.55 is an **assign** statement that sets the z output of the circuit. This statement is a concurrent statement and is independent of the **always** statement above it. When *current* or x changes, the right-hand side of this assignment is evaluated and a value of **0** or **1** is assigned to z. Conditions on the right-hand side of this assignment are according to values put in z in the state diagram of Figure 88.47. Specifically, the output is **1** when *current* is *got10* and x is **1**, otherwise it is **0**. This statement implements a combinational logic structure with *current* and x inputs and z output.

88.3.4.3 Huffman Coding Style

The Huffman model for a digital system characterizes it as a combinational block with feedbacks through an array of registers. Verilog coding of digital systems according to the Huffman model uses an **always** statement for describing the register part and another concurrent statement for describing the combinational part.

We will describe the state machine of Figure 88.43 to illustrate this style of coding. Figure 88.49 shows the combinational and register part partitioning that we will use for describing this machine. The *combinational* block uses x and *p_state* as input and generates z and *n_state*. The *register* block clocks *n_state* into *p_state*, and *reset p_state* when *rst* is active.

Figure 88.50 shows the Verilog code of Figure 88.43 according to the partitioning of Figure 88.49. As shown, parameter declaration declares the states of the machine. Following this declaration, *n_state* and *p_state* variables are declared as two-bit **reg**s that hold values corresponding to the states of the **101** Moore detector. The *combinational* **always** block follows this **reg** declaration. Since this is a purely combinational block, it is sensitive to all its inputs, namely x and *p_state*. Immediately following the block heading, *n_state* is set to its inactive or reset value. This is done so that this variable is always reset with the clock to make sure it does retain its old value. As discussed before, retaining old values implies latches, which is not what we want in our combinational block.

The body of the combinational **always** block of Figure 88.50 contains a **case**-statement that uses the *p_state* input of the **always** block for its **case**-expression. This expression is checked against the states of the Moore machine. As in the other styles discussed before, this **case**-statement has **case**-alternatives for *reset*, *got1*, *got10*, and *got101* states.

In a block corresponding to a **case**-alternative, based on input values, *n_state* and z output are assigned values. Unlike the other styles where *current* is used both for the present and next states, here we use two different variables, *p_state* and *n_state*.

The next procedural block shown in Figure 88.50 handles the register part of the Huffman model of Figure 88.49. In this part, *n_state* is treated as the register input and *p_state* as its output. On the positive edge of the clock, *p_state* is either set to the *reset* state (**00**) or is loaded with contents of *n_state*. Together,

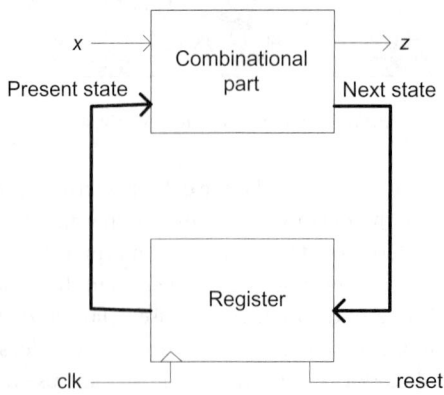

FIGURE 88.49 Huffman partitioning of 101 Moore detector.

```
module moore_detector (x, rst, clk, z);
  input x, rst, clk;
  output z;
  parameter [1:0]
    reset = 2'b00, got1 = 2'b01, got10 = 2'b10, got101 = 2'b11;

  reg [1:0] p_state, n_state;

  always @ ( p_state or x ) begin : combinational
    n_state = 0;
    case ( p_state )
      reset:  begin
            if( x==1'b1 ) n_state = got1;
            else n_state = reset;
          end
      got1:  begin
            if( x==1'b0 ) n_state = got10;
            else n_state = got1;
          end
      got10:  begin
            if( x==1'b1 ) n_state = got101;
            else n_state = reset;
      end
      got101:  begin
            if( x==1'b1 ) n_state = got1;
            else n_state = got10;
      end
      default: n_state = reset;
    endcase
  end

  assign z = (p_state == got101) ? 1 : 0;

  always @( posedge clk ) begin : register
    if( rst ) p_state <= reset;
    else p_state <= n_state;
  end

endmodule
```

FIGURE 88.50 Verilog Huffman coding style.

combinational and *register* blocks describe our state machine in a very modular fashion. As in the other coding style done for this machine, an **assign** statement handles the assignment of values to the circuit output.

The advantage of this style of coding is in its modularity and defined tasks of each block. State transitions are handled by the *combinational* block and clocking is done by the *register* block. Changes in clocking, resetting, enabling, or presetting the machine only affect the coding of the *register* block. If we were to change the synchronous resetting to asynchronous, the only change we had to make was adding ***posedge*** *rst* to the sensitivity list of the register block.

The examples discussed above, in particular, the last style, show how combinational and sequential coding styles can be combined to describe very complex digital systems.

88.3.5 Memories

Verilog allows description and use of memories. Memories are described as arrays of vectors of variables and are declared as **reg**. Verilog allows **reg** data types for memories. Figure 88.51 shows a **reg** declaration declaring *mem* and its corresponding block diagram. This figure also shows several valid memory operations.

```
reg [7:0] mem [0:1023];
reg [7:0] data;
reg [3:0] short_data;
wire [9:0] addr;
    .
    .
    .
data = mem [956];
    .
    .
short_data = data [7:4];
    .
    .
    .
mem [932] = data;
mem [addr] = {4 b0, short_data};
```

FIGURE 88.51   Memory representation.

```
module memory (inbus, outbus, addr, clk, rw);
    input [7:0] inbus;
    input [9:0] addr;
    output [7:0] outbus;
    input clk, rw;

    reg [7:0] mem [0:1023];

    assign outbus = rw ? mem [addr] : 8'bz;

    always @ (posedge clk)
        if (rw == 0) mem [addr] <= inbus;

endmodule
```

FIGURE 88.52   Memory description.

Operations shown here comply with Verilog 1999 and not Verilog 2001. More advanced features of Verilog 2001 are described in Chapter 90.

Square brackets that follow the **reg** keyword specify the word-length of the memory. The square brackets that follow the name of the memory (*mem*), specify its address space. A memory can be read by addressing it within its address range, e.g., *mem[956]*. Part of a word in a memory cannot be read directly, i.e., slicing a memory word is not possible. To read part of a word, the whole word must first be read in a variable and then slicing done on this variable. For example, *data[7:4]* can be used after a memory word has been placed into *data*.

With proper indexing, a memory word can be written into by placing the memory name and its index on the left-hand side of an assignment, e.g., *mem[932] = data;*, memories can also be indexed by **reg** or **net** type variables, e.g., *mem[addr]*, when *addr* is a 10-bit address bus. Writing into a part of the memory is not possible. In all cases data directly written into a memory word affects all bits of the word being written into. For example to write the four-bit *short_data* into a location of *mem*, we have to decide what goes into the other four bits of the memory word.

Figure 88.52 shows a memory block with separate input and output buses. Writing into the memory is clocked, while reading from it only requires *rw* to be **1**. An **assign** statement handles reading and an **always** block performs writing into this memory.

88.4 Writing Testbenches

Verilog coding styles discussed so far were for coding hardware structures, and in all cases synthesizability and direct correspondence to hardware were our main concerns. In contrast, testbenches do not have to have hardware correspondence and they usually do not follow any synthesizability rules. We will see that delay specifications, and **initial** statements that do not have a one-to-one hardware correspondence are used generously in testbenches.

For demonstration of testbench coding styles, we use the Verilog code of Figure 88.53 that is a **101** Moore detector, as the circuit to be tested. This description is functionally equivalent to that of Figure 88.44. The difference is in the use of condition expressions (**?:**) instead of **if-else** statements, and separating the output assignment from the main **always** block. This code will be instantiated in the testbenches that follow.

88.4.1 Generating Periodic Data

Figure 88.54 shows a testbench module that instantiates *moore_detector* and applies test data to its inputs. The first statement in this code is the **'timescale** directive that defines the time unit of this description. The testbench itself has no ports, which is typical of all testbenches. All data inputs to a circuit-under-test are locally generated in its testbench.

Because we are using procedural statements for assigning values to ports of the circuit-under-test, all variables mapped with the input ports of this circuit are declared as **reg**. The testbench uses two **initial** blocks and two **always** blocks. The first initial block {XE "**initial** block"} initializes *clock*, *x*, and *reset* to **0**, **0**, and **1**,

```
module moore_detector ( x, rst, clk, z );
    input x, rst, clk;
    output z;
    parameter [1:0] a=0, b=1, c=2, d=3;
    reg [1:0] current;

    always @( posedge clk )
        if ( rst )      current = a;
        else case ( current )
            a : current <= x ? b : a ;
            b : current <= x ? b : c ;
            c : current <= x ? d : a ;
            d : current <= x ? b : c ;
            default : current <= a;
        endcase
    assign z = (current==d) ? 1'b1 : 1'b0;
endmodule
```

FIGURE 88.53 Circuit under test.

```
`timescale 1 ns / 100 ps

module test_moore_detector;
    reg x, reset, clock;
    wire z;
    moore_detector uut ( x, reset, clock, z );
    initial  begin
        clock=1'b0; x=1'b0; reset=1'b1;
    end
    initial #24 reset=1'b0;
    always #5 clock=~clock;
    always #7 x=~x;
endmodule
```

FIGURE 88.54 Generating periodic data.

respectively. The next **initial** block waits for 24 time units (ns in this code), and then sets *reset* back to **0** to allow the state machine to operate.

The **always** blocks shown produce periodic signals with different frequencies on *clock* and *x*. Each block waits for a certain amount of time and then it complements its variable. Complementing begins with the initial values of *clock* and *x* as set in the first **initial** block. We are using different periods for *clock* and *x*, so that a combination of patterns on these circuit inputs is seen. A more deterministic set of values could be set by specifying exact values at specific times.

88.4.2 Random Input Data

Instead of the periodic data on *x* we can use the **$random** predefined system function to generate random data for the *x* input. Figure 88.55 shows such a testbench.

This testbench also combines the two **initial** blocks for initially activating and deactivating *reset* into one. In addition, this testbench has an **initial** block that finishes the simulation after 165 ns.

When the flow into a procedural block reaches the **$finish** system task, the simulation terminates and exits. Another simulation control task that is often used is the **$stop** task that only stops the simulation and allows resumption of the stopped simulation run.

88.4.3 Timed Data

A very simple testbench for our sequence detector can be done by applying test data to *x* and timing them appropriately to generate the sequence we want, very similar to the way values were applied to *reset* in the previous examples. Figure 88.56 shows this simple testbench.

Techniques discussed in the above examples are just some of what one can do for test data generation. These techniques can be combined for more complex examples. After using Verilog for some time, users form their own test-generation techniques. For small designs, simulation environments generally provide waveform editors and other tool-dependent test-generation schemes. Some tools come with code fragments that can be used as templates for testbenches.

An important issue in developing testbenches is external file IO. Verilog allows the use of **$readmemh** and **$readmemb** system tasks for reading hex and binary test data into a declared memory. Moreover, for writing responses from a circuit-under-test to an external file, **$fdisplay** can be used.

88.5 Synthesis Issues

Verilog constructs described in this chapter included those for cell modeling as well as those for designs to be synthesized. In describing an existing cell, timing issues are important and must be included in the

```
`timescale 1 ns / 100 ps

module test_moore_detector;
    reg x, reset, clock;
    wire z;
    moore_detector uut( x, reset, clock, z );
    initial     begin
        clock=1'b0; x=1'b0; reset=1'b1;
        #24 reset=1'b0;
    end
    initial #165 $finish;
    always #5 clock=~clock;
    always #7 x=~x;
endmodule
```

FIGURE 88.55 Random data generation.

```
`timescale 1ns/100ps

module test_moore_detector;
   reg x, reset, clock;
   wire z;

   moore_detector uut( x, reset, clock, z );

   initial begin
      clock=1'b0; x=1'b0; reset=1'b1;
      #24 reset=1'b0;
   end

   always #5 clock=~clock;

   initial begin
      #7  x=1;
      #5  x=0;
      #18 x=1;
      #21 x=0;
      #11 x=1;
      #13 x=0;
      #33 $stop;
   end

endmodule
```

FIGURE 88.56   Timed test data generation.

Verilog code of the cell. At the same time, description of an existing cell may require parts of this cell to be described by interconnection of gates and transistors. In contrast, a design to be synthesized does not include any timing information because this information is not available until the design is synthesized, and designers usually do not use gates and transistors for high-level descriptions for synthesis.

Considering the above, taking timing out of the descriptions, and only using gates when we really have to, the codes presented in this chapter all have one-to-one hardware correspondence and are synthesizable. For synthesis, a designer must consider his or her target library to see what and how certain parts can be synthesized. For example, most FPGAs do not have internal three-state structures and three-state busings are converted into AND-OR buses.

88.6   Summary

This chapter presented the Verilog HDL language from a hardware design point of view. The chapter used complete design examples at various levels of abstraction for showing ways in which Verilog could be used in a design. We showed how timing details could be incorporated in cell descriptions. Aside from this discussion of timing, all examples that were presented had one-to-one hardware correspondence and were synthesizable. We have shown how combinational and sequential components can be described for synthesis and how a complete system can be put together using combinational and sequential blocks for it to be tested and synthesized. This chapter did not cover all of Verilog, but only the most often used parts of the language.

Register-Transfer Level Hardware Description with SystemC

Shahrzad Mirkhani and
Zainalabedin Navabi

Nanoelectronics Center of Excellence
School of Electrical and Computer Engineering
University of Tehran

CONTENTS

Many details that are needed in register transfer level (RTL) designs are ignored in a system-level design. As discussed in Chapter 86, the SystemC language [1,2] is developed for system-level design. In spite of this, SystemC language has the necessary constructs for RTL and even gate-level designs. This chapter presents SystemC at the RTL for describing hardware components at this level of abstraction. The purpose is to familiarize readers with the main topics of this language and present it so that after reading this chapter, readers can start writing SystemC codes for simple components like arithmetic and logic units (ALUs) and state machines. The chapter first covers the basic definitions in SystemC. We will then discuss hardware examples in an incremental fashion starting with simple components progressing into more complex ones. In a later section after a general familiarity with the language is gained, more complex features of the SystemC language with emphasis on testbench development will be described. Finally, a comprehensive example will be presented.

89.1 The Role of RTL SystemC in Hardware Design

Although digital design technology has powerful hardware description languages like VHDL and Verilog, the need for RTL SystemC is still justified. Generally, a system specification, implemented at the system level of abstraction, must be mapped to hardware (such as an IC, SoC, or board). Therefore, after system-level design (for example in SystemC), there must be an easy way to feed designs to a synthesis tool for obtaining a netlist of gates or low-level components. Since most of the available synthesis tools only support

RTL designs, we have to convert our high-level system descriptions to their corresponding RTL models. There are two ways to perform this: (1) refining the system-level design in SystemC to RTL SystemC, and (2) redesigning the system-level design in SystemC with another HDL (like Verilog).

The first solution has certain advantages. Since refinements are performed in the same environment, the conversion process will be faster than recoding it in another language. Besides, it is more error-free to refine different parts of a design than to recode the whole design in another language. Another advantage of the first solution is that after developing the RTL model, the designer still has the opportunity to cosimulate the hardware and software parts together.

Therefore, to reach the synthesis process from the system-level design, RTL SystemC design plays an important role in the design flow. Today, there are high-level synthesis tools that convert a high-level design, developed in C language, to RTL SystemC. Furthermore, there are converter tools for converting RTL SystemC descriptions to VHDL and Verilog languages. These processes together enable designers to go from system-level down to a netlist of gates with transition through SystemC.

Another scenario in which knowledge of RTL SystemC becomes useful is in design reuse. To be able to use existing VHDL or Verilog components in design of systems at the system-level in C or C++, these designs must be translated to SystemC and then used along with higher level components described in C++. Whether this translation is done manually or automatically, knowing SystemC is useful for verification of the translated codes or manually performing the translation.

89.2   SystemC Constructs for RTL Design

In this section we will introduce the basic components in SystemC for describing hardware at RT level. This section uses a series of examples starting from a very primitive one to a more complex one.

Each example incrementally discusses SystemC constructs at the RT level. After discussing basic parts of the SystemC language, we start with combinational circuits and then cover several RT-level sequential circuits. Combinational circuits cover basic functions, adders and ALU functions, and the sequential examples cover basic flip-flops, functional registers, and finally, several ways of describing state machines.

The last part of this section shows SystemC description of testbenches. This part is also example oriented in which we will be using the examples of the earlier parts of this section.

89.2.1   Modules

The entity used in SystemC for description of hardware components is a module defined by the **SC_MODULE** keyword. A module can describe a hardware component as simple as a multiplexer, or a network of complex digital systems. As shown in Figure 89.1, the definition of modules begins with the **SC_MODULE** keyword. This is usually done in a header file (*.h* file). In the definition of a module, ports, member variables, and member functions are defined. The implementation of member functions is usually placed in a separate file (a *.cpp* file) as shown in the figure. In this section, most of the example

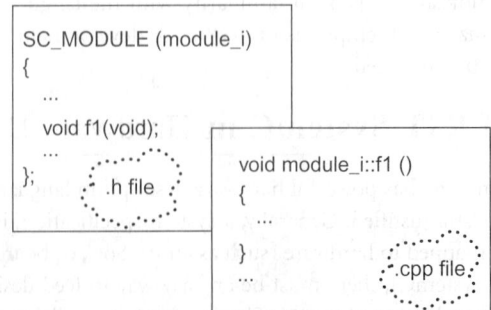

FIGURE 89.1   Module definition and implementation in SystemC.

codes are partitioned into two parts. The first part shows the header file (*.h* file) of the example code and the second part is the implementation of the member functions of the first part (*.cpp* file).

A design may be described in a hierarchy of other modules. The top-level module is the complete design, and modules lower in the hierarchy are the design's components. Module instantiation is the construct used for bringing a lower level module into a higher level one. Figure 89.2 shows a hierarchy of several nested modules. As shown in this figure, module instantiation is performed by defining a member variable of a lower level module inside a higher level module in hierarchy.

As shown in Figure 89.3, a module declaration includes several parts. These parts can be put in any order. They include module ports, local signal declarations, definition of concurrent processes of a module, member data declarations, and member function declarations.

89.2.1.1 Module Ports

A part of module declaration is a list of port declarations. This list includes inputs, outputs, and bidirectional input/output lines. Ports may be listed in any order. This ordering can only become significant when a module is instantiated and does not affect the way its operation is described. Usually, top-level modules used for testbenches have no ports.

In this part, the direction, type, and size of each port listed in the module header are specified. A port may be **sc_in**, **sc_out** or **sc_inout**. The latter kind is used for bidirectional input/output lines. A general declaration of ports is **sc_port** that is more useful in higher level designs than the RTL. The type and size of each port is defined after its direction. For example in Figure 89.4, the type of *a* and *b* input ports are **bool** and the type *w* output port is **int**. Size of vectored ports of a module is also declared in the module port declaration part. The size and indexing of a port is specified after its type declaration between < and > signs. Figure 89.4 shows an example circuit with scalar and vectored input, output and input/output ports. Ports named *a* and *b* are one-bit inputs. Ports *av* and *bv* are 8-bit logic vector inputs of *acircuit*. Port *w* of *acircuit* is declared as a 1-bit output, and *wv* is a logic vector bidirectional port of size 8.

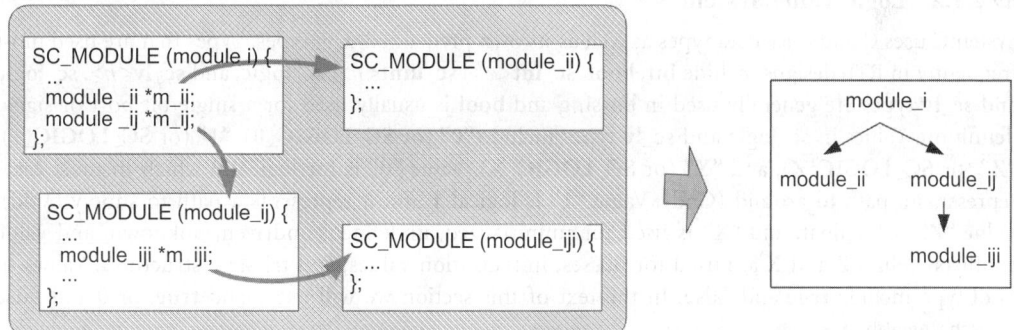

FIGURE 89.2   Module hierarchy.

```
SC_MODULE (name) {
    port declarations;
    internal signal declarations;
    concurrent part declarations;
    member data declarations;
    member function declarations;
};
```

FIGURE 89.3   Module outline.

```
SC_MODULE (acircuit) {
    sc_in <bool> a, b;
    sc_out <int> w;
    sc_inout <bool> c;
    sc_in <sc_lv <8> > av, bv;
    sc_out <sc_lv <8> > wv;
    sc_inout <sc_lv <8> > cv;
    sc_port <ud_type> p;
    ...
};
```

FIGURE 89.4 Module ports.

```
SC_MODULE (bcircuit) {
    sc_in <bool> a, b;
    sc_out <int> w;
    ...
    sc_signal <bool> d;
    sc_signal <int> e;
    sc_signal <sc_lv <8> > dv;
    ...
};
```

FIGURE 89.5 Module signals.

89.2.1.2 Logic Value System

SystemC uses C and C++ data types as well as its own predefined data types. Types that are used most frequently in RTL designs include **int**, **bool**, **sc_int**$<n>$, **sc_uint**$<n>$, **sc_logic**, and **sc_lv**$<n>$. **sc_logic** and **sc_lv** types are generally used in bussing and **bool** is usually used for a single-bit port or signal definition. Values in **sc_logic** and **sc_lv** types include "**0**" (or **SC_LOGIC_0**), "**1**" (or **SC_LOGIC_1**), "**Z**" (or **SC_LOGIC_Z**), and "**X**" (or **SC_LOGIC_X**). Value "**0**" is for logical **0** which in most cases represents a path to ground (Gnd). Value "**1**" is logical **1** and it represents a path to supply (Vdd). Value "**Z**" is for float, and "**X**" is used for uninitialized, undefined, undriven, unknown, and value conflicts. Values **Z** and **X** are used for busses, initialization values, and tri-state structures. Values of **bool** type include **true** and **false**. In the text of this section we will use **1** and **true**, or **0** and **false** interchangeably.

89.2.1.3 Module Signals

A portion of module declaration is a list of internal signals that are used inside the module (e.g., in its concurrent parts). Internal signals in SystemC are defined by **sc_signal** keyword.

Similar to ports, an internal signal has a type and it can be a single or a vectored signal. The acceptable data types for a signal are similar to those for a port. Figure 89.5 shows internal signals named *d*, *e*, and *dv* for the *bcircuit* module. As shown in this figure, signals *d* and *e* are defined as single signals, and signal *dv* is a logic vector of size 8.

89.2.1.4 Module Members

Similar to a *class* in C++, a module in SystemC has its own data and functions. These are called *members* of a module. Member definitions are similar to the definition of member data and member functions in C++.

```
SC_MODULE (ccircuit) {
    ...
private:
    int md;
public:
    void mf1 (bool arg1, bool arg2);
protected:
    void mf2 ();
};
```

FIGURE 89.6   Module members.

As shown in Figure 89.6, *md* is a member data of type **int**, *mf1* is a member function that returns **void** and accepts two arguments of type **bool**, and *mf2* is a member function with no arguments. Similar to C++, all members in a module must be **public**, **private**, or **protected**. Public members of a module can be accessed by other modules via the instance of that module. For example, *mf1* member function can be accessed in other modules, if they have an instance (a defined variable) of type *ccircuit*. Private members of a module can be accessed only by other members of that module. For example, *md* member data can only be accessed in other functions in *ccircuit*. Protected members of a module act as public members for the module's children and act as private members otherwise.

89.2.1.5   Module Concurrent Parts

A module in SystemC consists of one or more concurrent parts. These concurrent parts are called *processes* and they can be *methods* (defined by **SC_METHOD** keyword), *threads* (defined by **SC_THREAD** keyword), or *clocked threads* (defined by **SC_CTHREAD** keyword). These three kinds of processes are discussed in the examples of this section.

A method in SystemC is a light-weight process with a sensitivity list. Every time an event happens on one of the signals in its sensitivity list it is executed. A thread is like a method in that it also has a sensitivity list. But it can be suspended and resumed by the use of **wait** functions inside its body. A clocked thread is similar to a thread in that it can be suspended and resumed by **wait** functions. But a specified edge of a signal is defined as its clock in its registration, and it is sensitive to the defined edge of this signal.

A member function defined in a module can be defined as a process. These process definitions (also named as *process registration*) are performed inside the module *constructor*.

A mandatory part of a module is its constructor. Necessary variable initializations, memory allocations, and process registrations are performed in a constructor. Figure 89.7 shows two ways of constructor definition in SystemC.

A constructor of a module is a member function with the same name as the module name. In SystemC, the implementation of a constructor can be in a .cpp file (like other member functions) or it can be inside the module definition. In the former case, the **SC_HAS_PROCESS** keyword defines the member function that is a constructor, and in the latter case, the implementation of a module constructor begins with **SC_CTOR** followed by the module name and the body of the constructor.

In addition to member data initializations and memory allocations, process registration is done inside a constructor. Process registration specifies which member functions execute concurrently with other parts of the system and to which signals or ports they are sensitive. In Figure 89.8, the member function *mf2* defined in Figure 89.6 is registered as a *method* that is sensitive to port *a* and signal *ib*. As shown in Figure 89.8, a sensitivity list is defined by **sensitive** keyword and the separation between signals in a sensitivity list is done by the << operator.

```
SC_MODULE (dcircuit) {
   dcircuit (int num, char *name);
   SC_HAS_PROCESS (dcircuit);
   ...
}; //module definition
...
dcircuit::dcircuit (int num, char *name)
{
   ...
} //constructor implementation
```

```
SC_MODULE (dcircuit) {
   ...
   SC_CTOR (dcircuit) {
      ...
   } //constructor implementation
   ...
}; //module definition
```

FIGURE 89.7 Module constructor using SC_HAS_PROCESS and SC_CTOR.

```
SC_MODULE (ccircuit) {
   sc_in <bool> a;
   ...
   sc_signal <bool> ib;
   ...
   SC_CTOR (ccircuit) {
      SC_METHOD (mf2);
      sensitive << a << ib;
   }
public:
   void mf1 (bool arg1, bool arg2);
   void mf2 ();
};
```

FIGURE 89.8 Process registration.

Using the basic concepts presented above, the sections that follow show how SystemC can be used for describing combinational and sequential components. Utilization of SystemC operators and constructs for specifying functions will be described in the examples that are discussed next.

89.2.2   Combinational Circuits Using Equations

A combinational circuit can be represented by its gate-level structure, its Boolean functionality, or description of its behavior. At the gate level, interconnection of its gates are shown; at the functional level, Boolean expressions representing its outputs are written; and at the behavioral level, a software-like procedural description represents its functionality. This section shows examples of combinational circuits using expressions of procedural descriptions with SystemC constructs.

At a higher level than gates and transistors, a combinational circuit may be described by use of Boolean, logical, and arithmetic expressions. These operations are performed on variables of C++ or SystemC data types and they are predefined or overloaded C and C++ operations. In the following, there are several combinational circuits implemented with equations in SystemC. Table 89.1 shows the predefined operations that can be used or overloaded for several data types in SystemC.

89.2.2.1 Majority Example

We use the majority circuit of Figure 89.9 to illustrate how expressions are used in a design. The description shown in Figure 89.10 corresponds to this circuit. The module description has inputs and outputs according to the schematic of Figure 89.9.

TABLE 89.1   SystemC Operators

Bitwise operators	&	|	^	!	
Arithmetic operators	+	−	*	/	%
Logical operators	&&	||	!		
Compare operators	<	>	<=	>=	==
Shift operators	>>	<<			

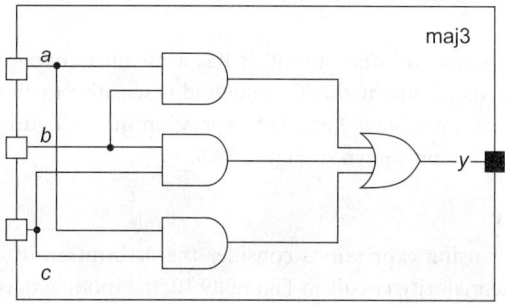

FIGURE 89.9   A majority circuit.

```
SC_MODULE (maj3) {
   sc_in <bool> a, b, c;
   sc_out <bool> y;

   SC_CTOR (maj3) {
      SC_METHOD (maj_func);
      sensitive << a << b << c;
   }
   void maj_func ();
}; //module definition
```

```
void maj3::maj_func()
{
   bool la, lb, lc;
   la = a.read();
   lb = b.read();
   lc = c.read();

   y = (la & lb) | (la & lc) | (lb & lc); //line 8
} //implementation of maj_func
```

FIGURE 89.10   SystemC code for the majority circuit.

```
SC_MODULE (xor3) {
   sc_in <bool> a, b, c;
   sc_out <bool> y;

   SC_CTOR (xor3) {
      SC_THREAD (xor_func);
      sensitive << a << b << c;
   }
   void xor_func();
};
```

```
void xor3::xor_func ()
{
   while (1) {
      y = a.read() ^ b.read() ^ c.read();
      wait(); //suspends this thread until the next event in the sensitivity list
   }
}
```

FIGURE 89.11 SystemC code for an XOR gate.

This module has three inputs and one output. It has a member function named *maj_func* that is defined as a method in the *maj3* constructor. This method is sensitive to all input signals. Figure 89.10 also includes the implementation of *maj_func*. The expression in this figure, at *line 8*, implements the functionality of the majority circuit shown in Figure 89.9.

89.2.2.2 XOR Example

As our second example for using expressions consider the description of an XOR gate as shown in Figure 89.11. Similar to the majority circuit in Figure 89.10, the module developed for XOR gate has a process that handles the functionality of a three-input XOR gate. The *xor3* module in Figure 89.11 uses a thread (SC_THREAD) process for its functional part. This thread is sensitive to the module's three inputs *a*, *b*, and *c*. Inside the thread, it waits for a new event on its sensitivity list and after each event it recalculates the output *y* and waits for the next event. The majority circuit in Figure 89.10 was developed by a method (SC_METHOD) process.

89.2.2.3 Full-Adder Example

After developing modules for a majority circuit and an XOR gate, we can develop our full-adder circuit using these two modules by instantiating them into the full adder. In SystemC, instantiating a module inside another module is performed by defining a member data of the instantiated module inside the module definition of the instantiating module. As shown in Figure 89.12, there are several options for instantiating a module inside another one. For example, in this figure, *xor_inst* and *maj_inst* are instances of *xor3* and *maj3* modules, respectively. *xor_inst* is defined as a reference to *xor3* module of Figure 89.11, while *maj_inst* is defined as a pointer to *maj3* module of Figure 89.10. The initialization of these two instances must be inside the *add_1bit* constructor, as shown in Figure 89.12.

After initializing the instances discussed above, their ports must be bound to proper ports or signals in the *add_1bit* module. The port binding of instantiated modules are done inside the *add_1bit* constructor. There are two ways for port binding in SystemC that are *instantiation by name* and *instantiation by position*. In Figure 89.12, ports *a*, *b*, *c*, and *y* of *xor_inst* are bound to *a*, *b*, *cin*, and *s* of *add_1bit*, respectively. These ports are bound in the same order they are defined in the *xor3* module. Also in this figure, ports *a*, *b*, *c*, and *y* of *maj_inst* are bound to *a*, *b*, *cin*, and *co* of *add_1bit*, respectively. These ports are bound *by name*.

As shown in Figure 89.12, by instantiating modules *xor3* and *maj3* inside the *add_1bit* constructor, a full adder is developed. Two processes inside *xor3* and *maj3* perform the function of full adder.

```
SC_MODULE (add_1bit) {
  sc_in <bool> a, b, ci;
  sc_out <bool> s, co;

  xor3 xor_inst;
  maj3 *maj_inst;

  SC_CTOR (add_1bit):
    xor_inst ("xor4fulladd") //xor_inst initialization
  {
    xor_inst (a, b, ci, s); //instantiation by position

    maj_inst = new maj3 ("maj4fulladd"); //memory allocation
    //instantiation by name
    maj_inst.y (co);
    maj_inst.a (a);
    maj_inst.b (b);
    maj_inst.c (ci);
  }
};
```

FIGURE 89.12 SystemC code for a full adder by instantiation.

```
SC_MODULE (add_1bit) {
  sc_in <bool> a, b, ci;
  sc_out <bool> s, co;

  SC_CTOR (add_1bit) {
    SC_METHOD (add_func);
    sensitive << a << b << ci;
  }
  void add_func ();
};
```

```
void add_1bit::add_func()
{
  bool la, lb, lci;

  la = a.read();
  lb = b.read();
  lci = ci.read();

  s = la ^ lb ^ lci;
  co = (la & lb) | (la & lci) | (lb & lci);
}
```

FIGURE 89.13 SystemC code for a full adder by expression.

We can also develop a full adder by directly describing it with two expressions. One expression for the XOR function and the other for majority function for the sum and carry-out outputs, respectively. This code is shown in Figure 89.13.

89.2.2.4 Adder Example

As another example for developing modules using expressions, we discuss a 4-bit adder. In this adder, shown in Figure 89.14, we use **sc_lv** data type and its methods for adding two 4-bit vectors. This adder

```
SC_MODULE (add_4bit) {
    sc_in <sc_lv <4> > a, b;
    sc_out <sc_lv <4> > s;
    sc_in <sc_logic> ci;
    sc_out <sc_logic> co;

    SC_CTOR (add_4bit) {
        SC_METHOD (add_func);
        sensitive << a << b << ci;
    }
    void add_func ();
};
```

```
void add_4bit::add_func ()
{
    sc_lv <5> tmp_result;
    int tmp_result_i;

    tmp_result_i = a.read().to_uint() + b.read().to_uint() + ci.read().value(); //line 6
    tmp_result = (sc_lv <5>) tmp_result_i;

    s = tmp_result.range (3, 0);
    co = tmp_result[4];
}
```

FIGURE 89.14   Four-bit adder with carry-in and carry-out.

has one bit carry-in (*cin* in Figure 89.14) and one bit carry-out (*co* in Figure 89.14). In this figure, *add_func* is a method that is sensitive to all of *add_4bit*'s input signals. In *line 6* of this function, the vectored inputs are converted to unsigned integers (using **to_uint** method) and added by the standard + operation in C++. Also in this line *ci*, which is an **sc_logic** type, is converted to **sc_logic_value_t** enumeration type that can be between **0** and **3** using **value** method and added to *a* and *b* vectors. *ci.value()* is converted to **0** when *ci* is **SC_LOGIC_0**, 1 when *ci* is **SC_LOGIC_1**, 2 when *ci* is **SC_LOGIC_Z**, and 3 when *ci* is **SC_LOGIC_X**. The total result is stored in an integer (*tmp_result_i*) that is converted to a 5-bit **sc_lv** type variable (*tmp_result*) using type casting. The leftmost bit of *tmp_result* is assigned to *co* and its first four bits are assigned to *s* using **range** function.

89.2.2.5   Multiplexer Example

Another example of a combinational circuit is a simple multiplexer that is shown in Figure 89.15. The *mux2_1* multiplexer uses a C style **if**-statement to implement a multiplexer functionality.

89.2.2.6   ALU Example

The **if**-statement, used in the previous example, is easy to use, descriptive, and expandable. However, when many choices exist, a **case**-statement that is more structured may be a better choice. The ALU description of Figure 89.16 uses a **case**-statement to describe an ALU with add, subtract, AND, and XOR functions.

The ALU has *a* and *b* data inputs and a 2-bit *f* input that selects its function. The SystemC code shown in Figure 89.16 uses *a*, *b*, and *f* on its sensitivity list for the *alu_func* method. The **case**-statement shown in the *alu_func* method uses *f* to select one of the case alternatives. The last alternative is the **default** alternative that is taken when *f* does not match any of the alternatives that appear before it. This is necessary to make sure that unspecified input values (here, those that contain **X** and/or **Z**) cause the assignment of the default value to the output and not leave it unspecified.

89.2.3   Sequential Circuits

As any digital circuit, a sequential circuit can be described in SystemC by use of gates, Boolean expressions, or behavioral constructs. While gate-level descriptions enable a more detailed description of timing and

```
SC_MODULE (mux2_1) {
    sc_in <sc_lv <4> > i0, i1;
    sc_in <bool> s;
    sc_out <sc_lv <4> > y;

    SC_CTOR (mux2_1) {
        SC_METHOD (mux_func);
        sensitive << i0 << i1 << s;
    }
    void mux_func ();
};
```

```
void mux2_1::mux_func ()
{
    if (s.read()) //s is true
        y = i1.read();
    else
        y = i0.read();
}
```

FIGURE 89.15 A multiplexer in SystemC.

```
SC_MODULE (ALU) {
    sc_in <sc_lv <4> > a, b;
    sc_in <sc_lv <2> > f;
    sc_out <sc_lv <4> > y;

    SC_CTOR (ALU) {
        SC_METHOD (alu_func);
        sensitive << a << b << f;
    }
    void alu_func ();
};
```

```
void ALU::alu_func ()
{
    switch (f.read().to_uint()) {
        case 0:   // add operation
                  y = (sc_lv <4>)(a.read().to_uint() + b.read().to_uint());
                  break;
        case 1:   // subtract operation
                  y = (sc_lv <4>)(a.read().to_uint() - b.read().to_uint());
                  break;
        case 2:   // and operation
                  y = a.read() & b.read();
                  break;
        case 3:   // xor operation
                  y = a.read() ^ b.read();
                  break;
        default: y = (sc_lv <4>) "0000";
    }
}
```

FIGURE 89.16 ALU SystemC code using case-statement.

delays, because of complexity of clocking and register and flip-flop controls, these circuits are usually described by the use of expressions inside the module processes. This section shows behavioral way of describing sequential circuits in SystemC. The following discusses primitive structures like latch and flip-flops, and then generalizes coding styles used for representing these structures to more complex sequential circuits including counters and state machines.

89.2.3.1 Basic Memory Elements

A clocked D-latch latches its input data during an active clock cycle. The latch structure retains the latched value until the next active clock cycle. This element is the basis of all static memory elements.

Although latches and flip-flops can be described by primitive gates and expressions, such descriptions are hard to generalize, and describing more complex register structures cannot be done this way. This section uses behavioral ways to describe latches and flip-flops. We will show that the same coding styles used for these simple memory elements can be generalized to describe memories with complex control as well as functional register structures like counters and shift-registers.

Latches. Figure 89.17 shows a D-latch described by a thread in SystemC. Latch clock and data inputs (c and d) appear in the sensitivity list of the *latch_func* thread, making this process sensitive to c and d. The reason for using a thread instead of a method in this latch is the delays used in the body of *latch_func*. Since we are modeling the delays between circuit inputs and its output, we have to use **wait** functions inside the process. As methods do not support **wait** functions, the use of threads is mandatory here. If this module had no delays, we could use methods as well as threads.

The **if**-statement enclosed in *latch_func* puts d into q when c is active. This means that if c is **1** and d changes, the change on d propagates to the q output after 4 ns. This behavior is referred to as transparency, which is how latches work. While clock is active, a latch structure is transparent, and input changes affect its output.

Any time *latch_func* is active, if c is **1**, it waits 4 ns and then puts d into q. It then waits another 3 ns and then puts the complement of d into q_b. This makes the delay of the q_b output 7 ns.

```
SC_MODULE (latch) {
    sc_in <bool> d, c;
    sc_out <bool> q, q_b;

    SC_CTOR (latch) {
        SC_THREAD (latch_func);
        sensitive << c << d;
    }
    void latch_func ();
};
```

```
void latch::latch_func ()
{
    while (1) {
        if (c.read()) {
            wait (4, SC_NS);
            q = d.read();
            wait (3, SC_NS);
            q_b = !d.read();
        }
        wait (); // waits for the next event on c or d
    }
}
```

FIGURE 89.17 A behavioral latch.

D Flip-Flop. While a latch is transparent, a change on the D-input of a D flip-flop does not directly pass on to its output. The SystemC code of Figure 89.18 describes a positive-edge trigger D-type flip-flop.

The sensitivity list of the *dff_func* thread, shown in Figure 89.18, includes only the positive edge of *clk* (using **pos** function). This process is activated when *clk* makes a **0** to **1** transition. When this process is active, the value of *d* is put into *q* after 4 ns and the inverted value of *d* into *q_b* after 7 ns.

Instead of **pos** function used in Figure 89.18, use of **neg** function would implement a falling-edge D flip-flop. After the specified edge, the flow into the *dff_func* thread begins. In our description, this flow is halted in 4 ns by the *wait (4, SC_NS)* statement. After this delay, the value of *d* is read and put into *q*. Following this transaction, the flow into the *dff_func* thread is again halted by 3 ns, after which *!d* is put into *q_b*. This makes the delay of *q* after the edge of the clock equal to 4 ns. The delay for *q_b* becomes the accumulation of the delay values shown, and it is 7 ns. Delay values are ignored in synthesis.

Synchronous control. The coding style presented for the above simple D flip-flop is a general one and can be expanded to cover many features found in flip-flops and even memory structures. The description shown in Figure 89.19 is a D-type flip-flop with synchronous set and reset (*s* and *r*) inputs.

The description uses a **SC_THREAD** that is sensitive to the positive-edge of *clk*. When *clk* makes a **0** to **1** transition, the flow into the *dff_func* thread begins. Immediately after the positive-edge, *s* is inspected and if it is active (**1** or **true**), after 4 ns *q* is set to **1** and 3 ns after that *q_b* is set to **0**. Following the positive edge of *clk*, if *s* is not **1**, *r* is inspected and if it is active, *q* is set to **0**. If neither *s* nor *r* is **1**, the flow of the program reaches the last **else** part of the **if**-statement and assigns *d* to *q*.

The behavior discussed here only looks at *s* and *r* on the positive-edge of *clk*, which corresponds to a rising-edge trigger D-type flip-flop with synchronous active high set and reset inputs. Furthermore, the set input is given a higher priority over the reset input. The flip-flop structure that corresponds to this description is shown in Figure 89.20.

Other synchronous control inputs can be added to this flip-flop in a similar fashion. A clock enable (*en*) input would only require inclusion of an **if**-statement in the last **else** part of the **if**-statement in the code of Figure 89.19.

Asynchronous control. The control inputs of the flip-flop of Figure 89.19 are synchronous because the flow into the thread is only allowed to start when the positive edge of *clk* is observed. To change this to

```
SC_MODULE (d_ff) {
    sc_in <bool> d, clk;
    sc_out <bool> q, q_b;

    SC_CTOR (d_ff) {
        SC_THREAD (dff_func);
        sensitive << clk.pos();
    }
    void dff_func ();
};
```

```
void d_ff::dff_func ()
{
    while (1) {
        wait (4, SC_NS);
        q = d.read();
        wait (3, SC_NS);
        q_b = !d.read();
        wait (); // suspends the process until next positive edge of clk
    }
}
```

FIGURE 89.18 A positive-edge D flip-flop.

```
SC_MODULE (d_ff) {
    sc_in <bool> d, clk, s, r;
    sc_out <bool> q, q_b;

    SC_CTOR (d_ff) {
        SC_THREAD (dff_func);
        sensitive << clk.pos();
    }
    void dff_func ();
};
```

```
void d_ff::dff_func ()
{
    while (1) {
        if (s.read()) {
            wait (4, SC_NS);
            q = true;
            wait (3, SC_NS);
            q_b = false;
        } else if (r.read()) {
            wait (4, SC_NS);
            q = false;
            wait (3, SC_NS);
            q_b = true;
        } else {
            wait (4, SC_NS);
            q = d.read();
            wait (3, SC_NS);
            q_b = !d.read();
        }
        wait (); // suspends the process until next positive edge of clk
    }
}
```

FIGURE 89.19   D flip-flop with synchronous control.

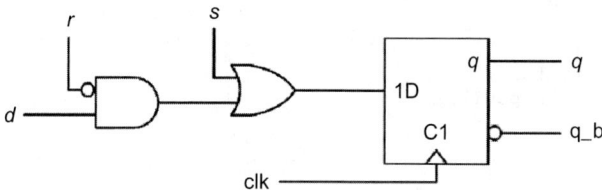

FIGURE 89.20   D flip-flop representation with synchronous control.

a flip-flop with asynchronous control, it is only required to include asynchronous control inputs in the sensitivity list of its process registration.

Figure 89.21 shows a D flip-flop with active high asynchronous set and reset control inputs. Note that the only difference between this description and the code of Figure 89.18 (synchronous control) is the inclusion of *s.pos()* and *r.pos()* in the sensitivity list of the *dff_func* thread. This inclusion allows the flow into the thread to begin when *clk* becomes **1** or *s* becomes **1** or *r* becomes **1**. The **if**-statement in this block checks for *s* and *r* being **1**, and if none is active (activity levels are high), then clocking *d* into *q* occurs. The graphic symbol corresponding to the flip-flop of Figure 89.21 is shown in Figure 89.22.

```
SC_MODULE (d_ff) {
   sc_in <bool> d, clk, s, r;
   sc_out <bool> q, q_b;

   SC_CTOR (d_ff) {
      SC_THREAD (dff_func);
      sensitive << clk.pos() << s.pos() << r.pos();
   }
   void dff_func ();
};
```

```
void d_ff::dff_func ()
{
   while (1) {
      if (s.read()) {
         wait (4, SC_NS);
         q = true;
         wait (3, SC_NS);
         q_b = false;
      } else if (r.read()) {
         wait (4, SC_NS);
         q = false;
         wait (3, SC_NS);
         q_b = true;
      } else {
         wait (4, SC_NS);
         q = d.read();
         wait (3, SC_NS);
         q_b = !d.read();
      }
      wait ();
   }
}
```

FIGURE 89.21   D flip-flop with asynchronous control.

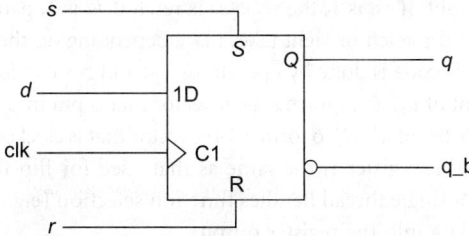

FIGURE 89.22   D flip-flop representation with asynchronous control.

89.2.3.2   Registers, Shifters, and Counters

Registers, shifter-registers, counters, and even sequential circuits with more complex functionalities can be described by simple extensions of the coding styles presented for the flip-flops. In most cases, the functionality of the circuit only affects the last **else** of the if-statement in processes of codes shown for the flip-flops.

Registers. Figure 89.23 shows an 8-bit register with synchronous set and reset inputs. The *set* input puts all 1s in the register and the *reset* input resets it to all 0s. The main difference between this and the flip-flop with synchronous control is the vector declaration of inputs and outputs.

```
SC_MODULE (reg) {
    sc_in <sc_lv <8> > d;
    sc_in <bool> clk, set, reset;
    sc_out <sc_lv <8> > q;

    SC_CTOR (reg) {
        SC_THREAD (reg_func);
        sensitive << clk.pos();
    }
    void reg_func ();
};
```

```
void reg::reg_func ()
{
    while (1) {
        if (set.read()) {
            wait (5, SC_NS);
            q = "11111111";
        } else if (reset.read()) {
            wait (5, SC_NS);
            q = "00000000";
        } else {
            wait (5, SC_NS);
            q = d.read();
        }
        wait ();
    }
}
```

FIGURE 89.23   An 8-bit register.

Shift-registers. A 4-bit shift-register with right- and left-shift capabilities, a serial-input, synchronous reset input, and parallel loading capability is shown in Figure 89.24. As shown, only the positive-edge of *clk* is included in the sensitivity list of *shiftreg_func* thread of this code, which makes all activities of the shift-register synchronous with the clock input. If *rst* is **1**, the register is reset, if *ld* is **1**, parallel *d* inputs are loaded into the register, and if none is **1**, shifting left or right takes place depending on the value of the *l_r* input (**1** for left, **0** for right). Shifting in this code is done by operations << and >> overloaded for **sc_lv**. For left-shift, *s_in* is concatenated to the right of *q[2:0]* to form a 4-bit vector that is put into *q* (*lines 14* and *15*). For right-shift, *s_in* is concatenated to the left of *q[3:1]* to form a 4-bit vector that is clocked into *q[3:0]* (*lines 18* and *19*).

The style used for coding this register is the same as that used for flip-flops and registers presented earlier. In all these examples, a single thread handles function selection (e.g., zeroing, shifting, or parallel loading) as well as clocking data into the register output.

Counters. The style described for the shift-register in the previous discussion can be used for describing counters. A counter counts up or down, while a shift-register shifts right or left. We use arithmetic operations in counting as opposed to shift operation in shift-registers.

Figure 89.25 shows a 4-bit up-down counter with a synchronous *rst* reset input. The counter has an *ld* input for doing the parallel loading of *d_in* into the counter. The counter output is *q*. We use *counter_func* process, which is a clocked thread (**SC_CTHREAD**), to implement the functionality of the counter. This clocked thread is sensitive to the positive edge of *clk*.

Discussions about synchronous and asynchronous control of flip-flops and registers also apply to counters. For example, if *counter_func* process is an **SC_THREAD** instead of **SC_CTHREAD** and *rst.pos()* is added to the sensitivity list of the *counter_func* of Figure 89.25, the counter resetting would be asynchronous.

```
SC_MODULE (shift_reg) {
    sc_in <sc_lv <4> > d;
    sc_in <bool> clk, ld, rst, l_r;
    sc_in <sc_logic> s_in;
    sc_out <sc_lv <4> > q;

    SC_CTOR (shift_reg) {
        SC_THREAD (shiftreg_func);
        sensitive << clk.pos();
    }
    void shiftreg_func ();
};
```

```
void shift_reg::shiftreg_func ()
{
    sc_lv <4> tmp_q;

    while (1) {
        if (rst.read()) {
            wait (5, SC_NS);
            tmp_q = "0000";
        } else if (ld.read()) {
            wait (5, SC_NS);
            tmp_q = d.read();
        } else if (l_r.read()) {
            wait (5, SC_NS);
            tmp_q = tmp_q << 1; //line 14
            tmp_q[0] = s_in.read(); //line 15
        } else {
            wait (5, SC_NS);
            tmp_q = tmp_q >> 1; //line 18
            tmp_q[3] = s_in.read(); //line 19
        }
        q = tmp_q;
        wait ();
    }
}
```

FIGURE 89.24   A 4-bit shift register.

89.2.3.3   State Machine Coding

Coding styles presented thus far can be further generalized to cover finite-state machines of any type. This section shows coding for *Moore* state machines. The example we will use is a simple sequence detector. This circuit represents the controller part of a digital system that has been partitioned into a data path and a controller. The coding styles used here apply to such controllers.

Moore detector. State diagram for a Moore sequence detector detecting **101** on its *x* input is shown in Figure 89.26. The machine has four states that are labeled, *reset, got1, got10,* and *got101*. Starting in *reset*, if the **101** sequence is detected, the machine goes into the *got101* state in which the output becomes **1**. In addition to the *x* input, the machine has an input named *rst* that forces the machine into its *reset* state. The resetting of the machine is synchronized with a clock.

The SystemC code of the Moore machine of Figure 89.26 is shown in Figure 89.27. After the declaration of inputs and outputs of this module, **enum** declaration for *states* declares four states of the machine as an enumeration type. Following **enum** declarations in the code of Figure 89.27, the *current* variable of type *states* is declared. This variable holds the current state of the state machine.

The *moore_func* thread used in the module of Figure 89.27 describes state transitions and output assignments of the state diagram of Figure 89.26. The main task of this thread is to inspect input

```
SC_MODULE (counter) {
    sc_in <sc_lv <4> > d_in;
    sc_in <bool> clk, rst, ld, u_d;
    sc_out <sc_lv <4> > q;

    SC_CTOR (counter) {
        SC_CTHREAD (counter_func, clk.pos ());
    }
    void counter_func ();
};
```

```
void counter::counter_func ()
{
    sc_lv <4> tmp_q;

    while (1) {
        if (rst.read())
            tmp_q = "0000";
        else if (ld.read())
            tmp_q = d_in.read();
        else if (u_d.read())
            tmp_q = (sc_lv <4>) (tmp_q.to_uint () + 1);
        else
            tmp_q = (sc_lv <4>) (tmp_q.to_uint () - 1);
        q = tmp_q;
        wait ();//suspends the process until next positive edge of clk
    }
}
```

FIGURE 89.25 An up-down counter in SystemC.

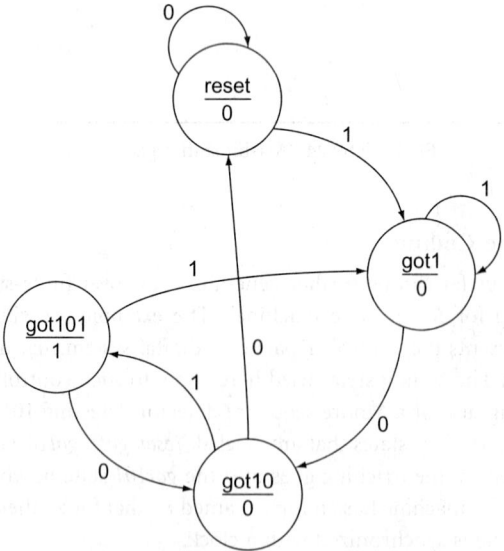

FIGURE 89.26 A Moore sequence detector.

conditions (values on *rst* and *x*) during the present state of the machine defined by *current*, and set values into *current* for the next state of the machine.

The flow into the *moore_func* thread begins with the positive edge of *clk*. Since all activities in this machine are synchronized with the clock, only *clk* appears on the sensitivity list of the *moore_func* thread.

```
SC_MODULE (moore_detector) {
    sc_in <sc_logic> x;
    sc_in <bool> rst, clk;
    sc_out <sc_logic> z;
    enum states {reset, got1, got10, got101};

    SC_CTOR (moore_detector) {
        SC_THREAD (moore_func);
        sensitive << clk.pos();
    }
    void moore_func ();
};
```

```
void moore_detector::moore_func ()
{
    states current;
    while (1) {
        if (rst.read())
            current = reset;
        else switch (current) {
            case (reset): if (x.read() == '1') current = got1;
                          else current = reset; break;
            case (got1): if (x.read() == '0') current = got10;
                         else current = got1; break;
            case (got10): if (x.read() == '1') current = got101;
                          else current = reset; break;
            case (got101): if (x.read() == '1') current = got1;
                           else current = got10; break;
            default: current = reset;
        }
        if (current == got101) z = SC_LOGIC_1;
        else z = SC_LOGIC_0;
        wait ();
    }
}
```

FIGURE 89.27 Moore machine SystemC code.

Upon entry into this block, the *rst* input is checked and if it is active, *current* is set to *reset* (*reset* is declared in the *states* enumeration and its value is **0**). The value put into *current* in this pass gets checked in the next pass with the next edge of the clock. Therefore, this assignment is regarded as the next-state assignment. When this assignment is made, the **if-else** statements skip the rest of the code of the *moore_func* thread, and this thread will next be entered with the next positive edge of *clk*.

Upon entry into the *moore_func* thread, if *rst* is not **1**, program flow reaches the **case**-statement that checks the value of *current* against the four states of the machine.

In this coding style, for every state of the machine there is a **case**-alternative that specifies the next state values. For larger machines, there will be more **case**-alternatives, and more conditions within an alternative. Otherwise, this style can be applied to state machines of any size and complexity. This same machine can be described in SystemC in many other ways. For example, we can use nested **if-else** statements instead of using **case**-statement.

Huffman coding style. The Huffman model for a digital system characterizes it as a combinational block with feedbacks through an array of flip-flops or a register. According to the Huffman model, SystemC coding of digital systems uses two concurrent processes; one for describing the register part and another for describing the combinational part.

We will describe the state machine of Figure 89.26 to illustrate this style of coding. Figure 89.28 shows the combinational and register partitioning that we will use for describing this machine. The *Combinational*

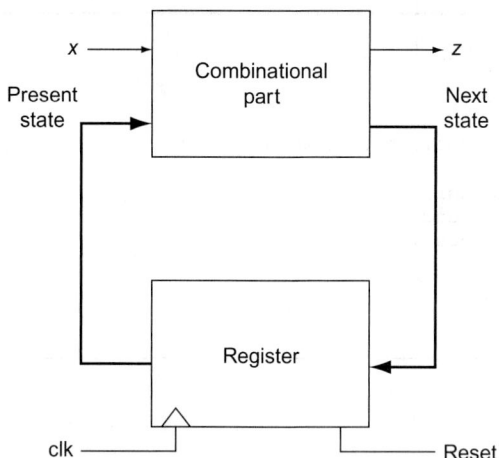

FIGURE 89.28 Huffman partitioning of a sequential circuit.

Part block uses *x* and *Present state* as input and generates *z* and *Next state*. The *Register* block clocks *Next state* into *Present state*, and resets *Present state* when *reset* is active.

Figure 89.29 shows the SystemC code of Figure 89.26 according to the partitioning of Figure 89.28. As shown, *n_state* and *p_state* variables are declared as variables of **enum** type *states*. These variables hold values corresponding to the states of the **101** Moore detector. The *comb_func* method is sensitive to *x* and *p_state*. In this method, *n_state* and *z* are set to their inactive or reset values. This is done so that these variables are always reset with the clock to make sure that they do not retain their old values. As discussed before, retaining old values implies latches, which is not what we want in our combinational block.

The body of the combinational method of Figure 89.29 contains a **case**-statement that uses the *p_state* input for its **case**-expression. This expression is checked against the states of the Moore machine. As in the other styles discussed before, this **case**-statement has **case**-alternatives for *reset, got1, got10,* and *got101* states.

In a block corresponding to a **case**-alternative, based on input values, *n_state* and *z* output are assigned values. Unlike the other styles where *current* is used both for the present and next states, here we use two different variables, *p_state* and *n_state*.

The next concurrent process (*seq_func* thread), shown in Figure 89.29, handles the *Register* part of the Huffman model of Figure 89.28. In this part, *n_state* is treated as the register input and *p_state* as its output. On the positive edge of the clock, *p_state* is either set to the *reset* state or is loaded with contents of *n_state*. Together, *comb_func* and *seq_func* processes describe our state machine in a modular fashion.

The advantage of this style of coding is in its modularity and well-defined tasks of each block. State transitions are handled by the *comb_func* method and clocking is done by the *seq_func* thread. Changes in clocking, resetting, enabling, or presetting the machine only affect the coding of the *seq_func* process. If we were to change the synchronous resetting to asynchronous, the only change we had to make was adding *rst.pos()* to the sensitivity list of the *seq_func* thread registration.

A more modular style. For a design with more input and output lines and more complex output logic, the *Combinational Part* may further be partitioned into a process for handling transitions and another for assigning values to the outputs of the circuit.

Figure 89.30 shows the coding of the **101** mealy detector using two separate processes for assigning values to *n_state* and the *z* output. In a situation like what we have in which the output logic is fairly simple, a simple assignment statement, like what we had in Figure 89.29, could replace the *outp_func* method.

```
SC_MODULE (moore_detector) {
   sc_in <sc_logic> x;
   sc_in <bool> rst, clk;
   sc_out <sc_logic> z;

   enum states {reset, got1, got10, got101};
   sc_signal <states> p_state, n_state;

   SC_CTOR (moore_detector) {
      SC_METHOD (comb_func);
      sensitive << p_state << x;
      SC_THREAD (seq_func);
      sensitive << clk.pos ();
   }
   void comb_func ();
   void seq_func ();
};
```

```
void moore_detector::comb_func ()
{
   n_state = reset;
   switch (p_state) {
      case (reset) :
            if (x.read() == '1') n_state = got1;
            break;
      case (got1) :
            if (x.read() == '0') n_state = got10;
            else n_state = got1;
            break;
      case (got10) :
            if (x.read() == '1') n_state = got1;
            else n_state = got10;
            break;
      case (got101) :
            if (x.read() == '1') n_state = got1;
            else n_state = got10;
            break;
   }
   if (p_state == got101) z = SC_LOGIC_1;
   else z = SC_LOGIC_0;
}
void moore_detector::seq_func ()
{
   while (1) {
      if (rst.read()) p_state = reset;
      else p_state = n_state;

      wait ();
   }
}
```

FIGURE 89.29 SystemC Huffman coding style.

The examples discussed above, and in particular the last two styles, show how combinational and sequential coding styles can be combined to describe very complex digital systems.

89.2.3.4 Memories

SystemC allows description and use of memories. Memories are variables that are declared as a two-dimensional array of a valid type in C or SystemC. These data types are usually **bool** and **sc_logic** types.

```
SC_MODULE (mealy_detector) {
    sc_in <sc_logic> x;
    sc_in <bool> en, clk, rst;
    sc_out <sc_logic> z;
    enum states {reset, got1, got10, got11};
    sc_signal <states> p_state, n_state;

    SC_CTOR (mealy_detector) {
        SC_METHOD (trans_func);
        sensitive << p_state << x;
        SC_METHOD (outp_func);
        sensitive << p_state << x;
        SC_THREAD (seq_func);
        sensitive << clk.pos ();
    }
    void trans_func ();
    void outp_func ();
    void seq_func ();
};
```

```
void mealy_detector::trans_func ()
{
   n_state = reset;
   switch (p_state) {
      case (reset): if (x.read() == '1') n_state = got1;
                    else n_state = reset; break;
      case (got1): if (x.read() == '0') n_state = got10;
                   else n_state = got11; break;
      case (got10): if (x.read() == '1') n_state = got1;
                    else n_state = reset; break;
      case (got11): if (x.read() == '1' ) n_state = got11;
                    else n_state = got10; break;
   }
}
void mealy_detector::outp_func ()
{
   z = '0';
   switch (p_state) {
      case (reset): z = '0'; break;
      case (got1):  z = '0'; break;
      case (got10): if (x.read() == '1') z = '1';
                    else z = '0'; break;
      case (got11): if (x.read() == '1') z = '0';
                    else z = '1'; break;
   }
}
void mealy_detector::seq_func ()
{
   while (1) {
      if (rst.read()) p_state = reset;
      else if (en.read()) p_state = n_state;
      wait ();
   }
}
```

FIGURE 89.30 Separate transition and output processes.

Also one-dimensional arrays of **sc_lv** type can be interpreted as memory. Figure 89.31 shows two different ways of memory declaration. This figure also shows several valid memory operations.

Figure 89.32 shows a memory block with separate input and output busses. Writing into the memory is clocked, while reading from it only requires *rw* to be **1**. The *read_func* method handles reading and the *write_func* thread performs writing into this memory.

```
sc_lv <8> mem [1024];
mem2 [1024][8];
sc_lv <8> data;
sc_lv <4> short_data;
sc_lv <10> addr;
...
data = mem2 [956];
...
short_data = data.range(7, 4);
...
mem [345] = data;
mem [addr] = "00100111";
```

FIGURE 89.31 Memory representation.

```
SC_MODULE (memory) {
    sc_in <sc_lv <8> > inbus;
    sc_in <sc_lv <10> > addr;
    sc_out <sc_lv <8> > outbus;
    sc_in <bool> clk, rw;

    SC_CTOR (memory) {
        SC_METHOD (read_func);
        sensitive << rw;
        SC_THREAD (write_func);
        sensitive << clk.pos ();
    }

    sc_lv <8> mem [1024];
    void read_func ();
    void write_func ();
};
```

```
void memory::read_func ()
{
    outbus = rw.read() ? mem [addr.read().to_uint()] : "ZZZZZZZZ";
}
void memory::write_func ()
{
    while (1) {
        if (!rw.read()) mem [addr.read().to_uint()] = inbus.read();
    }
}
```

FIGURE 89.32 Memory description and usage in SystemC.

89.2.4 Writing Testbenches

SystemC coding styles discussed so far were for coding hardware structures, and in all cases synthesizability and direct correspondence to hardware were our main concerns. In contrast, testbenches do not have to have hardware correspondence and they usually do not follow any synthesizability rules. We will see that delay specifications and initialization statements that do not have a one-to-one hardware correspondence are generously used in testbenches.

For demonstration of testbench coding styles, we use the SystemC code of Figure 89.33 that is a **101** Moore detector and generate a testbench for this circuit.

```
SC_MODULE (moore_detector) {
    sc_in <bool> x, rst, clk;
    sc_out <sc_logic> z;

    enum states {a, b, c, d};

    SC_CTOR (moore_detector) {
        SC_THREAD (moore_func);
    }
    void moore_detector::moore_func ();
};
```

```
void moore_detector::moore_func ()
{
    states current;

    while (1) {
        if (rst.read()) current = a;
        else switch (current) {
            case (a): current = x.read() ? b : a; break;
            case (b): current = x.read() ? b : c; break;
            case (c): current = x.read() ? d : a; break;
            case (d): current = x.read() ? b : c; break;
            default: current = a;
        }
        z = (current == d) ? SC_LOGIC_1 : SC_LOGIC_0;
        wait (clk.posedge_event ());
    }
}
```

FIGURE 89.33 Circuit under test: a 101 Moore detector.

This description is functionally equivalent to that of Figure 89.27. The difference is in the use of condition expressions (**?:**) instead of **if-else** statements. This code will be instantiated in the testbenches that follow.

89.2.4.1 Generating Periodic Data

Figure 89.34 shows a testbench module that instantiates *moore_detector* of Figure 89.33 and applies test data to its inputs. In SystemC, **sc_main** plays the role of the testbench for a design. There are several functions in SystemC that are useful for developing testbenches. The examples of these functions include **sc_time**, **sc_clock**, **sc_trace**, and **sc_simulation_time**, some of which we will use in our testbench examples.

One of the utility functions in SystemC is **sc_clock** that is used for generating periodic data on a specified signal. In Figure 89.34, signals *x*, *reset*, and *clock* are defined as periodic data. *x* is a periodic signal with 14 ns period and duty cycle of 0.5, which is the default value for duty cycle in **sc_clock**. The period of this signal and other periodic signals are defined by variables of kind **sc_time** (*line 5–7* in Figure 89.34). The period of the clock signal is 10 ns and its first transition is a **false** to **true** transition, which is the default value for transitions in **sc_clock**. In the testbench of Figure 89.34, *reset* is also defined as a periodic data. The period of this signal is 1000 ns, its duty cycle is 0.024, it starts from the beginning of the simulation (at **SC_ZERO_TIME**), and its first transition is the negative edge transition. These arguments cause the *reset* signal to change from **true** to **false** in the 24th ns of the simulation time and remains **false** until the simulation time reaches 1000 ns.

The reason that we define such a long period for *reset* signal is that we want to reset the *cut* only once in this example. As shown in Figure 89.34, the simulation time, defined as the argument of **sc_start**, is specified as 1000 ns. This simulation time causes the *reset* signal to change only once during the simulation time.

```
sc_main (int argc, char **argv)
{
    sc_signal <sc_logic> z;

    sc_time clk_period (10, SC_NS); //line 5
    sc_time x_period (14, SC_NS);
    sc_time rst_period (1000, SC_NS); //line 7

    sc_clock clock ("clock", clk_period);
    sc_clock x ("x", x_period);
    sc_clock reset ("reset", rst_period, 0.024, SC_ZERO_TIME, false);

    moore_detector cut ("circuit_under_test");
    cut (x, reset, clock, z);

    sc_start (1000, SC_NS);

    return 0;
}
```

FIGURE 89.34 Generating periodic data for Moore detector.

All data inputs to the *cut* are locally generated in this testbench. We are using different periods for *clock* and *x*, so that a combination of patterns on these circuit inputs is seen. A more deterministic set of values could be set by specifying exact values at specific times.

89.2.4.2 Timed Data

A simple testbench for our sequence detector can be done by applying test data to *x* and timing them appropriately to generate the sequence similar to values we wanted to observe on the *reset* signal in the previous example. Figure 89.35 shows this testbench.

Techniques discussed in the above example are just some of what one can do for test data generation. These techniques can be combined for more complex examples. After using SystemC for some time, users form their own test generation techniques. For small designs, simulation environments generally provide waveform editors and other tool-dependent test generation schemes. Some tools come with code fragments that can be used as templates for testbenches.

The testbench shown in Figure 89.35 uses a helper module (*apply_data*) to apply input data to *cut* at specified times. The *apply_data* module accepts a file name as one of its constructor arguments. This filename corresponds to a text file that contains values a signal should receive in specified times. An example of this file is shown in Figure 89.36. The first element in each line specifies the value of a signal and the second element specifies the relative timing of that value (relative to the previous time) that is to be applied to the signal. The *apply_data* module has an output port named *sig* that is bound to the signal to which the values in the text file should be applied. The *apply_data* module is shown in Figure 89.37. This module has a thread named *apply_func* and a helper function, which reads and analyzes each line of the specified data file, named *read_line*. The *apply_func* thread reads each line of the data file, waits for a proper time, and applies the proper value to its output *sig*.

With this helper module, the testbench in Figure 89.35 can easily be developed. The first part of this testbench (*lines 5* and 6) makes a clock signal with 10 ns period. The next part (*lines 8–12*) uses the *apply_data* module to read *reset.txt* and *x.txt* files and apply appropriate waveforms to *reset* and *x* signals, respectively.

Lines 14 and 15 of this testbench instantiates the circuit under test (*moore_detector* in this example).

Lines 18–20 of this testbench uses the predefined SystemC functions and data types to generate an output file for the result of *cut* instance (*z* in this example). To generate this output file **sc_trace_file**, **sc_create_vcd_trace_file, sc_trace**, and **sc_close_vcd_trace_file** data type and functions are utilized. The *zfile.vcd* output file is a file in VCD format that includes the values of the *z* signal for the first 100 ns of simulation.

```
sc_main (int argc, char **argv)
{
    sc_signal <bool> x, reset;
    sc_signal <sc_logic> z;

    sc_time clk_period (10, SC_NS); // line 5
    sc_clock clock ("clock", clk_period); // line 6

    apply_data apply_reset ("apply_reset", "reset.txt"); // line 8
    apply_reset.sig (reset);

    apply_data apply_x ("apply_x", "x.txt");
    apply_x.sig (x); // line 12

    moore_detector cut ("circuit_under_test"); // line 14
    cut (x, reset, clock, z); // line 15

    // creating an output file which stores z values in VCD format
    sc_trace_file *outp_file; // line 18
    outp_file = sc_create_vcd_trace_file ("zfile");
    sc_trace (outp_file, z, "z"); // line 20

    sc_start (100, SC_NS);

    sc_close_vcd_trace_file (outp_file);

    return 0;
}
```

FIGURE 89.35 Timed test data generation.

```
0  0
1  7
0  5
1  18
0  21
1  11
0  13

1  0
0  24
```

FIGURE 89.36 Text file for signal waveforms of *x* and *reset* signals of Figure 89.35.

89.3   Synthesizable Subset for SystemC

All examples (except the testbenches) presented in the previous section have direct hardware correspondence and therefore they can be manually synthesized to actual hardware. Since an RTL model is usually used as an input of a synthesis tool, there are constraints that must be followed when designing at this level. Therefore, the term "nonsynthesizable design" does not imply that there is no way to convert that design to a corresponding hardware. Rather, it denotes that the design conversion is not supported by most of the existing tools, or it is very complex to be performed automatically.

In May 2001, Synopsys released a subset for synthesizable SystemC. Many CAD tools dealing with RTL SystemC rely on this subset. However, with minor revisions, a number of its constructs that are

```
SC_MODULE (apply_data) {
   sc_out <bool> sig;

   apply_data (sc_module_name name, char *filename);
   SC_HAS_PROCESS (apply_data);

   FILE *fp;
   void apply_func ();
   void read_line (bool *sig_val, int *sig_time);
};
```

```
apply_data::apply_data (sc_module_name name, char *filename)
{
   fp = fopen (filename, "r");

   SC_THREAD (apply_func);
}
void apply_data::apply_func ()
{
   bool sig_val;
   int sig_time;

   while (!feof(fp)) {
      read_line (&sig_val, &sig_time);
      if (sig_time > 0) // to avoid delta time waits
         wait (sig_time, SC_NS);
      sig = sig_val;
   }
   fclose(fp);
   return; // stops the thread forever
}
```

FIGURE 89.37   Helper module for generating waveforms.

defined as nonsynthesizable can be used in special cases for automatic synthesis. In the following section, a list of constructs that can or cannot be used in a synthesizable design is shown. For more details, the reader can refer to Ref. [3].

89.3.1   Nonsynthesizable SystemC Constructs

The following constructs in SystemC are not supported for RTL synthesis:

- **SC_THREAD** and **SC_CTHREAD**: **SC_THREAD** and **SC_CTHREAD** are not synthesizable since they can contain **wait** functions. In the previous section, most of the threads and clocked threads used in examples can be easily converted to a method (**SC_METHOD**), which is synthesizable.
- **sc_main**: it can be used only as a testbench, not a part of the design.
- sc_start
- Global and local watchings
- Tracing: creating waveforms and trace files only can be performed in testbenches.

In addition, there are several C/C++ constructs that cannot be used in a synthesizable RTL design in SystemC. The major C/C++ constructs that are not supported are:

- Local and nested class declarations
- Derived classes: the only supported classes are SystemC modules and processes
- Dynamic memory allocation
- Exception handling

- Recursive functions
- Function and operator overloading: the only exception is the classes overloaded by SystemC
- **virtual** functions
- Inheritance and multiple-inheritance
- **public**, **protected**, **private**, and **friend** keywords are ignored during synthesis process. All the members of a class become **public**
- -> operator: the only exception is in module instantiation
- Static members
- * and **&** operators for dereferencing
- Commas in **for**-loops
- Pointers: the pointers are only allowed in the instantiation part of the hierarchical modules
- User-defined **template** classes
- Run-time type castings and type identifications
- **goto** statement
- **unions**
- Global variables
- Member variables shared by two or more processes
- **volatile** variables

89.3.2 Nonsynthesizable Data Types

The following list shows the data types which are not supported for RTL synthesis:

- Floating-point types
- Fixed-point types (e.g., **sc_fix**)
- Access types (e.g., pointers)
- File types and IO streams

Except the above list, other types of SystemC and C/C++ are synthesizable.

89.4   A Complete Example: An RTL Cache

In this section, a direct map cache module, developed in RTL SystemC, will be discussed. The header and implementation of this module are shown in Figure 89.38 and Figure 89.39, respectively.

As shown in Figure 89.38, the sample *Cache* module has a set of signals for connecting to a CPU (e.g., *read*, *write*, and *ready*) and a set of signals for connecting to a memory module (*mem_read*, *mem_write*, and *mem_ready*). The interface between this cache and CPU is byte-by-byte, while its interface with memory is block-by-block, using *blk_address* and *blk_size* ports.

The cache memory is modeled with an array of 128 logic vectors, as shown below. A logic vector in this array stores a 16-byte block of data from memory.

```
sc_lv <16*8> word [128];
```

The module developed here is actually a cache controller that is a state machine with seven states (*s0* to *s6*). This state machine consists of two methods named *cache_op* and *state_change* (as shown in the **SC_CTOR** part of Figure 89.38). The *state_change* method is sensitive to *clk* and *reset* signals. This method changes the *present_state* at the positive edge of *clk*.

The other method (*cache_op*) performs cache operations in each state. It is sensitive to the *read*, *write*, *mem_ready*, and *present_state* signals (Figure 89.38). As shown in Figure 89.39, it has a **case**-statement with all possible values of states. In each state, it performs specific tasks, asserts proper values for its memory and CPU side signals, and defines the value of *next_state*. This variable is used in the *state_change*

```
#include <systemc.h>
#include "sc_util.h"
#define BLOCK_SIZE 16

SC_MODULE (Cache)
{
    sc_in <bool> clk;
    sc_in <bool> reset;
    // interface with cpu
    sc_in <sc_uint <16> > address;
    sc_in <bool> read;
    sc_in <bool> write;
    sc_inout <sc_lv <8> > data;
    sc_out <bool> ready;
    // interface with memory
    sc_out <sc_uint <16> > mem_address;
    sc_out <bool> mem_read;
    sc_out <bool> mem_write;
    sc_out <sc_uint <16> > blk_address;
    sc_out <sc_uint <8> > blk_size;
    sc_in <bool> mem_ready;
    //Module Constructor
    SC_CTOR (Cache) {
        SC_METHOD (cache_op);
        sensitive << read << write << mem_ready << present_state;
        SC_METHOD (state_change);
        sensitive << clk << reset;
    }
private:
    void cache_op ();
    void state_change ();
    enum state {s0, s1, s2, s3, s4, s5, s6};
    sc_signal <state> present_state;
    sc_signal <state> next_state;
    sc_lv <5> tag [128];
    sc_lv <7> block [128];
    bool dirty_bit [128];
};
```

FIGURE 89.38 RTL cache declaration.

method. Predefined functions for logic vectors, like **range** and **to_unit**, are used in this method. The state machine, shown in Figure 89.39, is depicted in Figure 89.40.

As an example, consider state *s2* in cache controller state machine (Figure 89.39 and Figure 89.40). The state of state machine changes to *s2* if the CPU asserts the *write* signal. In this state, *data* and *address* values are read and the corresponding block of the given address is checked. If this block is in the cache memory (*word* variable), then the given data is written into the corresponding address of the block (*cache hit* happens). If a *cache miss* occurs, there are two situations that can happen.

One situation is that another block resides in place of the corresponding block of the given address (dirty bit is **1**). In this case, the current block must be written back to the memory, the corresponding block of the given address must be read from the memory, and the given data must be written to the given address of the newly read block. These operations are performed by going from *s2* to *s6*, *s6* to *s5*, and *s5* to *s0*, respectively.

Another situation is that the corresponding block of the given address is not occupied by any other block (dirty bit is **0**). In this case, this block must be read from memory, and the given data must be written to the given address of the newly read block. These operations are done by going from state *s2* to *s5*, and *s5* to *s0*, respectively.

```
...
void Cache::cache_op ()
{
  ...
  switch (present_state){
    case s0:
        if (read.event () && read.read () == true) {
          next_state = s1;
          ready = false;
        }
        else if (write.event () && write.read () == true) {
          ...
        }
      break;
      ...
    case s2:
      target_block = address.read ().range (10, 4).to_uint ();
      if (tag[target_block] == (sc_lv <5>) address.read().range(15, 11)) {// cache hit
        word[target_block].range (address.read ().range (3, 0).to_uint ()*8 + 7, \
              address.read ().range (3, 0).to_uint ()*8) = data.read ();
        ready = true;
        next_state = s0;
      }
      else {
          if (dirty_bit[target_block]) {
            mem_write = true;
            mem_address_temp = (tag[target_block], block \
                          [target_block], (sc_lv <4>)"0000");
            mem_address = mem_address_temp.to_uint ();
            blk_address = target_block;
            blk_size = BLOCK_SIZE;
            next_state = s6;
          }
          else {
            mem_read = true;
            mem_address = address.read ();
            blk_address = target_block;
            blk_size = BLOCK_SIZE;
            next_state = s5;
          }
      }
      break;
      ...
    case s5:
      mem_read = false;
      mem_write = false;
      if (mem_ready) {
          target_block = address.read ().range(10, 4).to_uint ();
          dirty_bit[target_block] = false;
          tag[target_block] = (sc_lv <5>) address.read ().range (15, 11);
          block[target_block] = (sc_lv <5>) address.read ().range (10, 4);
          word[target_block].range (address.read ().range (3, 0).to_uint ()*8+7, \
              address.read ().range (3, 0).to_uint ()*8) = data.read ();
          dirty_bit[target_block] = true;
          ready = true;
          next_state = s0;
      }
      break;
      ...
  }
}
void Cache::state_change()
{
  if (reset == true) {
      present_state = s0;
      for (int i = 0; i< 128; i++) // clear all dirty bits
        dirty_bit[i] = false;
  }
  else if (clk.event () && clk == true)
    present_state = next_state;
}
```

FIGURE 89.39 Part of RTL cache implementation.

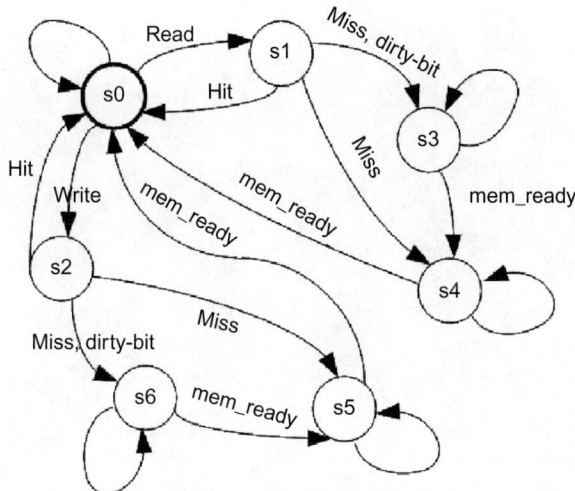

FIGURE 89.40   A state machine diagram to exemplify RTL cache.

89.5   Summary

This chapter presented constructs of the SystemC language that are used for RT-level descriptions. The chapter used complete design examples at various levels of abstraction for showing ways in which this language could be used for RTL designs. We used examples with clear one-to-one hardware correspondence. We have shown how combinational and sequential components can be described and how a complete system can be put together using combinational and sequential blocks. Testbench methods used in SystemC were also described in this chapter.

References

1. D.C. Black and J. Donovan, *SystemC: From The Ground Up*, Kluwer Academic Publishers, Dordrecht, 2004.
2. J. Bhasker, *A SystemC Primer*, 2nd edition, Star Galaxy Publishing, Allentown, 2004.
3. http://www.synopsys.com.

90
System Verilog

Saeed Safari

Nanoelectronics Center of Excellence
School of Electrical and
Computer Engineering
University of Tehran

CONTENTS

This section provides the basic concepts and techniques of hardware modeling using the System Verilog language. It is assumed that the reader is familiar with the Verilog HDL language and knows the basic hardware design concepts.

90.1 System Verilog Origin

Verilog HDL and several CAD tools, e.g., simulation tools, synthesis tools, and verification tools, were designed in 1984 by Gateway Design Automation. In 1990, the Open Verilog International (OVI) was formed and efforts for Verilog HDL popularization were started. These efforts led to the standard Verilog HDL as IEEE 1364 Verilog HDL (Verilog-95). In the late 1990s, the need for using higher level constructs for system level design increased. As a result some new features were added to Verilog-95 and the language was specified as the IEEE 1364 Verilog-2001. Since then several EDA companies and committees worked on Verilog-2001 and finally in May 2002 it was finalized as the System Verilog 3.0 standard [1,2,3].

90.2 Basic Concepts

90.2.1 Data Types

There are several built-in data types in Verilog-95, such as **net**, **reg**, and **integer**. In System Verilog, these data types are extended to new data types that are used to ease hardware description at RTL and system level. System Verilog also allows users to define their own data types.

90.2.1.1 Verilog-95 Data Types

There are two basic data types in Verilog-95 for hardware description. These types include **net** and **reg**. Both **net** and **reg** data types use a 4-logic value system including **0, 1, Z,** and **X**. The **net** data type is used to model an interconnection between two components (as an actual wire). The **reg** data type is used to model the behavior of the circuit (as a software variable). The usage of these types is restricted, which implies that you are only allowed to use the **net** type in module instantiation or in continuous assignments. In contrast, the **reg** data type should only be used in procedural assignments.

90.2.1.2 System Verilog Basic Data Types

System Verilog is backward-compatible with Verilog-95, which means that all data types in Verilog-95 can be used in System Verilog. In addition, System Verilog has several new data types, with two logic values (**0** and **1**) for RTL and system level description. These data types include **bit, byte, shortint, int,** and **longint** to show 1-bit, 8-bit, 16-bit, 32-bit, and 64-bit variables, respectively. Figure 90.1 shows two examples of variable declarations, including declaring a 1-bit variable named *carry*, and a 16-bit variable named *d_in*.

By default, variables declared as **byte, shortint, int,** and **longint**, are considered as signed numbers. However, they can be changed to unsigned numbers using the **unsigned** modifier. A variable defined as a vector of **net**s, **reg**s, or **bit**s is considered an unsigned number. The **signed** modifier enables us to declare a signed variable with arbitrary size. Figure 90.2 shows the declaration of an unsigned 8-bit variable *sum*, and a signed 12-bit variable *d_out*.

As described earlier, type **reg** is used to define a 1-bit variable. The keyword **reg** may mislead the user that the object corresponds to a hardware register. To avoid this misleading, a 4-logic value data type, called **logic**, is added in System Verilog. The **logic** data type has all features of type **reg** and can therefore be used instead of it.

Two other data types **void** and **shortreal** are used to show a function which has no return value and a 32-bit single precision floating point number, respectively.

As an example consider a signed adder with two signed 8-bit inputs *a* and *b*, a 1-bit input *c_in*, a signed 8-bit output *sum*, and a 1-bit output *c_out*. Figure 90.3 shows the System Verilog code for the adder that uses a continuous **assign** statement. In Verilog-95, ports are defined in two steps: first, module ports appear in the module definition's list of ports and second they are declared in the module body. As shown in Figure 90.3, these two steps can be performed simultaneously in System Verilog.

90.2.1.2.1 Variable Declaration

In Verilog-95, all variables and nets should be declared in the module body enclosed by keywords **module** and **endmodule**. However, declarations outside the module body are allowed in System Verilog. Such a variable is visible in all modules compiled at the same time with .the file containing the variable. In

```
Bit         carry;      // 1-bit variable
shortint    d_in;       // 16-bit variable
```

FIGURE 90.1 System Verilog **bit** and **shortint** variable declaration.

```
byte unsigned       sum;       // unsigned 8-bit variable
reg signed [11:0]   d_out;     // signed 12-bit variable
```

FIGURE 90.2 System Verilog signed and unsigned modifier.

```
module signed_adder_8( input  signed  [7:0] a,
                       input  signed  [7:0] b,
                       input  bit           c_in,
                       output signed  [7:0] sum,
                       output bit           c_out);

    assign {c_out, sum} = a + b + c_in;
endmodule
```

FIGURE 90.3 Signed 8-bit adder.

```
reg [2:0] alu_func;

module m1(input [3:0] a, b, output [3:0] y);
    always @(a or b or alu_func)
    begin
        case (alu_func)
            ...
        endcase
    end
endmodule

module m2(input  [3:0] a, b,
          input  [2:0] alu_func,
          output [3:0] y);
    always @(a or b or alu_func)
    begin
        case (alu_func)
            ...
        endcase
    end
endmodule
```

FIGURE 90.4 Variable declarations outside of the module body.

Figure 90.4, a 3-bit binary variable, called *alu_func*, is defined. To find the reference of a variable, the local variables in the scope are searched first. If the variable is not found, then it is searched outside the module. The *alu_func* in model *m1* refers to the *alu_func* declared outside the module, while *alu_func* in module *m2* refers to its *alu_func* input port

90.2.1.2.2 Net and Variable Assignment

In Verilog-95, a **net** type can be assigned from an output of a module or through a continuous assignment statement, while a variable of type **reg** can be assigned in procedural assignment statements. In System Verilog, these restrictions are removed from existing and new data types. This means that any variable of any type can be assigned by only one of the following ways:

- From any number of procedural blocks (i.e., initial and always block)
- From an output of a single module (or primitive)
- From a single continuous assignment statement

Figure 90.5 shows an example of variable assignment. In module *m1*, a variable of type **reg**, named *y1*, is assigned in a continuous assignment, which is a valid use of the **reg**-type in System Verilog. In contrast, a **reg**-type variable, named *y2*, is assigned in both continuous and procedural assignments in module *m1*, which is not valid (because of the mix of continuous and procedural assignment to a variable). In Figure 90.6, a variable of type **logic**, named *y1*, is assigned in two continuous assignments, which is

```
module m1 (input        a, b,
           output reg y1, y2);
   // Valid:
   assign y1 = a & b;
   // Invalid:
   assign y2 = a & b;
   always @(a or b) y2 = a | b;
endmodule
```

FIGURE 90.5 Assignment to a **reg** variable.

```
module m1 (input a, b, output logic y1, y2);
   // Invalid:
   assign y1 = a & b;
   assign y1 = a | b;
   // Valid:
   always @(a or b) y2 = a & b;
   always @(a or b) y2 = a | b;
endmodule
```

FIGURE 90.6 Assignment to a **logic** variable.

```
byte'(56.00)                     // Example 1
16'(a)                           // Example 2
signed'(b)                       // Example 3
shortint a; $cast (a, 2.0 * b);  // Example 4
```

FIGURE 90.7 Type casting.

not a valid use of the **logic** type. In the same figure, a valid use of type **logic** is shown. Variable *y2* of type **logic** is assigned in a procedural block.

90.2.1.2.3   *Type Casting*

System Verilog provides static (compile-time) and dynamic (run-time) type casting. There are three types of static casting used for casting a value to a specified data type, casting a value to a specified vector size, and casting a value to **signed** or **unsigned**. Dynamic casting is performed by using the **$cast** (*des, src*) system function which casts the *src* variable to the *des* variable.

Figure 90.7 shows several examples of type casting. In the first example, the value of 56.00 is cast to **byte**. The second example shows casting the value of *a* to a 16-bit vector size. The third example shows casting the value of *b* to a signed value and finally the last example, that consists of a declaration and a dynamic cast, shows casting the value of expression *2.0 * b* to the **shortint** data type.

90.2.1.3   User-Defined Data Type

System Verilog allows new data type definitions with the **typedef** keyword. The syntax of the **typedef** statement is:

   typedef *type name*;

where *type* is an existing data type and *name* is the new name for this type. Figure 90.8 shows an example of a user-defined type definition.

```
typedef signed [25:0] vec26
vec26 jump_adr;
```

FIGURE 90.8 User-defined data type.

```
enum {init, got1, got11, got110} present_state;

present_state = got1;                    // Valid
```

FIGURE 90.9 Enumerated data type.

```
enum { init   = 0,
       got1   = 2,
       got11  = 4,
       got110 = 8} present_state;
```

FIGURE 90.10 A legal code assignment.

```
// Valid:
enum [3:0] { init   = 4'b0001,
             got1   = 4'b0010,
             got11  = 4'b0100,
             got110 = 4'b1000 } present_state;

// Invalid:
enum { init   = 4'b0001,
       got1   = 4'b0010,
       got11  = 4'b0100,
       got110 = 4'b1000 } present_state;
```

FIGURE 90.11 Code assignment.

90.2.1.4 Enumerated Data Type

Enumerated data type is a set of user-defined names that specifies all the valid values that a variable of that type can obtain. Each user-defined symbol corresponds to an integer value one greater than the previous symbol. The value of the first symbol is **0**.

Figure 90.9 shows an example of enumerated data type definition, which is used for state machine coding. Variable *present_state* can have the values of *init*, *got1*, *got11*, and *got110*. The integer values of *init* and *got11* are **0** and **2**, respectively.

The default integer value assigned to a symbol can be modified explicitly by assigning the desired value to the symbol. Figure 90.10 shows an example of legal code assignments to the enumerated symbols. This feature can be employed to specify a desired code assignment in state machine implementations. Figure 90.11 shows an example of a one-hot code assignment to the states of Figure 90.9. As shown in Figure 90.11, the lengths of the enumerated symbols are specified and matched with the assigned codes. Figure 90.11 also shows an illegal code assignment due to the violation of this rule.

90.2.1.5   Array Type

In Verilog-95, only one-dimensional array declarations of data type **reg**, **integer**, and **time** are allowed. System Verilog extends array declarations to allow multidimensional array declarations of any data type, and calls it unpacked array. The syntax of an unpacked array declaration is as follows:

<type> <name> <dimension *1*> ⋯ <dimension *n*>;

Figure 90.12 shows several examples of unpacked array declarations. Variable *a1* is a one-dimensional array of 16 elements of a 1-bit wire type. Variable *a2* is declared as a one-dimensional array of 1024 elements of 8-bit **reg** vectors, and finally variable *a3* is declared as a two-dimensional array of bytes.

An array element can be accessed using indexes. Figure 90.13 shows how to access the fifth element of array *a2* and how to access the second row and the third column of array *a3*. As shown in this figure, you can assign **0**, **1**, **Z**, and **X** values to all bits of a vector using **'0**, **'1**, **'Z**, and **'X**, respectively. So the statement *a2[4]* = **'0**; sets all bits of the fifth element of the array *a2*, which is an 8-bit **reg** vector, to **0**. Bit select and part select access to an array are allowed in System Verilog, and multiple select that selects multiple elements of an array is not allowed.

The C-like style can be employed to initialize an unpacked array. Figure 90.14 shows an example of initializing an unpacked array. This statement assigns values **0**, **1**, ... to *a4[0][0]*, *a4[0][1]*, ... elements

```
wire      a1 [0:15];
reg [7:0] a2 [0:1023];
byte      a3 [0:2] [0:3];
```

FIGURE 90.12   Unpacked array declarations.

```
a2 [4]     = '0;
a3 [2][4] = 3;

// Valid: bit select is allowed
a2 [2][4] = 1'b1;
// Valid: part select is allowed
a2 [1][3:1]
// Invalid: multiple select is not allowed
a3 [1][1:2]
```

FIGURE 90.13   Access to the array elements.

```
// Unpacked array initialization:
byte a4 [0:2][0:3] = { {0, 1, 2, 3},
                       {4, 5, 6, 7},
                       {8, 9, 10, 11}};
// Unpacked array assignment:
// Assignment to full unpacked array
a4 = {{0, 1, 2, 3}, {4, 5, 6, 7}, {8, 9, 10, 11}};
// Assignment to full unpacked array using default operator
a4 = {default:5};
// Assignment to the multiple select
a4[1] = {'1, '1, '1, '1};
// Assignment to an element
a4[2][2] = 34;
```

FIGURE 90.14   Unpacked array initialization and assignment.

of *a4*, respectively. Unpacked array assignment is performed in the same manner (see Figure 90.14). System Verilog adds the **default** operator to fill all array elements with a specified value. As shown in Figure 90.14, statement *a4 = {default: 5};* fills all elements of *a4* with 5.

System Verilog also provides multidimensional vectors of the following types: **bit**, **logic**, **reg**, and **net**. This type of vectors is called packed arrays. Packed arrays are declared based on the following syntax:

<type> <dimension *1*> ··· <dimension *n*> <name>;

Figure 90.15 shows an example of packed array declaration. Variable *pa* is declared as a two-dimensional packed array. As shown in Figure 90.16, we can consider *pa* as a vector with two elements, where each element is a 4-bit **logic** vector. System Verilog allows access to multiple elements, bit select, and part select of a packed array as illustrated in Figure 90.15. As shown in Figure 90.16, packed arrays can be assigned using simple assignment statements. Assignments to bit select, part select, and also to a slice of an array are allowed in System Verilog.

90.2.1.6 Structure Data Type

Structures can be defined in System Verilog using a C-like syntax. As in software languages, a structure is a collection of various elements that are not necessarily of the same type. Figure 90.17 shows an example of structure definition. The variable *inst* has four fields including *opcode*, *rs*, *rt*, and *adr*. An element of a variable of type structure can be accessed using the C-like "." operator, for instance statement *inst.adr = 'b0* is used to fill in the *adr* element of the *inst* variable with zeros.

```
logic [1:0] [3:0] pa;

pa [0] [1]    // bit select of
              // packed array is allowed

pa [0] [2:1]  // part select of
              // packed array is allowed

pa [0]        // multiple select of
              // packed array is allowed
```

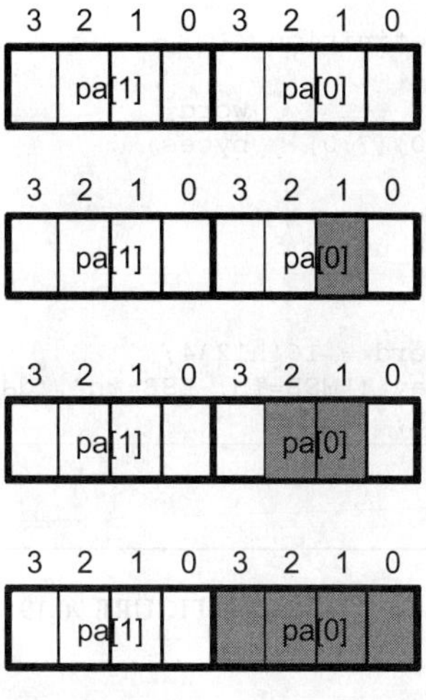

FIGURE 90.15 Packed array declaration.

```
// Assignment to full packed array
pa = 8'b0100_1100;
// Assignment to a slice
pa[0] = 'b1;
// Assignment to a part select
pa[0][2:1] = 2'b01;
// Assignment to a bit select
pa[0][1] = 1'b1;
```

FIGURE 90.16 Packed array assignment.

```
// Structure definition
struct {
   bit [5:0]  opcode;
   bit [4:0]  rs;
   bit [4:0]  rt;
   bit [15:0] adr;
} inst;

// Structure assignment
inst.adr = 'b0;
```

FIGURE 90.17 Structure definition and assignment.

```
// Structure assignment
inst = {6'b001000, 5'd21, 5'd17, 'b0};
// Structure assignment
inst = {default:0};
```

FIGURE 90.18 Structure assignment.

```
// Union definition
union {
   int             word;
   bit [1:0][7:0]  bytes;
} data;

module test_union;
  initial
  begin
    data.word = 16'h1234;
    $display ("MSB=%h, LSB=%h", data.bytes[1], data.bytes[0]);
    $finish;
  end
endmodule
```

FIGURE 90.19 Union definition and assignment.

A set of values enclosed in { } and separated by commas can be used to assign values to elements of a structure. As with unpacked arrays, all structure elements can be set by a default operator. Figure 90.18 shows two types of structure assignments.

90.2.1.7 Union Data Type

As in the C programming language, **union** in System Verilog is a memory location shared by two or more different variables. The union declaration is similar to structure declaration and the union elements can be accessed in the same way. Figure 90.19 shows an example of union declaration, mapping its two elements together. The elements are a 16-bit integer variable *word*, and a packed array called *bytes*. As described before, these two elements share memory locations (see Figure 90.20). In module *test_union* a 16-bit value is assigned to element *word* of the union and can then be accessed as two bytes through element *bytes*.

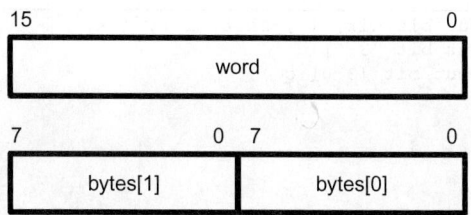

FIGURE 90.20 Memory sharing between union elements.

```
module counter (input bit clr, u_d, clk,
                output bit [3:0] q );
  always @(posedge clk)
  begin
    if (clr)
      q = 'b0;
    else if (u_d)
      q ++;
    else
      q --;
  end
endmodule
```

FIGURE 90.21 Increment and decrement operators.

TABLE 90.1 Shift Register Truth Table

Clr	ld	shft	l_r	clk	q3	q2	q1	q0
1	X	X	X	↑	0	0	0	0
0	1	X	X	↑	d3	d2	d1	d0
0	0	1	1	↑	q2	q1	q0	0
0	0	1	0	↑	0	q3	q2	q1

90.2.2 Operators

The increment and decrement operators (++ and − −) are added to System Verilog. The operator ++ (− −) adds (subtracts) 1 to (from) its operand. If the operator ++ (or − −) appears before the operand, the increment (or decrement) operation is performed before using the value of the operand. However, if these operators appear after the operand, the value of the operand is used before incrementing (or decrementing) it. These operators have a blocking behavior, which means that they block their following statements until they are completed. Figure 90.21 shows an example of increment and decrement operators used to describe a 4-bit up-down counter.

System Verilog also provides several C-like assignment operators including +=, −=, *=, /=, %=, &=, |=, ^=, <<=, >>=, <<<=, and >>>=. Assignment operators perform the specified operation on both left- and right-hand-side operands and then the result is assigned to the left-hand-side operand. The last two operators are used for arithmetic left and arithmetic right shift operators. Because of the blocking behavior of the assignment operators, special care is needed to be taken when using them. Consider a shift register with the truth table shown in Table 90.1. The module *shift_reg*, shown in Figure 90.22, describes the functionality of the shift register using a procedural always block. The assignment statement *q <<= 1;* means that the value of *q* is shifted 1 bit to the left and then the result is assigned to *q*.

```
module shift_reg ( input bit clr, ld, shft, l_r, clk,
                   input bit [3:0] d,
                   output bit [3:0] q );
   always @(posedge clk)
   begin
     if (clr)
        q = 'b0;
     else if (ld)
        q = d;
     else if (shft)
     begin
       if (l_r)
          q <<= 1;
       else
          q >>= 1;
     end
   end
endmodule
```

FIGURE 90.22 Shift register example using assignment operator.

```
// Equality operator
4'b0010 == 4'b001x             // return value is x
// Case equality operator
4'b0010 === 4'b001x            // return value is 0
// Wild equality operator
4'b0010 =?= 4'b001x            // return value is 1
```

FIGURE 90.23 Equality, case-equality, and wild-equality operators.

System Verilog also adds two new equality operators called *wild equality operator* (=?=) and *wild inequality operator* (!?=). The **X** and **Z** value in a bit position in any operand are treated as a don't-care bit and can be matched with any value. The wild-case equality operator returns **1**, if the operands are equal, and returns 0 otherwise. To check the equality of the operands all **X** and **Z** values in any bit position are considered as don't-cares and can be matched to all logic values (i.e., **0**, **1**, **X**, and **Z**). Figure 90.23 shows an example of using equality, case equality, and wild equality operators and illustrates their differences.

90.3   Hardware Design with System Verilog

System Verilog provides several design abstraction levels including switch level, gate level, RT level, and behavioral level. Here, we will focus on behavioral level hardware description.

90.3.1   Behavioral Modeling Constructs

Verilog as has several fundamental behavioral constructs, such as **if-else**, **case**, **for-loop**, and **while-loop**. This section describes how System Verilog enhances these constructs.

90.3.1.1   The Unique and Priority Modifier

System Verilog adds two modifiers, **unique** and **priority** that are used with **if-else** and **case** statements. These modifiers specify the way in which simulation and synthesis tools interpret the **if-else** (**case**) conditions. The **unique** modifier is used to make sure that only one condition in a series of **if-else-if** is true. This means that all conditions can be interpreted in parallel. However, the **priority** modifier is used to force the simulation or synthesis tools to interpret the condition as they appear in the code.

As an example, consider a barrel shifter with a 4-bit data input d, a decoded 4-bit shift amount input s, and a 4-bit data output y. Table 90.2 shows the truth table for this barrel shifter while its description is shown in Figure 90.24.

Figure 90.25 shows our barrel shifter using the **unique** modifier. In this case, if two or more conditions become true simultaneously (e.g., $s[2] == 1\text{'b}1$ and $s[0] == 1\text{'b}1$), a run-time error will be generated.

Figure 90.26 shows our *barrel_shifter* using the **priority** modifier. Here, if two or more conditions become true simultaneously (e.g., $s[2] == 1\text{'b}1$ and $s[0] == 1\text{'b}1$), the first condition that appears in the code (i.e., $s[2] == 1\text{'b}1$) will be met, and the corresponding statement (i.e., $y = \{d[1:0], 2\text{'b}00\}$) will be executed. These modifiers can be applied to **case** statements in the same manner.

TABLE 90.2 Barrel Shifter Truth Table

S_3	S_2	S_1	S_0	y_3	y_2	y_1	y_0
0	0	0	1	d_3	d_2	d_1	d_0
0	0	1	0	d_2	d_1	d_0	0
0	1	0	0	d_1	d_0	0	0
1	0	0	0	d_0	0	0	0

```
module barrel_shifter ( input  bit [3:0] d,
                        input  bit [3:0] s,
                        output bit [3:0] q );
  always @(d or s)
  begin
    if (s == 4'b1000)
       y = {d[0], 3'b000};
    else if (s == 4'b0100)
       y = {d[1:0], 2'b00};
    else if (s == 4'b0010)
       y = {d[2:0], 1'b0};
    else if (s == 4'b0001)
       y = d;
  end
endmodule
```

FIGURE 90.24 Barrel shifter example.

```
module barrel_shifter ( input  bit [3:0] d,
                        input  bit [3:0] s,
                        output bit [3:0] q );
  always @(d or s)
  begin
    unique if (s[3] == 1'b1)
       y = {d[0], 3'b000};
    else if (s[2] == 1'b1)
       y = {d[1:0], 2'b00};
    else if (s[1] == 1'b1)
       y = {d[2:0], 1'b0};
    else if (s[0] == 1'b1)
       y = d;
  end
endmodule
```

FIGURE 90.25 Barrel shifter example using unique modifier.

```
module barrel_shifter ( input  bit [3:0] d,
                        input  bit [3:0] s,
                        output bit [3:0] q );
  always @(d or s)
  begin
    priority if (s[3] == 1'b1)
      y = {d[0], 3'b000};
    else if (s[2] == 1'b1)
      y = {d[1:0], 2'b00};
    else if (s[1] == 1'b1)
      y = {d[2:0], 1'b0};
    else if (s[0] == 1'b1)
      y = d;
  end
endmodule
```

FIGURE 90.26   Barrel shifter example using priority modifier.

```
for (byte i = 0, byte j = 0; i+j < 15; i++, j++)
begin
   ...
end
```

FIGURE 90.27   For loop enhancements in System Verilog.

90.3.1.2   For Loop Enhancement

System Verilog enhances the **for** loop statement by allowing

- Local variable declaration in the initial part of a **for** loop
- Multiple statements in the initial and update part of a **for** loop

Figure 90.27 shows an example of a **for** loop that employs these enhancements.

90.3.1.3   Do-While Loop

System Verilog adds a C-like **do-while** loop. The loop's termination condition is tested at the end of the loop, which means that the loop body is executed at least once.

90.3.2   Combinational Circuit Design

Combinational circuits can be described using an **always** block. To do so, all inputs of the circuit should appear in the sensitivity list of the **always** block. In addition all combinational outputs should get a value for all possible combinations of the inputs. As an example, consider Figure 90.28 that shows a 2-to-1 multiplexer implemented with the **if-else** construct. As shown, all inputs (i.e., i0, i1, and s) are listed in the sensitivity list, and the y output requires some value for all input combinations.

Verilog-2001 allows the use of wild-card * in the sensitivity list of an always block as **always @ (*)** or **always @***. This implies that each variable used as an input of the **always** block, i.e., variables read by **always** block, should be implicitly considered in the sensitivity list. If there are several signals that the always block is sensitive to, the use of the wild-card * shortens the code and prevents the user from probable coding errors. Figure 90.29 illustrates a 2-to-1 multiplexer with an enable input. In this example **always @*** infers **always @ (i0 or i1 or s)**. Note that the wild-card * does not infer the variables used in the function called by the **always** block. For example, in Figure 90.29, function *enable* that uses variables g1 and g2_b, is called in an **always** block. As a result of using wild-card *, g1 and g2_b are not listed in the sensitivity list and therefore they cannot sensitize the **always** block for execution.

```
module mux2to1 ( input  bit i0, i1, s,
                 output bit y );
   always @(i0 or i1 or s)
   begin
     if (s)
       y = i1;
     else
       y = i0;
   end
endmodule
```

FIGURE 90.28 A 2-to-1 multiplexer.

```
bit g1   = 1'b1;
bit g2_b = 1'b0;

module mux2to1 ( input  bit i0, i1, s,
                 output bit y );

   function bit enable;
     if ((g1 == 1'b1) && (g2_b == 1'b0))
       enable = 1'b1;
     else
       enable = 1'b0;
   endfunction

   always @*
   begin
     if (s)
       y = i1 & enable ();
     else
       y = i0 & enable ();
   end
endmodule
```

FIGURE 90.29 A 2-to-1 multiplexer with enable input using wild-card.

System Verilog adds a new procedural block, named **always_comb**, to resolve the discussed problem. The **always_comb** procedural block is similar to **always @(*)**, with the difference that it infers all variables that the procedural block is sensitive to, including variables used in functions called in the procedural block. Figure 90.30 illustrates the multiplexer example using **always_comb**. In this case the procedural block is sensitive to variables $i0$, $i1$, s, $g1$, and $g2\_b$. The procedural **always_comb** block cannot contain any explicit event or timing control. In addition any other assignment to the variables assigned in the **always_comb** block is not permitted. The **always_comb** block should describe a combinational logic. In other words, outputs should never retain their value from one activation to the next.

90.3.3 Level-Sensitive Latch

An input combination in an **always** block, where the output is not assigned a value represents a level-sensitive latch. Consider the gated latch with two inputs g and d, and an output q described in Figure 90.31. Any change in d or q triggers the **always** block and if $g = \mathbf{1}$, q gets the value of d, otherwise q holds its previous value. This performance describes the latching behavior.

System Verilog provides a new procedural block, called **always_latch**, to describe latching behavior. As an **always_comb** block, the **always_latch** block infers all variables that the procedural block is sensitive to, including the variables used in functions called in the **always_latch** block. The **always_latch** should describe a latch, which implies a condition where the output does not get any value for at least one input

```
bit g1   = 1'b1;
bit g2_b = 1'b0;

module mux2to1 ( input  bit i0, i1, s,
                 output bit y );

   function bit enable;
     if ((g1 == 1'b1) && (g2_b == 1'b0))
       enable = 1'b1;
     else
       enable = 1'b0;
   endfunction

   always_comb
   begin
     if (s)
       y = i1 & enable ();
     else
       y = i0 & enable ();
   end
endmodule
```

FIGURE 90.30   A 2-to-1 multiplexer with enable input using **always_comb**.

```
module gated_d_latch ( input  bit d, g,
                       output bit q );

   always @(d or g)
   begin
     if (g)
       q = d;
   end
endmodule
```

FIGURE 90.31   Gated d-latch.

```
module gated_d_latch ( input  bit d, g,
                       output bit q );

   always_latch
   begin
     if (g)
       q = d;
   end
Endmodule
```

FIGURE 90.32   Gated d-latch using always_latch.

combination. As with **always_comb**, any other assignments to the latched output are not allowed outside the **always_latch** block. Figure 90.32 shows the gated d-latch using **always_latch**.

90.3.4   Edge-Triggered FF

If an **always** block is sensitive to the positive or negative edge of a signal, it can usually be synthesized as an edge-triggered sequential circuit. Verilog-95 provides two operators **posedge** and **negedge**, which can be used to show sensitivity to the edge of a signal. As an example, consider Figure 90.33 that shows an example of a positive edge-triggered *d-ff* with active low asynchronous *aclr* input.

System Verilog also provides an **always_ff** procedural block to show the behavior of an edge-triggered sequential circuit. All signals in the sensitivity list should be used by a **posedge** or **negedge** operator. Any

```
module d_ff ( input  bit d, aclr, clk,
              output bit q );

  always @(posedge clk or negedge aclr)
  begin
    if (aclr == 1'b0)
      q = 1'b0;
    else
      q = d;
  end
endmodule
```

FIGURE 90.33   Edge-triggered d-ff.

```
module d_ff ( input  bit d, aclr, clk,
              output bit q );

  always_ff @(posedge clk or negedge aclr)
  begin
    if (aclr == 1'b0)
      q = 1'b0;
    else
      q = d;
  end
endmodule
```

FIGURE 90.34   Edge-triggered d-ff using **always_ff** block.

other assignment to the sequential output is not allowed outside the **always_ff** block. Figure 90.34 shows the d-ff of Figure 90.33 using an **always_ff** block.

90.3.5   Interfaces

There are two types of port connections in Verilog-95: connection by position and connection by name. Using these methods in a hierarchical design makes the creating and maintenance of connection procedure difficult. For example, Figure 90.35 shows an example of an add-and-shift multiplier, which is partitioned into *data_path* and *controller* parts. These two parts are connected using several wires. As shown in the figure these wires should be declared as ports in both modules, as well as being declared in the top-level module. Finally, they are used in both module instances to make interconnections.

System Verilog provides a new port type called **interface**. Interface is a group of signals that can be used as a single port. As shown in Figure 90.36, an interface is declared using the **interface** keyword, followed by an interface name. Then a list of interconnection signals should be declared. Usually the interface signals are declared as inputs in some modules and declared as outputs in some other modules. Therefore, the **port** mode should be specified by using the **modport** construct, which is similar to the port declaration of a module excluding the port type. For example, in Figure 90.36 there are two types of **modport**s, *dp_mode* and *cu_mode*. The **endinterface** keyword shows the end of the **interface** declaration. Now the interface can be used in module port declarations by using the interface name followed by a **modport** name (see Figure 90.36). Then the interface should be used in a top-level module to interconnect the modules. To do so an instance of the interface is needed to be used in the port connection. Note that the **modport** cannot be used in both module declaration and module instantiation.

```
module data_path ( input  logic [3:0] inbus,
                   output logic [3:0] outbus,
                   input  logic       clk,
                   input  logic       clrA,
                   input  logic       ldA,
                   input  logic       ldX,
                   input  logic       shft,
                   input  logic       ldY,
                   input  logic       add,
                   input  logic       AOnDbus,
                   input  logic       XOnDbus,
                   output logic       X0      );

  ...
endmodule

module controller ( input  logic  clk,
                    input  logic  reset,
                    input  logic  start,
                    output logic  done,
                    output logic  clrA,
                    output logic  ldA,
                    output logic  ldX,
                    output logic  shft,
                    output logic  ldY,
                    output logic  add,
                    output logic  AOnDbus,
                    output logic  XOnDbus
                    input  logic  X0 );

  ...
endmodule

module multiplier ( input  logic [3:0] inbus,
                    output logic [3:0] outbus,
                    input  logic       clk,
                    input  logic       reset,
                    input  logic       start,
                    output logic       done );
  logic   clrA, ldA, ldX, shft, ldY,
          add, AOnDbus, XOnDbus, X0;

  data_path  dp (inbus, outbus, clk, clrA, ldA, ldX,
                 shft, ldY, add, AOnDbus, XOnDbus, X0);
  controller cu (clk, reset, start, done, clrA, ldA, ldX,
                 shft, ldY, add, AOnDbus, XOnDbus, X0);

endmodule
```

FIGURE 90.35 Module interconnection.

90.4 Complete Design Example

In this section we describe a 4-bit add-and-shift multiplier using System Verilog. The multiplier accepts two 4-bit numbers and multiplies them to generate an 8-bit product. The multiplier has a 4-bit input *inbus*, for inputting the multiplier and multiplicand, and a 4-bit output *outbus* for outputting the 8-bit product as two 4-bit numbers. The multiplier looks for a complete pulse on the start input and then in the next two clock pulses reads the multiplier and multiplicand and stores them into the internal registers X and Y, respectively. The next four clock pulses are used to perform multiplication. When the 8-bit product is ready, the *done* output becomes 1 for two clock pulses. During this period the least and most

```
interface bus;
  logic  clrA, ldA, ldX, shft, ldY, add,
          AOnDbus, XOnDbus, X0;
  modport dp_mode (input   clrA, ldA, ldX, shft, ldY, add,
                            AOnDbus, XOnDbus,
                   output X0);
  modport cu_mode (output clrA, ldA, ldX, shft, ldY, add,
                           AOnDbus, XOnDbus,
                   input  X0);
endinterface

module data_path ( input   logic  [3:0]    inbus,
                   output logic  [3:0]    outbus,
                   input   logic          clk,
                           bus.dp_mode    dp_bus);

   ...
endmodule

module controller ( input   logic         clk,
                    input   logic         reset,
                    input   logic         start,
                    output logic         done,
                            bus.cu_mode  c_bus );

   ...
endmodule

module multiplier ( input   logic [3:0] inbus,
                    output logic [3:0] outbus,
                    input   logic      clk,
                    input   logic      reset,
                    input   logic      start,
                    output logic      done );

  bus connection_bus;

  data_path  dp (inbus, outbus, clk, connection_bus);

  controller cu (clk, reset, start, done, connection_bus);

endmodule
```

FIGURE 90.36 Module interconnection using interface.

significant 4-bits of the product are put on the *outbus* output. Figure 90.37 illustrates the block diagram of the multiplier partitioned into data and controller parts.

The data path of the multiplier of Figure 90.37 is partitioned into functional units (adder), memory elements (registers and shift registers), and buses (tri-state buffers). Figure 90.38 shows the block diagram of this data path. The Y register holds the multiplicand and the X shift register holds the multiplier. The 8-bit multiplication result is stored in both A and X registers. In each step, the $X0$ bit is examined and if it is equal to 1 contents of Y will be added to A. Then both A and X are shifted one place to the right. Here, these two operations (addition and shifting) have been performed in a single clock pulse.

The multiplier controller, which is a finite-state machine, is shown in Figure 90.39. In the *idle* state, the controller searches for a complete pulse on start. In two consecutive states the required control signals are issued to load the multiplier and multiplicand in X and Y, respectively. In the next four states, based on the value of $X0$, the addition of A and Y is performed and both A and X are shifted right.

The adder used in the data-path is a 4-bit adder that adds the partial product with the multiplicand. Figure 90.40 shows the description of the adder.

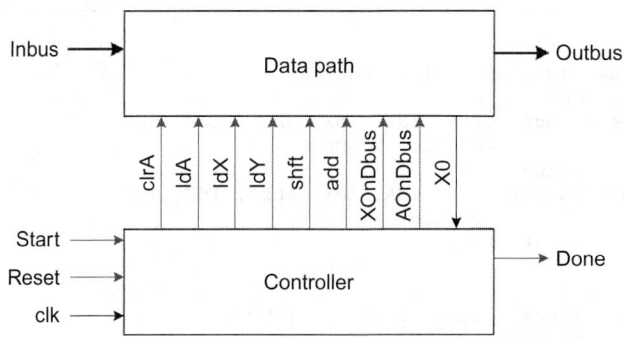

FIGURE 90.37 The add-and-shift multiplier block diagram.

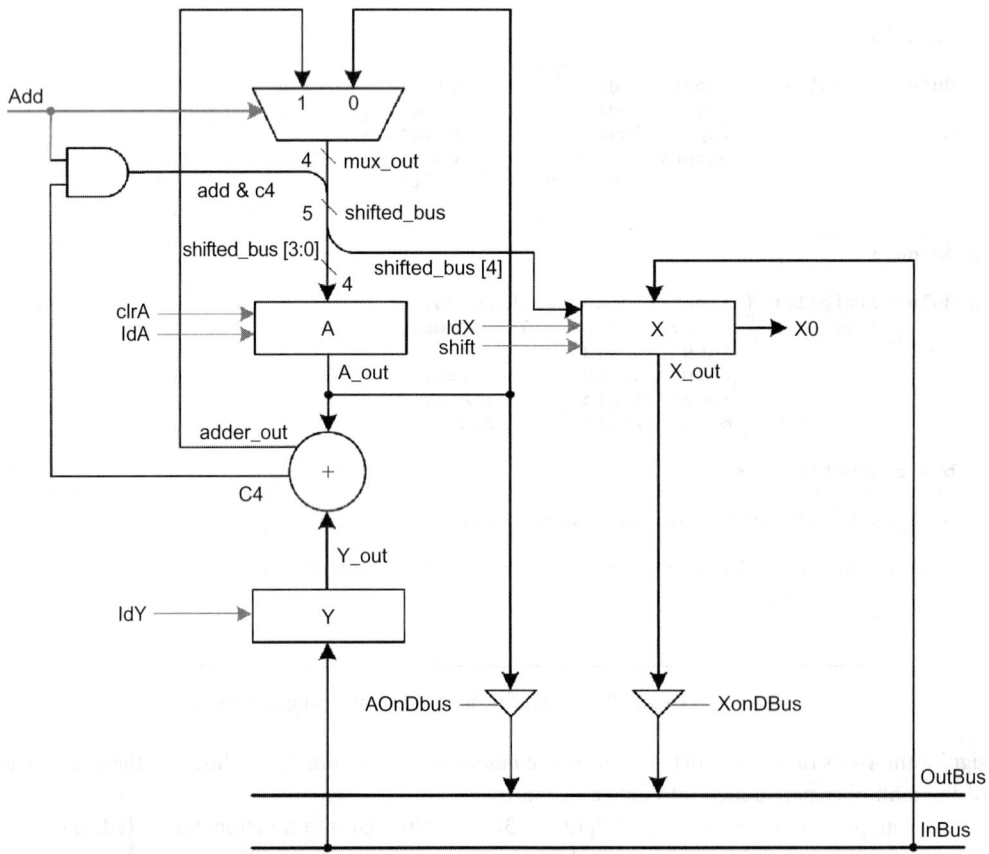

FIGURE 90.38 Add-and-shift multiplier data path.

Figure 90.41 shows the description of a 4-bit positive edge-triggered register with two synchronous inputs *clr* and *ld*. Two instances of this register will be used to store the multiplicand, i.e., *Y*, and to store the partial products, i.e., *A*, in the data path.

The next module is a 4-bit positive edge-triggered shift register with two synchronous inputs, *ld* and *shft*. With a positive edge of the *clk*, if *ld* = **1**, the 4-bit input *d* is loaded into the shift register, otherwise if *shft* = **1**, the content of the shift register is shifted right 1 bit and its left-hand bit is filled by *ser_in*. Figure 90.42 shows the description of the shift register. One instance of this module (labeled *X*) is used in the data path to hold the multiplier value.

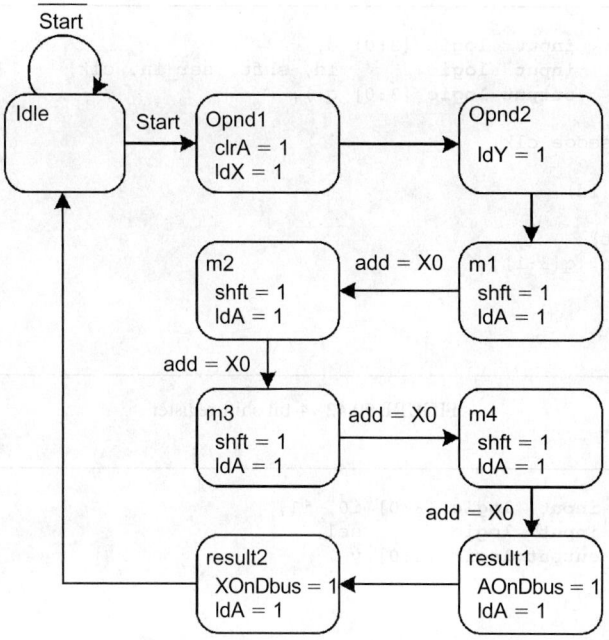

FIGURE 90.39 Add-and-shift multiplier controller.

```
module adder ( input   logic  [3:0] a, b,
               input   logic        c0,
               output  logic  [3:0] s,
               output  logic        c4 );

  always_comb
    {c4, s} = a + b + c0;

endmodule
```

FIGURE 90.40 4-bit adder.

```
module register ( input   logic  [3:0] d,
                  input   logic        clr, ld, clk,
                  output  logic  [3:0] q );

  always_ff @(posedge clk)
  begin
    priority if (clr)
      q = 'b0;
    else if (ld)
      q = d;
  end

endmodule
```

FIGURE 90.41 4-bit register.

```
module shift_reg ( input  logic [3:0] d,
                   input  logic       ld, shft, ser_in, clk,
                   output logic [3:0] q );

  always_ff @(posedge clk)
  begin
    priority if (ld)
      q = d;
    else if (shft)
      q = {ser_in, q[3:1]};
  end

endmodule
```

FIGURE 90.42 4-bit shift register.

```
module mux2to1 ( input  logic [3:0] i0, i1,
                 input  logic       sel,
                 output logic [3:0] y );

  always_comb
    if (sel)
      y = i1;
    else
      y = i0;

endmodule
```

FIGURE 90.43 A 2-to-1 multiplexer.

```
interface bus;
  logic  clrA;
  logic  ldA;
  logic  ldX;
  logic  ldY;
  logic  shft;
  logic  add;
  logic  AOnDbus;
  logic  XOnDbus;
  logic  X0;

  modport dp_mode (input  clrA, ldA, ldX, shft, ldY, add,
                   AOnDbus, XOnDbus, output X0);

  modport cu_mode (output clrA, ldA, ldX, shft, ldY, add,
                   AOnDbus, XOnDbus, input X0);

endinterface
```

FIGURE 90.44 The interface used to interconnect multiplier data path and controller.

As shown in Figure 90.43, the *mux2to1* module is a 4-bit 2-to-1 multiplexer with two 4-bit inputs *i0* and *i1*, a *sel* input and a 4-bit output *y*.

Figure 90.38 shows how the above components should be interconnected to each other to build the multiplier data path. We will use an **interface**, called *bus*, to interconnect the multiplier data path and controller. As illustrated in Figure 90.44, *bus* contains nine signals named *clrA*, *ldA*, *ldX*, *ldY*, *shft*, *add*,

AOnObus, XOnObus, and *X0.* The modes of the interface signals are specified by two **modport** constructs called *dp_mode* and *cu_mode.* For instance all signals are considered as inputs of the data path except *X0* which is an output.

Consider Figure 90.45 which shows the data-path description. As shown in this figure, an interface of type *bus* is declared as data-path port. Its mode, specified as *dp_mode,* declares all signals as input except *X0,* which is declared as output. Interface signals can be accessed using interface instance names followed by ".". together with the signal name. For instance expression *dp_bus.clrA* is used to access the *clrA* signals.

As shown in Figure 90.39, the multiplier controller is designed as a finite-state machine. The Huffman coding style is used to describe the controller shown in Figure 90.46. An interface of type *bus* is declared as the data-path port, and its mode is specified as *cu_mode.*

The interconnection of the multiplier and the data path to build the multiplier is shown in Figure 90.47. An instance of the **interface** *bus* is used to interconnect these two components.

Figure 90.48 illustrates a typical test-bench developed to test the multiplier. This module uses an instance of the multiplier called circuit under test (*cut*), and applies 10 pairs of random data to the multiplier with an **initial** block.

```
module data_path ( input   logic  [3:0]   inbus,
                   output  logic  [3:0]   outbus,
                   input   logic          clk,
                           bus.dp_mode    dp_bus );

   wire [3:0] Y_out;
   wire [3:0] X_out;
   wire [3:0] A_out;
   wire [3:0] adder_out;
   wire [3:0] mux_out;
   wire [4:0] shifted_bus;
   wire       c4;

   // A instantiation
   register A (shifted_bus[4:1], dp_bus.clrA, dp_bus.ldA,
               clk, A_out);

   // Y instantiation
   register Y (inbus, 1'b0, dp_bus.ldY, clk, Y_out);

   // X instantiation
   shift_reg X (inbus, dp_bus.ldX, dp_bus.shft, shifted_bus[0],
                clk, X_out);

   // Adder instatniation
   adder add_4bit (A_out, Y_out, 1'b0, adder_out, c4);

   // Multiplexer instatiation
   mux2to1 mux (A_out, adder_out, dp_bus.add, mux_out);

   // Bussing
   assign outbus = dp_bus.AOnDbus ? A_out :
                   dp_bus.XOnDbus ? X_out : 'bz;

   // Assign shifted_bus
   assign shifted_bus = {(dp_bus.add & c4), mux_out};

   // Assigning X0
   assign dp_bus.X0 = X_out[0];

endmodule
```

FIGURE 90.45   Add-and-shift multiplier data path.

```
module controller ( input   logic          clk,
                     input   logic          reset,
                     input   logic          start,
                     output  logic          done,
                             bus.cu_mode    cu_bus );

   typedef enum [3:0] {idle, opnd1, opnd2, m1, m2, m3, m4, result1, result2}
states;

   states ps, ns;

   always_ff @(posedge clk)
   begin
     if (reset)
       ps = idle;
     else
       ps = ns;
   end

   always_comb
   begin
     case (ps)
       idle:    ns = start ? opnd1 : idle;
       opnd1:   ns = opnd2;
       opnd2:   ns = m1;
       m1:      ns = m2;
       m2:      ns = m3;
       m3:      ns = m4;
       m4:      ns = result1;
       result1: ns = result2;
       result2: ns = idle;
     endcase
   end

   always_comb
   begin
     cu_bus.clrA    = 1'b0;
     cu_bus.ldA     = 1'b0;
     cu_bus.ldX     = 1'b0;
     cu_bus.ldY     = 1'b0;
     cu_bus.shft    = 1'b0;
     cu_bus.AOnDbus = 1'b0;
     cu_bus.XOnDbus = 1'b0;
     done           = 1'b0;
     case (ps)
       idle:    ;
       opnd1:   begin
                  cu_bus.clrA = 1'b1;
                  cu_bus.ldX  = 1'b1;
                end
       opnd2:   cu_bus.ldY = 1'b1;
       m1:      begin
                  cu_bus.shft = 1'b1;
                  cu_bus.ldA  = 1'b1;
                end
       m2:      begin
                  cu_bus.shft = 1'b1;
                  cu_bus.ldA  = 1'b1;
                end
       m3:      begin
                  cu_bus.shft = 1'b1;
                  cu_bus.ldA  = 1'b1;
                end
       m4:      begin
                  cu_bus.shft = 1'b1;
                  cu_bus.ldA  = 1'b1;
                end
```

FIGURE 90.46 The multiplier controller.

```
        result1: begin
                  cu_bus.AOnDbus = 1'b1;
                  done           = 1'b1;
               end
        result2: begin
                  cu_bus.XOnDbus = 1'b1;
                  done           = 1'b1;
               end
     endcase
  end

  assign cu_bus.add = cu_bus.X0 ? 1'b1 : 1'b0;

endmodule
```

FIGURE 90.46 (*Continued*).

```
module multiplier ( input         clk, reset, start,
                    input  [3:0]  inbus,
                    output [3:0]  outbus,
                    output        done );
  bus connection_bus ();

  data_path   dp (inbus, outbus, clk, connection_bus);
  controller  cu (clk, reset, start, done, connection_bus);

endmodule
```

FIGURE 90.47 Add-and-shift multiplier.

```
module test_multiplier;
  logic         clk, reset, start;
  logic [3:0]   inbus;
  logic [3:0]   outbus;
  logic         done;

  byte i;

  multiplier cut (clk, reset, start, inbus, outbus, done);

  initial
  begin
       clk   = 1'b0;
       reset = 1'b1;
    #10 reset = 1'b0;
  end

  always
    #5 clk = ~clk;

  initial
  begin
    for (i = 0; i < 10; i++)
    begin
      #10 start = 1'b1;
      #10 start = 1'b0;
          inbus = $random;
      #10 inbus = $random;
      #70;
    end
    $finish;
  end

endmodule
```

FIGURE 90.48 The multiplier test-bench.

90.5 Assertions

Assertions are used to validate that a design works correctly. Assertions can be checked dynamically by a simulation tool. An assertion or property language is designed to capture the design intent in an executable, formal, and unambiguous way. In System Verilog, there are two types of assertions: immediate assertion and concurrent assertion which are described here.

90.5.1 Immediate Assertion

The immediate assertion statement is used to test a nontemporal expression as a condition in an if statement. If the expression is evaluated to **1** then assertion passes and the first statement, called *pass statement* is executed. And if the expression is evaluated to **0, Z,** or **X** then the assertion fails and the statement associated with **else**, called *fail statement* is executed. Figure 90.49 illustrates an example of immediate assertion designed to verify the multiplier behavior. The use of pass statement and fail statement are optional.

The failure of an assertion has a severity associated with it. There are four severity system tasks that can be used in the fail statement to specify a severity level: **$fatal, $error, $warning,** and **$info.** By default, the **$error** is used as severity level of an assertion failure [1].

90.5.2 Concurrent Assertion

Concurrent assertions describe the clocked behavior of the design. Therefore a concurrent assertion is evaluated only at the clock edges. Figure 90.50 shows an example of using concurrent assertion to verify that the asynchronous *clr* and *pr* input of a flip-flop never be asserted simultaneously. Note that in this example both pass and fail statements are omitted.

Concurrent assertions are often constructed by sequential behavior and described using a list of System Verilog Boolean expressions in a linear order of increasing time. The implication construct (I->) allows to monitor sequences based on satisfying some conditions. The syntax of the implication is as follows:

s1 I-> s2;

where both *s1* and *s2* are sequences. The left-hand side operand of the implication is called *antecedent* sequence while the right-hand side operand is called *consequent* sequence. If the antecedent sequence matches then the consequent sequence should match. If there is no match of antecedent operand then the implication returns true. As an example consider a bus arbiter that receives a *req* input and provides a *gnt* output. Suppose that the *gnt* should occur no more than two clock cycles after the *req* is asserted. Figure 90.51 shows that the concurrent assertion describes this behavior. The ## operator followed by [1:2] is used to illustrate that *gnt* should occur in the next or in the following clock cycles after the clock cycle in which *req* is asserted.

```
assert (product == a * b) $display ("The multiplier works correctly.");
else $error ("The multiplier does not work correctly");
```

FIGURE 90.49 Immediate assertion.

```
assert property (! (clr && pr));
```

FIGURE 90.50 Concurrent assertion.

```
assert property (@(posedge clk) req |-> ##[1:2] gnt);
```

FIGURE 90.51 Bus arbiter.

90.6 Our Coverage of System Verilog

This chapter provides an overview of the new System Verilog language constructs, including new basic data types, user-defined data types, structures, unions, arrays, typecasting, enhancements in loops, new **do-while** loop, new operators, new procedural blocks (**always_comb**, **always_latch**, **always_ff**), and **interface**. In each case the construct was described and several examples were provided to clarify its usage. Then a complete example, an add-and-shift multiplier, was described and modeled by System Verilog. Finally, we had a brief overview of assertions in System Verilog.

References

1. System Verilog 3.1 Language Reference Manual, Accelera's Extensions to Verilog available at http://www.accelera.com.
2. S. Sutherland, S. Davidmann, and P. Flake, *System Verilog for Design*, Kluwer Academic Publishers, Dordrecht, 2004.
3. Z. Navabi, *Verilog Digital System Design*, 2nd Edition, McGraw-Hill, New York, 2005.

91

VHDL-AMS Hardware Description Language

Naghmeh Karimi and
Zainalabedin Navabi

Nanoelectronics Center of Excellence
School of Electrical and Computer
Engineering
University of Tehran

CONTENTS

With the growth of VLSI technology, the design specification is moving toward higher levels of abstraction. For many years, VHDL, Verilog, and other HDLs have been used to model digital components, while the behavior of analog circuits have been described using analog modeling languages such as SPICE.

Modeling and simulation of analog and digital parts of a design separately required individual modeling and simulation tools, while the interaction between these tools for exchanging data was a bottleneck in the design process.

IEEE Standard 1076.1 was presented in 1999 as an extension of the VHDL standard to develop a standard way for modeling and simulation of analog and mixed-signal designs. The IEEE standard 1076.1 along with the IEEE standard 1076 (VHDL standard) formed the VHDL-AMS standard [1,3].

Very high speed integrated circuit Hardware Description Language for Analog and Mixed Signal (VHDL-AMS) allows designers to model digital and analog parts of a design with a uniform description language. VHDL-AMS is a powerful language which supports multilevel and multidomain modeling as well as mixed-signal modeling.

The VHDL-AMS standard is the superset of the VHDL 1076 standard, meaning that a valid model in VHDL 93 is also valid in VHDL-AMS. Digital modeling by means of VHDL has been discussed in another chapter of this book. In this chapter we only deal with the analog and mixed-signal modeling.

91.1 VHDL Analog Constructs

VHDL-AMS supports various energy domains implemented by corresponding packages. The common energy domains presently supported by VHDL-AMS-related packages are *Electrical, Mechanical, Thermal,* and *Fluidic* [1]. An AMS description for analog electronic circuits begins with the specification of the energy domain of such circuits using the following statements:

```
library IEEE;
use IEEE.electrical_systems.all;
```

The *electrical-systems* package enables definition and declaration of domains, analog ports, quantities, and behavior of analog parts using constructs that are described below.

91.1.1 Natures

In VHDL-AMS the domain used in a description is specified by the **nature** construct. A **nature** construct has two floating-point aspects: **across** and **through**. These aspects are used to model force and flow of the related domain, respectively. In addition, a **nature** may represent a **reference terminal** with respect to which the quantity represented by the **across** aspect is introduced. For example, the **across** type, **through** type and **reference terminal** of the *electrical* nature are voltage, current, and ground, respectively. Figure 91.1 shows the definition of the *electrical* nature.

Natures can be scalar or composite. A composite **nature** is either an array or record of natures with the same base nature.

91.1.2 Terminals

An analog interface or internal node is referred to as a **terminal**. Terminals are of the **nature** type. Figure 91.2 shows two examples of **terminal** declaration.

91.1.3 Quantities

Quantities are used to represent continuous values in VHDL-AMS. This standard supports three types of quantities: free, branch, and source.

Free quantities, either scalar or composite, represent analog values. Branch quantities are analog values used to model the **across** or **through** aspects of terminals, and source quantities are used for response and noise modeling of small-signal spectral in the frequency domain [2].

```
nature electrical is
voltage across
current through
electrical_ref reference;
```

FIGURE 91.1 Electrical nature definition.

```
library IEEE;
use IEEE.electrical_systems.all;
use IEEE.math_real.all;

entity NonLinearResistor is
   port (terminal t1, t2: electrical);
end NonLinearResistor;

architecture functional of NonLinearResistor is
   quantity VR across IR through t1 to t2;
begin
   VR==50.0*IR+0.5*IR**3.0 tolerance "default_voltage";
end functional;
```

FIGURE 91.2 A model of a nonlinear resistor.

In an electrical system, **across** and **through** branch quantities represent the voltage and current flow between the terminals of the specified branch.

A branch is declared by its two endpoint terminals. If a branch is between a specified terminal and the **reference terminal** it can be represented by its single specified **terminal**, i.e., mentioning the **reference terminal** is not required in its declaration. Figure 91.2 shows the VHDL-AMS model of a nonlinear resistor. The energy domain of this description is *electrical_systems*. Terminals of this nonlinear resistor are *t1* and *t2* with *electrical* nature type. Branch quantities *VR* and *IR* are used to model **across** and **through** aspects of terminals *t1* and *t2*. The rest of the code shown in this figure defines *VR* in terms of *IR*.

91.1.4 Attributes

VHDL_AMS includes a number of predefined attributes for terminals, quantities, and signals. As in VHDL, attribute names begin with a single quote (').

'reference and **'contribution** are terminal attributes by use of which **across** and **through** quantities between the specified **terminal** and the reference are defined.

'tolerance, 'above, 'delayed, 'dot, 'integ, 'slew, 'ztf, 'ltf, and **'zoh** are predefined quantity attributes in VHDL-AMS. These attributes are discussed below.

- *Q'tolerance.* Q'tolerance string is used to represent the precision of the scalar quantity Q. VHDL-AMS also allows the designers to define tolerance groups whose elements are quantities with the same tolerance string. In Figure 91.2, "*default_voltage*" is defined as the **tolerance** of VR.
- *Q'delayed* (*T*). This quantity is a copy of quantity Q delayed by time *T*.
- *Q'dot.* Q'dot quantity specifies the derivative of quantity Q with respect to time.
- *Q'integ.* Q'integ quantity specifies the integral of quantity Q with respect to time.
- *Q'slew* (*max_rising_slope, max_falling_slope*). This quantity follows quantity Q, but with limited rising and falling slopes.
- *Q'above* (*E*). Q'above is a boolean signal that is true when the value of quantity Q is greater than that of expression *E*.
- *Q'ztf* (*num, den, t, initial_delay*). This quantity represents the *z*-domain transfer function of quantity Q.
- *Q'ltf* (*num, den*). This quantity represents the Laplace-domain transfer function of quantity Q.
- *Q'zoh.* Q'zoh quantity is the sampled version of the quantity Q.

The **'ramp** and **'slew** are signal attributes and are used in mixed-signal modeling. S'ramp (t_rise, t_fall) is a quantity that follows signal S with a specified rise/fall time. S'slew (rising_slope, falling_slope) quantity follows signal S with a specified slope.

91.1.5 Simultaneous Statements

Simultaneous statements are used for analog modeling. Using simultaneous statements, the behavior of an analog system can be modeled by a set of Differential and Algebraic Equations (DAEs). The left- and right-hand sides of each DAE equation are both either scalar floating-point-valued expressions or composite expressions with floating-point subelements. The expressions include any number of (including 0) quantities, constants, literals, signals, and function calls. A simultaneous statement always includes at least one quantity.

For solvability of the DAE equations, the number of equations and unknowns of the equation set must be equal, i.e., the number of simultaneous statements in an architecture must be equal to the sum of the number of free, through, and output interface quantities minus the number of quantities related to output quantities of the instantiated components in the architecture. Simultaneous statements can be used anywhere concurrent statements are allowed [2].

Figure 91.3 models the behavior of a time-variant resister using two simple simultaneous statements. In this model R is a free quantity, while VR and IR are branch quantities. Since this model includes one free and one through quantity, to make it solvable, its behavior is described using two simultaneous equations. The **NOW** function in this model represents the current time.

Figure 91.4 models the behavior of a capacitor with one DAE. This DAE includes the **'dot** attribute to show the direct relation of the capacitor current with the derivative of its voltage.

Simultaneous **if** and **case** statements are two other types of conditional simultaneous statements used for modeling the behavior of an analog design. These statements use keywords **use** and **end use** for bracketing their statements [4]. Figure 91.5(a) models a clipper using simultaneous **if** statements. In this circuit the lower and upper bounds of the output voltage are represented by V_{min} and V_{max}, respectively. The transfer function diagram of this circuit is shown in Figure 91.5(b). The **break** statement used in this model is discussed in the following sections. The terminal *electrical_ref* is for the ground terminal.

```
library IEEE;
use IEEE.electrical_systems.all;
use IEEE.math_real.all;

entity  TimeVariantResistor is
    port (terminal t1, t2: electrical);
end TimeVariantResistor;

architecture functional of TimeVariantResistor is
    quantity VR across IR through t1 to t2;
    quantity R:real;
begin
    R==NOW+1.0;
    VR==IR* R;
end functional;
```

FIGURE 91.3   A model of a time-variant resister ($R=t+1$).

```
library IEEE;
use IEEE.electrical_systems.all;
use IEEE.math_real.all;

entity Capacitor is
    generic (C:real);
    port (terminal t1, t2: electrical);
end Capacitor;

architecture functional of Capacitor is
    quantity VC across IC through t1 to t2;
begin
    IC==C*VC'dot;
end functional;
```

FIGURE 91.4   A model of a time-invariant capacitor.

```
library IEEE;
use IEEE.electrical_systems.all;
use IEEE.math_real.all;

entity Clipper is
    generic(Vmin,Vmax:real);
    port (terminal input,output: electrical);
end Clipper;

architecture functional of Clipper is
    quantity Vin across Ii through input to electrical_ref;
    quantity Vout across output to electrical_ref;
begin
    if (not(Vin'above( Vmin))) use
        Vout==Vmin;
    elsif (not(Vin'above( Vmax))) use
        Vout==Vin;
    else
        Vout==Vmax;
    end use;
    break on Vin'above(Vmin), Vin'above(Vmax);
end functional;
```

FIGURE 91.5(a)   A model of a clipper.

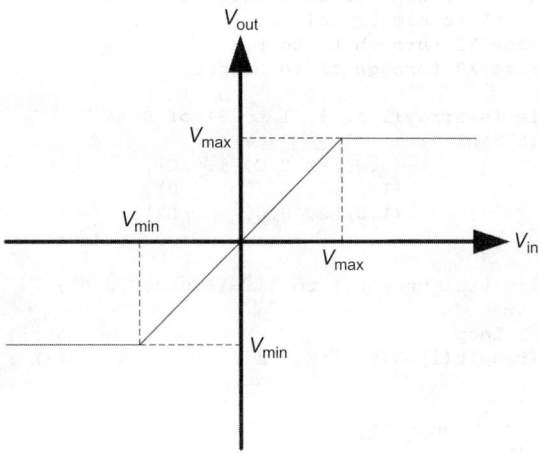

FIGURE 91.5(b)   Diagram of the voltage transfer function of Figure 91.5(a).

Another type of simultaneous statements is the procedural statement. This construct allows us to describe the behavior of an analog design using sequential statements enclosed in the pair of **procedural** and **end procedural** keywords. A procedural statement may be identified with a label.

As with other simultaneous statements, a procedural statement appears where a concurrent statement is allowed. This construct may include any sequential statement except **wait**, signal assignment, and **break** [1].

Figure 91.6(a) shows an example circuit and Figure 91.6(b) represents the circuit model using state-space method. In this model the state variables ($X1$, $X2$, and $X3$) correspond to the voltage of C capacitor, the current of $L1$ inductor, and the current of $L2$ inductor, respectively. This example illustrates the use of procedural statements. As shown, a procedural statement is similar to VHDL process, and allows use of variable declarations, sequential constructs (i.e., loop in this example), and variable assignments.

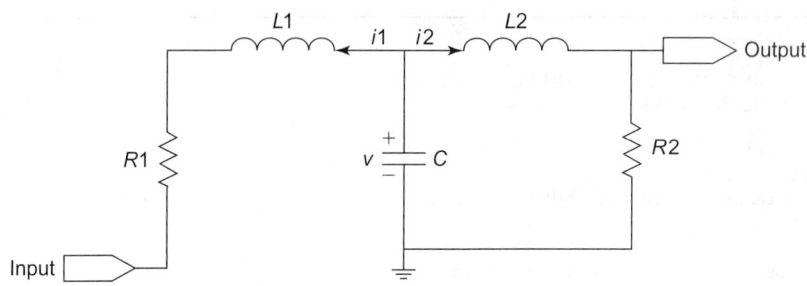

FIGURE 91.6(a)   A sample circuit.

```
library IEEE;
use IEEE.electrical_systems.all;
use IEEE.math_real.all;

entity Circuit is
    generic (L1,L2,R1,R2,C:Real);
    port (terminal input, output: electrical);
end Circuit;

architecture structural of Circuit is
    terminal t1, t2 : electrical;
    quantity Vinput across Iinput through input to electrical_ref;
    quantity X1 across t2 to electrical_ref;
    quantity X2_v across X2 through t2 to t1;
    quantity X3_v across X3 through t2 to output;

    type circuit_table is array(1 to 3, 1 to 3) of Real;
    constant A:circuit_table :=(
                             (0.0,  -1.0/C,-1.0/C),
                             (1.0/L1,-R1/L1,  0.0),
                             (1.0/L2,  0.0,  -R2/L2));
begin
    procedural is
        variable result: real_vector(1 to 3):=(0.0,0.0,0.0);
    begin
        for i in 1 to 3 loop
            result(i):=result(i)+A(i,1)*X1+ A(i,2)*X2+ A(i,3)*X3;
        end loop;
        X1'dot:=result(1);
        X2'dot:=result(2)-Vinput/L1;
        X3'dot:=result(3);
    end procedural;
end structural;
```

FIGURE 91.6(b)   VHDL-AMS model of circuit shown in Figure 91.6(a) based on state-space method.

91.1.6   Break Statements

Break statements are used to model the discontinuity in analog models, including discontinuity at the start time which is initialization. When a discontinuity occurs, the analog solver must be informed of the occurred discontinuity and solve the DAEs with new initial values. For an example see the last part of Figure 91.5(a).

   Break statements can be categorized into two different types: concurrent break statements, and sequential break statements. Both forms can be used in a conditional clause. Concurrent break statements appear anywhere a concurrent statement can be used, while sequential break statements appear within a process that is sensitive to the signal or conditions that cause the discontinuity [4].

Break statements may include one or more quantity assignment constructs. In this case, the current value of the specified quantity is replaced by the value of the represented expression whenever the break statement is executed. Figure 91.7 models an inductor with an initial current. In the body of this description, the inductor current (*IL*) is set to the integral of its voltage (*VL*) over its inductance. The break statement shown sets the initial value of *IL* to 2.0.

Figure 91.8 is the schematic of another example for the description of which we will use the break statement. The code for this example is shown in Figure 91.9. As shown in Figure 91.8, the circuit is composed of an inductor, a resistor, and a switch. The switch position can be altered between *A* (modeled by *Switch='1'*) and *B* (modeled by *Switch='0'*) positions. The behavior of this circuit is modeled in Figure. 91.9.

The **break** statement shown makes the architecture of *Circuit* sensitive to the *Switch* input. Occurring an event on *Switch* input results in describing the behavior of *Circuit* with the following equation provided that *Switch* value is **0**.

$$IL => IL'dot^*(-L)/R$$

This DAE replaces the original DAE (*IL'dot == IL*(−R)/L + Esource/L*) given that *Switch* value is **0**.

Another form of **break** is the **break for** statement. This statement is used for initializing quantities that are not directly in the integral form. A **quantity** is in the integral form if it is an integral **quantity** or if its **'dot quantity** appears in the analog expressions (e.g., *IL* in example of Figure 91.9). To show the discontinuity of a quantity for which its integral form does not exist, the break for statement is used and indicate which implicit equation must be replaced when the break statement is executed. The selector quantity is represented by a **for use** clause. Consider Figure 91.10 as an example for this situation. The

---

```
library IEEE;
use IEEE.electrical_systems.all;
use IEEE.math_real.all;

entity Inductor is
    generic (L:real:=1.0E-3);
    port (terminal t1, t2: electrical);
end Inductor;

architecture functional of Inductor is
    quantity VL across IL through t1 to t2;
begin
    IL==VL'integ/L;
    Break IL=> 2.0;
end functional;
```

FIGURE 91.7   A model of an inductor with initial current.

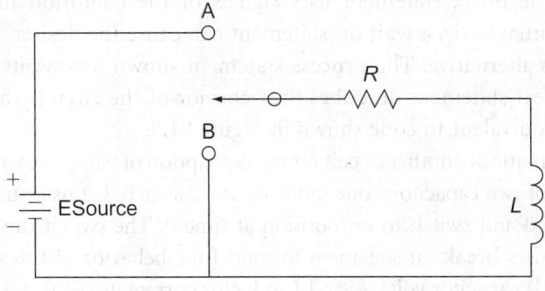

FIGURE 91.8   A sample circuit.

```
library IEEE;
use IEEE.math_real.all;
use IEEE.std_logic_1164.all;
use IEEE.electrical_systems.all;

entity Circuit is
    generic (R:real:=8.0; L:real:=4.0; Esource:real:=10.0);
    port(signal Switch:in std_logic);
end Circuit;

architecture con_break of Circuit is
    quantity IL: real;
begin
    IL'dot==IL*(-R)/L+Esource/L;
    break IL=>IL'dot*(-L)/R on Switch when Switch='0';
end con_break;
```

FIGURE 91.9   A behavioral model of Figure 91.8 circuit using break on statement.

```
library IEEE;
use IEEE.math_real.all;
use IEEE.std_logic_1164.all;
use IEEE.electrical_systems.all;

entity Circuit is
    generic (R:real:=8.0; L:real:=4.0; Esource:real:=10.0);
    port(signal Switch:in std_logic);
end Circuit;

architecture breakfor of Circuit is
    quantity IL,Flux:  real;
begin
    Flux==L*IL;
    Flux'dot==Flux*(-R)/L+Esource;
    break for Flux use IL=>IL'dot*(-L)/R when Switch='0';
end breakfor;
```

FIGURE 91.10   A behavioral model of Figure 91.8 circuit using break for statements.

DAE of this model does not include the derivative of *IL*. Thus to indicate which implicit equation must be replaced when a discontinuity occurs on *IL*, the **break for** statement is used. In this example *IL=>IL'dot*(-L)/R* equation replaces the *Flux'dot=0* implicit equation when a discontinuity occurs on *IL*.

A break statement can also be used in sequential statements. As discussed above, **break on** causes sensitivity to signals that appear after the **on** keyword. Furthermore, for a **break** without the **on** part, the condition part of the break statement uses signals of the condition in an implicit **break on**. Alternatively, in a sequential body, a **wait on** statement can cause the desired sensitivity. The example of Figure 91.11 uses this alternative. The **process** statement shown here waits for an event on *Switch*. When this occurs, the next statement describes the behavior of the circuit when *Switch='0'* condition holds. This example is equivalent to code shown in Figure 91.9.

Figure 91.12 is the schematic of another circuit for the description of which we will use the break statement. This circuit is composed of two capacitors; one inductor and a switch. Let us assume that the switch shown has been off before time 0 and switch to on position at time 0. The switch can be in different positions afterwards. Figure 91.13 uses **break on** statement to model the behavior of this circuit. In this circuit the initial values for *C*1 and *C*2 capacitor voltages and *L* inductor current are −1.0, 4.0 V, and 1.0 A, respectively. Figure 91.14 models the behavior of the circuit shown in Figure 91.12 using sequential **break** statement.

```
library IEEE;
use IEEE.math_real.all;
use IEEE.std_logic_1164.all;
use IEEE.electrical_systems.all;

entity Circuit is
   generic (R:real:=8.0; L:real:=4.0; Esource:real:=10.0);
   port(signal Switch:in std_logic);
end Circuit;

architecture seq_break of Circuit is
   quantity IL:  real;
begin
   IL'dot==IL*(-R)/L+Esource/L;
   process(Switch)
   begin
      break IL=>IL'dot*(-L)/R when Switch='0';
   end process;
end seq_break;
```

FIGURE 91.11   A behavioral model of Figure 91.8 circuit using sequential break statements.

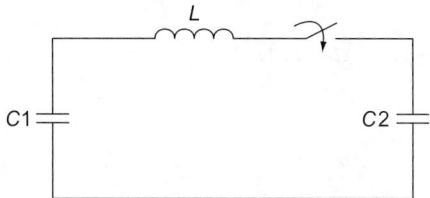

FIGURE 91.12   A sample circuit.

```
library IEEE;
use IEEE.electrical_systems.all;
use IEEE.math_real.all;
use IEEE.std_logic_1164.all;

entity Circuit is
   generic (C1,C2:real:=4.0; L:real:=2.0);
   port (signal SwitchOpen :in std_logic);
end Circuit;

architecture con_break of Circuit is
   terminal t1,t2,t3:electrical;
   quantity VC1 across IC1 through t1 to electrical_ref;
   quantity VC2 across IC2 through t3 to electrical_ref;
   quantity VL across IL through t1 to t2;
   quantity VSwitch across ISwitch through t2 to t3;
begin
   IC1==C1*VC1'dot;
   IC2==C2*VC2'dot;
   IL==VL'integ/L;
   if SwitchOpen='1' use
      ISwitch==0.0;
   else
      VSwitch==0.0;
   end use;
   break VC1=>-1.0, VC2=>4.0, IL=>1.0;
   break on SwitchOpen;
end con_break;
```

FIGURE 91.13   A behavioral model of Figure 91.12 circuit using break on statement.

```
library IEEE;
use IEEE.electrical_systems.all;
use IEEE.math_real.all;
use IEEE.std_logic_1164.all;

entity Circuit is
    generic (C1,C2:real:=4.0; L:real:=2.0);
    port (signal SwitchOpen :in std_logic);
end Circuit;

architecture seq_break of Circuit is
    terminal t1,t2,t3:electrical;
    quantity VC1 across IC1 through t1 to electrical_ref;
    quantity VC2 across IC2 through t3 to electrical_ref;
    quantity VL across IL through t1 to t2;
    quantity VSwitch across ISwitch through t2 to t3;
begin
    IC1==C1*VC1'dot;
    IC2==C2*VC2'dot;
    IL==VL'integ/L;
    if  SwitchOpen='1' use
        ISwitch==0.0;
    else
        VSwitch==0.0;
    end use;
    break VC1=>-1.0, VC2=>4.0, IL=>1.0;
    process(SwitchOpen)
    begin
        break;
    end process;
end seq_break;
```

FIGURE 91.14   A behavioral model of Figure 91.12 circuit using sequential break statement.

91.2   Mixed-Signal Design

A mixed-signal system is composed of digital and analog parts each of which has its own constructs, data formats, and timing models. What is important in a mixed-signal system is the interaction of these two parts to exchange the required data at the correct time.

To utilize continuous values in the digital part of a model, these values must be sampled and scaled to the appropriate values at discrete time instances. Sampling can be performed by utilizing a **process** statement that is sensitive to the sampling event. The **'above** attribute can be used to trigger the sampling event.

After sampling continuous values at discrete-time instances, these values must be converted into appropriate values using type conversions supported by VHDL-AMS. Figure 91.15 shows an example for the use of **'above** attribute.

In contrast, the signals defined in the digital part of a system can be used by the simultaneous statements of the analog part using VHDL-AMS type conversions. To consider the discontinuity that appears on the resulted analog parameters, break statements as well as **'ramp**, and **'slew** attributes can be used. Figure 91.16 shows an example for the use of **'ramp** attribute.

91.3   Frequency-Domain Modeling

VHDL-AMS supports quiescent, time, and frequency simulation domains identified by the domain signal. The domain signal is set to quiescent domain during the initialization phase of simulation by the simulator. While it is assigned to time domain or frequency domain when the system becomes stable at time zero depending on the kind of simulation we wish to perform.

```
library IEEE;
use IEEE.electrical_systems.all;
use IEEE.math_real.all;

entity Analog2Digital is
   generic(threshold:real:=4.0);
   port(terminal t1 :electrical; signal DataOut: out bit);
end Analog2Digital;
architecture simple of Analog2Digital is
   quantity Vin across t1 to electrical_ref;
begin
   DataOut<='1' when Vin'above(threshold) else '0';
end simple;
```

FIGURE 91.15   A model of an analog to digital converter circuit.

```
library IEEE;
use IEEE.electrical_systems.all;
use IEEE.math_real.all;

entity Digital2Analog is
   port(terminal t1 :electrical; signal input: in bit_vector(3 downto 0));
end Digital2Analog;
architecture simple of Digital2Analog is
   signal AnalogResult:real:=0.0;
   quantity Vt across It through t1 to electrical_ref;
begin
   process(input)
   variable result: natural:=0;
   begin
      for i in input'right to input'left loop
         if input(i)='1' then
            result:= 2**i + result;
         end if;
      end loop;
      AnalogResult<=real(result);
   end process;
   Vt==AnalogResult'ramp(1.0E-8);
end simple;
```

FIGURE 91.16   A model of a digital to analog converter circuit.

Time-domain simulation was discussed above. In this section, we deal with the frequency domain. Note that frequency-domain simulation is based on the small-signal model. In this domain, the behavior of a system can be modeled using Laplace- or *z*-domain transfer functions. The Laplace transform is a continuous transfer function, while *z* transform is a discrete one.

Several VHDL-AMS models for the high-pass filter of Figure 91.17 are shown in Figure 91.18. The structural model of this filter is shown in Figure 91.18(a). Figure 91.18(b) models the high-pass filter in time domain, while Figure 91.18(c) and Figure 91.18(d) model this filter in frequency domain using Laplace and *z* transforms, respectively. The latter two models are discussed below.

As shown in Figure 91.18(c), the Laplace transform of a scalar quantity is specified by using the 'ltf attribute. This attribute represents the numerator and denominator polynomials of the corresponding quantity using **real_vector** arrays. Note that the first element of the denominator array cannot be zero.

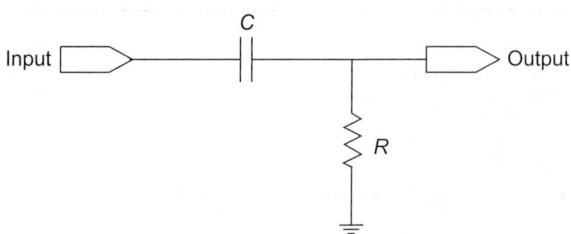

FIGURE 91.17   Schematic of an RC high-pass filter.

```
library IEEE;
use IEEE.electrical_systems.all;
use IEEE.math_real.all;

entity HighPassFilter is
   port (terminal input,output: electrical);
end HighPassFilter;

architecture structural of HighPassFilter is
begin
   comp1:entity work.Capacitor (functional) -Capacitor model of Fig. 4
      generic map(0.01E-6);
      port map(input,output);
   comp2:entity work.Resistor (functional)- A Linear Time-Variant Resistor
      generic map(1.0E+3);
      port map(output,electrical_ref);
end structural;
```

FIGURE 91.18(a)   Structural model of the high-pass filter of Figure 91.17.

```
library IEEE;
use IEEE.electrical_systems.all;
use IEEE.math_real.all;

entity HighPassFilter is
   port (terminal input,output: electrical);
end HighPassFilter;

architecture functional of HighPassFilter is
   quantity Vi across input to electrical_ref;
   quantity Vo across Io through output to electrical_ref;
   Constant tp:real:=0.01E-3;
begin

   Vi==Vo+Vo'integ/tp;-- tp=RC
end functional;
```

FIGURE 91.18(b)   Modeling the high-pass filter of Figure 91.17 using 'dot attribute.

Consider the high-pass filter shown in Figure 91.17 and its Laplace-domain model in Figure 91.18(c). The voltage transfer function of this circuit is

$$\frac{V_{\text{out}}}{V_{\text{in}}} = \frac{S}{S + w_{\text{p}}}$$

```
library IEEE;
use IEEE.electrical_systems.all;
use IEEE.math_real.all;

entity HighPassFilter is
    port (terminal input,output: electrical);
end HighPassFilter;

architecture LaplaceForm of HighPassFilter is
    quantity Vi across input to electrical_ref;
    quantity Vo across Io through output to electrical_ref;
    Constant wp:real:=1.0E+5;
    Constant num:real_vector:=(0.0, 1.0);
    Constant den:real_vector:=(wp, 1.0);

begin
    Vo==Vi'ltf(num, den);
end LaplaceForm;
```

FIGURE 91.18(c)   Modeling the high-pass filter of Figure 91.17 using Laplace transform.

```
library IEEE;
use IEEE.electrical_systems.all;
use IEEE.math_real.all;

entity HighPassFilter is
    port (terminal input,output: electrical);
end HighPassFilter;

architecture ZForm of HighPassFilter is
    quantity Vi across input to electrical_ref;
    quantity Vo across Io through output to electrical_ref;
    Constant wp:real:=1.0E+5;
    Constant T:real:=1.0E-5;--Sampling priod
    Constant a0:real:=2.0;
    Constant a1:real:=-2.0;
    Constant b0:real:=2.0+T*wp;
    Constant b1:real:=-2.0+T*wp;
    Constant num:real_vector:=(a0, a1);
    Constant den:real_vector:=(b0, b1);

begin
    Vo==Vi'ztf(num, den, T);
end ZForm;
```

FIGURE 91.18(d)   Modeling the high-pass filter of Figure 91.17 using *z* transform.

where $w_p = \frac{1}{R \times C}$. This transfer function is represented by *num* and *den* vectors in the model of Figure 91.18(c). In these constants, coefficients of the numerator and denominator polynomials starting with 0 power are presented in ascending order.

The behavior of a continuous object can also be modeled by the *z*-transform. To approximate the discrete transform of a continuous object, the values of that object must be sampled at discrete times.

VHDL-AMS provides two mechanisms for *z*-domain modeling. The first mechanism uses **'zoh** and **'delayed** attributes to perform the sample-and-hold process and the *z*-domain delay (z^{-1}) implementation. The second method combines the sample-and-hold process and the *z*-domain delay implementation by using **'ztf** attribute [2].

Figure 91.18(d) shows the *z*-domain model of the high-pass filter shown in Figure 91.17. In this figure, the *z* transform of the V_i quantity is specified by the **'ztf** attribute. The arguments of this attribute represent numerator and denominator polynomials of the corresponding quantity as well as the sampling period. Note that the *z*-domain transfer function of a circuit is created from its Laplace-domain transfer function by using bilinear transform. Bilinear transform replaces all occurrence of operator S in the given Laplace-domain transfer function with

$$\frac{2}{T}\left(\frac{1 - z^{-1}}{1 + z^{-1}}\right)$$

where *T* represents the sampling period. As discussed above the voltage transfer function of the circuit shown in Figure 91.17 in Laplace-domain is

$$\frac{V_{out}}{V_{in}} = \frac{S}{S + w_p}$$

Using bilinear transform, the voltage transfer function of this circuit in *z*-domain is represented by

$$\frac{V_{out}}{V_{in}} = \frac{a_0 + a_1 \times z^{-1}}{b_0 + b_1 \times z^{-1}}$$

where $\quad a_0 = 2$
$\qquad a_1 = -2$
$\qquad b_0 = 2 + T \times w_p$
$\qquad b_1 = -2 + T \times w_p$

As before, *num* and *den* of Figure 91.18(d) that are used with the **'ztf** attribute are **real_vector** array constants. These arrays represent numerator and denominator coefficients in ascending order.

As discussed, quantities have three categories of free, branch, and source. Free and branch quantities have already been discussed. In this section we will present source quantities. A source quantity is used for response and noise modeling of small-signal spectral in the frequency domain. A spectral source **quantity** is a stimulus for the frequency-domain simulation. To model a sinusoidal signal of the form $A = A_m \cos(wt + \alpha)$, a source **quantity** can be used. For example, a source **quantity** representing signal $A = 5\cos(wt + \pi/6)$ is defined as

quantity A: real **spectrum** 5, math_pi/6.0

As shown above, a spectral source quantity represents the magnitude and phase of the related source as well as the type of the quantity values. The value of a spectral source quantity is zero except during frequency-domain simulation.

91.4 Noise Modeling

For precise modeling of the behavior of a system, noise impact must also be considered. Noise can be modeled as a kind of source quantities in VHDL-AMS named as noise source quantities. To represent a noise source, its power as well as the type of the quantity values should be identified. The value of a **noise** source quantity is zero except during noise simulation. Both kinds of source quantities, spectral and noise quantities, allow the user to specify stimulus for small-signal frequency-domain simulation. Figure 91.19 shows an example of modeling the behavior of a resistor including the effects of thermal noise.

```
library IEEE;
use IEEE.electrical_systems.all;
use IEEE.math_real.all;

entity Resistor is
    generic(R:real; temp:real:=1.0; R0:real:=500.0E+3);
    port (terminal t1, t2: electrical);
end Resistor;

architecture noisy of Resistor is
    constant K:real:=1.0E-5;
    quantity VR across IR through t1 to t2;
    quantity ThermalNoiseSource:real noise 4.0*K*temp/R0;
begin
    IR==VR/R+ThermalNoiseSource;
end noisy;
```

FIGURE 91.19   A model of a nonlinear resister including noise effect.

91.5   Our Coverage of VHDL-AMS

VHDL-AMS is tied with the VHDL hardware description language. The presentation above can only be useful if it is considered in conjunction with VHDL. Chapter 87 of this book provided an introduction to this language. Like the VHDL chapter, the purpose of this chapter was to provide a general knowledge and overall structure of the VHDL-AMS language. VHDL-AMS is a very complex and complete language and readers are encouraged to refer to some of the references listed at the end of this chapter for a more complete presentation of the language and its applications.

References

1. *IEEE Standard VHDL Analog and Mixed-Signal Extensions,* IEEE Std 1076.1, IEEE, Inc., New York, 1999.
2. P.J. Ashenden, G.D. Peterson, and D.A. Teegarden, *The system Designer's Guide to VHDL-AMS: Analog, Mixed-Signal, and Mixed-Technology Modeling,* Morgan Kaufmann, Los Altos, CA, 2003.
3. S.A. Huss, *Model Engineering in Mixed-Signal Circuit Design: A Guide to Generating Accurate Behavioral Models in VHDL-AMS,* Kluwer Academic Publishers, Dordrecht, 2001.
4. U. Heinkel, M. Padeffke, W. Hass et al., *The VHDL Reference: A Pratical Guide to Computer-Aided Integrated Circuit Design including VHDL-AMS,* Wiley, New York, 2000.

92

Verification Languages

Hamid Shojaei

Nanoelectronics Center of Excellence
School of Electrical and Computer Engineering
University of Tehran

Zainalabedin Navabi

Nanoelectronics Center of Excellence
School of Electrical and Computer Engineering
University of Tehran

CONTENTS

Present-day systems are complex and proving the correctness of a design has become a major concern. Simulation-based methods do not provide enough capabilities for design validation and Correctness Checking of a design. In recent years, formal methods [1,2,3] and assertion-based verification (ABV) methods [4] are used as alternative approaches to prove the correctness of hardware designs. Since these methods overcome some of the limitations of traditional validation techniques such as simulation and testing, more designers are moving toward them to guarantee the correctness of their designs.

Formal verification constructs mathematical proofs for the behavior of design. Model checking is a powerful formal verification technique, which gets a model of design plus some desired properties and exhaustively verifies whether or not the model satisfies all the desired specifications under all possible input sequences. In this method the model characteristics are specified in a property specification language such as computation temporal logic [1,2,3] (CTL), which will be described in first section of this chapter. This section presents an introduction to model checking of hardware systems and discusses the basic elements required for understanding of model checking. We will then discuss the details of CTL for specification of properties.

In ABV method, assertions capture the design properties. They specify both legal and illegal behaviors of a circuit. The ABV method compares the implementation of a design against its specified assertions. Discrepancies found will be reported.

Three standards are proposed that are related to ABV: open verification library (OVL), property specification language (PSL [4]), and system verilog assertions (SVA). Of these standards, PSL, which is a powerful property language and a language for ABV, is discussed below. This section introduces the concept of ABV, and then covers an introduction to the PSL language. The introductory material provides the readers with writing the basic properties and assertions for their verilog or VHDL designs.

92.1 Computation Temporal Logic

In this section, theory and practice of model checking and related topics are first introduced and then CTL and its application in specification of properties will be described.

92.1.1 Formal Verification

The need for reliable hardware systems is an important issue and the involvement of these systems in our daily life is increasing day by day. The rapid growth of technology inquires into the methods that increase our confidence in the correctness of these systems. To increase the level of this confidence, we highly need robust techniques to verify these systems. Simulation is the traditional approach for verifying finite-state systems, but its problems are: working much slower than the real system, being expensive and having no guarantee for all possible input combinations to be simulated. For these reasons, the application of formal verification (FV) is increasing every day. Formal verification is the process of checking whether a design satisfies for requirements (properties) specified in a logical language or not.

Formal verification is a new emerging hardware validation method. It is an alternative to simulation in some designs, and complementary to simulation in many others. Different formal verification methods have been proposed that make it useful for hardware verification. Among these methods model checking is mostly used which is described in the next subsection.

92.1.2 Model Checking

Model checking [1,2,3] is the most popular formal verification technology for property verification. In this method, the verification problem is reduced to graph algorithmic problems that can be fully automated. This method, which is relatively easy to use, generates a counterexample if the property is not satisfied. A counterexample describes conditions in the design under which a property cannot be satisfied.

Model checking uses transition systems (Kripke structure) to model systems [1,3] and temporal logics to specify properties. To understand the term "model," we need to be familiar with transition systems and Kripke structures which are described as follows.

92.1.2.1 Transition System

A transition system [1,3] is a structure $TS = (S, S0, R)$, where S is a finite set of states, $S0 \subseteq S$ the set of initial states, and $R \subseteq S \times S$ a transition relation that should be total, i.e., for every s in S there exists s' in S such that (s, s') is in $R (\forall s \in S \; \exists \; s' \in S, (s,s') \in R)$.

92.1.2.2 Kripke Structure

A Kripke structure [1,3] models a system defined by $M = (S, S0, R, AP, L)$. In this structure, $(S, S0, R)$ is a transition system. An atomic proposition directly corresponds to a variable in the design being verified. AP is a finite set of atomic propositions. L is a labeling function that labels each state with a set of atomic propositions that are true in that state. Together, atomic propositions and L convert a transition system into a model.

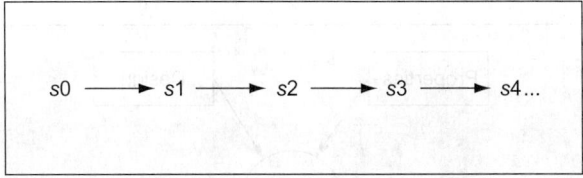

FIGURE 92.1 A computation.

A computation [3] in a model, as shown in Figure 92.1, is the basic object of reasoning. A complete computation consists of several time frame each expressed in a path or a tree. Linear time logic (LTL) expresses a computation in a path, while CTL is used for a computation tree with branches. In linear time we use LTL and consider individual computations starting from $s0$ (the first state). In contrast, in branching time we look at a computation tree (CTL) originating from $s0$.

As mentioned before, two main aspects of model checking are model representation and property specification. Models are represented by Kripke structures which are described here. For property specification we used temporal logics described later.

92.1.3   Temporal Logics

Properties are characteristics of a design that should hold if the design is to perform certain functions. For example, a property of a 4-bit BCD counter that rolls over is that the next count after 9 is always 0. The foremost step to verify a system is to specify the properties that the system should have. For example, if we want to prove that a system never deadlocks, we have to specify a set of properties for that system. These properties are represented by temporal logic. Temporal logic refers to representation of time-dependent information within a logical framework.

CTL is a version of temporal logic which is currently one of the popular frameworks used in verifying properties of concurrent systems.

In this section, we focus on temporal logic for model checking. Once we are familiar with the important properties, the second step is to construct a formal model for the system being verified. The model should capture properties that must be considered for the establishment of correctness of a system. Model checking includes traversing the state transition graph (Kripke structure) and verifying whether it satisfies the formula representing the property or not; or more concisely, whether the system is a model of the property or not.

92.1.4   CTL Model-Checking Steps

A CTL formula [3] in a given state of a Kripke structure is either true or false. Its truth is evaluated from the truth of its subformulas in a recursive fashion, until atomic propositions (variables of system being verified) which are either true or false in a given state, are reached. A formula is satisfied by a system if it is true for all the initial states of the system.

Mathematically, suppose a Kripke structure $K = (S, S0, R, AP, L)$ (system model) and a CTL formula Ψ (specification of the property) are given. We have to determine if $K \models \Psi$ is satisfied; this expression reads as K is a model of Ψ. It holds if $K \models \Psi$ holds for every initial state, $s0$, of K in the set of $S0$ states, i.e., $K, s0 \models \Psi$ for every $s0 \in S0$. If the property does not hold, the model checker will produce a counterexample, which is an execution path that cannot satisfy the specified formula (Figure 92.2).

92.1.5   CTL Formulas

Atomic propositions, standard Boolean connectives of propositional logic (e.g., AND, OR, NOT), and temporal operators are used all together to build CTL formulas.

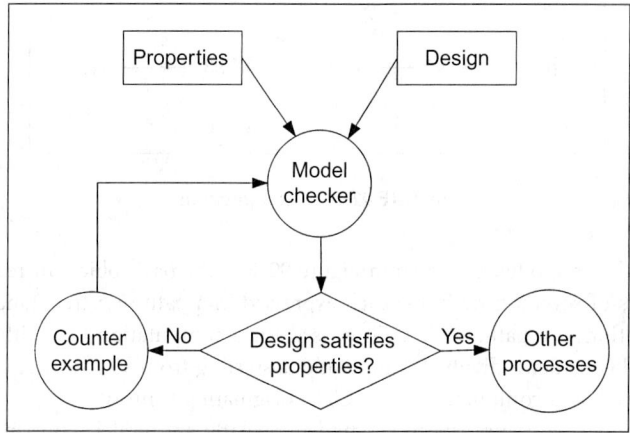

FIGURE 92.2   Model-checking process.

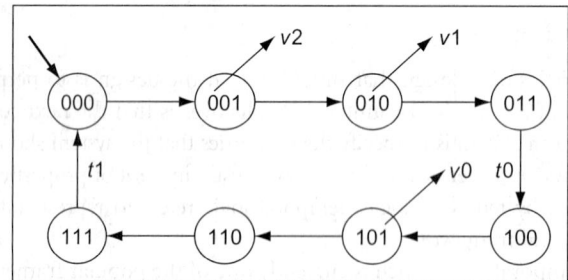

FIGURE 92.3   Transition system of a mod-8 counter.

A temporal operator is composed of two parts, a path quantifier (universal (**A**) or existential (**E**)) followed by a temporal modality (**F, G, X, U**), which are described in the next subsections. A CTL property is interpreted relative to an implicit "current state". There are generally many execution paths (sequences) of state transitions of a system starting from the current state.

A path is an infinite sequence of states $(s_0, s_1, s_2, \ldots)$ for $(s_i, s_{i+1}) \in R$ to hold for all i. Consider the modulo-8 counter shown in Figure 92.3, variables of system in this example are $v0$, $v1$, and $v2$. The domain of these variables is $\{0, 1\}$, which means that these variables can hold either 0 or 1. As mentioned above, a state is a function that assigns a value to each variable in its domain. For example, if $s(v0) = 0$, $s(v1) = 1$, and $s(v2) = 1$, the state is (1 1 0).

Since a set of states can be selected by a formula, we can predicate states using this state diagram. For example, $X = v2 \lor v0$ selects the set $\{100, 101, 110, 111, 001, 011\}$. Similar to states, since a set of transitions can also be selected by a formula, we can predicate the transition relation as well. For example, consider function $R3 = (v2 \neq v2')$ where $v2$ and $v2'$ are the values before and after a transition respectively. As shown in Figure 92.3, the only transitions in which $v2$ in one state is not equal to $v2$ of the next state are $t0$ and $t1$, therefore $R3 = \{t0, t1\}$.

92.1.5.1   Path Quantifiers

The path quantifier indicates if the modality, which defines a property, is true for all the possible paths or the property only holds on a few paths. The former case is denoted by the universal path quantifier, **A**, while the latter case is denoted by the existential quantifier, **E**. Properties that begin with **A** are called **ACTL** and those beginning with **E** are called **ECTL**.

92.1.5.2 Temporal Operators

The temporal modalities describe the ordering of events in time in an execution path. They can be **F, G, X,** and **U** described below.

- **F** ∅ (which is read as "'∅' holds sometime in the future") is true in a path if there exists a state in that path where formula '∅' is true.
- **G** ∅ (which is read as "'∅' holds globally") is true in a path if '∅' is true at every state in that path.
- **X** ∅ (which is read as "'∅' holds in the next state") is true in a path if '∅' is true in the state reached immediately after the current state in that path.
- ∅ **U** φ (which is read as "'∅' holds until 'φ' holds") is true in a path if 'φ' is true in a state in that path, and '∅' holds in all preceding states of that state.

92.1.6 CTL Syntax

The general syntax of CTL expressions including variable names, indexing, formulas, and combination of formulas are described below. Examples provided later in this chapter show utilization of these structures.

- **TRUE, FALSE,** and *var-name == value* are *CTL* formulas, where *var-name* is the name of a variable, and *value* is a legal value in the domain of that variable.
- *var-name1 == var-name2* is the atomic formula that is true if *var-name1* has the same value as *var-name2*. *var-name1[i:j] == var-name2[k:l]* can be used if the lengths of vectors are equal. These characters may be used for variable names and values: `A-Z  a-z  0-9  ^  ?  |  /  [  ]  +  *  $  <  >  ~  @  _  #  %  :  "  '`.
- If *f* and *g* are CTL formulas, the following combinations are also considered as formulas. In these formulas, symbols of Table 92.1 are used.
- (f), f ∧ g, f ∨ g, f ^ g, !f, f → g, f ←→ g,
- **AG** f, **AF** f, **AX** f, **EG** f, **EF** f, **EX** f, **A**(f **U** g) and **E**(f **U** g).

In the above, **AX**:*n*(f) is allowed as a shorthand for **AX**(**AX**(. . . **AX**(f). . .)), where *n* is the number of invocations of **AX**. Also, the term **EX**:*n*(f) is defined similarly. The operators used in the formulas and shown in Table 92.1 have the precedence shown in Table 92.2. A complete formula, the elements of which are shown above, should be followed by a semicolon. Text written from # to the end of a line is considered as a comment.

92.1.7 CTL Semantics

A brief overview of the semantics of CTL is given below. In this overview K is considered as a Kripke structure and s a typical state of this structure. The semantics of the CTL operators are stated below:

- $K, s \models \mathbf{EX}(\Psi)$: There exists s' such that $s \rightarrow s'$ ($R(s, s')$) and $K, s' \models \Psi$. It means that s has a successor state s' at which Ψ holds.

TABLE 92.1 Symbols in CTL

Symbols	Meaning
∧	AND
∨	OR
^	XOR
!	NOT
→	IMPLY
←→	EQUIV

TABLE 92.2 CTL Operator Precedence

High
!
AG, AF, AX, EG, EF, EX
$\wedge$
$\vee$
$\wedge$
$\longleftrightarrow$
$\rightarrow$
U
Low

- $K, s \mid= \mathrm{EU}(\Psi 1, \Psi 2)$ *iff* there exists a path $L = s_0, s_1, \ldots$ from s *and* $k >= 0$ such that $K, L(s_k) \mid= \Psi 2$, and if $0 \leq j < k$, then $K, L(s_j) \mid= \Psi 1$. It means that $\Psi 1$ should hold in all states of a path until we reach to a state in which $\Psi 2$ is satisfied.
- $K, s \mid= \mathrm{AU}(\Psi 1, \Psi 2)$ iff for every path $L = s_0, s_1, \ldots$ from s there exists $k >= 0$ such that $K, L(s_k) \mid= \Psi 2$, and if $0 \leq j < k$, then $K, L(s_j) \mid= \Psi 1$. It means that for all paths $\Psi 1$ should hold in all states until we reach to a state in which $\Psi 2$ is satisfied.
- ***AX*** (Ψ): There is no case where a next state exists at which Ψ does not hold, i.e., for every next state Ψ holds.
- ***EF*** (Ψ): There exists a path L from s and $k >= 0$ such that $K, L(s_k) \mid= \Psi$.
- ***AG*** (Ψ): There is no case where a path L from s exists and $k >= 0$ such that $K, L(s_k) \mid= \Psi$, i.e., for every path L from s and every $k >= 0$; $K, L(s_k) \mid= \Psi$.
- ***AF***(Ψ): For every path L from s, there exists $k >= 0$ such that $K, L(s_k) \mid= \Psi$.
- ***EG***(Ψ): There is no case where for every path L from s there is a $k >= 0$ such that $K, L(s_k) \mid= \Psi$. It means that there exists a path L from s such that, for every $k >= 0$: $K, L(s_k) \mid= \Psi$. In other words, there is path in which all states satisfy Ψ.

To clarify the concept, several basic CTL operators stated above are shown graphically in Figure 92.4 to clarify the concept. In this figure, states that satisfy the property are marked using $\sqrt{}$ and states that do not satisfy the property are marked using $\times$.

92.1.8   CTL Formula Conversions

After property specification and model representation, we should verify the model against each of its properties that are described in CTL. For this purpose, a CTL formula should be checked against the states of the model being verified. For a universal CTL formula (ACTL), all states in a design that are reachable from the initial states should be checked. However, for an existential CTL formula (ECTL), only one case from the initial states should be found that satisfies the formula. It is clear that algorithms of existential CTL formula can be implemented easier than that of universal CTL formula.

Universal formulas can be converted into existential formulas. That is, all universal path quantifiers can be replaced with the appropriate combination of existential quantifiers and Boolean negations. "*finally*" (**F**) operators are also converted into "*until*" (**U**) operators. This returns a new formula that may look different from the original one (even the strings are different), but has the same semantics. These conversions are shown in Table 92.3.

92.1.9   Building a Computational Tree

We usually use the CTL, which is a version of temporal logic, for specifying formulas for model checking. But usually, design representations are not in tree forms. They are usually in the form of a transition

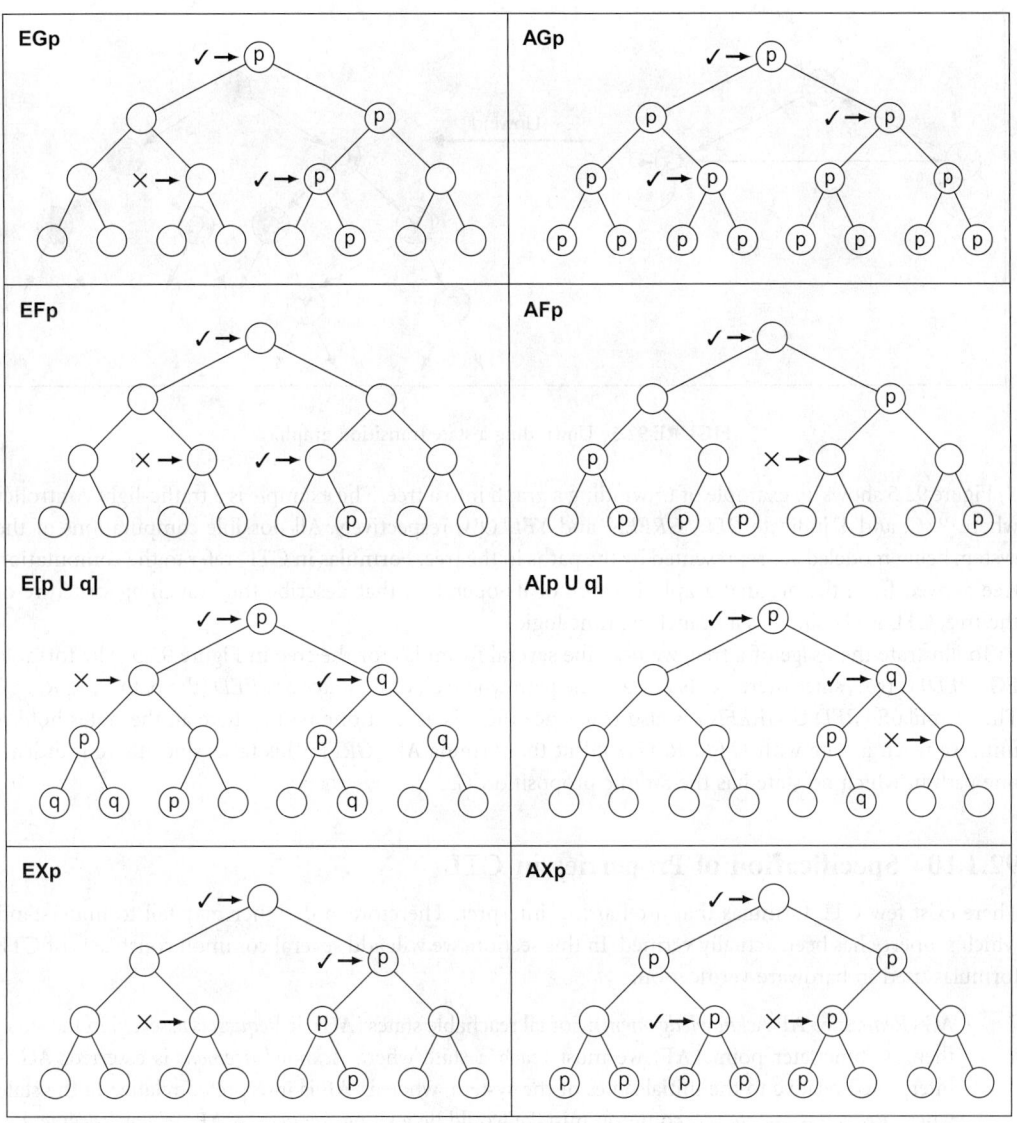

FIGURE 92.4 Basic CTL operators.

TABLE 92.3 Conversion of CTL Formulas

Formulas	Same As
AX f	~**EX** (~f)
EF f	**E** (True **U** f)
AG f	~**EF** (~f)
AF f	~**EG** (~f)
A (f **U** g)	~**E**[~g **U** (~f ∧ ~g)] ∧ ~**EG** ~g

graph called state transition graph (STG). In model checking, a STG is used to derive a computation tree that CTL formulas are applied to. In the following discussion, we show how a tree can be built from a given STG. For this purpose, the graph structure is unwounded into an infinite tree rooted at the initial state.

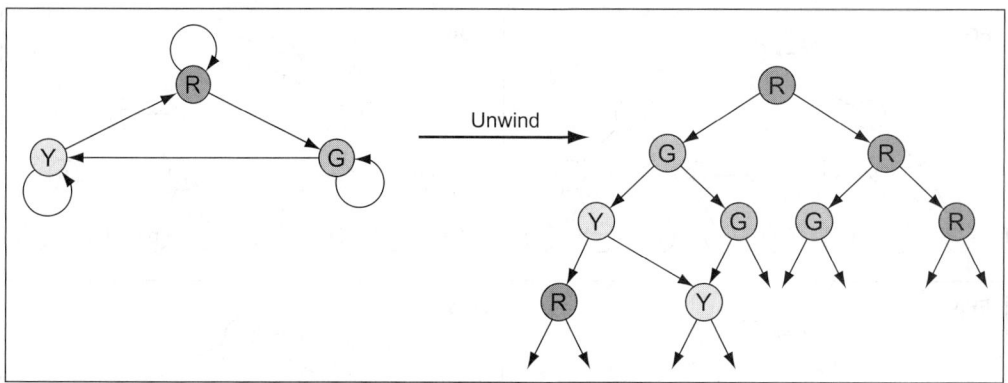

FIGURE 92.5 Unwinding a state transition graph.

Figure 92.5 shows an example of unwinding a graph into a tree. The example is a traffic-light controller where *R*, *G*, and *Y* indicate *RED*, *GREEN*, and *YELLOW*, respectively. All possible computations of the system being modeled are represented by the paths in the tree. Formulas in CTL, refer to the computation tree derived from the original graph. Because of its operators that describe the branching structure of the tree, CTL is classified as a branching time logic.

To illustrate the usage of a tree, we describe several formulas for the tree in Figure 92.5. The formula **EG** *(RED)* is true, since there exists at least one path where all of its states are *RED* (the path *R*, *R*, *R*, …). The formula **E** *(RED* **U** *GREEN)* is also true since there is at least one path, where all the states hold *R* until we reach a state with *G* (*R*, *R*, *G* …). But the formula **AF** *(GREEN)* is false, since there is at least one path in which no state has the atomic proposition *G*.

92.1.10 Specification of Properties in CTL

There exist few CTL formulas that are hard to interpret. Therefore, a designer may fail to understand which property has been actually verified. In this section, we will add several common constructs of CTL formulas used in hardware verification.

- **AG** (*Request* → **AF** *Acknowledgement*): For all reachable states (**AG**), if *Request* is asserted in the state, then at some later point (**AF**) we must reach a state where *Acknowledgement* is asserted. **AG** is interpreted relative to the initial states of the system, whereas **AF** is interpreted relative to the state where *Request* is asserted. A common mistake would be writing *Request* → **AF** *Acknowledgement* in the **AG** (*Request* → **AF** *Acknowledgement*). The meaning of the former is if *Request* is asserted in the initial state, it is always the case that eventually we reach a state where *Acknowledgement* is asserted. The latter requires that the condition is true for any reachable state where *Request* holds. If *Request* is identically true, **AG** (*Request* → **AF** *Acknowledgement*) reduces to *Acknowledgement*.
- **AG** (**AF** *DeviceEnabled*): The proposition *DeviceEnabled* holds indefinitely on every computational path, starting with the starting state.
- **AG** (**EF** *start*): From any reachable state, there must be a path starting from that state which reaches a state where *start* is asserted. In other words, it must always be possible to reach the start state.
- **EF** (*x* ∧ **EX** (*x* ∧ **EX** *x*)) → **EF** (*y* ∧ **EX** **EX** *z*): If it is possible for *x* to be asserted in three consecutive states, then it is also possible to reach a state where *y* is asserted; and from there to reach a state where *z* is asserted in two more steps.
- **EF** (~*Ready* ∧ *Started*): It is possible to reach a state where *Started* holds, but *Ready* does not hold.
- **AG** (*Send* → **A** (*Send* **U** *Receive*)): It is always the case that if *Send* occurs, then eventually *Receive* is true, and until that time, *Send* must continue to be true.
- **AG** (*in* → **AX** **AX** **AX** *out*): Whenever *in* goes high, *out* will go high within three clock cycles.

- **AG** (*~storage_coke*→ **AX** *storage_coke*): If the coke storage of a vending machine becomes empty, it is recharged immediately.
- **AG AF** ((*~storage_coke* ∨ *~storage_coffee*) ∧ (*storage_coke* ∧ *storage_coffee*)): The recharge transaction of a vending machine (of coke and coffee) often takes place indefinitely.

92.1.11 Model-Checking Example: SAYEH Controller

In this section, we illustrate model checking using CTL with a small example that is the controller part of a simple processor called SAYEH. This processor has also been described in Chapter 86. The architecture of this processor is simple, but it has a good number of hardware components for formal verification. The SAYEH processor has a 16-bit data bus and a 16-bit address bus. It has 8- and 16-bit instructions. Short instructions may contain shadow instructions, which effectively pack two such instructions into a 16-bit word. The state transition graph of its controller is shown in Figure 92.6.

The controller's Kripke structure is described below by stating the set of states *S*, the initial state *S0*, the atomic proposition *AP*, the labeling function *L* and the transition relation *R*. The details of this structure are:

- *AP* = {reset, halt, fetch, decode, exec}
- States are *S* = {1, 2, 3, 4, 5}
- Initial state is *S0* = {1}
- Labeling functions are:
  - *L*(1) = {reset, ~halt, ~fetch, ~decode~, exec}
  - *L*(2) = {~reset, halt, ~fetch, ~decode, ~exec}
  - *L*(3) = {~reset, ~halt, fetch, ~decode, ~exec}
  - *L*(4) = {~reset, ~halt, ~fetch, decode~, exec}
  - *L*(5) = {~reset, ~halt, ~fetch, ~decode, exec}
- The transition relations *R* are shown in Figure 92.6.

Let us take a specification to check if it is satisfied by this Kripke structure. Let *State (f)* be the set of all states labeled by the subformula *f*. Here, we first proceed with the atomic formulas and then with the more complicated ones. These atomic formulas or others, which do not cover the whole formula, are termed as subformulas.

Consider the CTL formula and its simplified form shown below for evaluation.

$$\Psi = \textbf{AG} \; (\texttt{reset} \rightarrow \textbf{AF} \; \texttt{decode})$$

$$\texttt{\~}\textbf{EF} \; (\texttt{reset} \; \wedge \; \textbf{EG} \; \texttt{\~decode})$$

The starting state is given as:

$$\texttt{State (reset) = \{1\}}$$

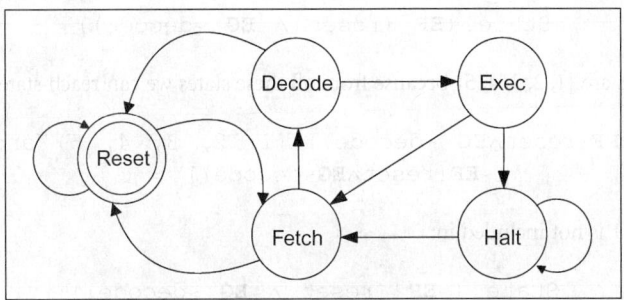

FIGURE 92.6 State machine of *SAYEH* controller.

To find all states that satisfy the formula,

$$\sim\mathbf{EF} \ \ (\text{reset} \ \wedge \ \mathbf{EG} \ \sim\text{decode})$$

We start with *reset*. Since *reset* is an atomic proposition, state {1} in which *reset* is active is the result of the evaluation of the formula.

The same can be said about the *~decode* that is also an atomic proposition. This leads to the following result of the evaluation of the formula:

$$\text{State} \ \ (\sim\text{decode}) = \{1, 2, 4, 5\}$$

Computing the above formula for states starting in **AG**, **EG**, etc. is more complex since they are not atomic propositions. For **EG**(*~decode*) we first find states where *~decode* holds, i.e.,

$$\text{State} \ \ (\sim\text{decode}) = \{1, 2, 4, 5\}$$

then we find

$$\text{State} \ \mathbf{EG}(\sim\text{decode}) = \{1, 2, 4, 5\}$$

Therefore the states for the formula,

$$(\text{reset} \ \wedge \ \mathbf{EG} \ \sim\text{decode})$$

will be the intersection of the following sets:

$$\text{State} \ (\text{reset}) \ \text{ and State} \ (\mathbf{EG} \ \sim\text{decode})$$

So the result will be as follows:

$$\text{State} \ (\text{reset} \ \wedge \ \mathbf{EG} \ \sim\text{decode}) \ = \ \{1\}$$

Here, we have to compute

$$\text{State} \ (\mathbf{EF} \ (\text{reset} \ \wedge \ \mathbf{EG} \ \sim\text{decode}))$$

as we have already found that

$$\text{State} \ (\text{reset} \ \wedge \ \mathbf{EG} \ \sim\text{decode}) \ = \ \{1\}$$

We need to include those states where there is at least one path to enter state 1 for computing

$$\text{State} \ (\mathbf{EF} \ (\text{reset} \ \wedge \ \mathbf{EG} \ \sim\text{decode}))$$

Clearly, these states are {1, 2, 3, 4, 5} because from all these states we can reach state 1. So

$$\text{State} \ (\mathbf{EF}(\text{reset}\wedge\mathbf{EG} \ \sim\text{decode})) = (1, \ 2, \ 3, \ 4, \ 5) \ \text{ and state}$$
$$(\sim\mathbf{EF}(\text{reset}\wedge\mathbf{EG}\sim\text{decode})) \ = \ \varnothing$$

As the initial state 1 is not included in:

$$\text{State} \ (\sim\mathbf{EF} \ (\text{reset} \ \wedge \ \mathbf{EG} \ \sim\text{decode}))$$

The system described by this Kripke structure does not satisfy this specification.

All other properties of this state machine are written in the CTL property language. These properties are in three classes. With these classes explained below, any state machine can be completely verified. The three classes are as follows:

- The first class of properties should be checked for all states. This class is divided into three sets of properties that are:
  - "There is no deadlock in any state." This property is expressed in CTL as:

    AG((Pstate = S) → **EX**(Pstate != S), S is from {reset,fetch, decode,exec}

  - "States are reachable from the initial state (*reset*)." This property is presented in CTL as:

    AF((Pstate = S), S is from {reset, fetch, decode, exec}

  - "Each state is reachable from any other state." This property is shown in CTL as:

    AG((Pstate = exec) → **EX**(Pstate = reset)

- The second class of properties verifies state transitions. In this class "*immediate states after each state*" are checked. The following is an example.

$$\textbf{AG}((\text{Pstate} = \text{exec}) \rightarrow \textbf{EX}(\text{Pstate} \mathrel{!=} \text{exec})$$

- The third class of properties check transitions between states with respect to the input signals and instructions. The following is an example.

$$\textbf{AG}((\text{Pstate} = \text{exec} \ \& \ \text{ExternalReset} = 1) \rightarrow \textbf{AX}(\text{Pstate} \mathrel{!=} \text{reset})$$

92.1.12 Our Coverage of CTL and Formal Verification

Formal verification replaces simulation in certain applications. For testing the correctness of a digital system that consists of FSMs, formal verification is efficient and easy to use. This is an exact method and does not require test data. This section convers the topics related to model checking of hardware systems, focusing on the CTL property language that is used for property specification. In addition, we covered basics of developing properties for components of a design. We also presented ways of verifying a typical system. Methods described here are general and are applied to most hardware systems. Another property language used for formal verification and assertion verification is PSL. This language will be discussed in the next section.

92.2 Property Specification Languages

Use of property based verification [4] is increasing among the design and verification community. As mentioned before, there are two major methods in property based verification: ABV and FV. CTL that was described in the pervious section is a property language which is used only in formal verification. On the other hand, PSL [4] is a standard assertion language and a language for description of properties. It is used by engineers to specify functional properties of logic designs.

A property is a Boolean valued fact about a design-under-test and PSL is designed to capture design properties in an executable, formal, unambiguous manner. It uses many of the underlying HDL operators and expression syntax to build necessary Boolean expressions in properties. On top of that, when required, PSL defines its own syntax, (separate from HDLs) to build complex temporal relationships among the Boolean expressions.

In contrast, an assertion is a conditional statement that checks for a specific behavior and displays a message if it occurs. Assertions are generally used as monitors looking for bad behaviors, but may be used to create an alert for a desired behavior. Assertions are added during verification to monitor conditions that are otherwise hard to check using simulation and sometimes, they are used to simplify the debugging of complex design problems. Assertion monitors can be considered as internal hardware test points that wait for a particular problem to occur and then alert the designer when it does. In these cases, assertions are used to improve the ability of observing bugs once they are triggered by a simulation vector.

Application of PSL for formal verification is similar to that for CTL described in the previous subsection. So in this section, we focus on PSL as an assertion language. It is important to note that all assertions can be used as properties for a formal verification tool. In this section that follows we first show all components of a typical PSL assertion to give an overview of this language. Then we will describe the details of PSL.

92.2.1 Components of a PSL Assertion

The general structure of a PSL assertion is shown in a simple property in Figure 92.7. The subsections that follow, describe various parts of this structure.

Various pieces that comprise a complete PSL assertion are described in the following subsections.

92.2.1.1 Label

Label is optional for every assertion. It is a recommended practice to use a meaningful name for assertions. It helps in identifying failures, success reports coming from a PSL tool.

92.2.1.2 Verification Directive

It is also possible for users to specify an action to be executed when an assertion passes or a possibly different action to be executed when the assertion fails. PSL has a set of constructs to build complex properties. A property itself is a declaration and a verification tool does not know what to do with it unless told otherwise. A verification directive sits on top of a property and instructs a tool whether the property should be checked to satisfy or it should never be true.

92.2.1.3 Occurrence Operators

Occurrence operators as part of the temporal layer are means to specify "when to check for a property." PSL supports the following occurrence operators:

- always
- never
- eventually!
- next

The **always** operator is the most frequently used one and it specifies that the property expression which follows it should be checked every clock.

92.2.1.4 Property to be Checked

The property to be checked forms the core of PSL which allows properties to be declared and then used in an assertion or simply specifies the complete property in the assertion itself as shown in Figure 92.7. Later, we will describe this part in more details.

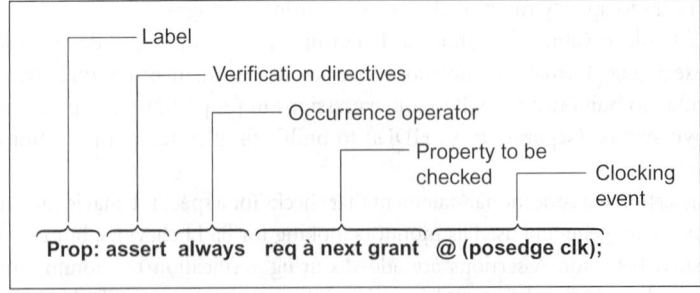

FIGURE 92.7 Components of a PSL assertion.

92.2.1.5 Clocking Event

Properties can be either clocked or unclocked. A clock for a property can be specified in the property definition as shown in Figure 92.7. The symbol @ is used to specify clocking while any Boolean expression can be used for it. PSL also allows a default clock specification as given below:

$$\textbf{default}\ \texttt{clock} = \texttt{(posedge clk);}$$

Using this, assertion and its equivalent, one can be specified as:

$$\textbf{assert always}\ \texttt{\{req}\ \texttt{|}\to\ \texttt{ack\};}$$

$$\textbf{assert always}\ \texttt{\{req}\ \texttt{|}\to\ \texttt{ack\}}\ \texttt{@(posedge clk);}$$

92.2.2 PSL Flavors

In the previous sections PSL assertions that we showed used syntax that was similar to Verilog. However, this is not necessarily a requirement of PSL. PSL is a multiflavored, multilayered language. A flavor is the syntax used in a property that may be of VHDL or Verilog. This syntax dictates the syntax for the Boolean expressions, clocking and in general, the way a property is written. Properties shown below have VHDL and Verilog flavors of the same.

```
VHDL_Prop: assert always not (busy1 and busy2) @(rising_edge(clk));
    Verilog_Prop: assert always ~(busy1 & busy2) @(posedge clk);
```

92.2.3 PSL Layers

In Figure 92.7, a PSL assertion is shown. Different layers are distinguished in this assertion. All properties in PSL are structured. They have Boolean, temporal, verification, and modeling layers. The details of these layers are described in a later section. Here, we will merely discuss the categorization of these layers.

The Boolean layer forms a Boolean expression that becomes part of a property. An example for the Boolean layer is "**not** (ack1 **and** ack2);". This example property states that *ack1* and *ack2* should not be asserted simultaneously. The second layer in PSL is the temporal layer that consists of sequential extended regular expression (SERE) and properties. An example for this layer is "{req;busy[*1:10];ack};", which is a sequence of three Boolean expressions. In this layer, the timing of the expressions is expressed. Using this sequence, a property can be formed such as "reset |=> {req;busy[*1 to 10];ack}". The next layer is the verification layer that constructs a property. An example for this layer is "**assert** (reset |=> {req;busy[*1 to 10];ack});", which uses the property we just formed. This layer applies the verification directive (**assert**) to the defined property.

The modeling layer is used for formal verification. This layer associates PSL properties with a hardware model we are verifying. Other alternatives such as file name may also be used for this association.

92.2.4 Property Expressions

Specification of properties in PSL can be as simple as a Boolean equation that is true for just one clock cycle or can be a complex temporal SERE that is constructed from multiple Boolean expressions in multiple cycles. So we only use *Boolean layer* and *temporal layer* to build a property and then a verification directive from the verification layer converts this property into an assertion. The next two subsections describe property expressions as *Boolean layer* and *temporal layer*.

92.2.4.1 Boolean Layer

The Boolean layer consists of Boolean expressions containing variables and operators from the underlying language. All expressions that can be used as a condition in the underlying language can be used for

Boolean expressions of the Boolean layer. Two simple Boolean expressions in Verilog and VHDL languages are shown below, respectively:

$$(a \; \& \; b) \; == \; 0$$

$$(a \; \textbf{and} \; b) \; = \; 0$$

As shown, at the Boolean layer, PSL looks very similar to the underlying HDL. This characteristic of PSL is often referred to as language neutrality.

92.2.4.2 Temporal Layer

The temporal operators of PSL sit on top of the LTL operators. These temporal operators include **always, never, next, until,** and **before,** among others, and are described later.

The Boolean layer forms the core of PSL and the real power of PSL comes from its temporal layer. The term temporal refers to the design behavior expressed as a series of Boolean expressions over multiple clock cycles. To support this, PSL has two major components in the temporal layer: *sequences* and *properties*. Sequences are built from basic Boolean expressions and sequence operators such as repetition operators. Properties are built on the top of sequences and can include Boolean expressions, sequences, and other properties.

92.2.4.2.1 Sequences

It is necessary for an assertion language to be able to express design behavior over multiple clocks. PSL supports SEREs to meet this requirement. PSL provides an easy and familiar way for engineers to capture sequential behavior. The syntax is derived from standard UNIX regular expressions and hence the name SERE. The first and foremost requirement of any temporal sequence is a quick way to move time forward. PSL uses SERE concatenation to achieve this. This operator is represented with a semicolon.

For example, this pseudo-code: {a;b} describes the following behavior:

- *a* being high in the current clock tick,
- wait until the next clock tick ($t + 1$) and
- check for *b* being high.

The curly brackets around the sequence mark the beginning and ending of a SERE. In real life, the delay between two such expressions can be

- More than one
- A range
- Not necessarily occurring in contiguous clock cycles

PSL supports all these requirements via its repetition operators. There are three types of SERE repetition operators, which are consecutive repetition operator, nonconsecutive repetition operator, and GOTO repetition operator. These operators are discussed below.

a. Consecutive repetition operator

The consecutive repetition operator is used to specify that a signal must be asserted continuously for *n* clocks. The following example states that *busy* signal is asserted for three clocks.

$$\{busy;busy;busy\}$$

Alternatively, the following shortcut can be used:

$$\{busy[*3]\}$$

We can also specify a range for a SERE with a MIN and MAX specifications as shown below:

```
{busy[*MIN:MAX]}
```

Few notes on the ranges:

- Both MIN and MAX have to be constants.
- Both have to be natural numbers (≥ 0).
- MIN can be set to 0.
- MAX can be set to the keyword *inf* to indicate infinity.

The following SEREs show several possible repetition operators. They all start when signal *req* is asserted. In the very next clock, *busy* is asserted. The number of clocks that this signal (*busy*) remains asserted is different in the properties shown. For all four properties, the sequence finishes when an *ack* is seen after a desired number of *busy*.

```
Prop_1:req; busy[*2]; ack;

Prop_2:req; busy[*0:100]; ack;

Prop_3:req; busy[*2:inf]; ack;

Prop_4:req; busy[+]; ack;
```

Property *Prop_1* states that when *req* is asserted, signal *busy* can be asserted two clocks later and then signal *ack* should be asserted. Property *Prop_2* states that when *req* is asserted, signal *busy* can be asserted between 0 to 100 clocks later and then, signal *ack* should be asserted. Property *Prop_3* states that when *req* is asserted, signal *busy* can be asserted between 2 and infinity clocks later and then, signal *ack* should be asserted. Property *Prop_4* states that when *req* is asserted, signal *busy* should be asserted at least 1 clock later and then signal *ack* should be asserted.

b. Nonconsecutive repetition operator

To have a repetition in which occurrences of the repeated expression or sequence needs not to be contiguous, a nonconsecutive repetition operator can be used. PSL uses the "=" symbol to denote nonconsecutive repetitions. The examples in the previous section with a nonconsecutive repetition are shown below:

```
Prop_1:req; busy[=2]; ack;

Prop_2:req; busy[=0:100]; ack;

Prop_3:req; busy[=2:inf]; ack;
```

c. GOTO repetition operator

The GOTO repetition operator is used to go to the *n*th repetition of the Boolean expression that it follows and immediately after the occurence of that last repetition, checks for the next expression in the sequence. The intermediate repetitions may be nonconsecutive. This is referred to as GOTO repetition in PSL and is represented with the "->" symbol. The examples in the previous section with a GOTO repetition are shown below:

```
Prop_1:req;  busy[->2];  ack;

Prop_2:req;  busy[->0:100];  ack;

Prop_3:req;  busy[->2:inf];  ack;
```

In the above, property *Prop_1* states that when *req* is asserted, we go to the second repetition of the *busy* signal and immediately after the occurrence of that signal, *ack* should be asserted.

92.2.4.2.2 SEREs within Operator

To construct a SERE in which one sequence's start and end points are fully contained within the other sequence, the SERE **within** operator can be used. In the following example,

```
Sere1  within  Sere2
```

sere1's start point should be after *sere2* and its end point should be before that of *sere2*.

92.2.4.2.3 Compound SERE Operators

Repetition operators give the ability to build basic SEREs in PSL. In contrast, compound SERE operators combine two or more sequences and describe complex sequences. PSL provides the following operators for building compound SEREs.

- Fusion operator (:)
- SERE nonlength matching AND (**&**)
- SERE length-matching AND (**&&**)
- SERE OR operator (|)

Table 92.4 shows all formats of SEREs and their descriptions.

92.2.4.2.4 Occurrence Operators

In addition to sequences, temporal (occurrence) operators are used in temporal layer to construct temporal properties. PSL provides several operators to specify when to check for a property. In the PSL LRM, these are referred to as "simple FL properties" (FL stands for foundation language) and in this section, they are termed as "occurrence operators" for clarity.

The temporal operators sit on top of the LTL operators. These temporal operators include **always**, **never**, **next**, **until**, and **before**, among others. The meaning of these operators is clear, but there are a few exceptions.

The **always** operator states that its operand holds in every single cycle, while the **never** operator states that its operand fails to hold in every single cycle. The **next** operator states that its operand holds in the cycle that immediately follows. Hence the assertion

```
assert always req -> next acknowledge;
```

means that whenever the HDL signal *req* is true, the HDL signal *acknowledge* must be true in the following cycle. The meaning of a cycle will typically be specified either by defining a default clock or by including the clocking operator @ within the property. Note that when *req* is true, this assertion says nothing about the value of *acknowledge* in any cycle other than the following cycle. It also says nothing about the value of *acknowledge* when *req* is false. In this case, true is returned as result. This assertion only says that whenever *req* is *true*, it must be the case that *acknowledge* is true in the very next cycle.

TABLE 92.4 Sequential Extended Regular Expressions

SEREs	Description	Example
<SERE>;<SERE>	Concatenation operator	`{req; busy; ack}`
<sequence1>:<sequence2>	Sequences2 comes after sequence1 and they overlap by one clock cycle	`{fetch; decode; exec} : {req; ack}`
<sequence1> | <sequence2>	Sequence1 or sequence2 are satisfied at a specific clock cycle	`{req; ack} | {req; busy}`
<sequence1> & <sequence2>	Sequence1 or sequence2 start at the same clock cycle, they need not be of the same length	`{fetch; decode; exec} & {req; ack}`
<sequence1> && <sequence2>	Sequence1 or sequence2 start at the same clock cycle and they need to be of the same length	`{fetch; decode; exec} && {lint[*]}`
<SERE>[*n]	SERE is repeated in *n* consecutive clock cycles	`{busy[*5]}`
<SERE>[*]	SERE is repeated for 0 or any number of clock cycles	`{busy[*]; busy; ack}`
<SERE>[+]	SERE is repeated for 1 or more clock cycles	`{busy[+]; ack}`
<SERE>[*n:m]	SERE is repeated for *n* to *m* number of clock cycles	`{busy[*2:5]} |=> {ack}`
<SERE>[=n]	SERE is repeated for *n* nonconsecutive clock cycles	`{req[=3]} |=> {ack}`

The **next** operator can take a number of cycles as an argument, enclosed in square brackets

$$\texttt{assert always } \texttt{req} \texttt{ -> } \textbf{next}\texttt{[2] (acknowledge);}$$

This assertion means that whenever *req* is true, *acknowledge* must be true two cycles later. It says nothing about the value of *acknowledge* one cycle after *req* is true. An interesting feature of this assertion is that it must hold in every single cycle, such that if *req* was true for three consecutive cycles, then *acknowledge* must be true for three consecutive cycles, but with a latency of two cycles.

For the meaning of the **until** operator consider the following example:

$$\texttt{assert always } \texttt{req} \texttt{ -> } \textbf{next}\texttt{ (busy until acknowledge);}$$

This property states that whenever *req* is true, *busy* is true in the following cycle and *busy* remains true until the first cycle in which *acknowledge* is true. When *acknowledge* goes true, the value of *busy* is not important. This is the same for any subsequent cycles. If *req* goes true and then *acknowledge* goes true in the following cycle, *busy* needs not to go true at all.

For the **before** operator consider the following example:

$$\texttt{assert always } \texttt{req} \texttt{ -> } \textbf{next}\texttt{ (busy } \textbf{before} \texttt{ acknowledge);}$$

This property states that whenever *req* is true, *busy* must be true at least once in the period starting in the following cycle and ending in the last cycle before *acknowledge* is true.

Table 92.5 shows all temporal operators in PSL and their descriptions.

92.2.5 Suffix Implication Operators

In a class of design properties, a property/sequence is expected to hold only after a condition occurs. This is also referred to as *an implication operator*. Suffix implication operators in PSL can either be of

TABLE 92.5 Temporal Operators in PSL

Temporal Operators	Description	Example
always <property or sequence>	The property or sequence should be always true	`always req -> ack`
never <property or sequence>	The property or sequence should never hold	`never busy1 and busy2`
next <property>	Property holds one clock cycle later	`always req -> next ack`
next[n] (<property>)	Property holds on the *n*th next cycles	`always req -> next[2](ack)`
next_a[<range>] (<property>)	Property holds at all clock cycles of a range of clock cycles	`always next_a[2:inf](req -> ack)`
next_e[<range>] (<property>)	Property holds at least once in the range of clock cycles	`always next_e[*2:5]( req -> ack)`
next_event (<bool>) [n] (<property>)	Property holds at *n*th occurrence of Boolean expression	`always next_event (rd)[1](req)`
next_event_a (<bool>) [<range>] (<property>)	Property holds at least once in the range of Boolean occurrences	`always next_event_a (req)[2:inf](next ack)`
<property> **until** <property>	Must hold until a certain event	`always (busy until ack)`
<property> **before** <property>	Must hold before a certain event	`always ack -> (req before ack)`
<property> **abort** <bool>	Terminate at certain event	`always (acknowledge -> (busy until done) abort reset)`
<sqeuence> |=> <sqeuence>	Suffix implication, precondition is followed by another sequence in the next clock cycle	`req |=> {busy;ack}`
<sequence> |-> <sequence>	Overlapping suffix implication, precondition is followed but another sequence starts in the last clock cycle	`busy1 |-> !busy2`
whilenot (<bool>) <sequence>	Sequence should hold until Boolean occurs	`whilenot (reset) {req;busy;ack}`

the overlapping suffix implication operator (I->) type or of the nonoverlapping suffix implication operator (I=>) type.

In the overlapping suffix implication operator, the condition is checked in the same clock as the precondition occurs. For example, to describe a property in an arbiter as "whenever there is a process using resource (indicated by *busy1*), the other process should be free (*busy2* should be low)," we can have this assertion:

```
busy1 |-> !busy2
```

In nonoverlapping suffix implication operator, the condition is checked one clock after the precondition occurs. A simple example would be to check for an arbiter:

```
req |=> ack;
```

The above property describes that once a process asserts the *req* signal, the next clock cycle shall make the arbiter to assert *ack* signal.

92.2.6 Verification Directives

Verification directives are used to direct a verification tool to check for the validity of properties. Without verification directives, a verification tool does not know what to do when an assertion passes or which action to be executed when the assertion fails. PSL supports the following directives:

- **assert**
- **assume**
- **assume_guarantee**
- **cover**
- **restrict**
- **restrict_guarantee**
- **fairness**
- **strong_fairness**

Among these, **assert** and **cover** are the most frequently used directives in simulation-based verification methods. Other directives are mostly used in formal verification. The **assert** directive instructs the tool to check whether the property is satisfied and if not, to report a failure. The **cover** directive checks if the sequence was satisfied during verification. PSL allows labeling of such directives, and it is a good coding style to use descriptive labels for them.

Other directives are described in Table 92.6.

92.2.7 Operator Precedence

The precedence of PSL operators, used above, is shown in Table 92.7.

92.2.8 Specification of Properties in PSL

In PSL, a property forms the top-level part of an assertion. A property can comprise sequences and Boolean expressions that are combined using various property operators described in this section.

The simplest form of property in PSL takes the form of a combinational Boolean condition that must be continuously true.

```
assert always CONDITION;
```

TABLE 92.6   Verification Directives in PSL

Verification Directives	Description	Example
assert <property>;	Verify that a property holds	`assert always` req -> ack;
assume <property>;	Assume that the property holds during verification	`assume never` busy1 && busy2;
assume_guarantee <property>;	Treated as **assume** if the **vunit** that links set of properties to a design and the directive is defined in, binds to the top level, and treated as **assert** otherwise	`assume_guarantee never` busy1 && busy2;
restrict <sequence>;	Constrain verification according to a specific sequence2	`restrict` {reset; !reset[*]};
restrict_guarantee <sequence>;	Treated as **restrict** if the **vunit** that the directive is defined in, binds to the top level, and as **assert** otherwise	`restrict_guarantee` {!wr[*]; rd; [*]};
cover <sequence>;	Check if the sequence was satisfied during verification	`cover` {state == BUSY; [*]; state == IDLE};

TABLE 92.7   Precedence of PSL Operators

Precedence	Operator	Description
Low	\<boolean\>	HDL operators
	@	Clocking operator
	; [*] [=] [->]	SERE construction operators
	:	& &&	Sequence implication operators
	|-> |=>	Foundation language implication operators
	always never next* within* whilenot* G F X	Foundation language occurrence operators
High	abort until* before *	Termination operators

However, this form of property is not particularly useful, since it leads to race hazards. It is more common to introduce a sampling event or clock. The following checks condition on every positive edge of clock.

```
assert (always CONDITION) @(posedge clk);
```

It is also possible to define a default clock and thus avoid the need to repeat the explicit clock operator @ in every single assertion.

```
default clock = (posedge clk);

assert always CONDITION;
```

It is more common for a property to take the form of an implication, with a precondition that must be satisfied before the main condition is checked.

```
assert always PRECONDITION -> CONDITION;
```

This property implies that whenever *PRECONDITION* holds, *CONDITION* must hold in the same cycle. The symbol -> denotes logical implication.

It is common for the precondition and condition within the assertion to each take the form of temporal sequences enclosed in braces.

```
assert always {a;b} |-> {c;d};
```

The sequence {a; b} holds if a holds in a cycle and b holds in the following cycle. The symbol | -> placed between the two sequences denotes suffix implication, meaning that if the first sequence holds, the second sequence must also hold, with the two sequences overlapping by a single cycle.

Finally, it is common for properties to include a terminating condition (such as a reset), which will cause the property to be abandoned through the matching of a temporal sequence.

```
assert (always ({a;b} |-> {c;d} abort reset)) @(posedge clk);
```

This property states that when the reset signal is asserted, condition evaluation is removed.

92.2.9   Verification Units

PSL is a property language that talks about the design and needs to be linked to a design unit for a tool to check if the design meets the requirements as described by PSL properties. PSL supports a set of verification units as containers of properties so that a set of properties can be linked to a design. Of these, **vunit** is the most commonly used and is described below.

92.2.9.1 Using *vunit*

A **vunit** is used as a container for PSL properties. In a property, there are various signals used in the basic Boolean expressions that are expected to exist in a design. The link to the design under test occurs via an argument to the *vunit* specification. These properties can be either bound to a design module or an instance of a module.

```
vunit <name> (<hierarchical_name>) {
        <psl_declarations>;
    <verification_directives>;
        }
```

92.2.10   Built-In Functions

PSL provides some of the most commonly used functions that can be used inside property expressions. Table 92.8 shows these functions and their descriptions.

The fourth layer in PSL is a modeling layer that is described in the next subsection.

92.2.11   Modeling Layer

The modeling layer is used for formal verification. This layer associates PSL properties with a hardware model that is being verified. For this association, we can use similar file names for the model and the property set. For example, a model's file name may be *File1.vhd* and its associated property set may be *File2.psl*. Other alternatives may also be possible.

As with the Boolean layer, the modeling layer comes in the underlying language format. PSL format in this layer defines design hierarchies and model correspondences. Different formats are used for VHDL and Verilog.

TABLE 92.8 Built-In Functions in PSL

Built-In Function	Description	Example
rose(<bool>)	This function returns a Boolean result that is true when there is a transition from 0 to 1. Else the result is false.	**rose** (start)} |=> {rerq; ack};
fell(<bool>)	Boolean was true at previous clock cycle and false at current.	**never fell**(working) && !done;
prev(<expression>)	A function, returns the value of <expression> in the previous clock cycle.	**always** ((!req -> next (**prev**(process_num) == process_num));
prev(<expression>, n)	This function returns the value of the argument in the previous clock. The argument can be any expression. The optional argument N specifies how many clock cycles to look back in the history and is defaulted to 1.	{(!req)[*3]; req} |=> {**prev** (process_num, 5) == process_num };

92.2.12 An Example: SAYEH ALU

In this section, we use SAYEH ALU as our example for verification. We will show various parts of properties that are needed for this example. This ALU is a 16-bit combinational logic. Its data inputs are 16-bit *A* and *B* and its data output is *aluout*. The ALU, shown in Figure 92.8, has control inputs that determine the function it performs. In addition, the *cin* input is its carry input used in arithmetic operations. Control outputs of this unit are *zout* and *cout* that are the zero and carry flags, respectively.

ALU function control inputs control its transparency, logic and arithmetic functions. For readability of Verilog code of the ALU, Verilog definitions shown in Figure 92.9 are used for functions of the ALU. The ALU has nine function control inputs, and only one is active at any one time.

Figure 92.10 shows the complete Verilog code of the ALU. Ports of the *ArithmeticUnit* module are defined as described above. The complete description of this ALU consists of an **always** block that is sensitive to the input ports of this module. To guarantee that this module is combinational and therefore avoid latches, all ALU outputs are set to their inactive values at the beginning of the **always** block of Figure 92.10.

Figure 92.11 shows several properties that are written in PSL for our ALU example. When this ALU is being used, we do not expect its control inputs to be asserted simultaneously. For example, *AandB* and *AorB* cannot both be 1 at the same time. The first property shown in Figure 92.11 checks this. The next seven properties shown in this figure check the output of the ALU output (*aluout*) to have the correct functionality when its corresponding function control input is active.

In the property set shown, two properties check that if *cin* is 0, then the ALU output is the sum of the two inputs if *AaddB* is 1 and subtraction of the two if *AsubB* is 1. Another property shown verifies that *cout* is affected when add or subtract operation is being done. The next to last property in Figure 92.11 says that if *aluout* is 0, *zout* must always be 0, which is of course the way this ALU has to function. The last property verifies that if add or subtract operation is being done and the data on *A* is 0 and data on *B* is all 1s (FFFF) and *cin* is 1, then we will definitely have a carry output.

Properties of Figure 92.11 are written with the Verilog flavor.

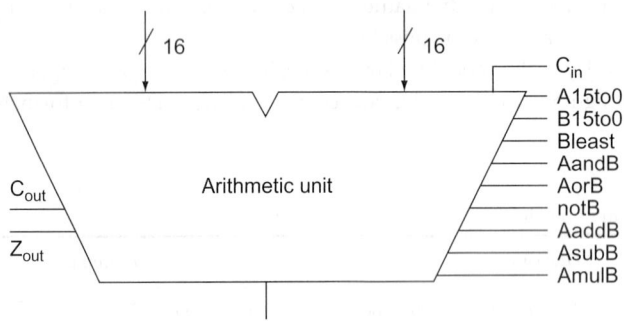

FIGURE 92.8   SAYEH ALU.

```
`define A15to0  9'b100000000
`define B15to0  9'b010000000
`define Bleast  9'b001000000
`define AandB   9'b000100000
`define AorB    9'b000010000
`define notB    9'b000001000
`define AaddB   9'b000000100
`define AsubB   9'b000000010
`define AmulB   9'b000000001
```

FIGURE 92.9   Verilog definitions.

```
//Arithmetic Logic Unit (ALU)
`include "ALUoperations.v"

module ArithmeticUnit (A, B,
            A15to0, B15to0, Bleast, AandB, AorB, notB, AaddB, AsubB, AmulB,
            aluout, cin, zout, cout);
input [15:0] A, B;
input A15to0, B15to0, Bleast, AandB, AorB, notB, AaddB, AsubB,  AmulB;
input cin;
output [15:0] aluout;
output zout, cout;
reg [15:0] aluout;
reg zout, cout;

always @(A or B or A15to0 or B15to0 or Bleast or AandB or AorB or notB or
            AaddB or AsubB  or AmulB  or cin)
begin
    zout = 0; cout = 0; aluout = 0;
    case ({A15to0, B15to0, Bleast, AandB, AorB, notB, AaddB, AsubB, AmulB})
        `A15to0: //Pass A[15:0] through unchanged
            aluout = A;
        `B15to0: //Pass B[15:0] through unchanged
            aluout = B;
        `Bleast: //Pass B[7:0], left-fill 0
            aluout = {8'h00, B[7:0]};
        `AandB:
            aluout = A & B;
        `AorB:
            aluout = A | B;
        `notB:
            aluout = ~B;
        `AaddB:
            {cout, aluout} = A + B + cin;
        `AsubB:
            {cout, aluout} = A - B - cin;
        `AmulB:
            aluout = A[7:0] * B[7:0];
        default: {zout, cout, aluout} = 0;
    endcase
    if (aluout == 0) zout = 1'b1;
    end
endmodule
```

FIGURE 92.10   ALU verilog code.

```
assume never (AandB & AorB);

assert always (A15to0 -> aluout == A);
assert always (B15to0 -> aluout == B);
assert always (Bleast -> aluout == B[7:0]);
assert always (AandB  -> aluout == (A & B));
assert always (AorB   -> aluout == (A | B));
assert always (notB   -> aluout == ~B);
assert always (AmulB  -> aluout == (A[7:0] * B[7:0]));

assert always ((cin==0) |-> (AaddB  -> aluout == (A + B)));
assert always ((cin==0) |-> (AsubB  -> aluout == (A - B)));
assert always (cout -> (AaddB || AsubB));

assert always ((aluout == 0) -> zout);
assert always ((AaddB || AsubB) |-> ((A == 0 && B == 16'hFFFF && cin) -> cout));
```

FIGURE 92.11   Properties of SAYEH ALU.

92.3 Conclusions

In this chapter we discussed formal verification, assertions, CTL and PSL. CTL is a mathematical form of expressing properties, while PSL which we discussed was for assertion verification. We talked about use of assertion languages for specification of properties. The ABV methodology compares the implementation of a design against its specified assertions. Assertions are written in a standard property language such as PSL. The PSL property language was described in this section.

References

1. M. Huth, *Logic in Computer Science: Tool-Based Modeling and Reasoning About Systems*, Proceedings of the International Conference on Frontiers in Education 2000, Kansas City, Missouri, pp.~T1C/1-T1C/6, October 2000.
2. E.M. Clarke, E.A. Emerson and A.P. Sistla, *Automatic Verification of Finite-State Concurrent Systems Using Temporal Logic*, ACM Trans. Programming Languages Syst., Vol. 8, pp. 244–263, 1986.
3. E.M. Clarke, Jr., O. Grumberg and D.A. Peled, *Book: Model Checking*, MIT Press, Second printing, 2000.
4. Accellera Property Specification Language, http://www.eda.org/vfv/docs/psl_lrm-1.1.pdf.

93

ASIC and Custom IC Cell Information Representation

Naghmeh Karimi and
Zainalabedin Navabi

Nanoelectronics Center of Excellence
School of Electrical and Computer
Engineering
University of Tehran

CONTENTS

Postsynthesis and postlayout information for ASIC and custom design deals with the timing, physical, and geometrical information of the cells as well as their interconnections. VITAL and GDS2 collectively provide such information. While VITAL is a representation for functionality and timing of ASIC cells, GDS2 provides physical and geometrical information of such cells. In contrast, EDIF is a neutral interchange format used to transfer data between different EDA tools. The transferring data can be in the netlist, schematic, and PCB/MCM domains.

This chapter covers the basics of VITAL, GDS2, and EDIF representations and familiarize the reader with the overall structure of these languages.

93.1 VITAL

Owing to the growth of VLSI technology, the specification and design of digital systems is moving toward higher levels of abstraction. VHDL, Verilog, and other hardware description languages have been established to define circuits independent of design tools and technology of the end product. To perform a precise simulation, timing information should be annotated to the simulation model. Originally, Verilog HDL used Standard Delay Format (SDF) to describe timing data for back annotation and specify timing constraint for forward annotation. Later on, VHDL and other HDL-based tools also used SDF for their detailed timing specifications. VHDL Initiative Toward ASIC Libraries (VITAL) is a VHDL-based standard that serves the same purpose [1].

VITAL standard was presented in 1995 to develop a standard way to perform back annotation in VHDL models and introduce ASIC libraries in the VHDL environment.

One of the key issues of VITAL is its ability to improve VHDL simulation performance at the gate level. To facilitate generating timing models applicable to both VHDL and Verilog HDL, the underlying data formats of VITAL are designed similar to that of the Verilog's SDF.

The VITAL standard includes three standard packages: **VITAL_Primitives**, **VITAL_Timing**, and **VITAL_Memory** [1]. These are discussed in the following paragraphs.

Improving the VHDL simulation performance at the gate level is one of the key issues of moving toward the VITAL standard. An important factor of this acceleration is utilizing a set of primitives and general-purpose truth, memory, and state tables. These components are included in the **VITAL_Primitives** package.

To be able to perform a precise simulation, the **VITAL_Timing** package provides different functions for delay selection, output scheduling, and timing violation checks. This package also includes generic timing parameter types.

The **VITAL_Memory** package is used for defining memory models. This package includes different data types and routines for functional modeling, corruption handling, delay specifications, etc.

Before reading details of VITAL discussed below, readers are encouraged to learn its overall structure by viewing the VITAL code of Figure 93.1.

93.1.1 Overall Structure of VITAL

VITAL defines three modeling levels (Level 0, Level 1, and Level 1 Memory) to meet its requirements, each of which has its corresponding rules. Level 1 and Level 1 Memory deal with the architecture of a design, while Level 0 pertains to the entity of a design as well as its architecture.

- *Level 0.* VITAL Level 0 mainly deals with generic parameters of a design and includes a number of rules for back annotation and modeling of timing checks. This level addresses external interface of a model and supports its portability and interoperability.
- *Level 1.* VITAL Level 1 rules apply to the architecture of a model. Level 1 allows an HDL compiler to optimize the compiled model for faster setup and simulation.
- *Level 1 Memory.* VITAL Level 1 Memory rules pertain to the architecture of a model. This modeling level accelerates the simulation of memory modules by allowing the compiler to optimize the compiled model for faster setup and simulation.

93.1.1.1 Level 0 Modeling

A Level 0 entity or architecture is specified by its corresponding attribute. As discussed above, this modeling level addresses external interface of a model and supports its portability and interoperability [1]. Figure 93.1 shows an example of a VITAL entity and its corresponding Level 0 architecture.

93.1.1.1.1 Level 0 Entity

Entity of a model is for specifying its input/output ports and generic parameters. The only allowable port types in Level 0 are **Std_ulogic** and its subtypes for scalar ports, and **Std_logic_vector** for array ports. Another restriction of this modeling level is that port names should not include the underscore character.

Level 0 generic parameters are classified into three different categories: timing generics, control generics, and other generic parameters.

Timing generics deal with the actual timing data associated with a cell for which a VITAL model is written. The values of these parameters can be set either during the back annotation or negative constraint calculation phase of simulation. Of these parameters, the values of propagation delay (**tpd**), input setup time (**tsetup**), input hold time (**thold**), input recovery time (**trecovery**), input removal time (**tremoval**), period (**tperiod**), pulse width (**tpw**), input skew time (**tskew**), no change setup time (**tncsetup**), no change hold time (**tnchold**), interconnect path delay (**tipd**), and device delay (**tdevice**) are specified during the back annotation phase of simulation. In contrast, values of internal signal delay (**tisd**), biased propagation delay (**tbpd**), and internal clock delay (**ticd**) are set during the negative constraint calculation phase of simulation.

A set of control generic parameters has been provided in the VITAL standard. These parameters have a large impact on the simulation result and are discussed below.

- **InstancePath**. Instance path is a string that is used for demonstrating an instance path.
- **TimingChecksOn**. This boolean parameter is for controlling the execution of timing checks.
- **XOn/XOnChecks**. XOn/XOnChecks boolean parameter is used to control the X value generation on certain output ports while a timing violation occurs.
- **XOn/XOnGlitch**. This boolean parameter is used for controling the X value generation on certain output ports while a glitch violation occurs.
- **MsgOn/MsgOnChecks**. This boolean parameter is used to control assertion message generation while a timing violation occurs.
- **MsgOn/MsgOnGlitch**. This boolean parameter is used for controlling assertion message generation while a glitch violation occurs.

Designers can define other generic objects to control the functionality of a design.

93.1.1.1.1.1   Delay Types — There are two types of VITAL delays: simple delays and transition-dependent delays. Simple delays include a single value (**VITALDelayType**), while transition-dependent delays introduce different delay values for their corresponding ports depending on the occurring transition on the ports. A transition-dependent delay includes a list of 2 (**VITALDelayType01**), 6 (**VITALDelayType01Z**), or 12 (**VITALDelayType01ZX**) values. In VITALDelayType01ZX, each value corresponds in sequence to occurring the following transitions on the related ports: $0{\to}1$, $0{\to}Z$, $0{\to}X$, $1{\to}0$, $1{\to}Z$, $1{\to}X$, $Z{\to}0$, $Z{\to}1$, $Z{\to}X$, $X{\to}0$, $X{\to}1$, $X{\to}Z$. For the two-valued (six-valued) transition-dependent delay constructs, only the first 2(6) delay values of the above transitions are presented. Delay values corresponding to other transitions can be specified according to Table 93.1.

Note that for the simple delays, the corresponding value is applicable to all occurring transitions.

Simple and transition-dependent delays have scalar and vector forms. The vector form allows representing different delay values for individual bits of a vector port.

As discussed above, values of timing generics can be set during the back annotation phase of simulation. For an SDF annotator to be able to map the timing generics of VITAL to their corresponding values, VITAL timing generic names follow a certain naming convention. For this, a VITAL generic name is composed of the following parts:

- A prefix that indicates the parameter type
- Signal(s) or path(s) to which the timing values are related
- A suffix that represents the probable conditions and edges of the corresponding signal(s) or path(s).

For example, *tsetup_D_Clk*, *tipd_D*, and *tpw_Clk_negedge* are three samples of timing generics. These generics specify in sequence setup time and interconnect delay of D input and the negative pulse width of the *Clk* signal.

TABLE 93.1   Deriving Delay Values

Transition	VITALDelayType01	VITALDelayType01Z
$0{\to}1$	$\mathrm{Delay}(0{\to}1)$	$\mathrm{Delay}(0{\to}1)$
$0{\to}Z$	$\mathrm{Delay}(0{\to}1)$	$\mathrm{Delay}(0{\to}Z)$
$0{\to}X$	$\mathrm{Delay}(0{\to}1)$	$\mathrm{Min}(\mathrm{Delay}(0{\to}1), \mathrm{Delay}(0{\to}Z))$
$1{\to}0$	$\mathrm{Delay}(1{\to}0)$	$\mathrm{Delay}(1{\to}0)$
$1{\to}Z$	$\mathrm{Delay}(1{\to}0)$	$\mathrm{Delay}(1{\to}Z)$
$1{\to}X$	$\mathrm{Delay}(1{\to}0)$	$\mathrm{Min}(\mathrm{Delay}(1{\to}0), \mathrm{Delay}(1{\to}Z))$
$Z{\to}0$	$\mathrm{Delay}(1{\to}0)$	$\mathrm{Delay}(Z{\to}0)$
$Z{\to}1$	$\mathrm{Delay}(0{\to}1)$	$\mathrm{Delay}(Z{\to}1)$
$Z{\to}X$	$\mathrm{Min}(\mathrm{Delay}(1{\to}0), \mathrm{Delay}(0{\to}1))$	$\mathrm{Min}(\mathrm{Delay}(Z{\to}0), \mathrm{Delay}(Z{\to}1))$
$X{\to}0$	$\mathrm{Delay}(1{\to}0)$	$\mathrm{Max}(\mathrm{Delay}(1{\to}0), \mathrm{Delay}(Z{\to}0))$
$X{\to}1$	$\mathrm{Delay}(0{\to}1)$	$\mathrm{Max}(\mathrm{Delay}(0{\to}1), \mathrm{Delay}(Z{\to}1))$
$X{\to}Z$	$\mathrm{Max}(\mathrm{Delay}(1{\to}0), \mathrm{Delay}(0{\to}1))$	$\mathrm{Max}(\mathrm{Delay}(1{\to}Z), \mathrm{Delay}(0{\to}Z))$

93.1.1.1.2 *Level 0 Architecture*

A Level 0 architecture starts with the attribute specification and is followed by declaration (constant or signal declaration) and architecture statement parts. Note that the only restriction applied to Level 0 architectures is that all functionality and timing parts must be in the VHDL format. Figure 93.1 shows a Level 0 model of an inverter. Constructs in this figure, which have not been discussed in the above discussions, will be explained in the following sections.

```
 1   Library IEEE;
     Use IEEE.STD_LOGIC_1164.all;
     Use IEEE.VITAL_Timing.all;
     Use IEEE.VITAL_Primitives.all;
 5
     Entity inv is
             Generic (
                             tpd_A_Y:            VitalDelayType01:=(1 ns, 1 ns);
                             tipd_A:            VitalDelayType01:= VitalZeroDelay01;
10                           TimingChecksOn:  Boolean := True;
                             InstancePath:     STRING := "*";
                             Xon:              Boolean :=True;
                             MsgOn:            Boolean := True
                     );
15           Port     (
                             A:                in    STD_ULOGIC;
                             Y:                out   STD_ULOGIC
                     );

20           Attribute VITAL_LEVEL0 of inv: entity is True;
     End inv;

     Architecture netlist of inv is
             Attribute VITAL_LEVEL0 of netlist:architecture is True;
25           Signal A_ipd:                STD_ULOGIC:='X';

     Begin
             WireDelay : Block
             Begin
30                   VitalWireDelay (A_ipd, A, tipd_A);
             End Block;

             VITALBehavior  :Process(A_ipd)
                     Variable Y_zd : std_ulogic:= 'U';
35                   Variable Y_GlitchData : VitalGlitchDataType;
             Begin
                     Y_zd:=VitalINV(A_ipd);

                     VitalPathDelay01 (
40                           OutSignal          => Y,
                             GlitchData         => Y_GlitchData,
                             OutSignalName      => "Y",
                             OutTemp            => Y_zd,
                             Paths              => (
45                               0 => (InputChangeTime=>A_ipd'last_event,
                             PathDelay=>tpd_A_Y,
                             PathCondition=>True))
                                     );
             End Process;
50   End netlist;
```

FIGURE 93.1   VITAL Level 0 model of an inverter.

93.1.1.2 Level 1 Modeling

VITAL Level 1 rules only apply to the architecture of a model and not to its entity. An entity is always described using Level 0 entity rules. Level 1 architectures are more restricted than Level 0. One such restriction is that in this level, signals have at most one driver. In addition, only subprogram calls and operators declared in the **Standard**, **Std_logic_1164** and VITAL standard packages are permitted in this level. The only allowable signal types are **Std_ulogic**, **Std_logic_vector**, and their subtypes [1].

A Level 1 architecture, an example of which is shown in Figure 93.2(b) starts with the attribute specification part and is followed by declaration (constant, signal, or alias declaration) and architecture statement parts. The following statements are allowed in the architecture statement part:

- Wire Delay Block Statement
- Negative Constraint Block Statement
- Process
- Primitive Concurrent Procedure Call

93.1.1.2.1 Wire Delay Block

A wire delay block is used for applying interconnect delays to input signals. This block, labeled with **Wire-Delay**, contains one or more concurrent calls to **VitalWireDelay** procedure (declared in **VITAL_Timing** package). This procedure can be called for each input (or inout) port of a design [2].

VITAL Level 1 code of Figure 93.2(b) corresponds to the seven-segment display circuit of Figure 93.2(a). Lines 52–59 of this figure show a sample utilization of the wire delay block construct. In this example signal *A_ipd* is driven by input port *A* after delay of *tipd_A*.

Generate statements can be used for applying the interconnect delays to an array port.

93.1.1.2.2 Negative Constraint Block Statement

A negative constraint block statement supports negative timing constraints in VITAL models. This construct, labeled with **SignalDelay**, includes calls to the **VitalSignalDelay** procedure (declared in **VITAL_Timing** package) for each timing generic that represents an internal clock delay or an internal signal delay. Using generate statements, negative timing constraints can be applied to a vector port. In line 48 of Figure 93.3, signal *Clk_dly* is driven by signal *Clk_ipd* after a delay of *ticd_Clk*.

93.1.1.2.3 Process

A VITAL process starts with the declaration section which deals with constants, variables, aliases, and attribute definitions and is followed in sequence by timing check, functionality, and path delay sections.

Allowed variable types in a VITAL Level 1 process are **Std_ulogic**, **Std_logic_vector**, boolean, and a number of restricted variable types including **GlitchData** (used for glitch detection), **TimingData** (used for timing violation checking), **PeriodPulseData** (used for pulse width and period specification), **PreviousDataIn** (used for specifying previous state information), and **SkewData** (used for skew violation checking). Line 78 of Figure 93.2(b) illustrates the use of restricted variables.

Timing check section of a VITAL model looks for timing violations, utilizing a number of specified procedures in **VITAL_Timing** package. **VitalSetupHoldCheck**, **VitalRecoveryRemovalCheck**, **Vital-PeriodPulseCheck**, **VitalInPhaseSkewCheck**, and **VitalOutPhaseSkewCheck** are procedures that are defined in the **VITAL_Timing** package for timing constraint checks. The last two procedures detect a skew violation between two signals. If after the specified skew timing constraints, values of signal parameters of **VitalInPhaseSkewCheck** (**VitalOutPhaseSkewCheck**) are different (identical) a skew violation will be issued.

In functionality section of a process, the behavior of a component is modeled by variable assignment and VITAL procedure call statements. **VitalStateTable** procedure and **VitalTruthTable** function can be called in this section. In addition, variable assignment may include function calls to functions defined in **Std_logic_1164** package and/or VITAL primitives. The seven-segment display example of Figure 93.2 uses a truth table, while the T-Type flip-flop of Figure 93.3 uses a state table.

Truth and state tables represent the functionality of combinational and sequential circuits, respectively. To reduce the size of these tables, a number of specified symbols (including -, ^, /, \, etc.) defined by **VitalTableSymbolType** can be utilized. The slash character is used for representing clock edge, and symbol 'S' in a state table represents the same next state as the present.

In the path delay section of a VITAL Level 1 process, a signal is driven by a value after the specified propagation delay, provided that a specified condition is met (see lines 90–101 of Figure 93.2(b)). Glitch handling and message-generation management are performed in this section.

A VITAL Level 1 process shall include at least one of the timing check, functionality, and path delay sections. The truth table used in modeling the circuit of Figure 93.2(a) is described by the code shown in lines 61–75 of Figure 93.2(b). This truth table is called in lines 85–88.

93.1.1.2.4 *Primitive Concurrent Procedure Call*

Primitive concurrent procedure calls form another part of a Level 1 model. A number of VITAL primitives have been defined in **VITAL_Primitives** package. These primitives are used in the functionality section of a process or the primitive concurrent procedure call section in a VITAL Level 1 model. VITAL primitives include logic primitives, truth tables, and state tables.

93.1.1.3 Level 1 Memory Modeling

Level 1 Memory modeling was introduced in VITAL 2000 to model the functionality and timing specifications of a memory structure. This modeling level accelerates the simulation of memory modules by allowing the compiler to optimize the compiled model for faster setup and simulation.

Level 1 Memory applies to the architecture of a memory model. As with other VITAL models, the corresponding entity of a Level 1 Memory architecture is a VITAL Level 0 model [1].

Similar to the Level 1 model, the Level 1 Memory architecture starts with the attribute specification part and is followed by declaration and architecture statement parts. The structure of declarative parts in these two modeling levels is identical, while the attribute specification and architecture statement parts are somewhat different. The following statements are allowed in the architecture statement part of a VITAL Level 1 Memory model:

- Wire Delay Block Statement
- Negative Constraint Block Statement
- Process
- Memory Output Drive Block

The first difference between Level 1 and Level 1 Memory architecture statements appears in the process statement describing them (third item above) [2]. As discussed in Section 93.1.1.2.3, the Level 1 architecture may include more than one process statement, while the Level 1 Memory architecture contains exactly one process statement labeled **MemoryBehavior**. In addition, the Level 1 Memory model can include at most one Memory output drive block to propagate internal signals to the output ports. As shown in Section 93.1.1.2 this part does not exist in Level 1 models.

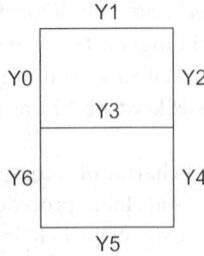

FIGURE 93.2(a)   Seven-segment display circuit.

```
 1   Library IEEE;
     Use IEEE.STD_LOGIC_1164.all;
     Use IEEE.VITAL_Timing.all;
     Use IEEE.VITAL_Primitives.all;

 5
     Entity seven_segment is
              Generic (
                        tpd_A_Y0:              VitalDelayType01:=(1 ns, 1 ns);
                        --repeat for Y1 to Y6
10
                        tpd_B_Y0:              VitalDelayType01:=(1 ns, 1 ns);
                        --repeat for Y1 to Y6

                        tpd_C_Y0:              VitalDelayType01:=(1 ns, 1 ns);
15                      --repeat for Y1 to Y6

                        tpd_D_Y0:              VitalDelayType01:=(1 ns, 1 ns);
                        --repeat for Y1 to Y6

20                      tipd_A:                VitalDelayType01:= VitalZeroDelay01;
                        --repeat for all inputs

                        TimingChecksOn:        Boolean := True;
                        InstancePath:          STRING := "*";
25                      XOn:                   Boolean :=True;
                        MsgOn:                 Boolean := True
                      );
              Port (
                        A:                     in    STD_ULOGIC;
30                      B:                     in    STD_ULOGIC;
                        C:                     in    STD_ULOGIC;
                        D:                     in    STD_ULOGIC;
                        Y0:                    out   STD_ULOGIC;
                        Y1:                    out   STD_ULOGIC;
35                      Y2:                    out   STD_ULOGIC;
                        Y3:                    out   STD_ULOGIC;
                        Y4:                    out   STD_ULOGIC;
                        Y5:                    out   STD_ULOGIC;
                        Y6:                    out   STD_ULOGIC);
40
              Attribute VITAL_LEVEL0 of seven_segment: entity is True;
     End seven_segment;

     Architecture netlist of seven_segment is
45           Attribute VITAL_LEVEL1 of netlist: architecture is True;

             Signal A_ipd:                     STD_ULOGIC:='X';
             Signal B_ipd:                     STD_ULOGIC:='X';
             Signal C_ipd:                     STD_ULOGIC:='X';
50           Signal D_ipd:                     STD_ULOGIC:='X';
     Begin
             WireDelay : Block
             Begin
                  VitalWireDelay (A_ipd, A, tipd_A);
55                VitalWireDelay (B_ipd, B, tipd_B);
                  VitalWireDelay (C_ipd, C, tipd_C);
                  VitalWireDelay (D_ipd, D, tipd_D);

             End block;
60           VITALBehavior: Process (A_ipd, B_ipd, C_ipd, D_ipd)
                  Constant seven_segment_table: VitalTruthTableType :=(
                  --A   B   C   D   Y0   Y1   Y2   Y3   Y4   Y5   Y6
                  -------------------------------------------------
                  ('0','0','0','0', '1','1', '1', '0', '1', '1', '1'),
65                ('0','0','0','1', '0','0', '1', '0', '1', '0', '0'),
                  ('0','0','1','0', '0','1', '1', '1', '0', '1', '1'),
                  ('0','0','1','1', '0','1', '1', '1', '1', '1', '0'),
                  ('0','1','0','0', '1','0', '1', '1', '1', '0', '0'),
```

FIGURE 93.2(b)   VITAL Level 1 model of the seven-segment circuit in Figure 93.2(a).

```
                          ('0','1','0','1', '1','1', '0', '1', '1', '1', '0'),
70                        ('0','1','1','0', '1','1', '0', '1', '1', '1', '1'),
                          ('0','1','1','1', '0','1', '1', '0', '1', '0', '0'),
                          ('1','0','0','0', '1','1', '1', '1', '1', '1', '1'),
                          ('1','0','0','1', '1','1', '1', '1', '1', '1', '0'),
                          ('1','0','1','-', '-','-', '-', '-', '-', '-', '-'),
75                        ('1','1','-','-', '-','-', '-', '-', '-', '-', '-') );

                Variable Yout : std_logic_vector(0 to 6);
                Variable Y0_GlitchData : VitalGlitchDataType;
                --repeat for Y1 to Y6
80
                Alias Y0_zd : std_ulogic IS Yout(0);
                --repeat for Y1 to Y6

        Begin
85              Yout:=VitalTruthTable(
                TruthTable => seven_segment_table,
                DataIn => ( A_ipd, B_ipd, C_ipd, D_ipd)
                );

90              VitalPathDelay01 (
                    OutSignal          => Y0,
                    GlitchData         => Y0_GlitchData,
                    OutSignalName      => "Y0",
                    OutTemp            => Y0_zd,
95                  XOn                =>XOn,
                    MsgOn              =>MsgOn,
                    Paths              => (
                    0 => (A_ipd'last_event, tpd_A_Y0, True),
                    1 => (B_ipd'last_event, tpd_B_Y0, True),
100                 2 => (C_ipd'last_event, tpd_C_Y0, True),
                    3 => (D_ipd'last_event, tpd_D_Y0, True) ) );
                    --repeat for Y0 to Y6

        End Process;
    End netlist;
```

FIGURE 93.2(b)   *(Continued)*.

93.1.1.3.1   *Process*

A Level 1 Memory process statement includes the declaration, timing check, functionality, and path delay sections.

The allowable variable types in a Level 1 Memory process are **Std_ulogic**, **Std_logic_vector**, boolean, time, and a number of restricted variable types including **MemoryData**, **PrevControls**, **PrevDataInBus**, **PrevAddressBus**, **PrevEnableBus**, **PortFlag**, **PortFlagArray**, **AddressValue**, **ScheduleData**, **Schedule-DataArray**, **InputChangeTime**, **.InputChangeTimeArray**, **TimingData**, and **PeriodPulseData**.

To detect timing constraint violations, a number of procedures defined in the **VITAL_Timing** and **VITAL_Memory** packages can be called in the timing check section of a process statement. **VitalSetup-HoldCheck**, **VitalRecoveryRemovalCheck**, **VitalPeriodPulseCheck**, **VitalInPhaseSkewCheck**, and **VitalOutPhaseSkewCheck** are allowed procedures that are in the **VITAL_Timing** package, and **VitalMemorySetupHoldCheck** and **VitalMemoryPeriodPulseCheck** are in the **VITAL_Memory** package.

The functionality section of a VITAL Level 1 Memory process addresses the behavior of the corresponding component utilizing variable assignment and VITAL memory procedure call statements. In this section, at least one of **VitalMemoryTable**, **VitalMemoryViolation** or **VitalMemoryCrossPorts** procedures (defined in **VITAL_Memory**) or **VitalStateTable** procedure (defined in **VITAL_Primitives**) should be used. The first two procedures deal with memory operations (read, write, and corrupt)

```vhdl
 1  Library IEEE;
    Use IEEE.STD_LOGIC_1164.all;
    Use IEEE.VITAL_Timing.all;
    Use IEEE.VITAL_Primitives.all;
 5
    Entity T_flip_flop is
                Generic (
                        tipd_Reset:             VitalDelayType01:= VitalZeroDelay01;
                        tipd_Clk:               VitalDelayType01:= VitalZeroDelay01;
10                      tpd_Reset_Q:            VitalDelayType01:=(1 ns, 1 ns);
                        tpd_Clk_Q:              VitalDelayType01:=(1 ns, 1 ns);
                        trecovery_Reset_Clk:    VitalDelayType:=1 ns;
                        tremoval_Reset_Clk:     VitalDelayType:=1 ns;
                        ticd_Clk:               VitalDelayType:=0 ns;
15                      tisd_Reset_Clk:         VitalDelayType:=0 ns;

                        TimingChecksOn:         Boolean := True;
                        InstancePath:           STRING := "*";
                        XOn:                    Boolean :=True;
20                      MsgOn:                  Boolean := True
                    );
                Port (
                        Clk:                    in   STD_ULOGIC;
                        Reset:                  in   STD_ULOGIC;
25                      Q:                      out  STD_ULOGIC);

                Attribute VITAL_LEVEL0 of T_flip_flop: entity is True;
    End T_flip_flop;

30  Architecture netlist of T_flip_flop is
                Attribute VITAL_LEVEL1 of netlist: architecture is True;

                Signal Clk_ipd:                 STD_ULOGIC:='X';
                Signal Reset_ipd:               STD_ULOGIC:='X';
35              Signal Clk_dly:                 STD_ULOGIC:='X';
                Signal Reset_dly:               STD_ULOGIC:='X';

    Begin
                WireDelay : Block
40              Begin
                    VitalWireDelay (Clk_ipd, Clk, tipd_Clk);
                    VitalWireDelay (Reset_ipd, Reset, tipd_Reset);
                End Block;

45              Signal_Delay : Block
                Begin
                    VitalSignalDelay (Reset_dly, Reset_ipd, tisd_Reset_Clk);
                    VitalSignalDelay (Clk_dly, Clk_ipd, ticd_Clk);
                End Block;
50
                VITALBehavior: Process (Clk_dly,Reset_dly)
                    Constant T_flip_flop_table: VitalStateTableType := (

                    --Violation    Reset    Clk    Q(previous)    Q
55              ------------------------------------------------------
                        ( 'X',      '-',     '-',        '-',       'X'),
                        ( '-',      '1',     '-',        '-',       '0'),
                        ( '-',      '0',     '/',        '1',       '0'),
                        ( '-',      '0',     '/',        '0',       '1'),
60                      ( '-',      '0',     '-',        '-',       'S'));

                    Variable violation : X01;
                    Variable viol_Reset_Clk : X01:='0';
                    Variable Q_zd : std_ulogic;
65                  Variable PrevData : std_logic_vector (0 to 2);
                    Variable Q_GlitchData : VitalGlitchDataType;
                    Variable Timingdata_Reset_Clk: VitalTimingDataType;
                Begin
```

FIGURE 93.3 VITAL Level 1 model of a T-type flip-flop.

```
                             IF (TimingChecksOn) Then
70                               VitalRecoveryRemovalCheck(
                                     TestSignal => Reset_dly,
                                     TestSignalName => "Res",
                                     TestDelay=>tisd_Reset_Clk,
                                     RefSignal => CLK_dly,
75                                   RefSignalName => "CLK",
                                     RefDelay=>ticd_Clk,
                                     Recovery => trecovery_Reset_CLK,
                                     Removal => tremoval_Reset_CLK,
                                     ActiveLow => FALSE,
80                                   CheckEnabled => TRUE,
                                     RefTransition => '/',
                                     HeaderMsg => InstancePath & "/T-flip-flop",
                                     TimingData => Timingdata_Reset_Clk,
                                     XOn => XOn,
85                                   MsgOn => MsgOn,
                                     Violation => viol_Reset_Clk
                                 );
                             End IF;

90                           VitalStateTable (
                                     StateTable => T_flip_flop_table,
                                     DataIn => (Violation, Reset_dly, Clk_dly),
                                     Result => Q_zd,
                                     PreviousDataIn => PrevData
100                              ) ;

                             VitalPathDelay01 (
                                     OutSignal      => Q,
                                     GlitchData     => Q_GlitchData,
105                                  OutSignalName=> "Q",
                                     OutTemp        => Q_zd,
                                     XOn            => XOn,
                                     MsgOn          => MsgOn,
                                     Paths          => (
110                                  0 => (Clk_dly'last_event, tpd_Clk_Q, True),
                                     1 => (Reset_dly'last_event, tpd_Reset_Q, True)) );
                    End Process;
      End netlist;
```

FIGURE 93.3   (*Continued*).

and timing constraint violation, while the third procedure pertains to implementation of multiport contention as well as cross port read. In addition, variable assignment may include function calls to VitalTruthTable function, functions defined in **Std_logic_1164** package and/or VITAL primitives. Note that only **VitalMemoryTable** call is required and calling other procedures in this section is optional.

In path delay section of a VITAL Level 1 Memory process, by calling the appropriate procedures of **VITAL_Timing** package, each signal is assigned the corresponding value after the specified propagation delay.

93.1.1.3.2 *Memory Output Drive Block*

Memory output drive block, labeled with **OutputDrive**, is used to propagate internal signals to output ports. In addition this block can be used for modeling tri-state's output behavior.

The block includes procedure calls to one of **VitalBUF, VitalBUFIF0, VitalBUFIF1,** or **VitalIDENT** primitives declared in the **VITAL_Primitives** package.

93.1.2 Our Coverage of VITAL

VITAL is tied with the VHDL hardware description language and is used for postsynthesis hardware representation. Because VITAL is a VHDL subset, the presentation above can only be useful when it is considered along with VHDL, discussed in Chapter 87 of this book. Like the VHDL chapter, the purpose of the preceding section was to provide a general knowledge and overall structure of VITAL. In this section, using an example-oriented flow we discussed the basics of VITAL. For a complete description of this language readers are encouraged to see the references at the end of this chapter [1,2].

While VITAL is a representation for functionality and timing of ASIC cells, GDS2 provides physical and geometrical information of such cells and their related interconnects. GDS2 format along with a layout example are presented in the following section.

93.2 GDS2

The Graphical Design System (GDS) format is a binary format developed by Calma Company in 1971 to convey IC layout information between IC designers and fabricators. Calma presented GDS2 format, the improved version of GDS, in 1978. Finally in 1991, when Cadence acquired Calma, GDS2 rights were transferred to it.

In the design process, after the completion of routing and placement, design rule check (DRC) and layout versus schematic (LVS) are performed to verify the final layout. Then the verified layout, dumped in a GDS2 file, is transferred to IC foundries for fabrication.

93.2.1 GDS2 File Format

GDS2 is a standard mask layout intermediate format containing a single cell or a library of cells each of which includes a number of geometrical objects such as boundaries, paths, boxes, etc. These objects correspond to different layers of a cell where each layer is specified by a layer number and data type [3].

The file format of GDS2 is in binary. A GDS2 file contains a number of variable length records each of which follows immediately the previous one. The first four bytes of each record, known as the record header, contain the information about the entire record, while the rest of the record (starting with byte 5) contains data. The first two bytes of the header represent the total record length, while its 3rd and 4th bytes introduce the record type and its current data types, respectively.

93.2.1.1 Data Types

Byte 4 of a GDS2 record indicates the type of data included in this record. Allowable data types in the GDS2 format are bit array, 2-byte signed integer, 4-byte signed integer, 4-byte real, 8-byte real, and ASCII string. These types are represented in sequence by the integers between 1 and 6. Value 0 for the data type is used for types of records with no available data.

A bit array data is a 2-byte data regarded as 16 individual bits of data. In a GDS2 file, signed integer and real number types use 2's complement and floating point formats, respectively. Each floating point number is composed of three different parts: sign, exponent, and mantissa, where its most significant bit represents the sign, the next 7 bits introduce the exponent, and the rest of the bits specify the mantissa of the real number. In this format, the actual exponent of a floating point number is 64 units less than the value presented by the related bits [4]. Thus the value of a floating point number is equal to

$$(-1)^{\text{sign}} \times (\text{mantissa}) \times 16^{(\text{actual exponent})}$$

ASCII strings include a number of ASCII characters each of which are one byte long. According to the GDS2 format the size of each record should be an even number of bytes. Thus if the ASCII strings includes an odd number of characters, a null character is appended to the end of the string.

Record	Mandatory
HEADER	√
BGNLIB	√
LIBDIRSIZE	
SRFNAME	
LIBSECUR	
LIBNAME	√
REFLIBS	
FONTS	
ATTRTABLE	
GENERATIONS	
FORMATTYPE*	
UNITS	√
STRUCTURE*	
ENDLIB	√

FIGURE 93.4   GDS2 stream.

Note that value 0 is allocated to the types of records with no available data. Obviously the number of data pieces in each record can be specified by considering the record length and its current data type. Each record can be between 4 and 65536 ($=2^{16}$) bytes long where a 4-byte record contains no data. An example of a 4-byte record is **TEXT** record which indicates the state of a text record.

93.2.1.2   Record Types

Byte 3 of a GDS2 record indicates the type of the record. A GDS2 file contains a number of ordered variable lengths records, i.e., the order in which records appear in a GDS2 file must be according to the GDS2 syntax. This file starts with the header record and is followed by other records in the order shown in Figure 93.4.

In Figure 93.4, items marked with an asterisk (FORMATTYPE and STRUCTURE) represent a group of records while other items are individual records. Of the records shown in Figure 93.4, a GDS2 file must include **HEADER, BGNLIB, LIBNAME, UNITS,** and **ENDLIB** records while it may or may not include the other records.

As discussed above, record type and data type of records are each 1 byte long. However, all combinations of record types and data types are not valid, i.e., for each predefined record type only one unique data type is valid.

Commonly used record types and their related data type values are discussed below. A 4-byte number represents record and data types: the first two digits show the hexadecimal value of the record type and the other two digits specify the data type. For example, the **HEADER** record is represented by 0002, where 00 shows that the record type is header and 02 represents that its corresponding data type is a 2-byte signed integer.

93.2.2   A GDS2 Example

To better understand the structure of GDS2 that will be described in the next section, a simple example and its record file will be presented here. We will refer to this example when describing records and data types of GDS2.

Figure 93.5 shows the physical layout of an inverter. Because of the size of the GDS2 file for this complete layout, only a section of this design along with its hexadecimal representation of the related GDS2 file are shown in Figure 93.6(a) and Figure 93.6(b) respectively. GDS2 records, which consist of a header and data, are shown in Figure 93.6(b). For each record, the header is shown in bold, and its

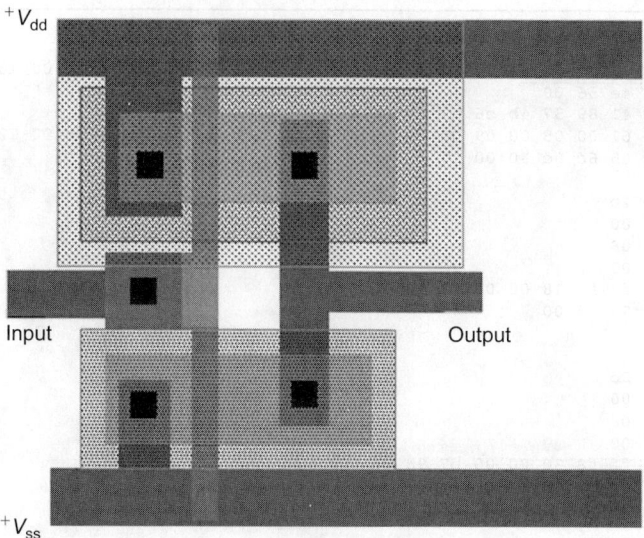

FIGURE 93.5 An inverter layout.

FIGURE 93.6(a) A part of layout shown in Figure 93.5.

corresponding data follows it. For clarity in the GDS2 information, we have also provided a symbolic representation of the GDS2 file of Figure 93.6(b). This representation (Figure 93.7) uses mnemonics for record types. The actual data of the records follow the mnemonics.

93.2.3 Structure of GDS2

The structure of a GDS2 file is as described here.

Header. A GDS2 file starts with the **HEADER** record (code 0002), which indicates the GDS2 version number. This is shown in line 1 of Figure 93.6(b) and in Figure 93.7.

Library specification. Specification of a library of cells in GDS2 begins with the **BGNLIB** record (line 2, Figure 93.6(b)) and if the optional records do not exist, it is followed by **LIBNAME** (line 3, Figure 93.6(b)) and ends with the **ENDLIB** record (line 42, Figure 93.6(b)).

```
1  00 06 00 02 00 03
   00 1c 01 02 00 69 00 0c 00 13 00 13 00 05 00 01 00 69 00 0c 00 13 00 13 00 05 00 01
   00 08 02 06 49 4e 56 00
   00 14 03 05 3e 41 89 37 4b c6 a7 f0 39 44 b8 2f a0 9b 5a 54
5  00 1c 05 02 00 61 00 05 00 09 00 0c 00 2c 00 37 00 61 00 05 00 0b 00 12 00 1c 00 2e
   00 0a 06 06 43 65 6c 6c 30 00
   00 04 0c 00
   00 06 0d 02 00 2b
   00 06 16 02 00 00
10 00 06 17 01 00 08
   00 06 1a 01 00 00
   00 0c 10 03 ff ff fc 18 00 00 13 88
   00 08 19 06 56 73 73 00
   00 04 11 00
15 00 04 0c 00
   00 06 0d 02 00 2b
   00 06 16 02 00 00
   00 06 17 01 00 08
   00 06 1a 01 00 00
20 00 0c 10 03 ff ff f8 30 00 00 b7 98
   00 08 19 06 56 64 64 00
   00 04 11 00
   00 04 08 00
   00 06 0d 02 00 2e
25 00 06 0e 02 00 00
   00 2c 10 03 00 00 32 c8 ff ff fc 18 00 00 3a 98 ff ff fc 18 00 00 3a 98 00 00 b7 98
                00 00 32 c8 00 00 b7 98 00 00 32 c8 ff ff fc 18
   00 04 11 00
   00 04 08 00
30 00 06 0d 02 00 31
   00 06 0e 02 00 00
   00 2c 10 03 00 00 0f a0 00 00 a4 10 00 00 da c0 00 00 a4 10 00 00 da c0 00 00 b3 b0
                00 00 0f a0 00 00 b3 b0 00 00 0f a0 00 00 a4 10
   00 04 11 00
35 00 04 08 00
   00 06 0d 02 00 31
   00 06 0e 02 00 00
   00 2c 10 03 00 00 13 88 00 00 00 00 00 00 de a8 00 00 00 00 00 00 de a8 00 00 0f a0
                00 00 13 88 00 00 0f a0 00 00 13 88 00 00 00 00
40 00 04 11 00
   00 04 07 00
   00 04 04 00
```

FIGURE 93.6(b)   Hexadecimal representation of GDS2 file of Figure 93.6(a).

The **BGNLIB** record is identified by 0102 and as data, it has the date of the library. The data included in the **BGNLIB** record includes twelve 2-byte data which are the corresponding year, month, day, hour, minute, and second of the last modification and access times, respectively (line 2, Figure 93.6(b)).

Following the **BGNLIB**, the **LIBNAME** record (code 0206, example: line 3, Figure 93.3) contains the library name. In our example the library name is *INV*. The **ENDLIB** record (code 0400, example: line 42, Figure 93.6(b)) contains no data.

A library specification may also include several optional records discussed here. When a new library is created, the number of pages in the library directory as well as the access control list data and the name of the sticks rules file can be introduced by the **LIBDIRSIZE**(code 3902), **LIBSECUR**(code 3B02), and **SRFNAME**(code 3A06) records, respectively. Furthermore, to attach reference libraries to a working library, the **REFLIBS** record (code 1F06) is used. This record specifies consecutively the name of all reference libraries allocating 44 byte for each library name.

Units. The geometrical dimensions of the elements of a GDS2 file are expressed using the **UNITS** record (code 0305, example: line 4, Figure 93.6(b)), which shows the relation between database and user units. The **UNITS** record contains two 8-byte real numbers to demonstrate in sequence the size of database unit in user unit and meters.

```
1   Header 03
    BGNLIB 2005 12 19 19 5 1 2005 12 19 19 5 1
    LIBNAME INV
    UNITS 1.0E-03 1.0E-09
5   BGNSTR 1997 5 9 12 44 55 1997 5 11 12 28 46
    STRNAME Cell0
    TEXT
    LAYER 43
    TEXTTYPE 0
10  PRESENTATION (font:00 vertical:10 horizental:00)
    STRANS (ref=0 angle=0 mag=0)
    XY (ff ff fc 18 00 00 13 88)
    STRING Vss
    ENDEL
15  TEXT
    LAYER 43
    TEXTTYPE 0
    PRESENTATION (font:00 vertical:10 horizental:00)
    STRANS (ref=0 angle=0 mag=0)
20  XY (ff ff f8 30 00 00 b7 98)
    STRING Vdd
    ENDEL
    BOUNDARY
    LAYER 46
25  DATATYPE 0
    XY (00 00 32 c8 ff ff fc 18 00 00 3a 98 ff ff fc 18 00 00 3a 98 00 00 b7 98
        00 00 32 c8 00 00 b7 98 00 00 32 c8 ff ff fc 18)
    ENDEL
    BOUNDARY
30  LAYER 49
    DATATYPE 0
    XY (00 00 0f a0 00 00 a4 10 00 00 da c0 00 00 a4 10 00 00 da c0 00 00 b3 b0
        00 00 0f a0 00 00 b3 b0 00 00 0f a0 00 00 a4 10)
    ENDEL
35  BOUNDARY
    LAYER 49
    DATATYPE 0
    XY (00 00 13 88 00 00 00 00 00 00 00 00 de a8 00 00 00 00 00 00 00 00 de a8 00 00 0f a0
        00 00 13 88 00 00 0f a0 00 00 13 88 00 00 00 00)
40  ENDEL
    ENDSTR
    ENDLIB
```

FIGURE 93.7 Symbolic representation of GDS2 file of Figure 93.6(a).

Structures and elements. A GDS2 file contains a number of structures, each of which includes an arbitrary number of elements. Mnemonic STRUCTURE* in Figure 93.4 represents a group of records for defining a structure. Each structure begins with **BGNSTR** and is identified by its name (using **STRNAME** record, code 0606).

The last modification and access date of the structure are included in the **BGNSTR** record (code 0502, example: line 5, Figure 93.6(b)). The end of a structure is marked by **ENDSTR** (coded with 0700, example: line 41, Figure 93.6(b)). Note that **BGNSTR** record includes twelve 2-byte data in the same format as the **BGNLIB** record.

GDS2 elements are geometrical objects assigned to different layers of a design. These objects have the following types:

- Boundary
- Path
- SREF

- AREF
- Text
- Box
- Node

Boundary. A boundary element is a closed polygon specified by the XY coordinates of its vertices in database units. The first and last vertex of a boundary must be the same. A boundary specification starts with the **BOUNDARY** record (code 0800, example: line 23, Figure 93.6(b)), is followed by **LAYER** (code 0D02, example: line 24, Figure 93.6(b)), **DATATYPE** (code 0E02, example: line 25, Figure 93.6(b)), and **XY** (code 1003, example: lines 26 and 27, Figure 93.6(b)) records, and terminates with the **ENDEL** record (code 1100, example: line 28, Figure 93.6(b)). **LAYER** and **DATATYPE** records specify the layer number and label of the specified element, respectively, while the **XY** record includes an array of XY coordinates related to the vertices of the specified object. Note that the number of specified XY coordinate pairs for a boundary element should be between 4 and 200.

Path. A path element is an open polygon specified by the XY coordinate pairs of its vertices. Each path specification starts with the **PATH** record (code 0900), is followed by **LAYER**, **DATATYPE**, and **XY** records, and terminates with **ENDEL** record. The minimum and maximum number of required XY coordinate pairs to specify a path element are 2 and 200, respectively. In addition to the above records a path specification may contain two other records to specify element width in database units and the type of path endpoints.

SREF. Utilizing a structure reference (SREF) element within a structure results in the instantiation of that structure and its placement in the referencing structure. Each SREF specification starts with **SREF** (code 0A00) record, is followed by **SNAME** (code 1206) and **XY** records, and terminates with **ENDEL** record. **SNAME** and **XY** records demonstrate in sequence the name of the referenced structure and its location within the referencing structure. In addition to the above records, an SREF specification may contain another record to deal with reflection, magnification, and rotation of the referenced structure in the referencing one.

AREF. An AREF element is used to instantiate and locate an array of specified structures within another structure. An AREF specification starts with **AREF** record (coded with 0B00) and is followed by **SNAME**, **COLROW** (coded with 1302), and **XY** records and terminates with **ENDEL**. **COLROW** specifies the size (number of rows and columns) of the instantiated array. The **XY** record in an AREF element includes three pairs of XY coordinates. The first pair introduces the array reference point while second and third pairs are used to specify the locations of other instantiated elements. Similar to an SREF element, an AREF element may contain another record to deal with reflection, magnification, and rotation of the referenced structures in the referencing one.

Text. Each text specification in a GDS2 file starts with the **TEXT** record (code 0C00, example: line 7, Figure 93.6(b)), is followed by **LAYER**, **TEXTTYPE**, **XY**, and **STRING** records, and terminates with **ENDEL** record. **TEXTTYPE** (code 1602, example: line 9, Figure 93.6(b)) represents the text-type specification, while the **STRING** record (code 1906, example: line 13, Figure 93.6(b)) includes the character string that is presented by this element. The location of each text element is specified by a single pair of XY coordinates. In addition to the above records, a text specification may contain other records to deal with the text font, alignment, and width as well as the format of its endpoints, reflection, magnification, and rotation of that text element.

Box. A box is a four-sided polygon represented by exactly five pairs of XY coordinates, where the first and last pairs must be exactly the same. Box is a special case of boundary. A box specification starts with **BOX** (code 2D00), is followed by **LAYER**, **BOXTYPE** (code 2E02), and **XY** records, and terminates with the **ENDEL** record.

Node. A node element specification starts with the **NODE** record (code 1500), is followed by **LAYER**, **NODETYPE** (code 2A02), and **XY** records, and terminates with the **ENDEL** record. Each node location can be introduced by at most 50 pairs of XY coordinates.

In addition to the discussed records, each element specification may contain two other records to introduce template and external data as well as Plex number and Plexhead flag.

Other specifications. GDS2 files can be categorized into two different groups: archive stream and filtered stream files indicated by the **FORMAT** record (code 3602). In the former, the elements of GDS2 structures can be located in all layers and data types. In the latter category, only the user-defined layers and data types may include the elements. In an archive stream file, **UNITS** must immediately follow the **FORMAT** record, while in a filtered stream at least one **MASK** record (code 3706) precedes the **UNITS** record. In this stream, the last **MASK** record must be immediately followed by **ENDMASKS** record (code 3800). The **MASK** record is used to represent the list of layers and datatypes specified by the user when creating the file. Mnemonic FORMATTYPE* in Figure 93.4 introduces a group of records representing the stream type of the GDS2 file as well as the list of related layers (if available).

In a GDS2 stream, the number of preserved copies and back-up structures is specified by utilizing **GENERATIONS** record.

93.2.4 Our Coverage of GDS2

GDS2 is a mask layout intermediate format used to convey layout information between EDA tools. We discussed the basics of GDS2 in the preceding section and provided a layout example along with its hexadecimal and symbolic representations of the related GDS2 file.

The constructs not included in this example follow the same basic patterns discussed here. For a more complete description of this language readers are encouraged to see the references at the end of this chapter [3,4].

The following section is on EDIF. This EDA format is a standard data interchange format used to transfer design data between different EDA tools. In the following section, we deal with the common constructs of the EDIF format.

93.3 EDIF

Electronic Design Interchange Format (EDIF) is a text format used to convey information between different EDA tools. The purpose of EDIF is to provide a solution for data transfer problem by defining a neutral intermediate format. EDIF was first developed in 1985. The first real public release of EDIF was EDIF version 2 0 0, which was published in 1987 and adopted as an ANSI/EIA standard in 1988 [6]. To handle busses, bus rippers, and busses across multipage schematics, EDIF version 3 0 0 was developed and approved in 1993. EDIF version 4 0 0 was developed in 1996 to transfer PCB/MCM layouts between EDA tools.

93.3.1 EDIF Structure

Since most of EDA tools use EDIF 2 0 0, instead of other more recent versions of EDIF, in this section we deal with the fundamentals of EDIF 2 0 0. EDIF is a LISP-like format including several constructs each of which is surrounded in a pair of parenthesis. The regularity of the EDIF format makes parsing the file very simple.

The overall structure of an EDIF file is shown in Figure 93.8. As shown in this figure, an EDIF file has a hierarchical nature and may include one or more library elements. Each library element consists of a technology section and one or more cell descriptions. The technology section of a library includes the information about the specified technology used for the related cells of that library. Each cell has a number of views used to specify different aspects or representations of that cell. The specification covers all aspects of electronic design including schematic, netlist, mask layout, PCB layout, documentation, etc. A view consists of an interface section and possibly a contents section where the interface section deals with port definitions of the corresponding cell and the contents section pertains to the internal structure of that cell [5].

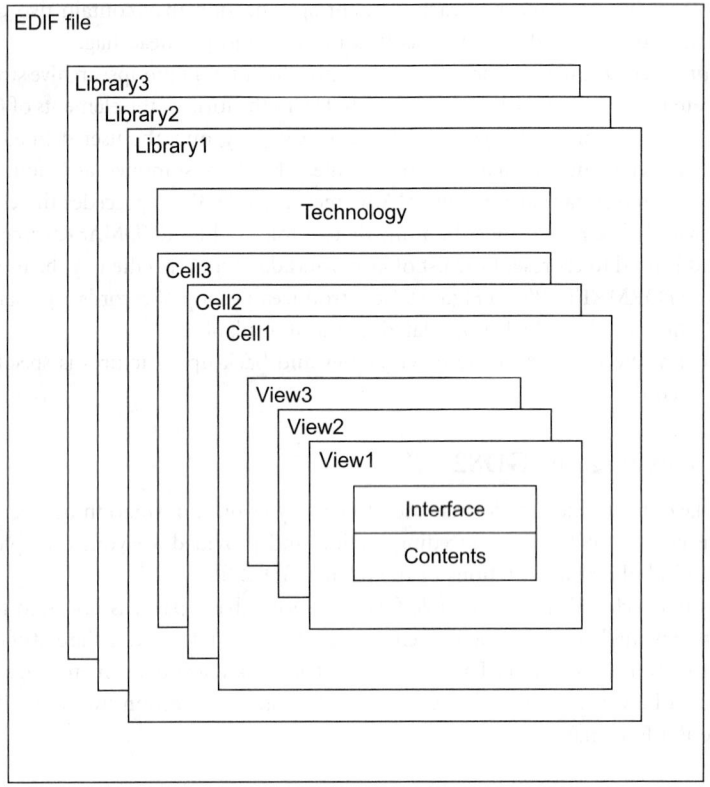

FIGURE 93.8   Overall structure of an EDIF file.

93.3.2   EDIF Syntax

EDIF defines three different levels of complexity, named as level 0, level 1, and level 2. Level 0 is the basic level of EDIF description and is more restricted than other levels. In this level only literal constants are permitted. Level 1 supports level 0 constructs as well as parameters, expressions, and frames while level 2 adds control flow constructs to level 1. The level of an EDIF file that includes several libraries is considered to be the maximum level of the corresponding libraries [6].

As discussed above an EDIF file includes a number of constructs each of which is surrounded in a pair of parenthesis. Each construct is composed of an EDIF keyword followed by a list of items. The items can be other constructs, identifiers, or data items, where a construct can be a nested construct.

The ability of keyword definition and abbreviation has been provided in EDIF files using **keywordMap** construct, where keyword level 0 construct specifies that no keyword definition and abbreviation is supported.

Identifiers are used to define objects in EDIF files. Each identifier is a string of up to 255 characters consisting of alphanumeric and the underscore characters. A string must be preceded with an ampersand unless its first character is an alphabetic character or underscore. This ampersand is not considered as a part of the object name. Identifiers and keywords are not case sensitive. Thus &a1, &A1, A1, and a1 strings are identical.

Numbers can be used to represent different aspects in EDIF files such as path widths, gate delays, etc. EDIF supports 32-bit signed integer numbers. To represent real numbers or integers, which cannot be specified by a 32-bit number, a scale factor is used. The triple (e mantissa exponent) format is used to represent an exponential number that is equal to mantissa $\times 10^{exponent}$. For example (e 8 $^{-2}$) represents 0.08.

The strings in EDIF structures are a sequence of legal characters enclosed in a pair of double-quote characters.

To represent the general location of a specified point in EDIF, the triple (pt x y) is used, where x and y are the x and y locations of the corresponding point.

As discussed above, EDIF can handle different representations of a design. Of these representations, **netlist** is the most common view used in HDLs. To give a general introduction to the syntax and semantics of EDIF language, we will present the netlist view of a T-type flip-flop in the following section.

93.3.3 An EDIF Netlist Example

Figure 93.9 shows an EDIF netlist for a T-type flip-flop composed of a D-type flip-flop and an inverter. As shown in this figure, each EDIF file is composed of a number of constructs surrounded in a pair of parentheses.

An EDIF file starts with the **edif** keyword, which marks the highest level of hierarchy in a design and is followed by the design name (line 1, Figure 93.9). **edifVersion**, **edifLevel**, and **keywordMap** constructs which were discussed above follow the design name (lines 2–4, Figure 93.9).

An optional construct (**status**) may be included in an EDIF file to present additional information about the creation date, creating program, and author of the current EDIF file. This is shown in lines 5–9 of Figure 93.9.

To describe a design in EDIF, the referenced library as well as the design library must be identified. To declare a library in an EDIF file, the **library** (**external**) construct is used. The **library** (**external**) construct describes a library that is internal (external) to the EDIF data. Line 10 of Figure 93.9 introduces the external library *cub* as the reference library.

Scaling information in an EDIF file is presented by using **numberDefinition** construct (line 12, Figure 93.9). This construct is included in each EDIF file but it is not applicabale and not utilized in netlist structures.

Lines 13–25 of Figure 93.9 show that the *cub* library contains a number of cells to be used in our design. This library includes *DFF* and *INV* cells (lines13–25, Figure 93.9).

The netlist view of the *DFF* cell has three inputs and two output ports (lines 15–20, Figure 93.9), while the netlist view of the *INV* cell has one input and one output port (lines 23–25, Figure 93.9).

To describe our complete design (*Tflipflop*) the same constructs used for *DFF* and *INV* are used. Lines 26–34 of Figure 93.9 introduce the design library (*work*), and its interface. In this example the design (*Tflipflop*) has *Reset*, and *Clk* input ports and *Dataout* output ports (lines 31–34, Figure 93.9).

After the instantiated cells of a referenced library are identified, ports of the design must be mapped to the ports of these cells. The **contents** construct shown in line 35 of Figure 93.9 is used to give instance names *ix44* and *ix45* to *DFF* and *INV* cells. The **cellref** constructs (lines 36 and 38, Figure 93.9) links the cell instances *ix44* and *ix45* to *DFF* and *INV* cells in library *cub*.

The **net** construct links the design nets with the input/output ports of the library cells. Lines 40–43 of Figure 93.9 show that the *Reset* port of this design has been tied to the port *A* of the *ix45* instantiated cell. The remainder of this code specifies the links between other design ports and input/output ports of the instantiated cells.

The EDIF file terminates with the **design** construct (line 60, Figure 93.9). This construct introduces the design name (*Tflipflop*) and places the design in the working library (*work*).

93.3.4 Our Coverage of EDIF

The description of EDIF presented here was an introduction to this format. We tried to cover the basics of this format and familiarize the reader with the overall structure of this language and its applications. For this purpose we used a small example. EDIF is a complete format and includes many constructs for design description. Dealing with all of these constructs is beyond the scope of this chapter. For a complete description of this format, readers are encouraged to see the references at the end of this chapter [5,6].

```
1(edif Tflipflop
     (edifVersion 2 0 0)
     (edifLevel 0)
     (keywordMap (keywordLevel 0))
5    (status
        (written
            (timestamp 2006 01 17 13 31 29)
            (program "Sample Program"(version "v2"))
            (author "ASIC Inc.")))
10   (external cub
        (edifLevel 0)
        (technology (numberDefinition ))
        (cell DFF (cellType GENERIC)
            (view NETLIST (viewType NETLIST)
15              (interface
                    (port C (direction INPUT))
                    (port D (direction INPUT))
                    (port RN (direction INPUT))
                    (port Q (direction OUTPUT))
20                  (port QN (direction OUTPUT)))))
        (cell INV (cellType GENERIC)
            (view NETLIST (viewType NETLIST)
                (interface
                    (port A (direction INPUT))
25                  (port Q (direction OUTPUT))))))
    (library work
        (edifLevel 0)
        (technology (numberDefinition ))
        (cell Tflipflop (cellType GENERIC)
30          (view INTERFACE  (viewType NETLIST)
                (interface
                    (port Reset (direction INPUT))
                    (port Clk (direction INPUT))
                    (port Dataout (direction OUTPUT)))
35              (contents
                    (instance ix44 (viewRef NETLIST (cellRef DFF (libraryRef cub)))
                    )
                    (instance ix45 (viewRef NETLIST  (cellRef INV (libraryRef cub )))
                    )
40                  (net Reset
                        (joined
                            (portRef Reset )
                            (portRef A (instanceRef ix45 ))))
                    (net Clk
45                      (joined
                            (portRef Clk )
                            (portRef C (instanceRef ix44))))
                    (net Dataout
                        (joined
50                          (portRef Dataout )
                            (portRef Q (instanceRef ix44 ))))
                    (net nx6
                        (joined
                            (portRef QN (instanceRef ix44 ))
55                          (portRef D (instanceRef ix44))))
                    (net nx44
                        (joined
                            (portRef Q (instanceRef ix45 ))
                            (portRef RN (instanceRef ix44 )))))))))
  (design Tflipflop (cellRef Tflipflop (libraryRef work )))
  )
```

FIGURE 93.9 EDIF file of a T-type flip-flop netlist.

References

1. *IEEE Standard for VITAL ASIC (Application Specific Integrated Circuit) Modeling Specification*, IEEE Std 1076.4-2000, IEEE, Inc., New York, 2001.
2. R. Munden, *ASIC and FPGA Verification: A Guide to Component Modeling*, Morgan Kaufmann, Los Altos, CA, 2005.
3. *GDSII Stream format Manual*, Release 6.0, Documentation No. B97E060, Cadence Design Systems, Inc./Calma., 1987.
4. H. J. Levinson, W. H. Arnold, *Handbook of Microlithography, Micromachining, and Microfabrication*, Vol. 1, SPIE Optical Engineering Press, Washington, 1997.
5. M.J.S. Smith, *Application-Specific Integrated Circuits*, Addison-Wesley, MA, 1997.
6. *Electronic Design Interchange Format*, Version 2 0 0, ANSI/EIA-548-1988, EIA, Inc., 1988.

94

Test Languages

Shahrzad Mirkhani and
Zainalabedin Navabi

Nanoelectronics Center of Excellence
School of Electrical and Computer Engineering
University of Tehran

CONTENTS

An essential process that must be performed on the result of each stage in the design and manufacturing of electronic systems, from design specification to design manufacturing (and even after the manufacturing), is the test process. This implies that the result of each stage of hardware manufacturing must be tested to detect probable bugs, problems, or faults.

Based on a general categorization, there are two kinds of tests: *design test* and *manufacturing test*. With the design test, the design is examined for detecting probable design errors. Design test is usually called design verification. It can either be simulation-based or it can be performed by verification methods.

Hardware or manufacture test is performed on a real hardware by automatic test equipment (ATEs). These equipments apply predefined test vectors to a Circuit Under Test (CUT). These test vectors can be obtained by several methods such as test generation algorithms applied on the model of the CUT, random generation, or a combination of both. The expected (good) responses to test vectors are then stored for further comparisons. Circuit outputs, obtained by the ATE after applying a test vector to the CUT, are compared with the prestored expected results. The ATE must indicate an error when these two responses do not match in a specified amount of time or interval. This process recognizes faulty CUTs from the good ones.

To be able to use test methods on ICs, there must be a strong link between design and test processes. Important points in testing a design or a real hardware are test vector application and response analysis. Accessing a good set of vectors and storing the expected responses in an efficient way are critical in reducing the test time and increasing test coverage. In simulation-based test processes, testbenches perform test vector application and response analysis tasks. These testbenches can also compare the design results with the expected ones. In manufacturing test process, testers get the necessary information about test vectors and expected results from testbenches with special formats and perform the test vector application and result comparison. A good data exchange format that can be used for both design and manufacturing tests helps the test process.

In recent years, a few specific languages and standards have been developed for both design and manufacturing test purposes. One of the major features of these languages is their ability to define input vectors with special values and timings. Other features of these languages include response storage and, in a few cases, result comparison with good responses of the design model or the real hardware.

In the sections that follow, we will describe WAVES, which is considered as a standard format more than a language, and the Standard Test Interface Language (STIL).

94.1 WAVES

WAVES stands for Waveform and Vector Exchange Specification. It is a standard for defining input vectors, collecting the outputs of a design, and comparing the stored results with the received outputs from the CUT. WAVES includes a number of standard libraries developed in the VHDL language. In these libraries, several useful concepts and functions, which help a testbench developer, have been implemented.

Only the sequential subset of VHDL is used in WAVES. Therefore, this standard is more similar to a programming language, like the C-language, than an HDL. This standard has been standardized under IEEE 1029.1-1998 standard. There are two levels defined for WAVES specification: WAVES level 1 and WAVES level 2. WAVES level 1 includes the basic utilities and definitions for testbench development, while in WAVES level 2 more complex constructs are utilized for testbench development.

To test a design with WAVES, one should develop a testbench with the utility functions in WAVES standard libraries. This testbench has several parts including one for test vector reading and applying, one for storing the expected results, and another part for analyzing the outputs (comparing the expected results with the CUT outputs). There are also several tools developed for automatic generation of the necessary files from the description of the CUT. Figure 94.1 shows the flow of using WAVES in testing a design. Note that the testbench generated using WAVES can be used by both simulators and hardware testers.

In the following sections, the basic definitions of WAVES will be described. Then, the functions in the WAVES built-in libraries and the accepted formats for waveforms in WAVES will be discussed. Finally, the WAVES test set will be discussed using a simple example. For further study on WAVES, the reader is encouraged to refer to Refs. [1,2].

94.1.1 Basic Definitions

In this section, we will describe a few expressions and data structures used in the WAVES standard.

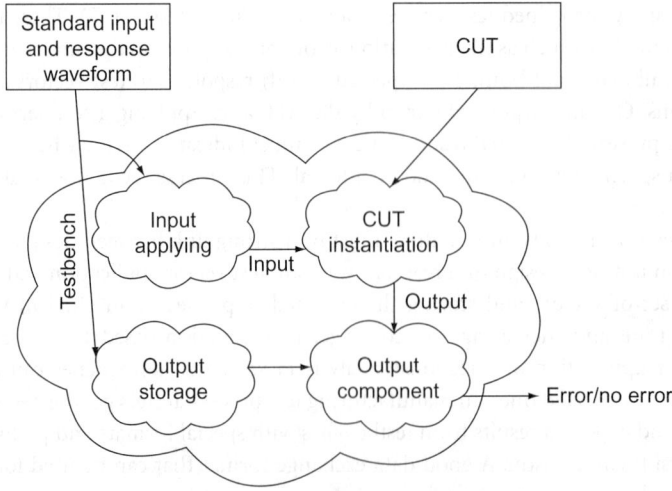

FIGURE 94.1 Test flow of a design using WAVES.

94.1.1.1 Logic Values

The input and output values used in WAVES are similar to the STD-1164 standard. There are two packages defined for logic values in WAVES. One package is *WAVES_1164_pin_codes* that defines input/output values as a STRING-type constant named *pin_codes*. In the other package (*WAVES_1164_logic_value*) an enumeration type named *logic_value* is defined for the input/output values. The values used in the *WAVES_1164_pin_codes* package are **X, 0, 1, Z, W, L, H**, and **-**. The "-" sign is for *don't care*. The other values are interpreted as in the STD-1164 standard.

In the *WAVES_1164_logic_value* package, the values used in the enumeration type are **dont_care, sense_x, sense_0, sense_1, sense_z, sense_w, sense_l, sense_h, drive_x, drive_0, drive_1, drive_z, drive_w, drive_l**, and **drive_h**. The default value of variables of this type is **dont_care**, since this is the first member in this enumeration. The *sense* group in this type is used to show output values, while the *drive* group is used for the input values.

A difference between STD-1164 and *pin_codes* **or** *logic_value* types is the **U** value. **U**, which is the default type in STD-1164, means un-initialized. Initialization is only used in simulations, and not in real designs. Therefore, this type is omitted in WAVES. It is assumed that every signal in a design has a **dont_care** initial value in WAVES.

94.1.1.2 Slice

A **slice** or time slice is simply the tester cycle. This cycle can be interpreted as the period in which a test vector is applied to the CUT.

94.1.1.3 Frame

The list of transitions on a signal, which occurs during a slice, is called a **frame**. The set of frames for all possible patterns which can be applied on a signal is called a frame-set. Collectively, all inputs and outputs of a CUT generate a frame-set array.

94.1.1.4 t0, t1, and t2

Symbols **t0, t1**, and **t2** represent predefined time markers. **t0** represents the time in which a test cycle begins. **t1** shows the time of leading-edge transition of a signal while **t2** represents the time of trailing-edge transition of a signal. These definitions are used in waveform specification functions and procedures.

94.1.1.5 WAVES Built-in Libraries

WAVES built-in libraries used in upper-level packages are *waves_standard, waves_interface, waves_port*, and *waves_object*. Instead of discussing these libraries, we will show the usage of libraries that are more at the front-end of user interfaces.

94.1.2 Waveform Formats

To test and verify a CUT, a number of predefined test vectors must be applied to the circuit at specified times, and the responses from the CUT must be compared with the expected responses. To perform these operations, we have to define a format for input and output waveforms. These waveforms must carry test vector and output data. They must also have information about their timing (transition points) and their format (shape). There are several predefined waveform formats and timings (frames) that are implemented in the *WAVES_1164_FRAMES* package.

Predefined waveform formats are categorized into two main groups. One group is suitable for defining input (drive) waveforms and the other is for comparing output waveforms. The first category includes **nonreturn, return high, return low, pulse high, pulse high skew, pulse low, pulse low skew**, and **surround by complement** formats. In the second category, **window** and **window skew**, which are a span of time, are defined. Comparing the output to the expected output is done in this specified span of time.

As stated above, these formats are implemented as functions in the *WAVES-1164_FRAMES* package. This package uses functions and types defined in the WAVES standard packages to develop its functions. The waveform frame formats are described below.

94.1.2.1 Non-Return

As shown in Figure 94.2, if a signal has a transition in **t1** and it holds this value until the next **t1** (the leading edge at next **t0**), it has nonreturn frame format. It is important to note that in this and other figures related to frame formats *Logic level 1* and *Logic level 2* are two general logic values that only show a transition occurrence on a signal. For example, in Figure 94.2 *Logic level 1* is not necessarily higher than *Logic level 2*. In the following figures, logic levels (high or low) are explicitly stated if necessary. In such cases, we show the logic levels with *high* and *low* labels.

94.1.2.2 Return High

In the return high frame format demonstrated in Figure 94.3, a signal has its data during a time interval between **t1** and **t2** (in one cycle) and it is held *high* between **t2** and the next **t1**.

94.1.2.3 Return Low

The return low frame format is similar to return high format. They differ in logic value between **t2** and the next **t1**. In this frame format, the signal value must be held *low* during **t2** to the next **t1**. Figure 94.4 shows this frame format.

94.1.2.4 Pulse High

The pulse high format is a fixed frame format that is useful for signals like clocks in a design. The signal is held high between **t1** and **t2** (in one cycle) and becomes *low* until the next **t1**. Figure 94.5 demonstrates this frame format.

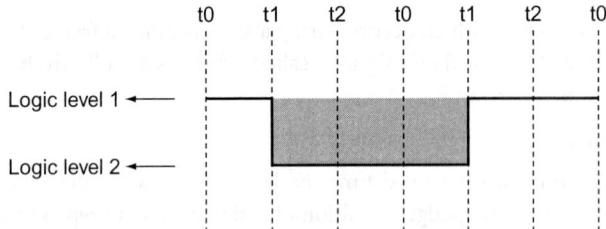

FIGURE 94.2   Nonreturn frame format.

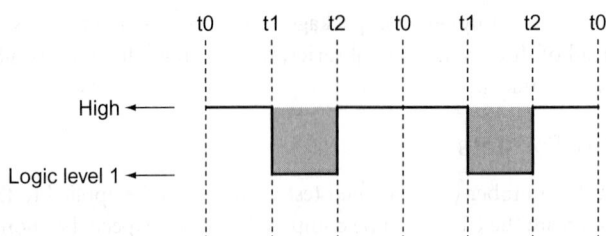

FIGURE 94.3   Return high frame format.

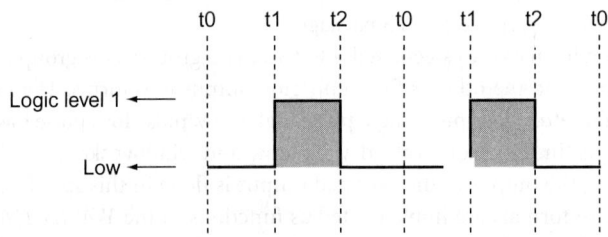

FIGURE 94.4   Return low frame format.

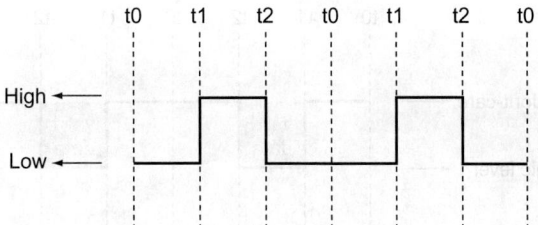

FIGURE 94.5 Pulse high frame format.

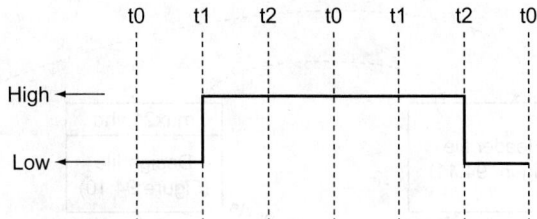

FIGURE 94.6 Pulse high skew frame format.

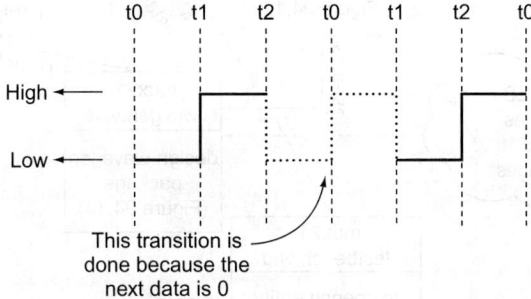

This transition is
done because the
next data is 0

FIGURE 94.7 Surround by complement frame format.

94.1.2.5 Pulse High Skew

The pulse high skew format is similar to the pulse high frame format except that in this format the signal must be held *high* between **t1** in one cycle and **t2** in its next cycle. This frame format is shown in Figure 94.6.

94.1.2.6 Pulse Low and Pulse Low Skew

The two frame formats, pulse low and pulse low skew, are similar to pulse high and pulse high skew (Figure 94.5 and Figure 94.6), respectively. Their difference is in their logic values. In these two formats the signal logic value must be held *low* between **t1** and **t2**, and *high* otherwise.

94.1.2.7 Surround by Complement

In the surround by complement frame format, a signal has its data between **t1** and **t2** (in the same cycle) and the complement of that data between **t0** to **t1** duration and **t2** to the next **t0** duration. Figure 94.7 shows an example of a surround by complement frame format. This figure indicates that the signal is *high* in a cycle and *low* in the next cycle.

94.1.2.8 Window

A window is a time duration in which an output signal must have a specified value. It begins at **t1** and ends at **t2** (in the same cycle). For a correct comparison, the signal is set to **dont_care** in time durations **t0** to **t1** and **t2** to next **t0**. Figure 94.8 shows an example of a window.

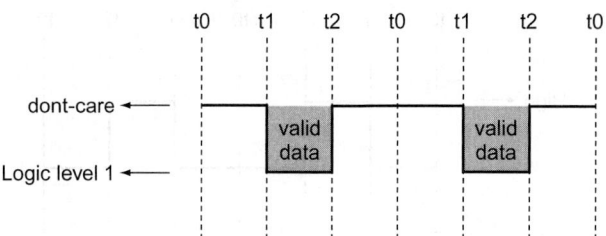

FIGURE 94.8   Window frame format.

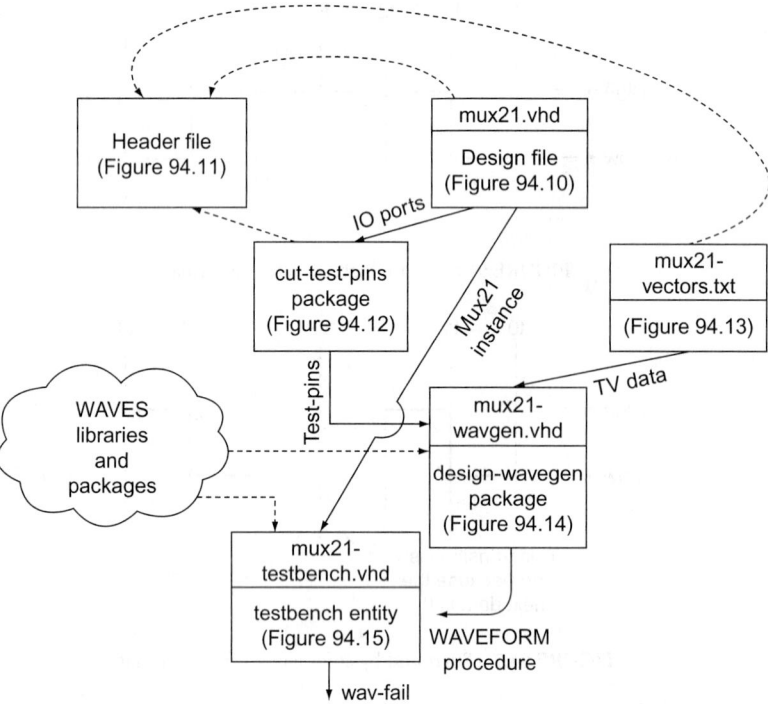

FIGURE 94.9   WAVES test set for a multiplexer.

94.1.2.9   Window Skew

The window skew frame format is similar to a window, except that the output data is valid between **t1** in a cycle and **t2** in the next cycle.

All the above frame formats are implemented by functions with proper names in the *WAVES_1164_FRAMES* package (e.g., *window_skew* function for window skew frame format, etc.). When called with proper values of **t0**, **t1**, and **t2**, they return a frame-set which has all the possible waveforms for all logic values. The testbench developer uses a returned frame-set to look up the proper waveform for a given logic value.

94.1.3   An Example: WAVES Testbench for a Multiplexer

To verify and test a design with WAVES, a *test set* must be defined. As shown in Figure 94.9, this test set includes several files that all together apply test vectors to a design, measure the outputs, and compare the outputs with the expected results. These files include:

- A WAVES header file containing information about the author, revision level, files dependencies, external files, waveform generator procedure, etc.

- A VHDL package including an enumeration-type definition for test pins. Enumeration elements of this type are names of input and output parts of the CUT.
- An input pattern file.
- A testbench file.
- A process for conversions between binary test data waveform formats.
- Standard and upper-level 1164 packages and libraries.

The above test set can either be generated manually or automatically in part or as a whole. There are several automatic testbench generator tools that use the design file as input and generate the header file (or at least its structure) and a test pin file.

In this section, we use a 2-to-1 one-bit multiplexer as our example CUT. We will show various parts of WAVES test set (files indicated in Figure 94.9) that are needed for this example. Figure 94.10 shows the VHDL code for the multiplexer example.

The header file for this example is shown in Figure 94.11. This file contains all test set information.

The necessary test pin file for the multiplexer example is shown in Figure 94.12. Note in this file that *test-pins* type is defined to include inputs and outputs of the multiplexer of Figure 94.10.

A test data file for our example test set is shown in Figure 94.13. This file shows input data (ordered according to enumeration elements of *test-pins*), expected data, and time increments at which data is

```
entity mux21 is
    port (i : in std_logic_vector (0 to 1); ctl : in std_logic; o : out std_logic);
end mux21;

architecture rtl of mux21 is
begin
    o <= i(0) after 10 ns when (ctl = '0') else
         i(1) after 50 ns when (ctl = '1') else
         'Z';
end rtl;
```

FIGURE 94.10 Multiplexer VHDL code.

```
-- Header file for entity mux21
TITLE              A design for a 2-to-1 multiplexer
DEVICE-ID          mux21
DATE               Sat Dec 10 12:54:39 2005
ORIGIN             CAD group
AUTHOR             S. Mirkhani
OTHER              This is a very simple design!
-- Data set construction information
VHDL_FILENAME      mux21.vhd                          WORK
WAVES_FILENAME     mux21-pins.vhd                     WORK
library            waves_1164;
use                waves_1164.waves_standard.all;
use                waves_1164.waves_interface.all;
use                work.cut-test-pin.vhd              WORK
VHDL_FILENAME      mux21-testbench.vhd                WORK
WAVES_UNIT         waves_objects                      WORK
WAVES_FILENAME     mux21_wavegen.vhd                  WORK
EXTERNAL_FILENAME  mux21-vectors.txt                  VECTORS
WAVEFORM_GENERATOR_PROCEDURE WORK.mux21.waveform
```

FIGURE 94.11 A sample header file for multiplexer.

```
-- This file can be generated automatically or manually
-- It is generated for mux21 entity
package cut_test_pins is
   type test_pins is (i_0, i_1, ctl, o);
end cut_test_pins;

package body cut_test_pins is
end cut_test_pins;
```

FIGURE 94.12   A sample test pin file for multiplexer.

```
%   DATA-IN         DATA-EXPECTED         TIME
    101                  0            :    200 ns;
    110                  1            :    200 ns;
    011                  1            :    200 ns;
```

FIGURE 94.13   A sample test data file for multiplexer.

used. As shown, data and timing are separated by a colon, and data slices are separated by a semicolon. The % sign in this figure marks a comment line. There are other formats for input data and timings that can be found in the IEEE WAVES document [1].

Another part of the multiplexer test set, shown in Figure 94.14, is a waveform generator file that includes a package containing the *WAVEFORM* procedure. This procedure opens the test vector file of Figure 94.13, reads input vectors and responses from this file, applies test data to the *WFS* port of the procedure, and reads the next slice after the given time increment (e.g., 200 ns in our example). In this procedure, we specify **nonreturn** frame format for the three inputs and the output. For each slide of data we use 150 and 180 ns for **t1** and **t2**, respectively. The testbench calls *WAVEFORM* procedure in its waveform-generation part. The underlined phrases show that phrase is defined in WAVES libraries.

The above files and packages are used in the testbench of the multiplexer example shown in Figure 94.15. Similar to Figure 94.14, the underlined phrases in Figure 94.15 are implemented in WAVES libraries. For reading data, this testbench uses the *WAVEFORM* procedure of Figure 94.14. Instantiation of the CUT also takes place in this testbench. The last process in this testbench compares obtained and expected test data.

94.1.4   WAVES Level 2

As stated at the beginning of this chapter, there are two levels defined for the WAVES standard. The above discussions applied to both WAVES *level 1* and *level 2*. In WAVES level 2, the WAVES user is allowed to build testbenches using more complex data structures and features of VHDL.

Constructs supported in WAVES level 2 include:

- Floating-point types
- Composite data types like arrays of records
- Access types
- Type castings and conversions
- Subtypes
- Variable assignments: in WAVES level 1, variables can only be assigned in the initialization phase
- Complex file operations: using predefined *TEXTIO* procedures, such as *read* and *write*, is allowed in WAVES level 2

```
-- Waveform generator file
-- mux21-wavegen.vhd
use std.textio.all;
library waves_1164;
use waves_1164.waves_1164_frames.all;
...

package design_wavegen is
   procedure WAVVEFORM (signal WFS : inout waves_port_list);
end design_wavegen;
package body design_wavegen is
   procedure WAVEFORM (signal WFS : inout waves_port_list) is
      file vect_file : text open READ_MODE is "mux21-vectors.txt";
      variable vect : FILE_SLICE := NEW_FILE_SLICE;
      constant inps : PINSET := i_0 + i_1 + ctl;
      constant outps : PINSET := o;
      variable timing : FRMAE_DATA := BUILD_FRAME_DATA ((
                                     (inps, non_return (0 ns)),
                                     (outps, window (50 ns, 180 ns)) ));
   begin
      READ_FILE_SLICE (vect_file, vect);
      while (not (vect.end_of_file) loop
         apply (WFS, vect.codes.all, timing);
         DELAY (vect.FS_TIME);
         READ_FILE_SLICE (vect_file, vect);
      end loop;
   end WAVEFORM;
end design_wavegen;
```

FIGURE 94.14 A sample waveform generator for multiplexer.

```
-- Testbench for mux21 design mux21-testbench.vhd
library ieee;
library waves_1164;
use ieee.std_logic_1164.all;
use waves_1164.waves_1164_utilities.all;
use waves_1164.waves_1164_interface.all;
use work.design_wavegen.all;

entity testbench is
end testbench;

architecture mux21 of testbench is
   signal wav_stim_data : std_logic_vector (0 to 2);
   signal wav_actual_o, wav_expect_o : std_logic;
   signal wav_fail : boolean;
   signal tb_wfs : waves_port_list;
begin
   wav_stim_data <= to_stdlogicvector (tb_wfs.signals (1 to 3)); --input vector
   wav_expect_o <= tb_wfs.signals (4); expected output read from file
   READVECT: WAVEFORM (tb_wfs); --wfs data structure is filled with file data
   CUT: entity work.mux21 (rtl) port map (wav_stim_data (0 to 1),
                                           wave_stim_data (2),
                                           wave_actual_o);
   --comparison section of the testbench
   compare_output: process (wav_actual_o, wav_expect_o)
   begin
      wav_fail <= Compatible (wav_actual_o, wav_expect_o);
      report "actual and expected outputs were compared at " & now() & ".";
   end process;
end mux21;
```

FIGURE 94.15 A sample testbench for multiplexer using WAVES.

Although there are VHDL features added to WAVES level 2, there are still several constructs that are not supported. These constructs include:

- Components and configuration specifications
- Signal declarations and assignments
- Wait statements
- Aliases, attributes, file, and physical types

94.1.5 Our Coverage of WAVES

In this section a library that is based on the VHDL language for generating efficient testbenches was discussed. Testbenches developed as such can be used for simulation-based verification of a design or they can be used by testers to perform a manufacturing test. The next section discusses another test language for hardware designs.

94.2 STIL

STIL is used as an interface between test tools and test equipments. Test tools generate a set of test vectors for a design with specified timings. They can be applied to the real design and their responses can be compared with the expected outputs using a STIL file loaded to an ATE. Unlike WAVES, which is a set of packages written in VHDL, STIL is a stand-alone language and not a library.

Several test tools that generate test vectors for design inputs or its scan chains have the option to save their generated test vectors in the STIL format. In contrast, STIL has become a popular language supported by most ATEs. Since STIL is easy to understand, a STIL file can also be generated or changed manually. As there can be a large number of test vectors to test a design, this standard also accepts GNU *gzip* format as its input file. This reduces the amount of data transferred to ATE.

The STIL standard was published in August 1999 under the IEEE standards for digital test vectors. In December 2002, the STIL working group released another version of this language identified as the IEEE 1450.2-2002. This version was an extension to STIL for DC-level specification. In June 2005, another extension to STIL for semiconductor design environments was published under the IEEE P1450.1 standard. Other extensions are also being developed by the STIL working group. An extension to STIL for supporting core test language (CTL) has been approved in 2005 under the P1450.6 standard. Extensions to STIL for tester target specification (P1450.3), test flow specification (P1450.4), semiconductor test method specification (P1450.5), and analog and mixed-signal (P1450.AMS) are the extensions that the STIL working group is working on.

In the next section the different parts of a STIL file, its constructs, and a simple example for each construct will be described. Then, a sample test vector file developed in STIL (IEEE 1450.0-1999) will be demonstrated to test Sayeh processor (this processor is discussed in Section 86.3). To study more details on STIL, see Refs. [3,4].

94.2.1 STIL Syntax

The main purpose of STIL is to develop a file containing the characteristics (IO pin information), related input test vectors and expected responses of a design in a proper timing. This file consists of several different parts and blocks. These parts must be ordered as defined in the STIL standard. These parts and their ordering are listed below.

1. **STIL** statement
2. **Header** block
3. **Signals** block
4. **SignalGroups** block
5. **ScanStructures** block

6. **Spec** block
7. **Timing** block
8. **Selector** block
9. **PatternBurst** block
10. **PatternExec** block
11. **Procedures** block
12. **MacroDefs** block
13. **Pattern** block

The above-mentioned parts will be discussed in the following subsections. There are also other blocks and definitions that can be written in any part of a STIL file. These parts are:

- **Ann.** This keyword shows an annotation statement and can be placed anywhere after the **STIL** statement.
- **UserKeywords.** The **UserKeywords** defines additional keywords to STIL. This construct can be written anywhere in a STIL file, not inside STIL blocks.
- **UserFunctions.** It defines additional timing expression functions. It can be written anywhere except inside other blocks.
- **Include.** This keyword opens a specified file in a STIL file (like **#include** preprocessor in C-language). It can be written anywhere after the **STIL** statement.

Several points about this language are worth mentioning. STIL is a case-sensitive language, e.g., *InputBus* is interpreted different from *inputbus*. The other point is that the user identifiers in STIL can be defined as both *quoted* identifiers and *nonquoted* identifiers. Nonquoted identifiers should begin with an alphabetical or an underscore sign and include only alphanumeric or underscores. In contrast, a quoted identifier can include any character. A quoted identifier should be referenced with its quotes. For example, if a quoted identifier named "*my-bus*" is defined, it must be referred to as "*my-bus*" everywhere, and not as *my-bus*. Identifiers in STIL should not exceed 1024 characters (including the double quotes in the quoted identifiers). In addition, an identifier longer than 1024 characters can be defined by partitioning it into 1024 character-length quoted identifier segments and concatenating the various segments by placing a period (.) between them.

The format of line comments in STIL is similar to those in C++ (// is used). For block comments, (*...*) is used.

94.2.1.1 STIL Statement

The **STIL** statement specifies the version of STIL used in the file. Its syntax and example are shown in Figure 94.16. *In all figures in this chapter, (a) and (b) parts are separated by a line.*

94.2.1.2 Header Block

The **Header** block comes after the **STIL** statement and contains file title, its creation date, and other file-related information. Figure 94.17 illustrates the **Header** block syntax and a related example. Note that **History** in this figure is a block which contains one or more annotations.

```
STIL stil_version_number;
```

```
STIL 1.0;
```

FIGURE 94.16 STIL statement. (a) Syntax. (b) Example.

```
Header {
    (Title "title";)
    (Date "date";)
    (Source "source";)
    (History { })
}
```

```
Header {
    Title "Sayeh TB";
    Date  "Sun Dec 11 9:13:17 EST 2005";
    History {
        Ann {* revision 1 - Sep 2005 - it didn't work! *}
        Ann {* revision 2 - Nov 2005 - bugs were fixed *}
        Ann {* revision 3 - Dec 1ˢᵗ - timings were completed *}
    }
}
```

FIGURE 94.17   Header block. (a) Syntax. (b) Example.

```
Signals {
    (sig-name <In|Out|InOut|Supply|Pseudo>;)* {
        (Termination TerminateHigh|TerminateLow|TerminateOff|TerminateUnknown>;)
        (DefaultState <U|D|Z|ForceUp|ForceDown|ForceOff>;)
        (Base <Hex|Dec> waveform_char_list;)
        (Alignment <LSB|MSB>;)
        (ScanIn (decimal-number);)
        (ScanOut (decimal-number);)
        (DataBitCount decimal-number;)
    })*
}
```

```
Signals {
    MemRead Out;
    MemWrite Out;
    Dbus InOut {
        Termination TerminateOff;
        Alignment MSB;
    }
    ExternalReset In {DefaultState ForceDown;}
    s_in  In  {ScanIn;}
    s_out Out {ScanOut;}
}
```

FIGURE 94.18   Signals block. (a) Syntax. (b) Example.

94.2.1.3   Signals Block

The input and output pins of a CUT are defined inside a **Signals** block shown in Figure 94.18. Only one **Signals** block is considered in a file and any other **Signals** block is ignored by the parser. The following characteristics of a signal can be specified in this block:

- **Signal type.** The signal type can be **In** for inputs, **Out** for outputs, **InOut** for input/outputs, **Supply** for power or ground (power supply signals), or **Pseudo** for a pseudo input/output signal (not a primary input/output).

- **Termination.** This attribute shows the value that an ATE applies to a signal when it finishes the testing process. **Termination** value is useful for increasing the fault coverage and speeding up the state transitions in a design. It can be **TerminateHigh, TerminateLow, TerminateOff,** or **Terminate Unknown** values.

  The **TerminateHigh** and **TerminateLow** switch the signal to high and low state, respectively. **TerminateOff** causes the ATE to set the voltage for the signal to float-state voltage. In **Terminate Unknown**, the ATE is programmed to set the load to float-state voltage level.
- **DefaultState.** This attribute indicates the default value assigned to a signal. It is used if the signal is not derived by a test vector.
- **Base.** A signal can be evaluated by hex (**Hex** option) or decimal (**Dec** option).
- **Alignment.** This attribute is useful when signal groups are used instead of individual signals, and the length of assigned data is larger than the number of signals in the group. It has two options: **MSB** and **LSB.** The **MSB** option maps the left-most bit of the data to the left-most signal of the group and other bits follow this pattern. The **LSB** option maps the right-most bit of data to the right-most signal in the group.
- **Scan Signals.** Scan signals include **ScanIn** and **ScanOut** and are used to define the scan chain scan-in and scan-out signals of a design. The length of these signals can also be defined.
- **DataBitCount.** This option is useful for multibit signals. It defines the number of data bits required for a waveform to be completed.

94.2.1.4   SignalGroups Block

The **SignalGroups** block, shown in Figure 94.19, defines a group of signals. This group can be empty or it can have only one member. The group of signals is treated as an individual signal in a STIL file. A **SignalGroups** block may have a domain name. Several **SignalGroups** blocks with different domain names can be defined in a STIL file, but only one global **SignalGroups** block can be defined in each file. The syntax and an example of **SignalGroups** block is shown in Figure 94.19.

94.2.1.5   ScanStructures Block

The **ScanStructures** block is an optional block and is mainly used for documentation. In this block, the characteristics of one or more scan chains in the design are specified. These characteristics include the name of scan chain input and output signals, the name of scan cells, the list of scan clocks, and the

```
SignalGroups (domain-name) {
  (group-name = sigref-expr;)*
  (group-name = sigref-expr {
    (Termination <TerminateHigh|TerminateLow|TerminateOff|TerminateUnknown>;)
    (DefaultState <U|D|Z|ForceUp|ForceDown|ForceOff>;)
    (Base <Hex|Dec> waveform_char_list;)
    (Alignment <LSB|MSB>;)
    (ScanIn (decimal-number);)
    (ScanOut (decimal-number);)
    (DataBitCount decimal-number;)
  })*
}
```

```
SignalGroups {
  DataBus = 'Dbus[0]+...+ Dbus[7]';
  AddBus = 'Abus[0]+...+ Abus[7]';
  MemCtl = 'MemRead + MemWrite';
  Scan_in = 's_in' {ScanIn 100; Base Hex LHX; Alignment LSB;}
}
```

FIGURE 94.19   SignalGroups block. (a) Syntax. (b) Example.

```
ScanStructures (scan-name) {
   (ScanChain chain-name {
       ScanLength decimal-number;
       (ScanOutLength decimal-number;)
       (ScanCells cell-list;)
       (ScanIn sig-name;)
       (ScanOut sig-name;)
       (ScanMasterClock sig-name-list;)
       (ScanSlaveClock sig-name-list;)
       (ScanInversion <0|1>;)
   })*
}
```

```
ScanStructures sayeh_scan {
   ScanChain chain1 {
       ScanLength 4;
       ScanCells dff1 dff2 ! dff3 ! dff4;
       ScanIn sc_in1;
       ScanOut sc_out1;
       ScanMasterClock clk;
       ScanInversion 0;
   }
   ScanChain chain2 {
       ScanLength 5;
       ScanCells ! msff1 msff2 ! msff3 msff4 ! msff5;
       ScanIn sc_in2;
       ScanOut sc_out2;
       ScanMasterClock mclk;
       ScanSlaveClock sclk;
       ScanInversion 1;
   }
}
```

FIGURE 94.20 ScanChain block. (a) Syntax. (b) Example.

inversion value of the scan path. In Figure 94.20, an example and the syntax of this block are shown. The design of the example in Figure 94.20 has two scan chains. The first scan chain has four D-type flip-flops, its input signal name is *sc_in1* and its output signal name is *sc_out1*. The clock signal for this chain is *clk*. The second scan chain has five master–slave flip-flops. Its input signal name is *sc_in2*, its output signal is *sc_out2*, the clock signal for master parts of the scan cells is *mclk*, and the clock signal for slave parts is *sclk*. The different parts of the block syntax are:

- **ScanChain.** This option indicates the name of a chain.
- **ScanLength.** The length of the scan chain is defined by this parameter. This parameter is mandatory in the block.
- **ScanOutLength.** This option specifies the number of observable cells in the scan chain. Its default value is the value of **ScanLength.**
- **ScanCells.** This parameter is a list of cell names that constitute this scan chain. The number of cells in the list should be equal to the **ScanLength** parameter. The left-most member of this list indicates the input cell of the scan chain, while the right-most member is the output cell of the scan chain. The order of the cells in the list must be identical to that in the scan chain. If there is an inversion before a cell, it must be defined by ! before the name of that cell in the list. The cells in the list are separated by a white space.
- **ScanIn.** This parameter defines the signal name through which scan data propagates into the scan chain.

- **ScanOut.** This parameter defines the output signal of the scan chain.
- **ScanMasterClock.** This parameter defines the list of signals that are connected to the clock ports of the master latches in a dual memory or the clock ports of single memories.
- **ScanSlaveClock.** This parameter defines the list of signals connected to the clock ports of the slave latches in a dual memory.
- **ScanInversion.** The value of this option can be **0** or **1**. Value **1** means that an overall relative inversion exists between the input of the first scan cell (before a scan cell) and the output of the last scan cell (after the last scan cell).

94.2.1.6 Spec Block

The **Spec** optional block, shown in Figure 94.21, is used for timing purposes. A signal in a design may have different delays in different situations. The kinds of delays defined in STIL are:

- **Min.** For minimum delay
- **Max.** For maximum delay
- **Typ.** For typical delay
- **Meas.** For measured delay. This delay is measured during run-time and its mechanism is not specified in STIL

In a **Spec** block, there can be one or more *categories* and *variables* defined. A **category** may contain several variables and each of them can have their own timings. In contrast, a **variable** may belong to several categories, and it can have different timings in each category. Therefore, **Spec** blocks can be defined as a list of categories or a list of variables, but not both. The categories can be used in the **PatternExec** block (see Section 94.2.1.10).

94.2.1.7 Timing Block

The **Timing** block, shown in Figure 94.22, is one of the main blocks in a STIL file. The **Timing** block specifies the waveform values and their timings for design signals. This block contains a number of waveform tables. In a waveform table (defined by **WaveformTable** block) a cycle time and the list of signals and their waveform shapes (defined in **Waveforms** block) are defined. All shapes in the **Waveforms** block use the same cycle (period) in a waveform table.

The **Timing** block is a large block of STIL. We will discuss each part and option of this block with simple examples.

- **SignalGroups.** This part is optional. In this part, the signal groups for which timings are to be defined are selected. In the absence of this part, the global **SignalGroups** block (a **SignalGroups** block with no domain name) will be used.
- **WaveformTable.** A **Timing** block can have several waveform tables. A waveform table has its own period (cycle) and waveform shape for the signal groups. The following options are defined inside a **WaveformTable** block.
- **Period.** This option indicates the period of signals defined in the **Waveforms** block. For example, the expression **Period** '500 ns'; indicates that each test vector is applied 500 ns after the previous one.
- **InheritWaveformTable.** This construct allows the use of definitions done in other waveform tables. A local definition always overwrites the inherited definition. The **Period** option can be removed from a waveform table if we inherit a waveform table that already has a period.
- **SubWaveforms.** Waveforms defined for different signals may have common parts. We can define common parts of several waveforms in the **SubWaveforms** block and use it in our waveform definitions. A **SubWaveform** is referenced in a waveform definition by its label (**SubWaveform** label in Figure 94.22). A **SubWaveform** can have a duration (defined by **Duration** keyword in Figure 94.22). This duration is used when a **SubWaveform** is repeated in a waveform. The rest of a **SubWaveform** block is an event list definition which is similar to the definition of an event list in **Waveforms** block and is described in the **Waveforms** part, described next.

```
Spec (spec-name) {
    (Category cat-name {
        (var-name = time-expr;)*
        (var-name = {
            (Min time-expr;)
            (Typ time-expr;)
            (Max time-expr;)
        })*
    })+
}
```

```
Spec sayeh_spec {
    Category bus_time {
        low_to_high = '2.0ns';
        high_to_low = {Min '1.0ns'; Type '3.0ns';}
    }
    Category signal_time {
        high_to_low = '2.0ns';
        low_to_high = '3.0ns';
    }
}
```

```
Spec (spec-name) {
    (Variable var-name {
        (cat-name = time-expr;)*
        (cat-name = {
            (Min time-expr;)
            (Typ time-expr;)
            (Max time-expr;)
        })*
    })+
}
```

```
Spec sayeh_spec {
    Variable low_to_high {
        bus_time = '2.0ns';
        signal_time = '3.0ns';
    }
    Variable high_to_low {
        bus_time = {Min '1.0ns'; Typ '3.0ns';}
        signal_time = '2.0ns';
    }
}
```

FIGURE 94.21 Spec block. (a) Syntax alternative 1. (b) Example for alternative 1. (c) Syntax alternative 2. (d) Example for alternative 2.

- **Waveforms.** In this block, which can only be defined once in a **WaveformTable** block, one or more waveform shapes can be defined for a signal in the design. A signal waveform shape indicates that a given event in a given time should be applied or measured on that signal. A waveform consists of the following parts:
  - ○ **Waveform Character.** This is a simple character (shown by *wf_char* in Figure 94.22) which is used in a vector (in the **Patterns** block). This character is the representative of a set of events that happens on a signal. For example, if we need to apply a periodic 10 ns **0** pulse to a signal,

```
Timing (domain-name) {
  (SignalGroups group-domain;)*
  WaveformTable name {
    (Period time-expr;)
    (InheritWaveformTable (tim.)wft;)*
    (SubWaveforms {
      (swf-label: Duration time-expr {
        ((event-label:) (time-expr)
                (<event-list ([event-num])|event>);)*
      })*
    })
    Waveforms {
      (sigref-expr (wav) {
        (InheritWaveform ((tim.)wft.)wav;)*
        (wf-char {
          (InheritWaveform (((tim.)wft.)wave.)wf-char;)*
          ((event-label:) (time-expr)(event);)*
        })*
        (wf-charlist {
          (InheritWaveform (((tim.)wft.)wav.)wf-char;)*
          ((event-label:) (time-expr)
                  (<event-list ([event-num])|event>);)*
        })*
        (<wf-char|wf-charlist> {
          ((event-label:) (time-expr) (\rN)
                  <sub-wf-lab;|sub-wf-lab[N];|sub-wf-lab[#];>)*
        })*
      })*
    }
  }
}
```

FIGURE 94.22 Timing block syntax.

we can define a waveform for this signal with waveform character *s*. Then, in the **Patterns** block, this character can be used to represent the defined pulse.

- ○ **Event label.** This is a label (shown by *event_label* in Figure 94.22) assigned to the time expression that follows it. We can use this label instead of its next time expression in the current block. This label can be interpreted as a local variable which is assigned once by a time expression and can be used several times in other timing expressions. Event labels are useful to calculate the timing of a signal from its durations. There is a special label @ which is used to store the last timing expression used in a waveform.

- ○ **Event.** There are several predefined values in STIL for deriving events (used for inputs) and comparing events (used for outputs). Some of these values have both short and long representations. Events used for inputs include:

  - **ForceDown or D.** Shows that a signal is forced to be logic low.
  - **ForceUp or U.** Shows that a signal is forced to be logic high.
  - **ForceOff or Z.** Shows that a signal is forced to be high impedance.
  - **ForcePrior or P.** Shows that a signal should preserve its last derived state.

  Values used for comparison are as follows:

  - **CompareLow or L.** Shows that a signal should be logic low.
  - **CompareHigh or H.** Shows that a signal should be logic high.
  - **CompareOff or T.** Shows that a signal should be high impedance.

- **CompareUnknown or x or X.** Indicates that the signal should not be compared with any values.
- **CompareValid or V.** Shows that a signal should be compared for a valid logic level. It is compared for a voltage lower than the low threshold or higher than the high threshold.
- **CompareLowWindow or l/CompareHighWindow or h.** Shows that a signal should be compared with logic low/logic high in a period of time and should be terminated by CompareUnknown.
- **CompareOffWindow or t/CompareValidWindow or v.** Shows that a signal should be high impedance/a valid logic level in a period of time and should be terminated by CompareUnknown.

The following values are expected values from the outputs:

- **ExpectLow or R.** The CUT output is expected to be logic low.
- **ExpectHigh or G.** The CUT output is expected to be logic high.
- **ExpectOff or Q.** The CUT output is expected to be high impedance.
- **Marker or M.** With this event, no activity from the CUT or the tester is expected.

There are also a few unresolved events defined in STIL as follows:

- **ForceUnknown or N.** Shows the driver is turned on, but the up or down state is not defined yet.
- **LogicLow or A.** Shows that the driver is low, but its direction is unknown.
- **LogicHigh or B.** Shows that the driver is high, but its direction is unknown.
- **LogicZ or F.** Shows that the driver is high impedance, but its direction is unknown.
- **Unknown or ?.** Shows that the driver has an unknown logic with an unknown direction.

Using the events described above, a waveform for a signal can be defined. The shape of this signal is shown in Figure 94.23(a). The STIL file corresponding to this waveform is shown in Figure 94.23(b). In this figure *p* and *n* characters are defined for pulsing and turning off the *clk* signal.

o **Waveform character list.** The various waveforms defined by a waveform character for a signal can be merged. The merged waveform is called a waveform list. For example, in Figure 94.23(b) we can merge two waveforms defined for the *clk* signal as shown in Figure 94.24.

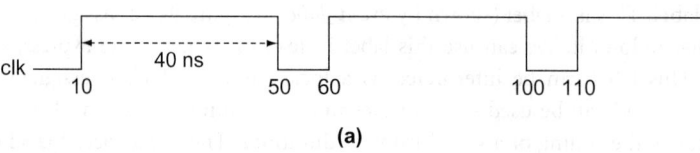

(a)

```
WaveformTable clk_wft {
    Period '50ns';
    Waveforms {
        clk {
            p {'0ns' ForceDown; RISETIME: '10ns' ForceUp;
               'RISETIME+40ns' ForceDown;}
            n {'0ns' ForceOff;}
        }
    }
}
```

(b)

Figure 94.23   (a) A sample waveform. (b) Waveforms block for (a).

```
...
clk {
    pn {'0ns' ForceDown/ForceOff;
         RISETIME: '10ns' ForceUp/ForceOff;
         'RISETIME+40ns' ForceDown/ForceOff;}
}
...
```

FIGURE 94.24   An example of waveform character list.

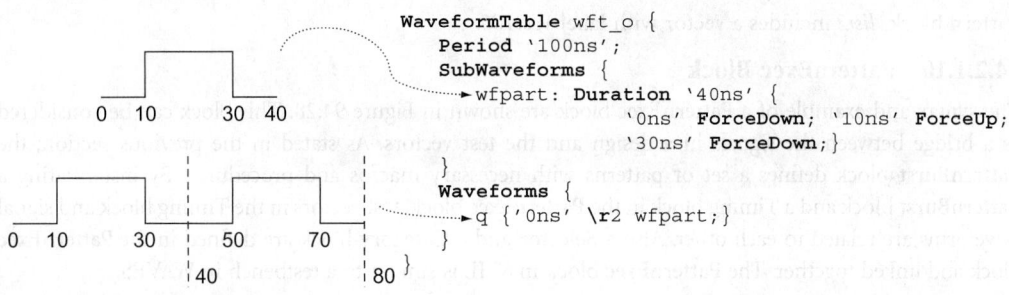

FIGURE 94.25   An example of SubWaveforms block.

```
Selector sel-name {
    (var-name <Min|Typ|Max|Meas>;)*
}
```

```
Selector sayeh-sel {
    low_to_high Typ;
    high_to_low Min;
}
```

FIGURE 94.26   Selector block. (a) Syntax. (b) Example.

- **Using subwaveforms.** Figure 94.25 shows an example subwaveform. A waveform can use another waveform as one of its parts. If a waveform is completely similar to another waveform, then it is best to use that waveform by the **InheritWaveform** option (see the last part of Figure 94.22). A subwaveform should be defined in a **SubWaveforms** block. This subwaveform can be repeated in a waveform by \rN option shown in Figure 94.22, where *N* is a positive integer. Also the two formats **sub_wf_lab[N]** and **sub_wf_lab[#]** in Figure 94.22 are related to the subwaveform event list with *event_num* index and are not discussed here.

94.2.1.8   Selector Block

A **Selector** block, shown in Figure 94.26, selects a delay type for each variable in a **Spec** block (see Section 94.2.1.6). These selected times are used in a **PatternExec** block. The example in Figure 94.26 is based on the **Spec** example shown in Figure 94.21. In this example, if *sayeh-sel* is used in a **PatternExec** block, the default delays for *low_to_high* will be 2 ns, while the default delay for the *high_to_low* variable will be 1 ns.

94.2.1.9 PatternBurst Block

Figure 94.27 shows the syntax and an example of a **PatternBurst** block. This block defines a set of patterns and defines certain characteristics for them. Application of test vectors defined in a **PatternBurst** block takes place in a **PatternExec** block. As shown in Figure 94.27(a) (the syntax part), the **PatternBurst** block has several parts. In this block a specified signal group, macro definitions, procedures, scan structures, and a termination state for signals can be defined. Also the first and the last test vectors in the burst can be specified (by **Start** and **Stop** options).

The main part of a **PatternBurst** block is the **PatList** block. In this block, a set of patterns or another previously defined **PatternBurst** block can be specified. Note that the termination state for a signal in a defined **PatternBurst** block overwrites the termination state defined in the **Signals** block. An example of this block is shown in Figure 94.27(b). In this example, *list1* and *list2* are pattern lists defined in the Pattern block. *list2* includes a vector with label *vect100*.

94.2.1.10 PatternExec Block

The syntax and example of a **PatternExec** block are shown in Figure 94.28. This block can be considered as a bridge between the signals in a design and the test vectors. As stated in the previous section, the **PatternBurst** block defines a set of patterns with necessary macros and procedures. By instantiating a **PatternBurst** block and a **Timing** block in the **PatternExec** block, test vectors in the **Timing** block and signal waveforms are related to each other. Also a **Selector** and a **Category** block are defined in the **PatternExec** block and linked together. The **PatternExec** block in STIL is similar to a testbench in WAVES.

```
PatternBurst pat-burst-name {
    (SignalGroups group-domain;)*
    (MacroDefs macro-domain;)*
    (Procedures proc-domain;)*
    (ScanStructures scan-name;)*
    (Start pattern-label;)
    (Stop pattern-label;)
    Termination {
        (sigref-expr <TerminateHigh|TerminateLow|TerminateOff|TerminateUnknown>;)*
    })*
    (PatList {
        (patt-or-burst-name;)*
        (patt-or-burst-name {
            (SignalGroups group-domain;)*
            (MacroDefs macro-domain;)*
            (Procedures proc-domain;)*
            (ScanStructures scan-name;)*
            (Start pattern-label;)
            (Stop pattern-label;)
            Termination {
                (sigref-expr <TerminateHigh|TerminateLow|TerminateOff|TerminateUnknown>;)*
            })*
        })*
    })+
}
```

```
PatternBurst sayeh_pb {
    Termination {Abus TerminateOff;}
    PatList {
        list1;
        list2 {Stop vect100;}
    }
}
```

FIGURE 94.27   PatternBurst block. (a) Syntax. (b) Example.

```
PatternExec (pat-name) {
   (Category cat-name;)*
   (Selector sel-name;)*
   (Timing tim-name;)
   (PatternBurst pat-b-name;)
}
```

```
PatternExec sayeh_pe {
   Category bus_time;
   Selector sayeh_sel;
   Timing sayeh_tim;
   PatternBurst sayeh_pb;
}
```

FIGURE 94.28   PatternExec block. (a) Syntax. (b) Example.

```
Procedures (proc-domain-name) {
   (procedure-name {
      (pattern-statement)*
   })*
}
```

```
Procedures {
   sayeh_proc1 {
      W wf1;
      V {sig = 10;}
      V {sig = 11;}
      V {sig = 01;}
   }
   sayeh_proc2 {
      W wf2;
      V {sig = 10;}
      V {sig = 01;}
   }
}
```

FIGURE 94.29   Procedures block. (a) Syntax. (b) Example.

94.2.1.11   Procedures Block

The syntax and example for **Procedures** block are shown in Figure 94.29. This block performs a given set of operations on its arguments. It can be called from inside a **Pattern** block, a **MacroDefs** block, or another **Procedures** block. When a procedure or macro calls another procedure, the caller procedure should be defined after the called procedure to prevent recursion. Procedures can also have domain names. Only one global procedure (a procedure without a domain name) can be defined in a STIL file.

94.2.1.12   MacroDefs Block

A macro in STIL is defined in a **MacroDefs** block (Figure 94.30). Macros are similar to procedures. A macro is *expanded* via the **Macro** statement. Unlike procedures, macros do not need to be meaningful portions of a block. For example, the procedure of Figure 94.29 must have a waveform statement, while in the macro of Figure 94.30 the waveform statement (**W** *wf1*) can be removed. This is because this statement can be defined before macro expansion.

```
MacroDefs (mac-name-domain) {
   (mac-name {
      (patt-statement)*
   })*
}
```

```
MacroDefs {
   sayeh_mac {
      V {sig = 10;}
      V {sig = 11;}
      V {sig = 01;}
   }
}
```

FIGURE 94.30   MacroDefs block. (a) Syntax. (b) Example.

```
Pattern patt-name {
   ((label:) TimeUnit 'time-expr';)
   (pattern-statement)*
}
```

FIGURE 94.31   Pattern block syntax.

```
(label:) V(ector) {(cyclized-data)* (non-cyclized-data)*}
```

```
v_label: Vector {Abus = 000110100;}
V {Dbus = 01100100;}
```

FIGURE 94.32   Vector statement. (a) Syntax. (b) Example.

94.2.1.13   Pattern Block

Test vectors and the expected output responses are defined in the **Pattern** block. As shown in Figure 94.31, this block contains an optional statement for time unit definition and several pattern statements.

   Pattern statements have been designed to define a set of test vectors for a design. There are several types of **Pattern** statements which we list below:

   • **Vector statement.** A test vector is applied by this statement. Its syntax is shown in Figure 94.32.

      As shown in the syntax part of Figure 94.32, there are two kinds of data for a test vector: *cyclized* and *noncyclized*. Cyclized data are the vectors that are applied in common period of time (defined in the **WaveformTable** block). V {*Abus* = 00011010;} defines data on *Abus* as cyclized data. On the other hand, a noncyclized data is a kind of data that is applied to the design or measured from a design in a specified time. Its syntax and an example are shown in Figure 94.33. In this figure, the *time_value* is a positive integer which shows the relative time (in the defined time unit) to the start of the vector (or a common *T0*).

```
@time-value sigref-expr = event-list;
@time-value {(sigref-expr = event-list;)⁺}
```

```
@30 clk = U; @70 clk = D; @110 clk = U;
@122 {Dbus = DDDUDDUD; ExternalReset = U;}
```

FIGURE 94.33 Noncyclized data. (a) Syntax. (b) Example.

```
rst: V {ExternalReset = 1; @100 ExternalReset = 0;}
V { } //Repeats the previous vector statement
```

FIGURE 94.34 An example of a vector with various data formats.

```
(label:) W(aveformTable) wft-name;
```

```
wf1: W sayeh_wft;
```

FIGURE 94.35 WaveformTable statement. (a) Syntax. (b) Example.

```
(label:) Call proc-name;
(label:) Call proc-name {(scan-data)* (cyclized-data)*
                         (non-cyclized-data)*}
```

```
Call sayeh_proc1;
pr_lab: Call sayeh_proc2
```

FIGURE 94.36 Procedure call. (a) Syntax. (b) Example.

In addition to cyclized and noncyclized data, there is a serial data construct defined in STIL. This is useful for applying and reading data to and from a scan chain. An example of two vector statements is shown in Figure 94.34.

- **WaveformTable statement.** A waveform table, shown in Figure 94.35, is defined in this statement. Signal values in the vector statements up to the next **WaveformTable** statement follow the timings and shapes defined for the waveform characters in this defined **WaveformTable**.
- **Procedure call statement.** A procedure can be called in a **Pattern** block, as shown in Figure 94.36. Arguments can be passed to the procedure being called.
- **Macro call statement.** Similar to a procedure, macros can be called inside a **Pattern** block. This is done by the **Macro** keyword, as shown in Figure 94.37.
- **Goto statement.** An unconditional branch can happen inside a **Pattern** block by the **Goto** statement. The syntax and an example of the **Goto** statement are shown in Figure 94.38.
- **Condition statement.** This statement, shown in Figure 94.39, causes an input vector or output value to be established, but its actual application is postponed until a vector statement is defined.

```
(label:) Macro mac-name;
(label:) Macro mac-name {(scan-data)* (cyclized-data)*
                         (non-cyclized-data)*}
```

```
mac1: Macro sayeh_mac1;
      W wft2;
mac2: Macro sayeh_mac2 {sig = 01;}
```

FIGURE 94.37 Macro expansion. (a) Syntax. (b) Example.

```
(label:) Goto label-name;
```

```
begin_1: V {ExternalReset = 0;}
...
Goto begin_1;
```

FIGURE 94.38 Goto statement. (a) Syntax. (b) Example.

```
(label:) C(ondition) {(cyclized-data)* (non-cyclized-data)*}
```

```
c1: Condition {MemRead = L;}
C {MemWrite = H;}
...
V { }
```

FIGURE 94.39 Condition statement. (a) Syntax. (b) Example.

```
(label:) Loop loop-cnt {(patt-statement)*}
```

```
lp1: Loop 20 {
   V {clk = 1;}
   V {ExternalReset = 0;}
   ...
}
```

FIGURE 94.40 Loop statement. (a) Syntax. (b) Example.

- **Loop/MatchLoop statements.** A **Loop** statement repeats a set of vectors for a specified number of iterations. Its syntax and an example are shown in Figure 94.40. A **Loop** statement can enclose another Loop statement. **MatchLoop** statements are conditional. They can be infinite loops (until that condition is met) or they can have an upper limit. Figure 94.41 shows the syntax and an example of a **MatchLoop** statement.

- **Breakpoint statement.** The **Breakpoint** statement, shown in Figure 94.42, is generally used for segmenting a large number of vectors (which may exceed the memory of the tester). A **Breakpoint** statement may also contain a statement. The signals in this statement run until the next vector after the **Breakpoint** statement is executed. The other signal values of the CUT do not change during this statement.
- **IDDQ testpoint.** This statement is used when an IDDQ measurement is to be performed on a design. Its syntax is shown in Figure 94.43.
- **ScanChain statement.** This statement defines the name of the scan chain on which the data is going to be applied. Figure 94.44 shows its syntax and an example.
- **Stop statement.** The **Stop** statement (its syntax is *(label:) Stop;*) stops the execution of patterns.

```
(label:) MatchLoop <loop-cnt|Infinite> {
   (patt-statement)⁺
   (b-label:) BreakPoint {(patt-statement)⁺}
}
```

```
ml1: MatchLoop <Infinite> {
   V {clk = 1;}
   V {ExternalReset = 0;}
   ...
   exit_1: BreakPoint {Abus = xxxxxxxx;}
}
```

FIGURE 94.41   MatchLoop statement. (a) Syntax. (b) Example.

```
(label:) BreakPoint;
(label:) BreakPoint {(patt-statement)*}
```

```
bp_1: BreakPoint {clk = 0; @50 clk = 1;}
```

FIGURE 94.42   BreakPoint statement. (a) Syntax. (b) Example.

```
(label:) IDDQ test-point;
```

FIGURE 94.43   The syntax of IDDQ statement.

```
(label:) ScanChain chain-name;
```

```
...
ch1: ScanChain chain1;
V { }
V { }
...
```

FIGURE 94.44   ScanChain statement. (a) Syntax. (b) Example.

```
STIL 1.0;
Header {
    Date "Tue Dec 13 11:15:20 EST 2005";
    Source "Generated in CAD group";
    History {
        Ann {* rev1 = Dec 13 - file created *}
    }
}
Signals {
    AddressBus0  InOut; AddressBus1  InOut; AddressBus2  InOut; …
    DataBus0 InOut; DataBus1 InOut; DataBus2 InOut; …
    ExternalReset In; IORead Out; IOWrite Out; MemRead Out; MemWrite Out;
    DataReady In; clk In;
}
SignalGroups {
    ABus = 'AddressBus7+AddressBus6+…+AddressBus0';
    DBus = 'DataBus7+DataBus6+…+DataBus0';
    Inps = 'ExternalReset+DataReady';    Outps = 'MemRead+MemWrite+IORead+IOWrite';
}
SignalGroups simpleG {
    simple_ios = 'Inps+Outps';    clocks = 'clk';
}
SignalGroups busses {
    bus = 'ABus+DBus';
}
Spec sayeh_spec {
    Category MemComm {
        tplh = {Min '1ns'; Typ '2ns'; Max '3ns';} tphl = {Min '2ns'; Typ '2ns'; Max '4ns';}
    }
}
Timing simple {
    SignalGroups simpleG;
    WaveformTable simp1 {
        Period '100ns';
        Waveforms {
            Clocks { P {'0ns' ForceDown; '20ns' ForceUp;  '40ns' ForceDown;}
            Inps { 01Z {'0ns' D/U/Z;}
            Outps {01Z {'0ns' CompareOff; '30ns' CompareLowWindow/ CompareHighWindow/
                        CompareOffWindow; '50ns' CompareOff;}
        }
    }
    SignalGroups busses;
    WaveformTable simp2 {
        Period '100ns';
        Waveforms {
            bus {01 {'0ns' Z; '25ns' D/U; '35ns' Z;}
            bus {LH {'0ns' T; '30ns' L/H; '40ns' T;}
        }
    }
    WaveformTable total {
        InheritWaveform simp1;
        InheritWaveform simp2;
    }
}
Selector min_sel {
    tplh Min; tphl Min;
}
Selector max_sel {
    tplh Max; tphl Max;
}
PatternBurst sayeh_pb {
    SignalGroups busses; SignalGroups simplG;
    Start st_lb; Stop sp_lab;
    Termination {bus TerminateOff;}
    PatList {
        sayeh_pat;
    }
}
```

FIGURE 94.45   STIL file for Sayeh example.

```
PatternExec sayeh_pe {
    Category MemComm; Selector min_sel;
    Timing simple; PatternBurst sayeh_pb;
}
Pattern sayeh_pat {
    W total;
    st_lab: V {clocks = P; ExternalReset = 1; @9 ExternalReset = 0;}
    V {clocks = P; ABus = LLLLLLLL; DBus = LHHLHHHL;}
    Lp: Loop 3 {
        V {clocks = P;}
    }
    V {ABus = LLLLLLHL; DBus = HHHHLLLL;}
    ...
    sp_lab: V {clocks = P; ABus = LHLLLHHL; DBus = HHHLLLHL;}
    Goto st_lab;
}
```

FIGURE 94.45 (*Continued*).

This section gave an overview of the STIL language. Other details and features not discussed here can be found in references listed at the end of this chapter [3,4].

94.2.2 An Example: A STIL File for Sayeh Processor

A simple example of a STIL file is demonstrated in Figure 94.45. Only a few of STIL features are used in this example. This example tests a processor called Sayeh by first resetting the processor (this is done by activating the **ExternalReset** signal), and then pulsing the clock in each tester cycle. For testing this processor, a specified program, which has known outputs, is loaded into the memory of the system. Therefore, after each cycle address and data busses (*Abus* and *Dbus* in Figure 94.45) have known values. The tester would stop applying the vectors to the CUT (Sayeh in this example) when it reaches the vector labeled *sp_lab* in the **Pattern** block. In the next cycle after applying this vector, data bus and address bus go to the high impedance state.

94.2.3 Our Coverage of STIL

In this section, STIL, which is a language used for testing a design, was discussed. This language is widely used for manufacturing test, and many commercial test tools and ATEs support this language. This section used several illustrative examples to discuss syntax and semantics of STIL. The last section in this section showed a complete example of STIL for testing a simple CPU.

This chapter discussed two test languages widely used in design and hardware testing. The first language was WAVES, which is actually a library based on the VHDL language. The second part of this chapter discussed the STIL language. STIL is a stand-alone language for testing devices and designs. Several examples were discussed in this chapter, which describe the features of these two languages.

References

1. *IEEE Standard for VHDL Waveform and Vector Exchange to Support Design and Test Verification (WAVES)*, IEEE Std 1029-1, Inst. of Elect. & Electronic, New York, 1999.
2. J.P. Hanna, R.G. Hillman, H.L. Hirsch, T.H. Noh, and R.R. Vemuri, *Using WAVES and VHDL for Effective Design and Testing*, Springer, Berlin, 1996.
3. *IEEE Standard Test Interface Language (STIL) for Digital Test Vector Data*, IEEE Std 1450, Inst. of Elect. & Electronic, New York, 1999.
4. G.A. Maston, T.R. Taylor, and J.N. Villar, *Elements of STIL: Principles and Applications of IEEE Std. 1450 (Frontiers in Electronic Testing)*, Springer, Berlin, 2003.

95

Timing Description Languages

Naghmeh Karimi and
Zainalabedin Navabi
Nanoelectronics Center of Excellence
School of Electrical and Computer
Engineering
University of Tehran

CONTENTS

In addition to languages for design and description, today's hardware design requires formats or languages for describing structures, timing constraints, physical layouts, and other hardware characteristics. On the timing side, SDF is a language (or standard format) for describing timing of hardware components, and VCD is a format for describing values and timings of signals of a hardware system. SDF and VCD are used for delay and signal value specification of hardware components.

This chapter covers the basics of SDF and VCD representations and familiarizes the reader with the overall structure of these formats.

95.1 SDF

Standard Delay Format (SDF) is an ASCII format used to convey timing information between EDA tools. SDF was first developed by Cadence in 1990. Open Verilog International (OVI) approved SDF version 2 in 1993 and SDF version 2.1 and 3 in the following years. Improving SDF version 3 features, resulted in IEEE Standard 1497 [1].

95.1.1 Role of SDF in Design Process

SDF can be used in different levels of a design process to annotate timing specifications, including timing data and constraints.

Timing constraints are used during the forward-annotation process while timing data are applied during the back-annotation process. Forward-annotation process deals with porting timing constraints to synthesis, floorplanner, layout, and routing tools to meet the required constraints.

During the design process, timing data can be back-annotated to analysis tools (including simulators, static timing analysis tools, etc.) to provide more accurate timing representations.

95.1.2   SDF Structure

An SDF file is in ASCII format with a structure as shown below.

```
(DELAYFILE
    .
    .          }  Header  Section
    .
    (CELL
       .
       .
       .
    )
    (CELL
       .
       .          Cell Section
       .
    )
       .
       .
       .
)
```

An SDF file starts with the **DELAYFILE** keyword and is followed by a header section and one or more cell descriptions. Each cell corresponds to a part of the design and can be an ASIC library primitive, a modeling primitive for a specific analysis tool or a user-created part of the design hierarchy.

95.1.2.1   Header Section

The header of a SDF file includes documentation, physical and formatting information about the entire file. The documentation information includes version of the current SDF file, related design and vendor name, date of creation, and generating tool version and generating tool name of the current SDF file. Physical information includes the operating voltage, temperature and process factor for which the corresponding timing parameters have been evaluated, and timescale for the corresponding timing parameters. In addition, formatting information that represents the symbol demonstrating the hierarchy in the SDF file is included in the header. Among the discussed items the only required entry is the SDF version, while other entries are optional. Note that the default value for timescale is 1 ns. The hierarchy divider symbol must be either a period (.) or a slash (/) with a default value of period (.).

The first part of Figure 95.1 (lines 1 to 12) shows the header of a sample SDF file. Following the header, the cell part begins on line 14 of this figure, which will be discussed in the following sections.

95.1.2.2   Cell Section

Each cell of a design is identified by its name (**CELLTYPE**), hierarchical location (**INSTANCE**), and corresponding timing parameters. As shown in Figure 95.1, each cell starts with the **CELL** keyword and is followed by its type, instance, and timing specifications. These fields are discussed below:

CELLTYPE is a string that indicates the name of an ASIC library primitive or a user-created region.

INSTANCE is a string that represents the hierarchical location of the current cell in the design.

```
 1    (DELAYFILE
        (SDFVERSION             "4.0")
        (DESIGN                 "T flip-flop")
        (DATE                   "Dec 12, 2005, 01:39:52")
 5      (VENDOR                 "ASIC Chips")
        (PROGRAM                "Sample Program)
        (VERSION                "v4.0")
        (DIVIDER                / )
        (VOLTAGE                5.5:5:4.5 )
10      (PROCESS                "typical")
        (TEMPERATURE               :35: )
        (TIMESCALE              1 ns)

        (CELL
15        (CELLTYPE "DFF")
          (INSTANCE ix57)
          (DELAY
            (ABSOLUTE
              (PORT C (0.01:0.01:0.02) (0.01:0.02:0.03)) )
20          (ABSOLUTE
              (PORT D (0.01:0.01:0.02) (0.01:0.02:0.03) ) )
            (ABSOLUTE
              (PORT RN(0.01:0.01:0.02) (0.01:0.02:0.03) ) )
            (ABSOLUTE
25            (IOPATH C Q (::0.19) (::0.18) (::0.18) (::0.18) (::0.18) (::0.18) )
            (ABSOLUTE
              (IOPATH RN Q (::0.00) (::0.13) ) )
            (ABSOLUTE
              (IOPATH C QN (::0.16) (::0.14) ) )
30          (ABSOLUTE
              (IOPATH RN QN (::0.12) (::0.00) ) )
          )
          (TIMINGCHECK
            (SETUP D (posedge C) (0.1) )
35          (HOLD D (posedge C) (0.01) )
          )
        )
        (CELL
          (CELLTYPE "INV")
40        (INSTANCE i x67)
          (DELAY
            (ABSOLUTE
              (PORT A (0.01:0.01:0.02) (0.01:0.02:0.02) ) )
            (ABSOLUTE
45            (IOPATH A Q ( (0.02:0.03:0.04) (0.01:0.02:0.03) (0.02:0.03:0.04) )
              ( (0.03:0.04:0.05) (0.02:0.03:0.04) (0.03:0.04:0.05) ) ) )
          )
        )
      )
```

FIGURE 95.1   SDF file of a T-type flip-flop.

Timing specification fields of a cell (starting on line 17 of Figure 95.1) contain the actual timing data associated with that cell. Four different types of timing specifications can be used for each cell. These timing specifications are **DELAY, TIMINGCHECK, TIMINGENV,** and **LABEL.**

CELLTYPE and **INSTANCE** are required for identifying each cell, but timing specification fields are optional, i.e., each cell in the SDF file can include zero or more timing specifications.

95.1.2.2.1 Delay Values.

Each delay construct includes a list of 1, 2, 3, 6, or 12 values (see, for example, line 19 of Figure 95.1). In the case of a 12-valued list, values specify in sequence the delays corresponding to transitions in column 1 of Table 95.1. For a 2-, 3-, or 6-valued list, delay values corresponding to each transition can

TABLE 95.1   Deriving Delay Values

| Transition | Number of Delay Values in a Delay List | | |
	2 value	3 value	6 value
0→1	0→1	0→1	0→1
1→0	1→0	1→0	1→0
0→z	0→1	0→z	0→z
z→1	0→1	0→1	z→1
1→z	1→0	0→z	1→z
z→0	1→0	1→0	z→0
0→x	0→1	min(0→1, 0→z)	min(0→1, 0→z)
x→1	0→1	0→1	max(0→1, z→1)
1→x	1→0	min(1→0, 0→z)	min(1→0, 1→z)
x→0	1→0	1→0	max(1→0, z→0)
x→z	max(0→1, 1→0)	0→z	max(0→z, 1→z)
z→x	min(0→1, 1→0)	min(0→1, 1→0)	min(z→0, z→1)

be specified according to columns 2 to 4 of this table. For example in a 6-valued list, the represented values specify in sequence the delays corresponding to the first 6 transitions in column 1 of Table 95.1. In this case the related delays of other transitions are determined according to column 4 of this table. If there is only one value in the list then this value is applied to all the transition delay values. If a delay value is null, the parentheses enclosing that value is empty. This null value acts as a placeholder and allows specifications of the corresponding parameters further down the list.

Figure 95.1 shows the SDF file of a T-type flip-flop composed of a D-type flip-flop and an inverter. Line 25 of this figure shows an example of utilizing a 6-valued delay list.

A delay construct includes a list of delay values each of which is composed of a list of 1, 2, or 3 values enclosed in a pair of parentheses. An example is shown in line 45 of Figure 95.1 and will be discussed later in this section.

If a delay value contains three values, the first value specifies the delay value, while the second and the third values represent rejection limit (r-limit) and filter limit (e-limit), respectively. In addition, if a delay value is composed of two values, the first value specifies the delay value while the second value represents both limits (r- and e-limit). Finally, in the case of having only a single value, this value represents delay value, r-limit and e-limit. r-limit and e-limit are the lowest allowable pulse widths, i.e., any pulse narrower than r-limit will be rejected (no pulse appears on the port) and any pulse narrower than e-limit will appear as the **X** value.

Each of these three values (delay value, r-limit, and e-limit) can be a single or a triple value. The triple values that are separated by colons represent minimum, typical, and maximum values computed at three different operating conditions of a design. Line 45 of Figure 95.1 represent min:typical:max r-limits and e-limits for the rise and fall transition delays (of a 2-valued list). In this construct rise delay, rise r-limit and rise e-limit values are specified by the first three triples, i.e., 0.02:0.03:0.04, 0.01:0.02:0.03, and 0.02:0.03:0.04, while fall delay, fall **r**-limit, and fall e-limit values are represented by the last three triples (line 46), respectively, i.e., 0.03:0.04:0.05, 0.02:0.03:0.04, and 0.03:0.04:0.05.

95.1.2.2.2   *Timing Specifications.*

As discussed above, there are four types of timing specifications for each cell that are identified by **DELAY, TIMINGCHECK, TIMINGENV,** and **LABEL** keywords representing delay, timing check, timing environment, and label, respectively. Delay and timing check specifications are used for back annotation while timing environment specification is used for forward annotation. Labels represent timing constants used for annotating behavioral timing or readability purposes.

95.1.2.2.2.1 Delays — The delay type of timing specification that is used for back annotation starts with **DELAY** keyword and identifies the delay values of interconnects, input ports, input to output paths, and device outputs. Delays can be classified into four different groups: **ABSOLUTE, INCREMENT, PATH-PULSE,** and **PATHPULSEPERCENT.**

Absolute (increment) delays specify delay values that replace (increment) the existing delay values in a design. PathPulse and PathPulsePercent represent the absolute and relative values of pulse propagation limits between an input and an output port of a device. These types of delays specify whether a pulse on an output port of a device is rejected, filtered to **X,** or remains unchanged.

Absolute and increment delays are described by **IOPATH, RETAIN, COND, CONDELSE, PORT, INTERCONNECT, NET,** and **DEVICE** constructs.

- **IOPATH delay:** The input–output delay introduces the delay from an input/bidirectional port to an output/bidirectional port of a device.
- **RETAIN delay:** Retain delay represents the time duration in which an output/bidirectional port remains unchanged after a change in the corresponding input/bidirectional port.
- **CONDITIONAL delay:** Conditional delay starts with the COND keyword. In this construct, the corresponding delay values are applied to the related ports only if the introduced conditions are met.
- **CONDELSE delay:** Default delays can be specified for conditional paths using the CONDELSE structures. In this case, if the introduced conditions are not met, the specified delay in the CONDELSE structure is utilized by the annotator.
- **PORT delay:** Port delay represents the interconnect delay of an input port.
- **INTERCONNECT delay:** Interconnect delay specifies the propagation delay of a net connecting an output/bidirectional port (driving module port) to an input/bidirectional port (driven module port).
- **NET delay:** Net delay represents the propagation delay from all sources to all loads of a net.
- **DEVICE delay:** Device delay specifies the delay of all paths through a cell to a specified output port. If the output port is not specified, then the device delay represents the delay of all paths through a cell to all output ports of that cell.

Figure 95.2(a) and Figure 95.2(b) show an example design and its corresponding SDF file. Timing check constructs of this SDF file are discussed in the following section.

(a)

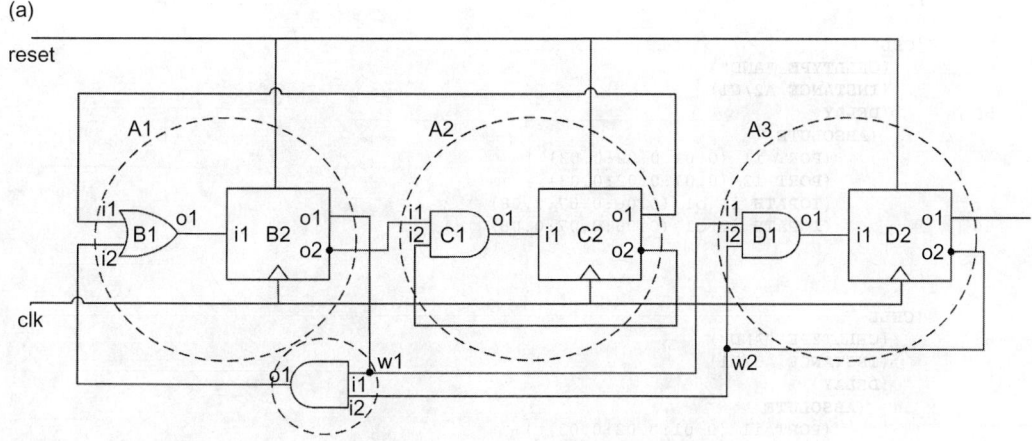

FIGURE 95.2 (a) A sample circuit.

(b)

```
 1  DELAYFILE
        (SDFVERSION          "4.0")
        (DESIGN              "Counter")
        (DATE                "Dec 13, 2005, 20:12:15")
 5      (VENDOR              "ASIC Chips")
        (PROGRAM             "Sample Program")
        (VERSION             "v4.0")
        (DIVIDER             / )
        (VOLTAGE             5.5:5:4.5 )
10      (PROCESS             "typical")
        (TEMPERATURE         :35: )
        (TIMESCALE           1 ns)

        (CELL
15         (CELLTYPE "Counter")
           (INSTANCE )
           (DELAY
             (ABSOLUTE
               (INTERCONNECT   A1/B1/o1   A1/B2/i1   (0.04:0.05:0.06) )
20             (INTERCONNECT   A1/B2/o2   A2/C1/i1   (0.04:0.05:0.06) )
               (INTERCONNECT   A2/C1/o1   A2/C2/i1   (0.04:0.05:0.06) )
               (INTERCONNECT   A2/C2/o1   A1/B1/i1   (0.04:0.05:0.06) )
               (INTERCONNECT   A2/C2/o2   A2/C1/i2   (0.04:0.05:0.06) )
               (NETDELAY       W1         (0.05:0.06:0.08) )
25             (NETDELAY       W2         (0.05:0.06:0.08) ) )
           )
        )

        (CELL
30         (CELLTYPE "OR")
           (INSTANCE A1/B1)
           (DELAY
             (ABSOLUTE
               (PORT i1 (0.01:0.02:0.03) )
35             (PORT i2 (0.01:0.02:0.03) )
               (IOPATH i1 o1 (0.05:0.06:0.07) )
               (IOPATH i2 o1 (0.05:0.06:0.07) ) )
           )
        )
40      (CELL
           (CELLTYPE "AND")
           (INSTANCE A4)
           (DELAY
45           (ABSOLUTE
               (PORT i1 (0.01:0.02:0.03) )
               (PORT i2 (0.01:0.02:0.03) )
               (DEVICE o1 (0.05:0.06:0.07) ) )
           )
50      )

        (CELL
           (CELLTYPE "AND")
           (INSTANCE A2/C1)
55         (DELAY
             (ABSOLUTE
               (PORT i1 (0.01:0.02:0.03) )
               (PORT i2 (0.01:0.02:0.03) )
               (IOPATH i1 o1 (0.06:0.07:0.08) )
60             (IOPATH i2 o1 (0.06:0.07:0.08) ) )
           )
        )

        (CELL
65         (CELLTYPE "AND")
           (INSTANCE A3/D1)
           (DELAY
             (ABSOLUTE
               (PORT i1 (0.01:0.02:0.03) )
70             (PORT i2 (0.01:0.02:0.03) )
```

FIGURE 95.2 (b) SDF file of the circuit shown in (a).

```
                     (IOPATH i1 o1 (0.06:0.07:0.08) ) )
                     (IOPATH i2 o1 (0.06:0.07:0.08) ) ) )
                  )
             )
 75     (CELL
             (CELLTYPE "DFF")
             (INSTANCE A1/B2)
             (DELAY
                (ABSOLUTE
 80              (IOPATH reset o1 (0.12:0.13:0.13) )
                 (COND (reset==0)
                     (IOPATH clk o1 (0.17:0.18:0.19) ) )
                )
 85          (TIMINGCHECK
                 (SETUPHOLD (COND ~reset i1) (posedge clk) (0.1) )
                 (RECREM (negedge reset ) (posedge clk) (0.07) )
                 (WIDTH (posedge clk) (1:1:1) )
                 (WIDTH (negedge clk) (1:1:1) )
 90              (WIDTH (posedge reset) (1:1:1) ) )
                )
         )
 95     (CELL
             (CELLTYPE "DFF")
             (INSTANCE A2/C2)
             (DELAY
                (ABSOLUTE
100              (IOPATH reset o1 (0.12:0.13:0.13) )
                 (COND (reset==0)
                     (IOPATH clk o1 (0.17:0.18:0.19) ) )
                )
         )
105          (TIMINGCHECK
                 (SETUPHOLD (COND ~reset i1) (posedge clk) (0.1) )
                 (RECREM (negedge reset ) (posedge clk) (0.07) )
                 (WIDTH (posedge clk) (1:1:1) )
                 (WIDTH (negedge clk) (1:1:1) )
110              (WIDTH (posedge reset) (1:1:1) ) )
                )
         )
         (CELL
115          (CELLTYPE "DFF")
             (INSTANCE A3/D2)
             (DELAY
                (ABSOLUTE
                 (IOPATH reset o1 (0.12:0.13:0.13) )
120              (COND (reset==0)
                     (IOPATH clk o1 (0.17:0.18:0.19) ) )
                )
         )
125          (TIMINGCHECK
                 (SETUPHOLD (COND ~reset i1) (posedge clk) (0.1) )
                 (RECREM (negedge reset ) (posedge clk) (0.07) )
                 (WIDTH (posedge clk) (1:1:1) )
                 (WIDTH (negedge clk) (1:1:1) )
130              (WIDTH (posedge reset) (1:1:1) )
                )
         )
```

95.2 (*continued*).

95.1.2.2.2.2 Timing Checks — The timing check timing specification, which is used for back annotation, starts with the **TIMINGCHECK** keyword and identifies timing check limits for a cell instance. There are 11 different types of timing checks that are discussed below.

- **SETUP (HOLD) timing check:** Setup (hold) timing check represents the time interval during which a specified signal must remain unchanged before (after) another signal transition.
- **RECOVERY timing check:** Recovery timing check introduces the minimum required time for active transition of clock after the release of an asynchronous control signal from the active state.
- **REMOVAL timing check:** Removal timing check specifies the minimum required time for release of an asynchronous control signal from its active state after an active transition of clock.
- **SETUPHOLD timing check:** Setup–hold timing check represents both setup and hold timing limits in an entry.
- **RECREM timing check:** Recovery–removal timing check specifies both recovery and removal limits in an entry.
- **SKEW (BIDIRECTSKEW) timing check:** Skew (bidirectional skew) timing check introduces the maximum permitted unidirectional (bidirectional) delay between two signals.
- **WIDTH timing check:** Width timing check represents the minimum acceptable pulse width of a specified signal.
- **PERIOD timing check:** Period timing check specifies the minimum allowable pulse period of a specified signal.
- **NOCHANGE timing check:** No-change timing check describes the minimum time for a specified signal to be stable before the start and after the completion of a control pulse.

A condition can be associated with all types of the discussed timing checks. For this purpose, the **COND** keyword is used and precedes the corresponding condition. The following is an example of a conditional setup timing check. In this example setup time between *din* and positive edge of *clk* is checked only when *enable* is active.

$$(\textbf{SETUP}\ din\ (\textbf{COND}\ enable\ (\textbf{posedge}\ clk))\ (2))$$

Two other forms of condition are **SCOND** and **CCOND**. In multiple signal timing checks, **SCOND** applies to the first signal that changes, while **CCOND** applies to the second. In the following example, the condition is applied to signal *clk* for setup check and to signal *din* for hold check.

$$(\textbf{SETUPHOLD}\ din\ (\textbf{posedge}\ clk)\ (2)\ (3)\ (\textbf{CCOND}\ enable))$$

Figure 95.2(b) shows a number of timing check constructs to specify timing check limits for the cell instances.

95.1.2.2.2.3 Timing Environment — The timing environment type of timing specification, used for forward annotation, starts with the **TIMINGENV** keyword and represents constraint values of critical paths in a design. This construct also specifies the timing information of the design's operating environment. There are two forms of timing environment specification. The first form represents constraint values for the design, while the second deals with the information about the timing environment in which the circuit operates. These constructs are described below.

Constructs for representing constraints include path, period, sum, diff, and skew constraints that are described below. Several of these constructs are used in the example of Figure 95.3(b).

- **PATH constraint:** The path constraint represents the maximum permitted delay of a path. A path is specified by its two ending ports. However, to unify the path, an intermediate port in the path can also be specified.

- **PERIOD constraint:** This constraint is used to introduce a path constraint value for a group of paths in a synchronous circuit. These paths lie between a common clock of several flip-flops and their related outputs. The starting point of all these paths is introduced by an output port of a device that drives the clocks of these flip-flops. Only flip-flops whose clock input is connected to this driving port directly or via a number of buffers are considered in the related period constraint structure. To be able to exclude a number of paths from the path group in this structure, an optional exception construct can be used. This construct starts with the **EXCEPTION** keyword and identifies the cell instances to be excluded from the path group.
- **SUM constraint:** The sum constraint includes two or more paths in a design and specifies the maximum allowable value for the sum of their related delays.
- **DIFF constraint:** This constraint includes two paths in a design and specifies the maximum allowable value for the absolute difference of their related delays.
- **SKEW constraint:** The skew constraint represents the maximum acceptable delay from a common driver to all driven inputs. These inputs must be driven directly by the common driver. In this constraint, only the driver port is specified.

As discussed above, the second form of timing environment constructs deal with information about the timing environment in which a circuit operates. These constructs are discussed below. Of what follows the waveform timing environment construct is used in the example of Figure 95.3(b).

- **ARRIVAL time:** The arrival time construct specifies the time in which a transition occurs on a primary input of a design. In this construct the primary input port along with four related values are presented. These values define the earliest rising, latest rising, earliest falling, and latest falling arrival times, respectively. In addition to these fields, another port edge can be specified in this construct. This port edge is considered as the time reference for arrival time specifications.
- **DEPARTURE time:** The departure time construct represents the time in which a transition occurs on a primary output of a design. The structure of this construct is the same as that of the arrival time construct.
- **SLACK time:** The slack time construct represents the maximum value that a path delay can vary without violating any design constraints. This construct includes a specified port along with four related values. To avoid specifying different paths in this structure, slack time structure considers all paths ending at the specified port. The four values in this construct are rising setup, falling setup, rising hold, and falling hold slack, respectively. Rising (falling) setup slack specifies the additional delay that can be tolerated by a design when there is a rising (falling) transition at the specified port. On the other hand, rising (falling) hold slack represents the reduction of delay that can be tolerated by a design when there is a rising (falling) transition at the specified port. An optional real number can be added to this construct to show the clock period on which the slack values are based.
- **WAVEFORM time:** The waveform construct introduces the specification of a waveform applied to a primary input of a circuit during its normal operation (see Figure 95.3(b) for an example). This structure includes the primary input port to which the waveform is applied, the period of the waveform, and its transition times during a given period. The transition times are specified by one or two pairs of transitions. Each pair consists of two edges: a positive edge followed by a negative edge or a negative edge followed by a positive edge. In the case of having two pairs of transitions, the order of edges in both pairs must be the same. Each edge is followed by one or two real numbers. If one real number is used, then this number defines the transition time and if two real numbers are used, they can define an uncertainty region in which the transition can occur (ambiguity area).

(a)

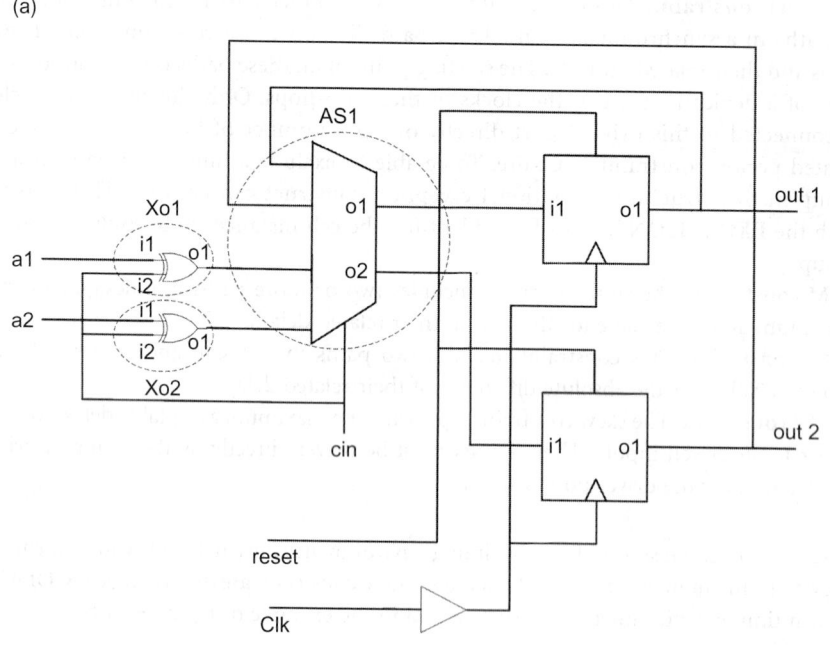

(b)

```
 1    (DELAYFILE
         (SDFVERSION             "4.0")
         (DESIGN              "AddSub")
         (DATE          "Dec 12, 2005, 01:39:52 ")
 5       (VENDOR            "ASIC Chips")
         (PROGRAM            "Sample Program)
         (VERSION            "v4.0")
         (DIVIDER            / )
         (VOLTAGE            5.5:5:4.5 )
 10      (PROCESS            "typical")
         (TEMPERATURE          :35: )
         (TIMESCALE           1 ns)

         (CELL
 15         (CELLTYPE  "AddSub")
            (INSTANCE )
            (TIMINGENV
               (WAVEFORM Clk 2 (negedge 0) (posedge 1) )
               (PATHCONSTRAINT Xo1/i1 AS1/o1 (1.5) )
 20            (PATHCONSTRAINT Xo2/i1 AS1/o2 (1.5) )
               (SKEWCONSTRAINT (posedge Clk) (0.1) ) )
            )
         )
```

FIGURE 95.3   (a) An incrementer/decrementer circuit. (b) SDF file of the circuit shown in (a).

Figure 95.3(a) shows a circuit and Figure 95.3(b) shows a part of its corresponding SDF file including a number of timing environment constructs. As shown in Figure 95.3(a), in each clock cycle the input signal values are added/subtracted to/from the contents of the registers. Let us assume that we are driving this circuit with a clock of 500 MHZ. Thus the propagation delay from primary input signals to register inputs (i1 of Xo1 to o1 of AS1 and i1 of Xo2 to o2 of AS1) must be less than 2 ns (1/500 MHZ). Lines 19 and 20 of Figure 95.3(b) illustrate this constraint.

95.1.2.2.2.4 Labels — Labels represent timing constants used for annotating behavioral timing or readability purposes. Using label constructs improves the annotation performance, since labels can be looked up faster than port names and conditions. This type of timing specification starts with **LABEL** keyword and is classified into two different groups, **ABSOLUTE** and **INCREMENT**, each of which includes a list of delay values. Absolute (increment) labels specify delay values which replace (increment) the existing delay values in a design.

95.1.3 Our Coverage of SDF

The description of SDF presented here was an introduction to this format. We tried to cover the basics of SDF to familiarize the reader with the overall structure of this language and its applications. This section provided complete examples, and SDF structure, in the context of these examples, were described. Other language structures not covered by our examples follow the same basic patterns discussed here. This section did not elaborate on some of the constructs that were not covered in the above examples. For a complete description of this language readers are encouraged to see the references at the end of this chapter [1,2].

Related to the SDF format which describes timing of hardware components is the VCD waveform format. This format describes waveforms of circuit signals and is used for displaying data and for moving data across various platforms. VCD, including its timing representation is described in the section that follows.

95.2 VCD

A Value Change Dump (VCD) file is used for representing waveforms in the ASCII format. VCD logs the value changes of selected variables in a design during a simulation session or waveform generation. VCD can also be created by inserting the VCD system tasks in the Verilog source file and simulating the Verilog file. As another example, SystemC has also a number of features for creating VCD files. This file format was first introduced in the IEEE 1364-1995 Verilog standard.

95.2.1 VCD Structure

VCD files can be categorized into four-state VCD files, and extended VCD files. In a four-state VCD file, variable changes are represented by **0**, **1**, **X**, and **Z** with no strength information while in an extended VCD file, variable changes are represented by all states and strength information. A VCD file is composed of sections for header specification, node information, and value changes.

95.2.1.1 Four-State VCD File Structure

A four-state VCD file is an ASCII file composed of three different sections discussed below.

95.2.1.1.1 Header Section.
The header of a VCD file consists of the information about the entire file. This information includes creation date and time of the current VCD file, the simulator version used to generate the VCD file, and the timescale for the corresponding parameters. To represent the creation date and time of a VCD file **$date** keyword is used. The VCD writer version and the timescales are specified using **$version** and **$timescale** keywords.

To specify the timescale, a timescale number and a unit are required. According to the IEEE Verilog standard, the timescale number should be 1, 10, or 100 and the timescale unit should be s, ms, us, ns, ps, and fs [3]. All header section constructs terminate with the **$end** keyword. A complete example of a VCD file is shown in Figure 95.4.

```
$date      Nov 15 2005, 17:24:10
$end
$version   Sample Simulator Version 3
$end
$timescale 1 ns
$end
$scope module    Inv        $end
$var wire 1    !    a        $end
$var reg  1    *    b        $end
$upscope $end
$enddefinitions $end

#0
$dumpvars
x!
x*
$end
#10
1!
0*
#15
0!
1*
```

FIGURE 95.4   VCD file of an inverter.

95.2.1.1.2 *Node Information Section.*

The node information section starts with **$scope** and terminates with **$enddefinitions $end** keywords. This section deals with scope definitions and the types of variables being dumped. **$scope, $upscope, $comment**, and **$var** keyword commands can be included in this section.

- **$scope:** $scope keyword command defines the scope of the variables being dumped. Scope types can be classified into five different categories as shown below:
  - **module:** This type represents the top-level module and module instances. The **$scope module** *mod1* command specifies that signals of *mod1* module are to be dumped.
  - **task:** This type is used to introduce tasks.
  - **function:** This type represents functions.
  - **begin:** This type specifies scope as a named sequential block.
  - **fork:** This type is utilized to define scope to be a named parallel block.
- **$upscope:** This keyword command indicates the change of the current scope to the next higher scope in the design hierarchy.
- **$comment:** Comments can be inserted in a VCD file using the **$comment** keyword command.
- **$var:** This keyword command is used to specify a VCD symbol for each variable being dumped. A VCD symbol can be any sequence of printable ASCII characters. Each dumped variable is mapped to a unique VCD symbol. In some cases more than one variable is represented by a symbol. This occurs only if two differently named variables reference the same value. **$var** command has four different fields. The first, second, and fourth fields specify the type, size, and name of the variable being dumped, while the third field specifies the corresponding VCD symbol. The acceptable variable types in this command are: *event, integer, parameter, real, reg, supply0, supply1, time, tri, triand, trior, trireg, tri0, tri1, wand, wire,* and *wor.*

Note that the individual bits of a vector can be dumped individually. Figure 95.5 shows the waveforms of a 4-bit 2-to-1 multiplexer (Figure 95.5(a)) and the header and node information sections of the corresponding extended VCD file (Figure 95.5(b)).

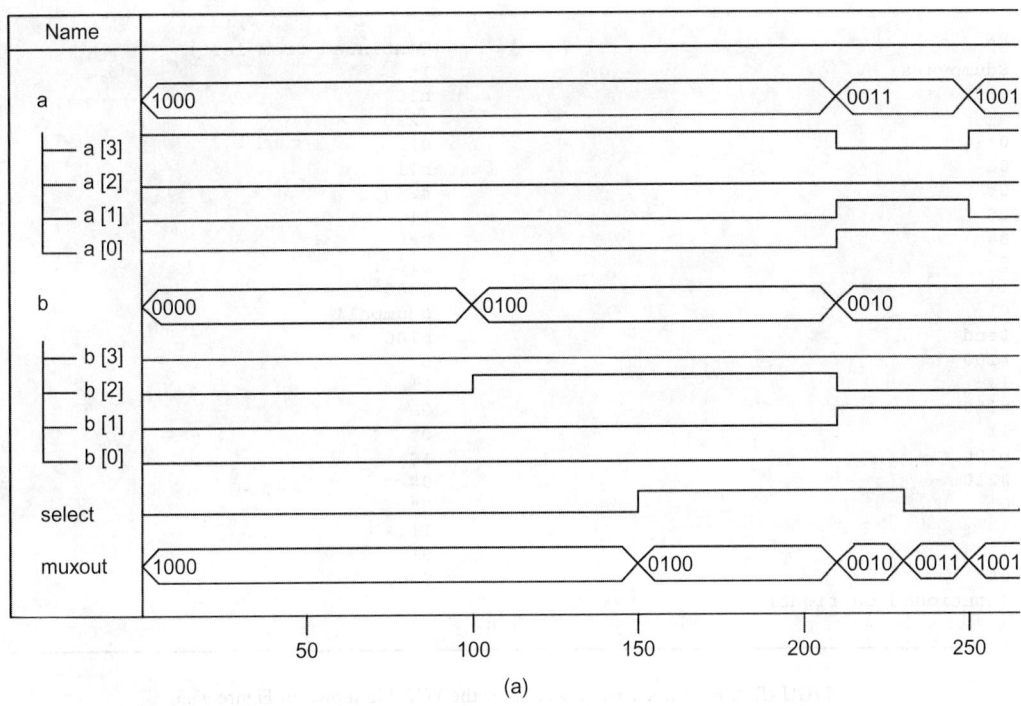

```
$date       Dec 10 2005, 8:14:10
$end
$version    Sample Simulator Version 3
$end
$timescale 1 ns
$end
$scope module    Mux        $end
$var wire 1    !    a [3]    $end
$var wire 1    "    a [2]    $end
$var wire 1    #    a [1]    $end
$var wire 1    $    a [0]    $end
$var wire 1    %    b [3]    $end
$var wire 1    &    b [2]    $end
$var wire 1    '    b [1]    $end
$var wire 1    (    b [0]    $end
$var wire 1    )    select   $end
$var reg 4 * muxout [3:0] $end
$upscope $end
$enddefinitions $end
```

(b)

FIGURE 95.5 (a) Waveform of a 4-bit 2-to-1 multiplexer. (b) A part of the VCD file corresponding to the waveform of (a).

95.2.1.1.3 *Value Change Section.*

This section deals with the value change of variables declared in the node information section.

95.2.1.1.3.1 Variable Values Format — Scalar and vector value changes are dumped according to their related formats. To introduce a scalar variable, the VCD symbol should immediately follow the corresponding value digits, i.e., no space character is permitted between the VCD symbol and the corresponding value digits.

```
#0                                    (Continued from left)
$dumpvars                             1'
b1000 *                               b10 *
0(                                    #230
0'                                    0)
0&                                    b11 *
0%                                    #250
0$                                    1!
0#                                    0#
0"                                    b1001 *
1!                                    #300
0)                                    $dumpall
$end                                  b1001 *
#100                                  0(
1&                                    1'
#150                                  0&
1)                                    0%
b100 *                                1$
#210                                  0#
0!                                    0"
1#                                    1!
1$                                    0)
0&                                    $end
(Continued on right)
```

FIGURE 95.6   Value change section of the VCD file shown in Figure 95.5.

To represent a vector variable, the corresponding value digits immediately follow a base character but a space character is required between the value digits and the VCD symbol. This base character is used to declare the type of the vector variable. 'B' and 'b' characters are utilized for binary vector representation while 'R' and 'r' characters are used to specify real vectors. Note that data in the VCD file is case-sensitive.

95.2.1.1.3.2   Keyword Commands — The value of the variables being dumped can be specified by utilizing (but not be limited to) the following commands.

- **$dumpvar:** This keyword command indicates the initial values of all variables being dumped.
- **$dumpoff:** $dumpoff keyword command shows that all variables are dumped with value **X**.
- **$dumpon:** This keyword command indicates the current values of all the variables being dumped. Between a **$dumpoff** and its next **$dumpon**, no value changes are dumped.
- **$dumpall:** $dumpall keyword command specifies the current values of all variables being dumped.

The value change section of the four-state VCD file corresponding to the waveform of Figure 95.5(a) is shown in Figure 95.6.

95.2.1.2   Extended VCD File Structure

In a four-state VCD file, variable changes can be represented by four different states: **0, 1, X**, and **Z** with no strength information. However, this set of states is not always sufficient for representing value changes, i.e., in some cases more than four states are required to provide the necessary information about bidirectional pins, such as the driving direction, driving strength, and collision detection. The format of the extended VCD file is similar to that of the four-state VCD file, i.e., an extended VCD file is an ASCII file composed of three different sections: header, node information, and value change sections.

95.2.1.2.1   Header Section.

The header of an extended VCD file is similar to that of a four-state VCD file. This section includes the creation date and time of the current VCD file, the version of the simulator that has been used to generate the VCD file, and the timescale for the corresponding parameters.

95.2.1.2.2   Node Information Section.

The node information section which starts with **$scope** and terminates with **$enddefinitions $end** keyword commands, deals with the definitions of ports being dumped. **$scope**, **$upscope**, **$comment**, **$var**, and **$vcdclose** keyword commands can be included in this section.

- **$scope**: The only allowable scope type in an extended VCD file is module. In this case, the module keyword is utilized to represent the top-level module and module instances.
- **$upscope**: This keyword command indicates the change of the current scope to the next higher scope in the design hierarchy.
- **$comment**: Comments can be inserted in an extended VCD file using **$comment** keyword command.
- **$var**: In the extended VCD file (similar to the four-state VCD file) **$var** keyword command specifies a VCD symbol for each variable being dumped. The **$var** keyword command has four different fields. The first, second, and fourth fields specify the type, size, and name of the variable being dumped, while the third field introduces the corresponding VCD symbol. Note that the only allowable variable type in an extended VCD file is **port**. If a port is a bus, the size field indicates its actual index; otherwise the size value is 1. The VCD symbol of each port is an integer number preceded by the <character. This integer value starts at zero and is increased one unit per port in the order found in the module declaration.
- **$vcdclose**: This keyword command which is not included in a four-state VCD file indicates the final simulation time.

Figure 95.7 shows the waveforms of a 4-bit counter (Figure 95.7(a)) and the header and node information sections of the corresponding extended VCD file (Figure 95.7(b)).

95.2.1.2.3   Value Change Section.

This section discusses the value changes of the ports identified in the node information section. The following keyword commands can be used to represent the values of the dumped ports.

- **$dumpportsoff**: $dumpportsoff keyword command shows that all variables are dumped with value **X**.
- **$dumpportson**: This keyword command indicates the current values of all variables being dumped. Between a **$dumpportsoff** and its next **$dumpportson** no value changes are dumped.
- **$dumpportsall**: This keyword command specifies the current values of all variables being dumped.
- **$dumpports**: This keyword command indicates the current values and strengths of the dumped ports and consists of the following five fields: the **p** character, port-value, 0-strength-componenet, 1-strength-component, and the corresponding VCD symbol. 0- and 1-strength-components represent 0 and 1 strengths of a port value.

An integer between 0 and 7 specifies the strength0 (1) specification of a port. Integers 0 to 7 represent high, small, medium, weak, large, pull, strong, and supply strength values, respectively.

The **p** character which begins a port value is followed by what is referred to as a state character. This character specifies port states for inputs, outputs, and bidirectional ports. Acceptable values for input ports are **D, U, N, Z, d**, and **u** that are used for low, high, unknown, tri-state, low (two or more drives active), and high (two or more drives active) states, respectively. Values for output ports are **L, H, X, T, l**, and **h** that are used for low, high, unknown, tri-state, low (two or more drives active), and high (two or more drives active) states, respectively. Values for bidirectional ports are **0, 1, F, ?, A, a, B, b, C, c**, and **f**. The first two values are used for low and high states, the third value represents tri-states and other values represent the unknown states (depending on the input and output values of the specified port).

Figure 95.8 shows the value change section of the extended VCD file corresponding to the waveform shown in Figure 95.7(a).

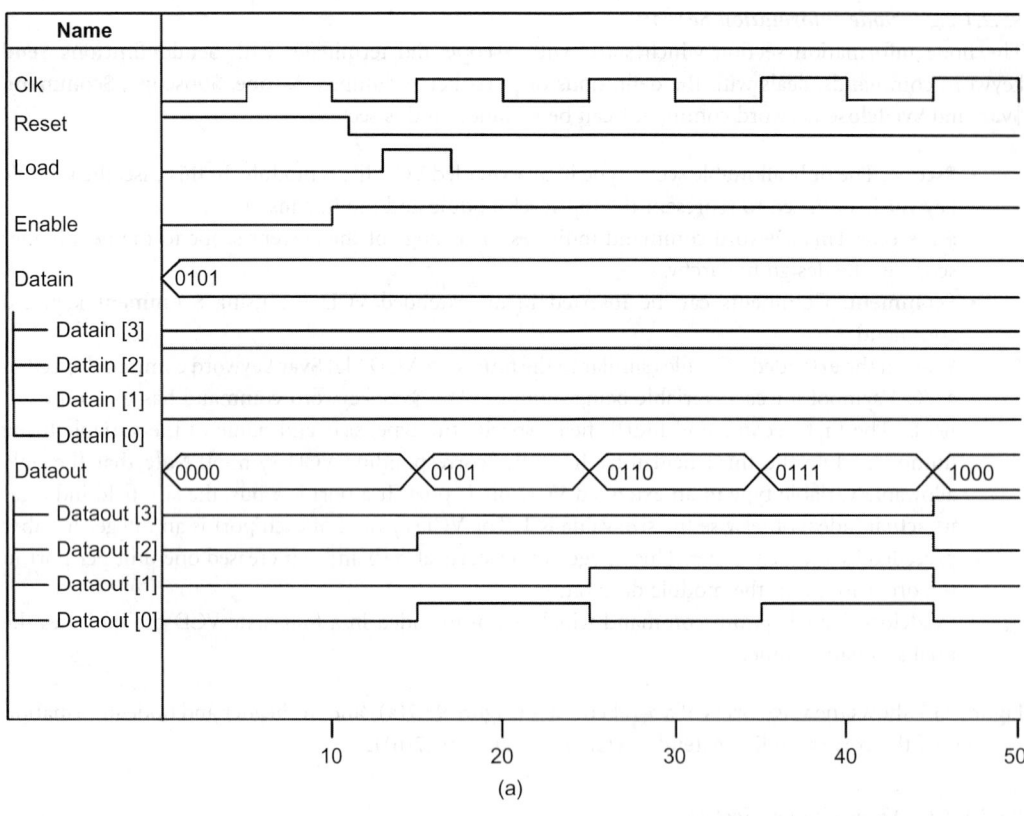

(a)

```
$date       Dec 09, 2005   22:26:28
$end
$version    Sample Simulator Version 3
$end
$timescale 1 ns
$end
$scope module   Counter                     $end
$var port       1 <0        Clk             $end
$var port       1 <1        Reset           $end
$var port       1 <2        Load            $end
$var port       1 <3        Enable          $end
$var port       1 <4        Datain [3]      $end
$var port       1 <5        Datain [2]      $end
$var port       1 <6        Datain [1]      $end
$var port       1 <7        Datain [0]      $end
$var port       1 <8        Dataout [3]     $end
$var port       1 <9        Dataout [2]     $end
$var port       1 <10       Dataout [1]     $end
$var port       1 <11       Dataout [0]     $end
$upscope $end
$enddefinitions $end
```

(b)

FIGURE 95.7   (a) Waveform of a 4-bit counter. (b) A part of the VCD file corresponding to the waveform of (a).

```
#0
$dumpports                          (Continued from left)
pU   0   6   <1                     #20
pD   6   0   <0                     pD   6   0   <0
pD   6   0   <2                     #25
pD   6   0   <3                     pU   0   6   <0
pU   0   6   <7                     pU   0   6   <10
pD   6   0   <6                     pD   6   0   <11
pU   0   6   <5                     #30
pD   6   0   <4                     pD   6   0   <0
$end                                #35
#5                                  pU   0   6   <0
pU   0   6   <0                     pU   0   6   <11
#10                                 #40
pU   0   6   <3                     pD   6   0   <0
pD   6   0   <0                     #45
#11                                 pU   0   6   <0
pD   6   0   <1                     pU   0   6   <8
#13                                 pD   6   0   <9
pU   0   6   <2                     pD   6   0   <10
#15                                 pD   6   0   <11
pU   0   6   <0                     #50
pU   0   6   <9                     pD   6   0   <0
pU   0   6   <11                    $vcdclose
#17                                 #50
pD   6   0   <2                     $end
(Continued on right)
```

FIGURE 95.8   Value change section of the VCD file of Figure 95.7.

95.2.2   Our Coverage of VCD

VCD is a general portable format for waveform display. In the preceding section, we discussed the basics of VCD and how it could be used for representation of waveforms. In the examples that we presented, constructs of this language and their utilization were discussed. Because VCD is a simple format, in the above discussion we were able to discuss most of its structures and capabilities.

References

1. *IEEE Standard for Standard Delay Format (SDF) for the Electronic Design Process*, IEEE Std 1497™, IEEE, Inc., New York, 2001.
2. M.J.S. Smith, *Application-Specific Integrated Circuits*, Addison-Wesley, Reading, MA, 1997.
3. *IEEE Standard Hardware Description Language Based on the Verilog™ Hardware Description Language*, IEEE Std 1364-2001, IEEE, Inc., New York, 2001.

96

HDL-Based Tools and Environments

Saeed Safari

Nanoelectronics Center of Excellence
School of Electrical and Computer Engineering
University of Tehran

CONTENTS

As IC process technology improves, the number of transistors that can be fabricated on a single IC increases. The design effort also increases with the circuit density. Progress in VLSI applications, with more than 10 million transistors per chip, is limited not only by the circuit technology but also by the capability of the design and validation work related to such complex circuits. Therefore, computer-aided design (CAD) tools and new design methods are needed to manage this complexity.

Figure 96.1 illustrates a typical ASIC design flow. The design flow begins with a design idea and continues with describing the design at the various levels of abstraction and ends with generating a layout. In the design entry step, the design can be described by an existing HDL (i.e., VHDL, Verilog, System Verilog, or SystemC) or a combination of them. The design is converted into an internal data structure (e.g., task graph, data flow graph, or BDD) based on the input abstraction level. Then the appropriate synthesis algorithms are applied on the internal data structure and the layout (GDS2, EDIF for layout or net-list, SDF and VITAL for timing) is generated. During the synthesis steps several circuit qualification parameters, such as area and delay are optimized. As shown in the figure, design validation should be performed (by a simulation or verification tool) in any design step to check for probable existing errors. A design error can occur due to several reasons, e.g., ambiguities in design specification or designer errors.

In this chapter, we describe CAD tools that are used in a typical VLSI design flow including simulation tools, synthesis tools, test tools (not shown in the figure), and verification tools.

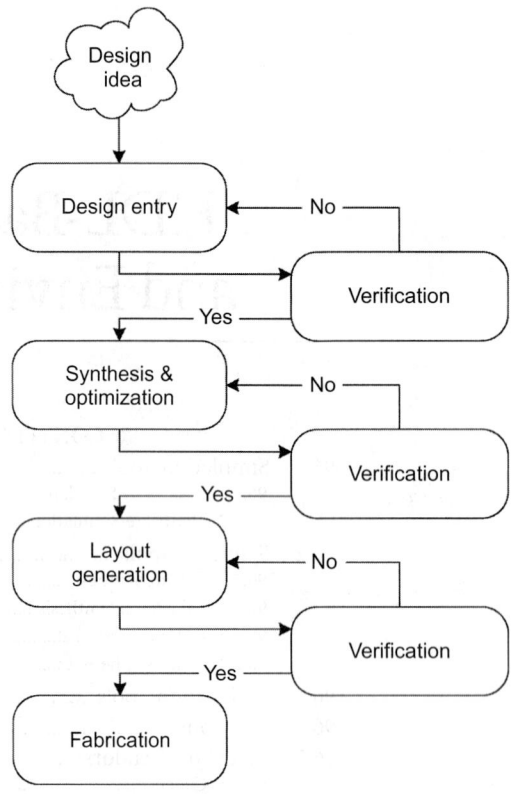

FIGURE 96.1 Typical VLSI design flow.

96.1   Simulation Tools

Simulation is probably the most stabilized method for verifying the correctness of a designed system. A simulation tool is a software tool that helps us apply test vectors to the inputs of a designed system and observe the circuit responses. Here we will introduce the basic concepts of the simulation and types of simulation tools.

96.1.1   History of Simulation Tools

In the late 1960s, the first circuits were manufactured by integrating devices manually. Owing to the lack of CAD tools, the necessary design validations were performed after the fabrication process, and this obviously increased the chip cost and time to market. In the 1970s the first CAD tools were designed and employed in design entry. The first event-driven digital circuit simulator was designed in the mid-1970s and since then simulators are enhanced day by day to meet the design methodology requirements.

96.1.2   Hardware Simulation Methods

Figure 96.2 shows the structure of a typical hardware simulator. A hardware simulator is supplied with a description of the design called model and a set of input data as test data for this model. Then the hardware simulator is run and test vectors are applied to the model and outputs are generated. The results depend on both the abstraction level of the model and the input data applied to the model. The best benefit of simulation is that later modifications to the model can easily and quickly be applied with low costs.

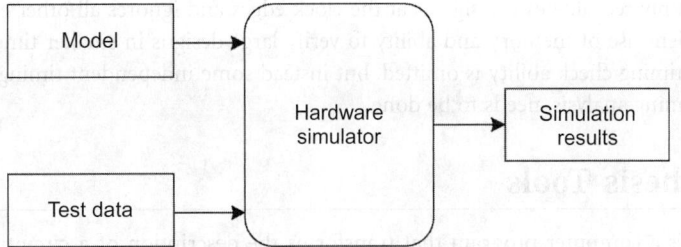

FIGURE 96.2   Hardware simulator.

Several implementations of hardware simulators exist in today's design automation technology. Of these, *interpreted* and *compiled* implementations of hardware simulators are the most common and typical of simulation engines.

- Interpreted implementation: In this method, the simulator interprets the input model based on the data structures. The simulator converts the input model to an internal data structure and then the simulation process is performed by applying the input test data to the internal data structure and generating the appropriate results. The advantage of this method is simple conversion of the input model into the data structure and also simple implementation of the simulator. In contrast, the simulator executes slowly.
- Compiled implementation: In this method, the simulator converts the input model into a compiled code. Then the simulator executes the generated compiled code to perform simulation. The disadvantage of this method is the high compilation time, which is required for the conversion of the input model into the compiled code. The advantage of this method is that the execution time of the simulation is low.

Regardless of the level of abstraction chosen for simulation, hardware simulators are classified into three main categories: *event-driven, oblivious,* and *cycle-based.*

- Event-driven simulators: When an event (any change in the value of a signal) occurs in the simulation model, the simulator only evaluates the signals affected by the event. The event-driven simulator engine holds a time-ordered priority queue of events. The simulator engine picks all events from the queue which have time stamp t (the next simulation time) and applies the events to change the corresponding signal values. Then the affected signals are reevaluated based on the new input values. New events are appended to the priority queue. This loop is repeated until the queue is empty or the simulation end time is reached. This method employs a complex data structure (disadvantage) to provide a reasonable simulation time (advantage).
- Oblivious simulators: The oblivious simulators run the simulation at fixed time points. These simulators advance the simulation engine clock a fixed amount of time. The values for all signals in the design are calculated at the end of each time step. The time step should be selected in a way that no events on the interested signals are missed. Sometimes, analog (or mixed-signal) simulators choose a variable time step controlled by the convergence properties of the equations. The simple implementation of this method can be considered as its advantage. The disadvantage of this method is that all signals should be reevaluated at each time step even if they have not changed since the last time step.
- Cycle-based simulators: As mentioned before, one disadvantage of an event-driven simulator is that the simulation models use a lot of memory (as a result of complicated data structure) to hold the timing data required for a signal. Thus, it is very difficult to simulate a large design in a reasonable amount of time by event-driven simulators. Cycle-based simulators are designed and provided to resolve the event-driven simulator problems only in sequential circuits. Consider a classical synchronous design as combinational logic sandwiched between blocks of registers. The cycle-based

simulator only reevaluates the signals at the clock edges and ignores all other events. This results in an efficient use of memory and ability to verify large designs in smaller time. In a cycle-based simulator timing check ability is omitted, but instead some independent timing verification, such as static-timing analysis, needs to be done.

96.2 Synthesis Tools

A synthesis tool is a computer program that transforms the description of a circuit into a lower level description which is closer to the hardware, while at the same time optimizes some objectives such as area, delay, testability, and power consumption [1]. Here we will introduce the different types of the synthesis tools, i.e., system synthesis, behavioral synthesis, RTL synthesis, and logic synthesis.

96.2.1 System Synthesis

Today, most designers implement embedded systems as core-based *systems-on-a-chip* (SOC) [2]. They use different implementation of cores provided by different core vendors to achieve the required functionality. This approach reduces system design time by reusing the existing cores. The set of existing cores defines the design space, while the set of tasks defines the system specification. The designer should select an implementation of the system for both achieving the system functionality and satisfying the design objectives (e.g., area, performance, price, testability, and power consumption) from the design space. This process is called *system synthesis*. Therefore, a system synthesis tool takes in a specification of the system and a set of cores as inputs and provides an architecture that is an interconnected network of cores as its output.

Specification of a system is described in a high-level language such as System Verilog or SystemC and then converted into a directed acyclic *task graph*. Each node of a task graph is associated with a task, while each edge, connecting two nodes, is associated with a data transfer link between the two nodes. Data dependency edges ensure the correct order of execution of tasks. This implies that each task can execute when all its predecessor tasks are completed. Each edge is associated with a scalar describing the amount of data that must be transferred between the two connected tasks, representing the communication time between these tasks if mapped on different cores.

There are four major steps in a system synthesis tool: *allocation, assignment, scheduling,* and *cost estimation.* In the first step, i.e., allocation, we determine the quantity (number) of each type of core and buses needed to be used. The next step is assignment, where we assign each task to a core (for execution), and each communication link to a bus (for data transfer). Then we perform scheduling to determine the start time of all tasks and communications. Cost estimation indicates the price, area, performance, testability, and power consumption of the solution. This step is required to select the optimal solution from the existing ones.

Several design parameters should be optimized during system synthesis steps including price, area, performance, testability, and power consumption. Here we will describe these qualification parameters.

- Price: The price of the final solution can be defined as the sum of the prices of all the cores on the chip plus the area-dependent price of the chip.
- Area: At system level, the area of a system can be estimated with the sum of the area of used cores and buses. This area should be minimized.
- Performance: The scheduler provides accurate information on the start and finish times of tasks in the allocated cores and finds the critical-path worst-case execution time. This value is used to estimate the performance of the synthesized system.
- Testability: Since embedded cores are not directly accessible via chip IO pins, a special *test access mechanism* (TAM) is required to test them after system integration. Testability optimization means that TAM designs should be integrated into the system synthesis process to find the best testable solution.
- Power consumption: The power consumption of the synthesized core equals the sum of the power consumed by all the allocated cores and buses.

96.2.2 Behavioral Synthesis

A behavioral synthesis tool usually translates the behavioral description from VHDL, Verilog, System Verilog, or SystemC into a suitable intermediate format, such as a data flow graph (DFG). In a DFG, each node is associated with an operation, while each edge is associated with a variable. Figure 96.3 shows a partial VHDL code and its corresponding DFG, called *ex1*. To generate the RTL architecture, the behavioral synthesis performs three major tasks: scheduling, resource allocation, and design optimization.

96.2.2.1 Scheduling

Scheduling assigns each operation to one or more clock cycles, and specifies cycle-by-cycle behavior of a circuit. The result of scheduling is called scheduled DFG (SDFG). For an operation o in a DFG, $o.earliest$ ($o.latest$) denotes the earliest (latest) cycle time, in which the operation can be executed. In an SDFG, the cycle time, at which a variable is first defined, is called the birth time of the variable, while the cycle time, at which a variable is used last, is called the death time of the variable. The lifetime of the variable is defined as the interval [birth, death]. Two variables are compatible if their lifetimes do not overlap. Such variables can be mapped on to a single register, and two variables are incompatible if their lifetimes overlap. For example, in *ex1* variables, a and f are compatible while variables e and f are incompatible. Scheduling algorithms are used for assigning operations to specific clock cycles.

96.2.2.2 Resource Allocation

Given an SDFG, resource allocation assigns modules to perform operations (module allocation), and registers to store variables (register allocation). In module allocation, operations in an SDFG are mapped to proper data-path modules available in the given library. Operations used in high-level synthesis can be classified as arithmetic operations (e.g., *add*, *subtract*, *multiply*, *divide*, and *compare*) and logical operations (e.g., *and*, *or*, *nand*, and *nor*). If a data-path library has modules with the same functionality but different characteristics, high-level synthesis can achieve better performance. For example, an "add" operation can be mapped to a ripple-carry adder, a carry look-ahead adder, or a carry-save adder. The trade-off among different characteristics enables the synthesized circuit to have smaller area, higher performance, or less power consumption. In register allocation, a register should be assigned to each module input or output variable. The left-edge algorithm (LEA) that is used for both module and register allocation finds the minimum-number modules and registers required for data-path implementation.

To illustrate scheduling and allocation, we apply some of their corresponding algorithms to the example of Figure 96.3. In this example, using a method of scheduling that is called, *force-directed*, the operations $+_1$ and $+_2$ are scheduled at time frame **0** and the operation $+_3$ is scheduled at time frame **1**. Using LEA

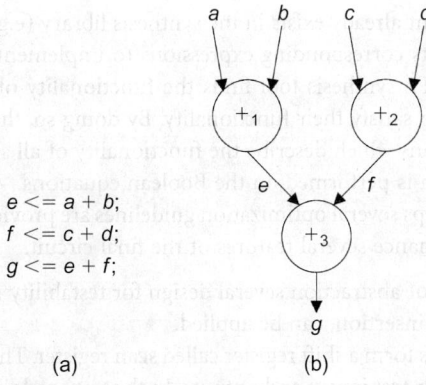

$e <= a + b;$
$f <= c + d;$
$g <= e + f;$

(a)   (b)

FIGURE 96.3 (a) VHDL code of *ex1*. (b) Its DFG.

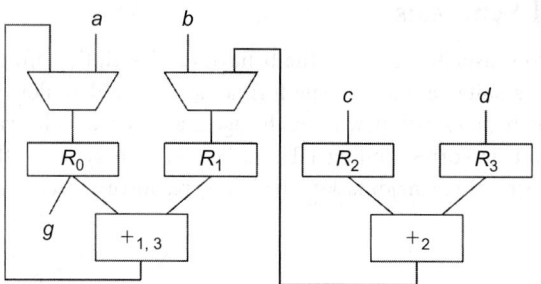

FIGURE 96.4   An RTL implementation of *ex1*.

for module allocation results in $M_{LEA} = \{(+1, 3), (+2)\}$ allocation, which means that operations $+_1$ and $+_3$ are mapped on to the same adder, and the operation $+_2$ is mapped to another instance of adder. Using LEA for register allocation leads to $R_{LEA} = \{(a, e, g), (b, f), (c), (d)\}$ allocation, which means that variables a, e, and g are mapped to register R_0, variables b and f are mapped to R_1, and variables c and d are mapped to registers R_2 and R_3, respectively. The RTL implementation of *ex1* is shown in Figure 96.4.

96.2.2.3   Behavioral Design Optimization

In behavioral synthesis several qualification parameters are optimized. These parameters are:

- Area: The area of the final circuit should be minimized. At the behavioral level the area of the circuit can be estimated with the sum of the area of functional units and interconnections. When we move to the deep submicron era, the wiring area should be considered.
- Delay: The delay of the resulted circuit should be optimized during the synthesis steps. The delay of the circuit is estimated with the sum of the delay of functional units on the critical path. In deep submicron, since wiring delays dominate gate delays, they should be considered.
- Testability: The architecture provided by a synthesis tool is ready to be fabricated. However, manufacturing test and debug remain as major problems, in which the test sequences provided for the system components should be justified and propagated. In addition, increasing the test time, results in increase in chip costs and time-to-market. Therefore, testability consideration during the early stages of behavioral synthesis generates an RTL circuit that is optimized for testability.
- Power consumption: Dynamic power consumption that is caused by switching of logic values is the main source of power consumption in CMOS designs. Scheduling and allocation algorithms can help optimize a circuit for reducing its power consumption.

96.2.3   RTL Synthesis

If the described RTL component already exists in the synthesis library (e.g., an adder) the RTL synthesis tool simply uses the cell and its corresponding expressions to implement the functionality of the RTL component. Otherwise, the RTL synthesis tool finds the functionality of all signals in the circuit and generates Boolean equations to satisfy their functionality. By doing so, the output of the RTL synthesis tool is a set of Boolean equations which describe the functionality of all signals. It should be noted that in this process no optimization is performed on the Boolean equations.

During the RTL synthesis steps several optimization guidelines are provided by different tools (e.g., test tools) to assist designers to enhance several features of the final circuit.

- Testability: At this level of abstraction several design for testability methods, e.g., scan design and built-in self test (BIST) insertion, can be applied.
  - In scan design, all FFs form a shift register called scan register. This scan register can be accessed from external pins for test inputs and outputs. In the test mode, the scan register can be loaded by shifting appropriate data into the scan register. Then the data applied to a specific part of

the circuit and the outputs are loaded into the scan register. And finally, the results are shifted out via the scan register.

○ In BIST, on-chip circuitry is included to generate test vectors and analyze the outputs. For test pattern generation and analyzing the output, the registers in BIST must be redesigned. Functions of a BIST normally involve random test pattern generation (RTPG), serial shifting of data into scan registers, collecting and compressing results using multiple-input signature register (MISR), and analyzing the results.

- Power consumption: The main concern for power reduction at the RT level is to select low-power cells from the library.

96.2.4 Logic Synthesis

A logic synthesizer parses the input design (which is described as Boolean equations) and builds an internal data structure (usually a graph represented by linked lists). Next, logic optimization uses a series of factoring, substitution, and elimination steps to simplify the equations that represent the synthesized network. To make logic optimization tractable, most tools use algorithms based on algebraic factors rather than Boolean factors. Logic optimization attempts to simplify the equations in the hope that this will also minimize area and maximize speed. Following the optimization pass, the technology-mapping pass decides on cells to use for the optimized representation of a circuit. In technology mapping, the algorithms attempt to minimize area (the default constraint) while satisfying other user constraints (delay, testability, and power constraints).

In logic synthesis process several qualification parameters of the circuit are optimized. These parameters are:

- Testability: The aim of testability optimization in logic synthesis is mainly to minimize the number of scan FFs and consequently reduce the design area overhead and cost.
- Power consumption: As mentioned before, the main source of power consumption in digital CMOS circuits is switching power that can be reduced at the logic optimization stage.

96.3   Verification Tools

As mentioned before, there are two ways to inspect the correctness of the design steps: simulation and verification. Simulation methods were described in Section 96.1 and this section is dedicated to the verification methods. Verification methods are classified into two categories:

- Assertion-based verification: In design validation using simulation, the simulation results should be inspected manually. In an assertion-based verification, assertion monitors are used to check the simulation results continuously. The designer provides several properties to show the correct behavior of a circuit and develops the assertion monitors to check that these properties are not violated. When a design property is violated, the assertion triggers and alerts the designer. A set of common assertion monitors are provided in the open verification library (OVL).
- Formal verification: After completing a design, the designer provides a set of properties to show the correct behavior of the design. The process of property checking is called formal verification. A formal verification tool tries to match a property with the specification of a design and if it is unsuccessful, it finds the input conditions that make the property fail.

96.4   Test Tools

Test tools used in design flow are divided into the following categories:

- Fault simulator: A fault simulator is a simulator that is capable to simulate a circuit in presence of faults. Because of the lack of fault model in the higher levels of abstractions, fault simulation is usually performed at the gate level. Fault simulation is normally done for the development of manufacturing tests, i.e., testing the manufacturing defects. A fault simulator receives the circuit gate-list and a set of test vectors and finds the fault coverage of a test vector. There are different implementations for fault simulators, such as *serial fault simulation, parallel fault simulation, deductive fault simulation,* and *concurrent fault simulation.*

- Automatic test-pattern generator (ATPG): ATPG is a program used to generate test patterns to test a circuit. An ATPG usually operates in conjunction with a fault generator program that creates the minimal collapsed fault list. There are many features that are important in test generation, including the cost of test generation, the quality of the generated tests, and also the cost of applying tests. Several implementations of ATPGs are exhaustive test generation, random test generation, and path-sensitized test generation.
- Test advisor: This program usually performs a testability analysis on the circuit being designed and based on this analysis, proposes testability techniques to increase the testability of the circuit. There are several types of the test advisor program including programs used to select scan register and programs used to insert a BIST architecture in the design. These tools are usually used in conjunction with synthesis tools.

96.5 CAD Tool Vendors

96.5.1 Cadence

Cadence Design Systems, Inc. was established in 1988 through the mergence of two electronic design automation (EDA) pioneers—ECAD, Inc. and SDA Systems. Since then, Cadence provides a series of CAD tools used in hardware design in different abstraction levels from layout to system level [3]. Cadence design technologies include:

- System-level design
- Functional verification
- Emulation and acceleration
- Synthesis/place and route
- Analog, RF, and mixed-signal design
- Custom IC layout
- Physical verification and analysis
- IC packaging
- PCB design

Cadence provides a series of simulators called *NC-Verilog, NC-VHDL,* and *NC-SC* used for simulating Verilog, VHDL, and SystemC designs. These simulators are employed in a verification platform, called *Incisive,* used for design verification in any abstraction level from gate level to system level. The *BuildGates* is the synthesis tool of Cadence that generates a Verilog net-list. Then *Encounter* does all the other tasks including floor-planning, placement, clock tree generation, routing, timing analysis, power analysis, and signal integrity. Encounter Test is used for ATPG and DFT design and analysis.

96.5.2 Magma

Magma design automation provides EDA software that enables chip designers to handle multimillion-gate designs. Magma's complete RTL-to-GDSII design flow includes design planning, prototyping, synthesis, place and route, and signal and power integrity chip design capabilities in an integrated environment [4]. Magma's software also includes products for advanced physical synthesis and architecture development tools for programmable logic devices (PLDs) and capacitance extraction.

Magma's *Blast Create* provides a predictable path from RTL to placed gates. It is an integrated environment for general logic and high-performance data-path synthesis, DFT analysis and insertion, physical synthesis, power optimization, and static timing analysis. *Blast DFT* provides test quality and yield management within Magma's design flow. It supports several DFT strategies such as full-scan ATPG, memory and logic BIST, and TAM and Boundary Scan. *Blast Power* provides a solution for power optimization and management.

96.5.3 Mentor Graphics

Mentor Graphics provides an event-driven simulator called *Modelsim* that supports assertions. This makes it a useful assertion-based verification tool. *Modelsim* can take Verilog, VHDL, System Verilog, or SystemC inputs [5].

Leonardo Spectrum of Mentor is an RTL synthesis tool that converts its Verilog, VHDL, and System Verilog input into net-lists for both ASIC and PLD targets [5]. The designer can adjust the synthesis process to optimize area, delay, or both.

DFT advisor is a testability analysis tool and test synthesis tool used to insert full scan into the design. It also provides several methods for partial scan selection [5]. *LBISTArchitect* is a BIST insertion program that analyzes the testability of the design and synthesizes test structures into the design and finds the design's fault coverage.

The ATPG tool for full-scan design is called *FastScan*. It uses an innovative compression method to generate compact test vectors. It supports extensive fault models, including stuck-at, IDDQ, transition, and path delay. *FlexTest* is an ATPG tool for nonscan or partial scan designs. It supports fault simulation for functional vectors and also supports stuck-at, IDDQ, transition, and path delay fault models.

96.5.4 Synopsys

Synopsys provides a simulator called *VCS* that takes standard HDLs including Verilog, VHDL, System Verilog, and SystemC as input. It supports System Verilog assertions that make it suitable to be used as a verification tool [6].

Synopsys *Behavioral Compiler* is a high-level synthesis tool that allows users to synthesize behavioral HDL input descriptions. *DFT Compiler* enables designers to quickly and easily implement high-quality test early in the design flow. *DFT Compiler* supports RT-level and gate-level scan design rule checking, constraint-optimized test synthesis, and fault coverage verification.

Synopsys also provides a power analysis tool. *PrimePower* is a dynamic full-chip power analysis tool for complex ASICs.

96.6 Summary

In this chapter, a typical VLSI design flow was presented and based on it several required CAD tools including simulation tool, synthesis tool, verification tool, and test tool were described. In each case the different types and different implementations were described. Finally, we introduced the four major CAD tool vendors and some of their products.

References

1. G.D. Micheli, *Synthesis and Optimization of Digital Circuits*, McGraw-Hill, New York, 1994.
2. S. Safari, A.H. Jahangir, SOC Test Synthesis using Test Access Mechanism Design, ECTI-CON, pp. 799–803, May 2005.
3. http://www.cadence.com
4. http://www.magma-da.com
5. http://www.mentor.com
6. http://www.synopsys.com

Index

N